英汉土木工程大词典

上

An English-Chinese Civil Engineering Dictionary

中交第四航务工程勘察设计院有限公司

罗新华 主编

人民交通出版社
China Communications Press

内 容 提 要

《英汉土木工程大词典》共收录词条约 82 万，内容涉及建筑工程、港口航道工程、道路工程、铁路工程、岩土工程、桥梁与隧道工程、市政工程、水利水电工程、城市轨道交通工程、工程机械、建筑材料、健康安全环保(HSE)等相关学科专业方面的词组、短语，可供从事土木工程方面相关工作的广大科技人员参考使用。

图书在版编目(CIP)数据

英汉土木工程大词典/罗新华主编. — 北京：人民交通出版社，2014.2
 ISBN 978-7-114-11134-1

Ⅰ. ①英… Ⅱ. ①罗… Ⅲ. ①土木工程—词典—英、汉 Ⅳ. ①TU-61

中国版本图书馆 CIP 数据核字(2014)第 011098 号

书　　名：	英汉土木工程大词典
著 作 者：	罗新华
责任编辑：	邵　江　杜　琛
出版发行：	人民交通出版社
地　　址：	(100011)北京市朝阳区安定门外外馆斜街 3 号
网　　址：	http://www.ccpress.com.cn
销售电话：	(010)59757973
总 经 销：	人民交通出版社发行部
经　　销：	各地新华书店
印　　刷：	北京市密东印刷有限公司
开　　本：	880×1230　1/16
印　　张：	233
字　　数：	17334 千
版　　次：	2014 年 2 月　第 1 版
印　　次：	2014 年 2 月　第 1 次印刷
书　　号：	ISBN 978-7-114-11134-1
印　　数：	0001～1500 册
定　　价：	868.00 元(上、下两册)

(有印刷、装订质量问题的图书由本社负责调换)

《英汉土木工程大词典》编委会

总 策 划：朱利翔

主　　编：罗新华

副 主 编：覃　杰　张丽君

编　　委（按姓氏笔画顺序）：

　　卢永昌　朱利翔　余巧玲　张丽君　张宏铨　李　刚
　　李　虹　李华强　李伟仪　杜　宇　沈力文　肖玉芳
　　陈策源　周晓琳　罗　梦　罗新华　徐少鲲　高月珍
　　曹培璇　覃　杰　廖建航

审　　定：林　鸣　王汝凯

校　　审（按姓氏笔画顺序）：

　　刘少珍　吕邦来　何康桂　何智敏　张　勇　张欣年
　　张冠绍　肖向东　邱创亮　麦若绵　周　野　林　琳
　　林吉兆　苗　辉　祝刘文　贾登科　郭玉华　梁　桁
　　曾青松　韩凤亭　廖先斌　蔡泽明　薛轩宇

顾　　问：蔡长泗　杨孟愚　王将克

电脑录入及排版：罗新华　余树阳　余树敏　张喜顺　张再嘉　张丽琼

《英汉土木工程大词典》
编 辑 组

责任编辑： 邵　江　杜　琛

编　　辑： 卢　珊　温鹏飞　富砚博　潘艳霞　卢俊丽　丁　遥
　　　　　　李　洁　杜　琛

复　　审： 刘　涛　陈志敏　王　霞　杜　琛　王文华　郑蕉林

终　　审： 赵冲久　韩　敏　周往莲　田克运　尤晓暐

特邀审稿： 李志荣（民族出版社）

序

随着国家"走出去"战略的实施,我国工程建设行业的海外业务也在不断扩展,施工企业承担的海外建设项目种类越来越多,而国内现有的有关土木工程专业方面的英语辞书已经不能适应这种新形势的要求,很有必要编写一部综合性的土木工程方面的英语辞书,以满足广大科技工作者在工作和学习中的需要。

20世纪90年代初,中交第四航务工程勘察设计院罗新华硕士等人即开始了该方面词条的收集工作,并于2000年4月出版了《汉英港湾工程大词典》(人民交通出版社出版)。在此基础上,罗新华等人又参考了大量的国内外相关辞书、现行规范和标准术语,进一步广泛收集了与土木工程相关的专业词汇,涉及建筑工程、港口航道工程、道路与桥梁工程、铁路工程、水利水电工程、岩土工程、建筑材料、环境保护和安全等相关学科专业。经过多年的不懈努力,终于完成了《英汉土木工程大词典》的编纂工作,共计收录土木工程类英语词条近82万条。该书是国内迄今为止最为完整的土木工程方面的英语辞书。

该大词典的出版,将在很大程度上满足从事土木工程方面工作的广大科技工作者阅读和翻译英文的需求。

中国工程院院士:谢世楞

2013年9月于天津

前　　言

20世纪80年代初的某个晚上，编者在广州中山大学图书馆晚自修，因学习需要进入外文工具书阅览室借阅有关地质方面的英汉词典。但看到的是，宽敞的书架上只放了少量仅有几百页（16开本）的非土木工程专业词典和两本厚厚的美国人编的韦氏词典（大16开本）。当时编者便感慨于韦氏词典编者的伟大；佩服之余，心生今后也要编出这样大部头综合性权威词典的理想。1990年，编者硕士毕业，其后一直在国内外从事岩土工程勘察类工作，经常接触到工程报告、招投标书的翻译，也经常因找不到可用的相关专业词典而苦恼，于是萌发了编写汉英土木工程方面词典的想法，并在王汝凯、麦土金等专家教授的鼓励下开始了资料搜集工作。

1997年初，编者完成了《汉英港湾工程大词典》的编撰工作，收录词汇约186000条，该书于2000年4月由人民交通出版社出版。在该词典编写过程中，偶然在一本有关土木工程的辞书中看到这样一句话："土木工程学科发展到今天，英汉土木工程词汇估计有三十多万条"。为了这个答案，在好奇心的驱使下，编者凭借兴趣和长期在国外及港澳地区的工作需要，尝试收集，结果发现，在实践中收集的词条远远超过这些。经过悉心整理和筛选，汇集了词条近82万，编写成本书。编者在本书收录词条的过程中，特别注意以下两方面：第一，高校和科研单位需要；第二，土木工程项目招投标、勘察设计以及施工过程中的需要，以期满足土木工程科技工作者的实际需要。

本词典的词条首先源自近年来土木工程专业大类相关出版单位出版的土木工程类词典和专业图书，并在收集时做了大量核查比对工作，力求对每个词条给出最为准确、常用和实用的解释；其次，编者参考了全国科学技术名词审定委员会公布的《土木工程名词》，并根据行业发展近况，对部分内容进行了修订。此外，据编者近二十年在国外和港澳地区参与国际土木工程建设的经验，目前的土木工程勘察设计与施工特别注重健康、安全、环保（HSE）方面问题，因此本词典也收集了这方面的词条，方便广大科技工作者在工作中使用。

在词典编排顺序上，一般有两种方式：一种是先列出基本词，再在基本词下按字母顺序分别列出相关词组；一种是不区分基本词和词组短语，按字母顺序依次排列。本词典采用了后一种方式，这种排序的优势是读者可按字母顺序快速地找到所需要的词条，节省时间，同时这种方式也符合一般科技工作者的查阅习惯。

本词典在编写过程中，先后有一百多位国内外专家、教授、勘察设计大师、博士、硕士、英语专业毕业的大学生、在校大学生、常年奋战在工程项目第一线的工程师、技术人员参与。同时，也得到了交通运输部科技司、中国交通建设集团有限公司、人民交通出版社、中交第四航务工程勘察设计院以及其他兄弟单位的大力支持。本词典的编纂可以说是凝聚了我国从事土木工程工作的广大科技人员的心血，在此表示衷心的感谢！

科技发展迅速,土木工程新词汇不断出现,且随时间的推移某些词的词义也有所变化,热切期望广大读者使用过程中发现问题给予批评指正,也欢迎随时提供该词典没有收集到的相关词条,为以后再版输送新鲜血液。相信有读者的热心参与,该词典将更加完善。

主编:罗新华

2014 年 1 月 13 日于广州

凡　　例

1. 本词典系按英文字母顺序排列，术语单词、复合词和短语一律顺排。对于含有逗号(,)和阿拉伯数字的英文词条，只考虑逗号(,)及阿拉伯数字前面的英文词的排序。

2. 单个英文单词或单条词组在有若干释义时(即一词/词组多义情况)，释义间用分号(;)隔开。如:saddle joint 阶形接榫;咬口接头;咬接头;鞍状关节;鞍形接头;鞍(形)接合

3. 括号的用法

(1) ( ) 圆括号

① 释义的补充，如 sacellum (露天的)小型罗马祭坛;

② 可省略的字母，如 sabin(e)。

(2) [ ] 方括号

① 可替换前面的单词，如 overhead railway [railroad] 高架铁路;高架铁道;

② 缩略语，如 uninterrupted power supply [UPS] 不间断(供电)电源;不中断电源;

③ 缩略语解释，如 FVT [field vane test] 现场十字板剪切试验;

④ 词源，如 passing point [美]越行站;相会点;

⑤ 单词的复数形式，如 palestra 体育场;希腊或罗马体育训练馆;[复] palestrae;

(3)【 】鱼尾括号，表示专业或学科，如【铁】，表示铁路工程;

(4)〈 〉尖括号，表示词语属性，如〈旧〉，表示旧时用。

4. 专业或学科代号

【数】数学	【港】港口航道工程	【建】房屋建筑工程
【物】物理	【道】道路工程	【航海】航海学
【化】化学化工	【铁】铁路工程	【航空】航空学
【无】无线电子学	【给】给排水	【岩】岩土工程
【地】地质学	【矿】矿物学	【水文】水文学
【测】测量学	【气】气象学	【机】工程机械
【疏】疏浚工程	【计】计算机	【天文】天文学
【救】救捞工程	【声】声学	【植】植物学
【动】动物学	【生】生物学	

目　录

序 …………………………………………………………………………… I

前言 ………………………………………………………………………… Ⅲ

凡例 ………………………………………………………………………… V

正文 …………………………………………………………………… **1～1879**

参考文献 ………………………………………………………………… 1880

目 次

A

A. A. Brodisky classification A A 布罗德斯基分类
aa channel 块熔岩通道
AA conductor 铝镁合金导线
AADT[Annual Average Daily Traffic]年平均日交通量
aa-field 块熔岩区
aa flow 渣状熔岩流
aagregation of capital 资本集聚
aa-lava 块熔岩
aa lava structure 块状熔岩构造
Aalenian 阿连阶
Aare massif 阿勒地块
Aaron's rod 亚伦丈(绕蛇杆形饰)
AASHO[American Association of State Highway Officials] 美国各州公路工作者协会
AASHO flexible pavement design method 美国各州公路工作者协会柔性路面设计法
AASHO method 美国各州公路工作者协会方法
AASHO road test 美国各州公路工作者协会道路试验
AASHO soil classification 美国各州公路工作者协会土分类法
AASHO system 美国各州公路工作者协会岩土分类法
AASHTO System 岩土的 AASHTO 分类系统
Aba basin 阿坝盆地
abac 算图;列线图;坐标格网;大圆改正量列线图
abaca 马尼拉制品;马尼拉麻;吕宋麻;麻蕉
abaca rope 白棕绳
abaciscus 用于马赛克的砖石;小柱冠;小型檐托;嵌饰
aback 使船后退;逆帆
abac scale 列线图标度线
abacterial 无细菌的;无菌的
abacterial operation 无菌操作
abactinal side 反辐射对称面
abaculus 嵌镶块;嵌饰
abacus 柱顶板;算盘;栏杆小柱顶;柱冠;冠板;顶板;[复]abaci 或 abacuses
abacus column 具有顶端托板的柱;带有冠板的柱
abacus conveyer 算盘式输送机
abacus flower 圆柱顶板花纹;柱顶托板花
abad 镇;村;城市
Abadan Refinery 阿巴丹炼油厂(伊朗)
abadging material bin 研磨材料储[贮]罐;研磨材料储[贮]仓
abaft 向船尾;在船尾
abaft the beam 在船的正横以后
abaiser 象牙黑
a balanced package 一揽子均衡守则(关贸总协定术语)
abalienate 让渡;让出
abalienation 让渡
abalone 鲍鱼
abalyn 阿巴林
abambulacral 非步带的
abampere 绝对安培;电磁系电流单位;电磁安培
abampere per square centimeter 电磁安/平方厘米
abamurus 扶梁;支墩(块);挡土墙;挡水墙;扶壁
abandon 弃权;抛弃;放弃;废弃
abandon an anchor 弃锚
abandon application 放弃申请
abandon a ship 弃船
abandon a well 将井报废
abandon catch basin 废弃截留井;废弃集水池
abandoned aerodrome 废弃停用的机场
abandoned assets 废弃资产
abandoned basin 废池;废水池
abandoned call 废弃呼叫;放弃呼叫
abandoned channel 抛弃废河道;(河流的)旧道;牛轭湖;故河道;废水渠;废(弃)河道;废河槽
abandoned channel deposit 废弃河道沉积
abandoned charges 废弃费用
abandoned city 废城
abandoned cliff 崩塌崖
abandoned coal pillar 残留煤柱
abandoned dam 废坝
abandoned dike 废弃的堤坝
abandoned ditch 废沟
abandoned dredge-cut 废挖槽
abandoned field succession 弃耕地演替
abandoned goods 废弃货物;被弃物
abandoned hole 报废孔
abandoned land 撂荒地;撩荒地
abandoned lease cost 废弃租约成本
abandoned levee 废堤
abandoned line 废弃线(路)
abandoned loop 废河网
abandoned manhole 废弃窨井
abandoned meander 牛轭湖;残余河曲;废曲
abandoned merchandise 无主货物
abandoned mine 废矿(井);废旧矿坑
abandoned mining area 禁止开矿地区
abandoned oil 废弃油
abandoned pit 废矿(井);废坑
abandoned property 委托财产;已放弃产权财产;废弃财产
abandoned river course 废河道
abandoned road 废路
abandoned ship 委弃的船;废弃船舶
abandoned shore line 旧岸线;废岸线
abandoned stream course 废河道
abandoned submarine cable 报废海底电缆
abandoned support 废弃的支柱
abandoned tender 废标
abandoned thalweg 旧深泓线;河流旧泓线
abandoned track 废线【铁】;废弃线(路)
abandoned tunnel 废隧道
abandoned vehicle 废弃车辆
abandoned vessel 委弃船只;被弃的船
abandoned well 废弃井;废矿井;废井;报废(油)井
abandoned working 报废巷道
abandoned Yellow River zero datum 废黄河零点
abandonee 废弃财物受领人;被委付者
abandoner 委托者;委弃者;委付者
abandonment 废弃;放弃(产权);抛弃;委付(书);拒绝受领;报废等值线
abandonment charges 废弃费用
abandonment clause 委托条款
abandonment contour 报废等值线
abandonment expenses 废弃费用
abandonment in extractive industry 采掘工业的废弃
abandonment loss 废弃损失
abandonment of action 放弃诉讼
abandonment of a patent 放弃专利权
abandonment of a patent application 放弃专利申请
abandonment of appeal 放弃上诉
abandonment of a ship 弃船
abandonment of cargo 放弃货载
abandonment of claim 放弃索赔(权利)
abandonment of homestead 弃宅文书;放弃产权文书
abandonment of option 放弃期权;放弃购货保留权
abandonment of property 房地产不租不买;财产放弃
abandonment of ship 船舶委付
abandonment of the voyage 放弃原定航次
abandonment of water rights 放弃水权
abandonment plug 封井水泥塞
abandonment pressure 排气压力;废弃压力
abandonment value 清算价值;废弃价值
abandon non-effect nodule 废弃无效根瘤
abandon patent 放弃专利权
abandon ship drill 弃船演习
abandon ship signal 弃船信号
abandon ship station 弃船部署
Aba old land 阿坝古陆
a bare contract 无条件契约
abas 诺谟图;列线图
abasciscus 着色拼花小块
abase 降下(帆或旗)
a basket of currencies 一篮子货币
a batch 一批
a batch of concrete 一拌混凝土

abate 双硫磷;减退;减弱;削弱;倒钝;废止
abate a price 降价;还价
abated amount of reserve 核销储量
abatement 扣减;减轻;减免;减低;免赔;压低;作废;除却;冲消;废弃
abatement and exemption from penalty 免除刑罚
abatement claim 减价要求
abatement cost 消除有害事物的成本
abatement noise 消除噪声
abatement of debts 废除债务
abatement of diesel exhaust 柴油机废气消除法
abatement of dust(ing) 消除灰尘;除尘
abatement of noise 减轻噪声;消除噪声
abatement of nuisance 排除障碍;消除公害
abatement of pollution 清除污染;污染消除;消除污染
abatement of smoke 消除烟雾法;消除烟尘;烟雾消除
abatement of smoke and dust 消烟除尘
abatement of swell 消除肿胀;消除隆起;消除涌浪
abatement of tax 减税
abatement of water pollution 水污染消除;消除水污染
abatement order 废除令
abatement pollution 污染防治
abating 作废;削阀;减少;撤销
abating pollution 消除污染
abatjour 亮窗;反射器;灯罩;窗上槛斜面;天窗;斜片百叶窗
abaton 礼拜堂;圣堂;圣所
abatsons 使声音向下传播的装置
A-battery 甲电池(组);A 电池组
abat(t)is 三角形木架透水坝;通风隔墙;鹿砦;拒木;风挡;鹿寨;杩杈(坝);障碍物
abattoir 屠宰场
abattoir wastes 肉类加工废水;屠宰场废物
abattoir wastewater 肉类加工废水;屠宰场废水
abat-vent 坡屋顶;金属烟囱;挡风斜板;致偏板;折转板;通风帽;通气帽;转向装置;障风装置;固定百页窗;挡风装置
abat-voix (讲台上的)止响板;声响反射板;吸声板;吸音板;单声反射板;反射板
abaxial 离(开)轴心;远轴的;轴外的
abaxial astigmatism 轴外像散
abaxial ray 轴外光线
abbatre 凹凸(花)纹
Abbaye 克吕尼修道院(十至十二世纪法兰西)
Abbe apertometer 阿贝数值孔径计;阿贝孔径计
Abbe apparatus 阿贝折射计
Abbe autocollimation measuring 阿贝自准直量度法
Abbe autocollimation measuring method 阿贝自动对准测量方法
Abbe camera lucida 阿贝化显微镜描图器
Abbe comparator 阿贝比长仪
Abbe condenser 阿贝聚光器;阿贝聚光镜
Abbe constant 阿贝常数
Abbe criterion 阿贝准则;阿贝判据
Abbe doublediffraction principle 阿贝双衍射原理
Abbe eyepiece 阿贝目镜
Abbe illuminator 阿贝照明器
Abbe invariant 阿贝不变式;阿贝不变量
Abbe number 阿贝值;阿贝数
Abbe photometric law 阿贝光度定律
Abbe prism 阿贝棱镜
Abbe radiance law 阿贝辐射率定律
Abbe refractometer 阿贝折射(率)计;阿贝折射仪
Abbe resolution criterion 阿贝分辨率判据
abbertite 黑沥青
Abbe's comparator principle 阿贝坐标仪原理;阿贝量测仪原理
Abbe's formula 阿贝公式
Abbe's spherometer 阿贝球径计
Abbe's principle 阿贝原理
Abbe's sine condition 阿贝正弦条件
Abbe's sine law 阿贝正弦定律
Abbe's sine rule 阿贝正弦定则
Abbe's theory 阿贝理论

Abbe's theory of image formation 阿贝成像原理;阿贝成像理论
Abbe's theory of imaging 阿贝透镜成像理论
Abbe's treatment 阿贝处理
Abbe's value 阿贝值
abbey 修道院;大寺院;寺院
abbey church 修道院教堂
abbey court 教堂墓地
abbeystead 教堂所在地;庙宇寺堂占地;修道院所在地
abbey tile 凹瓦
abbot's cloth 粗厚方平织物
Abbott's bearing curve 艾卜特支承曲线
abbozzo 底画
abbreuvoir 石块间接缝;石块间隙缝(拱石或石砌体的缝)
abbreviate 缩写;简略
abbreviated address 短缩地址
abbreviated address call(ing) 短缩地址调用;简缩寻址访问
abbreviated addressing 节略寻址;简缩编址;短缩寻址;短缩编址
abbreviated analysis 简项分析;简略分析
abbreviated and depayed ringing 压缩延迟振铃
abbreviated dial(l)ing 简略拨号;缩位拨号
abbreviated dial system 缩位拨号方式
abbreviated drawing 简图
abbreviated equation 简化方程
abbreviated expression 缩语
abbreviated flight plan routing 简略飞行计划路线
abbreviated form 简写形式;简体
abbreviated formula 缩写式;简(写)式;简化公式
abbreviated indication 缩写字样
abbreviated keyboard 略键盘
abbreviated ladder 小型爬梯
abbreviated list 简表
abbreviated name 略语名称
abbreviated scenery 缩景
abbreviated signal code 传输电码
abbreviated version 节本
abbreviated words 缩写词汇
abbreviation 缩短;缩写;省略;简略符号
ABC[Aggregate Base Course] 集料基层;骨料基层
ABC[Associated Builders and Contractors] 建造商及承包商联合会
ABCA[American Building Contractors Association] 美国建筑承包商协会
ABC analysis 分类控制法;分类管理法
ABC analysis method ABC 分析法
ABC control 分级存货管理
ABC inventory system ABC 存货控制制度
ABC management method of stock classification ABC 库存分类管理法
ABC of life-saving 救生基本知识
abcoulomb 绝对库仑
abcoulomb centimeter 电磁库厘米
abcoulomb per cubic centimeter 电磁库/立方厘米
abcoulomb per square centimeter 电磁库/平方厘米
ABC process 污水净化三级过程;三级过程(污水处理);污水化学沉淀法
Abcrrcombie, Fitch 阿伯克龙比和菲奇公司
ABC standard sampling system ABC 标准抽样制度
ABC system ABC 分类法
Abdel-Herrin model 阿蓓达汉合一赫林模型
Abderhalden's reaction 阿布德哈尔登反应
Abde's zero in variant 阿贝零不变量
abdicate 退位
abdominal mass 积聚
abdominal version 外倒转术
abduction 外展(法)
abduction plate 仰冲板块
abduction splint 外展夹板
abeam 横(梁)向;正横
abeam direction 正横方向
abeam distance 正横距离
abeam wind 侧风
abecedarian 按字母顺序排列
Abegg's rule 阿贝格规则
Abegg's rule of eight 阿贝格八值定则
Abel c.c. flash point apparatus (沥青的)阿贝尔闭杯闪点试验仪
Abel closed cup flash point apparatus (沥青的)阿贝尔闭杯闪点试验仪;阿贝尔闭杯闪点测定仪

Abel closed tester 阿贝尔闭式试验器
abele 杨树;白杨;银白杨
Abel flash point 阿贝尔闪点
Abel flash point apparatus 阿贝尔闪点(测定)仪
Abel flash point tester 阿贝尔闪点试验器
Abel heat test 阿贝尔耐热试验;阿贝尔加热试验
Abelian domain 阿贝尔域
Abelian field 阿贝尔域
Abelian group 阿贝尔群
Abelite 阿贝立特炸药
Abell richness class 阿贝尔富度
Abel-Pensky closed cup flash point tester 阿贝尔-宾斯基闭杯式闪点试验器
Abel Pensky flash point test 阿贝尔闭杯试验
Abel reagent 阿贝尔试剂
Abel's close test 阿贝尔密闭实验(一种测定液体燃料和润滑油闪点的方法);阿贝尔闭杯试验
Abel's generalization of binomial theorem 二项式定理的阿贝尔推广
Abel's inequality 阿贝尔不等式
Abel's integral equation 阿贝尔积分方程
abelsonite 紫四环镍(矿);卟啉镍石
Abel's problem 阿贝尔问题
Abel's reagent 阿贝尔金相浸蚀剂
Abel's test of convergence 阿贝尔收敛性判别法
Abel test 石蜡闪点测定仪;阿贝尔(闪点)试验
Abel tester 阿贝尔闪点试验器;阿贝尔测定仪;石蜡闪点测定仪
Abel testing apparatus 阿贝尔闪点试验仪
Abel theorem 阿贝尔定理
abend 异常结束;异常终止
aber 两河回流点;河口
Aberdeen granite 亚伯丁花岗岩(英国)
abernathyite 水砷钾铀矿;阿别纳提矿
aberrance 延误;停顿;越轨
aberrancy 越轨
aberrant conduction 差异传导
aberrant copy 残本
aberrant ductules 迷管
aberrant source 偏差源;像差源
aberrated optics 有像差光学系统
aberration 失常;偏差;误差;畸变像差;畸变;像差;过失;光行差
aberration angle 光行差角
aberration coefficient 像差系数
aberration constant 光行差恒量;光行差常数
aberration correction 光行差校正
aberration curve 偏差曲线;离异曲线
aberration day numbers 光行差日数
aberration defect 像差
aberration ellipse 光行差椭圆
aberration-free 无像差
aberrationless 无像差的
aberration mark 像差符号
aberration of fixed stars 恒星像差
aberration of light 光行差
aberration of needle 磁针偏差
aberration of reconstructed wave 重建波的像差
aberration of reconstruction wave 重现波像差
aberration of shift 光行差位移
aberration of stars 恒星光行差
aberration of wind 风动差
aberrometer 象差计
Aberson brick 艾波生砖
Aberta oil 阿贝他石油
a better option 较优方案
abeyance 潜态;未定;暂缓;暂搁
abfarad 电磁法拉;绝对法拉
abhenry 电磁亨利;绝对亨(利)
abherent 防粘剂
abhesion 脱粘
abidance by rules 遵守条例;遵守规章
abide 遵守;船被风浪阻止;持续
abide by state regulations 遵照州规则
abide by the agreement 遵守协议
abide by the contract 遵守合同
abiding place 寓所;住宅
abies oil 松香油;冷杉油
Abies pectinata 欧洲白冷杉
abies sachalinensis 库页冷杉
abietate 松香酸盐;枞酸酯
abietene 松香烯;枞烯
abietic acid 松香酸;枞酸
abietic amine 松香胺

abietic anhydride 松香(酸)酐
abietic resin 枞脂;松脂;松树脂
abietic-type acid 枞酸型酸
abietin 枞树脂
abietinal 枞醛
abietinol 松香醇;枞醇
abietyl 松香酸基;枞酸基
abietyl alcohol 松香醇
abietylamine 枞胺
ability 能力;效率;才能;本领
ability and earnings 能力与收入
ability and wisdom 才智
ability density of seawater 海水条件密度
ability factor 工作能力指标
ability in self-support 自承能力
ability of bearing taxation 负税能力
ability of equalization 均衡能力
ability of manager 管理者能力
ability of migration of elements 元素迁移能力
ability of mobilization 动员能力
ability of self-cleaning of aeration zone 包气带自净能力
ability of water and sediment conveyance 过水输沙能力
ability of water supply from groundwater reservoir 地下水库供水能力
ability rating 汽车动力定额;动力定额
ability test 能力测验
ability to accommodate social funds 社会融资能力
ability to bargain 讨价还价的能力;交易能力
ability to borrow 借债能力;借款能力
ability to climb gradients 爬坡能力;上坡能力
ability to communicate 通信能力
ability to compete 竞争能力
ability to design plans and strategies 策划能力
ability to dismantle 拆卸能力
ability to flow 流动性(指混凝土等)
ability to harden 硬化性能;硬化能力
ability to invest 投资能力
ability to pay 纳税能力;支付能力;购买力;负税能力
ability-to-pay approach 支付能力学说
ability to pay basis 负税能力基准
ability-to-pay principle 负担能力原则;付税能力原则
ability-to-pay principle of taxation 课税的支付能力原则;纳税的支付能力原则;负税能力原则
ability-to-pay theory 负税能力说
ability to penetrate 穿透能力
ability to provide the auxiliary items 配套能力
ability to react to market conditions 市场反应能力
ability to repay loan 借贷偿还能力;还款能力
ability to sell 销售能力
ability to swim 游动能力
ability to work 劳动能力
ability to work operation(al)capability 工作能力
ab initio wave-function 初始波函数
abiochemical pollutant 无机化学污染物
abiochemical pollution 无机化学品污染
abiochemistry 无生化学;无机化学
abiocoen 无机生境
abiogenesis 无生源说;自然发生说;自然发生(论);非生物起源
abiogenetic 非生物成因的
abiogenic gas 非生物成因气
abiogloph 无机物层面印痕
abiological 非生物学的
abiology 无生物学;非生物学
abioseston 非生物浮游物
abiotic 非生物的
abiotic biodegradation 非生物降解
abiotic components 非生物部分
abiotic control 非生物防治
abiotic degradation 非生物分解
abiotic environment 非生物环境
abiotic environment(al)factor 无机环境因素;非生物环境因素
abiotic factor 非生物因子;非生物因素
abiotic non-living 非生活的
abiotic reaction 非生物反应
abiotic substance 非生物物质
abiotrophy 营养不足
a bipolar type ion-exchange membrane electrolyzer 复极式离子膜电解槽
Abitibi belt 阿比提比带

abject poverty 赤贫
abjunction 脱离
ablated weight 烧蚀重量
ablaters 防腐蚀料
ablating rate 烧蚀率;消蚀率
ablating surface 烧蚀(表)面;消融面
ablation 切除;烧蚀;磨削作用;磨蚀;消蚀;消融(作用);冲蚀;冰面消融;冰面融化
ablation amount 消融量
ablation area 消融区
ablation belt 消融带
ablation breccia 溶蚀角砾岩;垮塌角砾岩;消融角砾岩
ablation coating 烧蚀涂料;烧蚀涂层;消蚀涂层
ablation cone 融蚀锥;冰锥
ablation cooling 烧蚀冷却
ablation drift 融蚀冰碛;消融冰碛;风化冰碛
ablation factor 烧蚀因数;消融系数;消融率
ablation form 融蚀形态
ablation insulating material 烧蚀绝热材料
ablation intensity 消融强度
ablation material 烧蚀材料
ablation mechanism 烧蚀机理
ablation moraine 融蚀冰碛;消融(冰)碛
ablation of dike body 坝体冲蚀
ablation of snow cover 积雪消融
ablation period 消融期
ablation polymer 烧蚀性聚合物
ablation rate 消融率
ablation resistance 耐消蚀性;耐烧蚀性
ablation season 消融期
ablation shield 烧蚀防护罩;防烧蚀屏蔽;蚀防护罩
ablation swamp 消融沼泽;冰融沼泽
ablation test 烧蚀试验
ablation till 消融冰碛
ablation velocity 融蚀速率;烧蚀速率
ablation zone 消融层;融蚀区;烧蚀范围;融蚀层带;消融区;消融带
ablative 烧蚀材料
ablative agent 烧蚀剂
ablative characteristic 烧蚀特性
ablative coating 消蚀涂料
ablative composite material 烧蚀复合材料
ablative cooling material 烧蚀冷却材料
ablative flashlamp 烧蚀闪光灯
ablative insulative material 烧蚀隔热材料
ablative layer 烧蚀层
ablative material 烧蚀剂;烧蚀材料
ablative mode of protection 烧蚀防护法
ablative performance index 消蚀性能指数
ablative plastics 烧蚀性塑料
ablative polymer 烧蚀聚合物;消蚀聚合物
ablative protection 烧蚀防护
ablative recording 冲蚀记录
ablative response 烧蚀反应
ablative shielding 烧蚀保护层
ablative thermal protection 烧蚀热防护
ablativity 烧蚀率
ablatograph 融化测量仪;消融仪
ablator 烧蚀(性)材料;烧蚀体;烧蚀剂
ablator ceramics 陶瓷烧蚀材料
ablaze 着火(的)
able-bodied 强壮的
able-bodied farm worker 全劳动力
able-bodied labo(u)rer 强劳动力
able-bodied seaman 一级水手;熟练水手;二等水手;二等水兵
able-minded 能干的
ablepsia 视觉缺失
Ablers projection 阿伯斯投影
able seaman 熟练水手
abletyl 松香
A-block A形砌块
abluent 清洗剂(清洗的;清净剂;洗涤剂
ablution 清洗法;洗净(液);吹掉;吹除
ablutionary 洗净的
ablution board 滴水板(洗涤用);泄水板
ablution fountain 洗涤泉;洗涤池;喷泉式洗涤槽(为水磨石环槽,同时可供六人使用)
ablykite 水合多水高岭石;阿布石
abmho 电磁姆
abmiralty gun metal 海军炮金属
Abney clinometer 阿布尼测斜器
Abney effect 阿布尼效应

Abney flicker photometer 阿布尼闪变光度计
Abney level 手水准(仪);手持水准仪;手持水平仪;阿布尼式测斜仪;阿布尼水准器;阿布尼水准仪
Abney mountin for concave grating 阿布凹面光栅装置
Abney mounting 阿布尼装法
Abney's colo(u)r patch apparatus 阿布尼色度装置
Abney's law 阿布尼定律
abnomal loss 特别损失
abnormal 异常的;反常的;不正常的;不平常的;变态的
abnormal addressing 异常寻址【计】
abnormal animal behavio(u)r 动物异常行为
abnormal anticlinorium 上敛复背斜;逆复背斜(层);倒换状复背斜
abnormal attribute 异常属性;异常表征
abnormal audibility zone 非标准可听度范围
abnormal behavio(u)r 异常行为
abnormal blooming of plankton 浮游生物异常繁殖
abnormal bridging 非正常桥
abnormal cathode fall 反常阴极电位降;反差阴极电位降
abnormal circumstance 异常情况
abnormal climate 异常气候
abnormal combustion 不正常燃烧
abnormal component 异常成分
abnormal condition 异常状况;非(正)常状态;反常条件;反常情况;不正常条件;不正常情况
abnormal condition interrupt 异常条件中断
abnormal contact 异常接触
abnormal cost 特别成本;异常成本;非(正)常成本
abnormal current 异常海流;异常电流
abnormal curvature 异常弯曲
abnormal curve 异常曲线;非正态曲线;非正规曲线;非正常曲线
abnormal density 异常密度;反常密度
abnormal depreciation 特别折旧;非常折旧
abnormal development 发育异常
abnormal deviation 异常偏差
abnormal differentiation 异常分化
abnormal dip 不正常倾斜
abnormal discount 不正常折扣
abnormal dispersion 反常色散
abnormal dispersion glass 反常色散玻璃
abnormal distribution 非正态分布
abnormal drainage 异常水系;反常水系
abnormal drop 不正常降低
abnormal dump 异常转储
abnormal echo 反常回波
abnormal E layer 异常电离层
abnormal end 异常终止;异常结束;不正常终止;不正常结束
abnormal end control table 异常结束控制表
abnormal end of task 任务异常终止;异常结束
abnormal erosion 畸形的侵蚀;异常侵蚀;异常冲刷;异常侵蚀;反常冲蚀
abnormal exchange 不正常交换;非正常行市;非正常交易
abnormal expansion 反常膨胀
abnormal fan-shaped fold 逆扇形褶皱;逆扇状褶皱
abnormal farms 不规则农场
abnormal fault 逆断层;异常断层;变态断层
abnormal fermentation 异常发酵
abnormal field 异常磁场
abnormal fixation 变态固着
abnormal flow 非常流
abnormal focal length 反常焦距
abnormal formation pressure 异常地层压力
abnormal function 异常函数;不正常工作
abnormal gains 特别收益;非常收益;非常收入
abnormal glow 不规则辉光放电
abnormal glow discharge 异常辉光放电;反差辉光放电;反差辉光放电
abnormal grain growth 晶粒反常长大;异常晶粒生长;反差晶粒生长
abnormal growth 异常生长
abnormal high water-level 异常高水位
abnormal indication 异常指示
abnormal information 异常信息
abnormal interference colo(u)r 异常干涉色
abnormal item 异常项目;非常项目
abnormality 畸形;不规则;异常情况;非正态性;反常(性);反常的性质和状态;不正常性;不正常现象

abnormality of force of labor 产力异常
abnormality of mark 货物标志不正规;航标异常
abnormal lag 反常滞后
abnormal landing 不正常着陆
abnormal load 异常荷载;不规则荷载;非常荷载
abnormal loading 异常荷载
abnormal loss 非正常损失
abnormal low-voltage arc 反差低压电弧
abnormally flat 反差过小
abnormal magnetic variation 异常磁差;磁差异常;反磁变化的
abnormal melting 反常熔化
abnormal noise 异常噪声
abnormal nuclear state 反常核态
abnormal observation 反常观察值
abnormal obsolescence 非正常陈旧
abnormal odo(u)r 异常气味;异常臭气;异臭
abnormal operating condition 异常工作条件;异常车行条件;反常操作情况;不正常的工作条件
abnormal operation 异常运行;操作异常;不正常运转
abnormal overload 异常过载
abnormal overpressure 高异常压力
abnormal overvoltage 事故过电压;异常过电压
abnormal phenomenon 畸形现象;异常现象;反常现象
abnormal place 反常采煤区
abnormal polarization 反常偏振;反常极化;反向偏振
abnormal pressure 异常压力;反常压力
abnormal pressure formation 异常压力下形成的岩层;高压岩层;超高压岩层;异压岩层
abnormal profile 畸形的剖面;异常剖面
abnormal profit 超额利润;非正常利润
abnormal program(me) termination 程序异常终止
abnormal propagation 反常传播;不规则传播
abnormal recrystallization 反差再结晶
abnormal redshift 反常红移
abnormal reflection 反常折射;反常反射
abnormal refraction 反常折光;异常折射;反常蒙气差;不正常折射
abnormal retirement 非正常报废
abnormal return 异常返回;异常收益
abnormal return address 异常返回地址
abnormal rise zone of fault 断层异升高度带
abnormal risk 异常风险;特殊风险;异常危险;非常风险
abnormal sale 异常出售
abnormal scattering 反常散射
abnormal scour 异常冲刷
abnormal sea current 异常电流;异常海流
abnormal sea level 异常潮位
abnormal seismic intensity 异常地震烈度
abnormal series 非正规列
abnormal setting 非正常凝结;反常硬化;反常凝聚;反常凝结;反常固化
abnormal shock 异常振动
abnormal shrinkage 非正常短缺
abnormal situation 异常情况
abnormal smell 异常气味
abnormal soil 不正常土;畸形的土;异常土(壤);非正常土(壤)
abnormal sound 异响
abnormal spoilage 非正常损失;非正常损耗;非正常破损;非常损坏
abnormal spoiled goods 非正常废品
abnormal state 异常状态;不正常状态
abnormal statement 异常语句;不规则语句
abnormal steel 非正常钢;反常钢
abnormal stress 异常应力;反常应力
abnormal structure 反常结构
abnormal structure steel 反常组织钢
abnormal swell 异常涌浪
abnormal synclinorium 逆复向斜
abnormal task termination 异常任务终结
abnormal temperature distribution 反常温度分布
abnormal termination 异常终止;异常终结
abnormal test condition 特殊试验条件
abnormal tide 异常潮;非常潮;稀遇潮
abnormal time 异常时间
abnormal traffic 反常交通
abnormal transfer 反常转移
abnormal value 异常值
abnormal viscosity 反常黏度

abnormal voltage 异常电压;反常电压
abnormal waste 非正常损耗
abnormal water halo 异常水晕
abnormal water level 异常水位;非常水位
abnormal wave pattern 异常波型
abnormal wear 反常磨耗;异常磨损;不正常磨耗
abnormal weather 反常天气
abnormal weight 反常重量
abnormity 异形
abnormous 变态的
abnoxious fishes 厌弃鱼类
aboard 靠近(他船)舷侧;在船上;在船内
abode 住宅;居住;营业所
abohm 绝对欧姆;电磁(制)欧姆
abohm centimeter 电磁欧厘米
aboideau 沼地挡潮闸;低地防潮(水)堤坝
aboil 沸腾的
A-boiler 用原子反应堆加热的锅炉
aboiteau 沼地挡潮闸;低地防潮(水)堤坝
abolish 废除
abolish commodity economy 取消商品经济
abolishment 取消;废止
abolition 消失;禁止;撤销;取消;废除
abolitionist 废止论者
abolition of ground-rent 取消地租
abolition of reflex 无反射;反射消失
a bond 轴向键
abopon 硼磷酸钠液体混合物
aboral 反口的
aboral organ 反口器
aboral pole 反口极
aboral porcelain tube 反口瓷管
aboriginal 土著居民;原始的;土产
aboriginal cost 原始成本
aboriginality 原始状态;本土性
aboriginal value 原始(价)值
aborigines 土著居民;土生植物;土产;本地居民
aborption spectrograph 吸收摄谱仪
aborption spectrum 吸收光谱
abort 截止;发生故障;半途终止
abort command 异常终止命令
aborted landing 落地未成(飞机)
abort engine 失事应急发动机
abort escape system 紧急逃逸系统;紧急疏散系统
abort handle 应急把手
aborting job 异常终止作业
abortion 失事
abortion statistics 流产统计学
abortion within the first month of pregnancy 暗产
abortive 失败的
abortive eruption 不完全喷发
abortive seed 败育种子
abort light 紧急故障信号
abortoscope 波状热凝集试验器
abort program(me) 异常终止程序
abort sensing 故障测定
abort sensing and implementation instrumentation system 故障传感和处理仪表系统
abort sensing control unit 故障传感控制装置
abort sequence 异常终止序列
abort signal 故障信号
abort situation 故障位置
abort station 止喷按钮
abort switch 止喷开关
abound 丰富;丰产
abounding 丰富的
abound resources 资源丰富
abound volume 装订成册
about 在周围;大约
about days after application 大约在施用后天
about equal to sample 同样品大致相同
aboutsledge 铁锤;大铁锤
about to depart 准备出发状态
above 超过
above and under 上和下
above average 中上;在一般水平之上
above center 上中心
above-critical 临界以上的;超临界的;超限差的
above-critical state 超临界(状)态
above curb 路边标高以上;侧石标高以上
above deck equipment 甲板(上)设备;舱面设备
above facility 地面设施;地面构筑物
above floor level 高于地面高程

above freezing 零上;冰点以上
above-ga(u)ge 飞边
above-grade 在一般等级以上;高于原定级别(指质量或技巧);地平面上;地面(高程)以上
above-grade hydrant 地面上给水消防龙头;地面上给水消防栓
above-grade pipeline 地上管线;地面(上)管线;地面管道
above-grade piping 地上管线;地上配管
above-grade subfloor 房屋建筑底层楼地面;首层楼面;首层楼层;底层楼地面
above-grade wall 地面以上的墙
above-ground 地上的;地面的;在地面以上;地面以上;地表以上
above-ground equipment 地面设备
above-ground facility 地上设施;地面设施
above-ground hydrant 地面消火栓;地面消防栓;地上消防栓
above-ground installation 地面敷设
above-ground level 高出地面;地面高程以上
above-ground masonry 地面上圬工
above-ground(pipe)line 地面上管线;地上管线
above-ground plant 地面装置
above-ground portion 地面部分
above-ground silo 地上青储[贮]塔
above-ground storage gas 地面储气
above-ground storage tank 地上储[贮]罐;地面储罐
above-ground structure 地面建筑物
above-ground subfloor 首层楼面;房屋建筑底层
above-ground tank 地上储[贮]罐;地面储[贮]罐;地面蓄水池
above groundwater table 地下水位以上
above-ground work 地上工作;地上工种;地面工作
above high water level 高水位以上的
above high water mark 在高水位以上;高水位以上的
above hydrant 地面式消火栓
above line accounts 国家预算平衡项目
above masonry 地面上的砖砌构筑物
above mean sea level 平均海面之上;平均海拔高度
above measure 上述措施
above-mentioned 上述的
above norm 高出平价;定额以上的;超标准(的);限额以上;超(定)额
above-normal loss 超正常损失
above-normal speed 超正常速度
above-norm project 限额以上的合同
above par 高出平价;超过面值;超出票面价值
above par value 平价以上
above(pipe)line 架空管道;地面管道
above quota 配额以上;限额以上;超额
above-quota loan 超定额贷款
above-quota purchase 超购
above sea level 海拔(高度);高出海面
above stair(case) 在楼上
above tank 地面储[贮]水池
above target profit 超额利润
above the average quality 一般水平以上的质量
above the erosion(al) basis 侵蚀基准面以上
above the line 线上(项目);线上预算(英国预算中的经常性项目)
above-the-line expenditures 经常预算支出;线上(项目预算)支出
above-the-melt polymerization 熔融聚合(作用)
above thermal 超热(的)
above-thermal neutron 超热(能)中子
above threshold 超阈(值)
above-threshold operation(al) method 超阈值运转法
above-threshold region 超阈值区
above tide 高出海面
above transmitted as receive 以上发送收讫
above water 水上的;在水面上;船体水线以上部分
above water cantilever 前伸臂(码头前沿装卸桥)
above-water craft 水面船舶
above-water damage 船体水线以上部分破损;水上部分破损
above-water fish lamp 水上集鱼灯
above-water torpedo 水上鱼雷发射管
above works 地面工程
ABPR[American Bureau of Public Roads] 美国公路局

ABR[Acrylate Butadience Rubber] 丙烯丁二烯橡胶
abra 岩洞
A-bracket 人字架;艉轴架
abradability 磨损性;磨损度;磨蚀性;磨耗性
abradant 磨蚀剂;磨料;磨粉;磨擦剂;研料;擦除的
abradant action 研磨作用
abradant surface 研磨面
abrade 磨蚀;磨掉;擦去
abraded action 磨蚀作用
abraded bedrock surface 浪蚀基岩面
abraded diamond 磨损的金刚石
abraded filament yarn 擦毛长丝纱
abraded glass 磨光玻璃
abraded parts 磨屑
abraded plane 磨蚀面
abraded platform 浪成台地
abraded quantity 损耗量;减量;磨损量
abraded stone 磨光石
abraded surface by blasting 用喷砂或喷丸清理过的表面
abraded test 磨蚀试验
abraded tester 磨蚀试验机
abraded yarn 擦毛丝
abrade platform 浪蚀台地
abrader 磨蚀(试验)机;磨光机
abradibity 可研磨性;磨损度
abradible ceramics 可磨蚀陶瓷
abradible sealing coating 可刮削封接涂层
abrading 磨光;研磨
abrading apparatus 磨蚀器;研磨设备
abrading device 磨耗试验器;打磨机
abrading material 研磨材料
abrading material bin 研磨材料储[贮]罐;研磨材料储[贮]仓
abrading powder 磨蚀粉;研磨粉
abrading section 磨损部位
abrading tool 磨光机
abradum 抛光氧化铝粉
Abraham's consistometer 亚伯拉罕稠度计
Abraham's law 亚佰拉罕(混凝土)水灰比定律
Abram's cone 阿布拉姆圆锥;混凝土强度试验锥
Abramsen straightener 阿布拉姆森型矫直机
Abram's fineness modulus 阿布拉姆细度模数;阿布拉姆细度模量
Abram's law 混凝土强度试验规则;混凝土强度试验规范;阿布拉姆定律(混凝土强度取决于水灰比)
Abram's method 混凝土强度试验法
Abramson code 阿布拉姆逊码
Abram's test (对混凝土骨料进行的)氢氧化钠检验
Abram's water/cement ratio law 阿布拉姆斯水灰比定律
abra position 溶洞的位置
abrased glass 磨光玻璃
abraser 研磨剂;磨料;耐磨试验仪;测定物质抗磨性的器械
abrasimeter 耐磨试验仪
Abrasin oil 七年生桐油;皱桐油;次桐油
abrasion 磨损;磨蚀;磨耗;研磨;冲蚀;擦伤;擦破;擦刮;磨掉;剥蚀
abrasion action 磨搓作用
abrasion alloy 耐磨合金
abrasion blasting 喷砂加工
abrasion by glaciers 冰川磨蚀
abrasion coast 剥蚀河岸;剥蚀海岸;海蚀岸
abrasion coefficient 磨耗系数
abrasion concealing coating 耐磨罩面涂层
abrasion cycle 磨损试验转数;磨蚀周期
abrasion damage 磨伤
abrasion degree 磨耗度
abrasion disc method 盘式磨耗(测定)法
abrasion drill 回旋钻
abrasion drilling 磨蚀式回旋钻进;冲蚀钻井
abrasion embayment 磨蚀河湾;磨蚀海湾;侵蚀河湾;侵蚀海湾
abrasion fatigue 磨损疲劳
abrasion finishing 磨光
abrasion geomorphy 海蚀地貌
abrasion groove 磨蚀沟;研磨槽
abrasion hardness 耐磨硬度;耐磨蚀度;磨损硬度;磨蚀硬度;磨耗硬度;研磨硬度
abrasion hardness test 研磨测硬法
abrasion index 磨损指数;磨蚀指数;磨耗指数
abrasion inspection 磨损检查

abrasion loss 磨损量;磨耗损失(量);磨耗量
abrasion loss of carbon 炭的摩擦损失
abrasion loss of gloss 磨失光泽
abrasion machine 磨损试验机;磨蚀试验机;磨耗(试验)机
abrasion mark 磨蚀刻痕;磨蚀痕迹;摩擦痕迹;擦伤痕
abrasion mechanism 磨损机理
abrasion number 磨蚀数
abrasion of aggregate 集料磨损(性)
abrasion of blown sand 风沙磨蚀
abrasion of cutter 铰刀磨损
abrasion of draghead 耙头磨损
abrasion of dredge pump 泥泵磨耗
abrasion particle 磨蚀微粒
abrasion pattern 磨纹
abrasion performance 耐磨耗性;耐磨性
abrasion plain 海蚀平原;冲蚀平原
abrasion plane 浪蚀崖面
abrasion platform 浪蚀平台;浪蚀平台;浪成台地;海蚀台地;海蚀平台
abrasion-proof 抗磨的;耐磨的
abrasion pulsator 间歇(式)耐磨试验仪
abrasion resistance 抗(磨)能力;耐磨阻力;抗磨损性;抗磨强度;耐磨耗性;耐用度;耐磨性;耐磨度;耐冲刷性;耐擦性;磨蚀抗力;磨耗阻力;耐冲蚀性;磨损阻力
abrasion resistance index 抗磨(耗)指数;耐磨指数;磨耗指数
abrasion resistance of plastic 塑料的抗磨性
abrasion resistance of wood 木材耐磨性
abrasion resistance test 抗磨损试验
abrasion-resistant 耐磨耗的;耐磨的
abrasion-resistant coating 耐磨涂层;耐磨层
abrasion-resistant filler 耐磨填料
abrasion-resistant refractory 抗磨耐火材料
abrasion-resistant steel 耐磨钢
abrasion-resistant steel alloy 耐磨合金钢
abrasion shore line 浪蚀岸线
abrasion strength 抗磨力
abrasion surface 磨耗层(路面);浪蚀崖面;海蚀面
abrasion tableland 剥蚀(基岩)台地
abrasion terrace 海蚀台地
abrasion test 耐磨试验;磨蚀度;磨损试验;磨耗试验;磨蚀测定
abrasion tester 磨损试验器;磨耗试验机;磨蚀试验机;磨耗试验机
abrasion testing machine 磨耗试验机
abrasion test of dredge pump 泥泵磨耗试验
abrasion value 抗磨值;磨损值;磨耗值;磨耗量
abrasion value of dredge pump 泥泵磨耗量
abrasion volume 磨耗量
abrasion wear 磨耗量
abrasion wear index 磨损指数
abrasion wear test machine 耐磨试验机
abrasion wheel 磨轮
abrasite 刚铝石
abrasive 磨损的;磨蚀的;磨蚀剂;磨料;擦伤的;擦除的
abrasive action 磨损作用;磨耗作用;研磨作用
abrasive agent 磨蚀剂
abrasive aggregate 耐磨集料;耐磨骨料;耐磨粒料
abrasive band 砂带;研磨带
abrasive band polishing 砂带抛光
abrasive belt 研磨带;砂布带;砂带(式);磨带
abrasive belt grinding lubricants 磨蚀油
abrasive belt grinding machine 砂带磨床
abrasive bin 研磨材料储[贮]罐;研磨材料储[贮]仓
abrasive blade 耐磨锯条(用于割断混凝土接头);磨轮
abrasive blast equipment 喷砂(磨光)设备
abrasive blasting 喷砂清理(法);磨料喷射清理;用磨料进行喷砂处理
abrasive block 研磨砧板;研磨块
abrasive brick 耐磨砖;磨块;研磨砖
abrasive capacity 剥蚀能力
abrasive cement 磨料黏结剂
abrasive cleaner 擦洗剂
abrasive cleaning 喷砂清理(法)
abrasive clearing 喷砂清洁法
abrasive cloth 擦光布;砂布
abrasive coating 磨料表层;砂卫表层
abrasive compound 研磨剂;研磨材料;复合磨料
abrasive cone 砂锥
abrasive-containing 含有磨料的
abrasive cut-off machine 砂轮切断机
abrasive cut-off wheel 切割砂轮
abrasive cutting drilling 磨蚀钻井
abrasive cutting-off 砂轮切断;研磨切断
abrasive deflashing 磨蚀修边
abrasive diamond 磨料级金刚石
abrasive disc 砂轮;磨轮;研磨砂轮;研磨盘;金刚砂磨盘;磨片;磨盘
abrasive-disc cutter 砂轮研磨切割器;砂轮锯
abrasive dresser 砂轮修整器
abrasive drilling 磨钻
abrasive dust 磨屑;磨料粉尘
abrasive erosion 磨损性剥蚀;磨蚀性侵蚀;腐蚀性侵蚀
abrasive fabric 金刚砂布
abrasive file 方头锉
abrasive filler 研磨剂
abrasive finishing 打磨修整
abrasive finishing machine 抛光机
abrasive floor 防滑地面;防滑地板
abrasive floor tile 防滑地(面)砖
abrasive force 磨削力
abrasive for cutting and grinding wheels 切削工具和砂轮用研磨料
abrasive formation 磨蚀性岩层;研磨性岩层;研磨性地层
abrasive grain 抛光粉;磨料粒度;磨粒
abrasive grain dispensing 上砂
abrasive grain size 磨料粒径
abrasive grinding wheel 砂轮;磨轮;磨油石
abrasive grit 研磨用砂粒
abrasive ground 研磨性岩层;研磨料
abrasive hardness 耐磨硬度;磨损硬度;磨蚀硬度;磨蚀硬度;研磨硬度
abrasive hardness of rock 岩石研磨硬度
abrasive hardness test 耐磨硬度试验
abrasive industry 研磨工业
abrasive ink 磨墨;研磨油墨
abrasive jet 喷射式钻头
abrasive jet drilling 磨蚀喷射钻井;冲蚀钻井
abrasive jet wear testing 喷砂磨损试验;喷砂磨蚀
abrasive-laden 含有磨料的
abrasive lapping 磨料研磨
abrasive lapping machine 磨料研磨机
abrasive machine 磨料机
abrasive machining 强力磨削;磨削加工;研磨加工
abrasive material 耐磨材料;磨料;磨平材料;研磨剂;研磨材料
abrasive material spreader 防滑料撒布机(冬季路面)
abrasive medium 研磨介质
abrasive mineral commodities 矿物磨料矿产
abrasive molding hydraulic press 磨料制品液压机
abrasive nature 磨损性质
abrasiveness 磨损性;磨耗度;磨蚀(性);磨料性;磨耗度;研磨性;研磨剂
abrasiveness factor 磨耗系数
abrasiveness of ink 研墨
abrasiveness of rock 岩石研磨性
abrasive nosing 防滑突沿
abrasive of road metal test 道砟磨耗试验
abrasive paper 砂纸
abrasive paper for wood 木砂纸
abrasive particle 磨料颗粒;磨料粒度;磨粒
abrasive paste 磨蚀膏;研磨膏
abrasive peeler 磨擦去皮机
abrasive pencil 橡皮头铅笔
abrasive planing 磨削
abrasive point 磨削针
abrasive points for chrome-cobalt alloy 铬钴合金用砂石针
abrasive powder 金刚(砂)粉;磨料(粉);研磨粉
abrasive product 磨料制品
abrasive property 研磨性
abrasive rate of road metal 道砟磨耗率
abrasive resistance 抗磨性;抗磨力;磨蚀强度;磨蚀抗力;磨耗阻力;耐磨强度;耐磨(能)力;耐磨阻力
abrasive resistant alloy 耐磨合金
abrasive-resistant material 抗磨材料
abrasive resistant steel liner 抗磨钢衬板
abrasive rubber 耐磨橡胶
abrasive sand 金刚砂;磨料砂;研磨砂
abrasive sawing machine 砂轮切割机;砂轮锯断机
abrasive segment 砂瓦
abrasives filter 磨料过滤器
abrasive slurry 磨粉浆;研磨性泥浆;研磨性矿粉
abrasive soap 磨蚀皂;擦洗皂
abrasive stair tread 防滑楼梯踏步板
abrasive steel shot 铁砂喷射处理;磨料钢珠;磨料钢球
abrasive stick 磨条;油石
abrasive substance 研磨材料
abrasive surface 磨损面;磨蚀面;磨耗层;研磨面
abrasive suspension 磨蚀液
abrasive terrazzo 防滑水磨石(地面)
abrasive tester 磨损试验机
abrasive tool 研磨工具
abrasive tumbling 滚擦
abrasive-type 研磨式的
abrasive-type blade 研磨式刀片
abrasive wear by vehicles (路面的)车辆磨耗
abrasive wear(ing) 磨损;磨蚀;磨耗
abrasive wheel 砂轮;磨(砂)轮;研磨(砂)轮
abrasive wheel cutting-off machine 砂轮切割机
abrasive wheel dresser 磨轮整型器
abrasive wheel with dust remover 除尘式砂轮机
abrasivity 磨蚀度;冲蚀度;研磨
abrasivity of ground 土壤冲蚀度
abrasor 打磨用具;打磨器械;刮除器;擦除器
abrator 喷丸清理装置;喷丸(清理)机;喷砂清理机;抛丸清理机
abrator head 喷丸喷头;喷砂头;喷嘴抛头丸
abraum 红赭石(一种颜料);硬红木着色用土红(颜料);层积石
abraum salt 层积盐
abrazite 水钙沸石
abreast 平齐
abreast milling 并列铣
abreast type stall 并列式挤奶台
abre(a)uvior 石块间细缝;(砌块中的)砂浆缝;圬工缝;石砌体灰缝;块石砌缝
abreuvage 机械黏给
abri 岩穴;防空洞;岩洞
abriachanite 镁铁青石棉
abridge 缩短;节略
Abridged Admiralty Chart Folio 英版海图图夹节本
abridged armilla 简仪
abridged drawing 略图;简图
abridged edition 缩写本;简版
abridged general view 示意图
abridged indication 简略字样
abridge division 简便除法
abridged life table 简略寿命表
abridged list 简表
abridged method of analysis 简化分析法
abridged monochromator 滤色单色仪
abridged multiplication 捷乘法
abridged nautical almanac 简本航海天文历
abridged notation 简记法
abridged spectrophotometer 滤色光度计;滤色分光光度计;简易型分光光度计
abridged trial load method 简化试载法(拱坝应力分析)
abridged version 节本
abridge trial load method 截面荷载分配法
abridgment of expenses 削减费用
abrim 满满地
abroad 海外;在国外;国外
abrogate 取消;废除
abrogate contract 废除合同
abrogate original contract 废除原合同
abrogate the agreement 取消协议
abrogation of agreement 废除合同
abroma fibre 昂天莲属纤维
Abros 阿布罗斯镍铬锰耐蚀合金
abrupt 生硬的;陡峭的
abrupt bend 急弯(段);突然弯曲
abrupt change 突然变化;突变;急剧变化
abrupt change in facies 岩相突变
abrupt change in thickness 厚度突变
abrupt change of cross-section 截面突变;断面突变
abrupt change of ship section 船体急剧变形
abrupt change of slope 坡度突然改变
abrupt change of voltage 电压突变
abrupt change point 陡变点

abrupt contraction 突然缩窄;突然缩小;突然收缩
abrupt contrast border 高反差界
abrupt curve 急转曲线;陡变曲线;急弯曲线;陡度曲线
abrupt deceleration vehicle 制动车辆
abrupt discharge 骤然排出(量);猝然排出(量)
abrupt ecotone 分离式生态过渡带
abrupt halt 突然停机
abruption 拉断;断裂
abruption test 断裂试验;爆破试验
abrupt junction 突变点;阶跃结
abruptly stressed 突加应力的
abrupt maneuver 急剧操纵
abrupt potential 突变势
abrupt relief 突变地形
abrupt slope 陡坡(度)
abrupt succession 急转演替
abrupt taper 锐锥度
abrupt transformation 突(跃)变换
abrupt translatory wave 推进陡波;移动陡波
abrupt turn 急剧转折
abrupt wall 陡壁
abrupt wave 突变波;陡浪;陡波
abrupt wind 急风
Absarokan orogeny 阿布萨罗克造山运动
absarokite 正边玄武岩;橄榄粒玄岩;橄榄安粗岩
abscess 气孔;泡孔;(金属中的)砂眼;夹渣内孔
abscisic acid 脱落酸;阿伯西斯酸
abscisin 脱落酸;脱落素
abscissa 横标距;脉转线;横坐标;横轴坐标;横线
abscissa axis 横轴【测】
abscissa-axis circular current 横轴环流
abscissa error of traverse 导线横坐标误差
abscissa of crippling point 拐点的横坐标
abscissa of principal point 像主点横坐标
abscissa of saturation point 饱和点的横坐标
abscissa of turning point 转折点横坐标
abscissa scale 横坐标尺度
abscissa value of flow-duration curve 径流期曲线的横坐标值
abscissing layer 离层
abscission 切除;截去
abscission joint 脱离节
abscission layer 脱离层
abscission of fruit 落果
abscission period 落叶期
abscission zone 脱离区
absciss phelloid 木栓离层
abscond 切断
ABS co-polymer 氰基乙烯
abscured aperture 遮拦孔径
absence 缺席;缺少
absence of business 无生意
absence of consideration 缺乏对价(指合同)
absence of convection 对流停止
absence of draft 不通风
absence of draught 不通风
absence of glare 无眩光;无闪光;无光线
absence-of-ground searching selector 无接地搜索选择器
absence of load(ing) 无荷载;无负载
absence of multicollinearity 非多重共线性
absence of restrictions 无约束
absence of stock 缺乏存货
absence of streaks 无条痕;无纹理;水泥和添加剂混合时无痕线
absence of style 风格上不协调的建筑物
absence of wear 无消耗
absence or incapacity 缺席或不能担任工作
absence rate 缺勤率
absentee 空号
absenteeism 缺勤;旷工;不遵守劳动契约
absenteeism rate 缺勤率
absentee management 由他人代理缺席业主进行经营管理
absentee owner 未在场物主;遥领地主
absentee ownership 缺席者所有权;遥领制
absente management 业主不在位管理
absent order 缺序
absent reflection 无反射
absent refractory period 绝对反应期
absent shagreen surface 无糙面
absent variability 绝对变异量
absent weight 绝对重量

absidiole 教堂中的小型半圆室
absiemens 电磁姆
absinthe green 苦艾绿(的);淡绿色(的)
absinthe yellow 灰绿黄色
absinthol 侧柏酮
absis (教堂的)小型半圆室;半圆形后室半圆壁(教堂)
absite 钍钛铀矿
absolute abundance 绝对丰度
absolute acceleration 绝对加速度
absolute acceptance 无条件承付;无条件承兑;单纯承兑
absolute accuracy 绝对准确(度);绝对精(确)度
absolute action 绝对行动
absolute activity 绝对活性;绝对活度;绝对放射性
absolute activity determination 放射性活度绝对测量
absolute activity method 绝对放射性法
absolute address 绝对地址
absolute addressing 绝对寻址;绝对选址;绝对认址【计】
absolute address in the absolute loader 绝对装入程序的绝对地址
absolute adjustment 绝对性调节
absolute advance 绝对超前
absolute advantage 绝对优势;绝对利益
absolute advantage theory 绝对利益论;绝对成本说
absolute age 绝对年龄
absolute age analysis 绝对年龄分析
absolute age dating 绝对年龄测定
absolute age determination 绝对年龄测定
absolute age of landform 地貌绝对年龄
absolute age of soil 土壤绝对年龄
absolute alcohol 无水酒精;无水乙醇;纯无水酒精;纯酒精
absolute altimeter 绝对航高测定仪;绝对高程计;绝对测高计
absolute altitude 标高;绝对海拔;绝对高度;绝对高程;绝对标高;海拔;拔海高度
absolute amount 绝对数量
absolute amplification 绝对放大系数;固有放大系数
absolute amplitude 绝对幅值
absolute and permissive staff instrument 绝对容许路签机
absolute angle of attack 绝对攻角;绝对冲角
absolute annual range of temperature 温度绝对年较差;绝对年温差
absolute anomaly 绝对异常
absolute assembler 绝对地址汇编程序
absoliute assembly 绝对汇编
absolute assignment 绝对转让
absolute asymmetric(al) synthesis 绝对不对称合成
absolute asymmetry 绝对不对称
absolute atmosphere 绝对大气压
absolute availability of import 绝对进口供应额
absolute average error 绝对平均误差
absolute azimuth 绝对方位角
absolute basin 绝对流域
absolute bearing 绝对方位
absolute binary 绝对二进制
absolute bireflectivity 绝对双反射率
absolute black body 绝对黑体
absolute blackness absolute element measurement method 绝对黑度绝对元素测量法
absolute block 绝对阻遏;绝对区截;绝对块;绝对断路;绝对闭塞
absolute block section 绝对闭塞区间
absolute block signal 绝对闭塞信号
absolute block signalling 绝对闭塞信号法
absolute block system 绝对闭塞制
absolute boiling point 绝对沸点
absolute bolometric luminosity 绝对热光度
absolute bolometric magnitude 绝对热星等
absolute bond 不附条件的债券
absolute bound 绝对(值)界限;绝对边界
absolute branch 绝对转移;绝对分支
absolute brightness 绝对亮度
absolute calibration 绝对校准;绝对标定
absolute calling 绝对呼叫
absolute capacity 绝对容量
absolute ceiling 绝对云幂高;绝对云底高度;绝对升限;绝对航高上限
absolute chronology 绝对年代(学)【地】

absolute closing error 绝对闭合差
absolute code 特种码;绝对码;绝对代码;基本码;代真码
absolute coding 绝对(地址)
absolute coefficient of expansion 绝对膨胀系数
absolute command 绝对命令
absolute complement of a set 集合的绝对补
absolute complement set 绝对余集;绝对补集
absolute compression 绝对压缩
absolute concept 绝对概念
absolute configuration 绝对构型
absolute constant 绝对常数
absolute construction 独立结构
absolute contour 绝对等高线
absolute contraband 绝对禁运品
absolute contract 不附加条件的合同
absolute control of the switch points 道岔的绝对控制
absolute convergence 绝对收敛
absolute conveyance 绝对让度
absolute coordinate data 绝对坐标数据
absolute coordinate dictionary 绝对坐标词典
absolute coordinates 绝对坐标(系)
absolute cost 绝对成本;纯粹成本
absolute cost advantage 绝对成本优势
absolute cost of production 绝对生产费(用)
absolute covariant 绝对共变式
absolute cover 绝对保险
absolute cumulative error 绝对累积误差
absolute damping 绝对阻尼
absolute data 绝对数据
absolute date 绝对年龄
absolute dating 绝对年代测定(法);绝对断代
absolute datum 绝对基面
absolute deflection 绝对变形;绝对偏移;绝对挠度;绝对垂线偏差;绝对垂度;绝对变位;绝对弯曲度
absolute deformation 绝对失真;绝对变形(量)
absolute delay 绝对延迟
absolute demand 绝对需要;绝对需求
absolute demineralization 绝对除盐(水的软化);绝对软化(水的)
absolute density 绝对密度;重度
absolute depth 绝对深度
absolute derivative 绝对导数
absolute desire 绝对需求
absolute detection limit 绝对探测极限
absolute determination 绝对测定(法)
absolute deviation 绝对偏差;绝对变差
absolute deviation of the vertical 绝对垂线偏差
absolute difference in cost 绝对成本差
absolute difference of component weight of index 标志分权绝对差
absolute digital number 绝对值数字
absolute dimension 绝对量;绝对尺寸
absolute dispersion 绝对离差;绝对差量
absolute displacement 绝对位移;绝对变位
absolute distortion 绝对畸变
absolute distribution 绝对分布
absolute draft 绝对压下
absolute drift 绝对漂移;绝对航差
absolute drought 绝对干旱;大旱
absolute dry 绝对干燥(的)
absolute dry-bulb temperature 绝对干球温度
absolute dry condition 绝干状态
absolute dry specific gravity 全干比重;绝干比重
absolute dry weight 绝干重量
absolute dullness 绝对浊音
absolute dullness area 绝对浊音区
absolute duty of water 低限水量
absolute dynamic(al) modulus 绝对动态模量
absolute efficiency 绝对效率
absolute efficiency inspection 绝对效率检验
absolute elevation 绝对高程;绝对标高;海拔(高程)
absolute elongation 绝对伸长(率)
absolute encoder 绝对编码器
absolute endorsement 无条件背书
absolute entry address 绝对入口地址
absolute environmental capacity 绝对环境容量
absolute equilibrium pressure 绝对平衡压力
absolute error 绝对误差
absolute error in magnetic survey 磁测绝对误差
absolute ether 无水醚

absolute ethyl alcohol 无水乙醇
absolute execution area 绝对执行区域
absolute expansion 绝对膨胀
absolute expansion coefficient 膨胀系数绝对值
absolute expression 绝对表达式
absolute extension 绝对延伸
absolute extract 纯净萃
absolute extreme 绝对极值;绝对极限
absolute fasting 绝对禁食
absolute figure 绝对数值
absolute filter 绝对过滤器(高性能空气过滤器)
absolute filtration rating 绝对过滤率
absolute fine aggregate percentage 绝对细集料率;绝对细骨料率
absolute fitting error of water table 水位绝对拟合误差
absolute fix 绝对定位
absolute fixed capital 绝对固定成本;纯粹固定资本
absolute fixed cost 绝对固定成本
absolute flow field 绝对流场
absolute flying height 绝对航高
absolute force 绝对力
absolute fore-and-aft tilt 绝对航行倾斜
absolute form-factor 绝对形数
absolute form quotient 绝对形率
absolute frequency 绝对频率
absolute frequency function 绝对频率函数
absolute function 求绝对值函数;绝对值函数
absolute galvanometer 绝对检流计
absolute geochronometry 地貌绝对年龄研究方法
absolute geopotential topography 位势高度图
absolute gradient 绝对梯度
absolute gravity 绝对重力
absolute gravity determination 绝对重力测定
absolute gravity measurement 绝对重力测量
absolute gravity point 绝对重力点
absolute gravity station 绝对重力网;绝对重力点
absolute gravity value 绝对重力值
absolute ground-rent 绝对地租
absolute ground stress measurement 绝对地应力测量
absolute growth 绝对生长(速度)
absolute growth curve 绝对生长曲线
absolute hardness degree 绝对硬度
absolute heating effect 绝对供暖效果;绝对(加)热效应
absolute height 绝对高度;绝对高程;海拔
absolute humidity 绝对湿度
absolute humidity of the air 空气绝对湿度
absolute idiosyncrasy 绝对特异反应性
absolute impoverishment 绝对贫困化
absolute inclinometer 绝对倾斜仪;绝对测斜仪
absolute income 绝对所得;绝对收入
absolute income hypothesis 绝对收入假设
absolute increase 绝对增长
absolute increment 绝对增(长)量
absolute index 绝对指标
absolute index of refraction 绝折射率;绝对折射率
absolute indicatrix 绝对指标
absolute inequality 非条件不等式
absolute inflation 绝对通货膨胀
absolute instruction 绝对指令
absolute instrument 绝对仪器;绝对测量仪表;绝对不稳定性;一级标准仪器
absolute intensity 绝对强度;绝对烈度
absolute intensity of pressure 绝对压强
absolute interest 绝对权益
absolute interferometric laser 绝对干涉测量激光器
absolute jump 绝对转移
absolute lag value 滞后绝对值
absolute land rent 绝对地租
absolute language 绝对语言;机器语言
absolute length 绝对长度
absolute lethal concentration 绝对致死浓度
absolute lethal dose 绝对致死量;绝对致死剂量
absolute level 绝对水平;绝对级;绝对电平;绝对标高
absolute liability 绝对责任;绝对赔偿责任
absolute limit 绝对界限
absolute linear misclosure 绝对线长闭合差
absolute line of flow 绝对流域
absolute load address 绝对装入地址;绝对写入地址
absolute loader 绝对装入程序
absolute loader routine 绝对装入程序
absolute load module 绝对装入模块
absolute location 绝对位置
absolute log 绝对计程仪
absolute luminance threshold 绝对亮度阈;绝对发光率阈
absolute luminosity 绝对亮度
absolute luminosity curve 绝对亮度曲线;绝对光视效率曲线;绝对发光度曲线
absolutely connective stability 绝对联结稳定性
absolutely continuous distribution 绝对连续分布
absolutely continuous function 绝对连续函数
absolutely dried 干透的
absolutely dried condition 绝干状态
absolutely dry 干透的
absolutely dry wood 全干木材
absolutely fatal wound 绝对致命伤
absolutely normal number 绝对范数
absolutely rigid structure 绝对刚性结构
absolutely symmetric(al) function 绝对对称函数
absolutely unbiased estimation 绝对无偏估计
absolutely unbiased estimator 绝对无偏估计量
absolute machine location 绝对机器单元
absolute magnetic permeability 绝对磁导率
absolute magnetometer 绝对磁强计
absolute magnification 绝对放大率
absolute magnitude 绝对值;绝对星等;绝对量;绝对大小
absolute magnitude effect 绝对星等效应
absolute majority 绝对多数
absolute manometer 绝对压力计;绝对压力表
absolute manual block 绝对人工闭塞
absolute marine combined magnetometer 海洋组合式绝对磁力仪
absolute mass unit 绝对质量单位
absolute maximum 绝对最大值;绝对最大极限;绝对极大值
absolute maximum fatal temperature 绝对最高致死温度
absolute maximum loss 实际最大损失
absolute maximum moment 绝对最大弯矩
absolute maximum rating 绝对最高额定值;绝对最大额定值
absolute maximum speed 绝对最大速度
absolute maximum stage 绝对最高水位
absolute mean deviation 绝对平均偏差
absolute measure 绝对度量
absolute measurement (物理量的)绝对计算(值);绝对度量;绝对测量
absolute measure of dispersion 离差的绝对度量
absolute measure of variation 变差的绝对测度
absolute methanol 无水甲醇
absolute method 绝对(值)法
absolute method of measurement 绝对测量(方)法
absolute minimum 绝对最小值;绝对极小值
absolute minimum error projection 最小绝对误差投影
absolute minimum fatal temperature 绝对最低致死温度
absolute minimum stage 绝对最低水位
absolute mobility (离子的)绝对淌度;绝对迁移率
absolute mobility of element 元素绝对活动性
absolute mode 绝对模式;绝对方式
absolute modulus 绝对模量
absolute moisture content 绝对湿度;绝对含水量
absolute moment 绝对矩
absolute monopoly 绝对垄断
absolute motion 绝对运动
absolute near point 绝对近点
absoluteness 绝对(性)
absolute net loss 绝对净损(保险)
absolute number 绝对数
absolute object 绝对目标
absolute object module 绝对目标模块
absolute object program(me) 绝对目标程序
absolute offset 绝对绿时差
absolute optic(al) frequency measurement 绝对光频测量
absolute optimal function 绝对最佳函数
absolute orbit 绝对轨道
absolute order 绝对指令;绝对命令
absolute order to pay 绝对支付指示;简单支付指示
absolute orientation 绝对定向;大地定向
absolute outflow of gas 瓦斯对出量
absolute overpopulation 绝对过剩人口
absolute owner 绝对所有人
absolute ownership 绝对所有权
absolute par 绝对平价
absolute parallax 绝对视差
absolute par value exchange 绝对汇兑平价
absolute path 绝对路径
absolute pathogenic bacteria 绝对致病菌
absolute pauperization 绝对贫困化
absolute periodic movement 绝对周期运动
absolute permeability 绝对渗透率;绝对浸水率
absolute permissive block 绝对容许闭塞
absolute permissive blocking 绝对容许闭塞法
absolute permissive block scheme 绝对容许闭塞方案
absolute permissive block signalling 绝对容许闭塞信号法
absolute permissive block system 绝对容许闭塞系统
absolute perturbation 绝对摄动
absolute perturbation method 绝对摄动法
absolute phase 绝对相位
absolute photoelectric(al) magnitude 绝对光电星等
absolute photographic(al) gap 绝对漏洞
absolute photographic magnitude 绝对照相星等
absolute photometry 绝对光度学;绝对光度测量;绝对测光
absolute photovisual magnitude 绝对仿视星等
absolute pitch 绝对音调
absolute plotter control 全值绘图机控制;绝对绘图机控制器
absolute pointing device 绝对指针设备
absolute polygon command 绝对多边形命令
absolute porosity 绝对孔隙性;绝对孔隙率;绝对孔隙度
absolute position 绝对位置
absolute potential 绝对电位
absolute potential vorticity 位势涡量
absolute pounds per square inch 绝对气压为……磅/平方英寸
absolute poverty line 绝对贫困线
absolute practical system of units 绝对实用单位制
absolute precision 绝对精密度;绝对精(确)度
absolute predominance 绝对优势
absolute pressure 绝对大气压力;绝对压力;绝对气压
absolute pressure controller 绝压控制器
absolute pressure ga(u)ge 绝对压力计
absolute pressure head 绝对压头
absolute pressure indicator 绝对压力指示器
absolute pressure(intensity) 绝对压强
absolute pressure vacuum ga(u)ge 绝对压力真空计
absolute price 绝对价格
absolute priority 绝对优先权
absolute probability 绝对概率
absolute profit 绝对利润
absolute program(me) 绝对程序
absolute programming 绝对程序设计
absolute prohibition 绝对禁止令
absolute promise to pay 绝对无条件付款
absolute proper motion 绝对自行
absolute quantity 绝对数量
absolute quota 绝对配额
absolute radiometer 绝对辐射表
absolute radiometric magnitude 绝对辐射星等
absolute rating 极限参数
absolute reaction rate theroy 绝对反应速率理论
absolute reading 绝对读数
absolute reference system 绝对参考系;基准参考系
absolute refraction factor 绝对折射系数;绝对折射率
absolute refractive index 绝对折射指数;绝对折射率
absolute refractory period 绝对不应期
absolute register 精确套合
absolute reliability 绝对可靠性
absolute rent 绝对房租;绝对地租
absolute representation 绝对值表示法
absolute rest 绝对静止
absolute rest precipitation tank 静止沉淀池(废水处理);绝对静止沉淀池(废水处理);间歇式沉淀池

absolute retention time 绝对滞留时间;绝对停留时间;绝对保留时间
absolute retention volume 绝对保留体积
absolute rolling 绝对横摇
absolute rotation 绝对转动
absolute roughness 绝对粗糙度;绝对糙率;绝对糙度
absolute rule 绝对规则;公认规则
absolute safe 绝对安全性
absolute sale 无条件销售;绝对销售额
absolute sale level 绝对销售水平
absolute salinity 绝对盐度
absolute scale 绝对温度范围;绝对尺度;绝对标度;绝对标尺
absolute scarcity 绝对稀缺
absolute scattering 绝对散射能力
absolute scattering power 绝对散射率;绝对散射力;绝对散射本领
absolute seasonal fluctuation 绝对季节变动
absolute (self-calibrating) pyrheliometer 绝对直接日射表
absolute sensitivity 绝对灵敏度
absolute sequence 绝对序列
absolute settlement 绝对沉降(量)
absolute signal 绝对信号
absolute signal delay 绝对信号时延
absolute signal device 绝对停止信号机
absolute size 绝对额
absolute solvent 绝对可溶量
absolute solvent power 绝对溶解力
absolute sound sensation 绝对音感
absolute specific gravity 绝对比重;真(正)比重;绝对密度
absolute specificity 绝对专一性;绝对特效性
absolute spectral response 绝对频谱反应;绝对光谱响应
absolute spectral sensitivity 绝对光谱灵敏度
absolute-spectral sensitivity characteristic 绝对光谱灵敏度特性
absolute spectral transmission 绝对光谱透射率
absolute spectrophotometric gradient 绝对分光光度梯度
absolute speed-drop 绝对速度降
absolute speed indicator 绝对速度指示器
absolute speed limit 规定最大速限
absolute stability 绝对稳定(性)
absolute stability constant 绝对稳定常数
absolute staff 绝对路签
absolute standard 绝对基准;绝对标准
absolute standard barometer 一级标准气压表
absolute standard entropy 绝对标准熵
absolute standard of living 最低生活水准
absolute star catalogue 绝对星表
absolute stereoscopic parallax 绝对立体视差
absolute stop aspect 绝对停车显示
absolute stop light 绝对停车灯号
absolute stop signal 绝对停车信号;红灯信号
absolute strength 绝对强度;绝对力量
absolute surplus value 绝对剩余价值
absolute symbol 绝对符号
absolute symmetric(al) balance 绝对对称平衡
absolute symmetry 绝对对称
absolute system of measurement 绝对测量制
absolute system of measures 绝对度量制
absolute system of unit 绝对单位制
absolute task set 绝对任务集
absolute temperature 热力学温度;绝对温度
absolute temperature scale 绝对温标
absolute temperature system 绝对温度制
absolute temperature zero 绝对零度
absolute tensor 绝对张量
absolute term 绝对项;常数项
absolute term of an equation 方程式常数项
absolute thermal efficiency 绝对热效率
absolute thermometer 绝对温度计
absolute thermometer scale 绝对温度标
absolute thermometric scale 开氏温标;绝对温标
absolute threshold 绝对阈
absolute time 绝对时间
absolute time code 绝对时间码
absolute titer 绝对滴定值;绝对滴定度
absolute title 绝对(产)权
absolute topography 绝对地形
absolute total loss 实际全损;绝对全损;绝对全部损失
absolute toxicity 绝对毒性
absolute track address 绝对道地址
absolute traffic capacity 绝对交通(容)量
absolute truth 绝对真理
absolute uncertainty 绝对不可靠性;绝对不精确性
absolute undertaking 绝对保证
absolute unit 绝对单位
absolute unit of current 绝对电流单位;电流绝对单位
absolute unit system 绝对单位制
absolute urgency 绝对紧急情况
absolute utility 绝对效用
absolute vacuum 绝(对)真空
absolute vacuum ga(u)ge 绝对真空计;绝对真空表
absolute valency 最高价【化】
absolute value 绝对值
absolute value computer 全值计算机
absolute value device 绝对值器件
absolute value error 绝对值误差
absolute value instruction 绝对值指令
absolute value machine 全值计算机
absolute value of a complex number 复数的绝对值
absolute value of a real number 实数的绝对值
absolute value of a symbolic expression 符号表达式的绝对值
absolute value of a vector 向量的绝对值
absolute value sign 绝对值符号
absolute value transducer 绝对值传感器
absolute vapo(u)r pressure 绝对蒸汽压力
absolute variability 绝对变异性;绝对变异;绝对变(化)率
absolute vector 绝对向量;绝对矢量
absolute vector address 绝对向量地址
absolute velocity 绝对速度;对地速度
absolute veto 无限制的否决权;绝对否决权
absolute viscosity 绝对黏度
absolute viscosity coefficient 黏滞系数绝对值;绝对黏滞系数
absolute visual threshold 绝对视觉阈
absolute volt 绝对伏特
absolute voltage level 绝对电压量规
absolute volume 净体积;绝对体积;绝对容积;净容积
absolute volume method 绝对体积法
absolute vorticity 绝对涡量;绝对涡度
absolute wage cost 绝对工资成本
absolute wages 绝对工资
absolute warranty of seaworthiness 绝对适航保证
absolute water 绝对水分
absolute water absorbing capacity of rock 岩石绝对吸水量
absolute water content 绝对含水量
absolute water content of slip 泥浆绝对含水率
absolute water content of snow 雪中绝对含水量
absolute watt 绝对瓦特
absolute wavemeter 绝对波长计
absolute weight 绝对重量;绝对权数
absolute white body 绝对白体
absolute worst case 绝对最坏情况
absolute yield 绝对产量
absolute zero 绝对零度
absolute zero of temperature 温度的绝对零度
absolute zero point 绝对零点
absolute zero temperature 绝对零度
absolution 解除责任;免除
absolutization 绝对化
absolve 解除责任;免除
absonant 不合理的
Abson recovery method 阿布松回收法;阿布松分离(萃取)法
absoprtion treatment of odo(u)r 臭气吸收处理
absorb 吞并;分摊转账
absorbability 可吸收状态;可吸收性;吸收性;吸收(能)力;吸收量;被吸收性
absorbable 可吸收的
absorbable cellulose 可吸收性纤维素
absorbable composite material 吸收的复合材料
absorbable gelatin sponge 明胶海绵;吸收性明胶海绵
absorbable suture 吸收性缝线
absorbance 吸收系数;吸收率;吸收性;吸收度;吸光率;吸光度;吸附量
absorbance for solar radiation 太阳辐射热吸收系数
absorbance index 吸光指数
absorbance of the colo(u)r 色素吸收率;色素强度
absorbance ratio 吸收度比值
absorbance spectrum 吸收光谱
absorbance unit of full scale 满刻度吸收度单位
absorbancy 吸光度;吸收系数;吸收率
absorbancy index 吸光指数
absorb anisotropy 吸收异向性
absorbate 吸物;吸收物;吸附物;被吸收物
absorbate density 吸附物密度
absorbate incineration 吸收物煅烧
absorbate volume 吸附物体积
absorbed bed 吸收床;吸收层
absorbed burden 已负担间接成本;已分配间接费用;已承担间接费用
absorbed cation 吸附性阳离子
absorbed cost 已吸收成本;已分配成本
absorbed declination 已吸收跌价;已分摊跌价差额
absorbed dose 吸收(辐射)剂量
absorbed dose calorimeter 吸收剂量量热器
absorbed dose distribution 吸收剂量分布
absorbed dose index 吸收剂量指标
absorbed dose rate 吸收剂量率
absorbed energy 吸收能;吸附能;被吸收的能量
absorbed expenses 已吸收费用;已负担的费用;已分配费用
absorbed film of water 吸附水膜
absorbed flux 吸收通量
absorbed fraction 吸收份额
absorbed gas 吸附气体
absorbed gas in coal 煤层吸附气
absorbed gas in sedimentary rock 沉积岩中吸附气
absorbed ground 吸水地层
absorbed heat 被吸收的热量
absorbed infracture energy 冲击软性(强度);冲击功
absorbed ions 吸附离子
absorbed isotherm 吸附等温线
absorbed layer 吸着层;吸水层;吸收层
absorbed manufacturing expenses 已吸收制造费;已分配制造费用
absorbed moisture 吸湿;吸附水;吸潮;吸收水分
absorbed orientation 吸附定位
absorbed overhead(cost) 已分配间接费用;已分配间接成本
absorbed phase 吸附相
absorbed power 吸收功率
absorbed ratio 吸收比
absorbed striking energy 被吸收的冲击能
absorbed surface film 吸收表面膜
absorbed torque 被吸收的扭矩
absorbed water 吸着水;吸收水;吸附水
absorbed water in wood 木材中的自由水分;木材的结合水
absorbed wave energy coefficient 波能吸收系数
absorbefacient 吸收性的
absorbency 吸光度;光密度;吸收率;吸收本领;吸墨性;被吸收状态;吸收能力
absorbency test 吸收性能试验
absorbent 吸收物;吸收的;吸附体;吸附的;中和剂
absorbent activity 吸附剂活性
absorbent aggregate 吸附设备;吸水集料;吸水骨料
absorbent backing 吸声背衬材料
absorbent bed 吸收床;吸收层
absorbent blanket 吸声毯毡;吸声垫;隔音垫
absorbent board 吸声板
absorbent brick 吸声砖;隔音砖
absorbent building unit 吸声建筑单元
absorbent capacity 吸收能力;吸声能力
absorbent carbon 碳吸收剂;活性炭
absorbent cassette 吸声(护)墙板
absorbent ceiling 吸声天花板;吸声顶板
absorbent ceiling sheet 吸声天花板板条;隔音天花板板条
absorbent chamber 吸声室
absorbent charcoal 活性炭;吸收性炭
absorbent coffer 吸声镶板
absorbent concentration 吸附浓度
absorbent construction material 吸声建筑材料
absorbent construction(method) 吸声(材料)施工法

absorbent cotton 脱脂棉
absorbent cover(ing) 吸声面层
absorbent deactivator 吸附减活剂
absorbent equipment 吸附装置
absorbent facing 吸声面层
absorbent felt(ed fabric) 吸声毡
absorbent fiber board 吸湿纤维板
absorbent filter 滤湿器;吸收(性)过滤器;吸附滤器;吸附滤池
absorbent filtering medium 吸收过滤介质;吸附过滤介质
absorbent filtration 吸收性过滤
absorbent foil 吸声(金属)箔
absorbent formation 吸水地层
absorbent gauze 脱脂纱布;吸水纱布
absorbent glass 吸声玻璃
absorbent gradient 吸附梯度
absorbent ground 吸水地层
absorbent hollow tile 吸声墙空心砖
absorbent hung ceiling 吸声平顶;吸声吊顶
absorbentia 吸收剂;吸附剂
absorbent lined barrier 吸声屏障
absorbent lining 吸声衬层;吸声衬板
absorbent masonry wall 吸声砖石墙
absorbent material 吸声体;吸声剂;吸声材料;吸声材料
absorbent metal 吸收型轴承合金
absorbent metal ceiling 吸声金属天花板;吸声金属顶棚
absorbent oil 吸收油
absorbent pad 吸声(隔)层;吸声垫
absorbent paint 吸声涂料;吸声漆
absorbent panel 吸声层;吸声镶板
absorbent paper 吸水纸
absorbent plaster 吸声粉刷层
absorbent plaster aggregate 吸声粉刷集料;吸声粉刷骨料
absorbent plaster ceiling 吸声粉刷天花板
absorbent powder 粉状吸声材料;吸声粉末;粉状吸声剂
absorbent pressure 吸附压力
absorbent pump 吸附泵
absorbent refrigerator 吸收式制冷机;吸收式冷冻机
absorbent sheet 吸声薄板
absorbent shutter 吸水模板;吸声模板
absorbents of carbon monoxide 一氧化碳吸收剂
absorbent solution 吸收(剂)溶液
absorbent sprayed-on plaster 吸声喷涂粉刷
absorbent surfacing 吸声面层
absorbent suspended ceiling 吸声平顶;吸声吊顶
absorbent system 吸声系统
absorbent temperature 吸附温度
absorbent tile 吸声瓦(板);吸声面砖
absorbent type filter 吸附型滤器;吸附型滤池
absorbent unit 吸声单元
absorbent wall 吸声墙;隔音墙
absorbent wall block 吸声墙板;隔音墙板
absorbent wall brick 吸声墙砖;隔音墙砖
absorbent wallpaper 隔音墙纸;吸声墙纸
absorbent wall tile 隔音墙空心砖;吸声墙空心砖
absorbent well 吸水井
absorbent wood fiber board 吸声纤维板;吸声木丝板
absorber 减振器;吸收装置;吸收体;吸收器(泵);吸收剂
absorber circuit 吸收器电路
absorber control 吸收控制
absorber cooler 吸收(塔)冷却器;吸收器冷却器
absorber of long wave radiation 长波辐射吸收剂
absorber oil 吸收油
absorber plate 太阳能吸热片;吸收板
absorber spring 减震(器)弹簧
absorber washer 吸收洗涤器
absorb foreign capital 吸收外资
absorb foreign investment 利用外资
absorb idle funds 吸收游资
absorbing ability 吸收能力
absorbing apparatus 吸收装置;吸收仪
absorbing backing 吸声背衬材料
absorbing band 吸收光谱带
absorbing board 吸声板;隔音板
absorbing boom 吸油栅
absorbing brick 吸声砖
absorbing building unit 吸声组合设备;隔音组合设备

absorbing capacity 吸收能力;吸收量
absorbing cassette 吸声镶面板;隔音镶面板
absorbing ceiling(board) 吸声天花板;吸声天花板;隔音天花板
absorbing ceiling paint 吸声天花板涂料;吸声顶棚涂料
absorbing ceiling tile 吸声天花贴面板;隔音天花贴面板
absorbing chamber 吸声室;隔音室
absorbing chemical 吸收试剂
absorbing coefficient 吸收系数
absorbing coffer 吸声(镶)板;隔音板
absorbing colloid flo(a)tation 吸附胶体浮选
absorbing column 吸收塔;吸收柱
absorbing complex 吸收性复合体
absorbing construction 吸声构造
absorbing construction material 吸声建筑材料;吸声构造材料;隔音结构材料
absorbing cover(ing) 吸声(镶)面层;吸声涂面层;吸声贴面层;隔音镶面层;隔音涂面层;隔音贴面层
absorbing duct 吸声管道
absorbing dust 吸收灰尘
absorbing dust mass 致尘物质
absorbing dynamometer 吸收功仪
absorbing facing 吸声面层
absorbing felted fabric 吸声毡;隔音毡
absorbing felted fabric ceiling 具有吸声毡的天花板
absorbing fiber board 吸声纤维板;隔音纤维板
absorbing filter 吸收滤光镜
absorbing foil 吸声(金属)箔;隔音箔
absorbing glass 吸声玻璃;隔音玻璃
absorbing hole 吸声孔
absorbing hollow tile 隔音空心砖
absorbing hung ceiling 吸声平顶;吸声吊顶;隔音吊顶
absorbing lining 吸声内衬;吸声涂面;吸声贴面;隔音涂面;隔音贴面
absorbing Markov chain 吸收马尔可夫链
absorbing masonry wall 吸声砖石墙;吸声砖砌墙;隔音砖砌墙
absorbing material 吸收性材料;隔声材料;吸声剂;吸声材料;吸声材料;隔音材料
absorbing medium 吸收(性)介质;吸声媒质;吸收媒介
absorbing membrane 吸收薄膜
absorbing metal ceiling 吸声金属天花板;吸声金属顶棚;金属吸声天花板;金属隔音天花板
absorbing pad 吸声垫;隔音垫
absorbing paint 吸声油漆;吸声油漆
absorbing panel 吸声镶板;隔音镶板
absorbing pipet(te) 吸收球管(气体的)
absorbing plaster 吸声灰泥;吸声灰膏;吸声粉刷(层);隔音灰膏
absorbing plaster aggregate 吸声粉刷灰浆骨料;隔音粉刷灰浆集料
absorbing plaster ceiling 具有吸声粉刷层的天花板
absorbing powder 吸声性粉末;吸附粉末
absorbing power 吸收本领
absorbing rod 控制棒;承受棒
absorbing selector 吸收选择器
absorbing septum 吸收隔板
absorbing set 吸收集
absorbing sheet 吸声板条;吸音板条
absorbing silencer 吸收消音器;吸声消音器
absorbing spectacle glass 吸光眼镜玻璃
absorbing sprayed on plaster 吸声喷涂灰泥;吸声喷涂粉刷;采用喷射施工的吸声粉刷层;采用喷射施工的隔音粉刷层
absorbing state 吸收(状)态
absorbing surface 吸收表面
absorbing surfacing 吸声面层
absorbing suspended ceiling 吸声平顶;吸声吊顶;隔音吊顶
absorbing system 吸声装置;吸声系统;隔音装置;隔音系统
absorbing tile 吸声面砖;吸声空心砖
absorbing tile ceiling 吸声空心砖天花板;吸声空心砖顶棚;隔音空心砖天花板;隔音空心砖顶棚
absorbing time 吸水时间
absorbing tower 吸收塔
absorbing trap 吸收阱
absorbing type gas air filter 除气用吸附式空气过滤器

absorbing waffle 吸声镶板;吸声方形块;隔音方形块
absorbing wall block 吸声砌块
absorbing wallpaper 吸声墙纸
absorbing wedge 吸收劈;吸收光楔;吸声尘劈
absorbing well 泻水井;污水井;补给井;渗水井;吸水井;吸收井;扬水试验的抽水井
absorbing wood fibre board 吸声木质纤维板;隔音木质纤维板
absorbing zone 吸收区(域);吸收层
absorb loss 吸收亏损
absorb more labo(u)r power 容纳更多的劳动力
absorb shock 减震;消震
absorb the price difference 分担差价
absorb these charges 承担此项费用
absorb workload 承担工作量
absorpion state 吸收状态
absorptance 吸收性;吸收能力;吸收度;吸收比;吸光率;辐射吸收系数
absorpting colo(u)r 吸收色
absorpting loss in optic(al) thin-film 光学薄膜的吸收损耗
absorptiometer 吸收率计;吸收测定器;透明液体比色计;溶气计(液体);调液厚器;调稠器;吸收比色计;吸收光度计;吸收计;液体吸光计;液体溶气计
absorptiometric analysis 吸光分析
absorptiometry 吸收测量(学);吸光分析;吸光测定法
absorption 吸附(作用);承运人间的承诺;分摊
absorption account 附属账(户);费用分摊账户
absorption affinity 吸附力
absorption air conditioning 太阳能空气调节系数;吸收式空调设备
absorption air inlet 吸气口
absorption applicability 吸附适用性
absorption approach 吸收理论;吸收分析法
absorption attenuation compensation 吸收衰减补偿
absorption attenuation compensation value 吸收衰减补偿值
absorption attenuator 吸收式衰减器
absorption axis 吸收轴
absorption band 吸收光(谱)带;吸收带;吸附带
absorption band intensity 吸收带强度
absorption basin 静力池;消力池
absorption bed 吸收床;吸收层
absorption bottle 吸收瓶
absorption brake 吸收制动器
absorption by impurities 杂质吸收
absorption capacitor 吸收电容器
absorption capacity 吸收性能;吸声性能;吸收量;吸收能力
absorption capacity of turbine 水轮机吸收能力
absorption capture 吸收俘获
absorption cell 吸收槽;耗能元件;吸收匣;吸收(试验)池;吸收杯
absorption chamber 吸收室
absorption circuit 吸收电路
absorption cleaning 吸收式净化
absorption coefficient 吸收率;吸收系数
absorption coefficient of gas 气体吸收系数
absorption coefficient of medium 媒质吸收系数
absorption coefficient section 吸收系数剖面
absorption coil 吸收盘管
absorption colo(u)r filter 滤光片
absorption colo(u)ring 吸收着色
absorption column 吸收柱;吸收塔;吸附塔
absorption complex 吸附性复合体
absorption compound 吸收化合物
absorption constant 吸收常数
absorption-contrast image 吸收衬度像
absorption control 吸收控制
absorption correction 吸收修正(量)
absorption cost 全额成本;全部成本
absorption costing 全额成本计算;完全成本法;摊配成本法;吸收成本(计算)法;吸纳成本法;分担成本计算
absorption cross-section 吸收截面
absorption curve 吸收曲线
absorption cycle 吸收循环
absorption cycle air conditioning 定期吸收式空气调节
absorption damper 减震器;消音器

absorption dehumidifier 吸附式减湿装置
absorption denitrification 吸收法脱氮
absorption deodo(u)rizing 吸附脱臭
absorption distribution 吸收分布
absorption dynamometer 吸收式测功器;吸收功率计;阻尼式测力计;制动测功器
absorption edge 吸收极限;吸收端;吸收带边缘;吸收边
absorption edge energy 吸收极限能量;吸收边能量
absorption effect 吸收效应
absorption effect correction 吸收效应修正
absorption efficiency 吸收效率;吸附效率
absorption electron image 吸收电子图像
absorption enhancement effect 吸收增强效应
absorption equilibrium 吸收平衡;吸附平衡
absorption equipment 吸附设备
absorption factor 吸收因数;吸收因子;吸收系数
absorption factor of solar radiation 太阳辐射吸收率
absorption fading 吸收衰落
absorption field 泻水场地;吸收(水)层;吸收场地;渗流场地;浸润区;浸润范围;吸水场地;下水道排放场地
absorption field trench 灌溉水沟;灌溉渠(道)
absorption filter 吸收滤光片;薄膜吸收滤光片
absorption flaw detector 吸收式探伤器
absorption function 吸收功能
absorption gas 吸附气体
absorption gasoline 吸收汽油;吸收法回收的汽油
absorption ga(u)ge 吸附计
absorption heat 吸收热
absorption heat pump 吸收性热泵;吸热泵
absorption hygrometer 吸收湿度计;相对湿度度计;吸收式湿度表;吸湿式湿度计
absorption impurity 吸收沾光
absorption in cavity 空腔吸声
absorption index 吸收率;吸附指数
absorption indicator 吸附指示剂
absorption isotherm 等温吸附线
absorption law 吸收(定)律
absorption layer 吸附层
absorption lens 吸收透镜
absorption level 吸收能级
absorption limit 吸着界限;吸收极限;吸收范围
absorption limiting frequency 吸收截止频率;吸收极限频率
absorption limit of absorbed flat 吸收片的吸收限
absorption limit of penetrated flat 透过片吸收限
absorption line 渗水管线(地下排水系统)
absorption loss 吸收损失(量);吸收损耗
absorption loss water 浸润损失;吸收损失(量)
absorption machine 吸收机
absorption maximum 最大吸收
absorption meter 调液厚器;吸收(光度)计
absorption method 吸附方法
absorptionmetric determination 吸收法测定
absorption modulation 吸收调制
absorption mud 吸浆量
absorption of ash 熟料对煤灰的吸收
absorption of corporate bond 公司债吸收
absorption of dyes 色吸收
absorption of energy 能量吸收
absorption of gas and vapo(u)r 气体吸收
absorption of gypsum method 石膏吸收法
absorption of idle capacity 闲散生产力吸收
absorption of labo(u)r power 吸收社会劳动力
absorption of light 光(线)吸收;光的吸收(作用)
absorption of moisture 吸湿;吸潮
absorption of nourishment 吸收营养;养分吸收
absorption of polar molecule 极性分子吸附;极化分子吸附;有极分子吸附
absorption of power 功率吸收
absorption of radiation 照射吸收;辐射吸收
absorption of shock 减震;减振;缓冲
absorption of sound 声(音)吸收
absorption of surplus funds 剩余资金吸收
absorption of vegetation 植被吸收
absorption of vehicle 载体吸收
absorption of water 吸水作用;吸水性
absorption paper 吸收纸
absorption peak 吸收峰(值)
absorption photometry 吸收光度法;吸光测定法
absorption pipe 吸收管
absorption plane 吸附面
absorption plant 吸收车间
absorption plate 吸附板
absorption point 饱和点
absorption pond 溢洪道的前池;消力塘;消力池
absorption pool 消力池
absorption potential 吸附势;吸附(电)位
absorption power 吸声能力;吸收能力;吸收功率;吸收本领
absorption process 吸收过程
absorption property 吸收能力;吸收性能;吸收特性
absorption pump 吸收泵
absorption pyrometer 吸收式光学高温计
absorption quality 吸声性能;吸收能力;吸声能力
absorption radiation 辐射吸收
absorption radon time 吸附氡时间
absorption rate 砖块吸收率;吸收率;吸水速率;吸收速率;吸附精制吸附速率;分摊率
absorption rate of rock 岩石的吸水率
absorption ratio 吸水比率;饱和系数(指建材吸水量)
absorption ray 吸收射线
absorption reaction 吸附反应
absorption refrigeration 太阳能冷却系统;吸附式制冷
absorption refrigerator 吸收式制冷机;吸收式冷冻机;吸热式冰箱
absorption region 吸收区(域);吸收范围;吸收层
absorption rim 吸收边
absorption shoulder 吸收肩
absorption solution 吸收溶液
absorption spectroanalysis 吸收光谱分析
absorption spectrograph 吸收摄谱仪;吸收分光摄像仪
absorption spectrometer 吸收光谱分析仪
absorption spectrometry 吸收光谱术
absorption spectrophotometer 吸收(式)分光度计
absorption spectrophotometry 吸收分光光度测定(法)
absorption spectroscopy 吸收光谱检查;吸收光谱分析;吸收分光学
absorption spectrum 吸收光谱;吸收波谱
absorption stage 吸收阶段
absorption strength 吸收强度
absorption stripping voltammetry 吸收溶出伏安法
absorption surface 吸附表面
absorption system 吸收系统
absorption tank 吸收池
absorption terrace 保水阶地
absorption test 陶瓷坯的吸水试验;吸水试验;吸收试验
absorption tinting strength 吸收着色力
absorption tissue 吸收组织
absorption tower 吸收塔;吸附塔
absorption trap 吸收陷阱;吸附阱
absorption tray 吸收塔板;吸收盘
absorption trench 灌溉渠;吸水管道沟槽;吸收沟槽
absorption tube 吸收管
absorption tube method 吸收管法
absorption type frequency meter 吸收式频率计
absorption type gas air filter 除气用吸附式空气过滤器
absorption type ga(u)ge 透过式测厚计
absorption-type liquid chiller 吸收式液体冷却装置
absorption type terrace 垄形阻流梯地;蓄水梯田;吸水式梯田;保水式地埂
absorption unit 单位面积吸声能力;吸收式调温机
absorption unit of carbon dioxide 二氧化碳吸收器
absorption value 吸收值
absorption vessel 吸收容器
absorption volume of grouting hole 注浆孔吸浆量
absorption water 吸着水;吸附水
absorption wave 吸附波
absorption well 扬水式抽水井;渗水井;吸水井;抽水井
absorption zone length 吸收带长度
absorptive 吸水性的;吸收(性)的;有吸收力的
absorptive action 吸附作用
absorptive attenuator 吸收式衰减器
absorptive attraction 吸收引力
absorptive backing 吸声垫板;吸声衬板;石膏吸声板
absorptive blanket 吸声毡;吸声隔板;隔音隔板
absorptive board 隔音板;吸声(隔)板
absorptive brick 吸声砖
absorptive building unit 吸声组装构件;吸声组装部件
absorptive capacity 吸着能力;吸水能力;吸收量;吸声能力
absorptive capacity for external assistance 吸收外援的能力
absorptive capacity for rain water 雨水滞留容量;雨水滞留能力
absorptive capacity for water 吸水能力;吸水性能
absorptive capacity of soil 土壤吸收能力
absorptive cassette 吸声镶(面)板
absorptive ceiling 吸声顶棚;吸声天花板
absorptive ceiling sheet 吸声顶棚薄板
absorptive chamber 隔音室;吸声室
absorptive character 吸声性;吸收特性
absorptive coffer 吸声镶(面)板;装饰吸音板
absorptive complex 吸收性复合体
absorptive construction material 吸声建筑材料;隔声建筑材料;吸声构造材料
absorptive control glass 吸声玻璃
absorptive cover(ing) 吸声面层;吸声护面;吸声涂层
absorptive extraction 吸收抽提
absorptive facing 吸声面层;吸声垫
absorptive felt(ed fabric) 吸声毡长织物;吸声毡
absorptive fibre board 吸声纤维板
absorptive foil 吸声金属箔;吸声箔
absorptive force field 吸附力场
absorptive form 吸声性模板
absorptive form lining 吸水性模板衬里
absorptive glass 吸声玻璃
absorptive index 吸声指数;吸收率
absorptive lens 滤光透镜
absorptive liner 吸水衬里
absorptive lining 吸音板;吸水(性模板)衬里;吸声板
absorptive masonry wall 吸声砖石墙;吸声砖砌墙
absorptive material 吸收性材料;吸声材料
absorptive metal ceiling 吸声金属天花板;吸声金属顶棚
absorptive method 吸收法
absorptive modulation 吸收调制
absorptive pad 减震垫;吸声垫
absorptive paint 吸声油漆
absorptive panel 吸声镶板;吸声(墙)板
absorptive plaster 吸声灰膏
absorptive powder 吸收性粉末;吸附粉末
absorptive power 吸收能力;吸收率;吸收本领
absorptive property 吸收性能;吸声性能
absorptive quality 吸收性
absorptive rate of radiation 辐射吸收率
absorptive sheet 吸声木条
absorptive sprayed on plaster 吸声喷涂灰泥;喷涂吸音粉刷;吸声喷涂粉刷
absorptive(sur)facing 吸声面层;吸声护面
absorptive system 吸声装置
absorptive tile 吸声砖
absorptive transition 吸收跃迁
absorptive type 吸收型的
absorptive type filter 吸收型(过)滤器;吸附型滤池;吸收式(过)滤器
absorptive waffle 吸声镶板;吸声薄板
absorptive wall 吸声墙
absorptive wavemeter 吸收波长计
absorptive wood fibre board 吸声木质纤维板
absorptivity 吸引率;吸引力;吸声率;吸热率;吸性;吸收能力;吸收率;吸光系数
absorptivity coefficient 吸收系数
absorptivity-emissivity ratio 吸收发射比
ABS plastic fittings ABS 塑料管件
ABS plastic pipe ABS 塑料管
abstain from pleading 放弃申辩权
abstatampere 绝对安培
abstatvolt 绝对伏特
abstention 弃权;自制
abstergent 洗涤器;洗涤剂;去污剂;去污粉;去垢剂;洗涤的
abstract 抽象概念;抽象(的);抽取
abstract account 结账户
abstract algebra 抽象代数
abstract alphabet 抽象字母
abstract and index 文摘和索引
abstract art 抽象艺术
abstract automata theory 抽象自动机理论

abstract automation 抽象自动机
abstract average 抽象平均数
abstract book of receipts and payments 收支摘要簿;收支摘要本
abstract code 理想码;抽象(代)码
abstract data 抽象数据
abstract data structure 抽象数据结构
abstract data type 抽象数据类型
abstract designs 局部花纹图案
abstract directed graph 抽象图
abstract drawing 抽象图
abstracted river 被夺河(流)
abstracted stream 被夺河(流)
abstract equation 不名方程
Abstract Expression 抽象派风格
abstract form 抽象形式
abstract individual 抽象单体
abstract invoice 简式发票
abstraction 提取;提炼;袭夺(河流);萃取;抽象(化);抽象观点;抽象概念;抽出
abstraction borehole 抽水孔
abstractionism 抽象派
abstractionist school 抽象派
abstraction loss 降雨损失;降水损失
abstraction number 不名数
abstraction of heat 减热
abstraction of observed values 观测值摘录表
abstraction of river 导流河
abstraction of stream 导流河
abstraction of utility 效用抽象
abstraction of water 排水;水的提取
abstraction reaction 提取反应
abstraction sequence 提取序列
abstraction tool 抽象工具
abstraction volume 引水量;抽水量
abstraction well 抽水井
abstraction well-water level 抽水井水位
abstraction works 引水设施;引水工程
abstractive use of water 水的提取应用
abstract labo(u)r 抽象劳动
abstract ledger 作业费用总账
abstract log 航海日志摘要
abstract machine 抽象机
abstract mechanism 抽象原理;抽象(模型)机构
abstract model 抽象模型
abstract norm 抽象范数
abstract number 无名数;抽象数;不名数
abstract object 抽象对象
abstract of account 对账单
abstract of bids 单价汇总表(工程);投标价单一览表
abstract of financial statement 财务报表摘要
abstract of invention 发明摘要
abstract of judgement 判决摘要
abstract of observations 观测记簿
abstract of posting 过账录
abstract of title 权利证明要约书;土地所有权简介;产权说明书
abstractor 萃取器;引水户
abstract painting 抽象画
abstract program(me) 抽象程序
abstract quantity 抽象量
abstract real field 抽象实域
abstract set 抽象集;电视演播室布景
abstract signal 抽象信号
abstract simplex 抽象单(纯)形
abstract space 抽象空间
abstract statement 摘要说明
abstract structure 抽象结构
abstract symbol 抽象符号
abstract syntax 抽象语法
abstract system 抽象系统
abstract term 抽象术语
abstract thinking 抽象思维
abstriction 断裂作用
absurd 荒唐的;不合理的
absurdity 荒唐的事情;不合理
abtes squamigerus 固定蜗壳
Abt-rack 齿轨铁路;齿轨铁道
Abtsbessingen faience 阿布茨百新根陶器
Abt system railway 瑞士爱伯特式山区齿轨铁路
Abt track 爱伯特式轨道
a bubblepoint reservoir 溶气驱动油藏
abuilding 正在建造的

abukumalite 硅磷灰石;铋磷灰石
Abukumatype facies 阿武隈钇硅磷灰石型相
abuminal plate making 蛋白板制版
abuminal press plate 蛋白版
abuminal sensitized plate 蛋白感光版
a bunch 一串
a bunch of 一束
abundance 富裕;丰富;丰度;分布量
abundance broadening 丰度致宽
abundance by atom 原子丰度
abundance by volume 体积丰度
abundance by weight 重量丰度
abundance clockwise foram 有旋有孔虫丰度
abundance commodity supplies 商品供应充足
abundance estimation method 丰度估计法
abundance indicator of organic matter 有机质丰度指标
abundance of bitumen 沥青丰度
abundance of capital 资金充足;资本充足
abundance of coldwater foraminifera 冷水有孔虫丰度
abundance of counter clock wise foram 左旋有孔虫丰度
abundance of element 元素的丰度
abundance of food 饵料量
abundance of foraminifera warm water 暖水有孔虫丰度
abundance of hydrocarbon 烃类丰度
abundance of labo(u)r 劳动力过剩;劳动力充裕
abundance of lightly exploited stock 轻度开发的群体资源量
abundance of minerals 矿物丰度;矿藏丰富
abundance of offers 出价过高;报价很多
abundance of organic matter 有机质丰度
abundance of remainder organic matter 残留有机质丰度
abundance of species 物种丰富;物种丰度
abundance of water 水的丰度;丰水
abundance ratio 丰度比(率)
abundance sensitivity 丰度灵敏度
abundant 富裕的;丰富的
abundant fund 雄厚资金
abundant in fish 鱼类丰富
abundant in water 水量丰富
abundantly 富裕地
abundant number 过剩数
abundant observation 多余观测
abundant of precipitation 降水丰富
abundant of water quantity 水量丰富
abundant precipitation 雨量充沛;过量降水;过冷降水;丰量降水
abundant proof 充分证据
abundant snowfall 过冷降雪;降雪丰富
abundant sunshine 充足日照
abundant watery overflow zone 强富水溢出带
a bundle 一捆
a bundle of laths 一捆板条
abura 硬木(产于尼日利亚的深棕色有光泽的);大叶帽柱木
aburton 横向装载;横排
abuse 误用;违反操作规程;滥用;机械损伤;不遵守运行规程
abuse in the granting of loans 任意滥发贷款
abuse of credit 滥用信贷
abuse of flag 旗帜的滥用
abuse of monopoly 滥用垄断权
abuse of process 不合法诉讼
abuse of rights 滥用权利
abuse of trust 滥用信用
abut 支托;支架;对头接合;端部
a-butalene 丁烯
abut equilibrium 基本均衡
abutilon 白(青)麻
abutment 桥基;接界;(架空雪道两端的)锚墩;支座;礅柱;墩台;顶柱;坝座;坝肩;坝岸连接结构;拱座
abutment anchorage 桥台锚固
abutment anchor bar 桥台锚固栓钉
abutment arch 桥台拱;岸墩拱;支墩拱
abutment backwall 桥台背墙
abutment bay 桥台横向间节;桥台横跨;临岸跨;近岸桥跨;岸墩孔
abutment block 人造坝座
abutment body 台身

abutment cap(ping) 台帽
abutment cheek 雄榫肩部受压表面;榫孔四周表面;榫孔周围与雄榫对抵的夹面;支座承压面;推压力承受面
abutment crane 台座起重机;高座起重机;高架起重机
abutment deformation 坝座变形;桥台变形;拱座变形
abutment drainage 桥台排水设施
abutment fill 桥台填土;岸墩填土
abutment for jacks 千斤顶垫座
abutment gallery 岸墩廊道;桥台廊道
abutment hinge 座铰;岸墩接合铰
abutment joint 桥台接缝;平接缝;对接接头;端接;坝肩接缝
abutment line 闭线
abutment load 支座负载
abutment location 桥台定位
abutment masonry 墩座圬工;支墩砖台;拱座圬工
abutment of arch 拱座
abutment of flare wing walls 斜翼墙桥台
abutment of framed structure 框架式桥台
abutment opening 岸墩孔
abutment pad 岸墩基座
abutment piece 支座;基石(木);隔墙底槛;(房屋构架的)底部构件;桥台;墩台;垫底横木;岸墩(支座)
abutment pier 靠岸桥墩;制动墩;墩式桥台;对接栈桥;边墩;岸墩
abutment pressure 拱脚压力;桥台压力;支座压力;支点压力;支承压力;支撑压力;拱座压力;墩台压力;坝座压力
abutment pressure zone 支承压力带
abutment rock 支墩基石
abutment seat back wall 桥台耳墙
abutment sidewall 岸墩边墙;堰岸墩
abutment span 桥台跨度;桥台边跨;近岸(桥)跨;边跨;岸墩孔;岸墩(间)距
abutment stability 坝肩稳定(性)
abutment stone 桥台石;拱座石
abutment system 支墩系统;拱座系统;岸墩设施
abutment thrust 坝肩推力
abutment toe wall 桥台前址墙;岸墩前址墙;岸墩背水面底部加围墙
abutment tool block 刀架挡板
abutment tooth 基牙
abutment wall 桥台壁;扶壁;桥台(翼)墙;桥台边墙;翼墙;拱座墙;墩台;边墙
abutment with cantilevered retaining wall 耳墙式桥台
abutment with cantilevered wings 耳墙式桥台
abutment with flare wing wall 八字形桥台
abutment with wing wall 翼墙式桥台
abutment zone 岸墩承压区
abuttals 境界;接界;地界;相邻房地产分界线;分界线[建]
abutted surface 贴合面;邻接面;相接面
abutter 邻人;接壤道路的土地所有者;邻屋地产业主;邻接房地产所有人
abutting 凸出的;毗连的;邻接的;对抵的;端接的;末端相接的
abutting building 毗连建筑物;毗连房屋;邻接建筑物;密集建筑;沿街建筑
abutting end 对接端
abutting flange 对接翼板;对接法兰
abut ting joint (管子等的)对接(焊)接头;对接缝;对接(缝);毗连接头;贴靠接合;对抵接头;端接头
abutting land 沿路用地
abutting lane 毗连车道;相邻车道
abutting lot 邻近地段;接近地段;邻接地段;相邻地段
abutting owner 毗邻房地产业主;邻业主
abutting pier 对接栈桥
abutting property 邻接产业;相邻地产
abutting tenon 对接榫;端接榫头
AB valve test device 空制阀类试验装置
aversion 外转
abvolt 电磁伏特;绝对伏特;电磁系电势单位
abvolt per centimeter[aV/cm] 电磁伏/厘米
abwatt 绝对瓦特;电磁瓦特
abysm 非常深的海
abyssal area 深海区
abyssal circulation 深层环流

abysmal clay 深海黏土
abysmal deposit 深海沉积(物)
abysmal facies 深海相
abysmal floor 深海底
abysmal injection 深部贯入
abysmal rock 深海岩(石);深成岩
abysmal sea 深海
abysmal sea bathometer 深海测深仪
abyss 深渊;深海;非常深的海
abyssal 深水的;深海的;深成环境;深成的
abyssal algae 深水藻类
abyssal algal 深海藻类
abyssal area 深海区
abyssal assimilation and contamination 深部同化混染作用
abyssal association 深海植物群丛
abyssal basin 深海盆地
abyssal benthic 深海海底的
abyssal benthic fauna 深海底栖动物
abyssal benthic zone 深渊底栖带;深海底栖带;最深海底带;大海深渊地带
abyssal cave 海底扇
abyssal circulation 深海环流
abyssal community 深海底栖群落
abyssal cone 深海锥
abyssal current 深海流
abyssal deeps 深海深处
abyssal deposit 深海沉积(物)
abyssal deposition 深海沉积作用
abyssal deposits association 深海沉积组合
abyssal depth 深海深度;海渊深度
abyssal differentiation 深部分异作用
abyssal dune 深海沙丘
abyssal environment 深海环境
abyssal facies 深海相;深成相
abyssal fan 深海扇;海底扇
abyssal fan deposit 深海扇沉积
abyssal fault 深断层
abyssal fauna 深海动物群;深海动物区系
abyssal fishes 深海鱼类
abyssal floor 深海(海)床;深海底
abyssal gap 深海狭缝;深海通道;深海山口;深海裂隙;海脊山口
abyssal heat recharge 深热源补给
abyssal hill 深海丘陵
abyssal injection 深成贯入
abyssal intrusion 深成侵入(体)
abyssall-semiabyssal succession 深海—半深海序列
abyssal mineral resources 深海矿物资源
abyssal Mn nodule 深海锰结核
abyssal mud ripple 深水淤泥质波痕
abyssal ooze 深海淤泥;深海软泥
abyssal pelagic 深海海面的;深海水域的
abyssal pelagic ecology 远洋深海生态学
abyssal pelagic fauna 远洋深海动物群;远洋深海动物区系
abyssal pelagic fishes 深海鱼类
abyssal pelagic organism 远洋深海生物
abyssal pelagic zone 深渊带;远洋深海带
abyssal phase 深海相
abyssal plain 深海平原;海洋平原
abyssal plain deposit 深海平原沉积
abyssal population 深海种群
abyssal region 深水区;深海区
abyssal rock 深海岩(石);深成岩
abyssal sea 深海;海渊
abyssal sediment 深海沉积(物)
abyssal sediment sequence 深海沉积物层序
abyssal structure 深海构造
abyssal temperature 深海温度
abyssal theory 深海说;深成理论
abyssal thermometer 深海温度计
abyssal tholeiite 深海拉斑玄武岩
abyssal valley deposit 深海谷地沉积
abyssal zone 深海区;深海底带
Abyssinian architecture 阿比西尼亚建筑
Abyssinian driven well 阿比西尼亚井
Abyssinian gold 阿比西尼亚假金
Abyssinian pump 深井泵;管井泵
Abyssinian well 阿比西尼亚(式)井
abyssobenthic 深海海底的
abyssobenthos 底栖生物
abyssopelagic 深渊的;深海水域的;深海海面的

abyssopelagic ecology 深海生态学
abyssopelagic plankton 深海浮游生物
abyssoplite 深海软泥
ac 粗紫胶
Acabijev classification 阿加比耶夫分类
acacia 洋槐;阿拉伯洋槐;阿拉伯(橡)胶;相思树
acacia ecosystem 金合欢属植物生态系统
acacia gum 金合欢树胶;合金欢胶;阿拉伯(橡)胶
acacia harpophylla 镰叶相思树
acacia senegal 阿拉伯胶树
acacin 合金欢树胶
academia 学术界
Academia Sinica 中国科学院
academic 学术(性)的;高等专科学校的
academic achievement 学术成就
academic activity 学术活动
academic authority 学术权威
academic body 学术团体
academic city 学院城市;学术城
academic community 学界
academic conference 学术会议
academic degree 学位
academic degree system 学位制度
academic discussion 学术讨论(会)
academic exchange 学术交流
academician 会员;院士
academic institute 学术机构
academic intelligence 学校智力
academi(ci)sm 学院式
academic journal 学报
academic meeting 学术会议
academic painting 学院派绘画
Academic Press 科学出版社
academic profession ability 学历和业务能力
academic rank 学衔
academic report 学术报告
academic research 学术研究
academic society 学术团体
academic style 学院派风格
academic style of building 建筑物的学院派风格;学院式建筑风格
academic trends 学术动态
academic writing 学术论文
academy 学院;学会;研究院;研究所;专科院校
academy blue 油画蓝
academy board 油画板
academy committee 学术委员会
academy of fine arts 美术学院
academy of management 经营管理研究院;管理研究院
academy of music 音乐学院
Academy of Sciences 科学院
Academy of Sciences of China 中国科学院
ACAD environment variable 制图 CAD 环境变量
Acadia Forest 阿卡迪亚森号【船】
acadialite 红斜方沸石;红菱沸石
Acadian geosyncline 阿卡德地槽
Acadian orogeny 阿卡德造山运动【地】
Acadian series 阿卡德统
ac adjustable voltage speed control 调压转速控制
acajou 木假如树;腰果;桃花心木
Acala 阿卡拉
a calcium compound 一种钙化物
acampsia 屈挠不能
AC [Automatic Control] and frequency speed control 调频转速控制
acanthicone 绿帘石
acanthiconite 绿帘石
acanthite 螺(旋)状硫银矿;螺硫银矿
acanthostegous 复棘的
acanthus 茛苕叶形装饰;卷叶饰;叶形装饰;叶纹装饰板;叶板饰
acanthus foliage 叶形板
acanthus frieze 叶形装饰檐壁;雕画带
acanthus leaf 叶形板;叶板
acanthus-mollis 阔叶饰
acanthus scroll 卷涡叶饰;涡卷形装饰
acanthus spinose 针叶装饰(科林思柱)
acapus 柱头凹线脚;柱座
acardite 二苯脲
acaria 阿卡胶
acariasis 壁虱病
acaricide 杀螨剂
acaroid(es) gum 禾木胶;草树胶

acaroid resin 草树树脂;禾木树脂
acaroid solution 禾木树脂溶液
acaroid varnish 禾木树脂清漆
acarpous 不结实的
a case-by-case basis 一种情况接情况的基础
acaustobiolith 非燃性生物岩;非(可)燃性有机岩
acaustophytolith 非可燃性植物岩
accede 继承
acceding state 参加国
accelerant 加速剂;催速剂;催化剂;促染剂;促凝剂;促进剂
accelerant aeration method 加速曝气法
accelerant aeration tank 加速曝气池
accelerant coating 速燃层
accelerate 催速;促进
accelerated accumulation 快速累积
accelerated action photography 快动作摄影
accelerated admixture 促凝外加剂
accelerated ag(e)ing 人工(加速)时效;加速时效;加速老化;加速陈化;加快陈腐
accelerated ag(e)ing of lake 湖泊加速老化
accelerated ag(e)ing of pond 池塘加速老化
accelerated ag(e)ing test 快速老化试验;加速老化试验
accelerated agent 促凝剂;促进剂
accelerated air drying 加速气干
accelerated alluvial deposits 加速冲积物;加速冲积沙
accelerated amortization 快速摊提;加速摊销
accelerated amortization method 加速摊销法
accelerated autoclave test 加速压蒸试验
accelerated binder content determination 胶结料含量的快速测定;胶合剂含量快速测定
accelerated breakdown test 加速破损试验;加速破坏试验
accelerated cement 速凝水泥;快凝水泥;促凝水泥
accelerated charging 加速充电
accelerated circulation 加速循环;加速环流
accelerated circulation by pump 用泵加速循环;泵力加速循环
accelerated clarification 加速澄清
accelerated clarifier 机械澄清池;加速澄清池
accelerated clause 加速偿付条款
accelerated combustion 加速燃烧
accelerated concrete test 混凝土耐蚀加速试验;混凝土快速试验
accelerated conditioning 加速调理
accelerated consistency test 稠度快速试验;稠度快速检验
accelerated consolidation 快速固结;加速固结
accelerated construction 快速施工;加速施工
accelerated construction method 快速施工法
accelerated cooling 加速冷却
accelerated correction 加速度校正
accelerated corrosion 加速腐蚀
accelerated corrosion test 加速锈蚀试验;加速腐蚀试验
accelerated course 速成班
accelerated creep 加速蠕动;加速蠕变
accelerated crucible rotation technique 加速坩埚旋转法
accelerated cure 加速硫化
accelerated curing (混凝土的)快速养护;加速养护
accelerated curing of concrete 混凝土快速养护
accelerated curing of concrete test cubes 混凝土试块(立方体)快速养护
accelerated curing tank 加速养护箱
accelerated curing test 快速养护试验
accelerated debt maturity 提前还债期
accelerated delamination test 快速层离试验;加速层离试验
accelerated deposit 加速沉积物;加速沉积沙
accelerated depreciation 加速贬值;快速折旧;加速折旧
accelerated depreciation allowance 加速折旧提存
accelerated depreciation method 快速折旧法
accelerated development 加速开发;加速发展;加速发育
accelerated dicvandiamide 促进剂双氢胺
accelerated drainage 加速排水
accelerated draught 加速通风
accelerated durability test 耐久性快速试验;快速疲劳试验
accelerated electron engine 加速电子发动机

accelerated erosion 加速侵蚀;加速冲蚀
accelerated fatigue test 加速疲劳试验
accelerated filter 加速滤池;加速过滤器
accelerated filtration 加速过滤
accelerated flow 加速(水)流
accelerated frame 加速参考架
accelerated goods traffic 快速货物运输
accelerated graphics port 图形加速端口
accelerated growth 加速增长
accelerated growth phase 加速生长期
accelerated gum 速成胶质
accelerated hardening 加速硬化
accelerated hydration 加速水化
accelerated iterative method 加速度迭代法
accelerated laboratory test 室内快速试验;实验室内快速试验
accelerated leaching 加速浸出
accelerated leaching rate 加速渗毒率
accelerated Liebmann method 加速列布曼(方)法
accelerated life 强化试验寿命;加速老化寿命
accelerated life test(ing) 加速寿命试验
accelerated load test 加速荷载试验
accelerated method 快速法;加速法
accelerated methods of depreciation 加速折旧法
accelerated motion 加速运动
accelerated motion of a fluid 加速流
accelerated movement 加速运动
accelerated multifamily processing 简化多户住房贷款的手续
accelerated nozzle 加速型喷管;加速喷嘴
accelerated outdoor exposure test 加速室外暴露试验
accelerated oxidation 加速氧化
accelerated oxidation test 加速氧化试验
accelerated paint weathering machine 加速油漆耐老化试验机
accelerated parameter 加速度参数
accelerated particle 加速粒子
accelerated periodd 加速期
accelerated polishing test 加速磨光试验
accelerated procedure 加速程序
accelerated random search 加速随机搜索
accelerated rate of depreciation 快速折旧率
accelerated rating process 加速腐化法
accelerated reaction 加速反应
accelerated return stroke 加速回程
accelerated roller conveyer 加速辊道
accelerated run 赶点
accelerated sediment 加速沉积物
accelerated seismograph 加速度地震仪;加速度地震计
accelerated service 加速运转
accelerated solvent extraction 加速溶剂萃取
accelerated soundness test 快速安定性试验;加速安定性试验;安定性快速试验
accelerated speed 加速速率;加速度
accelerated stock 加速胶料
accelerated storage 加速储(贮)存
accelerated stream bed erosion 河床加速冲刷
accelerated strength 快速试验强度
accelerated strength test 快速强度试验
accelerated sulfur system 促进剂硫黄体系
accelerated surface aeration 加速表面曝气;表面加速曝气
accelerated test 加速试验(法)
accelerated test for colo(u)r stability 颜色稳定性加速试验
accelerated trafficking test of road pavement 路面快速行车试验
accelerated valley deposition 河床加速沉积
accelerated velocity 加快的速度;加速度
accelerated water tower 加速给水塔
accelerated wear test 加速磨损试验
accelerated weathering 人工老化;露天试验(建筑材料和油漆等);风雨侵蚀试验;加速风化;加快风化
accelerated weathering machine 加速大气老化试验机
accelerated weathering test 加速风蚀试验;快速风化试验;加速耐候试验;加速气候试验;加速风化试验;加速大气老化试验
accelerate stop distance available 加速停止距离(飞机起飞时)
accelerate the combustion 加速燃烧

accelerate the failure 加速破坏
accelerate the maturity 提前到期日
accelerate the reform and opening up an economic development 加快改革开放和经济发展
accelerating ability 加速(运动)能力;加速性能
accelerating additive 添加剂(混凝土);促凝剂
accelerating admixture 速凝剂;催化剂;快速剂;加速(附加)剂
accelerating aeration method 加速曝气法
accelerating ag(e)ing 加速老化
accelerating agent 速凝剂;加速剂;催化剂;促染剂;促凝剂;促进剂
accelerating amplitude 加速度幅值;加速度幅度
accelerating and water reducing agent 促凝型减水剂
accelerating anode 加速阳极
accelerating chain 加速节
accelerating chamber 加速箱
accelerating charge 加速进气
accelerating coil 加速线圈
accelerating contactor 加速接触器
accelerating convergence 加速收敛
accelerating conveyer 加速输送机;加速输送带
accelerating creep test 加速蠕变试验
accelerating curve 加速度曲线
accelerating death phase 加速死亡期
accelerating device 加速装置
accelerating device for humps 丘陵路段的加速设备;驼峰路段的加速设备
accelerating effort 加速作用力
accelerating electrode 加速电极
accelerating factor 加速因子;加速因数;加速系数
accelerating field 加速(电)场
accelerating flow 加速流
accelerating fluid 加速流体
accelerating force 加速力
accelerating force curve 加速力曲线
accelerating grade 加速(缓)坡【铁】
accelerating grade of hump 驼峰加速坡
accelerating installation 加速装置
accelerating interval 加速周期
accelerating jet 加速喷嘴
accelerating lens 加速透镜;加速电子透镜
accelerating lever 加速操纵杆
accelerating load 加速载荷;加速荷载
accelerating machine 加速器
accelerating maturity 催熟
accelerating nozzle 加速型喷管
accelerating period 加速时期
accelerating potential 加速位差;加速势差;加速度势;加速电势
accelerating pressure gradient 加速时的压力梯度
accelerating pump 热加速泵(采暖系统);加速泵
accelerating pump lever 加速泵杠杆
accelerating reducing valve 快速减压阀
accelerating region 加速区
accelerating relay 加速转播;加速替续器;加速继电器
accelerating resistance 加速阻力
accelerating rocket engine 加速火箭发动机
accelerating slits 加速狭缝
accelerating slope 加速坡【铁】
accelerating space 加速空间
accelerating stage 加速级
accelerating system 加速系统
accelerating test 快速试验;加速试验
accelerating travel 加速行程
accelerating tube 加速电子管
accelerating unit 加速器
accelerating voltage 加速电压
accelerating weight 加速负载
accelerating well 加速油壶;补偿油井
accelerating yellowness 促进黄色度
acceleration 加速响应;加速时期;加速(度);加速到期
acceleration agent 促进剂
acceleration amplitude 加速振幅
acceleration analysis 加速度分析
acceleration and deceleration operation 加速和减速操作
acceleration area 加速区段
acceleration attenuation 加速度衰减
acceleration by powering 赋力加速
acceleration by reflection 反射加速

acceleration cam 加速凸轮
acceleration capacity 加速能力
acceleration centripetal 向心加速度
acceleration characteristic 加速运动特性
acceleration clause 提前支付条款;提前偿付条款;加速还款条款
acceleration consolidation with sand drains 用排水砂井加速固结
acceleration constant 加速常数
acceleration constraint 加速限制
acceleration control 加速度控制
acceleration-controlled system 加速度控制系统;加速驾驶装置;加速调节系统
acceleration controller 加速控制器
acceleration correction 加速度差的修正
acceleration creep curve 加速变曲线
acceleration creep test 徐变快速试验法
acceleration curve 启动特性曲线;加速(度)曲线
acceleration damper 加速度消震器;动力消振器
acceleration-deceleration delay 加减速延误
acceleration-deceleration fluctuation 加减速波动
acceleration depreciation 加速折旧
acceleration depreciation allowance 加速折旧扣除
acceleration diagram 加速度图
acceleration direction 加速(度)方向
acceleration disturbance 加速故障;加速干扰
acceleration due to attraction 引力加速度
acceleration due to gravity 重力加速度
acceleration effect 加速作用;加速效果
acceleration erosion of channel 航道加速冲刷
acceleration error 加速度误差
acceleration error constant 加速度误差常数
acceleration factor 加速(度)因子;加速(度)系数
acceleration feedback 加速(度)反馈
acceleration field 加速度场
acceleration/flow-time ratio 单位时间加速作用
acceleration force 加速力
acceleration governor 加速度调速器
acceleration grade 加速坡(度)【铁】
acceleration impedance 加速度阻抗
acceleration in curves 曲线加速率
acceleration in money-circulation 货币流通加快
acceleration in pitch 纵摇加速度;俯仰加速度
acceleration instrument 加速度仪表
acceleration in yaw 偏航加速度
acceleration lag 加速度延迟
acceleration lane 加速车道
acceleration limit 加速极限
acceleration limiter 加速限制器
acceleration load test 加速荷载试验
acceleration measurement 加速度测量
acceleration mechanisms 加速机制
acceleration meter 加速计
acceleration method 加速途径;加速方法
acceleration misalignment 加速角误差
acceleration mode 加速方式
acceleration motion 加速运动
acceleration motor 加速发动机;加速电动机
acceleration movement 加速运动
acceleration noise 加速(度)噪声;加速骚扰
acceleration nozzle 加速喷嘴
acceleration of base rock 岩面加速度
acceleration of earthquake 地震加速度
acceleration of falling body 落体加速度
acceleration of following 牵连加速度
acceleration of gravity 重力加速度
acceleration of hardening by heat 加热快凝;(混凝土的)加热快凝法
acceleration of inflation 通货膨胀加速
acceleration of ions 离子加速
acceleration of set(ting) (混凝土的)加速凝结;缩短硬化
acceleration of shuttle movement 货叉伸缩加速度
acceleration of sintering 快速烧结;促进烧结
acceleration of the earth 地心引力加速度
acceleration of tide (日潮先于月潮)先潮;潮时提前
acceleration of translation 平移加速度;移动加速度
acceleration parameter 加速(度)参数
acceleration peak 加速度峰值
acceleration pedal 加速踏板
acceleration performance 加速性能
acceleration period 加速周期;加速期
acceleration pick-up 加速度拾振器
acceleration pick-up sensor 测加速度传感器

acceleration potential 加速度势
acceleration profile 加速度分布图
acceleration pulse 加速(度)脉冲;加速冲量
acceleration pump 加速泵
acceleration pump well 加速泵油池
acceleration rate 加速度变化率
acceleration reaction period 加速反应期
acceleration relay 加速度继动器
acceleration resistance 加速阻力
acceleration response 过载反应;加速度反应
acceleration response factor 加速度反应系数
acceleration response spectrum 加速度反应谱
acceleration restrictor 加速度限制器
acceleration section 线路加速段
acceleration seismograph 加速度地震仪;加速度地震计
acceleration sensitive device 加速度敏感装置;加速度敏感元件;加速度敏感设备
acceleration sensitive drift rate 加速度敏感漂移率
acceleration sensitive trigger 加速度式触发器
acceleration sensor 加速传感器
acceleration setter 加速度给定器
acceleration space 加速空间
acceleration spectra 加速度谱曲线
acceleration spectrum 加速度谱
acceleration switching valve 快速开关阀
acceleration test 加速试验;加速度实验
acceleration test equipment 加速度试验设备
acceleration theory 加速(度)理论
acceleration time 启动时间;加速时间
acceleration time constant of unit 机组起动时间常数;机组加速时间常数
acceleration time data 加速度时间数据
acceleration time history 加速度时间历程
acceleration tolerance 加速容限;加速耐受度;加速度耐力
acceleration torque 加速转矩
acceleration transducer 加速度传感器
acceleration valve 加速阀
acceleration vector 加速度矢量
acceleration verse time curve 加速度时间曲线
acceleration voltage 加速电压
acceleration zone graph 加速度分区图
accelerative 加速的
acceleraton disturbance 加速失调;加速扰动
accelerator 速凝剂;速滤剂;加速器;加速剂;加速过程;加速澄清池;加速比;机械搅拌澄清池;催速剂;促进剂;风门
accelerator admix(ture) for hardening (混凝土的)快硬外加剂
accelerator anode 加速阳极
accelerator card 加速卡
accelerator cavity 加速器谐振腔
accelerator coefficient 加速系数
accelerator complex 加速器组合
accelerator conveyer 高速胶带运输机
accelerator cross shaft 加速器横轴
accelerator-decelerator 速度控制器;加速—减速器
accelerator dosage 促进剂用量
accelerator dossage ratio 促进剂用量比率
accelerator dynamic(al)test 加速器动态试验
accelerator for hardening 促硬剂
accelerator globulin 促凝血球蛋白
accelerator gun 加速器电子枪
accelerator intensity 加速器束流强度
accelerator jet 加速器喷口
accelerator lever 加速器杠杆
accelerator linkage 加速连杆机构
accelerator neutron source 加速器中子源
accelerator of charge particles 电荷质点加速器
accelerator of curing 流化加速器
accelerator of Ohkawa type 大川型加速器
accelerator pedal 加速器踏板
accelerator pedal bracket 加速器踏板支架
accelerator pedal hinge 加速器踏板铰链
accelerator pedal pad 加速器踏板垫
accelerator pedal rod 加速器踏板杆
accelerator plunger 加速器柱塞
accelerator pump 加速器泵
accelerator pump lever 加速器泵杠杆
accelerator pump outlet valve ball 加速器泵出油阀钢球
accelerator pump spray nozzle 加速器泵喷管
accelerator ratio 促进剂配比;促进剂比率

accelerator-retarder 加减速器
accelerator rod 加速器拉杆;加速杆
accelerator spectrometer 加速器分光计
accelerator spring 加速器弹簧
accelerator theory of investment 加速投资理论
accelerator treatment 加速器辐射处理
accelerator-type neutron generator 加速器型中子发生器
accelerator valve 加速器阀
accelerator well 具有加速泵的油井;加速泵油井
acceleratory collapse 加速度性虚脱
acceleratory reflex 加速反射
accelerogram 加速度图;加速度谱;自记加速(度)图
accelerogram of real earthquake 真实地震加速度图
accelerograph 地震力记录仪;加速(自)记录器;加速(度)自记仪;加速度(记录)图;自记加速(度)计
accelerometer 加速度仪;加速度计;加速度(度)表
accelerometer calibrator 加速计校准器
accelerometer chart 过载传感器图表;测震仪图表;加速计图表
accelerometer diaphragm 加速计膜盒;加速表膜盒;加速表隔膜
accelerometer drift 加速计的偏差
accelerometer falling sphere 加速计落球
accelerometer response 加速度计特性;加速计的灵敏度;加速表的灵敏度
accelerometer tube 加速度测量管
accelerometer type seismometer 加速计型地震计;加速度地震仪;加速度地震检波器;加速度地震计
accelerometric governor 加速度调速器
Accelerotor 埃克西来罗试验仪
accellular 非细胞组成的
accelofilter 加速过滤器;加速滤池
accendant 维修人员
accent colo(u)r 重点色彩
accented contour 指示等高线
accented letter 标音字母
accented term 重点项目
accent light 强光灯(标);塑形光;加强灯光
accent lighting 直接照明;突出照明;塑型照明;重点照明
accent plant 主景植物
accent stone 罕见纹理的青石
accentuate 加重;强调(指出)
accentuated contrast 加重对比度
accentuation 频率强化;加重;预增频
accentuation of attention 注意增强
accentuator 增强器;增强剂;选频放大器;促染剂;频率校正电路;加重器;加重电路;加强电路;振幅加强线路
accept 认付(支票)
acceptability 税收的可接受性;可接受性;可承兑性;合格
acceptability criterion 验收规范
acceptability for cyclic operation 对循环操作合格
acceptability limit 容许范围;认付范围;合格范围;承兑范围
acceptability of colo(u)r match 容许色差
acceptability of drinking water quality 接受饮用水水质
acceptability of risks 风险的可接受性
acceptability quality level 可接受的质量水平;满意的质量水平
acceptability test 可接受性测试;可接收性测试
acceptability when radiographed 当射线照相时的合格性
acceptability when sectioned 当切片时的合格性
acceptable 容许的;合意的;合格的
acceptable accounting principle 可接受的会计原理;公认的会计原理
acceptable ages 采用年龄
acceptable amplitude 容许振幅
acceptable cargo density 货物许可密度
acceptable cargo temperature 货物许可温度
acceptable close limit 允许趋近极限
acceptable concentration 容许浓度;容许集中度
acceptable concentration limit 容许浓度限度
acceptable container condition 认可的集装箱条件
acceptable currency 可接受的货币
acceptable daily intake 容许摄入量(日);日摄入容许量;日容许摄入量;污染物浓度的每日容许摄取量;每日许摄入量

acceptable date 可以接受的期限
acceptable defect level 容许缺陷标准;合格缺陷标准;允许缺陷标准
acceptable dose 容许剂量
acceptable drill hole 合格(钻)孔
acceptable emergency dose 容许紧急剂量
acceptable end-product 合格品
acceptable environment 验收环境
acceptable environment(al)impact 容许环镜影响
acceptable environment(al)limit 容许环镜限度;容许环镜极限
acceptable explosive 合格炸药
acceptable failure error 允许故障率
acceptable failure rate 容许故障率;允许故障率
acceptable flow 容许流速
acceptable function 可接受函数
acceptable gap 可接受空档
acceptable hazard rate 可接受损伤率
acceptable heat treatment procedure 验收的热处理工艺
acceptable hypothesis 可接受假设
acceptable indexing 可接受标记
acceptable input 容许输入
acceptable item 可接受的项目
acceptable level 容许水平;可接受水平
acceptable level of environmental protection 容许环境保护水平
acceptable level of radioactive material 容许放射性物质水平;放射性物质容许水平
acceptable level of reliability 可接受的可靠性水平
acceptable life 有效使用寿命
acceptable limit 容许限(度);容许极限
acceptable malfunction level 容许事故等级;容许故障水平;容许故障等级
acceptable malfunction rate 容许故障率
acceptable material 合格原料;合格材料
acceptable noise level 容许噪音级;容许噪声标准;可承受噪音级;允许噪声级
acceptable numbering 可接受配数
acceptable phosphorus loading 容许磷负荷
acceptable precision 容许精度
acceptable price 可接受价格
acceptable principle 可接受原理;公认原理
acceptable product 合格产品
acceptable productivity 可接受的生产率
acceptable product quality 可接受的产品质量
acceptable program(me) 可接受的方案;可接受程序
acceptable quality 合格的质量
acceptable quality criterion 合格质量标准
acceptable quality inspection 合格质量检查
acceptable quality level 容许质量级;容许质量标准;可通过质量标准;可接受质量水平;可接受品质水准;可接受的质量级别;可接受的质量标准;可接受的品质水平;合格质量水平;允许质量指标;验收质量水平;验收质量标准;质量合格标准
acceptable quality level test 可接受质量等级试验;可接受的质量标准测试
acceptable quality limit 可接受质量等级;合格质量标准
acceptable quality standard 验收质量标准
acceptable reliability level 容许可靠性水平;容许可靠性程度;可靠性合格标准;可接受的可靠性水平;合格可靠性水平
acceptable risk 可接受的危险度
acceptable risk level 可接受的危险水平
acceptable sample 合格样品
acceptable setting 容许沉降
acceptable standard of safety 安全容许标准
acceptable summary report 验收总结报告
acceptable term 可接受的条件
acceptable test 验收试验;验收测试
acceptable use policy 许可使用策略
acceptable value 容许值
acceptable velocity 容许速度
acceptable water supply 满意供水
acceptable water temperature 容许水温
acceptable wave condition 允许波况
acceptable weekly intake 周容许摄入量
accept a clam 接受索赔
accept a commission 接受佣金;接收委托
acceptancc test 法定试验
acceptance 容纳;认付(支票);试收;接受性;接受分保;接收;合格;承诺;承付;承兑

acceptance(accepting)house 承兑商业银行
acceptance agreement 承兑协议
acceptance and checkout 验收与测试
acceptance and delivery document of goods 货物交接单
acceptance and delivery document of wagon 车辆交接单
acceptance and guarantee 接受承付
acceptance and settlement of import bills 进口票据的承受与清算
acceptance and transfer 验收与移交
acceptance angle 可允许角;接收角;接近角
acceptance bank 承付行;票据承兑银行;承兑银行
acceptance bill 承兑票据
acceptance boundary 接收范围;接收边界
acceptance bribe 受贿
acceptance business 票据承兑业;承兑业务
acceptance by intervention 参加承兑
acceptance certificate 合格证(书);验收证(明)书;验收鉴定书;验收合格证(书)
acceptance certificate of contract 合同验收证书
acceptance certificate of the goods 货物验收证书
acceptance charges 承兑费
acceptance check 验收合格;验收检查;验收;工程验收
acceptance coefficient 合格系数
acceptance commission 接受佣金;承兑手续费
acceptance committee 验收委员会
acceptance condition 合格条件;验收条件
acceptance cone 接受光锥
acceptance contract 承兑合同
acceptance control chart 接受控制图
acceptance corporation 承兑票据公司
acceptance credit 票据承兑信贷;承兑信用;承兑信贷
acceptance criterion 合格标准;接受准则;验收准则;验收标准
acceptance date 承运日期
acceptance dealer 承兑商
acceptance declaration 申报承兑;承兑申报单
acceptance deviation 容许误差;验收偏差
acceptance documents 验收资料;验收文件
acceptance domain 接受域
acceptance draft 承兑票据
acceptance drawing 认可图(纸);验收图
acceptance efficiency 细粉率;产品选出效率
acceptance environment 验收环境
acceptance error 接受错误
acceptance fee 认付费;承兑费
acceptance for carriage 客车验收
acceptance for honour 信用承兑;荣誉承兑
acceptance for honour supra protest 已提出拒绝证书后承兑
acceptance functional test 验收性能试验;验收功能试验
acceptance ga(u)ge 试收量规;验收规
acceptance house 期票承兑行;票据承兑所;承兑所
acceptance in blank 空白承兑;不记名承兑
acceptance in intermediate construction stage 中间交工验收
acceptance inspection 接受检验;接收检验;验收检验;验收(检查)
acceptance insurance slip 承兑保险单
acceptance ledger 承兑票据分类账
acceptance letter of credit 远期信用证;承兑信用证
acceptance level 验收标准
acceptance limit 容许极限;合格极限;验收极限
acceptance line 承兑限额
acceptance margin 验收公差
acceptance market 票据承兑市场;贴现市场;证券市场;承兑市场
acceptance maturity record 承兑票据到期记录
acceptance number 验收数目
acceptance of a batch 批量验收
acceptance of a bill of exchange 承兑汇票
acceptance of an order at below-cost price 接受低于成本的订货
acceptance of a train 列车验收
acceptance of batch 分批接受
acceptance of bills 票据承兑
acceptance of carriage 受理
acceptance of checks 票据承兑
acceptance of completed project 竣工验收
acceptance of concealed works 隐蔽工程验收
acceptance of concrete 混凝土的验收
acceptance of constructional works 工程验收
acceptance of(construction)works 工程验收
acceptance of conveyance 承运
acceptance of delivery 交货验收
acceptance of divisional work 分部工程验收
acceptance of freight 货物承运
acceptance off-size 容许尺寸误差;公差;验收公差
acceptance of goods 承运货物
acceptance of grout 吸浆能力;吸浆量;吃浆情况;吃浆率
acceptance of hidden subsurface works 隐蔽工程验收
acceptance of luggage 行李收运
acceptance of luggages and parcels 行包的承运;行李包裹承运
acceptance of materials 材料验收
acceptance of offer 接受报价
acceptance of project 工程验收
acceptance of promissory notes 接受期票
acceptance of rail 钢轨验收
acceptance of risk 接受保险;承担风险
acceptance of service 接受送达
acceptance of subdivisional work 分项工程验收
acceptance of the bid 中标;得标
acceptance of the carcass 屋架验收;骨架验收;车架验收
acceptance of the tender 中标;得标;接受投标(书)
acceptance of wagons 货车验收
acceptance of work 工程验收;加工验收
acceptance of work subelements 分项工程验收
acceptance one-half angle 接受半角
acceptance on security 担保承兑
acceptance or rejection 合格或不合格
acceptance pattern 接受图;验收模式
acceptance payable 应付承兑票据
acceptance period 贴现期;验期;承付期;承兑期
acceptance permissible deviation 验收容许误差
acceptance permissible variation 公差;验收允许公差;验收容许偏差
acceptance potential 接收电位
acceptance probability curve 接收概率曲线
acceptance probability per batch 批接受概率
acceptance problem 接受问题
acceptance property 验收质量
acceptance qualified as to place 限定地点承兑;对付款地点有限制的承兑
acceptance qualified as to time 限定时间承兑;对付款时间有限制的承兑
acceptance quality 验收质量
acceptance quality level 允许质量指标
acceptance range 验收范围;作用范围
acceptance rate 票据贴现率;票据承兑(利)率;接收率;验收比率
acceptance rate of exchange 外汇承兑率
acceptance receivable 应收承兑票据
acceptance region 接受域;接受区(域);验收区域
acceptance register 承兑票据登记簿
acceptance report 接收报告;验收报告(单)
acceptance requirement 接收规格;验收要求
acceptance rule 验收规则
acceptances 分入再分保
acceptance sample 承认货样
acceptance sampling 取样验收;接受抽样;接收抽样;选样认可;验收抽样
acceptance sampling inspection 接受取样检验;验收取样检查
acceptance sampling plan 验收抽样计划
acceptance sampling scheme 验收取样方案
acceptance sheet 验收签认单
acceptance specification 验收规范
acceptance stamp 验收盖章;验收盖印
acceptance standard 检验验收标准;验收标准
acceptance standards for radiography 射线照相验收标准
acceptance summary report 验收总结报告
acceptance surface 接收面
acceptance survey 竣工测量;接收检验;验收
acceptance test 收料试验;接受验收;接收测试;合格试验;验证试验;验收试验;验收考核;验收检验;验收测试
acceptance testing at ambient temperature 常温验收试验
acceptance test manual 验收试验手册
acceptance test procedure 验收试验程序
acceptance tolerance 允许公差;验收允许量;验收容许偏差;验收(容许)公差
acceptance trial 接收试航;验收试车;验收试验;验收检验;验收测试
acceptance trial trip 验收试航
acceptance value 合格值;验收值
acceptance with reservation 有保留验收
accept an invitation 接受邀请
accept an offer 接受发盘;接受报盘;接受报价
accept an order 接受定货;接受订货;接受订单;接收订单;承接订单
accept a quotation 接受报价
accept bribe 受贿
accept delivery 接受交割;接收交货
accepted 已承兑;一般承认的;公认的
accepted bid 中选标;认可的(投)标书;接受投标;得标
accepted bill(draft) 已承兑(的)票据;已承兑汇票
accepted bond 已承兑债券
accepted chips 合格木片
accepted contract 默认合同
accepted depth 观测深度
accepted draft 已承兑汇票
accepted engineering practices 公认(符合工程)质量标准
accepted engineering requirements 认可的工程要求;公认(符合工程)质量标准
accepted flag 接受标志
accepted indicator 合格指示器
accepted letter of credit 已承兑信用证
accepted load 输入负荷;容许荷载;容许负荷;输入荷载;承载;承兑荷载
accepted practice 常例
accepted product 合格(产)品
accepted risk 容许危险性
accepted scale 采用的比例尺;采用比例
accepted set 接受集
accepted signal call 接受呼叫信号
accepted standard 通用标准;采用基(准)面;采用(的)标准
accepted stock 合格浆料
accepted tolerance 容许公差;验收公差;规定限差;规定差;规定公差
accepted value 允许值;常用值;采用值
accepter 认付汇票人;受款人;验收人;承人
accepter for honour 荣誉承兑人
accepter level 接受级;承受水平
accept for carriage 承运
accept forward shipment 接受远期装运
accepting bank 承兑行
accepting charges 承兑费用
accepting configuration 接受格局;接收配置;接收格局
Accepting Houses Committee 票据承兑所协会;承兑商行委员会
accepting of data 接收数据
accepting station 接收站;接收站;接收台
accepting station terminal 接收站码头
accepting unit of geologic(al) map 图幅验收单位
acception survey 竣工测量
accept membership 接收成员资格
accept offer 接受报价
accept one's order as per instruction 接受买方订单的指示
acceptor 通波器;受主;接受体;接受(电)器;接收器;谐振回路;带通电路;承付人;承兑人;被诱物
acceptor adjusted crystal 受主调整晶体
acceptor band 受主能带
acceptor bindinng energy 受主结合能
acceptor center 受主中心
acceptor circuit 接受器电路;接收器电路;接收机电路;迎收电路;带通电路
acceptor concentration 受主浓度
acceptor density 受主浓度
accept order 接受定货
acceptor dopant 受主掺质
acceptor doping 受主掺杂
acceptor impurity 受主杂质
acceptor impurity level 受主杂质能级
acceptor level 受主能级
acceptor material 受主物质;受主
acceptor reaction 受主反应

acceptor resonance 接受器谐振;电压谐振;串联谐振
acceptor site 受体部位
acceptor spectrum 受主光谱
acceptor state 受主态
accept the assignment 接受任务
accept the bid 接受投标
accept the challenge 接受挑战;接受任务
accept the conveyance of luggages and parcels 行李和包裹的承运
accept the tender 接受投标
accept your suggestion 接受你方建议
access 入门;入口(通道);入段线【铁】;取数;调整孔;市场销路;进入口;进路;捷径;接入;门路;选取;存取;出入隧道;出入口
access activity 存取活动
access address 存取地址
access adit 巷道口
access algorithm 存取算法;访问算法
access arm 取数臂;定位臂;存取臂;存储臂;磁头臂
access arrangement 通道布置;孔口布置;出入口布置
accessary 辅助部件用具
access attribute 存取属性;访问属性
access authority 存取权;访问权限
access balcony 阳台通道;入口通道;入口廊;进口楼厅
access bearer capability 接入承载能力
access bit 取数位;存取位;访问位
access board 搭板;入口跳板;入口铺板;跳板;便桥
access bridge 便桥;入口过桥;交通桥;引桥
access canal 进(出)港运河;引渠
access capability 接入能力;存取能力
access card 通达卡;赊购卡;信用卡
access category 访问类别
access chamber 清扫室;清扫间;出入舱
access channel 入网信道;入口通道;进出港航道;接入信道;引航道
access circuit 存取电路
access code 接续码;接续号;接入码;选取码;存取码;访问码
access code control 存取码控制
access coefficient 接近系数
access conflict 存取冲突;访问冲突
access connection 引道;路口;(可以出入干道的)车行道设施;进路连接;进口连接;公路干道出入口
access constraint 存取约束
access control 进路控制;入口控制(高速公路);进口控制;进出站控制;接入控制;选取控制;存取控制;访问控制
access control block 存取控制块;访问控制块
access control category 存取控制类别
access control code 存取控制码;访问控制码
access control field 存取控制字段;访问控制域
access control key 存取控制键
access control list 存取控制表;访问控制表
access control lock 存取控制锁
access control-logging and reporting 访问控制-录入和报告程序
access control mechanism 使用控制机构
access control procedure 存取控制过程
access control register 存取控制寄存器
access control scheme 计数定标电路;存取控制电路
access control section 存取控制部分
access control violation 访问控制违章
access control violation fault 访问控制违章故障
access control words 存取控制字
access corridor 进口走廊
access corridor block 通廊式大厦
access count 存取计数;访问计数
access counter 访问计数器
access cover 进入孔盖;进(出)口盖;检修盖
access cycle 存取周期;访问周期
access cylinder 存取圆柱体
access delay 接入延迟
access dimension (入口的)进口直径;出入口尺寸
access door 检修孔;人孔盖;投料孔;通道门;进入门;检修门;便门
access drift 交通坑道
access driveway 进入支路;进入干道的(专用)支路
access duct 进线管道;进出孔道
access environment 存取环境;访问环境
access excavation 通道挖掘

access exception 存取异常
access eye 管道检查孔;清扫口;掏堵;检查孔;管道清扫孔
access facility 通行设施;接入设施
access floor(ing) 活地板;出入过道地板;双层楼板;双层地板;架高活地板;活动地板
access flooring system 活动地板式;双层地板式;架高地板式;活动地板体系
access for fire-fighting 消防入口
access frequency 需用资料的次数
access front 入口正面
access function 存取函数;访问功能
access gallery 进水(管)道;入口廊道;进入廊道;进口走廊;交通廊道
access gap 存取间隔;存取差距;访问间距
access gradient 进站坡度
access guide 存取指南;查索指南;访问指南
access gulley 雨水口检查盖板;窨井检查盖板;排水沟检查盖板;雨水口
access gully 进入雨水沟;交通沟;检查窨井
access hatch 出入闸门;舱口;出入舱口
access hatch cover 出入舱口盖
access hole 人孔;取数孔;通道孔;进入孔;进出孔;交通孔;检修孔;检修口;检查口;检查孔;检查井;存取孔;出入孔
access hole of rudder 舵检修孔
access hook 入口铁爬梯;铁爬梯
access hook for chimneys 烟囱铁爬梯
accessibility 渗入性;可亲性;可接近性;可及性;可访问性;可达性;可存取性;易接近性
accessibility after installation 安装后的可及性
accessibility in stowage 积载性能
accessibility of marking 标记的可及性
accessible 可理解的;可接近的;可及的;交通方便的;易受影响的;易接近的;便于拆卸的
accessible address space 可存取地址空间
accessible boundary 可达边界
accessible boundary point 可过边界点;可达边界点
accessible canal 可通行管道
accessible coast 可通航(的)海岸;可靠近海岸
accessible compressor 易卸压缩机;半封闭式压缩机
accessible depth 探及深度
accessible duct 进入管道
accessible emission limit 可接受的发射极限
accessible flat roof 上人平屋顶(指屋顶之一部分)
accessible hermetic refrigerant compressor 半封闭式制冷剂压缩机
accessible location 交通可达地区
accessible permutation 可达排列
accessible pipe duct 进入管沟;进入管道
accessible pipe trench 进入管沟
accessible point 能达点;可达点
accessible radiation 可接受的辐射
accessible state 可接近(状)态;可达状态
accessible stationary points 可达稳定点
accessible trench 可通行地沟
access inconsistency 存取的非一致性
access interface 存取接口
access interrupt mark 存取中断标志
accession 接近;财产自然增益
accession number 存取号;登录号
accession rate 人事新进率;就业增长率
accession tax 财产增值税
access ladder 出入口爬梯;出入口扶梯;爬梯;通道梯;进入爬梯;便梯
access lane 小巷;便道;引入车道;便巷
access language 存取语言
access level 存取级;访问级
access level structure 访问的层次结构
access line 进站线(路);进场线路;接入线路;存取线路;访问线
access line at depot 车辆段出入线
access line for double track 双线出入线
access line for single track 单线出入线
access lock 出入闸门
access macro 存取宏(指令)
access macro instruction 存取宏指令
access management 存取管理
access manhole 通道人孔;进出入孔;检查井;出入人孔
access matrix 存取矩阵
access mechanism 存取机构

access memory 访问存储器
access ment 通路
access method 取得资料的方法;存取(方)法;访问(方)法
access method control block 存取方法控制块;存取法控制部件
access method module 存取法模块
access method routine 存取方法(例行)程序;存取方法例程;存取法程序
access method service 存取方法服务(程序)
access mode 取数方式;读出方式;存取方式;访问方式
access mode clause 存取方式子句
access module 存取模块
access monitoring system 出入口安全监视系统
access monument 入口界石
access network 接入网;存取网络
access number 存取号;访问号
access object 存取对象;访问对象
access of air 空气进出口;进气
access opening 人孔;出入口;通道孔;检修口;检修孔;检查口;检查孔
access opening for maintenance 维修用出入口
accessor 存取器
accessor control 存取器控制
accessorial 附属的
accessorial service 附加服务
access-oriented 面向存取的
accessories 附属物;附属设施;附属设备;附属品;附属结构物;辅助设备
accessories and parts 部件与零件
accessories clause 附加条款;附加款
accessories for steel rail 钢轨配件
accessories of a product 产品附件
accessories of engine 发动机附件
accessories of rail 钢轨附件
accessories of style 建筑风格构件
accessories of universal operation table 万能手术台附件
accessories per unit 每台附件
accessories section 附件部分;附加部分
accessories store 备品库
accessory 属具;次要;附属的;附体;附生的;附加的;辅助仪器;辅助仪表;辅助部件用具;辅料;补助物
accessory agent 补助剂
accessory attachment 附件;辅助装置
accessory base 附属基地
accessory block 配楼;附属房屋;附属建筑物;辅助装置;辅助设备;辅助厂房
accessory building 次要建筑;附属建筑;辅助建筑(物);辅助房屋
accessory camera attachment 摄影机辅助装置
accessory case 附件箱;备件箱
accessory channel 附属通道;辅助通道
accessory charge 附加指控
accessory circuit 辅助电路
accessory claim 附属要求;附加权利要求;附带权利要求
accessory constituent 副组分;副成分
accessory contract 附约;附加契约;附加合同
accessory design specification 辅助设计要求;附件设计要求;附件设计任务书
accessory diagnostic method 辅助性诊断法
accessory diving equipment 附属潜水设备
accessory drive 辅助传动(装置);控制驱动装置;副机传动
accessory drive casing 附件传动机匣
accessory drive gear 附件传动机构
accessory drive shaft 附件主动轴
accessory ejecta 次要抛出物;副喷出物
accessory element 痕量元素;附属元素
accessory equipment 附属设施;附属设备
accessory examination 辅助检查
accessory fibers 副纤维
accessory filter 辅助滤波器
accessory finder 附加取景器
accessory fruit 附果
accessory garage 附属的汽车间
accessory gas appliance 煤气灶及管道辅助用具
accessory gearbox 附件传动箱
accessory housing 合住住房
accessory ingredient 副成分;辅助物料;辅助剂
accessory living quarters 附属生活用房

accessory maceral 次要显微煤岩组分
accessory mineral 副矿物;附产矿物
accessory organ 附属器
accessory parking area 附属停车场
accessory parts 附件;配件
accessory pigment 辅助颜料;辅助色素
accessory plant 附属设施;附属设备
accessory plate 试板;附加片
accessory plume 副羽
accessory power supply 备用电源
accessory program(me) 附属程序【计】
accessory risk 附加(危)险;附加保险
accessory room 附属房间
accessory sample 副样
accessory shaft 辅助轴
accessory shaft hub 副轴衬套
accessory shock 副震
accessory shoe with electrical contact 触点式插座
accessory shoot 辅助枝
accessory sleeping quarters 附属卧房部分
accessory species 附生物种
accessory structure 附属结构;附属建筑物
accessory substance 副产物;附属物质;副产品
accessory system 辅助作业;辅助(设备)系统
accessory terminal 辅助终端
accessory use 附带用途
accessory vacuole 辅助浓泡
access panel 观察板;检修门;检修窗;检查口;观察检查盘;观测台
access path 存取通道;存取路径;访问路径
access path design level 访问路径设计级
access path selection 存取路径选择
access pit 检查井;清扫孔
access plate 清扫盖;检修孔盖板;检查孔盖板
access plug 检修插头;检查口堵塞
access point 入口处;进口点(道路立体交叉);接入点;存取点
access point identifier 接入点识别符
access port 通路孔;入口孔;进入孔;存取口;通路口;人孔口;调整孔
access prevention planting 禁止栽植;禁入栽植
access procedure 存取过程
access program(me) 存取程序;访问程序
access protection 存取保护
access protocol 接入协议;存取协议
access railroad 进场铁路;专用铁路;专用铁道
access railway 铁路专用线;进港铁路专线;专用铁路(线);专用铁道
access railway for airport 连接机场铁路
access ramp (立体交叉的)出入口坡道;入口匝道;入口坡道(道路交叉处);进入匝道;进口坡;接坡
access rate 存取速率;访问速率
access right 通路权;存取权;访问权;出入权
access-rights terminal 直接访问终端
access road 支路;辅助道路;入口通道;入境道路;临时道路;进路;进港航道;进港道路;进出(通)道;对外(交)通道;地区街道;出入口道路;便道
access roadway 进出通道
access roof 通行屋顶
access route 进出线路;交通线路
access routine 存取例程;存取程序;访问程序
access rule 存取规则
access scan 取数扫描
access screw 检修螺丝;清扫口丝堵;清扫口堵头
access selector 入口选择器;存取选择器
access sequence 存取顺序
access shaft 进入井;通道竖井;交通(竖)井;出入井
access shaft of the caisson 沉箱出入井;沉井人行井;沉箱出交通井;沉井出人行井;沉井出交通井
access shaft to draft tube 尾水管进入井
access side 进口端
access sile 进入井
access speed 数据存取速度;存取速度;存储速度;访问速度
access stair(case) 室内楼梯;普通楼梯;交通楼梯
access stairway 人行梯;外部楼梯通道
access stencil 存取模板
access strategy 存取策略
access street 出入道路
access switch 进入开关
access technique 存取技术
access time 取数时间;取出时间;数据选择时间;信息存取时间;存入时间;存取时间;访问时间
access time gap 存取时间间隙
access to a site 通往工地入口
access to certification system 认证体系的利用
access to court 向法院起诉
access to data 资料的查阅
access to embarkation 登艇甲板通道
access to information in short 信息不灵
access to market 进入市场;打入市场
access to modern technology 取得现代技术
access to plant 进入装置
access to railway premises 铁路站界入口
access to sea 出海权
access to sewer 沟管入口
access to site 进入现场
access to store 存储器存取
access to street 通街道出口
access to the sea 入海通道
access tower 入口塔;门塔
access to works 进入工程现场
access traffic 进入交通
access trestle 栈桥
access trunk 升降道
access trunk for free escape and free ascent 自由上浮脱险通道
access tube 检查管
access tunnel 进出支洞;横坑道;进出(口)隧道;交通隧洞;交通隧道;出入口隧道
access type 存取类型;访问类型
access value 存取值;访问值
access variable 访问变量;存取变量
access violation 访问违章;访问违例;访问破坏
access volume 存取卷(宗);访问卷
access way 进车道;入口引道;进港航道
access website 访问站点
access well 升降井;进入井;交通(竖)井
access width 存取位变;存取宽度;访问宽度
access zone 入口区;接近(区)段
accidence insurance 意外保险
accident 事故;失事;偶然事件;意外(伤害);故障;工作事故
accidental 偶然的
accidental accuracy 随机精度
accidental action 偶然作用
accidental aggregation 偶然聚积
accidental air 截留的空气(混凝土中);混凝土中空隙
accidental alarm 事故警报
accidental ambulance station 急救车站
accidental base point 偶然基点
accidental change 偶然变化
accident(al) collision 意外碰撞
accidental colo(u)ring 突变色;窑变
accidental combination for action effects 作用效应偶然组合
accidental consumption 意外消耗
accidental contact 事故性接触
accidental contamination 突发性污染;偶然污染;意外污染
accidental convergence 偶然收敛
accidental cost 偶然成本;意外费用;不可预见费(用)
accidental cover 临时掩蔽物
accidental current 偶发电流
accident(al) damage 意外损坏;意外损害;失事险;偶然破坏;意外损失
accident(al) death 事故死亡;死亡事故;意外死亡
accident(al) death rate 事故死亡率
accidental degeneracy 退化简并;随机简并度
accidental destruction 意外性破坏
accident(al) discharge 事故泄放;事故排放;突发性排放;意外泄放;走火
accidental discharge of oil 意外泄油
accidental displacement of wheel on axle 车辆在轴上的意外窜动
accidental eccentricity 偶然偏心;附加偏心距
accidental ejecta 外源喷出物;偶然喷出物;偶然抛出物;意外喷出物
accident(al) error 偶然误差;随机误差
accidental exposure 意外辐射
accidental factor 偶然因素
accidental failure 偶然破坏
accidental fluctuation 偶然变动;意外变动
accidental force 不可抗力
accident(al) hazard 意外事故
accidental host 偶栖宿主
accidental inclusion 外来包体(地);外源包体;捕房体【地】
accidental injury 负伤事故
accidental insurance 意外保险
accidental irradiation 事故性辐照
accidental lake 偶成湖泊
accidental lights 旁射光;横射光
accidental load(ing) 偶然荷载
accidental loss 意外损失
accidental maintenance 事故维修
accidental oil spill 意外漏油
accidental omission 意外遗漏
accidental operation 事故运行
accidental operation mode 事故运行方式
accidental phenomenon 偶然现象
accidental point 附加灭点
accidental pollution 突发性污染;偶然污染;意外污染
accidental pollution source 突发性污染源
accidental profit 意外的利润
accidental release 事故性排放;意外泄漏
accidental release of organisms 生物体的意外释放
accidental shutdown 事故停工
accidental species 附生物种
accidental spill 事故性溢漏
accidental spillage 偶然漏出;意外逸出
accidental state 事故状态
accidental torque 异常转矩
accidental torsion 偶然扭转
accidental variation 偶然变异;偶然变化
accidental whorl 杂形斗
accidental wound 意外伤
accidental xenolith 外源捕掳岩;外源捕掳体
accident ambulance station 急救车站;事故救护站
accident analysis 事件分析;事故分析
accident and death benefit insurance 意外及伤亡保险
accident and health plan 事故及健康保障计划
accident and indemnity 意外事故赔偿损失
accident at sea 海上事故
accident at work 操作事故
accident averting 事故预防
accident beyond control 不可抗力灾害
accident block 外源喷块
accident boat 救生艇;急救艇;应急救生艇
accident brake 紧急制动器
accident-cause code 事故发生原因及规程;工伤事故原因准则
accident cause factor 事故征候
accident circular telegram 事故速报
accident condition 事故情况
accident congestion 事故阻塞;交通事故拥塞
accident contamination 事故性污染
accident control 事故控制
accident control ratio 事故控制对比
accident cost 事故支出;事故损失费;事故费用
accident crane 救险起重机;救护起重机;吊车;救护吊车
accident damage 事故损坏;意外事故损坏
accident data-recorder 偶发记录器
accident death 交通死亡事故;意外事故死亡
accident death insurance 事故死亡保险;意外死亡险
accident death rate 事故死亡率
accident defect 事故损坏
accidentde to skidding 滑溜事故
accident disposal 事故处理
accident due to negligence 责任事故
accident due to skidding 滑溜事故(车辆)
accidented 凹凸不平的
accidented relief 崎岖地形
accidented topography 表面不平的地形;凹凸不平的地形
accident error 偶发误差
accident exposure 易出事故(地)点;事故照射;事故苗子;交通事故苗子
accident exposure index method 事故苗子指数法
accident failure 事故失效
accident fault 事故损坏;偶然缺陷;偶然缺点
accident forecast 事故预测
accident form 突变地形
accident frequency 事故频率;交通事故频率;意外事故频率

accident fund 意外事故准备金
accident hazard 危险事故;事故的危险;意外事故
accident hospital 急救医院;事故救护医院
accidential spillage 意外漏油
accident indemnity 事故赔偿
accident in factory 工伤事故
accident injury 交通负伤事故
accident in shunting operation 调车事故
accident insurance 意外保险;事故保险
accident investigation 事故调查
accident involvement 交通事故所牵连的人或物
accident involvement rate 车祸死亡率
accident involvement severity 事故严重度(死伤与非死伤事故之比);事故严重性
accident jam 事故阻塞;交通事故拥塞
accident liability 事故责任
accident lighting transfer board 事故照明切换盘
accident location 事故地点;出事地点
accident maintenance 事故检修
accident management 事故管理
accident management system 事故管理系统
accident map 事故现场情况图;事故图
accident mean square error of elevation difference 高差偶然中误差【测】
accident measures 安全措施
accident number of passenger casualties 旅客伤亡事件数
accident of falling 落物事故
accident of navigation 航行事故
accident of the ground 褶皱
accident of train derailment 列车脱轨事故
accident on board 船上事故
accident on duty 工伤事故
accident on the job 工伤事故
accident pattern 交通事故类型
accident points 事故记分
accident prevention 事故预防;事故防止;防止事故;安全技术;安全措施
accident prevention instruction 技术安全规程;防止事故说明;安全(技术)规程;安全规章;安全规则
accident prevention norms 技术安全规程
accident prevention officer 安全员;事故预防员
accident prevention program(me) 事故预防计划
accident prevention regulation 防事故规章;事故预防规程;安全规章;安全规则;安全规程
accident prevention tag 防止事故标记(警告性标记)
accident probability 事故可能性;偶然概率
accident-prone 事故敏感;事故多发的,易致事故的;事故多的;易发生事故的
accident-prone investigation 事故调查
accident proneness 事故征候;事故倾向性;事故多发性;易出事故
accident-prone section 事故多段
accident-proof 保险的;安全的
accident protection 事故预防;安全措施;事故防护
accident psychology 事故心理
accident rate 事故(发生)率;交通事故率
accident rate of drilling equipment 钻探设备事故率
accident rate of factory equipment 工厂设备事故率
accident record 事故统计;事故记录
accident reduction bonus 减少事故奖
accident reductive rate 事故降低率
accident report 事故报告
accident rescue 事故救援
accident risk 事故风险
accident signal 事故信号
accident site 事故发生地点
accident spot 事故地点
accident spot map 事故地点图;交通事故位置图
accidents resulting from poor quality of projects 工程质量事故
accident statistics 事故统计资料;事故统计学
accident survey(ing) 事故调查
accident threat 事故隐患
accident time 事故时间
accident toll 事故发生率;事故损失
accident to person(s) 人身事故
accident to workmen 人身事故;伤害工人事故
accident treatment 事故处理
accident water environmental risk 突发性水环境风险
accident water environment risk 事故性水污染风险
accident wildfire 偶然野火
acclimatation 气候适应;服水土
acclimate 服水土
acclimated activated sludge 驯化活性污泥
acclimated culture 驯化培养
acclimated microorganism 驯化微生物
acclimated strain 驯化菌种
acclimation 气候适应;水土适应;服水土;风土驯化
acclimation fever 水土病
acclimation of activated sludge 活性污泥驯化;活性污泥系的驯化
acclimation sludge 驯化污泥
acclimation to heat 热适应
acclimatization 气候驯化;气候适应;水土适应;适应期;适应能力;适量沉淀;服务土
acclimatization fro recovery 恢复性驯化
Acclimatization Institute 驯化研究所
acclimatization of activated sludge system 活性污泥系统驯化
acclimatization time 驯化时间
acclimatization to anoxia 缺氧适应
acclimatization to cold 冷适应
acclimatization with metallic ion 金属离子驯化
acclinal valley 顺斜谷
acclive 有坡度的;倾斜的;升坡的
acclivitous 倾斜的;升坡的;上斜的;向上倾斜的
acclivity 向上倾斜;上行坡;上斜坡;向上斜坡
acclivous 向上斜的;倾斜的;升坡的;上坡的;慢坡的
acclivous column 盘旋柱;柱头蜗卷式柱
acclivous valley 顺斜谷
accolade 门窗顶部双弯曲线装饰;葱形饰;门窗顶葱形线饰;拱门窗上的葱形饰;葱形饰脚
accolle(e) 缠绕在颈部的
Accoloy 阿科洛伊镍铬耐热合金;镍铬铁耐热合金
accommodate by accepting time letter of credit 通融接受远期信用证
accommodate positive tolerance 容许正偏差
accommodate the traffic 调节交通
accommodating accounts 调节账户
accommodating barge 住宿船
accommodating capital movement 调整资本移动
accommodating financial transaction 通融性金融交易
accommodating transactions 调节性交易
accommodating vessel 住宿船
accommodation 容纳能力;通融资金;通融贷款;调视;调节;宿营地;适应范围;设备;供应(设备);变通办法
accommodation acceptance 通融承兑(汇票)
accommodation acceptor 通融承兑人;贷款承兑人
accommodation allowance 膳宿津贴;住房津贴
accommodation area 暂居地段;备用地段;专用地段;住处面积
accommodation barge 居住驳船
accommodation berth 通融船席;预约码头泊位
accommodation bill 通融票据;食宿费用收据;空头票据
accommodation bill of lading 通融提单;空头提单
accommodation bills issued 已发出通融票据
accommodation bridge 特设桥梁;专用桥(梁)
accommodation bulkhead 住舱壁;船舱壁
accommodation by transverse structure 横向构造调节
accommodation check 通融支票
accommodation cheque 通融支票
accommodation coefficient 调节系数;适应系数;供应系数
accommodation craft 代雇船舶
accommodation crossing 平交道口
accommodation deck 居住甲板;舱室甲板
accommodation density 居住密度;住宿人员密度
accommodation draft 通融汇票
accommodation endorsement 通融背书
accommodation endorser 通融背书人
accommodation kite 欠据;欠单;通融票据;空头票据
accommodation ladder 爬梯;扒梯;舷梯;舷门
accommodation ladder davit 舷梯吊柱
accommodation ladder winch 舷梯绞车
accommodation lane 专用车道
accommodation line 通融(受保)业务
accommodation maker 通融票据出票人
accommodation net 有横木踏板的绳梯;(船用的)舷梯;木踏板舷梯;木踏板绳梯(大的有横木)
accommodation note 欠单;通融票据;空头票据
accommodation of eye 眼的适应
accommodation of human eye 人眼调节
accommodation outfit 舱室设备
accommodation paper 欠据;欠单;通融票据
accommodation party 通融票据关系人;票据上的通融人;通融有关方面
accommodation payment 索高价贿赂法
accommodation plan 舱室布置图
accommodation quarters 居住处所;居住舱室
accommodation rail 舷栏杆
accommodation ramp 调节坡
accommodation recording 产权保险登记
accommodation reflex 调节反射
accommodation road 专用公路;专用道路;房屋后街
accommodation ship 居住船;住宿船
accommodation space 居住处所;居住舱室
accommodation stairway 简易楼梯
accommodation taxes for the investing direction 投资方向调节税
accommodation trading 欠单交易
accommodation trailer 拖车住房;建筑工地的活动住房;居住拖车
accommodation train 普通旅客列车;慢车;管内旅客列车
accommodation vessel 居住船
accommodator 通融贷款人;调停人;调解人;调节装置;贷款人;贷款部门
accommodometer 调节仪
accomodation loan 无担保融资
accomodatory ability 远近适应力
accomodatory reflex 远近适应反射
accompanied by 附有
accompanied luggage 随身行李
accompanied vehicle 随行客车
accompanied with 带有
accompaniment 陪伴物
accompanist 伴随物
accompany 伴随;伴生
accompanying 并发的;伴发的
accompanying diagram 插图;附图
accompanying element 伴生元素
accompanying figure 插图;附图
accompanying gold deposit 伴生金矿床
accompanying information 附加情况
accompanying map 插图;附图
accompanying mineral 伴生矿物
accompanying sound 伴音
accompanying sound trap 伴音陷波器;伴音带阻滤波器
accompanying vessel 护送船
accompany movement 跟随运动
accomplish 实现;达到
accomplished 完成的
accomplished as planned 按计划完成
accomplished bill of lading 已提货的提单
accomplishment 技能;完成(量);实施;技艺;成就;成绩
accomplishment rate of working efficiency 工作效率完成率
accopanying passengers of parcel 带运
accordance 调和;一致;按照
accordance of height 高度平齐
accord and satisfaction 协议与补偿;一致和满意
accord and satisfaction clause 一致满意条款
accordant connection 匹配连接
accordant drainage 调和水系;协调水系
accordant fold 同向褶皱;协和褶皱【地】
accordant junction 平齐汇流;顺层汇流;协和汇流
accordant morphology 平齐地貌
accordant river 顺向河
accordant stream 顺向河
accordant summit(level) 平齐峰顶线
accordant to schedule 按预定计划
accordant unconformity 平行不整合
accordant value 符合值
according above stated 如上所述
according to 根据
according to actual condition 按实际情况;实际情况

according to a definite base 按切块的办法
according to drawing 按图
according to international practice 按照国际惯例
according to local condition 因地制宜
according to modular coordination 与模数协调相符的
according to plan 按计划
according to priority 按顺序
according to requirement 按要求
according to schedule 按预定计划;按计划进行
according to specification 按照说明书
according to specified quantity 按量
according to standard sample 根据标准样品
according to the chemical reaction of soil 根据土壤化学反应
according to the ISO standard 符合 ISO 标准
according to the local conventions 按当地习惯做法
according to the merits of each case 分别情况
according to the rule 按照规定
according to usual practice 按照惯例
according to value 按值
accordion 折式地图
accordion binding 折叠式装订本
accordion buoy 蛇腹状浮标
accordion cable 折垫电缆
accordion cam 提花三角
accordion coil 折状线圈;折叠式线圈
accordion contact 手风琴式触点簧片
accordion conveyer 折叠式输送机
accordion curtain 折门;折屏障
accordion door 折遮;折门;摺门;折叠门
accordion door furniture 折叠门家具
accordion fold 棱角褶皱;折叠式褶页
accordion folding door 手风琴式折叠门
accordion hood 折棚横顶罩
accordion like wrinkles 折叠状褶皱
accordion map 折叠式地图
accordion partition 折隔屏;折扇墙;折扇间断;折扇隔壁;折隔扇;折叠(式)隔断;折叠隔屏
accordion pipe 波纹管
accordion plate 折叠式图板
accordion plate roof 折叠式屋顶
accordion roller conveyer 折叠式辊道输送机
accordion shades 折叠式活动隔断;屏风
accord with the demands 符合要求
accoropode 扭王字块体
accoropode armo(u)r 扭王字块体护面
account 会计(指工作);科目;计账;计算书;户头;合计;账目;账户;报告书
accountability 可计量性;责任
accountability analysis 可计量分析
accountability cost center 责任成本中心
accountability cost system 责任成本制度
accountability information 说明性资料
accountable 可计量的;负有(会计)责任的
accountable officer 出纳员
accountable person 负责人
accountable receipt 正式收据
accountable warrants (指未经事前审核的)责任支付命令;负有责任的支付命令
account analysis 账目分析;账户分析
accountancy 会计学;会计师之职;会计(指工作)
account(ancy)law 会计法
accountancy profession 会计员
accountant 会计员;会计师;会计(指人员)
accountant and secretary 账房先生
accountant file 信用调查档案
accountant general 会计主任;会计长
accountant in bankruptcy 清理会计师;破产核算员;破产财产管理人
accountant in charge 主任会计师
accountant office 会计师室
accountant officer 会计主管;会计员
accountant room 会计师室
accountant's certificate 会计师证明书
accountant's department 会计部门
accountant's fee 会计师公费
accountant's firm 会计师事务所
accountant's office 会计事务所
accountant's report 会计师报告;查账报告
account as recorded in a ledger 分类账账目;分类账科目
account balance 账户余额;账户结余

account balanced 结平的账户
account bill 账单
account book paper 账册用纸;账簿纸
account books 会计账簿;账册;账簿
account category 账户分类
account chart 账户一览表
account classification 科目分类;账户分类
account clerk 账务员
account closing procedure 结算手续
account credit 赊账
account current 来往账(目);会计月报表
account current among senior administrative and subordinate unit 上下级往来账户
account current transport income 运输收入往来
account customer 顾客账户
account day 交割日;付欠款(结账)日
account due from customers 顾主欠账款
account due to custom 应付顾主款项
accountee 开证人;开证申请人(信用证)
account en route 未还账
account executive 业务员;业务经理;业务经管人
account file 记账文件
account flow chart 会计流程图;账户流程图
account form 账户格式
account form of balance sheet 账户式资产负债表
account form of income sheet 账户式收益表
account form of income statement 平衡式收益表;账户式收益表
account form of profit-loss statement 账户式损益计算表
account goods 承销清单
account heading 账户名称
account heading of capital construction 基本建设科目
account holder 顾客账户
account in arrears 积欠
account in dispute 有争议账目
accounting 清算账目;会计学;会计手册;会计(指工作);核算
accounting adjustment 账务调整
accounting alternatives 各种会计方法
accounting analysis 会计分析
accounting and business research 会计和商业研究
accounting and financial audit 会计及财务审计
accounting assistant 会计助理
accounting authority identification code 账务机构识别码
accounting based on shift 按班核算
accounting books 账簿
accounting by electronic data processing 会计工作电子计算机化;会计电算化
accounting by machine method 会计机器记账法
accounting by month 月份结账
accounting check 会计校验;会计检验
accounting classification 会计分类
accounting clerk 会计员;记账员
accounting code 会计条例
accounting concept 会计基础理论
accounting consultation 会计咨询
accounting control 会计监督
accounting cost 会计成本
accounting cost control 会计成本监督;成本管理
accounting data 会计数据
accounting day 结算日
accounting department 会计科;财务科
accounting device 计算装置
accounting division 会计部门
accounting documents 会计凭证
accounting equation 会计恒等式
accounting evaluation 会计估价
accounting evidence 会计凭证;记账凭证
accounting exchange rate 会计汇率
accounting exit-routine 记账退出例行程序
accounting fiduciary 财产信托
accounting file 会计文件
accounting firm 会计(师)事务所
accounting for a pooling 联营会计
accounting for assets expirations 资产消耗会计处理
accounting for branch office 分支机构会计;分店会计
accounting for branch operation 分店会计
accounting for discontinued operations 已停业务的会计处理

accounting for equities 产权会计
accounting for human resources 人力资源会计
accounting for idle time 停工时间的会计处理
accounting for inflation 通货膨胀会计;按币值调整的会计核算
accounting for internal reporting 内部报告会计
accounting for inventories 盘存会计;库存资产会计
accounting for joint venture 合资企业会计
accounting for lost unit 损失产品会计处理
accounting form 会计簿;账户簿
accounting for management control 控制管理会计
accounting for management planning 业务计划会计
accounting formula 记账公式
accounting for ownership equities 资本会计
accounting for planning 计划编制会计
accounting for public enterprise 公营企业会计
accounting general 会计主任
accounting head 会计主任
accounting house 会计室;会计科
accounting identity 会计恒等式
accounting income 账面收益
accounting information 会计情报
accounting information system 会计信息系统;记账信息系统;计算信息系统
accounting investment 国外业务会计
accounting item 会计科目;账项
accounting logging 报表记录
accounting machine 会计(计算)机;记账机;费率结算机
accounting manager 会计主任
accounting method 会计方法;记账方法;核算方法
accounting number 账号
accounting of business 营业会计
accounting of capital construction 基本建设会计核算
accounting of capital construction account 基本建设账户
accounting of railway construction enterprise 铁路施工企业会计
accounting of railway transport enterprise 铁路运输企业会计
accounting of railway transport income 铁路运输收入会计
accounting of subordinate units 附属单位会计
accounting on the accrual basis 权责发生制;应收应付制
accounting on the cash basis 收付实现制;现金收付制;现金付会计制
accounting organ 会计机构
accounting organization 会计组织
accounting period 会计年度;会计期;结算期
accounting price 账面价格;会计价格
accounting principle 会计原理
accounting procedure 记账程序;衡算计量程序
accounting profession 会计专业
accounting profession organization 会计职业团体
accounting profit 账面利润
accounting program(me) 记账程序
accounting rate 账面汇率
accounting rate of returns 会计期间资本回收率;投资报酬率
accounting recapitalization 会计上的调整资本
accounting record 财务记录
accounting regularity audit 会计常规审计
accounting report 会计报告;财务报告
accounting report of capital construction 基本建设会计报表
accounting research study 会计研究
accounting room 会计师室
accounting routine 记账程序;费用计算程序
accounting standard 会计准则
accounting statement 会计报表
accounting study 会计研究
accounting subject 会计科目
accounting supervision 会计监督
accounting system 统计报告制度;会计制度;金额换算系统;记账系统
accounting tabulating 会计制表
accounting technique 会计技术
accounting temperature 计算温度
accounting terminology 会计用语;会计术语
accounting transaction 会计经营业务;会计事项

accounting treatment 会计处理
accounting trend 会计趋势
accounting trends and techniques 会计发展与技术
accounting unit 会计单位;衡算(量)单位;核算单位
accounting voucher 会计专票;会计凭证;财物单据
accounting work order 计算工作指令
accounting year 会计年度
account in transit 未达账项;在途账
account in trust 信托账户
account item 账户项目
account level 核算标准
account name 账户名称
account note 账目单;账单
account not in order 账目不清
account number 账号;记账号码;账户编号
account of 记账
account of application of fund 资金运用账户
account of assets 资产台账
account of bankruptcy 破产账
account of business 营业科目
account of cash in vault 库存现金簿
account of credit sale 赊卖账户
account of disbursement 支出账
account officer 会计主管
account of finance 财产账(户)
account of goods purchased 购货账目
account of goods sales 销货账目
account of goods sold 销货账
account of payments 支出账户;支出表;付款账户
account of purchase 采购账户
account of receipts 收入账户;收入表;收入账目
account of receipts and payments 收支账户
account of sales 销货账目
account of the exchequer 国库账目
account of treasury 金库账;国库账目
account paid 已付账款;已付款;账项付清;付讫
account party 开证申请人(信用证);进口贸易商
account payable 赊购费;应付账目
account payable interline 同行业单位之间应付账款;同行间应付账款
account payable subsidiary ledger 应付款明细分类账
account payee 入账受款人;收款入账
account purchase 赊账
account receivable 应收账款
account receivable control 应收账款控制
account receivable financing 应收账款筹资
account receivable from installment 分期付款应收账款
account receivable insurance 应收账款保险
account receivable internal control system audit 应收账款内部控制制度审计
account receivable ledger 应收账款分类账
account-receivable program(me) 收款程序
account receivable register 应收账款登记簿
account rendered (指贷方交给借方审查和清算的)借贷细账;结账清单;结欠清单;付款账单
account sale 赊销;清单(材料账);销售账;销货清单
account sales book 承销货物账簿
account sales ledger 承销货物总账
accounts are not in order 账目不清
accounts due 应收到期账款
account section 账务科
account settled 清算;清迄;决算;账项结清;付讫
accounts for goods and material services 货物和物质性服务账户
accounts of railroad transportation 铁路运输会计账户
accounts of railway transportation 铁路运输会计账户
accounts open to public 经济公开
accounts payable 应付账款
accounts payable account 应付账款账户
accounts payable audit 应付账款审计
accounts payable authenticity and legality audit 应付账款真实性和合法性审计
accounts payable ledger 应付账款分类账
accounts payable others 其他应付账项
accounts payable register 应付账款登记簿
accounts payable stretching 应付账款增多
accounts receivable 列清单;应收账款
accounts receivable assigned 已转让应收账款
accounts receivable financing 应收筹资账款
accounts receivable from installment sales 分期销货的应收账款
accounts receivable others 其他应收款项
accounts system 会计制度
account stated 认可账额;借贷双方认可的细账;明细账单;已定账目;债务人认为无误的确定清单
account structure 账户结构
account title 会计科目;账户名称
account titles for annual expenditures 岁出科目
account to be credited(debited) 应列贷(借)方科目
account to give 支出账目
account transaction 估价;选价预算【建】;账目事务;造价预算
account transfer memo 转账账目;转账通知单
account valuation 估价;预算;造价预算;估价账户
account working 业务账户
account year 会计年度
accouple 偶合;连接
accouplement 匹配;配合;木条;对柱;并立柱式;系木;系材;成双构件
accouplement of columns 对柱
accouplement of pilasters 双壁柱
accoustic convolver 卷积器
accouterment 配备;装备
accredit 信任;归于;发行信用证;发给委任书
accreditation 确认;委派;鉴定(合格);信赖
accreditation body 认可机构
accreditation criterion 认可准则
accreditation of testing laboratory 测试试验室的认可
accreditation system 认可体系
accredited 可信任的;质量合格的;被认可的
accredited agent 信任代理人
accredited buyer 开证申请人(信用证)
accredited farm and land broker 田地交易注册经纪人
accredited party 开证申请人(信用证);被信托人
accrediting body 鉴定机构
accrete community 合生群落
accreted terrane 增生地体
accreting bank 沉积岸;淤积河岸;淤积岸;冲积岸
accreting beach 淤积滩
accreting continent 增生大陆
accreting convergent plate boundary 增生聚敛板块边界
accreting plate 增生板块;增长板块【地】
accreting plate boundary 增生板块边缘
accreting zone 增生带
accretion 粘连;外展作用;投落作用;添加物;结块;加速生长量;加积;吸积;淤积;撞冻;增涨作用;增生;增大作用;增长(作用);堆积物
accretional crustal consumption zone 地壳叠接消减带
accretional ice sheet 堆积冰盖
accretional prism 加积棱柱体
accretionary bar 增长沙坝
accretionary basin 增生盆地
accretionary boundary 增生边界
accretionary lava ball 外加熔岩球
accretionary limestone 加积石灰岩;堆积灰岩
accretionary meander 河曲的淤长砂嘴
accretionary meander point 河曲凸部淤长的沙嘴
accretionary phase 增长期
accretionary prism 增生(棱)柱体
accretionary ridge 外展滩脊
accretionary wedge 加积楔;增生楔
accretion beach 淤积滩;增长海滩;堆积海滩;冲积滩
accretion bed 加积层
accretion bedding 加积层理
accretion by neutron star 中子星吸积
accretion coast 外展海岸;堆积海岸
accretion deposit 淤积物;淤积泥沙;淤积层
accretion disc 吸积盘
accretion hypothesis 吸积假说
accretion line 生长线
accretion of beach 海滩堆积
accretion of bed levels 河床淤积;河床(的)淤高
accretion of bottom 河底淤高
accretion of channel 航道淤高
accretion of groundwater 地下水冲积层
accretion of population 人口增加率;人口增长率
accretion of sand 淤沙
accretion of the capital 资本加大
accretion ridge 古海滩脊
accretion surface 加积面
accretion theory 吸积理论
accretion topography 加积地形;堆积地形
accretion vein 加填矿脉;增生矿脉
accretion zone 加积地带
accretive crystallization 聚集结晶;增生晶体
Accrington brick 阿克林顿红砖
accroides resin 禾木胶;禾木树脂
accropode block 钩连块体
accroter 山墙装饰物
accrual 应计项目;增加物;权责发生额
accrual accounting 应收应付会计制;权责发生制
accrual basis 权责应计制;权责发生制;应收应付制
accrual basis accounting 权责应计制会计
accrual basis of accounting 会计权责发生制
accrual concept 权责发生概念
accrual convention 权责应计制惯例
accrual date 应收日期
accrual of interest 累计利息;应计利息
accrual parity principle 权责相当原则
accrual system 权责体制;权责发生制
accrual wage 应付工资
accrued 权责已发生的
accrued account 应计账户
accrued annual leave 应计年假
accrued assets 流动资产;应计资产
accrued basis 权责发生制
accrued bond interest 应计债券利息
accrued bond interest receivable 应计未收债券利息
accrued bonus 应计奖金
accrued charges 应计费用
accrued commission 应计佣金
accrued commission receivable 应计未收佣金
accrued cumulative dividends 应计累计股利;应计未收股利
accrued debit account 应收款项账户
accrued depreciation 累积折旧;累积贬值;应计折旧;增加折旧
accrued dividend 应计股息;应计股利
accrued dividend account 应计股息账
accrued expenditures 应计开支;应付开支
accrued-expenditures basis 应计支出制;支出应计制
accrued expenses 应计费用
accrued expense payable 应计未付费用
accrued income 应计收益
accrued income receivable 应计未收收益
accrued income tax 应计未付所得税款
accrued interest 待收利息;应计利息
accrued interest on bonds 应计(未付)债券利息
accrued interest on bond sold 应计未付售出债券利息
accrued interest on investment 应计投资利息
accrued interest on loan 应计未付借款利息
accrued interest on mortgage 应计抵押利息
accrued interest on notes payable 应计(未付)期票利息
accrued interest receivable 应计未收利息
accrued item 应计项目
accrued items of expenses 应计开支项目;应付费用项目
accrued liabilities 应计债务;应计负债
accrued of depreciation 计提折旧
accrued payable 应付款项;应计未付项目
accrued payroll 应计未付工资;应付工资
accrued profit 应计利润
accrued real 应计未付不动产税
accrued real estate taxes 应计不动产税(金)
accrued receivable 应计未收款项
accrued rent(al) 应计租金
accrued rent payable 应计未付租金
accrued rent receivable 应计未收租金
accrued revenue 应计营业收入;应计收益;应收收益
accrued rights 已产生的正当权利
accrued salaries 应计薪金
accrued salaries payable 应计未付薪金
accrued taxes 应计税金;应缴税金;应计未付税捐;应计税款
accrued taxes payable 应计未付税金
accrued tax on income 应计所得税款
accrued tax on value added 应计增值税款
accrued wages 应计工资
accrue payable 应付款

accruing amount 累计金额
accul 深弯;深河口
accumbent 斜倚的
accumbent posture 横卧姿势
accumulate 累加;累计;累积(量);聚集;积聚;蓄积
accumulate at interest 将利息滚入本金计算复利;复利计算
accumulated amortization 累计摊销
accumulated amount 累计额
accumulated amount of 1 每元的复利终值
accumulated amount of 1 dollar 每美元复利终值
accumulated amount of one 每元复利本利和
accumulated angle 总角
accumulated average grade of ore 累计矿石平均品位
accumulated capital 累计资本;累积资本
accumulated cost 累计成本
accumulated decrease of water table 水位累积下降值
accumulated deficiency 累积差值;累积缺量
accumulated deficit 累计亏损;累计赤字
accumulated deflection 累计耗竭
accumulated deformation 累积变形
accumulated degree day 累积度一日
accumulated depletion 累计损耗
accumulated depreciation 累计折旧
accumulated depreciation ratio 累积折旧率
accumulated discrepancy 累积误差;累积差(值);累积不符值
accumulated distance 累计距离
accumulated distribution 累积分配
accumulated distribution function 累计分布函数
accumulated divergence 累积发散;积累发散
accumulated dividend 累积股利;未发累积股利;累计股利
accumulated dose 累积剂量;总剂量
accumulated earnings 累计盈利;积累盈余;累积收益
accumulated energy 储蓄能
accumulated error 累积误差;积累误差
accumulated excess 累积超值
accumulated fatigue 蓄积疲劳
accumulated filth 积垢
accumulated fund 累计基金;公积金
accumulated funds of each enterprise 每一单位积累额
accumulated gabbro 堆积辉长岩
accumulated income 累计收益;累积收益;积累收益
accumulated inflow 累积入流量;累积进水量
accumulated island 堆积岛
accumulated loss 累积损失
accumulated net cash flow 累计净现金流量
accumulated net income 累计净收益
accumulated net loss 累计净亏损
accumulated outflow 累积furt水量;累积出流量
accumulated outlay 累计支出;累积支出
accumulated pitch error 周节累积误差
accumulated plastic deformation 累计塑性变形
accumulated profit 利润滚存;累计利润;累积收益;累积利润
accumulated public funds 公共积累
accumulated radio-nuclide 聚集的放射性核素
accumulated rate 积率;积差率
accumulated reserves of metal 累计金属量
accumulated reserves of ore 累计矿石量
accumulated(retained)earning 累积收益
accumulated round-off 累计舍入
accumulated roundoff error 累积舍入误差
accumulated running hours 累积运转小时
accumulated running time 累计运转时间
accumulated sediment 淤积物;沉积泥沙
accumulated speed 累积速度
accumulated stock 积压存货
accumulated storage 累积蓄水量
accumulated storm water on ground surface 暴雨积水
accumulated sum of differences between fore and back sighting distances 前后视距累积差
accumulated surplus 累计盈余;累积盈余
accumulated temperature 积温
accumulated total punch 累计穿孔机
accumulated total punching 累加总穿孔数;累计总穿孔数
accumulated total sediment volume 沉积泥沙总量

accumulated value 累加值;累积值
accumulated variance contribution 累积方差贡献
accumulated water 水库中蓄水量;积水;汇集流量;水库中蓄积的水量
accumulated wealth 累计财富
accumulated yearly quantity of production 本年累积产量
accumulate frequency value 累积频率值
accumulate output 累计量
accumulate texture 补堆积结构
accumulating 累加;聚积
accumulating biomass 停止累积生物量
accumulating condenser 聚集冷凝器;聚集冷凝器
accumulating conveyer 储运机
accumulating counter 累加计数器
accumulating curve 累积曲线
accumulating diagram of water demand 累积用水图;累积用水曲线
accumulating frequency of radon 氡累积频率值
accumulating level 积累水平
accumulating mileage 累计行驶里程
accumulating probability value 累积概率值
accumulating reproducer 累加复印机;累计复制机;累积重复装置;累积再现装置
accumulating sampler 累积式采样器
accumulating total 累计;累加总计
accumulating traffic counter 累积运输计数器;累积交通计数器;累积车辆计数器
accumulating type heat exchanger of revolution 旋转型蓄热式换热器
accumulation 累加;累积过程;累积(额);聚集作用;聚集物;聚积;积聚;积存;堆积作用;堆积(物)
accumulation account 累计账户;累积账户
accumulation action 累积作用
accumulational 堆积的
accumulational flood plain 堆积河漫滩
accumulation(al)form 堆积形态
accumulational geomorphy 堆积地貌
accumulational island 堆积岛
accumulational material anomaly 堆积物异常
accumulational mode of coal-forming material 成煤物质堆积方式
accumulation(al)platform 堆积台地
accumulation(al)relief 堆积地形
accumulation(al)terrace 堆积阶地
accumulation area 堆积区;补给区
accumulation benefit 集聚效益
accumulation capacity 积聚能力
accumulation coefficient 累积系数;聚集系数
accumulation curve 累计曲线;累积曲线
accumulation cycle 累加循环
accumulation data 积累数据
accumulation depth of harbor 港湾堆积物厚度
accumulation diagram 累积(曲线)图
accumulation diagram of water demand 用水积累图
accumulation distribution unit 累加分配器
accumulation energy 储能
accumulation equilibrium 累积平衡
accumulation factor 累积因素;累积系数
accumulation frequency figure of radon value 氡值累积频率图
accumulation fund 积累基金
accumulation fund draw by 提取积累
accumulation horizon 积累层;积聚层;淤积层;堆积层
accumulation hydraulic system 累积液压系统
accumulation hypothesis 积累假说
accumulation in body tissues 在人体组织中积累
accumulation inflow 累积来水量
accumulation lake 蓄水湖
accumulation layer 淤积层;存储层
accumulation line(冰川冻结渗透区中的)累积线;雪线
accumulation measuring method 累积测量法
accumulation method 累积法
accumulation mode 累加态;积模
accumulation mountain 堆积山
accumulation multiple 蓄积倍数
accumulation of capital 资本积累
accumulation of cold 蓄冷
accumulation of commodities 商品积累
accumulation of data 资料积累

accumulation of discount 折价累积;折价积累
accumulation of electric(al)energy 电能存储
accumulation of error 误差累积
accumulation of floating ice 浮冰积聚
accumulation of funds 资金累积;资金的积累
accumulation of harmful material 有害物质积累
accumulation of heat 热量蓄积;热力储存;蓄热(作用)
accumulation of impurities 杂质积聚
accumulation of money 货币积累
accumulation of mud 污泥淤积;泥浆淤积;泥浆堆积;淤积污泥
accumulation of one per period 每元定期复利累积值
accumulation of organic substance 有机质积累
accumulation of pressure 压力累积;压力积累
accumulation of radioactive waste 放射性废料积存
accumulation of risks 责任积累;风险积累
accumulation of rounding errors 舍入误差累积;凑整误差累积
accumulation of runoff 径流汇集
accumulation of salt in the surface soil 返盐;返碱
accumulation of salts 盐的积聚
accumulation of silver 银离子聚集
accumulation of sludge 污泥淤积
accumulation of snow 积雪
accumulation of strain 应变累积
accumulation of stress 应力累积;应力集中;应力积蓄;应力积累;应力积聚
accumulation of wagon flow 车辆积压
accumulation phase 堆积相
accumulation plain 堆积平原
accumulation plan 积聚方案
accumulation platform 堆积台地
accumulation point 聚集点;聚点;凝聚点
accumulation precipitation 累积降水量
accumulation principle 累加原理
accumulation process of oil and gas 油气聚集作用
accumulation-quotient register 累加商寄存器
accumulation rate 频度累积率;累积率;累积率;积聚率;增长率(冰川);堆积(速)率
accumulation rate of organic matter 有机质堆积速率
accumulation rate of sediment 沉积物堆积速率
accumulation rate of wave train 波列累积率
accumulation-regeneration theory 累积再生理论
accumulation register 累加寄存器;积累寄存器
accumulation rock 堆积岩
accumulation schedule 累积表;累积表
accumulation season(冰川的)储积期
accumulation sinking fund 累积偿债基金
accumulation slope 堆积坡
accumulations of quaternary period 第四纪堆积物
accumulation soil 堆积土
accumulation soil moisture 积累土壤湿度
accumulation temperature 积温
accumulation terrace 淤积阶地;堆积阶地
accumulation test 蓄气试验;蓄积试验;储[贮]气试验
accumulation toxicity 积累毒性
accumulation trust 储蓄信托
accumulation units 将投资利息滚入本金的积累
accumulation volume 累积(总)量;堆积量
accumulation zone 累积区;堆积区
accumulation zone of fault terrace 断阶聚集带
accumulation zone of growth fault-rollover anticline 同生断层—逆牵引背斜聚集带
accumulation zone of middle-strong folding anticline 中强褶皱背斜聚集带
accumulation zone of mudlump and diapir fold 泥火山—底辟背斜聚集带
accumulation zone of placanticline 长垣聚集带
accumulation zone of salt dome and anticline 盐丘和盐背斜油气聚集带
accumulation zone of slightly anticline 平缓背斜聚集带
accumulation zone of thrust-fold 逆掩断裂褶皱聚集带
accumulative 累积的;积聚的;堆积的
accumulative action 累积作用;蓄积作用
accumulative area 汇水区;积水区
accumulative carry 累加进位
accumulative coefficient method 聚集系数法

accumulative column 聚集柱
accumulative commission 累计佣金
accumulative comparison 累积比较
accumulative cost 累计成本
accumulative crystallization 聚основ结晶
accumulative deformation 积累变形
accumulative eluvial landscape 堆积残积景观
accumulative error 积累误差;累积误差
accumulative estimation 累积估计
accumulative facies 堆积相
accumulative factor 堆积因素
accumulative failure 积累破坏
accumulative formation 堆集形成;堆积(地)层
accumulative frequency 累积频率
accumulative frequency curve 累积频率曲线
accumulative frequency diagram 累积频数(曲线)图;累积频率曲线图
accumulative frequency distribution 累积频率分布
accumulative frequency function 累积频率函数
accumulative hours 累积时数
accumulative hydrocarbon productivity 累积产烃率
accumulative length of core 累计岩芯长度
accumulative method 累加法
accumulative numbering method 累计编号法
accumulative numbering method in advance 提前期累计编号法
accumulative percentage 累计百分数
accumulative pitch error 周节累积误差
accumulative process 积累过程
accumulative rain ga(u)ge 累积量雨器
accumulative rock 堆积岩
accumulative sampling 聚集采样;积累采样
accumulative sinking fund 积累偿债基金
accumulative timing 积累计时法
accumulative total 累计
accumulative total on the sieve percent 筛上累计百分数
accumulative toxicant 蓄积性毒物
accumulative weight blender 积蓄重量掺和机
accumulative weighing 累积称(重)量
accumulative zone 累积区
accumulator 受液器;收集器;累加器;累积器;积储[贮]器;积累器;蓄水区;蓄能器;蓄电池组;压缩空气筒;储[贮]液器;储[贮]料塔;储[贮]料坑;电瓶;低压储[贮]液筒;存储器;储压器;储蓄器;储能器;储能电路;储存器
accumulator acid 蓄电池酸液
accumulator addressing 累加寻址
accumulator battery 蓄电池组
accumulator blowdown valve 蓄热器排污阀
accumulator box 蓄电池箱;蓄电池容器;蓄电(池)槽;电瓶箱
accumulator capacity 蓄电池容量
accumulator car 储[贮]藏车;电瓶车;电力搬运车;电动汽车;电池搬运车
accumulator carriage 累加(载运)器
accumulator case 蓄电池箱
accumulator cell 蓄电池
accumulator charger 蓄电池充电器;蓄电池充电机
accumulator charging 蓄能器增压;蓄电池充电
accumulator charging pushcart 蓄电池充电小车
accumulator charging room 蓄电池充电间
accumulator charging valve 蓄加充电开关;蓄加充电阀门
accumulator container 蓄电池容器
accumulator drive 电池牵引
accumulator grid 蓄电池栅板
accumulator-head machine 储料缸式机头吹机塑机
accumulator house 蓄电池室
accumulator insulator 蓄电池绝缘子
accumulator jar 蓄电瓶;蓄电池容器
accumulator jump instruction 累加器转移指令
accumulator latch 累加器闩锁
accumulator loco(motive) 蓄电池牵引机车;蓄电池式电机车;电池机车
accumulator metal 蓄电池金属极板;蓄电池极板合金
accumulator operation 累加器操作
accumulator plant 聚集植物;聚积植物;积累植物;蓄电池站;蓄电池室;蓄电池充电间;蓄电池车间;储电站;充电室
accumulator plate 蓄电池极板
accumulator pocket 储气筒袋

accumulator pressure 储能器液体压力
accumulator railcar 蓄电池轨道车;电瓶轨道车
accumulator rectifier 蓄电池充电整流器
accumulator register 累加寄存器
accumulator roll 成型筒
accumulator room 蓄电池室;蓄电池舱
accumulator separator 蓄电池隔板;电池极板间的隔离板
accumulator shift instruction 累加器移位指令
accumulator spring 缓冲系统弹簧
accumulator stable instruction 累加器稳定指令
accumulator stage 累加计数器单元
accumulator still 累加器蒸馏室;缓冲釜
accumulator switch 蓄电池转换开关
accumulator switchboard 蓄电池配电盘
accumulator system 累加器系统;存储器系统
accumulator tank 集油罐;缓冲罐;储蓄槽
accumulator traction 蓄电池牵引
accumulator transfer instruction 累加器转移指令
accumulator tray 集油塔;集油盘
accumulator tube 储气筒管
accumulator turbine 蓄热器汽轮机
accumulator vehicle 蓄电池车
accumulator workshop 蓄电池间
accumulator workshop with resistance discharge 带电阻放电蓄电池间
accumulograph 累积曲线(图);累积线
accuracy 精确性;精密性;精密度;准确性;准确度
accuracy adjustment 角度平差
accuracy and correction of deviation 偏移的精度和校正
accuracy and source strength 准确度和能源强度
accuracy block 精确块;加强块
accuracy checking 精度检查
accuracy class 精确度级别;精度等级
accuracy constraint 精度约束
accuracy control 精确(度)控制;精度控制
accuracy control character 精确度控制记号;精度控制符号;准确度控制字符
accuracy control system 精度控制系统;准确度控制系统;差错控制系统
accuracy control time window 精度控制时窗
accuracy distribution 精度分配
accuracy error 精度误差
accuracy exponent 精度指数
accuracy factor of mapping 作图准确度
accuracy for horizontal control 平面控制测量精度
accuracy for vertical control 高程控制精度
accuracy grade 精确度等级;精度等级;准确度级(别)
accuracy in calculation 计算精确度
accuracy in computation 计算准确度;计算精(确)度
accuracy in measurement 量测精度;测量精(确)度
accuracy lasting 精度耐久性
accuracy life 精确度寿命;精度寿命
accuracy limit 精度界限;精度极限
accuracy map 精确地图
accuracy margin 精度余量
accuracy modulus 精度模
accuracy of adjusted base station network 基点网平差后精度
accuracy of adjustment 调整准确度
accuracy of analysis 分析准确度
accuracy of angle 测角精(确)度
accuracy of an instrument 仪表准确度
accuracy of base station network 基点网联测精度
accuracy of breaking through 贯通精度
accuracy of chaining 链测精度
accuracy of colo(u)r value 色泽准确性
accuracy of configuration 外形精度
accuracy of data 资料数据的准确性
accuracy of discharge observation 流量观测精度
accuracy of drawing 绘图精度
accuracy of equalization 均衡精度
accuracy of estimate 估算精度;估计精度
accuracy of exploration 勘探精度
accuracy of factor of mapping 作图准确度
accuracy of finish 最后加工精度
accuracy of flatness 压平精度
accuracy of focusing 对光精度
accuracy of foundation 基础的精确度
accuracy of ga(u)ge 尺寸的准确性
accuracy of geologic(al)surveying 地质调查精度

accuracy of gravity anomaly 重力测量精度
accuracy of groundwater resources evaluation 地下水资源评价精度
accuracy of horizontal control 平面控制精度
accuracy of instrument 仪器精确度;仪表精(确)度
accuracy of latitude correction 纬度改正精度
accuracy of latitude distance 纬向距精度
accuracy of level(s) 平整度;平滑度
accuracy of manufacture 制造精度
accuracy of manufacturing 制造精度
accuracy of map 地图精度
accuracy of measurement 测验精度;测量精度;量测准确度;量测精度;度量精度
accuracy of measuring 计量准确度;计量的准确性;测量精(确)度;计量精度
accuracy of mesh 啮合精度
accuracy of network 控制网精度
accuracy of observation 实测精度;观测精度
accuracy of presentation 表示精度
accuracy of ranging for laser range finder 激光测距机测距准确度
accuracy of reading 读数精度
accuracy of repetition 重复精度
accuracy of sample analysis 样品分析精度
accuracy of scale value determination 重力仪精度;格值测定精度
accuracy of separation 筛分的精细度;筛分率;分离精度;分割精度
accuracy of shape 形状的准确性;形状的精确度;成型的精确度
accuracy of ship's position 船位精确度
accuracy of sounder reading 测深仪读数精度
accuracy of terrain correction 地形改正精度
accuracy of terrain correction farther distance 远区地形改正精度
accuracy of terrain correction on medium distance 中区地形改正精度
accuracy of terrain correction on nearby distance 近区地形改正精度
accuracy of test glass 玻璃样板准确度
accuracy of the estimated ore reserves 估算储量精度
accuracy of time scale 时标的准确度
accuracy of topographic(al) map 地形图精度
accuracy of vertical circle 竖直度盘精度
accuracy of vertical control survey 高程控制测量精度
accuracy of water-level observation 水位观测精度
accuracy order 精确度等级
accuracy range 精度范围
accuracy ranging 测距精度
accuracy rating 准确率;额定精(确)度
accuracy requirement 精度要求
accuracy specification 精度要求
accuracy standard 精度标准
accuracy table 精度表
accuracy test(ing) 精度检验;准确度试验;精度试验
accuracy tolerance 精确限度;准确度限值
accuracy to size 尺寸精(确)度
accuracy user 准确度用户
accuracy verification package 精密检验组件
accurate 精确的;精密的;准确的
accurate adjustment 精确调整;精密校正;精密调整;精度调整
accurate aiming 精确瞄准
accurate beam-scanning method 精确波束扫描法
accurate colo(u)r value 精确色值
accurate complete equipment 精密成套设备
accurate control 精确控制
accurate cost 精确费用
accurate curve method 精确分光光度法
accurate data 精确数据;准确数据
accurate degree 精确等级;精确程度
accurate determination 精确测定
accurate diagnosis 确诊
accurate estimate 精确估算
accurate fix 准确定位
accurate forming 准确成型
accurate grinding 精密磨削
accurate interception 精确交叉截获
accurate mass measurement 精确质量测定
accurate measurement 精确计量;精确测量
accurate measuring instrument 精确测量仪表
accurate picture 准确图像

accurate pointing 精确定向;准确定向;点测
accurate position finder 精确定位仪;精确测位装置;精确测位仪;精密目标调整
accurate position indicator 精确位置指示器;精确位置显示器;精密位置指示器
accurate range display 精确距离显示器
accurate range marker 准确距离标志
accurate rapid dense method 高速度高密度精密压铸法
accurate rate 精密度
accurate reading 精确读数;准确读数
accurate record 精确记录
accurate shape 准确的外形
accurate sizing 精确筛选;精确分级
accurate solution 精确解
accurate sweep 精密扫描
accurate sweep generator 精密扫描振荡器;精密扫描发生器
accurate thread 精密螺纹
accurate to dimension 符合加工要求;符合加工尺寸
accurate to ga(u)ge 符合加工尺寸
accurate to within plus or minus five percent 精度在正/负百分之五以内
accurate tracking 精确跟踪
accurate value 准确值
accusation 谴责;诉讼;指控
accusatoru 非难的
accuser 控告者;控告方;原告
accustomed 按照风俗习惯的
accustomed average 小海损
accustom to climate 适应气候
accutron 电子手表
ACD[automatic closing device] 自动关闭装置
AC-DC general-use switch 交直流通用按钮开关
ac dump 掉电
acebrochol 醋布洛可
acebutolol 醋丁洛尔
acecarbromal 醋卡溴脲
acedapsone 醋氨苯砜
acediasulfone sodium 醋地砜钠
acedronoles 醋椿脑染料
aceglatone 醋葡醛内酯
aceko-black 醋黑
Acelan 阿塞纶聚丙烯腈短纤维
Acelba 阿策尔巴乙酰化黏胶纤维
Acele 阿西尔醋酯纤维
acenaphthenequinone 苊醌
acenaphthylene 苊烯;苊
acendrada 白泥灰岩
acenocoumarol 醋硝香豆醇
acentric 离心的;非中正的;无中心的;非中枢的
aceperinaphthane 醋代萘烷
acephatemet 甲胺磷
aceprometazine 醋异丙嗪
aceptable daily intake 每日允许摄入量
acequia 灌溉水渠(美西南部)
acerate 针状的;尖的
acerbity 涩度
acerdol 高锰酸钾;高锰酸钙
acerilla 方铅矿
acerose 针叶状的;针状的
acerous 无角的;针状
acerous tree 针叶树
acervation 堆积
aceryl cellulose 乙酸纤维素
acescency 微酸味
acescent 微酸的;变酸的
acess door 入检门
acet 次乙基
Aceta 阿策塔醋酯长丝
acetabular 碟状体的
acetabulum 臼
acetabutar 杯状的
acetal 醛缩醇类;乙缩醛;乙醛缩二乙醇
acetal copolymer 缩醛共聚物;乙缩醛共聚物
acetaldehydase 乙醛酶
acetaldehyde 乙醛;醋醛
acetaldehyde polymer 乙醛聚合物
acetaldehyde resin 乙醛树脂
acetaldol 丁间醇醛
acetaldoxime 乙醛肟
acetal homopolymer 缩醛均聚物
acetal plastics 缩醛塑料
acetal resin 缩醛树脂;聚甲醛树脂;乙缩醛树脂

acetamide 天然醋酸胺;乙酰胺;醋胺石
acetamide chloride 二氯代乙酰胺
acetamino 乙酰氨基
acetaminophen 对乙酰氨基酚;醋氨酚
acetanilid(e) 乙酰(替)苯胺;醋酰苯胺
acetarsol 醋酰胺胂
acetas (地板上或油桶内用以抗沥青的)专用酸
acetate 乙酸酯;乙酸盐;乙酸基;醋酸酯;醋酸盐;醋酸纤维透明片;醋酸纤维基
acetate base 醋酸盐磁基带;醋酸纤维(片)基;醋酸片基
acetate base film 醋酸基胶片
acetate cellulose adhesive 醋酸纤维素橡皮胶
acetate cellulose fibre 醋酯纤维
acetate chrome 乙酸法铅铬黄
acetate clear dope 乙酸纤维(素)蒙皮(清)漆
acetate coated polyethylene fabric 乙酸酯涂层聚乙烯织物
acetate colo(u)r 醋酸性染料
acetate disc 醋酸酯唱片
acetate dope 乙酸纤维酯涂布漆;乙酸纤维酯蒙皮漆
acetate dye 醋酯纤维染料
acetate fiber 醋酸纤维
acetate fibre paste 醋纤浆料
acetate film 醋酸纤维素膜;醋酸纤维胶片;醋酸纤维薄膜
acetate green 乙酸法铅铬绿
acetate hollow filament 醋酯中空长丝
acetate ink 乙酸酯油墨
acetate ion 乙酸根离子
acetate of copper 醋酸铜
acetate of lead 乙酸铅
acetate of lime 醋酸石灰;醋石
acetate process 醋酸纤维法
acetate rayon 醋酸人造丝;醋酰纤维
acetate rayon filament 醋酯人造丝
acetate rayon staple 醋酯短纤维
acetate sheet 醋酸盐胶片
acetate silk 醋酸纤维素丝;醋酸丝;醋酸人造丝
acetate spinning machine 醋酯纤维纺丝机
acetate staple fibre 醋酯短纤维
acetate visor 醋酸纤维护目镜
acetate wire 醋酸纤维绝缘电缆线;醋酸绝缘线
acetate wool 醋酸人造毛
Acetat Rhodia 阿策塔特罗迪阿醋酯短纤维
acetazolamide 醋唑磺胺
acetic 醋酸的
acetic acid 乙酸;醋酸
acetic acid bacteria 醋酸细菌
acetic acid ethanoic acid 乙酸
acetic acid fiber 醋酸纤维
acetic acid rubber 醋酸橡胶
acetic acid salt spray test 乙酸盐雾试验
acetic anhydride 乙酸酐;醋酸酐;醋酐
acetic ester 乙酸酯
acetic ether 乙酸乙酯
acetic fermentation 醋酸发酵
acetic oxide 氧化乙酰
acetic peracid 过乙酸
acetic starch 醋酸淀粉
acetification 醋(酸)化作用
acetificator 醋化剂
acetifier 醋化器
acetify 醋化
acetimeter 酸计;醋酸比重计
acetimetry 醋酸定量法;醋酸测定法
acetin 单乙酸甘油酯;醋精
acetine 醋精
acetin method 乙酸酯法
acetiromate 醋替罗酯
acetoacetate 乙酰乙酸酯;乙酸盐;乙酰醋酸盐
acetoacetic acid 乙酰乙酸;丁间酮酸
acetoacetic ester 乙酰乙酸(乙)酯
acetobutyl acid 乙酰丁酸
acetobutyl alcohol 乙酰丁醇
acetobutyryl cellulose 乙酸丁酯纤维素
aceto-carmine 醋酸洋红
acetoglyceride 乙酸甘油酯;醋酸甘油酯
acetol 乙酰甲醇;丙酮醇
acetolysis 乙酰解作用;醋酸水解
acetomeroctol 醋辛酚汞
acetometer 醋酸比重计
acetometry 醋酸比重测定法
acetonate 丙酮酸盐

acetonation 丙酮化作用
acetone 丙酮
acetone acid 醋酮酸
acetone alcohol 羟基丙酮
acetone amine 丙酮胺;氨基丙酮
acetone-benzol dewaxing plant 丙酮苯脱蜡装置
acetone-benzol process 丙酮苯脱蜡法
acetone-butanol bacteria 丙酮丁醇细菌
acetone chloroform 丙酮氯仿
acetone cyanohydrin 丙酮氰醇;丙酮合氰化氢
acetone-diacetic acid 丙酮二乙酸
acetone dicarboxylic acid 丙酮二羧酸
acetone dichloride 二氯代丙酮
acetone extract 丙酮提取物
acetone-formaldehyde resin 丙酮甲醛树脂
acetone-furfural resin 丙酮糠醛树脂
acetone-methylene chloride solvent 丙酮二氯甲烷溶剂
acetone number 丙酮值;丙酮数
acetone oil 丙酮油
acetoneoxime 丙酮肟
acetone peroxide 过氧化丙酮
acetone phenylhydrazone 丙酮苯腙
acetone poisoning 丙酮中毒
acetone pyrolysis 丙酮热解
acetone raffinate 丙酮提余物
acetone resin 丙酮树脂
acetone solution 丙酮溶液
acetone value 丙酮值
acetonic 丙酮的
acetonic acid 醋酮酸
acetonide 丙酮化合物;丙酮宁
acetonitrile 乙腈
acetonitril test 氰化甲烷试验
acetonyl 丙酮基
acetonyl acetone 丙酮基丙酮
acetonylamine 丙酮基胺;氨基丙酮
acetonyl urea 丙酮基脲
aceto-orcein 醋酸地衣红
acetophane 醋酸薄膜
acetophenone 甲基苯基(甲)酮;乙酰苯;苯乙酮
acetophenone oxime 苯乙酮肟
acetopropion cellulose 乙酸丙酸纤维素
Acetopurpurin 阿赛托红紫
acetorthotoluid 邻乙酰氨基甲苯
acetous 有酸味的;醋的
acetoxime 丙酮肟
acetoxytrimethylsilane 乙酰氧基三甲基硅烷
acetrizoic acid 醋碘苯酸
acetum 醋剂;醋
acetyl 乙酰
acetyl acetone 乙酰丙酮
acetylated cotton 乙酰化棉
acetylated dammar 乙酰达玛树脂
acetylated hydroxyethyl cellulose 乙酰化羟乙基纤维素
acetylating agent 乙酰化剂
acetylation 乙酰化(作用)
acetylator 乙酰化器
acetyl base 乙酰基
acetyl benzene 苯乙酮
acetyl benzoyl peroxide 乙酰过氧化苯甲酰
acetyl biuret 乙酰基缩二脲
acetyl brilliant blue 乙酰亮蓝
acetyl butyryl cellulose 乙酰丁酸纤维素
acetylcarbromal 对乙酰基二溴乙酰脲
acetyl cellulose 乙酰纤维素;醋酸纤维素
acetyl cellulose lacquer 乙酰纤维素漆;醋酸纤维素漆
acetyl-cellulose plastics 乙酰基纤维素塑料;醋酸纤维素塑料
acetyl cellulose sheet 醋酸纤维板
acetyl chloride 乙酰氯
acetylcholine iodide 碘化乙酰胆碱
acetyl colo(u)r 乙酰染料
acetyldigoxin 醋地高辛
acetylene 乙炔;电石气
acetylene acids 烃属酸
acetylene apparatus 乙炔气焊设备
acetylene black 乙炔(炭)黑
acetylene bottle 乙炔(气)瓶
acetylene burner 乙炔燃烧器;乙炔焊炬;乙炔炬;乙炔灯
acetylene carbon black 乙炔炭黑

acetylene chloride 乙炔基氯
acetylene cutter 乙炔烧割器;乙炔切割器
acetylene cutting 乙炔切割;乙炔气割;氧炔切割
acetylene cylinder 乙炔罐;乙炔(气)瓶
acetylene cylinder valve 乙炔瓶阀
acetylene dichloride 对称二氯乙烯
acetylene filled countor 乙炔计数器
acetylene flame 乙炔焰;乙炔火焰
acetylene flame carburizing 乙炔焰渗碳
acetylene flasher 乙炔闪光器
acetylene flashing lamp 乙炔闪光灯
acetylene flashing light 乙炔闪光灯
acetylene gas 乙炔;乙炔气(体)
acetylene gas lighted buoy 乙炔灯照明的救生圈;乙炔灯照明的浮标
acetylene gas-powered 以乙炔为能源的
acetylene generating station 乙炔发生站
acetylene generator 乙炔发生器
acetylene headlight 乙炔照明信号头灯;乙炔照明信号前灯;乙炔前灯
acetylene lamp 乙炔灯
acetylene light buoy 乙炔灯浮标
acetylene phenol resin 乙炔苯酚树脂
acetylene producer 乙炔发生器
acetylene reduction method 乙炔还原方法
acetylene removal of stress 乙炔(加热)解除应力
acetylene residue 电石渣
acetylene slag wastewater 电石渣废水
acetylene sludge 乙炔渣;电石渣
acetylene starter 乙炔起动机
acetylene torch 乙炔炬;氧乙炔炬;乙炔焰;乙炔喷灯;乙炔吹管
acetylene trichloride 乙炔化三氯
acetylene welder 乙焊机;乙炔气焊设备;乙炔(气)焊枪;乙炔(气)焊机;乙炔(气)焊工
acetylene weld(ing) 乙炔焊;气焊;乙炔焊接
acetylene welding torch 乙炔焊(接)吹管;乙炔焊炬
acetylenic acid 炔属酸
acetylenic ketone 炔属酮
acetylenyl benzene 乙炔基苯
acetylenyl carbinol 乙炔基甲醇
acetyl furan 乙酰呋喃
acetyl gasoline 含乙炔汽油;乙炔汽油
acetyl group 乙酰基
acetyl hydrate 醋酸
acetylide 乙炔化(合)物
acetyl iodide 乙酰碘
acetylization flask 乙酰化烧瓶
acetyl number 乙酰值
acetyl oxide 醋酸酐
acetylperchlorate 高氯酸乙酰酯
acetylperoxide 过氧乙酰
acetylphenol 乙酰苯酚
acetyl phosphate 乙酰磷酸
acetyl pure yellow 乙酰纯黄
acetyl pyrazine 乙酰吡嗪
acetyl pyridine 乙酰吡啶
acetyl-resorcin 乙酰苯间二酚;间乙酸基苯酚;间苯二酚一乙酸酯
acetyl ricinoleate 乙酰基蓖麻油酸酯
acetyl saponification number 乙酰皂化值
acetyl scarlet 乙酰猩红
acetyl value 乙酰值
acetyl wood 乙酰化木材
acetysalicylic acid phenobarbital 阿苯
acevaltralte 醋戊曲酯
acexamic acid 醋己氨酸
AC-generator 交流发电机
Achaemenid architecture 阿契美尼建筑(波斯)
a chain of undulating hills 丘陵起伏
Achamenian Persian architecture 阿昌门利亚建筑艺术(公元前六世纪到公元前四世纪)
a chance occurrence 偶发事件
achatine 并行色
achavalite 硒铁矿
acheb 短生植被区
a chemical element 一种化学元素
Acheson furnace 阿奇逊电炉
Acheson graphite 阿奇逊人造石墨
Acheulean 阿舍利文化
Acheulean age 阿舍利时代
achiasmate 无交叉的
achievable 可用成就的;可完成的
achieve 达到

achieved reliability 实际可靠性;工作可靠性
achievement 完成;效益;成就;成果(功绩)
achievement age 成就年龄
achievement analysis 成就分析
achievement figure table 成果图表
achievement function 达到函数
achievement quotient 成就商;成绩商数
achievements in scientific research 科研成果
achieves 档案库存储器
achieving maintenance history 完成修善历史
Achilles group 阿基里斯群
achiral reagent 非手性试剂
achloride 非氯化物
achloropsia 绿色色盲
achlusite 钠滑石
achmatite 绿廉石
achnakaite 黑云(倍)长岩
Achnanthes 曲壳硅藻属
achnath erum splendens 芨芨草
achondrite 无球粒陨石
achrodextrin 消色糊精
achroite 无色电气石;无色碧硒
achromat 消色差透镜
achromatic 无色的;不着色的
achromatic aerial 消色差天线
achromatic antenna 消色差天线
achromatic colo(u)r 无色;无彩色;消色(差颜色);非彩色
achromatic condenser 消色差聚光器;消色差聚光透镜;消色差聚光镜(头)
achromatic convex lens 消色差集光镜
achromatic doublet lens 消色差双合透镜
achromatic edition 单色版
achromatic eyepiece 消色差目镜
achromatic fringe 消色差条纹
achromatic image 消色差像
achromatic indicator 消色指示剂
achromatic interference fringe 消色差干涉条纹
achromatic interval 无色间隙
achromaticity 无色;消色差(性);非彩色(性)
achromatic lens 消色差透镜
achromatic light 消色光;消色差光
achromatic locus 消色差色线
achromatic magnetic mass spectrometer 消色差磁色谱计
achromatic map 单色图
achromatic objective 无色差物镜;消色差物镜
achromatic ocular 消色差目镜
achromatic phase plate 消色差位相版
achromatic pigment 非彩色颜料
achromatic point 消色差点;非彩色点
achromatic power 消色力
achromatic prism 消色差棱镜
achromatic region 中性色区域
achromatic sensation 消色差感觉
achromatic sheet 单色图
achromatic stimulus 无彩色刺激
achromatic system 消色差镜系
achromatic telescope 消色差望远镜
achromatic threshold 无色阈值
achromatin 非染色质
achromation 消色差
achromatism 无色;消色差(性)
achromatization 无色彩;色差消除;消色差
achromatize 使消色;使无色
achromatophil 不染体;不染色的
achromatophilia 不染色性
achromatophilic 不染色的
achromatopsy 全色盲;完全色盲
achromic method 消色测定法
achromic period 消色时间;消色期
achromic point 消色点
achromophil 不染色的
achromophilous 非嗜色的;不染色的
achromophilous smear examination 不染色涂片检查
Achucarro's stain 阿丘卡罗染剂
acicular 针状晶体;针状的;针形的
acicular blastic texture 针状变晶结构
acicular coke 针状石油焦
acicular constituent 针状组织
acicular crystal 针状(结)晶体
acicular ferrite 针状铁素体
acicular ice 纤维状冰;屑冰;针冰

acicular iron 针状铸铁;贝茵体铸铁
acicular-leaved tree 松柏类植物;针叶树
acicular martensite 针状马氏体
acicular pigment 针状颜料
acicular powder 针状粉末
acicular-prismatic blastic texture 针柱状变晶结构
acicular structure 针状构造
acicular texture 针状结构
acicular type zinc oxide 针状氧化锌
aciculate crystal 针状结晶
aciculate type zinc oxide 针状氧化锌
aciculifruticeta 针叶灌木群落
aciculilignosa 针叶植被;针叶木本群落;针叶树林和灌木
aciculisilvae 针叶乔木群落;针叶林
acid 酸性物;酸的;砷酸;膜酸
acid absorber 酸吸收器
acid acceptance 容酸性
acid acceptor 酸性中和剂
acid acclimated seed sludge 酸驯化接种污泥
acid accumulator 酸蓄电池;酸性蓄电池
acid activation 酸活化
acid-affected 受酸影响的
acid ager 酸气蒸化机
acid agglutination 酸凝集
acid alcohol 酸性酒精;酸醇
acid alizarine black 酸性茜素黑
acid alizarine blue 酸性茜素蓝
acid-alkali cell 酸碱电池
acid-alkaline measurement device 酸碱液计量箱
acid-alkaline pump 酸碱泵
acid-alkali titration 酸碱滴定(法)
acid alteration 酸性蚀变
acid alum spring 酸性矾泉
acidamide 酰胺
acid ammonium carbonate 酸式碳酸铵
acid analyzer 酸分析仪
acid and alkaline matter 酸碱物质
acid and alkaline resisting brick 抗酸碱砖
acid and alkaline wastewater 酸碱废水
acid and alkali of groundwater 地下水的酸碱性
acid and alkali-resistant 耐酸碱的
acid and alkali-resistant grout 耐酸碱水泥浆;耐酸碱泥浆
acid and alkali-resistant mortar 耐酸碱水泥砂浆;耐酸碱砂浆
acid and alkali treatment 酸碱处理
acid and base pollution 酸碱污染
acid anhydride 酸酐
acid annealing 酸蚀后退火
acid anthracene brown 酸性蒽棕
acid anthraquinone blue 酸性蒽醌蓝
acidation 酸化;酸处理;酰化(作用)
acid attack 酸性腐蚀;酸侵蚀;酸浸蚀
acidazo-colo(u)r azo-dye 酸性偶氮染料
acid badging 酸蚀打印;戳记刻蚀;酸蚀印记
acid-base analyzer 酸碱分析仪
acid-base balance 酸碱平衡
acid-base balance reaction 酸碱平衡反应
acid-base catalysis 酸碱催化
acid-base compexation reaction 酸/碱络合反应
acid-base conjugate pair 酸/碱共轭偶
acid-base degree of media 介质酸碱度
acid-base determination 酸碱测定法
acid-base equilibrium 酸碱平衡
acid-base equilibrium reaction 酸碱平衡反应
acid-base group 酸/碱基
acid-base indicator 酸碱指示剂
acid-base metabolism 酸碱代谢
acid-base pair 酸碱对
acid-base pollution 酸碱污染
acid-base ratio 酸碱性系数
acid-base reaction 酸碱反应
acid-base relationship 酸碱关系
acid-base titration 酸碱滴定(法)
acid-base titration detector 酸碱滴定检测器
acid-base treatment 酸碱处理
acid-base wastewater 酸碱废水
acid bath 酸浴镀样;酸浴;酸性熔池;酸洗柜;酸池
acid bating 酸性软化(法)
acid Bessemer brick 酸性转炉砖
acid Bessemer cast iron 酸性转炉铸铁
acid Bessemer furnace slag 酸性转炉钢渣
acid Bessemer iron 酸性转炉铁

acid Bessemerizing 酸性转炉冶炼
acid Bessemer pig 酸性转炉生铁
acid Bessemer process 酸性转炉(炼钢)法
acid Bessemer steel 酸性转炉钢
acid binding agent 酸性黏合剂
acid black 酸性黑
acid blast machine 喷酸液机
acid blow case 吹气扬酸箱;吹气箱
acid blue 酸性蓝
acid bottle (测定钻孔偏差测量用的)酸瓶
acid brass 耐酸黄铜
acid brick 酸性砖;耐酸砖
acid brilliant blue 酸性亮蓝
acid brittleness 氢化脆性;酸性脆裂;酸脆性;酸脆
acid bronze 耐酸青铜
acid bronze alloy 铜基合金
acid brown soil 酸性棕色土
acid carbonate 酸性碳酸盐;酸式碳酸盐
acid carboy 酸坛
acid carmoisine 酸性淡红(偶氮染料之一)
acid catalyst 酸催化剂
acid-causing substance 致酸物质
acid-chloride spring 酸性氯化物泉
acid chrome black 酸性铬黑
acid chrome dye 酸性铬媒染料
acid chrome salt 酸性铬盐
acid circulating pump 酸循环泵
acid clay 酸性黏土;酸性白土
acid cleaning 酸洗
acid cleaning waste(water) 酸洗废水
acid cleavage 成酸分裂
acid coagulation 酸性絮凝;酸性凝固;酸凝固法;加酸凝固
acid coke 酸性焦炭
acid colo(u)r 酸性颜料;示酸色
acid combining capacity 酸性化合能力
acid complex dye 酸性络合染料
acid concentration 酸浓度
acid concentrator 酸浓缩器
acid condenser 酸冷凝器
acid consumption 耗酸量
acid container 酸容器
acid-containing 含酸的
acid-containing soot 含酸煤烟
acid content 酸含量
acid converter process 酸性转炉炼钢法
acid cooking 酸蒸煮
acid cooler 酸冷却器;冷酸器
acid copper 酸性铜电解溶液
acid copper chromate 酸性铬酸铜
acid corrosion 酸(性)腐蚀
acid cracking process 酸性裂化法
acid cracking treatment 酸裂化处理
acid cresol red 酸性甲酚红
acid cupric chromate 铜铬木材防腐剂
acid cure 酸渍;酸性处理;酸固化
acid-cured epoxy resin 酸固化环氧树脂
acid-cured resin 酸凝树脂
acid-cured varnish 酸性漆;酸硬化清漆;酸固化清漆;酸凝清漆
acid curing 酸固化
acid curing acrylic resin system 酸固化丙烯酸树脂系统
acid cut 酸定
acid decomposition 酸性分解
acid decomposition method 酸分解法
acid deficient 缺酸的
acid degradation 酸(性)降解
acid deposition 酸沉降
acid deposition assessment model 酸沉降评估模型
acid deposition monitoring 酸沉降监测
acid deposition system 酸沉降体系
acid derivative 酸衍生物
acid detergent 酸性去污剂
acid detergent fiber 耐酸性去污剂纤维
acid developer 酸性显影液
acid development 酸显影
acid dewpoint 酸露点
acid dew point corrosion 酸露点腐蚀
acid diamide 酰二胺
acid digested sample 酸浸样品
acid digestion 酸性菌分解
acid digestion analysis 土壤酸解分析;酸煮分析

acid digestion analysis of soil 土的酸煮分析
acid dilution 酸液稀释
acid dip 酸洗
acid dip pickle 酸洗装置;沉浸式酸洗装置
acid dipping 酸浸渍;酸浸
acid dissociation constant 酸离解常数
acid dissociation reaction 酸离解反应
acid drain 酸性废水
acid drainage 酸性排水
acid drift 酸性矿石;变酸倾向
acid droplets 酸滴
acid dye lake 酸性染料色淀
acid dye(stuff) 酸性染料;酸性染剂
acid effluent 酸性排水;酸性流出物;酸性废水
acid electric(al) steel 酸性电炉钢
acid electrode 酸性焊条
acid elevator 升酸器;扬酸器
acid embossing 蚀刻;酸性压花(采用氟氢酸在玻璃表面形成闪光装饰花纹);酸蚀刻浮雕;酸蚀;酸浸析;玻璃酸蚀法
acid embossing of glass 玻璃酸蚀法
acid embrittlement 酸脆性
acid emulsion 阳离子乳液;酸性阳离子型乳浊液;酸乳剂;酸性乳液
acid environment 酸性环境
acid equivalent 酸当量
acid erosion index 酸性侵蚀指标
acid eructation 嗳酸
acid estimation apparatus 酸定量器
acid etch 酸浸蚀;浸酸;酸洗
acid-etched 酸蚀刻;酸浸蚀的
acid-etched nail 酸蚀铁钉
acid-etched plaster 酸蚀粉饰
acid etching 蚀刻的;酸蚀(刻);酸浸;玻璃蚀刻法
acid etching of glass 玻璃蚀刻法
acid etch method 酸蚀法
acid ethyl oxalate 草酸氢乙酯
acid exchange 酸性交换
acid extract 酸性萃取物;酸浸出物;酸浸提液
acid extraction 酸浸取;酸萃取
acid factor 酸度系数
acid fallout 酸性沉降物
acid-fast 抗酸的;耐酸的
acid-fast asphalt(ic)tile 耐酸沥青面砖
acid-fast bacillus 抗酸杆菌;耐酸杆菌
acid-fast bacteria 抗酸(细)菌;耐酸(细)菌
acid-fast chimney 耐酸烟囱
acid-fastess test 耐酸性试验
acid-fast joint 耐酸接头;耐酸接缝
acid-fast masonry(work) 耐酸圬工;耐酸砖石工
acid-fast mastic 耐酸玛琋脂;耐酸胶合料
acid-fastness 耐酸度;耐酸性
acid-fast paint 耐酸漆;耐酸涂料
acid-fast refractory concrete 耐酸耐火混凝土
acid-fast stain 抗酸染色
acid-fast staining 抗酸染色法
acid-fast varnish 耐酸凡立水;耐酸清漆
acid feeder 送酸器;加酸器
acid feed system 抗酸系统
acid fermentation 酸性发酵
acid firebrick 酸性耐火砖
acid fixer 酸性定影液
acid fixing bath 酸性定影液
acid flux(material) 酸性熔剂;酸性焊剂
acid foaming 酸泡沫
acid fog 酸雾
acid fog liquid 酸雾液
acid-formation stage 成酸阶段
acid former 酸化器;酸化剂;成酸物质
acid-forming food 成酸食物
acid-forming group 成酸基
acid-forming substance 成酸物质
acid fortification 酸增强处理
acid-free 无酸的
acid-free oil 无酸油;不含酸油品
acid free radical 酸自由基
acid frosting 酸蚀
acid fuchsin 酸性品红
acid fume 酸雾;酸烟(雾);酸性尘雾
acid-fume scrubber 酸雾涤气器;酸雾洗涤器
acid gas 酸性气体;酸气
acid gas absorption 酸气吸收
acid gilding 腐蚀金

acid granite rare element-bearing formation 酸性花岗岩含稀有元素建造
acid green 酸性绿
acid-ground glass 毛玻璃;酸蚀毛玻璃
acid group 酸根
acid-handling pump 酸液泵;酸液输送泵
acid hardening fixer 酸性坚膜定影剂
acid heat 酸性反应供暖;酸反应供热;耐酸;酸反应热
acid heat test 酸热试验
acid hot pool 酸性热水塘
acid humidification 酸性腐殖化作用
acid humus 酸性腐殖质
acid hydrolysate 酸解产物
acid hydrolysis 加酸水解(作用);酸(水)解
acid hydrothermal leaching 酸性水热淋滤
acid hypo 酸性定影液
acidic 酸性的;酸式
acidic activity 酸性活度
acidic aggregate 酸性骨料
acidic asphalt 酸性地沥青
acidic barrier 酸性障
acid(ic) bleaching 酸性漂白
acidic bleaching solution 酸性漂白液
acidic bridged hydroxyl 酸性桥接羟基
acidic coal-cleaning leachate 酸性洗煤浸沥液
acidic component 酸性组分
acidic compound 酸性化合物
acidic condition 酸性状况
acidic constant 酸性常数
acidic constituent 酸性组分
acidic content 含酸量;酸度
acidic cooling water treatment 酸化法冷却水处理
acidic deposit(e) 酸性沉积
acidic deposit(ion) 酸性沉积物
acidic dilute 稀酸
acidic dye(stuff) 酸性染料
acidic dystrophic lake 酸性贫营养湖
acidic electrolysis 酸性电解
acidic electrolysis bath 酸性电解液
acidic electrolyte bath 酸性电解质
acidic extractant 酸性萃取剂
acidic feed 酸性饲料
acidic fluoride-containing wastewater 含氟酸性废水
acidic functional group 酸性官能团
acidic functional group content 酸性官能团含量
acid(ic) ground 酸性土(壤)
acidic groundwater 酸性地下水
acidic groundwater plume 酸性地下水羽
acidic group 酸基
acidic jet 酸性喷气孔
acidic lava 酸性熔岩
acidic lava flow 酸性熔岩流
acidic method 酸性法
acidic oxide 酸性氧化物
acidic oxidizing agent 酸性氧化剂
acidic particle emission 酸性微粒散发
acidic precipitation 酸性降水
acidic property 酸性
acidic reaction 酸性反应
acidic refractory 酸性耐火材料
acidic resin 酸性树脂
acidic rock 酸性岩石;酸性岩
acidic rose red 酸性玫瑰红
acidic salt 酸性盐
acidic slag 酸性炉渣;酸性矿渣
acidic snow 酸雪
acidic sodium alumin(i)um phosphate 磷酸钠铝
acidic sodium chlorate 酸性氯酸钠
acidic soil 酸性土(壤)
acidic solution 酸性溶液;酸溶液
acidic surface 酸性面
acidic titrant 酸性滴定剂
acidic waste 酸性废物
acidic waste(water) 酸性废水
acidic water 酸性水
acidiferous 含酸的
acidifiable 可酸化
acidification 酸化(作用);加酸处理;变酸
acidification effect 酸化效应
acidification-extraction process 酸化萃取法
acidification hydrolysis 酸化水解
acidification of alkaline water 碱性水的酸化
acidification of natural waters 天然水体酸化

acidification of surface waters 地表水体酸化
acidification of water body 水体酸化
acidified milk 变酸乳
acidified silage 加酸青储[贮]料
acidified softwater 酸化软水
acidified water 酸化水
acidifier 酸化器;酸化剂
acidifying machine 浸酸机
acidifying oil 酸化油
acidifying reaction kinetics 酸化反应动力学
acidifying substance 酸化物质
acid igneous rock 酸性火成岩
acidimeter 酸液比重计;酸度计;酸比重计
acidimeter for battery 电池用测酸计
acidimetric analysis 酸量滴定分析
acidimetric estimation 酸量测定
acidimetric method 酸(量)滴定法
acidimetric standard 标酸基准;标定酸溶液的基准物
acidimetry 酸量滴定法;酸定量法;酸滴定法
acid industry 制酸工业
aciding 酸蚀;假石面轻微刻痕;酸蚀法
acid inhibitor 酸抑制剂
acid-in-oil emulsion 油包酸乳液
acid-insoluble lignin 酸不溶木质素
acid-insoluble residue 酸性不溶残渣
acid ion 酸性离子
acid iron waste 酸性铁废物
acidising 酸化
acidite 酸性岩;硅质岩
acidity 酸性;酸度
acidity-alkalinity reaction of soil solution 土壤溶液的酸碱度
acidity and alkalinity test 酸碱度试验
acidity classify of spring 温泉的酸碱度分类
acidity coefficient 酸性系数;酸度系数
acidity constant 酸度常数
acidity error 酸度误差
acidity index 酸度指数;酸度指标
acidity of residue 残渣酸度
acidity of water 水的酸度
acidity reaction 酸性反应
acidity rose red dyeing wastewater 酸性玫瑰红印染废水
acidity soil 酸性土(壤)
acidity test 酸度试验;酸度检定;酸度测定;酸测定
acidization 酸化(作用);酸处理
acidize 酸处理;使酸化;用酸处理
acidized wastewater 酸化废水
acidizing 酸化;酸处理;酸蚀法;酸化处理
acidizing of wells 酸洗井
acidizing technology 酸化工艺
acid jetting 喷酸作业
acid knock-out drum 酸中和容器
acid lake 酸性湖泊
acid lake reacidification 酸性湖泊再酸化
acid lake reacidification model 酸性湖泊再酸化模型
acid-leach 酸浸出
acid leaching 酸浸出;酸浸滤
acid leaching method 酸浸法
acid leach plant 酸性浸出设备
acid lead 耐酸铅
acid light yellow 酸性嫩黄
acid-lined converter 酸性转炉
acid lined cupola 酸性冲天炉
acid lining 酸性窑衬;酸性内衬;酸性炉衬;酸性衬里;耐酸衬砌
acid liquor 酸液
acid loss 酸减量
acidluoren sodium 杂草焚
acid maceration 酸浸渍
acid magmatism 酸性岩浆作用
acid mark 酸蚀痕;酸刻
acid material 酸性材料
acid medium 酸性介质
acid metal 耐酸铜合金
acid milling 酸磨法
acid milling black 酸性磨黑
acid milling green 酸性磨绿
acid mine drain 酸性矿井水
acid mine drainage 酸性矿物排放;酸性矿山排水;矿山酸性排水
acid mine drainage from coal mine 煤矿酸性废水

acid mine(drainage)waste 酸性矿山废物
acid mine drainage wastewater 酸性矿坑污水
acid mine pollution 酸性矿山废水污染
acid mine water 酸性矿水
acid mist 酸雾
acid mist eliminator 酸雾捕集器
acid moor 酸性沼泽(地)
acid mordant dyes 酸性媒染染料
acid mordant red 酸性媒介红
acidness 酸度;酸性
acid neutralizing capacity 酸中和能力
acid neutralizing unit 酸中和装置
acid nitrile 腈
acid number 酸值;酸数;酸价
acidogenic bacteria 产酸细菌
acidogenic digestion 酸发生消化
acidoid 酸性胶体;酸胶基;似酸;类酸物质;可变酸物质;变酸的
acid oil 酸性油
Acidol dye 阿齐多染料
acidolysis 酸—酯交换;酸解(作用)
acidometer 酸液比重计;酸比重计;测酸仪;pH计;酸度计
acidometry 酸定量法;酸液定量法
acid open-hearth furnace 酸性平炉
acid open-hearth process 酸性平炉法
acid open-hearth steel 酸性平炉钢;酸性马丁炉钢
acidophil(e) 适酸的
acidophilic 嗜酸的
acidophilic bacteria 嗜酸(细)菌
acidophilic microorganism 嗜酸微生物
acidophilic plant 嗜酸植物
acidophilous 嗜酸的;喜酸的
acidophobous 嫌酸的
acidophole 嗜酸生物
acidophytes 酸土植物
acid or alkali poisoning 酸或碱中毒
acid orange 酸性桔红;酸性橙
acid ore 酸性矿石
acidoresistant 抗酸的;耐酸的
acidosis 酸中毒;酸毒症
acid oxalate 酸式草酸盐
acid oxide 酸性氧化物;成酸氧化物
acid pasting 酸溶法
acid peat 酸性泥炭
acid peat soil 酸性泥炭土
acid permanganate oxidation method 酸性高锰酸盐氧化法
acid-phase 酸相
acid-phase anaerobic digestion 酸相厌氧消化
acid-phase upflow anaerobic sludge blanket system 酸相升流厌氧污泥层系统
acid phosphate 酸性磷酸盐;酸式硫酸盐
acid phosphatic manures 磷酸盐肥料
acid phosphat of lime 重过磷酸钙
acid pickle 废酸液
acid pickling 酸渍;酸洗;酸蚀;酸浸
acid pickling wastewater 酸浸废水
acid pig 酸性生铁
acid pigment 酸性颜料
acid pit water 酸性矿坑水
acid plagioclase 酸性斜长石
acid plant 酸土植物;制酸厂
acid poisoning 酸中毒
acid polishing 酸蚀(法)抛光;酸蚀法磨光;酸抛光;酸法制浆;酸处理抛光;酸处理法
acid pollution 酸污染
acid potassium acetate 酸式醋酸钾
acid potassium permanganate index method 酸性高锰酸钾指数法
acid potassium permanganate oxidation method 酸性高锰酸钾氧化法
acid potassium sulphate 硫酸氢钾
acid precipitation 酸雨量;酸雨;酸沉降
acid precipitation system 酸沉降体系
acid preservative method 酸防腐法
acid-processed 酸处理的
acid-producing bacteria 产酸细菌
acid producing material 产酸材料
acid-proof 防酸的;耐酸的
acid-proof alloy 耐酸合金
acid-proof asphalt(ic)tile 耐酸沥青面砖
acid-proof asphalt varnish 耐酸沥青清漆

acid-proof battery box 耐酸电池箱
acid-proof brick 耐酸砖
acid-proof casting 耐酸铸件
acid-proof cast iron 耐酸铸铁
acid-proof cement 耐酸水泥;耐酸胶泥
acid-proof ceramic pipe 耐酸陶(瓷)管
acid-proof ceramic products 耐酸陶瓷制品
acid-proof coating 耐酸涂料;耐酸面层;耐酸防腐层;防酸面层;防酸保护层
acid-proof concrete 耐酸混凝土;防酸混凝土
acid-proof enamel 耐酸搪瓷;耐酸瓷漆
acid-proof floor 耐酸地面;耐酸地板
acid-proof gloves 耐酸手套
acid-proof hose 耐酸软管
acid-proofing 防酸
acid-proofing agent 防酸剂
acid-proofing ceramic pipe 耐酸陶管
acid-proofing flooring 防酸地面
acid-proof ink 耐酸油墨;耐酸印剂
acid-proof lining 耐酸内衬;耐酸衬料;耐酸衬里
acid-proof mastic 抗酸胶;耐酸玛埃脂
acid-proof material 耐酸(防腐)材料;防酸材料
acid-proof mortar 耐酸砂浆;防酸砂浆
acid-proof paint 耐酸涂料;耐酸漆
acid-proof paper 耐酸纸
acid-proof refractory 耐酸耐火材料;防酸耐火材料
acid-proof refractory material 耐酸耐火材料
acid-proof spirit varnish 耐酸挥发清漆
acid-proof stoneware 耐酸陶器;耐酸炻器;耐酸缸器;耐酸缸瓷器
acid-proof sulphuric mortar 硫黄耐酸砂浆;耐酸硫黄砂浆
acid-proof test 耐酸试验
acid-proof trap 防酸存水弯;耐酸存水弯;耐酸衬料;耐酸衬里
acid-proof valve 耐酸阀
acid-proof varnish 耐酸清漆
acid protection 酸防护;耐酸;防酸
acid protection coating 防酸表层;防酸涂层
acid pump 酸液泵;酸泵
acid purification system 酸净化系统
acid radical 酸基;酸根
acid rain 酸雨;含酸雨
acid rain damage 酸雨危害
acid rain monitoring 酸雨监测
acid rain(monitoring)network 酸雨监测网
acid ratio 酸比
acid reaction 酸反应
acid receiver 储酸器;盛酸器
acid reclaiming plant 酸回收装置;酸回收设备
acid recovery 酸回收
acid-recovery plant 废酸回收厂;酸回收设备
acid recyling 酸回收
acid red 酸性红
acid-refined linseed oil 酸漂亚麻仁油
acid refined oil 酸漂油;酸精制油
acid refined tall oil 酸漂妥尔油
acid refractory 酸性耐火砖;高硅耐火材料
acid refractory brick 高硅耐火砖
acid refractory product 酸性耐火制品
acid-regenerated 酸再生的
acid regression stage 酸性消退阶段
acid-repelling 防酸的
acid residue 酸渣;酸性残渣;酸性残余物;酸残渣
acid-resistance 抗酸力;抗酸性;耐酸度;耐酸性;防酸(性)
acid-resistance casting 耐酸铸件
acid-resistance metal 耐酸合金
acid-resistance property 耐酸性能
acid-resistance rate 耐酸率
acid-resistance test 抗酸(性)试验;耐酸(性)试验
acid-resistant 抗酸的;耐酸的
acid-resistant brick 耐酸砖
acid-resistant castable 耐酸浇注料
acid-resistant cement mortar 耐酸水泥砂浆
acid-resistant ceramic jet pump 耐酸陶瓷喷射泵
acid-resistant ceramic pipe 耐酸陶瓷管
acid-resistant ceramic pipeline 耐酸陶瓷管道
acid-resistant ceramic pump 耐酸陶瓷泵
acid-resistant ceramic slurry pump 耐酸陶瓷砂浆泵
acid-resistant coating 耐酸涂层
acid-resistant flooring 耐酸地面;耐酸地板

acid-resistant glaze 耐酸釉
acid-resistant gull(e)y 耐酸勾缝
acid-resistant insulation 耐酸绝缘
acid-resistant intalox saddle packing 耐酸陶瓷矩鞍形填料
acid-resistant mastic 耐酸胶合料;耐酸玛琋脂
acid-resistant material 抗酸材料
acid-resistant mortar 耐酸砂浆
acid-resistant motor 耐酸电动机
acid-resistant paint 耐酸漆;耐酸涂料
acid-resistant paving 防酸砖地面(带环氧面层)
acid-resistant pipe 耐酸管
acid-resistant refractory brick 耐酸耐火砖
acid-resistant refractory concrete 耐酸耐火混凝土
acid-resistant steel 耐酸钢
acid-resistant stoneware blower 耐酸陶瓷鼓风机
acid-resistant stoneware container 耐酸陶瓷容器
acid-resistant stoneware pump 耐酸陶瓷泵
acid-resistant stoneware tower 耐酸陶瓷塔
acid-resistant stoneware valve 耐酸陶瓷阀门
acid-resistant system 耐酸设备
acid-resistant test 耐酸试验
acid-resistant thermotolerant ceramic brick 耐酸耐温砖
acid-resistant tile flooring 耐酸缸砖地面
acid-resisting 抗酸的;抗酸性;耐酸的
acid-resisting aggregate 耐酸集料;耐酸骨料
acid-resisting alloy 抗酸合金;耐酸合金
acid-resisting asphalt 耐酸沥青
acid-resisting bacteria 耐酸细菌
acid-resisting binder 耐酸胶结料
acid-resisting bituminous varnish 耐酸沥青清漆
acid-resisting black varnish 耐酸沥青黑清漆
acid-resisting brick 耐酸砖
acid-resisting casting 耐酸铸件
acid-resisting cement 耐酸水泥
acid-resisting ceramic tile 耐酸瓷砖
acid-resisting ceramic veneer 耐酸陶瓷板
acid-resisting ceramic ware 耐酸陶瓷制品
acid-resisting concrete 抗酸混凝土;耐酸混凝土;防酸混凝土
acid-resisting flooring 防酸地面
acid-resisting glass coil 耐酸玻璃蛇管
acid-resisting glazed tile 耐酸釉面砖
acid-resisting lining 耐酸衬料;耐酸衬里
acid-resisting material 耐酸材料;防酸材料
acid-resisting mortar 耐酸砂浆;耐酸灰泥;防酸砂浆
acid-resisting paint 抗酸涂料;耐酸涂料;耐酸漆
acid-resisting pump 耐酸泵
acid-resisting standard 耐酸标准
acid-resisting steel 耐酸钢
acid-resisting stoneware 耐酸缸瓦器;耐酸陶器
acid-resisting tile 耐酸面砖
acid resistivity 耐酸性
acid-restoring plant 废酸回收装置
acid-ripening stage 酸性发酵成熟期
acid river 酸性河
acid rock 酸性岩(石)
acid rocks 酸性岩类
acid salt 酸性盐;酸式盐
acid saponification 加酸皂化
acid seal paint 酸封漆;防酸封漆
acid separation air-floated 酸离气浮
acid separator 除酸器
acid settler 酸沉降器;酸沉降池;酸沉降槽
acid settling tank 酸淀池;酸沉降槽
acid silage 酸青储[贮]料
acid siphon 酸虹吸管
acid slag 酸性渣;酸性炉渣;酸性矿渣
acid sludge 酸渣;酸性淤渣;酸性污泥;酸性废渣
acid-sludge asphalt 酸渣(石油)沥青;酸渣(地)沥青
acid-sludge from petroleum refinery 炼油厂废酸渣
acid-sludge pitch 酸渣柏油脂;酸渣(煤焦)沥青
acid slurring 酸浆法
acid slurry 酸性稀浆;酸性水泥浆
acid smut 酸性污物;酸性碳黑;酸性片状炭黑;酸性煤尘;干硫煤尘
acid soaking 酸渍
acid sodium carbonate 酸式碳酸钠
acid soil 酸土;酸性土(壤)
acid soil neutralization 酸性土的中和
acid(soil)plant 酸性植物

acid-soluble 溶于酸液的;可溶于酸的
acid-soluble lignin 酸溶木质素
acid-soluble oil 酸溶油
acid-soluble rate 酸溶率
acid solution 酸溶液
acid soot 酸烟垢;酸性烟灰
acid spar 酸性萤石
acid splitting 酸分解法;加酸分解
acid spot resistance 耐酸浸蚀性
acid spring 酸性泉
acid stage oil 酸性油
acid stain 酸性沾污;酸性染(色)剂;酸斑
acid stamping 酸蚀商标
acid steel 酸性炉钢;酸性钢
acid stop bath 酸性停影液
acid storage battery 酸性蓄电池组
acid(storage)tank 储[贮]酸罐;储[贮]酸槽;储酸罐
acid store room 储酸室
acid streaking (在玻璃表面进行的)酸蚀条纹;(在玻璃表面进行的)酸性蚀斑
acid strength 酸强度
acid strengthening 酸增强
acid substance 酸性物质
acid sulfate-chloride spring 酸性硫酸盐—氯化物泉
acid sulfate soil 酸性硫酸盐土
acid sulfate spring 酸性硫酸泉
acid sulfite semichemical pulp 酸性亚硫酸盐半化学纸浆
acid sulphate chloride thermal water 酸性硫酸盐氯化物型热水
acid sulphate thermal water 酸性硫酸盐型热水
acid sulphite 酸式亚硫酸盐
acid suspension 酸性悬液
acid swelling 酸胀法
acid tank 酸槽
acid tanker 运酸船
acid tank truck 酸罐车
acid tar 酸渣焦油(沥青);酸渣柏油;酸性焦油沥青油渣;酸焦油
acid test 酸性试验
acid tester 耐酸试验器
acid test ratio 流动资产负债比;酸性测验比率
acid tolerant species 耐酸品种
acid tower 酸塔;储[贮]酸塔
acid transformer 绝缘油酸化变压器
acid-treated clay 酸化黏土
acid-treated finish (混凝土的)酸(性)处理整面法
acid-treated method 酸处理法
acid-treated oil 酸洗油
acid-treated plaster 酸处理粉饰
acid-treated sodium silicate 酸化硅酸钠
acid treatment 酸蚀刻;酸清洗;酸性处理;酸化处理;酸处理;加酸处理
acid tube 酸管
acidulae 碳酸矿水
acidulant 酸化剂
acidulate 酸化
acidulated rinsing 酸浴冲洗
acidulated water 酸化水
acidulating agent 酸化剂
acidulation 酸化
acidulous 微酸的;有酸味的;带酸味的
acidulous element 成酸元素
acidulous spring 微带酸性的泉;酸性矿泉
acidulous water 微酸性水;微带酸性的水;酸性水
acidum aceticum 醋酸
acidum benzoicum 苯甲酸
acidum carbolicum 碳酸
Acidur 阿西杜尔铁硅合金
aciduric bacteria 耐酸细菌;耐酸菌
acid value 酸值;酸价
acid vapo(u)r 酸性蒸气
acid violet 酸性紫
acid volatile sulfide 酸性挥发硫化物
acid wash colo(u)r test 酸性差色试验;酸洗消色试验;硫酸着色试验
acid-washed active carbon 酸洗活性炭
acid-washed darkened sand 酸洗黑砂
acid-washed plaster 酸蚀粉刷;酸蚀粉饰;酸洗粉刷;酸化粉刷处理
acid wash(ing) 酸洗
acid washing liquor 酸洗(废)液
acid washing test 酸洗试验

acid waste 酸性废水
acid waste assimilation 酸性废水的同化作用
acid waste liquid 酸性废液
acid waste liquor 酸性废液
acid waste sludge 酸性废污泥;废酸渣
acid waste utilization 废酸渣利用
acid water 酸水
acid water corrosion 酸性水腐蚀
acid water discharge 酸性水排放
acid water drainage 酸性水排放
acid water pollution 酸性废水污染
acid wood 干馏材
acidyl 酸基
acidylation 酰化(作用)
acierage 钢化;表面钢化;渗碳
acieral 铝基合金
acieration 碳化;金属镀铁硬化
acies 缘(端)
aciform 针状的;针形的
Acilan dye 阿西仑染料
aciniform 葡萄状的
aci-nitro 异硝基
acinoscope 放射量测定器
acinose 细粒状
acinose structure 粒状构造
acinous 细粒状
Acipenser sinensis 中华鲟
acisculis 石工小锤
acitinide 锕类
acitrin 苯基金鸡纳酸乙酯;阿齐特林
aciular structure 针状组织
acivinil alcohols 不饱和烯醇类
ack-ack ship 防空船
acker 怒潮
Acker cell 阿克电解池
Ackeret method 阿克莱特方法
Ackermann front axle 阿克曼车前轴
Ackermann ribbed floor 阿克曼肋形楼面
Ackermann's function 阿克曼函数
Ackermann steering 阿克曼转向;阿克曼车转向驾驶盘
Ackerman steering gear 阿克曼转向机构
Acker process 阿克法
ackey 硝酸硫酸混合酸浸渍液(电镀用)
acknoeledgment button 警惕按钮
acknowledge 确认;收到确认;应答;承认
acknowledge character 确认(字)符;肯定字符
acknowledge cycle 应答周期
acknowledged 确认的
acknowledged information transfer service 确认式信息传递服务
acknowledged rule of technology 公认的技术规则
acknowledged run flag 确认运行标志
acknowledged sequence number 确认序号
acknowledge enable 确认允许;允许应答
acknowledge input 确认输入
acknowledge interrupt 肯定中断;确认中断
acknowledgement 认可;收条;应答
acknowledgement confirmation 确认
acknowledgement of debt 确认债务;债务的承诺
acknowledgement of orders 接受订单;接受定货
acknowledge(ment)signal 接受信号;承认信号
acknowledge output 确认输出
acknowledger 瞭望装置
acknowledge receipt 证实收到
acknowledge sheet 验收回单
acknowledging contactor 认收接触器
acknowledging relay 认收继电器;接受继电器
acknowledging switch 认收开关
acknowledgment 承认书
acknowledgment and merger 承认与合并
acknowledgment lamp 证实指示灯
acknowledgment lever 警惕手柄
acknowledgment money 承认金
acknowledgment of a debt 承认债务;承认负债
acknowledgment of an obligation 承认债务
acknowledgment of liability 确认责任
acknowledgment of receipt 收妥通知;收据
acknowledgment signal 证实信号
aclantate 阿克兰酯
A class 甲级
aclastic 无折光的
acle (沉重坚硬耐久的)阿光木材
aclimatic soil formation 非气候性成土作用

aclinal 无倾角的;不倾斜的;水平的;无倾斜角的
acline 水平地层;无倾线;零等倾线
acline line 无倾线
aclinic 无倾斜角的;不倾角的;水平的
aclinic line 无倾线;地磁赤道(线);磁赤道
aclutter 拥挤的;乱糟糟的
acme 极度;极点;顶点;最高点;顶峰期
Acme Commodity and Phrase Code 爱克米商品及用语密码
acme harrow 阔刀齿耙;刀齿耙
Acme screw thread 平口螺纹;爱克米制螺纹
Acme thread 平口螺纹;梯形螺纹;爱克米螺纹
Acme thread ga(u)ge 爱克米螺纹规;梯形螺纹规
Acme thread tap 梯形丝锥;爱克米丝锥
acme-zone 顶峰带
acmite 锥辉石
acmite-augite 霓辉石
acmite-granite pegmatite 绿辉花岗伟晶岩
acmite trachyte 绿辉粗面岩
acmonital 意大利硬币铜
acnodal cissoid 弧点蔓叶线
acnode 极点;孤立点;孤立波;孤点
a coat of paint 一度漆
acock 向上卷
a' cockbill 悬吊;吊锚
A-coefficient 自发发射系数
acoin 阿可因
acolite 低熔合金
acomous 无发的
acone 无晶锥
aconiazid 阿克烟肼
aconitic acid 乌头酸;丙烯三羧酸
a connected sequence 一条龙
A-connection 三角形接线;三角形接法
a considerable depth 相当大的深度
a coordinated process 一条龙
acoradiene 菖蒲二烯
acorane 菖蒲烷
acorenone 菖蒲烯酮
acoretin 菖蒲脂
acoric acid 菖蒲酸
acorin(e) 菖蒲苦苷;菖蒲苷
acorite 锆土;锆石
acorn 桅顶风标帽;橡子;橡树子;橡树果;橡实夹;柱头圆球饰;柱头桅顶装饰;整流罩
acorn barnacle 藤壶
Acorn cell 通信机偏压用电池;橡实形电池
acorn cup 橡碗
acorn nut 圆(顶)螺母;圆顶螺帽;盖形螺母;盖帽式螺帽;半圆头螺帽
acorn shell 藤壶
acorn tube 橡实管
acorn worm 柱头虫
acorone 菖蒲酮
Acorus 菖蒲属
acorus aclamus 菖蒲
acosmia 违和
acoubuoy 声监听仪
acoumeter 听力计;听力测验器;测听计
acoumetry 测听(技)术
acount currency 记账货币
a course of bricks 一皮砖
acousimeter 听力计;测听计;测声器
acoustele 吸声天花板;吸声顶棚
acoustextile 吸声织物
acoustic(al) 声障;声学的;声波的
acoustic(al) absorbent 吸声(剂);消音材料;吸音材料;吸声材料;隔音材料
acoustic(al) absorbing device 吸声设施;吸声设施
acoustic(al) absorbing device of tunnel 隧道吸声设施
acoustic(al) absorbing medium 吸声介质
acoustic(al) absorption 声吸收
acoustic(al) absorption coefficient 声吸收系数
acoustic(al) absorption loss 声吸收损耗
acoustic(al) absorption wedge 吸声尖劈
acoustic(al) absorptivity 吸声力;声吸收系数;声吸收率;吸音率;吸声系数;吸声率
acoustic(al) absorptivity measurement 吸声测量
acoustic(al) admittance 声导纳
acoustic(al) adsorption coefficient 吸声系数
acoustic(al) adsorption factor 吸声系数
acoustic(al) adsorption loss 吸声损失
acoustic(al) alarm(unit) 声(响)警报器;音响警报器
acoustic(al) altimeter 声学高度计;声学测高计;声响测高计;回声测高计
acoustic(al) amplifier 声(波)放大器
acoustic(al) amplitude log 声波幅度测井
acoustic(al) amplitude log curve 声波幅度测井曲线
acoustic(al) amplitude logger 声波幅度测井仪
acoustic(al) analysis 声学分析
Acoustic(al) and Insulating Materials Association 美国声学及隔声材料协会
acoustic(al) anemometer 声学风速仪;声学风速计;声学风速表
acoustic(al) anomaly 声学异常
acoustic(al) array 声线列阵;声基阵
acoustic(al) attenuation 声的衰减
acoustic(al) attenuation coefficient 声衰减系数
acoustic(al) attenuation constant 声衰减常数
acoustic(al) backing 吸声背衬;隔音填料
acoustic(al) baffle 声障板;隔声板
acoustic(al) barrier 声屏障;隔音墙;隔声屏障;隔声层;隔声壁;隔声板
acoustic(al) beacon on seabottom 海底声标
acoustic(al) beam 声波束
acoustic(al) beam deflector 声束偏转器
acoustic(al) bearing 声(响)方位;声测向
acoustic(al) bed-load recorder 声学自记推移质测定器
acoustic(al) behavio(u)r 声学特性;吸声性能;声学性能
acoustic(al) blanket 吸声(隔)层;吸声垫
acoustic(al) blanking 消声作用
acoustic(al) block 吸声砌块
acoustic(al) board 共鸣板;隔声板;吸音板;吸声板;隔音板
acoustic(al) booth 隔声小室;吸声间;隔音间
acoustic(al) box 消声罩
acoustic(al) branch 声支路;声频支路;声频支
acoustic(al) brick 吸声砖;隔音砖;隔声砖
acoustic(al) bridge 声桥
acoustic(al) bulkhead 隔声舱壁
acoustic(al) buoy 声浮标;音响浮标;音频浮标;发声浮标
acoustic(al) burglar alarm 防盗警报器
acoustic(al) calibrator 校正声源
acoustic(al) capacitance 声容
acoustic(al) capacitance unit 声容单位
acoustic(al) cassette 吸声镶板;吸声护壁板;吸声贴面板
acoustic(al) cavitation 声空化
acoustic(al) ceiling 吸音平顶;吸音顶棚;吸声平顶;吸声吊顶
acoustic(al) ceiling board 吸声顶棚;吸声顶棚;吸音天花板;吸声天花板
acoustic(al) ceiling paint 吸声天花板涂料;吸声顶棚涂料
acoustic(al) ceiling panel 吸声天花板
acoustic(al) ceiling system 吸音顶棚体系;吸声顶棚系统
acoustic(al) ceiling tile 吸声顶棚砖;吸声吊顶砖
acoustic(al)-celotex board 纤维隔声板;隔音(纸)板;赛璐特克吸声板
acoustic(al) celotex(tile) 纤维隔声板;甘蔗纤维吸声板;赛璐特克吸声板
acoustic(al) channel 声信号通道;声道
acoustic(al) characteris impedance 声特性阻抗
acoustic(al) characteristic 声学特征;声学特性;音响特性
acoustic(al) cladding 隔音罩
acoustic(al) clarifier 传声明晰器;声干扰消除器
acoustic(al) clip 隔声垫夹
acoustic(al) cloud 声波反射云;回声测深时的隔音层
acoustic(al) coagulation 声波凝聚作用
acoustic(al) coating 吸声涂料
acoustic(al) coefficient 声响系数
acoustic(al) coffer 吸声镶板;吸声隔板;吸声夹层板
acoustic(al) colo(u)ration 声配置;室内声学条件
acoustic(al) comfort index 声舒适度指数
acoustic(al) compatibility 声相容性
acoustic(al) compliance 声顺;声容抗
acoustic(al) condition factor 听感因数;声学条件因素;听觉因数
acoustic(al) conductance 声导;传声性
acoustic(al) conduction 声导
acoustic(al) conductivity 声导率;传声性
acoustic(al) conductivity of wood 木材传声性
acoustic(al) conglomeration 声波凝聚
acoustic(al) constant of rock 岩石声阻抗
acoustic(al) construction 声学结构;声学构造;吸音构造;吸声建筑;吸声构造;隔音建造法;隔声构造
acoustic(al) construction material 声学处理建筑材料
acoustic(al) construction method 声学处理施工(法)
acoustic(al) control 声控
acoustic(al) control and telemetry system 声控遥测系统
acoustic(al) control glass 消声玻璃;隔声玻璃
acoustic(al) correction 声响改善;声响处理;音质校正
acoustic(al) corrector 测音诸元修正器
acoustic(al) correlation log 声相关测井仪
acoustic(al) countermeasure 声对抗
acoustic(al) coupler 音频调制解调器;声音耦合器;声耦合器
acoustic(al) coupling 声音耦合
acoustic(al) coupling agent 声耦合剂(用于材料的超声无损检验)
acoustic(al) cover(ing) 隔声衬砌层
acoustic(al) criterion 音质评价标准
acoustic(al) current meter 声响式流速计;音响流速计;音响海流计
acoustic(al) curtain 隔声幕
acoustic(al) curve acrossing boreholes 声波透视曲线
acoustic(al) damper 消声器
acoustic(al) damping 声阻尼
acoustic(al) dazzle 狂响聚声;声致眩晕;声干扰;异常聚声
acoustic(al) decoy 声响假目标
acoustic(al) deflection circuit 声偏转电路
acoustic(al) deflector 声偏转
acoustic(al) delay 声延迟
acoustic(al) delay line 声(音)延迟线
acoustic(al) delay-line memory 声延迟线存储器
acoustic(al) delay-line storage 声延迟线存储器
acoustic(al) depth finder 回声测深仪
acoustic(al) depth sounding 回声测深(法);声波测深(法)
acoustic(al) design 声学设计;音质设计
acoustic(al) detection 声(波)探测
acoustic(al) detector 声检测器;声检波器;声波探伤仪;声波探测器;声波定位器;音响探测器
acoustic(al) diffraction 声衍射
acoustic(al) diffraction grating 声衍射栅
acoustic(al) diffusion 声扩散
acoustic(al) dipole 声偶极子
acoustic(al) direction-finder 声响探向器;声响探向仪
acoustic(al) discontinuity 声不连续性
acoustic(al) dispersion 声频散射;声频散
acoustic(al) displacement detector 声位移探测器
acoustic(al) dissipative element 声耗能元件
acoustic(al) distortion 声畸变
acoustic(al) disturbance 声扰动;声干扰
acoustic(al) door 隔声门;吸声门;隔音门;吸音门
acoustic(al) Doppler current meter 声波多普勒测流计
acoustic(al) Doppler current profiler 声波多普勒水流断面测绘仪;超声波多普勒流量断面仪
acoustic(al) Doppler current recorder 声波多普勒测流计
acoustic(al) Doppler fixing system 声学多普勒定位系统
acoustic(al) Doppler positioning system 多普勒声学定位系统
acoustic(al) Doppler technology 声学多普勒技术
acoustic(al) doublet 声偶极子
acoustic(al) duct 声波导
acoustic(al) duct lining 吸声管道衬砌;风道隔音衬里
acoustic(al) echo repeater 回声重发器
acoustic(al) effect 声(学)效应;声响效果
acoustic(al) effects generator 音响效果发生器
acoustic(al) effects mixer 音响效果混合器
acoustic(al) efficiency 声效率

acoustic(al) element 声学元件
acoustic(al) embedded strain ga(u)ge 回声埋入式应变计
acoustic(al) emission 声发射
acoustic(al) emission analysis technology 声发射分析技术
acoustic(al) emission detection system 声发射探测系统
acoustic(al) emission monitor(ing) 声发射监测
acoustic(al) emission monitoring system 声发射监测系统
acoustic(al) emission rate 声发射率
acoustic(al) emission signal 声发射信号
acoustic(al) emission source 声发射源
acoustic(al) emission spectrum 声发射频谱
acoustic(al) emission technique 声发射技术
acoustic(al) emission technology 声发射技术
acoustic(al) emission test 声发射检测
acoustic(al) emission testing 声发射试验
acoustic(al) emission wave 声发射波
acoustic(al) emissive method 声发射法
acoustic(al) enclosure 隔声罩;隔声间;隔音罩;隔音间
acoustic(al) enclosure equipment 隔声设施
acoustic(al) energy 声能(量)
acoustic(al) energy density 声能密度(单位为 J/m^3)
acoustic(al) energy flow density 声能流密度
acoustic(al) energy ratio 声能比
acoustic(al) engineering 声学工程;声工程学
acoustic(al) environment 声(学)环境
acoustic(al) environment design 声环境设计
acoustic(al) environment laboratory 声环境试验室
acoustic(al) equipment 声学设备
acoustic(al) excitation 声激励
acoustic(al) experiment 声学试验
acoustic(al) exploration 声波勘探;回声勘探(法)
acoustic(al) facing 声学处理(面)层
acoustic(al) factor 吸声系数;声学因素
acoustic(al) Faraday rotation 声法拉第旋转
acoustic(al) fatigue 声(致)疲劳;噪声疲劳
acoustic(al) fatigue test 声疲劳试验
acoustic(al) feedback 声回授;声反馈
acoustic(al) feedback suppression 声反馈抑制;机震抑制
acoustic(al) felt 吸音毡;吸声毡
acoustic(al) fiber board 声学处理纤维板;隔声纤维板;吸声纤维板
acoustic(al) fiber building board 建筑用吸声纤维板
acoustic(al) fidelity 声音逼真度;声音保真度;声保真度;音质
acoustic(al) field 声场
acoustic(al) field fluctuation 声场起伏
acoustic(al) figure 声图形
acoustic(al) filter 滤声器;消声器
acoustic(al) filter board 吸音板;滤音板
acoustic(al) filtering 声的过滤
acoustic(al) fix 声响船位;声定位
acoustic(al) fixing and ranging 声定位和测距
acoustic(al) fix ranging 声定位
acoustic(al) flow meter 声响测流仪
acoustic(al) focus(ing) 声聚焦
acoustic(al) fog 声雾
acoustic(al) fog signal 音响雾号
acoustic(al) foil 消声金属箔;声学处理金属箔;吸声箔
acoustic(al) form board 声衬板
acoustic(al) free field 自由声场
acoustic(al) frequency 声频
acoustic(al) frequency amplifier 声频放大器
acoustic(al) frequency generator 声频发生器
acoustic(al) frequency signal 声频信号
acoustic(al) fuse 声引信;感应引信
acoustic(al) gas analyzer 声学气体分析器
acoustic(al) gas purification 超声净化气体(法);超声净化处理
acoustic(al) generator 发声器;发生器
acoustic(al) glass 隔音玻璃;声控玻璃;隔声玻璃
acoustic(al) grating 声栅;声光栅
acoustic(al) guidance 声学制导
acoustic(al) guiding 声制导
acoustic(al) guiding system 声制导系统
acoustic(al) hangover 声残余;声的迟滞

acoustic(al) heating 声加热
acoustic(al) hologram 声全息照片;声全息图
acoustic(al) holographic system 声全息系统
acoustic(al) holography 声学全息术;声全息照相术;声波全息照相术;声波全息学;声波全息术
acoustic(al) holography system 水声全息系统
acoustic(al) holography with detector array 布阵声全息图
acoustic(al) homing 声制导;音响修正
acoustic(al) homing device 声自导装置;声学寻的装置
acoustic(al) homing head 声自动引导头
acoustic(al) homing system 声自导系统;声制导系统;声引导系统;声导引系统
acoustic(al) horn 射声器
acoustic(al) illumination 声波照射
acoustic(al) image 声像
acoustic(al) image converter 声像转换器;声像转换管;声像变换器
acoustic(al) imaging 声成像
acoustic(al) impedance 声阻抗;声抗
acoustic(al) impedance factor 声阻抗因子
acoustic(al) impedance ratio 声阻抗比
acoustic(al) impulse 声脉冲
acoustic(al) indicator 声学指示器;声学显示器;声学测量仪
acoustic(al) inertance 声质量;声惯量;声感抗
acoustic(al) inertia 声惯量
acoustic(al) input device 声(音)输入装置
acoustic(al) instability 声学不稳定性
acoustic(al) instrinsic impedance 声特性阻抗
acoustic(al) instrument 声学仪器;声学仪表
acoustic(al) insulation 隔声(材料);声绝缘;隔音;隔声性
acoustic(al) insulation board 音响绝缘板;隔声板
acoustic(al) intensity 声强度;声强
acoustic(al) intensity level 声强级
acoustic(al) interaction 声互作用
acoustic(al) interference 声干涉;声波干扰
acoustic(al) interferometer 声(波)干涉仪
acoustic(al) intrusion detector 声入侵探测器;声干涉探测器;声干扰探测
acoustic(al) intrusion protection 声入侵防护;防声扰装置
acoustic(al) investigation 声学试验;声学探测;音响测量
acoustic(al) ionization 声致电离
acoustic(al) jamming 声干扰
acoustic(al) labyrinth 声迷宫
acoustic(al) lattice vibration 声学点阵振动
acoustic(al) lay-in panel 吸声镶板
acoustic(al) lens 声透镜
acoustic(al) lens guide 声透镜定向
acoustic(al) level 声级
acoustic(al) levitation 声悬浮
acoustic(al) line 声传输线
acoustic(al) line of position 声(响)位置线
acoustic(al) lining 吸声面层;吸声衬垫
acoustic(al) load 声负载;声负荷
acoustic(al) load cell 回声荷载盒
acoustic(al) locating device 声定位器;声定位仪;音响定位仪
acoustic(al) locating installation 声定位装置
acoustic(al) log 水声计程仪;声波测井;声速测井
acoustic(al) log curve 声波测井曲线
acoustic(al) logging 声波(法)测探;声波测井;声速测井
acoustic(al) loss 声损失
acoustically satisfactory auditorium 声响处理合适的会堂
acoustically transparent vessel 透声压容器
acoustically treated construction 声处理结构
acoustic(al) Mach meter 声马赫计
acoustic(al) magnetic mine 磁声水雷;声响磁性水雷;音响磁性水雷
acoustic(al) marine speedometer 声航速仪
acoustic(al) maser 声脉塞
acoustic(al) masonry wall 消声碎石墙;消声砖石墙;消声圬工墙
acoustic(al) mass 声质量
acoustic(al) mass reactance 声质量抗
acoustic(al) material 隔声材料;声学材料;吸音材料;吸声材料;音响材料;隔音材料
acoustic(al) maximum 最大音响

acoustic(al) measurement 声学量度;声学测量
acoustic(al) measurement of suspended solid 悬移质的超声测量
acoustic(al) measuring system 声学测量系统
acoustic(al) measuring unit 声学测量装置;声学测量单元
acoustic(al) mechanism 发声机构
acoustic(al) medium 声介质
acoustic(al) memory 超声波延迟线存储器;声音存储器;声存储器;声储存器
acoustic(al) metal ceiling 消声金属天花板;消声金属顶棚
acoustic(al) metal deck 隔声金属面层;隔声金属面板
acoustic(al) meter 测声仪;测声计;比声计
acoustic(al) method 声学(方)法
acoustic(al) method acrossing borehole 声波透视
acoustic(al) microscope 声显微镜
acoustic(al) mill feed control 磨声喂料控制
acoustic(al) mill load control 磨机负荷声音控制;磨声喂料控制
acoustic(al) mine 声引信水雷;声响水雷;音响水雷
acoustic(al) mirror 声镜;声反射镜
acoustic(al) mismatch 声失配
acoustic(al) mode 声模
acoustic(al) model 声学(建筑)模型
acoustic(al) model(l)ing 厅堂的声学模型分析
acoustic(al) model test 声学模型试验
acoustic(al) modem 声调制解调器
acoustic(al) mode scattering 声模散射
acoustic(al) modification 音响修正
acoustic(al) monitoring 声监听;声监控
acoustic(al) monitoring system 声监控系统;声监测系统
acoustic(al) navigation 声(学)导航;声响导航(法)
acoustic(al) noise 声噪声;声音干扰
acoustic(al) noise measurement 声噪声测量
acoustic(al) noise test 声振试验
acoustic(al) nozzle 声波喷嘴
acoustic(al) ocean-current meter 声洋流计
acoustic(al) oceanography 声学海洋学
acoustic(al) ohm 声欧姆
acoustic(al) organ 螺旋器
acoustic(al) oscillation 声波振荡;音波
acoustic(al) oscillogram 声示波器
acoustic(al) oscillograph 声(学)示波器;声示波器
acoustic(al) output 声输出
acoustic(al) output power 声输出功率
acoustic(al) pad 隔声垫
acoustic(al) paint 消声漆;吸音油漆;吸音涂料;吸音漆;吸声(油)漆;吸声涂料
acoustic(al) panel 吸音镶板;吸音板;隔声嵌板;吸音镶板;吸音(嵌)板
acoustic(al) paper 隔音纸
acoustic(al) parameter 声参数
acoustic(al) particle size analyzer 声学粒度分析仪
acoustic(al) partition 活动隔声板;活动隔音板;吸声隔墙
acoustic(al) pattern 声波图
acoustic(al) pay-in board 吸音天花板
acoustic(al) perforated gypsum board 石膏吸声板;穿孔石膏吸声板
acoustic(al) performance 吸声特性
acoustic(al) perfume 遮掩噪声的音响
acoustic(al) permeability 透声率
acoustic(al) permeability of wood 木材透声性
acoustic(al) perspective 声的立体感
acoustic(al) phase 声相(位)
acoustic(al) phone booth 隔声电话间
acoustic(al) phonon 声频声子
acoustic(al) pick-up 拾声器
acoustic(al) picture 声波图
acoustic(al) piezometer 声发射测压计;声发射井计;回声孔压计
acoustic(al) pile test 混凝土桩声学检验法
acoustic(al) plaster 吸声墙粉;吸音灰膏;吸声灰膏;吸声粉刷(层)
acoustic(al) plaster aggregate 消声粉刷集料;消声粉刷骨料;隔声粉刷集料;隔声粉刷骨料;消声粉刷材料
acoustic(al) plaster ceiling 隔声天花板;隔声顶棚
acoustic(al) plastics 吸声塑料

acoustic(al) plenum 防噪声室
acoustic(al) pollution 声(响)污染;噪声污染
acoustic(al) positioning 声学定位;声定位
acoustic(al) positioning and ranging 声学定位和测距
acoustic(al) positioning system 声(学)定位系统
acoustic(al) position reference system 声定位参考系统;音响定位参照系统
acoustic(al) power 声响能;声响功率;声功率
acoustic(al) power density 声功率密度
acoustic(al) power level 声功率级
acoustic(al) power output 声功率输出
acoustic(al) pressure 声压
acoustic(al) pressure cell 回声压力盒
acoustic(al) pressure field 声压场
acoustic(al) pressure level 声压级
acoustic(al) pressure meter 声压计
acoustic(al) pressure value 声压值
acoustic(al) probe 探声管
acoustic(al) profile 声剖面图
acoustic(al) proof ceiling 隔音天花板;隔音顶棚
acoustic(al) proof deck 隔音天花板;隔音顶棚
acoustic(al) propagation 声传播
acoustic(al) propagation coefficient 声传播系数
acoustic(al) propagation constant 声传播常数
acoustic(al) property (建筑物的)声学性能;声性质;声响性能;声特性;吸声性能
acoustic(al) property of rock and soil 岩土的声学性质
acoustic(al) property of sea floor 海底声学特性
acoustic(al) pulse 声脉冲
acoustic(al) quality design 声质设计;音质设计
acoustic(al) quartz 传声石英
acoustic(al) radar 声雷达
acoustic(al) radiation 声辐射;噪声干扰
acoustic(al) radiation pressure 声压
acoustic(al) radiometer 声辐射计
acoustic(al) range 声学测距
acoustic(al) ranger 声测距仪
acoustic(al) ranging 声音测距法
acoustic(al) ratio 声强比;声波比
acoustic(al) reactance 声抗
acoustic(al) receiver 声接受器;声接收机
acoustic(al) reciprocity theorem 声学互易定理
acoustic(al) recognition input 声识别输入
acoustic(al) reduction coefficient 隔声量;减噪系数
acoustic(al) reduction factor 隔声量;声响消减系数;声响消减系数;降音因数;消声因数;消声系数;隔音系数
acoustic(al) reference 声学基准系统
acoustic(al) reflection coefficient 声反射系数
acoustic(al) reflection factor 声反射系数
acoustic(al) reflectivity 声反射比;声反射率;声反射性;声反射系数;声反射率;音响反射性;音响发射性
acoustic(al) reflex 声反射
acoustic(al) reflex enclosure 反音匣
acoustic(al) refraction 声折射
acoustic(al) regeneration 声再生;声反馈
acoustic(al) relay 声中继器
acoustic(al) remote control 声学遥控
acoustic(al) remote measurement 声波遥测(法)
acoustic(al) remote sensing 声遥感
acoustic(al) reproduction 重放声
acoustic(al) research vessel 水声研究船
acoustic(al) resistance 吸声阻力;声阻;吸音阻力
acoustic(al) resistance ga(u)ge 声波应变仪;声阻应变仪
acoustic(al) resistance inspection 声阻检验
acoustic(al) resonance 声共振;声共鸣;音响共鸣
acoustic(al) resonance device 共鸣器
acoustic(al) resonator 声共振器;共鸣器
acoustic(al) responder 水声应答器
acoustic(al) room 隔声室
acoustic(al) saturation 声饱和
acoustic(al) scanner 声学扫描器
acoustic(al) scattering 声(波)散射
acoustic(al) scintillation 音响起伏
acoustic(al) screen 声屏(蔽)幕;声屏遮板;吸音屏幕;隔音遮板
acoustic(al) seal 声密封
acoustic(al) sealant 隔声密料;隔声密封剂;隔声密封膏;隔声料;隔声剂;隔音密封;隔音密封材料

acoustic(al) sensor 声敏元件
acoustic(al) septum 隔音声板
acoustic(al) shadow 声影区
acoustic(al) shed 隔声遮板;声屏
acoustic(al) sheet 隔声板
acoustic(al) shell 声罩
acoustic(al) shield 隔声罩
acoustic(al) shielding 声屏(蔽)
acoustic(al) ship positioning 船舶声定位
acoustic(al) shock 声震
acoustic(al) shock absorber 冲击声吸收器;冲击声吸音器;声响消震器;声冲击吸收器;音响冲击吸收器
acoustic(al) signal 声(音)信号;声响信号;声响标志;声频信号;音响信号
acoustic(al) signal device 音响信号装置
acoustic(al) signal(l)ing equipment 声信号设备
acoustic(al) signal processing 声信号处理
acoustic(al) signal recognition 声信号识别
acoustic(al) signature 声特征;声波标记图
Acoustic(al) Society of America 美国声学学会
acoustic(al) sounder 回声探深仪;回声探测器;回声测深仪;回声测深器
acoustic(al) sounder calibration 回声测深仪校准
acoustic(al) sounding 声学探测;声波测深(法);射波测深法;回声探测;回声测深(法)
acoustic(al) sounding of atmosphere 声波大气探测
acoustic(al) source 声源
acoustic(al) source strength 声源强度
acoustic(al) spectrograph 声谱仪
acoustic(al) spectrometer 声谱仪
acoustic(al) spectroscope 声谱仪
acoustic(al) spectrum 声谱
acoustic(al) speed 声速(度);音速
acoustic(al) sprayed-on material 吸音喷涂材料;吸音喷涂材料;隔声喷涂材料
acoustic(al) sprayed-on plaster 隔声喷涂粉饰
acoustic(al) stiffness 声劲
acoustic(al) stiffness reactance 声劲度抗
acoustic(al) storage 超声波存储器;声(音)存储器;声储存器
acoustic(al) strain ga(u)ge 声速应变仪;声学应变仪;声应变仪;声响传感器;声波应变仪;声波应变计;发声式应变计
acoustic(al) stratigraphy 声波地层学
acoustic(al) streaming 声流
acoustic(al) structure 隔声构造
acoustic(al) surface strain ga(u)ge 回声表面应变计
acoustic(al) surface wave delay line 声表面波延迟线
acoustic(al) surface wave transducer 声表面波换能器
acoustic(al) surveillance 声音监视
acoustic(al) survey(ing) 声学调查;音响测量;声音测量法
acoustic(al) susceptance 声呐
acoustic(al) suspension 声支撑
acoustic(al) sweep 水声搜索
acoustic(al) switch 声开关
acoustic(al) synchronizer 声响整步器;声响同步器
acoustic(al) system 声学系统;声系统
acoustic(al) tablet 声感图像输入板
acoustic(al) technique 声学技术
acoustic(al) technology 声学技术;吸声技术
acoustic(al) telemetering 声学遥测(技)术;声学遥测
acoustic(al) telemetry 声遥测(技)术
acoustic(al) telemetry bathythermometer 声遥测海水温度计
acoustic(al) telemetry system 声学遥测系统
acoustic(al) test 声学试验;敲击试验;音响试验
acoustic(al) tex 吸声板
acoustic(al) theodolite 声学流速廓线仪;声经纬仪
acoustic(al) thermometer 声学测温法
acoustic(al) tile 吸音砖;吸音板;吸声砖;吸声瓦(板);吸音贴砖;吸声面砖;吸声板;隔音(贴)砖;吸声墙面砖
acoustic(al) tile ceiling 吸声砖平顶;吸声砖天花板;吸声顶棚
acoustic(al) timing 声音计时
acoustic(al) timing machine 声音计时机构;音响计时机构

acoustic(al) tomography 声层析法
acoustic(al) torpedo 音响鱼雷
acoustic(al) tracking 声跟踪
acoustic(al) tracking system 声响跟踪系统
acoustic(al) transducer 声换能器;声波换能器;传声器
acoustic(al) transducer array 声换能器阵列;声波换能器阵列
acoustic(al) transfer 声变化器
acoustic(al) transformer 声变换器
acoustic(al) transmission 声传输;声传递
acoustic(al) transmission coefficient 透音系数;透声系数;传声系数
acoustic(al) transmission factor 声透射系数;传声系数;声音传递因数;声响传导系数;声透系数
acoustic(al) transmission system 声学传输系统;声传输装置;声传输系统;传声系统
acoustic(al) transmissivity 声透射比;透声系数;声透射性;声透射系数;声传输性;声波传输系数;传声系数
acoustic(al) transponder 声应答器;有声应答器;音响脉冲转发器;音响脉冲收发两用机
acoustic(al) transponder navigation system 声学应答器导航系统
acoustic(al) trauma 声损伤
acoustic(al) treatment 防声措施(处理);声学处理;音响处理;隔音措施
acoustic(al) tube 传声筒
acoustic(al) type strain ga(u)ge 声学应变仪
acoustic(al) underwater survey equipment 水下测声设备
acoustic(al) unit 声学单位
acoustic(al) vault 吸音穹隆;吸声穹隆
acoustic(al) vector averaging current meter 声学矢量平均海流计
acoustic(al) velocity 声速(度)
acoustic(al) velocity correlator 声速相关器
acoustic(al) velocity in rock 岩石声波传播速度
acoustic(al) velocity log 声波速度测井
acoustic(al) velocity log curve 声波速度测井曲线
acoustic(al) velocity logger 声速测井仪
acoustic(al) velocity logging 声速测井
acoustic(al) ventilating ceiling 吸声通风天花板;吸声通风顶棚
acoustic(al) vibration 声(学)振动;声波振动
acoustic(al) waffle 吸声隔板
acoustic(al) wallboard 隔声墙板;隔声壁板;隔音壁板
acoustic(al) wallpaper 吸声墙纸
acoustic(al) wall tile 吸声墙面砖;吸音墙面砖
acoustic(al) warning 音响报警
acoustic(al) warning device 发声警告装置
acoustic(al) wave 声波;音波
acoustic(al) wave analysis 声波分析
acoustic(al) wave filter 声滤波器
acoustic(al) wave ga(u)ge 声学测波仪
acoustic(al) waveguide 声波导
acoustic(al) wave hydrophone 声波仪
acoustic(al) wave method 声波法
acoustic(al) wave theory 声波理论
acoustic(al) wave velocity 声波传播速度
acoustic(al) wedge 吸声楔
acoustic(al) well logging 声测井
acoustic(al) well sounder 钻井液面声学测深仪
acoustic(al) window 声窗
acoustic(al) window wall 隔声玻璃幕墙;隔音玻璃窗墙
acoustic(al) wood fibre board 吸声木质纤维板
acoustic(al) zero indication signal 零点指示信号
acoustic-celotex board 木质纤维隔声吊顶板;纤维隔音板
acoustic-celotex tile 木质纤维隔声吊顶板;纤维隔音板
acoustician 声学家;声学工作者;声学工程师
acousticolateral system 侧听系统
acousticon 助听器
acoustic-pulse propagation 声脉冲传播
acoustic-ranging indicator 水声测距指示器
acoustics 声学;音质;音响装置;音响学;音响(效果);传音性
acoustics consultant 声学顾问
acoustics engineer 声学工程师
acoustics expert 声学专家
acoustics insulation 消声;隔声;隔音

acoustics of churches 教堂声学
acoustics of room 房间音质
acoustictooptical image converter 声光像转换器
acoustic-wave amplifier 声波放大器
acoustimeter 声测测量计;声强仪;声强计;声强测量器;声强测量计;声级仪;噪声仪;测声仪;测声计
acoustodynamic(al) 声动力的
acoustodynamic(al) effect 声动力效应
acousto elasticity 声弹性
acoustoelectric(al) coefficient 声电系数
acoustoelectric(al) effect 声电效应
acoustoelectric(al) index 声电指数;声电变换效率
acoustoelectric(al) material 声电材料
acoustoelectric(al) oscillator 声电振荡器
acoustoelectric(al) power ratio 声电功率比
acoustoelectric(al) pressure ratio 声压电压比
acoustoelectric(al) wave 声电波
acoustoelectronics 微声电子学
acoustolith tile 吸音贴(面)砖;吸声贴砖
acoustometer 声波测量器;声强测量计;测声仪;测声计;比声计
acoustomgneto-electric(al) effect 声磁电效应
acoustomotive pressure 声压
acoustooptic(al) 声光的;声与光的
acousto-optic(al) beam steering deflector 声光束控制偏转器
acousto-optic(al) Braggdiffraction 声光布喇格衍射
acousto-optic(al) buoy 声光浮标
acousto-optic(al) cavity 声光腔
acousto-optic(al) cell 声光盒;声光调制器;声光变换元件
acousto-optic(al) crystal 声光晶体
acousto-optic(al) deflection 声光偏转
acousto-optic(al) deflection device 声光偏转器
acousto-optic(al) deflector 声光偏转器
acousto-optic(al) effect 声光效应
acousto-optic(al) filter 声光滤波器
acousto-optic(al) glass 声光玻璃
acousto-optic(al) hydrophone 声光水听器
acousto-optic(al) imaging 声光成像
acousto-optic(al) interaction 声光(相互)作用
acoustooptically tuned laser 声光调谐激光器
acousto-optic(al) material 声光材料
acousto-optic(al) medium 声光介质
acousto-optic(al) modelocker frequency double 声光锁模倍频器
acousto-optic(al) modelocking 声光锁模
acousto-optic(al) modulation 声光调制
acousto-optic(al) modulation device 声光调制器
acousto-optic(al) modulator 声光调制器
acousto-optic(al) pulse modulator 声光脉冲调制器
acousto-optic(al) quantity switch 声光光量开关
acousto-optic(al) quality factor 声光品质因数
acousto-optic(al) scanner 声光扫描器
acousto-optic(al) scan(ning) 声光扫描
acousto-optic(al) signal system 声光信号系统
acousto-optic(al) tunable filter 声光可调滤波器
acousto-optics 声光学
acoustophotorefractive effect 声光折射效应
acquaint with the market condition 了解市场情况
acqualung 深水潜水衣
acquire a target 录取物标;捕捉物标
acquired 获得性
acquired company 出盘公司;被收购公司
acquired impurity 外物杂质
acquire display 申请显示
acquired right 取得的权利;既得权利;获得的权利
acquired surplus 已取得盈余滚存
acquire full legal force 具法律效力
acquirer 让受方
acquiring 瞄准
acquiring company 受盘公司
acquiring party 让受方
acquisition 征用;征收;探测;收集;勘测;接管企业;接办企业;征购;捕捉;捕获
acquisition adjustment 取得额调整数
acquisition adjustment of assets 对资产购置的调整
acquisition and tracking radar 捕获和跟踪雷达
acquisition and tracking system 捕获与跟踪系统
acquisition appraisal 不动产购置估价
acquisition approval 拨地
acquisition commission 收购手续费

acquisition cost 购置费;取得成本;获得成本;展业费;购置资本;购置成本
acquisition cost inventory 存货的购置成本
acquisition cost method 购置成本法
acquisition cost of assets 资产的购置成本
acquisition equipment 捕获装置
acquisition error 获取误差
acquisition expenses for buildings 房屋购置费
acquisition laser 搜索激光器
acquisition mode 探测方式;捕获方式
acquisition model 获得模式
acquisition of assets 购置资产
acquisition of building land 征用建筑用地;购买建筑用地;获得建筑用地
acquisition of concessions 让与权取得
acquisition of control 取得控制权
acquisition of debenture 债券购进
acquisition of development rights 获得扩界权
acquisition of exchange value 取得交换价值
acquisition of information 收集信息;获得信息
acquisition of land 征用土地;土地获得;土地探测;申请用地
acquisition of right 权益取得
acquisition of technology 取得技术;技术引进
acquisition of territory 大地探测
acquisition parameter 采集参数
acquisition phase 获取阶段;采购阶段
acquisition probability 占用概率;捕获概率
acquisition processing 探测进程;探测程序
acquisition processor 采集处理机
acquisition program(me) 获取程序
acquisition radar 搜索雷达;目标指示雷达
acquisition radar coverage 雷达作用区
acquisition range 录取物标距;接收范围
acquisition satellite 探测卫星
acquisition sensor 探测传感器
acquisition strategy 获取策略;采购策略
acquisition time 收集时间;采集时间
acquisition-without rehabilitation 不拟复兴整修购置房产
acquisition-with rehabilitation 拟复兴整修购置房产
acquisitive capital 期望获得的资本
acquisitive prescription 取得时效
acquit 尽责
acquittal 清偿债务;履行义务;解除责任
acquittance 清欠凭据;偿还
acquittance of a debt 免除债务;免偿债务
acraldehyde 丙炔醛
Acramin pigment colo(u)rs 阿克拉明涂料
acrasia 自控能力缺失
acratex 乳香棕色沥青漆
acratopega 冷泉
acratotherm(e) 温泉
Acrawax 阿克蜡
acre(age) 英亩(1 英亩=6.07 亩);土地面积;英亩数
acreage allotments 田亩分派
acreage estate company 地产公司
acreage meter 田间作业面积计数器
acreage property 未经开发的的大片地产
acreage rent 租地金
acreage under crop 播种面积(英亩)
acreage under cultivation 种植面积;耕种面积(以英亩计算的)
acreage yields 每英亩产量
acreage zoning 大面积用地开发区划
Acree's reaction 阿克里反应
acre-foot 亩;英亩一英尺
acre-inch 英亩一英寸
acre-inch day 英亩一英寸日;灌溉量英亩一英寸
acre yield 英亩产量
acrg 土地面积
acribometer 精微测量器
acrid 苛性的
acridine 氮蒽;吖啶
acridine dye 吖嗪染料;吖啶染料
acridine orange 吖啶橙
acridine orange count 吖啶橙直接计数
acridine orange flourescent staining 吖啶橙荧光染色法
acridine orange staining method 吖啶橙染色法
acridine red 吖啶红
acridine system dye 吖啶系染料
acridine yellow 吖啶黄
acridinic acid 吖啶酸

acridinyl 吖啶基
acridol 吖啶酚
acridone 吖啶酮
acrid poison 烈性毒药
acriflavine 吖啶黄(素)
acriflavine direct cell count 吖黄素直接细胞计数
acriflavinium chloride 吖啶黄
Acrilan 阿克里隆
acrisorcin 吖啶琐辛
acritical 无极期的
acritol 吖糖醇
Acrizane 吖啶氯
acrobatic 特技的
acrobatics and tumblings 运动技巧
Acrocel 阿克罗塞尔聚酯纤维
acrocephalia 尖头
acrocinonide 阿克缩松
acrogens 顶生植物
acrolein 丙烯醛
acrolein cyanohydrin 丙烯醛氰醇
acrolein dimer 丙烯醛二聚物
acrolein poisoning 丙烯醛中毒
acrolein polymer 丙烯醛聚合物
acrolein resin 丙烯醛树脂
acrolein test 丙烯醛试验
acrolite 酚与丙三醇合成的树脂
acrolith 石首石肢木身雕像;石木雕像
acrometer 油类比重计
acromorph 火山瘤;盐瘤
Acron 铝基铜硅合金(铝95%,铜4%,硅1%);阿克隆铝铜硅合金
acron bar 六角栅条;六角筛条
Acronol dye 阿克罗诺染料
acronym 首字母缩略词;简称
acropetal coefficient 器官系数
acrophobia 高空恐惧症;高处恐怖
acrophytia 高山植物群落
acropodium 高支座
acropolis (古希腊城市的)卫城
Acropolis of Athens 雅典卫城
a cropping 沿露头方向
acrose 吖糖
acrosomic system 顶体系统
across 交叉;横越
across back 背宽
across bulkhead 横向隔墙
across corners 对角
across cutting 横向切割
across earth-scraping process 跨铲法
across flats 对边
across grain 横纹;横行纤维;横木纹
across impact matrix 交叉影响矩阵;互影响矩阵
across rule 全域规则
across the board cut 全面削减
across-the-board increase 普遍增加工资
across-the-board reduction 每项费用中均衡缩减
across-the-board tariff changes 全面调整关税
across-the-board tariff reduction 全面降低关税
across-the-board tax cut 全面减税
across the bow 横越船首
across the grain 横纹;与木纹垂直;横过纤维
across-the-line 跨接在线路上;跨接线;并行线路
across-the-line motor 全压起动电动机;直接起动电动机;直接启动式电动机
across the line starter 直接起动器
across-the-line starting 直接起动
across the open 横过开阔地
across to the grain 横过木纹
acrostolion 船首尼花纹装饰
acrostolium 船首尼花纹装饰
acroter (山尖或山花端部的)小型雕塑座;建筑顶台座;人像或装修座
acroteria 山墙饰物;山花
acroteric 末梢的
acroterion 山墙顶饰;山尖饰;门窗梁上饰物台座;顶花
acroterium 山墙饰物衬托;山尖形装饰;山墙饰物底座;山墙顶饰
acrotheater 杂技与戏剧结合演出剧院
acrotherium 船首尼花纹装饰
acrotorque 最大扭力;最大扭矩
acrotrophic 端滋的
acrow 建筑施工设备的统称
acrox 强酸性氧化土;高风化氧化土

acrozone 极顶带
Acrybel 阿克里贝尔丙烯腈共聚物短纤维
Acrylafil 玻璃纤维增强聚丙乙烯酸系塑料
Acrylaglas 玻璃纤维增强聚丙烯酸系塑料
acrylaldehyde 丙烯醛
acrylamide 丙烯酰胺(一种化学灌浆剂)
acrylamide base(injecting paste) 丙凝(注浆材料);丙烯酰胺堵漏(一种化学灌浆剂)
acrylamide copolymer 丙烯酰胺共聚物
acrylamide dye 丙烯酰胺染料
acrylamide grouting 丙烯酰胺灌浆;丙凝灌浆
acrylamidine 丙烯脒
acrylate 丙烯酸酯(树脂);丙烯酸盐
acrylate-based 丙烯酸酯基;丙烯酸盐基
acrylate-butadiene rubber 丙烯酸丁二烯橡胶
acrylate coated fiber 丙烯酸酯涂覆光纤
acrylate copolymer 丙烯酸酯共聚物;丙烯酯聚合物
acrylate emulsion 丙烯酸酯乳液
acrylate modified PVC 丙烯酸酯改性聚氯乙烯
acrylate plastic paint 丙烯酸酯塑料涂料
acrylate resin 丙烯酸酯树脂
acrylate sealing paste 丙烯酸酯密封膏
acrylate-styrene-acrylonitrile resin 丙烯酸酯—苯乙烯—丙烯腈树脂
acrylate-vinyl chloride ethylene-vinyl acetate pressure polymer 丙烯酸酯—氯乙烯—乙烯—乙酸乙烯加压聚合物
Acrylene dye 阿克里利染料
acryl glass 有机玻璃;丙烯酸酯类有机玻璃
acrylic 丙烯
acrylic acid 丙烯酸
acrylic acid amide 丙烯酰胺
acrylic acid paint 丙烯酸酸性涂料
acrylic acid series 丙烯酸系
acrylic acid wastewater 丙烯酸废水
acrylic aldehyde 丙烯醛
acrylic anhydride 丙烯酸酐
acrylic baking enamel 丙烯酸烘干瓷漆;丙烯酸(树脂)系烘烤瓷漆
acrylic baking hammer finish 丙烯酸烘干锤纹漆
acrylic baking paint 丙烯酸烘漆
acrylic baking transparent paint 丙烯酸透明烘漆
acrylic baking varnish 丙烯酸烘干清漆
acrylic base 丙烯酸酯基;丙烯酸树脂基团
acrylic(based)emulsion 丙烯酸乳液;丙烯酸乳胶
acrylic binder for pigment printing 丙烯酸涂料印花黏合剂
acrylic brilliant paint 丙烯酸自干闪光漆
acrylic(building)mastic 丙烯酸(防潮)玛琋脂;丙烯建筑玛琋脂;丙烯建筑胶合料
acrylic(bulk)compound 丙烯酸表面密封材料;丙烯酯密封剂
acrylic coated fiber 丙烯酸涂覆光纤
acrylic coating 丙烯酸涂层;丙烯酸树脂清漆
acrylic coating for plastics 丙烯酸塑料漆
acrylic colo(u)r 丙烯酸染料
acrylic compound 丙烯酸化合物
acrylic concrete 丙烯酯混凝土
acrylic continuous filament 丙烯腈系长丝
acrylic copolymer 丙烯酸共聚物
acrylic dispersion 丙烯酯悬浮体;丙烯酯分散系
acrylic electro static spraying baking enamel 丙烯酸静电烘干瓷漆
acrylic electro static spraying baking varnish 丙烯酸静电烘清漆
acrylic emulsion 丙烯酸类乳液
acrylic emulsion binder 丙烯酸黏合(乳)剂
acrylic emulsion paint 丙烯酸系乳化漆
acrylic emulsion thickener 丙烯酸乳液增稠剂
acrylic enamel 丙烯酸瓷釉;丙烯酸瓷漆
acrylic epoxy baking paint 丙烯酸环氧烘漆
acrylic epoxy resin 丙烯酸环氧树脂
acrylic ester 丙烯酸酯
acrylic ester-based 丙烯酸酯基;丙烯酸盐基
acrylic ester emulsion 丙烯酸酯乳液
acrylic fiber 丙烯酸(类)纤维;丙烯纶;丙烯酸系纤维;丙烯腈系纤维
acrylic fiber carpet 腈纶地毯
acrylic fiber wastewater 丙烯纤维废水
acrylic flashing paint 丙烯酸闪光漆
acrylic glass 丙烯酸(有机)玻璃
acrylic glazing 丙烯酸瓷漆
acrylic glue 丙烯酸胶

acrylic lacquer 丙烯酸清漆;丙烯酸(挥发性)漆
acrylic latex 丙烯乳液;丙烯酸胶乳
acrylic latex coating 丙烯酸乳胶漆
acrylic latex polymer ca(u)lk 聚丙烯乳胶嵌缝料
acrylic lens 透明丙烯板;丙烯有机玻璃透镜
acrylic lithin coating 丙烯酸系彩色水泥(砂浆)涂料
acrylic low-temperature baking enamel 低温丙烯酸烘干瓷漆
acrylic low-temperature baking varnish 丙烯酸低温烘清漆
acrylic mastic 聚丙烯酯胶合铺料;丙烯酯胶合铺料
acrylic matte water soluble baking paint 丙烯酸无光烘干水溶性漆
acrylic metallic baking paint 丙烯酸金属烘漆
acrylic modified alkyd 丙烯酸改性醇酸树脂
acrylic modified amino-alkyd baking varnish 丙烯酸改性氨基烘清漆
acrylic modified nitrocellulose lacquer 丙烯酸改性硝基清漆
acrylic modified vinyl perchloride enamel 丙烯酸改性过氯乙烯瓷漆
acrylic monomer 丙烯酸单体
acrylic-nitrile rubber 丙烯腈橡胶
acrylic nitrocellulose lacquer 丙烯酸硝基清漆
acrylic orange-figured paint 丙烯酸橘纹漆
acrylic orange peel enamel 丙烯酸橘皮瓷漆
acrylic outdoor enamel 丙烯酸外用瓷漆
acrylic outdoor varnish 丙烯酸外用清漆
acrylic packing resin 丙烯酸填充性树脂
acrylic paint 丙烯酸涂料;丙烯酸漆
acrylic painting 丙烯画
acrylic panel 丙烯酸板;丙烯塑料板;丙烯塑胶板
acrylic plastering 聚丙烯酸涂料;丙烯酸涂灰;丙烯酸抹灰
acrylic plastic(bath)tub 丙烯塑料浴盆
acrylic plastic board 丙烯塑料板
acrylic plastic corrugated board 丙烯塑料瓦楞板;丙烯塑料波纹板
acrylic plastic cube 丙烯酸类塑料块
acrylic plastic domed rooflight 丙烯塑料圆顶天窗
acrylic plastic glaze 丙烯酸塑料膜玻璃贴面;丙烯酸系塑料涂膜;丙烯酸系玻璃塑料
acrylic plastic glazing 丙烯酸有机玻璃
acrylic plastics 丙烯酸(类)塑料
acrylic polyelectrolyte 丙烯酸类聚电解质;丙烯酸聚合电解质
acrylic polymer 丙烯(酸)聚合物
acrylic polymer based sealant 丙烯酸嵌缝膏
acrylic polyurethane 丙烯酸(改性)聚氨酯
acrylic polyurethane coating 丙烯酸聚氨酯漆
acrylic polyurethane enamel 丙烯酸聚氨酯瓷漆
acrylic polyurethane wall paint-exterior 丙烯酸聚氨酯外墙面漆
acrylic polyvinyl chloride 丙烯酸聚氯乙烯
acrylic primer 丙烯酸底漆
acrylic quick drying baking paint 丙烯酸快干烘漆
acrylic quick drying baking varnish 丙烯酸快干烘清漆
acrylic resin 丙烯酸系树脂;丙烯酸树脂;丙烯酸类树脂;丙烯树脂
acrylic resin base 丙烯树脂基
acrylic resin concrete 丙烯酸树脂混凝土
acrylic resin dispersion 丙烯树脂悬浮体;丙烯树脂分散系
acrylic resin emulsion 丙烯酸树脂乳胶
acrylic resin paint 丙烯酸树脂涂料;丙烯酸树脂漆
acrylic(resin)seal(ing)compound 丙烯酸酯密封料
acrylic roof lighting sheet 丙烯酸屋顶采光板
acrylic rubber 丙烯酸(类)橡胶
acrylic rubber paste 丙烯橡胶糊剂
acrylics 丙烯酸脂类
acrylic sealant 丙烯酸密封剂;丙烯酸树脂表面密封剂;丙烯酸酯密封膏;丙烯酸嵌缝膏;丙烯酯密封料
acrylic seal(er)丙烯酸密封剂
acrylic semi-gloss water soluble baking paint 丙烯酸半光烘干水溶性漆
acrylic sheet 丙烯酸(有机玻璃)板;丙烯酸片
acrylic size 丙烯酸浆料
acrylic stearate 丙烯酸硬脂酸盐
acrylics thermoplastic resin 丙烯酸类热塑性树脂
acrylic syrup 丙烯酸树脂浆
acrylic terpolymer sealant 丙烯酸酯三聚密封胶
acrylic tile 丙烯酸屋面采光瓦
acrylic varnish 丙烯酸清漆

acrylic veneer crown 丙烯酸酯罩冠
acrylic wall paint 丙烯酸外墙涂料
acrylic wall sealer 丙烯酸墙封
acrylic water based putty 丙烯酸水性腻子
acrylic water-soluble transparent baking paint 丙烯酸透明烘干水溶性漆
acrylic-wood flooring 丙烯酸浸渍木地板
acrylide maroon 甲苯胺红
Acrylite 聚丙烯酸塑料
acryl-nitrile rubber 丙烯腈橡胶
acryloid 丙烯酸树脂溶剂;丙烯酸剂
acryloid cement 丙烯酸黏合剂
acrylonitrile-acrylate-styrene copolymer 丙烯腈丙烯酸酯苯乙烯共聚物
acrylonitrile-butadiene rubber 丙烯腈丁二烯橡胶
acrylonitrile-butadiene-styrene 丙烯腈丁二烯—苯乙烯三元共聚物
acrylonitrile-butadiene-styrene copolymer 丙烯腈丁二烯—苯乙烯共聚物
acrylonitrile butadiene styrene resin 丙烯腈二乙烯丁二烯树脂;丙烯腈丁二烯苯乙烯树脂
acrylonitrile copolymer 丙烯腈共聚物
acrylonitrile poisoning 丙烯腈中毒
acrylonitrile waste 丙烯酯废物
acrylonitrille 丙烯腈
acrylonitrille-styrene copolymer 丙烯腈苯乙烯共聚物
acrylonitrille-styrene resin 丙烯腈苯乙烯树脂
acryloyl 丙烯酰
acryloyl chloride 丙烯酰氯
acryl resin 丙烯酸树脂
acryogenic 非他冻的;非冰川的
acrysol 丙烯酸树脂乳剂
acry-wood flooring 丙烯酸木地板
act 行为;法案
act and deed 物证
Actanium 镍铬钴低膨胀合金
act at one's discretion 酌情处理
act curtain 大幕
act drop 大幕;舞台口的幕(场间降下)
act for the encouragement of trade 贸易促进条例
actic 潮区
Actidione 放线酮
actification 再生作用
actified solution 再生溶液
actifier column 再生塔
acting 作用的;动作
acting agent 临时代理人
acting area 舞台的演戏面积;表演区
acting area light 舞台灯光;表演区灯
acting circle of blasting 爆破作用圈
acting current 动作电流
acting duration of force 外力作用持续时间
acting force 作用力
acting glacier 流动冰河
acting head 有效水头;作用水头;工作水头
acting in good faith 重义
acting level 舞台面;表演平台
acting management 代经理
acting manager 代理经理
acting surface 作用面;工作面
acting surface by force 外力作用面
acting time 动作时间
acting type of imposed force 外力作用方式
actinic 光化性的;光化的;光化
actinic absorption 光化吸收(作用)
actinic achromatism 光化消色差;日光谱线消色差
actinic balance 分光测热计
actinic chemistry 光化学
actinic effect 光化效应
actinic focus 光化焦点
actinic glass 闪光玻璃;护目玻璃;光化玻璃;光弧玻璃
actinic green 光化绿
actinic-green glass (毒物瓶用的)光化绿玻璃
actinicity 光化性;光化度
actinic lamp 光化灯
actinic light 光化性光;光化光
actinic melanogen 光色素原;光化性黑色素原
actinic opacity 光化密度
actinic radiation 光化辐射
actinic rays 光化射线
actinic stability 光化稳定性
actinide chalcogenide 锕系元素硫族元素化合物

actinide chloride 锕系元素氯化物
actinide contraction 锕系收缩
actinide element 锕系元素
actinide halide 锕系元素卤化物
actinidelanthanide searation 锕系—镧系元素分离
actinide metallide 锕系元素金属化物
actinide oxometallate 锕系元素的金属含氧酸盐
actinide oxyfluoride 锕系元素氟氧化物
actinides 锕类(元素);锕化物
actinide series 锕系
actinide thiocyanate 锕系元素硫氰酸盐
actinism 射线作用;射线化学;光化作用;光化性;光化力;辐射作用
actinium(Ac) 锕
actinium becay series 锕衰变系
actinium emanation actinon 锕射气
actinium family 锕系
actinium lead 锕铅
actinium series 锕系
actinium-uranium 锕铀
actinobiology 光化生物学
actinochemistry 射线化学;光化学
actinocytic 环状辐射型
actinocytic type of stomata 辐射型气孔
actinodielectric 光敏介电的
actinoelectricity 光化电
actinogram 日射曲线图;光能曲线图;光化记录
actinograph 自记曝光计;光强测定仪;日射计;日光强度自动记录器;光能测定仪;光化线强度记录器;光化线强度测定器;光化力测定器;感光记录器;辐射计
actinography 光强测定仪;光能测定术;光量测定法;光化线强度测定法;光化力测定术
actinoid 辐射状的
actinoid elements 锕系
actinolite 阳起石
actinolite glaucophane epidote chlorite schist 阳起蓝闪绿帘绿泥片岩
actinolite rock 阳起石岩
actinolite schist 阳起石片岩
actinolite slate 阳起石板岩
actinolitization 阳起石化
Actinolitum 阳起石
actinology 光化学
actinomere 辐节
actinometer 日照计;日射计;日光能量测定器;日光辐射(率)计;曝光计;曝光表;太阳辐射仪;太阳辐射计;露光计;光化线强度计
actinometer paper 曝光记正纸;光化线强度记录纸
actinometrical 曝光测定的
actinometrical time 曝光测定时间
actinometry 日射测量法;曝光测定;露光测定(术);光作用测量术;光能(强度)测定学;光能强度测定术;辐射测量(学);辐射测量术;辐射测量法
actinomorphic 放射型的
actinomycete 放线菌
actinomycosis 放线菌病
actinon 锕射气
actinopraxis 放射性物质应用
actinoscope 光强测定仪;光能测定器;辐射测定器
actinoscopy 放射检查
actinostome 辐状口
actinotherapy 射线疗法
actinouran 锕铀
act in violation of the stipulation 违反条约
action 行为;作用(量);动作
actionable course of action 可行行动
actionable fire 违禁火;违法火
action and reaction 作用力与反作用力
action at a distance 远距离作用;超距作用
action at a distance theory 超距作用理论
action bars 操作杆
action by contact 接触作用
action center 作用中心
action-centered leadership 以活动为中心的领导方式训练
action command 运行命令
action current 作用电流
action cycle 作用周期;工作周期;动作周期
action data automation computer 动作数据自动计算机
action directive 动作控制命令
action effect 作业效应
action element 执行元件

action ex contract 依据合同或契约的诉讼
action ex delict 根据侵权行为的诉讼
action factor of human 人类活动因素
action for compensation for loss 要求赔偿损失的诉讼
action for damages 损害行动;请求损害赔偿的诉讼
action for indemnity 损害赔偿诉讼
action for reimbursement 申请赔偿的诉讼;要求补偿的诉讼
action for restitution 退回原物诉讼;要求退还原物的诉讼
action for the recovery of damage 损害赔偿诉讼
action founded in contract 基于合同的诉讼
action founded in tort 基于侵权的诉讼
action in rem 物权诉讼;告船(法)
action in respect of damage to goods 就货物损坏的诉讼
action integral 作用量积分
action in the medium 介质间的作用
action lag 行动时滞
action length 作用长度
action level 作用水平;干预水平;动作水平
action limit 行动点;作用界限;处置界限;处理界限
action line 激励行;作用线
action message 请求干预消息;请求操作信息;动作信息;动作消息
action method 作用方法
action of bank suction 岸吸作用
action of frost 霜冻作用;冰冻作用
action of gravity 重力作用
action of light 光的作用
action of magnesia 氧化镁的膨胀作用
action of microorganisms 微生物作用
action of natural-electric(al)field 自然电场作用
action of pile group 群桩作用
action of points 尖端作用
action of rescission 解除合同诉讼
action of restitution 要求宣判契约无效的诉讼
action of rust 锈蚀作用
action of sheet erosion 片状侵蚀作用;薄层冲刷作用;表面侵蚀作用
action of silicium dust 矽尘作用
action of surge chamber 调压室功能;调压井功能
action of the wind 风力作用
action parameter 行动参数
action period 作用周期;作用期
action pitch 作用节距
action plan 实施计划
action plan for the biosphere reserves 生物圈保护行动计划
action plan for the program(me)of studies 气候对人类影响研究计划的行动计划
action plan for the protection of the atmosphere 保护大气行动计划
action potential 有效电位;有效电势;动作电位
action principle 作用(量)原理
action program(me)plan 行动方案计划
action pulse 动作冲量
action radius 有效半径;活动范围;活动半径;作用半径
action range 作用范围
action-reaction law 作用反作用定律
action report 执行报表
action research 行动研究
action roller 活动滚轮;动辊
action routine 活动程序;动作例行程序
action schedule 动作表
action sketch 动作草图
action skill 行动技能
action slow 慢动作
action space 作用范围
action specification 动作说明
action spectrum 作用光谱
action spot 作用点;工作点
action spring 撞针簧
action stretch 高弹性伸缩
action sub-coefficient 作用分项系数
action through medium 通过介质起作用
action time 动作时间
action token 作用特征;动作记号
action to quiet title 产权凝点消除
action to take 指定采取行动
action training 行动训练
action turbine 冲击式涡轮机;冲击式水轮机;冲动式涡轮机;冲动式透平;冲动式水轮机
actionuranium series 锕铀系
action valve 自动阀;投币阀
action variable 作用变数;作用变量
action variable in decision situation 决策形势中的行动变量
action wheel 主动轮
actithiazic acid 阿克唑酸
actium 海岩群落
activate 激化;活(性)化
activate business 扩展业务
activate button 启动按钮
activated 活性的
activated absorption 活性吸附
activated aeration 活性曝气;活化曝气
activated aeration process 活性曝气法;活性曝光法;活化曝气过程
activated agent 放射化物质;活化剂
activated agent concentration 活化剂浓度
activated algae process 活性藻类法;活化藻类法
activated alum 活性明矾
activated alumina 氧化铝干燥剂;活性氧化铝;活性铝土;活性矾土
activated alumin(i)um 活性铝
activated alumin(i)um oxide 活性氧化铝
activated atom 活化原子
activated bauxite 铝矾土滤质;活性矾土
activated bentonite 活性膨润土
activated biofilter 活性生物滤池
activated biofilter system 活性生物滤池系统
activated biological filtration 活性生物过滤
activated brilliant organe 活性艳橙
activated calcium carbonate 活性炭酸钙
activated carbide 活性碳化物
activated carbon 活性炭
activated carbon absorption 活性炭吸收
activated carbon adsorption bed 活性炭吸附床
activated carbon absorption column 活性吸附柱
activated carbon adsorption (工艺)
activated carbon adsorption process 活性炭吸附法
activated carbon adsorption separation-biological regeneration process 活性炭吸附分离生物再生工艺
activated carbon bed 活性炭床
activated carbon bed-depth service time 活性炭床厚度作用时间
activated carbon biodisk 活性炭生物转盘
activated carbon black 活性炭黑
activated carbon canister 活性炭罐
activated carbon-catalysed ozonation process 活性炭臭氧催化工艺
activated carbon cloth 活性炭布
activated carbon concentration 活性炭浓度
activated carbon denitrification 活性炭脱氮
activated carbon eductor 活性炭排放管;活性(炭)喷射器
activated carbon felt 活性炭毡
activated carbon fiber 活性炭纤维
activated carbon fiber electrode electrolysis system 活性炭纤维电极电解系统
activated carbon filter 活性过滤器;活性炭滤池
activated carbon filter process 活性炭过滤法
activated carbon filtration 活性炭过滤
activated carbon method 活性炭法
activated carbon molasses number 活性炭糖蜜值
activated carbon moving bed 活性炭移动床;移动性活性炭床
activated carbon particle swarm 活性炭颗粒群
activated carbon pneumoconiosis 活性炭尘肺
activated carbon polishing 活性炭后处理
activated carbon pond 活性炭池
activated carbon pore volume 活性炭孔隙容积
activated carbon process 活性炭法;活性炭处理
activated carbon process in wastewater 废水的活性处理法
activated carbon process in wastewater treatment 废水的活性处理法
activated carbon process of fume gas desulfurization 活性炭吸附法烟气脱硫
activated carbon reactivation 活性炭再生(作用);活性炭复活作用
activated carbon selecting 活性炭筛选
activated carbon surface chemistry 活性炭表面化学性质

activated carbon treatment 活性炭处理
activated carbon treatment system 活性炭处理系统
activated carbon tube 活性炭管
activated carbon types 活性炭类别
activated carrier 激活载体;活化载体
activated center 活化中心
activated chalk 活性碳酸钙
activated char 活性炭
activated charcoal 活性(木)炭
activated charcoal absorption 活性炭吸收;活性炭吸附
activated charcoal adsorption method 活性碳吸附法
activated chlorine 活性氯
activated clay 活性黏土;活性白土
activated coal plough 摆动式刨煤机
activated complex 活性(化)络合物;活化络合物;活化复体
activated diatomite 化学硅藻土
activated draghead 活动耙头
activated driver cab 激活端司机室
activated dye 活性染料
activated earth 活性土
activated fiber 活性纤维
activated form 活化型
activated gelatin 活性明胶
activated granular carbon 活性颗粒炭
activated group 活化基团
activated hot pressing sintering 活化热压烧结
activated indicator 活性指示器
activated ion 活性离子
activated magnesia 轻烧镁砂
activated magnesium oxide 活性氧化镁
activated manganese oxide process 活化氧化锰法
activated material 激活物质
activated microbial concentration 活性微生物浓度
activated mineral purified water system 化学矿物质净化水系统
activated montmorillonite 活性蒙脱石
activated montmorillonite clay 活性蒙脱土;活性微晶高岭土
activated parthenium carbon 活性单性炭
activated petroleum coke 活性石油焦煤
activated phosphate 活性磷酸盐
activated plant 活水处理厂
activated plough 动力刨煤机
activated powdered carbon 粉状活性炭
activated process 活化过程(法)
activated resin 活性树脂
activated rosin flux 活性(松香)焊剂;不腐蚀钎焊剂
activated sand filter 活性砂滤池
activated sewage 活性污水;活化污水
activated sewerage 活性污物;活性污水
activated silica 活性矽土;活性矿土;活性硅(土);活性二氧化硅
activated silica gel 活性硅胶;活化硅胶
activated silicate 活性硅酸盐;化学硅酸
activated silicate glass 激活硅酸盐玻璃;活化硅酸盐玻璃
activated silicon switch 硅光敏开关
activated silt 活性淤泥
activated sintering 活化烧结
activated sludge 活性淤泥;活性污泥;活化污泥
activated sludge adsorption 活性污泥吸附
activated sludge adsorption equation 活性污泥吸附方程
activated sludge aeration 活性污泥曝气
activated sludge aeration method 活性污泥曝气法
activated sludge aeration tank 活性污泥曝气池
activated sludge aeration travel 活性污泥曝气运动
activated sludge biofilm wastewater treatment 活性污泥生物膜废水处理
activated sludge biomass 活性污泥生物质
activated sludge bioreactor 活性污泥生物反应器
activated sludge biota 活性污泥生物群
activated sludge bulking 活性污泥膨胀
activated sludge bulking and foaming 活性污泥膨胀发泡
activated sludge bulking and foaming control 活性污泥膨胀发泡控制
activated sludge culture 活性污泥培养
activated sludge digestion 活性污泥消化
activated sludge disposal 活性污泥处置;活性污泥处理
activated sludge effluent 活性污泥出水;活性污泥废液
activated sludge filamentation bulking 活性污泥丝状膨胀
activated sludge filtration 活性污泥过滤
activated sludge floc 活性污泥絮凝物;活性污泥绒粒;活性污泥块
activated sludge floc structure 活性污泥絮凝物结构
activated sludge foam 活性污泥泡沫
activated sludge foam control 活性污泥泡沫控制
activated sludge inhibition 活性污泥抑制
activated sludge kinetics 活性污泥动力学
activated sludge loading 活性污泥负荷
activated sludge loading rate 活性污泥负荷率
activated sludge metabolic property 活性污泥代谢特性
activated sludge method 活性污渣法;活性污泥(处理)法
activated sludge mineralization 活性污泥矿化
activated sludge model 活性污泥模型
activated sludge nomenclature 活性污泥术语
activated sludge nutrient requirements 活性污泥营养需要
activated sludge operational difficulties 活性污泥运转问题
activated sludge oxidation 活性污泥氧化(作用)
activated-sludge package unit 装配式活性污泥单元
activated sludge particle 活性污泥颗粒
activated sludge phase 活性污泥相
activated sludge pilot plant 活性污泥试验厂
activated sludge plant 活性污泥(法处理)厂;活化污泥设施
activated sludge plant design 活性污泥厂设计
activated sludge-powdered activated carbon process 活性污泥粉末-活性炭工艺
activated sludge process 活性污泥工艺;活性污泥(处理)法(一种废水生物处理法);活化污泥法
activated-sludge process for continuous flow stirred tank 连续流搅拌池活性污泥处理法
activated sludge procession 活性污泥行列
activated sludge process microbiology 活性污泥法微生物学
activated sludge property 活性污泥特性
activated sludge reaction tank 活性污泥反应池
activated sludge reactor 活性污泥反应器
activated sludge respirometer 活性污泥呼吸计
activated sludge rising 活性污泥上浮
activated sludge separation 活性污泥分离
activated sludge settling curve 活性污泥沉降曲线
activated sludge settling tank 活性污泥沉淀池
activated sludge solubilization 活性污泥溶液化
activated sludge submerged hollow fibre membrane bioreactor 淹没式活性污泥中空纤维膜生物反应器
activated sludge system 活性污泥系统
activated sludge tank 活性污泥池
activated sludge treatment 活性污泥处理
activated sludge treatment plant 活性污泥处理设备
activated sludge treatment process 活性污泥处理工艺
activated sludge treatment system 活性污泥处理系统
activated sludge treatment technology 活性污泥处理技术
activated sludge treatment unit 活性污泥处理设备
activated sludge wastewater 活性污泥废水
activated sludge wastewater treatment plant 活性污泥废水处理厂
activated sludge with zeolite addition 沸石附加物活性污泥
activated sludge yield reduction 活性污泥产量减缩
activated solid 活性固体
activated state 激活状态;活性状态;活化(状)态
activated sulphur 活性硫黄
activated surface 激活面
activated temperature 活化温度
activated time 活化时间
activated waste 活化废物
activated water 受辐射作用的水;活化水
activated zeolite 活性沸石;活化沸石
activated zinc oxide 活性氧化锌
activate key 起动键;启键;启动键
activating agent 激活剂;活性剂;活化剂;放射化物质
activating analysis 活化分析
activating catalyst 活化催化剂
activating channel 活化槽
activating condition 活化状态
activating energy 活化能
activating isotope 活化同位素
activating oxide 活性氧化物
activating period 活化期
activating radiation 活化辐射
activating reaction temperature 活化反应温度
activating reagent 活化试剂
activating signal 起动信号
activating solution 活化剂
activating stress 活化应力
activating treatment 活化处理
activation 驱动;激活(作用);激活激励;敏化;活性化(作用);活化作用;活化(法);触发
activation accommodation 激活调节
activation adsorption 活化吸附(作用)
activation analysis 激活分析;激化分析;活化分析
activation centre 激活中心
activation coefficient 活化系数
activation control 活化控制
activation controlled reaction 活化控制反应
activation core 活化区
activation cross-section 活化截面
activation cycle 激活周期
activation detector 活化探测器
activation energy 激活能;活化能
activation energy barrier 活化能势垒
activation energy for fission 裂变临界时能量
activation energy of organic matter 有机质活化能
activation energy of reaction 反应活化能
activation enthalpy 活化焓
activation entropy 活化熵
activation experiment 激活试验
activation factor 活化因子
activation fiber 激活纤维
activation fragment 激活段;激活段
activation grade 活化度
activation heat 活化热
activation index 活化指数
activation level 激活水平
activation log 活化测井
activation log curve 活化测井曲线
activation mechanism 激活机制
activation nuclide 活化核素
activation number 活化值
activation of a procedure 过程的激活
activation of blocks 程序块启动;分程序的活动
activation of epithermal neutron 超热中子活化
activation of homing 接送自导系统
activation parameter 活化参量
activation pointer 激励指示字;激活指示字
activation polarization 活化极化
activation potential 活化电势
activation process 敏化过程;活化过程(法)
activation processor 激活处理机
activation product 活化产物
activation ratio 活化比
activation request 启动请求
activation sand 活化砂
activation stack 激活栈
activation state 激化状态
activation system 激活系统
activation table 激励表
activation tank 活化池
activation temperature 活化温度
activation time 激活时间
activation volume 活化体积
activation zone 活化区
activative mineral filler 活化矿粉
activator 污化剂;激励器;激活质;激活剂;活性剂;活化器;活化剂;催化剂
activator centre 激活中心
activator development 激活显影
activatory 活化的
active 激活的;积极的;活(性)的;主动的;工作的;操作的
active absorption 活性吸收;有源吸收;主动吸收
active account 流动账户;活动账户

active accumulated temperature 活动积温
active acetate 活性乙酸盐
active acidity 活性酸度
active addition 活性混合材料
active admixture 活性混合材料
active aerating zone 充气区
active aerodrome 现用机场
active agent 活性剂;活化剂
active age of active fault 活断层活动时代
active aggregate 活性集料;活性骨料
active air 活性空气
active alkali 活性碱
active alumina 活性氧化铝
active alumin(i)um oxide 活性氧化铝
active amount of air for combustion 燃烧所需的实际空气量
active amyl propionate 丙酸旋性戊酯
active analysis 活动分析
active and passive Rankine state 朗肯主动及被动土压力状态
active antenna 激励天线;有源天线;辐射天线
active anthracite 活性无烟煤
active antiroll system 主动抗摇系统
active arch 有效拱
active area 现用区;现役区;有效面积;工作面积
active area in West Pacific 西太平洋活动区
active area of new tectogenesis 新构造运动活动地区
active arm 通航汊道;活汊道
active assets 活动资产;活动财产;动产
active augmentation system 主动放大系统
active autofocusing 有效自聚焦
active balance 顺差;结算盈余;节余
active balance of payments 收支顺差
active bank 可使用数据库
active bank account 有存款的银行账户
active beam 激励光束
active belt 活动带
active biological film 活性生物膜
active biological sand 活性生物砂
active biological surface 活性生物表面
active block 有源组件;有效组件
active block upwarping 活动地块隆起
active bond 有效债券
active braking distance 实际制动距离
active bromine 活性溴
active business 推动业务
active calcium oxide(of lime) (石灰的)活性氧化钙
active call path 工作呼叫电路
active capacity 有效库容
active capital 流动资金;流动资本;能动资本;活动资本
active car 车场车辆
active carbon 活性炭黑;活性炭
active carbon adsorption 活性炭吸附法
active carbon filter 活性炭过滤器
active carbon model 活性炭型号
active carbon monitoring 活性炭检验
active card 现用卡片;常用卡片
active catch (阀门分配机构的)活动挡
active cavity 激活腔
active cell 动子
active center 活化中心;有效中心;活性中心
active chaff 有源干扰物
active chain 有效链
active channel 占线通道;占线沟道
active charcoal 活性木炭
active charcoal filter 活性炭过滤器
active chlorine 活性氯;有效氯
active circuit 有源电路
active circuit mesh equation 有功电路网络方程
active circulation 流行额
active cirque 活动冰斗
active clay 活性黏土;有效黏土
active cliff 活动悬崖
active coating 活性敷层;活性被覆层;激活层;活性涂层;活性面层;放射层
active coil 有效线圈
active commerce 以本国船只运输的贸易
active communication satellite 有源通信卫星
active component 活性组分;活性部件;有源元件;有效组分;有效元件;有功分量;有功电流;有功部分;主动部件
active compound 活性化合物

active computer 现用计算机
active condition 活性状态
active conductor 有效导体
active connector 有源连接器
active constituent 活性组分;有效分量;有效成分
active constraint 起作用约束;有效约束
active containment alternatives 主动抑压法
active content 流动储料量;活性浓度
active continental margin 活动大陆边缘;主动大陆边缘
active cooling surface 活性冷却面;有效冷却面积
active correlator 主动相关器
active cross-section 过水断面;水流断面;有效过水断面
active crystalline material 激活晶体物质
active current 有效电流;有功电流
active cycle of active fault 活断层活动周期
active data area 现用数据区;现役数据区
active day 地磁受扰日
active debt 活动债务;需付利息的债务;付利息的债券
active decoy 主动假目标
active defect 运行故障
active demand 迫切需要;畅销
active deposit 放射性淀积;活性淀积;活性沉积;活期存款;放射性淀质
active detergent 活性洗涤剂
active device 激活装置;能动元件;有源器件
active diluent 活性稀释剂
active display 主动式显示
active DO-loop 现用循环
active door 先启门扇;活络门
active Doppler velocity sensor 主动多普勒速度传感器
active draft 正阻力
active draghead 活动耙头;主动耙(吸)头(疏)
active drainage area 有效排水面积
active drawing area 有效绘图面积
active drift 正阻力
active dual-direction-transmitting transponder 有源双向传送应答器
active duty 现役
active dye 活性染料
active dyeing wastewater 活性染色废水;活性染料废水
active earth 漂白土;活性土;活化漂土
active earth pressure 主动土压力
active earth pressure factor 主动土压力系数
active earth pressure of cohesive backfill 粘性填土的主动土压力
active earth pressure redistribution 主动土压力重分布
active earth pressure rupture 主动土压力状态破坏面
active earthquake 主动地震
active earth thrust 主动土推力
active edge 主动边
active electric(al) network 有源网络
active electrode 有源电极;有效电极;作用电极
active electron 激活电子
active element 激活元素;能动元件;活动的元素;有源元件
active element memory 有源元件存储器
active emanation 放射性射气
active emitting material 活性发射材料
active-energy meter 瓦时计;电度表;瓦特小时计(通称电表)
active entry 生产煤巷
active erosion 强烈侵蚀
active exercise 主动运动
active extreme pressure lubricant 活性极压润滑剂
active face 刃面;掌子面;工作面
active face of flexible wall 柔性墙的主动面
active factor 活性因子
active failure 自行破坏;主动损坏;主动破坏
active failure wedge 主动破坏棱体
active fare table 在用票价表
active fault 活性断层;活断裂;活层;活动(性)断层
active fault block 活动断块
active fault in scale 活断层规模
active fault specificity 活断层特性
active fault system 活动断裂系
active fault trace 活动断层线
active fault type 活断层类型

active fault zone 活动断裂带
active feature of each movement 各次活动特征
active fertilizer 速效肥料;活性肥料
active fiber 活性纤维
active field 搅拌区
active field shielding 主动场屏蔽
active figure control 有效图像控制
active file 现用文件;现用文档【计】;有用资料;有用文件;有用档案;常用资料;常用文件
active filler 活性填料;补强剂
active filter 有源滤波器
active fire defense 积极防火;自动消防(包括发出火警信号和传感系统的装置)
active fissure 活动裂缝
active fleet 现役舰队;常备舰队
active floridin 活性白土
active flow zone 有效流动区;有效流动带
active fluid pressure 有效流体压力
active fold 活动褶皱
active following 主动追踪;主动跟踪
active force 活力;有效力;主动力
active frequency of active fault 活断层活动频率
active front 活跃峰
active furnace area 炉膛有效面积
active gas 活性气体;腐蚀性气体
active gas container 激活气体容器
active gas material 激活气体材料
active ga(u)ge 工作应变丝(电阻应变片的)
active gelatin 活性明胶
active geothermal area 活动的地热区
active geyser 活间歇泉;现间歇泉
active glacier 活(动)冰川;运动冰川
active glacier front 活动冰川前沿
active glassy material 激活玻璃材料
active glycolaldehyde 活性乙醇醛
active gradient 活动梯度
active granular carbon 颗粒活性炭
active grate 活动箅板
active gravitational mass 主动引力质量
active group 活性基团
active growth centre 活化生长中心
active growth stage 积极生长期;活性生长期
active guidance 主动制导;主动导航
active gull(e)y 活动冲沟;活冲沟
active highway grade crossing protection 主动式平交道口防护
active homing 主动导航
active humus 活性腐殖质
active hydrocarbon 活性烃;活性碳氢化合物
active hydrothermal area 活动水热区
active illumination 有源照明;主动照明
active infrared detection system 主动红外探测系统
active infrared equipment 主动式夜视仪
active infrared tracking system 主动式红外跟踪系统
active ingredient 活性成分
active instruction 活动指令;运行指令
active instrument 主动式仪器
active interference filter laser amplifier 有源干涉滤光激光放大器
active interferometer 有源干涉仪
active investment 能动投资
active ion 激活离子
active ion electrolyser 活性离子电解仪
active ion electrolysis unit 活性离子电解仪
active ion of transition(al) element 过渡元素激活离子
active iron 活动铁
active island arc 活动岛弧
active isolation 积极隔振;有效隔振;有效隔离;有隔绝
active jamming 人为干扰;积极干扰
active job 运行中的作业
active land mass 活动地块
active landslide 活滑坡
active laser tracking system 主动激光跟踪系统
active lattice 堆芯栅格
active layer 融冻层;激活层;活性层;活土层;活化层;活动层;冻胀层;冻融层;地面土层变动层;充气活化层
active leaf 活络门;主动门扇;双扇门中能先开启的一扇
active length 活动时间;有效长(度);工作长度
active level 激活能级;活动程度

active level transformation 有效电平变换
active life 有效寿命
active lifejacket 主动式救生(衣)服
active line 加电线;有效线(路);作用线;工作线(路);有效行
active link 有效链路
active liquid material 激活液体材料
active list 现用表;现役表
active load 活荷载;有源负载;有功负荷
active logic 有源逻辑
active loss 有功损耗;有功负失
active low-pass filter 有源低通滤波器
active low structure black 活性低结构炭黑
actively and prudently transform econonmic system 积极稳妥地转变经济体制
actively heated suit 主动加热式潜水服
active maintenance 有效维修
active margin 活动边缘
active margin basin 活动边缘盆地
active market 市场活跃
active maser oscillator 超高频量子振荡器
active mass 活性团;有效质量
active mass approximation 活性质量近似法
active mass fraction 活性质量部分
active master file 现用主文件;常用主文件;常用基本文件
active master item 现用主项;现役主项;常用字组;常用主文件项;常用基本项目
active material 激活材料;活性物质;活性材料;活化材料;放射性物质;放射(性)材料
active material host 激活材料基质
active matter 活性物
active matter content 活性物含量
active mechanism 作用机制
active medium 激活媒质;激活介质;活性介质;工作介质
active member 主动构件
active memory 立动式存储器;快速存储器;主动式存储器
active memory cell 有源存储元件
active metal 活性金属;活泼金属
active method 主动法
active microflora 活跃的微生物区系;活性微生物区系
active microorganism 活性微生物
active microwave 主动微波
active microwave system 主动微波系统
active model 主动式
active money 活动货币
active moraine 活动冰碛
active movement 主动运动;主动活动
active mycelium 活性菌丝
active negativism 主动违拗
active network 有源网络
active nitrogen 活性氮
active noise control 有源噪声控制
active noise reduction 有源降噪
active operator 现用操作符
active optic(al) component 有源光学元件
active optic(al) fibre 激活光纤
active optic(al) network 有源光网络
active optics 能动光学;主动光学
active organic matter 活性有机质;活性有机物
active organism 活性生物
active orientation 强制定向排列
active or passive thrust 主动或被动土压
active output 实际产量;有用功率;有功输出
active oxygen 活性氧;有效氧
active oxygen method 活性氧法
active oxygen test 活性氧试验
active page field 有效页面区
active page queue 有效(活动)页面队列
active part 有源器件
active particle 活性粒子;放射性粒子
active partner 普通股东;积极合伙人;执行业务股东;执行业务的合伙人
active permafrost 活动永冻层
active phosphorus 活性磷
active photospheric region 光球活动区
active pigment 活性颜料
active pile 主动桩
active pipeline 活动流水线
active plate 活动板块;移动板块
active plate margin 活动板块边缘

active pole 主极
active pollution 放射性污染;放射线污染
active pool 活性沉淀池
active population 激活粒子
active porosity 有效空际(度);有效孔隙度
active portion 活动部分
active position address 主动寻址
active posture 自动体位
active powder 活性粉末
active powdered carbon 化学粉末炭
active power 有效功率;有功功率
active power loss 有效功率损耗
active power relay 有功功率继电器
active pressure 实际压力;主动压力
active primary damping 有效一次阻尼
active primer 磷化底漆;金属底层处理用漆;活性底漆;洗涤底漆
active probe 有源探头;有源探测器
active process 活动进程
active processor module 活动处理机模块
active program(me) 活动程序;运行中的程序
active protection 主动保护
active pull up 有源上拉
active pure strategy 活动纯策略
active radar 有源雷达
active radar probe 主动雷达探测器
active radar sensor 主动雷达探测器
active radiator 有源辐射器
active radical 活性基
active range target 主动雷达物标
active ranging system 主动测距系统
active Rankine pressure 朗肯主动土压力
active Rankine state 朗肯主动土压力状态
active Rankine zone 主动朗肯区
active reabsorption 主动再吸收
active reaction 活性反映;活性反应;活化反应;有效反作用;有效反应
active recording device 主动记录设备
active recreational areas 动态游憩用地;有效游乐面积
active redundance 主动冗余
active redundancy 主动冗余;工作储[贮]备
active reflector 有源反射器
active region 激活区;活性中心;活性区域;活化区;活动区;有源区;作用区(域);非饱和区
active reinforcement 受力钢筋
active relation 能动关系
active relay station 有源中继站
active remote sensing 人工源遥感;有源遥感;主动式遥感
active repair time 实际修理时间;实际维修时间;有效修复时间
active reserve 动储量(地下水);活动准备金
active reservoir capacity 有效库容
active resistance 自动抵抗力
active resonator 有源共振器
active resorption 主动再吸收
active rest 积极休息
active return loss 回声衰减
active rift 活动裂谷;活动断裂
active rig 现用钻机;工作钻机
active rod grinding 激活材料芯子研磨;激活材料棒研磨
active roll damping facility 主动横减摇装置
active roller 动辊
active roll resisting fins 可操纵减摇器
active rudder 主动舵
active runway 使用中的跑道;现用跑道
active rupture 主动状态破坏面;主动土压力状态破坏面
actives 活性物
active salt marsh 活性盐沼
active sand dune 流动沙丘
active satellite 有源卫星;主动(人造)卫星
active section 活性段
active security 流通证券
active sediment 活性沉积物
active segment 活动段
active seismic pressure 主动地震压力
active seismic zone 活动地震带
active semiconductor material 激活半导体材料
active sensor 有源传感器;主动式传感器
active service 现役
active session 有效时间

active set 作用集
active shear stress 主动剪应力
active side of coil 线圈有效边
active silica 活性硅酸;化学硅酸;活性硅;活性二氧化硅
active silica-gel 活性硅胶
active site 激活部位;活性点;活性部位
active sleeve 活化套(阴极)
active slide area 活动性滑坡区域;主动滑动区
active sludge 活性污泥
active sludge process 活性污泥工艺
active sludge treatment 活性污泥处理
active soil force 主动土力
active soil formers 活性成土因素
active soil pressure 主动土压力
active soil pressure redistribution 主动土压力重分布
active solar energy system 主动太阳能系统;主动式太阳能装置
active solar heating system 主动式太阳能采暖系统
active solar house 主动式太阳房
active solid 活性固体;吸附剂气固色谱
active solute 活性溶质
active solvent 活性溶剂;有效溶剂
active sonar 主动(式)声呐
active sonar detection 主动声呐检测
active sonar system 主动声呐系统
active sound absorption 有源声吸收
active sound reduction 有源减噪
active source method 主动源法
active stabilized platform 有源稳定平台
active stabilizer 活性加固剂;主动减摇装置;主动安定面
active stack 活动栈;现役栈;操作栈
active stage 进行期
active standard 现行标准
active state 激活态;活性(状)态;活动状态;有效状态;工作状态
active state of plastic equilibrium 塑性平衡主动状态;主动塑性平衡状态
active station 活动站;工作站
active stope 生产工作面;生产采场
active storage 机动储备;有效库容;有效储备;主动式存储器
active stream sediment 活性水系沉积物
active stream sediment above water level 高于水面的岸边活性沉积物
active stream sediment near water surface 水面附近的岸边活性沉积物
active strength 有效强度
active structural type 活动构造类型
active structure 活动构造
active structure zone 活动性构造带
active substance 活性物质
active sulfur 活性硫
active superfine calcium carbonate 活性超细碳酸钙
active surface 有效表面
active surface of sliding 主动滑动面
active survival suit 主动式救生(衣)服
active system 有源系统;太阳能储存系统;主动系统;主动式
active target 主动式目标
active task list 任务表
active tectonic belt 活动区域构造带;活动大地构造带
active tectonic pattern 活动构造型式
active tectonic stress 活动构造应力
active tectonic system 活动性构造体系
active tectonic zone 活动构造带
active thermal seepage 活动性热渗捅
active thrust 主推力
active thrust of earth 土壤有效压力;主动土压力;主动土推力
active time 服役时间
active tonnage 在营运吨位;在航吨位
active tracking 主动跟踪
active tracking system 主动跟踪系统
active trade balance 贸易顺差
active trading 活跃贸易
active trading fleet 营运商船队
active transducer 主动换能器;有源能量转换器;有源换能器;有源变换器;主动式转换器;主动变换器

active transfer 活性转移;主动传递
active transition 活动变换
active transport 活性转移;自动运输;主动转移;主动运输;主动输送
active trust 自动信托
active unit 活性单位
active user 活动用户;现时用户;当前用户
active valence 有效化合价
active vector 活动向量
active vibration isolation 主动隔振
active vibration reduction 有源减振
active viewport 激活视口
active volcano 活火山
active voltage 有功电压
active volt-ampere 有效伏安
active volume 有效容量;有效容积
active wait 主动等待
active waste disposal site 现用废物处置场;现用废物处理场
active water 侵蚀性水;活性水;有腐蚀性的水
active water supply source 有效供水水源
active way of active fault 活断层活动方式
active wheel 工作轮
active window 激活窗口
active window against blasting 防爆活门
active workings 生产巷道;生产工作区
active zeolite 活性沸石
active zone 活动区;活动带;活动层;工作范围
active zone of well 井的补给区;井水水源区
activities not-adequately described 非标准经济活动;非标准活动
activities of a vessel 船舶活动范围
activities of project 项目业务
activity 激活度;机构;能动性;活性;活力;活化度;活度;活动性;活动(度);信号活动;作用活度
activity account 业务账户
activity accounting 作业会计
activity-activity diagram 活度—活度图
activity address code 单位地址代码
activity address directory 单位地址簿
activity analysis 活性分析;活度分析;活动分析;业务活动分析
activity arrow 作业箭头
activity-based cost(ing) 活动成本;作业成本
activity-based costing method 作业成本法
activity-based depreciation 产量折旧法
activity-based equilibrium constant 活度平衡常数
activity capability 实际能力
activity chart 活动示意图
activity classification 业务活动分类
activity classify of spring 温泉的活动状态分类
activity coefficient 激活系数;活性系数;活度系数
activity coefficient of adsorbed species 吸附物种活度系数
activity coefficient of hydrogen ion 氢离子活度常数
activity concentration 放射性浓度
activity constant 活度常数
activity cost 活动成本;作业成本
activity costing 作业成本计算
activity counter 活动计数器
activity curve 活性曲线;活度曲线;放射性曲线;放射曲线
activity cycle 活动周期
activity degree 活性度
activity duration 活动期限;作用期限;作业持续时间;有效期限
activity factor 活性系数;活度因子
activity finish time 作业完成时间
activity for defocus 散焦灵敏度
activity index 活性指数;活度指数
activity indicator 活动指标
activity infill 活动空间
activity in the model 模式中的活动
activity is middle strong 活动性中等
activity is strong 活动性强烈
activity is very strong 活动性极强烈
activity is weak 活动性弱
activity level 活动程度
activity life 实际寿命
activity line 作业线
activity loading 活动装入法;有效装入法
activity measure of activated sludge 活性污泥活性测定

activity median aerodynamic diameter 活度中值空气动力学直径
activity network 作业网络
activity number 活性指数
activity number of soil 土的活性指数
activity of bitumen 沥青活性
activity of catalyst 催化剂活性
activity of cement 水泥活性
activity of cohesive soil 黏性土的活动度
activity of component 组分活度
activity of faults 断层活动程度
activity of fore-shock 前震活动
activity of free electrons 自由电子活度
activity of fundamental fault 基底断裂活动
activity of microorganism 微生物活性
activity of pollutant 污染物活度
activity of soluble species 溶性物种活度
activity of surface species 表面物种活度
activity on arrow 以箭头代表作业
activity-on-arrow network 箭示活动网络
activity on branch 双代号网络
activity on node network 节点网络;单代号网格
activity product 活度积
activity product of ions 离子活度积
activity quota 作业定额
activity quotient 活性系数;活度系数;活度商
activity rate 单位劳动系数;单位劳动量;就业活动率;业务活动率
activity ratio 放射性比(率);激活率;活性比;活度比(率);活动率;业务活动比率
activity relation chart 作业相关图
activity room 活动室
activity sampling 工作的抽样检验;工作抽样检查
activity save area 活动保留区域
activity series 活度顺序
activity space 活动空间
activity stage 活动阶段
activity test 活性试验
activity time cost trade-off graph 工作的直接费与历时关系图
activity trace 活动追踪
activity unit 放射性单位
activity variance 业务活动差异;工作差异
activity vector 活动向量
activization of platform 地台活化作用
act of arbitration 仲裁条例;仲裁法
act of authorization 授权行为
act of bankruptcy 破产行为;破产法
act of business taxes 营业税法
act of conservation 土壤保持法
act of destroying forest 毁林行为
act of flood control 防洪法
act of force 暴力
act of foreign states 非本国行为
act of God 天灾;自然灾害
act of hono(u)r 信誉承兑;承兑
act of hostility 战争行为;敌对行为
act of nature 天灾;自然灾害;不可抗力
act of omission 不作为
act of pollution control 污染防治法
act of production 生产行为
act of purchase and sale 买卖行为
act of randomization 随机化行动
act of sale 卖契;销售记录
act of the seller 卖方的行为
act of tort 侵权行为
act of war 战争行为
act on 作用于
acton 铜解气
act on schedule 按计划进行
actor 操作符
actor formalism 动作者表示法
acts of God clause 天灾及海难除外条款
act tax 行为税
actual 实在的
actual aberration 实际像差
actual absorption 实际吸收;实际吸附
actual absorption density 实际吸附密度
actual achievement actual performance 实际成绩
actual acidity 实际酸度
actual acquisition cost 实际取得成本
actual acquisition cost basis 实际购置成本法
actual address 实际地址
actual advance 实际进度

actual age 实际年龄;绝对年龄;绝对年代【地】
actual aggregate breaking strength 骨料实际断裂强度
actual allowance 实际容差;实际间隙;实际偏差;实际公差
actual amount of money 实际金额
actual amount of rainfall 实际降雨量
actual angle of attack 实际攻角;实际冲角
actual annual interest rate 实际年利率
actual area 实际面积;有效面积
actual argument 实自变量;实元;实际变元;实际变量;实变元;实变数
actual argument array 实际变元阵列
actual argument value 实际变元值
actual array 实在数组
actual array argument 实在数组变元
actual array declarator 实在数组说明符
actual assets 实际资产
actual balance 实存
actual balanced budget 决算
actual balance of money 实际货币余额
actual banking assets 实际金融资产量
actual basis accounting 实际发生额会计
actual behavio(u)r 实际性状
actual block processor 实际数据块处理程序
actual boring time 钻孔实际时间
actual bottom 实际海底
actual boundary shear stress 实际边界剪切应力
actual breaking force 实际破断力
actual breaking load 实际破坏荷载;实际破断荷载
actual breaking stress 真实断裂应力
actual budget 决算
actual budget on completion 竣工结算
actual burden rate 实际制造费用分配率;实际负担率
actual calculating area 实际计算面积
actual capacitance 实际电容量
actual capacity 实际(生产)能力;实际容量
actual capacity to repay 实际偿还能力
actual capital 实际资本
actual capital coefficient 实际资本系数
actual cargo capacity 实际装货量;实际载货量
actual carrier 实际承运人
actual carrying capacity 实际运载能力
actual cash balance 实际现金余额
actual cash value 实际现金价值
actual challenges 现实挑战
actual change 实际变动
actual charges 实际费用
actual clearance 实际净空
actual code 实际代码
actual coding 实际编码
actual compression 实际压缩
actual concentration 实测浓度
actual condition 实际工况
actual construction procedure 实际施工程序
actual construction sequence 实际施工程序
actual construction time 实际施工期;实际建筑期限;实际(建筑)工期
actual consumption 实际消费
actual corpus 实际财产主体
actual cost 实际价格;实际造价;实际费用;实际代价;实际成本
actual cost accounting system 实际成本会计制度
actual cost calculation 实际成本计算
actual costing 实际成本计算
actual cost method 实际成本法
actual cost of living index 实际生活费指数
actual cost of producing export commodity 出口商品生产的实际成本
actual cost payment 实报实销
actual cost system 实际费用制度;实际成本制
actual cross-section area 实际横剖面面积;实际横断面面积
actual crystal 实际晶体
actual cut 实际切削
actual cycle 实际循环
actual damages 实际损害赔偿金
actual damping 有效阻尼
actual data transfer rate 实际数据传输率
actual deadweight 实际载重吨位
actual death rate 实际死亡率
actual debt 实际债务;实际债款
actual decimal point 实际(十进制)小数点;实际

存在的十进制小数点;有效小数点
actual declarator 实在说明符
actual declarer 实在说明词
actual degree of belief 实际置信度
actual degree of calcinations 实际煅烧率;实际分解率
actual degree of decarbonation 实际分解率
actual delivery 实际扬水量;实际交货;实际交付;实际财产的转移
actual delivery of pump 泵的实际送水量
actual delta 实在增量
actual demand 实际需要;实际需求
actual density 实际(疏)密度;真密度
actual density of soil 实际土壤密度
actual depreciation 实际折旧
actual depth 实际深度;实际水深
actual derived data item 实际任意数据项
actual deviation 实际偏差
actual device 实际设备
actual device coordinates 实际设备坐标
actual dimension 实际维数;实际尺度;实际尺寸
actual discharge 实际偿还;实际流量
actual discharge coefficient 实际流量系数
actual dishono(u)r 实际拒付;拒绝承兑
actual displacement 汽缸工作容积;实际位移;实际排水量【船】
actual distance 实距;实际距离;实际航程
actual diving depth 实际潜水深度
actual drilling time 纯钻进时间
actual drying 实干
actual drying process 实际干燥过程
actual earth fill 实际填土
actual eccentricity 实际偏心
actual economic condition 经济现状
actual effect 实效
actual efficiency 实际效率
actual element 实有元件
actual elevation 实际高程;实际标高
actual entry 实际入口;实际登记项
actual error 似真误差;实际误差
actual evapo(u)ration 实际蒸发(量);有效蒸发(量)
actual eviction 强行收回;实际收回
actual exchange rate 实际汇率
actual execution 实际执行
actual exhaust velocity 喷管实际流速;实际排气速度
actual expected standard cost 实际预算标准成本;实际预期标准成本
actual expenditures 实际支出
actual expenses 实际开支;实际费用
actual family size 实际子女数;实际家庭规模
actual fault and privity 实际过失和知情
actual feed 实际给料
actual figure 实际数字
actual filling depth 实际填充深度
actual finish time 实际完成时间
actual flow 实际流量;实际流动;真实流动
actual flow curve 实际车流曲线
actual fluid 实际流体
actual freight rate 实际运费率
actual frequency 实际频率
actual frog point 辙叉(心轨)实际尖端;辙叉实际尖点
actual fuel consumption of locomotive 机车燃料实际消耗量
actual full supply line 实际最大供水管路
actual gain 实际增益;有效增益
actual gap 实际间隙;实际缝隙
actual garbage 实际无用数据;实际废料
actual gas 真实气体
actual glass composition 实际玻璃组成;实际玻璃成分
actual goods 现货
actual grading curve (粒状材料的)实际筛分
actual grain size 实际粒径;实际晶粒度
actual gravity 真实重力
actual gross national products 实际国民生产总值
actual gross weight 实际总重量;实际皮重
actual groundwater velocity 地下水实际流速;地下水实在流速;地下水实地流速
actual growing situation 实际长势
actual handing over 实际交付
actual head 实际扬程;实际水头

actual height 实际高度;实际高程
actual height of groundwater 地下水实际高度
actual height of water-level 实际水位高度
actual holder 实际占有人
actual horsepower 实有功率;实际马力;有效马力
actual hydraulic gradient 实测水力梯度
actual image point 实际像点
actual implementation 实际实现
actual income 实际收益
actual indicator 实际示意图
actual indicator card 实际示功图
actual industrial effluent 真实工业污水
actual inspection 实际检查
actual instruction 实际指令;有效指令
actual interest income 实际利息收益;实际利息收入
actual interface 实际接合
actual inventory 实际盘存;实地盘存
actual investment 实际投资
actuality 现状;现实;实在
actual key 实际关键字;实际关键码;绝对关键字
actual labo(u)r rate 实际工资率
actual land form 实在地形
actual lead (打桩时的)实际导程
actual leakage 实测漏泄量
actual length 实际长度
actual length of turnout 道岔实际长度
actual level 实际标高
actual liability 实际负债
actual life 实际使用期限;实际(使用)寿命;实际使用年限;有效寿命
actual lime content 实际含灰量
actual limit 实际限度;实际极限
actual line 实线;实际行
actual liquid 真实液体
actual load factor 实际负载因素;实际负荷因素
actual load(ing) 实际荷载;实际负载;实际负荷;有效负载;有效负荷
actual loading method 真实负载法
actual loading test 实物试验;实际荷载试验;实际负载试验;实际负荷试验;真载试验
actual load rate 实载率
actual loss 实际损失;实际亏损
actual lower bound 实在下界;实际下界
actually 事实上;实际上
actually measured data 实测资料;实测数据
actually realized productive capacity 实际到达能力
actually tested self-weight 实测自重湿陷量
actually transmitted 实际透过的
actual man-days 实际工作人日数
actual man-days at work 实际作业工作日
actual man-days for loading/discharging 装卸实际工作日数
actual man-days in attendance 实际出勤人日数;实出勤工日数
actual man-hour 实际工时
actual manufacturing cost 实际制造成本
actual manufacturing overhead 实际间接制造费用
actual map 实地图
actual market 现货市场
actual market price 现货市场价格
actual market volume 市场实际容量
actual material map 实际材料图
actual maximum supply line 实际最大供水管路
actual mean pressure 实际平均压力
actual measured value 实测值
actual measurement 实物测量;实际测量;实测;有效测量
actual mechanism 真实机理
actual moment 实际弯矩;实际力矩
actual money holding 实际货币持有额
actual monitor 输出监听器;线路监视器
actual motion 实际运行;实际运动;对地运动
actual net weight 实际净重
actual no-par value stock 真正无面值股票
actual normal cost 实际正常成本
actual normal cost system 实际标准成本法
actual number 实数【数】
actual number analysis 实数分析法
actual numeric(al) value 实际数值
actual observation record 实测记录
actual observed value 实际观测值
actual occurrence time 实际发生时间
actual operation 实际运转;实际操作
actual order 最终任务;实际执行合同;实际(执行)

订货单
actual output 实际输水量;实际输出;实际出力;实际产量;实际产出;有效输出
actual output of pump 泵的实际出水量
actual oxygen requirement 实际耗氧量
actual pairing 实际并行;真并行
actual parameter 实在参数;实际参数
actual parameter association 实在参数结合
actual parameter display 实在参数显示
actual parameter list 实在参数表
actual parameter part 实在参数部分
actual parking volume 实际停放量
actual payload 实际载货量;实际收费载重;实际净载重量
actual peak load 实际峰荷
actual performance 实际性能;实际完成情况;实际履行;实绩;真实性能
actual performance measurement 实际工况测定
actual performance of a company 经营实绩
actual personnel 实际工作人员;实际人员(情况)
actual pitch 实际螺距
actual play 实际运转;实际间隙
actual plot 真实图形
actual point of frog 辙叉心轨尖端;辙叉实际尖端
actual point of switch 转辙器实际尖端
actual population 实际人口;现有人口
actual porosity 实际孔隙率
actual position 实际位置
actual possession 实际占有(权)
actual power 实际功率;有效功率;有效动力
actual power chart 实际功率图
actual power consumption 实际消耗功率
actual practice 实务
actual pressure 实际压力;实际气压
actual price 实际价格
actual production 实际产量
actual profile rate of funds 实际资金利润率
actual profit 实际利润
actual purchase 实际购买
actual purchase price 实际购入价格
actual quality 实物品质
actual quantity 实际(数)量
actual quantity issued 实发数量
actual quantity to be shipped 实际准备装船数量
actual query 实例查询
actual quotation 实际开价;实际交易值;行情;暗盘
actual range 射程
actual rate of exchange 实际汇率
actual rate of growth 实际增长率
actual rate of interest 实际利率
actual rate of obsolescence 实际陈废率
actual rate of wastage 实际耗蚀率
actual rate of wastage and obsolescence 实际损耗率及废弃率;实际耗损与陈旧率
actual ray trace 实际光路追迹
actual recognizer 实在识别算法
actual reflux 实际回流
actual register 现行寄存器
actual relay 实际延迟
actual rent 实际租金
actual requirement 实际需要量
actual reserve 实际储藏量;实际储[贮]量;可靠储量
actual result 实际效果;实际结果
actual routine data area 实际例行程序的数据区
actual row of rower 实在行标行;标行程序的实际行
actual run-off 实际径流量
actual run time 实际运行时间
actual runtime data area 实际运行时数据区;实际运行程序数据区
actuals 实际商品;实际货物;标准黄麻
actual safe load 实际安全荷载
actual safe load of pile 桩的实际安全荷载
actual sale 实际销售量
actual sample 现货样品
actual sea level 实际海平面
actual section 实际横截面;实际(横)断面
actual seepage velocity 实际渗流速度
actual sequence 实在序列
actual service 实际送达
actual service condition 实际使用条件
actual service life 实际使用寿命;实际使用年限
actual (service) load 有效荷载
actual service test 实际使用试验
actual setting 实际设置

actual settlement-time curve 实际沉降—时间曲线
actual shear force 实际切力
actual ship form 实际船型
actual shot number 实际炮数
actual size 真实尺寸;实际细度;实际规格尺寸;实际大小;实际尺度;实际尺寸
actual slip 实滑距
actuals market 实物市场
actual solid retention time 固体实际停留时间
actual sound position 实际声源位置
actual specification 实际规范
actual specific gravity 实际比重;真实比重
actual specific impulse 实际比推力
actual speed 实际速率;实际速度
actual stability 实在稳定性
actual staff 实际工作人员;实际人员(情况)
actual standard 实际标准;现行标准
actual state 实际状态;实际状况;现状
actual state of affairs 真相
actual steam consumption 实际汽耗
actual steam rate 实际耗汽率
actual stock 实际库存;实际储存(量)
actual strain 实际应变
actual strength 实际强度
actual stress 实际应力;有效应力
actual stress at fracture 实际破坏应力;断裂处的实际应力
actual structure 实际结构;真实结构
actual stuff 现货
actual supplementary head 实际附加水头
actual supply 实际供给
actual survey dip of foundation 地基实测变形值
actual surveyed water-levels over the years 实测历年水位
actual system 实际的系统
actual tare 实际皮重
actual taxpayer 实际付税人
actual technical strength 实际力学强度
actual temperature 实际温度
actual tender 实际承包人
actual test 实物试验
actual testing pressure 实际试验压力
actual thermal efficiency 实际热效率
actual thickness 实际厚度
actual throat 焊缝实际厚度
actual throat depth 实际焊喉深度
actual throughput 实际通过量;实际输量;实际产量
actual thrust 实际推力
actual tide 真潮
actual time 实时;实际时间;真实时间
actual time of arrival 实际到达时间
actual time of completion 实际完成时间
actual time of departure 实际起飞时间;实际开航时间;实际出发时间
actual time of interception 实际拦截时间
actual time of observation 实际观测时间
actual time over 实际飞越时间
actual tooth density 齿内的有效磁感应
actual total loss 全部损失;实际全损;绝对全损;绝对(全部)损失
actual track capacity 实际线路通过能力
actual trade-off curve 实在权衡曲线
actual transportation time 实际运输时间
actual travel direction 实际航向
actual(tube)length 实际长
actual upper bound 实在上界;实际上限
actual user 实际用户
actual utility 实际效用
actual value 实际值;实际价值;现实的价值;真值
actual value basis 实际价计算基础
actual value method 实际价值法
actual variable 实际变量
actual variable parameter 实际变量参数
actual velocity 实际速度;实际流速
actual velocity of groundwater 地下水实际流速
actual volume 实际容积
actual volume of consumption per unit 实际单耗
actual water depth 实际水深
actual water output 实际出水量
actual water temperature 实际水温
actual wave 实际波浪
actual weight 实重;实际重量;实际重力;实际权数;真实重量
actual weight of load 实际装载量

actual weld-throat thickness 焊缝厚度
actual working days 实际工作天数
actual working hours 实际工时;实动工时
actual working machine-days 实作台日数;实际工作台日数
actual working pressure 实际工作压力
actual work(ing)time 实际工作时间
actual worth 实值
actual yield 实际收益率;实际生息率
actual yield point 实际屈服点
actual yield stress 实际屈服应力
actual zenith distance 实际天顶距
actual zero point 绝对零点;基点
actual zoning 现状分区
actuarial 精算的
actuarial basis 保险业计算标准;保险统计计算标准
actuarial cost 精算费用
actuarial equivalent 精算等值
actuarial evaluation 精算估值
actuarial method 保险统计法
actuarial reserves in respect of life insurance 人寿保险精算准备金
actuarial statistics 保险统计
actuarial surplus 精算剩余;精确盈余
actuary 保险精算师;保险统计员;保险统计师;保险(公司)计算员
actuary of fixed assets 固定资产统计师
actuate 使动
actuated controller 感应式信号控制机
actuated controller assembly 感应式信号控制
actuated roller switch 促动滚子开关
actuated valve 控制阀
actuating 启动
actuating air 推进空气
actuating apparatus 启动装置;调节器;控制器;执行器;促动器;驱动器
actuating arm 驱动杆;驱动臂;力臂;开动臂;工作臂;操纵杆;推动臂
actuating battery 操作电池
actuating bellows 控制膜盒
actuating brake cylinder 制动器的作用缸
actuating cable 控制电缆;操作电缆
actuating cam 凸轮;推动凸轮;主动凸轮
actuating cavity 激励谐振腔
actuating code 执行码
actuating coil 激励线圈;工作线圈;动作线圈
actuating current 动作电流
actuating cylinder 作动筒;主动油缸;动力汽缸
actuating device 驱动装置;启动装置;调节装置;伺服装置;激励装置;感应装置;感应设备
actuating element 执行元件;驱动机构
actuating lever 起动杆;启动杠
actuating mechanism 调节装置;促动器;驱动机构;执行机构;促动机构;传动机构;操作机构
actuating medium 工作介质
actuating motor 起动电动机;驱动电动机;伺服电动机;执行电动机
actuating of relay 继电器动作
actuating pressure 驱动压力;作用压力;促动压力
actuating quantity 作用量
actuating rod 作用杆
actuating section 起动部分
actuating shaft 驱动轴
actuating signal 起动信号;启动信号;激励信号;作用信号;感应信号;动作信号
actuating solenoid 操纵电磁铁
actuating speed 驱动速度
actuating spring 促动簧
actuating strut 作用支柱;动作筒
actuating system 动作系统;传力系统;传动系统;操作系统
actuating time 起动时间;动作时间
actuating unit 动力传动装置;驱动机组;执行装置;动力机构
actuation 促动;冲动
actuation duration 激励时间;作用时间
actuation gear 作用机构;促动装置
actuation register 激励记录计
actuation time 作用时间
actuator 激励器;传动装置;调速控制器;调节器;螺线管;激振器;油缸;作动器;执行元件;执行机构;促动器;传动机构
actuator assembly 驱动器组件
actuator cylinder 促动器油缸

actuator disc 起动圆盘;促动盘
actuator housing 执行机构罩
actuator motor 驱动电动机;执行电动机
actuator piston ring 气环
actuator pole 连接杆
actuator port 执行器油缸通油口;工作油口
actuator power 激发器功率
actuator rod 促动杆
actuator switching pulse 激励器转换脉冲
actuogeology 现实地质学
actuopaleontology 实证古生物学;现实古生物学
act upon plan 按计划进行
act with hasty 草率行事
act without consideration 无偿行为
actynolin 阳起石
acuasto-phytogenic rock 非燃性植物岩
acuiry for defocus 散焦锐度
acuition 锐敏性
acuity 锐利;清晰度;尖锐分辨力;锐敏度;锐度;鉴别力;敏锐度;敏度
acuity for defocus 散焦灵敏度
acuity matching 锐度匹配;分辨率匹配
acuity of colo(u)r 色觉锐度;色分辨力
acuity of colo(u)r definition 色分辨力锐度
acuity of eye 人眼分辨能力
acuity of hearing 听敏度
acuity photometer 敏锐度光度计
aculei 翅刺
acumen 尖头
acuminate 渐尖叶
acuminated roof 尖顶屋顶
acuminate element 尖锐的构件
acumination 尖头;锋利;尖锐
acupuncture treatment room 针灸室
acusis range 正常听觉范围
acutance 明锐度
acutance curve 锐度曲线
acutance developer 高锐度显影液
acutance matching 锐度匹配
acutate 微尖的
acute 尖锐
acute action 急性作用
acute adaptation 急剧的适应
acute alcoholism 饮酒中毒
acute angle 锐角;尖角
acute-angle attachment 锐角附件
acute angle blade 锐角叶片
acute angle block 锐角断块
acute-angle crossing 锐角辙叉;锐角交叉
acute-angled triangle 锐角三角形
acute angle intersection 锐角交叉
acute arch 锐拱;外二心桃尖拱;尖拱
acute bisectrix 锐角等分线
acute cone 锐锥
acute corner 锐角墙角
acute crossing 锐角交叉
acute deficiency of housing 严重的住房短缺
acute disease 急性病
acute effect 急性效应
acute episode 急性(中毒)事件
acute exposure 急性照射;急剧曝光
acute fold 尖角褶皱
acute hammer 开槽锤
acute hazardous waste 急性有害废物
acute infectious disease 急性传染病
acute-intersection 锐角交会
acute irradiation 强烈辐射;锐辐照;强烈照射;急性辐照
acute leaf 急尖叶
acutely hazardous waste 急性有害废物
acuteness 锐利;锐度;敏锐;敏度
acuteness of vision 视觉敏锐度
acute poisoning 急性中毒
acute poisoning episode 急性中毒事件
acute radiation disease 急性放射病
acute radiation injury 急性辐射线损伤
acute radiation sickness 急性放射疾病
acute squint 锐角转角隅石
acute-squint corner 锐角墙角
acute toxic effect 急性毒性效应
acute toxic effect zone 急性毒作用带
acute toxicity 急性毒性
acute toxicity test 急性毒性试验
acute toxicity test of aquatic organism 水生生物

急性毒性试验
acute-toxic pesticide 剧毒农药
acute toxin 剧毒素
acute triangle 锐角三角形
acute turnout 锐角道岔
acute zee 斜腹板Z形钢
acut frog 锐角辙叉
acyclic(al) 单极的;非周期(性)的;无环的;非循环的;非轮列的;非环形的
acyclic(al) compound 无环化合物
acyclic(al) dependence relation 非周期相关性关系
acyclic(al) digraph 非循环有向图
acyclic(al) directed graph 非循环有向图
acyclic(al) dynamo 单极发电机
acyclic(al) feeding 非周期(性)馈送;非周期输送
acyclic(al) graph 非循环图(形)
acyclic(al) hydrocarbon 无环烃
acyclic(al) machine 单极电机;非周期性电机
acyclic(al) model 非循环模型
acyclic(al) motion 无旋流动
acyclic(al) network 非周期网络;非循环网络
acyclic(al) oscillation 非周期性振荡;非循环性振荡
acyclic(al) point 非循环点
acyclic(al) potential flow 无环量势流
acyclic(al) potential motion 无环量势流
acyclic(al) wave 非周期波
acyclovir 阿昔洛维
acylamide 酰胺
acylamino yellow 酰氨基黄
acylated polymer 酰化聚合物
acylation 增酸作用;酰化(作用)
acyl exchange 酸解(作用)
aczoiling 防腐;电杆防腐
aczol 铜锌氨酚防腐剂
adalert 补充警报
adamant 金刚石;硬石
adamantanamtne 氨基三环癸烷
adamantane 金刚烷
adamant bend 硬质金属弯管
adamant clinkers 金刚砖(具有斜削边缘的砖,铺路用)
adamantine 阿达曼铬锰钢;金刚石制的;金刚合金;冷铸钢粒
adamantine bit 金刚石钻头;金刚砂钻探
adamantine chain 极坚固的铁链
adamantine clinker 坚硬炼渣
adamantine compound 金刚化合物
adamantine drill 金刚石钻头;金刚砂钻;钻粒钻进钻机;钢球钻头
adamantine head 金刚石钻头
adamantine luster 金刚石光泽;金刚石光彩;金刚光泽
adamantine shot 钻粒
adamantine spar 刚玉
adamant metal 以锡为主的锡锑铜轴承合金
adamant plaster 块干石灰墁灰;速凝石灰料;速凝石膏粉刷;速凝灰泥;硬石灰膏
adamant steel 铬钼特殊耐磨钢
Adam architecture 亚当式建筑;古典建筑装饰的建筑形式
adamas 金刚石
adambulacral 侧步带的
adamellite 石英二长石;石英二长岩;二长花岗岩
Adamesque 从亚当风格派生的(指建筑和家具)
adamic earth 红黏土
adamite 羟砷锌石;水砷锌矿;氧化铝磨料
adamite roll 高碳铬镍耐磨铸铁轧辊
Adam's arm 铲(俚语)
Adams chromatic value 亚当斯色度值
Adams chromatic value colo(u)r difference equation 亚当斯色度值色差方程(式)
Adams chromatic value system 亚当斯颜色定量指示法;亚当斯色度值定量指示法
Adam's formula 亚当斯砖墩实用公式
adamsite 吩砒嗪化氯;暗绿云母
Adams method 亚当斯方法
Adams-Nickerson space 亚当斯—尼克尔逊彩色空间
Adams-Nickerson-Stultz colo(u)r difference equation 亚当斯—尼克尔逊—斯塔尔茨色差方程
Adamson joint 阿达姆松接头(锅炉)
Adamson's ring 炉筒补强环;阿达姆松联接环
Adams process 亚当斯玻璃砂除铁法
Adams projection 亚当地图投影

Adam's saw 长柄小锯
Adam's style 亚当风格
Adam's water bar 门底横木(用以排除雨水和穿堂风)
ada mud 添加泥浆
Adansonia fibre 阿丹森尼亚纤维
adapt 适应
adaptability 适用性;适应性;适应能力;适应力;灵活性;可塑性
adaptability of operation 加工适应性
adaptability of software 软件的适应性
adaptability of soil 适土性
adaptability to change 适应变化的能力
adaptable 可适应的
adaptable species 适应性强的物种
adaptable stage 灵活舞台
adaptation 适应现象;改用;采用;变通办法
adaptation brightness 适应亮度
adaptation function 适配功能
adaptation kit 成套配(合)件
adaptation level 适应亮度;适应电平
adaptation lighting 适应照明
adaptation luminance 适应亮度
adaptation of environment 环境适应
adaptation of methods 方法的适应
adaptation of product 产品适应性
adaptation of the budget 核定预算
adaptation period 适应阶段;适应时间
adaptation point 适应点
adaptation process 适应过程
adaptation product 反应质
adaptation study 适应(性)研究
adaptation theory 适应理论
adaptation to cold 冷适应
adaptation to high temperature 高温适应
adaptation tolerance 适应耐受性
adaptation to local condition 因地制宜
adaptation to short-day region 适合短日照的地区
adaptation to the ground 利用地形
adaptation unit 适配单元
adaptation zone (车辆在隧道口前使驾驶员适应光度变化的)适应区段
adaptative development strategy 变通发展战略
adapted breed 适应品种
adapted circumstance 适宜的环境
adapted form 适应形状
adapted types 适应类型
adapted varieties 适宜品种
adapter = adaptor
adapter amplifier 匹配放大器
adapter assembly 紧接套组件
adapter bearing 带固接套轴承
adapter bend 异径弯头
adapter-booster 传爆药管
adapter box 接头盒
adapter bush 接头衬套
adapter check 适配器检查
adapter connector 接合器;连接器(接头);接头
adapter control block 适配器控制块
adapter converter 附加变频器
adapter coupling 管套转接;套筒式联轴器;连轴齿套;接头
adapter flange 配接凸缘
adapter gear 齿轮接合器
adapter glass 玻璃接头;附接玻璃
adapter junction box 适配箱;接线盒;分线匣;分线盒
adapter kit 成套附件
adapter lens 接合器透镜;附加适配透镜
adapter path 连接路线
adapter plate 衬板;垫板
adapter plug 转接(器)插头;分接插头
adapter ring 适配环;配接环;过渡环;接圈;接合镜筒;接合环;中间环
adapter set screw 接头调节螺钉
adapter shifting coupling 接合器
adapter skirt 适配裙部;连接裙
adapter sleeve 连接套筒;连接套管;紧固套;紧定套(筒);接头套管;接合器套筒
adapter socket 卡盘
adapter transformer 附加变压器
adapter-type bearing 接合式轴承
adapting device 配合装置;连接装置
adapting flange 配接法兰;连接接盘;连接法兰;法兰

adapting form 承插式
adapting lighting 适应照明
adapting pipe 套管;连接管;异形管;承接管(套管)
adaption 配合;适合
adaptive 适应的
adaptive algorithm 适应算法
adaptive alternatives 适合的备择方案
adaptive array 自适应阵列
adaptive assessment 自适应评价
adaptive attenuation 自适应衰减
adaptive automatic equalizer 自适应式自动均衡器
adaptive autopilot 自适应自动驾驶仪;自适应自动舵
adaptive capacity 适应能力
adaptive colo(u)ration 适应赋色
adaptive colo(u)r shift 色适应位移
adaptive communication 适应式通信;自适应通信
adaptive control 适应(性)控制;适配控制;自适(应)控制
adaptive control action 自适应控制作用
adaptive control constraint 限制式适应控制
adaptive controller 自适应控制装置
adaptive control optimization 自适应控制优化
adaptive control procedure 自适应控制程序
adaptive control system 适应式控制系统;自适应控制系统
adaptive control unit 自适应控制装置
adaptive convergence 趋同适应
adaptive deconvolution 自适应消卷积
adaptive delta modulation 自适应增量调制
adaptive differential pulse code modulation 自适应差分脉冲编码调制
adaptive dispersion 适应性扩散
adaptive divergence 适应趋异
adaptive element 适应元件
adaptive equalization 自适应均衡
adaptive equalizer 自适应均衡器
adaptive evolution 适应(性)进化
adaptive faculty 适应(能)力
adaptive feedback control 适应反馈控制;自适应反馈控制
adaptive filter 适应滤波器
adaptive filtering material 自适应滤料
adaptive forecasting 自适应预测
adaptive forecasting coding 自适应预测编码
adaptive form 适应类型
adaptive linear dynamics and quadratic criterion 适应的线性动态及二次判据
adaptive linear filtering 自适应线性滤波
adaptive management 自适应管理
adaptive mechanism 适应机制
adaptive migratory 适应迁移
adaptive model 自适应模型
adaptive modelling 适应性建模
adaptive modelling approach 适应性建模方法
adaptive multiplexer 自适应复用器
adaptive optimal control 自适应最优控制;自适应最佳控制
adaptive optimal system 自适应最佳系统
adaptive optimization 适应最优化;自适应最优化
adaptive peak 适应峰
adaptive phase 适应阶段;自适应阶段
adaptive prediction 适应预报;自适应预测
adaptive prediction coding 自适应预测编码
adaptive process 自适应过程
adaptive radar-controlled autopilot 自适应雷达控制自动舵
adaptive radiation 适应(性)辐射;适应性放射
adaptive reception 自适应接收
adaptive response 自适应性响应
adaptive re-use 适应性再使用
adaptive routing 适应式路径选择;自适应路由选择;自适应路径选择
adaptive routing algorithm 自适应路径选择算法
adaptive servo 自适应伺服机构
adaptive servo-system 自适应伺服系统
adaptive signal control optimization technique 自适应信号控制优化技术
adaptive signal processing 自适应信号处理
adaptive sleeve 紧定套(筒)
adaptive smoothing method 拟合平滑法
adaptive stack 自适应叠加
adaptive steering module 自适应操舵模型
adaptive system 适应性系统;可适应系统;自适应

系统
adaptive system theory 自适应系统理论
adaptive technique 适应技术;自适应技术
adaptive test 自适应测试
adaptive testing 适应测试
adaptive traffic control optimization technique 自适应交通控制优化技术
adaptive traffic control system 自适应交通控制系统
adaptive transform coding 自适应变换码
adaptive value 适应值
adaptometer 匹配测量计;适应计;自适应计
adaptor 接头;适配器;拾音器;拾波器;联轴套管;联接装置;接续器;接套;接合器;衔接器;应接管;应接板;转接器;转(换)接头;针接头;多头插销;承接管;衬套
adaptor bend 注料嘴接头弯管;管子接头弯头
adaptor for product discharge 排料接口
adarce 泉渣;石灰华;石灰沉积;钙华
adatom 被吸附的原子
adaxial paratraceid parenchyma 近轴旁管薄壁组织
adaxial parenchyma 近轴薄壁组织
Adcock antenna 爱德考克天线
Adcock directional finder 旋转天线测向仪;爱德考克测向仪
Adcock direction finder 旋转天线探向器;爱德考克测向仪;爱德考克定向器
Adcock range 爱德考克无线电测距
Adcock system 爱德考克天线系统
adcumulate 补堆积岩
add 加重;加法;加多
add and drop multiplexer 分插复用器
add-and-subtract relay 增减继电器
addaverter 加强转换器;加法转换器
add bus 加法总线
add confirmation 加保兑
add control unit 加法控制部件
add-delete list 附加删除表
added burden 额外负担
added capacity 新增生产能力
added circuit 加法电路;相加电路
added edition 增订版;重版
added entry 副款目;附加著录;附加款目;被加项
added entry point 补充进入点
added fixed assets 新增固定资产
added items of expenditures 追加支出
added lane 附加车道
added loss 附加损耗
added martix 附加矩阵
added mass 附加质量;附加物质;附加块体
added metal 熔接金属;熔淀金属(熔接时);填充金属
added net income 附加纯收益
added protection 附加防护
added purchasing power 追加的购买力
added recording 附录
added resistance 扩量程电阻;附加电阻
added resistance gradient 附加阻力梯度
added return 附加报酬
added turning lane 外加转弯车道
added value 增值;附加价值
added value analysis 附加值分析
added value of railway transportation enterprise 铁路运输业的增加值
added value tax 增值税
added water 附加水
added water mass 附加动水质量
added weight 附加重量
added weight method 重量增加法
addehdum 补遗
addend 加数;被加数
addend address operand 加数地址操作数
addend length operand 加数长度操作数
addendpartial product register 被加数部分积寄存器
addend register 加数寄存器
addend string 加数串
addendum 追加;建筑合同附件;附加物;补遗;补充文件;补充通知;齿顶(高);附约;附录;[复] addenda
addendum angle (伞齿轮的)齿顶角
addendum bearing 齿顶承载区
addendum circle 齿顶圆
addendum coefficient 齿顶高系数

addendum cone 齿顶锥
addendum correction 齿高变位量;齿顶高修正
addendum envelope (指螺旋齿轮的)齿顶包络面
addendum factor 齿顶高系数
addendum flank 上半齿面;齿顶高齿面
addendum line 齿顶线
addendum modification 齿高变位量;齿顶高修正;变位量
addendum modification coefficient 齿顶修正系数
addendum to charter party 租约附录;租约补订书
adder 求和装置;求和器;加法器;加法电路
adder-accumulator 加法累加器
adder amplifier 加法放大器
adder gate 加法器门
adderss out 地址输出
adder stage 相加级
adder-subtractor 加减装置;加减器;加减法器;加法器
adder tree multiplication 加法器树形乘法
adder tube 加算管;加法管
adder unit 加法器单元
add file 加文件
add gate 加法门;相加门
add in 加入;加进
adding 抹灰脱层
adding additional area 增加辅助面积
adding additional space 增加辅助面积
adding alumina 加矾土
adding box 加法器
adding element 加法元件
adding gas mud 充气泥浆
adding limestone 加进石灰石
adding machine 求和器;加数器;加法器;加法机
adding mechanism 加法机构
adding molybdenum compounds 添加钼化合物
adding mortgage 加添抵押
adding operation 加法运算
adding operator 加减运算符;加法运算符;加法算子
adding rate 掺加速度;掺加率;掺加比
adding sand and organic matter 加入砂子和有机物
adding storage register 加法存储寄存器;求和存储计数器
adding tape 添加卷尺;附尺
adding up problem 相加问题
adding wheel 加法轮
Addis count 艾迪斯计数
additament 附加数法
addition 外加;添加;加入量;加建工程;加法;加成(作用);用量;增添;掺加物;附加(部分);补充
addition agent 添加剂;合金元素;触媒剂;外加剂;加合剂;加成剂;掺和剂
additional 附加费;附加的;额外(的)
additional acceleration 附加加速度
additional accelerator 附加促进剂
additional acid/base group 加成酸/碱基
additional adsorbate 加成吸附质
additional advance 追加垫款
additional air 补助空气
additional air governor 补给空气量调节器
additional air regulator 补给空气量调节器
additional allowance 额外津贴
additional amount 附加额
additional appropriation 追加预算拨款
additional article 增订条款;附文;附加条款
additional auditing procedure 附加审计程序
additional auditing standards 附加审计标准
additional audit procedures 补充审计程序
additional background radiation 附加的背景辐射
additional band 附加光带
additional bar 附加钢筋
additional base point 辅助基点
additional bending moment 附加弯矩
additional bending moment on derrick boom 吊杆附加弯矩
additional benefit 附加利益;额外利益
additional budget 追加预算
additional budget allocation 追加预算拨款
additional building 扩建房屋;附加房屋
additional burner 辅助燃烧器
additional call 追加保险费;增收保险费
additional call library 附加调用程序库
additional capital 追加资本
additional cargo 附加载货

additional cargo list 加载货清单
additional channel 辅助通道
additional character 特殊符号;附加字符;附加符号;辅助字符;辅助符号
additional charge 附加电荷;再充电;附加费用;额外费用;补充装料;补充充电
additional circuit 附加电路
additional circulating flow 循环附加流量
additional circulating flow rate 循环附加流率
additional clause 附加条款;补充条例
additional coating 追加涂层;附加涂层
additional code 附加码;辅助码
additional collateral 追加担保品
additional colo(u)r filter 附加滤色镜
additional combining 相加合成
additional component 附加元件
additional condition 附加条件
additional consolidation under traffic 交通运输荷载下的附加固结(沉降)
additional construction 增建工程
additional contract clause 附则;合同补充条款
additional control point 加密控制点;辅助控制点;额外控制点
additional coordinatograph 辅助坐标仪
additional core sampling 补取岩芯样品
additional cost 新增成本;附加费用;额外费用
additional cost assigned to by-product 副产品再加工费用
additional cover 追加保证金;追加保险额
additional coverage 增加承保范围
additional cultivation 补耕
additional damage 附加的破坏
additional damping 附加阻尼;补充阻尼
additional debt restriction 追加借款限制
additional demand 追加需求
additional deposit 附加定金;附加保证金
additional depreciation 追加折旧
additional depth 附加深度
additional details 补充细节
additional device of gravimeter 重力仪辅助设备
additional dialing 附加拨号
additional displacement 附加位移
additional disturbing force 附加扰动力
additional documents 补充文件
additional drag 附加阻力
additional drain 附加沟
additional duty 追加(关)税;附加税项
additional dynamic(al) load 附加动荷载
additional education taxes 教育附加费
additional electromotive force 附加电动势
additional element 添加元素;加入元素;附加要素
additional emergency bilge suction pipe 附加应急舱底水吸入管
additional end moment 附加端部力矩
additional equation 附加方程
additional equipment 附加设备;辅助设备;补充的设备
additional error 附加误差
additional examination 附加检查
additional expenditures 追加支出;追加费用;额外开支
additional expenses 追加费用;附加支出;额外费用
additional expense war risks 战争风险附加费
additional exploration 补充勘探
additional exposure 补充曝光
additional factor for exterior door 外门附加率
additional factor for intermittent heating 间歇采暖附加率
additional factor for wind-force 风力附加率
additional felting 附加油毡做法
additional fertilization 追肥作业;额外施肥
additional fertilizer 追肥
additional finance 额外资金
additional fire hazard 外加失火危险
additional first year depreciation 首年追加折旧费
additional flexural force 挠曲附加力
additional flexural load 挠曲附加力
additional force 附加力
additional for optional port of discharge 选择卸货港费
additional freight 附加运费;附加货运
additional fuel 附加燃料
additional function 附加功能;辅助功能
additional fund 增加资金;附加基金

additional gain 附加增益
additional girder 加强桁材;附加纵桁;附加桁材
additional halfspan loading 半跨径附加荷载
additional hanging(pipe)line 附加吊索管线;附加吊索管道
additional header information 附加报头信息
additional head loss 附加水头损失
additional heating 附加供热;补充加热;补偿加热（焊接中）
additional heat loss 附加耗热量
additional height 附加高度
additional income 附加收益;额外收入
additional income tax 追加所得税;附加所得税
additional income tax return 附加所得税申报表
additional indicator system 附加指标法
additional information 可加信息;附加信息;附加情报;补充信息
additional inner bow door 附加内门首
additional input source 附加输入源
additional instrument 辅助仪器;辅助仪表
additional insulation 附加绝缘;辅助隔声;辅助隔热
additional insurance 加保;追加保险;附加保险
additional insurance of works 对工程进行附加保险
additional investment 附加投资;额外投资;补充投资
additional investment in the means of production 追加生产资料投资
additional item 增列项目;附加项;补充项;补偿项（目）
additional labo(u)r cost 附加人工成本
additional lateral force 附加侧向力
additional lateral pressure 附加侧向压力
additional layer 附加层
additional layer of felt 附加油毡层;辅助油毡层
additional lending 额外放款
additional living expenses 追加生活费;附加生活费
additional load(ing) 附加负载;附加负荷;附加荷载;增加荷载
additional local stress 附加局部应力
additional local tax 地方附加税
additional loss 附加损失;附加损耗
additional lubrication of axle-boxes 轴箱补充润滑
additional main storage 附加的主存储器
additional margin 追加押金;追加保证金
additional mark 附加标志
additional mark on 额外加成
additional mark-up 再加价
additional mass 补充质量
additional matter 添加物
additional means of isolation 附加隔离设施
additional mirror 附加镜(头)
additional monitoring 附加监控
additional motion 附加运动
additional necessary condition 附加的必要条件
additional network 附加网络
additional nitration 补充硝化
additional noise 附加噪声
additional on heavy lifts 超重附加费
additional on long lengths 超长货物附加费
additional order 加定货单;追加定货;追加订单
additional outlay 额外开支
additional paid-in capital 额外实收资本;超面值缴入资本;附加资本;附加实缴股本
additional parameter 附加参数
additional parking space 附加停车车位
additional part of a bill 期票附件
additional passenger train 临时旅客列车
additional payment 附加费用;追加支付;追加付款;额外付款
additional peak 附加峰
additional percentage tax 附加比例税
additional perils 附加危险
additional perils clause 附加险条款;标准班轮附加险条款
additional period 延长期限;附加时间
additional perspective condition 附加透视条件
additional perturbation 附加摄动
additional phase 附加相
additional pipe 接长管;支管
additional plan of wage fund 工资基金追加计划
additional point 补充点
additional pollution 增添污染
additional pore water pressure 附加孔隙水压力
additional port of call 加靠港

additional potential 附加位势
additional premise 附加前提
additional premium 追加保险;附加额外保费;附加保险费;额外保险费
additional premium-reinsurance 再保险附加保费
additional pressure 附加压力
additional process 附加过程
additional producer 补充生产井
additional product 附加产品
additional property 附加性能
additional protection 附加防护
additional protocol 附加议定书
additional provision 附加条款
additional rate 附加比率;附加费率
additional rate for room height 高度附加率
additional reaction 附加反应
additional record 补充记录;附加记录
additional redundancy 附加冗余度
additional regulation 补充规定
additional reinforcement 附加钢筋
additional remark 补充说明
additional remittance 追加汇款;追加备用金
additional requirement 附加要求;补充技术条件
additional reserve 追加准备金
additional reservoir 附加传染源
additional resistance 附加阻力
additional resistance due to tunnel 隧道附加阻力
additional resistance for curve 曲线附加阻力
additional resistance for gradient 坡道附加阻力
additional resistance for tunnel 隧道附加阻力
additional return 追加报酬
additional revenue 附加收益;附加收入
additional risks 附加险;附加危险
additional safety factor 额外安全因素;附加安全系数
additional safety valve 附加安全阀
additional sampling 增加采样;补充采样
additional secondary phase factor 附加的二次相位因子
additional security 追加保证金
additional sediment 附加的泥沙;附加沉积物
additional service 附加责任事项;附加服务(项目);附加业务;附加职责
additional services of the architecture or engineer 建筑师或工程师的附加服务项目
additional shipment 加载【船】;中途加载
additional side girder 附加旁龙骨;附加边桁材
additional soil data 补充土壤资料
additional sounding 水深加密测量
additional source 补充水源;补充来源
additional special contribution 额外特别捐款
additional stake 加桩【测】
additional state 附加状态
additional steel 辅助钢筋
additional stop 附加光栏
additional storage 辅助存储器
additional storage capacity 附加库容
additional stream 附加水流
additional strength 补充强度
additional strengthening 加强结构
additional stress 附加应力
additional stress factor method 附加应力系数法
additional stress of foundation base 基底附加压力
additional subscription 再行增订
additional substantive test 附加的实质性测试
additional sum 附加金额
additional sun's correction 太阳总改正增订量
additional survey 附加检验;附加测量;补充勘测;补充测量
additional symbol 附加符号
additional tax 追加税;追加关税;附加税
additional tax for education 教育附加费
additional tax rate 附加税率
additional temperature force 伸缩附加力
additional temperature load 伸缩附加力
additional tensioning 附加张拉
additional test 附加检验
additional thermal resistance 附加热阻
additional thrust 附加推(进)力
additional ticket 附加票
additional time 附加时间
additional track 补充轨线
additional train 储备列车
additional unit 附加装置;新增机组

additional value of wall thickness 壁厚附加量
additional vapo(u)rization 附加蒸发
additional variable 附加变数;附加变量;补充变量
additional vehicle 储备车辆
additional voltage 补充电压
additional vulcanization 后硫化
additional wage 附加工资
additional water sources 补充水资源
additional water supply 增加供水;补充供水
additional water supply source 增加供水源;补充供水水源
additional wave 附加波
additional wave data 补充波浪资料
additional weight 配重;附加重量
additional well 补充井
additional winding 附加绕组;辅助绕组
additional work 新添工程;追加工作;附加工作;附加工程;追加工程
additional yard 辅助编组场
addition and betterment reserve 扩建及改造准备
addition and subtraction 加减法
addition by mixing 掺和料
addition by subtraction 采用减法运算的加法
addition command 加法命令
addition compound 加成(化合)物
addition constant 加常数
addition copolymerization 加成共聚合
addition dimerization 加成二聚合作用
addition displacement 附加位移
addition entry point 附加入口点
addition extension 增加
addition file 附加文件
addition formula 加法公式
addition instruction 加法指令
addition item 附加项;补充项
addition method 加入法
addition of assets 资产扩充
addition of a wagon 货车附加物
addition of cement 水泥外加剂;水泥混合材(料)
addition of clay 掺黏土
addition of complex quantities 复数的加法
addition of constraints 附加约束
addition of diffraction patterns 衍射图形叠加
addition of fixtures 添加固定设备
addition of matrices 矩阵加法
addition of modes 模叠加
addition of optic(al) fields 光学场叠加
addition of polynomial 多项式加法
addition of sand 掺砂
addition of series 级数的加法;序列的相加
addition of variables 加变量;变量相加
addition of vectors 矢量加法;矢量合成;向量加法;向量相加
addition of waveforms 波性叠加;波形叠加
addition of wavefronts 波前叠加
addition percentage 附加百分比
addition polymer 加聚物;加成聚合物
addition polymerization 加聚作用;加聚反应;加成聚合
addition polymerization reaction 加成聚合反应
addition principle 加法原理
addition procedure 掺加方法
addition process 加法工序;加成法
addition product 加成物
addition property 加法性质
addition rate 掺和率
addition reaction 加成反应
addition resin 加成聚合物
addition rule 相加规划
additions and alterations 扩建和改建
additions and betterments reserve 扩充及改良准备
addition solid solution 加成固溶体
addition speed 加法速度
addition-subtraction bookkeeping method 增减记账法
addition table 加法表
addition theorem 加法定理
addition to a building 房屋增建
addition to existing assets 现有资产增值;现有资产扩建
addition to property through surplus 由盈余增加得资产
addition to retained earnings 留存资产增加额

addition to standard 标准补充件
addition to structure 附属建筑物
addition to the contract sum 合同金额的追加部分
addition without carry 按位加(无进位加法)
additive 外加剂;添加料;添加剂;可加的;加性;加法;相加;助剂;掺加剂;掺和物;掺加料;附加剂;附加的
additive action 相加作用;叠加作用
additive agent 附加剂;掺加剂;配合剂;助剂
additive alternate 变更标价;投标商附加标价;附加标价;补充投标;补充比较方案;变更方案;备用方案
additive attribute 加属性;附加属性
additive cement 混合水泥;有掺和料水泥
additive channel 可加信道
additive colo(u)r 加色
additive colo(u)ration method 附加着色法
additive colo(u)r blending 加色(法)混合
additive colo(u)r compositing 加色合成
additive colo(u)r filter 附加滤色镜
additive colo(u)r matching 加色法配色
additive colo(u)r method 加色法
additive colo(u)r mixing 加色(法)混合
additive colo(u)r mixture 附加混色
additive colo(u)r photography 加色彩色摄影
additive colo(u)r process 加色法
additive colo(u)r system 混色系统;加色剂
additive colo(u)r viewer 加色法合成
additive complementary colo(u)r 加色法补色;相加合成补色
additive compound 加成化合物
additive concentrate 浓缩添加剂
additive concentration 添加剂浓度
additive congruential method 加同余法
additive constant 加(法)常数;相加性常数;附加常数;测绘工作修正系数
additive correction 附加校正
additive depletion 添加剂耗损;添加剂耗减
additive dimer 加成二聚物
additive disc 加色法混色盘
additive effect 累加效应;加性效应;加和效应;相加作用;相加效应;叠加作用;叠加效应;附加效应
additive effect of pesticide 农药的附加效应
additive engine oil 含添加剂的机油
additive error 相加误差;附加误差
additive evaluation model 辅助评价模型
additive expenses 额外开支
additive factor 累加因子;加性因素;附加因素
additive feeder 添加剂进料装置
additive file attribute 附加文件属性;补充的文件属性
additive filter 附加滤光片
additive for gasoline 汽油添加剂
additive for grinding 研磨添加剂
additive for improving the property 改性剂
additive for setting time control 控制凝结时间的外加剂
additive function 可加性函数;加性函数
additive group 加法群
additive halo 累加晕
additive identity 加法恒等元;加法单位元
additive independence 可加独立性
additive independence assumption 可加独立性假设
additive index 加和指数
additive in soill-cement and soill-lime 水泥土和石灰土中的掺和
additive interaction 累加相互作用
additive inverse 加性逆元
additive joint action 相加性的联合作用
additive law of irregularity 不匀率相加定律
additive layer 附加层
additive loss 添加剂损失
additive lubricating oil 含添加剂的润滑油
additively separable 加性可分的
additively separable function 加法可分函数;辅助可分函数
additive migration 添加剂迁移
additive mixing 相加混频
additive mixture 加色混合
additive mixture of colo(u)rs 加色混合色
additive mixture of colo(u)r stimuli 加色法刺激
additive motor oil 含添加剂的机油
additive nature of the laws of Lambert-Bouguer and Beer 朗伯·波格和比尔定律的加和性
additive noise 相加噪声;叠加噪声
additive of effect 效应可加性;效力可加性
additive offset plate 加色法胶印平版
additive operation 加法运算
additive operator 加法算符
additive order 加法阶
additive plasticizer 外加增塑剂;添加增塑剂
additive polarity 加极性
additive pollution 添加剂污染
additive preference 可加偏好
additive pressure 附加压力
additive primary 加色法基色;加色法的原色
additive primary colo(u)rs 加(色)法三原色
additive printer 加色光印片机
additive process 添加法;可加过程;加法过程
additive property 加和性;相加性
additive random walk process 可加随机游走过程
additive ratio 累加比值
additive reaction 加成反应
additive resin 填充树脂
additives and spacer fluid 添加剂和隔离液
additive seasonal effect 附加季节性影响
additive set function 加性集函数
additive soil stabilization 加有稳定剂的地基土;掺和物稳定土(法);水泥灌浆
additive structure 可加结构
additive term 补充项
additive theory 堆垒论
additive time 加法时间
additive to extraction 浸出添加剂
additive toxicity index 加和毒性指数
additive trichromy by approaching 渐近加色法三原色
additive trichromy by superposition 叠加加色法三基色
additive-type oil 含添加剂(类型)的油
additive utility function 可加效用函数;相加的效用函数
additive wire 附加导线
additivity 可加性;加性;加成性;相加性;叠加性
additivity assumption 可加性假设
additivity for a production set 生产集的可加性
additivity of mean 平均值的可加性
additivity test 可加性检验
additonal area 追加区
additon of forces 分力相加
additus maximus (古罗马圆形剧场的)大门;古罗马圆形剧场的主入口
addling 鼓胀;起壳;抹面剥落;抹灰剥落
add list 附加表
add mode 加方式
add network 加法网络
addometer 加算器
add-on 扩充;比例附加
add-on board 添加插件板
add-on hardware 添加硬件
add-on interest 利上加利;(分期付款的)附加利息
add-on memory 添加存储器;附加存储器
add-on method 加象法
add-on process 加象法
add ons 附加装置
add-on storage 添加存储器
add-on system 增加系统
add-on technology 增添技术
add operation 加法运算;加法操作;加操作
addorsal line 侧背线
addorsed (柱头上的)背对背人兽图案雕饰
add-or-subtract 加或减
add parity 加奇偶;加法奇偶
add pulse 加法脉冲;相加脉冲
address 通信录;收信地址
add rod ahead 孔口接钻杆
addrout file 地址输出文件
adds 扣除前总量;毛总量
add-subtract control unit 加减控制器
add-subtracter 加减装置;加减器
add-substract key 加减键
add the weight 增加重量
add time 加法运算时间
add to list 加到表中
add to set 锯齿顶部微折
add to the prices 提价

adduce evidence 引证证据
adduct 加合物
adduct curing agent 加合固化剂
adduct forming agent 加合物形成剂
adduction 加合(作用)
adductive crystallization 加合结晶;引晶作用
add up 累计
add up error 累加误差
add up to 合计(达);共计
add with carry 进位加
a decline in price 价格下降
adeem 废止
Adelaide brown colo(u)r 阿德莱德棕色料
Adelaide geosyncline 阿德莱德地槽【地】
Adelaide wool 阿德莱德羊毛
Adeline steel making process 阿德莱熔模铝热离心浇注法
adelite 砷钙镁石
adelomorphous 不显形的
Adenos cotton 阿登诺棉
adequacy 不充分
adequacy of fit 拟合适当
adequacy of insurance 保险的完备性
adequate 适当的;合适的
adequate and systematic service 配套服务
adequate cause 充分理由
adequate definition 精确定义
adequate design 稳妥可靠的设计
adequate disclosure 充分反映
adequate distribution 均匀分布
adequate habitat 适宜生境
adequate irrigation 水利化
adequate lubrication 适当的润滑作用
adequate management 适当管理
adequate measures 适当措施
adequate nutrition 合理营养
adequate preparation 充分准备
Adequate Public Facility Ordinance 充足公共设施法
adequate quality 优等品
adequate rent 适当的租金
adequate sample 足够样本;充足样本
adequate shielding 适当的屏蔽
adequate supply 充足的供应
adequate to the demand 适应于要求
adequate variability 适应变异性
adequate ventilation 充分通风
adequate verification 充分的核实
adequation 足够;调整
ader perthite 脉状条纹长石
A-derrick 动臂起重机;A形转臂起重机
ader wax 生地蜡;含硫地蜡;粗地蜡
a design of flowers 花形图案
adfluxion 汇集;汇流
adfreezing 冻硬;冻附;冰附过程
adfreezing force 冻结力;冻附力
adfreezing strength 冻附强度
adfrontal 旁侧片
adglutinate 烧结;凝聚;胶结
adhension 附着性
adhere 黏固;黏附
adhered glass 黏附尘埃的玻璃
adhered pollutant 黏着污染物
adhered residue 落渣
adherence 黏附;依附力;黏着;附着(力)
adherence of nappe 外层黏附;贴附水舌
adherence promoter 密着剂
adherence ratio of insulating gel 绝缘胶黏附率
adherence test(ing) 密着(强度)试验(橡胶);黏结强度试验;黏着试验
adherence to design 符合设计要求
adherence to specification 遵守技术规范;符合规范规定
adherence to specifications 符合规范要求
adherence to the agenda 遵照议程
adherency 黏附
adherend 黏附件;黏附物;黏附体;黏合体;被黏物;被黏体
adherend failure 黏附体破坏;被黏物破坏
adherent 胶黏体;黏着的;黏附的;附着的;防黏着材料
adherent nappe 黏附流态;附壁溢流;黏附溢流水舌
adherent point 依附点
adherent property 黏着性质

adherent surface 黏着面
adherent water 死水;静止水;吸附水
adhere to principle 坚持原则
adhering mo(u)lding material 涂膜材料;涂模材料;铸模涂料;铸型涂料
adhering nape 附着水舌
adhering nappe 贴附水舌
adhering root 附着根
adhering sand 黏砂
adhering to blade 黏附在叶片上
adhering zone 黏合物层;附着区域
adherography 胶印法
adherometer 漆膜附着力测定仪;黏附计
adhesion 贴生;胶黏;黏着作用;黏着物;黏着力;黏结作用;黏结力;黏合(力);黏附性;黏附(力);黏着物;黏连;船底寄生物;附着(作用);附着物
adhesion additive 黏结外加剂
adhesion agent 黏结剂;胶黏剂;黏着剂;增黏剂
adhesional energy 黏附能;黏合能
adhesional strength 附着力
adhesional wetting 黏合湿润;黏湿作用;黏湿;附着润湿
adhesional work 附着功
adhesion attraction 附着力
adhesion capacity 黏着性;黏着能力;黏着力;胶黏度
adhesion characteristic 黏附特性
adhesion coefficient 黏附系数;黏着系数
adhesion connection 胶连接
adhesion constant 黏结常数
adhesion contract 附意合同;附从合同;服从契约
adhesion due to electrostatic charge 静电黏结
adhesion energy 附着能
adhesion factor 黏着系数;黏附系数
adhesion failure 黏胶(力)破坏;黏附破坏
adhesion heat 黏结热;附着热
adhesion in peel 剥离黏结性
adhesion layer 黏附层
adhesionless soil 非黏性土(壤)
adhesion limit 黏着力强度极限;附着极限
adhesion limit between wheel and rail 轮轨间的黏着极限
adhesion locomotive 机车黏着(状态)
adhesion loss 黏着力消失;脱胶;附着力损失
adhesion meter 黏附力计;附着力测定仪;黏附强度表
adhesion of bacteria 细菌黏合
adhesion of clay 黏土附着力
adhesion of film 漆膜附着力;薄膜黏附
adhesion of granules 颗粒间黏着力
adhesion of particles 附着微粒
adhesion of slag 炉渣黏附;渣瘤
adhesion of thin-film 薄膜的附着力
adhesion phenomenon 粘附现象;黏连现象
adhesion power 胶黏度
adhesion-preventing film 防黏薄膜;黏附保护膜
adhesion preventive 防黏剂
adhesion promoter 黏结增强剂;增粘剂;底胶;促黏剂;附着力促进剂
adhesion promoting agent 黏附剂;防剥落剂;抗剥落添加剂;黏附强度增强剂
adhesion property 胶黏性(质);黏附性能;黏附特性
adhesion quality 黏附力
adhesion railway 黏着铁路
adhesion ratio 黏附率;黏着率
adhesion separation or spasm-relaxation 分筋手法
adhesion settling 吸持沉降
adhesion spot 黏合斑
adhesion strength 黏着力;黏附力;黏结强度;黏合强度;黏附强度;黏着强度;附着强度
adhesion stress 黏着力;黏附力;黏结应力
adhesion substance 黏合物质
adhesion system 黏着方式
adhesion tape 黏合小带
adhesion tension 黏合张力;黏附张力;附着张力
adhesion tension analyzer 附着张力分析仪
adhesion test 黏结试验;附着(力)试验
adhesion tester 附着力试验仪;附着力测定仪
adhesion testing 密着(强度)试验(橡胶);密着检验(橡胶)
adhesion traction 黏附牵引力;黏着牵引力
adhesion tractive effort 黏着牵引力
adhesion-type ceramic veneer 黏贴型陶瓷面板;黏接型陶瓷板;黏附型陶瓷饰面板;黏附型陶瓷面砖
adhesion value 黏结值
adhesion water 黏附水
adhesion weight 黏着重量;附着重量
adhesion wetting 黏润
adhesive 胶黏剂;黏着剂;黏结材料;黏合剂;黏附的;黏连性;黏合的;隔黏材料;附着剂;氟橡胶黏剂
adhesive action 胶黏作用;附着作用
adhesive agent 黏合剂
adhesive aggregation 黏聚
adhesive anchor 黏合锚固件;胶黏锚接;胶黏锚固
adhesive application 胶黏剂的涂敷
adhesive-assembly 黏结装配
adhesive attraction 黏力;黏吸力;黏吸作用;附着引力;附着力
adhesive backed 涂满黏合剂的
adhesive backed tape 胶带
adhesive backing 裱版
adhesive band 涂胶封条
adhesive bandage 胶布绷带
adhesive based on coal tar 煤焦油沥青黏着剂
adhesive binding 胶黏装订
adhesive bitumen primer 沥青黏着剂;沥青冷底子油;冷底子油;胶黏沥青打底料
adhesive bond 黏着键;黏接;黏合界面
adhesive-bonded fabric 无纺布
adhesive-bonded joint 黏结接头;胶结接头;黏结接合面;黏接接头;黏接接合
adhesive bonding 胶接;黏结结合;附着黏合
adhesive bond strength tester 黏合强度试验机
adhesive capacity 胶黏度;胶黏力;黏着能力;黏结能力;黏附能力
adhesive cement 黏胶剂;胶结剂;黏结(质)水泥
adhesive coat 黏合层
adhesive coating 涂胶;黏合层;黏附层
adhesive coefficient 黏着系数;黏附系数
adhesive colo(u)r 黏着色
adhesive composition 黏胶(合)剂
adhesive compound 黏胶(合)剂
adhesive dispersion 胶黏剂分散体
adhesive effect 黏着作用;黏结作用
adhesive emulsion 胶黏乳油液;黏着乳剂
adhesive factor 黏附系数;附着系数
adhesive failure 胶黏破坏;胶黏剂失效;黏接失效;黏合疵点;内聚破坏
adhesive film 胶膜;黏着膜;黏合膜;黏附膜;黏性膜
adhesive foil 烫金箔
adhesive for blocks 砌块黏结剂
adhesive force 黏着力;黏附力;附着力
adhesive for cold pressing 冷压胶合剂
adhesive for concrete 混凝土黏合剂
adhesive for laminated film 复合薄膜胶黏剂
adhesive for laying 敷设黏结剂
adhesive for metals 金属胶合剂
adhesive for plywood 胶合板用黏结剂;胶合板用胶黏剂
adhesive for wall tile 壁砖黏合剂
adhesive for wood grain film 木纹膜胶黏剂
adhesive-glue 黏合胶
adhesive industry 胶黏剂工业
adhesive in film form 黏膜;黏性薄膜
adhesive insulation 胶合绝缘材料
adhesive interface 胶黏界面
adhesive interlayer 胶浆夹层
adhesive joint 黏胶接合;胶接;黏结接缝;胶接接头;黏结结合;黏接结合面;黏接接头;黏接接合;黏接合缝
adhesive joint failure 胶黏接头破坏
adhesive jointing 黏接法
adhesive-laminated wood 胶合层积材;黏合叠层木;胶合叠层木料
adhesive latex 包封用胶乳
adhesive layer 黏胶层;胶层;黏性土层
adhesive line 胶黏接头
adhesive lubricant 胶黏润滑剂
adhesive masking tape 遮黏带;(装修技术中的)饰面胶带布
adhesive material 胶黏材料;黏结材料;黏合剂
adhesive measuring tape 胶黏卷尺;胶黏测尺
adhesive meter 黏着力计;黏度计
adhesive mixture 胶黏混合料
adhesive moisture 附着水分
adhesive mortar 胶泥;黏结砂浆;黏结灰泥;黏结灰浆
adhesive nail-on method 钉压黏固法
adhesiveness 胶黏性;胶黏度;黏着性;黏性;黏合性;黏合度;黏附性;黏附度;附着性
adhesiveness test 黏着性试验
adhesive paper 胶水纸
adhesive paper tape 胶纸带
adhesive paste 浆糊;糊状黏结剂
adhesive peel joint 剥离节
adhesive phase 黏合剂层
adhesive plaster 胶布(橡皮膏);橡皮膏
adhesive plastic strip 塑料胶带
adhesive powder 粉末胶
adhesive power 黏着(能)力;黏结力;黏合力;黏附能力;附着(能)力
adhesive property 胶黏性(能);胶黏剂性能;黏附性
adhesive property of beads 玻质球的黏合性
adhesive putty 胶结(质)水泥;黏性水泥;人工橡胶填充剂;胶接水泥
adhesive remover 黏胶清除剂
adhesive retention 黏滞性
adhesive rod 胶结型锚杆
adhesive rubber 黏合橡胶
adhesive seal 胶黏密封;胶泥密封;密封油膏
adhesive shale 黏页岩
adhesive slate 黏板岩;黏性板岩
adhesive solution 胶黏溶液
adhesive solvent 胶黏剂溶剂
adhesive sprayer 喷胶机
adhesive spreader 摊胶机
adhesive spreading 上胶机
adhesive spreading machine 涂胶机
adhesive starved area 贫胶区
adhesive strength 黏着力;黏附力;胶黏强度;胶接强度;黏接强度;黏合强度;黏附强度;附着强度
adhesive strength tester 附着强度试验仪
adhesive strip 黏合带
adhesive surface 黏结面
adhesive suspension 黏结悬胶液;黏性悬浮物
adhesive system 胶黏体系;黏合结构
adhesive tape 绝缘带;胶黏带;胶条;胶皮带;胶(布)带;胶黏带;黏黏布;黏合带
adhesive tape test 胶带试验
adhesive tar composition 焦油沥清黏性成分
adhesive tar compound 焦油黏结剂
adhesive tension 黏着强度;黏附张力;黏合张力
adhesive test 黏性试验;黏接试验;黏合试验
adhesive thickness 胶层厚度
adhesive tie 黏合锚固件
adhesive tire 胶口轮胎
adhesive treatment 黏结处理
adhesive type vibrator 附着式振动器
adhesive tyre 胶口轮胎
adhesive varnish 胶黏漆;胶性(清)漆;黏合清漆
adhesive water 吸附水;胶黏水;黏附水;附着水;薄膜水
adhesive wax 黏着蜡;胶黏蜡;黏蜡;封蜡
adhesive weight 黏着重量;附着重量
adhesive weihgt controller 附着重量控制器
adhesive zone 黏合带
adhesivity 黏合性;黏附能力;黏着性
adhint 黏连接头;胶接
Ad Hoc Advisory Committee of Experts 特设专家咨询委员会
ad hoc analysis 专门分析
ad hoc approach 特定设计法;特定方法
ad hoc arbitration 特设仲裁;特别仲裁;临时仲裁
ad hoc committee 特设委员会;专门委员会;特别委员会
ad hoc decision rule 特定决策规则
ad hoc expert group 特设专家小组;特别专家小组
ad hoc fashion 特定方式
ad hoc group 特别小组
ad hoc judge 专案法官
ad hoc logic 专门逻辑
ad hoc meeting 特别会议
ad hoc observer 特别观察员
ad hoc panel 特别专家小组
ad hoc program(me) 专门计划
ad hoc request 特别要求
ad hoc resources 特定资源;特别资源
ad hoc rules 特定规则;特别规则

ad hoc task 特别任务
ad hoc test 特设试验
ad-holism 特定主义
adiabat 绝热线
adiabatic 绝热的
adiabatic apparatus 绝热装置
adiabatic approximation 绝热近似法
adiabatic atmosphere 绝热大气
adiabatic calorimeter 绝热(式)量热器;绝热热量计;绝热卡计;绝热测热器;断热热量表
adiabatic change 绝热变化
adiabatic change of air 空气绝热变化
adiabatic chart 绝热变化图
adiabatic coefficient of compression 绝热压缩系数
adiabatic column 绝热式精馏柱
adiabatic combustion 绝热燃烧
adiabatic combustion temperature 绝对火焰温度
adiabatic compaction 绝热压密
adiabatic compressibility 绝热压缩性
adiabatic compression 绝热压缩
adiabatic condensation 绝热凝结;绝热冷凝
adiabatic condensation pressure 凝结气压
adiabatic condensation temperature 绝热凝结温度;绝热冷凝温度
adiabatic condition 绝热状态;绝热条件
adiabatic constant 绝热常数
adiabatic contraction 绝热收缩
adiabatic cooling 绝热冷却
adiabatic cooling curve 绝热冷却曲线
adiabatic cooling line 绝热冷却线
adiabatic cooling temperature of air 空气绝热冷却温度
adiabatic curing 绝热湿治(法);绝热养护;隔热养护;混凝土绝热养护(工艺)
adiabatic curve 绝热(过程)曲线
adiabatic decomposition temperature 绝热分解温度
adiabatic degasing 绝热放气
adiabatic demagnetization 绝热去磁
adiabatic diagram 绝热图(解)
adiabatic diaphragm 绝热隔膜
adiabatic drift wave 绝热漂移波
adiabatic dryer 绝热干燥器
adiabatic drying 绝热干燥
adiabatic efficiency 绝热效率
adiabatic elasticity 绝热弹性
adiabatic ellipse 速度椭圆
adiabatic engine 绝热发动机
adiabatic envelope 绝热包壳
adiabatic equation 绝热方程(式)
adiabatic equilibrium 绝热平衡
adiabatic evapo(u)ration 绝热蒸发
adiabatic expansion 绝热膨胀;绝热扩张
adiabatic exponent 绝热指数
adiabatic extrusion 绝热挤压;绝热挤出
adiabatic flame temperature 绝热火焰温度
adiabatic flow 绝热流(动)
adiabatic function 绝热函数
adiabatic gradient 绝热梯度;绝热递减率
adiabatic gradient plate 绝热梯度板
adiabatic head 绝热压头
adiabatic heat drop 绝热热降
adiabatic heating 绝热增温;绝热升温
adiabatic humidification 绝热加湿
adiabatic humidifying 绝热加湿;等焓加湿
adiabatic incompressibility 绝热不可压缩性系数
adiabatic index 绝热指数
adiabatic insulation 绝热隔层
adiabatic invariant 绝热(式)不变量
adiabatic inversion 绝热反转
adiabatic ionization energy 绝热电离能量
adiabaticity 绝热性
adiabatic lapse rate 绝热直减率;绝热衰减率;绝热递减率
adiabatic lapse rate of air temperature 气温垂直递减率
adiabatic law 绝热定律
adiabatic layer 绝热板
adiabatic line 绝热(曲)线;绝热过程曲线
adiabatic liquid water content 绝热液态含水量
adiabatic magnetization 绝热磁化
adiabatic method 绝热法
adiabatic mixing 绝热混合
adiabatic modulus of elasticity 绝热弹性模量

adiabatic optics 绝热电子光学系统
adiabatic path 绝热线路;绝热曲线
adiabatic phenomenon 绝热现象
adiabatic principle 绝热原理
adiabatic process 绝热过程
adiabatic property 绝热性能
adiabatic psychrometer 绝热干湿表
adiabatic pulsation 绝热脉动
adiabatic rate 绝热率
adiabatic recovery temperature 绝热恢复温度
adiabatic rectification 绝热精馏
adiabatic rectification column 绝热精馏柱
adiabatic relation 绝热比
adiabatic response 绝热响应
adiabatic reversible process 绝热可逆过程
adiabatic rigidity modulus 绝热刚性模量
adiabatic rise of temperature 绝热升温
adiabatics 绝热曲线
adiabatic saturated change 绝热饱和变化
adiabatic saturated temperature 绝热饱和温度
adiabatic saturation 绝热饱和
adiabatic saturation pressure 凝结气压
adiabatic saturation temperature 绝热饱和温度;凝结温度
adiabatic shell calorimeter 绝热壳量热器
adiabatic shift 绝热移位
adiabatic shrinkage 绝热收缩
adiabatic state 绝热状态
adiabatic storage condition 绝热温升状态
adiabatic strain 绝热应变
adiabatic switch-on 绝热收缩
adiabatic system 绝热系统
adiabatic temperature 绝热温度
adiabatic temperature change 绝热温度变化
adiabatic temperature gradient 绝热温度梯度
adiabatic temperature rise 绝热温度上升;绝热升温
adiabatic test 绝热状态的试验;绝热试验
adiabatic titration calorimeter 绝热滴定量热计
adiabatic turnon 绝热起动
adiabatic variation 绝热变化
adiabatic wall 绝热墙
adiabatic warning 绝热增温
adiabator 保温材料;绝热材料
adiactinic 绝光的;绝光化辐射的;不透射线的;绝光化;不透光化线的;不透光的
adiactinic glass 防放射玻璃
adiadochokinesis 更替运动不能
adiagnostic 非特征性的;不可判明的
adiathermal 绝热的
adiathermal body 绝热体
adiathermance[adiathermancy] 不透热(性);绝热性;不透红外线性
adiathermanous 不透红外线的
adiathermanous body 不透热体
adiathermic 不透热的
adiathermic body 不透热体
adiathermic membrane 绝热膜
adiathetic 非素质性的
adicillin 阿地西林
adicity 化合价
adijunct 附加的
ad infinitum 无限地
adinole 钠长英板岩
adinolite slate 钠长英板岩
ad interim 临时
adion 吸附离子;被吸附离子
adipamide 己二酰二胺
adipate 己二酸盐
adipic acid 己二酸
adipinketone 环戊酮
adiponitrile 己二腈
adipopexia 积脂
a-dipping 沿倾斜方向
Adiriatic plate 亚得里亚板块
Adiriatic Sea 亚得里亚海
A display 距离显示
adit 入口;平硐;平硐水平巷道;水平坑道;坑道;勘察坑道;交通坑道;横坑道;导坑开挖
a ditch to carry extra irrigation water 排出剩余灌溉水的水沟
adit collar 平硐口;支洞口
adit development 平硐开拓
adit digging survey 巷道掘进测量
adit end 平硐工作面

adit entrance 坑道(入)口
adit for draining 排水坑道
adit level 平硐水平
adit opening 排水坑道;坑道(出入)口;平硐口;水平坑道口
adit planimetric map 坑道平面图
adit portal 平硐口;支洞口
adit prospecting engineering survey 坑探工程测量
adit-shaft development 平硐-井筒联合开拓
adit test 试坑调查
adit to well (矿井的)井口
adjacency 相关项;邻近间距;邻近;邻接(性)
adjacency analysis 近邻性分析
adjacency effect 邻接效应;邻基效应
adjacency list 邻接表
adjacency matrix 连接矩阵;相邻矩阵
adjacent 毗连的;邻接的;邻;交界的;相邻的
adjacent accommodation 邻近建筑物;厢房;相邻设备;附属建筑物;附近可利用物;附近建筑物
adjacent angle 邻角
adjacent anomaly 毗邻异常
adjacent area 毗邻区
adjacent area inset 位置示意图
adjacent blocks 相邻街坊;相邻房屋;相邻浇筑块
adjacent building 相邻房屋;邻接建筑物;邻近建筑物;相邻建筑物
adjacent carbon couple 邻碳偶合
adjacent catchment 毗邻流域
adjacent channel 邻道;相邻信道;相邻通道
adjacent channel attenuation 邻信道衰减;相邻信道衰减
adjacent channel frequency 相邻信道频率
adjacent channel interference 邻近波道干扰;相邻信道干扰;相邻波道干扰
adjacent channel noise 邻近信道噪声
adjacent channel rejection 邻道抑制
adjacent channel selectivity 相邻信道选择性
adjacent character 相接字符;相邻字符
adjacent coil 毗连线圈;连接线圈
adjacent colo(u)r 邻近色
adjacent contract section 相旁地(段);交界收缩断面;邻接收缩断面;交接收缩断面
adjacent control region 邻接控制区
adjacent country 邻接区;相邻地区
adjacent curve in one direction 同向曲线
adjacent curves 相邻曲线
adjacent domain 邻近区域;相邻域
adjacent edge 相邻边
adjacent effect 相邻影响
adjacent effects chart 相邻影响曲线图【岩】
adjacent enhancement 邻区法增强
adjacent extreme point 相邻极值点
adjacent flight lines 邻航线;相邻航(空)线
adjacent flight line viewfinder 侧方瞄准器
adjacent flight strips 相邻航(空)线
adjacent habitat 毗邻生境
adjacent halo 毗邻晕
adjacent intersection 相邻交叉口
adjacent junction 相邻交叉口
adjacent land 毗邻地块;邻近地段;沿路用地
adjacent landowner 邻近土地所有者
adjacent layers 毗连层次
adjacent lift 相邻混凝土浇筑层
adjacent line 邻线;邻接线
adjacent link station 邻路路站;邻近链站
adjacent link station image 邻近链站映像
adjacent load 邻近荷载
adjacent location 邻接位置
adjacent map 邻接图
adjacent model 相邻模型
adjacent mullion 附加木杆
adjacent navigation channel 引航道
adjacent network control program (me) 邻近网络控制程序
adjacent node 邻近节点;邻接的节点;相邻节点
adjacent opening 相邻的桥跨;邻孔;邻跨(孔)
adjacent orthogonal 相邻(的)正交线
adjacent owner 相邻(房地产)业主
adjacent pair 相邻像对
adjacent peak 邻峰
adjacent periods 相邻期
adjacent photograph 相邻像片
adjacent picture carrier 邻道图像载波
adjacent picture carrier trap 邻道图像载波陷波器

adjacent pile 邻桩
adjacent pitch error 相邻周节误差
adjacent plank 门头线；踢脚板；门框梃；门框边挺；相邻板
adjacent position 相邻位置
adjacent premises 相邻地产
adjacent property 邻近地权；邻近产业（指房产地产等）；邻接房地产；周围房地产
adjacent rank 相邻等级
adjacent region 相邻区
adjacent rock 围岩；邻岩
adjacent room 邻室
adjacent sea 毗邻海域；邻海；近海；相邻海区；边缘海
adjacent section 相邻断面
adjacent segment 邻接块（管片）
adjacent sheet 邻接图幅；相邻图幅
adjacent shothole 邻孔
adjacent side 邻边
adjacent solution 相邻溶液
adjacent span 相邻的桥跨；邻孔；邻跨（度）
adjacent spread's number 毗邻排列数
adjacent square 相邻方格
adjacent stereomodel 相邻立体模型
adjacent strip 毗邻碾压带；邻接带；邻港；铁路联络线；沥青混凝土覆面邻接带
adjacent structure 毗连建筑物；相邻建筑物；附近建筑物
adjacent subarea 邻近子区
adjacent surface 比连面
adjacent teeth 邻接牙
adjacent tie 邻枕
adjacent to welds 邻近焊缝开孔
adjacent track 邻线；相邻线；相邻轨道
adjacent traffic lane 相邻车道
adjacent turns 相邻线匝
adjacent vertexes 邻顶
adjacent vertices of a graph 图的邻接顶点
adjacent video carrier 邻道图像载波
adjacent vision carrier 邻道图像载波
adjacent vortex 邻涡
adjacent wall 相邻的墙；邻墙
adjacent waters 毗邻水域；毗邻海域；毗连水域；相邻水域
adjacent wave 邻波
adjacent wayside stations 相邻小站
adjacent window 邻窗
adjacent zone 毗连（地）区
adjective dye 间接染料
adjective law 程序法
adjoin 邻接
adjoin boundary condition 伴随边界条件
adjoiner 邻近处；邻接处；接合部
adjoin functions 共轭函数
adjoining 邻接的；毗邻的
adjoining angle 邻角
adjoining benefit 伴随效益
adjoining blocks 毗连房屋；相邻房屋
adjoining building 邻屋；邻近房屋
adjoining carbon 邻接碳原子
adjoining concession 相邻特许（租借）地；邻接租借地；附近居民的许可；附近居民的同意（交通线）
adjoining construction 邻近施工
adjoining course 邻接（砖）层；结合层
adjoining district 相邻区段
adjoining flight strips 相邻航（空）线
adjoining hole 毗邻孔
adjoining land 邻接地；毗连土地；邻地
adjoining layers 毗连层次
adjoining lift 邻接浇注层；相邻混凝土浇筑层
adjoining lot 毗连土地
adjoining model 相邻模型
adjoining opening 相邻桥孔；邻跨（孔）
adjoining owner 邻近业主
adjoining photograph 相邻像片
adjoining plane 相邻面
adjoining premises 毗邻地产；相邻房地产
adjoining property 邻近产业（指房产地产等）；邻近财产；邻接房地产；邻产；周围房地产
adjoining railway 铁路引入站
adjoining rock 围岩
adjoining room 邻屋；邻室
adjoining section 相邻区间
adjoining sheet 邻接图幅；相邻图幅

adjoining sheet names 相邻图幅名称
adjoining span 邻跨（度）；邻接跨
adjoining square 邻接方格
adjoining stratum 毗邻岩层
adjoining survey 相邻航线测量
adjoining teeth 邻接牙
adjoining traces stack 相邻道叠加
adjoining wall 毗邻的墙；邻墙
adjoin linear transformation 伴随线性变换
adjoin matrix 相伴矩阵
adjoint 依附图；伴随
adjoint branch 毗连支路
adjoint curve 伴随曲线
adjoint determinant 伴随行列式
adjoint equation 伴随方程
adjoint function 伴随函数；共轭函数
adjoint mapping 伴随映射
adjoint matrices 伴随阵
adjoint matrix 共轭矩阵；伴随矩阵
adjoint network 伴随网络
adjoint of a matrix 附加矩阵
adjoint ointorthogonal system 共轭转置正交系
adjoint operator 伴随运算子；伴随算符；伴算符
adjoint orthogonal system 伴随正交系
adjoint problem 伴随问题
adjoint simulation 伴随仿真
adjoint space 伴随空间
adjoint terminal 半随终结符
adjoint variable 伴随变量
adjoint vector space 伴随向量空间
adjoint wave functions 伴波函数
adjoning tree 对照植株
adjourned sale 延期销售
adjudgement 裁决
adjudicating panel 评审小组；招标（评议）委员会；裁判委员会
adjudication 判决；裁决；裁定
adjudication fee 法庭费用
adjudication of bankruptcy 宣告破产；裁定破产
adjudication of tax 税收裁决
adjudication of water rights 水权判决
adjugate 伴随
adjugate determinant 转置伴随行列式
adjugate matrix 转置伴随矩阵
adjunct 添加剂；附件；附加物；附属品；辅助的
adjunct account 附加账户
adjunction 附益
adjunct pollution 添加剂污染
adjunct register 附加寄存器
adjunct register set 附加寄存器组
adjunct spring 辅助弹簧；弹簧自动闭门器
adjust 对准
adjustability 调整性能；可调性；可调整性；可调性
adjustability on release 缓解调节能力
adjustable 可校准的；可调整的；可调节的
adjustable air damper 风窗口调节门
adjustable air gap 可调气隙
adjustable air supplying fan 可调供气扇
adjustable anchorage 可调锚杆
adjustable anchorage bar 可调整的锚杆
adjustable angle square 可调角尺
adjustable angle aperture 可调孔径；可调光圈
adjustable-aperture grizzly 可调孔式格筛
adjustable array 可调节数组
adjustable array declarator 可调数组说明符
adjustable attachments 可调整附件；活支撑；建筑结构可调支撑
adjustable attenuator 可调衰减器
adjustable axle 可调节轴
adjustable ballast 可调压载
adjustable ballast tank 可调压载水舱
adjustable ball hinge 可调节球形铰链；可调节球形合页
adjustable bar 伸缩杆；可调杆
adjustable base anchor 可调门框座；活动底锚
adjustable basket technique 一揽子调整方法
adjustable beam saddle 可调梁座
adjustable bearing 可调支承；可调轴承（支）座；可调轴承支架；可调支座；可调整（的）支承；可调（式）轴承
adjustable bed 可调工作台；可调台面
adjustable bed press 工作台可调（式）压力机
adjustable bellows 可调伸缩软管；可调膜盒；可调波纹管

adjustable bench level 可调台（式）水准仪
adjustable binocular loupe 可调式双筒放大镜
adjustable blade 可调整叶片；可调整刀片；可调推进器叶片；可调导叶；变距桨叶
adjustable blade propeller pump 可调叶片式轴流泵；可调叶片桨叶泵
adjustable blade propeller turbine 转桨式水轮机
adjustable blade pump turbine 可调整叶片的的水泵—水轮机
adjustable blade reamer 可调叶片扩锥；可调节刃的铰床；刀片可调铰刀
adjustable blade unit 可调整叶片的水轮机组
adjustable bolt 调节螺栓；可调螺栓；调整螺栓
adjustable brake block 可调整闸瓦；可调（式）闸瓦
adjustable brake release device 可调制动器释放装置；可调节的制动机缓解装置
adjustable brush 可调电刷
adjustable bush 可调轴衬
adjustable calliper ga(u)ge 可伸缩内径规
adjustable cam 可调凸轮
adjustable capacitor 可调电容器；可变电容（器）
adjustable cast iron planes 精刨
adjustable circuit breaker 可调式断路器
adjustable circular plane 可调整的圆刨
adjustable cistern barometer 调槽式气压表；福丁气压计；福丁气压表
adjustable clamp 可调夹头；活络钳；活夹具；活动钳（夹）
adjustable clearance 可调（节）间隙
adjustable clip 可调位的夹捆
adjustable collar 调整环
adjustable compound tap 可调节丝锥
adjustable condenser 可调电容（器）；可变电容（器）
adjustable conductor rail support 可调导轨支撑
adjustable cone 可调锥体
adjustable contact 可调接点
adjustable contact plate 调整触片
adjustable counter balance 可调配重
adjustable coupling 可调联轴节
adjustable crank 可调曲柄；加减拐肘
adjustable crank pin 可调曲柄销
adjustable crest 可调整式堰顶
adjustable curtain wall 可调节吊墙
adjustable curve 校正曲线
adjustable cushion 可调缓冲器
adjustable cut nippers 活刃剪钳
adjustable cutter bush 调刀轴环；刀杆调整环
adjustable damper 可调减震器；可调风门；可调挡板
adjustable declarator subscript 可调说明符下标
adjustable deflector 可调闸门；可调导向板
adjustable delay 可调延迟（时刻）
adjustable delivery pump 可调输料泵
adjustable device 可调整的装置
adjustable dial 可调标度盘
adjustable diaphragm 可调膜片；可调光阑；可调隔仓板
adjustable die 活动螺丝钢板；可调扳牙；活络螺纹扳；活络扳牙；活动扳牙
adjustable dimension 可调维数；可调尺寸
adjustable discharge flow gate 卸料调节阀
adjustable discharge gear pump 可调卸料齿轮泵
adjustable divider 可调两脚规
adjustable dog 可调制动爪；可调行程限制器；可调挡块
adjustable dog hook 挽钩；可调节的伐木钩；带钩撑杆；可调爪形钩
adjustable door frame 可调门框；可按墙厚调整的门框
adjustable drawbar 可调拉杆
adjustable drawing table 可调绘图桌；活动制图桌
adjustable drum 可调转筒；可调绕线架；可调成型机头
adjustable eccentric wheel 可调偏心轮
adjustable end stop 升降挡板
adjustable engine 变速发动机
adjustable equatorial mount 可调赤道仪装置
adjustable error 可校正误差
adjustable extent 可调整的可扩充；可调（节）范围
adjustable eyepiece 可调目镜
adjustable face spanner 可调节的平面扳手；可调端面扳手；端面扳手；叉形螺母扳手
adjustable fitting 铰接接头；铰接接合；液压铰
adjustable fixing 调节支柱；可调支柱

adjustable flap 可调活板;可调风门片
adjustable flat wrench 扁平活动扳手
adjustable flow beam 可调喷油嘴
adjustable flume 可调整水槽;活动变坡水槽
adjustable for height 高度可调的
adjustable friction damper 可调摩擦减震器
adjustable front axle 可调前轴
adjustable front sight 可调整准星
adjustable fulcrum 可调节支点
adjustable gang-condenser 可变电容器组
adjustable gangway 活动引桥
adjustable gate 调节门;可调式闸门
adjustable ga(u)ge 可调卡规
adjustable gib 调整镶条;可调条;可调导轨
adjustable glazed louvers 活动通风玻璃百叶窗
adjustable grill(e) 可调栅格;可调整格栅
adjustable guide 可调导板
adjustable guide blade 可调导叶片
adjustable guide rail 可调导向器
adjustable guide vane 可调导叶
adjustable hand reamer 可调式手铰刀
adjustable head 可调进刀架;活动头丁字尺
adjustable head T-square 活头丁字尺;丁字尺可调头;活动丁字尺
adjustable heater switch 加热器调节旋钮
adjustable-height seat 高度可调座椅
adjustable hitch bar 可调式牵引杆
adjustable hollow cutter 可调空心刀具
adjustable hollow mill 可调整空心铣刀
adjustable hook rule 可调整钩头尺
adjustable index 可调整指标
adjustable indicator 可调(节)指示器
adjustable inductance 可调电感
adjustable inductor 可变电感线圈
adjustable in steps 可分级调整的
adjustable instrument table 可调式器械桌
adjustable integrated stabilizer 可调集成稳压器
adjustable interconnection block 选择单元
adjustable iron force plane 粗刨;粗长刨
adjustable jack 可调千斤顶
adjustable jack table 可调重机架
adjustable jaw flap 可调颚板
adjustable jet 可调(节)喷管
adjustable jib 可调挺杆;可调起重机臂;可调吊机臂
adjustable key 伸缩钥匙;可调键
adjustable knee tool 可调式膝形刀
adjustable lapping ring 可调精研圈
adjustable leak 可调漏孔
adjustable legs 伸缩式三脚架
adjustable lever 可调杠杆
adjustable lifter scoop 可调扬料勺
adjustable limit snap thread ga(u)ge 可调式极限螺纹卡规
adjustable lock rod 可调锁杆
adjustable louvers 调整活动百叶窗;活动百叶
adjustable luminaire 可调灯具
adjustable marker 可调校准器;可调曲线规;可调划线规;可调标识器
adjustable mask 可调掩模;可调屏障
adjustable micrometer 可调千分尺
adjustable mortgage loans 可调抵押贷款
adjustable mortgages 可调节的抵押契据
adjustable mouth 可调进气口
adjustable multiple-point suspension scaffold 多点调节吊脚手架
adjustable multiple spindle drill 可调式多轴钻床
adjustable nozzle 可调喷嘴
adjustable nut 松紧扣;可调螺母;可调螺帽
adjustable nut wrench 活络扳手;活动扳手
adjustable orifice 调节嘴;可调喷口
adjustable orifice scrubber 变径式除气器
adjustable outrigger collector 可调节外架总管
adjustable overload friction clutch 可调式过载摩擦离合器
adjustable pedal 可调式踏板
adjustable peg 可调固定汇率
adjustable peg rate of exchange 可调固定汇率
adjustable peg rate system 可调固定汇率制
adjustable peg system 可调整的关系汇率制;可调固定汇率制
adjustable pile head 活动桩顶法
adjustable pipe die 可调管子板牙
adjustable pipe for feed flow 送料调节管
adjustable pipe tongs 可调管钳;活动管钳

adjustable pitch 可调螺距;可变螺距
adjustable pitch airscrew 调距螺旋桨
adjustable pitch blade 可调节距叶片
adjustable pitch propeller 可调(螺)距螺旋桨;变距推进器;变距螺旋桨
adjustable plane 可调木工刨
adjustable plate 可调板
adjustable plough 可调清扫器;可调刮板
adjustable plow 可调的犁
adjustable plumb bob 可调垂球
adjustable pocket bench plane 钢刨
adjustable port 可调孔
adjustable port proportioning valve 可调节的空调进口阀
adjustable positioner 可调定位器
adjustable precentering device 调位式导口【船】
adjustable premium 可调整的保险费
adjustable probe 可调指示器
adjustable prop 可调支柱;可调的支柱
adjustable proscenium 可调整宽度的舞台口;假台口;活动台口
adjustable pulse width pacemaker 脉宽可调起搏器
adjustable pump vane 可调泵叶
adjustable rail brace 可动轨撑;可调式钢轨撑;可调轨撑
adjustable rail fastening system 可调式钢轨扣件
adjustable rail ga(u)ge 可调的轨距尺
adjustable range 调节范围;可调节范围
adjustable rate bond 可调整利率债券
adjustable rate mortgage 可调整抵押贷款利率
adjustable-ratio transformer 可调变比变压器
adjustable reamer 可调(节)铰刀
adjustable rear sight 可调整后准星
adjustable release 阶段缓解
adjustable resistance 可调电阻;可变阻力;可变阻抗
adjustable resistor 可调电阻器;可变电阻器
adjustable ridging body 可调起垄犁体
adjustable ring 控制环
adjustable riveting machine 可调的铆接机
adjustable rod 可调拉杆
adjustable roller 可调(整)滚轴
adjustable rope sling 活动绳扣
adjustable round die 可调圆扳手
adjustable round split die 开口圆扳牙;开缝环形板牙
adjustable round split die with adjusting screw 带调整螺钉的可调式圆板牙
adjustable rubber rest bar 可调橡皮停止杆
adjustable saw bow 可调整锯架;可调(节)的锯弓
adjustable saw frame 可调(节)的锯框
adjustable scoop 可调进气口
adjustable screw 调节螺旋;可调螺丝;可调螺钉
adjustable screw type steering gear 可调整式螺旋型转向装置
adjustable seat 可调节座位
adjustable sheave 可调滑车轮;变距槽轮
adjustable shelf 活动(支)架;活动书架
adjustable shell reamer 可调筒形铰刀
adjustable shelving 可调整高低的搁板
adjustable shock absorber 可调减震器
adjustable shoe sole 可调制动器底板
adjustable shore 可调(整)的支撑;可调顶柱
adjustable shore ramp 可调式岸用跳板
adjustable shoring 可调支撑
adjustable short-circuit termination 可调短路终端
adjustable shuttering 活动模板
adjustable sieve 可调筛孔筛
adjustable single-end wrench 活动单头扳手
adjustable siphon 活动虹吸
adjustable size array 可调数组
adjustable size wet suit 可调节湿式潜水服
adjustable slatted shutter 活动百叶窗
adjustable sleeve 可调套管
adjustable slide 可调导板;可调整导板
adjustable slit 可调(狭)缝
adjustable slit assembly 可调狭缝组件
adjustable snap ga(u)ge 可调整式卡规
adjustable sounder 可调发声器
adjustable spanner 活络扳手;可调扳手;活(动)扳手;活动扳手
adjustable spanner angle 弯头活络扳手
adjustable-span recorder 可调间隔记录器
adjustable speed 可调速度
adjustable-speed belt 无级变速器皮带

adjustablespeed drive 无级变速传动装置
adjustable speed motor 调速电动机;可调速电动机;变速电动机
adjustable speed pump 变速泵
adjustable speed wheel 调速皮带轮
adjustable spindle milling machine 可调主轴铣床
adjustable spring collar 可调整弹簧环
adjustable spring ga(u)ge 可调式弹簧定位装置
adjustable square 可调直角尺
adjustable stabilizer 可调整稳定面
adjustable stair ga(u)ge 调整式阶梯规
adjustable starter 可调起动器
adjustable starting rheostat 可调起动变阻器
adjustable stem 可调整基杆;校正杆
adjustable stop 可调行程限制器;可调光栏
adjustable stopper 调整挡铁
adjustable stop screw 可调整止动螺钉
adjustable stroke feeder 可调式冲程给料机
adjustable strut 可调整支柱
adjustable submerged orifice 可调潜孔
adjustable support 可调支架
adjustable suspension 可调节悬挂装置
adjustable switch operating rod 密贴调整杆
adjustable tap 可调丝锥
adjustable tap wrench 活动丝锥扳手;可调丝锥扳手
adjustable tax 调节税
adjustable thrust engine 推力可调的发动机
adjustable thrust journal 可调(整)的止推轴颈
adjustable tongs 可调管钳
adjustable track ga(u)ge 活动(轨)道尺
adjustable tractor seat 拖拉机可调式座位
adjustable transformer 可调变压器;可变比变压器
adjustable triangle 坡度尺;斜度尺
adjustable tripod 可调三脚架;活腿三脚架
adjustable turning head 可调车削头
adjustable valve 调整阀;调节阀
adjustable vane 可调(节)叶片
adjustable vane turbine 可调桨式水轮机;转桨式水轮机
adjustable varying-speed motor 可调速电动机;可调变速马达
adjustable velocity 可调速度
adjustable vent-flap 可调的通风口盖;可调整通风折翼
adjustable ventglazed louvers 活动玻璃百叶窗
adjustable vertical gate feeder 可调式垂直闸门给料器
adjustable viewing angle 可调视角
adjustable voltage divider 可调分压器
adjustable voltage motor 可调压电动机
adjustable voltage stabilizer 可调稳压器
adjustable washer 角形垫块
adjustable weight 可调压载
adjustable weir 可调节堰板
adjustable weir crest 活动堰顶
adjustable weir ring 溢流调节环
adjustable wheel set 可调节轮对
adjustable wooded legs 伸缩式木三脚架
adjustable wrench 活板子;可调扳手;活络扳手;活动扳手
adjust accounts 清理账目;复算账目;复核核算
adjust actual inventory 调整实际库存量
adjustage 喷射管;精整设备;管接头
adjust and control 调控
adjust and control function 调控职能
adjust angle 平差角
adjust a ship's draft 调整船舶吃水
adjust board 摇臂滑杆支架
adjust claim 评定赔偿要求
adjust data 订正资料
adjust downward 降低
adjusted acquisition cost 调整后采购成本
adjusted actual inventory 调整后实际库存量
adjusted angle 平差角
adjusted bank balance 调整后银行余额
adjusted bank balance of cash 调整后银行存款(余)额
adjusted base 调整的实际库存量
adjusted base cost 调整的基本造价;调整成本;变动单价
adjusted basis 已调整基准
adjusted bond 调整债券
adjusted book balance of cash 调整后账面现金额
adjusted claim 调态索赔金额

adjusted coefficient of determination 调整的测定系数
adjusted cost basis 调整后成本基础;调整成本制
adjusted data 校核资料;订正的资料
adjusted death rate 修正死亡率;订正死亡率;标准化死亡率
adjusted decibel 调整分贝
adjusted drainage 适应水系
adjusted elevation 平差高程
adjusted entry 调整分录
adjusted figures 调整后数字
adjusted frequency table 调整的频率表
adjusted geoid 调整大地水准面
adjusted gross income 调整后的总收益(房地产);调整后总所得;调整后毛收入所得额
adjusted historical cost 调整后历史成本
adjusted income 调整后所得;调整后收益
adjusted income statement 调整后损益计算表
adjusted indexes 调整指数
adjusted key 调整键
adjusted mean 算术平均数;校正均值;修正均值
adjusted mortality (rate) 调整后的死亡率
adjusted mortgage 调整后的抵押
adjusted multiple correlation coefficient 调整的复相系数
adjusted net fill 调整后的净填方
adjusted net profit 调整后净利润
adjusted nut 调整螺母;调整螺帽
adjusted original cost 调整原价
adjusted pension 调整后年金
adjusted position 平差位置;平差后的位置;已平差点位
adjusted price 调整后价格
adjusted pro-rata 按比例调整
adjusted quality 平差量
adjusted rate 修正率
adjusted retention time 调整保留时间;调焦保留时间
adjusted retention volume 调整保留体积
adjusted reverse trend 经过调整的翻转趋势
adjusted river 调整河流;适应河;经过整治的河流
adjusted sales price 调整后的售价
adjusted solution 调整溶液
adjusted spring 调整弹簧
adjusted stream 调整河流;适应河;经过整治的河流
adjusted total price of contract 调整合同总价
adjusted trial balance 调整后试算表
adjusted value 平差值;调整值;校正值
adjusted vertically 按高度调整
adjusted wedge 调整楔块
adjusted weight 校正(体)重
adjustement of exchange rate 调整汇价
adjuster 调准装置;调整器;调整工;调停人;理算师;精调装置;校准者;校准器;校正师;海损理算人;海损理算人;保险赔偿估定员
adjuster bar 调整杆
adjuster board 调节器板;调整器板
adjuster for windows 窗开关调整器;撑窗杆
adjuster's note 理算师附记
adjuster valve 调节器阀门;调整器阀门
adjust for wear 按磨损调整
adjust(ing) 调准
adjusting account 调整账户
adjusting accuracy 调整精度
adjusting and repairing work 调整和修补工程
adjusting angle 旋轮安装角
adjusting appliance 调整仪器;调节设备;调节工具;调整仪器;调整设备;调整工具;调焦设备
adjusting block 调整(垫)块;调节楔座
adjusting bolt 调整螺栓;调节螺栓;校正螺栓
adjusting chamber 调压室
adjusting cock 调节旋塞
adjusting colo(u)r 调色
adjusting cylinder 调节汽缸
adjusting damper 调节风门
adjusting device 调整装置;调整设备;调节装置;校准装置;校正装置
adjusting distance of track lining 拔距【铁】
adjusting ear 调整耳;拉线用复滑轮
adjusting element 细调摘纵叉
adjusting entries 调整记录;整理账
adjusting entry 调整分录
adjusting eyepiece 校正目镜
adjusting file 可调锉;校准锉刀

adjusting flume 活动变坡水槽
adjusting force 调节力
adjusting ga(u)ge 调整量规;整定仪表;整定卡规
adjusting gear 调整装置;调整齿轮;调节装置;调节齿轮;校准装置
adjusting gear for rudder 方向舵调整装置
adjusting gib 活动扁栓
adjusting handle 调整手把;调整柄;调节手轮
adjusting-head T-square 活头丁字尺
adjusting idler wheel 调整用惰轮;履带诱导轮
adjusting journal entry 日记账调整分录
adjusting key 调整键
adjusting knob 调整旋钮;调节(旋)钮;校正旋钮
adjusting lath 伸缩式标尺
adjusting lever 校正杆;调整后毛收入;调整(杠)杆
adjusting link 可调联结杆
adjusting lock nut 调节锁紧螺母;可调锁紧螺母;可调锁紧螺帽
adjusting machine 调整机;校准机
adjusting magnet 自差校正磁铁
adjusting mark 调整标记
adjusting mechanism 调整机构;校正装置
adjusting microinching 寸动(压力机上)
adjusting microscope 可调显微镜
adjusting needle 调准针
adjusting needle valve 调节针阀
adjusting nut 调整螺帽;调节螺母;调节螺帽;调整螺母
adjusting of cross level 整正水平
adjusting pin 配合销;调整销;校正针;校正销;固定销;改针;定位销;拨针
adjusting pipe tongs 活动管钳
adjusting plane 可调木工刨;木工校正刨
adjusting plate 机枪瞄准具;调节(垫)板;调整盘;调节板
adjusting precipitation records 调节降水记录
adjusting price 调剂价;价格调整
adjusting quantity meter 调量表
adjusting range 调整范围
adjusting resistance 调节电阻
adjusting ring 调整环;调节环
adjusting rod 调节杆;调整杆
adjusting roller 调整滚子;调整辊
adjusting scale 刻度调节
adjusting screw 调准螺旋;调准螺丝;调整螺旋;调整螺丝;调整螺杆;调整螺钉;调节螺钉;松紧螺旋扣;校正螺旋;校正螺丝;校正螺钉;定位调节器
adjusting screw rod 调节螺杆
adjusting screw for elevation 标高调整螺丝
adjusting screw for pan position 锅体水平调节螺杆
adjusting screw for transverse 横向调整螺丝
adjusting screw nut 调距螺母
adjusting shim 调整垫片
adjusting shop 调整车间;校准车间;装配(车)间
adjusting shutter 调节门
adjusting slider 调节滑块
adjusting spacer 调整垫圈
adjusting speed 调整速度
adjusting speed motor 变速电动机
adjusting spindle 调整轴
adjusting spring 调准螺丝;调整螺丝;调整螺杆;调整螺钉;调整弹簧;校正弹簧
adjusting spring case 调准簧盒
adjusting strip 调整片;调整衬条
adjusting stud 调整用双头螺栓
adjusting surface 调整面
adjusting tank 调整水柜;调整水舱
adjusting to a line 直线对准
adjusting tool 校正工具
adjusting unit 调节装置
adjusting valve 调整阀
adjusting variables 调整变数
adjusting washer 调整垫圈;调节垫圈
adjusting wedge 调整楔(块);调节楔块
adjusting well 调整井
adjusting wheel 调整轮
adjusting worm 可调整蜗杆;可调螺旋运送机;调节蜗杆
adjusting wrench 活络扳手
adjusting yard 校正场
adjusting yoke 调节拉杆
adjustment 平差;调准;调整;调节;顺应;适应;理算;校准;校正

adjustment account 调整账户
adjustment accuracy 调正精度
adjustment administrative setup 调整管理机构
adjustment assistance 调节性援助
adjustment bolt 校正螺栓;调整螺栓
adjustment bond 调整债券
adjustment buoyancy life jacket 浮力可调救生衣
adjustment by angles 角度平差
adjustment by condition 条件平差
adjustment by coordinates 坐标平差
adjustment by correlate 联系数平差
adjustment by directions 方向平差
adjustment by method of junction points 接点平差;结点平差;结点法
adjustment by method of polygon 多边形(平差)法
adjustment by parameter 参数平差
adjustment by partitioning 分区平差
adjustment by the method of least squares 最小二乘法平差
adjustment by variation of coordinates 坐标平差
adjustment charges 理算费用
adjustment clause 伸缩条款
adjustment coefficient 调整系数
adjustment compass 可调圆规;调整罗盘
adjustment computation 平差计算
adjustment continuation condition 调整连续条件
adjustment continuous condition 调整连续条件
adjustment controler 调节控制器
adjustment correction 平差校正;平差改正
adjustment cost 调整费用;理算费用
adjustment credit 调整性信贷
adjustment curve 调制曲线;调整曲线;校正曲线;缓和曲线
adjustment device 校正装置
adjustment diagram 调整图
adjustment economic texture 调整经济结构
adjustment entries 过账登记
adjustment entry 调整分录
adjustment equation 平差方程
adjustment error 校准误差
adjustment factor 调整因数;调整因素;调整系数
adjustment for altitude 高度校正;高度调整
adjustment for definition 调整清晰度
adjustment for horizon glass of sextant 六分仪动镜校正
adjustment for index glass of sextant 六分仪动镜校正
adjustment for length of stroke 调整行程长度装置
adjustment formula 平差公式
adjustment for over-withholding 超额扣缴调整
adjustment for parallax 视差改正
adjustment for position of stroke 行程位置调整
adjustment for price fluctuation 价格波动调整
adjustment for tilt 倾角校正;倾角改正
adjustment for trend 趋势调整
adjustment for turnover of staffs 职员更替调整
adjustment for wear 补偿磨损调整
adjustment free 全部调整完毕;全部调节完毕
adjustment function 调整函数;调节函数
adjustment indicator of techno-economics 调整技术经济指标
adjustment in direction 方向修正;方向调整
adjustment in flow direction 调整流向
adjustment in groups 分组平差(法);分层平差
adjustment in height 按高度调整
adjustment in one cast 整体平差【测】
adjustment in successive steps 逐次平差
adjustment lags 调整时滞
adjustment letter 海损理算书
adjustment mark 调整标记;安装标记
adjustment memo 调整凭单
adjustment memorandum 调整凭单
adjustment method 平差(方)法;调试方法
adjustment method in groups 分组平差法
adjustment method of variation of coordinates 坐标变化平差法
adjustment notch 调整槽口;安装记号;安装标记
adjustment of account 修正账单
adjustment of actual total loss 实际全损理算;实际全损调整
adjustment of aids 航标调整
adjustment of aids layout 调波
adjustment of air volume 空气体积调节

adjustment of alignment 定线调整
adjustment of angles in all combination 全组合角度平差
adjustment of assets value 资产价值的调整
adjustment of astrogeodetic network 天文大地网平差
adjustment of average 海损理算
adjustment of balance 余额调整
adjustment of bit load 钻头负荷调节
adjustment of business income 营业收入调整
adjustment of car flow 车流调整
adjustment of car loading 装车调整
adjustment of car loading by direction 方向别装车调整
adjustment of compass 罗盘校正
adjustment of condition equation with unknowns 带有未知数的条件平差
adjustment of condition observations 条件观测平差
adjustment of condition 条件平差
adjustment of constructive total loss 推定全损理算
adjustment of correlated observation 相关观测平差
adjustment of data 数据校正;数据改正;资料校正;资料订正
adjustment of dike face 坝面整理
adjustment of direct observation 直接(观测)平差
adjustment of environment 环境调整
adjustment of errors 平差;误差调整
adjustment of exchange rate 调整汇率
adjustment of figure 图形平差
adjustment of final design 最终设计修正
adjustment of general average 共同海损理算
adjustment of grading 颗粒级配调整
adjustment of gravity base station network 重力基点网平差
adjustment of gross average 共同海损理算
adjustment of ignition 发火调整
adjustment of images 像片镶嵌
adjustment of incomplete sets of observation 不完全观测组平差
adjustment of indirected observation with condition equations 带有条件方程的间接观测平差
adjustment of indirect observations 间接(观测)平差
adjustment of individual grievances 个人不满的调整
adjustment of instrument 仪器校正;仪表校正
adjustment of intermediate observations 间接(观测)平差
adjustment of inventory 库存调节
adjustment of inventory investment 库存投资的调整
adjustment of itself 自动调整
adjustment of joints 接头调节
adjustment of level(l)ing circuits 水准线路平差
adjustment of level(l)ing network 水准网平差
adjustment of light level 调光
adjustment of loaded car 重车调整
adjustment of loss 损失理算;损失调整
adjustment of mepty cars 空车调整
adjustment of mix 配合比调整
adjustment of mixture 配合比调整;混凝土配合比调整
adjustment of model 模型修正
adjustment of network 测量网平差
adjustment of network of triangulation chains 三角锁网平差
adjustment of observation 观测平差;测量平差
adjustment of observed equations 间接观测平差
adjustment of particular average 单独海损理算
adjustment of price 价格调整
adjustment of printing plate 上版
adjustment of quota 配额调整
adjustment of rate 调整费率
adjustment of reserved cars 备用车调整
adjustment of river 河流整治;河流调节
adjustment of salvage loss 施救费用理算
adjustment of seasonal variations 季节性变动调整;按季节变化调整
adjustment of sextant 六分仪校正
adjustment of slip 泥浆性质调整
adjustment of steel material 钢材矫正

adjustment of stream 河流整治;河流调节
adjustment of surplus 调整盈余额
adjustment of survey 测量平差
adjustment of track 轨道调整
adjustment of track ga(u)ge 轨距校正;轨距规的调整
adjustment of track level 轨距水准仪调整
adjustment of traverse 导线平差
adjustment of traverse network 导线网平差
adjustment of trigonometric(al) level(l)ing network 三角高程网平差
adjustment of trilateration network 测边网平差
adjustment of trim/heel 吃水差调整;倾斜度调整
adjustment of typical figures 典型图形平差
adjustment of valve clearance 调整阀隙
adjustment of wage rates 调整工资率
adjustment panel 调整板
adjustment parameter 平差参数
adjustment period 调整期(间)
adjustment phase 调整期
adjustment plan(ning) 计划调整;修正计划
adjustment plate 调整填片;调整垫片;调整板
adjustment process 调整过程
adjustment productive texture 调整产品结构
adjustment program(me) 平差程序
adjustment range 调整幅度;调整范围
adjustment rate 调整率
adjustment reaction 顺应反应
adjustment residual 平差残差
adjustment scale 刻度调节
adjustment sheet 调节片
adjustment stud 调整螺栓
adjustment tax 调节税
adjustment technologic(al)texture 调整技术结构
adjustment the compass 确定罗盘偏差
adjustment the letter of credit as requested 按要求修改信用证
adjustment to expenses 费用调整
adjustment to income 收入调整
adjustment tool 调整工具
adjustment value method 调整现值法
adjustment variable 整定参数
adjustment washer 调整垫圈
adjustment weir crest 活动堰顶
adjustment with fictitious observed quantities 虚拟观测量平差法
adjustment work expenses 调整费用
adjustment works of stream 河流整治工程
adjust on control point 根据控制点定向
adjustor 校准器
adjust point 调整点
adjust present value method 调整现值法
adjust price 调整价格
adjust prices upward 调高物价水平
adjust scale 调整比例尺
adjust spring 弹簧自动闭门器
adjust supply and demand for funds 调节资金供求
adjust supply to a current demand 调整供求关系
adjust the angles by curves 使曲线光滑
adjust the difference 理算差额
adjust the instrument to zero 把仪器调到零
adjust the price ratios 调整比价
adjust the time 调整时期
adjust the use of labor force 调剂劳动力
adjust to zero 调整至零点;调整到零,调到零(值);置零;归零
adjust valve 调节阀;制动调节阀
adjust vertically 高度调整;垂向调整
adjust work for position 调整工件位置
adjutage 排水筒,调节管;射流管;延长臂;管嘴;接;放水管
adjuvant 助手;辅助剂;辅助的
ad-layer 吸附层
Adler silk 阿德勒铜铵丝
Adler tube 阿德勒高速射线管
adlet 小广告
adlittoral 近岸的
Adlrey 阿尔德雷导线用铝合金
admeasure 量度;计量;测定;测量
admeasurement 量测;计量;丈量;尺度;测量
admeasuring apparatus 测像仪(器);测量仪
admicelle 吸附胶团;叠加胶束
adminiculum lineae albae 白线支座

administer 染毒;施行;管辖;管理;操纵
administer claim 解决索赔案件
administered price 垄断价格;控制价格
administered-price theory 操纵价格论
administered rate 控制价格
administered trade 控制贸易
administering authority 管理权限;管理当局
administering country 管辖区(域)
administering power 管理权
administering state 管理国
administrating system for work of nightsoil and sewage from boats and ships 船舶粪便污水作业管理系统
administration 行政部门;管理(局);给药
administration according to law 依法行政
administration account 管理科目
administration anchor 海军式锚
administration and supervision cost 行政管理费
administration and training 管理与训练
administration area 管理区
administration at different levels 分级管理
administration audit 行政审计
administration authority 行政管理部门
administration block 行政建筑物;行政办公楼
administration boundary 行政边界
administration branch 管理分局
administration budget 行政预算
administration building 办公大楼;行政办公楼
administration center 行政中心
administration complex 行政建筑群;行政办公建筑
administration cost 管理成本;管理费用
administration data processing 行政管理数据处理
administration duty 行政事务工作
administration expenses 管理费(用)
administration expense account 企业管理费账户
administration expense budget 一般管理费预算
administration fee 管理费
administration in bankruptcy 破产清理
administration in charge of marine affairs 国家海洋管理部门
administration launch 港口公务联络艇
administration law 行政法(令)
administration line 局界会议电话
Administration Management Society 行政管理学会
administration manager 行政事务主任
Administration Measures of Nature Reserves for Forest and Wild Animals 森林和野生动物型自然保护区管理
administration of accounts 账目管理
administration of budget 管理预算
administration of city 城市管理
administration of construction contract 施工合同管理
administration of estates of small value 小额财产的管理
administration office 码头管理所;行政(办公)室
Administration of Justice Act 司法管理法例
administration of law 法律的实施
administration of law in water 水行政执法
administration of power supply 供电所;供电局
administration of radiation 辐射处理
administration of research activities 研究管理
administration of the construction contract 施工合同管理
administration of treatment plant 处理厂管理
administration of water 水行政管理
administration of water rights 水权的管理
Administration Procedure Law of the People's Republic of China 中华人民共和国行政诉讼法
administration program(me) 管理程序
administration reconsideration of water 水行政复议
administration section 行政管理部门;事务部门;管理部门
administration staff 行政管理人员
administration tower 高层建筑行政办公楼
administration unit 行政办公楼;管理单元
administration unit group 管理单元组
administration unit pointer 管理单元指针
administration-welfare quarter 厂前区
administrative ability 经营才能
administrative account(ing) 经营管理会计
administrative action 行政诉讼

administrative action of water 水行政行为
administrative agency 行政机构;管理机构
administrative and financial services 行政和财务处
administrative and maintenance expenses 管理及维护费用
administrative and selling expenses 管理及销售费用
administrative approval 行政批准;正式验收
administrative area 行政区;管理区
administrative arrangement 行政管理
administrative audit 经营管理审计;行政审计;管理审计
administrative authorities 行政机关
administrative authority 行政管理当局;主管当局;管理权限;管理当局;管理部门
administrative autopsy 行政解剖
administrative backstopping 行政支持
administrative behavio(u)r 行政行为
administrative body 办事机构
administrative boundary 行政区划界;行政界
administrative budget 管理预算;行政费预算
administrative building 行政建筑;行政管理楼;行政办公室;行政办公楼;办公大楼
administrative bureau for industry and commerce 工商行政管理局
administrative census 行政调查
administrative center 行政中心
administrative classification of highway 公路分类
administrative command 行政指挥
administrative committee 管理委员会
administrative complex of a factory 厂前区
administrative control 行政控制;行政管制
administrative cost 行政管理费;行政费(用);管理费
administrative council 行政理事会行政管理费
administrative data 行政区划资料;管理资料
administrative data processing 事务管理的数据处理;行政管理数据处理;管理数据处理
administrative data processor 行政管理数据处理机
administrative decision 管理决策
administrative delimitation 行政区域;行政区划
administrative department in charge of grassland 草原主管部门
administrative detail 行政区划资料
administrative diagram 行政区划(略)图
administrative division 行政区域;行政区划;行政区;管理区段;行政组;管理组
administrative enclave 飞地
administrative engineer 管理工程师
administrative engineering 经营工程(学);管理工程
administrative expenditures 行政支出;行政管理费
administrative expenses 行政开支;行政管理费;行政费(用);管理费用;办公费
administrative expense audit 管理费用审计
administrative expense budget 管理费用预算
administrative expense statement 管理费用表
administrative facility of road information 道路情报管理设施
administrative fine 行政罚款
administrative function 行政功能
administrative guidance 行政指导
administrative hierarchy 行政等级
administrative improvement 经营管理改善
administrative index 行政区划示意图
administrative information 管理信息
administrative interference 行政干预
administrative intervention 行政干预
administrative jurisdiction of roads 道路的管辖权
administrative law 行政法(令);行政法规
administrative legislation 行政立法
administrative legislation of water 水行政立法
administrative litigation 行政诉讼
administrative machinery 行政机构;管理机构
administrative management service 行政管理处
administrative management theory 行政管理理论
administrative manager 行政经理
administrative map 市政管理图;行政区划图;政区地图
administrative means 行政手段;行政方法
administrative measure 行政手段;行政措施
administrative office 行政机关;办公室
administrative officer 管理员;管理人员
administrative of justice in water 水行政司法
administrative operator 管理操作员
administrative operator station 管理操作员站;管理操作台
administrative organ 行政机关;行政管理机构;办事机构
administrative organization 管理机构
administrative overhead 间接管理费;企业管理费(用);管理间接费
administrative penalty 行政处罚
administrative personnel 行政人员;行政管理人员;管理人员
administrative planning system 经营规划系统
administrative policy 管理政策;管理方针
administrative policy for environmental protection 环境保护行政管理政策
administrative price 管理价格
administrative procedure 管理程序
administrative procedure act 管理程序条例
administrative proceeding 行政诉讼
administrative process 管理过程;管理方法
administrative processor 管理处理机
administrative protection 行政性保护
administrative rate 规定的费率
administrative region 行政区域;专区
administrative regulation 管理制度;行政法规;管理规则;管理法规
administrative regulation for adopting international standards 采用国际标准的管理办法
administrative regulations of water 水行政法规
administrative regulations regarding wastewater 废水管理法
administrative responsibility 行政责任
administrative rule 行政规范;行政法规;管理规则;管理法规
administrative rules of water 水行政规章
administrative sanctions 行政处罚
administrative services 行政性业务
administrative set-ups 行政体制
administrative smuggling 行政走私
administrative software 管理软件
administrative spending 行政管理费
administrative staff 行政人员;行政管理人员
administrative standard 管理标准
administrative statistics 行政统计资料
administrative support 后勤支援;后勤工作;行政支援
administrative system of marine environment 海洋环境管理体系
administrative terminal system 事务终端系统;行政管理终端系统;管理终端系统
administrative time 修理准备时间;管理时间
administrative tribunal 行政法院
administrative unit 行政单位
administrative worker 行政人员
administrative zone 行政地区
administrator 行政人员;行政管理员;行政负责人;管理(人)员;管理人;财产经理人
administrator in a bankrupt estate 破产管理人
administrator of a home 住宅建设管理人;住宅建设管理员
administrator's deed 遗产管理人的契约
adminstrative division 行政部门
adminstrative regulations of environment 环境行政法规
admiral's bridge 司令台
admiralty 海事的;海军部
admiralty action 海事诉讼
admiral(ty)anchor 海军锚
admiralty brass 耐酸黄铜;海军黄铜;船用黄铜
admiralty bronze 海军青铜;船用青铜
admiralty chart 海事图;海图;英版海图
Admiralty Chart Datum 英版海图零点
admiralty cloth 海军呢
admiralty coal 船用煤(炭)
admiralty code 海事法典
admiralty coefficient 海军常数
admiralty constant 海军常数
admiralty court 海事法院;海事法庭
admiralty creeper 探海锚
admiralty dispute 海事纠纷
admiralty fuel oil 船舶燃料油
admiralty gun metal 海军炮铜
admiralty hitch 绳针结
admiralty jurisdiction 海事审判权;海事裁判权;英国海事审判管辖权
admiralty knot 节(航海单位);英国法定海里
admiralty law 海事法;英国海事法规
Admiralty List of Lights and Fog Signals 英版灯标雾号表
Admiralty List of Radio Signals 英版无线电航标表
admiralty measured mile 海军部海里;英制海里(1英制海里=6080英尺,折合1853.24米)
admiralty metal 耐海水金属;海军金属;海军黄铜;船用黄铜;铜锡锌合金
admiralty mile 英(制)海里(1英制海里=6080英尺,折合1853.24米)
admiralty mixture brass 海军混合黄铜
Admiralty Notices to Mariners 英版航海通告
admiralty(pattern)anchor 有杆锚;海军锚
admiralty port 海军要塞
admiralty proceedings 海事诉讼
Admiralty Salvage Organization 英国海军打捞组织
admiralty stockless anchor 海军锚
admiralty test 干湿陈化试验
Admiralty tide tables 英国海军潮汐表;英版潮汐表
admiralty type mooring buoy 英国海军型系泊浮筒
Admiralty unit 英国海军单位
Admiralty visco(si)meter 雷德伍德黏度计
Admiro 阿德密拉铜锌镍合金(48%铜,35%锌,10%镍,2%铝,3%锰,2%铁)
admissibility 可容许性;可采纳性
admissibility in evidence 接受作证据
admissibility of Bayes decision rule 贝叶斯决策规则的宜取性
admissibility of strategies 策略的容许性
admissible 容许;可容纳的;许用
admissible absolute random error 允许绝对偶然误差
admissible action 容许行动
admissible assets 可(课)税资产;认可资产
admissible average content of harmful impurity 有害杂质平均允许含量
admissible basis 可采基
admissible clearance 容许间隙
admissible concentration 容许浓度
admissible control 容许控制
admissible curve 容许曲线
admissible decision 可采纳的决策
admissible decision function 容许判决函数
admissible deviation 容许离差
admissible draught 允许吃水
admissible error 容许误差;允许故障率
admissible estimate 容许估计
admissible estimator 容许估计量
admissible evidence 可靠证据;可接受的证据
admissible failure mechanism 容许破坏机理
admissible function 容许函数
admissible hypothesis 容许假设
admissible interpolation constraint 容许内插约束
admissible length 容许长度
admissible load 容许荷载;容许负载;容许负荷
admissible mark 容许符号;允许数字
admissible maximum of ash content 最高可采灰分
admissible maximum of minable depth 最大可采深度
admissible maximum of outline area 最大轮廓面积
admissible minimum of average grade 最低可采平均品位
admissible minimum of average meter percent 最低米百分数
admissible minimum of caloricity 允许最低发热量
admissible minimum of minable thickness 最低可采厚度
admissible minimum of minable width 最低可采宽度
admissible number 容许数
admissible overload 容许超载
admissible parameter 容许参数
admissible parameter space 容许参数空间
admissible pollutant concentration 容许污染物浓度;允许污染物浓度
admissible prediction function 容许预测函数
admissible random error 允许偶然误差
admissible region 容许区域
admissible relative random error 允许相对偶然误差

admissible relief variation 容许地貌变化
admissible set 容许形状;容许集
admissible situation 宜取形势
admissible solution 容许解(法)
admissible statistical hypothesis 容许统计假设
admissible strategy 容许策略;可采纳的策略;宜取策略
admissible stress 容许应力;许用应力
admissible stress to rupture 容许断裂应力
admissible structure 容许结构
admissible surface 容许曲面
admissible test 容许检验
admissible transformation 宜取变换
admissible value 容许值
admissible value for a player 局中人的容许值
admissible variation 容许变异;容许变分
admission 容许进入;允许;准允;公差
admission cam 进气凸轮;吸进凸轮
admission cam shaft 进气凸轮轴
admission chamber 进入箱;进气室
admission delay 进场延迟
admission department 住院部
admission edge 进汽边
admission free 免费入场
admission gear 进汽装置;进气传动装置
admission intake 进气
admission lead 提前进气;进气导程
admission line 进气线;进气管路;进气管道
admission of goods free of duty 准许免税输入商品
admission of heat 输热;输入热量
admission of liability 认债
admission of light 光的输入;采光
admission of partner 入伙;合伙人入伙
admission of seaworthiness 承认适航
admission of steam 进汽
admission opening 进汽孔;进口
admission passage 进汽管路;进汽道;进气道;进路
admission pipe 进汽管;进气管
admission port 进汽口;进气口;进口;给水口
admission pressure 容许压强;容许压力;进气压力
admission section 进气口截面
admission space 进气空间;进气容积;进气空间;装填体积
admission stroke 进气冲积;进气冲程
admission tax 入场税
admission ticket 入场券
admission valve 进水阀;进入阀;进汽阀;进气阀(门)
admission velocity 入口速度;进入速度
admittance 容差;进入;进汽;公差;电纳;导纳;通道
admittance area 通导截面;流导面积
admittance bridge 导纳电桥
admittance circle 导纳圆
admittance circle diagram 导纳圆图
admittance comparator 导纳比较器
admittance diagram 导纳图
admittance function 通道作用;导纳函数
admittance gap 导纳隙
admittance matrix 导纳矩阵
admittance measuring instrument 导纳测量仪
admittance parameter 导入参数;导纳参量
admittance ratio 导纳比
admittance relay 导纳继电器;导纳继电保护装置
admittance wave 入口波
admitted assets 容许资产
admitted company 准许开业公司
admitted to dealings 准予成交
admitting pipe 输入管;进水管;进入管;进气管
admitting port 进气口
admix 调和物;掺和;废料回炉
admix dosaging equipment 混合配料设备
admixer 掺和器;混合器
admixture 外加剂;添加剂;混流;混合材;混质;掺和物;掺和料;掺和剂;附加剂
admixture agitator 外加剂搅拌器
admixture dosage 外掺剂量;单位掺和料量
admixture effect 外加剂效果
admixture heat 掺和热
admixture instability 混合不稳定性
admixture (of concrete) (混凝土的)掺加料
admixture of hard and soft layers 软硬互层
admixture result 外加剂效果
admixtures for concrete 混凝土掺和料

admixture stabilization 掺拌加固法;拌和加固(法)
admixture stabilization of skin 表层拌和法
admixtures to architectural concrete 装饰混凝土外加剂
admonish 警戒
admonition 警告
admontite 水硼镁石
Admos die casting 阿德莫斯排溢铸造(法)
adnate alage 联生藻类;并生藻类
adnate hydrophyte 固着水生植物
adnation 贴生
adnexa 附器
adnexed 附着的
Adnic 海军镍;阿德尼克铜镍合金
adobe 土筑墙;土坯(房屋);瘦黏土;灰质黏土;砖坯;风干砖(坯);风干土坯;粉黏土;土坯砖
adobe blasting 外部爆破;裸露爆破;泥盖(爆破)法;糊炮爆破
adobe block 土坯块
adobe brick 土坯砖;砖坯;风干砖;不烧的砖
adobe brick wall 土坯墙
adobe brown 土棕色
adobe building 土坯建筑;土坯房
adobe clay 灰质黏土;制砖黏土
adobe clay block 风干坯砖
adobe clay brick 风干土坯砖
adobe(clay)construction 土坯结构;土坯构造;土坯建筑;干打垒;土墙房屋
adobe flat 黏粉土沉积平原
adobe house 土坯房屋
adobe masonry 土砖圬工;土坯圬工;土坯砌体
adobe shooting 外部装药爆破(法)
adobe soil 灰质黏土;制砖土;龟裂土;龟裂地;冲积黏土
adobe wall 土坯墙
adohero ga(u)ge 漆膜附着力测定仪
adolescence 青年期
adolescent 青年期的;成材木
adolescent coast 少壮海岸
adolescent river 青年河;少壮河
adolescent stream 青年河
adolescent valley 青年期河谷;青春期河谷
Adonic 铜镍锡合金
adonic acid 阿东酸
adopt 采取
adopted base-line value 采用基线值
adopted datum 采用基(准)面
adopted latitude 纬度采用值
adopted longitude 经度采用值
adopted plan 拟定的计划;采用的规划
adopted scale 采用比例
adopted value 采用值
adopter 接受管
adopting by equivalent 等效采用
adopting by reference 参照采用
adopting safety measures in productive technologies 安全技术组织措施
adoption 选用;采用;采纳
adoption of agenda 通过议程
adoption of existing standards 采用现行标准
adoption of indigenous method 土法上马
adoption of innovation 采纳革新
adoption of international standards and advanced oversea standards 采用国际标准和国外先进标准
adoption of mew technology 采用新技术
adoption process 采购程序
adoption society 团体群聚
adoptive transfer 继承性转移
adopt thickness 采用厚度
adorn 装饰;修饰
adorned 装饰的;富于装饰的
adornment 饰物;饰品;盛饰;装饰物;装饰品
adorsed 背靠背动物图案雕饰
a dozen(of) 一打
adpedance 导纳阻抗
ad-radius 从辐管
adrasion resistance of thin-film 薄膜的抗磨强度
ad referendum 还要斟酌;有待核准;待进一步审议
ad referendum contract 暂定合同;草约;暂定契约
adret(to) 阳坡
adrift 漂浮
A-drill rod 外径 15/8 寸钻杆

adrollal 乙酸环己酯
adscape 广告景观
adsolubilization 叠加溶解作用
adsorb 吸附
adsorbability 可吸附性;吸附性;吸附能力
adsorbability material 吸附性材料
adsorbable dissolved organic carbon 可吸附溶解有机碳
adsorbable ion 可吸附离子
adsorbable molecule 可吸附分子
adsorbable organic halogen compound 可吸附卤素化合物
adsorbable organic halogenide 可吸附有机卤化物
adsorbable organic sulfur compound 可吸附有机硫化合物
adsorbable species 可吸附物种
adsorbate 吸附质;被吸附物
adsorbate accumulation 吸附质累积
adsorbate activity 吸附质活度
adsorbate concentration 吸附质浓度
adsorbate molecule 吸附质分子
adsorbed 吸附的
adsorbed amount 吸附量
adsorbed argon 吸附氩
adsorbed cations 吸附的正离子;吸附的阳离子
adsorbed charge 吸附电荷
adsorbed concentration 吸附浓度
adsorbed detector 吸附检测器
adsorbed film 吸附膜
adsorbed film of water 吸附水膜
adsorbed gas 吸附气体
adsorbed ions 吸附离子
adsorbed layer 吸附层
adsorbed material 吸附材料
adsorbed matter 吸附材料
adsorbed moisture 吸附水分;吸潮
adsorbed molecule 吸附分子
adsorbed molecule bind 吸附分子接合
adsorbed natural organic matter 吸附天然有机质
adsorbed oil 吸附油
adsorbed phase concentration 吸附相浓度
adsorbed phosphorus 吸附磷
adsorbed sodium 吸附性钠
adsorbed solute 吸附溶质
adsorbed species 吸附物种
adsorbed state 吸附(状)态
adsorbed substance 吸附物质
adsorbed surface film 吸附表面膜
adsorbed water 吸附水
adsorbed water with a meniscus 带弯月面吸附水
adsorbency 吸附性
adsorbent 吸附剂
adsorbent activity 吸附剂活性
adsorbent-adsorbate pair 吸附剂—吸附质对
adsorbent bed 吸附剂床
adsorbent coated glass strip 涂吸附剂的玻璃条
adsorbent concentration 吸附(剂)浓度
adsorbent deactivator 吸附减活剂
adsorbent dose 吸附剂剂量
adsorbent equipment 吸附装置
adsorbent filter 吸附剂过滤器
adsorbent filtering medium 吸附(剂)过滤介质
adsorbent gradient 吸附梯度
adsorbent-loaded paper 带吸附剂纸
adsorbent material 吸收材料
adsorbent-membrane system 吸附剂膜系统
adsorbent modifier 吸附改性剂
adsorbent particle 吸附剂粒子
adsorbent phase 吸附剂相
adsorbent pressure 吸附压力
adsorbent regeneration 吸附剂再生
adsorbent resin 吸附树脂
adsorbent selection 吸附选择
adsorbent surface area 吸附剂表面积
adsorbent temperature 吸附温度
adsorber 吸附剂;吸附装置;吸附器
adsorbing 吸收的
adsorbing agent 吸附剂
adsorbing capacity 吸附量
adsorbing clay 吸附黏土
adsorbing colloid flo(a)tation 吸附胶体浮选
adsorbing column 吸附柱
adsorbing group 吸附基
adsorbing impurity 吸附杂质

adsorbing layer 吸附层
adsorbing material 吸附剂
adsorbing medium 吸附剂
adsorbing performance 吸附性能
adsorbing substance 吸附剂
adsorptiometric analysis 吸收测量分析
adsorption 吸收作用;吸收(法);吸附作用
adsorption activity 吸附活度
adsorption affinity 吸附力
adsorption agent 吸收剂;吸附剂
adsorption air conditioning 吸收式空调设备
adsorption amount 吸附量
adsorption-anaerobic sequencing batch reactor 吸附厌氧序批间歇式反应器
adsorption analysis 吸附分析
adsorption apparatus 吸收试验仪
adsorption area 吸收面积
adsorption balance 吸附天平
adsorption balance time 吸附平衡理论
adsorption band 吸收谱带;吸收频带;吸收(光谱)带;吸收波段;吸附谱带
adsorption barrier 吸附障
adsorption bed 吸收床;吸附床
adsorption behavio(u)r 吸收性能
adsorption biodegradation activated sludge method 吸附生物降解活性污泥法
adsorption biodegradation process 吸附生物降解法
adsorption bio-oxidation process 生物吸附曝气氧化法;吸附生物氧化法
adsorption bubble separation 吸附气泡分离法
adsorption bulb 吸收瓶
adsorption capacity 吸收能力;吸收本领;吸附容量;吸附能力
adsorption capacity index 吸附容量指数
adsorption capacity of colloids 胶体吸附容量
adsorption catalysis 吸附催化
adsorption cell 吸收池
adsorption centre 吸收中心
adsorption characteristic 吸附特征
adsorption chromatography 吸收色谱学;吸附色谱分离(法);吸附色谱法;吸附色层分离法;吸附层析
adsorption clarifier 吸附澄清剂
adsorption cleaning 吸附式净化
adsorption coefficient 吸水系数;吸收系数;吸收率;吸附系数
adsorption colo(u)r 吸收色
adsorption column 吸收柱;吸附柱
adsorption column chromatography 吸附柱色谱法
adsorption compound 吸附化合物
adsorption constant 吸附常数
adsorption cooling 吸收(式)供冷
adsorption correlation 吸附容量相关
adsorption cross-section 吸收横断面
adsorption crystal spectrum 吸收晶体光谱
adsorption current 吸收电流(非完全介质中);吸附电流
adsorption curve 吸收曲线;吸附曲线
adsorption cycle 吸附周期
adsorption-decay steady state 吸附衰减稳态
adsorption denitrification 吸附法脱氮
adsorption density 吸附密度
adsorption-desorption cycle 吸附及脱附周期
adsorption-desorption equilibrium 吸附及脱附平衡
adsorption-desorption of catalyst 催化剂的吸附与脱附
adsorption-desorption-regeneration-readsorption 吸附—脱附—再生—再吸附
adsorption device 吸收装置
adsorption differential pulse voltammetry 吸附差示脉冲伏安法
adsorption dynamics 吸附动力学
adsorption dynamometer 吸收测功器
adsorption edge 吸收限;吸收边缘;吸附限
adsorption effect 吸收作用;吸附效应
adsorption efficiency 吸收效率
adsorption elution 吸附洗脱
adsorption energy 吸附能量
adsorption equation 吸附公式
adsorption equilibrium 吸附平衡
adsorption equilibrium constant 吸附平衡常数
adsorption equilibrium theory 吸附平衡理论

adsorption equilibrium time 吸附平衡时间
adsorption equipment 吸附装置
adsorption exponent 吸收指数
adsorption factor 吸收因子;吸收因素;吸收因数;吸收系数
adsorption film 吸附膜
adsorption filtration 吸附过滤
adsorption frequency 吸收频率
adsorption from solution 溶液吸附
adsorption function 吸收函数
adsorption gas 吸附瓦斯;吸附气体
adsorption gas chromatography 气相吸附色谱(法);气相吸附层析;吸附气相色谱法
adsorption gasoline 吸附汽油
adsorption gas refrigerator 吸收式燃气制冷机
adsorption heat 吸附热
adsorption heat detector 吸附热检测器
adsorption hologram 吸收全息图
adsorption hygrometer 吸收湿度计;吸收湿度表;吸附湿度表
adsorption index 吸收指数
adsorption inhibitor 吸附型抑制剂;吸附型钝化剂
adsorption isobar 吸附等压线;等压吸附线
adsorption isostere 吸附等容线
adsorption isotherm 吸收等温线;吸附等温线;等温吸附线
adsorption isotherm equation 吸附等温方程
adsorption isotherm isopotential 吸附等温等势线
adsorption kinetics 吸附动力学
adsorption kinetics curve 吸附动力学曲线
adsorption kinetics equation 吸附动力学方程
adsorption layer 吸附层
adsorption limit 吸着界限;吸附限
adsorption line 吸收(谱)线
adsorption liquid 吸收液
adsorption load curve 吸附负荷曲线
adsorption loss 吸着损失;吸收损失(量);吸收损耗
adsorption mechanism 吸附机理
adsorption membrane 吸附膜
adsorption method 吸收法;吸附法;表面吸附法
adsorption model 吸附模式
adsorption modeling 吸附模拟
adsorption-moisture 吸附水
adsorption of bacteria 细菌吸附作用
adsorption of detergent 洗涤剂吸附
adsorption of discrete ion 离散离子吸附
adsorption of foreign investment 吸收外资
adsorption of gas and vapo(u)r 气体吸收
adsorption of ion 离子吸附
adsorption of labo(u)r power 吸收劳动力
adsorption of pollutant in water body 水体污染吸附
adsorption of solute 溶质的吸附
adsorption oil 吸收油
adsorption pan 吸收盘
adsorption pattern 吸附非方程式;吸附方式
adsorption performance 吸附性能
adsorption phenomenon 吸附现象
adsorption photometry 吸光测定法
adsorption pipe 吸收管
adsorption pit 吸水坑;渗滤坑;吸收坑
adsorption plant 吸收装置
adsorption plate 吸附板
adsorption point 吸附点
adsorption potential 吸附势;吸附电位
adsorption potential theory 吸附势理论
adsorption power 吸收能力;吸收功率;吸附力
adsorption precipitant 吸附沉淀剂
adsorption precipitation 吸附沉淀
adsorption pressure 吸附压力
adsorption principle 吸附原理
adsorption process 吸附过程;吸附(处理)法;吸附法
adsorption process in wastewater treatment 废水吸附处理法
adsorption process of wastewater 污水吸附法
adsorption property 吸收性能;吸附性能
adsorption radon technique 吸附氡法
adsorption rate 吸收(比)率;吸收(速)率
adsorption rate constant 吸附速率常数
adsorption rate theory 吸附速率理论
adsorption ratio 吸附比
adsorption reaction 吸附反应
adsorption reactor 吸附反应器

adsorption refining 吸附精制
adsorption refrigerating machine 吸收式冷冻机
adsorption refrigeration 吸收式制冷
adsorption refrigeration cycle 吸收式制冷循环
adsorption refrigeration machine 吸收式制冷机;吸收式冷冻机
adsorption refrigeration system 吸收式制冷系统
adsorption refrigerator 吸收式冰箱
adsorption regeneration activated sludge process 吸附再生活性污泥法
adsorption regeneration aeration 吸附再生曝气
adsorption regeneration process 吸附再生法
adsorption resin 吸附树脂
adsorption screen 吸光屏
adsorption section 吸附段
adsorption selection 吸附选择
adsorption selectivity 吸附选择性
adsorption separation 吸附分离
adsorption silencer 吸收式消音器
adsorption site 吸附位置
adsorption solution 吸收溶液
adsorption space 吸附空间
adsorption spectrochemical analysis 吸收光谱化学分析
adsorption spectrogram 吸收光谱图
adsorption spectrograph 吸收摄谱仪
adsorption spectrometer 吸收光谱仪;吸附分光计
adsorption spectrometry 吸收光谱(测定)法
adsorption spectroscopy 吸收光谱学
adsorption spectrum 吸收频谱;吸收光谱
adsorption stage 吸收阶段
adsorption strength 吸附强度
adsorption stripping 解吸;吸附气提
adsorption substance 吸附物质
adsorption surface 吸收面
adsorption surface area 吸收表面积
adsorption swelling 吸水膨胀
adsorption system 吸收系统;吸附系统
adsorption tank 吸收池;吸附罐;吸附池;吸附槽
adsorption terminology 吸附术语
adsorption test 吸收试验;吸附试验
adsorption themodynamics 吸附热力学
adsorption theory 吸附理论
adsorption time 吸附时间
adsorption tower 吸收塔;吸附塔
adsorption trap 吸附罩
adsorption treatment 吸收处理
adsorption treatment of wastewater 废水吸附处理法
adsorption trench 吸收(沟)槽
adsorption tube 吸附管
adsorption type liquid chiller 吸收式液体冷却装置
adsorption-type refrigerating machine 吸收式制冷机
adsorption type terrace 吸水式梯田
adsorption unit 吸收装置
adsorption velocity 吸附速度
adsorption velocity theory 吸附速度理论
adsorption vessel 吸收器
adsorption volume 吸附体积
adsorption water 束缚水;结合水;薄膜水;吸附水;吸着水;附着水
adsorption water content 吸附含水量
adsorption wavemeter 吸收式波长计
adsorption well 吸水井
adsorption zone 吸收区(域);吸收带;吸附区
adsorptive 吸附的;被吸附物
adsorptive action 吸附作用
adsorptive agent 吸附剂
adsorptive attraction 吸附引力
adsorptive bubble technique 吸附气泡技术
adsorptive capacity 吸附(容)量;吸附能力
adsorptive catalyst 吸附催化剂
adsorptive displacement 吸附置换
adsorptive equilibrium constant 吸附平衡常数
adsorptive filter 吸附性过滤器
adsorptive filtration 吸附(性)过滤
adsorptive force 吸附力
adsorptive form 吸收性模板
adsorptive hull 吸附罩;吸收薄膜;吸附(薄)膜
adsorptive interaction 相互吸附
adsorptive load curve 吸附负荷曲线
adsorptive material 吸附性材料

adsorptive matter 吸附物
adsorptive media 吸附介质
adsorptive performance 吸附性能
adsorptive potential theory 吸附势理论
adsorptive power 吸附能力
adsorptive precipitation 吸附沉淀
adsorptive process 吸附法
adsorptive property 吸附性能
adsorptive region 吸附区
adsorptive resin 吸附树脂
adsorptive solid-solution interface 吸附固体—溶液界面
adsorptive support 吸附性载体
adsorptive surface 吸附表面
adsorptive system 吸附系统
adsorptive thrust force 吸附推动力
adsorptive time 吸附时间
adsorptive transition 吸收过渡
adsorptive value 吸收值;吸附值
adsorptivity 吸附性;吸附率;吸附度
ADT [Average Daily Traffic] 日平均交通量
adularescence 冰长光彩
adular(ia) 冰长石;低温钾长石
adularization 冰长石化
adult 成体;成年
adult age 成熟期
adult barnacle 藤壶成虫
adult bookstore 成人书店
adult death rate 成人死亡率
adulterant 掺杂用的;混杂料;掺杂物;掺杂剂;掺混物;掺和剂
adulterant process 掺杂过程
adulterate 低劣;掺假
adulterated 掺杂的
adulterated food 掺假食品
adulterated goods 冒充货
adulterated oil 污油;掺杂油
adulterated water 含杂质的水
adulteration 劣等货;伪造;劣质品;冒牌货;掺杂;掺假
adultered food 掺杂食品
adult fish 成鱼
adult form 成体型
adult nucleus 成年核
adult phase 成年阶段
adult population 成年种群;成年人口;成年群体
adult stage 成熟期;成年期
adult suite 套房
adult theater 成人戏院
adult ticket 成人票
adult wood 成年林
adumbral 阴暗的;遮阳的
adumbrate 画轮廓
adumbration 素描;草图;遮阳
adurol 氯对苯二酚显影剂
adustion 烘焦
adustion of timbering 木支架起火
ad valorem 贵重货物运费;从价;按价格(计税)
ad valorem duties 从价税
ad valorem duty 按价税
ad valorem freight 从价运费
ad valorem method 从价法
ad valorem rate 从价费率
ad valorem taxation 从价课税
advance 前移;前进;推进;升高;进展;回转纵距;预支;垫付;超前;分条移档
advance A 卷片方向
advance account 预付款账户
advance acquisition 先行获取
advance against collection 托收垫款
advance against goods 用商品抵押的预付款
advance agent 先遣人员
advance allocation 预付款项
advance and final arrival notice of train 到达火车预确报
advance angle 提前角
advance appropriation 提前拨款
advance arrival notice of ship 船舶到港预报
advance ball 滑动滚珠
advance beam 超前梁
advance bill 先开汇票;发货前开出的账单
advance billing 预开账单
advance block signal 预告比塞信号
advance business 贷款业务
advance buying 预先采购
advance by hand 人工掘进
advance by machine 机械掘进
advance by overdraft 透支
advance call 预付保险费
advance capital 垫(付资)本
advance change notice 更改的先期通知
advance charges 预收费用;预付款
advance chart 现势图
advance coefficient 前进系数;进速系数
advance collection 预收款项
advance collection account 预收货款账户
advance consignment-in 带销商品预付;带托销商品预付
advance consignment-out 托销商品预收款
advance constant 前进系数
advance control 步进控制
advance control mechanism 点火提前控制机构
advance control of cost 成本预先控制
advance copy 预印本;样本
advance corporation tax 预缴公司税
advance cut 超前沟
advanced 前置;提前的;先进的
advanced absorption 快速吸附
advanced administrative system 先进管理系统
advanced algebra 高等代数(学);大代数
advanced announcing post 预告车标(列车进站)
advanced architecture 先进建筑
advanced automatic compilation system 最新自动编图系统
advanced average 先进平均数
advanced average quota 平均先进定额
advanced ballistic missile defense agency 高级弹道导弹防御局
advanced battery 高级电池
advanced bill of lading 预借提单
advanced biochemical technique 高级生化技术
advanced biochemistry 高级生化化学
advanced biological enhanced nutrient removal process 强化营养去除高级生物工艺;高级强化生物除营养物工艺
advanced bore(hole) 超前钻孔;超钻
advanced calculator 高级计算器
advanced camera 高级照相机
advanced capital 预付资本
advanced ceramics 现代陶瓷;先进陶瓷;高级陶瓷
advanced charges 预付费用
advanced communication function 高级通信功能
advanced communication system 高级通信系统
advanced composite material 现代复合材料;先进复合材料
advanced computational processor 先进计算处理机;高级计算处理机
advanced computing environment 先进计算环境
advanced configuration and power interface 先进配置和电源管理界面
advanced control 先行控制;超前控制
advanced control system 高级控制系统
advanced converter 先进转换堆
advanced copy 试行本;样本
advanced cost accounting 高等成本会计
advanced country 先进国家
advanced course 高级课程;高级教程
advanced data base system 先进数据库系统;高级数据库系统
advanced data link controller 高级数据链路控制器
advanced data management system 高级数据管理系统
advanced data system 高级数据系统
advanced decay 深度腐朽;早期腐朽
advanced department 贷款部
advanced design 先进设计
advanced development 研制;试验研究;试制
advanced development objective 远景发展规划
advanced development of model machine 样机试制
advanced development of sample 样品试制
advanced diver 高级潜水员
advanced diving system 先进潜水系统;高级潜水系统
advanced drill hole 超前钻孔
advanced drinking water treatment process 深度饮用水处理工艺
advanced economy 发达的经济
advanced effluent treatment 深度污水处理
advanced engineering and design 超前工程设计
advanced engineering plastics 先进工程塑料
advanced environment(al) control system 先进环境控制系统
advance department 放款部;货款部
advance deposit 进口保证金;预缴押金
advanced equipment 先进设备
advanced estimate 事先估计值
advance development 超前掘进
advance dewatering 预先疏干
advanced experience 先进经验
advanced exploration 初勘
advanced facultative pond 高效兼性塘
advanced feed 前置导孔带
advanced feed-hole 前置导孔
advanced feed tape 前置导孔纸带
advanced fiber reinforced composite 高级纤维增强复合材
advanced financial programmable calculator 高级财务可编程序计算器
advanced flexible reusable surface insulation 高级柔性防热材料
advanced forecasting techniques 高级预测技术
advanced forecast observation system 高级预报观测系统
advanced forward looking infrared 高级红外前视仪器(系统)
advanced freight 预付运费
advanced fry 后期鱼苗
advanced function for communication 高级通信功能
advanced gallery 超前小平巷
advanced gas-cooled reactor 改进型气冷反应堆;先进气冷反应堆;高级气冷却反应器;高级气冷反应堆
advanced gear 进给机构
advanced glass 先进玻璃;高级玻璃
advanced graphic display controller 高级图形显示控制器
advanced headway distribution 改进型车间分隔分布
advanced house 先进房屋
advance diesel(road)roller 前行式柴油压路机
advanced igniter 提前点火器
advanced ignition 提前点火
advanced information society 高度情报化社会
advanced information system 高级信息系统
advanced information technology 高级信息技术
advanced in methodology for research 研究方法的进展
advanced inorganic material 无机新
advanced instrument 先进仪器
advanced integrated data system 高级综合数据系统
advanced integrated light meter 高级积分曝光表
advanced interactive debugging system 先进交互调试系统
advanced interactive executive 高级交互执行程序
advance direction(al)sign 前置指路标志;方向预告标志
advance distance 前置距离
advanced landing ground 前置飞机场;前方飞机场;军用机场;前置降落场;前方降落场
advanced light rapid transit 新型轻轨快速运输系统
advanced line of position 超前位置线
advanced lines 高世代
advanced map depot 辅助图库
advanced material 先进材料;高级材料
advanced maturation pond 高效熟化塘
advanced modeling extension 高级建模扩展
advanced network system architecture 先进网络系统体系
advanced nitrogen removal 深度脱氮
advanced norm 先进定额
advanced ocean drilling project 高级深海钻探计划
advanced optic(al)character reader 高级发光字符读出器
advanced oscilloscope 高级示波器
advanced over-burden 超额剥离量
advanced oxidation process 高级氧化工艺
advanced passenger train 高速旅客列车
advanced payment 预付款
advanced payment audit 预付款审计

advanced payment for construction works 建筑工程预付款
advanced photo oxidation 高级光氧化
advanced planning 制定远景规划
advanced plant 先进设备
advanced potential 提早势;超前势;超前电位
advanced power management 高级电源管理
advanced preparation 事前准备
advanced price 预付价格;已上涨的价格
advanced process 先进工艺
advanced programming 高级程序设计
advanced project 尖端计划;远景规划;研究计划
advanced proof 初样图
advanced qautic biology 高级水生生物学
advanced quantitative analysis 高等定量分析
advanced query 高级查询
advanced radar traffic control system 高级雷达交通管制系统
advanced reclamation treatment 先进的回收处理;高级回收处理
advanced record system 先进记录系统
advanced redemption 提前偿还
advanced reentry system 高级再入系统
advanced research 预研;远景研究
advanced research division 远景研究部
advanced research monograph 最新研究报告集
Advanced Research Projects Agency 美国高级研究项目机构
advanced reserves 高级储量
advanced resistry 高级注册
advanced school building 高等学校建筑
advanced science 尖端科学
advanced scientific computer 先进科学计算机;高级科学计算机
advanced scientific programmable calculator 高级科研可编程序计算器
advanced secondary treatment 二级深度处理
advanced sheet 样图
advanced shipboard command communication system 高级舰船通信系统
advanced ship's operation control system 先进船舶操纵控制系统
advanced sign 前置标志
advanced signal processor 高级信号处理机
advanced solid logic technology 先进固体逻辑技术
advanced spectroscope technique 高级光谱技术
advanced stage 晚期;深度裂化阶段
advanced stage of decay 后期腐朽阶段
advanced starting valve 预升起动阀
advanced statistical analysis program(me) 高级统计分析程序语言
advanced statistical analysis program(me) language 高级统计分析程序语言
advanced statistics 高级统计学
advanced strains 高级品系
advanced surveillance radar 高级监视雷达
advanced synchronous meteorological satellite 高级同步气象卫星
advanced technique 先进技术
advanced technique circuit module 先进技术电路模块
advanced technology 先进技术
advanced technology satellite 先进技术卫星;高级技术卫星
advanced text management system 高级文本管理系统
advanced thermal reactor 转换反应堆
advanced thinning 前期疏伐
advanced timing 超前成圈
advanced trade balance 入超
advanced training machine 高级教练机
advanced transport of products 物资提前运输
advanced treatment 完全处理;深度处理;高级处理
advanced treatment method 高级处理法
advanced treatment of domestic wastewater 生活污水深度处理
advanced treatment of secondary effluent 二级出水深度处理
advanced treatment of treated sewage 处理过的污水深度处理
advanced treatment process 深度处理过程;深度处理法
advanced unmanned search system 先进的无人调查系统
advanced user interface 高级用户接口
advanced version 改进型
advanced very high resolution radiometer 高级甚高分辨率辐射计;超高分辨辐射计
advanced VHF communication system 先进甚高频通信系统
advanced vidicon camera system 最新光导摄像管摄像机系统
advanced wage 预付工资
advanced waste(water)treatment 先进污水处理(法);废水深度处理;废水三级处理;废水高级处理;污水深度处理;深度废水处理
advanced water technology 先进废水处理技术
advanced water treatment 污水三级处理法;高级废水处理
advanced water treatment laboratory 废水高级处理实验室
advanced water treatment program(me) 废水高级处理方案
advanced wave 前进波;推进波;进行波;移动波
advanced ways of working 先进工作法
advance education center 预培训中心
advance estimate 事前估计(值);预先估值;预先估计
advance exploration 初步勘探;初步勘察
advance face 超前工作面
advance fee 预付金
advance fitting-out 提前舾装;预先舾装
advance for construction 施工预付款
advance for work in process and construction materials 预付工料款项
advance frame 超前帧
advance freight 预付运费
advance from customers 顾客预付款
advance fund 预付经费;超前经费
advance funding 基金预支;预拨资金
advance gallery 超前巷道
advance geologic(al)control 掘进地质指导
advance grouting 超前工作面灌浆
advance growth 前生树
advance heading 超前开拓道路;前置的开拓道路;超前平巷;超前工作面;前置的开挖道路
advance in path 光程提前量
advance(in)price 提价;涨价
advance in the world 发迹
advance irrigation 提早预灌;预灌溉;早期灌溉
advance item 前置项目;超前项;步进项目
advance key 前置键
advance line of position 船位线超前
advance load for mobilization 预付款
advance maintenance 超前维护
advance material order 预付款器材订单
advance material request 先期材料申请
advancement 先进;垫件;前进运动;提升;进展;进刀;预付
advance metal 前进铜镍合金;高比阻铜镍合金
advance mining 前进式开采
advance money on a contract 合同预付定金;预付定金
advance note 预支通知单
advance notice 预先通知;预告
advance of face 工作面推进
advance of flood wave 洪水波推进
advance of ice cover 冰层运动;冰层推进
advance of periastron 近星点进动
advance of sea 海侵;海进
advance of shoreline 滨线推进
advance of the face 工作面推进;工作面进度
advance of the perihelion 近日点进动
advance of tool 进刀;钻具给进
advance of wages 提高工资
advance on construction 预付建筑工程款
advance on maintenance contract 维修合同预付款
advance on sale 预收货款
advance order 预付单
advance pack 超前充填带
advance packing credit 货物装船前预付款
advance payment 预付(款);提前支付;预付项;预支
advance payment bond 预付金
advance payment for contract 承包预付款
advance payment for materials 预付材料款
advance payment guarantee 预付款保函
advance payment of premium 预缴保险费
advance payment on construction 预付建筑工程款
advance payments for future purchase 预购定金
advance per attack 每循环进尺;一个循环的进度
advance per round 每循环进尺;一个循环的进度
advance preparation 前期准备工作;提前准备;事先准备;预先准备
advance print 试样(指印刷);样本(指印刷)
advance pulse 推进脉冲
advance quantity 掘进量
advancer 超前补偿器;超前补偿机
advance ramp control warning sign 匝道控制前预告标志
advance range counter 提高标度计数器
advance rate 钻进速度
advance received 暂收款
advance received account 暂收款账户
advance received on contract 按合同预收款项
advance refunding 提前转期偿还(美国公债);提前转期
advance release 预泄(洪水)
advance remittance 预付汇款
advance rental payment 预付租金
advance repayment 预先清偿
advance replacement 预更换
advance report 先期报告;预报通知
advance report notice 预报通知单
advance report on available seating accommodation on trains by stations and train conductors 站车预报
advance ring 提前点火环
advance sample 货样;先发样品
advances from customers 客户预付款
advance sheets 样本
advance sign 前置(交通)标志;预告标志
advance signal 前置信号(机)
advance signal of computer 计算机推进信号
advance slope grouting 预置集料横向压力灌浆;斜向压力灌浆;预置骨料横向压力灌浆
advance slope method 混凝土斜面浇灌法
advances on materials 预收工料款;预付备料款
advance speed 推进速度
advances received on consignment-out 发出委托销售商品暂收款
advance starting signal 总出站信号机
advance staysail 预加支索帆
advances to subcontractors 预付转承(包)者款
advance street-name sign 前置路名标志
advance stripping 超前剥离
advance support 超前支护
advance surveying 高等测量学
advance the speed 增加速率
advance timbering 超前支架
advance time 全程航行历时
advance timing 提前定时;导前
advance to notice-to-mariners 航海预告
advance to officers and employees 职工预支
advance to subcontractor 分包者预支款;分包者预付款;分包商预支款;分包商预付款;分包人预支款;分包人预付款
advance to the supplies 预付货款
advance warning 提前报警;事前警告
advance warning device 提前警告装置
advance warning sign 前置警告标志
advance warning system 安全预警系统;安全警报系统
advance warning time 洪水预警时间
advance water-detecting hole 超前探水孔
advance wave 超前波
advance workings 前进式开采面;超前巷道;长壁前进式全面开采法
advancing 前进式开采法;进桩;正车;超前
advancing against current 逆流前进
advancing against stream 逆流前进
advancing blade 前进桨叶
advancing boring 超前钻探
advancing colo(u)r 暖色;前进色;似近色
advancing contact angle 前进接触角
advancing crack 缓慢开展中的裂缝
advancing delay factor 前移延迟系数
advancing drift 导坑
advancing drilling 超前钻探
advancing edge 前缘;前沿
advancing front 前沿

advancing gallery 前导坑
advancing glacier 前进冰川
advancing load stress 升高荷载应力
advancing longwall mining 前进式长壁开采法；前进式长壁采煤法
advancing longwall system 建筑物的突出部分（外伸悬臂梁等）；突出建筑
advancing shock front 前进激波前
advancing side 皮带张紧侧；受拉部分
advancing side of belt 皮带紧边
advancing slope grouting 横向灌浆法；斜坡推进灌浆（法）
advancing slope method 斜坡推进法（一种水平移动式浇注隧道衬里方法）；衬砌灌注法
advancing velocity front 峰面推进速度
advancing wave 前进波
advantage 利益；有利条件；优势；优点
advantage in competition 竞争优势
advantage of economic integration 经济联合优越性
advantage ratio 有利比
advantages and disadvantages 优缺点
Advastab 艾德发稳定剂
advection 平流；水平对流
advection fog 平流雾
advection heat 对流热
advection inversion 平流逆温
advection layer 平流层(指大气)
advection radiation fog 平流辐射雾
advective cooling 平流冷却
advective region 平流区
advantage of certain tax benefits 免税好处
adventitious 偶发的；不定株的
adventitious fiber root 不定须根
adventitious form 不定型
adventitious phase 不定期
adventitious root 不定根
adventitious shoot 不定枝
adventitious sound 附加音
adventitious variation 不定变异
adventive 不定株
adventive cone 寄生(火山)锥
adventive crater 寄生(火山)口
advent of freshet 洪水演进
adventure 投机事业；投机(买卖)；商业投机；由托运人自行承担危险的货物
adventure ground 探险游乐场
adventure playground 勇敢者游戏场；惊险游乐场
adversaria 杂记；备忘录
adversary 对手；对方当事人
adverse atmospheric condition 不利大气条件
adverse balance 贸易逆差
adverse balance of payment 支付逆差
adverse balance of trade 贸易逆差
adverse clearing 不利的票据交换
adverse climate change 不利的气候变化
adverse climatic condition 不良气候条件
adverse condition 不利条件
adverse current 逆流
adverse effect 逆效应；坏作用；副作用；反作用；反效应；不良影响；不利影响
adverse environment(al) effect 负面环境效应；对环境不利的影响
adverse environment(al) impact 负面环境影响
adverse exchange 逆汇
adverse factor 有害因素；不利因素
adverse field 反转场
adverse geologic(al) phenomenon 不良地质现象
adverse grade 逆坡；逆比降；倒坡；反向坡度；反倾斜；反坡；反比降
adverse grade for safety 反坡安全线
adverse gradient 倒坡
adverse health effect 对健康的不利影响
adverse impact 不利影响
adverse land use 有害的土地使用
adverse lighting condition 逆光条件
adverse operating condition 不利的营运条件；不利的运行情况
adverse opinion 反面意见
adverse opinion report 否定意见报告书
adverse physical condition 不利的自然条件
adverse physical congestion 自然条件障碍
adverse physical obstructions 不利的外界障碍
adverse physical obstructions or conditions 不利的外界障碍或条件
adverse possession 非法占有
adverse pressure 反压力
adverse pressure gradient 反(向)压力梯度
adverse pricing structure 逆反价格体系
adverse pricing system 逆反价格体系
adverse reaction 有害反应
adverse selection 逆选择；不利的选择
adverse separation practice 选矿后作业
adverse slope 逆坡；逆比降；倒坡；反坡；反比降
adverse slope terrace 倒坡梯田；反坡梯田
adverse trade balance 逆差；入超；贸易逆差
adverse visibility condition 不利的视见条件
adverse weather 坏天气；反常天气；恶劣天气；不良天气；不良气候；不利天气
adverse weather condition 不良气候条件
adverse weather lamp 防风灯；防雾前灯
adverse wind 逆风；迎面风
adverse witness 对方证人
advertise 通知
advertised depth 公布的水深
advertisement 通知；广告
advertisement board 广告牌
advertisement boarding 杂贴广告的临时棚栏
advertisement charges 广告费
advertisement curtain 舞台广告幕；广告幕
advertisement for bids 招标公告；公开招标(建筑工程)；招标广告；公开询价
advertisement lighting 广告照明
advertiser 信号装置；登广告者
advertising agency 广告公司
advertising board 广告牌
advertising board holder 公告牌支架
advertising colo(u)r 广告色
advertising encroachment 广告破坏沿路景色；广告侵占(公路用地)
advertising expenses 广告费
advertising for bid 招标公告
advertising kiosk 广告亭
advertising lamp 广告灯
advertising lamp box 广告灯箱
advertising lighting 广告照明
advertising map 宣传(地)图
advertising paper 广告纸
advertising sign 广告牌；广告标志
advertising structure 广告构筑物
advertising tower 广告塔
advertising wall 广告墙
advertize 通知
advice 通知(书)；建议
advice added to delineation 略图参考说明
advice and pay 通知交款；通知(并)付款
advice boat 通信船；联络桥；联络船
advice book 通知登记簿；汇票通知簿
advice for collection 托收委托书；托收通知书
advice note 邮包通知书；汇票通知书；通知单；发货通知书
advice note of surveyor 验船师建议书
advice notice 通知
advice of adjustment 调整通知
advice of arrival 货物运到通知；到货通知
advice of audit 审核通知书
advice of bill accepted 先付票据通知
advice of bill collected 票据付款通知
advice of bill paid 票据付款通知
advice of charges 索赔通知
advice of credit 信用证通知书
advice of damage 破损通知书；损坏通报
advice of debit 借记报单
advice of delivery 交付通知书；货到通知书
advice of dispatch 发送通知书
advice of documents 寄送单证通知书
advice of drawing 提款通知；汇票通知书
advice of fate 出险事故通知
advice of irregularity 违章通知书
advice of letter of credit 信用证通知书
advice of payment 付款通知书；付款通知单
advice of receipt 收领货物通知书
advice of settlement 结算通知书；结算通知书
advice of shipment 装运通知(单)
advice of total transfer of credit 全额转让通知书
advice of works in progress 工作进展报告；工程进度通知
advice on production for export 出口商品通知书
advice sheet 汇兑交易通知书
advice ship 通报船
advised line of credit 通知贷款额度；已通知信用额度
adviser 指导者；劝告者；顾问
adviser on financial affairs 财政顾问
advise shipping date 通知装运日期
advising bank 通知银行；通知行
advising charges 通知手续费；通知费用
advising commission 通知手续费
advisor = adviser
advisory 顾问的
advisory architect 顾问建筑师
advisory board 咨询委员会；顾问团
Advisory Board on Compensation Claims 索赔事项咨询委员会
advisory body 咨询机构；咨询机关；顾问团
advisory commission 顾问委员会
advisory committee 咨询委员会；顾问委员会
Advisory Committee for Safety 安全咨询委员会
Advisory Committee on Marine Pollution 海洋污染咨询委员会
Advisory Committee on Oceanic Meteorological Research 海洋气象研究咨询委员会
Advisory Committee on Pollution of the Sea 海上污染咨询委员会
Advisory Committee on Technical and Operation Matters 技术与操作事务咨询委员会
Advisory Committee on Weather Control 天气控制咨询委员会
advisory council 咨询委员会；顾问团
advisory data 参考数据
advisory department 咨询部
advisory engineer 咨询工程师；顾问工程师
advisory fee 咨询费
advisory for gale and rain 风雨警报
advisory for heavy snow 大雪警报
advisory group 咨询组；咨询(服务)小组；顾问(小)组
advisory opinion 咨询意见
advisory panel 顾问团
advisory panel on program(me) policy 计划政策咨询小组
advisory plan 纲要(性)规划
advisory service 咨询业务；咨询服务
advisory service of standard information 标准情报咨询服务
advisory signal 劝告信号
advisory speed 劝告车速；推进车速
advisory speed sign 推荐车速标志
advocacy planning 特殊利益保护计划；建议计划；建议规划
advocate 辩护人
advocate green activities 开展绿色活动
adynamic(al) soil 非动力土壤
adynamic(al) stage 无效期；无效力期
adytum 庙堂内室；(古代庙宇中的)密室
adz(e) 横口斧；刮刀；锛(子)；扁斧
adz(e) block 锛架；刨身
adzed work 斩琢工；刮刀工；扁斧工
adz(e)-eye claw hammer 斧眼拔钉锤
adz(e)-eye hammer 小铁锤；鱼尼锤；羊角锤；拔钉锤
adz(e) finish 斧削面
adz(e)-hewn 用斧斩削的；用扁斧加工的
adz(e) plane 刮刨
adz(e) work 砍削处理（用扁斧对木材表面进行处理后代替刨面）
adzing 砍平
adzing and boring machine 铲平钻孔机
adzing ga(u)ge 削平榫规；轨枕槽规
adzing machine 削枕机
Aebi kiln 埃毕顶烧式隧道窑
Aebl 埃伯尔
AE-cellulose 氨乙基纤维素
aecium 锈子器
aedelforsite 硅灰石
aedes 教堂或庙宇；大厦或神龛(古罗马)
aedicular architecture 神龛建筑；壁龛建筑
aedicule 神龛式门窗；神龛；壁龛；(用作庙宇的)小建筑物；(教堂内的)小祈祷处
aedicule altar 壁龛祭坛；壁龛隔墙
aedile 古罗马营造司官员
aeetylenic alcohol 炔属醇
aegagropilus 毛块

Aegean architecture 爱琴海(文化)建筑
Aegean art 爱琴海艺术
Aegean civilization 爱琴海文化
Aegean Sea 爱琴海
aegiapite 霓磷灰石
aegirine 霓石
aegirine alvikite 霓石方解石碳酸岩
aegirine aplite 霓细晶岩
aegirine augite 霓辉石
aegirine beforsite 霓石镁云碳酸岩
aegirine felsite 霓石霏细岩
aegirine rauhaugite 霓石白云(石)碳酸岩
aegirine sovite 霓石黑云碳酸岩
aegirinzation 霓石化
aegricane 锤头装饰；羊头饰
aegrit 涂饰墙板的墙粉；涂饰墙板的灰泥
A eliminator 灯丝电源整流器；代甲电池
Aelosomatidae 爱罗科
aemilogarithmic line chart 单对数坐标折线图
Aeneolithic Age 次新石器时代
aenigmatite 三斜闪石；钠铁非石
aeolation 风化作用
aeolian 风成的
aeolian abrasion 风蚀作用
aeolian accumulation 风力堆积；风聚集；风积；风沉积
aeolian-accumulation landform 风积地貌
aeolian basin 风蚀盆地；风成盆地
aeolian bedding 风成层理
aeolian clay 风积黏土
aeolian cross-bedding 风成交错层理
aeolian cross-bedding structure 风成交错层理构造
aeolian-demoiselle 风蚀蘑菇
aeolian deposit 风聚集；风积(物)；风积沉积；风成沉积(物)；风成沉积(地)层；风积土
aeolian dispersion 风成分散
aeolian dune 风成沙丘
aeolian dune deposit 风成沙丘沉积
aeolian dune facies 风成沙丘相
aeolian dune field 风成沙丘原
aeolian erosion 风蚀(作用)；风力侵蚀
aeolian facies 风成相
aeolian flat 风坪
aeolian harp 风奏琴
aeolianite 风成岩；风成沉积
aeolian landform 风蚀地貌；风成地貌
aeolian material 风积物
aeolian plain 风成平原
aeolian-residual hill 风蚀残丘
aeolian ripple 风成沙(波)
aeolian ripple mark 风成波痕；风成涟痕
aeolian rock 风成岩
aeolian sand 风积沙；风成沙
aeolian sand ripple 风成波纹
aeolian sandstone 风成砂岩
aeolian sediment 风积物
aeolian soil 风积土(壤)；风成土
aeolian soil type 风积土类
aeolian tone 风吹声
aeolian tower 风蚀塔
aeolian transportation 风力搬运
aeolian vibration 风激振动；风吹震动
aeolic 风成的
Aeolic capital 爱奥尼亚式柱头
aeolic soil 风积土(壤)
aeolipile 汽转球
aeolosphere 各向异性球
aeolotropic 各向异性的
aeolotropic crystal 各向异性晶体
aeolotropism 偏等性；各向异性(现像)
aeolotropism of rock 岩石各向异性
aeolotropy 偏等性；各向异性(现像)；非均匀性；不等方程
aeolotropy of rock 岩石各向异性
a. e. p. of cohesionless backfill 无黏性填土的主动土压力
aeradio 航空无线电台
Aeral 阿拉尔铝合金
aerarium 供气器；通风器；国库(古罗马)；公共金库(古罗马)
aerate 曝气；通气；松料；供风；吹风；充气
aerated 充气的
aerated activated carbon filter 曝气活性炭滤池
aerated annular base 环形充气库底

aerated apparatus 曝气装置
aerated bath nitriding 通气液体氮化；空气搅拌液体渗氮法
aerated biofiltration 曝气生物过滤
aerated biological treatment 曝气生物处理
aerated blending silo 气拌库；空气搅拌库；空气搅拌仓
aerated block 空心混凝土块；充气砌块
aerated bottom 充气炉底
aerated building 充气房屋
aerated burner 预混(空气)烧器
aerated calcium silicate 泡沫硅酸钙制品
aerated cascade photoreactor 阶梯式曝气光化反应器
aerated cement 充气水泥
aerated cement screed 加气水泥刮板
aerated channel 曝气渠；曝气槽
aerated channel air requirements 曝气槽需气要求
aerated column 曝气柱
aerated concrete 泡沫混凝土；加气混凝土；充气混凝土；掺气混凝土
aerated concrete block 加气混凝土(砌)块
aerated concrete compressive strength 加气混凝土抗压强度
aerated concrete density 加气混凝土容重
aerated concrete floor slab 加气混凝土楼板
aerated concrete glue-joint block partition 加气混凝土黏胶缝隔墙
aerated concrete member 加气混凝土构件
aerated concrete mixer 加气混凝土搅拌机
aerated concrete panel 加气混凝土板
aerated concrete partition 加气混凝土隔墙
aerated concrete pipe insulating section 加气混凝土管套
aerated concrete slab 加气混凝土板
aerated concrete wall panel 加气混凝土墙板
aerated concrete window panel 加气混凝土带窗墙板
aerated conduit 曝气管(道)；通气管道
aerated contact bed 曝气接触床
aerated conveying trough 空气输送斜槽
aerated discharge channel 空气卸料槽
aerated filler 疏松填料
aerated filter 曝气滤池
aerated flame 充气焰；富空气焰
aerated flow 曝气水流；充气流；掺气水流
aerated flow region 曝气水流区
aerated funnel shaped bottom discharge 充气漏斗形底部卸料
aerated geothermal fluid 受空气污染的地热流体
aerated grit chamber 曝气沉砂池
aerated grit removal 曝气除砂
aerated gypsum 加气石膏
aerated gypsum board 加气石膏板
aerated gypsum core 加气石膏芯
aerated gypsum sheet 加气石膏板
aerated lagoon 曝气塘；充气水池；曝气氧化塘；曝气湖
aerated lagoon kinetics 曝气塘动力学
aerated launder 曝气流槽
aerated material 充气材料
aerated mortar 加气砂浆；加气灰浆；加气灰膏
aerated mud 掺气泥浆；加气泥浆；混气泥浆；充气泥浆
aerated nappe 自由溢流水舌；瀑布；曝气水舌；掺气水舌
aerated pipe 曝气管(道)
aerated plastics 泡沫塑料；充气塑料；多孔塑料
aerated pond 曝气塘；稳定塘—曝气塘
aerated pond effluent 曝气池出水
aerated pool 曝气塘
aerated porosity 通气孔隙度；含气孔隙；(土壤的)充气孔隙
aerated prestressed concrete 加气预应力混凝土
aerated reaeration 曝气复氧
aerated reinforced concrete 加气钢筋混凝土
aerated ring 充气圈
aerated seawater 充气海水
aerated section 充气区
aerated sewage lagoon 曝气污水(氧化)塘
aerated sewage lake 曝气污水池；污水曝气湖；污水曝气池
aerated sheet of water 瀑布；自由溢流水舌
aerated silo floor 库底充气板

aerated sintered(concrete) aggregate 加气烧结(混凝土)集料
aerated skimming tank 气浮除油池；曝气撇油池
aerated solid 气溶胶；充气固体
aerated solid contact 曝气固体接触
aerated spillway 真空式溢流道
aerated spiral flow grit channel 曝气旋流沉砂槽
aerated spring 曝气处理泉水
aerated stabilization basin 曝气稳定池
aerated structure 孔隙结构；气孔结构
aerated submerged biofilm reactor 淹没式曝气生物膜反应器
aerated surface 充气面
aerated trickling filter 曝气滴滤池
aerated trough conveyer 空气输送斜槽
aerated upflow sludge bed 升流曝气污泥床
aerated water 气水；碳酸水；充气的水；掺气水
aerated water body 曝气水域；曝气水体
aerated water machine 汽水机
aerated weir 真空(式)溢流堰；掺气溢流堰
aerated yarn 充气纱
aerated zone 充气带；饱气带
aerating agent 掺气剂；充气剂；发泡剂
aerating apparatus 曝气装置；充气装置
aerating blending 充气搅拌
aerating block 充气板
aerating box 充气箱
aerating chemical 发泡剂；充气剂
aerating corrosion 通气腐蚀
aerating device 曝气装置
aerating filter 空气滤清器；曝气滤池
aerating grit chamber 曝气沉砂池
aerating grit removal 曝气除砂
aerating jet 曝气射流
aerating lagoon 曝气塘
aerating launder 曝气流槽
aerating panel 透气板；多孔板
aerating powder 发泡粉
aerating process 曝气法
aerating quadrant 扇形充气区；充气象限
aerating rate 曝气速率
aerating regulation 曝气调节
aerating root 通气根
aerating rotor 曝气转子
aerating sewage lagoon 曝气污水(氧化)塘
aerating spiral flow grit channel 曝气旋流沉砂槽
aerating stabilization basin 曝气稳定池
aerating submergence 曝气浸没
aerating system 曝气系统；通气系统
aerating tank appearance 曝气池外貌
aerating test 曝气试验
aerating test method 曝气试验方法
aerating time 曝气时间
aerating tissue 通气组织
aerating tower 曝气塔
aerating treatment 曝气处理
aerating unit 充气装置
aerating waste(water) 曝气废水；废水曝气
aerating zone 曝气区
aeration 溶液充气；气泡影响；起泡作用；曝气；透气；通气(性)；吹气；充气(法)；掺气
aeration air 松料空气
aeration amount 曝气量
aeration basin 曝气池；通气池
aeration basin loading 曝气池负荷
aeration-basin oxygen demand 曝气需氧量
aeration blending 充气搅拌
aeration blower 曝气鼓风机；通气鼓风机
aeration cell 氧气电池；充气电池
aeration chamber 充气室
aeration channel 通气渠；通气槽
aeration circulation pattern filter cell 曝气循环型滤池
aeration coefficient 曝气系数；氧转移系数；鼓风系数；充气系数；掺气系数
aeration column 充气塔
aeration concentrater 掺气浓度仪
aeration conduit 曝气管道
aeration cone 曝气漏斗
aeration cooling 曝气冷却；通风冷却；通风降温
aeration degree 含气度；掺气度
aeration device 曝气装置
aeration disc 曝气盘
aeration ditch 曝气沟；通气渠

aeration drainage 曝气排水
aeration drilling 充气钻进;充气泥浆钻井
aeration drying 通风干燥;风干法
aeration efficiency 曝气效率
aeration equipment 曝气设备;通气设备;通风设备
aeration experiment 曝气实验
aeration facility 曝气装置
aeration factor 掺气因素;曝气因素;通气因素;充气因子
aeration flo(a)tation 曝气上浮法;曝气气浮法
aeration flotator 充气式浮选机;气升式浮选机
aeration frequency 通气次数
aeration funnel 曝气漏斗
aeration grit settling tank 曝气沉砂池
aeration intensity 曝气强度
aeration irrigation 曝气灌溉
aeration jet 曝气射流;喷气发动机;通气喷嘴;进气喷管;含气射流;充气射流
aeration lagoon 曝气塘;氧化塘
aeration machine 曝气机
aeration method 曝气法
aeration methods of activated sludge process 活性污泥法曝气方法
aeration mud drilling 充气泥浆钻进
aeration need 通气需要
aeration nozzle 通气管嘴
aeration of cement 水泥风化
aeration of mo(u)lding sand 松砂
aeration of river water 河流水体复氧
aeration of sewage 污水曝气处理;污水曝气
aeration of soil 土壤通气(性)
aeration of the cell 浮选机充气
aeration period 曝气时间;曝气周期
aeration period of activated sludge process 活性污泥曝气时间
aeration period of sewage 污水曝气周期;污水曝气时间
aeration phase 曝气阶段
aeration pipe 曝气管(道);通气管
aeration plant 曝气机长;曝气设备;充气装置;曝气站
aeration porosity 通气孔隙度
aeration process 曝气法
aeration quantity 充气量
aeration rate 曝气(速)率
aeration reactor 曝气反应器
aeration rotor 曝气转子
aeration-sedimentation tank 曝气沉淀池
aeration skylight 通风天窗
aeration slot 通气槽;掺气槽
aeration sprinkler 航空人工降雨机
aeration strength 曝气强度
aeration submergence 曝气浸没
aeration system 曝气系统
aeration tank 曝气池;曝气槽;充气箱;充气槽
aeration tank appearance 曝气池外貌
aeration tank biological oxygen demand loading 曝气池生化需氧负荷
aeration tank cross-section 曝气池横断面
aeration tank detention time 曝气池停留时间
aeration tank diffuser 曝气池扩散器
aeration tank splitter box 曝气池分流管
aeration test 曝气试验
aeration test burner 验气燃烧器
aeration test method 曝气试验方法
aeration time 曝气时间
aeration time sewage 污水曝气时间
aeration tower 曝气塔;通风塔;洗气塔
aeration treatment 曝气处理
aeration unit 曝气设备;曝气单元
aeration vent scrubber 充气孔洗涤器
aeration volume 曝气量
aeration zone 通气带;通气层;含气层;掺气层;饱气带;通气区;掺气区
aeration zone water 包气带水
aerator 曝气装置;曝气设备;曝气器;曝气机;曝气池;透气机;通气机;松砂机;熏烟装置;液体充气装置;充气器;充气机
aerator fitting 充气管配件;曝气送风机;曝气设备;充气器
aerator nozzle 曝气喷嘴
aerator performance testing 曝气器性能试验
aerator pipe 曝气管(道)
aerator tank 充气槽
aeremia 气泡栓塞

aerenchyma 通气组织
aereous 青铜色的
Aeress 阿勒斯变性聚丙烯腈纤维
Aerex fan 艾瑞克斯型轴流式扇风机
aerial 气生的;天线;空中(的);架空的;航空的
aerial adapter 天线接头
aerial application of pesticides 空中喷药;飞机喷药
aerial archaroloy 航空考古学
aerial area coverage 航测成图区
aerial array 多振子天线;天线阵
aerial attack 空袭
aerial beacon 航空信标
aerial beam width 喇叭形天线张开角
aerial bell 空中警钟
aerial belt conveyer 架空式输送带
aerial bird 飞鸟
aerial bridging instrument 空中三角测量仪
aerial cable 天线电缆;架空电缆;高空电缆
aerial cable line 架空电缆线路
aerial cable spinning 架空钢索
aerial cableway 缆车道;空中索道;架空索道
aerial cableway survey 架空索道测量
aerial camera 航摄仪;航空照相机;航空(测量)摄影机;航测摄影机
aerial camera constant 航摄仪常数
aerial camera lens 航摄仪镜头
aerial camera mount(ing) 航摄仪座架
aerial camera operator 航空摄像员
aerial car 气球吊篮;高架铁道车
aerial cartographic(al) topography 地形测图航空摄影
aerial cascade 大气急流;大气暴流
aerial chart 航空图
aerial choke 天线扼流圈
aerial circuit 架空线路
aerial circuit breaker 天线断路器
aerial condenser 空气冷凝器;空气冷凝机
aerial conductor 空气导管;架空电线;架空导线;架空避雷针;航摄范围
aerial conduit 架空管道
aerial contact line 架空电车线
aerial contaminant 空气污染物(质)
aerial contamination 空气污染
Aerial Contamination Act 空气污染法
Aerial Contamination Code 空气污染法规
aerial contamination complaint 空气污染诉讼
aerial contamination concentration 空气污染浓度
aerial conveyer 架空运输设备;钢索运输装置;钢索运输机;吊运器;空中输送机;架空输送设备;架空式输送机;空气运输机;悬链输送机;悬空索道;高架轨道;高架轨道(用货明)
aerial coverage 航摄范围;航测范围
aerial craft 航空器
aerial current 气流
aerial data 航摄资料;航测资料
aerial defence 防空
aerial depth charge 空投深水炸药;空投深水炸弹
aerial detection 空中检测;空气检测
aerial discharge 气体放电;空气中放电
aerial dispersion 空气扩散
aerial drainage 泄流
aerial dump 空中卸料(架空索道起重机斗)
aerial duplicating material 航空摄影复制资料
aerial dust 大气灰尘;大气尘埃
aerial dust filter 空气滤尘器
aerial earth(ing) wire 架空接地线
aerial earth wire support 架空地线支撑
aerial earth wire suspension 架空地线悬挂
aerial earth wire termination 架空地线下锚
aerial engineer 航空测量工程师
aerial equipment 航摄装备
aerial exposure index 空中曝光指数
aerial extension 架空线(路)
aerial fairlead 天线引线管
aerial farming 航空农业
aerial ferry 列车轮渡;浮桥
aerial film 航摄软片;航摄胶片;航空胶片
aerial flight 空中飞行
aerial fog 气雾;空气灰雾
aerial for television transmitter 电视发射天线
aerial frog 空中索扣(吊桥)
aerial funicular 架空缆车
aerial gallery 天线廊道(广播塔);天线平台
aerial geology 航空地质(学)

aerial geophysics prospecting 航空地球物理调查
aerial growth 地上部分生长;地面生长
aerial growth plant 地上部分生长植物
aerial halyard 天线升降索
aerial image 空间(成)像;航空像片
aerial infection 空气传染
aerial information 航摄资料;大气资料
aerial infrared imagery 航空红外成像
aerial infrared scanner 航空红外扫描仪
aerial installation 天线安装;天线设备;天线装置;架空敷设
aerial insurance 空运保险
aerial ladder 架空消防梯;消防梯;云梯;云杉
aerial level(l)ing 空中三角测量法;航空水准测量;空中水准测量
aerial lift 升降工作平台
aerial lift vehicle 高处作业车
aerial light house 航空灯塔;航空标塔
aerial line 架空天线;空气管线;架空线(路)
aerial liner 定期飞机;航空班机;定期民航机;班机
aerial magnetometer 航空磁力仪
aerial map 航摄图;航空图;航测图
aerial mapping 航空制图;航空测量;航空测绘;航测制图
aerial mapping camera 绘图航空摄影机;航空绘图摄影机
aerial mapping photography 航测摄影(学)
aerial mast 天线柱;天线塔;天线杆
aerial material 航空材料
aerial metal 航空用金属
aerial method 航空测量方法;航测方法
aerial monorail tramway 架空单轨道
aerial mosaic 空中照片嵌拼地图
aerial multiband camera 航空多波段摄像机
aerial navigation 空运;航空运输;空中导航
aerial navigational equipment 导航设备
aerial navigational viewfinder 空中导航检信器
aerial negative 航摄底片
aerial negative number 航摄底片编号
aerial network 架空电力网
aerial night camera 夜间航空摄影机
aerial object 空中物体
aerial observation 航空观测
aerial optic(al) cable 架空光缆
aerial oxidation fog 空气氧化灰雾
aerial part 地上部分
aerial part management 地上部分作物管理
aerial perspective 立体透视;空中透视;鸟瞰透视图;大气透视
aerial photo coverage 航摄像片覆盖区
aerial photogrammetry 空中摄影测量
aerial photo(graph) 航空照片;航空摄影学;航摄像片;航摄图;航空像片
aerial photograph converter 航空像片转绘仪
aerial photograph fix 航迹摄影定位
aerial photographic(al) absolute gap 航摄绝对漏洞
aerial photographic(al) gap 航摄漏洞
aerial photographic(al) interpretation 航摄照片解译
aerial photographic(al) mapping 航空测绘;航空摄影制图
aerial photographic(al) relative gap 航摄相对漏洞
aerial photographic(al) survey 航空摄影测量
aerial photograph interpretation 航空照片解释
aerial photography 空中摄影;航空摄影;航测
aerial photography by movable camera mounting 摇摆航空摄影
aerial photo interpretation 航空照片解译;航摄像片判读
aerial photomap 空中摄影地图;航测地(形)图
aerial photomapping 航空摄影制图;航空测绘
aerial photomosaic 航摄照片拼接图
aerial photo pair 航摄像片对
aerial photo-taking organization 航摄机构
aerial picture map 空摄地图;航空摄影地图
aerial pipe crossing 管道桥架
aerial pipeline 架空管线
aerial plankton 空气浮游生物
aerial plug-in point 天线插座;公用天线接线点;公用天线插座
aerial pole 天线杆;架空杆
aerial pollutant 空气污染物(质)
aerial pollution 空气污染

Aerial Pollution Act 空气污染法
Aerial Pollution Code 空气污染法规
aerial pollution complaint 空气污染诉讼
aerial pollution concentration 空气污染浓度
aerial port 航空港
aerial print 航空照片;航空像片
aerial propeller 空气推进器
aerial propeller vessel 空气螺旋桨船
aerial prospecting 航空探矿
aerial radiological measurement and survey 大气放射性测量考察
aerial radiological measuring survey 大气放射性测量考察
aerial radiological measuring system 大气放射性测量系统
aerial radiological monitoring system 大气放射性监测系统
aerial radome 天线罩(雷达);整流罩
aerial railroad 架空路;架空铁道
aerial railway 架空铁路;架空铁道;架空索道;高架铁路;高架铁道
aerial reconnaissance 飞行勘察;空中探测;空中勘测;航空目测踏勘;航空勘察;航空勘测
aerial remote sensing 机载遥感;航空遥感
aerial root 气生根
aerial ropeway 空中缆道;空运索道;缆车道;架空索道;架空缆车道
aerial ropeway conveyer 架空索道运输机
aerial route selection 航测选线
aerial sediment 大气沉积(物)
aerial seeding 飞机播种
aerial sewer 露置下水管道;露明下水管道;架空污水管;架高排水管
aerial shoot 地上茎
aerial shot 航空摄影;空中摄影;航摄照片
aerial sickness 航空病
aerial signal 空中标志
aerial sketch master 航片转绘仪
aerial skidder 架空集材机;架空滑道
aerial skidding 架空式集材
aerial snow survey 航空测雪
aerial socket 天线插座
aerial sound ranging 空中声波测距
aerial speed 航速(指航空器)
aerial spray 空气喷雾
aerial sprayer 航空喷雾机
aerial spraying 空中喷药;飞机喷药
aerial spud 悬空缆
aerial surveillance 空中监视
Aerial Survey and Remote Sensing Company 航测遥感公司
aerial survey camera 航空测量摄影机;航测摄影机;航空测量摄影仪
aerial survey craft 航测飞机
aerial survey(ing) 航测;航空摄影测量;航空测量(法);航空勘测;航空调查
aerial surveying camera 航摄仪;航测摄影机
aerial survey team 航摄队;航测队
aerial telescope 航空望远镜
aerial topographic(al) analysis 航空摄影分析
aerial topographic(al) map 航测地形图
aerial tram 架空索道小车
aerial tramway 架空缆车道;架空索道;架空缆车道;架空电车道;高架电车道
aerial transit system 架空运输系统
aerial transmission 空中传递;空气传递
aerial transport(ation) 航空运输
aerial transporter 空中缆车;架空缆车
aerial transport gear 架空调度
aerial triangulation 空中三角测量;航空三角测量
aerial triangulation block 空中三角网区;空中三角测量区域网
aerial triangulator 空中三角仪;辐射三角仪
aerial velocity 航速(指航空器)
aerial view 空中景观;鸟瞰(图);俯瞰(图)
aerial volume table 航空材积表;航测材积表
aerial windscreen 航空风挡玻璃
aerial wire 天线;架空导线
aerial work basket 高级作业吊篮
aerial work carriage 高级作业车
aerial work platform 高级作业平台
aeriductus 气门;尾状管
aeriferous 含有空气的
aerification 气体化;掺入空气;掺气

aeriform 气样的
aerify 气化;充气
aerinite 青泥石
aeriscope 超光电摄像管
aeroaccelerator 加速曝气池
aeroaccelerator combined 曝气沉淀池
aeroaccelerator method 加速曝气法
aeroacoustics 空气声学
aeroallergen 空气(中)变应原
aeroaspiration 空气吸入
aerobacter aerogenes 产气杆菌
aerobacteria 需氧细菌
aerobacteria corrosion 好气菌腐蚀
aerobe 好氧生物;好气微生物;需氧(微)生物;需氧菌
aerobe bacteria 需氧细菌
aerobian 需氧微生物;需氧的
aerobic 需氧的;需气的
aerobic activated sludge method 好氧活性污泥法
aerobic activation 好氧活性化
aerobic aerated lagoon 好氧曝气氧化塘;好气曝气氧化塘
aerobic aerated oxidation pond 好氧曝气氧化塘
aerobically grown microbial granules 好氧生长微生物颗粒
aerobic ammonium oxidation 好氧氨氧化
aerobic ammonium oxidation bacteria 好氧氨氧化菌
aerobic-anaerobic oxidation lagoon 好厌氧氧化塘
aerobic-anaerobic process 好厌氧处理法
aerobic-anaerobic-repeated-coupling fixed bed bioreactor 好厌氧反复耦合固定床生物反应器
aerobic-anaerobic-repeated-coupling process 好厌氧反复耦合工艺
aerobic-anaerobic stabilization pond 好厌氧稳定塘
aerobic-anaerobic treatment 好厌氧处理
aerobic-anoxic process 好缺氧处理法
aerobic area 好氧区
aerobic atmosphere 需氧环境
aerobic bacteria 好氧(细)菌;好气(细)菌;需氧(细)菌;需气细菌;喜氧(细)菌
aerobic bacteria degradation 喜氧细菌降解
aerobic bacteria degradation zone 喜氧细菌降解作用带
aerobic bacteria treatment 需氧细菌处理
aerobic biodegradation 好氧生物降解
aerobic biofilm layer 好氧生物膜层
aerobic biofilm process 好氧生物膜法
aerobic biofilter 好氧生物滤池
aerobic biological disc 好氧生物转盘
aerobic biological filter 好氧生物滤池;需氧生物滤池
aerobic biological fluidized bed 好氧生物流化床
aerobic biological fluidized bed reactor 好氧生物流化床反应器
aerobic biological fluidized bed system 好氧生物流化床系统
aerobic biological formation 好氧生物转化
aerobic biological process in wastewater treatment 废水需氧生物处理
aerobic biological reactor 好氧生物反应器
aerobic biological treatment 好氧生物处理;需氧(性)生物处理
aerobic biological treatment method 好氧生物处理法
aerobic biological treatment of wastewater 污水好氧生物处理
aerobic biological treatment reactor of wastewater 污水好氧生物处理反应器
aerobic breakdown 好氧降解;需氧分解
aerobic catabolism 有氧分解代谢
aerobic cellulolytic bacteria 好气性分解纤维细菌;好气分解纤维素细菌
aerobic composting 好氧堆肥
aerobic condition 好氧条件;好气条件
aerobic culture 好氧培养;需氧培养
aerobic cycle 好氧循环
aerobic de-ammonification process 好氧除氨工艺
aerobic decomposition 好氧分解;好气分解;需氧分解;有氧分解
aerobic denitrification 好氧脱氮作用;好氧反硝化;好氧硝脱作用
aerobic denitrification bacteria 好氧脱氮菌;好氧反硝化菌

aerobic denitrification phenomenon 好氧脱氮现象;好氧反硝化现象
aerobic digester 好氧消化池;需氧消化池
aerobic digestion 好氧消化;好气消化;需氧消化(法)
aerobic digestion of refuse 垃圾好氧消化
aerobic domestic wastewater treatment 好氧生活污水处理
aerobic environment 不缺氧环境;好氧环境;充氧环境
aerobic filter 好氧滤池
aerobic fluidized bed 好氧流化床
aerobic granular sludge 好氧颗粒污泥
aerobic granular sludge reactor 好氧颗粒污泥反应器
aerobic granulation 好氧粒化作用
aerobic granule 好氧颗粒
aerobic groundwater 好氧地下水
aerobic heterotrophic bacteria 好氧异养菌;需氧异养菌
aerobic heterotrophic dominant bacteria 好氧异养优势菌
aerobic incubation 供气培养
aerobic lagoon 需氧菌塘
aerobic landfill 好氧土地填埋
aerobic media trickling filter 好氧介质滴滤池
aerobic membrane biological process 好氧膜生物工艺
aerobic membrane biological reactor 好氧膜生物反应器
aerobic membrane bioreactor 好氧膜生物反应器
aerobic membrane bioreactor treatment system 好氧膜生物反应器处理系统
aerobic metabolism 好氧代谢作用;需氧代谢;有氧代谢
aerobic microorganism 好氧微生物;好气性微生物
aerobic moving bed biofilm reactor 好氧移动床生物膜反应器
aerobic nitrification reactor 好氧硝化反应器
aerobic nitrifying granules sludge 好氧硝化颗粒污泥
aerobic nitrifying granules sludge sequence batch reactor 好氧硝化颗粒污泥序批式间歇反应器
aerobic nitrogen-fixing 需氧固氮细菌
aerobic nitrogen fixing bacteria 好气固氮细菌
aerobic organic sediment 好氧有机底泥;需氧有机沉积物
aerobic organism 好氧生物;好氧(细)菌【给】
aerobic oxidation 好氧氧化;需氧氧化(作用);有氧氧化
aerobic pollutant 好氧污染物
aerobic pond 稳定塘—好氧塘;好氧塘;需氧塘
aerobic pond with water plant 水生植物好氧塘
aerobic process 好氧工艺;好氧处理法;需氧过程;需氧法
aerobic purification 好氧净化;需氧净化
aerobic reactor 好氧反应器
aerobic respiration 需氧呼吸;有氧呼吸
aerobic respiration zone 喜氧呼吸带
aerobic secondary treatment 好氧二次处理;需氧二级处理;需氧二次处理
aerobic sludge digestion 好氧污泥消化;需氧污泥消化
aerobic sludge granule 好氧污泥颗粒
aerobic sludge treatment 好氧污泥处理
aerobic spore populations of soil 土壤需氧孢子体
aerobic stabilization pond 好氧稳定塘
aerobic treatment 需氧(细菌)处理;好氧处理;好气性处理
aerobic treatment efficiency 好氧处理效率;好气处理效率;需氧处理效率
aerobic wastewater treatment 好氧污水处理
aerobic water 有氧水
aerobic water body 有氧水域;有氧水体
aerobic waters 有氧水域;有氧水体
aerobiology 空气生物学;航空生物学;高空生物学;大气生物学
aerobiont 好氧生物;好气生物
aerobioscope 空气细菌计数器;空气细菌测定器
aerobiosis 需氧生活
aerobism 需氧性
aeroboat 飞行艇;水上飞机;飞船
aerobond 环氧树脂

aerobridge 登机桥
aerobronze 航空青铜
aerobus 客(班)机
aerocab 空中客机
aerocamera 航空测量摄影机;空中摄影机;航空照相机;航空摄影机
aerocar 飞行车;气垫车
aerocarburetor 航空汽化器
aerocartograph 摄影测量绘图仪;空中照片制图仪;航空测图仪;航空测量图;航空测绘仪;航测仪
aerocartography 航空摄影测量学
aerochart 航空图
aerochemistry 气体化学;空气化学
aerochlorination 空气氯化;加气氯化;压气氯化法
aerochlorination of wastewater 废水空气氯化处理
aerochronometer 航空精密时计
aeroclay 空气离析黏土;空气分选黏土
aeroclimatology 高空气候学
aeroclinoscope 天候信号器;风信计
aerocode 航空电码
aerocolloid 气溶胶;气凝胶体
aeroconcrete 多孔混凝土;加气混凝土;充气混凝土
Aerocor 埃罗柯尔(一种超细玻璃绝缘材料)
aerocraft 飞机
aerocraft noise level 飞机噪声水平
aerocrete 气孔混凝土;泡沫混凝土;加气混凝土;充气混凝土
aerocrete block partition 加气混凝土隔墙
aerocrete compressive strength 加气混凝土抗压强度
aerocrete density 加气混凝土容重
aerocrete floor slab 加气混凝土楼板
aerocrete panel 加气混凝土板
aerocrete slab 加气混凝土板
aerocrete unit 加气混凝土构件
aerocurve 曲翼面
aerocycle 空中自行车(一种垂直起落飞行器)
aerodiesel 狄塞尔航空发动机;柴油航空发动机
aerodifferential spectrum survey 巷空微分能谱测量
aerodiscone antenna 机载盘锥天线;航空盘锥形天线
aerodist 航空微波测距仪
aerodone 滑翔机
aerodreadnaught 特大飞行器;巨型飞机
aerodrome 航空站;飞机库;飞机场
aerodrome beacon 机场信标;机场灯塔;机场灯标;机场指向灯;机场指向标
aerodrome conical surface 机场锥面
aerodrome control radar 机场监视雷达
aerodrome control tower 机场管制塔台
aerodrome diagram 机场平面图
aerodrome elevation 机场标高
aerodrome identification sign 机场识别标记
aerodrome light beacon 飞机场标志灯
aerodrome lighting 机场灯光设备
aerodrome locating beacon 机场导航台
aerodrome marker 机场标志板
aerodrome markings 机场标志
aerodrome master planning 机场总体规划
aerodrome obstruction chart 机场净空图
aerodrome of entry 入境机场;国际机场
aerodrome reference point 机场位置;机场控制点
aerodrome station 机场电台
aerodrome surface movement control 机场地面活动管制
aerodrome taxi circuit 机场滑行路线
aerodrome traffic circuit 机场起落航线
aerodrome traffic zone 机场交通空域
aerodromometer 气流速度表;风速计;风速表
aeroduct 离子冲压式发动机;冲压式空气喷气发动机
aeroduster 航空喷粉器;喷粉器
aerodux 酚醛树脂黏合剂;酚醛树脂胶粘剂;酚醛树脂胶合剂
aerodynamic(al) 空气动力(学)的;气动的
aerodynamic(al) ablation 气动烧蚀
aerodynamic(al) analysis 空气动力分析
aerodynamic(al) balance 气体动力平衡;气动力天平;气动力补偿;空气动力天平
aerodynamic(al) bearing 空气动力轴承;气体动压轴承
aerodynamic(al) body 空气动力绕流体
aerodynamic(al) brake 气力制动器;空气制动器;空气动力制动装置
aerodynamic(al) braking 空气动力制动
aerodynamic(al) center 空气动力中心;空气动力焦点
aerodynamic(al) characteristic 空气动力特性
aerodynamic(al) chord 空气动力弦
aerodynamic(al) coefficient 气体动力系数;空气动力系数
aerodynamic(al) compensation 气动力补偿
aerodynamic(al) compressor 空气动力压缩机
aerodynamic(al) configuration 空气动力设计外形
aerodynamic(al) contact force 空气动力接触力
aerodynamic(al) control 空气动力控制
aerodynamic(al) controller 空气动力控制器;空气动力调节器
aerodynamic(al) controlling device 气动力操纵装置
aerodynamic(al) criterion 气动力准则
aerodynamic(al) damper 气动阻尼器
aerodynamic(al) damping 气动阻尼
aerodynamic(al) data 气动数据
aerodynamic(al) decelerator 空气动力减速器
aerodynamic(al) derivative 气动力导数
aerodynamic(al) diameter 空气动力学直径
aerodynamic(al) drag 气动阻力;空气阻力
aerodynamic(al) effect 空气动力作用
aerodynamic(al) efficiency 空气动力效应
aerodynamic(al) excitation 气动激振
aerodynamic(al) experiment 空气动力试验
aerodynamic(al) force 气动力
aerodynamic(al) form 气流形;流线型;流线形状;空气动力外形
aerodynamic(al) heating 气动加热;空气动力加热
aerodynamic(al) heat transfer 气动热传递
aerodynamic(al) influence 气动力影响
aerodynamic(al) instability 气动失稳性;空气动力失稳;空气动力不稳定性
aerodynamic(al) interference 气动力干扰
aerodynamic(al) laboratory 空气动力学试验室
aerodynamic(al) lift 气动升力
aerodynamic(al) lifting surface 空气动力升力面
aerodynamic(al) load 空气动力荷载;空气动力负荷
aerodynamic(al) load distribution 气动力荷载分布
aerodynamic(al) lubrication 气体动力润滑
aerodynamically equivalent particle 风成等级颗粒
aerodynamically induced noise 空气动力噪声
aerodynamically shaped 空气动力型的
aerodynamically stream-lined deck cross-section 风动流线型桥面断面
aerodynamic(al) measurement 空气动力测定
aerodynamic(al) model 空气动力模型
aerodynamic(al) moment 空气动力(力)矩
aerodynamic(al) noise 气动噪声;空气动力噪声
aerodynamic(al) noise test 气动噪声试验
aerodynamic(al) oscillation 气体动力摆动;气体动力(学的)波动
aerodynamic(al) phenomenon 空气动力现象
aerodynamic(al) properties of a vehicle 车辆的空气动力学特性
aerodynamic(al) property 空气动力特性
aerodynamic(al) quality 空气动力性能;空气动力升阻比
aerodynamic(al) resistance 气动阻力;空气动力(的)阻力
aerodynamic(al) roll control system 空气动力控制滚动系统
aerodynamic(al) roughness 空气动力学粗糙度
aerodynamic(al) rudder 空气动力舵
aerodynamic(al) shape 流线型车体外形设计
aerodynamic(al) size 空气动力学粒径
aerodynamic(al) smoothness 空气动力平滑度
aerodynamic(al) stability 气动稳定性;空气稳定(性);气体动力(学的)稳定性;空气动力稳定性;抗风稳定性
aerodynamic(al) surface 空气动力面;空气动力表面
aerodynamic(al) test 气动力试验;空气动力试验
aerodynamic(al) test model 气动力试验模型
aerodynamic(al) tool 气动力方法
aerodynamic(al) trimmer 气动力调整片
aerodynamic(al) turbine 空气动力透平;空气动力涡轮机;闭式循环透平
aerodynamic(al) turbulence 气动湍流
aerodynamic(al) vane 空气动力导流片
aerodynamic(al) wave drag 气动波阻
aerodynamics 气体动力学;气流动力学;空气动力学;航空动力学
aerodynamics laboratory 空气动力学试验室
aerodynamics of super-sonic flight 超音速空气动力学
aerodynamo 航空发电机
aerodyne 重飞行器
aeroelastic 气动力弹性
aeroelastic interaction 气动力弹性干扰
aeroelasticity 气动力弹性力学;气动弹性(学);空气弹性力学
aeroelastics 气体弹性力学;气动弹性力学
aeroelastic vibration 气动弹性振动
aeroelectromagnetic method 航空电磁法
aeroembolism 气压病;气泡栓塞;空气栓塞症
aeroengine 航空发动机
aeroengine oil 航空润滑剂
aerofall mill 气落磨;气落式磨矿机;气吹磨碎机;干法自磨机;干法无介质磨
aerofilter 空气滤床;空气过滤器;加气滤池;加气过滤器
aerofilter oil 空气过滤器(用)油
aerofloat 二硫代磷酸型浮选剂;二乙基二硫代磷酸盐;水机浮筒
aerofloated sulfur 风选硫磺
aerofloat reagent feeder 泡沫供给器
aerofloc 气凝剂
aerofluxus 排气;泄气
aerofoil 气翼;空气动力翼面;机翼;翼型
aerofoil boat 气翼船
aerofoil camber 翼形曲面
aerofoil contour 翼型
aerofoil fan 轴流通风机;轴流式(通)风机
aerofoil profile 翼型剖面
aerofoil section 机翼型截面
aerofoil strut 机翼支柱
aeroform 爆炸成型
aerogamma survey 航空伽马测量
aerogas turbine 航空燃汽轮机
aerogas turbine engine 航空燃汽轮机发动机
aerogel 气溶胶;气凝胶
aerogene 产气微生物
aerogenerator 航空发电机;风力发电机
aerogenesis 产气
aerogenic 产气的
aerogenic bacteria 产气细菌
aerogenous 产气的
aerogeography 航空地理学
aerogeologic(al) mapping 航空地质制图
aerogeologic(al) survey 航空地质测量
aerogeology 航空地质学
aerogradimeter 航空倾斜度计
aerogram 航空信件;高空图解;无线电报
aerograph 喷漆枪;喷画笔;喷雾器;喷彩;无线电报机;空中气象记录仪;空气喷枪;航空气象图;修版汽笔;高空气象记录仪;高空气象计;天线电报机
aerographer 气象员
aerographic(al) chart 高空气象图
aerographic(al) film 航摄胶片
aerographing paint 喷漆
aerograph plotter 航测制图仪
aerography 喷染术;描述性气象学;大气(状况)图;大气学
aerogravimeter 航空重力仪
aerogravity survey 航空重力测量
aerohaler 喷雾吸入器
aerohydromechanics 空气水动学;空气流体力学
aerohydrous mineral 包液矿物
aerohydrous water 空穴水
aerohypometer 高空测高仪;高空测高计
aeroidogram 膜盒气压曲线;空盒气压曲线
aeroidograph 膜盒(式)气压记录器
aerointegral spectrum survey 航空积分能谱测量
aeroionization 空气电离(作用)
aeroisoclinic wing 等迎角机翼
aerojet 喷气式飞机;空气喷射
aerojet engine 空气喷气发动机
aerojet network analyzer 航空喷气网络分析器
aerolead battery 航空用铅蓄电池
aerolight 航空灯(标)
Aerolineas Argentinas 阿根廷航空公司
aerolite 石陨石;陨石;航空铝合金;埃罗铝合金
aerolith 石陨石
aerolog 航行记录簿;航行计程器
aerologation 测高术航行

aerologic(al) 高空气象的
aerologic(al) and oceanographic(al) research ship 气象及海洋考察船
aerologic(al) ascent 高空探测;大气上层探测
aerologic(al) days 高空观测日
aerologic(al) instrument 高空仪器
aerologic(al) map 航空气象图
aerologic(al) observation 高空观测
aerologic(al) sounding 高空探测;大气上层探测
aerologic(al) station 高空气象(观测)站;高空观测站
aerologic(al) theodolite 高空经纬仪
aerologic(al) throttle valve 高空节流阀
aerologist 气象学家;高空气象学家;大气学家
aerology 航空气象学;高空气象学
aerolotropic 各向异性的
aeromagnetic 空中地磁探测的;航磁的
aeromagnetic exploration 航空测探
aeromagnetic gradiometer 航空磁力梯度仪;航磁梯度仪
aeromagnetic horizontal gradiometer 磁水平梯度仪
aeromagnetic map 航空磁测图
aeromagnetic survey(ing) 航空磁测
aeromagnetic vertical gradiometer 航磁垂直梯度仪
aeromagnetometer 航空地磁仪;航空磁强计;航空磁力仪
aeromammography 充气乳腺造影术
aeromancy 天气预告;天气预测
aeromap 航空地图
aeromarine 海空的;海上飞行;海上航空的
aeromarine light 海空兼用航标灯
aeromechanical 航空力学的
aeromechanics 气体力学;空力力学
aeromedicine 航空医学
aerometal 航空铝合金;航空合金;航空铅合金
aerometeorograph 航空气象记录仪;高空气象计
aerometer 气体(比重)计;气量计;量气计;空气流量计;浮秤
aerometric and emission reporting system 气体测量和排放报告系统
aerometric measurement 大气测量
aerometric network 高空气象网
aerometro 空中快速运输系统
aerometry 气体测定(法);气体测量学;气体比重测定法;量气学
aeromicrobe 需氧菌
aeromobile 气垫汽车;气垫(式)运载工具;气垫(式)运载车
aeromotor 航空发动机
aeromovel 风动(快速轨道)运输系统
Aeron 阿埃隆铝合金
aeronaut 浮升员
aeronautical 航空的
aeronautical amp 航空图
aeronautical anchorage buoy 水上机场锚地浮标;水上飞机锚地浮标
aeronautical beacon 航空信号灯;航空灯标;海空立标
aeronautical chart 航行图;航空图
aeronautical climatology 航空气候学
aeronautical data 空中领航数据
aeronautical edition 航空版
aeronautical engineering 航空工程
aeronautical engineering institute 航空学院
aeronautical environment 航空环境
aeronautical environment chart 机场进近区航空图
aeronautical facility 航空设备
aeronautical fixed circuit 导航用固定电路
aeronautical fixed service 航空定点通信服务
aeronautical flutter 航空颤振
aeronautical information 航向资料
aeronautical information overprint 图上加印的领航资料;附加航空信息
aeronautical instrument 航空仪器
aeronautical laboratory 航空实验室
aeronautical light (机场的)航行灯;航空灯(标);航空标灯
aeronautical Loran charts 罗兰航空图
aeronautical meteorology 航空气象学;航空气象测量
aeronautical meteorology planning 航空气象学计划

aeronautical mile 空里;航空英里
aeronautical national taper pipe threads 国际航空用锥形管螺纹标准
aeronautical pilotage chart 航空导航图
aeronautical planning chart 航空计划图;战略航空图
aeronautical plotting chart 航线标绘图;战术航空图
aeronautical radio 航空无线电设备;导航无线电设备
aeronautical radio altimeter 航空无线电高度计
aeronautical radio beacon 航空无线电指向标;航空无线电信标;航空无线电立标
aeronautical satellite system 航空卫星系统
aeronautical symbol 航空符号
aeronautic navigation 航空
aeronautics 航空学
aeronavigator 领航飞行员
aeronomy 天体大气学;星体大气物理学;高空大气(物理)学;高层大气物理学;超高层大气物理学
aeronomy satellite 高层大气科学卫星
aeroobstruction light 航空障碍灯光;航空障碍标志灯;航空碍航指示灯
aerooil 航空(润滑)油
aerooleopneumatic shock absorbing strut 空气油压减震支柱
aerootitis 气压性中耳炎;航空耳炎
aeropause 大气上界;大气航空边界
Aerophane 埃罗芬素色纱罗
aerophankton 大气浮游生物
aerophare 航空用信标
aerophile bacterium 好气性细菌
aerophilic 气生的;需气性
aerophilic algae 气生藻类
aerophilons 需气性
aerophone 对讲机;报话机
aerophore 通风面具;输气机;防毒面具吸呼器
aerophoto base(line) 航摄基线
aerophotogeodesy 航空摄影大地测量(学)
aerophotogeology 航空摄影地质学
aerophotogrammetric(al) map 航空摄影测量图;航测图
aerophotogrammetric(al) plan 航空摄影测量平面图;航测平面图
aerophotogrammetric(al) survey 航空摄影测量
aerophotogrammetric(al) survey team 航摄队;航测队
aerophotogrammetry 航空摄影测量(学);航测学
aerophoto(graph) 航空照片;航摄像片;航空摄影照片;空中摄影照片
aerophotographic(al) apparatus 航摄仪器;航空摄影仪器
aerophotographic(al) camera 航仪;航空摄影机
aerophotographic(al) map 航测平面图
aerophotographic(al) mosaic 航空像片镶嵌图;航摄像片拼图
aerophotographic(al) plan 航测平面图
aerophotographic(al) reconnaissance 航摄勘测;航空勘察
aerophotographic(al) sketch 航摄草图;航片略图;像片图
aerophotography 空中摄影术;航空摄影(学);高空摄影学
aerophoto interpretation 航摄像片判读
aerophoto topography 航空摄影测图
aerophysics 空气物理学;航空物理学;高空物理学
aerophyte 气生植物
aerophytobiont 需氧土壤微生物;好气土壤微生物
aeroplane carrier 航空母舰
aeroplane cloth 航空布料
aeroplane depot boat 航空母舰
aeroplane dope 航空涂料
aeroplane hangar 飞机库
aeroplane head-light 飞机起落灯
aeroplane mapping 航空制图
aeroplane oil 航空润滑油
aeroplane photograph 航摄像片;航空像片
aeroplane photography 航空摄影
aeroplane's call sign 飞机呼号
aeroplane shed 飞机库;飞机棚
aeroplane spotter 标绘观测孔
aeroplane station 飞机电台
aeroplane tender 飞机供应船
aeroplane view 高瞰图;高空俯视图;高空俯瞰图
aeroplankton 空气浮游生物

aeroplanktophyte 空中浮游植物
aeroplathysmograph 呼吸气量测量计
aeroplex 航空安全玻璃
aeropolygon 空中导线(测量)
aeropolygon procedure 空中导线法
aeroport 机场;航空港;飞机场
aeroprojection method 立体投影(测图)法;空中投影法
aeroprojector 航空制图仪;航空投影测图仪;立体投影器;航空投影仪;航测制图仪
aeroprojector multiplex 多倍仪;多倍投影测图仪;航空投影多倍测图仪
aeropropeller 空气推进器
aeropropeller vessel 空气螺旋桨船
aeroprospecting 航空勘探
aeropulse engine 脉冲式喷气发动机
aeropulverizer 气流粉碎机;喷气粉磨机;喷磨机;吹气磨粉机
aeroquay 停机码头
aeroradiator 航空散热器
aeroradioactivity 大气放射性
aerosal paint 喷雾罐装油漆;喷雾罐装涂料
aeroscope 空气微生物取样器;空气集菌器;空气纯度(检查)镜;空间观测器
aeroscopy 空中观测术
aeroseal 空气密封
aeroseal door 空气密封门
aeroseat 航空座椅
aeroshed 飞机库
aerosiderite 铁陨石
aerosiderolite 铁石陨石
aerosil 氧相二氧化硅
aerosimplex 简单投影测图仪;航空摄影测图仪
aerosinusitis 航空鼻窦炎
aerosketch master 航片转绘仪;航空像片转绘仪
aerosol 气雾喷涂;气雾剂;气溶胶;气喷制剂;雾化剂;悬浮微粒(空气中);烟雾剂;大气悬浮物;大气微粒;按钮式喷雾器
aerosol analyzer 气溶胶分析器;空气气溶胶分析器
aerosolator 喷雾除尘器
aerosolcan 气溶胶罐
aerosol chamber 气雾室
aerosol chemistry 气溶胶化学
aerosol coating 气雾(喷涂)涂料;气溶胶涂料
aerosol concentration 气溶胶浓度
aerosol container 喷雾剂瓶
aerosol contaminant 污染性悬浮微粒;污染性气溶胶;(空气中的)悬浮微粒污染物
aerosol detector 气溶胶检测器
aerosol development 干粉显影
aerosol dispenser 气溶胶喷雾器
aerosolether-xylene 气溶胶醚二甲苯
aerosol filter 气溶胶过滤器
aerosol filtration efficiency 气溶过滤效率
aerosol flow 气溶胶态流
aerosol for external use 外用气雾剂
aerosol inspiration scanning 气溶胶吸入扫描
aerosolization 烟雾化(作用)
aerosolize 雾化;成烟雾状散开
aerosolizing fan 喷雾风扇
aerosol lacquer 气雾喷漆;气溶胶漆
aerosol layer 烟雾层
aerosol load 气溶胶浓度;气溶胶含量
aerosol monitor 气悬监视器;气溶胶监测器
aerosol monitoring 气溶胶监测
aerosology 气溶胶学
aerosoloscope 气溶胶检测器;空气(中)微粒测量表
aerosol paint 喷漆;气溶胶涂料;气溶胶漆;喷雾涂料;罐装喷漆
aerosol particle 气溶胶微粒;气溶胶粒子;悬浮微粒
aerosol pesticide 气雾剂农药
aerosol pollutant 气溶胶态污染物
aerosol producer 烟雾发生器
aerosol propellant 气雾推进剂;气溶剂的挥发剂;气溶胶推进剂;气溶胶喷射剂;气溶胶发生剂
aerosol release agent 气溶胶脱模剂
aerosol sampling 气溶胶采样
aerosol sampling device 气溶胶采样器;大气微粒取样器;空气采样装置;空气采样器
aerosol size spectrum 气溶胶粒度谱
aerosol spectrometer 气溶胶谱仪;气溶胶分光仪
aerosol spray 气溶胶喷射制剂
aerosol spray-can 气溶胶喷罐;喷雾罐
aerosol sprayer 烟雾喷射机

aerosol spraying 气溶胶(式)喷涂;喷烟
Aerosol surfactant 阿罗索表面活性剂
aerosol test 烟雾试验
aerosol transmission 气溶胶传播
aerosonator 脉冲式喷气发动机
aerosowing 飞机播种
aerospace 航空和航天
aerospace automatic landing 航空航天自动着陆
aerospace cartography 空间制图学
aerospace craft 宇航飞船;航天飞行器;宇宙飞船
aerospace disease 宇宙病
aerospace engineering 航空和航天工程
aerospace environment 航天环境;航空航天环境
aerospace glass 航天玻璃
aerospace ground equipment 航空航天地面设备
aerospace industry 空间工业;航空和航天工业;宇航工业
aerospace material 航天材料;宇航材料
aerospace seal 气密;气封;航空密封
aerospace standards 宇宙标准
aerospace survey 航天测量
aerospace surveying 航天测量法
aerospace vehicle 航空航天器;宇航飞行器;空间飞行器
aerosphere 气圈;气界;空气圈;高空气层;大气圈;大气界;大气(层)
aerospring shock absorber strut 航空弹簧减震器支柱
aerostat 气球驾驶员;航空器;高空气球;飞艇
aerostatic bearing 气体静压轴承;空气静压轴承;空气静力轴承
aerostatic lubrication 气体静力润滑
aerostatic press 气压机
aerostatics 气体静力学;空气静力学
aerostation 浮空术
Aerostyle 手提式喷洒设备
aerosurvey 航空测量;航空勘测
aerosurveying 航空摄影测量;航空勘测
aerotank 曝气池
aerotar 航摄镜头
aerotaxis 趋氧性
aerotechnics 航空技术
aerothermoacoustics 湍流声学
aerothermochemistry 气动热化学;空气热化学
aerothermodynamic duct 冲压式喷气发动机
aerothermodynamics 气体热动力学;气动热力学;空中热动力学;空气热力学
aerothermoelasticity 气动热弹性;空气热弹性(理论)
aerothermopressor 气动热力压缩器
aerotire 飞机轮胎
aerotolerant 耐氧的
aerotolerant bacteria 耐氧细菌
aerotometer 气体张力计
aerotonometer 气体张力计
aerotonometry 气体张力测量法
aerotopography 航空摄影测图;航空地形测量
aerotorelant bacteria 耐氧菌
aerotow 空中牵引
aerotrack 简易机场
aerotrain 气垫列车;气垫火车;悬浮火车;飞行式无轨列车;气垫式超高速列车(法国)
aerotraverse 空中导线
aerotraversing 空中导线测量
aerotriangulation 航空三角测量;空中三角测量
aerotriangulation by independent model 独立模型法空中三角测量
aerotriangulator 空中三角仪
aerotron 三极管
aerotropism 向气性
aeroturbine 航空涡轮机
aerotympanal 空气鼓室的
aeroultra filter paper 高效空气过滤纸
aerovan 运货机
aerovane 旋翼;风速计;风车;方向计;风向风速仪;风向风速器;风向风速计
aerovelox 小型投影测图仪
aerovibro screen 悬吊振动筛
aeroview 鸟瞰(图);高空俯视图;高空俯瞰图
aerozine 航空肼
aeruginous 涂氧化铜的;铜绿色的
aerugite 块砷镍矿
aerugo 金属氧化物;腐锈斑;凹坑;铜锈;铜绿
aes 铜锡合金

aesar 蛇(形)丘
aeschynite 易解石
aeschynite-(Y) 钇易解石
Aesculus hippocastanum 马栗树;欧洲七叶树
aesenometric titration 酸酸滴定法
aestatisilvae 夏绿乔木林
aesthete 美学家
aesthetical 美学的
aesthetic appeal 美学作用
aesthetic appreciation 地图美学
aesthetic area 景色美丽的地区;观光区
aesthetic aspect 审美的问题;美学角度;美学观点;美学方面
aesthetic charm 美学魅力;美学作用;艺术吸引力
aesthetic concept 美学观念;美学设想;美学概念
aesthetic consciousness 审美意识
aesthetic consideration 审美考虑;美学观点
aesthetic control 景观控制;美观控制
aesthetic criterion 美术标准
aesthetic design 美学设计
aesthetic design of bridges 桥梁的美观设计
aesthetic education 美学教育
aesthetic effect 艺术效果;美学效果
aesthetic feeling 美感;艺术情调;艺术鉴赏力;艺术感受力
aesthetic forest 风景林
aesthetic idea 审美观念;美学思想;美学设想
aestheticism 唯美主义
aesthetic judgment 审美能力
aesthetic measure (色彩调配的)美观程度
aesthetic properties of water 水的感官性状
aesthetic quality 感官质量
aesthetic road 景色美丽的道路;风景道路
aesthetics 美学
aesthetic sense 艺术感觉;审美意识;美学理念;美感
aesthetics of bridges 桥梁建筑艺术
aesthetics of highway 公路美学
aesthetic square 城市艺术广场;城市美观广场
aesthetic standard 审美观
aesthetic theory 美学理论
aesthetic treatment 艺术处理
aesthetic value 美学价值;风景价值
aesthetic zoning 美化地区规划;景观区划;美观区划;风景区规划
aestidurilignosa 夏绿硬叶木本群落;夏绿硬叶林;落叶常绿硬混合硬木群落
aestilignosa 夏绿木本群落夏绿硬叶木;夏绿林夏绿硬叶林
aestival 夏季的
aestival aspect 夏季相
aestivation 花被卷叠式
aetheriastite 变柱石
aethoballism 阻击变质作用
aethylis chaul-moogras 大风子油酸乙酯
aetiology 病原学;病因学;病理学
aetite 矿石结核
aetoma 三角山墙楣饰
aetos 三角山墙檐饰
aetylsalicylic acid 醋柳酸
a fake 一盘绳索
afara 伦巴木(非洲榄仁树)
Afar depression 阿法尔洼地
Afar triple junction structure 阿法尔三向联结构造
afebrile 无热的
afercooling 后冷却
affect departure 偏差
affected area 受影响的区域
affected coefficient of mean settlement 平均沉降影响系数
affected degree of new usage 新用途影响程度
affected degree of substitution 代用品影响程度
affected head 影响水头
affected layer 损伤层;加工变质层;影响层
affected people 影响人口
affected population 影响人口
affected seedling 受害苗
affected zone 波及区
affecting area of collapse tunnel top 洞顶塌陷影响范围
affecting factor 影响因素
affection caused by summer-heat 冒暑
affectless polynomial 无偏差多项式
affect zone (焊接的)变质区
afferee 被发价人

afferent blockage 向中路遮断
afferent echo 进波
afferent hindrance 向中路遮断
afferent limb 传入支
afferent limb block 传入支封闭
afferent phase 传入期
affidavit 声明;具结书;宣誓书;正式书面陈述;公证书(银行);保证书
affidavit of non-collusion (投标人的)非串通证誓;(投标人的)非串通保证书;(投标人的)非串通誓言
affidavit of title 产权证明书;产权宣誓声明书
affiliate 联营;分(支)机构
affiliated agencies 分支机构
affiliated company 联营公司;联络公司;联号公司;附属公司;分公司
affiliated enterprise 联营企业;附属企业;分支企业
affiliated interests 联营股权;联络股权
affiliated middle school 附属中学
affiliated primary school 附属小学
affiliated retailer 附属零售店
affiliated school 附属学校
affiliated society 分会
affiliate member 从属构件
affiliation 亲缘关系;联营
affinage 精炼
affinant 亲和力
affination centrifuge 精制离心机
affine 仿射的
affine and projective model 仿形和投影模型
affine and projective transformation 仿形和投影转换
affine block cipher 仿射分组密码
affine cipher 仿射密码
affine collineation 仿射共线
affine connection 仿射联络
affine connection space 仿射联络空间
affine conversion 仿射变换
affine coordinates 仿射坐标
affine curvature 仿射曲率
affine deformation 仿射变形;仿射变形;对应变形;均匀变形;拟似变形
affine distortion 仿射畸变差
affined restitution instrument 仿射纠正仪
affined transformation 仿射变换(式)
affined transformer 仿射变换器
affine factor 仿射因数
affine film shrinkage 胶片仿射收缩
affine geometry 仿射几何
affine group 仿射群
affine hull 仿射包
affine image 仿射影像
affinely connection space 仿射联络空间
affine mapping 仿射映象
affine-plotter 仿射绘图仪;仿射测图仪
affine plotting 仿射绘图;仿射测图;变换光速测图
affine rectification 倾斜面纠正;仿射纠正
affine ring 仿射环
affine set 仿射集
affine space 仿射空间
affine stereoplanigraph 仿射精密立体测图仪
affine transformation 亲缘转换;仿射变换
affinitizing solution 亲合液
affinity 亲合性;亲合势;亲合力;仿射性
affinity chromatography 亲合色谱法;亲合层析法
affinity coefficient 亲合系数
affinity constant 亲合常量
affinity curve 亲合曲线
affinity diagram 亲和图法
affinity factor 变换光速系数
affinity for metal 对金属表面的亲合力
affinity for water 亲水性
affinity labelling 亲合标记
affinity law 相似定律
affinor 反对称张量
affirmation 确认
affirmation of contract 批准合同
affirmative 肯定的
affirmative acknowledgement 肯定应答;肯定回答
affirmative action 肯定行动
affirmative fair housing marketing 公平住房销售计划
affirmative field 确认字段
affirmative flag 同意旗

affirmative network diagram 肯定型网络图
affirmative proposition 肯定命题
affirmative sign 正号
affirmative vote 赞成票
affix 签署;小装饰物;附加物;附加(上);附标
affix a seal 加盖印章
affixing of seal 盖章;盖印
affixing trademark tags 刷贴商标标签
affix ion 添加
affix one's seal 盖章;盖印
affix one's signature 署名;签名
affix revenue stamp 贴用印花
affix stamps 贴邮票
affixture 粘上;贴上;添加物;结合物;加成物
affluence 集övers;汇入
affluent 流畅的;丰富的;富裕的;富饶的;支流;汇流的
affluent consumption structure 小康型消费结构
affluent level 上游水位;壅高水位;支流水位
affluent society 小康社会;富裕社会;丰裕社会
affluent stream 汇入的支流;补给河
afflux 群集;集流;汇入;汇集;堰水位差;补给河
afflux curve 汇流曲线;回水曲线
affluxion 群集;汇入;补给河
afflux mass 流入质量
afflux of waste water 废水的流入
afflux xion 流注
afford 提供
affordable housing 低价住宅;低价住房
affordable method 可供给额法
afford no relief 不能缓解
afford priority 给予优先权
afforest(ation) 造林;绿化(造林);荒(山)地造林;植树造林;造林计划
afforestation area 造林区;造林面积
afforestation committee 绿化委员会
afforestation fee 绿化费
afforestation for erosion control 防蚀造林;防沙造林;防冲刷造林
afforestation for sand control 造林治沙
afforestation of political science 政治学绿化
afforestation project 造林工程
afforestation relocation 绿化迁改
afforestation sand-fixation 固沙造林
afforestation station 植树站
afforested area 造林区;造林面积
afforested mountain closed off for protection 封山育林
afforesting at portal 洞口绿化
affreight 租船
affreightment 海运契约;海上运输契约;租船运输;租船运货;租船合同;租船契约;海运运货合同
affreightment contract 海运契约
affronte(d) 面对面人兽图案雕饰
afghanite 阿钙霞石
afield 在野外
a fifteen minutes frequency 每十五分钟测一次
A finish 木纹饰面
a five-flight shiplock system 五级的船闸系统
aflame 冒烟
afloat 漂浮;可流通;业经装船
afloat cargo 已装船货物;路货
afloat contact 浮动接点;浮动触点
afloat evapo(u)rimeter 漂浮(式)蒸发计
afloat goods 已装货物;业经装船之货
afloat repair 水上检修;不进坞检修
afocal 焦外;非聚焦系统
afocal lens 无焦透镜;远焦透镜
afocal system 无焦系统;远焦系统
afoliate 缺叶
afordability 可承受性
afore mentioned rules 上述规定
afoul of a ship 碰船
A-frame 三角支架;A 形构架;人字构架;金字顶式建筑物;A 形框架
A-frame dam A 形构架坝;A 架坝
A-framed derrick 人字起重机;人字架转臂起重机;A 形钢架式摇臂起重机
A-frame design A 形框架结构;A 形框架设计
A framed weir A 形构架(活动)堰
A-frame gantry 人字支架
A frame groyne A 形丁坝
A frame portainer crane A 形构架岸边集装箱起重机

A-frame timber dam A 形木构架坝
A-frame weir A 架堰
A-framing derrick 人字吊杆
A F ratio 铝铁比
Africa-Antarctica separation 非洲大陆与南极大陆分离
Africa-Eurasia collision 非洲大陆与欧亚大陆碰撞
African, Caribbean and Pacitic states [ACP states] 非加太国家
African cherry 西非樱桃木
African chess-board structure 非洲棋盘格式构造
African Development Bank 非洲开发银行
African Development Fund 非洲发展基金
African ebony 非洲黑檀木;非洲乌木
African Economic Community 非洲经济共同体
African Economic Unity 非洲经济一组织;非洲经济统一体
African Financial Community France 非洲金融共同体法郎
African Franc 非洲法郎
African incense 非洲油树脂
African Institute for Economic Development and Planning 非洲经济发展和规划研究所
African Intellectual Property Organization 非洲知识产权组织
African mahogany 非洲桃花芯木;非洲红木
African Ministerial Conference on Environment 非洲环境问题部长会议
African Non-government(al) Organization Environment Network 非洲非政府组织环境网
African ochre 非洲赭石
African plate 非洲板块
African platform 非洲地台
African Regional Center for Technology 非洲地区技术中心
African rosewood 西非荧檀木;西非青龙木;西非花梨木
African shield 非洲地盾
African transgression 非洲海浸
African walnut 非洲核桃棟的黄色木材
African whitewood 非洲轻木;非洲白木
African zebrawood 非洲条纹木;非洲斑马木
A frome timber dam A 形构架木坝
afrormosia 非洲红豆木
aft 在船尾
aft antenna 后部天线
aft-bay 堰下游建筑部分;尾水池;后池;下游堰(船闸);船闸下游建筑部分;下游河湾;闸下游建筑部分;石架间
aft breast line 船尾横缆
aft closure 后盖
aft cockpit 后座舱
aft draft 后吃水;船尾吃水(深)
aft end bulkhead 后端壁
aft end face 后端面
aft end face of boss 桨毂后端面
aft end stiffener 后端壁扶强材
aft-engine boat 尾机型船
aft-engine ship 艇后机型船;艉机船;尾机(型)船
aft-engine vessel 尾机型船
after acceleration 后(段)加速
after accommodation 后部舱室
after acquired property 经法庭裁定后债权获得的财产留置权
after-admission 补充进汽
after air filter 后置空气过滤器;次级空气过滤器
after anchor 后锚;船尾锚
after anchor light 后锚灯
after appraisal 后评法
after autumn 秋后
after back spring 后助倒缆
after-bake 二次焙烧;后期烘烤;烘
after baking 后期焙烧;后烘烤;二次焙烧;后热处理;后期烘烤
afterbay 后架间;后湾;后池;闸下游建筑部分;尾水池;下游塘;下游河湾;尾水池;石架间
afterbay dam 尾水池坝
afterbay reservoir 尾水水库
after bitt 船尾系缆桩
after-blocking 后粘连
after blow 二次吹风;后吹
afterbody 后体;物体后部;机尾;机身尾部;机身后部(飞机);后尾部;后段船体;弹体尾部;船体后部
afterbody effect 后体效应

afterboiler 后锅炉;二次锅炉
after boiler room 后锅炉舱
after-break 地面沉陷后的破坏
after breast(line) 后横缆;艉横缆;尾横缆
after breast rope 后横缆
after bridge 艉桥楼;船尾桥楼
after brow 船尾跳板
after bulkhead 后舱壁
afterburner 后燃装置;二次燃室;后燃烧器;补燃器;补热器
afterburner control 加力燃烧室控制
afterburner fuel control 加力燃烧室供油调节
afterburner nozzle 强化燃烧室喷嘴;加力燃烧室喷嘴
afterburner rating 加力燃烧室额定功率
afterburning 后燃烧(现象);后燃;后期燃烧;二次燃烧;补燃;迟燃;后烧;再烧;重烧;加力燃烧
afterburst (地下坑道岩石爆破后发生的)后期岩崩;突出后崩
after cabin 后舱室
after capstan 船尾绞盘
after-care 房屋竣工后维修;抚育
after-care of disposal sites 垃圾堆置场的使用后管理
aftercarriage 后车体
after castle 船尾楼
after charges 附加费率
after chemical reaction 经过化学反应
afterchine 后舷脊
afterchlorinate 补充氯化
after-chlorinated polyvinylchloride fibre 后氯化聚氯乙烯纤维
after-chromed dye 后铬媒染料
after-clap 意外变动
after closing 结账后
after closing trial balance 结账后试算表
after cofferdam 尾隔离舱
after-collector 二次集尘器;二次收集器;后(加)收集器
after-combination 复烧
after-combustion 二次燃烧;后燃烧;再次燃烧;复燃;补充燃烧
after-compaction 后压密;补压;补层
after compass 尾罗经
after-condenser 后冷凝器;再冷凝器;二次冷却器;二次冷凝器
after conning position 船尾操纵部位
after-contraction 残余收缩;后(期)收缩;再加热收缩;成型后收缩;附加收缩
after-cooler 后冷却机;后(级)冷却器;烟气冷却器;二次冷却器;终冷却器;减热去湿器;末级冷却器;后(置)冷却器;压缩空气冷却器;最后冷却器;再冷却器
after-cooler core 后冷却系统
aftercooling 后冷却;再冷却;二次冷却
after cooling zone 后冷却带
after-cost 后生成本;后续成本
after crank 后曲拐
after crank shaft 后曲柄轴
after-crop 第二次收获;晚季作物
after-culture 补栽补种;补播
after cure 后硫化;后固化;后处理
aftercurrent 残余电流;后电流
after damage 后遗灾害
afterdamp 爆炸后毒气;煤矿爆发后毒气;煤矿坑煤气;爆后毒气
after darkening 变深;变暗
after date 期后;出票后
after date bills 发票日后付款汇票
after date of draft 出票后
after deck 后甲板;后部甲板
after depolarization 后除极
after-discharge 后放
after draft 尾吃水;后吃水;船尾吃水(深)
after draught 尾吃水
after drawing 后拉伸
after-drying 后干燥;再次干燥
after earthquake 余震
after edge 后缘
after effect 后效(应);后果;余功;滞后效应;副作用
after end 后端;船尾端
after engine room 艉机舱
after engine ship 艉机船;尾机船
after-etching 残余腐蚀

after-expansion 后(期)膨胀;附加膨胀;残余膨胀;残存膨胀
after exposure 后(期)曝光;二次曝光
after fender 尾碰垫;尾护舷材
after file 锉过
after-filter 终端过滤器;后(过)滤器;二次过滤器;补充过滤器;后滤池
after-filtration 后滤;最后过滤
afterfiring zone 后烧段
after fixing 后固定
afterflaming 余火;烧尽;后期燃烧;二次燃烧;补充燃烧
after flow 蠕变;塑性变形;后期塑性变形;后吹;残余塑性流动
after-flush 水箱残渣;冲洗滞留;便器冲洗水;冲后水封
after flush cistern 抽水马桶蓄水箱
after-flush compartment 背水箱(水冲便池背后的水箱);冲后水室
after-flushing compartment 背水箱
after-fractionating 再蒸馏;二次分馏
after frame 补架;尾肋骨;后框(架)
after gallows 后吊架
after-gas 爆后气体
aftergases 灾后气体;爆炸后的有害气体
afterglow 后辉;余辉
afterglow correction 余辉校正
afterglow duration 余辉延续时间
afterglow eliminator 余辉消除器
afterglow light 余辉光
afterglow resistance 抗余辉性
afterglow screen 余辉屏
afterglow time 余辉延续时间
afterglow tube 余辉管
aftergrass 再生草
aftergrowth 后生树;再生
after guy 后张索;后稳索
after-hardening 后(期)硬化;发黏
after hatch 尾舱口;后舱口
after head spring 后助缆
after-heater 后热器;复燃室补燃器
after heat(ing) 后热;后加热;秋老虎;余热
afterhold 后货舱;后舱;船尾舱
after hood 尾端外板
after hours value 闭市后价格
after hyperpolarization 后超极化
after-image 后像;余留影像;残留影像;残留图像;残像
afterimage phenomenon 残像现象
after imbibition of water 水分吸胀后
after impulse 剩余脉冲
after inspection 检查后
after-irrigation moisture content of soil 灌水后土壤中含水量
after island 船尾上层建筑
after island poop 船尾楼
after landing quality terms 起岸后品质条件
after leech rope 后帆边绳
after-light 尾灯;余辉;夕照
after line 艉缆;尾缆;后缆;船尾缆
afterload 后负荷
after-loading 后负荷
after-loading technique 后装源技术
after loss realization 已有亏损后;变现损失
after market 零件市场;修配业市场;闭市后市场
aftermast 后桅
aftermast head light 后桅灯
aftermath 后果;再生草;次生岩(石);次生矿床;沉积岩
aftermath pasture 再生草地
aftermost breadth 尾端宽度;船尾宽度
after-movement 后继性运动
afternoon 下午
afternoon fever 潮热
afternoon watch 午后班
after of paper 纸张泛黄
after-only design 事后设计(企业管理)
after-only measurement 事后测定
after overhang 跨距;悬后
afterpart 后部;船体后部;尾部;后段船体
after peak 船尾尖舱
after peak bulkhead 尾尖舱舱壁
after peak(water)tank 后尾舱;尾尖舱
aft(er)perpendicular 尾垂线;后垂线;船尾垂线

after piece 舵踵;舵板根
after-polarization 后极化
after polymerization 后聚合(作用)
after poppet 尾支架【船】;后支架;船尾垫架
afterpotential 后电位
afterpower 滞后功率
after precipitation 再凝结;再沉淀(作用);二次沉淀;后沉积(作用)
after-precipitation fouling 后沉淀污垢
after-precipitation structure 后沉积构造
after pressure 后压
afterproduct 后产物;二次产品
afterpulse 剩余脉冲;残留脉冲
after purification 后净化;二次净化;最后净化;补充净化
after quake 余震
after range light 后叠标灯
after refuge compartment 后部脱险舱
after replacement 补植
after rib 后肋骨
after-ripening 后熟
after rope 尾缆
after run 去流段;船尾端部
after sails 后部帆
after-sale quality service 产品售后技术服务
after-sale service 返修服务;保修服务;售后服务;交工后技术服务;包修服务;包修保用技术服务
after screen derrick 舱尾壁吊杆
aftersection 尾部
after separation 裂开后
after separation cost 分离后成本
after service 售后服务;产品售后技术服务;包修服务;包修保用技术服务
after settler 后澄清器
aftershaft 副羽
after sheer 尾舷弧
aftershock 后震;后波;余震
aftershock activity 余震活动
aftershock sequence 后震序列;余震序列
aftershock wave 余震波
after-shrinkage 后(期)收缩;成形后收缩;残余收缩;二次收缩
after sight bill 见票后若干天付款的期票
after slip 余滑
after-sound 余音;残音
after spring(line) 后尾缆;艉倒缆;船尾倒缆
after stain 后染
after-stove post baking 后烘烤
after strain 后变形
after streaming light 后桅灯
after-stretch(ing) 后拉伸
after summer 秋老虎
after table 后工作台
after-tack 回黏(性);软化;返黏;不干漆;残余黏性;油漆过黏;发黏漆
after-tackiness 残余粘性;后效粘性
after tank 船尾水舱
after-taste 余味
after-tax cash flow 税后现金流量
after-tax deduction 纳税后扣除
after-tax income 税后收入;纳税后收入
after-tax profit 扣除所得税后的利润;税后利润;纳税后利润
after-tax profit rate 纳税后利润率
after-tax yield 税后收益;付税后收益
afterteeming 补浇
after telemotor 后部液压遥控操舵装置;操舵摇控传动装置
after-tensioning 后加张力
after-the-fact analysis 事后分析
after the same pattern 按同一方式
after the shell dries 外壳干燥后
after-thickening 后期增稠;后稠化;返稠
afterthreshold behavio(u)r 阈值后工作状态
aftertime 余辉时间
aftertossing 船尾浪;船尾波动;船尾余波;风暴后余波;艉浪
after-tow 后曳;艉曳
after transfer cost of material 材料二次搬运费
after transfer expenses 二次搬运费
after-transformation 后期转变
after-treatment 第二处理;二次处理;补充处理;后(期)处理
after vibration 后振;后期振动;余振动

after-vision 后视觉
after waist spring 后舷斜缆
after-wale 后隆起
after-wash(ing) 后水洗
after-wind 余风
after-working 后效(应)
after yard 后桁架
after-yellow(ing) 枯黄后;泛黄后;后黄变
aft fan 后风扇
aft foil hydraulic actuator 尾翼液压传动装置
aftfoot bulb 球鼻形尾柱底部
aft gate 尾水闸门;下(游)闸门
aft-hold 后舱
aft ice belt region 尾部冰带区
aft main anchor 船尾主锚
aft most 靠近船尾
Aftonian interglacial epoch 阿夫唐冰期【地】
Aftonian interglacial stage 阿夫唐间冰阶【地】
aft part 后部
aft perpendicular 艉垂线;后垂线
aft port side anchor 后左边锚
aft ramp 后调节板
aft section of hull 船尾分段
aft shaft 后轴
aft sidelight 尾舷灯
aft spheric(al) chamber 尾球形舱
aft spud 艇定位桩
aft starboard anchor 后右边锚
aft steering wheel 后操舵轮
aft superstructure 艉部上层建筑
aft thruster 后推力器
aft type ship 尾机型船
afunction 机能缺失
afwillite 硅钙石;水合硅酸钙;柱硅酸钙
afzelia 亚氏孢粉
Agache-Willot 阿加什—维约公司
agacol 沉香萜醇
agad 海滨植物
against 针对
against actual 以期换现
against all risks 全险;投保全险;保全险
against corrosion 抗蚀
against current 顶流
against documents 现金支付
against hydrostatic(al) uplift 抗静水扬压力
against stream 逆流
against the ear type earphone 触耳式耳机
against the force of the spring 克服弹簧阻力
against the reaction of the spring 克服弹簧阻力
against the stream 顶流
against the sun 反盘绳法
against the wind 顶风
against to the grain 垂直于纹理
against us 负债
against vote 反对票
against wall 靠墙
against weather 逆风顶浪
agalite 纤(维)滑石
agalloch 沉香木
agallochum 加罗木;沉香木
agalma 向神供品(古希腊);奉献给上帝的艺术品(古希腊)
agalmatolite 寿山石;蜡石;叶蜡石;冻石
agalmatolite(fire)brick 蜡石耐火砖
agalmatolite fireproofing material 蜡石耐火材料
agamete 拟配子
agamous plant 隐花植物
agar 琼胶
agar-agar 琼脂;冻粉
agar attar 东方沉香油
agar block method 琼脂块法
agar block technique 琼脂块技术
agar chromatography 琼脂色谱法
agar culture 琼脂培养
agar cup method 琼脂杯法
agar deep culture 琼脂深层培育
a garden patch 一块园地
agar diffusion method 琼脂扩散法
agar disc method 琼脂片法
agardite 砷钇铜石
agar double diffusion technique 琼脂双扩散测定法
agar electrophoresis 琼脂电泳
agar embedding material 琼脂包埋剂
agar gel 琼脂凝胶

agar gel diffusion test 琼脂扩散试验
agar gel electrophoresis 琼脂凝胶电泳(法)
agar gel reaction 琼脂凝胶反应
agar hanging block 琼脂悬块
agaric 蘑菇
agaric acid agaricin 琼脂酸
agaric mineral 伞形矿物；蘑菇状矿物；石菜花矿物
agar jelly strength tester 琼脂胶体强度计
agar medium 琼脂培养基
agaroid 类琼胶
agaropectin 琼脂胶
agarose coagulation 琼脂糖凝胶
agar photoelectric(al) photometer 琼脂光电光度计
agar plaque technic 琼脂空斑技术
agar plate 琼脂平皿；琼脂平板
agar plate culture 琼脂平面培养(法)
agar plate method 琼脂平面培养法
agar slant 琼脂斜面(培养)
agar slant culture-medium 琼脂斜面培养基
agar stab 琼脂针刺培养
agar stab culture 琼脂针刺培养
agar streak method 琼脂划线法
agar strength 琼脂强度
agar tube 琼脂管
Agassiz orogeny 阿格赛兹造山运动【地】
Agassiz trawl 阿格赛兹拖网
Agassiz Valleys 阿格赛兹海下谷
agate 玛瑙
agate artware 玛瑙工艺品
agate balance edge 玛瑙天平刀子
agate ball 玛瑙球
agate bead 玛瑙珠
agate bearing 玛瑙轴承
agate burnisher 玛瑙抛光机
agate cap 玛瑙尖顶
agate carving 玛瑙玉雕；玛瑙雕刻品
agate cup 玛瑙杯
agate end stone jewel 玛瑙端面轴承
agate glass 玛瑙玻璃
agate guide 玛瑙导丝器
agate hole jewel 玛瑙通孔轴承
agate jasper 玛瑙碧玉
agate knife-edge 玛瑙刀口
agate marble paper 仿大理石纹纸
agate mortar 玛瑙研钵；玛瑙乳钵
agate mortar and pestle 玛瑙臼和杵
agate opal 玛瑙蛋白石
agate pearl 玛瑙珍珠
agate pestle 玛瑙钵杵
agate plane 玛瑙面
agate plate 玛瑙片
agate spatula 玛瑙调刀
agate spheric(al) jewel 玛瑙球形轴承
agateware 玛瑙纹饰陶器；玛瑙器皿；仿玛瑙陶器(皿)；斑纹玻璃
agathenedi-carboxylic acid 贝壳松烯二(羧)酸
agathic acid 贝壳松烯二(羧)酸
agathin 阿加赛因
agatized wood 硅化木
Ag-Au ore 银金矿石
agave fiber 龙舌兰纤维
agawood 沉香木
agbar 非洲阿勃木
Ag-bearing copper ore 银铜矿石
age 朔望间隙；熟化；时代；龄期
age-adjusted death rate 年龄校正死亡率
age-adjusted mortality rate 年龄校正死亡率
age assessment 年龄鉴定
age at last birthday 实足年龄；足岁
age at loading 加载时龄期；加荷时龄期
age at retirement 退休年龄
age at test 试验时龄期
age at withdrawal 退休年龄
age-boundary method 年界法
age-bracket 年龄范围
age census 年龄普查
age characteristic 年龄特征
age class 同龄级木；龄级；年龄等级；分年龄组；年龄级(动物出生年)
age class distribution 龄级分布
age class period 龄级期
age class structure 龄级结构
age coating 老化覆层；老化变黑层
age coating of lamp 灯泡老化层

age cohort 年龄群组；年龄分组
age composition 年龄组成；年龄构成
age-conversion factor 年龄换算系数
age-corrected basis 年龄校正基准
age-crack 老化开裂
age cycle 龄排列
aged alloy powder 时效合金粉末
aged and rale tree 古树名木
age dating 年代测定；年龄测定
aged batches of solution 分批熟化溶液
aged cement 过性水泥；过期水泥；陈水泥
aged clay 老黏土；陈化黏土
aged clay layer 老黏土层
aged clinker 陈化熟料
aged column 老化柱
age determination 年龄鉴别；年龄测定
age determination of the earth 地球年龄测定
aged factor 年代因素；年代因数；时效因素；龄期因数；老化因数
aged flat 老年公寓
aged flexibility 时效柔韧性
aged floc 老凝聚体
aged ice ridge 老化冰脊
aged insulation value 老化后的热阻值
Age Discrimination in Employment Act 就业年龄差异法(美国)
age distribution 寿命分布；年龄分布(图)；账龄分类
age distribution function 年龄分配函数
aged people's apartment 老人公寓；老年公寓
aged people's community center 老年社区活动中心
aged people's dwelling 老人住宅；老年人住房
aged person's apartment 老人公寓；老年公寓
aged person's home 敬老院；养老院
aged property 老化后性能
aged-refuse fixed bed 矿化垃圾固定床
aged ridge 老化冰脊；多年的冰丘脉
aged seawater 老化海水
aged shore 成年海岸
aged soil 老年土
aged steel 时效钢
aged vessel 老船
aged viscosity 熟化黏度；老化黏度
age effect 年代效应
age embrittlement strain 老化脆性应变；陈化脆性应变
age-fast 抗老化的；抗时效的；抗老化的
age forming 时效成形
age gradation 龄阶
age group 龄期组；年龄组；年龄群
age growth 年龄生长
age hardening 时效硬化；经时硬化；经久硬化；龄期
age hardening stainless steel 时效硬化不锈钢
age hardness 陈硬
age incidence 年龄发生率
age indicator 年龄标记
ag(e)ing 熟化；时效化；时效；老熟；老化
ag(e)ing accounts receivable 应收账款的分期
ag(e)ing apparatus 老化试验器
ag(e)ing behaviou(u)r 时效效应；老化性能；老化特性；耐老化状态；时效行为
ag(e)ing blemish 老化污点
ag(e)ing bunker 陈化仓
ag(e)ing can 老化罐
ag(e)ing change 年龄变化
ag(e)ing characteristics of thermistor 热敏电阻衰老特性
ag(e)ing city 老龄化城市
ag(e)ing coefficient 老化系数
ag(e)ing condition 熟化条件；时效规范；老炼规范
ag(e)ing contraction 老化收缩
ag(e)ing crack 自然开裂；时效裂纹；自然裂纹；自然裂开
ag(e)ing effect 时效；老化作用；老化效应；迟后效应；陈化效应
ag(e)ing error 老化误差；老化错误
ag(e)ing factor 老化因素；龄期因数
ag(e)ing failure 老化故障
ag(e)ing hardening 时效硬化；经久硬化
ag(e)ing index 老化指数
ag(e)ing industry 落后工业；过时工业
ag(e)ing life 老化寿命
ag(e)ing maturing 陈化
ag(e)ing of asphalt 沥青老化

ag(e)ing of clay 土壤的陈化；黏土老化
ag(e)ing of column 柱老化
ag(e)ing of dam 坝的老化
ag(e)ing of insulation 绝缘老化
ag(e)ing of lake 湖泊老年期；湖泊老化
ag(e)ing of metal 金属的时效
ag(e)ing of pipe 管道老化
ag(e)ing of plastic material 塑料(的)老化
ag(e)ing of plastics 塑料老化
ag(e)ing of population 人口老(龄)化
ag(e)ing of roofing felt 屋面油毡老化
ag(e)ing of rubber 橡胶老化
ag(e)ing of solidified product 固化物老化
ag(e)ing of thermometer glass 温度计玻璃陈化
ag(e)ing of trees 树木老化
ag(e)ing of valve 电子管的老化
ag(e)ing oven 老化试验箱；老化炉
ag(e)ing period 熟化期；老化(时)期；成熟期；放置(时)期
ag(e)ing phenomenon 老化现象
ag(e)ing plaster 陈腐烧石膏
ag(e)ing population 老年人口
ag(e)ing printing 香味印刷
ag(e)ing process 熟化过程；时效过程；老化过程；老化法
ag(e)ing property 老化性质
ag(e)ing rack 老炼台；老化架
ag(e)ing rate 老化速度；老化率
ag(e)ing resistance 抗时效(性)；时效阻力；时效电阻；老化抗力；抗老化性(能)；抗老化
ag(e)ing-resistant 抗老化的；抗老化剂
ag(e)ing resistant agent 防老(化)剂
ag(e)ing-resistant grease 耐老化润滑脂
ag(e)ing retardation 防老化
ag(e)ing room 陈腐室
ag(e)ing schedule 客账分析表；分期表
ag(e)ing silo 陈化仓；陈腐室
ag(e)ing society 老年化社会
ag(e)ing stabile 时效稳定性
ag(e)ing stability 老化稳定性；抗老化稳定性；经时老化稳定性
ag(e)ing stage 陈化阶段
ag(e)ing steel 时效钢
ag(e)ing step 老化阶段
ag(e)ing strengthening 时效强化
ag(e)ing tank 老化容器
ag(e)ing test 时效试验；老化试验；陈化试验
ag(e)ing tester 老化实验仪
ag(e)ing theory 老化理论
ag(e)ing time 熟化时间；老化时间
ag(e)ing tower 陈化塔；陈腐仓
ag(e)ing treatment 时效处理
ag(e)ing unit 老机组
ag(e)ing vessel 老化器
age inhibiting 防老化的
age inhibiting addition 防老化添加剂
age inhibiting additive 抗老化添加剂
age involution 年老衰退
age limit 使用年限；年限
age limit of chart folios 换海图年限
age map 龄图
age mark 年龄标志
Agenaclass lunar orbiter 阿琴那型月球轨道飞行器
Agena control system 阿琴那控制系统
Agena rocket 阿琴那火箭
agency 所；经销处；媒介；代理机构；代理公司；代理(处)；代办所；办事处
agency account 代理账户；分经理账
agency agreement 通汇契约；代理协议；代理合同
agency arrangement 通汇契约
agency bank 分行
agency bills 代理期票
agency commission 代理手续费
agency contract 代理契约；代理合同
agency fee 代理费
Agency for International Development 国际开发署(美国)
agency fund 代理基金；分销处基金
agency in charge 主管机关
agency in charge of land 土地主管部门
agency of necessity 必需的代理
agency service 代理业务
agency-shop 代理商店
agenda 记事册；记录；待议事件；程序序列；备忘录

agendum 议事日程;议程;记事簿;执行规程;[复] agenda
agendum call card 记事调用卡片;执行规程卡片
agent 试剂;营力;作用剂;动因;代理行;代理商;代理(人);代表
agent banks 代理行
agent concentration 灭火剂浓度
agent contract 代理契约
agent fee 代理费
agent for bidding 投标代理商
agent for clearing customs 报关代办者;报关代办人
agent for collection 托收代理
agent middleman 居间商
agent of carrier 承运人代理
agent of deflocculation 反絮(凝)剂
agent of disease 病原体
agent of erosion 侵蚀力;侵蚀动因;侵蚀的原动力;冲刷力;冲刷动因;腐蚀剂
agent of fusion 熔剂
agent of mineralization 矿化剂
agent of necessity 紧急代理人
agent of transportation 搬运营力
agent's coupon 出票人联
agent service 代理业务
agent sign 代理人签署
agent's lien 代理人留置权
agents of erosion 侵蚀作用;冲刷的动因
agent's staff 代理人的工作人员
age of air-mass 气团存在期
age of apical plate of aquifer 含水层顶板地层时代【地】
age of aquifer 含水层时代【地】
age of aquifuge 隔水层时代
age of artificial recharge stratum 人工补给层时代
age of bottom layer of aquifer 含水层底板地层时代
age of cave sediments 洞穴沉积物年龄
age of concrete 混凝土年龄;混凝土龄期
age of container 箱龄
age of decline 衰朽阶段
age of diurnal inequality 日潮不等潮龄
age of diurnal tide 日潮不等潮龄
age of earth crust 地壳年代
age of earthquake 发震年代
age of entry into employment 就业年龄
age of fishes 鱼类时代
age of gliding tectonics 滑动构造时代
age of hardening 受荷龄期;结硬期龄;硬化龄期;龄期
age of ice 冰龄
age of impact-crater 撞击坑形成年代
age of intrusion 侵入体年龄
age of loading 受荷龄期
age of mammals 哺乳动物时代(早第三纪~晚第三纪)
age of menarche 初潮年龄
age of metamorphism 变质作用时代
age of mine 矿山可采期
age of minerals 矿石年龄
age of pal(a)eoshoreline 古海岸线年代
age of parallax inequality 视差不等潮龄
age of parallax tide 视差潮龄
age of phase inequality 月相不等潮龄;潮龄不等
age of plate 板块时代
age of rail 钢轨使用期
age of reservoir 储集层时代
age of retirement 退休年龄
age of sedimentary rock 沉积岩时代
age of sludge 污泥变陈
age of stand 林木年龄;林龄
age of surveying 施测年代
age of tectonic layer 构造层时带
age of tectonic migration 构造迁移的时代
age of tectonic movement 构造运动时带
age of the moon 月龄
age of tide 月相不等潮龄;潮龄
age of tide inequality 潮龄不等
age of vegetation sampling organ 植物采样器官年龄
age of vessel 船龄
age of volcanic activity 火山活动时代
age of waste disposal stratum 排污层时代
age of water-storing bed 储[贮]水层时代

ageostrophic wind 非地转风
ageotropism 无向地性
age period 年龄期
age premium 超龄保险费
age-proof 抗衰老的;抗时效的;抗老化的
age pyramid 年龄锥体;年龄金字塔(图);百岁图;人口百岁图
ager 熟化器;老化器
age-resistance 抗老化性(能);抗老化力
age-resistant 抗衰老的;抗时效的;抗老化的
age resistant steel 抗氧化钢
age-resister 抗氧化剂;抗老化剂
age sampling 时效抽样检验法
age-sex composition 年龄性别构成
age softening 时效软化
ages of various meanings 各种涵义的年龄
age-specific birth rate 年龄(性别)出生率
age-specific death rate 年龄死亡率;按年龄的死亡率
age spectrum 年龄谱
age stability 老化稳定性
age-strength relation 强度—龄期关系;龄期—强度关系(曲线)
age-strength relation of concrete 混凝土龄期强度关系曲线
age structure 年龄结构;年龄构成
AGE system AGE 方式【机】
age table 龄系列表
age test 时效试验
age tester 老化试验机
age theory 年龄理论
Agfacolo(u)r 阿克发彩色
Agfa-Gevaert Group 阿克发—古伐集团
Agfa polyamide 阿克发聚酰胺
agger 双潮;古罗马的道路;复潮
agger arenal 小石;泥沙;沙洲
agger of nose 鼻堤
agglaciation 加强冰川作用
agglomerant 熔结剂;凝聚剂;凝集剂;粘结剂;附聚剂
agglomerate 烧结(团)块;烧结矿;粒子附聚体;胶凝剂;集块岩;成团;附聚(物)
agglomerated cake 烧结块
agglomerated cork 烧结软木;石软木
agglomerated cork brick 烧结软木砖;黏结软木砖
agglomerated particle 团聚颗粒
agglomerated settlement 聚居村
agglomerate-foam 烧结矿渣泡
agglomerate-foam concrete 烧结矿渣泡沫混凝土
agglomerate lava 集块熔岩
agglomerate of powder 粉团
agglomerate tabling 团粒台浮
agglomeratic ice sheet 块集冰盾
agglomeratic lava texture 集块熔岩结构
agglomeratic texture 集块结构
agglomerating agent 烧结因素;凝结剂;烧结添加剂;胶凝剂
agglomerating factor 烧结因素
agglomerating force 团聚力
agglomerating index 烧结指数;结焦指数
agglomerating installation 烧结设备
agglomerating plant 烧结厂;烧结装置
agglomerating value of coal 煤凝结值
agglomeration 团矿;团集;烧结(作用);聚居区;聚集作用;聚集成团(现象);结团;结块;集块作用;集结;凝聚增长;住宅区;成团;附聚作用;附集作用
agglomeration aids 助凝剂;助滤剂
agglomeration economy 聚集经济
agglomeration effect 凝聚效应
agglomeration in mill 磨内结块
agglomeration of activated sludge 活性污泥成团
agglomeration of grains 颗粒成团
agglomeration of industry 工业密集;工业集结
agglomeration of population 人口聚居点;人口居住点
agglomeration of primary particles 胶凝成团;初级颗粒烧结
agglomeration rate 附聚率
agglomeration roast 烧结焙烧
agglomerative 附聚的
agglomerative factor 聚集因素
agglomerator 团结剂;结团剂;沉降电极
agglomeritic ice 附集冰

aggloporite 多孔熔渣
agglutinability 可凝集性;凝集能
agglutinant 烧结剂;胶着剂;胶凝剂;凝集剂;促凝物质
agglutinant for concrete 混凝土凝集剂
agglutinate 胶结(物);集结;黏合集块岩
agglutinate cone 寄生熔岩锥
agglutinating index 结焦指数;黏结指数
agglutinating property 烧结性
agglutinating test 烧结试验
agglutination 胶结(作用);集结;凝集(作用);凝集测验;黏合物;成胶状
agglutination lysis test 凝集溶解试验;凝溶试验
agglutination phenomenon 凝集现象
agglutination reaction 凝集反应
agglutination reaction test tube 凝集反应试管
agglutination test 凝集试验
agglutination titer 凝集价
agglutinative 集结的;凝集的
agglutinative absorption 凝集吸收作用
agglutinator 凝集物
agglutinin 凝集素
agglutinin absorption 凝集素吸收
agglutinin absorption test 凝集素吸收试验
agglutinogen 凝集原
agglutinogenic 产生凝集素的
agglutinophore 凝集簇
agglutinoscope 凝集反应镜
agglutinoscopy 凝集反应镜检查
agglutogen 凝集原
agglutogenia 产生凝集素的
agglutometer 凝集反应器
aggradated flood plain 逐年游离的潮漫滩
aggradated plain 填积平原;加积平原
aggradation 外展作用;填积(作用);加积(作用);淤积;淤高;(地下土层永冻深度面的)递升
aggradation above reservoir 水库前淤积
aggradational deposit 加积沉积
aggradation base level 填积基准面
aggradation beach 加积海滩
aggradation coast 加积海岸;淤积海岸
aggradation island 加积岛;淤积岛
aggradation of bed level 河床淤高;河床抬高
aggradation of ice level 冰位抬高
aggradation of level 河床淤积高度
aggradation of risks 风险增加
aggradation of riverbed 河床的淤高
aggradation of stream channel 河槽淤高
aggradation plain 填积平原;淤积平原;冲积平原;沉积平原;加积平原
aggradation shore 加积海岸
aggraded floodplain 洪积平原;淤高的洪泛平原
aggraded thickness 加积厚度
aggraded valley floor 加积谷底
aggraded valley plain 冲积平原
aggrading action 加积作用;淤积作用
aggrading channel 淤积水道;淤积河槽
aggrading continental sea 填大陆海
aggrading growth sequence 填积型生长层序
aggrading pattern 填积型
aggrading river 加积河(流);淤积河流;冲积河流
aggrading stream 加积河(流);淤积河流;冲积河流
aggrandizer 扩大器
aggravate 加重;加剧
aggreaasion 侵袭
aggreagtion community 集聚群落
agreement on general terms and conditions of business 一般交易条件协议书
aggregate 团粒;团块;团聚体;粒料;集料;集结;集合体;极薄片状团聚体;机组;凝集体;合计;原子团;总数;总计;总和;整套装置;整套设备;骨料;共计;成套装置
aggregate abrasion 集料磨损(性)
aggregate abrasion value 集料磨损值
aggregate adjustment option 综合调整方案
aggregate agent 腐蚀剂
aggregate-alkali reaction 集料—碱反应
aggregate amount 总金额
aggregate analysis 团集体分析;集料分析;总体分析;综合分析
aggregate angularity 集料棱角指数;集料棱角性
aggregate area of wires 钢丝总面积
aggregate assignment 集合体赋值
aggregate attrition machine 集料磨耗试验机

aggregate averaging 集料平均粒径(测定);骨料平均粒径测定;测定集料平均粒径;测定骨料平均粒径
aggregate averaging size 骨料平均粒径
aggregate balance sheet 综合资产负债表
aggregate barge 砂石驳
aggregate base 骨料底层;集料底层
aggregate base course 集料基层;骨料基层
aggregate batcher 集料配料器;集料量配器;集料计量箱;集料计量仓;集料分批计量器;集料配料器;骨料配料箱;骨料配料机(器);骨料计量器;骨料称量器
aggregate batcher bin 集料分批箱;集料配料仓
aggregate batching bin 骨料配料仓
aggregate batching plant 集料配料装置;集料配料设备;骨料投配器
aggregate-bed filter 集料床过滤器
aggregate bin 集料储(存)斗;集料仓;骨料仓
aggregate bituminizing plant 骨料沥青搅拌装置;骨料沥青拌和装置;沥青集料搅拌机;沥青骨料搅拌机
aggregate blending 集料混合;集料拌和;骨料混合
aggregate B/L freight 提单总运费
aggregate breakage 集料压碎量
aggregate breaking force 钢丝破断拉力总和;钢丝拉力总和
aggregate bridging 集料鼓泡
aggregate bucket elevator 骨料斗式提升机
aggregate capacity 聚集容器;聚集容量;机组功率;总容量;总功率
aggregate cash inflow 现金流入总额
aggregate cash outflow 现金流出总额
aggregate-cement ratio 集料水泥比;骨料水泥比;骨灰比;集灰比
aggregate chamber 集料室
aggregate chips 石屑;集料屑
aggregate cleanness 集料清洁度
aggregate-coated panel 石粒喷涂板;干粘石饰面板
aggregate coating 集料包裹
aggregate combination 集料混合;集料化合;集合结构
aggregate composition 集料颗粒组成;集料成分
aggregate concept 统计概念;集合概念
aggregate consumption 总消耗
aggregate consumption funds 累计的消费基金
aggregate content 集料含量
aggregate correction factor 集料修正系数;骨料修正系数
aggregate cost 总价;总成本
aggregate crushing index 骨料压碎指标
aggregate crushing test 集料抗碎试验;骨料压碎试验;骨料抗碎试验
aggregate crushing value 集料压碎值
aggregate decision 总体决定
aggregate demand 累积总需求;总需求(量)
aggregate demand function 累积总需求函数;总需求函数
aggregate demand price 累积总需求价格
aggregate deposit 集料沉积(层)
aggregate depth 集料沉积深度;总厚
aggregate depth of elements 构件总厚
aggregated error 累积误差;总误差
aggregated feed 骨料供给
aggregated filter 集中滤波器
aggregate dimension 总尺寸
aggregate discharge 汇合流量;总流量
aggregate discount 总合折扣
aggregated model 集合模型;总体模式;综合模式
aggregated momentum 总动量
aggregated particle structure 聚集粒子结构;次结构
aggregate drain 集料盲沟;骨料盲沟
aggregated rebate cartels 统一回扣卡特尔
aggregate dredge(r) 采砂船
aggregated dryer 集料烘干机
aggregated shipment 合并装运
aggregated structure 团聚构造;聚粒结构
aggregated subarea 群落分区
aggregated subregion 群落分区
aggregated texture 团聚结构
aggregate economics 总量经济学
aggregate effective demand 总合有效需求
aggregate efficiency 总效率
aggregate elevator 集料提升机;斗式集料提升机
aggregate envelopment 集料包裹(现象)

aggregate error 集合误差;合计误差
aggregate estimation of mineral resources 总合式定量预测
aggregate excavator 骨料开采设备;骨料采掘设备
aggregate expenditures 总支出
aggregate exposure 骨料外露(混凝土墙);集料裸露(以获得装饰效果)
aggregate expression 聚集表达式;聚合表达式;集合表达式
aggregate facing panel 外露骨料饰面墙板
aggregate feeder 骨料进料器;集料供给装置;集料供给器
aggregate feeding 供料;集料供给;集料给料;骨料供给
aggregate feeding unit 集料供给装置
aggregate filling 填缝料;结构填料
aggregate finish 骨料饰面;粗粒饰面;砂面装饰
aggregate floor area 楼面总面积
aggregate floor depth 楼面总厚度;地板总厚度
aggregate floor space 楼面总面积
aggregate formation 团集体形成
aggregate for reinforced concrete 钢筋混凝土集料;钢筋混凝土骨料
aggregate fracture 集料破裂;骨料破裂
aggregate frictional characteristics 集料摩阻特性
aggregate gradation 集料分级;骨料级配;骨料分级
aggregate grader 骨料级配机;骨料分级机
aggregate grading 选配骨料;骨料分配;选配集料;集料级配
aggregate grading curve 集料粒径(级配)曲线;骨料粒径(级配)曲线;集料筛分曲线
aggregate grain 集粒;骨料颗粒
aggregate grain-size distribution 集料粒径组成;集料粒径分布
aggregate granulosity 骨料粒质;骨料粒度;粒度
aggregate handling dock 砂石码头
aggregate heating 骨料加热(混凝土)
aggregate heating concrete 骨料加热混凝土
aggregate heating unit 骨料加热装置
aggregate height 总高(度)
aggregate hopper spreader 斗集料撒布机;斗骨料撒布机
aggregate impact value 集料冲击值;总冲击强度;冲击试验值
aggregate income of families 家庭总收入
aggregate interlock 集料嵌锁;混凝土集料的穿插锁结;骨料咬合
aggregate interlocking 骨料嵌锁(作用)
aggregate interlocking effect 集料嵌锁作用
aggregate interlocking force 集料咬合力;骨料咬合力
aggregate interlock(ing) modulus 集料嵌锁模量;骨料嵌锁模量
aggregate investment 投资总量;投资总额;总投资额
aggregate launder 集料槽
aggregate length 全长;总长(度)
aggregate limit (配额的)总限额
aggregate load 集中荷载;集中负载
aggregate losses 累计损失
aggregate machine-tool 组合机床
aggregate macroeconomic change 宏观经济的总合变化
aggregate making property 轧成碎石的能力(岩石);轧成骨料的性能;集料形成(岩石轧成碎石);集料生产能力;骨料的轧制性能
aggregate map 综合地图
aggregate market 综合市场;总体市场
aggregate market value 市场价格总额
aggregate matrix 集料母岩;集料母material;集结矩阵
aggregate measuring plant 骨料配料装置;骨料配料设备;集料配料装置;集料配料设备
aggregate meter 骨料计量器;集料称量器
aggregate mix(ture) 大小颗粒混合物;粒度成分;颗粒组分
aggregate model 集结模型
aggregate model of the economy 经济集结模型
aggregate motion 组合运动
aggregate notation 聚集记法
aggregate number 总数
aggregate of broken bricks 碎砖骨料
aggregate of broken concrete 碎混凝土骨料
aggregate of data 数据簇

aggregate of soil 土壤团块
aggregate of soil particles 土粒团;土粒聚集
aggregate output 输出量;总生产额;总产出
aggregate panel 外露集料墙板;外露骨料墙板
aggregate particle 集料颗粒;骨料颗粒;集料粒度
aggregate particle orientation 集料定位
aggregate paver 石料摊铺机
aggregate payment 综合支付
aggregate plan(ning) 总体计划;总体规划
aggregate plant 砂石(料)厂;集料加工场;集料(加工)厂;骨料轧制设备
aggregate plywood 石屑敷面胶合板
aggregate polarization 集偏振化
aggregate polish 集料磨光
aggregate polishing 集料磨光现象;骨料磨光现象
aggregate population measure 总体人口情况
aggregate precooling 骨料预冷
aggregate preparation machine 骨料制备机
aggregate preparation plant 砂石配备厂;沙石配备厂;集料制备厂;集料配备厂;骨料轧制设备;骨料加工厂
aggregate preparation system 骨料加工系统
aggregate processing 骨料的制备
aggregate processing plant 骨料制备
aggregate producer 轧石机;集料生产者;骨料生产者;集料制备机;骨料制备机
aggregate production 总生产
aggregate production function 总量生产函数
aggregate production planning 总生产计划;总生产规划
aggregate production plant 集料制备厂
aggregate property 骨料性能
aggregate proportioning plant 集料配料装置;骨料配料装置
aggregate pullout (from concrete pavement) (混凝土路面的)集料脱出
aggregate pump 砂石泵
aggregate quality 骨料质量
aggregate railroad 运送骨料铁路;(装运集料的)工地窄轨铁路
aggregate railway 集料供应列车
aggregate ratio 集料水泥比(混凝土中);骨料水泥比(混凝土中)
aggregate ray 聚合射线
aggregate realized savings 实际储蓄总额
aggregate reclaiming equipment 骨料采掘设备
aggregate reclaiming plant 砂石料厂;骨料开采场;骨料开采设备;骨料采掘设备;砂石料场;骨料开采场
aggregate recoil 集合反冲
aggregate recovery unit 骨料回采设备
aggregate reinforcement area 钢筋总面积
aggregate resistance 总抵抗力;总承载能力
aggregate resistance of support 支柱总承载力
aggregate retention 集料的滞留;集料保持能力
aggregate roughness 集料粗糙度
aggregate roundness 集料圆度
aggregate sample 混合样品;混合样(本)
aggregate scale 集料计量器;集料称
aggregate scheduling 总进度计划
aggregate segregation 集料离析
aggregate shape 集料形状
aggregate size 集料粒径;集料尺寸;骨料粒径;骨料尺寸
aggregate social product 社会总产品
aggregate social supply 社会供应总量
aggregate softness 集料松软度;集料柔软度;骨料柔软度
aggregate species 复合种
aggregate specific gravity under oven dry condition 骨料绝对干燥比重
aggregate sphericity 集料的圆球度
aggregate spreader 砂粒撒布器;集料撒布机;骨料撒布机
aggregates producer 骨料生产厂
aggregates stabilized soil base 粒料稳定土基层
aggregate state 聚集状态
aggregate steel area 钢筋总截面积
aggregate stock 砂石堆场
aggregate storage 总蕴藏量;总库容;集料堆场;集料仓库;总储(存)量;总储藏量;骨料储存场;骨料仓库
aggregate storage bin 骨料仓;集料仓
aggregate storage capacity 总库容

aggregate strength 骨料强度;集料强度
aggregate structure 集合体结构;团粒构造;团聚体构造;团集结构;集合体构造
aggregate subbase course 骨料底基层
aggregate suction dredge(r) 吸扬(式)采砂船
aggregate supply 集合供应
aggregate supply function 累积总供给函数;总供给函数
aggregate supply price 总供给价格
aggregate supply wharf 骨料供应码头
aggregate surface 集料表面;骨料表面
aggregate-surface-area method 集料表面积法(计算沥青混合料中沥青用量的一种方法)
aggregate-surfaced 碎石铺面的
aggregate-surfaced shoulder 铺集料的路肩;铺骨料的路肩
aggregate surface texture 集料表面结构
aggregate test(ing) 集料实验;骨料试验;集料试验
aggregate texture 集料结构;集合体结构
aggregate thickness 总厚度
aggregate-to-aggregate contact 骨料(颗粒)相互接触;集料(颗粒)相互接触;集料接触
aggregate tonnage 总(计)吨位;总吨数
aggregate trailer 集料拖运车;骨料拖运车
aggregate transfer 石屑敷面转移法
aggregate transparency 集料灰斑
aggregate type 集合类型
aggregate unit 组合单位
aggregate unloading dock 砂石卸货码头
aggregate use-values 使用价值总和
aggregate value 总值;总合价值
aggregate value method of weighting 总值加权法;累积加权法
aggregate value of an invention 发明的总价值
aggregate value of listed stock 时价总额;上市股票的总市值
aggregate value of net output 净生产量的总价值
aggregate velocity 集合速度
aggregate volume index 附聚体积指数
aggregate volume reduction 集料体积收缩
aggregate volume shrinkage 集料体积收缩
aggregate washer 集料洗涤机;集料清洗机
aggregate washing 骨料冲洗
aggregate washing equipment 骨料冲洗设备
aggregate washing plant 集料冲洗设备;洗石设备;骨料冲洗设备
aggregate weigh hopper 集料称量斗
aggregate weighing barrel 集料称筒
aggregate weighing batcher(with scale) 骨料重量配料机;集料称量斗;集料称料机
aggregate weighing equipment 骨料称量设备
aggregate weighing hopper 骨料称量斗
aggregate width 总宽(度)
aggregate yard 砂石堆场;骨料堆场
aggregation 群聚;团集作用;聚集作用;聚集(体);集团;集积;集合物;集合(体);积聚
aggregational model 集合模型
aggregational rock 集合岩
aggregational structure 集合体构造
aggregation analysis 聚集分析
aggregation bed 聚集层
aggregation body 凝聚体
aggregation density 群集密度
aggregation force 聚合力
aggregation function 集合函数
aggregation index 群落指数
aggregation moisture 团集化湿度
aggregation of capital 资本聚集
aggregation of clay 黏土的聚集(作用);粘土的成团(作用)
aggregation of individual preferences 个人偏好集合
aggregation of marine cargoes 集中海洋货物
aggregation of mineral particle 矿物颗粒集合体
aggregation of particles 颗粒团聚;颗粒结团
aggregation of soil particles 土团粒;土粒聚集
aggregation of variables 变量的归并
aggregation on furnace bottom 结底
aggregation plain 集积平原
aggregation process 凝聚过程
aggregation stability 聚集稳定性
aggregation state 团集体状态
aggregative company 集团公司
aggregative economic control 宏观经济调控

aggregative economics 宏观经济学
aggregative fluidized bed 聚式流化床
aggregative index(number) 综合指数
aggregative indicator 综合指示数;综合指标
aggregative indicatrix 综合指标
aggregative model 总量模型;综合模式
aggregative quantity index 综合量指数
aggregative study 综合研究报告
aggregative table 综合表
aggregative weighted index 综合加权指数
aggregometer 集合度计
aggremeter 集料计量器;集料秤量器;集料秤量计;骨料计量计;骨料称量器
aggressive action 腐蚀作用;侵蚀作用;侵蚀性;腐蚀性
aggressive agent 侵蚀物;侵蚀剂;腐蚀剂
aggressive atmospheric(al) condition 侵蚀性大气状况
aggressive behavio(u)r 攻击行为
aggressive carbon dioxide 侵蚀性二氧碳
aggressive carbonic acid 侵蚀性碳酸
aggressive characteristic 侵蚀性特性
aggressive condition 腐蚀性环境
aggressive device 主动装置
aggressive dosage 冲击剂量
aggressive fume 腐蚀性气体
aggressive liquid 侵蚀性液体
aggressive magma 侵进岩浆
aggressive matter 侵蚀性物质
aggressive medium 侵蚀介质
aggressiveness 侵蚀
aggressiveness of water 水的侵蚀性
aggressive power 侵蚀能力
aggressive salt 侵蚀性盐
aggressive selling 卖气旺盛
aggressive solution 侵蚀性溶液
aggressive substance 侵蚀性物质
aggressive tack 干黏(性)
aggressive tack hour 接触干燥时间(涂层)
aggressive tax 累积税
aggressive to concrete 混凝土侵蚀性;对混凝土的侵蚀
aggressive toward concrete 对混凝土有侵蚀性的;对混凝土有害的
aggressive toward metal 对金属具有侵蚀性的
aggressive toward structural materials 对结构材料具有侵蚀性的
aggressive volume of completed investment 累积完成投资额
aggressive water 侵蚀(性)水;侵入水;侵进水;有腐蚀性的水;腐蚀性水
aggressivity 侵蚀性
aggressivity water 侵蚀性水
aggrieved party 受害方
agiasterium 教堂设置祭坛部分
Agidisc filter 爱杰克斯型真空过滤机
agiepis film forming foam 水膜形成泡沫
Agil 阿吉尔聚乙烯纤维
agilawood 沉香
agility of frequency 频率快变
Agilon 阿吉纶
aging 分期
aging accounts receivable 应收账款的分期
aging error 衰化误差
aging schedule 账龄分析表
aging the receivable 分析应付款欠账的时间
agio 贴水(升水);扣头
agio account 贴水账目;贴水账户
agipanthus 百子莲
agistment area 放牧面积
agitate 搅动
agitate conveyer 搅拌式运输车(混凝土);搅拌式输送车(混凝土)
agitated batch crystallizer 搅动结晶器;搅拌式分批结晶器
agitated bed 搅拌床
agitated compartmented extractor 搅拌式分隔萃取塔;搅拌式萃取塔;搅拌间隔式萃取器
agitated cylinder dryer 圆筒搅拌干燥器
agitated dryer 搅动式干燥机;搅拌(型)干燥器
agitated film evapo(u)rator 搅动膜蒸发器
agitated horizontal autoclave 搅拌卧式高压釜
agitated kettle 搅动锅
agitated leaching autoclave 搅拌浸出高压釜

agitated tank 搅拌槽
agitated thin-film evapo(u)rator 搅拌薄膜蒸发器;搅拌薄膜浓缩器
agitated trough dryer 槽式搅动干燥器
agitate lorry 搅拌式运输车;搅拌式输送车(混凝土)
agitate truck 搅拌式运输车;搅拌式输送车(混凝土)
agitating 搅动;搅拌
agitating arm 搅拌杆;拌和杆
agitating auger 搅拌螺旋
agitating bar 搅拌杆;拌和杆
agitating conveyer 搅拌式运输车;搅拌式输送车(混凝土)
agitating cooker 回转式杀菌锅
agitating device 搅拌式输送车;搅拌装置
agitating heater 搅动加热器
agitating lorry 搅拌运料车;搅拌运货车;搅拌式运输车;搅拌式输送车(混凝土);混凝土搅拌(汽)车;混凝土拌和(汽)车
agitating pan 搅拌锅
agitating speed 搅拌速度
agitating sprocket 抖动链轮
agitating tank 搅拌箱;搅拌罐;搅拌槽
agitating truck 搅拌运料车;搅拌式运输车;搅拌式输送车(混凝土);混凝土搅拌运料车;混凝土拌和车;拌和(卡)车
agitating vane 搅动叶片;搅动叶轮;搅刀;搅拌桨叶
agitating vessel 搅拌桶
agitation 搅动作用;搅拌(作用)
agitation ball mill 搅拌球磨机
agitation cascade 高落搅拌
agitation characteristics 搅拌特性
agitation device 搅拌装置
agitation dredging 扰动疏浚;搅动挖土(法);搅动挖泥(法);搅动疏浚(法)
agitation dredging method 搅动疏浚法
agitation error 骚动误差
agitation-froth machine 机械搅拌式浮造机;搅拌(式)泡沫浮选机
agitation-froth process 搅动生泡法
agitation layer 混合层
agitation leach 搅动浸滤;搅动浸取;搅拌浸青;搅拌沥滤
agitation of the bath 搅动溶液
agitation subaeration 底充气搅拌
agitation type flo(a)tation machine 搅动式浮选机
agitation vat 搅拌瓮
agitation vessel 带搅拌器的容器
agitator 搅拌桶;搅拌器;振荡器
agitator arm 搅动机叶轮;搅动杆;搅拌机杆
agitator ball mill 搅拌球磨机;有搅拌器的球磨机
agitator bar 搅拌杆
agitator bath 搅拌槽;搅拌桶
agitator blade 搅拌叶(片);搅拌桨
agitator blender 搅拌机
agitator board 搅拌板;抖动机板
agitator body 搅拌器转筒;搅拌机机身
agitator body truck 搅拌汽车
agitator capacity 混合料容量
agitator car 搅拌汽车;拌和器车
agitator conveyer 搅拌送料车;带搅拌器的混凝土输送车;搅拌送料车
agitator dryer 搅动干燥器
agitator extension shaft 搅拌器伸出轴
agitator flo(a)tation cell 空气搅拌式浮选槽
agitator for stored pulp 储[贮]浆搅拌机
agitator jet 搅拌器喷管
agitator joint shaft 搅拌器连接轴
agitator kettle 搅拌锅
agitator machine 空气搅拌式浮选机
agitator mill 立式球磨机;搅拌磨
agitator nozzle 搅拌器喷嘴
agitator shaft 搅拌器转轴
agitator support 搅拌机支座
agitator tank 搅动(器)槽;搅拌箱;搅拌罐;搅拌槽
agitator treating 搅拌洗涤;搅拌器处理
agitator truck 搅拌运输车(混凝土);搅拌汽车;混凝土搅拌(汽)车;混凝土拌和(汽)车;有搅拌装置的(混凝土)运输车;带搅拌工具的汽车
agitatory crusher 搅拌式轧碎机
agium 海滨群落
a given area 一定地区
a given period 一定时期

aglaite 变锂辉石
Aglaophenia 羽螅属
A glass 高碱玻璃；A 玻璃
A glass fibre 高碱玻璃纤维
aglet 金属箍
aglime 农用石灰
aglite 烧结膨胀黏土；轻质膨胀黏土集料；轻质集料；轻质骨料；烧结黏土；陶粒；膨胀黏土骨料
aglomerator 沉尘极
agmatite 角砾(状)混合岩
agmatitic gneiss 角砾(混合)片麻岩
agmen 聚集；集合(物)
agnosia 失辨觉能
agnosticism 不可知论
agnostogenic rock 成因不明的岩石
Agnotozoic age 元古代
Agnotozoic erathem 元古界
agnus dei 神羔像(象征基督的一种标志)；神之羔羊饰
Ag-O-Cs photocathode 银氧铯光阴极
agoing 开动
agollissartog 霜冻土丘
Agomet E polymer resin 阿哥墨特 E 聚合树脂
agon 辅基
agonic 无偏差的
agonic line 无偏(差)线；无磁偏线；零磁偏线；零磁差线；坐标轴线
agonistic behavio(u)r 格斗行为
a good plot of tree 一块好林木
agora(e) 广场(古希腊贸易及集会用)；阿格拉
agoraphobia 广场恐惧症
agouni 宽急流冲沟；干冲沟
agpaite 钠质火成岩类
agpaitic nepheline syenite 纳质火成岩型霞石正长岩
A-grade wood 头等木材；甲级木材；甲级木材
agraf(f)e 铁箍；(石块相连结的)金属搭扣；铁夹子；钩子
agranular endoplasmic reticulum 非颗粒型内质网
agranular type 无颗粒型
agraphitic carbon 无定形碳；结合碳；非石墨碳；非结晶碳
agrarian 土地的；耕地的
agrarian law 土地法(规)
agrarian population 农村人口
agrarian reform 土地改良；土地改革
agrarian revolution 土地革命
agravic 无重力状态
agravity 失重(情况)
Agra work 阿格拉细工(印度一种镶贴石料细工)
agree 赞成；符合
agreed compensation 商定补偿
agreed-dimension slab 协议板
agreed-dimension stone 协议石料
agreed duty 议定关税
agreed frequency 约定频率
agreed insured value 约定保险价值
agreed payment 议定付款
agreed period 商定的期限
agreed price 议定的价格
agreed quantity of trade 议定贸易量
agreed rate 协定运费率；议定的费率
agreed stowage factor 议定积载因素
agreed sum 商定金额
agreed tariff 协定关税
agreed territory 协议区域
agreed text 议定文本；一致同意的文本
agreed-upon price 协议价格
agreed valuation 商定价
agreed value 协定保险价额；保险评定价值
agree life 模拟寿命
agreement 契约；同意；合同；协议；协商；协定；约定书；一致；承包合同；符合
Agreement Board 鉴定委员会
agreement button panel 同意按钮盘
agreement by cubic(al)meter 按方计货
agreement by meter 量方协议；量方契约；量方合同；按方计资；按方计售
agreement by piece 按件计资；计件协议；计件契约；计件合同
agreement certificate 鉴定证明书；官方证书
agreement corporation 协议公司
agreement effective 协议生效
agreement for deed 土地买卖合约

agreement for hire 租赁合同；租赁协议；租赁协定；租赁契约
agreement for hire purchase 租购合同
agreement form 协议格式；合同格式；协议书；协议表格
agreement of deed 契约协议
agreement of intent 意向协议书；意向性协议
Agreement of International Monetary Fund Organization 国际货币基金组织协定
agreement of international through passenger traffic 国际旅客联运协定
agreement of lease 租赁协议；租赁协定；租赁契约
agreement of sale 销售协议
agreement of understanding 非正式协议
agreement on buying option 优先购买权协议
agreement on commerce 通商协议
agreement on consignment 寄售契约
agreement on environmental protection 环境保护协议
Agreement on Exchange Guarantees 外汇保值协定
agreement on protection of investment 投资保护协定
agreement on reinsurance 分保险合同
agreement price 协议价格；协定价格
agreement tariff 协定税则
agreement to hire 租赁协议书
agreement to resell 转售协议
agreement to sell 经销协定
agreement to sell at valuation 评价出售的协议
agreement water rate 协议水费
agreement year 合同年度；协议年度；协定年度
agrellite 氟626钙钠石
agribusiness 务农活动；农业综合企业；农商联合企业
agric 农业的
agricat 履带式拖拉机
agricere 土漆膜
agrichemicals 农用化学品
agric horizon 耕作熟化层
agriclimatic map 农业气候图
agricorporation 农业综合公司
agricultural 农业的
agricultural accounting 农业会计
agricultural acreage 耕地亩数；耕地亩数；耕地面积
agricultural activity 农业活动
agricultural aircraft 农用飞机
agricultural amelioration 农业土壤改良
agricultural and mechanical 农业机械的
agricultural antibiotic 农用抗菌素
agricultural area 农业区
agricultural area source pollution 农业面源污染
agricultural association 农会
agricultural automobile 农用汽车
agricultural aviation 农用航空；农业航空
agricultural backwater pollution 农业回水污染；农业回流污染
agricultural bacteriology 农业细菌学
agricultural belt 农业地带
agricultural biometeorology program(me) 农业生物气象学规划
agricultural biotechnology 农业生物技术
agricultural building 农用建筑；农用房屋；农产品加工建筑
agricultural capital 农业资本
agricultural catchment 农业流域
agricultural change 农业变化
agricultural chemical 残留性农药
agricultural chemical contamination 农用化学污染
agricultural chemical pesticide 农药残留
agricultural chemical residuum 农药残留
agricultural chemicals 农用药剂；农用化学物；农业化学品；农药
Agricultural Chemicals Regulation Law 农业化学品管理条例；农药管理法
agricultural chemicals wastewater 农用化学物废水
agricultural chemistry 农业化学
agricultural city 农业城市
agricultural classification 农艺分类
agricultural climatology 农业气候学
agricultural commodity 农业商品
agricultural comphrehensive prevention and treatment 农业综合防治
agricultural conservation program(me) 农业保护纲要
agricultural contamination 农业污染
agricultural contract works 农业承包工程
agricultural control 农业防治
agricultural credit 农业贷款
agricultural crop 农作物
agricultural damage 农业损害
agricultural data 农业数据
agricultural development 农业发展；农业开发
Agricultural Development Advisory Service 农业发展顾问局
agricultural development zone 农业发展区
agricultural disposal 农业处理
agricultural district 农业(地)区
agricultural division 农业区划
agricultural drain 农用排水沟；农业排水沟；农业排水管；农田排水渠；农田排水(沟)
agricultural drainage 农业排水；农田排水
agricultural drainage system 农业排水系统
agricultural drain pipe 农用排水管；农田排水管(道)
agricultural economics 农业经济
agricultural economy 农业经济
agricultural ecosystem 农业生态系统
agricultural ecosystem forecasting 农业生态预测
agricultural effluent 农业废水
agricultural engineer 农业工程师
agricultural engineering 农业工程
agricultural enterprise 农业企业
agricultural entomology 农业昆虫学
agricultural environment 农业环境
agricultural environmental destroy 农业环境破坏
agricultural environmental management 农业环境管理
agricultural environmental monitoring 农业环境监测
agricultural environmental protection 农业环境保护
agricultural environmental science 农业环境学
agricultural equipment 农业机具；农具；农机设备
agricultural experiment station 农业试验站
agricultural exports 出口农产品
agricultural fertilizer 农业肥料
agricultural geography 农业地理
agricultural geology 农业地质学
agricultural hydraulic engineering 农田水利工程
agricultural hydrology 农业水文学
agricultural implement 农业工具
agricultural income 农业收入
agricultural industrial complex 农业工业结合体
agricultural industrial enterprise 农场企业
agricultural injury 农业损伤
agricultural insecticide 农用杀虫剂；农业杀虫剂
agricultural insect management 农业害虫的防治
agricultural insects 农业害虫
agricultural land 农业用地；农业土地；农田(水利工程)
agricultural law 农业法
agricultural legislation 农业立法
agricultural lien 农业留置权
agricultural lime 农用石灰
agricultural loader 农用装载机
agricultural machinery 农业机械
agricultural machinery design 农业机械设计
agricultural map 农业化学土壤图；农业地(质)图
agricultural meteorology 农业气象学
agricultural method 农业措施；农业方法
agricultural microclimate 农田小气候
agricultural mineral 农业矿物
agricultural miticide 农用杀螨剂；农业杀螨剂
agricultural nonpoint source of pollution 农业非点污染源；农业非点源污染
agricultural nonpoint source pollution 农业非点污染源污染
agricultural nonpoint source pollution control 农业非点污染源污染控制
agricultural nonpoint source pollution model 农业非点污染源污染模型
agricultural nonpoint source pollution simulation 农业非点污染源污染模拟
agricultural nonpoint source water quality model 农业非点污染源水质模型
agricultural output value 农业生产值
agricultural pastoral area 半农草牧区

agricultural pasture 农业放牧
agricultural pest 农业害虫;农业病虫害
agricultural pipe drain 农田排水管(道)
agricultural plan 农业计划
agricultural plot 农业用地;农田
agricultural poison 农业毒物
agricultural policy 农业政策
agricultural pollutant 农业污染物
agricultural pollution 农业污染
agricultural pollution of water body 水体农业污染
agricultural pollution sources 农业污染源
agricultural pollution treatment 农业污染处理
agricultural practice 耕作方法
agricultural prevention and treatment 农业防治
agricultural processing building 农业加工建筑
agricultural processing waste 农业加工废物
agricultural product 农产品
agricultural production 农业生产
agricultural program(me) 农业计划
agricultural protection forest 农田防护林
agricultural rear-tipping trailer 后卸农用挂车
agricultural region 农业区
agricultural regional environment 农业区域环境
agricultural regionalization 农业区划
agricultural regrouping 农业改组
agricultural remote sensing 农业遥感
agricultural research 农业研究;农业调查
agricultural research project 农业研究课题
agricultural research service 农业研究所
agricultural roller 农田用滚筒
agricultural runoff 农田径流
agricultural season 农作季节
agricultural service building 杂用建筑(如仓库、厨房、马厩或附属在田园的等)
agricultural service transformer 农业用变压器
agricultural side tipping trailer 侧卸农用挂车
agricultural soil 农业用地;农业土壤;农田土壤;农田土;种植土;耕作土(壤);耕种土(壤)
agricultural soil loss 农业土壤流失
agricultural soil loss rate 农业土壤流失率
agricultural solid waste 农业固体废物
agricultural source 农业污染源
agricultural source of water pollution 水污染农业源
agricultural sprinkler irrigation 农田喷(水)灌(溉)
agricultural storage 农业储藏
agricultural sulphur 农用粗硫粉
agricultural system 农业系统
agricultural systems engineering 农业系统工程
agricultural technique 农业技术
agricultural terminal market 农业品散市场
agricultural tools and equipment 农业机具和设备
agricultural traction engine 农用拖拉机;农业拖拉机
agricultural tractor 农用拖拉机;农业拖拉机
agricultural use 农业利用
agricultural utilization of sludge 污泥的农业利用
agricultural value of sludge 污泥的农业价值
agricultural waste 农业废(弃)物;农业剩余物
agricultural waste disposal 农业废物处置
agricultural wastewater 农业污水;农业废水
agricultural water 农业用水
agricultural watershed 农业集水区
agricultural weather forecast 农业天气预报
agricultural wrench 农用扳钳;农用扳手
agricultural zoning 农业区划
agriculture 农业;农学
agriculture aid funds 补农资金
agriculture and forest chemical industry 农林化工
agriculture division 农业分区
agriculture in greenhouse 温室农业
agriculture in transition 转变中的农业
agriculture residue 农业残渣;农业残余物
agriculture technique 农业技术
agriculture water supply 农田供水
agriculturist 农艺师;农学家
agrigeology 农业地质学
agrimotor 农用拖拉机;农用汽车;农用动力机;农业动力机
agrinoerite 钾钙锶铀矿
agriplastics 农用塑料
agrisulviculture 农林作业
agrium 栽培群落

agroamelioration 农业土壤改良
agroatomizer 农用喷雾器
agrobased industry 农基工业
agrobiology 农业生物学;农业生态学
agrobiology analysis 农业生物学分析
agrocement batcher 农业联合配料器;混合量配器(水泥)
agrochemicals 农业化肥;农用化学物
agrochemical soil map 土壤农化图;农业化学土壤图
agrochemistry 农业化学
agroclimatic 农业气候的
agroclimatic delimitation 农业气候区划;农业气候分区;农业气候分界
agroclimatic region 农业气候区(域)
agroclimatic zone 农业气候区;农业气候带
agroclimatological resources 农业气候资源
agroclimatology 农业气候学
agroeco-economic system 农业生态经济系统
agroeco-engineering 农业生态工程
agro-ecological environment 农业生态环境
agroecology 农业生态学
agroecosystem 农业生态系统
agroecosystem research 农业生态系统研究
agroecotype 农业生态型
agroforestry 农业林学
agrogeological map 农业地质图
agrogeology 农业地质学
agrohydrology 农业水文学
agroindustrial-commercial enterprise 农工商联合企业
agroindustrial complex 农工联合企业;农业工业结合体
agroindustrial enterprise 农工企业
agroindustry 农村工业;农用工业;农(业)工业;农产品加工(工)业
agrology 土壤学;农业土壤学
agromelioration 农业土壤
agrometeorological forecast 农业气象预报
agrometeorology 农业气象学
agromicrobiology 农业微生物学
agronatural resources 农业自然资源
agronomical survey 作物调查
agronomical traits 农业性状
agronomist 农艺师;农业专家;农学家
agronomy 农艺学;农学
agronomy farm 试验农场
Agronomy Journal 农学杂志
agropedic degradation 农业土壤改良
agropedology 农业土壤学
agrophilous 适耕地生物
agrophysical property 农业物理性质
agrophytocoenosium 农作物群落
agrosocial resources 农业社会资源
agrosol 氰化甲汞
agrosprayer 农用喷雾机
agrosterol 草本甾醇
agrotechnical level 农业技术水平
agrotechny 农产品加工(学)
agrotown 建在农业地区的城镇;农村城镇;村镇
agrotype 农业土壤类型;农作物类型
aground sweeping 拖地扫海
agstone 石灰石粉;农用石灰
aguada 集水洼地
aguaje 秘鲁赤潮
aguano 大叶桃花心木
aguaria 水族馆
agueweld 穿心佩兰
a guide to the structural design of flexible and rigid pavement for new roads 英国新建道路柔性和刚性路面结构设计指南
aguilarite 辉硒银矿
Agulhas basin 厄加勒斯海盆
Agulhas Current 厄加勒斯海流
Agulhas plateau 厄加勒斯海台
agustite 磷灰石
agysical 矽炭银;硅碳银
ahead cam 前进凸轮
ahead clutch line 前进离合器气管
ahead of front axle 前轴之前
ahead of schedule 提前完成计划;超越进度表;超前进度
ahead of time 提前(完成计划)
ahead one 前进一【船】

ahead one port 左进一
ahead one starboard 右进一【船】
ahead power 前进力;推进力
ahead pull 前进拉力
ahead running 正车
ahead steering test 前进舵效试验
ahead swell 先趋涌
ahead three 前进三【船】
ahead turbine 前进汽轮机;推进涡轮机;推进透平;推进汽轮机;顺车汽轮机
ahead two 前进二【航海】
a heavy bloom 一层厚厚的蜡被
a heigh dosage must be used 施用量必须加大
ahen employment 外国人的雇用
aheneus 灿铜色
ahermatypic 非造礁型;非造礁的
ahermatypic coral 非造礁珊瑚
a high-precision depth-measuring and sounding system 精密量深和水深测量系统
ahlfeldite 水硒镍石;复硒镍矿
A-hoe blade 箭形锄铲
ahold 贴风航行
ahooting plane 大型刨
A-horizon 淋溶层(A层);淋滤土层;表土层;A层位
Ahrens polarizing prism 阿伦斯偏振棱镜
Ahren's prism 阿伦斯棱镜
ah-set 埋置在混凝土中的支架或柱
ahu 用作边界、路标和纪念牌的土丘;用作边界、路标和纪念牌的石头堆
ahuehuete 尖叶落羽杉
a hundred million 亿
a hundred percent 百分之百
a hundred years 百年
aicd lead-containing wastewater 含铅酸性废水
Aich's alloy 艾希六含铁黄铜
Aich's metal 艾希合金;艾希铁黄铜
aid 辅动设备;补助税
aid abnormality 航标失常
aid agency 援助机构
aid agreement 援助协议;援助协定
aid changed 航标变动
aid-coagulant 助凝剂
aid cooperation agreement 援助合作协议
aid corrected 航标恢复正常
AID Credit 美援信用证
aid-cross section 辅助横断面图
aid decision making 帮助决策
aided 辅助的
aided design 辅助设计
aided laying 半自动瞄准
aided matching 辅助装定器
aided range gearing 半自动距离数据发送
aided recall 帮助回忆法
aided tach(e)ometer 半自动跟踪转速计;半自动转速表
aided tachometer generator 半自动测速发电机
aided tracking 辅助跟踪;半自动跟踪
aided-tracking mechanism 半自动跟踪装置;半自动跟踪机构
aided-tracking ratio 半自动跟踪系数
aided tracking system 半自动跟踪制
aide-memoire 备忘录
aid establishment 航标设置
aid given gratis 无偿援助
aid-giving agency 提供援助的机构
aid-hearing 助听器
aiding in keeping equipment records 协助保管设备档案
aiding planning for utilities 协助计划公共设施
aiding planning materials flow 协助计划材料流程
aid moved 航标移位
aid-off work 停工
aid project 援助项目
aids adjustment 调波
aids discrepancy 航标失常
aids failure 航标失常
aids failure due to improper maintenance 维护性失常
aids failure not due to improper-maintenance 非维护性航标失常
aid shifted 航标移位
aids-layout adjustment and fairing transferring 调标改建
aid station 卫生所;急救站

aids tending craft 航标工作船
aids tending man 航标员
aids tending station 航道站
aids-to-navigation 航路标志;航标;助航设施;助航设备;助航标志
aids-to-navigation course 航标课程
aids-to-navigation in lock area 船闸航标
aids-to-navigation maintenance 航标维护
aids-to-navigation manual 航标手册
aids-to-navigation on canal 运河航标
aids-to-navigation on inland waterway 内河助航标志;内河航标
aids-to-navigation on lake and reservoir waterways 湖泊水库航志
aids-to-navigation quota 航标定额
aids-to-navigation school 航标学校
aids-to-navigation seminar 航标研究班
aids-to-navigation system on inland waterway 内河航标制式
aid to foreign countries 对外援助支出
aid to navigation 导航辅助设备;航路标志
aid to operation 辅助操作
aid transmission 有偿转拨
aid tying 援助的附带条件;援助限制
aid withdrawal 航标撤除
aie injection starter 压缩空气起动机
aie oxidation desulfation tower 空气氧化脱硫塔
aiguille 钻石器;钻孔器;尖(山)顶;石钻头;尖峰;钻孔机;针状峰
Aiken code 阿肯码
aikinite 针硫铋铅矿
aile 剧院观众席走道;教堂侧廊;耳房【建】
aileron 翼墙(教堂侧廊,通道,耳堂等);副翼;半山墙
aileron actuator 副翼促动器
aileron cable 副翼操纵索
ailette 建筑物的侧翼;钢甲两翼护肩
ailing aircraft 有故障飞机
ailsyte 钠闪微岗岩
aimantine 磁铁矿
aim colo(u)r 目标色
aim corrector 瞄准仪校正器
aimed launcher 定向发射装置
aimed midwater trawling 瞄准中层拖网捕捞
aimed trawling 瞄准线拖网捕捞
aim(ing) 目标;目的;瞄准;照准;对准
aiming adjust screw 瞄准调节螺钉
aiming axis 照准轴;对准轴
aiming circle 罗针仪;瞄准环;引导圆;照准圆;测角器;测角盘
aiming device 瞄准器
aiming error 瞄准误差
aiming field 目标字段;引导域
aiming head 照准头
aiming light 标灯
aiming line 瞄准线;照准线;觇线
aiming mechanism 瞄准装置
aiming point 瞄准点;照准点;觇点
aiming position 瞄准位置
aiming post 标杆;瞄准标杆
aiming rule 瞄准尺;觇尺
aiming rule sight 瞄准镜
aiming screen 调试屏幕;大灯对光屏幕
aiming screw 瞄准调节螺钉
aiming sight 瞄准仪
aiming stake 标杆
aiming symbol 目标符号;瞄准符;引导符号
aiming telescope 水准仪上的望远镜
aimless drainage 紊乱水系
aimlessly drifting population 盲流人口
aim-off 瞄准点提前量
aim-oriented approach 实现目标的方法
aims of system(s) 系统目标
ainfall washing 雨水冲刷
aiophyllous 常绿的
aiphyllium 常绿乔木;常绿林
air 214Bi influence 大气铋214放射性影响
air absorption 空气吸收(率)
air accidental 偶然进气
air accumulation 空气蓄积
air accumulator 气柜;气罐;空气蓄电池;空气存储器;压缩空气储[贮]罐
air acetylene welding 空气乙炔焊(接)
air-activated bulk cement container 空气松料散装水泥罐

air-activated cement container 空气松料水泥罐
air-activated chute 空气槽
air-activated conveyer 气动输送机;空气输送机
air-activated gravity conveyer 气动动力输送机;气动动力传送带;气动重力运输机
air-activated sludge design 空气活性污泥法设计
air activation 空气活化系统
air activator 气动激励器;气动输送机
air activity monitor 空气发射性指示器;大气放射性污染监测器;大气活动监测器
air activity monitoring station 大气污染监测站
air-actuated 压缩空气驱动的;压缩空气控制的
air-actuated clamp 气动夹具
air-actuated jaw 气动卡爪;气动夹具
air admission 进气
air admission port 进气口
air admission valve 进空气阀;真空破坏阀
air admittance 进气
air-admitting surface 进风口面积
air aerated sludge system 空气曝气污染系统
air aeration 空气曝气
air after cooler 空气二次冷却器
air agitation 空气扰动;空气搅动;充气搅拌
air agitation system 空气搅动系统
air agitator 压气搅拌器;充气搅拌器
air alert 空中待机
air almanac 航空(天文)历
air alternator 空气振荡器
air analyser 空气分析器
air analysis 空气分析
air analysis test 风筛(分)析试验;风力筛分试验;风化筛分试验
air analyzer 空气分析器
air and air-conditioning equipment 气空调设备
air and gas 焊接用氧气和乙炔
air and gas mixer 煤气混合器
air and gas premixer 空气燃气预混器
air and light space 采光通风场所
air-and-screen cleaner 风筛谷物清选机
air and steam blast 空气蒸汽鼓风
air and vacuum valve 空气真空阀
air angle grinder 风动角度磨床
air-arc cutting 电弧气刨
air arc furnace 空气电弧炉
air assist assembly 空气助流组件
air assisted airless spraying machine 混合式喷涂机
air assisted electrostatic gun 空气辅助静电喷枪
air assisted pressure jet burner 空气助压喷烧器
air atomization 气压雾化;空气雾化
air atomizer 气力弥雾器;空气雾化器
air-atomizing 气力喷雾
air atomizing burner 空气喷雾燃烧器;空气喷雾喷灯
airator 冷却松砂机
air back valve 空气止回阀;止回气阀
air bacteria 空气细菌
air baffle 空气反射层;空气挡板;折流板(烟道)
air bag 气袋;空气囊
air bag buffer 空气胎缓冲器
air balance 空气平衡;风量平衡
air balanced pumping unit 空气平衡泵组
air balancing chamber 空气平衡室
air balloon 气球
air-bar 压气钻井
air barometer 显震气压表
air barrage 通风隔墙
air barrier 气障;气流阻挡层;气泡屏蔽
air base 空中基线;空军基地;航空基地
air based laser antisatellite system 空基激光反卫星系统
air base inclination 空中基线倾斜;空测基线倾斜
air baseline 航摄基线;空中基线
air base tilt 基线倾斜
air bath 空气(淋)浴
air-bath device 气浴装置
air-bath equipment 气浴装置;气浴设备;空气浴装置
air-bath unit 气浴装置
air battery 空气电池(组)
air beacon 航空信标
air bearing 气浮垫轴承;气垫;空气轴承
air-bearing gyro 气体轴承陀螺仪
air bearing kit 空气垫

air-bed 气褥;气垫床;气床
air bell (显影时乳剂层表面上产生的)气泡;气流罩;砂眼
air bellows 鼓风机;风箱
air belt 气带;空气带
air-bend die 弯曲冲模
air bends 气压病
air bill 空运单
air bill of lading 航空提单
air binding 气阻;气塞;气结;气缚;气堵(塞)
air binding of filter 滤池气阻
air blanket countercurrent regeneration 空气顶压逆流再生
air blanketing 空气夹层
air blast 气喷净法;气流;气吹;喷气(器);喷气除垢;空气冲击(波);鼓风喷射;鼓风机;吹喷净法
air blast atomizer 气压喷雾器;空气喷射喷雾器
air blast atomizing 气压喷雾
air-blast breaker 喷气流断路器
air blast burner 鼓风式燃烧器
air blast circuit breaker 空气灭弧断路器;空气吹弧式断路器;空气吹弧开关
air blast cleaner 鼓风清洗机
air blast connection pipe 高压空气导管
air blast cooled 空气冷却的
air blast cooling 强制空气冷却;强制风冷;通风冷却;空气喷射冷却;吹风冷却
air blast dusting machine 鼓风除尘机
air blaster 空气炮;空气鼓风机;压气喷砂机
air blast freezer 空气喷射冷冻机;吹风冻结
air blast freezer tunnel 隧道式鼓风冻结机
air blast gas burner 鼓风煤气喷灯
air blast gin 气流式轧花机;气流式打桩机;鼓风机
air blast goaf-stowing machine 风力充填机
air blast gun 空气炮
air blasting 空气喷净法;鼓风(喷射);空气爆破
air blasting wave 空气冲击波
air blast loading 气流冲击荷载
air blast nozzle 鼓风喷嘴
air blast pump 掺气泵
air blast quench 鼓风淬冷;风冷淬火
air blast saw gin 气流式锯齿打桩机
air blast separation 气流分离
air blast sprayer 鼓风喷雾机
air blast switch 空气吹弧开关;气吹弧开关
air blast transformer 风冷式变压器
air bleed 放气
air bleeder 通气小孔;放气阀;气眼;气泡管;通风罩
air bleed hole 放气孔
air bleed orifice 放气孔
air bleed piston 放气活塞
air bleed port 放气孔
air bleed restrictor 抽气限制器
air bleed set 抽气机(组)
air bleed valve 空气量孔阀;放气阀(门);排气阀
air blending silo 空气搅拌库;空气搅拌仓
air blister 气泡;砂眼
air block fender 充气(橡胶)碰垫;充气(橡胶)护舷;充气橡胶防撞装置
air-block fender unit 充气护舷部件
air blocking area 风阻面积
air blocking coefficient 风阻系数
air blocking length 风阻长度
air blocking thrust 风阻推力
air blocking vehicle 风阻车辆
air blow 吹练;吹气;吹风
air blow aeration tank 鼓风曝气池
air blow asphalt 喷射沥青
air blower 增压机;风扇;压气机;鼓风机;吹风机
air-blower noise 通风机噪声
air blow flo(a)tation device 鼓风浮选装置
air blow flo(a)tation unit 鼓风浮选装置;鼓风浮选装置
air blowing 吹制法;吹(炼油)法
air-blowing installation 鼓风设施
air-blowing oxidation 吹气氧化
air-blowing process 吹气(氧化)过程
air-blowing treatment 吹气处理
air-blown aeration tank 空气鼓风曝气槽
air-blown asphalt 吹制(地)沥青;吹气(地)沥青;加气沥青;吹气氧化沥青
air-blown bitumen 吹制(地)沥青
air-blown concrete 喷射混凝土

air-blown converter 鼓气转炉
air-blown core 吹制型芯
air-blown mortar 喷射砂浆;吹气砂浆
air-blown producer 空气鼓风发生器
air-blown rod 风冷盘条
air-blown steel 富氧吹制转炉钢;吹气炼钢;空气吹制钢
air blowout 喷气;空气灭弧
air blowpipe 砂浆喷器;喷射管;气流吹洗管;喷气管
air blow system 鼓风系统
air boat 飞艇;飞船;空气推进艇
air boiler 空气加热器
air-bond 气障
air-boost compressor 空气压缩机;增压式空压机;增压式压气机
air boosted 增能的;增加气压的;升压的
air-booster 空气升压泵;气力增器;压力助力器;压缩空气伺服装置;压缩空气伺服设备
air boring 风锥钻岩;空气钻进;干式凿岩
air-borne 空运的;机载
air-borne activity 大气中放射性
air-borne aircraft 空中飞机
air-borne alert 空中警戒
air-borne allergen pollution 空传变应原污染
air-borne altimeter 航空高度计
air-borne and marine vector measurements 航空海洋向量测量
air-borne astrographic camera 航空天体摄影机
air-borne bacteria 空中微生物;气载(浮游)细菌
air-borne beacon 机载信标
air-borne biological form 空气生物形态
air-borne camera 航空照相机
air-borne communication 空中通信
air-borne communicator 空中通信装置
air-borne contaminant 空气传播污染物;气载污染物;气传污染物
air-borne contaminant concentration 气载污染物浓度
air-borne contamination 空气放射性污染;空气发射性污染;空传污染;气载污染
air-borne control system 空中控制系统;机载控制测量系统
air-borne data processor 航空数据处理机
air-borne data terminal set 机载数据终端机
air-borne debris 大气中散落物
air-borne detector 机载探测器
air-borne direction finder 航空测向器
air-borne dirt 空气污尘
air-borne disease (经)空气传播(的)疾病
air-borne display system 空中显示系统
air-borne drill 风动骨钻
air-borne dust 空中浮游尘埃;空气挟带的灰尘;空气粉尘;空传尘埃;气载尘埃;大气尘埃
air-borne-dust elutriator 水洗法空气除尘装置
air-borne dust pollution 飘尘污染
air-borne early-warning station 机载预警雷达站
air-borne electrical method 航空电法
air-borne electromagnetic prospecting 航空电磁找矿;航空电磁性勘探
air-borne electromagnetics 航空电磁法
air-borne electronic computer 机载电子计算机
air-borne electronic survey control 机载电子测量控制
air-borne emission 空气中的排放物
air-borne expendable bathythermograph 航空投掷式海水测温仪
air-borne flux-gate magnetometer 航空磁通门磁力仪
air-borne fraction of carbon 空气中碳化合物留分
air-borne gamma ray spectrometry 航空能谱测量
air-borne gap 气隙
air-borne gas chromatographic analyzer 航空气相色谱分析仪
air-borne geochemical exploration 航空地球化学勘探
air-borne geochemical prospecting 航空地球化学勘探
air-borne geochemistry 航空地球化学
air-borne geographical surveying 航空地球物理测量
air-borne geophysical prospecting 航空物探;航空地球物理勘探
air-borne goods 空运货物
air-borne gradiometer 航空梯度仪

air-borne gravimeter 空中重力计;航空重力仪
air-borne gravity measurement 航空重力测量
air-borne gravity meter 机载重力仪;机载重差计;航空重力仪
air-borne hydrography technique 航空测量水道技术
air-borne infection 空气传染
air-borne infrared astronomy telescope 航空红外天文望远镜
air-borne infrared equipment 机载红外装置
air-borne infrared high energy laser 航空红外高能激光器
air-borne infrared mapper 机载红外成像仪;机载红外测绘仪;航空红外测绘仪
air-borne infrared measurement instrument 机载红外测量仪器
air-borne infrared radiometer 机载红外辐射计
air-borne infrared scanner 航空红外扫描仪
air-borne infrared surveillance set 机载红外监视装置
air-borne interrogator 机载询问器
air-borne IR imaging system 机载红外成像系统
air-borne IR transmissometer 机载红外透射仪
air-borne laser absorption spectrometer 航空激光吸收光谱仪
air-borne laser beacon 机载激光信标
air-borne laser-camera scanning system 航空激光摄影机扫描系统
air-borne laser illuminator 机载激光照明器
air-borne laser profile-graph 航空激光纵剖面测绘仪
air-borne laser profiler 航空激光剖面测绘仪
air-borne laser rangefinder 航空激光测距仪
air-borne laser sounding 机载激光测深
air-borne laser target tracker 航空激光目标跟踪器
air-borne laser terrain profiler 航空激光地形断面记录仪
air-borne laser tracker 机载激光跟踪器
air-borne lifeboat 随机救生船;空投救生艇
air-borne light optic(al) fibre technology 机载轻型光纤技术
air-borne lightweight optic(al) tracking system 机载轻型光学跟踪系统
air-borne locator-designator 机载定位器指示器
air-borne magnetic exploration 航空磁力探矿
air-borne magnetic gradiometer 航空磁力梯度仪
air-borne magnetic prospecting 航空磁法找矿
air-borne magnetic survey(ing) 航空磁测
air-borne magnetometer 机载磁强计;航空磁强计;航空磁力仪
air-borne magnetometric(al) gradiometer 航空磁力梯度仪;航空磁力梯度计
air-borne mercury 空气中汞
air-borne mercury survey 航空汞量测量
air-borne meteorological radar 机载气象雷达
air-borne method 航空测量
air-borne microbial 悬浮微生物
air-borne multispectral scanner 航空多谱段扫描仪
air-borne multispectral scanning image 航空多谱段扫描图像
air-borne noise 空气噪声;空气声;空传噪声;直接传播噪声
air-borne non-dispersive infrared monitor 航空非分光红外监测仪
air-borne obstruction plan 机场进近净空区平面图
air-borne oceanographic lidar system 机载海洋激光雷达系统
air-borne oil spill surveillance system 航空油溢视系统
air-borne oil surveillance system 机载油污染监测系统;机载水面油污染监视系统
air-borne operator 空勤作业员
air-borne particle 气传微粒
air-borne particle analyzer 航空粉尘分析仪
air-borne particle counter 航空粉尘计数器
air-borne particle mass monitor 大气粉尘物质浓度监测仪
air-borne particles 大气尘粒
air-borne particulate 空中悬浮微粒体;空中悬浮尘雾;气传颗粒物
air-borne particulate contaminant 气载颗粒污染物
air-borne particulate contamination 气载颗粒污染
air-borne particulate pollutant 气载颗粒污染物
air-borne particulate pollution 气载颗粒污染

air-borne pathogenic organism 空气传播致病生物;空传致病生物
air-borne photogrammetric(al) camera 航摄仪;航空摄影机;机载摄影机
air-borne photography 空中摄影
air-borne platform 机载平台
air-borne pollutant 空气传播污染物;气载污染物;气传污染物
air-borne pollutant concentration 气载污染物浓度
air-borne pollution 空气污染
air-borne pollution oil detection system 机载水面油污染探测系统
air-borne profile 空中断面记录图;航空剖面
air-borne profile recording 空中断面记录
air-borne proton-precession magnetometer 航空质子旋进磁力仪
air-borne radar 机载雷达
air-borne radar geologic(al) reconnaissance 空中雷达地质勘察
air-borne radar measurement 机载雷达测量
air-borne radar platform 机载雷达平台
air-borne radiation thermometer 机载辐射温度计
air-borne radioactivity 空(气)中的放射性;空(气)中的放射量;气载放射性
air-borne radioactivity prospecting 航空放射性找矿
air-borne radiosonde recorder 航空无线电探空记录仪
air-borne ranger 航空测距仪
air-borne remote sensing 航空遥感
air-borne remote sensing system 机载遥感系统
air-borne salt 空中悬浮盐类
air-borne scanner 航空扫描仪
air-borne sea and swell recorder 航空海浪记录仪
air-borne sediment 空气挟带的泥砂;风沙沉积(物)
air-borne sensor 机载传感器
air-borne sensor system 空中传感系统
air-borne set 空中台
air-borne single point and differential GPS navigation 航空单点与差分全球空中定位导航系统
air-borne sonar 机载声呐
air-borne sound 空气噪声;空气(载)声
air-borne sound insulation 空气声隔声;空气消音;空气声隔绝
air-borne sound intensity 空气声声强
air-borne sound transmission 空气声传播;空气传声
air-borne sound transmission loss 空气载声传声损失
air-borne spectral measurement instrument 航空波谱测量仪器
air-borne station 机载电台
air-borne strip mapper 机载航带成像仪
air-borne survey 航空测量
air-borne system 机载系统
air-borne target designator 机载目标指示器
air-borne target locating system 机载目标定位系统
air-borne telesensing 航空遥感
air-borne television system 机载电视系统
air-borne tellurometer 航空微波测距仪
air-borne thermal infrared scanner image 航空热红外扫描图像
air-borne trade 空运贸易
air-borne transmission 经空气传播
air-borne transport 大气输运
air-borne typochemical compound 气迁标志化合物
air-borne typochemical element 气迁标志元素
air-borne warning and control system 机载预警和控制系统
air-borne waste 空气中废物;飞机上的废气污染;气载废物
air bottle 空气瓶;空气罐;集气罐;压缩空气瓶
air bottle of pump 泵的空气(拱)室
air-bound 被气体阻塞;空气阻塞;气体障碍;气塞的
air-bound pipe 空气束缚管
air box 气箱;气室;通气箱;空气箱
air brake 气(动)制动;气动力制动装置;减速板;气闸;气压制动器;刹车;气力制动器;空气制动(器);空气式功率计;风闸
air brake distributor 空气制动分配器
air brake grease 气闸润滑脂

air brake hose 气闸软管;空气制动软管
air brake hose nipple 气闸软管螺纹接套
air brake-pipe feed 空气制动管
air brake switch 空气制动开关
air brake system 空气制动系统;风闸系统
air brake system pipe 空气制动系统管道
air brake system testing stand 空气制动装置试验台
air braking 空气制动
air braking tachometer 空气制动转速计;摩擦式转速计
air brattice 风帘
air break 漏气;空气间隙;空气断路
air break contactor 空气断路接触器
air breakdown 空气击穿
air breaker 风钻破碎机;风镐破碎机;空气断路器;通风装置;通风孔
airbreaking 空气爆破
airbreak switch 空气(断路)开关
air breakup 大气层破坏
air breakwater 喷气式消波防波堤;气泡防波设施;气动防波堤;空气防波堤
air-breather 空气吸潮器;透气装置;通气装置;吸潮器
air-breathing 吸气
air-breathing laser 吸气式激光器
air-breathing power plant 空气喷气动力装置
air-breathing power unit 喷气功率单位;喷气发电机组;喷气动力装置
air brick 穿孔砖;透式砖;通风空心砖;空心砖;空气风干砖;有孔砖;多孔通风砖;风干砖
air bridge 空中桥梁;空运线
air brine 充气盐水
air-bronchography 含气支管照相术
air broom 气帚
air brush 气压喷雾刷色器;气压喷漆器;气刷;气笔;喷雾刷色器;喷刷器;喷漆器;喷笔
air brush coater 气刷涂布机
air brush coating 气刷涂布
air bubble 空气泡;砂眼;气泡
air-bubble breakwater 空气帘防波堤;消波空气帘;气泡防波堤;气帘防波堤
air-bubble curtain 气泡帷幕
air-bubble eliminator (铸件中的)气泡排除装置
air-bubble injection 气泡喷射法
air-bubble level 气泡水准仪;气泡水准器;气泡水平仪
air-bubble level(l)ing 气泡水准测量
air-bubble parameter 气泡参数
air-bubble pitting 气泡点蚀
air-bubble plume 气泡束
air-bubbler system 气泡防冻系统
air-bubble screen 空气幕;空气帘;气泡屏;气泡帘幕
air-bubble size 气泡尺寸
air-bubble spacing 气泡间隙;气泡间距
air-bubble system 气泡喷射系统
air-bubble technique 气泡法
air-bubble visco(si)meter 气泡黏度计
air bubbling system 气泡喷射系统
air buffer 软靠枕;空气缓冲器;空气隔层
air buffering 空气缓冲
air buffing 空气喷磨
air bulking 空气胀;气胀
air bumper 空气阻尼器;空气减震器
air buoyancy 空气浮力
air buoyancy tank 空气箱;浮箱
air burner 喷灯
air burning 空气燃烧
airburst 空中爆炸
airbus 大型客机;客机;空中公共汽车
air caisson 气压沉箱;压缩空气沉箱
air calorimeter test 吹风发热试验
air camel 空气浮箱;打捞浮筒;充气浮筒;浮柜
air camera 航摄仪;航空摄影机;航测摄影机
air canal 通风管路
air cap 空气帽
air capacitor 空气电容器
air capacity 空气输能力;含气量;空气容量;容量;空气量
air capping 通风帽;气帽
air car 气垫汽车
air carbon arc cutting 吹气碳电弧切割
air carbon arc cutting machine 吹气碳电弧切割机
air cargo 空运货物;航空货物

air cargo terminal 航空航运站
air cargo terminal building 航空货运楼
air carrier 航空公司;运输飞机
air case 空气罩;气箱;浮箱
air casing 气套;(平屋顶的)气隔层;管外(隔气)套管;防热气套;隔气套管
air castle 空中楼阁
air catapult 气力弹射器
air cataract 气瀑
air catcher 吸气器
air cavity 气穴;气腔;气泡;气孔;气眼
air ceiling 通风天花板;通风顶棚
air cell (从水底向上冒的)气泡;气胞;空气室;空气电池;充气室浮选机;充气式浮选机
air cell engine 空气室发动机
air cell piston 带空气室的活塞
air cement gun 水泥喷枪;风动水泥喷轮
air cement handling 气动水泥装卸
air centrifuge 空气离心机
air chain 坯桶中呈链状气泡;链状气泡
air chamber 气压室;气室;气室(分配)室;空气包;储(贮)气罐;浮囊;气泡腔;通风房
air chamber of pump 泵的气(压)室
air chamber pump 气室泵;气包水泵;带空气室的泵
air chamber type of surge-tank 气箱式调压塔
air chamber water shock absorber 气箱式水击吸收器
air change 换气;通风换气流量(按小时计算)
air change rate 通风换气次数;空气换气率;换气率;换气次数
air-changing device 换气装置
air channel 风管;排气道;通气槽;风道;空气道;风道
air charged accumulator 充气蓄能器
air charging 充气
air charging apparatus 充气装置;充气设备
air charging machine 风送器
air charging system 充气装置
air cheaner oil cup 空气过滤器油杯
air checks 麻孔
air chemistry 空气化学;大气化学
air chest 配气箱;风箱
air chill 空气冷硬;空气冷淬
air chilling 空气淬冷
air chimney 通风道;通风烟囱;通风筒;通风管
air chipper 风凿;风铲
air choke 阻风门;空心扼流圈
air churning resistance 空气扰动阻力
air chute 通风道;降落伞;电弧隔板
air circle 弥散圆
air circuit breaker 空气(自动)断路器;空气开关
air-circulating blower 空气循环鼓风机
air-circulating chamber 空气循环室
air-circulating device 空气循环设备
air-circulating installation 送风设备
air circulation 空气循环;空气洗井;空气流通;空气环流
air clamp 气压夹具
air clamper 气动压紧装置;气动夹紧装置
air classification 空气风力分级;空气分类;空气分级;风选;风力分级
air classifier 风洗器;气流分级器;气力分级装置;空气选粒器;空气分级器;空气分级机;风选器;风筛机;风力选粉机;风力分级器;风力分级机
air cleaner 空调洁净器;滤气器;空气滤清器;空气滤净器;空气净化器;空气净化机;涤气器
air cleaner air seal 滤气器气封
air cleaner and silencer 空气净化器及消声器
air cleaner base 空气滤清器底壳
air cleaner body 滤气器壳体
air cleaner cartridge 滤气器滤筒
air cleaner clamp 滤气器夹钳
air cleaner element 滤气器元件
air cleaner hair 滤气器发卷
air cleaner inlet 空气净化进气管
air cleaner servicer 空气滤清器服务车
air cleaner spring 滤气器弹簧
air cleaner tray 空气净化器支架;空气净化器台座
air cleaning 空气过滤;空气净化;风力选矿;风力清洗;风力清除
air cleaning device 空气净化设备
air cleaning equipment 气体净化设备;空气净化装置;空气净化设备

air cleaning facility 空气净化设施
air cleaning load 污染负荷量
air cleaning plant 空气净化设备;空气净化厂
air cleaning system 空气净化系统
air cleaning tray 空气净化器支架;空气净化器台座
air cleaning unit 空气净化系统;空气净化装置;空气净化器
air cleanliness 空气洁净(程)度
air clutch 气胎离合器
air coach 客机
air cock 气阀;气嘴;气栓;圆锥形风标;排气旋塞;空气旋塞;空气阀;放气旋塞;放气活栓;放气塞;放气阀(门)
air code generator 航空电码信号发生器;航空编码发生器
Airco-Hoover sweetening 爱尔柯—胡佛脱硫法
air coil 气冷蛇(形)管;气冷盘管;空气冷却蛇管
air collection 集气
air collector 气瓶;气柜;气筒;空气收集器;集尘器;集气箱;集气器;集气瓶;集气罐;储(贮)气罐;储气罐
colo(u)red 无色的
air column 气柱
air-comatic welding 惰性气体保护电弧焊
air compartment 气室;通风室;浮箱
air compensating jet 空气补偿喷嘴
air composition 空气组成;空气成分;大气组成;大气成分
air compressing engine 空压机;空气压缩机
air compressing machine 空压机;空气压缩机
air compress ink spray 喷笔
air compression 空气压缩
air compressor 空气压缩机;压缩空气机;压气机;气压机
air compressor chamber 空气压缩机室
air compressor crank case 空气压缩机曲柄箱
air compressor crank shaft 空气压缩机曲柄轴
air compressor cylinder 空气压缩机缸体
air compressor for inflating tires 轮胎充气压缩机
air compressor for mo(u)lding machine 模压机压气具
air compressor installation 空压机站
air compressor mounting 空气压缩机安装
air compressor oil 空气压缩机油
air compressor piston ring 空气压缩机活塞环
air compressor plant 空压机站;空压机房;空压(机)房;缩(机)房
air compressor pulley 空气压缩机皮带轮
air compressor pump 空气压缩泵
air compressor pumping installation 空压机抽水装置
air compressor system 空气压缩系统
air compressor test bench 空气压缩机试验台
air compressor test stand 空气压缩机试验台
air compressor unloader 空气压缩机卸荷器
air compressor valve 空气压缩机阀门
air concentration 空气浓度
air concrete 加气混凝土
air concrete placer 混凝土喷射器;混凝土喷枪
air condenser 空气凝汽器;空气冷凝器;空气冷凝机;空气电容器;风冷式冷凝器
air-condition 通风;空气调节
air-conditional train 空调列车
air-conditioned 有冷气设备的;装有空气调节器的;装有空调设备的
air-conditioned building 空调房屋;装有空调的建筑;设有空调房间
air-conditioned cabin 空气调节舱;空调舱室
air-conditioned ceiling 通风天花板;通风顶棚
air-conditioned coach 空气调节客车
air-conditioned container 空气(调节)集装箱
air-conditioned rolling stock 空调车辆
air-conditioned space 经空气调节的空间;空调室
air-conditioned train 空调列车
air-conditioned wind tunnel 空气调节风洞
air-conditioned zone 空气调节区
air-conditioner 空气调节装置;空气调节器;空调设备;空调器;空调机
air-conditioner for vehicle 车辆用空调器
air-conditioner fresh air fan 空调新风机
air-conditioner room 墙间空调器;空调机房;单独空调房间
air-conditioner unit 空气调节设备
air-conditioner window 窗间单独空调器

air-conditioner workshop 空调装置检修车间
air-conditioning 空气调节;空调
Air-conditioning, Heating and Ventilating 空气调节采暖与通风(期刊)
air-conditioning and exhausting apparatus 空调及排风装置
air-conditioning chamber 空气调节室;空调室
air-conditioning comfort 舒适空调
air-conditioning condenser 空调冷凝器
air-conditioning condition 空调工况
air-conditioning conduit 空气调节管道
air-conditioning control device 空调控制设备
air-conditioning control panel 空调控制板
air-conditioning design 空调设计
air-conditioning discipline 空调专业
air-conditioning duct 空气(调节)管道;空调管道
air-conditioning engineering 空调技术;空调工程
air-conditioning equipment 空气调节设备;空调设备
air-conditioning equipment room 空调设备室
air-conditioning evapo(u)rator 空气调节蒸发器
air-conditioning inlet 空调口
air-conditioning installation 空调设施;空调机组;空气调节装置;空调装置
air-conditioning layout 空调平面图
air-conditioning load 空调荷载;空调负荷
air-conditioning lock 空调房间专用窗锁;空调气闸
air-conditioning luminaire 空调照明装置;空调照明设备
air-conditioning machine room 空调机房
air-conditioning machine room layout 空调机房平面图
air-conditioning machinery 空气调节机
air-conditioning measuring facility 空调测量装置
air-conditioning noise 空调噪声
air-conditioning package 空气调节组件
air-conditioning piping 空调管线
air-conditioning planning 空气调节计划;空调节设计
air-conditioning plant 空调设施;空调节装置;空气调节设备;空调设备;空调机房
air-conditioning shaft 空调管道井
air-conditioning space 空调空间
air-conditioning station 空调站;中心空调站
air-conditioning system 空气调节系统;空调系统;空调机组
air-conditioning system cooling load 空气调节系统冷负荷
air-conditioning tower 空调塔
air-conditioning unit 房间空(气)调(节)器;空气调节装置;空气调节设备;空气调节器;空调设备;空气(调节)机组
air-conditioning zone 空调区
air-conduction 空气传导;气导
air-conduction anti-noise microphone 气导式抗声传声器
air conductivity 空气电导率;大气导电性;透气性
air conduit 通风道;输气管;通风道管;空气管道;风管
air connection 风管接头;送风连接管
air conservation 空气保护;防止空气污染;防止大气污染
air consignment note 空运托运单;空运发货单
air consumption 空气用量;空气消耗(量);耗空气量;耗风量
air container 储气器
air contaminant 空气污染物(质)
air contamination 空气污染
air contamination analysis 空气污染分析
air contamination concentration 空气污染浓度
air contamination control 空气污染控制
air contamination control district 空气污染控制区
Air Contamination Control Law 空气污染防治法
air contamination controller 空气污染检测器;空气污染管制员
Air Contamination Control Regulation 空气污染防治条例
air contamination control system 空气污染防治系统
air contamination disaster 空气污染事故
air contamination emission 空气污染排放
air contamination emission factor 空气污染排放因素
air contamination episode 空气污染事件

air contamination exposure level 空气污染程度
Air Contamination Fund 空气污染研究基金会
Air Contamination Hearing Board 空气污染听证委员会
air contamination index 空气污染指数
Air Contamination Law 空气污染法
air contamination legislation 空气污染立法
air contamination measure-ment program (me) 空气污染测定规划
air contamination meteorology 空气污染气象学
air contamination model 空气污染模型
air contamination monitoring 空气污染监测
air contamination observation station 空气污染观测站
air contamination occurrence 空气污染偶发事件
air contamination potential 空气污染潜势
air contamination reduction device 空气污染减轻装置
air contamination region 空气污染区
Air Contamination Regulation 空气污染法规;空气污染条例
air contamination sensor 空气污染传感器
air contamination sources 空气污染源
air contamination standard 空气污染标准
air contamination surveillance 空气污染监视
air contamination survey 空气污染调查
air contamination zone 空气污染区
air content 空气(含)量;含气率
air content determination 含气量测定
air content of concrete 混凝土含气量
air content test 含气量试验
air content test for aerated concrete 混凝土含气量试验
air continuous and automatic monitoring system 大气连续自动监测系统
air control 气动控制;空气控制;压气控制;气压控制
air control butterfly 翻板式空气调节闸板
air control center 空中管制中心
air controlled isothermal lehr 空气控制等温退火炉
air control system 气压操纵系统
air control valve 控气阀;空气控制;空气调节阀
air convection 空气对流
air convection loss 空气对流损失
air conveyer 气动运输机;气动输送机;气力输送机;航空输送机;航空运输机;压缩气输送器;压缩空气输送机;空气输送器
air conveying pipeline 气动输送管路
air coolant 冷却空气;空气冷却剂
air-cooled 气冷式;空气冷却的;船用空气压缩机;风冷式
air-cooled air-compressor 空冷式空气压缩机
air-cooled air-conditioner 气冷式空调器;风冷式空调器
air-cooled blast-furnace slag 气冷高炉溶渣;气冷高炉炉渣;气冷高炉矿渣
air-cooled cargo 风冷货;气冷货
air-cooled cascade blade 气冷叶册
air-cooled ceiling 气冷式天花板
air-cooled chiller 空气冷却致冷器
air-cooled chiller unit 气冷式冷却机组
air-cooled collector 空气冷却整流子
air-cooled compressor 气冷式压缩机;空气冷却压缩机;风冷式压缩机
air-cooled condenser 空气冷却器;气冷冷凝器;空冷凝器;风冷式冷凝器;气冷式(空气)冷凝器
air-cooled conditioner 气冷式空调机
air-cooled diesel engine 风冷柴油机
air-cooled diffusion pump 空气冷却扩散泵
air-cooled engine 空气冷却发动机;气冷(式)发动机;风冷式发动机
air-cooled heat exchanger 空气冷却热交换器
air-cooled ignitron 气冷(式)引燃管
air-cooled jacket 气冷式套筒
air-cooled lamp 空气冷却照明器;气管
air-cooled mercury rectifier 气冷式水银整流器
air-cooled mill 气冷磨
air-cooled motor 风冷式电动机;气冷式电(动)机
air-cooled reactor 空气冷却反应器;空气冷却反应堆
air-cooled refrigerant condenser 气冷式制冷剂冷凝器
air-cooled rotor 气冷式转子
air-cooled slag 空气冷却渣;熔炉渣;气冷矿渣
air-cooled slag aggregate 熔炉渣集料;熔炉渣骨料
air-cooled steel 气冷钢

air-cooled system 气冷系统
air-cooled transformer 气冷式变压器
air-cooled turbo-generator 气冷式汽轮发电机
air-cooled two-stroke gasoline engine 风冷二冲程汽油发动机
air-cooled wall 空冷炉墙;气冷壁
air cooler 气冷器;空气冷却器
air cooler coil 空气冷却器盘管
air cooler sensitive cooling effect 空气冷却器冷却效能
air cooling 气冷;空气冷却;风冷
air cooling apparatus 空气冷却装置
air cooling by evapo(u)ration 蒸发气冷却
air cooling cargo 凉货(冷藏船运保持华氏40～60度的货物)
air cooling coil 空气冷却盘管
air cooling compressor 空气冷却压缩机
air cooling dehumidifier 空气冷却去湿装置
air cooling engine 气冷式发动机
air cooling fin 散热片;空气散热片;空气冷却片
air cooling refrigeration 风冷式制冷
air cooling system 气冷系统;气冷设备;空气冷却系统
air cooling tower 空气冷却塔
air cooling unit 空气冷却机组
air cooling valve 气冷阀
air coordinates 空中坐标
air coordinate system 空中坐标系
air core 空心
air core cable 空气纸绝缘电缆
air core coil 空心线圈;空气心线圈
air core deflection coil 空心偏转线圈
air cored instrument 空心电动仪表
air core focusing coil 空心聚焦线圈
air core reactor 空心扼流圈;空心电抗器
air core solenoid 空心螺线管
air core tire 半实心轮胎
air core transformer 空气心变压器
air core tyre 空心轮胎
air corridor 空中走廊
air corrosion 空气腐蚀
air coupling valve 空气连接阀
aircourse 空气通道;风洞;风道;通风巷道(工程)
air course board 通风道壁板
air cover 航测成图地区
air-crack 干裂
air-craft 航空器;飞机;飞艇
aircraft accessories 飞机附件
aircraft altimeter 航空测高计
aircraft assembly plant 飞机装配厂
aircraft axes 飞机坐标系
aircraft ball bearing 航空球轴承
aircraft-base observation 机上观测
aircraft bonding 飞机搭接
aircraft-borne scanner 航空扫描仪
aircraft cable 航空钢丝绳
aircraft camera 航摄仪;航空摄影机
aircraft camera mount(ing) 航摄仪座架;航空摄影机座架
aircraft carburetor 航空汽化器
aircraft carrier 航空母舰
aircraft ceiling 飞机升限
aircraft chart board 航空图板
aircraft coating 航空涂料;飞机涂料
aircraft compassI aeroplane compass 航空罗盘
aircraft cord 航空细钢丝绳
aircraft cord wire 航空钢丝
aircraft critical speed 单发临界速度
aircraft deicer 飞机除冰器
aircraft derived gas turbine 航空改装型燃汽轮机
aircraft direction frigate 航空指挥海防舰
aircraft dispersal area 疏散机场
aircraft early warning station 飞机早期警报站
aircraft electrification 飞机带电
aircraft engine 航空发动机
aircraft engine emissions 飞机发动机排放物
aircraft engine fuel 航空发动机燃料
aircraft engine oil 航空机油
aircraft exhaust 飞机排气
aircraft factory 飞机制造厂
aircraft ferry 飞机运输舰
aircraft fittings 飞机配件
aircraft fluid 航空液压油
aircraft flying 飞机航高

aircraft fuel 航空燃料;飞机燃料
aircraft gas turbine 航空燃汽轮机
aircraft grease 航空润滑脂
aircraft hangar 飞机库
aircraft identification sign 飞机识别标志
aircraft impactor 机载冲击采样器
aircraft instrument 航空仪表
aircraft lighting 飞机灯光设备
aircraft lubricating oil 航空润滑油
aircraft machine-gun turrent 航空飞机关炮机炮塔
aircraft machinery 航空加工机械
aircraft manoeuvring area 起落活动区
aircraft manufactory 飞机制造厂
aircraft manufacturing shop 飞机制造车间
aircraft marking 飞机标志
aircraft motor gasoline 航空汽油
aircraft movement 飞机运动
aircraft movement area 飞机活动区
aircraft noise 航空器噪声;飞机噪声
aircraft noise certification 飞机噪声验证
aircraft noise level 飞机噪声水平
aircraft observation 飞机观测
aircraft operation 飞机运动
aircraft overhaul waste 飞机检修废物
aircraft parking (area) 停机坪
aircraft parking guidance system 飞机停靠目视引导系统
aircraft parking system 停机坪系统
aircraft pen 飞机掩体
aircraft position chart 战术航空图
aircraft power vehicle 航空地面电源车
aircraft quality steel 飞机用优质钢
aircraft revetment 飞机掩体
aircraft rocket 航空火箭
aircraft sextant 航空六分仪
aircraft snow survey 航空测雪
aircraft sounding 飞机测探
aircraft speed 航速(指航空器)
aircraft stand taxilane 机位间滑行线
aircraft surveying 航测
aircraft tanker 空中加油飞机
aircraft tilt 飞机倾斜
aircraft towing tractor 飞机牵引用拖拉机
aircraft transparency 航空器镶玻璃部分;飞机玻璃窗
aircraft turbine 航空透平
aircraft varnish 航空清漆
aircraft vectoring 飞机导引
aircraft velocity 航速(指航空器)
aircraft wake 飞机尾流
aircraft warning light 飞机警告灯
aircraft washing area 飞机洗涤场
aircraft watch 航空表
aircraft windows 飞机摄像窗口
aircraft wing panel 飞机翼镶板
aircraft wire rope 航空钢丝绳
aircraft yaw 航偏角
air crossing 井下交叉风道;交叉气道;风桥
air cup 喷雾嘴
aircure 空气固化
air cure agent 空气硬化剂
air-cured 空气养护的
air-cured concrete 空气养护混凝土
air-cured specimen 空气养护试件
air curing 空气养生;空气养护;常温固化;空气固化
air curing binder 空气硬化黏结剂
air current 气流;空气流;地空传导电流
air current control 气流控制
air current mark 风成波痕
air current mill 气流磨
air current ripple 风成波纹
air curtain 气帘;空气屏障;空气幕;空气帘;风幕;气屏;气幕
air curtain doorway 门前热风幕
air curtain installation 气幕装置;气幕设备
air curtain method 气幕法
air curtain unit 空气幕装置;空气幕设备
air cushion 气垫;空气减震器;空气缓冲器;空气缓冲层;空气垫(层)
air cushion barge 气垫驳船
air cushion bearing 气垫轴承
air cushion belt conveyer 气垫(胶带)输送机
air cushion boat 气垫艇;腾空艇
air cushion brake 气闸;风闸

air cushion car 汽垫车;气垫车
air cushion conveyer 气垫输送机
air cushion craft 气垫艇
air cushion crash rescue vehicle 气垫应急救助船;应急救助气垫船
air-cushion culvert 气垫排水渠
air-cushion dredge(r) 气垫式挖泥船
air-cushioned 气垫式;气垫的
air-cushioned seat 气垫座
air-cushioned tempering 气垫钢化
air cushion high speed ground transportation 气垫式运输
air cushion ice breaking platform 气垫破冰平台
air cushioning 空气缓冲
air cushioning craft 气垫船
air cushioning machine 气垫汽车;气垫机
air cushion joint (混凝土路面的)气垫缝
air cushion kiln 气垫窑
air cushion landing system 气垫着陆系统
air cushion pad 气垫输送装置
air cushion pontoon 气垫方驳
air cushion railway 气垫铁路
air cushion seaplane 气垫式水上飞机
air cushion ship 气垫船
air cushion shock absorber 气垫减震器;空气弹簧隔振器
air cushion sprayer 气垫式喷雾器
air cushion surge chamber 气垫式调压室
air cushion train 气垫火车;气垫列车
air cushion transporter 气垫输送装置
air cushion vehicle 气垫船;气垫列车;气垫车(辆);气垫(式)运载工具;气垫(式)运载车;腾空艇
air cushion vessel 气垫船
air cushion vessel in non-displacement mode 气垫船非排水状态
air cut 气割
air cycle 空气循环
air cycle efficiency 空气循环效率
air cycle equipment 空气循环设备
air cyclone 空气旋流器
air cylinder 汽缸;空气汽缸;储气筒
air cylinder gasket 储气筒垫片
air cylinder pusher 气动堆料机
air cylinder valve 气筒阀
air cylinder valve bonnet 气筒阀帽罩
air cylinder valve gland 气筒阀压盖
air cylinder valve stem 气筒阀杆
air damped 空气减震;空气减幅
air damped pendulum 空气阻尼摆
air damper 气流调节器;气动阻尼器;空气阻尼器;空气节制器;空气调节板;空气挡板;节气门;压水板;挡风板
air damping 空气阻尼;空气制动;空气减震;空气减幅
air damping balance 空气阻尼天平
air data 气象数据;航空资料;大气数据
air data computer 空中数据计算机
air data sensor 空气数据传感器;大气数据传感器
air data system 大气数据系统
air defence 防空
air defence area 防空区
air defence command computer 防空指挥计算机
air defence computer 防空计算机
air defence equipment 防空设备
air defence measures 人防措施;防空措施
air defence of city 城市防空
air defence radar 防空雷达
air-defence sighting telescope 防空观测望远镜
air-defence station 防空站
air defence system 防空系统
air defence work 防空工程;防空工作
air deflector 空气折流板;空气偏导器;空气导流器;空气导流板;导风板
air dehumidifier 空气去湿器;空气干燥器
air deliverable anti-pollution transfer system 气送防污输送系统;海面防污(染)空投转移系统
air delivered anti-pollution transter system 海面防油污转移系统
air delivery 气力输送;风动输送
air delivery pipe 空气输送管
air delivery system 输气系统
air demand 需气量;供气量
air demolition pick 风镐需气量

air density 空气密度
air density coefficient 空气重度系数
air density correction factor 空气密度校正因数
air density difference 空气重度差
airdent 喷砂磨齿机
air depolarized battery 空气去极化电池
air deposited clay 风积黏土
air depot 飞机维修站
air-detraining admixture 除气外加剂
air detrainment 除气剂
air differential pressure transducer 空气差压变送器
air diffuser 空气散流器;空气扩散器;空气护散器
air diffusing equipment 空气扩散设备
air diffusing outlet 空气扩散(流)口
air diffusion 空气扩散
air diffusion aeration 空气扩散曝气
air diffusion aerator 空气扩散式曝气器
air diffusion coefficient 空气渗透系数
air diffusion method 压缩空气曝气法;空气扩散法
air diffusivity 空气扩散率
air digger 风动铲具;风铲
air discharge 排气;空中放电
air discharge cock 排气开关
air discharge compartment 排风室
air discharge equipment 排气设备
air discharge grill(e) 排气格栅
air discharge gyroscope 气体轴承陀螺仪
air discharge wagon 气动自卸车
air dispersion performance 布气性能
air displacement 排气量
air-displacement pump 空气排代泵;排气泵
air distance 空中距离
air distillation 常压蒸馏
air-distributing acoustic(al) ceiling 配气吸声顶棚;送风吸音吊顶
air-distributing ceiling 配气顶棚;配气吊顶
air-distributing valve 配气阀
air distribution 气流组织;通风量分配;空气分配;空气分布
air distribution bar 空气调节杆
air distribution equipment 空气分配设备;空气分布设备
air distribution line 空气分布管道
air distribution outlet 空气分布口
air distribution system 空气分配系统;空气分布系统
air distributor 空气分配器;空气分配阀;空气分布器
air dive profile 空气潜水方案
air dive project 空气潜水方案
air diver 空气潜水员
air dive schedule 空气潜水表
air diving 空气潜水
air diving breathing apparatus 空气潜水呼吸器
air diving decompression 空气潜水减压
air doctor blade 气体刮墨刀
air doctor coating 气刮刀涂布
air doctor of spreader 涂布机气刮刀
air dolly 气压顶铆器
air dome 半圆形空气室;气室;空气包;充气帐篷
air dome of pump 泵的气(压)室
air door 气门;风门;通气门
air door operation 气门操作
air dose 空气中辐射剂量;气剂量
air-douche unit with water atomization 喷雾风扇
air downtake 空气下降道
airdox 压气爆破筒
airdox system 压气爆破筒采景系统
air draft 气流;通气(管);空气通风;船高;抽风
air draft sensitivity 气流敏感性
air drag 空气阻滞作用;空气阻力
air drag perturbation 大气阻力摄动
air drag tach(e)ometer 气阻式转数计
air drain 通气管;气眼;气门;排气道;通气沟;出气口
air drainage 空气流泄;排气
air draught 气流;通风;水上高度;空气通风;气流
airdraulic actuator 气液压力转换缸
air drawing 吸气
airdreadnought 大型飞船;大型航空器
air-dried 风干的;空气干燥的
air-dried basis 气干基;风干基
air-dried brick 风干砖
air-dried coil varnish 空气干绕组清漆;空气干绕

圈清漆
air-dried condition 气干状态;风干状态
air-dried density 风干密度(混凝土管)
air-dried lumber 气干木材;自然干燥木材
air-dried material 空气干燥材料;风干材料
air-dried sample 风干样(品)
air-dried size 风干尺寸
air-dried soil 风干土
air-dried state 风干状态;空气干燥状态
air-dried strength 风干强度
air-dried timber 风干圆木
air-dried varnish 风干清漆
air-dried volumetric(al)specific gravity 风干容重
air-dried weight 风干状态重量;风干重(量)
air-dried wood 风干木材
air drier 空气干燥器
air-drifted drill 风动凿岩机;风动架式钻机
air drill 风锥;风镐;气钻;气动凿岩机;风钻
air drilling 风钻;空气钻进;空气洗井钻进;压缩空气冲井钻进
air drive 气压传动;气动装置;空气驱动
air-driven 气动的
air-driven generator 风动发电机
air-driven grout pump 气动灌浆泵
air-driven gyroscope 气动陀螺仪
air-driven hammer drill 风动气锤钻;风动冲击凿岩机
air-driven hydraulic jack 风动油压千斤顶
air-driven mechanical tube cleaner 风动管子清洁机
air-driven mine car loader 矿山风动装车机
air-driven pump 气泵;压缩空气泵
airdrome 飞机库;航空站
airdrome control tower 机场指挥塔
airdrome environment 机场环境
air drop hammer 气锤;蒸汽重力落锤
air droppable expendable ocean sensor 空投抛弃式海洋遥感器
air drum 空气罐[贮]气筒;储气器
air dry 气干;晾干;自干;常温干燥;风干(的)
air dry cell 空气干电池
air dry concentrates 风干精料
air dry density 风干密度(混凝土管)
air dryer 空气干燥器;风干器;风干机
air dry fiber 风干纤维
air drying 空气干燥;自干;空气干化;自然干燥
air drying enamel 风干瓷漆
air drying kiln 烘干房
air drying loss 风干失重
air drying machine 空气干燥机
air drying neoprene sealant 风干氯丁橡胶嵌缝膏
air drying paint 气干(色)漆;自干(色)漆
air drying sealant 风干嵌缝膏
air drying time 气干时间;自干时间;常温干时间
air drying varnish 气干清漆;空气干燥清漆;自干清漆;风干清漆
air dry matter 风干物质
air dry roughages 风干粗料
air dry sample 风干样(品)
air dry shrinkage 风干收缩
air dry strength 风干强度
air dry vent system 空气干燥通风系统
air dry weight 风干重量
air dry wood 气干木材
air duct 通气管;通气道;通风道;送气管道;空气通道;空气(流导)管;空气(导)管;进气管道;导气管;风管
air duct area 风渠面积
air duct burner 风道燃烧器
air duct design 风道设计
air duct detector 风管探测器
air ducting 通风管;导风装置;风管装置;风道
air duct installation 风道安装
air duct of tunnel 隧道通风风道
air duct riser 暖气管道竖井;送风立管;暖气竖管
air duct system 风道系统
air duct works 风道工程
air-dump car 气动翻卸车;气动自卸车;气动倾卸车;风动自卸车
air dunnage bag 空气袋
air duster 吹风拂尘器
air duster blow gun 吹尘器喷枪
air dynamic(al)noise 空气动力噪声
air dynamic(al) stability of suspension bridge 悬索桥空气动力稳定性
air-earth conduction current 地空传导电流
air earth current 地空电流
air earth hammer 风动打夯机;气动打夯机
aired bar 脱碳钢条
air eddy 空气涡流
aired ware 跑釉制品;跑釉
air-ejecting fan 抽气机;排气通风机
air ejection 气动脱模
air ejector 空气喷射器;空气抽出器;抽气器
air ejector fan 空气喷射风扇;抽气风扇;抽风机
air ejector for starting 起动弹射器
air electrode 空气电极
air elimination 去气;排气
air eliminator 排气装置;空气净化器;排气器;去气器;放气门
air elutriation 空气淘析;空气淘净;风筛
air elutriation apparatus 空气淘洗装置
air elutriation method 空气淘析法;空气离析法
air elutriator 风筛机;风力洗分器;风力离析机
air embolism 沉箱病;气栓症;气泡栓塞;空气栓塞症
air emission 空气散发
air end 通风口
air endway 副回风煤巷
air engine 风力发动机;气动机;空气发动机;航空发动机
air enrty 吸气口
air-entrained 加气的
air-entrained air content 引气量
air-entrained cement 加气水泥;引气水泥
air-entrained cement mortar 加气水泥砂浆
air-entrained concrete 掺气混凝土;无气孔混凝土;加气混凝土;引气混凝土
air-entrained light weight concrete 加气轻质混凝土
air-entrained mortar 加气灰浆;加气灰浆;加气砂浆
air-entrained slag concrete 加气矿渣混凝土
air entrainer 加气器
air-entrainer for concrete 混凝土加气剂;加气剂
air-entraining 加气(工艺);引气作用
air-entraining admixture 加气剂;加气掺和料
air-entraining agent 加气剂;混凝土加气剂;引气剂;携气剂
air-entraining and fibres concrete 加气纤维混凝土
air-entraining cement 加气水泥;加气剂;掺加气剂的水泥
air-entraining chemical compound 掺气剂
air-entraining concrete 加气混凝土
air-entraining fiber concrete 加气纤维混凝土
air-entraining fibrous concrete 加气纤维混凝土
air-entraining hydraulic cement 水硬性加气混凝土;加气水硬水泥
air-entraining mortar 加气砂浆;加气灰浆
air-entraining Portland blast furnace-slag cement 加气硅酸盐高炉矿渣水泥;加气波特兰高炉矿渣水泥
air-entraining Portland cement 加气硅酸盐水泥;加气波特兰水泥;引气硅酸盐水泥
air-entraining Portland pozzolan cement 加气硅酸盐火山灰水泥;加气波特兰火山灰水泥
air-entraining slag cement 加气矿渣水泥
air-entraining technology 加气工艺
air-entraining workability 加气增塑剂;和易性剂(用于混凝土、砂浆等)
air-entrainment 加气剂;加气(处理);引气;掺气
air-entrainment concentration 掺气量;加气量;掺气浓度
air-entrainment concrete 加气混凝土
air-entrainment light(weight)concrete 加气轻质混凝土
air-entrainment meter 引气量测定仪
air-entrainment mortar 加气砂浆;加气灰泥;加气灰浆
air-entrainment test 引气量试验
air entrap capacity 掺气能力
air entrapment 滞留在漆膜内的气泡
air-entrapping 加气
air-entrapping cement 加气水泥
air-entrapping structure 多孔构造;多孔结构
air entry 空气进口
air-entry permeameter 掺气渗透仪
air envelope 大气层
air environment 空电设施
air environmental parameter 空气环境参数
air equilibrium distillation 常压平衡蒸馏
air equipment 通风设备;空调设备;压缩空气装置
air equivalent 空气当量
air equivalent material 航空等效材料
airer 烘衣架;干燥装置
air escape 漏气;放气孔;放气管;放气(阀);排气孔;排气(出)口;泄气
air escape cock 排气嘴;排气阀;放气开关
air escape valve 排气阀;排气活门;放气阀(门)
air escaping 放气
air evapo(u)ration 空气蒸发
air evapo(u)ration valve 抽气阀
air excess 空气过剩
air excess factor 空气过剩系数
air exchange 空气交换
air exchange fan 换气扇
air exchange rate 气体交换速度
air exhaled 浊气
air exhaust 通风孔;排气;抽气;抽风;废气排除
air exhaust device 排气装置
air exhaust duct 排气管道
air exhaust duct network 排气(管)网
air exhauster 排气器;排气风机;抽风机
air exhaust hose 放气软管
air exhausting pump 排气泵
air exhausting tee 排气三通
air exhausting vent 排风口
air exhaustion optimal number 最佳抽气次数
air exhaust opening 排气口
air exhaust ventilator 排气通风机;排气风窗;排风机
air exhaust vent-pipe 排气通风管
air exit 排风口
air expansion chamber 空气膨胀室
air expansion lightning arrester 气膨胀避雷器
air express 特快空运
air external vibrator 气动表面振捣器;附着表面振捣器(混凝土)
air extracting 抽气
air-extracting pump 抽气泵
air extraction 抽气
air extraction device 抽空装置
air extraction main 抽气总管道
air extractor 排气机;抽风机;抽气设备;抽气器;抽气机
air face 空气表面;下游坝面;(坝的)背水面
air-faced tile 饰面砖
air-faced work 饰面工作
air failure 气源故障
air-fast 不透风的
air faucet 气旋塞
air feed 增加压缩空气;气力给料;进气;供气;供风;风力推进;风力牵引
air feeder 送气机;空气供给装置;进气管;吹气机
air-feeding and water-pumping 封舱压气抽水打捞法
air feed leg 压缩空气给进支腿
air feed leg drill 气腿式钻机
air feed mask 气罩
air feed motor 风动马达
air feed pump 供气泵
air-felting 气流铺装;气流交粘法
air-fence 导流栅
air ferry 渡运飞机
air ferry service 空运业务
airfield 机场;飞机场
airfield circuit 机场起落航线
airfield classification method 机场分类法
airfield clearance 机场净空
airfield complex 综合性机场
airfield cone pentrometer 机场贯锥测实计
airfield contours 机场地形等高线
airfield control communication 机场指挥通信
air field equation 气隙磁场方程
airfield first aid 机场救护
airfield flight pattern 机场起落航线
airfield hygiene 机场卫生
airfield layout 飞机场规划
airfield lighting 飞机场照明;机场照明
airfield meteorological station 飞机场气象站;机场气象站
airfield rescue 机场救护
airfield runway 机场跑道
airfield runway survey 机场跑道测量
airfield soil classification 机场土(壤)分类

airfield survey 机场测量
airfield traffic pattern 机场起落航线
air-filled pore space 充满空气的孔隙
air-filled porosity 充气孔隙度
air-filled thermocouple 充气热电偶;充气温差电偶
air-filled type 充气轮胎
air-filled waveguide 充气波导管
air-filling connection 充气连接
air film 气膜;空气膜
air film bearing 气膜垫
air film conveyer 气垫式输送机
air film interface 空气膜层界面
air film transfer coefficient 空气侧放热系数
air filter 气滤;滤气器;空气滤清器;空气过滤器
air filter apparatus 空气过滤器装置
air filter bank 过滤组合件
air filter casing 空气过滤器箱体
air filter chamber 空气过滤室
air filter cover 滤气器盖
air filter element 空气过滤元件
air filter module 过滤器组件
air filter oil 空气过滤器(用)油
air filter silencer 空气滤清器消声器
air filter tray 空气过滤盘
air filter unit 空气过滤器单元
air filtration 空气净化;空气过滤
air filtration paper 空气过滤纸
air filtration unit 空气滤清装置;空气过滤设备
air fin cooler 空气散热叶片冷却器;翅片管空气冷却器
air fixed capacitor 固定空气电容器
air fixture 通风装置
air flap 气阀瓣;气瓣;风门片
air float 气动抹灰工具
air flo(a)tation 气浮(法);气浮选(法);充气浮选
air flo(a)tation-anaerobic-aerobic process 气浮—厌氧—需氧工艺
air flo(a)tation classifier 气浮(法)分离器
air flo(a)tation dry separation 风选
air flo(a)tation-hydrolysis-contact oxidation process 气浮—水解—接触氧化工艺
air flo(a)tation physical-chemical treatment 气浮物化处理
air flo(a)tation powder 空气浮选粉料
air flo(a)tation process 气浮工艺
air flo(a)tation thickener 气浮浓缩池
air float conveyer 风力运输机
air floated 风分选的
air-floated clay 风选黏土
air-floated magnetic head 空气浮动磁头
air-floated powder 气溶微粒;气溶粉尘;空气浮选粉料;风分选粉料
air float fines 喷浆粉末;喷浆粉尘
air floating table 气浮桌
air floor heating 热风地板采暖
air flora 空气植物群落
air-flow 气流;空气流(量);气辉
air-flow balancing 气流平衡
airflowbed gasification 气流床气化
airflow breakaway 气流分离
airflow classification 分流式分选
airflow classifier 气流选粉机;粗粉分离器
airflow combustion 气流式燃烧
airflow differential 空气压差
airflow diffusion 气流扩散
airflow direction 空气流动方向;风流方向
airflow distribution 气流分布;气流分配
air flow dryer 气流式干燥器;气流式干燥机;热风烘燥机
airflow duct 空气流导管
airflow failure device 风压不足切断装置
airflow makeup 气流组织
airflow mechanics 空气动力学
air flow meter 气流计;空气流速计;空气流量计;气流流量计
airflow meter 空气流量表
airflow method 透气法
airflow model 气流模型
airflow noise 气流噪声
airflow organization 气流组织
airflow pattern 气流形式
airflow pipe 空气输送管道
airflow rate 空气流量;气流(速)率
airflow resistance 耐气流性

airflow resistivity 耐气流阻率
airflow sprayer 气流喷雾机
airflow through filter 自然通风滤池
airflow vane (管道转弯处设的)气流导向叶片
air flue 风道;气道;烟道;通风管
air flush drilling 空气钻探;空气洗井钻进
air foam 空气泡沫;泡沫橡胶;海绵
air foam fire extinguisher 空气泡沫灭火器
air foam fire extinguishing system 空气泡沫灭火系统
air foam fire extinguish method 空气泡沫灭火法
air foam nozzle 空气泡沫喷嘴
air foam rubber 气沫橡胶;泡沫橡胶
air foam system 空气泡沫(灭火)系统
air fog signal 空气雾号
airfoil 气翼;机翼;翼片;空气动力面;翼剖面
airfoil cascade 导向叶片条
airfoil fan 高效风机;翼型叶片
airfoil profile 水翼型;翼型
airfoil section 翼型;翼剖面
airfoil shape 翼型
airfoil surface 翼面
airfone 传话筒
air force 空气负荷;空气反作用力;气动力;风力
air for combustion 燃烧需空气量
air fork 气动拨叉
air former 风罩
air fountain 气泉;喷气器
air fractionation 空气分离;空气分馏;风动分级
airframe 弹体构架;飞机骨架;飞机构架;飞机机架
airframe dynamics 构架力学
air frame opening seal 飞机舱口密封胶
air free 无空气的
air-free concrete 真空处理混凝土;无气混凝土;密实混凝土
air-free dilution 空气稀薄
air-free geothermal fluid 未受空气污染的地热流体
air-free water 无空气水;除气水;无气水
air freezing 空气冷冻法;空气冻结
air freezing index 空气冻冻指数
air freight 空运(运)费;空运(货物);货运飞机;航空运输费;航空运送;运输机
air freight bill of lading 空运提单
air freight building 航空货运大楼;航空货运厅
air freight cost 空运成本
air-freighter 货运(飞)机
air freight list 空运清单
air friction 空气阻力;空气阻;空气摩擦;风摩擦
air friction dynamometer 空气摩擦测力计
air friction loss 空气摩擦损失;风摩耗
air froth equipment 空气泡沫设备
air fuel 航空油料
air-fuel gas mixture 空气燃料气混合物
air-fuel ratio 过剩空气系数;空气燃料比
airfuge ultra-centrifuge 超高速气动离心机
air fumigation plume spreading model 大气烟羽扩散模式
air funnel 空气管道;通风筒;通风孔;通风斗;货舱通风筒
air furnace 热风采暖炉;空气炉;自然通风炉;鼓风炉;反射炉
air gap 气隙;升台高度;空隙;空气(间)隙;火花放电隙
air-gap asymmetry protection 气隙不对称保护(装置)
air-gap breakdown 气隙击穿
air-gap commutator 气隙整流子
air-gap cylinder 气隙圆筒
air-gap diameter 气隙直径
air-gap field 气隙磁场
air-gap flux 气隙磁通;气隙通量
air gap flux density 气隙磁通(量)密度
air-gap ga(u)ge 气隙量规
air-gap induction 气隙磁密度;气隙磁感应
air-gap leakage 空隙泄漏
air-gap leakage flux 气隙漏磁通
air-gap line 气隙磁化线
air-gap measuring device 气隙测量装置
air-gap method 空隙法;气隙法
air-gap mmf 气隙磁动势
air-gap permeance 气隙磁导率
air-gap reluctance 气隙磁阻
air-gap reluctance motor 气隙磁阻电动机
air-gap separation 气隙间距

air-gap torque motor 干式力矩马达
air-gap torsion meter 气隙扭力计
air-gap voltage 气隙电压
air gas 空气气体;空气煤气;风煤气
air gas attack 毒气空袭
air-gas proportioner 空气煤气配比调节器
air gate 气眼;气门;排气口;阀门;通气道;空气门;出气孔
air ga(u)ge 气压计;气动量仪;气动量测仪表;气动微仪;空气压力表;空气量规;气压量仪
air getter 吸气剂
air glass block 通风玻璃砖;通风玻璃块
air glass brick 通风玻璃砖
air glass interface 空气玻璃界面
air glass reflection 空气玻璃反射
air glass surface 空气玻璃界面
airglow 夜天光;夜辉;大气辉光
airglow excitation 大气辐射激励
air goods 空运货物
air governor 送风调节器;空气调节器;压气调节器
air grain 气眼
air grate 通风花格;通风格栅;通风箅子
air grating 气力炉算;通风;箅子通风格栅
air-gravel 气干砾石
air-gravel concrete 气干砾石混凝土
air grease unit 气动油脂枪;气动黄油枪
air grid 通风栅;通风栅格;通风栅子板;通风格栅;通风算子
air grill(e) 通风口格栅;通风花窗;通风算子
air grinder 气动砂轮机;风动磨头
air grommet washer 空气密封垫圈
air ground communication 陆空通信;空对地通信
air ground correlation 空对地相关
air-ground detection 地空巡护
air guide section 导风段
air gun 喷雾器;喷枪;喷漆枪;风动铆枪;气枪;气铆钉枪;空气(喷)枪;压缩气喷嘴
air gun fire delay time 气枪激发延迟时间
airgun waveform 气枪波形
air hammer (打桩用的)气锤;气动冲击器;空气锤;钢坯气动定心机;风动锤
air hammer drill 风动冲击凿岩机;风锤钻
air hand grinder 风动手提砂轮机
air hand hammer rock drill 手持风镐(凿岩机)
air-handling 通风(竖)井;通风
air-handling area 通风面积;通风范围
air-handling arrange 通风面积;通风范围
air-handling block 通风空心砌块;通风砖
air-handling brick 通风砖
air-handling capacity 空调容量;空调能力
air-handling cavity 通风空腔;通风孔;通风道
air-handling ceiling 通风天花板
air-handling ceiling system 通风天花板系统;通风顶棚系统
air-handling ceiling system 通风天花板系统;通风天花板方式;通风顶棚系统;通风顶棚方式
air-handling connection 风管接头
air-handling connection pipe 通风管连接管
air-handling device 空气处理设施
air-handling duct 通风管道
air-handling ducting 管道通风
air-handling equipment 空气调节设备;空气处理设备
air-handling fan 送风机
air-handling glass brick 通风玻璃砖
air-handling grille 通风口格栅;通风格子窗
air-handling installation 通风装置;通风设备
air-handling lighting fixture 通风照明装置
air-handling line 通风管道
air-handling luminaire 兼照明的空调器;空调照明装置;空调照明设备
air-handling opening 通风孔
air-handling panel 通风镶板;通风板
air-handling piece 通风管道系统配件;通风管配件
air-handling pipe 通风管
air-handling pipe tray 通风管道支架
air-handling ridge 通风屋脊
air-handling shaft (隧道的)通风竖井
air-handling station 通风站;通风机房;通风测量站
air-handling system 带有空气处理的空调系统;空气调节系统;空气处理系统
air-handling tray 通风管支架
air-handling troffer 气调顶棚长槽形灯
air-handling tube 通风管

air-handling unit 空气输送设备;空气调节装置;空气处理机组;风柜
air-handling unit room 空气调节机房
air-handling window 通风窗
air harbo(u)r 空港;航空港
air-hardenable 气硬(性)的;风干的
air hardened quality 气硬性
air hardening 气硬性的;气冷硬化;空气硬化;空气淬火;气硬(性)
air-hardening binding material 气硬性胶凝材料
air-hardening cement 气硬性水泥
air hardening lime 气化石灰;风化石灰;气硬(性)石灰
air-hardening quality 气硬性(质)
air-hardening refractory cement 气硬性耐火胶结材料;气硬性耐火水泥
air-hardening steel 气冷硬化钢;空气硬化钢
air hatch 通风舱口
air hawser 大型运输滑翔机
air head 通风平巷;通风巷道;空降场
air header 空气立管;集合管
air heading 通风巷道;通风坑道
air heater 热风器;热风炉;热风机;空气减流器;空气加热器
air heating 热风供暖;热风采暖;空气加温法;空气加热;空气供暖法
air-heating apparatus 空气加热装置;空气加热设备
air-heating device 空气加热装置;空气加热设备
air-heating installation 热风设备;空气加热装置;空气加热设备
air-heating plant 空气加热装置;空气加热设备
air-heating radiator 热风散热器;热风供暖器;风放射器
air-heating system 空气加热系统
air-heating unit 空气加热机组
air hight equipment 气体密封装置
air hoist 气压提升起重机;气压提升绞车;气压提升机;气力起重机;气动提升机;气动起重机;气动卷扬机;气动葫芦;空气提升机;气动升液机;风动提升机;风动绞车
air holddown countercurrent regeneration 空气顶压逆流再生
air holder 气罐;气包;空气收集器;空气罐;空气储蓄器;储气罐
air hole 通气孔;空中陷阱;风窗;气眼;通气筒;风眼;通风联络孔
air-hole grate 通气孔格栅;气孔炉算
airhood 空气罩;空气室
air horn 气喇叭;气笛;空气操纵杆;导罩盖
air horsepower 空气马力;空气功率
air hose 气软管;通风软管;输气软管;空气软管;充气软管;风管
air hose for salvage pontoon 打捞气筒输气管
air hose inner diameter 气管内径
air hospital 空军医院
airhouse 充气房;吹气屋
air housing 喂料端机罩;冷烟室
air humidification 空气加湿
air humidifier 空气增湿器;空气加湿器
air humidifying 空气加湿;空气湿润
air humidifying equipment 空气加湿设备
air humidity 气湿;空气湿度
air humidity comfort curve 空气湿度舒适曲线
air humidity index 空气湿度指数
air humidity indicator 空气湿度指示仪;空气湿度指示器
air hydraulic accumulator 空气水压蓄力器
air hydraulic converter test bench 空油变换器试验台
air hydraulic jack 气动液压千斤顶;气液千斤顶
air hydraulic press 气动水压机
air hygiene 空气卫生
air illness 潜函病;高气压病
air immersed transformer 空气绝缘变压器;干式变压器
air impact hammer 气锤
air impeller 气动驱动器
air impermeability 气密性;不透气性
air impervious liner 气密层;内衬层
air impingement starter 气动力起动器
air-in 供气;送气;进气
air inclusion 气体夹杂物;夹气
air index 大气折射率
air indicated horsepower 空气指示马力

air induction 空气诱导
air induction valve 吸气阀
air infiltration 空气渗入;空气渗透
air infiltration coefficient 空气渗透系数
air infiltration resistance 空气渗透阻力
air inflatable jacket 气胀式救生衣
air inflatable raft 气胀式救生筏
air-inflated 充气的
air-inflated building 气承建筑;充气(式)房屋;充气(式)建筑;气胎式建筑
air-inflated structure 充气结构
air inflating 打气
air inflation 充气
air inflation indicator 充气表
air inflation jacket 气胀式救生衣
air inflation raft 气胀式救生筏
air information 航摄资料;航空情报;大气资料
air information symbol 航空符号
airing 气爆;起沫;通气;晾干;凉干;空气干燥法;鼓风;风干
airing of cargo hold 晾舱
airing system 供气系统
airing system of lifeboat 救生艇供气系统
airing valve 通风阀;换气阀
air inhibition 空气阻滞;空气阻聚;空气抑制
air injection 气压喷油;喷气;空气喷射;压气喷射
air injection burner 空气喷射式燃烧器
air injection engine 空气喷射式柴油机
air injection machine 低压铸造机
air injection nozzle 空气喷射喷嘴
air injection pressure 空气喷射压力
air injection system 空气喷油系统;空气喷射系统
air inleakage 空气渗入;空气漏入
air inlet 空气入口;进气口;进气孔;进气道;进气;进风口
air inlet cam 吸气凸轮
air inlet clack valve 进气活门;进气瓣阀
air inlet dish valve 进气圆盘阀
air inlet filter 进风口过滤器
air inlet grill(e) 空气入口;进风格栅;进气格栅
air inlet hanged trough cell 充气挂槽浮选机
air inlet hose 进气软管
air inlet louvers 进气百叶窗
air inlet pipe 进气管
air inlet screen 气道滤网;进气纱窗(口);进气栅
air inlet shaft (隧道的)通风竖井;进风竖井
air inlet shutter 进气节气门
air inlet sound absorber 进气消声器
air inlet sound attenuator 进气消声器
air inlet strainer 空气进口滤气器
air inlet tower 进风塔
air inlet valve 空气进气阀;进气阀(门)
air inlet valve lever 进气阀杆
air inlet wood shutters 进风木百叶窗
air input 风量;空气输入量;进气(量)
air-in screen 进气滤网
air installation 空气设备;航空设施;通风装置
air instrument 气动量测仪表
air insulated 空气绝缘的
air insulated roofing system 空气绝缘屋顶做法
air insulation 空气绝缘;空气绝热
air intake 通风管道;空气入口;进气(口);进气孔;进风;吸气(剂)
air intake adjusting screw 进气调节螺钉
air intake adjusting spring 进气调节弹簧
air intake branch 进气支管
air intake casing 进气道外壳
air intake cooler 进气冷却器
air intake duct 吸气管
air intake duct system 进气管系统
air intake equipment 进风设备
air intake filter 空气进气滤气器
air intake heater 进气加热器
air intake lever 进气杆
air intake opening 进气口
air intake pipe 进气口;进气管;进风管
air intake shaft 进气竖井;进风竖井;进气道;通风竖井
air intake sound absorber 进气消声器
air intake system 进气装置;进气系统
air intake tower 进风塔
air intake valve 空气进口阀;进气门;进气阀(门);进气口活门
air intake volume 进气量

air intelligence 航空情报
air interchanger 换气装置;换气器
air intermodal container 空陆水联运集装箱
air interrupted spark generator 空气吹断式火花发生器
air interstice 气隙;空气间隙
air introduction 空气的引入
air intrusion 气流侵蚀
air invasion 空气侵入
air in/water out concept 气进水出概念
air in/water out design 气进水出设计
air ion 空气离子;大气离子
air ionization 空气离子化
air isolating cock 空气割断旋塞
air jack 气压千斤顶;气压顶升器;气力千斤顶;气力起重器;气动起重器;气压起重器
air jacket 空气套;吹气救生衣;充气救生衣
air-jacketed condenser 空气(外)套冷凝器
air jet 喷射水流;喷射气流;喷砂喷枪;涂布喷气刀;空气喷射;气喷;喷气
air jet blower 空气喷嘴鼓风机
air jet bulking 喷气变形
air jet dispersion 喷气分散(法)
air jet lift 空气提升器;喷气提升;空气喷射提升机
air jet mill 气流磨;空气喷射喷磨
air jet noise 喷气噪声
air jet pump 空气射流泵
air jet screen 空气喷式筛分机
air jet system 喷口送风系统
air jet texturing 喷气变形
air jig 气压夹具;气动夹具;风力跳汰机
air junction manifold 压缩空气连接总管
air kiosk 风亭
air knife 气刀
air knife coater 气刀辊涂机;气刀刮涂机
air knife coating 气刀涂层;气刀刮涂法;喷气涂布
air knocker 气力抖动器
air lance 气压喷枪;空气(吹灰)枪;压缩空气吹风管;吹管
air-lance type cell 风吹管式浮选机
airlandable 可降落飞机的
air landing strip 机场跑道;机场内跑道区;机场内跑道带
air lane 气廊;空运线
air lateral 空气支管
air lay drying machine 悬浮烘燥机
air layer 空气层
air layering 空中压条
air lay-in ceiling 通风天花板;通风顶棚
air leak(age) 空气渗透;漏气(量);漏风(量);泄气
air leak(age) test 漏气试验;气密(性能)试验;空气渗漏试验;空气漏气试验
air leak(age) rate 漏气速率
air leak(age) tachometer 漏气式转速表
air leg 气腿;气动杆;风力锤(气动杆);气动伸缩式气腿;风动支架;风腿
air leg attachment 风动钻架
air leg mounted(rock) drill 风动支架凿岩机
air leg type rock drill 气腿式凿岩机
airless blast cleaner 抛丸清理机
airless blast cleaning 真空喷砂清理
airless injection 无气喷射;机械喷射
airless paint spraying 液压喷涂;无气喷漆
airless paint spraying machine 无气喷漆机
airless shot blasting installation 无空气抛丸清理装置;无空气抛丸清理机
airless shot blasting mchine 喷丸清理机
airless spray gun 真空喷枪;无气喷漆枪;无空气喷枪
airless spraying 无空气喷涂;液压喷涂;液压喷漆
airless spraying equipment 无空气喷涂装置
airless spray painting 无空气喷漆涂装
air level 水平尺;气泡水准仪;气泡(式)水准器
air level(l)ing 气泡水准测量
air lever 吸阀控制杆
air-life pump 风动抽水泵
airlift 气压提升机;气提;气升(举);气力起重机;气力提升;气力升降机;气举;气动起重器;气动提升机;用压缩空气扬水;压缩空气扬水泵;空中补给线;空运(航线);空运(运气);空气吸升;空气提升;空气升液机;空气升力;空气举;空气抽水器
airlift agitator 气动提升搅拌机;空气搅拌器;提升式气动搅拌机
airlift booster 空气提升升压器

airlift dredge(r) 空气提升挖泥船;气升式挖泥船
airlift drier 提升式气流干燥机
airlift drilling 空气吹除岩土的钻孔法
airlift drilling rig 空气吹除岩土的钻(井)架
airlift ejector 气动喷射器;气动喷射泵
air lifter 气动升降机构;气压泵
airlift flushing 空气升液洗井
airlift foot 空气升液管管脚
airlift gravity drop hammer 空气重力落锤;蒸汽重力落锤;单作用空气锤
airlift machine 气动提升机;气升式浮选机;气力提升机
airlift membrane bioreactor 气提式膜生物反应器
airlift method 气压提升法;气压扬水法
airlift oxidation ditch 气提式氧化沟
airlift pump 气压提升泵;气压泵;气升泵;气力提升泵;气泡泵;气动提水泵;气动提液泵;气动抽水泵;空气提升泵;空气升液泵;压缩空气扬水泵
airlift pump for fish 气力吸鱼泵
airlift reactor 气提式反应器
airlift reverse circulation drilling 气举反循环钻进
airlift system 气力升降机;气提升系统;空气升液器
airlift thermofor catalytic cracking 空气提升储热催化裂化
airlift type agitator 气升式搅拌器;空气搅拌器
airlift type circulating reactor 气提式循环反应器
airlift unit 空气升提升装置
airlift unloader 气动式升降卸载机
airlift well pump 井用气压泵
airlift well test 空气提升井试验
airlight 空气光;航空灯(标)
airlight formula 大气光学公式
air-light troffer 通风吸顶灯(兼有通风和照明功能);空调照明装置;空调照明设备;兼照明的空调器
airline 气泡线;送气管;空运(路)线;空气谱线;空气管路;空气管道;进气管线;架线路;航空线;细长条纹;空中直线距离;飞机航线
airline aerial photography 航线航空摄影
airline aerodrome 航线机场
airline airport 航线机场
airline breathing apparatus 气喉型呼吸器具
airline cells 空气管路单元
airline cleaner 风管式净棉器
airline company 航空公司
airline correction 空中线段校正;航线的修正【航空】;干绳改正;测深水上部分偏移校正;干线校正
airline distance 空中直线距离;空气管路距离;空间直线距离;架空线路距离;航空线距离;大圆航线距离
airline lubricator 管路给油器
airline main 主压气管道
airline map 航空线图
airline mask 蛇管式防毒面具
airline mile 空中直线里程
airline network 空运网
airliner 客机;班机
airline respirator 蛇管式防毒面具
air line-up clamp 气动对管器
air line valve 送气管阀门
air liquefier 空气液化器
air liquefying apparatus 空气液化装置
air-liquid interface 气液界面
air-load 气压负荷;气动力负荷
air-loaded accumulator 储气器
air-loaded roller 气压辊
air loader 风动装载机
air load-interruption switch 空气负荷中断开关
airlock 气闸;气阻;气窝;气锁;气塞;气门;气封;锁气室;锁风装置;双层风闸门;空气闸门;门斗;隔热室;风闸;气压过渡舱
air lock bell 气闸钟
air lock caisson 气压沉箱
air lock chamber 气闸室
air lock device 锁气装置
air lock door 气闸门;双层风门
air locked 不透气的
air lock feeder 气闸进料器;气密性喂料机;气闸喂料器
air lock gate 气闸门;气密性阀门
air locking 气密装置
air locking diving bell 气闸潜水钟
air locking material 气密材料

air lock of laminating film 胶合层气泡
air lock strip 挡风条;阻风带;密封条;防风条
air lock system 气闸系统;气塞系统;风闸门
air lock tank 气闸室
air lock type head 截止探头
air lock valve 气闸阀
air log 航空里程计
air-logged 管路被气阻阻塞不通的
air loop 空气环路
air louver 放气孔;空气调节孔;放气窗
air lubrication 空气润滑作用
air lubricator 压缩空气润滑装置;空气润滑装置;压缩空气润滑器
air luminaire(fixture) 通风照明装置
air machine 风扇;通风机
airmail 航空邮件;航空信;航寄;空邮
airmail bill of lading 航空提单
airmail paper 航空邮件纸
airmail receipt 空邮收据
airmail transfer 航空信汇
air main 空气总管;空气主管路;风管干线
air make-up 补给新风
air make-up unit 空气补偿装置
airman 飞行员
air management 大气管理
air manifold 压气管;输气管;风管;空气支管;集气管;集合管;压缩空气分配总管;汇气总管
air manometer 空气压力计;测大气压仪
airmanship 导航术;飞行技术
air-mapping 航空测绘
air-mapping plane 测绘飞机
air marking 麻孔;小气泡麻孔;小气泡
air marks 麻孔;小气泡麻孔
air-mass 气团
air-mass analysis 气团分析
air-mass and front analysis 气团及锋面分析
air-mass characteristic 气团特性
air-mass chart 气团图
air-mass classification 气团分级
air-mass climatology 气团气候学
air-mass cloud fog 气团云雾
air-mass flow 空气流量
air-mass fog 气团雾
air-mass identification 气团鉴别
air-mass modification 气团变性
air-mass number 气团号
air-mass precipitation 气团(性)降水
air-mass property 气团属性
air-mass rain 气团雨
air-mass shower 气团阵雨
air-mass source 气团发源地
air-mass-source region 气团源区;气团源地
air-mass temperature 气团温度
air-mass thunderstorm 气团雷暴
air-mass tragectory 气团轨迹
air-mass transformation 气团变性
air-mass transport 气团输送
air-mass type diagram 气团类型图解
air master 气动伺服制动器;气动加力制动器
air mat 气垫
air mat filter 孔垫式滤器
air mattress 空气垫;橡胶气垫
air-measuring device 供气量测定装置;供气量测定仪
air-measuring station 风流测量站
air mechanic foam 空气机械泡沫
air mechanics 空气力学
air melted 空气熔炼
air-melted alloy 敞熔合金
air merge system 空气搅拌系统
air meter 风速计;气流计;气体计;气流表;气量计;量气计;空气流量计;混凝土加气量测定仪;含气量测定仪;风速表
air metering device 空中测量设备
air micrometer 气压测深器;气动测微仪;空气测微计
air mile 空里
air mileage indicator 机载里程计;飞行航程指示器
air mileage unit 空中里程装置;空运里程测量计;气流速度计
air mineral adhesion 空气—矿物黏附
air mist sprayer 鼓风式弥雾机
air mixer 空气混合器
air mixing chamber 空气混合室

air mixing plenum 混合送气室;空气混合室
air mixture 空气混合物
air mode 空气模
air mode container 航空集装箱
air model study 气流模型试验;空气模型试验
air model test 气流模型试验
air moistening 空气加湿
air moisture 空气水分;空气含湿量
air moisture comfort curve 空气湿度舒适度曲线
air monitor 空气监测器;大气(污染)监测器
air monitoring 空中监测;空气污染监视;空气监测;大气监测
air monitoring analysis and prediction 空气监测分析和预测
air monitoring car 大气监测车
air monitoring centre 大气监测中心
air monitoring instrument 大气污染监测仪;大气监测仪器;大气监护警报器
air monitoring network 大气污染测定网;大气监测网
air monitoring procedure 大气监测规程
air monitoring regulation 大气监测规程
air monitoring station 大气监测站
air mortar 加气砂浆
air mortar gun 泥浆喷射机
air motion 空气流动;空气运动
air motor 气动马达;空气发动机;压缩空气电机;压气发动机;风动马达
air motor drill 风钻;风动钻机;风动钻车
air motor driven jumbo 风动钻车;风动台车
air motor hoist 气动起重机;气动卷扬机
air motor hoist with top hook 带顶钩的风动起重机
air motor powered 风力马达驱动的
air movement 空气流动;空气运动
air movement data 空中运动数据
air-moving device 气体输送机械;气动装置;风动装置
air-moving velocity in cave 洞穴空气运动速度
air natural oxidation 空气自然氧化
air navigation 空中导航;航空
Air Navigation Act 空中导航法案
air navigation chart index 航空图接图表
air navigation computer 领航计算机
air navigation information 领航资料【航空】
air navigation map 航空图
air navigation planning amp 航空计划图
air navigation plotting chart 战术航空图
air navigator 航空积算机;航空导航设备
air need for combustion 燃烧所需的空气量
air network 大气采样网
air noise 空气噪声
air non-conditioning space 非空调空间
air nozzle 喷雾嘴;喷枪喷嘴;喷气嘴;喷风口;空气喷嘴
air nuclear burst 空中核爆炸
air oasis 空气净区
of combustion 燃烧需要的空气;燃烧空气;燃烧需空气量
air off 断气
air oil actuator 气液压力转换缸
air oil separator 油气分离器
air oil shock strut 空气油压减震支柱
air oil strut 空气油液式支柱
air oil suspension 气—油悬浮体
airometer 量气计
air on 送入压缩空气
air-on-the-run system 车辆在行驶中调整轮胎气压的设备
air opening 通风口;通风孔;通风洞
air-operated 风动的;气动的;空气操作
air-operated bucket 气动卸料斗
air-operated ca(u)lking gun 气动嵌缝枪
air-operated clay spade 风动铲土器
air-operated control 气压操纵;气动操纵
air-operated controller 气动控制器
air-operated crawler type loader 风动履带式装车机
air-operated dump car 风动自卸车
air-operated equipment 空气操纵设备
air-operated expanding mandrel 气动胀心开轴
air-operated fixture 气动卡具
air-operated flying shears 气动飞剪
air-operated folding door 气动折门
air-operated grab 风动抓岩机

air-operated grease unit 气动油脂枪；风动油脂枪
air-operated impact wrench 气动扳手；风动扳手
air-operated mechanism 风动机构
air-operated portable grinder 风动砂轮机
air-operated pump 气动泵；压缩空气驱动泵
air-operated reversing valve 气动换向阀
air-operated rock breaker 风动凿岩机
air-operated spreader 气动扩胎机
air-operated take-up 气动张紧装置
air-operated thermostat 气动恒温器
air-operated throttle motor 风动节流器
air-operated tool 风动工具
air-operated valve 气动阀
air-operated valve grinder 气动阀门研磨机
air-operated vice 气动虎钳
air-operating switch 气动开关
air operation 空气推动
air-out 出气
air outlet 送风口；空气出口；出气口；风口；放气口
air outlet slit 空气出口
air-outlet valve 出气阀；排气阀
air output 排气量；空气输出量；出气量
air oven 空气烘箱；烘炉；热空气干燥炉
air oven ag(e)ing 热空气老化
air oven ag(e)ing test 热空气老化试验；空气加热老化试验
air over board dump mask 舱外排气面罩
air over hydraulic booster 气液增力器
air over hydraulic brake 气液制动器
air over hydraulic intensifier 气液增力器
air oxidation 空气氧化
air oxidation desulfation 空气氧化脱硫
air oxidation desulphation 空气氧化脱硫
air oxidation desulphation tower 空气氧化脱硫塔
air oxidation method 空气氧化法
air ozonizer 臭氧发生器
air pack 空气包
air packing 气密填料
air pad conveyer 气动输送机
air-painter 喷漆器；喷漆设备
air-painting equipment 喷漆设备；喷漆发生器；压缩空气喷漆设备
air-painting sprayer 气喷漆枪；压缩空气喷漆器
air parameter 空气参数
air parcel 气块；航空包裹
air parent base 主供给基地
air park 私人停机场；小(飞)机场；小型机场
air particle 气粒；空气质点
air particle detector 空气微粒检测器
air particle monitor 空气微粒监测器；空气粒子（放射性）监测器；大气粒子监测器；大气粉尘监测器
air partition 隔风墙
air parts 空运部件
air passage 气道；排气道；通气道；通风道；风道
air passenger 航空旅客
air passenger terminal 航空客运站
air patenting 空气淬火
air path 大气路径
air pavement breaker 气动路面破碎机；风动路面破碎机
air pavilion 风亭
air peak 空气峰
air peener 气力喷砂机；气动喷丸机；气动喷砂机
air percussion rotary drilling 空气冲击旋转钻进
air performance 空气动力性能
air permeability 透气性；透气率；透气度；空气渗透性；空气磁导率
air-permeability apparatus 透气仪；透气度试验仪
air-permeability fineness tester 透气法细度试验仪
air-permeability method 透气(性)法(测量比表面法)；空气渗透率法
air-permeability specific surface apparatus 透气法比表面积测定仪
air-permeability test 透气(性)试验
air-permeability value 透气法比表面积值
air-permeable 透气的
air permeable coating 透气性涂层
air pervious material 透气材料
air petroleum 航空煤油
air photo duplicate 复制航摄像片
air photogrammetric(al) original 航摄原图
air photogrammetry 航空摄影测绘学；航空摄影测量图(学)；航空摄影测量学

air photo(graph) 航摄照片；航摄像片；航空照片；航摄图；航空像片；航空摄影
air photographer 航空摄影员
air photographic(al) crew 航摄队
air photographic(al) mosaic 航空像片镶嵌图
air photographic(al) survey 航空摄影调查
air photography 航空摄影
airphoto intelligence 航摄资料
airphoto interpretation 航空照片解释；航空照片判读；航空照片识别
airphoto mosaic 镶嵌航片
airphoto stereoscopic instrument 立体摄像仪；立体航测仪器
air phototechnique 航空摄影技术；空中摄影法
airphoto transfer 航片转绘
air pick(er) 风镐；振捣器(混凝土)；空中警戒哨
air picnometer 空气比重计
air piece 通风系统配件
air pile hammer 铲煤风镐；气压桩锤；气动(打)桩锤；风压打桩机；风动打桩锤
air pillow 空气垫(座)；气枕；气垫
air pills 气丸
airpilot risks 机师险
air pilot valve 空气导向阀
air pipe 气管；空气管(道)；风管；压气管；通气管；通风管；空气旁通管
air pipe channel 风管槽
air pipe cock of pantograph 受电弓气路塞门
air pipe head 通气管头
air pipeline 空气管线；空气管路；风管路
air pipeline governor 风管路调压设备
air pipe strainer 气管滤器
air pipe tray 通风管支架；通风管托架
air piping 空气管道(布置)；风道
air piping compressed 压缩空气管道
air piping filter 空气管道过滤器
air piston 空气活塞；压气活塞
air pit 气坑；通气坑；通风坑；通风井；风井
air placed concrete 喷抹混凝土；气压浇注混凝土；气压浇注混凝土；气压浇注混凝土
air placer 气压机；喷枪；喷灌机(混凝土)；气力压送机；气力喷注机
air placing 喷射法；喷浆法(混凝土、水泥浆等)；气动浇灌(混凝土)
air placing machine 气压机；喷灌机(混凝土)
airplane 飞机
airplane altimeter 飞机高度计；机载高度计
airplane balance 飞机平衡
airplane handling winch 飞机绞车
airplane hangar 飞机库
airplane horizon 航摄地平线
airplane insulator 航空绝缘物
airplane maintenance 飞机维护
airplane mother ship 航空母舰
airplane paint 航空漆；飞机漆
airplane repair depot 飞机修理厂
airplane repair shop 飞机修理厂
airplane rigging 飞机装配
airplane shed 飞机库；飞机棚
airplane's magnetization compensation precision 飞机磁干扰场补偿精度
airplane structure 飞机结构
airplane type 飞机型别
air plankton 空气浮游生物
air plant 通风系统配件
air plasma torch 空气等离子体吹管
air plate 气门板；通风板；蜂窝板
air plenum 空气室
air plenum chamber 空气静压箱
airplot 航空标绘；空中测绘
airplot wind velocity 航空标绘风速
air plug 气塞；气孔塞；通气塞
air plug ga(u)ge 气动塞规
air pocket 气眼；气潭；气穴；气窝；气泡；气囊；气阱；气袋；砂眼；空中低压区；混凝土麻点；大气泡；残存空气
air pocket eliminator 砂眼排除装置；(铸件的)气泡排除装置；气泡消除器
air poise 空气平衡；衡气表
air pollutant 空气污染物(质)；大气污染物
air pollutant concentration 空气污染物浓度
air pollutant emission inventory 空气污染物排放清单
air pollutant matter 空气污染物质

air polluter 空气污染者
air pollution 空气污染；大气污染
air pollution abatement 消除空气污染
Air Pollution Act 空气污染法
air pollution agency 大气污染管理机构
air pollution agent 大气污染因子
air pollution alert 空气污染警报
air pollution analysis 空气污染分析
air pollution auto-monitoring system 大气污染自动监测系统
air pollution by automobile 汽车造成的大气污染
air pollution carcinogenesis 空气污染致癌作用；空气污染致癌性
Air Pollution Code 空气污染法规；防止空气污染规范
air pollution coefficient 空气污染系数
air pollution complaint 空气污染诉讼
air pollution concentration 空气污染浓度；大气污染浓度
air pollution concentration forecasting 空气污染浓度预报
air pollution control 空气污染控制；空气污染防治；大气污染控制
air pollution control activity 空气污染控制活动
Air Pollution Control Administration 空气污染管理局
Air Pollution Control Association 空气污染控制协会（美国）
air pollution control center[centre] 空气污染防治中心
air pollution control district 空气污染控制区；空气污染管制区；大气污染控制区
air pollution control engineering 大气污染防治工程
air pollution control law 大气污染防治法
air pollution controller 空气污染检测器；空气污染管制员
Air Pollution Control Office 空气污染控制处（美国）
air pollution control planning 大气污染控制规划
air pollution control policy 大气污染防治政策
Air Pollution Control Regulation 空气污染防治条例
air pollution control system 空气污染控制系统；空气污染防治系统
air pollution disaster 空气污染灾害；空气污染事故
air pollution effect 空气污染效应
air pollution emission 空气污染排放
air pollution emission factor 空气污染排放因素
air pollution engineer 空气污染工程师
air pollution environmental problem 空气污染环境问题
air pollution episode 空气污染事件；大气污染事件
air pollution episode game 大气污染事件
air pollution exercise 空气污染实施
air pollution exposure level 空气污染程度
air pollution focal point 空气污染集中点
air pollution forecasting 空气污染预报
air pollution forecasting by model 空气污染模式预报
air pollution forecasting system 空气污染预报系统
Air Pollution Fund 空气污染研究基金会
Air Pollution Hearing Board 空气污染听证委员会
air pollution index 空气污染指数；空气污染指标；大气污染(源)指数
air pollution information and computer system 空气污染情报和计算机系统
Air Pollution Law 空气污染法
air pollution legislation 空气污染立法
air pollution load 空气污染物浓度；空气污染物含量
air pollution measurement program(me) 空气污染测定规划
air pollution meteorology 污染气象学；空气污染气象学
air pollution model 空气污染模型
air pollution monitor 空气污染监测器；空气污染度监视器
air pollution monitoring 空气污染监测；大气污染监测
air pollution monitoring laser 空气污染监测激光器
air pollution monitoring laser radar 空气污染监测激光雷达
air pollution monitoring network 大气污染监测网
air pollution monitoring system 空气污染监测系统
air pollution monitoring van 大气污染监测车

air pollution observation station 空气污染观测站
air pollution occurrence 空气污染偶发事件
air pollution order classification 空气污染气味分类
air pollution potential 空气污染潜势
air pollution potential forecasting 空气污染潜势预报
air pollution prevention 空气污染防治；大气污染防治；防止空气污染
air pollution prevention and control law 大气污染防治法
air pollution reduction device 空气污染减轻装置
air pollution region 空气污染区
air pollution regulation 空气污染规章；空气污染条例
air pollution sensor 空气污染传感器
air pollution sources 空气污染源
air pollution standard 空气污染标准
air pollution statistical forecasting 空气污染统计预报
air pollution surveillance 空气污染监视
air pollution surveillance system 空气污染监视系统；空气污染监测系统
air pollution survey 空气污染调查
air pollution syndrome 空气污染综合症
air pollution system 大气污染系统
air pollution technical data 空气污染的技术数据
Air Pollution Technical Information Center 空气污染情报中心
air pollution trend 空气污染趋势
air pollution with lead particles 铅害；空气铅粒污染
air pollution zone 空气污染区
air porosity 孔隙率
airport 气门；通气口；机场；航空港；舷孔；舷窗；空运港；空气孔；航空站；现用机场；飞机场
airport activity 航空港运行
airport administration 空港管理；航空港管理
airport air pollution 机场空气污染
airport approach zone 飞机进场区
airport area 飞机场面积
airport beacon 航空灯塔；机场灯塔；机场灯标
airport city 航空港城市
airport construction plan 机场进近净空区平面图
airport construction work 机场建设工程
airport drainage system 机场排水系统
airport engineering 航空站工程学
airport facility 机场设施
airport fire fighter 机场消防人员
airport fireman 机场消防人员
airport glass 舷窗玻璃
airport ground traffic control 机场地面交通管制
airport hazard zone 机场危险区
airport identification mark 航空港标志
airport identification sign 空港标志；航空港标志
airport light beacon 航空港灯标
airport lighting 机场照明
airport location 空港位置；飞机场位置
airport lounge 候机厅；候机室
airport management 航空港管理
airport noise 机场噪声
airport of departure 起运机场；启运机场
airport of entry 入境机场
airport operation 机场运行
airport pavement 飞机场道路路面；飞机场跑道面；机场跑道路面
airport pavement loading 机场(跑道)路面荷载
airport platform area 机场地坪区
airport platform site plan 机场地坪位置图
airport runway 飞机跑道；机场跑道
airport runway pavement 机场跑道路面
airport runway pavement loading 机场跑道路面荷载
airport screen 舷孔纱窗；舷窗帘
airport surface detection equipment 机场地面侦察设备
airport surface movement indicator 机场地面活动目标显示器
airport surveillance radar 机场监视雷达
airport survey 飞机场测量
airport tax 机场税
airport taxiway 飞机场滑行道；机场滑行道
airport terminal 机场候机楼
airport terminal building 候机楼
airport traffic 航空港航运量

airport traffic control tower 航空港交通管制塔台
airport traffic zone 机场跑道
airport wastewater 机场废水；飞机场废水
airport zoning 航空港地区规划
air position 无风位置；投影位置；空中位置
air position indicator 空中位置指示器
air power 空中力量；空气动力
air-powered 气动的
air-powered equipment 气动设备
air-powered master cylinder 气动油制动总泵
air-powered pump 气力泵；气动泵
air-powered servo 气压伺服装置
air-powered steering unit 气动转向助力装置
air power noise 空气动力噪声
air pre-cleaner 空气粗滤器；空气预过滤器
air precooler 空气预冷器
air preheater 空气预(加)热器
air preheater of revolution 回转式空气预热器
air preheating 空气预热
air preheating chamber 空气预热室
air pressure 气压；空气压力
air pressure brake 气制动(器)
air pressure change 气压变化
air pressure conveyer 气力输送机
air pressure drainage system 气压排水系统
air pressure drive 空气压力动；气压驱动
air pressure drop 空气压降
air pressure duct 气压管道
air pressure engine 气压发动机；空气压缩机
air pressure ga(u)ge 气压表(测大气)
air pressure gun 压缩空气枪；风枪
air pressure probe 空气压力传感器
air pressure pump 空气压缩机
air pressure reducer 空气减压器；空气减压阀
air pressure-reducing valve 空气减压阀
air pressure regulating valve 气压调节阀
air pressure regulator 气压调节器；空气压力调节器
air-pressure release 气压快门开关
air pressure shield 气压盾构
air pressure switch 气动开关
air pressure tank 空气压力灌
air pressure test 气压试验；压缩空气试验
air pressure test for strength and tightness 气压强度及气密试验
air pressure valve 气压阀
air pressure vessel 储[贮]气筒；储[贮]气罐
air-pressure water(supply)system 气压给水系统
air-pressurized water extinguisher 加压水灭火器
air prilling 喷射造粒
air producer gas generator 空气煤气发生器
air-proof 不透气的；不漏气的；气密的
air-proof test 气密试验
air propeller 空气推进器；空气螺旋桨；鼓风机
air propelling 压缩空气驱动
air pulsated jig 脉动跳汰机
air pulse conveying 脉冲式气力输送；泄差输送
air pulse ga(u)ge 脉冲式气动量仪
air pulsing 气动脉冲
air pump 空气压缩机；打气筒；排气唧筒；排气泵；空气泵；抽气机；抽风泵；风泵
air pump governor 气泵调节器
air pumping method 气动扬水法；空气泵送法
air pump valve 气泵阀
air puncher 风镐；风动凿孔机
air purge 气洗；空气清洗；空气净化；空气吹扫；鼓风吹扫
air purged instrument 空气清洗式计量仪器
air purge unit 抽气装置
air purge valve 排气阀；空气净化阀
air purification 空气净化
air purification equipment 空气净化装置
air purification system 空气净化系统
air purifier 气体提纯器；空气净化器；空气净化机
air purifying respirator 空气净化(式)呼吸器
air purifying unit 空气净化机组
air purity 空气纯度；空气洁净度
air pyrorechnic 对空联络发光信号设备
air quality 空气质量；大气质量
Air Quality Act 大气质量法
air quality analysis 空气质量分析
air quality and emission standard 空气质量和排放标准；大气质量和排放标准
air quality classification 空气质量分级
air quality control 空气质量控制

air quality control region 空气质量控制区；大气质量控制区
air quality criteron 空气质量标准；空气质量准则；大气质量准则；大气质量基准；大气质量标准
air quality cycle 大气质量循环
air quality data handling system 大气质量影响模型系统；大气质量数据处理系统
air quality display model 大气质量显示模型
air quality forecast 大气质量预测
air quality index 大气质量指数
air quality maintenance 空气质量维护
air quality management 大气质量管理
air quality measurement 空气质量检测；大气质量检测
air quality meteorology 大气质量气象学
air quality mode 空气质量模式；大气质量模式
air quality model 气质模式；空气质量模型；大气质量模型
air quality modeling 大气质量模拟
air quality monitoring 空气质量监测；大气质量监测
air quality monitoring network 大气质量监测网
air quality region 大气质量区域
air quality simulation model 空气质量模拟模型；大气质量弥散模型
air quality standard 空气质量标准；大气质量标准
air quality surveillance 大气质量监视
air quality surveillance network 大气质量监视网
air quantity 空气量；风量
air quenching 空气淬冷；空气骤冷；空气冷却淬火；空气淬火
air quenching clinker 气淬熔渣
air quenching cooler 炉箅子冷却器
air quenching grate clinker cooler 空气骤冷箅式熟料冷却器
air quenching hardened glass 空气淬冷钢化玻璃；风冷钢化玻璃
air quenching nozzle 空气淬冷喷嘴；风冷口
air-rack-raking device 气动拨火耙
air radiator 散流器；空气散热器
air radon daughter corrected value 大气氡子体校正值
air raid 空袭
air-raid alarm 空袭警报；空气袭警板
air-raid precaution 防空(措施)
air-raid shelter 防空洞；人防地下室；防空掩体
air-raid sheltering basement 人防地下室
air-raid siren 防空警报器
air-raid underground shelter 人防地道
air rake agitator 气力耙式搅拌机
air ram jolt 气动柱塞震击；气垫微震
air ram(mer) 气动活塞；气动夯锤；气夯；气动顶杆；空气压机；风动锤
air rammer press 气锤捣打机
air ramming 气锤捣打成型；气锤捣打成形法；空气振捣
air rate 通风量；空气流量；混气率；含气率
air ratio 空气比
air reaction 空气阻力
air reamer 气压扩张机；气压扩孔机
air receiver 气包；空气罐；空气储蓄器；储[贮]汽箱；储[贮]气罐；储气筒；储气室；储气器；储气罐；储风筒
air recirculation 空气再循环
air reconnaissance 航空勘察；航空勘测
air recourses 空气资源
air-rectification 空气吹脱；吹氧(法)；吹气(法)
air-rectified bitumen 吹氧提炼地沥青
air recycling 空气二次循环
air reducer 气压减压器；空气减流器
air reducing agent 除气剂
air reducing valve 空气减压阀
air-refined asphalt 吹制精制地沥青；吹制精炼石油沥青
air-refined steel 精炼钢；富氧底吹转炉钢
air refining 吹炼；精炼
air refractivity effect 空气折射率效应
air refrigerating machine 空气制冷机；空气冷冻机
air refuelling 空中加油
air regenerating device 换气设备
air regeneration unit 空气再生装置
air regenerative exhaust 空气蓄热废气
air regenerator 空气蓄热室
air regenerator flue 空气蓄热室烟道
air regime 空气状况

air register 节气门;气体记录仪;(锅炉的)配风(空气调节)器;调气装置;调风器;空气调节器;空气挡板
air regulating valve 风量调节阀
air regulating valve for dewatering 压气脱水调节阀
air regulation 空气调节;风量调节
air regulator 气流调节器;送风调节器;空气调节器;风流调节装置;风量调节器
air reheater 空气再热器
air reheating 空气再热
air relay 气压替续器;气压继电器;气动转换;气动替续器;气动继电器
air release 排气
air release screen 筛网式通气塞
air release valve 气阀;泄气阀;放气阀(门)
air relief 排气;放气(门);空气排放
air relief cock 放气旋塞
air relief elbow 放空弯管
air-relief installation 排气设备;排气装置
air-relief shaft (隧道的)通风竖井;排气竖井(风筒)
air relief valve 通气阀;减压阀;放气阀;安全阀;排气阀
air reluctance 空气磁阻
air remote-sensing technique 航空遥感技术
air removal 通风;排气;换气;抽气;除气
air removal jet 抽气射流;抽气器
air removing roll 除气辊
air renewal 换气
air replenishing 补气
air requirement 空气需要量;空气量;需气量
air requirement according to least air-speed 按防尘最小风速计算风量
air requirement by diluting harmful gas 稀释有毒气体需风量
air requirement for every man 按人员计算风量
air reservoir 压缩空气储[贮]罐;储[贮]气器;储气筒;储气缸;储汽缸
air reservoir carrier 储气桶架
air resistance 空气阻力;抗气性;不透气度;气阻
air resister 空气调节器
air resisting liner 气密层;内衬层
air resource management 空气资源管理
air respirator 空气滤光器
air-retaining substance 加气剂;保持空气的物质;稳泡剂;保气剂
air retaining wall 气密层;内衬层
air retention time 空气保留时间
air return 回风
air return/exhaust 回排风机
air return fan 回风机
air return inlet 回风口
air return mode 回风方式
air return opening 回风口
air return system 回气系统;回风系统
air return through corridor 走廊回风
air return way 回风道
air reversal valve 空气交换器;空气换向阀
air revitalization 空气更新
air revitalization material 空气更新物质
air rich 最佳空气燃料比;充足空气燃料混合比;充足空气
air rig 风钻
air right 上空使用权
air ring 混凝土多孔喷嘴;环形(空)气囊;通气环;冷却环;环形垫
air rising duct 升气管
air risks 空险;空运险
air riveter 气压铆锤;气动铆枪;气动铆钉机;气头铆钉机;气压铆钉枪
air riveting hammer 气动铆钉机;气压铆钉机
air roasting 空气焙烧;氧气煅烧
air rock drill 风钻;风动凿岩机
air rotary drilling 空气旋转钻进
air rotary method 空气旋转法
air route 空气路线;航【航空】;航空线;飞机航线
air route for power transmission survey 架空送电线路测量
air route in use 通航里程
air route map 航线图;航空线图
air route surveillance radar 航线监视雷达
air route traffic control center 航线交通管制中心;空中交通控制中心

air sac 气袋;气囊
air sack 充气堵漏袋
air sample 气样
air sampler 空气取样器;空气采样器;大气采样器
air sampling 空气取样;空气采样(法);大气采样
air sampling device 大气采样装置
air sampling measurement 空气采样测量
air sampling method 空气采样法
air sampling network 空气采样网;大气采样网
air sampling rig 空气取样装置;空气采样装置;大气采样装置
air sampling system 空气采样系统
air sand blow 空气喷砂
air sand blower 气压喷砂器;喷砂器;喷砂机;空气喷砂器;空气喷砂机;吹砂器
air sander 气力喷砂机;气动打磨机;自动打磨机
air sanding 风动撒沙装置
air sand process 风沙分选法
air sanitary protection 大气卫生防护
air sanitation 空气卫生
air saturation 空气饱和
air saturation diving technique 空气饱和潜水技术
air saturator 溶蚀气罐
air saw 气锯
air scape 空中鸟瞰图
air scavenging 空气吹除
air scavenging gear 气动净化装置;换气装置;吹风装置(灰浆喷射机)
air science 大气科学
airscoop 进气口;通风罩;导气管;导风罩;风斗;自然通风吸入口;招风斗;进气喇叭口;进气道
airscoop drag 进气口阻力
airscoop heater 进气道加热器
airscoop shroud 进气罩
air scour 气体冲刷
air-scour concurrent with water 冲洗气水并流
air scouring 空气冲洗
air scouring manifold 空气冲洗总管
air-scour supply pipe 气洗供气管
air scout 航空侦察机
air screen 气帘;滤气网;空气幕;空气帘
air screw 空气螺旋桨
airscrew balancing apparatus 螺旋桨平衡器
airscrew blade 螺旋桨叶
airscrew boss 螺旋桨壳
airscrew brake 螺旋桨制动器;螺旋桨闸
airscrew circle 螺旋桨盘面圆
airscrew diameter 螺旋桨直径
air screwdriver 风动螺丝刀
airscrew engine 螺旋桨发动机
airscrew hub 螺旋(桨)毂
airscrew pitch 螺旋桨螺距
airscrew reduction gear 螺旋桨减速齿轮
airscrew thrust 螺旋桨推力
airscrew turbine system 涡轮螺旋桨系统
air scrubber 空气洗涤器;空气滤净器
air scrubbing 空气洗涤
air-sea boundary layer 大气海洋边界层
air-sea interaction 大气海洋相互作用
air-sea interface 气-海界面;大气海洋分界面
air seal (由隔声材料做成的)吹气封口;不透气;气密;气封;防气圈
air-seal dust-valve 密封排灰阀
air sealed discharging device 气密式卸料装置
air sealed discharging valve 气密式卸料阀门
air sealed rotary discharging device 气密式回转卸料装置;气密式分格卸料装置
air sealing apparatus 空气密封装置
air sealing parts 气密部件
air seal ring 气封圈;空气密封;密封圈
air search radar 防空雷达
air-sea rescue 空中海上救助;海空救助
air sea rescue boat 航空救生艇
air season 自然养护
air seasoned 露天时效的;气干的;风干的
air seasoned lumber 风干木材
air seasoned timber 风干木材
air seasoned wood 风干木材
air seasoning 通风干燥(法);空气干燥;自然干燥;风干
air seasoning method 空气干燥法
airsecond 时间标度
air section 气段
air sedimentation balance 微粒沉积分析天平

air seeding 飞机播种
air self-purification 大气自净(作用)
air separate 分解炉供风系统
air-separated sulfur 风选硫磺
air-separating mill 带选粉机的磨机;带选分器的磨机;气流分选器
air-separating plant 空气分离装置
air-separating tank 吹气分离箱;排气池
air-separating tray 除气盘
air separation 空气分离;吹气分离;风选;风力分离;用气分离
air separation tank 气水分离箱
air separation technology 空气分离技术
air separation unit 空气分离装置
air separator 分级机;空气选粉机;空气分离器;空气分离机;吹气分离器;风选器;风力选矿机
air separator mill 吹离磨
Air Service Agreements 民用航空运输协定
air servicer 充气车
air set 气凝;空气中固化;常温自硬;自然硬化;吸入硬化(的);气硬(性);空气中凝固(的);常温凝固(的)
air set core 自硬芯
air-set lamps of cement 水泥结硬块
air-set lump 结硬块
air set mo(u)ld 自硬型
air set mo(u)lding 自硬法造型
air set oil 常温凝固油
air set pipe 空气冷凝管
air setting 气硬的;气凝性;气凝的;空气冷凝;自凝的;自干;常温固化;气硬性胶泥;常温凝固;气硬(性)
air-setting binder 自凝黏结剂;自硬黏结剂
air-setting cement 气硬性水泥
air-setting jointing material 气硬性黏接材料
air-setting mortar 气硬性砂浆
air-setting oil 自硬油
air-setting refractory cement 气凝性耐火胶结材料;气硬性耐火水泥
air-setting refractory cement mortar 气硬性耐火水泥砂浆
air-setting refractory mortar 气硬性耐火砂浆;气硬性耐火泥
air-settling mortar 自凝泥浆
air sextant 航空用六分仪
air shaft 通风竖井;通风竖管;通风筒;通风井;风井;防火道;(隧道的)通风竖井
air shaker 气动振打器
air sharing 空气分配
air shears 风动剪断机
air shed 风干棚;空气污染区;飞机棚
air shed model 空气污染区模式;空气污染模型
air shield driving 气压盾构法开挖;气压盾构法掘进;压缩空气盾构开挖
airship 气球船;飞艇;飞船
airship dope 飞艇蒙布漆
airship envelope fabric 飞艇气囊织物
airship hangar 飞船库
airshipper 空运企业
air/shipping service suspending period 停航期
airship shed 飞机棚;飞船棚
air shock absorber 空气减震器
air shooter 气体输送管
air shooting 气垫爆破;空中爆炸(法);空中爆破(法);空气爆破;压气爆破
air shot 空隙爆破
air shower 空气淋浴;空气簇射;空气吹淋室;大气簇射;大气族射
air shrinkage 风干收缩;气干收缩
air shut-off cock 空气切断旋塞
air shut-off valve 空气关闭阀
air shutter 阻气器;调风门;空气闸门;关气机
air shutter lever 关气杆
air shutter shaft 阻风器轴
air shuttle service 空中短程往返服务
airsickness 航空晕;航空病;高空病
airside face 堰的下游面;坝的下游面;下风面
airside of terminal building 候机楼对空侧
airside scuttle 通风舷窗
air sifter 风选机;风力分级器;气流分离器;风筛机
air sifting 气流筛分;气流分选;空气分离
air silencer 空气消音器;空气消声器
air silk 空心丝
air siltometer 吹气式颗粒分析仪;气扰混沙分析

仪;空气尘埃颗粒分析测定仪
air sinker 风动钻孔器;风动凿岩机
air sinking 压缩空气下沉;气压沉箱下沉;气动凿井
air siphon (敷设在墙内的)防潮斜管
air skimmer 除气器
air slacking 气化;潮解;空气熟化
air slake 空气熟化;空气中水化(指石灰)
air slaked 空气中水气被水解石灰和水泥吸收;空气消和的
air-slaked lime 气化石灰;空气消化石灰;风化石灰;潮解石灰(粉末)
air-slaked mortar 潮解的砂浆
air-slaked quick lime 消石灰;风化石灰
air slaking 风化;潮解;风化潮解
air sleeve 风标
air slide 气动溜槽;气动滑板;气动传送机;气动活塞;气动滑槽;空气输送斜槽;空气活塞;空气调节板;压缩空气传送(槽);风动滑槽;气力输送;空气槽
air-slide bulk cement trailer 带有压缩空气输送槽的散装水泥拖车
air-slide classifier 气动分选机;气滑式分级机
air-slide conveyer 风送式输送机;气滑式输送机;空气输送溜槽;压缩空气输送器;空气输送斜槽;气力输送机;压风输送器
air-slide feeder 气滑式给料器
air sliding 气动活塞;压缩空气驱动
air sliding window 通风推拉窗
air slot 空气槽
air sluice 风闸门
airslusher 气动刮板卷扬机;气动扒煤机
air snap 气动卡规
air snap ga(u)ge 气动外径量规
air sniffer 气体臭探器;吸气机
air-soild ratio 气固比
air soil hammer 风动夯土机;气动夯土机
air solenoid valve 空气电磁阀
air/solid ratio 气固比
air sound attenuator 空气消音器
air sounding 探空(仪);大气探测
air sounding line 吊015测绳
air space 气域;空隙;空气间隔;空气间层;气隙;气道;领空;空域;大气空间
air-space above sea 海上空间
air-space cable 空气(纸)绝缘电缆;充气纸绝缘电缆
air-spaced coax 空气间隔同轴线
air-spaced coaxial cable 空气绝缘同轴电缆
air-spaced coil 大绕距电感线圈
air-spaced double anastigmat 双分离对称消像散镜头
air-spaced doublet 空气隙双合透镜;双分透镜;双不胶透镜;分离式双透镜
airspace for roof insulation 屋顶保温空气层
airspace insulation 空隙绝缘
airspace of soil 土中气隙
airspace paper core cable 空气纸绝缘电缆
airspace porosity 空气孔隙度
airspace ratio 气隙比;孔隙含水率;空气间隙比;空隙比
airspace reservation 空中禁区
airspace warning area 空间警告区域
air spade 风动铲土器;风(动)铲
air sparger 喷气器
air sparger ring 布气环
air spectrography 气化光谱分析
air speed 气流速度;空速;风速
air-speed computator 空速计算器
air-speed ga(u)ge 风速表
air-speed head 空速管;气压感受器
air-speed indicator 空速表;空气速度指示器;风速指示仪;风速指示器;空速指示器
air-speed integrator 风速积分器
air-speed mach indicator 风速马赫指示器
air-speed measuring sensor 空速测量传感器
air-speed meter 空气流速计;航速表;空速计;空速表
air speedometer 气流速度计
air sphere 气圈;气界
air spinning machine 干法纺丝机
air split 风流分支
air-split system 风流分支装置;风流分支系统
air spora 风播体
air spots 表面不平
air spout 喷嘴
air spray 空气射流

air spray coating 空气喷涂
air spray disinfecter and disinfectant 尘气雾化消毒器及消毒剂
air sprayer pistol 压缩空气喷射器
air spray finishing 空气喷涂法;压气喷涂法
air spraying 压力喷涂
air spray pistol 压缩空气喷枪
air spring(cushion) 空气减震器;气垫;(弹簧);空气弹簧
air spring level(l)ing valve 空气弹簧调平阀;气垫调平阀
air-spring suspension 空气弹簧悬架;气垫吊架
air spring vibration isolator 空气弹簧隔振器
air-spun acetate rayon 干纺醋酯人造丝
air squeegee 空气吹拂器
air squeezer 气动挤压器
air stack 通风道;风管;排风管;空中堆旋
air stagnation 空气滞留;空气停滞
air standard cycle 空气标准循环
airstart 空中起动
air-starter 空气起动机;压气起动机
air starting bottle 起动空气瓶
air starting station 压缩空气站
air starting system 空气启动设备;压缩空气起动设备
air starting valve 空气起动阀
air station 空军基地;空间站;空中摄影站;航空站
air stave 通风侧板
air steam 人工气流
airsteps 随机客梯
air sterilization 空气消毒
airstone 砂滤多孔石
airstop 空降场
air stopping wall 挡风(雨)墙
air storage 储气;露天堆放(储[贮]气);储[贮]气
air storage capacity 空气储存量
air storage chamber 储气室;辅助空气室
air storage tank 储气筒;储气罐
air stove 热风炉
air strainer 空气滤网;滤气器;空气滤清器;空气滤净器;空气过滤网;空气过滤器;空气粗滤器
air strainer and silencer 空气滤器及消声器
air strainer oil 空气过滤器(用)油
air strake 通气列板
air strangler 空气阻气门
air stratification 空气分层
air stream 气流;空气射流;空气流
air-stream contamination 气流污染
air-stream deflector 导风器;气流偏导装置
air-stream detector 气流探测器
air-stream noise 气流噪声
air-stream pattern 气流模型
air-stream pollution 气流污染
air strength 自然干燥强度
air strike 空中攻击
air strip 简易跑道;着陆带;小飞机场;飞机跑道;简易机场
airstrip and runway specification 机场跑道技术规范
air stripper 空气分离装置
air-stripping blade 气动剥岩铲
air stripping(method) 空气吹脱法
air stripping of ammonia 空气解吸法除氨
air stripping stage 气提阶段
air stuffer 气动填充机
air suction 吸入空气;吸气
air suction inlet 空气吸入口;吸气口
air suction opening with slide plate 插板式吸风口
air suction pipe 吸气管
air suction port 吸气口
air suction separator 吸气式分选机
air suction stroke 吸气冲程
air suction valve 吸气阀
air-supplied respirator 供气面罩
air supply 气源;送气;送风;空气供给;压缩空气量;供氧;供气(量)
air supply cock 供气开关
air supply duct 送风管(道);送风道;供气管(道);供气道
air supply equipment 供气设备
air supply fan 送风机
air supply fan room 送风机室
air supply hose 送风软管;送风胶管;供气软管
air supply manifold 气源总管

air supply opening 进气口;进气孔
air supply opening with slide plate 插板式送风机
air supply outlet 供气出口
air supply panel 多孔透风板;送风孔板
air supply pipe 送风管;供气管(道)
air supply rate 供气率
air supply register 空气供应记录器
air supply section 送风段
air supply system 送风设备;送风系统;恒压通风
air supply valve 供气开关;供气阀
air supply volume of ventilation 通风机供风量
air supply volume per unit area 单位面积送风量
air-supported building 气胎式建筑;气承建筑;充气建筑
air-supported fabric dome 充气胶布帐篷
air-supported fiber 空气间加光纤
air-supported grid 陆空协同战联络格网
air-supported gyroscope 气浮陀螺仪
air-supported head 浮动头
air-supported plastics 充气塑料
air-supported plastic still 充气塑料蒸馏器
air-supported puck 气垫圆盘
air-supported structure 气承结构;空气支承结构;充气结构(物)
air-supported tempering 气垫钢化
air-supported vessel 气垫船
air supremacy 制空权
air/surface container 空陆水联运集装箱
air surveillance 空中监视
air surveillance patrol 空中巡视管线
air surveillance radar 空中监视雷达;对空监视雷达
air survey 航空勘测;航空测量
air survey aircraft 航空摄影飞机
air survey camera 航摄机;航空测量摄影机;航测摄影机
air survey compilation 航测编图
air survey equipment 航测设备
air surveying 航空摄影测量
air survey map 航测地形图
air survey of damage 震害航测
air survey photo(graph) 航空测量摄影
air survey plan 航测平面图
air survey plotting machine 航测绘图仪器
air suspended 空气中悬浮的;大气中悬浮的
air suspension 大气悬浮(物);空气悬浮(物);空气垫
air suspension bellow 空气弹簧气囊
air suspension block 气垫块;空气悬浮台
air suspension preheater 空气悬浮预热器
air suspension reservoir 空气弹簧风缸
air suspension system 空气悬架系统
air sweetening 空气脱臭
air swept 空气冲刷
air-swept ball mill 风吹式球磨机;风扫式球磨机
air-swept coal plant 风扫式煤磨设备
air-swept conic(al) ball mill 风吹式锥形球磨机
air-swept dryer 风扫式烘干机
air-swept grinding plant 风扫式粉磨设备
air-swept hammer mill 风扫式锤碎机;风扫式锤磨机
air-swept mill 风动排料磨机;风扫磨
air-swept pulverizer 空气吹扫粉磨机;通风式磨煤机
air swept surface 空气磨刷面
air-swept system 风扫磨系统
air switch 气开关
air system 通风系统;空气系统;大气系统
air table 风动工作台
airtack cement 封气(粘胶)水泥;密封粘胶
air tacker 气动钉钉机
air tamper 风动夯(具);风动捣棒;风动打夯机
air tank 气罐;空气箱;空气罐;空气舱;储[贮]汽箱;储[贮]气罐;储汽罐;储汽缸;充气油箱;浮箱
air tanker 灭火飞机
air tap 气旋塞;空气旋塞;空气栓
air target data 空中目标资料
air target material 空中目标资料
air target mosaic 空中目标镶嵌图
air-taxi 小型客机;出租飞机
air temperature 气温;空气温度
air temperature automatic fire alarm 测温式火警自动报警器
air temperature correction 空中温度校正
air temperature ga(u)ge 空气温度规
air temperature gradient 气温梯度;空气温度梯度

air temperature inversion 气温逆增;逆温
air temperature measurement 气温检测
air temperature rise 空气升温
air tempered glass 风钢化玻璃
air tempering 空气中回火;风钢化
air tent 气垫帐篷;充气帐篷
air terminal 建筑物上部的金属杆端;空中终端;机场;航空(终点)站;航空港;风道末端;飞机场;避雷针;航空集散站
air terminal building 航空港大楼
air terminal hotel 航空港旅馆
air terminal unit 空调终端设备
air termination network 屋面避雷网
air test 空气试验;气压试验;气密试验;漏气试验;充气试验
air tester 大气碳酸计
air testing of pipe 管道空气试验
air thawing index 空气熔化指数
air thermometer 气温表;空气温度计
air thermometer milliammeter 空气温度毫安计
air thermostat 空气恒温器
air throttle 空气节流阀;节气门;节气阀
air through 分解炉风
air throw 气流射程
air-tight 气密(的);密封的;密闭的;不透气;不漏气(的);气封
air-tight access door 气密门
air-tight bulkhead 气密隔板;气密舱壁
air-tight cabin 气密舱;密封舱
air-tight chamber 气密室
air-tight coating 密封涂层
air-tight concrete 密实混凝土;气密混凝土;不透气混凝土
air-tight construction 气密结构;密封构造
air-tight container 气密容器;密封箱;密封容器
air-tight cover 气密盖;密封盖;密闭盖
air-tight door 密封门;密闭门
air-tight double glazing window 密闭双层玻璃窗
air-tight double tipping valve 双叶锁风翻板阀
air-tight gangway connection 气密通道式连接
air-tight hood 密闭罩
air-tight joint 紧密接合;气密连接;气密接头;气密接合;气密接缝;密封接头;密闭接头;密闭接缝
air-tight machine 气密型机器;密封式机器
air-tight material 气密性材料;不透气材料
airtightness 不漏气性;不透气性;气密性;气密(度);密封性;密封度;密闭(性)
air-tightness test 气密试验
air-tight packing 气密密封;密封件;不透气填料
air-tight partition 气密隔墙
air-tight quick acting door 气密速闭门
air-tight sash 密封窗
air-tight screw pump 密闭式螺杆泵
air-tight seal 气垫;空气密封;密封性密封;密封封(闭);气密(层);密封
air-tight separator 气密式封罐机
air-tight shaft seal 气密轴封
air-tight test 气密(性)试验;气密(性)实验
air-tight test pressure 气密性试验压力
air-tight vessel 气密容器
air-tight work 气密操作
airtime 广播时间
air tint 氧化色
air tinting method 回火颜色法
air tire 气胎;充气轮胎
air-tired roller 气胎压路机
air tired tractor 空气轮胎拖拉机
air tite 不漏气的
air-to-air after cooler system 空气—空气后冷系统
air-to-air heat exchanger 气冷式热交换器
air-to air-heat transmission coefficient 空气中热传导系数
air-to-air-homing 空对空寻的
air-to-air-laser ranging 空对空激光测距
air-to-air resistance 总热阻
air-to-air total heat exchanger 全热换器
air-to-cloth ratio 气布比
air-to-glass surface 空气玻璃面
air-to-glass transition 空气玻璃的传输
air-to-ground laser rangefinder 空对地激光测距仪
air-to-ground laser ranging 空对地激光测距
air tool 压缩机驱动的工具;气动工具;风动工具
air torrent 空气湍流
air-to-solids bond 气固结合力

airtow 机场用牵引车
air toxicant concentration 空气中毒物浓度
air trace system 空中微迹系统
air-track drill 履带式重型风动凿岩机;风动履带式凿岩机
air traffic 空中交通;空运;航空交通
air-traffic area 机场空中交通管辖区
air traffic control 空中交通指挥;空中交通管制;空中交通管理;航空控制
air traffic control clearance 空中交通管制许可证
air traffic control interlock 空中交通管制联锁装置
air traffic controller 空中交通管制员
air traffic control system 空中交通管理系统;航空交通管制系统
air traffic control zone 控制区
air traffic network 空中交通(运输)网
air tramping 不定期空运
air transformer 气心变压器;空心变压器
air transit 航空转运
air transit of goods 货物空运过境
air transmission pipe 送气管
air transportable sonar 机载声呐
air transport(ation) 空运;航空运输;风力运输
air transportation insurance 空运保险
air transport service division 空运服务组
air transport station 空运站
air trap 气封口;(提高混凝土和易性的)化学掺料;防气弯管;防气阀(门);气阱;空气收集器;空气井;隔气具
air tray 通风管支架
air treatment 空气处理
air treatment plant 空气净化厂
air trials map 航空线图
air triangulation 空中三角测量
air trough conveyer 空气输送斜槽
air trowel 抹灰喷浆器;气动抹灰工具
air truck 运货飞机
air truck drill 汽车式风动钻机
air trunk 通气总管;通风总管;通气筒;通风井;导气管;防热气套
air tube 通风管;气管;送气管;空气管;风管;氧气导入管
air tube cooler 管状空气辐射冷却器
air tube installation 压缩空气管道装置;风管装置
air tube radiator 空气管状散热器
air tube system 压缩空气管道系统
air tugger 气动拖式卷扬机
air tumbler 空气转筒;空气换向器
air tunnel 用压缩空气在软土中开挖的坑道;空气通道;风洞
air tunnelling 气挖隧洞
air turbine 汽轮机;气体透平机;气动涡轮机
air turbine dental engine 气动涡轮牙钻;风动牙钻机
air turbine pump 空气涡轮泵
air turbulence 乱流;空气涡流;空气湍流;大气湍流
air turbulence method 空气扰动法
air type shock absorber 气动型避振器
air union nut 气管联接螺母
air uptake 空气上升道
air valve 通气阀;排气阀;气门;气阀;空气阀
air-valve box 气阀箱
air valve cage 气阀盖座
air valve camshaft 空气阀凸轮轴
air valve chamber 气阀井
air valve control lever 气阀控制杆
air vane 通风叶片;通风格子板;空气舵;鼓风机叶;风向标;风量测速器;风车叶片
air vane guide 气流导向叶片
air vehicle 航空器
air velocity 气流速度;空速;空气流速;风速
air velocity at work area 作业地带空气流速
air velocity at work place 工作地点空气流速
air velocity vector 空速矢量
air velometer 气流速度计
air vent 排气孔;排气口;通气孔;通气口;通风口;通风孔;出气口;出气孔;气孔;气眼;通气孔
air ventilation 通风量;空气通风
air ventilator 通风器;送风扇
air vent manifold 多叉通气管
air vent of tunnel 隧道风风口
air vent pipe 通风管;通气管
air vent seat 排气管座
air vent valve 喷气孔阀;排气孔阀;排气阀;通气孔

阀;调压阀;通气阀
air vent window 通气窗
air vessel 储[贮]气容器;气室;空气室;储[贮]气器;导气管;储[贮]气罐;泵的气(压)室
airveyor 气流式输送机
air vibration 空气振动
air vibrator 气动振动器
airview 空瞰图;鸟瞰(图);航空照片
air viscosity 空气黏度
air vitiation 空气污染;空气恶化
air void 气穴;气泡;孔隙率;空隙;含气率;残留气泡;气孔
air-void characteristic (土壤的)空气孔特性
air-void content 气孔率;(加气混凝土的)气孔含量
air-void in concrete 混凝土中的气孔
air-void ratio 气隙比;空气孔隙比;空隙比;土中气隙比
air-void spacing 孔隙中气泡分布;空气泡分布
air-volcano 气火山
air volume 空气体积;空气量;风量
air volume calculation method 风量计算法
air volume flow rate 风量流率
air volume measuring apparatus 空气体积测量器
air volume requirement at face 工作面需风量
air wall 气幕;气壁;空气壁
air wall ionization chamber 空气壁电离室
air wand 气棍
air wash 空气清洗
air washer 空气滤清器;淋水系统;空气洗涤器;净气装置;净气器;洗气装置
air washing 空气洗涤;空气冲洗;洗尘
air waste feed ratio 供气量与污水之比率
air-water backwashing 气水反冲洗
air-water cooling 空气水冷却
air-water heat pump 气水热泵
air water interaction 气水相互作用;空气水相互作用;气水交替作用
air-water interface 气水界面;空气水分界面
air-water jet 汽水混合喷射器;气水混合射流;气水混合喷射器
air-water mixture 气水混掺;气水混合物;气水混合体
air-water pressure vessel 空气水压力器;气水压力容器
air-water ratio 气水比;水气比
air-water separator 气水分离器
air-water storage tank 气压式储水箱
air-water surface 气水结合面
air-water system 空气水系统
air-water vapo(u)r mixture 空气水蒸气混合物
air-water washing 气水冲洗
air wave 气浪;气波;空气波
airway 空气通道;风巷;通风坑道;空运路线;航路;航(空)线;氧气导入管;导气管;飞机航线
airway beacon 航路信标;航路灯标;航空标塔;航线指向灯;航线指向标
airwaybill 空运输提单;空运单;航空运输提单;航空货运提单
airway bill of lading 空运(货物)提单
airway lighthouse 航路灯塔;航空灯塔
airway lighting 航路灯光设备
airway map 航路线图
airway obstruction 气道堵塞
airway resistance 气道阻力
airways forecast 航空天气预报
air way slits in rotor 转子上的条形风力分级孔
airways observation 航空天气观测
airway station 航路电台
airway surveillance radar 航线监视雷达
airway tube 气道管
air weather station 航空气象台
air weighted surge tank 空气补偿器
air well (隧道的)通风竖井;通风井
air wet cell 空气湿电池
air wetting 空气湿润
air wheel 气轮;低压轮胎车轮
air whirl 气旋
air whistle 气笛
air winch 空气驱动绞车;气动卷扬机;气动绞车;风动绞车
air window 通风窗
air wing 风扇叶子板;风扇叶片;风车叶片
air wiper 风刷
air wire 明线;架空线;架空导线;天线

air working chamber 气压沉箱作业室；沉箱工作室
airworthiness 适航性(指飞机)
airy 通风的
airy appearance 虚幻现象；轻快外形
Airy differential equation 艾里微分方程
Airy disc 艾里斑
Airy disk radius 艾里半径
Airy function 艾里函数
Airy isostasy 艾里地壳均衡说
airy phase 浅海爆炸声
Airy point 艾里(支)点
Airy's bar 艾里杆
Airy's circle 艾里圆
Airy's compensation 艾里均衡补偿
Airy's diffraction dise 艾里衍射盘
Airy's diffraction integral 艾里衍射积分
Airy's diffraction pattern 艾里衍射图样
Airy's floating theory 艾里浮力理论
Airy's function 艾里函数
Airy's hypothesis 艾里假说
Airy's integral 艾里积分
Airy's mechanism 艾里机制；艾里机构
Airy's pattern 艾里图形
Airy's point 艾里支撑点(简支梁自重变形最小的支点)
Airy's projection 艾里投影
Airy's spot 艾里圆斑
Airy's stress function 艾里应力(作用)函数
Airy's system 艾里系统
Airy's theory of isostasy 艾里均衡理论
Airy stress function 艾里应力函数
Airy's type objective 艾里型物镜
Airy's wave 艾里波
Airy transit circle 艾里子午环
air zoning 分区送风；分段送风
aisle 剧院观众席走道；教堂侧廊；走道；侧廊；侧房；耳房【建】
aisle bay 侧房；耳房【建】
aisled 有耳房的；有侧房的
aisled church 有厢房教堂
aisled hall 教堂中厅(座位所在部分)；正厅
aisle gallery 外走廊；侧廊
aisle gradient (剧场的)通道坡度；通道陡度
aisle-man 自动服务商店工作人员
aisle passage 厢房过道；走廊过道
aisle pier 走廊支柱；侧房扶壁
aisle roof 教堂半圆屋顶；后殿屋顶；侧房屋顶；侧房房顶；走廊屋顶
aisle seat 靠走道坐位
aisle space 过道空间(用于仓库内操作及防火)
aisle stacker 巷道堆垛机
aisle stacking crane 巷道堆垛起重机
aisle-vault 厢房拱顶；侧房拱顶；假房穹隆；教堂半圆后殿屋顶；侧房房顶
aisle wall 桥台斜翼墙；过道墙
aisle(wall)window 过道墙上的窗
aisleway 通道(堆货用)；走道
ait 湖心岛；湖泊中小岛；核心岛；河洲；河心岛；小岛
Aitch metal 艾奇合金
aithalium 常绿植丛
Aithen counter 艾肯计尘器
Aithen nuclei 艾肯核
Aithen particle 艾肯微粒
aitificial brain 仿真脑
aitiogenic 产生反应的
aitiogenous 产生反应的
aitionastic 反应运动的
Aitken nuclei 艾特肯核
Aitken's dust counter 艾特肯尘计
Aitken's estimator 艾特肯估计量
Aitken's formula 艾特肯公式
Aitken's generalised least-squares estimator 艾特肯广义最小平方估计量
Aitken's least square 艾特肯最小平方
Aitken's linear least square method 艾特肯线性最小平方法
Aitken's method of interpolation 艾特肯插值法
Aitken's nucleus counter 艾特肯核子计数器；艾特肯尘计
Aitken's theorem 艾特肯定理
Aitoff equalarea map projection 艾托夫等积投影
Aitoff projection 艾托夫投影(图法)
ait thermometer 气温计
aiwan 客厅(古伊朗建筑)

Aix gold leaf 艾克斯假金叶
Aix-la-Chapelle cathedral 亚琛大教堂
ajacine 阿加新
ajar 微开着；不协调；半开(指门)
ajarcara 装饰砖雕；硅砂体(装饰)浮雕；砖浮雕
ajar door hook 半开门钩
ajar hook 门梃钩；半开(门)钩
Ajar metal 埃杰克斯合金
ajawa oil 香旱芹油
Ajax 艾杰克斯铜锡铅轴承合金
Ajax bearing alloy 艾杰克斯铅锑锡轴承合金
Ajax casting alloy 艾杰克斯铸造铜硅锌合金
Ajax furnace 艾杰克斯吹氧倾动平炉
Ajax induction furnace 艾杰克斯感应炉
Ajax metal 艾杰克斯铅青铜；艾杰克斯(轴承)合金
Ajax-Northrup furnace 艾杰克斯—诺斯拉普式高频感应炉
Ajax phosphor bronze 艾杰克斯磷青铜
Ajax plastic bronze 艾杰克斯塑性青铜
Ajax powder 艾杰克斯火药
Ajax standard bronze 艾杰克斯标准青铜
Ajax-Wyatt furnace 艾杰克斯—怀阿特(式感应)电炉
ajimez 细直棂分隔式双窗
ajkite 块树脂石
ajoite 斜硅铝铜矿
ajour 漏空花格；采风镂雕；雕孔透光；透光雕刻
ajowan oil 香旱芹油；阿育凡油
a judgment by default 缺席裁决
ajuin 蓝方石
ajutage 送风管；喷水管；承接管；排水筒；放水管
akaganeite 四方纤铁矿；正方针铁矿
akamatsu 日本赤松
akanthus 莨苕叶型装饰；阔叶饰
akanticone 绿帘石
Akaroa fiber 阿卡罗厄纤维
akatoreite 羟硅铝锰石；硅铝锰矿
akbarum 冰岩
akdalaite 六方铝氧石
akee oil 阿开木油
akenobeite 明延岩
akerite 英辉正长岩
akermanite 镁黄长石；钙镁黄长石
Akesai basin 阿克赛盆地
a key national network of communications optic(al)cables 全国通信光缆骨干网
akhet 洪水季(节)
akimbo span 肘间距
akin 等效于
akinesia 行动不能
Akins' classifier 阿金斯(型)分级机；螺旋分选机；螺旋分级机
Akitsu epeirogeny 秋津造陆运动
Akiyoshi orogeny 秋吉尔造山运动
Akiyoshi orogeny cycle 秋吉尔造山旋回
Akkadian architecture 阿卡德建筑
akle 暗棕色硬木(产于菲律宾)
akmite 锥辉石
akmolith 岩刃
akmon 铁贴构(块)(一种防波堤护面异形块体)
akmon block 铁砧状块体
a knowledge innovation project 知识创新工程
Akoh reversed polarity chron 阿科反向极性时
Akoh reversed polarity chron zone 阿科反向极性时间带
Akoh reversed polarity zone 阿科反向极性带
Akousticos 热绝缘及隔声材料
Akoustolith tile 墙面及天棚快速吸声材料
Akrit 钴铬钨工具合金；艾克里特钴铬钨镍合金；特硬耐磨合金
akrochordite 球砷锰石
Akron 阿克隆黄铜
Akron(abrasion)machine 阿克隆磨耗试验机
Akron abrasion tester 阿克隆磨耗试验机
akroter 山墙饰物
akroter(ion)山尖饰；船首花纹装饰
Aksa 阿克萨聚丙烯腈短纤维
aksaite 阿硼镁石
Aksaybashishan-Keziletag fold-fault belt 阿克赛巴什—克孜勒塔格褶断带【地】
aktashite 硫砷汞铜矿
aktian deposit 陆坡沉积(物)
Aktiebolaget Svenska Kullager-fabriken 瑞典滚珠轴承公司

Akulon 阿库纶
Akvaflex 阿克瓦弗莱克斯聚烯烃纤维
akwara fiber 阿克瓦拉纤维
akyld resin build paint 厚浆醇酸漆
Al2O3-coated cemented carbide 氧化铝涂层硬质合金
Al2O3 saturability 三氧化二铝饱和度
AL2O3-SiC whisker cutting tool 碳化硅晶须增强氧化铝切削工具
ala 耳房【建】
Alabama design method 美国亚拉巴马州(柔性路面)设计法
Alabama marble 美国亚拉巴马大理石
alabandite 硫锰矿
alabascote 用于织物表面的塑料
alabaster 细白石膏；蜡石；雪花石膏(色调)；细纹大理石
alabaster column 雪花石膏柱
alabaster glass 骨灰瓷；乳色玻璃；乳白(灰)玻璃；雪花玻璃
alabaster quarry 雪花石膏采石场
alabaster ware 雪花瓷器
alabastine 嵌补石膏裂缝的墙粉
alachlor 草不绿
Al activation log 铝活化测井
Aladdin 阿拉丁铝铜锌合金
alafosfalin 阿拉磷
Alagia 阿拉盖绸
alaglyph 线形浮雕
Ala Grecque 希腊回纹装饰
alalite 绿透辉石
al-alloy cartain rod 铝合金窗轨
alamalt 阿拉玛特
alambique 立式蒸馏釜
alameda 林荫(散)步道；地沥青；屋顶小花园；林荫(小)路；林荫小道
alamosite 铅辉石；硅铅石
alanate 铝氧化物
alanginm 八角枫【植】
alanine 丙氨酸；氨基丙酸
alanosin 丙氨菌素
alantic acid 阿兰酸
alanyl 丙氨酰
alapoupee 单向版多色凹印
Alar 阿拉尔铝硅(铸造)合金
alar alloy 铝硅合金
a large number of secondary roots 大量次生根
a large proportion of 一大部分
alarima 侧舌间裂
alarm 警告；警报(器)；报讯；报警
alarm acknowledged 警报获悉
alarm and surveillance 报警及巡检
alarm annunciator 事故指示装置；警报(信号)器
alarm apparatus 警报器
alarm bar 告警条
alarm bell 警钟；警铃
alarm bell signal 警铃信号
alarm box 警报装置；警报(信号)器；报警(信号)箱
alarm buoy 警戒浮标；警报浮筒；警报浮标；报警浮标
alarm buzzer 报警音响器；报警蜂鸣器
alarm call point 报警站
alarm check valve 报警器控制阀；报警单向阀
alarm circuit 警报线路
alarm clock 闹钟
alarm code 报警码
alarm communication system 警报通信系统
alarm concentration 报警浓度
alarm condition 报警条件
alarm control panel 报警控制面板
alarm device 警报装置；告警装置；报警装置；报警信号设备
alarm display 报警显示
alarm dosimeter 报警剂量计
alarm facility 报警设备
alarm float 警戒浮标；警报浮筒；警报浮标；浮子报警器；报警浮标
alarm for a false ground 接地报警
alarm for burn-out of a fuse 熔断器断丝报警
alarm for burnout of a main filament 主灯丝断丝报警
alarm for burnout of filaments 灯丝断丝报警
alarm for switching split 挤岔警示；挤岔报警
alarm for trailed switch 挤岔警示；挤岔报警
alarm gamma ray survey 报警器伽马测量

alarm ga(u)ge 警号气压计;锅炉警报器;报警压力计
alarm glass 报警器玻璃
alarm glazing 报警器玻璃
alarm hooter 报警笛
alarm horn 警报喇叭;警报号角;高频振频器
alarm horn assembly 警笛组件
alarm indicator 警报指示器;闹指示器;告警指示器;报警指示器
alarm indicator panel 报警指示控制板
alarming unit 警报装置
alarm input level 报警输入电平
alarm installation 警报装置;报警装置
alarm lamp 警报灯;报警灯
alarm light 警戒灯;警报灯;报警(灯)光
alarm logging 警报记录
alarm manometer 气压报警计;示警气压计;警报信号压力计;报警压力计;报警气压计
alarm message 报警信号
alarm monitoring system 警报监测系统
alarm of water leak 漏水警报
alarm panel 警报盘;报警器控制板
alarm post 紧急集合处;应变岗位
alarm pressure 警报压力;极限容许压力
alarm pressure ga(u)ge 报警压力表
alarm printer 报警打字机
alarm reaction 紧急反应
alarm receiver 报警接收机
alarm recorder 报警记录器
alarm relay 警报继电器;报警信号继电器
alarm-repeated transmission 重复传送报警
alarm response 报警响应
alarm rocket wheel 闹速动轮
alarm sensor 告警传感器
alarm service 报警设备
alarm signal 警报信号;报警信号
alarm signal handle 报警信号操纵杆
alarm signal system 事故信号系统;警报信号系统
alarm siren 警报器;报警器
alarm stage 警戒水位
alarm standard 警报标准
alarm switch 报警信号开关;报警(器)开关
alarm switchboard 警报信号系统控制板
alarm system 报警装置;报警系统;报警器
alarm thermometer 温度报警器;示警温度计;定温自动报警器;报警温度计
alarm thermostat 温度报警;定温自动报警
alarm transmitter 报警发射机
alarm unit 警报装置
alarm valve 警告阀;警报阀;告警阀;报警阀
alarm warner 报警器
alarm watch 带闹手表
alarm water level 警戒水位;报警水位
alarm whistle 报警笛;警笛
alarm-wire luminated safety glass 嵌报警线发光安全玻璃
alar part 翼部
alar plate 翼板
alar septum 侧隔壁
alar sloping settling tank 翼片斜板沉淀池
alarum 闹钟
alar wall 翼壁
alary 翼状的;翼形的
Alaska cedar 阿拉斯加扁柏
Alaska current 阿拉斯加海流
alaskaite 银辉铅铋矿
Alaska long period array 阿拉斯加长周期天线阵
Alaska type of ultramafic rock body 阿拉斯加型超镁铁岩体
Alaska warm current 阿拉斯加暖流
Alaska yellow cedar 阿拉斯加黄杉
alaskite 白岗岩
alaskite aplite 白岗细晶岩
alaskite deposit 白岗岩矿床
alate 不等角骨针
alaum 明矾
alautun 阿劳顿
alazimuth 地平经纬仪
alb 高山平地
Alba alloy 阿尔巴银钯合金
alba alloy soldering 白合金焊
Alba crack detector 阿尔巴磁力探伤仪
Albada finder 阿尔巴达取景器
Albagloss 白色搪瓷制品的注册名称

Albaloy 铜锡锌合金;电解沉淀用合金;阿尔巴洛伊电解淀积用铜锡锌合金
albani stone 胡椒粉色的石料
albanite 地沥青;白榴岩;阿型白榴岩
Albany clay 阿尔巴尼黏土
Albany grease 钙基润滑脂;含钙皂黄油
Albany slip 陶泥釉
alba red 阿尔巴红
albarium 大理石灰;大白(一种粉饰用的白石灰);白石粉
albata goods 德国银器皿
Albata metal 阿尔巴达黄铜
Albatra alloy 阿尔巴特拉铜合金
albatross deep 信天翁海渊
Alba velvet 阿尔巴提花丝绒
albedo 扩散反射系数(反射光强与入射光强之比);漫反射系数;反照率;反射率
albedometer 反照仪;反照率计;反射仪;反射率计
albedo neutron dosimeter 反照率中子剂量计
albedo neutrons 反照粒子
albedo of sea 海洋反照率
albedo of snow 雪的反射性;雪反照;雪白
albedo of the earth 地球反射率
albedo particles 反照粒子
albedo perception 反照率知觉
albedo radiation 反照辐射
albedo theory 反照率理论
albendazole 丙硫咪唑;阿苯达唑
Albene 阿尔本醋配纤维
alberanite 紫苏安玄岩;拉长安山岩
alberene stone 高级皂石
Alberger process 阿贝格制盐法
Alberta basin 阿尔伯达盆地
Alberta low 阿尔伯达低压
albert coal 黑沥青煤
Albert cottage 艾伯特馆(英国第一次伦敦国际博览会上展出的工人典型住宅)
Albert crepe 艾伯特绉
albertine 脉沥青
albertite 黑沥青;艾伯特柏油
Albert lay 顺捻;同向捻(法)
Albert lay wire rope 同向捻钢丝绳;艾伯特顺捻钢丝绳
Albertol 艾伯塔树脂
Albert-Precht effect 阿尔伯特—普雷切特效应
albery 壁龛;壁橱
albescent 浅白色的;带白色的;发白的
Albian 阿尔必阶
albic 漂白土;白土层
albicans 白色的
albic-bearing nepeline syenite 含钠长石霞石正长岩
albinism 白化现象
albino 白化体
albino asphalt 白沥青
albinobitumen 浅色(石油)沥青
Albionian 阿尔比翁期
Albion metal 夹铅锡箔
albite 钠长石
albite-actinolite schist 钠长阳起片岩
albite-akenite 钠长正长岩
albite-anorthite series 钠长石—钙长石系列
albite-Calsbad twin law 钠长石—卡斯巴双晶律
albite-chlorite phyllite 钠长绿泥千枚岩
albite-chlorite schist 钠长绿泥片岩
albite-clossite rock 钠长石青铝闪石岩
albite-diorite 钠长闪长岩
albite-dolerite 钠长粒玄岩
albite-epidote actinolite hornfels 钠长绿帘阳起角
albite-epidote hornfels facies 钠长石—绿帘石角岩相
albite granite 钠长花岗岩
albite law 钠长石律
albite leuco granoblastite 钠长浅粒岩
albite phyllite 钠长千枚岩
albite porphyrite 钠长玢岩
albite rhyolite 钠长流纹岩
albite trachyte 钠长粗面岩
albite twin 钠长石双晶
albite twin law 钠长石双晶律
albitite 钠长(石英)岩;钠长岩;钠长玢岩
albitization 钠长石化
albitophyre 钠长斑岩
albocarbon 萘

albofungin 白真菌素
albolite 艾尔波里特水泥
alboll 漂白软土
Albondur 纯铝包皮超硬铝板;阿尔邦杜尔包皮纯铝超硬铝板;纯铝包皮;超硬铝板
alboranite 拉长安山岩;紫苏安玄岩
Albrac 铝黄铜高强度黄铜;阿尔布拉克高强度铝黄铜
albraium 大白
albronze 铜铝合金;铝青铜
albucid 阿尔布西
albugineous 白膜的
album 画册;像片簿
albumen process 蛋白版法
albumin 清蛋白
albumin adhesive 蛋白胶
albumin(al)glue 蛋白胶;白月元胶
albuminoid 类蛋白
albuminoid nitrogen 蛋白氮
albuminous matter 蛋白物质
album of painting 画册
Albuna 阿尔本纳无光醋酯长丝
alburnum 边材;边材;白木质
albus 白色的
albutoin 阿布妥因
albylcellulose 丙烯基纤维素
alcamines 醇胺类
alcazar 宫殿(西班牙);城堡(西班牙)(西班牙或阿拉伯的)要塞
alchemy 炼金术
Alchlor process 氯化铝法
Alchrome 铁铬铝系电炉丝;阿尔克罗姆铁铬铝电阻合金
Alchrotal 阿尔克罗塔尔电阻合金
Alcian dye 爱尔新染料
Alclad 阿尔克拉德纯铝覆面的硬铝合金
alclad 铝衣合金;铝包的;铝合金双面复合硬铝板;覆铝层;包铝(的硬铝)合金;包铝
alclad alloy 镀纯铝合金
alclad pipe 镀铝管材
alclad plate 镀铝板
alclad sheet 镀铝板材
alclad tube 镀铝管材
Alcoa 耐蚀铝合金
Alcoa alloy 阿尔科阿耐蚀铝合金
Alcock's canal 阿尔科克管
alcogel 醇凝胶
Alco gravure process 多色凹印照相网目制版法;阿尔科凹版法
alcohol 酒精
alcohol adsorption method 乙二醇吸附法
alcohol amide 醇酰胺
alcohol amine 醇胺
alcoholate 烃氧基金属;乙醇化物;醇化物
alcohol base flexo ink 醇基凸板油墨
alcohol bath 酒精擦浴
alcohol-benzene solubility 醇—苯可溶性
alcohol blast burner 酒精喷灯
alcohol burner 酒精灯
alcohol-burnt method 酒精燃烧法
alcohol content 酒精含量;醇含量
alcohol distillation 酒精蒸馏
alcohol duplicating 酒精复制
alcohol enamel 酒精瓷漆
alcohol ester 醇酯
alcohol ether 醇醚;羟基醚
alcohol fastness 耐醇性
alcohol foam 酒精泡沫
alcohol-free 不含酒精
alcohol group 醇基
alcoholic 含酒精的;醇的
alcoholic acid 醇酸
alcoholic acid paper mica sheet 醇酸纸云母片
alcoholic acid paper mica tape 醇酸纸云母带
alcoholic acid putty 醇酸腻子
alcoholic acid varnish 醇酸清漆
alcoholic acid varnished glass cloth tape 醇酸玻璃漆布带
alcoholic beverage 酒精饮料
alcoholic coma 醇毒性昏迷
alcoholic enamel 醇酸磁瓷漆
alcoholic fermentation 酒精发酵
alcoholic hydroxyl 醇式羟基
alcoholic lye 醇碱液
alcoholic mass 酒

alcoholic poisoning 醇中毒
alcoholic potash 氢氧化钾酒精溶液
alcoholic solution 酒精溶液;醇溶液
alcoholic tincture 醇制酊剂
alcoholic varnish 酒精清漆;泡立水;凡立水
alcoholimeter 醇定量器
alcoholimetry 醇定量法
alcohol insoluble matter 乙醇不溶物
alcoholism 酒精中毒;醇中毒
alcoholization 酒精饱和;醇化(作用)
alcohol kali 钾碱醇溶液
alcohol ketone 醇酮
alcohol lamp 酒精灯
alcohol level 酒精水准仪
alcoholmetry 酒精测定
alcohol number 醇值
alcoholometer 酒精表;酒精比重计;醇定量器;醇比重计
alcoholometry 醇定量法
alcohol phenol 醇酚
alcohol phosphate 醇的磷酸酯
alcohol plant 酒精厂
alcohol poisoning 酒精中毒;醇类中毒
alcohol-precipitation test 酒精沉淀试验
alcohol-proof 耐醇性
alcohol purification 酒精净化
alcohol resin 醇树脂
alcohol resistance 耐醇性
alcohols 醇类
alcohol slug process 酒精处理方法(用以增加油的回收率)
alcohol-sol-resin 醇溶树脂
alcohol-soluble phenolic resin varnish 醇溶酚醛树脂清漆
alcohol-soluble polyamid terpolymer 醇溶三元共聚聚酰胺
alcohol-soluble polymer 醇溶性聚合物
alcohol-soluble resin 酒精溶解树脂;醇溶性树脂
alcohol stain 醇溶着色剂;醇溶染料
alcohol strength 酒精浓度
alcohol sulfate 硫酸醇
alcohol thermometer 酒精温度计;酒精温度表
alcohol torch lamp 酒精喷灯
alcohol trough 酒精槽
alcohol varnish 酒精清漆;醇溶(性)清漆
alcohol wastewater 含醇废水
alcoholysate 醇解物
alcoholysis 醇解(作用)
Alcomax 阿尔科马克斯永久磁铁
Alcomax alloy 阿尔科马克斯铝镍钴铜型永磁合金
Alco metal 铝基轴承合金;阿尔科基轴承合金
Alcor 阿尔科相干测量雷达
alcosol 醇溶胶
alcove 料道;河岸凹壁;小亭;供料槽;附室;壁龛;凹形卧室;凹室;岸壁
alcove lands 砂页岩切割阶梯地形
Alcres 铁铬铝耐蚀耐热合金;阿克里斯铁铬铝(耐蚀耐热)合金
alcrete 预制房屋(预制房屋中的一种)
Alcumite 铜铝铁镍耐蚀合金;阿尔丘迈特金黄色铝青铜
Alcunic 艾尔科尼克耐蚀铝黄铜
alcuronium chloride 阿库氯铵
alcyone days 平安时期;平安时间
Alda 阿尔达钢
Aldal 阿尔达尔铝合金
Aldan facies 阿尔丹相;阿尔苏花岗岩相
Aldanian period 阿尔丹纪【地】
Aldan nucleus 阿尔丹陆核【地】
Aldan old land 阿尔丹古陆【地】
Aldan shield 阿尔丹地盾
aldary 铜合金;阿尔达里铜合金
Aldecor 高强度低合金钢;阿尔迪科高强度低合金钢
aldehyde 醛;庚醛
aldehyde alcohol 醛醇
aldehyde group 醛基
aldehyde oxidation 醛氧化作用
aldehyde resin 聚醛树脂
aldehyde wastewater 含醛废水
Alden power brake 奥尔登动力制动器;奥尔登测功器
alder 赤杨
alder-fen peat 赤杨沼泽泥炭
aldermanite 阿磷镁钼石

Alder reaction 艾德勒反应
aldesulfone sodium 阿地砜钠
aldimine 醛亚胺
Aldip process 热浸镀铝法;热镀铝法
Aldis(signaling)lamp 阿尔蒂斯信号灯
aldochlor 草毒死
Aldoform 阿尔多仿
aldol 醇醛
aldol condensation 醇醛缩合
aldol resin 醇醛树脂
aldotriose 丙醛糖
Aldray 铝镁硅合金
Aldrex 艾五混剂
Aldrey 铝镁硅合金;阿尔维铝镁硅合金
aldrin 艾氏剂;阿尔德林
Aldur 阿尔杜铝锌铬合金
aldural 高强铝合金;包纯铝的硬铝合金
Aldural alloys 阿尔杜拉尔包铝硬铝合金
Aldurbra 阿尔杜布拉铝黄铜
aldzhanite 氯硼钙镁石
aleatory contract 投机性合同
aleatory operation 投机活动
aleatory variable 复杂变量
ale bench 啤酒馆前面长凳
alectorium 娱乐厅(古罗马建筑);赌场(古罗马建筑);赌博室(古罗马建筑)
alee 向下风;下风
alee basin 深海异重流盆地
Alee-tree 行道树
ale-house 酒店;小酒店
alembic 蒸馏器;净化器具;蒸馏瓶;蒸馏壶;蒸馏罐;蒸馏釜
alemobholite 铝铈磷灰石
Alencon lace 阿郎松细嵌花针绣织物(室内装饰织物)
aleph-naught 阿列夫零
alephnull 阿列夫零
alephzero 阿列夫零
Aleppo pine 地中海白松
Aleppo silk 阿勒颇生丝
aleprestic acid 环戊烯戊酸
alepric acid 环戊烯壬酸
aleprolic acid 环戊烯甲酸
aleprylic acid 环戊烯庚酸
alerce 山达木
alert 警戒
alert data 警报数据;报警数据
alerting antenna 警戒天线
alerting signal 报警信号
alert level 警戒级(别)
alertor 报警器
alert platform 待飞坪(飞机场)
alert system 警报系统
alert time 警觉时刻;警戒时间
alert water level 警戒水位
aletophyte 村边杂草
alette 耳房【建】;门边框;拱墩侧座;房屋建筑侧翼室(古罗马和新古典建筑)
aleurite 粉砂
Aleurites ardata oil 罂子桐油
Aleurites cordata(steud) 罂子桐
Aleurites fordii 油桐
Aleurites fordii oil 光桐油
Aleurites montana 千年桐;木油树
Aleurites montana oil 木油树油
aleuritic acid 紫胶桐酸
aleuron(e) 糊粉(粒)
aleuron(e) grains 糊粉粒;糊粉粒
aleuron(e) layer 糊粉层;糊层
aleurorite 粉砂岩
Aleutian abyssal plain 阿留申深海平原
Aleutian current 阿留申海流;阿留申洋流
Aleutian low 阿留申低压
Aleutian trench 阿留申海沟
A-level A式水平仪
alevin 初孵鱼苗
Alexanderson altimeter 反射高度计;反射测高计;回波测高计
Alexander tester 凝胶强度试验器;亚历山大凝胶强度试验器
alexandria beans 凤倍子
Alexandrian paper 亚当建筑纸(英乔治三世时亚当建筑师采用的罗马亚历山大纸)
Alexandrian work 彩色大理石块镶嵌的玛赛克

Alexandrinum work 亚历山大式大理石镶嵌铺面
alexandrite 变石;金绿宝石
alexandrite laser 变石激光器
alexandrite-sapphire 变蓝宝石
alexandrolite 铬黏土
alexia 失读
alexidine 阿立西定
alexin unit 补体单位
alexipharmica 解毒剂
alexoite 磁黄橄榄岩
aleyurolite 粉砂岩
Alfa 埃斯帕托叶纤维;(多产于北非的)芦苇草
alfa fiber 芦苇草纤维
alfalda 紫花苜蓿
alfalfa 苜蓿(草)
alfalfa dehydrator 苜蓿干燥机
alfalfa dust 苜蓿尘
alfalfa gate 金属板滑动门;螺旋式闸门
alfalfa valve 螺旋式阀;牧场阀;灌水栓;灌溉水阀
alfameter 阿尔法仪
alfa process 阿尔法过程
alfatoxin 草霉素
Alfenide 铜锌镍合金;阿尔芬尼德铜锌镍合金
Alfenide metal 阿尔芬尼德金属
Alfenol 铝镍高导磁合金;阿尔芬诺尔铝铁合金
Alfenol flake 阿尔芬诺尔铝铁合金粉
Alfer 阿尔费尔铁铝磁致伸缩合金
Al-Fe ratio 铝铁比
Alferium 阿尔费留姆铝合金
Alfer(o) 铝铁合金;阿尔费罗铁铝磁致伸缩合金
Alferon 阿尔费隆耐酸合金
alferric 含铝铁
alferric minerals 铝铁矿物
alfesil (具有火山性质的)磨细硅质材料
Alfin alloy 阿尔芬二元铝合金;阿尔芬轴承合金
alfin catalyst 醇钠烯催化剂
alfin polymer 醇(钠)烯聚合物
Alfin process 阿尔芬铸塑法
alfin rubber 醇烯橡胶
alfisol 淋溶土
Alfol 阿尔富尔
Alford antenna 阿尔福德天线
Alford loop 阿尔福德环
Alford loop antenna 阿尔福德环形天线
Alforgas 阿尔福加斯厚帆布
alfosite 钡磷灰石
Alframe 阿芙栏(一种预制房屋)
Alfven frequency 阿尔文频率
Alfven number 阿尔文数
Alfven speed 阿尔文速度
Alfven turbulence 阿尔文湍动
Alfven velocity 阿尔文速度
Alfven waves 阿尔文波
algae 水藻;藻类;[复]alga
algae aerated lagoon 藻类曝气氧化塘
algae bacteria 藻类细菌
algae-bacterial-clay wastewater treatment system 藻类—细菌—黏土废水处理系统
algae-bacterial symbiotic system 藻菌共生系统
algae-bearing micropolluted water 含藻微污染水
algae-bearing wastewater 含藻水
algae-bearing water treatment 含藻水处理
algae bloom 水华;藻华;藻类水面增殖
algaecide 杀藻剂;除藻剂
algae control 海藻控制;藻类控制
algae floc 藻类絮凝体
algae growth 藻类生长
algae growth curve 藻类生长曲线
algae growth potential 藻类生长势;藻类生长潜力
algae-laden water 富藻水
algae-laden wastewater 含藻废水
algae-lysing bacterium 藻类溶源菌
algae pond 藻塘
algae population 藻类群体
algae preventive type anti-fouling paint 防藻型防污漆
algae removal 水灭藻;灭藻;除藻
algae settling pond 藻沉淀池
algae turbidity 水藻混浊度
algae turf scrubber 藻类泥炭净化床
alga glomerules 藻堆
alg(a)inite 藻质体
algal 藻的
algal anchor stone 藻锚石

algal-bacteria type 藻菌型
algal ball 藻球
algal bank facies 藻滩相
algal biological index 藻类生物指数
algal biomass 藻类生物量
algal biscuit 海藻饼;藻饼
algal bloom 水华;藻类水华;藻华
algal bloom control 藻华控制
algal bloom phenomenon 藻华现象
algal bound sand flat 藻结沙坪
algal coal 藻煤
algal community 藻类群落
algal community structure 藻类群落结构
algal control 藻类控制
algal density 藻类密度
algal dolomite 藻白云岩
algal filament 藻类丝状体
algal growth potential 藻类生长势
algal growth potential test 藻类生长势测试
algal gungi 藻类真菌
algal layer 藻层
algal limestone 藻灰岩
algal lump 藻凝块
algal mat 藻席;藻群
algal mixing depth 藻类混合深度
algal nutrient 藻类营养
algal ooid 藻鲕
algal oxidation pond 藻类氧化塘
algal pellet 藻球粒
algal pelletic siliceous rock 藻粒硅质岩
algal phosphoraite 藻磷块岩
algal pisolite 藻豆粒
algal pit 藻渊;藻坑
algal population 藻类群体
algal productivity 藻类产量
algal reef 藻礁
algal removal 除藻
algal ridge 藻脊
algal rim 藻缘;藻环
algal sapropelites 含藻腐泥煤
algal stromatolitic siliceous rock 藻叠层石硅质岩
algal structure 藻结构
algal type 藻质型
algal vegetation 藻类植被
algal water 有大量藻类水
algal zone 藻层
algam 铁皮
algarite 高氮沥青
algebra 代数学
algebraic(al) 代数的
algebraic(al) adder 代数加法器
algebraic(al) addition 代数加法;代数和
algebraic(al) analysis 代数解析
algebraic(al) approach restoration 图像代数修复
algebraic(al) average error 代数平均误差
algebraic(al) closure 代数闭包
algebraic(al) closure of a field 域的代数闭包
algebraic(al) coding 代数编码
algebraic(al) coding technique 代数编码技术
algebraic(al) complement 代数余子式
algebraic(al) complex 代数复型
algebraic(al) complexity 代数的复杂性
algebraic(al) computer 代数计算机
algebraic(al) configuration 代数构形
algebraic(al) criterion 代数判据
algebraic(al) curve 代数曲线
algebraic(al) cylinder 代数圆柱面
algebraic(al) decoder 代数译码器
algebraic(al) decoding 代数解码
algebraic(al) developable surface 代数可展曲面
algebraic(al) deviation 代数偏差值
algebraic(al) difference 代数差
algebraic(al) difference between adjacent gradients 坡度差
algebraic(al) differential equation 代数微分方程
algebraic(al) differentiation 代数微分
algebraic(al) equation 代数方程(式)
algebraic(al) equation solver 代数方程解算器
algebraic(al) expression 代数(表达)式
algebraic(al) expression manipulation statement 代数表达式操作语句
algebraic(al) extension of a field 域的代数扩张
algebraic(al) function 代数(型)函数
algebraic(al) geometry 代数几何(学)

algebraic(al) homotopy 代数同伦
algebraic(al) identity 代数恒等式
algebraic(al) language 代数(处理)语言
algebraic(al) linear programming 线性规划问题的代数解法
algebraically dependent set 代数相关集
algebraically equivalent 代数等价的
algebraically independent set 代数无关集
algebraic(al) manifold 代数流形
algebraic(al) manipulation 代数操作
algebraic(al) manipulation language 代数操作语言
algebraic(al) method 代数(解)法;数解法
algebraic(al) method for proportioning 代数法配料计算
algebraic(al) notation 代数记数法
algebraic(al) number 代数数
algebraic(al) number theory 代数数论
algebraic(al) operation 代数运算
algebraic(al) product 代数积
algebraic(al) program(me) 代数规划
algebraic(al) set 代数集()
algebraic(al) sign 代数符号
algebraic(al) sign convention 代数符号约定;代数符号规定
algebraic(al) simplification 代数简化
algebraic(al) solution 代数解
algebraic(al) spiral 代数螺线
algebraic(al) structure 代数结构
algebraic(al) sum 代数和
algebraic(al) surface 代数曲面
algebraic(al) symbol 代数符号
algebraic(al) symbol conventions 代数符号规定
algebraic(al) system 代数系
algebraic(al) topology 代数拓扑
algebraic(al) transformation 代数变换
algebraic(al) value 代数值
algebra isomorphism 代数同构
algebra of propositions 命题代数
algebra of regular represent 正规表示的代数
algebra of sets 集(合)代数
algebra of two state logic 二值逻辑代数
algebra system 代数系统
algeldrate 水合氢氧化铝
Algeria's New Petroleum Act 阿尔及利亚新石油法
Alger's metal 阿尔格锡锑(轴承)合金
algicidal bacteria 溶藻细菌
algicide 杀藻剂;灭藻剂;除藻剂
algid stage 寒冷期
Alg(i)er alloy 锡锑系轴承合金
Algier metal 阿尔加锡锑系轴承合金
algin 褐藻胶;海藻胶;海草藻素;藻胶
alginate 藻酸盐
alginate thickener 藻酸盐增稠剂;海藻酸盐浆料
alginate thickening 海藻酸盐浆料
alginic acid 褐藻酸;藻酸
alginite 藻类体
alginite-collinite 藻类无结构体
alginite-telinite 藻类结构体
alignment 准直调整;定测
algite 微藻类煤
algo-clarite 微藻类亮煤
algodonite 微晶砷铜矿
Algoi blue 阿果蓝
algology 藻类(学)
Algoman orogeny 阿勾曼造山运动
Algonkian period 阿尔冈纪
Algonkian system 阿尔冈系
algorism 十进位计数法;阿拉伯数字系统
algorithm 算法;运算法则
algorithmic analysis 算法分析
algorithmic block 算法字组
algorithmic dispatching 算法转接;算法调度
algorithmic elaboration 算法详细描述;算法推敲
algorithmic flow chart 算法流程图
algorithmic language 算法语言;代数符号语言
algorithmic level 算法级
algorithmic method 算法
algorithmic model 算法模式
algorithmic pattern 算法图
algorithmic queue control 算法排队控制
algorithmic routine 算法程序
algorithmic simulation 算法模拟
algorithmic statement 算法语句

algorithmic subroutine 算法子程序
algorithm to compute prime number 计算质数的算法
algoscopy 冰点测定法(溶液)
algosol 藻溶胶
algovite 辉斜岩
algraphy 铝版制版法
Alhambra (13~14 世纪西班牙的)阿尔汗布宫
al. honey-comb landing mat 蜂窝孔锌板(机场临时跑道用)
Aliaf 阿莱夫尼龙长丝
alias 同义名;替换入口;交换点;假频;换接口;别名
alias description entry 别名描述项目;别名描述入口
aliased grating ring 混杂栅环
alias filter 假频滤波器
alias frequency 假频频率
aliasing 混淆(现象);别名使用
aliasing distortion 折叠失真
aliasing effect 混迭效应
aliasing error 混迭误差;重叠误差
aliasing frequency 混杂频率
aliasing problem 假信号问题
aliasing spectra 混淆频谱
alias name 别名(名字)
alias section 别名段
Ali Baba 阿利巴巴
alibate 铝护层
alicatado 彩釉镶嵌工;彩色釉面图案砖
Alicinenne crepe 阿利新条子泡泡纱
alicyclic(al) compounds 脂环族化合物
alicyclic(al) hydrocarbon(s) 脂环烃
alidade 平板测量器;指示规;指方规;照准仪;照准器;照准架;带望远镜的方位镜;测向仪
alidade circle 照准仪度盘
alidade clamp 上盘制动螺栓
alidade method 图解交会法
alidade protractor 照准仪分度器
alidade rule 照准规;照准尺
alidade stadia method 照准仪视距法
alidade target 照准标
ali-digital correlation 全数字相关
alienability of property 财产的可让渡性
alienable land 可转让的土地
alienate 让与;转让
alienation 异化;相疏;(财产或产权的)转让
alienation clause 让与条款
alienation coefficient 相疏系数;不相关系数
alienation economy 转让经济
alienation of commodities 商品转让
alienator 让渡人(权利或财产)
alien bank 外国银行
alien cofactor 不相关余子式
alien corporation 外国公司
alien crop 引进作物
alien duty 侨民税
alienee 受让者;受让人
alien filling 外部充填料
alien merchant 外商;外国商人
alienor 财产转让者
alien pilot 外籍引航员
alien property 外国产业
alien species 侨民种
alien structure 转让建筑;拆迁建筑(物)
alien tones 外加噪声
aliettite 滑间皂石
alifera 上侧片
aliform 翼状的;翼形的;翼接口;八字墙
alight 降落
alighting 下车
alighting area 降落区域
alighting carriage 起落架(飞机)
alighting deck 降落甲板
alighting gear 起落架(飞机)
alighting passengers 下车人数
alighting passenger volume 下车客流量
alighting time 下客时间;着陆时间
align 配比;排列成行;调直;直线对准;对准;成直线
alignability 可对准性
align baring 镗同心孔
aligned attribute 列残属性;列式属性;列式表征
aligned bundle 定位光纤束
aligned current structure 流线构造;定向构造
aligned dimensioning 平行标注
aligned fiber 定向纤维

aligned series 配套单据
aligned structure 线性构造
aligned track 方向拨正的轨道
aligned view 展开画法
align(e)ment 排列成行
align(e)ment chart 列线图解
aligner 前轴定位器;校准器;校直器;转向轮安装角测定仪;直线对准器;整平者;整平器;对准器;定线器
aligner feeder 整列进料机
aligning 统调;校整;对准;定线测量
aligning arm 对中臂;对位舌板;对位活扳;(集装箱的)吊具四周抓臂;导向翼板
aligning bar 调心销钉;定心棒
aligning bearing 径向定心轴承
aligning capacitor 微调电容器
aligning elevation gear 俯抑调整机构
aligning gear 校正机构;对中装置
aligning guide 对准机构
aligning jig 直线校准用夹具
aligning key 校准键
aligning level 准直水平仪
aligning microscope 对准用显微镜
aligning parts 对准部件
aligning pin 调心销钉;对准销
aligning plug 卡口插头
aligning pole 定线标杆
aligning punch 冲孔对准器;定线打孔针
aligning seat 调心球面
aligning sleeve 调心套筒
aligning strip 调通板
aligning stud 调心销钉;定心棒
aligning telescope 定线望远镜
aligninum 木纤维敷料
alignment 汽缸布置;排成行;调准;调整;调试机器设备;校直;线形;线向;线列;准线;直线对准;找正;共轴调整;对准;对中(点);对正;对齐;定心;定线;成线法;房基线;成直线
alignment adjustment 调准校正
alignment and grades for highway tunnels 公路隧道的排列和坡度
alignment apparatus 对准装置
alignment area 排列区
alignment array 轴向辐射天线阵;直线列
alignment bar 停机线
alignment boat 定位船
alignment by sight 目测准直法
alignment chart 列线图;计算图(表);准线图;周线图;直线图(形);贯线图;定线图
alignment clamp 瞄准器;对管器
alignment clip 装置夹
alignment code 定位码
alignment coil 校准线圈;校正线圈;校列线圈
alignment collimator 视准仪
alignment connector 定位连接器
alignment correction 准线修正;定线修正;定线改正
alignment curve 定线曲线
alignment design 路线设计;线形设计;总体设计;定线设计
alignment design of channel 航道定线设计
alignment diagram 算图;列线图;诺谟图;定线图
alignment disc 直线对准盘
alignment element 线形要素
alignment error 调准误差;校直误差;校正误差;对准误差;定线误差;安装误差
alignment feeler 对准测隙规
alignment field 校正场
alignment flag 定线(大)旗;大旗(道路定线时用)
alignment function 对齐功能
alignment ga(u)ge 铁路轨距样板;定位规;水平仪;校准仪;定位仪
alignment gear 校准机械
alignment guiding line 导向线
alignment jitter 定位抖动
alignment line 列线图
alignment map 定线图
alignment mark 调节标记;对准标记;对准标记;定位记号
alignment network 对准网络;定位网络
alignment nomogram 算图
alignment of bearings by taut wire 轴承的拉线找正
alignment of canal 渠道定线;渠道布置
alignment of coast 岸线取直

alignment of crystal 晶体排列
alignment of data 数据重新定位
alignment of navigation channel 航道定线
alignment of orientation 方位对准
alignment of pipeline 管道定线
alignment of regulation line 整治线布置
alignment of road 公路定线
alignment of runnel 隧道定线
alignment of sewer central line 沟管中心定线
alignment of shafts 大轴找正;大轴对中;主轴找中心
alignment of sounding 水道测深线;测深定线
alignment of three axes for laser range finder 激光测距机的三轴一致性
alignment of tuned circuit 回路的调谐
alignment of tunnel 隧道定线
alignment of well 井位定线
alignment oscillator 校准振荡器;校直摇频振荡器
alignment pattern 校准图形
alignment pile 定位桩
alignment plate 定位板
alignment pole 中线桩;定线桩
alignment pole of channel 航道定线桩
alignment precision 对准精度
alignment procedure 调准过程
alignment reserve 调准余量;调整裕量;对中裕量
alignment scale 对准标尺
alignment scope 调准用示波器;调准装置
alignment sensitivity 定向灵敏度
alignment sensor 定位传感器
alignment stake 路线桩;中线桩;定线桩;定位桩;方向桩
alignment stake of channel 航道定线桩
alignment target 对准标setVisible
alignment technology 对准技术
alignment telescope 调直望远镜;对准望远镜
alignment-telescope bracket 校准用望远镜托架
alignment test 精度检查;校直试验
alignment tolerance 组对公差;准直容差;对准公差;定位公差
alignment tool 吊线工具
alignment tower 定位塔
alignment treatment 调准处理
alignment uncertain boundary 未定界限;未定界线
alignment wire (喷射混凝土施工时的)基准钢丝;定位线;照准(地)线;定线钢丝
align reamer 对准铰刀;长铰刀
align reaming 配铰;铰同心孔
aligreek 希腊式的格子花
alike 相同的
Alimasillillm 阿尔马西铝镁硅合金
alimemazine 阿利马嗪
alimentary canal 消化道
alimentary crop 食用作物;粮食作物
alimentary deficiency 食物不足;营养缺乏;营养不良
alimentary system 消化系统
alimentary toxicosis 食物中毒;食品中毒
alimentation 补给
alimentation area 补给区;补给面积
alimentation facies 补给相
alimentation of glacier 冰川补给
alimentation of river 河水补给;河流补给
alimentation of stream 河水补给;河流补给
alimentation region 补给区
alinamin 优硫胺
alinating 胺化
A-line A 线(指塑性图)
aline 矫直;对准;定线;支承力影响曲线
alineation 直线定测(法)
alinement 校直;线向;对准;成线法
alinement chart 线解图;诺谟图;列线图
alinement test 校直试验
alinite cement 阿利尼水泥
Alin's filter 阿林过滤管
alinternal reflection angle 全内反射角
aliobar 同素异重体;异组分体;变压区
alipamide 阿立帕米
aliphatic amine 脂肪族胺
aliphatic compounds 脂肪族化合物
aliphatic diisocyanate 脂肪醇二异氰酸酯
aliphatic glycidyl ether 脂烃基缩水甘油醚
aliphatic hydrocarbon 脂肪族烃
aliphatic polyester 脂肪族聚酯

aliphatic solvent 脂族溶剂
aliphatic unsaturated hydrocarbon 不饱和脂族烃
alipterion 浴池油身室(古罗马)
aliquant 除不尽的(数)
aliquation 层化
aliquot 可分量;整分部分;整除数;整除的;等分(部分的);除得尽的数
aliquot number 能除数
aliquot part 整除部分;等分部分
aliquot sample 等分试样
aliquot scaling 共振调音
aliquot tuning 共振调音
aliso 阿里索杉木
alist 船体倾侧
alite A 水泥石(水泥熟料中的矿物成分)
alit(e)cement 阿里特水泥
alitizing 铝化(处理);表面渗铝
a little fissures 有少量裂缝
Aliva concrete sprayer 阿里瓦混凝土喷射器
alive 通有电流;活的
alive circuit 有源电路;有电压的电路;带电线路
alizarin(e) 茜素
alizarin(e) astrol 茜素鲜红
alizarin(e) black 茜素黑
alizarin(e) blue 茜素蓝
alizarin(e) brown 茜素棕
alizarin(e) colo(u)r(stuff) 茜素染料
alizarin(e) complexone 氟试剂
alizarin(e) dye(stuff) 茜素染料
alizarin(e) lake 茜素色淀;茜素沉淀色料
alizarin(e) madder lake 茜素红色淀
alizarin(e) orange 茜素橙
alizarin(e) pigment 茜素颜料
alizarin(e) red 茜素红
alizarin(e) yellow 茜素黄
alkalescence 微碱性
alkalescent 弱碱性的
absorbent 碱性吸收剂
alkali accelerator 碱性促进剂
alkali accumulation 碱累积
alkali acid extinguisher 酸碱灭火机
alkali activated cement 碱激活水泥
alkali ag(e)ing 碱老化;(水泥的)碱性时效
alkali-aggregate activity 碱—集料反应活性;碱—骨料反应活性
alkali-aggregate expansion 碱—集料膨胀;碱—骨料膨胀
alkali-aggregate expansion inhibitor 碱性集料膨胀抑制剂;碱性骨料膨胀抑制剂
alkali-aggregate reaction 骨料碱性反应;碱(性)集料反应;碱(性)骨料反应;碱—集料反应;碱—骨料反应
alkali-aggregate reaction inhibiting admixture 碱—集料反应抑制剂;碱—骨料反应抑制剂
alkali alumina silicate 碱性硅酸矾土
alkali amide 氨基碱金属
alkali and related elements 碱性及其相关元素
alkali-antimonides 碱锑化物
alkali balance 碱平衡
alkali basalt 碱性玄武岩
alkali basaltic magma 碱性玄武岩浆
alkali blue 碱性蓝
alkali blue lake 碱性蓝色淀
alkali burn 碱烧伤
alkali-bursting 碱胀裂
alkali by-pass 除碱旁路放风
alkali by-pass calculation 除碱旁路系统的计算
alkali-calcic rock series 碱钙质岩系
alkali-calcic series 碱钙系列
alkali carbonate 碱金属硼酸盐
alkali cellulose 碱(性)纤维素;碱化纤维素
alkali chloride 碱性氯化物;碱金属氯化物
alkali chlorination process 碱性氯化法
alkali circulation 碱循环
alkali cleaning 碱洗
alkali cleanser 碱洗涤剂
alkalic marsh 碱性草沼
alkali compound 碱性磨光抛光剂
alkali consumption 耗碱量
alkali-containing glass fiber 含碱玻璃纤维
alkali-containing wastewater 含碱废水
alkali content 碱含量;含碱量
alkali content of cement 水泥含碱量
alkali corrosion 碱性腐蚀

alkalic pollution 碱污染
alkali cracking 碱性破裂;碱裂
alkalic rock 碱性岩(石)
alkalic rock series 碱质岩系
alkalic washing wastewater 碱洗废水
alkalic waste(water) 碱性废水
alkali cycle 碱循环
alkali cycle factor 碱循环系数
alkali damage 碱害
alkali degradation 碱降解
alkali degreasing 碱法脱脂
alkali denaturation test 碱变性试验
alkali detergent 碱洗涤剂
alkali diabase 碱性辉绿岩
alkali digestion 碱性消化;加碱消化
alkali-dimer 碱二聚物
alkali disease 碱病
alkali earth 碱(性)土
alkali earth metal 碱土金属
alkali earth metal element analysis 碱土金属元素分析
alkali elements 碱金属元素
alkali elimination 去碱;碱性消除
alkali embrittlement 碱性变脆
alkali error 碱度误差
alkali expansion 碱性膨胀
alkali-expansivity (集料的)碱膨胀性
alkalifast 耐碱的
alkali fast cement 耐碱水泥
alkali fast concrete 耐碱混凝土
alkali fastness 耐碱度
alkali feeding 加碱
alkali feldspar 碱(性)长石
alkali feldspar gneiss 碱性长石片麻岩
alkali feldspar granite 碱性长石花岗岩
alkali feldspar granite group 碱性长石花岗岩类
alkali-ferrous 含碱的
alkali filter paper process 碱性滤纸法
alkali flame detector 碱性火焰检测器
alkali flame ionization detector 碱(性)火焰电离检测器
alkali flat 碱质平地;碱性平地;碱滩;盐碱滩
alkali flume 碱性尘雾
alkali fluoride 碱金属的氟氧化物;氟化碱
alkali-free 不含碱
alkali free chemical resistant glass fiber 耐酸无碱玻璃纤维
alkali fume 碱性烟雾
alkali fusion(method) 碱熔法
alkali fusion pan 碱熔锅
alkali fusion reaction gas chromatography 碱溶化反应气相色谱法
alkalify 加碱
alkali gabbro 碱性辉长岩
alkali glass 碱性玻璃;高碱玻璃
alkali glass fibre 高碱玻璃纤维
alkali granite 碱性花岗岩
alkali granite group 碱性花岗岩类
alkali granite porphyry 碱性花岗斑岩
alkali group 碱性岩类
alkali halide 碱金属卤化物
alkali halide crystal 碱卤化物晶体
alkali halide photocathode 碱卤光电阴极
alkali halide prism 碱卤化物棱镜
alkali hydrometer 碱液比重计
alkali increase test for tung oil varnishes 桐油清漆加碱后的增稠度试验
alkali incrustation 碱化硬表层;碱垢;碱斑
alkali ion 碱离子
alkali isomerization 碱法异构化
alkali laden dust 含碱粉尘
alkali lake 碱湖
alkali lamprophyre 碱性煌斑岩
alkali land 碱地
alkali lava flow 碱性熔岩流
alkali lignin 碱木质素
alkali-lime index 碱—灰质指数
alkali-lime series 碱钙岩系
alkali liquor 碱液
alkali melting pan 碱熔锅
alkali metal 碱(性)金属
alkali metal element analysis 碱金属元素分析
alkali metal grease 碱金属润滑脂
alkali metal salt 碱金属盐

alkali metal soap 碱金属皂
alkali metasomatism 碱质交代作用
alkalimeter 碳酸定量计;碱(量)计;碱度计;碱定量器
alkalimetric analysis 碱量滴定分析
alkalimetric estimation 碱量滴定测定
alkalimetric standard 标碱基准
alkalimetric titration 碱量滴定法
alkalimetry 碳酸定量法;碱(量)测定法;碱度测定;碱定量法;用碱性当量溶液滴定
alkali microsyenite 碱性微晶正长岩
alkaline 强碱的;碱性的
alkaline accumulator 碱性蓄电池
alkaline action 碱性作用
alkaline activation 碱性激发作用;碱性活化作用
alkaline agent 碱性剂;强碱剂
alkaline aggregate reaction 碱(质)集料反应
alkaline air 氨气
alkaline amphibolization 碱性角闪石化
alkaline arsenic-containing wastewater 碱性含砷废水
alkaline barrier 碱性障
alkaline basalt 碱性玄武岩
alkaline bath 碱浴;碱液池;碱液槽
alkaline-bicarbonate thermal water 碱性重碳酸盐型热水
alkaline-calcareous glaze 碱性石灰质釉
alkaline-calcareous soil 碱性钙质土
alkaline carbonate 碱性碳酸盐
alkaline casein 碱性酪蛋白
alkaline cast 碱斑
alkaline catalyst 碱催化剂
alkaline cell 碱蓄电池;碱性(蓄)电池
alkaline circulation 碱循环
alkaline cleaner 碱性清洗剂;碱洗装置
alkaline cleaning 碱洗;碱清洗
alkaline cleanser 碱性洗净剂
alkaline colo(u)r 碱色
alkaline common salt spring 碱性食盐泉
alkaline corrosion 碱腐蚀
alkaline degradation 碱解;碱降解
alkaline degradation studies 碱的降解研究
alkaline deposit 碱性淀积;碱性沉淀
alkaline deruster 碱性除锈剂
alkaline derusting 碱液除锈;碱法除锈
alkaline detergent 碱液洗涤剂;碱性去垢剂
alkaline determination 碱度测定
alkaline development 碱性显影
alkaline digestion 碱性消化
alkaline dropping corrosion resistance test 滴碱腐蚀试验
alkaline earth 碱(性)土
alkaline earth family 碱土族
alkaline earth metal 碱土金属(元素)
alkaline earth metal ion 碱性土金属离子
alkaline earth radioactive nuclide 碱土族放射性核素
alkaline electrolysis 碱性电解
alkaline environment 碱性环境
alkaline error 碱误差;钠误差;钠差
alkaline etching 碱液浸蚀;碱腐蚀法
alkaline excess 碱过剩
alkaline extracts of soils 土壤碱性提取物
alkaline fermentation 碱性发酵
alkaline filler 碱性填料
alkaline filter paper method 碱片法
alkaline filter paper process 碱性滤纸法
alkaline flux 碱性溶剂
alkaline frit 碱性熔块
alkaline functional group 碱性官能团
alkaline fusion 加碱熔化
alkaline glaze 碱釉
alkaline hardness 碳酸盐硬度;水的硬度;碱性硬度
alkaline hydrolysis 碱性水解;加碱水解
alkaline hydrothermal leaching 碱性水热淋滤
alkaline kaoline 碱性高岭土;碱性高岭土
alkaline lake 硼砂湖;碱湖
alkaline land 碱(性)地;盐碱地
alkaline leach 碱浸出
alkaline leaching 碱性浸出
alkaline liquid 碱液
alkaline marsh 碱性草沼
alkaline meadow soil 碱性草甸土
alkaline medium 碱性培养基;碱性介质
alkaline metal 碱金属

alkaline metasomatism 碱质交代作用
alkaline methylene blue solution 碱性美蓝溶液
alkaline microorganism bio-flocculant 碱性微生物絮凝剂
alkaline mine drainage 矿山碱性排水
alkaline mineral water 碱性矿泉水
alkaline mud 碱性(沉)渣
alkaline oversaturation 碱过饱和
alkaline oxide 碱性氧化物
alkaline paint remover 碱性脱膜剂;碱性去漆剂;碱性洗漆剂
alkaline paint stripper 碱性去漆剂;碱性洗漆剂
alkaline pegmatite deposit 碱性伟晶岩矿床
alkaline phase 碱性相
alkaline postassium permanganate index method 碱性高锰酸钾指数法
alkaline postassium permanganate oxidation method 碱性高锰酸钾氧化法
alkaline potassium persulfate 碱性过硫酸钾
alkaline potassium persulfate digestion 碱性过硫酸钾消解
alkaline potassium persulfate digestion-ultraviolet spectrophotometry 碱性过硫酸钾消解—紫外分光光度法
alkaline process 碱液电镀锡法;碱法
alkaline pulp 碱性纸浆
alkaline pulp meter 碱性纸浆计
alkaline pyroxenization 碱性辉石化
alkaline reaction 碱性反应
alkaline reagent 碱性试剂
alkaline regenerant 碱性再生剂
alkaline remover 碱性脱除剂;碱性去除剂
alkaline reserve 碱储量
alkaline residue wastewater 碱渣废水
alkaline resistance 抗碱性;耐碱(性)
alkaline resistance property 耐碱性能
alkaline resistant cement mortar flooring 耐碱水泥砂浆楼地面
alkaline resistant flooring 耐碱楼(地)面
alkaline resistant mastic 耐碱玛琋脂
alkaline-resistant mortar 耐碱砂浆
alkaline resistant sealer 耐碱封闭剂
alkaline resisting 耐碱的
alkaline resisting aggregate 耐碱集料;耐碱骨料
alkaline resisting cement 耐碱水泥
alkaline-resisting mortar 耐碱砂浆
alkaline rinse 碱性洗液
alkaline rock 碱性岩(石)
alkaline rock beryllium-bearing formation 碱性岩含铍建造
alkaline salt 碱金属盐
alkaline scale 碱性垢
alkaline serozem 碱化灰钙土
alkaline siliceous thermal water 碱性硅质热水
alkaline-silicon kaoline 碱质高岭土;碱性硅质高岭土
alkaline slag 碱性炉渣;碱性矿渣
alkaline soil 碱性土(壤);碱土;碱化土(壤);碱地;盐碱地
alkaline solution 碱性溶液
alkaline solvent 碱性溶剂
alkaline spot 碱斑
alkaline spring 碱(性)泉
alkaline steeping agent 碱性浸渍剂
alkaline storage battery 碱液蓄电池;碱性蓄电池
alkaline sulfite process 碱性亚硫酸盐法
alkaline sulphide 碱性硫化物
alkaline suspension 碱性悬液
alkaline temporary hardness 碱性暂时硬度
alkaline test 碱性试验
alkaline tide 碱潮
alkaline tourmaline 碱电气石
alkaline tower 碱洗塔
alkaline treatment 碱处理
alkaline treatment tower 碱处理塔
alkaline type alteration 碱性蚀变
alkali neutralisation number 碱中和值
alkaline waste liquid 碱性废液
alkaline waste(water) 碱性废水
alkaline water 苦咸水;碱性水
alkaline zinc-air battery 碱性锌空气电池
alkaline zone 碱性范围;碱性(地)区
alkalinimetry 碱量滴定法

alkalinity 碱性；碱化程度；碱含量；碱度；含碱量
alkalinity chlorosity 氯容比(值)；碱度—氯度；氯碱度
alkalinity chlorosity factor 碱氯比(率)
alkalinity criterion 碱度指标
alkalinity end-point 碱度终点
alkalinity factor 碱度系数
alkalinity of irrigation water 灌溉水碱度
alkalinity of water 水的碱度
alkalinity-pH relation 碱度—pH 关系
alkalinity test 碱度试验
alkalinity titration 碱度滴定
alkalinization 碱化(作用)
alkalinous 碱性的
alkalinous metal 碱性金属
alkali olivine basalt 碱性橄榄玄武岩
alkalipenia 碱度减少
alkali poisoning 碱中毒
alkali prevention 防碱
alkali process 碱处理
alkali-proof 耐碱的
alkali-proof concrete 耐碱混凝土
alkali-proof glass 耐碱玻璃
alkali-proof glass fibre 耐碱玻璃纤维
alkali proofness 耐碱性
alkali-proof paint 耐碱涂料；耐碱漆
alkali-proof steel 耐碱钢
alkali-proof test 耐碱性试验
alkali-proof varnish 耐碱清漆
alkali pulp effluent 碱性纸浆废水
alkali purification 碱处理；碱净化；碱法净化
alkali reaction 碱性反应
alkali-reactive aggregate 对碱反应集料；对碱反应骨料；碱活性集料；碱反应集料
alkali-reactive material in concrete 混凝土中的碱性反应材料
alkali-reactive rock 碱反应岩石
alkali reactivity 碱性反应率；碱反应(性)；对碱反应性
alkali reactivity of aggregate 集料—碱反应性
alkali reclaiming process 碱性再生方法
alkali-refined oil 碱漂精制油
alkali-refined seed oil 碱漂亚麻仁油
alkali remover 碱性脱膜剂；碱性脱除剂；碱性去除剂
alkali reserve 碱储量；碱储备；碱藏量
alkali resistance 抗碱性；耐碱性；耐碱度
alkali resistance rate 耐碱率
alkali resistance test 耐碱(性)试验
alkali-resistant 抗碱的；耐碱的
alkali-resistant cement concrete flooring 耐碱混凝土地面
alkali-resistant fired-clay brick 耐碱黏土砖
alkali-resistant flooring 耐碱楼(地)面
alkali-resistant glass fiber 耐碱玻璃纤维；抗碱玻璃纤维
alkali-resistant mineral wool 耐碱矿(物)棉
alkali-resistant mineral wool reinforced cement 耐碱矿(物)棉增强水泥
alkali-resistant primer 耐碱底油
alkali-resistant refractory 耐碱耐火材料
alkali-resistant vegetation 耐碱植被
alkali-resistant zirconia glass fiber 含氧化锆抗碱玻璃纤维
alkali-resisting cast iron 耐碱铸铁
alkali-resisting cement 耐碱水泥
alkali-resisting mortar 耐碱砂浆
alkali-resisting paint 耐碱涂料；耐碱漆
alkali-resisting primer 抗碱底漆
alkali-resisting property 抗碱性
alkali-resisting resistant paint 耐碱涂料；耐碱漆
alkali-resisting varnish 耐碱清漆
alkali rhyolite 碱性流纹岩
alkali rhyolite group 碱性流纹岩类
alkali rich glass 纯碱玻璃(冕牌玻璃)
alkali rock 碱性岩(石)
alkali rocks 碱性岩类
alkali saline soil 盐碱土
alkali salt 碱金属盐
alkali salt flame ionization detector 碱盐火焰离子化检测器
alkalised alumina process 碱化氧化铝法
alkali-sensitive 对碱敏感的
alkali series 碱性岩系
alkali silicate 硅酸碱金属盐

alkali-silicate gel 碱—硅凝胶
alkali-silicate reaction 碱—硅反应
alkali soil 碱性土壤；碱土
alkali soil plant 碱土植物
alkali-soluble 可溶于碱液的；碱溶性的
alkali-soluble resin 碱溶树脂
alkali-soluble test 碱性浸出试验
alkali solution 碱溶液
alkali source 碱源
alkali spot 碱斑
alkali stain 碱性着色剂
alkali substance 碱性物质
alkali syenite 碱性正长石；碱性正长岩
alkali syenite porphyry 碱性正长斑岩
alkali-syenitic aplite 碱性正长细晶岩
alkali test 碱式法
alkali tolerance 耐碱性；耐碱度
alkali tolerance test 碱耐量试验
alkali trachyte 碱性粗面岩
alkali treatment 碱处理
alkalitrophic 碱性营养的
alkali tungsten bronze 碱金属钨青铜
alkali vapo(u)r 碱蒸气
alkali volatility 碱挥发性
alkali washed darkened sand 碱洗黑砂
alkali wash water 碱洗液
alkali waste 碱渣；废碱
alkali waste liquid 废碱液
alkali waste(water) 碱性废水；废碱水
alkali water 碱性水
alkali ws added to the soil 土壤里加碱
alkalization 碱化(作用)；碱处理
alkalize 碱处理；加碱
alkalized alumina process 碱化氧化铝法
alkalized plot 碱化地段
alkalized soil 碱化土(壤)
alkalized wastewater 碱化废水
alkalized water treatment 碱化水处理
alkalizer 碱化剂
alkali zone 碱性区
alkaloid 生物碱；植物碱
alkaloid toxicant 生物碱毒素
alkalophilic microorganism 嗜碱微生物
alkalosis 碱中毒
alkalsite 阿克赛混合炸药
alkamine 氨基醇
alkane 烷烃；链烷
alkanization 烷基化
alkanol amine 链烷醇胺
Alkar process 阿尔卡耳法
alkarylamine 烷芳基胺
alkathene 聚乙烯
alkelescent 微碱性的
alkene 链烯
alki 掺水酒精
alkide resin 醇酸树脂
alkine 链炔
alkoxide 烃氧基金属；醇盐；醇化物
alkoxycarbonyl 烷氧碳酰
alkoxylate 烷氧基化
alkyd adhesive solution 醇酸胶液
alkyd-aerylate copolymer 醇酸—丙烯酸酯共聚物
alkydal 邻苯二甲酸树脂
alkyd baking arc resistance enamel 醇酸烘干抗弧漆
alkyd baking insulating varnish 醇酸烘干绝缘漆
alkyd baking insulating wire coating 醇酸烘干漆包线漆
alkyd baking primer 醇酸烘干底漆
alkyd coating 醇酸树脂涂料
alkyd emulsion 醇酸树脂乳化液
alkyd enamel 醇酸瓷漆
alkyd neulating varnish 醇酸绝缘漆
alkyd marking paint 醇酸标志漆
alkyd matte enamel 醇酸无光瓷漆
alkyd-modified urea resin 醇酸改性脲醛树脂
alkyd mo(u)lding compound 醇酸模压混合料
alkyd mo(u)lding material 醇酸树脂模压料
alkyd oil 醇酸树脂油
alkyd oil resistance paint 醇酸耐油漆
alkyd paint 醇酸(树脂)涂料；醇酸(油)漆
alkyd plastics 醇酸塑料
alkyd ratio 邻苯二甲酸含量比
alkyd ready mixed paint 醇酸调和漆

alkyd resin 聚酯树脂；醇酸树脂；醇酸聚脂树脂
alkyd-resin adhesive 醇酸树脂黏合剂
alkyd-resin base 醇酸树脂底层
alkyd-resin clear varnish 醇酸树脂清漆
alkyd-resin emulsion 醇酸树脂乳剂
alkyd-resin enamel 醇酸树脂瓷漆
alkyd-resin gloss paint 醇酸树脂光泽漆
alkyd-resin lacquer 醇酸树脂喷漆；醇酸树脂腊克
alkyd-resin medium 醇酸树脂介质
alkyd-resin paint 醇酸树脂涂料
alkyd-resin pigmented varnish 醇酸树脂加色清漆
alkyd-resin primer 醇酸树脂底漆
alkyd-resin priming 醇酸树脂底层漆
alkyd-resin solution 醇酸树脂溶液
alkyd-resin undercoat(er) 醇酸树脂底漆
alkyd-resin varnish 醇酸树脂清漆
alkyd-resin vehicle 醇酸树脂载色剂
alkyd semi-gloss enamel 醇酸半光瓷漆
alkyd ship cabin paint 醇酸船舱漆
alkyd ship hull paint 醇酸船壳漆
alkyd shop primer 醇酸保养底漆
alkyd varnish 醇酸清漆
alkyd varnish for water sand paper 醇酸水砂纸清漆
alky gas 酒精(汽油混合物)燃料；酒精和汽油混合燃料
alkyl-acrylate 丙烯酸烷基酯
alkyl alkanephosphonate 烷基膦酸烷基酯
alkylalkoxy silane 烷基烷氧基硅烷
alkyl alkylphosphonate 烷基膦酸烷基酯
alkyl alumin(i)um halide 卤化烷基铝氨基
alkylamine 烷基胺
alkyl arylpolyether alcohol 烷基芳基聚醚醇
alkyl aryl recinoleate 蓖麻酸烷基芳基酯
alkyl aryl sulfonate 烷芳基磺酸盐
alkylate 烷基化(物)
alkylate bottoms 烷基化油蒸馏残液
alkylated tar 烷基化焦油
alkylate polymer 烷基化油蒸馏残液
alkylation 烷基取代；烷基化
alkyl benzene sulfonate 烷基苯磺酸盐
alkyl benzene sulfonic acid 烷基苯磺酸
alkyl carbamate 氨基甲酸烷基酯
alkyl chlorofluorosilane 烷基氟氯硅烷
alkyl chlorosilane 烷基氯硅烷
alkylene 亚烃基；烯化；烯烃基
alkyleneimine 烯亚胺
alkylene oxide 环氧烷烃；烯化氧
alkylene oxide polymer 环氧烷聚合物；烯化氧聚合物；氧化烯类聚
alkyl epoxy carboxylic ester 环氧烷基羧酸酯
alkyl ether 烷基醚
alkyl-etherified resin 烷基醚化树脂
alkyl ether sulfate 硫酸烷基醚
alkyl fluoride 氟化烷
alkyl halide 卤烷；卤代烷
alkyl hydrosulfate 硫酸烷基氢酯
alkyl lead 烷基铅
alkyl magnesium halide 卤化烷基镁
alkyl mercuric salt 烷基汞盐；卤化烷基汞
alkyl mercury 烷基汞
alkylnaphthalene sulphonate 烷基萘磺酸酯
alkylogen 卤代烷
alkylol 烷基醇
alkylol amine 羟基胺；醇胺
alkyl phenolic resin 烷基酚醛树脂
alkylphenol resin 烷基酚醛树脂
alkyl phenoxy poly(ethyleneoxy) ethanol 聚氧乙烯烷基苯醚乙醇；烷基苯氧基聚氧乙烯醚(乙)醇；烷基酚环氧乙烷聚合物
alkylphosphate 烷基磷酸盐
alkyl quaternary ammonium salts 烷基季铵盐
alkylsilanol 烷基硅醇
alkyl sulfate 烷基硫酸酯
alkylsulfonate 烷基磺酸盐
alkyl sulfur polyfluoride 多氟化烃基硫化物
alkyltetrahydronaphyhalene 烷基四氢化萘
alkyl zinc halide 卤化烷基锌
alkyne 炔烃
all 全部的
allactite 砷水锰矿；斜羟砷锰石
all-addendum gear 全齿顶角小齿轮
all aft cargo ship 尾机船
all-aged 全龄的

all-aged forest 择伐林;多龄林
allagite 绿蔷薇辉石
all air system 全空气系统;全空气方式
allalinite 蚀变辉长石;蚀变辉长岩
all-allocated quota 全球配额
all along 始终;一贯
all-alumin(i)um beverage can 全铝饮料罐
all-alumin(i)um building 全铝建筑物
all-alumin(i)um conductor 全铝导体
all amount invested in capital 投入资本总额
Allan alloy 阿伦铜铅合金
all-angle bracket 可调支架
allanite 褐帘石
Allan metal 阿伦铜铅轴承合金
Allan's alloy 阿伦合金
Allan's metal 阿伦(铜铅)合金
all-anthracite unit 全烧无烟煤锅炉机组
Allan valve 蒸汽机滑阀;阿伦滑阀
Allan variance 阿伦方差
Allan wrench 阿伦扳头
all application digital computer 通用数字计算机
allapulgite clay 硅镁黏土
Allard lens 阿拉德透镜
Allard's law 阿拉尔定律
Allard's relation 阿拉尔关系式
allargentum 六方锑银矿
all-armo(u)r rock mound 全护面防波堤;全护面堆石堤;全垒大石堆石堤
all armo(u)r rock mound breakwater 全部铺有护面石防波堤;全护面堆石防波堤
all-around 四周;连同杂费,包括各项费用;全向;全能的;万向的;通用的
all-around loading 全面加荷;四周加荷
all-around traverse 全圆周回转
all-around weld 环焊缝;整周焊缝
allasotonic contraction 变张力性收缩
allautal 纯铝包皮铝合金板;铝合金板;纯铝包皮
all autotensioned catenary equipment 全补偿链形悬挂装置
all autotensioned messenger wire and contact wire with balance 全补偿接触网
all-axle drive 全轴驱动
all axles motored 全动轴
all axle steering 全轮转向
all band 全波段
all-bank account 全国银行账户
all bell wye branch 全承式Y形分叉管
all-bituminous unit 全烧烟煤锅炉机组
all black 纯黑
all-blast heating system 送风式暖气设备
all-bottom sound 低频声
all-brick 全砖的
all-brick building 全砖建筑(物);砖砌房屋
all-bronze drain 全黄铜泄水口
all-bronze pump 全青铜泵
all-call push-button 全呼按钮
all capacity 全容积
all capital earnings rate 资本总额收益率
all-carbon brush 全炭电刷
all-cargo airline service 空中货运
all cargo ship 多用货船
all case furnace 全能渗碳炉;全能淬火炉
all casing drill 全套管钻机
all-cast 全铸的
all channel antenna 全频道天线;宽频带天线
all channel decorder 全信道译码器
all-channel tuning 全频道调谐
all charge for exchanging load 换装包干费
all charges paid 全部费用付讫;全部费用已付
allcharite 阿勒沙赖特矿
all-chromatism 全色性
all circuit law 全电路定律
all clay body 陶器挂釉质料;挂釉素地
all clear 许可;去路通畅;放行;全清爽;全清零;解除警报
all colo(u)r 全色
all common control system 全集中控制方式
all-concrete 全混凝土的
all-concrete pavement 全混凝土路面
all container ship 全集装箱船
all-core mo(u)lding 组芯造型;铸造物;制模
all correct 合格
all cost method 完全成本法
all cover 全罩式

all crop drill 通用条播机
all current accounting 全部市价会计
all curvature 全曲率
all day calls 全天呼叫次数
all-day parking 全日停车
all-dedendum gear 全齿根高齿轮
all-dielectric(al)filter 全介质滤光片
all-dielectric(al)interference filter 全介质干涉滤光片
all dielectric(al)mirror 全介质多层高反射膜
all-digital display 全数字显示
all-digital simulation 全数字仿真
all-direction 全方位
all-directional 全向的
all-directional intersection 全定向立体交叉;全定向交叉口
all direction propeller 全向推进器
all drop 全部落锻
all dry 全干;完全干燥(的)
all-dry cement mill 全干法(工艺)水泥厂
all-dry cement plant 全干法(工艺)水泥厂
all-dry construction 装配式施工;装配式结构
all-dry construction method 装配式施工法
Alleanthus fibre 阿利恩撒斯纤维
allee 园中小径;林荫宽步道
Allee's principle 阿利原理
allegation 混合计算法
allege 上下层窗空间墙;陈述;凹墙
alleghanyite 粒硅锰石;粒硅锰矿
Alleghenian life zone 阿莱干尼生物区
Alleghenian orogeny 阿莱干造山运动
Allegheny electric(al)metal 阿勒格尼高导磁率镍铁合金
Allegheny orogeny 阿勒格尼造山运动
Allegheny series 阿勒格尼统
allegory 有象征意义的图案;表意的图案;薄墙
all-electric(al) 全电气化的;全电动的、全部电气化
all-electric(al)control 全电动控制
all-electric(al)drive 全电力驱动(的)
all-electric(al)furnace 全加热炉;全电熔容
all-electric(al)grain drier 电热谷物干燥机
all-electric(al)interlocking 全部电气联锁装置
all-electric(al)melting 全电熔;全电加热熔融
all-electric(al)operation 完全电控制
all-electric(al)receiver 通用电源接收机
all-electric(al)set 通用电源接收机
all-electric(al)signalling 全电信号装置
all-electric(al)signalling system 全电气号志系统
all-electric(al)steering gear 全电动舵机
all-electrlc(al)system 全电气化系统
all-electronic 全电子化的
all-electronic ignition system 完全电子化点火系统
all-electrostatic tube 全静电讯线管
all-electrostatic vidicon 全静电光导摄像管
allelocatalysis 生化相促作用
allelochemistry 变异化学
allelopathic substance 他感作用物质
allelopathy 植物相克(现象);生物干扰;化学相克作用;化感作用;异种化感
allelotrope 稳变异构体
all-embracing 包括一切;全部的
all-embracing term 全面的条件;包含一切的条款;包含一切的条件;全面条款
allemontite 砷锑矿
Allen cell 阿伦电解池
allene 丙二烯
allene polymer 丙二烯聚合物
allenes 二烯烃
allen head 六角头
allenic compound 丙二烯(系)化合物
allenic hydrocarbon 丙二烯系烃
Allenite 阿伦奈特碳化钨系列
allenolic series 丙二烯系
Allen red metal 阿伦赤色金属
Allen salt velocity method 阿伦盐液测流法
Allen screw 阿伦螺钉;六角固定螺丝;六角固定螺栓
Allen's law 阿伦定律
Allen's loop test 阿伦环路试验法
Allen's metal 阿伦铅青铜合金
Allen's rule 阿伦规则;阿伦定律
Allen's salt velocity method 阿伦盐液测(流)速法
all environmental impact 极度环境影响
allen wrench 六角弯管套筒扳手;六角孔扳手;六角扳手;方孔螺钉头用扳手

all-epitaxial phototransistor 全外延光电晶体管
all-equipped car 全装备车
aller 栎木
allergen 变应原
allergenic extract 变态反应原浸出物
allergic 过敏的
allergic dermatitis 赤面飞
allergic disease 变应性病
allergic property 变态反应性
allergic reaction 人体变态性反应;过敏性反应;变态反应
allergic reaction disease 变态反应病
allergic response 变态反应
allergization 变应化作用
allergology 变态反应学
allergometry 变应测定
allergy 变应性;变态性;变态反应(性)
Allerod interstade 阿尔露德间冰段【地】
al(l)ette 古罗马式建筑;门边框;拱墩侧座;房屋翼室
allevardite 钠硬石;钠板石
alleviate 减轻
alleviating measures 减轻措施
alleviation 减轻
alleviator 减轻装置;缓和装置;缓和物;缓冲装置;缓冲器
all expected benefits 全部预期效益
all expected cost 全部预期代价;全部预期成本
all expenses 全部费用
alley 小路;小径;球场地;背街;林荫小道;里弄【建】;胡同
alley collection 里弄收集
alley dwelling 小巷住房
alley house (下巷内的)内店铺
alley influence 邻旁小路的影响
alley planting 并植
alley stone 矶石
alley tree 行道树
alley way 胡同;小径;小巷(道);通道(堆货用);走廊
all fast 全部挽牢
all fillet weld 满角焊
all financial resources 全部财力
all financial resources concept 全部财力概念
all-flanged cross 全法兰四通
all-flanged tee 全法兰三通
all-flanged Y-branch 全法兰Y形支管
all-flo(a)tation process 全浮选法
all-flying tail 全浮动尾翼
all found 舾装齐备
all fours moorings 四向锚泊法
all frozen 全冻结
all-function 全能;通用
all funds-combined balance sheet 全部基金综合平衡表
all-gas burner 通用燃气炉;通用燃烧器
all gas turbine 全燃气轮机
all gas turbine propulsion 全燃气轮机推进装置
all-gear drive 全齿轮传动
all-geared clutch 全啮合离合器
all-geared headstock 全齿轮主轴箱;全齿轮床头箱
all-geared upright drill 全齿轮立式钻床
all gear system 全齿轮传动
all gear tested 全部航行设备已经试验完毕
all-glass 玻璃壳的;全玻璃
all-glass building unit 全玻璃建筑构件
all-glass ceiling luninaire 全玻璃顶棚采光
all-glass construction 全玻璃结构
all-glass door 全玻璃门
all-glass facade (建筑物的)全玻璃立面
all-glass ion ga(u)ge head 玻壳电离规管
all-glass kinescope 全玻璃显像管
all-glass light fitting 全玻璃顶棚装配;全玻璃顶棚采光
all-glass paper 全玻璃纤维纸;玻璃纤维纸
all-glass pneumatic cell 玻璃式气压浮选机
all-glass skyscraper 全玻璃摩天大楼
all-glass sliding sash 全玻璃推拉扇;全玻璃推拉窗
all-glass thermal insulation 全玻璃热绝缘
all-glass work 全玻璃制品
all goods and services 全部产品和劳务
allgovite 辉绿玢岩
all-graphite reflector 全石墨反射层
all-hair pad 全毛垫
all hallown summer 晚秋晴热天
all hands 全体船员

all hatch vessel 全舱口船
all-haydite concrete 陶粒混凝土;全陶粒混凝土
all heart 全心木材
all hours 总时限
all-hydraulic 全液压的;全液式
all-hydraulic backacter 全液压反铲挖土机;全液压反铲挖掘机
all-hydraulic back digger 全液压反铲挖掘机
all-hydraulic concrete mixer 全液压混凝土搅拌机;全液压混凝拌和机
all-hydraulic concrete pump 全液压混凝土泵
all-hydraulic core drilling rig 全液压岩芯钻机
all-hydraulic crane 全液压式起重机
all-hydraulic crawler-mounted excavator 液压履带(式)挖土机;全液压履带(式)挖土机
all-hydraulic ditcher 液压挖沟机;全液压挖沟机
all-hydraulic ditching shovel 全液压反铲(挖沟)机
all-hydraulic dragshovel 全液压拖铲挖土机;全液压拖铲挖掘机;全液压拉铲挖土机;全液压拉铲挖掘机
all-hydraulic drill 全液压钻机
all-hydraulic driven static cone penetration test 全液压传动静力触探
all-hydraulic loader 全液压装载机
all-hydraulic mobile excavator 全液压移动式挖土机;全液压移动式挖掘机
all-hydraulic press 全水压机
all-hydraulic pullshovel 全液压索铲挖掘机;全液压拖铲挖土机;全液压拖铲挖掘机;全液压拉铲挖土机;全液压拉铲挖掘机
all-hydraulic trencher 全液压挖沟机
all-hydraulic trench-forming shovel 全液压反铲挖沟成形机;全液压反铲挖掘成形机
all-hydraulic trench-hoe 全液压反铲挖沟机
all-hydraulic(trenching)hoe 全液压反铲挖掘机
all hydro system 纯水电电力系统
alliance 联合;近似
Alliance International de la Distribution 国际无线电通信波段分配联盟
allidochlor 草毒死
Allied Chemical Corporation 联合化学公司(美国)
allied city 集合城市
allied city planning 联合城市规划
allied company 联营公司;联号(业务经营上的联系)
allied organization 类似组织
allied reflexes 联合反射
allied subject 相关学科
alligation 混合法;和均性
alligation alternate method 混合交叉法
alligator 皮带钩;水陆两用坦克;水陆两用车;胶责扣;木材拉坡机;辊式压渣机;齿销(木结构用);齿键;鳄式压轧机;鳄式碎石机;鳄式破碎机;鳄口压排渣机;鳄口工具;颚式碎石机;颚式破碎机;表面鼓起
alligator bonnet 碎石机罩
alligator clip 弹簧夹;鳄鱼嘴夹;鳄鱼夹;鳄式夹;鳄齿夹
alligator closed-ring dowel 齿轮形暗销
alligator crack 鳄裂;龟裂
alligator cracking 路面龟裂
alligator effect 粗面现象;橙皮效应;鳄皮现象
alligator forceps 鳄牙钳
alligator hide 鳄皮斑点
alligator-hide appearance 鳄皮状;鳄牙状
alligator-hide crack 龟裂;鳄裂
alligatoring (轧制表面的)起桔面;网状裂纹;涂膜皱皮;涂膜鳄纹;皱裂;龟裂;鳄嘴裂口;鳄纹;鳄皮裂纹;表面裂痕
alligatoring fishmounthing 鳄口裂纹
alligatoring honeycomb cracking 龟裂
alligatoring lacquer 皱纹漆;裂纹漆;鳄纹漆
alligatoring ring 齿环
alligator jaw 碎石机颚板;桩靴;桩脚;鳄鱼牙式桩靴
alligator lace 机械扣钩
alligator oil 鳄鱼油
alligator pliers 长钳
alligator point 桩套管的颚口;颚口
alligator rolling set 鳄式压轧机
alligator shears 鳄式切断机;鳄牙剪;鳄口剪切机
Alligator sinensis 扬子鳄
alligator-skin 鳄鱼皮(状表面)
alligator skin effect 碎矿机表皮效应;压轧表皮效应
alligator-skin glaze 鳄皮釉

alligator squeezer 杠杆式弯曲机;鳄式压轧机
alligator-type bonnet 鳄式发动机罩
alligator wrench 管(子)扳手;斜口扳手;鳄式扳手;鳄齿板钳
Allihn condenser 阿连冷凝器
all-in aggregate 未筛分的天然骨料;同粒度集料;未筛(分)骨料;未经筛分的集料;统货集料;天然混合骨料;毛集料;毛骨料
all-in allowance 全部容差;全部公差
all-in ballast 未筛分碎石;未筛分道砟;统货碎石;统货道砟
all-in-bid 总报价;工程设计与施工投标
all-inclusive 全部的
all-inclusive deed of trust 全盘财产信托契据
all-inclusive income statement 总括式损益表
all-inclusive trust deed 全部财产信托契据;返回租赁
all-in contract 总包合同;整套承包合同;统包合同;全部承包合同;总体合同;总承包合同
all-in cost 总费用;总成本
all-inertial guidance 全惯性制导
all-inertial guidance system 全惯性导航系统
all-inertial navigation system 全惯性导航系统
all in fare 包括伙食等的票价
Alling grade scale 阿令粒级标准
all-in gravel 混合石子;混合砾石
all-in mix 拌成混凝土料
all-in-one cable 合用电缆
all-in-one computer 单体计算机
all-in-one pav(i)er 筑路康拜因;联合铺路机;联合筑路机
all-in-one pliers 家用钳
all-in-one program(me) 一揽子项目
all-in price 统包价格
all-in rate 全部利率;各费在内的费率
all-in resistance 全部投入电阻
all-in-service station 保养站
all-insulated 全绝缘的
all-insulated sheathed cable 全绝缘铠装电缆
all-insulated sheathed wiring 全绝缘铠装电线
all-insulated switch 全绝缘开关;防电击开关
all-insulated wiring 全绝缘电线
all integer problem 全整数问题
all integer programming 全整数规划
all-intertial guidance system 全惯性制导系统
all in the wind 太靠风
all-in use 总费用
all-invar cavity 全殷钢空腔谐振器
all investment 全部投资
all-iron pump 全铁泵
Allischalmers critical experiment facility 艾里斯夏曼临界实验设施
Allis-Chalmers grinding index 艾里斯—查尔默斯粉磨指数
Allison's anchor 爱立逊无杆锚
allite 铝铁土;富铝土
allite red 铁红;矾红
allitic 富铝的
allitic soil 富铝性土
allitization 铝铁土化(作用)
Allium sativum 大蒜
allivalite 橄榄钙长岩;橄长岩
all-jute ball 纯麻球
all kinds kept in stock 一应俱全
all-length anchorage 全长锚固
all-lens endoscope 全透镜内窥镜
all level sample 全级试样(取自各个水平位置);多点试样;各层位样品;多层样品
all-levels sampler 多层取样器
all-light(weight)-aggregate concrete 轻骨料混凝土;全轻骨料混凝土
all-liquid engine configuration 纯液体火箭发动机配置
all made fast 全部挽牢
all-magnetic tube 全磁射线管
all marine and war risks 一切海洋运输及战争险
all marine risks 一切海洋运输险
all mark 全穿孔;全标记
all-metal 全金属(的)
all-metal body 全金属车身
all-metal brake lining 全金属闸衬片
all-metal building 全金属建筑物
all-metal coach 全金属客车;全金属车厢
all-metal construction 全金属结构

all-metal facade 全金属立面
all-metal focal-plane shutter 全金属焦面快门
all-metal high capacity burner 全金属高容量烧嘴
all metallic construction 全金属建筑物
all-metal mirror 全金属反射镜
all-metal panel 全金属镶板
all-metal plant 全金属设备
all-metal tube 金属管
all-metal waveguide 全金属波导
all-mirror optics 全反射镜光学系统
all-movable stabilizer 可操纵稳定器
all-moving tail 全动式水平尾翼
all-night service 通宵检修;通宵服务
All Nipon Airways 全日本航空公司
all-number calling 全数码呼叫;全数字(号码)呼叫
Alloa wheeling 阿洛粗绒线
Alloa yarn 阿洛(粗)绒线
allobiocoenosium 外源生物群落
allocable fixed cost 可分摊固定成本;可分摊的固定费用
allocatable space 可分配空间
allocate 配置;配给;配备;指定;支配;分摊;分派;拨给;保留
allocate a device 分配一个设备
allocate cargo 配货
allocate cost 分摊成本
allocated berth assign a berth 指定泊位
allocated channel 指配信道;指配频道;分配信道
allocated cost 已分摊的成本;已分配成本;分配的成本
allocated cutting method 分区轮伐法
allocated expenses 分摊费用
allocated frequency band 分配频带
allocated impact zone 分配影响区
allocated investment credit 分配投资贷款
allocated locomotive 配属机车【铁】
allocated price 调拨价格
allocated quota 调拨配额;分摊配额;分配(的)配额
allocated repair system 包修制【铁】
allocated storage 已分配的存储器
allocate file 分配文件
allocate income ratio for enterprise 企业分配指标
allocate income ratio for local 地方分配指标
allocate income ratio for nation 国家分配指标
allocate income ratio for workers and staff members 职工分配指标
allocate investment credit 分配投资贷款
allocate materials and facility for a new construction program(me) 把材料和设备拨给新的建设项目
allocate memory 分配存储器
allocate profits 分配红利
allocate shares 分摊股分
allocate statement storage 存储分配语句
allocate storage 存储区分配;分配存储面;分配存储器
allocate tractor 配备拖拉机
allocating point of all-field 全野外布点
allocating point of network 网段布点
allocating subordinary investment 拔付所属投资
allocating task 分配任务
allocation 配置;定位置;地址分配;分摊;分配(配置);分配;分派(配置);分派;分布配置;部署
allocation algorithm 分配算法
allocation and loading program(me) 地址分配和装入程序;分配和装入程序
allocation area 采矿用地面积
allocation base 分摊基数;分摊基础;分配基数
allocation buffer 分配缓冲区
allocation built in function 内部函数的配置;内部函数的分配;分配内部函数
allocation criterion 分配标准
allocation effect 分配效果
allocation error 分配误差
allocation factor 分配因数
allocation for test 分配测试
allocation function 配置职能
allocation index 地址索引
allocation map 土地分配图
allocation market 配额市场
allocation mode 分配模式
allocation model 分配模型
allocation of berths 泊位安排
allocation of burden 负荷分配

allocation of charges 费用的分配
allocation of cost 成本分配;分摊成本;费用分摊;工程费用分摊
allocation of cost to revenue 成本对收益的分配
allocation of cost variance 成本差异分配
allocation of expenditures 费用的分摊
allocation of expenses 费用的分摊
allocation of frequency 频率分配
allocation of funds 利base分配
allocation of fundsallocation of profits 资金分配
allocation of hardware resource 硬件资源分配
allocation of industry 工业布局
allocation of investment 投资分摊;投资分配
allocation of items 项目分配
allocation of labo(u)r force 劳动力调配
allocation of landownership and land-use right 土地权属
allocation of materials 物资分配;物质分配
allocation of parameters 参数的分配
allocation of production capacities 生产容量的分派
allocation of profits 利润分配
allocation of raw materials 调配原材料
allocation of real devices 实际设备的分配
allocation of residual expenses 费用余款分摊
allocation of responsibility 责任分担
allocation of risks 风险的分担
allocation of sample 样本分配
allocation of service department 服务部门成本的分配
allocation of service department cost 服务部门费用分摊
allocation of shares 股份分配
allocation of ships 配船;船舶调度
allocation of storage 存储器的分配
allocation of temporary storage 工作单元区的分配
allocation of water 水的分配
allocation plan 配置图;地址分配方案;分配方案;布局规划
allocation principles 分配原则
allocation problem 分配问题
allocation process 分派过程
allocation proportion 分配比例;分派比例
allocation routine 分配例行程序;分配程序
allocation scheme 配线图
allocation sheet 分配表
allocation signatures 指定符号差;分配符号差
allocations of special drawing rights 特别提款权的分配额
allocation system 轮乘制【铁】;分配制;包乘制【铁】
allocation theory 分配理论;分派理论
allocative decision 分配方案
allocative effects 配置效率
allocative efficiency 配置效率;分配效率
allocator 配给器;连续编辑用具;分配者;分配器;分配符;分配程序;分变程序
allocatur 费用查定证
allochem 外源化学沉积;异化粒
allochem grain texture 颗粒结构
allochem grain texture carbonate rocks 碳酸盐岩颗粒结构
allochemical limestone 异化粒灰岩
allochemical metamorphism 他化学变质(作用);异化变质(作用)
allochemical rock 异化岩
allochetite 霞辉二长斑岩
allochite 绿帘石
allocholestanoe 粪甾烷酮
allochonous terrane 外来地体
allochroic silicagel 变色硅胶
allochroism 易变色
allochroite 粒榴石
allochromatic 带假色的;掺杂色的;非本色的;变色的
allochromatic colo(u)r 他色;假色;掺质色
allochromatic crystal 掺质色晶体
allochromatic photoconductor 掺质光电导体
allochromatic photoconductor crystal 掺质光电导晶体
allochromatism 他色性;掺质色性
allochromy 磷光效应;荧光再放射
allochthon(e) 外来岩块;外来体;移置体
allochthonic groundwater 外来地下水
allochthonous 外来的

allochthonous block 异地岩块
allochthonous breccia 异地角砾岩
allochthonous coal 异地生成煤
allochthonous deposit 移置;移积
allochthonous ground-water 移置地下水
allochthonous limestone 异地灰岩
allochthonous material 外来物质
allochthonous river 外源河
allochthonous soil 运生土;移积土
allochthonous stream 外源河
allochthonous theory (煤形成的)移置学说;漂移学说
allochthony 异地堆积
allochthony coal 异地生成煤
alloclasits 斜硫砷钴矿
alloclastic breccia 火山碎屑角砾岩
allocortex 不均皮质
allocs 艾洛陶瓷
allodapic limestone 异地浊积灰岩
allodapic turbidite 异地浊积岩
allodial system 自主地产制
allodium 自主地产
allogene 外源物;他生物
allogenic 他生的
allogenic agglomerate 异源集块岩
allogenic agglomerate lava 异源集块熔岩
allogenic breccia lava 异源角砾熔岩
allogenic change 他生演变
allogenic mineral 他生矿物
allogenic process 异源过程
allogenic river 外源河
allogenic stream 外源河
allogenic succession 异发演替
allogenic transformation 异源转化
allogenic tuff 异源凝灰岩
allogenic tuff lava 异源凝灰熔岩
allogenic volcanic breccia 异源火山角砾岩
allogenic welded agglomerate 异源熔结集块岩
allogenic welded breccia 异源熔结角砾岩
allogenic welded tuff 异源熔结凝灰岩
allogonite 磷铍钙石
allogyric birefringence 圆偏振双折射;异旋双折射
allo-isomerism 立体异构(现象)
allolysogeny 异溶原性
allomer 异分同晶质
allomeric 同晶体
allomerism 同质异晶;同型;异质同晶性;异质同晶(现象);异分同晶性
allometamorphism 他变质作用
allometry 异速生长
allomicrite 异地泥晶灰岩
allomorph 同质异象变体
allomorphism 同质异晶现象;同质异晶假象;同质假象
allomorphite 贝状重晶石
allomorphosis 不对称发育
allomorphous 同质异晶的
allomorphs 同源异构包体
allonge 票据的黏度;票据背书附条;黏单;附笺;补笺;背书贴单
allopatric 分布不重叠的
allopatric speciation 分区物种形成
allopatric species 异域物种
allophanamide 缩二脲
allophanate 脲基甲酸盐
allophane 水铝英石
allophanite 水铝英石;铝英岩
allophase metamorphism 异相变质作用
allophone 音值
alloplasm 异质
alloprene 氯化橡胶
all optic(al) communication 全光通信
all optic(al) computer 全光系计算机
all-organic compound chemicals 全有机化合化学物
all organisms of this type 所有这类生物体
all-or-none embargo 全有或全无的限制
all-or-none toxic effect 有或无毒作用
all-or-nothing method 全有或全无分配法
all-or-nothing piece 开启杆;报时开关片
all-or-nothing relay 全有或全无继电器
alloskarn 外来夕卡岩
allosteric control 别位调节
allosteric modulator 别构调节物

allosteric transition 别构转变
allosterism 别位现象
all other perils 一切其他风险
allothigene 外源物;他生物
allothigenic 他生的
allothimorph 他形体
allotment 配给;调配;建矿用地;采矿用地;份额;分配(数);分配额拨款;分配额;分派;拨款
allotment advice 拨款通知
allotment area 分配面积
allotment control system 分配管理制度
allotment for building and outfitting 建筑安装工程拨款
allotment garden 供分配使用的园地;私用园地
allotment issued 已拨款项
allotment ledger 预算分配表
allotment letter 分派红利通知书
allotment money 分配金
allotment note 分付请求书
allotment of berth 派定泊位
allotment of labo(u)r 配工
allotment of shares 分配股份
allotment period 分配期
allotment plan 土地分配图
allotment profit 分配盈利
allotment ratio 分配比率
allotment system 分派制
allotment warden 分配土地管理人
allotrioblast 他形变晶
allotriomorphic 上反角;他形的
allotriomorphic crystal 不整形晶体
allotriomorphic granular 似花岗石的;人造的;不整形晶体的;人造花岗岩石面;不整形粒状
allotriomorphic granular texture 他形晶粒状结构
allotriomorphic mineral 他形矿物
allotrope 同素异形体
allotrophic lake 他养型湖
allotropic 同素异形的
allotropic modification 同素异形体
allotropism 同素异形(现象)
allotropy 同素异形性;同素异形(现象)
allotted land 放领地(作价售于需求者之公用土地)
allotter 分配者;分配器
allotter relay 分配器继电器;分配继电器
allotype 他型;异型;别模标本
all-out 全面的
all-out alert 全面戒备(机场)
all-out development plan 全面开发计划;综合开发计划
all-out development program(me) 全面开发计划
all outstanding advance 全部尚未归还的放款
allover 循环图案;重复图案
all over colo(u)r 印花色浆
all-over design 连续花纹
all-over glaze 满地釉;盖地釉
allow 容许
allowable 容许的;可容许的;许用
allowable acceleration 容许加速度(值)
allowable amplitude 容许振幅
allowable approximation 容许近似值
allowable axle loading 轴上许用荷载
allowable background level 容许本底水平
allowable bearing capacity 容许支承力;容许承载力;许用承载
allowable bearing capacity of foundation 地基容许承载力
allowable bearing capacity of pile 桩的容许承载力
allowable bearing force of soil 土壤的允许承载压力
allowable bearing load 轴承容许荷载
allowable bearing of foundation 地基容许承载力
allowable bearing of single pile 单桩容许承载力
allowable bearing pressure 容许支承压力;容许承压力;许用承载力
allowable bearing stress 容许支承应力;容许承压应力
allowable bearing stress for concrete foundation 混凝土基础的容许承载应力
allowable bearing unit stress 容许单位支承应力
allowable bearing value 容许承载力;容许承载值
allowable bearing value of foundation soil 地基土的容许承载力值
allowable bending radius 容许弯曲曲率半径

allowable bond stress 容许黏着应力;容许结合应力
allowable bond unit stress 容许单位结合应力
allowable bucklling stress 容许扭曲应力
allowable cabin load 容许客舱容量
allowable cargo load 容许装货量
allowable catch 容许的捕捉量
allowable compression ratio 容许压缩比
allowable compressive stress 容许压应力
allowable compressive unit stress 容许单位压应力
allowable concentration 容许浓度;允许浓度
allowable concentration index 容许浓度指数
allowable concentration limit 容许浓度限度
allowable concentration of carbon monixide 一氧化碳允许浓度
allowable concentration of waste water disposal 允许污水排放浓度
allowable constant current 容许恒定电流;容许持续电流
allowable contact stress 容许接触应力;容许接触应力
allowable contact unit stress 容许单位接触应力
allowable contamination level 容许污染水平
allowable coordinates 容许坐标
allowable crack width 容许裂缝宽度
allowable crushing stress 容许压碎应力
allowable crushing unit stress 容许单位压碎应力
allowable current 容许电流
allowable cut 容许采伐量
allowable daily intake 日容许摄入量
allowable defect 容许缺陷
allowable deflection 容许弯沉;容许挠度
allowable deformation of ground 地基容许变形
allowable degree 容许程度
allowable density of harmful gas 有害气体容许浓度
allowable depth 容许深度
allowable design 容许设计
allowable design stress 设计许用应力;许用设计应力
allowable deviation 容许偏差;允许偏差;允许公差
allowable deviation of grouting hole 注浆孔允许偏斜率
allowable differential settlement 容许差异沉降量
allowable dimension variation 允许尺寸偏差
allowable discharge of waste water 允许污水排放量
allowable displacement 容许位移
allowable drift 容许位移
allowable earth pressure 容许土压力
allowable elongation 容许伸长度
allowable emission 容许排放
allowable emission rate 容许排污率;允许排放速率
allowable error 容许误差;允许故障率
allowable expenses 免税支出
allowable exposure 容许照射;容许曝光
allowable extinction coefficient 容许吸光系数
allowable factor of safety 容许安全系数
allowable failure probability 容许破坏概率
allowable flexural stress 容许弯曲应力;许用弯曲应力
allowable flexural unit stress 容许弯曲单位应力
allowable flooding 容许浸水深度
allowable flooding depth 允许淹没深度
allowable flutter 容许颤动
allowable force 许用加载
allowable for deformation 预留变形量
allowable for shrinkage 容许收缩量;许可收缩(量)
allowable gradient of bed base 基底允许坡度
allowable gross weight 容许毛重
allowable ground-level concentration 地面容许浓度
allowable heat limit 容许热限度
allowable heeling 容许倾斜度
allowable heeling moment 许用倾侧力矩
allowable impact load 容许冲击荷载
allowable leakage 容许泄漏量;容许渗漏;允许漏失量
allowable limit 容许限(度);容许界限;容许极限;允许极限
allowable limited amount of dangerous cargo 小量危险品限额
allowable load 容许载重;容许荷载;容许负荷;允许负荷;许可荷载;许可负载;允许荷载

allowable load capacity of wagon 货车容许装载量
allowable loading 容许负载;容许负荷
allowable mass load 容许质量负荷
allowable mating maximum power 允许配带最大功率
allowable maximum of structure 结构物容许最大值
allowable maximum settlement 容许最大沉降量
allowable maximum torque 容许最大扭矩
allowable maximum water level 允许的最高水位
allowable mining modulus of groundwater 地下水允许开采模数
allowable mining yield 允许开采量
allowable noise exposure time 容许噪声接触时间;容许噪声暴露时间;容许曝噪时间
allowable non-scouring velocity 容许不冲刷流速
allowable nutrient loading 容许营养负荷
allowable offset 允许偏心
allowable offset in circumferential joints 圆周焊缝的允许偏移
allowable offset in longitudinal joints 纵向焊缝的允许偏移
allowable opening diameter 允许开孔直径
allowable over-depth 容许超深;允许超深
allowable parameter 容许参数
allowable permanent twist 许用永久扭曲
allowable pile bearing load 容许桩荷载;桩的容许荷载;单桩容许荷载
allowable pollution load of river 河流容许污染负荷量
allowable pressure 容许压强;容许压力;许用压力
allowable pressure difference 容许压(力)差
allowable pressure drop 允许压降
allowable pressure of serous material 浆液容许压力
allowable pull out capacity 容许固着力
allowable pulsary torque 许用脉动扭矩
allowable quantity of pollutant discharged 容许污染物排放量;污染物允许排放量
allowable radioactivity 允许放射性
allowable range 容许范围
allowable rebound deflection 容许回弹弯沉(值)
allowable rebound deflection of surface pavement 铺面表面容许弯沉
allowable relative deformation 容许相对变形
allowable residue limit 容许残留量
allowable resistance 容许阻力
allowable sample size 容许试样量
allowable seepage criterion 容许渗漏准则
allowable settlement 容许沉降(量);许可沉降;地基容许变形量
allowable shear 容许剪力
allowable shearing stress 容许剪应力
allowable shear stress 许用切应力
allowable smoke concentration 容许烟雾浓度
allowable soil bearing strength 土壤的允许耐力
allowable soil pressure 容许土压力;土壤的允许承载压力;土的容许压力
allowable space between running cars 容许通行车间隔
allowable speed of non-silt 容许不淤流速
allowable storey deformation 容许层间变形
allowable strength 容许强度;许可强度
allowable stress 容许应力;许用应力
allowable stress design 容许应力设计
allowable stress design method 容许应力设计法
allowable stress method 容许应力法
allowable subgroup 容许子群
allowable subject moving speed 可容许物体运动速度
allowable subject moving velocity 可容许物体运动速度
allowable suct lift 允许吸上真空高度
allowable switching current 容许合闸电流
allowable temperature 容许温度;许用温度
allowable temperature difference 允许温差
allowable temperature differential 容许温差
allowable tensile stress 容许拉应力;许用拉应力
allowable tensile stress at design temperature 设计温度下的许用拉应力
allowable tension 容许拉力
allowable test pressure 允许试验压力
allowable time for ensuring safety in loading and unloading of radioactive goods 装卸放射线货物容许作业时间
allowable tolerance 容许公差
allowable tolerance of filament voltage 灯丝电压容许公差
allowable torque 容许转矩;容许扭矩
allowable torsional stress 容许扭应力
allowable torsional vibration stress 许用扭振应力
allowable transition 容许跃迁
allowable transmissible bandwidth 容许传输带宽
allowable twisting stress 容许扭转应力
allowable twisting unit stress 容许单位扭转应力
allowable tyre load 轮胎容许荷载
allowable unit pressure 容许单位压力
allowable unit soil pressure 容许单位土压力
allowable unit stress 容许单位应力
allowable unit stress for bending 容许单位弯曲应力
allowable unit stress for buckling 容许单位扭曲应力
allowable unit stress for compression 容许单位压应力
allowable unit stress for torsion 容许单位扭应力
allowable use 容许利用度
allowable value 容许值;许用值;允许值
allowable value of movement of ground surface in mined-up region 采空区的容许地表变形值
allowable value of static(al) water pressure 允许静水压力值
allowable variation 容许变异
allowable variation range of objective variable 目标变量的允许变化范围
allowable variation range of value variable 价值变量的允许变化范围
allowable velocity 容许流速;容许速度
allowable velocity of flow of rock 岩石允许流速
allowable vertical pull strength of single pile 单桩竖向抗拔容许承载力
allowable vibrating acceleration 容许振动加速度
allowable vibration acceleration 容许振动加速度
allowable voltage 容许电压
allowable water leakage 容许渗水量
allowable wear 容许磨损;许用磨损
allowable working load 容许使用荷载
allowable working pressure 容许工作压力;许用工作压力
allowable working stress 许用工作应力
allowably relative deformation 容许相对变形
allow a credit 给予信贷
allowance 容许误差修正量;容许误差;容许量;容限;容隙;容差;配给品;剩油量;留量;宽放时间;津贴;减让;加工余量;免税项目;免税额;允许量;裕度;余量;余地;折让;公差;纯劳动时间;补助;补贴
allowance above nominal size 尺寸上偏差
allowance and rebate 折让及回扣
allowance bearing capacity of subgrade 地耐力
allowance below nominal size 下差;尺寸下偏差
allowance clause 宽容条款
allowance compression ratio 容许压缩比
allowance concentration index 容许浓度指数
allowance curve 公差配合曲线
allowance daily intake 日容许摄入量
allowance deflection 容许挠度
allowance deviation 容许偏差
allowance draft 允许吃水
allowance error 容许误差;许用误差;允许误差
allowance factor 公差因素
allowance for abrasion 磨蚀裕量
allowance for amortization 摊销备抵
allowance for awkward work 笨重工种的津贴
allowance for bad debts 呆账备抵;备抵坏账
allowance for breakage 破损定额
allowance for building workers 建筑工人津贴
allowance for bulking 湿胀容许量
allowance for camber 预变形曲率;(模型的)假曲率;反挠度
allowance for collection cost 备抵收回成本
allowance for contingency 不可预见费
allowance for contraction 收缩留量
allowance for corrosion 腐蚀裕量
allowance for current 流压差修整量
allowance for curvature 曲率容许量
allowance for damage 货损折扣;损耗折扣;亏损

补贴
allowance for depreciation 折旧提存;折旧(率);备抵折旧
allowance for discount available 折扣备抵
allowance for erosion 由于侵蚀所需要的裕度
allowance for exchange fluctuation 外汇牌价允许波动范围
allowance for extra-heavy work 繁重的工作津贴;繁重的工种津贴
allowance for finish 完工留量;精加工余量;加工余量;光制留量
allowance for impact 冲击预留量;冲击荷载的预留值;冲击容许量
allowance for leeway 风压差修正量
allowance for length of moment 弯矩长度的裕度
allowance for lost time 时间损失津贴
allowance for machining 机械加工裕度;机械加工余量;机械加工留量;机械加工公差
allowance for moments at support 支点弯矩的裕度;支点弯矩的预留量
allowance for overvaluation of branch inventory 分店存货高估价价准备
allowance for price declines 降价备抵
allowance for profit 利润补贴
allowance for sale 销货补偿
allowance for sales discounts 备抵销货折扣
allowance for sedimentation 备淤深度
allowance for settlement 容许沉降(量);沉降留量
allowance for shear 剪切预留量
allowance for shrinkage 收缩允许量;收缩余量;收缩留量;许可收缩量
allowance for siltation 备淤深度
allowance for the loss 损失补偿
allowance for trim 截齐允许量
allowance for uncollectible accounts receivable 备抵坏账
allowance for under-pit work 矿山井下津贴
allowance for uplift 浮托力容许量;土墩隆起容许量
allowance for vacancy 空房率;空房比(一定地区中未使用房屋数占总住宅数的比率)
allowance for width 宽度留量
allowance for work in high temperature 高温津贴
allowance for work stoppage 停工津贴
allowance gap 允许通车间隔
allowance in kind 实物津贴
allowance level of concentration 容许浓度水平
allowance limit 容许极限;容许范围
allowance of credit 提供信贷
allowance of frequency 频率分配
allowance of overhead 管理费
allowance of track irregularity 容许轨道不平整度
allowance over-height 容许超高
allowance over-width 容许超宽;允许超宽
allowance payment 津贴支付
allowance production 允许开采量;允许产量
allowance range 容许范围
allowance reserve 折让准备;折让储备金
allowance smoke concentration 允许烟雾浓度
allowance strength 容许强度;允许强度
allowance stress 允许应力
allowance temperature 允许温度
allowance test 容差试验;公差配合试验
allowance ticket 废票转换券
allowance tolerance 允许公差;公差
allowance unit 公差单位
allowance variation 允许偏差
allowance velocity 容许速度;容许流速
allowance vibration 允许振动
allow a worker act as a cadre 以工代干
allow clearance 容许间隙;许用余隙
allowed band 允许能带;公差带
allowed by law 法律允许的
allowed cost 容许成本
allowed defect 容许疵病
allowed depreciation 折旧范围额度
allowed energy band 容许能带
allowed energy level 容许能级
allowed equivalent crack length 允许当量裂纹尺寸
allowed frequency 可用频率;许用频率;准用频率
allowed indoor fluctuation of temperature and relative humidity 室内温湿度允许波动范围
allowed orbit 容许轨道
allowed predecessor 容许前置符
allowed pumping amount 允许抽水量

allowed pumping drawdown depth 允许抽水降深
allowed pumping tie 允许抽水时间
allowed space for expansion 容许胀隙
allowed spectrum 容许光谱
allowed spectrum shape 容许能谱形状
allowed successor 容许后继者
allowed time 容许时间;完成工作的规定时间;预定完工标准时间;放宽时间
allowed transition 容许转变;容许跃迁
allowed value 容许值
allow for an eccentric set 消除测站偏心
allow for grinding 磨加工裕量
allow for wind 风压差修正量
allowing for extension 扩充余地
allow temporary credit 通融资金
allow to approach the humping signal 允许预推信号
alloy 铝硅合金;合金
alloyage 炼制合金
alloy amalgam 合金汞齐
alloy analysis 合金分析
alloy bar 合金棒
alloy casting 合金铸造;合金铸件
alloy cast-iron 合金铸件;铸铁合金
alloy cast steel 合金铸钢
alloy cladding 包合金
alloy constructional steel 合金结构钢
alloy content 合金含量
alloy design 合金设计
alloy-diffused photo transistor 合金扩散型光电晶体管
alloy diffusion technology 合金扩散工艺
alloy diode 合金二极管
alloyd oil 掺和油
alloy(drill)bit 合金钻头
alloy drill steel 合金钢钎杆
alloy ductile iron roll 合金球铁轧辊
alloyed grain roll 合金纹理轧辊
alloyed metal 合金化的金属
alloyed nickel steel 镍合金钢
alloyed pig iron 合金生铁
alloyed steel 合金钢
alloyed steel bit 合金钢车刀
alloyed steel roll 合金钢轧辊
alloyed steel strip 合金钢带
alloyed tool steel 合金工具钢
alloyed zinc 锌合金
alloy electrode 合金焊条;合金钢电焊条
alloy fiber 合金型熔合纤维
alloy firm 合金薄膜
alloy for conductor 导线用合金
alloy for die casting 模铸合金
alloy frame 合金镜架
alloy gearing 合金齿轮传动
alloy grade niobium 合金级铌
alloy hardening 合金硬化
alloy head-hardened rail 合金淬火轨
alloy housing 合金外壳
alloying 熔结;熔合;合金化(处理)
alloying addition 添加合金;合金添加剂
alloying agent 添加合金元素;合金元素;合金添加剂;合金剂
alloying component 合金成分
alloying composition 合金成分
alloying constituent 合金剂
alloying element 合金元素;合金元件;成合金元素
alloying ingredient 合金元素
alloying pellet 合金球
alloying process 熔合过程
alloying technique 冶炼技术;合金技术
alloying technology 合金工艺
alloy inlay 合金嵌体
alloy iron 合金铁;合金铁
alloy iron roll 合金铸铁轧辊
alloy junction 合金结
alloy junction diode 合金结二极管
alloy latch 合金弹簧锁
alloy layer 合金层
alloy-lead pipe 合金铅管
alloy-lead tube 合金铅管
alloy live center[centre] 合金活顶尖
alloy nuclear fuel 合金核燃料
alloy of high percentage 高合金
alloy of low percentage 低合金

alloy of mercury 汞合金
alloy phase 合金相
alloy photocathode 合金光电阴极
alloy pig 合金生铁;合金锭
alloy pin porcelain tooth 合金钉瓷牙
alloy pipe 合金钢管;合金管
alloy plate 合金电镀层
alloy plating 合金电镀;镀合金
alloy powder 合金粉末
alloy rail steel 合金钢轨钢
alloy regrowth junction 合金再生结
alloy reinforcing steel 特种钢筋;合金钢筋
alloys 金属互化物
alloy sheet metal 合金薄板
alloy shutter 合金百叶窗
alloy solder 合金焊料
alloy steel 合金钢
alloy steel anchor 合金钢锚固件
alloy steel bearing 合金钢支承
alloy steel bit 合金钢钻头
alloy steel blade 合金钢刀片
alloy steel casting 合金钢铸件
alloy steel cutter 合金钢刀具
alloy steel gas welding rod 合金钢气焊条
alloy steel pipe 合金钢管
alloy steel pipe clamp 合金钢制管卡
alloy steel plate 合金钢板
alloy steel powder 合金钢粉末
alloy steel rail 合金钢轨
alloy steel sheet 合金钢片
alloy steel strip 合金钢带
alloy steel tube 合金钢管
alloy steel tubing 合金钢管
alloy steel wire 合金钢丝
alloy sticks 铝合金棒材;合金棒材
alloy structural steel 合金结构(型)钢
alloy structure 合金结构
alloy system 合金系
alloy throw-away bit 不重磨合金钻头
alloy tin 合金锡
alloy tool steel bit 合金工具钢车刀;合金钢钻头;合金钢刀头
alloy transfer efficiency 合金元素过渡系数
alloy treated steel 处理过的合金钢
alloy treatment 合金处理
alloy-type 合金型的
alloy watch band 合金表带
alloy wire 合金线;合金钢丝
alloy-wire drawer 合金丝拉制机
alloy zener junction 合金齐纳结
all painting 全船油漆
all-paper laminate 纸质层压贴面板;层压构件;层压板
all parallel analog(ue)/digital converter 全并行模/数转换器
all-pass 全通
all-pass filter 全通滤波器;移相滤波器
all-pass lattice 全通格子网络
all-pass network 全通网络【计】
all pass transducer 全通转换器;全通换能器
all-paved 全铺的
all-paved crossing 全铺式交叉(口)
all-paved intersection 全铺式交叉(口)
all-pebble mill 全砾磨机
all personnel labo(u)r productivity 全员劳动生产率
all-phase transaction tax 全阶段交易税
all-plastic bearing structure 全塑承载结构
all-plastic cold-glazed wall coating 全塑冷釉墙面涂料
all-plastic road tanker 全塑油车;全塑水车
all-plastic sandwich panel 全塑夹层板;全塑多层板
all-plastic structure 全塑结构
all-port open 全部油口打开
Allport oven 奥尔波特瓶式窑
all-ports block (阀的)中立关闭
all-position electrode 全位置焊条;适用于各种情况的焊条
all-position pipeline automatic arc welding 管道全位置自动弧焊
all-position welding 全位置焊接
all possible combination 全组合
all price 各费在内的价格
all province 全省

all-pull mill 斯蒂克轧机;张力轧机
all-purpose 万能的;适宜各种用途;适宜于多种用途的;全能的;通用的;多功能的
all-purpose adhesive 万能黏合剂;万能胶
all-purpose balance sheet 通用资产负债表
all-purpose bank 经营全面业务的银行
all-purpose bit 通用钻头;通用刀头
all-purpose calculator 通用计算器
all-purpose camera 通用摄影机
all-purpose capillary viscosimeter 通用毛细管黏度计
all-purpose cargo ship 通用货船
all-purpose chopper mill 通用切碎碾磨机
all-purpose communication system 通用通信系统;全能通信系统
all-purpose computer 全用途电子计算机;通用计算机
all-purpose concrete aggregate 通用混凝土骨料
all-purpose detector 通用检测器
all-purpose engine oil 通用机油
all-purpose excavator 通用式挖土机;通用式挖掘机;多功能挖土机;多功能挖掘机
all-purpose explosive 通用炸药
all-purpose extinguisher 通用灭火器
all-purpose financial statement 通用决算表;通用财务报表
all-purpose furnace black 通用炉黑
all-purpose gasoline 通用汽油
all-purpose grease 通用润滑剂
all-purpose instrument 万能仪表;通用仪器
all-purpose insulation 通用绝热隔声材料;通用绝热材料;通用隔声材料
all-purpose knife 通用刀
all-purpose language 通用语言
all-purpose launch 通用艇
all-purpose loader 万能装料机;万能(式)装载机;通用装载机
all-purpose machine 万能工具机
all-purpose machinery 通用机械
all-purpose mineral mix 通用矿物质混合料
all-purpose nozzle 万能喷嘴
all-purpose primer 通用底漆;万能底漆;多功能底漆
all-purpose protective coating (material) 万能保护涂层材料
all-purpose road 多功能道路;混合交通(道)路
all-purpose room 多功能住房;多用途房间
all purposes 装卸共用时间
all-purpose sand 统一砂
all-purpose saw 通用式锯
all-purpose shovel-crane 通用式挖土机;通用挖掘机
all-purpose tractor 万能拖拉机;通用拖拉机
all-purpose vacuum cleaner 万能真空吸尘器
all-purpose vegetable cutter 多能切菜机
all-purpose wrench 多用扳钳
all quarters 四方
all-radiant furnace 辐射炉
all red 全红灯
all red period 全红信号时间
all red signal 全红信号
all-relay 全继电器式的
all-relay interlocking 继电式电气集中联锁;全部继电联锁
all-relay interlocking for medium station 中站电气集中联锁
all-relay interlocking for shunting area 平面调车电气集中【铁】
all-relay interlocking for way-station 小站电气集中联锁
all risks 全险;一切险;协会全险条款;综合险
all risks account receivable 全风险应收账款
all risks clauses 一切险条款
all risks covers 全险保(险)单;一切险保单
all risks coverhulls 船舶一切险
all risks erection insurance 建筑工程一切性保险;建筑工程全损性保险;建筑工程的综合性保险;建筑工程保险
all risks insurance 全能保险;一切险;全险
all risks policy 一切险保单;综合险保单
all-rock breakwater 全石料防波堤;堆石防波堤
all room air conditioning 全室空调
all-round 全向;全面的;共计;多功能的
all-round absorbent 万向吸附剂;万能吸附剂
all-round advance 一齐涨价
all-round anchorage 全面锚碇;万能锚碇;多方向锚碇
all-round automation 全盘自动化
all-round cab (in) 全视野座室;全视野座舱
all round computer center 综合计算机中心
all-round conveyer 万能输送机
all-round coverage 全视野;全视界
all-round crane 旋臂(式)起重机
all-round decline 一齐跌价
all-round development 综合开发;全面发展
all-round frame 框架肋骨;加强肋骨;船体主肋骨
all-round improvement in the eco-environment 生态环境综合治理
all-round industrial system 完整工业体系
all-round investigation 全面调查
all-round light 环照灯
all-round looking antenna 环视天线
all-round looking radar 全景雷达;环视雷达
all-round looking scanner 环视扫描器;环扫描器
all-round mechanization 全盘机械化
all-round pattern 满地图案
all-round port 综合性港口;多用途港口
all-round pressure 周围压力
all-round price 全面价格;包括一切费用的价格
all-round property 全面性能;综合性能
all-round reversing gear 全周倒车杆装置;全逆转装置
all-round social progress 社会全面进步
all-round the clock 昼夜不停的
all-round the horizon 水平四周
all-round turret 全向转动炮塔;全向六角转台
all-round vessel 多用途船(舶)
all-round view 全视野;全视界
all-round view cab (in) 全视野座舱;全视界座舱
all-round view windscreen 全视野风挡玻璃
all-round visibility 周围能见度
all-round vision 全视野(景像);全视界
all-rowlock 丁顺交界层立砌法
all-rowlock wall 通斗墙;全空斗墙;空斗墙
all-rubber 全用橡胶制成
all-rubber article 全胶制品
all savers' certificate 免税储蓄证书
all second hand drums 全部旧桶
allseed 多子植物
all serial analog/digital converter 全串行模/数转换器
all-service gas mask 防毒面具
all ship calls 全呼(对船)
all-silent body 不透声车身
all-silent gearbox 无声齿轮箱
all-silk goods 全丝织物
all-skid landing gear 全橇式起落架
all-sky camera 全天空照相机
all sliming 全微粒化;全泥浆化;全矿泥化;矿石细磨
all space continuation 全空间延拓
all-squared 款项全清;全部结清
all stages of the investigation 所有阶段勘察
all-stations address 广播地址
all-steel 全钢制的;全钢
all-steel box car 全钢棚车
all-steel carriage 全钢客车
all-steel chain 全钢链条
all-steel coach body 全钢薄壳车体
all-steel construction 全钢建筑物;全钢结构
all-steel cord 钢丝帘线
all-steel curtain wall 全钢墙幕;全钢幕墙
all-steel leader 全钢导杆;全钢导管
all-steel stair (case) 全钢楼梯
all-steel threshing machine 钢制脱粒机
all-steel tipper 全钢自动倾卸车;全钢倾卸车
all-steel wagon 全钢货车
all streamer fathom accuracy 全缆定深精度
all-stretcher bond 全顺砖砌合
all suitable water 完全适用的水
all-supersonic 纯亚音速的;纯超音速的
all synchro 全自动同步机;全部同步
all-synchromesh gearbox 通用同步齿轮箱
all taxable individuals 全体纳税人
all-terrain aircraft 任意场地起落飞机
all-terrain vehicle 越野载重车;高级试验车;全地形交通工具;全地形车;高通过性车辆
all text before specified point 上文
all the shipping afloat 在航船总数
all the year round 终年;全年
all-the-year-round housing 全年住宅供给
all-timber door 木门
all-time 专职的
all-time high 最高记录
all-time low 最低记录
all told 全船船员;全部载重量;总共
all trades and professions 各行各业
all-transistor 全晶体管
all-transistor camera 全晶体管照相机
all-transistor circuitry 全晶体管电路
all-transistor computer 全晶体管计算机
all treatments 所有的处理
all trunks busy 全(部)占线;中继线全忙
all-turned furniture 镟木家具
alluaudite 磷锰钠石
allume 明矾
allumen 锌铝合金
all-up weight 全重(量);总重(量);空中总重量;满载重量;最大重量;最大载载;总荷载
allure 通道;吸引(力)
alluring colo(u)ration 诱惑色
all usual risks 一切通常险
alluvial 冲积土(壤);冲积的
alluvial apron 冲积坪;山麓冲刷扇;山麓冲积扇;冲积扇;冲积裙;冰水冲积平原
alluvial area 冲积区
alluvial bench 山麓冲积洲;冲积台地
alluvial brown soil 棕色冲积土
alluvial channel 冲积(平原)水道;冲积河槽
alluvial channel bed 冲积性河床;冲积河床
alluvial clay 冲积黏土;二次黏土
alluvial coast 冲积河岸;冲积海岸
alluvial community 冲积带群落
alluvial cone 锥状冲积层;冲积锥;冲积圆锥体;冲出锥
alluvial cone sand trap 冲积锥沙体圈闭
alluvial dam 冲积阻塞;冲积堤;冲积坝
alluvial deltaic cycle 冲积三角洲旋回
alluvial deposit 冲积土;漂砂矿床;冲积物;冲积床;冲积层;冲积沉积
alluvial/diluvial fan 冲积/洪积扇
alluvial district 冲积区;冲击地区
alluvial dolina 冲积落水洞
Alluvial epoch 冲积期;冲积世
alluvial facies 冲积相;冲积扇相
alluvial fan 扇状冲积砂;扇状冲积扇;扇形冲积;冲积扇;冲击扇
alluvial fan at river mouth 河口冲积扇
alluvial fan deposit 冲积扇沉积
alluvial fan environment 冲积扇环境
alluvial fan prograding sequence 冲积扇推进层序
alluvial fan sedimentation model 冲积扇沉积模式
alluvial fan sequence 冲积扇层序
alluvial fan shrinking sequence 冲积扇退缩层序
alluvial fan system 冲积扇体系
alluvial filter 冲积式过滤器
alluvial flat 河漫滩;冲积平原;冲积平地
alluvial flood plain 冲积漫滩
alluvial formation 冲积地层
alluvial gold 砂金;冲积砂金
alluvial gravel 河卵石;冲积砾石;冲积砾石
alluvial groundwater 冲积层地下水
alluvial horizon 冲积层
alluvial island 冲积岛
alluvial land 冲积土(壤);冲积地
alluvial laterite 冲积铁矾土;冲积红壤;冲积红土
alluvial layer 冲积层
alluvial layer filter 冲积层式过滤器
alluvial loam 冲积壤土
alluvial mantle rock 冲积风化表层岩;风化(表皮)岩
alluvial meadow 冲积草原
alluvial meadow soil 冲积性草甸土;冲积草土
alluvial meander 冲积型曲流;冲积曲折河段;冲积河曲
alluvial mining 淘金;冲积河床开采
alluvial ore deposit 冲积矿床
alluvial placer 冲积砂金;冲积砂矿(床)
alluvial plain 冲积平原;沉积平原
alluvial plain coast 冲积平原海岸
alluvial platform 冲积台地
alluvial process 冲积过程
alluvial-proluvial fan 冲洪积扇
alluvial-proluvial plain 冲洪积平原
alluvial reach 冲积河段
alluvial ridge 冲积脊

alluvial river 冲积河(流)
alluvial river bed 冲积河床
alluvial river bench 冲积河流阶地
alluvial sand 冲积砂
alluvial sandy soil 冲积砂土
alluvial sediment 冲积沉积物
alluvial series 冲积系
alluvial slope 冲积坡
alluvial slope spring 冲积坡泉;扇坡泉
alluvial soil 淤积土;冲积土(壤)
alluvial sorting 冲积分选
alluvial splay 河滩淤积;河渠扩展淤积
alluvial stone 冲积石块
alluvial stream 冲积河流
alluvial surface 冲积面
alluvial talus 冲积山麓堆积
alluvial terrace 冲积台地;冲积阶地
alluvial tract 河口冲积区;冲积区;冲积河段
alluvial valley 冲积河谷;冲积谷
alluvial value 沉积矿回收值
alluvial water 冲积层水
alluviated 冲积物覆盖的
alluviating bank 淤积带
alluviation 冲积(作用)
alluviation of delta 三角洲淤积;三角洲淤长
alluvideserta 冲积荒漠群落
alluvion 冲积成的新地;海浪冲刷;海岸冲积地;冲积层;沙滩;泛滥;冲积物;波浪击岸;波浪冲击
alluvious 冲积的
alluvium 淤积层;冲积土(壤);冲积层
alluvium accumulation 河滩堆积
alluvium anchor 冲积层碇泊
alluvium anomaly 冲积物异常
alluvium deposit 冲积层
alluvium grouting 冲积层灌浆
alluvium period 第四纪
alluvium splay 河渠扩展淤积
all-veneer construction 全镶板构造;全胶合板结构;全单板结构
all-veneer plywood 全单板胶合板;普通胶合板
all vertical pile 全直桩
all vertical pile of large diameter 大直径全直桩
all walk phase 全步行相位;步行专用相位
all water system 全水系统;全水方式
all water transportation 全程水运
all-watt motor 全瓦特电动机;全功率电动机
all-wave 全波
all wave antenna 全波天线
all-wave band 全波段
all-wave oscillator 全波带振荡器
all-wave receiver 全波(段)接收机;全波带接收机
all-wave rectification 全波整流
all-way airfield 多跑道机场
all-way guidance 全程制导;全程导航
all-way navigation 全程导航
always fuse 起爆引信;起爆雷管;起爆信管
all weather 适宜各种气候;全天候
all-weather aircraft 全天候飞机
all-weather airfield 全天候(飞)机场
all-weather base 全天候路基
all-weather capability 全天候工作能力
all-weather coat 全天候保护层
all-weather equipment 全天候设备
all-weather ga(u)ging device 全天候测器;全天候测计
all-weather guidance system 全天候制导系统
all-weather highway 全天候公路;全天候道路
all-weather landing system 全天候着陆系统
all-weather lifeboat 全天候救生艇
all-weather measuring capacity 全天候测量能力
all-weather navigation 全天候导航
all-weather operations 全天候飞行
all-weather patrol boat 全天候巡逻艇
all-weather plane 全天候飞机
all-weather plywood 全气候胶合板;全天候胶合板
all-weather port 全天候航空港;全天候港口;不冻港
all-weather quay 全天候码头
all-weather road 常年通车路;全天候道路;晴雨通车路;晴雨通车道
all-weather runway 全天候机场跑道;全天候跑道
all-weather sea target acquisition system 全天候海上目标搜索系统
all-weather service 常年通车;全天候空运;晴雨通车
all-weather slatted blind 全天候百叶窗
all-weather stability 常年稳定性;常年气候稳定性
all-weather strip 全气候带材;全气候板条
all-weather subbase 全天候道路基层;全天候基层
all-weather telephone set 全天候电话机
all-weather terrestrial rangefinder 全天候测距仪
all-weather visibility 全天候能见度;全天候良好视度
all-weather wharf 全天候码头
all-welded 全焊接的
all-welded boiler 全焊接锅炉
all-welded construction 全焊(接)结构
all-welded floating dock 全焊接式浮船坞
all-welded hull 全焊接船体
all-welded ship 全焊接船舶
all-welded steel bridge 全焊钢桥
all-welded steel girder or truss 全焊接钢梁
all-welded steel structure 全焊接钢结构
all-welded storage 全焊接储料仓
all-welded structure 全焊接结构
all-weld hull 全部焊接船体
all-weld-metal tension test 全焊金属的焊缝拉力试验
all-weld metal test specimen 全熔质试样
all-wet cement plant 全湿法水泥厂;全湿工艺水泥厂
all-wheel(bull)dozer 全轮驱动的推土机;全轮推土机
all-wheel compressed-air brake 全轮气压制动;全轮气压刹车
all-wheel drive 全轮驱动
all-wheel drive excavator 全轮驱动挖土机
all-wheel drive vehicle 全轮驱动汽车
all-wheel dump car 全轮自卸汽车
all-wheel dumper 全轮自卸车
all-wheel planetary wheel drive 全轮行星齿轮驱动
all-wheel pneumatic brake 全轮气压制动;全轮气压刹车
all-wheel steer(ing) 全轮转向;全轮操纵;全轮驾驶
all-wheel tipping lorry 全轮自卸汽车
all-wheel tractor 全轮牵引车;全轮拖拉机
all-winch mooring system 全绞车锚泊系统
all-wood 全木的
all-wood door 木门
all-work 全工种;全线开工;全部工程
all working time saved both ends 装卸节省时间的合计
all wrench 通用扳手
ally 联合
all-year navigation 全年通航
allyl acetate 乙酸烯丙酯
allylacetone 烯丙基丙酮
allyl alcoh 丙烯醇
allylalcohol 烯丙醇
allyl alcohol ester 烯丙醇酯
allyl alcohol polymer 烯丙醇聚合物
allyl aldehyde 丙烯醛
allylamine 烯丙胺
allylation 烯丙基化
allyl benzene 苯丙烯
allyl benzoate 苯酸烯丙酯
allyl bromide 烯丙基溴
allyl carbinol 乙烯基乙醇
allyl chloride 烯丙基氯
allyl cyanide 烯丙基氰
allylene 丙炔
allylene oxide 丙炔化氧
allyl ester polymer 烯丙酯聚合物
allyl ether 烯丙基醚
allyl glycidyl ether 烯丙基缩水甘油醚
allyl group 烯丙基
allyl halide 烯丙基卤
allylic 烯丙基的
allylic halides type 烯丙卤型
allylic rearrangement 烯丙重排作用
allylic substitution 烯丙取代
allylin 甘油(单)烯丙基醚
allyl iodide 碘丙烯
allyl mercaptan 烯丙硫醇
allyl plastic 烯丙树脂;烯丙塑料
allyl polymer 烯丙基聚合物
allyl propionate 丙酸烯丙酯
allyl resin 烯丙树脂;烯丙基树脂;丙烯树脂
allyl resins 烯丙基树脂类
allyl starch 烯丙基淀粉
allyl sulfhydrate 烯丙基硫醇
allyl sulfide 烯丙基化硫
allyl sulphide 烯丙基硫醚
all-zero signal 全零信号
all-zero state 全零状态
almacantar 等高圈
Almag 阿尔马格铝镁合金
Almag alloy 阿尔马格合金
almagra 深红赭色
Almalec 阿尔马莱克
almamor 讲台(桌);读经台(犹太教堂)
almanac 日历;讲台(桌);天文历;历书;年历
almanac of economy 经济年鉴
almanac time 历书时
almanac yearbook 年鉴
al-manazil 亚纳吉尔
almandine 铁铝榴石;贵榴石
almandine-biotite schist 铁铝榴石黑云母片岩
almandine-muscovite-schist 铁铝榴石白云母片岩
almandine spinel 红尖晶石;贵榴石尖晶石
almandite 铁铝榴石;贵榴石
almary 储[贮]藏室;圣餐厨具;库房;存放圣餐用具的壁龛;壁橱
Almasil 阿尔马赛高硬度铝镁硅合金
Almasil alloy 铝镁硅合金
Almasilium 铝镁硅合金;阿尔马西林铝镁硅合金
almatol 阿马托混合炸药
almehrabh (清真寺院面向麦加的那道墙内的)壁龛
Almelec 阿尔梅莱克铝基合金
Almelec conductor 铝镁合金导线
almemar 读经台(犹太教堂)
Almen extreme pressure lubricant testing machine 阿尔门极压润滑剂试验机
Almen friction machine 阿尔门摩擦试验机
Almen machine 阿尔门试验机
Almen-Ny-lander test 阿尔门—奈兰德试验
Almen tester 阿尔门摩擦试验机
almery 储[贮]藏室;圣餐厨具;库房;存放圣餐用具的壁龛;壁橱;教堂放圣器的木橱;圣餐具橱
Almeter 阿尔纤维长度试验仪
almighty 全能的
almighty adhesive 万能黏合剂
Alminal 阿尔米纳尔铝硅系耐蚀合金
alminum ornamental section 铝装饰型材
alminum wallpaper 铝墙板
Almiria ware 阿尔美利亚陶器
Almit 铝钎料
almond(神像画中的)光环;杏仁
almond green 杏绿
Almon distributed lag model 阿尔蒙分布滞后模型
almond leaved willow 毛柳
almond oil 杏仁油;扁桃油;扁桃仁油
almond pink 浅粉红色;杏红色
almond(shaped)cut 椭圆形切割
Almon lag 阿尔蒙滞后
Almon lag scheme 阿尔蒙滞后形式
almonry 救济品分发处;施赈所
almonry house 贫民所;救济所
Almon scheme of polynomial lag 阿尔蒙多项式滞后形式
almost admissible decision rule 几乎容许决策规则
almost-atoll island 准环礁岛
almost cylindric(al) shell 有初变形的圆柱面壳;初变形圆柱面壳
almost degenerate amplifier 近简并放大器
almost everywhere convergent 几乎处处收敛
almost everywhere divergent 几乎处处发散
almost full 基本有保证;基本满足
almost hemispheric(al)dome 初变形半球穹壳
almost linear 迨线性
almost periodic(al) 迨周期的
almost periodic(al) function 殆周期函数;迨周期函数
almost plain 准平原
almost ring 准环
almost spherical 近似于球形状的;近似球形的
almost triangular matrix 拟三角矩阵;准三角矩阵
Almquist unit 阿姆奎斯特单位
alms 救济品
alms-box (教堂、庙宇的)捐献箱;救济金箱
almshouse 养老院;救济院;贫民院;施赈所
almucantar 高度方位仪;地平纬圈;等高圈
almucantur 地平纬圈;等高圈
Almulit 阿尔穆莱特产炉用硅砖
Alneon 锌铜铝合金;阿尔乃昂铝锌合金;阿尔内昂

铝锌铜合金
Alni 阿尔尼铁镍铝(永磁)合金;铝镍合金
Alnic alloy 阿尔尼克铝镍沉淀硬化型永磁合金
Alnico 阿尔尼铁镍铝钴永磁合金;铝钴合金;铁铝镍钴金;铝镍钴永磁合金;铝镍钴合金
alnico alloy 铝镍钴合金
Alnico alumin(i)um-nickel-cobalt alloy 阿尔尼科铝镍钴合金
alnico magnet 铝镍钴磁钢
Alnico permanent magnet alloy 阿尔尼科铁镍铝钴系永磁合金
Alnico sintered magnet 阿尔尼科烧结铝镍钴磁铁
alnico V magnet 铝镍钴 V 形磁铁
Alniflex solution 阿尔尼夫莱克斯溶液
Alnisi 阿尔尼西铁镍铝硅合金
alnoite 黄长煌斑岩;橄黄岩
alnusenone 赤杨酮
Alnus japonica 赤杨
Alnus maritima 东北赤杨
Alnus tinctoria 辽东枫木
alochol aldehyde 醇醛
alodin 冰片丹
Alodine process 阿洛戴法(一种铝表面处理化学氧化法)
aloe fiber 芦荟属纤维
aloe hemp 龙舌兰属大麻
aloes wood 芦荟木;沉香木
aloe wood oil 东方沉香油;沉香木油
aloft work 高空作业
aloisite 结灰石
aloite 水矾石
Alon 阿龙(乙酰化高强度黏胶短纤维)
alone 单独
alone signal unit 独立信号单元
along fault strike 沿断裂带走向
along-grade 顺坡
along order 进行的次序
alongshore 河岸的;沿岸的
alongshore current 顺岸流;沿岸(水)流;沿岸海流;滨海潮流
alongshore drift 沿岸漂沙;滨海漂积物
alongshore feature 岸线特征;海岸特征;海岸要素
alongshoreman 港口工人
alongshore transport 沿岸输移;沿岸输沙
alongshore variation 顺岸变化
alongshore water movement 沿岸水流
alongshore wave 顺岸波
alongshore wind 海岸风;沿岸风
alongside 停靠;靠岸
alongside against the tide 顶流靠泊
alongside a pier 靠码头
alongside a ship 并靠某船
alongside a vessel 泊靠它船边
alongside bill of lading 船边交货提单
alongside date 开始装船日期
alongside delivery 航边交货;船边提货;船边交货
alongside port 靠左舷
alongside starboard 靠右舷
alongside tally 航边理货;船边理货
alongside the wreck to benefit passing sling 打千斤顶洞
alongside towing 并列拖带
alongside wharf 靠码头
along the coast 沿着海岸线
along the edge 沿边
along the fiber 顺纹
along the grain 顺木纹
along the river 沿河
along the road 沿路
along the strike of abundant groundwater zone 沿富水带走向
along the strike of aquifer 沿含水层走向
along track 沿径
along-track error 沿轨误差
along-track resolution 纵向分辨(率)
alopecia areata 斑形脱发;斑秃
aloring 教堂护墙;教堂女儿墙
a lossing proposition 亏本生意
a lot of 大批;大量的
a lot of fissures 裂缝多
a lowering of temperature below the freezing point 下降到水点以下的温度
aloxide bearing 刚玉轴承
aloxite 熔融氧化铝;人造刚玉;铝砂

aloxite tube 铝砂管;铝砂管
Aloyco 阿罗伊科铁镍铬系耐蚀合金
alp 高山草地
alpaca 阿尔帕卡
alpage 高山夏牧场
Alpaka 镍白铜
Alpakka 阿尔帕卡锌白铜
alpax 阿派合金;铝硅合金;硅铝明合金
alp climate 高山气候
alpenglow 高山辉
Alperm 高磁导率合金;阿尔帕姆高磁导率铁铝合金
Alpert bakable valve 全金属耐烘烤阀
alpert-ionization ga(u)ge 阿尔珀特电离计
Alpeth cable 聚乙烯绝缘铝芯电缆;阿尔贝斯电缆
alpha 阿尔法
alpha activation 阿尔法活化
alpha-active condition 阿尔法放射性条件
alpha activity 阿尔法放射性
alpha and omega 收尾;始终;全体
alpha anomalous scattering 阿尔法射线的反常散射
alpha-beta brass 两相黄铜
alphabetic addressing 字母选址;字母定址;字母编址
alphabetic arrangement 按字母顺序排列
alphabetic code 字母代码
alphabetic coding 字母编码
alphabetic digit 字母数字数码
alphabetic list 按字母排列的索引
alphabetic numeric 字母数字的
alphabetic puncher 字母穿孔机
alphabetic sorting 按字母分类
alphabetic string 字母行;字母串
alphabetic tube 字母数字显示管
alphabetic variant 初等变式
alphabet laser 多掺激光器
alphabet of lines 线型
alphabet table 字母表
alpha brass 单相黄铜;阿尔法黄铜
Alpha brass alloy 阿尔法黄铜合金
alpha bronze 阿尔法青铜
Alpha card meter model 阿尔法卡仪型号
Alpha card reader 阿尔法射线卡片阅读器
Alphacca 贯索四
alphacellulose 阿尔法纤维素
alpha chalcocite 蓝辉铜矿
alpha code 字母代码;字母编码
alpha coefficient 阿尔法系数
alpha counter 阿尔法计数
alpha-cristobalite 阿尔法一方英石
alpha decay 阿尔法衰变
alpha detection 阿尔法射线探测
alpha display 字母数字显示
alpha diversity 小生境多样性;阿尔法变化
alpha emitter 阿尔法放射体
alpha energy loss 阿尔法放射线的能量损失
alpha factor 阿尔法系数
alpha factor in wastewater 污染的阿发因子
alphageometric 字图图解方式
alphageometric graphic 阿尔法几何图形
alpha gypsum 阿尔法石膏
alpha heat releasing light method 阿尔法热释光法
alpha-Laval centrifuge 连续式转体离心机
alphameric conversion code 字母数字转换码
alphameric field 字母数字信息组
alphameric graphic display 字母数字图像显示(器)
alphameric keyboard 字母数字键盘
alpha-meter buried reading method 阿尔法仪埋设读数法
alpha-meter model 阿尔法仪型号
alpha method of pile design 桩জ设计计算的方法
alpha monitor 阿尔法射线监测仪
alphanol 长链烷醇混合物
alphanu meric characters 字母数字字符
alphanumeric character set 字母数字字符集
alphanumeric code 字母数字(代)码
alphanumeric coded character set 字母数字字符集
alphanumeric digit 字母数字数(代)码
alphanumeric display(device) 字母数字显示器
alphanumeric fluorescent screen display equipment 荧光屏字符显示器
alphanumeric grid 位置指示格网;地图指示格网
alphanumeric information 字母数字信息
alphanumeric instruction 字母数字(型)指令

alphanumeric item 字母数字项目
alphanumeric literal 字母数字文字
alphanumeric terminals 字母数字式终端
alpha omega 阿尔法欧米伽
alpha order 阿尔法次幂
alpha particle 阿尔法粒子
alpha radioactivity 阿尔法放射性
alpha radioactivity measuring 阿尔法放射性测量
alpha radiography 放射摄影法
alpha ray 阿尔法射线
alpha-ray spectrometry 阿尔法射线谱法
alpha survey method 阿尔法测量法
alphatizing 渗铬;镀铬(钢材表面)
alpha-tridymite 阿尔法—鳞石英
alphatron ga(u)ge 管真空规
alpha value 阿尔法值
alphitite 岩粉土
Alphonsus 阿尔芬斯
Alpides 阿尔卑斯造山带
Alpine air-swept sieve 阿尔卑斯气吹式清洗筛
alpine-arctic community 高寒群落
alpine belt 高山地带
alpine climate 高山气候
Alpine crustal type 阿尔卑斯地壳类型
alpine cushion-like vegetation 高山垫状植被
alpine desert 高山荒漠
alpine diamond 黄铁矿
alpine disease 高山病
alpine ecology 高山生态学
alpine ecosystems 高山生态系统
alpine fir 高山冷杉
Alpine fold 阿尔卑斯(式)褶皱
alpine form 高山地形
alpine garden 奇石园;高山植物园;岩石庭园;高山花园
alpine geomorphology 高山地貌学
Alpine geosyncline 阿尔卑斯地槽【地】
Alpine glacier 阿尔卑斯型(高山)冰川
alpine glacier 山地冰川;高山冰川
alpine glow 高山辉
Alpine-Himalayan geosyncline system 阿尔卑斯—喜马拉雅地槽系【地】
alpine lake 高山湖(泊)
alpine landscape 山岳景观
alpine light 人工日光;紫外线
Alpine marine trough 阿尔卑斯海槽
alpine meadow 高山草甸;高山草地
alpine meadow soil 高山草甸土;高山草地土
Alpine movement 阿尔卑斯运动
Alpine orogen 阿尔卑斯造山带【地】
Alpine orogeny 阿尔卑斯造山运动【地】;阿尔卑斯式高山造山作用【地】
Alpine paleoclimate epoch 阿尔卑斯古气候期
alpine pasture 高山牧场
Alpine period 阿尔卑斯期【地】
alpine piedmont 山前地带;高山麓
alpine planation terrace 高山夷平阶地
alpine plant 高山植物
alpine region 高山区
Alpine regression 阿尔卑斯海退
alpine relief 高山地形;高山地貌
alpine road 山岭道路;山道;高山道路
alpine rose 杜鹃花属
alpine sack 登山背囊
alpine scenic spot 高山风景区
alpine scrub 高山放牧
alpine snowfield 高山雪原;高山雪地
alpine sparse vegetation 高山稀疏植被
alpine steppe soil 高山草原土
alpine-sun lamp 高山太阳灯
alpine survey 高山调查
alpine talus vegetation 高山稀疏植被
alpine tundra 高山冻原
alpine tundra landscape spot 高山草甸风景区
alpine type facies 阿尔卑斯式相
Alpine type(of) folding 阿尔卑斯式褶皱
Alpine type of ultramafic rock body 阿尔卑斯型超镁铁岩体
Alpine type ultramafic associations 阿尔卑斯型超铁镁岩组合
Alpine type veins 阿尔卑斯型矿脉
alpine valley 高山河谷
alpine vegetation 高山植被
Alpine white bark pine 美国白皮松

alpine zone 高山植物带;高山带
Alpinotype tectonics 阿尔卑斯(型)构造;阿尔卑斯式大地构造
Alplate process 铝锌法热镀锌
Alps 阿尔卑斯山脉
Alrak method 表面防蚀化学处理法(铝及铝合金)
alranite 羟水钒钼矿
Alray 阿尔雷铁铬镍耐热合金;镍铬铁耐热合金
already-issued bond 已发行债券
alrksite 硫碲铋铅矿
alsbachite 榴云细斑岩
Al secoo 干壁画
alseco painting 绘壁画;用胶水和蛋黄调颜料绘
Alsever's solution 奥尔塞溶液
alsex 铝赛克斯合金
alsgraffits painting 拉毛涂装法
Alsia 阿尔西阿铝硅合金
Alsifer 阿尔西菲铁硅铝(磁性)合金;铁硅铝合金;铝硅铁合金
alsifilm 铝硅片
alsimag 铝硅镁合金
Alsimin 铝硅铁合金;阿尔西明铁硅铝合金
Al-Si ratio 铝硅比
Alsiron 耐热耐酸铝铸铁;阿尔西隆耐热耐酸铝硅铸铁
Alsithermic alloy 阿尔西塞迈克合金
alsithermic reducing agent 铝硅热还原剂
alsnritolic acid 木油树酸
alstonia-leaf 灯台叶
alstonite 碳酸钙钡矿;三斜钡解石;钡霞石
Altai 阿尔泰山脉
Altaides 阿尔泰造山带
altaite 碲铅矿
altalet 小祭坛
Alta-mud 阿尔塔膨润土;阿尔塔泥;膨润土(商品名称)
altana 柱承式阳台
altar 梯壁;(坞壁的)台阶;祭坛;干船坞梯壁
altar baldachin 祭台龛室;祭坛华盖
altar carpet 圣坛地毡
altar ciborium 祭台龛室;祭坛华盖
altar course (侧壁有踏步的)地下船坞
altar facing 祭坛饰角;祭坛饰罩
altar fresco 祭坛壁画
altar front(al) 祭坛饰幕;圣坛前挂饰
Altar of Land and Grain 北京社稷坛
altar of repose 圣库;(罗马天主教堂中储存圣餐的)侧圣台;(罗马天主教堂中储存圣餐的)壁龛
altar of Zeus 宙斯祭坛
altar over sarcophagus 石棺上的神坛
altar piece 祭坛后面的装饰屏幕;祭坛上方及后面雕画饰物
altar platform 坞壁平台
altar rails 圣坛栏杆;祭坛四周栏杆
altar screen 圣台隔扇;祭坛屏风;祭坛上方及后面雕画装饰
altar shrine 祭坛神龛
altar slab 圣台顶上的石板;圣台台面
altar stair(case) 祭坛楼梯
altar steps 坞壁梯阶;干船坞台阶
altar stone 圣坛台
altar-tomb 纪念坛;坛式石棺;祭坛式墓
Altay Diwa region 阿尔泰地洼区【地】
Altay early Paleozoic subduction zone 阿尔泰早古生代俯冲带
Altay fold system 阿尔泰褶皱系
Altay geosyncline 阿尔泰地槽
Altay marine trough 阿尔泰海槽
Altay-north Tashan marine trough 阿尔泰—北塔山海槽
Altay tectonic segment 阿尔泰构造段
altazimuth (同时测量经度和方位角的)天体测量经纬仪;天体经纬仪;高度方位仪;地平经纬仪
altazimuth instrument 地平经纬仪
altazimuth mounting 地平装置
altazimuth reflector 地平式反射镜
altazimuth telescope 地平式望远镜;地平经纬望远镜
alter 涂改;改动;变动
alterability 可变(更)性
alterable 可变的
alter a design 修改设计
alterant 变质剂;变色剂
alteration 涂改;蚀变(作用)【地】;交互变化;更新;改造;改装
alteration coefficient of rock 岩石风化程度系数
alteration mineralogy 蚀变矿物学
alteration of building 房屋的改建
alteration of course 改变路线;改变航向;改变航线;改变方向
alteration of cross-section 截面改变
alteration of destination 目的港变更
alteration of mark 航标变动
alteration of name 换船名
alteration of policy 保险条款变动
alteration of port 改港
alteration of premise 房舍改建
alteration of private rights 民事权利的变更
alteration of strata 互层
alteration of works 工程变更
alteration on entries in the register 变更注册事项
alteration point 蚀变点
alteration product 蚀变产物;变质产物;变异产物
alteration rock 蚀变岩石
alterations in goods transport 运输变更
alteration switch 变换转移点;变换开关
alteration to a building 改建
alteration type of adjoining rock neat ore body 近矿围岩蚀变类型
alteration work 改造工程;改建工程
alterative 变质剂;变质的
alter course 转向;改变路线;改变航向;改变航线;改变方向
alter course to port 向左转向
alter display 更换显示
altered bill 经更改的票据
altered check 涂改的支票
altered design 变更设计
altered granite 变质花岗岩
altered mineral 蚀变矿物
altered rock 蚀变岩石
altered sample 蚀变样品
altered sand mineral 蚀变砂矿物
altered volcanic rocks 蚀变火山岩
alter error 变更误差;变更差(错)
alter image 改后图像;更新后图像
altering 变更
altering error 变更误差;变更错;变更差(错)
altering the thermal property of the ground 改变地温性质
alterite 蚀变砂矿物
alter mode 修改方式;变更方式
alternataly pinnate 互生羽状的
alternate 轮换;交替的;交互;互生的;更迭的;变更的;变的
alternate aerodrome 备用飞机场;备用机场;备降机场
alternate airfield 备用航空站;备用机场
alternate airport 备用着陆机场;备用机场;备用航空站
alternate angle 交替角度;交错角;错角
alternate area 替代区;交替区
alternate arm 相邻臂
alternate balance line 交替平衡线
alternate bar 交替沙坝;交错边滩;游荡性沙洲
alternate bay construction 间隔施工法;隔仓施工法(混凝土道路施工中的);隔段施工法;隔仓(指混凝土路面施工)
alternate bay construction method 建筑路面间隔施工法;建筑路面隔仓施工法
alternate bays (建筑混凝土路面的)隔仓
alternate beating and cooling 交替加热及冷却
alternate bending 重复弯曲;反复弯曲
alternate bending strength 反复弯曲强度;交变弯曲强度
alternate bid 变动标价;可变标价;交替标价;备用标价;取舍标价
alternate black-and-white blocks 黑白格图案
alternate blade cutter 双面刀盘
alternate block progressive(traffic)system 隔区段进行交通管制系统
alternate channel 替代信道;交替信道;变更信道;变更通道;备用信道
alternate channel interference 相隔信道干扰
alternate channel selectivity 交流频道灵敏度
alternate clera-strip system 交互带状皆伐式渐伐作业法
alternate climate 交替气候;交变气候
alternate code 备用代码
alternate concentration limit 交变浓度限度
alternate concrete by construction method 混凝土路面隔仓施工法
alternate copolymer 交替共聚物
alternate cost 交替成本
alternate course 选修课程;可选用的航道;可选用的道路;(地层中的)交互层;交变层
alternate crop 填闲作物
alternate cropping system 交替耕作系统
alternate curing 交替养护
alternate current[AC] 交流电
alternate current-battery power supply 混合电源
alternate current-battery power supply system 混合供电制
alternate current electromagnet for valve 阀用交流电磁铁
alternate deep and shallow 交替的深槽和浅水
alternate depth 共轭水深;共轭深度;交替水深;交替深度
alternate design 替代方案;设计比较方案;交互比较设计;比较设计;比较方案;备选设计方案;备选方案
alternate design basis 代用设计依据
alternate design layout 比较设计方案
alternate device 替代设备;备用装置;备份设备
alternate direction 交替方向
alternate disjunction 相间离开
alternate distributor 双触点配电器
alternated moving average method 反复移动平均法
alternate energy source 备择能源
alternate extended route 更替扩充路由;更替扩充的路径选择
alternate exterior angle 外角;外错角
alternate facility 备用设施
alternate file 更替文件
alternate firing 轮换烧成
alternate flashing light 交替闪光
alternate folding 交替折叠
alternate form 替换式
alternate freezing and thawing 交替冻融;冻融交替
alternate freezing thawing 交替冰融
alternate frequency 补充频率
alternate fuel 人造石油;替用燃料
alternate function 交替函数;交代函数
alternate function key 更替功能键;备用功能键
alternate furrow irrigation 轮流沟灌;交替沟灌;隔沟施灌
alternate gate 备用闸门
alternate governor 副理事
alternate grazing 交替放牧
alternate group flashing light 交替闪动光群
alternate harbo(u)r site 比较港址
alternate heading 交叉平巷
alternate host 轮换寄生
alternate humidity and temperature test 湿热交变试验
alternate immersion corrosion test 交替浸没(腐蚀)试验
alternate immersion test 反复腐蚀试验
alternate in a production 产生式中的替元
alternate index 辅助索引
alternate index cluster 辅助索引簇
alternate index entry 辅助索引表目
alternate index record 辅助索引记录
alternate index upgrade 辅助索引升级
alternate input library 替代输入库
alternate installation 备用装置
alternate interior angle 内错角
alternate joint 交错接缝;相错式接头(错接);错列接头;错列接缝;错缝接合;错缝;相互式接头
alternate key 交替关键字;辅助关键字;备用关键字
alternate key stroke 交替击键;交错击键法
alternate lake 交替湖
alternate landing ground 辅助飞机场;备用飞机场
alternate lane construction 路面交替浇筑;路面分块浇筑;交替车道行;隔车道施工法;隔仓施工法
alternate lay (钢丝左右交替捻合成钢绳的)交捻;交替编绳
alternate layers 交互层
alternate layout 比较设计(方案);平面布置比较

方案;另一种布置方式;比较布置方案;备用方案;备选方案
alternate layout of hole 另一种布孔方式
alternate leaves 互生叶
alternate library facility 更替程序库功能
alternate-line scanning 隔行扫描
alternate load(ing) 交替加荷;交替负荷;交替荷载;反复荷载;交变荷载;交变负载
alternate location 另一个位置;(道路选线中的)比较位置
alternate locations of route selection 比较路线
alternately hard and soft strata 软硬互层
alternately reversed load 交替反复荷载
alternately used 交替使用过的
alternate magnet 交置磁铁
alternate marker 交替路线指示标
alternate mark inversion signal 传号交替反转信号
alternate material 替用材料;交换材料;代用材料
alternate math package 交替数学软件包
alternate matrix 交错矩阵
alternate-matrix of environmental impact assessment 迭代矩阵环境影响评价法
alternate method 交错法
alternate motion 往复运动;交变运动
alternate movement 交变运动
alternate name 选用名;别名
alternate/no answer option 替换/无应答选择
alternate numbering 间隔编号
alternate one-way traffic 交替单向交通
alternate one-way traffic control 交替单向通行控制
alternate operation column 轮换操作塔
alternate operations 轮换操作
alternate optima 交替最优法
alternate order 二者择一的订单
alternate partial polarizer filter 交变部分偏振滤光器
alternate path 更替路径;更换通路;第二通路;并联通路
alternate path retry 更替通路再试;更替插入再试
alternate pathway theory 交替途径
alternate pattern 备选范围
alternate phyllotaxy 互生叶序
alternate picture dot 交替像素;交替像点
alternate plan 交替方案
alternate plant 备用装置
alternate planting 棋盘式栽植;间作;间植
alternate plow 双向犁
alternate power source 交替电源;交流电源;备用电源
alternate product 代制品
alternate production 轮番生产
alternate production methods 轮流生产方法
alternate progressive system 绿灯交替显示的推进式信号系统;交替式联动信号系统
alternate quartersawn 四分之一交错锯木
alternate record key clause 更替记录关键子句
alternate recovery 交替恢复
alternate return 交错返回
alternate route 迂回路由;更替通路;比较线(路);比较路线;备用路线
alternate routing 指定替代通路;更替通路的指定;更替通路;备用路由
alternate row driving pile 隔行打桩
alternate row mutching 隔行盖草
alternate scanning 交替扫描
alternate scour 交替冲刷
alternate sector 备份扇区
alternate sector cylinder 备份扇区柱面
alternate sector identifier 备份扇区标识符
alternate signal system 交通(信号)系统
alternate site 另一个地点;比较厂址
alternate solution 另一种办法
alternate spring 交替泉;间歇泉
alternate staff 替换路签
alternate stage 交替水位;交变级
alternate stages of flow 水流的共轭水位
alternate stages of open channel flow 明渠水流更迭段
alternate standard 替代标准
alternate station method 隔站观测法;隔站法
alternate strain 交替应变;交变应变;反复应变
alternate strata 互层;交替层
alternate strength 交变强度

alternate stress 交替应力;交变应力;重复应力;反复应力
alternate stress test 反复应力试验
alternate subcarrier 交替的副载波
alternate system (联动信号系统之一的)交替系统
alternate system residence device 替代系统常驻设备
alternate tape 备用带
alternate tension 交替拉力;交替接力
alternate test for variance heterogeneity 非齐性方差的替换检验
alternate timer 更替定时器
alternate tooth slot 交错齿槽
alternate track 更替磁道;备份磁道
alternate triples 三重替代路由
alternate valuation 比较估价
alternate vertical camera 交替垂直摄影机
alternate viewing 交替观察
alternate volume 更替卷宗
alternate web splice 间隔腹板拼接
alternate wetting and drying 干湿交替
alternate wire-feed system 变速送丝方法(焊接用语)
alternating 交替的;交错的;交变的;更迭的
alternating acceleration peak 交变加速度峰值
alternating and flashing(light) 互闪光
alternating axis of symmetry 更迭对称轴
alternating beds 互层
alternating bending 反复弯曲;交替弯曲;交替反向弯曲;交变弯曲
alternating bending test 交替弯曲试验;交变弯曲试验;反复弯曲试验;反复冷弯曲试验
alternating component 交流分量;交流成分;交变分量
alternating component of pressure 压力的交变分量;交变压力分量
alternating continuous current commutating machine 交流一直流转换机构;交流一直流变换装置
alternating coordination polymerization 交替配位聚合(作用)
alternating copolymer 交替共聚物;间聚物
alternating copolymerization 交替共聚合(作用);间聚
alternating coursed ashlar masonry 乱石琢石墙;乱石方石墙;交错层状琢石砌体
alternating curing 交替养护
alternating current 往复流;交流电流;交变电流;反复潮流
alternating current[AC] 交流电
alternating current adaptor 交流电压转换器
alternating current anodizing process 交流阳极氧化法
alternating current arc 交流电弧
alternating current arc generator 交流电弧发生器
alternating current arc source 交流电弧光源
alternating current arc welder 交流电弧焊机;交流电焊机
alternating current arc welding 交流电弧焊
alternating current arc welding machine 交流弧焊机
alternating current argon arc welding machine 交流氩弧焊机
alternating current automatic regulator 交流自动调压器
alternating current balancer 交流电压平衡装置
alternating current bias 交流偏压
alternating current biased recording 交流偏置录音法
alternating current biasing 交流偏磁
alternating current bias recording 交流偏磁记录
alternating current brake 交流制动器
alternating current bridge 交流电桥
alternating current capacity 交流载流能力
alternating current circuit 交流电路
alternating current circuit theory 交流电路理论
alternating current commutatorless and brushless motor 交流无刷无换向器电动机
alternating current commutator motor 交流换向器式电动机
alternating current component 交流分量;交流电流分量
alternating current computer 交流计算机
alternating current contactor 交流接触器

alternating current continuous waves 交流等幅波
alternating current control coil 交流控制线圈
alternating current converter 交流转换器;交流交换器;交流换流器
alternating current copper coated electrode 交流镀铜电极
alternating current counting code automatic block system 交流计数电码轨道电路
alternating current counting code track circuit 交流计数电码轨道电路
alternating current coupling 交流耦合
alternating current/direct current 交直流两用
alternating current/direct current arc welding machine 交直流电弧焊机
alternating current/direct current change over 交直流电源切换
alternating current/direct current inverter 交直流转换器;交直流逆变器
alternating current/direct current relay 交直流继电器
alternating current/direct current traction unit 交直流牵引装置
alternating current/direct current welding machine 交直流两用焊机
alternating current distribution 交流配电
alternating current dump 交流清除;交流断电
alternating current dynamo 交流发电机;交流电机
alternating current earth relay 交流接地继电器
alternating current electric(al) traction 交流电力牵引
alternating current electronic motor 离子变频式交流电动机
alternating current electronic tube 交流电子管
alternating current equipment 交流电设备
alternating current erase 交流擦除
alternating current erasing 交流清洗
alternating current excitation 交流励磁
alternating current exciter 交流励磁机
alternating current filter 交流滤波器
alternating current galvanometer 交流电流计
alternating current gas metal-arc welding process 交流熔化极气保护焊
alternating current gas tungsten arc welding 交流钨极气保护焊
alternating current generating set 交流发电机组
alternating current generator 交流发电机
alternating current impedance 交流阻抗
alternating current in connector 交流电源输入插座
alternating current induced polarization method 交流激发极化法
alternating current induction motor 交流感应电动机
alternating current input 交流输入
alternating current input jack 交流输入插孔
alternating current input socket 交流电源输入插口
alternating current interference 交流干扰
alternating current inverter 交流逆变器
alternating current locomotive 交流电力机车
alternating current machine 交流电机
alternating current magnet 交流磁铁
alternating current magnetic biasing 交流偏磁
alternating current main 交流干线;交流电源;交流电网
alternating current method 交变电流勘探法
alternating current motor 交流电机;交流电动机
alternating current moving iron rectangular instrument 交流电磁系矩形电表
alternating current moving iron round instrument 交流电磁系圆形电表
alternating current network 交流网络;交流电网
alternating current operation thermocouple 交流工作温差电偶
alternating current outlet 交流电插座
alternating current panel 交流盘
alternating current pantograph 交流受电弓
alternating current pattern 交变脉冲图形
alternating current pick-up 交流干扰
alternating current pilot relaying 交流辅助线继电保护
alternating current point machine 交流电动转辙机
alternating current polarography 交流极谱法
alternating current portion 交流部分
alternating current potentiometer 交流电位计
alternating current power cord 交流电源线

alternating current power input jack 交流电源输入插口
alternating current power supply 交流电源
alternating current power supply panel 交流电源屏
alternating current power supply system 交流供电制
alternating current pumped gas 交流激发气体
alternating current rectifier 交流电整流器
alternating current relay 交流继电器
alternating-current resistance 高频电阻
alternating current series (wound) motor 交流串绕电动机；交流串激电动机
alternating current servo speed indicating generator set 交流伺服测速发电机组
alternating current set 交流发电机组
alternating current shunt generator 交流并激发电机
alternating current shunt (wound) motor 交流分激电动机；交流并绕电动机；交流并激电动机
alternating current signal 交流信号
alternating current signal(l)ing 交流振铃；交流电信令；交流传信
alternating current socket 交流电插座；交流插座
alternating current spark source 交流火花光源
alternating current stabilized voltage supply 交流稳压电源
alternating current stabilizer 交流稳压器
alternating current static converter 交流静态换流器
alternating current static switch 交流静态开关
alternating current synchronous generator 交流同步发电机
alternating current system 交流式
alternating current to digital converter 交流数字转换器
alternating current torque motor 交流陀螺修正电动机；交流扭矩电动机
alternating current track circuit 交流轨道电路
alternating current track circuit with two element two position relay 交流二元二位轨道电路
alternating current transformer 交流变压器
alternating current transmission 交流传输
alternating current tripping system 交流断路方式
alternating current tube 交流管
alternating current two element two position relay 交流二元二位继电器
alternating current two speed elevator 二级速度交流电梯
alternating current voltage 交流电压
alternating current voltage convertor 交流电压变流器
alternating current voltage selector 交流电压选择器
alternating current welding machine 交流电焊机
alternating current welding set 交流焊接变压器；交流焊机
alternating damp heat atmosphere 交变湿热环境
alternating deformation 交替形变
alternating deposit 交替沉积
alternating device 交替设备；交替处理机；交换器；分流器
alternating direction implicit method 交替方向稳定法
alternating direction iterative 交替方向迭代法
alternating direction method 交替方向法；交替方法
alternating discharge 周期性放电
alternating displacement 交变位移
alternating double bond 更迭双键
alternating double filter 交替双重滤池
alternating double filtration 交替运行二级过滤法；交替双重过滤
alternating edge train 交错四边序列
alternating electric(al) wave 交替电波
alternating electromotive force 交变电动势
alternating fault 交错断层
alternating field 交变范围；交变场
alternating field demagnetization 交变磁场退磁
alternating field demagnetizer 交变磁场退磁仪
alternating fixed and flashing 定闪互光
alternating fixed and flashing light 交替定夹闪光；互定闪光
alternating fixed and group flashing 定联闪互光
alternating fixed and group flashing light 交替定夹联闪光；互定联闪光
alternating flashing(light) 互闪光；彩色闪光灯；变色闪光灯
alternating flexural stress 交变弯曲应力
alternating flexure test 交变弯曲试验；反复弯曲试验；反复弯曲实验
alternating flow hollow spool valve 交变流空心滑阀（风动工具用）
alternating flux 交变通量
alternating force 交变力
alternating form 交错形式
alternating freezing and thawing 交替冻融
alternating-gradient accelerator 交变梯度加速器（磁场）
alternating-gradient cyclotron 交变梯度聚焦回旋加速器
alternating-gradient doublet 交变梯度双合透镜
alternating-gradient focused accelerator 交变梯度聚焦加速器
alternating-gradient focused section 交变梯度聚焦节
alternating-gradient lattice 交变梯度格子结构
alternating-gradient linac 交变梯度直线加速器
alternating-gradient quadrupole 交变梯度四极子
alternating-gradient ring 交变梯度环
alternating-gradient synchrotron 交变梯度同步加速器
alternating graph 交错图形；交错图
alternating group flashing(light) 互联闪光；联闪互光；交替联闪光；彩色闪光灯组
alternating group occulting 联明暗互光；互联明暗光
alternating group occulting light 彩色闪光灯组
alternating humidity atmosphere 交变湿度环境
alternating immersion test 干湿交替浸渍耐腐蚀试验
alternating impact 交变冲击
alternating impact machine 交变振动试验机
alternating impact test 反复冲击试验
alternating induced polarization 交流激发极化法
alternating lake 交替湖
alternating lateral loading 交变侧向荷载
alternating layer structure 混层结构
alternating light 交替光；明暗交替灯光；明暗互光；变光灯
alternating light method 两光交换法
alternating load(ing) 交变载重；交替荷载；交变负荷；交变荷载；交变负载；反复荷载；变动荷载；变动负荷
alternating load test 反复负荷试验
alternating logic 交替逻辑
alternating long and two short dashes line 二点划线
alternating magnetic field 交变磁场
alternating matrix 交错矩阵
alternating motion 往复运动；交变运动
alternating notch bending test 刻槽反复弯曲试验；刻槽反复弯曲试验；刻槽交变弯曲试验；刻槽交变挠曲试验；切口交变挠曲试验
alternating notch flexure test 切口交变挠曲试验
alternating occulting(light) 互明暗光；交替联明暗光
alternating operating system 交替的操作系统；供选用的操作系统
alternating operation 交错运算
alternating oscillation motion technique 交替跟踪观测法
alternating path 交替路(线)；交错路(线)
alternating personality 交替人格
alternating photocurrent 交变光电流
alternating potential 交变电势
alternating pressure 交变压力
alternating pressure method 频压法
alternating pressure process 加压真空交替法
alternating process 交替法
alternating project 比较方案
alternating pulse 交替脉冲
alternating quantity 交变量
alternating renewal process 交替可更新过程
alternating repeated loading 交变重复荷载
alternating return trap 双间回水盒；交替回水弯管；反复回水湾；双间回水箱
alternating run 交变路段
alternating sand bank 交替边滩
alternating search 交替搜索
alternating sequence 交互层理【地】
alternating series 交错序列；交错级数；交叉级数
alternating shift 轮替班
alternating shock load 交变冲击荷载
alternating side flat 交错边滩
alternating stills 交替操作蒸馏釜
alternating strain 交替应变；交变应变；反复应变
alternating strain amplitude 交替应变幅度
alternating stress 交替应力；交变应力；重复应力；反复应力
alternating stress amplitude 交变应力幅度
alternating stress intensity 交变应力强度
alternating stress test 交变应力试验
alternating structure 聚合物交替结构
alternating sum 交错总和
alternating tidal current 往复潮流
alternating torque 交变转矩
alternating torsional test 交变扭曲试验
alternating torsion fatigue test 交变扭曲疲劳试验
alternating two-stage filtration 交替运行二级过滤法；交替两级过滤
alternating-variable search 交替寻优法
alternating voltage 交流电压；交变电压
alternating wetting and drying test 干湿交替试验
alternation 强度交变；交替（工作）；交互变化；交变工作；修改工程；变更
alternation activity 择一活动
alternation and remedy work 更改和修补工作
alternation charges 变更卸货港附加费
alternation cost 更改费用
alternation in the loading plan 变更装货计划
alternation law 更迭律
alternation mark inversion violation 极性交替破坏
alternation number 交变周期数
alternation of agricultural and forest crops 林粮间作；农林间作
alternation of area 面积变形
alternation of beds 交互层；岩层交互层；岩层交互变化；层的交互变化
alternation of cross-section 截面改变；横断面交替
alternation of culture 轮作；轮换栽培；轮耕
alternation of freezing and thawing 冻融循环；冰融交替
alternation of generation 世代交替
alternation of mud barges 泥驳轮换
alternation of multiplicities law 重复性变化定律
alternation of rocks 岩石变异
alternation of ship's ETA 到港时间有改变
alternation of signs 符号的变更
alternation of strata 交互层；岩层交互层；岩层交互变化；层的交互变化
alternation of stress 应力变化；应力交替
alternation of trim 纵倾修正
alternation of wetting and drying 干湿循环；干湿交替
alternation policy 可供选择的方针政策
alternation-pseudomorphism 蚀变假像
alternation sheet 更正通知书
alternation survey 改装检验
alternation switch 交替开关；换接开关；转换开关；更变开关；变更开关
alternation theory 替代学说
alternation to plan 变更计划
alternation wind 反复变向风
alternation work 修改工程
alternation zone 蚀变带
alternative 替换（物）；另一种；可替换的；可供选择或采用的方法；可供选择或采用的方案；可供选择的；交换物；选择性的；方案；比较方案；备选方案；备选的
alternative accounting methods 可供选择的会计方法
alternative agriculture 替代农业
alternative arrangement 间隔布置
alternative attribute 择一属性
alternative biasing 交替偏压
alternative bid 可供选择的(投)标价；比较(投)标价
alternative box 交替框
alternative budget 代用预算
alternative carriage 交替装载
alternative check list method 替换措施表列法
alternative choice 备选方案选择；替代方案选择
alternative clause 可供选择的条款

alternative closure scheme 截流比较方案
alternative code 交替码
alternative coefficient 交错系数
alternative cofferdam scheme 围堰比较方案
alternative comparison 方案比较
alternative construction 变化组织
alternative construction method 比较施工方法;备选施工方法
alternative control 选择控制
alternative cost 替换成本;替代成本;交替成本;选择成本;择机代价;变换成本
alternative current power supply network 交流电源点阵
alternative dam site 比较坝址;坝址比较方案
alternative decision 选择性决策
alternative demand 可替代需要;交替需求
alternative design 方案设计;比较设计
alternative design project 比较设计方案
alternative development 可选择的发展
alternative device 替换装置
alternative dispute resolution method 替代纠纷调解法
alternative drawee 可替换的付款人
alternative duty 选择税
alternative energy sources 替代能源
alternative factory site 比较厂址
alternative file attributes 候选文件属性;择一文件属性
alternative folding 交错褶皱
alternative frequency 交变频率;备用频率
alternative front or rear loader 前后换置式装载机
alternative fuel 代用燃料
alternative-fuel engine 可变燃料发动机;油气两用机
alternative future 未来发展趋势预测
alternative goods 可替换货物
alternative height position 可选择高度位置
alternative host 交替寄主
alternative hypothesis 择一假说;更迭假设;对立假设;备择假设
alternative hypothesis test 替换性假设测试
alternative importer 可替换的进口商
alternative inheritance 交替遗传
alternative interchange 互通立交(体)交(叉)
alternative investment 投资比较方案
alternative item 可选择项目;选择项目
alternative layout 比较布置方案;可供选择的布置
alternative line 选取线;比较线(路);比较路线
alternative location line 比较路线
alternative location strategy 可供选择的布局对策
alternative lock site 比较闸址
alternative manner 迭代方法
alternative measurement 校核计量
alternative method 替代法;交替法;代замеsh法;变异法
alternative model 可供选择的模式;备择模型
alternative modes of transportation 不同的运输方式;各种不同的运输方式
alternative motion 往复运动;交替运动;交变运动
alternative multi-fuel burner 油气两用燃烧器
alternative naming 交替命名
alternative net ground 变化网眼地组织
alternative obligation 选择性债务
alternative offset 交替时差;交替(式)绿时差
alternative operator service 交错操作员业务
alternative optima 择其最优
alternative optimization 方案优化
alternative optimum 择一最优值;方案优化
alternative optimum solution 备择方案最优解
alternative option 比较选择方案
alternative order 二者择一的订单
alternative pack 可换容器;选择部件
alternative pattern 可选择的模式
alternative pattern of development and lifestyles 发展和生活方式的可选择模式
alternative payee 可替换的收款人
alternative plan 可供采用的方案;可供选择的方案;规划比较方案;比较规划方案;备选计划;备选方案
alternative pointer 更替指针
alternative port site 比较港址
alternative power plant 替代电站;备选电站;电站比较方案
alternative power station 备选电站
alternative pressure 交替压力

alternative price 比较价格
alternative project 替代方案;比较设计方案;比较方案
alternative proposal 可供选择的方案;比较方案投标
alternative quarantine anchorage 备用检疫锚地
alternative route 迂回进路;变更进路;比较路线
alternative routine 备择程序
alternative routing 替换线路
alternative scanning pattern 交变扫描图形
alternative scenario 替代方案
alternative schedule 比较方案
alternative scheme 其他方案;替代方案;比较方案;备用方案
alternative set 后备机组
alternative share 可换犁铧
alternative side street 平行主干线的复线
alternative site 比较坝址;坝址比较方案
alternative slot winding 交叉槽式绕组
alternative solution 替换溶液;替代方案;另一种办法;可选择的解决方案;备用方案
alternative steering station 备用操舵台
alternatives to the flush toilet 选择冲洗厕所
alternative strategy 变更的对策
alternative substitution 可供选择的代用品
alternative supply 交替供给
alternative survey 变通的检验
alternative technology 替代技术;可供选择的技术;交替技术;可选择的技术
alternative test 替换试验;替换测验;交错检验
alternative text 备选条文
alternative to pesticide 代用农药
alternative to the flush toilet 选择冲洗厕所
alternative transport strategies 交通对策比较方案
alternative usage of soil 土壤交换使用
alternative use 选择使用
alternative vehicle position determination 车辆定位替换方法
alternative wastewater treatment 交联污水处理
alternative water supply source 备选水源
alternative water use 轮换用水
alternative wave 地震转换波
alternative ways 可供选择或采用的方法
alternative wrap around guided-bend test 交变缠绕定向弯曲试验
alternator 同步发电机;交流发电机;振荡器
alternator armature 交流发电机电枢
alternator brake 交流发电机制动闸
alternator transmitter 高频磁电机
alterne 群落交错
alternobaric vertigo 压力眩晕
Alter periodogram 阿勒特周期图
alter polarity 交替极性;变更极性
Alter shield 阿勒特遮罩(雨量计上用以防止风影响的罩)
alter shield 改变护罩
alter shots 变化镜头
alter statement 更改语句
alter the design 变更设计
alter the economic relations 变革经济关系
alter the letter of credit 修改信用证
alter the procedure 变更程序
althausite 羟磷镁石
altherbosa 高草群落
altichamber 气压试验室;高空试验(模拟)室
alti-electrograph 高空电位计
altigraph 自记测高仪;高度记录器;高度记录器;测高计
altimeter 气压测高仪;高度仪(表);高度计;高度表;高程计;测高仪;测高器;测高计
altimeter-aneroid observation 空盒气压计高程测量
altimeter calibrator 高度表校准器
altimeter coder 高度计编码器;高度编码器
altimeter correction 测高仪校正
altimeter corrector 高度校正器;高度表校正器
altimeter setting 高度计设置;高度表装置;高度表拨正;测高仪调定
altimetric compensation 高度改正;高度补偿
altimetric data 高程资料;高程测量数据
altimetric displacement 垂直位移
altimetric measurement 高程测量;测高
altimetric point 高程注记点
altimetric reading 高程读数
altimetric sensor 高度传感器
altimetric survey 高程测量

altimetry 测高学;测高术;测高法
altimininium orthophosphate 正磷酸铝
altiperiscope 对空潜望镜
altiplanation 高山地夷平作用;高山剥夷作用;高地夷平作用
altiplanation surface 高山剥夷面
altiplanation terrace 山上阶地;高山剥夷面
altiscope 高度望远镜
altithermal 高热期;冰后期的高温期;冰后高温期
altithermal soil 高温土
altitude 高度;高程;顶垂线;地平纬度;地平高度
altitude above sea level 海拔高度
altitude acclimation 高原适应;高山适应;高空适应;高地适应
altitude acclimatization 高原适应;高空适应性
altitude acquisition technique 高空探测技术
altitude air 高空空气;高处空气
altitude and azimuth instrument 地平经纬仪
altitude and check valve 水位控制和止回阀
altitude angle 高度角
altitude angle of satellite 卫星高度角
altitude anoxia 高空缺氧
altitude axis 高度轴
altitude azimuth 高度方位角
altitude azimuth method 高度求方位法
altitude below pole 下中天高度
altitude bubble 高程水准器
altitude by observation 观测纬度;观测高度
altitude capability 升限;可达高度
altitude capsule 高度计膜盒
altitude carburetor 高空化油器
altitude chamber 气压试验室;高空试验舱;高空试验(模拟)室;高空舱;负压工作舱;负压工作室
altitude circle 竖直度盘;高度圈;地平纬圈;等高圈
altitude classification of plain 平原高程分类
altitude coefficient 海拔修正系数
altitude control 高度控制
altitude controller 高度调节器
altitude control lever 高度操纵杆
altitude control valve 水位控制阀;高度控制阀;高程控制阀
altitude control valve in pipeline 管线中的高度控制阀
altitude converter 高度变换器
altitude corrected value 高度校正值
altitude correction 投影差校正;高度修正;高度校正;高度改正
altitude correction factor 高度校正因数
altitude correction value 高度改正值
altitude curve 天体高度曲线;高度曲线
altitude data unit 高度数据发送器
altitude datum 高度基准;高程基准(面)
altitude decompression sickness 高空减压症
altitude delay 高度延迟
altitude diagram of top of marked layer 标志层顶板高程图
altitude difference 高度差
altitude difference method 高度差法
altitude-direction indicator 高度方向指示器
altitude disease 高原病
altitude distance diagram 高度距离列线图
altitude distribution 高度分布;垂直方向分布
altitude diving 高低潜水
altitude effect 高山作用;高度效应
altitude engine 高空发动机
altitude environment 高原环境
altitude figure 高程注记
altitude for maximum endurance 久航高度
altitude gain 爬高
altitude ga(u)ge 水深计;高度计;高度表;测高仪;测高计
altitude grade gasoline 高级汽油
altitude gyroscope 陀螺地平仪
altitude hold 高度固定
altitude hole 高度空白
altitude hypoxia 高空缺氧
altitude indicator 高度指示器
altitude intercept 高度差
altitude interval 高度间隔;高程间距
altitude level 竖盘水准器;高度水准器;测高水准仪
altitude limit 高度界限;高度范围;高程界限
altitude lobe 高度瓣
altitude low temperature test 高空低温试验
altitude marking radar 高度指示雷达

altitude meter 高度表
altitude of apogee of satellite orbit 卫星轨道远地点高度
altitude of aquifer 含水层产状
altitude of artesian water head 自流水头高度
altitude of a triangle 三角形的高线
altitude of back sight 过天顶高度
altitude of borehole mouth 孔口高程
altitude of cave distribution 洞穴分布高程
altitude of ground surface 地面高程
altitude of groundwater level 地下水位高程
altitude of landslide tongue bottom 滑坡舌底部高程
altitude of lower transit 下中天高度
altitude of perigee 近地点高度
altitude of perigee of satellite orbit 卫星轨道近地点高度
altitude of piezometric head 承压水头高度
altitude of pit mouth 坑口高程
altitude of projection surface 投影面位置
altitude of sample 取样高度
altitude of slope top 坡顶高程
altitude of source 河源高程;源头高程
altitude of spring emerging 泉的出露高程
altitude of terrace surface 阶地面高程
altitude of the heavenly body 天体高度
altitude of the pole 天极高度
altitude of the prime vertical 东西圈高度
altitude of water bursting point 突水点高程
altitude of weathering rock surface 风化岩面高程
altitude parallax 高度视差
altitude parallel 等高圈
altitude plant 高空设备
altitude reaction 高山反应
altitude reading 高程读数
altitude receiver 高度接收器;测高计接收机
altitude recorder 自记测高仪;高度记录器
altitude scale 高度标尺;比例尺;高程分划尺
altitude selector 高度转接开关;高度选择器;高度波段开关
altitude selsyn 高空自动同步机;高度自动同步机
altitude sensing apparatus 高度传感器
altitude sickness 高空病
altitude signal 高度信号
altitude slide 高度滑尺
altitude stress 高原反应;高山反应
altitude supercharger 高空增压器
altitude switch 高空电开关
altitude test facility 高空试车台
altitude tetany 高空痉挛
altitude throttle 高空环流阀
altitude throttle valve 高度节流阀
altitude tint legend 分层设色高度表
altitude tints 分层设色法
altitude to natural areas 对自然区的看法
altitude training 高原训练
altitude transmitter 高度数据发射器
altitude transmitting selsyn 仰角自动同步传送机
altitude valve 高空空气活门
altitude variation 高度变化
altitude vertical velocity indicator 高度升降速率指示器
altitudinal belt 垂直分带
altitudinal gradient 海拔梯度
altitudinal vegetation zone 高程植被带
altitudinal zones 海拔高的植被区
altizide 阿尔噻嗪
Altmag 阿特马格铝镁合金
altocumulus(clouds) 高积云
alto cumulus duplicatus 复高积云
alto cumulus glomeratus 簇状高积云
alto cumulus informis 无定状高积云
alto cumulus precipitus 降水性高积云
alto cumulus translucidus 透光高积云
altocumulus undulatus 波状高积云
altogether 完全
altoherbiprata 高原草本群落
altoherbosa 高草草本植被
altoll 黑软土
altometer 经纬仪;高度仪(表);高度表
altonimbus 高雨云
alto-relievo 隆浮雕;高凸浮雕;高浮雕;意大利高(凸)浮雕
Alto steel 加铝镇静钢

altostrato-cumulus 高层积云
altostratus clouds 高层云
altostratus densus 浓密高层云
altostratus fractus 碎高层云
altostratus maculosus 斑状高层云
altostratus precipitus 降水性高层云
altostratus translucidus 透光高层云
altra-mizer 超微粉碎机
altrasonograph 超声波图像记录仪
altricial 晚成
altro 不滑动踏步及瓦片
Alubond method 铝化学防蚀薄膜法
alucol 氢氧化铝
aludip 钢板热渗铝;热浸镀铝钢板;热浸镀铝;热度铝钢板
Aludirome 铁铬铝系电炉丝;铁铬铝合金;阿卢迪罗姆铁铬铝系电热丝
Aludur 阿鲁杜铝镁合金;铝镁合金
Alufer 包层钢板;阿卢弗尔包层钢板
Aluflex 阿卢弗莱克斯导电铝合金;锰铝合金
alugel 氢氧化铝
alum 明矾;矾;白矾
Alumag 阿留马格铝镁合金;铝镁合金
Alumal 铝锰合金;阿卢马尔铝锰合金
Aluman 阿卢曼铝锰耐蚀合金
alumatol 硝铵—铝粉炸药
alum bath 明矾水浴;矾水浴
alumbre 明矾
alum cake 明矾块;矾块
alum-carmine 明矾卡红
alum-carmine stain 明矾胭脂红染剂
alum carryover 铝矾携带
alum clay 铝土;明矾(黏)土;矾土
alum cleaning bath 明矾洗涤液
alum coagulation treatment 矾凝结处理
alum coal 明矾煤
alum dresed leather 明矾鞣革
alum earth 明矾土
alumed leather 明矾鞣革
Alumel alloy 镍铝锰硅合金;镍基热电偶合金;阿卢梅尔镍合金
alumel chromel 镍铝—镍铬合金
alumel chromel thermocouple 镍铝镍铬热电偶
alumel chromel thermoelement 镍铝镍铬热电偶
alumetized steel 涂铝钢;渗铝钢
alum feed apparatus 加矾器
alum feeder 明矾投加器
alum flocculation 明矾絮凝
alum flour 矾粉
alum gate 铝栅
alum glass 明矾玻璃
alum hematoxylin 明矾苏木精
alumian 无水矾石;水钠矾石
alumigel 氢氧化铝
alumilite process 铝阳极氧化法;硬质氧化铝膜处理法;阳极氧化铝加工处理法
alumina 铝氧;铝土;氧化铝
alumina anomaly 铝反常
alumina balls 氧化铝质球磨用球
alumina base catalyst 氧化铝基催化剂
alumina-based cutting tool 氧化铝基切削工具
alumina-bearing material 含铝材料;含矾土材料
alumina blanc fixe 氧化铝—重晶石粉
alumina block 高铝砌块
alumina-borosilicate glass 硼硅铝玻璃;铝硼硅酸盐玻璃
alumina brick 氧化铝砖;高铝砖;矾土砖
alumina bubble brick 轻质高铝砖;泡沫矾土砖;空心氧化铝球砖;氧化铝空心球砖
alumina cement 快干水泥;高铝水泥;铝土水泥;矾土水泥;矾土水泥浆
alumina cement castable refractory 高铝水泥浇注料
alumina ceramic bite 矾土陶瓷车刀
alumina ceramics 矾土陶瓷;氧化铝瓷;高铝陶瓷
alumina-chrome brick 铝铬砖
alumina-chrome refractory 铝铬质耐火材料
alumina clay 矾土
alumina clay ore 含铝黏土矿石
alumina column 氧化铝柱
alumina cream 氢氧化铝;矾土霜
alumina electrolysis bath 氧化铝电解槽
alumina-ferrous oxide ratio 铝铁比
alumina fiber 氧化铝纤维;氧化铝耐火织物

alumina fire(clay)brick 矾土耐火砖
alumina gel 铝胶;氧化铝凝胶
alumina-graphite refractory 氧化铝石墨质耐火材料
alumina hydrate 水化氧化铝;水合氧化铝;含水氧氧;氧化铝水化物
alumina hydrogel 氧化铝水凝胶
alumina iron modulus 铝率
alumina iron ratio 铝氧比
alumina lap 氧化铝研磨
alumina modulus 铝率
alumina mortar 矾土砂浆
alumina plant 氧化铝厂;矾土厂
alumina porcelain 氧化铝瓷;矾土陶瓷
alumina porcelain sectional insulator 高铝陶瓷分段绝缘子
alumina powder 铝氧粉;氧化铝粉
alumina ratio 铝氧率;铝率
alumina ratio of cement 水泥的铝率
alumina refractory 氧化铝质耐火材料;氧化铝耐火材料;高铝砖;矾土耐火砖
alumina refractory products 矾土耐火制品
alumina-riched weathering 富铝化风化壳
alumina self-stressing cement 铝酸盐自(动)应力水泥
alumina-silica ceramics fiber 硅酸铝陶瓷纤维
alumina-silica fiber 铝硅纤维
alumina-silica firebrick 硅酸铝耐火砖
alumina-silica ratio 铝硅比值
alumina silica refractroy 硅酸铝耐火材料
alumina silicate 硅酸铝
alumina silicate block 硅铝砖
alumina silicate brick 硅铝砖
alumina silicate fiber 硅铝纤维
alumina silicate refractory 硅酸铝耐火材料
alumina-silicon carbide-carbon refractory 氧化铝—碳化硅—碳质耐火材料
alumina single crystal 氧化铝单晶
alumina sol 氧化铝溶胶
aiumina substrate 氧化铝基片
alumina sulphate 铝硫酸盐;矾土硫酸盐
alumina supporter 氧化铝载体
alumina tabular 片状刚玉
aluminate 铝酸盐;铝氧根
aluminate cement 高铝水泥;铝酸盐水泥;矾土水泥
aluminate concrete 铝酸盐混凝土;矾土混凝土
aluminate glass 铝酸盐贝利玻璃
aluminate refractory cement 铝酸盐耐火水泥
aluminate refractory concrete 铝酸盐耐火混凝土
aluminate-silicate refractory 硅酸盐质耐火材料
alumina thickening 矾土增稠过程
alumina titanate ceramics 钛酸铝陶瓷
alumina tool 氧化铝陶瓷刀具
alumina white(lake) 矾土白
alumina whiteware 氧化铝瓷
alumincoat 铝制热渗铝法
alumine 铝土;矾土
aluminex 屋面上釉材料
aluminide deffusion coating 铝化物扩散敷层
aluminiferous 铝土的;含铝土的;含铝的;含矾
aluminite 矾石;铝氧石
aluminite powder 铝粉
alumin(i)um 自然铝
alumin(i)um absorbent ceiling 铝板吸声吊平顶;铝板吸声天花板;铝板吸声顶棚
alumin(i)um acetate 乙酸铝;醋酸铝
alumin(i)um acoustic(al)ceiling 铝板吸声天花板;铝板吸声顶棚;铝板吸声吊平顶
alumin(i)um acoustic(al)tiled ceiling 铝板吸声天花板;铝板吸声顶棚
alumin(i)um activation 铝活化反应
alumin(i)um adjustable bracket, slotted standard 铝活动支架
alumin(i)um adjuvant 铝佐剂
alumin(i)um air battery 铝空气电池
alumin(i)um alloy 铝合金
alumin(i)um alloy arc welding electrode 铝合金焊条
alumin(i)um alloy bar 铝合金棒
alumin(i)um alloy blind slat 铝合金百叶窗板条
alumin(i)um alloy bridge 铝合金桥
alumin(i)um alloy casting 铝合金铸件
alumin(i)um alloy checkered plate 铝合金防滑板
alumin(i)um alloy door 铝合金门
alumin(i)um alloy doorlock with knobs 铝合金

球型执手门锁
alumin(i)um alloy drill pipe 铝合金钻杆
alumin(i)um alloy drill rod 铝合金钻杆
alumin(i)um alloy dust 铝合金粉尘
alumin(i)um alloy eggcrate ceiling 铝合金格片顶棚
alumin(i)um alloy extrusions 铝合金型材
alumin(i)um alloy flat wire 铝合金扁丝
alumin(i)um alloy for temper 锻用铝合金
alumin(i)um alloy hinge 铝合金铰链
alumin(i)um alloy hull 铝合金船体
alumin(i)um alloy ingots 铝合金锭
alumin(i)um alloy joist 铝合金龙骨
alumin(i)um alloy levels 铝合金水平尺
alumin(i)um alloy plate 铝合金板
alumin(i)um alloy radiator 铝合金暖气片
alumin(i)um alloy rod 铝合金条
alumin(i)um alloy sash 铝合金窗扇
alumin(i)um alloy section 铝合金型材
alumin(i)um alloy shape 铝合金结构型材
alumin(i)um alloy sheet 铝合金薄板;铝合金扣板;铝合金板
alumin(i)um alloy ship 铝合金船
alumin(i)um alloy sink 铝合金洗涤盆
alumin(i)um alloy space frame 铝合金空间框架
alumin(i)um alloy strip ceiling 铝合金条吊顶
alumin(i)um alloy structure section 铝合金结构型材
alumin(i)um alloy tower bolts 铝合金插销
alumin(i)um alloy window 铝合金窗
alumin(i)um alloy wires 铝合金线(材)
alumin(i)um/alumin(i)um silicon alloy coating 铝铝硅合金涂层
alumin(i)um amalgam 铝汞齐
alumin(i)um ammonium sulfate 硫酸铵铝;铵铝矾
alumin(i)um ammonium sulphate 硫酸铝铵
alumin(i)um and alumin(i)um alloy extruded rods 铝及铝合金挤压棒材
alumin(i)um and alumin(i)um alloy pipes 铝及铝合金管材
alumin(i)um and alumin(i)um alloy wires 铝及铝合金线材
alumin(i)um and bauxite industry 铝矾土工业
alumin(i)um anode 铝阳极
alumin(i)um anode cell 铝阳极电池
alumin(i)um anomaly 铝反常
alumin(i)um antimonide 锑化铝
alumin(i)um architecture 铝结构建筑学
alumin(i)um arrester 铝电池避雷器
alumin(i)um arsenide 砷化铝
alumin(i)um austral window 铝制推拉窗
alumin(i)um backing 铝敷层
alumin(i)um bar 铝棒
alumin(i)um barium chart 铝钡状态图
alumin(i)um barked wire 铝合金刺丝;铝刺丝
alumin(i)um bars 铝巴
alumin(i)um base alloy 铝基合金
alumin(i)um base grease 铝基润滑脂
alumin(i)um bauxite 铝土
alumin(i)um beads collector 铝珠集电极
alumin(i)um beam 铝梁
alumin(i)um bearing 铝轴承
alumin(i)um-bearing alloy 铝轴承合金
alumin(i)um-bearing formation 含铝建造
alumin(i)um-bearing rock 含铝质岩
alumin(i)um-bearing structure 铝支承结构
alumin(i)um bellows 铝波纹管
alumin(i)um bichromate 重铬酸铝
alumin(i)um blasting cap 铝壳雷管;铝制雷管
alumin(i)um body 铝制车身;铝车身(运货卡车)
alumin(i)um bonded roof covering 铝结合屋顶屋面
alumin(i)um bonding 铝的键合
alumin(i)um borate 硼酸铝
alumin(i)um boride 硼化铝
alumin(i)um borohydride 硼氢化铝
alumin(i)um brace band 铝支撑带材;铝合金支撑带材
alumin(i)um brass 铝黄铜
alumin(i)um brass alloy 铝黄铜合金
alumin(i)um bridge 铝桥
alumin(i)um bronze 铝青铜
alumin(i)um bronze paint 银色漆;青铜色的铝粉涂料;铝青铜涂料;铝金粉漆

alumin(i)um builders hardware 铝制建筑小五金;铝制建筑附件;铝建筑小五金;铝建筑附件
alumin(i)um building component 铝制建筑部件
alumin(i)um building entrance door 铝制(建筑)外门
alumin(i)um building member 铝制建筑构件;铝制(建筑)板材
alumin(i)um building product 建筑铝制品;铝制建筑制品
alumin(i)um building sheet 铝制(建筑)板材
alumin(i)um building unit 铝制建筑构件;铝制(建筑)板材
alumin(i)um busbar 铝母线;铝汇流排
alumin(i)um cabide 碳化铝
alumin(i)um cable 铝芯电缆;钢芯铝线
alumin(i)um cable sheath 铝制电缆护套;铝芯电缆包皮;铝心电缆甲套
alumin(i)um cable steel reinforced 钢芯铝绞线;钢芯增强铝线;钢芯铝电缆
alumin(i)um calcium silicate 铝硅酸钙
alumin(i)um capping machine 铝盖压盖机;铝盖卷封机
alumin(i)um carbide 碳化铝
alumin(i)um-carbon microelectrolysis 铝炭电解法
alumin(i)um cassette 铝制镶板
alumin(i)um casting 铝合金铸件;铝铸件;铸铝
alumin(i)um casting alloy 铸造铝合金;铸铝合金
alumin(i)um cast iron 高铝铸铁
alumin(i)um ceiling 铝制天花板;铝制顶棚;铝板顶棚
alumin(i)um cell 铝极电池;铝电解槽
alumin(i)um cell arrester 铝(管)避雷器
alumin(i)um cell rectifier 铝电解整流器
alumin(i)um cement 矾土水泥
alumin(i)um chanelled foil 槽形铝箔
alumin(i)um channel 槽形铝材
alumin(i)um checker(ed) foil 花纹铝箔
alumin(i)um chelate 螯合铝
alumin(i)um chlorate 氯酸铝
alumin(i)um chloride 氯化铝
alumin(i)um chlorohydrate 水合氯化铝
alumin(i)um chrome steel 铝铬铬钢
alumin(i)um chromite 铝铬铁矿
alumin(i)um chromium 铝铬钢
alumin(i)um chromium coating 铝铬镀层
alumin(i)um-clad alloy 包铝(的硬铝)合金
alumin(i)um-clad iron 铝皮铁板
alumin(i)um-clad steel reinforced alumin(i)um stranded wire 铝包钢芯铝绞线
alumin(i)um-clad steel stranded wire 铝包钢绞线
alumin(i)um-clad wire 包铝钢丝
alumin(i)um cleaning solution 铝清洗液
alumin(i)um coagulant 铝混凝剂
alumin(i)um-coated fabric 涂铝(钢丝)网;镀铝(钢丝)网
alumin(i)um coated glass fiber 镀铝玻璃纤维
alumin(i)um coated iron 覆铝铁
alumin(i)um-coated mirror 镀铝镜
alumin(i)um-coated sheet steel 涂铝薄钢板
alumin(i)um-coated steel 镀铝钢板;包铝钢
alumin(i)um-coated steel contact wire 钢铝接触线
alumin(i)um coating 热镀铝法;铝涂覆层
alumin(i)um coffer 铝制镶板
alumin(i)um coil and sheet stack annealing furnace 铝卷材退火炉
alumin(i)um cold mill 铝材冷轧机
alumin(i)um colo(u)red fabric 着色铝网
alumin(i)um component 铝制建筑构件
alumin(i)um compound 铝化合物
alumin(i)um conductor 铝芯导线;铝芯导体;铝线;铝导线
alumin(i)um conductor steel reinforced 钢芯铝线(的);钢芯铝绞线
alumin(i)um container 铝质集装箱
alumin(i)um-containing alloy 含铝合金
alumin(i)um copper alloy 铝铜合金
alumin(i)um copper-iron alloy 铝铜铁合金
alumin(i)um copper zinc alloy 铝铜锌合金
alumin(i)um correcting material 铝质校正原料
alumin(i)um corroding bacteria 铝腐蚀细菌
alumin(i)um corrugated (building) sheet 铝制波纹片

alumin(i)um corrugated profile 铝制波纹形
alumin(i)um corrugated section 铝制波纹型材
alumin(i)um corrugated trim 波纹铝材
alumin(i)um corrugated unit 铝制波纹形构件
alumin(i)um curtain wall 铝制幕墙;铝幕墙
alumin(i)um curtain wall system 铝幕墙系统
alumin(i)um deck (roof) 铝板屋顶
alumin(i)um deck unit 铺平台的铝制件;铝顶板构件
alumin(i)um decorative section 铝装饰型材
alumin(i)um deoxidized metal 铝脱氧的金属
alumin(i)um detonator 铝壳雷管;铝雷管
alumin(i)um diboride 二硼化铝
alumin(i)um die cast alloy 模铸铝合金
alumin(i)um die casting 铝压铸件;铝压模铸件;压铸铝
alumin(i)um die-casting alloy 压铸铝合金;压模铝合金
alumin(i)um dihydrogen phosphate 磷酸二氢铝
alumin(i)um dip-brazing 铝浸钎焊
alumin(i)um door 铝合金门;铝制门;铝门
alumin(i)um door anodised in bronze 电镀黄铜铝门
alumin(i)um double window 双层铝窗
alumin(i)um dross 铝碴
alumin(i)um dry cargo container 铝制干货柜
alumin(i)um dust 铝粉
alumin(i)um eave(s) gutter 铝制檐沟
alumin(i)um eave(s) trough 铝制檐沟
alumin(i)um electrochemical stabilization 铝极电化学稳定
alumin(i)um electrode method 铝电极法(加固地基用)
alumin(i)um eletrolytic capacitor 铝电解电容器
alumin(i)um enamel 铝搪瓷;银粉漆
alumin(i)um-enriched process 富铝化作用
alumin(i)um-enriched weathering 富铝化风化壳
alumin(i)um equivalent 铝当量
alumin(i)um erection 铝质上层建筑
alumin(i)um expanded sheet 铝板网
alumin(i)um explosive 铝粉炸药
alumin(i)um extruded sections 铝型材
alumin(i)um extrusion 铝材挤压(成型);挤压铝制品
alumin(i)um fabric 铝网;铝合金网
alumin(i)um facade 铝制结构正面;铝制结构立面
alumin(i)um face 铝制结构正面;铝制结构立面
alumin(i)um facing 铝饰镶面;铝folded贴面
alumin(i)um family 铝族
alumin(i)um fence 铝制围栏
alumin(i)um fillings 铝制小五金
alumin(i)um filter 铝制滤光板
alumin(i)um finger plate (门锁的)铝指板
alumin(i)um finish 铝材饰面;铝表面处理;铝面
alumin(i)um fittings 铝制配件
alumin(i)um flake 铝粉;铝薄片;高岭土;片状铝粉;铝片;薄铝片
alumin(i)um flake pigment 铝粉颜料;铝粉色素;薄铝片颜料
alumin(i)um flashing 铝制泛水板;防水板;防雨铝板
alumin(i)um flat sheet 铝平片
alumin(i)um flat wire 铝扁丝
alumin(i)um floor covering 楼面铝覆面层
alumin(i)um fluoride 冰晶石;氟化铝
alumin(i)um fluosilicate 氟硅酸铝
alumin(i)um foil 碎铝片;铝膜;铝箔;铝板
alumin(i)um foil backing 铝箔背衬
alumin(i)um foil board 铝箔
alumin(i)um foil faced fiberglass board 铝箔贴面玻璃棉板
alumin(i)um foil ink 烫印油墨
alumin(i)um foil insertion 铝箔绝热
alumin(i)um foil insulation 铝箔隔声;铝箔隔热
alumin(i)um foil isolation paper 防热防水纸
alumin(i)um foil lamp 铝箔灯
alumin(i)um foil maltoid 铝箔油毡
alumin(i)um foil mill 铝箔轧机
alumin(i)um foil production line 铝箔生产线
alumin(i)um foil scrap 废铝箔
alumin(i)um foil with paper lining 衬纸铝箔
alumin(i)um folded plate roof 铝制折板屋面;铝制折板屋顶
alumin(i)um forging 铝锻造

alumin(i)um form(work) 铝制模板;铝模板
alumin(i)um for wire drawing 铝线锭
alumin(i)um foundry 铸铝车间
alumin(i)um frame 铝制框架
alumin(i)um framing strips 铝镜条
alumin(i)um front 铝制结构正面;铝制结构立面
alumin(i)um garnet source 铝石榴石光源
alumin(i)um gasket 铝密封垫圈
alumin(i)um gate and sillicon gate differences 铝门与硅门的区别
alumin(i)um glazing 铝制天窗
alumin(i)um gold 铝金(含铝黄铜的)
alumin(i)um grid 铝制窗格
alumin(i)um grid ceiling 铝制网格天花板
alumin(i)um grid dome 铝网格穹隆
alumin(i)um grill(e) 铝制格栅
alumin(i)um guide shoe 铝制引鞋
alumin(i)um halide 卤化铝
alumin(i)um hand railing 铝制围栏;铝制扶手
alumin(i)um hardware 铝制小五金;铝制零件
alumin(i)um hinge 铝制铰链
alumin(i)um hollow section 铝制空心型材
alumin(i)um hull 铝材船体
alumin(i)um humate 腐殖酸铝
alumin(i)um hydroxide 氢氧化铝;铝水化物
alumin(i)um hydroxide gel 干凝胶
alumin(i)um impregnation 渗铝;铝化
alumin(i)um industry 制铝工业
alumin(i)um industry waste 炼铝工业废物
alumin(i)um industry waste(water) 炼铝工业废水
alumin(i)um ingot 铝锭
alumin(i)um ink 铝粉油墨;银粉(油)墨
alumin(i)um iodide 碘化铝
alumin(i)um ion 铝离子
alumin(i)um iron 铝铁
alumin(i)um isopropoxide 异丙醇铝
alumin(i)um jacket 铝套;铝壳
alumin(i)um kicking plate 铝踢板
alumin(i)um leaf 铝薄片;铝箔
alumin(i)um lining 铝衬层;铝衬里
alumin(i)um lintel (门窗的)铝过梁;(门窗的)铝制过梁
alumin(i)um load-bearing member 铝制承重构件
alumin(i)um-loaded Shirasu zeolite 加载铝的斯拉苏沸石
alumin(i)um louvre panel 铝百叶板
alumin(i)um magnesium alloy 铝镁合金
alumin(i)um magnesium alloy conductor 铝镁合金导线
alumin(i)um magnesium alloy magaluma 铝镁合金
alumin(i)um manganese alloy 铝锰合金
alumin(i)um manufacturer 铝厂
alumin(i)um mesh 铝网;铝筛;铝制网
alumin(i)um metal 铝金属
alumin(i)um metallurgy 炼铝
alumin(i)um mill sheets in coils 铝卷片
alumin(i)um mirror coating 铝反射膜
alumin(i)um mist 铝雾;铝金属雾
alumin(i)um mixing varnish 调铝粉漆用漆料
alumin(i)um monostearate 单硬脂酸铝
alumin(i)um mo(u)ld 铝模
alumin(i)um mo(u)ld paint 铸铝件砂型涂料
alumin(i)um mo(u)ld plate 铝模板
alumin(i)um multicompound bronze 多种化合物铝青铜
alumin(i)um nail 铝钉
alumin(i)um naphthenate 环烷酸铝
alumin(i)um naphthenate varnish 环烷酸铝粉清漆
alumin(i)um nickel 铝镍合金
alumin(i)um nickel-cobalt alloy 阿尔尼科合金
alumin(i)um nickel steel 铝镍钢
alumin(i)um nitrate 硝酸铝
alumin(i)um nitride 氮化铝
alumin(i)um nitride ceramics 氮化铝陶瓷
alumin(i)um normal 铝正常的
alumin(i)um octahedron 铝氧八面体
alumin(i)um oleate 油酸铝
alumin(i)um ore 铝矿(石)
alumin(i)um ornamental section 铝制装饰型材
alumin(i)um outrigger base 铝支腿底板;铝支柱底板

alumin(i)um oversaturation 铝过饱和
alumin(i)um oxide 矿物陶瓷;三氧化二铝;铝土;氧化铝;矾土
alumin(i)um oxide abrasive 氧化铝研磨剂
alumin(i)um oxide adsorption 氧化铝吸附
alumin(i)um oxide ceramic insert 氧化铝陶瓷刀片
alumin(i)um oxide ceramic insert chip 氧化铝陶瓷镶装刀片
alumin(i)um oxide ceramic membrane 氧化铝陶瓷膜
alumin(i)um oxide cloth 氧化铝砂布
alumin(i)um oxide cloth in roll 卷筒砂布
alumin(i)um oxide flocculation 氧化铝絮凝
alumin(i)um oxide paper 氧化铝砂纸
alumin(i)um oxide tool 氧化铝陶瓷刀具
alumin(i)um oxygen bond 铝氧键
alumin(i)um paint 铝粉漆;铝涂料;铝漆;铝粉涂料;银灰漆;银粉漆;古铜色铝粉漆
alumin(i)um paint coating 银灰漆涂层
alumin(i)um palmitate 十六(烷)酸铝;棕榈酸铝
alumin(i)um panel 铝镶板;铝制镶板
alumin(i)um panel ceiling 铝板天花板;铝板顶篷;铝板顶板;铝板吊顶
alumin(i)um partition 铝板隔墙
alumin(i)um paste 铝涂料;铝粉浆;银灰漆;银粉浆
alumin(i)um patent glazing bar (不用油灰镶玻璃的)铝窗格条;(用于玻璃采光窗的)铝玻璃格条
alumin(i)um perforated ceiling 多孔铝制顶棚;多孔铝制天花板
alumin(i)um permanent mo(u)ld alloy 硬模铝合金
alumin(i)um permanent mo(u)ld casting alloy 硬模铸造铝合金
alumin(i)um phenylsulfonate 苯磺酸铝
alumin(i)um phosphate 磷酸铝
alumin(i)um phosphate binder 磷酸铝胶结料
alumin(i)um phosphate gel 磷酸铝凝胶
alumin(i)um phosphide 磷化铝
alumin(i)um phosphide poisoning 磷化铝中毒
alumin(i)um pig 铝锭
alumin(i)um pig casting machine 铝锭铸造机
alumin(i)um pigment 铝颜料;铝粉颜料;银粉颜料
alumin(i)um pigmented asphalt coating 铝颜料沥青涂料
alumin(i)um pipe 铝管;铝制管
alumin(i)um piston 铝质活塞
alumin(i)um plank runway 铝板拼装的跑道
alumin(i)um plant scrubber system 铝洗涤系统
alumin(i)um plate 铝版;铝板;厚铝板;铝元片
alumin(i)um-plated steel 镀铝钢;包铝钢
alumin(i)um plated steel wire 镀铝钢丝
alumin(i)um-plating 镀铝
alumin(i)um pneumoconiosis 铝尘肺
alumin(i)um polychloride 多氯化铝
alumin(i)um post 铝支撑;铝杆;铝柱;铝合金柱
alumin(i)um pot 铝罐
alumin(i)um potassium sulphate 硫酸铝钾
alumin(i)um powder 铝粉;银粉
alumin(i)um powder for painting 绘画用铝粉
alumin(i)um powder paint 铝粉漆;银粉漆
alumin(i)um primer 铝粉打底涂料;铝粉打底漆;铝粉船底漆;铝材底漆;银粉底漆;银粉底漆
alumin(i)um product 铝制品;铝材
alumin(i)um profile(d) panel 铝型材镶板;成型铝板
alumin(i)um profile(d) sheet 成型铝板
alumin(i)um profile polishing machine 铝型材抛光机
alumin(i)um profile sheet 成型薄铝板
alumin(i)um pull(handle) 铝门拉手
alumin(i)um purlin(e) 铝檩条
alumin(i)um rail coupling 铝栏杆连接器;铝合金栏杆连接器
alumin(i)um railing 铝栏杆;铝扶手
alumin(i)um rate 铝率
alumin(i)um recovery 铝回收
alumin(i)um resinate 树脂酸铝
alumin(i)um ridge capping 铝片脊瓦
alumin(i)um ring pull end 铝制易拉罐
alumin(i)um rivet 铝铆钉
alumin(i)um rivet head 铝铆钉头
alumin(i)um rod 铝杆;铝棒

alumin(i)um rolling grill(e) 卷开式铝格栅;滚动式铝格栅
alumin(i)um rolling plant 轧铝机;铝压延机;铝辊式破碎机
alumin(i)um rolling shutter 铝制卷帘门(窗);铝卷百页;卷帘铝质百叶窗(门)
alumin(i)um rolling slat 铝制卷板条
alumin(i)um roof cladding 铝屋面板;铝制屋面板
alumin(i)um roofing 铝制屋顶
alumin(i)um roofing roll 铝箔油毡
alumin(i)um roofing sheet 铝制屋面板;铝屋面板
alumin(i)um rosinate 松香酸铝
alumin(i)um salt coagulant 铝盐混凝剂
alumin(i)um salt coagulation method 铝盐混凝法
alumin(i)um salt residue 铝盐残渣
alumin(i)um salts 铝盐类
alumin(i)um sand-casting alloy 砂型铸造铝合金
alumin(i)um sandwich mat 铝夹芯飞机跑道板(铺临时跑道用)
alumin(i)um sandwich panel 夹层铝板
alumin(i)um sash 铝框条;铝窗框
alumin(i)um scrap 废铝
alumin(i)um screen window 铝合金纱窗
alumin(i)um screw 铝制螺丝;铝制螺钉;铝螺丝
alumin(i)um seal(ing) sheeting 铝密封皮;密封铝皮
alumin(i)um section 铝制型材;铝结构型材
alumin(i)um shape 铝制型材
alumin(i)um sheath 铝包皮
alumin(i)um-sheathed cable 铝铠电缆;铝管密封电缆;铝包电缆
alumin(i)um sheet 铝片;铝箔板;铝板;薄铝板
alumin(i)um sheet circle 铝圆片
alumin(i)um sheet facing 铝皮镶面;铝皮贴面;铝皮饰面;铝皮包面
alumin(i)um sheet lining 铝皮镶面;铝皮贴面;铝皮饰面;铝皮包面
alumin(i)um sheet mill 薄铝板轧机;铝板轧机
alumin(i)um sheet-rolling wastewater 铝箔板滚轧废水
alumin(i)um sheet roofing 铝板屋面
alumin(i)um sheet surface 铝皮镶面;铝皮贴面;铝皮饰面;铝皮包面
alumin(i)um shelf 铝搁板
alumin(i)um shielding 铝屏蔽
alumin(i)um shot 铝珠;铝豆
alumin(i)um shutter 铝制遮光窗;铝制百叶窗
alumin(i)um shuttering 铝制模板
alumin(i)um siding 铝制护墙板;铝墙板
alumin(i)um silicate 硅酸铝
alumin(i)um silicate fiber alumin(i)um silicate fibre 硅酸铝纤维
alumin(i)um silicate pigment 硅酸铝颜料
alumin(i)um silicate pneumoconiosis 硅酸铝肺
alumin(i)um silicofluoride 氟硅酸铝
alumin(i)um silico-fluoride 硅氟化铝
alumin(i)um silicon alloy 铝硅合金
alumin(i)um-silicon bronze 铝硅青铜;硅铝青铜
alumin(i)um silicon ingots 铝硅合金锭
alumin(i)um silo 铝仓筒(存放散装水泥)
alumin(i)um sink 铝合金洗涤盆;铝合金冲洗台
alumin(i)um skin 铝包皮
alumin(i)um skyscraper 超高铝制烟囱
alumin(i)um slatted blind 遮光铝百页;铝制百叶窗;铝板防护板
alumin(i)um slatted roller blind 铝制卷帘式遮光百叶;铝板卷升式百叶窗
alumin(i)um sliding door 铝制推拉门;推拉式铝门
alumin(i)um sliding folded shutterdoor 铝制折叠式百叶门
alumin(i)um smelting 炼铝
alumin(i)um smelting industry 炼铝工业
alumin(i)um smelting plant 铝冶炼厂;铝熔炼设备;铝熔炼厂
alumin(i)um soap 铝皂
alumin(i)um soap grease 铝皂润滑脂
alumin(i)um sodium fluoride 氟化铝钠
alumin(i)um sodium sulfate 硫酸钠铝
alumin(i)um sodium sulphate 硫酸钠铝
alumin(i)um soft wool 软铝绒;软铝棉
alumin(i)um solder 铝钎料;铝焊(料)
alumin(i)um soldering 铝钎焊
alumin(i)um sound control ceiling 铝制声控顶棚;铝制声控天花板

alumin(i)um spirit level 铝水平尺
alumin(i)um stabilizer base 铝支腿底板;铝制支柱底板;铝制支架底板
alumin(i)um stain 铝污斑
alumin(i)um steady arm 铝定位器
alumin(i)um stearate 硬脂酸铝
alumin(i)um stearic acid 硬脂酸铝
alumin(i)um steel 铝钢
alumin(i)um-steel cable 钢芯铝电缆;钢铝电缆
alumin(i)um sterilizing drum 铝质消毒桶
alumin(i)um street lighting column 铝制路灯柱;铝路灯柱;街道照明灯铝柱
alumin(i)um strip 铝板条
alumin(i)um sulfate 硫酸铝
alumin(i)um sulfonate foamer 石油磺酸酸铝泡沫剂
alumin(i)um sulphate 硫酸铝
alumin(i)um sunblind 铝制遮阳板;铝制百叶窗
alumin(i)um sunbreaker 铝制遮阳板
alumin(i)um surfacing 铝板饰面;铝板铺面
alumin(i)um swing door 双开式弹簧铝门
alumin(i)um tanker 铝槽车
alumin(i)um T(ee)bar T形铝杆
alumin(i)um tension band 铝张拉带材;铝合金张拉带材
alumin(i)um tension bar 铝张拉杆;铝合金张拉杆
alumin(i)um tension board 铝张拉板材;铝合金张拉板材
alumin(i)um tension wire 铝张拉丝;铝合金张拉丝
alumin(i)um tin bearing 铝锡合金轴承
alumin(i)um titanate 钛酸铝(一种低膨胀陶瓷)
alumin(i)um titanate ceramics 钛酸铝陶瓷
alumin(i)um titanate porcelain 钛酸铝瓷器
alumin(i)um titanate product 钛酸铝制品
alumin(i)um track 铝轨
alumin(i)um trapezoidal section 梯形截面铝材
alumin(i)um triacetate 三乙酸铝
alumin(i)um trim 铝型材;铝装饰线;铝质边框;铝结构型材
alumin(i)um truss brace 铝桁架支撑
alumin(i)um tube 铝管
alumin(i)um tube brackets 铝制管座
alumin(i)um turnings 铝刨丝
alumin(i)um unit 铝结构构件
alumin(i)um vacuum injection type matched die 铝真空注射配合膜
alumin(i)um vapo(u)r barrier 铝气封层;铝汽封;铝隔汽层;铝防潮层
alumin(i)um vessel 铝制容器
alumin(i)um waffle 铝格板;铝隔板
alumin(i)um wallpaper 铝墙纸
alumin(i)um wall rail 靠墙铝扶手
alumin(i)um ware 铝制器皿
alumin(i)um weight-carrying structure 铝承重结构
alumin(i)um welding rod 铝焊条
alumin(i)um welding wire 铝焊条
alumin(i)um window 铝合金窗;铝窗
alumin(i)um windowdoor and sections 铝门窗料
alumin(i)um window fittings 铝窗五金配件
alumin(i)um window furniture 铝窗装置;铝窗配件
alumin(i)um window hardware 铝制窗配件;铝窗小五金;铝窗配件
alumin(i)um wing type solar energy water heater 铝翼式太阳能热水器
alumin(i)um wire 铝线;铝丝
alumin(i)um wire cloth 铝丝布
alumin(i)um wire netting 铝丝网;铝窗纱
alumin(i)um wire splint 铝丝夹板
alumin(i)um wood composite system 铝木组合结构(体系)
alumin(i)um wool 铝绒;铝棉;铝纤维
alumin(i)um wrought alloy 轧制铝合金;熟铝合金;锻制铝合金
alumin(i)um yellow brass 铝黄铜合金
alumin(i)um zinc 锌铝(合金)
alumin(i)um zinc alloy 铝锌合金
alumin(i)um zinc coated sheet 铝锌涂层钢板
alumin(i)un aeschynite 含铝易解石
aluminize 镀铝;敷铝
aluminized apron 铝箔围裙
aluminized cathoderay tube 敷铝阴极射线管
aluminized coat 镀铝上衣
aluminized coated fabric 铝涂层织物
aluminized coated sheet steel 铝涂层钢板
aluminized coating 铝涂层;镀铝层
aluminized insulated garment 铝箔防护服
aluminized mirror 镀铝镜
aluminized phosphor 铝化磷光体
aluminized screen 铝荧光屏
aluminized steel 涂铝钢
aluminizing 涂铝;铝化;敷铝;渗铝;镀铝
aluminoferric 铝铁剂
aluminoferrite mineral 铁铝酸矿物
aluminography 铝版制版法
aluminonickel 铝镍合金
aluminoscorodite 铝臭葱石
aluminosilicate 铝硅酸盐;硅酸铝;硅酸盐
aluminosilicate brick 硅酸铝砖;硅硅酸盐耐火砖;硅酸盐耐火砖
aluminosilicate clay 硅酸铝黏土
aluminosilicate fiber 铝硅酸盐纤维;硅铝酸盐玻璃纤维;硅酸铝纤维
aluminosilicate firebrick 矾土硅酸盐耐火砖
aluminosilicate glass 铝硅酸盐玻璃
aluminosilicate refractory 硅酸铝耐火材料;铝硅酸盐耐火材料
aluminosilicophosphate 硅铝磷酸盐
aluminosis 铝尘肺;矾土肺;矾土沉着病
aluminosis pulmonum 铝质沉着病;铝尘肺
aluminothermic 加铝剂的;铝热的
aluminothermic method of reducing 铝热还原法
aluminothermic reaction 铝热反应
aluminothermics 铝热剂;铝热法
aluminothermic welded joint 铝热焊接头
aluminothermic welding 铝热剂焊接;铝热焊(接);铸焊;热铝焊接
aluminothermy 铝热法
aluminotype 铝凸版
aluminous abrasive 铝土磨料
aluminous aggregate of clinder type 熔渣型高铝集料;矾土熔渣集料;矾土熔渣骨料;熔渣型高铝骨料
aluminous barbed wire 铝合金刺线
aluminous cement 铝土水泥;高铝水泥;高铝矾土水泥;矾土水泥
aluminous cement concrete 高铝水泥混凝土
aluminous clinker 高铝熔渣;高铝水泥熟料;高铝炉渣;高铝矿渣
aluminous coated glass fiber 铝涂层玻璃纤维
aluminous coefficient 铝质系数
aluminous fire(clay)brick 高铝水泥砖;矾土火泥砖;矾土耐火砖;高铝耐火砖;铝质耐火砖
aluminous laterite 铝质红土
aluminous montmorillonite 铝蒙脱石
aluminous refractory 高铝耐火材料
aluminous refractory product 高铝耐火制品;矾土耐火制品
aluminous rock facies 铝质岩相
aluminous rocks 铝质岩类
aluminous sediments 铝质沉积物
aluminous shale 矾土页岩
aluminous silicate 硅酸铝;铝硅酸盐
aluminous silicate brick 高铝硅酸盐砖
aluminous slag 高铝炉渣;高铝矿渣
alumiseal 铝密封
alumite 明矾土;明矾石;氧化铝膜处理法;防蚀铝;防蚀钝化铝
alumite effluent 铝厂酸性废水
alumite process 阳极氧化铝加工法
alumite tile (吸声、隔音的)氧化铝板
alumite wire 氧化铝膜铝线;防蚀铝线
alum leather 矾制皮革
alum mordant 明矾媒染剂
alumo-calcium fluoride 铝钙氟化物
alumo-chrome slag 铝铬渣
alumochromite 铝铬铁矿
alumochromite ore 铝铬铁矿矿石
alumohydrocalcite 水碳铝钙石
alumopharmacosiderite 铝毒石
alumophosphate ore 铝磷酸盐矿石
alum ore 明矾石
alumotantite 钽铝矿
alumotungstite 铝钨华
alum potash 钾矾
alum potassium sulphate 钾矾
alum powder 明矾粉
alum precipitation 明矾沉淀作用
a lump sum 一次总付金额;一笔整数
alum recovery 明矾回收
alum rock 明矾岩
alum saturated class D(gypsum)plaster 德国大理石水泥经矾土处理石膏
alum schist 明矾页岩;明矾片状岩;明矾片岩
alum shale 明矾页岩;明矾片岩
alum shale concrete 明矾片岩混凝土;矾石板混凝土;明矾页岩混凝土
alum slate 明矾石板岩;明矾板岩;矾石板
alum slate concrete 明矾片岩混凝土;矾石板混凝土
alum sludge 明矾污泥
alum soil 明矾土
alum solution 明矾(溶)液;明矾水
alum stone 明矾石
alum storage tank 明矾液储[贮]存罐
alum tannage 明矾鞣
alum tannery 矾土鞣工厂
alum tanning liquor 矾土鞣液
alum tan solution 矾土鞣溶液
alum treatment 明矾处理
alum-vitriol spring 矾硫酸盐泉
alumyte 铝土矿
alundum 人造刚玉;三氧化二铝;铝氧粉;刚铝石
alundum aggregate 烧结刚玉骨料
alundum cement 氧化铝水泥;刚铝黏合剂;刚铝石水泥
alundum furnace 刚玉管炉
alundum insulator 刚玉绝缘体
alundum powder 刚玉粉
alundum tile 铝石砖
alundum tube 刚玉管
alundum wheel 人造刚玉砂轮
Aluneon 阿留尼翁锌铜镍铝合金
alunite 钠硼矾石;明矾石;氧化铝膜处理法
alunite cement 明矾石水泥
alunite deposit 明矾石矿床
alunite expansion agent for concrete 明矾石混凝土膨胀剂
alunite expansive cement 明矾石膨胀水泥
alunite high strength cement 明矾石高强水泥
alunite rock 明矾岩
alunitization 明矾化;明矾石化
alunitized 受含硫水作用而蚀变的
Alunize 阿卢奈兹焊药
alunogen 毛矾石
alupag 菲律宾大载木
alur 沿栏杆或女儿墙过道或走廊
alure 廊道;院廊
alurgite 淡云母
a lush growth of trees 繁茂的树木
Alusil 阿卢西尔(高硅耐热)铝合金
alusil alloy 铝硅合金
Alusirion 阿尔西隆高铝耐热铸铁
alustite 蓝高岭土
alutile (吸声、隔音的)氧化铝板
alvar 阿尔瓦乙烯树脂;矮化植被
Alvarez accelerator 阿尔瓦雷兹型直线加速器
alveated 蜂房式的;槽形的
alveolar 气泡状的;细胞状的;齿槽;肺泡的
alveolar-air measuring apparatus 残气量测定装置
alveolar crest 齿槽嵴
alveolate 有小窝的;有小孔的;蜂窝状的
alveolation 泡化
alveolina limestone 蜂窝状石灰岩
alveolus 小泡;蜂窝;[复]alveoli
alveolusity 蜂窝
alverine 阿尔维林
alveus 浴池(古罗马时嵌入楼层的);室内浴池;河床
alvikite 方解碳酸岩;方解石碳酸岩
alvikite group 方解石碳酸岩类
Alvinokaffric realm 南方区系
always afloat 始终保持浮泊
always open 长期有效
always-ready wrench 双面活动钳
always safe geometry 恒安全几何条件
always safely afloat 始终保持安全浮泊;经常安全浮动
alychn(e)零发光线
alyphite 轻沥青
alypine 阿里品
Alzen 阿尔曾铝铜锌合金
amadou 火绒

amafolone 阿马夫隆
amagat 阿马伽
Amagat density unit 阿马伽密度单位
Amagat diagram 阿马伽图
Amagat-Le-duc rule 阿马伽-勒杜克定则
Amagat system 阿马伽制
Amagat volume unit 阿马伽体积单位
amagmatic 非岩浆的
amagmatic water 非岩浆水
amaiosis 不减数分裂
a majority of the strains 大多数品系
amakinite 羟铁矿
amakusa 日本的一种瓷石
amalaka 馒头形顶(印度和耆那教建筑中高塔顶部的)
amalgaline plumbing 利用低熔点金属带将管子连接起来;金属带焊管法
amalgam 汞齐;汞合金;汞膏
amalgamable 可合汞的
amalgamate 混合物;汞合金;涂汞;混汞;汞齐化
amalgamated balance sheet 合并资产负债表
amalgamated dwelling(-house) 合住住宅;多户合住住宅;集合住宅
amalgamated metal 混汞金属
amalgamated pipe transmission 合管输送
Amalgamated Society of Engineers 工程师联合会
amalgamating-pan 汞合金盘
amalgamating plant 混汞设备
amalgamating table 混汞摇床;混汞台;混汞板
amalgamation 拼合;混合作用;混合法;合并;汞齐作用;汞合金化
amalgamation method 汞齐化法
amalgamation of contractors 承包商协会;承包商联合;建筑联合企业
amalgamation of the data sets 数据组混合
amalgamation pan 混汞盘
amalgamation point 混合点
amalgamation process 企业联营协议;联营协议;汞齐化过程;汞齐法
amalgamator 混汞器;汞合金调制器;拌和机
amalgam barrel 混汞桶
amalgam burnisher 汞合金研光器
amalgam catcher 汞齐捕集器
amalgam cell 汞齐电池
amalgam electrode 汞齐电极
amalgam exchange 汞齐交换
amalgam fluorescent lamp 汞齐荧光灯
amalgam matrix 汞合金型片
amalgam metallurgy 汞齐冶金
amalgam mortar 汞合金研钵
amalgam pestle 汞合金研棒
amalgam pot retort 汞膏蒸馏罐
amalgam press 汞齐挤压机
amalgam retort 汞齐蒸馏罐
amalgam squeezer 汞齐压榨机
amalgam treatment 混汞法
amalgam vapo(u)r lamp 汞齐灯
Amalog 阿马洛格镍铬钨合金
Amaloy corrosionresistant alloy 阿马洛伊耐蚀合金
Amamee 阿马米平纹织物
amanation 放射
amandone bordeaux 阿曼董枣红
amantadine 氨基三环癸烷
amaranth 紫红色;深紫色
amarantite 红铁矾
amargosite 膨土岩;膨润土
amarillite 黄铁钠矾
amaroidal 苦味质的
amaryllis 孤挺花
amassment 堆积
A-mast A 形轻便井架
amateur 业余者
amateur audit 非专业审计;非职业性审计
amateur earthquake prediction 业余地震预报
amateur gardener 业余园艺家
amateurish 浅薄的;业余的;不完善的;不熟练的
amateurism 业余活动;业余性质
amatol 阿马托炸药
a matter of urgency 紧急事件
amaurosis fugax 一时性黑蒙
amausite 燧石;奥长石
amazon 有图案轧制的暗色玻璃
Amazonas abyssal fan 亚马逊深海扇
Amazon Bore 南美亚马逊河口暴涨潮
Amazon-Colombian brachiopod region 亚马逊—哥伦比亚腕足动物地理大区
Amazon geosyncline 亚马逊地槽
Amazonia 亚马逊古陆
amazonite (装修用的)较次石料;微斜长石;天河石
amazonitization 天河石化
Amazon River 亚马逊河
amazon stone 天河石
amazu 塑料防潮材料
Ambam 代森铵
ambary 红麻
ambassador's residence 大使馆邸
ambeny 壁橱
amber 琥珀(色);淡黄色
amber acid 琥珀酸
amber blanket 琥珀色胶片
amber brown 琥珀棕;琥珀褐色;淡褐色
amber colo(u)r 琥珀色颜料
amber glass 琥珀玻璃;褐色玻璃
amber glaze 琥珀色釉
Amberg's line 安伯格线
amberina 艺术玻璃
amberiod 人造琥珀;合成琥珀
amberite 琥珀炸药;阿比里特炸药
amber lamp 黄色信号灯
amber light 琥珀色灯;黄色灯
amberlite ion exchange resin 琥石离子交换树脂;琥珀离子交换树脂
Amberlyst 大孔树脂(一种建材制品)
amber mica 琥珀云母;金云母
amber mutation 琥珀突变
Amberoid alloy 安伯罗依德镍银合金
amber oil 琥珀油
amberol resin 马来改性松香树脂
amber period (交通灯的)黄灯时间
amberplex A 阴离子交换膜
amber ray filter 黄色滤色镜;琥珀色滤光器
amber resin 琥珀树脂
amber sugar coated walnut meat 琥珀核桃仁
amber tar 琥珀焦油;琥珀柏油
amber varnish 琥珀清漆
amber white 琥珀白
amber wood 硬化层积材;酚醛树脂胶合板
amber-yellow 威尼斯黄;琥珀黄
amber-yellow glass 茶色玻璃
ambetti 龙涎香脂;晶粒装饰玻璃;装饰窗玻璃;装饰玻璃窗
ambience 周围;布景;气氛;环境
ambient 环境的;环境;背景的;包围物
ambient air 外界空气;环境空气;周围空气
ambient air level 环境空气级;环境空气水平
ambient air monitor 环境空气监测仪
ambient air monitoring 环境空气监测
ambient air quality 环境空气质量;周围空气质量
ambient air quality mode 空气质量模式
ambient air quality model 空气质量模型
ambient air quality standard 环境空气质量标准;周围空气质量标准
ambient air sampler 环境空气取样仪;环境空气取样器
ambient air sampling instrument 环境空气取样仪;环境空气取样器;环境空气采样仪
ambient air standard 环境空气标准
ambient air temperature 环境气温
ambient atmosphere 环境大气;周围大气
ambient background radiation 环境背景辐射
ambient CO-level 环境一氧化碳浓度
ambient concentration 环境浓度
ambient condition 介质条件;环境条件;环境情况;周围情况;周围条件
ambient cure 室温固化
ambient current 环境水流
ambient density 周围介质密度
ambient dive bell 开式潜水钟
ambient environment 周围环境
ambient exterior luminance 洞外亮度
ambient exterior luminance level 洞外亮度水平
ambient field 背景场
ambient fluid 外围流体
ambient gas analyzer 大气气体分析仪
ambient humidity 环境湿度
ambient level 环境水平;背景电平
ambient levels of toxic pollutant 环境有毒污染物水平
ambient light 外来光;环境光照度;环境照度;环境光线;周围光(线);侧面光;背景光
ambient light filter 环境保护滤光器;周光滤光器;背景光滤光器;保护滤光器(阴极射线管)
ambient light illumination 环境光照明;环境光照度;周光照明;背景光照明
ambient liquid water content 环境液态水含量
ambient medium 周围介质
ambient microseismic noise 环境脉动噪声
ambient noise 外界噪声;环境噪声;周围噪声
ambient noise level 环境噪声级
ambient operating temperature 环境工作温度
ambient operation 室温操作
ambient particulate concentration 环境微粒浓度;环境颗粒物浓度
ambient pollution burden 环境污染负荷
ambient pressure 外界压力;环境压力;周围压力
ambient pressure diving 直接潜水
ambient pressure effect 围压效应
ambient quality standard 环境质量标准
ambient radiant power 环境辐射功率
ambient receiving waters 环境纳污水体
ambient rock 峒室周岩
ambient standard 环境标准
ambient stress 包围应力
ambient stress condition 环境应力条件
ambient stress field 环境应力场
ambient temperature 外界环境温度;室温;环境温度;周围温度;周围介质温度;背景温度
ambient temperature and pressure 环境温度和气压
ambient temperature ferrite process 环境温度铁氧体法
ambient temperature seawater 常温海水
ambient test 大气环境下的试验
ambient test under surrounding media conditions 周界介质条件下的试验
ambient toxic chemical concentration 环境有毒化学物质浓度
ambient velocity 天然流速;环境流速
ambient vibration 环境振动
ambient water criterion 环境水质准则;环境水质标准
ambient water quality 环境水质
ambient water quality criterion 环境水质基准
ambient water quality monitoring network 环境水质监测网
ambient water quality standard 环境水质标准
ambient water temperature 环境水温
ambiguity 双值性;模糊点;含糊;多值性;多义性;二重性;二义性;不定性
ambiguity degree 模糊度
ambiguity delay 模糊延迟
ambiguity diagram 含糊图;分歧图
ambiguity error 模糊误差;模糊错误;多义性错误;二义性错误
ambiguity function 双值性函数;模糊函数;模糊度函数
ambiguity indicator 多值指示器
ambiguity mathematics 模糊数学
ambiguity of power and duties 权责不明
ambiguity problem 二义性问题
ambiguity relation fuzzy relation 模糊关系
ambiguity sensor 双值性传感器
ambiguous 模糊不清的;多义的;二义性的
ambiguous case 多值情况;分歧情况;含糊情况
ambiguous display 非单值显示
ambiguous expression 歧义表示式
ambiguous name of ship 暂定船名
ambiguous sign 重号
ambiophonic system 立体混响系统
ambi(o)phony 立体混响;环境立体声
ambiphasma 双等离子体
ambiphasma instability 双等离子体不稳定性
ambipolar 双极的
ambipolar diffusion 双极扩散
ambipolar effective mobility 双极性有效迁移率
ambipolar field 双极场
ambit 境界;区域;界限;范围
ambitus (教堂周围的)圣地;(地下墓中存放骨灰或棺材的)壁龛
Amble 阿姆勃尔方法
Amble's method 阿姆勃法
amblygonite 磷铝钾石;锂磷铝石

amboceptor 介体
amboceptor paper 介体试纸
amboceptor unit 介体单位
ambo(n) 高座;读经台;颂经台
ambon ciborium 祭坛华盖;读经台顶盖;读经台华盖
Amboyna 装饰用木材(产于东印度)
Amboyna wood 安波拿木材
ambra 琥珀
Ambrac 安布拉克铜镍耐蚀合金
ambrain 人造琥珀
ambrain cement 人造琥珀胶
Ambraloy 铜铝系耐蚀合金;铜合金;安布拉洛伊耐蚀铜合金
ambrettolic acid 黄葵酸
ambrices 屋檐与屋瓦之间安置的交叉板条
ambrite 灰黄琥珀
ambroid 人造琥珀树脂
ambroin 绝缘塑料;人造琥珀
ambroin cement 假琥珀胶
Ambrose alloy 安布罗斯铜镍耐蚀合金
Ambrose draghead 安布罗斯耙头
ambrosia 豚草病
ambrosin beetle 木菌甲虫
ambrosine 褐黄琥珀
ambrotype 玻璃干版照相法
ambry 食品室;教堂圣台上的碗柜或壁龛;餐具室;壁橱
ambucaine 氨布卡因
ambucetamide 氨由醋胺
ambu facepiece 安布式面具
ambulacral grooves 步带沟
ambulacralia 步带板
ambulacrum 步带
ambulance 救护船;野战医院
ambulance aircraft 救护飞机
ambulance boat 救护艇
ambulance car 救护车
ambulance/clinic boat 医务艇
ambulance coach 救援车
ambulance corps 流动救护队
ambulance party 救护班
ambulance station 救护站
ambulance train 救援机车
ambulant 不卧床的
ambulante (18世纪的)轻便茶桌
ambulator 测距离器;测距计;测距器;测程轮
ambulatorium 门诊所
ambulatory 测距计;回廊;不卧床的
ambulatory aisle 回绕通道(教堂中);回廊;通道;走廊
ambulatory bin 活动槽;活动料槽;移动式料仓;流动谷仓
ambulatory business 流动营业
ambulatory chattel 动产
ambulatory church 十字形布置圆顶教堂;十字形圆顶教堂
ambulatory clinic 流动诊所
ambulatory health care centers 非卧床病人中心
ambulatory laboratory 流动实验室
ambulatory vault 回廊拱顶;走道拱顶
ambulet 流动救护车
Ambursen-type dam 平板支墩坝;安布森式坝
Ambursen-type weir 平板支墩堰;安布森式堰
Ambursen weir 安布森溢流堰;安布森量水堰
ambuside 安布赛特
ambutonium bromide 安布溴铵
ambutte-seed oil 黄葵油
Amcide Ammate 氨基磺酸铵
amcinonide 安西缩松
Amco soaking pit 阿姆柯蓄热式均热炉;蓄热式均热炉
amco type tank furnace 双碹顶池窑
Amcrom chromium copper 阿姆克龙含铬无氧铜合金
ameba 变形虫;阿米巴
ameghinite 阿硼钠石
ameiosis 不减数分裂
ameliorated triple oxidation ditch 改进型三行式氧化沟
ameliorated wood 改良木料
ameliorating inferior groundwater 改良劣质地下水
amelioration 修正;土壤改良
amelioration of soil 土壤改良
amelogenic organ 造釉器

amemolite 分支钟乳石
amenability 可控制性
amenability of treatment 易处理性
amend 修正;修改;改正
amended draft 修正草案
amended focal length 改正焦距
amended items 更正的项目
amended plan 修正图;修正计划
amended standard 修正标准
amended subject 更正的项目;更改项目
amending agreement 修正条约的协定
amending clause 修正条款
amendment 调理剂;修正值;修正(量);修正案;更正;改正(量);改善;订正;修改
amendment advice 修改通知(书);修改通知单
amendment commission 修改手续费
amendment file 改正文件
amendment of bidding documents 投标文件修改
amendment of contract 修正合同;修改合同
amendment of tendering documents 投标文件修改
amendment of the agreement 合同条款的变更
amendment of the contract or other appendices 对合同及其附件的修改
amendment record 改正记录
amendments group 修订小组
Amendments of the Clean Air Act 洁净空气修正案
amendment tape 改正带
amendment to authorized signatures 更改签字;修正印鉴
amendment to forecast 预报改正;预报订正
amendment to standard 标准修改件
amend the amount of the letter of credit 修改信用证金额
amend the letter of credit 修改信用证
amend the terms of the letter of credit 修改信用证条款;修改信用证条件
amenities 舒适设施
amenities of sea coast 海洋的优美环境
amenities of ship 船上文娱活动
amenity 舒适性
amenity area 美化市容地带
amenity design 舒适性设计
amenity facility 舒适设施
amenity forests 风景林
amenity garden 观赏植物园
amenity grass land 宜人的草原
amenity planting strip 美化市容种植地带
amenity plot 美化市容地带
amenity right 舒适权
amenity room 更衣梳洗室
amenity standard 舒适性标准
amenity strip 美化市容地带
amenity value 环境舒适价值
amentities of the countryside 乡村的舒适
amentotaxus 穗花杉
Amerada Hess Corporation 阿姆达—赫斯公司
amercement 罚款;罚金
America caisson 箱形沉箱
American Academy of Arts and Science 美国科学技术研究院
American Academy of Sciences 美国科学院
American Accounting Association 美国会计学会
American-African-Antarctic triple junction 美洲—非洲—南极洲三向联结构造
American architecture 美国建筑
American Association for Contamination Control 美国污染控制协会
American Association for the Advancement of Science 美国科学发展协会;美国科学促进会
American Association of Port Authorities 美洲港务协会;美国港务协会
American Association of State Highway Officials 美国各洲公路及运输工作者协会
American Association of Textile Chemists and colo(u)rists 美国纺织化学师和着色师协会
American Automobile Association 美国汽车协会
American bank note 美国钞票
American Bank Note Company 美国钞票公司
American basement 半地下室;美式地下室
American beech 美洲大叶山毛榉
American black walnut 美国黑核桃木
American bond 普通砌式;普通砌砖法;美式砌法;美国式砌墙法

American boreal faunal region 美洲北方动物区
American brick 美国砖
American Building Contractors Association 美国建筑承包商协会
American Bureau of Shipping 美国船级社;美国船舶局
American caisson 美国开口沉箱;美式沉箱
American Ceramic Society 美国陶瓷协会
American Chemical Abstract 美国化学文摘
American Chemical Society 美国化学学会
American cherry 美国黑果稠李
American cloth 彩色防水台布;美国油布;彩色油布
American Concrete Institute 美国混凝土学会
American dewberry 美国悬钩子
American disc filter 美国式多盘过滤器
American Documentation Institute 美国文献资料研究学会
American dollar 美元
American Engineering Standard Committee 美国工程标准委员会
American Express 美国快递卡
American Federal Paint Specification 美国联邦油漆规格
American Federal Specification 美国联邦规范
American Federal Specification Board 美国联邦规格委员会
American Federation of Technical Engineers 美国技术工程师联合会
American filter 圆板过滤器
American foraminifera realm 美洲有孔虫地理区系
American gallon 美制加仑(1美制加仑=3.785升)
American Gas Association 美国煤气协会
American ga(u)ge 美国量规
American Geological Institute 美国地址学会
American Geophysical Union 美国地球物理协会;美国地球物理联合会
American hemlock 美国铁杉
American hotel china 美国旅馆瓷
American household china 美国家用瓷
American humid tropics 美洲湿热带地区
American-Indian architecture 美洲印第安人建筑;印第安人建筑
American Institute of Architects 美国建筑师学会;美国建筑师协会
American Institute of Decorators 美国室内装饰家学会
American Institute of Electrical Engineers 美国电气工程师学会
American Institute of Housing Consultants 美国住房顾问学会
American Institute of Planners 美国城市规划工作者学会
American Institute of Real Estate Appraisal 美国房地产估价师协会
American Institute of Real Estate Urban Economics Association 美国房地产及城市经济协会
American Institute of Steel Construction 美国钢结构学会;美国钢结构研究所
American Institute of Timber Construction 美国木结构学会
American Iron and Steel 美国钢铁协会
American jade 玉符山石
American Joint Intelligence Center 美国联合情报中心
American Joint intelligence Committees 美国联合情报委员会
American Journal of Science 美国科学期刊
American larch 美洲落叶松
American leather 油布
American leech 美洲水蛭
American linden 椴树;美国菩提树
American Institute of Crop Ecology 美国农作物生态学研究所
American lumber standard 美国木材标准
American Machine-tool Builder's Association 美国机床制造协会
American Machinist 美国机械师
American Marine Limited 海上限制(美国)
American Maritime Cases 美国海事案例
American material standard 美国材料标准
American melting point 美洲熔点
American Meteorologist Society 美国气象学会
American method of tunnel driving 美国式隧道法

American Military Agency for Standardization 美国军用标准化局
American Military Specification 美国军用规格
American Military Standard 美国军用标准
American Mobile Satellite Corporation 美国移动卫星公司
American National Standard Institute 美国国家标准学会
American National Standards 美国国家标准
American Office of Water Policy 美国水政策局
American Oil Chemists Society 美国油脂化学家学会
American Oil Company 美国石油公司
American open caisson 美国开口沉箱
American option 现货期权
american paraffin(e)oil 煤油
American Parquet Association Standard 美国镶木地板协会标准
American Patent 美国专利
American pattern pipe pliers 美式管钳
american pennyroyal oil 海地油
American Petroleum Institute 美国石油协会
American Petroleum Institute drill pipe thread 美国石油协会标准钻杆丝扣
American Petroleum Institute gravity 燃油比重度数;美国石油学会重度
American Petroleum Institute gravity index 美国石油协会容重指标
American Petroleum Institute Gravity Scale 美国石油协会标准比重度;美国石油协会标准比重表
American Petroleum Institute Process 美国石油学会工艺过程
American Petroleum Institute specification 美国石油协会技术规范
American pitch pine 美国油松(含脂松木)
American pitch pine strip floor(ing) 美国多脂松夹条地板
American plan 美式旅馆计价制
American plane 美国悬铃木
American plate 美洲板块
American Plywood Association 美国胶合板协会
American process zinc oxide 美国法氧化锌;直接法氧化锌
American Public Health Association 美国公共卫生协会
American Public Health Association's System 美国公共卫生协会色标
American Public Transit Association 美国公共交通运输学会
American Public Works Association 美国市政工程协会
American pump 美制捞砂筒
American Railway Engineering Association 美国铁路工程协会;美国铁道工程协会
American Real Estate and Urban Economics Association 美国房地产及城市经济协会
American sack 美国袋装水泥
American Society for Testing and Materials 美国材料试验学会
American Society Journal of Civil Engineers 美国土木工程师学会会刊
American Society of Agricultural Engineers 美国农业工程师学会
American Society of Appraissers 美国资产评估员协会
American Society of Civil Engineers 美国造船工程师学会;美国土木工程师学会;美国土木工程师协会
American Society of Civil Engineers and Architects 美国土木工程师和建筑师学会
American Society of Consulting Planners 美国规划咨询学会
American Society of Engineers and Architects 美国工程师及建筑师学会
American Society of Heating 美国供暖、制冷及空气调节工程师学会
American Society of heating and Ventilating Engineers 美国暖气通风工程师学会
American Society of Industrial Designers 美国工业设计师学会
American Society of Lubrication Engineering 美国润滑工程师学会
American Society of Mechanical Engineers 美国机械工程师学会
American Society of Real Estate Appraisers 美国房地产估价协会
American Society of Refrigeration Engineers 美国冷藏工程师学会
American Society of Safety Engineers 美国安全工程师学会
American Society of Sanitary Engineers 美国卫生工程学会
American Standard Association 美国标准协会
American standard beam 美国标准梁
American Standard cable system 美国标准钢绳钻进方法
American standard channel 美国标准槽钢
American Standard Code for Information Interchange[ASCII] 美国信息交换标准代码
American Standard Elevator Codes 美国标准电梯规范
American Standard fittings 美国标准配件
American Standard of Perfection 美洲家禽品种标准
American Standard of Testing Material 美国材料试验标准
American Standards 美国标准
American Standards Associations 美国标准委员会
American standard steel sections 美国标准型钢
American Standards Test Manual 美国标准试验手册
American Standard Straight Pipe Thread 美国标准直管螺纹
American Standard Taper Pipe Thread 布氏锥管螺纹;美国标准锥管螺纹
American steel and wire ga(u)ge 美国线径规
American Steel Structures Painting Council 美国钢结构涂装委员会
American Stock Exchange 美国股票交易所
American style cutter 美国式切刀
American sycamore 美国枫树
American system 美制
American system drill 顿钻
American system of drilling 美洲钻井方法;钢索冲击钻探法(美式)
American timber 美国木材
American ton 短吨(美国)
American turbine 混流反击式水轮机;美国水轮机
American type 美式白铁剪
American type bench vice 美式台虎钳
American type claw hammer 美式羊角锤
American vermilion 美洲朱砂(鲜红颜料)
American Water Environment Federation 美国水环境联合会
American Water Quality Association 美国水质管理协会
American Water Quality Research Council 美国水质研究理事会
American Water Research Council 美国水研究理事会
American Water Resource Association 美国水资源协会
American Water Resources Council 美国水资源理事会
American Water Works Association 美国自来水协会;美国给水工程协会
American Welding Society 美国焊接学会
American wheel 美国水轮机
American white pine 美国白松
American whitewood 美制白木
American wire ga(u)ge 美国线规
America Online 美国在线
America Petroleum Institute Scale 美国石油协会比重计标度
America Projection 第三象限投影法
americium 锔
Amerindian architecture 印第安人建筑
ameripol 人造橡皮;人造耐油橡胶
amerism 不分节
ameristic 不分节的
Amersfoort interstade 阿美尔斯福特间冰阶
amesdial 千分表;测微仪;测微表
amesiality 不正中
amesite 镁绿泥石;镁铝蛇纹石
Ames Laboratory 爱姆司实验室
a-mesosaprobic zone 甲型半污水生物带;甲型半腐生生物带
Ames surface ga(u)ge 爱姆司平面量规;爱姆司表面量规
Ames' test 爱姆司试验
amethocaine 阿美索卡因
amethyst 水碧;紫水晶;紫石英;紫晶色
amethyst colo(u)r 紫水晶色颜料
amethystine 紫石英色的
amethystine quartz 紫晶
amethystum 紫水晶
ametoecism 单主寄生
Amfion (均质的)离子交换膜
amggdaloidel 杏仁体
amherstite 反纹二长岩
Amherst sandstone 美哈斯特砂岩(一种产于美国俄亥俄州的浅灰色或浅黄色或杂色石料)
amiable composition clause 友好调解条款
Amianite 阿米奈特;(一种用石棉作填料的)塑料
amiant 细丝石棉
amianth 石棉绒
amianthine 石棉绒
amianthoide 石棉
amianthus 石棉绒;石麻;高级石棉;纯洁石棉
amiantos 细丝石棉;石棉
amiantus 石棉(绒);高级石棉;细丝石棉
amiathine 石棉绒的
amiatite 玻璃蛋白石
amicable allowance 让价
amicable number 亲和数
amicable settlement 调解书;友好解决
Amici Bertrand lens 阿米西—伯特兰透镜
amicibone 阿米西酮
amicinafide 安西非特
Amici objective 阿米西物镜
Amici prism 阿米西棱镜
amicite 斜碱沸石
Amick 阿米克尼龙丝
amicron 亚微粒;次微(胶)粒;超微料;超微粒;超视粒
amictic lake 永久封冻湖;永冻湖
amidation 酰胺化(作用)
amide 氨化物
amide-ester polymer 酰胺—酯聚合物
amide-imide resin 酰胺—酰亚胺树脂
amidine dye stuffs 脒染料
amido 胺基
amido aldehyde 氨基醛
amido black 胺黑
amidocarbonic acid 氨基甲酸
amido colo(u)r 酰胺染料
amidocyanogenphosphonic acid resin 胺基膦酸树脂
amidogen ether 胺醚
amidol 二氨基苯酚显示剂
amido naphthol red 酰胺萘酚红
Amidosulphuric acid 氨基磺酸
amidoxalyl 氨基草酰
amidoxime 偕胺肟
amidpulver 硝胺钾;酰胺粉;炭末炸药;硝铵
amidrazone 氨基腙
amid-river 在河道中间
amidship 在中纵线上;在船舶中段;船中央
amidship frame 船中肋骨
amidship ice belt region 船中冰带区
amid ship section 横中剖面【船】
amidships-engined ship 舯机型船
amidship weather deck 船中露天甲板
amid-stream 在中流;在河道中间
amid-stream entrance 流内进口(与流向正交的进口)
Amiens cathedral 亚眠大教堂(为 1200~1269 年法国哥特式建筑)
Amiesite 阿米赛脱(一种冷铺沥青)
Amilan 阿米纶;聚酰胺
amilomer 阿米洛姆
Amilon 阿米隆聚酯纤维
aminate 胺化产物
aminated 胺化了的
aminating agent 胺化剂
amination 胺化(作用);氨基化(作用)
amine 胺
amine absorption process 胺吸收法
amine acetate process 醋酸铵法
amine-acid complex inhibitor 胺酸络合阻蚀剂;胺络酸阻蚀剂
amine aldehyde resin 胺醛树脂
amine anti-oxidants 胺类抗氧化剂

amine black green 胺黑绿
amine blushing 胺致发白
amine carbonate 碳酸胺
amine cellulose 胺基纤维素
amine cured agent 胺固化剂
amine cured epoxy coating 胺固化环氧涂料
amine cured epoxy resin 胺固化环氧树脂
amine effect 胺效应
amine equivalent(weight) 胺当量
amine formaldehyde resin 胺醛树脂
amine-furfural resin 胺—糠醛树脂
amine group 胺基
amine hardener 胺固化剂
amine hydrochloride 盐酸胺类
amine oxide 胺化氧
amine perchlorate 胺高氯酸盐
amine pump 胺泵
amine ratio 胺比
amine resin 胺基树脂;氨基树脂
amine ring 胺基环
amine salt 胺盐
amine seed gas 胺基活化气体
amine stabilizer 胺稳定剂
amine surfactant 胺型表面活性剂
amine toluene 甲苯胺
amine treatment 胺处理
aminetrisulfonic acid 次氮基三硫酸
amine type curing agent 胺类固化剂
amine value 胺值
amine vapo(u)r cure 胺蒸气熟化
aminex 多孔性阴离子交换树脂
aming corrector 瞄准校正器
aminimide 胺化酰亚胺
aminitrozole 醋胺硝唑
aminized cotton 胺化棉
amino 氨基的
aminoacetal 氨基乙缩醛
aminoacethydrazide 氨基乙酰肼
aminoacetic acid 氨基乙酸
aminoacetone 氨基丙酮
aminoacetophenone 氨基苯乙酮
amino acid 氨基酸;胺酸
amino-acid in sludge 污泥中的氨基酸
amino-acid racemization 氨基酸消旋作用
amino-acid racemization age method 氨基酸消旋测年法
amino-acid racemization dating method 氨基酸外消旋测年法
amino-acid turnover rate 氨基酸转换率
amino-acid wastewater 氨基酸废水
aminoacyl 氨酰基
aminoacyl site 氨酰基部位
aminoadipaldehyde 氨基己二酸半醛
amino alcohol 氨基醇
aminoaldehexose 氨基己醛糖
aminoaldehyde resin 氨基醛树脂;胺醛树脂
amino alkoxydioxaborinane 氨代烷氧基二氧一硼杂环己烷
aminoalkyd baking surfacer 氨基烘干二道底漆
aminoalkyd resin 氨基醇酸树脂
aminoalkyd resin coating 氨基醇酸树脂涂料
aminoalkyd resin quick drying baking varnish 氨基醇酸树脂快速烘干清漆
aminoalkyl phosphate 氨烷基磷酸酯(或盐)
aminoalkyl sulfate 氨烷基硫酸酯(或盐)
aminoalkyl sulfonate 氨烷基磺酸酯(或盐)
aminoanisole 氨基茴香醚;氨基苯甲醚
amino anthracene 氨基蒽
aminoanthraquinone 氨基蒽醌
amino-arsenoxide 氨基胂化氧
aminoazabenzol 氨基偶氮苯
aminoazobenzene 氨基偶氮苯
aminoazo-compound 氨基偶氮化合物
aminoazonaphthalene 氨基偶氮萘
amino baking electro static hammer paint 氨基烘干静电锤纹漆
amino baking enamel 氨基烘干瓷漆
amino baking hammer paint 氨基烘干锤纹漆
amino baking insulating paint 氨基烘干绝缘漆
amino baking primer 氨基烘干底漆
amino baking putty 氨基烘干腻子
amino baking transparent paint 氨基烘干透明漆
amino baking varnish 氨基烘干清漆
amino baking water soluble primer 氨基烘干水溶性底漆
aminobenzaldehyde 氨基苯(甲)醛
aminobenzaldoxime 氨基苯(甲)醛肟
aminobenzene 氨基苯
aminobenzenesulfonic acid 氨基苯磺酸
aminobenzidine 氨基联苯胺
aminobenzoic acid 对氨苯甲酸;氨基苯甲酸
aminobenzonitrile 氨基苄腈
aminobenzophenone 氨基二苯甲酮;氨基二苯酮
aminobenzothiazole 氨基苯并噻唑
amino-benzoyl acetic acid 氨基苯酰醋酸
amino-benzoyl formic acid 氨基苯酰甲酸;氨基苯甲酰甲酸
aminobenzyl alcohol 氨基苄醇
aminobenzylation 氨苄基化
aminobiphenyl 苯基联苯胺
aminobphenyl 氨基联苯
aminobutyric acid 氨基丁酸
aminocaproic acid 氨基己酸
aminocapronitrile 氨基己腈
aminocaprylic acid 氨基辛酸
aminocarbamic acid 氨基甲酸
amino carboxyl chelating 氨羧络合剂
aminocellulose 氨基纤维素
amino-chloro-benzene 氨基氯苯
aminocidin 氨基杀菌素
amino-cinnamic acid 氨基肉桂酸
amino-complex 氨络物
amino-compound 氨络物;氨基化合物
amino-cumene 氨基异丙苯
aminocyclitol 氨基环醇
aminodiphenylamine 氨基二苯胺
aminodiphenylmethane 氨基二苯甲烷
amino electrostatic baking enamel 氨基烘干静电瓷漆
amino electro static spraying baking varnish 氨基静电喷涂烘干清漆
aminoethane 氨基乙烷
aminoethanol 氨基乙醇
aminoethy-lacetanilide 氨基苯乙酰乙胺
aminoethyl alcohol 氨基乙醇
aminoethylated cotton 氨乙基化棉
aminoethylbenzene 氨基乙苯
aminoethylcellulose and fixed quotas 乙氨基纤维素
aminoethyl iso-thiourea 氨基乙基异硫脲二溴氢酸盐
aminoethyl mercaptan 氨乙基硫醇
aminoethyl nitrate 氨乙硝酸
aminoffite 铍黄长石
amino formaldehyde resin 氨基甲醛树脂
aminoformic acid 氨基甲酸
amino-gas desulphurizing plant 氨法气体脱硫装置
aminoglutaric acid 氨基戊二酸
amino group 氨基
aminoguanidine 氨基胍
aminoguanidine bicarbonate 氨基胍重碳酸盐
aminoguanidine carbonate 氨基胍碳酸盐
aminoguanidine sulfate 氨基胍硫酸盐
amino halide 氨基卤化物
aminohexose 氨基己糖
amino-humic acid 氨基腐植酸
amino hydroxy acid 氨基羟基酸
amino hydroxy propionic acid 氨基羟基丙酸
aminoisobutyric acid 氨基异丁酸
amino-isovaleric acid 氨基异戊酸
aminolevulinic acid 氨乙酰丙酸
aminolipid 氨基脂
amino matt baking enamel 氨基无光烘干瓷漆
amino matt baking water soluble enamel 氨基无光烘干水溶性瓷漆
aminomercuric chloride 氨基汞化氯
aminomethane 甲胺
aminomycetin 氨基菌素
aminonaphthol 氨基萘酚
amino-naphthol sulfonic acid 氨基萘酚磺酸
amino-nitrile 氨基腈
amino nitrogen 氨基氮
amino-octane 氨基辛烷
aminoothyl mercaptan 氨基乙硫醇
aminooxamide 氨基草酰胺;氨基草酰肼
aminopentane 氨基戊烷
aminopeptodrate 卡米诺伊德
aminophenol 氨基苯酚
aminophenol sulfonic acid 氨基苯酚磺酸
aminophenyl acetic acid 氨基苯乙酸;氨基苯醋酸
aminopheny-lacetonitrile 氨基苯乙腈
aminophenyl arsine sulfide 氨基苯胂化硫
aminoplastic resin 氨基树脂
aminoplastics 氨基塑料
aminopolycarboxylic acid 氨基聚羧酸
aminopromazine 氨丙嗪
aminopropanol 氨丙醇
aminopropylbenzene 氨基丙苯
aminopropylsilica 氨丙基硅
aminopropyltriethoxysilicate 氨丙基三乙氧基硅酸盐
aminopterin 氨蝶呤
aminopyridine 氨基氮杂苯;氨基吡啶
aminopyrine test 氨基比林试验
amino quick dry baking paint 氨基快干烘漆
aminoquinoline 安纳晶;氨基喹啉
aminoquinoxaline 氨基喹恶啉;氨基喹酸铵
amino resin 氨基树脂
amino resin water repellent 氨基树脂抗水剂
aminoresorcinol 氨基间苯二酚
aminorosin 氨基松香
aminorthyl pyrazine 氨乙基对二氮杂苯;氨乙基吡嗪
amino rubber 氨基橡胶
aminosalicylate 氨基水杨酸盐
aminosalicylic acid 氨基水杨酸
amino semi-gloss baking enamel 氨基半光烘干瓷漆
amino sodium ferrocyanide spectrophotometry 氨基亚铁氰化钠分光光度法
aminosuccinamic acid 氨基琥珀酰胺酸
aminosuccinic acid 氨基丁二酸
amino-sumene 氨基异丙苯
amino terminal 氨基末端
aminothiazole 氨基噻唑
aminothiophene 氨基噻吩
aminothiourea 氨基硫脲
aminotoluene 氨基甲苯;苄胺
aminotoluene sulfonic acid 氨基甲苯磺酸
aminotransferation 氨基移作用
aminotriazole 氨基三唑
aminovaleric acid 氨基戊酸
amino-xylene 氨基二甲苯
aminoxytriphene 胺氧曲芬
amion acetic acid 氨基乙酸
amion formic acid 氨基甲酸
amiosis 不减数分裂
Amirante trench 阿米兰特海沟
amisometradine 阿米索美啶
amissum quod nescitur non-amittur 不知道的损失
amiton 胺吸磷
amitosis 直接分裂
amitraz 阿米曲拉
amitriptyline 阿米曲替林
amitrole 氨三唑
a mixing bowl 搅拌缸
Amman-Dead Sea transform fault 安曼—死海转换断层【地】
Ammanian 安曼阶
ammania oxidation converter 氨气氧化炉
Ammate 氨基磺酸铵除草剂
ammelide 氰脲酰胺
ammersoaite 钙伊利石
ammersooite 阿米水云母
ammeter 电流计;电流表;安培计;安培表
ammeter shunt 安培计分流器;安培表分流器
ammeter switch 安培计转换开关;安培表转换开关
ammine 胺络物;氨络物;有机肥料
ammino 氨合
ammino-complex 氨合物
ammiolite 铵锑汞矿
ammmonoid extinction 菊石绝灭
ammnonification rate 氨化速率
ammochrysos 白云母
ammonal 硝铵炸药;阿芒拿尔炸药
ammonation 氨合作用
ammon-dynamite 铵炸药
ammongelatine 铵胶
ammongelatine dynamite 铵胶炸药
Ammonia 阿摩尼亚
ammonia 氨
ammonia absorbent refrigerator 氨吸附制冷机
ammonia absorber 氨吸收器

ammonia absorption amchine 氨气吸收机
ammonia absorption refrigerating machine 氨吸收式制冷机;氨吸收式冷冻机
ammonia absorption refrigerator 氨吸收式制冷机;氨吸附制冷机
ammonia-air fuel cell 氨空气燃料电池
ammonia-alkali process 氨碱法
ammonia alum 铵明矾;氨矾
ammonia amalgam 氨汞齐
ammonia ash 氨法碳酸钠;氨法碱
ammonia bacteria 氨化菌
ammonia base sulfite waste(water) 亚硫酸铵废水
ammonia beam maser 氨束微波激射器;氨分子束微波激射器
ammonia blasting gelatin(e) 铵胶炸药
ammoniac 氨树胶
ammoniacal brine 氨盐水
ammoniacal copper aesenite 亚砷铜氨液(木材防腐用)
ammoniacal emulsion 氨乳剂
ammoniacal liquor 氨液
ammoniacal liquor plant 氨液工厂
ammoniacal nitrogen 氨(性)氮
ammoniacal salt 氨性盐
ammonia carburizing 氨气渗碳
ammonia carmine 氨洋红
ammonia centrifugal liquid chilling unit 氨离心制冷机
ammoniac gum 氨草胶
ammonia checker 氨测定器
ammonia chloride 氯化铵
ammonia circulating pump 氨循环泵
ammonia clock 氨钟
ammonia collector 氨气采集器
ammonia colo(u)rimeter 氨比色计
ammonia combustion catalyst 氨燃烧制氮催化剂
ammonia complex 氨络物
ammonia compressed refrigerator 氨压缩制冷机
ammonia compressing unit 氨压缩机组
ammonia compression refrigerating machine 氨压缩冷冻机;氨气压缩制冷机
ammonia compression refrigerator 氨气压缩冷冻机
ammonia compressor 氨(气)压缩机
ammonia compressor set 氨压缩机组
ammonia compressor unit 氨压缩机组
ammonia condenser 氨气冷凝器;氨冷凝器
ammonia-containing wastewater 含氨废水
ammonia converter 氨转化塔;氨合成塔
ammonia cooler 氨气冷却器;氨冷却器;氨冷凝器
ammonia cooling plant 氨制冷厂
ammonia cooling type 氨冷式
ammonia corrosion 氨腐蚀
ammoniacum 氨脂;氨草胶
ammonia dissociation separator 氨离解分离器
ammonia dissociator 氨离解器
ammonia distiller 氨蒸馏器
ammonia dynamite 硝铵炸药;铵炸药;铵爆炸药;氨代那买特炸药;氨爆炸药
ammonia dynamite wastewater 硝铵炸药废水
ammonia elimination process 除氨法
ammonia evapo(u)rator 氨蒸发器
ammonia filter 氨过滤器
ammonia formation 氨的形成
ammonia gas development 氨显影
ammonia gas maser 氨气微波激射器
ammonia gas mask 防氨面具
ammonia-gelatin(e) dynamite 氨胶质炸药;铵硝化钾炸药
ammonia gelatin(e) 硝酸铵胶质炸药;氨凝胶
ammonia gelatin(e) dynamite 胶质硝铵炸药;铵胶炸药
ammonia generator 氨发生器
ammonia group 氨基
ammonia gum 氨树胶
ammonia hydrate 氨制碱
ammonia intoxication 氨中毒
ammonia inversion transition 氨反演跃迁
ammonia is composed of nitrogen and hydrogen 氨由氮和氢合成
ammonia laser 氨激光器
ammonia leaching 氨浸滤
ammonia leak detector 氨探漏器
ammonia leak test 氨渗漏试验

ammonia liquor 煤气液;液体氨;粗氨水;氨液;氨水
ammonia machine 氨制冷机;氨制冰机
ammonia manometer 氨压计
ammonia maser 氨微波激射器
ammonia-maserspectrum analyzer 氨微波光谱分析仪
ammonia molecular laser 氨分子激光器
ammonia nitrogen 氨态氮;氨基氮;氨氮
ammonia nitrogen concentration 氨氮浓度
ammonia nitrogen degradation bacteria 氨氮降解菌
ammonia nitrogen removal 除氨氮
ammonia nitrogen wastewater 氨氮废水
ammonia oil 氨油;氨压缩机润滑油
ammonia oxidation 氨氧化
ammonia-oxidizing 氨氧化
ammonia oxidizing archaea 氨氧化古菌
ammonia oxidizing bacteria 氨氧化细菌
ammonia oxidizing bacterial community 氨氧化菌群
ammonia piping 氨管道
ammonia plant 氨植物
ammonia plant retrofit 合成氨厂改造
ammonia poisoning 氨中毒
ammonia pollution 氨污染
ammonia process 氨法;氨处理
ammonia process of desulfurization 氨法脱硫
ammonia pump 氨水泵;氨泵
ammonia purifier 氨净化器
ammonia reactor 氨合成塔
ammonia recirculation 氨再循环
ammonia recovery plant 氨回收厂
ammonia recovery process 氨回收法
ammonia recovery unit 氨回收装置
ammonia recycle system 氨循环系统
ammonia refrigerant 氨制冷剂;氨冷冻剂
ammonia refrigerating compressor 氨制冷压缩机
ammonia refrigerating machine 氨冷冻机
ammonia refrigeration 氨制冷
ammonia refrigerator 氨制冷机;氨冷冻机
ammonia removal 脱氨
ammonia removal and recovery process 除氨回收工艺
ammonia removal by air stripping 汽提法除氨
ammonia resisting mortar 耐氨砂浆
ammonia resol 氨催化甲阶酚醛树脂
ammonia saturator 氨饱和器
ammonia scrubber 洗氨器;氨洗涤器;氨气洗涤器
ammonia separator 氨分离器
ammonia soda 氨法苏打
ammonia soda ash 氨法苏打灰
ammonia soda process 氨碱法
ammonia solution 氨液;氨溶液
ammonia spirit 氨水(溶液)
ammonia spirit desulfurization 氨水法脱硫
ammonia still 氨气塔
ammonia still waste 氨蒸馏废物
ammonia strainer 氨滤器
ammonia stripping 氨气提;氨吸;氨吹脱;氨溶出
ammonia stripping process 脱氨法;氨解吸法
ammonia sulfate concentration 硫酸氨浓度
ammonia sulphate 硫酸铵
ammonia synthesis 氨合成(法);氨的合成
ammonia synthesis catalyst 氨合成催化剂
ammonia synthesis gas 氨合成气
ammonia synthesizer 氨合成装置
ammonia system 氨系统
ammoniate 有机氨肥;氨化物;氨合物
ammoniated beet pulp 氨化甜菜渣
ammoniated brine 氨盐水
ammoniated coal 氨化煤
ammoniated fertilizer 氨化肥料
ammoniated latex 充氨胶乳
ammoniated mercuric chloride 氨类汞化氯
ammoniated mercury 白降汞;氨基汞化氯
ammoniated peat 氨化泥炭
ammoniated silver nitrate 氨化硝酸银
ammoniated silver solution 氨银液
ammoniated superphosphate 氨化过磷酸钙
ammoniating agent 氨化剂
ammoniating vat 充氨瓮
ammoniation 氨化(作用)
ammoniation facility 氨化设备

ammoniation reaction 氨化反应
ammoniator 加氨器;氨熏机
ammonia treatment 氨处理
ammonia turbocompressor turbine 氨透平压机
ammonia valve 氨阀
ammonia volatilization 氨挥发
ammonia washer 洗氨器
ammonia waste liquor 含氨废液
ammonia water 氨水
ammonia water distributor 氨水撒洒机
ammonia water poisoning 氨水中毒
ammonibacteria 氨细菌类
ammonical silver staining technique 氨银染色技术
ammonification 加氨(作用);氨形成;氨化(作用)
ammonification rate constant 氨化速率常数
ammonificator 氨化菌
ammonifiers 氨化菌
ammonifying bacteria 成氨菌;氨化菌
ammonifying capacity 氨化强度;氨化能
ammonio 氨溶的
ammonioborite 水硼铵石
ammoniojarosite 黄铵铁矾;铵黄铁钾矾;铵黄铁矾
ammoniometer 氨量计
ammonionitrometry 氨氮测定法
ammonistion 氨化
ammonite 硝铵二硝基萘炸药;鲕状岩;阿芒炸药
ammonite faunal province 菊石动物地理区
ammonite scroll 菊石纹
ammonium 铵基;铵
ammonium absorption 吸铵量
ammonium acetate 乙酸铵;醋酸铵
ammonium acid fluoride 氟化氢铵
ammonium aldehyde 铵醛
ammonium alginate 藻酸铵
ammonium alum 铵铝矾;铵矾
ammonium alumin(i)um sulfate 铵铝矾
ammonium and ammonia in soils 土壤中铵和氨
ammonium bacteria 铵细菌;氨细菌
ammonium benzoate 苯氨酸铵;苯甲酸铵
ammonium bicarbonate 碳酸氢铵
ammonium bichromate 重铬酸铵
ammonium bifluoride 氟化氢铵
ammonium bisulfate 硫酸氢铵
ammonium bisulfite process 亚硫酸氢铵法
ammonium borate 硼酸铵
ammonium calcium phosphate 磷酸钙铵
ammonium carbonate 氨基甲酸铵;碳酸铵
ammonium caseinate 酪蛋白酸铵
ammonium chloride 氯化铵
ammonium chloride wastewater 氯化铵废水
ammonium chlorostamate 氯锡酸铵
ammonium chromate 铬酸铵
ammonium chromic alum 硫酸铬铵;铵铬矾
ammonium chromic sulfate 硫酸铬铵;铵铬矾
ammonium citrate 柠檬酸铵
ammonium compressing unit 铵压缩机组
ammonium cuprate 铵铜合物
ammonium cyanate 氰酸铵
ammonium dihydrogen phosphate 磷酸二氢铵
ammonium dithionate 连二硫酸铵
ammonium diuranate 重铀酸铵
ammonium explosive 硝铵炸药(爆破用)
ammonium ferrie oxalate 草酸铵
ammonium ferrous sulfate 硫酸亚铁铵
ammonium ferrous sulfate titration 硫酸亚铁铵滴定法
ammonium ferrous sulfate waste pickle liquor 硫酸亚铁铵废浸酸液
ammonium fixation 铵离子吸收;铵离子吸附;铵离子固定
ammonium fixation characteristics 铵固定特性
ammonium fluoride 氟化铵
ammonium fluoroborate 氟硼酸铵
ammonium fluosilicate 氟硅酸铵
ammonium halide 卤化铵
ammonium hydrogen fluoride 氟化氢铵
ammonium hydrogen sulfate 硫酸氢铵
ammonium hydrogen sulphite 亚硫酸氢铵
ammonium hydroxide 氢氧化铵
ammonium hypophosphite 次磷酸铵
ammonium iodate 碘酸铵
ammonium iodide 碘化铵
ammonium ion 铵离子
ammonium ion removal 除铵离子

ammonium iron alum 铵铁矾
ammonium magnesium phosphate 磷酸镁铵
ammonium magnesium sulfate 硫酸镁铵
ammonium manganum phosphate 磷酸锰铵
ammonium metaperiodate 偏高碘酸铵
ammonium metavanadate 偏钒酸铵
ammonium molybdate 钼酸铵
ammonium molybdate spectrophotometry 钼酸铵分光光度法
ammonium nickel sulfate 硫酸镍铵
ammonium nitrate 硝酸铵;多孔粒状硝酸铵
ammonium nitrate explosive 硝铵炸药
ammonium nitrate-fuel oil mixture 硝铵油炸药;铵油炸药
ammonium nitrate gelatin(e) 硝铵胶质炸药;胶质硝铵炸药
ammonium nitrate hose 硝铵炸药装药软管
ammonium nitrate permissible explosive 硝铵安全炸药
ammonium nitrate powder 粉状硝铵炸药
ammonium nitrate prill 粒状硝氨炸药
ammonium nitrite 亚硝酸二环己基铵
ammonium oxalate 草酸铵
ammonium perchlorate 高氯酸铵
ammonium perchlorate explosive 高氯酸铵炸药
ammonium permanganate 高锰酸铵
ammonium peroxoy-chromate ammonium perchromate 过铬酸铵
ammonium persulfate 过硫酸铵;高硫酸铵
ammonium persulfate test 高硫酸铵试法
ammonium persulphate 过硫化铵
ammonium phosphate 磷酸铵
ammonium phosphate fertilizer 磷铵肥料厂
ammonium phosphate sulfate 磷硫酸铵
ammonium phosphomolybdate 磷钼酸铵
ammonium phosphotungstate 磷钨酸铵
ammonium picrate 苦味酸铵
ammonium pitrogen ammoniacal nitrogen 氨态氮
ammonium polyphosphate 聚磷酸铵;多磷酸铵
ammonium polysulfide 聚硫化铵
ammonium polysulfide solution 多硫化铵溶液
ammonium pyrrolidine dithiocarbamate 吡咯烷二硫代氨基甲酸铵
ammonium removal 除铵
ammonium-rich leachate 富氨沥滤
ammonium-rich wastewater 富氨废水
ammonium salt 铵盐
ammonium salt attack 铵盐侵蚀
ammonium salt resisting polymeride binder 耐氨聚合物胶结料
ammonium silicofluoride 氟硅酸铵;氟硅化铵
ammonium soap 铵皂
ammonium sodium oxalate 草酸铵钠
ammonium soil 铵态土
ammonium stearate 硬脂酸铵
ammonium succinate 琥珀酸铵
Ammonium sulfamate 氨基磺酸铵
ammonium sulfate 硫酸铵
ammonium sulfate flocculation 硫酸铵絮凝
ammonium sulfate solution 硫酸铵溶液
ammonium sulfocyanate 硫氰酸铵
ammonium sulphate 硫酸铵
ammonium superphosphate 过磷酸铵
ammonium-syngenite 铵钾石膏
ammonium thiocyanate 硫氰化铵
ammonium thiosulfate 硫代硫酸铵
ammonium trinitrophenolate 苦味酸铵
ammonium vanadate 钒酸铵
ammonization 氨化作用
ammonizator 氨剂;氨化剂
ammonla maser clock 氨微波激射钟
ammonobase 氨基金属
ammonolysis 氨解(作用)
ammono-system 氨体系
ammonpulver 铵炸药
ammon-rosin-wax explosive 铵松腊炸药
Ammon's law 安蒙定律
ammo-phos 铵磷粉
ammoxidation 氨氧化(作用)
ammoxy-gen-nitrogen mixture 氨氧氮混合气
ammualization of expenses 费用的年度化
ammunition 弹药;弹药存放处
ammunition depot 军火(仓)库;子弹库;弹药库
ammunition disused 废弹药

ammunition dumps 弹药弃沉区
ammunition storehouse 弹药库
ammyonia absorption refrigerator 氨吸收冷冻机
annualization of expenses 费用年度化
Amobam 代森铵
Amoco International Oil Co. Ltd. 阿莫科国际石油公司
amodiaquine 氨酚喹
Am(o)eba 变形虫属
amoeba 阿米巴
am(o)ebiasis 变形虫病
amoeboid fold 不定形褶皱;不定形褶曲
amoeboid glacier 变形冰川
amoeboid movement 变形运动
amoeboid tapetum 变形绒毡层
amoire 衣柜;大食物柜
amolanone 阿莫拉酮
a moment 片刻
amonifying bacteria 氨化细菌
a monthly plan 月度计划
Amoora oil 阿莫拉油
Amor 阿莫尔
amorce 引爆药
Amor object 阿莫尔天体
amorphic soil 幼年土;初成土
amorphism 无定形性;无定形现象;非晶性
amorphous 无定形的;非晶质的;非晶态的;非晶、非定形的
amorphous binder 无定形黏合剂
amorphous body 非晶质体;非晶形体;非晶物体
amorphous boron 无定形硼
amorphous carbon 无定形碳;非晶质碳
amorphous cellulose 无定形纤维素;非晶纤维素
amorphous diamond 非晶金刚石
amorphous dust 无定形硅石粉
amorphous fiber 非晶态纤维
amorphous film 非晶形膜;非晶体膜;非晶膜
amorphous form 非晶态
amorphous frost 非晶质霜
amorphous graphite 无定形石墨;土状石墨;非晶质石墨
amorphous graphite in powder 非晶质石墨粉
amorphous ice 非晶质冰
amorphous inclusion 非晶包裹体
amorphous iron oxyhydroxide 非晶态氧氢氧化铁
amorphous laser 玻璃激光器
amorphous layer 非晶形层
amorphous magnetic bubble 非晶磁泡
amorphous material 无定形材料;非晶材料;非晶质;非晶态材料;非结晶材料
amorphous memory array 非晶存储器阵列
amorphous metal 非晶态金属
amorphous metallic coating 非晶态金属涂层
amorphous mineral 非晶质矿物
amorphous mixture 非晶混合物
amorphous nickel phosphide 非晶化镍
amorphous opaque matter 无定形不透明物质
amorphous optic(al) material 无定形光学材料
amorphous oxide film 非晶氧化薄膜
amorphous peat 无结构泥炭;无定形泥炭
amorphous phase 无定形相;非晶相
amorphous phosphate 无定形磷酸盐;非晶态磷酸盐
amorphous plastics 非晶态塑料
amorphous polymer 无定形聚合物;非晶态聚合物
amorphous precipitate 非晶态沉淀
amorphous precipitation 非晶态淀析
amorphous quartz 非晶质石英
amorphous refractory 不定型耐火材料;不定形耐火材料
amorphous region 无定形区;非晶区
amorphous rock 无定形岩
amorphous scanning signal 无定形扫描信号
amorphous sediment 无定形沉淀
amorphous selenium 无定形硒
amorphous semiconductor 非晶(态)半导体
amorphous semiconductor memory 非晶半导体存储器
amorphous silica 无定形氧化硅;无定形硅石;无定形二氧化硅;非晶态石英;非晶硅
amorphous sky 碎乱天空
amorphous snow 非晶质雪
amorphous solid 无定形体;无定形固体;非晶质固体;非晶形固体;非晶态固体;非晶固体

amorphous state 非晶形状态;非晶态
amorphous structure 非晶质结构;非晶态结构
amorphous substance 无定形物(质);非晶态物质
amorphous sulfur 无定形硫
amorphous switch 非晶半导体开关
amorphous type 无定形型
amorphous variety 非晶质变种
amorphous wax 软蜡;无定形蜡
amortise 提折旧费
amortisseur 减震器;缓冲器;消音器;阻尼器
amortisseur winding 阻尼绕组
amortization 清偿;提成;摊销的开办费;摊销;摊还;摊提;摊偿债款;零还储款;阻尼缓冲;阻尼;折旧;分期摊付;分期偿还
amortization allowance 备抵摊销
amortization charges 折旧提成;摊销费(用);分期偿还费;清偿费
amortization cost 偿还资金;折旧费;偿还费;分期偿还费
amortization expenses 摊还费
amortization factor 清偿因子;摊还率;折旧率;分期摊销系数
amortization fund 清偿费;偿债基金;偿资金(分期)
amortization idler 缓冲托辊
amortization loan 摊还借款;分期摊还债款
amortization method 摊销法;摊派法;分期偿还法
amortization money table 分期还款表
amortization of debt 分期偿还债务
amortization of discount 摊提贴水
amortization of fixed assets 摊提固定资产;固定资产摊销;固定资产摊提
amortization of government bond 政府公债分期偿还
amortization of initial expenses 开办费摊销;开办费摊提
amortization of loan 分期摊销的借款
amortization of organization expenses 开办费摊销
amortization payment 摊提款;摊派款;分期偿付的款项
amortization period 摊销期;清偿期(限);摊提期间;摊还期;偿还期;偿还年限
amortization plan of payment 分期摊还法
amortization quota 折旧定额;分摊额
amortization rate 分期偿还率;偿还率
amortization ratio 分摊比例
amortization rent 摊还租金;分期偿付租金
amortization schedule 分期偿付计划表;摊销表;摊提表;分期摊还日程表
amortization table 摊提表
amortization term 分期偿还期限
amortize 提折旧费;缓冲;折旧;分期注销费用;分期偿还债务
amortized cost 分期偿还费;摊余成本;摊提成本;已摊销成本;折旧费
amortized depreciation 分期偿还;折旧
amortized installation cost 设备折旧费
amortized load 分期偿还贷款
amortized loan 已摊还的借款;分摊的借款;摊还的借款
amortized mortgage 分期偿还抵押贷款
amortizement 扶壁的斜压顶;缓冲器;扶壁的斜坡顶;柱墩的斜压顶
amort winding 制动线圈
amosa asbestos 铁石棉
amosan 阿膜散
amosite 铁石棉;长纤维石棉;阿木基特石棉
amosite asbestos 长纤维石棉
amount 数目;金额;总值;总数;总金额;总计;共计
amount appraised 评定金额
amount assessed 分摊额
amount at risks 人寿保险的风险额;(保险中的)风险额
amount brought forward 前期转来金额
amount carried forward 结转金额
amount ceded 分出金额
amount declared 申报金额
amount deducted 扣除金额
amount due 应付款
amount due from somebody 欠款
amount forecasted 预测额
amount guaranteed 担保额
amount in arear 欠款金额;拖欠金额
amount in figures 小写金额

amount insured 投保金额
amount invested 投资额
amount in words 大写金额
amount limit 工作量(的)极限
amount made good 补偿量
amount needed by subsidiary materials 辅助材料需要量
amount not taken up 未用金额
amount of 1 每元本利和
amount of abiogenetic gas 无机成因气量
amount of acid 泡酸量；酸量
amount of acidizing fluid 酸液总量
amount of additional deviation of goods 货物的附加偏差量
amount of additive treating agent 处理剂加量
amount of a draft 汇票金额
amount of adsorbate 吸附质量
amount of adsorbed charge per unit surface area 单体表面积吸附电荷量
amount of agitation 搅拌强度；搅拌量
amount of aids maintenance 航标维护量
amount of air 空气量
amount of air-entrainment 掺气量
amount of aquatic habitat 水生生境度
amount of arrears 拖欠款
amount of attached table 附表数
amount of base 碱量
amount of bend(ing) 挠曲量
amount of biogenetic gas 生物成因气量
amount of blast 送风量
amount of business 贸易周转额
amount of camber 拱度；拱度
amount of capital invested 投资额
amount of capital invested by the owner 业主投资额
amount of cash inflow 现金流入量
amount of cash outflow 现金流出量
amount of casing compression 套管压缩距
amount of cement 水泥用量
amount-of-change scale 等差尺度；等差比例尺
amount of charge 电荷量
amount of claim 赔偿额
amount of coal-derived gas 煤成气量
amount of coal-formed gas 煤成气量
amount of coal-related gas 煤成气量
amount of coal seam 含煤层数
amount of combustion gas 烟气量
amount of compaction 压实度
amount of compression 压缩量；压缩度
amount of consumption 消耗量；消费额
amount of contract 合同金额
amount of contraction 收缩量
amount of contribution to capital 出资额
amount of cracking genetic gas of oil 油裂解气量
amount of crack offset 地裂缝位错量
amount of credit side 付项金额
amount of crown 路拱量；路拱高度；拱高(度)
amount of cure 熟化程度；固化程度
amount of cut 挖方量
amount of debit side 借方金额
amount of deflection 偏差值；偏差数值；挠度；垂度(量)；变位度；弯曲度
amount of deposit 预存费总数
amount of detail 地图负载量
amount of a deviation 偏差值；偏差量；离差量
amount of deviation of goods 货物偏差量
amount of discriminatory variant 判别变量数
amount of dispersion 偏差量；离差量
amount of displacement 变位量
amount of distortion 畸变量
amount of divisional map 图幅数量
amount of dredging 超挖深度【疏】
amount of drift 漂移量；偏航程
amount of drift of the gyro 陀螺漂移值
amount of drill working 钻探工作量
amount of drill working-month 钻月数
amount of eccentricity 偏心度；摆动量
amount of electric(al) energy per mole 每摩尔电能量
amount of energy 能量(值)
amount of energy-saving 节能量
amount of evapo(u)ration 蒸发量
amount of excavation 挖方量
amount of exercise 运动量

amount of explosive 炸药量
amount of exports 出口额
amount of feed 进给量；进刀量
amount of fill 填方量
amount of fishing 捕捞量
amount of forecast 预测额
amount of foreign exchange outflow 外汇流出量
amount of forward displacement 前移量
amount of forward shift 前移量
amount of fracturing fluid 压裂液总量
amount of fresh water 清水用量
amount of funds available for distribution 可支配的资金
amount of gas 气体量
amount of goods purchased 货物购入额
amount of groundwater 地下水超采量
amount of gyroscopic drift 陀螺漂移值
amount of handling 装卸费(用)；搬运费(用)
amount of heat 热量
amount of heat absorption 总吸热量
amount of heavy metals 重金属量
amount of hoisting 提升工作量
amount of ice 冰量
amount of illustration 插图数
amount of imports 进口额
amount of inclination (井身的)斜率；(地层的)倾角；倾斜量；倾斜度
amount of infiltration 渗透量；渗滤量
amount of information 信息量
amount of insurance 保险金额
amount of insurance carried 实保金额
amount of investment 投资额；出资额
amount of kick 井涌量
amount of labor in a certain form 一定形态的劳动量
amount of labo(u)r 劳动人数定额用工；劳动量
amount of labo(u)r used 用工量
amount of lap 搭接量
amount of leakage 渗漏量；漏水量；漏气量；泄漏量
amount of limitation 限额
amount of loan 贷款数量
amount of local unworkable coal seam 局部可采煤层数
amount of looseness 松动；公隙值
amount of loss 损失额；亏损额
amount of lost circulation material 堵漏料用量
amount of magensium applied 施镁量
amount of material 材料用量
amount of mixing 搅拌量；搅拌度
amount of money 金额
amount of monthly rainfall 月降雨量
amount of movement 位移量；移动量
amount of mud displacement 替泥浆量
amount of mud loss 漏失量
amount of negative charge 负电荷量
amount of net cash flow 净现金流量
amount of net foreign exchange 净外汇流量
amount of normal strain 标准应变量
amount of obligation 负债额
amount of offset 错动量
amount of offset in each time 每次错动量
amount of oiling 泡油量
amount of one 每元欠款到期偿还的款额(包括复利)
amount of one yuan 一元的本利和
amount of operating rigs 开动钻机台数；工作钻机台数
amount of ordinary annuity 普通年金终值
amount of overflowing 溢流水量
amount of overhaul depreciation of fixed assets 固定资产大修折旧额
amount of overlap 搭接量
amount of oxygen consumption 氧消耗量
amount of paper money in circulation 纸币流通量
amount of parting 夹矸层数
amount of percolation 渗入量；渗滤量
amount of photometry 光度值；测光值
amount of pollutant accumulation 污染物累积量
amount of porosity 孔隙度
amount of positive charge 正电荷量
amount of precipitation 降雨量；降水量；沉淀量
amount of precipitation per unit area and time 单位时间单位面积上的降水量
amount of prestress 预应力值

amount of profit 利润额
amount of propants 支承剂总用量
amount of rainfall 降雨量；暴雨量
amount of raw mix 生料量
amount of reinforcement 含钢量
amount of reservoir leakage 水库渗漏(量)
amount of reservoir seepage 水库渗漏(量)
amount of residual parallax 残余视差量
amount of rig-month 台月数
amount of rig-shift 台班数
amount of river load 可流荷载量
amount of runoff 径流量
amount of salary 薪额
amount of sales 销售额
amount of saturation 饱和量
amount of secondary crop of gas 二次生气量
amount of sediment inflow 来沙量
amount of sediment transport 输沙量
amount of seepage 渗漏量
amount of self-purification 自净量
amount of sensible information processing 可感信息处理量
amount of settlement 清算金额；沉降量
amount of setup 增水量
amount of shares 出资额
amount of shrinkage 收缩量
amount of sludge produced 污泥生成量
amount of slurry 水泥浆用量
amount of snowfall 降雪量
amount of spacer 隔离液数量
amount of standard strain 标准应变量
amount of stoichiometric(al) 化学计量反应量
amount of storm sewage 暴雨水量
amount of strength 强度值
amount of strong acid 强酸量
amount of surface evapo(u)ration 水面蒸发量
amount of target nuclear 靶核数量
amount of tax to be paid 税额
amount of the credit 赊账金额
amount of tidal rise 潮升量
amount of traffic 交通量运输量；交通量；货运量；运输量；运量
amount of transport 运输工作量
amount of unfinished products in circulation 流动在制品占有量
amount of unworkable coal seam 不可采煤层数
amount of used explosive 炸药用量
amount of variability 变异量
amount of variation 变异量
amount of ventilation 通风量；换气量
amount of vote 通过金额
amount of waste heat 废热总量
amount of waste material 废蒸汽总量
amount of waste water 废水总量
amount of water 水分含量
amount of water required 需水量
amount of wear 磨耗量
amount of wind 风量汽流量
amount of work 工作量
amount of workable coal seam 可采煤层数
amount of work completed 完成工作量
amount of work done 工程完成量；工程完成额
amount of work given out 发出的功
amount of work of exploration 调查工作量
amount of work of loading 装岩工作量
amount of works 工程量
amount of yaw 偏航角；摆艏角
amount or % per kg. ration 每公斤日粮含量或百分率
amount overcharged 多收款项
amount overpaid 多付款项
amount paid in 实缴金额
amount paid to subcontract 已付分包款金额
amount paid to subcontractors 支付给分包人的金额
amount payable 应付金额
amount purchased 采购量
amount receivable 应收入金额
amount received 已收入金额
amount retroceded 退还金额
amounts by weight 重量份
amounts differ 金额不符
amounts not recovered 未能收回的金额
amounts written off 冲销数

amount tendered 金额货币;偿付量
amount to 合计
amount under-collected 少收款项
amount used 消耗量
amount wear 磨损量
amount withheld 扣缴税额
amoxapine 阿莫沙平
amoxecaine 阿莫卡因
amozonolysis 氨解臭氧化作用;氨臭氧化反应
ampacity 安载流容量;安培容量;载流量
amparo blue 鲜蓝色
Ampco 铝铁青铜;耐蚀耐热铜合金;安普科耐热耐蚀铜合金
Ampcoloy 耐蚀耐热铜合金
ampelite 黄铁炭质页岩
amperage 电流强度;安(培)数
ampere 安(培)
ampere-balance 安培平衡;安培秤
ampere-capacity 安培容量
ampere-conductors 安培导体
ampere-current 安培电流
ampere density 电流密度
ampere-feet 安培英尺
ampere-hour 安时;安培(小)时
ampere-hour capacity 安时容量;安培小时容量
ampere-hour efficiency 安时效率;安培小时效率
ampere-hour meter 安时计;安时计;电度表
ampere meter 电流表;电流计;安培米;安计计;安培表
ampere/meter 安/米(磁场强度单位)
ampere meter squared 安培米平方
ampere-minute 安培分钟
ampere per meter 安培每米
ampere-second 安培秒;安秒
ampere's law 安培定律
ampere square meter per joule second [Am²/JS] 安培平方米每焦耳秒
Ampere's right-handed rule 安培右手定则
ampere's right handed screw rule 安培右手螺旋定则
ampere's rule 安培定则
ampere theorem 安培定理
ampere titration 安培滴定法
ampere-turns 安匝;安培匝数
ampere-turns factor 安匝系数
ampere-volt 伏安
ampere-voltage meter 电流电压表
ampere-winding 安培匝数;安匝
ampere-wires 安培导线;安培导体
amperite 平稳灯;限流器;镇流管
amperometer 安培计
amperometric analyser 电流分析仪
amperometric detection 电流检测;安培检测
amperometric determination 电流测定
amperometric electrode construction 电流电极结构
amperometric instrument 电流测定仪
amperometric titration 电流滴定;安培滴定
amperometric titration at constant voltage 定电压电流滴定
amperometric titration method 电流滴定法
amperometric titration test 电流滴定试验
amperometric titrator 电流滴定器
amperometric titrimeter 电流滴定计
amperometric zirconia oxygen sensor 安培米制氧化锆氧传感器
amperometry 电流分析(法)
ampferer-type subduction zone A 形俯冲带
amphenol connector 线夹;电缆接头;接线端子;电缆插销
amphenol plug 电缆插头
amphenone 氨苯丁酮
amphetamine 苯异丙胺;苯丙胺
amphi-antis temple 墙角墩前列柱神庙
amphi-backhoe dredge(r) 两栖反铲挖泥船
amphibarge 两栖方驳
amphibian 水陆两用飞机;水陆两用(的);两栖动物;两栖的
amphibian automobile 水陆两用汽车
amphibian barge 两栖方驳
amphibian dredge(r) 水陆两用挖泥船;两栖挖泥船
amphibian jeep 水陆两用吉普车
amphibian truck 水陆两用载重车;水陆两用车
amphibian vehicle 两栖船

amphibious 水陆两用的;两栖的
amphibious aircraft 水陆两用飞机
amphibious animal 两栖动物
amphibious assault ship 两栖攻击舰;直升(飞)机登陆运输舰
amphibious barge 两栖方驳
amphibious boat 两栖作战登陆艇
amphibious bulldozer 水陆两用推土机
amphi bious car 两栖汽车;水陆两用汽车
amphibious cargo ship 登陆物资运输舰
amphibious carrier 两栖运输车
amphibious container 水陆联运集装箱
amphibious craft 两栖船
amphibious crane 水陆两用起重机
amphibious dredge(r) 水陆两用挖泥船;两栖挖泥船
amphibious hovercraft 水陆两用气垫船
amphibious hover platform 两栖气垫平台
amphibious landing craft 两栖作战登陆艇
amphibious landing gear 水陆两用起落架
amphibious machine 水旱地两用拖拉机
amphibious mechanical shovel 水陆两用机械铲
amphibious plane 水陆两用飞机
amphibious plant 两栖植物
amphibious population 两栖人口
amphibious research craft 两栖研究船
amphibious shovel 水陆两用单斗挖掘机;两栖单斗挖掘机;两栖铲土机
amphibious site 水陆建筑场地
amphibious tank 水陆两用坦克
amphibious tracked carrier 履带式水陆两用输送车
amphibious tractor 水陆牵引车;水陆两用牵引车
amphibious truck 水陆两用载重车
amphibious vehicle 水陆两用汽车;水陆两用车
amphibious vessel 登陆艇
amphibious warfare ship 两栖战舰
amphibious warships 两栖舰艇
amphibious tank 水陆坦克
amphibole 闪石;角闪石
amphibole-ariegite 角闪尖榴辉岩
amphibole asbestos 闪石石棉;角闪石石棉
amphibole-chlorite schist 角闪绿泥片岩
amphibole dacite 角闪石英安岩
amphibole diopside granulite 角闪透辉麻粒岩
amphibole-gneiss 角闪石片麻岩
amphibole granite 角闪石花岗岩
amphibole granular poprhyrite 角闪安山玢岩
amphibole hornfels 角闪石角岩
amphibole hypersthene pyroxenite 角闪紫苏透辉岩
amphibole magnetite 闪石磁铁矿(石)
amphibole magnetite rock 角闪磁铁岩
amphibole magnetite serpentinite 角闪磁铁蛇纹岩
amphibole-mica schist 角闪云母片岩
amphibole rhyolite 角闪石流纹岩
amphibole schist 闪片岩;角闪石片岩
amphibolia 不稳定期
amphibolic 闪石的
amphibolic gneiss 角闪石片麻岩;闪石片麻岩
amphibolide 闪石岩
amphibolite 闪岩;角岩岩;角闪岩;斜长角闪岩
amphibolite facies 闪岩相;角闪岩相
amphibolization 闪岩化;闪石化
amphibololite 角闪石岩;火成闪岩
amphicelous 两面凹陷的;两边凹的
amphicentric 起止同源的
amphichrome 两色花
amphiclinous hybrids 双倾后代
amphiclinous progeny 双倾后代
amphicryptophyte 两栖植物
amphid 化感器
amphidetic 两面伸展的
amphi-dredge(r) 两栖挖泥船
amphidromic center 潮位不(稳)定点;潮流旋转中心
amphidromic current 旋转潮流
amphidromic point 无潮点
amphidromic region 无潮区
amphidromic system 无潮系统;无潮体系
amphidromic tide 旋转潮
amphidromic wave 无潮波
amphidromos 转风点;转潮点
amphigene 白榴石
amphigenite 白榴石岩

amphigenyte 白榴石岩
amphikaryon 倍数核
amphimixis 两性融合;两性混合
amphimorphic 二重
amphimorphic rock 二重岩
amphinucleus 中心核
amphioxus 文昌鱼
amphipathic compound 两亲化合物
amphipathic substance 两亲性物质
amphi-permeability 两透性
amphiphile 两亲物
amphiphyte 两栖植物
amphiplatyan 双平型;两端平的
amphi-position 跨位
amphiprostyle 前后排柱建筑;前后列柱式
amphiprostyle temple 前后排柱廊神庙
amphiprostyle tetrastyle temple 前后四柱廊神庙
amphiprostylos 前后有排柱而两边无柱的建筑(两排柱式建筑);前后列柱式
amphiprotic compound 两性化合物
amphiprotic reaction 两性反应
amphiprotic surface active agent 两性表面活性剂
amphirostylos 古典庙宇建筑
amphisapropel 半腐泥
amphisarca 瓜果
amphiscian region 日影双向区
amphistylar 两侧或两端有列柱式的;两排柱式的(建筑);两端侧柱廊式
amphitheater 倾斜看台;三楼楼厅;露天剧场;圆形竞技场;圆形凹地;圆形(露天)剧场(古罗马);圆形大剧场;野外圆形剧场;(古罗马时代的)斗技场;大围谷;大冰斗;冰围谷;手术观摩室
amphithura 希腊教堂圣壁入口的门幕
amphitriaene 双三叉体
amphodelite 钙长石
amphogneiss 混合片麻岩
ampholine 两性分子的
ampholine electrophoresis 两性电解质电泳
ampholyte 两性电解质
ampholytic detergent 两性洗涤剂
ampholytic surfactant 两性表面活性剂
ampholytoid 两性胶体
amphora 双耳长颈瓶;双柄细颈瓶
amphotalide 氨苯酞胺
amphoteric 两性的
amphoteric character 两性性质;两性特征
amphoteric chondrite 两性球粒陨石
amphoteric colloid 两性胶体
amphoteric compound 两性化合物
amphoteric copolymer 两性共聚物
amphoteric detergent 两性离子洗涤剂
amphoteric electrolyte 两性电解质
amphoteric elements 两性元素
amphoteric emulsifier 两性乳化剂
amphoteric flocculant 两性絮凝剂
amphoteric ion(ic) exchange resin 两性离子交换树脂
amphoteric isoelectric(al) compound 两性等电化合物
amphoteric organic compound 两性有机化合物
amphoteric oxide 两性氧化物
amphoteric polyacrylamide 两性聚丙烯酰胺
amphoteric polyelectrolyte 两性型聚电解质
amphoteric polymer 两性聚合物
amphoteric reaction 双性反应;两性反应
amphoteric resin 两性树脂
amphoteric salt 两性盐
amphoteric soil 两性土壤
amphoteric solute 两性溶质
amphoteric solvent 两性溶剂
amphoteric sulfide 两性硫化物
amphoteric surfactant 两性表面活性剂
amphtrac 水陆履带牵引车
amph-trk 水陆两用载重汽车
amphyl 酚衍生物
ampiclla 坛囊
ample aeration 自由掺气;充分掺气
ample area 足够的面积
ample clearance 充分间隙
ample evidence 充分证据
ample flow 夹水;充裕水量;丰水
ample power 大功率
ample rainfall 充沛雨量;充沛降雨;充沛降水
ample supply 供应充足

ample warning 提前报知;预先报知
amplexiform 贴接
ampli-aerobic pond 稳定塘—兼性厌氧塘
amplidyne 交磁放大机
amplidyne control unit 电机放大器控制部件
amplidyne generator 微场电流放大机;交磁放大器;交磁放大机;电机放大器;(旋转式的)磁场放大机
amplidyne servomechanism 电机放大器伺服系统;电机放大器伺服机构
amplification 扩大;增益;增幅系数;放大
amplification circuit 放大电路
amplification coefficient 放大系数
amplification constant 放大恒量;放大常数
amplification degree 扩大率;放大倍数
amplification factor 增幅因子;增幅率;放大因素;放大因数;放大系数
amplification factor bridge 测量放大系数的电桥
amplification factor of controlled plant 调节对象放大系数
amplification factor of wave 波浪放大系数
amplification frequency 放大频率
amplification gain 放大增益
amplification limit frequency 放大极限频率
amplification matrix 放大矩阵
amplification of ga(u)ge 轨距加宽
amplification of pulse 脉冲放大
amplification power 放大率
amplification range 放大范围
amplification ratio 扩大率;制动倍率;放大(比)率
amplification stage 放大级
amplification system 放大电路系统
amplifidyne generator 放大发电机
amplified action 放大作用
amplified automatic level control 放大式自动电平控制
amplified back bias 放大后负反馈偏压
amplified profile 放大纵断面图
amplified spont-aneous emission 放大的自发辐射
amplified surge 扩大的涌浪
amplified trial load method 扩大试载法
amplified verification 扩大(的)核查
amplifier 扩音器;扩大器;增音器;增强器;增强剂;增幅器;放大器
amplifier bandwidth 放大器带宽
amplifier block 放大器功能块
amplifier card 放大器电路插板
amplifier card connector 放大器电路板接插件
amplifier-chain 放大器线路
amplifier channel 放大器信道
amplifier detector 放大器检测器;放大检波器
amplifier deviation equalizer 放大器偏差均衡器
amplifier-filter 放大滤波器
amplifier gain 放大器增益
amplifier gain characteristic 放大器增益特性
amplifier installation 放大器装置
amplifier kit 放大器的整套零件
amplifier modulation 放大器调制
amplifier noise 放大器噪声
amplifier output 放大器输出
amplifier plug 放大器插件
amplifier-pulse analyzer 脉冲放大分析仪
amplifier recorder 放大器的纪录器
amplifier-rectifier 放大整流器;放大器—整流器
amplifier-rectifier trolly 放大—整流装置
amplifier stage 放大级
amplifier system 放大器系统
amplifier transformer 放大器变压器
amplifier tube 功率管;放大器电子管;放大管
amplifier-type meter 放大器式测量仪
amplifier type voltage regulator 放大器式电压调节器
amplifier unit 放大器部件
amplifier using discrete components 分离元件放大器
amplifier valve 放大器电子管;放大管
amplifier with time lag 带时滞的放大器
amplifilter 放大滤波器
amplify 增幅
amplifying 放大的
amplifying coefficient of dynamic(al) force 动力放大系数
amplifying delay line 放大延迟线
amplifying device 放大装置

amplifying gate structure 放大门电路结构
amplifying head 放大头
amplifying klystron 速调放大管;放大速调管
amplifying lens 放大透镜
amplifying medium 放大媒质
amplifying power 放大能力;放大功率;放大本领
amplifying ratio 放大比
amplifying relay 放大继电器
amplifying signal converter 放大信号变换器
amplifying system 扩声系统;放大系统
amplifying technique 放大技术
amplifying technique for light pulse 光脉冲放大技术
amplifying transformer 放大(用)变压器
amplifying triode 放大(用)三极管
amplifying tube 放大管;放大电子管
amplifying valve 放大阀
amplifying vibrograph 放大式示振器
amplifying winding 放大绕组
ampligraf 摆幅仪
ampligraph 摆幅记录仪
amplimeter 指针式摆幅仪
ampliscope 摆幅观测仪
ampliscript 摆幅仪
amplistat 内反馈式磁力放大器;自反馈式磁放大器
amplitrans 特高频功率放大器
amplitron 特高频功率放大器;增幅管
amplitud distortion 幅度畸变
amplitude 调幅;振幅;出没方位角;辐角;幅值;幅度;幅;变幅
amplitude adder 振幅加法器;幅度加法器
amplitude adjustment 振幅调整
amplitude analysis 振幅分析
amplitude and phase shift keying 振幅和相移键控
amplitude anomalous survey 振幅异常普查
amplitude anomaly 振幅异常
amplitude attenuation 振幅衰减
amplitude attenuation constant 振幅衰减常数
amplitude-change signalling 变幅信号;变幅通信
amplitude characteristic 振幅特性
amplitude circuit 振幅电路
amplitude code 振幅码
amplitude comparator 振幅比较器
amplitude compressor 振幅压缩器
amplitude contour 等幅线
amplitude contraction 缩小振幅;减幅
amplitude contrast 振幅反差
amplitude curve 振幅曲线
amplitude decrement factor 波幅衰减率
amplitude delimiter 增益压制器
amplitude depth diagram 幅值深度图
amplitude detection 幅度检测;幅度检波
amplitude discrimination circuit 振幅鉴别电路
amplitude discrimination technique 幅度鉴别技术
amplitude discriminator 脉冲振幅鉴别器;振幅译码器;振幅鉴别器;振幅辨别器;幅度甄别器;幅度鉴别器
amplitude distance diagram 幅值距离图
amplitude distortion 振幅失真;振幅畸变;幅度失真;非线性失真;波幅畸变
amplitude distribution 振幅分布
amplitude element 测振幅单元
amplitude encoder 幅度编码器
amplitude envelope 幅值包线
amplitude envelope section 振幅包络剖面
amplitude equalization 振幅均衡
amplitude excursion 振幅偏移
amplitude fading 振幅衰落;幅度衰落
amplitude filter 振幅过滤器
amplitude fluctuation 振幅摆动;幅值浮动
amplitude-frequency 幅频
amplitude-frequency characteristic 振幅频率特性;幅频特性
amplitude-frequency curve 振幅—频率特性曲线
amplitude-frequency distortion 振幅—频率失真;幅频失真
amplitude-frequency distribution 振幅频谱
amplitude-frequency response 振幅—频率响应;振幅—频率特性曲线;幅频响应
amplitude-frequency response characteristics 幅频响应特性
amplitude function 振幅函数
amplitude half adder 幅值半加法器;半振幅加法器
amplitude hologram grating 振幅全息摄影册

amplitude independent 振幅无关
amplitude indicator 幅度指示器
amplitude instability 振幅不稳定度
amplitude interferometry 振幅干涉测量
amplitude limit 限幅;波幅限制
amplitude limiter 限幅器;振幅限制器
amplitude limiter circuit 限幅电路
amplitude log 声幅测井;振幅测井
amplitude magnification factor 振幅放大因素;振幅放大系数;振幅放大图
amplitude measurement 振幅测定
amplitude meter 振幅计
amplitude-modulated indicator 调幅指示器
amplitude-modulated pulse 已调幅脉冲
amplitude-modulated signal 调幅信号;已调幅信号
amplitude-modulated sound transmitter 调幅伴音发射机
amplitude-modulated transmitter 调幅发射机
amplitude-modulated wave 调幅波;已调幅波
amplitude modulation 调幅;振幅调制;幅度调制;波幅调制
amplitude modulation communication system 调幅通信系统
amplitude modulation continuous wave laser ranging 连续波调幅激光测距
amplitude modulation frequency modulation 调幅—调频
amplitude modulation measurement 调幅测量
amplitude modulation mode locking 调幅锁模
amplitude modulation noise 调幅噪声
amplitude modulation phase modulation conversion 调幅调相转换
amplitude modulation radio 调幅无线电设备
amplitude modulation receiver 调幅接收机
amplitude modulation rejection 调幅抑制
amplitude modulation reticle 调幅分度线
amplitude modulation signature 调幅标志
amplitude modulation system 调幅系统
amplitude modulation telegraphy double side-band 双边带调幅电话
amplitude modulation telephony 减载波单边带电话
amplitude modulation transformer 调幅变压器
amplitude modulation with vestigial sideband 残留边带调幅
amplitude modulator 调幅器
amplitude of accommodation 调节幅度
amplitude of a heavenly body 天体出没幅角
amplitude of back-scatter(ing) sound 反向散射声振幅
amplitude of beat 拍振振幅
amplitude of component tide 分潮振幅
amplitude of constituent 分潮振幅
amplitude of drag 牵引幅度
amplitude of erosion and deposition 冲淤幅度
amplitude of fading 衰落幅度
amplitude of first harmonic 基波振幅
amplitude of flood 洪水波幅;洪水幅度
amplitude of fluctuation 波动幅度;变幅;变动幅度;摆动幅度
amplitude of meander 曲流幅度;蜿蜒幅度
amplitude of noise 噪声幅度
amplitude of oscillation 振幅;振荡幅度;摆幅
amplitude of partial tide 分潮振幅
amplitude of pitch 纵摇幅度
amplitude of pressure 压力振幅
amplitude of rolling 横摇幅度
amplitude of scouring and silting 冲淤幅度
amplitude of secondary magnetic field 二次磁场振幅
amplitude of seiche 假潮差;驻波幅
amplitude of setup 增水幅度
amplitude of stage 水位变幅
amplitude of stress 应力幅度
amplitude of swell 涌幅
amplitude of swing 摆动幅度
amplitude of the molecular vibration 分子振动幅度
amplitude of the variation 日变幅
amplitude of tidal current 潮流振幅
amplitude of tide 潮汐振幅;潮汐幅度;潮幅;半潮差
amplitude of total magnetic field 总磁场振幅
amplitude of variation 变化幅度;变幅
amplitude of vibration 振幅;振动幅值;振动幅度

amplitude of vibration method 振幅法
amplitude of water-level 水位变幅
amplitude of water-level of reservoir 水库水位变化幅度
amplitude of wave 波幅
amplitude of waveform 波形振幅
amplitude of wind tide 风壅水幅度
amplitude peak 最大幅度;幅度峰值
amplitude permeability 振幅磁导率
amplitude phase 幅度相位
amplitude-phase characteristic 幅相特性
amplitude-phase method 振幅相位法
amplitude process after stack 迭后振幅处理
amplitude pulse 振幅脉冲
amplitude quantization 振幅量化
amplitude quantized control 量化幅度控制
amplitude quantizing 振幅量化;幅度分层
amplitude range 振幅范围
amplitude ratio 振幅比;幅比
amplitude ratio of survey channel 测量道的幅度比
amplitude recovery 振幅恢复
amplitude reflectance 振幅反射率
amplitude residual curve in time domain 时域振幅剩余曲线
amplitude resonance 振幅共振;波幅共振
amplitude response 振幅频率特性
amplitude response curve 振幅响应曲线;振幅特性曲线;振幅反应曲线
amplitude sampler 振幅取样器
amplitude selector 脉冲振幅选择器;振幅选择器;振幅分离器
amplitude sensing 幅度检测
amplitude sensitivity 振幅灵敏度
amplitude separation 振幅区分
amplitude separation data detection 幅度分离数据检测
amplitude separator 振幅选择器;振幅分离器
amplitude shading 幅度束控
amplitude shift keying 振幅键控;振幅偏移键控法;变幅调制
amplitude sine-wave train 等幅正弦波列
amplitude spectrum 振幅谱;振幅频谱;振幅光谱
amplitude spectrum curve 振幅谱曲线
amplitude spectrum density 振幅密度谱
amplitude spectrum factor 振幅谱因子
amplitude-splitting interference 分振幅干涉
amplitude splitting 振幅剖分法
amplitude step time 台阶延迟时间;阶跃波时间;振幅阶跃时间
amplitude suppression ratio 幅调抑制比
amplitude taper 振幅锥度
amplitude to period ratio 振幅周期比
amplitude transformer 聚能器;变幅器
amplitude transmissivity 振幅透射率
amplitude transmittance 振幅透过率
amplitude value of log 测井曲线幅值
amplitude variation 幅度变动;幅变
amplitude velocity 幅速
amplitude versus distance curve 振幅距离曲线
amplitude versus frequency characteristic 幅度—频率特性曲线
amplitude-versus-frequency curve 振幅—频率特性曲线
amplitude wave 幅波
amplituhe modulation factor 调幅度
amporphous 非晶形
ampo(u)le 安瓿;小玻璃瓶
ampo(u)le forming process 安瓿成型法
ampulla 壶腹;(古希腊罗马的,有双耳的)细颈瓶
ampullaceal 坛形的
ampullaceous 坛形的
ampullaceous sensilla 罐形感器
ampulla vitrea 玻璃安瓿
amputating saw 切断锯;截断锯
amputation 剪枝;切断木;修剪树枝
amputation knife 截断刀
Amrad gum 阿姆拉特树胶
Amsco G 阿姆斯科合成芳烃油溶剂
Amsier's universal tester 阿姆勒万能材料试验机
Amsil silver copper 阿姆西尔银铜合金
AMSL[above mean sea level] 平均海平面以上
Amsler abrasion test 阿姆斯勒磨耗试验
Amsler abrasion tester 阿姆斯勒磨耗试验机
Amsler cupping test 阿姆斯勒杯形拉延试验

Amsler planimeter 阿姆斯勒求积仪
Amsler's universal tester 阿姆斯勒万能材料试验机
Amsler tester 阿姆斯勒试验机
Amsler vibraphone 阿姆斯勒(高频)扭转试验机
Amsterdam Bosh 阿姆斯特丹大公园
Amsterdam Exchange 阿姆斯特丹交易所(荷兰)
Amsterdam group 阿姆斯特丹学派
Amsterdam school 阿姆斯特丹学派
Amsterdamsch Peil 阿姆斯特丹高程基准面
Amsulf copper 阿姆萨尔弗无氧铜
Amtel tellurium copper 阿姆特尔含碲无氧铜
amtipodal points 对极点
amtrack 水陆两用车辆;履带式登陆车
AM transmitter with pulse duration modulation 脉宽调制式调幅发射机
amuminize 铝化
Amundsen Gulf 阿蒙森湾
Amundsen Sea 阿蒙森海
Amundsen trough 阿蒙森海槽
amusement center 娱乐中心
amusement city 娱乐城
amusement device 娱乐设施
amusement district 娱乐区
amusement expenses 交际接待费用
amusement facility 娱乐设施;娱乐设备
amusement hall 娱乐厅
amusement park 露天游艺场;公共游乐场;露天游乐场;娱乐园地;娱乐场;娱乐公园
amusement parlor 娱乐厅
amygdale 杏仁体
amygdaline 扁桃形的
amygdaloid 杏仁岩;扁桃样的
amygdaloidal 杏仁状
amygdaloidal basalt 杏仁状玄武岩
amygdaloidal diabase 杏仁状结构辉绿岩
amygdaloidal lava 杏仁岩
amygdaloidal texture 杏仁状结构
amygdaloidal Zechstein dolomite 蔡希坦统的杏仁状结构白云岩(二叠纪)
amygdaloid body 杏仁体
amygdoloidal structure 杏仁状构造
amygdule 杏仁子;杏仁孔
amylaceous 淀粉状的;淀粉质的
amyl acetate 乙酸(正)戊酯;醋酸(戊)酯
amyl acetate lamp 乙酸戊脂灯
amylaceum 葡萄糖
amylalcohol 戊醇
amyl benzoate 苯甲酸异戊酯;苯甲酸戊酯
amyl borate 硼酸戊酯
amyl carbamate 氨基甲酸戊酯
amyl cellulose 戊基纤维素
amyl citrate 柠檬酸戊酯
amylene 戊烯
amylene oxide 氧杂环己烷
amyl ether 戊醚;戊基醚
amyl formate 甲酸戊酯
amyl furylacrylate 呋喃基丙烯酸戊酯
amylin 糊精
amyl lactate 乳酸戊酯
amyl laurate 十二烷酸戊酯;月桂酸戊酯
amylodextrin 淀粉糊精
amylolysis 淀粉分解
amyloniazid 淀烟腙
amylopectin 胶淀粉
amylose 糖淀粉
amylose acetate 乙酸淀粉
amylphenol resin 戊酚树脂
amyl propionate 丙酸戊酯
amylsalicylate 水杨酸戊酯
amyl stearate 硬脂酸戊酯
amyl tartrate 酒石酸盐;酒石酸戊酯
amylum processing factory 淀粉加工厂
amyrol 白檀油醇
an abandoned pig house 废猪舍
Anabar nucleus 阿纳巴尔陆核
Anabar old land 阿纳巴尔古陆
Anabar shield 阿纳巴尔地盾
anabasine 阿那巴辛
anabasis sp 盐生段木贼
anabatic 谷风
anabatic wind 上升风;上坡风
anabergite 镍华
anaberrational reflector 消像反射望远镜
anaberrational telescope 消像差望远镜;消色差望远镜

anabohitsite 铁橄苏辉岩
anabolic metabolism 组成代谢
anabolic process 组成过程
anabolic substance 组成代谢物质
anabolism 合成代谢;有机质合成;组成代谢
anabranch 交织支流;河岔;小河道网;再流入主流的支流;再汇合支流;重汇支流;汊河
an abrupt descent 陡峭的斜坡
an abrupt turn 急转弯
anacampsis 弯曲;折射
anacamptics 反射光学
anacamptometer 反射计;反射测定计
Anacardiaceae 漆树科
anacardic acid 漆树酸
anacardol 腰果酚
anached documents 所附证件;所附单据;附件
anachronism 记时错误
anacidity 酸缺乏
an acid solution 一种酸性溶液
anaclinal 逆四周地层的倾斜方向而下降的
anaclinal river 逆斜河
anaclinal stream 逆斜河
anacline 正倾型
anacmesis 成熟受阻
anacom 分析计算机
anaconda 安钠康达(以三氧化砷为主要成分的木材防腐剂)
anacrobictank 加水分解池
anacrotic pulse 升线波脉
Anadarko basin 阿纳达科盆地
anadiagenesis 后生成岩作用;变生成岩作用
anadiagenetic 后生成岩期的
anadiagenetic stage 变生成岩期
anadiagensis 前进成岩(作用)
anadicrotic pulse 升线二波脉
anadromous 溯河(产卵)的;上行的;向上的
anadromous fish 溯河性鱼类
anadromous migration 溯河洄游
anaerobe 嫌气细菌;嫌气微生物;厌氧微生物;厌氧生物;厌氧菌
anaerobia 嫌气微生物
anaerobic 厌氧污泥消化池作用;厌气的
anaerobic acdogenesis 厌氧产酸
anaerobic acidification 厌氧酸化
anaerobic acidification method 厌氧酸化法
anaerobic activated sludge 厌氧活性污泥
anaerobic activated sludge method 厌氧活性污泥法
anaerobic activity 厌氧活性;厌氧活度
anaerobic adhesive 厌氧黏合剂;厌氧性黏结剂;厌氧胶黏剂;厌气性黏合剂
anaerobic-aerobic integrated biological aerated filter 一体式厌氧—好氧曝气生物滤池
anaerobic-aerobic treatment 厌氧—需氧处理
anaerobically digested sludge 嫌气消化污泥;厌氧消化污泥
anaerobic ammonia oxidation 厌氧氨氧化
anaerobic ammonia oxidation activity 厌氧氨氧化活性
anaerobic ammonia oxidation bacteria 厌氧氨氧化菌
anaerobic ammonia oxidation body 厌氧氨氧化体
anaerobic ammonia oxidation process 厌氧氨氧化工艺
anaerobic ammonia oxidation rate 厌氧氨氧化速率
anaerobic ammonia oxidation reactor 厌氧氨氧化反应器
anaerobic ammonia oxidation sludge 厌氧氨氧化污泥
anaerobic ammonia oxidation technology 厌氧氨氧化技术
anaerobic-anoxic/oxic process 厌氧—缺氧/好氧工艺
anaerobic-anoxic/oxic system 厌氧—缺氧/好氧系统;生物除磷脱氮系统
anaerobic-anoxic sequencing batch reactor 厌氧/缺氧序批间歇式反应器
anaerobic attached-film expanded bed 厌氧附着膜膨胀床法
anaerobic attached microbial film expanded bed 厌氧附着微生物膜膨胀床
anaerobic autotrophic bacteria 厌氧自养菌

anaerobic bacillus 厌氧杆菌
anaerobic bacteria 嫌气细菌；嫌气菌；厌氧细菌；厌氧菌；厌氧性细菌
anaerobic bacteria degradation 厌氧细菌降解
anaerobic bacteria degradation zone 厌氧细菌降解作用带
anaerobic baffled reactor 厌氧折流板反应器
anaerobic baffled sludge blanket reactor 厌氧折流板污泥床反应器
anaerobic biochemical treatment 厌氧生化处理
anaerobic biodegradation 厌氧生物降解能力
anaerobic biodegradation velocity 厌氧生物降解速度
anaerobic biofilm expansion bed 厌氧生物膜膨胀床
anaerobic biofilm method 厌氧生物膜法
anaerobic biofilter 厌氧生物滤池
anaerobic bioflocculation 厌氧生物絮凝
anaerobic biological rotating disc 厌氧生物转盘
anaerobic biological treatment 厌氧生物处理
anaerobic biological treatment of wastewater 废水厌氧生物处理
anaerobic biological treatment process 厌氧生物处理法
anaerobic biotransformation 厌氧生物转化
anaerobic catabolism 厌氧分解代谢
anaerobic-chemical oxidation-coagulation-air flo(a)tation process 厌氧—化学氧化—混凝—气浮法
anaerobic composite bed reactor 厌氧复合床反应器
anaerobic condition 缺氧状态；缺氧情况；嫌气条件；厌氧状况；厌氧条件
anaerobic contact digester 厌氧接触消化池
anaerobic contact pond 厌氧接触池
anaerobic contact process 厌氧生物膜法；厌氧接触过程；厌氧接触法；厌气接触法
anaerobic contact reactor 厌氧接触反应器
anaerobic corrosion 厌气(性细)菌腐蚀
anaerobic culture 厌氧培养
anaerobic culture apparatus 无氧培养器
anaerobic culture medium 厌氧培养基
anaerobic cycle 厌氧循环
anaerobic decay 嫌气腐烂；厌氧腐烂
anaerobic decolo(u)rization 厌氧脱色
anaerobic decomposition 厌氧降解；无氧分解；嫌气分解；厌氧分解
anaerobic decomposition of sludge 污泥的厌氧分解
anaerobic degradation 厌氧降解
anaerobic denitrification 厌氧脱氮过程
anaerobic denitrification process 厌氧脱氮法；厌氧反硝化过程
anaerobic dephosphorization 厌氧除磷
anaerobic dephosphorization bacteria 厌氧除磷菌
anaerobic digested process 厌氧消化法
anaerobic digester 厌氧消化池；厌氧消化器
anaerobic digesting sludge 厌氧消化污泥
anaerobic digestion 厌氧消化超滤；厌氧消化；厌气消化
anaerobic digestion kinetics 厌氧消化动力学
anaerobic digestion of refuse 垃圾厌氧消化
anaerobic digestion tank 厌氧消化池；厌气消化池；厌氧消耗罐
anaerobic ecosystem 厌氧生态系统
anaerobic environment 嫌氧环境；嫌气环境；厌氧环境
anaerobic expanded bed 厌氧膨胀床
anaerobic expanded granular sludge bed 厌氧膨胀颗粒污泥床
anaerobic-facultative-aerobic lagoon 厌氧—兼性—好氧塘
anaerobic-facultative lagoon 厌氧—兼性塘
anaerobic fermentation 厌氧发酵
anaerobic film expanded bed 厌氧膜膨胀床
anaerobic filter 厌氧滤池；厌氧过滤器
anaerobic filtration-contact aeration process 厌氧过滤—接触曝气工艺
anaerobic fixed bed reactor 厌氧固定床反应器
anaerobic fixed film reactor 厌氧固定膜反应器
anaerobic flocculation 厌氧絮凝(作用)
anaerobic fluidized bed 厌氧流化床
anaerobic fluidized bed biofilm reactor 厌氧流化床生物膜反应器
anaerobic fluidized bed reactor 厌氧流化床反应器
anaerobic granular sludge 厌氧颗粒污泥
anaerobic granular sludge reactor 厌氧颗粒污泥反应器
anaerobic habitat 缺氧生境
anaerobic hydrolysis 厌氧水解
anaerobic hydrolysis acidification 厌氧水解酸化
anaerobic hydrolysis acidification oxidation 厌氧水解酸化氧化
anaerobic hydrolysis basin 厌氧水解池
anaerobic hydrolytic acidification 厌氧水解酸化
anaerobic hydrolytic acidification-anoxic denitrification-oxic process 厌氧水解酸化—缺氧反硝化—好氧工艺
anaerobic hydrolyzation and biological contact oxidation process 厌氧水解—生物接触氧化工艺
anaerobic hydrolyzation high loading biological filter 厌氧水解高负荷生物滤池
anaerobic incubation 无氧培养；厌氧培养
anaerobic inhibition kinetics 厌氧抑制动力学
anaerobic inner loop reactor 厌氧内循环反应器
anaerobic jar 厌氧罐
anaerobic lagoon for wastewater treatment 废水处理厌氧塘
anaerobic landfill 厌氧土地填埋
anaerobic layer 厌氧层
anaerobic liquor 厌氧液体
anaerobic membrane bioreactor 厌氧膜生物反应器
anaerobic metabolism 缺氧代谢；厌氧代谢；不需氧代谢
anaerobic method 厌氧性处理法
anaerobic microbial 厌氧微生物
anaerobic micro-ecostructure 厌氧微生态结构
anaerobic microorganism 厌氧(性)微生物
anaerobic migrating sludge blanket reactor 移动式厌氧污泥床反应器
anaerobic moving bed bio-film reactor 厌氧移动床生物膜反应器
anaerobic nitrogen fixing bacteria 厌氧固氮细菌
anaerobic organism 嫌氧生物；厌氧菌；厌氧生物
anaerobic/oxic activated sludge process 厌氧/好氧活性污泥法
anaerobic/oxic alternating cycle system 厌氧/好氧交替循环系统
anaerobic/oxic/bioactivated carbon process 厌氧/好氧/生物活性炭工艺
anaerobic/oxic biological fluidized bed 厌氧/好氧生物流化床
anaerobic/oxic contact oxidation process 厌氧/好氧接触氧化法
anaerobic/oxic couple strengthening pond 厌氧/好氧偶合强化池
anaerobic/oxic process 厌氧/好氧工艺
anaerobic oxidation 厌氧氧化；不需氧氧化
anaerobic partion reactor 厌氧挡板式反应器
anaerobic phosphate release 厌氧释磷
anaerobic pond 稳定塘—厌氧塘；嫌氧塘；厌气塘；厌氧塘
anaerobic pre-denitrification 厌氧预反硝化
anaerobic pre-fermentation-intermittent aerated biofilm system 厌氧预发酵—间歇曝气生物膜系统
anaerobic pretreatment pond 厌氧预处理塘
anaerobic process 嫌气过程；厌氧过程；厌氧工艺；厌氧法
anaerobic reaction 厌氧反应
anaerobic reactor 厌氧反应器
anaerobic respiration 厌氧呼吸
anaerobic respiration zone 厌氧呼吸带
anaerobic rotating biological contactor 厌氧生物转盘
anaerobic sediment 厌氧底泥；缺气沉积
anaerobic sequencing batch bioreactor 序批式厌氧生物反应器
anaerobic sewage treatment 污水厌氧处理
anaerobic single component adhesive 厌氧单组分胶黏剂
anaerobic sludge 厌氧污泥
anaerobic sludge bed reactor 厌氧污泥床反应器
anaerobic sludge blanket reactor 厌氧污泥床反应器
anaerobic sludge digestion 厌氧(性)污泥消化
anaerobic sludge digestion tank 厌氧污泥消化池
anaerobic sulfide nitrate removal 厌氧除硫化物和硝酸盐
anaerobic supernatant 厌氧上层溶液
anaerobic suspended bed reactor 厌氧悬浮床反应器
anaerobic swine lagoon liquid 厌氧猪粪池液
anaerobic system 厌氧系统
anaerobic tank 化粪池
anaerobic toxicity 厌氧毒性
anaerobic toxicity assay 厌氧毒性检测
anaerobic toxic organics removal 厌氧去除有毒有机物
anaerobic toxic wastewater treatment 厌氧有毒废水处理
anaerobic treatment 嫌气处理；厌氧处理
anaerobic treatment process 厌氧处理法
anaerobic trickling 厌氧滴流
anaerobic trickling filter 厌氧滴滤池
anaerobic type 厌氧型
anaerobic ultrafiltration membrane reactor 厌氧超滤膜反应器
anaerobic waste treatment 废弃物厌氧处理
anaerobic wastewater 缺氧废水
anaerobic waste(water)treatment 废水厌氧处理
anaerobic water 缺氧水
anaerobic zone 厌氧区
anaerobiosis 厌氧生活
anaerobious preservation 缺氧保藏法
anaerogenic bacterium 非产气细菌
anaerophyte 厌氧微生物
anaerophytobiont 嫌气土壤微生植物
anaesthesia induction room 麻醉室
anaesthesia table 麻醉台
anaesthetic room 麻醉室
anaflow 上升气流
anafront 上滑锋
anagenesis 前演化；前进演化
anagenesis advancing coast 前进海岸
anaglacial period 初冰期
anaglyph 立体彩色照片；粗条浮雕；浅型浮雕；立体照片；互补色；浮雕装饰；补色立体相片；补色立体图
anaglyph board 浮雕装饰板
anaglyph ceiling 浮雕装饰吊顶
anaglyphic 互补色立体的
anaglyphic film 补色滤光纸
anaglyphic image 互补色图像
anaglyphic image map 互补色影像立体地图
anaglyphic lantern slide 立体幻灯片
anaglyphic map 互补色立体(地)图；互补色地图；补色立体(地)图
anaglyphic method 互补色法
anaglyphic picture 互补色立体像片
anaglyphic plotter 互补色立体测图仪
anaglyphic plotting instrument 补色法测图仪
anaglyphic principle 互补色法则
anaglyphic projection 补色立体投影
anaglyphic spectacles 补色立体眼镜
anaglyphic stereoscopic viewing 互补色立体观察
anaglyphic strip 条形浮雕
anaglyphic viewing system 补色立体观察系统
anaglyph map 全息(地)图；立体影片；浮雕画；浮雕图
anaglyphoscope 互补色镜；补色立体镜
anaglyph picture 浮雕图；彩色浮雕像
anaglyph topographic(al)map 互补色立体(地)图
anaglyphy 互补色立体法；浮雕艺术；补色立体法
anaglyptic 浮雕装饰的；浮雕的
anaglyptics 浮雕艺术
anaglyptic wallpaper (厚纸压花制成的)压花墙纸；浮雕墙纸
anaglytics 浮雕装饰术
anaglytographic(al)map 补色立体(地)图
anakinesis 高能化物合成
anakmesis arrest of maturation 成熟受阻
anakrom 硅藻土型色谱载体
analbite 歪长石；低温歪长石
analcime 方沸石
analcime basalt 方沸石玄武岩
analcime olivine basalt 方沸橄榄玄武岩
analcime syenite 方沸正长岩
analcimite 方沸岩
analcimization 方沸石化
analcimolith 方沸岩
analcite 方沸岩；方沸石

analcite basalt 方沸玄武岩
analcitization 方沸石化
analemma 升高的支座；日晷；罗马剧院的挡土墙；地球正投影仪；地球仪8字形曲线；赤纬时差图
anale setting scale 尺度标尺
analgene 安纳晶
anallatic 光学测距机
anallatic centre 准距中心【测】
anallatic correction 准距校正
anallatic distance 视距离
anallatic lens 测距镜；光学测距镜；视距镜；消加常数透镜；测距透镜
anallatic point 准距点
anallatic telescope 视距望远镜；消加常数望远镜
anallatism 准距性
anallobar 气压上升区；增压区
anallobaric center 升压中心
analmatic 自动检查分析装置
analog 类似物
analog/digital and digital/analog(ue) conversion 模/数与数/模转换
analogic(al) design 类比设计
analogic(al) method 类比法
analogic(al) method of coefficient water content 富水系数比拟法
analogic(al) pattern 类比方式
analog-intensity modulation 模内强度调制
analogous 模拟式；模拟的
analogous area key 类似地区对照判读样板
analogous articles 同类货；类似制品；类似货物
analogous circuit 相似电路
analogous column 模拟柱；比拟柱
analogous element 类似元素
analogous language 模拟语言
analogous navigation set 模拟式导航设备
analogous pole 热正极；模拟极
analogous structure 同功结构
analog scaling 定模比比例因子
analog(uc)类比；模拟(的)；同型物；模拟电视电话
analog(ue) accelerator 模拟加速器
analog(ue) adder 模拟加法器
analog(ue) aerial triangulation 模拟法空中三角测量
analog(ue) aerotriangulation 模拟法空中三角测量
analog(ue) amplifier 模拟放大器
analog(ue) answer 模拟应答
analog(ue) approach 模拟求解法；模拟法求解
analog(ue) assignment of variables 模拟变量指定；变量的模拟赋值；变量的模拟测定
analog(ue) auto pilot 模拟自动驾驶仪
analog(ue) backup 模拟备份；模拟后援
analog(ue) basin 相似流域；参证流域
analog(ue) board 模拟板
analog(ue) buffer 模拟缓冲器
analog(ue) bus 模拟母线
analog(ue) calculator 模拟计算器
analog(ue) channel 类比信道；模拟信道；模拟通道
analog(ue) chromatogram 模拟色谱图
analog(ue) code encryption unit 模拟代码加密器
analog(ue) column 模拟柱
analog(ue) communication 模拟通信
analog(ue) communication system 模拟通信系统
analog(ue) commutator 模拟换向器；模拟分配器
analog(ue) comparator 模拟比较器
analog(ue) compiler 模拟编译程序
analog(ue) compiler system 模拟编译程序系统
analog(ue) compounds 相似化合物
analog(ue) computation 模拟计算
analog(ue) computer 模拟计算机
analog(ue) computer system 模拟计算机系统
analog(ue) computing device 模拟(式)计算装置
analog(ue) computing system 模拟计算装置；模拟计算系统
analog(ue) control 模拟控制
analog(ue) control equipment 模拟控制设备
analog(ue) controller 模拟控制器
analog(ue) conversion 模拟转换
analog(ue) copier 模拟复印机
analog(ue) copying 模拟复印
analog(ue) correlator 模拟相关器；模拟相关函数分析仪
analog(ue) curve plotter 模拟曲线描绘器
analog(ue) data 模拟资料；模拟数据
analog(ue) data channel 模拟数据信道

analog(ue) data computer 模拟数据计算机
analog(ue) data digitizer 模拟数据数字转换器
analog(ue) data recorder 模拟数据记录器
analog(ue) data sink 模拟数据接转器
analog(ue) decoder 模拟解码器
analog(ue) deep seismograph in three-azimuth 三分量模拟深层地震仪
analog(ue) delay unit 模拟时延装置
analog(ue) device 模拟装置；模拟设备
analog(ue)-differential analyzer 模拟微分分析器；模拟微分分析机
analog(ue)-digital 模拟数字(的)
analog(ue)-digital adapter 模拟数字适配器
analog(ue)-digital-analog(ue) converter 模数模转换器；模拟数字模拟转换器
analog(ue)-digital and digital-analog(ue) conversion 模数与数模转换
analog(ue)-digital commutator 模数字转换装置
analog(ue)-digital computer 模拟数字计算机；模拟数字计算机
analog(ue)-digital computing system 模拟数字计算系统
analog(ue)-digital control system 模拟数字控制系统
analog(ue)-digital conversion 模数转换；模拟数字转换
analog(ue)-digital converter 模数转换器；模拟数字转换器；模拟数字变换器
analog(ue)-digital element 模拟数字元件
analog(ue)-digital integrating converter 模拟数字综合转换器
analog(ue)-digital integrating translator 模拟数字综合转换器
analog(ue)-digital recorder 模拟数字记录器
analog(ue)-digital recording equipment 模拟数字记录设备
analog(ue)-display unit 模拟显示装置；模拟显示部件
analog(ue)-distribution 模拟分布
analog(ue)-distributor 模拟量分配器；模拟分配器
analog(ue)-divider 模拟除法器
analog(ue) drawing machine 模拟绘图机
analog(ue) earth quake 模拟地震
analog(ue) electronic computer 模拟电子计算机
analog(ue) electrophotographic technology 模拟式电摄影技术
analog(ue) element 模拟元件
analog(ue) encoder 模拟编码器
analog(ue) equation solver 模拟方程解算器；方程模拟解算机
analog(ue) equipment 模拟(式)设备
analog(ue) experiment 模拟试验；模拟实验
analog(ue) extension 模拟扩展
analog(ue) feedback system 模拟反馈系统
analog(ue) filter 模拟信息滤波器；模拟滤波器
analog(ue) form 模拟形式
analog(ue) formation 制作模型
analog(ue) formatter 模拟格式器
analog(ue) function generator 模拟函数发生器
analog(ue) function switch 模拟功能转换开关
analog(ue) functuion 模拟功能
analog(ue) gate 模拟门
analog(ue) graphic(al) display 模拟图像显示
analog(ue) ground 模拟接地
analog(ue) horizontal deflection 模拟水平偏转
analog(ue) image 模拟图形
analog(ue) image processing 模拟图像处理
analog(ue) image scanner 模拟图像扫描器
analog(ue) indicator 模拟指示器
analog(ue) information 模拟信息
analog(ue) input 模拟输入
analog(ue) input channel 模拟输入通道；模拟量输入通道
analog(ue) input expander 模拟输入扩展器
analog(ue) input module 模拟输入组件；模拟输入模件
analog(ue) input operation 模拟输入操作
analog(ue) inputoutput unit 模拟输入输出装置
analog(ue) input sensor 模拟输入传感器
analog(ue) input voltage 模拟输入电压
analog(ue) instrument 模拟仪器；模拟式仪表
analog(ue) instrumentation system 模拟仪器系统
analog(ue) integrated circuit 模拟集成电路
analog(ue) integration 模拟积分

analog(ue) integrator 模拟积分仪；模拟积分器
analog(ue) interpolation 类比插值法
analog(ue) light deflector 模拟光偏转器
analog(ue) light spot seismograph 模拟光点记录地震仪
analog(ue) limit check 模拟量限值验算
analog(ue) line driver 模拟总线驱动器；模拟线路激励器
analog(ue) loading 模拟加载
analog(ue) logging system 模拟记录测井仪
analog(ue) logic 模拟逻辑
analog(ue) loopback 模拟环回
analog(ue) loopback testing 模拟环回测试法
analog(ue) machine 模拟机
analog(ue) map 模拟地图
analog(ue) matrix addressing 模拟矩阵寻址
analog(ue) memory(device) 模拟存储器
analog(ue) meter 模拟仪；模拟式仪表
analog(ue) method 模拟法；相似法
analog(ue) method of photogrammetric mapping 模拟法测图
analog(ue) microwave relay system 模拟微波接力通信系统；模拟(式)微波中继系统
analog(ue) model(ing) 模拟模型；类比模型；模拟模式；相似模型
analog(ue) model study 模拟模型研究
analog(ue) modem 模拟调制解调器
analog(ue) modulation 模拟调制
analog(ue) modulation system 模拟调制装置；模拟调制系统；模拟调制方式
analog(ue) multiplex equipment 模拟多路复用设备
analog(ue) multiplexer 模拟转换开关；模拟量多工器；模拟多路转换器；模拟多路复用器；模拟多路调制器
analog(ue) multiplication 模拟乘法
analog(ue) multiplier 模拟乘法器
analog(ue) nest unit 模拟组合装置
analog(ue) network 模拟网络
analog(ue) operational unit 模拟运算部件
analog(ue) out(put) 模拟输出
analog(ue) output channel 模拟输出通道；模拟量输出通道
analog(ue) output scanner 模拟输出扫描器
analog(ue) output submodule 模拟输出分组件
analog(ue) pattern memory 模拟图像存储器
analog(ue) photogrammetric mapping 模拟摄影测量制图
analog(ue) photogrammetric plotting 模拟摄影测量制图；模拟测图
analog(ue) photogrammetry 模拟摄影测量
analog(ue) plotter 模拟式立体测图仪；模拟绘图仪；模拟绘图机
analog(ue) plotting instrument 模拟测图仪
analog(ue) point 模拟点
analog(ue) pressure indicator 模拟压力指示器
analog(ue) procedure 模拟法
analog(ue) processing 模拟处理
analog(ue) processing equipment 模拟处理设备
analog(ue) processor 模拟处理器；模拟处理机
analog(ue) processor controller 模拟处理机控制器
analog(ue) profiler 模拟断面记录仪
analog(ue) projection 模拟摄影
analog(ue) protection 模拟保护(装置)
analog(ue) pulse power 模拟脉冲功率
analog(ue) pyrometer 模拟高温计
analog(ue) quantity 模拟量；模拟过程的物理量
analog(ue) radar absorber 模拟雷达吸收装置
analog(ue) radar image 模拟雷达图像
analog(ue) random process test 模拟随机过程试验
analog(ue) readout 模拟读出
analog(ue) readout device 模拟读出装置
analog(ue) receiver 模拟接收机
analog(ue) recorder 模拟记录仪；模拟记录器
analog(ue) record(ing) 模拟记录；模拟存储
analog(ue) rectification 模拟纠正
analog(ue) repeater 模拟中继器；模拟增音机
analog(ue) representation 模拟量表示；模拟表示(法)
analog(ue)-responsive pump 模拟信号泵
analog(ue) result 比拟结果；模拟结果
analog(ue) sampling 模拟采样
analog(ue) scaling 模拟定标

analog(ue) scan converter 模拟扫描变换器
analog(ue) scanning 模拟(式)扫描
analog(ue) scrambler 模拟置乱器
analog(ue) sensor 模拟传感器
analog(ue) servo system 模拟伺服系统
analog(ue) shift register 模拟移位寄存器
analog(ue) signal 模拟信号
analog(ue) signal demodulation equipment 模拟信号解调设备
analog(ue) signal encryption scrambler 模拟信号加密器
analog(ue) signal generator 模拟信号发生器
analog(ue) signal(l)er 模拟信号机
analog(ue) signal(l)ing 模拟信令
analog(ue) signal(l)ing device 模拟信号设备
analog(ue) signal processing transponder 模拟信号处理转发器
analog(ue) simulation 仿真模拟;类比模拟;模拟(式)仿真
analog(ue) single-crystal diffractometer 模拟单晶衍射计
analog(ue) slope detection 模拟斜率检测
analog(ue) slope detector 模拟斜率检测器
analog(ue) solution 模拟解
analog(ue) spectrophotometer 模拟分光光度计
analog(ue) spectrum analyzer 模拟谱分析仪;模拟光谱分析器
analog(ue) storage 模氦存储[贮]
analog(ue) storage expander 模拟存储扩张机构
analog(ue) storage module 模拟存储模块
analog(ue) strip 模拟航线
analog(ue) study 模拟研究
analog(ue) subset 模拟子集
analog(ue) sweep 模拟扫描
analog(ue) switch 模拟开关
analog(ue) system 模拟系统
analog(ue) tape seismograph 模拟磁带记录地震仪
analog(ue) technique 模拟技术;模拟工艺
analog(ue) technology 模拟工艺
analog(ue) telemeter 模拟遥测仪;模拟遥测计
analog(ue) test 模拟试验
analog(ue) tester 模拟测试器
analog(ue) test result 模拟试验结果
analog(ue) theory 模拟理论
analog(ue) timing pulse 模拟定时脉冲
analog(ue)-to-digital 模拟变数字(的)
analog(ue)-to-digital-analog(ue) converter system 模拟数字模拟转换系统
analog(ue)-to-digital conversion 模数转换;模拟数字转换
analog(ue)-to-digital conversion accuracy 模拟数字转换准确度
analog(ue)-to-digital conversion pulse 模拟数字转换脉冲
analog(ue)-to-digital conversion rate 模拟数字转换速度
analog(ue)-to-digital converter 模数变换器;模拟数字转换器;模拟数字转换程序;模拟数字变换器;模数转换器
analog(ue)-to-digital converter check program(me) 模拟数字变换器检验程序
analog(ue)-to-digital converter error 模拟数字变换器误差
analog(ue)-to-digital encoder 模数编码器
analog(ue)-to-digital programmmed control 模拟数字程序控制
analog(ue)-to-digital sensing 模拟数字式读出;模拟数字传感
analog(ue) to frequency converter 模拟频率变换器
analog(ue)to time to digital 模拟时间数字转换
analog(ue) translator 模拟译码器
analog(ue) transmission 模拟传输
analog(ue) transmission facility 模拟传输设备
analog(ue) transmitter 模拟发送器
analog(ue) treatment 模拟处理
analog(ue) type 模拟型;模拟式;模拟类型
analog(ue)type ferrite shifter 模拟铁氧体移相器
analog(ue) value 模拟值
analog(ue) variable 模拟量;模拟变量
analog(ue) variable assignment 模拟变量指定
analog(ue) visibility computer 模拟可见度计算机;模内可见度计算机

analog(ue) voltage 模拟电压
analog(ue) voltage signal 模拟电压信号
analog(ue) voltmeter 模拟伏特计
analog(ue) wave spectrum analyzer 模拟波谱分析器
analog(ue) weather map 模拟天气图
analog(ue) year 相似年份
analogultra-violet spectrophotometer 模拟紫外分光光度计
analogy 类似;类比;模拟学;相似性;比喻;比拟
analogy analysis 模拟分析
analogy approach 类比法;模拟法
analogy calculation 模拟计算
analogy calculator 模拟计算器
analogy computation 模拟计算
analogy computer 模拟计算机
analogy computing device 模拟计算装置
analogy decode 模拟译码
analogy decoder 模拟译码器
analogy display 模拟显示
analogy experiment of groundwater movement 地下水运动模拟试验
analogy filter 模拟滤波器
analogy machine 模拟机
analogy method 模拟方法;模拟法
analogy method for beam on elastic foundation 弹性地基梁比拟法
analogy method of drawdown 降深比拟法
analogy method of drawdown coefficient 下降系数比拟法
analogy method of drawdown coefficient of precipitation 降雨入渗系数比拟法
analogy method of electric(al) network 电网络模拟法
analogy method of hydrogeology 水文地质比拟法
analogy method of minimum spring discharge 泉水最小流量比拟法
analogy method of mining modulus 开采模数比拟法
analogy method of resistance-capacitance network 电阻—电容网络模拟法
analogy method of resistance network 电阻网络模拟法
analogy method of resistance-resistance network 电阻—电阻网络模拟法
analogy model 模拟模型
analogy of law 法律的推理
analogy of movement principle 运动原理模拟
analogy of narrow aperture trough 窄缝槽模拟
analogy of net model 网络模型模拟
analogy playback system 模拟回放系统
analogy procedure 模拟方法
analogy response computer 模拟反应计算机
analogy simulation 相似模拟
analogy tape 模拟带
analyptic wallpaper 有凸出花纹的墙纸
analyse 剖析;分析研究
analyser = analyzer
analyser on harmonic principle 和谐原理分析器(潮汐分析用)
analyser plate 检偏振片
analyser tube 分析管
analyses and handles 分析和处理
analysette 光学沉淀扫描仪
analysing 分析
analysing prism 分析棱镜
analysing semi-variable overhead 分析半变动间接费
analysi of data 资料分析
analysis 解析法;分析
analysis area 分析区域
analysis block 分析组;分析块
analysis by absorption of gases 吸气分析
analysis by finite differences 有限差分析法;有限差分法
analysis by measure 容积分析;量测分析
analysis by sedimentation 沉降分析(法)
analysis by sieving 筛分颗粒分析
analysis by synthesis method 综合分析法
analysis by titration 滴定分析
analysis by wet way 湿法分析
analysis cell 分析池
analysis certificate 化验证(明)书;分析证明书
analysis component 组分分析

analysis controller 分析员
analysis date of water sample 水样化验时间
analysis detected limit 分析准确度;分析检出限
analysis filter 分析滤光片;分光滤色片
analysis for credit 信用分析
analysis in time domain 时域分析
analysis item 分析项目
analysis item of water quality 水质分析项目
analysis life 分析年限
analysis life of economy 经济分析年限
analysis list 分析单
analysis meter 分析仪;分析计
analysis method of critical points in economy 经济临界点分析法
analysis method of developing regime 开采动态分析法
analysis method of economic safety coefficients 经济安全系数分析法
analysis method of mercury content 汞含量分析法
analysis method of satellite picture 卫片解译法
analysis methodology 分析方法论
analysis methods of exploratory profile 勘探剖面精度分析方法
analysis mode 分析式;分析方式
analysis of activated sludge 活性污泥分析
analysis of a factorial experiment 析因试验分析
analysis of a fuzzy function 模糊函数的分析
analysis of agglomeration 聚集分析
analysis of agricultural data 农业数据分析
analysis of agricultural drugs 农药分析
analysis of bed evolution 河床演变分析
analysis of business profit 企业利润分析
analysis of canonic(al) correlation 正规相关分析
analysis of capacity and load 能力与负荷分析
analysis of capital changes 资本变动分析
analysis of chemical composition of natural gas 天然气化学成分分析
analysis of coal ash 煤灰成分分析
analysis of complex system 复杂系统的分析
analysis of continuous girders by successive approximation 逐次渐近法分析连续梁
analysis of cost 费用分析
analysis of covariance 协方差分析;共变数分析
analysis of covariance model 协方差分析模型
analysis of covariance table 协方差分析表
analysis of cyanide ion 氰离子分析法
analysis of data 分析资料
analysis of data collected 采集数据分析
analysis of deviation 自差分析
analysis of dispersion 离差分析
analysis of dynamic(al) system behavio(u)r 动态系统行为分析
analysis of effect 效率分析
analysis of elasticity 弹性分析
analysis of environment(al) statistics 环境统计分析
analysis of existing conditions 现状分析
analysis of expenses 费用分析
analysis of fatigue 疲劳分析
analysis of field testing 现场实验分析
analysis of financial conditions 财务状况分析
analysis of financial liquidity 资产变现能力分析
analysis of financial statements 财务报告分析;财务报表分析
analysis of fluvial morphology 河流地貌分析
analysis of fluvial process 河床演变分析
analysis of frequency distribution 频率分布分析
analysis of fuel gases by Hempel method 亨佩尔燃料气体分析法
analysis of fumes 炉气分析
analysis of functional relationship 中函数关系分析
analysis of fuzzy functions 模糊函数分析
analysis of gamma spectrum method 伽马能谱分析
analysis of ground vibration 地面振动分析
analysis of hydrologic data 水文资料分析
analysis of ignited sample 煅烧试样分析
analysis of intense rainfall 暴雨分析
analysis of market links 市场环节分析
analysis of material price variance 材料价格差异分析
analysis of materials' placement 停放分析
analysis of materials quantity variance 材料数量差异分析

analysis of measure 量测分析
analysis of ocean wave spectrum 海浪谱分析
analysis of oscillogram 波形图分析
analysis of parameters 参数分析
analysis of pharmaceutical dosage forms 制剂分析
analysis of polluted water 污水分析
analysis of precipitation 雨量分析
analysis of price decline 跌价分析
analysis of prices 造价分析
analysis of program(me) 农业计划分析
analysis of range 极差分析
analysis of relationship 相关分析
analysis of resource trends 资源趋势的分析
analysis of results 成果分析
analysis of rigid frame 刚架分析
analysis of riverbed development 河床演变分析
analysis of river geomorphology 河流地貌分析
analysis of rodent poison bait 灭鼠药分析
analysis of scouring and sedimentation 冲淤分析
analysis of sewage effluent 污水处理后出水分析;污水出水分析;污染出水分析
analysis of sewage sample 污水水样分析
analysis of shells 薄壳(结构)分析
analysis of shifting products 搬运分析
analysis of shoal evolution 浅滩演变分析
analysis of shoal process 浅滩演变分析
analysis of simulation result 仿真结果分析
analysis of slide 滑坡分析
analysis of slope stability 边坡稳定性分析
analysis of stable isotope of natural gas 天然气同位素分析
analysis of statistic(al) data 统计数据分析
analysis of step test 分级试验分析
analysis of strain 应变分析
analysis of structural layers demarcated by well data 经钻井标定的构造层分析
analysis of structural member 结构构件计算
analysis of surface on Asia 亚洲地面天气分析图
analysis of tenders 投标审查;投标书分析;标书分析
analysis of the affairs of business 经营情况分析
analysis of the truss joints 桁架节点分析
analysis of time series 时间序列分析
analysis of transfer operation 装卸作业分析
analysis of transfer path 装卸作业分析路线
analysis of truss 桁架分析
analysis of truss stresses 桁架应力分析
analysis of uncertainty 不确定性分析
analysis of underground water level 地下水位分析
analysis of variance 离散分析;方差分析;变异数分析;变异分析;变量分析
analysis of variance analog(ue) 方差类比分析
analysis of variance diagram 方差分析图
analysis of variance for correlation ratio 相关比方差分析
analysis of variance model 方差分析模型
analysis of variance table 方差分析表
analysis of wastewater effluent 污水处理后出水分析
analysis of wastewater quality 污水水质分析
analysis of water content in inclusion 包裹体水含量分析法
analysis of water resources system 水资源系统分析
analysis of water sample 水样分析
analysis of weather map 天气图方法
analysis on acceptance 验收分析
analysis on investment decision 投资决策分析
analysis phase 分析阶段
analysis process(ing) 分析处理
analysis program(me) 分析程序
analysis report 分析报告
analysis room 分析室
analysis routine 分析程序
analysis sample 分析煤样
analysis sample of appointing item 指定项目分析水样
analysis sample of contaminated water 污水分析样
analysis sheet 分析表
analysis situ 拓扑
analysis stage 分析阶段
analysis statement 分析表
analysis technics 分析技术

analysis technique 分析法
analysis technique of X-ray fluorescence X射线荧光分析技术
analysis unit of water sample 水样化验单位
analysis upon entry 验收分析
analysis value calculation method 分析值计算法
analysor 分析器
analyst 化验员;化验师;分析(人)员;分析工作者
analyst backlog 分析者积累报告
analyst programmer 程序分析员
analyt 分析化学
analyte 分析物
analyte addition method 分析物添加法
analyte subtraction method 分析物减量法
analytic(al) 分析的
analytic(al) accounting 分析会计学
analytic(al) acoustics 分析声学
analytic(al) aerial triangulation 解析法空中三角测量
analytic(al) aerotriangulation 解析法空中三角测量
analytic(al) aerotriangulation in an individual strip 单航带解析空中三角测量
analytic(al) analysis 解析分析
analytic(al) and interpretive report 分析和解释性报告
analytic(al) anthropology 分析人类学
analytic(al) apparatus 分析仪器
analytic(al) approach 分析方法
analytic(al) atomic absorption spectroscopy 分析原子吸收光谱学
analytic(al) atomic spectroscopy 分析原子光谱学
analytic(al) audit file 分析性审计档案
analytic(al) auditing 分析(性)审计
analytic(al) balance 精密天平
analytic(al) balance weight 分析天平砝码
analytic(al) basis of composition 分析基组分
analytic(al) bias 分析偏倚
analytic(al) biochemistry 分析生物化学
analytic(al) calculation 分析计算法
analytic(al) calorimetry 分析量热学
analytic(al) center 分析中心
analytic(al) centrifugation 分析离心分离
analytic(al) characteristics 分析特性
analytic(al) chemistry 分析化学
analytic(al) chemistry in nuclear fuel reprocessing 核燃料回收分析化学
analytic(al) chemistry in nuclear technology 核技术分析化学
analytic(al) chemistry laboratory 分析化学试验室
analytic(al) chromatograph 分析用色谱仪;分析色谱仪
analytic(al) column 分析柱;分析栏
analytic(al) combustion 分析燃烧
analytic(al) computation of carrying capacity at station 车站通过能力分析计算法
analytic(al) computer equipment 解析计算机设备
analytic(al) concentration 分析浓度
analytic(al) condition 测试条件
analytic(al) continuation 解析延宕(光谱函数);解析开拓
analytic(al) control 分析控制;分析检验
analytic(al) costing 分析成本计算
analytic(al) curve 分析曲线
analytic(al) data 分析数据
analytic(al) data of oil field water 油田水分析数据
analytic(al) data processor 解析数据处理
analytic(al) data source 分析数据来源
analytic(al) date 分析日期
analytic(al) demonstration 解析证明(法);分析论证
analytic(al) detection limit 分析检测限度
analytic(al) determination 分析测定
analytic(al) distillation 分析蒸馏
analytic(al) dynamics 分析动力学
analytic(al) electrofocusing meter 分析聚焦电泳仪
analytic(al) electron microscope 分析用电子显微镜;分析电子显微镜
analytic(al) engine 解析机;分析机(器)
analytic(al) equation 解析方程
analytic(al) equipment 分析设备
analytic(al) equivalence 解析等价
analytic(al) error 分析误差
analytic(al) esthetics 分析美学
analytic(al) evidence 分析性证据

analytic(al) experimental physics 分析实验物理学
analytic(al) expression 分析式
analytic(al) expression of the law of distribution 分布律的解析表达式
analytic(al) extraction 分析萃取
analytic(al) extrapolation 解析外推法;分析外推法
analytic(al) facility 分析设备
analytic(al) factor 分析因数
analytic(al) findings 分析成果;分析结论;分析结果
analytic(al) forecast 分析性预测
analytic(al) form 解析形式
analytic(al) formula 解析式
analytic(al) fracture mechanics 分析断裂力学
analytic(al) framework 分析框架
analytic(al) function 解析函数;分析函数(式)
analytic(al) functional group 分析官能团
analytic(al) function generator 分解函数发生器;解析函数发生器
analytic(al) gap 分析间隙
analytic(al) geochemistry 分析地球化学
analytic(al) geometry 解析几何(学)
analytic(al) geometry of plane 平面解析几何
analytic(al) geometry of space 立体解析几何(学)
analytic(al) geometry of three dimensions 立体解析几何(学)
analytic(al) geomorphology 动力地貌学;分析地貌学
analytic(al) glassware 分析用玻璃器皿
analytic(al) grammar 解析文法;分析文法
analytic(al) graph 分析图形
analytic(al) grouping 分析分组
analytic(al) hierarchy process 层次分析法
analytic(al) image 解析像片
analytic(al) index 分析指数
analytic(al) index for sample 样品分析指数
analytic(al) industry 分析工业
analytic(al) inertial navigation 分析式惯性导航
analytic(al) instrument 分析仪器
analytic(al) instrumentation 分析仪器;测试设备
analytic(al) integration 解析积分
analytic(al) interim audit 分析性中期审计;分析性期中审计
analytic(al) interior orientation 解析内定向
analytic(al) intersection method 解析交会法
analytic(al) investigation 分析研究
analytic(al) isotachophoresis 等速电泳分析法
analytic(al) item 分析项目
analyticality 解析性
analytic(al) job evaluation 分解式作业评价
analytic(al) life 分析年限
analytic(al) line 分析线
analytic(al) line pair 分析线对
analytic(al) liquid chromatograph 分析液相色谱仪
analytic(al) liquid chromatography 分析液相色谱法
analytic(al) location 解析法定线
analytically pure 分析纯
analytically pure agent 分析纯试剂
analytic(al) manifold 解析流形
analytic(al) manufacturing process 分解制造过程
analytic(al) map 分析地图
analytic(al) mapping 解析测图
analytic(al) mapping control point 解析图根点
analytic(al) measurement 分析测量
analytic(al) measurement electrode 分析测量电极
analytic(al) mechanics 分析力学;分比力学
analytic(al) method 解析法;分析方法;分析法
analytic(al) method of groundwater resources evaluation 地下水资源评价解析法
analytic(al) method of photogrammetric mapping 解析法测图
analytic(al) method of profile accuracy 勘探剖面精度分析法
analytic(al) method of sample 样品分析方法
analytic(al) method of selection 分析式人员选择法
analytic(al) method of source rock 烃源岩分析方法
analytic(al) methodology 分析方法论
analytic(al) mode 分析模式
analytic(al) model(ing) 解析模型;分析(性)模型
analytic(al) molecular spectroscopy 分析分子光谱学
analytic(al) nadir point triangulation 天底点解析

三角测量;解析天底点辐射三角测量
analytic(al) nucleopore filter 分析核孔过滤器
analytic(al) number 分析编号
analytic(al) number theory 解析数论
analytic(al) orientation 解析定向
analytic(al) paper 分析滤纸
analytic(al) percentage 分析百分比
analytic(al) percents 百分比分析法
analytic(al) philosophy 分析哲学
analytic(al) photogrammetry 解析摄影测量(学)
analytic(al) photography 分析摄影(学)
analytic(al) photometry 分析光度学;分析光度测量法;光度分析法
analytic(al) phototriangulation 分析摄影三角测量;解析像片三角测量;解析摄影三角测量
analytic(al) picture 解析像片
analytic(al) plate number 分析塔板数
analytic(al) plotter 解析绘图仪;解析测图仪
analytic(al) plotting 解析测图
analytic(al) plotting instrument 解析测图仪
analytic(al) point of view 解析观点
analytic(al) precision 分析精确度
analytic(al) prediction 分析预测
analytic(al) principles 分析性原则
analytic(al) probability model 解析概率模型
analytic(al) procedure 分析方法;分析步骤
analytic(al) process 分析制造程式;分析过程;分解制造过程;分解式制造程序
analytic(al) product 分析产品;分解(的)产品
analytic(al) program(me) 分析程序
analytic(al) projective geometry 解析摄影几何学
analytic(al) proofness 解析证明
analytic(al) pure 分析纯
analytic(al) quadrupole mass spectrometer 分析四极质谱计
analytic(al) quality control 分析质量控制
analytic(al) quality control laboratory 分析质量控制实验室
analytic(al) quantity 分析量
analytic(al) radar prediction 解析雷达预测
analytic(al) radial triangulation 解析法辐射三角测量
analytic(al) radiobiology 分析放射生物学
analytic(al) radio chemistry 分析放射化学
analytic(al) reaction 分析反应
analytic(al) reagent 分析试剂;分析纯试剂
analytic(al) reagent grade 分析纯;分析试剂级
analytic(al) record 分析记录
analytic(al) rectification 解析纠正
analytic(al) reduction software 分析处理软件
analytic(al) region of resources situation 资源形势分析范围
analytic(al) regression 解析回归
analytic(al) relationship 解析关系
analytic(al) restitution 解析纠正
analytic(al) result 分析结果
analytic(al) review 分析(性)审核;分析(性)检查
analytic(al) review procedures 分析性审核程序
analytic(al) sample 分析样品;分析试样
analytic(al) sampling 分析取样;分析抽样
analytic(al) sampling error 分析取样误差
analytic(al) sampling survey 分析性抽样调查
analytic(al) scale 分析天平;分析度盘;分析标尺
analytic(al) schedule 分析程序表;分析表
analytic(al) sequence 分析序列
analytic(al) set 解析集;分析装置;分析校准设备;分析集;分析机
analytic(al) signal 解析信号;分析信号
analytic(al) smoothing 分析修匀;分析平滑
analytic(al) solution 解析解(法);分析解
analytic(al) solution of water quality model 解析解水质模型
analytic(al) specimen 分析标本
analytic(al) spectra 分析谱
analytic(al) spectrometer 频谱分析仪
analytic(al) spectroscopy 分析光谱学
analytic(al) standard 分析标准
analytic(al) statement 分析声明
analytic(al) statics 解析静力学;分析静力学
analytic(al) station 分析站
analytic(al) statistics 分析统计(学)
analytic(al) statistics method 分析统计法
analytic(al) stereophotogrammetry 分析立体摄影测量学

analytic(al) stereoplotter 解析立体测图仪;分析立体绘图仪
analytic(al) stereoplotting 解析立体测图
analytic(al) stereoplotting system 解析立体测图系统
analytic(al) stereo-triangulation 解析法立体三角测量
analytic(al) stratigraphy 分析地层学
analytic(al) study 分析研究(法)
analytic(al) surface 解析曲面;分析曲面
analytic(al) survey 分析性调查
analytic(al) system 分析系统
analytic(al) tacheometry 解析法视距测量
analytic(al) technique 解析技术;解析法;分析技术;分析法
analytic(al) term of sample 样品分析项目
analytic(al) test 分析试验;分析检验
analytic(al) three-point resection radial triangulation 三交点交会法分析径向三角测量
analytic(al) time 分析时间
analytic(al) topology 分析拓扑学
analytic(al) traffic forecasting methods 交通预测解析法
analytic(al) transformation 解析变换
analytic(al) treatment 解析法处理
analytic(al) tree-point resection radial triangulation 三点交会分析径向三角测量
analytic(al) trend 分析趋势
analytic(al) triangulation 解析三角测量
analytic(al) trigonometry 分析三角学
analytic(al) type 分析类型
analytic(al) type of sample 样品分析类型
analytic(al) ultracentrifugation 分析超速离心
analytic(al) ultra centrifuger 分析超速离心机
analytic(al) uncertainty 分析不确定性
analytic(al) unit 分析组件;分析单元;分析单位
analytic(al) variance 分析方差
analytic(al) vocoder 解析声码器
analytic(al) weights 分析砝码
analytic(al) work sheet 分析用细算表
analyticity 分析性;可分析性;解析性
analytics 分析学
analytics analysis 解析学
analyzability 可分析性
analyzable 可分析的
analyze case 分情形分析
analyzed area 分析区域
analyzed chart 分析图
analyzed pattern 分析图形
analyzer 上偏光镜;检偏器;检偏镜;测定器;分析员;分析仪;分析器;分析机;分析程序的程序;分析程序
analyzer-controller 分析控制器
analyzer for nitrogen oxides 氮氧化物分析仪
analyzer for supplying water for boiler 锅炉给水分析仪
analyzer magnet 分析器磁铁
analyzer of Ca and Fe by radioisotope 放射同位素钙铁分析仪
analyzer of residual chlorine 残氯分析仪
analyzer on harmonic principle 和谐原理分析器(潮汐分析用)
analyzer polaniscope 分析旋光镜
analyzer polariscope 分析偏振镜
analyzer pot 分析罐
analyzer pumping system 分析器抽气系统
analyzer system 分析器系统
analyzer tube 分析管
analyze the behavio(u)r of systems 分析系统特性
analyze the data 数据分析
analyzing 分析
analyzing adapter 检验接头
analyzing crystal 分光晶体
analyzing filter 分光滤色片
analyzing magnet 分析磁铁
analyzing mirror 分析反射镜
analyzing pal(a)eo-structure 古构造分析
analyzing rod 分析尺
anamalous heating process 异常加热过程
anamesite 细玄岩;中粒玄武岩
anamigmatism 重岩浆化作用
anamigmatization 深熔混合岩化
anammox 厌氧氨氧化

anammoxosome 厌氧氨氧化体
anamnesis 记忆力
anamorphetic lens 变形透镜
anamorphic compression 变形压缩
anamorphic cylindrical surface 变形的柱面
anamorphic effect 变形效应
anamorphic lens 艾奈莫尔费像变透镜;畸变透镜;变形透镜;变形镜头
anamorphic magnification 畸变放大
anamorphic prism 变形棱镜
anamorphic stretching 变形伸长
anamorphic zone 合成变质带
anamorphic zoom lens 变形变焦距透镜
anamorphism 合成变质
anamorphoscope 歪像校正镜
anamorphose 图像变形
anamorphoser 变形系统;失真透镜;失真光学仪;成像变形器;变形系数;变形器
anamorphosis 像变;失真;畸形;变形;歪像;图像变形
anamorphote 变形透镜
anamorphote lens 歪像透镜
anamorphotically squeezed image 畸形压缩图像;变形压缩图像
anamorphotic attachment 变形镜装置
anamorphotic copying objective 变形复制物镜
anamorphotic lens 变形物镜
anamorphotic optical system 变形光学系统
anamorphotic system 变形系统;变形系数
anandite 钡铁脆云母
anangenesis 前进性演化
anapaite 三斜磷钙铁矿
anapeirean 太平洋套
anaphalanx 暖锋面
anaphase 后期;分裂后期
Anaphe silk 阿纳菲野蚕丝
anaphoresis 阴离子电泳
anaphylactic 过敏的
anaphylactic disease 变应性病
anaphylactic reaction 过敏反应
anaphylaxis 过敏反应
anaplasia 间变
anaplerotic reaction 补给反应
anapophysis 副突
an area of 100m × 100m 面积为 100m × 100m
anarrhan 逆流
a narrow fissure running lengthwise 一条纵长形的狭窄裂缝
a narrow strip on each side of a fence 篱笆两侧的狭长地带
anasarca with shortness of breath 正水
anascope 正像镜
anaseism 背震中
anaseismic 背震中的
anaseismic onset 背震初动
anastatic printing 凸版印刷
anastatic printing process 凸版印刷法
anastatic process 锌版转印法
anastatic transfer 凸版印刷
anastatic water 毛管上限水;毛管上升水
anastigmat 去像散透镜组
anastigmatic 去像散的
anastigmatic beam 消像散射束
anastigmatic de-flection system 去像散偏转系统
anastigmatic lens 消像散透镜
anastigmatic objective 消像散物镜
anastigmatic reflector 消像反射望远镜
anastigmatism 消像散性
anastigmator 消像散射器
anastomose 联合;接通;接合;交叉合流
anastomosed river 网状河(流)
anastomosed stream 网状河(流);网结河
anastomosed stream sequence 网结河层序
anastomosing 交织网状的
anastomosing branch 支汊;副河汉
anastomosing cleavages 交织劈理
anastomosing deltoidal branch 交织三角洲汊河
anastomosing drainage 排水网;交织水系;辫河水系
anastomosing fault 交织网状断层
anastomosing river 交织河流
anastomosing stream 网状河(流);交织河流
anastomosis 网结;联合;接通;接合;交叉合流;水网;联结现象

anastomotc mode 交织水系模式
anastomotic cave 网状洞穴;交织洞穴
anastomses 水网
anastomsing river 网状河(流)
anatase 锐钛矿
anatase configuration 锐钛型
anatase modification 锐钛型
anatase titanium dioxide 锐钛型钛白粉;锐钛型二氧化钛
anatase variety 锐钛型
anatectic earthquake 深源地震
anatectic granite 深熔花岗岩
anatectic magma 深熔岩浆;重熔岩浆
anatectic process 重熔作用
anatectite 重熔混合岩
anatexis 深熔作用;重熔作用
anatexis way 深熔作用方式
anatexite 重熔混合岩
anathermal 增温期
Anatolian massif 安纳托里地块【地】
Anatolian old land 安纳托里古陆【地】
Anatomical alloy 阿纳多易熔合金
anatomical model 剖析模型
anatomic model 分析模型
Anatorlian carpet 安那托利亚地毯(土耳其)
anatricrotic pulse 升线三波脉
anatriptic 揉擦剂
anauxite 蠕陶土;富硅高岭石
an average annual increase of 每年递增
Anavor 安纳沃尔聚酯纤维
Anblock method 安卜洛克法
Ancaster stone 棕灰色粗纹石灰石
ancaustic decoration 烧入装饰
ancestral hall 祠堂
ancestral rivers 古河系;古代河系
ancestral task 世系任务;原始任务
ancestral temple 祠堂
ancharmonic ratio 交比
anchi-eutectic rock 近共结岩
anchimeric assistance 邻位协助;邻位促进
anchimeric effect 邻位效应
anchimetamorphism 近变质作用;近地表变质作用
anchimonomineralic rock 近单矿物岩
anchoic acid 壬二酸
anchor 锚固件;锚碇物;锚碇器;锚点;锚(件);锚钟;方爬器;铰钉;簧片;系紧;窗籍立柱;固定金属件;电楞衔铁
anchorable 可锚泊的
anchorage 停泊所;停泊区;停泊地;停泊处;停泊(指泊作业);水管支墩;拉车;铰钉固定;锚基;锚固体;锚固钢筋;锚固;锚杆支护;锚碇(作用);锚碇装置;锚固区(域);锚泊地;锚件;支抗;固着性;地牛;地锚;支抗
anchorage abutment 锚碇墩台;锚碇支座
anchorage and gunite support 喷锚支护
anchorage area 锚泊地;锚碇区;锚地
anchorage bar 锚碇钢筋;紧固件;紧固杆
anchorage basin 停船池;停泊池;锚泊地;系留港池
anchorage beam 锚梁
anchorage bearing 锚座;锚拉支座;锚固支座
anchorage bend 锚碇钢筋;(钢筋末端的)弯钩;锚碇弯曲
anchorage blister 锚碇齿板
anchorage block 锚碇块;镇墩;锚墩;锚块
anchorage bolt 系紧螺栓
anchorage bond 锚固黏结(预应力钢丝端部)
anchorage bond stress 握裹力;锚固黏结应力
anchorage buoy 锚地浮标;系船浮筒;系船浮标;系泊浮筒;标志锚泊区的浮标
anchorage by bond 握裹锚碇
anchorage by friction 摩阻力锚固;摩阻固定
anchorage cable 锚索;锚缆
anchorage capacity 锚固能力
anchorage chain 锚链
anchorage chamber (吊桥的)锚碇舱;(吊桥的)锚碇室
anchorage chart 停泊地海图;锚地图
anchorage clip 紧固件;紧固夹
anchorage cone 锚锥(头)
anchorage deadman 锚桩
anchorage deformation 锚碇变形;锚固件位移;锚固件滑脱;锚固件变形;锚头变形;锚具变形;锚杆变形;锚碇变形;预应力混凝土钢筋锚具应力损失;锚件变形

anchorage deformation of slip 锚固件(滑动变形)应力损失
anchorage deformation slip 锚固件位移;锚固件滑动
anchorage device 锚固设备;锚具(预应力混凝土);施锚设备;锚碇装置;锚碇设备;锚泊装置
anchorage device of cable 钢索锚具;钢绞索锚头
anchorage distance 锚固距离
anchorage dues 停泊税;锚泊费
anchorage element 锚固构件;锚碇构件
anchorage failure 锚碇失效;锚座破坏;锚固破坏
anchorage fixing 锚固安装;锚碇
anchorage force 锚固力;锚碇力
anchorage for deep draught vessel 深吃水船锚地
anchorage for explosion 爆炸品锚地
anchorage for quay wall 码头墙的锚碇
anchorage for seaplanes 水上飞机系留区;水上飞机停泊区;水上飞机锚地
anchorage for small vessels 小船锚地
anchorage foundation 锚碇基础;锚碇基础
anchorage insulator 锚碇绝缘子
anchorage length 锚着长度;锚固长度
anchorage length for reinforcement 钢筋锚固长度
anchorage limit 锚地界限
anchorage limit buoy 锚地界限浮标
anchorage location 锚座位置
anchorage loss 预应力钢筋的应力损失;锚固件滑动变形应力损失;锚固损失;锚具预应力损失
anchorage mast 锚柱;锚碇柱
anchorage of boats in a line 多船并排系泊
anchorage of hauling 拉索锚固
anchorage of rail 钢轨锁定
anchorage of reinforcement 钢筋锚固
anchorage of structure 结构物的锚碇
anchorage on pipeline 管线中的锚基
anchorage picket 路线标桩;锚桩;锚柱
anchorage pier 锚墩
anchorage place of mooring 锚泊地
anchorage point 预埋联结盒
anchorage-prohibited 禁止抛锚
anchorage-prohibited area 禁止抛锚区
anchorage pull 锚碇拉力
anchorage rope 锚链;索索
anchorage section 锚固段
anchorage shaft 锚碇竖井(吊桥)
anchorage shifting 移锚位
anchorage shoe 锚座;锚靴
anchorage slip 锚固滑移;锚碇滑动;锚具变形;锚固变形;锚碇滑移
anchorage space 锚地
anchorage spud 锚碇柱;锚碇支柱;系船柱;固船柱
anchorage steel 锚碇钢筋;锚钢
anchorage stone 底砾
anchorage stress 锚固应力
anchorage support 拉牢支柱
anchorage system 锚碇系统;锚固系统;锚碇体系
anchorage tower 锚碇塔
anchorage type 锚座类型
anchorage unit 锚碇装置;锚泊装置
anchorage zone 预应力混凝土锚固区(先张法或后张法);锚件区;锚固区;锚碇影响区;锚碇区
anchorage zone of fanned-out wires (预应力混凝土的)扇形钢丝锚固区
anchor agitator 锚固式搅拌器;杆住搅拌器
anchor and chain certificate 锚锚链证书
anchor and collar (门上的)铰链;锚环(闸门上的金属铰);导向轴承(水轮机)
anchor anemometer 锚碇风速表
anchor angle 锚碇角钢;锚碇角度;拉杆角度;锚固角度
anchor apeak 立锚
anchor arm 拉线支持腕臂;锚杆;锚臂
anchor arrangement 锚设备
anchor as a lead 用锚测深
anchor assembly 锚系总成
anchor awash 锚露出水面;锚出水
anchor away 锚离底
anchor ball 抛锚信号球;锚球;锚杆圆头;锚泊球
anchor band 固定带环
anchor bar (钢筋末端的)弯钩;拉线杆;锚着钢杆;锚筋;锚固钢筋;锚杆
anchor barge 起锚船;锚泊驳船;下锚船;移锚船
anchor bar of sheet piling abutment 板桩桥台锚固栓钉

anchor beam 锚碇梁
anchor bearing 锚方位
anchor bearing plate 锚碇座板;锚碇支承板
anchor bed 锚座;锚床
anchor behaviour 固结状态
anchor bend 锚结;渔人结
anchor berth 锚地泊位
anchor bill 锚爪尖部;锚式摘纵叉爪
anchor blade 锚叶
anchor block 墙内木砖;锚墩;锚碇体;锚枕;锚块;锚碇座;管道固定支座;地下横木;地锚
anchor board 锚床
anchor boat 起锚艇;起锚船;抛锚船;绞锚船;锚艇
anchor bolster 锚碇孔唇口
anchor bolt 锚碇基础螺栓;固定螺栓;地脚螺栓;锚栓;锚固螺栓;锚碇螺栓;地脚螺钉;底脚螺栓
anchor bolt and nut 锚碇螺栓螺母
anchor bolt box 地脚螺栓预留孔
anchor(bolt)hole 地脚螺栓孔
anchor bolt length 地脚螺栓长度
anchor bolt nut 固定螺栓螺母;地脚螺栓螺母
anchor bolt pitch 地脚螺栓间距
anchor bolt slot 地脚螺栓长孔
anchor bolt support 锚杆支护
anchor bolt with grout 注浆锚杆
anchor boom 抛锚杆;锚臂;移锚吊杆
anchor boom winch 抛锚杆绞车
anchor boom winch speed controller 抛锚杆绞车速度控制器
anchor bower 首锚
anchor box 燕尾形生铁盒
anchor box type 地脚螺栓预留孔式
anchor break out 锚出土
anchor brick 受钉木砖;木砖;木砧;锚砖;有钉木砖
anchor broken out 锚出土
anchor brought up 锚已抓牢
anchor bulkhead 锚碇挡土墙;锚碇岸壁(码头)
anchor buoy 系泊浮筒;锚用浮标;锚位浮标;锚位标;锚浮标;锚船浮筒;锚标;下锚浮标;系船浮标;标示下锚点的浮筒
anchor buoy rope 锚标绳
anchor button 锚钮
anchor by the stern 尾锚锚泊
anchor cable 锚链;锚缆;锚碇拉索
anchor cable drum 锚缆卷筒
anchor cable indicator 锚链示数器
anchor cable stopper 锚链制动器
anchor capstan 起锚铰盘;起锚绞盘;起锚机绞盘;起锚车
anchor casing packer 锚式套管封隔器
anchor cast on downward slope 下坡锚
anchor cast on upward 上坡锚
anchor catcher 锚捕捉器
anchor cell 锚箱
anchor certificate 锚证书
anchor chain 锚链
anchor chain meter 锚链拉力表
anchor chain shackle 锚链卸扣;锚链卡环
anchor chair 锚支座
anchor channel 锚碇铁件;锚固槽钢
anchor charge 簧片装药
anchor chock 锚楔;锚架
anchor clamp 夹具;拉线夹
anchor clip system 卡子固定法
anchor clutch 锚爪
anchor coat 结合层;打底胶浆;初层;锚固涂层;中间涂层
anchor column 竖旋桥支柱;锚柱
anchor cone 锚头;锚锥
anchor connector 锚连接器
anchor crane 起锚起重机;起锚吊车;吊锚机;吊锚吊杆
anchor crown 锚冠
anchor crown shackle 锚冠卡环
anchor cup 锚圈套
anchor cylinder 锚碇筒
anchor dart 锚尖装饰;锚饰箭头;锚箭饰;锚固箭头饰;锚尖;锚镖装饰
anchor davit 起锚吊杆;锚吊杆;吊锚柱;吊锚杆
anchor disc 固定盘
anchor ditcher 锚式挖沟机
anchor drag 探锚
anchor dragging 锚在拖;锚没有抓底
anchor dredge(r) 锚碇位挖泥船;锚泊挖泥船

anchor dues 锚泊费
anchor ear 桩环；柱箍；抱箍；抱钩
anchored and jacked pile 锚杆静桩
anchored bearing 锚固支承座；锚固支座
anchored bolt 锚固式锚杆
anchored bolt retaining wall 锚杆挡墙
anchored bulkhead 锚固式挡土墙；锚固式驳岸；锚碇崖壁；锚碇挡土墙；锚碇挡墙；有拉секжан的岸壁
anchored bulkhead abutment 锚碇板式桥台
anchored bulkhead retaining wall 锚碇板(式)挡土墙；锚杆挡土墙
anchored buoy 锚碇浮标
anchored buoy station 定泊浮标站
anchored buoy system 锚碇浮标系统
anchored cable 锚碇缆索
anchored cellular sheet-pile structure 有锚格形板桩结构
anchored compound 黏附的化合物
anchored concrete pavement with wire mesh reinforcement 锚挂钢筋混凝土护坡
anchored dune 固定沙丘
anchored earth 锚碇土
anchored end 锚固端（预应力混凝土）
anchored foundation 锚拉基础
anchored in the ground 锚固在土中；锚碇在土中
anchored location 锚碇位置
anchored mine 锚雷
anchored peg 锚桩
anchored pier 锚固墩
anchored point 锚碇点
anchored position observation of current 定点测流
anchored pretensioning （预应力混凝土的）锚碇先张法
anchored quaywall 锚碇墙式码头
anchored radio-sono-buoy 系泊无线电声呐浮标
anchored recording station 定泊自记（海洋水文）测站
anchored retaining wall 锚碇式挡土墙；锚杆挡土墙
anchored retaining wall by tie rod 锚杆式挡土墙
anchored sheet pile wall 锚碇板桩墙；锚系板桩墙；锚碇板桩墙
anchored sheet piling 锚系板桩；锚碇式板桩；打锚碇板桩；有拉杆的板桩
anchored sheet piling wall 锚碇板桩墙
anchored single-wall 锚碇单排桩墙
anchored single-wall sheet-pile structure 有锚单排板桩墙结构
anchored station 锚碇位置
anchored steel trestle 锚碇式钢栈桥；锚碇钢筋栈桥
anchored suspension bridge 锚固式悬索吊桥；锚碇式悬索桥
anchored tied back shoring system 后拉锚碇系统
anchored tower 锚碇钻塔
anchored tree 梢捆；沉树（将砍下的树锚碇于岸边以护岸防冲）；沉排（护岸用）
anchored-type ceramic veneer 锚碇型陶瓷板；用金属件固件固定的陶瓷面砖
anchored wall 锚碇式墙
anchor effect 固着效果
anchor efficiency 锚效率
anchor embedded in concrete 锚固在混凝土中；锚碇在混凝土中
anchor end of a bed 机架锚固端
anchor engaged 锚已收妥
anchor escapement 锚形擒纵机；锚式擒纵机构
anchor eye 锚眼；锚孔
anchor female cone 锚杯（预应力）
anchor fitting 锚固装置
anchor fixing angle steel 下锚固定角钢
anchor flags 锚旗
anchor fluke chock 锚爪垫
anchor flukes 锚爪
anchor force 锚固拉力
anchor-foul 锚纠缠
anchor fouled by the flukes 锚链缠住锚爪
anchor fouled by the stock 锚链缠住锚杆
anchor foundation 锚基础
anchor frame 锚碇架
anchor gap 火花（间）隙
anchor gate 锚柱闸门；锚支闸门；锚式闸门；闸门；闸；锚碇门（重闸门用）
anchor gear 起锚装置；起锚设备
anchor grip 锚固黏着力；楔形夹块（预应力锚具）

anchor ground 锚地；泊定地
anchor group 锚碇基团
anchor grout 锚碇灌浆
anchor grouting 锚固灌浆；锚碇灌浆；地脚灌浆
anchor handling tug 锚作拖船
anchor-handling vessel 抛锚船
anchor handling winch 移锚绞车
anchor hawse 锚链孔
anchor head 锚头；锚冠
anchor head and stern 抛首尾锚
anchor head by cold upsetting 冷镦锚头；镦头锚具
anchor hinge 锚杆铰链；锚栓铰链
anchor hoist 起锚绞车
anchor hoisting winch 移锚绞车
anchor hold 抓牢
anchor holding power 锚抓力
anchor hole 锚栓孔；锚杆孔；锚孔
anchor hole depth 地脚螺栓孔深度
anchor hole pitch 地脚螺栓孔间距
anchor hook 锚形钩
anchor hoops 锚木档铁箍
anchor hoy 起锚船；运锚船；港内运送长期系泊设备的小船；港口运送长期系泊设备的小船
anchor hydrographic(al) station 锚碇水文站；定泊水文（观测）站；定泊水道测量站
anchor ice 潜冰；锚冰【水文】；底冰
anchor ice sluice 底冰冲泄闸
anchor impeller 锚式搅拌器
anchoring 锚着；锚接；锚固；锚杆支撑；锚碇装置；锚碇环；锚碇；锚泊；装拉线；（位错的）钉扎；单锚泊
anchoring abutment 锚碇支座；锚碇墩
anchoring accessories 地脚钢筋；锚件附件；锚（碇）件（的）加固钢筋
anchoring agent 锚固剂；增黏剂
anchoring agent for glassine 玻璃纸用锚固剂
anchoring and guniting 锚喷
anchoring and mooring system 锚泊设备
anchoring basin 锚地
anchoring berth 锚地（泊位）；锚泊泊位
anchoring berthing 抛锚
anchoring block 锚墩；碇系块
anchoring bond 锚固黏结（预应力钢丝端部）
anchoring bracket 锚固联板
anchoring by friction 摩阻锚压
anchoring by the head and stern 抛首尾锚
anchoring chain 限位链
anchoring chamber （吊桥的）锚碇室；（吊桥的）锚碇舱
anchoring clamp 拉线夹
anchoring clip 拉线夹
anchoring collar 固定环
anchoring cone 锚锥；锚碇锥（头）
anchoring device 锚固装置
anchoring effect 锚固效应；锚碇效应；糙面效应
anchoring element 锚固构件；锚碇构件
anchoring equipment of crane 起重机锚碇装置
anchoring facility 锚碇设施
anchoring failure 锚固失效；锚固破坏
anchoring fascine 锚碇柴捆
anchoring fiber 固定纤维
anchoring filament 锚丝
anchoring force 锚栓力；锚固力
anchoring foundation 锚固基础
anchoring ground 锚地
anchoring hook 固定钩
anchoring jack 锚碇桩；锚碇柱；锚碇支架
anchoring lamp 锚泊灯
anchoring length 锚固长度
anchoring light 锚泊灯
anchoring loop of reinforced steel 钢筋锚固环
anchoring material 结合剂
anchoring method 锚固方法
anchoring of bedplate 座板锚碇
anchoring of dislocation 位错的钉扎
anchoring of prestress 预应力锚固
anchoring of sheet pile 板桩锚碇结构
anchoring of stonework 石作锚碇
anchoring of track 线路锁定
anchoring picket 路线标桩；锚柱
anchoring pier （桥梁的）锚墩
anchoring pipe 地脚螺丝套管；地脚螺栓套管
anchoring place 锚泊地
anchoring point 锚碇点
anchoring rail 锚固槽钢

anchoring resin 锚型树脂；碇系型树脂
anchoring rod 锚碇杆
anchoring screw 锚碇螺丝；止动螺钉
anchoring space 锚地；锚位
anchoring spud 锚碇桩；锚碇支柱
anchoring station 锚地
anchoring strength 锚接强度；锚固强度；锚碇强度
anchoring support 固定支撑
anchoring system 锚碇方法；锚碇方法；锚碇系统
anchoring tower 锚碇塔
anchoring tube 基础螺栓套管；地脚螺丝套管；地脚螺栓套管
anchoring wire 锚碇拉线
anchor in roadstead 在港外锚地抛锚
anchor insulator 拉桩绝缘子；拉杆绝缘子
anchor iron 锚铁
anchorite 脉状闪长岩；带状闪长岩
anchor jack 锚杆；支柱；撑柱
anchor key 固定锚销
anchor knot 锚结；渔人结
anchor lamp 锚灯
anchor lantern 锚灯
anchor leg mooring 锚腿式锚泊装置
anchor level ratio 锚碇高度比
anchor lift 抓钩提升器
anchor light 停泊灯；锚灯
anchor line 拉索；锚线；锚索；锚绳；锚链
anchor line guide spud 锚缆导桩
anchor line scow 抬缆方驳
anchor line string 系线
anchor link roller 锚链滚子；锚链转盘
anchor log 锚碇；锚木；锚固圆木；锚碇件
anchor loop 锚环；锚固圈；锚环；马蹄形螺栓
anchor loss 锚固件（滑移变形）应力损失
anchor male cone 锚塞
anchor mast 接触网支柱；锚柱
anchor method of stonework 石作锚碇法
anchor mixer 锚式混合器；锚式搅动器；锚式搅拌机；锚式混合机
anchor mooring 锚系定位
anchor moving delay factor 移锚延迟系数
anchor nut 锚碇螺帽；系紧螺母；地脚螺母
anchor of steel penstock 压力钢管镇墩
anchor opening （索桥、悬臂桥的）锚跨；锚孔
anchor operation 抛锚作业
anchor orders 锚令
anchor packer 锚碇式封隔器
anchor paddle mixer 锚桨式搅拌器
anchor palm 锚掌；锚齿
anchor pattern 锚系布置
anchor pea(k) 锚爪尖部
anchor pendant wire 吊锚缆
anchor picket 锚桩；锚固桩
anchor pier 锚墩
anchor pile 抗拔桩；拉杆桩；锚桩；锚固桩；系缆桩
anchor pin 固定销；连接销；锚销；锚式擒纵叉销；锚碇销；固定锚销；定位销；带动销
anchor pin cam 锚碇螺栓凸轮
anchor pin locknut 锁销防松螺母
anchor pin slot 带动销槽
anchor pintle 固定枢轴
anchor plate 锚碇（垫）板；锚垫板；锚板；螺栓垫圈；锚碇钢板；系定板；地脚板；地基板；带锚铺板；锚垫板
anchor plate in prestressed concrete 预应力钢筋混凝土锚板
anchor plate retaining wall 锚碇板挡土墙
anchor pocket anchor recess 锚窝
anchor point 锚碇点；绞接点；锚固点；锚段点；阻火位置；定位点；带缆点；防火位置
anchor point of planimeter 求积仪极点
anchor policy 船锚保险单
anchor position 锚位
anchor post 路线标桩；锚柱；固定柱；撑柱
anchor(post)hole 锚桩孔
anchor pressure 锚固压力
anchor prohibited mark 禁止锚泊标
anchor prop 锚栓支柱；锚杆支柱；固定支柱
anchor pull 锚固拉力；锚碇拉力
anchor raking pile 拉杆锚桩
anchor reinforcement 锚筋
anchor relay 簧片继电器
anchor rim 锚缘；锚环

anchor ring 锚碇环;锚环
anchor rod 锚筋;锚杆;锚碇杆;地脚螺栓;板桩拉杆
anchor rod jumbo 锚杆台车
anchor rod retaining wall 锚杆挡土墙
anchor root 固定根
anchor rope 锚链;锚索;锚碇拉索
anchor rope bend 锚缆结
anchors and chains proved 锚和锚链检验合格
anchor's aweigh 锚离底
anchor screw 基础螺栓;基础螺钉;锚栓;固定螺丝;固定螺钉;地脚螺钉;锚固螺栓
anchor section 锚段
anchor-section liner 下锚段衬砌
anchor-section lining 下锚段衬砌
anchor shackle 锚卸扣;锚链卡环;锚环;锚钩环【船】;锚穿链扣
anchor shaft 锚杆;锚柄
anchor shank 锚杆;锚柄
anchor shaped 锚形的
anchor sheeting 锚碇板
anchor sheeting pile 锚碇式板桩
anchor sheeting pile wall 锚碇式板桩墙
anchor sheet piling 锚固板桩
anchor shifting 移锚
anchor shoes 锚爪鞘;锚床;锚式鞋
anchor shop 锚工场;锚工厂
anchor short stay 抛近锚
anchor shoulder 锚肩
anchors in series 串连锚
anchor slab 锚碇板;锚板
anchor slab abutment 锚碇板桥台
anchor slab retaining wall 锚碇板挡墙
anchor slab structure 锚碇板结构
anchor slip 锚固滑移
anchor slot 锚碇缝隙;锚碇槽;锚缝;锚槽
anchor socket 锚碇插座;锚座
anchor span 锚柱跨距;锚跨(孔);下锚跨
anchor-spike hole 防爬道钉孔
anchors proved 锚已检验合格
anchor stake 拉紧桩;锚楔;锚橛;锚碇桩;锚柱
anchor stamping 锚检验钢印记
anchor station 水文观测站;定泊站;定泊点
anchor stay 锚索;锚碇拉索
anchor stock 锚座;锚杆;锚叉
anchor stock hoop 锚杆箍
anchor stone 底砾;固定石
anchor stopper 活砧式掣链器;掣链器
anchor store 商场中通中廊和外部的商店
anchor strap 系锚板;锚索;锚板;锚碇索;锚碇带
anchor strip 锚条
anchor strop 锚用绳环
anchor strut 拉线支柱;拉桩支栓;拉线撑杆;系泊桩
anchor stud 锚柱;锚固螺栓
anchor support 耐张支持物;分段支撑物
anchor swivel 锚链旋转接头;锚链旋转环
anchor system 锚碇系统;锚杆系统
anchor tackle 吊锚滑车组
anchor telegraph 起锚传令钟
anchor tenant (位于商场中部或端部商店的)承租人;关键租户
anchor testing 锚试验
anchor throat 锚喉
anchor tie 拉杆;锚条;锚拉杆;锚杆;锚碇拉条;锚碇杆;地锚拉索
anchor to check sheering 止
anchor tower 拉线塔;锚塔;锚碇塔;高架塔
anchor trial 锚设备试验
anchor tripper 锚板机
anchor trip(ping) line 浮锚拉索;弃锚拉索
anchor trumpet 锚碇喇叭管
anchor tube 锚固管
anchor type 锚固类型
anchor(type) agitator 锚式搅拌器
anchor-type ceramic veneer 铰接式陶瓷饰面板
anchor(type) mixer 锚式搅拌器
anchor-type stirrer 锚式搅拌器
anchor underfoot 脚下锚
anchor up 锚出水
anchor up and down 锚链垂直
anchor wall 锚墙;锚碇墙
anchor warp leader 锚索导口
anchor washer 锚碇板
anchor watch 锚更【船】;值锚更
anchor wedge 锚楔;锚键

anchor-weighing barge 绞锚驳
anchor winch 起锚绞盘;起锚绞车;锚杆收紧器
anchor windlass 起锚机
anchor windlass room 锚机间
anchor wire 桩索;锚索;锚缆;支持线
anchor with a spring 带尼缆抛锚
anchor work 抛锚作业;锚泊作业
anchor yoke 锚索环;锚轭
anchosine 硅铝石
Anchutz gyrocompass 安秀兹陀螺罗经
ancient 古代的
ancient animals 古动物
ancient architecture 古建筑(物);古代建筑
ancient art 古代艺术
ancient bridge 古代桥梁
ancient bronze colo(u)rs 古铜彩
ancient capital 故都;古都
ancient ceramics 古代陶瓷
ancient character 古代特点;古代人物;古代特色
ancient Chinese architecture 中国古代建筑
ancient Chinese garden 中国古代园林
ancient Chinese pottery 古陶器
ancient Chinese ware 古容器
ancient city 古代城市;古城
ancient civilization 古代文化
ancient coin 古钱
ancient colo(u)r 古彩
ancient colo(u)red drawing 古彩
ancient crucible 古代坩埚
ancient cultural 古代文物
ancient culture relic 古文化遗迹
ancient custom 古风(气)
ancient decoration 古式装饰
ancient deposited cohesiveness soil 老黏性土
ancient driveway 驰道
ancient Egyptian gardens 古埃及园圃
ancient erosion 古代侵蚀
ancient furniture 古式家具
ancient geologic(al) gorge 老河床;古峡谷
ancient glaciation 古代冰川作用;古冰川作用
ancient glass 古代玻璃;古玻璃
ancient imperial road 辇道
ancient Japanese perimmon juice 陈柿汁;陈柿油;陈柿漆
ancient karst 古岩溶;古喀斯特
ancient lakes 古代湖泊
ancient landform 古老地形;古地形;古地貌
ancient light 古式窗;采光权(英国);二十年以上的老窗户(别人无权遮挡其光线)
ancient map 古地图
ancient mausoleum 古墓
ancient meteorological observatory 古代气象台
ancient monument 古代纪念性建筑;历史遗迹;历史纪念物;古址;古迹
Ancient Mouments Acts 古迹条例
ancient object 古董
ancient painted pottery 古代彩陶
ancient palace 古宫
ancient picture 古画
ancient planation surface 古夷平面
ancient plank way 栈道
ancient plate 古板块
ancient porcelain 古瓷
ancient post road 驿道
ancient relics 古迹;古代遗物
ancient rent 过去租金
ancient river channel 古河槽
ancient river course 古河道
ancient rock slide 古滑坡
ancient Roman 古罗马
ancient Roman garden 古罗马花园
ancientry 古风
ancient sediment 古沉积物;古代沉积物
ancient slide 古滑坡
ancient slip plane 古滑移面
ancient soil 古土(壤)
ancient strand line 古滨线
ancient tomb 古墓
ancient tray support 托子(窑具)
ancient tree 古树
ancient tunnel 古隧道
ancient valley 古河谷
ancient volcano 古火山
ancient wall 古城墙

ancillary 从属的;辅助的;辅料
ancillary accommodations of stage 舞台附属用房
ancillary apparatus 辅助设备
ancillary attachment 特殊附件;附件
ancillary bill 附加起诉状
ancillary block 辅助部件
ancillary building 辅助房屋;辅助建筑(物)
ancillary cable 辅助电缆
ancillary control process 辅助控制进程
ancillary control processor 辅助控制处理机(器)
ancillary craft for dredging works 疏浚工程辅助船舶
ancillary credit 辅助信用
ancillary credit business 辅助信贷业务
ancillary customer 用户辅助设施
ancillary customer service 为用户服务的附属工作
ancillary device 辅助装置
ancillary documents 辅助文件;辅助的单据;附属文件
ancillary equipment 外围设备;外转设备;外部设备;辅助设备;补充设备
ancillary estimator 辅助估计式;辅助估计量
ancillary facility 辅助设备
ancillary information 辅助信息
ancillary investment 辅助投资
ancillary land use 土地的附属用途
ancillary lens system 附加透镜系统
ancillary measurement 辅助量测
ancillary process 辅助过程
ancillary profit 附业利润
ancillary revenue 工附业收入;副业收入;附业收入
ancillary right 附属权益
ancillary road 辅助道路
ancillary shoring 辅助支撑
ancillary species 附属树种
ancillary statistic 辅助统计量
ancillary telescope 辅助望远镜
ancillary volume 补充卷
ancillary works 辅助工作;附属工程;辅助工程
ancipital 二头的;剑形的
ancon 河渠弯道;悬臂托梁;肘状河曲
ancona 胡桃木
Ancona diagram 安科纳图
ancona ruby 红水晶
ancon(e) 肘托
anconeal 肘的
Ancor iron powder 安科海绵铁粉
ancylite 碳锶铈矿
ancyroid 锚状的;钩状的
anda-assu oil 石栗果油;蜡烛果油
andacite group 英安岩类
andalusite 红柱石片麻岩;红柱石
andalusite biotite quartz schist 红柱石黑云母石英片岩
andalusite biotite schist 红柱石黑云母片岩
andalusite dimicaceous schist 红柱石二云母片岩
andalusite-gneiss 红柱石片麻岩
andalusite hornfels 红柱石角岩
andalusite mica schist 红柱石云母片岩
andalusite muscovite quartz schist 红柱石白云母石英片岩
andalusite muscovite schist 红柱石白云母片岩
andalusite quartz schist 红柱石英片岩
andalusite slate 红柱石板岩
andalusite two mica quartz schist 红柱石二云母石英片岩
Andaman padauk 安达曼紫檀(木材)
Andaman Sea 安达曼海
andanite 硅藻土
Andean ammonite region 安第斯菊石地理大区
Andean bivalve subprovince 安第斯双壳类地理亚区
Andean Common Market 安第斯共同市场
Andean marine trough 安第斯海槽
Andean metallogenic belt 安第斯成矿带
Andean orogenesis 安第斯山型造山作用
Andean orogeny 安第斯运动
Andean-type orogeny 安第斯型造山运动
andelatite 二长安山岩
andemarcated boundary 未标定界
anderite 麻布基底上加沥青(防潮用)
Anderraa current meter 安得拉海流计
Andersin's classification statistic 安德森分类统计量

Anderson bridge 安德森电桥
Anderson-Daring statistic 安德森—达森统计量
andersonite 碳钠钙铀矿
Anderson's disease 安德逊病
Anderson's sedimentation pipet 安德森沉降管
Anderson's formula 安德森公式(用于计算大梁重量的公式)
Anderton's shearer-loader 安德登(滚筒式)联合采煤机;碎煤装载机
Andes 安第斯山脉
Andes glow 安第斯闪电
andesilabradorite 安山拉长岩
andesine 中长石
andesine andesite 中长安山岩
andesine anorthosite 中长斜长岩
andesine diopside amphibolite 中长透辉角闪岩
andesine labradorite anorthosite 中长拉长斜长岩
andesinite 中长岩
andesinite basalt 中长玄武岩
andesite 瓜子玉;安山岩
andesite ash 安山岩灰
andesite basalt 安山玄武岩
andesite block 安山岩块体
andesite lava 安山岩熔岩
andesite line 安山岩线
andesite paving sett 铺路安山石毛石;安山岩铺砌石块
andesite porphyry 安山玢岩;安山斑岩
andesite sett 安山岩石块
andesite-tuff 安山凝灰岩
andesitic agglomerate 安山质集块岩
andesitic glass 安山质玻璃
andesitic lava 安山质熔岩
andesitic line 安山岩线
andesitic magma 安山岩浆
andesitic porphyrite 安山玢岩
andesitic tuff 安山凝灰岩
andesitic volcanic breccia 安山质火山角砾岩
andesitic welded agglomerate 安山质熔结集块岩
andesitic welded breccia 安山质熔结角砾岩
andesitic welded tuff 安山质熔结凝灰岩
Andes lightning 安第斯闪电
Andeson air sampler 安得森空气取样器
Andeson cascade impactor 安得森多级撞击取样器
Andeson pipette 安得森移液管
Andeson sampler 安得森取样器;安得森采样器
Andes type geosyncline 安第斯型地槽
andhi 对流性尘暴
andible ringing tone 可闻振铃音
andiron (老式的)炉内炉条;炉壁内柴架;壁炉柴架
andlusite deposit 红柱石矿床
andorite 硫锑银铅矿
andosol 暗色土
Andrade's creep law 安得拉德蠕变定律
andradite 钙铁榴石
andradite garnet 钙铁榴石金刚砂
Andreasen pipet 安德列森门砂粒度测定仪
Andreasen pipet apparatus 安德烈森沉降管装置
Andreasen pipette 安德列森型砂粒度测定仪;安德烈森移液管;安德烈森型粒度测定仪
Andreassen apparatus 安德烈森装置
andremeyerite 硅钡铁石
Andre-Venner accumulator 银锌蓄电池;安德烈维内尔蓄电池;安德烈—弗纳蓄电池
Andrews elutriator 安德鲁斯淘析器
Andrew's formula 安德鲁公式
Andrew's Fourier-type plot 安德鲁的傅立叶型点图
andrewsite 羟磷铜铁矿
andron 罗马祭祖室旁走道;男用房间(古希腊);古希腊男宾室
androsphinx 人面狮身像;男性人头狮身像;狮身人面像
and so on 等等
andtostane 雄甾烷
anduoite 安多矿
anechoic 消声的;无回声的;无回音的
anechoic chamber 吸声室;无回音室;消音室;消声室;吸音室
anechoic hall 无回音厅
anechoic paint 吸声漆;吸音漆
anechoic room 无回声房间;无回响房间;消音室
anechoic studio 短混响播音室
anechoic test 消音试验
anechoic trap 消声槽

an eight-hour day 八小时工作制
anelastic 滞弹性的
anelastic attenuation 滞弹性衰减
anelastic behavio(u)r 滞弹性性能
anelastic deformation 滞弹性形变
anelasticity 内摩擦力;滞弹性
anelastic property 滞弹性质
anelastic strain 塑性应变
an elbow 锚链半绞花
anelectric 非电化体
anelectrotonus 阳极电紧张
anemoarenyte 风成砂(岩)
anemobarometer 风速气压表;风速风压仪;风速风压计
anemobiagraph 风速风压记录器;压力式风速计;压管风速计;风速记录仪;风速记录器
anemochorous 风播的
anemochory 风布
anemocinemograph 电动风速计
anemoclast 风成碎屑(岩)
anemoclastical rock 风成碎屑岩
anemoclastics 风成碎屑岩
anemoclinograph 铅直风速计;风斜计
anemoclinometer 铅直风速表;风斜表;垂直风速表
anemodispersibility 风力分散率
anemo-electric(al) generator 风力发电机
anemogenic curl effect 风生涡流效应
anemogram 风力记录图;自记风力计;风速自曲线;风速曲线;风速记录表;风力自记曲线
anemograph 自记风速计;自记风速表;自记风力计;自记风力表;测风仪;风速风压记录仪;风力记录仪
anemography 测风学
anemolite 弯钟乳石;不规则石钟乳
anemology 风学
anemometer 流速表;测风仪;风压计;风速仪;风速计;风速表
anemometer detector 风速探测器
anemometer mast 测风杆
anemometer of wind mill type 风车式风速计;风车式风速表
anemometer tower 测风塔
anemometer with stop watch 带停表的风速表
anemometrograph 风力记录仪;自动风速表;风速风向记录器
anemometry 测风(速)法;风向风速测定法;风速风向测定法;风速测定法
anemophilous flower 风媒花
anemophobe 嫌风植物;避风植物
anemoplankton 风浮生物
anemorphic distortion 不规则畸变
anemorumbograph 风速风向自记仪
anemorumbometer 风向风速表
anemoscope 风速仪;测风器;测风仪;风向(风速)仪;风向风速表;风向(风速)计;风速风向指示器
anemostat 风动起动器;风动启动器
anemostat 稳流管;扩散管
anemotachometer 风速转速表
anemotaxis 趋风性
anemo-thermometer 风速风温计
anemovane 接触式风向风速计;接触式风向风速器;风向风速器;风向风速计;风向标
anergia 无变应性
anergic 无变应性的
anergy 无反应性;无变应性
aneroid 无液的;膜盒
aneroid altimeter 膜盒高度表;无液气压测高计;无液高度计;空盒气压计;膜盒高度计
aneroid barograph 自记气压计;无液气压计;膜盒式气压记录仪;膜盒式气压记录器
aneroid barometer 无液气压计;无液气压表;空盒气压计;空盒气压表;膜盒式气压计;膜盒气压计;膜盒式气压表;高原空盒气压计
aneroid battery 干电池
aneroid calorimeter 膜盒式热计;膜盒量热计
aneroid capsule 无液压力传感器;膜盒压力传感器
aneroid chamber 真空膜盒;气压计盒
aneroid diaphragm 无液气压计膜片;膜盒感压膜片
aneroid flowmeter 膜盒式流量计
aneroid height control 气压计高程控制
aneroid height control point 气压计高程控制点
aneroid liquid-level meter 膜盒液面计
aneroid manometer 无液压力计

aneroid mixture controller 膜盒式混合比调节器
aneroidograph 无液气压器;无液式气压自记仪;空盒气压计;空盒气压记录仪;膜盒气压计;无液气压计
aneroid pressure meter 空盒气压计
aneroid pressure sensing device 空盒压力敏感装置
aneroid sensor 膜盒压力传感器
aneroid surveying barometer 膜盒气压测量计
aneroid thermometer 空盒温度计
aneroid valve 膜盒阀
an essential aspect 本质方面
an essential distinction 本质差别
anesthesia needle 封闭针
anesthetizing area 使用麻醉区(域)
anesthetizing room 麻醉室
Anethum graveolens 草茴香
anetic 弛缓的
anetiological 病原不明的
aneugamy 非整倍配合
aneuploid 非整倍体
aneuploid study 非整倍体的研究
aneuploidy 非整倍性
a new chemical method 一种新式化学方法
an exceptional bumper harvest 特大丰收
an exceptional case 特殊情况
anfo explosive 铵油炸药
anfractuosity 弯曲的通路
Angara floral realm 安加拉植物地理区系
Angaraland 安加拉古陆;安加拉地盾
angaralite 安加拉石
Angara paleo-island 安加拉古陆
Angara sea basin 安加拉海盆
Angara shield 安加拉地盾
angel 寄生目标;异常回波
angel beam 天使梁(在梁的尼雕刻人形);天使雕饰梁
angel choir 唱诗班圣坛
angel echo 杂散回波
angelellite 脆砷铁矿
angelic acid 当量酸
angelica-tree 楤木
angel light (窗花格中开的)三角形小窗
Angelrest 安吉尔雷斯特醋酯丝束
angel triforium 教堂拱门上入天使像
anghirol 朝鲜蓟酸
angica 巴西木(一种坚硬和沉重的装饰木材)
angico gum 巴西树胶
angina pectoris 心绞痛
angiografin solution 安吉格那芬溶液
angioneurosin 硝化甘油
angiosperm 被子植物
angiospermous forest 阔叶树林
angiportus 小路(古罗马)
angite porphyry 黑斑岩;安吉特斑岩
ang-kary 夹钳(用于百叶窗、拔松钉子、宽松包装等)
Angkor vat 吴哥寺(柬埔寨);吴哥窟(柬埔寨)
anglastique 安格拉斯提奎呢
angle 角度;角
angle abrader 角式磨耗实验机;角式磨耗机
angle addendum 齿端角高(指圆锥齿轮)
angle adjustable spanner 弯头活络扳手
angle admittance 角导纳
angle at angle 极角
angle at center 中心角
angle at eye of observer 在观测者眼角处的张角
angle at point of switch 尖轨尖端角
angle at spheric(al) center 球心角
angle attack of wind force 风力攻角
angle at the center 圆心角
angle at which the cranks are fastened 曲柄角
angle at zenith 天顶角
angle auger 挠性转角螺旋钻
angle back-pressure valve 背压角阀
angle back-to-back 角背间距;背靠背(组合)角接
angle backwards welding 后倾焊
angle-balancing method 角变位平衡法(框架结构分析方法);转角平衡法
angle bar 窗帘竖杆;三角铁;角型铁;角铁;角钢
angle bar joint 角铁连接;角铁板连接
angle bar strap 衬角铁
angle bead (剖面为直角形的)角接缝条;墙角护条;角条;护角(条)

angle bead tile 半圆形护角瓷砖
angle bead tile fitting 半圆形护角水泥板
angle beam 天使雕饰梁;对角支撑(杆);角光;角钢檩条;角梁;斜束;隅梁;转角梁;L形梁
angle beam probe 斜探头
angle beam searching unit 斜探头
angle bend 角形弯管;角形接头
angle bender 弯管机;弯板机;角度折弯机;钢筋折弯机;钢筋弯折机;钢筋弯曲机
angle bending machine 钢筋折弯机;钢筋弯折机;钢筋弯曲机
angle betvaxis long axis of finite strain ellipse and X 有限应变椭圆长轴与X轴夹角
angle between envelope and axial-plane 包络面与轴面夹角
angle between foliation and kink plane inside 内带夹角
angle between foliation and kink plane outside 外带夹角
angle between half power points 半功率点间的角
angle between lines of position 位置线交角
angle(a) between long axis and reference line 长轴与参考线夹角
angle between marker line and reference line 标志线与参考线夹角
angle between marker plane and reference plane 标志面与参考面夹角
angle between observation line and ground water flow direction 观测线与地下水流向夹角
angle between observation line and surface water flow direction 观测线与地表水流向夹角
angle between position lines 位置线交角
angle between principal strain axis and direction of shear 主应变轴与单剪方向夹角
angle between section line and bed dip 剖面线与岩层倾向夹角
angle between section line and bed strike 剖面线与岩层走向夹角
angle between teeth 齿间角
angle between wave crest and coast line 波峰线与海岸线夹角
angle between wave crest and contour line 波峰线与等深线夹角
angle between wavefront and bed contour 波峰与等深线的夹角
angle bisecting plane 角平分平面
angle bisection 二等分角
angle blade 角形平铲;斜铲推土板;斜板填土机;单刃平切锄铲
angle blade dozer 斜板推土机
angle blade scraper 斜铲(式)平地机
angle blanking 角坐标照明;角向照明
angle blasting 九十度内喷射
angle block 弯板;角铁;角砌块;角垫块;削角垫块;转向滑车;小木楔;三角木块;节点支块;胶接(木)块
angle block ga(u)ge 角度块规
angle board 角导板;角尺(板);斜薄板(木板墙)
angle bond 砖墙外角砌合;墙角连接件;墙角拉筋;墙勒脚外角砌合;墙角砌法;外角砌分(墙);角隅接合;角砌合;啮合隅角
angle book 角钢手簿
angle brace 对角撑;水平角撑;角铁撑;角撑;隅撑
angle bracing 隅撑;角撑;角铁撑
angle bracket 角铁托;角形托架;角形撑铁;角铁托架;角撑架;斜托座
angle brake 角钢制动器
angle branch 肘管;弯管
angle brick 斜角砖;角砖;异形砖;大小头砖
angle brush 弯柄漆刷
angle bulb 球缘角钢;圆趾角钢
angle bulb frame 球边角材肋骨
angle bulb iron 球头角钢
angle bulldozer 可调整角度的推土铲
angle butt joint 角度对接接头
angle buttress 角扶壁;转角扶垛
angle butt strap 斜口平接板;对接角材
angle butt weld(ing) 斜(口)对焊
angle capital 角柱顶;转角柱头;转角柱帽
angle capital decoration 转角柱头装饰
angle catamaran 八字双体浮座(浮式钻台的);八字双体船
angle catch 窗插销;暗窗闩
angle centrifuge 斜角离心机

angle change 角变位;角变化
angle chapel 小教堂;角形小教堂
angle check 直角单向阀
angle check-valve 直角止回阀;直角单向阀;直角型单向阀;直形单向阀;角部单向活门;角部止回阀;角部回阀
angle chimney 转角烟囱
angle clamp 角夹;角钢夹
angle cleat 短角钢;连结角钢;连接角钢;角钢(加强)夹板;角钢隅撑;角钢连接(在结构构架中用以支撑或搁置预构件)
angle clip 连接角铁;角铁系;角钢连接件;角钢夹;短角钢
angle closer 墙角砍砖;墙角半砖;镶边砖石;角接封闭砖石;交角镶边砖石
angle-closing error of traverse 导线角度闭合差
angle cock 弯头旋塞;弯管旋塞;转角管塞;直角旋塞;折角塞门
angle coefficient 角系数
angle collar 环套角钢;铸铁承插管弯头
angle collision 斜向相撞;直角碰撞
angle column 角柱;角钢柱
angle compressor V形汽缸排列风压机
angle coordinate 角坐标
angle corbel 角形梁托;角钢托座
angle correct 角度校正
angle count method 角测计算法
angle coupling 角铁连接器;直角接头
angle crane 三角架起重机;斜座起重机
angle cross-ties 角钢横系杆
angle curvature 弯曲角
angle cut 斜切
angle cut blasting 角锥形掏槽爆破
angle cut-off valve 角式截止阀
angle cutter 角铣刀;切角机;角铁切割机;角铁切断机;角钢切割机;角钢切断机;角钢剪床;斜切机
angle cutting tool 倒角铣刀
angled alumin(i)um steady arm 角形铝定位器
angle data transmitter 角度数据发送器
angle datum mark 角度基准标记
angled blade ear knife 横切口刀
angle defining wave direction 波向角
angle deformation 角变形
angle dekko 测角仪;测角器
angle derrick 角铁钻塔;角铁井架
angled field 斜向场
angled hole 斜钻孔
angle diagram 正形投影图
angle diameter (螺纹的)平均直径
angled inducer 角形导流器
angle discrimination 角分辨率
angle dispersion 角色散
angle displacement 角位移
angle display 角度指示器
angle diversity 角度分集
angle divider 分器器;分角仪;分角规
angle division tester 角分划检查仪
angled loop 角形环匙
angled nozzle 倾斜喷管
angle domain 角域
angle domain topography 角域地形
angledozer 角铲推土机;斜角推土机;斜铲(式)推土机;斜板推土机;侧铲推土机;倒铲式推土;铲土机;万能推土机
angle-dozing 侧铲推土;用斜铲推土机推
angle-drafted margin 转角琢边;转角蘑菇石剁边
angled rampway 斜角跳板【船】
angle drift 斜铰刀;锥形冲头
angle drill 角钻
angle drill attachment 锥形冲头联结;角钻附件
angle drive 转角传动法
angle drive grinder 转角传动研磨机;斜向传动碾磨机;斜向转动碾磨机
angle dropping tool assembly 减斜钻具
angled scissors 膝状剪
angled snubbing hole 斜掏槽(炮)眼
angled stair(case) 角形楼梯;转角楼梯
angled stern ramp 船首斜跳板;船尾斜引桥
angled stud 梯级高链节
angled type rampway 斜跳板
angle dumping 卸载角
angle dunting 石料粗加工
angled wheel 倾斜轮
angle-end of breakwater 堤根段(防波堤的)

angle equation 角方程
angle expression 角度表达式
angle factor 角系数
angle feedback 角反馈
angle file 三角形锉刀;三角锉(刀)
angle filler 斜板填土机
angle fillet 三角形嵌条;三角盖板;三角形盖板;三角焊缝;节点板
angle fire place 墙角壁炉;角形壁炉
angle fishplate 角型鱼尾板;角铁接合板;角铁鱼尾板;L形接合板
angle fitting 弯头配件
angle fixing pedal 角度固定踏板
angle flange 角铁凸缘
angle float 角抹子;角镘;直角镘刀
angle for maximum righting lever 最大复原力臂对应角
angle forwards welding 前倾焊
angle frame 角材肋骨
angle framing 角钢框架
angle gaining (building-up) tool assembly 增斜钻具
angle gate valve 弯头闸阀;角闸阀
angle ga(u)ge 倾斜计;量角器;量角规;角规;角度计
angle ga(u)ge block 倾斜计;量角器;量角规;角度块规;角度计;量角块规
angle gear 斜交轴伞齿轮;锥齿轮
angle generator 角度发生器
angle grinder 角磨;直角研磨机
angle guard 三角保护(器)
angle guide 角形导架
angle guide pin 定角位销
angle handle 开关拐柄
angle head 角条;弯头
angle head-wall culvert 进出口缩进岸坡
angle hinge 角铰链;角铰;角合页;角料撑
angle hip tile 人字形脊瓦;戗脊脊瓦;角形戗脊瓦;角形坡脊瓦;角形波脊瓦
angle hole 斜孔
angle impedance relay 角阻抗继电器
angle in a circular segment 圆周角
angle incidence indicator 倾角指示器
angle increment 角增量
angle indicator 角指示器;转角指示器
angle insulator 转角绝缘子
angle intersection method 角度交会法
angle iron 角钢;三角铁;角铁
angle iron bar 角铁条;角钢条
angle iron bending machine 角铁折弯机
angle-iron boiler ring 锅炉角铁环
angle iron frame 角铁框架;角钢构架
angle iron of the chords 弦杆角钢;桁架角钢;桁架角钢;弦杆角铁
angle iron of the flanges 梁翼角铁;梁翼角钢
angle iron purlin(e) 角钢檩条
angle iron ring 角铁环
angle iron rotor 角铁转子;角钢转子
angle iron shearing machine 角铁剪切机
angle iron smith 角钢锻工
angle iron stay 角铁撑条
angle iron stiffening 角钢加劲;角铁支撑;角铁加固;角钢支撑
angle iron with equal legs 等边角钢
angle jamming 角坐标干扰
angle jaw tongs 弯嘴钳
angle joint 角接(头);角部接合;隅接
angle joint bar L形鱼尾板
angle knife 角形刀
angle lacing 角铁缀条;角铁联系
angle lap 磨角
angle lapped specimen 磨角样品
angle lap stain method 磨角染色法
angle lead 导前角
angle leaf (柱脚方石板角上的)叶形装饰;柱座叶形角饰(欧洲中古建筑)
angle level 测角水准仪(斜孔控制角度用)
angle lever 曲柄杠杆;斜杠杆;肘节杆
angle lever shears 杠杆式角铁剪床
angle light 角花格中开的三角形小窗
angle lighting fitting 斜照型照明器
angle lighting luminaire 光线不对称分布的灯具;斜照型照明器;斜角采光光源
angle lintel 角钢过梁(门窗);角形过梁(门窗)

angle load 角荷载
angle loading chute V 形装载槽
angle loads method 角荷载法
angle locking pin 角铁锁销
angle lug 短角铁;短角钢
angle maintenance tool assembly 稳斜钻具
angle mark 角度符号;角度标记
angle measurement 角度测量;测角
angle measurement an ex-centre 偏心测角
angle measurement method in all combination 全组合测角法
angle measurement of satellite navigation system 卫星测角导航系统
angle measuring device 测角器
angle measuring equipment 角度测量装置;角测量装置;测角设备
angle measuring grid 量角格网
angle measuring instrument 角度测量仪器
angle meter 倾计计;量角器;量角计;测角器
angle method of adjustment 角度平差法
angle milling cutter 角铣刀
angle mirror 角镜;测角镜
angle modillion 檐角托饰;飞檐下悬臂石
angle moding press 直角形平板机
angle modulation 调角;角度调制;角调制
angle mo(u)lding press 压角机
angle mount 斜装
angle mounting 带角度安装
angle needle valve 弯角针形阀
angle newel (楼梯的)转弯扶手柱
angle niche 角钢壁龛
angle noise 角度噪声
angle notched disc 尖角缺口圆盘
angle number of pass azimuth 传递方位角角度个数
angle of aberration 光行差角
angle of acceleration 加速度角
angle of action 作用角
angle of advance 前置角;前进角;提前角;进程角;超前角
angle of alternation 偏离角
angle of altitude 仰角;高度角
angle of anisotropic(al) apparent rotation 非均质视旋转角
angle of anterior chamber 前房角
angle of apparent internal friction 视内摩擦角;视在内摩擦角
angle of application 施力角;作用角
angle of approach 靠船角;接近角;渐近角
angle of approach of vessel 船舶靠泊角
angle of arrival 入射角;到达角;出现角度
angle of ascent 上升角;螺旋角
angle of aspect 搜索角;方向角;反弦角
angle of attachment 联接角;连接角
angle of attack 迎角;冲击角;风的攻角
angle of attack detector 迎角探测器;迎角传感器
angle of attack indicator 迎角指示器;攻角指示器
angle of attack of minimum drag 最小阻力攻角
angle of avertence 偏角
angle of azimuth 方位角
angle of back of tooth 齿后角
angle of backstay 边索倾角
angle of balance 平冲角;均衡角
angle of bank 倾斜角;倾侧角
angle of bar 角铁
angle of base friction 基底摩擦角
angle of bedding 垫层倾角;地层倾角
angle of bend 屈折角;弯头角度;弯管角度;挠曲角;转折角;转弯角度
angle of bevel (焊接的)坡口角(度);削角;单边坡口角度
angle of bifurcation 分叉角
angle of bite 挟角
angle of boom 起重机的机臂转角;机臂(的)转(动)角;动臂回转角;吊机臂转角
angle of boom swing 吊机臂旋转角
angle of bosh 炉腰角
angle of bottom 齿根角
angle of bow 船首角
angle of bow rake 首柱倾角;船首柱倾角
angle of bracket 角形托座;角形撑铁
angle of branch 分叉角
angle of break(ing up) 崩裂角;破碎角;破坏角;断裂角

angle of brush lag 电刷移后角
angle of brush lead 电刷移前角
angle of cant 喷管摆角
angle of capillarity 毛细角
angle of chamfer 坡口斜角;斜切角;斜面角;削角;边缘斜角;凹线角
angle of chord 弦角
angle of circular segment 弓形角
angle of circumference 圆周角
angle of clearance 超越角
angle of climb 上升角
angle of collimation 瞄准角;准直角
angle of commutation 太阳行星角距;换向角
angle of connecting rod 连杆摆动角
angle of connection 连接角
angle of contact 接触角;包围角;包角
angle of contingence 切线角
angle of convergence 交向角;收敛角;视差角;聚合差;交汇角;辐合角
angle of convergency 收敛角;视差角;聚合差;交向角;交汇角;辐合角
angle of conversion 大圆改正量;半收敛差;半聚合差
angle of coverage 视场角;像场角
angle of crab 偏移角;偏流角;流压角;洗流角;侧航角
angle of crater 自然倾角
angle of crossing 交会角度
angle of current 水流流向角;水流角(度);水流方向角;流向角
angle of current conduction 电流通角
angle of curvature 曲率角;曲度角
angle of cut 方位角差
angle of cut-off 遮去角
angle of cutting 松动角;切割角;开挖角
angle of cutting edge 刃角(刀具);前角(刀具)
angle of deadrise 死角角;侧缘角;船底斜度角;船底横向升角
angle of declination 倾角;偏斜角;下倾角
angle of deflection 偏转角;偏向角;偏角;方向角
angle of deformation 角变位;变形角
angle of delay 滞后角
angle of departure 偏离角;离去角;出射角;出发角;分离角;发射场地角
angle of depression 倾角;低降角;俯角;伏角
angle of descent 下降角
angle of deviation 偏斜角;偏向角;偏差角(度);自差角
angle of diffraction 绕射角;衍射角
angle of diffusion 漫射角
angle of dig or drag 超前或拖后角
angle of dilatancy 剪胀角
angle of dip 磁偏角;倾向;倾角;眼高差;伏角
angle of direction 指向角
angle of discharge 卸料角;出口角(度)
angle of disintegration 松散角;松动角;崩裂角
angle of dispersion 色散角;散射角;弥散角;应力扩散角
angle of displacement 偏位角;位移角
angle of distortion 歪角;时变角;畸变角;扭转角;扭角
angle of distribution 分布角
angle of divergence 离向角;展开角;发散角
angle of diversion 分流角
angle of downwash 下洗角;冲淘角
angle of drag 牵引角
angle of drain 倾泻角
angle of draw 极限角;牵引角;临界角;陷落角;移动角;崩落角
angle of drift 漂流角;偏航角;偏摆角;风压差;风流压差
angle of dump 倾卸角
angle of dynamic(al) inclination 动倾角
angle of eccentricity 偏心角;离心角
angle of effluxion 射束角度
angle of elevation 仰角
angle of embrace 抱角
angle of emergence 射出角;出射角
angle of emergency 出射角
angle of emission 发射角
angle of encounter of a ship in waves 船与波的相会角
angle of engagement 啮合角
angle of entrance 水线角;进水角;进口角(度)

angle of entry 进水角;进入角;进口角(度)
angle of equilibrium 平衡角
angle of external friction 外摩擦角
angle of extinction 消光角
angle off 偏角
angle of fall 入射角;落下角
angle of fill slope 填土边坡角(路堤);路堤边坡角度
angle of flange 边缘弯角
angle of flare 喇叭口扩张角;扩张角;管扩角;承口角(管口扩张角)
angle of flexure 屈挠角度
angle of flooding 进水角
angle of flow 水流流向角;导通角
angle of flying trace 航迹角
angle of forking 分叉角
angle of friction 摩擦角;自然倾斜坡度
angle-offset (method) 夹角法;差角法
angle of funnel 烟囱后倾角
angle of geodesic contingence 短程切线角
angle of geodetic contingence 大地线切线角
angle of glide chute 溜泥槽倾角
angle of gradient 坡度倾角;坡度角;仰角;倾斜角
angle of hade 伸角;断层余角
angle of half field-of view 半视场角
angle of heel 倾斜角;倾侧角;横倾角;踵角;侧倾角
angle of helix 螺旋斜角;螺旋角
angle of helm 舵角
angle of hole 钻孔角度
angle of horizontal flare 水平喇叭口扩角
angle of horizontal swing 偏角;水平偏角;水平摆角
angle of ignition 燃弧角;点火角
angle of illumination 照明角度
angle of image 图像角;像角;成像角
angle of import 碰撞角;攻角;冲击角
angle of incidence 入射角;投射角;迎角;射入斜角
angle of incidence of maximum lift 最大升力冲角
angle of incidence of minimum drag 最小阻力冲角
angle of incidence of zero lift 无升力冲角
angle of inclination 倾斜角;斜角;伏角
angle of inclination of flame jet 火焰喷射角
angle of incurvature 凹角
angle of indraught 引入角
angle of inlet 入口角
angle of internal friction 内摩擦角(土壤)
angle of internal shear strength 抗剪内摩擦角
angle of intersection 交角;交会角;相交角
angle of isocline 消光角补角;等斜角
angle of isoline 等值线角
angle of jib swing 吊杆旋转角;吊杆悬臂角
angle of keenness 锐利角
angle of lag 落后角;移后角;滞后角;迟滞角
angle of landform plane 地形面角
angle of lay 布线角度
angle of lead 前置角;移前角;导移角;导程角;超前角
angle of lean 倾角(井架)
angle of leeway 漂流角;风压差;风流压差
angle of lift 升力角
angle of light 光射角;光入射角
angle of lighting 照明角度
angle of loading 装载角;吊车臂倾角
angle of loading distribution 荷载分布角
angle of lock (车轮的)转向限位角
angle of loosening 松动角(开挖);(松动的)开挖角
angle of loss 损失角
angle of maximum lift 最大升力角
angle of maximum righting arm 最大稳性力臂角
angle of minimum deviation 最小偏向角
angle of minimum resolution 最小分辨角
angle of natural heaping 自然堆积角
angle of natural repose 自然休止角
angle of natural stacking 自然堆积角
angle of needle insertion 针刺角度
angle of nip (碎石机的)扎入角;箝入角;钳入角;钳角;咬入角;颚式破碎机的啮角;(颚式碎石机的)颚板夹角
angle of no lift 无升力角
angle of nonslip point 临界角
angle of oblique 倾斜角
angle of obliquity 倾斜角;倾角
angle of obliquity of action 倾斜压力角
angle of obstruction 阻碍角;遮光角;遮断角
angle of opening 孔径角

angle of optic(al) axis 光轴间夹角
angle of orientation 取向角;定向角
angle of oscillation 摆动角
angle of outlet 出口角(度)
angle of overturn 倾翻角(度);倾覆角
angle of parallax 位差角;视差角
angle of penetration 隔层角
angle of penetrometer 锥头角度
angle of phase displacement 相位移角
angle of photographic(al) coverage 摄影视场角
angle of pinching 夹紧角
angle of pitch(ing) 俯仰角;坡度角;螺距角;桨叶安装角
angle of polarization 偏振角
angle of polygon 导线角
angle of position 星位角
angle of preparation 坡口面角度;焊接坡口角度
angle of pressure 压力角
angle of projection 投影角;投射角
angle of protection 保护角
angle of pull 牵引角
angle of radiation 辐射角
angle of rake 倾斜角;偏斜角
angle of rebound 回弹角
angle of recess 渐远角
angle of reflection 反射角
angle of refraction 折射角
angle of repose 稳定角;静止角;休止角;息角;粉体休止角;安息角;(从静止到滑动的)极限角;材料堆坡度角
angle of repose of natural slope 天然坡面休止角
angle of repose of the earth 土体休止角;土壤休止角
angle of residual shear 残余剪切角
angle of resistance 抗滑角;阻抗角
angle of rest 自然倾斜角;自然静止角;堆角;(松散体的)天然坡角;静止角;休止角;粉体休止角;安息角
angle of retard 落后角;减速角
angle of rifling 切螺纹角;切槽角;螺旋角;螺纹角;来复线角
angle of rise 上升角
angle of roll 横摇角;横倾角;旋转角度;辗压角;侧倾角;侧滚角
angle of rolling 轧入角
angle of roof 屋顶坡度;屋面倾角;屋顶斜角;屋顶倾斜角
angle of roof pitch 屋顶坡度角;屋顶高跨比角度
angle of roof slope 屋顶坡度角
angle of rotation 旋转角;旋光角;转角;转动角
angle of rudder 舵角
angle of ruling 网线角度
angle of run 去流角;尾尖角
angle of run off 铲背角
angle of rupture 破裂角;破坏角;断裂角;断角;粉体断裂角
angle of safety 安全角
angle of satellite altitude 卫星高度角
angle of saw-tooth 锯齿角
angle of scattering 散射角
angle of screen 银幕倾角
angle of shear 切变角;剪切角;滑移角
angle of shear blades 剪床开角
angle of shearing resistance 抗剪强度角;抗剪角;剪阻角;剪切阻力角;抗剪力角
angle of shearing strength 剪切强度角
angle of shift 位移角
angle of ship approach 靠泊角度
angle of ship departure 离泊角度
angle of shock 冲击角
angle of sides 侧壁斜角(孔型);侧壁斜度
angle of side slip 侧滑角
angle of sight 视线角;瞄准角;俯仰角
angle of sight instrument 高低机瞄准角
angle of silence 寂静角;哑点角
angle of site 位置角
angle of site knob 俯仰角转螺
angle of site level 俯仰角水准器
angle of site scale 俯仰角读数标尺
angle of situation 位置角
angle of skew 削角;斜砌石角度;桥对河流的斜角
angle of skew-back 侧倾角;拱脚
angle of slide 活动角;滑坡角;滑动角;自溜角
angle of sling 吊索与滑轮接触角

angle of slip 位移角;滑移角
angle of slope 坡度;倾斜角;坡度(度);坡度角;边坡角
angle of soil internal friction 土的内摩擦角
angle of soil shear resistance 土的抗剪角度
angle of solar incidence 太阳入射角
angle of spatula 刮铲角
angle of spiral 螺旋线升角
angle of spray 喷射角(度)
angle of spread 扩散角;展开角
angle of squint 斜视角
angle of stall(ing) 失速角;临界攻角
angle of static inclination 静倾角
angle of steepest slope 最陡坡度
angle of stern rake 尾柱倾角;船尾柱倾角
angle of stream junction 合流角度
angle of strike 岩层走向方位角;走向角
angle of sweep back 后掠角
angle of swing 旋转角度;旋角;摆角
angle of switch 转辙角
angle of tangent 切线角
angle of tangent on the arch axis 拱轴切线角
angle of taper 斜角;锥度
angle of tear 撕裂角
angle of the grain 木纹斜度
angle of the slope of the terrain 地面坡度
angle of the vertical 地理纬度订正量
angle of the V-groove 轮槽角度;槽角(度)
angle of thread 螺纹牙形角;螺纹截面角;螺纹角度
angle of throat 入口张开角;入口张角;刃角;前端角;喉角;楔角
angle of tilt(ing) 倾(斜)角
angle of tip 倾翻角(度)
angle of toe-in 前束角;内倾角
angle of tooth point 齿楔角
angle of torque 扭转角
angle of torsion 扭转角;扭角;转矩角
angle of total reflection 全反射角
angle of traction 牵引角
angle of traverse 射界角
angle of trim(ming) (船的)纵倾角;配平角;修整角
angle of true internal friction 天然内摩擦角;实际内摩擦角;真内摩擦角
angle of turn 转弯角度
angle of turning flow 水流转角
angle of twist 扭转角;扭角;捻转角;反转角
angle of twist per unit length 单位长度扭转角
angle of twist per unit length of shaft 轴的单位长度扭转角
angle of unconformity 不整合角【地】
angle of underlay 偃角
angle of unequal legs and thickness 不等边不等厚角钢
angle of valve seat 阀座角
angle of vanishing stability 稳性消失角
angle of vee V形(砌)沟;(焊接的)坡口角(度)
angle of view 视角;视场角;景观角
angle of visibility 视界角(度);可视角
angle of visible sky 可见天空角
angle of vision 视角
angle of visual field 视野角;视场角
angle of wall friction 墙摩阻角;墙面摩擦角;墙背摩擦角;外摩擦角;壁摩擦角;壁面摩擦角
angle of wave approach 来波角;迎浪角;波行角
angle of wedge 楔角
angle of widening 扩张角;扩散角;扩大角
angle of wipe 刮墨刀角度
angle of wound 创角
angle of wrap 包角
angle of yaw 偏转角;偏流角;偏航角;首摇角;船首偏荡角
angle on the bow 舷角
angle paddle 捋角器(抹灰用)
angle paint brush 弯柄漆刷
angle parked 斜列停放的
angle parking 倾斜式停车场;斜列停放;斜列停车;斜列停车
angle parking lane 斜列停车道
angle pavilion 角亭;休息室;耳房【建】
angle pedestal 角形轴架;角形支座
angle pencil of ray 宽光束
angle phase-digital converter 角度—相移—数字转换器
angle phase matching 角相位匹配

angle pier 角垛;角墩
angle pile 斜桩
angle pin 双角度针;角度针
angle pipe 肋管;曲管;弯管;角管
angle plane 角刨
angle planing 角刨;斜面刨削;斜刨法
angle plate 角型板;角板;L形板
angle plate jig 角铁钻模;角铁样板;角铁夹具;角板机架
angle plug 弯曲插座;弯插头
angle ply 角铺设层合板
angle pole 转角电杆
angle polisher 角抛(磨)光机
angle post 露明木架;角钢檐柱;转角柱;角柱
angle press 角压力机;角压机;角式模压机
angle prism 角棱镜
angle probe 斜探头
angle protractor 斜角规;分角规;量角器;量角规;分度规
angle pulley 换向滑轮;转向滑轮;变向滑轮
angle purlin(e) 角檩;斜角檩条;角钢檩条
angle quoin 屋角石(块);突屋角;楔形石;楔块
angle rafter 斜沟椽;角椽(子);角铁椽;角钢椽(条);斜角椽
angle range 可能角度
angle reading 角读数
angle reamer 斜铰刀
angle reciprocating compressor 角式往复压缩机
Angle red 恩格尔红
angle reed 斜齿箔
angle reflector 角形反射器
angle reflector antenna 角形反射天线
angle resolution 角分辨率
angle resolved photoelectron spectroscope 角分辨光电子能谱
angle resolver 分角器
angle rib 拱顶中间的弯曲拱梁;角肋
angle ridge (四坡屋顶的)面角椽;角脊;角椽;戗脊
angle ring 角钢圈
angle ring rod 角铁环杆
angle ring rod pin 角铁环杆销
angle ring spring 角铁环簧
angle riser 角度升降器
angle roller 角度矫正机
angle roof truss 角钢屋架
angle rotation of cantilever beam 悬臂梁转动角
angle rule 角尺;折尺;测斜仪
angle sander 角钢打磨器
angle sanding machine 角钢磨光机
angles back-to-back 角背间距;背靠背组合的角钢;背对背角钢组合;背对背角钢
angle scale 摄像角标度;角度盘;角度标尺
Angle's classification 安格尔分类法
angle scraper 弯头刮刀;带角度刮刀
angle seam 角缝;转角接缝
angle seat 角钢座;角钢支座;支座角铁
angle section 角钢截面;L形角铁截面;角形截面;角钢截面;角钢;角材;L形断面
angle section of equal legs 等边角形截面
angle section ring 角钢箍
angle separator 角钢加劲杆;(钢梁的)角形连接构件
angle setting of blade 叶片角度调整
angle shaft 装饰性转角小柱;角柱
angle-shaped arch 三角碹
angle shears 角钢剪切机;角铁剪床;剪角铁机;型材剪切机
angle sheet iron 角状扁铁;角扁铁
angle shot 角度拍摄;斜侧面镜头
angle shut-off valve 直角截流阀
angle side bracket 角托板;角撑架;角托趾
angle sieve 阶梯筛;角型筛;角型滤网;角度筛
anglesite 铅矾;硫酸铅矿
angle size 角大小
angle spider 角型鱼尾板;角型辐条
angle splice 鱼尾板;角型鱼尾板;角铁鱼尾板;角铁接合;角钢拼接件;角形连接板
angle splice bar 鱼尾板;(制造连接板用的)异形钢材;角钢连接板;角型鱼尾板;角形鱼尾板
angle splice joint 角形拼合接头;角钢拼接
angle splice plate 角形鱼尾板
angle square 角尺
angle staff 墙角木椽;角标杆;角标尺;角形柱;角饰;护角线(条);护角条

angle stair(case) 曲尺楼梯
angle staple 角肘钉;凸缘环;角扒钉
angle station (索道的)弯道站;(架空索道的)转角位置
angle steel 角铁;角钢
angle steel edged joint 角钢镶边接缝
angle steel log (洗矿机的)角钢支架
angle steel purlin(e) 角钢檩条
angle steel ruler 钢角尺
angle steel section 角钢截面
angle steel table 角撑架
angle steel with equal legs 等边角钢
angle steel with unequal legs 不等边角钢
angle-stem thermometer 曲杆温度计
angle step 转角梯级;转角踏步
angle stern ramp 船尾斜引桥
angle stiffener 角状加强材;角钢加强筋
angle stiffening 角钢加劲杆;角支撑;角加劲
angle stile 转角盖缝(木)条;三角形嵌条
angle stone 角石;隅石
angle stop 角形销;角形档;角钢制动器;角钢停车器;角钢挡铁;角钢挡板
angle stopper 角制动器
angle stop valve 角形断流阀
angle straightener 角钢矫直机
angle straightening machine 角钢矫直机
angle strap 木舵后边板;舌材
angle stringer 角形纵梁;角形桁条
angle stripper 剥取辊
angle strut 角材支柱;角铁支柱;角支柱;角撑
angle subtended by arc at center 弧所对的圆心角
angle suspension tower 角形铁塔
angle sweep 角扫描;角刮板;斜铰刀
angle-table 托座;加胰;角钢托座;三角桌(工具机上);角撑架;牛腿【建】;承托
angle tee[T] 斜角三通;斜角分支接头;分路
angle tester 测角计
angle test tube 氢氟酸测斜试管
angle thermometer 角式温度计;直角温度计
angle threshold 角阀;L形阀门;角钢门槛
angle tie 角钢拉杆;角形拉杆;水平斜撑;角(铁)撑;支撑脊椽梁
angle tile 转角塔楼;瞭望塔楼;脊瓦;人字形脊瓦;屋角石块;角砖;角瓦;屋脊盖瓦;饯脊盖瓦
angle-to-digit conversion 角度数字转换
angle-to-digit converter 角度数字转换器
angle toe 角材跟
angle to left 左转角
angle tolerance 角度限差;角度容许误差
angle to meet formation 隅层角
angle tooled 角形加工修整
angle to right 右转角
angle-to-right traverse 右转角导线【测】
angle tower 角楼;转角塔(架);转角电线塔
angle tracker 角跟踪器
angle tracking 角跟踪
angle tracking unit 角跟踪部件
angle trowel 角边抹子;阴角抹子;抹灰边抹子
angle truss 角钢桁架
angle turret 角楼;瞭望塔楼
angle type axial flow pump 弯管轴流泵;半贯流式轴流泵
angle type axial piston motor 斜轴式轴向活塞马达
angle type axial piston pump 斜轴式轴向柱塞泵
angle type joint bar 角形鱼尾板;L形接合板
angle type reciprocating compressor 角式复往复活塞压气机
angle type step (楼梯的)扇步
angle unconformity 角度不整合
angle unit 测角组件;测角部件
angle valve 角阀;直角形气门嘴
angle variation 角变化
angle washer 卷角垫圈;角形垫块;斜垫圈;斜垫块
angle web 转角挡板;角形排架肋;角腹板
angle weights method 弹性荷载法;角变位荷载法
angle weld(ing) 角焊;贴角焊缝;L形焊缝
angle with equal legs 等肢角钢
angle with equal sides 等边角钢
angle with sharp corners 锐边角型材;锐边角钢
angle with unequal legs 不等肢角钢;不等边角钢
angle wrench 斜口扳手
angling 偏角调节;安装角调节;斜角部件
angling blade bulldozer 角刃推土机
angling clip 钓夹

angling dozer 侧铲推土机;斜铲(式)推土机;斜板推土机
angling hole 斜炮眼;斜炮孔;斜炮洞;斜孔
angling well 斜钻孔
Anglo-Chinese style garden 中英混合式园林
Anglo-classic architecture 英国古典建筑
Anglo-classic style 安格鲁古典式
Anglo-Norman architecture 安格鲁—诺尔曼式建筑
Anglo-Norman style 安格鲁—诺尔曼式
Anglo-Palladian architecture 安格鲁—巴拉迪恩式建筑
Anglo-Saxon architecture 安格鲁—撒克逊式建筑
Anglo-Saxon masonry(work) 安格鲁—撒克逊式砖石建筑
Anglo-Saxon style 安格鲁—撒克逊式
Anglo-Saxon tower 安格鲁—撒克逊塔
Anglo-Tudor architecture 英国都铎式建筑
Angola abyssal plain 安哥拉深海平原
Angola basin 安哥拉海盆
Angola copal 安哥拉柯巴脂
Angola mending 安哥拉接补线
Angolan Escudo 安哥拉埃斯库多
Angola process 安哥拉工艺
angola yarn 安哥拉纱
Angosta bark 安格斯塔树皮
Angoumian 安古米阶【阶】
angrite 钛辉无棱粒陨石
angry pulse 弦脉
angry river 波涛汹涌的河流
angry stream 波涛汹涌的河流
angry wave 狂浪;巨浪;怒涛
Angström's coefficient 埃斯特朗系数
Angström's compensation pyrheliometer 埃斯特朗补偿日射热量计
Angström's formula 埃斯特朗公式
Angström's resolution 埃斯特朗分辨率
Angström's unit 埃斯特朗单位
anguclast 角碎屑;角斑晶
Anguilla 安圭拉岛
angular 棱角状的;角状的;角形的;角的
angular acceleration 角加速度
angular acceleration in roll 滚动角加速度
angular accelerator 角加速器;角加速计
angular accelerometer 角加速度测量仪;角加速度表
angular accuracy 角精(确)度;测角准确度;测角精(确)度
angular addendum 角齿顶高
angular adjustment 角度调整
angular advance 提前角;进程角
angular aggregate 棱角粗骨料;棱角粗集料;棱角(状)集料;棱角(状)骨料;有棱角骨料;粗制集料;粗制骨料
angular alignment 角对准
angular alternation 角度变形
angular altitude 高低角;地平纬度
angular aperture 天线张角;天线角开度;开角;方向图宽度;孔径角;角孔径;斜视孔;张角(天线)
angular asymmetry 角不对称
angular ball bearing 径向推力球轴承
angular beam misalignment 射束角偏调
angular beam width 束的角宽
angular bearing 止推轴承;枢日轴承;向心止推轴承
angular belting 三角胶带传动装置;转角皮带装置;转角皮带运输机构
angular berthing 斜向靠泊
angular bevel gear 斜交轴锥齿轮;斜交伞齿轮;非正交共轭圆锥齿轮副
angular bisector 角的平分线;角的等分线;等角分线;分角线
angular bit 角钻
Angolan bit stock 弯把手摇钻
angular blocky structure 角块状结构
angular boulder 块石
angular breadth 角宽度
angular brightness 角亮度;按角亮度分布
angular brightness distribution 角亮度分布
angular brush 倾斜电刷
angular calibration 角度核定
angular calibration constant 角度校准常数
angular capital 转角爱奥尼亚式(柱头;角柱的涡卷式柱头;角锥形柱头;角形柱头
angular change 角变位;角向变化;角度变形
angular characteristic function 角特性函数
angular coarse aggregate 棱角状粗集料;棱角状粗骨料;角状粗集料;有棱角的粗集料

angular cobble 碎石
angular column 角(形)柱;转角柱
angular compressor 角式压缩机
angular concrete aggregate 有棱角混凝土滑料
angular condition 角条件
angular contact 角接触;斜接触;斜接;斜角连接
angular contact ball bearing 角接触球轴承;向心止推滚珠轴承;向心推力球轴承
angular contact bearing 向心止推滚动轴承
angular conversion 角度换算
angular coordinate 角坐标
angular coordinate resolution 角坐标分辨能力
angular correction 角度校正
angular crack 梯段裂缝(露天开挖);斜裂缝
angular curvature 角度曲率
angular cutter 角铣刀;角度刀具;斜角铣刀
angular data 角数据
angular defect 角缺陷
angular deflection 角偏转;角挠度;角变位
angular deflection sensitivity 角偏移灵敏度;角偏差灵敏度
angular deformation 角度变形;角变形
angular degree 角度
angular dependence 角相关
angular detrusion sensor 角位移传感器
angular deviation 角度偏离;角度偏移;偏角;角偏移;角偏差;角离差
angular deviation loss 角偏离损失
angular deviation sensitivity 偏航敏感度;偏航灵敏度;角偏转灵敏度
angular diameter 角直径
angular diameter-red-shift test 角径—红移检验
angular difference 角差
angular dimension 角量;角度大小
angular discordance 角度不整合
angular discrepancy 角度不等值
angular dispersion 角向色散;角色散;角散布;角分散
angular displacement 角变位;失调角;角位移
angular displacement component 角位移分量
angular displacement of bogie 转向架角位移
angular displacement sensor 角位移传感器
angular distance 角距离;角距
angular distance of stars 恒星的角距离
angular distortion 角失真;角挠曲;角度畸变;角度变形;角变形
angular distribution 角向分布;角分布
angular distribution of input 输入角分布
angular distribution of output 输出角分布
angular divergence 开展角度;角散度;角分发散;角发散
angular diversity 角分集
angular drainage mode 角状水系模式
angular eccentricity 偏心角
angular epicentral distance 震中角矩
angular equation 角方程
angular error 角(度)误差
angular error of closure 角度闭合差;闭合差【测】
angular error signal 角误差信号
angular field 视场(角);角视场;视野;视界
angular field of view 视角范围;角视场
angular fish plate 角口接板;转角鱼尾板;角口尾板接板
angular flux 角流(量)
angular force 偏向力;角向力;转矩
angular foreland 角形岬
angular fracture 斜面断口
angular fragment 有棱角颗粒;棱角碎屑;棱形角碎片;角状石片;有棱角碎片
angular frequency 角频(率)
angular gap 角形裂隙
angular gear 人字齿轮;斜交轴伞齿轮
angular gradient 角梯度
angular grain 棱角颗粒;角粒;尖角颗粒;有棱角粒;多角形颗粒
angular grate 斜炉箅
angular gravel 角砾
angular gravel soil 角砾土
angular harmonic motion 角谐运动
angular height 高低角
angular hip tile 角形饯脊瓦;角形坡脊瓦;饯脊瓦
angular hole 斜孔
angular impulse 角转冲击;角冲量

angular impulse turbine 斜击式水轮机
angular inclined retaining wall 倾斜式挡土墙
angular indexing 角分度(法)
angular industrial thermometer 工业用曲管温度计
angular instrument 测角仪;测角器
angular intersection(method) 角度交会法
angular interval 间隔角
angularity 有角性;曲线度;曲率;倾度;翘曲度;弯曲角;棱角度;棱角;斜度;弯曲角
angularity cage antenna 角笼型天线
angularity chart 棱角图;棱角度对比图
angularity coefficient 棱角度系数
angularity correction 偏斜校正;角修正;角度校正
angularity factor 棱角性因素;棱角度因素;棱角度系数
angularity index (颗粒的)棱角(性)指数
angularity measurement 角因素测定
angularity number 棱角数
angularity of connecting rod 连杆斜度
angularity of direction finder 测向仪灵敏度
angularity of sounding line 测线斜度;测线倾斜度;测深线倾斜度
angular jitter 角度不稳定性
angular jointing 斜节理
angular Larmor frequency 拉莫尔角频率
angular lever 交角杠杆
angular lever with rolling surfaces 带滚动表面的肘节杆;带滚动表面的角杆
angularly dependent mode 随角度变化模式
angularly independent mode 不随角度变化模式
angular magnification 角放大率;角度放大率
angular magnification of pupils 瞳孔角放大率
angular measure 角度计算;角度计量;角度法;角度测量;角测度
angular measurement 量角;角测;测角
angular methyl 角上甲基
angular milling 角度铣
angular minute 角分
angular misalignment 角位移;角偏差;角度失准;管子接偏
angular modulation 角调制
angular modulation IR system 角调制红外系统
angular momentum 动量矩;角转动量;角动量
angular momentum density 角动量密度
angular motion 角运动;角动;转动
angular motion sensor 角运动传感器
angular mount 斜角架
angular movement 角运动;角度运动
angular movement pick-up 角变动传感器
angular notch 角切迹
angular observation 角度观测
angular orientation 角取向;角定向
angular oscillation 角摆动
angular overlap 重叠角
angular parallax 视差角;角视差
angular particle 棱角颗粒;有棱角颗粒
angular path length 轨迹角长度
angular pebble 角砾
angular perspective 斜透视;成角透视
angular phase difference 角相位差
angular pipe union 弯管接头
angular pitch 倾斜角;角距
angular planing 斜面刨削
angular point 角顶点尖顶;角顶点;角顶;角点;尖顶
angular polar coordinates 角极坐标
angular position 角坐标;角位置;角位
angular position control 角位控制
angular position indicator 角位置指示器
angular position pick-up 角位置传感器
angular prism 角棱镜
angular projection 角投影
angular prolate spheroidal function 角扁长球体函数
angular radiator valve 角形散热器阀
angular radius 角半径
angular rate 角速率
angular rate sensor 角速率传感器
angular readout 角坐标读出;角读出
angular reamer 斜角铰刀
angular reduction 角度归算
angular region 角区;角范围
angular resolution 角坐标分辨率;角分辨率;角分辨本领;角度分辨率;角分辨力
angular resolving power 角分辨率;角分辨本领

angular restraint 转角约束;角变位约束;角约束
angular revolution frequency 绕转角频率
angular ridge tile 脊瓦;角形脊瓦
angular rolling 角轧
angular rotation 角转动
angular rotation of flange 法兰转角
angular sand 棱角状砂;角粒砂;有棱角的砂;多角砂
angular scan 角扫描
angular scan(ning)rate 扫描角速度;角扫描速率
angular scattering function 角散射函数
angular scattering property 角散射特性
angular screw type liner 角螺旋衬板
angular second 角秒
angular section 角形(截面);斜剖面
angular sector 扇形角
angular semi-major axis 角半长径
angular semi-minor axis 角半短径
angular sensitivity 角灵敏度
angular sensitivity of gravimeter 重力仪角灵敏度
angular sensor 斜探头
angular separation 角距;角间距;方向夹角
angular shaft rotation 角轴旋转
angular shear 角剪切
angular shear mechanism 角度剪切机构
angular shear strain 角剪应变
angular sheen 斜向光泽
angular shift 漂移角度
angular solution 角度解算
angular spectrum 角谱;角频谱
angular speed 角速度
angular spheric(al)aberration 角球差
angular spiral liner 角螺旋衬板
angular spotting (集装箱的)偏斜对位;斜向对位
angular spread 角分散;角分布
angular spread of ion beam 离子束的角发散
angular standard 角度标准
angular stone 棱角石料;角形石块;有棱角石料
angular strain 角应变
angular subtense 角距
angular surface 斜面
angular surface grinding 斜面磨削
angular surveying 角度测量;测角工作
angular table (工具机上的)三角桌
angular templet 角度样板
angular tenon 斜口榫
angular test 弯曲试验
angular thread 三角螺纹
angular threshold of eye 视角阈值
angular tolerance 角容限
angular tool 弯头车刀
angular top 角顶
angular transducer 角位移传感器;角度传感器
angular transformation 角度变换;角变换
angular travel computer 角程计算机
angular trisector 角的三等分线
angular turn 角旋转
angular unconformity 角度不整合;斜交不整合;倾斜不整合
angular value 角值
angular variable 角变量
angular variation 角向偏移
angular velocity 角速度
angular velocity between plates 板块间的角不整合
angular velocity indicator 角速度器
angular velocity of precession 进动角速度
angular velocity of the earth's rotation 地球自转角速度
angular vernier 角游标
angular vertex 角顶点;角顶
angular wagon flow 折角车流
angular wheel 伞齿轮;锥形齿轮
angular wheel slide 斜置砂轮座
angular width 角宽度;角幅
angular width of arch at crest 拱顶角宽度
angulated mode 陵角状水系模式
angulate drainage 角状水系
angulated roping 角悬吊索;斜吊索
angulated sail 角帆
angulateration 边长—角度测量
angulation 形成角度;扭曲;成角度安装;量角;角度测量;成角
angulation deformity 成角畸形
angulation displacement 成角移位
angulator 角投影器;角度转换仪;变仪;变角器

angulometer 量角仪;量角器;测角器;测角计
anguloso-undulate 波角状
angulus 角
Angus-Smith process 安古斯—史密斯防腐蚀法
Angus-Smith's solution 安古斯—史密斯溶液
angustifoliol 安哥树酚
angustione 安哥树酮
anharmonic 非谐
anharmonic coupling 非谐耦合
anharmonic curve 非调和曲线
anharmonic force 非谐力
anharmonic interaction 非简谐作用
anharmonicity 非谐性
anharmonicity constant 非谐性常数
anharmonic oscillator 非谐振子;非谐荡振器;非简谐波振荡器
anharmonic ratio 非谐(性)比(例);非调和比
anharmonic temperature factor 非简谐温度因子
anharmonic theory 交比理论
anharmonic thermal vibration 非简谐热振动
anharmonic vibration 非谐振动
anharmonic wave 非谐波
anhedral 他形的
anhedral angle 上反角
anhedritite 硬石膏;无水石膏
anhedron 他形晶;劣形晶
an-hours for loading/discharging 装卸作业工时
anhtdrokainite 无水钾盐镁矾
anhydration 脱水(作用);干化
anhydric 无水的
anhydride 无水物;脱水物
anhydrite 无水石膏;硬石膏;酐
anhydrite band 无水带
anhydrite bander 硬石膏绷带;无水石膏绷带;硬石膏胶结料
anhydrite block 硬石膏砖;硬石膏砌块
anhydrite block partition 硬石膏砖块隔墙
anhydrite board 硬石膏板
anhydrite cement 无水石膏水泥;无水石膏胶凝材料;硬石膏水泥;硬石膏胶结物;硬石膏胶结料;硬石膏灰泥
anhydrite curing agent 酸酐类固化剂
anhydrite deposit 硬石膏矿床
anhydrite evaporite 硬石膏蒸发盐
anhydrite-gypsum ore 硬石膏—石膏矿石
anhydrite jointless floor(ing) 硬石膏无缝面层;硬石膏无缝地板
anhydrite lime mortar 无水石膏石灰砂浆;硬石膏石灰砂浆
anhydrite mortar 硬石膏砂浆
anhydrite ore 硬石膏矿石
anhydrite plaster 硬石膏灰泥;硬石膏灰浆
anhydrite process 石膏制酸法
anhydrite rock 无水石膏石
anhydrite screed 硬石膏无缝地板;硬石膏灰浆泥抹
anhydrite sheet 硬石膏板;无水石膏板
anhydrite tile 硬石膏砖;硬石膏瓦;无水石膏(砌)块
anhydrite tile partition 无水石膏砌块隔墙
anhydrobiotite 无水黑云母
anhydrock 硬石膏岩
anhydroferrite 赤铁矿
anhydrokaolin 无水高岭土
anhydromuscovite 无水白云母
anhydrone 无水高氯酸镁
anhydro-sorbite 脱水山梨(糖)醇
anhydrous 无水的
anhydrous acid 无水酸
anhydrous alcohol 无水乙醇;无水酒精
anhydrous alumin(i)um of silicate 无水硅酸铝
anhydrous ammonia 无水氨(液)
anhydrous calcium sulfate 无水硫酸钙;硬石膏
anhydrous calcium sulphate 无水硫酸钙
anhydrous calcium sulphate plaster 无水石膏灰泥;无水硫酸钙粉饰;烧石灰浆
anhydrous calcium sulpoaluminate 无水硫酸铝钙
anhydrous cement 无水石膏水泥
anhydrous copper sulfate 无水硫酸铜
anhydrous gypsum 无水石膏;硬石膏
anhydrous gypsum plaster 无水石膏粉饰;无水石膏灰泥;无水石膏胶泥;硬石膏灰泥;硬石膏灰浆
anhydrous gypsum plastering 无水石膏灰浆;无水石膏粉饰
anhydrous hydrogen chloride 无水氯化氢
anhydrous lime 生石灰;无水石灰;无水生石灰

anhydrous material 无水建筑材料
anhydrous period 无水时期
anhydrous phosphoric acid 无水磷酸
anhydrous plaster 无水粉饰;干灰膏;无水石膏粉饰
anhydrous saltcake 无水芒硝
anhydrous silicate 无水硅酸盐
anhydrous sodium carbonate 无水碳酸钠;苏打灰
anhydrous sodium sulfate 无水硫酸钠
anhydrous solvent 无水溶剂
anhydrous sulphate of calcium 无水硫酸钙
anhydrous sulphate of potassium 无水硫酸钾
anhydrous sulphate of silicate 无水硅酸钙
anhydrous sulphate of sodium 无水硫酸钠
anhydrous wool fat 无水羊毛脂
anhydrous wool wax 无水羊毛蜡
anhydrox 抗石膏(污染的泥浆处理剂)
anhydryitite 硬石膏
anhyetism 缺雨性;缺雨区
Anhyster 铁镍磁性合金;安西斯特铁镍合金
anhysteretic 非滞后
anhysteretic remanent magnetization 非磁滞剩余磁化强度
anianthus 白色短纤维石棉
a nice tilth 土壤耕性好
anicut 滚水坝;小坝;滚水堰
anideation 理想缺失
anidex 阿尼迪克斯纤维
anidoxime 阿尼多昔
anil 缩苯胺;靛蓝缩本胺
Anilana 阿尼拉纳聚丙烯腈系纤维
anilic sulphate 硫酸苯胺
anilinate 苯胺金属
aniline 苯胺(化);阿尼林
aniline acetate paper 醋酸苯胺试纸
aniline-aldehyde resin 苯胺甲醛树脂
aniline black 颜料黑;苯胺黑
aniline black byestuffs 苯胺黑染料
aniline blue 苯胺蓝
aniline brown 苯胺褐
aniline cloud point 苯胺浊点
aniline coating 苯胺涂层
aniline colo(u)r 苯胺染料;苯胺类颜料
aniline compound 苯胺类化合物
aniline dye 苯胺色素;苯胺染料
aniline dyeing 苯胺染料染色
aniline dye plant 苯胺染料厂
aniline dyestuff 苯胺染料
aniline equivalent aniline number 苯胺当量
aniline finish 苯胺涂饰剂
aniline foils 苯胺箔片
aniline-formaldehyde plastics 苯胺甲醛塑料
aniline formaldehyde resin 苯胺甲醛树脂
aniline-furfural 苯胺糠醛
aniline-furfural resin 苯胺糠醛树脂
aniline green 苯胺绿
aniline hydroxylation 苯胺羟基化
aniline ink 苯胺油墨
aniline leather 苯胺(皮)革
aniline modified phenolic mo(u)lding compound 苯胺改性酚醛模塑料
aniline nitrate 硝酸苯胺
aniline number 苯胺数
aniline oil 苯胺油;阿尼林油
aniline orange 苯胺橙
aniline pigment 苯胺颜料
aniline point 苯胺点
aniline poisoning 苯胺中毒
aniline press 苯胺印刷机
aniline printing 弹性版印刷;苯胺印刷(术);阿尼林印刷
aniline-p-thiocyanate 对硫氰基苯胺
aniline purple 苯胺紫
aniline-pyridine-insoluble matter 苯胺吡啶不溶解物质
aniline red 洋红;品红;苯胺红
aniline resin 苯胺树脂
aniline scarlet 苯胺猩红
aniline sulfonic acid 氨基苯磺酸
aniline sulphate 硫酸苯胺
aniline test apparatus 苯胺点测定仪
aniline violet 苯胺紫
aniline wastewater 苯胺废水
aniline yellow 苯胺黄
anilinium ion 苯胺离子

anilinoplast 苯胺塑料
anilite 斜方蓝辉铜矿;液态二氧化氮汽油炸药;安尼炸药
Anilo 阿尼洛韧皮纤维
animal 兽图;动物
animal adhesive 动物黏合剂;动物胶黏剂;动物胶
animal agriculture 畜牧业
animal and human excreta 人畜排泄物;人畜粪便
animal and plant life 动植物
animal balance 动物秤
animal black 兽炭黑;兽骨炭
animal borne disease 动物传染的疾病
animal brush 动物刷
animal by-products 畜牧副产品;畜产品;牲畜副产品
animal carbon preparation 动物碳制剂
animal chamber 动物舱
animal charcoal 兽炭;骨炭;动物炭;巴黎骨炭
animal checker 斗兽棋
animal climatology 家畜气候学
animal community 动物群落
animal container 动物集装箱
animal density 动物密度
animal disembarkation 动物上陆
animal dispersion 地动物分散
animal distemper 动物色胶
animal drawn traffic 畜力交通;兽力车;畜力车运输量
animal drying oil 动物干燥油(油漆用)
animal dung as fuel 动物粪(便作)燃料
animal dye 动物染料
animal ecology 动物生态学
animal excrement 动物排泄物
animal excreta disposal 动物粪便处置
animal farm 畜牧场
animal fat 动物脂
animal-fat emulsion 动物脂肪乳液
animal feces 动物粪便
animal feed yard 动物饲养场
animal fiber 动物纤维
animal flower 形似动物之花
animal gelatin 动物明胶
animal geography 动物地理学
animal glue 动物胶
animal glue and gelatin manufacture 动物胶制造
animal glue distemper 动物胶水粉涂料
animal glue manufacturing waste 动物胶生产废物
animal hair felt 兽毛毡
animal hospital 动物医院
animal house 牲畜棚;动物园兽禽舍
animal husbanding 畜牧业
animal husbandry 牲畜饲养;畜牧业;畜牧学
animal industry 畜牧经营
animalized cellulose fibre 动物质化纤维素纤维
animalized cotton 动物质化棉
animalized fibre 动物质化纤维
animalized viscose rayon fibre 动物质化黏胶纤维
animal judging 家畜鉴定
animal kingdom 动物界
animal labo(u)r 畜力
animall-shaped ornament 兽形装饰
animal manure 牲畜粪
animal manure odo(u)r 厩肥臭
animal matter 动物的有机残留物质
animal migration 动物移栖
animal monitoring 动物监测
animal motif 兽纹;动物花纹图案
animal oil 骨油;动物油
animal ornament 动物装饰(品)
animal painting 动物画
animal parchment 兽皮纸;动物羊皮纸
animal pattern 兽纹;模型动物
animal pigment 动物颜料;动物色素
animal plankton 浮游动物
animal population 动物种群
animal population assessment 动物总数估价
animal power 兽功率
animal products 畜产品;动物产品
animal production 畜牧生产;畜产品
animal quarantine 动物检疫
animal quarters 动物圈;动物棚;动物厩
animal quarter's annex 动物圈的附属建筑;动物棚的附属建筑;动物厩的附属建筑
animal relief 动物浮雕

animal remains 动物遗体;动物化石;动物残骸
animal resin 动物树脂
animal resource 动物资源
animal sample 动物样品
animal science 畜牧学
animal sculpture 动物雕塑
animal-shaped ornament 垂兽饰
animal shelter 畜圈;畜棚;畜厩;厩
animal skin 兽皮
animal symbolism 动物象征主义
animal tar 动物焦油
animal tolerance 动物耐受性
animal unit 牲畜单位
animal-unit month 牲畜单位月
animal waste 动物废料
animal waste disposal 动物粪便处置;动物粪便处理
animal wastes 动物粪便
animal wax 动物蜡
animal world 动物世界
animal yield 畜产量
animate cartoon 卡通电影
animated film 动画片
animated map 动画地图
animated or moving sign 活动招牌
animated perspective 动态鸟瞰
animation 生动(建筑艺术);活动图形;动画
animation camera 特技摄像机
animation controller 动画控制器
animation desk 特技合成桌
animation dissolve 特技慢转换
animation equipment 特技设备
animation film studio 美术电影制片厂
animation stand 特技台
animation system 动画制作系统
animation timing 特技同步
Animikean 安尼米基系
Animikie series 安尼米系统
animikite 铅银砷镍矿
animi resin 硬树脂
an impulse for new technologic(al) developments 新技术开发的推动力
animus 阿妮姆斯
an instant 片刻
anion 阴离子;负离子
anion activator 阴离子活化剂
anion active agent 阴离子活化剂
anion active detergent 阴离子活性洗涤剂
anion adsorption 阴离子吸附
anion analysis 阴离子分析
anion-cation balance 阴阳离子平衡
anion conductor 阴离子导体
anion-containing heavy metal 含阴离子重金属
anion defect 阴离子缺陷;阴离子亏损
anion detergent 阴离子洗涤剂
anion exchange 阴离子交换(体)
anion exchange bed 阴离子交换床
anion exchange capacity 阴离子交换容量;阴离子交换能力
anion exchange chromatography 阴离子交换色谱(法)
anion exchange coefficient 阴离子交换系数
anion exchange column 阴离子交换柱
anion exchange filter 阴离子交换滤池;阴离子交换过滤器
anion exchange material 阴离子交换物质;阴离子交换材料
anion exchange membrane 阴离子交换膜
anion exchanger 阴离子交换器;阴离子交换剂
anion exchange resin 阴离子交换树脂
anion exchange tower 阴离子交换塔
anion gap 负离子缺额
anion generator 阴离子发生器;负离子发生器
anion grouping 阴离子分组
anionic 阴离子型;阴离子的
anionic active agent 阴离子型活性剂
anionic asphalt emulsion 阴离子乳化沥青;阴离子(地)沥青乳液
anionic catalytic polymerization 阴离子催化聚合(作用)
anionic cellulose treister 阴离子纤维素三酯
anionic charge 阴离子电荷
anionic clay 阴离子黏土
anionic clay adsorption 阴离子黏土吸附
anionic complex 络阴离子;阴离子络合物

anionic conductor 阴离子导体
anionic coordinate polymerization 阴离子配位聚合(作用)
anionic cyclopolymerization 阴离子环化聚合(作用)
anionic detergent 阴离子洗涤剂;阴离子去垢剂
anionic dye 阴离子染料
anionic electric(al) mobility 阴离子电迁移率
anionic electrodeposit coating 阳极电泳漆
anionic emulsification bitumen 阴离子乳化沥青
anionic emulsifier 阴离子乳化剂
anionic emulsion 阴离子乳液
anionic exchange 阴离子交换
anionic exchange filter 阴离子交换滤池;阴离子交换过滤器
anionic exchange material 阴离子交换材料
anionic exchange tower 阴离子交换塔
anionic grafting 阴离子接枝
anionic initiation 阴离子引发(作用)
anionic membrane 阴离子膜
anionic organic contaminant 阴离子有机污染物
anionic permeable membrane 阴离子透膜;阴离子渗透膜
anionic polycrylamide 阴离子聚丙烯酰胺
anionic polyelectrolyte 阴离子聚合电解质;阴离子高分子电解质
anionic polymer 阴离子聚合物
anionic polymerization 阴离子聚合(作用)
anionic position coating 阴离子电沉积漆
anionic regeneratant tank 阴离子再生柜
anionic road emulsion 路用阴离子乳(化)液
anionic semipermeable membrane 阴离子半透膜
anionic site 阴离子部位
anionic surface 阴离子界面
anionic surface active agent 阴离子表面活化剂
anionic surfactant 阴离子型表面片性剂;阴离子(型)表面活化剂
anionic surfactant micelle 阴离子表面活化剂胶团
anionic surfactant solution 阴离子表面活化剂溶液
anionic synthetic(al) detergent 阴离子合成洗涤剂
anionic synthetic(al) surface-active pollutant 阴离子合成表面活性污染物
anionic transference number 阴离子迁移数
anionic type polycrylamide 阴离子型聚丙烯酰胺
anionic type polycrylamide flocculant 阴离子型聚丙烯酰胺絮凝剂
anionic type polyelectrolyte 阴离子型聚电解质
anionic type surface active agent 阴离子型聚表面活性剂
anionic wetting agent 阴离子润湿剂
anionite 阴离子交换剂
anionoid recombination 类阴离子重组;阴离子催化聚合(作用)
anionotropy 阴离子移变;负离子转移
anion penetration 阴离子渗入(作用)
anion precipitant 阴离子沉淀剂
anion radical 阴离子团
anion resin exchange column 阴离子树脂交换柱
anion retention 阴离子吸持(作用)
anion vacancy 阴离子空穴
anisallobar 等正变压线
aniseed-star oil 八角茴香子油
aniseikon 探测缺陷光电装置;侦疵光电装置;电子照相仪;电子探伤器
aniseikonia 物像不等
anisetree bark 地枫皮
Anisian 安尼西阶
anisic aldehyde 大茴香醛
anisobaric 不等压的
anisochronous digital signal 间隔可变的数据信号
anisochronous transmission 间隔可变的传输
anisodactylus 不等趾
anisodesmic 不均匀连锁
anisodesmic radical 不均匀连锁离子团
anisodimensional particle 不对称形粒子
anisoelastic 非等弹性的
anisoelasticity 各向异性弹性;非等弹性
anisole 甲氧基苯;茴香醚;苯甲醚
anisomeric 非异构的
anisomerous 不同数的
anisomery 不同数现象
anisometric(al) 非等轴的;不等轴的;不等大的
anisometric(al) breccia 不等粒角砾岩
anisometric(al) crystal 不等轴晶体;非对称晶体

anisometric(al) fraise 不对称角度铣刀
anisometric(al) growth 不等量生长
anisometric(al) particle 不等轴颗粒
anisophyllous 不等叶的
anisosmotic 不等渗的
anisosthenic 力量不等的
anisothermal 不等温的
anisotonic 不等渗的
anisotonic solution 不等渗溶液
anisotropic(al) 各向异性的;非均质的
anisotropic(al) absorption 各向异性吸收
anisotropic(al) apparent rotation 非均质视旋转
anisotropic(al) apparent rotation angle 非均质旋轴角
anisotropic(al) aquifer 各向异性含水层
anisotropic(al) astigmatism 各向异性像散现象
anisotropic(al) bed 非均质床
anisotropic(al) body 各向异性体;非均质体
anisotropic(al) ceramics 各向异性陶瓷
anisotropic(al) characteristic of rocks and minerals 岩矿石的各向异性特征
anisotropic(al) coefficient 各向异性系数
anisotropic(al) consolidation 不等向固结;各向异性固结(作用);各向不等压固结;非均质等固结;非等向固结
anisotropic(al) consolidation stress ratio 非均质等固结应力比
anisotropic(al) crystal 各向异性晶体
anisotropic(al) crystal quartz 各向异性石英晶体
anisotropic(al) dielectric 各向异性电介质
anisotropic(al) disc 暗板;横盘
anisotropic(al) dispersion 各向异性色散;非均质椭圆色散
anisotropic(al) distribution 各向异性分布
anisotropic(al) effect 各向异性效应
anisotropic(al) elastic body 各向异性弹性体
anisotropic(al) elasticity 各向异性弹性
anisotropic(al) emission 各向异性发射
anisotropic(al) energy 各向异性能
anisotropic(al) error 各向异性误差
anisotropic(al) etching 各向异性刻蚀
anisotropic(al) fabric 各向异性组构
anisotropic(al) filter 非均质滤池
anisotropic(al) fluid 各向异性流体
anisotropic(al) force system 各向异性力系
anisotropic(al) gain 各向异性增益
anisotropic(al) hardening 各向不等硬化
anisotropic(al) hardening model 各向异性硬化模型
anisotropic(al) index 各向异性指数
anisotropic(al) layer 非均质层
anisotropically consolidated drained test 各向不等压固结排水试验
anisotropically consolidated undrained test 各向不等压固结不排水试验;不等压固结不排水试验
anisotropic(al) magnetohydrodynamics 各向异性磁流力学
anisotropic(al) material 各向异性材料
anisotropic(al) medium 各向异性媒质;各向异性介质
anisotropic(al) membrane 各向异性膜;非均质膜
anisotropic(al) metal 非均质金属
anisotropic(al) mineral 非均质同性矿物
anisotropic(al) model filament 各向异性模型丝
anisotropic(al) oriented media 各向异性取向媒质
anisotropic(al) orthotropic body 正交各向异性体
anisotropic(al) plasma jet instability 各向异性等离子喷流不稳定性
anisotropic(al) plate 非各向同性板;各向异性(木)板
anisotropic(al) pressure 各向异性压力
anisotropic(al) propagation 各向异性传播
anisotropic(al) property 各向异性
anisotropic(al) refraction 各向异性折射
anisotropic(al) rock 各向异性岩石
anisotropic(al) rock mass 各向异性岩体;非均质岩体
anisotropic(al) rotation 非均质旋转
anisotropic(al) rotation dispersion 非均质旋色散
anisotropic(al) scattering 各向异性散射;不匀散射
anisotropic(al) section 各向异性剖面
anisotropic(al) seepage 各向异性渗流

anisotropic(al) shell 各向异性薄壳;各向异性层
anisotropic(al) soil 非均质土;各向异性土
anisotropic(al) sound bearing medium 各向异性传声介质
anisotropic(al) steel 各向异性钢片
anisotropic(al) stress 各向不均应力;不均应力;各向异性应力
anisotropic(al) structure 各向异性组织
anisotropic(al) substance 各向异性物质;重折光质;非均质物质
anisotropic(al) surface 各向异性曲面
anisotropic(al) surface wave 各向异性表面波
anisotropic(al) valley 各向异性河谷
anisotropisation 各向异性化作用
anisotropism 各向异性;非均质性
anisotropous disc 横盘;暗板
anisotropous 异向性;各向异性;非均质性
anisotropy constant 各向异性常数
anisotropy energy 各向异性能
anisotropy factor 各向异性因数;不对称性因数
anisotropy field 各向异性磁场
anisotropy instability 各向异性不稳定性
anisotropy of conductivity 传导各向异性
anisotropy of seismic wave 地震波速异向性
anisotropy of variogram 变差函数的异向性
anisotropy silicon steel 各向异性硅钢片
anisotropy surface 各向异性曲面
anisyl alcohol 大茴香醇
anjan 木砖(印度,有黑色条纹,颜色由红到暗棕色)
Anka 安卡不锈钢
ankar 木纤维热绝缘墙板
ankaramite 橄榄辉玄岩;富辉橄玄岩
ankaratrite 橄榄霞岩
ankerite 铁白云石;富铁白云石
ankerite rock 铁白云石岩
ankh 带圆环十字架
ankle boot 中统靴
ankle-deep mud 深及踝部的泥泞
ankylomele 弯探子
ankylosis 长合关节
ANLAB colo(u)r difference equation 亚当斯·尼克尔逊 LAB 色差方程(式)
Anlan Bridge 安澜桥
an-line plot 联合绘图
an-line plotter 联合绘图机
annaline 沉积硫酸钙
annals 史册;年刊;年表;编年史
annals of local history 地方志
annatto 胭脂树(红)
annealed 退火的
annealed alumin(i)um 退火铝;软铝
annealed alumin(i)um wire 软铝线;退火铝线
annealed casting 锻铸;退火铸件;退火浇铸
annealed cast iron 韧(性)铸铁
annealed clinker 缓冷熟料
annealed condition 退火状态
annealed copper 韧铜;退火软铜
annealed copper sheet 锻铜片
annealed copper wire 软铜线;软铜丝;韧铜线;炼铜线
annealed glass 退火玻璃
annealed in process 中间退火
annealed-inprocess wire 中间退火钢丝
annealed metals 退火金属
annealed sheet steel 退火薄钢板
annealed steel 韧钢;退火钢
annealed structure 退火组织
annealed tensile 退火后的拉力
annealed tensile strength 退火后的拉力强度
annealed wire 退火金属线;软铁丝;退火钢丝
annealed zone 重结晶区
annealer 退火炉
annealing 韧化(作用);退火(处理);转色试金
annealing box 退火箱
annealing breccia 退火角砾岩
annealing can 退火箱
annealing colo(u)r 退火色;烧色
annealing container 退火箱
annealing crack 退火裂(纹)
annealing crystallization 退火结晶;退化结晶作用
annealing curve 退火曲线
annealing cycle 退火周期
annealing effect 退火效应
annealing for workability 改善加工性退火

annealing furnace 退火容;退火炉
annealing grade 退火程度
annealing hearth 退火敞炉
annealing heat treatment 退火热处理
annealing increment of refraction 折射度的退火增加
annealing in process 中间退火
annealing kiln 退火炉;退火窑
annealing line 退火作业线
annealing of lattice disturbance 晶格破坏退火
annealing oil 退火油
annealing operation 退火工序
annealing oven 退火炉
annealing pig iron 可锻铸铁用生铁
annealing point 退火温度;退火点
annealing pot 退火罐
annealing process 退火过程;回火处理;二次煅烧过程
annealing range 退火温度范围
annealing recrystallization 退火重结晶(作用)
annealing region 退火温度范围
annealing room 退火工段;退火车间
annealing schedule 退火制度;退火程序
annealing slack 退火不完全
annealing sticker 退火黏结板
annealing stove 回火炉
annealing temperature 退火温度;退火点
annealing temperature range 退火温度范围
annealing treatment 退火处理
annealing tunnel 退火隧道
annealing twin 退火双晶;退火孪晶
annealing welding wave 退火焊波
annealing welds 退火焊条;耐火焊条
annealing wire 退火丝
annealing zone 徐冷区
anneal pickle line 退火酸洗作业线
annexa 附层;附轨
annexation 合并物;归井;附件;连接关系
annex(e) 附加建筑;附属建筑;附加房屋;辅助建筑;随附;附录;附件(附属建筑);分室干燥炉
annexed budget 附属预算
annexed building 群体建筑
annexed level(l)ing line 附合水准路线
annexed table 附表
annexed traverse 附合导线【测】
annexed triangulation method 附合三角法
annexed triangulation net 附连三角网;附合三角网
annexed type vibrator 附着式振动器
annex(e) foil 水翼衬翼
annex(e) memory 相联存储器;附加存储器
annex(e) point 过渡点;附加点
annex(e) storage 内容定址存储器;相联存储器;附加存储器
annicut 堰
annihilation 湮没作用;湮没现象
annihilation γ-ray 湮没伽马射线
annihilation of electron pair 电子对湮没
annihilation operator 湮灭算符;湮没算符
annihilation radiation 湮没辐射
annihilator 阻尼器;零化子;减弱器;灭火器;消除器;熄灭器;吸收器
annihilator of energy 消能槛
annite 铁云母
anniversary clock 周年钟
anniversary wind 季节风;年周风;定期风
anno domini 纪元;公元
annonation classification of picture 相片调绘
annotate a photograph 像片判读
annotated bibliography 附有说明的资料目录
annotated list of items 附有说明的项目表
annotated mosaic 有注记镶嵌图
annotated photograph 调绘像片;已注释像片;已校正像片;已调绘像片
annotation 注解;注记;附注
announce 通报;宣布
announce booth 播音员室
announced depth 公布的水深
announce in advance 预告
announcement 通告;声明;宣告;布告
announcement day 宣布日
announcement for open tender 招标公告
announcement of tender 招标通知;招标公告
announcer 表示器
announce room 广播室

announcer's booth 广播室
announcing removal 迁移通告
announcing room 广播室;播音室
annoyance value 干扰值
annoying pulse 扰动脉冲
annual 全年;年刊;年鉴;年度的;每年(的)
annual aberration 周年光行差
annual abstraction volume 年引水量
annual abstract of labo(u)r statistics 劳动统计年报表
annual account 年度账;年度决算;年度结算
annual accumulation of earthquake 年地震累积数
annual accumulation of sediment 年淤量;年淤积量;年泥沙淤积量;年泥沙年淤积量;泥沙年沉积量
annual actual mortality 每年实际死亡率
annual advance 年推进量;年推进度;年进度;每年垫付
annual advances 每年垫款
annual age determination 年代测定
annual aircraft movements 年飞机起落架次
annual allowance 每年免税额
annual amortization 每年分摊
annual amortization factor 每年分期偿还数
annual amount 年额
annual amplitude 年较差;年变幅
annual anomaly 年距平
annual apparent motion 周年视运动
annual appropriation 年度拨款
annual artificial lake 年调节水库
annual audit 年度审计
annual authority 限当年有效的(预算或拨款)授权
annual automated control system survey 自控系统年检
annual average 年平均(量)
annual average concentration 年平均浓度
annual average cost 年平均费用
annual average daily traffic 年平均日交通量;年平均交通量
annual average daily traffic volume 年平均日交通量
annual average flow 年平均流量
annual average productive capacity 年平均生产能力
annual average rate of increase 平均年增长率
annual average sediment yield 年平均淤积量;年平均沉积量
annual balance 年终结余
annual balance sheet 年终决算书;年终决算表;年度决算书;年度资产负债表
annual balancing reservoir 年调节水库
annual basis 年度基准
annual bearer 年年结果树
annual-benefit function 年收益函数
annual borrowing program(me) 年度贷款计划
annual budget 年度预算
annual bulletin 统计年报
annual business expenses 年度经营费用
annual capacity 年生产能力
annual capacity factor 年利用系数
annual capital cost 年总成本
annual cargo tonnage 年货运量
annual cargo turnover 年货运周转量
annual cash ceiling 年度支付最高限额
annual cash flow 年度现金流量
annual cash inputs 每年现金投入
annual change 年差;年变异;年变化;周年变化
annual change of magnetic variation 磁差年变化
annual change of spring discharge 泉水流量的年变幅
annual closing 年终结算;年度决算;年度结账;年度结算
annual coldest month 年最冷月份
annual compositive depreciation rate 年综合折旧率
annual conference 年会
annual construction organization design 年度施工组织设计
annual construction plan 年度建设计划
annual consumption 年消耗量;每年耗用数
annual consumption at percentage 消耗量占百分比
annual convention 年会
annual cost 年费用;年度费用;年度成本;年成本;常年费用

annual cost capital recovery 资金回收年成本
annual cost function 年费用函数
annual cost method 年度成本法
annual cost method of comparison 年成本比较法
annual cost of heavy repair 大修年度费用
annual cost of upkeep 年度维修费(用)
annual crop 年收获量;一年生作物
annual cropcycle 每年收获循环
annual cumulative environmental carrying capacity 年积累环境承载容量;年积累环境承载能力
annual cut 年收获量;年伐量
annual cutting 年筏面积
annual cycle 年周期;年循环
annual dam 挡水墙
annual damping depth 年衰减深度
annual data 年度资料
annual debt constant 年债务偿还常数
annual debt service 年还本付息
annual decrease 周年减量
annual deficit 年度亏损
annual depletion 年亏耗量
annual depletion rate 年亏耗量(地下水);年退水量
annual depreciation 年(度)折旧
annual depreciation allowance 每年折旧免税额
annual depreciation cost 年折旧费用
annual depreciation rate 年折旧率
annual depreciation reserve 年折旧回收
annual discharge 年流量
annual distribution 年内分配
annual distribution of runoff 年径流分配
annual dose 年剂量
annual dose-rate 年剂量率
annual drainage 年排水量
annual earnings 年收益;每年收益
annual earthquake extreme 年地震极限
annual efficiency 年效率
annual electric(al) supply 环形供电
annual energy output 年发电量
annual energy production 年发电量
annual environmental capacity 年环境容量
annual epact 年闰余
annual equation 周年差
annual erosion 年侵蚀量;年冲刷量
annual erosion rate 年土壤流失率
annual evapo(u)ration 年蒸发量
annual expected loss ratio 年期望损失率
annual expenditures 全年支出额;岁出;年度支出
annual expenses 年费用
annual export 年出口量
annual export at percentage 出口量占百分比
annual feed and supply of mineral commodities 矿产品供需量
annual financial statement 年度决算书;年度决算表;年度财务报表
annual fixed charges 年固定支出;年固定费用
annual fixes cost 年固定支出;年固定费用
annual flood 年最大流量;年量大洪水量;年洪水(量)
annual flood flow 年最大流量
annual flood peak series 年洪峰序列
annual-flood-series method 年洪序列法
annual flow 年流量;年径流(量)
annual fluctuation of water-level 水位的年变幅
annual frost zone 年冻层;年霜带
annual funding program(me) 年度拨款计划
annual grass 一年生牧草
annual groundwater regime 地下水年内动态
annual growth 全年增长;年生长量
annual growth layer 年生长层;年轮层
annual growth rate 年增长率;年生长率
annual growth ring 年轮(树木)
annual handling capacity 年装卸能力
annual harvest 年收获量
annual heat consumption 年热能消耗量
annual holding cost 年储存成本
annual holidays 例假
annual hotest month 年最热月份
annual housing survey 年度住房调查
annual hydrologic balance 年水文平衡;年水量平衡
annual import 年进口量
annual impounding reservoir 年调节水库
annual improvement coefficient 年增长系数
annual improvement factor 年增长系数;每年整修费;每年按生产率调高工资的条款

annual income 全年所得额;岁入;年收入
annual increase 周年增加
annual increase rate 年增长率
annual increment 年增(长)量;年生长量;年度增长
annual index 年度索引
annual inequality 年差异(量);年不均衡性;年不等;周年不等性
annual inflation rate 通货膨胀年率
annual inspection 年度检查
annual installment 分期付款的每年付款数;按年支付
annual interest 年息
annual interest rate 年利率
annual investment 分年度投资
annualization of expenses 费用按年折算
annualized expenses 费用按年折算
annual layer 年层
annual leave 年度假;每年一次的休假
annual load constant 年贷款偿还率
annual load curve 全年负载曲线;年负载曲线;年负荷曲线
annual load diagram 年负荷曲线(图)
annual load duration curve 年负荷历时曲线
annual load factor 年负荷系数;全年负载因数;年平均负载因素;年负载因数;年负荷率
annual loss 年度损失;年度亏损
annually 年年
annual magnetic change 周年磁变化;磁周年变化
annual magnetic variation 磁周年差
annual maintenance 岁修;年度养护;年度维修;年度维护;年度检修;年度保养维修
annual maintenance charges 年度维修费(用)
annual maintenance cost 年维护费用;年度养护费
annual march 年过程
annual marketing 市场销售量
annual maximum 年最高量;年最大(值)
annual maximum earthquake 年最大地震
annual maximum flood flow 年最大洪水流量
annual maximum flow 年最大流量
annual maximum level 年最大水位
annual maximum permissible dose equivalent 年最大容许剂量当量
annual maximum runoff 年最大径流
annual maximum series 年最大系列
annual maximum stage 年最大水位
annual meadowgrass 早熟禾
annual mean 年平均值;年平均量;年平均
annual mean daily flow 年平均日交通量;年平均日(车)流量
annual mean sea level 年平均海平面
annual mean sediment discharge 年平均输沙量
annual mean temperature 全年平均气温;年平均温度
annual mean tidal cycle 年平均潮汐周期
annual mean value 年均值
annual mean(water)discharge 年平均(水)流量
annual mean water level 年平均水位
annual meeting 年会
annual minimum 年最小(值)
annual minimum series 年最小系列
annual mining output 年开采量
annual mining output at percentage 开采量占百分比
annual mortality rate 年死亡率
annual mortgagor statement 抵押人年度报告
annual nebula 环状星云
annual normal flow 年正常径流(量)
annual normal runoff 年正常径流
annual objectives 年度目标
annual operating charges 年运行费;年经营费
annual operating cost 年运行费
annual operating hours 年运转小时
annual operating report 年运行报告
annual operation days(of port)年营运天数【港】
annual operation plan of railway construction enterprise 铁路施工企业年度经营计划
annual ordering cost 年订货成本
annual output 年排出量;年产量;每年产量
annual output concentration at percentage 产量占百分比
annual output of by-products 副产品年产量
annual output of concentrates 精矿年产量
annual output of metals 金属年产量
annual output of new mining 新增矿山生产能力

annual output of ores 原矿年产量
annual overhaul 岁修;年修;年度检修
annual parallax 日心视差;恒星视差;周年视差
annual passenger movements 年旅客吞吐量
annual pay 年薪
annual payment 年补偿;年支付;年投资回收额
annual peak load 年最大负荷;年峰荷
annual per capita gross national product 年人均国民生产总值
annual percentage rate 年百分利率
annual percent of increase 年增长率
annual period 年周期
annual phytoplankton cycle 浮游植物年循环
annual plan 年度计划
annual plan of capital construction 基本建设年度计划
annual plan of passenger transportation 旅客运输年度计划
annual plant 一年生植物
annual plant output 电厂年发电量
annual power generation 年发电量
annual power output 年发电量;年输出功率;年出力
annual power production 年发电量
annual precession 年岁差;周年岁差
annual precipitation 年雨量;年降雨量;年降水量
annual precipitation factor 年降雨量因素;年降水量因素
annual premium 年度保险费;年保险费
annual premium costing 养老金交费逐年计算法
annual production 年生长量;年产量
annual productivity of lake algae 湖泊藻类年产量
annual productivity of lake aquatic organism 湖泊水生生物年产量
annual productivity of lake bacteria 湖泊细菌年产量
annual productivity of marine algae 海洋藻类年产量
annual productivity of marine organism 海洋有机体的年产量
annual productivity of marine plankton 海洋浮游生物的年产量
annual profit 年收益;年利润
annual program(me)年度计划
annual program(me)of work 年度工作计划
annual progress report 年度进展报告
annual proper motion 年自行
annual quota 年度配额;年度定额
annual rainfall 年雨量;年降雨量;年降水量
annual rainfall factor 年降雨量因素;年降水量因素
annual range 年全距;年较差;年变幅
annual range of temperature 气温年变幅;温度年较差
annual rate 年率;周年变化率
annual rate method 年率法
annual rate of change 周年变化率
annual rate of deposition 年淤积量;年沉积速率
annual rate of profit 每年利润率
annual rate of return 年度盈利率
annual ratio of runoff 年流量率;年径流率
annual reckoning 年度决算;年度结算
annual regulating reservoir 年调节水库
annual regulation 季调节;年调节
annual rental 年租
annual repair 年度维修;全年需修量;岁修
annual report 年报;年终报告;年度报告
annual residual 年富余量
annual returns 年度统计表;每年报税表格
annual returns of emoluments paid to staff 每年支付给雇员之薪酬报税表
annual revenue 岁入;年收益;年度收入
annual review 年度审查
annual ring density 年轮密度
annual ring(of timber)木材年轮;树木年轮;年环
annual risk 年危险性
annual river inflow 年河流入流
annual runoff 年径流(量)
annual runoff depth 年径流深度
annual salable expenditures 年度推销费
annual sale volume 年营业额
annual sediment deposition 年淤积量;年泥沙积量;年落淤量;年沉积量
annual sediment discharge 年输沙量;年排沙量
annual sediment transport 年输沙量
annual sediment yield 年淤积量;年产沙量

annual series 年系列
annual session 年会
annual short 年短缺量
annual siltation 年淤积量
annual snowfall 年降雪量
annual soil loss 年土壤流失量;年水土流失
annual solar motion 太阳周年视运动
annual special survey of hull 船体每年特别检验
annual statement 年度报表
annual statistic(al)table 年度统计表
annual storage 年蓄水量;年蓄能量;年堆存量;年调节量;年调节库容
annual-storage plant 年调节水库电厂
annual-storage reservoir 年调节水库
annual-storage station 年调节水库电站
annual summary 年度汇编
Annual Summary of Notices to Mariners 航海通告年度摘要
annual surplus or deficit 岁入余缺
annual symposium 年会
annual target 年度指标;年度目标
annual tax 年税额
annual technical progress report 年度技术进展报告
annual temperature range 气温年较差;温度年较差;年温(度)差
annual temperature variation 年温度变化
annual thermal wave 周年热波动
annual tide 年周(期)潮
annual tonnage capacity 年通过能力;年货运量
annual traffic 年交通量;年运输量
annual traffic ability 年通过能力
annual traffic capacity 年通过能力
annual transporting capacity required by the state 国家要求的年运输能力
annual transport plan 年度运输计划
annual trend 年趋势
annual turnover 年周转量;年周转额;年营业额;年逆转
annual unbalance 年不平衡
annual upkeep 年度保养;年度维修
annual utilization rate 年利用率
annual value 年值;周年值
annual values of by-products 副产品年产值
annual values of concentrates 精矿年产值
annual values of ores 原矿年产值
annual values of other products 其他产品年产值
annual values of products 成品年产值
annual valve 环形阀(门)
annual variation 年差;年变异;年变化;周年偏差;周年变化;磁差年变化
annual variation in soil temperature 土温的年变化
annual variation of runoff 径流年变化
annual variation of shoal 浅滩年内变化
annual variation of shoal channel 浅滩河槽的年变化
annual variation of temperature 气温年变化
annual variation rate 年变率
annual volume of carried silt 年输沙量
annual volume of traffic 年度运量
annual volume of water abstraction 年引水量;年抽水量
annual waste water discharge 年废水排放量
annual wave 周年波动
annual weed 一年生杂草
annual working program(me)年度工作计划
annual workship 年度讨论会
annual workshop 年度研讨会
annual yield 年收获量;年伐量;年产量
annual yield by area 单位面积年产量
annual zone 年轮(树)
annuation 周年变异
annuitant 享受年金权利者
annuity 年金;每年;养老金
annuity agreement 年金契约
annuity certain 确定固定年限的年金
annuity due 期初应付年金
annuity fund 年金基金
annuity in advance 预付年金保险;期初年金;预付年度保险
annuity in arrears 拖欠年金
annuity method 年金法
annuity of compound interest 复利年金
annuity table 年金表

annuity trust account 年金信托账
annul 废止;取消
annul a concession agreement 中断租让合同
annular 轮状;环状的;环形的
annular amplifier 环形放大器
annular arch 环形拱
annular auger 环孔钻
annular ball bearing 环形滚珠轴承;径向球轴承;径向滚珠轴承;向心球轴承
annular ball mill 环球磨
annular basin 环形水池
annular bearing 环形轴承
annular bit 圆锯;环状钻头;环形钻(头);环孔锥
annular blast pipe 环形吹管
annular blowout preventer 环形防喷器
annular borer 套环钻;环孔镗床
annular burner 环状喷灯;环形喷灯
annular bushing 环状漏板
annular cathode 环状阴极
annular cell construction method 环形构体施工法
annular channel 环状通道;环形沟道
annular classifier 环式分级机
annular conductor 环状绞合线
annular contact 环形接触
annular contact thrust ball bearing 推力向心球轴承
annular corrugation 环形波纹
annular counter 环状计数器
annular crack 环形裂纹
annular cross-section 环形断面;环形截面;环形接触
annular cutting 环形切割
annular cylinder type servomotor 环形(油缸式)接力器
annular dam 环形闸;环形坝;胸墙
annular diaphragm 环状光圈
annular disc 环形盘
annular distance 环孔
annular drainage 环状水系;环形排水系统
annular drainage mode 环状水系模式
annular drill 环孔钻
annular eclipse 环食;日环食
annular element 环形单元
annular fin 环形散热片
annular float 环形浮子;环形浮标
annular flow 环状流;环形流
annular furnace 环形炉
annular gasket 环形密封垫;环形垫片;环形垫密片
annular gear 内齿轮;内齿环;环形齿轮
annular girder 环形梁
annular grate kiln 环形焙烧机
annular green space 环状绿型
annular grid 环形格栅
annular groove 环形槽
annular guard ring 环形保护圈
annular holder 环形夹具
annular hydraulic jump 环形水跃
annular intake limit 年摄入量限值
annularity 环状
annular jet 环喷口;环状喷口;环形喷射流
annular-jet nozzle 环形喷嘴
annular joint 环形接头;环形接合
annular kiln 轮窑;环窑;环形窑
annular knurl 滚花
annular largest event 年最大事件
annular limit of intake 年摄入量限值
annular longitudinal force 环形纵向力
annular magnet 环形磁铁
annular mark 环形测标
annular membrane 环形薄膜;环形膜
annular micrometer 圆径千分尺
annular mill 环形磨
annular nail 环纹钉;刺钉
annular nozzle 环形喷嘴;环状喷嘴
annular opening 圆孔
annular packing 环状填料;环形密封垫
annular passage 环形通道
annular pervious boundary 环形透水边界
annular pipe network 环状管网
annular piston 环形活塞
annular piston meter 环形活塞式仪表;环形活塞表
annular piston valve 环形活塞阀门;环形活塞阀
annular plate 环形板
annular plate valve 环形阀(门)

annular plug ga(u)ge 空心柱形测孔规
annular pressure drop in collar 钻铤环空压降
annular pressure drop of drill pipe 钻杆环空压降
annular pressure loss 环空泵压损失
annular preventer 囊式防喷器;环形防喷器
annular producer 圆形发生炉
annular radiation 环形放射
annular recess 环形槽
annular reduced cost 年合算费用
annular return velocity 环状空间返流速度;环空偏速
annular ring 木材年轮;树木年轮
annular ring(ed) nail 环形钉
annular ring of timber 木材年轮
annular ring valve 环形阀(门)
annular roller mill 环状辊式粉碎机
annular saw 圆锯
annular screw plate 环状板牙
annular seal 密封环;环状密封
annular section 环形截面;环形段;环形部分;环形区域;圆断面
annular shaft kiln 环形立窑
annular shake 环裂
annular shape basin 环形水池
annular slab 环形板
annular sleeve valve 环形套管式阀门
annular slot nozzle 环槽式喷嘴
annular space 环形空间;环形缝隙;环状间隙
annular spillway 环形溢洪道
annular strain 环形应变;环变形
annular structure 环状构造
annular surface 环形(表)面
annular tank 环形池;循环水槽
annular tapered roller bearing 向心圆锥滚子轴承
annular thickening(of cell wall) 环纹加厚
annular tube 环管;套管
annular-type stripper 环形冲孔模板
annular valve 节制器阀门;滑动柱形阀;环状阀;针形阀;针舌阀
annular vault 筒形拱顶;环形穹顶;圆形穹顶;圆拱顶
annular void 环间孔隙
annular wheel 内齿轮;内齿环
annular yield 年收获率
annular zone 环状地区;环状地带;环形区
annulated 有环纹的;用环做记号的;环纹的
annulated bit 环形钻头
annulated column 有环纹的柱;绕花饰群柱;环箍柱;环柱;环饰柱
annulated shaft 环饰柱
annulate lamella 环饰板
annulation 环状物;环状纹;环状结构;环状构造
annulet 小轮;小环;轮缘;轮状平缘;环形线脚;环形活线脚;圆箍线;柱环饰
annuli fibrosi 纤维环
annulment 注销;取消;废止
annulment of contract 取消合同
annul null 无效
annuloid 环状的
annulospiral ending 螺旋环终末
annulus 环状裂缝;环形体;环形套筒;环形件;环节;环带;圆环
annulus chamber 环形室
annulus conjecture 圆环猜想
annulus pressure at surface 井口环空压力
annum 年度
annum running 年息
annunciation 报警
annunciator 通告者;回转拨号机;呼铃器;呼唤器;号码箱;车钟【船】;报讯器
annunciator bell 信号铃
annunciator electric(al)clock 报信号电钟
annunciator system 警报信号系统;预报信号系统
anochromasia 不染色性
anodal 阳极的;正极的
anodal opening contraction 阳极断电收缩
anode 阳极;正极;板极
anode bag 阳极罩
anode bar 阳极棒
anode battery 乙电池组;阳极电池组;高压电池组
anode bend rectification 利用屏栅特性弯曲部分检波
anode brightening 阳极电抛光;电抛光
anode button 阳极帽
anode cap 阳极帽

anode capacity 阳极电容量
anode cap inserting machine 阳极帽封接机
anode casting machine 阳极电镀机
anode-cathode distance 极间距离
anode cell 阳极罩
anode chamber 阳极空间
anode circuit 阳极回路;阳极电路
anode coating 阳极镀层
anode column 阳极光柱
anode compartment 阳极空间;阳极室
anode conductance 阳极电导
anode converter 阳极变流器
anode copper 电解铜;阳极铜
anode corrosion 阳极腐蚀;阳极侵蚀
anode current 阳极电流
anode current fluctuation 阳极电流起伏
anode dark space 阳极暗区
anode detection 阳极检波
anode detector 阳极检波器
anode disc 盘形阳极
anode dissipation 板极耗散
anode dissipation power 阳极耗散功率
anode drop 阳极电压降;阳极电势降;阳机压降
anode effect 阳极效应
anode efficiency 阳极效率
anode film 阳极膜
anode finish 阳极处理保护层
anode follower 阳极跟随器
anode glow 阳极辉光
anode half-cell 阳极半电池
anode high voltage 阳极高压
anode inductance 阳极电感
anode keying 阳极键控法
anode light 阳极射线;阳极辉光;阳极发光;板极辉光
anode liquor 阳极液;阳极电解液
anode loop 阳极回路;阳极电路;板极回路
anode loss 阳极损耗
anode metal 阳极金属
anode modulation 阳极调制
anode mud 阳极残渣;阳极泥
anode oxide film 阳极氧化膜
anode oxide film streak 阳极氧化膜花纹
anode oxidizing 阳极氧化
anode passivation 阳极钝化
anode plate 阳极板
anode plate charging machine 阳极板充电机
anode ray 阳极射线;正射线
anode reaction 阳极反应
anode reactor 阳极电抗器
anode resistance 阳极电阻
anode saturation 阳极饱和
anode scrap 废阳极;阳极残铜;残阳极
anode slime 阳极淀渣;阳极沉积层
anode's optimization 阳极优化
anode stub 阳极棒
anode supply 阳极电源
anode tap 阳极端
anode tapping point 阳极活接点
anode voltage 阳极电压
anodic area 阳极区
anodic attack 阳极腐蚀
anodic backfill 阳极填充料
anodic battery 阳极电池组
anodic beam 阳极射线束
anodic behavio(u)r 阳极行为
anodic brightening 阳极抛光
anodic-cathodic electrocatalysis 阳极—阴极电催化
anodic-cathodic wave 换极连续波
anodic chamber 阳极室
anodic cleaning 阳极酸洗;阳极清洗;反向电流清洗
anodic coating 阳极氧化镀层;阳极涂层;阳极敷层;阳极镀层;电极氧化罩面;电镀饰面
anodic colo(u)r 阳极氧化着色
anodic control 阳极制约;阳极控制
anodic corrosion 阳极腐蚀
anodic corrosion control 阳极腐蚀控制
anodic corrosion efficiency 阳极腐蚀效率
anodic corrosion inhibitor 阳极腐蚀抑制剂
anodic corrosion protection 阳极防腐蚀(法)
anodic corrosion protector 阳极防腐板
anodic current density 阳极电流密度
anodic current intensity 阳极电流强度
anodic depolarized wave 阳极去极化波
anodic depolarizer 阳极去极剂

anodic deposition 阳极沉积
anodic diffusion current 阳极扩散电流
anodic dissipation 阳极耗散
anodic dissolution 阳极溶解
anodic dissolution current 阳极溶解电流
anodic dissolution wave 阳极溶解波
anodic drop pressure 阳极降压
anodic earth 阳极接地
anodic electrodeposition 阳极电沉积
anodic etching 阳极蚀刻
anodic film 阳极镀层
anodic finish 阳极氧化镀层;阳极镀层;阳极处理保护层
anodic finish of alumin(i)um component 铝电极氧化饰面
anodic inhibitor 阳极缓蚀剂;阳极钝化剂;阳极(溶解)抑制剂
anodic load 阳极负荷
anodic metal 阳极金属
anodic oxidation 阳极氧化
anodic oxidation colo(u)r 阳极氧化着色
anodic oxidation method 阳极氧化法
anodic oxidation treatment 电极氧化处理
anodic oxygen-transfer reaction 阳极氧—转移反应
anodic passivation 阳极钝态
anodic passivity 阳极钝态
anodic pickling 阳极酸洗;阳极腐蚀
anodic polarization 阳极极化
anodic process 阳极法;阳极处理
anodic protection 阳极防腐蚀(法);阳极保护(法)
anodic ray 阳极射线
anodic scouring 阳极浸蚀
anodic sensitivity of photomultiplier tube 光电倍增管的阳极灵敏度
anodic shield 阳极护罩
anodic slime 阳极附着物;阳极沉渣;阳极残渣
anodic stripping 阳极溶出法;阳极解吸
anodic stripping polarography 阳极溶出极谱法
anodic stripping voltammetry 阳极溶出伏安法
anodic treatment 阳极氧化处理;阳极化;阳极处理
anodic treatment method 阳极处理法
anodic tripping voltammetry with collection 聚集阳极溶出伏安法
anodise 阳极氧化;阳极化处理;阳极电镀
anodised alumin(i)um 电化铝;阳极氧化铝;阳极化铝
anodised finish 阳极化表面处理;阳极化抛光
anodising 阳极化处理;阳极化
anodization 阳极化;阳极化防腐法;阳极电镀;阳极处理
anodize 阳极电镀;阳极化处理;阳极防腐
anodized alumin(i)um 阳极化铝
anodized coating 阳极化镀层
anodized finish 阳极化抛光
anodized mirror 阳极化镜
anodized surface of alumin(i)um 电极氧化铝面
anodizing 阳极氧化(处理);阳极化(处理);阳极防腐法;阳极电镀;阳极处理;电极氧化
anodizing electrolytic colo(u)ring 阳极氧化电解着色
anodizing plant 阳极化处理厂;阳极氧化工厂
anodizing process 阳极化过程
anodizing sheet 阳极氧化板
anodizing waste 阳极氧化废物
anodizing waste(water) 阳极氧化废水
anoganotrophic 非有机营养的
anogene 喷出的
anohoteric ion 两性离子
anohoteric surfactant 两性表面活性剂
anol 对丙烯基苯酚
anole 变色蜥
anolyte 阳极电解液
anomalistic 近点的
anomalistic distance 近点距离
anomalistic drift 不规则漂移
anomalistic inequality 月角差
anomalistic mean motion 近点平均运动
anomalistic month 近日点月;近点月
anomalistic motion 近点运动
anomalistic period 近点周期
anomalistic revolution 近点周;异常转动
anomalistic Schottky effect 反常肖特基效应
anomalistic tidal cycle 近点潮汐周期

anomalistic tide 近点潮
anomalistic tide cycle 近地点潮汐周期
anomalistic year 近点年
anomalon 反常子
anomaloscope 色觉比较镜
anomalous 距平(的);异常的;反常的;不规则的
anomalous absorption 反常吸收
anomalous abundance 异常丰度
anomalous coefficient 异常系数
anomalous colo(u)r vision 异常色觉;反常色视觉
anomalous coupling 反常耦合
anomalous diffusion 反常扩散
anomalous dispersion 反常色散;不规则弥散
anomalous dispersion phase-matching 非常色散相位匹配
anomalous dispersion profile 反常色散分布图
anomalous earthquake 异常地震
anomalous effect 反常效应
anomalous expansion 反常膨胀
anomalous fading 反常日晒褪色
anomalous felt zone 异常震区
anomalous field 剩余磁场;异常磁场;异常场
anomalous film 反常膜
anomalous flocculation 反常絮凝(作用)
anomalous flow property 反常流动性
anomalous fluctuation 反常起伏;反常波动
anomalous fluorescence scattering 反常荧光散射
anomalous growth 畸形生长
anomalous high water-level 异常高水位;异常高潮(位)
anomalous inclusion of chemical origin 化学成因异常包裹体
anomalous intensity 烈度异常
anomalous interference colo(u)r 异常干涉色;反常干涉色
anomalous lag 异常滞后;反常滞后;不规则滞后
anomalous lead 异常铅
anomalous magma 异常岩浆
anomalous magnetic moment 不规则磁矩;反常磁矩
anomalous magnetic variation 局部异常磁变
anomalous magnetization 异常磁化;反常磁化
anomalous material 异常物质
anomalous melting 反常熔化
anomalous metallic element 异常金属元素
anomalous mixed crystals 反常混晶;反常和晶
anomalous mode 异常态
anomalous numbers 反常数
anomalous peak 反常峰
anomalous potential 异常位
anomalous propagation 异常传播;反差传播
anomalous propagator 异常传播子
anomalon refraction 异常折射;反常折射
anomalous scattering 异常散射;反差散射
anomalous sea level 异常潮位;反常潮位
anomalous secondary thickening 异常次生加厚
anomalous singularity 反常奇异性
anomalous site 变态点
anomalous site colo(u)r particle 颜料粒子上的反常点
anomalous spectral broadening 反常光谱增宽
anomalous structure 异常构造;反常结构
anomalous succession 不规则演替
anomalous thermal conductivity 反常热导率
anomalous thermal expansion 反常热膨胀
anomalous tide 反常潮汐
anomalous tide level 异常潮水位;异常潮位
anomalous transfer 反常转移
anomalous transport 反常输运
anomalous trichromat 反常三原色视觉
anomalous trichromatism 色弱
anomalous upper mantle 异常上地幔
anomalous value 异常值
anomalous value of rock 岩石异常值
anomalous viscosity 反常黏滞性;反常黏度
anomalous watershed 异常分水岭
anomalous weight 反常重量
anomalous Zeeman effect 反常塞曼效应;反差塞曼反应
anomalous zone 异常区;异常带
anomaly 距平(的);近点距离;近点角;异态;异常;反常
anomaly attenuating curve 异常衰减曲线
anomaly characteristic analysis method 异常特征分析法
anomaly characteristics 异常特征
anomaly classification 异常的分类
anomaly composition zoning 异常组份分带
anomaly concentration zoning 异常浓度分带
anomaly contour 距平线;异常等值线
anomaly contrast 异常衬度
anomaly curve of aeroelectromagnetic method 航空电磁法异常曲线
anomaly curve of frequency sounding with a source 人工频率测深异常曲线
anomaly curve of ground EM method 地面电磁法异常曲线
anomaly curve of magnetotelluric sounding method 大地电磁测深异常曲线
anomaly curve of mise-a-la-masse method 充电法异常曲线
anomaly curve of self-potential method 自然电场法异常曲线
anomaly cutoff 异常截止点
anomaly decay pattern 异常衰减模式
anomaly deduction 异常推断
anomaly delineation 异常的圈定
anomaly dispersion pattern 异常分散模式
anomaly distribution map of survey area 测区异常分布图
anomaly electric(al) field 异常电场
anomaly evaluation 异常评价
anomaly explanation 异常解释
anomaly file 不规则形锉刀
anomaly finder 异常测定器
anomaly grade 异常级别
anomaly homogeneity 异常均匀性
anomaly horizontal displacement curve in repetition line 重复线异常水平错动图
anomaly inspection 异常检查
anomaly in stratification 层理不规则;不规则层理
anomaly intensity 异常强度
anomaly map 距平线图;异常图;异常等值线图
anomaly map of geochemistry classification 地球化学分类异常图
anomaly map of geophysics 地球物理异常分类图
anomaly mean value 异常平均值
anomaly nature of electric(al) prospecting 电法勘探异常性质
anomaly not related to mineral deposit 非矿物异常
anomaly of alkaline metasomatite 碱性交代岩异常
anomaly of alumyte 铝土矿异常
anomaly of anisotropy 各向异性反常
anomaly of antimony ore 锑矿异常
anomaly of carbonatite 碳酸盐岩异常
anomaly of crevice-water 裂隙水异常
anomaly of diamond 金刚石异常
anomaly of electric(al) prospecting 电法勘探异常
anomaly of geopotential difference 动力高度偏差
anomaly of gold ore 金矿异常
anomaly of granite 花岗岩异常
anomaly of gravitational field 重力场异常
anomaly of greisen 云英岩异常
anomaly of heat bearing structure 储热构造异常
anomaly of heavy spar 重晶石异常
anomaly of hot-water alteration 热水蚀变异常
anomaly of karst water 岩溶空隙水异常
anomaly of mechanic dispersion hole 机械分散晕异常
anomaly of mercury ore 汞矿异常
anomaly of molybdenum ore 钼矿异常
anomaly of phosphorite 磷块岩异常
anomaly of rare metals ore 稀有金属(矿)异常
anomaly of Rn-Ra water 氡—雷水异常
anomaly of secondary enrich 次生富异常
anomaly of strontium ore 锶矿异常
anomaly of tine ore 锡矿异常
anomaly of tungsten ore 钨矿异常
anomaly of underground water behaviou(u)rs 地下水动态异常
anomaly of uranium ore 铀矿异常
anomaly of volcanic rock 火山岩异常
anomaly of wave velocity 波速异常
anomaly peak form map of survey area 测区异常峰形图
anomaly peak type 异常峰类型
anomaly peak value 异常峰值

anomaly per centage 百分异常
anomaly ranking 异常评序
anomaly related to mineral deposit 矿异常
anomaly rich potassium shale 富钾页岩异常
anomaly seasonal variation 异常季节性变
anomaly shape 异常形状
anomaly size 异常大小
anomaly standard deviation 异常标准离差
anomaly value of mercury 汞的异常值
anomaly while drilling 钻进中的异常现象
anomaly width 异常宽度
anomer 反构体
anomie 无价值状况；无法纪的社会状态；称名不能；反常状态
anomite 褐云母
anomocytic type of stoma 不规则型气孔
anomy 反常状态
anona oil 依兰油
anonymous 无名的
anonymous block 匿名块
anonymous bonus 模糊奖金
anonymous dimensionless group 无名无因次群
anonymous file transfer protocol server 匿名 FTP 服务器
anonymous literature 不具名资料
anonymous processor pool 均一处理机群
anonymous server 匿名服务器
anophelifuge 防蚊剂
anophorite 钠钛闪石
anorak 防水衣；防水布；防风衣
anorogenic 非造山的
anorogenic province 非造山省
anorogenic time 非造山期
anorthic 三斜的
anorthic crystal 三斜晶体
anorthic system 三斜系；三斜晶系
anorthite 钙长石；钙长石
anorthite anorthosite 钙长石斜长岩
anorthite basalt 钙长玄武岩
anorthite diorite 钙长石闪绿岩
anorthite troctolite 钙长橄长岩
anorthitite 钙长岩
anorthoclase 歪长石
anorthoclasite 歪长岩
anorthopia 不正视
anorthose 歪长石；斜长石
anorthosite 歪长岩；斜长岩
anosovite 黑钛石
Anosov's theorem 安诺索夫定理
another infection 另次侵染
anothermic 大洋平流层
another mistake (in treatment) 再逆
anotron 辉光放电管
anoxia 缺氧症；乏氧症
anoxia condition 缺氧条件
anoxic 缺氧的
anoxic activated sludge 缺氧活性污泥
anoxic-anaerobic-Carousel oxidation ditch 缺氧—厌氧—卡鲁塞尔式氧化沟
anoxic area 缺氧区
anoxic baffled reactor 缺氧折流板反应器
anoxic basin 缺氧区
anoxic biodegradation 缺氧生物降解
anoxic bottom condition 海底缺氧状态
anoxic condition 缺氧状态
anoxic degradation 缺氧降解
anoxic denitrification 缺氧反硝化
anoxic deposit 缺氧沉积
anoxic environment 缺氧环境
anoxic fluid bed reactor 缺氧流化床反应器
anoxic inactivated sludge 缺氧灭活污泥
anoxic length 缺氧期
anoxic/oxic/oxic process 缺氧/好氧/好氧工艺
anoxic/oxic process 缺氧/好氧工艺
anoxic/oxic system 缺氧/好氧系统；生物脱氮系统
anoxic phosphorus uptake 缺氧吸磷
anoxic pond 缺氧池
anoxic process 缺氧过程；缺氧工艺
anoxic reactor 缺氧反应器
anoxic sediment 缺氧底泥
anoxic sludge 缺氧污泥
anoxic time 缺氧时间
anoxic waters 缺氧水域；缺氧水体
anoxic zone 缺氧区；缺氧地区；缺氧地带

anoxybiotic 缺氧的；绝氧的
anoxybiotic bacteria 嫌氧菌
anoxycausis 无氧燃烧
anoxygenous environment 缺氧环境
anoxyphotobacteria 无氧光细菌
anoxyscope 实示需氧器
anphtha distillation 石脑油蒸馏
Anping Bridge 安平桥
anprolene 氧丙环
A-N radio range 分区式无线电导航设备
anroral zone disturbance 极光带扰动
ansa lenticularis 豆状核袢
ansamacrolide 桥环大环内酯
ansate 有柄的；蹄系状的；环状的
Ansbacher unit 安斯巴彻尔单位
Anschutz gyrocompass 安秀兹陀螺罗盘
anse de panier 三心拱
Anson unit 安森单位
Anstett test 安斯台特试验法
Anstie's iimit 安斯提极限
anston (美国约克郡的) 奶油色白云石
anstu 增水【水文】
Answer aback 船首开始反转
answer back 回答信号；应答；对答
answerback code 回答码；应答码
answer-back device 应答装置
answer-back unit 应答机构
answer extraction 解答抽取
answer extraction process 解答抽取过程
answer for a crime 负有罪责
answering 应答
answering back mechanism 应答机构
answering circuit 反响线路
answering cord 应答塞绳
answering delay 应答延迟
answering detection pattern 应答检测码型
answering equipment 应答设备
answering jack 应答塞孔
answering line 反响线路
answering pendant 回答旗；答旗
answering plug 应答插头
answering process 应答进程
answering signal 回答信号；应答信号
answering system 反响系统
answering time 应答时间
answering tone 应答音
answer interval 应舵时间间隔
answer in writing 书面答复
answer lamp 应答指示灯；应答信号灯；应答灯
answer list 应答表
answer only 只应答
answerphone 应答电话
answer print 校正拷贝；标准校正拷贝
answer too slow 舵反应太慢
anta 壁角柱；墙角墩
anta cap 壁角柱顶
anta capital 壁角柱柱头
antacid 解酸剂；制酸剂
Antaciron 安塔西隆耐蚀耐磨铁硅合金；硅铁合金
antae 墙角墩；副柱；端头墙墩；壁角柱
antae capital 壁角柱柱头
antaefix (ae) 瓦檐饰
antagonism 拮抗作用；对立；对抗作用
antagonism game 对抗性对策
antagonism of interest 利益的对抗性
antagonist 拮抗剂；对抗剂；对合牙
antagonistic (al) 对抗的；相反的
antagonistical symbiosis 对抗共生
antagonistic effect 拮抗效应；对抗效应；对抗效果
antagonistic function 拮抗作用
antagonistic interest 相互对立的利益
antagonistic spring 复原弹簧；反作用弹簧
antagonizing screw 调置螺旋
antal 墙墩；壁柱；半露柱；墙柱
antalgol 逆大陵变量
antalkali 解碱剂
antapex 背点
ant-arching device 防结拱装置
Antarctica 南极洲；南极地区；南极大陆
Antarctic air 南极 (圈) 气团
Antarctic anti-cyclone 南极反气旋
Antarctic Circle 南极圈
Antarctic circum polar current 南极绕极 (海) 流
Antarctic Continent 南极洲

Antarctic convergence 南极辐合带；南极汇聚
Antarctic ecosystem 南极生态系统
Antarctic exploration ship 南极探险船
Antarctic faunal region 南极动物区
Antarctic floral realm 南极植物地理区系
Antarctic front 南极锋
Antarctic ice sheet 南极冰源
Antarctic Interme diate water 南极中层水
antarcticite 南极石
Antarctic mapping 南极制图
Antarctic Ocean 南冰洋
antarctic ozone hole 南极臭氧层空间
antarctic penguin 南极企鹅
Antarctic plate 南极洲板块
Antarctic platform 南极洲地台
Antarctic Pole 南极
Antarctic realm 南极洲界
Antarctic region 南极 (地) 区；南极植物区
Antarctic source water 南极源水
antarctic spring time ozone depletion 南极春季臭氧消耗
antarctic stratospheric circumpolar vortex 南极平流层绕极涡旋
Antarctic treaty 南极条约
Antarctic waters 南极水域
Antarctic west wind drift 南极西风洋流；南极冷流
Antarctic whiteout 南极乳白天空
Antarctic zone 南极 (地) 带；南极区
Antarctogea 南极地区
ant attack 蚁侵蚀；蚁害
antecabinet (通住密室的) 前室
antecedent 前项；前述；先例的；比例前项
antecedent condition 前期条件
antecedent consequent river 顺向先成河
antecedent discharge 前期流量
antecedent drainage 前期排水；先成水系；先成河
antecedent engineering 前期工程
antecedent flow 前期流量
antecedent gorge 先成峡
antecedent moisture 前期含水量；前期水分；前期湿度；降水前土壤的湿度
antecedent moisture condition 原有湿度条件
antecedent moisture content 前期含水量
antecedent money 押金
antecedent party 提前背书人
antecedent platform 先成地台
antecedent precipitation 前期降水 (量)
antecedent precipitation index 前期降水指标
antecedent rainfall 前期降雨
antecedent regulation 始动调节作用
antecedent river 先成河
antecedent rule 先行规则
antecedent runoff 前期径流
antecedents 履历 (表)
antecedent soil moisture 前期土壤水分；土壤前期含水量
antecedent soil water 土壤前期含水量；灌前土壤含水量
antecedent stream 先成河
antecedent temperature 前期温度
antecedent temperature index 前期温度指数
antecedent valley 先成谷
antecedent wetness 前期湿度；前期水量；原有湿度
antecedent year 上年 (度)
antechair (教堂的) 唱诗班前厅
antechamber 前厅；前室；接待室；小方厅；预燃室；沉沙室；沉积室；沉淀室
antechamber with built-in wardrobe 带衣帽间的前厅；带衣帽间的接待室
antechapel (殡仪馆的) 小礼堂；教堂前厅
antechoir 唱诗班前室 (教堂)；唱诗台
ante christum 公元前
antechurch 前部 (古教堂)；教堂前厅
anteclise 台背斜
anteconsequent stream 顺向先成河
antecourt 前庭【建】
antedate 填早日期；早填日期；倒填日期
ante date cheque 填早日期支票
antedated bill 日期提前票据
antedated bills of lading 日期提前提单
antedated check 日期提前支票
antedating 日期填早
antediluvial 前洪积世

antediluvian 前洪积世的
antedorsal 前背部
antefix(ae) 檐口饰；瓦檐饰；滴水瓦；瓦顶饰；瓦挡；脊吻
antefixal tile 瓦檐；端部装饰瓦
antefix tile 檐瓦；瓦檐饰；瓦挡
antehall 前厅
anteisoalkane 反异构烷烃
anteklise 台背斜；陆背斜；台拱；陆梁
antemarginal 近边缘的
ante meridian 上午；子午圈；午前的
antemural (城堡的)外墙
antenatal-clinic 产前检查所
antenave 前门廊；教堂前殿
antenna 天线方位；天线
antenna aperture 天线孔径
antenna array 天线阵
antenna beam 天线射束
antenna beam direction 天线波束方向
antenna cable 天线电缆
antenna capacitance 天线电容
antenna characteristic impedance 天线特性阻抗
antenna circuit breaker 天线断路器
antenna condenser 天线电容器
antenna connection 天线接线
antenna connector 天线馈线连接管
antenna counterpoise 天线地网；地网
antenna coupler 天线耦合器
antenna current 天线电流
antenna curtain 天线幕；天线屏幕
antenna depression angle 天线俯角
antenna diplexer 天线双工器；天线共用器
antenna directional angle 天线指向角
antenna directivity 天线方向性
antenna directivity diagram 天线方向性图
antenna down-lead 天线引下线
antenna effective area 天线有效面积
antenna efficiency 天线效率
antenna elevation pawl 升高天线用摇柄
antenna eliminator 假天线；天线抑制器；等效天线
antenna feeder 天线馈电线；天线馈线
antenna field 天线场
antennafier 天线放大器
antenna flabellate 扇形触角
antenna frame 天线架
antenna gain 天线增益
antenna gallery 天线廊道；天线平台
antenna height 天线高度
antenna impedance 天线阻抗
antenna input impedance 天线输入阻抗
antenna input power 天线输入功率
antenna installation 天线设备；天线装置
antenna insulator 天线绝缘子
antenna iron tower 天线铁塔
antenna jack 天线插座
antenna laser modeling 激光天线模型
antenna lead(-in) 天线引(入)线
antenna lead-in insulator 天线引入绝缘子
antenna length 天线长度
antenna lens 透镜天线
antenna load 天线负载
antenna loading 天经加载
antenna loss 天线损耗
antenna main lobe 天线主瓣
antenna mast 天线杆；天线塔
antenna matching 天线匹配
antenna matching network 天线匹配网络
antenna matching unit 天线匹配装置
antenna mine 触线水雷
antennamitter 天线发射机
antenna mount 天线安装机械；架设天线机构
antenna optic(al) modeling 光学天线模型
antenna pattern 天线方向性图；辐射方向图
antenna pedestal 雷达天线座
antenna pick 天体摄影
antenna pick-up 天线电器中产生的起伏电压；天线噪声
antenna pole 天线杆
antenna positioning control unit 天线定位控制装置
antenna ramosa 分枝触角
antenna reflector 天线反射器
antenna repeat dial 旋转天线位置指示刻度盘
antenna rotation number 天线转数

antenna rotator 天线旋转器
antenna scanning center 天线扫描中心
antenna servo system 天线伺服系统
antenna side lobe 天线旁瓣
antenna socket 天线插座
antenna spike 天线销；天线杆；鞭状天线
antenna splitting device 天线共用器
antenna standing wave 天线驻波比
antenna support 天线杆
antenna switching device 天线交换器
antenna system 天线系统
antenna tilt mechanism 天线仰角机构
antenna tower 天线塔
antenna tracking system 天线跟踪系统
antenna trailer 拖曳天线
antenna tuner 天线调谐器
antennaverter 变频天线
antenna windshield 天线风挡玻璃
antenna wire 天线用线
antennifer 支角突
antepagment 门窗洞口装饰线脚；加在房屋上的镶边饰
antepagmenta (装修门用的)装饰石
antepagmentum 门窗洞口装饰线脚
antependium 祭坛前部覆盖物；祭台前的帷幕；祭台前的屏物；帷幔；屏饰
antepodium 教堂唱诗班后面的一个座位
anteporch 外门廊
anteporlico 教堂前门廊
anteport 外门；门道；外门槛；外大门
antequire 唱诗班前室(教堂)
anterides (古建筑墙体的)支墩；扶垛；扶壁
anterior aspect 前面观
anterior border 前缘
anterior part 前部
anterior surface 前面
anterior wall 前壁
anterio-superior to the ear 曲周
alternatives analysis 替代方案分析
anteroom 前室；前厅；前厅休息室；门斗；接待室
anteroposterior diameter 前后径
anteroposterior position 前后位
ante-solarium 朝日阳台；朝阳阳台；阳台
antetemple 前厅(寺庙)；前殿(寺庙)
antetheca 前壁
ante-venna 遮阳幕；遮阳板
anteversion 前倾
anteverted 前倾的
ant-friction material 耐磨蚀材料
anthanthrene 蒽嵌蒽
anthanthrone 蒽嵌蒽醌
anthead 白蚁堆
anthelate 长侧枝聚伞花序
anthelic arcs 反九弧
anthelion 日映云辉(幻日)；幻日；反假日
anthemion(mo(u)lding) 棕叶(花)饰；棕榈叶饰(古罗马和古希腊建筑)；忍冬饰；花状平纹；[复] anthemia
anthemion ornament 棕叶装饰(古希腊建筑)；棕叶饰(古希腊建筑)
anthill 蚁冢
anthiolimene 安锑锂明
anthoblast 珊瑚体
anthocaulus 珊瑚茎
anthochroite 青辉石
anthocyan 花色素类
anthocyanidin 花色素
anthocyathus 珊瑚杯
anthodite 石膏花
anthoinite 水钨铝矿
anthonyite 水氯铜矿
Anthony's capsule staining 安东尼荚膜染色法
Anthony's model 安东尼模型
anthophore 花冠柄
anthophyllite 直闪石
anthophyllite asbestos 直闪石石棉
anthophyllite leuco granoblastite 直闪浅粒岩
anthophyllite-mica schist 直闪云母片岩
anthophyllite schist 直闪石片岩
anthopoclimatology 人类气候学
anthostele 反口区
anthracene 蒽；并三苯
anthracene black 粗蒽炭黑
anthracene blue 蒽蓝

anthracene brown 蒽棕
anthracene cake 蒽坯；蒽饼
anthracene carboxylic acid 蒽甲酸
anthracene chrome yellow 蒽铬黄
anthracene crystal scintillation counter 蒽晶体闪烁计数器
anthracene green 蒽绿
anthracene nucleus 蒽环
anthracene oil 蒽油
anthracene oil tar 蒽焦油；蒽油渣
anthracene ring 蒽环
anthracene scintillation counter 蒽闪烁计数管
anthracene single crystal 蒽单晶
anthracene sulfonic acid 蒽磺酸
anthracene tetrone 蒽二醌
anthracene yellow 蒽黄
anthracenol 蒽酚
anthracenone 蒽酮
anthraciny 蒽化作用
anthracite 无烟煤
anthracite capping 白煤覆盖
anthracite coal 无烟煤
anthracite coal base refractory 碳质耐火材料
anthracite duff 无烟煤粉
anthracite filter media 无烟煤滤池介质
anthracite fines 无烟煤细粒
anthracite mine 无烟煤矿
anthracite-sand filter 煤气砂双层滤池
anthracite stove 温室(英国温室的一种)；无烟煤炉
anthracitization 无烟煤化(作用)
Anthracolithic period 大石炭纪(即石炭二叠纪)【地】
Anthracolithic system 大石炭系(即石炭二叠系)【地】
anthracolitization 煤化作用
anthracology 煤岩(石)学；煤炭学
anthra colo(u)rs 蒽素染料
anthracometer 二氧化碳计
anthracometry 二氧化碳定量法；二氧化碳测定法
anthraconite 沥青灰岩；黑沥青灰岩；黑方解石
anthra copper 蒽素铜
anthracoses 炭肺病
anthracosilicosis 炭末石末沉着病；煤矽肺
anthracosis 炭末沉着病；炭肺病；炭尘肺；煤肺(病)；煤尘肺
anthracotic deformity 炭末沉着性变形
anthracoxene 碳沥青质；炭沥青质
anthracoxenite 醚不溶树脂
anthracyl 蒽基
anthradiamine 蒽二胺
anthradiol 蒽氢醌；蒽二酚
anthradiquinone 蒽二醌
anthrafilt 过滤用无烟煤
anthrafilter 炭过滤器
anthrahydroquinone 蒽氢醌
anthraldehyde 蒽甲醛
anthramine 蒽胺
anthranil 氨茴内酐
anthranilate 氨基苯甲酸盐；氨茴酸盐
anthranilic acid 邻氨基苯甲酸；氨茴酸
anthranilo 氨茴基
anthranilo nitrile 氨基苯甲腈
anthraniloyl 氨茴酰
anthranol 蒽酚
anthranol chrome blue 蒽酚铬蓝
anthranol colo(u)r 蒽酚染料
anthranone 蒽酮
anthranoyl 氨茴酰
anthranoyl-anthranilic acid 氨茴酰氨茴酸
anthrao-silicosis 煤气矽肺
anthrapurpurin 蒽紫红素
anthraquinol 蒽二酚
anthraquinone aldehyde 蒽醌甲醛
anthraquinone blue 蒽醌蓝
anthraquinone diazonium zinc chloride compound 蒽醌重氮氯化锌复合物
anthraquinone dye 蒽醌染料
anthraquinone glycoside 蒽醌甙
anthraquinone pigments 蒽醌色素
anthraquinone sulfonic acid 蒽醌磺酸
anthraquinonic acid 蒽醌酸
anthrasol 蒽焦油
anthratetrol 蒽四酚
anthratriol 蒽三酚

anthrax 煤质;古宝石
anthraxolite 碳沥青(含固定碳97.2%)
anthraxylon 细屑镜煤;纯木煤
anthrinoid 高变质镜质体
anthrinoid group 炭质组
anthroecology 人类生态学
anthrogenics 合成有机化合物
anthroic acid 蒽甲酸
anthrone dye 蒽酮染料
anthrophyllite 云母
antropic activity 人为活动;人类活动
antropic epipedon 熟化层;人为表层;耕作表层
antropic factor 人为因素
antropic soil 人造土壤;熟土;耕作土;表层土(壤)
antropic zone 人为区
anthroplithic age 石器时代
anthropobiology 人类生物学
anthropochore 人为传布植物
anthropochory 人为散布;人为传播
Anthropogene 人类纪(灵生纪)
anthropogenetic 人类发生的
anthropogenetic force 人类营造力
anthropogenic 人为的;人类发生的
anthropogenic acidification 人为酸化
Anthropogenic age 人类时代
Anthropogenicalluvial soil 灌淤土
anthropogenic association 人为植物群丛
anthropogenic climax 人为顶极
anthropogenic contaminant 人为污染物
anthropogenic contaminant concentration 人为污染物浓度
anthropogenic discharge 人为排放
anthropogenic environmental effect 人为环境影响
anthropogenic eutrophication 人为富营养化
anthropogenic factor 人为因素
anthropogenic influence 人为影响;人类影响
anthropogenic non-point pollution source of water body 水体人为非点污染源
anthropogenic organics 人造有机物
anthropogenic pollutant 人为污染物
anthropogenic pollution sources 人为污染(来)源
anthropogenic process 耕作土壤发生过程
anthropogenic slope 人工坡
anthropogenic sources of pollution 人为污染(来)源
anthropogenic succession 人为演替
anthropogenic wastes 人为废弃物
anthropogentic form 人工地物
anthropography 人类分布学;人口地理分布学
anthropokinetics 人类活动学
anthropology 人类学
anthrometric data 人体测量数据
anthropometry 人体测量学
anthropomorphic diving suit 拟人型常压潜水服
anthropomorphic robot 人形机器人
anthropomorphous phantom 人体伦琴当量
anthroposphere 人类圈
Anthropozoic era 灵生代
anthryl 蒽基
anthryl carbinol 蒽甲醇
anti-abrasion additive 抗磨蚀添加剂
anti-abrasion coating 耐磨涂层
anti-abrasion layer 防磨擦保护层;保护层
anti-abrasive 耐磨损的
anti-abrasive material 耐磨损材料
anti-acid 抗酸剂;抗酸(的);制酸剂
anti-acid additive 抗酸添加剂
anti-acid bronze 抗酸青铜
anti-acid cement 抗酸水泥;耐酸水泥
anti-acid ceramic equipment 防腐陶瓷设备
anti-acid ceramic mechanical sealing 耐酸陶瓷机械密封
anti-acid coat(ing) 防酸涂层;耐酸涂层
anti-acid flooring 防酸地面
anti-acidic paint 耐酸油漆
anti-acid metal 抗酸金属
anti-actinic glass 隔热玻璃;防射线玻璃;反光玻璃;阻光玻璃;吸热玻璃
anti-activator 活化阻止剂
anti-adhesiveness 抗粘连性
anti-adsorption 抗吸附
anti-aeration plate 阻气板
anti-ag(e)ing agent 抗老化剂
anti-ag(e)ing dope 防老(化)剂

anti-ag(e)ing protective 抗老化(剂)
anti-ager 抗老化剂;防老(化)剂
anti-agglutinin 抗凝集素
anti-aircraft 防空的
anti-aircraft control 防空管制
anti-aircraft defence 防空设备;防空工事
anti-aircraft defence system 防空系统
anti-aircraft lookout 对空监视
anti-aircraft protection 防空
anti-aircraft range finder 防空测距计
anti-aircraft searchlight 防空探照灯
anti-aircraft tower 对空监视哨;对空观测台;防空观察塔
anti-air pollution 防止空气污染
anti-air pollution system 防止空气污染系统;空气防污系统;防空气污染系统
anti-alias filtering 去假频滤波
anti-aliasing 反混淆
anti-amarillic serum 黄热病血清
anti-atom 反原子
anti-atomic defence 原子防御;原子防护
anti-attrition 减少磨损;减磨;耐磨损
anti-automorphism 反自同构
anti-axial growth type 反向生长型结晶纤维
anti-backlash device 防后冲装置
anti-backlash gear 无齿隙齿轮
anti-backlash spring 防止齿隙游移的弹簧;消隙弹簧
anti-bacterial 抗细菌的;抗菌的
anti-bacterial action 抗菌作用
anti-bacterial agent 制菌剂
anti-bacterial cement 抗菌水泥
anti-bacterial paint 抗菌涂料;抗菌漆
anti-balance action 反平衡作用
anti-balloon plate 隔纱板
anti-bandit glazing 防盗玻璃
anti-bank movement 反银行运动
anti-baric flow 反压流;反气射流
anti-barreling 反桶形畸变;桶形失真校正
anti-baryon 反重子
anti-biosis 抗生作用;抗生现象
anti-biosis cement 抗菌水泥
anti-biotic fertilizer 抗菌废料
anti-biotic formulation wastewater 含抗生素成分废水
anti-biotic wastewater 含抗生素废水
anti-bleed sealer 防渗腻子;防渗封闭剂
anti-block 开段
anti-blocking 抗阻塞
anti-blocking agent 抗粘连剂;抗黏剂;抗结块剂
anti-blocking brake 防抱死制动器
anti-blooming agent 防潮剂;防白剂
anti-blowing agent 消泡剂
anti-blushing agent 抗混浊剂;防发白剂;防变色剂
anti-body 抗体;反物质
anti-bomb 防弹的
anti-bonding 反键(作用)
anti-bonding orbital 反键轨;反成键轨道
anti-bonding orbital function 反键轨函数
anti-boreal faunal region 南寒带动物区
anti-borer 防虫蛀
anti-borer carpet 防蛀地毯
anti-borer plywood 防蛀胶合板;防虫蛀胶合板
anti-boson 反玻色子
anti-bouncer 减振器;防跳装置;防回跳装置
anti-breaker 防碎装置
anti-bronzing agent 防青铜色剂
anti-bubble effect 反汽泡效应
anti-bubbling agent 消泡剂
anti-bucking 抗屈曲
anti-bugging 防错法
anti-bumping stone 防爆沸石块
anti-bunch 反聚束;反聚束
anti-bunching effect 反聚束效应
anti-burglary glazing 防盗玻璃
antic 怪状动物雕像
anti-caking agent 防颜料沉淀结块剂
anti-capillary 抗毛细作用
anti-carburizer 渗碳阻止剂;防渗碳剂
anti-carburizing paint 渗碳防护涂料;防渗碳涂料;防渗碳漆
anti-carburizing paste 防渗碳膏
anti-carcinogen 抗致癌物
anti-carrier 反载体

anti-catalyst 抗催化剂;催化毒物;负催化剂;反催化剂
anti-catalytic property 负催化性
anti-catalyzer 抗催化剂;负催化剂;反催化剂
anti-cathode 对阴极;对负极
anti-cavitation 防穴蚀;防气穴
anticenter 震中对点;反中心;反震中;反银心(天空中与银河系中心相距180度的方向)
anti-center of earthquake 震中对点
anti-centrifugal splash 抗甩性
anticer 防冰装置
anti-chain 反链
anti-chatter spring 抗震弹簧
anti-checking agent 防龟裂剂
anti-checking iron 扒钉;防裂钩
anti-chemical 防化
anti-chill 防白口涂料;防白口镶块
anti-chipping paint 抗石击漆;抗片落漆
anti-chirping 反线性调频
anti-chlor 除氯剂;脱氯剂
anticipant 预期的
anticipate 抢先
anticipated acceptance 提前支付的承兑汇票;提前偿还的承兑汇票;预文的承兑汇票
anticipated accuracy 预期精(确)度
anticipated action 预期行动
anticipated area of drift 预测漂移区
anticipated buying 预先购买
anticipated cost 提前成本;预期费用;预计成本
anticipated differential settlement 预计差异沉降量
anticipated discharge 预期流量;预计流量
anticipated discount 预计折扣
anticipated economic index 先行经济指数
anticipated effect 预期影响;预期效果;预期收益
anticipated environmental policy 预期环境政策
anticipated expenditures 预期开支
anticipated final field stress 预计最终原位应力
anticipated freight 预支运费;预期运费
anticipated ground water yield 预计地下水产量
anticipated import 提前进口
anticipated input 抢先输入
anticipated load(ing) 预期荷载;预期负荷
anticipated operation(al) event 预期运行事件
anticipated payment 提前付款
anticipated population 预测人口
anticipated population increase 预计人口增长
anticipated price 预期价格
anticipated probability 预测概率
anticipated profile 预期纵断面
anticipated profit 预期利润
anticipated redemption 提前兑付
anticipated redemption of bonds 提前兑换债券
anticipated request 先行请求
anticipated revenue 预期收入
anticipated risk 预计风险
anticipated seepage 预计渗涌量;预计渗透量;预计渗漏量;预计渗流量
anticipated service life 预计使用期限;预计使用年限;预测使用期限;预测使用年限
anticipated settlement 预计沉陷量
anticipated shipping date 预计装货时间
anticipated stock 预计存量
anticipated structural life 预测运行结构使用期限
anticipated traffic 预期交通量
anticipated use or development expenditures 预期开发费
anticipated volume of traffic 预期交通量;预计交通量
anticipate price 预计价格
anticipater 超前预告器
anticipate shipment 提前装运
anticipating control 提前控制;预调制
anticipating of marketing competition 市场竞争前景
anticipating signal 预告信号;超前信号
anticipation 预先考虑;预先处理;提前出现;提前付款;预期;预料;早期出现;超前作用
anticipation fee 申请费
anticipation mode 先行形式;先行方式
anticipation network 超前网络
anticipation of mineral reserve 资源储备前景
anticipation rate 提前付款利率
anticipation redemption of a bond 提前兑换债券
anticipation survey 前景调查

anticipator 预感器;预测器;超前预防器
anticipatory control 超前校正;超前控制
anticipatory control of aeration 曝气预控制
anticipatory data 提前处理的资料
anticipatory environmental action 防患未然的环保行动
anticipatory index 预期性指标
anticipatory staging 先行传送
anti-clamp function 防夹功能
anti-clash key 防碰撞键
anti-clastic 一面凸的;抗裂面;互反曲面;鞍形面的;互反的
anti-clastic shell 鞍形薄壳;互反曲面薄壳;鞍形面薄壳
anti-clastic shell system 鞍形面薄壳系统
anti-clastic surface 抗裂面;互反曲面;鞍形面
anti-climber 防攀登装置;防攀登金属网
anti-climbing 防攀登的
anti-climbing metal 防攀登的金属
anti-climbing metal mesh 防攀登金属网
anticlinal 垂周的;背斜的
anticlinal axis 背斜轴;鞍轴
anticlinal belt 背斜带
anticlinal bend 背斜曲部
anticlinal bowing 构造隆起;构造鼻
anticlinal bulge 背斜隆起
anticlinal closure 背斜闭合度
anticlinal deposit 背斜矿床
anticlinal division 垂周分裂
anticlinal fault 背斜断层
anticlinal fissure 脊缝;背斜裂隙
anticlinal fold 背斜褶皱
anticlinal form structure 背斜型构造
anticlinal layer 背斜层
anticlinal limb 背斜翼
anticlinal meander 背斜河曲
anticlinal mountain 背斜山
anticlinal nose 背斜鼻
anticlinal oil-gas field 背斜油气田
anticlinal pivot 背斜枢
anticlinal pool 背斜油藏
anticlinal ridge 背斜岭;背斜脊
anticlinal river 背斜河
anticlinal spring 背斜泉
anticlinal stratum 背斜层
anticlinal theory 背斜理论
anticlinal trap 背斜圈闭
anticlinal trap by differential compaction 差异压实背斜圈闭
anticlinal trap by differential compaction over buried hill 潜山差异压实背斜圈闭
anticlinal trap by differential compaction over reef 礁差异压实背斜圈闭
anticlinal trap by differential compaction over sand lens 砂岩透镜体差异压实背斜圈闭
anticlinal trap by differential compaction over uplifted block 上升断块差异压实背斜圈闭
anticlinal trap by igneous plug 由火成岩栓侵入造成的背斜圈闭
anticlinal trap by mud diapir 由泥底群造成的背斜圈闭
anticlinal type of hydrodynamic(al) trap 背斜型水动力圈闭
anticlinal-unconformity combination trap 背斜—不整合复合圈闭
anticlinal-up-dip edge out combination trap 背斜—上倾尖灭复合圈闭
anticlinal valley 背斜壳;背斜谷
anticlinal wall 垂周壁
anticlinal zone 背斜带
anticline 背斜(层)
anticline core 背斜核部
anticlinorium 复背斜(层)
anti-clock circuit 倒钟形电路
anti-clockwise 逆时针方向旋转;反时针(方向)的
anti-clockwise direction 逆时针方向
anti-clockwise motion 反时针方向运动
anti-clockwise movement 反时针方向运动
anti-clockwise polarized electromagnetic wave 反时针偏振电磁波
anti-clockwise reading 反时针方向读数
anti-clockwise rotation 逆时针旋转;反时针旋转;反时针方向转动;反时针方向旋转
anti-clockwise vorticity 反时针方向涡轮

anti-clogging agent 防阻塞剂
anti-clogging separator 防堵塞分离器
anti-cluster sea 抗海浪干扰
anti-clutter 抗杂波(干扰);抗干扰;抗地物干扰系统;抗本地干扰;杂波抑制;反干扰(系统)
anti-clutter circuit 抗本地干扰电路;反杂乱回波电路
anti-clutter gain control 减干扰增益控制;海浪干扰抑制;反杂波增益控制
anti-clutter rain 抗雨雪干扰
anti-clutter sea 海浪干扰抑制
anti-coagulant 抗凝素;抗凝剂;阻凝剂
anti-coagulant action 阻凝作用
anti-coagulant property 抗凝剂性质
anti-coagulant separator 防堵塞分离器
anti-coagulant step 防堵塞措施
anti-coagulation 防凝;阻凝
anti-coaxial magnetron 反同轴磁控管
anti-code 反密码
anti-codon 反暗码子
anticoherer 散屑器
anti-coincidence 反重合;反符合
anti-coincidence circuit 防止重合线路;反重合电路;反符合电路
anti-coincidence counter 反符合计数器
anti-coincidence detector 反重合探测器
anti-coincidence element 反符合元件
anti-coincidence gate 按位加门
anti-coincidence technique 反符合技术
anti-collineation 反直射变换
anti-collision 防撞(击)
anti-collision beacon 防撞灯
anti-collision cushion 防碰垫
anti-collision device 防撞装置;防撞装置
anti-collision display unit 避碰显示装置;避碰雷达显示装置
anti-collision gear 防撞设备
anti-collision indicator 防撞指示器
anti-collision pier 抗撞墩
anti-collision property 抗撞性能
anti-collision radar 防撞雷达;避碰雷达
anti-collision safety device 防撞装置
anti-collision system 防碰系统;避碰系统
anti-collision wall 防撞墙
anti-colodal 金属防腐剂
anti-commutation relation 反对易关系
anti-commutator 反换位子;反对易子
anti-commute 反交换;反对易
anti-compound generator 反复励发电机;反复激发电机
anti-condensation 防凝;防冷凝;防凝结
anti-condensation ceiling 防凝水顶棚;防结露顶棚;防凝水天花板
anti-condensation coating 防凝结涂层
anti-condensation lining 防凝(水)内衬
anti-condensation paint 防冷凝作用漆;抗凝聚漆;防凝水油漆;防结露漆
anti-condensation plaster 防结露粉刷;防凝水石膏
anti-condensation protective measure 抗结露保护措施;抗凝水保护措施
anti-configuration 反型;反式构型
anti-consequent river 逆向河
anti-consequent stream 逆向河
anti-contamination 防沾污;防污染;抗沾污性
anti-contamination clothing 防沾污服
anti-convolution divide layer interpretation method 反褶积分层解释法
anti-corodal 高强(度)耐蚀铝合金;铝基硅镁合金;安奇可罗矣达耐蚀铝硅镁合金
anti-corona 反日冕
anti-corona coating 防晕漆;防晕(涂)层
anti-corona collar 电晕环;防晕环
anti-corona protection collar 电晕保护环
anti-corona ring 防晕环
anti-corona tape 防晕带
anticorrodant 防锈剂;防蚀剂
anti-corrosion 耐蚀;防锈(蚀);防蚀;防腐(蚀)
anti-corrosion additive 防腐添加剂;防腐蚀添加剂
anti-corrosion agent 防腐蚀添加剂;防腐蚀剂;伤蚀
anti-corrosion alloy 耐蚀合金;防蚀合金
anti-corrosion anti-corrosive paint 防腐蚀漆
anti-corrosion cable 防腐电缆
anti-corrosion coat(ing) 防锈面层;防蚀面层;防锈蚀涂料;防蚀涂料;防蚀涂膜;防蚀涂层;防腐涂层;

防腐蚀涂层;防锈层
anti-corrosion composition 防蚀锈油漆;防蚀化学剂;防蚀油漆;防蚀化合物
anti-corrosion film 耐蚀薄膜
anti-corrosion grease 防腐蚀滑酯
anti-corrosion insulation 防蚀绝缘(层);防蚀层;防腐蚀绝缘层
anti-corrosion lining 防腐衬砌
anti-corrosion method 防蚀法;防腐蚀法
anti-corrosion paint 防锈蚀漆;防蚀涂料;防腐蚀涂料
anti-corrosion paper 防腐蚀纸
anti-corrosion promenade tile 防腐蚀缸砖
anti-corrosion protection coat 防蚀涂层
anti-corrosion test 耐腐蚀试验
anti-corrosion treatment 防腐蚀处理;防腐处理
anti-corrosive 防蚀的;防腐的
anti-corrosive additive 防腐蚀添加剂
anti-corrosive agent 防蚀剂;防腐蚀添加剂;防腐蚀剂
anti-corrosive agent for synthetic(al) fibres 合成纤维防腐剂
anti-corrosive alloy 防蚀合金
anti-corrosive alloy coating 防蚀合金涂层
anti-corrosive alumin(i)um-coated steel wire 防蚀镀铝钢丝
anti-corrosive blanket 防腐面层;防腐层;防腐覆面层
anti-corrosive coat(ing) 防腐涂层;防锈蚀涂料;防腐涂饰;防蚀涂层;防腐层;氧化防腐膜
anti-corrosive coating with oxides 氧化防护膜
anti-corrosive composition 防腐蚀成分;防锈漆;防锈剂;防蚀油漆;防腐蚀涂料;防腐蚀剂
anti-corrosive equipment lined with Teflon 衬聚四氟乙烯防腐设备;衬里防腐设备
anti-corrosive foil 防腐箔
anti-corrosive grout 防锈蚀灌浆;防腐水泥灰浆
anti-corrosive insulation 防锈绝缘;防蚀绝缘(层);防蚀绝缘层;防腐层
anti-corrosive layer 防腐层
anti-corrosive lining 防腐衬里
anti-corrosive material 防腐材料
anti-corrosive mortar 抗蚀砂浆;防腐砂浆
anti-corrosive oil 防腐油
anti-corrosive paint 防锈油漆;防锈涂料;防锈蚀漆;防锈漆;防蚀油漆;防蚀涂料;防腐蚀油漆;防腐涂料;防腐蚀涂料;防腐漆;通用防锈漆
anti-corrosive pigment 防蚀锈颜料;防腐蚀颜料;防锈颜料;防腐颜料;防腐涂料
anti-corrosive precaution 抗侵蚀性措施
anti-corrosive preparation 防腐蚀制剂
anti-corrosive prime coat 防锈蚀打底层;防腐蚀打底层;防腐打底层
anti-corrosive primer 防锈底漆;防腐打底层
anti-corrosive priming paint 防锈打底漆
anti-corrosive slurry 防蚀灌浆;防腐水泥灰浆
anti-corrosive treatment 防锈处理;防蚀处理;防腐处理
anti-corrosive valve 防蚀阀门;防腐阀门
anti-corrosive varnish 防锈清漆
Anticosti Island 安提科斯提岛
anti-coustic 反聚光线
anti-crack 抗裂的
anti-crack bar 抗裂钢筋
anti-crack coating 抗裂(涂)层;抗裂涂料层
anti-cracking agent 防裂剂橡胶;防裂剂
anti-crack reinforcement 抗裂钢筋;抗裂缝钢筋;防裂钢筋
anti-crack steel 抗裂钢筋
anti-crash device 防撞装置;防破裂装置
anti-cratering agent 防缩孔剂
anti-creaming agent 防膏冻剂
anti-crease 防皱纹
anti-creep 逆蠕动;防蠕动
anti-creepage 防漏电
anti-creep baffle 防蠕爬板
anti-creep barrier 防蠕爬障栅
anti-creeper 防爬行装置;防滑动装置;钢轨防爬器;防蠕动装置;防爬器;防漏电设备
anti-creeper baffle 防爬障板
anti-creeper barrier 防爬障栅
anti-creeper clip 防爬夹钉
anti-creeper device 防爬设备
anti-creeper of cover plate 路面板防爬装置

anti-creeper shield 防爬挡板
anti-creeper stake 防爬桩
anti-creeper strap 防爬带
anti-creeper strut 防爬器支撑
anti-creep heap 挡车堆(厂矿道路)
anti-creeping capacity 防爬能力
anti-creeping device 防爬装置;防爬器
anti-creeping iron clop 防爬扣铁;防爬卡铁
anti-creeping plate 防爬板
anti-creeping resistance 防爬阻力
anti-creeping stake 防爬桩
anti-creeping steel angle 防爬角铁
anti-creeping strut 防爬(木)撑
anti-creeping switch 防漏电开关
anti-creep link 防爬连杆
anti-creep shield 防蠕爬挡板
anti-creep strap 防爬链条
anti-creep strut 防爬支撑
anti-crepuscular rays 反曙暮辉
anti-crisis measures 反危机措施
anti-crustator 表面沉垢防止剂;防水锈剂;防垢剂;表面硬化防止剂
anticryptic colo(u)r 侵隐色
anti-crystallinic pipe 防晶管
anticum 门廊(寺庙);前门廊;壁角柱
anti-curl 防卷曲
anti-curl coating 防卷曲涂层;防卷曲层
anti-curl cutter fax 防卷曲切纸传真机
anti-cyclical policy 反周期政策;反衰退政策
anti-cyclogenesis 反气旋生成;反气旋发生;反气旋的形成
anti-cyclolysis 反气旋消散
anti-cyclone 反气旋;反低气压;高气压
anti-cyclonic 反气旋的
anti-cyclonic area 反气旋区
anti-cyclonic circulation 反气旋式环流
anti-cyclonic current gyre 反气旋式环流
anti-cyclonic curvature 反气旋曲率
anti-cyclonic divergence 反气旋辐散
anti-cyclonic eddy 反气旋式涡流
anti-cyclonic gloon 反气旋阴天
anti-cyclonic inversion 反气旋逆温
anti-cyclonic shear 反气旋式切变
anti-cyclonic storm 反气旋风暴
anti-cyclonic subsidence 反气旋下沉
anti-cyclonic swirl 反气旋(型)漩涡
anti-cyclonic system 反气旋系统
anti-cyclonic vorticity 反气漩涡度
anti-cyclonic wind 反气旋风
anti-cyclotron tube 反回旋加速管
anti-damping 抗阻尼;抗衰减
anti-damping agent of concrete mass 大体积混凝土防堵剂
anti-darkening agent 防黯色剂
anti-dated bill of lading 倒签提单
anti-dazzle 防眩
anti-dazzle device 防目眩装置
anti-dazzle device for vehicle 车辆防眩装置
anti-dazzle glass 遮光玻璃;防眩玻璃
anti-dazzle hedge 防眩绿篱
anti-dazzle lamp 静光灯;防眩灯
anti-dazzle lighting 防眩照明;防眩灯光
anti-dazzle mirror 防眩镜
anti-dazzle pedal 防眩踏板
anti-dazzle screen 遮阳光器;遮光片;防眩屏;防眩挡板
anti-dazzle shade 防眩遮光板
anti-dazzling screen 遮阳板;遮阳屏;遮光片;防眩屏;防眩挡板
anti-decaying paint 防腐漆
anti-derailing rail 护轨;防止脱轨轨条;防止出轨的轨道
anti-derivative 反式衍生物;反导数【数】;不定积分
anti-desertification program(me) 防止沙漠化计划
anti-desiccant 抗平解剂;保湿剂;防干剂
anti-deteriorant 防老(化)剂;防坏剂
anti-detonant 汽油抗爆剂;抗爆剂
anti-detonating fluid 抗爆液
anti-detonating fuel 防爆燃料
anti-detonating quality 抗爆性
anti-detonation 抗震;抗燃;防爆
anti-detonation fuel 抗爆燃料;抗爆汽油
anti-detonation timbering 抗震加强支架
anti-detonator 防震剂;抗震剂;抗震化合物;抗爆剂;防爆剂

anti-diazo compound 反偶氮化合物
anti-diazosulphonate 反重氮磺酸盐
anti-diazotate 反重氮盐
anti-diffusion 反扩散
anti-dilution provision 反稀释条款
anti-dilutive 反稀释化的
antidim(compound)(防止水分集结于玻璃上的)保明剂
antidimmer 抗膜剂;保明剂
anti-dip river 反倾斜河;反倾角河
anti-dip stream 反倾斜河;反倾角河
anti-direction(al)finding 反向定位;反定位;反测向
anti-dirt 防污;防尘(的)
anti-disaster access 救灾通道
anti-discolo(u)ration 反褪色;防变色
anti-dislocation 反位错
anti-distortion ring 加固圈
anti-distortive transition 反畸变相变
anti-disturbance 反扰动
anti-dote 解毒剂
anti-dotum 解毒剂
anti-downdraught terminal 防倒灌烟囱帽
anti-drag 减阻(力)的;减阻;逆牵引;反阻力;反阻
anti-drag cap 减阻帽
anti-drag ring 减阻环
anti-drag speed ring 减阻增速环
anti-drawing mask 带氧气呼吸器的轻型潜水器
anti-drier 干燥延缓剂
anti-drip 防滴漏;防滴
anti-drip device 防滴装置
anti-dripping agent 防流挂剂
anti-dromic 逆行的;逆向行的;逆向的
anti-dromic conduction 逆向传导
anti-dromic fibers 逆行纤维
anti-dromic impulse 逆行冲动
anti-dromic volley 逆行排放
anti-dromy 反旋
anti-dropping device 防坠装置
anti-drowning mask 抗溺水面罩
anti-drumming coat(ing) 防震耳涂层;防噪声涂料;隔音涂刷(层)
anti-dryer 干燥延缓剂
anti-drying surface 防干裂面
antidugout 防空壕
anti-dumping 抗倾覆(作用);抗翻倒作用
anti-dumping duty 反倾销税
anti-dumping unit 反倾单位
anti-dune 背转沙丘;逆行沙丘;逆波痕;反向沙波;反沙丘;逆沙丘
anti-dune bed 反向沙波河床
anti-dune cross-bedding 反沙丘交错层
anti-dune cross bedding structure 逆行沙丘交错层理构造
anti-dune motion 逆沙丘运动
anti-dune movement 逆沙丘运动
anti-dunes on bottom 河底逆行沙波
anti-dust 防尘
anti-dust binder 去污斑胶合剂
anti-dust compound 去污斑化合物
anti-dust gun 防尘喷枪
anti-dusting 抗尘作用;抗尘性(的)
anti-dust spraying system 防尘喷洒系统
antiedrite 钡沸石
anti-elastic bending 反弹性变形弯曲
anti-electron 正电子
anti-electro shoes 防静电工作鞋
anti-element 反元素
anti-epicenter 震中对点;反震中
anti-epicentrum 震中对点
anti-epidemic 防疫的;防止传染的
anti-epidemic center 防疫中心
anti-epidemic measures in a focus 疫区处理
anti-epidemic station 防疫站
anti-equivalence 反等价
anti-eroding ability of soil 土壤抗冲(刷)能力
anti-evaporant 蒸发抑制剂;防蒸发剂
anti-explosion 防爆
anti-explosion automatic air switch 防爆自动空气开关
anti-explosion distribution equipment 防爆配电设备
anti-explosion fuel 防爆燃料
anti-explosion magnetic starter 防爆磁力起动器

anti-explosion manipulator 防爆机械手
anti-explosion part ventilating fan 防爆局部通风机
anti-explosion sheet 防爆板
anti-export statute 反输出法令(调水等);反输出法regulation(调水等)
anti-fading antenna 防衰落天线
anti-fading device 衰落补偿装置
anti-fading varnishing 防褪色漆
anti-fan complex fold 反扇形复式褶皱
anti-fatigue 抗疲劳剂;耐疲劳
anti-fatique agent 防疲劳剂
anti-feedback 回授消除
anti-feerromagnetics 反铁磁质
anti-ferment 防酵剂
anti-fermentative 防发酵剂;防酵(处理剂)
anti-fermite oil 防蚁油
anti-ferroelasticity 反铁弹性
anti-ferroelectric(al)ceramics 反铁电陶瓷
anti-ferroelectric(al)crystal 反铁电体;反铁电晶体
anti-ferroelectric(al)distortion 反铁电畸变
anti-ferroelectric(al)energy storage ceramccapacitor 反铁电储[贮]能电容器
anti-ferroelectric(al)material 反铁电材料
anti-ferroelectric(al)phase transition 反铁电体相变
anti-ferroelectric(al)state 反铁电态
anti-ferroelectricity 反铁电(性);反铁电现象
anti-ferroelectrics 反铁电体
anti-ferromagnet 反铁磁体
anti-ferromagnetic 反铁磁性的
anti-ferromagnetic compound 反铁磁性化合物
anti-ferromagnetic crystal 反铁磁晶体
anti-ferromagnetic domain 反铁磁畴
anti-ferromagnetic exchange integral 反铁磁交换积分
anti-ferromagnetic material 反铁磁性材料
anti-ferromagnetic non-gyro tropic crystal 反铁磁非回转晶体
anti-ferromagnetic resonance 反铁磁谐振;反铁磁共振
anti-ferromagnetic spin wave 反铁磁自旋波
anti-ferromagnetic state 反铁磁态
anti-ferromagnetic substance 反铁磁性物质
anti-ferromagnetic susceptibility 反铁磁化率
anti-ferromagnetism 反铁磁性;反铁磁现象
anti-ferromagnetism mineral 反铁磁性矿物
anti-ferromagnon 反铁磁振子
Antifier block 四面开槽方块
anti-fix 反曲线(状)挑檐饰
anti-flaming 防燃
anti-flammability agent 防燃剂
anti-flash 防闪
anti-flashing equipment 防溶化装置
anti-flex cracking 抗折裂;抗弯裂;抗弯开裂
anti-flexure system 反弯曲系统
anti-floating agent 防发花剂
anti-flocculating 防絮凝的
anti-flocculating agent 防絮凝剂
anti-flocculating property 防絮凝性质
anti-flocculation 反絮凝(作用)
anti-flood 防洪(的)
anti-flooding agent 防浮色剂;防浮剂
anti-flooding devie 防溢流装置
anti-flooding interceptor 防洪截流渠;止回阀
anti-flooding valve 满水保护阀
anti-flood interceptor 防泛滥截流管
anti-flood measure 防淹措施
anti-flood platform 防淹平台
anti-flood valve 防洪阀
anti-flood wall 防洪墙
anti-fluctuator 稳压器;缓冲器
anti-fluorite 反萤石
anti-fluorite structure 反萤石(型)结构
anti-fluting mill 防折轧机
anti-flux 抗熔渣;抗焊媒;高温熔剂;防溶剂
anti-fly glass 防蝇玻璃
anti-foam 抗起沫剂;防沫剂;消沫剂;阻沫剂
anti-foam additive 防起泡添加剂;消泡添加剂
anti-foamer 防沫剂;消沫剂
anti-foaming additive 防泡沫(添加)剂;防沫添加剂
anti-foaming agent 防泡(沫)剂;防沫剂;消沫剂;抗泡沫剂;除沫剂
anti-foaming aid 防沫添加剂
anti-foam package 破沫装置

anti-foam plate 破沫板
anti-fog 防灰雪
anti-fog antismog 消除烟雾设施
anti-fog device 雾中航行设备
anti-foggant 灰雾抑制剂;防雾剂;防灰雾剂
anti-fogging 防雾
anti-fogging agent 防雾剂;防结露剂;防灰雾剂
anti-fogging coating 防雾涂层;防结露涂层
anti-fogging compound 防雾剂
anti-fogging resistance 防结露性
anti-fog insulator 抗雾绝缘子;防雾绝缘子;防雾绝缘器
antifooding interceptor 止回阀
anti-forgery ink 防伪油墨
antiform 背形;背斜型构造;背斜形态;反式
antiformal 背形的
antiformal anticline 背形背斜
antiformal stack duplex 背形叠加双冲构造
antiformal syncline 背形向斜
antiformal trap 背形圈闭
antiform structure 背斜型构造
anti-foulant 船底防污剂;防污剂;防污底漆
anti-foulant additive 防污添加剂
anti-foulant agent 防垢剂
anti-foulant formulation 防垢剂成分
anti-foulant treatment 防垢处理
anti-fouling 防污着;防污塞;防污;防塞
anti-fouling agent 防臭剂;防污着剂;防污剂;防垢剂;防腐臭剂
anti-fouling bottom paint 船底防污漆
anti-fouling coating 船底防污涂料;防污涂层;防污漆;防污底涂层
anti-fouling composition 防污剂;除臭剂;船底防污组分;船底防污漆;船底防污漆
anti-fouling compound 船底防污剂;防污剂;防臭剂
anti-fouling electrolyze sea water 电解海水防污法
anti-fouling hawse 防止链纠缠
anti-fouling marine coating 船用防污涂层
anti-fouling motor 防污电动机
anti-fouling paint 船底防污漆;防藻漆;防污油漆;防污涂料;防污染涂料;防污漆;防垢油漆;防腐油漆;防腐涂料;防腐漆;防虫漆
anti-fouling property 抗附着性能
anti-fouling rubber 防污橡胶
anti-fouling technology 防污着技术
anti-fraying 抗磨损
anti-freeze 阻凝(剂);抗凝(剂);防冻液;不冻液;不冻剂;防冻的;阻冻剂
anti-freeze additive 防冻添加剂
anti-freeze admixture 防冻外加剂;防冻添加剂;防冻附加剂
anti-freeze agent 抗冻剂;防冻剂
anti-freeze compound 抗冻剂;防冻剂
anti-freeze extinguisher 防冻灭火器
anti-freeze fluid 防冻液
anti-freeze irrigation 防冻灌溉
anti-freeze layer 防冻层
anti-freeze liquid 防冻液
anti-freeze mixture 防冻混合物;防冻混合剂;防冻合剂
anti-freeze operation 防冻作业
anti-freeze paint 防冻漆
anti-freeze pin 防黏螺栓
anti-freeze powder 防冻粉末
anti-freezer 抗冻剂;防冻剂
anti-freezer for gasholder 储气罐防冻装置
anti-freeze solution 防冻液;防冻溶液;阻冻溶液
anti-freeze substance 防冻物质
anti-freezing 防冻(的)
anti-freezing action 防冻作用
anti-freezing admixture 防冻外加剂;防冻掺和物
anti-freezing agent 防冻剂;防冻掺和剂;阻冻剂
anti-freezing agent for concrete 混凝土防冻剂
anti-freezing aid 防冻添加剂
anti-freezing brick 防冻黏土砖
anti-freezing case 防冻箱
anti-freezing clay brick 防冻黏土砖;抗冻黏土砖
anti-freezing coat 防冻层;防冻涂层;防冻面层
anti-freezing compound 防冻涂料;防冻混合剂
anti-freezing cover 防冻覆盖物;保温罩
anti-freezing extinguisher 防冻灭火器
anti-freezing fluid 防冻液
anti-freezing hydrant 防冻消火栓;不冻防火栓
anti-freezing liquid 防冻液

anti-freezing lubricant 防冻润滑剂
anti-freezing material 防冻材料
anti-freezing measure 防冻措施
anti-freezing mixture 防冻剂;防冻膏;防冻掺和料
anti-freezing oil 防冻油
anti-freezing painting 阻冻涂料
anti-freezing process 防冻法
anti-freezing protection 低温防护
anti-freezing solution 防冻溶液
anti-fretting compound 防腐蚀剂
anti-friction 抗摩阻;抗摩(擦);耐磨;防摩擦
anti-frictional 抗摩擦力的;抗摩擦的;减轻摩擦的
anti-frictional bearing 抗磨轴承;抗摩轴承
anti-frictional cast iron 抗摩铸铁
anti-frictional composition 抗摩制品
anti-friction alloy 抗磨合金;减摩合金;耐磨合金
anti-friction axle pulley 减摩轴滑轮
anti-friction ball bearing 减摩滚珠轴承
anti-friction bearing 滚珠轴承;减摩轴承;耐磨轴承;滚动轴承
anti-friction bearing grease 减摩轴承润滑脂
anti-friction bearing pillow 减摩轴承垫座
anti-friction block 减摩滑轮
anti-friction bolt 减摩螺栓
anti-friction box 减摩轴箱
anti-friction brick 耐磨砖
anti-friction coat 减摩涂层
anti-friction coating 增滑涂层
anti-friction composition 减摩制品;减摩剂
anti-friction grease 减摩脂;减摩润滑脂;轴承润滑脂
anti-friction latch bolt 减摩插销
anti-friction material 润滑油;润滑剂;减摩材料;耐磨材料;耐磨材料
anti-friction metal 减摩金属;耐磨金属
anti-friction pivot 减摩枢
anti-friction plate 耐磨板
anti-friction property 抗磨性;减摩性;耐磨性;增滑性
anti-friction reducing roller 减摩滚柱
anti-friction ring 减摩圈
anti-friction ring and holder 减摩环及托
anti-friction ring holder screw 减摩环托螺钉
anti-friction roller 减摩滚柱
anti-friction slide 减摩溜板
anti-friction (of tapered bearing) 锥形滚柱轴承的减摩
anti-friction thrust bearing 减摩推力轴承
anti-friction wheel 减摩轮
anti-friction worm conveyer 抗磨蜗杆输送机;抗磨螺旋运输机;抗磨螺旋输送机;抗磨蜗杆运输机
anti-frost 防冻(的);防冻(冻)的
anti-frost additive 防冻添加剂;防霜添加剂
anti-frost additive for concrete 混凝土防冻添加剂
anti-frost device 防霜冻装置
anti-frosting agent 防霜花剂
anti-frost layer 抗冻层;防冻层;防霜冻层
anti-frost smoke projector 防霜冻烟雾喷射器
anti-frother 防起泡添加剂;防起泡剂
anti-fuei 抗爆燃料
anti-fungal property 抗霉性能
anti-fungal property of malthoid 油毡防霉性
anti-fungicide paper 防霉纸
anti-fungin 硼酸镁
anti-fungus 防霉油漆
antigalling compound 丝扣(润滑)油
anti-galvanic paint 防电解(油)漆
anti-gas 防毒气
anti-gas defence 防毒
anti-gas kit 防毒气装置
anti-gas mask 防毒面具
anti-gassing agent 防放气剂
anti-gelling agent 防胶凝剂;防胶化剂
anti-gen 抗原
anti-genicity 抗原性
anti-glare 防眩的;遮光
anti-glare coating 防耀眼涂层;防眩涂层;防反光膜
anti-glare device 防眩设施;防眩装置
anti-glare fence 防眩栅栏;防眩屏
anti-glare lamp 防眩灯
anti-glare paint 无光漆;防眩涂料
anti-glare planting 防眩栽植
anti-glare screen 防眩屏
anti-glare shading 遮光屏;遮光帘;防眩屏;防眩帘
anti-glare shield 防眩挡板;防眩板

anti-glow 对日照;反日照
anti-gold policies 反黄金政策
antigorite 蛇纹岩;叶蛇纹石
anti-gradient 逆梯度
anti-graffiti paint 抗涂划漆
antigraphy 契据副本
anti-gravitation 防重力;反重力;抗重力;耐重力
anti-gravity 抗重力;反重力;反引力
anti-gravity device 抗重力装置;防过载装置
anti-gravity filtration 抗重力过滤;抗重过滤
anti-gravity machine 抗重力机;反重力机
anti-gravity screen 抗重力筛(筛料自下而上筛过);抗重力筛;反重力式筛分机
anti-gravity valve 抗重力阀
anti-ground 消除地面影响
anti-ground noise 消除背景噪声
anti-growth movement 反增长运动
Antigua 安提瓜岛
anti-gum inhibitor 防胶剂
anti-g valve 防超重活门;抗重力阀
anti-hadron 反强子
anti-halation 抗光晕层;消晕作用;防光晕;反光晕的
anti-halation backing 背面防光晕层
anti-halation backing film 防光晕背层胶片
anti-halation layer 防光晕层
anti-halation material 防光晕材料
anti-halation protection 防光晕层
anti-halation undercoat 防光晕底层
anti-halo 防光晕
anti-hard caking agent 防结(硬)块剂
anti-hazard classification 防爆等级
anti-hazing agent 防潮剂;防白剂
anti-heave measure 防胀措施
anti-helium 反氦
anti-Hermitean operator 反厄密算符
anti-hermition function 反厄米函数
anti-homomorphism 反同态
anti-horse-fly oil 防马蝇油
antihum 抗哼声;哼声抑制器;哼声消除器;静噪器;交流声消除;消声的
antihum capacitor 消除噪声电容器
antihum condenser 去噪声电容器;声抑制电容器
anti-hunt 反振荡;反搜索
anti-hunt action 阻尼作用
anti-hunt circuit 防振电路;阻尼电路
anti-hunt device 防振器件
anti-hunting 防止摆动;防摆动阻尼
anti-hunting protection 防震保护装置
anti-hunt means 防振设备;稳定器;稳定方法;阻尼器
anti-hydro 憎水剂;疏水剂;憎水
anti-hygienic 不卫生的
anti-hyperon 反超子
anti-hypo 高碳酸钾
anti-ice measures 防冰措施
anti-ice paint 防冰漆
anti-icer 防冰装置;防冰设备;防冰器
anti-icing 防冰
anti-icing additive 防冻添加剂;防冰添加剂
anti-icing admix 防冻掺和剂
anti-icing duct 防冰系统导管
anti-icing equipment 防冰装置
anti-icing fluid 防冻液
anti-icing gasoline 防冻汽油
anti-icing system 防冰系统
anti-incrustation 防垢
anti-incrustator 锅炉防垢剂
anti-induction 防感应(作用)
anti-industry 反工业
anti-infection apparel 防感染服
anti-inference method 反算法
anti-inflation policy 反通货膨胀政策
anti-injunction act 反禁止令法
anti-interference 抗干扰;反阻塞干扰
anti-interference antenna 抗干扰天线
anti-interference circuit 反干扰电路
anti-interference equipment 抗干扰设备
anti-interference filter 抗干扰滤波器
anti-interference for power grid 抗电网干扰
anti-interference installation 抗干扰装置;抗干扰设备
anti-interference measure 抗干扰措施
anti-interference technique 反干扰技术
anti-intrusion device 防渗透装置

anti-inversion 逆反演
anti-ionization 反离子作用
anti-isobar 反同量异位素
anti-isomorphism 反同形性;反同构
anti-jackknife attachment (铰接式列车、汽车的)防折褶装置;防褶皱装置
anti-jam 抗干扰
anti-jamming 抗干扰;防咬缸
anti-jamming blanking 抗干扰遮隐
anti-jamming unit 抗干扰装置
anti-jump baffle 防跃挡板
anti-kickback attachment 防回弹装置;防反向安全装置;反冲安全装置
anti-kinesis 逆向运行
anti-klystron 反速调管
anti-knock 汽油抗爆剂;抗震(的);抗爆剂;抗爆
anti-knock additive 抗爆添加剂
anti-knock agent 防爆剂;抗爆剂
anti-knock blending value 抗爆掺和值
anti-knock block 防震挡块
anti-knock component 抗爆组分;高辛烷值成分
anti-knock compound 抗震油剂;抗爆剂;抗爆化合物
anti-knock device 防爆设备
anti-knock dope 抗爆添加剂
anti-knock dope for diesel fuel 柴油抗爆添加剂
anti-knock fuel 抗爆燃料
anti-knock gasoline 抗爆汽油;高辛烷值燃料
anti-knock performance 抗爆性能;防爆性能
anti-knock petrol 抗爆汽油;高辛烷值燃料
anti-knock petrol additive 抗爆汽油添加剂
anti-knock quality 抗爆性(质);抗爆燃性
anti-knock rating 抗爆值;抗爆率
anti-knock substance 防爆剂
anti-knock test 抗爆试验
anti-knock value 抗震值;抗爆值
anti-laser beam coating 抗激光束涂层
anti-laser goggles 激光防护目镜
anti-laser protection 反激光防护
anti-leakage factor 阻越流系数
anti-leak cement 防漏水泥
anti-leak ring 密封环;防漏环
antileprol 大风子油酸乙酯
anti-lift rope 固定索
anti-lift wire 固定索
anti-linear 反线性
anti-linear mapping 反线性映射
anti-linear operator 反线性算符
anti-lithic 去垢剂;抗石的
anti-livering agent 防胶化剂
Antillean bivalve subprovince 安替列双壳类地理亚区
Antillean marine trough 安替列海槽
Antilles current 安德列斯海流
anti-loga 反性曲线
anti-logarithm 反对数
anti-log(arithmic) amplifier 反对数放大器
anti-log(arithmic) circuit 反对数电路
anti-log(arithmic) multiplier 反对数乘法器
anti-log(arithmic) voltage converter 反对数电压转换器
antilogous pole 热负极
anti-loosing washer 防松垫圈;防松垫片
anti-lowering device 防下降装置;防松吊装置;防减压装置
anti-lubricant 防滑剂
anti-lubricating agent 防润滑剂
anti-macassar 防污方巾;椅背硬套
anti-magnet 防磁
anti-magnetic 抗磁(性)的;防磁(性)的
anti-magnetic bearing 防磁轴承
anti-magnetic glass 逆磁玻璃
anti-magnetic iron 防磁铁
anti-magnetic shield 防磁屏蔽;防磁屏蔽
anti-magnetic shift 反磁性位移
anti-magnetized coil 消磁线圈
anti-matter 反物质
antimer 反映体
antimere 体辐
anti-meridian 逆子午线;反面子午线
anti-metabolite 抗代谢物
antimetric 非对讲的
anti-microbial agent 抗微生物剂;抗菌剂;防细菌生长剂

anti-microbial substance 抗微生物物质
anti-microbial wastewater 抗微生物废水;抗菌废水
anti-microbic substance 抗微生物质
anti-microbin 灭菌质
anti-microphonic 抗噪声的;反颤噪声的
anti-migrant superclear 防泳移剂
anti-migration 反式迁移作用
anti-migration shield 防蠕爬挡板
anti-mildew agent 防霉(菌)剂
anti-mist cloth 防雾织物
anti-mode 反众数
anti-moist 防潮湿的
anti-moist battery 防潮电池
anti-molecule 反分子
antimonate 锑酸盐
antimonial glass 锑玻璃
antimonial lead 锑铅合金;含锑铅;硬铅;锑铅
antimonial lead pipe 硬铅管;锑铅合金管
antimonial nickel 锑镍矿;红锑镍矿
antimonial silver 含锑银
antimonial tin solder 锑锡焊条
antimonian rustenburgite 含锑等轴锡铂矿
antimoniate of lead 锑酸铅
antimonic fluoride 氟化锑
antimonic oxide 氧化锑
antimonide 锑化物
antimonite 亚锑酸盐;辉锑矿
anti-monopoly law 反垄断法
anti-monotone 反单调
antimonous fluoride 氟化亚锑
antimonpearceite 锑硫砷铜银矿
anti-Monson curve 反莫森曲线
anti-monsoon 反季(节)风
antimony acetate 醋酸锑
antimony alloy 锑合金
antimony-bearing formation 含锑建造
antimony black 锑黑
antimony blende 红锑矿
antimony bloom 锑华
antimony bronze 锑青铜
antimony chloride paint 氯化锑涂料
antimony containing alloy 锑基合金
antimony detector 锑检波器
antimony-electrode 锑电极
antimony flint glass 锑火石玻璃
antimony glance 辉锑矿;辉锑矿
antimony glass 锑玻璃
antimony halides 卤化锑
antimony hydride 氢化锑
antimony iode-sulphide type structure 碘硫化锑型结构
antimony lead 锑铅
antimony orange 锑橘黄;锑橙
antimony ore 锑矿(石)
antimony oxide 锑白;氧化锑
antimony oxide-coated silica 包覆氧化锑的二氧化硅
antimony oxysulfide 氧硫化锑
antimony pentachloride 五氯化锑
antimony pentasulfide 五硫化二锑
antimony pigment 锑白颜料
antimony pneumoconiosis 锑尘肺
antimony poisoning 锑中毒
antimony potassium oxalate 草酸氧锑钾
antimony red 五硫化二锑;锑红
antimony regulus 锑块;炼锑
antimony ruby glass 锑红玻璃
antimony slab 锑锭
antimony sodiate 锑酸钠
antimony spot 锑斑
antimony sulfide 硫化锑
antimony sulphide 硫化锑
antimony sulphide ore 硫化锑砂
antimony tetroxide 四氯化二锑
antimony trichloride 三氯化锑(堵塞岩石空隙用的化学药剂)
antimony trioxide 三氧化锑;三氧化二锑
antimony trisulfide 硫化锑
antimony vermilion 锑朱
antimony white 锑白
antimony yellow 锑黄
anti-morph 反形体;反效等位基因
anti-mosquito oil 防蚊油
anti-moth agent 防蛀剂

anti-motoring device 防止倒拖装置
anti-multiplet 反多重态
anti-mycin 抗菌素
anti-neutrino 反中微子
anti-neutrino beam 反中微子束
anti-neutrino helicity 反中微子螺旋性
anti-neutrino spectrum 反中微子能谱
anti-neutron 反中子
anti-niacin 抗烟酸
anti-nicotinic acid 抗烟酸
anti-nodal plane 反节面
anti-nodal point 腹点;反交点;波腹点
anti-node 对交点;腹点;反波节;波腹
anti-noise 抗噪声;防止噪声;反噪声
anti-noise measure 防噪声措施
anti-noise microphone 抗噪声传声器;防噪声送话器
anti-noise paint 吸音油漆;吸音涂料;吸音油漆;吸音涂料;消声漆;隔音漆
anti-noise screen 防噪声屏
anti-nose-dive leg 箱体支架
anti-nucleon 反核子;反方位
anti-nucleus 反原子核;反核
Antioch process 安蒂奥奇方法
anti-offset 防粘脏的
anti-oil pollutant 抗油污染物
anti-oscillation device 止摆装置
anti-oscillation mounting 防振支座
anti-osmosis purification 反渗透净水
anti-overload device 防过载装置;防超载装置
anti-overloading 防止过载;防过载的
anti-overloading amplifier 防过载放大器
anti-overloading performance 抗超载特性
anti-overshoot 反过冲
anti-overturning 抗倾
anti-oxidant 抗氧(化)剂;抗老化剂;防氧化剂;防老(化)剂
anti-oxidant additive 抗氧添加剂;防氧化添加剂
anti-oxidant flux 防氧焊剂
anti-oxidant for lube oil 润滑油抗老化剂
anti-oxidation 抗氧化作用;抗氧化性
anti-oxidation coating 抗氧化涂层;防氧化层
anti-oxidizing agent 抗氧化剂
anti-oxygen 抗氧剂;抗氧化剂
anti-oxygenic activity 抗氧活性
anti-ozonant 抗臭氧剂
anti-ozone compound 抗臭氧化合物
anti-packing chemical 除积垢剂
anti-pairing effect 反成对效应
antiparabema 拜占庭教堂入口处附属教堂
anti-parallax 反视差
anti-parallax mirror 防视差镜;反视差镜
anti-parallel 逆平行;反平行的;反并联
anti-parallel coupling 反并联接法
anti-parallel crank 不平行曲柄
anti-parallel domain 逆平行畴;反平行畴
anti-parasitic 抗寄生物剂
anti-parasitic resistor 寄生振荡抑制电阻器
anti-particle 反质点;反粒子
anti-pendulum 防摆动的
anti-percolator 防渗装置
anti-perforation layer 抗穿孔层
anti-periodic 反周期
anti-periodic(al) function 反周期函数
anti-period style 无时代性形式;反时代性形式
anti-periplanar 反迫
anti-peristaltic 逆蠕动的
anti-peristaltic wave 逆蠕动波
anti-permeability 抗渗透性
anti-peroxide additive 抗过氧化物添加剂
anti-perspirant 防汗剂
anti-perthite 反纹长石;反条纹长石
anti-petalous 对瓣的
anti-phase 逆相;反相(位)
anti-phase boundaries 反相界面
anti-phase boundary 反相边界
anti-phase dipole 反相偶极子
anti-phase domain 反相畴
anti-phase effect 反相位效应
anti-phase nuclei 反相核
anti-phase region 反相区
anti-phase relay 反相继电器
anti-phase splitting 反相分裂
anti-phlogistic theory 反燃素学说

anti-phone 防音器
anti-pinking fuel 抗爆燃料；高辛烷值汽油
anti-piping compound 冒口防缩剂；防缩（孔）剂；缩孔防止剂
anti-piracy guard 护航队；船上公安队
anti-pirating agreement 反挖墙脚协议
anti-pitching device 减纵摇装置
anti-pitching fin 抗纵摇鳍
anti-pitting agent 麻面防止剂；防点蚀添加剂
anti-plague 抗鼠疫的
anti-plane shear mode 3 crack 反平面剪切Ⅲ型裂纹
anti-plane strain 非平面变形
anti-plasma 反等离子体
anti-plastering agent 阻黏剂
anti-plastering aid 抗黏剂
anti-plasticization 反增塑作用
anti-plasticizer 反增塑剂
anti-pleion 负偏差中心；欠准区
antiplug relay 电流跃变平缓继电器
anti-pode 对跖点
anti-podal 对映体；正反对
anti-podean day 跨界日（增或减一日）；过日界线日期
antipoisoning 消毒
anti-poison respirator 防毒口罩
anti-polarizing winding 反极化绕组
anti-polar phase 反极性相
anti-pole 相反极；相对极
anti-pole diagram 反极图
anti-pole effect 反极点效应
anti-pollutant 抗污染剂；防污染剂
anti-pollution 抗污染（作用）；防污染（作用）；反污染（作用）
anti-pollution act 防污染法令
anti-pollution barrier 防污染栅栏
anti-pollution campaign 防污染运动
anti-pollution cost estimates 反污染费用估计值
anti-pollution device 抗污染装置；防污染装置
anti-pollution heat treatment 无公害热处理
anti-pollution incentives 抗污染物质的刺激；反污染的物质鼓励
anti-pollution industry 反污染工业
anti-pollution investment 消除污染投资
anti-pollutionist 反污染者
anti-pollution law 防污染法令
anti-pollution legislation 防污染立法
anti-pollution loan 消除污染贷款
anti-pollution measure 防污染措施
anti-pollution plant 防污染装置
anti-pollution plantation 防污染绿化；防污绿化
anti-pollution plantation in urban industry districts 城市工矿区防污绿化
anti-pollution standard 防污染标准
anti-pollution system 防污系统；防污染系统
anti-pollution transfer and storage system 防污染转移与储[贮]存系统；防污染传递与储[贮]存系统
anti-pollution tree species 抗污染树种
anti-pollution type insulator 防污型绝缘子
anti-pollution unit 防污染装置
anti-pollution vessel 反污染船
anti-polymerizer 反聚剂
anti-port 反向运输
anti-position 反位
anti-precipitant 阻垢剂；抗氧化剂；反沉淀剂
anti-precipitin 反沉淀素
anti-pressure diving 抗压潜水
anti-preventing paint 防锈涂料
anti-priming 止水装置
anti-priming pipe 汽水隔离管；多孔管；防止汽水共腾管；筛孔管
anti-principal plane 负主平面
anti-principal point 负主点；反主点
anti-profiteering ordinance 取缔暴利法令
anti-profiteering tax 暴利税
anti-propagation 反向传播
anti-proton 反质子
anti-proton accumulator ring 反质子累积环
anti-protonic atom 反质子原子
anti-proton injection 反质子注入
anti-pump device 重合闸闭锁装置
anti-putrefactive 防腐的
antipyrine 氨替比林
antiquarian 大幅图纸；大裁画纸

antiquarium 古物室
anti-quark 反夸克
antiquated 已废弃的；过时的；陈旧的
anti-quated idea 过时观念
anti-quated sand 废砂
antique 素压凸印；古物；古玩；古董
antique and curio dealer 古玩铺
antique-and-curio shelves 博古架
antique architecture 古风建筑
antique art 古代艺术
antique bevel 古式斜边（小于7.5°）
antique brass 古铜色
antique brick 古砖；古式砖
antique bronze 古青铜色
antique bronze colo(u)r 古铜色
antique brown 古褐色
antique building 古老房屋；古建筑；古老建筑；古风建筑
antique character 古老的特点；古式特征
antique colo(u)r 古彩
antique copper and bronze ware 古铜器
antique-copper iron square hinge loose pin 古铜色轴心铁方铰链
antique crown （散射光芒的）头冠纹章；古冠；东方冠；古皇冠
antique drawn glass 古法拉制玻璃
antique effect 古色古香效果
antique famille rose enterprise 古粉彩业
antique finish 仿古处理（家具）；仿古装饰；仿古加工
antique finished carpet 仿古地毯
antique finished ivory carving 仿古牙雕
antique finished wood carving 仿古木雕
antique finishing 仿古髹饰法；仿古涂装（法）
antique flat glass 古法平板玻璃
antique glass 中世纪式的玻璃（有色玻璃窗）；古式（彩色）玻璃；古镜；古玻璃
antique gold 古金色
antique lace 粗线亚麻梭结花边
antique making 仿古制作
antique mirror 古式镜；古色镜；古镜
antique motif 古式的主题；古老的主题
antique ornament 古老装饰品
antique paper 仿古纸
antique reproduction 仿古玩
antique rug 仿古地毯
antique satin 仿古缎；悬挂织物；毛织物
antique shelf 博古架
antique shop 古玩店；古玩商店
antique site 古迹
antique structure 古建筑；古老结构
antique temple 古寺庙；古神殿；古庙
antiquing 仿古漆法；打光；仿古涂装；仿古装修；仿古处理；打光
antiquing finish 古彩涂饰剂
antiquing paint 仿古漆
antiquity 古代习俗；古迹；苍古；古物；古代
antiquum opus 乱石墙；毛石墙
anti-rabic 防狂犬病药
anti-racer 防空转装置
anti-radar coating 防雷达涂层
anti-radar paint 防雷达涂料；防雷达漆
anti-radiation 抗辐射；防辐射；反辐射
anti-radiation coating 防辐射涂层
anti-radiation effect 抗辐射效应
anti-radiation material 防辐射材料
anti-radioactive contamination 抗放射性污染性
anti-rads 抗射线（老化）剂
anti-rattle clip 防振线夹
anti-rattler 减声器；消声器；防震器
anti-rattler spring 防震弹簧
anti-reaction 回授消除
anti-reaction coil 反再生线圈
anti-recession program(me) 反经济衰退政策；反衰退计划
anti-reciprocal circuit 反可逆电路
anti-redeposition agent 抗再沉积剂
anti-reflecting coating 消反射敷设；增透涂层
anti-reflecting film 减反射膜；消反射膜
anti-reflection 减反射
anti-reflection coated 防反射膜的
anti-reflection coating 抗反射敷层；抗反射涂层；抗反射层；减反射敷设；增透膜；防反射膜
anti-reflection diaphragm 拦杂光光栏
anti-reflection film 抗反射涂膜；抗反射涂层；增透

膜；防反射膜；防反射胶片
anti-reflection glass 防反射玻璃
anti-reflection layer 减反射层
anti-reflective 减反射
anti-reflective coating 增透膜层；抗反射敷层；增透涂层
anti-reflective film 减反射膜
anti-reflexion 减反射
anti-reflexive relation 非自反关系
anti-regular 反正则的
anti-representation 反表示
anti-resistant paint 防腐蚀涂料
anti-resonance 电流谐振；防共振；防共鸣；谐振；反共振；并联谐振
anti-resonance frequency 反谐振频率；反共振频率；并联振频率
anti-resonant circuit 并联谐振电路
anti-reverse-rotation device 防逆转装置
anti-ricochet device 防回跳装置
antiripple 不对称小波痕
anti-rivelling paint 防起条纹漆
anti-rodent 防噬齿动物
anti-roll 防滚
anti-roll bar 抗侧倾杆；车体角位移横向平衡杆
anti-roll device 转向架防侧滚装置
anti-rolling 抗横摇；防滚动
anti-rolling apparatus 减摇装置
anti-rolling device 减摇装置
anti-rolling fin 抗横摇鳍
anti-rolling guy 防滚索
anti-rolling gyroscope 抗横摇陀螺仪；减摇陀螺仪
anti-rolling stabilizer 抗横摇装置；减摇装置
anti-rolling tank 抗横摇水舱；稳定水舱；抗摆水舱；减摇液体舱；减摇液体舱；消摇液体舱
anti-roll pump 减摇泵
anti-roll stabilizer 减横摇装置
anti-root 反根【数】
anti-Rossi circuit 分离线路；分离电路
anti-rot 防腐的
anti-rotating rope 抗转绳
antirotation door control decoupling method 反转门控去耦法
anti-rotation resume method 反转恢复法
anti-rot material 防腐材料
anti-rot solution 防腐液
anti-rot substance 防腐剂；防腐材料
anti-rumble 防闹；防嘈杂；减声器；防噪声
anti-rumble room 防噪声室
anti-running agent 抗流动剂；防流挂剂
anti-rust 耐腐蚀的；耐锈的；防锈的
anti-rust action 防锈作用
anti-rust coat(ing) 防锈涂层；防锈层
anti-rust composition 防锈组分；防锈配方；防锈混合漆；防锈混合剂；防锈配料
anti-rust compound 防锈剂；防锈混合漆；防锈混合剂
anti-rust enamel 防锈珐琅；防锈瓷釉；防锈搪瓷
anti-rust grease 防锈油脂；防锈润滑脂；防锈脂
anti-rust hard gloss paint 防锈硬质有光漆
anti-rusting agent 防锈剂
anti-rusting coat 防锈面层
anti-rusting flat steel 防锈扁钢
anti-rusting paint 防锈颜料；防锈漆
anti-rusting pigment 防锈颜料
anti-rust metal coat(ing) 金属防锈涂层
anti-rust oil 防锈油
anti-rust oil paint 防锈油漆
anti-rust paint 防锈漆；防锈油漆；防锈涂料
anti-rust paper 防锈纸
anti-rust protection 防锈保护
anti-rust solution 防锈溶液
anti-rust steel 耐锈蚀钢
anti-rust treatment 防锈处理
anti-rust varnish 防锈清漆
anti-sag agent 防流挂剂
anti-sag bar 防垂（吊）杆；支垂杆；桁架中心抗垂杆件；桁架顶（脊）竖直杆
anti-sagging agent 抗流动剂；防流挂剂
anti-salinity intrusion lock 防咸船闸
anti-saprobic 防腐生的
anti-saprobic zone 防污染带；防腐生带
anti-saturated logic(al) circuit 抗饱和逻辑电路
anti-saturation 抗饱和
anti-saturation amplifier 防饱和放大器

anti-scalant 防垢剂;阻垢剂
anti-scald device 防烫伤设备
anti-scale 除垢剂;防垢
anti-scale boiler fluid 锅炉防垢液
anti-scaling 防垢
anti-scaling composition 防垢剂
anti-science 反科学
anti-scientism 反科学主义
anti-scorching 抗焦作用
anti-scorching agent 抗焦化剂
anti-scour 抗冲刷;消能的;防冲刷的
anti-scour apron 抗冲刷护坦
anti-scour lip 防冲槛;消力槛
anti-scour prevention 防冲槛
anti-scour prism 消能棱体
anti-scour protection 防冲护坦;防冲措施;防冲层
anti-scour sill 防冲槛;消力槛
anti-scour wall 防冲墙
anti-scratch 抗划痕;抗划道
anti-scuff agent 抗磨剂
anti-scuff coating 防擦伤涂层
anti-scuffing paste 抛光膏
anti-sediment agent 防沉剂
anti-seep(age) 防渗漏;防渗
anti-seepage cofferdam 抗渗围堰
anti-seepage diaphragm 截水环
anti-seepage mark number 抗渗标号
anti-seep collar 防渗环
anti-seep diaphragm 防渗环;防渗层;截水环;刚性防渗心墙;防渗心墙
anti-seismic 防震的;抗震的;抗地震的
anti-seismic calculation 抗震计算
anti-seismic computation 抗震计算
anti-seismic construction 抗震构造
anti-seismic design 抗震设计;防震设计
anti-seismic engineering 抗震工程
anti-seismic joint 抗震缝;防震缝
anti-seismic strength 抗震强度
anti-seismic structure 抗震结构;抗震建筑;防震建筑(物)
anti-seize 防黏;防卡塞
anti-seize compound 抗扯裂化合物;防黏剂;防凝添加剂
anti-seize lubricant (螺纹结合部分的)防止过热(卡孔)润滑剂
anti-seizing property 抗黏性
anti-selene 反假月;幻月
anti-sense strand 反意义链
anti-sensitizer 减感剂
anti-sepsis 防腐法
antiseptic(al) 防腐剂;防腐的;防腐药
antiseptic agent 防腐败剂;抗菌剂
antiseptic concrete 防腐混凝土
antiseptic effect 防腐作用;防腐效应
antiseptic gauze 消毒纱布
antiseptic oil 防腐油
antiseptic plaster 防腐膏
antiseptic room 消毒房间;无菌房间
antiseptics 防腐剂;防腐材料
antiseptic substance 防腐剂
antiseptic treatment 防腐处理
antiseptic varnish 防腐清漆
antiseptic wash 防腐洗涤剂
antiseptic wooden sleeper 防腐枕木
anti setoff 防黏脏的
anti setoff powder 吸墨粉
anti-setting 防固着的
anti-settling 防沉
anti-settling agent 抗沉降剂;防沉降剂;防沉剂;防沉淀剂
anti-settling property 悬浮性;防沉淀性
anti-shielding effect 反屏蔽效应
anti-shipping activity messages 反航运活动电文
anti-shock 防震
anti-shock gasoline 抗爆汽油
anti-shock mounting 防冲装置
anti-shock substance 抗爆剂
anti-shoe rattler 闸瓦减震减声器;闸瓦减振器
anti-shrink 抗缩的;耐缩的;防缩的
anti-shrinking medium 防收缩剂
anti-shrinking mortar pad 抗缩砂浆垫层
anti-shunt field 反旁路场;反分流场
anti-sidetone circuit 消侧音电路
anti-silking agent 防丝纹剂

anti-siltation installation 防淤设施
anti-sine 反正弦
anti-singing device 振鸣抑制装置
anti-sing lamp 无声白炽灯
anti-siphonage 反呼吸作用
anti-siphon(age) device 防虹吸装置
anti-siphon(age) pipe 反虹吸管;防虹吸管
anti-siphon(age) valve 防虹吸阀
anti-siphon ball cock 防虹吸球形阀
anti-siphoning device 防虹吸装置
anti-siphoning pipe 防虹吸管
anti-siphon trap 防虹吸存水弯
anti-skid 抗滑(的);防滑(的)
anti-skid brake 防抱死制动器
anti-skid braking device 防滑制动装置
anti-skid chain 防滑链
anti-skid device 防滑装置
anti-skid drive 防滑传动
anti-skid factor 防滑系数;抗滑系数
anti-skid finish 防滑面层
anti-skid floor 防滑地面;防滑地板
anti-skid groove 防滑花纹沟
anti-skid heap (厂矿道路的)防滑堆
anti-skiding 防滑的
anti-skid material 防滑材料
anti-skid paint 防滑涂料;防滑漆
anti-skid pile 防滑桩
anti-skid plate 防滑板
anti-skid property 防滑性能
anti-skid quality 防滑特性
anti-skid rib(bed)tile 防滑条纹地砖;防滑花纹瓷砖;防滑(肋)板
anti-skid road surface treatment 路面防滑处治;路面防滑处理
anti-skid tread 防滑轮胎纹;防滑轮胎路面;防滑梯级
anti-skid treatment 防滑处理
anti-skid tyre 防滑轮胎
anti-skid unit 防滑器
anti-skinning 防结皮的
anti-skinning agent 防结皮剂
anti-slide pile 抗滑桩;防滑桩
anti-slide rack 防止滑动架
anti-sliding device 防滑设备
anti-sliding key 防滑齿坎
anti-slip 抗滑;防滑(转)
anti-slip agent 增磨剂;防滑剂
anti-slip aggregate 防滑集料;防滑骨料
anti-slip device 防空转装置;防滑装置
anti-slip finish 防滑装修;防滑饰面;防滑面层
anti-slip metal 防滑金属
anti-slip paint 防滑漆
anti-slip pile 抗滑桩
anti-slip retaining wall 抗滑挡土墙
anti-slip rib(bed)tile 防滑花纹瓷砖
anti-sludge 抗淤沉;抗淤淀;抗淤积;去垢
anti-sludge additive 抗淤沉添加剂
anti-sludging 去垢的;抗淤沉的;抗淤积的
anti-smog 防烟的
antismoke agent 消烟剂
anti-smudge ring 防污垢环
anti-smuggling 反走私
anti-softener 防软剂
anti-solar 防太阳(光)的
anti-solar glass 吸热玻璃
anti-solar point 对日点;避光点
anti-somorphism 反同形性
anti-sonar coating 防声呐涂层
anti-sound 反声
anti-sound technology 反声技术
anti-spalling agent 抗散列剂;抗剥落剂
anti-spalling brick 抗剥蚀砖
anti-spark 消火花;防火花的
anti-speaker 消声器
anti-spin direction 反螺旋方向
anti-spin engine 防尾旋发动机
anti-spin parachute 反螺旋伞
anti-splash 挡溅;防溅污的
anti-splash device 防溅装置;雨水导板;防溅水龙头;防溅沟道
anti-splash nozzle 防溅喷嘴
anti-spliting bolt 防裂螺栓
anti-spray 防溅的
anti-spray film 隔沫薄膜;隔沫层;防喷沫膜
anti-spray guard 防油渍溅护板

anti-spray guard plate 防油喷溅护板
anti-spurion 反假粒子
anti-squeak 消音器;消声器;减声器;防尖声
anti-stain 防污;防变色
anti-stain agent 防沾污剂
anti-stain treatment 防变色处理
anti-stall 防止熄灭;防止失速
anti-stall gear 防止失速装置
anti-stat 抗静电剂
anti-static 防静电的
anti-static additive 抗静电添加剂;静电添加剂
anti-static agent 抗静电物;抗静电剂;静电剂
anti-static antenna 抗静电干扰天线
anti-static backing 防静电层
anti-static brush 抗静电(毛)刷
anti-static cleaning 防静电清洗
anti-static coating 抗静电涂料
anti-static device 抗静电装置;静电设备
anti-static dirt 防静电吸尘
anti-static dust 防静电吸尘
anti-static effect 抗静电效果
anti-static electricity property 抗静电性能
anti-static fabric 抗静电织物
anti-static fiber 抗静电纤维;抗电纤维
anti-static finish fabric 防静电整理织物
anti-static floor 抗静电地板
anti-static fluid 防静电液;防干扰涂液
anti-static layer 抗静电层
anti-static material 抗静电材料
anti-static plastic plate 防静电胶板
anti-static polishing cloth 抗静电擦布
anti-static prevention 防静电
anti-static resistance 抗静电
anti-static rubber 抗静电橡胶;防静电橡胶
anti-statics 防静电干扰
anti-static shoes 防静电工作鞋
anti-static-shunt 防静电分流器
anti-static tile 抗静电墙地砖
anti-status detector 反状态检测装置
anti-step 反阶步
anti-sticking 抗黏着;防黏的
anti-sticking agent 抗黏剂;防黏着剂;防黏剂
anti-sticking lease coating 防粘涂料
anti-stickoff voltage 反粘电压
anti-stick-slip device 防滑装置
anti-Stokes component 反斯托克斯分量
anti-Stokes displacement 反斯托克斯位移
anti-Stokes fluorescence 反斯托克斯荧光
anti-Stokes frequency 反斯托克斯频率
anti-Stokes law 反斯托克斯定律
anti-Stokes light 反斯托克斯光
anti-Stokes line 反斯托克斯线
anti-Stokes radiation 反斯托克斯辐射
anti-Stokes Raman scattering 反斯托克斯—拉曼散射
anti-Stokes Raman spectroscopy 反斯托克斯—拉曼光谱学
anti-Stokes scattering 反斯托克斯散射
anti-Stokes transition 反斯托克斯跃迁
anti-stone-bumping sealing paint 密封防石击涂料
anti-stop list 非禁用词表
anti-storm glazing 耐风压玻璃
anti-stress 反应力
anti-stress mineral 反应力矿物
anti-strip performance 抗剥落性(能)
anti-strip(ping) 抗剥落;抗剥离(沥青及充填物间的黏结力)
anti-stripping additive 抗散落添加剂;抗剥落添加剂;抗剥落掺和料
anti-stripping agent 抗剥落剂
anti-stripping force 抗剥落力
anti-structure 反结构
anti-structure defect 反结构缺陷
anti-structure disorder 换位无序;反结构无序
anti-submarine 反潜艇;反潜
anti-submarine barrier 反潜屏障
anti-submarine cabinet 反潜舱
anti-submarine detector 反潜探测器
anti-submarine early warning system 反潜预警系统
anti-submarine laser system 反潜激光系统
anti-submarine reconnaissance 反潜搜索
anti-submarine search 反潜搜索
anti-submersible detection indicator 探潜器

anti-subsidy 反津贴税
anti-subsidy duties 反补贴税;反补贴关税
anti-suction valve 负压防止阀
anti-suffocation valve 防窒息活门
anti-suggestion 反暗示
anti-sulfate casing 防硫套管
anti-sun 抗日光;防止日光照射的
anti-sun cantilever roof 遮阳挑檐;抗日光悬臂式屋顶
anti-sun checking 耐晒裂;耐日光晒裂
anti-sun coating 遮日涂层
anti-sun cracking agent 防晒裂剂;防日光龟裂剂
anti-sun glass 遮光玻璃;抗日光玻璃;反光玻璃
anti-sun material 抗日光照射材料;抗光物质
anti-superheating 防止过热
anti-surge mechanism 阻increasingly 扰办法
anti-sway device 减摇装置;防摆装置
anti-sway winch 防摇绞车;防摆绞车
anti-sweat covering 防露罩;防露覆盖层
anti-sweep means 反扫描器材
anti-swing crane 防摆动起重机
anti-swing system 防摆系统
anti-swirl baffle 防涡流挡板
anti-symmetric(al) 反对称的
anti-symmetric(al) anomaly 不对称异常
anti-symmetric(al) bending vibration 反对称弯曲振动
anti-symmetric(al) dyadic 反对称并向量
anti-symmetric(al) function 反对称函数
anti-symmetric(al) ket 反对称刃
anti-symmetric(al) law 反对称律
anti-symmetric(al) load 反对称荷载;反对称载重
anti-symmetric(al) matrix 反对称矩阵;非对称矩阵
anti-symmetric(al) mode 反对称模式
anti-symmetric(al) mode of vibration 不对称振型
anti-symmetric(al) potential 反对称势
anti-symmetric(al) relation 反对称关系
anti-symmetric(al) state 反对称态
anti-symmetric(al) stretching vibration 反对称伸缩振动
anti-symmetric(al) tensor 反对称张量
anti-symmetric(al) tensor operator 反对称张量算子
anti-symmetric(al) vibration 反对称振动
anti-symmetric(al) wave function 反对称波函数
anti-symmetry 非对称(性);反对称(性)
anti-symmetry postulate 反对称性假设
anti-synbiosis effect 反共生效应
antisynchronism 异步性
anti-syphon 防虹吸存水弯
anti-syphonage 反虹吸
anti-syphonage pipe 倒虹吸管;反虹吸管
anti-tack agent 抗黏着剂;抗黏附剂;抗黏剂
anti-tail 逆向彗尾
anti-tank ditch 防坦克壕;反坦克壕
anti-tarnish 防失辉;防晦暗
anti-tarnishing agent 去防锈剂
anti-tarnishing chemical for window 窗户防退色化学品
anti-tarnish paper 防锈纸
anti-template 反模板
anti-termination factor 反终止因子
anti-theft alarm 防盗报警器
anti-theft alarm for vehicle 车辆防盗警报器
anti-theft device for vehicle 车辆防盗装置
anti-thermal stress coefficient 抗热应力系数
antithesis 正相反
antithesis between manual and mental labor 体力劳动和脑力劳动的对立
antithetic 正相反;正反对
antithetic(al) block fault 对偶块状断层;反向块状断层
antithetic dominance 独亲显性
antithetic fault 反向对偶断层;反向断层;断层
antithetic variable 对偶变量
anti-thixotropy 抗摇溶现象;反摇溶现象;抗触变性;反触变性
anti-thrust 止推
anti-thrust bearing 推力轴承
anti-tipping moment 稳定力矩;抗倾覆力矩;反翻倒力矩
anti-tone 反序
anti-tone mapping 反序映射;反序变换
anti-tonic regression 反序回归

anti-torque 反作用转矩;反向转矩
anti-torque moment 抗扭(力)矩
anti-torque pulsation device 防扭振装置
anti-torque rotor 反扭矩旋翼
antitower 防空观测塔;对空监视塔
anti-toxic filtration material 防毒过滤材料
anti-toxic reaction 抗毒反应
anti-toxin 防毒
anti-tracking 反跟踪
anti-tracking lacquer 防炭化漏电漆
anti-tracking varnish 防炭化漏电清漆
anti trades 反信风(带);反季(节)风
anti trade wind 反信风;反季(节)风
anti-transform 反变换
anti-transmit-receive tube 收发管
anti-transmitting receiving switch 反收发开关
anti-transpirant 防散发剂;抗蒸腾
anti-transpiration 防蒸腾;防蒸发;防逸散(作用)
anti-trigonometric function 反三角函数
anti-triptic wind 摩擦风;减速风
anti-trochanter 反转节
antitropous 倒向的
anti-trust act 反托拉斯法
anti-trust law 反托拉斯法;反垄断法;反独占法
anti-trust legislation 反托拉斯立法;反托拉斯法规
anti-trust policy 反托拉斯政策
anti-turbidity system 防混浊系统
anti-tussin 二氟二苯膏
anti-twilight 反辉
anti-twilight arch 反曙暮光弧
antium (古建筑的)南向门柱;门前阶梯;门廊;柱廊
anti-universe 反宇宙
anti-urban 反城市化
anti-urban attitude 反城市状态
anti-vac trap 反虹吸作用排臭阀
anti-vacuum 非真空;反压力
anti-vacuum pipe 反压力导管
anti-vacuum profile 非真空剖面
anti-vermicular 除蠕虫的
anti-verminous 除蠕虫的
anti-vertex 奔离点
anti-vibrating stability 抗震稳定性
anti-vibration 抗振;防振;防颤
anti-vibration clamp 防振线夹;防振夹具
anti-vibration cushioning material 减振防振垫料
anti-vibration damp 减振器;阻尼器;振动阻尼器
anti-vibration device 抗振装置;防振装置
anti-vibration handle 防振手柄;防振拉手
anti-vibration mounting 抗振托架;防振台;抗振装置;防振机座;防振装置;防振台
anti-vibration pad 缓冲垫;防振填料;防(振)衬垫
anti-vibration strength 抗振强度
anti-vibration supports 防振支撑
anti-vibration track slab 防振轨枕板
anti-vibration trench 防振沟
anti-vibrator 阻尼器;减振器;防振器
anti-vignetting effect 防晕映效果
anti-vignetting filter 防晕滤色镜
antivoid valve 真空安全阀
anti-vortex baffle 导流片;导流环;防漩涡板
anti-warp wire 反翘线
antiwarthships bulkhead 横舱壁
anti-water hammer 防水锤
anti-water logging 除涝;防涝;防止木头泡胀
anti-wear 抗磨损的;减磨;耐磨;防磨损的
anti-wear additive 抗磨添加剂;抗磨剂
anti-wear agent 抗磨剂;耐磨剂;防磨剂
anti-wear performance 抗磨损性
anti-weeds cement 防藻水泥
anti-welding 抗焊接的
anti-wheel locking device (汽车的)车轮防抱死装置
anti-whitening agent 防白剂
anti-wind 防缠绕
anti-wind cap 防塞罩;防缠(绕)罩
anti-wind shield 防缠罩
anti-wist batten 防绞杆
anti-wrap shield 防缠罩
anti-wrap strip 防缠板条
anti-wrinkling 防皱
anti-wrinkling agent 防皱剂
anti-zigzagging 反锯齿现象
antler 鹿角
antlerite 块铜矿;块铜矾

Antler orogeny 安特勒造山运动;安特勒运动
antlophobia 洪水恐怖
ant nest 蚁巢
ant net 蚁穴
antodyne 苯氧丙二醇
anto-flood terrace 防淹平台
antogenous grinding 无介质粉磨
antogenous mill 无介质磨
antogenous welding 氧炔焊
Antoine equation 安托尼方程
antomobile sprayer 自动喷雾器
Antonaldi scale 安东纳迪标度
antonite 白云母
Antonoff's rule 安东诺夫定则
Antonucci test 安托努齐试验
ant plant 好蚁植物
ant queen 移动式可伸缩卷扬机;移动式可伸缩装置
antracen 树脂沥青
antre 洞窟
Antron 安特纶
antrorse 向上的
ant salt of copper 蚁酸铜;甲酸铜
antu 安妥
antunesite 黄钾铁矾
Antwerp blue 安特卫普蓝
Antwerp lace 安特卫普花边
Antwerp Port 安特卫普港(比利时)
an urgent demand 迫切需要
anvil 铁砧;碎矿板;压砧;锤砧;测砧
anvil base 砧座
anvil beak 铁砧嘴;砧嘴;砧角
anvil bed 砧座
anvil block 砧台;铁砧台;(凿岩机活塞的)撞针;砧座;砧块;吊锤[岩];垫锤
anvil cap 砧枕;锻模座
anvil chip 测砧片
anvil chisel 砧凿
anvil cinder 锻渣
anvil cloud 砧状云
anvil cushion 铁砧垫;砧枕
anvil cutter 砧凿
anvil-cylinder type 顶锤一缸式
anvil-cylinder ultra high pressure apparatus 压砧一缸式超高压装置
anvil dross 锻渣
anvil face 砧面
anvil faced rail 抗磨钢轨;耐磨钢轨
anvil horn 砧角
anvil jolt 砧座震击
anvil jolter 震击平台
anvil pallet 砧面垫片
anvil piece 千分尺可换测砧;可换测砧
anvil pillar 砧座柱
anvil plate 砧面垫面
anvil ratio 砧锤重量比
anvil scale 锻屑
anvil seat 砧座
anvil stake 桩砧
anvil stand 砧座
anvil swage 砧铁(型)模;下型模
anvil synchronizing 压砧同步
anvil test 纵轴压顶试验
anvil tool 下锻模
anvil type percussion drill 凿岩机;冲击钻
anvil vice 铁砧虎钳;台用虎钳;台虎钳;砧台虎钳
anvil vise 铁砧虎钳;砧钳
anvil with an arm 鸟嘴砧
anxo-action 促进作用
anxybiotic bacteria 厌氧生活细菌
Anyemagen anti-clinorium belt 阿尼玛卿复背斜带
anyone event 任一事件;所有事件
anyone loss 任一损失;所有损失
anyone risk 任一保险;所有保险
anyone vessel 任一船舶;所有船舶
any other perils 其他一切危险;其他一切风险
any selection 任意选择
anywhere carpet 室外通用地毯;通用地毯
anywhere rule 随所规则
a-oligosaprobic zone 甲型寡污水生物带;甲型寡腐生生物带
apa (一种产于尼日利亚红褐色有多种纹理的)硬木
apachite 闪辉响岩;蓝水硅铜石
a pack 一包
apad(h)ana 古波斯皇宫中柱式宫殿大厅

A pair of neighboring massives 相邻二地亮块体
a pair of stipules 一对托叶
Apapa Port 阿帕帕港(尼日利亚)
apaque 不透光的
A-parameter 孔隙压力系数 A;系数 A(孔隙压力)
a parcel 一宗;一包
apart check 取样检验
apart continuous layout 间隔连续观测系统
apartment 合住公寓;成套房间
apartment and lodging house combined 公寓与分租房屋合一
apartment and rooming house combined 公寓与分租房屋合一
apartment area 公寓式住宅
apartment block 街坊;公寓建筑;公寓大厦;公寓大楼
apartment building 公寓大楼;公寓建筑;公寓房屋
apartment combining shop and dwelling units 底层为集体公寓
apartment complex 公寓建筑;公寓大厦;公寓大楼
apartment dweller 居民;住户
apartment entrance 楼层住宅入口
apartment floor 公寓楼层;住宅楼层
apartment for the aged 老人公寓
apartment for the elderly 老年(人)公寓
apartment for the single 单身公寓
apartment hotel 公寓饭店;公寓(式)旅馆
apartment house 公寓(房屋);公共住宅;公寓式住宅
apartment house for employees 职工住宅
apartment house keying 公寓钥匙系统
apartment house of corridor access 廊式公寓
apartment house of direct access 服务性空间共用的公寓
apartment house of employees 职工住宅
apartment housing 多户住宅;公寓楼(房)
apartment in clogs 底层设有公共建筑的住宅
apartment kiln 分室干燥窑
apartment kitchen 住宅厨房
apartment metered 装有水表的公寓
apartment of building 公寓楼(房)
apartment project 公寓建设项目
apartments combining shop and dwelling units 底层为商店等的集体宿舍;底层为商店等的公寓
apartment service area 公寓服务面积
apartment skyscraper 公寓摩天大楼;公寓大楼
apartment storey 公寓楼层
apartment tower 公寓大楼塔楼
apartment unit 公寓套房;公寓居住单元
apartment unit entrance 公寓居住单元入口
apartment unit entrance door 公寓居住单元大门
apartment unit floor space 公寓居住单元楼板面积;公寓居住单元面积
apartment unit kitchen 公寓居住单元厨房
apartment using 多户住宅
apartotel 公寓酒店
aparture card 隔离卡片
apastron 远星点
apathy 淡漠
apatite 磷石灰;磷灰石
apatite deposit 磷灰石矿床
apatite muscovite quartz schist 磷灰石白云母石英片岩
apatite-muscovite-schist 磷灰石白云母片岩
apatite ore 磷灰石岩矿石
apatite-rich rock 富磷灰石岩
apatite-rich rock with chromite 含铬铁矿富磷灰石岩
apatite-rich rock with ilmenite 含钛铁矿富磷灰石岩
apatite-rich rock with magnetite 含磁铁矿富磷灰石岩
apatite schist 磷灰石片岩
apatite sovite 磷灰石黑云碳酸岩
apatitolite ore 磷灰岩矿石
Apco 有电木面层的墙板
apeak 竖立;立锚
apear pyrites 矛白铁矿
aperient 苦矿水
aperient water 泻性水
aperiodic(al) 非周期(性)的;不定期的
aperiodic(al) antenna 非周期天线;非调谐天线
aperiodic(al) chain 非周期链
aperiodic(al) compass 无振荡罗盘;速示罗盘;非周期罗经;不摆罗经
aperiodic(al) component 非周期部分
aperiodic(al) damped motion 非周期的阻尼运动
aperiodic(al) damping 非周期(性)阻尼;非周期(性)衰减
aperiodic(al) damping motion 非周期性阻尼运动
aperiodic(al) detector 非调谐检波器
aperiodic(al) discharge 非周期放电
aperiodic(al) element 非周期摆动的可动部分
aperiodic(al) flood 不定期洪水
aperiodic(al) function 非周期(性)函数
aperiodic(al) galvanometer 不摆检流计;不摆电流计;直接指示电流计;非周期电流计
aperiodic(al) instrument 不摆式仪表
aperiodic(al) irreducible chain 非周期不可约链
aperiodic(al) mode of motion 非周期运动形式
aperiodic(al) motion 非周期(性)运动
aperiodic(al) oscillation 非周期性振荡;非周期的振荡
aperiodic(al) pendulum 无周期摆
aperiodic(al) phenomenon 非周期现象
aperiodic(al) regeneration 非周期性再生
aperiodic(al) state 非周期状态
aperiodic(al) strain 非周期应变
aperiodic(al) stretch 非周期延伸
aperiodic(al) transition 非周期性过渡
aperiodic(al) transitional condition 非周期性过渡条件
aperiodic(al) type compass 非周期性罗盘
aperiodic(al) variation 非周期(性)变化
aperiodic(al) vibration 非周期性振动
aperiodic(al) voltmeter 速示电压表;不摆(式)伏特计;直指伏特计;大阻尼伏特计
aperiodic(al) wave 无周期波;非周期波
aperiodicity 非周期性
a period of ten days 旬
apertometer 数值口径计;数值孔径计;孔径仪;孔径计
aperture 球面镜口径;口径;孔口;孔洞;开口;门窗孔;光圈;地震裂缝;窗口;窗孔;壁孔
aperture aberration 孔径像差
aperture adhesive 窗孔黏合剂
aperture admittance 孔径透力
aperture and stops 孔径与光阑
aperture angle 孔径角
aperture angle of imaging lens 成像透镜孔径角
aperture antenna 孔径天线;开口天线;隙缝天线
aperture area 开孔面积
aperture area efficiency 开口面积效率
aperture blocking 孔径遮光
aperture board 窗孔板
aperture card 窗孔卡片;穿孔卡;方孔卡
aperture card window 卡片窗孔
aperture colo(u)r 孔径色;孔观颜色;光孔色
aperture compensation 孔径失真补偿;孔径畸变补偿
aperture compensator 孔径校正器;孔径失真补偿器;孔径调准器
aperture conductivity 孔的传导率
aperture control-grid 控制栅孔径
aperture correction 孔径校正
aperture corrector 孔径校正器
aperture correlation function 孔径相关函数
apertured disc 开孔圆盘;穿孔圆盘
aperture delay 间隙延迟
aperture diaphragm 孔径光阑
aperture dimension 孔径
aperture disc 光圈盘
aperture distortion 孔径失真;孔径畸变;孔径变形;小孔畸变;光阑小孔畸变
apertured mask 荫罩
apertured shadow mask 影孔板;荫罩
aperture effect 孔径效应
aperture efficiency 孔径效率;孔径面积率;开口面积效率;开口比
aperture equivalent 孔径当量
aperture error 孔径误差
aperture factor 孔径因子
aperture flux density 孔径通量密度
aperture function 孔径函数
aperture gap 孔隙
aperture grill 孔径栅;荫栅;障栅
aperture illumination 照度分布;孔径照明;孔径面照明
aperture image 孔径像
aperture lens 孔隙透镜;孔径透镜;膜孔透镜;针孔透镜;针孔电子透镜
aperture lever 光圈拨杆
aperture loss 孔径失真
aperture-magnetic confinement 孔径磁约束
aperture mask 孔障板;孔眼掩模;显像管荫罩;多孔障板;彩色显像管阴罩
aperture number 孔径数
aperture of a mirror 反射镜的口径
aperture of beam 射束孔(径)
aperture of bridge 桥孔;桥洞
aperture of diaphragm 光阑孔径
aperture of door 门口;门孔;门洞
aperture of mirror 镜口径
aperture of screen 筛孔
aperture of sight 观测孔
aperture of the diaphragm 光圈
aperture of window 窗口;窗洞
aperture opening ratio 开口率
aperture plate 栅网;光圈挡片
aperture-priority automatic exposure camera 光圈优先式曝光量自动控制照相机
aperture-priority mode 光圈优先式
aperture radiance 孔径辐射强度
aperture ratio 孔率;孔径比;镜头直径与焦距之比
aperture ratio number 孔径比值
aperture response 孔径响应;孔径特性曲线
aperture restrictor 孔径限制器
aperture setting 孔径定位
aperture setting ring 光阑装置环;光阑调节环
aperture sight 孔径瞄准器;孔径准具
aperture size 筛眼孔径;筛眼尺寸;筛隙宽度;筛孔度;孔径尺寸
aperture slot 缝隙光圈
aperture stop 孔阑;孔径遮光;孔径光阑;隔膜
aperture surface 孔径表面
aperture synthesis 孔径综合;综合孔径
aperture synthesis radio-telescope 综合孔径射电望远镜
aperture time 孔径时间;穿孔时间;缝隙时间
aperture type fluorescent lamp 狭缝式荧光灯
aperture vignette 孔径渐晕
aperture width 筛网孔净径;孔径宽度;隙缝宽度
aperturing 孔径作用
apetalae 不完全花群
apex 褶皱顶点;顶尖;底底排出口;反射点;奔向点
apex 尖;[复]apexes 或 apices
apex angle 位置角;孔径角;尖角;顶角
apex arch 拱顶;顶石
apex block 顶石;拱顶(砖)石
apex dentis 齿突尖
apex distance 钻尖偏移距离;顶距
apex hinge 拱顶铰;顶点铰;顶铰
apex hog 拱顶隆起;顶点隆起;拱曲顶
apex inductor 顶点电感器
apex joint 拱顶节点;顶节点;顶部节点;链接合
apex linguae 舌尖
apex load 顶点载重;顶点荷载
apex mo(u)ld 顶尖造型;拱顶线脚;顶模板;分水岭地形
apex of arch 拱冠;甑顶;拱顶
apex of bend 弯头顶点;弯曲顶点;曲线顶点;弯顶;弯道顶点;河弯顶点;变头顶点
apex of concave bank 凹岸顶点
apex of cusp 海滩嘴脊梁
apex of deposit 沉积层的顶点
apex of dry dock 干船坞头顶部
apex of earth motion 地球向点
apex of grade 坡顶
apex of nose 鼻尖
apex of prism 棱镜脊
apex of theoretic(al) triangle 理论三角形顶点(坝体断面)
apex of vein 矿脉顶;脉尖
apex partis petrosae 岩部尖
apex piece 拱顶石;顶端部件
apex point 顶点;钻尖
apex radicis dentis 根尖
apex sag 拱顶下垂;顶点下垂
apex seal 径向密封片
apex stone 山墙顶石;拱顶石;房脊石
Apgar score 阿普伽评分
aphaniphyric 显微隐晶质的;非显晶基斑状
aphanite 隐晶岩;非显晶岩

aphanitic 隐晶质的；非显晶质的
aphanitic earthy graphite deposit 隐晶质土状石墨矿床
aphanitic limestone 隐晶质灰岩
aphanitic texture 隐晶结构；隐晶质结构
aphanocrystalline texture 隐晶结构
aphanophyre 显微隐晶斑岩
aphanophyric 非显晶基斑状
aphantobiont 超视微粒
aphelion 远日点；远核点
aphelion distance 远日点距离
a-phenyl phenacyl 二苯乙酮基
Aphidan 丰丙磷
aphlebia 无脉羽叶
aphlogistic 不能燃烧的
Aphloia theaformis 茶红木
aphodus 排水沟
aphotic layer 无光层
aphotic marine environment 无光海洋环境
aphotic zone 无光深水区；非生化带；不透光带
aphotometric 非光度计的
aphototropism 反趋光性
aphrite 鳞方解石
aphrizite 泡沸电气石；黑电气石
aphrodite 镁泡石
aphroid 互嵌状
aphrolite 镁沫岩
aphrosiderite 铁绿泥石；铁华绿泥石
aphryzite 黑电气石
aphthitalite 钾芒硝
aphyactic projection 不等角不等积的投影
aphylactic map projection 任意投影
aphyllous plant 无叶植物
aphyric 无斑隐晶的；无斑非晶质
aphytal 深水湖盆；深水带；湖渊
aphytal zone 湖渊区
apiary 蜂房；养蜂场
apiary alarm 蜂场报警器
apical 顶端的；顶尖的；峰顶的
apical angle 顶切角；顶角
apical anomaly 顶部异常
apical axis 顶轴
apical branch 尖支
apical dominance 顶端优势（度）
apical grafting 顶接
apical growth 顶端生长
apical organ 顶器
apical plate altitude of aquifer 含水层顶板高程
apical pore 顶孔
apical rosette 顶玫板
apical segment 尖段
apical sense organ 端感器
apical spikelet 顶小穗
apical string 顶带
apical tooth 顶齿
apical yellowing 顶端变黄
apiciform 尖形的
API/cm API/厘米
apiculus 细尖；顶端
a piece(of) 一件；一块
a piece of drawing 一幅图纸
a piece of land 一块土地
a piece of scene 园林小品
Apiezon 阿匹松真空泵油；阿匹松
Apiezon grease 阿匹松脂；阿匹松密封腊
Apiezon oil 阿匹松真空泵用油
Apiezon wax 封蜡；阿匹松蜡
API filtration API 失流
API hydrometer API 比重计
API method API 法（桩工）
apisin 蜂毒
apisin poisoning 蜂毒液中毒
apitong 大花龙脑树（装饰用木材）；大花腊树（装饰用木材）
API unite mark 美国石油组织 API 组合标记
apjohnite 锰明矾；锰铝矾
Apjohn's formula 阿浦约翰公式
Aplacophora 无板纲
aplanat 消球差透镜
aplanatic 等光程；不晕的
aplanatic condenser 等光程聚光镜
aplanatic condition 不晕条件
aplanatic foci 齐明点
aplanatic focus 等光程焦点

aplanatic image formation 不晕成像
aplanatic lens 消球差透镜
aplanatic mirror pair 不晕反射镜对
aplanatic objective 不晕物镜
aplanatic points 不晕点
aplanatic refraction 等光程折射
aplanatic telescope 消球差望远镜
aplanatism 等光程；不晕(性)
aplanogamete 不动配子
aplastic 非塑性的
a plastic chart for use in the Unified Soil Classification system 土的统一分类塑性图
Aplataer process 阿普拉特尔热镀锌法
a play pen for babies 幼儿园游戏场
Apl gravity API 比重度
apline marsh 高山沼泽
apline road 高山道路
apline valley 高山河谷
aplite 细晶岩；半花岗岩
aplite-granite 细晶花岗岩
aplite group 细晶岩类
aplitic dike 细晶岩脉
aplitic texture 细晶结构
aplitic trachyte 细晶质粗面岩
apllo glass 透紫外线玻璃
aplogranite 淡云花岗岩
aploid 梨形体
aplowite 四水钴矾
Apl scale API 比重计标度
AP meter 碱性纸浆计
apoanalcite 变方沸石
apoandesite 脱玻安武岩
apoapsis 远拱点
apob 飞机仪器观测
apobasalt 脱玻玄武岩
apobasidium 分离担子
apocenter 远主焦点；远心点
apochromat 消多色差的；复消色差；复消色差透镜
apochromatic 复消色差的
apochromatic correction 复消色差校正
apochromatic lens 复消色差透镜
apochromatic micro objective 复消色差显微物镜
apochromatic objective 消多色差物镜；复消色差物镜；复消色差透镜
apochromatic system 复消色差系统
apochromatism 复消色差
a pocket-handkerchief lawn 小块草地
a pocket-handkerchief of land 小块土地
Apocrenic acid 阿波克连酸
apocynein 磁麻素
apocynthion 远月点
apodisation apodization 变迹法
apodization filtering 切践滤波
apodization transducer 变迹换能器
apodized aperture 变迹孔径
apodized transducer 变迹指长换能器
apodizer 切践器；变迹器
apodyterium （古罗马、古希腊的）庭院；（古罗马、古希腊的）浴场更衣室
apo-fenchene 从衍小茴香烯
apofocus 远主焦点；远心点
apogean current 远地点潮流
apogean(tidal)range 远日潮潮幅；远地点潮差
apogean tide 远月潮；远地点潮
apogee 远地点；最远点；极点
apogee motor 遥控电动机
apogee rocket 远地点火箭
apogee tide 远月潮；远地点潮
apogeoesthetic 背地弯曲的
apogeotropism 无向地性
apogranite 变花岗岩
apogrit(e) 杂砂岩
apohomotypic meiosis 无同型减数分裂
apolar 无极的；非极的
apolegamic 选配的
apollinaris spring 碳酸泉
apollinaris water 碳酸矿泉水
Apollo alpha spectromter 阿波罗阿尔发能谱仪
Apollo application 阿波罗应用
Apollo applications program(me) 阿波罗应用计划
Apollo bioenvironmental information system 阿波罗生物环境信息系统
Apollo Data Bank 阿波罗数据库
Apollo docking test device 阿波罗飞船对接试验装置
Apolloniu's circle 阿波罗尼圆
Apolloniu's theorem 阿波罗尼定理
Apollo object 阿波罗天体
Apollo panoramic camera 阿波罗全景摄象机
Apollo program(me) 阿波罗计划
Apollo simple penetrometer 阿波罗简单透度计
apolune 远月点
a-polysaprobic zone 甲型多污水生物带；甲型多腐生生物带
apomagmatic hydrothermal mineral deposit 中岩浆热液矿床
apomecometer 视距仪；测远器；测距仪；测角测距仪
apomeiosis 不完全减数分裂
apomorph 离态
a poor risk for insurance 风险大的保险户
apophorometer 升华仪
apophyge 柱座凹线；凹线脚；柱头凹线脚；（古典建筑物的）蜗牛形柱墩
apophyllite 鱼眼石
apophysis 岩枝
apoporphyry 变斑岩
Apo-Ronar lens 阿波罗纳尔镜头
aport 向左舷；在左舷
A portainer A 形门架集装箱起重机（装卸桥）
aposafranone 阿朴藏红酮
aposandstone 石英岩
aposedimentary 沉积后生的
aposelene 远月点
aposhere 过渡球面
a posterior 凭经验的；后验的（拉丁语）；由结果追溯到原因的
a posteriori estimate 事后估值；事后估计
a posteriori probability 事后概率
aposthesis （柱身与柱头或柱脚的）凹圆形弯曲线脚
apostilb 阿熙提
apostrophe 上撇号；向下转形
apotectonic 造山后的；构造后的
apotheca 储存室；藏酒窖
apothecaries system 药衡制（英国）
apothem 浸剂沉淀物；边心距；远心距
apothem of a regular polygon 正多边形的垂辐
apotropine 阿朴托品
apotype 补型
Appalachia 阿帕拉契亚
Appalachian 阿帕拉契亚造山运动【地】
Appalachian basin 阿帕拉契亚盆地【地】
Appalachian fault-fold belt 阿帕拉契亚断裂褶皱带【地】
Appalachian geosyncline 阿帕拉契亚地槽【地】
Appalachian marine trough 阿帕拉契亚海槽【地】
Appalachian Mountain 阿帕拉契亚山脉
Appalachian orogeny 阿帕拉契亚造山运动【地】
Appaachian relief 阿巴拉契亚式地形
Appalachian revolution 阿帕拉契亚变革
apparatus 器械室；器械；器具；用具；仪器；仪表；装置
apparatus analysis of petroleum 石油仪器分析
apparatus capacity 设备容量；设备能力
apparatus casing 综合仪器箱
apparatus constant 仪器常数；仪表常数
apparatus cutting 切割器
apparatus dew point 机器露点
apparatus dew point temperature 机器露点温度
apparatus engineering 仪表工程(学)；设备工程(学)；仪器设备工程(学)
apparatus error 仪器误差
apparatus-fixed height 仪器安置高度
apparatus floor 设备楼层；设备层；仪器设备层
apparatus for absolute pitch 绝对音调器
apparatus for air conditioning 空气调节设备
apparatus for butt welding 平接压焊夹具
apparatus for dam observation 大坝观测仪器
apparatus for degasification 脱气器
apparatus for determination of ammonia 氨测定器
apparatus for determining mortar consistency 砂浆稠度(测定)仪
apparatus for determining normal consistency and setting time 标准稠度与凝结时间测定仪
apparatus for determining stratification of mortar 砂浆分层度测定仪
apparatus for determining water retention 测定保水性仪器

apparatus for directional solidification 定向凝固设备
apparatus for inclusion study 包裹体实验设备
apparatus for lathe centers 车床顶尖磨具
apparatus for locating pipe 管道定位仪
apparatus for nitrogen examination 氮测定仪
apparatus for orientation 定向仪
apparatus for rectification 纠正仪器
apparatus for semi-micro elementary analyzer 半微量元素燃烧分析装置
apparatus for suspension 滑车架
apparatus for testing rail profile 钢轨断面测定器
apparatus for transformation by drawing 转绘仪
apparatus for water softening 水的软化装置
apparatus function 仪器函数
apparatus glass 仪器玻璃;仪表玻璃
apparatus insulator 电器绝缘子
apparatus load factor 设备负荷系数
apparatus of neutron activation analysis 中子活化法装置
apparatus of oxygen supply 供氧装置
apparatus of photonuclear analysis 光核子分析装置
apparatus of resistance 测电阻仪
apparatus of X-fluorscenco survey X 荧光测量装置
apparatus porcelain sleeve 电器瓷套
apparatus room 设备房
apparatus spectrum 仪器谱
apparatus stor(e)y 设备层;设备楼层
apparatus to remove iron 除铁装置
apparatus with homing position 复原式装置
apparatus without homing position 非复原式装置
apparatus with several arm wippers 多弧刷旋转选择器
apparel 船上用具;外表;舾装;船具;船舶舾装
apparel and tackle 舾装;船具
apparent 外表的;视在;明显的;表观的
apparent ablation 视消融
apparent absorption 视吸收;表观吸收
apparent abundance 视丰度
apparent accumulation 视累积
apparent acidity 表观酸度
apparent activation energy 表观活化能
apparent activity 表观活度
apparent additional mass 表观附加质量
apparent adsorption density 表观吸附密度
apparent adsorption equilibrium constant 表观吸附平衡常数
apparent adsorption isotherm 表观吸附等温线
apparent age 视年龄;表面年龄
apparent altitude 视高度;视高程;视地平纬度
apparent amplitude 外视振幅
apparent angle 视角
apparent angle of attack 视迎角
apparent angle of dip 视倾角
apparent angle of friction 视摩擦角;表观摩擦角;视在摩擦角
apparent angle of internal friction 表观内摩擦角;视在内摩擦角
apparent angle of shearing resistance 似抗剪强度角
apparent anomaly 视近点角
apparent approximation 表观逼近
apparent arc 明弧
apparent area 视面积;名义面积;表观面积
apparent attenuation degree 视衰减度
apparent availability 市场供应率
apparent available area 表观可用面积
apparent azimuth(angle) 视在方位角;视方位角
apparent bed-thickness 视地层厚度
apparent black 视在黑色
apparent bolometric magnitude 视热星等
apparent bond strength 表观黏结强度
apparent bottom 视海底
apparent brightness 视在亮度;视亮度;表观亮度
apparent bulk density 表观体密度;表观松密度;表观容重;表观堆积密度
apparent bulk modulus 表观体积模量;表观压缩系数
apparent candle power 表观烛光;视烛光
apparent capacity 视在容量;视在电容;视容量;表观容量;表观电容
apparent charge 视电荷;表观电荷

apparent chargesability 视极化率;视充电率
apparent coagulation 明显聚沉
apparent coastline 视海岸线
apparent coefficient 表观系数
apparent coefficient of compressibility 表观压缩系数
apparent coefficient of friction 表观摩擦系数
apparent coefficient of heat transmission 表现传热系数
apparent coefficient of viscosity 视黏滞系数
apparent cohesion 似黏聚力;视内聚力;假黏结;假内聚力;表观黏力;表观黏聚力;表观黏结;表观凝聚力;表观内聚力
apparent colo(u)r 视在颜色;表色;表观颜色;观色;视在色
apparent compressibility 视压缩性
apparent compressional wave 视压缩波
apparent condition of goods 货物的表面状态
apparent conductivity 表观电导率
apparent consistency 表观稠度
apparent constant 表观常数
apparent contrast 表观对比度
apparent conversion 表观转化率
apparent coordinates 视坐标
apparent course 视航向
apparent crater 视环形火山
apparent crystallization constant 表观结晶速率常数
apparent damage on discharge 卸货时外表明显损坏
apparent decay time 视衰减时
apparent declination 视赤纬
apparent defect 外观缺陷;表观缺陷
apparent defect of refractory product 耐火制品外观缺陷
apparent degree of calcination 表观分解煅烧率
apparent degree of decarbonation 表观碳分解率
apparent density 散装密度;松密度;视密度;堆积密度;堆积比重;表观密度;松装密度;容重
apparent density of hydrocarbon 烃类的视密度
apparent density of matrix 岩石骨架视密度
apparent density porosity of shale 泥质的视密度孔隙度
apparent depression of the horizon 视地平俯角;地平俯角
apparent depth to the magnetic body 磁性体视深度
apparent deviation component 视偏差分力
apparent diameter 视直径
apparent diffusion coefficient 表观扩散系数
apparent digestibility 表观消化吸收率;表观消化率
apparent digestible energy 表观可消化能
apparent dip 视倾斜;视倾角;表观倾斜度
apparent direction 视方向
apparent direction of wind 视风向
apparent distance 视距离;表观距离
apparent diurnal motion 周日视动
apparent diurnal path 视周日路径
apparent downthrow 视下落断距
apparent drag 视阻力
apparent drag coefficient 视阻力系数;表观阻力系数
apparent easement 外观附属建筑物
apparent effective field 视有效场
apparent efficiency 视在效率;视效率;表观效率
apparent elastic limit 视弹性极限;视弹性限度;显似弹性极限;表观弹性极限
apparent elastic modulus 表观弹性模量
apparent elevation angle 视在仰角
apparent elongation 视拉长
apparent energy 表观能量
apparent energy consumption unit 单位视在能耗【铁】
apparent energy meter 全功电度表;伏特安培小时计;表观能量计
apparent equatorial coordinates 视赤道坐标
apparent equilibrium constant 表观平衡常数
apparent error 视在误差;视误差;直观误差;表观误差
apparent expansion 视膨胀;表观膨胀
apparent exterior contact 视外切
apparent fault displacement 视断距
apparent field 外视场;表观视场
apparent field of view 表观视场

apparent fineness 表观细度
apparent fixity 表观嵌固点
apparent flattening 视扁率
apparent fluidity 视流质;表观流动性
apparent focus 视焦(点)
apparent force 视在力;表观力
apparent formation constant 视地层常数;表观形成常数
apparent formation factor 视地层因数
apparent free nape (水流出口的)自由射流;表观溢洪水舌
apparent free space 表面自由空间;视观自由空间
apparent gap area 空隙表面面积
apparent good order and condition 外表情况良好;表面完好
apparent gravity 视比重
apparent groundwater table 视地下水面
apparent groundwater velocity 视地下流速;地下水表观流速
apparent half-life 表观半衰期
apparent hardening curve 表观硬化曲线
apparent hardness 表观硬度
apparent head 视在落差
apparent heave 视平错;视横断距
apparent heave slip 视地层位移幅度
apparent height 视在高度;视高度;视高程;有效高度
apparent horizon 视水平(线);视地平(线)
apparent horizontal overlap 视平超距
apparent horizontal separation 视水平离距
apparent image sharpness 影像视觉清晰度
apparent impedance 视阻抗;视在阻抗;表观阻抗
apparent initial deformation 表观初始变形
apparent initial softening temperature 荷重软化点
apparent interior contact 视内切
apparent intersection 虚交点
apparent ion exchange capacity 表观离子交换容量
apparent ionization constant 表观电离常数
apparent ionization yield 表观电离率
apparent isochron age 视等时年龄
apparent latitude 视黄纬
apparent level 视水平
apparent libration 视天平动
apparent life 视年龄
apparent lifetime 表观寿命
apparent lignin 表观木素
apparent lineation 视线理
apparent load 视在负载;视荷载;近似荷载
apparent longitude 视黄经
apparent loss 视损失;毛估损失;表观损失
apparent luminance 视亮度
apparently drowned 溺水假死
apparently higher than 明显高于
apparent magnitude 视星等
apparent mass 视(在)质量;虚质量;有效质量;表现质量
apparent material 透明材料
apparent maximum altitude 视最大高度
apparent metal factor 视金属因素
apparent modulus 视模数;视距离模数;表观模数
apparent modulus elasticity 表观弹性模量
apparent molal volume 表观摩尔体积
apparent molar adsorptivity 表观摩尔吸附系数
apparent molecular weight 表观分子量
apparent molecular weight distribution 表观分子量分级
apparent mortality 表观死亡率
apparent motion 视运动;视动;表观运动
apparent movement 视运动;视动
apparent movement of faults 视断层活动;断层视运动
apparent neutron porosity of fluid 流体的视中子孔隙度
apparent neutron porosity of matrix 岩石骨架视中子孔隙度
apparent neutron porosity of shale 泥质的视中子孔隙度
apparent noon 视准正午;视正午;视午
apparent normal fault 视正断层
apparent onlap 视上超
apparent optic(al) axial angle 视光轴角;假光轴角
apparent orbit 视轨道
apparent order and condition 外表状况
apparent output 视(在)输出;视(在)功率;表观

输出
apparent overconsolidation 表观超固结
apparent oxygen utilization 表观耗氧量
apparent partition coefficient 表观分配系数
apparent path 视轨道
apparent permeability 视在渗透系数
apparent phase velocity 视相速度;表观相速度
apparent photored magnitude 视仿红星等
apparent photovisual magnitude 视仿视星等
apparent pitch 视在螺距;视螺距
apparent place 视位置
apparent plunge 视倾伏
apparent polarity 视极性
apparent polarization 视激电率
apparent polar-wandering 视极移
apparent polar-wander(ing) path 视极移曲线;视极移轨迹
apparent porosity 视孔隙率;显气孔率;表观气孔率;表观孔隙率;表观孔隙度
apparent position 视位置
apparent powder density 粉末表观密度
apparent power 视在功率;视功率;表观功率
apparent precession 视岁差;视漂移;视进动
apparent preconsolidation pressure 表观先期固结压力;表观前期固结压力;视在预固结压力;表观预固结压力
apparent pressure 表观压力
apparent profile 视纵断面
apparent proof 表观度
apparent purity 表观纯度
apparent quality 外观质量
apparent quality of paper 纸张外观质量
apparent radiant 视辐射点
apparent radius 视半径;表观半径
apparent range 视在距离
apparent range rate 视在距离变化率
apparent rate 视速率;名义费率;表面费率
apparent rate constant 表观速率常数
apparent rate of motion 表观速率
apparent rediometric magnitude 视辐射星等
apparent red magnitude 视红星等
apparent reflectance 表观反射比
apparent relative movement 视相对运动
apparent relief 视突起
apparent reluctance 表观磁阻
apparent resistance 视阻力;视在电阻;视电阻;表观阻力;表观电阻
apparent resistivity 视电阻率;视比阻;表观电阻力
apparent resistivity curve 视电阻率曲线;表观电阻率曲线
apparent resistivity log curve 视电阻率测井曲线
apparent resistivity of formation water 视地层水电阻率
apparent resistivity of mud filtrate 视泥浆滤液电阻率
apparent resolution 视(在)清晰度;视(在)分辨率;可见分辨率
apparent retention time 表观保留时间
apparent retention volume 表观保留体积
apparent revolution 视公转
apparent right ascension 视赤经
apparent rolling 视摇摆;表观摇摆
apparent rotation 视自转
apparent salinity 表观盐度
apparent scanner temperature 扫描器记录的表观温度
apparent seepage velocity 视渗透速度
apparent seismic coefficient 表观地震系数
apparent semidiameter 视半径
apparent servitude 外观附属建筑物
apparent setting 视设置
apparent shear(ing) strength 表观抗剪强度;视在抗剪强度
apparent shear-strength 视抗剪强度;似抗剪强度
apparent shear-strength parameter 表观剪切强度参数
apparent shear stress 似剪应力;表观剪(切)应力
apparent shoreline 视海岸线
apparent sidereal-time 视恒星时
apparent size 视尺寸;表观尺寸
apparent sky brightness 视天空亮度
apparent slip 视滑脱;视滑距;视断距
apparent slope 视坡度;表观坡度
apparent sludge production coefficient 表观产污泥系数
apparent soil parameter 表观岩土参数
apparent solar day 视太阳日
apparent solar time 视时;视太阳时;真太阳时
apparent solar year 视太阳年
apparent solid density 表观固体密度
apparent solid volume 固体表面体积;固体表观体积
apparent solubility 表观溶(解)度
apparent solubility product 表观溶度积
apparent sonic porosity of shale 泥质的视声波孔隙度
apparent sonic porosity of wet clay 湿黏土的视声波孔隙度
apparent specific gravity 外表比重;体比重;视比重;假比重;表观比重
apparent specific heat 表观比热
apparent specific resistance 视电阻率;表观电阻系数;表观电阻率
apparent specific resistivity 视电阻率
apparent specific volume 表观比容
apparent speed 视速率;表观速度
apparent spot diameter 视光点直径
apparent strain 视应变;名义应变;公称应变
apparent stratigraphical gap 视地层距
apparent stratigraphical interval 视地层间隔
apparent stratigraphic(al) separation 视地层离距
apparent strength 视强度;表观强度
apparent stress 视应力;表观应力
apparent sun 视太阳
apparent sunrise 视日出
apparent sunset 视日没
apparent superluminal motion 视超光速运动
apparent superluminal velocity 视超光速
apparent superposition 视迭覆
apparent surface 视场表面;视表面;相对表面
apparent surface velocity 视地面速度;视表速
apparent survey 目视检验
apparent susceptibility 视磁化率
apparent susceptibility chart 视磁化率图
apparent susceptibility value 视磁化率值
apparent target luminance 表观目标亮度
apparent tax 名义税
apparent tax rate 名义税率
apparent temperature 视温度;表观温度
apparent temperature drop 表观温度降
apparent thermal conductivity 表观导热系数;表观传热系数
apparent thermal resistivity 表观热阻率
apparent thickness 视厚度;假厚度;表观厚度
apparent thickness of ore body 矿体视厚度
apparent threshold 表观阈
apparent throat 焊缝计算厚度
apparent throw 视纵断面;视落差
apparent tilt 视倾斜
apparent title 虚假所有权;表面所有权
apparent tooth density 视在齿密度;表观齿磁束密度
apparent to the naked eye 肉眼可见的
apparent unconformity 视不整合
apparent valence 外表价
apparent value 表观值
apparent variable 视变数;视变量;约束变项;约束变量
apparent velocity 视速度;表观速度;视速流速
apparent velocity correlation 视速度对比
apparent velocity of propagation 表观传播速度
apparent vertex 视奔赴点
apparent vertical 视垂线;动垂线
apparent viscosity 视黏度;表观黏(滞)性;表观黏(滞)度
apparent visual angle 表观视角
apparent visual magnitude 视目视星等
apparent volume 松装体积;松装容量;视体积;视容积;表观体积;表观容积
apparent volume of distribution 表观分布容积
apparent volumetric(al) efficiency 表观容积效率
apparent wander 视漂移;视进动
apparent water-level 视水位
apparent water stage 视水位
apparent water table 视水位;上层滞水位;上层水面
apparent wave 表观波
apparent wave frequency 表观波频率
apparent wave height 视波高;表观波高
apparent wavelength 表观波长
apparent wave period 波浪表观周期;表观波周期
apparent weight 视重量;毛重;表观重量
apparent wellbore radius 视井径
apparent wellbore ratio 视井径比
apparent width (地层的)视厚度
apparent wind 视风;相对风
apparent zenith distance 视天顶距
apparition 初现;初显;出现期
appeal committee 仲裁委员会
appeal for compensation 赔偿要求;要求赔偿
appeals board 上诉庭(房地产税);上诉部门
appearance 外形;外貌;外观
appearance age of continental ice-sheet 大陆冰盖出现年代
appearance and dimension check 外观尺寸检查
appearance defect 外观缺陷;外表缺陷
appearance failure 严重外表损伤
appearance fracture test 断口外观检验;断口外观试验
appearance grade 外观等级
appearance height 出现点高度
appearance inspection test 外形检查
appearance of drinking water 饮用水外观
appearance of forests 林相
appearance of fracture 破碎现象;破碎外观;破碎特征;裂面外貌;断口形状;断口外观
appearance of potable water 饮用水外观
appearance of slush 初冰
appearance of treatment plant 处理厂外貌
appearance of wastewater 废水外观
appearance of weld 焊缝成形
appearance point 出现点
appearance potential 外观电势;出现电位;表观电势
appearance quality 外观质量
appearance rating 外观质量评级
appearance ratio 夸大立体感比值;超体视
appearance retention 外观保持性
appearance standard 外观标准
appearance surface 外表
appearance test 外观检查;外部检查
appearant volume 外观容积
appeared depth of crater 漏斗可见深度
appeerence potential spectroscopy 出现电势谱
appearing 版面高度
appearing and subsiding 出现和消失
appellant 上诉人
append 增补;附上
appendage 属具;船体附属体;附属部分;附器;附加物;附加程序;备用仪表
appendage pump 附属泵;备用泵
appendage resistance 附体阻力
appendage routine 附属例行程序;附加例行程序
appendage task 附属任务;附加任务
appendage vacuum pump 高真空维持泵;备用真空泵
appendant 附属的;附加的;附属物;附属权利
appended description of standard 标准附加说明
appended documents 附加文件;附件
append file 附加文件
appendix 随附;输气管;增补;附录;补遗
appendix man hole 输气管进入孔
appendix of regional geologic(al) survey report 区域地质调查报告附本
appendix of the bill of quantities 工程量表附件
appendix to bid 投标书附录
appendix to contract 合同附件
appendix to the tender 投标书附件
append macros 附加宏指令
appentice 屋顶房间;坡屋;门缝;耳房;门前之雨篷;厢房;雨罩
Apperleen 阿珀利恩起绒呢
appertaining documents 有关证件
appinite 富闪深成岩
applanate 扁平的
applauseograph 噪声录音机;采声记录仪
Appleby-Frodingham process 阿普尔比—福罗丁翰法
apple coal 沥青煤
Applegate diagram 阿普尔盖特图
apple green 苹果绿;豆青
Appleman gumbo bit 底喷式鱼尾钻头
apple oil 腰果油

apple orchard 苹果园
apple ring dowel 楔形暗销
apples and pears 梯级(俚语);楼梯(俚语)
applet 小应用程序;小程序
Appleton layer 阿普顿层
apple tree 苹果树
applewood 苹果木
appliance 器械设备;器具;适用;矫正器;用具;仪表
appliance circuit 仪表线路;仪表用电路;仪表电路
appliance clock 仪表钟
appliance computer 应用计算机
appliance control system 设备控制系统
appliance finish 仪表器具用涂料
appliance flue 灶具烟道;适用烟道
appliance load 民用负荷
appliance mains switch 电器电源开关
appliance outlet 电源插口;设备接头;设备插口
appliance panel 设备检修配电板
appliance saturation 燃具饱和率
appliance ventilation duct 用具设备通风管;仪器设备通风管
applicability 可贴性;可适用性;贴近度;适用性;适用范围
applicability parameter 适用性系数
applicable 可适用的
applicable codes and standards 现行规范和标准
applicable condition 适用条件
applicable for setstock grazing 适用于定区放牧
applicable JIS standard 现行日本工业标准
applicable law 适用的法律;实用法律;可适用的法律
applicable material specification 现行材料规格
applicable material standard 现行材料标准
applicable notes 适用条款
applicable provision 适用条款
applicable safety procedures 现行安全规程
applicable surface 可贴曲面
applicable water quality standards 适用的水质标准
applicant 请求人;投保人;投保方;申请人;应征人员
applicant applicant for the credit (信用证的)开证申请人
applicant block position exploitation 申请开采区域位置
applicant country for exploitation 申请开采的国家
applicant for a discount 贴现人
applicant for collection 托收人
applicant for import or export credits 押汇人
applicant for insurance 投保人;要保人
applicant for letter of credit 信用证申请人
applicant for payment 托付人
applicant for shares 入股申请人;应募股票人
applicant proposing to build 建筑申请者;建造中申请人
applicant state 起诉国
applicant time 申请日期
applicant to do exportation 出口许可证申请人
applicate 贴于表面
application 涂装;适用;施加;申请;应用;要求书;敷用;敷贴;敷剂
application(al)life(span) 使用寿命
application amnagement application process 应用过程管理应用过程
application area 应用领域
application blank 空白申请书
application bond 黏结力
application builder 应用建立程序;应用程序的构造程序
application butyl 丁基胶黏剂
application by brushing 用刷子刷(油漆涂料)
application cement 胶合料
application consistency 操作连续性;操作稠度
application control block 应用控制块
application control code 应用控制代码;应用程序控制代码
application control language 应用控制语言
application control language instruction 应用控制语言指令
application control table 应用控制表
application control valve 作用控制阀
application cylinder 控制筒
application cylinder cover 控制缸盖
application data base 应用数据库
application definition record 应用定义记录
application development cycle 应用开发周期

application development system 应用开发系统
application development tools 应用开发工具
application drawing 申请用图纸;应用图;操作图
application entity address 应用实体地址
application equipment 填料填缝用设备
application factor 应用因子;应用因素;应用系数
application failure 黏合失效
application fee 申请费
application fields 应用范围
application file 应用文件
application for advice of transfer of L/C 信用证转让申请书
application for alteration of credit 信用证修改申请书
application for alteration of export permit 出口许可证修改申请书
application for amendment of import permit 进口许可证修改申请书
application for amendment of letter of credit 修改信用证申请书
application for a patent 申请表项专利;专利申请表
application for bail 请求担保
application for bill of lading endorsement 提单背书申请书
application force 作用力
application for code stamp 规范印章的使用
application for credit 申请程序
application for documentary bills for collection 出口托收申请书
application for drawback 申请退款
application for employment 求职申请书;求职申请表
application for exchange 外汇申请书
application for export 出口报单
application for export permit 出口许可证申请书
application for extrawork 要求办公时间外提货
application for foreign exchange 申请外汇
application for import of foreign goods 外国货物进口报单
application for import permit 进口许可证申请书
application for import quota 进口配额申请书
application for instituting legal proceedings 起诉申请书
application for insurance 投保单;保险申请书;保险申请
application for letter of credit 信用证申请书
application for licence 许可证申请书
application form 求职申请书;施工方式;申请书;申请表格;空白申请书
application form for marine insurance 投保海上保险申请书;海上保险申请书
application form for a banking account 开立银行账户的申请书
application for negotiation of drafts under letter of credit 出口押汇申请书
application for partial transfer of letter of credit 信用证部分转让申请书
application for patent 申请专利
application for payment 请求付款;付款申请(书)
application for registration 申请登记;申请注册
application for shares 认股书
application for shipment 申请装船单;货运舱位申请书;装船申请书;订舱单
application for space 货运舱位申请书;订舱单;舱位申请书
application for special commitment 申请特别承诺
application for tenders 投标书
application for transfer of letter of credit 信用证转让申请书
application for treasury paying-in 解库申请书
application for withdrawal 提款申请
application function routine 应用函数例行程序;应用功能例行程序;应用操作例行程序
application graduating spring 控制递开弹簧
application graduating stem 递开操纵杆
application group name 应用组名
application guide 使用导则
application image 应用图像;应用目标的形态
application importing of protective coating 运用导入保护涂层
application information services 应用信息服务程序;应用信息服务
application in planting hole 穴施法
application instituting proceeding 起诉申请

application integrated module 应用集成模块
application interface module 应用接口模块
application in two-dressings 分两次施用
application knifer 追肥铲刀
application layer 应用层
application level protocol 应用级协议
application library 应用程序库
application life 使用寿命
application limit 最高使用极限
application load balancing 应用负载均衡;应用程序均衡装入
application load list 应用装入表;应用程序装入表
application management program(me) 应用管理程序
application manual 实用手册;应用手册
application method 应用方法;操作方法;(油漆技术中的)涂刷方法;使用方法;施药方法;敷贴法
application mode 应用方式
application moment 施加力矩
application money 认股金
application monitor 应用监督程序
application mortar 粘贴用灰泥
application neoprene 氯丁橡胶涂层
application note 应用注释
application number 申请号;应用号
application of artificial intelligence 人工智能的应用
application of a surface 曲面的贴合
application of binder 施加黏结剂;施加黏合剂
application of brake 制动
application of building mastics 嵌缝膏施工法
application of ca(u)lking compound 嵌缝膏施工法
application of enamel 涂搪
application of equipartition theorem 均分定理的应用
application of flame 喷火焰;火焰处理
application of force 施力;加力
application of funds statement 请款书
application of geosynthesis 合成材料应用
application of insecticide 施用除虫剂
application of linear programming 线性规划的应用
application of load 施加荷载;施加负荷;加载(荷);加荷(载)
application of mortar 铺砂浆;铺灰浆;抹灰;打底;粉刷
application of normative documents 标准文件的应用
application of optimization 优化技术的应用;优化方法的应用;最优化的应用
application of pesticide 农药施用
application of plaster 用粉饰
application of repayment of an overcharge 要求退还多收费用
application ofr retiring bill 赎票申请书
application of sample 加样;样品的使用或申请
application of sewage and sludge in agriculture 污水和污泥在农业中的应用
application of stress 施加应力
application of the brake 施行制动;施加制动
application of water 供水
application of wax resist 蜡抗蚀的应用(玻璃刻花);用蜡保护层
application-oriented 面向应用的
application-oriented protocol 面向应用的协议
application-oriented software 应用软件
application-oriented system 面向应用系统
application package 应用程序包
application piston 操纵活塞
application piston cotter 作用活塞扁销
application plot program(me) 应用绘图程序
application point 作用点
application position 制动位置
application printing 直接印花
application procedure 申请程序
application process 施用方法;应用进程;应用过程
application process group 应用过程组
application processing function 应用处理操作
application program(me) 使用程序;应用(服务)程序;操作程序
application program(me) exit routine 应用程序出口例行程序
application program(me) identification 应用程序

application program(me) image 应用程序映像
application program(me) interface 应用程序接口
application program(me) major node 应用程序主节点
application program(me) manager 应用程序管理程序
application program(me) output limit 应用程序输出范围
application programmer 应用程序员；应用程序设计员
application program(me) stub 应用程序子例程集
application programming 应用程序设计
application property 施工性能
application property at high temperature 高温使用性能
application protocol data unit 应用协议数据单元
application rate 应用率；使用率；施加速率；投加率；单位面积材料用量；施用量
application reference manual 应用参考手册
application-required language 应用要求语言；要求应用的语言
application reservoir 作用风缸
application right 应用权；认购权
application routine 应用例行程序；应用程序
application services 应用服务
application slip 货运舱位申请书；订舱单
applications of meteorology program(me) 气象学应用计划
application software 应用软件
application software engineering 应用软件工程
application software language 应用软件语言
application software package 应用软件包
application solid 施工固体分
application source program(me) 应用源程序
application specific integrated circuit 专用集成电路
application spreader 铺胶镘刀
application standard 应用标准
application study 应用研究
application subsystem 应用子系统
application support 应用支援(说明书)；应用支持
application support package 应用支援程序包
application technology satellite 应用技术卫星
application temperature 泼油温度(沥青材料)；使用温度
application temperature range 施工稳定范围
application terminal 应用终端
application thickness 涂抹厚度；涂布厚度；撒布厚度；使用厚度
application to application services 应用到应用的服务程序
application to prepare for export 出口准备申请书
application to seismic study 地震研究应用
application to set aside an award 申请撤销仲裁裁决
application to unload 卸货申请书
application tuning 应用调整
application valve 控制阀
application valve body 控制阀体
application valve bracket 控制阀托架
application valve conver 控制阀盖
application valve covergasket 操纵阀盖垫密片
application valve pin 操纵阀销
application valve spring 控制阀簧
application wall clips 贴合墙支架；贴合墙板夹
application ware 应用件
application ways of pesticide 农药施用方法
application weight 坯面施釉重量
application-with irrigation water 随水灌溉
applicator 漆膜涂布器；喷射枪；涂药器；涂膜器；涂膜机；申请者；洒施器；撒放机；加热电感器；消防喷头；高频加热电极；高频发热电极；敷贴器；敷料器；布粉器
applicator blade 追肥铲
applicator boot 施肥开沟器
applicator roll 上漆辊(子)
applieate 紧贴于表面的
applied 贴花的
applied accounting 应用会计
applied addition 添加物
applied administrative expenses 已分配管理费
applied aerodynamics 应用空气动力学
applied analysis 应用分析

applied analysis of mineral commodities 矿产品用途分析
applied anthropology 应用人类学
applied art 实用艺术；应用美术
applied chemistry 应用化学
applied classification of minerals 矿物实用分类
applied climatology 应用气候学
applied cloud physics 应用云物理学
applied code 适用标准
applied colo(u)r label 贴花；贴彩色标签
applied column 墙柱；壁柱；外加圆柱
applied computer science 应用计算机科学
applied cost 实用价格；应用价格；已分配成本
applied current 外加电流
applied database 应用数据库
applied data research 应用数据研究
applied design 贴花
applied divisional plan 实用目的区划
applied driving energy 作用锤击能量
applied ecology 应用生态学
applied economics 现实经济学；应用经济学
applied ecosystem analysis 应用生态系统分析
applied elasticity 应用弹性学
applied electromotive force 外加电动势
applied environmental geochemistry 应用环境地球化学
applied expenses 已分配费用；分配费用
applied experimental psychology 工程心理学
applied factory overhead expenses 已分配工厂间接费用
applied fine arts 实用美术
applied fluid dynamics 应用流体动力学
applied fluid mechanics 应用流体力学
applied force 外加力；作用力
applied forecasting 应用预测
applied forward voltage 正向外加电压
applied geography 应用地理学
applied geology 应用地质学
applied geomorphologic(al) map 应用地貌图
applied geophysics 应用地球物理学
applied geothermal 应用地热学
applied graphics 应用图学
applied hydraulics 应用水力学
applied hydrology 应用水文学
applied information management system 应用信息管理系统
applied linings 应用衬里
applied load 所施荷载；外施荷载；外加荷载；外加负载；使用负载；施加荷载；施加负荷；作用荷载
applied loading 作用荷载
applied manufacturing expenses 已分配制造费用
applied manufacturing overhead account 已分配间接制造费用账户
applied map 专用地图
applied material handling expenses 已分配材料管理费用
applied mathematician 应用数学家
applied mathematics 应用数学
applied mechanics 应用力学
Applied Mechanics Reviews 应用力学评论(期刊)
applied meteorology 应用气象学
applied metrology 应用计量学
applied microbial ecology 应用微生物生态学
applied microbiology 应用微生物学
applied mineralogy 应用矿物学
applied moisture 应用水分
applied moment 所施力矩；外施力矩；外加力矩；施加力矩；施工力矩；作用力矩
applied mo(u)lding 贴附线脚；贴附饰；(门窗、家具的)木条板；应用造型；镶板或护板分格线脚；墙板分隔线条
applied naval architecture 应用造船学
applied occurrence 应用的出现
applied oceanography 应用海洋学
applied optics 应用光学
applied overhead 已分配间接费用
applied overhead rate 间接费用分配率
applied physics 应用物理(学)
applied plate 防水贴板
applied-potential test 加压试验
applied power 外加功率
applied pressure 外加压力；外加电压；作用压力
applied probability 应用概率

applied probability theory 应用概率论
applied proceeds swap 利用收益的调换
applied pruning 人为修剪
applied range 应用范围
applied relief 捏雕
applied research 实用研究；应用研究
applied resource image exploitation system 应用资源图像开发系统
applied science 应用科学
applied science laboratory 应用科学实验室
applied seismology 应用地震学
applied shock 外施震动；外施冲击；外加冲击；作用冲击波
applied soil biology 应用土壤生物学
applied spectroscopy 应用光谱学
applied statics 应用静力学
applied statistics 应用统计学
applied stratigraphy 应用地层学【地】
applied stress 外施应力；外加应力；使用应力；施加应力；作用应力
applied tectono-geochemistry 应用构造地球化学
applied tension 外施张力
applied thermodynamics 应用热力学
applied thread 加线装饰法(玻璃上)；贴玻璃线
applied thrust 外施推力；施加的推力；作用推力
applied to 施加
applied torque 施加的扭力矩
applied train 运用车
applied trim 外加贴面；现制贴面
applied voltage 外加电压
applied weathering 防水(件)；防风化措施
applied work 外施的功
applique 热合镶嵌；贴花织物；镶刻
applique armor 装甲披挂
applique carving 贴雕
applique circuit 附加电路
applique decoration 贴花(镶饰)
applique design with exposed body 露胎贴花
applique work 补花装饰
apply 涂；提出申请；施行；运用
apply a header coat 露头砖盖面
apply a mist coat 施加薄雾涂层
apply at the same rate 施用同样剂量
apply boring test 普查钻探
apply brake board 施闸牌；踩下制动板
apply by brushing 用刷子涂上
apply by letter 通信申请
apply colo(u)r 敷色
apply fertilizer 施肥
apply for berth 泊位申请
apply force 作用力
apply for collection 托收
apply for insurance policies 投保
apply for loan secured 申请抵押借款
apply for remittance 托汇；托付
applying cost ratio to ending inventory 应用成本比例计算期末存货
applying for loans 申请贷款
apply oil 上油；加润滑油
apply sanction 实行制裁
apply the ax to nonproductive expenditures 削减非生产性开支
apply to the customs 向海关申请出口或进口；报关
apply work 做功
appoint 任命；约定；指定
appointed bank 指定银行
appointed berth 指定泊位
appointed cost 分摊成本
appointed day 指定日期
appointed import licensing bank 进口签证银行
appointed licensing bank 签证银行
appointed project manager 派任项目经理
appointed store 特约商店
appointed surveyor 指定检验人
appointee 被任命者
appointing authority 指定仲裁员的机构
appointment 位置；设备；家具；车身内部装饰；任命；委派；指派；指定；职位
appointment and performance 委托办理
appointment call 定人定时呼叫；约访；约电话
appointment of an adjuster 委任理算人
appointment of assistants 任命助理
Apollo 太阳神(希腊神话)
apportion 分配；按计划分配；按比例分配；摊派；分列

apportionable part 可分摊部分
apportion design 分配设计;分配法
apportioned by volume 按体积分配(的)
apportioned by weight 按重量分配
apportioned charges 摊派费用;摊派成本;分摊费用;分派费用
apportioned cost 分派成本;分派费用
apportioned fixed cost 分摊固定成本
apportioned purchase price 派购价格
apportioned tax 摊派税捐;摊派税款;分配税款;分配赋税
apportioning cost 分摊费用;分摊成本;分摊值;分配成本
apportioning depreciation 分摊折旧;分配折旧
apportionment 拨款;土地或资金分配;分摊;分配
apportionment charges 分摊费用
apportionment clause 按比例分摊条款
apportionment cost 摊派成本
apportionment exercise 分派法
apportionment method 分摊法
apportionment of cost of grade separation 立交造价分派
apportionment of general average 共同海损分摊额
apportionment of losses 亏损分摊
apportionment of valuation 价值分摊额
apportionment ratio 分配率
apportionment tax 摊派税
apportion the total conversion cost 分解过渡费用
apposed glacier 汇合冰川;合流冰川
apposition 归并;对接;并置;对向;并列
apposition beach 归并海滩;附生海滩;并列海滩
apposition fabric 并生组构
appraisal 评价书;鉴定估价;鉴定;鉴别;估价
appraisal account 估价账户
appraisal adjustment 估价调整
appraisal basis 估价基础
appraisal capital 评估资本;增值资本;估定资本额
appraisal clause 估价条款
appraisal company 评估公司;估价公司
appraisal cost 鉴定成本;评价费用;鉴定费(用)
appraisal criterion 估价标准
appraisal curve 估计曲线
appraisal date 鉴定日期;估价日期
appraisal drilling 评价钻探
appraisal entry 估价分录
Appraisal Foundation 资产评估基金会
appraisal increment 重估增值
appraisal institute 资产评估学会
appraisal inventory 估价财产目录
appraisal item 估价项目
appraisal meeting 评议会
appraisal method 鉴定方法;评定方法;估价法
appraisal method of depreciation 折旧估价法
appraisal method with the aid of contour graph 轮廓图形评价法
appraisal mission 估价工作组
appraisal of assets 资产估价
appraisal of business 经营估价;营业估价
appraisal of construction project 建设项目评价
appraisal of damage 估损
appraisal of equipment 设备鉴定
appraisal of mine long-range 矿山远景估价
appraisal of new technology 新技术鉴定
appraisal of plant assets 设备资产估计;工厂资产估价;厂场资产的估价
appraisal of project from the view of natural resources 项目自然资源评价
appraisal of property and funds 清产核资
appraisal of quality 品质鉴定;质量鉴定
appraisal of real assets 不动产估价
appraisal of real estate 房地产估价;不动产估价
appraisal of scientific and technical achievements 科技成果鉴定
appraisal of scientific and technologic(al) achievements 科技成果鉴定
appraisal operation 估价作业
appraisal person 估价师
appraisal price 估价
appraisal principles 资产评估原则
appraisal procedure 估价程序
appraisal process 资产评估过程
appraisal purpose 估价目的
appraisal report 评估报告书;评定报告;鉴定报告
appraisal reserve 估价准备

appraisal survey 评价勘查;估价调查
appraisal tenet 估价原理
appraisal value 估定价值
appraise 鉴别;评价;鉴定;估价
appraise and examine 评审
appraise assets 评估资产
appraise at the current rate 变价
appraised price 估价
appraised value 评定价值;估定价值
appraisement 鉴定;估价;评价;鉴别
appraiser 评价人;鉴认人;鉴定员;鉴定人;验估人;估价人
appraiser in chief 主审人
appraiser's surveyor 验光库
appraising date 评审时间
appraising grade 评审级别
appraising property 评估资产
appraising unit 评审单位
appraising unit in chief 主持单位
appreciable 值得重视的;可观的;可估计的;可观的
appreciable assets 应升值资产
appreciable attack 重侵蚀
appreciable error 可估误差;可察觉误差;显著误差;粗差
appreciable rebound 可估测回弹
appreciate 升值;估价
appreciated currency 增值货币;增值通货
appreciated value 增值价值
appreciation 市场涨价;升值;涨价;增值
appreciation in asset value 资产增值
appreciation of a project 对工程项目的评价
appreciation of arts 艺术鉴赏
appreciation of depth 景深鉴别力
appreciation of fixed assets 固定资产增值
appreciation of land value 土地增值
appreciation of market price 市场上涨
appreciation surplus 涨价盈余
apprehend 扣押
apprehension 理解
apprentice 徒工;实习生;见习生;学徒;帮工
apprentice draughtsman 见习绘图员;绘图学员
apprentice-driller 钻探徒工
apprentice engineer 实习轮机员
apprentice officer 实习驾驶室
apprentice pilot 实习引航员
apprentice sailor 实习水手
apprentice seaman 实习水手;见习船员
apprenticeship 做学徒;训练期;学徒期间;学徒制
apprenticeship act 学徒条例
apprenticeship system 学徒制度
apprenticeship training 学徒培训
apprentice's school 技工学校
apprentice system 徒工制
apprentice training 学徒训练
apprentice(work)shop 学徒实习工场;学徒实习车间
appression 被压紧状态
approach 入门;趋近;驶近;进口航道;进口道;进场;近岸;接近;渐近;门路;逼近;引道
approachability 可接近性
approachable 易接近的
approachable subset 可接近子集
approach adit (隧道、坑道施工的)平巷;(隧道、坑道施工的)导洞;施工导流隧洞
approach aid(s) 引航标志;导标;着陆导航设备;导航设备
approach alignment 桥头引道接线;桥头引线定线;引道接线
approach-and-landing 进近着陆
approach and landing chart 机场进近区起降航空图
approach angle 前进角;(工作部件的)碎土角;接近角;渐近角
approach announcing in cab 机车接近通知
approach announcing lamp 接近预告灯
approach-avoidance conflict 趋避冲突
approach bank 引桥;桥梁引道路堤;护岸;引堤;引道
approach beam 临场引导波束
approach block 接近闭塞区段
approach bridge 引桥
approach channel (气体的)引入通道;引槽;进口航道;进港口航道;进港航道;进出港航道;引水渠;引航道;引渠

approach channel chart 进港航道图
approach chart 进港引水图;进出港海图;进场图;机场进近区航空图
approach coding 接近发码
approach cone 入口锥(体);桥头锥体
approach contact 渐近接触;啮入接触
approach control zone of air station 航空站进场控制区
approach conveyer 连接输送机
approach course 进场路线
approach curve 涨前段
approach cut 不完全切削
approach cutting 桥头挖方;隧道两端挖方;进洞引道挖方;引桥挖方;引道开挖
approach cutting quantity 引桥挖方量
approach delay 进口道延误
approach embankment 桥头引线;桥头路堤;引线路堤;引堤;引道(路)堤
approach end 尖端【道】;引道尽头;鼻端
approach end runway 跑道进近端
approaches and problems 方法和问题
approaches to architecture 建筑创造途径
approach fill 桥头填方;引桥填方;引道填筑;引道(路)堤
approach flap (浮码头的)铰接板;(移动式桥的)引道(折)板
approach flow 行近流;迎面气流
approach flume 引水槽
approach gate 进场门
approach grade 引线坡度;引道坡度
approach gradient 引道坡度
approach grafting 贴接
approach hole 导孔
approach indication 接近表示
approach indication system 预告显示制度;预告复示式显示制度
approaching announcing 列车接近通知
approaching channel 引槽
approaching flow 临近水流;行近(水)流
approaching side 进入侧
approaching speed 驶近(靠码头)速度;靠码头速度;靠泊速度
approaching vehicle 到达车辆
approaching velocity 驶近(靠码头)速度;靠码头速度;靠泊速度;临近速度;接近速度;接近流速;行近速度
approaching velocity before the pier 墩前行进流速
approaching vessel 开来的船舶
approach jetty 引道突堤
approach junction terminal tracks de-crossing 枢纽进站线路疏解
approach light 指示灯;信号灯;降落灯;进近灯;着陆灯
approach lighting 接近点灯
approach lighting system (of airfield) 飞机场夜航降落灯光系统
approach limb 涨前段
approach line 进站线(路);进出站线路;进出线路
approach locking 进场锁定;接近锁定;接近锁闭(完全锁闭)[铁]
approach-marker-beacon transmitter 机场信标发送机
approach method 趋近法
approach nose 临靠两条公路的土地;路岛端;接近端
approach of bridge 通往桥的道路;引桥路
approach parapet 翼墙上的防浪墙
approach path 渐近轨迹;啮入轨迹
approach pavement to the road decking 铺设过渡段路面
approach phase 渐近相位;啮入相位
approach pit 导坑
approach point 接近点
approach rail 引轨
approach ramp 建筑物坡道;进出口斜坡段;引桥坡道;引坡匝道;引道斜坡段;引道坡道;公路引道
approach relay 接近继电器
approach road 引路;引道
approach route 进出港航道
approach sea lane 进出港航道
approach section 进入段;(车辆在接近隧道时光度变化的)接近区段
approach section of a highway crossing 道口接

近区段
approach segment 进场段;涨前段
approach shoulder 引道路肩
approach sign 接近标志;引线标志;引道标志
approach signal 列车近站信号;接近信号;行近信号;港口信号
approach slab 桥头搭板;引道板
approach span 引桥跨(度);引跨;岸跨
approach speed 接近速度;驶进速度;进入车速;进口车速;进车车速;进场速度;驾近速度
approach spillway channel 岸边溢洪道
approach stream velocity 迎面气流速度
approach structure 引道结构
approach surface 机场跑道进入面
approach system 导进机场系统
approach table 输入辊道
approach the advanced world level 接近世界先进水平
approach the orebody 接近矿体(钻井)
approach the topmost level 接近最高水平
approach time 临场时间;逼近时间
approach track 进站轨道;进站线(路);接车线
approach transition 入口渐变段
approach trench 交通濠;交通沟
approach trestle 引道栈桥;引道(排架)栈桥
approach trestle pier 引桥式码头
approach tunnel 引线隧道
approach velocity 驾近速度;行进速度;行近流速
approach velocity head 行近流速水头
approach vessel 驾来的船
approach viaduct 引道高架桥;高架引桥
approach visibility 倾斜能见度;接近能见度
approach wall 导墙(船闸、进水渠)
approach way 引道;进场段
approach zone (车辆在接近隧道时光度变化的)接近区段;接近区;引道区
approach zone district 引道区域
approach zone luminance 洞外照明;洞外亮度
approbation 批准;核准;委员会结论;品种鉴定
approcynum hendersonü hook 大花罗布麻(一种固砂植物)
appropriate 适当的;合适的;合理的;拨用;拨给
appropriate accumulation rate 适度积累率
appropriate A for B 把 A 拨给 B
appropriate a fund 拨款
appropriate assistance 必要的协助
appropriate authority 相关当局
appropriate berth 专用泊位
appropriate body 有关机关;有关机构;有关部门;主管机关
appropriate chart 专用图
appropriated berth 提供专用泊位;专用码头
appropriated berth scheme 专用泊位计划
appropriated materials 调拨材料;拨定材料
appropriated profit 拨定利润
appropriated retained earnings 指定用途留存收益;拨定的留存盈余
appropriated right 专权权;用水专有权
appropriated surplus 指定用途盈余;拨定盈余
appropriated wharf 专用码头
appropriate economic growth 适度的经济增长
appropriate edge matrix 合理边缘矩阵
appropriate goods 溢收货
appropriate interest rate 适当利率
appropriate level of detail 适当的细化程度
appropriately stringent financial policy 适度从紧原则
appropriate money 拨款
appropriateness 适合程度
appropriate package 合适包装
appropriate price 调拨价格
appropriate public authority 适当的公共机关
appropriate right of wage 工资分配权
appropriate sanction 适当制裁
appropriate scale of operation 规模经营
appropriate season of giving birth 最佳生育季节
appropriate shed 专用货棚
appropriate speed for economic growth 经济适度增长
appropriate technology 适应性技术;适用技术;适应的技术;适当的技术
appropriation 私用;经费;占用;拨款
appropriation account 拨款账户;分配账户
appropriation allotment 经费分配;支出分配款

appropriation analysis 支出明细账
appropriation budget 经费预算;支出预算
appropriation change to loan 拨改贷
appropriation encumbrance 支出保留数
appropriation expenditures 经费支出
appropriation expenditures ledger 经费支出分类账
appropriation for addition of property 添置财产拨款
appropriation for pollution control 治理三废拨款
appropriation for three categories of scientific use 科技三项费用拨款
appropriation from state treasury 国家预算拨款
appropriation language 经费说明
appropriation ledger 支出拨款分类账
appropriation limitation 拨款限额
appropriation of charges 费用的划分
appropriation of economy 经济占用;经济的占用;经济的拨款;经济拨款
appropriation of land 土地的占用
appropriation of net income 净盈余占用
appropriation of payment 支付指定
appropriation of profit 利润分配
appropriation of the funds 提取基金
appropriation of water right 水权的专有
appropriation period 拨款期限
appropriation plan 调拨计划
appropriation reserves 经费储备金
appropriations for additions of property 扩展产业的拨款;扩充事业的拨款;发展产业拨款
appropriations reserve 经费准备金
appropriation warrant 年度拨款支付书
appropriative rights 使用权
appropriative water rights 专用水权;专利水权;专有水权
appropriator 拨款者
appropriator of water rights 水权拥有者;水权享用者
approral test 认可试验
approval 认可;批准;核准;许可
approval authority 批准权;批准机关;审批机构;核准权
approval authority for projects 项目核准权
approval by an authorized foreign exchange bank 经指定外汇银行认可
approval container 批准集装箱
approval drawing 认可图(纸);批准图
approval in writing 书面批准
approval notice of the contract 合同批准通知
approval of casting repairs 铸件修补的批准
approval of defects in material 材料中缺陷的认可
approval of draft 草图核准
approval of drawings 批准图纸
approval of individual container 个别批准(指集装箱等)
approval of material repairs 对材料修补的批准
approval of materials 材料的审批
approval of new materials 新材料的批准
approval of parts forgings repairs 锻件修补的批准
approval of plans 批准规划;批准计划;批准方案
approval of project 工程立项
approval of sample 认可样品
approval procedure 审批手续
approval process 审批过程
approval sales 试销
approval stage 批准阶段
approval standard 合格标准;许可标准
approval test 控制试验;控制实验;鉴定试验;检查试验;核定试验;合格性检验;效能试验;验收试验;验收测试
approve 批准;许可;赞成
approved 审定;已批准的
approved accountant 认可特许会计师;批准会计师
approved apparatus 合格设备;防爆装置
approved attorney 经保险公司认可的律师
approved budget 核定预算;法定预算
approved cable 防爆电缆
approved capital 核定资金
approved collection area 经核准的采种区
approved combustible plastics 被认可的可燃塑料;被批准的可燃塑料
approved continuous examination program (me)decal 连续检验计划标记
approved data 认可资料
approved design 批准设计

approved drawings kept at building 存于施工场地的批准图样
approved equal 认可同等替换;同意代用品;准许代用品;同意换用合格;互换合格材料;公认互换相等
approved equal substitution 同意代用品;准许代用品
approved equivalent method 认可的等效方法
approved estimate 核定概算
approved facility 合格设施
approved first-aid outfit 认可的急救包
approved for construction 批准施工
approved investment 核定投资总额;核定投资
approved marking 认可的标志;合法的标志;安全标志(放炮器)
approved masonry 核准圬工
approved materials 许用材料
approved method 经认可的方法;经批准的方法;批准方式;准许的方法
approved norm 核定定额
approved officer 签核人;核付员
approved parts list 批准的部件清单
approved pattern 批准的型号
approved plan 核准的图纸;核准的计划;批准的图纸
approved plastic filler 许用塑料填料
approved polyvinyl chloride 改性聚氯乙烯
approved product 定型产品
approved program(me) 核定计划
approved rule 经认可的规则;经批准的规则;实行的规则(建筑部门、权威机构)
approved sample 批准样品;同意样品
approved shot-firing apparatus 安全放炮器;防爆型放炮器
approved signatory 批准签署人
approved sod 合格草皮
approved symbol 审定图号
approved vendor list 批准的供应商清单
approved wagon requisition plan 核定的要车计划
approving authority 有批准权的权威单位
approximability 可近似性;可逼近性
approximant 近似值;近似式
approximate 近似的;近似;约;概略的;逼近;约计
approximate adjustment 近似平差;约略调整
approximate alignment 概略定线
approximate amount 概算
approximate analysis 近似分析
approximate analysis method 近似分析(方)法
approximate analytic(al)method 近似分析(方)法
approximate answer 近似回答
approximate approach 近似解;近似法;近似解法;逐次逼近法
approximate assumption 近似假设;近似假定
approximate boundary 近似边界;概略界线
approximate calculation 概算;近似算法;近似计算法;近似计算
approximate characteristic equation 近似特征方程
approximate circle 近似圆
approximate coefficient of deviation 自差近似系数
approximate colo(u)r harmony 近似色调
approximate computation 近似算法
approximate condition 近似条件
approximate configuration 大致的尺寸形状
approximate construction 近似组成;近似作图
approximate continuity 近似连续
approximate contour 假想构造等值线;近似等高线
approximate contour line 未精测等高线
approximate convergence 近似收敛
approximate coordinates 近似坐标;概略坐标
approximate cost 近似成本;约计成本;估算成本
approximate data 近似数据
approximated cost 概算成本
approximate depth contour 未精测等深线
approximate derivative 近似导数
approximate diameter 近似直径
approximate differentiation 近似微分
approximate dimension 概略尺寸
approximate distribution 近似分布
approximate drawing 近似作图
approximate ellipse 近似椭圆
approximate ellipse arch 近似椭圆拱
approximate equation 近似公式;近似方程
approximate equilibrium 大致平衡
approximate equilibrium condition 近似平衡条件
approximate error 近似误差;逼近误差

approximate estimate 粗略估算;粗略估计
approximate estimate on cubic(al) meter basis 单立方概算
approximate estimate on square metre basis 单方概算
approximate estimate sum 近似估算额
approximate estimation sheet 概算估价单
approximate evaluation 近似计值;近似计算;近似估值;近似估算;近似估计
approximate expansion 近似展开式
approximate expression 近似(公)式;近似表达式
approximate factor values 近似要素价值
approximate formula 近似(公)式
approximate integration 近似求积;近似积分
approximate lethal dose 大致致死量
approximate-limit order 不确切限额指专
approximate loaded displacement 近似满载排水量
approximate location 近似位置
approximately 大约的;大约
approximately axial 近似同轴的
approximately continuous 近似连续
approximately equal 近似相等
approximately equal to 近似等于
approximately equidistance conic(al) projection 近似等距圆锥投影
approximately equidistance minimum error conic(al) projection 近似等距最小误差圆锥投影
approximately equidistance projection 近似等距投影
approximately estimated cost 概算造价
approximately normal distribution 近似正态分布
approximately vertical air photograph 近垂直摄影航片
approximate match 近似符合
approximate measure 近似值测定数;近似测度
approximate measurement 近似测量;概略尺寸
approximate method 逐渐接近法;近似法;逐次逼近法
approximate mode 近似众数
approximate normality 近似正态性
approximate number 近似数;概数
approximate numerical measures 近似数值度量
approximate packed weight 近似总重量
approximate parallel motion 近似平行运动
approximate plotter 低精度测图仪
approximate position 近似位置;概位;概略位置
approximate quantity 约略数量;近似数量;近似量;概略工程量
approximate reading 近似读数
approximate reasoning 近似推理
approximate relief 草绘地貌
approximate result 近似结果
approximate rule 近似律
approximate sampling analysis 近似抽样分布
approximate sampling distribution 近似抽样分布
approximate shrinkage 适量收缩
approximate sight 概略瞄准器
approximate solubility 近似溶解度
approximate solution 近似解;逐次逼近法
approximate surveying 概略测量
approximate theory 近似理论
approximate treatment 近似计算;近似处理
approximate unit weight 近似容重
approximate valuation 近拟赋值
approximate value 近似值;近似价值
approximate value measure 近似值度量
approximate variance of the ratio estimate 比率估计的近似方差
approximate vertical interval 近似等高线
approximate water course 草绘河流符号
approximate weight 约计重量
approximating function 逼近函数
approximating optimum allocation 近似最优配置
approximating variance 近似方差
approximation 近似化;近似;逼近
approximation approach 近似法
approximation assumption 近似假设;近似假定
approximation by least squares 用最小二乘方的近似法;最小二乘法逼近
approximation by power 用幂公式逼近
approximation by semi-infinite slopes 半无限斜率逼近
approximation characteristic 近似特征
approximation composition 大致成分
approximation construction 近似作图
approximation curve 近似曲线
approximation equation 近似公式
approximation equation for migrating interface between salt and fresh water 移动咸淡水界面近似方程
approximation equation for steady interface between salt and fresh water 稳定咸淡水界面近似方程
approximation equation of interface movement 界面移动近似方程
approximation error 近似误差
approximation expression 近似公式
approximation formula 近似(值)公式
approximation for swing 旋角近似法
approximation for tilt 倾角近似法
approximation hypothesis 近似假设;近似假定
approximation method 近似方法
approximation of 1st degree 一次近似
approximation of function 函数近似;函数逼近
approximation on the average 一次近似
approximation polynomial 近似多项式
approximation problem 逼近问题
approximation reading 近似读数
approximation solution 近似解
approximation theorem 近似定理;逼近定理
approximation theory 逼近理论;近似理论
approximation theory of price index number 近似值物价指数理论
approximation to distribution 近似分布
approximation value 近似值
approximity warning indicator 逼近警报指示器
appulse 表观逼近;半影月食
appurtenance 配件;零件;管件;辅助机组;属具;附属物;附属设施;附属设备;附属建筑;附件;辅助设备;辅助工具;附属部件
appurtenance material 附属材料
appurtenant 附属物;附属的
appurtenant easement 附属通行权;附属使用权
appurtenant structure 附属结构;附属建筑物;附属工程
appurtenant works 附属设施;附属设备;附属工程
a pressing matter of the moment 当务之急
a previous representative period 以前有代表性的时期
Aprey faience 阿普列陶瓷
apricot 杏黄色
apricot kernel oil 杏仁油
apricot tree 杏树
aprindine 安搏律定
apring main leaf camber 钢板主片拱度
a prior estimate 事前估计
a priori 先验的(拉丁语)
a priori distribution 事前分布;先验分布
a priori error bound 事前误差界;先验误差界
a priori estimate 先验估值;先验估计
a priori probability 事前概率;先验概率
apriorism 先验论
a priori standard deviation 先验均方根差;先验标准差
apron 前方堆场;舞台前唇;坞头槛;台唇;散水(房屋外墙脚坡);码头前沿作业地带;码头前沿;护柱;护坦;护脚板;护底;海漫;选矿槽析流板;底脚护坡;底板;堤岸码头边缘;导脂层;垂层羊毛;窗台;副船首材;防冲铺砌;防爆坡;冰水沉积平原;冰川前砂砾层;前沿作业地带
apronal 丙戊酰脲
apron applicator 带式上浆装置
apron area 停机坪
apron arm 裙臂(轮动式铲运机)
apron belt 锭子传动带(俗称龙带)
apron block 护坦块体;护脚块体;镇块;挡板座
apron board 裙板
apron bolt joint 平板带接头
apron cable (推土机的)铲索;(推土机的)铲缆;裙索(轮式铲运机)
apron control valve 铲动控制阀(轮式铲运机)
apron conveyer 槽形链板输送机;槽板式输送机;裙板式输送机;链板式输送机;裙式运输机;鳞板输送机;链板式输送带;板式输送机
apron conveyer feeder 链板式输送机进料器
apron crane rails 码头前沿起重机轨道
apron crib 护坦笼框
apron distributor 带式自动分配器
apron dryer 输送器式干燥炉;帘式干燥机;罩式干燥器;帘式烘干机
apron eaves piece 檐口披水板;遮阳板;封檐板;遮檐板
apron elevator 平板提升机;链板式提升机
apron extension 护坦延伸段;护坦延长段;海漫
apron facing 窗肚墙;小梁侧封板;窗肚墙面装饰
apron feed distributor 带式撒肥机
apron feeder 制砖传送带;裙式给料器;裙板式喂料机;带式加料机;带式给料器;带式给料机;槽形链板给料机;板式喂料机;板式给料机
apron feeder spillage 裙板式喂料机的漏料
apron flashing 引水板;屋面泛水;挡雨板;披水板;遮檐板
apron flood lighting 前场(舞台)泛光照明;停机坪泛光照明
apron gallery 码头前沿廊道
apron guide 帆布输送带导向板
apron hay loader 传送带式装干草机
apron jack 铲斗液压缸(轮式铲运机)
apron lathe 拖板箱车床
apron lever 溜板手把;铲板操纵杆
apron lift 扩大舞台口升降装置
apron lift arm 铲斗提升臂(轮式铲运机)
apron lines 铲斗绳(轮式铲运机)
apron lining 楼梯井的竖向套筒;裙板线脚;楼梯装修中的护墙裙板;楼梯平台裙板;小梁侧封装
apron lip 铲斗唇板(轮式铲运机)
apron marking 停机坪标志
apron mo(u)lding 门中帽板装饰线脚;门挡板上部的楣饰;柜底线脚;裙板线脚;门中档装饰
apron of a dock 船坞底板
apron of dike 堤前护坦
apron opening (大小)(轮式铲运机的)铲斗斗门开口;(轮式铲运机的)铲斗斗门口径
apron panel (窗户的)拱肩镶板
apron picking conveyer 板式拣矿输送机
apron pickup 带式检拾器;带式检拾器
apron piece (楼梯平台的)支承小梁;遮檐板;承台梁
apron pivot 闸门枢轴
apron plain 冰前平原;冰川沉积平源
apron plate 闸门;挡板;裙板;前护板;围护板;大型混渗板
apron plate of foundation 基础裙板
apron rail 有挡板的围栏;护坦栏杆;有挡板的护栏
apron railway 码头前沿铁路
apron ring 裙圈
apron rolls 裙板式输送机托辊;运输机皮带滚轴;运输带的托辊
apron rope 拉铲缆(铲土机);拉铲索
apron sequence valve (轮式铲运机的)铲斗顺序阀
apron slab 护坦(底)板;防冲刷底板;防冲底板
apron slab of overflow 溢流道护坦
apron slab of spillway 溢流堰顶护坦;溢流道护坦
apron slope 护坦坡度;迎面坡;迎波面坡
apron space 码头前沿区;码头前沿作业地带
apron stage 舞台前附台;固定舞台口的外延部分
apron stone 护坦石料;护脚石
apron strip 盖缝条
apron tank 胶带式显影罐
apron track 前沿区轨道;码头前沿区轨道;轻便轨道;码头(沿边)铁路线;码头前沿铁路(线)
apron-type 裙式的;板式的
apron-type feeder 链板式加料器;链板式加料机;带式进料器;板式进料机
apron-type mechanical feeder 链板式输送机进料器
apron wall 水下窗间墙;裙墙;(水下窗间的)护墙板;前护墙;窗心墙
apron wall facing 窗台下墙饰面;窗台墙饰面;前护墙修饰
apron wall lining 窗台下墙衬筑;窗台下墙衬砌;窗台墙衬砌;窗台墙衬砌;前护墙衬砌;前护墙封板
apron wall panel 窗台下墙板;窗台墙板;前护墙镶板
apron washer 传送带式清洗机
apron wharf surface 码头前沿
apron wheel lever
aprotic media 疏质子介质;对质子惰性的介质
aprotic solvent 疏质子溶剂;非质子传递溶剂
aprotogenic solvent 非给质子溶剂
aprous 耐火的
aps 欧洲山杨
apsacline 倾斜型;斜倾型
apse 近星点;教堂半圆形后殿;拱点;半圆形室;半

圆形龛
apse aisle 后殿外廊;教堂内由唱诗班通道伸向半圆室或周围的通道;教堂唱诗班通道
apse arch (教堂的)半圆后堂拱
apse arch impost (教堂的)半圆后堂拱墩
apse-buttress 半圆形状扶垛;半圆形扶垛;半圆室扶垛
apse-buttressed 用半圆壁龛支撑的
apse chapel 半圆形小教堂;辐射式小教堂
apse in echelon 梯(形排)列拱点
apse line 拱线
apse node terms 拱交项
apses in echelon 梯形排列拱点;梯列拱点
apse window 教堂半圆形窗户;(教堂的)半圆室窗
apsidal 半圆形建筑的;半圆建筑;多角形建筑;半圆室的
apsidal angle 毗拱拱角距;横心角;拱心角
apsidal chapel 辐射式小教堂;半圆形小教堂;唱诗班礼拜堂
apsidal choir 半教堂圆形歌唱半席位
apsidal constant 近星点运动常数
apsidal distance 拱点力心距;拱距
apsidal entrance hall 半圆形后堂门厅;半圆形门厅
apsidal line 拱线
apsidal motion 拱线转动;拱线运动
apsidal period 近星点转动周期;拱线转动周期;拱线运动周期
apsidal rotation 近星点转动;拱线转动
apsidal rotational motion 拱线旋转运动
apsidal surface 长短径曲面
apsides 拱点
apsidiole 教堂小型半圆形后殿;附属半圆形建筑;教堂中小型半圆室
apsilate 有纹饰的
apsis 极距点;拱顶;顶点;半圆室;半圆壁龛;近星点;教堂半圆形后殿;拱点
apsis arch (教堂的)半圆后堂拱
apsis arch impost 教堂半圆形后殿拱(墩)
apsis window 教堂半圆形后殿窗户;(教堂的)半圆室窗
apteral 无侧柱的(古希腊,古罗马);两侧无柱式的
apteral temple 无侧柱的寺院;无侧柱的神殿;无侧柱的寺庙(古希腊)
apterium 裸域
Apt faience 阿珀特陶器
Aptian 阿普第阶
aptitudal station 恰当布置;适当规范
aptitude factor 适应系数
aptitude for stereoscopic vision 立体观测能力
aptitude test 性能试验;适应性试验;性能测验;才能测验
aptitude to rolling 可轧性
apuanite 硫氧锑铁矿
apyre 红柱石
apyretic 无热的;不发热的
apyrexia 热歇期;无热期
apyrexial 无热的
apyrite 红电气石
apyrogenetic 不致热的
apyrogenic 不致热的
apyrous 不易燃的;抗火的;防火的
aqua 溶液;[拉]水液体;[拉]水凝胶;水剂
aqua alta (尤指意大利威尼斯城的)洪水
aqua ammonia 氨水(溶液)
aqua ammonia applicator 氨水注施机
aqua ammonia pump 氨水泵
aqua bidestillata 重蒸馏水
aqua blue 水蓝色
aquaboard 滑水板
aqua bulliens 开水;沸水
aqua calcis 石灰水
aqua communis 普通水
aqua complex 含水复合物
aqua compound 含水化合物
aqua condensate pump 溶液冷凝液泵
aquacrete 防水波特兰水泥
aquaculture 水产养殖;水中栽培
aquaculture plant 水生栽培植物
aquaculture water 养殖用水
aquadag 石墨沉淀;胶态石墨;碳末润滑剂;水石墨;石墨滑水涂料;胶体石墨;导电敷层
aqua distillate 蒸馏水
aquafact 海滩巨砾
aquafalfa 地下水位高的土地
aquafarm 海洋牧场

aqua fervens 热水
aqua fluvialis 河水
aqua fontana 泉水
aqua fortis 浓硝酸;硝酸
aqua frigida 冷水
aqua garden 水景园
aquage 水路
Aquagel 艾阔杰尔土(美国产的一种膨润土);水凝胶
aquagene tuff 玻质碎屑岩
aquagraph 导电敷层
aqua landscape 水景观
aqualf 潮湿淋溶土
aqualite 冰岩
aqualithic 混凝土地面液体硬化剂
Aqualon 阿கேலon 阿奎仑高吸水性纤维;阿奎仑丙烯腈短纤维
aqualung 轻型的水下勘探装置;潜水呼吸器;潜水肺;水中呼吸器;供氧潜水罩
aqualunger 蛙人
aquamarine 海水;水蓝宝石;蓝绿玉;蓝绿石;蓝晶;海绿色;海蓝色;海蓝宝石
aquamarine chrysolite 海蓝贵橄榄石
aquamarine glass 海蓝玻璃;海蓝宝石玻璃
aquamarinus 海绿色
aquamarsh 沼泽;水沼地
aquamarsh soil 水沼土
aquameter 水力计
aquametry 滴定测水法;测水法
aquamotrice 挖泥器具(其中的一种)
aquanaut 潜水员;潜航员;海底观察员
aquanaut work 潜水作业
aquanote apparatus 海中居住设施
aquanote planning 海中居住计划
a quantitative index 定量指数
a quantity of seeds 一定数量的种子
aquaphone 听漏器;漏水探知器;检漏器
aquaplaning 液滑现象
aquaplant 水生植物
aqua pluvialis 雨水
aquapolis 水上城市
aquapulper 水力碎浆机
aqua pure 纯水
aqua regia 王水
aquarelle 水彩画(法);具有透明水彩画
aquarium 培养缸;水族馆;玻璃缸
aquarium disinfection 水族馆消毒
aquarium reactor 水池式反应堆
Aquaroll mangle 阿夸罗水压式轧液机
aquarter 大约四分之一
aquaseal 密封的;密封剂(电缆绝缘涂敷用);水封
aqua-solution method 水溶液法
aquastat 水温自动调节器
aqua sterilisa 灭菌水
aqua storage tank 储水池;储水槽
aqua system 水压储[贮]存系统;充水式储油罐
aquatard 密封焦油;防水层
aquated ion 水合离子
aquatel 水上旅店
aqua tepida 温水
aquathermal pressure 水热增压
aquathermal pressuring theory 火热增压说
aquathruster 气压扬水机
aquatic 水生的
aquatic acidification 水的酸化
aquatic adjustment 水中调整
aquatic adventitious roots 水生不定根
aquatic and terrestrial ecosystem 水陆生态系统
aquatic animal 水动物
aquatic bacteria 水生细菌;水生菌
aquatic bioacoustic(al) 水生生物声
aquatic bioda 水生生物相
aquatic biologic(al)test(ing) 水生生物试验;水生生物测试
aquatic biology 水生生物学
aquatic biota 水生生物群;水生生物
aquatic bird 水禽;水鸟
aquatic blue oak 水青榈
aquatic building(s) 游泳建筑物;水上运动建筑物;浴建筑物
aquatic chemistry 水生化学;水化学
aquatic community 水生群落
aquatic crops 水生作物
aquatic earthworms 水蚯蚓
aquatic eco-environment 水生生态环境
aquatic eco-environment environmental risk as-

sessment 水生生态环境风险评价
aquatic eco-environment quality 水生生态环境环境质量
aquatic ecological comprehensive index 水生生态综合指数
aquatic ecologic environmental quality 水生生态环境环境质量
aquatic ecology 水生生态学
aquatic ecosystem 水生生态系(统)
aquatic ecosystem element 水生生态系统要素
aquatic ecosystem pollution 水生生态系统污染
aquatic ecotoxicology 上生态毒理学;水生生态毒理学
aquatic environment 水生环境
aquatic environmental chemistry 水环境化学
aquatic environment restoration 水环境修复
aquatic farm 水产养殖
aquatic fauna 水生动物(区系)
aquatic feed web 水生食物网
aquatic flora 水生植物(区系)
aquatic flower 水生花卉
aquatic food chain 水生食物链
aquatic fulvic acid 水生富里酸
aquatic fungi 水生真菌
aquatic green manure 水生绿肥
aquatic growth 水生植物;水草;水生生长物;水草生长
aquatic habitat 水生生境;水生环境
aquatic habitat resources 水生生境资源
aquatic habitat suitability 水生生境适宜性
aquatic herbicide 杀水草剂
aquatic humic substance 水生腐殖质
aquatic insect 水生昆虫
aquatic invertebrate 水生无脊椎动物
aquatic life 水生植物;水生物;水生生物
aquatic life toxicity 水生生物毒性
aquatic livestock 水产动物
aquatic macrophyte 水生植被;水生大型植物
aquatic macrophyte vegetation 水生大型植被
aquatic macrophyte vegetation restoration 水生大型植被修复
aquatic mammal 水中哺乳动物;水栖哺乳动物;水生哺乳动物
aquatic microbial ecosystem 水生微生物生态系统
aquatic microbiology 水生微生物学
aquatic microorganism 水生微生物
aquatic migration capacity 水迁移能力
aquatic migration coefficient 水迁移系数
aquatic mobility coefficient 水可流动性系数
aquatic model ecosystem 模型水陆生态系统
aquatic moss 水中苔藓
aquatic oligochaeta 水栖寡毛类
aquatic organic matter 水生有机物
aquatic organism 水生生物
aquatic organism determination 水生生物检定;水生生物测定
aquatic pest 水生害虫
aquatic pest control 水生害虫控制;水生害虫防治
aquatic pesticide contamination 水中农药污染;水生生物农药污染;水的农药污染;农药水污染
aquatic plant 水生植物;水产植物;水草
aquatic plant biology 水生植物生物学
aquatic plant community 水生植物群落
aquatic plant garden 水生植物园
aquatic pollutant 水污染物
aquatic pollution 水质污染;水污染
aquatic pond 水塘
aquatic pool 水塘
aquatic product processing factory 水产品加工厂
aquatic products 水产(品货);海产(物)
aquatic products industry 水产业
aquatic recreation 水上娱乐
aquatic reptiles 水生爬行类
aquatic resources 水资源;水产资源
aquatic respiration 水中呼吸
aquatic science and fisheries information system 水生科学和渔业资料系统
aquatic soil 水生土;水成土
aquatic sport area 水上运动场
aquatic sports 水上运动
aquatic toxicant 水生毒素
aquatic toxicity 水生毒性
aquatic toxicity test 水生毒性试验
aquatic toxicology 水生毒理学

aquatic vascular plant 水生维管束植物
aquatic vegetation 水生植物;水生植被
aquatic weed 水生杂草;水草
aquatic weed control 水草控制;防除水生杂草
aquatic weed cutter 水草切除器
aquatint 镂蚀;凹蚀镂法;凹版腐蚀制版法
aquatint grain 腐蚀颗粒
aquation 水化作用;水合作用
aquatolysis 陆解作用
aquatone 照相平版
aquatorium 水族馆
Aquazur filter 阿奎左重力式滤池
aqueduct 输水管(道);输水道;架空水渠;引水管;沟渠;高架水道;高架渠(道);导水管;导管
aqueduct and sewer 导水管下水道
aqueduct arch 输水道孔;输水道拱
aqueduct bridge 渠桥;高架水渠;渡槽
aqueduct diversion 渡槽引水;渡槽导流
aqueduct of cochlea 蜗水管
aqueduct trough 高架水渠;渡槽
aqueductus cochleae 蜗水管
aque-glacial deposit 冰川沉积(物)
aque motrice 挖泥船;匙斗式挖泥船
aquent 潮新成土;潮湿新成土
aqueoglacial deposit 冰川沉积(物);冰水沉积
aqueo-igneous 水火成的;含水岩浆成的;岩浆水成的
aqueo-igneous rock 水火成岩
aqueo-residual sand 水蚀残沙
aqueous 水(成)的;含水的
aqueous alcohol 含水酒精
aqueous alkali 碱水溶液
aqueous ammonia 氢氧化铵;氢氧化氨
aqueous caustic 苛性碱液
aqueous chemistry 水化学
aqueous chlorination 水加氯处理
aqueous chlorination kinetics 水氯化动力学
aqueous coating 水性涂料
aqueous concentrate 浓水剂
aqueous corrosion 水腐蚀;潮湿腐蚀
aqueous cross-bedding structure 水成交错层理构造
aqueous deposit 水成沉积(物)
aqueous desert 海底荒漠
aqueous desizing 水洗除浆处理;洗涤处理
aqueous dispersion 水(性)分散液;水分散系;水分散体
aqueous distillate 水馏分
aqueous electron 水合电子
aqueous emulsion 乳剂;水乳液;水乳剂
aqueous emulsion paint 水乳胶漆;水乳化漆
aqueous ferric sulphate solution 含水硫化铁溶液
aqueous film 水膜
aqueous fuel 水作燃料
aqueous fuel-oil extract 含水燃油萃取
aqueous fusion 水融
aqueous homogeneous reactor 水均匀反应堆
aqueous horizon 水成层
aqueous humor 水样液;水样
aqueous hydrogen peroxide solution 过氧化氢水溶液
aqueous-injection 水液注入的
aqueous lava 泥熔岩;泥流岩
aqueous layer 水层
aqueous leaching 水浸出
aqueous liquid 水剂
aqueous medium 水介质;含水介质;液体介质
aqueous metamorphism 水变质作用
aqueous organic pollutant 水成有机污染物
aqueous ozonation 水臭氧化(作用)
aqueous ozone 含水臭氧
aqueous phase 液相;水相
aqueous phase ionic equilibrium of water 水的液相离子平衡
aqueous phase mole fraction 水相摩尔分数
aqueous phase solar photocatalytic detoxification process 水相日光催化消毒工艺
aqueous phase solar photocatalytic disinfection process 水相日光催化消毒工艺
aqueous phenol 含水酚
aqueous pollutant 含水污染物
aqueous potassium ferrate solution 含水高铁酸钾溶液
aqueous resin emulsion 水合树脂乳化液;水合乳化树脂

aqueous resources building 供水建设
aqueous rock 水成岩
aqueous sample 水样;含水试样
aqueous slurry 含水泥浆
aqueous soil 水成土;含水土(壤);饱气水;饱和土壤
aqueous soil water-saturated soil 饱水土(壤)
aqueous soil zone 饱和土壤带
aqueous solubility 水溶度;水的溶解度
aqueous soluble paint 水溶性漆
aqueous solution 水溶液
aqueous solution cooling method 水溶液降温法
aqueous solution co-polymerization 水溶液共聚合(作用)
aqueous solution electrolytic cell 水溶液电解槽
aqueous solution evapo(u)ration method 水溶液蒸发法
aqueous solution growth method 水溶液生长法
aqueous solution polymerization 水溶液聚合(作用)
aqueous solution temperature differential method 水溶液温差法
aqueous species 溶解态组分
aqueous stratum 蓄水层
aqueous stream 含水液流
aqueous suspension 水悬剂;水悬浮;含水悬浮物
aqueous tissue 储水组织
aqueous titanium dioxide dispersion 含水二氧化钛分散
aqueous treatment 水处理
aqueous two-phase polymer separation 水的两相聚合分离
aqueous vapo(u)r 水蒸气;水汽;空气水汽
aqueous vapo(u)r pressure 水汽压(力)
aqueous vehicle 水性漆料;水彩溶液;水媒介物;水车
aqueous wood preservative 木材防腐水剂;木材防水剂
aquept 潮湿始成土
aquert 潮湿变性土
aquex (水泥灰浆及混凝土的)防火粉末
Aquia Greek stone (一种产于美国弗吉尼亚州的浅灰色和浅黄色的)白垩纪砂岩
aquic 饱水缺氧的
aquiclude 滞水岩层;滞水层;透水性微弱的含水层;阻水层;隔水层;半含水层;弱透水层
aquic taxa 水成土
aquiculture 水产养殖
aquiculture and poultry 养殖
aquifer 蓄水层;含水层
aquifer air binding 含水层气封
aquifer basin 含水层的集水区;含水层补给区
aquifer boundary 含水层边界
aquifer coefficient 含水层系数
aquifer contaminantion 含水层污染
aquifer contaminantion hazard map 含水层污染危害图
aquifer description 含水层描述
aquifer eous stratum 含水层
aquifer exploration 含水层勘探
aquifer group 含水岩组
aquifer loss 含水层损失
aquifer of gradual change in hydraulic conductivity along flow direction 渗透性沿流向渐变含水层
aquifer of sudden change in hydraulic conductivity along flow direction 渗透性沿流向突变含水层
aquiferous 水成成因的
aquifer permeability 含水层渗透能力
aquifer response time 含水层响应时间
aquifer restoration 含水层恢复
aquifer sand 含水层砂层
aquifer storage 含水层储[贮]量;含水层储水量;含水层储气
aquifer storage of gas 含水层气体蓄积
aquifer system 含水岩系
aquifer test 含水层试验
aquifer thickness 含水层厚度
aquifer transmissibility 含水层输水能力;含水层输水率
aquifer volume 含水层体积
aquifer with changeful thickness 厚度变化的含水层

aquifer with fixed thickness 厚度不变的含水层
aquifer with limited thickness 有限厚度含水层
aquifer with semi-limited thickness 半无限厚度含水层
aquifer yield 含水层储水量;含水层储水量;含水层产水量
aquifuge 滞水(岩)层;无水层;不透水(岩)层;不透水层
aquiherbosa 水生草本群落
aquiline 钩状的;弯曲的;似鹰
Aquilonian 阿基隆阶
aquiprata 水中草原;水生草本群落
Aquitaine basin 阿奎特因盆地
Aquitanian 阿启坦阶
aquitard 滞水层;弱透水性岩体;弱含水层;隔水层;弱透水层
aqult 潮湿老成土;潮老成土
aquo-acid 水系酸;水溶剂酸
aquo-base 水系碱;水溶剂碱
aquo-complexed ion 水络合离子
aquod 潮湿灰土;潮灰土
aquogel 水凝胶
aquo-hydroxo complex ion 水羟水络离子
aquo ion 水合离子;含水离子
aquoll 潮湿软土;潮软土
aquolysis 水解作用
aquometer 蒸汽吸水机;鼓动车水机
aquosity 水性;潮湿
aquosol 水性溶胶
aquo-system 水系
aquox 潮氧化土;潮湿氧化土
Ar39/Ar40 dating method 氩39—氩40 测年法
Arab Accounting Dinar 阿拉伯记账第纳尔
araban 阿拉伯树胶
arabate 阿拉伯酸盐
Arab Bank for Economic Development in Africa 阿拉伯非洲经济开发银行
Arab Common Market 阿拉伯共同市场
arabescato 大花白石
arabesque 奇异的;蔓藤花纹;精致的;花叶饰;阿拉伯图案;阿拉伯式花纹;阿拉伯式花饰
arabesque decoration 阿拉伯风格装饰;阿拉伯式花纹装饰
arabesque ornament 阿拉伯(风格)装饰;阿拉伯式花饰品
arabesquitic 蔓藤花纹;花纹结构
arabesquitics of crystal face 晶面花纹
Arabeva 阿拉贝瓦缝编机
Arab Gas Symposium 阿拉伯天然气讨论会
Arabian American Oil Company 阿美石油公司
Arabian architecture 阿拉伯(式)建筑
Arabian basin 阿拉伯海盆
Arabian capital 阿拉伯式柱头;阿拉伯式柱顶
Arabian crepe 阿拉伯绉
Arabian cypher 阿拉伯数码
Arabian luster 阿拉伯光泽彩
Arabian millet 高粱
Arabian oil 阿拉伯石油
Arabian old land 阿拉伯古陆
Arabian Platform 阿拉伯地台
Arabian satellite 阿拉伯人造卫星
Arabian Sea 阿拉伯海
Arabian style 阿拉伯式
Arabia plate 阿拉伯板块
Arabic acid 阿拉伯酸
Arabic arch 马蹄形拱(圈);阿拉伯拱
Arabic figure 阿拉伯数字
Arabic gum 阿拉伯树胶;金合欢胶;阿拉伯胶
Arabic number(al) 阿拉伯数字
Arabic numerals 阿拉伯数码
Arabic style 阿拉伯式
arabin 阿糖胶;阿拉伯酸
arabinal 阿醛
arable 可开垦的
Arab League 阿拉伯国家联盟
arable area 可耕(地)面积
arable land 适于耕种的地;适耕地;可耕地;耕地
arable layer 耕种(土)层;耕作层
arable soil 可耕土
Arab Monetay Fund 阿拉伯货币基金组织
Arabo-Nubian Shield 阿拉伯努比亚地盾【地】
arachidic acid 花生酸
arachnoid 蛛形的;蛛网状的;蛛丝状
aradi system 旱谷型渠系

aradi type depression 旱谷型洼地
araeometer 比浮计;液体比重计
araeo-picnometer 比重测定仪
araeostyle 四柱径式;疏柱式建筑物(柱距等于柱径的四倍或四倍以上)
araeostyle temple 疏柱式寺庙;疏柱式建筑寺庙
ar(a)eosystyle 对柱式建筑(柱距以柱径两倍和四倍交替排列的成对柱列式建筑)
araeosystyle 对柱式柱廊
aragonite 文石;霰石
aragonite compensation depth 文石补偿深度
aragonite law 文石律
aragonitic limestone 霰石灰岩;文石灰岩
Arago's disc 感应涡流盘;阿拉戈圆盘
Arago's distance 阿拉戈距离
Arago's point 阿拉戈点
Arago's rotation 阿拉戈旋转
aragotite 美国加里福尼亚州产天然沥青;黄沥青
a rain ga(u)ge installed at the experimental site 在实验地设置的量雨表
arake 倾斜于垂直线的
Aralac 干酪素塑胶纤维;阿腊拉克
Araldite 合成树脂黏结剂;阿拉尔代特环氧树脂;阿拉地胶
aralkyl 芳烷基
Araloop machine 阿拉卢普缝编机
Aral Sea 咸海
aramayoite 硫铋锑银矿
aramid 芳族(聚)酰胺
Aramidae fiber rope 阿拉米德纤维缆索
aramid fabric 芳香族聚酰胺织物
aramid fiber 芳香族酰胺纤维;芳纶纤维
aranga (产于菲律宾的淡黄~淡咖啡色的)硬木
Arany's number 阿兰尼数(土料开始流动时的含水量)
arapahite 磁玄岩
a rapid method 一种快速法
a rapid method of measuring the CBR by the ball drop test 快速测定 CBR 的落球试验法 (CBR 指加州承载比)
a rapid prototype 一种快速原型
Arathene 高密度聚乙烯合成纸
a rather large pulvinus 较大的叶座
araucaria 南洋杉
Arberdeen granite 阿伯丁花岗岩
arber press 压装压力机
arbitary average 假定平均数
arbitary selection 任意选择
arbite 安全炸药(其中的一种)
arbiter 抛丸清理机;判优程序;判别器;调停器;仲裁员;仲裁人
arbiter juris 依法仲裁员
arbiter of the dispute 争端的仲裁人
arbiter speed 判优速度
arbitrage 套汇;仲裁
arbitrage account 套汇账;套购账
arbitrage bond 套汇股票
arbitrage business 套购业务
arbitrage dealer 套汇人;套购人
arbitrage house 套利公司
arbitrager 套汇人
arbitrageur 套汇商;套汇人
arbitral agreement 仲裁协议;仲裁协定
arbitral award 仲裁裁决
arbitral award for which no reasons are given 不说明理由的仲裁裁决
arbitral clause 仲裁条款
arbitral court 仲裁法庭
arbitral decision 仲裁裁决
arbitral institution 仲裁机构
arbitral instruction 仲裁机构
arbitral procedure 仲裁程序
arbitral proceedings 仲裁程序
arbitral rules of procedure 仲裁程序规则
arbitral settlement 仲裁解决
arbitrament 调停;仲裁(决定)
arbitrarily sectioned file 随机分段文件;任意分段文件
arbitrary 任意的;比例附加运价
arbitrary access 任意存取;随意存取
arbitrary angle 任意角
arbitrary assignment point 任取点
arbitrary assumption 任意假设;任意假定
arbitrary average 任意平均数

arbitrary axis meridian 任意轴子午线
arbitrary central meridian 任意中央子午线
arbitrary circulation distribution 任意环流分布;不规则环流分布
arbitrary concentration 任意浓度;随意浓度
arbitrary constant 任意常数;假定常数;泛定数
arbitrary course computer 任意航向计算机;航线计算机
arbitrary data point 任意数据点
arbitrary datum 任意基面;假定基(准)面
arbitrary datum line 任意基准线
arbitrary deformation 任意变形
arbitrary depreciation method 任意折旧法
arbitrary duty 任意征税
arbitrary flow 任意流
arbitrary fraction 任意分数
arbitrary function 任意函数
arbitrary function generator 任意函数发生器;通用函数发生器
arbitrary grid 任意格网;(地质图上的)假定网
arbitrary ground point 任意地面点
arbitrary hue 任意色调
arbitrary inspection 任意抽查
arbitrary Lagrangian-Euleuan method 任意拉格郎日—欧拉方法
arbitrary lateral load 任意横向荷载
arbitrary level 任意水位
arbitrary load 任意荷载
arbitrary map projection 非等角等积投影
arbitrary moment diagram 任意弯矩图
arbitrary number 任意数
arbitrary orientation 假定方位标定
arbitrary origin 任意原点
arbitrary parameter 任意参数
arbitrary phase 任意相位
arbitrary phase-angle power relay 任意相角功率继电器
arbitrary point 任意点
arbitrary point method 任意点法
arbitrary point of origin 任意原点
arbitrary polygon 任意多边形
arbitrary pricing 人为定价
arbitrary projection 任意投影
arbitrary proportioning 习用配料法;习用配合法
arbitrary proportion method 体积比配料法(混凝土);经验配合法;经验配比法;习用配料法;习用配合法;指定配合法
arbitrary reference value 任意基准值
arbitrary sample 任意样本
arbitrary scale 任意刻度;任意标度
arbitrary selection 任意选择
arbitrary sequence 任意顺序;任意次序;可变时序
arbitrary sequence computer 任意顺序计算机;可变时序计算机
arbitrary surface 任意面
arbitrary taxation 任意征税
arbitrary three-colo(u)r density 任意三色浓度
arbitrary traverse 任意联测导线
arbitrary unit 任意单位
arbitrary value 任意值
arbitrary variation 任意变化
arbitrary weight 任意权数
arbitrary-weight lag scheme 任意权数滞后型式
arbitrary zone 任意带
arbitrate 仲裁的;裁定
arbitrate a dispute 仲裁争端
arbitrate between two parties 在两方向进行仲裁
arbitrate by default 缺席仲裁
arbitrated exchange 裁定外汇
arbitrated exchange rate 裁定汇率;裁定汇兑比例价;比例外汇汇率
arbitrated house 套利公司
arbitrated interest rate 裁定利率
arbitrated loan 裁定借款
arbitrated par of exchange 裁定外汇平价;比例外汇平价
arbitrated rate of exchange 汇兑比例价;裁定汇兑比价;比例外汇汇率
arbitration 判优法;仲裁;公断;裁定
arbitration act 仲裁法
arbitration agency 仲裁机构
arbitration agreement 仲裁协议;仲裁协定
arbitrational 有关仲裁的
arbitration award 仲裁裁决

arbitration award on appeal 仲裁上诉裁决
arbitration bar 抗弯试棒
arbitration board 仲裁委员会
arbitration by summary procedure 简易仲裁
arbitration clause 仲裁条款
arbitration commission 仲裁委员会
arbitration committee 仲裁委员会
arbitration court 仲裁法庭
arbitration decison 仲裁裁决
arbitration expenses 仲裁费(用)
arbitration fee 仲裁费(用)
arbitration in disputes between labor and capital 调停劳资纠纷
arbitration in law 依法仲裁
arbitration law 仲裁法
arbitration legislation 仲裁立法
arbitration logic 判优逻辑
arbitration notice 仲裁通知书
arbitration organization 仲裁机构
arbitration procedure 仲裁程序
arbitration proceedings 仲裁程序
arbitrationrate of exchange 比例外汇汇率
arbitration rules 仲裁规则
arbitration sample 检验用样品;查对用样品
arbitration tribunal 仲裁法庭
arbitrator 仲裁员;仲裁人;公断人
arbitrator de jure 仲裁员;依法仲裁员
arbond 粘氯乙烯薄膜钢板
arbor 林荫步道;凉亭;边框;乔木;藤架;技编棚架;花架;芯骨;心轴;刀轴;刀杆;柄轴
arbor adapter 柄轴连接;柄轴接头
arbor bearing sleeve 柄轴轴承衬套
arbor bolt 圆盘刮把方轴
arbor collar 轴环
arboreal 树状的;树上生活的;乔木的
arboreous 树木茂盛的;树状的
arborescence 枝状;树状;树质;乔木状
arborescent 树枝状的;枝状
arborescent crystal 树枝状晶体;树枝形晶体;枝晶
arborescent drainage 树枝状水系
arborescent powder 枝状粉末
arborescent river network 树枝状河网
arborescent structure 枝晶组织
arboret 小树;灌木
arboretum 树木园;林园;植物园
arbor flange (铣刀杆上的)盘式刀架;柄轴凸缘
arbor for drill chuck 钻头夹盘轴
arbor for shell reamer 套状铰刀轴
arbor for spacing piece 定距片轴
arbor holder 刀杆支座
arbor hole 心轴孔
arboricity 荫度
arboriculture 造林(学);树木栽培;树艺学
arbo(u)r 螺孔刮面;棚架;凉亭;刀杆
arbo(u)ring tool 螺孔刮面刀具
arborist 树木学家;林学家
arborization 树枝分枝;树枝状
arbor nut 刀杆螺母
arboroid 树状的
arbor press 手动平板机;手扳压机;手扳冲床;矫正机
arbor Saturni 萨图尼树
arbor support 柄轴支架
arbor-type cutter 套式铣刀
arborvitae 香柏;侧柏
arbor walk 花棚下步道;蔓棚架下步道
arbor wheel 轴轮
arbovirus 树木病毒
Arbuckle orogeny 阿尔布克尔造山运动
arbuscle 灌木
arc 弧形;拱圈
arc absorber 电弧吸收器
arca custodiae 牢房
arcade 连拱廊;连拱;有拱廊街道;拱形建筑物;拱廊
arcade apex 连拱道;拱廊顶;连拱顶
arcade cornice 拱廊墙顶线脚;连拱檐;拱式挑檐;拱式飞檐;连拱式飞檐
arcade crown 拱廊顶(部);连拱顶
arcaded cornice 连拱式飞檐
arcaded court 带回廊庭院;带拱廊庭院;连环拱廊
arcaded facade 带拱廊建筑(立)面;连拱立面;拱形建筑物正面
arcaded gallery 连拱走廊;连拱楼座;连拱廊道;连拱画廊

arcaded ground floor 拱形底层
arcaded lobby 连拱形走廊
arcaded sidewalk 有拱廊的人行道;骑楼式人行道
arcaded tribune 拱形看台;带拱顶讲坛
arcaded window 连拱窗;拱窗
arcade impost 连拱墩
arcade key 拱廊顶石;连拱顶石
arcade lobby 连拱形走廊
arcade of shops 商店廊;商店街
arcade pier 连拱支墩;连拱墩;拱支墩
arcade quay wall 连拱式驳岸墙
arcade rib 拱肋;连拱肋
arcade sidewalk 廊式人行道;有顶棚的人行道;拱廊人行道
arcade top 连拱顶
arcade type wall 拱廊式墙;拱廊式岸壁
arcade vertex 拱廊顶(部);连拱顶
arcading 连拱饰;柱间拱
arcae 古罗马内庭水沟
arcafure 盲拱
arcagraph 划弧器
arc air cutting 电弧气割
arc air gouging 压缩空气电弧气刨;电弧气刨;电弧气焊;电弧气动钻孔;电弧气动割槽;电弧空气气刨
arc air gouging clearage 电弧气刨清理
arc air gouging method 电弧气割法
arc air process 电弧喷气切割法
arc air torch 电弧气刨枪
arcanal 防腐漆
arc and spark stand 电弧电花架
arcanite 单钾芒硝
arc arrester 火花消灭器;火花熄灭器;弧形避雷器;消弧器;电弧避雷器;活化熄灭器;放电器;熄弧器
arc atmosphere 电弧气氛;电弧炉内气氛
arcatron 冷阴极管
arcature 假拱廊;小型连拱;小拱廊;(不开洞的)装饰拱廊
arc-back 逆弧
arc bearing(plate) 弧形支座
arc belt type 弧形条带式
arc blasting 弧形爆破法
arc blow 电弧偏吹;磁偏吹
arc blow out 灭弧;消弧
arc booster 起弧稳定器(焊接)
arc-boutant 扶拱(垛);拱门式扶壁;飞拱
arc braze welding 电弧钎焊
arc brazing 弧铜焊;电弧硬焊;电弧铜焊;电弧钎接;电弧钎焊
arc break 断弧;电弧断流
arc breaker 电弧断流器
arc carbon 电弧碳精棒
arc cast 弧熔铸锭;电弧熔铸
arc cast metal 弧熔金属锭;电弧熔铸的金属
arc catcher 消弧器
arc cathode 弧光放电阴极;电弧阴极
arc cell 连拱段(钢板桩);弧形格仓;圆弧形格体
arc chain 弧形锁
arc chamber 放电室
arc characteristic 电弧特性(曲线)
arc chord 弓弦
arc chute 消弧栅;灭弧沟;电弧隔板;灭弧罩;灭弧栅;弧形槽;熄弧沟;灭弧室
arc circuit 电弧电路
arc concentration technique 电弧浓缩法
arc control 控制电弧;灭弧;消除火花;电弧控制
arc-control device 灭弧装置;电弧控制器件
arc converter 电弧振荡器;电弧变流器
arc core 电弧中心;电弧心
arc correction 摆幅改正
arc cosecant 反余割
arc cosine 反余弦
arc cosine transformation 反余弦变换
arc cotangent 反余切
arc cracking 电弧裂化法
arc crater 弧坑(焊接时);电弧陷口
arc crest 弧形堰顶
arc current 电弧电流
arc cutter 绞链式导弧器;弧式导弧器
arc cutting 电弧切割(法)
arc cutting machine 电弧切割机
arc deflector 灭弧器;熄弧隔板;电弧转向器;电弧偏转器
arc descaling machine 弧轮去锈机

arc detector 电弧检验器
arc de triomphe 凯旋门;教堂大拱门
arc discharge 电弧放电
arc discharge coating equipment 电弧放电蒸镀装置
arc discharge tube 电弧放电管
arc dissociation 电弧断开
arc dissociation unit 电弧离解设备
arc distance 弧距
arc doubleau 大跨拱;巨拱
arc-dozer 弧板板推土机
arc drilling 电弧钻进
arc drive characteristic 电弧激励起动特性
arc drop 弧降;电弧电压降
arc duration 电弧持续时间
arc dynamo 电弧发电机;弧光灯用直流发电机
arc east 弧熔的
arceite 单钾芒硝
arc elasticity of demand 需求的弧弹性
arc electrode 电弧极
arc element 元弧;弧元
arcella 干酪室;中世纪奶酪房
arc end 引弧端
arc-energy 电弧能(量)
arcer 高功率火花震源;电火花激发器
arc erosion 电弧腐蚀
arc excitation 弧激发
arc extinction 灭弧;消弧;电弧熄灭
arc extinguish chamber 灭弧室;消弧室
arc extinguish device 灭弧器
arc-extinguishing 灭弧;消弧
arc-extinguishing coil 灭弧线圈;消弧线圈
arc-extinguishing equipment 灭弧装置
arc-extinguishing medium 灭弧介质
arc-extinguishing plastics 灭弧性塑料
arc fault 电弧故障
arc flame 电弧焰;电弧火焰
arc flame heater 电弧火焰加热器
arc flash welding 电弧闪光焊
arc force 电弧力
arc formeret 侧向拱肋;墙拱;(拱结构中的)墙肋
arc form hull 拱形肋骨船体
arc form ship 弧形肋骨船
arc form vessel 弧形肋骨船
arc furnace 直接电弧炉;电弧炉
arc furnace transformer 电弧炉变压器
arc fusing welder 电弧熔焊机
arc fusion splicing machine 电弧熔接接头机
arc gap 弧光间隙
arc generator 弧光发生器;电弧振荡器;电弧发生器
arc gouging 电弧凿槽;电弧气刨;电弧刨削
arc gouging torch 电弧切割炬
arc guide 水银蒸汽阻隔筒;弧导
arch 穹起;炉顶;拱;弓状物;弓形;半圆形
arch-abdomen dam 腹拱坝
arch abutment 圈拱座柱;圆拱墩柱;拱座;拱台;拱墩;拱端
arch action 拱架;起拱作用;拱作用;拱圈作用;成拱作用
archaeal ammonia oxidation 古菌氨氧化
archaean and eo-algonkian megacycle 太古及老元古巨旋回
archaean basement 太古代基底
Archaean era 太古代【地】
Archaean group 太古界【地】
Archaean rock 太古岩
archaeocyathid extinction 古杯绝灭
archaeo-hydrology 考古水文学
Archaeoid 太古
archaeologic(al) chemistry 考古化学
archaeologic(al) dig 考古发掘
archaeologic(al) diving 考古潜水
archaeologic(al) evidence 考古学证据
archaeologic(al) exploration 考古勘察
archaeologic(al) map 考古图
archaeologic(al) method 考古法
archaeologic(al) photogrammetry 考古摄影测量
archaeologic(al) site 考古地点;古文化遗址
archaeology 考古学
archaeology data analysis 考古资料分析
archaeomagnetism 古地磁(学)
archaeomorphic rock 原型侵入岩
Archaeozoic era 太古代【地】
Archaeozoic erathem 太古界【地】

archaic 古代遗存;古代的
archaic architecture 古建筑(学)
archaic map 古地图
archaic sepulcher 古墓(葬);古代墓葬;拱形古墓穴
Archaic temple at Ephesus (古希腊城市的)以弗所的古神殿;以弗所古庙(小亚细亚)
archaism 古风(气)
archaist 古物研究者
arch analysis 拱静力学分析
arch and buttress blockwork 拱和扶壁块体
arch and crown cantilever method 拱冠梁法(拱坝受力分析)
arch and pier system 拱墩系统
archangel pitch 松焦油沥青
arch apex 拱顶
arch aqueduct 拱式渡槽
arch arching 拱作用
arch axis 拱轴线;拱轴
arch axis coefficient 拱轴系数
arch back 拱背
arch band 拱肋;横向拱;拱箍;拱带;与拱面相连的一条砌体
arch bar 弯曲杆件(窗扇中);拱形铁条;拱木条;拱筋;拱杆;拱板;弓形杆
arch bar column bolt 拱板撑柱螺栓
arch bar drilling machine 弓柄钻机
arch barrel (横截面高度小于宽度的板拱桥的)拱圈;拱形筒壳;拱筒
arch bar truck 菱形转向架;菱形车架
arch bay 拱跨
arch beam 拱形梁;拱梁
arch bearing 拱支承;拱结构支承;拱支座
arch block 拱旋块;拱楔块;楔形拱石;拱圈砌块;拱圈块;拱圈;(楔形的)拱面石;拱块;木拱楔块
arch bond 拱砌合
arch bound 拱镶(边)
arch boutant 飞拱垛;飞扶壁
arch brace 拱形撑架;拱支撑;拱形支撑
arch braced roof 拱支屋顶;拱支架屋顶;拱支桁架
arch braced truss 拱支桁架;拱撑桁架
arch bracing 拱支撑
arch breaker 料拱排除器
arch brick 窑顶砖;砌拱用砖;人孔(楔形)拱砖;砌拱砖;窑砖;楔形砖
arch brick for manholes 砌窨井扇形砖;人孔形拱砖
arch bridge 拱桥
arch bridge with flat hinges 平铰拱桥
arch bridge with variable cross-section 变截面拱桥
arch buttress 下拱壁;拱扶垛;拱柱;拱支墩;拱壁;拱形扶壁
arch buttress dam 拱式支墩坝
arch camber 起拱;拱势;拱矢
arch canopy 拱形罩篷
arch cantilever bridge 拱式悬臂桥
arch cantilever method 拱梁法
arch casting 浇筑混凝土拱
arch cave 拱洞
arch ceiling 拱顶棚;拱形顶棚;拱形天花板
arch center 拱鹰架;拱模架;拱中心;拱形中心;拱架
arch center for crest 坝顶拱圈中心(拱坝);拱的中线
arch centering 券模;拱胎;拱鹰架;拱架;拱(顶)脚手架
arch center line 拱轴线;拱中心线;拱的中线
arch centrum 弓形锥体
arch chord 拱弦
arch clouds 弓形云
arch cofferdam 拱形围堰
arch compression 拱压力
arch concrete 拱部混凝土
arch concreting 浇筑混凝土拱
arch construction 拱构造;拱券结构;拱结构
arch core 拱心;背斜核部
arch corner bead 拱形墙角装饰线条;拱角护条;拱边圆线条
arch cornice 拱檐;拱形檐
arch cover(ing) 拱形铺装;拱盖;拱上铺装;拱板;拱上填平层
arch crown 拱顶;拱冠
arch crown section 拱冠断面
arch culvert 券涵;拱形涵洞;拱涵(洞)
arch curvature 拱曲率

arch dam 拱坝
arch dam pad 拱坝垫座;拱坝坝座
arch dam with peripheral hinge 边铰拱坝
arch deck 梁拱甲板;龟背甲板
arch-deck buttress dam 拱面支墩坝
arch depth 拱结构高度
arch door 拱门
Archean 太古代【地】;始生代
archean basement 弧形基底
Archean nucleus 太古代陆核
archean rock 原生岩石
arc heating 电弧加热
arc heating furnace 电弧加热炉
archecentric 建筑中心的
arched 曲拱的;拱形的;弓形的;弓形
arched abutment 拱形支座;拱形桥台;拱式桥台;拱桥台;拱桥墩
arched area 背斜顶部;隆起地带;背斜地带
arched axis 拱轴线
arched bar 拱形条铁
arched barrel roof 筒壳屋顶;筒形屋顶
arched beam 拱主梁;拱(形)梁;拱副梁;筒形梁
arched beam bridge 拱梁桥;拱形梁式桥
arched bent 拱形排架
arched bond 拱砌合
arched boom 弧形弦杆
arched boom angle iron 弧形角铁弦杆
arched brace 拱形撑架
arched brick 拱形砖
arched brick roof covering 砖砌窑洞屋盖;砖拱屋盖
arched bridge 拱形桥;拱桥
arched bridge construction 拱桥建筑
arched building 拱形建筑;拱形房屋;弧形房屋
arch(ed) buttress 拱形扶跺;拱式支墩;拱式扶垛;拱扶垛;飞扶垛;飞扶壁
arched camber 拱矢
arched cantilever bridge 悬臂式拱桥
arched ceiling 拱形天花板;拱形顶棚
arched center 拱心
arched chord 拱形弦杆;弓形弦杆
arched concrete block 拱形混凝土砌块
arched concrete block dam 拱形混凝土砌块坝
arched concrete dam 拱形混凝土坝;混凝土拱坝
arched concrete roof 混凝土拱形屋顶
arched conduit 拱形管道
arch(ed)construction 拱形结构;拱形建筑(物);拱形构造;拱式结构;拱式构造;拱式建筑;拱架施工(方)法
arched corbel table 弯托梁挑檐;连拱饰带挑檐
arched core 拱心
arched cornice 拱形飞檐
arched cover(ing) 拱形屋盖;拱盖;拱板;拱形屋面
arched crown 拱顶
arched culvert 拱形涵洞;拱涵;弓形排水渠;弓形涵洞
arched culvert pipe 拱形涵管
arched dam 拱坝
arched diagonal 弓形斜(支)撑;弓形斜杆
arch(ed) dome 拱形圆屋顶;拱形;圆屋顶式拱;圆顶穹隆
arched door 拱形门;拱门
arched element 拱形结构
arched entrance 拱形入口
arched falsework 拱架;拱形脚手架;拱模
arched fault 穹断
arched flange 弧形弦杆
arched floor 拱形屋盖;拱形楼盖;拱形楼板
arched flume 拱形水槽
arched fold 穹褶
arched form 拱形模板
arched formation 拱桥构造
arched foundation 拱形基础
arched frame 拱形框架;拱形构架
arched frame construction 拱形框架结构
arched framework 拱形模板
arched frieze 拱形檐壁
arched furnace roof 拱形炉顶
arched gallery 拱形走廊;拱形廊道
arched girder 拱主梁;拱形主梁;拱形大梁;拱梁
arched girder chord 弓形大梁翼缘;弓形大梁弦杆
arched girder hinged at the abutment 支座铰接的拱(梁);两脚拱(梁);与支座铰连的拱梁
arched girder length 拱形梁长度
arched girder load 拱形梁荷载

arched girder of constant cross-section 等截面拱形梁
arched girder parabola 拱梁抛物线
arched girder plane 拱梁平面
arched girder with braced spandrels 有支撑拱肩的拱形桁架
arched girder with diminished horizontal thrust 有衰减水平推力的拱形桁架
arched girder with intermediate tie 有中间系杆的拱形桁架
arched girder with invariable horizontal thrust 有不变水平推力的拱形桁架
arched girder without horizontal thrust 有拉杆的拱;无水平推力的拱形桁架
arched girder with parabolic chord 有抛物线型弦杆的拱形桁架
arched girder with polyonal outlines 有多边形状的拱形桁架
arched girder with tie 有拉杆的拱形大梁;带拉杆拱梁
arched girder with tieback 有系杆的拱梁
arched gravity dam 拱式重力坝
arched hunch 拱腰
arched iceberg 拱形冰山
arched keystone 拱心石
arched mole 拱形突堤;拱形防波堤;弓形防波堤;弯防波堤
arched mo(u)lding 拱饰
arched opening 拱形门洞;弧形洞
arched outline 拱形曲线
arched piling bar 拱形(钢)板桩
arched pipe 管拱;拱形管道;拱形管
arched plate 筒形板;弧拱形板
arched portal 拱形门;拱形桥门
arched principal 拱形主梁;拱形大桁梁
arched principal girder 拱形主梁
arched rail set 钢轨拱形支架
arched recess 拱形壁龛
arched retaining wall 拱形挡土墙
arched rib 拱肋
arched ring laid by cantilever method 拱桥悬装施工法;拱桥拱砌施工法
arched rise 拱矢
arched roof 拱形屋面;拱形屋顶;拱形筒顶
arched roof in advance of wall tunnel(l)ing method 先拱后墙法
arched roof truss 拱形屋架;拱形屋顶桁架
arched scaffold(ing) 拱形脚手架;拱模支架
arched sheeting 拱形板
arched shell 拱形薄壳
arched shuttering 拱形模板
arched solid 拱斜石;拱楔石
arched spillway 拱形溢洪道
arched squall 拱状云飑;拱形云;弓形黑云风暴
arched stair(case) 拱形楼梯
arched stone 拱斜石;拱楔石
arched stone bridge 曲拱石桥
arched structure 弧形结构;拱形结构;拱形建筑物
arched style(of architecture) 拱形建筑风格
arched system 拱形体系
arched system for absorption of thrust 推力吸收拱系统;推力衰减拱系统;推力缓冲拱系统
arched timbering 拱形支撑;拱模支架;拱模
arched truss 拱形桁架
arched truss bridge 拱形桁架式桥
arched trussed girder 拱形桁架式大梁
arched tunnel 拱形隧洞;拱形隧道
arched type piling bar 拱形板桩
arched type sheet pile 拱形(钢)板桩
arched up folds 穹形褶皱;隆起褶皱
arched vault 拱顶地下室;拱顶
arched viaduct 拱形栈桥;拱形高架桥;拱形旱桥
arched wall 拱墙
arched window 拱窗
arched work 拱形构造
arc height 弧高
archeion 古希腊或罗马档案库
arch element 拱形组件;拱的组成部分
archeological area 古文化遗址
archeological site 遗址
Archeozoic 太古生代
archeria (城墙上开的)箭洞
archerite 磷钾石
archery 射箭场

archery ground 靶场
arches 钢拱
archetype 模型标本;原型(的);原始模型
arch excavation 拱部开挖
arch expansion joint 拱桥伸缩缝
arch extrados 拱背线;拱背
arch face 弧面;拱面
arch falsework 拱形支撑;拱形脚手架
arch filament 拱状暗条
arch filaments system 拱状暗条系
arch-flat 平拱
arch floor 拱形底板
arch foot bridge 人行拱桥
arch form 拱型
arch formation 拱桥现象;拱桥结构
arch form carrier 拱模移运器
arch formed on filled earth 土牛拱胎
arch form traveler 拱形起重行车;拱模起重行车
arch formwork 拱架
arch frame 拱形框架;拱形结构
arch frame construction 拱形框架结构
arch framing 拱形肋骨
arch furnace 拱顶窖;拱顶炉
arch-gravity 重力拱
arch-gravity dam 重力(式)拱坝;拱形重力坝
arch hand 拱带
arch haunch 拱肩石;拱背圈
arch height of a natural bridge 天生桥拱高
arch hinged at ends 双铰拱
arch hinged at the abutments 铰连在支座上的拱形桁架
archibenthic zone 半深海底带
archibenthos 深海底栖生物
archicenter 原始中心
archiepiscopal cross 主教十字架;两横臂十字架;大主教带的十字架
archifoglio 方铅矿
archigenesis 非生物起源
archigony 自然发生
archil 海石蕊
Archimedean axiom 阿基米德公理
Archimedean drill 螺旋钻;取水用螺旋钻;阿基米德钻;阿基米德螺旋钻
Archimedean principle 阿基米德原理
Archimedean pump 螺旋泵
Archimedean screw 阿基米德螺旋(钻);螺旋升水器;阿基米德螺线
Archimedean screw elevator 阿基米德螺旋提升机;垂直螺旋式输送机
Archimedean screw pump 螺旋泵;阿基米德螺旋抽水机;阿基米德螺旋泵
Archimedean screw water lift 螺旋提扬水机;阿基米德螺旋提扬水机;螺旋提水器
Archimedean solid 阿基米德体;阿基米德多面体
Archimedean spiral 阿基米德螺线
Archimedean valuation 阿基米德赋值
Archimedes' axiom 阿基米德公理
Archimedes' law 阿基米德定律
Archimedes number 阿基米德数
Archimedes' principle 阿基米德原理
Archimedes' problem 阿基米德问题
Archimedes' screw 阿基米螺(旋)桨
Archimedes' screw pump 阿基米德螺泵
Archimedes' screw vessel 阿基米德螺旋桨船
Archimedes' spiral 阿基米德螺线
Archimedes spiralfield 阿基米德漩涡场
arch impost 拱墩;拱座
archine 阿尔申(俄尺等于71.12cm)
arching 扫船底;挠度;中部起拱;拱作用;拱形支护;拱圈作用;拱起;船首尾下垂;成拱作用;粉体架桥现象;背斜作用
arching action 弯拱作用
arching effect 拱效应;成拱作用
arching factor 弯拱因素;拱作用系数;拱度
arching layer 拱枝压
arching of cylinder 滚筒中高度
arching of material 物料架拱
arching theory 拱理论
arch intrados 拱腹线
arch in trellis work 格形拱
arch invert 仰拱;倒拱;倒拱形沟底
archipelagian 群岛的
archipelagic 群岛的
archipelagic apron 列岛裙

archipelagic baseline 群岛基线
archipelagic waters 群岛水域
archipelago 群岛海区；群岛；列岛；多岛海
archiplasm 初质
architect 设计师；建筑师
architect and designer 建筑设计师
architect assistant 建筑师助理
architect engineer 建筑师兼工程师；建筑工程师
architect-engineer, architectural engineering firm 建筑师—工程师联合事务所
architect errant 无定住所的建筑师
architect in charge 总建筑师；主持建筑师；主任建筑师
architect in private practice 开业建筑师；私营建筑师
architect-in-training 实习建筑师；见习建筑师
architective 关于建筑的；建筑的
architect office 建筑师事务所
architectonic(al) 建筑学的；建筑师的；地质构造的；大地构造的；成体系
architectonic division 大地构造区划
architectonic feature 建筑特征
architectonic geology 构造地质学
architectonics 体系论；建筑原理；建筑学；大地构造
architect partnership 建筑师合作关系；建筑师合伙关系
architect's approval 建筑师认可证明
architect's basic service 建筑师基本服务项目
Architect's Co-partnership 建筑师联盟
architect's fee 建筑师设计费；建筑师费用；建筑师酬金
architect's inspection certificate 建筑师验收证书
architect's office 建筑师事务所
architects' registration ordinance 建筑师注册法
architect's scale 三棱尺；比例尺
architect's sketch 建筑师的草图
architect's table 绘图桌
architectural 建筑学的
architectural acoustics 建筑声学
architectural aesthetics 建筑美学
architectural aggregate 建筑用混凝土集料；建筑用混凝土骨料
architectural alloy 建筑用合金
architectural alumin(i)um 建筑用铝
architectural alumin(i)um profile production line 建筑铝型材生产线
architectural analysis 建筑分析
Architectural and Exploring Designing Institute 建筑勘察设计院
architectural and industrial ceramics 建筑工业陶瓷
architectural and sanitary ceramics 建筑卫生陶瓷
architectural appearance 建筑外形；建筑外观
architectural appearance grade 建筑外观等级
architectural area 建筑面积
architectural area of building 房屋建筑面积
architectural art 建筑艺术
architectural assistant 建筑助理；建筑描图员；建筑绘图员
architectural Association 建筑协会
architectural award 建筑设计奖；建筑奖；建筑获奖
architectural barriers 建筑障碍
architectural bid 建筑投标
architectural brass 建筑用黄铜
architectural bronze 铜锌铅合金；建筑用青铜；建筑青铜
architectural builder's fitting 建筑施工五金配件
architectural builder's hardware 建筑施工小五金
architectural building material 装饰建筑材料
architectural career 建筑事业
architectural cast concrete product 建筑装饰用混凝土产品；建筑混凝土产品；装饰预制混凝土制品
architectural cast(ing) 建筑组件浇注；装饰用预制混凝土
architectural casting concrete element 装饰预制混凝土构件
architectural characteristic 体系特性；结构特征；结构特性
architectural climatic zoning 建筑气候分区
architectural coating 建筑涂料
architectural comparison 结构比较；总体比较
architectural compatibility 建筑协调性；建筑调和性
architectural competition 建筑图案评比；建筑设计竞赛；建筑方案评比；建筑方案竞赛

architectural complex 建筑总体；建筑群
architectural composition 建筑构造(方式)；建筑构图；建筑布置；建筑布局
architectural compression seal 建筑用压缩密封件
architectural concept 建筑思想；建筑构思；建筑设想
architectural conception 建筑构思
architectural concrete 艺术混凝土；装饰用混凝土；装饰(性)混凝土
architectural concrete aggregate 装饰用集料
architectural concrete casting 装饰用预制混凝土
architectural concrete construction 建筑艺术化的混凝土建筑
architectural concrete element 装饰用预制混凝土构件
architectural concrete finishing 混凝土一次抹面
architectural concrete product 装饰用预制混凝土制品
architectural conservation 建筑文物保护
architectural constructional materials 建筑材料
architectural construction company 建筑工程公司
architectural context 建筑文脉
architectural control 建筑管理规则；建筑管理
architectural coordinate system 建筑坐标系(统)
architectural course 建筑研究室
architectural creation 建筑创作
architectural critic 建筑评论(家)
architectural criticism 建筑评论
architectural damage 建筑破坏
architectural decoration 建筑装饰(学)
architectural definition 结构定义
architectural design 建筑设计
architectural design and research division 建筑设计研究所
architectural design criterion 建筑设计准则
architectural designing documents 建筑设计文件
architectural design institute 建筑设计院
architectural design methodology 建筑设计方法学
architectural design standardization 建筑设计标准化
architectural detail 建筑详图；建筑施工详图；建筑设计细部；建筑大样；反映建筑风格构造；反映建筑风格构件；建筑细节；建筑细部
architectural device 反映建筑风格构造；反映建筑风格构件
architectural difference 体系差异
architectural difinition 体系定义
architectural discipline 建筑专业
architectural division 建筑布置；建筑分隔
architectural door 装饰门
architectural drafting 建筑制图
architectural draughtsman 建筑绘图员
architectural drawing 建筑制图；建筑图
architectural education 建筑学教育
architectural effect 建筑效果
architectural element 建筑构件；装饰构件
architectural elevation 建筑立面图
architectural enamel 建筑搪瓷
architectural enameled glass panel 建筑用油彩玻璃板
architectural engineering 建筑工程(学)
architectural engineering drawing 建筑工程图
architectural engineering firm 建筑师—工程师事务所；建筑工程事务所
architectural engineering institute 建筑工程学院
architectural ensemble 建筑总体效果
architectural expert 建筑专家
architectural extruded section 建筑挤压型材；装饰用挤制叶型
architectural faience 建筑彩陶
architectural fashion 建筑模型
architectural feature 体系特征；结构特征；建筑特征；建筑特色；建筑风格
architectural fee 建筑业务费；建筑酬金
architectural finish(ing) 建筑表面装饰；建筑物涂漆；建筑涂装
architectural firm 建筑公司；建筑师事务所
architectural fitting 建筑小五金；建筑零件；建筑施工装置；建筑施工器具
architectural flute (建筑物柱子的)槽纹；(建筑物柱子的)凹槽；柱体凹槽
architectural form 建筑形式
architectural fountain 艺术喷泉；装饰性喷泉
architectural furniture 建筑小五金；建筑配件；建筑零件；建筑施工装配器具；建筑施工五金配件

architectural garden 建筑构图式庭院；建筑庭园
architectural geometry 建筑几何学
architectural glass 建筑玻璃；装饰玻璃
architectural glass article 装修玻璃制品；建筑装修用玻璃制品；建筑装修用玻璃成品
architectural glass plant 建筑用玻璃生产厂
architectural grade concrete form panel 符合设计要求的混凝土模板
architectural granite 建筑装修用花岗石；建筑装修用花岗石
architectural grid 建筑方格网
architectural grill(e)work 建筑花格；建筑格架结构
architectural hardware 建筑(小)五金
architectural heritage 建筑遗产
architectural historian 建筑史学家；建筑历史学家
architectural history 建筑(历)史
architectural hygiene 建筑卫生学
architectural idea 建筑设想
architectural illumination 建筑照明
architectural image 建筑隐喻
architectural implement 建筑设备
architectural impost 建筑拱墩
architectural inspection 建筑检查
architectural inspector 建筑检查员
architectural institute 建筑学院
architectural instruction 建筑指示；建筑学细则；建筑学课程；建筑学规程
architectural ironmongery 建筑五金器具；建筑五金
Architectural Journal 建筑学报；建筑杂志
architectural journalism 建筑新闻事业(包括编辑、出版、刊物和管理等)
architectural laminated glass 建筑用叠层(安全)玻璃；建筑叠层(安全)玻璃
architectural lamp 建筑灯；装饰灯
architectural layout 建筑布局
architectural league 建筑联合会
architectural lighting 建筑照明；建筑采光
architectural limit 结构限制；体系限制
architecturally 建筑学上
architecturally beautiful squares 具有建筑特色的地方
architecturally enframed window 镶有装饰框的窗；建筑上配的框架窗户
architecturally treated 建筑上经过处理的；具有建筑艺术的
architectural management 建筑管理
architectural marble 建筑用大理石
architectural masonry(work) 建筑砖石工；砖石建筑
architectural mechanics 建筑力学
architectural metal 小五金；建筑五金材料；建筑用金属材料
architectural metalwork 建筑装修五金；建筑大五金
architectural metaphor 建筑造型
architectural millwork 定做木活；建筑装饰木制品
architectural modelling 制作建筑模型；建筑造型；预制建筑构件生产(混凝土)
architectural module 建筑模数
architectural monument 纪念性建筑物
architectural morphology 建筑形态学
architectural motif 建筑风格主题；建筑风格特征；建筑特色；建筑设计主题思想
architectural movement 建筑运动；建筑思潮
architectural operation 建筑施工
architectural order 建筑圆柱式样；古典建筑柱式；建筑柱型；建筑(圆)柱式
architectural organism 建筑整体；建筑机体；建筑组织
architectural ornament 建筑装修；建筑装饰
architectural paint 建筑涂料
architectural panel 建筑墙板
architectural paradigm 建筑范例
architectural parts 建筑配件
architectural perspective 建筑透视(图)
architectural philosophy 建筑哲理
architectural photogrammetry 建筑装饰线条；建筑摄影测量
architectural plan 建筑平面(图)
architectural planning 建筑规划；建筑设计
architectural planting 整形种植
architectural porcelain 建筑陶瓷；建筑陶器
architectural pottery 建筑釉陶；建筑陶瓷
architectural practice 建筑实践
architectural precast concrete product 预制建筑混凝土制品

architectural preservation 建筑保护
architectural principle 建筑原理
architectural product 建筑用制品
architectural profile 建筑轮廓;建筑分布图
architectural program(me) 建筑计划;建筑规划
architectural project 建筑项目
architectural projected window (一种较高级的)凸窗;(按建筑学原理设计的)窗户
architectural rendering 建筑渲染
architectural safety glass 建筑(用)安全玻璃
architectural scale 建筑尺度
architectural science academy 建筑科学院
architectural sculpture 建筑雕塑;建筑雕刻
architectural section 建筑剖面图;建筑型材
architectural seminar 建筑研究班
architectural semiotics 建筑符号学
architectural services 建筑业务
architectural shape 建筑用型材;建筑形状;建筑形式
architectural shatterproof glass 建筑安全玻璃
architectural sketch 建筑草图
architectural slate 建筑石板;建筑用石板
Architectural Society 建筑学会
Architectural Society of China 中国建筑学会
architectural specifications 建筑设计说明书;建筑规范
architectural stoneware 建筑炻器
architectural structure 结构体系;总体结构
architectural structure material 装饰用建筑材料
architectural style 建筑形式;建筑式样;建筑风格
architectural style garden 建筑式庭园
architectural surfaces 符合建筑要求的(模浇混凝土)表面
architectural survey(ing) 建筑测量
architectural system 建筑体系
architectural team 建筑师小组;建筑设计小组
architectural terra cotta 建筑用陶砖;建筑用琉璃砖;建筑用琉璃瓦;建筑用红土陶器;建筑陶板
architectural theorist 建筑理论家
architectural theory 建筑理论
architectural tradition 建筑习惯;建筑惯例;建筑传统
architectural training 建筑培训;建筑训练;建筑教育
architectural treatment 建筑艺术处理;建筑(术的)处理
architectural treatment and finish 艺术装修
architectural trend 建筑发展趋势;建筑动向
architectural trim 建筑装修
architectural trimming unit 建筑装修元件
architectural unit 建筑型材
architectural volume 建筑体积;建筑容积
architectural work 建筑设计作品;建筑设计成果;建筑作品;建筑成果
architectural worker 建筑工人
architectural working drawing 建筑施工图
architectural works 建筑工程
architecture 体系结构;结构格式;结构;建筑学;建筑设计;建筑
architecture adaptable to conversions 建筑重建;建筑改造
architecture adaptable to extension 建筑物扩建
architecture appearance 建筑形状
architecture assemble 建筑群组合
architecture complex 建筑群
architecture composition 建筑构图;建筑布局
architecture coordinate system 建筑坐标系(统)
architecture design 体系结构设计;建筑设计
architecture design model 建筑设计模型
architecture elect 选用建筑(推荐采用的建筑形式)
architecture element 建筑构件
architecture engineering and construction 建筑工程和结构
architecture engineering drawing 建筑工程图
architecture evaluation 体系结构评价
architecture free machine 无结构式计算机
architecture free processor 无结构式信息处理机
architecture function 体系结构功能
architecture layout 建筑布局
architecture of Asia minor 小亚细亚建筑
architecture of Australia 澳大利亚建筑
architecture of Austria 奥地利建筑
architecture of minority nationalities 少数民族建筑
architecture of the Dai nationality 傣族建筑
architecture of the Hui nationality 回族建筑
architecture of the Korean nationality 朝鲜族建筑
architecture of the Miao nationality 苗族建筑
architecture of the Mogul Empire 莫卧儿帝国建筑(印度)
architecture of the Qiang nationality 羌族建筑
architecture of the United States of America 美国建筑
architecture of the Zang nationality 藏族建筑
architecture ornament 建筑装饰
architecture planning 建筑规划
architecture proper 狭义建筑学(不包括铺管道,雕刻装饰等)
architecture realistic graphics 建筑逼真图形;建筑真实感图形
architecture requirement 建筑要求
architecture simulation 体系结构模拟
architecture sketch 建筑草图
architecture treatment 建筑艺术处理
architrave 贴脸板;贴脸;框缘;门头线;门贴脸;门窗周围线条板;线脚;下楣;柱顶过梁;额枋【建】
architrave block 门磴座;贴脸墩;门贴脸墩子;门贴脸座;门窗线条板座块;线条板座块
architrave cornice 门窗框上槛檐板
architrave jamb 门窗边框线脚;门窗线条;门窗线条板边框
architrave of a door 门缘饰;门头线条饰
architrave of proscenium 舞台口边框
archival bag paper 档案袋纸
archival film 档案胶片
archival memory 数据库存储器;档案库存储器
archival paper 档案纸
archival quality 档案质量;存档质量
archival-storage condition 档案储存条件
archive 编档保存;保密归档;公文;档案馆;档案;案卷
archive administration 档案管理
archived file 归档文件;存档文件
archive establishment 档案馆
archive of standardization 标准化档案
archive repository 档案库
archive sample 编档保存样品
archives building 档案馆;档案房屋;档案大楼
archives keeping 档案保管
archives of environmental protection 环境保护志
archives office 档案室
archiving 档案;存档
archiving process 归档过程;归档处理
archivist 档案保管员
archivium 古希腊及罗马档案库
archivolt 穹隆门上饰;拱门侧缘饰;拱边装饰线脚;穹隆形;拱缘装饰;拱门缘饰;拱门饰;拱门墙侧装饰线条
archivoltum 中古期阴沟管式垃圾箱
arch jack 等厚拱
arch key 拱冠石;拱顶石;龙口石;拱石
arch keystone 龙口石;拱趾石;拱顶石
arch length 拱长
archless 无拱的
archless continuous kiln 无固定盖连续窑
archless furnace 开式无拱炉膛
archless kiln 无拱窑;无拱炉
arch limb 穹翼;拱翼;倒转褶皱上翼;伏卧褶皱上翼;背斜翼;背斜翼;顶翼
arch line 拱曲线
arch lining 拱圈衬砌
arch lining of tunnel 隧道顶拱衬砌
arch lining with curved wall 顶拱曲墙衬砌
arch lining with straight wall 顶拱直墙衬砌
arch load 拱荷载
arch mast 拱门桅
arch material 拱材料
arch method 拱架法
arch motor 弓形电动机
arch mo(u)lding 拱饰;拱线脚
archo 群岛海区
arch of anticline 背斜鞍部
Arch of Augustus 奥古斯塔斯拱(位于意大利佩鲁贾)
arch of brickwork 砖拱;砖砌拱
arch of Constantine at Rome 罗马康士坦丁拱门
arch of elevation 仰拱
arch of five center 五心拱
arch of furnace 窑拱
Arch of Janus Quadrifrones 雅努斯四面凯旋门
Arch of Septimus Severus at Rome 罗马塞弗拉斯凯旋门
Arch of Tiberius at Orange 奥兰治蒂比略凯旋门
Arch of Titus 罗马泰塔斯凯旋门
Arch of Titus at Rome 罗马泰塔斯凯旋门
Arch of TRajan at Ancona 安科纳图拉真凯旋门(意大利)
arch of triumph 凯旋门
arch of vault 穹拱
archoplasm 初原生质
archoplasmic loop 初浆袢
archoplasmic vesicle 初浆泡
arch order 拱柱式;中古期拱和柱层层内叠;由附墙柱和檐头框起的拱
arc horn 角形避雷器
arch oval 卵形拱
arch parabola 拱抛物线
arch pattern 弓形纹
arch peak shear plate 拱顶承剪板
arch piece 尼框顶部;船尼拱架
arch plane 拱平面
arch ponor 拱石拱门落水洞
arch pour(ing) 浇筑混凝土拱
arch press 拱形压力机;拱门式压力机;拱门式冲床
arch pressure 拱压力
arch radius 拱半径
arch reinforcing 拱的加劲
arch rib 拱肋;拱肋
arch rib footing 拱肋基
arch ring 拱圈;拱环;拱圈;拱环
arch rise 拱高;拱高
arch rise-span ratio 拱高与跨度比
arch roof 拱形屋顶
arch roof truss 拱形屋架
arch saw 拱锯
arch scaffolding 拱架;拱膺架;拱脚手架
arch section 拱的断面;弧形部分;弧段
arch set 拱形支架;拱架;钢拱形棚子
arch-setter 脚手架工
arch-shaped step 拱形踏步
arch sheeting 拱板
arch skewback 拱座
arch skew block 拱脚斜块
arch slice bridge 拱片桥
arch soffit 拱腹
arch span 拱跨;拱宽;拱跨
arch spandrel 拱肩;拱侧墙;拱臂
arch springer 起拱点;起拱石;拱脚
arch springing 起拱点;拱趾;拱脚
arch springing height 起拱高度
arch springing line 起拱线;拱的起拱线;拱门上升线
arch stair(case) 拱形楼梯
arch stay 拱的加劲杆
arch-stern ship 拱尼形船
arch stiffener 拱的加劲杆
arch stiffening 拱的加劲
arch stone 楔形拱石;拱石
arch stress 拱应力
arch striking 拆除拱架
arch structural analysis 拱结构分析
arch structure 拱结构;弓形结构
arch-subsidence structure 穹起断陷式构造
arch support 拱座;拱座;拱支撑;拱形支架
arch system 拱系统
arch thrust 拱应力;拱推力;拱的轴向压力
arch tie 拱拉杆;拱拉焊
arch timbering 拱形支撑;拱形板桩;拱架
arch top 拱顶
arch truss 拱桁;拱形桁架
arch twilight 曙暮光弧
arch-type 弓形的
arch-type culvert 拱形涵洞
arch-type fender 拱型橡胶护舷
arch-type press 拱式压力机
arch-type sheet pile 拱形(钢)板桩
arch-type steam hammer 导杆汽锤;拱式自由锻锤
arch-type structure 拱式结构
arch-type sub press 拱式小压力机
arch-type support 拱形支撑;拱形支撑
arch-type truss 拱式桁架;弓形桁架
arch vertex 拱冠(石);拱顶
arch viaduct 拱形栈桥;高架拱桥
arch wall 拱墙
arch walling 拱形支护

arch way 拱廊;拱道;拱门
arch web type sheet pile 拱腹板桩
arch weir 拱形围堰
arch width of a natural bridge 天生桥拱跨
arch window 拱形窗
archwise 拱廊似的;成弓形;拱状(的);拱似地
arch with 带啮合接头的拱
arch with apex hinge 具有顶铰链的拱,顶铰链的拱;单铰拱;带顶铰的拱
arch with built-in ends 固端拱
arch with crown hinge 具有顶铰链的拱;顶铰链的拱;带顶铰的拱
arch with fixed ends 固端拱
arch with joggled joints 错缝拱
arch with key hinge 带顶铰的拱
arch with order 同心渐扩踏步式卷
arch without articulation 无铰拱
arch with three-articulations 三铰拱
arch with thrusting force 推力拱
arch with tie 带拉杆拱
arch with top hinge 具有顶铰链的拱;顶铰链的拱;带顶铰的拱
arch with vertex hinge 带顶铰的拱
archy 拱形的
arc-hyperbolic 反双曲的
arc hyperbolic function 反双曲(线)函数
arciform 拱状的,拱形的;成弓形的
arciform density 弧形致密区
arc ignition 电弧点火
arcilla 粗酒石
arc-image furnace 弧成像炉;电弧反射炉
arc image growth 弧成像法
arc incision 弧形切口
arcing 起弧;跳火;击穿;形成电弧;构成电弧;电弧闪击,电弧放电;飞燃弧;飞弧;发生电弧;发弧光
arcing back 回弧
arcing chamber 灭弧室;灭弧腔
arcing contact 电弧触头;灭弧触点
arcing current 超弧电流
arcing device 电弧装置
arcing distance 火花(间)隙;放电距离
arcing fault 闪络故障
arcing gap 火花(间)隙
arcing gas 电弧气体
arcing ground 接地弧;电弧接地
arcing-ground suppressor 接地弧遏制器;电弧遏制器
arcing horn 防闪络角形件;角形避雷器;消弧角
arcing over 电弧击穿
arcing ring 环形消弧器;消弧环
arcing shield 电弧屏蔽环
arcing time 燃弧时间;电弧燃烧时间;飞弧时间
arcing time factor (焊机的)暂载率
arcing tip 弧尖;电弧接(触)点
arcing voltage 跳火电压;电弧电压
arc interlocking relay 电弧联锁继电器
arc interruption 断弧
arc into unit conversion 度数时间换算
arc-jet engine 点弧喷射发动机;电弧喷气式发动机
arc lamp 弧光发生器;电弧灯
arc lamp illumination 弧光照明;弧光灯照明
arc lamp lighting 弧光灯照明
arc leakage power 电弧漏过功率
arc length 弧长
arc length of meridian quadrant 子午线象限弧长
arc light 弧光发生器;弧光(灯);电弧火光
arc light carbon 弧光碳棒
arc light diffuser 弧光扩散器
arc lighting 弧光照明;弧光灯照明
arc lighting dynamo 电弧照明电机
arclight projector 弧光投影器
arc line 弧线;弧光谱线
arc loader 弧线装船机
arc loading 弧体荷载
arc loss 电弧损失
arc magnetism 电弧磁特性
arc maintenance 维弧
arc manipulation 运条
arc measure 弧度法
arc measurement 弧度测量;弧度测量
arc melting 弧熔化;电弧熔化;电弧熔;电弧炉熔炼
arc melting furnace 电弧熔炼炉
arc melting metal powder 电弧熔化用的金属
arc melting method 电弧熔炼法

arc metal 弧焊电极
arc modulation 电弧调制
arc navigation 弧形导航;圆弧导航
arc nodal 弧点的
arc nodal cissoid 弧点蔓叶线
arc noise 电弧噪声
arc of action 作用弧
arc of approach 渐近弧
arc of a sextant 六分仪刻度弧
arc of circumference 圆弧
arc of contact 接触弧
arc of excess 负弧
arc of folding 褶皱弧
arc of lighting 发光扇面
arc of oscillation 振动弧;摆弧
arc of parallel 平行圆弧;纬圈弧
arc of recess 渐远弧
arc of rotation 回转度;旋转弧
arc of swing 摆幅
arc of visibility 视界弧;能见弧(度);能见光弧;光弧;观察扇面
arcogen welding 电弧氧乙炔焊
arcogeny 弧曲运动
arcograph 圆弧规
arcola 小锅炉
arcolite 阿尔科列特酚醛树脂
arcological city 仿生城市
arcology 不用汽车的城镇;生态建筑学
Arco microknife 阿尔科显微刀
arconograph 电弧稳定性测定仪
arc oscillator 电弧振荡器
arcose 长石砂岩
arcosic grit 长石砂岩磨料;长石砂岩
arcosolium 罗马地下墓穴中的拱顶小室
arcotron 显光管
arc-over 跳火;电弧放电
arc-over voltage 电弧放电电压
arc-oxygen cutting 电弧氧气切割
arc-oxygen welding 氧弧焊
arc penetration 电弧穿透
arc perimeter 弧形视野计
arc pile 弧形桩
arc piling 弧形桩
arc pistol 电弧喷枪
arc plasma 电弧等离子体;电弧的等离子区
arc plasma gun 电弧等离子体喷枪;等离子体电弧枪
arc plasma spraying 电弧等离子体喷涂;等离子体电弧喷涂
arc point welding 电弧电焊
arc preventive coveralls 防电弧工作服
arc process 电弧法
arc production furnace 电弧熔炼炉
arc protect circuit 打火保护电路
arc radial curvature 弧弓曲率
arc rectifier 弧光放电整流阀;电弧整流器;电弧振荡器
arc regulator 电弧调节器
arc resistance (绝缘材料的)抗电弧性(能);耐电弧性;弧阻;弧电阻;电弧电阻
arc resistance meter 弧阻计
arc ring 弧环;引弧环;分弧环
arc sag 弧垂
arc scale 弧度数;弧形尺;弧度
arc-seam weld 电弧缝焊
arc secant 反正割
arc second 弧秒
arc self-regulation 电弧自身调节
arc shaped collar-pressing machine 弧型压领机
arc shape cutting machine 弧形切割机床
arc shaped releasing plate 弧形放松板
arc-shaped tile 筒瓦
arc shears 弧剪机
arc shooting 弧形爆破法;弧线爆破
arcsine 反正弦
arc-sine distribution 反正弦分布
arcsin function 反正弦函数
arc-sine law 反正弦定律
arc sine transformation 反正弦变换
arcs of visibility 能见范围
arc source 弧光源;电弧离子源
arc-spark direct reading spectrometer 电弧火花直读式光谱仪
arc spectrograph 弧光摄谱仪;电弧摄谱仪
arc spectrography 弧光摄谱学

arc spectrometer 弧光分光计
arc spectrum 弧光谱
arc spectrum analysis 电弧光谱分析
arc splitter 分弧器
arc spotlight 弧光聚光灯
arc-spot welder 电弧点焊机
arc-spot welding 电铆焊;电弧点焊
sprayed coating 电弧喷镀层
arc spraying 电弧喷涂;电弧喷射;电弧喷镀
arc spray metallizing system 电弧喷涂金属系统
arc stability 电弧稳定性
arc stabilizer 稳弧剂
arc starting 起弧;电弧接通
arc stiffness 电弧稳定度;电弧挺度
arc stream 电弧气流
arc strike 弧光放电;电弧击打;电弧放电;电弧触发
arc striking mechanism 电弧触发器
arc stud welding 螺柱电弧焊;柱钉电弧焊
arc subtended by a chord 对弦弧
arc suppressing 灭弧;消弧
arc suppressing transformer 灭弧变压器
arc suppression coil 灭弧线圈;消弧线圈
arc suppressor 灭弧器;熄弧器;电弧遏制器
arc suppresssing reactor 灭弧电抗器
arc surfacing 电弧堆焊
arc system 电弧通信系统
arctalpine flora 北极高山植物区系
arctan 反正切
arctan function 反正切函数
arc tangent 反正切
arctation 孔道狭窄
arc thickness 弧线厚度
arc-through 弧穿;电弧击穿;电弧穿过
arctic 北极地方;北极的
Arctic air mass 北极气团
arctic-alpine 高寒的
Arctic-Antarctica crustal-wave system 北冰洋—南极洲地壳波系
Arctic anti-cyclone 北极高压
Arctic belemnite region 北极箭石地理大区
Arctic bottom water 北极底层水
Arctic brown earth 极地棕壤;北极褐土
Arctic cap 防寒帽
Arctic Circle 北极圈
Arctic climate 极地气候;北极气候
Arctic climate of ice field 极地冰原气候
Arctic construction 寒冻区建筑
Arctic current 北极海流
Arctic drainage 北极水系
Arctic easterlies 北极东风带
Arctic eastern wind current 北极东风流
Arctic ecology 北极生态学
Arctic ecosystem 极地生态系统;北极生态系统
Arctic fox 北极狐
Arctic front 北极锋;北极边缘
Arctic glass (有金属丝加固的)毛玻璃
Arctic grassland 极地草地
Arctic haze 北极霾
Arctic high 北极高压
Arctic ice 北极多年冰
Arctic ice breaker 北极破冰船
Arctic ice cap 北极冰盖
Arcticization 低温准备
Arctic Mid-Oceanic Ridge 北极海中央海岭
Arctic mist 北极雾
Arctic navigation 极地航行
Arctic Ocean 北冰洋
Arctic offshore 北极外洋
Arctic ozone hole 北极臭氧层空洞
Arctic-Pacific realm 北极太平洋区系
Arctic plant 极地植物
Arctic Pole 北极
Arctic pollution prevention certificate 极地防污染证书
Arctic province 北极地区
Arctic realm 北冰洋带
Arctic region 北极地区;极地
Arctic Sea 北冰洋;北极海
Arctic sea smoke 北冰洋蒸发雾
Arctic slope basin 北极陆坡盆地
Arctic smoke 北冰洋蒸发雾
Arctic soil 极地土壤
Arctic source water 北极源水
Arctic suite 北极岩套

Arctic tree line 北极树木线
Arctic tundra 极地冻原
Arctic tundra climate 极地苔原气候
Arctic vegetation 北极植物
Arctic vessel 极区航行船;极区船
Arctic waters 北极水域
Arctic weather 极冷的气候
Arctic whiteout 乳白辉
Arctic zone 北极区;北极区
arc tight 耐弧的
arc timber 弧形木
arc time 拉弧时间;弧焊开动时间;弧光发生时间;发弧时间
arc timer 燃弧时间测定装置
arc tip 电弧接(触)点
arctite 北极石
arc to chord correction 转换角;半收敛差;半聚合差
arc-to-chord correction in Gaussian projection 高斯投影方向校正
arc-to-chord reduction (投影的)方向改正
Arcto floral realm 泛北植物地理区系
arc torch 弧焊炬;电弧焊炬
arc tracking 电弧径迹
arc transmitter 弧式发送机;电弧(式)发射机
arc trench couple 弧沟对
arc trench gap 弧沟间隙
arc trench system 弧沟体系
arc trench tectonic 弧沟构造
arc triangulation 三角测量;弧三角测量
arcual construction 拱形构造;弓形构造
arcualia 弓片
arcual structure 拱形结构;拱形建筑物;弓形结构
arcual style 拱形建筑风格;弓形建筑形式
arcual system 拱系统
arcual system for absorption of thrust 推力吸收拱系统;推力衰减拱系统;推力缓冲拱系统
arcuate 弧状;拱形的
arcuate architecture 拱式建筑
arcuate collecting tubule 弓状集合小管
arcuate construction 拱形结构;拱建筑
arcuated 拱式的
arcuated architecture 拱式建筑
arcuate dark point 弧形暗点
arcuated building 拱式建筑
arcuated construction 拱形建筑物;拱形构造;拱式结构;拱式建筑;拱式构造
arcuate delta 阔叶三角洲;弧形三角洲;弓形三角洲
arcuated structure 拱式结构
arcuated style 拱式建筑风格
arcuated system 拱系统;拱式系统
arcuate echelon type 弧形斜列式
arcuate eminence 弓状隆起
arcuate fault 弧形断层
arcuate fold 弧形褶皱
arcuate linear 弧线状的
arcuate meander 弧形曲流
arcuate scotoma 弓形暗点
arcuate zone 弓形带
arcuatio 古罗马拱建的构筑物;古罗马拱构筑物
arcuation 弯曲;拱工;成弧作用
arcuation construction 拱式构造
arcuayite 硫铋铜银矿
arcus 弧状云;弧形云;弓
arcus ecclesiae (中世纪教堂中殿听众席与圣台间的)拱券;中世纪教堂中殿与圣台的拱券
arcus pedis transversalis 横弓
arcus presbyteri 中世纪教堂东端半圆室上的拱券
arcus toralis 教堂中殿和唱诗班室间的格子屏
arc voltage 起弧电压;电弧电压;弧电压
arcwall coal cutter 弧形截槽截煤机
arc wall face 弧形面
arcwalling 弧形掏槽
arc-weld 电弧焊
arc welded 电弧熔接的;电弧焊接的
arc welded pipe 弧焊管
arc welded steel pipe 电弧焊接钢管
arc welder 弧焊机;电弧焊机;电焊机
arc-welder's disease 铁尘肺
arc-welder's gloves 电焊手套
arc welding 弧焊;电弧焊接;电焊;电弧焊
arc welding electrode 弧焊电极;电弧焊条;电弧焊用焊条
arc welding equipment 弧焊设备

arc welding generator 弧焊发电机;电弧焊接用发电机
arc welding generator with independent excitation 他激电焊发电机;自激弧焊发电机
arc welding generator with self-excitation 自激电焊发电机
arc welding machine 弧焊机;电弧焊机;电焊机
arc welding plant 电焊厂;弧焊机
arc welding process 电弧焊接工艺过程
arc welding rectifier 弧焊整流器
arc welding robot 弧焊机器人
arc welding set 弧焊装置;弧焊机;电(弧)焊机组
arc welding transformer 电弧变压器;弧焊变压器;电焊变压器
arc without contact 无切弧
ardaite 氯硫锑铅矿
ardal 阿达铝合金
Ardamu silk 阿达穆生丝
Ardand type polygonal roof 阿尔丹德多角形屋顶(后哥特式屋顶);木拱脚悬臂托梁屋顶
ardealite 磷石膏
Ardelt process 阿达特砂型离心铸管法
Ardennes massif 阿登尼斯地块
Ardennian orogeny 阿登造山运动
ardennite 砷硅铝锰石;锰硅钒铝矿
ardometer 光测高温器;光测高温计
ardor 铝箔状薄板
Ardran-Crooks cassette 阿德兰—克鲁克斯储[贮]片盒
arduinite 发光沸石;安沸石
arduous 险峻的
are 公亩(1公亩=100平方米)
area 区;领域;面积;地下室前凹地;地区;场地
area adjustment 区域网平差;面积平差
area aerial photography 面积航空摄影
area aerial survey 区域航空测量
area affected 涉及地区
area age 面积令
area allocation diagram 区域分配图
area allotted for the construction 建设项目所在地
area-altitude curve 面积高度曲线
area analysis 面分析
area analysis intelligence 区域分析信息
area and yard drains 区域和堆场排水
area approach center 区域进近管制中心
area array 面积组合
area array X-ray sensor 分区阵列X射线传感器
area assessment 区域评价;分区评价
area assignment 区域赋值
area assist action 区域供电临时改进措施
area attribute 区域属性;区域表征
area available for reinforcement 有效加强面积
area averaging method 面积平均法
area-balanced current 等量面积电流;面积平衡电流
area-balanced waveform 面积平衡波形
area bar-chart 面积条图
area bare-cultivated 未耕过地区
area based matching 基于灰度影像匹配
area boundary 区域界限;地区界(线)
area-capacity curve 面积—库容曲线;面积—容积曲线
area chart 面积图
area classification 区域类别;按地区分类
area cleared of buildings 已拆除房屋的场地
area closed sign 封锁区标志
area closed to fishing sign 禁渔区
area code 区域(代)码;区代码;地区(代)码
area coefficient 面积系数
area composition machine 区域作图机
area computation 面积计算
area condition 区域状况;区域条件
area contact 平面接触
area control 面控制;地区管制
area controller 区域控制员
area control radar 地区控制雷达
area coordinates 面积坐标
area coordinator 工区协调员
area correction 面积校正
area correlation 面相关
area cover (为查明未来矿区地质水文条件的)井网控制面积
area coverage 区域范围
area covering structural element 平面承重结构;覆盖结构构件

area covering structure 覆盖结构;平面承重结构建筑物
area cultivation 耕作区
area curve 断面积曲线;面积曲线;断面曲线;断面积曲线
area data 区域数据;面积数据
area defence 区域防空
area delimiting line 区域境界线
area density 布面密度
area-depth curve 同深度面积曲线;面积—深度曲线
area-depth curve for rainfall 面积—深度降雨曲线
area-depth distribution curve 面积—深度分配曲线
area-development 地区开发
area dispatcher 工地调度员
area distance curve 面积距离曲线
area distance method 面积距离法
area distortion 面畸变
area-distortion coefficient 面积变形系数
area-distortion ratio 面积变形比
area distribution 区域分布
area-distribution curve 面积分配(关系)曲线
area-distribution ratio 面积变形比
area district (土地利用规划的)面积分区
area dosimetry 区域剂量测定法
area drain 露天排水斗;地区排水;地面排水(沟);场地排水
area echograph 面积回声曲线(图);面积回声测深仪
area effect 区域效应;面积效应
area efficiency 面积效率
area-elevation curve 面积—高程曲线;同高程面积曲线
area-elevation distribution curve 面积—高程分配曲线
area-elevation graph 面积—高程曲线图
area emission sources 区域污染源
area emission standard 地区排放标准
area error 面积误差
area exaggeration 面积夸大
area exchange 区域电话局;区交换
area factor 面积因数;面积系数
area fine measurement 面积细测
area flowmeter 截面流量计;面积式流量计;面积流量仪;面积流量计
area forecast 区域预报
area forecast center 区域预报中心
area forecast system 区域预报系统
area format 区域形式
for new settlements 新居民区
area free from defect 无缺陷区域
area frequency coordinator 地区频率协调者
area function 面积函数
area geology 区域地质学
area graph 面积图
area grating 阴盖;阴井盖;场地雨水口算盖
area grating cover(ing) (阴井的)格栅盖;格栅保护层
area gravity survey 面积重力测量
area grouting 铺盖灌浆;固结灌浆
area heating installation 局部加热装置
area histogram 面积直方图
area identification 区域标识符
area identifying number 区域识别码
area ignition 全面点火
area illumination 区域照明
area image sensor 面图像传感器
area increment 面积增长
area increment method 面积增长法
area in storage 存储区(域)
area in the front of yard 厂前区
area kill 单位面积杀量
areal acceleration 掠面加速度
areal analysis 面积分析
area landfill 地区性卫生填地
areal average pressure 区域平均压力;面积平均压力
areal average rainfall 面平均雨量
area law 面积定律
areal category 按地区分类
areal center 分布区中心
areal change 区域性变化
areal coordinates 面坐标;重心坐标
areal coverage of forest 森林面积;森林覆盖度
areal deformation 表面变形

areal density 面密度;表面密度
areal differentiation 区域差异;区域差距;地域分异
areal distribution 面上分布
areal division of power 电力区划
areal dose monitor 照射面剂量监视器
areal effect of colo(u)r 装饰色彩的面积效应
areal element 面积元(素)
areal eruption 区域(性)喷发
areal level(l)ing 面测准(测量)
areal feature 面积要素
areal forecast 区域预报
areal geochemical anomaly 区域性地球化学异常
areal geologic(al) map 区域地质图
areal geologic(al) structure 区域地质构造
areal geology 区域地质学
area light 地下室窗采光;场地灯光
area lighting 场地照明
area lighting source 面光源
a realistic plan 切实可行的方案
areal level(l)ing 面水准测量
areal limit 传播范围;分布范围
areal limits of oil sand 油层含油范围
areal map 区域图;地区图
areal mean precipitation 面平均降水量
areal mean rainfall 面平均雨量;地区平均(降)雨量
areal metric 面积度规
areal non-point source 面非点污染源
areal non-point source watershed environment response simulation model 面非点污染源流域环境响应模拟模型
area load(ing) 面荷载;面积荷载;均布荷载
areal oceanography 区域海洋学
arealometer 纤维细度气流测定仪
areal packing density areal storage density 面存储密度
areal pollution 区域污染;面污染
areal pollution source 面污染源
areal power substation 区域变电所
areal precipitation 区域降雨量;区域降水(量);面降水量
areal productivity 面金属量
areal rainfall 面雨量
areal rainfall depth 地域平均雨量
areal rate 面积速率
areal sampling 面积抽样(法)
areal settlement 面状沉降
areal source 面震源;面污染源
areal source pollution 面源污染
areal stockpiling 平层堆料
areal structure 区域构造
areal study 区域研究;区域勘探;区域勘察;区域勘测
areal suction effectiveness 表面吸收效能
areal symbol 面状符号
areal transfer 区域转移
areal variation 区域性变化
areal variation in rainfall 区域性雨量变化
areal variation of intense precipitation 降水强度面积变化
areal velocity 掠面速度;面积速度
area magnification 面积放大;面放大率
area mask 套色区
area mean pressure 表面平均压力
area measurement 面积测定
area meteorological watch 区域气象监视
area meter 面积仪;面积计量仪;面积计量器;面积测速计
area method 区域法;求面积法;面积法;面波法;范围法;面积估价法
area method of estimating cost 按面积估算造价法;面积估价法
area method of landfill 区域填冲法
area mismatch loss 面积失配损失
area moment 力矩面积;面积矩;面积矩
area moment method 力矩面积法;几何面矩法;面法;弯矩面积法
area monitor 区域监察器;特定范围放射线检测器
area monitoring 区域监测;局部放射量探测
area-name 区域名
area navigation 区域导航
area navigation computer 区域导航计算机
area navigation station 区域导航站
area normalization 区域正常化
area normalization method 面积归一化法
area occupied by building 建筑物占地面积

area of abnormal variation 磁差异常区
area of absorption device 吸附器面积
area of a building 一栋房屋面积;房屋面积
area of accumulation 堆积区
area of activity 客厅;(在一幢公寓楼内的)起居室;活动区
area of a curved surface 曲面面积
area of adhesion 黏附面积
area of air survey 航测面积
area of a karst hall 岩溶洞厅的面积
area of ambiguity 不定面积
area of application 应用范围
area of aquifer intake 含水层进水面积
area of artesian flow 承压水流出面
area of artificial recharge engineering 回灌工程面积
area of a structure 建筑物面积
area of audibility 能听范围
area of balance calculation 匀衡计算区面积
area of ball imprint 球凹面积
area of base 基底面积;地基面积;底面积
area of beam 束截面
area of bearing 支承面积;支承面;承载面积;负荷面积
area of building 房屋面积;建筑面积
area of calculated region 计算区面积
area of capillary rise 毛细上升区
area of catchment 流域面积;集水面积;汇水面积
area of chemical reagents 药品房面积
area of circle 圆面积
area of circular features 环形体面积
area of close depression 封闭洼地面积
area of cohesion 聚结面积;凝聚面积;黏着面积
area of comprehensive redevelopment 大规模重建区
area of concentrated emphasis 重浓度区
area of contact 接触面积
area of contour 投影面积;等高线面积;等高线包围的面积
area of control region 控制区面积
area of cooling 受冷面积
area of cooling surface 冷却面积
area of cool or temperate climate 冷凉或温暖气候地区
area of counter pressure 反压面积;反压力面积
area of coverage 涉及范围;观测区;覆盖区(域);覆盖面积
area of cross-section 横断面(面)积
area of cup 压痕面积(试验硬度时用)
area of cut 切削面积
area of deposition 喷镀区;喷镀面积;喷镀范围;淤积区;堆焊区;堆焊面积;堆焊范围;沉积区;沉积面积
area of depression 沉降地区;洼地区;沉降截面;沉降地段
area of diagram 图形面积
area of discovery 发现面积
area of dispersion 扩散面积
area of dissipation 散热面积;消融区;消融面积;消融范围;消能区;消能面积;消能范围
area of diversion 分水面积
area of double of triple cropping 复种面积
area of draw 放矿面积(崩落采矿法)
area of dwelling unit 居住单位面积
area of efflux 出流面积
area of environmental stress 环境受到威胁地区
area of evapo(u)ration 蒸发区
area of evasion 避航区
area of exceed the flight elevation 超高面积
area of explosion 炸裂面积
area of extraction 影响面积;作用面积
area of face per hole 单位钻孔表面积
area of factory site 厂区占地面积
area of falling-head standpipe 变水头竖管面积
area of fallout 落下灰沉降区
area of faulting 断层区
area of feasible solutions 可行解区域;可行区域
area of fire division 防火面积
area of fire grate 炉算面积
area of flood basin 汇水盆地面积
area of floor 底面积
area of flow 水流横截面
area of flow rate diagram 流率图面积
area of flue 烟道工作段;烟道表面

area of force 外力图面积
area of fracture 断口面积
area of gamma logging curve 伽马测井曲线面积
area of gas cap 气顶面积
area of geothermal manifestation 地热显示区
area of geothermics anomaly 地热异常区
area of getting 开采面积;影响面积
area of glazing 装配玻璃面积
area of grate 算子面积
area of grate diagram 火床面积;火算面积
area of ground contact 地面支承面积
area of ground subsidence 地面沉降面积
area of groundwater discharge 地下水溢出带;潜水溢出带;地下水流出区;地下水出流区
area of groundwater reservoir 地下水库面积
area of heating surface 加热面积
area of heat transfer 热转换面积;热交换面积;导热面积
area of heavy rainfull 多雨地区
area of high nuclear radiation 高核辐射区域
area of high shear 高切力区;高剪切区;高剪力区
area of high stress 高应力区
area of high temperature and rainfall 高温多雨地区
area of house trailer spaces 拖车住房停放面积
area of hydrofoil 水翼面积
area of hysterisis loop 磁滞曲线面积
area of inclusion 包裹体面积
area of increased radioactivity 放射性增升区
area of indentation 压痕面积(试验硬度时用)
area of infection 受影响面;影响区;感染区
area of influence 影响区;影响面积;影响范围;作用面积
area of influence line 影响线面积
area of influence of extraction 采空影响区
area of influence of well 水位降落区;深井抽水影响面积;水井影响区;水井影响面积
area of injection orifice 喷口面积
area of investigation 调查范围
area of investigation region 周查区面积
area of klippe 飞来峰面积
area of land 土地面积
area of lateral surface 侧面积
area of leaky region 越流区面积
area of liquid-gas interface 气液界面面积
area of load distribution 荷载分布面积
area of loading 受荷面积;荷载面积
area of lower contour 下等高线面积
area of maximum rainfall 最大降雨区;最大降雨面积
area of meander belt 迂回区面积
area of memory 存储区(域)
area of mesh 筛孔面积
area of mice sheet 云母片面积
area of midship section 船中剖面面积
area of mill site 厂区占地面积
area of mining 开采面积;影响面积(抽水)
area of mining funnel 开采漏斗面积
area of mobile bottom 海底易变区
area of moment 力矩面积;弯矩面积
area of nappe 推覆体展布面积
area of natural reserves 自然保护区
area of nature reserves 自然保护区
area of non-point source pollution control 非点源污染控制区
area of non-point sources of pollutant 非点污染源区
area of opening 筛孔面积;开孔面积
area of ore block 块段面积
area of orebody 矿体面积
area of ore district 矿区面积
area of outstanding natural beauty 天然风景区;自然风景区
area of passage 通路面积;通道面积;过流面积;出入口面积
area of penetrometer 锥底面积
area of perceptibility 有感区
area of permanent high pressure 常定高压区
area of perpetual shadow 永久阴影面(建筑物常年没有日照部分)
area of per unit volume 比面
area of pictorial surface 图面积
area of pile head 桩头面积;桩顶面积
area of plane 平面面积

area of possible collision 可能碰撞区
area of predicted district 预测面积
area of prediction region 预测区面积
area of pressure 受压面积;加压面积;承压面
area of principal building 主要建筑物面积
area of probe bottom 探头锥底面积
area of propeller 螺旋桨桨叶面积
area of prospective district 远景区面积
area of pumping depression 抽水漏斗;抽水降落区;抽水下降面积
area of quay 码头面积
area of rain 雨区;降雨面积
area of recharge 补给区面积
area of reference 参考面积
area of reinforcement 钢筋强面积;钢筋面积;钢筋截面积;钢筋加强面积
area of reinforcing steel 钢筋面积
area of reservoir 储集层面积
area of rivet shaft 铆钉杆面积
area of road and place 道路及广场面积
area of rudder 舵面积
area of safe operation 安全工作区;安全操作区
area of sample 试样面积
area of seawater intrusion 海水入侵面积
area of section 剖面面积;截面面积;截面积
area of sedimentary rock 沉积岩面积
area of seismographic(al) disturbance 有震区;地震区
area of shelter belt 防护林面积
area of site 场建面积
area of slack water 憩流区;平潮区;缓流区;憩流段
area of sluggish flow 缓流区
area of snowmelt 融雪区
area of soil sample 土样面积
area of solid angle projection 立体角投射面积
area of source 来源地区
area of special biological significance 特种生物显著区
area of steel 钢筋面积;钢筋断面积;钢材面积
area of storage yard 露天堆场面积;堆场面积
area of structural steel 型钢截面积
area of structure 建筑区;建筑面积;建筑结构面积;构筑物面积;构造面积
area of subsidence 下沉面积
area of supply 供给范围;服务区(域);服务面积;补给区;补给面积;补给范围
area of suspected pollution 可疑污染区
area of target 靶面积
area of tectonic region 构造区的面积
area of thrust surface 推力面面积
area of unit occupancy 使用单元面积
area of upper contour 上等高线截面积
area of use 使用面积
area of visibility 能见范围
area of volcanic products 喷出物面积
area of water development 水资源开发
area of water drenching 淋水系统
area of water infiltration test 灌水入渗试验面积
area of water plane 水线面面积;吃水线面积
area of water quality over standard 水质超标区面积
area of waterway 水流截面积;水道截面积;河道断面面积;过流断面
area of wave generation 波浪发生区;波发生区
area of well influence 井影响面积;井影响范围
area of wetted cross-section 浸水断面积;过水横截面面积;过水横断面积;过水断面面积
area of wetted surface 浸水面积
area of wetter cross-section 水下断面面积
area of wharf 码头面积
area of window 构造窗面积
area of woods 森林面积
area of work 工作面积
area of working 开采面积;影响面积
area on vertical projection 垂直投影面积
area opaca 暗区
area opening of filter 滤池空隙面积
area pattern 面状模式
area pellucida 明区
area picture 区域像片
area planning 土地规划;地区规划
area ploughed by tractors 机耕面积
area pollution 区域污染
area pollution source 面污染源

area postrema 极后区;最后区
area precipitation 区域降雨量;区域沉降(量)
area-preserving map 等积投影地图
area projection method 立体投影(测图)法
area ratio 面积比(率)
area ratio of earth fracture 裂缝面积率
area ratio of the sampler 取土器面积比
area reconnaissance 地区草测;区域勘探;区域勘察;区域勘测;面积普查
area redevelopment act 地区再发展条例
area redevelopment program(me) 区域重新发展规划
area redistribution 面积的再分配
area reduction 面积缩小;断面减缩率
area-reduction method 面积减少法
area regulation 面积调节法
area requirement 面积要求;需要面积;需要工作面
area requiring soil improvements 需要加固土壤的地区;需要改良土壤的地区
area research 区域调查
area retrieval 区域检索
area rug 小地毯(房间用);装饰性小块地毯
area rule 面积律
area rule concept 面积律概念
area safe for surface navigation 水面安全航行区
area sample 区域样本
area sampling 区域抽样(法);面积抽样(法);地区抽样(法)
area scale 面积比(率)
areas covered with forests 森林覆盖地区
areas dangerous due to mines 水雷危险区
area search 区域检索;领域检索
area sensitivity 面灵敏度
area separation wall 分隔墙;风火墙
area separator 区域分隔带
area sketch 地区性略图
area-solid-angle product 面立体角乘积
area source 面污染源;面震源;地区性(空气)污染源
area source of air pollution 空气污染面源
area source of sound 面声源
area source of water pollution 水污染面源
area source pollution characterstc 面污染特征;面源污染特征
area source survey 面污染源调查
area spray 面施药;地面施药
area standard 地区标准
area standardization 地区标准化
area strain 区域应变
area striata 纹状区
area substation 地段变电站
area summation method 面积总值法
area supplementary control 区内辅助性控制;区内辅助性调节
area supported 支承表面
area survey 区域测量
area sweep 面积扫油
areas with gently dipping schistosity 具缓倾片理的地区
area system 断面系统
area take-off 面积的粗略估算;粗略估计(面积);粗略估算
area target 面积目标;面积目标
area thermal substation 小区热力站
area tie line 地区联络线
area to avoid by ships of certain classes 某一类型船避航区
area to be avoided 避航区
area to be dredged 挖土区;挖泥区;挖泥范围;疏浚区
area-to-mass ratio of satellite 卫星面质比
area traffic control 地区交通控制
area traffic control system 区域交通控制系统
area triangulation 三角网测量;面积三角测量
areatus 簇状的
area type meter 面积型流量计
area under a curve 曲线下的面积
area under canopy 覆盖面
area under crops 种植面积;播种面积
area under glass 保护区
area under reclamation 填筑区
area under residual static stability curve 剩余静稳性面积
area under the jurisdiction 管辖区(域)
area unit 区域单位;面积单位

area utilization factor 面积利用系数
area variable 区域变量;区变量
area-velocity ga(u)ging method 面积—流速测流法
area-velocity method 面积速度(测流)法
area-volume curve 面积—容积曲线
area-volume ratio 面积—体积比
area wall 四周挡土墙(地下室);空地围墙;工地围墙;地下室前空地周围的挡土墙;窗井墙;建地围墙
areaway 过道(地下室);地下室前空地;地下室窗前的采光井;地下室采光井;窗井
areaway grating 采光井格栅盖
areaway wall 窗井墙;采光井墙
area weighted average resolution 面积加权平均分辨率
areawide bargaining 全区域谈判
areawide control 区域控制
areawide count 地区交通量观测;区域交通观测
areawide water quality planning and management 广域水质规划与管理
area within the jurisdiction 管辖区(域)
area with main services 具有管道设施的地区;主干管服务面积
area yield value 区域产量值
area zoning (城市规划的)面积分区制
areca betel nut 槟榔
arecadeine 槟榔因
arecaidine methyl ester 槟榔碱
arecaine 槟榔因
areca nut 槟榔
areca red 槟榔红
a recession in price 价格暴跌
Arecibo ionospheric observatory 阿雷西波电离层观测台
arecin 槟榔红
arecoline 槟榔碱
arecolone 槟榔酮
arefaction 除湿;干燥
areflexia 无反射;反射消失;反射缺失
A register 运算寄存器
arena 竞技场;圆形剧场中表演区;圆形竞技场;动界
arenaceous 砂的;沙质的;含砂的;多砂质的
arenaceous cement 砂质胶结物;沙质胶结物
arenaceous clay 砂质黏土
arenaceous-conglomeratic facies 砂质砾岩相;沙质砾岩相
arenaceous deposit 砂屑岩;粗砂质
arenaceous limestone 砂质石灰岩;砂质石灰石;沙质石灰岩;沙质石灰石
arenaceous limestone chip(ping)s 砂质石灰岩碎屑;砂质灰岩碎屑;石灰岩碎屑;石灰砂岩碎片
arenaceous-petitic facies 砂泥质相
arenaceous quartz 石英砂
arenaceous region 砂质地区;砂质地段
arenaceous rock 砂质岩;沙质岩
arenaceous sediment 砂质沉积物;砂岩;沙质沉积物
arenaceous sedimentary rock 砂质沉积岩;沙质沉积岩
arenaceous shale 砂质页岩;沙质页岩
arenaceous texture 松散结构;砂质结构;沙质结构
Arena Chapel at Padua 帕多瓦圆形大教堂(意大利)
arena stage 三面临观众的舞台;中心式舞台
arena theater 舞台设在观众席中央的剧院;中心舞台剧场
arena type stage 圆形剧场式舞台;突伸式舞台;圆形剧场
arena vomitory 通向中央舞台的进出口
arendalite 暗绿帘石
arene 砂质黏土;风化粗砂;芳烃
arenicolite 砂栖石
Arenigian 阿仑尼阶【地】
arenite 砂质岩;砂屑岩;砂碎屑岩;砂粒碎屑岩;粗砂碎屑岩
arenose 含砂砾的;粗砂质
arenosol 红砂土
arenous 砂质的;多砂的
arenturine 耀水晶
arenyte 砂质岩;砂碎屑岩;砂粒岩;屑岩
areodesy 火星测量学
areodetic 火星测量的
areographic 火星表面的

areographic geography 分布区地理学
areographic spectrum 分布区谱
areola 细隙
areolar 细隙的
areolar tissue 蜂窝状组织
areolat(ed) 小空隙的
areolation 网眼状结构;形成网眼状空隙
areology 火星学
areometer 浮秤;液体比重计;比重计;比浮计
areometer analysis 液体比重分析
areometry 液体密度测定(法);液体比重测定法;稠液比重测定法;比重测定法
areopycnometer 黏液比重计;稠液比重计;稠度比重计;比重瓶比重计
aerospace map 宇航地图
areostyle 对柱式建筑;疏柱式建筑物(柱距等于柱径的四倍或四倍以上)
arepycnometer 液体密度计
arete 刃岭
arfvedsonite 钠铁闪石;亚铁钠闪石
Arfwedson distribution 阿尔夫威德森分布
argand 管状灯
argand burner (具有空心管子的)煤气燃烧器
Argand diagram 阿尔干图
Argand lamp 阿尔干灯
Argan oil 阿干树油
Arga polarity hyperchron 阿尔加极性巨时
Arga polarity hyperchron zone 阿尔加极性巨时间带
Arga polarity hyperzone 阿尔加极性巨时带
Argasoid 阿加索依德铜合金
Argelander method 阿格兰德法
argemone seed oil 蓟罂粟油
argent 银白色
argental 银质的;银汞膏
argentalium 银铅
argental mercury 含银汞;银汞膏
Argentea 阿根泰有光黏胶长丝;银膜
argentian auricupride 含银的金铜矿
argentic 高价银的
argentic oxide 低氧化银
argentic salt 银盐
argentic sulfide 硫化银
argentiferous 含银
argentiferous galena 银方铅矿
argentiferous lead 含银铅;银铅
argentimetry 银液滴定法
Argentina abyssal plain 阿根廷深海平原
Argentina basin 阿根廷海盆
Argentina hook 阿根廷钩
argentine 银色页状方解石;银色金属;珠光石
argentine compound 银化合物
argentine finish 银光面漆
Argentine metal 宽赛银锡锑合金;银色锡锑合金
argentite 辉银矿
argentojarosite 银铁矾
argentometer 银盐定量计;电量计;测银比重计
argentometry 银量法
argentopentlandite 银镍黄铁矿
argentopyrite 阿硫铁银矿
argic water 通气层水;含气带水;包气带的水(土壤中)
argid 黏化旱成土
argil 陶土;矾石;白黏土;白土
argilla 陶土;泥土;铝氧土;高岭土
argillaceous 黏土质的;泥质的;含黏土的
argillaceous bauxite 泥质铝矾土
argillaceous bottom 黏质底土;黏土地基
argillaceous breccia 泥质角砾岩
argillaceous cement 黏土水泥;泥质胶结物
argillaceous conglomerate 泥质砾岩
argillaceous desert 黏土荒漠
argillaceous dolomite 泥质白云岩
argillaceous earth 黏土
argillaceous earthenware 泥质陶器
argillaceous facies 泥质岩相
argillaceous fracture 泥质组分
argillaceous gall 黏土团;泥质薄片
argillaceous grain 黏土颗粒
argillaceous gravel 泥质砾石
argillaceous gypsum 含黏土石膏;泥质石膏
argillaceous hematite 土状赤铁矿
argillaceous ice layer 含土冰层
argillaceous inclusion 泥质岩包裹体

argillaceous ingredient 黏土质混合物;泥质混杂物
argillaceous intercalated bed 泥土夹层
argillaceous intercalation 泥质夹层
argillaceous iron ore 黏土质铁矿;泥质铁矿
argillaceous iron stone 黏土质铁矿石;泥质铁矿
argillaceous limestone 黏土质石灰岩;黏土质灰岩;泥质石灰岩;泥质灰岩
argillaceous marl 泥质泥灰岩;黏土灰岩
argillaceous material 黏滞性粘质
argillaceous mud 黏土泥浆;泥浆
argillaceous odo(u)r 土腥味;泥土气味
argillaceous particle 黏土颗粒
argillaceous raw material 黏土质原料
argillaceous rock 黏土岩;泥质岩;泥质岩
argillaceous rock facies 泥质岩相
argillaceous sand 黏土质砂(土);泥质砂;含黏土的砂
argillaceous sand ground 泥质砂地
argillaceous sandstone 黏土质砂岩;泥质砂岩
argillaceous schist 陶土片岩;泥质片岩
argillaceous sediment 泥质沉积(物)
argillaceous shale 黏土质页岩;泥质页岩
argillaceous shale formation 泥质页岩建造
argillaceous siltstone 泥质粉砂岩
argillaceous slate 陶土板岩;泥质板岩
argillaceous smelt 泥土气味
argillaceous soil 黏土泥土
argillaceous texture 黏土质结构;泥质结构;泥土质结构
argillaceous zone 黏土化带
argillation 黏土化;泥质化
argillation zone 泥化带
argillazation 泥质化
argile scagliose 碎片泥层;鳞片状泥页岩;叠瓦黏土
argillic 黏土的;泥质的
argillic alteration 泥岩蚀变;泥蚀变
argillic horizon 黏化层
argillic intercalated layer 泥化夹层
argilliferous 泥质的;黏土的;含黏土的
argillious 黏土似的;泥质的;含黏土的
argillite 厚质板岩;泥质岩;泥质板岩;泥板岩;厚层泥岩
argillite shale 泥质页岩
argillization 黏土化;泥化作用
argillo-arenaceous 泥沙质的;泥砂质的
argillo-arenaceous ground 泥砂地
argillo-arenaceous melange 泥沙质混杂体
argillo-calcareous 泥砂质的;泥灰质(的)
argillo-calcareous chalk 泥砂质白垩
argillo-calcite 泥灰方解石;泥砂质石灰岩
argilloid 泥岩类
argillolith 泥质层凝灰岩
argillous 黏土似的;含黏土的;泥质的
argillutite shale 泥页岩
argil powder 白土粉
Argo basin 阿尔戈海盆
Argo-Bond alloy 阿尔戈一邦德易熔合金
argodromile 缓流河流
Argofil alloy 阿戈菲尔铜锰硅合金;铜锰合金
Argo-Fio alloy 阿尔戈一弗洛四元合金
argoflex camera 反光摄影器
Argoflon fiber 阿戈弗纶聚四氟乙烯纤维
argol 粗酒石
argon arc 氩弧
argon-arc cutting 氩弧切割
argon-arc torch 氩弧焊炬
argon-arc welder 氩弧焊机
argon-arc weld(ing) 氩弧焊(接)
argon atmosphere 氩气氛
argonaut 船蛸
Argonaut welding 自动调节电弧焊
argon cast 氩气保护铸造
argon detector 氩检测器
argon-filled lamp 氩光灯
argon-filled tube 充氩管
argon filling 充氩
argon flash 氩气闪光灯
argon gas 氩气
argon gas laser 氩气体激光器
argon glow lamp 氩辉光灯
argon ionization detector 氩离子化鉴定器;氩电离检测器
argon ion laser 氩离子激光
argon lamp 氩(气)灯

argon laser 氩激光(器)
argon laser photocoagulation 氩激光凝固
Argonne high flux reactor 阿尔贡高通量反应堆
Argonne National Laboratory 阿尔贡国立实验室
Argonne reactor computation 阿尔贡反应堆计算
Argonne three-roll tube reducer 阿尔贡三辊式冷轧管机
argon oxygen decarburization 氩氧脱碳
argon plasma 氩等离子体
argon-potassium method 氩钾法
argon shield 氩气保护
argon tungsten-arc welding 钨极氩弧焊
argon welder 氩弧焊机
Argo-Swift alloy 阿尔戈一斯维夫特四元合金
argosy 大商船
Argovian 阿尔高夫亚阶【地】
argtum 橡皮
argtum eraser 软橡皮
arguerite 银汞齐
Arguerite alloy 阿格莱特银矿物型合金
argulite 沥青砂岩
argument 位相;论证;引数;自变数;自变量;复角;幅角;幅度
argument association 变元结合
argument back and forth 前后论证
argument byte 自变量字节
argument count 变元计算;变元计数
argument descriptor 变元描述器;变元描述符
argument function 辐角函数;幅角函数
argument list 自变量表;变元表
argument of a complex number 复数的辐角
argument of function 函数自变量
argument of latitude 升交角距
argument of perigee 近地点角距;近地点辐角
argument of perigee of satellite orbit 卫星轨道近地点角距
argument of perihelion 近日点辐角
argument of resistance 阻力增加
argument of subroutine 子程序的自变量
argument of vector 矢量幅角;矢幅角
argument pointer 变元指针
argument principle 幅角原理
argument register 变元寄存器
argument segment 变元段
argument subscript 变元下标
argument table 变量表
argument value of well function at matching point 配合点的井函数自变量值
argutite 锗石
Arguzoid alloy 阿盖佐依德铜镍锌焊剂合金
argyria 银质尘着病
Argyris force method 阿吉利斯方法(又称柔度矩阵法)
argyrism 银中毒
argyrodite 硫银锗矿
argyrol 含银的防腐剂
argyrophilia 嗜银性
Arhat 光罗汉
arhbarite 阿砷铜石
arheic region 无流区;无河区
arheism 无流区;无河区
arheol 白檀油烯醇
arhitect's scale 三棱尺
ariadne system 艾里亚德尼系统
ari contaminator 空气净度指示器
arid 干旱的
arid and half arid desert soil zone 干旱半干旱荒漠土带
arid and semi-arid area 干旱和半干旱地面
arid and semi-arid land 干旱和半干旱地
arid area 干燥地区;干旱地区
arid area underground water area 干旱地区潜水区
arid belt 干旱带
arid biogeographic zone 干旱生物地理区
arid climate 干燥气候;干旱气候
arid cycle 干燥周期;干旱周期
arid desert 干旱荒漠
arid desert region 干旱沙漠地区
arid ecosystem 干旱生态
arid erosion 干旱侵蚀
aridextor 产生侧向力的操纵装置;产生侧向力的操纵机构;侧向力的操纵机构
arid homoclimate 相同干旱气候
aridisol 旱成土;干燥土

aridity 干燥性;干燥度;干旱性;干旱程度
aridity coefficient 干燥系数
aridity index 干燥指数;干旱指数
aridized plaster 盐处理石膏
arid land 贫瘠地;旱地;干旱地
arid land ecosystem 干旱地区生态系统
arid landform 干旱地形
arid lithogenesis 干成岩石成因论
aridness 干旱程度
arid period 旱季;干季;干旱周期;干旱期
arid plain 干旱平原
arid region 干燥地区;干燥区域;干燥区;干旱区;干旱地区
arid region karst 干旱区岩溶
arid region relief 干燥区地貌
arid regions in tropic(al) and temperate zones 热带和温带的干旱地区
arid soil 旱境土壤;干燥土;干土
arid tract 干旱地带;干旱带
arid transition life zone 干旱过渡生物带
Aridye pigment colo(u)rs 阿里代涂料
arid zone 旱地区;干旱带;干旱带;赤道干旱带
ariegite 尖榴辉岩;尖榴辉石
Arikareean 阿里卡里阶【地】
arildone 阿立酮
Ariloft 阿里洛夫特醋酯长丝
Arimina 阿里米纳韧皮纤维
ariness 通风;空气流通
Ariron 耐酸铸铁;阿里龙耐蚀高硅铸铁
aris 棱角;棱边;尖脊
arisarainite 硼钠镁石
ariscope 移像光电摄像管
arise 上升
arising of expansion joint 胀缝拱起
arising of joint edges 缝边上升(指混凝土路面);缝边起拱(指混凝土路面)
arising period 旱季;干季
arising region 干旱区;干燥地带
arising soil 干燥土
aristiform 芒状
arithemtic 算术;四则运算;计算
arithlog paper 半对数纸
arithmetic(al) addition 算术加法
arithmetic(al) and control 运算控制装置
arithmetic(al) and control unit 运算控制器
arithmetic(al) and logic(al) unit 运算与逻辑;运算器;运算逻辑单元;运算逻辑部件
arithmetic(al) array 算术数组
arithmetic(al) assignment 算术赋值
arithmetic(al) assignment statement 算术赋值语句
arithmetic(al) attribute 算术属性
arithmetic(al) average 算术平均数
arithmetic(al) average grade 算术平均品位
arithmetic(al) average index 算术平均指数
arithmetic(al) average pressure method 算术平均压力法
arithmetic(al) average thickness 算术平均厚度
arithmetic(al) average value 算术平均数
arithmetic(al) average volume weight 算术平均体重
arithmetic(al) binary computer 二进制计算机
arithmetic(al) built-in function 算术(型)内部函数
arithmetic(al) capability 运算能力
arithmetic(al) check 运算校验
arithmetic(al) circuit 运算电路
arithmetic(al) complement 余数
arithmetic(al) computation 算术计算
arithmetic(al) computer 算术计算机
arithmetic(al) constant 算术常数
arithmetic(al) continuum 算术连续统
arithmetic(al) controller 运算控制器
arithmetic(al) control unit 运算控制部件
arithmetic(al) conversion 运算转换
arithmetic(al) coordinate 普通坐标
arithmetic(al) data 运算数据
arithmetic(al) debug 运算调试
arithmetic(al) device 运算装置;运算器
arithmetic(al) difference 算术差
arithmetic(al) element 算术元素;运算元素;运算元件
arithmetic(al) error 算术错误
arithmetic(al) example 算例
arithmetic(al) exception 算术异常;算术例外

arithmetic(al) exception interrupt 运算例外中断
arithmetic(al) expression 算术式;算术表达式;运算式
arithmetic(al) facility 运算设备;运算部件
arithmetic(al) fault mode 运算故障模型
arithmetic(al) field 运算字段
arithmetic(al) filter 算术递增滤色镜
arithmetic(al) for network optimization 网络优化算法
arithmetic(al) function 算术函数
arithmetic(al) growth 算术增长;线性增长
arithmetic(al) identifier 算术标识符;运算标识符
arithmetic(al) invariant 算术不变数
arithmetic(al) logic register stack 算术及逻辑寄存器栈
arithmetic(al) logic unit 算术逻辑单元;运算逻辑部件;运算器
arithmetic(al) mean 均数;等差中项;算术中项;算术平均值;算术平均数;算术平均
arithmetic(al) mean diameter 平均直径;算术平均直径
arithmetic(al) mean error 算术平均误差
arithmetic(al) mean method 算术平均法
arithmetic(al) mean of population 总体算术均数
arithmetic(al) mean of water level 平均水位
arithmetic(al) mean sound pressure level 算术平均声压级
arithmetic(al) mean temperature difference 算术平均温(度)差
arithmetic(al) mean value 算术平均数
arithmetic(al) mode 运算形式;运算模式;运算方法
arithmetic(al) operation 算术运算;算术操作;四则运算
arithmetic(al) operator 算术运算符;算术算子
arithmetic(al) overflow 算术运算溢出;算术溢出;运算溢出
arithmetic(al) picture data 算术图像数据
arithmetic(al) pipeline 运算流水线
arithmetic(al) plotting paper 算术坐标纸
arithmetic(al) point 训数点
arithmetic(al) portion 运算部分
arithmetic(al) primary 算术初等项
arithmetic(al) probability 算术概率
arithmetic(al) product 乘积;算术乘积
arithmetic(al) progression 等差数列;算术级数;等差级数
arithmetic(al) quadruple 算术运算四元组
arithmetic(al) quartile deviation 算术分位数差
arithmetic(al) quartile kurtosis 算术分位数峭度
arithmetic(al) reactive factor 算术无功因数
arithmetic(al) register 运算寄存器
arithmetic(al) scale 等差尺度
arithmetic(al) section 运算装置
arithmetic(al) sentence 算术语句;算术句子
arithmetic(al) series 算术级数;等差级数
arithmetic(al) series of higher order 高阶算术级
arithmetic(al) shift 算术移位;运算移位
arithmetic(al) shift left 算术左移
arithmetic(al) shift right 算术右移
arithmetic(al) sign 运算符号
arithmetic(al) software 算术软件
arithmetic(al) solution 数值解
arithmetic(al) speed 运算速率
arithmetic(al) statement 算法语句;算术语句;运算语句
arithmetic(al) subroutine 算术子程序;算术例行子程序;运算子程序
arithmetic(al) system 运算系统
arithmetic(al) trap 算术自陷;算术俘获
arithmetic(al) trap enable 算术自陷赋能;允许算术自陷;允许算术俘获
arithmetic(al) trap mask 算术自陷陷阱;算术俘获屏蔽
arithmetic(al) unit 运算器;算术单元;运算单位;运算部件
arithmetic(al) verb 算术动词
arithmetic-geometric(al) series 算术几何级数
arithmetic-geometric(al) mean 算术几何平均
arithmetization 算术化
arithmograph 运算图
arithmometer 四则算术机;四则运算机;四则计算机
Arius sinensis 中华海鲇
Arizona marble 亚利桑那大理石
Arizona ruby 红榴石

arizonite 红钛铁矿;正长脉岩
ark 高处壁橱;保险箱;高处壁柜;大平底船;避难所;庇护所;平底船;救生吊车
Arkansas marble 阿肯色大理石
Arkansas River Navigation System 阿肯色河航行系统
Arkansas soft pine 阿肯色软松
Arkansas stone 均密石英岩;阿肯色岩;阿肯色磨石
arkansite 黑钛矿
arkite 霞白斑岩;白榴霞斑岩;阿克斯炸药
Arkon (抗热及绝缘的)浅色脂环饱和烃树脂
arkose 长石砂岩
arkose deposit 长石砂岩矿床
arkose quartze 长石石英岩
arkose quartzite 长英岩
arkosic 长石砂岩的
arkosic bentonite 长石斑脱岩
arkosic conglomerate 长石硬岩;长石砾岩
arkosic limestone 长石灰岩
arkosic sandstone 长石砂岩;花岗质砂岩
arkosic wacke 长石玄武
arkosite 长英岩;长石石英岩
arksutite 锥冰晶石
arm 上臂;机械臂;锚臂;港汊;电极臂;辐条;辐;分支;防波堤或狭长陆地伸出部分;臂状物;臂杆
Arma-Brown gyrocompass 阿马—勃郎罗经
armada 舰队;机群
armalcolite 镁铁钛矿
Armaleolite 阿马科月球石
armal sign 悬挑标志牌
armament 军械
armament error 兵器误差
armament factory 军械长;装备生产厂
armament training camp 靶场
Armand Pine 华山松
armangite 砷锰矿
armarium 储藏室;备餐室;(教堂圣台上的)碗柜或壁龛
armar wood 有装饰面的金属板
armature 铠装;铠甲;加强料;加劲材料;旋转圆板;电枢;电缆铠装;磁舌
armature band 电枢带
armature bar 电枢条
armature bearing 电枢轴承;电枢支座
armature bearing brass 电枢轴承铜衬
armature bearing shield 电枢轴承护片
armature binder 电枢束带
armature binding wire 电枢包扎钢丝
armature body 电枢体
armature characteristic curve 电枢特性曲线
armature chatter 衔铁振动
armature coil 衔铁线圈;电枢线圈;电枢绕组
armature conductor 电枢导体
armature contact 活动触点
armature copper 电枢绕组
armature core 衔铁铁芯;电枢铁芯;电枢铁芯
armature core disc 电枢(铁)芯片
armature core head 电枢铁芯压板
armature core lamination 电枢芯叠片
armature core length 电枢铁芯长度
armature detector 电枢检查仪
armature diameter 电枢直径
armature dropout overtravel 衔铁释放超行程
armature drum 电枢芯子
armature duct 电枢风道
armature end 衔铁端;磁舌端
armature end connexions 电枢端部接线
armature end plate 电枢端板
armature excitation curve 电枢励磁曲线
armature field 电枢磁场
armature flange 电枢端板
armature flux 电枢磁通(量)
armature gap 衔铁间隙;引铁间隙;电枢间隙
armature head 电枢压板;电枢端板
armature hesitation 衔铁滞缓
armature inductor 电枢感应线圈
armature iron 衔铁;引铁
armature iron loss 电枢铁损
armature key 电枢轴键
armature keyway 电枢轴键槽
armature lamination 电枢铁片
armature lead bushing 电枢引线套管
armature leakage reactance 电枢漏(磁电)抗
armature lifter 衔铁推杆

armature lock washer 电枢锁紧垫圈
armature loop 转子绕组；电枢绕组元件
armature loudspeaker 舌簧扬声器；舌簧喇叭
armature of cable 电缆的铠装
armature of relay 继电器衔铁
armature ohmic loss 电枢铜损
armature pickup overthrow 衔铁吸合过度
armature pinion 电枢小齿轮
armature play 衔铁游隙；电枢游隙
armature projection 电枢的齿
armature ratio 衔铁比
armature reactance 电枢电抗
armature reaction 电枢反作用力；电枢反应
armature reaction-excited machine 反应式电机
armature relay 衔铁继电器；电枢继电器
armature residual gap 剩余气隙；衔铁（防粘）余隙
armature rotor 电枢转子
armature shaft 电枢轴
armature shaft oil retainer 电枢轴护油圈
armature sleeve 电枢轴套
armature slot 电枢槽
armature smooth core 电枢平滑铁芯
armature spider 电枢十字支架；电枢辐式机架
armature spider pin 电枢十字架销
armature stamping 电枢冲片
armature stop 磁舌挡
armature stroke 衔铁行程
armature structure 电枢构造
armature terminal insulating bushing 电枢电极绝缘衬套
armature terminal stud 电枢线头接柱
armature tester 电枢检查仪；电枢测试器
armature testing apparatus 电枢试验仪（器）
armature tooth 枢齿
armature travel 衔铁行程；衔铁动程
armature type magneto 电枢式磁电机
armature winding 电枢绕组
armature winding machine 电枢绕组机
armature with closed slots 闭槽电枢
armature with ventilation 有通风（槽的）电枢
arm axle 臂轴
armband 臂带
arm base pin 悬臂支承销
armboard 起纹板；搓花板
arm brace 撑臂；臂形拉条；臂拉杆；撑脚；附加支柱
arm bushing 臂衬套
arm-chair 钢筋支座；单人沙发；扶手椅
arm cleat 单挽钩
Armco 波纹白铁管
Armco 48 alloy 阿姆柯48 软磁性合金
arm coal-cutter 杆式截煤机
Armco aluminized steel 阿姆柯渗铝钢；表面浸镀铝钢
Armco iron 阿姆柯磁性铁
Armco magnetic iron 磁性铁
arm contact breaker 臂板接触（断路）器
arm conveyer 悬臂输送机
Armco process 阿姆柯冶炼法
Armco stabilized steel 阿姆柯稳定化钢
Armco steel 阿姆柯软钢；不硬化钢
arm crane 挺杆式起重机；挑杆起重机；横臂起重机；悬臂式起重机；悬臂吊机；悬臂吊车
armed beam 加强梁
armed boarding vessel 武装检查船
armed horseback figure 刀马人
armed interrupt 待命中断；待处理中断
armed mast 拼合桅；二部组合桅
armed merchant 辅助巡洋舰
armed merchantman 武装商船
armed state 待命状态；待处理状态
armed vessel 武装商船
arm elevator 臂式升降机
Armenian bole 亚美尼亚红土
Armenian red 亚美尼亚红
Armenian style 亚美尼亚式
armenite 钡钙沸石；钡钙大隅石
armeria maritima 海石竹
arm extension 电极臂伸出长度
arm file 手锉；粗齿方锉
arm for theodolite 悬挂经纬仪横托架
arm for transit 悬挂经纬仪横托架
arm fulcrum pin 臂支销
arm hydraulic cylinder 斗机液压缸
armillary sphere 浑天仪

arm indicator 转向指示器
arming 水砣黏样剂
arming hole 填脂孔
arming pin 保险销
armistice line 停火线
armlak 电枢用亮漆
arm length 臂长
armless chair 无扶手沙发
armlet 小湾；护臂；小海湾；臂圈
arm lock magnet 臂连锁磁铁
arm microscope 带柄放大镜
arm mixer 桨叶式搅拌机；桨叶式拌和机；叶片式搅拌机；叶片式拌和机
arm nail 臂骨钉
arm of anchor 锚臂
arm of angle 角边
arm of balance 秤杆；平衡臂
arm of canal 运河支叉
arm of couple 力偶臂
arm of crane 起重机悬臂；吊车臂
arm of delta 三角洲河汊；三角洲分港汊
arm of flywheel 飞轮辐
arm of force 力臂
arm of lake 湖湾；湖汊
arm of lever 杠杆臂
arm of mixer 搅拌机浆叶；混合机浆臂
arm of pivoting floor 转动底板臂
arm of ratchet 棘轮爪臂
arm of resisting couple 抗力偶臂
arm of stability 稳定力臂；稳定臂
arm of the sea （狭长深入陆域的）海湾；狭长海港
arm of wheel 轮辐
armoire 大衣柜
armored bubble 带矿粒的气泡
armored gravel bed reach 粗化砾底河段
armored layer 粗化层
armored mud ball 贴沙砾泥球
armorial decoration 纹章装饰
Armorican 阿莫尔力克运动【地】
Armorican massif 阿莫尔力克地块【地】
Armorican orogeny 阿莫力克造山活动【地】
armoring concrete block 混凝土护面块体
armoring effect 粗化作用
armoring of bed material 床沙粗化
armoring phenomenon 粗化现象
armoring process of bed material 床沙粗化过程
armory 武器库；军械库
armo(u)r 外部铠装；铠板；铠甲；装甲；防具；防弹板；包封；人工防冲铺盖
armo(u)r belt 装甲带
armo(u)r block 护面块体；保护板
armo(u)r block of breakwater 防波堤护面块体
armo(u)r casting 装甲铸件
armo(u)r-clad 装甲的；复块的
armo(u)r coat 沥青护面层；保护层；护面层（油漆等）；护层；保护涂料；沥青保护面层
armo(u)r concrete block 混凝土护面块体
armo(u)r course 保护层；保护面层
armo(u)r-course slope 斜坡式防波堤面坡度
armo(u)r debris 护面碎石
armo(u)r door 装甲门；防火门
armo(u)red 铠装的；包镀的；增强的
armo(u)red automobile 装甲汽车
armo(u)red bed 有护面的河床
armo(u)red bubble 矿化气泡
armo(u)red bulkhead 装甲舱壁
armo(u)red cable 铠装电缆；装甲电缆
armo(u)red cable wire 铠装电缆钢丝
armo(u)red carrier 装甲运输车
armo(u)red cassete 装甲箱
armo(u)red cast iron 钢筋铸铁
armo(u)red coat （油漆的）防护层；保护层
armo(u)red concrete 钢筋混凝土
armo(u)red concrete paving flag 钢筋混凝土铺路砖；钢筋混凝土铺路板
armo(u)red concrete screed 加筋混凝土找平层；钢筋混凝土找平层
armo(u)red concrete slab 钢筋混凝土板
armo(u)red conduit 包皮管道
armo(u)red corner 铁皮包角；包角；铠装棱角；铠装帮
armo(u)red course 护面层（块石、碎石等）；防护层；保护层（道路路面）
armo(u)red cruiser 装甲巡洋舰

armo(u)red curb 镶铁路缘；包铁缘石
armo(u)red deck 装甲甲板
armo(u)red diving apparatus 铠装潜水设备
armo(u)red diving dress 铠装式潜水服
armo(u)red diving suit 铠装式潜水服
armo(u)red door 铁皮门；装甲门；防火门；包金属板门
armo(u)red equipment 装甲装备
armo(u)red fabric 防护织物
armo(u)red face conveyer 有铠装承载面的输送机
armo(u)red faceplate 护面板；金属护面板
armo(u)red fire proof door 装甲门；防火门；装甲防火门
armoured flexible conveyer 铠装挠性运输机
armo(u)red front 铠装面；防撬护面板
armo(u)red glass 嵌有铁丝网的玻璃；铠装玻璃；钢化玻璃；防弹玻璃
armo(u)red hose 铠装软管；缠丝软管；包铁软管；包皮（软）管
armo(u)red joint 包铁接头；包铁接缝
armo(u)red layer 抗冲保护层；铠装层；护面层（块石、碎石等）；防护层
armo(u)red lead cable 铠装铅包电缆
armo(u)red loader 装甲装载机
armo(u)red material 铠装材料
armo(u)red motor 带防护套电动机
armo(u)red oxygen rubber hose 编织氧气胶管
armo(u)red paving slab 装甲铺面板（工业厂房地坪）
armo(u)red paving tile 钢筋混凝土铺路砖；钢筋混凝土铺路板；装甲铺面砖（工业厂房地坪）
armo(u)red plate 装甲板
armo(u)red plate glass 钢化玻璃（板）；装甲玻璃；铠装玻璃板
armo(u)red plywood 铠装胶合板；金属面胶合板
armo(u)red pump 铠装泵
armo(u)red rolling contact joint 铠装球形活动接头
armo(u)red rope 铠装绳索；钢丝绳
armo(u)red sedan 装甲轿车
armo(u)red ship 装甲船
armo(u)red steel rope 钢缆保护外层
armo(u)red thermocouple 铠装热电偶
armo(u)red thermometer 铠装温度计；带套温度计
armo(u)red tubular floor 铠装管状地板；小梁空心管楼盖
armo(u)red vessel 装甲舰
armo(u)red vest 防护背心
armo(u)red wire 铠装电线
armo(u)red wood 加金属箍木材；包铁皮木材；包铁皮防火门
armo(u)r excavation method 插管支护开挖法
armo(u)r glass screen 防弹玻璃板
armo(u)ring 装甲；套；壳；铠装
armo(u)ring cable wire 铠装电缆钢丝
armo(u)ring concrete block 混凝土护面块体
armo(u)ring material 铠装材料
armo(u)ring steel tape 铠装钢带
armo(u)ring wire 铠装钢丝；铠装电缆用金属线；铠装电缆
armo(u)rite rubber lining 橡胶衬砌
armo(u)r layer 护面层；面层
armo(u)r layer fail 护面层失效
armo(u)r particle 护面小卵石；护面颗粒
armo(u)r-piercing 穿甲的
armo(u)r plate 铠装板；护面板；护铠板；装甲钢板；防踢板；护防层；防弹钢板；保护板
armo(u)r plated 装甲的
armo(u)r plate glass 铠装玻璃板；钢化玻璃板；防爆玻璃
armo(u)r plate mill 装甲钢板轧机
armo(u)r plate rolling mill 装甲板轧制厂；装甲板轧机
armo(u)r plating 装甲板；包铁板
armo(u)r ply 包铁皮胶合板；装甲胶合板
armo(u)r plywood 金属贴面板
armo(u)r rock 护面石；护面块石
armo(u)r slate 护面板
armo(u)r stability coefficient 护面块体稳定系数
armo(u)r stone 护面石料；护面块石；护面大块石；防护石（料）
armo(u)r strake 装甲列板
armo(u)r tubing 保护导管
armo(u)r unit 护面块体；阿穆尔单位

armo(u)ry 军械库;兵工厂
armo(u)ry wire 铠装线
arm path 桥臂支路
arm phenomenon 臂现象
arm pin 支杆销
arm ratio 臂的比例;两臂比例
arm rest 悬臂支承;搁臂架;扶手;臂橐;靠手
arm revolving gear 转动齿轮;臂架旋转装置
arms 兵器
arm saw 手锯;摇臂锯
arm saw blade 手锯条
Arms bronze 特殊铝青铜;阿姆斯铝青铜
arms depot 军械库
arm's-length bargaining 普通讨价还价;正常商业谈判
arm's-length deal 正常商业交易
arm's-length price 正常交易价格
arm's-length sale 彼此独立的买卖
arm's-length transaction 正常交易;公平交易;正常商业交易
arm sling 悬臂吊索;臂悬;臂吊带
arm spider 臂辐射支架
arms plant 兵工厂
arm spool pin 插线灯
arm sprue cut 冒口切割
arm stirrer 桨叶式搅拌器;桨臂搅拌机
arm straight paddle mixer 直臂旋桨拌和机;直臂桨式混合机
Armstrong circuit 超再生式接收电路;阿姆斯特朗电路
armstrongite 水硅钙锆石
Armstrong motor 阿姆斯特朗活塞液压马达;活塞液压马达
Armstrong oscillator 阿姆斯特朗振荡器
Armstrong process 阿姆斯特朗双金属轧制法
Armstrongs axiom 阿姆斯特朗公理
Armstrong system 阿姆斯特朗调制方式
arm structure 悬臂结构
arm support 交叉支架;臂架
arm support bracket 刀杆支座
arm supporter 搁臂架
arm-suspended belt scale 悬臂式电子皮带秤
arm tie 支撑杆;斜撑;连结臂;拉板;横臂拉条;连臂板;交叉撑
arm-to-arm vaccination 臂臂接种法
arm-tongue circulation time 臂舌循环时间
arm type of locomotive routing system 肩回交路
armure 小卵石花纹
army air defence system 陆军防空系统
army camp 军营
Army Corps of Engineers 陆军工程兵(美国)
army data system 陆军信息编码系统
army depot 集团军仓库
army field data code 陆军信息编码
army hospital 陆军医院;军医院
army munitions plant 军火工厂
army role in water pollution control 军队在水污染控制的作用
army service boat 陆军用船
army specification 军用规范
army transport 陆军运输舰
armyworm 行军虫;黏虫
Arnaudon's green 阿诺东绿
Arnd alloy 阿恩合金
Arndt-Eistert reaction 阿尔登特—埃斯特尔特反应
Arndt-Eistert synthesis 阿尔登特—埃斯特尔特合成
Arnel 阿尼尔三醋酯长丝和短纤维
arnentoflavone 阿曼伯黄素
Arneth's classification 阿尔内斯分类
Arneth's count 阿尔内斯指数
Arneth's cunt 阿尔力斯计数
Arneth's formula 阿尔内斯公式
Arneth's index 阿尔内斯指数
Arneth's scale 阿尔内斯计
arnidiol 阿里二醇
Arnild motor 安诺德电动机
arnimite 块铜矿
Arnold distribution 阿诺德分布
Arnold steam sterilization 阿诺德通蒸汽灭菌法
Arnold steam sterilizer 常压蒸汽灭菌器;阿氏锅;常压灭菌器;阿诺德灭菌器
Arnold test 阿诺德试验
arnott 榧木(地板用木材)
Arnott's bed 水褥

Arnott valve 阿诺特式阀门;阿诺特式截门(近天花板处的逆止开关,用以排出油气)
Arnu-Audiberts dilatometer test 奥亚膨胀度试验
Arny solutions 阿尼溶液
arochlors 芳氯物
aroclor (密封料中使用的)增塑剂;氯化三联苯
arohebient 生命起源
arolium 中垫
aroma 芳香
aromatic 芳族的;芳香的
aromatic acid 芳酸;芳香酸
aromatic alcohol 芳族醇;芳香醇
aromatic aldehyde 芳族醛;芳香醛;芳醛
aromatic amine 芳香族胺;芳香胺;芳族胺
aromatic amino compound 芳香族氨基化合物
aromatic-asphaltic oil 芳香—沥青型石油
aromatic cedar 东奥红杉木;香杉木
aromatic chemicals 合成香料
aromatic compound 合成香料;芳族化合物;芳香族化合物;芳烃化合物;芳香化合物
aromatic content 芳族含量
aromatic copolyamide microfibre 芳族共聚酰胺超细纤维
aromatic cyclodehydration 芳化成环脱水作用
aromatic diamine 芳香二胺
aromatic epoxide 芳烃环氧化物
aromatic ester 芳香酯
aromatic ether 芳香醚
aromatic-free 不含芳烃
aromatic-free white oil 不含芳烃石蜡油
aromatic fuel 芳烃燃料
aromatic group 芳族基
aromatic halide 芳族卤化物
aromatic herb 芳香草本植物
aromatic heterocyclic(al) polymer 芳香族杂环聚合物
aromatic hydrocarbon 芬方烃;芳香烃;芳香碳氢化合物;芳烃
aromatic hydrocarbon oil 芳香烃油
aromatic hydrogenation 芳香氢化
aromatic-intermediate oil 芳香中间型石油
aromatic ketone 芳族酮
aromatic mercuration 芳烃汞化作用
aromatic-naphthenic oil 芳香—环烷型石油
aromatic nature 芳香本性
aromatic nitration 芳烃硝化作用
aromatic nitro compound 芳香硝基化合物
aromatic nucleus 芳香环;芳香核
aromatic nylon fibre 芳族尼龙纤维
aromatic odo(u)r 芳香气味
aromatic oil 石油芳烃;芳香油;芳烃油类
aromatic oxide 芳香醚
aromatic peroxide 芳族过氧化物
aromatic petroleum naphtha 芳族石脑油
aromatic petroleum residue 芳族石油(残渣)树脂
aromatic polyamide 芳族聚酰胺;芳香族聚酰胺
aromatic polyamide fibre 芳族聚酰胺纤维;芳香族聚酰胺纤维
aromatic polybenzoxazole fibre 芳族聚苯并恶唑纤维
aromatic polyimide fibre 芳族聚酰亚胺纤维
aromatic polymer 芳族聚合物
aromatic polymerization 芳香聚合作用;芳香族聚合作用
aromatic property 芳香性
aromatic radical 芳族基;芳基
aromatic recovery device 芳香族回收器
aromatic red cedar 芳香红雪松;芳香红杉木
aromatic ring 芳族环;芳香环
aromatics 芳族;芳香族化合物;芳香剂
aromatic series 芳族;芳香系
aromatic sheet 芳香页片
aromatic smell 芳香气味
aromatics of asphalt 沥青芳香烃;地沥青的芳香份
aromatic solvent 芳族溶剂;芳香族溶剂;芳烃熔剂
aromatic substance 芳香物
aromatic substitution 芳香族取代
aromatic sulfinic acid 芳族亚磺酸
aromatic sulfonic acid 芳族磺酸
aromatic sulfuric acid 芳香硫酸
aromatic tar 芳香焦油
aromatic vinegar 芳香醋
aromatization 芳构化
aromatize 芳香化

aromatizer 芳香剂
aromatous 芳香的
aromoline 阿莫灵
arone 芳酮
Aron meter 阿龙计
aronotta 胭脂树(红)
Arons chromoscope 阿龙斯验色器
Aronson's culture-medium 阿龙森培养基
arope creep 钢丝绳砂浆锚
around 围绕;周围
around-corner sliding shutter 转角推拉遮光窗
around opening 环绕开孔
around openings for welded attachments 环绕焊接附件孔口
around rhythm 昼夜节律
a round sum 一大笔钱
around synchronization 循环同步
around-the-city expressway 绕城高速公路
around-the-city road 绕城公路
around-the-clock 昼夜不停的;三班倒的;连续24小时的
around-the-clock job 日夜施工;昼夜施工
around-the-world 环绕世界的;环球世界的
arovane 风车
Arpenamef formula 阿勃拉莫夫公式
arpent 阿尔品(法国旧时土地面积单位)
arrange 排列;支配
arrange cutted piece 分片
arranged rate 协议运价
arranged total loss 协议全损
arrange form of pollution sources 污染源分布形状
arrange in rows along the taproot 沿主根排成行
arrangement 配置;排列(法);整理;部署;布置;安排
arrangement and method for construction 施工方案
arrangement and method statement 施工方案
arrangement at dry dock cope 干坞坞墙顶的布置
arrangement diagram 布置图
arrangement drawing 配置图
arrangement for despatching empty wagon trains by specified train paths and number and types of wagons 定线、定辆数、定车种的配空列车工作安排
arrangement for isostatic pressing 等静压制装置
arrangement for raising a wreck 起浮布置
arrangement for settlement 清算布置
arrangement for stowage of grain cargo 谷物积载安排
arrangement in parallel 并列;并联布置
arrangement in series 串联
arrangement map of water withdrawal engineering 取水工程布置图
arrangement nomenclature of tunnel 坑道装置名称
arrangement of axes 注记轴配置
arrangement of bars 钢筋排列;布筋;配筋图
arrangement of building 建筑布局
arrangement of clause 条款的议定
arrangement of coordinate axis 坐标轴配置
arrangement of curve 曲线测设
arrangement of diamonds 金刚石排列
arrangement of electrodes 电极的排列
arrangement of exploratory work 勘探工程布置
arrangement of framework 桁架排列;桁架布置
arrangement of goods 放置货物
arrangement of groundwater water behavio(u)r observation 地下水动态观测工作布置
arrangement of harbo(u)r and port spacing 港口布局规划
arrangement of holes 钻孔布置
arrangement of lanterns 路灯布置;道路照明布置
arrangement of levers 控制杆布置;操作杆布置;操纵杆装置
arrangement of materials 材料安排
arrangement of mirrors 组镜;反射镜组合;反射镜排列方式
arrangement of mosaic 镶嵌排列
arrangement of operating sequence 加工顺序安排
arrangement of piles 桩位图;桩的布置
arrangement of plan 配置图
arrangement of props 支柱布置(图);支架布置;支撑布置
arrangement of re-fleeting 重新编队
arrangement of reinforcement 配筋图;钢筋排

列;钢筋布置;布筋
arrangement of rhombohedron single particle 单粒菱形体排列
arrangement of rights 权利安排
arrangement of seats 座位安排
arrangement of soil particles 土壤颗粒排列;土粒排列
arrangement of spring mouth 泉口排列方式
arrangement of square body single particle 单粒正方体排列
arrangement of station tracks 站线布置
arrangement of terminals 端子布置
arrangement of tetrahedral single particle 单粒四方体排列
arrangement of the cells 电解槽配置
arrangement of trees and shrubs 树木配植
arrangement of trussed beam 桁梁排列
arrangement of wells 钻井布置
arrangement of wires 布线;安线
arrangement pf station line 车站线路布置
arrangement plan 配置图;布置图
arrangement rule 排列规则
arrangements for attaching and detaching of wagon within the district 区段内甩挂作业安排
arrangement track 编组线【铁】
arrangement with system 按规则排列
arranger 传动装置
arranging routine 整理程序
arras 挂毯;壁毡;饰墙花毡;阿拉斯挂毯
arrastra 矿石粗磨;阿赖斯塔式研磨
array 排列;级数;阵列;装置
array algebra 数组代数
array algorithm 数组算法
array allocation 数组分配
array antenna 天线阵;阵列天线
array architecture 数组结构;阵列结构
array-array operation 数组间运算
array assignment 数组赋值
array bound 数组界
array box 数组块
array component 数组成分;阵列组件
array computer 数组计算机;阵列计算机
array copy subroutine 数组复写子程序
array curtain 天线阵台;阵帘
array declaration 数组说明
array declarator 数组说明符
array declarator statement 数组说明语句
array defining 数组定义
array detector 阵列探测器
array dimension 数组维数
array effect 组合效应
array element 数组元素;阵列单元
array element name 数组元素名
array element reference 数组元素引用
array element subscript 数组元素下标
array element successor function 数组元素后继函数
array expression 数组表达式
array extent 数组宽度
array factor 排列因数
array file 数组文件
array for combined profiling 联合剖面装置
array for combined symmetric(al) quadripole 复合四极装置
array for dipole-dipole profiling 偶极剖面装置
array for longitudinal mid-gradient 纵向中梯装置
array for seismic monitoring 地震监视台阵
array for symmetric(al) quadripole 对称四极装置
array for transverse mid-gradient 横向中梯排列
array for two electrodes 二极装置
array geophone number 组合检波器数
array identifier 数组标识符
array item 数组项
array length 组合基距
array linkage field 数组连接字段
array list 数组表
array logic 数组逻辑;阵列逻辑
array manipulation 数组处理;数组操作
array manipulation built-in function 数组处理内部函数
array module 数组模块;阵列模块
array multiplier 阵列乘法器
array name table 数组名字表
array of cores 磁芯阵列
array of data 数组;数据阵列
array of difference 差分格式
array of structures 结构布置;结构数组;构件数组;结构数列
array operation 数组运算;阵列运算;阵列操作
array partitioning 数组划分
array pass-band width 组合通放带宽度
array pattern 组合形式
array pitch 排距;行距
array printer 阵列打印机;点阵打印机
array processing 数组处理;阵列处理
array processing computation 数组处理计算
array processor 数组处理机;阵列处理机
array redimension 数组重新定义
array reduction analysis circuit 阵列简化分析电路
array response 台阵响应;台阵反应
array segment 数组段
array seismology 台阵地震学
array sensitivity 组合灵敏度
arrays of seismometers 地震仪阵列
array sonar 阵列声呐
array spectrum 台阵谱
array station 台阵站
array stream 数组流
array subscript 数组下标
array symbol table 数组符号表
array tester 阵列测试器
array test for asbestos spinning fiber 石棉纺丝堆放试验;石棉纺丝阵列试验
array transpose 数组转置
array-type computer 阵列计算机
array unit 阵列单元;阵列处理机
array variable 数组变量
array vector 数组向量
array verify 数组校验
array water gun 组合水枪
Arrbenius plot 阿雷纽斯作图法
arrear 未完成的工作;拖延;余债
arrearage 拖欠款;拖欠;保留物;欠款
arrear of interest 滞纳利息
arrear of work 剩下未做完的工程
arrears 欠款;未付尾数;尾数;拖欠;收尾工作;收尾工程
arrears of dividends 拖欠红利
arrears of interest 拖欠利息
arrears of road maintenance 养路延误工程
arrears of wages 拖欠工资
arrested anticline 平缓背斜;不明显背斜
arrested crushing 有限性破碎(法)
arrested decay 抑制分解阶段
arrested dune 稳定沙丘;固定沙丘
arrested evolution 滞留演进;滞留演化
arrested failure 制止损坏
arrested property 被扣押财产
arrester 限动器;限程器;停机装置;制止器;制动装置;过压保险丝;避雷放电器;放电器;捕集器;避雷针;避雷器;避电器
arrester carbon 避雷器用炭精
arrester catch 避雷器锁挡;避雷器掣子
arrester gear 着陆拦阻装置
arrester hook 停机钩;制动钩
arrester hook actuation gear 制动钩操纵机构
arrester switch 避雷器开关
arrester tester 避雷器测试器
arrest inflation 限制通货膨胀
arresting device 卡子;止动装置;止挡;掣子
arresting disc 止动盘
arresting efficiency 捕集效率
arresting gear 停车装置;行程限制器;制动装置;制动器
arresting lever 止动杆
arresting pin 止动销
arresting stop 止动块挡铁
arrest mark 停止痕
arrestment 停止设备;扣押;扣留;制动;财产扣押
arrest of ship 扣船
arrestor = arrester
arrest point (加热或冷却的)停止点;临界点;转变点;驻点
arrests (加热及冷却的)临界点
arrest sweating 敛汗
ar retarder material 隔气材料
Arrhenius equation 阿伦纽斯方程式
Arrhenius-Guzman equation 阿伦纽斯—古兹曼方程
Arrhenius law 阿伦纽斯定律
Arrhenius theory of dissociation 阿伦纽斯电离理论
Arrhenius viscosity formula 阿伦纽斯黏滞性公式;阿伦纽斯黏度公式
arrhythmia 无节律
arrhythmic pulse 不规则脉
arriere-voussure 支承厚墙体的拱心;砖墙内的辅助拱
Arrindy silk 阿林迪丝
arris 棱角;尖棱;脊梭;边棱(角);圆角
arris compression 边缘压缩(力)
arris cover angle 棱边角铁;棱边角钢;棱边保护角铁;棱边保护角钢;边缘保护
arris cover strip 棱边铁皮;棱边保护铁皮;边缘保护带
arris crack 棱边裂缝;边缘裂缝
arris edge 尖棱(角)
arrises of joint (混凝土路面的)接缝圆角
arris fillet 檐口垫瓦条;垫条;三角垫木;棱角线条;棱角嵌块;尖嵌条;反水尖角块
arris gutter V形檐沟;V形檐槽
arris hip tile 人字形脊瓦;四坡顶(直)脊瓦;角形饿脊瓦;角形波脊瓦;回坡顶垂脊脊瓦
arrish tile 棱瓦
arris knot 边棱节
arris length 边缘长度
arris of joint 接缝(圆)角
arris of quay 码头边棱
arris of slab 板肋;板棱
arris-piece 棱形楔
arris pressure 边缘压力
arris protection 边棱保护
arris rail 三角木条;三角轨;三角嵌条;三角栏杆
arris rounded 圆阳角
arris rounding 棱角圆;棱角圆削;棱边修圆;边磨圆;边缘倒棱
arris section V形截面;棱边截面;边棱断面
arrissed corner 拼V形缝的单边
arrissed edge 斜边;拼V形缝的单边
arrissing tool 边缘抹子;新浇混凝土棱边圆角修整器;磨圆器(新浇混凝土边缘);抹棱工具;抹角镘;新浇混凝土边缘圆角抹刀
arris stress 棱边应力;边缘应力
arris tile 饿脊盖瓦;屋脊盖瓦;棱角瓦;角形瓦
arris trowel 阳角抹灰泥刀;棱角抹灰泥刀;棱边抹灰泥刀;边角抹子
arris ways 六角形瓦铺屋面;锯齿形铺砌;屋面石板;盖瓦对角铺砌法
arris wire 对角线
arris-wise 锯齿形铺砌;对角式(铺砌砖瓦或切割材料);拼如锯形;成对角方向
arris-wise paving 对角铺砌;成对角方向铺砖;对角方向铺砌
arrival-and-departure sidings 到发线【铁】
arrival-and-departure track 到发线【铁】
arrival ballast 到港压载
arrival bearing 到达方位
arrival card 更改地址通知单
arrival clause 到港付款条款
arrival conditioning 进场准备过程
arrival contract 期货合同;目的地契约
arrival current curve 输入电流曲线;终端电流曲线
arrival curve 输入曲线;到达曲线
arrival distribution 到达分布
arrival draft 已到汇票;到港吃水
arrival end 入口端
arrival flow 到达流量;到达车流量
arrival flow rate 到达流率
arrival interval 到达间隔
arrival level 行近平面
arrival line 入口线
arrival list 到达船旅客名单
arrival lounge 休息厅(机场)
arrival manifold 入口汇管
arrival notice 到货通知(书);到达通知书;到港通知书
arrival notice of container 集装箱到达通知单
arrival notice of ship 船舶到港通知
arrival notice of vessel 船舶到港通知
arrival of cargo 货物到达
arrival of vessel 货船到达
arrival operation expenditures 到达作业支出

arrival pattern 到达状态；到达图形；到达图式；到达方式
arrival platform 末站月台
arrival point 目的地；到达地
arrival port 到达港
arrival process 到达过程
arrival quality 到岸品质
arrival rate 到达速率；到达率；到达车率；到达比率
arrival route 进场线路
arrival shaft 到达竖井
arrival ship 到达船
arrival signal 进站信号(机)
arrival time 到达时间；波至时间
arrival time difference 到时差
arrival time distribution 到达时间分布
arrival time indicator 到达时间表示器
arrival track 到达线
arrival wave 到达波
arrival weight terms 卸货重量条件
arrival yard 到达场【铁】
arrive at 达到
arrived ship 到达船
arrived weight 到达时重量；到岸重量
arriving draft 到港吃水
arriving passenger 到站旅客；到港旅客；到场旅客
arriving train 到达列车
arriving yard 到达场【铁】
arrojadite 磷碱铁石；钠磷锰铁矿
arrondi 曲线形的
Arrou trough 阿鲁海槽【地】
arrow 箭线；指针
arrow antenna 矢形天线；箭形天线
arrow bamboo 箭竹
arrow bamboo grove 箭竹林
arrow block 箭头块实体
arrow diagram 网络图；网格图；矢线图；矢量图；箭图表(进度计划)；箭头图；箭式网络图；向量图；指示图；双代号网络图
arrow diagramming 箭头图示
arrow direction exclusive lane sign 专用车道分流标志
arrowed 标有箭头的
arrow engine 剪形发动机
ArrowEnthoven sufficicy theorem 阿罗·艾索文充分性定理
arrowhead 箭头(饰)
arrowhead clasp 箭头形钩子
arrowhead depression 箭头形坑
arrowheaded 箭状的；箭头(形)的
arrowhead method 运动线法
arrow head twills 人字斜纹
arrow height 矢高；箭头所示高度；臂高
arrow-like 箭形的
arrow loop 炮眼；中古期要塞墙上的竖射箭孔；中古期要塞墙上的瞭望孔
arrow mark 指向标志；指向标记；刮痕
arrow network 双代号网络图；箭形网络图；箭头网络
arrow notation 箭头表示法
arrow pass 山间狭道
arrow plot 蝌蚪图；箭头图
arrow point bracing (桁架梁的)双斜杆支撑系统；(桁架梁的)双斜杆系统；测量觇标支撑
arrowroot paper 淀粉复制纸
arrow shaped share 箭形铲
arrow side 箭头侧
Arrow's Paradox 阿罗反论
arrow stone 燧石；箭石
arrow stripe 箭线状箭头花纹
arrow symbol 箭头符号
arrow-type network 箭头式网络
arroyo 细流；旱谷干涸的沟壑；旱谷；小河；干涸河道；干河；干谷
ar ruling rate 现行折合率
arsanilic acid 对氨苯基胂酸；阿散酸
arse 滑车底部
Arsem furnace 碳粒发热体电炉；螺丝状硬质炭精管式电炉
arsenal 武器库；军火库；兵工厂
Arsenal process 阿塞诺炮筒离心铸造法
arsenate 砷酸酯；砷酸盐；次胂酸盐
arsenate concentration 砷酸盐浓度
arsenate dinitrophenol 砷酸二硝基酚(木材防腐用)
arsenate ion 砷酸根离子

arsenate ion removal 除砷酸根离子
arsenblende 雌黄
arsenbrackebuachite 砷铁铅石
arsendescloizite 羟砷锌铅石
arseniasis 砷中毒；慢性砷中毒
arsenic 砷酸；信石
arsenical 砷制剂；砷剂；砷化物
arsenical antimony 砷锑矿
arsenical bronze 砷青铜
arsenical copper 砷铜合金
arsenical lead 砷铅合金
arsenic alloy 砷合金
arsenical nickel 红砷镍矿
arsenical paralysis 砷毒性麻痹
arsenical poisoning 砷中毒
arsenical pyrite 毒砂
arsenical wastewater 含砷废水
arsenic and mercuric iodides solution 碘化汞砷溶液
arsenic apparatus 砷检出装置；测砷仪
arsenic bloom 砷华；毒石
arsenic-caustic desulfuration 砷碱法脱硫
arsenic compound 砷化(合)物
arsenic contaminant 砷污染物
arsenic-contaminated wastewater 含砷废水
arsenic contamination 砷污染
arsenic cycle 砷循环
arsenic dip 砷浸液
arsenic disulfide 二硫化二砷
arsenic hydride 砷化氢
arsenic ion 砷离子
arsenic measuring apparatus 测砷器
arsenicorange 雌黄
arsenic oxidation 砷氧化
arsenic oxide 三氧化二砷
arsenic pentasulfied 五硫化二砷
arsenic pentoxide 五氧化二砷
arsenic pigment 含砷颜料
arsenic poisoning 砷中毒
arsenic pollutant 砷污染物
arsenic removal 脱砷；除砷
arsenic sulfate 硫化砷
arsenic sulphate 硫化砷；雄黄
arsenic trifulfide 三硫化二砷
arsenic triiodide 三碘化砷
arsenic trioxide 三氧化二砷；白砒
arsenic trisulfide 雌黄
arsenic vacuum coating 砷真空镀层
arsenic yellow 砷黄
arsenide 砷化物
arseniopleite 红砷铁矿；红砷钙石
arseniosiderite 砷铁钙石；菱砷铁矿
arsenious ion 砷离子
arsenious oxide 白砒
arsenite 亚砷酸盐
arsenite pesticide 亚砷酸盐农药
arsenobismite 羟砷铋石
arsenoceptor 砷受体
arsenoclasite 羟砷锰石；水砷锰矿
arsenocrandallite 纤砷钙铝石
arsenogoyazit 砷锶铝石
arsenohauchecornite 硫砷铋镍矿
arsenoklasite 水砷锰石
arsenolamprite 斜方砷
arsenolite 砷华
arsenomiargyrite 砷辉锑银矿
arsenopallasinite 砷钯矿
arsenopyrite 砷黄铁矿；毒砂；毒砂(含量)
arsenosulvanite 等轴硫砷铜矿
arsenous acid 亚砷酸
arsenous acid anhydride 亚砷酐
arsenpolybasite 砷硫锑铜银矿
arsenthorite 砷钍石
arsenuranospathite 铝砷铀铀云母
arsenuranylite 砷钙铀矿
arshinovite 胶锆石
arsine 砷化氢
arsine poisoning 砷化氢中毒
arsinic acid 次胂酸
Ar-sintering 氩气保护烧结
arsoite 辉橄粗面岩
arson 放火
arso-trachyte 橄榄粗面岩
art and craft room 美术工艺室

Art and Crafts Export Corporation 工艺品出口公司
art anotomy 艺术解剖
Artbond 黏聚氯乙烯薄膜钢板(商品名)
art book cover paper 美术书皮纸
art brown 美术棕色
art bulb 美术灯泡
art carpet 美术地毯
art ceramics 美术陶瓷；艺术陶(瓷)
art collection 艺术收藏(品)
art college 艺术学院
art connoisseur 艺术品鉴赏家；艺术品爱好者；艺术鉴赏家；艺术鉴定家
art corner 美术相角
art cover paper 美术装饰纸
art creation 艺术作品；艺术中心
art critic 艺术批评家
art criticism 艺术批评
art deco 装饰艺术(风格)
art designer 美工
art designing department 美工车间
art director 艺术指导(者)
art education 艺术教育
artefact 人为现象；人工制品；人工产物；伪迹
artefact pollution 人为污染
art effect 艺术效果
artemiseion 阿尔特弥斯神庙(供奉月亮女神的建筑)
artemisia annua 臭蒿【植】
artemisia argyi 艾蒿【植】
Artemisia leaf 艾叶【植】
Artemission 月神庙(古希腊)
artenkreis 种种
artensian flow 承压水流
arterial 干线的
arterial canal 总渠；运河干河；干线运河；干渠
arterial control 干道控制
arterial drain 脉状排水干渠；排水干沟
arterial drainage 排水干管；干渠；脉状排水系统；有支渠的排水设施；主要排水；干渠排水系统；干渠排水孔道
arterial drainage canal 排水干渠
arterial drainage pipeline 排水干管
arterial drainage system 排水干渠系统
arterial duct 博塔洛管
arterial grid 干线网
arterial highway 公路干线；干线公路；干线道路；干道
arterial hypertension 高动脉压
arterial intersection control 干线相交控制；干道交叉口控制
arterial-local type intersection 主次道路交叉口
arterial pair 双线干道
arterial pattern 干道类型
arterial primary main 主干线
arterial railway 铁路干线；干线铁路
arterial road 主干道；干线道；主干路
arterial road of port 港口主干道
arterial street 干线街道；主要街道；城市主干路；城市干道
arterial system 配水干线系统；配水干管系统
arterial system of distribution 干线分布系统
arterial traffic 干路交通；干线交通；干线道路；干道交通
arteries of communication 交通主干线；交通网；干线网
arteriole 微动脉
arterite 脉状混合岩
arteritic migmatite 脉状混合岩
artery 动脉；大路；运输线；干线；干道
artesian aquifer 自流水层；自流含水层；承压水层；承压含水层
artesian aquifer power 承压出水动力
artesian area 自流水(地)区；承压水地区
artesian basin 自流水盆地；自流泉盆地；自流盆地；自流井盆地；地下水流域；承压水盆地
artesian bored well 钻出自流井；自流管井
artesian capacity (井的)自流量；自流产水率
artesian capacity of well 井的自流水生产量；自流井出水量
artesian condition 承压条件；自流水情况；地下水承压状态；承压状态；承压水条件
artesian discharge 自流流量；泉水流量；自流水排泄；自流排水；自流水出量；地下水流量
artesian drainage method 自流式排水法

artesian flow 自流流量;地下水流量
artesian-flow area 自流水面积;自流区域
artesian flowing well 自流水井;自流水井
artesian formation 有压含水层;自流含水层;承压含水层
artesian fountain 自流泉;自流喷泉
artesian groundwater 有压地下水;自流地下水;承压地下水
artesian head 自流(井)水头;承压水(水)头
artesian hydrostatic(al) pressure 自流水静压力
artesian land 自流地带
artesian layer 承压(含水)层
artesian leakage 自流渗漏;承压渗漏
artesian level 承压水位
artesian loss 承压水水头损失;承压地下水水头损失;自流水损失
artesian monoclinal stratum 自流水斜地
artesian pressure 自流压力;自流井压力;地下水压力;承压水压力
artesian pressure gradient 自流压力坡度
artesian pressure head 自流水水压头;地下水压力水头;承压水(水)头
artesian pressure surface 自流水压面
artesian province 泉水区;自流区
artesian recharge 自流回灌
artesian relief well 自流减压井
artesian seepage 承压渗透
artesian slope 自流斜坡;承压含水层坡降
artesian slope and basin 自流坡度及流域
artesian sources 自流水源
artesian spring 涌水池;涌泉;自流泉
artesian spring tank 涌水池
artesian surface 自流井水头;自流水水面;自流井压力水位
artesian system 自流系统
artesian table 承压水位
artesian waste 自流排水;自流井水量损耗
artesian water 自流水;自流井水;自流水池
artesian water circulation 承压水循环;自流水循环
artesian water flow 自流水流
artesian water head 承压水头
artesian water leakage 承压水渗漏
artesian water power 自流水动力
artesian water struck 初见地下水位;初见承压水
artesian well 涌泉井;自流井;承压水井;承压井
artesian well capacity 自流井产水率;自流井出水量
artesian well pump 深井泵;自流井用抽水机;自流井泵
art felt 美术呢
art form 艺术形式
art gallery 美术陈列室;画廊;美术馆
art glass 美术玻璃;艺术玻璃;工艺玻璃
art glassware 艺术玻璃器皿
art history of ceramics 陶瓷美术史
arthrophyte 有节植物
arthropleure 侧板
arthropod 节肢动物
Arthrospira 节旋藻
arthurite 水砷铁铜石
Arthur remote-indicating thermometer 阿瑟遥测温度仪
Arthur unit 阿瑟单位
Arthus phenomenon 阿瑟斯现象
Arthus reaction 阿瑟斯反应
article 物品;物料;条款;论文;制品;章程;产品
articled clerk 契约学徒;(见习生(英国)
article for sale 卖品
article from the store rooms 库存货
article in custody 保管品
article not for sale 非卖品
article of association 合股营业章程
article of consumption 消费项目
article of contraband 违禁物品
article of trade 商品
article prohibited for use in harbo(u)r 港内禁用品
articles accompanying passengers 旅客携带品
articles for daily use 日用品
articles for ship 船用品
articles free 免税商品
articles from the storeroom 仓库货物
articles in great demand 流行货
articles not for sale 非卖品
articles of antiquity (地下的)古典文物

articles of association 公司章程
articles of consumption 消费品
articles of co partnership 合资条款;合资公司章程
articles of corporation 公司章程
articles of export 出口商品
articles of handicraft art 手工艺品
articles of value 有价值物品;珍贵物品
articles on free market 市售商品
articular branches 关节支
articulate 环接
articulated 铰接的
articulated arch 铰接拱
articulated arch(ed) girder 铰接拱梁
articulated arm 软节臂;蛇节杆
articulated arm-type stone grinder 软节臂式磨石机;软节臂式石头打磨机
articulated axle 铰接轴;活动关节式车轴
articulated bar 铰接杆件
articulated barge 铰接式驳船
articulated beacon 活节式灯桩
articulated beam 联接梁;连接梁;铰接梁;铆接梁
articulated bed plate 铰接底板
articulated boom 铰接悬臂;铰接桁架式吊臂
articulated boom platform 铰接吊杆(起重)平台;多节式构架高空平台
articulated bridge 铰接桥
articulated bus 铰接式公共汽车
articulated buttress dam 铰接支撑扶垛坝;带活接支撑的扶壁式坝;带活接支撑的垛坎
articulated car 路平车
articulated carrier 活节运载车
articulated chamber floor 铰接式闸室底板
articulated chute 溜管;活节卸槽;铰接溜管;(浇混凝土用的)活接卸槽;象鼻管
articulated coach 铰接车辆
articulated concrete 铰接式混凝土;铰接混凝土;活节混凝土
articulated concrete block 铰接混凝土块
articulated concrete matting 活节混凝土罩面;活节混凝土块盖面;活节混凝土块护坡
articulated concrete mattress 活节混凝土块护坡;铰接混凝土沉排;活节混凝土褥垫
articulated concrete pavement 铰接混凝土路面
articulated concrete slab 活节混凝土板
articulated conduit 铰接管;分节管道
articulated connecting rod 铰接式连杆;活节(式)连杆
articulated connection 活节接合
articulated construction 铰接式结构;活节式构造;装配式构造
articulated coupling 铰接车钩;铰链式联轴器;铰接接头;铰接接合;活节联结器
articulated deep-water atmospheric diving system 铰接式深水常压潜水系统;关节式深水潜水系统
articulated diving suit 关节式潜水服
articulated(drop)chute 溜管;溜槽;象鼻管;相连接的漏斗车(浇注混凝土用)
articulated dumper 铰接式翻斗车
articulated dump truck 铰接翻斗车;有铰接的翻斗车
articulated fender frame 铰接防撞构架
articulated flat-slab buttress type dam 活节平板支墩坝
articulated fork lift truck 转叉式装卸车
articulated frame 铰接框(架);铰接构架;铰接车架
articulated frame chassis 铰接车架底盘
articulated frame steering 铰接车架转向
articulated gate shoe 座铰;支铰;闸门制动器;闸门铰座
articulated gear 链式二级减速齿轮
articulated go-devil 铰接清管器
articulated hydraulic crane 关节液力起重机
articulated index 挂接索引
articulated jack 铰接千斤顶
articulated jib 铰接起重臂;铰接臂;活节臂
articulated joint 关节接合;铰接接合;铰接接头;铰接接合;铰接;活节接合;活接结合;活动接头;活接合;分节连接
articulated landinng gear 摇臂式起落架
articulated link chain 铰接链
articulated loader 铰接装载机;活节装载机;折腰转向装载机
articulated loading arm 铰接装油臂;关节式装油臂

articulated locomotive 关节式机车
articulated lorry 铰接式货车
articulated manned submersible 关节式载人潜水器
articulated mattress 活节混凝褥垫
articulated mat type concrete pavement 活节铺块式混凝土铺面
articulated mirror 万向转镜
articulated mobile crane 铰接流动式起重机
articulated operation arm 铰接装油臂;关节式装油臂
articulated pig 铰接清管器
articulated pin 活节销
articulated pipe 铰接管;旋转绞接管
articulated pivot pins 铰链心轴
articulated plate 铰接板;活接板
articulated plate method 铰接板法
articulated portal frame 铰支门形架
articulated purlin(e) 铰接檩条;活接檩条
articulated rail vehicle 铰接式有轨车辆
articulated rear axle 铰接后桥
articulated robot 多关节型机器人
articulated rocker 铰接摇座(桥梁)
articulated rod 活节连杆
articulated roller 铰接式碾压机;铰接式压路机
articulated rotor 活节式回转器
articulated scraper 铰接清管器
articulated shaft 万向转轴;活动关节轴
articulated shield 铰接盾构
articulated shield machine 铰接盾构机
articulated slab 蛇节连接板;铰接板
articulated spindle 铰接主轴
articulated steel plate 铰接钢板
articulated steel plate conveyer 铰接钢板运送带
articulated steering 活节转向
articulated stock anchor 铰接有杆锚
articulated structure 栓接结构;活节式构筑物
articulated support 活节支座
articulated system 铰接结构系统;铰接系统
articulated traffic 拖挂车运输
articulated trailer (用铰链连接的)拖车;半拖挂车;活接拖车;铰接拖车;铰接挂车;活接拖车
articulated train 铰接车辆;(直达的)固定编组列车
articulated train set 铰接式列车车组
articulated train unit 关节动车组
articulated trolley-bus 铰接式无轨电车
articulated truck 铰接式自卸车
articulated truss 铰接桁架
articulated tube 铰接管;旋转绞接管;象鼻管
articulated turntable 转车台【铁】;铰接转盘;铰转车台
articulated type dam 块接式坝;铰接式坝
articulated type diesel hydraulic locomotive 关节式内燃液力传动机车
articulated unit substation 单元组合式变电站
articulated vehicle 拖车;铰接式车辆;活接车辆
articulated vibrating trough conveyer 铰接振动槽(式)运送机;铰接振动槽式输送机
articulated wheel tractor mounted shovel 活节轮式拖拉铲土机
articulated wood(en)slate 铰接木板运送带;铰接木板
articulated yielding(roadway)arch 可变形铰接拱
articulate latex duct 有节乳液管
articulating 铰接的
articulating boom 铰接起重臂;活接头弦杆
articulating boom platform 多节式构架高空平台
articulating paper 咬合纸
articulating paper pliers 咬合纸夹
articulating pin 关节销
articulation (声音的)清晰度;铰链轴;铰链;活接头;咬合;关节;传声清晰度
articulation by ball-and-socket 球窝关节;球铰接合
articulation efficiency 清晰效率
articulation equivalent 等效清晰度
articulation index 清晰指数;清晰度指数
articulation joint 连接头
articulation of financial statement 财务表中的结合关系
articulation point 关节点
articulation reduction 清晰度降低
articulation reference 清晰度参考当量
articulation reference equivalent 等效清晰度衰减
articulation score 清晰度百分数

articulation statement 棋盘式账目分析表;棋盘式对照表
articulation test 清晰度试验;联结检验;联接检验;连接检验
articulation trochoidea 车轴关节
articulator 联接器;铰接车;咬合架
articulatory 关节的;发音清晰的
articulite 可弯砂岩
artifact 人造制品;人为现象;人工制品;人工效应物;人工产物;伪迹;大型仿古制品;仿制品
artifact during analysis 分析期人造物品
artifactitious 人工产物的
artificer 设计者;技术员;技工;发明者;发明人
artificer diver 潜水技师;潜水员
artificer's knot 技工结
artificial 人造的;人为的;人工的
artificial abradant 人造研磨料;人工金刚砂(磨料)
artificial abrasive 人造磨料;人工金刚砂
artificial activity 仿真活动
artificial adhesive 人工合成胶粘剂
artificial adverse 人为障碍
artificial aeration 人工曝气
artificial aeration basin 人工曝气池
artificial aerial 仿真天线
artificial aerosol tracer technique 人造气溶胶示踪技术
artificial afforestation 人工林
artificial ag(e)ing 人工硬化;人工老化;人工陈化
artificial aggregate 人造集料;人造骨料;人工团聚体;人工集料;人工混凝土骨料(如煤渣、矿渣、陶粒);人工骨料
artificial aging 人工时效硬化;人工时效;回火硬化
artificial allocation of funds 人为分配资金
artificial almond oil 人造杏仁油
artificial amphibole 人造角闪石
artificial anhydrite 人造硬石膏;人造无水石膏;人工无水石膏
artificial antenna 假天线;仿真天线
artificial anthracite 人造无烟煤
artificial aquatic mat 人工水草
artificial asbestos 人造石棉
artificial asphalt 人造石油沥青;人造(地)沥青
artificial asphalt-aggregate mix(ture) 人造(地)沥青骨料混合料
artificial asteroid 人造小行星
artificial atmosphere 人工气候;人工空气;调节空气;空气调节
artificial attenuation of recharge sources 补给源人为衰减
artificial background 人造背景
artificial barrier 人工障碍物;人为障碍
artificial base 人工基地
artificial basin 人工港池
artificial basis 人工基底
artificial beach 人工海滩
artificial bed 人工床层
artificial biological method 人工生物法
artificial birefringence 人工双折射
artificial bitumen 人造焦油沥青;人造柏油;人造(地)沥青
artificial black signal 模拟黑色信号
artificial black white 仿黑白色信号
artificial block 人造块体
artificial block dike 人造方块堤
artificial blowing 人工吹制
artificial bonding adhesive 合成黏性胶;合成树脂胶;树脂基黏结料;资脂基胶粘料;人造黏结料;人造胶结料
artificial bonding agent 合成黏合剂;合成树脂胶;树脂基黏结料;资脂基胶粘料;人造黏结料;胶结料
artificial bonding medium 合成黏结介质;合成树脂胶
artificial boundary 人为边界线;人工边界
artificial cable 仿真电缆
artificial calculation 人工计算
artificial canal 人工运河;人工渠道
artificial capacitor 模拟电容
artificial capital 人造资本;人为资本
artificial carbon 人造碳
artificial carbonation 人工碳化处理;碳化处理
artificial carborundum 人造金刚砂
artificial cavern 人工洞室
artificial caving 人工放顶

artificial cement 人造水泥;仿制水泥
artificial cementation 压浆处理;人工浆胶结
artificial cementing 人工压浆处理;人工灌浆胶结
artificial cementing agent 人工胶凝材料;人造水泥
artificial cementing method 人工压浆处理法
artificial channel 人工河道;人工渠道;人工渠槽;人工河槽;人工航道
artificial characteristic 人为特性
artificial circuit 模拟电路
artificial classification 人为分类(法)
artificial climate 人造气候;人工气候
artificial climate chamber 人工气候室
artificial cloud 人造云;人工云
artificial coal 煤球
artificial coalification experiment 人工煤化试验
artificial coarse aggregate 人造粗骨料
artificial coast 人工岸
artificial cognition 人工识别
artificial cold 人工制冷;人工冷却
artificial collodion 人造胶木
artificial colo(u)ring 人工着色;人工着色
artificial community 人工群落
artificial commutation 强制整流
artificial compaction 人工夯实
artificial compost 人造堆肥;人工堆肥
artificial computation 人工计算
artificial concrete aggregate 人造混凝土骨料;人造混凝土集料
artificial concrete block 人工混凝土块体
artificial concrete coarse aggregate 人造混凝土粗骨料
artificial condition 人为条件
artificial consolidation 人工密实法;人工加固;人工地基密实
artificial constraint 人为限制;假限制
artificial contaminant 人工污染物
artificial control 人为防治;人工控制;人工防治
artificial cooling 人工冷却;强制冷却
artificial corundum 人造金刚砂;人造刚玉
artificial cotton 人造棉
artificial counting 人工计数
artificial course 人工航道
artificial cover 人工覆盖物
artificial crater 人造环形山
artificial crown 人造冠
artificial cryolite 人造冰晶石
artificial crystal 人造晶体;人工晶体
artificial culling 人工淘汰
artificial culture 人工养殖;人工培养
artificial culture medium 人工培养基
artificial cutoff 人工截弯;人工裁弯
artificial damping 人工阻尼
artificial data 人工数据
artificial daylight 人造日光;人工采光;太阳灯
artificial decorative board 人造饰面板
artificial defect 人工缺陷
artificial defence 人工防护
artificial delay line 延迟线
artificial deposition 人工淤填
artificial dewatering 人工降低地下水位
artificial diamond 人造金刚石
artificial dielectrics 人造介质
artificial digestion 人工消化
artificial discharge 人为排放;人工排泄
artificial discharge of groundwater 地下水人工排出量
artificial discharging 人工卸货
artificial disintegration 人为蜕变;人工分解
artificial disturbance 人为扰动;人工破坏程度
artificial diversion of river 人工改道
artificial divide of groundwater 地下水人工分水岭
artificial draft 人工通风;机械通风;机力通风
artificial drain 人工下水道;人工排水沟
artificial drainage 人工消耗;人工水系;人工排水;人工河网
artificial drainage system 人工排水系统
artificial draught 压力气流;人工通风;机械通风
artificial dredging 人工挖泥
artificial drying 人工烘干;人工干燥(法)
artificial drying out 人工干燥;人工干燥
artificial drying oven 人工干燥室;人工干燥炉
artificial ear 仿真耳
artificial earthed body 人工接地体
artificial earth fill 人工填土

artificial earthing 人工接地
artificial earthquake 人造地震;人工地震
artificial earthquake wave 人工地震波
artificial earth satellite man-made earth satellite 人造(地球)卫星
artificial echo 人造回波;假回波
artificial echo unit 人造回声器;人工混响器
artificial ecological system 人工生态系统
artificial emery 黑刚玉
artificial enamel 合成树脂瓷漆
artificial environment 人造环境;人工环境
artificial environmental anomaly 人为环境异常
artificial error signal 人为误差信号
artificial excavation shield 人工开挖盾构
artificial exposure 人工露头
artificial eye 仿真眼
artificial feature of terrain 人工地物;非自然地貌
artificial feeding 人工饲养
artificial ferrite 人造铁氧体
artificial fertilizer 人造肥料;人工造肥
artificial fertilizer distributor 人造肥料撒播机
artificial fever 人工发热
artificial fiber[fibre] 人工纤维
artificial field 人为场;人工场
artificial field method 人工场法
artificial fill 人工填方;人工回填
artificial filter media 人工滤料
artificial filtrable membrane 人工滤膜
artificial fine aggregate 人造细颗粒;人造细集料;人造细骨料
artificial fine grain 人造细(颗)粒;人造细集料;人造细骨料
artificial fissure 人工裂隙
artificial flagstone 人造石板;人工石板
artificial flooding 漫灌
artificial flood wave 人造洪水波;人为洪水波
artificial flow control 人工水流控制
artificial flow field 人工流场
artificial fluctuation 人工波动
artificial flushing 人工冲洗
artificial foam 人造泡沫
artificial footing 人工基础
artificial forest 人造林
artificial formation 人造结构
artificial forming 手工成形;人工成型;手工成型
artificial foundation 人工基础
artificial fracturing 人工致裂
artificial freezing of ground 人工冻结地基;地基冻结工艺;地基冻结法;地基冰结过程
artificial freezing of soil 土壤人工冻结法
artificial gas 人造煤气
artificial ga(u)ging 人工计量
artificial gem 人造宝石
artificial geothermal fluid reservoir 人造地热流体储层
artificial gilsonite 人造硬沥青;人造黑沥青
artificial glue 人工合成树脂胶;合成树脂胶
artificial grade 人工坡降
artificial graphics 人造图形;人工图形
artificial graphite 人造石墨
artificial grass 人工草皮
artificial gravel material 人工砾料
artificial gravel-pack well 人工填砾井
artificial gravity 人造重力
artificial grind stone 人造磨(刀)石
artificial grit 人造磨石
artificial ground 人造土地;人为接地;人工基地;人工地基
artificial ground freezing method 冻结施工法
artificial ground motion 人工地面运动
artificial ground noise 人工地面噪声
artificial groundwater 人工地下水
artificial groundwater recharge 人工地下水回灌;人工地下水补给;人工补给地下水
artificial guard a slope 人工护坡
artificial gum 人造树胶;人造胶质
artificial hand 模拟手
artificial harbo(u)r 人造港湾;人工港湾;人工港
artificial hillock 假山
artificial hologram 人工全息照片;人工全息图;仿真全息照片
artificial horizon 假定地平线;人造地平;人为水平;人为地平;人工地平;水银盘地平;水平镜地平;假地平(线);仿真地平仪

artificial hot-water circulation system 人造热水循环系统
artificial hydraulic mortar 人工水凝灰浆
artificial ice 人造冰
artificial Iceland spar 人造冰洲石
artificial ice nucleus 人工冰核
artificial ice rink 人工溜冰场；人造滑冰场；人工冰场
artificial identification 人工识别
artificial illumination 人工照明
artificial immobilized engineering bacteria 人工固定化工程菌
artificial indoor illumination 室内人工照明
artificial indoor lighting 室内人工照明
artificial inducement(al) of rainfall 人工降雨；人工致雨；人工催雨；人工催化降水
artificial induction of hereditary changes 人工引变
artificial inertia 虚惯性
artificial infection 人工传染
artificial inorganic dust 人造无机粉尘
artificial intelligence 人工智能；智能模拟
artificial intelligence approach 人工智能法；智能模拟法
artificial intelligence languages 人工智能语言
artificial intelligence system 人工智能系统
artificial ionization 人造电离层；人为电离
artificial iron oxide 合成氧化铁
artificial irrigation 人工灌溉
artificial island 人工筑岛；人工海岛；人工岛(屿)
artificial island construction method 筑岛施工法
artificial island method 人工筑岛法(深基础施工时下沉井,沉箱用)；人工岛法
artificial isotope 人造同位素；人工同位素
artificial ivory 人造象牙
artificial labo(u)r cost 人工费(用)
artificial lake 人工蓄水池；人工湖；水库
artificial landslide 人为滑坡
artificial language 人工语言
artificial latex 人造胶乳
artificial law 制定法
artificial leather 人造革
artificial leather hanging 人造革帷幕；人造革挂幕
artificial leg 假腿
artificial lens 人工晶体
artificial levee 防洪堤
artificial lift 人工提升；人工举升
artificial light 人造光；人工日光
artificial light filter 仿真滤光器
artificial lighting 人工照明；人工闪电；人工光照；人工采光
artificial lighting generator 人造闪电发生器
artificial-lighting stair(case) 暗楼梯
artificial light source 人造光源；人工光源
artificial light(weight) aggregate 人造轻集料；人造轻骨料
artificial limestone 人造石灰石
artificial line 人工线路；模拟线；仿真线
artificial line duct 仿真线通道
artificial load 模拟负载；虚负载；仿真荷载；仿真负载；仿真负荷
artificial loading 人工装货
artificial lubricating oil 人造润滑油
artificially aerated trickling filter 人工曝气滴滤池
artificially aged alumin(i)um 人工老化铝
artificially blunted 人工钝化处理
artificially created water resources 人工水力资源；人工水利资源
artificially crushed aggregate 人工碎石集料；人工集料；人工骨料
artificially dried 人工干燥的
artificially excavated harbo(u)r 人工开挖的港口
artificially excavated port 人工开挖的港口
artificially flattened bed 人工整平的基床；人工整平的河床
artificially generated acceleration 人工加速度图
artificially graded aggregate 人工筛分的骨料；人工级配集料；人工级配骨料
artificially improved soil 人工加固土
artificially lighting 人工照明的
artificially low price 人为低价
artificially low tender 故意压低的报价
artificially marked point 人工标志(点)
artificially precipitated calcite 人工淀积方解石
artificially recharged groundwater 地下水回灌(量)；人工补充的地下水
artificially recharged underground water 人工补充的地下水
artificially refractory construction material 人工耐火建筑材料
artificially sheltered harbo(u)r 筑防波堤防护的港口
artificiallysis 人工裂解
artificially sown pasture 人工草地
artificially synthesized 人工合成
artificial made land 人工填土
artificial magnet 人造磁铁
artificial magnet field 人造磁场
artificial magnetization method 人工磁化法
artificial mains network 仿真电源网络
artificial manure 人造肥料；人工厩肥；堆肥
artificial marble 仿云石；人造大理石；假大理石
artificial masonry unit 人造砖石；人造圬工用砖石
artificial material 人造材料
artificial maturation 人工成熟
artificial measures 人造措施
artificial media 人工介质
artificial membrane 人工膜
artificial meteor 人造流星
artificial meteorology 人工气象
artificial mica 人造云母
artificial mill-stone 金刚砂磨石
artificial mineral 人造矿物
artificial monument 人造标石；人工标石；人造古迹
artificial mound 土山
artificial mudflow 人为泥石流
artificial mulch 人工覆盖物
artificial naturalized plant 人为归化植物
artificial navigable waterway 人工航道；人工通航水道；人工可航水道
artificial navigation canal 通航运河；人工航道
artificial neroli oil 氨茴酸甲酯
artificial network 模拟网络；仿真网络
artificial nourishment 人工围堤填土；人造食物；人工淤滩；填砂护滩
artificial nuclear reaction 人工核反应
artificial nuclear survey method 人工核法
artificial nucleus 人工核；人工成核
artificial object 人造天体
artificial observation installation 海上观测站
artificial of everyday use 日用品
artificial oil 合成油
artificial outcrop 人工露头
artificial-packed 人工包轧
artificial parchment 仿羊皮纸
artificial pare a slope 人工削坡
artificial pasture 人工牧场；人工草地
artificial perception 人工识别
artificial person 法人
artificial petroleum 人造石油
artificial placer sample 人工重砂样品
artificial placer sampling 人工重砂采样
artificial planet 人造行星
artificial point 人工标志(点)
artificial polarization 人工极化
artificial pollution 人为污染；人工污染
artificial pollution test 人工污染试验
artificial pond 人工池(塘)
artificial porcelain 软质瓷
artificial port 人工港
artificial pozzolana 人造火山灰(质材料)
artificial pozzolanic material 人造火山灰质材料
artificial precipitation 人造雨；人工雨；人工降水
artificial precipitation enhancement 人工增雨
artificial precipitation stimulation 人工致雨；人工催雨；人工催化降水
artificial propagation 人工繁殖
artificial quartz crystal 人造石英晶体
artificial radiation 人工辐射
artificial radiation belt 人造辐射带；人工辐射带
artificial radiation dose 人工辐射剂量
artificial radiation isotope 人工放射性同位素
artificial radioactive element 人工放射性元素
artificial radioactive energy 人工辐射能
artificial radioactive isotope 人造放射性同位素
artificial radioactive product 人工放射性产物
artificial radioactive source 人工辐射源
artificial radioactive tracer 人工放射性示踪剂
artificial radioactivity 人工放射性
artificial radioactivity product 人工放射性产物
artificial radio aurora 模拟无线电极光
artificial radio element 人工放射性元素
artificial radionuclide 人工放射性核素
artificial rain device 人工降雨装置
artificial rain(fall) 人工降雨；人造雨
artificial rainmaking process 人工造雨法
artificial rainmaking rocket 降雨火箭
artificial rain method 人工降雨法
artificial range 人工建筑叠标
artificial realignment 人工改道
artificial recharge 人工回灌；人工灌注；人工补水；人工补给量；人工补给地下水；人工补给；回灌
artificial recharge data of groundwater 地下水人工补给数据
artificial recharged groundwater 人工补给地下水
artificial recharge of groundwater 人工地下水回灌；人工地下水补给；地下水人工补给(量)
artificial recharge resources of groundwater 地下水人工补给资源
artificial recharge through well 人工竖井回灌
artificial recharge well 人工回灌井
artificial recharging 人工灌注；人工充电
artificial reef 人造礁石
artificial refilling 人工填土
artificial reforestation 人工造林(法)
artificial regeneration 人工更新
artificial regulation 人工整治；人工调控
artificial renourishment 人工淤滩；填砂护滩
artificial replenishment 人工填充；疏浚排泥填岸
artificial rerouting 人工改道
artificial reservoir 人工水库
artificial reservoir fluid 人工热储流体
artificial resin 合成树脂；人造树脂
artificial resin-based bonding adhesive 合成树脂黏性胶；合成树脂胶；合成树脂胶黏剂
artificial resin-based bonding agent 合成树脂胶；合成树脂胶粘剂
artificial resin-based cement(ing) 合成树脂凝结剂
artificial resin-based cementing agent 合成树脂胶凝剂
artificial resin-based glue 合成树脂胶
artificial resin exchanger 人工树脂胶交换器
artificial resistance 模拟电阻
artificial respiration 人工呼吸
artificial reverberation 人工混响
artificial rockwork 假山
artificial roof 人工假顶；人工顶板；假顶
artificial roughness 人为粗糙率
artificial rubber 人造橡胶
artificial runoff field 人工径流场
artificial salt 人工盐
artificial sample 人工样本；人工抽样
artificial sampling 人工抽样
artificial sand 人造砂；人工砂
artificial sand island 人工砂岛
artificial sandstone 人造砂石
artificial satellite 人卫
artificial satellite geodesy 卫星大地测量学
artificial scarcity 人为的稀缺性
artificial scheelite 钨酸钙
artificial seasoning 人工陈化；人工时效；人工干燥(法)；人工干材法；烘干
artificial seawater 人造海水
artificial seismic source 人工震源
artificial selecting 人工选择
artificial selection 人为淘汰；人工选择
artificial series 人工数列
artificial shating rink 人造溜冰场
artificial sheltered harbo(u)r 筑堤防护的港口
artificial shift erosion 人为侵蚀
artificial shipway 人工浮道；人工船道
artificial silk 人造丝
artificial silk fabrics 人造丝织物
artificial silting 人工淤填
artificial simulator 人工降雨装置
artificial skating rink 人造滑冰场；人工溜冰场
artificial sky 人造天空；人工天空
artificial slate 人造石板
artificial slipway 人工浮道
artificial slope 人工边坡
artificial sludge 人工污泥
artificial sludge drying 污泥人工脱水；污泥人工干燥
artificial smoke 烟幕

artificial snow-making 人工造雪
artificial snowmelt 人工融雪
artificial soil 人工土壤
artificial soil freezing method 冻结施工法
artificial solidification 压浆处理；人工压浆处理
artificial source of illumination 人工光源
artificial spacecraft 人造飞船
artificial spawning ground 人工产卵场(鱼类)
artificial stability 人为稳定
artificial stabilization 人为稳定
artificial stable isotope 人工稳定同位素
artificial star-field calibrator 人字星物检定器
artificial stereoscope 人造立体镜
artificial stone 铸石；斩假石；假石
artificial stone block 人造石块；人造石材
artificial stone coating 斩假石墙面
artificial stone floor cover(ing) 人造石花砖饰面
artificial stone floor dressing 人造石楼面修整
artificial stone floor(ing)finish 人造石楼板面层；人造石地板面层；人造石花砖地面装修
artificial stone pavement 人造石铺面；人造石铺砌层
artificial stone shop 人造石工场
artificial stone skin 人造石面层
artificial stone stair(case) 人造石楼梯
artificial stone tile 人造地砖；人造石砖
artificial stone tile floor cover(ing) 人造楼面砖；人造地面砖；人造花砖地面装修
artificial stone waterproofer 人造石防水层
artificial stone work 人造石工艺；人造石构件
artificial storage 人工蓄水；人工储存量
artificial stream 人工河流；人工河道
artificial strengthening 人工加固
artificial structure 人工结构(物)；人工建筑物；人工构筑物
artificial subgrade 人工地基；人工路基
artificial sub-irrigation 人工地下灌溉
artificial subsoil 人工地基
artificial sun light 人工太阳光
artificial sunlight lamp 人造太阳灯
artificial suppression 人工抑制
artificial swarm 人工分群
artificial sweetener 人工增甜剂
artificial synthesis 人工合成
artificial system 人工分类法
artificial tamping 人工夯实
artificial target 人工目标；人工标志(点)
artificial temperature stress 人工温度应力
artificial terrace 人工梯田；人工阶地
artificial traffic 模拟通信量
artificial transmission line 仿真传输线
artificial transmitting boundary 人工透射边界
artificial travertin(e) 人造云石；人造大理石；人造凝灰石
artificial turbulence flume 人工紊流水槽
artificial turf 人造草皮
artificial ultramarine 合成群青
artificial value 人为值
artificial variable 人为变量；人工变量；仿真变量
artificial variable method 人工变量法
artificial vaseline 人造凡士林
artificial vector 仿真向量
artificial ventilation 人工通风；机械通风
artificial venting 人工通风
artificial vibration 人工振动
artificial vibration source 人工振源
artificial voice 仿真口声
artificial watercourse 人工水道
artificial waterfall 人工瀑布
artificial water flooding 人工注水
artificial watering 人造雨；人工浇水；人工灌溉
artificial water injection 人工注水
artificial water tank 人工水槽
artificial waterway 人工水道；人工开挖的水道；人工航道
artificial wealth 人为财富
artificial weathering 人工风蚀；人工老化；人工风化；人工大气老化
artificial weathering test 人工气候试验
artificial wetland 人造湿地
artificial works 人工构筑物
artificial zeolite 分子筛；人造沸石
artifitial modification of microclimates 小气候的人工调控

artillery 火炮
artillery cart 双轮炮车
artillery director 炮兵罗盘
artillery gun 大炮
artillery map 炮兵地图
artillery panorama picture 炮兵照相写景图
artillery survey 炮兵测地
artillery survey service 炮兵测地勤务
art infection 艺术感染
artinite 水纤菱镁矿；纤水碳镁石
Artinskian 阿尔丁斯克阶【地】
artisan 手艺(工)人；技工；工匠
artisanal inshore fishery 沿海个体渔业
artist 美术家；画家；艺术家
artist-craftsman 手工艺工作者；艺术家工匠
artistic adviser 美术顾问；艺术顾问
artistic anatomy 美术解剖学；艺术解剖学
artistic attainments 艺术造诣
artistic carving 雕刻工艺
artistic ceramics 美术陶瓷
artistic characteristics 艺术特征
artistic circle 艺术界；艺术馆中心；文化界；文化馆中心；艺术范围
artistic conception 意境
artistic conception in gardening 园林意境
artistic creation 艺术创造
artistic criterion 艺术标准
artistic design 艺术造型设计；美术设计；艺术设计
artistic door opening 洞门（如月亮门壶门等）
artistic drawing set 美术用品
artistic effect 艺术效果
artistic engraving of papers 艺术剪纸
artistic exaggeration 艺术夸张
artistic expression 艺术造型；艺术表现
artistic form 艺术形式
artistic formation 艺术表现形式；艺术造型
artistic form of architecture 建筑艺术形式
artistic form of expression 艺术表现形式
artistic glass 艺术玻璃
artistic glassware 美术玻璃器皿；艺术玻璃制品
artistic glaze 艺术釉
artistic glaze tile 精美琉璃瓦
artistic greeting card 美术贺卡
artistic handicraft 美术手工艺品
artistic intuition 艺术直观
artistic lamp 灯具
artistic layout of garden 园林艺术布局
artistic maturity 艺术上成熟
artistic modelling 艺术塑造品
artistic monument 不朽的艺术作品；艺术纪念牌
artistic object in ivory 象牙艺术雕刻品
artistic porcelain 美术瓷；艺术瓷
artistic pottery 美术陶器
artistic presentation 美术装潢
artistic quality 艺术性
artistic style 艺术风格
artistic tapestry 艺术挂毯；工艺美术壁毯
artistic terrazzo 美术水磨石
artistic terrazzo flooring 美术水磨石地面
artistic theatre 艺术剧院
artistic treatment 美化处理；艺术处理
artistic treatment in architecture 建筑艺术处理
artistic type 艺术型
artistic value 艺术价值
artistic ware 艺术器皿
artist lounge 演员休息室
artist paint 油画漆
artist restorer of ancient paintings 古画修复艺术家
artistry 艺术手法；艺术技巧；艺术性；艺术效果
artists' colo(u)r 绘画颜料
artists' impression 透视图
artists' oil 美术油
artists' paint 绘画漆
artists' primary 美术三原色
artists' studio 艺术家工作室
artizan 手艺(工)人
art layout 艺术设计
art-lover 艺术鉴赏家；艺术鉴定家；艺术爱好者
art marble 人造大理石；艺术大理石
art material 美术用品
art metal 艺术金属物
artmobile 流动艺术展览车
art museum 美术馆；艺术博物馆
Art Nouveau 新技艺；(19 世纪末法国和比利时的)新艺术运动；新艺术派
Art Nouveau faience 新艺术风格的(上釉)陶瓷
art of architectural modeling 建筑造型艺术
art of architecture 建筑艺术
art of building 建筑艺术
art of design 装饰艺术
art of dynamic(al)programming 动态规划的技巧
art of fortification 要塞建筑技艺；城堡建筑技艺；要塞建筑术；城堡建筑术
art of garden colo(u)rs 园林色彩艺术
art of garden-making 造园艺术
art of inlaying 镶嵌艺术
art of navigation 航海技术
art of pottery 陶瓷艺术
art of pyrography 烙画法
art of sculpture 雕刻艺术
art of the individual 个性艺术
art of vaulting 拱顶建造艺术；拱顶建造技艺；圆顶建筑艺术
artogravure 照相丝网印刷法
artotype 明胶板(照相制板)
art paper 铜版纸；美术纸
art parchment 仿羊皮美术纸
art porcelain 艺术瓷
art poster board 美术广告纸板
art poster paper 美术广告纸
art post paperboard 白卡纸
art pottery 艺术陶(瓷)
art pottery and porcelain 美术陶瓷
art printing paper 美术印刷纸
art room 美术教室
arts and crafts 工艺美术
arts and crafts college 工艺美术学院
arts and crafts factory 工艺美术工厂
arts center 艺术中心；艺术厅
art school 美术学校；艺术学校
arts faculty 艺术学院；艺术系
art shade 雅致色泽
art silk 刺绣丝线
art square 图案方毯
art stoneware 艺术炻器
art store 美术商店
art theatre 艺术剧院
art-title 美术字幕
art treasure 文物；艺术珍品
art vellum 美术充皮纸
artware 艺术制品；实用艺术工艺品；工艺品
artware glaze 美术釉；艺术釉
art weave 艺术花纹
artwork 原图；艺术品；工艺图；布线图；工艺品；图形；画稿拼版
artwork generation 图形发生
artwork master 照相原图
artwork system 原图系统
artwork tape 原图信息带
Artype 阿尔太普印字传输系统
Artz press sheet 特殊薄钢板
aruhuesiru (预填集料灌浆混凝土的)掺和剂；(预填集料灌浆混凝土的)外加剂
Arundel formation 阿伦德组【地】
Arundel method 阿伦德法
Arusha Agreement 阿鲁沙协定
Arusha Convention 阿鲁沙协定
Arvee 游乐汽车
aryl 芳烃基；芳基
aryl acid 芳基酸
arylamine 芳基胺
arylated alkyl 芳脂基
arylating 芳基化作用
arylating agent 芳基化剂
arylation 芳基化作用
aryl chloride 芳基氯
aryl chlorofluorosilane 芳基氟氯硅烷
aryl chlorosilane 芳基氯硅烷
aryl compoundiarylide 芳基化合物
aryl diazo compound 芳基重氮化合物
aryle 芳基金属(化合物)
aryl halide 芳基卤
arylide 芳基化合物(化合物)
aryl lithium 芳基锂
arylmercurial 芳基汞的
arylmercury 芳基汞
aryl olefin 芳基烯烃
aryl oxide 芳族醚

aryloxy 芳氧基
aryloxy acetic acid 芳氧基乙酸
aryloxy compound 芳氧基化合物
aryloxy siloxane 芳氧基硅烷
aryl silazane 芳基硅氨烷
arylsulfamic acids 芳代氨基磺酸
aryl sulfonyl chloride 芳基磺酰氯
aryne 脱氢芳烃;芳炔
arythmia 无节律
arzrunite 氯铜铅矾
as agreed 依照协定
Asahi process 对辊法
Asahi process drawing roller 对辊辊子
as a matter of experience 按照实践经验
as a matter of fact 事实上
Asam Kirche 阿赛姆教堂(巴洛克式建筑的代表作品)
as analyzed basis 分析基
Asarco lead 高耐蚀铅合金;阿萨科耐蚀铅合金
Asarco-Loy alloy 阿萨科－洛伊镉镍合金
Asarco method 连续铸造法
asarite 粗细辛脑
asarotum 油彩地面;拼花路面;彩绘人行道(古罗马);[复]asaroria
asbecasite 砷铍钙石;砷铍硅钙石
asbestic 石棉的
asbestic half-round tile 半圆形石棉瓦
asbestic tile 石棉瓦
asbestification particle 石棉颗粒
asbestiform 石棉构造的;似石棉;石棉状
asbestiform half-round tile 半圆形石棉瓦
asbestiform mineral 石棉状矿物
asbestiform tile 石棉瓦
asbestine 纤滑石;滑石粉;石棉(状)的;石棉质的;滑石棉
asbestinite 微石棉;角闪石棉
asbestoid 似石棉的;石棉状的
asbestone 石棉水泥材料;防火布
asbestonite 石棉制绝热材料;石棉绝缘材料;石棉绝热材料
asbestophalt 石棉地沥青
asbestos 石绒;石棉
asbestos apron 石棉盖板;石棉底板;石棉挡板
asbestos article 石棉制品;岩棉制品
asbestos ash 石棉灰
asbestos asphalt products 石棉沥青制品
asbestos base asphalt paper 层压石棉沥青油纸板
asbestos based 石棉基的
asbestos-based asphalt bitumen felted fabric 石棉沥青油毛毡
asbestos-based asphalt felt 石棉基沥青油毡;石棉沥青油毡
asbestos-based asphalt felted fabric 石棉基(地)沥青油毛毡
asbestos-based asphaltic-bitumen paper 石棉基沥青油毛纸
asbestos-based asphalt paper 石棉基(地)沥青油毛纸
asbestos-based bitumen felt 石棉基沥青油毡;石棉沥青油毡
asbestos-based bitumen paper 石棉沥青油毛纸
asbestos-based felt 石棉基油毡
asbestos based laminate 层压石棉板
asbestos-bitumen 石棉沥青
asbestos-bitumen sheet 石棉沥青板
asbestos-bitumen thermoplastic sheet 石棉沥青热塑板
asbestos blanket 石棉保温层;石棉毡
asbestos block 石棉块
asbestos board 石棉板
asbestos boarding 铺钉石棉板
asbestos board shield 石棉保护板
asbestos body 石棉小体
asbestos brake belt 石棉制动带
asbestos brake facing 石棉制动摩擦片
asbestos brake lining 石棉闸衬;石棉制动衬片
asbestos brick 石棉砖
asbestos building board 建筑用石棉板;建筑石棉板
asbestos building sheet 建筑用石棉板;建筑石棉板
asbestos calcium silicate board 石棉硅酸钙薄板
asbestos calcium silicate sheet 石棉硅酸钙薄板
asbestos canvas 石棉布
asbestos cap 石棉帽
asbestos carcinogenesis 石棉致癌(性)

asbestos cardboard 石棉纸板(垫层)
asbestos card liner 石棉衬圈
asbestos cement 石棉水泥
asbestos-cement article 石棉水泥制品
asbestos-cement board 石棉水泥板;石棉胶合板
asbestos-cement board ceiling 石棉水泥板平顶;石棉水泥板顶棚
asbestos-cement board structural roofing 石棉水泥板结构屋面
asbestos-cement box roof gutter 石棉水泥屋顶箱形天沟;箱形石棉水泥天沟
asbestos-cement building component 石棉水泥建筑构件
asbestos-cement building member 石棉水泥建筑构件
asbestos-cement building unit 石棉水泥建筑构件
asbestos-cement cellulose 石棉水泥纤维素
asbestos-cement cistern 石棉水泥蓄水池;石棉水泥水槽
asbestos-cement cladding 石棉水泥板覆面;石棉水泥填充墙;石棉水泥外墙护板;石棉水泥外挂板
asbestos-cement closure 石棉水泥封口;石棉水泥封闭
asbestos-cement composition 石棉水泥制品
asbestos-cement conductor 石棉水泥落水管;石棉水泥雨水落水管
asbestos-cement corrugated board 石棉水泥波纹板
asbestos-cement corrugated panel 石棉水泥瓦楞板;石棉水泥波纹板
asbestos-cement corrugated roof cladding 石棉水泥瓦楞屋顶;石棉水泥波纹屋顶
asbestos-cement corrugated roof(ing) board 石棉水泥瓦楞屋面板;石棉水泥波纹屋面板
asbestos-cement corrugated sheet(ing) 石棉水泥波形瓦;石棉水泥瓦楞(薄)板;石棉水泥波纹板
asbestos-cement corrugated sheet roofing 石棉水泥波形瓦屋面
asbestos-cement discharge pipe 石棉水泥排水管
asbestos-cement distance piece 石棉水泥隔板
asbestos-cement downcomer 石棉水泥落水管;石棉水泥泄水管
asbestos-cement downpipe 石棉水泥落水管
asbestos-cement downspout 石棉水泥落水管
asbestos-cement drain(age)pipe 石棉水泥排水管
asbestos-cement duct 石棉水泥管;石棉水泥导管
asbestos-cement eaves gutter 石棉水泥檐沟
asbestos-cement eaves trough 石棉水泥檐沟
asbestos-cement electric(al)insulating board 石棉水泥电气绝缘板
asbestos-cement extract(ion)ventilation unit 石棉水泥抽风装置;石棉水泥抽气装置;石棉水泥通风装置
asbestos-cement extractor 石棉水泥通风器;石棉水泥抽风器;石棉水泥通风机;石棉水泥抽气机
asbestos-cement facade 石棉水泥门面;石棉水泥饰面;石棉水泥铺面
asbestos-cement facade slab 石棉水泥饰面板
asbestos-cement face 石棉水泥饰面;石棉水泥面
asbestos-cement facing 石棉水泥饰面层;石棉水泥面层;石棉水泥饰面;石棉水泥粉面
asbestos-cement factory 石棉水泥厂
asbestos-cement fall pipe 石棉水泥落水管
asbestos-cement fascia board(for flat roof)(用于平屋面的)石棉水泥檐口板;(用于平屋面的)石棉水泥挑口板
asbestos-cement fat siding 石棉水泥平板
asbestos-cement fence 石棉水泥围栏;石棉水泥栅栏
asbestos-cement filler strip 石棉水泥填充片;石棉填充片
asbestos-cement fitting 石棉水泥管配件
asbestos-cement flat board 石棉水泥平板
asbestos-cement flat(run)panel 石棉水泥镶面板
asbestos-cement flat(run)sheet 石棉水泥;石棉水泥平板
asbestos-cement flexible board 石棉水泥柔性板
asbestos-cement floor(ing) 石棉水泥面层;石棉水泥地板
asbestos-cement flooring material 石棉水泥地板材料
asbestos-cement flower container 石棉水泥花盆
asbestos-cement flue 石棉水泥暖气管;石棉水泥烟道

asbestos-cement flue liner 石棉水泥烟道衬管
asbestos-cement fluted board 石棉水泥瓦楞板;石棉水泥波纹板;石棉水泥槽纹板
asbestos-cement fluted sheet 石棉水泥槽纹板
asbestos-cement form board 石棉水泥模板;石棉水泥型板
asbestos-cement formwork 石棉水泥模板
asbestos-cement foul water pipe 石棉水泥污水管
asbestos-cement fountain basin 石棉水泥喷水池;石棉水泥喷泉池
asbestos-cement front 石棉水泥饰面;石棉水泥粉面
asbestos-cement gutter 石棉水泥檐沟;石棉水泥天沟
asbestos-cement header 石棉水泥集管
asbestos-cement insulating sheet 石棉水泥绝缘薄板
asbestos-cement insulation board 石棉水泥保温板;石棉水泥绝缘板
asbestos-cement joint 石棉水泥接头
asbestos-cement lateral 石棉水泥支管
asbestos-cement leader 石棉水泥排水管;石棉水泥雨水落水管
asbestos-cement lining 石棉水泥衬里;石棉水泥衬垫
asbestos-cement lining sheet 石棉水泥封檐板;石棉水泥衬板
asbestos-cement member 石棉水泥构件
asbestos-cement mortar 石棉水泥砂浆
asbestos-cement mo(u)ld 石棉水泥模型
asbestos-cement panel 石棉水泥板
asbestos-cement partition 石棉水泥隔墙
asbestos-cement pipe 石棉水泥管
asbestos-cement pipe corrosion 石棉水泥管腐蚀
asbestos-cement pipe fitting 石棉水泥管配件
asbestos-cement pipe joint 石棉水泥管接头
asbestos-cement pressure pipe 石棉水泥压力管
asbestos-cement product 石棉水泥制品
asbestos-cement profile(d)board 石棉水泥定型板;石棉水泥异型板
asbestos-cement rain water conductor 石棉水泥雨水管
asbestos-cement rainwater gutter 石棉水泥雨水檐沟
asbestos-cement refuse water pipe 石棉水泥污水管;石棉水泥废水管;石棉水泥排水管
asbestos-cement ridge capping 石棉水泥脊瓦
asbestos-cement ridge capping tile 石棉水泥压脊瓦
asbestos-cement ridge covering tile 石棉水泥压脊瓦
asbestos-cement ridging tile 石棉水泥脊瓦
asbestos-cement roof cladding 石棉水泥屋面覆盖层
asbestos-cement roof covering 石棉水泥屋面覆盖层
asbestos-cement roof gutter 石棉水泥屋面檐沟
asbestos-cement roofing 石棉水泥屋顶;石棉水泥屋面
asbestos-cement roofing board 石棉水泥屋面板
asbestos-cement roofing cover 石棉水泥屋面覆盖层
asbestos-cement roofing panel 石棉水泥屋面板
asbestos-cement roofing sheet 石棉水泥屋面板
asbestos-cement roofing shingle 石棉水泥(平)瓦;石棉水泥屋面板;石棉水泥屋面瓦
asbestos-cement roofing slate 石棉水泥屋面平瓦
asbestos-cement roofing tile 石棉水泥屋面瓦
asbestos-cement roof sheathing 石棉水泥屋面板;石棉水泥望板;石棉水泥衬板
asbestos-cement rubber tile 石棉水泥橡胶瓦
asbestos-cement semi-corrugated shingle 石棉水泥半波瓦
asbestos-cement separator 石棉水泥隔板
asbestos-cement septic tank 石棉水泥化粪池
asbestos-cement sewage pipe 石棉水泥污水管
asbestos-cement sewer pipe 石棉水泥污水管
asbestos-cement shake 石棉水泥盖屋瓦
asbestos-cement sheet 石棉水泥墙板;石棉水泥板
asbestos-cement sheeting 石棉水泥(镶面)板
asbestos-cement shingle 石棉水泥平瓦;石棉水泥墙板;石棉水泥瓦
asbestos-cement siding 石棉水泥挂墙板;石棉水泥墙面板;石棉水泥防雨墙板;石棉水泥防雨壁

板;石棉水泥防雨板
asbestos-cement siding shake 石棉水泥墙板
asbestos-cement slate 石棉水泥板瓦;石棉水泥板(条);石棉水泥(屋面石板)瓦
asbestos-cement slate roofing 石棉水泥平瓦
asbestos-cement solid board 石棉水泥实板;石棉水泥实心板
asbestos-cement spacer 石棉水泥垫块
asbestos-cement structural roofing 石棉水泥板结构屋面
asbestos-cement sur facing 石棉水泥饰面
asbestos-cement sur facing sheet 石棉水泥饰面板
asbestos-cement tile 石棉水泥砖;石棉水泥瓦
asbestos-cement tile roofing 石棉水泥瓦屋面
asbestos-cement valley board 石棉水泥斜沟底板
asbestos-cement valley gutter 石棉水泥斜沟槽
asbestos-cement vent(ilating) pipe 石棉水泥通风管
asbestos-cement ventilator 石棉水泥材料通风装置
asbestos-cement wallboard 石棉水泥墙板
asbestos-cement wall panel 石棉水泥墙板
asbestos-cement wall shingle 石棉水泥墙板
asbestos-cement ware 石棉水泥器皿;石棉水泥制品
asbestos-cement window sill 石棉水泥窗台板
asbestos center gauze 石棉心铁丝网
asbestos clay 石棉泥
asbestos cloth 石棉织物;石棉布
asbestos clothing 石棉罩;石棉保温层
asbestos composition 石棉泥
asbestos concrete 石棉混凝土
asbestos concrete pipe 石棉混凝土管
asbestos concrete slab 石棉混凝土板
asbestos conduit 石棉管
asbestos cord 石棉绳
asbestos cord covering 石棉绳绝缘层;石棉绳防潮层;石棉绳缠层
asbestos-corrugated board 石棉波纹板
asbestos-corrugated panel 石棉波纹板
asbestos cover 石棉盖
asbestos-covered metal 石棉包覆金属
asbestos-covered wire 石棉包线;包石棉线
asbestos covering 石棉覆盖(层)
asbestos curtain 石棉幕帘;安全幕帘;防火幕(帘)
asbestos cushion 石棉垫
asbestos deferring machine 开棉机
asbestos deposit 石棉矿床
asbestos diaphragm electrolytic cell 石棉隔膜电解槽
asbestos-diatomaceous earth 石棉硅藻土
asbestos-diatomite 石棉硅藻土;防火石棉板
asbestos dressing 石棉选矿;石棉加工
asbestos duct 石棉管道
asbestos dust 石棉粉尘;石棉尘
asbestos encapsulation 石棉密封罩
asbestos encasement 石棉外罩
asbestos fabric 石棉织物;石棉织品;石棉布
asbestos felt 石棉毡;石棉油毡;浸油石棉毡
asbestos fiber 石棉纤维
asbestos fiber board 石棉纤维板
asbestos fiber filling 石棉纤维填缝
asbestos fiber in combination 分解棉
asbestos fiber insulation 石棉纤维绝热;石棉纤维保温
asbestos fiber plaster 石棉纤维灰浆
asbestos fiber-reinforced 石棉纤维加筋的
asbestos fiber reinforced polyester 石棉纤维加劲聚酯
asbestos fiber-reinforced sheet(ing) 石棉纤维加筋(墙)板
asbestos fiber tile 石棉纤维砖
asbestos fibrillation 石棉松解(度)
asbestos filler 石棉垫片;石棉填料;石棉填充料
asbestos filter 石棉滤纸;石棉滤(油)器;石棉过滤器
asbestos filter cloth 石棉滤布
asbestos firebrick 石棉耐火砖
asbestos fire curtain 石棉防火幕
asbestos flat board 石棉平板
asbestos flat building board 石棉建筑平板
asbestos flat building sheet 石棉建筑平板
asbestos float 石棉绒
asbestos flour 石棉粉末
asbestos foam 泡沫石棉
asbestos foam(ed) concrete 石棉泡沫混凝土

asbestos formboard 石棉模板
asbestos-free calcium silicate 无石棉微孔硅酸钙
asbestos-free calcium silicate board 无石棉硅酸钙板
asbestos-free friction product 无石棉制动制品
asbestos friction material 石棉摩擦材料
asbestos friction product 石棉制离合器摩擦片;石棉制动制品
asbestos friction sheet 石棉摩擦片
asbestos gasket 石棉垫片;石棉填料
asbestos gasket joints for pipes 管道接缝—石棉垫圈接缝
asbestos gland packing 石棉压盖填料
asbestos gloves 石棉手套
asbestos gutter 石棉雨水天沟
asbestos heat-insulation material 石棉绝热材料
asbestos hose 石棉软管
asbestos insert(ion) 石棉嵌料;石棉插件;石棉垫
asbestos insulated cable 石棉绝缘电缆
asbestos insulating board 石棉绝热板;石棉保温板
asbestos insulating sheet 石棉绝热薄板;石棉保温板
asbestos insulation 石棉隔热层;石棉保温层;石棉绝缘;石棉绝热体
asbestosis 石棉肺;石棉沉着病
asbestos jointing 石棉填料
asbestos joint runner 填缝石棉绳;石棉填缝浇口;石棉接头浇道;石棉接口浇道
asbestos lagging 石棉外套
asbestos laminate 石棉层压板
asbestos layer 石棉层
asbestos liner sheet 石棉垫板
asbestos lining 石棉衬垫
asbestos manufacturing 石棉加工
asbestos mat 石棉毯;石棉垫
asbestos mica 石棉云母
asbestos millboard 石棉书皮纸板;石棉覆盖板
asbestos milling 石棉碾磨
asbestos mine 石棉矿(井)
asbestos mineral 石棉矿物;石棉矿石
asbestos mining 石棉矿开采;石棉开采
asbestos mitten 石棉手套
asbestos mortar 石棉砂浆;石棉灰浆
asbestos mortite 石棉纤维粉
asbestos-organic 石棉—有机物的
asbestos packing 石棉填料;石棉填充;石棉盘根;石棉密封;石棉垫(料)
asbestos-packing gasket 石棉橡胶垫片
asbestos pad 石棉垫
asbestos pad bearing 石棉垫层支座
asbestos paint 石棉涂料
asbestos paper 石棉纸
asbestos paper gasket 石棉纸垫片
asbestos pipe 石棉管
asbestos plaster 石棉灰泥;石棉灰浆;石棉灰;石棉粉饰
asbestos plastic floor(ing) 石棉塑料楼面板;石棉塑料地板
asbestos plywood 石棉胶合板;石棉层压板
asbestos pollution 石棉污染
asbestos powder 石棉粉(末)
asbestos product 石棉制品
asbestos product factory 石棉制品厂
asbestos-protected metal roofing 石棉阻燃金属屋顶;石棉护层金属屋顶;石棉保护金属屋顶
asbestos protection 石棉防护层;石棉保护;瓦冷石棉屋面防锈层;波纹石棉屋面防锈层
asbestos PVC 石棉聚氯乙烯
asbestos PVC floor(ing) 石棉—聚氯乙烯塑料地板
asbestos PVC floor(ing) tile 石棉—聚氯乙烯塑料地砖;石棉—聚氯乙烯塑料楼面砖
asbestos quarry 石棉开采;石棉采矿场
asbestos rainwater gutter 石棉雨水沟;石棉屋面雨水天沟
asbestos reel 石棉卷筒
asbestos-reinforced panel 石棉加强板;石棉增强板
asbestos-reinforced plastics 石棉增强塑料
asbestos removal 石棉清除
asbestos ribbon 石棉带
asbestos ring 石棉垫圈;石棉圈;石棉箍
asbestos rock 石棉岩
asbestos roll 石棉辊
asbestos roof gutter 石棉屋面檐沟;石棉屋面雨水天沟

asbestos roofing 石棉屋面材料;石棉瓦屋顶;石棉水泥屋顶板
asbestos roofing material 石棉屋面材料
asbestos roofing sheet 石棉屋面板
asbestos roof shingle 石棉耐火屋顶板;石棉屋面瓦;石棉屋顶瓦
asbestos rope 石棉绳
asbestos roving 石棉粗纱
asbestos rubber gasket 石棉橡胶密封垫片
asbestos rubber sheet 石棉橡胶板;石棉胶板
asbestos rubber tile 石棉橡胶瓦;橡皮层石棉盖板
asbestos runner 填缝石棉绳(承口处)
asbestos sealing product 石棉密封制品
asbestos sheet(ing) 石棉纸;石棉瓦;石棉片;石棉板;石棉镶板
asbestos sheet packing 石棉密封垫片
asbestos sheet with insertion rubber 石棉橡胶板
asbestos shingle 石棉屋面板;石棉瓦
asbestos shoes 石棉鞋
asbestos siding 石棉壁板;石棉墙板
asbestos silk 石棉丝
asbestos slag wool 石棉矿渣棉
asbestos slate 石棉岩板
asbestos sliver 石棉条
asbestos standard testing machine 石棉标准检验筛
asbestos steel slice 石棉钢片
asbestos string 石棉线
asbestos strip 石棉条
asbestos structural roofing 石棉结构屋面
asbestos suit 石棉衣;石棉服
asbestos tailing 石棉尾矿
asbestos tape 石棉带
asbestos tester 石棉试验机
asbestos texolite 石棉胶布板
asbestos textile fabric 石棉纺织品
asbestos thread 石棉线
asbestos tile 石棉瓦;石棉板瓦
asbestos tile roofing 石棉瓦屋顶
asbestos tile works 石棉瓦厂
asbestos tube 石棉管
asbestos vein of slip fibre 纵相位石棉脉
asbestos-veneer plywood 石棉饰面胶合板;石棉夹心胶合板
asbestos-vinyl comound 石棉乙烯制品
asbestos-vinyl composition 石棉乙烯制品
asbestos-vinyl floor cover(ing) 石棉乙烯楼板面层;石棉乙烯地板面层;柔性地板复面层
asbestos-vinyl floor(ing) finish 柔性楼面整修;柔性楼板整修
asbestos-vinyl mass 石棉—乙烯(制品)材料
asbestos-vinyl material 石棉—乙烯(制品)材料
asbestos-vinyl tile 石棉乙烯饰面砖;石棉乙烯饰面板
asbestos wadding 石棉絮;石棉填料
asbestos wallboard 石棉墙板;高级石棉板;高级石棉墙板;高档石棉镶板
asbestos wall sheet 石棉墙板
asbestos wart 石棉疣
asbestos washer 石棉缓冲垫;石棉垫圈
asbestos waterproof felt 防水石棉毡
asbestos waterproofing 石棉防水
asbestos waterproofing cloth 石棉防水布
asbestos waterproofing slab 石棉防水板
asbestos wire 石棉线
asbestos wire gauze 石棉衬网
asbestos wood 防火石棉板;石棉墙板
asbestos wool 石棉绒
asbestos woven fabric 石棉织物;石棉布
asbestos yarn 石棉绒;石棉线
asbestous 石棉的
asbestumen 石棉沥青
asbestus 石绒;石棉
asbolane 钴土
asbolane ore 钴土矿石
asbolite 珠明料;钴土
as-brazed 硬钎焊态
as-built description 竣工说明
as-built drawings 竣工图
as-built drawing for settling accounts 结算竣工图;决算竣工图
as-built for settling the accounts 竣工图;结算图
as built profile 竣工断面
as-built return plan 竣工平面图

as-built return plan data 竣工平面图资料
as-built return plan information 竣工平面图资料
Ascalloy 阿斯卡罗伊铬钼耐热钢
ascanite 蒙脱石
ascarite 苏打石棉;烧碱石棉剂
as cast 铸造的
as-cast condition 铸态
as-cast finish 清水混凝土面;毛混凝土面
as-cast-finish concrete 清水面混凝土
as-cast metal 铸态金属
as-cast state 铸态
as-cast structure 铸态组织;铸态结构
ascend 上升;升高
ascendancy 优势
ascendant quantity 抬升量
ascending 起浮;上升顺序;向上的;递升
ascending ability 上升能力
ascending aestivation 上升花被卷迭式
ascending air 上升空气
ascending air current 上升气流
ascending angle 爬坡角;爬高角
ascending a rapid 上滩
ascending arrangement 升序排列
ascending barge 上行驳船
ascending belt 上行输电带
ascending boat 上行船(舶)
ascending branch 上行分支;上行段;上升段;向上分支
ascending buoy 救生浮标
ascending chain 升序列
ascending chromatogram 上行色谱图
ascending chromatography 上行色谱法
ascending current 上行气流;上升水流;上升流;涌升流
ascending curve 上升曲线
ascending degeneration 逆行性变性
ascending-descending chromatograph 升降色谱分离
ascending-descending chromatography 升降色谱(法);上下行色谱
ascending development 上行展开(法)
ascending development method 上行展开法
ascending eddy 上升漩涡;上升涡流
ascending face 上升割面
ascending flow 上升流;上升径流
ascending grade 上坡(度);升坡;上升坡度
ascending gradient 上坡道;上坡
ascending indicated rod 指示起浮的标尺
ascending inflorescence 上升花序
ascending key 升序标号;递升键
ascending kiln 阶梯窑;阶级窑
ascending method 爬高法;上行法
ascending ming 上行开采
ascending mode of satellite orbit 卫星轨道升交点
ascending motion 上升运动
ascending node 升交点;上升节点
ascending of reservoir head 库尾上翘
ascending order 顺序由小而大;升序(排列);递增次序;递升次序
ascending pipe 泵的压入管;上行管;注入管;直管;增压管
ascending posterior branch 后升支
ascending power 升力;上升功率;升幂;上升力
ascending power series 升幂级数
ascending sequence 升序序列;升序(排列)
ascending series 递增序列;递升级数
ascending sort 升序排序;递升排序;递升分类;比较分类
ascending spring 上升泉
ascending system 向上分类法
ascending tractive force 上坡牵引力
ascending trajectory 上升轨道
ascending tube 上行管;直管
ascending velocity 提升速度;上升速度;上升流速;上坡速度
ascending vertical angle 仰角
ascending vessel 上行船(舶);浮升船舶
ascending water 上升水
ascensional 上升的
ascensional air current 上升气流
ascensional difference 赤经差
ascensional velocity 上升速度;上升流速
ascensional ventilation 上行通风;上升通风
ascension force 升力

ascension pipe 上行管
ascension spring 上升泉
ascension theory 上升说
ascension verse 赤经共轭量
ascensor 竖式缆索铁路
ascent 爬高;升起
ascent calculation 上浮计算
ascent curve 上升曲线
ascent decent weighting system 上浮下潜重量系统
ascent of elevation 拔起高;克服高度
ascent rate 上浮速度
ascent resistance 坡道阻力;坡道附加阻力
ascent time 上浮时间
ascertain 探明;查明
ascertainability 可确定性
ascertainable 可查明的
ascertainable limit 可确定界限
ascertainable production 小批生产
ascertainable profit centre 可确定利润中心
ascertained weight 确认重量;确定的重量
ascertaining a ship's position 定船位
ascertaining well 探井(探查井深与井径的工序)【岩】
ascertain loss 确定损失
ascertainment 探知;探明
ascertainment of damage 确认损失;确定损失;查明损害情况
ascertainment of loss 确定损失
ascham 存放滑车及设备仓库
aschamalmite 阿硫铋铅矿
ascharite 纤硼镁石
ascharite porcelain 硼镁瓷
Ascheim-Zondek test 阿沙姆—聪德克试验
aschistic 非片状的
aschistite 未分异岩
Ascidiacea 海鞘目
ascidian 海鞘状动物
ASCII keyboard 美国信息交换标准(代)码健盘
as-cold 处于冷却状态的
as-cold reduced 冷轧成的
Ascoli's reaction 阿斯科利反应
Ascoli's test 阿斯科利试验
Ascoloy 阿斯科洛伊高温合金;阿斯科洛伊镍合金钢;镍铬铁(防锈)合金
as-completed drawing(setting the accounts) 竣工图;结算(竣工)图
ascon grade 单沟型
as-constructed density 建成时的密实度
as-constructed drawings 竣工图
as-construction drawings 竣工图
as contracted 按照合同
ascribe 归于
Ascu 铜砷铬防腐剂
as customary 照例;按照惯例
asdic 探器器;水下探测仪;回声定位声呐;超声波水下探测器;测潜艇仪
asdic bearing 水声方位
asdic control room 水声设备控制室
asdic gear 声呐;阿斯迪克水下声波探测仪;水下探测器
asdic method 水声法;声呐探测法;声呐法;超声波水下探测法
asdic range 声呐作用距离
asdic repeater 声呐复示器
asdic search 声呐搜索
asdic spotting 声呐定位
as-drawn 拉拔状态
as-dug gravel 原状砾石
asecertain 探查
a second time 再次
aseismatic 耐地震的
aseismatic design 抗震设计;耐震设计
aseismatic joint 抗震接头
aseismatic planning of city 城市抗震规划
aseismatic region 无地震区
aseismatic stability of bridge 桥梁抗震设计
aseismatic structure 抗震结构;耐震结构
aseismatic structure with shear wall 抗震剪力墙结构
aseismatic structure with shear wall core 抗震剪力墙筒体结构
aseismatic structure with shear wall frame interaction system 抗剪力墙框架相互作用体系

(的)抗震结构
aseismatic structure with shear wall system 抗剪力墙体系(的)抗震结构
aseismic 耐震的;耐地震的
aseismic base isolation 抗震用的基底隔振
aseismic bearing 抗震支座
aseismic brace 抗震支撑
aseismic calculation 抗震计算
aseismic capacity 抗震能力
aseismic code 抗震规范
aseismic computation 抗震计算
aseismic construction 抗震构造
aseismic decision analysis 抗震决策分析
aseismic deformation 无震形变
aseismic design 抗震设计;耐震设计;防震设计
aseismic design code 抗震设计规范
aseismic design decision analysis 抗震设计决策分析
aseismic detailing 抗震细部
aseismic engineering 抗震工程(学)
aseismic fault displacement 无震断层位移
aseismic front 无震前沿
aseismic ground deformation 无震地面变形
aseismic hardening 抗震加固
aseismic joint 抗震缝;防震缝
aseismic joint cover 抗震缝盖板
aseismic lake 无震湖
aseismic measures 抗震措施
aseismic plate 无震板块
aseismic policy 抗震政策
aseismic protection 抗震保护
aseismic protective measures 抗震措施
aseismic region 无震区;无地震区;非震区
aseismic requirement 抗震要求
aseismic ridge 无震海岭;无震海脊
aseismic rise 无震隆起
aseismic safety 抗震安全性
aseismic shear wall 抗震剪力墙
aseismic slip 无震滑动
aseismic stability 抗震稳定性
aseismic stability of bridge 桥梁抗震设计
aseismic strength 抗震强度
aseismic strengthening 抗震加固
aseismic structure 抗震结构;抗震建筑;抗震构造
aseismic support 抗震支承
aseismic technique 抗震技术
aseismic test 抗震试验
aseismic upgrading 抗震加固
aseismic wall 抗震墙
Asellus aquatics 水豚虱;木节水虱
asepsis 无菌
aseptic 无菌的;起净化作用(的);无菌性;防腐剂
aseptic culture 无菌培养
aseptic distillation 防腐蒸馏
aseptic filling 无菌填充
aseptic filtration 无菌过滤;除菌过滤
aseptic hole-towel 无菌洞巾
aseptic manipulation 无菌操作法
aseptic manipulation cabinet 无菌操作柜
aseptic manipulation room 无菌操作室
aseptic production 无菌生产
aseptic region 无地震地
aseptic technics 防腐技术
aseptic technique 无菌技术;防腐技术
a series of 一系列
a set 一套
a set of apparatus for road making materials 路面材料测试仪
a set of sieves 一套筛子
as-extruded 挤出来的
as fast as possible 尽快交货
as fast as steamer can deliver 按船舶交货速度
as fertilizer sources 作为肥料
as fired basis met calorific value graduation 应用基低位发热量分级
as fires basis 应用基
as-forged 黑皮锻件;锻压状态
as-forged state 锻后状态
as found 全数好质量;凭样(品)销售;凭样(品)买卖
asgrown crystal 刚长成的晶体
ash 水曲柳;灰烬;灰尘;白蜡树
ash absorption 灰吸收
ash absorption rate 灰吸收率
ash agglomeration 灰渣团聚

ash air 含灰空气
ash analysis 灰分分析;灰的分析
ashanite 阿山矿
Ashanti-Goldfields Corporation Ltd. 阿散蒂金矿公司
A-shaped pylon A 形塔架
A-shaped pylon inward leaning legs 柱腿内倾的 A 形塔架
A-shaped pylon with inward leaning legs 内倾柱 A 形塔架
a sharp rebound in price 价格猛涨
ash attack 灰渣侵蚀
ash barge 装煤灰船驳
ash basement 出灰间
ash bed 凝灰岩层;火山灰(层)
ash-bed diabase 凝灰辉绿岩
ash bin 出煤灰桶;出灰桶;垃圾桶;灰箱;灰坑;灰仓
ash boat 煤渣船;装灰船
ash box 灰箱
ash box door 灰箱门
ash-box trap-door 炉灰箱卸渣门
ash breaker 碎渣机
ash breeze 平静海面;无风
ash bucket 灰斗
ash bunker 灰斗;灰槽;灰仓
ashburton marble 硬石灰石
Ashbury metal 阿什贝利锡合金
ash-can 垃圾桶
ash car 清洁车;土车;灰车
ash cart 垃圾车
ash cellar 灰坑;灰槽
ash channel 出灰道;灰道
ash characteristic curve 灰分特性曲线
ash chute 灰槽
ash coal 高灰分煤;多灰分煤
ash collector 集尘室;除尘器
ash composition 灰分组成;灰的成分
ash concrete 炉灰混凝土;煤灰混凝土;灰渣混凝土
ash cone 凝灰火山锥;火山灰锥
ash constituent 灰分
ash content 灰分(含量);含灰量
ash content graduation 灰分产率分级
ash content test 灰分含量测定
ash conveyer 灰渣输送机;灰渣输送带;出灰输送机;炉渣运输带;运灰机;出货输送机;出灰运输机
Ashcroft cell 阿希克罗夫特电池
ashcroftine 钾杆沸石;硅碱钙钇石
ash crusher 碎渣机
ash crusher machine 碾灰机
ash determination 灰分测定
ash discharge 出灰
ash discharge hopper 排灰斗;排灰仓
ash discharging gear 出煤灰装置;出灰器
ash disposal 除灰
ash disposal area 储[贮]灰场;除灰场
ash door 灰门
ash drop 炉灰道
ash dump 卸灰口;除灰翻斗;灰堆
ashed 灰化了的
ashed or mulch 草木灰或覆盖
ashed sample 灰化样品
ash ejector 排灰器;煤灰吹除器;吹灰器;出灰器
ashen 灰色的;灰白色的
ashen-grey soil 灰色土
ash entering the clinker 进入熟料的煤灰
ash erosion 炉内结渣
ashery 烧灰场;堆灰场
ashes of roses 灰青红
ash exhauster 排灰装置;排灰器
ash expeller 除灰器
ash fall 炭尘沉降;火山灰下降;火山灰沉降(层);灰尘沉降;烟灰降落;烟尘降落
ash field 火山灰原
ash fire 慢火;余烬
ash flow 火山灰流
ash flow tuff 火山灰流凝灰岩
ash formation 灰渣层;灰层
ash-forming impurity 成灰杂质
ash-free 不含灰的;无灰的
ash-free basis 无灰基;除灰计算
ash-free coal 不含灰分煤;无灰煤;低灰煤
ash-free fuel 无灰燃料
ash from incineration 烧却灰
ash fusibility 灰分熔度

ash fusion temperature 灰溶温度;灰分熔点
Ashgillian 阿石极阶【地】
ash glaze 灰釉
ash gray 灰白色;淡灰色的
ash-gray soil 灰化土
ash gun 排煤渣灰器;排灰器;吹灰枪
ash-handling 除灰;灰尘处理;灰分处理;粉尘处理;尘灰处理
ash-handling device 除尘装置
ash-handling equipment 除灰装置;除尘装置
ash-handling pump 粉尘处理泵;除灰泵;灰尘处理泵
ash-handling system 灰控制系统;除灰系统
ash hoist 起灰装置;煤灰吊车
ash hoist engine 起灰机
ash hole 出灰孔;灰坑;出灰口;出灰洞
ash hopper 底卸式灰斗;灰渣斗;灰斗
ashing 除灰;成灰;推光;湿式粉抛光;灰磨;灰化
ashipboard 在船内
a shipper's option 由托运人选择
ash-laden gas 含灰烟气;含尘烟气;含尘气体
ash-lagoon 灰分处理池;灰池
ashlar 石板;琢石;琢方石
ashlar arch bridge 料石拱桥
ashlar bond 条石砌合;方石砌合;琢石砌合
ashlar bonder 琢石砌体露头石;琢石砌体露头锚石;琢石砌体露头顶石;琢石砌体砌合丁砖
ashlar bonding header 琢石砌体露头石;琢石砌体露头锚石;琢石砌体露头顶石
ashlar bond stone 琢石束石
ashlar brick 石面砖;面砖;琢石砖;琢面砖;仿石砖;琢细面砖
ashlar buttress 琢石扶壁
ashlar-faced rubble wall 琢石面毛石墙
ashlar facing 黏土、油页岩或天然石块墙面;方石镶面;琢石镶面;琢石贴面;琢石饰面;琢石面面;琢面饰面
ashlar finish 琢石饰面
ashlaring 琢石镶面;砌琢石墙;砌琢石;贴琢石;用琢石砌面
ashlar joint 琢石接缝
ashlar line 石砌外墙面线;琢石线;外墙面线
ashlar masonry 方石砌筑工程;石砌圬工;琢石砌体
ashlar masonry work 琢石砌筑;条石砌筑;琢石砌体建筑
ashlar pavement 琢石铺装;琢石路面;琢石层
ashlar paving 琢石路面;方石路面
ashlar piece 石板;石片
ashlar pier 琢石柱;琢石桥墩;琢石墩
ashlar slab 琢石(石)板
ashlar structure 琢石结构;琢石建筑物
ashlar vault 琢石拱顶;琢石穹隆
ashlar veneer 琢石饰面;琢石板
ashlar wall 琢石墙
ashlar walling 砌琢石墙面
ashlar window 琢石窗
ashlar work 琢石工程
ash layer 灰层
ashler 琢石(砌体);细方石;(屋顶层中楼板至屋檐的)立柱;方石;顶棚短吊柱;粗陶砌块;(琢石墙中的)板墙筋
ashlering 琢石;贴墙面;砌琢石;板墙筋;阁楼立柱砌琢石;阁楼主柱;阁楼立柱
ashless 无灰的
ashless filter paper 无灰滤纸
Ashley type groin [groyne] 艾什莱型丁坝;艾什莱式不透水丁坝(新西兰的一种丁坝)
ash lift 煤灰提升机;炉灰提升机;提灰机
ash light 灰光(灯)
ash lighter 煤渣船;装煤灰船驳
ash-lime block 灰渣石灰砖(砌)块;粉煤灰—石灰砌块
ash-lime tile 粉煤灰—石灰面砖
ash lock 排灰锁斗
ash loss 煤灰损失;灰分损失
ashman 清灰工
Ashmolean Museum 阿斯麻林博物馆
ash mortar 毛石砂浆
Ashmouni cotton 阿什穆尼棉
ash on dry coal 干煤灰分
ashore 向岸上;在岸上;到岸上;被冲上岸
ashore navigation centre 陆上导航中心
a short while 片刻
ash paint 灰油涂料

ashpan 灰盆;灰盘;灰盒;炉灰膛
ash pan damper 炉灰挡板;灰盘调节板
ash pan door 灰坑门;除灰门
ash picker 灰扒
ash pipe 出灰管;灰管
ashpit 灰坑;储灰槽;灰仓;除渣井;灰池
ashpit damper 灰坑调节板
ashpit door 煤渣池门;煤渣池挡板;灰坑门;除渣门;出灰门
ash pit road 卸灰道(蒸汽机车的)
ashpit track 灰坑线
ash plain 火山灰原
ash pocket 聚灰点
ash pond 灰沉淀池
ash pool ash pond 灰池
ash production rate of coal 煤的灰分产率
ash pump 灰浆泵;除灰泵
ash pumping 水力除灰(法)
Ashraf-Lotfi's Suggestion 阿什拉夫—卢特菲建议
ash rain 火山灰沉降
ash receptacle 烟灰盒
ash removal 清灰;脱灰;除灰
ash removal crane 除灰吊车
ash removal from raw coal 煤炭脱灰
ash removal system 除灰系统
ash removal system by slurry pump 水力除灰系统
ash-rich fuel 多灰燃料;多灰分燃料
ash ring 前结圈;灰圈
ash rock 凝灰岩;火山灰岩
ash screen 滤灰尘网
ash separator 除灰器;除尘器
ash shoot 出渣槽;出灰槽
ash shower 火山灰下降;火山灰沉降
ash slate 灰岩板(火山);凝灰质灰岩;灰板岩
ash sluice 水流输渣道;水流输灰道;灰槽
ash sluice gate 出灰闸门
ash sluice way 水力冲灰沟
ash-sluicing water 冲淡灰水;冲灰排水
ash specification 灰分规格;灰分
ashstone 凝灰岩;火山灰岩
ash structure 火山灰结构;火山灰构造
ash test 灰分试验
ash tile 灰渣石灰(砌)块
ashtonite 发光沸石
ash transport water 冲灰水
ash tray 烟灰盒;烟灰碟;煤灰缸;炉盆;灰盆;灰盘;烟灰缸
ashtree 水曲柳
ash trench system 水力除灰系统
ash-tuff 火山凝灰岩;火山灰凝灰岩;灰质凝灰岩
ash valve 排灰阀
ash washing 洗灰
ash water 灰水
ash water glass 灰水玻璃
ash water pump 灰水泵;冲洗泵;冲灰泵
ash whip 吊灰桶绞辘
ash with high temperature 煤的高温灰
ash with low temperature 煤的低温灰
ash wood 木炭灰
ashy 似灰;灰色的
ashy grit 火山灰砂砾
ash yield of cleaned coal 精煤灰分产率
ash yield of raw coal 原煤灰分产率
ashy shale 凝灰质页岩;灰页岩;火山灰泥板岩
ashy substance 似灰物质;含灰物质
ash zone 灰层
Asia-Europe seismotectonic map 亚欧地震构造图
Asian-African Legal Consultative Committee 亚非法律咨询委员会
Asian Association of Management Organization 亚洲管理组织协会
Asian-Australian-Pacific realm 亚澳太平洋大区系
Asian Clearing Union 亚洲清算联盟
Asian Development Bank 亚州开发银行
Asian Development Center 亚州开发中心
Asian Development Fund 亚州开发基金
Asian Development Institute 亚州开发研究所
Asian golden cat (东南亚产的)金猫
Asian Industrial Development Council 亚洲工业发展理事会
Asian-Pacific Council 亚洲太平洋理事会
Asian Payment Union 亚洲清算同盟
Asian Productivity Organization 亚洲生产力组织

Asian Rice Trade Fund 亚洲大米贸易基金
Asian white birch 桦树
Asia-Pacific economic circle 亚太经济圈
Asia-Pacific International Trade Fair 亚太地区国际贸易博览会
Asia Reinsurance Company 亚洲再保险公司
asiatic acid 积雪草酸
Asiatic base 亚洲式柱座
Asiatic clam 河蚬
Asiatic closet 亚洲式厕所;蹲式大便器
asiatic relapsing fever 亚洲回归热
Asiatic sweetleaf 白檀
Asiatic WC pan 蹲式便桶;蹲式便池
asiderite 无铁陨石;陨石
asif 干河谷
as inspected 依据验货付款
asisculis 石匠锤
a six loaned nut 六角螺母;六角螺帽
Askania 液压自动控制装置
Askania gravimeter 阿斯卡尼亚重力仪
Askania optical tracker 阿斯卡尼亚光学跟踪器
Askania pressure regulator 阿斯卡尼亚压力控制器
Askania sea gravimeter 阿斯卡尼亚海洋重力仪
Askania theodolite camera 阿斯卡尼亚经纬仪摄影机
askanite 蒙脱石
askanite clay 蒙脱土
Askarel 爱斯开勒电介液体
asked 卖方要价
asked price 叫价;卖主的开价;卖价;卖方报价;要价
asked quotation 卖价
askew arch 歪曲拱;斜拱
askew bridge 斜桥
ask for a loan 借债;借款
ask for an advance on one's pay 借支
ask for a ruling 要求作出裁决
ask for loan 告贷
ask for payment 催付款
ask for samples 索样
asking for correction 请求更正
asking price 讨价;卖主的开叫价;卖价;卖方报价
ask price 卖方开价
ask the price 询价
ask to make a bid 要求递价
aslant 倾斜地;成斜角地
asleep 瘦风帆
a slight heel 轻微倾侧
a slip of the pen 笔误
aslope 倾斜地
as low as possible 尽可能低
as low as reasonable achievable principle 合理可行尽量低原则
as low as reasonably achievable 可以合理做到的最低水平
as maintained 依据验货付款
a small area of ground 一小块地
a small circle of the evolutive 小周天
as many as 多至
asmission ticket 门票
as-mixed concrete 新鲜混凝土;新拌混凝土
as of date 截止日期
Asohoner's reflex 阿施内反射
aso lava 阿苏熔岩
Asonkobi 阿桑科比野蚕丝
A-spacer A 形钢筋定位件
asparagolite 黄绿磷灰石
asparagus fern 文竹
asparagus stone 黄绿磷灰石
aspartame 阿斯巴特
aspartic acid 天(门)冬氨酸;丁氨二酸;氨基丁二酸
aspatria (产于英国坎伯兰郡的)暗红色砂岩
aspect 物标舷角;视方位;色灯显示;季相(变化);信号方式;侧面;反弦角
aspect angle 视线角;视界角(度);扫描角
aspect camera 空间稳定摄影机
aspect card 特征卡(片);标号卡片
aspect card system 标号卡片系统
aspect indicator 物标反舷角指示器
aspection 季相(变化);周期变化
aspect of approach 目标地貌
aspect of drainage basin 流域形态
aspect of slope 坡向
aspect photocell 方位光电管
aspect ratio 图像纵横比;图像宽高比;宽深比;宽高比;宽长比;画面比;型数比;形状比;形态比;影像比;纵横比;展弦比;高度比;高度与直径之比;长宽比(率);长径度
aspect ratio of blade 桨叶纵横比
aspect ratio of bridge 桥梁宽跨比
aspect ratio of fibers 纤维形状比
aspect ratio of weld 焊缝成形系数
aspect sensor 方位传感器
aspect source flags 特征源标记
aspect system 标号系统
aspen 白杨(树)
aspen wood 山杨木
as per advice 依照通知
as per charter party 按租船合同规定
aspergillic acid 曲霉酸
asperity 粗糙性;粗糙度;不平滑
asperity contact 粗糙面接触
asperity ratio 粗糙度比
asperity-surface interaction 粗面与平面相互作用
as per list 按照表列
asperromagnetism 散反铁磁性
as per sample 见样品;与样品相符;按样品
asperses 有粗点的
aspersorium 圣水器
asphalite 沥青矿
asphalt 涂柏油;沥青;地沥青
asphalt addition 沥青掺配;沥青掺和;地沥青掺和;石油沥青掺和剂
asphalt adhesion-preventing agent 沥青防黏剂
asphalt adhesive 沥青胶合剂;沥青胶黏剂
asphalt adhesive compound 地沥青胶黏剂
asphalt aggregate Marshall analysis system 沥青集料马歇尔试验分析系统;沥青骨料马歇尔试验分析系统
asphalt-aggregate mixture 石油沥青骨料混合料;沥青集料混合物;沥青集料混合料;沥青集料拌和物;沥青骨料混合物;沥青骨料混合料;沥青骨料拌和料;地沥青集料混合料;地沥青骨料混合料
asphalt-aggregate ratio 沥青集料比;沥青骨料比;油石比【道】
asphalt alligator crack 沥青龟裂缝
asphalt and coated macadam finisher 沥青与黑色碎石路面整修机
asphalt and coated macadam finisher without hopper 不带料斗沥青与黑色碎石路面整修机
asphalt and coated macadam mixing plant 沥青与黑色碎石路面拌和厂
asphalt and tar (melting) kettle 沥青及煤焦油熔锅;沥青与煤焦油熔锅
asphalt-asbestos 沥青石棉黏膏;沥青石棉玛琋脂;沥青石棉胶泥
asphalt-asbestos composition 沥青石棉混合物;沥青石棉拌和料
asphalt-asbestos compound 沥青石棉混合物;沥青石棉拌和料
asphalt-asbestos fiber cement 沥青石棉纤维水泥
asphalt-asbestos material 沥青石棉材料
asphalt-asbestos roof coating 沥青石棉屋面涂料
asphalt-asbestos sealant 沥青石棉嵌缝膏
asphalt-asbestos sheeting 沥青石棉瓦
asphalt-asbestos tile 沥青石棉砖
asphalt-base 沥青打底的;沥青基
asphalt-base coat(ing) 沥青基层涂层;沥青底涂层
asphalt-base course 石油沥青底层;沥青基底层;沥青底层
asphalt-base crude oil 沥青地原油
asphalt-based rust protective paint 沥青基防锈漆;沥青防锈涂料
asphalt-base emulsion 沥青基乳液
asphalt-base mastic joint 沥青基玛琋脂填缝料
asphalt-base oil 地沥青基石油;沥青冷底子油;石油沥青冷底子油;沥青底油
asphalt-base paint 沥青基油漆;沥青基涂料;沥青打底子涂料
asphalt-base paste 地沥青基涂料;沥青基涂料
asphalt-base waterproofing paint 沥青基防水涂料
asphalt batcher 沥青配料计量器
asphalt batching pump 沥青配料泵;沥青装料泵
asphalt bearing 含沥青的;含沥青的
asphalt bearing petroleum 含沥青石油;含地沥青石油
asphalt bearing residue 含沥青残渣;含地沥青残渣
asphalt bearing shale 含沥青页岩;含地沥青页岩
asphalt bearing stock 含沥青原料;含沥青油料
asphalt bend test 沥青柔性试验
asphalt binder 石油沥青基料;沥青粘合料;沥青结合料;地沥青结合料
asphalt binder course 沥青连结层;沥青结合层
asphalt block 沥青砌块;地沥青块;沥青块材;沥青混凝土块材
asphalt block pavement 沥青块铺砌路面;沥青块路面;浸沥青木块铺面;地沥青块铺砌路面;沥青砌块路面
asphalt block surface 地沥青预制块面层
asphalt blowing 沥青空气氧化;沥青吹制
asphalt blowing catalyst 沥青氧化催化剂
asphalt board 防潮纸板
asphalt board strip 沥青填缝条
asphalt-bonded 沥青胶结的;地沥青胶结的
asphalt-bonded felt 沥青黏毡
asphalt bonding adhesive 沥青胶结料;沥青胶黏剂
asphalt bonding agent 沥青胶结料
asphalt bonding medium 沥青胶黏料
asphalt bottom 沥青渣油;地沥青渣油
asphalt brick 沥青砖
asphalt bridge carriageway pavement 桥梁车行道沥青铺面层
asphalt briquet(te) 沥青试块;沥青标准试块;标准沥青试块
asphalt brown 褐色沥青颜料
asphalt bucket 沥青铲斗
asphalt building paper 建筑用沥青油毛毡;沥青油毡
asphalt built-up roof(ing) 地沥青组合屋面;沥青组合屋面
asphalt cake 沥青砂胶块;地沥青砂胶块
asphalt calking compound 沥青防水油膏
asphalt canal lining 沥青渠道村砌;渠道沥青衬砌
asphalt cap sheet 沥青面层油毡
asphalt carpet 沥青油毡层;沥青毡层;沥青铺面层;沥青覆盖层
asphalt carpet coat 沥青油毡层;地沥青毡层
asphalt carrier 沥青运载船;沥青运输船
asphalt cauldron 沥青熔锅
asphalt cement 沥青胶结料;沥青膏;沥青水泥;沥青冷底子油;沥青胶(泥);黏结地沥青(美国);膏体地沥青;地沥青胶结料
asphalt cement admixture 沥青水泥混合物
asphalt cemented ballast bed 沥青道床
asphalt cement(ing) agent 沥青里底子油膏
asphalt cement mortar 沥青水泥砂浆
asphalt cement with road tar addition 含筑路焦油的沥青胶泥
asphalt chip(ping)s 沥青石屑
asphalt clinker 地沥青熔渣;石油沥青熔渣;沥青熔渣
asphalt coal 石油沥青煤;沥青煤;地沥青煤
asphalt coat 沥青涂层
asphalt-coated 涂沥青的
asphalt-coated aggregate 涂沥青的骨料;沥青拌制的集料;沥青拌制的骨料;地沥青拌制的集料;地沥青拌制的骨料;用沥青拌制的骨料
asphalt-coated ballast 沥青拌制的道砟
asphalt-coated chip(ping)s 涂沥青的铺路石屑;涂沥青的铺路片石
asphalt-coated chip(ping)s carpet 沥青拌石屑面层;涂沥青片石铺面层
asphalt-coated glass fiber venting base sheet 沥青玻纤排气基层油毡
asphalt-coated gravel 涂沥青砾石
asphalt-coated metal pipe 涂沥青金属管
asphalt-coated pasteboard 沥青胶纸板;涂沥青硬纸板
asphalt-coated sand 沥青拌制砂;涂沥青砂
asphalt coating 沥青面(层);沥青涂层
asphalt coating compound 沥青胶涂布料
asphalt coating material 沥青覆盖胶料;沥青涂料
asphalt coating roof(ing) 涂沥青的屋面
asphalt colo(u)r coat 沥青色面层;沥青彩色面层;有色沥青封层;有色地沥青面层;彩色地沥青封层
asphalt composition 沥青组成;沥青成分;地沥青组成;沥青拌和料
asphalt composition roofing 地沥青屋面油毛毡;沥青屋面油毡
asphalt compound 沥青混合料;沥青化合物;复合地沥青
asphalt concrete 沥青混凝土;石油沥青混凝土;地

asphalt concrete base 沥青混凝土防水层;沥青混凝土垫层;沥青混凝土底层;沥青混凝土基层
asphalt concrete carpet 沥青混凝土(铺)面层
asphalt concrete core 沥青混凝土心墙
asphalt concrete core dam 沥青混凝土心墙坝
asphalt concrete mineral skeleton 沥青混凝土矿物骨架
asphalt concrete mixer 沥青混凝土拌和机
asphalt concrete mixing plant 沥青混凝土拌和设备
asphalt concrete mixture 沥青混凝土配合料;沥青混凝土混合料;沥青混凝土料
asphalt concrete pavement 沥青混凝土路面;地沥青混凝土路面
asphalt concrete plant 沥青混凝土拌和厂;沥青混凝土拌和机
asphalt concrete surface 沥青混凝土地面
asphalt concrete surfacing 沥青混凝土面层
asphalt content 沥青含量;地沥青含量
asphalt cooker 沥青加热锅;地沥青加热锅
asphalt cord filler 沥青麻绳填料
asphalt cork surface 沥青毡层;沥青软木贴面
asphalt cover(ing) 沥青盖层;沥青罩面;沥青涂层;沥青砂浆面层;沥青铺面;沥青面层;地沥青盖层
asphalt curb 沥青路缘(石);地沥青路缘石
asphalt cushion course 沥青垫层
asphalt cutback 溶于石油馏出油的沥青;低熔点沥青混合物;轻质石油沥青;轻制沥青;轻制地沥青
asphalt cutter 沥青面层凿缝斧;沥青面层切割器;石油沥青面层切割机
asphalt damp course 沥青防水层
asphalt damp-proof course 石油沥青防潮层;沥青防潮层
asphalt deck pavement 桥面沥青铺面层
asphalt deposit 沥青沉积;沥青矿;地沥青沉积
asphalt dip 沥青浸液
asphalt dispersion 沥青分散液
asphalt distributor 石油沥青喷洒机;路油洒布机;沥青洒布机;沥青洒布车;沥青撒布机;沥青铺洒机;沥青铺路机;沥青洒布机;沥青喷布机;地沥青洒布机
asphalt dry penetration surfacing 干贯式(地)沥青碎石路面;表面经沥青处理的碎石
asphalt dry process penetration macadam 干贯式(地)沥青碎石路面
asphalt ductility testing machine 沥青延性试验仪;沥青延(伸)度试验仪;地沥青延度试验仪
asphalt ductilometer 沥青延(伸)度仪
asphalt earth pavement 沥青土路面
asphalted cast iron pipe 沥青涂层铸铁管
asphalted paper 沥青纸
asphalted pipe 涂沥青管(子)
asphalted tube 涂沥青管(子)
asphalt emulsion 沥青乳胶;沥青乳(浊)液;地沥青乳液
asphalt emulsion slurry seal 乳化沥青密封膏
asphalt emulsion system 乳化沥青系统;乳化沥青设备
asphaltene 沥青烯;沥青精;地沥青质
asphaltene of asphalt 沥青的沥青质;地沥青的沥青质
asphaltenic concentrate 沥青质浓缩物
asphalter 沥青摊铺机;沥青铺路机;沥青工人
asphaltex 多种颜色的地沥青塑性屋面材料
asphalt expansion joint 沥青伸缩缝
asphalt facing 沥青涂面层
asphalt felt 石油沥青毡;沥青毡;沥青油毛毡;沥青浸渍毡;油毛毡
asphalt felt coat 沥青护面;沥青毡层
asphalt-felt-grade talc 油毡级滑石
asphalt felt roof 油毛毡屋面;沥青毡屋面
asphalt felt roof cladding 沥青毡屋面覆盖层
asphalt felt roof covering 屋顶覆面油毛毡
asphalt felt with punched holes 带孔油毡
asphalt filler 沥青填缝料;石油沥青填料;地沥青填缝料
asphalt-filler mix(ture) 沥青填料混合物;沥青填料混凝土
asphalt fillet 沥青填角
asphalt film 沥青膜;地沥青膜;沥青薄膜
asphalt finisher 沥青整修机;沥青熨平机;沥青修整机;沥青路面整修机;沥青滚平机;地沥青整修机
asphalt firepipe melter 火管式沥青溶化装置

asphalt fish oil ca(u)lk 沥青鱼油油膏
asphalt flashing cement 涂泛水用沥青胶结材料;沥青泛水胶泥
asphalt floor cover(ing) 沥青楼面;沥青地面
asphalt floor cover(ing) tile 沥青地板砖;沥青楼面砖;沥青地面砖
asphalt floor finish 沥青楼面层;沥青地面层;沥青地面
asphalt floor(ing) 沥青楼面;沥青地面;石油沥青地面;沥青楼地面
asphalt floor tile 沥青地面砖
asphalt flow meter 沥青流量计
asphalt flux 沥青助熔剂;沥青稀释剂;沥青软化油;残渣油沥青
asphalt fluxing oil 地沥青溶剂;沥青稀释油;沥青软制油;沥青溶剂;地沥青软制油
asphalt fog 沥青雾
asphalt fog coat 沥青雾层;地沥青雾层(极薄的表面处治层,不加撒石屑);薄沥青面层
asphalt fog seal 沥青雾密封;沥青喷雾密封
asphalt for damp-proof coatings 地沥青防潮涂层;沥青防潮涂层
asphalt fork 沥青耙
asphalt-free oil 不含沥青油料
asphalt fuel oil 地沥青基燃油;沥青基燃油
asphalt furnace 沥青炉
asphalt grade 沥青种类;沥青等级
asphalt gravel 沥青拌砾石;沥青砾石
asphalt-gravel roofing 沥青拌碎石屋面;沥青碎石屋面
asphalt groove joint 沥青槽接
asphalt ground 沥青地坪;沥青地面
asphalt-grounded surfacing 沥青灌浆碎石路(面)
asphalt grout 石油沥青灌注料;沥青砂胶;沥青胶泥;沥青浆;地沥青砂胶
asphalt-grouted 沥青灌浆的
asphalt-grouted macadam 灌沥青碎石路;沥青灌浆碎石
asphalt-grouted stone 沥青灌浆块石
asphalt-grouted surfacing 地沥青贯碎石面层;地沥青贯碎石路
asphalt grouting 石油沥青灌浆;沥青灌浆;地沥青灌浆
asphalt gun 沥青喷射器;沥青喷枪
asphalt gunite 沥青喷浆
asphalt gunite equipment 沥青喷浆装置;沥青喷浆机具
asphalt gunite process 气动沥青喷浆法;沥青喷浆法
asphalt gutter 沥青面(排)水沟;沥青面天沟
asphalt hardness(degree) 沥青硬度
asphalt heater 石油沥青加热锅;沥青加热器;地沥青加热器;柏油锅
asphalt heating kettle 沥青加热器;沥青加热锅;地沥青加热锅
asphalt heating plant 地沥青加热厂
asphalt heating pot 沥青锅
asphalt highway 沥青公路
asphalt highway emulsion 公路用乳化沥青;道路用乳化沥青;沥青公路用乳化沥青
asphalt hot-mix recycling 沥青热拌再生利用
asphalt hot oil meter 导热沥青熔化装置
asphaltic acid 石油沥青砖;沥青酸;地沥青酸
asphaltic acid anhydride 沥青酸酐;地沥青酸酐
asphaltic adhesive 沥青黏结剂;沥青胶合料
asphalt(ic) adhesive mixture 沥青胶黏剂混合料
asphaltic base(course) 沥青基层;沥青底层
asphaltic base crude 沥青基原油
asphaltic base crude oil 沥青基原油
asphaltic base crude petroleum 沥青基原油
asphaltic base oil 沥青冷底子油;沥青基石油
asphaltic base petroleum 沥青基石油;地沥青基石油
asphaltic belt 油毡
asphaltic binder 沥青结合料
asphaltic binder course 沥青联结层;沥青结合料(路基)层
asphaltic bitumen 天然沥青;沥青结合料;沥青胶;地沥青(英国)
asphaltic-bitumen addition 沥青掺和料
asphaltic-bitumen adhesion preventing agent 沥青防黏剂
asphaltic-bitumen aggregate 沥青拌料;沥青拌骨料;沥青集料;沥青骨料

asphaltic-bitumen aggregate mixture 沥青集料混合物;沥青集料混合料;沥青骨料混合物;沥青骨料混合料
asphaltic-bitumen asbestos composition 沥青—石棉纤维复合物;沥青石棉混合料
asphaltic-bitumen asbestos compound 沥青石棉混合料
asphaltic-bitumen asbestos fiber cement 沥青石棉纤维胶泥
asphalt(ic-bitumen) asbestos mastic 沥青石棉玛琋脂;沥青—石棉砂胶
asphaltic-bitumen base 沥青基;沥青打底
asphaltic-bitumen based 沥青基的;沥青打底的
asphaltic-bitumen-based building mastic 建筑用沥青基玛琋脂
asphaltic-bitumen-based mastic joint sealer 沥青基玛琋脂嵌缝料
asphaltic-bitumen-based paste 沥青基黏膏
asphaltic-bitumen-based rust protection paint 沥青基防锈漆;沥青防锈层
asphaltic-bitumen-based rust protective paint 沥青基防锈漆;沥青防锈涂层
asphaltic-bitumen batching pump 沥青装料泵;沥青配料泵
asphaltic-bitumen binder 沥青胶结料
asphaltic-bitumen blanket 沥青铺(面)层
asphaltic-bitumen bonding adhesive 沥青胶黏剂
asphaltic-bitumen bonding compound 沥青胶合料
asphaltic-bitumen briquet(te) 沥青试块;标准沥青试块
asphaltic-bitumen bucket 沥青料斗
asphaltic-bitumen building paper 建筑用沥青油纸
asphaltic-bitumen calcium silicate brick 沥青硅酸钙砖
asphaltic-bitumen carpet 沥青油毡(层);沥青铺面层
asphaltic-bitumen coated 涂沥青的
asphaltic-bitumen-coated chip(ping)s carpet 涂沥青碎石磨耗层;涂沥青石屑磨耗层
asphaltic-bitumen-coated gravel 沥青拌制砾石
asphaltic-bitumen-coated(road)metal 涂沥青碎石(筑路用)
asphaltic-bitumen-coated sand 沥青拌制砂
asphaltic-bitumen coat(ing) 沥青涂层
asphaltic-bitumen concrete 沥青混凝土
asphaltic-bitumen concrete base 沥青混凝土(路)基层
asphaltic-bitumen concrete carpet 沥青混凝土铺面层
asphaltic-bitumen concrete mineral skeleton 沥青混凝土矿物质骨架
asphaltic-bitumen concrete pavement 沥青混凝土路面
asphaltic-bitumen content 沥青含量
asphaltic-bitumen cooker 沥青加热器;地沥青加热器
asphaltic-bitumen deck surfacing 桥梁行车道沥青铺(面)层
asphaltic-bitumen dispersion 沥青分散剂
asphaltic-bitumen distributor 沥青洒布机
asphaltic-bitumen emulsion 沥青路用乳液
asphaltic-bitumen emulsion with colloidal emulsifier 含胶态乳化剂的沥青乳浊液
asphaltic-bitumen facing 沥青涂(面)层;沥青铺层
asphaltic-bitumen felt roof 沥青油毡屋面;沥青油毡屋面
asphaltic-bitumen filler mix(ture) 沥青填缝料;沥青混合填料
asphaltic-bitumen film 沥青薄膜
asphaltic-bitumen for damproof coating 沥青防潮层
asphaltic-bitumen for road purposes 筑路用沥青
asphaltic-bitumen globule (乳浊液中的)沥青珠;(乳浊液中的)沥青球
asphaltic-bitumen grade 沥青等级
asphaltic-bitumen gravel 沥青拌制砾石;沥青砾石
asphaltic-bitumen gravel roofing 沥青拌制砾石屋面
asphaltic-bitumen grouted macadam 灌沥青碎石路
asphaltic-bitumen gun 沥青乳液喷枪
asphaltic-bitumen gunite 沥青乳液喷浆枪

asphaltic-bitumen gunite process 沥青乳液喷浆法
asphaltic-bitumen gunite quipment 沥青乳液喷浆机
asphaltic-bitumen heater 沥青加热器;沥青加热炉
asphaltic-bitumen-impregnated 沥青乳液浸渍的
asphaltic-bitumen-impregnated calcium silicate brick 沥青浸渍的硅酸钙砖
asphaltic-bitumen-impregnated lime-sand brick 沥青浸渍的灰砂砖
asphaltic-bitumen-impregnated paper 沥青纸
asphaltic-bitumen-impregnated sand 沥青拌制砂
asphaltic-bitumen-impregnated strip 沥青浸渍条;沥青浸渍的卷材
asphaltic-bitumen injection control valve 沥青喷射控制阀
asphaltic-bitumen insulating coat 沥青绝缘涂层
asphaltic-bitumen insulating slab 沥青绝缘板
asphaltic-bitumen joint 沥青嵌缝;沥青接缝
asphaltic-bitumen joint pouring compound 沥青接头灌缝料
asphaltic-bitumen joint runner 沥青接头嵌缝绳;沥青接头嵌缝料;沥青封口油麻丝
asphaltic-bitumen laminated 沥青层压的;沥青叠层的
asphaltic-bitumen-latex emulsion 沥青橡浆乳剂
asphaltic-bitumen layer 铺沥青工人;沥青层
asphaltic-bitumen lime-sand brick 沥青灰砂砖
asphaltic-bitumen liner 沥青衬砌;沥青衬垫
asphaltic-bitumen macadam surfacing 沥青碎石路面层
asphaltic-bitumen measuring pump 沥青计量泵
asphaltic-bitumen melting boiler 沥青融锅
asphaltic-bitumen melting tank 沥青融罐
asphaltic-bitumen membrane 沥青膜
asphaltic-bitumen metering pump 沥青计量泵
asphaltic-bitumen mortar 沥青砂浆
asphaltic-bitumen pavement 沥青路面
asphaltic-bitumen pavement mix(ture) 沥青路面材料
asphaltic-bitumen penetration madacam 沥青灌注碎石(路面)
asphaltic-bitumen perforated sheet(ing) 沥青多孔板
asphaltic-bitumen pouring compound 沥青浇注复合料
asphaltic-bitumen pouring rope 沥青浇注嵌缝绳
asphaltic-bitumen pre-impregnating 沥青预浸渍处理
asphaltic-bitumen prepared roofing 沥青屋面料;沥青层防水油毡
asphaltic-bitumen pre-saturating 沥青预浸渍处理
asphaltic-bitumen pre-saturation 沥青预浸渍
asphaltic-bitumen primer 头道沥青;沥青打底涂层;沥青打底涂料
asphaltic-bitumen proportioning pump 沥青计量泵
asphaltic-bitumen protection coat(ing) 沥青保护(涂)层
asphaltic-bitumen pump 沥青泵
asphaltic-bitumen putty 沥青油灰;沥青腻子
asphaltic-bitumen rag felt 沥青油毛毡;沥青屋面毡;建筑用沥青粗制毡
asphaltic-bitumen ready roofing 沥青油毛毡;沥青屋面毡
asphaltic-bitumen road pavement 沥青路面
asphaltic-bitumen road surfacing 沥青路面
asphaltic-bitumen rolled-strip roofing 沥青屋面卷材
asphaltic-bitumen roofing cement 屋面用沥青胶泥
asphaltic-bitumen roofing felt 沥青屋面油毡
asphaltic-bitumen roof(ing) sheet 沥青屋面层;沥青屋面板
asphaltic-bitumen-rubber compound 沥青—橡胶复合料
asphaltic-bitumen-rubber mass 沥青橡胶块
asphaltic-bitumen-rubber material 沥青橡胶材料
asphaltic-bitumen-rubber strip 沥青橡胶条
asphaltic-bitumen-saturated 浸透沥青的
asphaltic-bitumen-saturated felt(ed fabric) 浸透沥青的毛毡织物
asphaltic-bitumen-saturated felted mat 浸透沥青的毛毡编织物
asphaltic-bitumen-saturated felted mat with cork 浸透沥青的混有软木的毛毡编织物坐垫
asphaltic-bitumen-saturated felted pad 浸透沥青的毛毡织物衬垫
asphaltic-bitumen-saturated felted pad with cork 浸透沥青的混有软木的毛毡衬垫
asphaltic-bitumen-saturated paper 浸透沥青的纸
asphaltic-bitumen-saturating 浸透沥青的
asphaltic-bitumen sea(ing) paste 沥青密封膏
asphaltic-bitumen seal(ing) 沥青密封的
asphaltic-bitumen sealing compound 沥青密封混合物
asphaltic-bitumen sealing sheet(ing) 沥青密封层
asphaltic-bitumen sheet(ing) 沥青多孔片;沥青地毯
asphaltic-bitumen sheet roofing 沥青毡屋面
asphaltic-bitumen shingle 沥青屋面板;沥青墙面板
asphaltic-bitumen slurry 沥青软膏
asphaltic-bitumen slurry seal 沥青胶浆填缝
asphaltic-bitumen solution 沥青溶液
asphaltic-bitumen spray bar 沥青喷油管
asphaltic-bitumen stabilization 沥青稳定化(土壤);沥青加固(土壤)
asphaltic-bitumen subseal(ing) 沥青嵌缝;沥青挤嵌
asphaltic-bitumen surface coating 沥青表面涂层
asphaltic-bitumen surfacing 沥青面(层)
asphaltic-bitumen tape 沥青带
asphaltic-bitumen test 沥青试验
asphaltic-bitumen top dressing 起来表面整修
asphaltic-bitumen trailer 拖挂式沥青车
asphaltic-bitumen-type wood fibre board 沥青基木纤维板
asphaltic-bitumen waterproofer coat 沥青防水层
asphaltic-bitumen wearing course 沥青路面磨耗层
asphaltic-bitumen weighing batcher 沥青计量分批箱
asphaltic-bitumen with rock flour 混有石粉的沥青
asphaltic-bitumen wool felt 沥青羊毛毡
asphaltic block 沥青块(材)
asphaltic bonding composition 沥青黏结成分;沥青胶合材料
asphaltic briquet(te) 沥青模制试块;沥青标准水泥试块
asphaltic cardboard roof 沥青纸板屋顶
asphaltic cement 沥青冷底子油膏;沥青胶粘剂;沥青胶泥;沥青膏;地沥青胶泥
asphaltic cementing agent 沥青胶结剂;沥青黏结剂
asphaltic cementing composition 沥青黏结成分
asphaltic cementing compound 沥青黏结混合料
asphaltic clinker 地沥青熔渣;石油沥青熔渣;沥青熔渣
asphaltic coal 地沥青煤;沥青煤
asphaltic coat 沥青涂料面
asphaltic coating 沥青涂层;沥青涂料面层
asphaltic composition 沥青成分
asphaltic compound 沥青混合物;地沥青化合物
asphaltic concrete 沥青混凝土;地沥青混凝土
asphaltic concrete base 沥青混凝土路基层;沥青混凝土基础;沥青混凝土基层
asphaltic concrete batching plant 沥青混凝土配料装置
asphaltic concrete block 沥青混凝土块
asphaltic concrete carpet 沥青混凝土(铺)面层;沥青混凝土路面
asphaltic concrete corewall 沥青混凝土堤坝心墙;沥青混凝土防渗心墙
asphaltic concrete cutoff wall 沥青混凝土截水墙
asphaltic concrete-faced rockfill dam 沥青混凝土面板堆石坝
asphaltic concrete facing dam 沥青混凝土面板坝
asphaltic concrete flooring 沥青混凝土楼地面
asphaltic concrete mineral skeleton 沥青混凝土矿质骨架
asphaltic concrete mixer 沥青混凝土拌和机;沥青混凝土搅拌机
asphaltic concrete mixing plant 沥青混凝土搅拌装置;沥青混凝土拌和站;沥青混凝土搅拌设备;沥青混凝土拌和装置;沥青混凝土拌和设备
asphaltic concrete mix(ture) 沥青混凝土配合料;沥青混凝土混合料
asphaltic concrete pavement 沥青混凝土路面;沥青混凝土铺面
asphaltic concrete paver 沥青混凝土铺路机;沥青混凝土铺料机
asphaltic concrete paver without hopper 无料斗的沥青混凝土摊铺机;无料斗的沥青混凝土铺路机
asphaltic concrete road 沥青混凝土路;地沥青混凝土路
asphaltic concrete runway 沥青混凝土跑道
asphaltic concrete skeleton 沥青混凝土骨架
asphaltic concrete spreading machine 沥青混凝土摊铺机
asphaltic concrete surfacing 沥青混凝土(路)面层
asphaltic crude 沥青原油
asphaltic cutback 溶于石油馏出物中的沥青;沥青稀释
asphaltic dispersion 沥青耗散
asphaltic dispersion phenomenon 沥青扩散现象
asphaltic emulsion 乳化沥青;沥青乳(浊)液;地沥青乳液
asphaltic emulsion with colloidal emulsifier 用胶体乳化剂乳化沥青
asphaltic facade slab 正面沥青胶块;沥青贴面板
asphaltic facing 沥青饰面;沥青铺面
asphaltic felt 油毡;石油沥青油毡
asphaltic felt panel 沥青油毡板;沥青毡板
asphaltic filler 沥青填(缝)料
asphaltic gravel 沥青拌制砾石;沥青砾石
asphaltic grouted macadam 灌沥青碎石路面;沥青灌浆碎石
asphaltic insulating slab 沥青绝缘板
asphaltic joint 沥青缝
asphaltic lac(quer) 沥青漆
asphaltic layer 沥青层
asphaltic limestone 沥青(质)石灰岩;沥青(质)石灰石;沥青涂层;地沥青石灰石
asphaltic limestone powder 沥青质石灰岩粉;沥青质石灰石粉
asphaltic limestone quarry 沥青石灰岩矿
asphaltic lining 沥青衬砌;沥青衬层
asphaltic macadam 沥青贯入式碎石铺路法;沥青碎石路
asphaltic macadam pavement 沥青碎石路面
asphaltic macadam surfacing 沥青碎石路面
asphaltic mastic 沥青砂胶;沥青玛琋脂;石油沥青玛琋脂
asphaltic mastic pavement 沥青砂胶路面
asphaltic mattress 地沥青毡;沥青隔板
asphaltic mattress laying vessel 沥青垫层铺设船;沥青毡铺设船
asphaltic membrane 沥青膜
asphaltic mixture 沥青混合料
asphaltic mixture joint filler 沥青混合料填缝料
asphaltic mortar 地沥青砂浆;沥青砂混合料
asphaltic mortar masonry(work) 沥青砂浆砖石建筑;沥青胶泥圬工
asphaltic mulch 沥青盖料
asphaltic nature 沥青属性
asphaltic oil 沥青油;含沥青的石油
asphaltic overlay 沥青面(层);沥青护面层;沥青涂料
asphaltic pavement 沥青路面层;沥青铺面
asphaltic paving-mix(ture) 沥青铺路材料;沥青铺面混合料
asphaltic penetration macadam 沥青浸透的碎石路;沥青灌浆碎石
asphaltic petroleum 沥青质石油
asphaltic plank 沥青板
asphaltic plank wearing course 沥青板(混凝土)磨耗层
asphaltic plant-mix surface 厂拌地沥青混合料面层
asphaltic pouring rope 沥青浇口油麻丝
asphaltic primer 沥青透层
asphaltic product 沥青制品;沥青产品
asphaltic pump 沥青泵
asphaltic pyrobitumen 石油(焦)沥青;沥青焦沥青;地沥青质焦沥青
asphaltic pyrobituminous shale 沥青焦沥青页岩
asphaltic rake 沥青齿耙
asphaltic residual oil 沥青质残油;沥青油渣;重油中沥青膏渣滓油
asphaltic residue 沥青残渣
asphaltic resin 沥青树脂;地沥青树脂
asphaltic road 沥青路
asphaltic road-mix 路拌沥青混合料
asphaltic roadway 沥青公路
asphaltic rock 含沥青石块;沥青岩
asphaltic roofing 沥青屋面

asphaltic roof(ing) cement 沥青屋面黏结剂；沥青屋面胶结料
asphaltic roof(ing) felt 沥青油毛毡；沥青屋面毡
asphaltic roof(ing) shingle 沥青屋面板
asphaltics 沥青质成分
asphaltic sand 沥青砂；(含大量矿物质的)天然沥青
asphaltic sandstone 沥青砂石；沥青砂岩
asphaltic saturated felt 沥青浸渍毡
asphaltic sealing 沥青密封；沥青堵漏
asphaltic seal(ing) paste 沥青密封膏；沥青密封浆料
asphaltic sediments 沥青质沉积物
asphaltic sheet 沥青板；地沥青板
asphaltic slab 沥青板；地沥青板
asphaltic slag 含沥青炉渣
asphaltic surfacing 沥青面(层)
asphaltic surfacing work 沥青涂面工作；沥青路面工种；沥青覆面工作；沥青表面加工
asphaltic terrazzo tile 沥青磨石子砖(板的下层为沥青材料,上层为石子面)；沥青水磨石砖
asphaltic tile 沥青砖；沥青板
asphaltic tile floor covering 沥青瓷砖地面
asphaltic tile flooring 沥青砖铺地
asphaltic tile gutter 沥青砖雨水沟
asphaltic tile made under high-pressure 高压模制沥青砖；高压模制沥青瓦
asphaltic varnish 沥青漆；沥青清漆
asphaltic wall tile 沥青贴墙板；沥青贴墙砖
asphaltic water-proof coat 沥青防水面层；沥青防水涂层
asphaltic water-proof course 沥青防水层
asphaltic wearing course 沥青磨耗层
asphaltic work 沥青工种
asphalt-impregnated calcium silicate brick 沥青浸渍的硅酸盐砖
asphalt-impregnated joint runner 沥青浸渍的填缝浅口
asphalt-impregnated lime-sand brick 沥青浸渍的灰砂砖
asphalt-impregnated pouring rope 沥青浸渍的嵌缝油麻丝
asphalt-impregnated protective felt 沥青浸渍的保护毡
asphalt-impregnated sand 浸沥青的砂；沥青浸渍砂
asphalt-impregnated sand-limestone brick 沥青浸渍的灰砂砖
asphalt-impregnated sealing rope 经沥青处理的嵌缝绳；经沥青处理的封口绳；沥青浸渍的密封绳
asphalt-impregnated stone 经沥青处理的板条；沥青浸渍的板条
asphalt-impregnated strip 经沥青处理的板条；沥青浸渍的板条
asphalt impregnating 沥青浸渍的；沥青灌注
asphalt impregnation 沥青处理；沥青浸渍
asphalt individual coat 沥青单独涂层
asphaltine 沥青质
asphaltine cement 地沥青胶
asphalting 沥青防水；沥青表面处理；浇沥青；浇灌沥青
asphalting tool 浇沥青工具；沥青摊铺工具
asphalt injection control valve 沥青浇注控制阀
Asphalt Institute 沥青学会(美国)；地沥青学会(美国)
Asphalt Institute Method 地沥青学会法(美国)
Asphalt Institute(Smith) triaxial method of mix-design 地沥青学会(或史密斯式)混合料配合三轴试验设计(美国)
asphalt insulating coat 沥青绝缘涂层
asphalt insulating paste 沥青绝缘胶；沥青绝缘膏
asphalt insulating slab 沥青绝缘板
asphalt intermediate course 沥青结合层
asphalt ironing plate 沥青路面熨袋
asphaltite 石油沥青岩；沥青石；沥青岩；沥青石；高熔点天然沥青；地沥青石
asphaltite coal 黑沥青
asphaltization 沥青化作用
asphalt joint 地沥青缝；沥青缝；地沥青灌缝
asphalt jointed 沥青填缝的
asphalt jointed pitching 沥青砌石护坡
asphalt joint filler 沥青填缝料；沥青接缝填料
asphalt jointing 沥青填缝
asphalt joint pouring compound 沥青接头灌缝混合料；灌缝沥青混合物
asphalt joint runner 沥青填缝浅口
asphalt jungle 大城市；柏油丛林

asphalt jute 沥青浸黄麻；柏油麻布；柏油布
asphalt jute pipe 沥青黄麻管；沥青黄布管
asphalt kerb 沥青路边石；沥青处理路边石
asphalt kettle 沥青锅；柏油壶
asphalt lac 沥青清漆
asphalt lake 沥青湖；地沥青湖
asphalt-laminated 沥青叠层的；沥青层的
asphalt-laminated paper 层压防潮纸
asphalt lamination 沥青压层；沥青叠层压合纸；沥青叠层压合毡；沥青叠层压合板；沥青垫层材料；沥青层压
asphalt-latex emulsion 沥青橡胶浆乳化液
asphalt layer 沥青层
asphalt level(l)ing course 沥青找平层
asphalt limestone 沥青石灰岩；沥青石灰石
asphalt lined pipe 沥青衬里管道
asphalt liner 沥青嵌条
asphalt lining 沥青防水层；石油沥青衬里；沥青衬里
asphalt macadam 沥青灌入碎石铺路法；地沥青碎石路；沥青碎石路
asphalt macadamix 沥青混凝土与碎石灌沥青面层混合铺设法
asphalt macadam pavement 沥青碎石铺面；黑色碎石路面；地沥青碎石路面
asphalt macadam road 沥青碎石路
asphalt macadam surfacing 沥青碎石路面；沥青碎石铺面
asphalt mark 沥青标号
asphalt mastic 石油沥青砂胶；石油沥青玛琋脂；沥青砂胶；沥青胶黏剂
asphalt mastic block 沥青胶泥块
asphalt mastic board dummy joint 嵌填假缝用(地)沥青砂胶板；沥青玛琋脂模制假缝
asphalt mastic finisher 地沥青砂胶磨光机
asphalt mastic floor(ing) 沥青砂胶铺面；沥青玛琋脂地面；沥青砂胶地面
asphalt mastic pointing 沥青胶泥勾缝
asphalt mastic seal coat 沥青玛琋脂封层；沥青砂胶封层
asphalt mastic sidewalk 地沥青砂胶人行道
asphalt mat 沥青面层；沥青垫层；地沥青面层
asphalt material 石油沥青材料；沥青材料；地沥青材料
asphalt matrix 沥青乳胶香
asphalt mattress 石油沥青垫层；沥青垫层；地沥青毡
asphalt mattress revetment 沥青垫层护岸；地沥青垫层护岸
asphalt measuring pump 沥青计量泵
asphalt measuring system 沥青秤量系统
asphalt measuring unit 沥青计量装置
asphalt melting kettle 沥青加热炉；沥青熔锅
asphalt melting tank 沥青熔解罐
asphalt membrane 石油沥青膜；沥青防水膜；地沥青膜
asphalt membrane waterproofing 防水沥青膜；沥青防水薄膜；沥青薄膜防水
asphalt metering pump 沥青计量泵
asphalt-mix-design method 地沥青混合料(配合比)设计(方)法
asphalt mixer 沥青拌和机；沥青搅拌机；沥青混合料拌缸
asphalt mixer silo 沥青混合料储仓
asphalt mixing plant 沥青混合料拌和设备；沥青拌和装置；沥青拌和厂；地沥青(混合料)拌和厂；沥青混合料搅拌设备；沥青混凝土搅拌设备
asphalt mixture 石油沥青混合料；沥青拌和物；地沥青混合物
asphalt mortar 石油沥青砂浆；沥青砂浆
asphalt mortar finish 沥青砂浆面层
asphalt mortar flooring 沥青砂浆楼地面
asphalt mortar plaster 沥青砂浆粉刷
asphalt mulch 沥青盖料；地沥青盖料
asphalt mulch top 沥青盖顶
asphalt nature 地沥青属性
asphalt odo(u)r 地沥青臭味
asphaltogenic 成沥青的
asphalt oil 地沥青油
asphalt(os) bitumen 地沥青
asphaltous acid 地沥青酸
asphaltous acid anhydrite 地沥青酸酐
asphalt oven 沥青炉
asphalt overlay 沥青罩面层；沥青盖面层；沥青覆盖层；沥青补强层

asphalt oxidation 沥青氧化；沥青空气氧化
asphalt paint 石油沥青涂料；沥青油漆；沥青涂料；沥青漆；地沥青涂料
asphalt panel 沥青板
asphalt paper 沥青油纸；油毡纸；柏油纸
asphalt pavement 沥青路面；地沥青路面
asphalt pavement recycling 沥青路面再生作用
asphalt pavement sealer 沥青路面保护层；沥青面封缝料
asphalt pavement structure 沥青路面结构
asphalt paver 沥青摊铺机；沥青铺洒机；沥青铺路机；沥青路面摊铺机；沥青混凝土摊铺机
asphalt paving 沥青铺砌；沥青铺面；地沥青路面
asphalt paving block 沥青铺面砌块；沥青砖块；地沥青铺砌块
asphalt paving equipment 沥青铺设机械；沥青路面设备
asphalt paving machine 沥青摊铺机；沥青路面机械
asphalt paving plant 沥青铺料设备；沥青铺料机；沥青铺料厂；地沥青铺料厂
asphalt penetration 沥青针入度(指数)
asphalt penetration macadam 地沥青贯碎石面层；地沥青贯碎石(路)
asphalt perforated sheet(ing) 沥青多孔板
asphalt petroleum 含沥青石油
asphalt pipe 沥青输送管
asphalt piping 沥青输送管
asphalt pit 沥青坑
asphalt pitch 沥青脂
asphalt plank 沥青板
asphalt plank floor 沥青预制板地板
asphalt plant 沥青制备厂；沥青混合料拌和厂；沥青工厂；沥青拌和装置；沥青拌和厂
asphalt plant central control station 沥青混合料拌和设备中心控制室
asphalt plant mix 沥青厂拌合料；地沥青厂拌混合料
asphalt plate 沥青板；地沥青板
asphalt pliability test 沥青柔性试验
asphalt pollution 沥青烟雾污染；沥青污染
asphalt pouring compound 沥青嵌缝(填充)料
asphalt pouring ground 浇灌沥青场地
asphalt pouring rope 沥青嵌缝用油麻丝；沥青铅封口用油麻丝
asphalt powder 沥青粉(末)；石油沥青粉末；地沥青粉
asphalt pre-impregnation 沥青预浸渍(处理)
asphalt prepared roofing 预制沥青屋面板；沥青屋面料；沥青卷材屋面；沥青处理的屋面材料
asphalt pre-saturating 沥青预浸渍(处理)
asphalt prime coat 沥青头道涂层；路面头道沥青；石油沥青透层；沥青透层；沥青打底；地沥青透层
asphalt-primed 浇过地沥青透层的
asphalt-primed base 浇过石油沥青透层的基层；浇过沥青透层的基层；浇过沥青透层的基层
asphalt primer 石油沥青透层；沥青涂层；沥青透屋；沥青冷底子油；沥青底层；地沥青透层
asphalt processing 沥青制造法
asphalt-proof coating 沥青防水膜层
asphalt proportioning pump 沥青泵；沥青计量泵
asphalt protection coat 沥青防护层
asphalt protective coating(material) 沥青保护层(材料)；沥青防护涂料
asphalt pug mixer paddle arm 沥青混合料拌缸浆柄
asphalt pug mixer plates 沥青混合料拌缸衬板
asphalt pug mixer shaft 沥青混合料拌缸轴
asphalt pug mixer tips 沥青混合料拌缸叶片
asphalt pump 石油沥青泵；沥青泵
asphalt putty 沥青腻子；沥青灰；沥青封泥；地沥青油灰
asphalt pyrobitumen 沥青质焦性沥青
asphalt rake 沥青齿耙
asphalt raker 铺刮沥青工具；铺刮沥青工人
asphalt ready roofing 沥青屋面料；预制沥青屋面板
asphalt ready roofing with asphalt coat(ing) on both 预制双面沥青涂层的沥青屋面板
asphalt reclaimed rubber roofing 再生胶油毡
asphalt reclaimed rubber waterproofing paint 再生胶沥青防水涂料
asphalt recycler 沥青再生机
asphalt recycling 沥青再生利用；沥青回收
asphalt recycling plant 沥青旧料再生设备；沥青混合料旧料再生设备

asphalt refinery 沥青加工厂
asphalt rejuvenating agent 沥青复苏剂
asphalt remixer 复拌沥青混合料摊铺机;复拌沥青混合料拌和机
asphalt residual oil 沥青残渣油
asphalt revetment 沥青护面;沥青护墙;沥青护岸
asphalt road 石油沥青路;沥青柏油路;地沥青路;柏油路
asphalt road burner 烫沥青路面机
asphalt road mix 沥青路拌混合料;地沥青路拌混合料
asphalt road oil 地沥青铺路油
asphalt road roller 熨沥青路面机
asphalt road surfacing 沥青路铺面
asphalt rock 沥青岩(石);块沥青
asphalt roll 沥青卷材
asphalt roll(ed-strip) roofing 屋面沥青卷材;沥青油毡;沥青屋面卷材;沥青卷材屋面;油毡屋面
asphalt roof cement 沥青屋面黏结剂
asphalt roof coating 沥青屋面涂料
asphalt roofing 沥青屋面
asphalt roofing cement 沥青屋面胶泥
asphalt roofing felt 沥青屋面毛毡;屋面沥青油毡;沥青屋面防水毡
asphalt roofing sheet(ing) 沥青屋面薄板
asphalt roofing shingle 沥青屋面板
asphalt roof paint 沥青屋面漆
asphalt rubber 沥青橡胶
asphalt-rubber ca(u)lk 沥青橡胶胶油膏
asphalt-rubber mass 沥青胶料
asphalt-rubber material 沥青胶料
asphalt-rubber strip 沥青胶条;沥青防水胶带
asphalt runway 沥青跑道
asphalt sand 沥青浸渍灌浆砂;浸沥青的砂;沥青砂
asphalt sand lime brick 沥青灰砂砖
asphalt(sand) mastic 地沥青砂胶;沥青砂胶
asphalt sandstone 沥青砂岩;沥青砂石
asphalt saturated 沥青饱和的
asphalt saturated and coated asbestos felt 沥青石棉油毡;沥青石棉卷材
asphalt-saturated and coated felt 沥青毡
asphalt saturated and coated jute fabric 麻布沥青油毡
asphalt saturated and coating organic felt 沥青纸胎油毡
asphalt-saturated asbestos felt 沥青浸渍石棉油毡;石棉纸油毡;浸沥青棉毡
asphalt-saturated felt 沥青油纸;沥青岩棉毡;沥青浸润毡;地沥青浸润毡;沥青浸汁矿棉纸油毡;纸胎油毡
asphalt-saturated felted fabric with cork 具有软木的沥青饱和毡织物
asphalt-saturated felt mat 沥青油毡隔层;沥青油毡垫层
asphalt-saturated glass cloth felt 玻璃布油毡
asphalt-saturated glass fiber felt 沥青浸渍玻璃纤维油毡;玻璃纤维油毡
asphalt-saturated loose fiber felt 沥青浸汁网状屋面板
asphalt-saturated mineral wool felt 矿棉纸油毡
asphalt-saturated organic felt 石油沥青纸毡
asphalt-saturated paper 沥青饱和(的)纸
asphalt-saturated rag felt 沥青浸渍粗制毡;沥青浸渍粗制毡
asphalt-saturated rock wool felt 浸沥青石棉毡
asphalt-saturated roofing felt 沥青浸渍屋顶油毡
asphalt-saturated woven fabric 浸沥青防水布(屋面用)
asphalt saturating 沥青饱和的
asphalt saturation 沥青浸透
asphalt scale 沥青计量器;沥青秤
asphalt screen 沥青罩面
asphalt seal 沥青止水;沥青封闭
asphalt seal coat 沥青密封涂层;沥青封层;沥青封闭层;地沥青封层
asphalt-sealing compound 沥青封塞料;沥青防水料;沥青填缝料
asphalt sealing rope 沥青封口油麻丝
asphalt sealing sheet(ing) 沥青封塞薄片
asphalt seepage 沥青苗
asphalt sheet 沥青纸毡;沥青片(材);油毛毡
asphalt sheet mattress 薄沥青垫层
asphalt sheet roofing 片状铺的沥青屋面;沥青纸毡屋面

asphalt shingle 组合屋面盖片;石油沥青片瓦;沥青油毡瓦;沥青油毡盖片;沥青屋面板;沥青瓦;沥青墙面板;沥青木瓦;沥青浸渍卷材
asphalt shingle roofing 沥青瓦屋面;沥青卷材屋面
asphalt ship 沥青船
asphalt skirting 沥青勒脚
asphalt slab 沥青板;地沥青板
asphalt slurry 沥青软膏;沥青稀浆
asphalt slurry seal 沥青浆塞缝;沥青浆封缝
asphalt smoother 沥青平整器;沥青地面平整机;沥青平整机;沥青刮面机
asphalt softening point apparatus 沥青软化点测定仪
asphalt softening point test 沥青软化点试验
asphalt soil stabilization 沥青土稳定(法);土的沥青稳定法
asphalt spray bar 沥青喷管
asphalt sprayer 沥青洒布器;沥青洒布机
asphalt spray nozzle 沥青喷嘴
asphalt spray pump 沥青喷射泵;沥青喷洒泵
asphalt spreader 沥青摊铺机;沥青铺装机;沥青路面摊铺机;地沥青(混凝土)摊铺机
asphalt stabilization 石油沥青稳定;沥青稳定处理;沥青稳定;地沥青稳定
asphalt steam pipe melter 蒸汽式沥青熔化装置
asphalt stone 地沥青块;沥青岩石
asphalt storage and melting plant 沥青储[贮]存熔化装置
asphalt storage tank 沥青储[贮]料箱
asphalt storage unit 沥青储[贮]存装置
asphalt street 城市柏油路;柏油道路
asphalt strip 沥青条;沥青板条
asphalt subbase 路面沥青层
asphalt subseal(ing) 沥青基层处理;沥青封底;沥青基封底
asphalt surface coat(ing) 沥青表面涂层;沥青面层
asphalt surface course 沥青砂浆面层;沥青磨损层;沥青面层;地沥青面层
asphalt surface heater 沥青循环加热器;沥青路面加热器;地沥青路面加热器
asphalt surface heating 沥青路面加热
asphalt surface heating drum 沥青路面加热滚筒
asphalt surface treatment 沥青浇面处理;沥青敷面处理;地沥青表面处治
asphalt surfacing 沥青面(层)
asphalt tack coat 石油沥青黏层;沥青黏结层;沥青黏层;地沥青黏层
asphalt tamper 沥青夯;沥青捣棒
asphalt tank 沥青桶;沥青罐
asphalt tank car heater 沥青加热车
asphalt tanker 沥青船
asphalt tape 沥青胶带;沥青带
asphalt tar 沥青焦油;沥青柏油;地沥青柏油
asphalt-tar pitch 硬焦油沥青
asphalt test 沥青试验
asphalt tile 沥青瓦;沥青砖;沥青铺面砖;地沥青砖
asphalt tile base 沥青砖基;沥青预制块地基
asphalt tile floor 沥青砖地面
asphalt tile flooring 沥青面砖楼地面
asphalt tile sealer 沥青砖封闭剂
asphalt top coat 沥青面层
asphalt top dressing 沥青浇面(料)
asphalt topping 石油沥青浇面;沥青面层;沥青浇面;地沥青表面处治
asphalt transfer pump 沥青转运泵
asphalt-treated 沥青处治过的;沥青处理过的;地沥青处治过的
asphalt-treated base 沥青处治的底层;沥青处理的基层;沥青处理的基底
asphalt-treated layer 地沥青处治层
asphalt-treated material 沥青处理的材料
asphalt truck distributor 沥青摊铺车
asphalt tube 沥青输送管
asphalt tubing 沥青输送管
asphalt-type wood fiber board 沥青型木纤维板
asphaltum 防腐沥青油漆;石油沥青;地沥青;沥青
asphaltum oil 柏油;沥青油;地沥青油
asphalt underseal(ing) 沥青封底
asphalt underseal-work 沥青封底工程;铺筑底层的沥青混凝土
asphalt upstand 沥青勒脚
asphalt varnish 沥青油漆;沥青漆
asphalt vermiculite plate 地沥青蛭石板
asphalt viscosity 沥青黏(滞)性;沥青黏(滞)度

asphalt-void ratio 沥青空隙比
asphalt volatile constituent 沥青挥发分
asphalt wall tile 沥青贴面砖;沥青墙砖
asphalt water-proofing 沥青防水
asphalt water-proofing paint 地沥青防水涂料
asphalt water-proofing membrane 沥青防水膜
asphalt wearing course 沥青磨耗层
asphalt weighing batcher 沥青称量配料机
asphalt weighing bucket 沥青称量斗
asphalt with rock flour 石粉沥青
asphalt wood fiber-board 沥青木纤维板
asphalt wool felt 沥青羊毛毡;沥青毛毡
asphalt work 铺沥青工作;铺沥青工程
aspheric(al) 非球面的
aspheric(al) coefficient 非球面系数
aspheric(al) condenser 非球面聚光镜
aspheric(al) correcting lens 非球面校正镜
aspheric(al) correction 消球差校正
aspheric(al) corrector 非球形校正器;非球面校正镜
aspheric(al) figure 非球面形
aspheric(al) lens 消球差透镜;非球面透镜;非球面镜头
aspheric(al) magnifier 消球差放大镜
aspheric(al) mirror 非球面镜;非球面反射镜
aspheric(al) optic(al) system 非球面光学系统
aspheric(al) particle 非球形颗粒
aspheric(al) polishing machine 非球面抛光机
aspheric(al) reflecting surface 非球面反射面
aspheric(al) reflector 非球面反射镜
aspheric(al) surface 非球面
aspheric(al) telescope 消球差望远镜
asphericity coefficient 非球面系数
aspheric-mirror 非球面反射镜
aspherics 非球面仪;非球面光学
aspherism 消球差性
aspherizing 非球面形结构
asphxiator 排水管道漏水试验;二氧化碳灭火器;二氧化碳灭火机
asphyctic 窒息的
asphyctous 窒息的
asphyxia 窒息
asphyxial 窒息的
asphyxiant 窒息剂;发生窒息的
asphyxiant action 窒息作用
asphyxiant gas 窒息性气体
asphyxiant poisoning 窒息性毒剂中毒
asphyxiat carbonica 煤气窒息
asphyxiating gas 窒息性气体;窒息性毒气
asphyxiation 窒息
asphyxiator 碳酸气灭火器;窒息装置;下水管漏泄试验器
aspirate 吸出物;抽出物
aspirated-air grain conveyer 气吸式谷物输送器
aspirated dry-(and-) wet bulb thermometer 通风干湿球湿度计;通风干湿球湿度表
aspirated hygrometer 吸气湿度计
aspirated pit pair 闭塞纹孔对
aspirated psychrometer 通风干湿表
aspirating 除尘
aspirating burner 吸入煤气式烧嘴;吸气燃烧器
aspirating chamber 抽气室;吸气室
aspirating cylinder 抽吸机汽缸;抽气机汽缸;吸汽缸
aspirating engine 抽吸机;抽气机;吸气发动机
aspirating hole 入气口;吸气孔;吸气口;抽气口
aspirating mouth 入气口;吸气孔;吸气口
aspirating needle 吸液针
aspirating pressure 吸气压(力)
aspirating propeller pump 真空螺旋泵
aspirating psychrometer 吸气湿度计
aspirating pump 抽吸泵;抽气泵;吸气泵
aspirating pyrometer 吸气式高温计
aspirating screen 振动吸力筛;抽气筛
aspirating stroke 进汽行程;进汽冲程;进气冲程
aspirating syringe 吸引注射器
aspirating valve 吸出阀
aspiration 气吸;吸引术;吸取;吸气;吸出;抽吸
aspiration channel 吸风管
aspiration column 吸气筒;吸气塔
aspiration condenser 吸入冷凝器;吸气冷凝器;吸尘凝集器
aspiration drainage 吸引引流法;吸引导液法
aspiration inlet 吸气口
aspiration leg 吸风塔;柱形吸风筒
aspiration meteorograph 通风气象计

aspiration method of asbestos processing 石棉吸选法
aspiration phychrometer 通风干湿机
aspiration piping 抽吸管;抽气管;吸气管道
aspiration pressure control 吸气压力控制
aspiration probe 抽气式探针
aspiration psychrometer 通风湿度计;通风干湿球湿度计;通风干湿球湿度表;吸气球湿度计;吹风式干湿球湿度表
aspiration pump 吸扬式泵;抽吸泵
aspiration temperature 通风温度
aspiration temperature meter 通风温度计;通风温度表
aspiration thermograph 通风温度记录器
aspiration thermometer 通风湿度计;通风温度表
aspiration ventilation 排气通风(法)
aspirator 气吸管道;水流抽气管;流水抽气器;流水抽气机;流水抽气管;吸引器;吸液器;吸气器;吸出器;抽吸器;抽气器;抽风扇;抽风器;防尘呼吸面具
aspirator bottle 吸气瓶
aspirator cleaner 气吸式风力清选机
aspirator combined with dust collector 吸气积尘器
aspirator filter pump 吸气过滤泵
aspirator pump 抽气泵;吸气泵
aspirator with dust collector 吸气集尘器
aspiring pump 抽吸泵;抽风机;抽风泵
aspirochyl 氨基苯胂酸汞
aspite 大口火山
as-placed concrete 未凝固的混凝土;新浇混凝土;新拌混凝土
Asplumd process 阿斯普兰德式木丝板制造法
AS position AS 位置
as-quenched 淬火状态
as-raised gravel 未洗砾石;未经筛洗的天然砾石
as received basis 收到基
as required 按要求
as-rolled 轧制状态
ass 滑车底部
Assama silk 阿萨马丝
Assam cotton 阿萨姆棉
Assam fever 阿萨姆热
assart 垦伐;开荒
assault aircraft 突击运输机
assault air cushion vehicle 攻击型气垫船
assault of waves 浪击
assay 确定成色;试验;试料;试金;鉴定;检定;化验;验定;测定
assay balance 试金天平;化验天平
assay bar 鉴定棒;检定金
assay certificate 鉴定证明书
assay determination 分析测定
assayer 试金者;化验员;分析者
assay foot 分析英尺
assay for aquatic organism 水生生物鉴定
assay furnace 试样炉;化验炉
assay grade 分析品级
assay inch 分析英寸
assaying date analytic(al) date of sample 样品分析日期
assaying of mineral 矿物鉴定;矿物分析
assay laboratory 试金室;试金实验室;分析实验室
assay mill 试验研磨机
assay period 鉴定期
assay plan 分析图
assay pound 化验磅
assay procedure 化验程序
assay sample 分析用试样
assay size 检验粒度
assay ton 检定吨;化验吨
assay ton weight 试金吨砝码
assay value 试金值;分析值
assay walls 矿体可采界限
as seen 看货买卖
asselbornite 砷铋铅铀矿
Assel elongator 阿塞尔辗轧机
Asselian 阿舍尔阶【地】
Assel mill 阿塞尔轧管机;阿塞尔轧管机
assemblage 聚集体;集合艺术(品);集合;系综;组合(品);装配;安装
assemblage body 集合体
assemblage business 来件装配
assemblage characteristic 群落特征

assemblage cost 合并邻近土地的购置费用
assemblage form of discontinuity 结构面组合形式
assemblage index 图幅接合索引(表);图幅接合表;接图表
assemblage of curves 曲线组合;成组曲线
assemblage of forces 力系
assemblage point 连接点;接点;汇集点;(桁架拉杆用交叉的)结点
assemblage point connection 点连接
assemblage technology 集群技术
assemblage tent 集会帐篷
assemblage zone 组合带
assemble 集中;汇编;合模;装配
assemble-and-go 边汇编边执行
assemble by welding 铆合
assemble carbonic acid fat 聚碳发酯
assembled 装配的
assembled battery 电池组
assembled car 组装的车辆
assembled die 拼合式模具;拼合模;镶块模
assembled double-flanged hollow shaft 组装式双法兰空心轴
assembled fitting 组装连件
assembled frog 组合(式)辙叉;钢轨组合辙叉
assembled micromodule 组装微型模件
assembled milling cutter 装配式铣刀
assembled part 装配部件
assembled parts list 装配部件清单
assembled plate column structure 装配式板柱结构
assembled product 装配式成品;装配产品
assembled reinforced concrete structure 装配式钢筋混凝土结构
assembled structure 装配式结构
assemble error 汇编错误
assemble in the field 就地拼装
assemble level 集计水平
assemble list 汇编(程序)表
assemble operator 汇编算符
assemble program(me) 汇编程序
assembler 汇编程序;装配工
assembler advantages 汇编程序优点
assembler code 汇编码
assembler command 汇编命令;汇编程序命令
assembler compiler 汇编编译程序
assembler control 汇编控制
assembler development system 汇编程序编制系统
assembler directive 汇编指令;汇编程序指令;汇编程序控制指示字
assembler directive command 汇编程序引导指令
assemble reinforcement concrete conver plated culvert 拼装式钢筋混凝土盖板涵
assembler error code 汇编错误码
assembler error message 汇编错误信息;汇编程序错误信息
assembler generator 汇编生成程序
assembler handler 汇编控制程序
assembler instruction 汇编指令
assembler language 汇编语言;汇编程序语言
assembler language program(m)ing 汇编语言程序设计
assembler level 汇编程序级
assembler listing 汇编列表
assembler loader package 汇编装入程序包
assembler machine 汇编机
assembler operation code 汇编程序控制指令
assembler operator 汇编程序算符
assembler option 汇编选择
assembler output 汇编输出
assembler output listing 汇编程序列表输出
assembler pseudo-operation 汇编语言伪操作;汇编程序伪操作
assembler source code 汇编程序源程序码
assembler source program(me) 汇编语言源程序
assembler statement 汇编语言语句
assembler system 汇编系统
assembler translator 汇编翻译程序
assemblies 层积木;层积构件
assembling 汇编;合模
assembling bolt 组合螺栓;装配螺栓;安装螺栓
assembling cost 安装费;总成本
assembling department 装配车间;装配部(门)
assembling die 组合模(具)
assembling factory 装配厂
assembling form(work) 拼装式模板

assembling hall 会场
assembling jig 装配机架;装配夹具;工作夹具
assembling line 装配线;装配生产线;装配流水线
assembling list 汇编(程序)表
assembling machine 装配机器
assembling mark 装配记号
assembling market 聚集市场
assembling method 装配方法;安装方法
assembling mo(u)ld 装配铸型
assembling of axle-boxes 轴箱装配
assembling of steel bridge 钢桥拼装
assembling parts 组装部件
assembling parts supplied by clients 来件装配
assembling plant 装配厂;装配(车)间
assembling process 装配过程;装配工艺;装配程序
assembling rate 装配定额
assembling reinforcement 绑扎钢筋
assembling segmentally 分段拼装(法)
assembling site 装配现场;安装位置
assembling speed 装配速度
assembling stand 装配座;装配台
assembling support 装配式支撑
assembling table 成型操作台
assembling technology 装配工艺
assembling time 汇编时间
assembling unit 装配(部)件
assembling with customer's parts 来件装配
assembling work 装配工种;安装工作
assembly 联合;集群;汇总图;汇编;组装;组坯;合;组对;总成;装置
assembly adhesive 装配黏胶剂;装配黏合剂;装配用胶黏剂;装配胶黏剂
assembly aids 装配用辅助装置
assembly and checkout 装配与检验;装配与检查;装配和测试
assembly and disassembly 汇编与反汇编;装配与拆卸
assembly and repair 装配与维修;装配和修理
assembly and test 装配与测试
assembly area 编组场(机场);集装区;装配场;安装场地
assembly area of wheel pair 存轮场【铁】
assembly average 统计平均值;集合平均值;汇集平均值;系集平均值
assembly battery 干电池组
assembly bay 装配间;装配场
assembly belt 流水装配线
assembly belt technique 总装配带技术
assembly bolt 组合螺栓
assembly buffer 汇编缓冲区;汇编缓冲器
assembly building 装配式建筑;装配式房屋;装配间
assembly chaining 汇编连接;装配系列;装配链接
assembly code 汇编码
assembly cost 装配费;装配成本
assembly crew 装配组
assembly data 汇编数据
assembly debug program(me) 汇编排错程序;汇编调试程序
assembly department 装配部(门);装配(车)间
assembly diagram 组装图;装配图(样)
assembly dimension 组装尺寸
assembly drawing 组装图;总图;装配图(样);安装图
assembly drill 钻机机组;成套钻具
assembly drum 成型鼓
assembly element 供装配构件
assembly equipment 装配设备
assembly fitting 供装配配件
assembly fixture 装配架
assembly floor 装配台
assembly for installation 安装图
assembly frame 装配架
assembly glue 装配胶
assembly gluing 装配胶合
assembly hall 礼堂;会议厅;会堂;会场
assembly handler 汇编控制程序
assembly holding block 安装木架
assembly industry 加工装配工业;装配工业
assembly jig 拼装台;拼装点;组装模具;装配夹具;装配夹架;组装胎型
assembly key 紧固键;装配楔块;装配键
assembly language output 汇编语言输出
assembly language processor 汇编语言加工程序;汇编语言处理程序

assembly language programing vs high level 高级汇编语言程序设计
assembly language program(me) 汇编语言程序
assembly layout (室内的)平面装配图
assembly level 汇编级
assembly line 生产线;流水作业(装配)线;流水(装配)线;装配(作业)线;装配流水线
assembly line balancing 装配线平衡
assembly line flow process 装配流水作业
assembly line method 装配流水作业法
assembly line operation 流水作业;装配作业线生产
assembly line process 装配流水作业法
assembly line processing 汇编行处理
assembly line production 装配线生产
assembly line technique 流水作业装配法
assembly list 汇编列表;汇编(程序)表
assembly machine 装配机
assembly mark 装配记号;装配标记;组装记号
assembly method 装配方法
assembly occupancy 集体使用的房屋
assembly of members 杆件组装
assembly of nozzle necks to vessel walls 容器壁同接管的装配
assembly of platefin elements 板翅元件的组装
assembly of purification furnace 净化炉装置
assembly of reduction furnace 还原炉装置
assembly of thin-films 薄膜组合件
assembly operation 编组作业
assembly or unit overhaul 装配成部件大修
assembly output language 汇编语言输出
assembly packaging 集合包装
assembly parameter 汇编参数
assembly parts 组合零件;组装(部)件;装配件
assembly phase 汇编阶段
assembly phase of a compiler 编译程序汇编阶段
assembly pit 装配地坑
assembly place 会议地点;会场
assembly plant 小群落植物;装配(工)厂;安装场地;装配车间
assembly plate 接合板;装配板
assembly platform 装配台
assembly procedure 装配程序
assembly process 汇编过程;装配程序
assembly processing 组装加工
assembly processing step 组合加工步骤;组装加工步骤
assembly program(me) 组装程序;组合程序;综合程序;汇编程序;安装程序
assembly quality 安装质量
assembly rack 装配架
assembly rake 装配架
assembly rate 装配速度
assembly rig 装配架
assembly rod 组装螺栓
assembly room 会议室;装配室;装配间;成型车间
assembly routine 汇编程序;汇编常规
assembly schedule 装配时间表;装配程序表
assembly screw 安装螺旋;安装螺丝
assembly section 装配区
assembly sequence 装配次序
assembly shop 总装车间;装配(车)间
assembly siding 集结线
assembly site 安装场地
assembly space 观众厅部分;集合空间
assembly speed 装配速度
assembly stage 装配阶段
assembly statement 汇编语句
assembly station 装配站
assembly stress 装配应力;安装应力
assembly subprogram(me) 汇编子程序
assembly subroutine 汇编子程序
assembly system 汇编系统;装配系统
assembly testing 成组测试
assembly time (黏接的)晾置时间;积压时间;组合时间;装配周期;装配时间
assembly track 集结线;车辆集结线
assembly trolley rails 装配场运输轨
assembly twister 并线初捻机
assembly type 装配式
assembly type structure 装配式建筑物;装配式结构
assembly unit 可汇编的;汇编单元;装配单元;装配单位;装配(部)件;供装配构件
assembly with foreign custom's goods 国外来件装配

assembly work 装配工作;装配工种
assembly yard 装配工场;装配场
assent and consent 一致通过
assented bond 同意债券
assented stock 同意股票
asser 古木作
assert a maritime claim 海事索赔
assert a set-off 提出债权作为抵销
assertion 推断;断定
assertion statement 断定语句
assertive colo(u)r 醒目色;鲜明色
assess 核计;征收;估价;估定
assessable 可征收的
assessable capital stock 可催缴成本
assessable income 可纳税收入;应评税之收入
assessable land 可征地
assessable stock 可征税股票;可估值股票
assessable value of property 财产评税值
assess a tax 评定税额
assessed budget 摊派预算
assessed contribution 分摊的会费
assessed cost 摊派费用
assessed program(me) cost 摊派的计划费用
assessed tax 估定税额
assessed valuation 估税价值;估税价格;估定价值
assessed value 课税估定财产价值;政府估定价值;估价价值;房地产课税价值
assessed value of land 估定地价
assessee 财产被估定者
assess effectiveness 评价效率
assessing capital 核定资本
assessing the treatability of wastewater 废水可处理性评价
assess market opportunities 判断市场机会
assessment 评价;评估;摊派税款;摊派税捐;课征;价格的评定;估征额;估定;查定
assessment and computational method of resource 资源量预测和计算方法
assessment assess 评定
assessment base 房地产估值
assessment basis 评价依据;评价基础
assessment categories 评价等级
assessment center 人员评价选择中心
assessment district 估值区(域)
assessment drilling 矿体评价钻进
assessment endpoint 评价终点
assessment for ambient environment 环境现状评价
assessment fund 摊派税捐基金;摊派捐税基金;赋税基金
assessment index of pollution 污染评价指数;污染评价指标
assessment index system 评价指标体系
assessment language 估价语言
assessment method 评价方法
assessment method of resource 资源量预测方法
assessment model of water body pollution sources 水体污染源评价模型
assessment of additive utility function 可加效用函数的评估
assessment of claim 要求权评定
assessment of conformity 评定合格
assessment of depreciation 折旧估算
assessment of exposure 照射评价
assessment of function 功能评定
assessment of geochemistry anomaly 地球化学异常评价
assessment of geologic(al) environment 地质环境评价
assessment of groundwater condition 地下水状况评价
assessment of loss 评估损失;损失估算;估损
assessment of offers and tenders 报价与投标的估定
assessment of performance 性能估计
assessment of project 工程评定
assessment of property value 财产价值评定
assessment of radiation protection 辐射防护评价;放射防护评价
assessment of regional geophysical anomaly 区域地球物理异常评价
assessment of results 成果评定
assessment of reviewable environment 环境回顾评价

assessment of risks 风险评估
assessment of scaling constants 度量常数的评估
assessment of social effect 社会影响评价
assessment of the consequences of technology 技术的社会效益评价
assessment of the productivity of plutons 岩体含矿性评价
assessment of the productivity of sedimentary horizons 地层含矿性评价
assessment of the productivity of structure 构造含矿性评价
assessment of time preference 时间偏好的评估
assessment of toxicological state of water body 水体毒性生物状况评价
assessment of waste water 废水估计量
assessment of water policy 水政策评价
assessment of water quality 水质评价
assessment of water resources 水资源评价
assessment of work done 计工
assessment of yieldcapacity 定位测产
assessment paid 税款付讫;已完税;已纳税
assessment panel 评估小组
assessment period 征税期;房地产估价期
assessment procedure 评价程序
assessment process 评价程序
assessment receipt 摊派捐税收入;摊派捐税收入
assessment rolls 摊派税款清册
assessment valuation 摊派的税捐估价
assessment work (美国法律规定每年的)最低钻探工作量
assessor 鉴定者;鉴定器;鉴定管;估税员;公估人;财产估价人
assess the cost 核计成本
assess toxicity 估计毒性
asset-backed securities 具资产保证的证券
asset-backed-securitization model ABS 模式
asset-income ratio 资产收益比(率)
asset life 设备使用年限;可用年限
assets 资产;财产;贵重器材;财富
assets account 资产账户
assets acquired for non-cash assets 非现金取得的资产
asset acquisition and retirement budget 资产购置和退废预算
assets acquisition budget 资产购置预算
assets alteration 资产更迭
assets and liabilities 资产与负责;资产负债表
assets appraisal 资产重估价
assets audit 资产审计
assets control 资产控制
assets cover 资产抵偿;资产担保
assets coverage 资产担保率
assets depreciation 资产折旧
assets depreciation range 资产折旧(年限)幅度
assets form 资产形态
assets guideline period 资产折旧基准期
assets held as collateral 作为抵押品的资产
assets income 资产所得;资产收益
assets-income ratio 资产收益比(率)
assets in general 一般资产
assets in kind 实物资产
assets inventory shortage 资产盘亏
assets inventory surplus 资产盘盈
assets liabilities 资产负债
assets-liabilities ratio 资产负债比(率)
assets life-time 财产寿命
assets management reponsibility system 资产经营责任制
assets manager 资产管理者
assets of railway transport unit 铁路运输单位资产
assets out of account(book) 账外资产
assets qcquisition budget 资产购置预算
assets reduction 资产减项
assets reduction account 资产抵减账户;资产减项
assets reserves 资产储备金
assets retirement 资产报废
assets revaluation 资产重估价;资产重估
assets revaluation tax 资产重估税
assets settlement 资产决算
assets stock 资产股份
assets structure 资产结构
assets subject to lien 留置资产;扣押资产
assets transfer ledger 资产转让分类账
assets turnover 资产周转

assets turnover of operating unit 运营单位资金周转额
assets turnover ratio 资产周转率
assets valuation 资产评估;资产计价
asset-transmutation effect 资金转形效应
asset-utilization ratios 资产利用情况类比率
assign 交割;转让;指定;赋值;分派
assignability 可转让性;财产的可转让性
assignable 可指定的;可分配的
assignable cause 可指出的原因;指明原因;不可忽视的原因
assignable claim 转让债权
assignable contract 可转让合同
assignable credit 可转让的信用证
assignable indirect charge 可分摊的间接成本;可分配间接费用
assignable instrument 可过户证券
assignable variation 可指定变差
assign a buffer 指定一个缓冲区;赋值一个缓冲区
assign a channel 指定一个通道;赋值一个通道
assign a contract 转包;转让
assignation 转让;指定;委托;分配
assign a value 赋值;分配一个值
assigned 定岗位
assigned address 赋值地址
assigned amount unit 指定的总量单位
assigned branch 赋值转移;赋值分支
assigned channel 指配频道
assigned cost 转让成本
assigned frequency 指配频率;指定频率;规定频率;工作频率
assigned frequency band 指配频带;分配频带;分配的频带
assigned hospital 合同医院
assigned in blank 空白转让书;空白过户单
assigned position 指定位置
assigned priority 指配优先
assigned risk 转让危险;分派风险;分担风险;分出风险
assigned shed 指定货棚
assigned state 赋值状态
assigned voltage 规定电压
assigned volume 分配量;分配交通量
assignee 让受人;受让者;受让人;接受转让人;被指定人;受委托者;受委托人;代理人
assignee in insolvency 破产清算人
assigner 转让者;转让人;出让人;出让方
assign in blank 不记名转让
assigning crew system of passenger train 旅客列车包乘制
assign letter of credit 让购信用证
assignment 任务书;任务;让渡;利益转让;课题;转让;指派;定岗位;赋值;分配的工作;分配
assignment allowance 任务津贴;外派津贴
assignment analysis 任务分析
assignment charges 转让费(用)
assignment clause 让与条款;让渡条款;转让条款
assignment command 分配指令
assignment component 赋值分量
assignment key 呼叫键
assignment lamp 联络灯;呼叫灯
assignment of a claim 债权转让
assignment of contract 合同签订;合同签定;转包;转让
assignment of copyright 版权转让
assignment of debt 债务转移
assignment of deferred compensation 延期补偿款转让
assignment of freight 货运转让
assignment of interests 转让利益
assignment of leases 租借的转让
assignment of loadline 载重线勘定
assignment of marketing right 经销权转让
assignment of mortgage 抵押贷款的转让
assignment of policy 保险单的转让
assignment of property 转让产权
assignment of rents 租金过户;租金转让
assignment of rights 权让权
assignment of title 过户
assignment operation 赋值运算
assignment operator 赋值算符
assignment phase 赋值阶段
assignment priority 指定优先;赋值优先级
assignment privilege 转让权利

assignment problem 指派问题;分配问题;分派问题
assignment procedure 赋值过程
assignment program(me) 分配程序
assignment right 转让权
assignment statement 计算语句;赋值语句
assignment switch 呼叫开关
assignment symbol 赋值符号
assignment test 标定测验
assignment-type model 分派模型
assignor 让与者;委托者;委派人;转让者;转让人
assimilable organic carbon concentration 可同化有机碳浓度
assimilable organic carbon content 可同化有机碳含量
assimilable organic carbon nutrient bioassay 可同化有机碳营养生物测定
assimilate into new working evironment 同化于新工作环境
assimilation 同化(作用);同化代谢
assimilation ability 同化能力
assimilation colo(u)r 拟色
assimilation effect 同化效应
assimilation layer 同化层
assimilation products 同化产物
assimilation quotient 同化物;同化商
assimilation system 同化系统
assimilation type 同化作用类型
assimilative capacity 同化容量;同化能力;吸收污染度;吸收污染量
assimilative capacity of water body 水体同化容量;水体同化能力
assimilative reaction 同化反应
assimilatory coefficient 同化系数
assimilatory colo(u)ration 保护色
assimilatory depletion 吸收衰竭;同化衰竭
assimilatory efficiency 同化效率
assimilatory metabolism 吸收代谢(作用);同化代谢
assimilatory power 吸收能力;同化能力
assimilatory quotient 同化商
assimilatory root 光合根
assimilatory tissue 同化组织
assist 支援;增加推力;扶助;帮助
assistance 协助
assistance and salvage at sea 海上救助
assistance on a refundable basis 有偿援助
assistance with local regulation 协助查明当地规章
assistant 助手;助理(人)员;助剂;副手;辅助的
assistant accountant 助理会计师
assistant chemist 助理化学师
assistant chief engineer 副总工程师
assistant compiled unit 协编单位
assistant director 副主任;副校长;副社长;副理事;副董事;副厂长;副经理
assistant driller 副司钻
assistant editor 协编人
assistant engine 辅助机车
assistant engineer 轮助【船】;助理轮机员;助理工程师;副工程师
assistant fitter 安装助手
assistant foreman 领班;工长助理
assistant index 辅助指标
assistant locking force 辅助锁闭力
assistant locomotive 辅助机车
assistant manager 经理助理;副经理
assistant mate 驾助;助理驾驶员
assistant officer 助理;助理驾驶员
assistant of the engineer 工程师助理
assistant pilot 助理引航员
assistant purser 助理事务长;助理管事
assistant research fellow 助理研究员
assistant statistician 助理统计员
assistant superintendent 助理监督
assistant supervisor 助理监督员;助理管理员
assistant town planner 助理城市设计师
assistant track foreman 养路副工长
assistant vent stack 副通气立管
assist devices 辅助医疗装置
assisted access 加速存取
assisted area 可获资助地区
assisted charges 附加费
assisted circulation 强制循环;辅助循环
assisted-circulation boiler 强制循环锅炉
assisted circulation pipe 辅助循环管

assisted control 辅助操纵
assisted draught 人工通风;辅助通风
assisted-draught cooling tower 辅助通风冷却塔
assisted housing 公助住房
assisted instruction 加速指令
assisted operation 补机运行
assisted running 双机运行
assisted take-off rocket 火箭助推器
assisted ventilation 辅助通气
assist exercise 助力运动
assisting diver 援救潜水员
assisting grade 辅助坡(度);推送坡度;加力牵引坡度
assisting locomotive 补机【铁】
assisting locomotive for pushing 推送补机
assisting pusher grade 加力牵引坡度
assisting run of a locomotive 补机运行
assisting vehicle 补机
assistor 助推机;加速器;加力器;助推器
assitant master 助理值班员
assize 圆柱形区块;砌石工程;材积测定;圆柱体石块;石砌块层;(对商品价格规定的)法定标准
as skavl 雪面波纹
Assmann aspiration psychrometer 阿斯曼通风干湿球湿度计;阿斯曼通风干湿表
Assmann psychrometer 阿斯曼干湿表;阿斯曼干球湿度计
associable design 可相伴设计
associate 同事;联合;结合;缔合子
associate architect 协作建筑师;副建筑师
associate architect or engineer 建筑师或工程师临时事务所
associate broker 经纪人助手
associate charges 附加费
associate curve 伯特兰曲线
associated 配套的;联合的;结合的;关联的
associated architect 合作建筑师
associated array 相伴阵列
associated Au ore 伴生金矿石
associated bank 联营银行
associated bilateral aid 联合双边援助
associated bridge structure 跨桥结构
Associated Builders and Contractors 建造商及承包商联合会
associated center 缔合中心
associated circulation boiler 辅助循环锅炉
associated cleavage 伴生劈理
associated company 联营公司;联号(业务经营上的联系)
associated conductor 组合导线;组合导体;组合导管
associated contaminant 伴生污染物
associated contractor 副承包者;副承包商;副承包人
associated corpuscular 伴生微粒
associated cost 投资追加费;联合费(用);联合成本;联带费用;附带费用
associated craft for dredging works 疏浚工程配套船舶
associated data 关联数据
associated deposit 共生矿产
associated diagram 关联图
associated digraph 伴随有向图
associated directed graph 相伴有向图
Associated Electrical Industries 英国联合电气工业公司
associated element 伴生元素
associated elliptic integral 连带椭圆积分;可结合的椭圆积分
associated emission 伴生发射
associated en echelon veins 伴生雁列脉
associated engineer 合作工程师
associated enterprise 联营企业;联合企业;组合企业
associated environmental effect 伴生环境影响
associated equipment 配套设备;相关联设备
associated facility 配套设施
associated file 相关文件
associated flow rule 相关流动准则
associated folds 伴生褶皱
associated function 联带函数
associated gas 伴生气
Associated General Constructors of America 美洲总承包商联合会
associated graph 相伴图

associated ion 缔合离子
associate director 副董事长
associated Laguerre function 连带拉盖尔函数；可结合的拉盖尔函数
associated Laguerre polynomial 连带拉盖尔多项式；可结合的拉盖尔多项式
associated Legendre function 连带勒让德函数；可结合的勒让德函数
associated Legendre polynomial 连带勒让德多项式；可结合的勒让德多项式
associated line 共用线路
associated liquid 缔合液体
associated material 配套材料
associated matrix 结合矩阵；相伴矩阵；转置矩阵；共轭矩阵；关联矩阵
associated member 辅助构架
associated metal 共生金属；伴生金属
associated mineral 组合矿物；共生矿物；伴生矿物
associated model 伴随模型
associated molecule 缔合分子
associated movement 联合运动；伴随运动
associated natural gas 缔合气；伴生天然气
associated nitrogen fixation 联合固氮
associated operation 联营作业
associated ore 伴生矿
associated outfit 相关联设备
associated parking area 毗邻的停车场地；毗邻停车场
associated particle monitor 伴生粒子监测器
associated phenomenon 伴生现象
associated pollutant 伴生污染物
associated reaction 缔合反应
associated rock 伴生岩石
associated scalar type 关联标量类型
associated series 连带级数
associated set type 关联集合类型
associated shear joints 伴生剪节理
associated ship 联营船
associated signal(l)ing 关联信令；对应信号方式
associated structure 伴生构造
associated structures of diapir 底辟伴生构造
associated structures of fold 褶皱伴生构造
associated support program(me) 连带辅助程序
associated symmetric directed graph 相伴对称有向图
associated tensor 相伴张量；关联张量
associated transformer 联合变压器；附属变压器
associated undertaking 附属子公司
associated undirected graph 相伴无向图
associated useful component 次要有用组分；伴生有用组分
associated useful mineral 次要有用矿物
associated variate 相伴变量
associated variation 相联变动
associated water 共生水；伴生水
associated wave 缔合波
associated wind 联合风
associated word 关联词
associated works 辅助工程；相关工程；联合工程
associate engineer 副工程师
associate expert 助理专家
associate member 副会员；准会员
associate members of conference 班轮公会准会员
associate participant 非全额参与人
associate professor 副教授
associate research fellow 副研究员
associate thickener 缔合增稠剂
association 群丛；配对；联ої；联合会；连带；学会；协会；相联；组合；植物群丛；缔合作用
association analysis 关联性分析；缔合分析
association area 联络区
association between variables 变量间的联系
association centre 联系中心
association coefficient 相依系数
association complex 植物群丛复合体
Association Computer Machinery 计算机协会
association constant 缔合常数
association control service element 连续控制服务单元；相关控制服务单元
association diagram 关联图
association Economic for Europe 欧洲经济学会
association factor 关联因子
association fiber 联合纤维
Association for Anticontamination 污染控制协会

Association for Applied Solar Energy 太阳能利用协会(美国)
Association for the Purification of Sewage 污水净化协会
association headquarters 协会总部；协会本部
association in time 时间相关；时间关联
association list 关联表
association mechanism 联合机理
Association of American Engineer 美国工程师协会
Association of America for Quaternary Research 美国第四纪研究协会
Association of American Geographers 美国地理学家协会
Association of America of Scientific Workers 美国科学工作者协会
Association of American Pesticide Control Occifials 美国农药管理委员会
association of architects 建筑师协会
Association of Asphalt Paving Technologists 沥青铺面专家协会；沥青铺路专家协会
association of average adjuster 海损理算师协会
association of computer users 计算机用户协会
Association of Consulting Engineers 顾问工程师协会
association of deposits 沉积物共生组合
association of fisheries 渔业协会
Association of French Normalization 法国标准化协会
association of house lessor 房主协会
association of ideas 概念的组合
Association of Iron-Ore Export(ing) Countries 铁矿砂出口国协会；铁砂出口国联盟
Association of Maritime Engineering 水运工程协会
association of maturity and yield 成熟期与产量的相关性
association of mountain-shape 山地形态组合
Association of National Pipe Thread 全国管道螺纹协会
Association of Natural Rubber Producing Countries 天然橡胶生产国协会
association of organisms 生物群丛
association of plants and animals 动植物群落
association of sedimentary rock 沉积岩岩石组合
Association of Ship Brokers and Agents 船舶经纪与代理协会
association of soil 土壤组合
association of source reservoir and cap rock 生储盖组合
association of tectono-sedimentary rocks 构造岩石组合
Association of Tungsten Producers 钨生产商协会
Association of West European Shipbuilders 欧洲造船商协会
association pair 相联对
association rate 结合速率
association scheme 结合方案
associations of laborers 劳动者联合体
associative algebra 结合代数
associative and parallel computation 相联并行计算
associative chemisorption 缔合化学吸附
associative collar 结合环
associative data processing 相联数据处理；相关数据处理
associative dimensioning 相关尺寸标注
associative flow rule 相适应流动法则
associative key 启发式判读样式；相联关键码
associative law 结合律
associative memory 联合存储器；内容定址存储器；相联存储器；关联内存储器；伴随存储器
associative memory match 相联存储符合
associative operation 结合运算
associative perception 整体感
associative polyurethane thickener 缔合聚氨酯增稠剂
associative processor 相联处理机
associative property of hologram 全息图的组合性质
associative retrieval 相联检索
associative ring 结合环
associative search 联系查找；联想检索；相联检索
associative storage 内容定址存储器；关联存储器
associative storage register 相联存储寄存器

associative structure 联想结构
associative table looking 相联查表
associativity 结合性
associator 联系文件
assommoir 投掷矢石台；(要塞的)战台
Assorcebunder silk 阿索斯本德丝
assort 相配；配齐；分类；配套；分级
assortative mating 选配
assorted 配套的；混合的；各种各样的；分类的
assorted brass 杂铜
assorted colo(u)red buoy 杂色浮标
assorting 整理；分配；分类
assorting effect 分选作用
assorting process 分类过程
assorting room 分选工段
assorting tolerance 选别公差；分选公差
assortment 品质；无分选；各种各类的聚合；材种；分类；分级
assortment charges 分类费
assortment of particle 颗粒分级
assortment table 材种表
assseets appraisal 资产重估
assume 设想；采取
assumed azimuth 假方位角；假定方位角
assumed bonds 保付证券
assumed computation load 假定计算荷载
assumed condition 假定条件；假设条件
assumed coordinate system 假定坐标系(统)
assumed cost 假定成本
assumed datum 假定起算值；假定基(准)面
assumed decimal point 假设的小数点；假定(十进位)的小数点
assumed design speed of a highway 公路假定设计车速
assumed design stress-strain 假定设计应力应变
assumed duration 假定持续时间
assumed elevation 假定高程
assumed formula 假定公式
assumed frame line 假设肋骨线；理论肋骨线
assumed ground elevation 假定地面高程
assumed ground plane 假定高程起始面
assumed height system 假定高程系
assumed interest rate 预定利率
assumed latitude 假设纬度；假定纬度；选择纬度
assumed liability 假定负债
assumed load(ing) 设计荷载
assumed load 假设载荷；假设荷载；假设负载；假定载荷；计算荷载
assumed longitude 假设经度；假定经度；选择经度
assumed mean 假平均值；假定平均值；假定平均数；假定均数
assumed mean sea level 假定平均海平面
assumed origin 假定原点
assumed plane coordinates 假定平面坐标
assumed position 假定位置；假定方位；选择位置；选择船位
assumed principal-point 假定主点
assumed priority 假定优先级
assumed radial centre 假定辐射中心
assumed(ship) position 假设船位；假定船位
assumed spheroid 假定椭球
assumed statistic(al) inspection 统计假设检验
assumed stress approach 设计应力法；应力估计
assumed swing assumption of swing 假定旋角
assumed symbol 常用符号
assumed tilt 假定倾角
assumed unit weight method 假定容重法
assumed value 假定值
assumed working plane 进给剖角面
assume no risk 不承担风险
assume responsibility 承担责任
assumption 假设；假定
assumption agreement 承担协议书；承担协议
assumption diagram 理论线图；结构图
assumption fee 承担费(用)
assumption of isotropy 各向同性假设
assumption of mortgage 承担产权抵押
assumption of plane cross-section 平截面假设
assumption of tilt 假定倾角
assumptions economic analysis 经济分析中的假定
assumption value 假定值；采用值
assurance 确信；担保；保证
assurance coefficient 保证系数；保险系数；安全系数
assurance company 保险公司

assurance factor 保证系数;保证度;安全因素;安全系数
assurance mutuelle 互保险
assured 已保险;被保险人
assured against fire 火灾保险
assured mineral 可靠储量;开拓储量
assured power 保证功率
assured system capacity 系统保证出力
assured with average 担保单独海损
assure factor 保证系数
assurer 保险商;保险人;保险公司
assurgent 向上升起的;向上浮起的
assurin 二氨基双磷酯
assuring rate of depth datum 深度基准面保证率
assymmetric(al) motion 不对称运动
Assyntian orogeny 阿森特造山运动【地】
Assyrian architecture 亚西利亚建筑(公元前1275~前538年)
Assyrian ornament 亚西利亚装饰
astable 不稳定的
astable balance 不稳定平衡
astable blocking oscillator 不稳定间歇振荡器
astable circuit 不稳定电路
astable equilibrium 不稳定平衡
astable multioscilator 非稳(定)多谐振荡器
astable multivibrator 单稳多谐振荡器;非稳态多谐振荡器;非稳多谐振荡器;非稳(定)多谐振动器;不稳定多谐振荡器
astable operation 非稳态运动
A-stage 第一阶段;初期状态
A-stage phenol-formaldehyde resin 可熔阶段酚醛树脂
A-stage resin 可熔阶段树脂;可溶酚醛树脂
a standing funny monk in blue-and-white 青花站济公
astarboard 向右舷;在右舷
Astarte 阿斯塔特(腓尼人所崇拜的丰饶和爱的女神)
astatic 无定向的
astatic balance 无定向平衡
astatic buckling load 不稳定的屈曲荷载;非静止压曲临界荷载
astatic coil 不定向线圈
astatic constant 无定向常数
astatic control 无定向控制;不定向控制
astatic drainage 无定向水系
astatic elastic pendulum 无定向弹性摆
astatic galvanometer 无定向电流计;不定向电流计
astatic governing 无定向调节
astatic governor 无定向调节器;恒速调速器
astatic gravimeter 无定向重力仪;无定向重力计;助动重力仪
astatic gyroscope 无定向陀螺仪
astatic instrument 无定向仪表
astaticism 无定向性
astatic lake 无定湖;内陆湖;不定湖
astatic magnetometer 无定向磁力仪;无定向磁力计
astatic magnetometer method 无定向磁力仪法
astatic meter 助动重力计
astatic microphone 无定向传声器;不定向传声器
astatic needle 无定向磁针
astatic pair 无定向对
astatic pendulum 无静差摆;无定向摆
astatic regulation 无定位调节
astatic regulator 无静差调节器
astatide 砹化物
astatine analyzer 砹含量分析仪
astatism 无定向性
astatized gravimeter 无定向重力计
astatotyrosine 砹代酪氨酸
astay 同支索平行;与锚时链与睡眠成锐角的状态
astay freight 超载货物
astel 平巷护顶板;平巷顶板背板;屋面顺槽中的拉条;隧道平巷支撑板
astelic 无中柱的
astely 无中柱式
aster 碎件;星体
asteria 星彩宝石
asterion 星点
asteriscus 星状耳石
asterisk 加星号于;星状物;星形标记;星号;星标;注星号
asterisk notation 星号记法
asterisk picture character 星号图像字符
asterisk protection 星号保护
asterisk with based variables 关于基本变量的星号用法
asterism 星芒;星彩性;星彩
astern 向船尾;在船尼;船尼后
astern cam 倒车凸轮
astern clutch line 后退离合器气管
astern flag 倒车旗号
astern guardian valve 倒车辅助阀
astern motion 向后运动
astern navigating bridge 船尼部驾驶台
astern of one's reckoning 误期
astern power 倒车功率
astern reach 倒车惯性滑动距离
astern running 倒车【船】
astern speed 后退速度
astern stage 倒车级
astern steering test 后退操舵测验;后退操舵测试
astern trial 倒驶试车;倒车试验
astern turbine 倒车透平;倒车汽轮机;倒车轮机
astern wheel 倒车轮
Asterococcus limneticus 湖生绿星球藻
asteroid 星状的;星形的;小行星
asteroid belt 小行星带
asteroid-shaped curve 星形曲线
aster venustus 美丽紫苑
asthenobiosis 不活动生活
asthenolith 熔岩浆体
asthenolith hypothesis 放射性热熔化假说;放射热成因岩浆假说
asthenopia 视力疲劳
asthenosphere 软流圈;软流层
asthernosphere bump 软流圈凸起
as the season progresses 随着季节的推移
as they are 现状条款
Astian 阿斯蒂阶【地】
a stiff penalty 严厉处罚
astigmatic 散光的
astigmatic aberration 像散差;像散
astigmatic accommodation 像散调节
astigmatic bundle 像散光束
astigmatic correction 像散校正
astigmatic corrector 像散校正器
astigmatic difference 像散差
astigmatic focus 像散焦点
astigmatic image 像散成像
astigmatic lens 像散透镜
astigmatic mounting 像散装置
astigmation 像散性;像散;像差
astigmation correction 像散校正
astigmatism 散光;像散(现像)
astigmatism against the rule 反规性散光;不正常像散
astigmatizer 像散装置;像散镜;像散器;夜间测距仪
astigmat lens 像散透镜
astigmatometer 散光计;像散计
astigmator 像散校正装置
astigmatoscope 散光镜;像散镜
astigmatoscopy 散光镜检查
astigmia 散光
astigmometer 散光计;像散计
astigmoscope 散光镜
astillen (坑道中的)过梁;岩脉间层;(平碉的)隔墙
astite 红柱云母角岩
astler 琢石(古英语)
ASTM colo(u)r standard 美国材料试验学会颜色标准
Ast-Molin rib(bed) floor 阿斯特—莫林加肋楼板;阿斯特—莫林肋形楼板
ASTM slope 美国材料试验学会黏度—温度特性曲线图
astoli canvas 车篷防水帆布(爱尔兰)
astomatal 无气孔的
astomous 无开口的
Aston 阿斯顿
Aston dark space 阿斯顿阴极暗区;阿斯顿暗区
astonomical theory of business cycles 天文经济周期理论
Aston process 阿斯顿方法
Aston spectrum 阿斯顿光谱
Aston whole number rule 阿斯顿整数定则
astoop 在倾斜的位置上的
Astoria Bridge 阿斯托里亚桥(美国)
Astoria seamer 阿斯托里亚式封罐机
astos 石棉沥青屋面材料
ast-pulley 定滑轮
astraddle 跨
Astrafoil 阿斯特拉富尔塑料片;透明箔
astragal 门扇盖缝条;(柱头或柱脚的)圆剖面小线脚;半圆装饰;半圆形挡水条;半圆线脚;半圆饰
astragal cornice 小凸装饰线脚;圆线型飞檐
astragal frieze 小珠壁缘;半圆雕带
astragal front 串珠饰面;半圆饰面;圆弧锁面
astragal joint 半圆凸凹线脚接头
astragal plane 圆角刨;圆缘刨;圆线刨
astragal tile 筒瓦;半圆形瓦;半圆装饰砖
astragal tool 半圆刀具
astragal window 横木窗;玻璃格条
astrakanite 白钠镁矾
astrakhan 阿斯特拉罕羔
astrakhan cloth 毛圈织物
astrakhanite content 白钠镁矾含量
astrakhanite rock 白钠镁矾岩
astral 观察天窗
astral crown 星冠
astral dome 天文观测窗;天文观测舱
Astralene 阿斯特雷涤纶弹力丝
astral fiber 星丝
astralite 星字炸药;奥司脱拉特
astral lamp 无影灯
astral movement clock 夜光钟
astral oil 变质精制石油
Astralon 阿斯特拉龙尼龙假捻变形纱
Astrasil 阿斯特希尔(一种夹层材料)
astration 物质改造
astray freight 过剩货物
Astrazon dye(stuff) 阿斯特拉宗染料
astreated 星状装饰
astre fictif 参考恒星
a stretch of high land from which water drains 河水流走的一片高地
astriction 限制;收缩
astride 跨越
astringe 收敛
astringency 收敛作用;收敛性;黏滞性
astringency taste 涩味
astringent 收敛性的;收敛剂;黏结的
astringent clay 涩黏土;含明矾的黏土
astringent substance 收敛性物质
astrionics 天文电子学
astrium 房屋内庭
astroarchaeology 天文考古学
astroaz 天文方位角
astroballistics 天文弹道学
astrobiology 天体生物学
astrobleme 天体碰撞坑;古陨石坑
astrobotany 天植物学
astrocamera 天体照相机
Astrocarb 阿斯特罗卡布
astrochanite 白钠镁矾
astrochemistry 天体化学
astrochronology 天文年代学
astroclimate 天文气候
astroclimatology 天文气候学
astrocompass 天文罗盘;天文罗经;星象罗盘
astrodome 天文圆顶;天文观测窗;天文观测舱
astrodynamics 天文动力学
astrofix 天文定位
astrofixation network 天文定点网
astrogation 天文导航
astrogeodesy 天体测量
astrogeodetic data 天文大地测量数据
astrogeodetic datum orientation 大地基准的天文大地定位;天文大地基准定向
astrogeodetic deflection 天文大地偏差;天文大地测量偏差
astrogeodetic deflection of the vertical 天文大地垂线偏差
astrogeodetic level 天文大地水准测量
astrogeodetic method 天文大地测量方法
astrogeodetic net adjustment 天文大地网平差
astrogeodetic net(work) 天文大地网
astrogeodetic undulation 天文大地(水准面)起伏
astrogeodynamics 天文地球动力学
astrogeography 天体地理学
astrogeology 天文地质学;天体地质学;大地天文学
astrogeophysics 天文地球物理学
astrograph 天文高度图;天文定位器;天体照相仪;

天体摄影仪
astrographic(al) catalogue 照相星表
astrographic(al) chart 照相星图
astrographic(al) doublet 天体照相双合透镜
astrographic(al) objective 天体照相物镜;天体摄影物镜
astrographic(al) position 天体照相位置;天体摄影位置
astrograph mean time 天体照相平时
astrography 天体照相学;天体摄影学
astrogravimetric 天文重力测量的
astrogravimetric level (ling) 天文重力水准(测量)
astrogravimetric points 天文重力测量点
astrogravimetry 天文重力测量
astroid 星形线;星形的
astrolabe 决定维度的检测仪;棱镜等高仪;圆圈测天仪;观象仪;等高仪;测星盘
astrolithology 陨石学
Astroloft 阿斯屈罗夫脱捻假捻变形丝
astrology 超耐热镍合金
Astroloy 阿斯特罗伊镍基超耐热合金
astromagnetism 天体磁学
astrometer 天体测量仪
astrometric aspect 天体测量方位
astrometric baseline 天体测量基线;测星基线
astrometric binary 天测双星
astrometric declination 天体测量赤纬
astrometric expectation 天体测量期望
astrometric method 天体测量法
astrometric orbit 天体测量轨道
astrometric right ascension 天体测量赤经
astrometry 天体测量学;天体测量学;天体测定学
astrometry satellite 天体测量卫星
astron 仿天器
astronaut 宇宙航行员;宇宙飞行员;宇航员
astronautics 宇宙航行学
astronavigation 天文领航;天文航海(学);天文航法;天文导航;天体导航;航天;星际航行;宇航
astronavigation system 天文导航装置;天文导航系统
Astron barker 阿斯特朗链式剥皮机
astronegative 天长底片
astronomer 天文学家
Astronomer Royal 皇家天文学家
astronomic(al) 天文学的
astronomic(al) aberration 天文光行差
astronomic(al) almanac 天文年历
astronomic(al) azimuth 天文方位角
astronomic(al) azimuthal station 天文方向点
astronomic(al) calculation 天文计算
astronomic(al) camera 天文照相机;天文摄影机
astronomic(al) chart 天文图
astronomic(al) chronicle 天文历书
astronomic(al) climate 天文气候
astronomic(al) clock 天文钟;天文时计
astronomic(al) compass 天文罗盘;天文罗经
astronomic(al) constants 天文常数
astronomic(al) constant system 天文常数系统
astronomic(al) constituent 天文分量
astronomic(al) control 天文控制
astronomic(al) coordinate measuring instrument 天文坐标测量仪
astronomic(al) coordinates 天文坐标
astronomic(al) coordinate system 天文坐标系
astronomic(al) date 天文日期
astronomic(al) datum 天文大地基准
astronomic(al) day 天文日
astronomic(al) determination 天文测定
astronomic(al) distance 天文距离
astronomic(al) electron coupler device spectrometer 电子耦合器件天文光谱仪
astronomic(al) electronics 天文电子学
astronomic(al) ephemeris 天文历;星历表
astronomic(al) equator 天文赤道
astronomic(al) eyepiece 倒像目镜
astronomic(al) figures 天文数字
astronomic(al) fix 天文定向;天文定位;天测船位
astronomic(al) fixation 天文定位
astronomic(al) fixation network 天文定点网
astronomic(al) geodesy 天文大地测量(学)
astronomic(al) geography 天文地理
astronomic(al) geology 天文地质学;天体地质学
astronomic(al) gravity level 天文重力水准

astronomic(al) horizon 天文地平(线);天球地平(圈)
astronomic(al) inertial navigation aids 天文惯性导航设备
astronomic(al) instrument 天文仪器
astronomic(al) latitude 黄纬;天文纬度
astronomic(al) lens 天文透镜
astronomic(al) level(l)ing 天文水准(测量)
astronomic(al) line of position 天文定位线
astronomic(al) longitude 天文经度;黄经
astronomic(al) map 天体图
astronomic(al) measurement 天文观测;天文测量
astronomic(al) meridian 天文子午线;天文子午圈;天球子午线
astronomic(al) meridianal plane 天文子午面
astronomic(al) method 天文学方法
astronomic(al) navigation 天体导航;天文航海(学);天文航法;天文导航
astronomic(al) navigation system 天文导航系统
astronomic(al) navigation tables 天文航海表
astronomic(al) nomenclature 天文命名
astronomic(al) number of artificial heavenly bodies 人造天体天文代号
astronomic(al) nutation 天文章动
astronomic(al) objective 天文物镜
astronomic(al) observation 天文观察;天文观测;天体观测
astronomic(al) observatory 天文台;天文馆
astronomic(al) optics 天文光学
astronomic(al) orientation 天文定向;天文定位
astronomic(al) parallel 天文纬圈;天文纬度;天球纬圈
astronomic(al) perturbation technique 天文摄动技术
astronomic(al) photography 天文摄影学;天文摄影术;天体照相学
astronomic(al) photometry 天体光度学;天文光度测量;天体测光学
astronomic(al) plate 天文底片
astronomic(al) point 天文点
astronomic(al) position 天文位置;天文点;天测位置
astronomic(al) position circle 天文船位圆
astronomic(al) position determination 天文定位
astronomic(al) position finding 天文定位
astronomic(al) position line 天文位置线;天文船位线;天测位置线
astronomic(al) pyrometer 天体测温计
astronomic(al) reflector 天文反射望远镜
astronomic(al) refraction 大气折射;大气差;天文折光差;天文蒙气差
astronomic(al) research centre 天文研究中心
astronomic(al) satellite 天文卫星
astronomic(al) scintillation 天文闪烁
astronomic(al) sight 测天;天文观测
astronomic(al) sign 天文符号
astronomic(al) spectrograph 天体摄谱仪
astronomic(al) spectrophotometry 天体分光度测量
astronomic(al) spectroscopy 天体光度学
astronomic(al) station 天文点;天文台
astronomic(al) sun 平太阳
astronomic(al) surveying 天文测量
astronomic(al) synchronome 同步天文钟
astronomic(al) telescope 天文望远镜;天体望远镜
astronomic(al) test 天文检验
astronomic(al) theodolite 地平经纬仪;天文经纬仪
astronomic(al) tidal constituent 天文分潮
astronomic(al) tide 天文潮
astronomic(al) tide level 天文潮位
astronomic(al) time 天文时(期)
astronomic(al) time clock 日光自动开关
astronomic(al) time scale 天文时间标度
astronomic(al) transit 子午仪;中星仪
astronomic(al) traverse 天文测量用导线
astronomic(al) triangle 球面三角形;定位三角形;天文三角形;天文船位三角形
astronomic(al) twilight 天文曙暮光;天文晨昏蒙影
astronomic(al) unit 天文单位
astronomic(al) vertical 天文垂线
astronomic(al) watch 测天表
astronomic(al) year 回归年
astronomic(al) year book 天文年历
astronomic(al) zenith 天文天顶

Astronomische Gesellschaft 德国天文学会
astronomy 天文学
astronomy of position 位置天文学
astrophile 天文爱好者
astrophotocamera 天文摄影仪;天体照相机
astrophotogram 天文底片
astrophotograph 天体照相
astrophotography 天文摄影学;天体照相学;天体照相术
astrophotometer 天体光度计;天文光度计
astrophotometry 天体光度学;天文光度测量;天体测光学
astrophyllite 星叶石
astrophysics 天文物理学;天体物理学
Astroplax 抹灰用水泥(基层为白色硬水泥)
astropolarimeter 天体偏振计
astro-position line 天文位置线
astroscope 天文仪;星宿仪
astrospectrograph 天体摄谱仪
astrospectrometer 天体分光计
astrospectroscopy 天体光谱学
astrostereogram 天体立体照片
astrotracker 星象跟踪仪
Astroturf 阿斯特罗特尼龙草皮
astrovehicle 宇宙飞行器;航天器
astrovelocimeter 天体视向速度仪
Asturian architecture 阿斯图里建筑(西班牙)
Asturian orogeny 阿斯图造山运动【地】
asty 苔星体
astylar 无柱式的
astylar back 无柱(式)背面
astyllen 阻水小坝;挡水槛(坑道)
A-subduction basin A 式俯冲盆地
a sudden 暴洪
a suit 一套
a suite of racks 一套机架
asummable theorem 不可和定理
Asuncion Bridge 亚松森桥(巴拉圭)
Aswan High Dam 阿斯旺高坝
as-welded 焊态;焊后状态
as-welded condition 焊接状态
asylum 收容所;避难所;庇护所
asylum for aged 敬老院
asylum for lunatic 精神病院
asylum for the aged 敬老院;养老院
asylum switch 锁盖开关
asymbiotic 非共生的
asymbiotic system 非共生系统
asym-dipheny hydrazine 不对称二苯肼
asymmeter 非对称计;不对称计
asymmetric(al) 不对称的
asymmetric(al) absorption band 不对称吸收谱带
asymmetric(al) A. C. trigger switch 不对称交流触发开关
asymmetric(al) amplitude modulation 不对称调幅
asymmetric(al) anastigmat 非对称性消像散透镜;不对称肖像散透镜
asymmetric(al) antenna 不对称天线
asymmetric(al) anticline 不对称背斜(层)
asymmetric(al) arch 不对称拱
asymmetric(al) arch dam 不对称拱坝
asymmetric(al) array 不对称组合
asymmetric(al) atom 不对称原子
asymmetric(al) balance 不对称平衡
asymmetric(al) bars 高低杠
asymmetric(al) beam 不对称梁
asymmetric(al) bedding 不对称层理
asymmetric(al) bending 不对称弯曲
asymmetric(al) branch 非对称支路
asymmetric(al) branching process 非对分法
asymmetric(al) building 非对称建筑物;不对称建筑物
asymmetric(al) carbon 不对称碳
asymmetric(al) carbon atom 不对称碳原子
asymmetric(al) cell 不对称管;不对称电池
asymmetric(al) chevron 不对称人字形
asymmetric(al) circuit 不对称电路
asymmetric(al) circuit element 单向导电性元件
asymmetric(al) class 不对称晶族
asymmetric(al) climbing ripple 不对称上攀波纹;不对称爬升波痕
asymmetric(al) coefficient 不对称性系数
asymmetric(al) conductivity 不对称导电性
asymmetric(al) conductor 不对称导体

asymmetric(al) configuration 非对称配置；不对称形状；不对称形体；不对称形态；不对称配置；不对称构造
asymmetric(al) current 不对称电流
asymmetric(al) curvature 非对称弯曲；非对称曲率
asymmetric(al) curve 非对称曲线；不对称曲线
asymmetric(al) deflection 非对称偏转；不对称偏转
asymmetric(al) device 非对称器件
asymmetric(al) digital subscriber line 非对称数字用户线路
asymmetric(al) diopter 不对称屈光度
asymmetric(al) dipole 不对称偶极子
asymmetric(al) dispersion 非对称色散
asymmetric(al) distortion 非对称畸变；不对称失真；不对称畸变
asymmetric(al) distribution 非对称分布；不对称(性)分布
asymmetric(al) distribution curve 不对称分布曲线
asymmetric(al) double-chamber surge tank 不对称(式)双室调压井
asymmetric(al) double vee groove 非对称V形槽；不对称V形坡口
asymmetric(al) drainage 不对称水系；不对称排水(系统)
asymmetric(al) effect 不对称效应
asymmetric(al) elliptic conduit 不对称椭圆形通道
asymmetric(al) error 不对称误差
asymmetric(al) estimation method 非对称估计法
asymmetric(al) facade 不对称立面；不对称正面
asymmetric(al) face 不对称面
asymmetric(al) factorial design 不对称阶乘设计
asymmetric(al) fault 不对称故障
asymmetric(al) fission 非对称裂变；不对称裂变
asymmetric(al) flight 不对称飞行
asymmetric(al) flow 不对称水流
asymmetric(al) flower 不对称花
asymmetric(al) flow of water 不对称水流
asymmetric(al) focusing 不对称聚焦
asymmetric(al) fold(ing) 不对称褶皱
asymmetric(al) foundation 不对称基础
asymmetric(al) four terminal network 不对称四端网络
asymmetric(al) frequency curve 不对称频率曲线
asymmetric(al) fringe 非对称条纹
asymmetric(al) girdle 不对称环带
asymmetric(al) half disc 不对称圆盘
asymmetric(al) heating 不对称热变形工艺
asymmetric(al) high voltage impulse track circuit 不对称脉冲轨道电路；不对称高压脉冲轨道电路
asymmetric(al) image 不对称的像
asymmetric(al) impulse response 非对称脉冲响应
asymmetric(al) induced copolymerization 不对称诱导共聚作用
asymmetric(al) induction 不对称诱导
asymmetric(al) input/output 非对称输入输出设备；非对称输入输出
asymmetric(al) interchange 不对称互换
asymmetric(al) interface 不对称接口
asymmetric(al) ionic atmosphere 不对称离子氛
asymmetric(al) joint 不对称接合
asymmetric(al) karyokinesis 不对称有丝分裂
asymmetric(al) kink-band 不对称膝折带
asymmetric(al) laccolith 不对称岩盖
asymmetric(al) laminate 不对称层板
asymmetric(al) lens 不对称透镜
asymmetric(al) light distribution 不对称光分布
asymmetric(al) lighting 非对称照明
asymmetric(al) load(ing) 不对称荷载；不对称负载；不对称负荷
asymmetric(al) luminaire 非对称配光型灯具
asymmetrically distributed 不对称分布的
asymmetrically placed 不对称布置的
asymmetrically split anode magnetron 不对称分辨阳极磁控管
asymmetric(al) meander loop 不对称河环
asymmetric(al) membrane 不对称膜
asymmetric(al) migration 偏对称迁移
asymmetric(al) mineralization belt 不对称成矿带
asymmetric(al) mitosis 不对称有丝分裂
asymmetric(al) mode 不对称振型
asymmetric(al) molecult 不对称分子
asymmetric(al) multiprocessing system 非对称多处理系统

asymmetric(al) multiprocessor 非对称型多处理机
asymmetric(al) multivibrator 不对称多谐振荡器
asymmetric(al) network 不对称网络
asymmetric(al) operation 不对称运行(状态)
asymmetric(al) parafocusing arrangement 不对称仲聚焦装置
asymmetric(al) particle 不对称粒子
asymmetric(al) peak 不对称峰
asymmetric(al) periodic(al) orbit 非对称周期轨道
asymmetric(al) point 非对称点；不对称点
asymmetric(al) polarization 不对称极化
asymmetric(al) polyvinylidene fluoride membrane 聚偏氟乙烯不对称膜
asymmetric(al) potential 不对称电位；不对称电势
asymmetric(al) pulse 不对称脉冲
asymmetric(al) pulse depth modulation 不对称脉冲宽度调制
asymmetric(al) radiation 非对称辐射
asymmetric(al) ratio 不对称比率
asymmetric(al) reaction 不对称反应
asymmetric(al) refractor 不对称折射器
asymmetric(al) relation 非对称关系
asymmetric(al) response 非对称响应；不对称反应
asymmetric(al) ripple mark 不对称波痕
asymmetric(al) riverbed 不对称河床
asymmetric(al) roller bearing 非对称滚子轴承
asymmetric(al) rotating disc 不对称转盘
asymmetric(al) rotor 不对称转子
asymmetric(al) second division 不对称第二次分裂
asymmetric(al) selection polymerization 不对称选择聚合作用
asymmetric(al) shape 非对称形状
asymmetric(al) shedding 不对称梭口
asymmetric(al) sideband 不对称边带
asymmetric(al) sideband transmission 残余边带传输
asymmetric(al) sideband transmitter 不对称边带发射机
asymmetric(al) spreading 不对称扩张
asymmetric(al) state 不对称状态
asymmetric(al) step function 非对称阶梯函数
asymmetric(al) stretch 不对称伸缩
asymmetric(al) stretching vibration 不对称伸缩振动
asymmetric(al) structure 不对称结构
asymmetric(al) surface ridge 不对称表面皱纹
asymmetric(al) synase 不对称突触
asymmetric(al) synthesis 不对称合成
asymmetric(al) system 非对称系统；不对称系统
asymmetric(al) taper ends cheese 不对称双锥筒子
asymmetric(al) terminal voltage 不对称端子电压
asymmetric(al) terrace 不对称阶地
asymmetric(al) test 不对称检验
asymmetric(al) three phase short circuit current 不对称三相短路电流
asymmetric(al) top 不对称陀螺
asymmetric(al) transcription 不对称转录
asymmetric(al) transducer 不对称换能器
asymmetric(al) transmission 不对称传输
asymmetric(al) treatment 不相称的处理方法
asymmetric(al) valley 不对称谷
asymmetric(al) vein 不对称矿脉
asymmetric(al) vibration 非对称振动；不对称振动
asymmetric(al) vibrator 非对称振捣器；不对称振子
asymmetric(al) wave 不对称波
asymmetric(al) wave ripple 不对称波痕
asymmetric(al) well pattern 不对称井网
asymmetry 偏位；非对称(性)；反对称(性)；不对称(性)；不对称现象
asymmetry angle 非对称角
asymmetry coefficient 偏度系数；非对称性系数
asymmetry control 不对称性控制
asymmetry distribution 非对称性分布
asymmetry error 不对称误差
asymmetry flow 不对称水流
asymmetry of distribution 分布的不对称现象
asymmetry of peak 峰不对称性
asymmetry of slope 不对称坡
asymmetry parameter 非对称(性)参数
asymmetry potential 非对称电位
asymmetry spectrum 不对称谱
asymptote 渐近线
asymptote circle 渐近圆
asymptote convergence 渐近收敛

asymptote equation 渐近方程
asymptote expansion 渐近展开(式)
asymptote expression 渐近式
asymptote formula 渐近公式
asymptote integration 渐近积分
asymptote line 渐近线
asymptote method 渐近法
asymptote of convergence 辐合线
asymptote of electric(al) sounding curve 电测深曲线的渐近线
asymptote of the first part of the curve 首枝渐近线
asymptote of the last part of the curve 尾枝渐近线
asymptote relation 渐近关系(式)
asymptote solution 渐近解
asymptote value 渐近值
asymptotic(al) 渐近的；渐近线的
asymptotic(al) analysis 渐近线分析；渐近分析
asymptotic(al) approximation 渐近性近似；渐近近似(法)；渐近逼近(法)
asymptotic(al) behavio(u)r 渐近特性；渐近状态
asymptotic(al) branch 渐近支
asymptotic(al) characteristic 渐近特性
asymptotic(al) cone 渐近锥(面)
asymptotic(al) curvature 渐近曲率
asymptotic(al) curve 渐近曲线；主切曲线
asymptotic(al) developable 渐近可展曲面
asymptotic(al) direction 渐近方向
asymptotic(al) direction of arrival 到达的渐近方向
asymptotic(al) distribution 渐近分布
asymptotic(al) distribution of eigenvalue 特征值的渐近分布
asymptotic(al) efficiency 渐近效率
asymptotic(al) equal 渐近相等
asymptotic(al) error 渐近误差
asymptotic(al) error constant 渐近误差常数
asymptotic(al) expansion 渐近展开(式)
asymptotic(al) expansion formula 渐近展开式公式
asymptotic(al) flux 渐近通量
asymptotic(al) frequency equation 渐近频率方程
asymptotic(al) giant branch 渐近巨星支
asymptotic(al) integration 渐近积分
asymptotic(al) LC30 起始半数致死浓度
asymptotic(al) line 渐近线
asymptotic(al) lower bound 渐近下界
asymptotically efficient estimate 渐近有效估计(值)
asymptotically normal process 渐近正态过程
asymptotically normal random variable 渐近正态随机变量；渐近正态随机变量；渐近正规随机变量
asymptotically stable in the large 全局渐近稳定
asymptotic(al) method 渐近法
asymptotic(al) model 渐近宇宙模型
asymptotic(al) normality 渐近正态性；渐近正态法
asymptotic(al) orbit 渐近轨道
asymptotic(al) point 渐近点
asymptotic(al) polarization 渐近偏振
asymptotic(al) population 饱和人口
asymptotic(al) progressive sampling 渐近取样
asymptotic(al) radiance 渐近辐射率
asymptotic(al) relation 渐近关系(式)
asymptotic(al) representation 渐近表示
asymptotic(al) series 渐近级数
asymptotic(al) shortest confidence interval 渐近最短置信区间
asymptotic(al) solution 渐近解
asymptotic(al) stability 渐近稳定性；渐近稳定度
asymptotic(al) standard error 渐近标准误差
asymptotic(al) surface 渐近面
asymptotic(al) theory 渐近理论
asymptotic(al) threshold concentration 渐近阈值浓度
asymptotic(al) unbias(s)ed test 渐近无偏检验
asymptotic(al) value 渐近值
asynapsis 不联会
asynchronism 时间不同；异步性；不同时性；不同步
asynchronism automatic reclosing device 非同期自动重合(闸)装置
asynchronization 时间不同；异步；非同步化
asynchronized synchronous motor 异步化同步电动机
asynchronous 非同步的；不同期的

asynchronous balanced mode 异步均衡方式
asynchronous bus interface 异步总线接口
asynchronous bus system 异步总线系统
asynchronous character 异步字符
asynchronous circuit 异步电路
asynchronous communication 异步通信;异步传送
asynchronous communication control adapter 异步通信控制适配器
asynchronous communication control attachment 异步通信控制附件
asynchronous communication controller 异步通信控制器
asynchronous communication interface 异步通信接口
asynchronous communication interface adapter 异步通信接口适配器
asynchronous computer 异步计算机
asynchronous condenser 异步调相机;异步电容器;异步补偿机
asynchronous control 异步控制
asynchronous counting 异步计数
asynchronous data 异步数据
asynchronous data channel 异步数据信道
asynchronous data collection 异步数据收集
asynchronous data transfer 异步数据传送
asynchronous data transmission 异步数据传输
asynchronous development 不同步发展
asynchronous device 异步装置
asynchronous disconnected mode 异步断开方式
asynchronous display 异步显示
asynchronous dynamo 异步电动机;异步电机
asynchronous entry point 异步入口点
asynchronous event 异步事件
asynchronous exit 异步退出;异步出口
asynchronous exit routine 异步出口例行程序
asynchronous finite state machine 异步有限自动机
asynchronous flow 异步数据流;异步(径)流
asynchronous frequency changer 异步变频器
asynchronous generator 异步发电机;不同步发电机
asynchronous impedance 异步阻抗
asynchronous input 异步输入
asynchronous interface 异步接口
asynchronous interrupt request 异步中断请求
asynchronous link 异步链路
asynchronous machine 异步机;异步电机
asynchronous measurement 异步测量
asynchronous memory 异步存储器
asynchronous memory capability 非同步存储能力
asynchronous modem 异步调制解调器
asynchronous modem controller 异步调制解调控制器
asynchronous motor 异步电动机;非同步电动机
asynchronous multiplexer 异步多路转换器
asynchronous network 异步网(络)
asynchronous opening and closing 6-jaw grab 异步启闭式六颗板抓具
asynchronous operation 异步运行;异步工作;异步操作
asynchronous operator 异步算子
asynchronous output 异步数据输出;异步输出
asynchronous parallel algorithm 异步并行算法
asynchronous phase modifier 异步整相器
asynchronous phase modulator 异步调相机
asynchronous photograph 非同步像片
asynchronous processing 异步处理
asynchronous quenching 异步遏制
asynchronous record operation 异步记录操作
asynchronous request 异步要求;异步请求
asynchronous response mode 异步应答方式
asynchronous sending mode 异步发送方式
asynchronous sequential network 异步时序网络
asynchronous serial data 异步串行数据
asynchronous servomotor 异步伺服电动机
asynchronous shift register 异步移位寄存器
asynchronous signal 异步信号
asynchronous signalling 异步信号发送
asynchronous spark gap 异步电花隙
asynchronous speed 异步转速;异步速率
asynchronous starting 感应起动;非同步起动
asynchronous structure 异步结构
asynchronous system 异步系统;异步方式;非同步系统
asynchronous terminal 异步终端

asynchronous terminal support 异步终端支持
asynchronous termination 异步终结
asynchronous time division multiplexer 异步时分多路器
asynchronous time division multiplexing 异步时分多路方式
asynchronous timer 异步计时器
asynchronous transfer mode 异步传输模式
asynchronous transmission 异步发送;异步传输;非同步传输
asynchronous transmission output 异步输出;异步传送输出
asynchronous transmit mode 异步传输模式
asynchronous travel(l)ing wave 异步行波
asynchronous working 异步工作;异步操作
asynchrony 异步;不同时性
asynergia 协同不能;动作失调
asynergy major 步行协同不能
A system 甲级道路【道】
aszyzygetic bitangents 不合冲双切线
ata 阿泰
atacamite 氯铜矿
ATA carnet 临时进口证
atactic polypropylene 无规聚丙烯
atactic polypropylen modified asphalt 无规聚丙烯改性沥青
atactose 无规溶胶
at-a-glance 简表
atala (产于尼日利亚的红棕色～紫红色的)硬木
at all level 各级水位
at all stages of water 各级水位
at all time 随时
at anchor 锚泊的;锚泊;下锚
at and from insurance 在港及出航时期的保险
at any stage of development 在任何一个发育阶段
at any time 随时;不时
at a premium 超过票面以上的价格(股票)
a task of top priority 当务之急
atavism 返祖性
ataxiameter 协调功能测试器
ataxite 角砾斑杂岩;镍铁陨石
ataxitic 角砾斑杂状
ataxitic microdolerite 角砾斑杂显微粒玄岩
ataxitic prophyrite 角砾斑杂玢岩
Atbas alloy 阿特巴斯镍铬钢
at carrier's option 由船东选择
Atchison 艾奇逊—托皮卡—圣菲铁路公司
at current cost 按现时费率计算
at current price 按当时价格
Atdabanian 阿特达板阶
at discount 贴水;按折价
atectonic 非构造的
atectonic dislocation 非构造错位;非构造变位
atectonic pluton 非造山运动深成岩体
atectonite 非构造岩
atelectasis 不张
atelene 不完全晶形
atelestite 砷酸铋矿;板羟砷铋石
atelier 工作室;雕刻室;画室
at equal spacing 等间距
Aterite 阿特赖特铜镍锌合金
aterrimus 深黑色
a testing program(me) 一种试验方法
atestor 证人
at every identifiable change of strata whichever is met earlier 遇变层(岩土工程勘察术语)
at factory 在工厂交货;工厂交活
at fair price 价格公道
at full capacity 全部开工
at given speed 达到给定速度
at godown 在货栈交货
at grade 在同一平面上
at-grade expressway 平交快速干道
at-grade highway 平交公路
at grade intersection 平面交叉
at-grade station 地面站
at ground level 底层的
at-ground transport system 地面运输系统
athabascaite 斜方硒铜矿
athalassohaline water (内陆的)咸水湖水
at hand 手边
Atha's alloy 阿瑟无磁性耐热耐蚀钢
athen(a)eum 图书馆;阅览室;雅典娜神殿
atheneite 砷汞钯矿

Athenian window 雅典式窗(希腊);阁楼窗(希腊)
Athens Charter 雅典宪章(1933年确定城市规划原则的,法国)
athermal 冷却变态
athermal solutions 无热溶液
athermal transformation 非热转变
athermancy 不透热性;不透辐射性;不透辐射热性
athermanous 不透热的;不透辐射热的
athermic 不发热的
athermic effect 绝热效应
athermous 不透辐射性的
athermous body 不透辐射热体
atheromatous plaque 粥样斑块
athey wagon 平板卸底拖车;载重拖车;载重翻斗车;平板拖车
Athey wheel 履带轮
a thin coating on ceramics 陶衣
athletes' village 运动员村
athletic back-stop fence 运动场的后阻篱障
athletic facility 体育设施
athletic field 运动场;操场
athletic surfacing 人造运动场地
athodyd 冲压式(空气)喷气发动机
athodyd combustion chamber 冲压式空气喷气发动机燃烧室
a thousand-year flood 千年一遇的洪水
a three-step process 一种三步骤过程
athwart 船的横向
athwart-hawse 横过他船船首
athwart sea 舷侧浪
athwartship 横向的【船】;与龙骨线直交;横向地【船】
athwartships corrector 横向校正器
athwartships inclination 横倾【船】
athwartships magnet 横向自差校正磁铁;横向磁铁
athwartship permanent magnetism 横向永久船磁
athwartships propeller 侧推器
athwartships stability 横稳性【船】
athwartships trim list 横倾【船】
athwart the bow of a ship 横越对方的船首
athwart the tide 横流航行
athwart thrust 横向推进;横向推动
a tight(easy)money market 银根紧(松)的金融市场
atilt 倾斜的
a timesaving technique for measuring 一种测样省时技术法
ation 增强辐射
Atkinson cycle 艾金森循环
Atkinson motor 艾金森电动机
Atkinson unit 艾金森通风阻力单位
atlantes 男像柱
Atlantic 大西洋的
Atlantic arctic front 大西洋极锋
Atlantic cedar 北非雪松;大西洋雪松
Atlantic coast area 大西洋沿岸地区
Atlantic current 大西洋海流
Atlantic deep water 大西洋深水层
Atlantic epoch 大西洋期
Atlantic floral realm 大西洋植物地理区系
Atlantic fracture zone 大西洋破裂带
Atlantic Free Trade Area 大西洋自由贸易区
Atlantic greyhound 大西洋航线快速定期船
Atlantic Indian basin 大西洋—印度洋海盆
Atlantic international geophysical year program (me) 大西洋国际地球物理年计划
Atlantic Intracoastal Waterway 大西洋沿岸水道
Atlantic line 大西洋航线
Atlantic liner 大西洋班轮
Atlantic Merchant Vessel Report System 大西洋商船报告系统
Atlantic Ocean 大西洋
Atlantic ocean circulation 大西洋环流
Atlantic Ocean region 大西洋地区
Atlantic polar front 大西洋极锋
Atlantic ports 大西洋沿岸各港
Atlantic province 大西洋区
Atlantic ridge seismotectonic zone 大西洋海岭地震构造带
Atlantic route 大西洋航路
Atlantic sardine 大西洋沙丁鱼
Atlantic series 大西洋岩系
Atlantic Standard Time 大西洋东岸标准时间;大西洋标准时间;大西洋时间

Atlantic suite 大西洋岩组;大西洋岩系;大西洋岩套
Atlantic time 大西洋时间
Atlantic tracking ship 大西洋遥测跟踪船
Atlantic type coast 大西洋型海岸
Atlantic type coastline 大西洋型海岸线
Atlantic type continental margin 大西洋型大陆边缘
Atlantic type geosynclines 大西洋型地槽
atlantite 暗霞碧玄岩
Atlas 阿特拉斯
atlas 图集;图册;地图集;地图册
Atlas bronze 阿特拉斯青铜
atlas cartography 地图集制图学
Atlas cedar 大西洋雪松
Atlas Computer Laboratory 阿特拉斯计算机实验室
Atlas geosyncline 阿特拉斯地槽
atlas grid 图集位置指示格网;地图指示格网
atlasite（表面呈大理石花纹或木纹的）石棉薄板
Atlas-Johnson tubing joint 阿特拉斯—约翰逊接头
atlas leaf 地图集图页
atlas maps of oceans 海洋图集
Atlas mooring 阿特拉斯式系泊装置
atlas of cloud 云级图
atlas of map 地图集地图
atlas of stars 恒星图
atlas of surface waters 地面水图册
atlas of the living resources of the seas 海洋生物资源分配图
atlas of the world 世界地图集
atlas of tidal stream 潮流海图
atlas of tides 潮汐海图
atlas page 地图集图页
atlas paper 印图纸
atlas sheet 地图集图幅
atlas type 地图集类型
Atlas-Werk system 阿特拉斯—沃克方式
atlas white 美国白硅酸盐水泥
Atlon 阿特纶醋酯纤维
at market price 按市场价
atmidometer 蒸发器;蒸发表;蒸发计;汽化计
atmidometrograph 描记式蒸发计
atmidometry 蒸发率测定法;蒸发测定法
atmocautery 蒸气烙器
atmoclast 大气碎屑;风成碎屑
atmoclastic 大气碎屑岩
atmoclastic rock 气碎屑;大气碎屑岩;风成碎屑岩
atmogenic deposit 大气沉积(物)
atmogenic rock 气生岩
atmogeochemical anomaly 气圈地球化学异常
atmolith 气成岩;风积岩
atmology 水蒸汽学;蒸汽学;大气学
atmolysis 微孔分气法
atmometer 蒸发器;汽化计;蒸发计
atmometry 水蒸气测定法;蒸发量测定术;蒸发测定法
atmophile element 亲气元素
atmoseal 气封(法)
atmosphere 大气圈;大气层
atmosphere air 大气
atmosphere circulation 大气环流
atmosphere composition 大气成分
atmosphere controller 气氛控制器
atmosphere density 大气密度
atmosphere dispersion model 大气弥散模型
atmosphere drag 大气阻力
atmosphere equipment 保护气氛供应设备
atmosphere gas 气圈气;保护气体
atmosphere gas converter 保护气体转化器
atmosphere gas generating plant 常压煤气发生设备
atmosphere gas generator 常压气体发生器
atmosphere ga(u)ge 气压计
atmosphere humidity condensate 大气水汽凝结物
atmosphere influence 气ం影响
atmosphere laser 大气激光器
atmosphere melting 控制气氛熔炼法
atmosphere moisture 大气水分
atmosphere monitoring satellite 大气监测卫星
atmosphere-ocean dynamics 大气海洋动力学
atmosphere-ocean-land-ice system 大气—海洋—陆地—冰雪体系
atmosphere-ocean system 大气海洋系统
atmosphere pollution 大气污染
atmosphere pressure 大气压(力)

atmosphere pressure affected coefficient of groundwater 地下水的气压影响系数
atmospherepurifying equipment 保护气体净化设备
atmosphere resistance of malthoid 油毡大气稳定性
atmosphereretaining sinter box 保护气体烧结箱
atmosphere revitalization section 大气活化区域
atmosphere sampler 气体采样器
atmosphere science 大气科学
atmosphere sensing and maintenance system 大气传感与维护系统
atmosphere sintering 气氛烧结
atmosphere suit 常压潜水服
atmosphere temperature transducer 大气温度传感器
atmosphere transfer temperature 气氛转变温度
atmosphere transparency 大气透明度
atmosphere turbidity 大气浑浊度
atmosphere venting 静压孔;大气压力孔
atmospheric 大气的
atmospheric absorption 大气吸收(作用)
atmospheric accumulation 大气累积(作用)
atmospheric acidity 大气酸度
atmospheric acid precipitation experiment 大气酸沉降试验
atmospheric acoustics 大气声学
atmospheric action 大气作用
Atmospheric Administration 大气管理局
atmospheric adsorbed water 大气吸附水
atmospheric aeration coefficient 大气复氧系数
atmospheric aerosol 大气气溶胶
atmospheric ag(e)ing 大气作用老化;大气老化
atmospheric agitation 大气抖动
atmospheric air 大气
atmospheric and vacuum distillation unit 常减压装置;常减压蒸馏装置
atmospheric anomaly 大气异常
atmospheric arc rectifier 大气电弧整流器;大气弧整流器
atmospheric area 大气污染物滞留区
atmospheric argon 大气氩
atmospheric argon correction 大气氩校正
atmospheric armo(u)red suit 常压铠甲式潜水服
atmospheric assimilation 大气同化
atmospheric attenuation 大气衰减
atmospheric backscatter 大气反向散射
atmospheric billows 大气波
atmospheric-biologic(al) system 大气—生物系统
atmospheric blast 大气鼓风
atmospheric boiling poit 常压沸点
atmospheric boundary layer 地表边界层;大气边界层
atmospheric braking 大气制动
atmospheric burner 低压燃烧器;大气式燃烧器;常压燃烧器
atmospheric capacity 大气湿度
atmospheric carbon dioxide concentration 大气二氧化碳浓度
atmospheric carcinogen 大气致癌物(质)
atmospheric cascade 大气急流
atmospheric chamber 常压舱
atmospheric change 大气变化
atmospheric chemical kinetics 大气化学动力学
atmospheric chemistry 大气化学
atmospheric chemistry of troposphere 对流层大气化学
atmospheric chlorine 大气氯
atmospheric chlorine build-up 大气氯累积
atmospheric chlorine detector 大气含氯检测仪
atmospheric circulation 大气循环;大气环流
atmospheric circulation pattern 大气环流模式
atmospheric circulation type water cooler 大气循环式水冷却器
atmospheric cleanes itself 大气自净(作用)
atmospheric cleansing 大气净化
atmospheric climate 大气气候
atmospheric cloud 大气干扰云
atmospheric colo(u)ration 大气着色;大气染色
atmospheric colo(u)ration effect 大气着色效应;大气染色效应
atmospheric column 常压塔
atmospheric commotion 大气震荡
atmospheric compartment 常压舱

atmospheric compartment dryer 常压间隔干燥室;常压式干燥器;常压干燥室
atmospheric component 大气成分
atmospheric composition 大气组成;大气成分
atmospheric composition monitor 大气成分监测仪
atmospheric concentration 大气浓度
atmospheric condensation 大气凝结;降雨;降水
atmospheric condenser 淋雨式冷凝器;空气冷凝机;大气压式凝水柜;常压冷凝器
atmospheric condition 大气状态;大气状况;大气条件;大气情况;气候条件
atmospheric constituent 大气组元
atmospheric contaminant 大气污染物
atmospheric contaminant analyzer 大气污染物分析仪
atmospheric contaminant sensor 大气污染物传感器
atmospheric contamination 大气沾染;空气污染;大气污染
atmospheric contamination burden 大气污染负荷
atmospheric contamination potential 大气污染潜在性
atmospheric control 常压控制;气动力操纵
atmospheric controller 常压控制器
atmospheric cooler 空气冷却器;大气冷却器
atmospheric cooling rate tester 大气冷却率测试仪
atmospheric cooling tower 开放式冷却塔
atmospheric core 通气芯
atmospheric correction 大气修正;大气校正;大气改正
atmospheric corrosion 大气侵蚀;大气腐蚀(作用)
atmospheric corrosion factor 大气腐蚀因素
atmospheric corrosion resistant 大气腐蚀防护剂
atmospheric corrosion test 大气腐蚀试验
atmospheric crack(ing) 大气作用开裂;大气开裂;风化裂纹;风干开裂;老化裂缝
atmospheric curing chamber 常压养护箱
atmospheric cycle 大气循环
atmospheric damping 大气衰减
atmospheric densitometer 大气密度计
atmospheric density 大气密度
atmospheric deposit 大气沉积(物)
atmospheric deposition 大气沉降
atmospheric depth 大气深度;大气厚度
atmospheric diffusion 大气扩散
atmospheric diffusion equation 大气扩散方程
atmospheric diffusion measuring system 大气扩散检测系统
atmospheric diffusion model 大气扩散模式
atmospheric dilution 大气稀释;排入大气
atmospheric discharge 天电放电;大气放电
atmospheric discharge coil 天电放电线圈;大气放电线圈
atmospheric discharger 大气放电器
atmospheric dispersion 大气色散;大气弥散;大气扩散;大气分散
atmospheric dispersoid 大气弥散胶体
atmospheric distillation 常压蒸馏
atmospheric disturbance 天电干扰;大气扰动;大气干扰
atmospheric diving 常压潜水
atmospheric diving suits 充气潜水服
atmospheric diving suit system 常压潜水服系统
atmospheric diving system 常压潜水系统
atmospheric drag 大气阻力
atmospheric drier 空气干燥器
atmospheric drought 天气干旱;大气干旱
atmospheric drum drier 鼓风式常压干燥器
atmospheric duct 大气中的飞尘;大气含尘量;大气波导(层)
atmospheric dust 大气尘埃
atmospheric dynamics 大气动力学
atmospheric eclipse 大气食
atmospheric effect 大气影响;大气效应
atmospheric effect of cooling system 冷却系统的大气效应
atmospheric electric(al) field 大气电场
atmospheric electricity 天电;大气电学;大气电
atmospheric electricity measurement 大气电学检测
atmospheric emission standard 空气排放标准;大气排放标准
atmospheric engine 大气机

atmospheric entry trajectory 进入大气层轨道
atmospheric environment 大气环境
atmospheric environmental capacity 大气环境容量
atmospheric environmental impact assessment 大气环境影响评价
atmospheric environmental quality assessment 大气环境质量评价
atmospheric environmental quality map 大气环境质量图
atmospheric evapo(u)ration 大气蒸发
atmospheric explosion 大气层爆炸
atmospheric exposure test 室外暴露试验；空气曝露试验；耐候性试验；大气暴露试验；暴试验；暴露试验
atmospheric extinction 大气消光；大气削弱
atmospheric fading 烟气褪色；大气褪色
atmospheric fallout 大气沉积物
atmospheric feedback process 大气反馈过程
atmospheric feeder 大气压力冒口
atmospheric filter 常压失水仪；常压过滤仪
atmospheric flash tower 常压闪蒸塔
atmospheric flow pattern 大气流模式
atmospheric fluctuation 大气起伏；大气波动
atmospheric fluidized bed combustion 常压流化床燃烧
atmospheric fluorescence 大气荧光
atmospheric forcing 大气作用；大气外力作用
atmospheric free radical 大气游离基
atmospheric front 大气锋
atmospheric gas 大气气体；大气成因；大气
atmospheric gas-burner 自燃通风煤气燃烧器
atmospheric gas-burner system 自动通风煤气燃烧器系统
atmospheric gas-oil ratio (换算大气条件下的)地面油气比
atmospheric gas phase reaction 大气气相反应
atmospheric general circulation 大气总循环；大气总环流
atmospheric general circulation model 大气环流模型
atmospheric geology 气界
atmospheric haze 大气雾霾
atmospheric HF analyzer 大气中氟化氢浓度分析仪
atmospheric humidity 大气湿度
atmospheric hydrocarbon analyzer 大气含烃量分析仪
atmospheric impurity 大气杂质；大气污染物
atmospheric induction burner 自动进气燃烧器
atmospheric influence 大气影响
atmospheric injection-type burner 空气注入式燃烧器
atmospheric input 大气输入
atmospheric interaction 大气相互作用
atmospheric interface 大气界面
atmospheric interference 天电干扰
atmospheric inversion 大气逆温
atmospheric ion 大气离子
atmospheric ionization 大气电离
atmospheric jet 空气喷气发动机
atmospheric laser return 激光大气反射
atmospheric layer 大气壳层；大气层
atmospheric life of a substance 物质在大气中的存在时间
atmospheric light 大气光线
atmospheric line 大气压力线；大气谱线
atmospheric liquid phase reaction 大气液相反应
atmospheric loading 大气负载
atmospheric-marine-petrology system 大气—海洋—岩系
atmospheric mass density 大气质量密度
atmospheric measurement 大气测量
atmospheric metamorphism 大气变性
atmospheric microphysics 大气微观物理学
atmospheric mixed-layer 大气混合层
atmospheric model 大气模型
atmospheric moisture(capacity) 大气湿度；空气湿度；大气水分
atmospheric monitoring 大气监测
atmospheric monitoring satellite 大气监测卫星
atmospheric monitoring station 大气监测站
atmospheric monitoring system 空气监测系统；大气监测系统
atmospheric movement 在大气中运动
atmospheric nitrogen 大气氮

atmospheric noise 天电噪声；天电干扰；大气噪声
atmospheric noise measurement 大气噪声检测
atmospheric observation 大气观测
atmospheric observation bell 常压观察钟
atmospheric observation manipulator bell 带机械手常压观察潜水钟
atmospheric observation network 大气观测网
atmospheric opacity 大气不透明度
atmospheric optics 大气光学
atmospheric origin of reaction 大气应变源
atmospheric oscillation 大气震荡；大气涛动；大气波动
atmospheric over flow dyeing machine 常温溢流染色机
atmospheric overvoltage 大气过电压
atmospheric overvoltage protection 大气过电压保护
atmospheric oxidant 大气中氧化性污染物；大气氧化剂
atmospheric oxidation 大气氧化
atmospheric oxygen 大气中的氧气；大气氧
atmospheric ozone 大气臭氧
atmospheric ozone column 大气臭氧气柱
atmospheric ozone detector 大气臭氧检测仪
atmospheric ozone layer 大气臭氧层
atmospheric parameter 大气参数
atmospheric particles 大气粒子
atmospheric particulate loading 大气悬浮颗粒物量；大气微粒负荷
atmospheric particulates 大气(中)微粒；大气悬浮微粒；大气颗粒物
atmospheric pause 大气上界
atmospheric penetration capability of radar 雷达大气穿透力
atmospheric perturbation 大气扰动
atmospheric phenomenon 大气现象
atmospheric photochemistry 大气光化学
atmospheric photolysis 大气光解作用；大气光解
atmospheric physicist 大气物理学家
atmospheric physics 大气物理学
atmospheric pipe 放空管路；通大气管线；通大气管路；常压管线
atmospheric planetary motion 大气行星运动
atmospheric polarization 大气偏振；大气波动
atmospheric pollutant 大气污染物
atmospheric polluting material 大气污染物
atmospheric pollution 大气污染
atmospheric pollution agency 大气污染管理机构
atmospheric pollution assessment 大气污染评价
atmospheric pollution auto-monitoring system 大气污染自动监测系统
atmospheric pollution burden 大气污染负荷
atmospheric pollution control 大气污染防护
atmospheric pollution control engineering 大气污染防治工程
atmospheric pollution control equipment 大气污染控制设备
atmospheric pollution control planning 大气污染控制规划
atmospheric pollution effect 大气污染效应
atmospheric pollution forecasting 大气污染预报
atmospheric pollution index 大气污染指数；大气污染(源)指数
atmospheric pollution map 大气污染图
atmospheric pollution model 大气污染模型
atmospheric pollution monitor 大气污染监测器
atmospheric pollution monitoring 大气污染监测
atmospheric pollution monitoring instrument 大气污染监测仪
atmospheric pollution monitoring network 大气污染监测网
atmospheric pollution monitoring station 大气污染监测站
atmospheric pollution monitoring van 大气污染监测车
atmospheric pollution policy 大气污染防治政策
atmospheric pollution potential 大气污染潜在性
atmospheric pollution sensor 大气污染传感器
atmospheric pollution simulation model 大气污染模拟模式
atmospheric pollution sources 大气污染源
atmospheric pollution system 大气污染系统
atmospheric pool 大气库
atmospheric precipitation 降水；大气降水量；大气降水；降雨
atmospheric press bed counter current regeneration 无顶压逆流再生【给】
atmospheric pressure 大气压(力)；气压
atmospheric-pressure boiler 低压蒸汽锅炉
atmospheric-pressure chamber 常压舱
atmospheric-pressure chemical ionization mass spectrophotometer 气压式化学电离质谱分光光度计
atmospheric-pressure consistometer 常压稠度仪；常压稠度计
atmospheric-pressure distribution 大气压分布
atmospheric-pressure diving suit 常压潜水服
atmospheric-pressure equipment 低压设备；低压仪器
atmospheric-pressure gas 低压煤气
atmospheric-pressure gas system 低压煤气系统
atmospheric-pressure head 大气压力冒口；大气冒口
atmospheric-pressure kiln 常压炉
atmospheric-pressure leaching 常压侵出
atmospheric-pressure level 气压高程测量
atmospheric-pressure method 抽气加速固结法；真空预压法(软基加固)
atmospheric-pressure saturated steam 低压饱和蒸汽
atmospheric-pressure spraying 低压喷洒
atmospheric-pressure steam boiler 低压蒸汽锅炉
atmospheric-pressure steam cured 低压蒸汽养护的
atmospheric-pressure steam curing 常压蒸汽养护
atmospheric-pressure steam heating 低压蒸汽供暖；低压蒸汽保暖
atmospheric-pressure steam pipe 低压蒸汽管(道)
atmospheric-pressure storage tank 常压储[贮]油罐
atmospheric-pressure system 低压蒸汽系统
atmospheric-pressure test 气压试验
atmospheric-pressure transfer module 常压人员转运舱
atmospheric-pressure wet steam 低压饱和蒸汽
atmospheric prevention and control law 大气污染防治法
atmospheric probing 大气探测
atmospheric process 大气过程
atmospheric propagation 大气传播
atmospheric propagation of high energy laser 高能激光大气传输
atmospheric pump 大气泵
atmospheric purification 大气净化
atmospheric quality and modification 大气质量改进
atmospheric quality assessment 大气质量评价
atmospheric radiation 大气辐射
atmospheric radioactivity 大气放射性；大气反射性
atmospheric radio noise 大气无线电噪声
atmospheric radio wave 大气无线电波；大气发射波
atmospheric radio window 大气无线电窗口
atmospheric reaction 大气反应
atmospheric recovery time 大气复原时间
atmospheric refraction 蒙气差；大气折射
atmospheric refraction correction 空气折射校正
atmospheric refraction displacement 大气折射位移
atmospheric refraction distortion 大气折射畸变
atmospheric refraction error 大气折射误差；大气折光差
atmospheric refraction index 大气折射率
atmospheric refrigerant condenser 大气式制冷剂冷凝器
atmospheric regenerant 大气更新剂
atmospheric region 大气壳层
atmospheric relief valve 向空(安全)泄放阀；大气安全阀
atmospheric rerun 常压再蒸馏
atmospheric research 大气研究
atmospheric research and remote sensing plane 大气研究和遥感研究专机
atmospheric researcher 大气研究者
atmospheric residence time 在大气中停留时间
atmospheric residue 常压渣油
atmospheric resistance 大气阻力
atmospheric riser 大气压冒口；大气压力冒口；大

气冒口
atmospheric riser core 冒口通气芯
atmospheric rock 气成岩
atmospherics 大气噪扰;大气干扰
atmospheric sampling 空气取样;大气取样;大气采样
atmospheric sanitation 大气卫生
atmospheric scattering 大气散射
atmospheric science 大气科学
atmospheric scientist 大气科学家
atmospheric scintillation 大气闪光
atmospheric sea salt 大气海盐
atmospheric sediment 大气沉积(物)
atmospheric sedimentation 大气沉降
atmospheric seeing 大气宁视度
atmospheric shell 大气圈;大气壳层;大气壳
atmospheric shimmer 大型闪烁;大气闪烁
atmospheric shower 大气簇射
atmospheric side 大气侧
atmospheric siphon 大气虹吸
atmospheric sound absorption 大气声吸收
atmospheric sounder 大气探测器
atmospheric sounding 大气探测
atmospheric sounding projectile 高层大气探测火箭
atmospherics suppressor 大气干扰抑制器
atmospheric stability 大气稳定度
atmospheric stamp 气锤
atmospheric state 大气状态
atmospheric statics 大气静力学
atmospheric steam 常压蒸汽
atmospheric steam cured concrete 蒸汽养护混凝土
atmospheric steam curing 混凝土蒸汽养护;常压蒸汽养护
atmospheric sterilizer 常压灭菌器
atmospheric still 常压管式加热炉
atmospheric storage tank 常压储[贮]柜;无压储[贮]箱;常压储[贮]槽;常压储油罐;常压储罐
atmospheric stratification 大气层结
atmospheric structure 大气结构
atmospheric submersible 常压潜水器
atmospheric substance 大气物质
atmospheric suit 常压潜水服
atmospheric sulfer 大气硫
atmospheric sulfur dioxide analyzer 大气中二氧化硫浓度分析仪
atmospheric sulfur dioxide detector 大气中二氧化硫含量检测仪
atmospheric surveillance 大气监视
atmospheric suspensoids 大气悬胶体
atmospheric syphon 大气虹吸
atmospheric system 大气系统
atmospheric tank 常压储罐
atmospheric temperature 大气温度;气温
atmospheric temperature inversion 气温逆转;气温逆增;气温倒布
atmospheric temperature of mine 矿井气温
atmospheric test 大气层试验
atmospheric thermodynamics 大气热力学
atmospheric tide 大气潮;气压潮
atmospheric top(ping) 常压拔顶;大气顶层
atmospheric trace gas 大气痕量气体
atmospheric transformation product 使大气变性的产物
atmospheric transmission 大气透射;大气传输;大气传递
atmospheric transmissivity 大气透射率;大气透过率
atmospheric transmittance 大气透射系数;大气透射率;大气透明率
atmospheric transparency 大气透明度
atmospheric transport 大气输运;大气输送
atmospheric transport-chemistry model 大气输运和化学反应模型
atmospheric tsunami 暴风海啸
atmospheric turbidity 大气混浊度
atmospheric turbulence 大气湍流;大气骚动
atmospheric type water cooler 空气循环式水冷器
atmospheric vacuum distillation process 常减压蒸馏过程
atmospheric vacuum distillation unit 常压真空蒸馏装置

atmospheric valve 放空阀;空气阀
atmospheric vapo(u)r 空气水汽
atmospheric vapo(u)rizer 常压汽化器
atmospheric ventilation 通风
atmospheric vessel 大气层容器;无压容器
atmospheric visibility 大气能见度
atmospheric vortex 大气漩涡;大气涡旋
atmospheric water 大气水;大气降水;降水
atmospheric water cooler 大气水冷器
atmospheric water cooling tower 大气水塔;大气式冷却塔
atmospheric water vapo(u)r 空气的水蒸气;大气水蒸气
atmospheric wave 大气波
atmospheric waveguide duct 大气层波导管
atmospheric weathering 大气风化
atmospheric window 大气窗(口)
atmospherium 大气馆;大气层次
atmos-valve 空气阀;大气阀;放空阀
Atokan 阿托克统
atokite 锡铂钯矿
atoleine 液体石蜡
atoline 液体石蜡
atoll 环状珊瑚岛;环形珊瑚岛;环礁
atoll island 环礁岛
atoll lagoon 环礁泻湖
atoll lake 环礁湖
atollon 小环(形)礁;大环礁圈;复环礁圈
atoll pool trend 环礁油气藏趋向带
atoll reef 环状珊瑚礁;环礁
atoll reef coast 环礁海岸
atoll reef lagoon deposit 环礁泻湖沉积
atoll ring 环礁圈
atoll structure 环状构造;环礁状构造
atoll texture 环形结构
a-toluic acid 苯基乙酸
atom 原子;最小识别单位;操作数;表中原子
atom-absorption mercury analyzer 原子吸收型测汞仪
atom engine 原子发动机
atom-fluorescence mercury analyzer 原子荧光型测汞仪
atomic absorption 原子吸收
atomic absorption flame spectrometer 原子吸收火焰光谱仪
atomic absorption method 原子吸收法
atomic absorption photometer 原子吸收光度计
atomic absorption photometry 原子吸收光度法;原子吸收光测法
atomic absorption spectrographic analysis 原子吸收光谱分析
atomic absorption spectrography 原子吸收光谱法
atomic absorption spectrometer 原子吸收分光计
atomic absorption spectrometry 原子吸收光谱法;原子吸收光谱测定
atomic absorption spectrophotometer 原子吸收分光光度计;放射性光谱吸收仪
atomic absorption spectrophotometry 原子吸收光谱法;原子吸收分光光谱法;原子吸收分光光度法
atomic absorption spectrum 原子吸收光谱
atomic accelerator 原子加速器
atomic acid 四价酸
atomic age 原子时代
atomic aircraft carrier 核动力航空母舰
atomic arc welding 原子氢焊(接)
atomic battery 核能电池;核电池
atomic beam 原子束
atomic binding forces 原子结合力
atomic blast 原子爆炸
atomic blast cloud 原子爆炸烟云
atomic blast excavation 原子爆破开挖
atomic boiler 原子锅炉
atomic bomb 原子弹
atomic bomb-proof shelter 原子弹掩蔽室
atomic bond 原子键
atomic bunker 防原子弹地下掩蔽室;防原子地堡
atomic charge 原子电荷
atomic chemistry 原子化学
atomic clock 原子钟
atomic cloud 原子烟云
atomic cluster 原子团
atomic concentration 原子浓度
atomic crystal lattice 原子晶格

atomic data element 基本数据单元
atomic decomposition 核衰变
atomic defect absorption 原子缺陷吸收
atomic defence 原子防御;原子防护;防核
atomic density 原子密度
atomic diffusion 原子扩散
atomic dipole moment 原子偶极矩
atomic disintegration 核分裂;原子衰变
atomic disruption 原子破裂
atomic draught strip 通风门用的磷铜合金密封带;通风窗用的磷铜合金密封带
atomic electric plant 原子能发电站;核电站
atomic emission 原子发射
atomic emission line 原子发射谱线
atomic emission spectrometry 原子发射光谱法
atomic emission spectrometry with inductively coupled plasma 电感耦合等离子体原子发射光谱法
atomic energy 核能;原子能
atomic energy battery 原子能电池
Atomic Energy Commission 美国原子能委员会;原子能委员会
atomic energy industry 原子能工业
atomic energy installation 原子能工业企业
atomic energy level 原子能级
atomic energy plant 原子能发电厂
atomic energy valve 核电站闸板;核电站挡板
atomic engineering 核电力工业
atomic fallout 沉降物
atomic fission 原子核裂变
atomic fissure 原子分裂
atomic fluorescence spectrophotometry 原子荧光光谱法
atomic force microscopy 原子力显微镜
atomic formula 结构式【化】;原子式
atomic frequency standard 原子频率标准
atomic fuel 核燃料;原子燃料
atomic ground state 原子基态
atomic group 原子团
atomic gyroscope 原子陀螺仪
atomic heat 核热
atomic heat capacity 克原子热容量
atomic H welding 氢原子焊接
atomic hydrogen arc welding 氢原子电弧焊接
atomic hydrogen maser 氢原子微波激射器
atomic hydrogen torch 原子氢焰
atomic hydrogen welding 原子氢焊(接);氢原子焊接
atomic hydrogen welding apparatus 原子氢焊机
atomic hypothesis 原子假说
atomic ice breaker 原子破冰船
atomic irradiation 核放射型辐照
atomicity 原子性;原子数;原子价
atomic laser 原子激光器
atomic link 原子链
atomic locomotive 原子机车
atomic marine plant 船用原子能动力装置;船用核动力装置
atomic mass 原子质量;原子量
atomic mass determination 原子质量测定
atomic mass unit 原子质量单位
atomic model 原子模型
atomic moderation ratio 原子慢化比
atomic modular lattice 原子模格
atomic moisture meter 原子湿度仪
atomic nucleus 原子核
atomic nucleus fission 原子核裂度;原子核裂变
atomic number 原子序数;原子序
atomic number correction 原子序数修正
atomic orbit 原子轨道
atomic orbital 原子轨函数;原子轨道
atomic ordering 原子有序化
atomic orientation 原子取向
atomic oscillation 原子振荡
atomic oxygen 原子氧
atomic oxygen layer 原子氧层
atomic paramagnetism 原子顺磁性
atomic parameter 原子参数
atomic particle 原子粒子
atomic photodissociation laser 光解原子激光器
atomic photoelectric(al) effect 原子光电效应;光化电离
atomic physics 原子物理学
atomic physics in nuclear experiments 核试验原

子物理学
atomic pile 核反应堆;原子·反应堆;原子堆
atomic pile reactor 原子反应堆
atomic plane 原子面
atomic polarizability 原子极化率
atomic polarization 原子极化强度
atomic power 核动力;原子动力
atomic-powered air-craft carrier 原子动力航空母舰
atomic-powered ship 原子动力船;原动力船;核动力船舶;核动力船
atomic power plant 原子能发电厂;原子能电站
atomic power reactor 原子动力反应堆
atomic power station 核能发电站;原子能发电站;原子能电站
atomic-propelled ship 原动力船
atomic propulsion nuclear engine 核发动机
atomic radiation 原子辐射
atomic radiation absorbing glass 防原子辐射玻璃
atomic radiation shielding coating 防原子辐射涂层
atomic radius 原子半径
atomic range 辐射范围
atomic ratio hydrogen/carbon of bitumen 沥青的氢碳原子比
atomic ratio of carbon to hydrogen 碳氢原子比
atomic ratio of hydrogen to carbon 氢碳原子比
atomic ray 原子射线
atomic reaction 原子反应
atomic reactor 核反应堆;原子反应堆
atomic reactor containment structure 原子反应堆建筑物;原子反应堆构筑物
atomic recombination 原子复合
atomic refraction 原子折射度
atomic research 原子能研究
atomic resonance line 原子共振谱线
atomics 原子学
atomic scale 原子标度
atomic scattering 核扩散;原子散射
atomic scattering factor 原子散射因子
atomic scattering power 原子散射本领
atomic separation 原子间距(离)
atomic shell 原子壳层
atomic shelter 原子弹掩蔽室
atomic spectrophotofluorometer 原子荧光光度计
atomic spectroscopy 原子光谱学
atomic spectrum 原子光谱
atomic standard 原子基准
atomic station 原子能发电站
atomic structure 原子结构
atomic submarine 核潜艇
atomic susceptibility 原子磁化率
atomic symbol 原子符号
atomic time 原子时
atomic time scale 原子时标;原子地质年代表
atomic unexcited state 原子基态
atomic vibration 原子振动
atomic volume 原子体积;原子容积
atomic waste 原子能工业废物;原子能废物;放射性废料
atomic wastewater 原子能工业废水
atomic weapon 原子武器
atomic weight 原子量
atomic weight scale 原子量标度
atomic weight unit 原子量单位
atom-inmolecule method 分子中单原子法
atomisation 原子化;粉化作用
atomise 雾化;喷雾
atomising concentrate 雾化剂;浓雾剂
atomistic competition 纯粹竞争
atomite 阿托迈(一种烈性炸药)
atomization 喷雾;雾化;原子化
atomization application 雾化运用
atomization burner 雾化喷嘴
atomization by pressure air 压气喷雾
atomization device 雾化装置;雾化器
atomization efficiency 原子化效率
atomization fuel spray 雾化燃料喷洒
atomization glazing 雾化上釉
atomization in graphite crucible 石墨坩埚原子化法
atomization in graphite tube furnace 石墨管炉原子化法
atomization medium 雾化介质
atomization(method) 喷雾法;雾化法
atomization nut 雾化喷嘴套管
atomization of water 喷水

atomization pattern 扩散图像
atomization period 雾化期
atomization plant 雾化装置;雾化设备
atomization steam 雾化蒸汽
atomize 喷雾;雾化粉化;雾化;弥雾
atomize by compressed air 压缩空气喷雾
atomized air 雾状空气
atomized aluminiun 细铝粉
atomized alumin(i)um powder 雾化铝粉
atomized by compressed air 压气喷雾
atomized dryer 雾化干燥器;雾化干燥剂
atomized fuel 雾化燃料
atomized fuel spray 雾化燃料喷洒
atomized liquid 雾化液体
atomized lubrication 喷油雾润滑法;喷油误养护装备;雾化润滑
atomized oil 喷成雾状的燃油
atomized particle 雾化粒度
atomized powder 雾化粉(末)
atomized spray injectior 雾化式喷嘴
atomized suspended oxidation technique 悬浮废液雾化技术
atomized suspension 雾化悬浮体
atomized suspension(al) technique 悬浮液雾化技术;悬浮雾化技术;雾化悬浮技术
atomized suspension technique of combustion 雾化悬浮物燃烧技术
atomized water 喷出水;雾化水;水雾化
atomized water jet 水雾喷嘴;雾化水喷嘴;水雾喷射
atomized water nozzle 雾化水喷嘴
atomized water spray 水雾喷洒;水雾雾;水喷洒
atomizer 喷雾器;喷丸器;喷笔;喷雾;雾化器;雾化头;微雾碎机;油雾化器;超雾粉碎机;超微粉碎机
atomizer airblast 空气喷油嘴
atomizer aperture 喷雾嘴
atomizer burner 燃烧喷嘴;喷射燃烧器;雾化燃烧器
atomizer by compressed air 压气喷雾器
atomizer chamber 雾化室
atomizer combustion chamber 雾化式燃烧室
atomizer cone 喷雾锥;雾化锥
atomizer device 喷管
atomizer lubrication 油雾润滑
atomizer mill 细磨机
atomizer passage 喷雾管路
atomizer retraction gear 喷射器缩回装置
atomizing 喷雾的;喷雾;雾化的;雾化;吹制珠粒
atomizing air 雾化空气
atomizing air blower 雾化风机
atomizing air intake 喷雾式进气口
atomizing application 雾化运用
atomizing burner 喷雾燃烧器;雾化燃烧器
atomizing carburet(t)er 喷雾式汽化器
atomizing column 喷雾塔
atomizing concentrate 雾化剂
atomizing cone 喷锥;雾锥;雾化锥
atomizing core 雾喷心
atomizing deaerator 喷雾式脱氧器;喷雾式除气器
atomizing device 喷雾装置;喷雾器
atomizing disc 雾化盘
atomizing dryer 喷雾烘干机
atomizing fineness 雾化细度
atomizing humidifier 喷雾加湿器
atomizing jet 喷雾嘴;水枪喷嘴
atomizing machine 喷雾机
atomizing medium 雾喷媒介;雾化介质
atomizing nozzle 喷雾喷嘴
atomizing nut 雾化喷嘴套管
atomizing of liquid fuel 液体燃料的雾化
atomizing pipe 喷管
atomizing pressure 雾化压力
atomizing pump 喷雾泵;雾化式泵
atomizing sprayer 弥雾机
atomizing spraying 雾化喷涂
atomizing-type humidifier 喷雾式加湿器
atomizing-type oil burner 喷雾式油燃烧器
atomizing unit 雾化设备
atom line 原子谱线
atom logic(al) expression 原子逻辑式
atomology 原子学
atomometer 气化计
atom polarization 原子极化
atom powered submarine 原子潜艇
atomseal 气封

atom section 原子截面
atom shell 原子壳层
atom site 原子位置
atoms of inert gas 惰性气体原子
atom-stricken 受原子爆炸污染的
ato-muffler 减声器
a ton by measurement 一吨位置
at-once-payment 立即付款
at one's own risk 自担风险
atony 弛缓
atopy 特异反应性
atoquino 苯基辛可宁酸
at or better 特定价格或更优价格
a total of tubes 总管数
ato unit 起动发动机
at owner's risks 风险由货主负担
Atox mill 阿托克斯辊式磨
Atoxyl 阿托益;氨基苯磺酸钠
atoxylate 氨基苯胂酸
Atozyarn 阿托兹纱
at par 平价;依股票面价格;按票面额
ATP on-board unit 车载列车自动保护装置
at premium 升水;按溢价
Atra-bor 阿硼混剂
a tractor serves several purposes 一机多用
atractozonic projection 分带等积投影
atractylis ovata 姜黄
atracurium besilate 阿曲库铵苯磺酸盐
a train cost of passenger transport 客运单列成本
ATR alloy 锆铜钼合金
atrament 十分黑的物质
atramental 墨水色的
atrament process 磷酸盐处理法;抗蚀磷化法
at random 无规律地
atraton 阿特拉通
atrazin 阿托拉辛
atrazine 阿特拉津
at regular interval 每隔一定时间
atrepsy 营养不足
atresia 闭锁畸形
at-rest earth pressure 静止土压力
at-rest pressure 静压力
at-rest pressure test 静压力试验
atretic body 闭锁体
atrial defect 房缺损
atrial septum 房间隔;房隔
a trickle irrigation system 一种滴灌系统
atrio crater floor 火口基座
atrio lake 火山原湖
atriolum 小中庭(古罗马);小正厅(古罗马);坟墓入口小室
atrionodal region 房结区
atrioseptoplasty 房间隔修补术
atrip 锚离底
at risk limitation 风险限额;风险极限
at risk rules 风险规定
atritor 烘干粉碎机
Atritor flash dryer 快速干燥器
atrium 内庭;天井;四季厅;露天庭院;(古罗马建筑物的)中庭;[复]atria
atrium architecture 内底式建筑;中庭式建筑
atrium house 内院式住宅
atrium lobby 中庭式门厅
atrochal 无原球的
atroglyceric acid 阿卓甘油酸
atrolactamide 苯乳胺
atronene 苯基二氢萘
atronic acid 阿窗酸
at room temperature 在室温中
at room temperature in daylight 室温光照条件下
atrophy 退化;减缩现象
atropic(al) acid 阿卷酸
atropine 阿托品
atropinization 阿托品化;阿托品处理法
atropoyl 阿托酰
atrous 纯黑色;暗灰色
atrovirens 墨绿色
A-truss 人字形桁架;A形桁架
at sea 在航海途中;在海上
at second hand 间接
at set interval 每隔一段时间
at ship owner's risk 由船东负责
at sight 见票时
at-site construction 现场施工

at site two 在第二试验点
at stake 插标志
at staked periods 在规定期限
at stump 在山场
attach 随附;加添;加上;附证件;附上;附加;绑上
attached account 法院查封账户
attached biofilm reactor 附着生物膜反应器
attached boat 附属船(舶)
attached bubble 黏附气泡
attached building 配套建筑物;附属建筑物;附属房(屋);附联式房屋;附属建筑物
attached chart 附图;附表
attached column 壁柱;嵌墙柱;附柱;附墙圆柱;半柱
attached decoration 附加装饰
attached documents 添附证件
attached dune 连岸沙丘;附着沙丘
attached dwelling 毗连住房;毗邻住房
attached edger 附设的立辊轧机
attached-film expanded bed 生物膜膨胀床;附着膜膨胀床
attached foundry 附属铸造车间
attached gable 饰山墙;附加山墙
attached garage 毗邻式汽车房;附属车库;毗连式汽车房
attached green space 附属绿地
attached ground water 连结地下水;附着地下水;吸着水;固定地下水
attached-growth biological process 附着生物生长法
attached-growth nitrification process 附着生长硝化处理法
attached-growth process 附着生长处理法
attached-growth reactor 附着生长反应器
attached-growth waste stabilization pond 附着生长污水稳定塘
attached hospital 附属医院
attached house 毗连住宅
attached housing 相连住宅
attached hydrobiont 附着水生物
attached information processor 附属信息处理机
attached list 附表
attached map of exploratory report 勘探报告图件
attached mass 附加质量
attached microbial film expanded bed process 附着微生物膜膨胀床工艺
attached middle school 附属中学
attached particle 黏附颗粒
attached pent-house 附联披屋;靠近主房的披屋
attached pier 堵墙;壁柱;支墩;扶墙
attached pier capital 附加的墩顶;附墙方柱头
attached plant 附着植物
attached primary school 附属小学
attached processor 附属处理机
attached pump 辅助泵;附备泵
attached sheet 附页
attached ship 附属船(舶);辅助船
attached shock 附体激波
attached shock wave 附体激波
attached sign 附设标志
attached sunspace 附加阳光间
attached support processor 增援处理机;附加增援处理机
attached task 被隶属任务
attached test bar 附铸试棒
attached test coupon 附铸试棒
attached thermometer 附装温度计
attached to 附属于
attached tower crane 附属塔式起重机
attached type 附着式
attached water 束缚水;结合水;吸附水;附着水;薄膜水
attach floating discharge pipeline 接排泥管
attaching clamp 固定夹
attaching creditor 行使扣押权的债权人
attaching label 附加标记
attaching nut 配合螺母
attaching organism 附着生物
attaching plug 小型电源插头;电源插头
attaching rubber to metal 橡胶与金属相结合
attaching task 隶属任务;归属任务;辅助性任务
attachment 起期;具具;连接件;连接法;连接;扣押财产;扣押;扣留;黏结物;查封;附着物;附着体;附着;附属装置;附属品;附录;附件;附加装置
attachment base 附加控制板;附加基板

attachment bolt 联接螺栓;连接螺栓
attachment chain 联接链;爪接链
attachment clamp 卡钉
attachment clip 轴环;铁箍;卡钉;夹子;夹线板
attachment coefficient 结合系数;附着系数
attachment constriction 附着缢痕
attachment driving shaft 辅助传动轴
attachment effect 附壁效应
attachment end 固定端
attachment feature 附加功能部件;附加电缆连接电路
attachment flange 连接法兰;接合凸缘
attachment for cranes 起重机附加设备
attachment for excavators 挖土机附加装置
attachment link 联接杆;连接杆
attachment objective 辅助物镜
attachment of a task 附加任务
attachment of connections 附件的连接
attachment of debtor's property 扣押债务人的财产
attachment of debts 债务扣押;扣押债款
attachment of devices 设备的连接
attachment of interest 起息期
attachment of non-pressure parts 非承压部件
attachment of nozzle 接管附件
attachment of policies 保险开始
attachment of risk 危险责任的起期;保险责任开始
attachment of the current collector 集电器联结法
attachment optic(al) system 辅助光学系统
attachment order 扣押令
attachment paper 随机文件
attachment plug 连接插头;插头;插塞
attachment point 固定点;安装点
attachment ring 联结环;接合圈
attachment screw 装配螺钉;连接螺钉;装合螺钉;止动螺丝;止动螺钉;定位螺丝;定位螺钉
attachments of forklift 叉式装卸车附属装置
attachments of forklift truck 叉式装卸车附属装置
attachments of fork truck 叉式装卸车附属装置
attachment table 连接表
attachment to a wall 墙的凸出物;墙的凸出部分
attachment tools 随机工具
attachment to steel blank 与钢体黏合
attachment to three point linkage 固定于三点悬挂装置上
attachment unit interface 附加装置界面
attachment weld 连接焊缝;连接件焊缝
attach to 附着于
attack 侵蚀;化学侵蚀;迎角;攻击;冲击
attack a fire 扑火
attack angle 迎角;冲角;冲击角
attack by groundwater 地下水作用
attack collision 撞击
attack cut 钻孔定位
attack dialing 突击拨号
attack director 打击指挥仪
attack drill 旋转钎头;钻孔器
attacked concrete 受侵蚀混凝土
attacked plant 受侵蚀植物;受保护植物
attack heading 攻击航向
attacking power 攻击能力;腐蚀性能力;腐蚀性能
attacking wave 来袭波
attack periscope 攻击潜望镜
attack polishing method 浸蚀抛光法;腐蚀抛光法
attack rate 开采强度;冲击速率;发病率
attack resisting glazing 防盗窗玻璃
attack sonar 攻击声呐
attack time 启动时间;增高时间;工作开始时间;出动时间;冲击时间
attack wave 发作波
attaclay 镁质黏土
attacolite 红橙石;红橙色
Attagel 硅镁土(可湿性粉剂掺和剂)
attainability 可及
attainable cost 可达到的成本
attainable high standard 可达到的高标准
attainable production set 可达生产集
attainable speed 可达车速
attainable standard 可达到的标准
attainable standard cost 实现标准成本;可达到标准成本
attainable state 可达状态
attainment of super-elevation 超高渐变段;超高缓和段

attainment of widening 加宽缓和段
attakolite 红橙石
attal 矸石
attapulgite 硅镁土(可湿性粉剂掺和剂);绿坡缕石;活性白土;凹凸棒石;阿塔普吉特镁质黏土
attapulgite catalytic oxidant 凹凸棒催化氧化剂
attapulgite clay 凹凸棒石黏土
attapulgite clay deposit 凹凸棒石黏土矿床
attar 挥发性油
attar of roses 蔷薇油
attemper 调温
attemperater 恒温器
attemperation 人工冷却;温度控制;温度调节(作用);减温
attemperator 温度调节器;调温器;减温器;减热器;恒温箱;保温水管
attemperment 调节温度;回火;调温
attempt 攻击
attempt frequency 尝试频率
attendance 养护;维护;保养;出席;出勤
attendance book 签到簿
attendance rate 出勤率
attendant 维护人员;话务员;值机员;值班人员;出席者;乘务员;附属品;伴随物
attendant board 转接台
attendant cabin 值班室
attendant claim 附带要求
attendant conference 话务员会议
attendant-controlled lift 服务员控制的电梯;值班员控制的电梯
attendant craft 附属船(舶)
attendant desk 转接台
attendant equipment 附带设备
attendant monitor 话务员监视器
attendant-operated control 服务员操作控制;值班员操纵控制
attendant-operated goods lift 服务员操作的货物升降机
attendant park 代理停车场
attendant parking 有服务员的停车场(代客停车,取车);服务停车处
attendant seat 乘务员座椅
attendant ship 服务船;保障船
attendant's platform 维修平台;维护平台
attendant's switchboard 人工交换台
attendant tug 辅助作业拖轮
attended 有人值守的
attended mode 连接方式
attended operation 连接操作;伴随操作
attended repeater 有人值班的增音站
attended station 有人值守的电站;有人观测站;值班台
attended telephone substation 人工打电话支站
attended time 值班时间;服役时间
attending personnel 维修人员;照管人员;操作人员
attention 引起注意;注意
attention code 通知码
attention device 维设设备;监护设备;引注器件;引起注意装置;注意装置
attention display 注目显示
attention interrupt(ion) 联机中断;注意中断
attention key 联机键;引起注意键;终端联机键
attention list 引起注意列表
attention program(me) 提示程序
attention sign 注意符号
attention signal 注意信号
attention to hygiene 讲究卫生
attentive response 注意反应
attent-unattent switch 自动转换开关
attenuant 稀释剂;冲淡剂
attenuate 稀释;衰减;弄细;减低;减弱
attenuated total reflectance 衰减全反射比;衰减全反射
attenuated total reflectance spectroscopy 衰减全反射光谱
attenuated total reflection 衰减全反射
attenuated total reflection spectroscopy 衰减全反射谱
attenuated type 减弱型
attenuater = attenuator
attenuate residual moveouts 衰减剩余时差
attenuating 拉细
attenuating blast 牵伸气流
attenuating cable 衰耗电缆

attenuating draft 变细拉伸
attenuating material 衰减材料;吸收性材料
attenuating medium 衰减媒质
attenuating mica ceramic 衰减云母陶瓷
attenuating plate 衰减板
attenuating zone 拉细区
attenuation 衰减作用;衰减;衰变减量;减毒作用;信号衰减;板宽收缩
attenuation band 衰减频带;阻带
attenuation by absorption 吸收衰减
attenuation by barrier 屏障衰减
attenuation characteristic 衰减特征;衰减特性
attenuation characteristic curve 衰减特性曲线
attenuation coefficient 衰减系数;衰变系数
attenuation coefficient value 衰减系数值
attenuation comparator 衰减比测仪
attenuation compensation 衰减补偿
attenuation constant 衰减常数;衰减常量;减幅常数
attenuation corrected value 衰减修正值
attenuation criterion 衰减标准
attenuation cross section 强度减弱截面;衰减截面
attenuation curve 衰减曲线
attenuation curve of relative anomaly 相对异常衰减曲线
attenuation decrement 减缩率
attenuation distance 衰减距离
attenuation distortion 衰减失真;衰减畸变;振幅失真
attenuation duct 声音衰减管道;隔音管道
attenuation equalization 衰减均衡
attenuation equalizer 衰减均衡器;衰减补偿器
attenuation equation 衰减方程
attenuation factor 衰减因子;衰减因素;衰减因数
attenuation frequency distortion 衰减频率失真
attenuation ga(u)ge 衰减测量计
attenuation index 衰减指数
attenuation limited operation 衰减限制作用
attenuation measurement 衰减测量
attenuation measuring device 衰减测定装置
attenuation motion 衰减运动
attenuation multiple slant stacking processing 衰减多次倾斜叠加处理
attenuation of combination 组合波衰减
attenuation of crust 地壳减薄
attenuation of energy 能量衰减
attenuation of flood wave 洪水波的削弱;洪峰降低
attenuation of flow 流量减少
attenuation of harmonic product 谐波衰耗
attenuation of noise 噪声衰减
attenuation of sound 声衰减;声波衰减
attenuation of spring discharge 泉流量衰减
attenuation of the first kind 第一种衰减
attenuation of the flood 洪水下退
attenuation of the second kind 第二种衰减
attenuation of wave 波浪衰减
attenuation pad 衰减器;缓冲垫
attenuation period 衰减周期
attenuation process 衰减过程
attenuation property 衰减性质
attenuation rate 拉丝速度(玻璃纤维);衰减率
attenuation ratio 衰减比
attenuation region 衰减区
attenuation step pilot lamp 衰减分级指示灯
attenuation term 衰减项
attenuation test 衰减测试
attenuation time 衰减时间
attenuative effect 衰减作用;阻尼效应;钝化(作用)
attenuator 阻尼器;消声器;衰减器;衰耗器;减压器;增益控制器;增益调整器
attenuator tube 衰减管
atteration 冲积土(壤)
atteration tank 泥沙池
Atterberg consistency 阿太堡稠度;阿氏稠度
Atterberg consistency limits 阿太堡稠度极限值;阿氏稠度极限值
Atterberg (grade) scale 阿太堡粒级标准;阿氏粒级标准;阿太堡土粒分级(标准);阿氏土粒分级(标准)
Atterberg limit grade scale standard 阿太堡极限分级标准;阿氏极限分级标准
Atterberg limits 缩限;阿太堡极限(土壤含水界限);阿氏极限(土壤含水界限);稠度极限;阿太堡限度

Atterberg limits of soil 土的阿太堡极限
Atterberg limit test 阿太堡极限试验;阿氏极限试验
Atterberg limit value 阿太堡极限值;阿氏极限值
Atterberg number 阿太堡指数;阿氏指数
Atterberg plasticity index 阿太堡塑性指数;阿氏塑性指数
Atterberg's scale 阿太堡土粒分组标准;阿氏土粒分组标准
Atterberg test 阿太堡黏土塑性测定法;阿太堡试验
attestation 认证;签证;作证;证明书;证明;证词
attestation clause 确认有效条款;签证条款;见证条款;验证条款;验签条款
attested copy 验证誊本;验签副本
attested documents 经公证或核证的文件;证书
attesting witness 契约见证人
attestor 证人;签证人
at the construction field 在施工现场
at the construction site 在施工现场
at the current market price 按现行市场价格
at the earliest possible date 在最短期限内
at the end of this period 在此阶段的末期
at the first sampling 在第一次取样时
at the international level 在国际范围
at the market 照市价
at the national level 在国内范围
at the opening 开盘价(格)
at the regional level 在区域范围
at the stump 按伐根茎计
attic 屋顶室;钻塔顶层平台;阁楼;顶楼
attic base 雅典式柱基座;(古希腊柱的)座盘
attic cladding element 阁楼骨架
attic fan 顶楼风机;阁楼(排)风机;顶棚风机
attic floor 屋顶层;阁楼层
attic furnace 阁楼炉
attic habitable 可住的屋顶阁楼
attic hand 塔上工人(钻探时)
attic joist 阁楼层格栅;阁楼层格栅
attic main system 上行下给式系统
attic order 雅典列柱式;顶层柱式;顶层角柱式
Attic orogeny 阿提克造山运动【地】
attic room 阁楼房间;屋顶房间
attic stair(case) 阁楼楼梯;屋顶楼梯
attic stor(e)y 挑檐墙以上的楼层;顶楼;屋顶层;阁楼层
attic tank 屋顶水箱;屋顶水池
atticurge 梯形门洞
attic vent block 屋顶通风空心砖;屋顶通风空心砖
attic ventilation 阁楼通风
attic ventilator 屋顶通风器
attic ventilator tile 屋顶通风装置
attic vent opening 屋顶通风孔;阁楼通风口;阁楼通风孔
attic window 阁楼窗;屋顶通风窗
attika 顶层(古典建筑)
attitude 空间方位角;产状【地】;层态
attitude control 姿态控制
attitude control subsystem 姿态控制子系统
attitude control system 自动驾驶仪;姿态控制系统
attitude director indicator 姿态指示器
attitude evolution 观点进化
attitude gyro 姿态回转仪
attitude gyro package 姿态陀螺组
attitude gyroscope 姿态陀螺仪
attitude indication 姿态显示
attitude jet 姿态喷射器
attitude of axial-plane 轴面产状
attitude of bed 岩层产状
attitude of bedded structure 岩层的位态
attitude of echelon plane 雁列面产状
attitude of enveloping surface 包络面产状
attitude of fault surface 断层面产状
attitude of fault zone 断层带产状
attitude of foliation 面理的产状
attitude of inclined surface 倾斜面产状
attitude of intact 接触面产状
attitude of kink band 膝折带产状
attitude of limbs 两翼产状
attitude of line 线的产状
attitude of main glide plane 主滑面产状
attitude of rock formation 岩层产状
attitude of rocks 岩层产状
attitude of stratum 地层产状
attitude sensor 位置传感器;(车身的)姿态传感器
attitudes scale 态度衡量

attitudes toward reuse of wastewater 对废水回用的各种看法
attitude studies 态度研究
attitude survey 态度调查
attitude test 不同姿态试验
attitude to labo(u)r 劳动观点
attitudinal reflexes 状态反射
attle 采石场石屑;矿石废料;采石石屑;矿屑;废屑
atto 阿托
attoc (由黏土烧结的)硬地板砖和砌砖
attorn 过户
attorney 律师;代理人
attorney at law 律师
attorney fee 律师费
attorney-in-fact 代理人
attorneyship 代理人职务;代理人身份;代理权
attorney's opinion of title 律师对房产所有权的判定意见书
attornment agreement 过户协议
attractant 诱引剂;引诱剂
attract armature relay 衔铁吸合式继电器
attracted continental sea 引缩大陆海
attracted-disc electrometer 吸盘静电计
attracted-iron motor 吸铁式电动机
attracted traffic volume 吸引交通(量)
attracted water 吸着水
attract foreign pound 吸收外国资金
attracting attention signal 引起注意信号
attracting foreign funds 吸引外国资金
attracting groin 引流水坝;引流丁坝
attracting groyne 引流防波堤
attracting mass 吸引质量
attracting voltage 吸起电压
attraction 吸引
attraction at equator 赤道上引力
attraction energy 日月引力能
attraction force 引力
attraction force field 吸引阻力场
attraction fore 吸引力
attraction meter 引力计
attraction of gravitation 地心引力;地心吸力;地球引力
attraction of the center of the earth 地心引力
attraction of topographic(al) masses 地形质量引力
attraction pole 吸引中心
attraction radius 吸引半径
attraction resistance 吸引阻力
attraction type linear suspension motor 吸引型直线悬浮电动机
attraction water 诱鱼水流
attraction water valve 诱鱼水流控制阀
attractive circle 吸引圈
attractive colo(u)r 醒目色
attractive dam site 条件优越的坝址
attractive force 吸引力;引力
attractive interaction 互相吸引
attractiveness 吸引
attractive potential 引力位
attractive power 吸引力
attractive radius 吸引半径;服务半径
attractive site 可取坝址;优越坝址
attractive sphere 吸引范围(设施服务)
attractive uniaxial crystal 正单光轴结晶
attractive wave dynamics in marine 海洋引力波动力学
attractor 吸引点
attributable 可归因于
attributable cost 可归成本;归属成本
attributable profit 可分配利润
attributable risk 特异危险性;归因危险度
attributable segment cost 可归核算单位成本
attribute 属性;标志
attribute-based model 属性设计模型
attribute data 品质资料;特征数据
attribute factoring 属性因子分解
attribute field 标志字段
attribute for processing 加工性能
attribute information 特征信息
attribute listing method 特性列举法
attribute of a relation 关系属性
attribute of discrete type 离散类型属性
attribute of file 文件属性
attribute of floating-point type 浮点类型属性

attribute of identifiers 标识符的属性
attribute of item 项目的属性
attribute of scenery 地形性状
attribute property 属性特征
attribute sampling 特性抽样(法);属性抽样;按属性抽样
attribute set 表征组
attribute statistics 品质统计
attribute test(ing) 品质测试;特性试验;质量分类试验
attributive classification 按属性分类;品质分类
attributive indicant 品质标志
attributive indication 品质标志
attributive sampling 按属性抽样
attrinite 细屑体
attrital anthrxylon 细屑丝炭
attrital coal 杂质煤
attrita terra 劣地
attrite 摩擦;擦去
attrited black 球质炭黑
attrition 损耗;磨耗(作用);磨擦;擦除术
attrition grinding 碾磨
attrition hardness 耐磨硬度
attritioning 精碎
attrition loss 磨耗;磨损
attrition medium 研磨介质
attrition mill 圆盘磨;双盘式磨碎机;立式搅动磨机;碾磨机;磨粉机;磨盘式磨粉机
attrition mixing 磨碎混合;磨擦搅拌
attrition rate 损耗率;磨损率;磨损程度
attrition resistance 耐磨性;抗磨性;抗磨损性;抗磨耗性
attrition resistant 耐磨的
attrition scrubber 擦洗机
attrition test 耐磨试验;磨损试验
attrition tester 磨损试验机
attrition testing machine 磨损试验机;磨耗试验机
attrition value 磨耗值;磨损值
attritive powder 微细粉末
attritor 立式球磨机;磨粉机;超微磨碎机;碾磨机
attritor mill 碾磨机
attritus 暗煤;碎屑组分;碎集煤;杂质煤;暗煤质
Attwood's formula 爱特伍德公式
at utmost 最大程度
at weld edge 在焊缝边上
at wholesale price 批发价(格)
at will 随意
Atwood machine 阿特伍德机械
a two-wheel vehicle drawn by man 人力车
A-type chromium door handle 镀铬A形门拉手
A-type graphite 均匀片状石墨
A-type lineation A线理
A-type mast A形井架
A-type steel A类钢
atypical 非典型的;不典型的
atypical characteristic 不规则特征
atypical estimation procedure 非典型估计程序
atypical form 非典型型
atypical recurrent fever 非典型回归热
Au-Ag ore 金银矿石
Au-bearing copper ore 金铜矿石
aubergine 乌紫色
aubergine purple 茄皮紫
Aubert gate 奥伯特闸门;奥伯特门闸
aubertite 水氯铝铜矾
Aubert's phenomenon 奥伯特现象
aubleu 变成蓝色
aubrite 无球粒顽辉陨石;顽火无球粒陨石
auburn 栗色;赭色;暗棕色;枣红色;深褐色
aubusson carpet 奥步松毛圈地毯
aucillary detergent 辅助洗涤剂
aucillary documents 辅助证据
auction 竞卖
auctionable emission rights 可拍卖的排放权利
auction by candle 蜡烛竞卖法(到蜡烛点完为止,决定成交)
auction by tender 不公开拍卖
auction catalogue 拍卖目录
auction charges 拍卖费(用)
auctioneer 拍卖商;拍卖人
auctioneer's commission 拍卖佣金;拍卖人佣金
auctioneer's fee 拍卖费(用)
auctioneer's gavel 拍卖商的小槌
auctioneer's hammer 拍卖人的小槌

auctioneer vendee 拍卖买主
auction hall 拍卖所;拍卖(大)厅
auction house 拍卖行
auction land price 拍卖地价
auction market 拍卖市场;拍卖场
auction off 拍卖(掉)
auction of ship 船舶拍卖
auction of timber 木材拍售
auction price 拍卖价(格)
auction put up for 付与拍卖
auction room 拍卖行
auction sale 拍卖
auction sale notice 拍卖通告
audibility 可闻度;可听度;能听度
audibility factor 可闻系数;可听度系数
audibility meter 听力计;听度计;听度表
audibility range 可闻度范围;可听距离;可听距离;可听范围
audibility test 听力测定
audibility value 可听值
audibility zone 可听区
audible acoustic(al) alarm unit 音响警报装置
audible aids 音响航标
audible aid to navigation 声响助航系统
audible alarm 听觉警报;声音警报;声音报警;声频报警器;音响警报;音响报警设备;音频报警设备
audible alarm unit 音响警报器;声音报警装置
audible call 可闻信号;听得到的信号;可听到的信号
audible data 可听数据
audible frequency 可闻声频;声频
audible frequency amplifier 声频放大器
audible frequency generator 声频发生器
audible frequency signal 声频信号
audible indicator 音响指示器
audible leak detector 音响检漏器
audible limit 可闻限度;可听限制(声音强弱和频率高低的上下界限值)
audible navigation signal 音响航标
audibleness 可闻度
audible noise 声频噪声
audible range 可听到的范围;声达距离;可闻范围
audible region 声频区;音频频段
audible ringing tone 可听振铃声
audible signal 有声号志;听觉信号;声响信号;可闻振铃音;可听信号;听到的信号;有声信号;音响信号
audible signalling 音响信号装置
audible sound 可闻声;可听声音;可听声
audible spectrum 声谱
audible test 音频信号试验
audible threshold 可听阈值
audible tone 可听声调
audible-type transmitter 可听型发送机
audible value 可听值
audible-visual aid 光声辅助方法
audiblity 成音度
audience 用户;观众座椅
audience chamber 接见室;接待室;会见室
audience hall 听讲堂;会堂
audience room 会见室;接见室
audifier 声频放大器
audiga(u)ge 携带式超声波测厚仪;超声波测厚仪
audio 声响的;可听声信号;音频
audio amplifier 声音放大器;音频放大器
audio and video crosstalk 伴音图像串扰
audio and visual signal system 声光信号系统
audio attenuator 音频衰减器
audio band 音频波段
audio cable 音频电缆
audio carrier 音频载波
audio chanalyst 音频电路测试器
audio channel 声频通道
audio circuit 声频电路;音频电路
audio coder 音响信号编码器;声频信号编码器
audio communication 声音通信
audio-communication equipment 音频通信设备
audio-communication line 音频通信线路
audio-communication system 音频通信系统
audio control 音频控制
audio cue channel 声频插入通道
audio current meter 音频流速仪;音频海流计
audio cut off switch 音频断路开关
audio demodulator 伴音信号调解器;伴音解调器
audio detector 音频检波器

audio deviation limiting 频移限制
audio discriminator 伴音鉴频器
audio disc 声盘
audio engineering 声频工程(学)
audio equipment 声频仪
audio fader amplifier 声频控制放大器
audio fidelity 声频保真度;音频保真度
audio-fidelity control 声频保真控制
audioformer 声频变压器
audio frequency 声响频率;音频;成声频率;声频
audio-frequency amplifier 声频放大器;音频放大器
audio-frequency apparatus 声频设备;音频设备
audio-frequency choke 声频扼流圈
audio-frequency circuit 音频通信电路
audio-frequency coder 音频编码器
audio-frequency current 声频电流
audio-frequency generator 声频发生器;音频发生器
audio-frequency jointless track circuit 音频无缝轨道电路
audio-frequency meter 声频率计;音频频率计
audio-frequency noise 声频噪声;音频噪声
audio frequency oscillator 音频振荡器;声频振荡器
audio-frequency output 低频输出;声频输出
audio-frequency output level of receiver 接收机音频输出电平
audio-frequency pointing system 音频照准系统
audio-frequency power 声频功率;音频功率
audio-frequency range 音频范围
audio-frequency shift modulated track circuit 移频轨道电路
audio-frequency signal 声频信号
audio-frequency signal-to-interference ratio 音频信扰比
audio-frequency spectrometer 音频频谱计
audio-frequency transformer 声频变压器;音频变压器
audio-frequency variation factor 声频波动因数
audio generator 声频发电机
audiogram 听力图;声波图;声频图
audiograph 闻视图;声波图
audio image 声频图像
audio indicator 音频指示器;声频指示器
audio input device 声音输入装置
audio input system 音频输入系统
audio inquiry 声音询问;音音回答;声频询问
audio interpolation oscillator 声频内插振荡器;音频误差振荡器
audio line 声频线路;声频线;传声线
audiolloy 铁镍透磁合金
audiolloy Hypernic 铁镍透磁合金
audio-locator 声波定位器
audio-mail system 声邮系统
audio man 伴音工作人员
audiometer 听度器;听度计;听度表;声音测量器;音波计;测听器;测听计
audio method 声频法
audiometric earphone 测听耳机
audiometric laboratory 测听室
audiometric rooms 试听室
audiometry 听力测定法;测听(技)术
audio mill feed control 声控磨机喂料
audio mill loading device 磨机装料声控装置
audio mixer 调音台;声频混频器;混频控制板
audio modulating voltage 调制声压
audio modulation 音频调制
audiomonitor 监听设备;监听器
audiomonitor desk 监听台
audion 三极管
audio navigation 音频导航
audio noise meter 噪声计
audio-oscillator 声频振荡器
audio output limiter 音频输出限制器
audio patch bay 音频信号接线台
audio peak chopper 声频斩峰器
audiophone 助听器
audio piloting 音频导航
audio power 声频功率
audio preamplifier 前置声频放大器
audio response 声音应答;音频响应
audio response calculator 声音回答计算器;答话计算器
audio response control 声音应答控制
audio-response device 声音应答装置
audio response message 声音应答信息;声频响

应信息
audio response unit 声音应答器；声音应答装置；声音响应装置；声音响应器；声频响应装置；声频回音装置；音频应答器；答话终端
audio routing switcher 声频视频程序转换器
audio scanner 声频扫描器
audio signal 声频信号；音频信号
audio signal detector 音频信号指示器
audio signal generator 音频信号发生器
audiosonometry 声振动频率法
audio spectrometer 声谱仪
audio subcarrier 音频副载波
audio system 声频方式；放声系统
audio telecommunication line 声频远程通信线路
audio terminal 声音终端；声频终端；音频终端；答话终端
audio tone decoder 音频单音解码器
audio track 伴音声迹
audio transformer 音频变压器
audio tube 声频管
audio tunning 音频调谐
audio tweeter 音频扬声器
audio-vernier method 耳目切拍法
audio-video center 影视音像中心
audio-visual 声像的；视听的；直观的；直感的
audio-visual aids 直观教具；直感教学法
audio-visual book 视听图书；直感图书
audio-visual center 视听教育中心
audio-visual classroom 设有视听设备的教室
audio visual documents 视听文献
audio-visual education program(me) 电化教育
audio-visual information 视听情报；音像情报
audio-visual instruction 直观教学
audio-visual material 视听材料；直观教具；直感资料
audio-visual system 视听系统
audio-visual ultrasonic testing equipment 视听超声波检验设备
auditability 可审性
audit account 查账；盘账
audit by comparison 比较审计法；比较检查；比较稽查
audit capital sources of construction project 建设项目资金来源审计
audit certificate 审计证明书
audit conclusion 审计结论
audit contract 审计合同
audit coverage 审计范围
audit department 审计部门
audited accounts 审计决算；已审定账目
audited by test 抽审
audited entity 被审单位
audited net sales 已审查过的净销售额
audited organization 被审单位
audited voucher 已审查凭单
auditee 受审核方；被审单位
audit engagement 审计契约
audit entry 检查输入
audit expenses 查账费用
audit extent of financial and economic law and discipline 财经法纪审计的内容
audit fee 审计费
audit for railway assets and liabilities 铁路资产负债审计
audit-in-depth 分层检查
auditing 审计；审计(学)
audit 审查；会计检查；检查数据；查账
auditing bodies 审计机构
auditing by comparison 比较审计
auditing by rotation 轮流审计制
auditing by test and scrutiny 抽查账目
auditing offices 审计机关
auditing routine 检查例行程序；检查例程
auditing system 检查系统
audit interest and amortization charges of bonds 债券还本付息情况审计
audition 听力；试听；声量检查
audition studio 试听播音室
audit law 审计法
audit of accounting basis 会计基础审计
audit of account receivable 应收账款审计
audit of accounts received in advance 预收账款审计
audit of a shot in the locker 备用金审计

audit of assets management reponsibility system 资产经营责任制审计
audit of cash 现金审计
audit of charges for amortization 待摊费用审计
audit of completion settlement 工程结算审计
audit of construction plan feasibility 建设方案可行性的审计
audit of construction project 建设项目审计
audit of cost per product 单位产品成本的审计
audit of decrease of fixed assets 固定资产减少的审计
audit of economic efficiency 经济效益审计
audit of economic reasonableness of capital construction project 基本建设项目经济合理性审计
audit of estimation completeness of capital construction project 基本建设项目概算完整性审计
audit of estimation legality of capital construction project 基本建设项目概算合规性审计
audit of expenses 费用审计
audit of feasibiilty research of construction project 建设项目可行性研究的审计
audit of financial and economic law and discipline 财经法纪审计
audit of financial efficiency of construction project 建设项目财务效益审计
audit of financial expenditures 财务支出审计
audit of financial revenue 财务收入审计
audit of financial statement 财务报表审计
audit of fixed assets 固定资产审计
audit of increase of fixed assets 固定资产增加的审计
audit of increment of legal estate through revaluation 对法定财产重估增值的审计
audit of industrial production, mechanical work and auxiliary production 工业生产、机械作业、辅助生产审计
audit of internal control system 内部控制制度审计
audit of investment efficiency during productive period of construction project 基本建设生产阶段投资效益的审计
audit of investment expenses that should be cancelled after verification 应核销投资性费用审计
audit of investment result coefficient and payback period 投资效果系数与投资回收期的审计
audit of last account 竣工决算审计
audit of leaving office 离任审计
audit of legality of using surplus funds 盈余公积金使用的合法性审计
audit of liabilities 负债审计
audit of microcosmic investment efficiency of construction project 建设项目微观投资效益的审计
audit of national economy efficiency of construction project 建设项目国民经济效益审计
audit of profiteer loss and impairment of fixed assets 固定资产盘盈盘亏和毁损的审计
audit of putting estate into service 交付使用财产的审计
audit of rate of construction 建设速度的审计
audit of ratio of occupancy of unfinished project 未完成工程占用率的审计
audit of ratio of putting estate into service 交付使用财产率的审计
audit of reaching design capacity and level 达到设计能力和水平的审计
audit of returns 决算审计
audit of standard calculation and distribution method of cost and fee 成本费用开支标准和计算、分配方法审计
audit of total investment of construction project 建设项目总投资审计
audit of treasury 库存审计
audit of using budget quota 预算定额套用的审计
audit of working drawing budget 施工图预算审计
audit of working drawing budget of construction project 建设项目施工图预算的审计
auditor 审计员；审计委员会；审计师；查账员
auditor in charge of the audit 负责该项审计的审计人员
auditorium 正厅；观众席；听众席；礼堂；讲演厅；大讲堂；大会堂
auditorium acoustics 会场声学

auditorium area 会堂区域；会堂面积
auditorium-gymnasium 集会一体育两用会堂
auditorium seating 厅堂座椅排列；观众座椅；观众席位
auditorium space 观众席；观众厅
auditor's report 审计报告
auditor's standard report 标准审计报告
auditory 听众席
auditory acuity 听敏度；听力；听觉敏锐度
auditory area 听觉范围；可听范围
auditory direction-finding 听觉测向
auditory fatigue 听觉疲劳
auditory field 听野
auditory impression 听觉
auditory localization 声源定位
auditory perspective 声景象
auditory response 听觉效应；听觉响应；听觉反应
auditory sensation area 听觉区(域)
auditory sense modality 听觉势态；听觉力
auditory threshold 听阈
auditory threshold shift 听阈变化
audit program(me) 审查程序；审查程度
audit programming 审查程序设计
audit report 审计报告；审核报告
audit-review file 审查文件；审查评价文件
audit standard of investment efficiency of capital construction 基本建设投资效益审计的标准
audit team 审计小组
audit trail 审计跟踪；审核跟踪；审查追踪；检查追踪；检查跟踪；逐位跟踪；跟踪检查
audit trial 试查
AuDomac 奥多麦克精纺自动落纱机
Au-doped 掺金的
audrite 顽火辉石无球粒石陨石
Auer burner 奥厄灯
Auer metal 奥厄(火石)合金；(稀土金属65%，铁35%)
aufeis 层层结冰
aufhallung 浅化
aufwuch 附着生物
auganite 辉安岩
augelite 光彩石
augen 眼球状体【地】
augen amphogneiss 眼球状混合片麻岩
augen-chert 眼球状燧石
augend 加数；被加数
augend register 被加数寄存器
augen gneiss 眼状片麻岩
augen migmatite 眼球状混合岩
augen structure 眼状构造；眼球状构造
augen superposed fold 眼球形叠加褶皱
auger 螺旋钻头；螺旋推运器；螺旋取土钻；挤泥机；麻花钻机；长木钻
auger backfiller 万能螺旋钻式回填机
auger bin sweep 仓库卸粮用螺旋推运器；仓库卸粮用回转螺旋
auger bit 木螺钻；螺旋钻头；木螺锥；麻花钻头；短木钻
auger board 钻架
auger boring 螺旋钻探；螺旋钻孔；螺旋钻进；麻花钻钻探；麻花钻孔；钻孔【岩】
auger brick machine 带式制砖机
auger cleaner 勺形(钻孔)除渣器
auger concave 螺旋输送器底壳
auger conveyer 螺钻输送机；螺旋式输煤机
auger conveyor box 螺旋输送器外罩
auger core 螺旋推运器轴
auger cover 螺旋推运器外壳
auger delivery 螺旋输送
auger divider 螺旋式分配器
auger drill 螺钻；螺旋钻；回转式钻岩机
auger drill head 螺旋钻机头；螺旋钻机回转器；麻花钻头
auger drilling 螺旋钻孔；螺旋钻进；机械钻进
auger drive 螺旋推运器传动装置；旋旋传动
auger drive wheel 螺旋推运器传动链轮
augered pile 螺旋钻桩
Auger effect 俄歇效应
Auger ejection 俄歇发射
Auger electron 俄歇电子
Auger electron image 俄歇电子像
Auger electron spectrometer 俄歇电子能谱仪
auger elevator 螺旋升运器；螺旋升运机
Auger emission spectroscopy 俄歇发射光谱学

auger extension 螺旋接柄;钻杆
auger feed blower 螺旋输送式吹送机
auger feeder 螺旋送料器;螺旋送料机;螺旋进料器;螺旋加料器;螺旋加料机;螺旋给进器
auger-feeder cascading blender 螺旋给料阶式掺和机;螺旋给料级联掺和机
auger-feed granulator 螺旋供料成粒机
auger flight 螺旋叶片
auger fork 螺旋钻垫叉
Augergne transept 奥弗涅内型交叉甬道(十字形教堂翼部)
auger grain loader 螺旋装粮机
auger ground 软土层
auger hole 螺钻孔;螺旋钻孔;钻孔
auger hole charge 钻孔炸药
auger hole injection 钻孔注射法
auger hole test 钻孔试验;抽水试验
augering 螺旋钻进法
augering by hand 手摇螺旋钻
augering by power 动力螺旋钻
auger-lever 螺旋操纵杆
auger machine 螺旋钻孔机;螺旋制砖机;螺旋式练泥机;螺旋加料机;螺旋挤出机;带式制砖机
auger miner 钻采机
auger mining 螺旋钻孔开采薄煤层法;大直径钻孔法
auger packer 螺旋装料器
auger pile 螺旋桩
auger pile wharf 螺旋桩码头
auger pipe 螺旋钻杆
Auger rate 俄歇速率
Auger recombination 俄歇复合
auger reduction unit 齿轮减速器
auger rig 钻机
auger rod 螺旋钻杆
auger sampler 螺旋取样器
auger screw 手钻;螺旋钻土器
auger shaft 推运螺旋轴;钻井
auger shell 螺钻
Auger shower 俄歇簇射
auger-sinker-bar guide 冲击钻杆导向器
Auger spectrometer 俄歇能谱仪
Auger spectroscope 俄歇分光镜
auger spindle 钻轴
auger spout 螺旋卸粮管;螺旋喷卸管
auger sprocket 螺旋推运器传动链轮
auger steel 螺旋钎钢
auger stem 钻棒;螺旋钻杆;钻杆
auger string 螺旋钻杆柱
auger stripper 螺旋卸料器;螺旋刨煤机
auger system 螺旋钻进方法
Auger transition 俄歇跃迁
auger twist bit 螺旋式钻头;螺旋钻;麻花柱孔钻
auger-type 螺旋式
auger-type bit 螺旋式钻头
auger-type distributor 螺旋式分配器
auger-type extrusion(unit) 螺旋式挤压机;钻式挤出器
auger-type grain cleaner 螺旋式清粮机
auger-type grain elevator 螺旋式谷物升运器
auger-type grete concave 螺旋式栅格凹板
auger-type machine 螺旋式挤压机
auger unloader 螺旋式卸载机
auger vane test (土壤的)十字板试验
auger with hydraulic feed 水压推进式钻机;液压钻孔机;液压给进式钻机;水压钻进式钻机
auger with valve 舌阀捞砂筒;带阀土钻;带阀螺旋钻具(钻浅井用);阀式抽泥器
auget 雷管;爆破管
augetron 高真空电子倍增管
augite 普通辉石;辉石;斜辉石;雷管
augite albit pegmatite 辉石钠长伟晶岩
augite andesite 普通石安山岩
augite anorthite-rich gabbro 富普通辉石钙长石岩
augite diorite 辉石闪长石
augite granite 辉石花岗石
augite granophyre 辉石花斑岩
augite melaphyre 辉石暗玢石
augite noritegabbro 普通辉石苏长辉长岩
augite peridotite 普通辉石橄榄岩;辉橄岩
augite picrite 普通辉石苦橄岩
augite porphyrite 辉石玢岩
augite porphyry 辉石斑岩
augite propylite 辉石青盘岩
augite pyroxenite 普通辉石岩

augite-rich olivine basalt 富辉橄玄岩
augite rock 辉石岩
augite syenite 辉石正长石
augitite 玻基辉石岩;玻辉岩
augitophyre 辉斑玄武岩
augmentability 扩展性
augmentable 可增大的;可扩张的
augmentation 天体视增大;增加物;增加率;增大
augmentation coefficient 放尺率
augmentation correction 增量修正
augmentation distance 外推长度
augmentation of semi-diameter 半径增订量
augmented 增广;扩张的
augmented car length 延伸车长
augmented code 增信码
augmented core print 加大芯头;增大型芯头
augmented data manipulator 强化数据变换网络
augmented Lagrangian function 增广拉格朗日函数
augmented matrix 增广矩阵
augmented operator 扩充算子
augmented parity check code 扩大奇偶校验码
augmented Phillips curve 扩大的菲利普斯曲线
augmented product 引伸产品
augmented resistance 附加阻力
augmented thrust 增大推力;辅助推力
augmented transition network 扩充转移网络
augmented unblock 延伸开段
augmenter 增强压器;增量
augmenter tube 喷气短管
augmenting 使扩张
augmenting duct 加速引射器
augmenting factor 扩大因素;改正因数;补正系数
augmenting mining yield 扩大开采量
augmenting path 可扩路
augmentor 替身机器人;增强压器;增强器
augment tube 加速排气管;增速排气管
Augustaeum 奥古斯都神殿
Augustin process 奥古斯丁提银法
auiliary conductor 辅助线
auina 蓝方石
au jour le jour 每日贷款利率
aul 胶椴木;欧洲枫木;胶枫木
aula 院落(古建筑中);大学礼堂;广场
aulacogen basin 拗拉槽盆地
aulacogen trough 拗拉槽
aulodont dentition 管齿式
aulophyte 寄居植物
Aumann-Perles theorem 奥曼—柏雷定理
a(u)mbry (教堂的)下室;储[贮]藏室;圣餐具橱;库房;(存放圣餐用具的)壁龛;小室(教堂);壁橱
Aumund faience 奥蒙德陶器
a unit of cargo 一批货(物)
a unit of land measurement 土地面积的计算单位
Au placer ore 砂金矿
aura 电风
auragreen 碱式碳酸铜
aural alarm 声响警报
aural beacon 音响指向标;音响信标
aural carrier 声频载波;伴音载波
aural condition 听觉情况;伴音情况
aural course 可听方向;音响配向
aural detector 声响测察器;声波检波器;音频检波器
aural direction finder 音响测向仪
aural impression 听觉
auralite 褐块云母
aural modulator 伴音信号调制器
aural null 无声
aural perception 听觉
aural probe 探音管
aural radio beacon 音响无线电信标
aural radio range 可听无线电航标
aural receiver 伴音接收机
aural sensation 听觉
aural signal 声信号;音响信号;伴音信号
aural transmitter 伴音发射机
aural type beacon 发声式信标
aural warning 音响警报
aural zero 无声
auramine 金胺;碱性槐黄
aurantia 金橙黄
aurantiacus 橙黄色
aurantine 橙浸膏
auraptene 橙皮油素

auratus 金黄色
Aurelia 海月水母属
aureola 接触变质带
aureole 接触变质晕;华盖;光轮;电弧光环
aureolin(e) 钴亚硝酸钾;钴黄
auric 含金的
auric bromide 溴化金
aurichalcite 绿铜锌矿
auric oxide 氧化金
auric sulfate 硫酸金
auriculate bath 耳池
auricupride 斜方金铜矿
auriferous 含金的
auriferous gravel 含金砾石
auriferous pyrite 含金黄铁矿
auriferous quartz vein 含金石英脉
auriform file 半圆底小三角锉
aurin 蔷薇色酸
aurin red 金精红
auripigment 金色料
auripigmentum 雌黄
aurones 橙酮
aurora 极光
aurora australis 南极光
Aurora Babbitt metal 奥罗拉·巴比特合金
aurora borealis 北极光
aurora brightness 极光亮度
auroral absorption 极光吸收
auroral absorption event 极光吸收事件
auroral caps 极光帽
auroral corona 极光冕
auroral electrojet 极光电喷流
auroral excitation 极光激发
auroral fading 极光衰落
auroral forms 极光形状
auroral frequency 极光频率
auroral hiss 极光嘘声
auroral isochasm 极光等频数线
auroral latitude 极光纬度区
auroral line 极光线;极光谱线
auroral oval 卵形极光;极光椭圆区
auroral photometer 极光光度计
auroral poles 极光极
auroral proton flux 极光质子流量
auroral region 极光区域
auroral spectrum 极光谱
auroral storm 极光暴
auroral substorm 极光亚暴
auroral zone 极光区;极光(地)带
auroral zone blackout 极光地带中断
aurora polaris 北极光;极光
aurora yellow 镉黄
aurorite 黑银锰矿
aurosol 胶体金
aurostibite 方锑金矿
aurum foliatum 金箔
ausannealing 奥氏体等温退火
ausdrawing 奥氏形变拔丝法;奥氏体变拔丝法
ausforging 锻氏淬火;奥氏体锻造
ausform-annealing 奥氏体形变退火
ausformed steel wire 形变热处理钢丝
ausform hardening 奥氏体形变淬火
ausform(ing) 奥氏体形变;形变热处理(法)
Ausfuhrkredit GmbH 出口信贷股份公司(德国)
Au-silicon dioxide 掺金乳胶
auspices 主办;赞助
ausrolling 压延形变热处理;轧制形变热处理;奥氏体等温轧制形变热处理
aussage experimental 报告精确度实验
austausch 气团交换(量);紊流交换(量)
austausch coefficient 湍流交换系数
austemper 奥氏体回火;奥氏体(等温)淬火
austemper case hardening 等温淬火表面硬化
austempering 等温淬火;奥氏体回火
austemper stressing 加工应力等温淬火;等温淬火前加应力;等温淬火表面应力
austenag(e)ing 奥氏体等温时效
Austen-Cohen formula 奥斯丁—科恩公式
austenilic Ni-Cr stainless steel 奥氏体镍铬不锈钢
austenite 奥氏体
austenite actuargrain size 奥氏体实际晶粒度
austenite crystal 奥氏体晶粒
austenite former 奥氏体形成元素
austenite grain boundary 奥氏体晶界

austenite grainsize determination 奥氏体晶粒大小测定法
austenite inherent grain size 奥氏体本质晶粒度
austenite matrix 奥氏体基体
austenite stabilization 奥氏体稳定化处理
austenite stabilizer 奥氏体稳定剂
austenite welding 不锈钢焊接
austenitic 奥氏体的
austenitic alloy steel 奥氏体合金钢
austenitic area 奥氏体区
austenitic cast iron 奥氏体铸铁;奥氏体生铁
austenitic chromium-nickel steel 奥氏体铬—镍钢
austenitic clad steel 奥氏体复合钢
austenitic grain size 奥氏体晶粒度
austenitic grain-size classification 奥氏体晶粒度分级
austenitic heat-resistance steel 奥氏体耐热钢
austenitic iron 奥氏体铁
austenitic manganese steel 高锰钢;奥氏体锰钢
austenitic matrix 奥氏体基体
austenitic range 奥氏体区;奥氏体范围
austenitic region 奥氏体区域
austenitic stainless steel 奥氏体不锈钢
austenitic steel 奥氏体钢
austenitize 奥氏体化
austenitizing 奥氏体化
austenization 奥氏体化
austenize 奥氏体化
austenomartensite 奥氏马氏体
auster 奥斯特风
austere budget 紧缩预算;紧缩财政
austerity budget 紧缩预算;紧缩财政
austerity measures 财政紧缩措施
austerity program(me) 经济紧缩方案;紧缩计划;财政紧缩方案
Austin chalk 奥斯汀白垩层【地】
Austin dam 奥斯汀土坝
Austin data recorder 奥斯汀数据记录器
Austinian 奥斯汀阶【地】
austinite 砷锌钙矿;砷钙锌石
austral 南方的
Australasia 大洋洲
Australasian 大洋洲的
austral axis pole 南轴极
austral bivalve province 南方双壳类地理区
Australia-Antarctica separation 澳洲大陆与南极大陆分离
Australian cinnabar 澳洲朱砂矿
Australia current 澳大利亚海流
Australian gum 澳洲胶
Australia jade 澳洲玉
Australian kino 澳大利亚吉纳树胶
Australian leech 澳洲水蛭
Australian offset 盲目布井
Australian-Pacific coral realm 澳太珊瑚地理系
Australian pearl 银白珍珠
Australian plate 澳洲板块
Australian platform 澳洲地台
Australian sandal wood oil 澳洲檀香油
Australian ship reporting system 澳大利亚船舶报告系统
Australian soft eye splice 澳式插接软琵琶头
Australian Standard 澳大利亚国家标准
australis aurora 南极光
australite 澳洲玻璃陨体
Australorp 澳洲黑
austral region 南方区
austral spring 南半球春季
austral window 滑动窗;推拉窗;南向窗;南窗
austral winter 南半球冬季
Austrian black wattle gum 澳洲黑树胶
Austrian cinnabar 碱式铬酸铅
Austria drape 奥地利垂帘
Austrian method 全断面分块开挖法;上下导坑先墙后拱法
Austrian method of timbering 奥地利式隧道支撑法
Austrian method of tunneling driving 奥地利施工法(隧道掘进)
Austrian orogeny 奥地利造山运动【地】
Austrian school 奥地利学派
Austrian shade 奥地利遮帘
Austrian shade cloth 奥地利粗支阔条窗帘布
Austroriparian life zone 南方针叶林带
auswittering 铸件自然时效

autallotriomorphic 原生他形
Autaram 污水提送设备
autarky 自给自足;封闭化
autecology 个体生态学;单种生态学
authalic condition 等面积条件
authalic conic(al) projection 等积圆锥投影
authalic longitude 等积投影经度
authalic map projection 等积投影;等积地图投影
authalic projection 等面积投影
authalic projection latitude 等积投影纬度
authalic radius 等积半径
authalic sphere 等积球体
authentic 真实的
authenticate 出立证据
authenticated documents 鉴定证明书
authenticated protest 经公证人证明的拒付
authenticating code 押码
authentication 认证;鉴定
authentication agent 认证代理人
authentication code 鉴别符号;鉴别代码;标识代码
authentication of documents 文件的真确证明
authenticator 鉴定符;证明人
authentic communication 可靠信息
authentic data 可靠数据
authenticity 可靠性;真实性
authentic language 规定文字
authentic sample 可靠试样;真实样品
authigene 自生矿物
authigenesis 自生作用
authigenic deposits 自生矿床
authigenic material 自生物质
authigenic mineral 自生矿物
authigenic sediment 自生沉积物
authigenous mineral 自生矿物
authineomorphic rock 再造岩
authorise 审定
authorised agent 正式委托代理人
authorised pressure 容许压力;规定压力;法定压力
authoritarianism 专制制度;独裁主义
authoritarian theory 权力主义管理理论
authoritative inspection organization 权威检验机构
authority 权力;威信;职权;管理局;当局;官方
authority credential 权限值;权限凭证
authority engineer 权威工程师;管理工程师
authority file 规范文件
authority for operation 运转可靠性
authority for project 工程项目的批准
authority for the interpretation 解释权
authority for the report 报告的根据
authority having jurisdiction 有管辖权的机关;主管机关
authority of agency 代理权
authority of remote connection 远程连接审核
authority of sign 签署权
authority oriented 权力导向
authority-owned 官方所有的
authority pattern 权力模式
authority to delegate 权力委托
authority to draw 授权开立汇票;出票授权书
authority to negotiate 授权议付
authority to pay 委托支付证;委托付款证;托付;授权付款;支付权
authority to purchase 委托购买证;采购授权书
authority to sign 授权签字
authorization 授权;核准
authorization code 特许代码;授权代码
authorization control 授权控制
authorization data 特许数据
authorization exit-routine 授权出口例行程序
authorization file 授权文件;规范文件
authorization for project 项目的审定;项目的批准;项目的核准
authorization for shipment of goods 商品启运许可证
authorization message 授权信息
authorization of resources 财源核定
authorization of the investment 核准投资
authorization sheet for purchase of exchange 外汇购买核订单
authorization sheet for sales of exchange 出售外汇核订单;外汇出售订单
authorization to use 委任使用
authorization use 使用的委任

authorize 授权;审定
authorized 核准的;指定的
authorized access 特许取用;特许存储
authorized agency 授权机构;指定代理
authorized agency for foreign currency exchange 外币兑换指定代理处
authorized agent 全权代理人;授权代理人;指定代理人
authorized amount 核准金额
authorized auditor 法定审计师
authorized bank 授权银行;指定银行
authorized boiler pressure 认可的锅炉压力
authorized bonds 批准发行的债券
authorized buyer 认可的买主
authorized by 审核
authorized capital 认可资本;核定资本;法定资本;额定资本
authorized capital stock 法定股本
authorized capital stock account 法定股本账户;额定股本账户;额定股本账
authorized controlled material 经审定控制的物资
authorized data 特许数据
authorized data list 技术数据资料一览表;技术数据和技术情况一览表
authorized dealer 特许商人
authorized deferred payment 批准延期撤纳的税款
authorized depository 法定保管人
authorized documents 正式文本
authorized edition 审定版本
authorized extension 认可扩建
authorized foreign exchange bank 指定外汇银行
authorized full supply 核定的最大供水量;规定的最大供水量;批准的最大供水量
authorized inspector 授权检查员
authorized investment 授权投资;核定投资总额
authorized language 规定语言
authorized limit 管理限制;管理限值;特准极限
authorized mortgage bonds 批准发行的抵押债券;额定抵押债券
authorized officer 授权人员
authorized person 被派人员;授权人
authorized personnel 被授权人员
authorized pressure 容许压强;容许压力;审定压力;允许压力;规定压力
authorized program(me) 公认程序
authorized program(me) analysis report 授权程序分析报告
authorized project 已批准的工程项目
authorized representative 授权代表;经授权代表;被授权的代表
authorized senior member 全权高级人员
authorized shares 认可的股份;核定股份;额定股数
authorized signature 认可的印鉴;授权人签字;授权签署;授权的权力
authorized station 正规观测站;正规测站
authorized stock 核定股份;额定股份
authorized street parking 可停车街道
authorized surveyor 授权的检验人
authorized tariff 法定税则
authorized value 额定价值
author language 编辑语言
author of the invention 发明人
author's draft map 作者原图
authrone 蒽酮
autical star 导航星
autistic parasite 固有寄生物
auto 自动
autoabstract 自作文摘;自动摘录;自动抽样;自动抽取;自动抄录
autoacceleration 自加速(度);自动加速度
auto accident 行车事故
autoacquisition 自动采集(数据)
autoactivation 自身活化;自动活化
autoagglutination 自体凝集作用;自身凝集作用;自身凝集反应
autoagglutinin 自体凝集素
autoair heater 自动空气加热器
autoalarm 自动警报(器);自动报警接收机;自动报警
autoalarm device 自动警报装置;自动报警装置
autoalarm signal 自动警报信号
autoalarm system 自动报警系统
autoalignment 自动准直
autoallergy 自身变态反应

autoamputation 自身断离
autoanalyzer 自动分析仪;自动分析器
autoanalyzer flow chart 自动分析流程图
autoanalyzer flow diagram 自动分析流程图
autoanalyzer procedure 自动分析程序
auto and remote 自控与遥控
auto-answer 自动应答
autoarchitect 自动建筑师
autoatic strake 无极带式洗矿槽
auto axhaust analyzer 汽车排气分析仪
autobahn 超级公路;高速道路(德国);高速公路
autobalance 自动平衡机构;自动平衡
autobalance manometer 自动平衡压力计
autobalancer crane 自动平衡起重机
autobar 棒料自动送进装置
autobarotropy 自动正压状态
autobias 自偏压
autobicycle 机动自行车;摩托车
autobike 机器脚踏车;摩托车;机动脚踏车
auto-bin indicator 料仓储量自动指示器
autobiology 个人生物学
autoboat 汽艇;机动艇;摩托艇
autobond 自动链接;自动接合;自动焊接
autobrake 自动制动器;自动刹车
autobreccia 同生角砾岩
autobridge-factory 自动化桥梁厂
autobulb 汽车灯泡
autobus 机动车;公共汽车
autoby pass 自动旁路
autocade 汽车车列;一长列汽车
AutoCAD program(me) files AutoCAD 程序文件
AutoCAD software package AutoCAD 软件包
autocall 自动呼叫;自动调用
autocalorimeter 自动冷热试验计
autocar 汽车;机动车
autocarrier 汽车运输船;装运小汽车的大型载重汽车;装运汽车的专用船
autocartograph 自动制图仪;自动测图仪
autocatalysis 自动催化作用
autocatalytic reaction 自动催化反应
autocatheterism 自插导管
autocementation 自生胶结作用
autocentering 自动置中
autocentral computer system 自动中心计算机系统
autochangeover 自动接通
autochangeover unit 自动转换器
autochart 自动图表程序;自动流程图程序;自动画图;自动化流程图程序
autochromatic plate 彩色板
autochromatic printing 彩色印刷
autochrome 投影底片;天然色照片;彩色照片;彩色照片;彩色底片
autochrome plate 彩色照像干片;奥托克罗母微粒彩屏干版
autochrome printing 彩色印刷
autochrome process 彩色照像
autochthonal rocket 土火箭
autochthon(e) 原地岩
autochthonous 本地的
autochthonous bauxite 原地铝土矿
autochthonous breccia 原地角砾岩
autochthonous coal 原地生成煤;原地煤
autochthonous deposit 原地淤积;原地沉积
autochthonous granite 原地花岗岩
autochthonous groundwater 原地生成的地下水
autochthonous idea 自发观念
autochthonous limestone 原地石灰岩
autochthonous material 当地物质;本地物质
autochthonous microorganism 固有微生物
autochthonous parasite 固有寄生物
autochthonous stream 本地河
autochthony 原地堆积
autoclast 自碎岩;自生碎屑;蒸压器
autoclastic 自碎的【地】
autoclastic agglomerate lava 自碎集块熔岩
autoclastic breccia lava 自碎角砾溶岩
autoclastic fragmental facies 自成碎屑岩相
autoclastic phenocryst 自碎斑晶
autoclastic rock 自碎岩;自生凝结岩
autoclastic schist 自碎的片岩
autoclastic texture 自碎结构
autoclastic tuff lava 自碎凝灰熔岩
autoclave 热压器;压蒸釜;压热器;压力蒸汽锅;蒸压器;蒸压釜;蒸气灭菌器;高压蒸汽消毒器;高压蒸锅;高压消毒器;高压灭菌器;高压锅;高压釜;反应釜
autoclave body 釜体
autoclave brick 蒸压砖
autoclave-cured concrete 压蒸养护混凝土;压蒸混凝土
autoclave curing 高压釜养护;压蒸养护;蒸压养护;高压蒸汽养护(法)
autoclave curing cycle 蒸压养护周期
autoclave cycle 蒸压循环(时间);压蒸养护周期;高压蒸汽周期
autoclaved aerated block 压蒸加气砌块
autoclaved aerated concrete 压蒸加气混凝土;高压蒸汽加气混凝土;高压蒸汽养护的加气混凝土
autoclaved asbestos cement calcium silicate board 热压处理的石棉水泥硅酸钙板
autoclaved cellular concrete 高压蒸汽养护的蜂窝状混凝土;高压加气混凝土
autoclaved concrete 高压蒸汽养护混凝土
autoclaved concrete article 高压蒸汽养护混凝土制品
autoclaved concrete pile 蒸压混凝土桩;蒸汽养护混凝土桩
autoclaved concrete product 高压蒸汽养护混凝土制品
autoclaved concrete slab 压蒸混凝土板
autoclaved gypsum 高压蒸汽养护石膏
autoclaved light-concrete slab 蒸汽养护轻质凝土板
autoclaved light-weight concrete 压蒸轻质混凝土;高压轻质混凝土
autoclaved lime 蒸压石灰;水化石灰
autoclaved material 蒸压材料
autoclaved mo(u)lding 压蒸釜成型;蒸压罐模制法
autoclaved test 压蒸试验
autoclaved unit 压蒸构件;
autoclave expansion 热压膨胀(率);水泥蒸养膨胀(率)
autoclave expansion test (水泥安定性的)蒸压膨胀试验;体积安定性试验;压蒸膨胀试验;压蒸法;(水泥的)蒸煮试验;(用于测定水泥安定性)蒸压膨胀试验;蒸汽压力膨胀试验(用于水泥安定性测定)
autoclave expansion value 压蒸膨胀值
autoclave fly-ash brick 压蒸粉煤灰砖
autoclave leg 釜支座
autoclave lime 蒸压水化石灰
autoclave mo(u)lding 高压釜模塑;高压釜成型
autoclave plaster 压蒸石膏
autoclave press 压热硫化锅
autoclave process 压热釜法
autoclave-setting 高压釜定形
autoclave sterilizer 热蒸汽灭菌器;热压灭菌器;高压蒸汽消毒器;高压蒸汽灭菌器;高压消毒器
autoclave-style beck 高压釜式染色机
autoclave test 压热试验;蒸压试验;高压釜试验
autoclave test for cement 水泥高压釜试验(测定水泥安定性)
autoclave test for soundness 安定性压蒸试验
autoclave treated 高压釜处理的
autoclave treatment 压蒸处理
autoclaving 高压蒸汽养护法;热压灭菌法;压蒸养护;压蒸处理;高压蒸汽消毒灭菌法;压蒸养护
autoclaving cycle 高压蒸汽周期;蒸养制度
autoclaving equipment 高压消毒设备
autoclawing 自动抓紧
autoclear input 自动清除输入
autoclipping apparatus 自动清丝装置
autoclosing explosion flap 自关闭防爆阀门
autocoacervation 自动凝聚;自动分解脱水凝聚
autocoagulation 自动凝聚;自动凝结
autocoarse pitch 桨距自动调整装置;自动增大螺距装置
autocode 自动代码;自动编码
autocoder 自动编码语言;自动编码器;自动编码机;自动编码程序
autocoding 自动编码
autocoherer 自动粉末检波器
autocollimate 自动准直;自动对准
autocollimatic spectroscope 自准光谱仪
autocollimating eyepiece 自准直目镜
autocollimating spectrograph 自准直摄谱仪
autocollimating spectrometer 自准直分光仪;自准直分光计
autocollimating theodolite 自动对准经纬仪
autocollimation 自动对准;自准直;自动准直
autocollimation accessory 自动准直附件
autocollimation angle 自动准直角
autocollimation device 自动准直装置
autocollimation eyepiece 自准直目镜
autocollimation method 自动视准法
autocollimation spectrograph 自准直摄谱仪
autocollimator 自准直望远镜;自准直管;自动准直仪;自动照准仪;自动视准仪;自动瞄准仪;反射式望远镜
autocolo(u)rimeter 自动比色计
autocombustion system 自动燃烧系统
autocommand signal 自动指令信号;指令机构信号
autocompensation 自动补偿
autocompounded current transformer 自复绕式电流互感器
autocondense 自动压缩
autoconduction 自体导电法;自动传感
autoconglomerate 自生砾岩
autoconsequent fall 自顺向瀑布
autoconsequent river 自顺向河
autoconsequent stream 自顺向流;自顺向河
auto-container 汽车货柜;汽车集装箱
autocontrol 自动控制;自动调整
autocontrol valve 自动控制阀
autoconvection 自动对流
autoconvective gradient 自动对流直减率
autoconvective lapse rate 自动对流直减率
autoconverter 自动变换器
autoconvolution function 自动卷旋函数
Autocopser 奥托科泊自动卷纬机
autocorrection 自动校正
autocorrection coefficient 自动校正系数
autocorrelation 自相关作用;自相关
autocorrelation analysis 自相关分析
autocorrelation coefficient 自相关系数
autocorrelation computer 自相关式计算机
autocorrelation function 自相关函数
autocorrelation of data 数据自相关
autocorrelation receiver 自相关接收器
autocorrelation spectrometer 自相关频谱计
autocorrelation type radiometer 自相关型辐射计
autocorrelator 自相关器
autocorrelogram 自相关图
autocorrelogram computer 自相关式计算机
auto-counting 自动计数
autocoupler 自动车钩
autocoupling 自动耦合
autocovariance 自协方差
autocovariance function 自协方差函数
auto crack 热裂
autocrane 汽车吊;汽车式起重机;汽车起重机
autocrat 水质油彩或水粉画颜料;汽车阶级
auto-crazy 热爱汽车交通的;汽车狂
autocross 越野塞车
autocut-off 自动断路
autocut-out 自动阻断(器);自动断路
autocycle 机器脚踏车;机动自行车;摩托车;自动循环
autocycle control 自动循环控制
autocytolysis 自体溶解
autodata gather 自动数据收集器
autodata processor 数据自动处理机
autodecomposition 自体分解;自动分解
autodecrement 自动减(量);自动递减
autodecrement addressing 自减型编址
autodeflectometer 自动弯沉仪
autodegauss 自动消磁
autodemo(u)lding polymer 自动脱模聚合物
autodeposition 自动沉积
autodeposition paint 自泳涂装;自沉积漆
autodesign 自动设计
autodestructive fuse 自爆引信
autodetector 自动探测仪;自动检波器
autodevice 自动装置
autodial 自动拨号;自动标刻度盘
autodigestion 自体溶解;自身消化;自溶作用
autodispatcher 自动调度装置
autodistress signal apparatus 自动报警信号接收机
autododge 自动匀光印象机
autodoor (光电控制的)自动开关门
autodope 自动掺杂
autodraft(ing) 自动制图;自动绘图

autodrafting system 自动绘图器;自动绘图机
autodrinker 自动饮水器
autodrome 赛车场
autodynamograph 强力自动记录计
autodyne 自差
autodyne circuit 自差路;自差电路
autodyne oscillator 自差振荡器
autodyne receiver 自差式接收机;自差接收机
autodyne reception 自差接收法
autoecology 自我生态学;个体生态学
autoelectronic 场致发射
autoelectronic current 冷发射电流;场致发射电流
autoelevator device 自动找平装置
autoemission 汽车排气;冷发射
autoenlarging apparatus 自动放大器;自动放大机
autoenzyme 自溶酶
autoequipment 自动设备
autoexcitation 自激励;自激发
auto exhaust 汽车排气
auto exhaust oxidation catalyst 汽车排气氧化催化剂
autofeed 自动进料;自动进给;自动进刀;自动加料
autofeed drifter 自动推进架式钻机;自动推进架式凿岩机
autofeed drill 自动给进凿岩机
autofeed drilling 自动给进钻法
autofeeder 自动送料器;自动馈给装置
autofeed selector 自动给进换挡把手
autofermentation 自发酵(作用)
autoferry 汽车轮渡;汽车渡轮
autofettle 机械修坯;自动修坯
autofining 自动精炼
autoflare 自动拉平
autoflocculation 自絮凝(作用)
autoflow 自动绘流程图程序
autoflow chart 自动流程图
autofluorescence 自发体荧光
autofluorograph 自身荧光图
autofluoroscope 自身荧光镜
autofocus 自动聚焦
autofocus apparatus 自动对光设备
autofocusing 自动对焦
autofocus mechanism 自动调焦机构
autofocus radar projector 自动聚焦雷达投影仪
autofocus rectifier 自聚焦整形放大镜;自动对光纠正仪
autofocus reflecting projector 自动对光投影仪
autofollowing 自动跟踪
autofollowing laser radar 自动跟踪激光雷达
autoformer 自耦变压器
autofracturing expulsion mechanism 自裂开排驱机理
autofrettage 内膛挤压硬化法;自增强;自紧法
autofrettaged cylinder 自增强圆筒
autofrettage high-pressure cylinder 自增强高压圆筒
autogain control 自动增益控制;自动增益调整
autogardener 手扶园艺拖拉机
autogeneous grinding 自动研磨
autogeneous hardening 自动硬化
autogeneous rivet cutter 铆钉气割机
autogeneous welded steel pipe 焊接钢管
autogenesis 无生源说
autogenetic drainage 利用冲蚀而地形排水;河道本身构成的排水系统;自然排水系统;自成水系
autogenetic topography 自成地形
autogenic 自然的;气焊的;气割的;自生的
autogenic change 自生演变
autogenic mineral 自生矿物
autogenic movement 自发运动
autogenic river 内源河;自生河
autogenic stream 内源河;自生河
autogenic succession 自发演替
autogenor 自动生氧器
autogenous 气焊的;自生的
autogenous combustion 自体燃烧
autogenous curing 自然养护;自养护
autogenous cutting 乙炔熔化;乙炔气割;气割
autogenous electrification 自起电
autogenous fusing 气割
autogenous gouging attachment 自动气刨附加装置
autogenous grinding 自磨;自粉磨
autogenous grinding mill 自磨机;自动研磨机

autogenous growth of concrete 混凝土自生体积增长
autogenous healing 自愈性;自愈合;自生愈合;自动强化
autogenous healing of concrete 混凝土自行愈合
autogenous heating 自然发热
autogenous ignition 自然点火;自动着火
autogenous ignition temperature 自动着火温度
autogenous mill 气落磨
autogenous mineral 自生矿物
autogenous pressure welding 自动压合熔焊
autogenous shrinkage 自生收缩;自然收缩
autogenous smelting 自热熔炼
autogenous soldering 熔接(法);气焊(法);氧铁软焊
autogenous tumbling mill 自磨机
autogenous volume change 水泥由于水化作用的体积变化;自生体积变化
autogenous-welded steel pipe 气焊的钢管
autogenous weld(ing) 乙炔焊;氧炔焊;熔焊;气焊(法);熔接法
autogenous welding shield glass 气焊用护目镜玻璃
autogentic process 自发过程
autogeny 自生
autogeosyncline 平原地槽;自地槽;独立地槽
autogiration 自转
autogiro 旋翼飞机;直升(飞)机
autoglass 汽车玻璃;安全玻璃
autogonal projection 等角投影
autogram 压印
autograph 亲笔签名;签署;题辞;手稿;自动绘图仪
autographic(al) 自记的
autographic(al) apparatus 自动图示记录仪
autographic(al) oedometer 自录(式)固结仪
autographic(al) pyrometer 自动记录式高温计;自动记录高温计
autographic(al) rain ga(u)ge 自记录雨量计
autographic(al) record 自记录;自记录;自记器记录
autographic(al) recorder 自记仪;自动记录表
autographic(al) recording apparatus 自动记录仪;自记记录仪
autographic(al) strain recorder 自记式应变记录器
autographometer 自动图示仪;地形自动记录仪;地形自动记录器
autography 题辞;石版复制术;真迹版
autogravure 照相版雕刻法
autogrinding machine 自动研磨机;自动磨床;自动精磨机;自动粉碎机
autoguider 自动导星装置
autogyration 自旋转
autogyro 自动陀螺仪
autohand 自动手
autohandler 自动装入器
autohatch 自动舱盖
autoheading for minimum thrust 最小推力自动首向调整
autoheated thermophilic aerobic digestion 自热高温好氧消化
autohelm 自动舵
autohelm adjustment panel 自动舵调整面板
autohemagglutination 自身凝集反应
autohesion 自黏作用;自黏性;自黏力
autoheterodyne 自拍;自差
auto-highway 汽车公路
autohoist 汽车吊;汽车式起重机;汽车起重机;起重汽车;机动起重机
autohub 自动轮毂
autohyphenation 自动加连字符
autoidentify 自动认址【机】
autoignite 自燃;自动点火
autoigniter 自动点火器
autoignition 自燃;自动点火
autoignition conflagration 自燃火灾
autoignition point 自燃点
autoignition temperature 自动着火温度;自燃温度;自动点火温度
autoimpulse water meter 自动脉冲水表
autoincrement 自动增(量);自动递增
autoincrement addressing 自增型编址
autoincrement decrement 自动增量减量
autoincrement flag 自动增量标志
autoincrement mode 自动增量式

autoindex(ing) 自动索引;自动变址;自动编索引
autoindex register 自动变址寄存器
autoinduction 自动感应
autoinductive coupling 自耦变压器耦合
autoinductive door 自动感应门
autoinductive hand drier 自动感应烘手器
auto industry 汽车工业
autoinhibition 自动阻尼作用
autoinjection 自侵入(作用)
autoinjection panel 自动注射板
Autoinserter 自动插入
autointerference 自身干扰
autointrusion 自侵入(作用)
autoionization 自电离(作用)
autoist 汽车驾驶员
auto jack 车用千斤顶
autojig 自动卷染机
autojigger 自耦变压器
auto junkyard 旧汽车场
autokeyer 自动拍发器
autokinetic illusion 自发运动错觉
autokinetic phenomenon 自动现象
autoladle furnace 自动加料炉
autoland system 自动着陆系统
auto laundry 汽车清洗间
autolay 自动敷设;自动开关
autolayout 自动设计程序
auto-leak transformer 耦合自耦变压器
autolean mixture 自动贫化燃烧混合物
autolevel 自动找平;自动调节水准仪;自动安平水准仪
autoleveller bicoil draw box 自调匀整双圈条练条机
autoleveller draw box 自调匀整练条机
autoleveller gill box 自调匀整针梳机
autolevel(l)ing assembly 自动校平装置
auto-leveling-lifting-lining-tamping machine 自动整平-起道-拔道-捣固车;自动液压大型捣固车
autolevel mode 自动校平状态;自动校平方式
autolift 自动升降机
autolift trip scraper 铲斗自动起落刮土机
autolift with axle support 车轴支架式自动升降机
autoline 高速公路;高速道路;自动线
autoline hauler 自动起钓机
autolink 自动连接
autolith 同源包体;同源包体
autolithification 自生石化作用
autolithography 直接平版印刷法
autoload cassette 自动装片暗盒
autoloader 自动装载机;自动装卸机;自动装填器;自动装填机;自动装入程序;自动装货车
autoloading 自动加载
autoload preformer 自动装料预成形机
autolog-on 自动登录
autologous 自身的
autol red 阿脱红(色淀偶氮染料)
autoluminescence 自发光
autoluminescent paint 自发光涂料
autolysate 自溶产物
autolysis 自体溶解;自身溶解;自溶作用
autolyze 自溶解;自溶
automadesign 自动设计
automaker 汽车制造商
automan 自动手动开关
automan switch 自动手动开关
automanual 自动手动;半自动的
automanual exchange 半自动交换台;半自动交换台
automanual switch 自动—手动按钮;自动手动开关
automanual system 半自动制;自动手动式;半自动系统;半自动式;半自动化系统
automanual telephone exchange 半自动电话局
automanual telephone switchboard 半自动电话交换器;半自动电话交换机
automanual transfer 半自动传输
automanual transfer valve 半自动输送阀
automask 自动彩色校正
automason 自动砌墙机
automat 自动装置;自动售货机
automata 自动机;自动装置;自动开关;自动监控器
automatable 可自动化的
automatable machine 自动化机械
automat and automatic control 自动装置与自动控制

automata theory 自动控制理论;自动化理论
automated 自动化的
automated aids course 航标自动控制课程
automated analysis 自动化分析
automated analyzer 自动分析仪;自动分析器
automated answering question 自动回答问题
automated astronomic(al) positioning device 天文自动定位装置;自动化天文定位装置
automated astronomic(al) positioning system 天文自动定位系统
automated banking 自动化银行业务
automated bibliography 自动化书刊目录
automated cargo expediting system 自动化货物快速处理系统
automated cargo release and operations system 自动卸货与搬运系统
automated cartographic(al) system 自动制图系统
automated cartographic(al) system of thematic map 专题地图自动制图系统
automated cartographic(al) system of topographic(al) map 地形图自动制图系统
automated cartography 制图自动化
automated cartography development team 制图自动化研究小组
automated cartography software 自动制图软件
automated command system 自动化指令系统
automated communications and message processing system 自动通信和信息处理系统
automated component selection 自动部件选择
automated computer controlled electronic scanning system 计算机控制的自动电子扫描系统
automated control system of dredging process 挖泥操作程序自动控制系统
automated coordinate digitizer 自动坐标数字化器
automated coordinate plotter 自动坐标绘图机
automated coordinate recording 自动记录坐标
automated cost estimates 自动成本估计
automated data acquisition and control system 数据自动采集与控制系统
automated data extraction 数据自动抽出
automated data interchange system 数据自动交换系统;自动数据互换系统
automated data management information system 自动数据管理情报系统
automated data medium 数据自动传送介质;自动数据介体
automated data system 自动化数据系统
automated design engineering 设计自动化工程;自动设计技术;自动设计工程(学)
automated design tool 自动设计工具
automated detailing 自动详图制作
automated digital cartographic(al) system 自动数字制图系统
automated digital data assembly system 自动数字数据汇编系统
automated digital encoding system 自动数字编码系统
automated digital identification 自动数字识别
automated digital input/output system 自动数字输入—输出系统
automated digital message switching centre 自动数字信息交换中心
automated digital recording and control 自动数字记录与控制
automated digital test unit 自动数字测试装置
automated digital tracking 自动数字跟踪
automated digital tracking analyzer computer 自动数字跟踪分析计算机
automated door 自动开启的门;自动门
automated drafting 自动制图
automated drafting system 自动绘图系统
automated driverless metro 自动化无人驾驶地铁
automated drug identification 自动毒品识别
automated environment(al) chamber 自动化环境模拟室
automated equipment identification 自动设备识别
automated fibre optic biosensor 自动纤维光导生物传感器
automated flaw detector 自动故障检测器
automated forged system 自动化锻造装置
automated forging line 锻造自动线
automated gear 自动传动装置
automated generalization in cartography 自动制图综合

automated guided transit 自动导向交通(系统)
automated guideway transit 自动导向交通(系统)
automated highway 自动化公路
automated highway system 自动化道路系统
automated hollow fibre ultrafiltration method 自动空心纤维超滤法
automated imagery processing 自动成像处理
automated interpretation 自动解译
automated inventory control system 自动化存货管理系统
automated irrigation 自动灌溉
automated irrigation system 自动灌水系统
automated key-stop truck loading 槽车自动装载台(液化气卸用)
automated light rapid transit line 自动化轻型快速交通线路
automated logic diagram 自动化逻辑图
automated management 自动化管理
automated management control centre 自动化管理控制中心
automated manifest system 自动货单系统;自动舱单系统
automated map lettering 图面自动片字
automated mapping system 自动制图系统
automated marine propulsion system 船舶自动推进系统
automated material handling equipment 自动装卸材料的设备
automated measurement 自动量算;自动测量
automated measuring equipment 自动化量测设备
automated merchant vessel rescue system 商船自动救助系统
automated method 自动方法
automated mining 自动化开采
automated monitor 自动监测仪
automated monitoring system 自动监测系统;自动化监测系统
automated mutual-assistance vessel rescue system 船舶自动互救系统
automated mutual vessel rescue center 自动互助船舶营救系统中心
automated nautical chart index file 自动化海图索引夹
automated office equipment 自动化办公室设备
automated onboard gravimeter 船用自动重力仪;船上自动重力仪
automated particle size analyzer 自动粒径分析仪
automated paster 自动接纸装置
automated pre-concentration sampler 自动预选采样器
automated press line 压力机自动生产线;冲压自动线
automated production management 自动生产管理
automated program(me) 自动程序
automated program(me) control 自动程序控制
automated program(me) control unit 自动程序控制单元
automated radar plotter 雷达自动绘图器
automated radar plotting aids 雷达自动标绘仪
automated radar plotting system 雷达自动绘图系统
automated radar terminal system 自动雷达终端系统
automated radar tracking system 自动化雷达跟踪系统
automated railroading 自动化铁路管理
automated rapid transit 自动化高速运输
automated reasoning 自动推理
automated replenishment 自动补充存货
automated retrieval system 自动检索系统
automated route management 航路自动管理
automated seat reservation system 自动化定座系统
automated sensor 自动传感器
automated service station 自动服务站
automated ship 自动化船
automated shipboard aerologic(al) program(me) 自动化船载高空探测计划
automated shipboard forecasting system 船川自动预报系统
automated ship location and attitude measuring system 自动船舶定位和姿态测量系统
automated sliding door 自动滑门

automated stock control 自动库存控制
automated storage 自动储[贮]藏
automated subway train 自动化地下铁道列车
automated survey 自动测量
automated survey system 自动测量系统
automated system 自动系统;自动化系统
automated system control 自动系统控制
automated tariff filing and information system 自动运货单填与信息系统
automated test system 自动测试系统
automated tracing system 自动绘图系统
automated track inspection 自动化轨道检查
automated traffic control 交通自动控制;自动化交通控制
automated verification system 自动验证系统
automated water-quality control system 自动水质监测控制系统
automated water-quality monitor 自动水质监测仪
automated water-quality monitoring 自动水质监测
automated water-quality monitoring station 自动水质监测站
automated water-quality monitoring system 自动水质监测系统
automated workshop 自动化车间
automatic 自动化的;自动的
automatic abstract 自动编制
automatic abstracting 自动编制文摘
automatic acceleration 自动加速
automatic acquisition 自动探测和跟踪
automatic additional adjust 自动补偿测量
automatic addressing 自动选址
automatic addressing system 自动寻址系统
automatic address modification 自动改变地址;自动地址修改
automatic adjust 自动调节
automatic adjusting 自动调整的
automatic adjusting expansion chamber 自动调节膨胀室
automatic adjusting mechanism 自动调整机理
automatic adjusting system 自动调节系统
automatic adjustment 自动调整;自动调节
automatic advance 自动提前
automatic advanced breaker 自动提前断路器;自动提前断电器
automatic advanced element 自动提前点火装置;自动点火提前设备
automatic aerial film numbering machine 航空胶片自动编号机
automatic agitation 自动搅拌
automatic aiming 自动瞄准
automatic air bleed valve 自动排气阀
automatic air brake 自动风闸;自动气闸;自动气阀;自动空气制动机
automatic air-break switch 自动空气开关
automatic air circuit breaker 自动空气断路器
automatic air-compressed depth ga(u)ge 自动压缩空气测探仪
automatic air compressor 自动空气压缩机
automatic aircraft reporting system 机载观测资料自动发送系统
automatic air filter 自动滤气器
automatic air heater 自动空气加热器
automatic air release 自动排气
automatic air traffic control 自动空中交通管理
automatic air valve 自动气阀;自动浮球阀
automatic air vent 自动通风孔;自动排气口
automatic air vice 自动气压虎钳
automatic alarm 自动警报(器);自动告警装置;自动报警(器)
automatic alarm device 自动报警装置
automatic alarm equipment 自动报警设备
automatic alarm in pumping station 泵站自动警报
automatic alarm receiver 自动报警接收机
automatic aligning 自动对准的
automatic alignment 自动对准
automatically adjusting counter weight 自动调整平衡锤
automatically changing winder 自动换筒抽丝机
automatically closing door 自闭门
automatically closing fire door 自闭式防火门
automatically controlled continuous process 自动调整连续过程
automatically controlled fire pump 自动控制消

防泵
automatically controlled gate 自动栏木
automatically controlled rolling mill 自动控制的轧机
automatically controlled screw feed 自动控制螺旋给进
automatically damped suspension 自动减震悬挂装置
automatically engaging catch 自动连接轮挡
automatically flour banding plant 面粉自动包装机组
automatically lighted boiler 自动点燃锅炉
automatically operated damper 自动操纵的风门
automatically operated inlet valve 自动操作进给阀
automatically operated retarder 自动缓行器
automatically operated signal 自动号志
automatically operated valve 自动阀;自动操纵阀
automatically operating gas analyzer[analyser] 自动操作气体分析器
automatically plotted display 自动绘图显示
automatically positioned switch 自动转换道岔
automatically programmed tool 自动程序控制工具;自动数控程序
automatically recording magnetic balance 自动记录式磁秤;自动记录天平
automatically regulated feed speed 自动调节给进速度
automatically restored to operation 自动恢复运转
automatically rotated stopper drill 自动旋转冲缩式凿岩机
automatically shutting pressure of needle valve 针形阀自动关闭压力
automatically synchronous 自动同步的
automatically tared 自动配衡的
automatically tensioned carrying cable 自动张紧缆索
automatically tensioned catenary 自动张紧承力索;张力重力自动补偿装置
automatically tensioned equipment 自动张紧装置;张力补偿装置
automatically tensioned simple catenary 简易自动悬挂
automatically variable governor 自动转换调节器
automatic alternate voice/data 声音数字自动交替
automatic altitude control valve 高程自动控制阀
automatic alumin(i)um cap fitting machine 自动铝盖卷边机
automatic amino acid analyzer 氨基酸自动分析仪
automatic amplifier 自动放大器
automatic amplitude control 自动振幅控制;自动幅度控制;自动调幅控制
automatic amplitude control circuit 自动隔离控制电路
automatic analysis 自动分析
automatic anchoring device 自动抛锚装置
automatic and continuous air monitoring system 自动连续空气监测系统
automatic and continuous monitoring system 自动连续监测系统
automatic and hand-operated changeover switch 自动手动转换开关
automatic announcement device of approach of a train 列车进站自动通知
automatic answer 自动应答
automatic answering 自动回答;自动对答
automatic answering back device 自动应答器
automatic answering modem 自动应答调制解调器
automatic answering unit 自动应答器
automatic anti-collision aids 自动避碰设备
automatic anti-collision system 自动避碰系统
automatic antijam circuit 自动抗干扰电路
automatic aperture control device 光圈自动调节装置
automatic appliance 自动化用具;自动化设备
automatic approach 自动进场
automatic approach coupler 自动进场耦合器
automatic approval 自行生效
automatic arc welded tube 自动电弧焊缝管
automatic arc welding 自动电弧焊
automatic arc welding head 自动电弧焊(接机)头
automatic arc welding machine 自动电弧焊机;自动电弧焊接机;自动电焊机
automatic area meter 自动面积计
automatic ash remover 自动清灰装置

automatic assembly 自动装配
automatic assembly machine 自动装配机
automatic assembly technique 自动装配技术
automatic astronavigation system 自动天文导航系统
automatic astronomic(al) positioning system 自动天文定位系统
automatic atmospheric-pollution analyser 大气污染自动分析
automatic attribute 自动化属性;自动表征
automatic autocartograph 自动制图仪
automatic avoidance 自动失效
automatic back bias 自动反偏压
automatic background control 自动背景调整
automatic background noise control 背景噪声自动调整
automatic backup 自动备份
automatic back valve 止回阀;单向阀
automatic backwashing of rapid sand filter 快速砂滤池反冲洗
automatic bag filling and closing machine 自动装袋封口机
automatic bailer 自动排水装置
automatic balance 自动平衡;自动结算机
automatic balance instrument 自动平衡仪表
automatic balance manometer 自动平衡压力计
automatic bale sledge 草捆自动拖运器;包件自动拖运器
automatic bale stacker 自动堆包机
automatic ball valve 自动球阀;自动排气阀
automatic band-saw sharpener 带锯自动刃磨机
automatic bandwidth control 自动带宽控制
automatic barring gear 自动盘车装置
automatic bar type machine 自动杆式车床;杆式自动车床
automatic batch column dryer 竖筒式自动分批干燥机
automatic batcher 自动配料器;自动配料计量器;自动化配料器
automatic batch(er) plant 自动分批拌和设备;大型混凝土生产设备(混凝土生产厂内的)
automatic batching system 自动按量分配装置;自动配料装置
automatic batch mixing 自动配料拌和
automatic batch-type grain drier 分批式谷物自动干燥机
automatic batch weighing 自动配料称量
automatic baud rate detection 自动波特率检测
automatic beam changer 自动变光开关
automatic bearing indicator 自动方位指示仪
automatic belt sander 自动带式砂光机
automatic bending machine 自动弯机
automatic bias 自动偏压
automatic bias compensation 自动偏压补偿
automatic blade control 平土铲自动控制装置;铲刀自动控制装置
automatic block district 自动闭塞区段
automatic block equipment 自动闭塞装置;自动闭塞设备
automatic block(ing) 自动闭塞
automatic block installation 自动闭塞装置
automatic block line 自动闭塞电线路
automatic block operation 自动闭塞运用
automatic block power line 自动闭塞供电线路
automatic block project 自动闭塞计划
automatic block signal 自动闭塞信号;自动闭塞号志
automatic block signal in both directions 双向自动闭塞号志
automatic block signal in one-direction 单向自动闭塞号志
automatic block signal(l)er 通过信号机;自动闭塞信号机
automatic block signal line 自动闭塞信号线
automatic block signal(l)ing 自动闭塞信号法
automatic block signal(l)ing protection 自动闭塞信号防护
automatic block signal(l)ing rules 自动闭塞行车规则
automatic block signal system 自动闭塞信号系统
automatic block signal territory 自动闭塞信号区域
automatic block system 自动闭锁系统;自动区截制;自动闭塞装置;自动闭塞制;自动闭塞系统
automatic block system with alternating current

counting code track circuits 交流计数电码自动闭塞系统
automatic block territory 自动闭塞区域
automatic block with alternating current counting code track circuits 交流计数电码自动闭塞
automatic block with asymmetric(al) high voltage impulse track circuits 不对称高压脉冲自动闭塞
automatic block with audio frequency shift modulated track circuits 移频自动闭塞
automatic block with coded track circuit 电码自动闭塞
automatic block with double direction running 双向运行自动闭塞【铁】
automatic block with impulse track circuits 脉冲跟踪电路自锁
automatic block with polar frequency coded track circuits 极性频率脉冲自动闭锁;极频电码轨道电路自动闭塞
automatic block with single direction running 单向自动闭塞【铁】
automatic blowdown system 自动排污系统
automatic blowdown valve 自动排污阀
automatic blow off valve 自动放气阀
automatic boiler 自动锅炉
automatic boiler control 自动锅炉控制;锅炉自动控制
automatic bolt 自动插销
automatic bonding unit 自动捆扎装置
automatic booksewing machine 自动锁线机
automatic boost control 自动增压调节器
automatic booster control 自动升压控制
automatic bootstrap loader 自动引导装入程序
automatic bourette stripper 落棉自动剥取装置
automatic bowl latch 铲斗自动闭锁
automatic box splitter 自动分箱机
automatic brake 自动制动(器)
automatic brake adjuster 自动制动调节器
automatic brake application 自动制动
automatic brake rod 自动闸杆
automatic brake valve 自动制动阀
automatic braking 自动制动(作用);自动制动的
automatic branch exchange 自动交换分机
automatic break 自动切断;自动断路
automatic bridge 自动电桥
automatic bridge control 驾驶室自动控制
automatic brightness contrast control 自动亮度反差调整;自动对比度控制
automatic brightness control 亮度自动控制;自动亮度控制;自动亮度调整
automatic broadcast system of passenger station 客运站自动广播系统
automatic bucket 自动斗;自动吊桶
automatic bucket level(ling) device 铲斗自动调平装置(斗式装载机的);链斗自动调平装置(斗式装载机的)
automatic bucket positioner 链斗自动定位装置(装载机);铲斗自动定位装置;铲斗自动定位器
automatic buffering and draw coupler 自动缓冲牵引车钩;自动式牵引缓冲装置
automatic buoy 自动浮标
automatic burette 自给滴定管
automatic burette with reservoir 带储液瓶的自动滴定管
automatic burner 自动燃烧器
automatic burning appliance 自动燃烧装置;自动燃烧设备
automatic bus communication system 公共汽车自动通信系统
automatic bus location system 公共汽车自动定位系统
automatic bypass valve 自动溢流阀;自动旁通阀
automatic cab signal 列车自动信号装置;机车内自动信号
automatic cab signaling 机车自动信号
automatic cake discharge type filter press 自动卸料压滤机
automatic calculation 自动计算
automatic calculator 自动计算器
automatic calibration 自动校准
automatic callback 自动回叫
automatic call back-busy 遇忙自动回叫
automatic call device 自动报警器
automatic call(ing) 自动呼叫

automatic calling and answering unit 自动呼叫和应答装置
automatic calling equipment 自动呼叫装置;自动呼叫器;自动(调度)呼叫设备
automatic calling unit 自动呼叫装置;自动呼叫设备;自动呼叫器
automatic call library 自动调用程序库
automatic call point 自动烟雾报警装置;自动报警器;自动烟雾报警器
automatic call sign unit 自动呼叫装置;自动呼叫设备;自动呼叫器
automatic camera 自动摄影机
automatic camp-on 自动预占
automatic capping machine 自动压盖机
automatic carburetor 自动气化器
automatic car classification 车辆自动编组
automatic car count distance-to-coupling system 计算车辆到峰底、连挂点距离的自动计算系统
automatic car coupler 自动车钩
automatic card programmed rolling 穿孔带程序控制自动轧机
automatic card programmed rolling mill 程序控制自动轧机
automatic card reader 自动读卡机
automatic card transmitter 自动读卡机
automatic car identification 自动车辆识别;自动车号识别;自动抄车号;车辆自动识别
automatic car identification computer 车辆自动识别计算机
automatic car identification scanner 车辆自动识别扫描器
automatic car identification system 车辆自动识别系统
automatic carriage 自动送纸机构;自动滑架;自动滚轮
automatic carrier control 自动载波控制
automatic cartographic(al) program(me) 自动制图程序
automatic cartography 自动制图学;自动制图法;自动化制图
automatic car tracking system 车辆自动选路系统
automatic casting machine 自动化铸造机
automatic catch 自动挡
automatic categorization 自动分类
automatic cathead 自动套
automatic celestial navigation 天文惯性导航;自动天体导航
automatic celestial navigator 自动天体导航仪
automatic centering 自动置中;自动归心;自动对准中心;自动对中
automatic centering tripod 自动对中三角架
automatic center punch 自动中心冲床;自动冲床
automatic central buffer coupling 中央缓冲自动钩
automatic central grease station 自动油脂中心站
automatic centralized grease system 自动油脂集中润滑系统
automatic central mixing plant 自动中心拌和厂
automatic change-over 自动转换开关;自动转向机构
automatic change-over damper 自动变向阻尼器;自动变向器
automatic channel selection 自动信道选择
automatic charging 自动口料;自动加料
automatic charging equipment 自动加料机
automatic chart line follower 自动航线追随装置
automatic checker 自动检验器
automatic check(ing) 自动检验;自动检核;自动校验;自动检测
automatic checking and sorting machine 自动检验分选机
automatic check nozzle 自动关闭喷嘴
automatic checkout and control system 自动测试与控制系统
automatic checkout and evaluation system 自动检查与鉴定系统;自动检测和估算系统
automatic checkout and readiness equipment 自动检查和准备装置
automatic checkout and recording equipment 自动检测记录仪
automatic checkout device 自动检测装置
automatic checkout equipment 自动检测仪;自动检测设备
automatic checkout evaluation system 自动检测与鉴定系统

automatic checkout system 自动检出系统;自动检测系统
automatic checkout test equipment 自动检查试验装置;自动测试装置
automatic check-point restart 自动检查点再启动
automatic check-valve 自动止回阀
automatic chill casting-machine 自动硬模铸造机
automatic choke 自动阻气门;自动吸入器;自动节气门
automatic choke to manifold pipe 自动阻气门至歧管连接管
automatic chroma control 自动色度控制
automatic chroma control error 自动色度控制误差
automatic chromatogram scanner 自动色谱扫描器
automatic chrominance control 自动彩色控制
automatic chuck 自动钻头夹盘;自动卡盘
automatic chucking lathe 自动卡盘车床
automatic chucking machine 自动卡盘车床
automatic chute 自动装料槽
automatic circuit analyser 电路自动分析器
automatic circuit analyser and verifier 电路自动分析器和检验器
automatic circuit breaker 自动开关;自动断路器;自动断电器
automatic circuit-breaker with magneto-thermic protection 具有碰热保护的自动断路器;带有碰热保护的自动断路器
automatic circuit recloser 自动启闭电路装置;自动电路重合开关
automatic cistern 自动冲水箱
automatic clamping device 自动夹紧装置
automatic clapper 自动放水阀拍板
automatic classification 自动分类
automatic classification station 自动化编组站
automatic classification yard 自动化调车场【铁】;自动化编组场
automatic classifier 自动分类机
automatic cleaner 自动清洁器
automatic cleaning 自动清洗
automatic cleaning equipment 油轮自动化洗舱装置;自动化洗舱装置【船】
automatic cleaning flexible filter 自动清洗软管过滤器
automatic cleaning machine 自动清洗装置
automatic clear 自动清除
automatic clearing apparatus 自动清除装置;自动交换装置
automatic clearing indicator 自动折线指示器
automatic clearing key 报终信号键
automatic clipper 自动剪板机
automatic clock switch 时控开关;自动钟表开关;自动定时开关
automatic closed-loop control 闭环自动控制
automatic closed-loop process control 自动闭环过程控制
automatic closer 自动合箱机
automatic closing 自动关门;自动关闭
automatic closing device 自动封口装置;防火门自动开关装置;自动关门装置;自闭装置
automatic closing door 自关风门;自闭水密门
automatic closing gear 自动关闭装置
automatic closure machine 自动封缄机
automatic clutch 自动离合器
automatic coating machine 自动喷涂机
automatic cocking lever 自动扳机
automatic code 自动编码
automatic coder 自动编码器;自动编码机
automatic coding 自动设计程序;自动编码(技术);程序设计自动化
automatic coding language 自动编码语言
automatic coding machine 自动编码器;自动编码机
automatic coding system 自动编码系统
automatic coiling of tape 自动卷带
automatic coining 自动精压
automatic cold former 自动冷锻机
automatic collection 自动收集
automatic collector 自动收集器
automatic collimating telescope 自动准直望远镜
automatic collimation 自动准直
automatic collimation accessory 自动准直附件
automatic collision avoidance system 自动防撞系统;自动避碰系统
automatic colo(u)r compensation 自动光色补偿

automatic colo(u)r control 自动彩色控制
automatic colo(u)r control circuit 彩色自动控制电路
automatic colo(u)r dispersing machine 自动混色机
automatic colo(u)rimeter 自动比色计
automatic colo(u)r light block 色灯自动闭塞
automatic colo(u)r separation device 自动分色机
automatic combination boiler control system 锅炉自动燃烧控制装置
automatic combustion control 燃烧自动控制;自动燃烧控制
automatic communication device 自动通信装置
automatic communication monitor 通信自动监听器
automatic communication network 自动通信网
automatic compactometer 压实度自动检测装置
automatic compactor 自动内燃夯土机
automatic compensation 自动补偿
automatic compensator 自动补偿器
automatic component assembly 元件自动装配
automatic composing machine 自动排字机
automatic compressor-adjuster 压缩机自动调压器
automatic computation 自动计算
automatic computer 自动计算机
automatic computer-aided design 自动化计算机辅助设计
automatic computing 自动计算
automatic computing equipment 自动计算装置
automatic concrete block-making machine 混凝土(方)块自动浇制机
automatic concrete block mixing plant 混凝土(方)块自动拌和设备
automatic concrete column pourer 混凝土柱自动浇铸机
automatic concrete pipe machine 混凝土管自动浇铸机
automatic connection 自动联接;自动连接;自动接线
automatic connector 自动接线器
automatic consolidation 自动固结
automatic constant current charger 自动横流充电机
automatic container crane 自动化集装箱起重机
automatic contour digitizer 自动等深线数字转换器
automatic contouring 自动描绘等高线
automatic contouring roll lathe 自动仿形轧辊车床
automatic contouring system 等高线自动描绘系统
automatic contour liner 自动等高线仪;等高线自动描绘器
automatic contrast control 对比度自动控制;反差自动控制
automatic contrast selection function 反差自动选择功能
automatic control 自动控制;自动化控制
automatic control appliance of main engine 主机自动控制装置
automatic control assembly 自动控制装置
automatic control block diagram 自动控制框图
automatic control board 自动控制盘
automatic control circuit 自动控制电路
automatic control concentrating 自动调度集中
automatic control damper 自动控制气闸
automatic control device 自动控制装置
automatic control engineering 自动控制技术;自动控制工程(学)
automatic control equipment 自动控制装置
automatic control equipment for lock 船闸自动控制设备
automatic control error coefficient 自动控制误差系数
automatic control frequency response 自动控制频率响应
automatic control gear 自动控制装置
automatic controller 自动控制器;自动调节装置;自动调节器
automatic controller for spud carriage 定位桩台车自动控制装置
automatic controlling system of cement plant 水泥厂自动控制系统
automatic control machine tool 自动控制机床
automatic control of pumping station 泵站自动控制
automatic control on electric(al) power system

电力系统自动化
automatic control panel 自动控制台
automatic control rod 自动控制杆
automatic control servo valve 自动控制伺服阀
automatic control stability 自动控制稳定性
automatic control switches and routing of car 道岔和车辆进路的自动控制
automatic control system 自控系统;自动控制系统
automatic control system certified 持证的自动控制系统
automatic control system for blast furnace 高炉自动控制系统
automatic control system for industrial boiler 工业锅炉自控装置
Automatic Control System for Unattended Engine-room Certificate 无人机舱自动控制系统许可证
automatic control theory 自动控制理论
automatic control transient analysis 自动控制暂态分析
automatic control valve 自动控制阀
automatic control valve body 自动控制阀体
automatic control with feedback 反馈式自动控制
automatic convection 自动对流
automatic conveying 自动传输
automatic conveyer 自动输送机
automatic conveyor scale 自动运输秤
automatic coordinate recording device 坐标自动记录装置
automatic coordinate recording unit 自动坐标记录装置
automatic core-breaker 岩芯自动卡断器;自动卡芯器
automatic correction 自动校正
automatic correction coefficient 自动校正系数
automatic correction device 自动校正器
automatic correction system 自动校正系统
automatic correlator 自动相关器
automatic counter flow distributer 自动逆流分配器
automatic counter for tablets 自动数片机
automatic counterpart 自动计算部件
automatic counterweight gate 有平衡器的自动阀门
automatic counting 自动计数
automatic coupler 自动(式)车钩;自动联轴节;自动挂钩
automatic coupler detached 自动摘钩
automatic coupler head 自动车钩头
automatic coupling 自动联结器;自动联接;自动挂钩
automatic coupling device 自动耦合装置;自动耦合设备;自动联结装置;自动挂接装置;自动车钩
automatic coupling screwing on unit 管接头自动旋拧装置
automatic course keeping and depth keeping controls 航向深度自动控制仪
automatic course-keeping gear 自动操舵装置
automatic course stabilizer 自动稳定航向装置;自动航线稳定器
automatic cover 自动承保
automatic crab 自动起重小吊车
automatic crack-off machine 自动爆口机
automatic crane 自动起重机
Automatic cross-cut chain saw 自动横切割链锯
automatic crossing gate 自动栏木;交叉口自动启闭栅;自动启闭栅
automatic crossing gate with flashing light of bells 装有闪光灯或电铃的道口自动门栏
automatic crust breaker 自动打壳机
automatic crystal slicing machine 自动晶体切片机
automatic cup forming line 杯类自动成形线
automatic curing machine 自动养护机(混凝土)
automatic current recording meter 自动记录电流计
automatic current regulator 电流自动调整器
automatic curve follower 自动曲线跟踪装置
automatic curve plotter 自动曲线绘算器
automatic custody transfer 密闭式自动输送
automatic cut-out 自动断路;自动断流器;自动断电器
automatic cut-out valve 自动断流阀
automatic cutter 自动切割器;自动切割机
automatic cutter controller 铰刀自动控制器
automatic cutter for electric(al) tapping machine 电动铰牙自动切断机
automatic cutter grinder 自动工具磨床
automatic cutting machine 自动切割机;自动气割机
automatic cutting off 自动断路
automatic cycle 自动循环
automatic cycle control 自动循环控制
automatic cycling control 自动循环控制
automatic cylinder loader 自动灌瓶机
automatic dam 自动放水闸
automatic damper 自动挡板
automatic damper regulator 自动风挡调节器
automatic damping 自动阻尼
automatic data accumulator and transfer 自动数据储[贮]存与传输装置
automatic data acquisition 自动数据收集;自动数据录取;自动数据获取装置
automatic data acquisition and control system 自动数据采集与控制系统
automatic data acquisition system 自动数据录取系统
automatic data analysis system 自动数据分析系统
automatic data collection buoy 自动数据搜集浮标
automatic data digitizing system 自动数据数字化系统
automatic data editing and switching system 气象资料自动编辑中继系统
automatic data exchange system 自动数据交换系统
automatic data handling system 自动数据处理系统
automatic data plotter 数控绘图机
automatic data plotting 数控绘图;自动绘图
automatic data processing 自动资料处理;自动(化)数据处理
automatic data processing centre 自动数据处理中心
automatic data processing equipment 自动数据处理设备
automatic data processing machine 自动数据处理机
automatic data processing program(me) 自动数据处理程序
automatic data processing program(me) reporting system 自动数据处理程序报告系统
automatic data processing resources 自动数据处理方法
automatic data processing system 自动数据处理系统
automatic data processor 自动资料整理机;自动数据处理机
automatic data recording device 数据自动记录装置
automatic data reducer 自动数据简化器
automatic data reduction 自动数据简化(处理)
automatic data reduction equipment 自动数据简化设备
automatic data retrieval system 自动资料检索系统
automatic data service centre 自动数据服务中心
automatic data set 自动数据设备
automatic data switching center 自动数据转接中心;自动数据交换中心
automatic data switching system 自动数据转换系统
automatic data translator 自动数据转换器;自动数据译码器
automatic debugging 自动查错
automatic decal machine 自动贴花机
automatic decimal point arithmetic(al) 小数点自动定位运算
automatic decimal point computer 小数点自动定位计算机
automatic decrystalization device 自动除霜器
automatic defining file 自动定义文件
automatic deflected burning control system 偏火自动控制系统
automatic defrosting 自动除霜
automatic defrosting machine 自动除霜器
automatic demarcating device 自动划分装置
automatic depot 自动混凝土运送工厂;混凝土自动送运工厂
automatic depuration 自净(化)
automatic design 自动设计
automatic design center[centre] 自动设计中心
automatic design engineering 自动设计工程(学)
automatic design technique 自动设计技术
automatic desk computer 台式自动计算机
automatic de-sludger 自动除渣器
automatic desprucing 自动切除浇口
automatic detecting system 自动检测系统;自动查找系统
automatic detection 自动探测;自动检测;自动检波
automatic detection and integrated tracking 自动检测和综合跟踪;自动保护和联合跟踪
automatic detection finding system 自动查找系统
automatic detection system 自动探测系统
automatic detonator 自动雷管
automatic developing film 自动显影胶片
automatic developing machine 自动显影机;自动冲洗机
automatic device 自动装置
automatic dewatering 自行疏干
automatic diagnosis 自动诊断;自动识别;自动确定;自动发现
automatic diagnostic program(me) 自动诊断程序
automatic dial alarm system 自动拨号报警系统
automatic dialer 自动拨号机
automatic dialing unit 自动拨号装置
automatic dial-up two-way telephone 自动拨号双向电话
automatic diameter control 自动等径控制
automatic diaphragm 自动光圈
automatic diaphragm cone 自动膈膜式圆锥分级机
automatic dictionary 自动化检索词典
automatic die clamps 自动夹紧模具装置
automatic die head 自动板牙头
automatic differential count machine 自动分类计数机
automatic digit 自动化数字
automatic digital calculator 快速数字计算机
automatic digital computer 自动数学计算机
automatic digital drafting 自动数控绘图
automatic digital encoding system 自动数字编码系统
automatic digital network 自动数字网络
automatic digital tracking analyzer computer 自动跟踪分析数字计算机
automatic digitization 自动数字化
automatic direction(al) finder 自动测向仪;自动测向器;自动探向器
automatic direction finder bearing indicator 自动测向仪方位指示器
automatic direction finder remote controlled 遥控自动测向仪
automatic direction finder reversal 自动测向仪指针反转
automatic direction finding 自动测向
automatic direction finding system 自动测向系统
automatic direction plotter 航向自绘仪;航向自绘器
automatic discharge 自动卸料;自动泄水;自动排放;自动放电
automatic discharge device 自动放电器
automatic discharge ga(u)ge 自动流量计
automatic discharge pipe 自动排水管
automatic discharge unit 自动卸料装置
automatic discharge valve 自动排水阀
automatic discharge wagon 自动卸料车
automatic discharging 自动卸料的
automatic disconnecting device 自动切断装置;自动解列装置;自动拆卸装置
automatic disconnection 自动切断
automatic discrimination 自动判别
automatic dispatch system 自动调度系统
automatic display 自动显示
automatic display and plotting system 自动显示与绘图系统;自动显示标绘系统
automatic display call indicator 自动显示呼叫指示器
automatic dissemination 自动传播
automatic dissolver 自动溶解器;自动高速分散机
automatic distillator 自动蒸馏器
automatic distributor 自动洒布机;自动分配器;自动扬谷机(码头谷物卷扬)
automatic dividing head 自动分度头
automatic dock leveller 自动调节站台梯板
automatic document classification 自动文献分类
automatic document request service 自动文献请

求服务处
automatic document retrieval system 文献自动检索系统；自动文献检索系统；资料自动检索系统
automatic document storage and retrieval 文件自动存取与检索
automatic dodge 自动匀光印象机
automatic dodging 自动遮光
automatic dog 自动挡铁
automatic door 自动门
automatic door bottom 自动门底防风隔声附件；自动门槛；自动门底防风隔声设施
automatic door closer 自动关门装置
automatic door control 车门自动控制
automatic door hook 自动门钩
automatic doorlock 自动门锁栓
automatic door operator 自动开门装置；自动门控制器
automatic door reopening device 自动门再启装置
automatic door seal 自动密封垫
automatic door sensing equipment 自动门感应设备
automatic door stay 自动门钩
automatic door switch 自动门开关
automatic door unit 自动开门装置
automatic dose dividing apparatus 自动分量机
automatic dose rate controlled X-ray unit 剂量自动控制X线机
automatic dosing 自动投量
automatic double trap weir 双阱式活动堰；自动双叠门堰
automatic down feed 自动向下给进
automatic draft control 牵引力自动调节
automatic drafting 自动制图；自动绘图
automatic drafting machine 自动绘图器；自动绘图机
automatic drafting system 自动绘图系统
automatic draft regulator 自动通风调节器
automatic draft-tube fill line 自动尾水管充水管路；自动尾水管充水管线
automatic draghead winch controller 耙头绞车自动控制器
automatic drainage device 自动倒空装置
automatic drain cock 自动放水塞
automatic drain separator 自动排液分离器
automatic drain valve 自动放泄阀
automatic draught controller 吃水自动控制器
automatic draughting 自动绘图
automatic drawing stage 自动绘图阶段
automatic drawing system 自动绘图系统；自动绘图机
automatic drencher system 自动水幕系统
automatic drifting balanced circuit 漂移自动平衡电路
automatic drill 自动钻
automatic drilling 自动钻孔
automatic drilling machine 自动钻床
automatic drilling rig 车装钻机
automatic drilling robot 自动钻孔机器人
automatic drilling system 自动钻进系统
automatic drinking bowl 自动饮水器
automatic drinking cup 自动饮水器；自动饮水杯
automatic drive 自动传动
automatic drive control 传动自动控制
automatic driving 自动驾驶
automatic driving equipment 自动驾驶设备
automatic driving highway 自动驾驶道路
automatic drop bottom car 自动卸底车
automatic drop indicator 自动下落指示器
automatic dryer 自动控制烘干机；自动干燥器；自动干燥机
automatic drying line for welding electrode 电焊条自动烘焙线
automatic dry-pipe sprinkler system 干式自动喷水系统
automatic DSC operation(al) at ship station 船舶数字选呼自动操作
automatic dump 自动卸料
automatic dumping batch scale 自动卸料配料称
automatic dump truck 自卸式载重卡车；自动卸底车；自动倾卸卡车；自卸货车
automatic dye jig 自动卷染机
automatic earthquake detector 自动地震检测
automatic earthquake processing 自动地震数据处理；地震数据自动处理

automatic elastic plastic analysis 自动弹性分析；自动弹塑性分析
automatic electric(al) disappearing stair(case) 电动隐梯
automatic electric(al) ignition 自动电点火
automatic electrically heated waterer 电热自动饮水器
automatic electric(al) oven 自动电热烘箱
automatic electric(al) water heater 自动电热水器；自动电热水加热器
automatic electric(al) welding 自动电焊
automatic electrode assembly 电极自动装配
automatic electron compensation instrument 电子自动补偿仪
automatic electronic computer 自动电子计算机
automatic electronic data-switching center 自动电子数据交换中心
automatic electro-pneumatic brake 自动电力气动制动机
automatic elevator 自动升板机；自动电梯；自动升降机
automatic empty-and-load brake 自动空一重车制动器
automatic energy control 自动能量控制
automatic engineering 自控工程；自动工程
automatic engineering design 自动工程设计
automatic engine lathe 自动普通车床
automatic engine speed synchronizer 发动机转速自动同步器
automatic equalization 自动均衡
automatic equalizer 自动均衡器
automatic equation solver 方程式自动解算装置
automatic equipment 自动设备
automatic error correction 误差自动校正；自动误差校正；自动纠错；自动错误校正法
automatic error correction system 自动误差校正系统
automatic error detection 误差自动检测(程序)
automatic error request equipment 误差自动校正装置
automatic escape 自动退水闸；自动排气管
automatic exchange 自动交换(机)；自动电话局
automatic excitation cut off device 自动灭磁装置
automatic exclusion 自动排除
automatic exhaust air pipe 自动排气管
automatic exhaust air valve 自动排气阀
automatic exhaust gas analyzer 自动排气分析器
automatic exhaust gas filter 自动排气过滤器
automatic exhausting window 自动排风活门
automatic exhaust steam valve 自动排气阀
automatic expansion valve 自动调节阀；自动膨胀(滑)阀
automatic explosion-proof device 自动防爆装置
automatic exposure compensation 自动曝光补偿
automatic exposure control 曝光自动控制；自动曝光装置；自动曝光控制
automatic exposure equipment 自动曝光设备
automatic extraction 自动抽出
automatic extraction of radar information 雷达信息自动提取
automatic extraction turbine 自动调整抽汽式汽轮机
automatic fare collection 自动售检票
automatic fare collection machine 自动售票机
automatic fare collection system 自动售票系统
automatic fast return 自动快速返回
automatic fault detection 自动故障探测
automatic fault finding 自动故障寻找
automatic fault isolation 自动故障隔离
automatic fault signal 自动故障信号
automatic fault signal(l)ing 自动事故信号装置
automatic feed 自动进料；自动进给；自动进刀；自动供料；自动给料
automatic feed control 自动给料控制；自动进给控制
automatic feed drifter 自动推进架式钻机；自动推进架式凿岩机
automatic feed drill 自动给进凿岩机
automatic feed drilling 自动给进钻进
automatic feeder 自动续纸装置；自动投料器；自动投料机；自动送料器；自动进料装置；自动进刀装置；自动加料器；自动供料机；自动给料器
automatic feed grate 自动喂料算子；自动给煤机算盘
automatic feed grinder 自动进料式粉碎机

automatic feeding 自动供料；自动送料装置；自动给料的
automatic feeding and catching table 自动进出料辊道
automatic feeding and sorting maching 自动送料分类机
automatic feeding device 自动给料装置
automatic feeding mechanism 自动进给机构
automatic feed of electrode 电极自动送进
automatic feedoff 自动调正机构(转盘钻进，钢丝绳下放时)
automatic feed punch 自动送卡穿孔机
automatic feed system 自动给电系统；自动进料系统；自动给水系统；自动给料系统
automatic feed tray 自动给料盘
automatic feed unit 自动推进装置(钻机)
automatic feed valve 自动给气阀
automatic feed water control 自动进水控制；自动给水控制
automatic feed water pump 自动供水泵；自动给水泵
automatic feed water regulator 自动给水调节器
automatic fidelity control 自动逼真度控制；自动保真度控制
automatic field forcing 强行励磁；强迫励磁；自动强行励磁
automatic field suppressing 自动灭磁
automatic field weakening 自动减磁；磁场自动削弱
automatic file transmission 自动文件发射
automatic filing 自动文件编制
automatic filling and sealing machine 自动灌封机
automatic filling machine 自动装填机；自动填料机；自动加料机
automatic fill-up float collar 自动油浆浮箍
automatic fill-up float shoe 自动灌浆浮鞋
automatic film advance 自动卷片
automatic film advance mechanism 底片自动传输装置
automatic film developing equipment 自动显影设备
automatic filter press 自动压滤机
automatic filtration 自动过滤
automatic fine control 自动微调
automatic finetuning control 自动微调控制
automatic fine turning 自动微调
automatic fire alarm 失火自动报警器；火灾自动报警器；火警自动报警器；自动火灾报警系统
automatic firedamp alarm 自动瓦斯报警器
automatic fire damper 自动挡火闸
automatic fire detecting device 火灾自动检测设备
automatic fire detecting system 火灾自动检测系统；自动火灾警报装置；自动火灾警报系统；自动火警警报装置；自动火警报警系统
automatic fire detection device 火灾自动检测设备
automatic fire detection system 自动火灾警报系统
automatic fire detector 自动火警警报器
automatic fire device 自动燃烧装置
automatic fire door 自动防火门
automatic fire door parts 自动炉门零件
automatic fire-fighting system 自动消防系统
automatic fire protection system 自动消防水系统
automatic fire pump 自动消防水泵；自动消防泵
automatic fire shutter 自动断火闸；自动燃烧风门片；自动燃烧风门板
automatic fire sprinkler 自动灭火喷头；自动喷水灭火装置；自动洒水灭火装置；自动喷洒灭火装置；自动灭火装置；自动灭火喷淋器；自动火灾
automatic fire-sprinkler system 自动喷水灭火装置
automatic fire suppression system 自动灭火系统
automatic fire vent 自动出烟口；自动排烟口；自动排烟天窗
automatic fire warning device 自动报火警设备；火警自动报警装置
automatic fixed telecommunication network 自动固定电信网络
automatic flame photometer 自动火焰光度计
automatic flap gate 平衡式舌瓣闸门；自动活瓣闸门
automatic flash board 自动启闭闸门
automatic flash light 自动闪光灯
automatic flask machine 自动制瓶机
automatic flat grinding machine 自动磨盖板机
automatic flexible strapping machine 自动捆扎机
automatic float 自动浮标
automatic floater-controlled bleeder tap 自动控

制浮子放水阀(塞)
automatic floating station 自动漂浮测站
automatic flocculation 自动絮凝(作用)
automatic flow controller 自动供电调节器
automatic flow control system 流量自动控制系统
automatic fluid analyser 液体成分自动分析仪
automatic flushback 自动反冲洗
automatic flushing cistern 自动冲洗(水)箱;自动冲洗储水器;自动冲洗储[贮]水器;自动冲水储[贮]水器;自动冲水箱
automatic flushing device 自动清洗装置
automatic flushing out 自动冲洗
automatic flushing system 自动排水设施;自动冲水系统
automatic flushing tank 自动冲洗(水)箱
automatic flushing trough 自动冲洗槽
automatic fluviograph 自记水位仪;自记水位计
automatic focus 自动聚焦
automatic focusing 自动调焦
automatic focusing action 自动聚焦作用
automatic focusing apparatus 自动对光装置;自动调焦装置
automatic focusing system 自动调焦系统
automatic focus projector 自动对光投影仪
automatic focus rectifier 自动对光纠正仪
automatic fog bell 自动雾钟
automatic fog control 自动雾号控制
automatic fog gun 自动雾炮
automatic fog signal 自动雾号
automatic fog signal control 雾号自动控制器
automatic fog whistle 自动雾笛
automatic folding and thread-sealing machine 自动折页塑料线烫订机
automatic folding leaf door 自动折页门
automatic folding type air filter 自动更换折叠式空气过滤器
automatic following 自动跟踪
automatic following control 自动跟踪控制
automatic following microscope 自动跟踪显微镜
automatic following water level meter 自动跟踪水位计
automatic follow-up tuning 自动跟踪调谐
automatic food vending machine 自动食品售货机
automatic forecasting system 自动预报系统
automatic fork lift 自动叉车
automatic formatting 自动格式化
automatic forming 自动成形;自动成型
automatic foundry 自动化铸造车间
automatic fraction collector 自动馏分收集器;自动分选机
automatic freight handling car 自动装货车
automatic frequency analyzer[analyser] 自动频率分析器
automatic frequency control 频率自动控制;自动调频
automatic frequency correction 自动频率校正
automatic frequency fine control 自动频率微调
automatic frequency regulating device 自动调频装置
automatic frequency stabilization 自动频率稳定
automatic frequency tuner 自动频率调谐器
automatic friction skate retarder 自动摩擦铁鞋缓行器
automatic fuel drain valve 自动放油阀
automatic fuel economizing device 自动节省燃料设备
automatic fuel saving device 燃料节省自动调整器;自动节省燃料设备
automatic fuel shut off 自动停止进入燃料
automatic fuel spray valve 自动喷油阀
automatic fuel supply 自动加油;自动供给燃料
automatic full beam stop motion 满轴自停装置
automatic furnace 自动燃烧控制炉子;自动炉子;自动燃烧装置
automatic fusarc/CO_2 welder 二氧化碳保护自动弧焊机
automatic gain control 自动增益控制;自动增益调整
automatic gain regulator 自动增益调节器
automatic garden tractor 自走式园圃拖拉机
automatic gas analyzer 自动气体分析器
automatic gas chromatography 自动气相色谱仪
automatic gas control valve 煤气自动控制阀
automatic gas cutting 自动气割

automatic gas producer 自动气体发生器
automatic gas ratio control 气体比例自动调节
automatic gas sampling valve 自动气体进样阀
automatic gas shutoff device 煤气自动切断装置
automatic gas shutoff valve 煤气自动截止阀
automatic gas suspension device 煤气自停装置
automatic gate 自动大门;自动闸门;自动门;自动快门
automatic gate lock 自动门锁栓
automatic ga(u)ge 自记水位仪;自记水位计;自动水位计;自动计量;自动计量仪
automatic ga(u)ge classifier 厚度自动分选机
automatic ga(u)ge control 自动厚度控制
automatic ga(u)ge controller 厚度自动控制器;厚度自动调整器
automatic ga(u)ge register 自动计量器
automatic ga(u)ging 自动检验
automatic ga(u)ging and detecting machine 自动测验机
automatic gear 自动传动装置;自动装置
automatic gear change 自动换档
automatic gear cutting machine 自动切齿机
automatic gear shift(ing) 自动变速;自动换排档
automatic generating plant 自动发电装置;自动发电厂
automatic generating station 自动化发电站
automatic generation 自动生成
automatic generation chamber 自动发芽室
automatic generation control 出力自动控制;发电机组功率输出自动控制
automatic generator 自动控制的发电机
automatic geyser 自动喷泉
automatic gland packing 自动密封垫
automatic glass level control 玻璃液面自动控制
automatic governor 自动调速器;自动离合器;自动调节器
automatic grab 自动抓斗;自动夹具;自动抓岩机
automatic grab beam 自动抓梁
automatic grade control system 级配自动控制系统
automatic grader 自动分级机
automatic graphic(al) instrument 自动记录仪表
automatic graphic(al) pyrometer 自动记录高温计
automatic graph plotter 自动绘图仪
automatic grapnel for marine purpose 船用机械锚
automatic grapple fork 自动干草叉
automatic gravity classification yard 自动化驼峰编组场【铁】
automatic gravity die casting machine 自动硬模铸造机
automatic gravity marshalling yard 自动化驼峰编组场【铁】
automatic gravure etching machine 照相凹版自动腐蚀机
automatic greasing cycle 自动润滑循环
automatic grinder 自动研磨机
automatic grinding mill 自动式磨粉机
automatic grouter 自动调ät灌浆器;混凝土钢模压力灌浆机;自动灌浆器;自动灌浆机;自动调节灌浆机
automatic grouting machine 自动灌浆器;自动灌浆机
automatic guided system 自动导向运输系统
automatic guided vehicle 自动堆装机(库场用)
automatic guider 自动导星装置
automatic guiding 自动导星
automatic gun charger 自动装弹器
automatic gyropilot 自动陀螺驾驶仪
automatic gyroscope 自动驾驶仪
automatic hammer with two opposite working rams 双头对击锤
automatic handling equipment 自动装卸设备
automatic handshake 信号自动同步交换
automatic hardness tester 自动硬度测定仪
automatic hatch cover 自动舱口盖;自动舱盖
automatic head block signal 出站第一具自动闭塞信号机
automatic heading reference system 自动首向基准系统;自动船首基准系统
automatic heat detector 自动检热器;温感自动报警器
automatic heating 自动供暖
automatic heating tank 自动热水箱
automatic heat regulation 热力自动调节器
automatic heat regulator 温度自控器

automatic helmsman 自动操舵机
automatic hidden line removal 自动消除隐藏线
automatic high speed lathe 自动高速车床
automatic high voltage control 自动高压控制
automatic high voltage regulator 自动高压稳压电路
automatic high voltage safety switch 高压自动保险开关
automatic highway crossing gate 公路与铁路交叉口自动起闭栅
automatic highway crossing signal 道口自动信号(机);道口自动号志
automatic hillshading 自动晕渲
automatic hinge assembling and double-sided hinge pin spinning machine 铰链组合式自动化离心浇注(混凝土管)成形机
automatic hoist 自动升降机;自动卷扬机;自动绞车
automatic hold 自动夹紧
automatic holding 自动夹紧的
automatic holding device 自动夹具;自动保持装置
automatic hook 自动挂钩
automatic horizon 自动水平
automatic humidifier 自动加湿机
automatic humidity controller 湿度自动控制器;自动湿度调节器
automatic hump 自动化驼峰【铁】
automatic hump yard 自动化驼峰编组场【铁】
automatic hydraulic brake device 自动液压制动装置
automatic hydraulic sprayer 自动液压喷雾器
automatic hydraulic station 自动水力发电站
automatic hydraulic transmission 自动液力传动
automatic hydroelectric(al) plant 自动化水力发电厂
automatic hydroelectric(al) station 自动化水力发电站
automatic hydropower station 自动化水电站
automatic hysteresis loop recorder 磁滞回线自动记录器
automatic identification 自动识别
automatic identification equipment 自动识别装置;自动识别设备
automatic identification of outward dialing 外线拨号自动识别
automatic ignition 自动点火
automatic image processing equipment 图像自动处理设备
automatic image retrieval system 自动图像检索系统
automatic image scanner 图像自动扫描仪
automatic immersion heater 浸渍式自动加热器
automatic import quota 自动进口限额
automatic index(ing) 自动索引;自动变址;自动编索引;自动检索;自动加索引;自动分度(法);自动标引
automatic indexing device 自动示数装置
automatic indication pyrometer 自动指示高温计
automatic indicator 自动指示器
automatic induction coordinated system 自动感应联动(信号)系统
automatic inductive train control 列车自动保护感应控制
automatic information retrieval system 信息自动检索系统;自动信息检索系统
automatic infrared facility 红外(线)自动装置
automatic inherent storage 自动固有存储器
automatic injection advanced device 自动提前喷射装置
automatic input control valve 自动输料控制阀;自动给气控制阀
automatic input data 自动输入数据
automatic inspection 自动检查
automatic inspection and ga(u)ging machine 自动检查测量机
automatic inspection diagnostic and prognostic system 自动检验诊断与预报系统
automatic inspection line for glass container 玻璃瓶罐自动检验线
automatic inspection of data 数据自动检查
automatic instrument 自控仪表
automatic intake device 自动进水装置;自动进气装置
automatic integrator 自动积分仪
automatic intelligence data system 自动情报数

据系统
automatic intercepting valve 自动遮断阀
automatic interlock 自动联锁
automatic interlocking 自动联锁法
automatic interlocking device 自动联锁装置
automatic interlocking for hump yard 驼峰自动集中
automatic interpolation 自动内插
automatic interrupt 自动中断
automatic interrupter 断路器;自动开关;自动断续器
automatic interruption 自动中断
automatic iron carrier 铁件自动装卸机
automatic irrigation 自动灌溉
automatic isochronous staining machine 自动等时染色机
automatic isolated fixed cycle signal (独立的)固定周期自动信号
automatic isolating valve 自动隔离阀(门)
automaticity 自律性;自动性;自动化程度
automatic jet control valve 自动喷射调整阀
automatic key 自动电键
automatic keying device 自动键控装置
automatic knapsack sprayer 背负式自动喷雾机
automatic knotter 自动打结器
automatic knuckle coupler 自动钩舌车钩
automatic labeling machine 自动贴商标机
automatic lacing machine 自动穿连纹板机
automatic lamp changer 航标灯自动交换器;换灯泡机;自动航标灯交换器
automatic landing system 自动着陆装置;自动着陆系统
automatic language data processing 自动语言数据处理
automatic lap former 自动成卷装置
automatic large scale production 自动化大量生产
automatic laser grading system 自动激光控制坡度系统
automatic laser ranging and direction instrument 自动激光定位仪
automatic lathe 自动车床
automatic layboy 自动折叠机
automatic layout 自动布局
automatic leak control 自动漏量控制
automatic letter facing 自动信件盖销装置
automatic lettering on map 地图自动注记
automatic level 自动水准仪;自动找平水准仪;自动安平水准仪
automatic level compensation 自动电平补偿
automatic level control 液面自动控制;自动电平控制
automatic level(l)ing 自动取平的
automatic level(l)ing compensator 自动找平补偿器
automatic level(l)ing device 自动找平装置
automatic level regulation controlled by electromechanical element 机电式自动电平调节
automatic level-regulator 电平自动调节器
automatic library 自动化图书馆
automatic lift 自动电梯
automatic lift trip scraper 铲斗自动起落平地机
automatic light control 自动光量调节
automatic lighting 自动点灯
automatic lighting regulator 自动照明控制器
automatic light mixture overboard-discharging device 稀泥浆自动排出船外装置【疏】
automatic light mixture overboard installation 低浓度泥浆舷外自动排放装置
automatic light signal 自动灯光信号
automatic line 自动线
automatic linear positioning system 自动线性定位系统
automatic line feed 自动换行
automatic line-follower automatic line tracer 线划自动跟踪器
automatic line-following 线划自动跟踪
automatic line plotter 自动线绘图机
automatic line sectionalizer 自动线路分段开关装置
automatic line selection 自动送行
automatic line switching 线路自动转换
automatic line tracing 线划自动跟踪
automatic listing of constants 常数自动列表
automatic load balancing device 荷载自动平衡装置
automatic load control 自动荷载控制

automatic load curtailment 自动减(负)荷
automatic loader 自动装载机;自动装填机;自动装入程序
automatic loader and unloader 自动装卸机
automatic loading mechanism 自动装料机构
automatic loading regulator 自动变荷调压器
automatic loading ship 自动装载船
automatic load limitation 自动减载装置
automatic load limitation device 自动限载装置
automatic load maintainer 自动荷载稳定器
automatic load regulation 负荷自动调节
automatic load regulator 负荷自动调节装置;负荷自动调节器
automatic load sustaining brake 棘轮制动器
automatic lock 自动闭锁
automatic locking 自动锁
automatic lock system 自动闭锁系统
automatic log(ging) 自动记录
automatic logistrip 油轮自动装卸压载系统
automatic log-on 自动联机;自动注册
automatic long-range sonar 自动远程声呐
automatic louvers 自动百叶窗
automatic lowerator 自动降板机
automatic lubrication 自动润滑(法)
automatic lubricator 自动润滑器
automatic machine 机床自动操纵;自动校验;自动机(床)
automatic machine process 自动机械加工法
automatic machine tool control 机床的自动操纵
automatic machining 自动化机械加工
automatic magazine 自动装片暗盒
automatic magazine camera 自动装片摄影机
automatic magazing loader 自动储存放料装置
automatic making-up siding 自动编组线
automatic mallet 电动锤
automatic manual 自动手动
automatic map 自动化地图
automatic mapper 自动绘图器;自动绘图机
automatic mapping 自动制图;自动绘图;自动成图
automatic mapping system for marine map 海图自动制图系统
automatic marble feeder 自动加球机
automatic marine polarimeter 自动海洋旋光测定仪
automatic marker 自动划行器
automatic marshalling controller 自动编组控制设备
automatic marshalling station 自动化编组站
automatic marshalling yard 自动化编组场
automatic matching 自动匹配
automatic measurement 自动量测;自动测量
automatic measurement technology 自动检测技术
automatic measurer 自动量
automatic measuring device 自动计量装置;自动计量器
automatic measuring instrument for liter-weight of clinker 熟料升重自动测量仪
automatic measuring machine 自动量度仪
automatic measuring plant 自动计量设备
automatic measuring unit 自动测量器
automatic mechanical feed 自动机械送料器
automatic mechanical filter 自动过滤器
automatic mechanical rapping system 自动振打装置
automatic media renewing mechanism 滤材自动更换机构
automatic melt-sealer 自动熔封器
automatic merchant vessel report system 商船自动报告系统
automatic message counting 自动信息计算
automatic message distribution system 自动信息分配系统
automatic message recording 自动信息记录
automatic message switching centre 自动报文交换中心;自动信息交换中心
automatic metal feeding furnace 金属自动浇铸炉
automatic meteorological observation 自动气象观测
automatic meteorological observation station 自动气象观测站
automatic meteorological observation system 自动气象观测系统
automatic meteorological observing station 自动气象观测站
automatic meteorological oceanographic (al) buoy 海洋气象自动浮标(站)
automatic meteorological station 自动气象站;自动气象台
automatic meteorological system 自动气象系统
automatic metering tap 自动定量关闭龙头
automatic metering valve 自动定量关闭阀门
automatic microfilm information system 自动缩微胶片信息系统
automatic micrometer 自动测微计
automatic microphotographic(al) camera 自动显微照相机
automatic microtome knife sharpener 切片刀自动研磨器
automatic milling machine 自动铣床
automatic mixer 自动搅拌器
automatic mixing valve 自动混合阀(门)
automatic mixture controller 混合比自动控制器
automatic mixture regulator 混合比自动调节器
automatic mode 自动化模式
automatic modem selection 自动调制解调器选择
automatic modulation control 自动调制控制
automatic moisture control 湿度自动调节
automatic moisture meter 自动湿度计
automatic moisture titrator 自动湿度滴定器
automatic monitor 自动检查器;自动监视器;自动监控器;自动监测器;自动(化)监测仪
automatic monitor button 司机室自动控制按钮
automatic monitored control system 自动监控系统
automatic monitoring 自动监控;自动监测
automatic monitoring device 自动监测装置
automatic monitoring system 自动监测系统
automatic monkey 自动桩锤
automatic mosaicker 像片自动镶嵌仪
automatic motor starter 发动机自动起动器
automatic mo(u)ld 自动压模;自动塑模;自动模
automatic mo(u)lding line 自动造型线
automatic mo(u)lding machine 自动造型机;自动成型机
automatic mo(u)lding plant 自动造形设备
automatic mo(u)lding unit 自动造型机组
automatic moving target indicator 运动目标自动指示器
automatic muller 自动平磨机
automatic multiple channels cordless telephone 自动接入多信道无绳电话
automatic multiscrewdriver 多用自动起子
automatic multispanner 多用自动扳手
automatic multispindle lathe 多轴自动车床
automatic nature 自动性
automatic navigation 自动导航
automatic navigational computer and plotter 自动导航计算机和航迹描绘仪
automatic navigation and data acquisition system 自动导航和数据收集系统
automatic navigation control equipment 自动导航控制装置
automatic navigation device 自动导航仪
automatic navigation system 自动导航系统
automatic navigator 自动导航仪
automatic network analyzer 自动网路分析器
automatic new line 自动换行
automatic night service 自动夜间服务
automatic noise controller 自动噪声控制器
automatic noise figure indicator 自动噪声图形指示器;自动噪声数字指示器
automatic noise level(l)er 噪声自动限制器
automatic noise limiter 自动噪声限制器
automatic noise suppressor 自动噪声抑制器;自动噪声控制器
automatic nonreturn valve 自动逆向阀;自动止回阀
automatic notice to mariners system 自动航海通告
automatic nozzle 自动喷嘴
automatic number identification 自动号码识别
automatic numbering machine 自动编号机
automatic numbering transmitter 自动编号发射机
automatic number key 自动数字键
automatic number normalization 数值自动规格化
automatic nut tapping machine 自动攻螺母机
automatic object recognition and classification 自动标志识别和分类
automatic observatory 自动观测站;自动观测仪;自动观测台

automatic observer 自动观测仪
automatic oil 自动机润滑油
automatic oil cock 自动加油旋塞
automatic oil cup 自动加油杯
automatic oiler 自动加油器
automatic oiling 自动加油
automatic oil pressure ga(u)ge 自控油压表
automatic oil switch 自动油开关
automatic onboard gravimeter 船上自动重力仪
automatic onboard optic(al) system 机上自动光学系统
automatic one speed control 自动单速调整
automatic on load tap changer 带负荷自动切换开关;带荷自动抽头变换开关
automatic on-off type burner 自动开关式燃烧器
automatic open cycle control 开环自动控制
automatic opener 自动开门器
automatic opening 自动断开
automatic open loop control 开环自动控制;自动开环控制
automatic operating and schedule program(me)s 自动操作与调度程序
automatic operation 自动运行;自动运算;自动操作;自动操纵作业
automatic operations panel 自动操作台;自动操作板
automatic operation system 自动运算系统
automatic operator 自动运行设施;自动操作器;自动开关装置
automatic oscillograph 自动示波器
automatic output control 自动输出控制
automatic overboard-dump system for light mixture 稀泥浆自动排出船外装置【疏】
automatic overload circuit 自动过载电路
automatic overload control 超载自动控制;自动超载控制
automatic oxidation 自动氧化(作用)
automatic oxygen cutting 自动氧气切割
automatic oxygen generator 自动制氧机
automatic oxygen safety valve 自动氧气安全阀
automatic packaging 自动装配;自动包装
automatic packaging unit 自动包装机
automatic packer 自动包装机
automatic packing machine 自动打包机
automatic padding machine 自动拭浆机
automatic palletizer 自动摞包机
automatic pallet loader 自动托盘装载机
automatic paper interleaving machine 自动铺纸机;自动垫纸机
automatic parallax detection 视差自动检测
automatic paralleling device 自动同期装置;自动并行装置;自动并列装置
automatic parking ticket issuing machine 停车票证自动发放机
automatic particle counter 粒子自动计数器;自动粒子计数器
automatic particle size analysis 自动粒度分析
automatic particle size analyzer 自动粒度分析器
automatic passing button circuit 自动通过按钮电路
automatic paste filler 自动装膏机
automatic paste packaging machine 膏剂自动包装机
automatic paster 自动接纸装置
automatic patching system 自动编排系统
automatic paying gear 自动放索装置
automatic paying out gear 自动松缆装置
automatic peak limiter 自动峰值限制器
automatic performance 自动执行
automatic phase control 自动相位控制
automatic phase control loop 自动相位控制回路
automatic phase lock 自动锁相
automatic phase shifter 自动移相器
automatic phase synchronization 自动相位同步
automatic photocomposer 自动照相排字机
automatic photoelectric(al) titrater 自动光电滴定器
automatic photometer 自动光度计
automatic photosetter 自动照相排字机
automatic-pickup baler 自动捡拾压捆机
automatic picture control 自动图像控制
automatic picture transmission 图像自动发射;图像自动传输;自动图像传送;自动图像传输
automatic picture transmission system 自动图像发送系统;自动传真系统
automatic pile driver 自动打桩机
automatic piler 自动堆垛机
automatic pilot 自动驾驶仪;自动舵;自动导航;自动操舵装置
automatic pilot computer and plotter 自动导航计算机和航迹描绘仪
automatic pilot device 自动点火装置;自动导向装置;自动导航装置
automatic pilot system 自动驾驶系统
automatic pipe(making)machine 自动制管机
automatic pipet(te) 自动吸(移)管
automatic pitman 自动连杆
automatic plant 自动化发电厂;自动装置;自动设备;自动化工厂;自动工厂
automatic plate casting machine 自动浇版机
automatic plateforming line 盘类自动成形线
automatic plating 自动电镀(工艺)
automatic plotter 自动绘图仪;自动描绘器;自动绘图器;自动绘图机
automatic plot(ting) 自动绘图
automatic plotting board 自动标定图板
automatic plotting instrument 自动测图仪
automatic plotting system 自动绘图系统
automatic plug 自动塞
automatic plug mill process 自动轧管法;自动扎管法
automatic plug rolling mill 自动芯棒轧管机
automatic pneumatic control 自动气动控制
automatic point control 道岔自动控制
automatic point marking 控制点自动标定
automatic point setting 道岔自动转换
automatic point switching 道岔自动转换
automatic polymerization 自动聚合
automatic pontoon 自动浮船式闸门;自动浮桥
automatic pontoon lock gate 自动沉箱式闸门
automatic portable test set for cable 便携式自动测缆器
automatic position controller 自动位置调节器
automatic position indication 位置自动显示系统
automatic position indicator 自动位置指示器
automatic positioning 自动对位;自动定位
automatic positioning device 自动定位装置
automatic positioning equipment 自动定位装置
automatic position setting 自动定位
automatic position tracer 航迹自绘仪
automatic potentiometry 自动电位滴定法;自动电位测定
automatic pouring device 自动浇注装置
automatic powder spray gun 自动粉末喷枪
automatic power compensator 自动功率补偿器
automatic power control 自动功率控制
automatic power plant 自动发电厂;自动动力厂
automatic power regulator 自动功率调节器
automatic power slips 自动机械卡瓦
automatic power station 自动发电站
automatic premium loan 自动提缴保费贷款
automatic preset 自动程序调整
automatic press 自动印刷机;自动压力机
automatic pressing 自动压制;自动压形
automatic pressure and flow control valve 压力流量自动控制阀
automatic pressure control 自动压力控制
automatic pressure controller 自动压力调节器;自动调压器
automatic pressure filter 自动压滤器
automatic pressure jet burner 自动压力喷灯
automatic pressure-reducing valve 自动减压阀(门)
automatic pressure regulator 自动压力调节器
automatic pressure vent 自动排气门
automatic primer 自动灌水器;自动起爆装置;自动起爆器;自动雷管
automatic priming 自动起动;自动充水
automatic printing machine 自动印刷机
automatic probe test 自动探针测试
automatic process 自动程序
automatic process control 过程自动控制
automatic process controller 过程自动控制器;过程自动调节器
automatic process cycle controller 过程周期自动控制器
automatic process distillator 过程自动蒸馏分析仪
automatic processing equipment 自动处理设备
automatic processor 自动信息处理机;自动处理机
automatic production 自动生产;自动化生产
automatic production line 自动生产线;自动化生产线;自动化工厂
automatic production line of bean product 豆制品自动生产线
automatic production planning system 自动生产计划系统
automatic production record system 自动化生产记录系统
automatic profiling 自动生成剖面
automatic program(me)control 自动程序控制
automatic programmed checkout equipment 自动程序检查装置
automatic programmed control system 自动程序控制系统
automatic programmed tool 自动编程工具;刀具控制程序自动编制系统;程序自动化方法
automatic program(me)interruption 自动程序中断
automatic programmer 自动程序设计器
automatic programmer-and-test system 自动程序设计试验系统
automatic program(me)unit 低速自动程序装置
automatic programming 自动程序设计;自动编制程序;自动编程序;程序自动化;程序设计自动化
automatic programming machine 自动程序设计机
automatic programming tool 工具机控制程序;机床自动程序设计语言
automatic programming unit 自动编程序装置
automatic propeller 自动调距螺旋桨
automatic proportional water sampler 水自动比例采样器;自动比例水样采集器
automatic proportioning control 自动配料控制
automatic protection 自动保护
automatic protection of train running 列车运行自动保护
automatic protection switching 自动保护切换;自动保护开关
automatic protective system 自动保护系统
automatic protector 自动保护装置
automatic puller 自动拉出器;自动拔出器
automatic pump 自动水泵;自动泵
automatic pump controller 泥泵自动控制器
automatic pumping station 自动化泵站
automatic pumping system 自动抽水系统
automatic pump station 自动化泵站;自动泵站
automatic punch 自动穿孔
automatic puncher 自动穿孔机
automatic punching machine 自动冲床
automatic purifier 自动净水器;自动净水机
automatic push-button control 自动按钮控制;按钮自动控制
automatic push button control lift 按钮自动控制电梯
automatic pyrometer 自动指示高温计
automatic quadder 自动填空铅机
automatic quality control 自动质量控制
automatic radar 自动雷达
automatic radar beacon 自动雷达信标
automatic radar beacon sequencer 自动雷达信标定序器
automatic radar control and data equipment 自动雷达控制与数据设备
automatic radar measuring equipment 自动雷达数据测量设备
automatic radar plotter apparatus 自动雷达绘图仪
automatic radar plotting aids 自动雷达标绘仪
automatic radar terminal system 自动雷达终端系统
automatic radar tracking system 自动雷达跟踪系统
automatic radial check gate 自动弧形节制阀
automatic radiation monitoring system detection test program(me) 辐射自动监测系统测试计划
automatic radio alarm 无线电自动报警器
automatic radio compass 自动无线电测向仪
automatic radio direction finder 无线电自动测向仪;自动无线电测向仪
automatic radio direction finding system 自动无线电定向系统
automatic radiometeorograph 自记无线电气象仪
automatic rail trimming machine 钢轨自动研磨

机;钢轨自动平整机
automatic rail washer 自动轨条洗净器
automatic railway 自动化铁路
automatic railway control 铁路运输自动化控制
automatic railway control center 铁路运输自动化控制中心
automatic railway management 铁路运输自动化管理
automatic railway management center 铁路运输自动化管理中心
automatic railway telephone system 自动化铁路电话系统
automatic rainfall recorder 自动雨量计
automatic rain sampler 自动雨水采样器
automatic rammer 自动夯土机;自动捣固机
automatic ram pile driver 自动锤击打桩机;自动冲锤打桩机
automatic range control 自动距离控制
automatic range finder 自动测距仪
automatic range selector 自动量程选择器
automatic range tracking 自动测距跟踪
automatic range tracking system 距离自动跟踪系统
automatic range tracking unit 自动距离跟踪器
automatic range unit 自动测距装置
automatic ranging 自变换量程
automatic rapid transit 自动化快速交通系统
automatic raster plotter 光栅绘图机
automatic rate-and-price indicator 自动估价及价值指示器;估价及价值自动指示器
automatic rate-price indicator 估价及价值自动指示器;自动估价及价值指示器
automatic ratio control 自动比率控制
automatic reading 自动读数
automatic reading machine 自动读出器
automatic readout 自读器
automatic readout distance meter 自读测距仪
automatic ready mix(ed) concrete fabricator 自动化预拌混凝土工厂
automatic real time processing 自动实时(数据)处理
automatic rear signal control 尾部信号自动控制
automatic receptor 自动接收机
automatic reckoning plotter 船位自动绘迹仪
automatic reclosing 自动重合闸
automatic reclosing relay 自动重合继电器
automatic reclosing switch 自动重合开关
automatic reclosure 自动重合闸
automatic recognition 自动识别
automatic recorder 自记器;自动记录器
automatic recording 自动记录(的)
automatic recording altimeter 自动记录高程计
automatic recording and telemetering buoy 自记遥测浮标
automatic recording apparatus 自动记录装置;自动记录仪
automatic recording audiometer 自动记录测听计
automatic recording calculating machine 自动记录式计算机
automatic recording ga(u)ge 自动记录仪表;自动记录计
automatic recording indicator 自动记录指示器
automatic recording infrared spectrograph 自动记录红外线光谱仪
automatic recording instrument 自动记录装置;自动记录仪
automatic recording of mass spectra 质谱的自动记录
automatic recording oscillograph 自记示波仪;自记示波器;自动录波器
automatic recording pointer 自动记录针
automatic recording spectrometer 自记分光仪;自记分光计
automatic recording spectrophotometer 自记式分光光度计
automatic recording tide ga(u)ge 自记验潮仪;自动潮位仪;自动潮位记录器;自动潮位计
automatic recording titrator 自动记录滴定器
automatic recording titrimeter 自动记录滴定器
automatic record stereo-coordinator 自动记录立体坐标仪
automatic recovery 自动恢复;自动复原
automatic recovery system 自动回收系统
automatic recruitment control 自动重振控制

automatic rectifier 自动纠正仪
automatic reducing valve 自动减压阀(门)
automatic reduction 自动归算
automatic reduction tach(e)ometer 自动归算速测仪;自动归算视距仪
automatic registration 自动记录
automatic regression analysis 自动回归分析
automatic regression process 自动回归过程
automatic regulating apparatus 自动调整器
automatic regulating device 自动调节装置
automatic regulating system 自动调节系统
automatic regulation 自动调节
automatic regulation dam 自动调节坝
automatic regulation device 自动调节装置
automatic regulation mechanism 自动调整机理
automatic regulation valve 自动调节阀
automatic regulator 自动控制器;自动调整器;自动调节器
automatic reinstatement clause 自动补足保额条款
automatic rejection 自动选废
automatic relative plotter 自动相对运动作图器
automatic relay 自动中继
automatic relay calculator 自动继电器式计算机
automatic relay computer 自动中继计算机
automatic release 自动释放;自动切断;自动缓解
automatic release heating coupling 自动切断加热连接
automatic releasing hook 自动摘钩;自动解缆钩
automatic relief shading 地貌自动晕渲
automatic remote control 自动遥控;自动距离控制
automatic removable air filter 自动更换式空气过滤器
automatic rendering winch 自动收紧绞车
automatic renewable air filter 自动再生式空气过滤器
automatic reperforator switching 自动复凿机转换开关
automatic repetition system 自动重复发送方式
automatic report of wrongly shunted system 车辆误入另道的自动报告
automatic request for repeat 反复重复
automatic request for repetition 自动请求重复发送;自动请求重发;自动请求检索
automatic request for retransmission 自动重发请求
automatic request method 自动检索法
automatic rescue beacon 自动救生信标
automatic reseau measuring equipment 格网自动量测设备
automatic reset 自动重调;自动复位
automatic reset relay 自动复原继电器
automatic resolution control 自动清晰度控制
automatic respirator 自动呼吸器
automatic restart 自动再启动
automatic restoring 自动恢复
automatic retardation control 自动减速控制;自动缓行控制器
automatic retardation equipment 自动减速设备
automatic retarder classification yard 自动减速器编组场;自动缓行器驼峰编组场
automatic retarder control 减速器自动控制;缓行器自动控制
automatic retarder system 车辆自动减速系统
automatic retraction device 自动提动钻具设备(钻孔收缩时)
automatic return mechanism 自动返回机构
automatic return of points 道岔自动返位
automatic return of water 自动水环流;自动回水管路
automatic return trap 自动回流闸门;汽水分离器(蒸汽采暖);回水盒
automatic reversal film 自动反转片
automatic reverse 自动换向
automatic reverse current cutout 逆流自断器
automatic reverse function 自动折返功能
automatic reverser 自动逆转器
automatic revolving table 自动回转工作台
automatic rheostat 自动变阻器
automatic rhythmicity 自律性
automatic rice cooker with warming system 带保温系统的自动电饭煲
automatic ringing 自动振铃
automatic rise temperature 自动升温
automatic riveter 自动铆(钉)机

automatic riveting gun 自动铆(钉)枪
automatic road-builder 自动筑路机
automatic rock grab 自动抓岩机
automatic rod holder 钻杆自动夹持器
automatic rod lifter 钻杆自动提升器
automatic rod magazine 自动送料杆
automatic roller dropping device 自动落辊装置;自动落辊器
automatic roller float device 自动架辊装置;自动架辊器
automatic roll filter 自动卷绕式过滤器
automatic rolling down speed control 溜放速度自动控制
automatic rolling mill 自动轧钢机
automatic roll-up awning 自动卷帘遮篷
automatic rotating pan enamelling 转盘式自动搪瓷
automatic rotation stopper 自动旋转伸缩式凿岩机
automatic route release 进路自动解锁
automatic route restoration 进路自动复位解锁
automatic route selection 进路自动选择;自动路由选择
automatic route setting 自动排路;自动路由设置
automatic route setting system 进路自动排列系统
automatic routine 自动例行程序
automatic routine apparatus 自动例行测试器
automatic routineer 自动定期操作器
automatic routing 自动选路;自动路径选择
automatic routing control 自动选路控制
automatic routing system 自动列车进路
automatic run 自动运行
automatic run device 自动操纵装置
automatic sack applicator 自动插袋机
automatic safe load indicator 自动安全负载指示器
automatic safety belt release 安全带自动解脱
automatic safety device 自动保护装置;自动安全装置
automatic safety valve 自动安全阀
automatic sample changer 样品自动更换器;自动换样器
automatic sample handling system 自动进样系统
automatic sampler 自动选样器;自动取样器;自动采样器
automatic sampler analyzer 自动采样器分析仪;动采样分析仪
automatic sampler-counter 自动取样计数器
automatic sampling 自动取样;自动采样
automatic sampling analyzer 自动取样分析器
automatic sampling-counter 自动取样计数器
automatic sampling device 自动取样装置
automatic sampling equipment 自动取样设备
automatic sampling system 自动取样系统;自动采样系统
automatic sand distributor 自动砂料撒布机
automatic sanding 自动撒砂
automatic sand plant 自动化砂处理装置
automatic satellite communication control system 自动卫星通信控制系统
automatic satellite computer aid-to-navigation 卫星自动导航
automatic saws for cutting laminated and bullet-proof glass 夹层和防弹玻璃自动切割锯
automatic scale 自动天平;自动轨道衡;自动定标;自动秤
automatic scale control 自动比尺控制
automatic scan(ning) 自动扫描;自动搜索
automatic scanning telespectroradiometer 自动扫描远距离光谱辐射计
automatic scarfing 自动火焰清理
automatic scheduling 自动调度;自动编运行图
automatic scraper 自动刮面器
automatic screed control unit 自动整平控制装置
automatic screw lathe 自动螺纹车床
automatic screw machine 自动螺丝车床
automatic scribing 自动刻图
automatic seal 自紧密封
automatic sealing line 自动密封生产线
automatic search(ing) 自动搜索
automatic search jammer 自动搜索干扰器
automatic segmentation and control 程序自动分段及控制
automatic segregator 自动分离器
automatic selecting machine 自动拣选机
automatic selection 自动选择

automatic selection control 自动选择控制
automatic selective overdrive 自动换挡的超速传动
automatic selective system 自动选择系统
automatic selectivity control 自动选择性控制
automatic self-aligning roller 自动调心滚柱
automatic self-cleaning wash bowl 自动清洁洗毛机
automatic self-closing valve 自动关闭阀
automatic self-verification 自动核对
automatic selling 自动售货
automatic send and receive 自动收发
automatic sender 自动发送器
automatic send receiver 自动收发设备
automatic sense 自动定边
automatic sensibility control 自动灵敏度控制
automatic sensitivity correction 自动灵敏度校正
automatic sensor light 猫眼灯
automatic separation device 自动解列装置；自动分离装置
automatic sequence 自动顺序
automatic sequence computer 自动顺序计算机
automatic sequence control 自动顺序控制；自动时序控制
automatic sequence-controlled calculator 自动顺序控制计算器；自动程序控制计算器
automatic sequence-controlled computer 自动程序控制计算机
automatic sequence operation 自动时序操作
automatic sequences 自动序列
automatic sequencing equipment 自动顺序设备
automatic sequent(ial)operation 自动顺序操作；自动时序操作
automatic serial reporting system 自动连续汇报制
automatic service 自动业务
automatic setting 自动镶嵌(钻头的)；自动调整
automatic settlement plotter 自动沉降仪
automatic setup 自动设置
automatic sextant 自动六分仪
automatic shaft kiln 自动化立窑；自动化立窑
automatic sheet counter 薄板自动计数器
automatic sheet counting machine 薄板自动计数器
automatic sheet feeder 续纸器
automatic sheet stacker 自动堆纸机
automatic shift 自动移位
automatic shift position data encoder 自动移位数据编码机
automatic ship 自动化船
automatic shipboard forecasting system 海上自动预报系统；海上气象自动预报系统；船载自动天气预报系统
automatic shorebased acceptance checkout equipment 自动岸基接收检测设备
automatic short-circuiter 自动短路器
automatic shot blasting rotating table 自动喷净回转台
automatic shunting 自动溜放
automatic shutdown 自动停闭；自动停止；自动断开
automatic shut-off 自动切断；自动关闭；自动停带开关
automatic shut-off device 自动关闭装置；自动断路装置
automatic shut-off valve 自动切断阀(门)；自动断流阀
automatic shutter 自动快门；自动防火帘
automatic shutter door operator 自动卷帘式门控制器；自动门开启装置
automatic shuttle valve 自动关闭阀
automatic side-dump muck car 自动侧倾出渣车
automatic side tipping wagon 自动侧倾铁路货车
automatic sieving machine 自动振筛机
automatic signal 自动信号
automatic signal lamp 自动信号灯
automatic signal light 自动信号灯
automatic signal(l)ing 自动信号法；自动信号发送
automatic signal(l)ing apparatus 自动信号器
automatic signal(l)ing device 自动信号装置
automatic signal(l)ing section 自动信号区段
automatic signal(l)ing territory 自动信号区域
automatic signal locking 自动信号锁闭
automatic signal protection 自动信号保护(设备)
automatic signal recognition unit 自动信号识别装置

automatic signal relay 自动信号继电器
automatic single daylight press 自动单层压机
automatic siphon 自动虹吸(管)
automatic size control 自动尺寸控制
automatic sizing 自动测量
automatic sizing device 自动测量装置
automatic skidding device 自动制动机；自动制动滑行装置
automatic skip 自动跳越
automatic skip scale 自动翻斗秤
automatic slab and tile grinder 自动磨板及磨瓷砖机器；自动石板瓷砖研磨机
automatic slack adjuster 自动松紧调整器；闸瓦间隙自动调节器
automatic slag pool welding 自动电渣焊
automatic slide changer 幻灯自动变换器
automatic sliding door 自动滑门
automatic slip elevator 滑式自动提引器；自动吊卡
automatic sludge discharge pipe 自动排污管
automatic sluice 自动放水闸
automatic slusher control 电耙绞车自动控制
automatic smoke alarm service 烟气自动报警设备
automatic smoke detector 烟雾自动探测器；自动烟感器
automatic smoke(or fire)vent 自动排烟天窗
automatic smoke sampler 烟雾自动取样器；烟尘自动采样器；自动烟尘取样器
automatic smoke vent 自动排烟口
automatic snow ga(u)ge 自记量雪器；自记量雪计
automatic softening installation 自动软水装置
automatic softening plant 自动化软化(水)厂；自动软化水厂
automatic soldering 自动焊料焊接；自动焊接
automatic sorter 自动分类机
automatic sorting machine 自动分类机
automatic sound buoy 自动发声浮标
automatic sounding buoy 自动音响浮标；自动声响浮标
automatic spacing table 自动定距钻孔台
automatic spark advance 自动火花提前
automatic spark control 自动提前点火机
automatic spark extinguisher 自动灭火花机
automatic spark mechanism 火花迟早自节器
automatic spark timer 自动点火定时器
automatic spectrometer 自动分光计
automatic spectrophotometer 自动分光光度计
automatic spectrum analyzer 自动频谱分析仪
automatic speech recognition 自动语言识别【计】
automatic speed change valve 自动喷射速度调节阀
automatic speed control 速度自动控制；自动转速控制；自动速度控制；自动调速
automatic speed controller 速度自动控制器；自动调速器
automatic speed control system 速度自动控制系统
automatic speed governor 速度自动调节器；自动调速器
automatic speed measuring device 自动测速仪
automatic speed regulation 自动调速
automatic speed run-up equipment 自动升速装置
automatic spider 自动夹紧十字叉；自动夹持器(钻杆的)
automatic spiker 自动道钉机
automatic spillway 自动泄水道；自动溢洪道
automatic spinning lathe(machine) 自动旋压成形车床
automatic spot weld 自动点焊法
automatic spot welding 自动点焊
automatic spray apparatus 自动喷涂装置
automatic sprayer 自动喷雾器
automatic spraying 自动喷涂
automatic spraying machine 自动喷漆机
automatic spring loaded valve 自动弹簧阀
automatic sprinkler 自动喷灌机；自动洒水器；自动洒水机；自动喷水器
automatic sprinkler head 自动灭火喷头
automatic sprinkler riser 自动喷水器立管
automatic sprinkler system 湿式自动喷水系统；自动洒水系统；自动撒水灭火装置；自动喷水灭火系统；自动喷洒管系统
automatic squeezer machine 自动压榨机
automatic S/R machine 自动有轨巷道堆垛机
automatic stability 自动稳定性

automatic stabilization and control system 自动稳定和控制系统
automatic stabilization equipment 自动稳定装置；自动稳定设备
automatic stabilization system 自动稳定系统
automatic stabilizer 自动稳定器
automatic stack 自动搁架
automatic stacker 自动堆料机
automatic staff exchanger 路签自动授受机
automatic stamper 自动打印机
automatic stamping press 自动模锻压力机
automatic start 自动起动
automatic starter 自动起动装置；自动起动器；自动动机；自动启动器
automatic starting 自动起动的
automatic starting control 自动起动控制
automatic starting device 自动起动装置；自动启动装置
automatic starting equipment 自动启动设备
automatic starting pump 自动启动水泵
automatic starting pump station 自动启动泵站
automatic start-up 自动启动
automatic station 自动观测站
automatic station finding 自动对位；自动定位
automatic station keeping 台站自动管理；自动管理站；自动管理电站；自动管理电台；自动对位；自动定位
automatic station keeping system 自动化台站管理系统
automatic steam generator 自动蒸汽发生器
automatic steam temperature control 自动蒸汽温度控制
automatic steam trap 凝水自动排除器
automatic steel 易切削钢
automatic steering 自动控制航向；自动操舵
automatic steering device 自动操舵装置
automatic steering gear 自动操舵装置
automatic steering ship 自动操舵船
automatic steering system 自动驾驶系统
automatic stencil etching machine 自动蚀版机
automatic step adjustment 自动步长调整；步长自动调整
automatic stepping 自动步进
automatic step restart 自动步进再启动
automatic stereocompilation 自动立体测图
automatic stereo compiler 自动立体测图仪
automatic stereomapping 自动立体测图
automatic stereomapping instrument 自动立体测图仪
automatic stereomapping system 自动立体测图系统
automatic stereo plotter 自动立体测图仪
automatic stereo profiler 自动立体断面记录仪
automatic sterilizer 自动消毒器
automatic stirrer 自动搅拌器
automatic stoker 自动加煤器；自动加煤机
automatic stoking 自动加煤
automatic stoking boiler 自动加煤锅炉
automatic stop arrangement 自停装置；自动停止装置
automatic stop block 自动阻车器
automatic stop control 自动制动控制；自动停止控制
automatic stop(ping) 自动停车；自停；自动停止(器)；自动停机
automatic stopping device 自动停止装置；自动停车装置
automatic stop sleeve 自动停止套筒
automatic stop valve 自动停止阀；自动制动阀；自动止水阀；自动断流阀
automatic storage 自动存储区；自动存储(器)
automatic storage allocation 自动存储配置；存储器自动分配
automatic storage area 自动存储区
automatic storage class 自动存储类
automatic storm observation service 自动风暴观测站
automatic straight air brake 自动直通空气制动机
automatic strake 自动洗矿机
automatic strength control 自动强度控制
automatic stress scanner 自动应力扫描仪
automatic strip feed press 钢带自动进料冲床
automatic stripper 自动抄针器
automatic strip-straightening machine 带材自动

平整机
automatic strobe pulse 自动选择脉冲
automatic stroke adjuster 自动冲程调整器;冲程自调器
automatic stuff box 自动调浆箱
automatic submerged-arc welding 自动埋弧焊
automatic submerged-arc welding machine 埋弧自动焊机
automatic submerged slag welding of rail 钢轨自动埋弧电弧焊
automatic substation 自动变电站;自动变电所
automatic suction tube controller 吸管自动控制器
automatic supermarket 自动化超级市场
automatic supervisory control system 自动监控系统
automatic surfacer 自动刨床(木工用)
automatic surfacing 自动堆焊
automatic survey 自动测量
automatic suspended particulate sampler 悬浮颗粒物自动采样器
automatic suturing machine 自动缝合器
automatic swinging chain cleaner 自动摇摆链式清除器
automatic switch 自动转辙器;自动转换器;自动开关
automatic switchboard 自动开关板;自动交换机
automatic switching 自动交换
automatic switching control 道岔自动控制
automatic switching equipment 自动转换设备;自动交换设备;道岔自动转换设备
automatic switching machine 道岔自动转辙机
automatic switching of point 道岔自动扳动
automatic switching(over) 自动转换
automatic switching panel 自动转换板;自动开关板;道岔自动集中控制盘
automatic switching system 自动交换机;道岔自动集中系统
automatic switch of stand-by power supply 备用电源自动投入
automatic switch over 自动转接;自动切换;自动换用
automatic switch unit 自动开关组合
automatic synchronization 自动同步
automatic synchronizer 自同步机;自动同步器
automatic synchronous system 自动整步系统;自动同期系统;自动同步系统
automatic syphon 自动虹吸(管)
automatic system 自动装置;自动(化)系统
automatic system control 自动系统控制
automatic system for boiler operation 锅炉运行自动装置
automatic system for selection of receiver and transmitter 接收机和发射机自动选择系统
automatic system generation 自动系统生成
automatic system initiation 自动系统起始功能
automatic system of hydrologic(al) data collection and transmission 水文自动测报系统
automatic systems workshop 自动化车间
automatic system trouble analysis 自动系统故障分析
automatic tableting press 自动压片机
automatic tach(e)ometer 自动视距仪
automatic tachymeter 自动视距仪
automatic take-up device 自动张紧装置
automatic tally order 自动结算的指令
automatic tandem working 自动转接作用
automatic tank battery 自动油罐站
automatic tapchanging equipment 自动抽头切换设备
automatic tape punch(er) 自动纸带穿孔器;纸带自动穿孔机;自动纸带穿孔机
automatic tape transmitter 自动读带机
automatic tap perforator 纸带穿孔机
automatic tapping machine 自动攻丝机
automatic target selection 自动目标选样
automatic telecommunication centre 自动电信中心
automatic telemetering weather system 自动遥测气象仪
automatic telemetry tracking system 自动遥测跟踪系统
automatic telephone 自动电话
automatic telephone call 自动电话呼叫
automatic telephone exchange 自动电话交换台;自动电话交换机
automatic telephone service 自动电话业
automatic telephone set 自动电话机
automatic telephone switchboard 自动电话交换机
automatic telephone system 自动电话系统
automatic telephone trunk network 长途自动电话网
automatic telerecorder 遥测自动记录仪
automatic teleswitch 自动遥控开关
automatic telex test 自动电传测试
automatic teller machine 自动取款机;自动柜员机
automatic telling status 自动报警状态
automatic temperature control 温度自动控制;自动温度控制
automatic temperature-controlled electric(al) furnace 自动温控电炉
automatic temperature controller 温度自动调节器;自动温度控制器
automatic temperature recorder 温度自动记录器;自记温度计;自记温度表
automatic temperature recorder controller 自动温度记录控制器
automatic temperature recording controller 自动温度记录调节器
automatic temperature regulator 自动温度调节器
automatic tension balancer 自动张力平衡器
automatic tension control 自动张力控制
automatic tension controller 自动张力调节器
automatic tensioner 自动补偿器
automatic tensioning device 张力自动补偿装置
automatic tensioning equipment 自动拉紧设备
automatic tension mooring winch 自动调节绞缆机
automatic tension recorder 自动张力记录仪
automatic tension regulator 自动张力补偿器;自动电压调整器
automatic tension winch 自动调缆绞车
automatic terminal information service 自动终端信息业务;自动终端信息服务
automatic termination clause 自动终止条款
automatic termination cover 自动销保(当战争爆发时)
automatic termination of cover clause 保险自动中止条款
automatic test and checkout equipment 自动测试与检查设备
automatic test(ing) 自动测试
automatic test(ing) equipment 自动测试装置;自动试验设备;自动检验设备;自动测试仪;自动测试设备
automatic theorem proving 机械证明定理
automatic thermohygrograph 自动温湿度记录计
automatic thermostat 恒温器;自动恒温器
automatic thickness ga(u)ge 自动测厚规
automatic thin layer spreader 自动薄层涂布器
automatic third camera 第三镜箱投影器
automatic threading 自动装片
automatic thresher 自动脱粒机
automatic threshold closer 门底自动防风隔声设施;门底自动防风隔音设施
automatic threshold door reopening closer 自动门槛
automatic threshold variation 自动阈值变更
automatic throttle control 节流阀自动控制
automatic throttle valve 自动节流阀
automatic throw in device of emergency supply 备用水源自动投入装置;备用热源自动投入装置;备用供应源自动投入装置;备用电源自动投入装置
automatic throw-off of roll 滚筒自动脱开
automatic ticket dispenser 自动售票机
automatic tidal ga(u)ge 自动潮汐仪;自动潮位计
automatic tide ga(u)ge 自记验潮仪;自记潮汐计;自记潮位计;自动验潮仪;自动潮位仪;自动潮位计
automatic tide ga(u)ge station 自动验潮站
automatic tide-meter 自记潮汐计
automatic tide recorder 自记潮位仪;自动潮位仪;自动潮位记录器;自动潮位计
automatic tilting gate 自动倾侧闸(门)
automatic timed gate feeder 定时自动开闭给料机
automatic time difference circuit 自动时差电路
automatic timed magneto 自动定时磁电机
automatic timed save feature 自动时序存储特征
automatic time marking 自动时标;自动计时
automatic timer 自动定时器;自动计时器
automatic time relay compensator 自动延时补偿器
automatic time release 自动限时解锁
automatic time switch 自动计时开关;自动定时开关
automatic timing control 自动定时控制
automatic timing corrector 自动时间校准器;自动时间校正器
automatic timing device 自动记时装置;自动定时装置;自动调时器
automatic timing gear 自动定时齿轮
automatic tinning equipment 自动镀锡设备
automatic tipper 自动倾卸器;自动倾卸机构;自动倾卸槽;自动倾倒槽;自动倾侧槽
automatic tipping 自动倾倒式
automatic tire pump 自动汽泵;自动轮胎打气泵
automatic titrimeter 自动滴定计
automatic toll dialing 长途自动拨号
automatic toll switching system 长途电话自动交换机
automatic toll swtiching repeater 长途电话自动中继器
automatic tool lifter 自动抬刀装置
automatic tool pick-up 自动抬刀装置
automatic tool retracting unit 自动退刀装置
automatic tracing head 自动跟踪头
automatic tracker 自动跟踪器
automatic track follower 航线自动跟踪器
automatic tracking 自动跟踪
automatic tracking and anti-collision warning 自动跟踪避碰报警装置
automatic tracking antenna 自动跟踪天线
automatic tracking control 自动跟踪控制
automatic tracking controller automatic following controller 自动跟踪控制装置
automatic tracking device 自动跟踪装置
automatic tracking laser radar 自动跟踪激光雷达
automatic tracking lidar 自动跟踪激光雷达
automatic tracking radar 自动跟踪雷达
automatic tracking system 自动跟踪系统
automatic track plotter 自动描迹仪
automatic track shift 自动声道转换装置
automatic traction control equipment 自动牵引控制设备
automatic traction equipment 自动控制牵引装置
automatic traffic control 交通自动控制
automatic traffic counter 交通量自动计数器
automatic traffic recorder 交通量自动记录仪;自动车辆记录器
automatic traffic signal 自动交通信号
automatic traffic surveillance 自动交通监视;自动交通监测
automatic train control 列车自动控制(装置);机车色灯信号;自动化列车控制
automatic train control system 自动序列控制系统
automatic training idler 自动调节导向轮
automatic train loop feed unit 列车自动运行环形馈电装置
automatic train movement printer 行车自动打印机
automatic train operation 列车自动运行;列车自动操纵
automatic train operation system 自动化列车运行系统
automatic train operation unit 列车自动运行单元
automatic train protection 列车自动保护;自动化列车防护
automatic train protection overspeed alarm 列车自动保护超速警号
automatic train protection track feed unit 列车自动保护轨道馈电装置
automatic train protection trackside unit 轨旁列车自动保护装置
automatic train protection unit 列车自动保护单元
automatic train regulation 列车自动调节
automatic train regulation system 列车自动调整系统
automatic train speed regulation 列车自动调速;列车自动限速
automatic train stop 列车自动停车装置;自动停车
automatic train stop equipment 列车自动停车装置
automatic train stop lever 列车自动停车手柄
automatic train stopping device 列车自动停止装置
automatic train supervision 列车自动监测
automatic transfer 自动变压器

automatic transfer car system 自动搬运方式
automatic transfer equipment 自动切换设备;自动传送设备;自动传感设备
automatic transfer switch 自动转换开关;自动切换开关
automatic transformer switch 变压器自动开关
automatic transistor classifier 晶体管自动分类机
automatic transit circle 自动子午环
automatic translation 机器翻译
automatic translator 自动转换器
automatic transmission 自动传输;自动传动装置;自动变速装置;自动变速器;自动变速(箱)
automatic transmission of data 资料自动传输
automatic transmitter 自动发报机
automatic transport system 自动导向运输系统
automatic transverse welding 横向自动焊
automatic traverse computer 自动偏移计算机
automatic trigger 自动触发器
automatic triggering 自动触发
automatic trimmer 自动整修机
automatic trip 自动脱扣(器)
automatic trip circuit breaker 自动跳闸断路器
automatic tripping 自动跳闸
automatic tripping car 自动倾卸车
automatic trolley reverser 杆形集电路的自动反向器
automatic trouble-locating arrangement 自检装置;自动找寻故障装置;自动巡查故障装置;故障自动探测装置
automatic trouble locator 自动故障寻查器
automatic tube filling machine 自动装管机
automatic tuning 自动调谐
automatic tuning system 自动调谐系统
automatic turning 自动换档
automatic turning sander 自动可转动打磨机
automatic turntable 自动转台
automatic turret lathe 自动转塔式六角车床
automatic type belt tensioning device 皮带自动张紧机构
automatic type setting 自动排字;自动排版
automatic typewriter 电传打字机;自动打字机
automatic ultrasonic cleaning 超声波自动清洗
automatic ultrasonic wave recorder 自记超声测波仪
automatic uncoupling and air brake hose 自动摘钩设备
automatic unload 自动去载
automatic unloader 自动卸载器;自动卸荷器
automatic unloader for coating plate 涂装板的自动卸件装置
automatic unloading 自动卸料
automatic unloading machine 自动卸料机;自动卸货机
automatic vacuumcan sealing machine 自动真空封罐机
automatic vacuum valve 自动真空阀
automatic valve 自动阀(门);差压阀
automatic valve bag packer 自动封口水泥包装机
automatic valveless gravity filter 自动无阀重力滤池
automatic variable 自动变量
automatic variable delivery pump 自动变量输送泵
automatic vehicle for bulk cement 散装水泥汽车
automatic vehicle location system 自动车辆定位系统
automatic vehicle monitoring 自动车辆监控
automatic vehicle monitoring system 车辆自动监控系统
automatic vehicle tracking system 自动车辆跟踪系统
automatic vent 自动放气阀
automatic ventilating side scuttle 自动通风舷窗
automatic ventilation 自动通风
automatic ventilator 自动通风器;自动通风机
automatic vertical index 自动垂直指标
automatic vertical kiln 自动化立窑
automatic vibration monitor 自动振动探测计
automatic vigilance device 自动报警装置
automatic viscometer 自动黏度计
automatic viscous controller 自动黏度控制器
automatic viscous filter 自动黏液过滤器
automatic voice network 自动电话网
automatic voltage regulation 自动调压;电压自动调节
automatic voltage regulator 自动稳压器;自动稳压器;自动电压调整器;自动电压调节器;电压自动调节器;电压稳定器
automatic volume batching plant 自动体积计量设备
automatic volume compressor 声量压缩器
automatic volume control 自动音量控制
automatic volume expander 声量扩展器
automatic wagon control 矿车限速自动控制器
automatic warning device 自动报警装置
automatic warning system 自动报警系统
automatic washing machine 自动洗涤机
automatic waste-gas purifying plant 自动废气净化厂
automatic watch 自动手表
automatic water bowl 自动饮水器
automatic water closet 自动冲洗大便器
automatic water control of granulator 成球水分自动控制器
automatic water discharger 自动排水器
automatic waterer 自动饮水器
automatic water ga(u)ge 自动水位表(混凝土搅拌机的);自动量水表(混凝土搅拌机的)
automatic water heater 自动热水器
automatic watering system 自动洒水装置
automatic water(level)ga(u)ge 自记水位仪;自记水位计
automatic water measuring device 自动量水装置
automatic water-quality control system 自动水质监测控制系统
automatic water-quality monitor 水质自动监测器;自动水质监测仪
automatic water-quality monitoring 自动水质监测
automatic water-quality monitoring station 自动水质监测站
automatic water-quality monitoring system 水质自动监测系统;自动水质监测系统
automatic water sampler 自动水样采集器
automatic water softener 自动水处理装置;自动软水器
automatic water sprinkler system 自动喷水系统
automatic water stage recorder 自记水位仪;自记水位计
automatic water stage recording station 自记水位站
automatic water supply source 自动供水水源
automatic water system 自动洒水系统
automatic water treatment plant 自动(化)水处理厂
automatic water-wheel 自动水车
automatic wave(height) recorder 自记波高仪;自记波高计;自记测波仪;自记波浪仪
automatic wayside signal(l)ing 线路自动信号设备
automatic weather forecasting 自动天气预报
automatic weather station 自动气象站;自动气象台
automatic weather system 自动气象系统
automatic weigh batcher 自动重量称量器;自动配料秤
automatic weigher 自动衡器;自动秤
automatic weighing and feeding machine 自动定量喂给机
automatic weighing apparatus 自动称量装置
automatic weighing batching 自动称量配料
automatic weighing device 自动称量装置
automatic weighing equipment 自动称量装置;自动称量设备
automatic weighing filler 自动秤量填装机
automatic weighing lysimeter 自动称重式(土壤)渗漏仪;自动称重式(土壤)测渗仪
automatic weighing machine 自动秤
automatic weighing plant 自动称重设备;自动称量设备
automatic weighing scale 自动本重秤;自动计量秤
automatic weighing system 自动称量系统
automatic weighing unit 自动称量装置
automatic weight classifier 重量自动分选机
automatic weight control (钻头压力的)自动控制
automatic weighting plant 自动秤
automatic weir 自动堰;自动溢流堰
automatic weld deposit 自动焊缝
automatic welder 自动焊机;自动电焊机
automatic weld(ing) 自动焊(接);自动电焊
automatic welding head 自动焊头
automatic welding machine 自动焊接机
automatic welding procedure 自动焊接程序;自动焊接方法
automatic welding process 自动焊接工艺规程
automatic wet-pipe sprinkler system 湿式自动喷水系统;自动备水消防系统
automatic wheel defects detection system 车轮缺损自动探测系统
automatic wheel truer 自动砂轮整形器
automatic wheel truing device 自动砂轮整形装置
automatic wheel wear compensator 车轮摩损自动补偿器
automatic winch 自动升降机;自动卷扬机;自动绞车
automatic winder 自动化提升机
automatic window 自动开闭窗;自动开关窗
automatic wing tip slot 自开翼梢缝
automatic wire stripper 镀铬自动剥线器;柄绝缘自动剥线器
automatic wire tying device 自动式铁丝打结装置
automatic work cycle 自动工作循环
automatic working 自动工作
automatic working of a route 进路自动排列
automatic wrapping machine 自动包装机
automatic zero-point correction 自动调零
automatic zero setting 自动置零;自动调零
automation 自动装置;自动机械;自动化(作用)
automation and remote control of lock 船闸的自动化及遥控
automation assistant 设计自动化辅助工具
automation center 自动化中心
automation design 自动化设计
automation in cartography 地图制图自动化
automation instrument 自动化仪表
automation level 自动化水平
automation of a marshalling yard 编组场自动化
automation of blast furnace 高炉自动化
automation of cargo-handling 装卸作业自动化;装卸工作自动化
automation of dredging process 疏浚过程自动化
automation of information storage and retrieval 信息存储与检索自动化
automation of rolling 轧钢自动化
automation of rolling stock inspection and repair 车辆检修自动化
automation of route selection 选线自动化
automation of station for purification of sewage 污水净化站自动化
automation of stereo-compilation 立体测图自动化
automation of train regulation 调查调整自动化
automation of welding 焊接自动化
automation of wiring design 布线设计自动化
automation operation 自动化作业
automation paging system 自动寻呼系统
automation priming 水泵启动水
automation procedure 自动化作业;自动化工序
automation process 自动化过程
automation production 自动化生产
automation remote control of locks and bridges 船闸及开合桥自动化遥控系统
automation shaft kiln 机械化立窑
automation transportation 自动化运输
automatism 自动性
automatization 自动化(作用)
automatization of service 服务自动化
automatograph 自动记录器
automaton 自动机
autometamorphism 自变质作用
autometasomatic process 自交代作用
autometasomatism 自交代作用;自变质作用
autometer 汽车速度计;汽车速度表;速度表;自动测量计
automicriteic rock 自生泥晶灰岩
automicro densitometer 自动测微密度计
automicrometer 自动千分尺
automixture control 自动混合气控制
automobile 汽车;机动车;车辆
automobile accident 汽车事故
automobile camp 汽车旅游营地;汽车营地
automobile carburet(t)or 汽车汽化器
automobile chassis 汽车底盘
automobile container 汽车集装箱
automobile court 汽车游客旅馆
automobile crane 汽车吊车;汽车起重机
automobile diesel engine 汽车柴油发动机
automobile differential gear 汽车差速齿轮
automobile dimension 汽车尺寸

automobile duster 自动喷粉器
automobile dynamic(al) quality 汽车动力性能
automobile dynamics 汽车动力学
automobile effluent 汽车废气
automobile electronic igniter 电子汽车点火器
automobile emission 汽车排气;汽车废气排放
automobile engine 汽车发动机
automobile engineering 汽车制造业
automobile exhaust 汽车排气
automobile exhaust analysis 汽车排气分析
automobile exhaust catalytic cleaning cartridge 汽车废气催化净化器
automobile exhaust converter 汽车排气转化器
automobile exhaust gas 汽车废气
automobile exhaust gas cleaner 汽车废气净化器
automobile exhaust gas converter 汽车排放转换器
automobile exhaust gas purifier 汽车废气净化器
automobile factory 汽车制造厂
automobile ferry 汽车渡轮;汽车渡船
automobile ferry-steamer 汽车轮渡
automobile filling station 汽车加油站
automobile finish 汽车(罩面)漆
automobile frame 汽车车架
automobile headlight 汽车前灯
automobile highway 汽车道(路)
automobile horn 汽车喇叭
automobile improvement 汽车改进
automobile industry 汽车工业
automobile injure 机动车损伤
automobile insurance 汽车保险
automobile lacquer 汽车用(喷)漆
automobile laundry 汽车清洗间
automobile leaf-spring 汽车钢板
automobile map 公路图
automobile mechanic 汽车修理工
automobile mechanics 自动车机械学
automobile noise 汽车噪声
automobile oil 车用润滑油
automobile parking lift 汽车停放升降机
automobile parking space 汽车停车场
automobile parking structure 汽车停放建筑物;停车库
automobile pollution 汽车污染
automobile race course 赛车跑道
automobile race track 赛车跑道
automobile repair coating 汽车修补漆
automobile repair station 汽车修理站
automobile road 汽车专用道
automobile sales lot 汽车展销店
automobile service station 汽车服务站
automobile sling 汽车吊具;小汽车吊具
automobile test ground 汽车试验场
automobile test track 汽车试验跑道
automobile tower wagon 梯架汽车
automobile traffic 汽车交通
automobile traffic control system 汽车交通控制系统
automobile trailer 汽车拖挂的活动房屋;汽车拖车;汽车挂车
automobile tyre noise 汽车轮胎噪声
automobile wagon 运货汽车
automobile weight tax 汽车重量税
automobile with electric(al) transmission 电传动汽车
automobile(work) shop 汽车修理店;汽车修理(车)间
automobilism 汽车运行;汽车驾驶
automobilist 汽车驾驶员
automodulation 自调制
automonitor 自动监听器;自动监视器;自动监控器;自动监督程序;自动监测仪;自动监测器;自动程序监控器
automonitoring 自动监测
automonitoring system 自动化监测系统;自动监视系统
automonitor routine 自动监督程序
automorphic 自形的
automorphic-granular texture 自形粒状结构
automorphism 自同构
automorphosis 自变质作用
automorphous 自形的
automotive 机动的;自航的;自动车
automotive air conditioning 汽车空气调节;自动车空气调节

automotive air pollution 汽车空气污染;机动车对空气污染
automotive brake 汽车制动器
automotive coating 汽车涂料;汽车漆
automotive emission 汽车发散物;汽车排气;汽车废气
automotive engine 汽车发动机;自动车发动机
automotive engine combustion process 汽车发动机燃烧过程
automotive engineering 汽车技术;汽车工程
automotive frame 汽车大梁
automotive fuel 马达燃料
automotive glass 汽车玻璃
automotive ignition system 汽车点火系统
automotive industry 汽车制造业;汽车工业
automotive laundry 汽车清洗间
automotive lift 汽车提升机
automotive mount 汽车装配
automotive refinish paint 汽车修补漆
automotive service 汽车服务站
automotive steering 汽车转向系
automotive supplies 汽车器材;汽车备件
automotive suspension 汽车悬架
automotive top coat 汽车面漆
automotive transmission 汽车传动系
automotive truck 运货汽车;载重汽车;载货汽车
automotive type 汽车车型
automotive type steering 汽车式的转向机构
automotive vehicle 机动车;汽车;自动飞行器
automtitrator 自动滴定器
automuffler 自动消声器
automutagenicity 自发突变性
autonavigator 自动领航仪;自动导航仪
autonomic 自主的
autonomic movement 自发运动
autonomics 自调系统程序控制研究
autonomous 自主的;自治的
autonomous car 自动压道车
autonomous channel 自主通道;独立通道
autonomous channel operation 独立通道操作
autonomous city 自治城市
autonomous county 自治县
autonomous device 独立设备
autonomous federal state 自治联邦
autonomous financing 财务自主
autonomous group approach 小组自主生产法
autonomous investment 自发投资;自主性投资
autonomous navigation technology 自主导航技术
autonomous network of threshold elements 自治的阈值元件网络
autonomous oscillation 自振;自动振荡
autonomous port 自治港
autonomous prefecture 自治州
autonomous region 自治区
autonomous remotely controlled submersible 自主式遥控潜水器;自给式遥控潜水器
autonomous shift register circuit 自激移位寄存器电路
autonomous state 自治州
autonomous state boundary 自治州界
autonomous submersible 自主式潜水器
autonomous system 自主系统;自治系统;自控系统;独立系统
autonomous tariff 自主税则;法定税率;固定税则
autonomous temperature line acquisition system 自激式温度线数据采集系统
autonomous transaction 自发交易
autonomous underwater vehicle 自主式水下船;自激式水下船
autonomous unit 自主部件
autonomous working 独立工作
autonomy 自主性;自发性
autonomy in management 经营自主权
autonomy of functions 操作的自治权
autonomy of the will 意思自主
auto number plate 汽车号牌
autoodometer 自动固结仪
auto off control in condensation 结露自动关机控制
autoomatic wagon identification 货车自动识别
autoophthalmoscope 自身检眼镜
auto-oriented 机动车为主的
autooscillation 自振;自动振荡
autooxidation 自动氧化(作用);自身氧化

autooxidation of sludge 污泥自动氧化
autooxidation phase 自氧化期;自动氧化相;自动氧化期
autopack 自动填塞
autopack ring packing 自填塞环形填密件
autopact 自动定值
autopagination 自动加页码
autopallet swinger 自动集装箱装卸机构
autopanel 自动控制指示板
auto park or market 汽车停放展销市场
auto parts 汽车零件
autopatrol 巡路修路机
autopatrol grader 养路平路机;养路平地机;自动巡路平地机;自动巡路平路机
autoped 踏乘自动车;双轮机动车
autopelagic plankton 深海自养浮游生物;上层浮游生物
autophasing of wall rock 围岩自稳能力
autophoresis 自泳
autophoretic coating 自泳涂装
autopia 汽车社会;汽车专用区
autopicking section 自动拾取剖面
autopiler 自动化编译程序装置;自动编译程序装置
autopilot 自动驾驶仪;自动舵;自动导航仪;自动操舵装置
autopilot computer 自动驾驶仪计算机
autopilot controller 自动驾驶仪操纵器
autopilot control system 自动舵控制系统
autopilot coupler 自动驾驶仪耦合器
autopilot driver 自动舵驱动装置
autopiloted synchronous motor 自动驾驶同步电动机
autopilot engage and trim indicator 自动舵接通与调整指示器
autopilot navigator 自动驾驶领航仪
autopilot rate control 自动舵速率控制
autopilot servo unit 自动驾驶随动装置;自动舵伺服装置
autopilot steered ship 自动操舵船
autopiracy 本流袭夺
auto pista 高速公路(西方)
autoplaceme 自动绘图
auto-placement 自动布局
autoplant 汽车工厂;自动装置;自动设备
autoplotter 自动绘图仪;自动绘图器;自动绘图机
autoplugger 自动充填器
autopneumalolysis 自气化作用
autopneumatic circuit breaker 自动压缩空气传动开关
autopneumatic cylinder 气压水罐
autopneumatolysis 自气化
autopolarity 自动极性变换
auto-pollution 汽车污染
auto-pollution source control 汽车污染源控制
autopolymer 自聚物
autopolymerization 自动聚合
autopore 大孔
autopositioning 自动对位;自动定位
autopositive 直接正像材料;直接正片
autopositive film (一次显影的)直接正象胶片
autopour 自动开塞机
autopower spectrum 自功率谱
autoprecipitation 自沉淀(作用)
autoprobing machine 自动探针检测机
auto-production line for electronic facsimile 电子传真机自动生产线
autoprogram(ming) 自动程序设计;自动编程序
autoprotective tube 自动保护X射线管
autopsy 实地勘察;验尸室
autopsy room 尸(体)解剖室
autopulse 自动脉冲;振动式电压调节器
autopunch 自动冲压硬度试验机
autopurification 自净作用;自净作用;自净(化)
autopurification power 自净化能力
auto put 高速公路(西方)
auto-query 自动查询
autoradar plot 雷达自动绘图器;雷达标图板
autoradiograph 射线自显迹;自体放射照片;自动射线相;放射自显影照片
autoradiography 射线显迹法;自体放射相术;自动射线照相术;放射自显影法;放射性自显影术
autorail 轨路两用车
autoranging 自变换量程

autoranging digital display 自动量程选择数字显示
autoreclose breaker 自动重合闸断路器
autoreclose circuit breaker 自动重合闸;自动重合断路器
autorecloser starting relay 自动重合闸起动继电器
autorecording photometer 自动记录光度计
autoreducing tachymeter 自动简化视距仪
autoreduction 自动归算
autoreduction electronic distance measuring alidade 自动归算电子测距照准仪
autorefinish 汽车修补(涂漆)
autoregistration 自动重合
autoregression 自回归
autoregression analysis 自动回归分析
autoregression process 自回归过程
autoregressive coefficient 自回归系数
autoregressive formula 自回归公式
autoregressive integrated moving average 自回归累积移动平均(模型)
autoregressive model 自回归模型
autoregressive moving average 自回归流动平均数
autoregressive representative 自回归表达式
autoregressive series 自回归级数
autoregulation 自我调节;自体调节;自身调节;自动调节
autoregulation filtering method 自调滤波法
autoregulator 自动调节器
autoregulatory feedback mechanism 自动调节反馈机理
autorelay 自动继电器
autoreleasing hook 自动解缆钩
autoreleasing lifeboat 自释救生筏
autoreleasing lifecraft 自动放出救生筏
auto repair(work)shop 汽车修理(车)间
autorepeater 自动重发器
autoresidual static correction 自动剩余静校正
autorestart 自动启动
autorestarter 自动再启动
autoreversal film 自动反转片
autoreversing electric(al)motor 自动转向电动机
autorhythmicity 自动节律性
autoroad 自动行车道
autoroom 自动机房
autorotate 自转
autorotation 自转(动)
autorotation speed 自转速率
auto route 多车道高速公路;汽车行驶线
auto safety 汽车安全性
auto sales lot 汽车出售场所
autosampler 自动取样器
autoscaler 自动定标器
autoscan teleradiometer 自动扫描望远辐射计
autoscope 点火检查示波器
autoset level 自动水准仪;自动安平水准仪
autosetter 自动定形机
autoshaver 自动刨版机
autosignal(l)ing 自动信号
autosilo (具备升降机的)塔式汽车库;自动停车场
autosizing 自动尺寸监控
autosizing device 自动测量装置
autosled 机动橇
autosledge 自卸式拖运器
autosling 车辆吊具;小汽车吊具
autoslot 自动翼缝隙
autosorter 自动分类机
autospray 自身喷雾器
autosprinkling system 自动喷水灭火系统
autostability 自稳定性(能);自动稳定性
autostabilization system 自动稳定系统
autostabilized 自动稳定的
autostabilizer 自动稳压器;自动稳定装置;自动稳定器
autostabilizer unit 自动稳定装备
autostable 自动调节的;自动稳定的
autostable wall 自动稳定岸墙
autostacker 多层车场(大城市中心的);多层停车场
autostair 自动电梯
autostart 自动起动
autostart delay 自动起动延迟
autostarter 自耦起动器;自动起动器;自动启动器
autosteerer 自动操纵装置;自动转向装置
autostereo-profiler 自动立体断面记录仪
autosterization 自灭作用
autostop 自动停止;自动停机;自动停车

autostopper 自动制动器;自动停止装置
autostrada (多车道高速的)公路干线
autosurveyor 自动测量仪(器)
autosuspending turbidity flow 自浮浊流
autosuspension 自悬浮作用
autoswitch 自动开关
auto switch on 自投
autoswitch over 自投;自动转接;自动切换;自动换用
autosyn 远距传动器;自整角机;自动同步器;自动同步机
autosynchronous motor 自动同步电动机
autosyndesis 同源联会
autosyn instrument 自同步仪表
autosyn receiver 自动同步感受器
autosyn system 自动整步系统;自动同期系统;自动同步系统
Autotape 奥托塔普测距系统
autotare 自动去皮重
autotelegraph 电传真机;书画电传机
autotensiometer 自动张力计
autotensioned conductor 补偿导线
autotension mooring winch 自动绞缆机
autotension winch 自动张紧绞车;自动松紧绞车
auto-terminal 汽车终点站
autotest(ing) 自动测试;自动检测程序
autothermal packed bed reactor 自热填料床反应器
autothermal thermophilic aerobic digestion 自热高温好氧消化
autothermal thermophilic aerobic waste(water)treatment system 自热高温好氧废水处理系统
autothermic cracking 自裂化
autothermic piston 热自动补偿活塞;防热变形活塞
autothermoregulator 自动温度调节器
autothermostat 自动恒温器
autothreshold control 自动阈限控制
autothrottle 自动油门;自动节流活门
autotilly 自分
autotimer 自动计时器
autotrace 电气液压靠模仿型铣床
autotracing system 自动描图系统
autotracker 自动跟踪装置
autotrack(ing) 自动跟踪;自动位置对准
autotracking unit 自动跟踪器
autotrain 汽车列车
autotransductor 自饱和电抗器
autotransformer 自耦变压器
autotransformer coupled oscillator 自耦变压器反馈振荡器
autotransformer feeder 自耦变压器供电线
autotransformer feeding system 自耦变压器供电方式
autotransformer post 自耦变压器所
autotransformer starter 自耦变压器式起动器
autotransmitter 自动传送机
auto-tricycle 三轮卡车
autotrim clutch 自动配平离合器
autotronic system 电子式自动升降机系统
Autotron scanner 奥托特朗扫描器
autotrophic bacteria 自养细菌
autotrophic component 自养组分
autotrophic denitrification 自养反硝化
autotrophic index 自养指数
autotrophic microbe 自养微生物
autotrophic microorganism 自养微生物
autotrophic nitrogen 自养氮
autotrophic nutrition 自养营养;自养
autotrophic organism 自养生物
autotrophism 自养
autotropism 向自性
autotruck 运货汽车;卡车;载重汽车
autotruck control flat bed plotter 自动控制平底绘图仪
autotune 自动调谐
autoturret winder 自动转台式换卷机
autotype 影印(术);自型;照相印刷术;感光树脂制板
autovac 真空箱;真空罐;真空供油装置
autovalve 投币阀;自动阀;阀式放电器
autovalve arrester 阀型避雷器
autovalve lightning arrester 自动阀型避雷器;阀式放电器;阀式避雷器
autovalve type transformer 阀型避雷器
autovariance 自方差

autoverify 自动检验
autoveyor 装运汽车的平车
autovon 自动电话网
auto-vortex bowl classifier 浮槽自动调节式漩涡分级机
auto-vulcanization 常温硫化;自动硫化
auto-watch 自动值守
autoweigh disperser 自动称重分散器
autoweighing 自动计量
autoweigh totalizer 自动称重累加器
autowindow 自动窗孔
autoword wrap 自动文字切换
auto-wrecking yard 废车堆积场
autowrench 自动扳手
autoxidation 自动氧化作用
autoxidator 自动氧
autozero cloth raising machine 自调起绒机
autozero loop 自动零回路
autozeroset 自动调零
autrometer 自动多元素光谱仪
autumn 秋天;渐衰期
autumnal circulation 秋季环流
autumnal circulation period 秋季循环期
autumnal equinoctial period 秋分期
autumnal equinox 秋分(点)
autumnal equinox tide 秋分潮
autumnal fever 秋季热
autumnal overturn 秋季对流
autumnal turnover 秋季翻转
autumn aspect 秋季相
autumn colo(u)r 秋色
autumn construction 秋季施工
autumn crops 秋庄稼;秋季作物
autumn equinox tide 秋分潮
autumn flood 秋汛;秋洪
autumn garden 秋景园
autumn harvest 秋收
autumn ploughing 秋耕
autumn plowgh 秋耕
autumn plumage 秋植
autumn refuse 秋季垃圾
autumn timber 晚木;秋材
autumn variety 秋季品种
autumn wood 秋材;晚材;成熟木材
autumn wood ratio 秋材率
Autunian 奥顿阶【地】
autunite 钙铀云母
Auversian 奥维尔斯阶【地】
auxanograph 生长仪;生长谱
auxanometer 生长计
auxetic 促诱的
auxiliary 从属(的);次要;副臂;附属的;辅助剂;辅助的
auxiliary abrasive 辅助磨料
auxiliary absorber 辅助吸收塔;辅助吸收器
auxiliary accessories 辅助性零配件;附生副矿物;附属品
auxiliary account 辅助账
auxiliary accounting unit 附属会计单位;附属单位会计
auxiliary activator 辅助激活剂
auxiliary activity 辅助业务;附属作业
auxiliary aerodrome 辅助机场
auxiliary agent 助剂
auxiliary aiming mark 装配记号;辅助瞄准点
auxiliary air 辅助空气;二次空气;二次风;补给空气;补充空气
auxiliary aircraft transport 辅助飞机运输舰
auxiliary airfield 辅助飞机场
auxiliary air heater 辅助热风炉;备用热风炉
auxiliary air intake 辅助进气口
auxiliary air port 辅助风口;辅助机场
auxiliary air pump 辅助空气泵
auxiliary air receiver 辅助储气器
auxiliary air reservoir 副风缸;储气筒
auxiliary air valve 辅助气阀
auxiliary alarm 辅助警报器
auxiliary altitude lettering 辅助高程注记
auxiliary anchor 辅助锚锭
auxiliary angle (刀具的)副偏角;辅助角;补角
auxiliary anode 辅助阳极
auxiliary apparatus 辅助装置;辅助设备
auxiliary application 辅助申请
auxiliary arch 泄压拱

auxiliary assembly 辅助装置;辅助装配
auxiliary attachment 附属装置;辅助附件
auxiliary axis 辅助轴线;副轴
auxiliary balance 辅助平衡
auxiliary ballast high-pressure mercury lamp 高压汞灯外镇流器
auxiliary ballast tank 辅助压载水舱
auxiliary bar 辅助钢筋;辅助杆(件)
auxiliary base 辅助基准
auxiliary base-line 辅助基线
auxiliary battery backup 辅助电池后备
auxiliary battery power supply 辅助动力供应电池
auxiliary beacon 辅助指向标;辅助导航灯;辅助导航标;辅助指向灯;辅助信标
auxiliary beam 辅助横梁
auxiliary blade 辅助叶片
auxiliary blade wear strip 辅助叶片防磨板
auxiliary block 辅助部件
auxiliary block house 辅助所【铁】
auxiliary block post 辅助线路所;辅助所【铁】
auxiliary block station 辅助所【铁】
auxiliary blower 辅助鼓风机;辅助风扇;局部扇风机
auxiliary board 厂用配电屏;厂用配电盘
auxiliary boat 辅助船
auxiliary body 附属机构
auxiliary bogie 辅助转向架
auxiliary boiler 辅助锅炉;备用锅炉
auxiliary boiler equipment 辅助锅炉设备
auxiliary boiler house 辅助锅炉房
auxiliary bond 副键
auxiliary books 辅助簿册;辅助账簿
auxiliary booster 助推器
auxiliary box 辅助报警装置;辅件盒
auxiliary brake 副制动器;附加制动器
auxiliary brake pipe 辅助制动管
auxiliary brake shoe 辅助闸鞋
auxiliary breakwater 辅助防波堤
auxiliary bridge 临时桥;临时便桥;辅助桥;便桥
auxiliary bucket ladder 副斗桥
auxiliary building 附属建筑
auxiliary building equipment 备用的营造工具;辅助建筑设备
auxiliary buildings of cold store 冷库辅助建筑
auxiliary buildings for living 生活辅助建筑物
auxiliary buildings for port operation 港口生产辅助建筑物
auxiliary buildings for production 生产辅助建筑物
auxiliary burner 辅助燃烧器
auxiliary bus-bar 辅助母线;备用汇流条;备用母线
auxiliary capacitor 辅助电容器
auxiliary capital 辅助资本
auxiliary carrier(cable) 辅助承力(缆)索
auxiliary carry bit 辅助进位位
auxiliary cartographic(al) documents 辅助制图资料
auxiliary catenary(wire) 辅助承力(缆)索;加强线;辅助缆索
auxiliary cathode 辅助阴极
auxiliary cathode ray display tube 辅助阴极射线显示管
auxiliary center[centre] 辅助中心
auxiliary changeover valve 辅助切换阀
auxiliary channel 辅助渠道;辅助波道
auxiliary characteristics 辅导指标
auxiliary chart 辅助图
auxiliary circle 辅助圆
auxiliary circuit 辅助回路;辅助电路
auxiliary circulating pump 辅助循环泵
auxiliary civil defence 辅助性人防工程
auxiliary classification yard 辅助调车场【铁】
auxiliary cofferdam 副围堰;辅助围堰
auxiliary coil 辅助线圈
auxiliary column 辅助柱
auxiliary combustion equipment 辅助燃烧设备;助燃设备
auxiliary commerce 辅助商业
auxiliary compass 辅助罗盘;辅助罗经
auxiliary complexing agent 辅助络合剂
auxiliary component 辅助原料
auxiliary compressor 辅助压缩机
auxiliary compressor motor 辅助压缩空气电动机
auxiliary computer power unit 计算机辅助电源
auxiliary computing system 辅助计算系统
auxiliary condenser 辅助冷凝器

auxiliary conditioning 辅助调节
auxiliary condition 附加条件
auxiliary connecting rod 辅助连杆
auxiliary console 辅助控制台
auxiliary construction 附属构筑物;附属建筑物
auxiliary constructional appliance 辅助施工机具
auxiliary construction equipment 辅助施工设备
auxiliary contact 辅助触点;辅助触头
auxiliary contact cable 辅助接触电缆
auxiliary contactor 辅助接触器
auxiliary contactor block 辅助触头组
auxiliary contact strip 辅助触点簧片
auxiliary contour(line) 辅助等高线;间曲线;半距等高线间曲线
auxiliary control 辅助控制(机构)
auxiliary control box 辅助逆变器箱
auxiliary control panel 辅助控制盘;辅助控制面板;辅助操纵板
auxiliary control rod 辅助控制杆
auxiliary convection section 辅助对流段
auxiliary coolant system 辅助冷却系统
auxiliary counter 辅助计数器
auxiliary crab 副小车;辅助起重绞车
auxiliary craft 附属船(舶)
auxiliary craft for dredging works 疏浚工程辅助船舶
auxiliary crane 辅助起重机
auxiliary cross girder 副横梁
auxiliary cruiser 辅助巡洋艇
auxiliary crushing 辅助破碎
auxiliary current transformer 辅助电流互感器
auxiliary curve 辅助曲线
auxiliary cut 辅助切割;辅助裁切
auxiliary cutter bit 辅助切力
auxiliary cylinder 辅助唧筒
auxiliary dam 副坝;辅助坝
auxiliary data 辅助资料;辅助数据
auxiliary data processing equipment 辅助数据处理设备
auxiliary data translator unit 辅助数据转换装置
auxiliary dead latch 固死锁舌的附加锁闩
auxiliary deflection angle 分转向角【测】
auxiliary department 辅助车间
auxiliary department expenses 辅助部门费用
auxiliary detonator 传爆管
auxiliary device 辅助设备
auxiliary device of cage 罐笼附属装置
auxiliary diesel engine 辅助柴油机;备用柴油机
auxiliary directory 辅助目录
auxiliary documents 辅助单据
auxiliary drilling device 备用钻孔工具;辅助钻探装置
auxiliary drive 辅助驱动(装置);辅助传动(装置)
auxiliary drive shaft 辅助传动轴
auxiliary drive turbine 辅机驱动汽轮机
auxiliary drum 辅助卷筒
auxiliary ejector 辅助喷射器
auxiliary electric(al) drive 辅助电力传动装置
auxiliary electrode 辅助电极
auxiliary elevation 辅助正视图;辅助视图
auxiliary embankment 辅助堤
auxiliary emulsifying agent 辅助乳化剂
auxiliary energy dissipator 辅助消能工
auxiliary energy flow 补助能流
auxiliary engine 辅助发动机;辅机
auxiliary engine room 辅机舱
auxiliary enlarging lens 辅助放大镜
auxiliary enterprise 附属企业
auxiliary entrance 辅助入口
auxiliary equation 辅助方程
auxiliary equipment 附属设备;外围设备;附属装置;附带设备;辅助设备;备用设备
auxiliary equipment of transmitter 发讯机附属设备
auxiliary equipment power supply 辅助供电设备
auxiliary estimator 辅助预算员
auxiliary exciter 副励磁机
auxiliary exciting winding 辅助励磁绕组
auxiliary exhaust reservoir 排气副储筒
auxiliary expenses 辅助费用
auxiliary exploration line 辅助勘探线
auxiliary eyepiece 辅助目镜
auxiliary face labo(u)r 工作面辅助工
auxiliary facility 辅助设备
auxiliary factory for construction 施工附属企业

auxiliary fastenings 扣件附属配件
auxiliary fault 副断层;分支断裂;伴生断层
auxiliary feeder 辅助馈线
auxiliary feed pump 辅助进给泵;辅机给水泵
auxiliary feed valve 辅助进给阀
auxiliary feed water pump 辅助给水泵
auxiliary fiber 副纤维
auxiliary field 辅助磁场
auxiliary field stop 辅助视场光栏
auxiliary finder aperture 辅助探测孔径
auxiliary firm 附属公司
auxiliary fish entrance 鱼道辅助入口
auxiliary fleet 辅助船队
auxiliary float 辅助浮体
auxiliary floating dock 辅助浮船坞
auxiliary flocculent 辅助絮凝剂
auxiliary floor area 辅助面积
auxiliary fluid 辅助流体
auxiliary fluid ignition 起动燃料点火
auxiliary flux 辅助溶剂
auxiliary flying 辅助飞行
auxiliary follow up piston 辅助随动活塞
auxiliary force 临时工;辅助人员;随动传动力;辅助力
auxiliary formula 附加公式;辅助公式
auxiliary fracture 次要裂缝;伴生断裂
auxiliary frame 副车架;附加框架;辅助架
auxiliary friction 附件摩擦力
auxiliary fuel 辅助燃料
auxiliary fuel tank 辅助燃料箱
auxiliary function 辅助功能
auxiliary function symbol 辅助函数符号
auxiliary furnace 辅助炉
auxiliary fuse 辅助熔断器
auxiliary gain control 辅助增益控制
auxiliary gallery 辅助坑道
auxiliary ga(u)ge 辅助水尺;辅助量器
auxiliary gearbox 辅助齿轮箱
auxiliary generator 辅助发电机
auxiliary generator set 辅助发电机组
auxiliary geologic(al) map 提供辅助地质图件
auxiliary girder 辅助横梁
auxiliary gland 辅助填料压盖
auxiliary governor 辅助调速机;辅助调节器
auxiliary governor spring 副调节弹簧
auxiliary grade 辅助坡(度)
auxiliary great circle 辅助大圆
auxiliary grounding 辅助接地
auxiliary guide wall 辅助导航墙
auxiliary haulage 辅助运输
auxiliary head 辅助头
auxiliary heading 辅助异洞
auxiliary heater 辅助加热器
auxiliary heater switch 辅助加热器开关
auxiliary heating system 辅助加热系统
auxiliary hoist 辅助提升设备;辅助起重机
auxiliary hoisting mechanism 副提升机构;辅助提升机构;辅助起重机
auxiliary hologram 辅助全息图
auxiliary hook 辅助钩;辅助钩
auxiliary horizontal projection 辅助水平投影
auxiliary ignition 利用起动燃料点火
auxiliary illuminator 辅助照明器
auxiliary inductive parameter 辅助诱导参数
auxiliary information 附带资料;辅助资料
auxiliary input 辅助输入
auxiliary instruction buffer 辅助指令缓冲器
auxiliary instrument 辅助仪器;辅助仪表
auxiliary intersection point 副交点【测】
auxiliary inverter test equipment 辅助逆变器试验装置
auxiliary jack 辅助塞孔
auxiliary jet 辅助喷管
auxiliary jet carburetor 双嘴汽化器
auxiliary labo(u)r 辅助劳动
auxiliary lamp 辅助灯
auxiliary landing gear 辅助着陆装置;辅助起落装置
auxiliary landing ground 次要的飞行着陆场;辅助停机坪
auxiliary lane 辅道;附加车道;辅助车道;加宽车道
auxiliary ledger 辅助分类账
auxiliary lens 附加镜头;辅助透镜
auxiliary lens system 前镜系统;辅助透镜系统
auxiliary letter of credit 附属信用证

auxiliary level 辅助水准器
auxiliary liftmotor 辅助电梯电动机
auxiliary light 副灯;辅助光;辅助灯
auxiliary lighting 辅助照明
auxiliary line 辅助线
auxiliary line of sight 辅助瞄准线
auxiliary lintel 辅助过梁
auxiliary live steam admission valve 辅助新汽进汽阀
auxiliary live steam valve 进汽副阀
auxiliary locking force 副锁闭力
auxiliary locomotive 辅助机车
auxiliary locomotive moving kilometres 补助机车走行公里
auxiliary longitudinal member 辅助纵构件
auxiliary lubricating oil pump 辅助润滑油泵
auxiliary machine 辅助机(器);辅机;备用机器
auxiliary machinery 副机;辅助机械;补助机器
auxiliary machinery equipment 辅助机械设备
auxiliary machinery room 辅机舱
auxiliary machine shop building 机修厂建设
auxiliary magnet correction 辅磁改正
auxiliary magnet correction value 辅磁改正值
auxiliary manpower 半劳动力;辅助劳动力
auxiliary marker 路线指示标;辅助标示
auxiliary marshalling station 地方性编组站【铁】;辅助编组站【铁】
auxiliary mass curve 副积线
auxiliary material 副原料;附属材料;辅助原料;辅助材料
auxiliary material cost 辅助材料费
auxiliary material expenses 辅助材料费
auxiliary means 辅助工具
auxiliary measure 辅助措施
auxiliary measuring system 辅助测量系统
auxiliary memory 辅助存储器
auxiliary meridian 辅助子午线
auxiliary messenger wire 辅助挂索
auxiliary method 辅助方法
auxiliary microscope 辅助显微镜
auxiliary mill 辅助磨机;备用磨机
auxiliary mineral 次要矿物;副矿物
auxiliary mixer 辅助混合器
auxiliary module 辅助模数
auxiliary moment 辅助矩
auxiliary motor 辅助电动机;备用电动机
auxiliary motor test stand 辅助电机试验台
auxiliary needle valve 辅助针阀
auxiliary nest actuator 辅助连身齿轮传动器
auxiliary network address 辅助网络地址
auxiliary nozzle 辅喷嘴
auxiliary observation hole 辅助观测孔
auxiliary observation station 辅助观测站
auxiliary oil pump 辅助油泵
auxiliary operating rod 辅助操纵杆
auxiliary operation 外部操作;辅助作业;辅助运算;辅助操作
auxiliary operation time 辅助工作时间
auxiliary optic(al) axis 辅助光轴
auxiliary optic(al) system 辅助光学系统
auxiliary optics 辅助光学系统;辅助光学器件
auxiliary packer 辅助封隔器
auxiliary packing 辅助填料
auxiliary parameter 辅助参数
auxiliary penal measure 辅助惩罚措施
auxiliary pendulum 辅助摆
auxiliary person 辅助人员
auxiliary phase 辅助相
auxiliary pier 辅助墩
auxiliary pilot lamp 辅助信号灯
auxiliary pilot relay 辅助领示继电器
auxiliary piping 辅助管路
auxiliary piston 辅助活塞
auxiliary plane 辅助平面;辅助面
auxiliary plane method 辅助平面法
auxiliary plant 副厂房;辅助装置;辅助设备;辅助工厂;辅助厂房
auxiliary point 辅助点;补充点
auxiliary point orientation 辅助点定向【测】
auxiliary potential transformer 辅助电压互感器
auxiliary power 厂用动(力);厂用电(力);辅助动力;备用电源
auxiliary-powered vessel 机帆船
auxiliary power elevator 辅助动力驱动电梯;动力驱动升降机
auxiliary power plant 辅助发电厂;辅助动力装置;辅助电源设备;辅助电厂;备用电厂
auxiliary power requirement 辅助设备用电量
auxiliary power source 辅助电源
auxiliary power station 辅助电站;备用发电厂;备用电站
auxiliary power supply 备用电源;辅助供电;辅助电源
auxiliary power supply equipment 自备供电设备
auxiliary power supply system 厂用配电系统;辅助电源系统
auxiliary power unit 辅助动力装置;辅助动力机组;辅助电源设备
auxiliary prestressing 辅助预加应力
auxiliary prestressing jack 辅助预压千斤顶
auxiliary problem 辅助问题
auxiliary product 次要产品;附属产品;辅助物料
auxiliary production 辅助生产
auxiliary production building 辅助生产建筑
auxiliary production expenses 辅助生产费用
auxiliary production expense apportionment 辅助生产费用分配
auxiliary production subsidiary ledger 辅助生产明细表
auxiliary production unit 辅助生产部门
auxiliary profession 辅助职业
auxiliary project 附属工程;辅助工程;配套工程
auxiliary projection 线、平面或立体在另一平面上的投影;辅助视图
auxiliary projection drawing 辅助投影图
auxiliary projection plane 辅助投影面
auxiliary project plant 工程辅助设备
auxiliary propeller 辅助螺旋桨;辅螺旋浆
auxiliary property 次要性能
auxiliary propulsion 辅助推进
auxiliary propulsion unit 备用推进装置
auxiliary protection 辅助防护措施
auxiliary pump 辅助水泵;辅助泵;备用泵
auxiliary quarter-wave plate 辅助四分之一波片
auxiliary quay 辅助码头
auxiliary rack 辅助横移
auxiliary rafter 垫椽;辅助椽(子);便椽
auxiliary ram 辅助活塞
auxiliary raw material 辅助原料
auxiliary reactor 辅助反应器
auxiliary reagent for negative photoresist 负性胶配套试剂
auxiliary rear spring 副后弹簧
auxiliary receiver 辅助接收机;辅助集液器
auxiliary record 辅助记录
auxiliary recorder unit 辅助记录装置
auxiliary reference section 辅助参考剖面
auxiliary reflector 辅助反射器;辅助反光镜
auxiliary register 辅助寄存器
auxiliary regression 辅助回归
auxiliary regulator 辅助调节器;备用调节器
auxiliary reinforcement 副筋;辅助钢筋;预应力混凝土附加钢筋
auxiliary relay 中间继电器;辅助继电器
auxiliary repair 辅修
auxiliary repair dock 辅助修船坞
auxiliary repair drydock 辅助修理干船坞
auxiliary resistance 附加阻力;辅助电阻
auxiliary resources 辅助资源
auxiliary retarder 副缓行器
auxiliary reverser 反转机轮
auxiliary road 辅路;辅助道路
auxiliary rod 辅助杆
auxiliary rope 辅助绳;辅助索
auxiliary rope-fastening device 辅助系绳装置;钢丝绳辅助扣紧装置
auxiliary rotor 尾桨;辅助螺旋浆
auxiliary routine 辅助例行程序;辅助程序
auxiliary rudder 辅助舵
auxiliary sail 辅助帆
auxiliary scaffold(ing) 辅助脚手架;辅助钢筋手架
auxiliary scale 辅助尺
auxiliary scour 辅助冲洗
auxiliary screening device 辅助筛分装置
auxiliary screw 辅螺旋浆
auxiliary screw arc 辅助螺钉弧形板
auxiliary seat 副座
auxiliary sectional plane 副断面
auxiliary selector 辅助选择器
auxiliary service 附属服务设施
auxiliary servomotor 辅助伺服电动机;辅助接力器
auxiliary set 辅助用具;辅机
auxiliary shaft 辅助竖井;副轴;副井;辅助轴
auxiliary shears 辅剪
auxiliary shelter belt 副林带
auxiliary ship 辅助舰艇;辅助船舶;辅助船
auxiliary ship station 辅助船舶观测站
auxiliary shop 修配车间
auxiliary shoring 辅助支撑
auxiliary sign 辅助标志
auxiliary signal 辅助信号
auxiliary sluice valve 辅助蒸汽闸阀
auxiliary smoke bell 辅助烟火报警钟
auxiliary sound carrier unit 伴音载波设备
auxiliary source 辅助电源
auxiliary spar 副梁;辅助梁
auxiliary spark excitation 辅助火花激励
auxiliary spark gap 副火花隙;辅助电花隙
auxiliary species 附属树种
auxiliary sphere method 辅助球面法
auxiliary spillway 辅助溢洪道;副溢洪道
auxiliary spring 副簧
auxiliary spring bracket 副弹簧托架
auxiliary spring clamp 辅助弹簧夹
auxiliary spring nut 副簧螺母;辅簧螺母
auxiliary spring nut pin 副簧螺母销
auxiliary spring plate 辅助弹簧垫片
auxiliary spud 辅助钢桩【疏】
auxiliary stabilizer 辅助稳压器
auxiliary staff 辅助职员
auxiliary staff ga(u)ge 辅助水尺;辅助标尺
auxiliary staff instrument 辅助路签机
auxiliary stair(case) 辅助楼梯;便梯
auxiliary stand 辅助支架
auxiliary standard parallel 辅助标准纬圈
auxiliary standard pipe 副标准管
auxiliary statement about transport revenue plan finished 运输收入进款计划完成情况附表
auxiliary station 辅助站;辅助台站;辅助车站;辅助测站;副厂房
auxiliary steam 厂用汽
auxiliary steam cylinder 副汽缸
auxiliary steam pipe 辅汽管;副出汽管
auxiliary steam valve 副汽阀;辅助蒸汽阀
auxiliary steam valve spindle 副汽阀柄
auxiliary steel 附加钢筋
auxiliary steel road 辅助钢板路面
auxiliary steering gear 太平舵机;副舵机
auxiliary stop valve 辅助断流阀
auxiliary storage 后备存储器;辅助性储存;辅助存储器
auxiliary storage management 辅助存储器管理
auxiliary storage manager 辅助存储器管理(程序)
auxiliary store 辅助存储器
auxiliary straight line 辅助直线
auxiliary stressing method 附加应力法
auxiliary stretching 附加伸长
auxiliary structure 附属建筑物;辅助建筑(物);附属结构;附属构筑物
auxiliary submarine rescue ship 辅助潜水救生船
auxiliary substation 备用变电所
auxiliary suction 附属吸水管
auxiliary suction pipe 辅助吸管
auxiliary supply 辅助水源;辅助水渠
auxiliary support 辅助支架
auxiliary/support equipment 辅助/支缓设备
auxiliary survey 辅助测量
auxiliary survey vessel 辅助测量船
auxiliary switch 辅助开关
auxiliary switchboard 辅助配电盘;辅助配电板;厂用配电屏;厂用配电盘;厂用开关盘
auxiliary switchgear 厂用配电装置
auxiliary syntan 辅助性合成鞣剂
auxiliary system 辅助系统;辅助警报系统
auxiliary tablet instrument 辅助路牌机
auxiliary tangent 副切线;辅助切线
auxiliary tangent method 辅助切线法
auxiliary tank 辅助容器;副油箱;备用罐
auxiliary telescope 辅助望远镜
auxiliary template method 辅助量板法
auxiliary tensioning 附加的张拉;辅助预加应力

auxiliary test unit 辅助测试装置
auxiliary thermometer 辅助温度计
auxiliary tie-back 辅助锚碇拉杆
auxiliary tilted photocoordinate system 辅助倾斜像片坐标系统
auxiliary time of production 辅助生产时间
auxiliary token instrument 辅助凭证闭塞机
auxiliary track fastening 辅助轨道扣件
auxiliary traction system 辅助牵引系统
auxiliary transformer 厂用变压器;辅助变压器
auxiliary transmission 副变速箱
auxiliary transmitter 辅助发射机
auxiliary transom 辅助横梁
auxiliary traverse 辅助导线
auxiliary trolley 辅助吊运车
auxiliary truss member 辅助桁杆
auxiliary turbine 辅助涡轮机;辅助水轮机;辅助汽轮机
auxiliary unit 辅助总成;辅助装置;辅助单位
auxiliary unit accounting 附属单位会计
auxiliary unit budget 附属单位预算;附费单位预算
auxiliary utility model 辅助性实用新型
auxiliary valency 副原子价;副价
auxiliary value 校正系数;修正系数
auxiliary valve 备用阀;辅助阀
auxiliary valve seat 辅助阀座
auxiliary valve spring 副阀弹簧
auxiliary variable 辅助变量;辅助变数
auxiliary vault 附加的穹顶;辅助穹顶
auxiliary ventilator 辅助通风机
auxiliary vent piping 辅助通风管路
auxiliary vertical aerial 垂直辅助天线;辅助垂直天线
auxiliary vertical antenna 辅助垂直天线
auxiliary vertical plane 辅助垂直面
auxiliary vertical projection 辅助垂直投影
auxiliary vessel 辅助艇;辅助舰船;辅助船
auxiliary view 辅助视图
auxiliary wages 辅助工资
auxiliary washing 辅助冲洗
auxiliary washing unit 辅助冲洗设备
auxiliary water 诱鱼水流;辅助水流;备用水
auxiliary water discharge pumping station 辅助炉
auxiliary water meter 辅助水表
auxiliary water sources 辅助水源
auxiliary water supply 辅助供水;备用给水
auxiliary weather chart 辅助天气图
auxiliary wharf 辅助码头
auxiliary wheels 辅助轮
auxiliary winch 辅助绞车
auxiliary winding 辅助绕组
auxiliary wing of breakwater 防波堤的辅助翼堤
auxiliary wire rope system 辅助钢丝绳系统
auxiliary worker 辅助工
auxiliary work groups 辅助工组
auxiliary works 附属工程;辅助工程;辅助劳动;辅助工作
auxiliary work shop 辅助车间
auxiliary yacht 机帆游艇
auxiliary yard 子场;辅助调车场【铁】;辅助车场
auxilytic 促溶解的
auxiometer 透镜放大率仪;廓度计;放大率计
auxochrome 助色团;助色基团
auxochromic group 助色团
auxochromous group 助色团;助色基团
auxograph 体积变化自动记录器;生长记录器
auxometer 放大率计
avail 有用;有益于;有效;可用金额
availability 利用度;可用性;可用率;可靠性;可获量;性能最佳性;系统有效性;有效性;有效利用率;工作效率
availability date 生效日期
availability doctrine 信用的可供应情况学说
availability effect 可获性效应
availability factor 时间利用系数;时间利用率;利用因素;利用因数;利用率;可用因数;有效系数;有效利用系数;使用系数
auxiliary factor of aids 航标可用率
availability in market supplies 市场可供商品量
availability of aids 航标可用性
availability of client fund 可利用的委托人基金
availability of computer 计算机可靠性
availability of inputs 可能得到的投入
availability of inventory 可分配存货量

availability of labo(u)r 可利用的劳动力
availability of materials 可利用的材料
availability of notice to mariners 能取得航行通告
availability of oil 可采石油
availability of penetrating time 钻进时间利用率;纯钻进时间利用率
availability of service 服务有效性;服务效力
availability of soil water 土壤水分有效性
availability of water 水的有效性
availability period 支用贷款有效期
availability ratio 利用系数;可用率;有效度比;有效比
availability ratio of aids 航标可用率
availability risk 现有风险
availability space 有效舱容
availability system 系统利用率;系统利用度;有效工作系统
available 可用的;有效的
available accuracy 实际准确度;可及准确度;可达精度;可达到的精度;有效准确度;有效精度
available act 可取行为
available act of bankruptcy 可宣告破产的行为
available analysis 有效分析
available and latent superiorities 现有和潜在的优越性
available arsenic 有效砷
available assets 可用资产;可利用资产;可供清理债务的资产;可动用资产
available balance 可用余额;可动用结余
available base 符合要求的底层;符合要求的基层
available berth 可利用(的)泊位;现有泊位
available bit rate 可用比特率
available buoyancy 有效浮力
available capacity 有效库容;可用容量;可用功率;现有能力;有效容量;有效舱容
available car area 轿箱有限面积
available cash 可用现金;可动用现金
available channel capacity 河道有效行洪能力;现有航道通过能力
available characteristic 可用特性
available chlorine 有效氯
available-chlorine method 有效氯法
available coefficient 可利用系数;有效系数
available component 有效成分
available construction suplus 可动用建筑公积金
available cost 可控成本
available credit 可用信用证
available current surplus 可动用的经济盈余;可动用本期盈余
available data 现有数据;已有资料
available date 可动用日期
available depth 有效水深;资用水深
available designable reserve 保有可设计储量
available device table 可用设备表
available dilution 有效稀释(度)
available discharge 可用流量;有效流量
available discount 可能贴现额
available draft 可用风量;有效通风
available draught of flue 烟囱有效吸力
available ductility 可利用延性
available earning surplus 可用营业盈余
available energy 可用能(量);有用能;有效能(量)
available energy output 可用发电量;有效输出能量;有效发电量
available equipment 现有设施;现有设备
available estimate 可用估计值
available facility 现有设施;现有设备
available factor 利用系数;可用率;可用系数
available factors of reserves 储量利用系数
available field capacity 有效田间持水量
available field of view 有效视场
available flow 可用流量;有效流量
available for prompt delivery 可立即供水的;可立即供货的;可立即供电的
available frequency 可用频率
available from stock 可迅速交货;可从库存中取用的;可从库存中拨给的;仓库中有现货
available fund 可用基金;获得的资金
available gain 可用增益
available green time 有效绿灯时间
available groundwater 可用地下水;地下水开采储量;自由地下水;有效地下水
available head 可用压差;有效水头;可用水头

available heat 有效热
available heating value 有效热值
available height for storage 有效蓄水深度;有效蓄水高度;有效堆放高度
available horse power 可用功率
available hydraulic head 可用水头;有效水头
available income 可利用收益;可动用收入
available information 可用资料;可用信息;可得到的信息;现有资料;已有资料
available inventory 可调库存
available iron 可吸收铁
available length 有效长(度)
available length of short casing thread 短扣套管丝扣有效长度
available life 可用期;有效寿命
available line 可用线路
available machine time 可用机器的时间;开机时间;机器工作时间;(机器的)有效工作时间
available materials 可用材料;可取得的材料;现有材料
available memory 可用内存储量;现有内存储量
available metal 可用金属;速效金属
available method 有效方法
available moisture 可用水分;有效温度;有效水分
available moisture capacity 有效含水量;有效持水量
available moisture capacity of a soil 土壤有效浓度;土壤有效含水量
available moisture content 有效含水量
available national income 可使用国民收入
available natural resources 可利用的自然资源
available net head 可用有效水头;净有效水头
available nitrate 有效硝酸盐
available nutrient 有效养分
available nutrient of soil 土壤有效养分
available office space 有效办公空间;办公使用面积
available output 有用功率
available oxygen 有效氧;有效含氧量
available part 可用部分;可用部分
available period 适用期;有效期
available phosphorus 有效磷
available point 有效点
available pore space 有效孔隙
available potential energy 有效位能
available power 匹配负载功率;可用出力;有效功率;有效出力
available power efficiency 有效功率效率
available precipitation 可用降水量;有效降水(量)
available pressure 可用压力;有效压强;有效压力
available pressure head 有效压头;有效水压头
available productive capacity 可利用生产能力
available profit 可用利润;可动用利润
available proton concentration 有效质子浓度
available pump pressure 泵的有效压力
available quantity 可用量
available rainfall 可用雨量;有效雨量
available rate 可用率
available relief 有效地势
available reserves 有效储量;保有储量
available residual chlorine 有效余氯
available resources 可利用资源;现有资源
available runoff 可用径流(量)
available size 有效尺寸
available soil moisture 土壤有效湿度;有效土壤成分
available soil water supply 土壤有效供水量;土壤可供水量
available space list 可利用的空间表
available stock 可用货车库存
available storage(capacity) 有效容量;有效库容;可动用储[贮]存量;资用储量;有效储量
available sulfur 有效硫
available supply 可供水量
available supply of water 可用给水
available surface 可用表面;有效表面;(海的)自由表面
available surface area 有效表面积
available surface area of checker 格子体有效面积;格子体受热面积
available surface of evapo(u)ration 有效蒸发表面
available surplus 可动用盈余;可动用公积金
available temperature 有效温度
available temperature drop 有效温度降

available temporary surplus 可动用临时盈余
available thickness 有益厚度
available thrust 有效推力
available time 可用时间
available to plant 对植物有用
available uncultivated land 可垦荒地;可耕荒地
available volume 有效体积
available wagon 备用货车
available water 可用水量;可用水分;可供水;有效水(分)
available water carrying section 有效过水断面
available water consumption 可用水补给
available water depth 可用水深
available water holding capacity 有效含水量;有效持水能力
available water resources 可用水资源
available water resource survey 有效资源勘察
available water supply 可供水量;有效供水(量);有效给水量;可利用水量
available waterway traffic ability 现有航道通过能力
available work 可用功;有用功;有效工作;资用功
available workable reserve 保有可开采储量
available working hatch 可进行装卸舱口
availiability 有效使用率
avail length of long casing thread 长扣套管丝扣有效长度
availment of credit 信贷使用
avails 出售资产收入
avails of the sale 销货收入
avalanche 土崩;坍方;山崩;崩坍;崩塌;崩落
avalanche alarm 崩坍警报
avalanche baffle 坍方防御建筑物;场方防御建筑物;雪崩阻挡物;防坍构筑物;防坍挡板
avalanche baffle wall 坍方防护墙;山崩防户墙
avalanche blast 崩坍气浪
avalanche brake 防止滑坡的结构;排除落崩用设备
avalanche brake structure 山崩支017结构
avalanche breakdown 雪崩击穿
avalanche breccia 岩崩角砾岩
avalanche chute 雪崩槽沟
avalanche cone 崩落锥
avalanche control works 防崩工程;防坍工程
avalanche dam 崩坍形成的坝;崩坍坝
avalanche damage 崩坍破坏
avalanche debris 山崩废砾堆;崩坍堆积物
avalanche defence 崩落防御设施;坍落防御设施;场方防御栅;防雪栅;防雪崩栅栏;防坍设施;防坍工程;防坍;防场;崩坍防护
avalanche defence gallery 防雪崩廊道;防雪崩廊
avalanche-development constant 雪崩扩展常数
avalanche diode 雪崩二极管
avalanche diode oscillator 雪崩二极管振荡器
avalanche fence 坍方防御栅
avalanche gallery 防雪崩廊道;坍方防御廊
avalanche ice 雪崩冰;堆冰
Avalanche index 阿瓦兰切指数
avalanche-in-duced migration 雪崩感应徙动
avalanche observation 崩坍观测
avalanche observation station 崩坍观测站
avalanche of rock 石崩;岩崩
avalanche of sand and stone 砂石流
avalanche oscillator 雪崩振荡器
avalanche photodetector 雪崩光电检测器
avalanche preventing forest 坍方防止林;坍防护林;雪崩防止林;雪崩防护林
avalanche prevention 防坍
avalanche prevention works 防坍工程
avalanche protection 崩落保护
avalanche protection dike[dyke] 崩坍防护堤
avalanche protection works 防护崩落的工程;坍崩滞工程;崩坍防护工程
avalanche protector 防坍挡板;防崩落挡板;崩落防护装置
avalanche shed 坍方防御棚;雪崩防护板;场方防御板;雪崩防护棚;崩坍挡板
avalanche source 雪崩源
avalanche stoppage 防雪崩措施
avalanche test 崩坍试验
avalanche transistor 雪崩晶体管
avalanche trigger zone 崩坍触发区
avalanche voltage 雪崩电压
avalanche wall 坍方防护墙
avalanche wind 雪崩气浪;雪崩风

avalanche zone 崩塌区;崩坍区
avalanching 崩塌;磨球泻落
avalanching of grinding ball 磨球崩落
avalanchologist 崩落学家
avalite 钾铬云母;铬云母;路伊利石
Avalokitesvara with coiffure of Maku 麻姑观音(瓷器名)
avant-corps 前亭
avant-garde(style of) architecture 创新式建筑;建筑学独特风格;建筑学创新风格
avant port 前港;外港池
avant project 前期工程
a variety of dredging activities 各种挖泥业务
avasite 硅褐铁矿
avast hauling 停拖
avast veering 停松
Avco molecular test system 阿弗科分子试验系统
avehess 阿芙齐裁
aven 溶井;穹顶深坑;竖井;落水洞;岩溶坑
Avena unit 艾文纳单位
aventurin(e)feldspar 日长石;太阳石
aventurin(e)glass 茶色金星玻璃;砂金石;洒金玻璃(含金色细粒的不透明褐色玻璃)
aventurin(e)glass 金星玻璃;砂金石
aventurin(e)glaze 砂金釉;金星釉
aventurin(e)green 金星绿
aventurin(e)quartz 星彩石英
avenue 林荫大道;马路;大路;大街;大道
avenue of approach 解决途径
avenue of infection 传染途径
avenue of Sphinxes 斯芬克斯林荫路;斯芬克斯林荫大道
avenue planting 植树成行;道旁种植;行道栽植;道路两边树木
avenue planting system 街道栽植法
avenue to success 成功途径
avenue tree 路树;行道树
averagable income 可计入平均数的收入
average 平均数;平均(的);均分;海损
average absolute error 平均绝对误差
average absolute relative deviation 平均相对离差绝对值
average absolute relative error 平均相对误差绝对值
average abundance 平均丰度
average acceleration 平均加速度
average acceleration rate 平均加速度率
average acceleration response spectrum 平均加速度反应谱
average access time 平均存取时间
average accuracy 平均准确度
average accustomed 照常规办理单独海损免赔额
average adjuster 海损理算师;海损理算人
average adjustment 海损理算;海损检定书
average age 平均年龄
average agent 理赔代理人;海损代理人
average age of groundwater 地下水平均年龄
average age of receivables 应收账款平均账龄
average agreement 海损协议(书);海损分担保证书
average aircraft 一般性能飞机
average allowable amount of harmful components 有害组分平均允许含量
average altitude 平均海拔高度
average amount 平均量
average amount of current funds possessed 平均占用的流动资金
average amount of information 平均信息量
average amount of inspection 平均检验数
average amount of operating rigs 平均开动钻机数
average analysis 平均分析
average angle method 均角法
average annual 年平均
average annual abstraction 平均年抽水量
average annual accumulation 平均年淤积量;平均年沉积量
average annual capacity factor 年平均设备利用率
average annual cost 平均年度费用
average annual daily volume 年平均日交通量
average annual discharge 平均年流量;年平均流量;多年平均流量
average annual dredging quantity 平均年疏浚量
average annual duration curve 年平均历时曲线
average annual equivalent damage 平均年灾害损失

average annual erosion 平均年冲刷量
average annual flood 平均年洪水(流量);年平均洪水流量
average annual flow 年平均流量;年平均径流量;多年平均流量
average annual ground-rent 平均年地租
average annual growth 年均增长
average annual growth rate 年平均增长率
average annual output 年平均发电量;年平均产量
average annual precipitation 平均年降水量;年平均降水(量);多年平均降水(量)
average annual rainfall 平均年雨量;年平均雨量;平均年降雨量
average annual rate of earthquake occurrence 年平均发震率
average annual runoff(volume) 年平均径流量
average annual sedimentation rate 平均年沉积率
average annual sediment concentration 年平均含沙量
average annual sediment deposition 平均年淤积量
average annual sediment discharge 平均年输沙量;年平均输沙量
average annual sediment yield 平均年泥沙沉积量
average annual soil loss 平均年土壤流失量
average annual stand depletion 平均年疏伐度;年平均疏伐量
average annual trip rate of inhabitant 居民年平均出行率
average annual working capacity 年平均工作量
average annual yield 多年平均产量
average anomalistic motion 平均近点运动
average approximation 平均近似
average area of one passenger occupied in a carriage 客车车辆人均占用面积
average area precipitation 面平均降水量
average arrival rate 平均到达率
average assay 平均成分
average astronomic(al)north pole 平均天文北极
average atmospheric(al)condition 普通大气条件
average atmospheric(al)refraction 平均大气折射
average available discharge 平均可用流量
average award 海损裁决书
average background value 背景平均值
average balance 平均余额;平均差额
average balance of current fund 流动基金平均余额
average bearing loss 平均听力损失
average bearing stress 平均承载应力
average bed level 平均河底高程
average bit footage 平均钻头进尺
average bit weight 平均钻压
average block anomaly 平均区域异常
average block length 平均块长
average blockness 平均块度
average boiling point 平均沸点
average bond 海损协议(书);海损分担保证书
average bond stress (混凝土与钢筋的)平均黏结应力;平均黏结应力;平均握裹力
average brightness 平均亮度
average bulk modulus 平均体积弹性模数
average buried depth of basement 基ün的平均埋深
average by series 按系列计损
average candle 平均烛光
average capacity standard 平均能量标准
average capital 平均资本
average capital ratio method 平均资本比率(法)
average cargo payload 货物平均净载重
average cargo transportation density in a section 区段平均货物运输密度
average cargo transportation speed 货物平均运送速度
average cargo transportation time 货物平均运送时间
average carkilometers in one turnround 货车周转距离
average carkilometers per car-day 客车平均日车公里
average carrying intensity of line network 线网平均负荷度
average car study technique 平均车辆调查法
average cash balance 平均现金余额
average centre of the moon 平均月球中心

average charge 平均电荷
average charge density fluctuation 平均电荷密度起伏
average chart 平均值等深线图
average clause 海损条款;海损赔偿条款;分摊条款;分担条款
average clock 平均钟
average closure 平均闭合差
average coefficient of cubic expansion 平均体积膨胀系数
average coefficient of linear expansion 平均线膨胀系数
average coherence 平均相干
average collected balance 平均托收余额
average collection period 平均托收期;平均收账期间;平均收款期
average collection period of receivables 应收账款平均收账期;应收账款平均收款期
average collection period ratio 平均托收期比率
average combined cost 平均综合成本
average commissioner 海损检定员
average composite sample 平均拼合样品;平均并合样品
average composition 平均组成
average compressive strength 平均抗压强度
average concentration 平均浓度
average concrete 规范混凝土;普通混凝土
average condition 一般条件
average consistency 平均稠度
average constant cost 平均不变成本
average construction period 平均施工工期;平均建房工期
average consumer 一般消费者
average consumption 平均用(水)量;平均消费量
average content of mercury 汞的平均含量
average contour area method 平均等高线面积法
average contribution 海损分担
average core recovery 平均岩芯收获率
average cost 平均费用;平均成本
average cost curve 平均成本曲线;平均费用曲线
average costing 平均成本计算
average cost method 平均成本法
average cost per unit 单位平均成本
average cost profit rate 平均成本利润率
average cross section 平均截面
average current 平均流;平均海流;平均电流
average current of charging train 列车带电平均电流
average current of feeding section 供电臂平均电流
average current of train 列车平均电流
average current pulse responses 平均电流脉冲响应
average curvature 平均曲率
average daily balance 日均收付差额;每日平均存款余额
average daily cargo handling rate per ship in operation 船舶平均作业艘天装卸量
average daily cargo handling rate per ship in port 船舶平均在港艘天装卸量
average daily car loadings 日均装车数
average daily change 平均日变化
average daily collected balance 日均已收票据差额
average daily consumption 平均日消费量;平均日耗量;平均日用(水)量
average daily consumption of water 平均日用水量
average daily discharge 平均日流量
average daily dredging quantity 日平均疏浚量
average daily flow 日均流量
average daily gain 平均日增重
average daily locomotive kilometres 机车日车公里
average daily output 平均日供水量
average daily output freight locomotive 机车平均日产量
average daily output of wagon 货车日产量
average daily range 平均日温差
average daily solar temperature 日平均综合温度
average daily tonnage of cargo in storage 平均每天堆存货物吨数
average daily traffic 日平均运量;平均日交通量
average daily transport output of serviceable wagon 货车生产量
average daily variation 平均日变差
average daily water consumption 平均日耗水量
average daily water supply output 平均日供水量
averaged annual evapo(u)ration discharge over years 多年年平均蒸发量
averaged annual precipitation over years 多年年平均降水量
averaged annual runoff over years 多年年平均径流量
average data 平均数据
average data transfer rate 平均数据传输速度
average day consumption 平均日用水量
average day permissible concentration 日平均容许浓度
average days of supplying interval 平均供应间隔天数
average deceleration 平均减速度
average deficit 平均亏空;平均赤字
average definition 平均清晰度
average deformation 平均变形
average degree of consolidation 平均固结度
average degree of polymerization 平均聚合度
average demand 平均需要量
average density 平均容重;平均密度
average departure 平均偏差;平均距(平)
average deposit 海损备用金;海损保证金
average depth 平均高度;平均水深;平均深度
average depth of mining area 矿区平均水深
average detention time 一次货物作业平均停留时间
average detention time of car in transit 中转车平均停留时间
average deviation 平均偏差;平均离差
average device 平均机
average diameter 平均直径;平均粒径
average diameter of base face thread 基面丝扣平均直径
average diameter of grinding ball 磨球平均球径
average diameter of grinding media 平均球径;研磨介质平均直径
average diameter of handtight face 手旋紧面平均直径
average diameter of pore 孔隙平均直径
average diameter of well 平均井径
average difference 平均差(值)
average difference of flying altitude 飞行高度平均差
average difference of second order 平均二级差
average diffuse light transmission 平均漫射光穿透率
average dimension 平均尺寸;平均尺度
average direction 平均方向
average discharge 平均排出量;平均流量
average discharge modulus 平均流量模数;平均流量模量
average discharge power 平均放电功率
average discharge voltage 平均放电电压
average discount rate 平均贴现率
average discriminatory value 平均判别值
average dislocation 平均位错
average displacement 平均位移
average displacement response spectrum 平均位移反应谱
average distance 平均距离
average distance between stations 平均站间距离
average distance of freight transportation 铁路运程
average distance of passenger 乘客平均乘行距离
average distance of passengers carried 旅客平均旅程
average distribution 平均分布
average diurnal change 平均周日变化
average diurnal high water inequality 平均日高潮不等(值)
average diurnal low water inequality 平均日低潮不等(值)
average diurnal motion 平均周日运动
average diurnal temperature 日平均温度
average diurnal variation 平均周日变化
average divergence 平均散度;平均扩散度;平均发散度
averaged monthly evapo(u)ration discharge over years 多年月平均蒸发量
averaged monthly precipitation over years 多年月平均降水量
averaged monthly runoff over years 多年月平均径流量
average dosage 平均剂量
average down 以低于平均价格买进
average down grade of hump 驼峰平均下坡度
average downed proportional scale coefficient 平均比标系数
average draft 平均牵引阻力;平均吃水(深度)
average draught 平均吃水(深度)
average drift 平均漂移
average drift velocity 平均漂移速度
average driving speed 平均行车速度
average dry weather flow 平均旱天流量
average due date 平均支付期;平均到期日
average duration 平均延停时间
average duration curve 平均历时曲线
average duration hospitalization 平均住院日数
average duration of life 平均寿命长度
average duty on specified category of product 某几类产品的平均关税
average dwelling area per capita 平均每人居住面积
average dynamic(al) load of car 货车平均动载重
average dynamic(al) load of loaded wagon 重车动载量
average earnings 平均收入
average earnings per unit of time 平均工资率
average earth ellipsoid 平均地球椭球
average ebb discharge 平均落潮流量;落潮平均流量
average-edge line 平均边缘线
average effect 平均效应
average effective pitch 平均有效螺纹;平均有效螺距
average effective porosity 平均有效孔隙度
average effective pressure 平均有效压力
average effective stress 平均有效应力
average effective thickness 平均有效厚度
average effective thickness of source rock 有效母岩平均厚度
average effective value 平均有效值
average effective weight 平均有效容重
average efficiency 平均效率
average efficiency index 平均效率指数
average efficient ratio 平均效率比率
average electric(al) axis 平均电轴
average elevation 平均高程
average elevation of mine area 矿区平均标高
average elevation of terrace 阶地平均高度
average emergency braking deceleration 紧急制动平均减速
average empty wagon kilometers in one complete turnround of wagon 空周转距离
average empty wagon kilometres 空车周距
average end area 平均端面积
average end area formula 平均端面积(计算)公式
average end area method 平均端面积(计算)法
average energy 平均能量
average environmental capacity 平均环境容量
average epoch 平均历元
average equatorial moment 平均赤道惯性距
average erosion(al) depth 平均冲刷深度
average erosion(al) velocity 平均冲刷速度
average error 平均误差;平均差(值)
average error rate 平均出错率
average establishment 平均朔望潮高潮间隙;平均朔望潮
average estimated variance 平均估计方差
average evapo(u)ration 平均蒸发率
average excavation round 平均开挖半径
average excavation speed 平均推进率
average excess 平均排队超长
average exposure level 平均暴露水平
average external volume 平均外部体积
average extra defective limit 平均额外缺陷界限
average factor method 平均系数法
average failure rate 平均故障率
average fall velocity 平均沉落速度
average field intensity 电场强度平均值
average fineness 平均细度
average fissure length 平均裂隙长度
average fissure width 平均裂隙宽度
average fixed cost 平均固定成本;平均不变成本
average floor area per household 平均每户建筑面积
average floor area per suite 平均每套建筑面积

average floor area per unit 平均每套建筑面积
average flow 平均流量;平均径流
average flow duration curve 平均流量历时曲线
average flow rate 平均流率
average flow rate of carrier gas 载气平均流速
average flow time 平均流程时间
average flow velocity 平均流速
average footage per day 日平均进尺
average footage per hour 小时平均进尺
average free-air anomaly 平均空间异常
average free path 平均自由程
average freeze-up date 平均封冻日期
average freight rate assessment 评定平均运费率;运率平均估价;运费平均估计
average frequency 调频中心频率
average frequency section 平均频率剖面
average fuel (electric power) consumption of locomotive per 10000-gross weight-ton-kilometer 机车每万总重吨公里平均燃料(电力)消耗量
average functionality 平均官能度
average fund in transit 平均在途资金
average future life 平均预期寿命
average future time 平均预期寿命
average global temperature 全球平均温度
average grade 平均纵坡
average graded coefficient 平均等级系数
average grade of mineral deposit 矿床平均品位
average grade of ore 平均品位
average grade of ore block 块段平均品位
average grade of orebody 矿体平均品位
average grade of ore deposit 矿床平均品位
average gradient 平均纵坡;平均坡降;平均坡度;平均比率;平均比降
average gradient of river 河流平均比降
average gradient of track leading to crest of hump 驼峰推送线平均速度
average grading 平均粒径级配;平均粒度组成;平均级配
average grain diameter 平均粒径
average grain size 平均粒径;平均粒度
average gross haul tonnage of locomotive 机车平均牵引总重
average gross money wages 平均货币总收入
average gross tonnage of train 列车平均牵引总重
average ground elevation 平均地面高程
average ground level 平均地面高程
average ground surface 平均地面高程
average groundwater velocity 地下水平均流速
average growth speed 平均发展速度
average guarantee 平均海损担保函;海损保证书
average haul(distance)平均运程(土石方的);平均运距
average haul of goods 货物平均运程
average head 平均水头
average headway 平均车时距;平均车间(时)距
average heat radiation intensity 平均热辐射强度
average heavy swell 中狂涌;道氏七级涌浪
average height 平均高度
average height difference 平均高差
average height map 平均高度图
average height of the highest wave 最大波高均值
average higher high water tide 平均高高潮位;平均高高潮面;平均高高潮
average highest high tide 平均最高高水位;平均最高高潮面
average highest high water 平均最高高水位;平均最高高潮面
average high swell 中狂涌
average highway 一般公路
average highway speed 平均道路车速
average horizontal candle power 平均平面烛光
average hourly earnings 平均小时收益;每小时平均工资;小时平均收入
average hourly flow 平均时流量
average hourly wage 平均小时工资
average hour of work 平均工时数
average hours worked per week 平均周工作时数
average hour variation 平均小时变化
average household size 户均人口
average hydrocarbon productivity 平均产烃率
average igneous rock 平均火成岩
average illumination 平均照(明)度
average inaccuracy 平均不准确度

average income 平均收入
average income of nonagricultural workers 职工平均收入
average increase-decrease trend method 平均增减趋势法
average increasing rate 平均递增率
average increment 平均增量;平均生长量
average increment speed 平均增长速度
average index 平均指标
average index number 平均指数;平均指标数
average indicatrix 平均指标
average individual vehicle delay 平均每车延误(时间)
average infiltration 平均渗透率;平均渗透量;平均渗入率;平均渗入量
average infiltration capacity 平均下渗能力;平均入渗能力
average infiltration rate 平均入渗速率
average inflow capacity 平均截流量
average information content 平均信息量
average information rate mean information rate 平均信息率
average initial acceleration 平均初始加速度
average initial offering yield 平均开盘利率
average instruction execution time 平均指令执行时间
average instructions per second 每秒平均指令数;每秒平均执行指令数
average integrated demand 平均累计最大需量
average integrated demand meter 平均累计最大需量计
average interest rate 平均利率
average interest rate on total contracted 平均订约利率
average inventory 平均库存;平均存货
average inventory level 平均库存量;平均存货水平
average investment 平均投资数;平均投资额
average investment criterion 平均投资标准
average isopleth 平均等值线
average isopleth map 平均值等深线图
average issuing price 平均发行价格
average journey speed 平均行程速度
average kilometers of wagons in transit loaded at technical stations per transit 货车中转距离
average kilometre of a locomotive per day 机车平均日车公里
average labor-time expended in 平均耗费的劳动时间
average labo(u)r rate 平均工资率
average lake depth 平均湖深
average lake level 平均湖面
average least dimension 平均最小尺寸(骨料集料的)
average length 平均长度
average length of commuters' journey 平均通勤行程距离;月票乘客平均行程距离
average length of life 平均寿命
average length of paid holidays 平均有酬假日天数
average length of passenger journey 平均乘距
average length of swell 一般长涌
average length of working life 平均工作寿命
average level 平均水平(面);平均水面;平均能级
average liabilities per failure 每个破产公司的平均债务
average life expectancy 平均寿命预期值
average life expectancy of urban population 城市人口平均预期寿命
average life(period)平均使用年限;平均寿命;平均年限;平均满期日;借款平均偿付期限
average life rate 平均生长率
average life span 平均预期寿命
average life time 平均使用期限
average lift(force)平均升举力
average light curve 平均光度曲线
average limit of sea ice 平均海冰界线
average living floor area per capita 人均居住面积;平均每人居住面积
average loaded wagon kilometers in one complete turnaround of wagon 重周转距离
average locomotive drawing gross weight 机车平均牵引总重
average locomotive drawing rolling stocks 机车平均牵引辆数

average locomotive of check and repair per day 检修机车台日数
average locomotive technic speed 机车平均技术速度
average locomotive travel(l)ing speed 机车平均旅行速度
average log 中等原木
average log deviation 平均对数偏差
average longitudinal gradient 平均纵度
average long-term yield 多年平均产量
average loss 平均损失
average loss of energy 平均能量损失
average loss settlement 海损计算方法
average lower limit 平均下限
average lowest discharge 平均最小流量
average lowest water level 平均最低水位;平均最低潮位;平均最低潮面
average low flow 平均枯水流量
average low swell 短轻涌
average low water 平均低水位;平均低潮面
average low water flow 平均低水流量
average low water level 平均低水位;平均低潮面
average luminance 平均亮度
average luminous intensity 平均光强
average macroeconomic performance 宏观经济平均执行情况
average magnetic well 平均磁阱
average magnitude 平均量
average magnitude of earthquake 平均震级
average marked loading capacity per wagon 货车标记平均载重
average mark-on 平均成本加成
average maturity 平均到期日
average maximum demand 平均最大需要量
average maximum wave height 平均最大波高
average mean cushion length 平均气垫长度
average measured value 平均量测值
average melting point 平均熔点
average mileage 平均英里;平均里程
average moderate swell 中中涌
average modulus 平均模量
average modulus of deformation 平均变形模量
average moisture content 平均含水量
average molecular weight 平均分子量
average money wage rate 平均货币工资率
average monthly balance 平均月末余额
average monthly dredging quantity 月平均疏浚量
average monthly inventory 平均月库存
average monthly rainfall 月平均降雨量;月平均雨量
average monthly runoff 月平均径流量
average motion 平均运动
average movement 平均运动
average navigation period 平均通航期
average neap(tide)平均小潮
average net head 平均净水头
average net penetration per blow for the last five blows 打桩最后五击平均每击贯入度
average net profit 平均净利润;平均纯利润
average net weight hauled by locomotive 机车牵引平均净重
average night flow 平均夜流量
average noise 噪声电压平均值
average norm 平均定额
average normal ear 平均标准耳
average normal gravity 平均正常重力
average number 平均数
average number index 平均数指数
average number of cars sorted per hour 一小时平均编解车数
average number of cars unloaded per day 平均一日卸车数
average number of days to turnover 平均周转天数
average number of inhabitants per building 平均居住数;居住密度
average number of job 平均工作数量
average number of passenger carried per train 列车平均载客人数
average number of passengers per carriage 每辆客车平均乘客人数
average number of persons per habitable room 每一居室的平均居民数

average number of persons per household 平均每户人口
average number of wagon loadings and unloadings per unit of work in goods traffic 管内装卸率
average observatory 平均天文台
average occupied amount of current capital 平均占用流动资金额; 平均占用的流动资金
average of averages 平均数的平均
average of expected incomes 平均预期收入
average of high water stages 平均洪水位
average of low water stages 平均枯水位
average of normal water levels 平均常水位
average of operating cycle time 平均作业循环时间
average of relatives 相对平均数
average operating speed 平均运行速度
average operation time 平均作业时间; 平均运算时间; 平均工作时间
average orbit 平均轨道
average orders per second 每秒平均指令数
average ordinary water level 平均水位
average or mean shipped and landed quality terms 装卸平均品质条件
average out 盈亏两抵; 得到平均; 达到平均数
average outage time of equipment 设备平均停运时间
average outgoing quality 平均检出质量; 平均抽检质量; 平均抽查质量; 实际平均质量
average outgoing quality limit 平均质量检查最低限; 平均抽检质量界限
average outlays of material per day 每日材料平均支出数
average outlays per tourist per day 每一旅游者日平均花费
average output 平均效率; 平均生产率; 平均出力; 平均产量; 平均产出
average output per day of rail freight locomotive 货运机车平均日产量
average output power of pulse 脉冲平均输出功率
average outstanding balance of deposits 平均库存余额
average overall efficiency 平均总效率
average overall rate 平均总速率
average overall speed 平均路段速度
average oxidation number 平均氧化数
average paid-in capital 平均实交资本
average paleointensity 平均古地磁场强度
average parameter 平均参数
average particle diameter 平均颗粒直径; 平均颗粒粒径
average passenger change factor 乘客平均换乘系数
average passenger train technical speed 旅客列车技术速度
average passenger travel time 乘客平均乘行时间
average payable 可补偿海损
average penalty 平均罚款
average penetration rate 平均机械钻速
average per capita consumption 平均每人用(水)量; 平均每人消费量
average percentage deviation 平均百分离差
average percentage recovery of core 平均岩芯采取率
average per day 每日平均值
average performance 平均生产率
average period 平周期; 平均周期
average periodicity of stick-slip 平均黏滑周期
average period of production 平均生产期
average permeability 平均渗透率; 平均渗透量; 平均渗入率; 平均渗入量
average permeability coefficient 平均渗透系数
average permeability of oil bed 油层平均渗透率
average person trip rate 人均出行率
average plate height 平均塔板高度
average policy 海损保险单
average pore diameter 平均孔径
average pore size 平均孔径
average porosity 平均孔隙度
average port time of ship 船舶平均在港时间
average position action 摆动作用
average potential energy 平均位能
average potential water power 平均水力蕴藏量
average power 平均功率
average power demand 平均需电量
average power output 平均功率输出
average power spectrum method 平均功率谱法
average-practice coefficient 平均实行系数
average precipitation 平均沉淀; 平均降雨量; 平均降水量
average precipitation intensity 平均降雨强度
average precipitation over area 平均面积降雨量
average preference 平均偏好; 一般优惠
average premium 平均保(险)费; 海损保险费
average pressure 平均压力
average pressure grade line 平均压力梯度线
average pressure gradient mean pressure gradient 平均压力梯度
average price 平均价(格)
average price of production 平均生产价格
average pricing method 平均计价法
average prime cost 平均直接成本
average probability 平均概率
average product 平均产量
average production capacity 平均生产量
average production cost 平均生产成本
average production period 平均生产期
average productive capacity 平均生产能力
average productivity 平均生产率
average profit 平均利润
average profit margin 平均资金利润率
average profit rate 平均利润率
average profit rate on funds 平均资金利润率
average pulse amplitude 脉冲平均幅度
average pump pressure 平均泵压
average quadratic error 均方差
average quality 平均质量; 中等质量
average quality factor 平均质量因素
average quality protection 平均质量保护; 一般质量保护措施
averager 平均器; 海损理算师; 海损理算人; 中和器; 中和剂
average radiant temperature 平均辐射温度
average radius 平均半径
average rain catch (雨量器的)平均载雨量; (雨量器的)平均承雨量
average rainfall 平均雨量; 平均雨量; 平均降水量
average rainfall intensity 平均雨量强度; 平均降雨强度
average rainfall of the drainage area 流域平均雨量
average range 平均范围; 平均差(值); 平均变幅
average rank 平均等级
average rate 平均速率; 平均速度; 平均(保险)费率; 时价率
average rate of convergence 平均收敛速度
average rate of decrease 平均降低率
average rate of dividend 平均股息率
average rate of growth 平均增长率
average rate of increase 平均增长率; 平均递增率
average rate of irrigation 平均灌溉率
average rate of profit 平均利润率
average rate of profit on investment 平均资金利润率
average rate of progressive increase 平均递增率
average rate of sediment discharge 平均输沙率
average rate of tax 平均税率
average rating factor 平均工作等级因素
average ratio 平均比值
average reading 平均读数
average reading meter 平均值读数表
average receipts per person-night 人均每夜住宿收费
average recession curve 平均退水曲线
average recovery rate 平均回收率
average recovery velocity of water level 平均水位恢复速度
average recurrence interval 平均重现间距
average reflectance 平均反射系数; 平均反射率
average regional temperature 区域平均温度
average region anomaly 平均区域异常
average regressed value 平均回归值
average relative error 平均相对偏差
average relative price 平均相对价格
average remaining durable years 平均剩余耐用年限(建筑物)
average repair 海损修理
average repair time of locomotive 机车平均修车时间
average reservoir pressure 平均库容压力
average residence time 平均滞留时间; 平均滞留期
average response rate 平均回答率
average response spectrum curve 平均地震反应谱曲线
average return 平均收入
average return period 平均重现期
average revenue 平均收益; 平均收入
average revenue curve 平均收入曲线
average revenue product 平均收入产品
average revenue rate method 平均收入率法
average revenue rate of freight traffic 货运平均收入率
average revenue rate of passenger traffic 客运平均收入率
average revolutions 平均转数
average riding distance 平均乘距
average rise 平均潮位升高
average rise of tide 平均潮升
average risks 平均风险; 一般风险
average river flow 平均河流流量; 平均河川流量
average river level 平均河面高程
average road carrying intensity 线路平均负荷度
average road time of a freight train 货物列车平均在沿线时间
average room condition 平均室内条件
average root diameter 根平均直径
average rotary speed 平均转速
average roundness 平均圆度
average running time 平均行驶时间
average runoff 平均径流(量)
average runoff depth 平均径流深(度)
average runoff rate 平均径流率
average salary pension scheme 平均工资养老金方案
average sale expectancy 平均预期销售额
average salinity of seawater 海水平均盐度
average sample 平均样品; 平均样本; 平均试样
average sample number 平均值抽样数; 平均样品数; 平均样本数; 平均取样数; 平均检查量; 平均抽样(数)量; 平均抽查数
average sample number curve 平均抽样数量曲线
average sample size 平均抽样检验个数
average sample size in sequential tests 序贯检验的平均样本容量
average sampling 平均取样; 进口货指定检查
average sand content 平均含砂量
average saturation degree 平均饱和度
average saturation flow 平均饱和流量
average scale 平均比例尺
average sea depth 平均海深
average sea level 平均海平面高程
average sea level pressure 平均海平面气压
average search time 平均查找时间
average seasonal river 平均季节性河流
average seasonal runoff 平均季节性径流
average sediment concentration 平均泥沙浓度; 平均含沙量(指油、水、空气中含沙量)
average sediment particles 一般泥沙颗粒
average seeking time 平均查找时间
average service able life 平均使用年限
average serviceable years 平均使用年限
average service braking deceleration 常用平均制动减速度
average service conditions 一般使用条件; 正常使用条件
average service rate 平均使用率; 平均服务率
average set of final blows in pile driving 最后几击平均贯入量
average settlement 平均沉降
average settlement velocity (泥沙的)平均沉降速度
average sewage 一般污水; 中等程度污水
average sewage rate per day 平均日污水量
average shading 平均明暗度
average shear modulus 平均剪切模量
average shear strain 平均剪应变
average shear stress 平均剪应力
average shift 平均移位; 平均位移; 平均偏移; 平均变化; 平均变幅
average shift value of step 台阶的平均漂移值
average shipping weight 平均装载重量
average shortfall 平均欠额
average shunting speed in tranferring the train stock 车列转线平均速度

average side length of chains 锁网平均边长
average site condition 平均场地条件
average size 平均粒径;平均规模;平均尺度;每户平均人数
average size of aggregate 骨料平均粒径
average size of sand 砂平均粒径
average sky 平均天空
average slip direction 平均滑动方向
average slope 平均坡降;平均坡度;平均比降
average slope of the travel time curve 时距曲线的平均斜率
average smoothed spectrum 平均平滑谱
average snowline height 平均雪线高度
average soil loss 平均土壤流失量
average sorting interval per train 每列车平均分解间隔时间
average sorting time per train 每列车平均分解时间
average sound absorption coefficient 平均吸声系数
average sound absorptivity 平均吸声率
average sounding velocity 平均声速
average sound level 平均声级
average sound pressure 平均声压
average sound pressure level 平均声压级
average space headway 平均车头间距
average spacing of cleavage domain 劈理域平均间隔
average specific budget 平均比收支
average specific heat 平均比热
average specific train resistance 列车平均单位阻力
average spectrum 平均谱;平均频谱
average speed 平均速率;平均速度;平均航速
average speed between stops 在两个停车站间的平均速度
average speed difference 平均速度差
average speed of development 平均发展速度
average speed of growth 平均增长速度
average speed of train in departing from station 列车出站平均速度
average speed of train in entering into station 列车进站平均速度
average speed of train set leaded 车列牵出平均速度
average sphere level 平均地表面
average spheric(al) candle power 平均球面烛光
average sphericity 平均球度
average stage 平均水位;一般阶段
average stage in reservoir region 库区平均水位
average standard 平均定额
average standard of living 平均生活标准
average starting acceleration 平均起动加速度
average statement 海损理算书
average stater 海损理算师;海损理算人
average static load of wagon 货车平均静载重
average station efficiency 电站平均效率
average station interval 平均站间距
average stock 平均库存;平均存货
average stopped-time delay 平均停车延误
average storage time 平均堆存期
average storm sediment discharge 平均暴雨输沙率
average story number 平均层数
average strain 平均应变
average stray of concrete quality 混凝土质量平均偏差
average stream flow 平均河流流量;平均河川流量
average stream velocity 平均流速
average strength 平均强度
average stress 平均应力
average stripping ratio 平均采剥比
average subsidence rate 平均沉降速率
average supply 平均供水量
average surface diameter 平均表面直径
average surface temperature 平均地表温度
average taker 海损理算师;海损理算人
average tank method 平均秩法
average tare 平均皮重;平均包装重量
average taxation 平均课税
average tax rate 平均税率
average technical and vocational level 平均技术熟练水平
average technical speed 平均技术速度
average technical speed of passenger trains 客列车平均技术速度
average temperature 平均温度
average temperature value 平均温度值
average tempering 中温回火
average temporal velocity gradient 平均瞬时速度梯度
average tensile strength 平均拉力强度;平均抗拉强度
average terrestrial pole 平均地极
average test 平均测试
average test-car run 平均试验车行车
average thickness 平均厚度
average thickness of cumulative snow 平均积雪厚度
average thickness of ore block 块段平均厚度
average thickness of orebody 矿体平均厚度
average thickness of the layer 平均层厚
average threshold shear stress 平均临界剪应力
average throat width 毛细管半径
average tidal curve 平均潮位曲线
average tidal flow 平均潮流
average tide level 平均半潮水位
average time 平均时间
average time between maintenance 平均维修时间(间隔);平均维护时间(间隔);平均保养时间;维修平均间隔时间
average time for repair of breakdowns 平均故障修复时间
average time headway 平均车头时距
average time in transit per ton of freight 每吨货物平均运送时间
average time of detention per goods operation 一次货物作业平均停留时间;中转车平均停留时间
average time of turnover of locomotive 机车平均全周转时间
average time of turnround per wagon 货车平均周转时间
average time per complete turn-round of locomotive 机车平均全周转时间
average time to repair 平均维修时间
average time used for discharging a wagon 货车一次作业平均在港停留时间
average time used for loading a wagon 货车一次作业平均在港停留时间
average time used for loading (discharging) a train 火车一次作业平均在港停留时间
average-to-good 中等以上的;中上等的
average tonnage 平均吨位
average tonnage capacity of cars 货车平均载重
average topsoil loss 平均表土流失量
average total amount of ozone 臭氧平均总量
average total cost 平均总成本
average total inspection 平均检验数
average total ozone 臭氧平均总量
average tow 一般驳船队
average traffic density 平均交通密度
average traffic (volume) 平均交通(流)量
average train mile cost 列车英里平均成本
average transfer coefficient of line network 线网平均换乘系数
average transfer rate 平均输送率;平均传输率;平均传导率
average transit speed of passenger trains 旅客列车平均旅行速度
average transmission rate 平均输送率;平均传输率;平均传导率
average transmitted light intensity 平均透光强度
average transportation distance of goods 货物平均运程
average transporting passengers of coach 客车平均载运人数
average transport revenue payable per day 平均每日应缴运输进款
average travel(l)ing speed of freight 货运的平均旅行速度
average travel speed 平均行程速度
average travel time 平均行程时间
average trip 平均行程
average trip bit speed 平均行程钻速
average trip length 平均行程里程
average trip mileage 平均里程
average trip numbers 平均出行次数
average trip time by public traffic 公交平均出行时间
average trouble time of vehicle mechanism 车辆机件平均故障时间
average turnover time fo wagon 货车平均周转时间
average turnround time of loaded wagons to be delivered 移交车周转时间
average turnround time of local wagons to be unloaded 管内工作车周转时间
average unit cost 平均单位成本;平均单价
average unit hydrograph 平均单位(水文)过程线
average unit price 平均单价
average unit stress 平均单位应力
average up 以高于平均价格卖出
average upper limit 平均上限
average uranium content 平均铀含量
average useful life 平均有用寿命;平均有效寿命;平均有效使用年限
average value 平均值;平均数
average value indicator 平均值指示器
average value memory oscilloscope 均值记忆示波器
average value method 平均值法
average value of a function 函数的平均值
average value of a probability distribute 一种概率分布的均值
average value of a probability distribution 概率分布均值
average value of log 测井曲线平均值
average value of organic carbon 有机碳平均值
average value of subsidence 平均沉降量
average value of zero shift 零漂平均值
average value over years 多年平均值
average value process 平均值法
average value sample 平均值抽样
average value sampling 平均值抽样法
average value theorem 平均值定理
average variability 平均变异性;平均变化率
average variable cost 平均可变成本
average variable cost curve 平均可变成本曲线
average velocity 平均速度;平均流速
average velocity curve 平均流速曲线
average velocity in cross-section 断面平均流速;断面平均风速
average velocity of nature water level drawdown 天然水位的平均降速
average velocity of pore water 平均孔隙水速度
average velocity point 平均流速点
average velocity response spectrum 平均速度反应谱
average vertical pressure 平均垂直压力
average viscosity 平均黏度
average voltage 平均电压
average volume 平均体积
average volume of grain 平均颗粒体积
average volume weight 平均体重
average wage 平均工资
average wage earnings 平均工资收入
average wage index 平均工资指数
average wage plan 平均工资计划
average wage rate 平均工资率
average wage scale coefficient 平均工资等级系数
average wagon kilometers in one complete turnround of wagon 货车全周转距离
average wait per approaching 每进口道平均等待时间
average wastewater flow 平均污水流量
average water assumption per day 平均每日用水量
average water consumption 平均耗水量
average water consumption per capita per day 平均每人每日用水量
average water head 平均水头
average water(level) 平均水位
average watershed rainfall 流域平均雨量
average water velocity 水中平均波速
average water year 平水年
average wave 平均波
average wave height 平均波高
average wave period 平均波(浪)周期
average wear rate 平均磨损率
average weekly wage 平均周工资
average weight 平均重(量)
average-weighted 加权平均的
average-weighted index 加权平均指数

average weight statistics method 平均重量统计法
average white noise spectrum 平均白噪声谱
average width of fracture 裂缝平均宽度
average width of pore throat 喉道平均宽度
average width of watershed 流域平均宽度
average wind speed 平均风速
average wind velocity 平均风速
average winter level 冬季平均水位
average work load 平均工作负载
average workweek 平均工作周
average yaw 平均偏航角
average year 平年;平均年;一般年分
average yearly concentration of total phosphorus 年平均总磷浓度
average yearly flow 年平均流量
average yearly loading capacity 年平均负荷(容)量
average yearly rainfall 平均年降雨量;平均年降水量
average yield 平均收益(率);平均收率;平均生息;平均产量
average yield of single well 单井平均出水量
average zone velocity 平均区域速度
averaging 平均的;取平均值;求平均值;求平均数;分摊
averaging circuit 平均电路
averaging clause 海损条例;海损条款
averaging device 平均仪;平均器
averaging filter 平均滤波器
averaging method 求平均值法;均值法
averaging of accounts 计算账户平均余额
averaging of multiple image 多图像平均(法)
averaging of stock 库存量平均计算
averaging operator 求平均数算子
averaging out 平衡;更正;抵消;整平
averaging over a group 分组求平均值
averaging regulation 海损条例
averaging stock 平均库存
averaging taxation 平均课税
averaging time 平均次数
averaging unit 平均单元
averaging up 平均上涨
averaging wind 平均风
averge cost pricing 平均成本定价法
avermectin wastewater 除虫菌素废水
aversion 移转
aversion response 保护性反应
aversion to change 变化的恶感
averted angle of photographic(al) axis 主光轴偏角
averted hemi-sphere 背面半球
avertence 偏转
avertence of camera 摄影机偏转
avert the stall 防止失速
a very coarse sand layer 极粗砂层
Avery cupping test 艾弗里拉延试验
a very fine sand layer 极细砂层
a very old science 一门古老的科学
Avesta 阿维斯塔碳钢
avgas 航空汽油
Aviar 物镜类型
Aviar lens 阿维阿尔镜头
aviary 鸟舍;养禽所;飞禽饲料所;鸟屋
aviation 航空(学)
aviation accident 航空事故
aviation alkylate 航空烃化汽油
aviation astronomy 航空天文学
aviation barometer 航空气压表
aviation beacon 航空灯(标)
aviation channel 航空通信频道
aviation chart 航空图
aviation climatology 航空气候学
aviation clock 航空时钟
aviation decompression disease 航空减压病
aviation diesel oil 航空柴油
aviation dynamotor 航空发电机
aviation field 飞机场
aviation forecast 航空预报
aviation forecast zone 航空预报带
aviation fuel 飞行用燃料;航空燃料
aviation fuel fraction 航空燃料馏分
aviation fuel installation 航空燃料装置
aviation gas(oline) 航空汽油
aviation gas turbine 航空燃汽轮机
aviation gas turbine engine 航空燃汽轮机发动机
aviation grease 航空润滑脂

aviation industry 航空工业
aviation insurance 航空保险
aviation kerosene 航空煤油
aviation landing lamp 飞机着陆灯
aviation lifesaving 航空救生
aviation lubricating oil 航空润滑油
aviation lubrication grease 航空润滑脂
aviation map 航空图
aviation meteorology 航空气象学
aviation method 航空辛烷值测定法
aviation mineral exploration 航空探矿
aviation mix 航空汽油抗爆液
aviation obstruction light 航空障碍灯标;航空障碍标志灯
aviation oxygen supply system 航空供氧系统
aviation petrol 航空汽油
aviation radio 航空无线电台
aviation recorder 航空记录仪
aviation safety information 航空安全信息
aviation school 航空学校
aviation sickness 航空晕;航空病
aviation spectacle glass 航空风镜玻璃
aviation spirit 航空汽油
aviation sprayer 航空喷雾器
aviation starter 航空起动机
aviation toxicology 航空毒理学
aviation turbine 航空涡轮机
aviation weather forecast 航空天气预报
aviation weather observation 航空天气观测
aviator 飞行员
avicennite 褐铊矿
Avicron 阿维克纶黏胶丝束
avidity 亲和力强度
aviette 小型滑翔机;小型飞机(体育运动用)
avigation 空中领航;空中导航;航空术
avigation easement 航空通行权
avigator 领航员
avigraph 速度三角形机械计算器;航行计算仪;导航仪
Avila 阿维拉黏胶草杆纤维
Aviogon lens 阿维冈镜头
aviolite 堇云角岩
Avional alloy 阿维纳尔(硬质铝)合金
avionic 航空电子学的
avionic device 航空电子设备
avionic goggles 航空电子眼镜
avionics 航空电子学;航空电子设备;航空电子技术
Avior 海石一
aviotronics 航空电子学
avirulent 无病毒的
aviso 通信舰;侦察通信船
aviso boat 通信船
Avitene 阿维烯
Avlin polyester fibre 阿芙林聚醋纤维
avodire 乳油色硬木(产于加纳);白桃花心木
avogadrite 氟硼钾石
Avogadro's hypothesis 阿弗伽德罗假说
Avogadro's law 阿弗伽德罗定律
Avogadro's number 阿弗伽德罗数
avogram 阿伏克
avoid 废止;避免
avoidable cost 可避免成本
avoidable downtime 可避免的时间损失
avoidable error 可避免的误差
avoid agreement 废止协议
avoidance 避免
avoidance behavio(u)r 回避行动
avoidance of contract 宣告合同无效
avoidance of cracking 防止裂缝;防止开裂;避免破裂
avoidance of danger 避免危险
avoidance of dangerous region 危险区的回避
avoidance of delay 防止延迟
avoidance of double taxation 避免双重征税
avoidance of policy 保险失效
avoidance of repetition 避免重复
avoidance of tax 避税
avoidance reaction 回避反应
avoidance reaction test 回避反应试验
avoidance rendered by default 在一方缺席时作出的仲裁裁决
avoidance test 免毒试验;洄游试验;回避试验;回避实验
avoidance upon settlement 根据和解作出的仲裁裁决
avoid clauses 废止条款
avoid contract 废止合同
avoid creditors 避债
avoid damage 防止损坏
avoid delay 避免延误
avoiding angle 规避角
avoiding line 迂回线
avoiding reaction 回避反应
avoid obsolescence 防止过时
avoid regulations 废止条例
avoid rules 废止法规
avoirdupois 常衡
avoirdupois ounce 常衡盎司
avoirdupois pound 常衡制磅
avoirdupois system 常衡制
avoirdupois weight 英国常衡制;常衡制
avometer 万用电表;万用表;万能电表;三用电表;安伏欧计;安伏欧电表
avon 河流
Avonian 阿翁阶【地】
avulsio 土地转位
avulsion 撕脱;撕裂;河道裁弯;改道;抽出术;冲裂(作用);冲浪;冲决
avulsion cut-off 冲裂割断
avulsion of water course 水路改道
avulsive cutoff 自然穿透取直;自然穿孔取直
await 期待
awaiting parts 维修用备件;备件
awaiting repair 待修
awaiting repair time 等待修理时间;等待修复时间
awaiting transit 待运
awaiting transport 待运
await order 待命
award 判给;判标;授标;裁决书;裁定额;发包;认可书
award a damage 同意赔偿破损
award and signing of contract 发包订约
award at tender opening 现场决标
award containing reasons 说明理由的仲裁裁决
award contract 签订合同
award criterion 授标准则
award decision 授标决定
award fee 奖励费用
award for invention 发明奖
award-giving meeting 颁奖大会
award in final 最后裁决;终结裁决
awarding agency 合同主方或其代理人
awarding contract 签定合同;签订合同;授予合同
award mandate 决标委任
award meeting 决标会议
award notification 授标通知
award of bid 投标裁定;决标
award of contract 授标;签定合同;签订合同;决标;中标;投标
award of the contract 授予合同
award rendered by default 缺席仲裁;缺席情况下作出的仲裁裁决
award-winning 获奖的
awareness building 建立意识
awarf cherry 灌木樱
awaruite 铁镍矿
awash 齐水面的;适淹;适被淹没的;与水面齐平的;被水漫过的
awash at low-water 低潮时适淹
awash rock 浪刷岩
a water bath system 一种淹水装置
awave 在波浪上;波浪式的
away the collected water 疏通积水
a weak link 薄弱环节
aweather 向上风;向风;迎风
a week postponement 延期一周
awheft 卷系旗;小旗;小长旗
a wild scheme 轻率的计划
A-wire 正线
awja 藻类遗骸泥
awkward 有毛病的;难办的;笨拙的;粗劣的
awkward bend 危险弯段
awkward cargo 笨重货物
awl 钻子;尖锥;木工锥子;锥子
awl haft 钻子柄把(木工、皮革工用);尖锥
awn chaff 芒屑
awn cutter 除芒器
awner 除芒器

awner agitator 除芒器刀辊
awning 篷(养护混凝土用);天篷;天幕;天盖;遮阳篷;遮篷;罩棚;船缝;帆布篷
awning blind 篷式遮帘;篷式百叶窗;遮阳百叶窗
awning boom 天篷桁;天幕桁
awning brace 天幕撑架
awning curtain 天篷侧檐;帆布天幕;帆布幔
awning deck 天篷甲板;天幕甲板;上层轻甲板;小艇甲板;遮阻甲板
awning deck vessel 天篷甲板船;遮阳甲板船
awning fabric 遮篷用织物
awning jackstay 天幕边索
awning light 篷式天窗
awning rafter 天篷脊梁;天幕纵木
awning ridge 天幕纵木
awning sash 上旋窗;铰接式窗扇
awning sash window 篷式天窗
awning side stops 张幕索
awning spar 天幕横木
awning stanchion 天篷支柱;天幕柱
awning stops 天幕捆绳
awning stretcher 天幕缘材
awning type berth 开敞船台
awning type window 篷式天窗
awning window 上旋窗;篷式窗;遮阳窗;翻窗组
awn length 芒长
axe 石工斧;斧锤;斧槌;削减;斧头
axe blade 斧刃;斧刀刃
axed arch 斧琢砖拱;斧斩拱面;斧砍砖拱;斧拱
axed artificial stone 斩假石;剁假石
axed brick 琢石面;斧琢砖;斧斩砖;斧砍砖
axed face 斧斩面
axed work 凿石;工斩;琢石
axe for coaches 客车救险用斧;客车(救险)斧
axe hammer 剁斧槌;斧锤;斧槌
axe handle 斧柄
axe head 斧头
axeman 斧工
axenic culture 无菌培养
axes 轴心
axes of abscissa 横坐标轴
axes of an aircraft 飞机坐标轴
axes of major stress 最大主应力轴
axes of mean stress 中间主应力轴
axes of minor stress 最小主应力轴
axes of ordinate 纵坐标轴
axe stone 钺石;硬玉
axhammer 石工斧;石工锤;斧锤;斧槌
axial 轴向(式)的;轴的;中轴的
axial aberration 轴向像差
axial acceleration 轴向加速度
axial action 轴向动作
axial adjustment 轴调整
axial admission 轴向进气;轴向供给
axial air 轴流风
axial air section 轴向风区
axial ampere turn 轴向安匝
axial angle 轴向角;轴角;光轴角
axial armature 轴向电枢
axial back pressure turbine 轴流式背压汽轮机
axial bearing 轴向轴承;止推轴承
axial blower 轴流式风机;轴流
axial bond 轴向键
axial box cover 轴箱盖
axial box seating 轴箱座
axial bracket 船尾轴架
axial bracket arm 正心拱;泥道拱
axial brass 铜制轴衬
axial bundle 轴光速
axial cable 中轴索;中心索
axial cam 凸轮轴;圆柱凸轮;轴向凸轮
axial carrying capacity 轴向承载(能)力
axial cell 中轴压力盒(三轴压力仪用的)
axial channel 深泓线;河槽线
axial character 轴性
axial chromatic aberration 轴向色差
axial clearance 纵向间隙;纵间隙;轴向空隙;轴向间隙
axial clutch 轴向离合器
axial column 轴向柱;复中柱
axial component 轴向分量;轴向分力;轴向部分
axial compression 轴心受压;轴向压力;轴向压缩;轴向受压;轴挤压
axial compression ratio 轴压比
axial compressive strength 轴心抗压强度
axial compressive force 轴向压力
axial compressive stress 轴向压缩应力
axial compressor 轴向式压缩机
axial condenser 轴向布置凝汽器
axial cone 轴锥
axial consolidation 轴向固结
axial constraint 轴向约束
axial contact ball bearing 推力球轴承
axial contact bearing 推力轴承
axial cord 轴线
axial cross 轴交叉
axial cross-section 轴向剖面图
axial crystal 轴晶体
axial culmination 轴褶升区
axial curvature 轴向曲度;轴线曲度
axial curve 轴曲线
axial cyclone 卧式旋风筒;轴式旋流器
axial cylinder swashplate pump 轴向圆盘隔板泵
axial defocusing 轴向离焦
axial deformation 纵向变形;轴向形变;轴向变形
axial depositing 轴向堆料
axial deposition 轴向沉积
axial depression 轴陷
axial deviation 轴向偏移
axial diffusion 轴向扩散
axial diffusion casing 轴向扩散式外壳
axial dipole-dipole array 轴向偶极装置
axial dipole vector 轴偶极矢量
axial direction 轴向
axial discharge hopper wagon 轴向自卸漏斗车
axial dispersion 轴向渗透;轴向扩散;光轴色散
axial displacement 轴向位移
axial distribution 轴向分布
axial eddy 轴向漩涡;轴向涡流
axial electric(al) oil pump 轴向式电动油泵
axial element 晶体常数;结晶常数;轴向单元
axial elongation 轴向伸长
axial engine 轴向式发动机
axial entry impeller 轴向进口叶轮
axial error 轴向误差
axial exducer 轴向导流器
axial exhaust 轴向排气
axial expansion 轴向膨胀
axial extension test 轴向拉伸试验
axial fabrics 轴组构
axial fan 轴流式风扇;轴流式(通)风机;轴流风机
axial feather 轴羽
axial feed 轴向进刀
axial feed method 轴向进刀法
axial field 轴向场
axial figure 轴像
axial fixity 轴向约束;轴向嵌固;轴向固定
axial flame 轴流火焰
axial float 轴向窜动
axial flow 轴向流(动);轴流;轴对称流(动)
axial-flow action turbine 轴向冲击式水轮机
axial-flow aeration machine 轴流式通风机
axial-flow air compressor 轴流式空气压缩机
axial-flow blow 轴流通风机
axial-flow blower 轴流鼓风机;轴流式(通)风机;轴流风机;轴流式增压器;轴流式压缩机;轴向式压缩机
axial-flow compressor 轴流(式)压缩机;轴流式压气机;轴流式压缩机;轴流式压缩机
axial-flow exhaust fan 轴流式排气扇
axial-flow fan 轴流风扇;轴流送风机;轴流式风扇;轴流式(通)风机
axial-flow fan rotor 轴流式风扇转子
axial-flow gas separator 轴流式气体分离器
axial-flow gas turbine 轴流式燃气透平
axial-flow high-pressure fan 轴流式高压风扇
axial-flow hydroelectric(al) unit 轴流式水力发电机组
axial-flow impeller 轴流式叶轮;轴流风机叶轮
axial-flow impulse turbine 轴流式冲击汽轮机;轴流冲击式涡轮机
axial-flow jet engine 轴流式喷气发动机
axial-flow oil pump 轴流式油泵
axial-flow propeller 轴流螺旋浆
axial-flow pump 轴流式泵;轴流泵
axial-flow reaction turbine 轴流反击式透平
axial-flow reversing gas turbine 轴流式反转燃气轮机
axial-flow steam turbine 轴流式汽轮机
axial-flow turbine 轴流式汽轮机;轴流式涡轮(机);轴流式透平;轴流式水轮机
axial-flow turbo compressor 轴流式涡轮压缩机
axial-flow turbomachine 轴流式透平机
axial-flow type 轴流式
axial-flow type air compressor 轴流式空气压缩机
axial-flow type wheel 轴流冲击式水轮机
axial-flow vane 轴流式风机叶片
axial-flow ventilating fan 轴流式(通)风机
axial-flow ventilator 轴流式(通)风机
axial-flow wheel 轴流式水轮
axial flux 轴向通量
axial flux motor 轴向磁通电动机
axial force 轴(向)力;轴线力
axial force diagram 轴力图;轴向力图
axial force of rock bolt 锚杆轴力
axial force wave 轴力波
axial fracture 轴断裂面
axial freedom 轴向自由度
axial gasket 轴向密封垫
axial gas turbine 轴向燃气轮机
axial geocentric dipole field 轴向地心偶极子场
axial girder 轴梁
axial glide plane 轴向滑移面
axial gradient 轴级度
axial grease 轴用滑脂
axial grinding force 轴向磨削力
axial heat conduction 轴向传热
axial hydraulic thrust 液力轴向推力
axial image point 轴向像点
axial-inlet cyclone 轴向进口气旋分离器
axiality 同轴度
axial jet 轴向射流
axial jet velocity 轴向喷流速度
axial joint 轴向缝;轴节理
axial knock 撞击震颤声
axial labyrinth seal 轴向曲路密封
axial lead 轴心线;轴向引线
axial length 轴长
axial limit deposit velocity 轴向极限沉淀速度
axial line 轴线
axial line of fold 褶皱轴线
axial load carrying core 轴向负荷型芯
axial loaded column 轴向受力柱;轴向负荷柱
axial loaded member 纵载构件
axial load(ing) 轴向荷载;轴向加载;轴心荷载;轴向负载;轴向负荷
axial loading A-frame 轴向荷载A形井架
axial loading test of pile 桩轴向荷载试验
axial load test 轴向荷载试验
axially directed velocity 轴向速度
axially expansive seal 轴向膨胀密封
axially increasing pitch 轴向递增螺距
axially load 有轴向负载的
axially loaded column 轴向受力柱
axially loaded hollow cylinder 轴向受力空心圆柱体
axially symmetric(al) accelerator 轴对称加速器
axially symmetric(al) consolidation 轴对称固结
axially symmetric(al) deformation 轴对称变形
axially symmetric(al) flow 轴对称流动;轴对称水流;轴对称流(动)
axially symmetric(al) gravimeter 轴对称式重力仪
axially symmetric(al) jet 轴向对称射流
axially symmetric(al) load 轴向对称荷载
axially symmetric(al) motion 轴向对称运动
axially symmetric(al) radiator 轴对称辐射体
axially symmetric(al) shape 轴向对称型
axially symmetric(al) shell 轴向对称薄壳
axially symmetric(al) stress distribution 轴对称应力分布
axially varying pitch 轴向变螺距
axial magnet 磁棒
axial magnification 轴向放大率
axial mixing 轴向混合
axial mode 轴向波型
axial modulus 轴向模量
axial moment of inertia 极惯性矩;轴向惯性矩
axial motion 轴向运动
axial mount 轴向安装
axial movement 轴向运动;轴向移动;轴向窜动
axial movement of rotary kiln 回转窑轴向移动;回转窑轴向窜动
axial normal stress 轴向正应力

axial notch 轴向凹槽
axial object point 轴向物点
axial opening region of anticline 背斜轴部张开地段
axial organ 轴器
axial oscillation 轴向摆动
axial parallel beam 轴向平行光速
axial pencil 轴向光束
axial piston engine 轴向活塞机器;轴向活塞引擎
axial piston hydraulic equipment 轴向式活塞液压设备
axial piston hydraulic system 轴向式活塞液压系统
axial piston machine 轴向活塞机
axial piston motor 轴向柱塞马达;轴向活塞马达
axial piston pump 轴向活塞泵
axial piston transmission 轴向柱塞式传动装置
axial piston unit 轴向活塞装置
axial pitch 轴向螺距;轴向节距
axial plan 轴向布置;沿轴线布置
axial plane 轴平面;轴面体;轴面
axial plane cleavage 轴面劈理
axial plane foliation 轴面叶理
axial plane of fold 褶皱轴面
axial plane of symmetry 对称轴面
axial plane separation 轴面间距
axial play 轴向游隙;轴向游动;轴向间隙
axial plunger motor 轴向柱塞马达
axial plunger pump 轴向柱塞泵
axial point 轴点
axial polarization ratio 轴向极化比
axial pore 轴向孔隙
axial porosity 轴向孔隙率
axial position indicator 轴向位移指示器
axial power 轴功率
axial power of fan 风机轴功率
axial preload 轴向预加负荷
axial pressure 轴向压力
axial prestressing 轴向预张应力;轴向预加拉力
axial pretensioning 轴向预张力;轴向预加拉力
axial principal stress 轴向主应力
axial profile 轴向齿廓
axial profile angle 轴向齿形角
axial projection 轴投影
axial pump 轴流泵
axial quadrupole 纵向四极
axial rake (轴的)偏位角;轴向前角
axial rake angle 轴向前角
axial rate 轴向速率;轴率
axial ratio 极化轴比;轴率;轴比;长短轴比
axial-ratio of finite strain ellipse 有限应变椭圆轴比
axial ray 近轴射线;近轴光线
axial region 轴面
axial relief 离角
axial relief angle 轴向后角
axial resolution 纵向分辨(率)
axial response 轴向响应度;轴响应
axial restraint 轴向约束;轴向嵌固;轴向固定
axial rib 轴向肋;脊肋
axial rift 轴向裂谷;轴向断裂
axial rigidity 轴向刚度
axial road 辐射式道路;辐射道路;放射状道路;轴向道路
axial rotation 绕轴旋转;自转;轴向转动
axial run-out 轴向摆差;轴向摆动
axial seal 轴向密封
axial seal energized by internal pressure 轴向自紧密封
axial secondary 轴向次级
axial section 轴向剖面;轴向截面;轴向断面;轴断面
axial self-seal 轴向自紧密封
axial sensitivity 轴向灵敏度;正向灵敏度
axial slip 轴向滑脱
axial source level 轴向声源级
axial spotting 轴向找正
axial steel 轴钢
axial stockpiling 纵向堆划;轴向堆料
axial strain 轴应变;轴向应变
axial strained-body theory 轴向变形体理论
axial strand 轴束
axial stream 山谷主要水道;山谷主要河流;轴线水道
axial strength 轴向强度
axial stress 轴向应力
axial stretching 轴向延伸;轴向伸展;轴向伸长
axial subspace 轴向运动子空间

axial surface 轴面体;轴面
axial-surface fold 轴面褶皱
axial symmetric(al) diffusion 轴对称扩散
axial symmetry 轴向对称性;轴向对称;轴对称
axial system 轴向系统
axial-tag terminal 轴端
axial tectonic belt 轴向构造带
axial tensile force 轴心拉力;轴向张力;轴向拉力
axial tensile strength 轴向张拉强度;轴向抗拉强度
axial tensile stress 轴向拉应力
axial tension 轴心受拉;轴向张力;轴向拉力;中心拉力
axial tensioning 轴向张拉;轴心预加拉力
axial theory of growth 轴向发展理论
axial thickness 轴向厚度
axial thrust 轴向推力;轴推力
axial thrust balancing 轴向推力平衡
axial thrust balancing apparatus 轴推力平衡装置
axial thrust bearing 轴向止推轴承;轴向推力轴承;止推轴承
axial tilt 轴面刃倾角
axial tooth thikness 轴向齿厚
axial torque 旋转力矩;轴向转(力)矩;轴向扭矩
axial trace 轴线
axial translation 轴向位移;轴向平移
axial trough 轴槽
axial turbine 轴流式涡轮(机);轴流式透平;轴流式水轮机
axial turbo-blower 轴流式涡轮增压器
axial turbomachinery 轴流式透平机械
axial type 轴向式
axial vector 轴向向量;轴向矢量;轴量;轴矢量
axial vector coupling 轴矢量耦合
axial velocity 轴向速度
axial velocity ratio 轴向速度比
axial vibration 轴向振动
axial vortex 轴向漩涡;轴向涡流
axial wall 轴向的壁;轴壁
axial wave 轴面波
axial whirl 轴向涡流
axial winding 轴向绕组;轴向缠绕
axial wire bead 绝缘垫珠
axial withdrawal 轴向抽出
axial wobble 轴向摆动
axial zoning 轴向分带
axiation 体轴形成
axiator 轴向量
axi-compressor 轴流式压气机
axicon 轴棱镜;展像镜
axicon lens 锥镜;旋转三棱镜
axile placentation 中轴胎座式
axillary root 侧生根
axin 虫漆脂
axinite 斧石
axinitization 斧石化(作用)
axiolite 椭球粒;放射状椭球粒
axiolitic 椭圆状的
axiom 公理【数】
axiom as guide for consistency 当作相容性指南的公理
axiomatic 公理的
axiomatic approach 公理探索;公理方法
axiomatic devel-opment of Boolean algebra 布尔代数的公理化发展
axiomatic quantum theory 公理化量子理论
axiomatics 公理学
axiomatic semantics 公理语义学
axiomatic S matrix theory 公理化S矩阵理论
axiomatic theory 公理论
axiomatization 公理化
axiomatize 公理化
axiom based on utility difference 以效用差为基础的公理
axiometric 轴法测;投影图法;三向图(的)
axiom for dependency 依赖性公理
axiom for probability 概率的公理
axiom for social welfare function 社会福利功能的公理
axiom of aggregation 紧合公理;集合的公理
axiom of alignment 关联公理
axiom of connection 关联公理
axiom of continuity 连续性公理
axiom of convexity 凸性公理
axiom of perspective rotation 透视旋转定律

axiom of selection 选择公理
axiom of separation 分离公理
axiom of specification 分类公理
axiom of superposition 叠加公里
axiom of symmetry 对称公理
axiom schema 公理格式
axiom system 公理系统
axiopetal 向轴的
axiotron 磁控管
axipetal 向轴的
axiradial compressor 轴流式压缩机;轴流式压气机
axis 枢椎;机体轴;轴心;中轴;中心线
axis angle 轴线角
axis based error 轴线误差
axis bearing 轴向负载;轴承
axis congruence 轴汇
axis cylinder 轴突
axis dead in line 轴线重合
axis deviation 轴心偏斜
axis direction 轴线方向
axis distance of lithologic(al) flow 岩性流轴距
axis error 轴向误差
axis law 轴律
axis lentis 晶状体轴
axis line 通轴线;列柱中心线;两穿堂或门厅中心线
axis malalignment 轴线不对准
axis of abscissa 横坐标轴
axis of a cone 锥轴
axis of a conic 二次曲线的轴
axis of affinity 亲合力轴
axis of angular momentum 角动量轴;动量矩轴
axis of an orbit 轨道轴
axis of a quadric 二次曲面的轴
axis of arch 穹起轴;背斜轴
axis of a weld 焊缝中心线
axis of bank 纵轴
axis of bent 排架轴线
axis of binary symmetry 二次对称轴
axis of boiler 锅炉轴线
axis of bore 镗孔轴线
axis of buoyancy 浮心轴
axis of centres 中央轴
axis of channel 水流动力轴线;河道中泓线;河槽(轴)线;航道轴线;航道中心线
axis of cofferdam 围堰轴线
axis of collimation 视准轴;校准轴
axis of column row 柱列轴线
axis of commutation 中性线
axis of conic(al) scan 锥形扫描轴
axis of constant moments 常力矩轴
axis of convergence 收敛轴
axis of coordinates 坐标轴线;坐标轴
axis of couple 力偶轴线
axis of curvature 曲率轴
axis of dam 坝轴线
axis of deflection 偏转轴
axis of depression 低压轴
axis of dilatation 膨胀轴;展开轴
axis of direct elasticity 直弹性轴
axis of earth 地轴
axis of earth rotation 地球自转轴
axis of ecliptic 黄轴;黄道坐标轴
axis of elasticity 弹性轴
axis of elongation 伸长轴
axis of equilibrium 平衡轴
axis of expansion 膨胀轴
axis of extension 延伸轴
axis of figure 图形轴
axis of flo(a)tation 浮力轴线;漂心轴
axis of flow 水流轴线;流轴;流动轴线
axis of folding 褶皱轴
axis of freedom 自由度轴(线)
axis of gravity 重心轴(线)
axis of groove 轧槽中线
axis of guide 导向轴
axis of gyroscope 陀螺仪(主)轴
axis of heights 水标高轴线;水位轴(潮汐曲线的)
axis of hexagonal symmetry 六次对称轴
axis of high-pressure 高压轴
axis of homology 透视轴;透射轴
axis of imaginaries 虚数轴
axis of imaginary 虚轴
axis of inertia 惯性轴;惯量轴

axis of leading marks 导标轴线
axis of least compression 最小压缩轴
axis of lens 透镜轴;晶状体轴
axis of level 水准器轴
axis of level(l)ing bubble 水准器轴;水准气泡轴线
axis of level tube 水准管轴
axis of low 低压轴;低气压轴
axis of low-pressure 低压轴
axis of magnet 磁轴线
axis of magnetic needle 磁针轴
axis of maximum compression 最大压缩轴
axis of maximum inertia 最大惯性轴
axis of optic(al) system 光具组轴
axis of ordinate 纵坐标轴;纵坐标;纵轴
axis of oscillation 摇摆轴线;摆动轴线
axis of pencil 光锥轴
axis of perspective 透视轴
axis of perspectivity 透视轴
axis of pitch 横向轴线;纵摇轴
axis of plant site 厂址轴线
axis of pole 极轴
axis of precession 进动轴
axis of principal strain 主应变轴
axis of principal stress 主应力轴
axis of projection 投影轴;射影轴
axis of railway track 铁路轨道轴线
axis of reference 参照轴;参考轴(线);坐标轴;基准线
axis of revolution 旋转轴;绕转轴;回转轴(线)
axis of ridge 脊轴;脊线;海岭轴
axis of river 河流轴线
axis of riverbed 河床轴线
axis of roll 横摇轴
axis of rolling 轧线;滚动轴
axis of rotation 旋转轴;旋扭轴;自转轴;转动轴
axis of rotation of ecliptic 黄道回转轴
axis of rudder 方向舵轴
axis of screw 螺旋轴线
axis of shiplock 船闸轴线
axis of sideslip 侧滑轴
axis of sight 视轴;光轴;(望远镜的)照准轴
axis of sighting 瞄准轴线;瞄准轴
axis of spindle 立轴中线;轴的中线
axis of strain 应变轴
axis of stream 河流轴线
axis of stream bed 河床轴线
axis of stress 应力轴
axis of swing 旋转轴;摆动轴
axis of symmetry 对称轴(线)
axis of symmetry of basic rack 基准齿条对称轴线
axis of tetragonal symmetry 四次对称轴
axis of the couple 力偶轴线
axis of the fold 褶皱轴
axis of the gimbal 平衡环轴;常平轴
axis of the indicatrix 变形椭圆轴
axis of the laser gyroscope 激光陀螺轴
axis of thread 螺纹轴线
axis of thrust 推进线
axis of tilt 倾斜轴
axis of time 时间轴线;时间轴(潮汐曲线)
axis of torque 力矩轴
axis of total symmetry 全对称(轴)
axis of trigonal symmetry 三次对称轴
axis of tunnel 隧道纵轴
axis of visual cone 视锥轴;视轴;锥形视轴
axis of weld 焊接轴线;焊缝轴线
axis of welding 焊缝中心线
axis out of line 非重合轴线
axis rotation 旋转轴
axis shaft line 轴线
axis system 轴系
axistyle 轴线
axis unit distance 轴单位
axisymmetric(al) 轴对称的
axisymmetric(al) analysis 轴对称分析
axisymmetric(al) bending 轴线对称弯曲
axisymmetric(al) body 轴对称物体
axisymmetric(al) determinant 轴对称行列式
axisymmetric(al) expansion 轴对称扩大
axisymmetric(al) finite element method 轴对称有限元法
axisymmetric(al) flow 轴对称水流;轴对称流(动)
axisymmetric(al) forging 轴对称锻造

axisymmetric(al) jet 轴对称射流;轴向对称射流
axisymmetric(al) load 轴对称荷载
axisymmetric(al) problem 轴对称课题
axisymmetric(al) seismic response 轴对称地震响应;轴对称地震反应
axisymmetric(al) shell 轴对称壳体
axisymmetric(al) solid 轴对称体
axisymmetric(al) stress 轴向对称应力;轴对称应力
axisymmetric(al) stress and deformation 轴对称应力与变形
axisymmetric(al) stress condition 轴对称应力条件
axisymmetric(al) turbulent jet 轴对称紊动射流
axisymmetric(al) vibration 轴对称振动
axisymmetry 轴对称
axite 硝酸棉炸药;硝酸甘油炸药;石油炸药;无烟炸药
axle 轮轴;心棒;的;车轴
axle adjuster 轴调整器
axle arm 驱动桥定位臂
axle arrangement 布轴;轴向布置;轴式
axle attached by gluing 胶结轴
axle bar 铁轴棒
axle base 轴间距(离);轴距
axle beam 前轴梁;轴梁
axle bearing 轴承;车轴轴承
axle bearing clearance 车轴轴承间隙
axle-bearing in two parts 双半轴承
axle bed 轴座
axle body 轴身
axle bolt 轴螺栓
axle-box 车轴箱;轴箱;轴套
axle-box acceleration 轴箱加速度
axle-box bearing 轴箱轴承
axle-box body 轴箱体
axle-box bracket 轴箱座;轴箱托板
axle-box cage 轴箱罩
axle-box case 轴箱导框
axle-box cover 轴箱盖
axle-box cover joint 轴箱盖连接
axle-box dust-guard 轴箱防尘罩
axle-box fastened to the bogie frame by means of small articulated rods 用小连杆与转向架构架相连的轴箱
axle-box guard 轴向导板
axle-box guide 轴箱导框;轴箱导架
axle-box lid 轴箱盖
axle-box liner 轴箱轴瓦
axle-box lubricated as required 要求润滑的轴箱
axle-box lubrication check-up 轴检【铁】
axle-box lubrication check-up of vehicles detached from train 摘车临修【铁】
axle-box seating 轴箱底座
axle-box wedge 轴箱斜铁
axle brass 铜轴衬
axle breakage 车站切断
axle bush(ing) 轴衬
axle cap 轴帽
axle casing 轴箱;轴套;轴管
axle center 轴心
axle centering machine 车轴定心机
axle changing installation 车轴换装
axle clamp 轴夹
axle clearance 车轴净空
axle collar 轴领;轴环
axle concentration 轮轴集重
axle construction 轴结构
axle count 轴计数;轴向计数
axle-counter 计轴器;车轴计数器
axle-counter block 车轴计数装置
axle-counting installation 车轴计数器
axle-counting magnet 车轴计数器磁铁
axle-dead 轮轴;固定轴
axle detector 车轴探伤器
axle drive bevel gear 主传动锥齿轮;轴传动伞齿轮
axle drive bevel pinion 车轴传动的小伞齿轮
axle drive generator 动轴发电机
axle driven compressor 车轴带动的压缩机
axle driven dynamo 轴驱动发电机
axle driven generator 车轴驱动发电机;车轴发电机
axle drive pinion adjusting sleeve 车轴传动小齿轮调整套
axle drive pinion shaft 车轴主动小齿轮轴
axle drive shaft 车轴传动的轴
axle driving motor 车轴驱动的电动机
axle end 轴颈

axle fairing 车轴减阻装置
axle finishing lathe 光轴车床
axle for vehicle 车辆轴
axle fracture 轴折断
axle friction 轮轴摩擦
axle ga(u)ge 轴规
axle gear box 车轴齿轮箱
axle generator regulator 车轴发电机调节器
axle grease 轴用脂;车轴润滑脂;车轴滑脂
axle guard 车轴护挡
axle guard stay 车轴护挡撑条
axle guide 导向轴
axle guide fitting 导轴零件
axle guide stay 导轴固定
axle head 轴端
axle hole 轴孔
axle housing 车轴壳体;轴向住房建筑;轴箱;轴套;轴壳
axle housing bracket cap 轴向住房建筑伸臂柱头
axle housing cap 轴套帽
axle housing trunnion 轴壳万向节;轴壳十字头;轴向住房建筑枢轴
axle-hung motor 抱轴式悬挂电动机
axle-hung roller bearing suspension 抱轴式滚动轴承悬挂
axle I beam 工字梁
axle journal collar 轴颈挡环
axle journal grinder 轴颈磨床
axle journal head 轴颈头
axle journal(neck) 轴头
axle journal turning and burnishing lathe 轴颈车削及抛光车床
axle key 轴键
axle-kilometer 轴公里;车轴公里
axle king pin 轴向主销;轴向中心销
axle lathe 车轴车床
axle-live (水平旋转的)活动轴
axle-load adjusting device 轴荷载调节装置
axle load(ing) 轴荷载;轴重;轴载重(重);轴压
axle loading force 轴荷载
axle load limitation 轴载限制
axle load meter 轴重仪
axle load scale 轴载重刻度(尺);轴载重刻度(尺);轴荷载天平
axle load weight group 轴载重组
axle lock nut 轴锁紧螺母
axle lock ram 车轴锁塞
axle misalignment 轴安装误差
axle nut 车轴螺母
axle of drop hammer 落锤心棒;穿心锤轴
axle offset 轴偏移
axle oil 轮轴润滑油;轴用油;轴心油;重油
axle outer bearing 半轴外端轴承
axle packing 轴孔填密
axle pad 轴垫
axle pin 轴销
axle pinion 车轴小齿轮
axle power 轴功率
axle pulley 滑车轴;提拉窗滑轮;滑轮开关窗;轴开滑轮;(上下推拉窗开关的)轴滑轮
axle reduction gearbox 车轴减速齿轮箱
axle rod 主轴
axle saddle 轴鞍
axle-scope 方位—高度指示器
axle seat 轴座
axle section 轴截面
axle shaft 车轴;车辆轴;车后轴
axle shaft gear 驱动轴齿轮;半轴齿轮
axle shaft steel 车轴钢
axle sleeve 轴套
axle spacing 轴间距;轴距
axle spline bushing 车轴花键套
axle spur wheel 车轴正齿轮
axle stand 车轴修理台
axle steel 轴用钢;轴钢;车轴钢
axle steel bar 轴用钢棒;车轴钢粗钢筋
axle steel reinforcing 轴用钢配筋
axle steel reinforcing bar 车轴钢钢筋
axle straightener 轴校直机
axle strut 轴支柱
axle stub end 轴联杆端
axle suspension 轴支承;轴悬置;轴吊架
axle swivel 轴转向销

axle temperature 轴温
axle testing ga(u)ge 测轴规
axle torque 车轴扭矩
axle track 轴道
axle tree 轮轴;轴杆;车轴;车轮轴
axle trunnion 车轴万向节;车轴十字头
axle tube 轴管
axle turning 车轴车床
axle turning lathe 制轴车床
axle turning shop 车轴车间
axle weight 轴重;轴载(重);车桥荷载
axle weight limit 轴载重极限;轴荷载限度;轴荷载极限
axle weldment 车轴焊件
axle wire 中轴线
axle with lateral play 横动轴
axle yoke 轴轭;轴叉
Axline tester 阿克斯林弹性织物试验仪
axman 标桩工【测】;刀斧工;斧工;伐木工
Axminster carpet 阿克斯明斯特(式)地毯
axode 瞬轴面
axometer 调整轴器;光轴计;测轴计;测光轴计
axomometry 测晶学
axon 体轴
axon hillock 轴丘
axonometer 镜轴计
axonometric(al) drawing 不等角投影图
axonometric(al) chart 立体投影图
axonometric(al) perspective 三向图;轴测透视图;轴测投影三向图;不等角透视图
axonometric(al) projection 三向(投影);三向投影法;三角图;轴测投影;不等角投影(图)
axonometry 三向图;三面正投法;均角投影图法;镜轴测量法;晶轴测定;轴测法
axopetal 向轴的
axoplasm 轴浆
axoplasmic flow 轴浆流
axoplasmic transport 轴浆运输
axostyle 轴柱;轴杆
axostylus 轴杆
axotomous 立轴解理的
Ayer's test 脊管阻塞实验
Aylesford laboratory beater 艾勒斯福特实验室搅拌器
Ayre method 艾尔法
Ayres T-piece 艾尔斯 T 形管
Ayrshire beauxitic clay 艾俞里铁矾土
Ayrton-Jones balance 艾尔顿—琼斯秤
Ayrton-Mather shunt 艾尔顿—梅则分路
Ayrton-Mother galvanometer 艾尔顿—梅则检流计;艾尔顿—梅则电流计
Ayrton-Mother ring test method 艾尔顿—梅则环路测试法
Ayrton-Perry winding 艾尔顿—佩里绕组
Ayrton shunt 艾尔顿分路;艾尔顿分流器
azacosterol 阿扎胆肖
azalea 杜鹃花
azamethonium bromide 阿扎溴铵
Azamgar 阿赞加尔缎
azanidazole 阿扎硝唑
azanol 羟胺
azarin 亮红
azatadine 阿扎他定
azel 方位和高度
azelaic acid 壬二酸
azelain 壬二酸甘油酯
azelate 壬二酸酯
azel display 方位—高度显示(器)
azel drive 方位—仰角传动装置
azel scope 方位角和高低角显示器
azeotrope 共沸化合物;共沸(混合)物
azeotrope tower 共沸蒸馏塔
azeotropic 共沸的
azeotropic copolymer 同单体组分聚物;共沸共聚物
azeotropic distillation 共沸蒸馏
azeotropic mixture 共沸(混合)物
azeotropic point 共沸点
azeotropic process 共沸过程
azeotropic solution mixture refrigerant 共沸溶液制冷剂
azeotropic transformation 共沸点变换
azeotropism 共沸作用
azeotropy 共沸性
azepine 吖庚因

azetidine 吖丁啶
azide 叠氮化物
azide of lead 铅的叠氮化物
azidine dye 叠氮染料
azido compound 叠氮化合物
azidosulfonyl dyes 叠氮硫酰染料
azimethane 重氮甲烷
azimido-benzene 苯并三唑
azimino compound 叠氮撑化合物
azimuth 地平经度;方位
azimuth accuracy 方位角准确度
azimuth adjusting screw 方位调整螺钉
azimuth adjustment ruler 方位校正尺
azimuth adjustment slide rule 方位角修正计算尺;方位调节计算尺
azimuthal 方位角的;方位的
azimuthal angle 方位角
azimuthal angle error 方位角误差
azimuthal anomalous degree analysis of lineament 线性体方位异常度分析
azimuthal chart 天顶投影海图;方位投影(地)图
azimuthal compass 方位罗盘
azimuthal component 方位分量
azimuthal control 方位控制;方位角检核
azimuthal coordinates 方位坐标
azimuthal correction 方位校正
azimuthal determination 方位角测定
azimuthal deviation analysis of lineament 线性体方位偏差分析
azimuthal displacement 方位位移
azimuthal distribution 方位分布
azimuthal effect 方位效应
azimuthal equal-area equatorial projection 等积赤道方位投影
azimuthal equal-area oblique projection 等积斜方位投影
azimuthal equal-area polar projection 等积极方位投影
azimuthal equal-area projection 等积方位投影
azimuthal equatorial projection 赤道方位投影
azimuthal equidistant projection 等距(离)方位投影;方位等距投影
azimuthal equivalent projection 等面积方位投影
azimuthal error 方位误差
azimuthal gyro(scope) 航向陀螺仪
azimuthal indicating goniometer 方位指示器
azimuthal influence 方位影响
azimuthal lever 方位杆
azimuthal map 方位投影(地)图
azimuthal map projection 方位投影(法)
azimuthal mark 方位标志;方位标(记)
azimuthal mode number 方位角模数
azimuthal orientation 定方位;方向;方位定向
azimuthal orthomorphical projection 方位正形投影;等角方位投影
azimuthal orthomorphic projection 球面投影;正形方位投影
azimuthal projection 方位投影;天顶投影(法)
azimuthal quantum number 角量子数
azimuthal rotation 方位角旋转
azimuthal seismograph 方位地震仪
azimuthal shift 方位移动
azimuthal shooting 方位爆破
azimuthal table 方位角表
azimuth and altitude instrument 地平经纬仪
azimuth and elevation scope 方位角高低角显示器
azimuth and elevation tracking unit 方位及仰角跟踪装置
azimuth and speed imdicator console 方位和速
azimuth angle 方位角
azimuth angle coupling unit 方位角耦合装置;方位角耦合器
azimuth at future position 位置方位角
azimuth auxiliary division rotating wheel 方位辅助分划转轮
azimuth axis 方位轴线
azimuth bar 方位瞄杆;方位杆
azimuth bearing 方位角方向
azimuth bearing angle 轴承方位角
azimuth blanking 方位消隐
azimuth blanking tube 方位消隐管;方位角信号消隐管
azimuth by-pass 左右测管
azimuth calibrator 方位角校准器;方位定标器

azimuth card 方位盘
azimuth changing rate 方位角变化率
azimuth circle 地平经圈;方位分划盘;方位圈;方位刻度盘;方位角度盘;方位测定器
azimuth circle instrument 有方位度盘的测角仪
azimuth circle of compass 罗盘方位圈
azimuth closure error 方位角闭合差
azimuth commutator 方位角换向器
azimuth comparator 方位比较器
azimuth compass 方位罗盘
azimuth computer 方位计算机
azimuth condition 方位角条件
azimuth constant 方位角常数
azimuth control 方位角控制
azimuth-control housing 方位角控制架
azimuth-control system 方位控制系统
azimuth-control voltage 方位控制电压
azimuth correction 方位校正;方位改正数;方位改正(量)
azimuth counter 方位(角)测量器
azimuth coverage 方位视界
azimuth deflection yoke 方位(角)偏转线圈
azimuth determination 方位测定
azimuth determination over-long line 长边方位角测量
azimuth deviation 方位偏差
azimuth diagram 方位图
azimuth dial 日规;方位日晷仪;方位刻度盘
azimuth disc 方位盘
azimuth discrimination 方位分辨率;方位分辨力
azimuth-distance positioning system 极坐标定位系统
azimuth drive motor 方位驱动电动机;方位角传动电动机
azimuth drive pinion 方位角驱动小齿轮
azimuth drive shaft 方位角传动轴
azimuth elevation 方位和高度;方位—高程
azimuth elevation display 方位—仰角显示器
azimuth elevation indicator 方位—仰角指示器
azimuth equation 方位角方程;方位方程;方位差
azimuth error 方位误差;平经误差
azimuth error of closure 方位角闭合差
azimuth feature 方位物
azimuth finder 方位仪
azimuth following amplifier 方位跟踪放大器
azimuth followup amplifier 方位角跟踪放大器;方位随动放大器
azimuth followup system 方位随动系统;方位跟踪系统
azimuth-frequency histogram analysis of lineament 线性体方位—频效直方图分析
azimuth frequency plot 方位频率图
azimuth gain reduction 方位增益降低
azimuth gating 方位选示
azimuth gear 方位角传动装置;方位齿轮
azimuth gleiche 等方位曲线
azimuth gyroscope 方位陀螺仪
azimuth hand wheel 方位操纵盘
azimuth indicating device 方位指示器
azimuth indicating goniometer 方位指示测角仪;方位角指示器
azimuth indicating meter 方位指示计;方位角指示器
azimuth indicator 方位(角)指示器
azimuth indicator dial 方位指示度盘
azimuth information 方位信息
azimuth instrument 方位仪;方位角测量仪;方位杆
azimuth laying reticle 方位安置分划板
azimuth line 方位角线
azimuth lock 方位角制动锁
azimuth magnetic recording 方位磁记录
azimuth mark 方位标
azimuth marker 电子方位标记;方位角标识器
azimuth marker generator 方位角标志发生器
azimuth measurement 方位角量测
azimuth mechanism 方位机构
azimuth method 方位角法
azimuth micrometer 方位角千分尺;方位角测微器
azimuth micrometer knob 方位角微动手柄
azimuth mirror 方位仪;定向器;测向仪;方位镜
azimuth misclosure 方位角闭合差
azimuth motor 方位电(动)机
azimuth mounting 地平装置;方位式装置
azimuth navigation system 方位导航系统

azimuth observation 方位角观测;方位角测定
azimuth of axis structure 建筑方位角
azimuth of borehole dip 钻孔倾斜方位角
azimuth of distant point 远点方位角
azimuth of formation inclination 地层倾斜方位角
azimuth of heavenly body 天体方位
azimuth of hole 钻孔方位角
azimuth of light ellipse 光椭圆方位角
azimuth of lineament 线性体方位
azimuth of photograph 像片方位角
azimuth of principal plane 主垂面方位角
azimuth of principle structural line 主要构造线方位
azimuth of seismic line 地震测线方位角
azimuth of shoot array 炮排方位角
azimuth of station 台站方位角
azimuth operator 方位测量员;方位测定员
azimuth pedestal 测量方位角机械底座
azimuth planetary gear 方向行星轮传动装置;方位行星轮传动装置
azimuth plate 方位图板
azimuth point 方位点
azimuth polarimeter 方位偏振仪
azimuth position 方位角位置
azimuth potentiometer 方位角分压器;方位电位计
azimuth prism 方位棱镜
azimuth probe orientation 测斜仪定方位角
azimuth quadrant 地平象限仪;方位象限仪
azimuth-range 方位距离;方位和距离
azimuth-range potentiometer 方位距离电位计
azimuth rate 方位角变化率
azimuth rate circuit 方位角速率测量电路
azimuth reading 方位角读数
azimuth recording 方位记录
azimuth reference 方位参考标
azimuth resolution 方位角分辨能力;方位角分辨率;方位分辨率;方位分辨力
azimuth scale 方位分度器;方位度盘
azimuth scan 方位扫描
azimuth scanner 方位扫描器
azimuth scanning sonar 方位扫描声呐
azimuth scanning sonar equipment 方位扫描声呐设备
azimuth screw 方位螺旋
azimuth search sonar 方位搜索声呐
azimuth sensor 方位角传感器
azimuth servo 方位伺服
azimuth servomechanism 方位伺服机构
azimuth setting knob 方位安置钮
azimuth shift 方位移动;方位变动
azimuth sight 方位照准器
azimuth signal amplifier 方位信号放大器
azimuths of celestial bodies 天体方位角
azimuths of the sun 太阳方位角
azimuth stabilization 方位稳定
azimuth-stabilized indicator 方位稳定指示器
azimuth-stabilized plan position indicator 方位稳定平面位置指示器
azimuth star 方位星

azimuth station 方位点
azimuth stop 方向制动器
azimuth surveying 方位测量
azimuth sweep 方位扫描
azimuth sweep generator 方位扫描振荡器
azimuth synchro drive gear 方位同步驱动装置;方位同步传动装置
azimuth table 方位角表;方位表
azimuth telescope 方位仪
azimuth test 方位测量
azimuth tool orientation 钻具定方位角
azimuth tracker 方位跟踪器
azimuth tracking cursor 方位跟踪指示器;方位跟踪游标
azimuth tracking telescope 方位跟踪望远镜
azimuth transfer 方位转移
azimuth transmitting synchro 方位角传送同步机
azimuth traverse 方位角导线
azimuth vane 罗盘照准器
azimuth versus depth graph 方位—深度图
azimuth with telescope 望远方位镜
azimuth worn knob 炮队镜量角器
azine 连氮;吖嗪
azine dye 氮杂苯染料;吖嗪染料
azines 氮杂苯类;吖嗪染料
azinphos 谷硫磷
azinphose methyl 保棉磷
azipramine 阿齐帕明
azlactone 吖内酯
azlon 再生蛋白质纤维
azo 偶氮
azoamide 偶氮酰胺
azoaryl ether 偶氮芳基醚
azobenzene 偶氮苯
azobisisobutyronitrile 偶氮二异丁腈
azo blue 偶氮蓝
azocarmine 偶氮胭脂红;偶氮卡红
azocompound 偶氮化合物
azo derivative 偶氮衍生物
azo dyestuff 偶氮染料
azo-dye wastewater 偶氮染料废水
azoflavine 酸性偶氮黄
azoform colo(u)r 偶氮型染料
azofuchsine 偶氮品(红)
Azogeranine B 阿佐杰拉宁 B
azo-humic acid 偶氮腐殖酸
Azoic 前寒武系【地】;无生代【地】
azoic diazo component 不溶性偶氮染料重氮组分
Azoic era 无生代【地】
Azoic erathem 无生界【地】
Azoic group 无生界【地】
Azoic period 无生纪【地】
azoic(printing)composition 不溶性偶氮染料组分
Azoic system 无生系【地】
azoimide 三氮化氢;偶氮亚胺;叠氮酸;叠氮化氢
azoisobutyl cyanide 偶氮异丁基腈
azole 氮杂茂
azole series 吡咯系
azolimine 阿佐利明

azolite 硫酸钡硫化锌混合颜料
azomethane 偶氮甲烷
azomethine pigment 甲亚胺颜料;偶氮甲碱颜料
azonal 不分带的
azonal soil 原生土(层);非分带土;非地带性土壤;泛域土
azonal vegetation 非地带性植物
azonal water 不分带水
azone 氮酮
azophenylene 吩嗪
azophoska 氮磷钾肥
azopigment 偶氮颜料;偶氮涂料;偶氮色素
azoproite 硼镁铁钛矿
Azores anticyclone 亚速尔反气旋
Azores high 亚速尔高压
azorite 锆英石;锆土;锆石
Azor sea 亚速海
Azorubin 偶氮玉红
azosemide 阿佐酰胺
azotate 硝酸盐
azote 氮
azotea 建筑顶上平台;在住宅或其它建筑顶上的平屋顶或平台
azotic 含氮的
azotic acid 硝酸
azotification 固氮作用
azotine 艾弱丁炸药
azotize 渗氮;氮化硝化
azotometer 定氮仪;氮气测定仪;氮量计
azotorrhea 氮溢
azo violet 偶氮紫
azoxy compound 氧化偶氮化合物
azoxy dye 氧化偶氮染料
azoxytoluidine 二氨基偶氮甲苯
azo yellow 偶氮黄
azran 方位距离;方位和距离
Aztec architecture 古墨西哥建筑;阿兹台克建筑
Aztectron 阿兹泰克纶
azulejo 彩色釉面图案砖;西班牙彩砖
azulite 蓝菱锌矿
azure 蔚蓝色;天青色;天蓝(色);淡青色
azure black 黑天蓝色(的);暗天蓝色
azure blue 天青蓝;石青蓝(灰绿蓝色)
azure copper ore 石青
azure deep 暗图蓝色(的)
azure pigment 大青
azure quartz 蓝石英
azure spar 天蓝石
azure stone 石青;青金石;天蓝石;琉璃
azurite 铜蓝;石青;蓝铜矿
azurite blue 天青蓝
azurlite 蓝玉髓
azurmalachite 蓝孔雀石
azygosperm 非接合子
azygospore 非接合子
azygous 无对的
azyloxy 胺氧基
azylthio 胺硫基

B

Baad's copper test 巴德铜试验
Baad's window 巴德窗
bababudanite 镁铁钠闪石；紫纳闪石
babackck levee 戗堤
babakite 粗面粒玄岩
Babal glass 巴贝玻璃；钡硼铝玻璃
Babassu oil 巴巴苏棕(仁)油
Babassu palm kernel oil 巴巴苏棕(仁)油
Babbage 巴贝奇
Babbitt(alloy) 巴比合金(一种减磨合金)；巴氏合金
Babbitt bearing 巴比合金轴承
Babbitt bronze 巴比青铜
Babbitt bushing 浇铅轴承；浇铅轴衬
Babbitted fastening 巴比合金固定件
Babbitter 巴比合金镶嵌工
Babbitting 镶嵌巴比合金
Babbitting jig 轴承装配工具
Babbitt layer 巴比合金层
Babbitt-lined 浇有巴比合金层的；巴比合金衬垫的；衬巴比合金的
Babbitt-lined bearing 巴比合金衬管轴承
Babbitt lining 巴比合金内衬
Babbitt melter 巴比合金熔炉
Babbitt metal 巴比合金(一种减磨合金)
Babbitt metal bearing 巴比合金轴承
Babbitt packing 巴比合金衬垫
Babbitt packing ring 巴比合金填密圈
babble 集扰；多路串扰；串线杂音
babble signal 迷惑信号
Babcock and Wilcox boiler 巴韦锅炉；巴布科克及威尔科克斯锅炉
Babcock and Wilcox mill 巴布科克及威尔科克斯碾磨机
Babcock apparatus 巴布科克乳脂测定仪
Babcock coefficient of friction 巴布科克摩擦系数
Babcock magnetograph 巴布科克磁像仪
Babcock manila-rope socket 麻绳绳卡
Babcock socket 绳卡
Babcock's test 巴布科克试验；巴氏测定法
Babcok and Wilcox type luffing crane 巴韦型俯仰起重机
babefphite 氟磷铍钡石
Babel quartz 塔状石英
Babinet goniometer 巴比测角器
Babinet's compensator 巴比补偿器
Babinet-Soleit compensator 巴比涅—索累补偿器
Babinet's principle 巴比涅原理
babingtonite 硅铁灰石
baboon 建筑物或装饰品上风格奇特的图案；风格奇特图案
babosil 钡硼砂
Babo's law 巴博定律
Babo's psammism 巴博浴
Babo's sand bath 巴博氏沙浴
Babshkin approximation formula 巴布什金近似公式
Babson chart 巴布森预测图
Babuyan channel 巴布延海峡
Babuyan strait 巴布延海峡
baby air compressor 小型空气压缩机
baby battery 小电池
baby bessemer converter 小型酸性转炉
baby blue 亮蓝；淡蓝色；浅蓝(色)
baby bond 小额债券；小债券
baby budget 小额预算
baby bulldozer 小型推土机
baby-burst 出生率急剧下降
baby bus 小型公共汽车
baby can 小型聚光灯
baby car 微型汽车；小型汽车
baby-care unit 育婴堂
baby compass 小罗盘；小罗经
baby compressor 小型压缩机；小功率压缩机
baby concrete mixer 小型混凝土搅拌机；小型混凝土拌和机
baby container 小型集装船
baby cot 儿童床
baby drill 小钻；钻床

baby farm 育婴院；托儿所
baby knife switch 小型闸刀开关
baby light 小型聚光灯
Babylonian architecture 巴比伦建筑
Babylonian hanging gardens 巴比伦空中花园
Babylonian quartz 巴比伦塔状石英
baby pink 亮桃红；淡粉红色
baby rail 小钢轨
baby roller 小型压土器；小型压路机
baby room 婴儿室；儿童病房；保育室
baby spot(light) 小型聚光灯
baby square 小方材；小枋；小方木
baby switch 小型安全开关
baby tower 小型蒸馏塔；小塔
baby track 小轨道；工地小轨道(美国)；工地轻便轨道
baby track-layer 小型履带式拖拉机
baby truck 坑道运输车；小型运货卡车；小型货车
bac 平底渡船，桶；水槽
bacalite 淡黄琥珀
Bacat barge 巴卡特驳；双体载驳货船子驳
Bacat ship 巴卡特船
bacca 浆果
bacca box smoother 球形镘刀
baceous kerogen 草木干酪根
bachelor 学士(学位)
bachelor apartment 单身公寓(美国)
bachelor chest 五斗橱；有抽屉的柜
bachelor dwelling unit 单身居住单元
bachelor flat 单身公寓(英国)
bachelor hostel 单人宿舍
bachelor kitchen 单人用五斗橱；单人用厨房
bachelor of architecture 建筑学士
bachelor of civil engineering 土木工程学士
bachelor of science 科学学士
bachelor quarters 单身宿舍；未婚人寓所
bachelor's chest 单用五斗橱
bachelor's degree 学士学位
bachelor's dwelling unit 单身居住单元；单身公寓
bachelorship 学士学位
bachelor's quarters 单身宿舍；单人宿舍
Bachman diagram 巴切曼图
Bachman test 巴克曼氏试验
Bacho dust classifier 贝乔粉末分级器
bacile 碗形深碟
bacillar fabric 纤维状组织
bacillary 杆状(细菌)的；杆菌状的，杆菌性的
bacillary cement texture 串珠状胶结物结构
bacillary dysentery 杆菌(性)痢疾
bacillary layer 杆体锥体层
bacillicide 杀杆菌剂
bacilli culture 杆菌培养
bacillus 杆菌
bacillus anthracis 炭疽杆菌
Bacillus circulans 环状芽孢杆菌
bacillus coli 大肠杆菌
Bacillus lactis viscosus 黏性产碱杆菌
Bacillus lentus 缓慢芽孢杆菌
back 顶头盖；倒车风向左转；刀背；斧背；书背；背面；背脊；背材；背部；片基；退回；推进器车叶背面；书脊
back a bill 负责兑换票据
back abutment pressure 桥台后压力
back access 背径小路
back a check 负责付款；背书支票
back a cheque 负责付款
backacter 倒铲；反向铲；反铲挖土机
backacter dipper 反铲斗
backacter shovel 反向机械铲
back-acting excavator 反铲挖土机
back-acting hay tedder 反向作用式摊草机
back-acting shovel 反向铲；反铲(挖土机)；反铲挖掘机；反铲铲土机
back action 倒挡；反作用
back action lock 反作用锁
back-action shovel 反向铲；反铲挖掘机
backactor 反铲挖掘机
backactor shovel 反向机械铲

backaged fuel 包装燃料
back altitude 反向天体高度
back ampere-turn 逆向安匝数；反(作用)安匝
back analysis 逆分析；反(演)分析；事后分析
back an anchor 增挠串联锚；抛串联锚
back anchor 副锚；反力座
back-and-forth 来回地；前后来回；前后地
back-and-forth bending test 反复弯曲试验；正反弯曲试验
back-and-forth method 往返法
back-and-forth mode 前后扫描方式
back-and-forward method 往返法
back and forward motion 往复运动
back-and-forward observation 往返观测
back and stopping trial 倒车及停车试验
back angle 后视角；后视方位角；反向角；反方位角；背垫短角铁；背垫短角钢；切削后角
back-angle counter 记录逆向散射粒子的计数管
back-arc area 弧后区
back-arc basin 弧后盆地
back-arc downwarped extracontinental basin 弧后陆外下陷盆地
back arch 八字砖；内拱；内衬拱；拱背；墙内暗拱
back-arc spreading 弧后扩张
back-arc subduction 弧后俯冲
back arm 后伸臂
back around 风向反转
back astern 挡水
back axle 后车轴；后轮轴；后车轴
back azimuth 后视角；反方位角
back-back porch 水平同步信号后延时间
back balance 轮子坡；拉紧装置；后平衡；平衡重；平衡块；平衡锤
back band 背箍；背箍(门窗外边部件)；贴脸板
backbar 壁炉横梁；支承梁
back barrier complex 障壁后复合体；障壁岛复合体【地】
back bay 后湾
back beach 后滩；后滨；滨后；尾滩
back bead 封底焊道；背面焊道；填角焊缝
back-bearer 织机后梁
back bearing 后轴承；后方位角；反象限角；反方位角
back bed 密封垫层；密封料垫层(嵌玻璃或墙板槽中)
back bedding 嵌封垫料
back bench 后座舱
backbend 门窗框外缘装饰带；贴脸板
back-bent cement occlusion 后曲锢囚【气】
back bevel 后斜面
back bevel angle 后斜角
back bias 回授偏压；反偏压；反馈偏压；背面照明
back bias circuit 反馈偏压电路
back-biased barrier 反向偏压阻挡层
back-biased resistance 反馈偏压电阻
back-biased silicon diode 反向偏压硅二极管
back bias voltage 负偏压；反向偏压
back-blading 倒车叶片；倒刮
back blast 废气冲击；反向爆炸
backblast area 炮后风锥形区
back-blending 回调
back block 边远地区；内部街区
back blocking 背板粘贴
backblow 后座(力)
backblowing 倒冲洗；反吹(法)
back-board 靠背(板)；背板(打桩)；(脚手架外部的)安全背板；后挡板；垫板；底板；板背
back boiler (安装在明火或火炉后面的)热水炉；备用煮水器；家用热水炉
back bolting 顶板杆柱支护
back bolt 加固螺栓
back-bombardment 反轰击
back bond 反向键；反担保；退回证券；退回担保；限制财产处理权证书
backbone 脊柱；主脊；主干；骨干；构架；供给中心；干道；书脊
backbone artery 主干道
backbone chain 主链
backbone closet 主干线配线柜

backbone curve 中轴线;骨干曲线
backbone enterprise 骨干企业
backbone frame 脊梁架
backbone network 主干网(络);中框网络
backbone of reflex arc 反射弧脊柱
backbone pipeline 骨干管线;干线
backbone range 主脉
backbone road 主要干路;主干道;公路干线;干道
backbone route 基干路由
backbone subsystem 主干子系统
back boxing 背衬板;后挡板;窗盒围板
back brace 后撑;反斜撑;背板
back bracing 反斜撑
back brake 柱式制动器;倒闸;冲击钻机大轴刹车
back brake support 制动支架
back break 超爆;超挖
back breaker 反向断路器
back bridge relay 反桥接继电器
back bridge wall 挡烟桥
back-brush 反向刷
back brushing 挑顶
back-burn 逆风用火
back-by-back display 双向距离显示器
back calculation 反演计算;反演分析
back-calculation method 反算法
back cap 后盖
back cap and quadrant tap bolt 后盖及齿节弧形板
back cargo 回运货(物);回航货(物);回程货(物);归程货物
back casing 临时木棚子
back cast 回测
backcast stripping 倒堆剥离
back catch 门后扣钩;门后挂钩;门后钩;门后搭钩
back center 后顶尖;(车床的)尾顶尖
backchain 护舵链
back charges 决算后各种费用
back check 反向减速制动器
back chian (轮船的)倒车制动的链
back chipping 背琢;背面錾平
back choir 唱诗班后座;主祭坛后空间
back clamping 反向钳位
back-cleaned rack 背清渣格栅
back clearance 背隙
back clevis 后连接环;后夹具;后挂钩
back clip 背夹
backcloth 天幕;舞台天幕;彩画幕布;背景幕
back-coated mirror 背涂层反射镜
back coat(ing) 底面涂层;底涂;底面涂层;背涂层;屋顶板背面沥青涂层
back cock 反向板机
back coil 后圈
back coming 后退式回采
back concentrator 后浓缩器
back conductance 反向电导
back conductance ratio 反向电导率
back cone 背锥
back cone angle 背锥角
back cone distance 背锥距离
back connection 内侧连接;反接;背面连接;盘后接线
back connection diagram 背面接线图;盘后接线图
back connection type instrument 背面连接式仪器
back contact 静合接点;后接点;后触点;背面触点
back contact spring 后接点弹簧
back-contact transistor 背面接触式晶体管
back cooler 后冷却器;后冷却机
back coordination 反馈配位
back cornice 门窗框上槛檐板
back corona 反电晕
back counter (饭店服务柜台后面的)柜台;后柜台
backcountry 边远地区;偏僻村镇;边缘地区
backcountry district 边远地区
back-coupled generator 反馈振荡器;回授振荡器
back-coupled Hall generator 反馈耦合霍尔发生器
back coupling 回授;反馈耦合;反馈;反向耦合
back-coupling condition 反馈耦合条件
back coupling oscillator 反馈耦合振荡器
back court 后天井;后院
back cover 后壳
back crank pumping units 双井联动抽油设备
back-cross 反交;回交
back crossing 里层交错板
back cupping 背面受脂
back current 回流;反向流;反向电流

back current brake 反向电流电制动
back current relay 反向电流继电器
back current step 逆流级
back cushion 靠垫
back cut 顶部切槽;上口
back cut-off power recorder 反向截止功率记录仪
back cutting 必要的超挖部分;回挖
back cylinder cover 后缸盖
back cylinder head 后缸盖
backdate 倒填日期;回溯;追溯(到)
back dated bill of lading 倒签提单
back dead center 死顶尖
back deals 孔壁临时衬板;井壁临时衬板
backdeep 后渊
back diffusion 返扩散;反行弥漫;反(行)扩散;反(向)弥散;背面扩散
back digger 反向铲;反铲挖土机;反铲挖掘机;倒铲
back direction 后视方向
back discharge 反向放电
back dividend 退回股利
back door 后门;非途径;非法途径
backdoor bell 后门铃
backdoor financing 秘密融资;秘密理财
backdoor interference 后门干扰
backdoor operation 后门交易
backdoor selling 后门销售
back draft 反向气流;背попу通风;逆流通风;回通风;反斜度;反拔模斜度;反风流
back-draft damper 反向气流调节器;回风风挡;反向通风调节风门
back drain(age) (堤或墙的)背面排水;背排水;墙背排水(管)
back drain of wall 墙背排水设施
back draught 逆通风;反斜度;逆流通风;反向气流;回程;倒转
back drilling attachment 反钻附件
back drop 衬景;交流声;跌水井(竖管);跌落井;背景;窨井;舞台吊幕;干扰;跌水检查井;彩画幕布;天幕
back duty 追税;欠交税款
back dwelling-building 后面的居住房屋
back eccentric 倒行偏心轮;反向偏心轮
back eccentric wheel 倒行偏心轮
back echo 后回波;后瓣回波
back echo reflection 障碍物回波反射
backed bill 保证票据
backed bond 有担保的债券;已抵押债券;已背书债券
back eddy 船尾漩涡
back eddy area 回流区
backed frictional material 衬背摩擦材料
back edge 后缘;后沿;下降沿;叶片出汽边;后脊
backedge of impulse 脉冲后沿
back edge of slope 边坡后沿
back edge of tool 切削具的后缘
back edging 砖瓦清边(法);凿环切割陶管法;瓦管清边
back edging for brick and tile 砖瓦清边
backed-off(milling) cutter 铲齿(铣)刀
backed-off tap 铲齿丝锥
backed porcelain 烘瓷
backed stamper 衬板压模;复制模
backed-up bit 后备位
backed-up masonry(work) 支撑性圬工
backed-up type roller leveller 带支承辊的辊式矫直机
backed-up value 备用值
backed-up water level 回水位
backed-up water table 回水水位
backed-up-weld 后托焊接
backed with block 块体衬砌
back electromotive force coefficient 反电势系数
back element 后组
back elevation 后视图;后立面;背视图;背面图;背面立视图;背立(面)图
back emf 反电动势
back emission 反向放射
back emission electron radiography 反向发射式电子射线照相
back end 底;尾端;推力轭;后端
backend chamber 冷烟室;降尘室
backend computer 后端计算机
backend database processor 后端数据库处理机
backend loader 尾端装载机

backend machine 后端机
back end processor 后端处理器;后端处理机
back-end sill 后部端梁
backend wastes 尾端废水
backening 延迟
back entrance 侧门;背入口
backer 支持物;垫座;垫衬材料;背书人;背衬材料;背衬;石板瓦
backer board 背衬板;衬垫板
backer brick 背衬砖;墙心砖
backer pump 备用泵
backer strip 背衬条
backery 面包房
back-extract 反萃取
back extraction 返提取;反提取;反萃取
back face 冷面;后端面;背(割)面;反面
backfall 滑落;斜坡;山尾部;山坡
backfeed factor 反馈因数
backfeed loop 反馈回路
back fence 后篱笆
backfile 过期报刊合订本;备用文件;过期案卷
backfill 回填层;后面充填物;回填土;采空区充填;复土;反填充;填土;填坑;填充砌体
backfill around 在周围回填
backfill blade 刮板回填机
backfill coarse-gravel sand 回填粗砾砂
backfill compaction 回填土压实
backfill compactor 回填压实机
backfill concrete 回填混凝土;填补用混凝土
backfill consolidation 回填土压实;回填土固结;回填(土)夯实
backfill crown 回填土堆
backfill dam 填石坝
backfill density 回填密度
backfilled 反填充的;回填的
backfilled bulkhead 回填式岸壁
backfilled region 回填区
backfiller 回注管;回土机;回填物;回填机;充填器;覆土机;填土机;填土工;填沟机
backfiller blade 回填机刮板;回填铲;填沟机平铲
backfiller tamper 回填土捣固器
back fillet 平嵌饰线【建】
backfill grouting 回填注浆;衬后灌浆;回填灌浆;围填灌浆
backfilling 回填;再填(充);再充气
backfilling area 充填区
backfilling ballast machine 回填道砟机
backfilling blade 回填刮板
backfilling bucket 回填料斗;斗式回填机
backfilling grouting 衬后压浆
backfilling in layers 分层回填
backfilling materials 衬背注浆材料
backfilling medium-coarse sand 回填中粗砂
backfilling of a structure 结构的回填土
backfilling of boring 钻孔回填
backfilling of trench 沟槽回填
backfilling tamper 填土夯实机;回填捣固机
backfilling with grout 回填压浆
backfilling with rubble 回填乱石
backfilling zone 回填区
backfill line 回填线
backfill(material) 回填料;回填物质;回填材料
backfill pressure 回填土压力;回填压力
backfill pressure of soil 土的回填压力
backfill rammer 回填壕沟夯具;回填夯实机
backfill sand 回填砂
backfill side of trench 壕沟回填侧面
backfill soil 回填土
backfill surface 回填面
backfill tamper 回填夯(实机)
backfill under pressure 加压围填;加压回填
backfill wedge 回填楔
backfill with mortar 灰浆回填
backfill work 回填工
back filter 反回过滤
back fin 后脊;轧疤;裂纹
back financing 不可追溯的资金供应
back-fire 回火;逆燃;反焰;火箱回火;反燃;逆火;逆弧;迎面火;点迎面火(烧);倒焰
backfire antenna 背射天线
back-fire arrangement 回火装置
back-fire arrester 回火制止器
back-fire check valve 逆火止回阀
back-fire torch 点迎面火枪

back-flame boiler 回焰锅炉
back-flap 里扉;里垂帘;里百叶门窗;匣式软百叶里衬
back-flap hinge 明合页;明铰链
backflash 火焰反冲;反闪
back flashover 反击雷闪络
backflooding 地下水逆流
backflow 倒灌;返料;倒流;返流
backflow barrier 回流活门
backflow channel 回流水道
backflow circuit 回流线路
backflow connection 回流连接;回流管;倒流连接;相互连接;交叉连接
backflowing water 抽水
backflow obstruction 回流受阻
backflow of sewage 污水倒灌
backflow pipe 回流管
backflow piping 回流管
backflow pollution 回流污染
backflow pressure valve 逆止阀;止回阀;单向阀
backflow preventer 倒流防止器;逆止阀;回流防止器;防止回流设施;防回流装置
backflow prevention device 倒流防止设备;防回流装置;回流平方装置
backflow region 回流区
backflow ripple 回流波痕
backflow siphonage 回流虹吸作用
backflow stop valve 止回阀
backflow valve 逆止阀;倒流阀门;止回阀;单向阀
backflow velocity 回流速度
backflow ventilation 回流通风
backflush 逆流洗涤;回洗;反向冲洗;逆流冲洗
backflush chromatogram 反冲色谱
backflush device 反冲器
back-flushed 倒灌的;反冲的
back-flushed waste disposal 倒灌污水处治;倒灌污水
back-flushing 反冲洗;背面冲洗
back-flushing chromatography 反冲色谱法
back-flushing technique 反冲技术
back-flush peak 反冲锋
back-flush unit 反洗装置
back-flush valve 反冲阀
back focal distance 后焦距
back focal length 后焦距
back focal plane 后聚焦面;反聚焦面
back focus 后焦点;反焦点
back fold 折叠(窗)扇;里百叶门窗;里扉;摺叠(窗)扇
back-folding 折叠的;折合
backform 背模;底模;顶模
back freight 回运货(物);回航货(物);回程运费;回程货(物);回舱货运费;不足运费;退货运费
back-front 房屋背面;房屋背部
back furrow 蛇形丘
back garden 后花园;后园
backgate 底座阀门
back ga(u)ge 后支挡;车轮内距;反向行程限位器;背缘尺;边距(铆钉或螺栓中心至角钢或槽钢边缘的距离)
back gear 减速齿轮;跨轮;后齿轮;倒档齿轮;背轮;跨轮;慢盘(减速)齿轮;背齿轮
back-geared 背齿轮的;后齿轮的
back-geared motor 带减速齿轮的电动机
back-geared type motor 带减速器马达
back-geared upright drill press 带齿轮立式钻床;背齿轮立式钻床
backgearing 背轮
backgear ratio 背齿轮比
backgear shaft 背齿轮轴
back geosyncline 后地槽
back gluer 胶背机;书脊上胶机
back goods freight 退货运费;退货费用
back gouging 背刨;刨焊根
background 后台;腹地;本底;背景
background absorption 底层吸收;本底吸收;背景吸收
background activity 本底活动性;本底放射性
background air 本底空气
background air pollution 本底空气污染
background air pollution monitoring network 空气污染本底监测网;气象组织本底空气污染监测网
background area 本底区域
background atmosphere composition network 本底大气成分监测网
background block error 背景块差错
background block error ratio 背景块差错率
background briefing 背景情况介绍
background brightness 本底亮度;背景亮度
background brightness control 背景亮度控制
background camera 背景照相机;背景摄影机
background charge 背景电荷
background clutter 本底杂波;背景杂波
background collimating mark 后景标定点
background collimating point 后光标
background colo(u)r 背景色
background compiler 后台编译程序
background Compton scattering 背景康普顿散射
background concentration 本底浓度
background conductivity 本底传导性
background contamination 背景污染
background content 本底含量
background content of mercury 汞的背景含量
background contrast 背景反差
background control 背景控制
background corrected value 本底修正值
background correction 扣除背景;本底校正;本底改正;背景修正;背景校正;背景订正
background count(ing) 本底计数;背景计数
background counting rate 本底计数率
background current 基值电流;本底电流;维弧电流
background cycle 基础周期
background data 背景资料;背景信息;背景数据信息
background density 本底密度;背景密度
background determination 本底测量;本底测定
background discrimination 背景鉴别能力;背景鉴别
background discrimination technique 背景鉴别技术
background dispersion pattern 背景分散模式
background display image 背景显示图像
background doping 本底掺杂
background doping level 本底掺杂能级
background effect 背景效应
background electrolyte effect 本底电解质效应
background exposure 本底照射;本底接触;本底暴露
background fade-in 背景淡入
background flow 后台流
background fluctuation 背景起伏;背景波动
background fog 本底雾
background for plastering 抹灰底面
background gradient spectrum 背景陡度光谱
background heater 隐蔽式供暖器;附加采暖设备
background heating 局部供暖;背景供暖
background illumination 本底照明;本底照度;背景照明
background image 背景反映
background impurity 本底杂质
background indensity 背景光密度
background information 依据资料;背景资料;背景信息;背景材料
background ink 底色墨水;衬底印色;背景墨水
background intensity 背景强度
background interference 本底干扰
background ion 本底离子
background irradiation 本底辐射
background luminescence 本底发光
background job 后台作业
background level 本底值;本底水准;本底水平;背景值
background light 背景光
background lighting 背景照明
background-limieed infrared photography 背景限红外摄影
background limited infrared detectivity 背景限红外光电探测率
background-limited infrared photoconductor 背景限红外光电导体
background loop 背景循环
background loudspeaker 背景扬声器;背景声扬声器
background luminance 背景亮度
background luminance control 背景亮度控制
background mass spectrum 本底质谱图
background material 背景资料;背景材料
background measurement 本底测量
background measurement experiment 背景测量实验
background measurement radiometer 背景测量辐射计
background microbiology 本底微生物
background mixture 本底混合物
background mode 后台方式
background modulation noise 本底调制噪声;背景调制噪声
background monitor 本底监测器
background monitoring 本底监视;本底监察;本底监测;背景监测
background monitoring program(me) 本底监测规划
background monitoring station 本底监测站
background music 本底谐音;背景音乐
background noise 本底噪声;背景噪声;背景杂音;暗噪声
background noise control 本底噪声控制;背景噪声控制
background noise criteria 本底噪声标准
background noise level 背景噪声水平;背景噪声电平
background noise of channel 通路固有杂音
background of experience 经验积累;经验背景
background of information 经验信息背景;所积累的资料
background of mineralization 矿化背景
background of(navigation) mark 航标背景
background of region 区域背景
background of the project 工程项目背景
background of water environment 水环境本底值
background organics 本底有机物
background ozone 本底臭氧
background paper 背景文件;私自拨款
background partition 后台区
background patent 基础专利;背景专利
background planting 背景栽植;背景种植
background pollution 本底污染;背景污染
background pollution observation 本底污染观测;本底污染测定
background print 后台打印
background processing 基本处理;后台处理;背景处理
background processing interrupt 后台处理中断
background program(me) 后台程序
background projection 背景投映;背景投影
background projector 背景放映机
background radiation 本底辐射;背景辐射
background radiation intensity 本底辐射强度;背景辐射强度
background radiation noise 背景辐射噪声
background radiation value 本底辐射值
background radioactivity 本底放射性
background random process 背景随机过程
background range 背景范围
background reader 后台阅读器
background reflectance 基底反光能力;背景反射(性能);背景反光率;背景反差
background reflection 背景反射
background regional forecast 区域预报
background rejection 本底扣除
background removal for lettering 注记空位
background return 地物回波;背景反射信号
background sample 本底样本
background scattering 背景散射;背底散射
background science 基础科学
background screen 背景幕布
background seismicity 背景地震活动性
background seismic wave 本底地震波
background signal 本底信号;背景信号
background sound 背景声(音)
background spectrum 本底谱线图
background subtract(ion) 本底扣除
background suppression 本底抑制
background survey 本底调查;本底测量
background survey on economic structure 经济结构背景调查
background survey on meteorologic(al) environment 气象环境背景调查
background task 后台任务
background temperature 背景温度
background terminal 后备终端
background transmissivity 背景透射比

background value 背景值
background value of radiometer 辐射仪背景值
background value of soil 土壤背景值
background value of soil region 土壤区域背景值
background value of tracer 示踪剂本底值
background value survey 背景值调查
background vapo(u)r pressure 本底蒸气压
background voltage noise 基底电压噪声
background wallpaper 墙纸的底子
background water 背景水
background water quality 本底水质
background wave 背景波
back-grouting 二次灌浆
back guide 后导板；倒车导板
back gutter 烟囱排水边沟；背部雨水沟槽；泛水槽；烟囱雨水槽；包檐天沟
backguy 拉线；拉条；拉索；牵索；拉绳；拉缆；后拉索；支撑；绑缆；尾绳
backhand drainage 倒转水系
back handed rope 左搓绳
backhand method 右焊法
backhand technique 基值
backhand welding 后退焊(接)；退焊；反手焊(接)；逆向焊；后退式气焊；后焊法；向后焊法；反向焊(接)
backhaul 回运货(物)；回运(土方)；回航货(物)；回程运输；回运货(物)；后曳；返航载运；反向运输；空载传输；载货返航
backhaul cable 后曳缆；回运缆索；后曳索
backhaul drum 回程运输用圆筒
back haunch fillet of arch 护拱
back head 机车外门板；后尾部(凿岩机)
back head brace 外门板拉条
back heading 回风巷道；副平巷；通风巷道
back hearth 炉灰腔；壁炉床；壁炉背
backheating 反加热；回热；逆热
back hitch 止滑结
backhoe 反向锄；反向铲(土机)；反铲
backhoe boom 反铲臂
backhoe bucket 反铲挖土斗；反铲铲斗；反铲斗
backhoe digger 反铲挖土机
backhoe dredge(r) 反铲挖泥机；反铲挖泥船
backhoe excavator 反铲挖掘机
backhoe front end loader 后挖前卸式装载机
backhoe loader 反铲装载机；单头挖土机
backhoe-pushshovel 反铲正铲挖土机
backhoe shovel 反铲挖掘机
backhoe tamping 反铲夯法；反铲捣实
backhoe trenching machine 反铲挖沟机
back hole 峒顶爆孔；顶眼；顶炮眼；倒眼(爆破用一种炮眼)
back holes 后爆炮眼组
back hook 里钩；内挂钩；门后扣钩；门后钩；门后搭钩
back house 后屋；后房
backhousing 降尘室
back incline 后斜坡
back induction 去磁作用
backing 里壁；基材；逆转；后面充填物；垫底；底座；风向逆转；反接；背衬；墙背衬；窝托横梁；
backing a bill 在票据上背书
backing a chain 锚链上加绑缆
backing an anchor 增抛辅联锚；抛串联锚
backing angle 衬垫角铁
backing a warrant 对凭单背书担保
backing away 堵墨
backing away from fountain 供墨不匀
backing bar 撑杆；衬垫板条；反向角铁；背垫短角铁；背垫角钢；水平吊杆
backing belt 倒车皮带
backing block 墙心砖；靠枕；止动块；支撑块；衬垫；衬板
backing blowing 压气反洗
backing board 载模板；背衬衬板；底衬板；石膏底板
backing brick 顶砖；充填砖；衬边砖；墙心砖；外帮砖
backing brick wall 背衬砖墙
backing calcium silicate brick 填充灰砂砖
backing cloth 背衬布
backing coal 炼焦煤
back(ing) coat 底面涂层；抹灰层；底涂层；卷钢背(面)涂层；打底层；背面涂层
backing coat mixed plaster 底层混合灰泥
backing coat mix(ture) 底层混合物

backing coat plaster 底层灰泥
backing coil 反接线圈
backing compound 背衬胶粘剂
backing concrete 填充料混凝土；垫层混凝土
backing condenser 前级冷凝器
backing deal 顶衬板；背板
backing eccentric rod 后行偏心杆
backing fabric 衬布
backing field 反方向磁场
backing-filling material 砌体填充材料
backing for abrasives 砂布底布
backing for wall covering 贴墙衬材
backing frame 安片框
backing gas 背面保护气体；背衬气体
backing groove 焊缝；反面坡口
backing groove of weld 焊缝反面坡口
backing gypsum 石膏衬板
backing hip rafter 斜脊椽角的边棱
backing insulation 里绝热层；绝热背衬
backing layer 背层
backing light 背景灯光(舞台)
backing line 前级管道
backing log 挡轮槛(码头前沿)；挡边木
backing masonry(work) 墙心圬工；支持性圬工
backing material 基衬；衬底物；背衬料；底子料；衬里材料；衬底材料；背衬材料
backing memory 后援存储器；后备存储器；备用存储器
backing metal 底层金属；金属垫块
backing mixed plaster 底层混合灰泥
backing mixture 底层混合料
backing movement 回位
backing of drift direction 漂移方向的逆转
backing-off 凿(应力的)消除；铲；退绕
backing-off attachment 铲磨工附件；铲齿附件
backing-off cone clutch 反转离合器
backing-off lathe 铲工车床；铲齿车床
backing off stuck drilling tools 提出被卡钻具
backing of rafters 端支托；椽托
backing of smoke 推回火灾烟雾
backing of veneer 胶合板内衬；胶合板内层；贴面板衬背
backing of wall 墙托；墙背衬
backing of window 窗托
backing-out 调整需量；列车退行；反回
backing-out punch 打孔器；钉的冲子；顶出器；冲床顶出杆；冲床；退钉器
backing oven 烤箱
backing paint 镜背保护涂层；底层涂层；底漆；烤漆
backing paper 底层纸；垫纸；裱糊纸
backing pass 底层焊道；封底焊；背面焊
backing pin 支销；挡销
backing plank 衬板
backing plaster 底层灰泥
backing plate 背垫板；支撑板；后板；垫模板；托模板
backing power 后退功率；倒车功率
backing pressure 背压；前级压强；托持压力
backing printing 背面印刷
backing pulley 回行皮带轮
backing pump 增压泵；初级抽气泵；初级泵；前级泵；升压泵
backing register 后备寄存器；后援寄存器
backing ring 垫圈；衬圈；衬环；托圈
backing roll 支承辊；承压滚筒
backing rudder 倒车舵
backing run(weld) 底层焊接；打底焊道；背焊接；反向旋转；封底焊缝
backings 背衬木
backing sand 填充砂；填砂；背砂
backing sand-lime brick 砂灰砖墙
backing sheet(ing) 底基薄板；支撑板；副份；衬板；衬纸
backing side trap 前级冷井
backing soil 回填土
backing space technique 积累法；前级空间技术
backing stage 前级
backing stone 衬里石
backing storage 后援存储器；后备存储器；备用存储器
backing store 后备存储
backing strap 背衬；条状卷板
backing strip (搪瓷金属板的)背衬；背垫条；垫板
backing stuff 回填材料；底层涂料
backing system 前级真空系统；前级系统

backing tape 带基
backing tier (砖墙的)背层墙；内粗外细砖砌体
backing turbine 倒车透平；倒车汽轮机
backing-up 倒砌；衬砌；封底焊；里壁砖(墙内部用的瓷砖)；回流；回冲；背衬；背裱；刨刀护铁
backing-up brick 里壁砖
backing up of sewage 污水倒灌
backing-up screw 止动螺旋；止动螺钉
backing vacuum 前级真空
backing varnish 烤清漆
backing varnish finish 烤清漆饰面
backing vessel 里壁容器
backing wall 后墙；支撑墙
backing weld 底焊(缝)；封底焊(缝)；用垫环焊接；底焊焊缝
backing welding 打底焊；底层焊接
backing wind 逆转风；回旋风；风向反转
back injection 反注入
back injury 背部损伤
back inlet 后进水口
back-inlet gull(e)y 排水管支管进口；后部进水集水井
back-in parking 后退泊车
back iron 整修刀；刨刀护铁；护铁(木工刨刀)；背铁；刨刀护铁
back issue 过期期刊
back jamb 背衬；侧樘；背衬木条
backjoint 背衬；留槽待填；石阶踏步背榫；待填接口；平行解理走向的节理面【地】
backkick 倒转；反转；反向放电；反冲；冲火；逆转
back kiss roll 反面接触舔液辊
back lacquer 黑back 漆
back lacquer finish 黑光漆面
back laid rope 左搓绳
backland 腹地；后方；后陆；后方地带；海岸后面陆地；海岸后面地区；堤后滩地；堤后泛滥地；背地
backland deposit 堤后淤沙；堤后淤沙；堤后淤泥；堤后淤积物；堤后淤积层
backland of straight-forward 直接腹地
backland trackage 后方铁路线(码头)
backland trackage of dock 码头后方铁路线
back lane 后巷
back lapping 底印重叠错位
back lash 轮齿间隙；啮合间隙；后座；后冲；侧隙；木材反跳(锯木时)；空回；啮合间隙；后冲；齿隙(游移)；齿间隙；侧向间隙；反向爆炸；松动
backlash adjusting screw 齿隙调整螺钉
backlash allowance 侧隙公差
backlash circuit 间断电路；齿隙式电路
backlash compensation 间隙补偿
backlash current 间断电流
backlash eliminator 间隙消除器；后冲消除器；齿隙消除器
backlash error 侧隙误差
backlash nonlinearity 间隙非线性
backlash of rope 钢丝绳后滑
backlash phenomenon 回差现象
backlash potential 反栅极电位
backlash unit 反撞部件
backlayer 背面层
back lead weight 后压重物
back ledge 后横档
back leg 后床腿；斜支柱；背面支柱
back lens 后透镜
back letter 补充函件；认赔书
back levee 支流堤；内导水堤；内导流堤；支堤；戗堤；背水堤
back-levee marsh 漫滩一天然堤沼泽
back-level release 后层释放
back lever 后伸臂
back light 背光；逆光线；后灯；背景光；暗闪；背部照明
backlighted 背光的
backlighted plotting surface 反光绘图面
backlighted shot 逆摄影
backlighting 逆光照明；背面照明；背面补光；背景光；背光(照明)
backlight scribing 透光刻图
backlimb 背翼
backlining 衬脊纸；里衬；内衬；衬板；背衬(木条)；书脊衬
back link 后连杆
back lintel 隐蔽过梁；内衬过梁；暗过梁；隐蔽暗梁

backlit vinyl fabric 背打灯乙烯织物
back loader 反铲挖掘机
back loading 退装货物
back lobe 后瓣
back locking 回闭锁;反闭锁
backlog 订单积压;巨木;积压未交订货;积压任务;积累;储备物;储备;未交付订货积累;尚未用完的拨款
backlog control 储备管理
backlog demand 未满足要求
backlog document 积压文件
backlogging 积压
backlog of business 现有定货量
backlog of demand 未得到满足的需求
backlog of orders 现有定货量;已接受而尚未运出的定货总数;已接受订货总数
back magnetization 反磁化
backman 杂工;辅助工
back mark 反标记
back marsh 后背湿地;河漫滩沼泽;腹地湿地
back masonry wall 后面圬工墙
back measurement 后测
back method 后视法
back migration 反迁移;反向移动
back mill table 轧机后辊道
back mirror 后望镜;后视镜;后镜;后反射镜;望后镜
backmix-flow reactor 返混流反应器;反向混流反应器
backmixing 重新搅拌;逆向混合;返混;反混;回混
backmix reactor 反混合反应器
back money 过期未付款;拖欠款;拖欠分期款项
back-mop 背面涂镐
back mortaring 抹里皮;抹里衬;背面抹灰;砂浆衬里
back-most 最后面的
back motion 倒转;倒车
back mo(u)lding 底饰
back mo(u)ld joint 模缝线
back mounted 从背面安装的;背面安装的
back-moving spring 拉紧弹簧;延缓弹簧;复位弹簧
back mutation 反突变
back nail 平钉
backnailing 防水层钉合;钉暗钉;填塞钉孔
back noise 底噪声
back nut 底座螺母;锁紧螺帽;限动螺帽;支持螺帽;支承螺帽;后螺母;限动螺母;支持螺母;支承螺母;抵座螺母;背紧螺母;螺帽背
Back oars 倒桨【船】
back observation 后向测量;后向观测
back observational back sight 过天顶测天
back of arch 拱背
back of a wall 墙背
back of a window (窗框底部与楼板之间的)护墙板;窗下护壁板
back of bearing 轴瓦背面
back of beyond 内地(英国);偏远地区
back of board 板背
back of bridge seat 桥座背
back of cam 凸轮后侧边
back of dam 坝的上游;堰的上游;坝的背水面
back of dike 坝的背水面
back-off 倒扣;倒转后解松;铲背;补偿;去锐边;退下;凹进
back(-off)angle 后角
back office 后勤部门
back office force 内勤人员;后勤人员
back-off shooting 松动性放炮
back off system 补偿系统
back-off-tool 解松工具
back of levee 堤防背(水)面;堤背
back of panel mounting 在面板后面的装配
back of piston ring 活塞环内侧边
back of the saw 锯背
back of tool 刀背;工具背
back of tooth 齿背
back of vault (拱顶室的)外边缘线;拱背(穹隆)
back of wall 墙的受压面;墙的后面
back of weir 堰的上游
back of weld 焊缝背面
back of wind 风向反转
back order 暂时无法满足的订货;延期交货;延迟交货;待交订货单;返回次序;欠交订货;留待将来交货的订单
back order cost 延期交货费用;延迟交货成本
back ordering 延交订货;延迟交货

back order memo 延期交货通知单;欠交订货通知单
back order sales 欠拨销售量
back oscillation 回程振荡
backosmotic pressure 反渗压力
back-out 反回;逆序操作;旋出;保留原有现场;拧松
back-out by hand 用手拧紧
back-out of contract 背弃合同义务
back-out of the contract 不履行合同
back-out punch 退出穿孔机(铆钉)
back-out routine 回退程序
backpack 背包
backpack camera 背负式摄像机;背负式电视摄像机
backpacking 后背衬垫;回填(料);背法
backpack pump 背负式喷雾器
backpack transmitter 背负式发射机;便携式发射机
back pad 小鞍
back-paint 背面涂料
back painting 背面涂刷;背面涂漆;(金属构件的)底漆
back panel 后面板
back panel wiring 板后布线
back part of stage 舞台后方
Back-Paschen effect 巴克一帕邢效应
back patio 后院子;后平台
back pay 补发工资;欠薪;欠付工资
back payment 逾期支付
back-pedal 背弃
back pedaling brake 后踏制动器
back piece 舵板;反向角钢;背角材
back-pinched saw 夹背锯
back piston 后活塞
backpitch 铆钉背距;后节距;背螺旋;反螺旋;背节距
backplan 底视图
backplane 后挡板;护板;后板;底板;后连线板;背板(打桩)
back-planing 刨背
back plaster 底灰(指油漆)
backplastered 背面抹灰的;板条抹灰的
backplastering 抹里衬;在暗处抹灰;背面抹灰;板条背面抹灰
backplate 靠背板;后板;支点离合器圆盘;支承板;背板;背板(打桩);门上金属扣件;底板;护板;后插板;后挡板;背面电极;垫板
backplate lamp holder 带底板的灯座
back play 空挡(位置);空程
back poppet 后顶针座
back porch 后沿;后肩
back-porch effect 后沿效应
back pouring 补浇注;补充浇筑;补充浇注;补充浇灌
back-pressing method 压背法
back-pressure 回压;吸入压力;反压力;反压;背压力;背压;反压源
back-pressure control 反压控制
back-pressure control station 回压控制站
back-pressure curve 回压曲线
back-pressure evapo(u)rator 背压式蒸发器
back-pressure extraction turbine 背压抽汽式透平;抽气背压式汽轮机;背压抽气式汽轮机
back-pressure manometer 背压压力计;后压测压计
back-pressure of steam trap 凝结水背压力
back-pressure operation 背压操作
back-pressure regulating valve 背压调节阀
back-pressure regulation 吸入压力调节
back-pressure regulator 反压力调节器;吸入压力调节阀;蒸发压力调节阀;反压调节器;背压调节器
back-pressure relief port 反压泄料孔
back-pressure relief valve 背压安全阀
back-pressure return 余压回水
back-pressure test 回压试验
back-pressure testing 回压试井
back-pressure turbine 背压式水轮机;背压式汽轮机
back-pressure turbine refrigerator 背压式水轮制冷机;背压式汽轮制冷机
back-pressure turbine set 背压式机组
back-pressure valve 单向阀;回压凡尔;回压阀;回向阀;背压阀;止回阀;逆止阀;反压阀
back primed 背头;底漆;底涂;三面涂层;欠底漆的
back priming 背面底漆涂装
back-projection 后方投影;背面投影;幕后投影;背景放映
back-projection operation 反投射运算
back-projection slide 幻灯片
back prop 脚手支撑架(深沟槽);后支柱;后撑
back propagation network 反向传播网络

back pull 后张力;后拉力;反拉力
back pulley 支托滑轮;后背滑轮
back pullout 后开门
back pull wire drawing-machine 反拉力拉丝机
back pulse 反脉冲
back-pumping 反输
back pumping equipment 逆向抽水设备;回抽设备
back pumping method of injection well 回灌井回扬方法
back-pumping with air absorption 吸气回扬
back-pumping with flow back 回流回扬
back purge 反冲
back-purge disc valve 反冲洗圆盘阀
back-purge system 反冲系统;反冲洗装置;反冲洗系统
back putty 玻璃腻子;面灰(嵌门窗玻璃的);垫层油灰;打底油灰;嵌玻璃用油灰
back radiation 逆辐射;反向辐射
back radiator 反向辐射体;反辐射体
back rake 后倾角;底耙角(金刚石钻头切削刃)
back rake angle 纵向前角
back range 向后串视;尾向叠标;尾距
backreach 后伸距
back reaction 逆反应;反向反应
back readig 反向读数
backreed 后箅
backreef deposit 礁后沉积
back-reef lagoon 礁后泻湖
backreef lagoon deposit 礁后泻湖沉积
back-reef moat 礁后沟
back-reef zone 礁后带
back reflection 后反射;背反法;背反射
back-reflection photography 背反射照相
back reflection X-ray camera 背面反射X射线照相机
back reinforced knife saw 夹背刀锯
back rent 欠租
back residence block 后面的住宅;偏僻住宅
back residential block 后面的居住区;后面的住宅;偏僻住宅
back residential building 后面的居住房屋;后面的住宅;偏僻住宅
back resistance 逆向电阻;反向电阻;背部阻力
back resistance meter 反向电阻测量计
backrest 后梁;固定中心架;背靠;靠背;后刀架;床靠背
backrest block (镗床的)后立柱移动块
back-ridge 洋脊外侧
back rigidity 脊强
back ripping 二次挑顶
back river 河道壅流部分;后河;河道壅水部分
back road 乡村道路;地方性道路
back-roll 后滚筒;回原地再起动;倒卷;反绕;重新运行;重算
back roller 固定辊
back-rolling 反绕;倒卷;重复滚压
back room 里屋;密室;商店的辅助仓库
back room stock 商店辅助仓库的储存品
backrope 锚钩索;上索;斜桁撑杆后支索;反牵杆的反支索
back router 反向刻模机
backrun 底焊缝;反转;背面焊缝;逆行;逆向;逆行制气
back running 向后推进;倒车【船】
backrun process 逆行制气法
backrush 回卷浪;回卷;回溅;反流
backs 衬条
back sailing 顶风驶船;顶风帆
back salary 欠薪;欠付工资
back sand 背砂
back saw 弦锯;夹背锯;脊锯;镶边短锯;镶边手据;弓锯;背镶金属的锯
backsawing 弦锯
back saw with wood(en)handle 木柄手锯
back scarp 外堤岸
back-scatter 回压散射仪;反向散射体;逆散射
back scattered beta ga(u)ge 反散射型贝塔射线测厚计
back-scattered current 反散射电流
back scattered densitometer 同位素发射密度仪
back-scattered density robe 背散射式密度探测器
back-scattered dose 反散射辐射剂量
back-scattered electron 背散射电子
back-scattered electron detector 反散射电子检测器

back-scattered electron image 背散射电子像;背散射电子图像
back-scattered factor 反散射因子
back-scattered problem 反散射问题
back-scattered radiation 后向散射的辐射
back-scattered ultraviolet radiation 后向散射的紫外辐射
back-scattering 后向散射;反散射;反向散射;反射离散;向后散射;背散射
back-scattering beta ga(u)ge 反向散射贝塔测厚计
back-scattering coefficient 反向散射系数;背散射系数
back-scattering cross section 反向散射截面
back-scattering densitometer 反向散射密度计;反射密度仪
back-scattering density method 反向散射密度法
back-scattering factor 反向散射因子;反散射因子
back-scattering ga(u)ge 反向散射测量计
back-scattering image 反向散射像
back-scattering laser Doppler velocimeter 反向散射激光多普勒测速仪
back-scattering loss 反向散射损失
back-scattering peak 反向散射峰
back-scattering process 反向散射过程
back-scattering radiation dose 反向散射辐射剂量
back-scattering spectrometer 反向散射能谱仪
back-scattering thickness ga(u)ge 反向散射厚度计
back-scattering ultraviolet radiometer 反向散射紫外线辐射仪
back-scattering wave 反向散射波
back-scatter radar 后向散射雷达
back-scatter-type ga(u)ge 后向散射测厚仪
back-scatter ultraviolet radiometer 后向散射紫外辐射计
back scheduling 反向调度;反向安排日程
backscour 反向冲刷
back scraper 反向刮板
back-scratcher 背挠
backscrubbing 回洗
back seam 地毯背面接缝;背缝
backseat 后座;底座;阀后座;次要位置
back-seated valve 后座阀;强迫就位阀
back seat gasket 底座垫圈
back-selling 返销,逆销售
back service road 背街服务性道路
backset 倒流;逆行;回水;制动装置;锁挡;重耕;锁面至钥匙孔中心的距离
back-set bed 逆流层;回流层;后积层;涡流层
backset bedding 逆流层理;回流层理;后积层层理;涡流层理
backset current 逆流;回流
backsetting 圆缘凿石;收紧(砌墙等工作);缩进;收进(砌墙等工作)
backsetting effect 逆流效应
backshaft 后轴;后轮轴
back sheet 后板
back shift 交叉班;倒挡;二次变换
back ship 使船后退
back shooting 逆向爆炸;回程爆炸
backshop 修理厂;辅助车间;修理车间;大修厂;机车修理(车)间;大修车间
backshore 滨后;上滩;顶撑;背撑;尾滩;后海岸地带;后滨;滨后岸;水浸岸
backshore beach 滨后
backshore deposit 后滨沉积;滨后沉积(物)
backshore dune 后滨沙丘
backshore sediment 滨后沉积;滨后沉积(物)
backshore terrace 后滨阶地;海滨小丘;滨后阶地;滩丘
backshot 消声器内爆音;排气管内噪声
back-shot wheel 反射水轮;反射式水轮;反击叶轮;反击式水轮
backshunt 矿车自动折返装置;后端调车【铁】;反向推车器
back-shunt keying 后分路键控
back shu points 背俞穴
back shutter 里扉;里百叶门窗
backside 住宅后院(美国方言);背面;背部;后方;后部
backside printing 背面印刷
backside pull-apart fault 后缘拉离断层
backsight 瞄准口;后视
back sight angle 过天顶高度

backsight hub 后视标杆
back sight leaf 表尺板
backsight reading 后视读数
back sight slide 后视测滑座
back signal 回原信号
back silting 回淤
backsiphon(age) 回吸;逆虹吸(作用);倒虹吸作用;反虹吸;虹吸倒灌
backsiphonage pipe 反虹吸管
back-siphonage preventer with leakage water fitting 装有渗漏水配件的阻止反虹吸作用设备
backsite 不临街的建设基地;背面建筑基础;住宅后院(英国方言)
backsizing 背面涂胶
back slab 引板;岸板
back slagging 炉后出渣
back slope 反坡;内坡;后坡;(堤的)背水坡;背坡;倾向坡
back sloper 量坡规;内坡机;整坡器;(筑路机的)刮沟刀
backsloper 量坡规;推土机的推板
back sloping 切削边沟坡面
back slum 贫民窟
backspace 退位;回退;返回空格;反绕;反回;退格
backspace character 回车字符;退格(字)符
back space command 退回命令
backspace file 退回文件;回退文件
backspace key 后移键
backspace statement 回退语句
backspace subroutine 返回子程序
backspacing 反绕;后移;返回
back speed 退速
backspill 倒料;倒浆
backspin 回旋;倒旋
backspin rolls 传动托辊
back spin timer 反转时间继电器
backsplash 后挡板;防溅挡板(柜台)
backsplasher 防溅挡板(柜台);后挡板
backsplashing 后挡板
back splice 索尼插结
back splicing 索尼插结
back spray rinsing 反喷清洗(法)
back spread 不同市场间的差价
backspring 倒缆;反向弹簧
back square 矩尺;定线器
back stacking space 后方堆场
back stacking yard 后方堆场
back stackyard 后方堆场
backstage 后台;在后台的
back stage area 后台区
backstage deal 幕后交易
backstage entrance 后台入口
backstage flow 后台流
back stagger 后向斜翼
back stair(case) 后楼梯间;后楼梯;内部楼梯
back stamp 底印(器皿标记);底款(器皿底标记)
backstand 后座;支撑结构;张紧器
backstay 拉索;固定中心架;(帆船的)后支索;后牵;后缆;后拉索;后拉缆;后拉杆;后拉条;定位杆;背撑
backstay anchor 拉索锚碇;拉索锚固;拉索锚碇
backstay cable 斜缆;牵索;后拉索;后拉缆
back stay of pile driver 打桩架拉索
backstay rope 拉索
backstay stool 后牵座
backstein Gothic 哥特式建筑(14世纪在德国北部发展的)
backsteingotik 砖石特式(建筑)
backstep 退层;后部台阶;折标
backstep marks 折标
backstep sequence 分段后法次;分段退焊次序
backstep welding 分段退焊;逆向焊;逐步退焊法;分段逆焊;反向焊(接);反手焊(接)
back stitch 倒缝
backstop 逆行停止;后止件;止回器;止挡;防反转装置;托架;回复停止装置
back stopper 后退定程挡块
backstopping 支持
backstopping clutch 倒车排挡
backstopping porter bar 后定位送料杆
backstop tone 后定位夹钳
back strap 制动带;对接贴板;背垫条
back strapped 倒拉船
backstream 后河;回流;返回

back streaming 返流;逆流;回流
back street 后街
back strength 背力
backstroke 回程;倒划桨;逆行程;返回行程;返回冲程;反向冲击
back strut 后支柱;后撑;反撑
back strut prop 后支撑架
back-stuffing and sealing in part of well 部分回填封闭
back-stuffing and sealing in whole well 全孔回填封闭
back-stuffing with cement 水泥回填
back-stuffing with clay 黏土回填
back-substitute 回代
back substitution 逆代法;回代;倒转代换
back suction 反吸
back support 后背
back surface 后表面;底面;背面;热面
back surge 反涌波
backsurge tool 反冲洗工具
backswamp 漫滩沼泽;漫滩见泽;漫灌沼泽;河漫滩沼泽;河漫最低部分;堤后池沼;堤后沼泽
backswamp area 低沼泽区域
backswamp deposit 堤后泥沙
backswamp depression 漫滩洼地
backswept 后掠角
backswept wing 后掠翼
backswing 后摆(程);回转;回程;倒转;反向摆动
backswing voltage 回程电压;反向电压
back tabulation 反向制表
backtalk 工作联络电话
back taper 倒锥
back taper tap 机用丝锥;倒锥丝锥
back tax 退税;退缴税
back teeth 臼齿
back tension bridle 带材张紧装置
back-tensioned drawing 反张力拔丝
back terrace 后阳台
back test 弹性复原试验;加载卸载鉴定试验(弹簧称)
back the line of the hoist 绞车卷筒倒绳
back thrust fault 背冲式中断层
backthrusting 背逆断层作用
back tie bracket for steel mast 钢柱拉线角钢
back tilt 后倾
back timber 顶梁
back timbering 顶板支护
back titrating 反滴定
back titration 回滴定;返滴定;反滴定
back-to-back 反面对反面相叠;背靠背;背间距;背对背;自发自收测试;对头拼接
back-to-back account 对开账户
back-to-back agreement 背靠背协议
back-to-back angles 背贴角钢;背靠背拼合角钢;背靠背(组合)角钢;角钢背距离;T形组合角钢(背对背设置的角钢)
back-to-back circuit 背对背电路
back-to-back connection 背对背连接;交叉连接
back-to-back contract 背靠背合同
back-to-back(counter) 加倍计数器
back-to-back credit 转开信用证
back to back credit export first 输出为先的对开信用证
back to back credit import first 输入为先的对开信用证
back-to-back diode 背对背二极管
back-to-back duplex bearing 成对双联轴承
back-to-back escrow 先进后出交易
back-to-back grate 背对背设置的炉栅;背对背设置的炉条;背对背设置的炉算
back-to-back houses 背靠背房屋;背对背房屋
back-to-back L-C 背对背信用证
back-to-back letter of credit 对开信用证;背对背信用证
back-to-back loan 背对背贷款
back-to-back method 反馈法;背靠背法
back-to-back running 反馈运行;背靠背运行
back-to-back starting 反向起动
back-to-back test 成对流子片试验;反馈试验;背靠背试验
back-to-back testing 反馈试验;背靠背试验
back to face 反面对反面相叠
back-to-front ratio 反正比
back tool rest 后刀架
back-to-school goods 学生用商品

back to the engine 备用发动机
back trace 历史航迹;逆程;回扫描;尾迹
back trace technique 向后追踪技术
backtrack 历史航迹;回溯;向下顶替;反向跟踪
backtracking 以较高级职工代替低级职工;沿原路折回;倒向追踪
back tracking method 追踪法;倒向追踪法
back tracking operation 回溯操作
backtrack process 回溯法
back transient 反向瞬变过程
back transient procedure 反向瞬变过程
back traverse 返测导线;反向导线
back-trough 背槽
back tube sheet 后管板
back turn 逆向匝;反作用匝
backturn device for drilling string 反管器【岩】
back-turning section 折返段
back turn pulley 倒带轮
back-twist 反捻
back twisting 反向加捻
back-type life support system 背包生命保护系统
backup 截流措施;背撑;填背砂(铸造用语);后援;后备;支援;支承;备用品;备用(的);备份【计】
backup aircraft 备用飞机
backup alarm 倒退警钟
backup and resume 备份和恢复
back up area 备用堆置场
backup area of port 港口陆域
backup battery 后备电池
backup bearing 支承轴承
backup belt 支承皮带;支撑皮带
backup block 衬块;支撑块
backup breaker 备用断路器
backup brick 衬里砖;墙心砖
backup chain tong 链条扳手;固定链钳
backup chock 支承辊轴承座;支承辊的轴承座
backup coat 加固层
backup commitment 备用委托
backup computer 备用计算机
backup control data set 备份控制数据集
backup copies saving 备份存盘;备份操作员
backup copy 后备副本;副本;备份拷贝
backup cost 支助费用
backup data 复查资料;备查资料;复制资料;备份资料
backup data set 备用数据集
backup each other 互相支持;相互备用
backup equipment 后续设备
backup facility 后续设备
backup file 备份文件
backup filter 辅助过滤器
backup flame 辅助火焰
backup flange 支持法兰;支撑法兰
backup fuse 保险丝
backup gear (螺旋回转的)反向给进齿轮
backup heel 侧支承反
back upholstering 靠背
back up in parsing 分析的回溯
backup insulation 隔热填料
backup intercept control 备用截击指挥系统
backup item 备份项目;备用项目
backup lamp 倒车灯
backup land 后方陆域;后方场地
backup light 倒退信号灯;倒车灯
backup(lining) brick 背衬砖;衬砌砖
backup man 辅助工人;扶铆钉工人;(卸钻杆时的)下管钳操作工;支持的人
backup masonry(work) 填背圬工
backup material 填料(接缝用);背衬材料;填充材料;填衬材料
backup memory 后备存储器
backup operator 备份操作员
backup oscillator 后备振荡器
backup overspeed governor 备用超速保安器
backup path 备用通路
backup plate 靠板;垫片;垫板;撑板
backup post 支柱;增力柱
backup power source 备用能源;备用电源
backup procedure 后备程序
backup process 回报过程
backup program(me) 支援方案
backup protection 后备保护;辅助保护装置;备用保护装置
backup punch 反向凸模

backup radar 辅助雷达
backup relay 后备继电器
backup ring 嵌入圈;支承圈;支承环;支撑环;挡圈
backup roll 支承辊;固定辊;支撑轧辊
backup roll arbor 组合支承辊的心轴
backup roll bearing 支承辊轴承
backup roller 支承轧辊
backup roll extractor 支承辊换辊装置
backup roll sleeve 支承辊的辊套
backup sand 充填砂;填砂
backup satellite 备用卫星
backup shaft 支持辊辊轴
backup spring 支承簧
backup strip 衬板;背垫条;嵌入箍;背垫木条;支撑板条
backup system 备用系统;后援系统;后备系统;后备设施
backup system library 备用系统库;备用系统程序库;备分系统程序库
backup tape 后备带
backup tile 空心垫砖;砌混水墙的空心砖
backup tongs 下管钳;固定大钳
backup unit 支撑装置
back up value 倒推值
backup wall 后衬墙;保护墙
backup washer 支撑垫圈;保护(垫)圈
backup water level 回升水位;壅高水位
backup wrench 下管钳;固定大钳
Backus-Naur form 巴克斯—诺尔范式
Backus-Naur notation 巴克斯—诺尔范式
Backus normal form 巴克斯范式
back valve 止回阀;回动阀;止逆阀;单向阀
back-veneer 背板;背里层胶板;背面胶合单板;背面单板;胶合板背衬
back vent 净气进口;背部通气管;虹吸排气管;(虹吸管的)顶部排气管;风巷;背通气管;通气管
backventing 背部通气
back vent pipe 背部通气管
back view 后视图;背视图;背面图;背面立视图;背立(面)图
back view mirror 后视镜;望后镜
back voltage 反电动势;反电压
back wall 挡土墙;后墙;背墙(防弹片杀伤而在堑壕或掩体内构筑的土垛);前脸墙;投料端墙
backwall echo 背墙回波
back wall in jack-in method 顶进后座墙
back wall of abutment 桥台的顶墙
backward 落后(的);向后的;倒行;倒向
backward acting decoder 反向译码器
backward acting regulator 反向作用式调整器;反馈作用调节器
backward aerofoil blade 反向流线型叶片
backward and bidirectional 反向与双向的
backward and forward bending test 正反弯曲试验;反复弯曲试验
backward and forward extrusion 正反挤压
backward and forward integration of material 原材料供应和产品销售的一体化
backwardation 交割延期;现货溢价;现货升水
backwardation contango 交割延期费
backward attribute 反向属性;反向表征
backward bending supply curve 向后弯曲供给曲线
backward bent vane 后弯式叶片
backward-bladed aerodynamic(al) fan 后弯叶片风机
backward bladed impeller 后弯叶轮
backward brush lead 电刷后向超前
backward busying 反向占线信号
backward chaining 反向链接
backward channel 返回通道;反向信道;反向通道
backward channel ready 反向通道传送准备就绪
backward counter 减法计数器;反向计数器
backward country 落后国家;经济落后国家;不发达国家
backward creep 金属后滑
backward crosstalk 后向串扰
backward curve centrifugal fan 后叶片离心风机
backward-curved blade 后弯叶片
backward-curved blade fan 后向曲片式风机
backward-curved vane 后弯叶片;后弯式风机
backward decay 反向衰变
backward difference 后向差分;反向差分;退差(值)
backward difference operator 向后差分算子

backward-differnece formula 反向差分公式
backward diffusion 反向扩散
backward diode 反向二极管
backward dip 船尾下沉
backward drainage 逆向水系;逆向排水
backward-dumping(wheel) scraper 向后卸料刮土机
backward echo 反向重复
backward effect 逆流效应
backward emission 反向发射
backward entropy 后熵
backward erosion 向源侵蚀;反旋冲蚀;反漩冲蚀;反向冲刷;溯源侵蚀;溯源冲刷
backward extrusion 反向挤压;反向挤出;反挤压
backward failure 向后衰竭
backward feed 逆流送料法;反向进刀;反进给
backward field impedance 反向磁场阻抗
backward field torque 反向磁场转矩
backward flashover 反向闪络
backward flow 逆流;回流;倒流;反向流;反流
backward folding 背向褶皱
backward forward counter 正反向计数器;加减计数器
backward gain 反向增益
backward gear 倒退装置;倒挡齿轮
backward going 倒退
backward hold 反向锁定
backward impedance 反向阻抗
backward inclination angle 后倾角
backward inclined type of impeller 后弯式叶轮
backward integration 向后整合;向后合并;反向企业合并
backward interpolation 后向插值;向后内插;退插(值)
backwardization 延期交割金
backward jet 逆向射流;反向射流
backward lead 后移
backward leaning vane 后弯叶片
backward letter 赔偿保证书
backward link 向后连接;反向链接;反向连接
backward linkage 后向联系;向后连锁率
backward linkage effect 后向联系效应;向后联系效应;向后连锁效果
backward linking 后向信号联结;后向联结
backward migration 倒向迁移
backward motion 倒退
backward moving average 后向移动平均
backward operator 滞后算子
backward option 反向选择
backward pass computation 后退计算法
backward path 回程通路
backward phase component 反向相分量
backward photodiode 反向光电二极管
backward polyphase sort 反多相分类
backward position 回程位置;反转位置
backward power 反向功率;反波功率
backward prediction 反向预测
backward processing 反向处理
backward pruning procedure 反向修剪过程
backward pull 后张力;反张力
backward reaction 逆反应
backward read 回读;倒读;反读
backward reading 反向读出
backward reasoning 反向推理
backward recovery 反向恢复
backward recovery time 反向恢复时间
backward reference 向后引用;返回访问;反向引用;反向参照
backward repositioning 反向再定位
backward resistance 反向电阻
backward rolling 后退运行
backward running 反转
backward scatter 向后散射
backward sequence 逆序;反向序
backward shift 后移位;反移;反向移刷
backward shift operator 后向移(位)算子
backward sight 后视
backward signal 返回信号
backward slip 后滑
backward-sloping lettering 左斜体注记
backward-sloping supply curve 向后倾斜供给曲线
backward spatial harmonic 反向空间谐波
backward spinning 反旋压
backward spring 回程弹簧;延缓弹簧;反向弹簧

backward stagger 后向斜罩
backward station method 反向测点法
backward steps 回代过程
backward stroke 回冲程;向后划桨;返回冲程
backward substitution 后退代入
backward supervision 反向监控;反向管理
backward surging 逆涌
backward sweep 后掠角
backward swept vane 反弯式叶片
backward take 倒拍
backward tension 反张力
backward thrust 反推力
backward tilting of the bucket 后倾铲斗
backward tracing 反向追踪
backward transfer admittance 反向转移导纳
backward transfer characteristic 反向转移特性;反馈特性
backward transter character 反向传输特性
backward travel 后退行程
backward type centrifugal fan 叶片反曲离心式通风机
backward unloading elevator 后卸升运器
backward vane 反向叶片
backward vision 后视
backward voltage 反向电压
backward wave 反向波;反射波;回波;返波
backward-wave amplifier 逆行波放大器;回波放大器;反向波放大器
backward-wave crossed-field amplifier 返波正交场放大管;反向波正交场放大管
backward-wave magnetron 回波磁控管;返波式磁控管
backward-wave oscillation 逆波摆动;返波振荡
backward-wave oscillator 反向波振荡器;回波振荡器
backward-wave parametric(al) amplifier 回波参量放大器;反射波参量放大器
backward-wave power 回波功率
backward-wave power amplifier 回波功率放大器;回波参量放大器
backward-wave tube 回波管
backward welding 右向焊;后退焊(接);后倾焊;向右焊;反手焊(接)
backwash 击岸回浪;排水流;浪涛;逆吹;回流;回卷(流);回卷浪;回卷波;回冲;后洗流;反响;反洗;反激浪;尾流;退水浪花
backwash coagulant optimization 反凝聚剂优化
backwash controller 返洗液控制器
backwash cycle 回洗周期;反冲洗周期
backwash drain gullet 反洗排水管
backwash effect 出口畸形发展效应;反溅效果
back washer 复洗机
back wash extractor 反萃器
backwashing 回洗;回流冲刷;反溅;反冲(洗);水泉消退
backwashing filter 反洗滤池
backwashing intensity of filter 滤池反洗强度;滤池冲洗强度
back washing machine 复洗机
backwashing method 反洗法
backwashing of filter 滤池反冲洗
backwashing water 反冲洗水
backwash limit 回洗界线
backwash line 反循环冲洗管路;反冲洗管
backwash liquor 反洗液
backwash marks 退水浪花痕
backwash pipe 回洗管道;反冲洗管
backwash piping 回洗管道;后洗管道
backwash procedure 反洗程序
backwash rate 冲洗强度;回洗率;反冲速率
backwash-regeneration cycle 反冲洗再生循环
backwash ripple mark 回流波痕;回卷波痕
backwash tube 反冲洗管
backwash valve 反冲洗阀
backwash waste outlet 返洗液废水出口
backwash water 反洗水;反冲水
backwash water requirement 反冲洗水需要量
backwater 浆挡水;循环水;挡水;背射波;回水;壅水(回水);再用水;潮汐壅水;背水
backwater area 回水区;回水面积
backwater at bridge pier 桥墩处滞水
backwater computation of reservoir 水库回水计算
backwater curve 壅水线;回水曲线;回水线;回水曲线;壅水曲线
backwater deposit 回水沉积(物);回水沉积的泥沙
backwater distance 回水距离;回水长度
backwater effect 回水作用;回水影响;回水效应;壅水效应
backwater effect due to bridge piers 墩柱壅水作用
backwater effect of bridge pier 桥墩处滞水作用
backwater envelope curve 回水包络(曲)线
backwater formula 回水公式
backwater from ice jam 冰壅回水
backwater function 回水函数;壅水函数
backwater gate 逆流防止门;回水闸;挡水闸门;防回门;防倒灌闸门
backwater head 回水头
backwater height 壅水高(度)
backwater height in front of bridge 桥前壅水高度
backwater in open channel 明渠中的回水
backwater in reservoir 明渠中回水;水库中的回水
backwater in sewage duct 污水管道中壅水
backwater jump 壅水高度;水跃
backwater length 回水距离;回水长度;回收长度
backwater levee 壅水堤
backwater level 回水高程;壅水高程
backwater limit 回水界限;回水极限;回水范围
backwater mark 滞水标志
backwater pressure 回水压力
backwater profile 回水纵剖面;回水曲线;壅水曲线
backwater pump 回水泵
backwater quality 回水水质
backwater range 回水范围
backwater region 回水区
backwater sediment 回水沉积(物);回水沉积的泥沙
backwater silting 回水区淤积
backwater slope 回水坡度;回水比降
backwater storage 动库容;回水蓄水量;回水库容(量);回收库容;壅水库容
backwater structure 壅水结构;回水结构
backwater suppressor 回水制止设备;回水减低设备
backwater trap 回水存水弯
backwater valve 回水逆止阀;回水阀;倒流阀门;止回阀;逆止阀
backwater wave 回波
backwater zone 回水区
back wave 回波;反波
backway 后退距离
back wear 背向磨损
backweight method 锤回单杆吊货法
back weld 封底焊(缝);底焊缝;背面焊;背焊缝
back welding 底焊;背焊(接);退焊法
back wheel 后轮;反向截止阀门
backwind 倒片
backwinding 逆风的帆;不受的风帆
back window 后窗
back wiring 反面布线;背面布线;背景布线
backwoods 偏远的森林地区;森林地带;半开垦地带;边远林带;林区人;处女林
backwoodsman 林区人
back work 井下工作面以外的工作;后退式回采;后勤(辅助)工作;辅助工作
backyard 后院;后庭;后天井;后备场地
back yard garden 后花园
back yard industry 小型工业
bacon 火腿红(陶瓷);带状滴石;薄层方解石
Bacon fuel cell 培根燃料电池
bacon peat 含油泥炭土
Bacor 巴科尔刚玉锆英石(耐火材料)
bactard 白琥珀
bacteria 细菌
bacteria action 细菌作用
bacteria-algae system 菌—藻系统
bacteria amount 细菌总数
bacteria bearing 带菌的
bacteria bed 菌床
bacteria composition quality of water 水细菌成分分析
bacteria contamination 细菌污染
bacteria corrosion 细菌腐蚀
bacteria disinfection 灭菌消毒
bacteria ether 醋醚
bacteria filter 生物滤器
bacteria free 无菌的
bacterial action 细菌作用
bacterial analysis 细菌化验;细菌分析
bacterial and algae control 菌藻控制
bacterial attack 细菌侵蚀
bacterial balance index 细菌平衡指数
bacterial bed 细菌滤池;细菌床
bacterial colony counter 菌落计数器
bacterial community 菌群
bacterial concentration 细菌浓度
bacterial contaminant 细菌污染物
bacterial contamination 细菌污染
bacterial content 细菌含量
bacterial conversion 细菌转化
bacterial corrosion 细菌腐蚀
bacterial count 细菌计数;细菌繁殖数
bacterial cultivation 细菌培养法
bacterial culture 细菌培养物;细菌培养
bacterial culture medium 细菌培养基
bacterial decay 细菌衰减
bacterial decomposition 细菌分解(作用)
bacterial degradation 细菌降解法
bacterial degumming method 微生物脱胶法
bacterial density 细菌密度
bacterial dewaxing 细菌脱蜡
bacterial digestion 细菌消化
bacteria leaching 细菌浸出
bacterial equilibrium 细菌平衡
bacterial examination 细菌检验
bacterial examination of water 水的细菌检验
bacterial exudation 病原细菌的渗出块
bacterial fertilizer 细菌肥料
bacterial filter 滤菌器;细菌滤器
bacterial growth curve 细菌生长曲线
bacterial indicator 细菌指示物;指示细菌
bacterial insecticide 细菌性杀虫剂
bacterial leaching 细菌沥滤
bacterial metallurgy 细菌冶金
bacterial mortality 细菌死亡
bacterial organic carbon 细菌性有机碳
bacterial oxidation 细菌氧化(作用)
bacterial pollution 细菌污染
bacterial proof filter 防细菌滤器
bacterial prospecting 细菌勘察
bacterial purification of water 水的细菌净化
bacterial reduction 细菌减少
bacterial resistance 细菌抗力;细菌抗药性
bacterial slime 细菌黏液
bacterial species 菌种
bacterial survey 细菌测量
bacterial test 细菌检验
bacterial tracer 细菌示踪物;示踪细菌
bacterial treatment 细菌净化;细菌处理
bacterial vaccine 细菌菌苗
bacterial variation 细菌变异
bacteria pollution 细菌污染
bacteria reduction 细菌还原(作用)
bacteria resistance 抗菌能力
bacteria survey 细菌测量方法
bactericidal 杀菌的
bactericidal action 杀菌作用
bactericidal action of chlorine 氯的杀菌作用
bactericidal agent 杀菌剂
bactericidal coefficient 杀菌系数
bactericidal effect 杀菌效应
bactericidal efficiency 杀菌效率
bactericidal lamp 杀菌灯
bactericidal paint 杀菌漆
bactericidal rays 灭菌射线
bactericide 杀菌剂
bacterila population 菌群
bacteriogene 细菌成因
bacteriological aftergrowth 细菌再生
bacteriological analysis 细菌学分析;细菌化验;细菌分析
bacteriological count 细菌计数
bacteriological counting apparatus 细菌计数器
bacteriological examination 细菌试验
bacteriological examination of water 水的细菌检验
bacteriological examination of water quality 水质细菌学检验
bacteriological filtration 滤菌
bacteriological index 细菌学指标;细菌指数
bacteriological laboratory 细菌检验室
bacteriological pollution index 细菌污染指数
bacteriological process 使用细菌的过程
bacteriological purification 使用细菌进行净化

bacteriological standards for water body 水体细菌标准
bacteriological water quality 含菌水质
bacteriological water sampler 微生物采水器
bacteriology 细菌学；微生物学
bacteriology laboratory 细菌化验室
bacteriolytic reaction 溶菌反应
bacterioplankton 浮游细菌
bacteriostasis diameter 抑菌圈
bacteriostat 抑菌剂；制菌剂
bacteriostatic 制菌剂
bacteriostatic action 抑菌作用
bacteriostatic(al) agent 抑菌剂
bacteriostatic(al) plasticizer 抑菌(性)增塑剂
bacteriotoxin 细菌毒素
bacterium filter 除菌过滤器
bacteroid 假菌体
Bacticin 混酚灵杀菌剂(商品名)
bactrian camel 双峰驼
baculiform 杆状的；棒状的
baculite 暗杆雏晶
bad 不良的
bad accident 险性事故
bad account 坏账；呆账
bad anchorage 不良锚地
bad and doubtful account 坏账与呆账；坏债务与呆滞债务
bad audibility 可听度差
bad bargain 亏本生意
bad batch 一批次货
bad bearing sector 不良扇形区
bad break 错断
bad cast 钢丝盘卷不匀
bad check 空头支票
bad cheque 空头支票
bad claim 不良债权
bad colo(u)r 墨色不匀；颜色不正
bad conductor 不良导体
bad contact 接触不良；不良接点；不良接触；不良触点
bad debt 坏账；呆账
bad debt allowance 坏账限额
bad debt audit 应收账款坏账处理
bad debt expenses 坏账费用
bad debt insurance 坏账保险
bad debt loss 坏账损失
bad debt ratio 坏账比率
bad debt recovery 收回坏账
bad debt written off 勾销呆账
baddeckite 含有铁质的白云母；赤铁黏土
baddeleyite 斜锆石
bad delivery 运交不合格商品；不当交割
bad delivery of securities 违反有价证券规定的选择权
bad earth 接地不良；虚接地
bad economic benefit 经济效益差
bad faith 恶意
bad fishing job 复杂的打捞工作
badge 徽章；符号；象征；标记
badge card 标记卡
badge card reader 标记卡阅读器
bad geometry 不利几何条件
badger 管道清理器；大号企口刨；清洗器；排水管清扫器；刷子；刨边刨；宽槽刨
badge reader 标记阅读器
badger-hair softener 软毛宽刷
badger plane 宽槽推刨；槽刨；宽刨推刨
Badger rule 巴杰尔定则
badger softener (油漆工用的)软毛刷；软毛宽刷
badges of trade 贸易象征
badging 印商标
bad grade goods 劣品
bad grading 劣级配
bad ground 难钻地层；复杂地层；不稳定岩层；不稳定地层；不良锚地；不良地层
bad hadit 不良习惯
bad havest 歉收
bad holding ground 锚抓力不好的底质；不良锚地
badian essence 八角茴香香料
badigeon 补洞填料；嵌填膏泥(用于木石孔隙中)；腻子；嵌填灰膏；油灰
badijum 嵌填膏泥；嵌填膏泥
Badin metal 巴丁合金
bad-i-sad-o-bistroz 十二旬风

Badisch converter 巴迪氏转化器
badius 栗褐色
bad joint 不良接合；半圆凸缝
badland 荒地；恶劣地；劣地；瘠地；荒原；崎岖地区；崎岖地(带)；重蚀地；断绝地
bad loan 过期而未偿还的贷款；呆账；迟还贷款
badly annealed 退火不良的
badly bleeding 冒顶；跑钢
badly bleeding ingot 冒了顶的钢锭；冒顶钢锭
badly broken 破损严重
badly defined boundary 不明显分界
badly faulted 严重断裂(的)；被断裂严重破坏的
badly graded 级配不良的
badly graded sand 级配不良的砂
badly rigged 不良的安装
badly stowed 装舱不当
badly tracked surface 严重破坏的路面
badly weathered 强风化的
badly weathered granite 强风化花岗岩
badly weathered layer 强风化层
badly weathered mudstone 强风化泥岩
badly weathered rock 全风化岩
badly weathered sandstone 强风化砂岩
badly weathered shale 强风化页岩
badly weathered soil 强风化土
badly weathered stratum 强风化层
badminton court 羽毛球场
bad mo(u)ld joint 合缝线；模缝线
badness of fit 拟合劣度
bad oil 劣质油
bad oil tank 不合格油料储罐
bad order 包装不良
bad-order track 待修车存放线；破损车辆停留线
bad quality 劣等质量；质量低
bad repute 坏名誉；不守信用；无信用
bad result 恶果
bad risk 凶险；风险大(的保险)
bad rock 破碎岩石
bad rock mass 坏的岩体
bads 石膏糊团
bad sharpness 各种标志清晰度差；清晰度差
bad slip 严重滑坡；强烈滑坡；强滑坡
bad smell 臭气
bad soil management 土壤管理不良
bad-soluble salt test 难溶盐试验
bad start 不良起动
bad stowage 装载不良；堆装拙劣
bad time 停工时间；萧条；不景气
bad top 不稳固顶板
bad track 不良磁道
bad trade 贸易不振
bad trapping 套印性能不良(油墨)
bad usual practice 不良习惯
bad visibility 能见度低；坏能见度
bad-weather 坏气候；恶劣天气；不良气候；恶劣气候条件
bad weather condition 不良气候条件
bad weathering 严重风化
bad-weather landing 恶劣气候条件着陆
bad weld 劣质焊件
bad white 白色不白
badwill 负信誉；负商誉；商誉坏
bad year 凶年；歉收年
Baer's law 拜尔定律
Baeyer strain theory 拜尔张力学说
Baeza method 贝泽热压硬质合金法
bafertisite 硅铷铁钛矿；铷铁钛石
Baffin Bay 巴芬湾
baffle 拢海器；闷头；缓冲板；消力；阻隔板；助声器；折流板；遮护物；障板；栅板；导流叶板；导流体；导流板；挡雨板；挡浪板；挡光板；分流堰；分流墩；反射体；无定向风
baffle aerator 隔板曝气池；砥磴式通气器；挡板曝气器
baffle area 阻隔区
baffle arrangement 挡板装置；折流板排列式；挡板排列法
baffle beam 隔声梁
baffle blade 折流叶片
baffle blanket 吸音毡；吸声毡
baffle block 消能块；消能墩；消力块；消力墩；障块；障板
baffle board 截流挡板；烟道风帽；烟囱风帽；折流板；隔音板；隔声板；导流板

baffle board of sedimentation tank 沉淀池反射板
baffle box 消能箱；导料槽
baffle chamber 障板室；隔板室；隔板集尘区；挡板室
baffle cloth 喇叭布
baffle collar 带封挡圈的接箍(注水泥用)
baffle column 挡板塔
baffle-column mixer 挡板混合塔
baffle core 挡渣芯
baffle crusher 冲击反射式破碎机；冲击发射式破碎机
baffle cut 折流板弓形缺口
baffled 带有挡板的
baffled biologic(al) aerated filter 折流板曝气生物滤池
baffled decant line 拦水线
baffled evapo(u)rator 折流蒸发器
baffled exchanger 折流热交换器
baffled mixer 搅拌机
baffle(d) mixing chamber 隔板式混合池
baffled mixing tank 隔板混合槽；挡板混合槽
baffled photometer 反射式光度计
baffle(d) reaction chamber 隔板式反应室；隔板式反应池
baffled rotating packed bed 折流板旋转填料床
baffled settler 挡板式澄清器；挡板式沉降罐
baffled side weir 溢流侧堰；分流侧堰
baffled spray column 挡板喷雾萃取塔
baffle facility 消力设施
baffle feed heater 挡板式进料加热器；挡板进料加热器
baffle flame holder 挡板式火焰稳定器
baffle gate 单向门
baffle gear 板牙齿轮
baffle groin 消能丁坝
baffle heater 板式加热器
baffle junction 缓冲式交叉
baffle loudspeaker 障板扬声器
baffle mark 闷头印(玻璃制品缺陷)
baffle paddle 折流桨叶
baffle paint 保护色涂料；掩护色漆；伪装掩护色用涂料；伪装涂料；伪装迷彩漆
baffle painting 盖色涂装；伪装涂色；伪装迷彩涂装；涂保护色
baffle pier 挡墩；消能墩；消力墩；砥磴
baffle plate 火墙；模底板；缓冲板；消力板；折流板；遮护板；隔板；导向板；挡热板；分布板
baffle plate convertor 挡板变换器
baffle plate type separation 挡板分离
baffle plate type separator 挡板分离器
baffler 减音器；消能池；折流器；反射器
baffle reaction chamber 隔板反应池；隔板反应室
baffle reflection test 超声波反射试验
baffle ring 挡环
baffler module 消能构件
baffler prong 消能丁坝
baffler teeth 消能齿；消力齿
baffler wall 消力墙
baffles 导向环
baffle scrubber 折流洗涤器；挡板式洗涤器
baffle separation 折流分离
baffle separator 挡板式分离器
baffle shield 隔板屏蔽
baffle sill 门槛；消力槛
baffles in aeration tank 曝气池中挡板
baffle spray tower 挡板喷雾塔；挡板喷淋塔
baffle(s) spacing 折流板间距
baffle still 门槛
bafflestone 生物捕积岩
baffle structure 消力设施；消力结构
baffle threshold 消能槛；消力槛
baffle tower 挡板塔
baffle-type 百叶窗式的；有叶窗式的；挡板式的
baffle type collector 挡板式收尘器；百叶窗式收尘器
baffle type scrubber 折流式洗涤器
baffle vane of spiral case 蜗壳舌板
baffle wall 消能墙；折流墙；隔声墙；隔墙；砥墙；挡墙；挡火墙；分水墙
baffle wall in sedimentation tank 沉淀池隔墙
baffle washer 折流清洗器；折流洗涤器
baffle weir 砥堰
baffling 节流阀调节；加隔板；活门调节；阻遏；折流
baffling wind 无定向风
baftstone 障积岩
bag 料袋；满风；袋状物；袋

bag and package loading plant 袋物及小件货物装载机
bag and spoon dredge(r) 袋勺式挖泥船;袋匙式挖泥船;人工挖泥船
bag and spoon dredging 人工挖泥
bag applicator 插袋机
bagasse 蔗渣;甘蔗渣;干蔗渣
bagasse baler 甘蔗渣压捆机
bagasse board 甘蔗渣板;甘蔗板
bagasse dust 甘蔗渣尘
bagasse fiber 甘蔗渣纤维
bagasse fibreboard 蔗渣纤维板
bagasse sheet 甘蔗渣薄板
bagasse wastewater 甘蔗渣废水
bagassosis 甘蔗渣肺
B-A ga(u)ge B-A 规
bag auger 袋装式螺旋钻
bag barker 小型鼓状剥皮机
bag bonding 袋压黏结
bag breaker 拆袋器
bag broken 包破
bag bunch-sealing machine 米袋机
bag bundle elevator 成捆纸袋提升机
bag burst 包破裂
bag cargo 袋装货
bag carrier member 传袋装置
bag cart 行旅车;推货小车
bag chair 纸袋托架
bag clasping machine 扣夹封袋机
bag cleaner 清袋器
bag cleaning 清袋
bag closer 缝袋机
bag closing machine 封袋机
bag collector 集料袋;货袋收尘器;货袋除尘器
bag concrete 袋装混凝土
bag container 集装袋;袋式集装箱
bag conveyer 袋装货输送带;粮袋输送器;送袋机
bag counter 袋子计数器
bag cutter 刮袋器;袋装刮土器
bag dam 袋装土(壤)或混凝土筑成的坝;土袋埝坝
bag deviator 袋子转向板
bag dredge 水底采泥机;捞网;水底采样袋
bag dust collector 滤袋除尘器;袋式集尘器
bag dust filter 袋式滤尘器;袋式过滤除尘器;袋滤器
bag elevator 举袋机;货袋提升器;货袋升降机
bag filling 灌包
bag-filling machine 装包机;装袋机
bag filling plant 装袋车间
bag film 薄膜袋
bag filter 滤袋;过滤袋;袋式收尘器;袋式滤尘器;袋式过滤器;袋式除尘器;袋滤器
bag filter fabric 圆筒过滤布
bag filter installation 袋式集尘装置
bag filter process 袋式过滤法;袋滤法
bag-filter-type collector 袋滤式收尘器;袋滤式集尘器
baggage allowance 免费行李;行李重量限度
baggage arrival operation income 行李到达作业收入
baggage cabin 行李舱
baggage carrier 行李架
baggage car(t) 行李车
baggage cellar 地下行李房
baggage check 行李票
baggage(check)room 行李房
baggage claim area 行李认领处
baggage compartment 行李舱
baggage conveyance system 行李输送系统
baggage conveyer 行李传送带
baggage corridor 行李走廊
baggage declaration 行李申报单
baggage declaration form 行李报关单
baggage elevator 行李电梯;运行李升降机;运行李电梯
baggage entrance 行李入口处
baggage fee 行李费
baggage flow route 行李流程路线
baggage grid 行李格架
baggage hall 行李间
baggage handling 行李管理;行李搬运
baggage handling counter 行李处理柜台
baggage handling facility 行李搬运设施
baggage hold 行李舱
baggage holder 行李支架
baggage insurance 行李保险
baggage lift 行李升降机
baggage load and unload track 行李装卸线
baggage office 行李房;行包房
baggage platform 行李站台
baggage rack 行李架
baggage revenue 行李收入
baggage room 行李室
baggage roundabout 行李环形道
baggages and parcels flow paths 行包流线
baggage shed 行李棚
baggage shelf 行李架
baggage space 行李处所
baggage sufferance 行李通关单
baggage ticket 行李票
baggage tractor 行李拖车
baggage transshipment income 行包中转作业收入
baggage truck 行李小车;行李推车;行李卡车
baggage van 行李篷车;行李车
baggage wagon 行李车
baggage way 行李地下通道
bagged aggregate 袋装集料
bag(ged)cargo 袋装货物;袋物
bagged cement 袋装水泥;成袋水泥
bagged concrete 袋装混凝土
bagged concrete aggregate 袋装混凝土骨料;袋装混凝土集料
bagged conveyer 袋运送器
bagged grain 袋装谷物
bagged gypsum 袋装石膏
bagged hydrated lime 袋装水化石灰
bagged material 袋装料
bagged rice 袋装水稻
bagged salt 袋装盐
bagger 掘沟机;泥斗;装袋器;装袋机;多铲斗;挖泥机;勺斗;挖泥船
bagger combine 自动装袋式联合收获机
bagger elevator 装袋升运器
bagger weigher 计重装袋器;装袋计量秤
bagging 装包;打包布;装袋;排散气孔的水泥袋
bagging and stitching plant 灌包缝包设备
bagging and weighing machine 装袋秤重机
bagging apparatus 装袋装置
bagging attachment 装袋装置
bagging auger 装袋螺旋推运器
bagging-bag 厚黄麻袋布
bagging chute 袋装滑运道
bagging department 装袋部门
bagging hopper 袋装漏斗
bagging machine 灌包机;打包机;装填机;装袋机;装袋机;袋装机
bagging machinery 装袋机械
bagging(-off) 装袋;用气袋堵管
bagging-off by gravity 重力装袋
bagging-off chute 装袋滑槽
bagging-off spout 装袋口
bagging packer 灌包机
bagging platform 装包平台
bagging point 装袋装置
bagging pot 皮囊壶
bagging scale 装袋秤
bagging screw 袋装螺钉
bagging spout 装包机插嘴
bag goods 袋装货
bag grouting 袋装注浆
baggy 宽松下垂的;袋状的;膨胀如袋的
Baggy bed 巴吉岩层【地】
baggy fabric 袋状织物
baggy wrinkle 松皱结
bag-hanger 挂袋架
bag heat-sealing machine 袋热合机
bagheera 巴格希拉毛圈丝绒
bag hides 袋皮
bag holder 麻袋夹持器
bag holed by cargo hooks 包有货钩洞
bag hook 吊袋钩
baghouse 袋式过滤器;袋滤器;袋式除尘器;沉渣室;集尘室;集尘袋室;袋(滤)室;大气污染微粒吸收器
baghouse precipitator 滤袋除尘器;袋式除尘器
baghouse with mechanical shake cleaning 机械振动清灰袋式除尘器
bag job 非法搜查
bag lanyard 水手袋系绳
bag lift truck 举袋卡车
bag-like cave 袋状洞
bag loader 码包机
bag loading machine 水泥袋装载机
bag lock 袋锁
bagman 行商;推销员
bag mo(u)lding 袋形线脚;袋压成型;袋塑成型;袋模成型
bag net 袋网;吊兜
bagnette 小凸圆体花饰;小圆线脚;细小线条
bagnio 拘留所(土耳其);浴堂;妓院;(意大利、土耳其式的)浴室
Bagnold number 伯格诺德数
bag of cement 袋装水泥
bag of foulness 高压瓦斯包
bagon 沼泽林
bag opening machine 撑袋口机
bag-out 散装(货)袋卸系统
bag overturning machine 翻袋机
bagpack 背包
bag package 袋包装
bag packer 装袋机
bag-packing department 装袋部门
bag-packing machine 打包机
bag-packing plant 打包工厂
bag patched 包已缝补
bag pile 袋堆;袋包堆垛
bag piler 粮袋码垛机
bag piling machine 堆袋机;堆包机
bag placer 套袋机
bag plug 袋式堵头;袋栓
bag process 袋室除尘法;布袋收尘法
bag pump 风箱泵
bag rack 袋行李架;小包裹加
Ba-grease 钡基润滑油
bag reel 成卷纸袋
bag reep 缩帆带
bag reloading machine 叠包机
bag retainer 挂袋器
bag room 袋式集尘室
bag rope 袋索
bags 燃烧室空间
bag saddle 纸袋托架
bag sampler 袋式采样器
bag sampling 袋式取样;袋式采样
bag sand well 吸砂井
bags badly chafed 包擦破
bags chafed 包擦花
bag sealer 袋封口机;封袋机
bag sealing machine 纸袋封口机;袋封口机
bag seeding 袋播
bag sewer 缝袋机
bag sewing and dismantling 缝拆包
bag sewing machine 缝袋机
bag(s)for safe stowage 安全积载所需的袋装货
bag shaker 布袋震动器
bags mouths open 包开口
bag spiral slide 袋装货螺旋滑板;袋物螺旋溜槽
bag split 包撕裂
bag splitter 袋分裂器;拆袋器
bag squeezer 袋式挤压器
bags seam open 包开缝
bag stacker 摆包机;粮袋码垛机;码袋机
bag stopper 袋栓;袋阀
bag storage 纸袋库
bag swinging 振动滤袋法
bag system 口袋法(发给每个司钻一批钻头,每天检查钻头情况,以计算金刚石损耗的方法)
bag tank 软油箱
bag torn 包撕破
bag trap 袋形存水弯;袋式存水弯
bag truck 运袋卡车
bag turner 转袋装置
bag-type cell 软油箱
bag type cloth filter 袋式布滤器
bag type collection system 袋式集尘装置
bag-type collector 袋式集尘器
bag-type dust collection system 袋式除尘装置
bag type dust collector 袋式集尘器;袋式收尘器;袋式除尘器;布袋除尘器
bag-type duster 袋式除尘器
bag-type dust remover 布袋除尘器
bag-type filter 袋式过滤器
bag-type strainer 布袋滤器

baguette 小圆线脚;半圆饰;细小线条
baguio 碧瑶风
bag wall 火箱挡墙;砖窑隔火墙
bag weigh(ing) machine 称袋机
bag with hook 包有手钩洞
bagwork 袋装干拌混凝土;装袋作业;袋装混凝土堆层;砂包
bahada 山麓冲积扇;山麓冲积平原
bahamite 巴哈马石(细粒浅海灰岩)
Bahco air centrifuge 巴科型干式超微粒空气离心机
bahiaite 橄闪紫苏岩
bahianite 羟铝锑矿
bahnaluminium 巴恩铝铜合金
Bahnmetal 巴恩铅基轴承合金;巴恩合金
bahr 深水泉;[复]bahar
bahut 实心女儿墙;墙的圆形压顶;(挑檐上支持屋顶的)矮墙
bai 梅;黄沙雾
Baical old land 贝加尔古陆
baignoire 戏院最下层包厢;厅座
Baihuanshu 白环俞
Baikalian marine trough 贝加尔海槽
Baikalian orogeny episode 贝加尔构造作用幕
Baikal movement 贝加尔旋回
Baikal polarity hyperchron 贝加尔极性巨时
Baikal polarity hyperchronzone 贝加尔极性巨时间带
Baikal polarity hyperzone 贝加尔极性巨带
Baikal rift seismotectonic zone 贝加尔裂谷地震构造带
Baikal rift zone 贝加尔裂谷带
baikerite 贝地蜡
baikorite 钛镁尖晶石
bail 夹紧箍;戽斗水;戽;小勺;保证人;保释人;保释金;半圆形提环;卡圈;绳套;绳圈
bailable period 委托期
bail bond 保释保证书
bail bond lien 保释契约留置权
bail down 用捞砂筒清理钻孔;提捞
bailee 发运人;受委托者;受托人;受寄人;受托人
bailee clause 受托人条款
bailee's customers' insurance 受托人代客保险
bailer 捞筒;捞砂筒;泥浆泵;灰桨桶;戽斗;钻具提取管;抽筒【岩】;抽泥筒;抽泥机;财物委托人;排水吊桶;委托人;掏泥筒;水瓢戽斗;勺形钻
bailer bail 捞筒提环;捞砂筒钩环
bailer bit 勺形钻头
bailer-boring 用泥浆泵钻土
bailer conductor 钻孔加深(用)捞砂筒
bailer crownblock 捞砂(筒用)天车轮
bailer dart 捞砂筒下部活门;捞砂筒底阀
bailer dump 捞砂筒倒砂门
bailer grab 捞砂筒捞钩;打捞爪
bailer line 捞砂(筒)绳索
bailer link 捞砂筒连接环
bailer method of cementing 捞砂筒注水泥法
bailer sample 捞砂筒岩样
bailer sheave block 卷扬机滑车
bailer swivel 泥浆泵旋轴
bailer valve 捞砂筒阀门
bailer well 抽筒井井孔
bailey 城廓;城郭;城堡外墙
Bailey bridge 贝雷式桥;贝雷桥;活动便桥
Bailey flow meter 贝雷流量计
bailey form construction 城廓形建筑
Bailey meter 贝雷流量计
Bailey span 贝雷式桁架桥跨
Bailey test for sulphur 贝雷硫试验
Bailey truss 贝雷(式桁)架
bailey wall 水冷外壁;水冷耐火壁
bail handle 半圆形金属拉手;半圆形金属把手
bailiff 测量员;法警
bailing 泥泵清孔;排水;挖泥沙;提捞
bailing bucket 戽水斗;捞筒;捞砂筒
bailing depth 提务筒下入深度
bailing ditch 捞砂筒排水槽
bailing drum 捞泥沙用卷筒;捞砂卷筒;抽筒绞车卷筒
bailing experiment 扬水试验;抽水试验
bailing machine 捞砂机;自动吊罐机
bailing pulley 挖泥沙用滑轮
bailing reel 捞砂卷筒;抽筒绞车卷筒
bailing rope 捞砂(筒)绳索;挖泥沙有绳
bailing sheave 挖泥沙用滑车

bailing skip 排水箕斗
bailing test 抽水试验;挖泥沙试验;试抽
bailing times 提捞次数
bailing tub 捞泥沙用桶;捞砂桶
bailing tube 捞砂筒
bailing up 捞砂【岩】;用捞砂筒清理钻孔
bailing well (sampling) 挖泥沙井取样
bailiwick 本行范围;权限范围
Baillarger's external line 柏亚尔惹氏外线
Baillarger's internal line 柏亚尔惹氏内线
Baillarger's lines 柏亚尔惹氏日线
baillierod 水瓢罐
bailment 保释;寄托;财物委托;委托
bailment document 寄存依据
bailment lease 分期租让
bailor 财物委托人;委托人
bail out 保证偿还
bailout bottle 备用气瓶
bailout capsule 弹射座舱
bailsman 保证人;保释人
Baily furnace 贝利炭粒炉
Baily's beads 倍里珠
Bainbridge reflex 班布里奇反射
Bainbridge's mass spectrograph 班布里奇质谱仪
bainite 贝茵体;贝氏体
bainite ductile iron 贝氏体铸铁;贝氏法球墨铸铁
bainite finish point 贝氏体转变终了点
bainite hardening 贝氏体淬火
bainite start point 贝氏体转变开始点
bainite transformation 贝氏体转变
bainitic steel 贝氏体钢
bain-marie 蒸锅;水浴器;水浴
Baire function 贝尔函数
bait 加料(用铲子);有钩的长杆;自备食品(矿工);粗金刚石
bait box (管线工的)午餐盒
baiting 下料
bai-u 梅雨
bai-u rainy period 梅雨期
bai-u season 梅雨季
baize 粗呢;台面呢
bajada 山前冲积平原;山麓冲刷扇;山麓冲积扇;山麓冲积平原
bajada breccia 荒漠泥流角砾岩;山麓冲积平原角砾岩
bajir 沙丘长湖
Bajocian 巴柔阶【地】
bake 烘烤;烘焙;焙烧
bake and UV irradiation test 耐热和耐紫外辐射试验
bake cycle 烘烤周期
baked alumin(i)um paint 烘干的铝涂料
baked anode 预焙烧阳极
baked brick 欠火砖;焙干砖;晒干砖
baked carbon 炭精电极;碳电极;炭极
baked clay 烘干土;烧干土
baked contact test 烘烤接触检验
baked core 干型芯;干砂型心
baked cork 烘干成团的软木
baked cork brick 烧结软木砖;黏结软木砖;烘干软木砖
baked corkcard 烘过的软木板
baked enamel 烧搪瓷;烧珐琅
baked enamel coating 烧结搪瓷涂饰;烤漆饰面
baked finish 烘漆;烤漆饰面;烤漆处理;烘干饰面
baked flux 烧结焊剂;陶质焊剂
baked mo(u)ld 干型
baked(-on) enamel finish 烤漆饰面
baked-on synthetic(al) resin enamel 烧结合成树脂搪瓷
baked permeability 干透气性
baked plaster 烧石膏
baked resin 烘干的树脂
baked sand 烘干砂
baked strength 烘干强度;烘干强度
bakefill 磕包填埋
bake hardness 烘烤硬度
bake house 面包房;食品铺
bakelite 绝缘电木;胶木;电木;酚醛塑料
bakelite A 甲阶酚醛树脂;可溶酚醛树脂
bakelite adhesive 胶木人造树脂黏结剂;酚醛树脂胶黏剂;酚醛树脂胶合剂;树脂胶凝材料
bakelite B 不溶可熔酚醛树脂;酚醛树脂 B
bakelite bobbin 电木线圈管

bakelite bonded wheel 酚醛树脂结合剂砂轮
bakelite bowl 电木碗
bakelite C 不熔不溶酚醛树脂;丙阶酚醛树脂;酚醛树脂 C;不溶酚醛树脂
bakelite cement 酚醛树脂黏结料;酚醛树脂胶粘剂;酚醛树脂胶合剂;树脂胶凝材料
bakelite coating 酚醛塑料涂层
bakelite douche fittings 胶木灌洗配件
bakelited wood 绝缘胶木;电木
bakelite glue 胶木粘胶
bakelite-impregnated wood 用树脂浸渍的木料
bakelite lacquer 酚醛塑胶漆
bakelite mo(u)lding 电木造型;电木模型
bakelite paint 胶木漆
bakelite paper 胶木带;电木纸;电木带
bakelite paste 胶木膏;电木膏
bakelite powder 胶木(电木)粉;酚醛塑料粉
bakelite resin 胶木树脂
bakelite varnish 胶木清漆;胶木漆;电木漆;酚醛树脂清漆;清烤漆
bakelized paper 胶木纸;电木纸
baken peat 密实泥炭(形成泥炭沼泽地底基)
Bake-Nunn camera 贝克—隆摄影机
bake-on 焙干(砖头等);烘烤
bakeout 退火;烘烤;烘干
bakeout furnace 烘烤炉;烘烤箱;烘炉;烘箱
bakeout oven 烘箱;烘烤炉
bakeout temperature 烘烤温度
bake oven 烤箱;烘烤;烘炉;烘炉;恒温器;干燥箱
baker 烘干器;干燥炉;高温焙烘机
Baker bell dolphin 钟形系船墩;贝克系船墩;贝克式钟形系船墩;贝克式钟形系船柱
Baker cement float shoe (带止回阀的)贝克水泥浮鞋
Baker clamp 贝克钳位
Baker dolphin 贝克系缆桩
bakerite 纤硼钙石;瓷硼钙石
Bakerly normalized value 贝克莱标准数
Baker-Nunn camera 贝克—纳恩照相机;贝克—纳恩摄影机
baker oven 烤面包炉
Baker-Perkin mixer 贝克—帕金混合机
Baker Schmidt mirror system 贝克—施密特反射镜系统
Baker-Schmidt telescope 贝克—施密特望远镜
Baker shoe 贝克管鞋(方形齿刃)
Baker's rules 贝克挡土墙设计规则
baker's table 厨师案桌
Baker's water jet pump 贝克喷水抽气泵
bakery 烤面包室;食品店;面包厂
bakery and confectionery 食品厂
bakery room 面包房
bake sand bake process 烘烤—打磨—烘烤法
bake shop 面包店
bake system 烘烤系统
bakie 中间包
baking 烤干;烘烤(法);烘;低温干燥
baking black varnish 沥青黑烘漆;黑沥青(清)烘漆
baking coal 烤焦煤;结焦煤
baking enamel 烘烤瓷釉;烤干瓷漆;烘烤搪瓷;烘干瓷漆
baking enamel coating 烘烤搪瓷涂料
baking epoxy ester paint 环氧酯烘漆
baking filler 烘干填料
baking filling composition 烘干填料成分
baking finish 考漆;烘烤涂料;烤漆饰面;烘漆
baking furnace 烘烤炉;焙烧炉
baking hot 极热的
baking industrial enamel 烘干工业瓷釉
baking japans 沥青烘漆
baking machine 焙烘机
baking of resin 烘烤树脂
baking of varnish 烘烤油漆;漆的烘烤
baking oven 焙烧;烘箱;烘炉;烘炉;高温焙煤箱;干燥炉
baking paint 烘干油漆;烤干涂料;烘漆;烘烤涂料
baking pan 烘盘
baking powder 发酵粉;焙粉
baking quality 烘烤质量;烘干质量
baking schedule 烘烤(作业)规程
baking soda 小苏打;碳酸氢钠
baking stopper 烘干泥塞头
baking temperature 烘烤温度
baking test 烘干试验

baking time 烘烤时间
baking tunnel 烘干隧道炉
baking varnish 烤漆;烤干清漆;烘用清漆;烘干涂料;烘干清漆;热干燥漆;清烤漆;清烘漆
baking vehicle 烘干(型)漆料
baking zone 烘干层
Bakker-Nun camera 巴克—纳恩摄像机
baktun 白克顿
Baku 巴库地毯
Baku basin 巴库盆地
Baku olive oil 巴库橄榄油
bal 矿山;矿群;井上女工
Balaerti geodome system 巴拉额尔齐斯地穹系
balance 均势;均衡;结余;结算差额表;结存;衡算;相抵;余量;余额;秤;差额;权衡;配重平衡块;尾数;天平;收付平衡
balance a budget 平衡预算
balance account 决算(账户);结账;结余账目;结算账户;差额账;差额清算;清结账户
balance account book 余额登记簿
balance account with 与…结清账目
balance agitator 摆式搅拌器;摆式抖动器
balance amount 结算余额;差额
balance analysis of profit and loss 盈亏临界点分析
balance anchor 平衡重下锚
balance approach 均衡办法
balance arc 摆弧;平衡弧
balance area 均衡区
balance arm 均衡梁;秤杆;摆梁;平衡式窗扇托臂;平衡杆;平衡臂;天平臂
balance arm file 小三角锉
balance at the beginning of the period 期初余额
balance band 重心箍
balance bar 平衡杆(件);天平臂
balance based on an oversupply of products 按长线平衡
balance based on a shortage of products 按短线平衡
balance beam 成对的杆件;平衡杆件;平衡梁;天平梁
balance betweeen supply and demand of commodities 商品供求平衡
balance between cash receipts and payments 现金收支平衡
balance between credits and payments 信贷收支平衡
balance between imports and exports 进出口平衡
balance between income and expenditures 收支平衡
balance between pastures and livestock 畜草平衡
balance bob 平衡器;平衡锤;平衡臂
balance book 总账余额簿;分类账余额簿;分类账差额簿;分类余额簿
balance box 均衡箱;衡重盒;平衡箱;平衡锤;吊车起重臂平衡箱
balance bridge 开启桥;平衡桥;竖旋桥;(衡重式的)仰开桥
balance bridge with fixed axis 有固定轴线的开启桥
balance brought down 余额结转;承转余额
balance brought forward 结转(的)余额;转来的余额
balance brought forward from last year 上年结转余额
balance brought over from the last account 前期结转余额
balance brow 踏板
balance budget 平衡预算
balance budget with surplus 收支平衡并有盈余的预算
balance bunker 平衡仓
balance bush 平衡盘套筒;平衡衬套
balance calculation 平衡计算
balance calculation area 均衡计算区
balance calibration 天平校准
balance cantilever construction 平衡悬臂施工
balance cantilever segmental construction 平衡悬臂分段施工
balance car 平衡车
balance carried down 余额移后;余额结转;转次页;差额移下
balance carried forward 结转下期;余额承前
balance carried over 余额结转下期

balance chart 平衡图
balance check 零位检查;零点校验;对称校验;平衡校验;平衡检查
balance check mode 检衡式;平衡检验式
balance cock 摆轮夹板;摆夹板;平衡开关
balance coil 平衡线圈
balance column 天平架
balance cone 锥式液限(贯入)仪;平衡锥
balance contrast 平衡反差
balance converter 平衡换能器;平衡变换器
balance cost 均衡成本
balance crane 均衡式起重机;有平衡重的起重机;平衡(式)起重机
balance cuts and fills 均衡挖填;挖填土方平衡;填挖(上方)平衡
balance cylinder 平衡室;平衡汽缸
balance cylinder cover 平衡汽缸盖
balanced 有补偿的;均衡性的;对称的
balanced action 平衡作用
balanced addition 已抵除增加额
balanced allocation 均衡布局
balanced allocation of risk 均衡分摊风险
balanced ambient room type calorimeter 平衡环境房间型量热计
balanced amplifier 补偿放大器;平衡放大器
balanced arch 平衡拱
balanced area 平衡面积
balanced area method 面积平衡法
balanced armature 平衡衔铁
balanced-armature pick-up 平衡式衔铁拾声器
balanced armature unit 平衡式舌簧单元
balanced balloon 平衡气球
balanced base foundation 平衡底板基础
balanced biaxial tension 二向平衡拉伸
balanced bit 找中钻头
balanced block design 平衡区组设计
balanced box 平衡锤
balanced brake 平衡制动器
balanced bridge 平衡电桥
balanced bridge circuit 平衡电桥电路
balanced budget 自平预算;平衡预算;平衡的预算;收支平衡
balanced budget multiplier 平衡预算乘数
balanced budget rule 平衡预算规则
balanced budget with surplus 节余预算
balanced bus 均衡母线
balanced cable 对称电缆
balanced cable crane 平衡缆式起重机
balanced cantilever erection 平衡悬臂施工;平衡悬臂架设(法)
balanced cantilever method 平衡悬臂法
balanced capacity 平衡能力
balanced circuit 平衡电路
balanced clamper 平衡钳位电路
balanced code 平衡码
balanced condition of flame 稳定火焰
balanced configuration 均衡配置;对称配置
balanced connecting rod 平衡连杆
balanced construction 均衡施工;均衡结构;平衡型结构;平衡结构;平衡构造
balanced construction plywood 平衡结构胶合板
balanced control 平衡控制;平衡调整
balanced core 悬臂芯;平衡型心
balanced core barrel 平衡岩芯管
balanced crank 平衡曲柄
balanced crew 均衡工班
balanced cross-section 平衡剖面;平衡横断面(厚边混凝土路面)
balanced current protection 电流平衡保护(装置)
balanced cuts and fills 均衡挖填;平衡挖填;平衡填挖方
balanced cutting 平衡切削
balanced deflection 对称偏移
balanced demodulator 平衡解调器
balanced-density slurry technique 平衡密度泥浆法
balanced density solvent 平衡密度溶剂
balanced design 均衡设计;平衡设计
balanced detector 平衡检波器
balanced development 均衡发展
balanced diet 平衡膳食
balanced differences 平衡差
balanced distribution 均衡布局;平衡布局
balanced door 卸载门;平衡式上下推拉门;平衡门
balanced double mechanical seal 平衡式双端面机械密封

balanced draft 平衡通风;平衡送风
balanced draft sampling 平衡吸气取样
balanced draught 平衡式通风
balanced drawbridge 上开桥
balanced dynamometer 平衡功率计
balanced earthwork 均衡土方工程;平衡土方量;平衡土方工程;挖填平衡的土方工程;填挖平衡的土方工程
balanced eccentricity 界限偏心距
balanced economic development 平衡的经济发展
balance-delay cost 滞延成本
balanced electro-paint sprayer 静电喷漆机
balanced engine 平衡式发动机
balanced environment 平衡环境
balance(d)equation 平衡方程(式)
balanced erection 吊挂装配
balanced error 比较误差;摆轮误差;平均误差;平衡误差
balance des paiements courants 经常项目
balanced excavation 挖土平衡;平衡开挖
balanced factor 钻进规程参数
balanced feeder 平衡馈线
balanced filter 衡消滤波器;对称滤波器
balanced flow system 等惯性输水系统;水流分配均衡的输水系统
balanced flue 平衡烟道
balanced-flued appliance 密闭式(气用)具;咽气平衡炉具
balanced flued gas heater 平衡烟道式采暖炉
balanced flue nozzle 平衡烟道出口
balanced full trailer 拖带挂车;台车
balanced fund 平衡法的投资;平衡资金
balanced gasoline 平衡汽油
balanced gate 平衡闸门;自动闸门;自动阀门
balanced grade 均衡坡度;平衡坡度;平衡比降
balanced grading 挖填均衡的坡度;挖填平衡的坡度设计;土方平整;挖填平衡的坡度
balanced growth 均衡增长
balanced guard 平衡卫板;平衡防护装置
balanced gyroscope 平衡陀螺仪
balanced handle 球形手柄
balanced hoist 平衡升降机
balanced hole 均压孔;垂直炮眼
balanced housing 带平衡装置的机架
balanced impedance 平衡阻抗
balanced incomplete block design 平衡不定全区组设计
balanced incomplete block experiment 平衡不定全区组试验
balanced ingot 平衡钢锭
balanced-in-plane contour 平面封头曲面
balance district 均衡地段
balanced divisor 平衡因子
balanced jacking method 平衡顶进法
balanced laminate 均衡层合板
balanced lever-type bell ga(u)ge 杠杆平衡式钟形压力计
balanced line 对称双线线路;对称传输线;平衡线;平衡传输线
balanced linkage 平衡衔接
balanced load 平衡受力;均衡荷载;对称荷载;对称负载;平衡荷载;平衡负载;平衡负荷
balanced load meter 平衡负载仪表;平衡负载电度表
balanced load stress 均衡荷重应力;均衡荷载应力
balanced loop 平衡环;平衡回路
balanced-loop antenna 对称环形天线
balanced merge 对称合并;平衡合并;平衡归并
balanced method 对称法;平衡法;天平法
balanced mill 二辊式轧板机
balanced mixer 平衡混频器
balanced modulation 平衡调制
balanced modulator 平衡调制器
balanced moment 均衡弯矩;平衡弯矩;平衡力矩
balanced multivibrator 平衡多谐振荡器
balanced needle valve 平衡针阀
balanced network 平衡网络;平衡管网
balanced noise limiter 平衡噪声限制器;平衡式限噪器
balanced null point 稳定零点
balance dock 平衡(式)浮坞;平衡(式)船坞
balanced opposed compressor 对称平衡式压缩机
balanced opposed reciprocating compressor

对称平衡往复式压缩机
balanced optic(al) mixer 平衡光学混频器
balanced oscillator 平衡振荡器
balanced output 平衡输出
balanced overall development planning 平衡的全面发展规划
balance down 结转(金额);滚存余额
balanced polymorphism 平衡多态现象
balanced polyphase load 多相平衡荷载
balanced population 稳定的种群
balanced potential 平衡电势
balanced-pressure blow pipe 同压吹管
balanced-pressure rotameter 压力补偿转子流量计
balanced pressure torch 等压式焊(割)炬
balanced print 平衡芯头
balanced production line 均衡生产线
balanced product mixer 平衡乘积混频器
balanced profile 平衡剖面;平衡纵断面;填挖平衡的纵断面
balance draft 均匀通风
balanced reaction 平衡反应
balanced reaction rudder 平衡反应舵
balanced receipts and payment 收支平衡
balanced reduction of protective measures 均衡削减保护措施
balanced reinforcement 平衡钢筋;平衡配筋
balanced reinforcement beam 界限配筋梁;平衡配筋梁
balanced relay 差动继电器;平衡继电器
balanced repeater 对称式增音器
balanced reservoir 平衡水箱
balanced rigid 均衡刚架
balanced rigid frame 平衡构架
balanced ring modulator 平衡环形调制器
balanced risk 平衡危险性
balanced rock 坡栖岩块
balanced rotary kiln 平衡回转窑
balanced rudder 平衡舵;平衡舱【船】
balanced running 均衡运转
balanced salt solution 平衡盐液
balanced sash 平衡式窗扇;上下推拉窗;框格窗;滑窗
balanced sash window 竖向活动框格窗
balanced schedule 平衡计划进度表
balanced section 平衡截面
balanced sensitivity 平衡灵敏度
balanced shutter 平衡式百叶窗
balance(d) slide valve 平衡滑阀
balanced sluice 平衡水闸
balanced solution 缓冲液;平衡溶液
balanced solvent 平衡的溶剂
balanced sort 平衡分类
balanced sorting 对称排序;平衡分类法
balanced state 摆轮状态;平衡状态
balanced station 对称站
balanced steel 半镇静钢;半脱氧钢
balanced steel ratio 经济含钢率;平衡配筋率
balanced steps 扇形踏步宽度平衡法(螺旋形楼梯的);均衡踏步;扇形踏步
balanced stock 平衡储存
balanced strain 平衡品系
balanced superelevation 平衡超高(度)
balanced supply and demand of energy 能源供需平衡
balanced surface 平衡面;平衡操纵面
balanced system 对称系统
balanced system of ventilation 均衡制通风系统;通风平衡系统
balanced tank 平衡水柜
balanced taper roller bearing 平衡式滚锥轴承
balanced termination 平衡端接;平衡闭合电路
balanced thermometer 平衡式温度计
balanced three phase circuit 平衡三相电路
balanced three-phase system 平衡三相制
balanced threewire system 平衡三线制
balanced time 平衡时间
balanced to ground 对地平衡;对称接地
balanced top roll 非传动上辊
balanced trade 贸易平衡
balanced traffic 平衡交通运输
balanced transmission line 平衡传输线
balanced transport 均衡运输
balanced transportation problem 均衡运输问题
balanced tray thickener 平衡式多层增稠器;盘式增稠器;平衡多层增稠器
balanced tree 平衡树;平衡树
balanced twist 平衡加捻
balanced type 均衡型
balanced type carburet(t)or 平衡式汽化器
balance due 结欠(金额);不足额;欠款
balance due from 人欠
balance due from you 你结欠我方
balance due to 欠人
balance due you from us 你方结欠;你方结存
balance(d) valve 平衡阀;均衡阀
balanced vane relay 平衡叶片式继电器
balanced ventilation 均匀送风;均衡通风;平衡通风
balanced ventilation system 平衡通风系统;均衡通风系统;均衡制通风系统
balanced vertical lift 平衡直升装置
balanced vertical ship-lift 平衡重式升船机;衡重式垂直升船机
balanced voltage 平衡电压
balanced weight 均衡锤;平衡块;平衡重;砝码;配重
balanced weight lever 衡重杆
balanced weir 自动倾动式堰
balanced welding 对称焊
balanced wicket 活叶闸门;自动翻板闸门
balanced winders 转弯踏步
balanced yarn 平衡纱
balance elements 均衡要素
balance equation 均衡方程(式)
balance equation of confined water 承压水均衡方程
balance equation of phreatic water 潜水均衡方程
balance equation of total groundwater quantity 总的水量均衡方程
balance error 对称误差
balance error method 均衡误差法
balance exercise 平衡运动
balance expenditures with income 收支平衡
balance floating dock 平衡式浮坞
balance for specific gravity 比重天平
balance forward 结转余额;结转下页;余额移下;滚存余额
balance frame 船平行中体处(或船的中部的)肋骨;平衡机架
balance gate 均衡闸门;平旋门
balance gear 差动传动;补偿器;平衡齿轮
balance hanger 平衡吊架
balance height 平衡高度
balance held on deposits 存款余额
balance hoist 平衡起重机
balance hole 平衡孔
balance indicator 平衡指示器;天平指针
balance in hand 库存余额
balance in ledger 总账余额
balance in our favour 结欠我方;我方余额;我方受益的余额;我方结存
balance in your favour 你方(受益的)余额;你方结余;你方结存
balanceless 不平衡的
balance level 水准仪;水准器
balance lever 平衡杠杆
balance life 均衡寿命;等寿命
balance-like oscillations 天平式摆动
balance line system 平衡管道系统
balance magnetometer 电磁秤
balance mark 平衡标志
balance mass 平衡质量
balance matching 平衡配板
balance measurement 重量试验;天平法测量
balance mechanism 平衡机构
balance method 平衡方法;平衡表法;天平法(测密度)
balance method of stores accounting 材料核算余额法
balance mill 平衡铣
balance momentum 平衡动量
balance of accounts 账面上的余额
balance of activity 活度平衡
balance of amount 款项余额
balance of appropriation 经费结余
balance of assembly 装配程序
balance of births and deaths 出生和死亡的差额
balance of buying contract 购买合同余额
balance of capital account 资本账户结余
balance of capital movement 资本流动差额
balance of clearing 清算余额
balance of contract 现有契约限额
balance of current account 经常项目收支;经常项目差额;往来账户收支情况
balance of cut-and-fill 挖填平衡;土方平衡
balance of debt 负债余额
balance of estate 财产余额
balance of exchange 汇兑尾数
balance of external payments 对外支付差额
balance of finance and materials 财政和物资平衡
balance of finance credits an materials 三大平衡(指财政、信贷和物资平衡)
balance of foreign exchange 外汇收支结算
balance of foreign trade 贸易收支差额;外贸收支情况
balance of freight 运费余额
balance of fuel 燃料平衡表
balance of groundwater quantity 地下水水量均衡
balance of heat 热平衡
balance of income and outlay 收支平衡
balance of indebtedness 债权差额
balance of international payments 国际收支平衡;国际收支差额;对外收支结算
balance of manpower resources 人力资源平衡表
balance of materials 物料衡算;物料平衡
balance of matters 物料衡算;物料平衡
balance of migration 移入/移出的差额
balance of money loan over stock loan 货币贷款对股票贷款的差额
balance of mutual accounts 交互账户差额
balance of national economy 国民经济平衡
balance of nature 自然平衡;生态平衡
balance-of-payment assistance 收支差额补助
balance of payments 收支差额;支付平衡表
balance of payments credits and debits 借贷收付平衡
balance of payments deficit 收支逆差
balance of payments disequilibrium 收支不平衡
balance of payments on current account 流动账目收支差额;流动项目收支差额
balance of potential 电势均衡
balance of power 均势;动力平衡
balance of power level 电平平衡
balance of precision 精密天平
balance of preparation and winning work 采掘平衡
balance of profit 利润结余
balance of retained earnings 留存收益余额
balance of right and obligations 权利与义务的平衡
balance of roof 拱脚
balance of solar radiation 太阳辐射平衡
balance of stock 永续盘存制;存货余额
balance of stock system 成品盘存制
balance of stores 库存余额
balance of stores records 材料差额记录;库存余额记录;库存平衡记录;材料记录余额
balance of stores sheet 库存平衡表
balance-of-stores system 仓储盘存制
balance of supply and demand 供需平衡
balance of trade 贸易收支;贸易差额
balance of transaction 交易平衡
balance of voltage 电压平衡
balance of water and land resources 水土资源平衡
balance on current account 经常项目收支;经常项目平衡
balance on current account and long-term capital 经常项目和长期资本往来的平衡
balance on goods and services 货物和服务的平衡
balance on hand 库存余额;库存余数;现有余款
balance out 抵消
balance outstanding 余额
balance pan arrest 天平托盘
balance parameter 均衡参数
balance pipe 平衡管
balance piston 假活塞;平衡活塞
balance pivot 杠杆支点;天平支点
balance plate 平衡板
balance plate unit 平衡板组
balance plot of groundwater 地下水均衡场
balance point 平衡点
balance point detector 平衡点检测器
balance poising 摆轮平衡器

balance pressure reducing valve 平衡减压阀
balance process 平衡过程
balance quality 平衡品质
balance quality of rigid rotor 刚性转子平衡品质
balance quantity 阶段平衡产量
balancer 均衡器;平衡装置;平衡台;平衡器;平衡发电机;平衡(试验)机;无线电测向仪上消除天线效应的装置
balance ratio 平衡率
balance reading 天平示数;天平读数
balance reinforcement 平衡钢筋
balance rider 天平游码
balance rim 摆轮轮缘
balance ring 平衡圈
balance room 天平室
balance rope 平衡绳;平衡钢绳;尼绳
balancer set 平衡机组
balancer-transformer 差接变压器;均压变压器;平衡变压器
balance sash 平衡窗扇
balance screw 摆螺钉
balance set 均压机组
balance shaft 平衡轴
balance sheet 借贷对照表;决算书;借贷平衡表;资金平衡表;资产平衡表;资产负债表;收支平衡表
balance sheet account 资产负债表账户
balance sheet analysis 资产负债表分析
balance sheet audit 资产负债表审计
balance sheet date 结账日期;资产负债表日期
balance sheet for all purposes 全能资产负债表;通用资产负债表
balance sheet for anchor 下锚平衡板
balance sheet hedge 资产负债表项目对冲
balance sheet of investment 投资平衡表
balance sheet of transport economy 运输经济平衡表
balance sheet of transport revenue 运输收入总表
balance spring 弹簧秤;游丝
balance-spring chronometer 摆轮游丝精密记时计
balance staff 摆轴
balance statement 余额明细表
balance statistics 平衡统计
balance storage 平衡储水池;平衡库容;调节库容;日调节蓄水量;日调节库容;平衡蓄水池
balance subarea 均衡亚区
balance surplus with deficiency 余缺调整
balance test 差额测试;平衡试验
balance tester 平衡试验机
balance test of groundwater 地下水均衡试验
balance the books 结算
balance the profit and loss 平衡损益
balance ticket 决算单;结算单;余额证书;余额票证;余额对账单
balance tolerance 公差对照表
balance to surplus 结余盈余
balance to your credit 你方结余;你方结存
balance to your debit 你方结欠
balance transferred 余额过入
balance transferred to new account 余额转入新账
balance type potentiometer 补偿式电位差计
balance vapo(u)r recovery system 平衡管道蒸气回收系统
balance water content 平衡含水率;平衡含水量
balance weight 均衡锤;均衡器;坠砣;摆锤;平衡重(块);平衡重量;平衡(重)锤;天平砝码
balance weight abutment 衡重式桥台
balance weight bar 双强坠陀杆
balance weight chair 重力补偿托架
balance weight device 张力补偿装置
balance weight guide 平衡重导轨
balance weight retaining wall 衡重式挡土墙
balance weight rod 坠砣杆
balance weight tensioner 缓砣补偿器【电】;坠砣补偿器
balance weight tensioning 平衡重块绷紧
balance weight termination 张力补偿装置
balance-weight-type overhead garage door 平衡锤式车库翻门
balance wheel 均衡轮;飞轮转动;摆轮;平衡轮;平衡摆
balance-wheel watch 摆轮游丝表
balance zero 天平零件
balancing 平衡(法);平差;调质平衡
balancing account 结出账户余额;平衡账户

balancing action 平衡作用
balancing adjustment 平衡调整
balancing aerial 平衡天线
balancing apparatus 平衡仪
balancing arm 平衡臂;天平臂;秤杆
balancing a survey 平衡测量
balancing axial thrust 平衡轴向力
balancing band 锚吊箍【船】
balancing battery 缓冲电池组;补偿电池
balancing beam 平衡木
balancing bin 缓冲仓;调节料仓;调节仓
balancing calculation 平衡计算
balancing capacity 平衡容量
balancing cell 附加电池
balancing chain 平衡链
balancing chamber 平衡室
balancing coefficient 平衡系数
balancing components 平衡元件
balancing condenser 补偿电容器
balancing control 零位校正;平衡控制;平衡调整;调零装置;控制平衡装置
balancing delay 平衡延误;平衡延迟
balancing device 平衡装置;平衡器
balancing drum 平衡盘;平衡活塞;平衡鼓
balancing dynamo 平衡发电机
balancing dynamometer 平衡测力器
balancing effect 平衡效应
balancing factor for increasing live load 活载发展均衡系数
balancing fault 平衡误差
balancing flow 平衡流量
balancing force 平衡力
balancing gate pit 压力室
balancing generator 平衡发电机
balancing gyroscope 平衡陀螺仪
balancing hole 均压孔;卸荷孔;平衡孔
balancing illumination 均匀照明
balancing in locomotive 机车平衡
balancing item 差漏项目;平衡项目
balancing ladder 杠梯
balancing ledger 余额分类账
balancing lever 补偿杆;平衡杠杆
balancing line system 平衡管道系统
balancing link 锚杆环;平衡环
balancing load 平衡荷载;平衡负载
balancing loss 平衡损耗
balancing machine 平衡(试验)机
balancing margin 平衡的边际效用
balancing method 重心平衡确定法;重心平衡法;平衡方法
balancing moisture content 平衡水分
balancing moment 平衡弯矩法;平衡力矩
balancing network 平衡网络
balancing of building system 建筑物设备装置调整
balancing of construction volume 施工量平衡
balancing of cross-section 横断面土方量的平衡
balancing of cut and fill 挖土平衡
balancing of flow 平衡流量
balancing of sediment transportation 输沙平衡
balancing of stresses 应力补偿;应力平衡
balancing of tyre 轮胎平衡
balancing out 补偿
balancing period 均衡期
balancing piston 平衡活塞
balancing plane 修正面
balancing plate 平衡板
balancing plug 平衡销
balancing plug cock 平衡旋塞
balancing point 平衡点
balancing pressurizing tank 平衡压力罐
balancing process 平衡过程
balancing quantities 平衡土方
balancing rate 平衡速率
balancing repeating coil 平衡转接线圈
balancing reservoir 反调节水库;反调节库容;补偿调节水库;平衡水柜;平衡水池;平衡储槽;平衡池;调节水池
balancing resistance 平衡电阻
balancing resistor 平衡电阻器;配平电阻
balancing rheostat 平衡变阻箱;平衡变阻器
balancing rig 平衡试验装置
balancing ring 锚杆环;平衡环;平冲圈
balancing rotor 平衡转子
balancing sector thrust bearing 平衡块止推轴承

balancing sheet 补偿薄板;平衡薄板
balancing side 平衡侧
balancing speed 均衡速度;定常速度;平衡速率;平衡速度
balancing spring 平衡弹簧
balancing storage 调节库容;平衡库容
balancing survey 调整测量
balancing tank 平衡油槽;平衡油罐;平衡(水)箱;平衡水库;平衡水柜;平衡罐;平衡储槽;均衡槽;中和池;调节水池
balancing technique 平衡技术
balancing test 平衡试验
balancing tester 平衡试验机
balancing the books 结清各账户
balancing time 平衡工作时间
balancing transformer 平衡变压器
balancing trolley 平衡小车
balancing unit 平衡单元
balancing valve 平衡阀;平衡旋塞屏蔽阀(便于暖气片拆修)
balancing washer 平衡垫圈
balancing water(content) 平衡含水量
balancing weight 均衡锤;平衡重量;平衡重;天平砝码
balancing work station 平衡工作站
balaneion 澡盆;浴池(希腊语)
balangerite 羟硅铁锰石
balanid 小藤壶
balanus 藤壶
balas(ruby) 玫红尖晶石;浅红宝石;浅红晶石
balata 巴拉塔树胶
balata belt 巴拉塔胶带
balation shield 防融蚀的屏蔽
balatte 石灰石板
balaustre 栏杆装饰木
Balawangniduo basin 巴拉望尼多盆地
balayage method 扫除法
Balbach process 巴尔巴赫法
Balbiani ring 巴尔比尼氏环
balck cable 黑色电缆
Balco 巴尔可镍铁合金
balconet 阳台式窗栏;装饰横档;眺台(式)窗栏
balconette (在建筑物正面的)阳台栏杆
balconied 有阳台的
balconied access apartment 有阳台入口的成套房间
balconied access flat 有阳台入口的一套房间
balconied access living unit 有阳台入口的居住单元
balconied access type of block 有阳台入口的房屋群
balcony 楼厅;凉台;阳台;杆上工作台;船尾眺台;二楼厅;包厢;凸阳台;眺台;挑阳台;三层楼厅
balcony access 眺台入口;挑廊式入口
balcony-access apartment 有阳台入口的成套房间
balcony-access flat 有阳台入口的一套房间
balcony-access living unit 有阳台入口的居住单元
balcony-access type of block 有阳台入口的房屋群
balcony balustrade 楼厅扶手;阳台栏杆
balcony beam 楼厅梁
balcony beam-column 阳台梁柱
balcony column 阳台柱
balcony door 阳台门
balcony door fittings 阳台门上配件
balcony door hardware 阳台门上小五金
balcony drainage 阳台排水
balcony exterior exit 阳台外部通气管道
balcony facing 阳台饰面;阳台面层料
balcony front 眺台前沿
balcony girder 楼厅大梁
balcony greening 阳台绿化
balcony lift 平台升降机
balcony lifting door 阳台提升门
balcony-like intake 阳台(式)进水口
balcony like intake structure 平台式进水口结构
balcony lining 阳台饰面;阳台表面涂层
balcony outlet 阳台雨水口;阳台排水口
balcony parapet 阳台女儿墙
balcony partition(wall) 阳台隔墙
balcony plate 阳台板
balcony rail(ing) 阳台栏杆
balcony rainwater outlet 阳台雨水口
balcony slab 阳台板
balcony soffit 阳台底面
balcony stage 剧场舞台的表演挑台

balcony tile 阳台砖;阳台瓷砖
balcony truss 眺台(式)桁架
balcony-type 外廊式
balcony window 阳台窗
bald 无装饰的;秃山顶
baldachin 祭坛华盖;华盖
baldachin altar 有华盖的祭坛
baldachin tomb 有华盖的墓
baldacohino 无缝顶盖;墓上盖;龛室;祭坛华盖
baldanfite 铁水磷锰矿
baldaquin 织物罩盖;龛室;祭坛华盖
Baldet-John-son band 鲍尔德特—约翰逊谱带
bald-headed anticline 秃顶背斜
bald-headed derrick 无顶台钻塔;无顶台井架
bald mountain 秃山;童山
Bald pine 澳洲柏
bald spot 秃斑
bald tongue 光舌
Balduzzi's reflex 巴尔杜齐氏反射
Baldwin creep tester 鲍德温蠕变试验机
Baldwin solar chart 鲍德温太阳图
Baldwin's test 鲍德温氏试验
bale 捆(包);卡板;席水;大包;草捆;包
bale accumulator 草捆收集车
bale a hole dry 汲干井穴
bale aligner 整捆器;草捆整平器
bale band 叉座圈
bale block 分捆架
bale blower 草捆气力输送器
bale bogie 集捆车
bale breaker 拆包器;拆包机;松包机
bale broken 破包
bale buncher 草捆堆垛机
bale burst 破裂包
bale buster 草捆切碎机
bale capacity 件货容积;包装容量;包装货容积
bale cargo 件货;包装货
bale cargo capacity 包装货容积
bale carrier 运草捆车
bale chamber blade 压捆室的定切刀
bale channel 压捆室
bale collector 草捆收集器;拾草捆机
bale conveyer 草捆输送器;包裹件货输送机
bale cotton 成包原棉
bale counter 草捆计数器;包件计数器
balection mo(u)lding 企口护墙板装饰线条;盖缝线脚;镜框饰;突出嵌线的装饰线条
bale cubic 包装容积;包装货容积
bale cubic capacity 载货容量;包装货容积;载货容积
bale cutter 草捆切碎机
bale density 草捆密实度
bale density adjuster 压捆密度调节器
bale density control handle 草捆压实密度调节手柄
baled hay drier 草捆干燥机
baled material 包装货(物)
bale dog 草捆止退板
bale dumping device 捆包货倾卸设备
bale elevator 草捆升运器
balefire 火堆;烽火
bale fork 草捆叉
bale ga(u)ge 件货卡尺;件货长度
bale grab 草捆爪钩
bale handle 提把
bale handler 草捆堆垛机
bale hoop cutter 拆捆钳
bale loader 草捆装包机
bale measure 包装货容积
Balency system 巴朗西构造体系(采样预制墙板、现浇楼板)
bale of cotton 一包棉花
bale off 货物散包;散捆;散包
bale of wire 线束
bale oil stained 包装油污
bale out 勺取
bale-out furnace 勺取炉
bale-out pot furnace 固定式保温炉
bale packaging 捆包装
bale packing 捆包;麻包装
bale pickup 草捆捡拾机
bale pickup elevator 草捆捡拾升运器
bale piler 草捆码垛机
bale press 压捆机;打包机;包装机;填料压机
baler 捆扎机;汲水斗;压密机;压捆机;打包机;打包工;水瓢席斗;水瓢

bale retainer 草捆止退板
bale retarder 草捆止退板
baler plunger 捡拾压捆机柱塞
bale's catch 弹簧门碰头;自动扣;碰珠
bales cover torn 包皮破裂
bales disco lo(u)red 包装污染
bale separator 分捆器
bale slicer 分捆切刀;分捆器
bale sling 吊货绳套;(装卸件货的)绳扣
bale space 包装货容积
bales stained 包装污染
bale stacker 草捆码垛机;草捆堆垛机
bale stooker 草捆码垛机;草捆堆垛机
bale stook lifter 草捆堆提升机
bale stook mover 草垛运输车
bale tack 打包针;铅平头钉
bale thrower 草捆抛掷器
bale ties 打包绳
bale tie wire 打包钢丝
bale trailer 草捆挂车
bale transporter 包裹件货输送机
bale trip mechanism 草捆捆结接通机构
bale truck 载包车
bale valve 排水阀(门)
bale wagon 草捆挂车
bale wire 成捆线材
Balfour's law 鲍尔弗定律
Bali Declaration 巴厘宣言
baliki 风化残积黏土
baline 棉经马鬃纬衬里;粗麻布
baling 捆包装;压捆;打包;废物压实;废渣压块法
baling and printing wastewater 包装印刷废水
baling band 打包窄钢带;打包铁皮
baling chamber 压捆室
baling equipment 打包设备
baling faunal province 杷椰动物地理区
baling hay 干草压捆
baling hoop 打包铁皮;打包环箍
baling press 打包机;包装机;填料压机
baling press for waste 废料打包机
baling strip 打包带钢;打包钢皮带
baling wire 打包钢丝
baling wooden strips 打包木条
balipholite 纤钡锂石
Bali Sea 巴厘海
balistraria 城堡射箭孔;射箭孔(城堡上);弓箭室
Bali trough 巴厘海槽
Bali wind 巴厘风
balk 煤层尖灭;煤层包裹体;舷材;顶板突出岩石;地界;大木;粗木方(指大木);(以墩分隔的)桥跨;田埂
balkacchite 藻沥青;弹沥青
Balkan frame 巴尔干架
Balkan frame Balkan grippe 巴尔干夹板
balkanite 硫汞银铜矿;硫汞铜银矿
Balkan peninsula 巴尔干半岛
balk board 障碍板;隔板
balk cargo 散料
balk ring 摩擦环;阻尼环;阻环
balk roofing 重屋面
balk timber 大方木
balk-up photomap 背面影像地图
ball 结成团块混合料;耐火土块;线团;钻孔堵塞;滚珠;球形物;球端;气象风球;丝锥接套;沙囊
ball action 球磨机球磨作用
ball adaptor 球形连接器
ball and burlap planting 土球包扎装置
ball-and-burlap transplanting 带土栽植
ball-and-chain crawler 球形刮管器
ball and flower 滑球装饰
ball-and-lever valve 杠杆球阀
ball-and-line float 锚索浮标;水下浮子;水下浮标;深水浮标;悬浮浮子
ball-and-pillow structure 砂球和砂枕构造;球枕构造
ball-and-race mill 球环磨
ball-and-race type pulverizer mill 中速钢球座圈式磨煤机;钢球座圈式煤机;钢球座圈式粉碎机
ball and ring apparatus 球环软化点测定器
ball-and-ring method 环球(测定)法;(沥青软化点试验的)球环法;球环软化点测定法
ball and ring test 球环软化点测试
ball and roller bearing 滚珠和柱轴承;滚珠柱滚轴承
ball-and-seat (深井泵的)球和座

ball and sheet hinge 球状和片状铰链
ball-and-socket 滚珠和承窝;球窝
ball-and-socket attachment 球窝连接
ball-and-socket base 球轴座;球窝座
ball-and-socket bearing 球窝轴承;球窝支座;球窝承座;球铰承座
ball-and-socket coupling 球窝连接器;球形万向联轴节;球形万向节;球形万向接头;球窝联结器
ball-and-socket gear shifting 球窝式调档;球窝式变速杆
ball-and-socket head 球窝节头;球窝接头
ball-and-socket joint 球形万向联轴节;球形万向节;球形万向接头;球形接头;球窝连接;球窝接头;球窝关节;球铰接合;杵臼关节
ball-and-socket jointing 球窝节;关节状节理
ball-and-socket reamer 弧形造斜器;肘节扩眼器(弧线钻进用钻具)
ball and socket structure 关节构造
ball-and-socket type 球窝式
ball-and-socket type insulator 球窝式悬式绝缘子
ball-and-socket valve 球阀
ball-and-spigot 球销式;球塞式
ball-and-stick model 球棒模型
ball anode 球形阳极
Ballantine voltmeter 巴氏伏特计
ball arm 球头操纵杆
ballas 工业用金刚石;巴拉斯金刚石
ballast 加压载;压载【船】;压顶料;压舱物;镇流器;道砟;安定;平稳器;沙囊
ballastage 压舱费
ballast aggregate 道砟;碎石集料;碎石骨料;石渣
ballast air-dump breaker 风动卸碴车
ballast and sleeper 轨枕道基础
ballast arrival condition 压载到港状态
ballast bag 压舱袋
ballast bar 压载杆
ballast bed 道砟床;道床【铁】;道砟路基;碎石基床;碎石道床;石渣路基;石渣(道)床
ballast bed finisher 平整道砟机
ballast bed structure 铁路道床结构
ballast blown out 压载水排尽
ballast boat 有压载的船
ballast body 道砟床
ballast bolt 压载栓
ballast bonus 空放津贴
ballast box 压载箱;道砟箱
ballast boxing 为摊铺道砟而建筑的路槽(土路基)
ballast bridge floor 道砟桥面;铺道砟桥面;铺渣桥面
ballast car 道砟车;道床;铺渣车;石渣车
ballast cargo 压载货;压舱物;压舱货(物);底货
ballast chamber 压载舱;平衡重室
ballast channel plate 道砟槽板
ballast cleaner 道砟清洗机;道砟清筛机;清砟器
ballast cleaning 道床清筛;道砟清筛;清筛道砟
ballast cleaning machine 道砟清洗机;道砟清筛机;道床清筛机
ballast cleaning machine with removed track panels 大揭盖清筛机
ballast coefficient 道床系数
ballast coil 镇流线圈;负载线圈;平衡线圈
ballast compacting machine 道砟压实机
ballast compactor 道砟夯实机(械)
ballast compartment 道砟分隔间;压载舱
ballast compression 道砟压缩
ballast concrete 压重混凝土;压载混凝土;镇重混凝土;碎石混凝土;石渣混凝土;石混凝土;石碴混凝土
ballast condenser 压载凝结器
ballast configuration 压载布置
ballast connection 压载水管接头
ballast consolidating 夯实道床
ballast consolidating machine 道砟夯实机械
ballast consolidator 道砟夯实机械
ballast container 道砟容器
ballast crusher 道砟破碎机;道砟轧碎机
ballast curb 挡渣块;挡渣
ballast density 道砟密度;道床密实度;道床宽度;道砟密度
ballast departure 压载出港
ballast depth 道床厚度
ballast displacement 压载排水量
ballast distributing and regulating machine 道床配砟整形机

ballast distributing wagon 分碴车
ballast drag 拖拉道砟机
ballasted aeroplane 压重飞机
ballasted bridge floor 铺渣桥面
ballasted condition 空船压载状态;有压载状态;压载状态
ballasted deck 铺碎石砟桥面;铺道砟桥面;铺砟桥面(铁路桥);铺渣(上承)桥
ballasted deck bridge 铺渣上承式桥
ballast distributing and trimming 道床配碴整形
ballasted draft 压载吃水
ballasted floor 铺道砟桥面;铺碴桥面
ballast edge compactor 道床肩夯拍机;道砟肩夯拍机
ballasted pontoon 压载浮船;压载趸船
ballasted track 有砟轨道
ballasted track structure 有砟轨道结构
ballast engine 运渣机车
ballast factor 镇流系数
ballast filling device 垫渣装置
ballast filter 石渣滤水床;石渣滤水池;石渣滤水层
ballast fin 拔水板
ballast floor 道砟桥面
ballast for cruising 航行压载
ballast fork 道渣叉
ballast fouling index 道床脏污系数
ballast-free track 无砟轨道
ballast-free track structure 无砟轨道结构
ballast grading 颗粒尺寸级配;道砟级配
ballast guard 道砟围挡
ballast hammer 道砟锤;碎石锤
ballast harrow 道砟耙;扒渣机
ballast hoe 道砟铲
ballast hole 压载舷孔
ballast hopper wagon 石渣漏斗车
ballastic crusher 碎石机
ballast impedance 道砟漏泄电阻
ballasting 压载物;镇压荷重;加压舱物;镇流;铺渣;铺(道)砟
ballasting and sinking of mattress 压排沉排;柴排压载下沉
ballasting instruction 压载说明
ballasting lead weight 压铅
ballasting machine 运渣机;铺碴机
ballasting material 道砟(材)料;压载物;压舱物料;压舱材料
ballasting of floating unit 浮动装置压载
ballasting of mattress 柴排压载
ballasting oil storage tank 储油压载舱
ballasting palm 垫渣撑
ballasting plough 平渣犁
ballast(ing) system 压载系统
ballasting tanker 空放油轮
ballasting up 压载调整;铺完石渣;施加压重
ballast jettisoning arrangement 抛压载装置
ballast keel 压载龙骨
ballast lamp 镇流管;镇流灯;平稳灯
ballast leakage resistance 镇流漏泄电阻
ballast deck 无砟桥面
ballastless slab track 无砟板式轨道
ballastless track 无砟板式轨道
ballastless track structure 无砟轨道结构
ballast level(l)ing device 平渣器
ballast lighter 压载物驳船
ballast line 压载管系;压载吃水线;压载水管线
ballast load-line 压载吃水线
ballast main 压载水总管;压舱水总管
ballast mat 道砟(垫)层
ballast mattress 垫渣;沉砟垫层
ballast modulus 道床系数
ballast movement arrangement 压载调整装置
ballast mucking 出渣
ballast noise rating 镇流器噪声的分级
balast packing 垫砟
ballast particle 道砟块度
ballast passage 空放航次;压载航次;压舱货航程
ballast pipe 压载水管
ballast pipeline 压舱水管线
ballast piping system 压载管系
ballast pit 道砟坑
ballast plan 压载图
ballast plough 道砟犁
ballast plow 道砟犁
ballast pocket 压载舱;道砟窝(混凝土中);道砟囊;蜂窝
ballast porosity 道砟空隙比
ballast port 压载物舷门
ballast pressure 道砟压力
ballast pump 衡重水泵;压载水泵;压载泵;压舱水泵;镇定泵;平衡泵
ballast pumping system 压载水泵系统
ballast pump room 压载水泵舱
ballast rake 道砟耙;碎石耙
ballast raking equipment 扒渣机
ballast ramming 道床夯实;夯实道床
ballast regulating machine 道床整形机
ballast regulator 道砟整形机
ballast release 压载投放
ballast removal 清除道砟
ballast renewal 道砟更新
ballast replacement 道砟更换
ballast resistance 电阻箱;道床阻力;道砟电阻
ballast resistor 镇定电阻
ballast road 道砟路;碎石路;石渣路
ballast road bed 道砟路基
ballast-roller 道砟碾压机
ballastron 镇流管
ballast run 空载航次;空放航次;压载航次
ballast sailing 空放航行;压载航行
ballast sand 压舱砂
ballast scarifier 翻松道砟机;扒碴机;扒渣机
ballast screen 渣筛;道砟筛
ballast screening machine 道砟筛分机;筛渣机
ballast section 道砟断面;道床断面
ballast self-unloader 自动卸道砟
ballast shoulder 道砟(路)肩;道床砟肩;砟肩
ballast shoulder consolidating machine 道床边坡夯实机
ballast sinking 道砟陷槽
ballast slope 道砟边坡;道床边坡坡度
ballast specific weight 道砟比重
ballast spreader 道砟撒铺机;道砟撒铺车;铺渣机;撒石渣车;撒道砟车
ballast stress 道床应力
ballast suction pipe 压载舱吸水管
ballast surfaced 麻面的
ballast sweeper 道砟清扫设备
ballast tamper 道砟夯实机;道砟压实机;枕木捣固器;枕木捣固机;砸道机;捣砟机
ballast tamping 道砟捣固
ballast tamping machine 夯实道砟机;道砟捣固机;捣砟机
ballast tank 衡重水箱;压重箱;压载箱;压载水箱;压载水柜;压载(水)舱;压载双层底舱;压舱柜;潜水艇沉浮箱
ballast temping 道床捣固
ballast template 道砟肩板
ballast templet 道砟肩板
ballast thickness 道床厚度
ballast tightening device 重锤式张紧装置
ballast track bed 道砟道床
ballast train 运渣列车;石渣列车
ballast tray 镇气分馏塔盘
ballast trial 空放试航;压载试航
ballast trimming 压载调整;道床整形
ballast trough 道砟槽;桥梁道砟槽
ballast truck 石渣车;砂石料车
ballast tube 镇流管
ballast type hand roller 压重式手推滚筒
ballast undercutting-cleaner 道床破底清筛机
ballast unloader 卸除道砟机;卸渣器
ballast unloading 卸道砟
ballast volume 道砟体积
ballast voyage 空放航次;压载航次
ballast wagon 道砟车;石渣车
ballast wall 子墙;雉墙;桥台台帽上填土与上部构造间挡墙
ballast water 压载水;压舱水
ballast waterline 压载吃水线;压载水线
ballast water tank 压载水柜;压载水舱
ballast weight 压重;压载物;镇重;配重
ballast weight release test 压载重量释放试验
ball attachment 球形头
ball band 滚珠带
ball bank inclinometer 球式倾斜指示器
ball bank indicator 滚珠倾斜指示器;球式示倾器;球形倾斜仪
ball bank unit 滚珠倾斜装置

ball bearing 泵体底轴承;球座支承;球轴承;球支座;滚珠轴承;滚珠轴承
ball-bearing action (路面磨损微粒的)滚动作用
ball-bearing axle-box 滚动轴承轴箱
ball-bearing brass washer 滚珠轴承铜垫圈
ball bearing butt 滚珠轴承铰链
ball-bearing butt hinge 滚珠轴承铰(链);装有滚珠轴承的对接铰链
ball-bearing cap screw 滚珠轴承盖螺钉
ball-bearing center 滚珠轴承中心
ball-bearing core barrel 滚珠轴承式岩芯管(单动双层岩芯管)
ball-bearing coupling side 联轴节侧轴承
ball-bearing cover 滚珠轴承盖
ball-bearing drive spring 滚珠轴承主动簧
ball-bearing felt packing 滚珠轴承阻油毡
ball-bearing fit 滚珠轴承配合
ball-bearing grease 滚珠轴承润滑脂
ball-bearing head 滚珠轴承装置
ball-bearing hinge 滚珠轴承铰(链);滚珠轴承合页;滚球轴承铰
ball-bearing holder 滚珠轴承隔离圈
ball-bearing housing 滚珠轴承座
ball-bearing luboil 滚珠轴承润滑油
ball-bearing mill 离心球磨机;滚珠轴承式球磨机
ball-bearing motor 滚珠轴承电机
ball-bearing noise 滚珠轴承噪声;滚珠轴承噪声;球轴承噪声
ball-bearing nut 滚珠轴承螺母
ball-bearing pillow block 滚珠轴承台
ball-bearing plumber block 滚珠轴承台
ball-bearing puller 滚珠轴承拆卸工具
ball-bearing retainer 球轴承保持架
ball-bearing seat 滚珠轴承底座
ball-bearing shielded 带防尘盖球轴承
ball-bearing slewing crown 滚珠轴承转车台
ball-bearing slewing joint 滚珠轴承转动联接
ball-bearing steel 滚珠轴承钢
ball-bearing steel strip 滚珠轴承钢带
ball-bearing thrust cap 滚珠轴承抵盖
ball-bearing trolley of chain 链条滑轮滚珠轴承
ball-bearing tube 滚珠轴承管
ball-bearing washer 轴承座垫片
ball-bearing wheel 滚珠轴承轮
ball-bearing wire 滚珠轴承钢丝
ball-bearing with clamping sleeve 带套滚珠轴承
ball-bearing with filling slot 带滚珠沟槽的轴承
ball-bearing with seals 带密封圈的球轴承
ball-bearing with taper bore 锥孔滚珠轴承
ball blast 喷丸;抛丸
ball blasting 喷丸清理
ball-bolt 球头螺栓
ball bond 球压焊;球形接头
ball bonding 球形接头;球形接合;球铰节;球焊
ball breaker 撞锤;重锤
ball-burnishing 钢珠抛光;钢珠滚光;钢球抛光;球滚光
ball burst testing 钢球式顶破强力试验
ball bushing 直线运动球轴承;球轴套
ball cage 滚珠轴承架;滚珠隔离圈;滚珠槽;滚珠轴承罩
ball cal(l)ipers 球径量规
ball cam 球凸轮
ball case cover 滚珠轴承压盖
ball-casting machine 球铸机
ball catch 弹簧门碰头;门碰珠;自动扣;球掣;碰珠
ball catcher 承球短节
ball change 球形铰链换挡杆
ball charge 球磨机装球量
ball check 球形逆止阀;球形单向阀;球阀
ball check nozzle 球形止回喷嘴
ball check valve 钢球单向阀;球上回阀;球形止回阀;止回球阀;球形逆止阀
ball chuck 球夹
ball clack 球阀;球瓣
ball clay 泥球;陶土;黏土块;球状黏土;球黏土;球土
ball cleaner 小球清洗器
ball-clevis 球头挂板
ball closure member 球形关闭件
ball coal 球状煤
ball coating 糊球;包球(现象)
ball cock 浮球旋塞;浮球阀(门);球阀;球旋塞
ball-cock assemble 球阀组件
ball-cock assembly 球阀装置

ball-cock device 浮子装置;浮球旋塞装置;球阀装置
ball cold header 钢球冷镦机
ball collar 滚珠环
ball collar thrust bearing 环形滚珠止推轴承;(环形的)滚珠止推轴承;滚珠环推力轴承
ball condenser 球形冷凝器
ball consumption 磨球消耗量
ball contact tip 球测头
ball conveyer 球式输送机
ball coupling 球状连接器;球节;万向联轴器;万向联轴节;万向接头
ball crank 球状曲柄
ball crank handle 带球端摇把
ball crusher 磨球机;球磨碎石机;球磨机
ball cutter 球形铣刀;球面刀
ball delivery valve 球形排出阀
ball distance ring 滚珠轴承座圈
ball door knob 球形门把手
ball drive rotary head 球驱动旋转喷洒头
ball-driving contact-type air plug ga(u)ge 钢球传动型接触式气动塞规
ball durometer 球形硬度计
balled 再包土的
balled and burlapped 土球包扎(植物移植)
balled iron 铁坯
balled plant 打包厂
balled plant 带土栽植;再包土栽植
ball elbow 球面弯接头
ball-end eye-end rod 杵环杆
ball end mill 圆头槽铣刀;圆头雕刻铣刀
Ballentine hartester 鲍兰丁硬度计
bailer 搓球机;成圆器
bailerina check 斜交方形花纹
ballet house 芭蕾舞剧院
balleting 颠簸运动
balletini 路标玻璃球
ball-eye 球头挂环
ball faced hexagonal nut 球面六角螺母
ball feeder 球状冒口
ball finishing 钢球挤光
ball flashing machine 光球机
ball float 球状浮子;球状浮体;球形浮标;浮球(阀)
ball-float level controller 浮球液面控制器;浮子式液位控制器
ball float liquid level detector 浮子式液位检测器
ball-float(liquid-)level meter 浮球液面计;球形浮子液面计
ball float meter 浮球仪
ball-float trap 浮球式泄水活门;浮球式疏水器
ball-float type trap 浮球式泄水活门;浮球式疏水器
ball float valve 浮子阀;浮球阀
ballflower 圆球饰;球心花饰;球形花饰
ball flow meter 球形流量计
ball forming machine 制球机床
ball forming mill 成球机
ball forming rest 车球刀架
ball-frame carriage 球形支座;滚珠支座
ball-fringe 球形穗饰
ball-game ground 球场(总称)
ball-games hall 球类运动厅
ball gate 挡渣内浇口;球顶补缩内浇口
ball gatherer 球形挑料机
ball ga(u)ge 球形量规
ball go-devil 清管球
ball governor 离心调速器;球调节器
ball grading 磨球分选
ball granulator 成球机
ball grid array 球状矩阵排列
ball grinder(machine) 球磨机;磨球机;球磨床
ball grinding 球研磨
ball grinding machine 球磨机床;球磨床
ball grinding mill 球磨床;球磨机
ball grip 球状手柄
ball gudgeon 球体耳轴
ball guide 滚珠导筒;球状引导物
ball hammer 球形锒头;圆头锤
ball handle 球形手柄
ball hardness 钢球硬度;布氏硬度;布氏钢球硬度;布氏球测硬度;布氏钢球压入硬度;球印硬度
ball hardness machine 钢球硬度试验机
ball hardness number 布氏硬度值;球印硬度值
ball hardness test 球印硬度试验
ball hardness testing machine 球印硬度试验机
ballhead 飞球头;球形头

ball headed conic(al) bolt 球头锥形螺栓
ball header 钢球模锻机;球头模锻机
ball head governor 飞球调速器;飞球调节器
ball heading 钢球镦锻
ball head insulator 球头状绝缘子;球头形绝缘子
ball head stay 球头撑杆
ball heater 球形加热器
bail hinge 球形铰
ball holder 滚珠夹圈
ball-hook 球头挂钩
ball hooting 上滑道集材
ball house 球形房屋
ball hydrant 球形消防栓
ball hydrophone 球形水听器
ballicatter 冰脚
ball ice 浮冰球;球状冰
ballic pole 伊朗式脚手架
ball impact test 落球试验
ball impression (硬度试验的)球凹
ball impression of hardness test 硬度试验球凹
ball inclinometer 球形倾斜仪;球式倾斜计
ball indentation 钢球压痕
ball indentational hardness 球压硬度
ball indentation test 布氏球印硬度试验;球压硬度试验;球压试验
ball indenter 钢球压头
ball index 钢球分度装置
balling 泥团;糊钻;成球现象;成球;熟铁成球
balling disk 制粒盘;造球盘
balling drum 鼓形制球机;成球筒;球磨机滚筒
balling formation 易糊钻地层
balling furnace 搅炼炉
balling gun 投药器
balling machine 切包卷取机
balling of clinker 熟料成球
balling of the bit 泥包钻头
balling pan 成球盘
Balling scale 巴灵秤
balling up 成团儿;成球作用;糊钻
balling up of cement 水泥团;水泥结团;水泥成团
balling-up of particles 粉粒成球
ball in hand 手球法
ball insulator 拉紧绝缘子;球形绝缘子
ball iron 粒铁
ballista 投石器
ballistic 弹道的;冲击
ballistic arm 水银器连杆
ballistic breaker 快速作用断路器
ballistic cap 风帽
ballistic constant 冲击常数
ballistic curve 弹道
ballistic damping error 阻尼差;第二类冲击误差;冲击阻尼误差
ballistic deck 装甲甲板;防弹甲板
ballistic dedusting 冲击式除尘
ballistic deflection 惯性误差
ballistic deflection error 第一类冲击误差;冲击位移误差
ballistic electrometer 冲击静电计
ballistic equation 弹道方程
ballistic error 冲击误差
ballistic fabric 防弹织物
ballistic factor 冲击因素
ballistic factor of measuring instrument 测量仪表的阻尼因数
ballistic flow prover 冲击式流量校验器
ballistic galvanometer 冲击(式)电流计;冲击(式)检流计
ballistic impact performance 弹道冲击性能
ballistic instrument 冲击式仪器
ballistic kick 急冲
ballistic limit 冲击界限
ballistic link bearing 水银器连杆轴承
ballistic magnetometer 冲击磁强计
ballistic measurement 弹道测量;冲击式测量
ballistic method 冲击法
ballistic missile defence committee 弹道导弹防御委员会
ballistic missile office 弹道导弹局
ballistic missile test vessel 弹道导弹试验舰
ballistic mortar test 冲击(式)砂浆试验
ballistic motion 冲击运动
ballistic movement 冲击运动
ballistic pendulum 弹道摆;冲击摆

ballistic photogrammetry 弹道摄影测量学
ballistic pivot bearing 水银器轴承
ballistic precession 冲击进动
ballistic quality ratio 冲击强力比值
ballistic research laboratory 弹道研究实验室
ballistic separator 冲击式分离机;抛掷式分级机
ballistic test 冲击试验;冲击强力试验
ballistic testing machine 冲击式(强力)试验机
ballistic throw 冲击摆幅
ballistic work 冲击功
ballistite 无烟火药
ballistraria 塔楼墙上十字形孔
ballium 中世纪堡垒中的空地
ball jet 含球气动测头
ball joint 万向接头;球形接头;球窝接头;球合;球头节;球节(点);球接头;球形连接;球承
ball joint clamp 球节夹
ball joint cover 球节盖
ball joint cover latch 球节盖闩
ball jointed rocker bearing 球轴颈铰链支承
ball jointed screw 万向铰螺杆
ball joint manipulator 球节机械手;球承(式)机械手
ball joint net frame 球接点网架
ball joint prop 球铰接支承
ball joint roof bar 球铰接屋架杆件
ball journal 球轴颈
ball journal bearing 球颈轴承
ball knob 门球;球状捏手;球状把手;球形把手
ball lap 浮球开关
ball lapping machine 钢球精研机床
ball lathe 制球车床
ball launching 滚球下水法
ball(lead)screw 滚珠丝杠
ball lever 浮球杆;球隔流阀
ball lightning 球状闪电;球状电闪
ball load 球上荷载
ball lock 弹子锁;球形闸门
ball loss 磨球损耗;磨球磨损
ball lubricator 钢球式注油嘴
ball manipulator 球状铰接机械手;球窝机械手
ball method 压球法;球印法;球印(硬度)试验法
ball method of testing 钢球压印试验法
ballmill 用球磨机磨碎;球磨机
ball milling 球磨(机研磨)
ball mill lining 球磨机衬层
ball mill pulverizer 球磨粉碎机
ball mill refiner 球磨精制机;球磨精研机
ball mirror 球反射镜
ball non-return valve 单向球阀
ball nozzle 球形喷嘴;球形管嘴;球管嘴
ball nut 滚珠螺母;球状螺母
ballon 穹形圆山
ballonet 小气球;(气球、汽艇的)空气房;副气囊
ballonnement 气胀术
balloon 柱冠球【建】;短颈圆玻璃瓶;分量轻体积大的货物;球形大玻璃瓶;球形瓶;球形玻璃容器;气球;气瓶;漂浮式偿付款
balloon altimeter 气球高度
balloon astronomy 气球天文学
balloon-based astronomy 气球天文学
balloon-borne infrared telescope 气球载红外望远镜
balloon-borne instrument 球载仪器;气球运载仪器
balloon-borne IR telescope 球载红外望远镜
balloon-borne photoflash triangulation 气球闪光三角测量
balloon borne sensor 气球传感器
balloon borne telescope 球载望远镜
balloon camera 气球摄影机
balloon catheter 气囊导管
balloon construction 轻型结构;轻型构造;轻构造;轻捷构造
balloon densi(to)meter 囊式密度计
balloon down payment 成交后分批付定金
balloon dynamics 气球动力学
balloon equipment 气球仪
balloon flask 球形烧瓶
balloon frame 轻型木构架;轻捷构架
balloon frame construction 轻型木构架结构
balloon framed construction 轻捷木骨架构造;轻捷骨架构造
balloon frame house 轻捷构架房屋
balloon framing 轻型木构架;轻型骨架;轻捷木骨架;轻捷构架;轻骨构造

balloon hangar 气球库
ballooning 飞涨;气胀术;气球操纵
ballooning degeneration 气球样变性
ballooning over the runway 落地前平飘(飞机)
balloon loan 漂浮式货款
balloon loop 港内环行铁路
balloon maturity 全部或大部分的债券或欠款的到期日
balloon mortgages 短期抵押货款;特大尾数抵押货款;特大房屋抵押货款
balloon note 分期付款票据
balloon observation 气球观测
balloon ozone sounding 气球臭氧探测
balloon payment 漂浮式偿付款
balloon-payment load 最后一笔付款数额大的分期贷款
balloon photography 气球照相术;气球摄影
balloon platform 球载平台
balloon repayment 最后一笔金额特大的分期付款法;漂浮式偿付款
balloon satellite 气球卫星;气球式卫星
balloon sonde 高空测候气球
balloon sounding 气球探测
balloon structure 轻型结构
balloon survey 气球测量
balloon technique 气球技术
balloon telescope 气球望远镜
balloon test 漫画法(市场调查)
balloon theodolite 测风经纬仪;气球用经纬仪
balloon tire ga(u)ge 车胎低气压表;低压大轮胎
balloon transit 测风经纬仪
balloon-type rocket 增压式火箭
balloon tire 低压轮胎;低压大轮胎
balloon vessel 气球母船
balloon volumeter 囊式体积仪
ballot 小包货
ballotini 小玻璃球(在路面上反光用)
ballot papers null and void 废票
ballot problem 抽签问题
ball packing 球形填充物;球形填料
ball-pane hammer 球头锤
ball park 球场
ball path 磨球运动路线;磨球路线
ball peen (锤的)圆头;圆锤头
ball peen hammer 圆头锤;奶子榔头;圆头手锤
ball pen 圆珠笔
ball pendulum hardness of rock 岩石摆球硬度
ball pendulum test 球摆试验
ball penetration test (测定混凝土稠度的)球体贯入试验
ball penetrator 钢球压头
ball permeameter 球形渗透计
ball pig 清管球
ball pin 球轴颈
ball piston motor 钢球式发动机;球塞式电动机
ball piston pump 钢球活塞式油泵;球塞泵
ball piston type hydraulic motor 球塞式液压马达
ball pivot 球支枢;球铰;球节;滚珠支枢;球枢
ball-plant 土球包扎植物
ball planting 带土栽植;包土栽植
ball plug 球形塞;球塞
ball plug ga(u)ge 球塞规
ball plunger 球塞
ball plunger groove 球塞槽
ball pocket 球状囊
ball point 球形冲头端
ball point setscrew 球端定位螺钉
ball point water demand 球点吸水量
ball press 钢球冲压机
ball pressure test 球印硬度试验;球印试验;球压式(硬度)试验
bali-proof 防弹的
ball-proof clothes 防弹衣
ball pulverizer(mill) 钢球磨碎机
ball pump 球形泵
ball punch impact test 圆球冲击试验
ball race 座圈;滚珠座圈;滚珠座圈;滚球轴承座圈;球道
ball-race bearing 滚珠轴承
ball-race fit 滚珠座圈配合
ball race mill 钢球盘抱磨煤机;辊球立磨;球盘磨
ball race steel 滚珠轴承圈钢
ball rationing 磨球配量;球磨机球配量
ball reamer 菊形铰刀;球面铰刀

ball receiver 转播用接收机
ball reception 中继接收系统
ball reciprocating bearing 可后移滚珠轴承
ball relief valve 减压阀;安全阀;球形安全阀;释放球阀
ball resolver 球坐标分解器;球形解算器;球体解算器
ball retainer 滚珠护圈;球护圈;球阀座
ball retaining spring 滚珠回位弹簧
ball return spring 阀球返回弹簧
ball ring 滚珠环;球环
ball ring mill 辊球磨
ball roller bearing grease 滚珠滚柱轴承润滑脂
ball roller mill 回转球磨机
ball rolling mill 钢球轧机
ballroom 社交厅;舞极;舞厅
ballroom foyer 舞厅休息室
ball rotation speed 球磨机转速
ball safety valve 球形安全阀
ball scraper 清管球
ball screw 滚珠螺杆
ball sealer 密封球
ball seat(ing) 球座;球形阀座;球面座;球阀座
ball-shaped chemical vapo(u)r deposit diamond crystal 球形化学气相沉积金刚石晶体
ball-shooting 喷丸清理
ball sieve 圆筛筛
ball sizing 钢球挤孔;钢球挤光孔法;球校准
ball sleeve 球套
ball sleeve tubing union 球套管节
ball socket 杆座双耳;球座;球状支点;球形支座;球窝
ball socket base 球窝基座
ball socket bearing 球铰轴承
ball socketed bearing 耳轴轴承
ball spark gap 球形火花隙
ball spindle 球轴
ball spinning 滚珠旋压;钢球旋压
ball spline 滚珠花键
ball spray treatment 喷丸处理
Balls's sledge sorter 包氏棉纤维长度分析仪
ball stanchion bed 鹅头台;鹅头床
ball stone 褐铁矿;球石
ball stop 球形止逆阀;球阀止逆器件
ball structure 球状构造
ball structure parting 球状节理;球状剥离
ball stud 球枢轴;球头螺栓
ball stud for centershift 中心移动球头螺栓
ball subga(u)ge 球分规值
ball suction valve 球吸入阀
ball support 球支承
ball swage 钢球模具;钢球冲模
ball swager 钢球挤光机
ball table 球支承的桌子;球式工作台
ball tap 球阀;球形旋塞;球塞
ball terminal 球状端点
ball test 球印硬度试验;球印试验;球压试验;落球试验;滚球抗阻塞试验(排水系统中)
ball tester 球压式(硬度)试验;球式硬度试验机
ball test for hardness 洛氏硬度试验
ball-thrust bearing 球压轴承;止推球(滚珠)轴承;滚珠止推轴承;推力滚珠轴承
ball track 轴承滚道
ball transfer table 球式转运台
ball transplanting 带土包扎移植;土球移植
ball trap 球形存水弯
ball tube 滚珠循环导管
ball tube mill 球管磨机;钢球滚筒式磨煤机
ball tube-type mill 管式球磨机
ball-turning rest 车球刀架
ball type check valve 球型防逆阀;球型单向阀
ball type compensator 球形补偿器
ball-type corporation valve 球型总阀
ball-type handler 球状铰接机械手
ball type holding dog 球夹式提引器
ball-type interceptor 球形截水器
ball type printer 球型打印机
ball type pulling dog 球夹式提引器
ball-type valve 钢珠活门;球型瓣膜
ball up 形成球状;糊钻;成球;钻具为泥包住;阻塞;黏结成球;滚成球
ball valve 弹子阀;球形止逆阀;球阀(门);球形单向阀
ball valve copper float 球阀铜浮子
ball valve sampler 球阀式取土器

ball variometer 球形可变电感器
ball warping machine 整经机
ball washer 球形垫块
ball wear 球耗
ball welding 球焊
ball winder 绕球机;团绒机
ball wire nail 球头圆铁钉
ball with eye 球形挂环
balm 凹形崖;香脂;香油
balmaiden 女矿工
Balmer band 巴尔末光谱带
Balmer continuum 巴尔末连续区
Balmer discontinuity 巴尔末跳变
Balmer formula 巴尔末公式
Balmer jump 巴尔末跳变
Balmer limit 巴尔末系限
Balmer lines 巴尔末谱线
Balmer progression 巴尔末渐近
Balmer series 巴尔末(线)系
balmy breeze 和风
balnea 公共大浴场(罗马)
balnearium 私人浴室(古罗马时)
balneology 矿泉浴疗养学
balneotherapy 浴疗(法)
balneum pneumaticum 空气浴
balnstone 顶板岩石
baloney dollar 劣质金元
balop 反射式放映机
balopticon 立体放灯机;投影放大器
balsa 轻质木材
balsam 香脂;香胶;凤仙花
balsam fir 香脂冷杉;胶冷杉;胶枞
balsaming lens 胶合透镜
balsam of fir 枞香脂
balsam turpentine 香脂松节油
balsam wool 吸声毛毯
balsa wood 软木;轻木;轻筏木
balter 筛分机
balteus (古罗马竞技场上的或两讲堂之间的)通道;扁带饰
Baltic bow 破冰型船头;破冰型船首
Baltic old land 波罗的古陆
Baltic redwood 波罗的红杉
Baltic Sea 波罗的海
Baltic shield 波罗的地盾
Baltic-tow 破冰船型船头
Baltimore groove 巴尔蒂摩型凹形槽
Baltimore groove contacts 巴尔蒂摩型凹槽接触
Baltimore rule 巴尔摩规则(一种地产评估方法)
Baltimore truss 巴尔蒂摩桁架;平行弦再分式桁架
baltimorite 叶硬蛇纹石;叶蛇纹石
balun 巴伦仪
baluster 栏柱;栏杆柱;栏杆支柱;栏杆小柱
baluster capital 栏杆柱头
baluster column 短粗圆柱;栏柱;栏杆形柱;望柱
baluster cover flange 栏杆柱法兰底座
baluster railing 立柱栏杆
baluster seating 栏杆底座和压顶的冲模
baluster shaft 短粗圆柱;栏杆形柱;望柱
baluster side 栏杆边;(爱奥尼亚式卷形柱头的)侧面
baluster supporting a railing 小柱栏杆
balustrade 楼梯栏杆扶手;栏杆;阳台栏杆;扶手(装置)
balustraded caged-in window 勾栏槛窗
balustrade stay 栏杆支撑
balustrade wit glass panel 玻璃栏板
balyakinite 巴碲铜石
bambollaite 碲硒铜矿
bamboo 原竹;竹材
bamboo basket 竹篓;竹笼
bamboo blind door 竹帘门;篱笆门
bamboo blinds 竹帘
bamboo bolt 竹制螺栓;竹销
bamboo bridge 竹桥
bamboo broom 竹扫帚
bamboo cable bridge 竹索桥
bamboo cane 竹茅竹;竹基
bamboo carrying pole 竹杆子
bamboo carving 竹刻;竹雕
bamboo ceiling 竹材顶棚;竹材天花板
bamboo concrete 竹筋混凝土
bamboo concrete bridge 竹筋混凝土桥
bamboo corridor 竹制回廊;竹材回廊
bamboo curtain 竹帘

bamboo cylinder 竹笼
bamboo drain 竹暗渠
bamboo earth pan 竹土箕
bamboo enclosure 编竹围墙
bamboo engraving 竹刻
bamboo fence 竹篱
bamboo fence wall 竹围墙
bamboo fender 竹制碰垫
bamboo fiber 竹纤维
bamboo filament 竹丝
bamboo forest 竹林
bamboo frame 竹材构架
bamboo framing 竹châssis;竹材构架;竹结构
bamboo furniture 竹制家具;竹家具
bamboo gabion 竹笼;填石竹笼
bamboo gabion cofferdam 竹笼围堰;填石竹笼围堰
bamboo garden 竹园
bamboo groves 竹林
bamboo house 竹房子;竹材房子
bamboo ladder 竹梯
bamboo lath 竹板条
bamboo lathing 竹编条板
bamboo lath screen 竹栅
bamboo-lath transom 竹条气窗
bamboo lofty restaurant 竹轩(餐厅)
bamboo mat 竹席;竹凉席
bamboo mattress 竹茎沉排
bamboo mosaic 竹片马赛克
bamboo nail 竹钉
bamboo-notch tile 竹节瓦
bamboo ornament 竹饰
bamboo over-porcelain 瓷胎竹编
bamboo pavilion 竹制亭;竹亭;竹材亭
bamboo plantation 竹园
bamboo plywood 胶合竹板;竹胶合板;竹制胶合板
bamboo-pole 竹杆
bamboo-pole scaffold(ing) 竹脚手架;竹杆脚手架
bamboo pulp 竹浆
bamboo purlin(e) 竹檩条
bamboo raft 竹排;竹筏
bamboo rafter 竹椽
bamboo rail 竹制轨槽(推拉门用)
bamboo rail fence 竹篱
bamboo rapter 竹制
bamboo reeds 竹棍
bamboo-reinforced concrete 竹筋混凝土
bamboo reinforcement 竹筋
bamboo roof 铺竹屋顶
bamboo roof truss 竹屋架
bamboo scaffold(ing) 竹脚手架;竹材脚手架
bamboo scale 竹尺
bamboo screen 竹帘
bamboo shape hat stand with yellow glaze 黄釉竹形帽架
bambooshoot 竹笋
bamboo spatula 竹压片;竹刮刀
bamboo splits 竹篾
bamboo steel 竹节钢(筋)
bamboo stone cage 填石竹笼
bamboo tally 竹筹理货;发筹理货(法)
bamboo tape 竹尺
bamboo tile 竹瓦
bamboo walling 竹墙
bamboo ware 竹节形拓器;竹器
bamboo-wood inlay 竹木镶嵌
bamboo work 竹材细工;竹材工作
bamboo worker 竹木工;竹工
bamboo worker and scaffolder 搭棚工
bambusaceae 竹材
Bambusa ventricosa 佛肚竹
bamethan 丁酚胺
ban 禁止;禁令
Banach algebra 巴拿赫代数
Banach space 巴拿赫空间
Banach-Steinhaus theorem 巴拿赫—斯坦豪斯定理
Banach theorem 巴拿赫定理
banak (一种可以做材用的)美洲乔木
banalsite 钠钡长石;钡钠长石
banana apple 香蕉苹果
banana carrier 香蕉运输船
banana cluster 香蕉串
banana elevator 香蕉卸船机
banana jack 香蕉插头;香蕉插孔

banana oil 香蕉油;香蕉水
banana plug 香蕉插头
banana republic 香蕉国(指只有单一经济作物的拉丁美洲国家)
banana tube 长筒形单枪彩色显像管
banana unloader 香蕉卸船机
banburying 密炼
banbury mixer 班伯里重型双轴混合机;生胶混炼机
Banbury mixer 密炼机;密闭式混炼器;密闭式混炼机;封闭式混炼器;班伯里密炼机;班伯里密闭式混炼机;班伯里混炼机
Banbury mixing 密闭式混炼
bancal (黄色~橙色的)菲律宾硬木
banc(o) 法官席;牛轭湖;弓形湖
Banco di Napoli 那波里银行(意大利)
bancoul nuts 油桐籽
band 加双布条;能带;窖箍;箍;带(子);带状物;带环;带箍;波段;波带;扁带饰;区;嵌带;谱带;平线脚;条
banda 茅草屋(中非)
band absorption 光带吸收;谱带吸收
band adapter 置圈器
bandage 轮箍;纽绳;箍带;绷带;绑带;绑带
bandage for dressing 包扎绷带
bandage roller 卷绷带机器
bandage scissors 绷带剪
bandage shears 绷带剪
bandage treatment 绷带处理;包扎处理
bandage-winder 卷绷带机器
bandaging 绷带包扎;包扎;绑扎法
bandaging strapping 包扎法
bandal 竹框架工程(河道束水用)
bandaletta diagonalis 斜角小带
bandalling system 桩席屏系统
bandalls 桩席屏(河道束水用)
band analysis 带宽分析
band analyzer 带宽分析器;频带分析器
band and gudgeon 闸门活页;大门门铰;大门活页;大门合页;长铰;扁担合页
band and gudgeon hinge 大门门铰链;束带式铰链
band and hook 大门合页;扁担合页
band and hook hinge 大门门铰链;束带式铰链
band and peak 谱带及峰
band and strip rolling mill 带材轧机
band application 带施
Bandar Khomeyni 霍梅尼港(伊朗)
band-armo(u)red cable 钢带铠装电缆
band-armo(u)ring 钢带铠装
band articulation 频带清晰度
band attachment 接合圈
band belt 传送皮带
band bending 频带偏移
B and better lumber 二级以上的木料;二级以上的木材
bandbox 纸盒式小建筑
band brake 带闸;带形制动器;带式制动(器);带式刹
band bridge 薄带桥
band carrier 传送带
band centre frequency 频带中心频率
band chain 钢卷尺;卷尺;带链;平链
band chart 记录纸带;带状图;带形图
band cirrus 带状卷云
band clamp 带式固定夹;带夹;管箍;扁夹
band clutch 带式离合器
band combination 波段组合
band compensation 频带补偿
band compression 谱带压缩
band compressor amplifier 频带压放大器
band conduction 频波传导
band constant 谱带常数
band conversion 频带转换
band conveyer 皮带输送机;皮带输送器;皮带运输机;皮带运输器;带式运送机;带式运送器;带式输送机;带式传送器;带式传送机;传送带
band coupler 带形耦合器
band coupling 带形联轴节;带式连接器;平接联结器
band course 扁带饰;腰线;带形线脚;带饰;带;层;扁带层
band coverage 波段覆盖
band cramp 钢带夹
band curve chart 带状曲线图
band decoration 扁带饰;带饰
band dendrometer 带状测树计

band display 跑道标志
band division 波段划分
band drier 带式干燥器;带式干燥机;带状干燥器
band drive 带状传动装置
banded agate 带状玛瑙
banded appearance 带状;层状的;条状的
banded architrave 带线脚柱顶过梁;带饰门头线
banded bedding structure 带状层理构造
banded-bordered matrix 带状加边矩阵
banded cable 带扎电缆
banded cement texture 带状胶结物结构
banded clay 缟状黏土;带状黏土;纹泥(又称季候泥)
banded cloud 带状云系
banded coal 带状煤;条纹煤;条带状煤
banded collar 加筋圈;加箍圈;加箍环;轴环
banded colo(u)r filter 条状滤色镜
banded column 带饰柱;箍柱;箍形柱;分块柱
banded core 穿带芯板;封边板芯
banded differentiate 带状分异
banded elevator 带式升降机
banded fire 热备用锅炉
band edge 能带边缘;带缘
band-edge energy 带边能量
band-edge tailing 能带尾伸
banded glacial clay 有条纹的冰川黏土
banded gneiss 带状片麻岩;层状片麻岩;条带状片麻岩
banded granite 带状花岗岩;带纹花岗岩
banded impost 带线脚拱座;带线脚拱基;带线脚拱墩;拼合拱墩;有水平带形线脚的拱基
banded injected body 带状贯入体
banded iron-bearing formation 条带状含铁建造
banded iron formation 条带状铁建造
banded iron ore 条带状铁矿
banded ironstone 带状铁矿石
banded lace 条形花边
banded limestone 带纹石灰岩;带状石灰岩;薄层石灰岩
banded lode 带状矿脉
banded marble 条带状大理岩
banded matrix 带状矩阵
banded mo(u)lding 带状饰线条
banded one side 一侧镶边
banded ore 带状矿
banded pattern 带状样板
banded peat 带状泥煤;条带状泥炭
banded penstock 加箍压力水管;箍管
banded pilaster 带线脚壁柱;用突出宽带作装饰的半露柱体
banded porphyry 带状斑岩
banded pressure pump 加箍压力水泵
banded rustication 光毛相间条带状砌层
banded sediment 带状沉积
banded shaft 箍柱
banded steel pipe 加箍钢管
banded structure 加箍结构;带状组织;带状结构;带状构造;带纹构造;条带状结构;条带状构造
banded texture 条带状影纹
banded vein 带状脉;带状矿脉
band electric(al) cable connective-socket 带状电缆接头座
band electrode 带状电极
band(e)let 细带饰;带线脚壁柱;装饰细条线;素装饰线条;环柱扁饰线
band elevator 带式升降机;带式提升机;带式电梯
band-elimination 带阻
band-elimination filter 带除阻滤波器;带阻滤波器
Bandelkhand nucleus 本德尔康陆核【地】
Bandeloux's bed 班德劳氏气褥
band emission 能带发射
band envelope 谱带包络
bander 打捆工;箍工;打捆机
banderize 对钢铁施以防防腐的磷酸盐溶液涂料
bandero(ce) 扁带形雕饰
band(e)rol 带状雕饰
banderole 扁带雕刻饰;带状雕饰;扁带状雕刻装饰
banderoling machine 贴封条机
banderolle 花杆;标杆【测】
band-exclusion filter 带除滤波器
band expansion factor 频带展宽系数
band extruder 铸带机;带式挤压机
band feeder 带式喂料器;带式喂料机
band filter 带宽滤波器;带式过滤器;带式过滤机;

波段滤波器
bandforce 手动转向力
band-form 杆状的
band for wheel hub 车轮箍
band fulling machine 带式漂洗机
band fuse 条形保险丝
band fusing machine 带式热黏合机
band gap 禁带宽度;能带隙;带隙
band-gap energy 带隙能
band gap shrinkage 带隙收缩
band girdle 环剥
band girdling 带状环割
band guide 带材导板
band head 谱带头
band head rig saw 带锯主锯
band hinge 扁铰链;扁担式铰链
band horizontal dipole 宽频带水平偶极子
band hot mill 带材热轧机
bandicoot 害鼠
band identification 带站标识
band impurity 谱带杂质
band indicator 频带指示器
banding 打箍;加带环;加标记;木条镶边;描线;化学环割;箍紧;带状(的);封边木料;包箍;条状化;条状的
banding bedding 缩状层理;带状层理
banding conveyer 打捆运输机
banding insulation 绑带绝缘
banding machine 金属带捆箍机;胶粘带捆箍机;打锭绳机
banding plane 线脚刨
banding ring 绑环
banding steel 箍钢
band-interleaved-by-lines 陵行波段交叉
band interleaved by pixel pairs 按象元对波段交叉
band inversor 带状控制器
band iron 铁箍;扁钢;带铁;带钢;扁铁
band-iron tightener 扁钢张紧工具
band jaw tongs 锻工钳
bandknife 带形刀
bandknife machine 带式刀剖皮机
bandknife splitting machine 带刀片皮机
band leading peak 谱带伸前
bandlet 细带;装饰细线条;带饰细线条;扁带
band level 谱带级;频带级
band lightning 带状闪电
band-like foundation 条形基础
band-limited 带限
band-limited channel 有限带宽信道
band-limited filtering 带限滤波
band-limited frequency spectrum 限带频谱
bandlimited function 有限带宽函数
band-limited random process 限带随机过程;有限带宽随机过程
band-limited signal 限带信号;有限带宽信号
band-limited spectrum 有限带宽频谱
band-limited statistic(al) deconvolution 带限统计反褶积
band-limited white noise 有限带宽白噪声
band loss 绑线损耗
band loudspeaker 薄带扬声器
Bandl's ring 班都氏环
band matrix 带状矩阵;带形矩阵
band measurement 谱带测量
band merit 带宽指标;带宽优值
band meter 波长计
band mill 带式锯床
band modal 频带模
band model 能带模型;带模型
band mo(u)lding 带饰;装饰线条板;带形线脚
band number 带数
band number of image 图像波段号码
band of colo(u)r 色条(指漆于航标,测流标杆等上)
band of column 柱饰;柱环饰;柱环
band of error 船位误差带
band off 脱带
band of firn 永久积雪
band of hardenability 可淬硬地带
band of investment method 投资分段评估法
band of paint 油漆喷涂幅度
band of position 船位误差带;位置区域
band of rotation-vibration 转振谱带
band of westerlies 西风带

bandoleer 布子弹带
band origin 谱带起始线;谱带基(原)线
band oven 带式炉
bandpass 通带;带通
bandpass amplifier 带通放大器
bandpass channel filter 带通滤光片;带通滤光镜;带通滤波器
bandpass circuit 带通电路
bandpass filter 带通滤光片;带通滤光镜;带通滤波器;克里斯琴森滤光器
bandpass filtering 带通滤波
bandpass flatness 带通增益均匀性
bandpass frequency width 带通频率宽度
bandpass half-width 带通半宽
bandpass limiter 带通限制器
bandpass network 带通网络
bandpass of reflective coating 反射膜的通带
bandpass response 带通响应;带通频率响应特性
bandpass shape 带通频率特性
bandpass signal 带通信号
bandpass transformer 带通变压器
bandpass tuner 带通调谐器
bandpass tuning 带通调谐
band pickling machine 带材酸洗机
band pipet 谱带点样管
band plate 带板
band platform (舞台上的)伴奏席
band polishing machine 带式抛光机
band polymer 带密聚合物
band pressure level 频带压强级;频带压力级
band profile 谱带轮廓
band pulley 带动滑轮;带轮;带动滑车
band ratio 波段比值
band rationing 求频带比
band-rejection filter 带阻滤波器;带除滤波器
band resaw 剖分带锯机
band rope 扁绳;扁钢丝绳;带(绳)
band roughing mill 带材粗轧机
bands 索线
band sandstone 层状砂岩
band-saw 带锯机;带锯
band-saw blade 带锯条;带锯片
band-saw cutter 带锯铣刀
band-saw filing machine 带锯锉齿机
band saw for metal 金属带锯条
band saw for wood 木工带锯
band-saw(ing) machine 带锯床;带锯机
band-saw machine with autofeed carriage 自动送料带锯机
band-saw mill 带锯制材厂
band-saw mill with autofeed carriage 自动送料带锯机
band-sawn 带锯的
band-saw ring 带动皮盘
band-saw roll stretcher 辊子式带锯校正机
band-saw sharpener 带锯刃磨机;带锯磨齿机
band-saw sharpening grinding machine 带锯刃磨机
band-saw stretcher 带锯校正机
band scheme 能带图式
band screen 回转式滤网;链带形格栅;转筛;带式筛;带状格栅;带形屏蔽栅;带条筛;带式筛;带筛(机)
band screw 带式螺旋
band seeder 带播机
band seeding 带状播种
band selector 波段选择器;波段开关
band separation 能带间距
band-separation network 分带网络;频带分离网络
band separator 带式清洗机;带式分选机;带式除杂机;频带分隔器
band-sequential 按波段顺序的
band series 谱带系
band setting tuning condenser 波段选择用调谐电容器
bands for band-sawing machine 带锯机用带锯条
band-shaped prefabricated drain 带状预制排水板
band shaping circuit 频带形成电路
band sharing 频带共用;频带分割
band sharpening machine 磨锯机
band shell 狭长薄壳墙;(装有壳状反射板的)音乐台;(露天乐台后)回声墙;壳形演奏台;壳体结构音乐台
band shift 带迁移;带漂移;频带改变

band shoe 管箍座
band slpitting 频带分割
bands method 条带法
band spectrum 带状光谱图;带状光谱;带光谱
band spectrum analysis 带状光谱分析;带谱分析
band speed 带速度;通过带速度
bandspike 杠秤
band spinning machine 带式熔融纺丝机
band spitting equipment 分频设备
band splitter 频带分割器
band splitting 谱带分裂
band splitting equipment 频带分裂设备
band sprayer 带状喷雾器
bandspread 波段展宽;波段展开;频带展宽;波带扩展;调谐范围扩展
bandspreader 带式撒布机
band spread(ing) 频带扩展;谱带扩展
band spread receiver 展带接收机;带展接收机
band spread system 频带展开制
bandspread tuning control 频带展开调谐控制
band spring 钢板弹簧;板簧;弹簧压条
bandstand 音乐台;演奏台
band steel 扁钢;条钢
band stop 带阻
bandstop filter 带阻滤波器
band structure 带结构
band-suppression filter 带阻滤波器
band suspension meter 带悬挂仪表
band switch 波段转换开关;波段选择器;波段开关;换波带
band switching 波段转换;波段变换
bandtail 能带尾
band tailing 谱带拖尾
band tape 带钢;卷尺
band theory 能带理论;带理论
band theory of solid 固体能带论
band tire 载重轮胎
band-to-band transition 带间跃迁
band-to-band tunnelling 带间隧道效应
band tolerance 公差带
band tool 带式机床
band tubing 软韧橡皮管
band-type brake 抱闸
band type bucket elevator 皮带斗式提升机
band-type cutoff saw 带形切割器;带式切割锯;带形切割机
band-type magnetic separator 带式磁力分离机
band-type papering machine 带式砂纸打磨机
band-type screen 带型格栅
band visco(si)meter 带式黏度计
band wagon technique 潮流技术
band wander 矿物带偏移
band wheel 带动滑轮;带闸皮带轮;带轮;皮带轮
band wheel crank screw 带轮曲柄螺丝
band wheel crank screw nut 带轮曲柄螺丝帽
band wheel power unit 带轮动力单元
band wheel set screw 带轮主头螺丝
band wheel shaft 钢绳冲击钻机传动轴;带轮主轴
bandwidth 绿波(带);带宽;频带宽度;能带宽度;带宽;波带宽度
bandwidth assignation control protocol 带宽分配控制协议
bandwidth control 带宽调整;带宽控制
bandwidth correlator 带宽相关器;带宽环形解调电路;带宽乘积检波器
bandwidth effect 带宽效应
bandwidth expansion ratio 频带展宽比;频带扩展比
bandwidth in gradient elution 梯度洗脱中谱带宽度
bandwidth-limiting amplifier 带宽限制放大器;有限带宽放大器
bandwidth-limiting white noise 有限带宽白噪声
bandwidth method 带宽法
bandwidth of an antenna 天线宽度
bandwidth of sound absorption 吸声带宽
bandwidth of waveguide fiber 光波导纤维的带宽度
bandwidth on demand 按需分配带宽
bandwidth parameter 带宽参数
bandwidth reservation 带宽预约
bandwidth rule 带宽法则
band window 带形窗
band wrench 带式扳手
bandy 曲线;带状的

bandy clay 带状黏土；薄层黏土
bandy knees 弯腿
bandylite 氯硼铜矿
bang 冲击
Bangalore 爆破筒
bang-bang 继电器式控制
bang-bang action 继电器式动作；双位动作
bang-bang circuit 继电控制电路
bang-bang control 开关(式)控制；继电器(式)控制；舵面快速振动控制；起停控制；闪动控制
bang-bang device 随动开关装置
bang-bang robot 继电式机器人
banger 老爷车(俚语)
Banghum formula 斑厄姆公式
banging feeder 吊挂饲槽
banging the market 冲击市场
Bangkok Bank Ltd 曼谷银行
Bangkok Port 曼谷港
Bangor Series 班戈统【地】
bang-zone 飞机噪声区；(喷气式飞机的)声爆影响区
banian 榕树
Banish seine 丹麦拖网
banister 栏杆小柱；栏杆(立柱)；楼梯栏杆；楼梯扶手；扶手
banister-back chair 扶栏式靠背椅
banjo 机匣；短把铲
banjo axle 整体式桥壳；班卓轴
banjo bar U 形铆顶棍；U 形杠
banjo barometer 斑卓琴式气压表
banjo bolt 空心螺栓
banjo case 不可分箱体
banjo clock 班桌琴式挂钟
banjo fittings 鞍座系板配件
banjo fixing 对接接头
banjo frame 曲线规(放样用)
banjo housing 不可分箱体
banjo lubrication 离心式润滑；径向管子润滑；曲柄销离心润滑
banjo oiler 长嘴油壶；长嘴油壶；长油嘴加油器
banjo union 鼓形管接头
bank 井口区；江岸；海岸浅滩；线组；组；储[贮]料器；堆；存储[贮]单元【计】；存储体；存储单元；触排；采煤工作区；边坡；土脊
bankable bill 可通过银行承兑的票据；可贴现票据
bankable project 银行肯担保的项目
bank acceptance 银行承兑
bank account 银行账户；银行往来账户；往来存款
bank accounting 银行会计
bank address register 存储地址寄存器
bank address system 存储地址式
banka drill 班加钻；砂矿钻机；冲积层勘探钻；班克式人力勘探钻机
bank advance 银行来往账户
bank a fire 压火
bank against 填筑成滩；填筑成坡；填筑成堤
bank alignment 堤岸线定向
bank-and-climb gyro control unit 倾斜升降回转控制器
bank-and-turn indicator 转弯指示器；倾斜与转向指示器；倾斜转弯仪
bank-and-wiper switch 触排及弧刷转接器；触排和寻排开关；双动选择器
bank angle 超高角；倾斜角
bank armo(u)ring 堤岸护面
Banka tin 邦加锡
bank atoll 浅滩环礁
bank audit 银行审计
bank balance 银行往来账余额；银行结存
bank bar 坡地切割器；孔壁衬板；井壁衬板
bank barn 斜坡谷仓
bank barrier 有滩陡岸
bank bill 银行纸币；银行票据
bank bit 存储位
bank blasting 阶段爆破
bank block 银行大厦
bank bond 河岸稳固
bank book 银行存折；存折
bank breach 决堤；堤岸溃决
bank breaching 堤岸冲毁
bank building 银行房屋；造岸
bank buying rate 银行购买汇率
bank cable 线弧电缆
bank call 要求银行提交财务报表通知
bank capacity 线弧容量；触排容量

bank card 银行信用卡
bank caving 河岸淘空；河岸崩坍；堤岸坍陷；堤岸崩坍；崩坍；岸塌；坍岸；塌岸
bank cavity 崩岸
bank chain 海岸山链
bank charges 银行手续费
bank charter act 银行特许条例
bank check 银行支票；空白支票；支票；本票
bank check deposit 本票存款
bank checking account 银行支票户头
bank cheque 银行支票
bank cleaner 线弧清拭器；触排清洁器
bank clearance 离岸宽度；离岩净空；离岸净空；船舶离岸距离；船岸距；岸边间隙
bank clearing 银行票据交换
bank coast 岸
bank cohesiveness 堤岸土体黏聚性
bank collapse 河堤崩塌；岸塌
bank collapse investigation 坍岸调查
bank commission 银行佣金；银行手续费
bank consortium 银团
bank construction 护岸施工；河岸建造；堤坝工程
bank contact 触排接点
bank control 分组控制；倾斜操纵
bank controller 存储体控制器
bank copy 银行留的底本
bank correspondent 代理银行；同业银行
bank crane 岸边起重机
bank credit 银行信用证；银行信贷
bank credit fund 银行信贷基金
bank credit proxy 美联邦储备会员行存款平均额
bank creel 毛球架
bank crisis 银行危机
bank crown 单坡路拱
bank cubic(al) meter 立方米土方
bank cubic(al) yard 立方码土方；实体立方码(爆破岩石)；实体方位码
bank cushion 河岸缓冲力；岸推
bank cutter bit 光身车头刀
bank cutting 开挖岸坡；河岸侵蚀；河岸切割；岸坡开挖；岸边冲刷
bank debenture 金融债券
bank deformation 河岸变形；堤岸演变
bank deposit 河岸沉积物；银行存款；岸边沉积
bank deposit journal 银行存款日记账
bank deposit of major repair fund 大修理基金银行存款
bank deposits subject to withdrawal by check 凭支票取款的银行存款
bank discount 银行贴现
bank discount rate 银行贴现率
bank distance 离岸距离
bank draft 银行票据；沙洲吃水
bankruptcy court 破产法庭
banked 积起的
banked battery 并联电池组
banked bend 筑有护岸的河湾；筑成从内侧至外侧向上倾斜的弯道；超高弯道
banked boiler 备用锅炉；热备用锅炉
banked bowl 超高碗形曲面；碗形超高(用于高速弯道)
banked concrete 夹板现浇混凝土墙
banked concrete wall 夹板现浇混凝土墙
banked crown 单坡路拱
banked crown on curves 曲线超高；弯道上的单坡路拱
banked curve 超高曲线；横向倾斜曲线
banked eddy 岸边涡流；河边涡流
banked earth 储存土
banked eddy 岸边涡流
banked fire 压火
bank edge 岸边
banked relay 遥控继电器
banked turn 超高弯道；倾斜转弯
banked-up water 壅水；回水
banked-up water level 壅水水平；回水水位；壅水(水)位；壅高水位
banked winding 叠绕线圈；重叠绕法
bank effect 河岸效应；侧壁效应；岸吸岸推现象(吸船尼推船首)；岸壁影响
bank endorsement 银行背书
banker 井口工；满排流；银行业者；银行家；筑埋器；筑埋机；造台架；雕刻转台；堤防土工；挖土工人
banker mark 石工标记

banker mason 雕坏工；琢石工
bank erosion 河边侵蚀；河岸侵蚀；河岸冲刷；堤岸冲刷；岸坡侵蚀；岸坡冲刷；岸边侵蚀
bank erosion rate 河岸冲蚀率
bank error 倾斜度误差
banker's acceptance 银行承兑
banker's bill 银行票据
banker's card 银行信用卡
banker's check 银行支票
banker's clean bill 不抬头的银行票据
banker's clearing house 票据交换所
banker's invoice 银行发票
banker's letter of credit 银行信用状
banker's long draft 银行长期汇票
banker's margin 银行存款保证额
banker's rate 银行汇率
banker's rate selection method 银行等级选择法
banket(te) 护坡道；弃土堆；护脚；含金砾石层；致密石英砾岩
bank evolution 堤岸演变
bank exchange memo 银行兑换单
bank exchanges 通过交换托收的银行票据
bank face 阶段工作面
bank failure 银行破产；河岸淘空；河岸崩坍；堤岸溃决；堤岸崩坍；坍岸；塌岸
bank favo(u)rable rate 银行优惠利率
bank-filtered water 河岸渗滤水
bank filtration 岸边过滤；沙滩过滤
bank financing 银行投资
bank fire 封火；压火
bank fittings 银行柜台装置
bank fixation 河岸固定
bank flora 河滩植物
bankfull 满槽的
bankfulla 齐岸的(水位)
bankfull channel capacity 河道满槽过水能力
bankfull discharge 漫滩流(量)；满槽流量；齐岸流量；平滩流量；平岸流量
bankfull flow 漫滩流(量)；满流；满槽水流；满槽流量；满槽流；平岸流
bankfull stage 满流水位；平岸水位；漫滩水位；满库水位；满槽水位；满岸水位；齐岸水位；平岸水位
bankfull(water) level 满槽水位；漫滩水位
bankfull water stage 漫滩水位
bankfull width 齐岸宽度
bank grading 河岸修整；河岸修坡；削坡；坡岸修整
bank gravel 河岸砾石；采石坑砾石；河滩砾石；岸砾；未过筛砾(石)；河卵石
bank gravel-sand mix(ture) 天然砾石和砂混合料
bank guarantee 银行担保(书)；银行保证；银行保函
bank guarantee for bid bond 银行出具的投标保证书
bank guarantee for performance bond 银行出具的履约保证书
bank hall 银行大厅
bank hall floor 银行大厅地面
bankhead 坑口；岸首；坑口装卸平台；井口；丁坝；岸垛；挑水坝
bank height 露天矿阶段高度；岸高(度)
bank high flow 高岸流量
bank holding company 银行控股公司；银行持股公司
bankia 蛀船虫
Bankia spengler 船底附生物
bank improvement 岸坡处理
bank in 筑堤围入
bank indicator 触排指示器；倾斜转弯仪；倾斜指示器；倾斜指示表；倾斜显示器
bank infiltration 岸渗漏
banking 银行业；压坡度；筑堤；定位的；堤防；道路斜度；侧倾；封炉；填高
banking agreement 银行协议书
banking angle 侧倾角
banking body 培土犁体
banking business 银行业务
banking center 金融中心
banking curve 超高曲线
banking degree of freedom 侧倾运动自由度
banking file 齐头三角锉
banking fund 银行营业资金
banking hall 银行大厅
banking house 银行
banking inclinometer 倾角计
banking indicator 倾角计

banking institution 金融机构
banking interest 银行利率
banking locomotive 推进热机车
banking loss 机组热备用损失；埋火消耗量；压火损失；封火损失
banking machine 覆土机；培土机
banking materials 筑堤材料
banking pin 限位钉；限动销；制动箱
banking power 银行投资能力
banking practice 银行业务
banking process 堆积过程
banking room 出纳办公室
banking screw 限位螺钉；止动螺钉
banking structure 填土结构物
banking supervision 银行监督
banking transaction 银行交易
banking-up 压火
bank-inset reef 滩缘礁
bank intake 河岸进水口
bank interrupt processor 中断处理机
bank investment 银行投资
bank invoice 银行发票
Banki turbine 班克式水轮机；班基式水轮机；双击式水轮机
bank kiln 阶级窑
bank land 河岸平台
bank levee 井口出车台；河岸堤；岸式码头；岸堤
bank level(l)ing 平整岸线
bank light 泛光灯组；聚光灯；排灯
bank line 路肩线；河岸线；道路边线；岸线；岸边线
bank line of road 路基边缘
bank-line profile 路基边线纵断面；岸线纵断面；岸线轮廓图；路基边缘线纵断面
bank-line survey 岸线测量
bank liquidity 银行资产流动性
bank loading 河岸上的荷载
bank loan 银行贷款
bankman's handle 罐座操纵手把
Bank Markazi Iran 伊朗国家银行
bank material 河岸质；岸积物；土石方工程材料；土石坝填筑材料
bank measure (原地的)土石方量；堆层尺度；堤岸土方测量；未扰动土(石)方量；土石方量测定；填方数量；填方量
Bank Melli Iran 伊朗梅里银行
bank method of attack 梯段开采法；台阶采矿法
bank money 银行票据
bank money order 银行拨款单；银行汇票
bank moratorium 法定银行延期偿付期
bank multiple 触排变接盘
bank multiple cable 线弧复式电缆
bank note 空白支票；银行承兑的票据
bank note in circulation 流通的钞票
bank of a cut 剖线边沿
Bank of America 美洲银行
bank of bedroom closets 一排卧室壁橱
bank of capacitors 电容器组
bank of capacity 电容器组
bank of cells 槽组；浮选室
Bank of China 中国银行
bank of circulation 发行银行
bank of closets 一排壁橱
bank of commerce 商业银行
bank of condensers 冷凝器组；电容器组
bank of contacts 接点排；触头排
bank of cylinders 汽缸组；汽缸排
bank of deposit 开户银行；存款银行
bank of discount 贴现银行
bank of ditch 明沟斜度
bank of elevators 一排电梯；相对排列货物起卸口系统
bank of emission 发行银行
bank of firn 粒雪堆
bank of information 信息库；资料库
bank of international settlements 国际清算银行
bank of issue 发行银行
Bank of Japan 日本银行
bank of lamps 灯泡架
bank of material 资料库
bank of muddy sand 泥沙坝
bank of oil 油带
bank of ovens 炉组
bank of pier 桥台
bank of resistors 电阻器组

bank of river 江岸
bank of settlement 结算银行
bank of silt 淤沙沙洲
bank of slope 土坡
bank of soil 土坡；土堤
bank of staggered pipes 错列排管
Bank of Sweden 瑞典银行
bank of terraced field 梯田地坎
Bank of Thailand 泰国银行
Bank of Tonga 汤加银行
bank of transformers 变压器组
bank of tubes 管排
bank overdraft 向银行透支；银行透支
bank paging call 寻呼电话【船】
bank paper 银行承兑的票据
bank particle 堤岸土料颗粒
bank pass-book 银行存折
bank pavement 岸坡铺面
bank paving 河岸铺砌；护岸铺面
bank pier 岸墩；岸边桥墩；桥墩
bank plant 岸堤栽植
bank planting 堤岸种植；河岸堤树造林；堤岸栽植
bank portfolios 银行有价证券
bank post bill 银行汇票
bank prime rate 优惠利率；银行最优惠利率
bank process 岸坡冲碛法
bank protection 护岸；河流护岸；河岸保护；堤岸防护；防护岸
bank protection by short spur dike 短丁坝护岸
bank protection dam 护坡堤坝；护岸堤坝
bank protection structure 河岸保护结构
bank protection works 护坡工程；护岸设施；护岸工程
bank protective forest 护岸林
bank protector 护岸工程
bank pump 供水泵
bank pumping station 岸边式泵站
bank rate 银行利率
bank rate of discount 银行贴现率
bank rate of rediscount 银行再贴现率
bank reef 不规则水下礁；岸礁；滩礁；珊瑚礁
bank reference 银行资信书；银行证明书
bank refund guarantee 银行偿付保证书
bank regulation 河岸整治；平整岸线
bank reinstatement 河岸恢复原状；堤岸修复；复堤；复岸工程
bank reinstatement method 复岸法；复堤法
bank remittance fee 银行汇款手续费
bank renovation 堤岸修复
bank reserve 银行储备金
bank resistance 河岸阻力
bank restoration 堤岸修复
bank return 银行收益
bank revetment 护坡；护岸；堤岸护坡
bank roll 货币储备；有效基金；资助；资金
bank room 出纳办公室
bank roughness 堤岸粗糙度；堤岸糙率
bank ruin of reservoir 水库坍岸
bank-run 河滩砂石；河岸边的；岸边的
bank-run aggregate 河岸边集料；河岸边骨料；岸滩集料；岸滩骨料；岸料；岸骨料；岸边集料；岸边骨料
bank-run gravel 岸滩砾石；河滩砾石；河卵石；河岸砂砾；河岸砾石；河流砂砾；岸砾石
bank-run sand 河沙；岸沙；河砂；岸砂；原砂
bankrupt 破产者；破产人；破产的；无力还债的
bankrupt account 破产账
bankruptcy 破产
Bankruptcy Act 破产法例
bankruptcy administrator 破产财产管理人
bankruptcy court 破产审理法庭
bankruptcy debtor 破产债务人
bankruptcy law 破产法
bankruptcy notice 破产通知(书)；破产公告
bankruptcy petition 破产申请；申请破产
bankruptcy proceeding 破产程序
bankrupt landlord 破产地主
bankrupt's assets 破产资产
bankrupt's estates 破产财产
bankrupt stock 因破产而拍卖的货物
bank sand 河岸沙；滩砂
bank savings account 存折
bank scale 倾斜标度盘
banks clarifier 系列澄清器；系列澄清池；班克斯澄

清池
bank scouring 岸坡冲刷
bank's demand draft 银行即期汇票
bank's disposal credit 银行偿付信用
bank's draft 银行汇票
bank seat 碇桩；岸上碇桩；桥的岸座
bank selling rate 银行卖出汇票
bank service charges 银行手续费
bank settlement 河岸沉陷；河岸沉陷；堤的沉降；堤的沉陷
bank's guarantee for bid bond 投标保函
bank shear 河岸剪力
Banksian pine 北美短叶松
bankside 河岸斜坡；河岸；岸边的
bank sided 内倾
bankside dune 岸边沙丘
bankside erosion 岸边冲刷
bankside filtration 岸边过滤
bankside scouring 岸边冲刷
bankside storage 岸边原水；岸边储[贮]水；岸边储水
bank sill 岸基；岸坡底槛；梁梁；槛
bank-slag 炉层；炉堆；炉渣
bank sliding 河岸滑坡
bank slip 银行兑换单
bank slope 河岸斜坡；河岸边坡；边坡倾斜角；岸坡；路堤边坡
bank sloping 岸斜面的不稳定；岸坡筑料
bank slough 河岸坍塌；岸坡表层脱落
bank sloughing 河岸崩坍；堤岸坍塌
bank sluice 堤岸泄洪闸；河闸
bank(s)man 井口把钩工；塔机助手；吊车助手；井外监工
banks of ejectors 射流器组
bank soil 岸土
bank soil storage 河岸土壤蓄水量
bank's order 银行本票；本票
bank span 岸跨
bank stability 河岸稳定性；岸坡稳定性
bank stabilization 河岸稳定；河岸加固(作用)；堤岸加固；岸坡稳定(作用)；岸坡加固；岸坡处理；岸边加固
bank stage 满岸水位；齐岸水位；平岸水位
bank stamp 银行背书
bank statement 银行结账单；对账单
bank stock 银行股份
bank storage 河岸地下水储[贮]量；河槽蓄水量；河岸蓄水(量)；河岸间槽蓄；河岸调蓄量；河岸储量；岸储[贮]水
bank storage discharge 河岸调蓄流量
bank strengthening 护坡；河岸加固；堤岸加固；岸坡加固
bank strongroom 银行贵重物品保险库
bank subsidence 堤的沉陷
bank suction 河岸吸引作用；岸吸作用；岸吸现象；岸吸；偏贴向岸作用；贴岸吸力作用；贴岸吸力
bank suction force 岸吸力
bank switching 存储库开关
bank telegraphic(al) transfer 银行电汇
bank terrace 岸坡阶地；岸坡台地；岸坡阶地
bank the fire 封火
bank tip 河岸上挖出的泥堆
bank top width 堤顶宽度
bank transfer 银行汇款；银行转账
bank transference 银行划拨
bank trust fund 银行信托基金
bank tube 栅管
bank-type barn 双层畜舍
bank up 堵截；封窑；封炉
bank vegetation 河岸植被
bank wall 堤墙；挡墙；无窗墙
bank widening 路堤加宽
bank width 堤宽
bank-winding(coil) 交叠多层绕组；叠绕线圈
bank-wire 触排导线
bank wiring 触排布线
bankwise channel 沿岸航道
bankwise mark 沿岸标
bank with credits opened 往来银行
bank with whom the credits opened 开证(银)行
bank wound coil 叠层线圈
bank yards (土或岩石开挖前的)原方；未扰动土(石)方量
banlieu(e) 城郊住宅区；郊区

banned catching method 禁用捕捞法
banned fishing gear 禁用渔具
banned products 禁制品
banner 宽显线;旗帜;旗号
banner bead winding machine 钢丝圈卷成机
banner bracket 标识托座;旗帜托座
banner cloud 旗状云
bannering 修整边缘
banner land 旗地
banner line 旗帜线;标识线
bannermanite 碱钒石
bannerol 小旗(葬礼用);飘带或长卷形花纹;扁带形雕饰
banner word 标志字;标题字;标签字
Bannisdale Slates 班尼斯达尔板岩
bannister 栏杆;扶手;栏杆柱
bannister harness 棒刀吊综装置
bannisterite 硅锰矿;斑硅锰石
bannock 褐灰色耐火黏土
ban on export 禁止出口
ban on import 禁止进口
ban on maritime voyages 海禁
Banque Bruxelles Lambert 布鲁塞尔—朗贝银行
Banque Centrale de Tunisia 突尼斯中央银行
Banque Centrale du Mali 马里中央银行
Banque de Bruxelles 布鲁塞尔银行
Banque Nationale Suisse 瑞士国家银行
banquet cart 布宴车
banqueting hall 宴会厅
banqueting kitchen 备宴厨房
banqueting room 宴会室
banquette 护坡道;后戗;背水坡护道;戗台;弃土堆;凸出部分;台坎
banquette construction 建造护坡道
banquette of levee 堤戗;填出堤
banquette seating 火车站
banquette slope 踏垛坡
Bansen 班森(透气率单位)
bantam 小型设备
bantam car 轻便越野汽车
bantam mixer 非倾倒式拌和机;非倾侧式搅拌机;非倾侧式拌和机
bantam stem 短茎;小型管管脚;矮小心柱
banto faro 半沼泽
banyan 榕树
banyan tree 榕树
baobab oil 木棉油
baotite 包头矿
baotou beam 抱头梁
Baplorfstz function of back-water curve for positive inclination 正坡回水曲线的巴氏函数
Baplorfstz function of depression curve for positive inclination 芒坡下降曲线的巴氏函数;正坡下降曲线的巴氏函数
baptismal room 洗礼室;洗礼堂
baptisoid 野靛蓝素
baptistery 洗礼池;浸礼教的洗礼堂;洗礼堂
bar 拦门沙;拦江沙;卖酒柜台;河口浅滩;栅门;柜台;杆;挡杆;棒料;棒(材);巴(压强单位);沙坝
baraboo 重现残丘
Baraboo quartzite 巴拉波石英岩
barachois 弯河湾
baramite 菱镁蛇纹岩
bar anchorage device 帮条锚具
bar anchoring system 棒固定系统
bar-and-dot generator 点划发生器
bar and key grate 链条炉排
bar and shape mill 型材轧机
bar and sill method 上导坑法;上导沟法
bar and soda-fountain sink 酒吧间及冷饮间洗盆
bar and tube machine 制棒和制管机
bar and tube straightening machine 管棒矫直机
bar antenna 棒状天线;棒形天线
Barany's box 巴拉内箱
Barany's chair 巴拉内椅
Barany's test 巴拉内试验
bar apparatus 杆状基线尺
bararite 氟硅铵石
bar armo(u)r 线棒保护层
bar arrangement 钢筋布置;配筋布置
bar arrangement drawing 布筋图;配筋图
baras 粗袋布织物
barat 巴拉特风
baratovite 硅钛锂钙石

barb 毛口;毛刺;倒钩;倒刺(钩);风羽
barbacan 碉堡;城堡
Barbados 巴巴多斯岛
Barbados earth 巴多斯放射虫石
Barbary asphalt 柏伯尔沥青
Barba's law 巴尔拜定律
barbatol 坝巴醇
barbatol-carboxylic acid 坝巴醇羧酸
barbatolic acid 坝巴醇酸
barb bolt 棘螺栓;基础螺栓;倒钩螺栓;带刺螺栓
bar beach 堤洲;沙坝滩
barbed dowel 雌雄榫连接销钉
barbed dowel pin 有倒钩的接合销(木工用);带刺销钉;带刺合缝销;刺销钉
barbed drainage 倒钩状水系
barbed drainage pattern 倒钩状水系;倒刺状排水系统;上游羽状水系
barbed filament 毛刺丝
barbed mode 倒钩水系模式
barbed nail 带齿钉;倒刺钉;刺钉;不规则钉
barbed needle 带钩的针
barbedos lily 朱顶红
barbed spike 棘钉
barbed spindle 倒刺式纺锭;倒刺纺车轴
barbed steel wire 刺钢丝
barbed tributary 倒钩支流
barbed wire 刺丝;有刺铁丝;有刺钢丝;制网铁丝;带刺铁丝网;刺铁丝
barbed wire electrode 刺铁丝电极
barbed wire entanglement 刺铁丝网
barbed wire fence 倒刺铁丝围栏;刺铁丝围篱;刺铁丝围栏
barbed-wire machine 制刺钢丝机;制刺线机
barbed wire nail 带刺铁钉
barbed wire nippers 带刺铁丝剪钳
barbed wire reinforcement 带刺铁丝网;带刺钢筋;刺铁钢筋
barbed-wire strainer 倒刺铁丝拉紧器
barbell 杠铃
barbell set 举重器
bar bench 棒材拉拔机
bar bender 钢筋弯折机;钢筋弯曲器;钢筋弯曲机;钢筋弯具;钢筋挠曲器;弯钢筋机
bar bender and cutter 钢筋弯曲切割机
bar bending 弯钢筋
bar bending machine 弯钢筋机;钢筋弯折机;钢筋弯曲机;棒料弯曲机;棒材弯曲机;弯条机
bar bend table 弯钢筋台
barber 冷风暴;寒冷蒸汽;海上蒸汽雾;大风雪(美国和加拿大地区)
barber chair 椅状劈裂;打拌子
Barber-Colman automatic spooler 巴伯考尔门自动络纱机
Barber-Greene 巴伯格林铺路机
Barber Greene finisher 巴伯格林修整机
Barberio's test 巴伯里欧氏试验
barberite 铜镍锡硅合金
barber pole 理发店招牌
Barber pyrometer 巴伯隐线式光测高温计;巴贝尔高温计
barberry 小檗;刺檗;伏牛花
barbershop 理发店
barber shop and beauty parlor 理发店与美容店
barber's pole 理发店招牌
barber's shop 理发店
Barber's warp tying machine 巴柏氏自动接经机
barbertonite 水碳铬镁石;水镁铬矿
barbette 露天炮塔;炮座;炮台
Barbey orifice viscometer 巴比锥孔黏度计
barbican 碉堡;城堡;外堡
barbican entrance 望楼门
barbierite 钠正长石
barbing 柱;竿;棒
barbital 巴比土
barbital sodium 麦地那;巴比妥钠;巴比土钠
barbitone 巴比土
barbitone sodium 巴比妥钠
barbiturate 巴比妥酸盐;巴比妥酸盐
barbituric 巴比妥酸盐;巴比土酸盐
barbituric acid spectrophotometry 巴比妥酸分光光度法
barbosalite 复铁天蓝石
barbotage 起泡作用

barbotine 粗陶浮雕泥浆;装饰粗陶浮雕用的料浆
bar bound 船被拦门沙拦阻
barb radiator 芒刺式散热器
barb breaker 棒式破碎机
barb system 交错式布置
bar buckling 杆的压曲;杆的屈曲
Barbuda 巴布达岛
barbule 内齿层
bar buoy 拦江沙浮标;浅滩指示浮标;浅滩浮标;滩上浮标
bar burner 棒式燃烧器
barbus tie cell 母线联络间隔
bar button 杆state纽扣
barbwire 带刺铁丝网;制网铁丝;刺铁丝;刺钢丝
barb wire fence 有铁蒺藜的围栏;带刺铁丝围栏
bar by-passing 沙坝型转移
bar calibration 杆校准
bar capacity 棒料最大直径
bar capstan 人力绞盘;人力绞磨;推杆绞盘
Barcelona 巴塞罗那(西班牙东北海港)
Barcelona chair 巴塞罗那式椅子;镀铬钢材皮垫椅
bar centre line 杆的中心线
bar chain 杆链
bar chair 钢筋垫块;钢筋座;钢筋支座;钢筋支架;钢筋垫
bar chamfering machine 棒材倒角机
barchan 新月形沙丘
barchan chain 新月形沙丘链;沙丘链
bar chart 横道图;横线工程(计划)图表;线条图(表);线条进度表;柱状图;直条图;直方图;条形图表;条形图;条线图;条线表(统计用);条图;条件图表
bar chart current meter 直方图海流计
bar check 校准杆;杆校准
bar chuck plate 棒料卡盘
bar clamp 杆夹;棒夹;钢筋夹;钢夹
bar claw 闩爪
bar coach 酒吧车
bar coal cutter 杆式截煤机
bar coater 刮涂机;刮涂棒
bar code 条型码;条形码;条纹密码
bar code label 条形码标记;条件码标记
bar-code reader 条形码阅读机;条形码读出器;条码阅读器
bar-code scanner 条形码扫描器;线代码扫描器;条线代码扫描器
Barcol hardness 巴柯尔硬度
Barcol impressor 巴柯尔硬度计;巴柯尔压痕器硬度计
bar commutator 铜条整流子
bar connection 杆的连接
bar conveyer 杆式输送机;推杆运输机
bar copper 棒铜
barcoque 过分雕刻和怪诞的(巴洛克艺术风格)
bar corrosion activity evaluation 钢筋锈蚀活动性评定
bar coupler 钢筋连接套筒
bar coupling 杆形连接器
bar cramp 杆夹
bar crane 杆式起重机
bar cropper 线棒扭弯器;钢筋切断机;钢筋截断器;钢筋截断机;钢筋剪断机
bar-cropping machine 圆钢剪切机
bar cross-section 杆的截面
bar crusher 棒式破碎机
bar cut-off machine 棒料切断机
bar cutter 截条机;钢筋切割机;钢筋切断机;钢筋截断机;钢筋切断机;钢筋剪断机;切条机
bar cutting 钢筋切断
bar cutting machine 棒材切割机;棒材切断机
bar cutting shears 切割杆件机
bar cylinder 击杆式脱粒滚筒
Bardach's test 巴尔达赫氏试验
Bardac process 巴达克法
bar davit 圆形吊艇柱
bar deck 钢棒筛板
Barden 高岭土稀释剂
Bardenpho process 巴顿埔工艺
bardepot 现金存款
bar depth 沙槛水深
bar detector 钢筋探测器
bar diagram 直方图;图表;直条图;线条图;条图
bar diameter 杆的直径;钢筋直径
Bardine process 巴丁法(消除钻杆金属疲劳应力

方法)
bar disintegrator 棒式粉碎机
bar display 杆式显示装置
bar distance piece 杆的间隔件
bardolite 黑硬绿泥石;纤灰蛭石
bar door 酒吧间门;板条门
bardotine decoration 色浆装饰法
bar dowel 接合短销;接合销筋;接合短钢销;合缝接钢筋;传力杆;叉筋
bar down 撬整工作面;撬落矿石
bar draft 拦江沙(低潮)水深;浅滩水深;闩沙水深;沙槛水深
bar draught 拦门沙洲吃水;拦门沙堤深;过滩吃水;低潮沙槛水深
bar drawing 芯杆拔管法;棒材拉拔
bar drawing inspection 拉拔检验;拉出试验
bar drill 架式钻机;架式凿岩机;横梁架式钻机;横梁架式凿岩机;杆钻
bar dropper 投棒短节
bare 裸露;露出水面;露出盖片;光秃;微亏
bare alumin(i)um stranded wire 裸铝绞股线
bare and barren land 不毛(之)地
bare and line chart 划线图表
bare area 裸区;裸地;空地
bare-base support 无覆盖层地基
bare board 裸板
bare boat charter 光船租船合同;空船租赁
bare boat charter party 让租契约;空船租船契约
bare boat charters on annual contracts 空船年度租约
bare boat clause 光船条款
bare boat form charter 裸船租船合同
bare bones 梗概
bare busbar 裸汇流条;裸母线
bare cable 裸线;裸缆;裸电缆
bare channel 裸露渠道(没有植被的);裸露河槽
bare charge 裸装药;裸药
bare computer 裸计算机;硬件计算机;未配软件的计算机
bare conductor 裸电线;裸导线
bare contract 不附条件合同;无条件契约;无条件合同;无担保契约;无担保合同
bare cooling tower 开放式冷却塔
bare copper connection 裸铜接头
bare copper wire 裸铜线;裸铜丝
bare core 裸堆心
bare cut slope 新开挖边坡
bare deck 裸甲板;裸船体;光甲板
bar edge 条信号边沿
bared river mouth 有拦门沙的河口
bared-shaped window 圆筒形窗
bare electrode 裸甲板;裸焊条;裸电极;无药焊条
bare electromagnet 条形电磁铁
bare-engine 无辅助设备发动机;不带附属设备的发动机
bare face(d)tenon 裸面榫(头);半肩榫;侧肩榫;单面榫舌
bare face(d)tongue 裸面榫舌;裸面雄榫(榫舌);半肩榫
bare fallow 休闲地;绝对休闲地;无草休闲地
bare fiber 不含处理剂的纤维
bare filler rod 裸电焊条;裸焊条
bare foot 裸眼;无榫骨架;未上套管钻孔;未下套管的井段
barefoot completion 裸眼完井;裸眼成井
bare footed 无套管的钻孔
bare-foot(ed)well 裸眼井;裸底井;裸井
barefoot finishing 裸眼终孔;裸眼完成
barefoot interval (未下套管的)裸眼井段;裸井段
bare frame 空框架
barege 巴雷格纱罗
baregin 胶素;粘胶质
bare glass 原玻璃(纤维)
bare glass fiber 原纤维;不含浸润剂的玻璃纤维
bare grizzly 支架筛;格子筛
bare ground 露天采矿区;已采地区;采空区;不毛(之)地;无遮蔽的土地;无矿地层
bare ground wire 裸接地线
bare-headed 光着头的
bare hole 裸孔
bare hole cement plug 裸眼水泥塞
bare hull 裸船体;光船(体)
bare-hull charter 光船租船合同
bare ice 裸冰

bare ion 裸离子
bare karst 裸露型岩溶;裸露型喀斯特
bare lamp 明露灯
bare land 裸露地;基岩地区;出露地区;非耕地;不毛(之)地;白地
Barelattograph 巴氏仪(坯体收缩,失重记录仪);坯体收缩失重记录仪
bar element 条元
bare L-square 直角尺
barely enough moisture to produce crops 水分不足以生长作物
barely flow 明流
bare machine 裸机;硬件计算机
bare mass 裸质量
bare maximum 绝对最大值
bare metal 金属拷铲出白
bare metal arc weld 裸焊条电弧焊
bare-metal arc welding 裸焊条电弧焊
bare mill 桶式研磨机
bare minimum 绝对最小值
bare motor 无配件电动机
baren 压印垫板
bareness 露底
barentsite 氟碳铝钠石
Barents Sea 巴伦支海
bare optic(al)fiber 裸光纤
bare ownership 空头产权;不具名财产
bare particle 裸粒子
bare pavement maintenance (使路面露的)除雪养护
bare phace 不毛(之)地
bare pig 裸清管器
bare pipe 裸露管;裸管;光管;不绝缘管;无螺纹管
bare pipeline 裸露管线
bare place 非耕地
bare pole charter 光船租船合同
bare poles 不挂帆的桅;不挂帆的帆船
bare pulley 光面滑轮
bare radiator 裸露散热器;裸露散热片;明散热器
bare reactor 裸反应堆
bare rock 裸岩;裸岩石;裸礁石;明礁
bare rod 无药焊条
bare-rooted planting 裸根栽植
bare root plant 露根栽植
bare sleeve 白坯管
bare soil 裸土;裸露土(壤);裸地;不毛地
bare spot 裸点;光秃点
bare steerageway 舵效航速
bare stone 山墙凸石
bare terminal 焊条夹持端
bare terminal end 夹持端
bare terminal end of an electrode 电焊条裸头
bare the metal 出白
bare thermocouple 裸露热电偶
bare trustee 被动受托人
bare tube 裸管;露天管路
bare turbine 开式透平;开式涡轮机
bare universal joint 开式万向联轴节
bare value 裸值
bare weight 净重;空重;皮重
bare welding rod 光焊条
bare wire 裸线;裸铜线;裸导线;光丝
bare wire arc welding 光丝弧焊;光焊丝电弧焊
bare wire electrode 裸焊条
bar fabric 圆钢筋条组成的网
bar fagoting 小方形开口缝;梯形开口缝
bar feed 钢筋进给
bar feeder 栅式进料器;刮板给料器;杆式给料器;棒式进料器;棒材进给器
bar feed mechanism 棒材送进机构
bar fender 防撞杆
barffing 蒸汽(发蓝)处理
Barff's process 巴尔夫方法
bar field 杆件结构构造
bar-finger sand 指状沙坝;指状大透镜砂体
bar finish 酒吧装修
bar flight feeder 链条刮板喂料机
Barfoed's test 巴福德试验
bar folder 钢筋弯曲器;弯折机
bar force 杆力
bar for grooved springs 有槽弹簧用圆钢
bar fork 长叉
bar formation 钢筋组架;钢筋构成;沙洲生成
bar frame 杆架;巢框

bargain 廉价品;交易;买卖合同;成交条件;商定价格
bargain and sale deed 财产买卖契约;附条件的土地转让契约
bargain away 廉价脱手;卖脱
bargain basement 廉价部;地下廉价商场
bargain center 廉价市场;买卖中心;买卖市场;平价市场;廉价中心
bargain counter 廉价货品柜
bargainee 买主
bargainer 卖主;讨价还价者
bargain hunting 买廉价货
bargaining 契约;成交
bargaining group 交易谈判小组
bargaining policy 互惠通商政策
bargaining position 讨价还价的地位
bargaining power 议价能力;讨价还价的能力
bargaining process 议价过程
bargaining right 谈判权
bargaining strategy 讨价还价买卖策略
bargaining tariff 互惠关税协定;协议关税
bargaining transaction 议价交易
bargaining unit 议价单位;谈判单位
bargain money 定金;保证金
bargain offer 廉价优待
bargain on the spot 现款交易
bargainor 卖主;买主
bargain over the price 讨价还价
bargain price 廉价;特价
bargain purchase 廉价收购
bargain theory 谈判决定工资论
bargain transaction 廉价交易
Bargate beds 坝口层
bar ga(u)ge 钢筋规格;杆规;棒规
barge 座艇;浮吊;驳(船)
barge aboard catamaran 双体载驳货船
barge aboard catamaran barge 双体载驳货船子驳
barge aboard catamaran vessel 载驳双体船
barge above catamaran 载驳双体船
barge basin 驳船港池
barge bed 驳船停泊滩;驳船停泊区
barge berge 停泊处;浮码头
barge berth 浮码头;驳船码头;驳船泊位
barge berthing area 驳船靠泊区
barge board 挡板;山墙封檐板;博风板
barge canal 内陆运河;驳船运河;驳船渠
barge carrier 运驳母船;载驳(货)船;大型平板驳船
barge carrier ship 载驳船
barge carrier terminal 载驳货船码头
barge carrier transport 载驳运输
barge-carrying catamaran 载驳双体船;双体载驳船
barge-carrying ship 载驳(货)船
barge chain 船队
barge clipper system 载驳快船方式;载驳货船系统(浮船坞式)
barge commerce 驳运量
barge conveyance 驳运;驳船运输
barge couple 檐口人字木;山墙檐口椽;山墙挑檐椽
barge course 墙檐;山墙顶层;山墙顶部侧砌砖;山墙檐压顶;山墙檐瓦
barge cover 驳船舱盖
barge crane 船式起重机;浮式起重机;浮吊;驳船起重机;起重船
barge derrick 浮船;船式起重机;浮式起臂吊车;浮式桅杆起重机;浮式人字起重机;起重船
barged-in fill 吹填土
barge dock 驳船坞;驳船码头
barge drilling 船上钻探
barge elevator 平底船卸货用升降机
barge excavator 浮式挖掘机
barge fender 驳船护舷
barge fleet 驳船队
barge fleeting area 驳船集结区;驳船编队区
barge for hot cargoes 危险品运输船
barge hauler 驳船系结装置
barge hauling unit 驳船系结装置
barge haul system 驳船系结装置
barge hopper 开底泥驳;开底船;开底驳船
barge in tow 拖驳
barge job 船上钻进
barge lashing 驳船系结装置;驳船编队系结装置
barge lift 升船设备;升船机
barge lighterman 驳船船员
barge line 内海航运线;船只航道

barge line operator 驳运业者；驳运行；驳运公司
barge line system 驳船拖带法
barge-loading 驳船装货
barge loading belt conveyer[conveyor] 驳船装货皮带运输机；驳船装货皮带输送机
barge-loading runway 船下水滑道
bargeman 驳船水手；驳船工人；驳船船员
barge measure 上方（指船放泥量）
barge mooring area 驳船系泊区；驳船停泊区
barge-mounted 装在驳船上的
barge-mounted concrete plant 安装在船上的混凝土（搅拌）厂；安装在船上的混凝土（搅拌）设备
barge-mounted tower crane 浮式塔形起重机
barge of lighter aboard ship system 载驳船的船
barge on board ship 运驳母船；载驳（货）船
barge on board ship system 驳船运输船方式
barge operation system simulation 驳船操作系统模拟
barge pole 撑篙（驳船）；驳船钩杆；驳船撑篙
barge pool 驳船水域
barge port 内河港口；驳船港；驳船港；浅坝港埠
barger 绞驳设备；绞驳机
barge rafter 山墙檐口椽
barge ramp 船的斜坡道
barge sealing charges 驳船封仓费
barge shifter 绞驳设备；绞驳机；驳船系结装置
barge slip 驳船港址
barge spike 船用方钉；船钉
barge stone 压檐石；封檐石；山墙封檐石
barge stowage hold 载驳舱
barge string 驳船队
barge sucker 平底船上进气管；平底船上进油管；泥驳吸卸泵；非自航式吸泥船
barge system 驳运制；驳船队
barge terminal 驳船码头
barge tile 山墙封檐瓦
barge to rail transfer 驳船转铁路
barge tow 驳船队
barge train 平底船列；一列式拖带驳船队；驳船队；拖驳船队
barge-train 船列
barge train connecting ropes 拖驳船队系统
barge-train lock 原队过闸船闸；整队过闸船闸；船队船闸
barge train lockage 原队过闸
barge train towed alongside 绑拖驳队
barge train transportation 拖驳船队运输
barge-train unit 拖驳队
barge transport 驳运船
barge tug 拖轮；港作拖轮
barge-tug train 拖驳船队
barge type 驳船类型
barge unit 海洋测井船
barge unloading elevator 卸船用斗式提升机
barge-unloading suction dredge(r) 吹泥船
barge warping winch 移驳绞车
barge wharf 驳船码头
barge work 驳运工作
bargh 矿山企业
bargianing power 协议力量
barging 驳运
barging liquid sludge 用驳船运液体污泥
barging of sludge 污泥驳（船装）运；污泥驳
barging port 内河港口；驳载港
bar girder 型钢桁架
bar grading 直楞炉算
bar graph 线条图（表）；线条图形；柱状图；柱状表；直方图；棒状图；条形图；条线图
bar graph current meter 直方图海流计
bar graph generator 彩条信号发生器
bar graph memory 条形图存储器
bar graph schedule 横道图形进度表
bar shoeing chemical composition of oil field water 油田水化学组成成方图
bar grate 铁条格栅；条杆节；算条；条杆筛
bar grating 直冷；炉算；（机舱等的）棒格栅；直冷炉算
bar gravel 河滩砂石；河滩砾石；沙洲砾石
bar grinding attachment 棒料磨光附件
bar grit 条筛
bar grizzly 筐子筛；铁栅筛；棒栅筛；栅条式拦污筛；辊铁花格；格筛；粗条筛；棒条筛
bar handle 把手杆；把手梗；长条把手
bar head 轴端；杆端；带枢轴的螺钉头；沙洲头

bar hold 钳夹头
bar hooked at both ends 两端带弯钩的钢筋
baria 重晶石
bariable-length block 变长字组
bariandite 巴水钒矿
baric 气压的
baric area 气压区
baric center[centre] 气压中心
baric flow 气压流；气压流
baric gradient 压强梯度；压力梯度；气压梯度
baric gradient correction factor 压力梯度校正因子
baricite 镁蓝铁矿
baric minimum 最低气压
baric system 气压系统
baric topography 高度型式；气压形势
baric wind law 白贝罗定律
barie 微巴（声压压强单位）；巴列（旧时气压单位）
barilla 灰碱；苏打灰
bar in compression 受压杆
barines 巴林风
baring 出白；覆盖层；挖开岩层
bar insulation 线棒绝缘；排间绝缘
bar in tension 受拉杆
bariomicrolite 钡细晶石
bariopyrochlore 钡烧绿石
bar iron 型钢；条钢；钢筋；条状铁；钢条；铁条；条铁
bar iron crane 铁条起重机
barisal guns 震声
Bari-Sol process 巴瑞苏尔脱蜡法
barite 重晶石
barite aggregate 重晶石集料；重晶石骨料
barite aggregate concrete 重晶石集料混凝土；重晶石骨料混凝土
barite breccia 重晶石角砾岩
barite-calcite-fluorite ore 重晶石—方解石—萤石矿石
barite cement 重晶石水泥；重晶石胶结物
barite concrete 重晶石混凝土
barite deposit 重晶石矿床
barite fluorite composite mineralizer 重晶石萤石复合矿化剂
barite fluorite ore 重晶石萤石矿石
barite glaze 钡釉
barite grain mix(ture) 重晶石颗粒混合料
barite mortar 重晶石砂浆
barite ore 重晶石矿
barite ore aggregate 重晶石矿石集料；重晶石矿石骨料
barite ore concrete aggregate 重晶石矿石混凝土集料；重晶石矿石混凝土骨料
barite powder 重晶石粉
barite rosettes 重晶石玫瑰花结
barite slab 重晶石板
barite wall slab 重晶石墙板
barite-weighted mud 重晶石加重泥浆
barite yellow 重晶石黄色
baritic cement 重晶石水泥
baritite 重晶石
baritization 重晶石化
baritosis 钡中毒；钡尘肺；钡尘沉着病
barium 钡
barium acetate 醋酸钡
barium additive 钡添加剂
barium additives in diesel fuel 柴油燃料中的钡添加剂
barium alloy 钡合金
barium aluminium alloy 钡铝合金
barium-alumopharmacosiderite 毒铝钡石
barium amide 氨基钡
barium and strontium sulfate 硫酸钡锶
barium azide 叠氮化钡
barium based plaster 含钡灰浆
barium-base grease 钡基润滑脂；钡基润滑油
barium benzosulfonate 苯磺酸钡
barium binoxide 过氧化钡
barium bioxalate 草酸氢钡
barium bioxide 过氯化钡
barium borate 硼酸钡
barium carbide 二碳化钡
barium carbonate 毒重晶石；沉淀碳酸钡；碳酸钡
barium carbonate poisoning 碳酸钡中毒
barium cement 钡化水泥；含钡水泥；钡水泥
barium chloride 氯化钡；二水氯化钡
barium chloride poisoning 氯化钡中毒

barium chloroplatinate 氯铂酸钡
barium chromate 铬酸钡（钾）；钡黄
barium chromate indirect atomic absorption spectrophotometry 铬酸钡间接原子吸收分光光度法
barium chromate indirect spectrophotometry 铬酸钡间接分光光度法
barium chromate spectrophotometry 铬酸钡分光光度法
barium chrome 钡黄
barium colloidal suspension 钡胶浆
barium compound 钡化合物
barium concrete 含钡混凝土；钡混凝土
barium crown 钡黄
barium crown glass 钡冕玻璃
barium crystal glass 钡晶质玻璃
barium cyanide 氰化钡
barium dichromate 重铬酸钡
barium dihydrogen phosphate 磷酸二氢钡
barium dioxide 过氧化钡
barium diphenylaminesulfonate 二苯胺磺酸钡
barium discharger 钡放电器
barium dregs 钡渣
barium extended titanium dioxide 硫酸钡填充的二氧化钛
barium feldspar 钡长石
barium ferrate 高铁酸钡
barium ferrite 钡铁氧体
barium flint glass 钡质玻璃；钡燧石玻璃；钡火石玻璃；铅玻璃
barium fluochloride 氟氯化钡
barium fluoride 氟化钡
barium fluoride laser 费化钡激光器
barium fluoride substrate 氟化钡衬底
barium fluosilicate 氟硅酸钡；钡氟硅酸盐
barium fluxing agent 钡熔剂
barium fuel cell 钡燃料电池
barium getter 钡吸气剂
barium glass 钡玻璃
barium glaze 钡釉
barium hydrate 氢氧化钡
barium hydride 氢化钡
barium hydrogen phosphate 磷酸氢钡
barium hydroxide 钡氢氧化物；氢氧化钡
barium hypochlorite 次氯酸钡
barium hypophosphite 次磷酸钡
barium hyposulfite 硫代硫酸钡
barium ion 钡离子
bariumism 钡中毒
barium kitchen 钡餐室；调钡室
barium lithol toner 钡基立索调色料
barium manganate 锰酸钡
barium mercuric iodide 碘化汞钡
barium metaborate 偏硼酸钡
barium meta carbonate 低碳酸钡
barium metasilicate 硅酸钡
barium (mixed) plaster 含钡灰泥（医院特殊抹灰用）；钡混合灰泥
barium molybdate 钼酸钡
barium monohydrate 单水氢氧化钡
barium monosulfide 一硫化钡
barium monoxide 氧化钡
barium mortar 钡砂浆
barium naphthenate 环烷酸钡
barium nonatitanate ceramics 九钛酸钡陶瓷
barium octahydrate 八水氢氧化钡
barium oleate 油酸钡
barium ore 钡矿
barium orthosilicate 原硅酸钡
barium oxalate 乙二酸钡；草酸钡
barium oxide 结晶氧化钡；氧化钡
barium perchlorate 高氯酸钡
barium permanganate 高锰酸钡
barium peroxide 过氧化钡
barium peroxydisulfate 过二硫酸钡
barium persulfate 过二硫酸钡
barium-pharmacosiderite 毒铁钡石
barium phosphate 磷酸钡
barium phosphate crown glass 磷酸钡冕玻璃
barium plasma 钡等离子体
barium plaster 含钡砂浆；含钡灰浆；防射线抹灰；防射线灰浆；钡胶浆；钡灰泥；钡灰浆
barium platinate 铂酸钡
barium poisoning 钡中毒

barium polysulfide 多硫化钡
barium pyrovanadate 焦钒酸钡
barium rhodanid 硫氰酸钡
barium ricinoleate 蓖麻油酸钡
barium salt 钡盐
barium salt sedimentation method 钡盐沉淀法
barium silicate 硅酸钡
barium silicate cement 硅酸钡水泥
barium silicide 硅化钡
barium silicofluoride 氟硅酸钡
barium soap 钡皂
barium sodium niobate 铌酸钡钠
barium sodium niobate crystal 铌酸钡钠晶体
barium stannate 锡酸钡
barium star 钡星
barium stearate 硬脂酸(钡)
barium strontium cement 钡锶水泥
barium strontium niobate 铌酸钡锶
barium sulfate 硫酸钡
barium sulfate gravimetry 硫酸钡重量法
barium sulfate pigment 硫酸钡颜料
barium sulfate suspension 硫酸钡混悬液
barium sulfate turbidimetry 硫酸钡比浊法
barium sulfophenylate 苯代硫酸钡
barium sulphate 硫酸钡
barium sulphate concrete 硫酸钡混凝土
barium sulphate test 硫酸钡试验
barium sulphide 硫化钡
barium superoxide 过氧化钡
barium tetraiodomercurate 碘化汞钡
barium thiosulfate 硫代硫酸钡
barium titanate 钛酸钡(用于振动探测仪探头)
barium titanate ceramics 钛酸钡陶瓷
barium titanate piezoelectric(al) ceramics 钛酸钡压电陶瓷
barium titanate porcelain 钛酸钡瓷器
barium titanate thermistor 钛酸钡热敏电阻器
barium titanium oxalate 草酸氧钛钡盐
barium tungstate 钨酸钡
barium vacancy 钡空位
barium white 钡白;钨酸钡
barium wolframate 钨酸钡
barium yellow 钡黄
barium-zinc stabilizer 钡—锌稳定剂
barium zinc tantate 钽锌酸钡
barium zirconate 锆酸钡
bar joist 钢格栅;轻钢构架;小钢梁;钢桁条
bark 脱碳薄层;树皮鞣革;树皮;三桅帆船
bark allowance 树皮扣除率
bark background 暗背景
bark beetle 小蠹虫
bark blazer 刮皮器
bark borer 蛀虫
bark-bound 皮封的
bark breaker 树皮碎裂机
barkcloth 树皮布
bark coal 树皮煤
bark crepe 人丝树皮皱
bark durite 微树皮暗煤
bark dust 树皮粉尘
barked arrow 风矢
bar keel 立龙骨;矩形龙骨;方龙骨
barker 剥树皮机;剥皮器;剥皮机;剥皮工
barker chair 可调节椅
Barker index 巴克指数
Barker method 巴克法
Barker sequence 巴克序列
barker shop and beauty parlor 理发店与美容室
Barker's mill 巴克水磨机;巴克磨
Barker's point 巴克点
barkevikite 棕闪石
bark ga(u)ge 树皮规
barkhan 新月形沙丘
Barkhausen effect 巴克好森效应
Barkhausen interference 巴克好森干扰
Barkhausen-Kurz oscillator 巴克好森—库尔茨振荡器
Barkhausen oscillation 巴克好森振荡
barking 剥树皮;剥皮
barking drum 鼓式剥皮机;筒式去皮机;筒式剥皮机
barking iron 树皮剥刀
barking machine 剥(树)皮机
barkite 微树皮煤
bark knife 剥皮刀

bark liptobiolith 树皮残植煤
Barkman's reflex 巴克曼氏反射
bark mantle 皮层
bark mill 碾皮机
bark of paper mulberry 构树皮
bark of tree of heaven 椿皮
barkometer 鞣液比重计
bark peeling machine 剥皮机
bark picture 树皮画
bark pocket (木材的)夹皮;树穴
bark press 压皮机
bark scraper 刮树皮器
bark seam (木材的)夹皮
bark slip 树皮剥脱
bark slipping 树皮脱离
bark tan 树皮鞣料
bark tanner 树皮鞣料
bark tannery 树皮鞣革厂
Bark tile shake 搭扣屋面瓦
bar lathe 加工棒料车床
bar lattice boom 笼格(式)吊杆
barley 大麦
barley coal 大麦粒级无烟煤
Barley sugar 螺旋状柱子;麦芽糖
barley-sugar column 螺旋状柱子
bar lift 提物手柄;提手柄
bar lifter 杆式升降器
bar line 杆形线
bar-linear shrinkage test 钢筋线收缩试验;钢筋线收缩试验
bar liner 压条衬板
barling 柱;杆;脚手杆;木棒
bar link 型钢连接
bar linkage 连杆机械;连杆机构
bar list 配筋明细表;钢筋表;配筋表
bar loading 杆受荷载;杆荷载
bar lock 插销门锁
bar longwall macthine 链式截煤机
Barlow rod 巴罗水准尺
Barlow's equation 巴罗方程
Barlow's rule 巴罗规则
barlux lamp (表面温度较低的)管形真空灯泡
barmac 运送砖头的手推车
bar magnet 杆磁铁;磁棒;条形磁铁
bar magnetic compass 磁杆罗盘
barman 撬石工
bar mark 钢筋标牌
bar marker 杆式划行器
bar-master 矿山经理;矿长
bar mat 钢筋网;钢筋网片
bar mat reinforcement 钢筋网
bar mill 轧钢条厂;小型轧机;小型轧钢厂;棒磨机;棒材轧机;轧条机
bar mining 河滩金砂矿开采法
bar moment 杆受弯矩;杆弯矩;杆件弯矩
bar mo(u)ld 多穴压模
bar mo(u)lding 钢筋线条;柜台边棱线脚;钢筋装饰线条;凹凸线脚
Barmouth grit 巴尔茅斯粗沙岩
bar movement 杆状机芯
barn 粮食仓;粮仓;库房;农仓;畜棚;烟棚;谷仓;堆房;车库;仓;靶恩(1 靶恩=10⁻²⁴平方厘米);牲口棚
Barnach stone 巴纳克石
barnacle 藤壶
barnacled 盖满藤壶的
Barnard's satellite 巴纳德卫星
Barnard's loop 巴纳德圈
Barnard's star 巴纳德星
barn door 仓库大门;谷仓大门
barn-door hanger 仓库门吊架;(库房的)推拉门吊架;悬挂门滑机;推拉门吊轨
barn-door stay (库房的)推拉门吊架滑轮
barn-drying method 仓内干燥法
Barneby instrument 巴纳放射性测井仪
Barnes formula 巴尔纳斯水流公式
barnesite 水钒钠石
Barnesite polishing powder 氧化钍抛光粉
Barnett effect 巴涅特效应
Barnett method 巴涅特法
barney 小卡车;拉曳器;矿车用推车;小矿车;隔音套;保护套;平衡锤;上坡牵引车
barn hay drying plant 牧草人工干燥设备
barn hoist 厩舍起重机

barn lantern 风提灯
barnminutor 条状切碎机
bar No. 钢筋号
bar nodule 棒槌状结核
barn quay 粮仓码头
barn red 谷仓红
Barns formula 巴恩斯公式
barn siding 农仓板墙
barn terminal 粮仓码头
barn truss 三铰构架式桁架
barn wastewater 厩棚废水;畜圈废水
barn wharf 粮仓码头
barn yard 农仓空场;仓库前空地;畜棚场;深潜器
barnyard manure 动物粪便;圈粪;圈肥
barocepter 气压传送机
baroceptor 压力感受器;气压传感器
barochamber 压力舱
barock 畸形怪状的;变态式的;特饰建筑
baroclinic condition 斜压情况
baroclinic eddy 斜压涡流
baroclinic eddy transform 斜压涡动输送
baroclinic field 斜压场
baroclinic flow 斜压流
baroclinic fluid 斜压流体
baroclinic instability 斜压不稳定
baroclinicity 斜压性
baroclinic model 斜压模式
baroclinic ocean 斜压海洋
baroclinic process 斜压过程
baroclinic vortex 斜压涡旋
baroclinic wave 斜压波
barocliny 斜压性
Baroco 造浆黏土粉
barococo 过分装饰的
barocyclonometer 气压风暴表;旋风位置测定仪
baroduric bacteria 耐压细菌
barodynamics 重型结构力学;重型建筑动力学;重结构力学
bar of arch 拱式杆件
bar of cement 水泥试条
bar of continental shelf outer edge 大陆架外缘坝
bar off 挡去;排斥;阻挡
bar of flat 窄厚扁钢;扁平棒;板片
bar of muddy sand 泥砂洲
bar of rack and pinion jack 齿条给进齿筒
bar of statute of limitation 消灭时效
bar of uniform cross section 等截面棒
bar of uniform strength 等强度棒
barogram 自记压力记录;气压自记曲线;气压图;气压记录图
barograph 膜盒式气压记录器;膜盒气压记录器;自记气压计;自记气压表;气压自记器;气压自动记录仪;气压描记器;气压记录器;气压计
barograph trace 气压自记曲线
barogyroscope 气压陀螺仪;气压回转仪
Baroid wall-building test instrument 巴罗德造壁式试验仪(测定失水量用);巴罗德压滤仪(测定失水量用)
barokinesis 压力动态
barometer 大气压力计;晴雨计;晴雨表;气压计;气压测高仪;(测大气的)气压表
barometer altitude 气压计高度
barometer cistern 气压计杯;气压计水银槽
barometer condenser 气压式冷凝器
barometer constant 气压测高常数
barometer height 气压高度
barometer methods 景气预测法
barometer observation 气压计观测
barometer of business cycle forecasting 经济周期预测晴雨表
barometer reading 气压读数
barometer scale 气压计刻度
barometer stocks 标准股票
barometer table (测大气的)气压表
barometer tube 测压管
barometric(al) 气压计的;气压的;气压表的
barometric(al) admittance 气压导纳
barometric(al) altimeter 气压高度计;气压高度表;气压高程计;气压测高仪;气压测高计
barometric(al) altimetry 气压测高法
barometric(al) altitude 气压高程
barometric(al) area 气压区
barometric(al) change 气压变化
barometric(al) coefficient 气压系数

barometric(al) compensation 气压补偿
barometric(al) condenser 大气冷凝器；大气压凝汽器
barometric(al) condensing pipe 大气冷凝管
barometric(al) damper （减少烟道中上升气流的）挡板；挡风器；气压调解门；气流调节阀；烟囱空气阻滞器
barometric(al) day change 气压日变化
barometric(al) depression 低气压；气压下降
barometric(al) determination 气压测定
barometric(al) determination of altitude 气压测高
barometric(al) discharge pipe 大气排泄管
barometric(al) disturbance 气压扰动
barometric(al) diurnal range 气压日差
barometric(al) draft regulator 大气气流调节器
barometric(al) efficiency 气压效应；气压效率
barometric(al) efficiency of the aquifer 含水土层的气压效应
barometric(al) elevation 气压高程
barometric(al) equation 大气方程
barometric(al) fluctuation 气压扰动；气压变动
barometric(al) fuel control 按气压变化的燃料调节
barometric(al) gradient 气压梯度；压力梯度
barometric(al) gradient force 气压梯度力
barometric(al) gravimeter 气压重力仪
barometric(al) height 高气压；气压计高度；气压高度
barometric(al) height formula 利用气压差测高公式；气压高度公式
barometric(al) heighting formula 气压测高公式
barometric(al) height level(l)ing 气压高程测量
barometric(al) height measurement 气压高程测量；气压测高
barometric(al) high 高气压
barometric(al) hypsometry 气压测高法
barometric(al) information 气压情况
barometric(al) leg 气压排液管；气压柱；气压表管
barometric(al) leg condenser 气压柱冷凝器；大气压喷射冷凝器
barometric(al) level 气压高度计；气压高度表；气压高程计
barometric(al) level(l)ing 气压计测高程；气压水准测量；气压测高
barometric(al) level(l)ing point 气压测高点
barometric(al) low 低气压
barometric(al) maximum 气压最高值
barometric(al) millimeter of mercury 气压单位（以毫米水银柱计）；毫米水银气压表
barometric(al) minimum 气压最低值
barometric(al) non-periodic(al) changes 气压非周期变动
barometric(al) observation 气压计观测
barometric(al) pipe 气压排液管
barometric(al) pressure 大气压（力）；大气压；气压
barometric(al) pressure formula 大气压力公式
barometric(al) pressure sensor 气压传感器
barometric(al) pressure switch 气压继电器
barometric(al) price leadership 晴雨表式价格领导
barometric(al) rate 气压升降率
barometric(al) reading 气压读数
barometric(al) reduction 气压归正；气压订正
barometric(al) relay 气压计继电器
barometric(al) scale 气压计标度
barometric(al) scale factor 气压计标度因素
barometric(al) station 气压测高站
barometric(al) statoscope 气压计高差仪
barometric(al) surveying 气压水准测量
barometric(al) tank 气压罐
barometric(al) tendency 气压趋势；气压倾向
barometric(al) traverse 气压测高导线
barometric(al) tube 气压计管
barometric(al) unit 气压单位
barometric(al) variation 气压变化
barometric(al) waves 气压波
barometrograph 膜盒气压记录器；自记气压计；气压自动记录仪；气压自动记录器；气压描记器；气压记录器；气压计
barometry 气压测定法
baromil 毫巴（气压单位）
barong-barong 临时住房（菲律宾）
barophilic bacteria 适压细菌
barophoresis 压泳现象
Baroque 巴洛克式

Baroque architect 巴洛克建筑师
Baroque architecture 变态式建筑；巴洛克式建筑；巴洛克式建筑
Baroque art 巴洛克艺术
Baroque building 巴洛克房屋；巴洛克建筑
Baroque castle 巴洛克式城堡
Baroque cathedral 巴洛克大教堂
Baroque church 巴洛克教堂
Baroque fountain 巴洛克式喷泉
Baroque fresco painting 巴洛克壁画
Baroque garden 巴洛克式庭园
Baroque machine style 巴洛克机械型
Baroque master 巴洛克（风格艺术）大师
Baroque mirror 巴洛克美术镜
Baroque palace 巴洛克式宫殿
baroque pearl 异形珍珠
Baroque period 巴洛克（式）建筑时期
Baroque sanctuary 巴洛克寺院；巴洛克式圣堂
Baroque sculpture 巴洛克雕塑
Baroque square 巴洛克式广场
Baroque statue 巴洛克雕像；巴洛克塑像
Baroque style 巴洛克式
Baroque style garden 巴洛克式园林
baroreceptor 压力感受器；气压感受器
baroresistor 气压电阻器
baroscope 气压测定器；敏感气压计；验压器；大气浮力计；气压测验器；气压测深器
baroscu 硼砂
barosinusitis 航空鼻窦炎；气压性鼻窦炎
barosphere 气压层
barostat 气压计；恒压器；气压调节器；气压补偿器
baroswitch 气压转换开关；气压开关
barothermogram 气压温度图；气压温度曲线
barothermograph 自记气压温度计；气压温度自动记录仪；气压温度记录器；气压温度计；气压测温计
barothermohydrograph 压温湿记录器；自记气压温度湿度计
barothermohygrogram 气压温度湿度曲线
barothermohygrograph 气压温度湿度记录器；气压温度湿度计
barothermohygrometer 气压温度湿度表
barotor machine 高温高压卷染机
barotrauma 气压伤
barotraumatic otitis 气压性耳炎
barotropic atmosphere 正压大气
barotropic fluid 正压流体
barotropic gas 正压气体
barotropic gravity wave 正压重力波
barotropism 向压性
bar oven 条炉
bar past 榫焊接
bar pattern 线条形图形；线条测试图；条形图样
bar peel 简易杆式推料机
bar peeling machine 棒料粗车机床
bar pile 扁钢桩
bar placer 钢筋工
bar placing 布筋
bar placing drawing 布筋图
bar plain 沙坝平原
bar platform 沙洲台地；沙坝台地
bar pliers 杠钳
barpoint 棒钢铧尖
barpoint bottom 棒钢铧尖犁体
bar pointing machine 棒材轧尖机
barpoint plow 棒钢铧尖犁；伸出凿尖犁
barpoint share 棒钢铧尖犁铧；伸出凿尖铧
bar polishing machine 棒材抛光机
bar port 候潮港；有拦门沙的港口
bar post 形成外门的门栏；围栏上有榫眼的横木；榫接柱；榫接栏；榫接桩
bar pressure 大气压（力）
bar primary 条形一次绕组
bar printer 杆式打印机；棒式印刷机
barque 三桅帆船
barrabkie 爱斯基摩人茅舍
barrabora （爱斯基摩人住的）泥土住房
bar race mill 盘辊磨
barrack 栅门齿条；工棚；格栅池；粗条栅；临时工房；临时板房；营房；工房；兵营
barracks block 营房建筑
barracks building 活动房顶构筑物
barracks construction 兵营建造
barracks room 工棚房；兵营房间
barracks ship 仓库船；兵营船；水上营房

barracoon 集中营
barrage 拦阻射击；拦阻；拦河闸堰；拦河闸；拦河坝；拦河坝；拦坝；堰；烟幕；阻塞；挡水建筑物
barrage balloon 拦阻气球
barrage gate 堰顶闸门
barrage jamming 抑制干扰；阻塞干扰；全波段干扰
barrage lake 水库
barrage lasher 拦河坝
barrage mobile 活动堰坝；活动堰
barrage power station 堰坝式电站
barrage receiver 电池式接收机
barrage reception 多方向选择接收
barrage steps 梯级坝群；水坝梯级
barrage type spillway 堰式溢洪道；溢流堰
barrage width 抑止频带宽度
barrage with frames 构架式拦河堰
barrage with lock 有闸门的溢流堰
barrage with stop planks 插板式拦河堰；插板式拦河坝
bar-rail mo(u)lding 凹凸线脚；柜台边棱线脚
barranca 深峡；深山谷；深切沟
barras 毛松香；含油松脂；含油树脂；软树脂
barrator 有不法行为的船员
barratron 非稳定波型磁控管
barratry 海员的非法行为；恶意行为
Barratt-Halsall firemonth 瓶式陶瓷窑加煤口
barre 练功扶手；纬向条子织物；纬向条花
barred 被拦住
barred-and-braced gate 有斜框的横木大门
barred basin 海湾盆地；孤立盆地；隔绝盆地；受限盆地；沙槛盆地；沙坝盆地
barred body 棒体
barred claim 已过时效的债权；失去时效的债权
barred code 受阻码
barred debt 受时效限制的债务
barred door 横闩门；(不带框的)拼板门；实拼门；栅门
barred gate 栅门
barred obligation(debt) 失去时效的债务；受时效限制的债务
barred right 失去时效权利
barred river mouth 堵塞河口；分布有拦门沙的河口；沙洲拦阻的河口
barred speed range 禁制转速范围
barred spiral galaxy 棒旋星系
barred window 带铁栅的窗
bar reinforcement 增强钢筋；钢筋棒；钢筋；粗钢筋
barrel 镜头筒；挤泥机泥缸；圆筒；油桶；岩芯管；柱体；抵抗面；气包；炮管；筒体；桶状体；桶；条盒
barrelage 桶数
barrel amalgamation 桶式混汞法
barrel antenna 桶形天线
barrel arbor 条盒心轴
barrel arch 筒形拱；筒式拱桥；筒拱
barrel arch dam 筒形拱坝
barrel arched girder 筒形拱式大梁
bar relay 杆接多触点继电路；棒式多触点继电器
barrel bearing 圆筒轴承
barrel blockage 岩芯管自卡；岩芯管堵塞
barrel boiler 筒形锅炉
barrel bolt 圆筒形插销；圆筒插销；管形插销；管销；筒形插销
barrel bridge 浮筒桥
barrel bulk 一桶（松散物料体积计量单位，约合0.14立方米）
barrel buoy 筒形浮标；桶形浮标；桶式浮标
barrel-burnishing 圆筒体抛光；圆筒体滚光
barrel cam 可调式圆柱凸轮；凸轮鼓；筒形凸轮
barrel camber 圆筒顶
barrel camber of road 路面弧形拱度
barrel cargo 桶装货
barrel carrying trailer 载桶拖车
barrel cased turbine 桶壳式水轮机
barrel casing feed pump 筒式给水泵
barrel ceiling 半圆筒形天花板；筒形顶棚
barrel chair 半圆软垫椅；筒背椅
barrel cleaning 旋转滚筒清洗法；滚筒清理
barrel colo(u)ring 滚筒混色
barrel converter 筒式吹风转炉
barrel convertor 卧式吹炉
barrel copper 手选块铜
barrel core 套管
barrel crankcase 筒式曲轴箱
barrel culvert 筒形涵洞

barrel distortion 桶性畸变;桶形畸变
barrel door 栅门
barrel drain 筒形排水渠
barrel drain canal 筒形排水渠
barrel drain pipe 筒形排水管
barreled asphalt 桶装沥青
barreled bitumen 桶装沥青
barreled cement 桶装水泥
barreled goods 桶装物
barrel elevator 圆筒提升机;圆筒升降机;桶升降机
barrel enamelling 转鼓涂漆
barreler 高产油井;高产井
barrel exhausting pump 圆筒形排水泵;圆筒形排气泵
barrel filler 装桶机;灌桶机
barrel finish 抛光;滚磨
barrel finishing 滚筒抛光;磨光;转鼓涂漆(法);滚筒涂漆(法);滚筒清理;滚磨
barrel fitting 套筒接头;管式配件;管式接头;筒式接头;套筒式接头;套管
barrel frame 筒形构架
barrel goods 桶装货
barrel grab 圆筒形抓斗
barrel gravity method 桶装重量法
barrel handrail 筒形扶手
barrel head 桶底
barrel heater 圆筒形加热器
barrel heating pocket 加热筒形坑;圆筒加热器
barrel hoist 圆筒升降机;桶升降机
barrel hook 桶钩
barrel hoop 桶箍
barrel hopper 圆筒式装料斗
barrel-house 廉价小商店;油桶仓库
barreling 滚动涂漆;转鼓滚涂(法);滚研磨;滚筒清洗;滚磨
barrel key 管状钥匙
barrel lapping machine 精研缸机
barrel leaking at plugs 桶塞处漏
barrel length 辊身长度
barrel lid loose 桶盖松
barrel light 月牙老虎窗;筒形天窗;弯曲形老虎窗;天窗
barrel liner 桶衬
barrel lock 圆柱形锁
barrel lung off 桶塞脱落
barrel man 前桅瞭望员
barrel mast 桅柱;筒形桅杆
barrel mile 桶哩运费
barrel mill 鼓式研磨机
barrel mixer 混合筒;转筒式混合机;滚桶式混砂机;鼓筒式搅拌机;鼓式拌和机;鼓式拌和器;桶形搅拌机;桶式拌和机
barrel mooring buoy 筒形系船浮筒
barrel newel 旋转楼梯筒柱
barrel nipple 双外螺丝;带螺母外丝接头;筒形螺纹接套
barrel number 桶号
barrel of a pump 泵的汽缸
barrel of capstan 绞盘滚筒
barrel of cement 桶装水泥
barrel of oil per day 每日石油桶数
barrel of pipe 套管;管筒;管膛
barrel of pump 泵壳;泵体
barrel of puppethead 随转尾座套筒
barrel of roof 筒形屋顶
barrel oil 桶装油
barrel oil pump 油桶泵;油桶手摇泵
barrel oozing 桶渗漏
barrel organ 鼓形风琴
barrel outlet 管筒出水口
barrel packing machine 自动装桶机;装桶机
barrel per acre method 每英亩产油桶数
barrel pier 筒形支墩;筒形桥墩
barrel piston 活塞筒
barrel pitching machine 筒形涂沥青机;桶堆排机
barrel plating 滚镀;筒式电镀;筒镀;桶镀
barrel polishing 滚筒抛光(法);滚筒清砂法
barrel printer 鼓式打印机
barrel printing 鼓式印刷
barrel pump 筒形排水泵;筒形排气泵;筒形水泵;筒形气泵;桶式喷雾泵;桶泵
barrel punctured and leaking 桶穿洞并渗漏
barrel purlin(e) 管状檩条;筒檩条
barrel raft 浮筒筏

barrel railing 管状栏杆
barrel railing fitting 管状栏杆配件
barrel resoldered 桶重焊
barrel re-welded 桶重焊
barrel roll 横辊;筒形辊
barrel roof 筒壳;薄壳屋顶;筒形屋顶
barrel roof rib 筒形屋顶肋
barrel saw 筒形锯
barrel screen filter 圆筒形滤器
barrels daily 每日桶数
barrel shape 横向凸出;桶形
barrel-shaped 圆筒形的
barrel-shaped boudin 酒桶状石吞肠
barrel-shaped distortion 桶形失真
barrel-shaped roll 桶形轧辊
barrel-shaped roller bearing 筒状滚动轴承;筒状滚动支承
barrel shell 筒形屋顶;筒形薄壳;筒壳
barrel shell roof 筒形薄壳屋顶;筒壳屋顶;筒形屋顶;筒形穹顶;筒壳
barrel shutter 筒状快门;筒形遮光器
barrel skeleton 筒形骨架
barrel sling 桶吊具
barrels per calendar day 每日历日桶数;每月历日桶数
barrels per day 每日桶数;每日……桶
barrels per month 每月筒数
barrels per stream day 每个生产日筒数
barrel switch 鼓形开关;筒形开关
barrel theory 圆筒理论
barrel throttle 筒形节流阀
barrel-tumbling 滚筒抛光(法)
barrel-type airless shot-blasting machine 滚筒式抛丸清理机
barrel-type casing 筒型汽缸
barrel-type crankcase 筒形曲轴箱
barrel-type mixer 滚筒式混砂机
barrel-type pump 鼓形泵;筒式泵
barrel-type roller 鼓形辊
barrel-type roller bearing 鼓形滚柱轴承
barrel-type roll piercing mill 带桶形轧辊的穿孔机
barrel-type root cutter 滚筒式块根切碎机
barrel-type shot blasting machine 喷丸滚筒;筒式喷丸机
barrel vaulted roof 筒形拱顶屋顶
barrel vaulted with trelliswork 有格子细工的筒形圆屋顶
barrel vault form 筒拱模板
barrel vault formwork 筒拱模板
barrel vault(ing) 筒形拱顶;筒壳;半圆形拱顶;筒形穹隆;筒形穹顶;筒拱
barrel-vault roof 筒形穹顶;筒壳屋顶
barrel vault shell 筒壳;筒拱壳体
barrel vault with intersecting vault 交叉拱顶;筒形交叉穹顶
barrel wall lamp 筒形壁灯
barrel washing device 筒形清洗设备
barrel wear 缸套磨损
barrel winding 桶形绕组
barrel wrench 开油桶扳手
Barremian 巴雷姆阶
bar removal 清除沙槛
barren 裸荒地;荒芜地;不含矿物的;贫瘠地;无矿物的
barren gap 无油地段
barren ground 裸地;荒地
barren hill 荒山
barren hole 无矿钻孔
barren intrusive body 不合矿岩体
barren island 荒岛
barren land 荒芜土地;裸露地;瘠地;荒地;不毛(之)地
barren liquor 贫液
barren of oil 不产油
barren of practical value 无实际价值的
barren ore 贫矿
barren plateau 荒芜高原;不毛的高原
barren pluton 无矿岩体
barren rock 废石
barren sand 荒砂;不毛沙地
barren shale 无油页岩
barren spots 无矿地带
barren trap 空圈闭
barren waste 荒滩
barren waters 贫瘠海区

barren well 无油井
barren zone 无矿地带
barrerite 钠红沸石
bar response 条响应
Barret burette 巴瑞特氏滴定管
barrette file 扁三角锉
barretter 稳流灯
barrettes 连续墙基础;短垛【港】
barricade 路障;护栏;障碍物;防护栅;屏蔽墙;屏蔽板
barricade lantern 路障号志灯
barricade shield 可动屏蔽
barricading 设路障
barrican 碉楼
barrico 小桶
barrier 栏施;拦障;截水墙;海关关卡;堰(洲);障阻;障壁堤;障壁;障碍(物);栅障门;关卡;隔离物;隔离节;多儿隔板;垫条;堤(坝);挡堰;分车栅栏;防撞栏;妨碍因素;壁垒;人工堤;潜堰;潜坝;屏障
barrier action 屏障作用
barrier arm 栏木
barrier bar 离岸沙洲;拦门沙;沿岸沙坝;障壁砂坝;障壁坝;堤岛;滨外沙坝;潜坝型沙洲
barrier bar deposit 障壁沙坝沉积
barrier bar facies 障壁沙坝相
barrier-bar trap 障壁沙坝圈闭
barrier basin 堰塞盆地;堰塞湖盆;堤爆水池
barrier beach 离岸沙洲;沿岸沙埂;沿岸滩沙埂;沿岸沙洲;沿岸沙埂;障壁滩;滨外滩;滨外沙埂;围岸沙滩
barrier beach deposit 障壁滩沉积
barrier beach facies 障壁滩相
barrier block 路障
barrier breaker 障碍突破舰船
barrier centerline stripe 栏式中央分隔带
barrier chain 长串沙洲;岸外岛链;砂群岛;沙岛群;沙洲链;沙岛列
barrier clutch 安全矿柱
barrier coast 沙垣海岸
barrier coating 隔离涂料;隔离涂层;防蚀涂层;防渗涂料;不透气层;屏蔽涂料
barrier code 隔离码
barrier complex 障壁复合体
barrier curb 栏式缘石
barrier dam 拦鱼坝;拦污坝;拦沙坝
barrier ditch 设障沟
barrier draught 拦门沙洲吃水;拦门沙吃水深
barrier effect 遮帘作用;栅栏效应;屏障效应;势垒效应
barrier energy 势垒能
barrier field 势垒场
barrier film 阻挡膜;阻挡层;隔离薄膜;屏蔽膜
barrier film photocell 阻挡层光电管
barrier filter 吸收滤光片;遮光滤光器;隔紫外滤光片;屏状滤光片;屏障滤光片
barrier flashing light 海岸固定闪光灯
barrier flat 堰洲坪
barrier flood plain 堰垣式河漫滩
barrier-free 无障碍的(对残疾人)
barrier-free design 无障碍设计
barrier-free environment 无障碍环境
barrier-free environment design 无障碍环境设计
barrier frequency 截止频率
barrier function 障碍函数;闸函数
barrier function inversion 障碍函数反换式
barrier gate 势垒栅
barrier grid 阻挡栅极;制动栅;障栅
barrier grid mosaic 阻挡栅极马赛克
barrier-grid storage tube 雷德康管
barrier height 势垒高度
barrier ice 陆架冰;冰障(碍);冰堡;冰岸
barrier iceberg 平顶冰山
barrier island 礁岛;障壁岛;堤岛;大沙洲;滨外岛;滨海岛;堡礁岛;岸洲;沙垣岛
barrier island deposit 障壁岛沉积
barrier island facies 障壁岛相
barrier lagoon 泻湖;障壁泻湖
barrier-lagoon deposit 障壁岛—泻湖沉积
barrier-lagoon environment 障壁岛—泻湖环境
barrier-lagoon sequence 障壁岛—泻湖层序
barrier-lagoon system 障壁岛—泻湖体系
barrier lake 堰塞湖;堰湖;人工湖
barrier layer 栏障;耗尽层;阻挡层;屏障;屏蔽层;

阻隔层
barrier-layer cell 阻挡层光电池;障层电池;光生伏打电池
barrier-layer photocell 阻挡层光电管;光生伏打电池
barrier light 海岸探照灯
barrier line 拦阻线;警戒线;限制线;车道(分)隔线
barrier marsh 障阻沼泽
barrier material 不透水材料;隔绝材料;挡板材料;防潮材料;屏障材料
barrier mechanism 屏障机制
barrier membrane 不透水层
barrier net 拦网;木栅网
barrier on highway 陡边线
barrier paper 遮763纸;隔离纸
barrier penetration 势垒穿透
barrier pillar 间隔柱;间隔矿柱;岩柱;栅栏柱;隔离矿柱;边界矿柱;安全矿柱
barrier planting 栏式栽植
barrier platform 障壁台地
barrier post 栅栏柱
barrier potential 势垒
barrier power station 堤堰式电站
barrier property 隔离性能;防渗性能;屏蔽性能;防护性能
barrier railing 护栏
barrier reef 可折叠铁栏;堤礁;堡礁(指与海岸平行的珊瑚礁)
barrier reef coast 堡礁海岸
barrier reef of pool trend 障壁礁油气藏趋向带
barrier region 势垒区;势垒层
barrier-retreat carpet 障壁后退毯状沉积
barrier screw 屏障螺杆
barrier separation 膜分离
barrier sheet(ing) 防水板桩;挡水板桩;防水墙板
barrier shield 隐蔽墙;屏蔽墙
barrier shielding 掩体;阴障隐蔽;阻障隐蔽;阻碍物防护
barrier spit 堰洲嘴;障壁沙嘴;滨外沙嘴;滨海沙嘴;沙嘴;沙岬
barrier spring 堰塞泉;断层泉;堤泉
barriers to trade 贸易壁垒
barrier stripe 栏式分隔带;分隔栅
barrier to entry 进入市场壁垒
barrier to infrared radiation 红外(线)放射的障碍
barrier to trade 贸易障碍
barrier traffic 封锁交通
barrier vapo(u)r 隔汽层
barrier wall 阻隔墙;止水墙;防水墙
barrier width 势垒宽度
bar-rigged drifter 柱架式风钻;架式凿岩机;架式风钻
bar rigging 铁牵条
barring 慢速转动;盘车
barring device 盘车装置
barring down 撬落
barringerite 磷铁矿
barring gear 曲轴变位传动装置;曲柄移位装置;盘车装置
barring hole 盘车孔;盘车孔
barring motor 盘车电动机
barring-on 移圈
barring the entail 解除限定继承权
barringtonite 二菱镁矿;水碳镁矿
barring traffic 封锁交通
barring unit 慢转变装置
barrio-orthojoaquinite 斜方硅钡钛石
bar river mouth 拦门沙
barroisite 冻蓝闪石
bar rolled stock sorting unit 型钢分选机组
bar roller 棒条碾压机
Barrol system 抽气式通风系统;巴罗尔通风系统
barron bend 弯管
Barronia 高温耐蚀铅锡黄铜
Barron ligation 胶圈套扎法
Barron's consolidation theory 巴隆固结理论
bar room 酒吧间
Barrovian metamorphism 巴罗夫变质作用
Barrovian zone 巴罗式带
barrow 麻车;小车;担架;双轮手推车;手车
barrow area 吸泥机作业区
barrow excavation 露天开采(作业);手车采矿
barrow gang 手推车队
barrow-hoist 运料手推车提升机
barrowing 手推车运输;手推运输;手推车运输

barrow line profile 窄谱线轮廓
barrow men 手车工人
barrow pit 取土坑;材料坑;采石坑;借土坑
barrow pulse 窄脉冲
barrow run 手推车马道;手推车道
barrow runner 手车道板;手推用跳板;手推车跳板;手车跳板
barrow truck 小型运货车;手推运料车;手推车
barrow-type chip(ping)s spreader 手车式撒布碎屑机
barrow-type spreader 手车式撒布机
barrow-way 输送段
Barr Series 巴统【地】
Barr's method 巴氏反转图绘制法
bar sash lift 吊窗拉手
bar scale 海图线比例尺;线比例尺;直线比例尺
bar schedule 杆的一览表;钢筋表
bar screen 格筛;条行筛;铁栅筛;铁条格;炉条格算;栅条筛;粗筛格;分离筛;算子筛;算条筛;棒条筛;铁栅筛;条栅
barscreen 铁条格;铁栅算
bar section 杆剖面
bar separator 钢筋分隔物;钢筋分隔器;钢筋定位器
bar setter 钢筋工
bar sewage screen 污水铁格栅
bar shallow 拦江浅滩
bar shape 型钢;棒形型钢
bar shear 棒料剪床
bar shear by motor 机动钢筋切断机
bar shearing machine 剪条机;棒材剪切机
bar shear ring 钢筋剪切环
bar shears 剪条机;小型钢材剪断机;钢筋切割机;钢筋剪切机;棒料剪断机
bar sheet 钢筋网
bar shoal 拦门沙
bar sieve 栅栏筛;格筛;铁栅网
bar sight 表尺
bar signal 河口潮水(位)信号;通过拦江沙信号
bar sink 条形污水沟;条形污水槽
bar size 钢筋尺寸
bar size section 小断面型钢
bar slope 杆件斜率
bars of foot 蹄支
barsowite 钙长石
bar spacer 钢筋定位隔间(水泥)块;钢筋分隔器;钢筋定位卡
bar spacing 钢筋间距;钢筋分布;筛条间距
bar spade 杆叉铲
bar splicing 钢筋搭接
bar stand 棒料架
bar steel 异形型钢;六角型钢;矩型钢;扁型钢;圆型钢;条钢;棒钢;型钢
bar stock 棒材;异形型材;六角型材;矩型材;扁材;方型材;圆型材;钢筋储备;棒形材料;条材;棒料
bar stock cutting capacity 棒料切割容量
bar stock lathe 两脚车床
bar-stock needle valve 杆控针形阀
bar stock turnover device 棒料翻料机
barstool 酒吧高脚凳
Barstovian 巴斯图阶
bar straighten-cutter 钢筋调直切断机
bar straightener 棒材矫直机;棒杆调直机;棒材矫直机;调直机
bar straightening 钢筋调直
bar straightening machine 钢筋调直机;棒杆调直机;棒材矫直机
bar strainer 铁格栅;铁栅筛
bar stress 杆件应力;杆构件应力
bar strip 钢带;杆条;薄板坯
bar structure 杆系结构
bar subject to buckling 杆件屈曲
bar support 钢筋支架;钢筋坐垫;钢筋(支)座;钢筋垫块
bar sweep 铁杆扫床【疏】
bar sweeping 硬杆扫测【疏】
bar system 杆件系统
bar target 杆状目标
bartelkeite 铅钡锗矿
Bartel's diagram 巴特尔图
barter 做交易;换货;换算法;易货;物物交换;实物交易
barter agreement 易货协议;易货协定
barter arrangement 易货协定

barter away 亏本贱卖;亏本出售
barter business 实物交易
barter contract 易货合同
barter deal 易货交易
barter economy 实物交换经济
barter exchange 易货汇兑
barter peculiar to a semi-natural economy 半自然经济的物物交换
barter system 易货制度;以货易货制
barter terms of trade 纯交易条件;物物交易条件
barter trade 易货贸易
barter transaction 易货交易
barter treaty 换货协定
barter versus 易货贸易
bar test 钢棒测温法
bar theory 沙坝阻隔理论
bar thermometer 棒状温度计
Barth plan 巴斯计划
bar tie rod 圆钢拉条
bar timbering 横木加固;顶梁支护法;水平枝撑
bar tin 锡条(白铁皮)
bartizan 墙上凸出小塔楼;(墙上凸出的)小塔楼;墙外吊楼;(城堡上的)箭塔
Bartlett force 巴特勒脱力
Bartlett window 巴特勒窗口
bar-to-bar test 片间试验
bar-to-bar voltage 片间电压
barton 庄园中农场
Bartonian 巴尔顿阶【地】
Bartonian series 巴尔顿统
bartonite 巴硫铁钾矿
bar tracer 线条示踪器
bar tracery 铁棱窗花格;铁楞窗格;条饰花格窗
bar transformation 杆件位移
bar transition 彩条跃变
bar transposition 杆件位移
bar trap 砂坝圈闭
Bart reaction 巴尔特反应
bar turning machine 棒料车床
bar turret lathe 棒料转塔车床
bar type 钢筋类型;杆件类型
bar type crack(ing)test 圆棒抗裂试验
bar-type current transformer 棒式电流互感器;汇流条式电流互感器
bar type cylinder-gauge 卡尔玛斯缸径规
bar type grate 棒条筛
bar-type grating 条式格栅
bar-type multipoint tester 杆式多点检验器;棒式多点试验器
bar-type transformer 棒形铁芯变压器
bar-type trash rack 栅条式拦污栅
Barus effect 熔体突然膨胀
barverine 巴维林
bar visco(si)meter 杆式黏度计;杆型黏度计
bar way (有横路横木的)场内小路;有栏路横木的场内小路
bar weir 沙洲渔坝
bar welding 钢筋盖
bar winding 条形绕组
bar wire 裸电线
bar with hooked ends 带弯钩钢筋;带钩钢筋
barwood 条木;暗红色木材(产于非洲,装饰用)
bar work 棒料加工;棒料工件
bar-wound armature 杆式绕组电枢
barycenter 引力中心;质心;质量中心;重心
barycenter of curvature 曲率重心
barycentre coordinates 重心坐标
barycentric 质心的
barycentric coordinates 重心坐标;质心坐标
barycentric coordinate system 重心坐标系
barycentric dynamical time 质心力学时
barycentric element 质心要素
barycentric energy 重心能量
barye 巴列(气压力单位);微巴(声压压强单位)
barylite 硅钡铍矿
barysilite 硅铅矿;硅锰铅矿
barysphere 重圈;地心图;地心圈;地球核心
baryta 氧化钡;钡氧重土
baryta coated paper 大光钡地纸
baryta coated paper card 绸纹钡地纸
baryta coated paper card glossy smooth 厚光钡地纸
baryta coating 钡涂层;钡地涂层;钡地层
baryta crown 钡冕(玻璃)

baryta feldspar 钡长石；钡冰长石
baryta flint(glass)钡火石玻璃；含钡火石玻璃
baryta glass 钡铅玻璃
baryta green 钡绿
baryta light flint 含钡轻火石玻璃
baryta paper 钡涂纸；钡白纸
baryta phosphate 磷钡盐
baryta spar ceramics 钡长石陶瓷
baryta water 钡氧水
baryta yellow 钡黄；铬酸钡
baryte 天然硫酸钡
barytes aggregate 重晶石骨料；天然硫酸钡骨料；重晶石集料；天然硫酸钡集料
barytes aggregate concrete 重晶石骨料混凝土；重晶石集料混凝土
barytes breccia 重晶石角砾岩
barytes concrete 重晶石混凝土
barytes grain mix(ture) 重晶石混合颗粒
barytes mortar 重晶石砂浆
barytes mortar finish 重晶石砂浆饰面；重晶石砂浆抹面
barytes ore 天然硫酸钡矿石；重晶石矿石
barytes ore aggregate 重晶石集料
barytes ore concrete aggregate 重晶石矿石混凝土集料
barytes slab 重晶石板
barytes wall slab 重晶石墙板
barytes yellow 重晶石黄色
barytite 重晶石混凝土
barytocalcite 钡文解石；钡解石
barytolamprophyllite 钡刀叶石
barytosis 钡尘肺；钡尘沉着病
barytron 介子
basad 朝底地
basal 基础的
basal area 断面积；底面积
basal area of a stand 林分底面积
basal-area regulation 断面积调节法
basal arkose 基底长石砂岩
basal body 基体
basal body temperature 基础体温
basal cell 基室
basal cementation 基底胶结
basal cement(ing material) 玄武岩黏结材料
basal cleavage 基解理；底面解理【地】
basal coil 基底圈
basal complex 基底杂岩
basal conglomerate 底砾岩
basal contact 底部接触
basal coplane 基线共面；像对共面
basal crack 底面裂缝；底部裂缝
basal culture medium 基本培养基
basal diabasic texture 基底辉绿结构
basal disc 基盘
basal edge 底棱
basal erosion 底部侵蚀
basal face 底面
basal fertilizer 基肥
basal film 带基薄膜
basal granule 基粒
basal heave 基坑底隆起
basalia 基底；基板
basal(ic)sandstone 玄武岩砂石
basal infolding 基底褶
basal instability 坑底不稳定性
basal lamina 海底纹理叠层
basal layer 底层
basal layer of epidermis 表皮基层【生】
Basalloy 巴萨洛伊安全系统合金
basal matrix 玄武岩黏结料；玄武岩基质
basal metabolic rate 基础代谢率
basal metabolic test 基础代谢试验
basal metabolism 基础代谢
basal metabolism apparatus 基础代谢仪
basal moraine 底碛层；底冰碛
basal nerved 基础脉的
basal orientation 基向；基线定向
basal part 基底部
basal parts of the dome 穹隆底部
basal pinacoid 底轴面
basal plane 底面；基面；核面；底平面
basal plate 基层板
basal principle 基本原理
basal ration 基本定量

basal reflex 基础反射
basal ring 底足圈
basal sandstone 玄武岩砂石
basal sapping 基部掏蚀；基蚀；下切；底部掏蚀；挖掘
basal segment 基底段
basal slip 底面滑移
basal striation 基底纵纹
basal surface 基岩顶面；基底面；底面
basalt 黑色炉器；玄武岩
basalt-agglomerate tuff 玄块凝灰岩
basalt-cement tile 玄武岩水泥板；玄武岩水泥砖
basalt chip(ping)s 玄武岩碎屑
basalt chip(ping)s concrete 玄武岩碎屑混凝
basalt clay 玄武黏土；玄武土
basalt concrete 玄武岩混凝土
basalt concrete slab 玄武岩混凝土板
basalt depletion mechanism 玄武岩亏损机制
basalt face 玄武岩墙
basalt fiber 玄武岩纤维
basalt floor tile 玄武岩地板
basalt flow 玄武岩流
basalt-gabbro rock 玄武岩—辉长岩石
basalt glass 黑色玻璃；玄武(岩)玻璃
basalt group 玄武岩类
basaltic 玄武岩的
basaltic achondrite 玄武无球粒陨石
basaltic agglomerate 玄武质集块岩
basaltic chip(ping)s concrete 玄武岩碎屑混凝
basaltic clay 玄武黏土
basaltic concrete 玄武岩混凝土
basaltic crustal layer 玄武质地壳层；玄武岩地壳层
basaltic debris 玄武岩屑堆
basaltic dome 盾状火山
basaltic flood eruption 玄武岩泛滥喷发
basaltic floor tile 玄武岩楼面瓷砖
basaltic flow 玄武岩流
basaltic glassy substratum 玄武玻璃质底层；玄武波浪质地层
basaltic hornblende 玄武岩角闪石；玄闪石
basaltic iron ore 玄武岩铁矿石
basaltic joining 玄武岩节理
basaltic komatiite 玄武质科马提岩
basaltic lava 玄武岩熔岩；玄武熔岩；玄武熔岩
basaltic layer 玄武岩层；硅镁层
basaltic magma 玄武岩浆
basaltic meal 玄武岩粉末
basaltic parting 玄武岩节理
basaltic pea gravel 玄武岩小砾石
basaltic powder 玄武岩粉末
basaltic rock 玄武质岩
basaltic rocks 玄武岩类
basaltic sand 玄武岩砂
basaltic sandstone 玄武岩砂石
basaltic scree 玄武岩屑堆
basaltic sett 玄武岩铺路小方石
basaltic shell 玄武岩壳
basaltic slab 玄武岩板
basaltic stamp sand 玄武岩压碎砂
basaltic stone 玄武石
basaltic stratum 玄武岩层
basaltic substratum 玄武岩质底层
basaltic tuff 玄武凝灰岩；玄武火山凝灰岩；玄武质凝灰岩
basaltic volcanic breccia 玄武质火山角砾岩
basaltic wacke 玄武岩土；玄武土
basaltic wool 玄武岩的羊毛状物
basaltiform 玄武岩状
basaltine 辉石；玄武岩角闪石；玄武岩的
basalt magma 玄武岩浆
basalt meal 玄武岩石粉
basalt obsidian 玄武玻璃
basalt paving sett 玄武岩(铺路用)小方石
basalt pea gravel 玄武岩细砾；玄武岩豆砾石
basalt plateau 玄武岩高原
basalt-porphyry 玄斑岩
basalt powder 玄武岩粉末；玄武岩石粉
basalt quarry 玄武岩采石场
basalt sett 玄武岩铺路小方石
basalt sett paving 玄武岩小方石路面
basalt slab 玄武岩石板
basalt stoneware 黑炻器；玄武炻器
basalt stratum 玄武岩层
basalt structure 玄武岩构造
basalt substratum 玄武岩质底层

basalt wacke 玄武土
basalt ware 黑陶器；黑色无釉炻器
basalt water 玄武岩水层
basalt wool 玄武岩渣棉；玄武岩矿棉
basaluminite 羟铝矾
basal wall 基壁
basal water 最低地下水层；主要含水层；底水
basal water level 最低地下水位
basamite-tephrite group 碧玄岩—碱玄岩类
basan 书面羊皮
basanite 碧云岩；碧玄岩；试金石
basanitoid 玄武岩类
baschtauite 石英角斑岩
basculating fault 扭转断层
bascule 开合桥架；活动桁架；吊桥活动桁架
bascule barrier 竖旋式路栏
bascule bridge 竖升开启桥；开启桥；开合桥；活动桥；衡重式仰开桥；旋转吊桥；吊桥；竖旋桥；上开桥
bascule door 吊门
bascule gate 鱼腹式门；竖旋闸门；舌瓣闸门；吊门
bascule leaf 竖旋翼；竖旋翅
bascule pier 开启桥桥墩；仰开桥桥墩
bascule span 开启桥跨；竖旋孔；仰开跨(度)
base 路面基层；基座；基质底；基托；基地；基础；基部；轴距(指车)；底座；底数；底材；墙脚板供暖装置；瓶底；胎体
base address 基准地址；基地址【计】；基本地址
base address field 基地址字段
base addressing 基本地址
base address register 假定地址寄存器；基地址寄存器
base address relocation 基地址再定位
base adjuster 基址调整器；底座调整装置
base altitude 基地高度
base-altitude method 基线高度法；底高求积法
base-altitude ratio 基线与高度比；基线航高比
base amount 基础金额
base analog 碱类似物
base anchor 门销；门框底锚；地脚板；地板脚
base angle 基础边缘角铁；底角
base apex line 棱底线
base apparatus 基线器械；基线测量器械；基础装置；基本仪器；基本设备
base area 基底面积；支座面积；底区；底部断面积
base area of roof cavity 崩落面
base asphalt 底层沥青
base asphalt coating 沥青底子
base asphaltic bitumen 底层沥青
base assembly 基本汇编
base attribute 基属性；基本属性
baseball 棒球
baseball bar 棒球棒
baseball court 棒球场；垒球场
baseball diamond 棒球场
baseball field 棒球场
baseball finger mallet finger 棒球指
baseball stitch 垒球缝法；对缝法
baseband 基带；基本频带
baseband bank 基带组合
baseband branching station 基带分支站
baseband channel 基带信道
baseband coaxial cable 基带同轴电缆
baseband combiner 基带合并器
baseband digital transmission 基带数字传输
baseband distribution unit 基带分配单元
baseband equalization 基频带均衡化
baseband frequency repeating system 基频中继系统；基频接转制
baseband frequency response 基带频率响应
baseband local area network 基带局域网
baseband modem 基带调制调解器
baseband network 基带网络
baseband noise ratio 基带噪声比
baseband pulse 基带脉冲
baseband response 基带响应
baseband response function 基带响应函数
baseband signal 基频带信号；基带信号；基本信号
baseband signal(l)ing 基带信令
baseband signal(l)ing transmission 基带信号传输
baseband system 基带系统
baseband transmission 基带传输
base bar 夹角尺；基线杆(尺)；杆状基线尺
base barometer 基点气压计
basebead 金属刮平板(具有凸缘的，抹灰用)；冲

筋;勾泥尺
base beam lead 基极梁式引线
base bearing 基础轴承;主轴承;底轴承
base bed 微粒石灰石基层
base bend 支座弯头
base bias 基极偏压
base bias circuit 基极偏置电路
base bid 基本标价
base bid price 标底;投标底价
base bid specifications 基本标价项目说明书;基本标价清单;基本标价明细表;基本投标标价说明书
base bitumen 底层沥青
base bleeder circuit 基极分压电路
base block 洞口基石;踢脚板;脚石;基块;基底块体;门贴脸座;柱脚板;柱脚块;柱基石;踢脚墩(堵头);门基石
baseboard 粉刷的踢脚板;基线板;踢脚板;基线段
baseboard convector 踢脚板散热器
baseboard heater 护壁板散热器;底板加热器;踢脚板式取暖器;踢脚板散热器
baseboard heating 护壁板供暖
baseboard member 踢脚板构件
baseboard radiator 沿踢脚板铺设的散热器;踢脚板式暖气片;墙脚散热器;踢脚板散热器;踢脚板内暖气管
baseboard radiator unit 踢脚板散热器;踢脚板式暖气设备
baseboard register 底板节气门;踢脚板内调温装置;踢脚板调温装置
baseboard type unit 沿踢脚板铺设的管道设备
baseboard unit 踢脚板构件;踢脚板散热器
base box 基准箱;基盒;基本箱(镀锡薄板的商业单位);标准盒
base brick 碱性砖;底层面砖
base budding 底接
base bullion 不纯的金属锭;含银粗铅锭;粗金属锭
base bullion lead 粗铅锭
base buret 碱式滴定管
base burner 自给暖炉;底燃(烧)火炉
base camp 扑火营地;扑火野外指挥部
base cap 踢脚板顶条;踢脚板上缘线脚
base capacity 基本容量
base cargo 压舱物;垫底货(物)
base carriage 基线架;基线滑架
base carrier 下部行走部分
base case 基本情况
base-catalysed rearrangement 碱催化重排
base catalysis 碱催化(作用)
base catalyst 碱催化剂
base cation 碱性阳离子
base centered diameter of anchor(foundation) bolts 地脚螺栓中心圆直径
base-centered lattice 底心晶格;底心格子
base channel 基极沟道
base charge 雷管炸药;雷管药包;底部装药;扣除折旧费的净收益
base chart 原印海图;底图
base chart making 地图投影坐标底线
base circle 基圆;坡底圆
base circle diameter 基圆直径
base circular thickness 基圆齿厚
base class 基本类
base clip 门销;门框底锚
base cloth 基布
base cluster 基组;基体组
base coat 抹灰基层;底釉;底涂层;底色漆;底漆腻子;底胶;打底层;头道底漆
base coat/clear metallic system 金属闪光底涂层罩清漆涂装法
base-coated welding rod 碱性药(包)焊条
basecoat floating 底层抹灰
basecoating machine 底涂机
basecoat(mixed)plaster 底涂灰泥
basecoat mix(ture) 底层混合料
basecoat of paint 底漆
basecoat plaster 底涂灰泥;底层粉刷
basecoat stuff 底层混合涂料;底涂料
base code 瓶底厂家密码
base collector 基极集电极
base colo(u)r 底漆
base compilation 基本原图
base complement 基数补数;补码
base component 底部组成的单元;基线分量;基本部件

base concordance 底整合【地】
base concrete 承重混凝土;基础混凝土;混凝土受力层
base condition 基准条件
base cone 基圆锥;底锥
base configuration 基地布局
base contact 基极接点;基础接触;基部接触
base contact diffusion 基极接触扩散
base content 盐基含量
base control 基点控制
base correction 基线校正;基点改正
base course 墙基;路面下层;路面底层;路基;勒脚层;基底;基层;下垫层;垫层;底层;主航道线
base course concrete 底层混凝土
base-course constant 基层常数
base-course drainage 铺面的底层;路槽排水;基层排水
base course material 基层材料;基础材料
base course of air field runway 机场跑道垫层
base course of cement mortar 水泥砂浆底层
base course spreader 基层撒布机;基础撒布机
base-court 城堡外庭后面的院子;里庭;房屋后院
base cup 下盖;底盖
base current 基极电流;基流;基极电流;基电流
base curve 基础曲线
base cut 底部切削
base cutter 底切割机
base cylinder 基圆柱
based 向基的
base data 基数据;基本资料;基本数据组;原始资料;原始数据
base deficit 碱缺失
base density 片基密度
base density plus fog 胶片灰雾
base depot 基本机务段
base-depth-altitude ratio 基线景深比
base depth of wall 墙基深度
base desaturation 盐基脱饱和
base design 原图
based heave 基坑底隆起
base diameter 底座直径;底座对径
base diffusion 基极扩散
base diffusion layer 基极扩散层
base direction 基线方向;基线定向;起始方向
base discharge 基底流量
base displacement addressing system 基数位移定址系统
base distance 基距
base-distance ratio 基距比
based matching program(me) probability 基于匹配程序概率
based number 带基数
base domain 基畴
based on 根据
based on fuel 根据燃料计算
base drag 底部阻力
base drawing 底图
based reference 基址参照
base drive 基极激励
base-driven antenna 底部馈电天线
based storage allocation 基本变量存储区分配
based variable 基底变量;基本变量
base earth 基极接地
base elbow 支座弯头
base(electrode) 基极
base electrolyte 基底电解质
base element 成碱元素
base elevation 基线标高;基底高程
base ell 支座弯管
base emitter cut off current 基极发射极截止电流
base emitter saturation 基极发射极间饱和电压
base enamel 搪瓷底釉
base engine 后置发动机;后动发动机
base equation 基线方程
base excess 碱过剩;过剩碱
base exchange 离子交换;碱性离子交换;碱离子交换法;碱交换;盐基交换;阳离子交换
base exchangeable ion 碱交换离子
base-exchange capacity 碱交换量;盐基交换容量;盐基代换量
base exchange complex 盐基交换络合物
base exchange material 盐基交换物质
base-exchange process 碱交换过程

base-exchanger 碱性离子交换器;阳离子交换器
base exchange softening 盐基交换软水法
base-exchanging compound 碱交换剂;阳离子交换剂;碱性离子交换剂
base expansion 基线扩大
base expenditures 基本建设费(用)
base extension net 基线网
base fabric 打底的织物;基布;布基
base face 基准面
base failure 基塌;基底破坏;基础破坏;地基破坏;坡底塌陷
base failure calculaiton 剪切破坏计算
base failure formula 剪切破坏公式
base failure load 剪切破坏荷载
base failure of slope 坡脚坍塌;土坡基底破坏
base fee 基本费用
base felt 底层毛毡
base field 基域
base film 带基薄膜
base flashing 基础防水板;基层泛水;立墙泛水;接缝泛水
base flashing of parapet shirting 女儿墙勒脚泛水
base flight aircraft 基地建制飞机
base flow 基流;基本流量;基本径流;底部流动
base-flow depletion curve 基流退水曲线
base-flow hydrograph 基流过程线;地下径流过程线
base flow of river 河川基流量
base flow separation 基流分割
base flow storage 基流储量
base fog 乳剂灰雾
base foil 底层金属薄片
base for drop-hammer 落锤基础
base-forming food 成碱食物
base for plaster(ing) 抹灰的底
base for solar probe spacecraft 太阳探测宇宙飞船基地
base for truck farming 商品蔬菜栽培基地
base frame 基座;基架;下车支架;主架;底座;底台;底架;支架
base frame tilt correction value 基座倾斜校正值
base frequency 基本频率;固有频率
base fuel 基准燃料;基础燃料
base fuze 弹底引信
base gas 垫层气
base glass 基础玻璃
base glaze for colo(u)ration 配色基釉
base grid 基线格网;基本网格;坐标格网
base group 基群;基本组
base group rate 基群速率
base heater 踢脚板处供暖器;沿踢脚板铺设的供暖管道设备
base heating 踢脚板处供暖气;护壁板供暖
base-height ratio 底高比;基线与高度比;基线航高比;基线高度变化;基一高比
base helix 基圆螺旋线
base helix angle 基圆柱螺旋角
base hinge 底座铰接头
base hospital 后方医院
base hydrometric(al)station 水文基本(测)站
base identifier 基标识符
base impedance 基本阻抗
base-in 基线向内
base index 基本索引;指数基
base initiation 底起爆
base injection 基极注入
base ink 原墨;印刷原墨;印刷母墨
base insertion point 插入基点
base installation 基准信号机
base insulator 支座绝缘子;托脚绝缘子
base inventory method 基本盘存法
base iron 原铁水;原生铁
base isolation 基底隔振
base item 基本项
base joint 底部连接;底缝;底部接缝
base karst 裸露岩溶
base key 基键;基本键
base knob 门碰
base lacquer 底漆
base layer 基层;垫层;底层
base lead bullion 粗铅锭
base leaf 底部叶片
base leakage 管基漏泄
base length 基线长度;基距;基本长度
base length on photo 像片基线长度

baseless 没有基础的
base level 基准平面;基准面;基准高程;基数电平;基本水位;宁静辐射电平;起始水平
base-leveled plain 基准面平原
base level of abrasion 磨蚀基准面
base level of corrosion 溶蚀基准面;侵蚀基准面;切蚀基准面
base level of denudation 裸露基准面;剥蚀基准面
base level of deposition 沉积基准面
base level of erosion 侵蚀基准面;侵蚀基面
base level of stream 河流最低侵蚀面
base level of stream erosion 河流侵蚀基准面
base-level peneplain 基面准平原
base light 基础照明;基础光;衬底照明
base lighting 基本照明
base line 建筑场地主导线;准线【测】;计时基线;底线;扫描斤;基准线;基线;初始期间
baseline adjustment 基线调整
baseline adjustment technique 基线调整技术
baseline approach 基线设定途经
baseline-approved methodology 基准线已批准的方法
baseline break 基线中断
base line budget 基线预算
baseline case 基线案例
baseline check 地面校正
baseline chemical environment 基线化学环境
baseline concentration 基线浓度
baseline condition 基线条件
baseline configuration 基本构造形式;原始构造形式
baseline configuration management 基准配置管理
base line conversion method 基线换算法
baseline correction 基线校正
base line cost 基线费用;原始费用
baseline curvature 基线弯曲
baseline data 基准数据;基线资料;基线数据
baseline delay 基线延迟;基线时间延迟
base line dimensioning 基线尺度;基线尺寸法;基线尺寸
baseline direction 基线方向
baseline drift 基线漂移
baseline ecosystem monitoring 基线生态系统监测
baseline error 基线误差
baseline extension 基线扩大;基线延伸线;基线延长线
base line extension monitor 基延监测站
baseline fitting 基线拟合
baseline fluctuation 基线波动
baseline flying 基线飞行
baseline guide-channel 基线导槽
base line hologram 基线全息图
baseline information 基线资料;基线信息
baseline in shore station 岸台基线
baseline measurement 基线测量
baseline measuring apparatus 基线测量器械
baseline meteorological station 基线气象站
baseline method 基线法
baseline methodology 基准线方法
baseline monitoring 基准监测;本底值监测
baseline-new methodology 基准线—新方法
baseline noise 基线噪声
baseline observation station 基线观测站
baseline of diagram 图形基线
baseline of excavation 开挖基线
base line of projection 投影基线
baseline of section 断面基线
baseline of transducer in echo sounder 回声测深仪换能器基线
baseline projection 基线预测
baseline shift 基线位移;基线偏移
baseline site 基线场
baseline stability 基线稳定
baseline stake 基线桩
baseline stepping 基线颤动
baseline stream assessment 基线河流评价
baseline study 基线研究;基本研究
baseline survey 基线测量;背景值调查
baseline systematic concept 基线系统概念
baseline tape 基线尺
baseline terminal points 基线端点
baseline terminal stations 基线端点
baseline tilt 基线倾边
baseline value 基线值

base line value of regional environment 区域环境基线值
baseline wander 基线漂移
baseline water quality assessment 基线水质评价
baseline water quality condition 基线水质状况
baseline water quality monitoring network 基线水质监测网
base(line)year 基准年
base lining 基线定线
base-linked revenue 基数挂钩收入
base-load 基本荷载;基极负载;基底负载;基底负荷;最低负荷(发电厂)
base-loaded antenna 基载天线;底部负载天线
base-load electric(al)station 基底负载发电站
base-load electricity 基底负载电力;基底负荷电力
base-load gas 基底负载气;基底负荷气
base-load generator 基底负载发电机
base-load heat source 基本热源
base-load hydroelectric(al)project 水电计划的基底负载;水电计划的基底负荷
base-load hydroplant 基荷水电站
base load(ing) 基本负载;基本负荷
base-load operation 基荷运行;基本负荷运行
base-load power 基本负荷动力
base-load(power)station 动力站的基底负载;动力站的基底负荷;基荷电站
base-load source 基底负载源
base-load station 基底负载发电站
base-load turbine 基本负荷汽轮机
base-load unit 基荷机组;基本负荷机组
base machine 基型机床
base magnification 基线放大率
base map 基础图;基本图;基本地图;工作底图;工作草图;底图;三角网布设略图;索引图;基准图
base map of city 城市基本地图
base map of topography 地形底图
base mark 基础标志【测】
base mass storage volume 基本海量存储器卷
base material 基体材料;母料;基本材料;基层材料;基材;底层材料
base measurement 基线测量
base measurement party 基线组
base measuring equipment 基线测量器械
base measuring pressure 校正压力;测压的基准
base measuring tape 基线尺;带状基线尺
base measuring temperature 校正温度;测温的基准
base measuring wire 线状基线尺;钢丝基线尺
base medium effect influence 基медиум效应影响
base member 基础构件
basement 基座;基底;地下室;地窖
basement age 基底时代
basement air shaft 地下室通风(采光竖)井
basement area 地下掩蔽范围;地下掩蔽部
basement areaway 地下室采光井
basement boiler room 地下室锅炉房;地下锅炉房
basement car park 地下停车场;地下室停车;基层停车场
basement complex 基底杂岩
basement concourse 地下楼层大厅;地下中央大厅
basement contour map 基底等深线图
basement contours 基底轮廓线
basement cullet 炉底碎玻璃
basement door 地下室门
basement drain 地下室排水
basement drainage 地下室排水系统
basement drainage sump pump 地下室排除集水坑的水泵
basement dwelling 地下住宅
basement dwelling unit 地下室居住单元
basement entrance 地下室入口
basement excavation 基础开挖;地下室开挖
basement exit 地下室出口
basement extension 地下室延伸部分;基础延伸部分
basement fault 基底断层
basement flat 地下公寓
basement floor 地下室地面;地下室层
basement fracture 基底断裂
basement garage 地下室汽车库;地下车库
basement high 基底隆起
basement house 有地下室房屋
basement kitchen 地下室厨房
basementless 无地下室的
basementless space 非地下空间;爬行空间(地板下铺设管道设备所留出供人爬行的空间)
basement level 地下室层;底层
basement light well 地下通气竖井;地下采光井;窗井
basement main system 地下室干管系统
basement masonry 地下室圬工
basement membrane 基膜;基底膜
basement of basin 盆地的基底
basement of platform 地台基底
basement park(ing)garage 地下停车场;地下停车库
basement pit 地下窖
basement plan 地下室平面
basement retaining wall 地下室挡土墙
basement rock 基岩;基底岩石;基底岩
basement rock contour map 基岩等深线图
basement rock pool 基岩油藏
basement rock reservoir 基岩储集层
basement rock series 基底岩系
basement room 地下室房间
basement service road 地下辅助道路
basement shelter 地下避难室;地下避难所
basement slab 地下室地板
basement soil 基土;基础土(壤);土路基;路基
basement stair(case) 地下室楼梯
basement stor(e)y 地下楼层;(楼的)底层;地下室层面
base(ment)structure 基底构造;地下室结构
basement sump pump 地下室深水泵
basement tanking 地下室防水层
basement tectonic map 基底构造图
basement tectonics 基底构造
basement uplift 基底隆起
basement vault 地下拱顶室
basement velocity 基底的波速
basement wall 抗倾覆L形悬臂墙;基础墙;地下室墙
basement waterproofing 地下室防水
basement window 地下室窗
base metal 金属坯;金属底座;金属底层;碱金属;基本金属;基料;基焊料;基底金属;母体金属;母材;母板;主体金属;底金属;底层金属;底材;非贵重金属
base metal commodities 贱金属矿产
base metal couple 基金属热电偶
base metal fusion zone 母材熔合区
base metal test specimen 基底金属试样;母材试件;受焊金属试件
base-metal thermo couple 碱金属热电偶
base minus ones complement 反码【计】
base mix 原始混合料
base modulation 基极调制;基极调幅
base moment 基底弯矩;基底力矩;底部弯矩
base money 基准货币
base mortar 底座灌浆
base mo(u)lding (护墙板的)线脚装饰;踢脚线压顶条;踢脚板上缘线脚;底座线脚;底座饰
base net 基线网;基网
base network 基线网
base notation 基数制;基数记数法;基数表示法;基本记数法;基本记号;基本符号
base number 碱值;基数;底数
base number of a tender 标底
base number of emanation 射气底数
base of accounting period 以会计期间为基础
base of a column 柱基座
base of a cup decorated with carved designs 雕纹杯托
base of administrative control 以管理是否能控制为基础
base of a formation 岩层底基
base of a prism 棱镜底面;棱镜底座
base of a structure 结构底面
base of bottleneck 瓶颈下部
base of calculation 计算标准
base of column 柱基座;柱底
base of dam 坝脚;坝基
base of energy resources 能源基地
base of excavation 开挖基线;开挖基础;挖方基土
base of exponential function 指数函数的底
base offer 底标价
base of fill 填土的底
base of foundation 基底;基础底;地基
base of karstification 岩溶作用基准面;岩溶作用基底;溶蚀基准面

base of levee 堤基
base of logarithm 对数底(数);对数的底
base of logarithmic function 对数函数的底
base of natural logarithm 自然对数的底
base of notation 基本记号
base of producing goods for minority nationalities 民族用品生产基地
base of quay wall 岸壁基础
base of regularity of occurrence 以成本发生之经常与否为基础
base of relationship to the cost of unit 以成本与产品间的关系为基础
base of road 路面基层
base of sea mark 潮汛基线
base of slope 大陆坡底;边坡脚;坡脚(底部);坡底
base of slope sampling 沿山脚采样
base of source oil window 生油窗底界
base of taxation 课税标准;纳税基础
base of the rail 轨基
base of tool 刀具座
base of verification 控制基线;校准基线;基准面
base of wall 墙基
base of wet gas generation 湿气生成底界
base ohmic contact 基极欧姆接触
base oil 基础油;原油
base-one peak voltage 基极—峰值电压
base operand 基操作数
base operand specifier 基操作数说明符
base out 负基线;基线向外
base outcropping 基岩露头
base packer 底板垫塞物
base page 基页
base pair 碱基对
base pairing 碱基配对
base pair substitution 碱基对取代作用
base paper 原纸;纸胚
base parameter 基参数
base pav(i)er 石头铺路工人
base pay 基本工资
base peak 基峰
base period 基准期;基期;基本周期;地下径流时期
base period value 基期值
base pigment 基本颜料
base pin 底销;基础销子;管脚;底座销钉;插脚
base pitch 基圆节距;基圆齿距
base pitch error 基节误差
base pit engineering 基坑工程
base pit illustration 基坑图样
base plane 底平面;底板;基准平面;基面
base plate 基础底板;基板;护座(板);支承板;底座;底盘;底面;底板;壁脚板
base plate component 脚板部分
base plate cracking 基托折裂
base plate element 底板单元
base-plate engine 底座发动机
base plate for right angle and T-crank 直角及T角型拐肘用的垫板
base plate heater 踢脚板处的暖气管;底板散热器
base plate heating 踢脚板处暖气;底板供暖
base plate member 踢脚板单元;底板单元
base plate of stanchion 支柱底板
base plate radiator 踢脚板处暖气片;底板散热器;底板暖气片
base plate unit 踢脚板单元;底板单元
base plate wax 基托蜡
base plug (偏斜楔支座用的)下木塞
base plywood 底层胶合板
base point 基线点;基点;小数点;原点
base-point net 基点网
base pollution 碱污染
base polymer 聚合物基料;原始聚合物
base pressure 本底压强;地基压力;基准压强;基层压力;基础压力;底部压力
base price 基价;基本价格;地基费用;标底
base price on quality 按质论价
base priority 基优先数;基优先级
base projection 基线投影
base property 基本地产
base proposal 基本报价
base pump station 总泵站;泵总站
base quadrilateral 底四边形
base quantity 基本量
base radiator 踢脚板处暖气片
base radius 基圆半径

base rate 基本工资率;基本费率;基本单价;保证工资
base rate area 免费服务区;基本费率区
base rate of freight 基本运费率
base ratio 碱基比;基础航高比
base record 基本记录
base record slot 基本记录槽
base reducing tach(e)ometer 定基线自动归算视距仪
base region 基区;基极区;基本区
base register 基址寄存器;变址寄存器
base relief 浅浮槽
base relining 基托垫底
base rent 基本租金
base reservoir temperature 基底温度
base resin 基本树脂
base resistance 基区电阻
base ring 基座圈;基区环;基区环;垫环;底座圈;底环
base rock 基岩;底岩
base rock acceleration 基岩加速度
base rock motion 基岩运动
base rotation 基底转动
base runoff 枯水流量;基本径流
base salary 基薪
base saturability 盐基饱和度
base saturation 碱饱和(作用);盐基饱和作用;盐基饱和
base-saturation percentage 阳离子吸附饱和度;盐基饱和率
base screed 抹灰整平板;金属分格条;基准条;抹灰钢刮板;刮杠
base screw 底座螺钉
base segment 基段;基础砌块;基本段
base sequence 碱基顺序
base set 基本集
base setting 基线安置
bases for producing export commodities 出口商品生产基地
base shear 基底剪力;底面切力;底面剪力;底部剪力
base shear coefficient 基底剪力系数
base shear force 结构底部剪力;总地震荷载
base sheet 底图;基板;垫片;底层油毡;底层板;基层油毡;基本原图
base sheet(ing) 背衬护墙板
base shield 管座屏蔽
base-ship 基地船舶
base shoe 支座座板;地面周边饰条;踢脚压条;踢脚板压条
base shoe corner 墙角饰条;阴角处线条
base shoe mo(u)lding 地毯压条;地面周边饰条;踢脚板底缝压条
base-sill 底槛
base site 基线场
base situation 基本情况
base size 终拔前尺寸;钢丝终拔前尺寸
base slab 基础板;垫板;底板;承台
base slag 碱性炉渣;碱性矿渣
base slate 底层页岩
base slope 底坡(度)
base slot 槽子砖槽口
bases of well arrangement 布井依据
base soil 基土;底土
base-source of supply 基本供水源
base speed 基本速度
base spray 后飞破片
base spreader 底部材料撒布人;底部材料撒布机
base-spreading resistance 基区扩展电路;基极扩展电阻
base square 底正方形;塔底框
base stability 基底稳定性
base stabilization 地基稳定(作用)
base stacking 碱基堆积
base stanchion 基座标志柱
base standard 基础标准
base standard cost 基本标准成本
base station 基准站;基准台;基站;基线点;基台(指台站);基地址【计】;基地(电)台;基本站;基本测站;固定台;岸台
base station interface unit 接口单元
base status 盐基性质
base stock 基本组分;底纸
base stock bin 最低库存;安全存仓
base stone 底部基石;基石
base stream 主要河流

base structure 箱底结构
base subsistence density 基本生存密度
base substitution 碱基置换
base surface 基层;基础面;底面
base surge 底涌云;底涌;底涌云;水下爆破形成的大浪
base-surge deposit 底涌云沉积
base system 基本体系
base T 支座三通
base table 踢脚压顶条;踢脚板上缘线脚
base tangent length 公法线长度
base tankage 罐基本容量
base tape 基线尺
base-tar 硬沥青
base tee 支座三通
base temperature 基础温度
base tender 底标价
base test 基本方案试验(水工模型)
base thickness 基本轴向厚度
base thickness of wall 墙基厚度
base tile 基层花砖;底层面砖;底层花砖
base tilt factor 基底倾斜因数;基底倾斜系数
base time 基额定时间;基本工时;正常时间;额定时间;(劳动定额的)本工时基
base-timing sequencing 基站定时程序;时分定序
base titanium dioxide 基本二氧化钛;粗二氧化钛
basetone 低音
base track 地面滑轨(用于推拉隔断)
base transmission factor 基极传输因数
base transport factor 基区迁移率;基极传输因数
base-tray 基底托盘
base trim 基座线脚装饰;底座线脚
base tunnel 导坑
base turbine 后置汽轮机
base type 底盘式
base-type construction 底盘式结构
base union 底接结合点
base unit 下部结构;下部建筑
base unit weight 基量;基本单位重量
base value 基准数;基值
base variable 基变量
base vector 基准矢量;基向量;基矢量
base vehicle 生产成本最低的车型
base-vented hydrofoil 底面开孔水翼
base volume 基卷
base wad 底垫
base wafer assembly 基片装置
base wall 底基墙;底层墙
base weather station 基地气象台
base weight 基准重
base widening 基层加宽
base width 基准宽度;基区宽度;基层宽度;底宽
base width modulation 基区宽度调制
base word 基址字
base year 基年;基准年
base-year cost 基年费用
base year traffic volume 基本年交通量
bash 填充废巷道
basha 棚屋(印度)
Bashi channel 巴士海峡
bashing 充填采空区;采空区填充;封闭火灾区;废料回填废巷道
Bashore resiliometer 巴肖氏弹性试验机
bashrah 巴拉地毯
basic 基性的;基价数;基本的
basic access 基本接入
basic access method 基本访问方法
basic accounting records 基本的会计事项记录
basic accounting unit 基本会计单位;基本核算单位
basic acetate 碱式醋酸盐
basic active 碱活性
basic address 基本地址
basic agreement 基础协议
basic air diving 基础空气潜水
basic allowance 基本扣除额
basically controlled 达到基本控制
basically stable 基本稳定的
basically stable area 基本稳定地区
basic aluminium acetate 碱式醋酸铝
basic alumin(i)um chloride 碱式氯化铝;碱性氯化铝
basic amplification stage 基本放大级
basic amplitude 基本振幅
basic andesite 基性安山岩

basic and non-basic components of the economy 基础经济部门和非基础经济部门
basic angle 底角
basic apparatus 基本仪器
basic assembler 基本汇编程序
basic astronomic(al) point 基本天文点
basic balance of payments 基本收支平衡
basic benchmark 基本水准点;水准基点
basic Bessemer cast iron 碱性转炉铸铁
basic Bessemer converter 碱性转炉
basic Bessemer process 碱性转炉法
basic Bessemer steel 碱性贝氏转炉钢;碱性转炉钢
basic Bessemer steel converter 碱性转炉钢转炉
basic bismuth carbonate 碱式碳酸铋;次碳酸铋
basic black 碱性黑
basic blast furnace slag 碱性的高炉炉渣;碱性高炉熔渣
basic block 基本块;基本分程序;基本程序块
basic boom 主吊杆;基本臂(起重机)
basic bore 基孔
basic bottom 碱性炉底
basic brick 碱性砖
basic building codes 基本建筑规范;基本建设规范
basic building engineering geology 基建工程地质
basic building exploration 基建勘探方法
basic building material 基本建筑材料
basic capacity 基本生产率;基本通过能力;基本容量;基本能力;基本功率
basic capsule 基裂
basic carbonate 碱性碳酸盐;碱式碳酸盐
basic carbonate lead 碱性碳化铅
basic carbonate of white lead 碳酸铅白;白铅粉
basic cartographic(al) data 基本地图资料
basic cartographic(al) document 基本制图资料
basic case 基本情况
basic catalyst 碱催化剂
basic celerity 基本波速
basic cement 碱性水泥
basic cerise 碱性樱红
basic characteristic of river 河道基本特性
basic character set 基本字符集
basic chromate silico-chromate 碱式硅铬酸铅
basic circle 基准圆
basic circle table 基本度盘表
basic circuit 基本线路
basic circuit diagram 基本线路图
basic clause 基本条款
basic clinker 碱性溶渣
basic cobaltous carbonate 碱式碳酸钴
basic code 绝对码;绝对代码;基本码;基本代码;机器(代)码;代真码
basic coding 基本编码
basic colony system 基本群体制
basic colo(u)rs 基色;基本色
basic component 基础组分;基本成分;基本部件
basic computer 计算机主体
basic concept(ion) 基本概念;本义
basic configuration 基本形状
basic construction(al) material 基本建筑材料
basic content of mapping work 测绘工作基本内容
basic contour interval 基本等高距
basic contradiction 基本矛盾
basic control 基本控制
basic controller 基本控制器
basic control mode 基本控制方式
basic control point 基本控制点
basic control system 基本控制系统
basic converter 碱性炉衬吹炉;碱性吹炉
basic converter steel 碱性转炉钢
basic copper acetate 碱式醋酸铜;铜绿
basic copper arsenate 碱式砷酸铜
basic copper chloride 碱式氯化铜
basic copper tartrate solution 碱性酒石酸铜溶液
basic cost 基本费用;单位成本
basic counter 主计数器
basic course 基础课程;基本过程
basic cover 基本覆盖
basic credit line 贷款限额
basic creep 基准蠕变
basic crown gear 基准冠齿轮
basic cycle 基本周期;基本循环
basic data 基础资料;基本资料;基本数据;原始资料;原始数据
basic data manipulation function 基本数据操作功能
basic data of evaluation 评价基础数据
basic data type 基本数据类型
basic decision model 基本决策模型
basic density 基本密度
basic deposit 碱性沉积物
basic depreciation 基本折旧
basic depreciation expenses 基本折旧费
basic depreciation fund 基本折旧基金
basic depreciation rate 基本折旧率
basic depreciation rate of fixed assets 固定资产基本折旧
basic depth factor 基准齿高系数
basic design 基本设计;原图
basic design consideration 基本设计思想
basic design criterion 基本设计准则
basic design data 基本设计资料;设计基本数据
basic design information 基础设计资料;基本设计资料
basic design load 基本设计荷载
basic design principle 基本设计原则
basic design scheme 基本设计方案
basic deviation 基本偏差
basic device unit 基本设备单元
basic diagram 基本线图
basic difference 根本区别
basic dimension 基本尺度;基本尺寸
basic diorite 基性闪长岩
basic direct access method 基本直接访问方法;基本直接存取方法
basic display unit 基本显示装置
basic distance 基距
basic document 原始单据;基本资料;基本文献;基本文件
basic documentation 原始编література
basic dollar 基元
basic dose equivalent limit 基本剂量当量限值
basic drawings 基本图纸
basic dye 碱性颜料;碱性染料;碱染料
basic dyeing and printing wastewater 碱性印染废水
basic dye-phosphomolybdic heteropolyacid spectrophotometry 碱性染料—磷钼杂多酸分光光度法
basic dyestuff 碱性颜料;碱性染料
basic earnings share 每股基本盈利
basic edit 基本编辑
basic effect 基本效应
basic electric(al) furnace 碱性电炉
basic electric(al) furnace steel 碱性电炉钢
basic electrode 碱性焊条
basic element 基本元素;基本要素;基本动作
basic employment 基本就业
basic engineering 基础工程;基本建设工程
basic engineering design data 工程设计基本数据
basic environmental capacity 基本环境容量
basic environment(al) concept 基础环境概念
basic environment(al) science 基础环境科学
basic equalization 基本均衡
basic equation 基本方程(式)
basic equation of hydrodynamics 水动力学基本方程
basic equation of hydrostatics 水静力学基本方程
basic equation of two-dimensional steady water quality model of river 二维河流稳态水质模型基本方程
basic equation of water quality model 水质模型基本方程
basic equipment 基本设备
basic equipment list 基本设备表
basic excavator 主挖土机
basic exchange rate 基本兑换率
basic exemption 基本免税额
basic expenditures 基本支出
basic expenses 基本费用
basic exploration line 基本勘探线
basic external function 基本外部函数
basic facade unit 基本立面单元
basic factor 基本因素
basic failure rate 基本失效率
basic fare 基本票价
basic farmland conservation area 基本农田保护区
basic feasible solution 基础容许解;基础可行解;基本可行解
basic ferric acetate 碱式醋酸铁
basic fiber 贝西克纤维(聚碳酸酯纤维);原纤维
basic field attribute 基本字段属性
basic field description 基本字段说明;基本域说明
basic field descriptor 基本字段描述符;基本域描述符
basic figure 基价
basic filter 标准滤光片
basic filter value 基本滤光值
basic floor area ratio 建筑面积密度;建筑面积与基地面积比率(容积率)
basic flow 基本流量
basic flowchart 基本框图;基本流程图
basic flow sheet 基本流程图
basic flux 碱性熔剂;碱性溶剂
basic food 碱性食品;基本食品
basic food requirement 基本食品需求
basic for application of overhead 分配间接费的根据
basic form 基本形(式);基本式样
basic formula 基本配方
basic four colo(u)r system 四原色法
basic frame 基本框架
basic freight 基本运费
basic freight rate per ton for a stay in port 吨停泊基价
basic freight rate per ton for sailing 吨航行基价
basic frequency 基频;基本频率;主频率;主频
basic front 基性锋
basic fuchsin 碱性品红;碱性复红
basic fuchsin magenta 碱性洋红
basic fuel 基本燃料
basic function 基本功能
basic functions of commercial enterprises 商业企业基本职能
basic ga(u)ge cross-section 基本水尺断面
basic geochemical map 基本地球化学图
basic geodetic survey 基本大地测量
basic geologic(al) research to diminish in strength 基本地质研究工作削弱
basic geologic(al) research to increase in strength 基本地质研究工作增强
basic geologic(al) working 加强基础地质工作
basic geomorphic(al) element 地貌基本要素
basic gneiss 基性片麻岩
basic grade 基本级别;基本等级
basic group 碱性基;碱基;基群
basic group alarm 基群警报
basic group allocation 基群分配
basic group index method 基本分组指数法(设计柔性路面厚度的一种方法)
basic group link 基群线路;基群链路
basic hand tool 随车工具
basic harmonic 基波
basic health standards for radiological protection 放射卫生防护基本标准
basic heating 主加热
basic heat loss 基本耗热量
basic hole 基孔
basic horizontal control 基本平面控制
basic hornfels 基性角页岩
basic hydrograph 基本水文过程线
basic hydrography 基础水文学
basic hydrologic(al) data 基础水文资料;基本水文资料;基本水文数据
basic hydrologic forecast 水文趋势预报
basic hydrolysis 碱水解
basic hydrometric station 基本水文站
basic hypothesis 基本假设
basic igneous rock 基性火成岩
basic impulse insulation level 基准脉冲绝缘水平
basic impulse level 基本脉冲电平
basic increment of hour angle 时角基本量
basic indexed sequential access method 基本索引顺序存取法
basic induction variable 基本归纳变量
basic industry 基础工业;基本工业;制碱工业
basic information for design 设计基础资料
basic information of geology 基础地质资料
basic information of technologic(al) economic 技术经济基础资料
basic information unit 基本信息单元;基本信息单位
basic information unit segment 基本信息单位段
basic input-output system 基本输入/输出系统

basic instruction 基本指令
basic instruction set 基本指令表
basic instrument 基本仪表
basic insulation level 基本绝缘标准
basic insurance 基本险
basic intake rate 基本吸入速率;基本渗吸率
basic intelligence 基本情报
basic intensity 基本强度;基本烈度
basic interests 根本利益
basic interface 基本界面
basic internal operation(al) of lathe 车床基本内圆车削
basic international standard 基础国际标准
basic investigation 基本调研;基本调查
basic ion 碱离子;阳离子
basic iron 碱性铁
basicity 碱性;碱度;盐基度
basicity constant 碱度常数
basicity factor 碱度因素;碱度系数
basicity reaction 碱性反应
basicity stability 盐基度稳定性
basic jib 基本臂(起重机);最短主臂;主吊杆
basic lake 碱性色淀
basic landing runway length 降落跑道的基本长度
basic language 基本语言
basic language project 基本语言设计
basic language software 基本语言软件
basic lava 基性熔岩
basic lava flow 基性熔岩流
basic law 基本法
basic law and governing formula of seepage 渗流基本定律与公式
Basic Law for Environment Pollution Control 环境污染管制基本法
basic law of environmental protection 环境保护基本法
basic law of price change 价格变化基本规律
basic law of seepage 渗流基本定律
basic laws of river channel evolution 河演基本规律
basic lead acetate 碱式乙酸铅;碱式醋酸铅
basic lead arsenate 碱式砷酸铅
basic lead carbonate 碱式碳酸铅
basic lead chromate 碱式铬酸铅
basic lead chromate anti-corrosive paint 碱式铬酸铅防锈漆
basic lead nitrate 碱式硝酸铅
basic lead silicate 碱式硅酸铅
basic lead silicate white 碱式硅酸铅白
basic lead stabilizer 盐基性铅盐稳定剂
basic lead sulfate blue 碱式硫酸铅蓝
basic lead sulfate white 碱式硫酸铅白
basic lead sulphate 碱式硫酸铅
basic lead white 碱式碳酸铅白;铅白
basic length of basic runway 主跑道的基本长度
basic level(l)ing origin 水准原点
basic-level market 初级市场
basic-level production 初级生产
basic life-related facility 生活基础设施(住宅、交通、上下水道、电气、煤气、道路、公园绿化等)
basic limit 基本极限
basic line 基数;基础直线;底线;基线
basic-lined converter 碱性炉衬吹炉;碱性炉衬吹炉
basic line equalizer 基本均衡器
basic line of sight 基准视线
basic line space 基本行间距
basic lining 碱性衬里
basic link 基本链路
basic linkage 基量连接;基本连接指令;基本连接
basic link unit 基本链路单元;基本连接部件
basic list 基本表
basic loader 基本装入程序
basic load(ing) 静力荷载;基本荷载;基本负载;基本负荷;主要荷载;主要负载
basic logic 基本逻辑
basic loop 基本循环
basic machine 主要机器
basic machine time 基本周期
basic magma 主要岩浆
basic magnesium carbonate 碱式碳酸镁
basic major diameter 大径基准尺寸
basic manipulation 基本操作
basic manufacturing cost 基本制造成本
basic map 基本地图

basic map series 基本地图系列
basic maps of hydrogeology 水文地质基础性图件
basic market 基本市场
basic master group 基本主群
basic material 原料;碱性材料;基础材料;基础资料;基本材料;原材料
basic mathematics 数学基础
basic matrix 基矩阵
basic maxim 基本准则
basic mechanical design feature 主要技术性能
basic mesh size 网目尺寸;筛号
basic metal 碱性金属
basic metal fatigue strength 基本金属疲劳强度
basic method 碱性法
basic minor diameter 小径基准尺寸
basic mode 基模;基本方式
basic mode control procedure 基本方式控制过程;基本型控制规程
basic model 基本模型
basic mode link control 基本型链路控制(规程);基本方式链路控制
basic model of ideal body 理想物体的基本模型
basic module 基本组件;基本模块
basic modulus 基本模数
basic monitor 基本监督程序
basic motion 基本动作
basic multiplier 基础系数(总就业与基本就业的比率)
basic multiprogramming 基本多道程序设计
basic multiprogramming system 基本多道程序设计系统
basic national policy 基本国策
basic need for shelter 住房基本需求
basic neighbo(u)rhood 基本邻域
basic network 基本站网
basic neutralizer 碱性中和剂
basic noise 基本噪声;基本杂波;本底噪声
basic notation 基本符号
basic number 基本数;底数
basic nutrient 基本营养物
basic of distribution 分配根据
basic offshore survival 基础海上救生
basic offshore survival training 基础海上救生训练
basicole 碱生植物
basic open-hearth furnace 碱性平炉
basic open-hearth process 碱性平炉法
basic open-hearth steel 碱性平炉钢
basic operating system 基本操作系统
basic operational sea training 基本操作海上培训
basic optimal solution 基础最优解
basic options 基本期权
basic order code 主指令码
basic ore 碱性矿石
basic overlay 基本覆盖
basic oxide 碱性氧化物
basic oxygen furnace 氧气顶吹炉
basic oxygen furnace steel making 氧气顶吹炉炼钢
basic oxygen steel 碱性氧吹钢;氧气顶吹转炉钢
basic paint coding 涂料基本名称代号
basic parabola 基本抛物线
basic parameter 基本参数;基本参量
basic parameter for reserve calculation 储量计算基本参数
basic partitioned access method 基本划分访问法
basic parts list 基本零件清单
basic passenger ticket 基本客票
basic pattern 基本图式;基本格局;缠绕标准线
basic pay 基本工资
basic peak 基峰;主峰
basic period 基期;基本周期
basic physical infrastructure 有形基本设施
basic physical quantities and units 基本物理量及单位
basic pig iron 炼钢生铁;碱性生铁
basic pigment 碱性原料
basic pitch 基节
basic pitch diameter 基本中径
basic plagioclase 基性斜长石
basic plan 底图
basic plane 基本平面;基础面
basic plan generating system 基本规划生成系统
basic plate 坩埚漏板
basic plot subroutine 基本绘图子程序

basic plumbing facility 基本给排水设备
basic point 基地址(计);基础点
basic policy 基本政策
basic population 基本种群;基本人口;基本群体
basic porphyrite 基性玢岩
basic port 基地港
basic power source 基本能源;主要能源
basic practice 基本操作
basic premium 基础保险费
basic price 基价;基本价格
basic price date 基本价格数据
basic primary operator control station 基本主操作员控制台
basic principle 基本原理
basic principle of environment(al) law 环境法的基本原则
basic probability theory 基础概率论
basic process 碱性炼钢法;基本制法;基本过程
basic processing unit 基本信息处理单元
basic products 主要产品
basic project 基本项目
basic projection plane 基本投影面
basic property 碱性;基本特性;基本存储能力
basic property of crystal 晶体基本性质
basic propositional function 基本命题函数
basic pulse 基准脉冲;基本脉冲
basic pulse generator 基准脉冲发生器;标准脉冲发生器
basic pulse level 基本脉冲电平
basic pulse recurrence rate 基本脉冲重复频率
basic pulse repetition interval 基本脉冲重复周期
basic pulse repetition rate 基准脉冲重复频率;基本脉冲重复频率
basic Q factor 空载Q值
basic quantitative analysis 基本数量分析
basic quota 基本配额
basic rack 基本齿条;标准齿条
basic rack tooth profile 基准齿条齿廓
basic radiance 基本辐射度
basic range network 基本测距网
basic rate 基础费率;基本运费率;基本单价;基本比率
basic rate access 基本速率接入
basic rate interface 基本速率接口
basic rate of exchange 基本汇率
basic rate of income 基本收益率
basic raw material 基本原料
basic reaction 基本反应
basic real constant 基本实常数
basic reason 根本原因
basic recipe 基本配方
basic reflection mode 基本反射模式
basic refractory 碱性耐火材料
basic refractory brick 碱性耐火砖
basic regularity of channel evolution 河演基本规律
basic repetition frequency 基本重复频率
basic representation 基本表达式
basic requirement 基本要求
basic requirements for drafting standard 编写标准的基本要求
basic requirements for tendering 投标基本要求
basic research 基础研究;基础理论研究;基础科学研究;基本研究
basic research and development 基础研究与发展
basic resistance 基本阻力
basic resistance of car 车辆基本阻力
basic resistance of locomotive 机车基本阻力
basic resistance of train 列车基本阻力
basic return rate 基准收益率
basic risk 基本危险性
basic road way capacity 道路基本行车能力;道路基本通车能力
basic rock 碱性岩(石);基性岩
basic rocks 碱性岩类;基性岩类
basic rolling resistance 车辆基本溜放阻力
basic roof 碱性炉顶
basic rotary-hearth steel 碱性转炉钢
basic route 基本进路(铁)
basic rule 基本规则;基本规定
basic runoff 基本径流
basic runway 主要飞机跑道
basic runway length 主要跑道长度
basic safety training 基础安全训练

basic salary 基本薪金;基本工资
basic salt 碱式盐;基性盐
basic sample 正样
basic scale 基本刻度;基本比例(尺)
basic scale topographic(al) map 基本比例尺地形图
basic scarlet 碱性猩红
basic scheduling strategy 基本调度策略
basic scheme 基本方案
basic schist 基性片岩
basic science 基础科学
basic sealing width of gasket 垫片基本密封宽度
basic section 基础剖面
basic sediment 底部沉积物;碱性冲击物;碱性沉积物
basic sediment and water 水杂淀积
basic seismic intensity 基本地震烈度;地震基本烈度
basic seismic wave 基本地震波
basic sequential access method 基本按序存取法
basic series 基本级数
basic service port 基地港
basic service rate 基本服务费率
basic services 基本服务项目;基本服务程序;基本任务
basic set 基本集
basic set of solutions 基本解法组
basic shade 主底色
basic shaft 基轴;主轴
basic shaft system 基轴制
basic shearing resistance 基本抗剪力
basic shop cost 基本制造成本
basic shovel 主要挖土机;主要挖掘机
basic signal 基本信号
basic similarity 基本相似
basic size 名义直径;公称尺寸;标称直径;基准尺寸;基本尺度;基本尺寸;规定尺寸;标准规格
basic skill 基本技能;基本功
basic slag 碱性溶渣;碱性炉渣;碱性矿渣
basic sludge 碱性渣;碱性沉渣
basic soap 碱性皂
basic software 基本软件
basic software library 基本软件库
basic soil 碱性土
basic solution 碱液;碱性溶液;基解;基本解
basic solvent 碱性溶剂;基本溶剂
basic space 基本空间
basics parameters of earthquake 地震基本参数
basic-stable landslide 基本稳定滑坡
basic-stage flood 基点洪水
basic stain 碱性着色剂;碱性染剂
basic standard 基准;基础标准;基本标准
basic standard cost 基准成本;基本标准成本
basic standard cost system 基本标准成本制度
basic state 基态
basic statement 基本语句【计】;基本说明;基本论点
basic static capacity rating 额定静负荷
basic station 基本站;基本测站
basic statistics 基本统计资料;基本统计数字;基本统计数据
basic status 基本状态
basic status register 基本状态寄存器
basic steel 碱性钢
basic step 基本步骤
basic stimulus 基色;基本刺激
basic stipulation 基本规定
basic stock 碱性炉料;基本原料;正常储备
basic stress 主应力;基本应力
basic stress state 基本应力状态
basic structural index 基本结构指数
basic structural module 基本结构模数
basic structure 基本结构;主桁架;主构架;骨架
basic study on exploration of deep sea mineral resources 深海矿物资源勘查基础研究
basic sulfate white lead 碱性硫酸铅白
basic superphosphate 碱性过磷酸钙;沉淀磷酸钙
basic support software 基本支持软件
basic surface 基础面
basic switching term 基本转换项
basic symbol 基本符号
basic system parameter 基本系统参数
basic take-off runway length 起飞用跑道的基本长度
basic tankage 罐基本容量

basic tariff 基本运价表;基本价目表;基本关税
basic taxable year 基准纳税年度
basic technical data 基本技术数据;主要技术资料;主要技术数据
basic technique 基本技术
basic tectonic framework description 基本构造格架描述
basic term 基本项;基本术语;主项
basic term and definition for reliability 可靠性基本术语和定义
basic test 碱性试验
basic testing 基本测试
basic testing board 基本测试台
basic test on soil 土的基本试验;土的基本测试
basic theory 基础理论
basic thread overlap 螺纹接触基准高度
basic thread profile 螺纹基准牙形
basic through capacity 基本通过能力
basic time 基本时限;基本时间;基本工时
basic time factor 基时系数
basic time of production 基本生产时间
basic time table of trains 列车基本时刻表
basic timing 基本计时
basic timing cycle 基本计时周期
basic titrant 碱滴定液
basic token 基本记号;基本标记
basic tolerance 基本公差
basic toleranc table 基本公差表
basic tool 主要工具
basic trafficability 基本通过能力
basic traffic capacity 基本通行能力;基本通过能力;基本交通量
basic training 基本功
basic transmission unit 基本发送单位
basic travel(l)ing gear 主要着陆架
basic triangulation network 基本三角网
basic trig data 基本三角测量
basic type 基本类型
basic-ultrabasic pegmatite deposit 基性超基性伟晶岩矿床
basic unit 基层单位;基本单位;主要单元(机器)
basic unit budget 基本单位预算
basic unit price 基础单价
basic unprospected area 基本尚未勘探区
basic utility 基本实用程序
basic value 基本值
basic variable 基准变数;基准变量;基变量;基本变量
basic variance 基本差异
basic vegetative period 基本营养生长期
basic version 主要方案
basic vertical control 基本高程控制
basic vertical control data 基本高程控制数据
basic vibration mode 基本振型
basic violet 碱性紫罗兰
basic wage rate 基本工资率
basic wages 基本工资
basic water chemistry principle 基本水化学原理
basic water content 基本含水量
basic water ga(u)ge 基本水尺
basic water monitoring program(me) 基本水质监测程序
basic waveform 基本波形
basic weight 基准重量;基本重量
basic welding rod 碱性焊条
basic window length 基本窗长
basic window length correlation method 基本窗长对比法
basic wind pressure 基本风压
basic wind speed 基本风速
basic wood increment 定期木材生长率
basic work of economic calculation 经济核算基础工作
basic works of electric(al) prospecting 电法勘探基础工作
basic zero 基本零点
basic zinc carbonate 碱式碳酸锌
basic zinc chromate 碱式铬酸锌
basic zinc sulfate 碱式硫酸锌
basifacial 面基的
basification 碱化;基性岩化
basifier 碱化剂
basifixed 底着的
basifuge 嫌碱植物;避碱植物

basify 使碱化
basil 斜刃面;刀口;刃角
basilar 基底的
basilemma 基膜
basiliacal church 长方形教堂
basilica 长方形建筑物(尤指教堂);早期基督教堂;王宫;会堂
basilica julia 小庙
basilican cross-section 长方形会堂剖面图;长方形会堂断面图
basilican(ground)plan 长方形会堂平面图
basilican style 长方形会堂式
basilicon 松脂蜡膏;松香蜡膏
basil oil 罗勒油;丁香罗勒油
basin 港池;漫灌塘;海底盆地;洗脸盆;港湾内的停泊场;船坞;船模试验池;池(子);承盘;防潮船坞;盆地
basin accounting 流域水量平衡;算流域水账
basinal lake 内流湖;盆地湖;无(出)口湖
basin analog 流域相似
basin and range province 盆岭省;盆地山脉省
basin basic data 盆地基本数据
basin building 造盆运动
basin cell 流域单元
basin characteristic 流域特征(值);盆地特征
basin-check irrigation 畦灌
basin circularity 流域圆形度
basin classification 盆地分类
basin configuration 流域形状;流域外形
basin dam 漫水坝;谷口凹顶坝
basin desert 沙漠盆地
basin divide 流域分水线;流域分水岭
basin dry dock 干船坞;池式干船坞
basin due to damming 筑坝形成的流域
basin due to erosion 由涡流侵蚀形成的盆地
basin ecosystem 流域生态系统
basined flower bed 盆栽花坛
basin elongation ratio 流域伸长比
basin end period 结束期
basin entrance 港池口门;泊船池入口
basin facies 盆地相;盆地边缘相
basin-fill pattern 盆地充填样式
basin-fill sequence 盆地充填序列
basin fish pass 池式鱼道
basin flooding 淤灌;淹灌
basin-flooding irrigation 畦田漫灌
basin fold 盆地褶曲
basin for calibration 率定池;率定槽;校准池;检定池
basin for shipping 货运港池
basing 固结于基座
basing diagram 电子管管底接线圈
basing index 基本指数
basing machine 管底安装机
basing-point freight 基点运费
basing-point pricing 基点价格;基点定价
basing point system 基点制
basin grid 盆式炉栅
basing sales on production 以产定销
basin hydrography 流域水文地理(学);流域水道测量(学)
basining 盆地形成作用
basin into which the dock opens 干船坞前港池
basin investigation 流域调查
basin irrigation 漫灌;小区灌溉;淤灌;池灌;畦田灌溉;坑灌法
basin lag 流域滞时;流域滞留;流域汇流时间;流域迟滞
basin length 流域长度
basin line 流域界线
basin listing 沟槽作畦
basin lock 厂室船闸;港池闸门;池形船闸
basin map 流域图
basin margin deposit 盆地边缘沉积
basin mean precipitation 流域平均降水量
basin mean rainfall 流域平均雨量
basin mean runoff 流域平均径流量
basin method 淤灌法
basin method for irrigation 坑灌法
basin model 盆地模式
basin mouth 盆地出口
basin of aquifer 含水层补给区
basin of deposit(ion) 沉积盆地
basin of groin 丁坝间防波堤间水区;丁坝间水区
basin of internal drainage 内流盆地;内流流域

basin of the groin 丁坝间坝田
basin order 流域等级
basin peat 潜水滋育泥炭
basin perimeter 流域周长;流域范围
basin plain 盆地平原
basin planning 流域规划
basin plug 洗手盆排水栓
basin precipitation 流域降水量
basin range 断块山岭;断块盆岭;不连续山脉
basin-range landform 盆岭(相间)地貌
basin range structure 断块山岭构造;盆地山脉构造;断块构造;盆一岭构造
basin recession 流域退水
basin recharge 流域再补给(量);流域补给
basin-ridge system 洋盆一洋脊体系
basin room 洗餐具室
basin sewage 流域排水工程;流域下水道
basin shape 流域形状;盆地形状;盆地的形态
basin shape crucible 半圆形坩埚
basin-shaped strata 盆状层【地】
basin shape factor 流域形状系数
basins in range 排列着的盥洗盆;排列着的洗手盆
basin siol cultivation 盆地土壤栽培
basin slope 流域坡降;流域坡度
basin soil 盆地土壤
basin stand 脸盆架
basin storage 流域蓄水量
basin structure 盆地构造
basin-structured 盆形构造的
basin swamp 湖泰沼泽
basin tilting furance 半圆形炉缸倾转炉
basin topography 流域地形(学)
basin trial 系泊试验
basin valley 河谷盆地
basin(wall) block 池壁砖
basin water environment 流域水环境
basin water environment protection 流域水环境保护
basin water quality target management 流域水质目标管理
basin-wide 流域规模的;全流域的
basin-wide hydrological regime 全流域性水情
basin-wide program(me) 流域规划;整体流域规划;(整体的)流域计划
basin-wide water quantity and quality planning 广域流域水量水质规划
basin wrench 螺丝扳手;螺母扳手
basipodite 基节
basis 基底;底;玻基
basis box 基准箱
basis brasses 铜基合金
basis circuit matrix 基回路矩阵
basis contract 基本合同
basis cut matrix 基础割矩阵
basis for application of overhead 制造费用分配基础
basis for dating fault 断层时代确定依据
basis for dating fold 褶皱时代确定依据
basis for depreciation 折旧基础
basis for determining relative displacement of fault 确定相对位移的依据
basis for dividing active stages 活动分期依据
basis for establishing 规定的基础
basis for invoicing 会计学基础;开发票的根据;货单依据
basis for recording sales 销售计价基准
basis for rectification 纠正底图
basis function 基底函数
basis incidence matrix 基关联矩阵
basis ionization 基始电离电流
basis material 基础材料
basis matrix 基阵;基底矩阵
basis measurement 基本体积吨;丈量标准
basis meridian 本初子午线;首子午线
basis metal 基体金属
basis of allocation 分摊基数;分摊基础
basis of apportionment 分摊基础;分配基础
basis of assessment 课征基准;课税根据;课税对象;课税标准;摊额基准
basis of a vector module 矢量模的基底
basis of calculation 计算标准
basis of coal analysis 煤分析基准
basis of compilation 编绘底图
basis of credit 信用基础

basis of design 设计的基本原则;设计依据
basis of determination 确定依据
basis of distribution 分配基础
basis of hydro-geochemical division 水文地球化学分区依据
basis of integers 整数基底
basis of payment 支付的基础;付款方式
basis of price 买卖价格
basis of quotation 行市标准
basis of reference 参照基准
basis of satisfactory wages 满意工资基础
basis of valuation 估价标准
basis order 指定差距跨期单
basis period tax 课税基期
basis rate swap 基本利率互换交易
basis reserve 基本准备金
basis ship 母型船
basis soap 基皂
basis sulfate white lead 硫酸盐铅白
basis triangulation network 国家三角网
basis vector 基底向量
basis wages 基本工资
basis weight 基本重量
basis weight of paper 纸张重量
basite 粗麻布沥青防潮层;基性岩类
basked plant 悬篮植物
basked planting 悬篮栽培
basket 挂篮;笼形信号球;漏板纱网;篓;蓝(筐);转鼓;中间罐;吊篮;单管岩芯筒;触媒筐;挖斗机铲
basket arch 三心拱;花篮拱(形同花篮提手)
basket arm 捕集器拖臂
basketball 篮球
basketball court 篮球场
basketball field 篮球场
basketball nets 篮球网
basketball stands 篮球架
basket barrel 带岩芯爪的岩芯管;打捞筒【岩】
basket bent 篮状排架
basket bit 带取样筒钻头;齿状钻头
basket boom 铲斗吊杆
basket bottom 转鼓底
basket braid 方平组织饰带
basket capital 篮状柱头;花篮状柱头
basket central rate 一揽子中心汇率
basket centrifuge 篮式离心机;筐式离心机;过滤式离心机
basket-chair 柳条椅
basket charging 炉顶装料;料篮装料
basket coil 笼形线圈;篮形线圈;篮形绕组
basket container 篮状种植器
basket core 捞抓土样
basket(core) lifter 爪簧式岩芯提断器
basket crib 篮式排架;篮式木笼;叠木框架;木笼
basket curb (混凝土桥墩的)木笼护墙框架
basket dam 篮形充填坝
basket derrick 吊篮式起重机
basket drier 篮式干燥器;篮式干燥机
basket evapo(u)rator 悬篮式蒸发器
basket filter 篮式(过)滤器;筐式过滤器
basket for centrifuge 离心机转筒;离心机筛篮
basket grate 篮式栅;筐式炉栅
basket guard (灯的)篮状护罩
basket handle 篮状扶柄;蓝型扶柄
basket handle arch 篮筐拱;椭圆拱;三心拱
basket hitch 篮式套;吊篮索结;吊篮悬挂装置
basket hub 转鼓毂
basket jig 动篮式跳汰机
basket junk 捞爪;打捞蓝
basket layering 篮压
basket lid 棉箱顶盖
basket lip 转鼓唇缘
basket loan 灵活贷款;一篮子贷款;一揽子贷款
basket mast 笼形桅;格子桅
basket of currencies 一揽子货币
basket-of-eggs topography 雁列丘地形
basket of exchange rate 一揽子汇率
basket pot 篮罐
basket pouring 中间罐浇注;中间包浇注
basket purchase 一揽子购买;一揽子采购;一次总购;综合采购;整套购买;整批购买
basket rod 柳条
basketry 编篮细工
basket signal 篮状旗标
basket skirt 转鼓裙

basket strainer 莲蓬头;篮式滤网;篮式粗滤器
basket strainer wastes 花篮渣(污水池中固体废物)
basket tube 打捞爪(打捞小物件用)
basket tube combustion chamber 联筒式燃烧室
basket type 提篮式
basket type centrifugal separator 底部卸料离心机
basket type centrifuge 底部卸料离心机
basket-type core barrel 简单岩芯筒;简单岩芯管
basket type core lifter 篮状岩芯提取器
basket type evapo(u)rator 篮式蒸发器
basket type sampler 笼式采样器
basket type trash rack 笼形拦污栅;围笼式拦污栅
basket ware 网篮式格子体
basket(wave) coil 篮式线圈;篮形线圈
basket weave 席纹织物;席纹图案;粗厚方平织物(质地疏松,做家具装饰布、窗帘等);编篮纹装饰;方平组织
basket weave bond 席纹砌合;席纹砌法;棋盘式砌合
basket weave checkers 编篮式格子砖砌体
basket weave mosaic 席纹马赛克
basket weave packing 编篮式码法
basket weave parquetry flooring 席纹地板
basket weave pattern 编篮纹;席纹(图案)
basket weave pattern brick 席纹铺砖
basket weave pattern brick wall 席纹砖墙;席纹图案的磨砖对缝墙
basket-wheel excavator 斗轮式挖土机
basket winding 笼形绕组;链形绕组;篮形绕组;篮形绕线;篮式线圈
basket with spring steel fingers 带弹簧片的岩芯抓
basket work 编制;柳条编织品;编织物
basket work panel 编制面板
basket-work pattern 席纹图案
basofor 沉淀硫酸钡
basoid 碱性胶液
Basolan chrome dye 巴佐兰铬媒染料
Basolan dye 巴佐兰染料
basonym 有效名
basoolah 短柄斧(一种东方国家独有的斧头)
basophilla 嗜碱性
basophilous plant 好碱植物
basque 炉缸内衬
basquet 桶;蓝
Basrah 巴士拉地毯
bas-relief 半浮雕;底浮雕;浅浮雕
bas-relief frieze 浅浮雕带
bass 夹煤炭质页岩;黏土岩;硬黏土;椴属树木;椴木;低音;贝司;巴斯页岩
bassanite 烧石膏
bass boost 低音增益;低频音增强电路
bass-broom 韧皮纤维扫帚;级木扫帚
bass compensation 低频音补偿
bass control 低频音控制
basse-court 低级法院(英国);房屋后院;农家养禽场地;城堡外庭后面的院子
basse-taille 浅浮雕
basse-taille enamel 浮雕珐琅;浅浮雕珐琅
bassetite 铁铀云母
basset 岩层露头;(地层的)露头;地层露头部分;出露
basso 电解凹印版
bassoon 大管;巴松管
bassora gum 黄蓍胶;刺槐树脂;巴索拉树胶
bassoran 巴索兰
basso-relief 浅浮雕
basso-relievo 浅浮雕;浅浮雕槽;浅浮雕;半浮雕
bass reflex 低声频反射
bass reflex baffle 低频反音匣
bass response 低频响应
Bass Strait 巴斯海峡
basstonite 棕黑蛭石
basswood 级木树;级木材;美洲椴;菩提树;椴木
basswood oil 椴树油
Basszonoff's reagent 贝斯聪诺夫试剂
bast 韧皮;碳质黏土
bas-taille 浅浮雕
bastard 劣质(砂石);坚硬巨砾;假货;粗纹;粗齿;非标准的
bastard acacia 刺槐
bastard amber 杂质琥珀
bastard asbestos 变种石棉
bastard ashlar 贴面薄石板;毛石墙面琢石;琢面毛

bastard box 石(墙);粗琢石;粗饰琢石
bastard box 非标准母扣
bastard cedar 北美崖柏;香肖楠
bastard cement 粗粒水泥
bastard coal 劣质煤;煤矸;硬煤
bastard connection 非标准接头
bastard cop 中型管纱
bastard cut 粗切削
bastard-cut file 粗齿锉
bastard element 非标准件
bastard equaling file 粗齿扁平头锉
bastard fallow 短期休耕地
bastard file 毛锉;半圆锉;粗齿锉
bastard flat rasp 粗扁木锉
bastard flat wood rasp 粗扁木锉
bastard free stone 劣质毛石;劣质建筑毛石
bastard ganister 粗纹致密硅岩
bastard-grain shingle 不合标准的木纹瓦
bastard granite 片状花岗岩;片麻岩;假花岗岩;片麻状花岗岩
bastard halfround file 粗齿半圆木锉
bastard joint 假缝;瞎缝
bastard knife file 粗齿刀锉
bastard lumber 弦面材
bastard machine tool 联合机床
bastard masonry 仿琢石墙;粗琢石;乱石圬工;混杂圬工;琢面毛石(墙);石片面墙圬工;毛石块面层墙
bastard mortar 硬质块砂浆
bastard oak 美国栎
bastard pointing 假勒缝;假勾缝;粗嵌缝;凸勾缝
bastard quartz (岩石中的)石英巨砾
bastard rasp 劣质粗锉
bastard round straight file 粗齿直圆锉
bastard round wood rasp 粗圆木锉
bastard sawed 弦锯材
bastard(sawed)board 弦锯板;粗锯板
bastard sawed lumber 弦锯材
bastard sawing 粗锯;顺纹锯木;顺纹锯开;顺纺锯木;弦锯
bastard sawn 粗锯木料;粗锯材;弦锯材
bastard sawn lumber 弦锯材
bastard size 非标准尺寸
bastard size paper 等外品纸张
bastard spruce 美国松
bastard square blunt file 粗齿直方锉
bastard stucco 毛灰泥粉刷;粗粒水泥拉粉刷;粗粒水泥粉刷;粗拉毛水泥粉刷
bastard thread 粗螺纹;非标准螺纹
bastard trass mortar 混杂的火山灰水泥砂浆;粗面凝灰岩灰浆
bastard triangular file 粗三角锉
bastard tuck pointing 粗嵌灰缝;简便勾凸缝;凸勾缝
bast-bass 级木绳索;级木韧皮;椴木韧皮
baste 粗缝;衍缝
bastel house 设防农舍;宝塔建筑(局部设防御工事,底层常为穹结构)
Basteur effect 巴斯德效应
bast-fiber 麻类纤维;韧皮纤维
Bastianelli's method 巴斯蒂阿内利氏法
bastide (几何平面的)防御性房屋;小住宅(法国南部的农村)
bastil(l)e 城镇的防御高塔;巴士底狱;城堡;堡塔
bastil(l)e house 宝塔建筑(局部设防御工事,底层常为穹结构);堡塔建筑
bastion 阵地工事;设防地区;棱堡;堡垒
bastiste 细薄毛织物;细麻布
bastite 绢石
bastle 设防农舍
bastle house (苏格兰边境的)设防房屋
bast mat 韧皮坐垫
bastnaesite 氟碳铈矿
bastnaesite ore 氟碳铈矿矿石
ba(s)ton 半圆线脚;柱座盘;椭圆环;圆环;柱脚圆盘线脚;环状半圆线脚饰;座盘饰凸圆线脚
bastose 木质纤维素;黄麻的纤维素
bast pulling 拔麻纱
B. A. Sulin classification B. A. 苏林分类
baswood 美洲椴木
bat 泥塞;砖头;断头砖;短棒;封口砖;半砖;铅楔;棚板;推敲;湿土块
Bataan mahogany 红柳安木(菲律宾)
Batac jig 筛下气室跳汰机

Batalbra 巴塔尔布拉铜基合金
batalum getter 钡钽消气剂
batardeau 堤;围堰;围堤;坝
batatic acid 巴他酸
Batavia dammar 巴达维亚达玛(树)脂
bat batten 夹板;夹条;板条
bat bolt 棘螺栓
batch 炉批;每批容量;一批;一炉;一份;一拌(混凝土);装炉量;分批拌;批拌;批量;批次;生产批
batch accumulator 累计
batch aeration 间歇曝气
batch agglomeration 配合料结块;配合料成团
batch agitator 间歇式搅拌器;间歇式搅拌机;间歇式拨火板;分批配料搅拌器;分批搅拌器;分批搅拌机
batch analysis 分批试验
batch and interactive network 成批交互式处理网;成批和交互式处理网
batch annealing 箱式炉退火;分批退火
batch area 成批区域
batch asphalt mixing plant 间歇式沥青混合料搅拌设备
batch asphalt plant 间歇式沥青混凝土拌和设备
batch ball mill 间歇式球磨机
batch barrow 配合料车
batch bin 料槽;分批箱;分批投配斗;配料箱;配料舱;配料仓
batch blanket 配合料薄层;生料层
batch bleaching 分批漂白
batch blender 分配拌和机;间歇式拌和机;分批搅拌机;分批混合机;分批拌和机
batch blending 调整配方
batch blending control 成批混合控制
batch blending silo 间歇式搅拌库
batch box 投ުു;量箱;量斗;计量桶;混凝土配料箱;分批箱;分批投配斗;按体积配料器;配料量斗
batch briquet 配合料块
batch briquetting 配合料压块
batch bucket 配合料仓斗;配合料斗
batch budgeting 分批预算
batch-bulk 成批
batch by batch and stage by stage 分批分期
batch cake 料团
batch caking 配合料结块
batch calculation 配料计算
batch can 料罐;配合料罐
batch canister 料罐
batch capacity 搅拌机容量;分批容量
batch car 配合料车
batch card reader 成批卡阅读器
batch carry-over 飞料;配合料飞散
batch cart 配料车;运料车
batch centrifuge 间歇式离心机
batch change 配合料组成变化
batch changing into melt rate 熔成率
batch charger 加料器;加料工;投料机
batch charging 投料
batch charging end 投料端
batch checking 成批检查
batch coke still 分批焦化蒸馏器
batch command 成批命令
batch comminution 配料渐减
batch component 配合料组分
batch composition 泥料组分;配料成分;配合料组成
batch container 配料容器;配合料容器
batch control 间歇控制;分组控制;批量控制
batch controller 分批控制器
batch control sampling 批量控制采样
batch control total 分批控制总数
batch conveyer 仓式输送器
batch conveyor belt 配料运送带
batch cost 分批成本
batch costing 整批成本计算法;分批摊派成本法;分批成本计算法
batch costing method 分批成本计算法
batch counter 拌和计数器;配料记录器;配料计数器
batch counting 配料记录
batch crystallization 分批结晶
batch crystallizer 分批结晶器
batch cultivation 分批培养
batch culture 分批培养
batch culture medium 间歇培养基
batch cure 分批硫化;分批烘焙

batch cycle 间歇循环;断续循环;分批循环
batch cycle vacuum filter 分批循环真空过滤器
batch data exchange 成批数据交换
batch data exchange services 成批数据交换服务程序
batch data processing 成批数据处理;分批数据处理
batch demineralization 间歇式脱矿质;分批除盐
batch deodorizer 分批除臭机
batch depth 投料深度
batch desublimer 分批凝华器
batch dissolution 间歇溶解;分批溶解
batch dissolver 间歇溶解器;分批溶解器
batch distill 分批蒸馏器
batch distillation 分批蒸馏
batch dryer 分批干燥器;分批(式)干燥机;间歇干燥器;干燥炉
batch duct 母线导管
batch dust 配合料粉尘
batch dyeing 分批染色
batched aggregate 配制拌和的集料;配制拌和的骨料
batched communication 成批通信
batched job 成批作业
batched job processing 成批作业处理
batched layer 料层
batched materials 配制拌和的材料
batched question 成批提问
batched telecommunication 成批远程通信
batched water 搅拌水;一次拌和用水(混凝土);(混凝土的)拌和用水;一次拌量所需要的水
batched water volume 拌和水量
batch elevator 配料升降机
batch elutriation 分批陶析
batch end point 管线中石油产品分批点
batch-entry mode 成批输入方式
batcher 料斗;量配器;卷布机;进料量斗;混凝土分批搅拌机;给料机;定量器;分批箱;分批量具;分批加料器;分隔塞;拌和加料器;拌和机的配料仓;配料箱;配料量斗;配料计量器;配料称量器;送料计量器
batcher box 重量容器
batcher bucket 重量容器
batcher hopper 装载漏斗;装料斗;重量容器;定量装料仓
batcher installation 配料器装置
batcher mixture scale 自动式混料计量斗
batcher plant 混凝土搅拌厂;分批投配设备;分批加功装置
batcher plant barge 浮式搅拌楼;水上拌和厂
batcher scale 配料秤;进料量斗;自动式计量器
batcher weigher 间歇式计量设备
batch execution 成批运行
batch extraction 间歇萃取;分批提取;分批萃取
batch extruder 一次投料压出机
batch fed 分批给料的
batch-fed incinerator 间歇进料焚化炉
batch fed pugmill 间歇供料的叶片拌和机;分批给料叶片式搅拌机
batch feed 间歇喂料;间歇投料;间歇进料;批量喂料
batch feeder 料箱;料仓;给料器;配料;投料机
batch feeding end 池窑加料端;投料端
batch fermenter 间歇式发酵池
batch file 成批文件
batch filer press 间歇式压滤机
batch filler 玻璃加料工
batch filling 投料
batch filling end 池窑加料端
batch filter 间歇式滤池;间歇式过滤器;分批式过滤器;分批式过滤机
batch filtration 间歇式过滤;间歇式过滤作用;间歇过滤(法);分批式(过)滤
batch finishing 分批整理
batch firing system 间歇式燃烧系统
batch flo(a)tation 开路浮选;间歇浮选;分批浮选
batch floating 分批浮选
batch(-flow)production 成批生产
batch-flush valve 间歇冲洗阀
batch flux curve 分批通量曲线
batch flying 飞料
batch foam ion flo(a)tation technique 间歇泡沫离子浮选法
batch formula 料方;配料公式;配合料配方;配方
batch-free 生料熔尽的;玻璃液中无不熔物;玻璃生料熔尽的

batch-free time 配合料熔尽时间
batch furnace 分批炉；分层式烘炉
batch gas 配合料气体
batch gravity sludge thickener 间歇式重力污泥浓缩池
batch grinding 断续研磨；分批磨矿；分批粉磨
batch gun 喷浆枪
batch handling 配合料储存；配合料运输；分批处理（法）；配合料装卸
batch hauling 间歇处理
batch-header document 成批首部文件
batch heater 分批加热器；配料加热器（混凝土集料）
batch holding bin 储存配料的料仓；配料储仓
batch homogenization 间歇式均化；分批均化
batch hopper 装料漏斗；配料箱；配料斗；配合料斗
batch house 料房；配料房；配料车间
batching 配制拌和物；选择混凝土配合比；分配输送；卷取；卷布；剂量；计量；分组；分批（投配）；分批数据处理；配料；成批
batching and mixing plant 配料及搅拌设备
batching apparatus 卷取装置；配料装置
batching asphalt plant 间歇式沥青制备设备；间歇式沥青混合料搅拌设备
batching bin 材料储[贮]仓
batching box 配料箱
batching by conveyor belt 用运输带配料
batching by volume 按体积投配；按体积配料；按体积配合；容积配料法；体积配合法
batching by weight 重量配料（法）；重量配；按重量投配；按重量配料（法）；体积配料法
batching cart 配料轻便车
batching container 配料容器
batching conveyor belt 配料运输带
batching conveyor belt scale 运输带配料用的秤
batching counter 选组计数器；计量计数器
batching cycle 配料周期
batching department 配料部门
batching desk 配料台
batching drum 配料圆筒
batching equipment 分配配料装置；分批配料设备；分批配料秤重装置；分批配料秤重设备
batching equipment 分批投配设备；配料设备
batching frame 配料箱；配料架
batching go-devil 分隔塞
batching hopper 批量喂料斗；配料喂料斗
batching hygrometer 配料湿度表
batching lock 配料锁定器
batching machine 配料计量器
batching mixer 拌和机
batching of body and glaze 坯釉配方
batching out unit 排胶装置
batching pig 分隔塞
batching plant 混凝土搅拌厂；分批投配设备；分批配料装置；分批配料设备；分批配料秤重装置；分批配料秤重设备；拌和厂（混凝土）；搅拌设备；搅拌站；搅拌工厂；拌和楼（混凝土）；配料站；配料工厂；配料场
batching plug 分隔塞
batching redient 配料组分
batching roller 卷布辊
batching screw conveyer 配料螺旋计量输送机；配料螺旋计量运输机
batching section 配料部门
batching selector 配料选择器
batching set-up 配料装置
batching silo 定量配料仓；配料筒仓
batching sphere 充气硬橡胶球
batching star 配料星号
batching tower 配料塔楼
batching trough conveyer 配料槽输送机
batching unit 配料装置
batching water quality monitoring 间歇水质监测
batching weigher 分批配料装置；分批配料设备
batching weigh gear 重量配料装置
batching worm conveyer 螺旋式运输机配料
batch initiation 成批初启
batch input reader 成批输入读出器
Batchinsky relation 巴钦斯基关系
batch ion exchange 间歇离子交换；周期离子交换
batch irradiation 分批辐照
batch job 成批作业；分批作业
batch job foreground partition 成批作业前台区

batch job queue 成批作业等待序列
batch kiln 间歇式窑；间歇式加热炉；分批处理炉；分层式烘炉
batch kinetic experiment 间歇动力学试验
batch kinetic test 间歇动力学试验
batch kneader 批量搅拌机
batch layer 料层；配合料层
batch leaching 分批浸出
batch lease real estate 批租地产
batch-leasing landed property 批租地产
batch level controller 料位控制器
batch loader 分批装料机
batch loader unit 配料装料机提升装置；配料装料装置
batch loading 分批装料
batch lorry 配料运货卡车；配料车
batch lorry compartment 配料卡车间隔
batch lump 配合料块；料团；配合料块
batch manufacturing 成批制造
batch material 批料；配合料
batch melt-out line 池窑泡界线
batch message process(ing) program(me) 成批文电处理程序；成批报文处理程序
batchmeter 混凝土配料计量器；分配计；分批称量器；量箱；量料斗；量斗；计量给料器；给料漏斗；给料量斗；定量给料器；分批量料器；分批计（量器）；分批给料器
batch-meter control 分批配料控制
batch-meter controller 分批配料控制器
batch method 批处理法
batch method of operation 分批作业法
batch method of treatment 分批处理法
batch mixer 间歇式拌和机；间歇混合器；混料机；周期式拌和机；分批搅拌机；分批混合器；分批拌和机；批次式搅拌机
batch mixing 配料的断续拌和；间歇混合；分批混合；配合料混合
batch mixing plant 配料断续拌和厂
batch-mixing system 批料混合系统
batch mixing time 配料断续拌和时间；配料断续时间
batch mix(ture) 分批混合（物）；分批拌和
batch-mode data processing 成批式数据处理
batch monitor 分组方式监督程序
batch mortar mixer 间歇式灰浆搅拌机
batch number 产品批号；批号；批次号
batch of concrete 分批混凝土；分批拌和混凝土
batch of mortar 分批砂浆；分批拌和砂浆
batch of products for inspection 交验批
batch of treating agent 处理剂配比
batch oil 翻砂用油；绳缆用油
batch-operated computer system 分批作业计算机系统
batch operating system 成批操作系统
batch operation 间歇操作；间歇法；间歇操作；间歇运转；分批式操作；分批生产；分批操作；分配操作；配料作业
batch operation method 分批作业法
batch oven 分批热处理炉
batch paddle mixer 配料桨式拌和机
batch patenting 配料专利
batch(pebble) mill 间歇操作磨
batch pellet 配合料球粒；配合料粒
batch pelletizing 配合料造粒；配合料粒化
batch pile 料堆（在玻璃面上的未熔化的料）；配合料料团；料团；材料储备（量）；分批料堆；分批堆料；配合料杯
batch plant 搅拌设备；搅拌站；搅拌厂（混凝土）；分配配料装置；间歇式（混凝土）搅拌厂；拌和楼（混凝土）；配料车间；配料厂
batch-plush valve 定量冲洗阀
batch pneumatically 气动投料
batch polymerization 间断聚合；分批聚合
batch powder 配合料粉
batch precipitation 分批沉淀
batch preparation plant 配料车间
batch pricing 整批定价法
batch print function 成批打印功能
batch process 间歇过程；间歇法；间歇操作法；分批生产；批量制法；批量生产
batch processing 成批加工；成批处理；分批处理；批处理；批量处理；程组处理；分批加工；分批法
batch process(ing) interrupt 程序组处理中断；成批处理中断

batch process(ing) mode 成批处理方式；分批处理方式；批量处理方式
batch process(ing) monitor 成批处理监督程序
batch process(ing) system 成组处理系统；成批处理系统；批处理系统；批量处理系统
batch process(ing) technic 批处理技术
batch process(ing) terminal 成批处理终端；批处理终端
batch processor 分批显影机
batch production 连续生产；间歇生产；成批生产；分批（式）生产；批量生产
batch production process 分批生产法
batch program(me) 成批程序
batch pugmill mixing 卧式筒叶拌和机拌和配料
batch purification 间歇式净化；分批提纯；分批式精制；分批净化
batch pusher 玻璃加料工
batch quantity 每批配料量；每盘配料量；每盘混凝土数量；批量
batch quantity analysis 批量分析
batch query 成批质问；成批询问
batch queue 成批排队
batch reactor 分配反应器；间歇反应器；间歇反应器；分批处理反应器
batch record 配料记录
batch recorder 配料记录器；拌和记录器
batch recovery compressor 间断回收压缩机
batch rectification 分批精馏
batch reduction 配料缩减
batch refuse incinerator 分批垃圾焚化法
batch roller 卷取辊；分批辊轧机
batch run 成批运行
batch sample 试样（道路材料）；一次拌和量；批量试样
batch sampling 分批进样
batch sand mixer 分批式混砂机
batch save-restore 成批保留恢复
batch scale 分批称量秤
batch scheduler 成批调度程序
batch scouring 分批煮练
batch scraper 刮料器
batch segregation 配合料分层
batch session 成批对话
batch settling 分批沉淀
batch settling flux 分批沉降通量；间歇沉降通量
batch sheet 配料单
batch silo 配合料料仓
batch sintering 间歇烧结；分批烧结
batch size 批量
batch sizebulk 批量
batch sludge thickener 间歇式污泥浓缩池
batch steamer 分批蒸煮器
batch still 分批蒸馏器
batch stoker 批量加煤机
batch stone 配合料结石
batch stream 分批流量
batch system 间歇式系统；成批系统；分批系统；分批式系统
batch table 成批表；配料表
batch task 分批任务
batch terminal 成批处理终端
batch terminal controller 成批终端控制器
batch test 配料试验；成批试验；分批试验
batch timing device 定时秤料装置
batch-to-batch standard deviation 批间标准偏差
batch total 选组总数；程序组总计；成批总数；分批总数；分类总数
batch tower 搅拌塔；塔式配合料仓
batch transaction file 分批事务文件
batch treatment 间歇处理；分批处理
batch treatment method 分批处理法
batch treatment of wastes 废物分批处理；废料分批处理
batch truck 移动料仓；移动式料仓；分批运料车；拌和卡车；批运料车；配料车
batch turning 倒料
batch-type 分批式；分批的；间歇式；拌和式
batch-type asphalt plant 分批拌和式沥青（混合料）厂
batch-type concrete pugmill(mixer) 分批配料式混凝土筒叶拌和机
batch-type condensed milk cooler 分批式炼乳冷却器
batch-type controller 定量型控制器

batch-type counter current rinsing 间歇式逆流清洗
batch-type drying oven 间歇式烘炉;周期式烘干炉
batch-type dust destructor 分批式垃圾焚化炉;间歇式垃圾焚化炉
batch-type freezer 间歇式冻结器
batch-type furnace 间歇式加热炉;分层式烘炉;室式炉
batch-type heater 间歇式加热炉
batch(-type) mill 间歇式磨(碎)机;分批装料磨矿机
batch-type mixer 间歇式搅拌机;分批式配料和机
batch-type muller 间隙式混砂机
batch-type paving plant 分批式配料铺路拌和装置;分批式配料铺路和设备;分批式配料铺路拌和机;分批式配料铺路拌和机
batch-type production 分批式生产;分批混合式生产
batch-type pugmill mixer 分批式捣捏拌机;分批式叶片搅拌机
batch-type raw mix homogenization silo 间歇式生料均化库
batch-type seed mixer 分批式拌种器
batch-type sintering 分批式烧结
batch-type vacuum filter 间歇式真空过滤机;分批式真空过滤机
batch unit 配料器
batch up gear 卷取装置
batch user 成批用户
batch valve 计量阀
batch vapo(u)rization 分批蒸发;分批汽化
batch variation 批变异
batch volume 配料体积
batch water 拌和水;批量用水
batch water meter 分批式水表
batch weigh blender 分批称量掺和机
batch weighed 分批过秤的
batch weigher 称量配料装置;分批秤重器;批次式配料秤;配料秤;配料量器
batch weighing 断续量重;分批量重;分批称重
batch weighing equipment 分批量重设备
batch weighing feeder 称量喂料机
batch weighing plant 分批称重配料装置
batch weighing scale 配料秤
batch weight 批料重量;每批重量;每拌毛重;分批重量;批量
batch weight method of mo(u)lding 分批装模压型法
batch wetting 配合料润湿
batchwise 分批的;分批发酵法
batchwise dyeing 分批染色
batchwise homogenizing 分批均化
batchwise operation 分批操作
batchwise polymerization 分批聚合
batch word queue 分批字队列
batch zone refiner 分批区域精炼炉
bate 假节埋;脱灰液
batea 晶体收集器;尖底淘金盘;淘砂盘;淘金木桶
bateau 搭浮桥用的船;浮舟;浮桥墩;平底小船;平底船
bateau bridge 浮桥
batement light 跛窗(随楼梯斜度的窗);斜窗;坡窗
baten path 惯例;常规;熟路
bate pits 脱灰;脱斑
bateque 泉水沉积
baterred wall 收分墙
Batesian mimicry 贝茨拟态
Bates interferometer 贝茨干涉仪
bate stains 脱斑
Bates wavefront shearing interferometer 贝茨波前剪切干涉仪
batete 红棕色硬木(菲律宾)
bat faggot 柴排;粗枝捆;柴束;柴捆
bath 浴室;浴器;浴槽;池
bath accessory 室内游泳池;浴池附件
bath analysis 炉中分析;炉前分析;熔池分析
bat handle 球棒形手把
bat-handle switch 手柄开关;铰链式开关;棒球柄开关
bath and W.C. 浴室及厕所
bath brick 巴斯磨砖;(用于磨刀或打磨金属面的)砂砖
bath cabinet 浴箱
bath carburizing 液体渗碳
bath casing 池炉壳体
bath chair 行动椅;有篷轮椅

bath closet 沐浴更衣室;洗浴更衣室
bath-cock 浴用龙头;澡盆旋塞;澡盆龙头
bath component 熔体组分
bath composition 电解液成分;熔体组成;熔体成分
bath control 电解槽控制
bath delivery stacker 槽式堆叠收纸机
bath enamel 浴缸瓷漆
bath faucet 浴池水龙头
bath-fed tube 凝固浴管
bath filter 空气滤油器;浸油式空气滤清器
bath grip 浴盆上抓手
bathhouse 浴室;公共浴室
bathile 深湖底(25米以下)
bathing beach 海水浴场;海滨浴场;游泳场
bathing box 更衣棚
bathing laundry water 洗浴水洗衣水
bathing load (公共浴室的)浴负荷
bathing machine 活动更衣车;更衣车;活动更衣室
bathing place 海滨浴场;浴场
bathing-pool 游泳池
bathing process waste 鞣化理废物
bathing shed 沙滩泳屋;沙滩泳屋
bathing tub 浴盆
bathing waste(water) 洗澡污水;洗浴废水;洗澡废水
bathing water 洗澡水;洗浴水;洗涤用水
bathing water treatment 洗澡水处理
bath-laundry water 洗浴—洗衣粉;洗浴水洗衣水
bath life 电解槽寿命
bath lubrication 油浴润滑
Bath metal 巴思脆性白铜;电镀槽用金属
bathochromatic shift 向河移
bathochrome 向红(基)团;深色团
bathochromic effect 向红(增色)效应
bathochromic shift 向红移
bath of acceleration 促进剂溶液
bath of cell 电解槽熔体
bath of glass 玻璃水;玻璃池
bath of molten solder 熔焊锡浴槽
bath of molten zinc 熔锌池
batholite 岩基;岩盘
batholith 岩盘;岩基
batholith hypothesis of mineralization 岩基成矿说
batholithic granite 岩基花岗岩
batholith wall 岩基壁
batholyth 岩基
bathometer 测深器;测深计;测距计;水深测量器;深海测深仪
bathometric(al) chart 海洋深度图
Bathonian 巴通阶
Bathonian series 巴通统
bathothermograph 深度自记温度计
bath plug vortex 飓风涡
bath resistance 电解质电阻
bathroclase 水平节理
bath roof 池炉顶
bathroom 浴室
bathroom accessories 浴室附属设备;浴室附属件
bathroom and lavatory unit 成套的浴室及盥洗室单元
bathroom article 浴室用物品
bathroom building block module 规格化浴室单元
bathroom cabinet 浴室用品箱;(医药、梳洗、化妆品的)储[贮]藏柜;洗澡间
bathroom closet 浴室壁橱
bathroom curtain 浴室窗帘
bathroom door 浴室门
bathroom equipment 浴室设备
bathroom fitments 浴室装饰小五金件
bathroom fittings 浴室用具;浴室配件
bathroom fixtures 浴室内装置
bathroom gull(e)y 浴室排水沟
bathroom heater 浴室供暖器
bathroom heating 浴室取暖
bathroom heating appliance 浴室供热设备
bathroom hook 浴室挂钩
bathroom installation 浴室设备
bathroom layout 浴室布置
bathroom(lighting) fixture 浴室照明装置
bathroom product 浴室用品
bath(room) trap 浴室汽水分隔器
bathroom warning 浴室取暖
bath sample 电解质试样;熔体试样
bath shower 淋浴

bath solution 电镀槽液
Bath stone 石灰岩石;奶色石;巴斯石;巴思石;巴思鲕石
bath surface 熔体表面;熔池液面
bath towel holder 浴巾架
bath trap 浴盆弯管;浴室汽水分离器;浴盆存水弯
bathtub 机下浴缸形突出物;浴盆;浴缸
bath tub cock 浴盆塞字
bathtub construction 槽形结构
bathtub curve 浴缸形曲线;浴盆式曲线;故障率曲线
bath tube 浴用管子
bathtub faucet 浴缸龙头
bathtub hand grip 浴缸手握把
bathtub height 浴缸高度
bathtub lead trap 浴缸铅隔汽具
bathtub ledge 浴缸突出部分;浴缸边缘
bath tub plug 浴缸排水塞子
bathtub rim 浴缸凸边缘
bathtub/shower installation 浴缸或淋浴装置
bathtub truck 浴缸式卡车(运水泥混凝土用)
bathtub type 槽形
bathtub-type truck 浴盆式卡车(运水泥混凝土用)
bathvillite 黄褐块碳
bath voltage 电解槽电压;电镀槽电压;槽电压
bath waste 洗澡污水
bath water 洗澡水
bath water heater 大型(浴用)热水器
bath water treatment 洗澡水处理
bath weir 槽沟扫堰;槽沟挡块
bathwillite 暗光琥珀
bathyal 半深海底;半深海(的)
bathyal belt 半深海带
bathyal deposit 次深海沉积;半深海沉积(物)
bathyal deposition 半深海沉积作用
bathyal environment 次深海环境;半深海环境
bathyal facies 次深海相;半深海相
bathyal region 半深海区
bathyal region of lake 湖泊深水区;深水湖区
bathyal sediment 半深海沉积(物);深海沉积(物)
bathyal zone 半深海区;半深海带
bathybic 深海的
bathybic organism 深海层生物
bathyconductograph 海水电导仪;深度电导仪
bathycurrent 深处水流;海底暗流
bathyderm 深硅铝层;深部地壳变动
bathydermal 深壳的
bathydermal type of gravitative tectonic 深皮型重力构造
bathygenesis 陆沉运动;下降运动;造渊运动;沉降构造运动;负相运动;负构造运动
bathygram 海深图;测深自记图;水深图
bathygraph 自记测深仪
bathygraphic(al) chart 表示深浅的海洋图;海洋水深图;标示水深的海图
bathylic 半深海的
bathylimnetic organism 栖息湖底的生物
bathylite 岩基
bathylith 岩基
bathymeter 海洋测深器;海深测量仪;测深计;测深器;水深测量器;深海测深仪;深度计
bathymetric(al) 等深的;水深的;深海测量(学)的
bathymetric(al) and geotechnical survey 海底地形地质调查
bathymetric(al) biofacies 海深生物分相;深海生物相
bathymetric(al) chart 等海深线图;海洋水深图;海洋深度图;海底地形图;等深线图;等深海(底地形)图;水深图;世界大洋水深图
bathymetric(al) chart of the oceans 大洋深度图
bathymetric(al) contour 海底等深线;等深线
bathymetric(al) curve 等海深曲线
bathymetric(al) data 等深数据
bathymetric(al) detail 水深测量详细程度
bathymetric(al) line 等海深线;海洋生物深度分布线;等深线
bathymetric(al) low 海底低地
bathymetric(al) map 海底地貌图;等深线图
bathymetric(al) measure 水深测量
bathymetric(al) measurement 测深
bathymetric(al) navigation system 深海洋深导航系统
bathymetric(al) sounding system 测深系统
bathymetric(al) survey 测深;水下地形测量;水深

测量
bathymetric(al) transducer 测深换能器
bathymetry 海洋生物分布学;海洋测深学;海深测量法;测深学;测深法;深海测量法
bathyorographic(al) 大地高低分布的
bathyorographic(al) map 海底地形图;洋底地形图;水深地形图;水底地形图
bathyorographic(al) surveying 水下地形测量
bathypelagic 较深的深海区;海洋深处
bathypelagic belt 海洋深层带
bathypelagic fauna 海洋深层动物区系;深海动物群;深层动物区系
bathypelagic fishes 深海层鱼类
bathypelagic organism 深海生物;深海层生物
bathypelagic plankton 深海浮游生物
bathypelagic zone 海洋深层带;深海带;深层带
bathyphotometer 水下光度计;深水光度计
bathyphytia 低地植物群落;海洋植物群落
bathy plankton 深海层浮游生物
bathysalinometer 深海盐度计
bathyscaph(e) 探海艇;深潜水器;深海潜水器;深海勘探器
bathyscope 深海潜望镜
bathyseism 深震;深源地震;深海地震
bathysphere 海底观测球;测深球;球形潜水器;潜水球;探海球;(能载人和仪器供研究深海动物的)深海球形潜水器;深海潜水球
bathythemometer 深水温度计
bathythermogram 水温水深曲线;水温图;深海温度记录图
bathythermograph 海洋测温仪;海水温度计;温深仪;探水温器;深温仪;深温计;深度自记温度计;深度温度仪;深度温度计;海水深度温度自动记录仪;深海测温仪
bathythermograph recorder 海水温度深度自动记录仪
bathythermograph trace 温深记录曲线轨迹
bathythermograph winch 温深仪绞车
bathythermometer 海洋深水温度表
bathyvessel 深海潜水器
batice 井筒支架的倾斜
batik 蜡防印花
batik and waxprinting article 蜡染和拔染制品
batik fabrics 蜡染布
bating 井筒延深
bating material 脱灰剂
bating process wastewater 鞣化过程废水
bating wind 逆风
bat insulation 柔性毡垫状隔音层;柔性毡垫状保温层
batisite 硅钡钛石
Batnickoin 巴特尼科英铜镍铁合金
Batoan mahogangy 红柳安木(菲律宾)
bat of glass wool 玻璃棉毡片
batonet 接棍;系索棍
batons rompus 罗马建筑之字形线脚直段;(诺曼底或罗马式建筑风格的)波浪花饰中短直段凸圆线脚
batoon 覆墙窄条;挂瓦条;抹灰板条;拉结木板条;盖缝木板条
batoreometer 测厚计
bat printing 棒印法(陶瓷印坯装饰法)
batrachite 绿钙镁橄榄石
bat ribbon 帽带
bat rivet 锥头铆钉
bat's dung 夜明砂
batswing (排成扇形喷头的)煤气燃烧器
batswing burner 蝙蝠翼式灯
batt 黏质页岩;毛层;硬黏土;页岩沥青毡;页岩块;顶板下薄煤夹层;巴斯页岩;碳质页岩与煤线互层
batted surface 受击面;凿琢面
batted work 阔粗凿石
Battelle environmental evaluation system 巴特尔环境评价系统
battement light 跛窗
batten 警戒孔;木条(子);木板条;货舱壁护木;小圆材;小木条;小方材;狭板;压条;板条;外墙条板;凸条;头部桁条;同位穿孔;顺水条
batten and button 木板接合(法)
batten and corner trim 边角料;边角修整
batten-and-space bulkhead 格子舱壁
batten and space partition wall 格子分隔(墙)

batten bar 压条
Battenberg braid 巴藤贝克编带
batten board 板条芯胶合板;拼板心胶合板;夹芯板;条板蕊细木工板
batten ceiling 木铺板;木覆板;护舱板
batten check 同位穿孔检验;同位穿孔校验法
batten cleat 舱口楔耳
batten column 缀板柱;缀合柱
batten door 直板门;木栅门;板条门;(不带框的)拼板门;实拼门
batten down 封舱
battened column 缀板柱;双肢柱;缀合柱
battened grating 板条格栅
battened marconi sail 撑木条三角帆
battened panel 有压缝条的木墙板
battened partition 板条间壁;板条隔墙;板条隔断
battened plate column 缀合支柱
battened ply frames 板条分层框架
battened plywood 板条胶合板
battened sail 撑木条的帆
battened steel column 双肢钢柱
battened strut 缀合支柱;钉板条的支柱
battened wall 板条墙;板壁
batten ends (铺地板用的)短板条
batten fence 板条栅栏
batten floor 木条地板
batten flooring 狭木条楼面
battening 封舱;补隙板;用板条钉住;墙筋板条;砌木砖;钉板条
battening arrangement 固紧装置;封舱装置
battening bar 舱围压条
battening device 封舱装置
battening iron 舱围压条
battening wedge 封舱(木)楔
batten lighting fixture 用板条钉住照明装置
batten nails 板条钉
batten of roof truss 屋架水平撑条
batten plate 缀合板;缀板
batten plate column 空腹柱;缀合柱
batten roll 锥形接缝条(铅皮屋面);加固木条
batten roof cladding 狭条板屋面覆盖
batten seam 锥形压制缝;木条加固咬口(金属薄板表面);木楞式接缝;棒状折叠缝
batten seam metal 木楞式金属面
batten seam roofing 狭条板屋面;(金属板屋面的)棒状折叠缝式铺法
batten seat 压条架;板条座架
batten sheet piling 打条板桩
battens in the filling 填充料中木条
battens spaced for roof tiles 挂瓦条间距按瓦材尺寸决定
batten strip 撑条;压条;挂瓦条;板条
Batten system 同位穿孔(检索)系统
batten underdrain 木板暗沟
batten wall 倾斜墙壁
batten wall partition(ing) 板条隔断;板条隔墙
batter 连续打击;锤打;垂向倾斜;侧脚【建】;边坡;凹印;倾斜度;倾度;敲碎;墙身收分;坡度;收孔;收分(指墙)
batter board 龙门板;定斜度板;定斜板;槽板(施工放线用);施工放线槽板
batter brace 斜顶撑(隧洞);对角支撑;斜撑;桁架斜撑
batter chamber wall 斜面闸室墙
batter-curved wall 倾斜墙
batter drainage 斜向排水沟;斜水沟;斜坡排水
battered bank system 倾斜护岸工程
battered joint 钢轨低接头
battered joint of rail 低接头
battered masonry wall 倾斜圬工墙
battered pilaster 斜面壁柱
battered pile 斜桩
battered post 斜面柱;锥形变截面柱
battered upstream face 上游倾斜面;堰体上游倾斜面
battered wall 收分墙;挡土墙;抹灰板条墙;斜墙;斜面墙;倾斜墙
batter finder 倾斜仪
batterfly clack 蝶形瓣
batter ga(u)ge 斜坡样板;定斜规;土堤斜坡样板
battering 收分;倾斜;收孔
battering level 测斜器
battering ram 横向撞锤;斜撞锤;冲击夯
battering rule 定斜尺;测斜器;定斜规;坡规

battering wall 斜面墙;倾斜墙
Batterium 巴特里姆铜铝镍合金;铜铝镍合金
batterium alloy 铜铝镍合金
batter leader pile driver 斜桩机;斜桩打桩机;斜导架打桩机;有斜导架打桩机
batter leg tower 斜柱塔架
batter level 测坡仪;倾斜仪;测斜器
battern plate column 缀合柱
batter of column 柱侧斜
batter of facing 面坡
batter of pier body 桥墩墩身斜度
batter of wall 墙身收分
batter peg 内倾桩;定斜桩;放线木楔;坡脚桩
batter pier 斜面式墩;斜面桥墩;斜面墩
batter pile 角撑桩;支撑桩;斜桩
batter pile cluster 斜桩群
batter post 门道防护柱;圆锥形柱;斜撑;斜柱;斜支柱;斜杆;门道一边的立柱;房屋转角立柱
batter rule 定斜度规;测斜器;定斜器
Battersea gas washing process 巴特西烟气洗涤法
batter sheet piling 斜板桩
battersoable grommet 电池线护环
batter stick 靠尺;测斜竿
batter template 定斜样板
batter templet 定斜样板;斜坡样板;定斜样板
batter tension pile 斜拉桩
battery 溜口挡板;选矿场;电池(组);锤组;兵器群;炮台
battery acid 蓄电池酸液
battery acid level 电池酸位
battery assay 破碎试样分析
battery board 干电池用纸板
battery boiler 并列汽包式锅炉
battery box 蓄电池箱;电池箱;电池盒
battery box cover 电池箱盖
battery bracket 电池组托架
battery breast system 仓柱式开采法
battery bus 蓄电瓶汇流条
battery cable clamp 电池线夹头
battery capacity 电池组容量
battery car 蓄电池车;电瓶车;电池汽车
battery carbon 电池碳棒
battery cart 电瓶车
battery cartridge 电池套管
battery case 电池外壳;电池箱
battery casting 成组注浆
battery cell 蓄电池单位;原电池;电瓶;电池盒
battery cell tester 电池试验器;单电池测验表
battery charge and discharge box 充放电箱
battery charger 蓄电池充电器;蓄电池充电机;电池充电器;充电器;充电机;充电工
battery charging 蓄电池充电;充电
battery charging equipment 蓄电池充电设备
battery charging generator 电池充电发电机
battery charging isolation switch 电池充电隔离开关
battery charging regulator 电瓶充电调节器
battery charging set 电池充电成套设备;电池充电设备
battery charging switch 充电转换开关
battery charging system 电池充电系统
battery circuit 蓄电池电路
battery clamp 电池夹
battery clip 电池线夹;电池夹
battery clock 电池钟
battery coking 炉组炼焦
battery commander's telescope 炮队镜
battery commutator 电池互换器
battery-condenser exploder 电容器放电式爆破机
battery condition meter 电池电能指示表
battery-conserving switch 节电开关
battery contactor 电池接触器
battery container 电池箱
battery coupling 电池耦合
battery cover 蓄电池盖;电池盖
battery crane 蓄电池起重机
battery cubicle 电池柜
battery cupboard 蓄电池箱
battery cutout 电池电路自动断路器;自动开关;自动断箱
battery cyclone 旋风除尘器组
battery de-polarizer 去极剂
battery discharge 蓄电池放电
battery discharge indicator 蓄电池放电指示器

battery disposal 电池处理
battery distribution board 蓄电池配电盘
battery drive 蓄电池驱动；电池传动
battery-driven 电池驱动的
battery-driven movie camera 电池驱动摄影机
battery-electric 蓄电池电动车
battery electrolyte 蓄电池溶液；蓄电池电解质；电池电解液
battery eliminator 电池代用器；等效电池；代电池（组）
battery exchanger 电瓶交换器
battery excited generator 电池励磁发电机
battery failure alarm 电池故障信号设备
battery fault indicator 蓄电池故障指示器
battery-fed clock 电动钟
battery fork lift truck 电瓶叉车；电池叉车
battery form 成组立模（生产大墙板用）
battery fuse 电池保险丝
battery garages 一排汽车库
battery gassing 电池电液泡
battery ga(u)ge 袖珍电流计；电池电流计
battery generator 电池励磁发电机
battery grid 蓄电池铅板；电池铅板
battery ground cable 蓄电池地线
battery heating 电池组加热
battery holddown 电瓶夹板
battery holddown cover 电瓶夹板盖
battery house 蓄电池室；蓄电池充电间
battery hydrometer 电池液体比重计
battery ignition 蓄电池发火
battery ignition system 蓄电池点火系统
battery in quantity 并联电池组
battery insulator 电池绝缘子
battery jar 蓄电池容器；蓄电池壳；电瓶壳；电池外壳；电池瓶；电池槽
battery lamp 低压白炽灯
battery lead 电池引线
battery lead plate 蓄电池铅板；电极板
battery lighting 蓄电池照明
battery limit 界区
battery locomotive 蓄电池式机车；蓄电池机车；电瓶机车
battery meter 蓄电池充放电安时计；电池充放电用的安时计
battery motor 电池电动机
battery mo(u)ld 一组线脚
battery mo(u)lding 一组装饰线条
battery of autoclaves 高压釜组
battery of boilers 锅炉组；一组锅炉
battery of capacitors 电容器组
battery of coke ovens 炼焦炉组
battery of filters 滤池组
battery of filter wells 井点阵；井点群
battery of fixtures 成组器具
battery of formwork 一套模板
battery of hauling equipment 一组运输工具；成套运输设备
battery of lens 透镜组
battery of presses 压滤机组
battery of saws 锯组
battery of screens 一套筛
battery of shutters 一套模板
battery of sieves 一套筛
battery of silos 一排筒仓；谷仓组
battery of siphons 虹吸组
battery of tanks 罐群
battery of tests 成套测验
battery of three-flues 三孔烟道
battery of wells 井组；井群
battery-operated 电池供电的
battery-operated car 电动汽车
battery-operated conductivity recorder 电导率自记仪
battery-operated counter 电池式计数器
battery operated industrial truck 蓄电池操作的工业卡车
battery-operated motor cycle 电动摩托车
battery-operated radio receiver 电池式接收机
battery operated recorder 用电池工作的记录器
battery-operated toy 电动玩具
battery-operated toy ship 电动式玩具轮船
battery-operated train 电动火车
battery overcharge 电池过量充电
battery overload relay 电池过载继电器

battery pack 蓄电池组；电池组
battery panel 电池盘
battery paper 干电池用纸
battery plan 同类房间成组布置的平面
battery plane 同类房间成组布置的平面
battery plant 电池厂
battery pliers 长钳
battery pod 电池盒；电池舱
battery-powered 电池供电的
battery-powered haulage 蓄电池机车运输；蓄电池机车牵引
battery powered stacker 电瓶码垛车
battery radio set 电池无线电接收机
battery railcar 电瓶机动车；电瓶轨道车；电池轨道车；铁轨上蓄电池操作的小车
battery rearing 笼养(法)
battery receiver 电池式接收机
battery recharge 电池再充电
battery recharge room 充电站
battery resistor 电池电阻器
battery reverse switch 电池转向电闸
battery room 蓄电瓶室；蓄电池室
battery saver 电池保护元件
battery section 电池组
battery separator 蓄电池隔板；蓄电池薄片；电池隔板
battery set 电池组
battery solution 蓄电池溶液；电池溶液
battery starter 电池起动器
battery supply 电池电源
battery supply bridge 馈电电桥
battery supply coil 电池馈电线圈
battery supply set 电池供电设备
battery switch 电池组开关；电池转换开关
battery terminal 蓄电池接线柱；蓄电池电极；电池接线柱；电池极
battery tester 蓄电池实验器
battery traction 蓄电池牵引
battery tractor 蓄电池牵引车
battery triode 电池供电式三极管
battery trolley locomotive 双能源机车
battery truck 蓄电池机车；蓄电池车；电瓶车
battery truck garage 电瓶车库
battery type 成组立模制作方式(混凝土板)
battery unit 电池组
battery voltmeter in driver's cab 驾驶室电池电压计
battery wall (焚化炉的)炉膛隔墙
battery wastewater 蓄电瓶废水
battery welder 蓄电池式焊机
batting 棉絮；凿石面；撞击；打击
batting block 石膏硬块
batting machine 弹絮机
batting out 压制泥片
batting out device 压泥片装置
batting tool 打击工具；阔凿
batt insulation 牛皮纸包的玻璃纤维或矿棉毯(用于木墙筋和平顶格栅之间的隔热材料)；沥青隔热层
batt joint 平接口
battle against pollution of environment 反环境污染斗争
battle axe 灰泥名称(广泛用于熟石膏板面层)
battle deck 加劲钢板(用作甲板、桥面板)
battle deck (bridge) floor 战舰上承桥楼面；(房屋和桥梁的)焊接面板系；双向钢结构(楼)面板(房屋和桥梁用)；桥梁钢面板；双向钢格构(楼)面板系统
battledore 方头扁平长刀；板羽球球板
battle field 战场
battle light 舷窗铁盖
battle map 作战图
battlement (有城垛的)城堡女儿墙；锯齿形墙；雉堞；城墙垛
battlemented 城垛
battlemented bridge parapet 有城垛的桥梁栏杆
battlemented tower 防御塔楼
battlemented wall 防御墙
battlement tower 雉堞塔楼
battlement turret 雉堞角塔
battle plate 避火板
battle ship 军舰；主力舰；战舰
battleship gray 军舰灰色
battle short 保险短路开关；保安短路器

battle sight 表尺
battle-sight range 直射距离
battle tracer 航运自描器
batt printing 泥片印花法
batture 河洲；河滩高地；河滩地；心滩；淤高河床；冲积地
batture land 河洲地；心滩地
Batu gum 巴土树脂
bat wash 耐火泥浆
bat wing 望板
batwing aerial 超绕杆天线
bat-wing antenna 蝙蝠翼形天线
baty 炮台
bauble 小摆设
baud 波德(信号速度单位，1 波德 = 1 脉冲/秒)
Baud base 波特基准制
Baud base system 波特基准制
baudelot cooler 喷淋(表面)冷却器
Baudelot type cooler 膜式冷却器；波德洛特式冷却器
Baud generator 波特发生器
Baudoin's alloy 鲍多英合金
Baudot code 博多码
Baudot distributor 搏多分配器
Baudot multiplex speed 博多多工速度
Baudouin reaction 博杜安反应
Baudran expansion apparatus 鲍德兰高温膨胀仪
baud rate 波特速率；波特率；波特比
baud rate factor 波特速率系数
Baud rate generator 波特速率发生器
Baud rate switch 波特速率选择开关
bauerit 片石英
Bauer refiner 鲍尔双动盘磨机
bauge 柴泥
Bauhaus 包豪斯学派【建】
baulage tractor 搬运用拖拉机
ba(u)lk 梁木；木材堆放稳定木条；大方木；方木(材)
baulk 界埂；大木；粗主方(指大木)；田埂
Baulot cooler 表面液体冷却器
Baumann exhaust 鲍曼式排汽
Baumann hardness meter 鲍曼式硬度计
Baumann print 鲍曼硫印
Baumann sulfur printing 鲍曼试验
Baumbach formula 鲍姆巴赫公式
Baume 波美液体比重计
Baume degree 波美度
Baume gravity 波美比重(度)
Baume gravity instrument 波美液体比重仪
Baume hydrometer 波美液体比重计；波美(液体)比重计
Baume hydrometer scale 波美比重计标度
Baume liquid gravimeter 波美液体比重计
Baume scale 波美刻度；波美标度；波美比重瓶；波美比重计；波美比重标
Baume scale of specific gravity 波美比重标度
Baume's law 波美氏定律
Baume's scale 波美氏比重标
baumetester 波美测试仪
baumhauerite 褐硫砷铅矿
baumite 锰镁铁蛇纹石；波美石
Baum jig 鲍姆跳汰机；筛侧空气跳汰机
Baum wash box 鲍姆跳汰机；包姆洗煤机洗槽
bauranoite 钡铀矿
Baur-Leonhardt system 鲍尔—列昂哈尔特式预应力混凝土后张系
Bauschinger effect 包辛格效应；包氏效应
Bauschinger extensometer 包辛格伸长计
Bauschinger's expansion tester 包辛格式膨胀测定器
Bautz-Morgan type 鲍茨—摩根型
bauxite 铝土岩；铝土矿；铁铝氧石；铁矾土
bauxite based fused corundum 高铝刚玉；矾土基电熔刚玉
bauxite brick 铝土砖；高铝砖；矾土砖；土砖
bauxite cement 高铝水泥；矾土水泥；矾土泥浆
bauxite chamot 高铝矾土熟料；矾土火泥
bauxite clay 铝矾土；高铝黏土；铝质黏土
bauxite drier 矾土干燥器
bauxite fire brick 铝土耐火砖
bauxite fireclay 铝质耐火黏土
bauxite-ironstone formation 铝土—铁质岩建造
bauxite refractory 铝矾土耐火材料
bauxite refractory product 高铝耐火制品

bauxitic cement 高铝水泥
bauxitic clay 铝质黏土；铝土质黏土；铝铁黏土；铝铁矾土
bauxitic laterite 铝土质红土；水矾性砖红壤
bauxitic rocks formation 铝土岩亚建造
bauxitic shale 铝土页岩
bauxitite 铝质岩；铝土质岩
Bavarian Alps 巴伐利亚阿尔卑斯山脉
Bavarian splint 巴伐利亚夹
Bavarian trass 巴伐利亚火山灰(德国)
bavenite 硬沸石；硅铍钙石
Baveno law 巴温诺律
Baveno twin 斜坡双晶；斜面双晶；巴温诺双晶
Baveno twin law 巴温诺李晶律
bavin drainage 梢捆排水沟
bawk 水平拉杆；联系梁
bawke 料罐；吊桶
bawn 灰色泥炭；城堡围墙；蓄栏
bawneen 劳动服
bax ventilator 架在墙上的通风机
bay 跨度；开间；间档【港】；架间距；海湾；月桂(色)；隔舱；港湾；船闸附近河段；(集装箱船的)排(架)；台座
bayadere 横彩条花型；巴亚德横彩条绸；横条装饰
bay and cantilever system 跨度及伸臂系统
Bayard-Alpert ga(u)ge 巴雅—阿尔培特真空计
Bayard-Alpert ionization ga(u)ge 巴雅—阿尔培特电离真空计
bay area 海湾地区
bay area rapid transit 海湾地区高速公路
bay bar 河口沙洲；海湾沙洲；拦海坝；湾内沙坝；湾口洲；湾口沙嘴；湾口沙坝；湾内沙坝
bay barrier 湾口洲；湾口沙嘴；湾口沙洲；湾口沙坝
bay bear 支架
bayberry oil 月桂子油；月桂果脂
bayberry tallow 月桂脂
bay-bolt 基础螺栓；地脚螺栓
bay bridge 海湾桥；港湾桥
baycid 倍硫磷
Bay City 海湾城(美国旧金山的俚称)
bay crossing 海湾穿越
bay delta 海湾三角洲；湾头三角洲
bay deposit 湾积物
Bayer alumina 拜尔氧化铝
bayerite 拜三水铝石；拜尔石
Bayer process 拜尔工艺；拜尔(方)法
Bayes decision rule 贝叶斯判决规则；贝叶斯决策规则
Bayes detection 贝叶斯检测
Bayes detection theory 贝叶斯检测理论
Baye's estimation 贝叶斯估算(法)；贝叶斯估计(法)
Bayessian approach 贝叶斯分析法
Bayesian classification 贝叶斯分类
Bayessian decision rule 贝叶斯决策定则
Bayesian decision theory 贝叶斯决策理论
Bayesian method 后验概率法；贝叶斯法
Bayesian model 贝叶斯模型
Bayesian posterior analysis 贝叶斯后验分析
Bayesian probability decision method 贝叶斯概率决策法
Bayesian statistics 贝叶斯统计
Bayesian uncertainty 贝叶斯不确定性
Bayes law 贝叶斯定律
Bayes risk 贝叶斯风险
Bayes rule 贝叶斯规则
Bayes's classification 贝叶斯分类
Bayessian decision theory 贝叶斯决策原理
Bayes test 贝叶斯检验；贝叶斯检测
Bayes theorem 贝叶斯定理
bay harbo(u)r 海湾港(口)；泻湖港；海港；湖湾港
bay head 湾头
bayhead bar 湾头坝；湾顶坝；湾头沙洲；湾头沙埂；湾头沙坝；湾尖沙洲
bayhead barrier 湾头堰洲
bayhead beach 湾尖(沙)滩；湾头海滩；湾头滩
bayhead delta 湾头三角洲
bay ice 海湾冰；湾冰
bay joint 跨间接缝
bay lake 海湾湖；湾状湖；椭圆形小湖
baylanizing 钢丝连续电镀法
bay layout 建筑混凝土路面的隔仓施工法
bayldonite 乳砷铅铜矿
bayle 城堡围墙间空地

bay leaf 月桂叶(用于半圆形线脚装饰)
bay leaf garland 桂冠花饰【建】
bay leaf swag 桂冠花饰【建】
bayleaves oil 桂叶油
bayleyite 碳镁铀矿
bay-line 尽头式站台线；铁路专用线；专用支线
baylissite 水碳镁钾石
Baylis turbidimeter 培氏浊度计
baymouth 海湾口(沙洲)；湾口(洲)；湾口(沙)坝
baymouth bar(rier) 湾口沙嘴；湾口(沙)洲；湾口沙坝；湾口堰洲
bay number 集装箱船的排号；行号(集装箱等)
bay of a shaft 轴的分隔距离
Bay of Bengal 孟加拉湾
Bay of Biscay 比斯开湾
bay of joist 格栅间距
bay of lake 湖湾
bay oil 香叶油；月桂(叶)油
bayonet 接合销(钉)；刺刀；卡口
bayonet attachment 卡口连接装置
bayonet base 卡口底座；卡口灯座；卡口灯头；卡口灯泡；刺刀座；插口式灯座
bayonet cap 卡口帽；卡口灯头；插头盖；台灯插头盖
bayonet catch 卡口式连接；卡口连接；锁梢
bayonet clutch 插销式离合器；卡口离合器
bayonet connection 卡口连接
bayonet coupling 卡口式联结器；插栓式管接头；卡口接头
bayonet fastening 卡口连接
bayonet fitting 卡口式组装件
bayonet fixing 卡口式固定；管脚固定
bayonet flushed socket 插口
bayonet flush socket 平灯座
bayonet ga(u)ge 插入式油面计；插入式测量仪器；插入式表；机油表
bayonet ga(u)ge stick 插入式计量棍
bayonet harbo(u)r 海湾港
bayonet joint 卡口接头；卡口接合；插销节；插销接头；插接式接头；插刺接合
bayonet(lamp) holder 插销式灯座；卡口灯座
bayonet lamp-socket 卡口灯头；卡口灯座
bayonet lock 插销接头；卡住；卡口连接；插销栓
bayonet mount 卡口架；卡口固定件；插刀式插座
bayonet peg 灯头插脚
bayonet saw 机动竖锯；轻便电锯
bayonet socket 卡口插座；插口灯座；卡口灯座
bayonet-socket bulb 卡口灯泡
bayonet spindle nose 锁紧盘式主轴端部
bayonet stack 卡口式排气管
bayonet-tube exchanger 插管式热交换器
bayonet-type quick coupling 插栓式快速联结管接头
bayonet unit 卡口式连接
bayou 泥沼湖口；泥沼河口；泥沼海湾；漫流河段；湖口；河流入口；淤塞河流；长沼；浅滩海湾
bayou lake 旧废河道；牛轭湖；长沼
bayou platform 尽头式站台；分支站台
bay port 海湾港口
bay quoin of masonry wall 断裂的直角圬工墙角
bay salt 海盐；粗粒盐
bayshon 挡风墙
bayshore freeway 沿海岸的高速公路
bayside beach 湾侧滩；湾边海滩
bay span 支柱间跨距
bay stall (礼拜堂内的)凸窗前空间座位；凸墙处坐位；凸窗座；凸窗处坐位
bay stone 表面基石
Bayston Group 贝斯顿岩群
bay stop 灌水池门口
bay system 光栅系统
bayt 帐篷式住宅
bay termination module 间隔终端组件
baytex 倍硫磷
bay tree 月桂树
bay-tree oil 月桂子油
bay type 海湾型
bay-type collection 节间式集中荷载；柱距式集中荷载
bay wall 池墙；河湾边墙
bay water quality model 海湾水质模型
bay width 开间
bay window 凸窗；挑出窗；开窗壁槽；八角窗
baywood 洪都拉斯桃花心木；大叶桃花心木
baza(a)r 百货商店；市集；商场；商品陈列所；[复]

bazar
baza(a)r bill rate 商场证券利率
Bazar metal 巴札镍银合金
Bazett index 巴泽特氏公式
Bazin formula 巴金公式
Bazin-type sharp-crested weir 巴金式锐缘堰
Bazin weir formula 巴金堰公式
bazirite 硅锆钡石
bazooka 火箭发射架；活动螺栓运送器；螺旋运送器；火箭筒；超高频转接变换器；平衡变换装置
bazzite 钪绿柱石
B-battery 乙电池组；阳极电池组
BB-block brazing 加热块钎焊
bbey 修道院
B-B fraction 丁组分
B-box 变址数寄存器；变址寄存器
B-coefficient 受激发射系数
B delay 副延时
bdellium 芳香树胶
bdelygmia 恶心
Be10 dating method 铍—10 测年法
beach 海滩；海滨；岸滩；使船搁浅
beach accretion 海滨砂淤积；海滩淤积；海滩加积；海滩堆积；海滩冲积(物)
beach advance 海滩推进
beach angle 海滨倾斜(角)度
beach approach 岸坡；岸边；滩沿
beach area 河滩地；海滩区
beach at ebb tide 潮漫滩
beach barrier 滨外砂障；滨外沙坝；沿海沙埂；海滩沙堤；滨海沙堤；海滩暗礁；滨海沙坝
beach berm 海滩阶地；海滨滩肩；滩肩
beach berm crest 海滨滩肩外缘
beach berm scarp 滩肩崖
beach boat 海滨小划子；小渔船
beach bottom 滩底
beach breccia 海滩角砾岩
beach bridge 港湾桥；湾滩桥
beach building 海滨房屋；海滩形成过程；造滩(的)；成滩过程(的)
beach building material 造滩泥沙
beach-building phenomenon 造滩现象
beach-building process 海滩造床过程
beach chute 海滩流槽
beach city 海边城市
beach cloth 海滨薄呢；海滨布呢
beach comber 巨浪；击岸波；滚浪；拍岸浪
beach combing 海滩开采砂矿；海滩淘砂
beach concentrate 海滩砂矿
beach contamination 海滩污染物
beach crest 海滨顶点
beach cusp 海滩嘴；滩嘴；滩角；沙嘴
beach cycle 海滩旋回
beach deposit 海滩沉积；海滨沉积；滩沉积
beach detector 海滩探测器
beach development cycle 海滩发育循环
beach drift 海滨砂；海滩沿岸流
beach drifting 海滩漂移物；海滩漂积物；海滩漂浮物；海滩冲刷；沿滩漂移沙；沿滩漂沙；沿滩漂移；沿海滩沙；海滩漂积物；海滨漂流
beach dune 海滩沙丘
beached bank 铺石护岸；有砌石岸的岸坡；砌石岸坡；铺石护岸的岸坡；铺护岸
beach erosion 海滩侵蚀；海滩冲刷
Beach Erosion Board 海滩侵蚀局(美国)；海岸工程局
beach erosion control 海滩侵蚀控制；海滩防冲措施；滩地侵蚀防治
beaches and sea shores 海滨和海岸
beach excavator 海滩挖泥车
beach extension 海滨延伸；滩地外延；滩地后延
beach face 滩面
beach facies 海滩相
beach features 海滩地形
beach fill 海滩填护；海滨填土；填砂护滩；海岸淤填土；海岸淤积土
beach formation 海滩形成；海滩造建；海滨建造
beach formation by waves 浪积滩地；浪成海滩；海浪冲积滩地
beachfront 靠海滨的；海滨地区
beach fulls 滩脊
beach gear 两栖登陆装备；搁浅救助设备
beach gradient 海滩梯度；海滩坡度
beach grass 海滩水草

beach gravel 滩砾石;海滩砾石;海滨砾石;滩地砾石
beach-head 滩头(滩地)
beach house 海滨别墅
beach hut 海滩棚屋
beaching 海滩堆积;海岸堆积;搁滩;搁浅;船只搁浅;冲滩;抢滩;砌石护岸
beaching accommodation 登陆设备
beaching chassis 承船(底盘)车
beaching gear 陆上起落装置;登陆用轮架;登陆轮架
beaching of a distressed vessel 难船抢险
beaching of bank 护岸工程;堤岸护坡
beaching of raised wreck 沉船搁浅
beaching plan 比钦计划【铁】
beaching site 搁滩地点
beaching wedged chassis 斜架下水车【船】
beach lamination 海滩纹理
beach line 海滨线;海岸线;岸线
beach maintenance 护滩;海滩养护;海滩维护
beach marker 海滩航标;海滩标志器
beach markings 海滩状花纹
beach material 海滩物质;海滩泥沙;滩地(物)质
beach mining 海滩采矿
beach nourishment 海滨淤涨;海滨淤填;海滩填护;淤滩;补沙护滩;填砂护滩
beach of storm profile 风暴后形成的浪蚀滩地
beach ore 海滩砂矿
beach pipe 水力冲泥管;水力输泥管
beach placer 海滩砂矿;海滨砂矿
beach plain 海滩平原;滨平原;岸滩平原
beach plain coast 海滩平原海岸
beach plain spit 海滨平原沙嘴
beach pollutant 海滩污染物
beach pollution 海滩污染;海滨污染
beach pool 滩池
beach price 海边价格
beach process 海滩演变过程;岸滩演变过程
beach profile 海滩剖面;岸滩剖面
beach profile of equilibrium 海滨平衡剖面
beach profile of lake 湖滩剖面
beach profile survey 海滩剖面测量
beach protection 护滩;海滩保护
beach protection and accretion promotion 促淤保滩
beach race 滨海水道
beach reflection 海滩反射;消波
beach rehabilitation 人工滩淤
beach renourishment 补沙护滩;填砂护滩
beach replenishment 海滩护护;补沙护滩
beach resort 海滩度假村
beach retreat 海滩后退;滩地后退
beach retrogression 海滩后退
beach ridge 海滩脊;滩脊;滩埂
beach ridge deposit 海滩脊沉积
beach roadway 海滩(汽车)道路
beach rock 贝壳砂石;海滩岩;滩岩
beach sand 海滩砂;海滩泥沙;海滨砂;滩砂
beach sandstone 海滩岩;海滩砂岩
beach sand trap 滩砂圈闭
beach scarp 滩岩;海滩陡坎;海滨陡坎;滩崖;滩坎
beachscope 海滨风景
beach scour 海滩冲刷
beach sediment 海滩泥沙
beach seine 向岸拖的浅水围圈
beach seismic belt 海岸地震带
beach shack 海滩茅屋
beach-shallow sea 滩海
beach she-oak 木麻黄
beach shingle 海滩砾石
beach slope 海滨坡度;海滨比降
beach stability 海滩稳定
beach strand 滩岸
beach strand erosion 滩岸侵蚀
beach swale 沼泽地
beach turning 搁浅掉头
beach umbrella 太阳伞(海滨)
beach vehicle 海滩作业车
beach wagon 客货两用汽车
beach width 海滩宽度
beachy 有沙滩的;岸边浅滩的
beach zone 海滨区
beacon 立标;信号台;信标导航;信标;指向标;灯信号;灯塔标志;瞭标;标桩(作航标用);标向波;标灯;岸标

beaconage 航标税;岸标系统;设置航标;设置灯塔
beaconage and buoyage 航标费
beacon antenna 指向标天线
beacon antenna equipment 指向标天线设备
beacon boat 灯标船;警标艇;航标船;小灯船;灯标艇
beacon buoy 警告浮标;柱形浮标;灯标浮标;灯标浮筒;导航浮标;浮标
beacon-code 信标码
beacon course 灯标程;标程;无线电导标的航线
beacon delay 信标延迟
beacon direction finder 信标测向仪;导标测向仪
beacon flasher 闪光信号装置
beacon for compass adjustment 偏差指向标
beacon identification data 信标识别数据;示位标识别数据
beaconing 设标
beacon lamp 信标灯;标向灯
beacon lantern 航标灯
beacon light 探照灯;立标灯;航标灯;信标灯;灯标;岸标灯;无线电航标
beacon light tower 导航灯塔;航标桩;航标灯塔;探照灯塔
beacon point 海岸信标
beacon presentation 指向标显示
beacon processing system 信标处理系统
beacon radar 信标雷达
beacon range 信标有效距离
beacon receiver 标向波接收机;无线电导标接收机
beacon signal 航标信号;灯标信号
beacon skipping 指向标失答
beacon station 信标电台;标向电台
beacon stealing 信标遗失;指向标失迹
beacon tower 信标塔;烽火台
beacon tracking 信标跟踪
beacon trigger transmitter 指向标触发发射机
beacon turret 灯塔
be acted upon by temperature changes 受温度变化的影响
bead 泪滴(一种漆病);绝缘珠;卷边;金属丝;木压条;焊珠;小圆线脚;圆线条;有孔小珠;压缝条;叠角焊缝;磁珠;串珠饰【建】;串珠;车轮圆缘;玻璃珠;玻璃液滴;玻璃厚边;熔敷焊道;球边;胎边;水珠(一种漆病);泪珠;压条
bead-and-batten work 圆线板条工作
bead and butt 平圆接口;(板接缝的)圆线脚平头
bead and flush panel 串珠镶边板
bead and quirk 圆口饰;凹槽圆线脚
bead and reel 珠盘饰凸圆线脚;珠饰;珠链饰
bead and reel enrichment 珠卷饰
bead angle 阳角转角
bead apex(core) 三角胶带
bead bend test 串珠焊缝受弯试验
beadboard 粒状板
bead break 珠断
bead building machine 撑轮圈机
bead butt 圆线脚平头接;带盖缝条的平接接合;平圆接合(作业)
bead butt and square 平圆方角接(合)
bead chain 珠链
bead core 叶轮心
bead covering machine 包装机
bead crack 焊珠裂纹;焊缝裂纹;焊道裂纹
bead curtain 串珠帘
bead cut 切边;斜削
bead cutter 切边机
beaded 珠状饰
beaded bevel 椭圆面斜边
beaded board 串珠板
beaded border 串珠纹
beaded ceiling 珠饰天花板;串饰天顶;珠饰平顶
beaded centre 珠饰环中心
beaded coaxial cable 加垫圈的同轴电缆
beaded covering 波纹蒙皮
beaded drip 铁皮泛水圆形卷边
beaded eccentric reducing socket 偏心带边异径管箍
beaded edge 泪滴边;墙角护条紧固凸缘
beaded equal cross 带边同径四通(管)
beaded equal side outlet elbow 带边同径三向弯头
beaded equal side outlet tee 带边同径三向三通(管)
beaded esker 连珠蛇形丘
beaded ferrule 扩口箍
beaded flat 墙角护圆形扁条

beaded flat steel 半圆形扁钢
beaded glass 粒状玻璃;玻璃珠
beaded insulation 串珠绝缘
beaded intrusions 串珠状岩体
beaded joint 圆凸缝
beaded lakes 串珠湖
beaded lakes and depressions 串珠状湖泊洼地
beaded mo(u)lding 串珠线脚;墙上圆线条
beaded paint 油漆;涂料
beaded pearlite 球状珠光体
beaded reducing cross 带边异径四通(管)
beaded reducing elbow 带边异径弯头
beaded reducing tee 带边异径三通(管)
beaded river 串珠状河流
beaded screen 粒状荧光屏
beaded section 凹凸断面
beaded stream 串珠状河流;深切融冻河(冻土区)
beaded stripe 玻璃珠形镶条
beaded structure 带状结构
beaded support 球形支座
beaded tire 紧嵌式轮胎;嵌合轮胎
beaded volcanoes 串珠状火山口
beader 卷边器;卷边工具;弯边装置
bead esker 成串蛇形丘
bead fluorescence analysis 珠球荧光分析
beadflush 串珠饰滚边;串珠镶边
beadflush panel 串珠镶边板
bead foiming ring 撑轮圈
bead former 钢丝圈卷成机
bead glazing 压头镶玻璃;压条装配玻璃法;压条镶玻璃法
bead glue 珠形苏格兰胶
beading 压筋;作成细粒;珠缘;直接焊接;滚轮压花;串珠状装饰;串珠状缘饰;除去剩釉;成珠状;成细粒;玻璃熔接;半圆形嵌条;起泡;弯边;涂边;泪珠;压条
beading design 串珠状花样
beading die 压筋模
beading-down 挂护
beading enamel 边釉;搪瓷边釉
beading fillet 半圆饰边;凸圆线饰;半圆线脚
beading hammer 轮胎拆卸锤;弯板锤
beading machine 卷边机;压边机
beading nodules 串珠状结节
beading of glass 制成玻璃珠;玻璃成珠
beading of ribs 串珠状肋骨
beading plane 线角刨【建】;圆线条刨;圆边刨;圆线刨;线脚刨
beading roll 波纹轧辊
beading tool 线角刨【建】;圆线条刨;圆线刨;线脚刨
beading wall 珠饰壁
beading weld(ing) 珠焊;轻连续焊接;凸焊;叠珠焊缝
bead joint 圆凸勾缝
bead-jointed 圆凸勾缝
beadless tire 直边式胎;无翻边轮胎
beadless tyre 履带轮胎;直边式胎;无翻边轮胎
bead lightning 珠状闪电
beadlike 珠状的
bead lock ring 轮胎侧边撑圈
bead mill 珠磨(机);砂磨机
bead mo(u)ld(ing) 半圆饰;圆线条;串珠线条;串珠线脚;半圆线脚;圆带线条(柱头或柱脚凸起的)
bead-on-plate weld 堆焊焊缝
bead plane 型刨;圆线条刨;圆角刨
bead plate 侧边盖板
bead pointing 凸圆勾缝
bead polymer 粒状聚合物;珠状聚合物
bead polymerization 悬浮聚合(法);珠状聚合(法);成珠聚合(法)
bead reaction 珠球反应;熔珠反应;溶珠反应
bead resistance 珠状热敏电阻
bead saw 刻槽锯;镶边手锯
bead seal 珠状封接
bead seat band 圆线条带饰
bead sequence 焊道顺序
bead sleeker 修边饰器
bead slicker 小圆刮刀
beads-shaped 串球状
beads-shaped distribution 串珠状分布
beads-shaped drainage 串珠状水系
beads-shaped pluvial fan 串珠式洪积扇

beads-shaped structure 串珠状构造
beads string saw 串珠锯
beads structure 串珠状构造
bead-supported line 绝缘珠支持线路
beads with pottery body 陶胎琉璃珠
bead test 滴珠试验法;熔珠试验
bead thermister 珠状热敏电阻器;热电偶联结珠
bead tile 串珠贴面板
bead tool 卷边工具;圆头镘刀;平圆墁刀;旋边工具
bead weld 堆焊;焊道;珠焊
bead weld cracking test 焊缝开裂试验
bead welding 窄焊道焊接;堆焊;填角焊;珠焊
bead winder 钢丝圈卷成机
bead wire 轮胎钢丝;卷边嵌线
bead wire ring 侧边金属丝制胎圈
bead wire wrap 侧边金属丝制胎圈外壳
beadwork 半圆形凸圆线脚;珠状花边;串珠状橡饰
be affected by the heat 受热
be affected with damp 受潮
be afflicted with a disease 害病
be afflicted with a severe drought 旱情严重
beak 嘴状物;圆口灯;柱的尖头;壳尖;鸟嘴物;嘴;船首破浪材
beaked key 勾扳子
beaker 量杯;有倾口的烧杯;烧杯
beaker in low form 低型烧杯
beaker in tall form 高型烧杯
beaker sampler 烧杯式取样器
beak head 海角;鸟嘴头饰;岬
beakhead mo(u)lding 鸟嘴头饰线脚;鸟嘴头饰;鸟嘴饰;鸟嘴线脚
beakhead ornament 鸟嘴(头)饰(诺尔曼门道上的富丽装饰)
beaking joint 尖口接缝;尖口接头;尖口接合;削榫
beakiron 角砧;砧尖;尖嘴形铁砧;鸟嘴铁;丁字砧;铁角砧;鸟嘴砧
beak mo(u)lding 鸟嘴饰(线脚)
Beallon 比阿隆铜铍合金;铍铜合金
Bealloy 比阿洛伊铜铍中间合金;铜铍中间合金
bealock 垭口;分水岭山口
be always afloat 保持浮泊
beam 梁;犁辕;犁梁;扩音器有效范围;横梁;波束;束;射线;射束;射入辐射
beam accelerating voltage 射束加速电压
beam-accessible memory 光束存取存储器
beam action 梁作用;梁的作用
beam action hypothesis 梁作用假设
beam-action shear 梁的剪切作用
beam adjustment 犁辕调节
beam alignment 光束调整;束校正;射线校直;射束校正;射束校正装置
beam alignment assembly 射束校正装置
beam analogue 梁比拟法
beam analyser 分束器;波束分析器
Beaman arc 毕门视距弧
beam anchor 梁锚件;梁端锚栓;梁端锚固
beam anchorage 梁的锚碇;梁端锚固
beam anchorage in masonry 砌体中的锚梁
beam-and-column construction 梁柱结构
beam-and-column work 梁柱联合作用
beam-and-crank mechanism 横梁曲柄机件
beam and filler floor 梁和填充式楼板
beam-and-foil system 梁板体系
beam and girder construction 双向梁格结构;主次梁格结构;井字梁结构;交梁结构;交叉结构;主次梁式结构
beam and girder floor 井字梁楼板;交梁楼板;交叉梁楼板;主次梁式楼板
beam and girder framing 交叉梁构架;主次梁构架
beam and girder structure 井字梁构造
beam-and-rail 轨梁
beam-and-rail bender 轨梁弯曲压力机
beam and slab construction 梁板构造
beam-and-slab floor 梁式板楼;梁板式楼面;梁式楼盖;梁板式楼板结构;梁板楼盖;梁板式楼板
beam-and-slab-floor construction 楼板式楼板结构;梁板式楼板结构
beam and slab foundation 梁板基础
beam and slab raft 梁板筏式基础
beam-and-slab structure 梁板结构
beam and sling 吊链天平梁
beam and stringer 横梁与纵梁
beam angle 光束孔径角;光束角;波束宽度;波束角;射束孔径角

beam-angle-between-half-power-points 半功率点间的束夹角
beam angle of scattering 散射束锥角
Beaman stadia arc 毕门视距弧;比曼视距弧
beam antenna 定向天线
beam aperture 梁口;波束孔径;射孔口(径)
beam approach beacon system 盲目进场信标系统
beam approach seeker 波束引导进场寻的器
beam aqueduct 梁式渡槽
beam area 波束面积;射束面积;射束横截面
beam arm 叉形梁臂;横梁叉肋材
beam array 定向天线阵
beam attenuation 射束衰减
beam attenuation meter 波束衰减计
beam attenuator 光束衰减器;波束衰减器;波束减计;射束衰减器
beam at water line 水下线船宽
beam axis 梁轴线;光束轴;射束轴
beam axle 梁形横轴
beam balance 杠杆式天平;杠杆式秤
beam balance ratio 光束平衡比
beam barrel 轴筒
beam bearing 正横方位
beam-bearing block 垫梁块
beam bearing plate 承梁板;梁端支承板;梁垫板
beam bed 梁的加应力床;预应力梁张拉台
beam bender 压梁机;弯梁机
beam bending 梁弯曲;梁拱;射束弯曲
beam bending machine 弯梁机
beam bending press 弯梁压力机;弯梁机
beam bending test 梁弯曲强度试验
beam bending viscometry 梁杆法粘度测定
beam bifurcation 光束分叉
beam blank (扎制工字梁用的)异形钢坯
beam blanker 电子束熄灭装置
beam blanking 束消隐
beam block 结合梁;辕杆垫板
beam blocking 梁套
beam bolster (支梁的)铁马凳;梁垫;支梁钢杆
beam bolt 舱梁栓
beam bottom 梁底面;梁底模板;梁腹(模板);梁下皮;梁底(模)
beam bottom board 梁底模
beam bottom elevation 梁底标高
beam boundary 注流边界;光束界限;波束边界
beam box 墙上梁穴;墙托架;墙上梁支座;墙内电器盒;穿墙梁框套
beam brace 犁辕斜拉杆
beam bracket 梁肘板
beam breadth 梁宽(度)
beam breaking test 梁破坏试验;梁的破坏试验;梁的断裂试验
beam breakup 束流消失
beam brick 过梁面砖;砖过梁
beam bridge 梁式桥;梁桥
beam bridge with variable cross-section 变截面梁桥
beam brightness 光束亮度;束亮度
beam broadening 波束展宽;波束变宽
beam building 梁式锚杆支护法
beam building effect 组合梁效应
beam bunching 聚束
beam burst 束流脉冲
beam butt joint 梁的对接头;梁接头;梁接口
beam cage 梁用预制钢筋骨架
beam calculation 梁的计算
beam cal(l)ipers 卡尺;大卡尺
beam camber 梁上拱度;梁拱
beam cancellation 光线对消
beam candlepower 射束烛光
beam carling 短纵梁
beam carving machine 电子束雕刻机
beam casing 梁套;梁护面(混凝土);钢梁外壳层;钢梁耐火材料覆盖层
beam casting 梁的浇注(混凝土);梁浇捣(混凝土)
beam catcher 束捕集器;射线收注栅
beam cathode 集射阴极
beam ceiling 假梁顶棚;露梁平顶;露梁顶棚
beam center 束流中心;射束中心
beam centering 定束流中心;射束中心调整
beam center line 梁的中心线;梁轴线
beam centroid 束流重心
beam channel 槽钢;槽形梁;槽开差

beam characteristic 波束特性
beam clamp 钢大梁吊环;钢大梁吊钩;梁模紧固件;梁模夹具;副承梁材;悬吊耐火钢模夹紧装置;梁模夹紧件;(钢梁或钢大梁的)吊环
beam clipper 束流限制器
beam coaming 舱口活动梁
beam coaming shoe 舱口活动梁座
beam collimation 光束准直
beam collimation optics 平行光束光学装置
beam collimator 光束准直仪;光束平行光管;光束准直仪;束平行光管
beam column 梁柱;射束柱;梁型柱
beam column connection 梁柱连接;梁柱联结;梁柱结合;梁柱接合
beam column construction 梁柱结构
beam column panel zone 梁柱接点区
beam combiner 光束组合器
beam compasses 横切圆规;横杆圆规;横臂圆规;长径规;长脚圆规;长臂分规;长臂(圆)规
beam component 束流输送系统组件;束流成分
beam compression factor 波束压缩因数
beam-condensing unit 光束聚光部件
beam conductance 电子注电导
beam configuration 射束组态;射束排列
beam configurations of Doppler radar antenna 多普勒雷达天线的波束形状
beam confining electrode 聚束极;集射屏
beam connection 梁的联结
beam constant 梁常数
beam construction 梁结构系统
beam contact 船舷触碰
beam control 亮度控制;光束控制;光亮度控制;波束控制;束控;射束控制
beam-control(led)system 束流控制系统;调束系统
beam convergence 束集;束会聚;射束会聚
beam converter 光束转换器
beam coordinate 束流坐标
beam corridor 束流输运线
beam counter scale 杠杆式衡器
beam coupling 光束耦合;电子注相互作用;电子束耦合
beam coupling coefficient 光束耦合系数
beam crane 单梁起重机;梁式起重机
beam crossing 交的叉梁;交叉梁
beam crossing angle 束流交叉角
beam cross-section 梁横断面;光束截面
beam current 电子束电流;束电流;射束电流;束流
beam current amplifier 束流放大器
beam current control 射束电流控制
beam current gain 射束电流增益
beam current lag 束流残像
beam curvature 梁的弯曲曲率;梁的曲率
beam cushion block 梁垫块
beam cut-off 闭束
beam cutter 冲切机
beam-defining aperture 限束小孔
beam-defining clipper 限束器
beam-defining jaw 限束光阑
beam-defining slit 限束光阑
beam deflecting crystal 光束偏转晶体
beam deflecting mirror 射束偏转镜
beam deflection 梁弯曲;梁挠度;梁的弯曲;梁的挠度;梁的挠度;光束偏转;束偏转;射束偏转;射束偏转
beam deflection system 光束偏转系统
beam-deflection tube 电子束偏转管
beam deflection valve 电子束偏转管;波束偏转管;束流偏转管;射束偏转管
beam deflector 光束偏转器;电子束偏转器
beam deflexion 梁的挠度;束偏转
beam density 光束密度;射束密度
beam depth 梁的高度;梁深;梁高(度)
beam design 梁的设计
beam design formula 梁的设计公式
beam destructure 束流破坏
beam deviation 波束偏转
beam deviation factor 射束偏移因子
beam deviation loss 波束偏移损耗
beam diameter 光束直径;波束直径;射束直径
beam diaphragm 束光阑
beam direction 射束方向
beam directionality 光束方向性
beam displacement 束位移
beam distance 梁的间隔距离;梁间间距

beam divergence 光束散度;光束发散
beam divergence angle 光束发散角;束发散角
beam diverging 射束发散的
beam diversity 波束分散
beam divider 分束器;射束分配器
beam doffer 落轴装置
beam dosimetry 束流计量学
beam drill 摇臂钻床
beam droop 波束下垂
beam drop 波束降落
beam drying machine 经轴烘燥机
beam dump 束流收集器
beam-dump magnet 泄束磁铁
beam duty cycle 束流负载周期
beamed radiation 成束辐射
beam effect 指向效应
beam efficiency 波束效率
beam-emergence direction 射束出射方向
beam encasement 混凝土梁的外壳;梁的外壳(混凝土);梁护面(混凝土)
beam encasure 混凝土梁的外壳;梁的外壳(混凝土);梁护面(混凝土)
beam end 梁的端部;梁端
beam end face 梁端截面;梁端面
beam ends out-of-square 梁端(加工)歪斜;梁端(加工)不正
beam energy 电子束能量;射束能
beam engine 横梁式发动机;立式蒸汽机;横梁机;横梁发动机
beam engine pump 横梁发动机泵
beam envelope 线束包线;射束包线
beam equation 梁的方程(式)
beamer 卷轴机
beam evening device 卷经匀整装置
beam expander 光束扩展器
beam-expanding telescope 光束扩展望远镜
beam fabrication 梁式结构
beam fill 梁间墙
beam filling 梁端填充墙;(格栅间的)填充墙;梁间墙;梁间充填;梁头填充;梁墙;梁端间隙填塞物;填隙货
beam filling factor 注填充因数
beam finder 寻线器;寻迹器(示波器);射束探测器
beam fixed at both ends 两端固定梁
beam fixed at one end 一端固定梁
beam flange 梁翼(缘)
beam flanges out-of-square 梁翼(加工)歪斜;梁翼(加工)不正
beam flexural theory 梁的弯曲原理;梁的弯曲理论
beam flexure theory 梁的弯曲原理;梁的弯曲理论;梁弯曲理论
beam floor 梁楼面;小梁楼板;单层楼板
beam flume 梁式渡槽
beam flux 束通量
beam flying 光束扫描
beam focusing 光束聚焦;束聚焦;射线束聚焦;焦点相合;对光;射束聚焦
beam-focus voltage 束聚焦电压
beam-foil spectroscopy 束箔光谱学
beam-foil system 梁板体系
beam-folding 光束折叠
beam for air duct 风渠横梁
beam for dismounting bogie 转向架卸取梁
beam for exterior wall 外墙托梁;外墙托架
beam form 梁模(板)
beam forming 波束形成
beam-forming arrangement 成束装置;射束形成装置
beam-forming electrode 电子束形成电极
beam forming optics 聚束光学系统;射束形成光学;波束形成光学装置
beam foundation 梁式基础
beam frame 梁架
beam generating system 射versaillesbeam generating system 射束发生系统
beam geometry 波束几何形状;束几何
beam grab 抓梁钩
beam grid 交叉梁;梁格栅;格排(式)梁
beam grill(ag)e 格排梁;梁格式基础;梁格;格排梁式基础;交叉梁
beam grillage theory 梁排理论
beam guard rail 梁式护栏
beam guidance system 波束制导系统
beam guide 束流输运系统
beam half power angle 波束半功率角

beam handling 束流控制
beam hanger 游梁吊架;梁托;梁模支承架;梁模吊架;游梁悬支;游梁托架
beam haunch 梁腋(梁加厚部分)
beam haunching 梁托臂;梁加腋
beam head 梁端部截面;梁端;游梁头
beam height 梁高;射束高度
beam hinge 梁铰
beam holder 支梁杆;叉杆
beam holding 电子束存储复原
beam hole 引束孔道;束孔
beam hollow block 空心梁砌块
beam hooks 舱口梁吊钩
beamhouse 浸灰间;标准车间;准备车间
beam-house waste 鞣革工厂废物
beam impedance 电子束阻抗
beam index colo(u)r picture tube 电子束指引彩色显像管
beam indexing tube 电子束指引管;电子束引示显像管
beam indicator 灯光指示器;波瓣测定器
beaminess 粗大;放光
beaming 聚束;集束;整经;倒轴;辐射
beaming effect (声、光等的)束效应;束流效应
beam in prestressed clay 钢泥中梁
beam instability 二束不稳定性;束不稳定性
beam intensity 光束强度;光的强弱;束强度
beam intensity ratio 光束强度比;光强比
beam interferometer 射束干涉仪
beam iron 承重铁;梁铁
beam joint 梁的接合
beam knee 梁柱隅铁;梁肘材;梁肘板;梁尾接铁
beam knife 刮皮刀
beam lamp 光射束灯
beam-landing screen error 束着屏误差
beam layout 梁的布置
beam lead 梁式引线
beam-lead bonder 梁式引线焊接机
beam-lead bonding 梁式引线连接(法);梁式引线接合法
beam-lead technique 梁式连接法
beam length 梁长(度)
beamless 无梁
beamless hall 无梁殿
beamless slab 无梁板
beam level deflectometer 杠杆弯沉仪
beam lifters 梁式起重机(厂房中桥式行车)
beam light 光束
beam-limiting aperture 束限制孔径
beam line 梁线
beam load 横梁负荷;电子束负荷;梁载荷
beam loading 梁上荷载;梁荷载
beam loading conductance 电子注负载电导
beam loading impedance 电子注负载阻抗
beam lobe indicator 波瓣测定器
beam lobe switching 波束瓣转换法;波瓣旋转
beam lock 结合梁;梁箍
beam locking 束流锁定
beam luminous flux 光束光通量
beam machine 刨皮机
beam made of precast hollow blocks 预制空心块组成的梁
beam magnetron 电子束磁控管
beam-making machine 梁模压机器
beam maser 射束脉塞;射束量子放大器
beam masking 射束遮拦
beam-measuring transformer 束流监测变量器
beam mechanism 梁机构
beam method of rolling 梁形材轧制法
beam micrometer 可换尺杆千分尺
beam mill 钢梁轧机
beam misconvergence 光束失聚
beam model 束流模型
beam-modulating disk 光束调制盘
beam-modulating memory 射束调制存储器
beam modulation 光束调制;射束调制
beam modulation percentage 束调制度;束调制百分数
beam modulator 光束调制器
beam moment 梁中弯矩;梁力矩
beam monitor 束流监测器;射束监测器
beam mo(u)ld 梁压模;梁模样板
beam mo(u)lding machine 梁线脚压制机
beam mount(ing) 梁架

beam noise 束电流杂波;射束噪声
beam non-uniform section 变截面梁
beam null line 射束零线
beam number 梁号(指工程图纸中梁的编号)
beam of constant depth 翼缘平行梁;等高度梁
beam of constant strength 等强度梁
beam of electron 电子束
beam of hull 船体最大宽度
beam of light 光线;光束
beam-of-light-transistor 光束晶体管
beam of non-uniform section 变截面梁
beam of one bay 单跨梁
beam of one span 单跨梁
beam of ship 最大船宽;船幅;船舶最大宽度
beam of sound 声波束
beam of three-spans 三跨梁
beam of two spans 两跨梁
beam of uniform depth 等高梁
beam of uniform strength 等强度梁
beam of variable cross-section 交接截面梁;变截面梁
beam of vessel 船宽
beam on 舷侧相向;船侧相向;船侧靠
beam on alternate frame 间肋梁
beam on discrete elastic supports 连续单性点支承梁
beam on elastic foundation 弹性基础连续梁;弹性地基梁
beam on elastic subgrade 弹性基础(上)梁;弹性地基(上)梁
beam on elastic support 弹性支承梁
beam on multiple elastic supports 连续单性点支承梁
beam-on sea 横浪
beam on waterline 水线面宽
beam opening 梁上开洞
beam optics 束流光学
beam-optics arrangement 束流输运系统;束流光学装置
beam orifice 透射孔;束孔径;射线孔
beam oscillation 梁的振动
beam overall 总宽(指船)
beam over deck 甲板宽度
beam over fenders 包括护舷材的船宽
beam pad 梁垫
beam pass 梁形孔型
beam path 光路
beam pattern 指向性图(案);定向图;波束图;波束方向图
beam pencil 电子束
beam pitman 梁连接杆;锚筋;拉杆
beam pitman bearing 梁连接杆支座;锚筋支座
beam plan 梁布置图;梁平面图
beam-plasma amplifier 等离子束放大器
beam-plasma instability 二束不稳定性
beam-plate assemblage 梁式板组件;梁板组件
beam pocket 梁槽;梁模槽
beam polarization 光束偏振
beam post 游梁柱
beam pourer 混凝土浇筑工具;梁的混凝土浇筑工
beam power 波束功率
beam power amplifier 集射管功率放大器
beam-power tube 电子束功率管;电子注功率管
beam power valve 电子束功率管
beam practice 束流试验
beam problem 梁的课题
beam profile 梁的剖面图;梁的断面;光束轮廓;束剖面图
beam pulse rate 束流脉冲重复率
beam pump 摇臂泵
beam pumping unit 抽油机
beam rail brake 梁式轨道
beam ratio 光束强度比;光束比;束比
beam reactor 中子束反应堆
beam rear suspension 后悬架梁
beam reception 定向接收
beam recombination prism 光束组合棱镜
beam recording 电子束记录;射束记录
beam redundant 超静定梁
beam reflector 波束反射器;射束反射器
beam reinforcement 梁的配筋
beam relay 平衡杆式继电器
beam resisting horizontal wind load 防风梁
beam restrained at one end 固端梁

beam reversal 光束翻转;束反向
beam reversing lens 光束返回透镜
beam rib 梁肋
beam rider 波束制导导弹
beam rider guidance 波束引导;射束制导
beam rider mode 波束制导方式
beam rider radar 波束制导雷达
beam rider system 波束制导系统
beam right agreement 架ához权协议
beam rigidity 束流刚度
beam-rolling mill 钢梁轧机
beam roof 梁屋顶
beam rotation attachment 光束偏转装置
beam roughing method 平轧法
beam runner 梁下纵材
beam saddle 梁托;梁模吊架;梁座
beam sagging 梁下垂
beam scale 杠杆秤;杆式秤;杆式磅秤;杆秤
beam scanner 束流扫描器;射束扫描器
beam scanning 波束扫描
beam-scanning method 射束扫描法
beam sea 横向浪;横浪
beam search 定向搜索
beam seat 梁垫;梁座
beam selector circuit 波段选择电路
beam sensing system 束流显示装置
beam sensitivity 射束灵敏度
beam sensitivity of gravimeter 重力仪光线灵敏度
beam sensor 光束传感器
beam separation 光束间距
beam separator 梁间隔材;束流分离器;束分离器
beam-shaper 波束成形器
beam shaping 束成形
beam sharpening 光束锐化;束锐化
beam sheath coat 混凝土梁外壳
beam sheet coat 梁护面(混凝土)
beam shelf 承梁纵材
beam shoe 活动梁支座;舱口活动梁座
beam shutter 射束光闸
beam shuttering 梁模(板)
beam side 梁的侧面;梁模边线;梁侧模;梁侧面
beam side panels 梁侧模
beam sideway mechanism 梁侧倾机理
beam sizing 聚束上浆
beam slab 梁式板;梁板
beam-slab bridge 梁板式桥;梁板桥
beam-slab construction 梁板结构
beam-slab raft foundation 梁板筏式基础
beam-slab structure 梁板结构
beam slenderness ratio 梁的长细比
beam sling 梁吊索
beam socket 舱口梁座
beam soffit 梁下皮;梁拱腹
beam source 辅助源
beam space 梁的间距;波束空间
beam spacing 射束间距
beam span 梁跨距;梁跨(度)
beam spectrometer 光束分光计
beam splice 梁的接头
beam split camera 分光摄影机
beam splitter 集束分裂器;光束分裂器;光束分离器;分束器;分光分束镜;部分发射镜;射束分裂器;分光板
beam splitter mirror 分光镜;分束镜
beam splitter prism 分束棱镜
beam splitting 光束分离;分束;分光;射束剖析法
beam splitting coating 分束膜
beam-splitting cube 分光立方体
beam splitting dichroic mirror 二向色分束镜;双色分束镜
beam splitting element 分束元件
beam splitting layer 射束分离层
beam-splitting lens 分光透镜
beam-splitting prism 分光棱镜
beam splitting system 分光系统
beam spot 射束点
beam spraying 束流喷射
beam spread 光束发散度;波束分散
beam spread function 射束扩展函数
beam square 角尺;横梁角尺
beam stabilization 束稳定化
beam stanchion 梁支柱
beam steel 梁用钢筋;梁用钢材
beam steel plate 大梁钢板

beam steering 光束转向;光束控制;波束转向;波束扫描;波束控制;波束转向;束偏转;束流控制
beam stirrup 梁箍筋
beam sticking 射束黏附
beam storage 电子束存储器;射束存储器
beam storage tube 射束存储管
beam straightener 工字梁矫直机;工字梁矫直机;钢梁矫直机
beam strength 梁的抗拉强度;梁强度
beam strength of tooth 齿的弯曲强度
beam support 梁支座;犁辕支撑架
beam supported at both ends 两端支承梁;端托梁
beam-supported bushing 托梁漏板
beam suppression 波束抑制;束消隐;射束抑制
beam surface 梁表面;射束面
beam swinging 射束摆动
beam switch 束开关
beam switching 波束转换;射束转换法;射束变向
beam switching frequency 射束开关频率
beam-switching tube 射线开关管;射束转换管;电子注开关管;射束开关管
beam switchyard 束流分配装置
beam-swithing tube 电子注开关管
beam system 梁构造系统;定向通信系统
beam test 梁试验;梁抗弯试验;棱柱体试件抗折试验;抗弯试验
beam tetrode 电子束功率管
beam texture 梁结构;梁状结构
beam theory 梁的理论;梁理论
beam thermal velocity 束热速度
beam tide 横向潮流
beam tie 梁上系杆;锚梁
beam tilt 波束倾斜
beam tilt angle 束倾角
beam timber 梁材
beam-to-beam connection 梁与梁的连接
beam-to-beam moment connection 梁的刚性连接
beam-to-beam weld 引线焊接;梁间焊接
beam-to-column connection 梁柱联结;梁柱结合;梁柱接合
beam-to-girder connection 小梁和大梁的连接;主次梁连接
beam to length ratio 船舶宽长比
beam trace 电子束踪迹;束径迹;束道
beam trammel 长臂(圆)规
beam transformation matrix 光传输矩阵
beam translator 光束平移器
beam transmission 定向发射;定向传输
beam transmitter 定向发射机;波束发射机
beam transportation car 运梁车
beam trap 电子束阱;射线收注册;射束收集器
beam trawl 底层定向拖网
beam tube 集射管;辐射管;喷嘴;喷射管;束管
beam type 梁的类型
beam-type guard rail 梁式护栏
beam type tube 注型管
beam vibration 梁振动;梁的振动
beam vibrator 梁式(表面)振捣器
beam waist 光束腰(收敛部分)
beam-warping 整经
beam wave 横向波浪;横浪;横波;横状波
beam waveguide 光束波导
beam web 梁腹(板);梁的腹板
beam well 游梁抽油井
beam wheel 卷取辊齿轮
beam width 梁宽(度);波束宽度;束宽;射束宽度
beam width of antenna 天线波束宽度
beam wind 横向风;横风;侧风
beam with anchored end 固定端梁
beam with both ends built-in 砌端梁;两端固定梁;固端梁
beam with central prop 单筋梁;双跨连续梁;三支点梁;三托梁
beam with compression steel 复筋梁;双筋梁
beam with constant cross-sections 等截面深
beam with double reinforcement 双筋梁
beam with fixed ends 固端梁
beam with free ends 两端自由梁
beam with haunchs 加腋梁
beam with high depth span ratio 深梁
beam with one overhanging end 悬臂梁
beam with overhanging ends 悬臂梁;两端悬挑梁;两端悬臂梁
beam with overhangs 两端有悬臂的梁

beam with simply supported ends 简支梁
beam with single reinforcement 单面配筋梁;单筋梁
beam with uniform section 等截面梁
beam with variable cross-sections 变截面梁
beam wobbler 射束摆动器
beam wood 梁木料
beamy ship 宽型船
beamy vessel 宽型船舶
beam zone 波束区
bean 油嘴
bean bag chair 豆袋椅
bean blister beetle 豆芫菁
bean cake 豆饼
bean curd leaf 百页
bean dregs 豆渣
beanery 廉价饭店;经济餐厅;素餐厅
be an extremely great advantage 极大优越性
bean joint 喷嘴接头;喷嘴接合
bean noodle mill 粉坊
bean oil 豆油
bean ore 豆状铁矿(褐铁矿);豆铁矿
beanstalk 可移动式液压升降平台
bean stick 豆蔓架木
bean up 放大油嘴(直径)
bean vermicelli 粉丝
be anxious for success 急于求成
bear 利空;空头业者;空头;结瘤;小型冲机;小冲床;底部打孔;带有;承担;负担;轻便打孔器
bear abeam 在正横方向;保持正横方位
bearable load 可支承荷载;可承荷载
bear account 空头账户;卖主账(户)
bear a loss 负担损失
bear and bull 买空卖空
bear astern 在正后方【船】;保持正后方向
bear away 改变航道;避开逆风
bear building-up 底结
bear cat 难开采的油井
bear clique 空头集团
bear covering 补空;买回
beard 锯齿状缺痕
bearded needle 钩针
bearded needle warp knitting machine 钩针经编机
bearded needle warp loom 钩针经编机
bearder 除芒器
beard hair 发毛
Beard-Hunter protective system 补偿控制线保护系统
bearding 船舷填角楔;樑条的内边线
bearding angle 船首柱与龙骨接合的角材
bearding line 嵌接线;樑条的内边线
beard of a cereal spike 谷穗的须
Beard protective system 皮尔德发电机保护装置
bear drive 卖空
bearer 支座;支架;支承物;载体;持有者;持有人;持票人;承托;承木;拖枕;托架;筒枕
bearer arch 承重甑
bearer bar 支承梁;托梁
bearer bill 执票人票据;无记名票据
bearer bill of lading 不记名提单;无记名提单
bearer blade 刃形支承
bearer blade bearer 刀口支撑
bearer bond 无记名债券
bearer bracket 承重牛腿
bearer cable 受力绳(索);受力缆索;支持钢索;吊索;承载(钢)索
bearer carrier 承载器架;托架座
bearer channel 荷载信道
bearer channel connection 承载通路连接
bearer check 执票人支票;不记名支票;无记名支票
bearer cheque 不记名支票;无记名支票
bearer debentures 不记名公司债券
bearer depository receipts 无记名信托证券;无寄名受托保管收据
bearer draft 无记名汇票
bearer for accumulator 蓄电池架
bearer for the grate bars 炉条托架
bearer frame 支承框架
bearer frame separation (indoor) 摘架分隔(室内)
bearer instrument 持票人证券;无记名证券;无记名有价证券;无记名票据
bearer of a gutter 挑檐托梁;沟槽支撑

bearer of gutter 檐槽托
bearer of tax 负税人
bearer paper 无记名票据
bearer plate 支承板;底板;承重板
bearer securities 无记名证券
bearer service 承载业务
bearer shares(stock) 无记名股票
bearer stock certificates 无记名股票
bearer supporting bracket 垫块;托座
bearer-type timber dam 叠小坝;堆木场
bear expenses 承担费(用);负担经费;负担费用
bear forward 在前方
bear frame 支架
bear fruit 结果实
bear-garden 嘈杂场所
bearing 支座矿脉走向;向位;支承;支撑点;承重(的);承载;承压;承托部分;方向角;方位
bearing abutment 支撑桥台
bearing accuracy 方位准确度;定位精度;方向角精度
bearing adjustment 支承调整;轴承调整
bearing ainer 轴承支圈
bearing alloy 轴承合金
bearing amplitude 天体出没幅角
bearing anchor 轴瓦固定螺钉
bearing anchorage 承压锚固
bearing and distance 方位与距离
bearing and range indicator 方位距离指示器
bearing and sounding 方位与水深
bearing angle 象限角;方位角
bearing annular seal 轴承环形密封
bearing apparatus 支承设备
bearing application 轴承应用
bearing arc 象限弧
bearing area 接地面积;挤压面积;支座面积;支承面(积);支撑面积;承底面积;承底面积;承压区
bearing area of a foundation 基础的支承面积
bearing arrangement 轴承布置
bearing assembly 轴承装配
bearing axis 轴承轴线
bearing axle 轴承;支承车轴;承压车轴
bearing axle-box 滑动轴承轴箱
bearing axle-box lubricated as required 油杯润滑轴承
bearing babbit 轴承巴氏合金
bearing backing 轴承衬
bearing ball 轴承滚珠;轴承钢滚珠
bearing ball cover 轴承端盖
bearing band 弹带
bearing bar 罗针;指南针;支承梁;支承杆;支撑钢筋;支撑杆;承重杆;承压条;承受江;方位瞄针;方位杆;受力筋
bearing bar centers 承重棒的中心距;支承杆中心距
bearing base 承压垫层
bearing beam 支撑梁;支推梁;受力梁
bearing bed 承载(地)层
bearing block 煤矿柱;轴承座;支座垫块;支承块;支承垫块;定向滑块;垫木;承重砌块;承重块;承压块
bearing block connection 垫板接合
bearing blocks for ladder 斗架轴承座【疏】
bearing blue 蓝铅油
bearing body 轴承体
bearing bolt 轴承螺栓;支承螺栓
bearing bolt connection 承压螺栓连接
bearing bond 轴承黏合剂
bearing bore 轴承腔;轴承内孔
bearing box 轴承座;轴承箱;轴承体
bearing bracket 轴承架;承重隅撑;承受托
bearing bracket stand 轴承架
bearing branch 结果枝
bearing brass 轴承铜套;轴承黄铜合金;轴承黄铜;轴承巴氏合金;铜轴瓦
bearing brick 承受荷载的砖;承重砖
bearing brick wall 承重砖墙
bearing bridge 轴承支架
bearing bronze 轴承青铜
bearing burning out 轴承烧熔
bearing bush(ing) 轴瓦;轴承衬(料);套筒轴承
bearing by stars 星位导航
bearing cable 承重索
bearing cage 轴承座;轴承笼;轴承隔圈
bearing calibration 校正方法;方位校准
bearing cap 轴承盖

bearing capability 承载能力;承载量
bearing capacity 轴承允许荷载;支承能力;地基支承力;承重能力;承重量;承载能力;承载力
bearing capacity apparatus 承载能力测量仪;承载能力测定仪
bearing capacity factor 载重因数;载重系数;承载系数;承载力因子;承载力系数
bearing capacity of anchor 锚杆容许承载力
bearing capacity of a pile 单桩承载力
bearing capacity of driven pile 打入桩的承载力
bearing capacity of foundation soil 基土承载能力
bearing capacity of foundation soil for pile 地基对桩支承力
bearing capacity of group piles 群桩荷载力
bearing capacity of loess with subsidability 湿陷性黄土的地基承载力
bearing capacity of permissible soil 容许土(壤)承载力
bearing capacity of pile foundation 桩基承载能力
bearing capacity of pile shaft 桩身承载力
bearing capacity of single pile 单桩承载力
bearing capacity of soil 土壤承载力;土壤承载(能)力
bearing capacity of urban resources 城市资源承载力
bearing capacity under earthquake 地基抗震承力
bearing cap oil seal remover 轴承盖油封拆卸器
bearing cap snap ring replacer set 轴承盖弹簧圈拆装工具
bearing carrier 结构受力构件;支点;承重构件;承载构件;受力构件;支架
bearing case 轴承套;轴承壳
bearing cast-iron 轴承铸铁
bearing center 轴承中心;承载中心
bearing chair 支座;承重块
bearing characteristic 负荷特性
bearing chatter 轴承振颤
bearing chock 轴承座
bearing circle 方向圆;方位度盘;方和盘
bearing classification 方位分级
bearing clearance 轴承空隙;轴承间隙
bearing coefficient 轴承工作系数;方向系数
bearing collar 轴领;轴承环;推力头
bearing column 承重柱
bearing compass 定位罗盘;方位罗盘;探向罗盘
bearing cone 锥形轴承内环;轴承锥形内圈
bearing cone puller 锥形轴承内圈拔出工具
bearing cone replacing tool 锥形轴承内圈更换工具
bearing conformability 轴承与轴面的配合性
bearing constant 轴承常数
bearing construction 承力构造;承力结构
bearing cooling water 轴承冷却水
bearing correction 方位修正量
bearing counter pressure 轴承反压力
bearing course 承重层;承载层;承压层
bearing cover 轴承套;轴承盖
bearing cup 轴承(锥形)外圈;轴承杯;滚动轴承杯
bearing cup puller 轴承外圈拔出工具
bearing cursor 方位游标
bearing deformation 承压变形
bearing designation 轴承型号;轴承标志
bearing deviation 方位偏差
bearing deviation indicator 航向偏差指示器;方位偏差指示仪;方位偏差指示器
bearing diagonal 斜支撑
bearing diagram 方位图
bearing disc 支承盘;顶压板;止推轴承板
bearing discrimination 方位分辨率;方位分辨力;方位辨别
bearing disintegration 轴承面碎裂
bearing distance 方位距离;支承距(离)
bearing distance computer 航线计算机;方位距离计算机
bearing distance heading indicator 方位距离船首向指示器;方位航程航向指示器
bearing does not appreciably 方位没有明显变化
bearing door 通风大门
bearing drag 轴承摩擦阻力
bearing driver set 装轴承用的全套工具
bearing edge 支承边;支梁侧
bearing end leakage 轴承端部漏油
bearing end pressure 轴承端部油压
bearing enelosure 轴承油封

bearing error 方位误差
bearing extractor 轴承拆卸工具
bearing face 轴承支承面;轴承工作面;支承面;端部宽度(接摘);承载面;承压面
bearing factor 承载因数;支承系数;承载因素
bearing failure 轴承损坏;轴承失灵;承载破坏;承压破坏
bearing fatigue point 轴承疲劳强度极限
bearing feed pipe 轴承润滑油管
bearing felt 轴承毡圈
bearing film 轴承油膜
bearing finder 方位仪;方位测定器
bearing fit 轴承配合
bearing flaking 轴承面剥落
bearing flange 轴承凸缘;轴承法兰(盘)
bearing floor block 承重楼板面砖
bearing floor brick 承重楼板面砖
bearing floor clay brick 承重楼板泥砖
bearing force 支承力;承载力;承压力
bearing force of soil 土壤耐压力;土壤承压力
bearing for cover close 机盖压紧轴承
bearing for cover close shaft 筒体紧闭轴轴承
bearing formation 轴承构造;承载岩层
bearing for ring lubrication 环润滑用的轴承;环润滑用轴承
bearing for screw conveyer 螺旋输送机轴承
bearing for washing 洗涤用轴承
bearing frame 承重的框架;轴承架;支承框架;支承架;承重框架
bearing friction 轴承摩擦;支承摩擦
bearing friction loss 轴承摩擦损失;支承摩擦损失
bearing gasket 轴承垫
bearing ga(u)ge 同心度量规
bearing glass block 承重玻璃砖
bearing graph 反应曲线
bearing grease 轴承脂;轴承油脂
bearing groove 轴承油槽
bearing guard 轴承罩;轴承偏转限制器
bearing half 半底盘轴承
bearing hanger 轴承吊架
bearing holder 轴承套
bearing housing 轴承座;轴承箱;轴承体;轴承套;端罩;端盖
bearing hub 轴承毂
bearing-in 掏槽深度
bearing indicator 方位指示器;方位角指示器;偏位指示器
bearing in line 串视方位
bearing in longitudinal direction 纵向支承荷载
bearing inner race 轴承内圈
bearing inner spacer 轴承的内隔圈
bearing in pump 泵用轴承
bearing in reduction gears 减速机轴承
bearing insert 轴瓦;轴承衬套
bearing in shell 带外罩轴承
bearing installing tool 轴承安装工具
bearing instrument 方位仪
bearing insulation 轴承绝缘
bearing interest 负担利息
bearing in traverse direction 横向支承荷载
bearing jacket 轴承外套
bearing journal 滑转子;滑动轴承;轴颈(轴承);支承颈;支承轴承;轴承暗销
bearing keep 轴承盖
bearing labyrinth 轴承迷宫密封
bearing layer 支承层;承载(地)层
bearing layer supporting course 持力层
bearing life 轴承寿命
bearing line 基线;方位线
bearing liner 轴瓦;轴套;轴承衬(料);轴衬
bearing lining 轴承衬
bearing lip 轴瓦凸起
bearing load 轴承负荷;承载荷载;轴承负载
bearing loads and stress 承压载荷及应力
bearing location 轴颈
bearing lock 轴承锁圈
bearing locking collar 轴承锁圈
bearing lock nut 轴承锁紧螺母
bearing lock sleeve 轴承罩套
bearing luboil 轴承润滑油
bearing lubricant 轴承润滑剂
bearing lubrication 轴承润滑
bearing mark 方位标志;方位标(记)
bearing mark control 方位标志控制

bearing masonry(work) 承重的圬工墙;支承圬工;承重圬工(结构)
bearing material 轴承材料
bearing measurement 方位测量
bearing mechanism 支承机构
bearing member 承重构件
bearing metal 轴承合金
bearing metal wear 轴承金属磨损
bearing meter 方位指示器;方位计
bearing module 轴承系数;轴承模数
bearing monitor 轴承报警器
bearing neck 轴承颈
bearing nonscoring characteristics 轴承抗磨伤特性
bearing nose 鼻式托座
bearing number 轴承编号
bearing nut 轴承压紧螺母;轴承螺母;轴承螺帽
bearing nut lock washer 轴承锁紧螺帽垫圈
bearing of base line 基线方位角
bearing of beam 梁支承
bearing of bow 弓轴承
bearing of bridge 桥梁支座;桥梁支承
bearing of cutter head 刀盘轴承
bearing of die 模具的成形部分;模具成形部分
bearing of exposing axis 摄影轴方向
bearing off side 轴承泄油面
bearing of journals 轴颈轴承
bearing of line 测线方位角
bearing of lower tumbler shaft 下导轮轴承
bearing of motor 电动机轴承
bearing of optic(al) axis 光轴方向
bearing of station 电台方位
bearing of tangent 切线方位角
bearing of the connecting rod 连杆轴承
bearing of the crankpin 曲柄轴承
bearing of the cross head pin 十字头销轴承
bearing of the trend 矿床开发的方向
bearing of trend 走向【地】
bearing oil 轴承油;轴承润滑油
bearing oil cooler 轴承油冷却器
bearing oil hole 轴承油孔
bearing oil interface 轴承与润滑油的接触面
bearing oil pipe 轴承润滑油管
bearing oil pressure trip device 轴承低油压脱扣装置
bearing oil pump 轴承油泵
bearing oil seal 轴承油封
bearing oil sump 轴承油槽
bearing on vessel 海上方位
bearing operator 测向员
bearing outer race 轴承外座圈
bearing outer ring 轴承外圈
bearing overloading 轴承过载
bearing pad 支持垫板;轴承垫;支座垫;承压垫板;乌金轴瓦
bearing pad-stone 支承垫石
bearing panel system 预制承重墙板式建筑;大板(建筑)体系
bearing partition(wall) 支承荷载的隔墙;承重隔墙;承重壁
bearing pedestal 轴承座;轴承架
bearing pedestal movement indicator 轴承座位移指示器
bearing performance 轴承性能
bearing per gyrocompass 陀螺罗经方位
bearing per standard compass 标准罗经方位
bearing per steering compass 操舵罗经方位
bearing picket 花秆【测】;测量花秆;方位桩【测】;标杆
bearing pile 支桩;重柱(基础用);宽底盘型钢支座;支柱;支承桩;支撑桩;承压桩
bearing pile through structure 贯通支承桩
bearing pile wall 承重桩墙
bearing piling 重桩;支桩;承重桩
bearing pin 轴承销
bearing pintle 舵针;舵栓
bearing plane 支承荷载的面;支承面;承重面;承载面
bearing plant 轴承厂
bearing plate 基础底板;荷载板;支座垫板;支承垫板;支承板;支撑板;垫板;底垫;承重垫板;承压板;承载板;衬托板;托板
bearing plate bar 钢轨垫板
bearing plate bed 支承垫石

bearing plate process 承压板法
bearing plate stiffness 承压板刚度
bearing plate test 承压板试验
bearing platform 承台
bearing plummer block unit 立式轴承箱组
bearing point 支承点;方位点
bearing position 支承位置
bearing post 支承柱;支承柱
bearing potentiometer 斜度分压器
bearing power 承重能力;承载能力;承载力;负载功率
bearing power of ground 地面承载能力
bearing power of soil 土的承载力
bearing preload indicator 检验轴承预示功器
bearing preloading 轴承预先受载;预应力承载
bearing pressure 承压应力;接地比压;轴承压力;支座压力;支力;支承压力;承载压强;承载压力;承压力
bearing pressure distribution 承压力分布;支承压力分布
bearing pressure of foundation 地基承压力
bearing pressure on foundation 基础压力;基础承压力;地基承压力;土壤压力
bearing property 支承荷载能力;承载性能
bearing puller 轴承拉出器;轴承拉拔器
bearing pulley 轴承滑轮
bearing pump 承压泵
bearing quality 承载能力
bearing quality steel 滚珠钢
bearing race 轴承座圈;轴承套圈;轴承滚道
bearing race puller 轴承座圈拉出器
bearing race snap 轴承固定环
bearing rate 方位变化率
bearing ratio 承重比;持力比;承载比
bearing ratio test 载重比试验;承重系数试验
bearing reaction 轴承反作用(力);轴承反力;支承反力
bearing release lever 轴承放松杠杆
bearing remover 轴承拉出器
bearing repeater 方位分罗经
bearing replacer 轴承换装器
bearing resistance 抗压力;轴承阻力;轴承摩阻力;支承阻力;承压强度
bearing resolution 方位角分解度
bearing retainer 轴承托;轴承护圈;滚动轴承保持器
bearing retainer puller 对中心式轴承拆卸工具
bearing retaining ring 轴承护圈
bearing rib 承重肋;轮辐
bearing ring 井框支承圈;主要井框;轴承套圈;轴承环;支承圆环;支承环;方位圈
bearing rod 支柱;支杆;承重杆;标杆
bearing roller 轴承滚柱;滚柱支座
bearing running hot 轴承过热
bearing rust 轴承生锈
bearing saddle 轴承稳管夹;轴承座;支座锚固
bearing saddle bore 轴承座孔
bearing scraper 柳叶刮刀;轴瓦刮刀;轴承刮刀
bearing screw 支承螺钉;支承螺栓
bearing seal 轴承密封
bearing seat 轴承柱;支承座;支承垫石
bearing segment 轴衬
bearing seizure 轴承咬死
bearing selector 方位选择器
bearing-sense switch 定向转换开关
bearing series 轴承系列
bearing servo amplifier 方位伺服放大器
bearing set 支承压定;支承装置;大齿轮组
bearing setting plate 轴承固定板
bearing shaft 支承柱;支承轴
bearing shell 轴瓦;轴承壳(套)
bearing shim 轴承垫片
bearing shoe 轴承瓦;支梁靴(桥梁)
bearing shoot 结果枝
bearing signal 方位信号
bearing skeleton 承受荷载的骨架;承重框架
bearing skeleton construction 承重框架构造;承重框架结构
bearing skeleton member 承重框架构件
bearing skeleton structure 承重框架结构
bearing slackness 轴承净空;轴承间隙;轴承公差
bearing sleeve 轴承座套;轴承套筒
bearing sleeve cover 轴承座套盖
bearing sleeve gasket 轴承座套垫片
bearing sleeve pin bushing 轴承座套销衬套

bearing sleeve seal 轴承座套式密封
bearing socket 轴承承窝
bearings of pumps 水泵轴承
bearing soil 底土
bearing sole plate 支承底板
bearing spacer 轴承间隔器;轴承隔离圈
bearing spacer ring 轴承隔环
bearing spigot 轴承座孔
bearing spring 承压弹簧;悬架弹簧;承重板簧;片簧;托簧
bearing stand 轴承台
bearing steel 轴承钢
bearing stiffener 支承加劲肋;承重加劲杆
bearing stiffener of steel web 钢腹板支承加劲肋
bearing stone 石头支承面;基石;底座石;门枕石
bearing stool 轴承座架
bearing stop ring 轴承止环
bearing strain 交变应变;挤压应变;支承应变;承载应变;承压应变;承压变形
bearing stratum 持力层;承重层;承载(地)层;地基
bearing strength 挤压强度;轴承强度;支承强度;承载强度;承压强度
bearing strength of ballast bed 道床承载力
bearing strength test 负荷应力
bearing stress 挤压应力;支承应力;承载压强;承压应力;支撑应力
bearing strip 轴瓦垫片;衬垫
bearing structure 支承结构;承重结构
bearing structure of plain web girders 支承荷载的平腹板大梁结构
bearing structure of solid web girders 支承荷载的实腹板大梁结构
bearing stud 支承螺栓
bearing support 轴承支架;支座
bearing support of agitator 搅拌器轴承座
bearing surface 轴承支承面;轴承面;支承面;支撑面;持力面;承重面;承载面;承载表面(积)
bearing surface area 支承面积
bearing surface of a ramp 跳板受力面
bearing surface of foundation 基础支承面;基础承压面
bearing surface of ground 地基支承面
bearing surface of rail 钢轨支承面
bearing suspension 轴承吊架
bearing symbol 轴承代号
bearing synchro 方位同步机
bearing system 支承体系;承载体系
bearing take up 轴承拉紧
bearing telescope 方位望远镜
bearing temperature alarm 轴承温度报警器
bearing tension indicator 轴承预载指示器
bearing test 轴承试验;荷载试验;持力试验;承重试验;承载试验;方位测定
bearing test of soil 土壤承载(力)试验;土的承载试验
bearing thrust 轴承止推(作用);轴承推力
bearing tracking 方位角跟踪
bearing transmission unit 方位测定装置
bearing transmitter 方位发送器
bearing tree 定向树
bearing tree witness 方位树
bearing type 轴承类型
bearing-type connection 承压型连接
bearing-type mix 轴承粉料
bearing unit 定向装置;方位测定装置
bearing unit capacity 单位承载力
bearing unit stress 承压单位应力
bearing upper 上轴承
bearing up pulley 紧轮
bearing value 承载能力;支承值;支承量;载重量;承载值
bearing value of soil 土壤承载(能)力;土的承力
bearing wall 结构墙;持力墙;承重墙;承重炉墙
bearing wall and frame structure 内框架结构;承重墙内框结构
bearing wall construction 承重墙构造;承重墙结构
bearing wall panel construction 承重墙板结构
bearing wall structure 支承荷载的墙结构;承重墙结构
bearing wear 轴承磨损
bearing wearing mode 轴承磨损模式
bearing wearing rate 轴承磨损速度
bearing white metal 轴承白合金
bearing with an outer aligning ring 带球面衬圈

的轴承
bearing with extended inner ring 宽内圈轴承
bearing with segments 分开模制式橡胶轴承
bearing with side plate 带侧板的轴承
bearing with taper bore 带锥孔轴承
bearing zone 测向范围
bear interest 承担利息;行息;负担利息
bearish market 利空市场;空头市场;市场疲软;市场疲跌
Bearium bearing alloy 比里昂姆轴承合金
bear loss 承担损失
bear market 空头市场;卖空市场
bear on 依靠;依靠;压在……上;支承在……上
bear operation 卖空行为
bear pool 卖方联合;卖方联(合经)营
bear position 空头行情;空头头寸;空头部位
bear raid 空头猛跌;卖空浪潮
bear reputation 享有盛誉
bear risk 承担风险
bears 卖空者
bear sail 张帆
bear sales 抛售
bear's breech 叶形装饰(古希腊科林思柱头)
bear's den gray (产于美国新罕布什尔州的)深色花岗岩
bear seller 空头;卖空的证券交易投机商
bear-shaped lamp 熊形灯
bearsite 水砷铍石
bear speculation 空头投机
bear squeeze 空头轧平
bear the test 试验合格
bear transaction 空头交易
bear trap 熊陷堰
bear-trap dam 开合坝闸坝;斜撑闸板坝;浮体闸;开合式闸坝;熊陷式坝;浮体活动坝;人字式活动坝
bear trap drift-chutes 斜撑闸筏道
bear-trap gate 熊陷式闸门;斜撑闸板门;卧式斜撑闸门;开合式闸门(又称熊陷闸门);熊陷闸门;浮体闸门;屋顶式闸门
bear-trap weir 斜撑闸板堰;开合式闸坝(又称熊陷堰)
bearty in form 形式美
bear up for 驶向下风
bear up to 驶向下风
bear weight 承担重量
bear witness 作证
beast column 兽形柱
beast of burden 力畜;驮畜
beat 节拍;巡逻区;音差;捶;差拍;敲击;敲打;偏幅;偏摆;拍音;拍打;搅打
beat about 掉抢;迎风斜驶【航海】
beataway 斜开采面
beat carrier 差频载波
beat-chambersite 贝塔锰方硼石
beat cob work 夯土工作;夯土工程;捣土工作;夯土工程
beat counter 差频式计数器
beat cutter 平压切断机
beat down 跌价
beat-down method 逐次差拍法
beat down price 压价
beat effect 差频效应;拍效应;拍频效应
beaten alumin(i)um 铝片;铝箔(片);铝板
beaten-cob construction 夯土建筑;干打垒(建筑);版筑
beaten-coben gold 金箔
beaten-coben path 走出来的小路
beaten copper 薄铜片;薄铜皮
beaten gold 金叶;金箔
beater 搅拌器;击碎轮;夯土机;夯实机;捣碎轮;轮;轮锯;打碎机;打浆机;锤;捶捣棒;炮棍;拍打器;脱粒滚筒
beater blade 轮叶;击轮叶片
beater colloid mill 打浆胶体磨
beater grate 漏栅格;漏谷条
beater grinder 锤式粉碎机
beater head 锤头
beat-ericaite 贝塔铁方硼石
beater mechanism 冲击机构;拍打机构
beater mill 锤式研磨机;冲击磨;冲击式研磨机;冲击磨磨机;冲击磨机
beater movement 间歇机构
beater paddle 击送轮叶片
beater pick 夯土镐;捣土镐;捣固道砟镐
beater pulverizer 锤式粉碎机;锤式研磨机

beater pulverizer beater mill 锤式粉碎机
beater-refiner 打浆精研机
beater roll(er) 打浆辊;回转器;打浆机轴
beater stuff tester 打浆浆料检测仪
beater-type agitator 冲击式搅拌机
beat frequency 节拍频率;差频;拍频;拍频
beat frequency amplifier 拍频起放大器;拍频放大器
beat frequency change 差频变化
beat frequency detector 拍频检波器
beat frequency meter 差频频率计;差频测试计
beat frequency oscillator 差频振荡器;拍频振荡器
beat frequency receiver 拍频接收机;外差式接收机
beat gargain 还价
beat generator 节拍脉冲发生器
beatify the environment 美化环境
beat indicator 差频指示器
beating 击打法;锻伸;打浆;打扁;补充造林;跳动;搅打
beating crusher 打碎机;锤碎机
beating generator 差拍信号发生器
beating-in 拍入;进入同步;合拍;拍同
beating into leaf 锻打成片
beating machine 打浆机
beating of optic(al) frequencies 光学频率的拍频
beating of waves 波浪拍岸;击岸波
beating the gun 抢先成交
beating tower 打浆塔;塔式碎解机;塔式破碎机
beating transport 加温运输
beating-up 打纬
beating wave 拍频波
beat interference 交调干扰;相拍干扰;差频干扰;拍频干扰
beatl 尿素
beat mass 拍频质量
beat off 乘风破浪
beat of pointer 指针摆动(测量仪器)
beat out 锤薄(金属)
beat output 拍频信号输出
beat patten 拍频波形图
beat period 拍频周期
beat phenomenon 拍现象
beat pin 节拍调整销
beat pump 热力泵
beat radiation sound 拍频辐射声
beat rate 拍频率
beat receiver 超音频外差式接收机;外差式接收机
beat reception 差拍接收
beat screws 节拍螺钉
beat telephone 调度电话
Beattie and Bridgeman equation 比特和布里奇曼方程
beat tone 拍音
beat up 掉抢;年久失修的
beatway 用楔锤破地
beaufet 简便食堂;小吃店
Beaufort number 蒲福数
Beaufort scale 蒲福风级
Beaufort sea 蒲福海(浪)
Beaufort('s)(wind)scale 蒲福风级(表)
Beaufort(wind)force 蒲福风力
beaulate 用天然页岩作室内装修
beaumontage 蜂蜡和虫胶制成的填孔腻子;蜂蜡和虫胶制成的填孔料;填孔;填孔腻子;填孔料;密孔剂
beaumontite 黄束沸石
beautification 美化;城市美化运动;装饰
beautify 装饰;美化
beauty defects (木材的)表面疵病
beauty of art 艺术美
beauty parlour 美容厅;美容院
beauty rest mattress 甜眠型床垫
beauty salon 美容室;美容院
beauty shop 美容院
beauty-spot 风景点;风景区;风景瞭望点
beauvais tile (法国北方常用的)巧克力红色瓦片
beaux arts style 建筑艺术学院式建筑风格
beauxite cement 铝土水泥
beauxite mine 铝土矿山
beauxite ore 铝土矿石
beauxite refractory 铝土耐火材料
beauxite slurry injection pump 铝土泥浆注射泵
beave 河狸
beaver 头盔上所附的活动脸罩
beaver board 木纤维热绝缘墙板;木纤维板;纤维板;人字纤维板;轻质木材纤维板
beaverite 铜铅铁矾
beaver-tail 强力千斤顶;测高天线;扇形雷达束;扇形雷达波束
beaver-tail tile 海狸尾形瓦
beaver type 颏甲式;斜锥式
beaver-type timber dam 木笼填石坝;堆木坝;梢木坝
beavy oil engine 重油机
bebeerilene 贝比烯
be behind in payment 支付误期;拖欠
be booked up 预定一空
Bebout wicket dam 贝包特旋转桁架上木板坝;贝包特式栅孔活动坝;贝包特式门坝;活动撑板坝
be built by contract 包建;包盖
be buried 入土
becanthone 胺甲噻吨酮
be cased up for transport 装箱待运
Becchi-Millian test 贝奇一米利安试验
Bechamp reduction 贝尚还原
be charged extra 额外征收
beche 打捞母锥;取杆器(用以取出钻井时断掉的钻杆);挖井机具
beche-de-mer 海参
beche-de-mer without spike 光参
Bechstein photometer 贝克斯坦光度计
beck 页岩(中的一种);小溪;细流;大桶;山涧
Beck apertometer 贝克数值口径计
beck arc 静弧
beck biologic(al)index 贝克生物指数
Becke line 贝克线
beckelite 方钙铈镧矿
Becker and Kornetzki effect 贝克尔与科尔内兹基效应
beckerite 酚醛琥珀
beckern 鸟嘴砧;丁字砧;铁砧
beckern mill 扩径机
becket 环索;滑轮下吊圈;吊绳;带环短绳;尼眼;绳环
becket bend 单编结
becket block 带眼环滑车
becket bridle 带环短绳
becketed life line 半连环状救生索
Becke test 贝克试验
becket eye 单编眼环琵琶头
becket hitch 环索结;单编结
becket rowlock 桨圈
Beck hydrometer 贝克比重计
becking 碾轧;扩孔锻造
becking mill 碾轧厂;轮箍粗轧机
beck iron 砧角
Beckley ga(u)ge 贝克莱雨量计
Becklin-Neugehaver object 贝克林一诺伊格鲍尔天体
Beckmann apparatus 贝克曼仪器
Beckmann bolometer 贝克曼辐射计
Beckmann coulometric SO_2 analyzer 贝克曼二氧化硫库仑分析仪
Beckmann electronic analyzer of paramagnetic type 贝克曼顺磁型电子分析仪
Beckmann flow colo(u)rimeter 贝克曼流动测色仪
Beckmann molecular transformation 贝克曼分子转换
Beckmann process oxygen monitor 贝克曼流程氧量监测仪
Beckmann rearrangement 贝克曼重排
Beckmann SO_2 analyzer 贝克曼二氧化硫分析仪
Beckmann spectrophotometer 贝克曼分光光度计
Beckmann thermometer 贝克曼温度计
Beckmann total oxidant analyzer 贝克曼总氧化剂分析仪
Beckmann trace moisture analyzer 贝克曼微量水分分析仪
Beckmantown limestone 贝克曼镇石灰碳
Beckwith series 贝克威思统
beclamide 贝克拉胺
be clearly better 明显好转
becloforte spray 必可复气雾剂
beclomethasone 倍өөe松
become annulled 作废;废止
become due 满期;到期;期满
become effective 生效
become lodge in a hole 留在孔内
become operative 生效

become partner 入股
become shareholder 入股
become stuck in a hole 卡在孔内
becoming foggy 起雾
becoming mat(t) 变成暗淡无光
be compelled to cancel the order 被迫取消订货
be computed and taxed separately 分别计算纳税
be concise and to the point 简明扼要
be considerably subnormal 明显低于正常
be contrary to the fact 违背事实
Becorit system 贝氏里悬吊式单轨运输设备
becotide spray 必可酮气雾剂
Becquerel 贝克勒耳
Becquerel cell 贝克勒耳电池
Becquerel effect 贝克勒耳效应
becquerelite 黄钙铀矿；深黄铀矿
Becquerel's ray 贝克勒尔射线
be credited 放账
becueing 防锚咬底法
be cut hard back 重剪
bed 路床；地层；底盘；底；道床；床；衬；层【地】；伐木平垫；版台；台座
bed accretion 河床抬高；河床淤高
bed and board 兼包伙食的宿舍
bed-and-joint width 面缝宽度；砌石接缝宽度；砌块接缝宽度
bed armo(u)ring 河床面层；河床护面；河床粗化；河床保护
bed-a-tree 砍树筑路
Bedaux plan 贝杜工资奖励方案
bed bench 底层平台
bed block 底层地体
bed bolt 底脚螺栓
bed box 机床脚；台座
bed box bed 箱形底座
bed bracket 床铺托
bed brush 床刷
bed-building discharge 造床流量
bed-building material 床沙质；造床泥沙
bed-building process 造床过程
bed-building sediment 造床泥沙
bed-building(wear)stage 造床水位
bed-built water ga(u)ge 造床水位计
bed-built water level 造床水位计
bed-built water stage 造床水位；造船水位
bed cage 床上支架；床护架
bed capacity 床交换能力【化】
bed center 病床工作中心；寝具消毒整理中心
bed chair 卧椅；躺椅
bed chamber 卧室
bed characteristics 河床特征；河床特性
bed charge 底料
bed clearance 船底以下的水深
bed closet 吊折床壁橱；藏床壁橱；卧具间
bed clothes 床上用品
bed coat 底涂层
bed coke 底焦
bed coke height 底焦高度
bed composition 河床组成；河床底质成分；底质成分；底沙组成
bed condition 河床状况
bed conditions of ore body surrounding rock 矿体围岩技术条件差
bed configuration 河床形状；河床形态；河床地形
bed contour 河床等高线；河床等高线
bed course 垫层；矿层走向；底层
bed cover 床帷；床罩
bed cradle 床上支架；床护架
bed crop-rotation 温床轮作
bed current 底层流
bed curtain 床帷
bedded 层状；成层的；层状的
bedded agglomerate 层状集块岩
bedded aquifer 层状含水层
bedded bar 埋置钢筋
bedded chert 层状燧石
bedded clay 层状黏土
bedded deposit(ion) 层状沉积；层状矿床
bedded fissure 层裂
bedded fissure water 层状裂隙水
bedded formation 层状岩系；层状建造；层状地层
bedded media 成层介质
bedded rock 成层岩(石)；层状岩体；层状岩石
bedded rockfill 成层填石；成层干砌块石；分层抛石；分层堆石
bedded sandstone 层状砂岩
bedded soil 层状构造土
bedded structure 夹层结构；垫衬结构；层状构造
bedded tuff 层状凝灰岩
bedded vein 层状脉
bedded volcanic breccia 层状火山角砾岩
bedded volcano 层状火山
bedded water 夹层水
bed degradation 河床刷深；河床加深
bed density 床层密度
bed depth 床深
bedder 制垫片的模型；渣饼；垫石；垫饼；观赏苗圃
bedder lister 开沟作垄器
bed detector 油层检测装置
bed development 河床演变
bed die 阴模
bed dimension 工作台尺寸
bedding 基床；密封膏条；埋藏；管道垫层；垫褥；垫料；底货；打底(指油漆)；成层；层状排列；层理【地】；寝具；铺垫；卧具
bedding a core 稳芯垫砂
bedding angle 层面角；层面方向
bedding armour rock 垫底防护块体(防波堤)
bedding article 床上用品
bedding bar 作垄器安装梁
bedding bottom 作垄型体
bedding cement 底层胶泥
bedding cleavage 层面劈理；顺层劈理
bedding coat 打底涂层
bedding composition 垫层成分；基层结构
bedding compound 底层胶；密封剂填料；垫层成分
bedding concrete 垫层混凝土
bedding condition 成层条件；层面状况
bedding constant 层面常数
bedding corrosion zone 顺层溶蚀带
bedding course 垫层(材料)；垫层(层)；垫底层
bedding course material 垫层材料
bedding course mortar 坐浆；砂浆垫层
bedding course thickness 垫层厚度
bedding dot 点贴短板；灰饼
bedding ductile shear zone 顺层韧性剪切带
bedding error 端热误差；端基误差
bedding face 坐浆面；大面
bedding factor 基床反力系数
bedding failure 基床病害
bedding fault 模压缺陷；层面断层；顺层断层
bedding fissility 层面易裂性；层面裂开性；层面开裂性；层面断开性；顺层裂开性
bedding foliation 层面叶理
bedding for immersed tubes 管段基础处理
bedding frame 铺垫层框架
bedding glide 层面滑移；层面滑裂；层面冲断层；顺层滑动；顺层冲断层
bedding gravel 垫层砾石
bedding-in 研配；研磨；刮研；配合；挖地造型
bedding-in of bearings 轴承的配合
bedding-in period 走合期；走合时间
bedding in soil 土壤中垫层；土路基
bedding into 嵌入；砌入
bedding into stone or concrete 砌入石块或混凝土中
bedding-in zone 分级区
bedding joint 层面节理；层间接缝；顺层节理；接缝
bedding lamellae 层面壳层
bedding layer 基床垫层；垫层
bedding layer of mound type breakwater 斜坡堤基床垫层
bedding layer of soil 土壤垫层；土垫层
bedding magnetization 顺层磁化
bedding material 成层材料；垫底材料
bedding mix 垫层配合比；垫层混合料
bedding mortar 砂浆层；浆层；垫层砂浆；砂浆垫层
bedding of a furnace 分层铺炉底
bedding of bearing 支承刮削
bedding of brick 砖铺基床；砖铺垫层
bedding of land 土地平整
bedding of pipes (ion)管层；管道垫床；管道垫层
bedding of rubble mound type 明式基床
bedding on rubber cushions 装在橡皮垫上
bedding out 货物搁置方法
bedding plane 岩层层面；垫层面；层理面；顺层面；层面
bedding plane fault 层面断层
bedding plane fracture 层面裂缝
bedding plane landslide 顺层滑坡
bedding plane slip 挠褶滑动；层面滑动；曲滑
bedding plant 花坛花草；花床植物；储料场；花坛植物
bedding plate 层理面
bedding putty 油灰样可移动的底层；打底腻子
bedding recumbent folds 顺层掩卧褶皱
bedding reinforcement 底层配筋
bedding sand 垫层用砂；垫层砂；地基砂层；砂垫层
bedding schistosity 层面片理；顺层片理
bedding sheath 电缆内套
bedding slip 层面滑动；顺层滑动
bedding slip fault 顶层滑动断层
bedding slip structure 层滑构造
bedding stone 平置石；基石；垫石；验平石；座石；扁平石
bedding stones 石垫层
bedding structure 层状结构；层状构造
bedding surface 层面；岩层层面
bedding surface structure 层面构造
bedding system 衬垫方法
bedding thrust 层面冲断层；顺层冲断层
bedding value 地基系数；基床值；基床系数
bedding void 层间空隙
bedding wood material 垫木
bed disturbance 河床扰动；底沙扰动；床沙扰动；床面扰动
bed dowel 灰缝销；砌体中的暗销；石砌体定位销；石砌体暗销
bed dredge(r) 海底挖泥船
bed dunes 河底沙丘；床面沙纹
bed dwelling unit 住宅居住单元
be decided on job-site 现场决定
be declared bankrupt suspend 宣告破产
be deficient in resources 缺乏资源
bed elevation 基床高程；河底高程；河床高程
bed elevator (摇动病床的)活动装置
bed erosion 岩层风化；河床冲刷；河床刷深；河床加深；河床冲刷
be determined on job-site 现场决定
be development 河床演变
bed expansion 滤层膨胀
bed factor 河床系数
bed-factor equation 河床因数方程；河床因素方程
bed filter 滤床式过滤器
bedflow 底流
Bedford cloth 贝德福呢
Bedford cord 经向凸条织物；经条灯芯绒；厚实凸条布
bed form 河底形状；河床形状；河床形态；海底形状；底形；床面形态
bed formation 河床形成；海底形成
bed-forming discharge 河道造床流量；造床流量
bed-forming material 造床泥沙
bed-forming process 河床造床过程；河床演变过程；河床形成过程；造床过程
bed-forming stage 造床水位
bed form roughness 底形糙度；床面形态糙率
bed form superimposition 底形叠加作用
bed for structural testing 结构试验台座
bed frame 床宽；基架；机座底架；座架；床架；支承结构；底座框架
bedframe ga(u)ntry 支承结构
bedframe of pile driver 打桩架基座；桩架基座
bed friction coefficient 河床摩擦系数
bed friction factor 河床阻力系数；河床摩擦系数；底摩擦系数
bed friction stress 基底摩擦应力
bed-generating flow 造床流量
bed geometry 河床形状；河床形态
bed glazing 安玻璃的油灰层
bed gradient 河底坡降；河床比降
bed groin 河底潜堤；河床潜(没式)丁坝；河床丁坝；填积堤
bed groyne 河底潜堤；河床丁坝
bed hedgehopping 垫高；垫层厚度
bed-holding 河床容纳装置；河床容纳能力
bed hook 底座凹槽
bed house 简易旅馆
bediasite 贝迪阿熔融石
bedim 使模糊不清
bed in 卧模(地坑造型)
bed-in-series 串联床

bed interchange 河床泥沙交换
bed irregularity 河床不平整度
bed irrigation 灌床;床灌;污水畦灌
be divided into three classes 分为三类
be divided into three types 分为三种型号
bed joggle 底面榫接
bed joint 横缝;底层接缝;成行(砖)缝;层面节理;层间接缝;平层节理;水平砌缝;水平灰缝;拱的放射形接缝
bed knife 固定刀;底刀(刀片)
bed lake 河床湖
bedlam 精神病院
bed lamp 床灯;床头灯
bed layer 河床底层;基层;推移质运动层
bed level 基床高程;河底高程;河床高程
bed leveler 河床平整器;底部整平器
bed lift 病床摇高装置
bed light 床头灯
bed load 底沙;底负载;底负荷;床沙荷载;推移质
bed-load calibre 河床推移质输送能力
bed-load deflecting apron 挑砂坝
bed-load deflecting sill 拦沙底槛;导沙底槛;挑砂底槛
bed-load deflection apron 拦沙护坦;挑砂护坦;拦沙槛;导沙护坦
bed-load discharge 底沙输送量;推移质输移量;底质输移率;底质输送量;推移质输沙量
bed-load discharge measurement 推移质输沙率测验
bed-load discharge per unit width 单宽推移质输沙量;单宽输沙能力
bed-load diversion 底沙分流
bed-load ejector 底沙排射器
bed-load feeder 模型试验用加砂器
bed-load formula 床沙公式;底沙公式;推移质公式
bed-load function 床沙函数;底沙函数
bed-load intensity 推移质输移强度
bed-load measurement 推移质测验;推移质测定
bed-load model test 推移质泥沙模型试验
bed-load motion 底沙运动;推移质运动;底质运动
bed-load movement 底质运动;底沙运动;推移质运动
bed-load movement above competence 河床碎屑移动超限度
bed-load movement below competence 河床碎屑移动小于限度
bed-load moving strip 推移质移动带
bed-load of pebbles 卵石推移质
bed-load production 河床碎屑产量
bed-load rate 河床碎屑产生率;推移质输沙率
bed-load sampler 河床碎屑取器器;底沙取样器;底沙取样器;推移质取样器;推移质采样器
bed-load sampling 底沙取样;底质取样
bed-load sediment 底质;推移质泥沙
bed-load sweep 底沙横向输沙;横向输沙
bed-load trajectory 推移质轨迹
bed-load transport 河床碎屑移动;底沙输移;推移质搬运
bed-load transportation 推移质输移
bed-load transport intensity 推移质输移强度
bed-load transport test 推移质输移试验;推移质输沙试验
bed-load trap 河床碎屑截除器;推移质集沙槽
bed-making 铺床;寝具整理工作
bed material 河床质;河床物质;底质;底沙;床沙;河砂
bed material armo(u)ring 河床质粗化
bed material density chart 底质密度图
bed-material discharge 河床质输移量;底质输移率;底质输送量;床沙输移量;推移质输沙量;推移质输沙量
bed material exploration 底质勘察
bed-material load 床沙质;底负载;底负荷;推移质
bed material measurement 河床质测验
bed material sampler 河床质采样器;底沙取样器
bed material sampling 河床质采样;底沙取样
bed material size 河床质粒径;底沙粒径;床沙粒径
bed material size distribution 河床质颗粒分布
bed measure 下方【疏】
bedment 垫板(压缩试验用)
bed mortar 坐浆
bed mo(u)ld(ing) 底线脚;粉刷线脚层;檐板下的线脚;底层线脚;深凹饰
bed movement 河床运动;河床移动

bed mud 底泥
bed necessaries 床上用品
bed of a slate 石板底层
bed of boiler 锅炉座
bed of brick 砖铺底层;运砖框
bed of clay 黏土下卧层
bed of land preparation 播前土壤准备
bed of material 料层;物料层
bed of mortar 灰浆打底层;灰浆层
bed of nodules 料球层
bed of pitching 护坡垫层;砌石护坡基床
bed of river 河底;河床
bed of slate 石板底层
bed of the high sea 公海海床
bed of the paving 护坡基床
bed of the pitching 护坡基床
bed of vein 矿脉层;层状矿脉
bed oil paint 油质底层涂料;油性底层涂料
bed on the grate 算上物料层
Bedoulian 贝杜尔阶
bedpan 垫盘;便盆
bed pan closet 便盆间
bedpan washer 大便器;便盆洗涤器
bedpan washer hose 冲洗便盆软水管;便器冲洗软管
bed particle size 底沙粒径;底沙粒度
bed perimeter 床边视野计
bed perturbation 河床扰动;底沙扰动;床沙扰动;床面扰动
bed piece 堆垛垫木;床身;座板;垫板;底座;台板
bed plane 层面
bed plant 花坛植物
bedplate 座板;台板;炉底基台;道岔垫板;(安装机器的)床板;基础底板;基板;支座板;垫板;底(座)板;底刀盘
bed-plate foundation 底板式基础;板式基础;座板基础;底板基础
bedplate knife 底刀刀片
bed plug 床边(电)插头;层栓
bedpost 床柱
bed process 河床演变过程;河床形成过程
bed process observation 河床演变观测
bed production 河床泥沙产生;推移质泥沙形成
bed profile 河底纵断面;河床纵剖面;河床纵断面;河床剖面
bed puddle clay 夯实黏土层;夯实夯土层
bed putty 油灰底层(玻璃);打底油灰;嵌玻璃用油灰
bedrail 床栏
bed recess 底座凹槽
bed reinforcement 河床加固
bed resistance 河床阻力
bedrest 床支架
bed ripple 河底高低不平;河底波痕;河底沙纹;河底沙波;沙纹;沙坡;沙波
bedrock 基岩(层);基石(层);岩床;根底;底岩
bedrock aquifer 基岩含水层
bedrock estuary coast 基岩港湾海岸
bedrock geologic(al) structural element 基岩地质构造要素
bedrock geology 基岩地质(学)
bedrock intensity 基岩强度
bedrock landslide 基岩滑坡
bedrock lithologic(al) type 基岩岩性类型
bedrock map 基岩地质图
bedrock mark 基岩标
bedrock motion 基岩运动
bedrock outcrop 基岩露头
bedrock price 最低价(格)
bedrock rapids 基岩急滩
bedrock seated flood plain 基座河漫滩
bedrock seated terrace 基座阶地
bedrock soil 基岩土壤
bedrock-soil layer model 基岩土层模型
bedrock spur 基岩露头;暗礁
bedrock surface 基岩顶面
bedrock surveying 基岩测定;基岩标定
bedrock test 确定浮土厚度和基岩性质的钻探
bedrock topography 基岩地形
bedroom 寝室;卧室
bedroom basin 卧室水盆;卧室洗涤盆
bedroom block 住宅区
bedroom building 宿舍房屋;住宅区
bedroom chair 卧室(用)椅(子)
bedroom city 卧城

bedroom closet 卧室壁橱
bedroom closet bank 衣橱式卧室壁橱;卧室壁橱台架
bedroom closet for children 小孩卧室壁橱
bedroom closet for men 男人卧室壁橱
bedroom closet for women 女人卧室壁橱
bedroom community 郊区;住宅区;卧城
bedroom cupboard 卧室内壁橱
bedroom decoration 卧室装饰
bedroom door 卧室门
bedroom floor 卧室楼层
bedroom house 住宅区
bedroom lamps 卧室灯具
bedroom stor(e)y 卧室楼层
bedroom suburb 卧城郊区
bedroom town 中等住宅区;卧城
bedroom unit 住宅区;宿舍楼
bedroom window 卧室窗(子)
bed roughness 河床粗糙度;河床糙率;沟床糙度
bed sample 河床沙样
bed sampler 底沙取样器
bed sand 底沙
bed scour 河床冲刷;冲刷;河床冲刷;河床刷深;河床加深;海底冲刷
bed sediment 底质;底沙;推移质
bed sedimentation 河床沉积(作用)
bed sediment discharge 底沙输送量;推移质输移量;推移质输沙量
bed sediment load 沉积底沙
bed sediment size 底沙粒径;底沙粒度
bed separation 层状剥落;层间分离;分层
bed separation cavities 层间脱离成穴
bed set 层组
bed shear 河床剪切力;河床剪力;层面剪切
bed shear stress 河床剪切应力
bed shear velocity 河床摩阻流速
bedside board 床边挡板
bedside cabinet 床头柜;床头(小)桌
bedside card 床头卡
bedside cupboard 床头柜
bedside lamp 床头灯
bedside monitor 床边监护器
bedside table 床头(小)桌
bed-sill 底槛
bed silt 底沙;床沙;河底淤泥
beds-in-parallel 并联床
bed sintering 料层结块
bed sit(ter) 起卧室;卧室兼起居室;卧室兼起居(间;租用起居室
bed-sitting flat 单间宿舍;卧室及起居两用的单间宿舍
bed-sitting room 起卧室;卧室兼起居室;卧室兼起居间
bed-sitting room dwelling unit (卧室兼起居室共有的)居住房屋;住宅住宅单元
bedskirt 床帷
bed slide 刀架底板
bed slope 河底坡降;河底坡度;河底比降;河床坡降;河床底坡;河床比降;底坡(度)
beds of passage 过渡层
beds of precipitation 沉淀层
bedsoil 支承土壤
bedspace 床位总数(酒店、医院、宿舍等)
bedspread 床罩
bedspring 弹簧床
bedspring array 横列定向天线阵
bed stability 河床稳定性;床的稳定性
bed stability factor 河床稳定因素;河床稳定系数
bed stability index 河床稳定性指标
bed stabilization 河床加固
bedstand 试验台
bedstead 床架;试验装置;试验台
bedstone 垫石;基石;底石
bed stressing 应力工作台
bed surface 河床表面;床面
bed surround 底层覆盖
bed sweep(ning) 扫海;扫床(指探测河床有无障碍物)
bed-sweep(ning) boat 扫床船
bed system 以床为寝具的居住方式【建】;畦植法
bed table 诊察台
bed-terrace 阶地;梯田;台地
bed thickness correction 地层厚度校正
bed timber 枕木;垫木

bed town 卧城
bed town suburb 卧城郊区
bed trap 推移质拦截井
bed-type milling machine 高刚性铣床；床型卧式铣床；床身式铣床
bed-type pneumatic concentrator 床层式风选机；床层式风力选矿机
bed-type vertical miller 立柱移动工作台不升降铣床；床型垂直式铣床
be dull of sale 滞销
bed under-locker 床下储[贮]柜
bed urinal 便壶
bed vein 层状脉
bed velocity 河底流速；海底声速；底流速；底部流速
bed vibrator 床用振荡器
bed volume 床体积；床层体积
bedward 病房
bed wave 沙浪；沙波；河底沙波；河底波纹状起伏
bed way grinder 床身导轨磨床
bedways 机床导轨；床身导轨(机床)；视层状
bedye 施彩；加彩；着色
bee block 船首斜桅桅牵槽板
beech 山毛榉
beechnut 地空通信系统
beechnut oil 山毛榉实果油
beech parquet 山毛榉镶木地板
beech shingle 山毛榉墙面板
beech strip floor covering 山毛榉板条楼面
beech strip floor(ing) 山毛榉板条楼面
beech wood 山毛榉木(材)；榉木焦油
beechwood mallet 山毛榉木锤
beechwood parquet(ry) 榉木条拼镶地板
beechwood shingle 榉木瓦
beechwood strip floor cover(ing) 山毛榉木条地板
beechwood tar 山毛榉木焦油
beechwood tread 榉木楼梯踏步板；山毛榉木踏步板
bee colony 蜂群
beedle 伸出；突出；木夯；木槌
bee-escape 脱蜂器
bee farm 养蜂场
beef stearin 油硬脂
beef tallow 牛脂
beef up 充实
beefwood 木麻黄；硬红木；土麻黄
beefwood extract 木麻黄栲胶
beehive 蜂箱；蜂房
beehive coke 蜂窝状炉焦炭
beehive coke oven 蜂房炼焦炉
beehive cooler 蜂房式冷却器
beehive house 蜂窝式房屋
beehive kiln 多孔窑；倒焰圆窑；蜂窝式窑；蜂巢式窑；蛇窑
beehive oven 蜂窝式炼焦炉；蜂巢炼焦炉
beehive tomb 地下蜂窝式墓室；蜂窝型地下墓(古希腊)
beehive type radiator 蜂窝式散热器；蜂窝散热器；蜂巢式散热器
bee-hole 蜂孔
bee-hole bore 蜂孔蛀虫
bee honey 蜂蜜
bee house 养蜂场
beekeeping 养蜂(业)
Beek index 比克指数
beekite 玉髓燧石
be eliminated 被淘汰
beeline 直线；空中距离；捷径；最短距离；最短路径；两点间直线
beeling boom 顶木架
beep 高频笛音
beep box 操纵台
beeper 遥控员；雷达遥控装置
beerachite 辉长细晶岩
beerbachite 辉长岩红钴银矿；微晶辉长岩
Beer-Bouer law 比尔—鲍格定律
Beer-Bouguer law 比尔—布格定律
beer brewing 啤酒酿造
beer cellar 啤酒地窖
beer garden 啤酒店花园；屋外花园酒店
beer hall 啤酒店
beer house 啤酒店(英国)
Beer-Lambert-Bouguer law 波格—朗伯—比尔定律
Beer Lambert's law 比尔—朗伯(吸收)定律
Beer-Lambert's equation 比尔—郎伯方程
beer-money 小费

beer mug 带柄啤酒杯
beer shop 啤酒店
Beer's law 比尔定律
beer stone 无鲕石灰石
beer wagon 啤酒车
bee's homenycomb brick 蜂窝砖
bee's wax 蜂蜡
beeswax 黄蜡；蜂蜡(色)；漂白黄蜡；上蜡；蜜蜡
beeswax tar 蜂蜡焦油
beeswax yellow 蜜蜡黄
bee's wing 蜂翅纹
beet 亚麻束
beet(cancer) 甜菜
beet cleaner loader 甜菜清理装载机
beet collecting station 甜菜收购站
beet cutter 甜菜切碎机
beet digger 甜菜挖堆机
beetle 搅打(机)；木夯；捣打机
beetle attack 虫蛀
beetle back 标志牌背后
beetle calender 锤打轧光机；捣打轮压机
beetle cement 尿素树脂胶；脲醛树脂黏合剂
beetle faller 捣布锤
beetle head 桩头落锤；汽锤；打桩锤；送桩锤
beetle socket 可上下冲击的打捞筒
beetle stone 龟背石
beetling 捣布
beetling cliff 悬崖；峭壁
beetling machine 搅打机；捣布机
beetling wall 绝壁；峭壁
beet loosenet 甜菜松土器
beet mallet 甜菜木锤
beet piler 甜菜堆藏机
beet pulp 甜菜渣(堵漏材料)；甜菜浆
beet root 甜菜块根
beet silo 甜菜地下储[贮]藏室
beet slicer 甜菜切片机
beet sugar factory 甜菜糖厂
beet sugar process wastewater 甜菜糖厂废水
beet sugar waste 甜菜制糖废渣
beet sugar waste liquour 甜菜制糖废液
beet top chopping machine 甜菜顶切碎器
beet top silage 甜菜顶青储[贮]
beet washing machine 甜菜清洗机
be evaluated for salt tolerance 耐碱性鉴定
bee venom 蜂毒
bee-venom poisoning 蜂毒液中毒
beewax 蜜蜡
beewax jade 蜜蜡黄玉
be exempted from income tax 免征所得税
be exempted from individual income tax 免征个人所得税；免纳个人所得税
be exempted from taxation 免予征税
bee yard 养蜂场
beffry 教堂尖塔
Beflast roof truss 贝氏弓形屋架
be flexible 变通
before and after 前后效果对比
before and after analysis 前后对比分析
before-and-after contrast approach 前后对比法
before and after method 前后法
before and after study 前后对比分析；前后对比调查
before-and-after test 辐照前后试验
Before Christ[BC] 公元前
before evaluating for yield 测定产量之前
beforehand 预先；事前
before jacketed 安装夹套前
before planting 播前
before-separation cost 分离前成本
before shipment 装船前
before tax 纳税前
before-tax cash flow to equity 税前现金流量财产价值
before-tax income 税前收入
before test 事先测试
before the beam 正横前
before-the-lens shutter 镜前快门
beforsite 镁云碳酸岩
beforsite group 镁云碳酸岩类
befouled water 污水
befouling environment 污染环境
be found in alluvium 在潮湿冲积层中发现
be found in several kinds of rocks 在若干种岩石中找到
beggar-my-neighbor import barriers 以邻为壑的进口壁垒
beggar-my-neighbour policy 以邻为壑政策
beggar weed 荒地杂草
Beggiatoa 贝氏硫菌属
begin 开始
begin block 开式子程序；开始块
begin chain 链头；开始链
begin column 开始行；开始列；开始栏；首列
begin construction 动工
beginning 开端
beginning and ending 开始与结尾
beginning and end of water way 通航起讫点
beginning average method 期初平均法
beginning balance 期初余额；期初差额
beginning block 开始分程序
beginning capital 期初资本
beginning character 开始字符
beginning character position 字符开始位置
beginning correlation time of each trace 每道起始相关时间
beginning event 始点事项
beginning height 出现点高度
beginning inventory 期初存货
beginning lamp 开始信号灯
beginning node 始结点
beginning of a month 上旬
beginning of astronomic(al) morning twilight 天文晨光始
beginning of Besselian fictitious year 贝塞尔假年岁首
beginning of break up 解冻开始(期)
beginning-of-chain 链头；链首(部)；开始链
beginning of circular curve 圆曲线起点
beginning of construction 施工开始
beginning of curve 曲线起点；曲线的起点；平曲线起点；始曲点
beginning of curve sign 曲线起点标志
beginning of data handling 数据处理开始
beginning of design 设计起点
beginning of ebb 初落(潮汐)
beginning of fall 开始退水；退水开始(期)
beginning-of-file label 文件开始标号
beginning of floating 通航河段起点
beginning of flood 初涨(潮汐)
beginning of freeze up 封冻开始(期)
beginning-of-information marker 信息开始标志
beginning of month 月初
beginning of morning twilight 晨光始
beginning of nautical morning twilight 航海晨光始
beginning of navigation 通航河段起点
beginning of partial eclipse 初亏
beginning of period 期初
beginning of rise 开始涨水；涨洪开始(期)；起涨
beginning of scale 标度起点
beginning of shipping 通航河段起点
beginning of snow cover 积雪开始(期)
beginning of storage 开始蓄水
beginning-of-tape marker 带开始标记
beginning of the bai-u 梅雨开始
beginning of the fiscal year 财政年度初
beginning of the route 进路起点
beginning of transition curve 缓和曲线起点
beginning-of-volume label 卷头标；卷开始标号；卷标的开始
beginning of year 年初；岁首
beginning point 出现点；始点
beginning point number of migration 偏移起始点号
beginnings 开始阶段
beginning tape label 带起始标记
beginning temperature 初温
beginning threshold value of energy 能量起始阈值
beginning time of migration 偏移起始时间
begin statement 开始语句
begin symbol 开始符号
begin time of subsidence 沉降起始时间
begohm 千兆欧(姆)
begonia 秋海棠
be good enough to reply early 请早日答复
be hard to cultivate 耕作起来费劲
behaved wood 改性木材
behavio(u)r 功效；性状；性能；行为；行动；特性

behavio(u)r abnormality 行为异常
behavio(u)ral decision function 行为决策函数
behavio(u)ral ecology 行为生态学
behavio(u)ral engineer 行为工程师
behavio(u)ral geography 行为地理学
behavio(u)ral science 行为科学
behavio(u)ral system 行动系统
behavio(u)r approach 行为研究法
behavio(u)r at sea 抗浪性
behavio(u)r constraint 状况约束
behavio(u)r decision making 行为决策
behavio(u)r equations 行为方程
behavio(u)r in a seaway 抗浪性
behavio(u)r in fire 火中性能
behavio(u)r in natural environment 自然环境中的特性
behavio(u)r in service 运转状况;运转性能;运用状态;使用状况;使用性能
behavio(u)r model 行为模型;外态模式
behavio(u)r modification 行为修正
behavio(u)r observation of spring 泉水动态观测
behavio(u)r observation of surface water 地表水动态观测
behavio(u)r observation of surface water level 地表水位动态观测
behavio(u)r of average cost 平均成本的行为
behavio(u)r of boundary layer 附面层特性
behavio(u)r of cross-section 截面变化(过程)曲线
behavio(u)r of dam 大坝特性
behavio(u)r of electricity 电气性能
behavio(u)r of fluids dynamics 流体动力学特性
behavio(u)r of ground deformation 地面变形
behavio(u)r of ozone 臭氧特性
behavio(u)r of profit 利润动态
behavio(u)r of rail joint 钢轨接头工作状态
behavio(u)r of river 河流特征;河流生状;河流特性
behavio(u)r of structures 构造性能;结构工作状况;建筑物性能;结构性能
behavio(u)r of surrounding rock 围岩特性
behavio(u)r of the economy 经济状况的变化
behavio(u)r of variogram 变差函数性状
behavio(u)r of well 井的状态
behavio(u)r on heating 加热时的行为
behavio(u)r on storage 储[贮]藏时状况
behavio(u)r pattern 行为模式
behavio(u)r rating scale 行为量表
behavio(u)r science 行为科学
behavio(u)r setting 行为背景
behavio(u)r theories of the firm 厂商的行为理论;企业行为科学
behavio(u)r under fire 着火时状况
behavio(u)r under heat 受热时状况
behead 河流的夺流;断头;夺流
beheaded river 夺流河;断源河;断头河
beheaded stream 断源河;断头河
beheaded valley 断头谷
beheading 夺流;河流袭夺
beheading of river 河流的尽头;河道夺流;夺流
beheading of stream 河道夺流
be heated 受热
be heavier than air 比空气重
behen 贝昂
behen oil 贝昂油
behierite 硼钽石
behind completion 没有按期完成
behind lining 衬砌背面
behind propeller efficiency 螺旋桨船后效率
behind schedule 跟不上进度;误期;拖期
behind test 船后(螺旋桨)试验
behind the lens 镜头后测光标志
behind-the-lens shutter 镜面快门;镜后快门
behind-the-scene master 幕后操纵者
behind the schedule 落后于预定计划;晚点
behind-the-scheming 幕后策划
behind time 过期
Behmer method 勃赫曼法
Behm lot 柏姆型回声测深仪
behoite 羟铍石
Behrens-Fisher problem 贝伦斯-费希尔问题
Behrman's projection 贝尔曼投影
Beibu Bay 北部湾
Beibu Bay basin 北部湾盆地
Beibu Bay depression 北部湾拗陷地带
beidellite 贝得石

beige 米色;灰褐色;驼色
Beijing coordinate system 北京坐标系
Beijing Geodetic Coordinate System 1954 1954年北京坐标系
Beilby layer 贝尔伴层;拜尔比层
Beiley flow meter 贝雷流量计
Beilstein flame detector 贝尔斯坦火焰电离探测器
Beilstein test 贝尔斯坦试验
beilupeimine 白炉贝碱
be impatient for success 急于求成
be in account with 与……有账务来往
be in arrears 拖欠
be in arrears with tax payment 拖欠税款
be in debit 亏空
be in default 不履行(契约)
being build area 建设矿区
being in arrears in handing over profits 拖欠上交利润
being poisoned by agricultural chemicals 农药中毒
being produced area 生产矿区
being to the hammer 送去拍卖
Bein kep gear 贝恩罐笼托座
be in low water 缺少资金
be in terms 在交涉谈判中;在交涉商量中
be interred 入土
be in the hole 亏空
Beishouling type pottery 北首岭类型陶器
beit hilani 古叙利亚宫殿
beiyinite 白云矿
be kept perfectly 保持完整
be kept upright 保持正直
Bekk's viscometer 贝克黏度计
bel 贝(尔)(音强单位)
Belady algorithm 贝莱迪算法
Belain plate atmometer 贝兰板式蒸发计
Belanger's critical flow 伯朗格临界流
Belanger's critical velocity 伯朗格临界速度
belated 过期的;误期的
belated claims 迟索的赔偿;逾期索赔
belay 牢拴绳子;系绳处
belaying cleat 加强板;索系耳;系绳栓;导缆钳
belaying pile 套管桩
belaying pin 紧缆棍;索索栓;挽缆插栓;套索桩;系索桩;索索柱
belaying pin rack 挽索插栓架
belch 猛烈爆发;冒烟;冒火;一阵阵冒出
belching 喷出;嗳气
belching well 间歇自喷井
belection 凸出平面的线脚
belemnite 箭石
belemnite faunal province 箭石动物地理区
Belfast roof 贝式弓形盖;贝式弓弦屋盖;贝尔法斯特弓形屋顶
Belfast roof truss 贝尔法斯特弓形屋架
Belfast sink 贝尔法斯特水槽
Belfast truss 弧形桁架;贝式弓形屋架;贝式弓弦桁架;贝尔法斯特桁架
belfried 有钟楼的
belfry 钟楼;船钟架
belfry cloth 粗厚方平棉织物
belge 米色
Belgian architecture 比利时式建筑
Belgian biotic index 比利时生物指标
Belgian block 比利时式砌石块;比利时大理石
Belgian block pavement 比利时式砌石块路面
Belgian (block) road 比利时(矩形块石)路
Belgian blue marble 比利时蓝色大理石
Belgian excavation method 比利时(开挖)法
Belgian flexible pavement design method 比利时柔性路面设计法
Belgian kiln 比利时窑
Belgian looping mill 比利时式线材轧机
Belgian method 先拱后墙法;支承顶推法;比利时先拱后墙法
Belgian method of timbering 比利时式隧道支撑法
Belgian method of tunnel(l)ing 比利时式隧道法;隧道施工比利时法
Belgian mill 比利时式轧机
Belgian rod mill 比利时式线材轧机
Belgian roof truss 斜腹方三角形桁架;比利时水屋架
Belgian sandwich cable method 比利时预应力混凝土法
Belgian truss 法国式桁架;比利时式桁架;比利时式桁架

Belgian-type layout 横列式布置
Belgian wire mill 交替二辊横列式线材轧机;比利时式线材轧机
Belgian zinc distilling furnace 比利时式锌蒸馏熔炉
belgite 硅锌矿
Belgium architecture 比利时建筑
Belgium truss 比利时式桁架
belgravia 富裕住宅区;中上流社会
Belimer bleacher 贝尔麦式漂白机
Beliminator 阳极电源整流器;屏电源整流器
be limited to temperate climates 限于温带气候
Belisha beacon 倍利夏警标(穿越道路标志);交通指示柱
belite 二钙硅酸盐
belite cement 二钙硅酸盐水泥;贝利特水泥
belith 斜方硅灰石
Belitzski's reducer 比利兹基减薄剂
beljankite 铝氟石膏
bell 炉盖;漏斗状接口;料钟;号钟(避碰规则);旋杯;钟形物;钟盖;高炉盖;承口;起落架舱
belladonna lily 孤挺花
Bellamy drift 贝拉米偏航角
bell and bell reducer 双承大小头
bell and flange piece 承口法兰短管
bell and flange reducer 承盘大小头
bell and hopper 炉口装料斗和盖;进料器
bell and hopper arrangement (高炉的)钟斗装置
bell-and-plain end joint 套管接头;平接
bell and reducer 承插式大小头
bell and socket joint 承插接头
bell and spigot 承插式接口;承插三通;承插口
bell and spigot bend 承插弯管;承插式弯管
bell-and-spigot cost-iron pipe 承插口铸铁管
bell and spigot cross 承插四通
bell and spigot eccentric reducer 承插偏心渐缩管
bell and spigot joint 承插口钟口(式)接头;承插盒;管端套筒接合;承插式接头;承插接头;承插接合
bell and spigot offset 承插式乙字管
bell and spigot pipe 套筒接管;承插管
bell and spigot pipe joint 管道承插接口
bell and spigot reducer 承插式大小头
bell-and-spigot stoneware pipe 承插式粗陶管
bell and spigot wye branch 承插Y形分叉管
bell arch 钟状拱;钟形拱
Bellarmine 盐釉陶罐
bell beam 炉钟梁;料钟平衡杆
bell bearing 钟杆推力轴承
bell brass 钟黄铜
bell bronze 钟青铜
bell buoy 警钟浮标;响铃浮标;钟形浮标;钟浮标
bell button 电铃按钮
bell cage 钟形构架;钟架
bell caisson 钟形沉箱
bell canopy 钟形天篷;钟顶天篷
bell cap 钟帽;泡帽
bell capital (早期英国建筑的)柱顶;倒钟形柱头;钟形柱头;钟状柱帽
bell cast 滴水槽;钟口式滴水槽
bellcast caves 嵌齿扣合屋檐;钟口屋檐
bell cell 钟形电解池;钟式电解池
bell center punch 钟形中心冲头;钟形中心冲孔器
bell chamber 钟楼
bell character 报警字符;报警符(号)
bell chime steam whistle 鸣声汽笛
bell chuck 钟形卡盘;带螺钉钟壳形夹头
bell cistern 虹吸式自动冲洗水箱;钟罩式虹吸水箱;高水箱;钟盖虹吸水箱
bell clapper 电铃锤
bell collar 锥形承口套管
bell cord 铃绳
bell cot 罩子;小床;钟状小屋;钟架;屋顶上小钟楼
bellcote 吊钟构架
bell counterweight 料钟平衡锤
bell coupling 钟形联接器
bell course 环饰【建】
bell cover 炉盖
bell crank 钟形曲柄;曲柄杠;钟锤杠杆;直角杠杆;曲拐;双臂曲柄;摇臂吊
bell crank governor 直角杠杆调节器
bell crank lever 直角形杠杆;双臂曲柄杠杆
bell crank shaft collar 直角杠杆轴环
bell crowned 钟形冠的

bell damper 钟状挡板;钟形气流调节器;钟式闸板
bell design 钟形结构设计
bell diving technology 钟潜水技术
bell dolphin 钟形系船墩;钟形靠船墩
Bell dresser 贝尔型整形砂轮器;贝尔型修整砂轮器
belled 承插口的;套接的;有钟形口的;钟形口的
belled-and-flange joint 插塞式凸连接
belled excavation 钟形开挖
belled mouth 喇叭口
belled-out cylindric(al) pile 大头圆柱桩
belled-out pier 整体基础的码头
belled(-out) pile 扩底桩;大头桩
belled-out pit 扩大竖井;扩大截面
belled-out section 扩大段
belled pier 扩底墩;钟形墩
belled shaft 扩底墩
Belleek ware 贝利克瓷
bell end 扩大端;漏斗口;母锥开口端;圆形头;(管道的)承插端
bell end bored pile 扩底钻孔灌注桩
bell end pipe 管承插端;承插管
Belleville spring 贝氏弹簧;盘形弹簧
Belleville spring washer 贝氏弹簧垫圈
bell exhauster 钟形排气机
bell fender 钟形护舷木
bell flower 吊钟柳
bell frame 钟形构架
bell furnace 罩式炉
bell gable 吊钟山墙;吊钟尖塔;(屋顶上的)吊钟构件
bell glass 玻璃钟罩;钟形玻璃制品
bell guide 导向漏斗;钟形口管
bell hammer 棱结花边杆;铃锤
bell hanger 吊钟钩
bell hanging (机械、电气设备用的可移动的)挂钟线
bellhole 矿核冒落空穴;钟形坑;钟形坑
bellhole bucket 钟穴形(挖掘)铲斗;钟式穴坑挖掘铲斗
bellhole housing 钟形罩
bellhole welder 焊接坑焊工
bellhole welding 冠顶塞焊;钟形焊接
bell house 钟楼
bell housing 漏斗状罩;离合器壳(体);钟状罩;钟形罩;钟形壳
bell idles 空闲码
bellidoite 灰硒铜矿
bellied 凸圆曲线;有突出部分的板;凸起的
bellied file 凸锉
bellied jaw 凸形腭板
bellied jaw crusher 凸腹形颚板破碎机
bell indicator 电铃指示器
belling 制造管子的喇叭口
belling bucket 扩口铲斗;扩底铲斗;钟形钻孔锥;钟形铲斗
bellingerite 水碘铜矿
belling expander 扩管口器;扩管器
belling in 向内凸胀
belling out 向外凸胀
Bellini-Tosi antenna 贝里尼一托西天线
bell installation 钟的安装
bell insulator 铃形绝缘子;单裙绝缘子
Bell integrated optical device 贝尔集成光学器件
Bell interpretive system 贝尔解释系统
bellite 贝里特混合炸药;砷铬铅矿
bell jar 取杆器;取管器;压入头;钟(形)罩;钟形烧结炉;钟形缸;玻璃(钟)罩
bell-jar exhaust 钟罩排气
bell jar high straight form 高直型钟罩
bell jar high straight form with knob 高直型有圆鼻钟罩
bell jar testing 钟罩试验
bell joint 钟口接头;承插式接头;喇叭口管;承口管;插接接合;套管接合
bell kiln 钟形窑
Bell laboratory 贝尔实验室
bell man 钟人
bell manometer 浮钟压力计
Bellman's equation 贝尔曼方程式
Bellman's principle of optimality 贝尔曼最优原则;贝尔曼最优性原则
Bellmer bleacher 贝尔默漂白机
bell metal 铸钟铜;钟青铜
bell-mounted 有承口的
bell-mounted pipe 承插管
bell-mouth 喇叭口;钟形口管;漏斗口;喇叭形的;钟形套管;钟形口;钟形孔;钟形接头;钟形的
bell-mouth area 喇叭口区
bell-mouth chamber 大沟管连接井
bell-mouthed 喇叭口的;带托叶鞘的;钟形的;承口
bell-mouthed entrance of lock 喇叭形船闸入口
bell-mouthed opening 喇叭口;漏斗口;钟形;钟形孔
bell-mouthed orifice 喇叭形孔口
bell-mouthed overflow 钟形溢流
bell-mouthed pipe 承插管
bell-mouthed spillway 钟口溢流(坝);钟口溢流堰
bell-mouth entrance 喇叭形进口;喇叭形入口
bell-mouth estuary 喇叭形河口
bell-mouth estuary regulation 喇叭形河口整治
bell-mouthing 喇叭状扩大开口;喇叭口
bell-mouth inlet 喇叭形进水口;喇叭形入口
bell-mouth intake 喇叭形进水口;钟形进水口
bell-mouth nozzle 喇叭形喷嘴
bell-mouth of central tube in sedimentation tank 沉淀池中心管喇叭口
bell-mouth of pipe 管子大头
bell-mouth orifice 喇叭孔;钟形孔(口);喇叭形孔口;钟形口
bell-mouth overflow 喇叭口溢流;钟形溢水口;钟形口溢流
bell-mouth overflow spillway 漏斗式溢洪道;喇叭形溢洪道
bell-mouth pipe 喇叭形管
bell-mouth pond 喇叭口排水塘
bell-mouth socket 喇叭口打捞器
bell-mouth ventilator 喇叭形风斗
bell nipple 外径接头;套管顶端
bell nozzle 钟形喷管
bell of capital 钟形柱头
bell of die 凹模拉入的圆角部分
bell of pipe 管子承(接)口;管承插口
bell operating rigging 料钟操纵装置
bell or gong 号钟或号锣
bell-out bored pile 扩底钻孔灌注桩
bellow 膜盒压力传感器
bellow form contact 膜盒形接点
bellow-framed door 折门
bellow-framed door fitting 折门零件
bellow-framed door furniture 折门装配件
bellow gas meter 膜盒式气量计
bellows 膜盒组件;膜盒;皱纹管;真空膜盒;感压箱;吹风器;风箱;波纹管;皮老虎;膨胀节;手用吹风器;手动吹风器;伸缩软管
bellows attachment 皮腔
bellows blow pipe 风箱吹管
bellows compensator 风箱式补偿器
bellows differential flowmeter 膜盒式差压流量计;膜盒式差动流量计;波纹管差(压)流量计
bellows differential ga(u)ge 波纹管差动压力计
bellows expansion joint 波形膨胀接头;波纹管膨胀接头;风箱式伸缩缝
bellows expansion piece 皮腔式膨胀接合件
bellows extension 皮腔延伸
bellows flowmeter 膜盒式流量计;波纹管流量计;波纹管式流量计
bellows frame 风箱式框架
bellows-framed door 折扇门;摺扇门
bellows ga(u)ge 膜盒压力计;膜盒式压力计
bellows instruments 膜盒式仪表
bellows joint 膜盒连接;风箱形接头;波纹管连接
bellows manometer 膜盒式压力计;波纹管压力计;波纹管式压力计
bellows mechanical seal 波纹管机械密封
bellows meter 波纹管式差压流量计
bellows operated pilot valve 膜盒控制阀
bellows pressure ga(u)ge 波纹管压力表
bellows pump 隔膜泵;风箱泵
bellows radiator 风箱式散热器
bellows seal 皮腔式密封;膜盒密封;波形密封;波纹管式密封;波纹管密封
bellows sealed gate valve 波纹管密封闸门阀
bellows-sealed valve 风箱密封阀;波纹管式密封阀
bellows seal gland 伸缩箱式封盖
bellows seal valve 风箱
bellows suspension 波纹管悬挂
bellows-type 膜盒式;风箱式的;波纹管式
bellows-type expansion joint 风箱式伸缩接头;风箱伸缩缝;风箱型伸缩缝
bellows-type gas flowmeter 膜盒式气体流量计;波纹管气体流量计
bellows-type manometer 皱纹管压力计;波纹管式压力计
bellows-type mechanical seal 风箱型机械密封
bellows-type meter 膜盒式仪表;波纹管式计量计
bellows-type pneumatic suspension 波纹管气动悬架装置
bellows-type regulator 波纹管式调节器
bellows-type steam trap 波纹管式阻汽排水器
bellows unit 波形补偿器
bellows valve 波纹管阀
bellow-type expansion joint 波纹管补偿接头
bell pepper 灯笼椒
bell pier 钟形桥墩
bell pipe 承接管接头管
bell press 电铃按钮
bell protecting tube 承口保护管
bell prover 校表气罐
bell pull 敲钟拉绳;铃扣;门铃拉索
bell punch 钟形冲孔器
bell purchase 两个动滑车四单轮滑车组
bell push 电铃按钮
bell recorder 浮钟计压器
bell rigging 警钟装置
bell ringing transformer 电铃变压器
Bellrock hollow plaster slab 空心石膏板
bell rod 钟杆
bell roof 钟形屋顶
bell room 钟室
bell rope 敲钟索
bell rope hand pile driver 拉绳式打桩机;拉绳打桩机;人工拉索式打桩机;人工拉绳式打桩机;人工打桩机
bells 排钟
bell screw 钟形螺钉;打捞母锥;丝锥接套
bell set 编钟
bell shade pendant lamp 枫叶罩吊灯
bell shape 喇叭皮带线盆
bell-shaped 钟形
bell-shaped anchorage 钟形组合锚头
bell-shaped capital 钟形柱头
bell-shaped chain wheel 杯形链轮
bell-shaped crank 钟形曲柄
bell-shaped crown gasket 钟形柄头防水垫圈
bell-shaped curve 钟形曲线
bell-shaped curve model 钟形曲线模型
bell-shaped distribution 铃形分布
bell-shaped dome 钟形穹顶
bell-shaped insulator 裙式绝缘子;碗形绝缘子
bell-shaped nozzle 钟形喷嘴
bell-shaped shell 钟形壳
bell-shaped sprocket wheel 杯形链轮
bell-shaped stupa(mound) 钟状印度塔墓
bell-shaped turbine 钟形涡轮
bell signal 铃声信号;振铃信号
bell silencing switch 止铃开关
bell socket 母锥;承插接口;套接;丝锥接套;钟形套筒(打捞工具)
bell-spigot-bell wye branch 承插式Y形分叉管
bell-stoneware pipe 承插式粗陶瓷接头管
bell stupa(mound) 钟形印度塔墓
bell system 钟的安装
bell tap 打捞母锥;丝锥接套
bell tent 钟形帐篷
bell test 电铃式通路试验;电铃式导通试验;电铃测验
bell top kiln 钟罩窑
bell tower 钟楼;古代钟楼
bell transformer 电铃用变压器;电铃变压器
bell trap 钟形凝汽器;钟形存水弯
bell-turret 钟楼;钟角楼
bell type 带外锥复合中心钻
bell-type accumulator 钟型储蓄器
bell type chest piece 钟型胸件
bell-type distributing gear 钟式旋转布料器;高炉配料钟
bell type flushing cistern 钟形冲洗水箱;钟式冲洗水箱
bell-type furnace 罩式炉
bell type gasholder 钟罩式储气罐
bell-type generator 浮筒式发生器;钟形乙炔发生器;浮筒式乙炔发生器
bell type governor 钟罩调压器
bell-type manometer 浮钟式压力计;钟式压力计
bell type pulsator 钟罩式脉冲器

bell type trap 钟式存水弯
bell valve 钟形阀；钟开阀
bell-weevil hanger 套管挂
bell whistle 钟形汽笛
bell whistle buoy 铃哨浮标；声响浮标
bell wire 电铃线
belly 炉腰；隆腹形；桁腹；鼓风的帆；弯板内面；拖网下部
belly band 补漏管箍
belly board 塔内台板（下管时用）
belly brace 曲柄钻
belly buster 安全绳（横向拉钻塔的）
belly container 机腹集装箱
belly core 内芯型
belly discharge semi-trailer 底部倾卸的半挂车
belly dumper 底卸载重汽车；底卸车
belly-dumping 底部倾卸
belly-dump semi-trailer 底部倾卸的半挂车
belly-dump wagon 底部倾卸的运输车
bellying 鼓出部
bellying in 向内凸胀
bellying out 向外凸胀
belly of blast furnace 高炉炉腰
belly pipe 直吹管
belly-rod truss 桁架式梁；倒单柱桁架
belly sling 艇腰挂锚索
belly stay （机车锅炉的）腹撑；桅中部支索
belly strap 艇腰挂锚索
belly tank 机腹油箱
belly-up 中途关停（房地产开发项目）
belly wagon 活底卸料车
belongings 附属物；所有物
belonite 针雏晶
belonospasis 针导法
Belotti straddle crane 贝洛梯跨式起重机
belovite 锶铈磷灰石
below average 中下；不及平均值
below bearing type generator 伞式发电机
below bridge 船桥以下
below-center offset 下偏置
below chart datum 海图零点以下；海图基准面以下
below curb 路缘石标高以下；路缘石高程以下
below day 在地面以下
below deck equipment 下层甲板设备；舱内设备
below freezing 零下；冰点之下
below grade 基准标高以下；基线以下的；地面以下；不合格的；不合格；标线以下的
below-grade application 地下应用
below grade iron bar 不合格的铁条
below grade masonry 地面标高以下的圬工
below-grade water 不合格的水
below ground 在地面以下；地下；地面以下
below-ground installation 地下装置
below-ground loading 地下装载
below-ground masonry 地面以下的圬工；地下圬工；地下砌体
below-ground silo 地下青储[贮]窖
below low-water 低水位以下；低潮位以下
below minimums 低于最低气象条件
below-norm 限额以下
below normal 水平之下
below-order benchmark 等外水准点
below-order level(l)ing 等外水准测量
below par 低于票面价值；低于面值
below par stock 低于票面价值的股票
below proof 废品；不合格品；不合格（的）
below proof bar iron 不合格的铁条
below resonance balancing machine 谐振平衡试验机
below sea floor 海底
below sea level 海平面以下
below-sea level contour 负向地貌等高线
below stair(case) 楼下层；底层内；地下室内
below standard performance 效率低于标准
below the aleurone layer 糊粉层下面
below the average 平均以下
below the average quality 一般水平以下的质量
below the face of a mountain 在山的阳面下部
below-the-line 栏外会计；线下项目
below-the-line expenditures 线下预算支出；投资预算支出
below-the-line item 线下项目；线下收支
below the mark 标准以下
below-threshold laser behavio(u)r 阈值前激光工作状态
below-threshold operation 在阈值以下运转
below velocity 河底流速
below zero 零下；零度以下
belt 引带；地带；带子；带状物；带；垂直带；皮带；同心带
belt adjustment 皮带松紧度调节
belt alignment switch 胶带调正开关
belt and bucket elevator 带头式升降机；带式提升机；皮带斗式提升机
belt-and-chain driven drill 皮带及链条传动的钻机
belt applicator 带式上浆装置；带式浸润器
belt-apron feeder 铠装胶带给料机
belt balance 皮带称量计
belt balancer 皮带称量计
belt batcher 带式配料（计量）器
belt-bedding 带状花坛
belt bench 皮带定长器；皮带测长台
belt brake 带闸；带式制动（器）；皮带制动器；皮带刹车
belt bridge 皮带运输桥；皮带输送机桥；皮带桥
belt-bucket elevator 斗链式升运器；带式链斗提升机；斗式提升机；斗带式提升机
belt buckle 带扣
belt building machine 黏带机
belt canal 环形运河
belt carrier 传动皮带轮
belt car unit 皮带装卸车
belt caster 带式连铸机
belt casting machine 带式连铸机
belt channel 带形槽
belt city 带形城市
belt clamp 传送带联结卡；皮带联结卡；皮带扣；皮带夹
belt clasp 带子扣钩
belt cleaner 皮带擦拭器；皮带机除尘器
belt clip 皮带夹
belt cloud system 带状云条
belt composition 传动皮带润滑剂；带皮结构；皮带润滑剂
belt concreting tower 皮带运输浇灌混凝土塔；带式混凝土浇注塔
belt cone 皮带塔轮
belt contact 皮带包角
belt conveyance 带式输送；皮带运输；皮带输送；皮带传送
belt-conveyed concrete 皮带运输的混凝土
belt conveyer 搬运者；传送带；皮带运输机；皮带运输带；运输机；带式运送机；皮带运输器；带式运送机；带式输送机；带式传送机；传动运输机；胶带输送机
belt-conveyer bridge 带式输送机桥
belt conveyer scale 传送带秤
belt conveyer takeups 带式运输机收紧装置；带式运输机收紧器；皮带运输机收紧器
belt-conveyer type furnace 传送带式电炉
belt conveyer type trencher 皮带输送式挖掘机
belt conveyer with lateral discharger 带侧向卸料器的带式输送机
belt-conveyer 搬运者
belt conveyor scale 皮带输送秤
belt conveyor stacker 带式堆垛机；带式堆垛机
belt conveyor transfer point 带式输送机转运点
belt course 垫层；腰线；带状层；腰带层（墙上沿窗台凸出的装饰层）；束带层（墙上凸出的装饰层）；圈梁；带状板
belt cover 皮带运输机外罩
belt creep 皮带轮缘滑动；皮带打滑
belt deviation 皮带跑偏
belt dressing 皮带装置；皮带油；皮带涂料
belt drier 带式条形干燥器
belt drive 皮带传动
belt drive double housing planer 皮带传动龙门刨床
belt drive implement 皮带传动作业机
belt-driven 皮带传动的；带式传动
belt driven blower 皮带传动鼓风机
belt-driven elevator 皮带传动升降机；皮带传动电梯
belt driven lathe 皮带车床
belt-driven lift 皮带传动升降机；皮带传动电梯
belt driven mixer 皮带传动混合器
belt driven pump 皮带传动泵
belt driven riveting machine 皮带传动铆接机
belt driven tach(e)ometer 皮带传动转速计
belt drive unit 皮带传动装置
belt drive winch 皮带传动绞车
belt driving over 带上张紧传动；上侧拉紧的皮带
belt driving under 下侧拉紧的皮带；带下张紧传动
belt drop hammer 皮带落锤
belt dryer 带式条形干燥器
belt-dynamometer 传动式测力计
belted 带状的；带式传动的
belted-bias tyre 带束斜交轮胎
belted cable 铠装电缆
belted coastal plain 带状沿海平原
belted electronic balance 电子皮带秤
belted outcrop plain 高低地相间准平原
belted plain 分带平原
belt electronic weigher 皮带电子秤
belt elevator 带式升降机；皮带提升机；皮带斗式提升机
belt experiment 传动带试验
belt extractor 胶带出料机
belt failure 皮带损坏
belt fastener 皮带连接器；皮带扣；皮带接头固紧件；皮带带扣
belt fastener wire 皮带扣件钢丝
belt feeder 带式喂粉器；带式喂料机；带式送料器；带式进料器；带式进料机；带式给料器；带式给料机；皮带喂料机；皮带给料器
belt feed(ing) 皮带喂料
belt feed launcher 弹带式发射装置
belt feed lever 拨弹带杆
belt feed lever pivot 拨弹杆轴帽
belt-feed planter 带式播种机
belt feed slide 带传滑板
belt filled 带状上料
belt filler 皮带注油口；皮带油
belt filter 带式过滤器；带式过滤机
belt filter press 带式压滤机
belt filter pressure 带式过滤器压力
belt filter screen 带式过滤筛
belt finish （水泥混凝土路面施工时的）皮带拖平
belt finishing 皮带拖平
belt fish 带鱼
belt float 带墁
belt fork 移带叉
belt friction(al) elevator 摩擦带式提升机
belt furnace 直通炉；带式炉
belt gear 皮带传动装置；皮带传动机构
belt-gear reduction type 皮带－齿轮减速型
belt generator 带式静电发电机
belt grader 皮带式分级机
belt grain conveyer 带式谷物输送器；带式谷物输送机
belt graph 带形图
belt gravity separator 带式比重分选机
belt grease 皮带润滑脂
belt grinder 带式磨光机；砂带磨光机；皮带传动机构
belt grinding 带式磨光；砂带磨削；砂带磨光
belt grinding lapping machine 带式磨光机
belt grinding machine 带式磨床；带式磨边机
belt guard 皮带罩；皮带护挡
belt guidance 皮带导槽
belt guide 导带器
belt guide pulley 皮带导向轮
belt hammer 皮带落锤
belt hay loader 带式装干草机
belt heading 运输巷道
belt highway 带状公路；环形公路；环形公路；环行公路；环路【道】；环城公路
belt hoist 皮带提升机；皮带起重机
belt holding pawl 弹带扣勾
belt hold roller 压带辊
belt hook 带钩；皮带扣
belt hook clipper 皮带扣钳压器
belt house 皮带罩
belt housing 皮带罩
belt humidifier 带式加湿器
Beltian body 贝尔特体
Beltian orogeny 倍尔特造山运动（前寒武纪末）【地】
Beltian series 倍尔特统【地】
belt idler 张紧轮；皮带托辊；皮带空转轮；皮带紧轮
belting 小艇护舷材；装置传动带；扎线；带装置；带料；带类；传动带装置；传动带材料；传动带；包带；皮带装置
belting course 带层；垫层

belting duck 传动带帆布
belting fabric 传动带的纤维织物
belting leather 传动带皮
belting loom 传动带织机
belting tightener 皮带张器器;皮带张紧轮
belting wax 传动带用蜡
belting wire 线带装置
belt insulated cable 皮带绝缘电缆
belt insulation 铠包绝缘
belt joint 皮带结合;皮带拉合;皮带接头
belt kiln 带式窑
belt lace 皮带扣;皮带接头固紧件
belt lacer 皮带扣;皮带卡子;皮带接头
belt lacing 引带接头;皮带扣;皮带卡子;皮带结头;皮带结合;皮带接头
belt lacing machine 皮带接头机
belt leather 带皮
belt lifter 带式起重机
belt lift hammer 带式落锤
belt lifting arrangement 皮带起重装置
belt line 流水线;环形线(路);环形线;皮传送带;枢纽区域环行线
belt line city 带状城市
belt-line highway 环路;环行公路;带状公路
belt-line production 传送带式流水生产
belt-line railroad 环形铁路;环行铁道;临港铁道;临港(带)铁路
belt-line railway 环形铁路;环行铁道;带形铁道
belt-line theory 按已钻出油井延线布井法
belt loader 带式装载机;带式装载机;带式挖土平路机;皮带装料机
belt loading station 皮带运输机装载站
belt lubricant 皮带油
belt magnetic separator 带状磁性分离器;带式磁选机
belt mark 链带痕
belt material 带材料;带衬;带材;皮带衬料
belt material thrower 带式抛料机
belt meander 带状河曲
belt mill 带式锯床
belt mo(u)ld board 带式犁壁
belt mo(u)ldboard plow 带形犁壁犁
belt mo(u)lding 车窗上部镶条
belt mounted conveyer 吊运带式输送机
belt of cementation 胶结带
belt of convergency 辐合带
belt of demorphism 风化作用带
belt of discharge 泄水带
belt of fault 断层带
belt off-centre 皮带走偏
belt of fluctuation 变动带
belt of fluctuation of water 水位变动带
belt of fluctuation of water table 水位波动带
belt of folded strata 褶皱带
belt of fortesses 要塞地带
belt of Gaussian projection 高斯投影分带
belt of ice 冰带
belt of land 带状土地
belt of ore 矿石条纹
belt of phreatic fluctuation 地下水位波动带;潜水波动带
belt of seismic activity 地震活动带
belt of soil moisture 土壤水带
belt of soil water 土壤水分带;土壤水带
belt of totally 全蚀带
belt of transition 过渡(地)带
belt of wandering 河流摆动带;河道蜿曲带
belt of water table fluctuation 地下水位波动带
belt of wave loop 波腹带
belt of wave mode 波节带
belt of weathering 风化(地)带
belt-operated 皮带传动的
belt package conveyer 带式包装件输送机
belt pipe wrench 带式管子钳
belt planting 带播;防护带栽植;带植
belt polishing 皮带抛光;砂带抛光
belt polishing machine 带式抛光机;皮带抛光机
belt press 压带机
belt printer 带式打印机
belt production 流水作业;皮带传动生产作业
belt prover 带运机
belt pull 皮带张力
belt pulley 皮带盘;皮带轮;皮带传动滑轮
belt pulley clutch 皮带轮离合器

belt pulley drive 皮带滑轮传动
belt pulley flywheel 皮带滑轮
belt punch 皮带冲压机;皮带冲头;皮带冲孔器;皮带冲孔机
belt puncher 皮带冲孔机
belt punch pliers 接合带冲钳
belt quarter-turn drive 皮带直角转弯传动
belt rail 带形轨条;带式横挡
belt railroad 环形铁路;带式铁道;带形铁道
belt railway 环形铁路;环行铁道;环路;环行铁道
Beltrami's theory of failure 贝尔塔密强度理论
belt respirator 带式呼吸器
belt reversing device 传动带反转装置
belt reversing drive 皮带回动机构
belt reversing gear 皮带回动装置
belt rivet 皮带铆钉
belt road 带状道路;环路;环线
belt roller 皮带轮;皮带滑车;滚轴;皮带运输机;胶带托辊;皮滚柱;皮滚轴;皮滚筒;皮辊;皮带轴
belt sander 带式磨光机;带式砂光机;带式打磨机;砂带抛光机;砂带磨光机
belt sanding 砂带打磨
belt sanding machine 带式磨光机;带式磨床
belt saw 带锯
belt saw blade 带锯条;带锯片
belt scale 胶带秤机;计量运送带;带秤;皮带电子秤;皮带秤
Belt's corpuscle 贝尔特体
belt scraper 皮带擦拭器;皮带刮板
belt screen 钢丝布;链带形格栅;带式筛;带状拦污栅;带式格栅;带式筛分机;带式筛;带筛机;带筛
belt seeder 带式播种机
belt separator 带式分离器
Belt series 倍尔特统【地】
belt sewage screen 污水钢丝布
belt shifter 移带器;皮带移动装置
belt shifting mechanism 皮带移动机构
belt shift lever 皮带移动手把
belt sizer 带式分级机
belt sling 皮吊带
belt slip monitoring system 皮带防滑装置
belt slip(page) 皮带打滑
belt slope lifter 倾斜皮带提升机
belt sorter 皮带式分级机
belt span 相带宽度
belt spanner 平带拉紧装置
belt speeder 皮带变速装置
belt speed transducer 带速传感器
beltstower 抛掷式胶带填充机
belt stowing machine 带式充填机
belt stress 圆屋顶上的水平压力;穹顶中的水平应力
belt stretcher 带式伸长器;皮带张紧装置;皮带伸张器
belt stretching machine 带式拉幅机;皮带展幅机
belt stretching roller 皮带轮;皮带紧轮
belt strike 移带叉
belt striker 移带器
belt system 皮带系统;输送带系统
belt take-up 紧带装置
belt tearing 胶带撕裂
belt tension 皮带张力
belt tension clutch 皮带张紧式离合器
belt tensioner 胶带张紧装置;胶带张紧轮;皮带松紧调整器
belt-tensioning hinged support 受拉皮带的铰支座
belt tension(ing) idler pulley 皮带惰轮
belt tensioning pulley 皮带张紧轮
belt tension release lever 皮带松紧杆
belt tightener 紧带器;皮带张紧装置;皮带拉紧轮;拉紧轮;紧带轮;皮带绞轮
belt tightening 皮带张紧度;实行紧缩
belt tightening pulley 带式张紧滑轮;游轮;带式拉紧滑轮
belt topographic map 带状地形图
belt training 胶带位置校正;皮带位置校正
belt transect 样带
belt transmission 带式传动;皮带传送;皮带传动
belt transom 输送机架
belt tripper 胶带卸料口;带式卸料器;带式输送机卸料器;皮带运输机的卸料器;输送机卸料
belt-trough elevator 斗带式升运器
belt troughing idler 胶带机槽形托辊
belt truck loader 带式装载机

belt tunnel dryer 带式隧道干燥器
belt-type 皮带式
belt-type apparatus 带式装置
belt-type batching unit 带式配料机;带式配料器
belt-type bucket elevator 皮带斗式提升机
belt-type bucket elevator loader 皮带斗式提升加载机
belt-type concrete placer 混凝土浇灌带
belt-type distribution 带式分布
belt-type dry oven 带式烘干炉
belt-type electromagnetic separator 带式电磁分离机
belt-type fertilizer distributor 带式撒肥机
belt-type filter 带式过滤器;带式过滤机
belt-type measuring unit 皮带式量料装置
belt-type moving walk 带式活动人行道
belt-type(oil)skimmer 带式刮油器
belttype pickup 带式捡拾器
belt-type proportioning unit 皮带式分配装置
belt-type sheet machine for dry process 干法带式制板机
belt-type sinterer 带式烧结机
belt-type sludge filter 带污泥过滤器
belt-type suspended magnetic separator 皮带式悬挂磁选机
belt-type vacuum filter 带真空过滤器;带真空过滤机;带式真空过滤器;带式真空过滤机
belt-type wrench 带式扳手
belt veneer dryer 带式原片干燥器
belt-wagon 胶带车
belt wax 皮带蜡
beltway 围绕道路;环行路;环城路;环城公路;带运输机
belt weigher 胶带称量器;带式计量秤;皮带计量机;皮带秤量器
belt weigh feeder 皮带式称量喂料机
belt weight 皮带秤
belt weight meter 胶带秤;皮秤;皮带秤
belt wheel 皮带轮
belt wiper 皮带擦拭器;胶带清扫刷;皮带刮土机
beltwork 皮带传动
belt workline 传送带工作线
beluga 白鲸
belvedere 八角窗;瞭望台;观景楼;瞭望塔楼
Belvedrer Wien 贝尔维德雷宫(18世纪奥地利维也纳)
belyankinite 锆钛钙石
bema 讲坛;圣坛
be markedly increased 明显增加
bematis 教堂安放圣器处
Bemberg 本伯格铜铵丝
Bemberg sheer 本伯格铜铵丝双面绉
Bemberg's stretch spinning apparatus 本伯格式拉伸纺丝机
Bembridge beds 本布里奇层
bementite 蜡硅锰矿
be meticulous in design 精心设计
be mingled with 夹杂
be modelled on 仿造
be much more likely to suffer injury 易遭受危害
ben 小山;内厅;高地;住宅内室;高山丘
benab 棚屋;小屋(几内亚)
Benacon 一种木材运输标准合同格式
Benard cell 贝纳尔德窝;贝纳尔德单体
Benard circulation motion 贝纳尔德环流(运动)
Benares hemp 贝拿勒斯麻
benavidesite 硫锰铅锑矿
bench 露天台阶;拉拔机;阶地;护道;河岸马道;海蚀平台;下部半断面;下半断面;狭长台地;光学台;光具座;长椅;长凳;爆破阶面;板凳;锐道;钳工台;钳工工作台;台阶;台架;舢板横座板
bench abutment 张拉座;张拉台
bench and stool cover 凳套
bench angle 台阶坡面角
bench anvil 台砧
bench apron 工作台挡板;工作台面板
bench arc welding machine 台式弧焊机
bench axe 木工用斧;木工斧
bench belt 工作台
bench blasting 爆破成踏步;阶梯式爆破;阶梯爆破;阶段爆破;台阶式爆破;台阶爆破
bench blasting with multiple rows 多排踏步爆破;多列踏步爆破
bench block 校正平台

bench blower 台式吹芯机
bench board 斜操纵台;操纵台;台式配电盘
bench border irrigation 梯田灌溉
bench brake 台式弯板机;台式制动机
bench calibration 试验台校验
bench charge 台架充电
bench check 工作台检验
bench chisel 钳工錾
bench clamp 台钳
bench cloth 铺放绒布
bench construction 建造护坡道;护坡道构造
bench cut 阶梯式开挖;正台阶法;台阶式开挖
bench cut method 台阶开挖法;台阶法(隧道开挖)
bench cutting 台阶式挖土法;台阶式开挖
bench cutting on side slope 边坡台阶开挖
bench dog 工作台挡块;台轧头
bench drill 台钻;台式钻床
bench drilling 阶梯钻探;阶地钻探;梯段凿岩;台阶穿孔
bench drilling machine 台钻
benched 陡坎的
benched excavation 阶梯式开挖
benched footing 阶状基础;台式底脚
benched foundation 阶状基础;阶形基础;阶梯形基础;梯式基础;台阶式基础;踏步形基础
bench edge 坡顶线
benched joint 阶状接缝
benched pit sinking 阶梯形沉井
benched section 台口式断面
benched subgrade 台口式路基
bench-end 凳端围屏(教堂信徒席的)
bench engine lathe 台式普通车床
bench excavation 台阶式挖土法;台阶式挖掘;台阶式开挖
bench face 台阶工作面
bench filing and sawing machine 台式锉锯机
bench floor 阶段底板;平盘
bench flume 阶地水槽;支架渡槽;支架渡槽;低渡槽;山坡护坡旁的截水沟
bench gravel 阶地砾石;河滩砾石;滩地砾石
bench grinder 磨床;台式手摇砂轮;台式磨床
bench group 组合台阶
bench hammer 钳工小锤;台用锤;台锤
bench height 露天矿阶段高度;台阶高度
bench holdfast 台座夹钳
bench hole 阶地钻孔
bench hook 木工工作台挡头木;挡木头;缝帆钩;台卡
benching 阶梯式开挖法;阶地;河岸阶地;管道定边坡;定边坡;钳工加工;梯段开挖法;台阶式挖土(法);台阶式开挖;台级采矿法
benching cut 台阶式开挖;台阶式挖土;台阶式回采
benching iron 标尺垫【测】;测杆铁垫;测杆铁座;测杆铁插头
benching method 分段开挖法
benching tunnel(l)ing method 台阶法(隧道开挖);分层开挖法(指隧道)
bench inkometer 台式油墨仪
bench investment-earnings ratio 基准收益率
bench investment yield 基准收益值
bench joint 阶状接缝
bench knife 木工台刀挡头
bench land 河滩阶地;滩地;沙洲
bench lathe 台式车床
bench lava 台阶熔岩
bench level 台用水准仪;水准点水平面;水准点高程
bench life of a sand mix 砂子存放期
bench-like form 阶状地形
bench lumber 台面材
bench magma 台阶熔岩
bench mark 基准;基准桩;基准水位标志;基准标记;测定应变时地基标志;标准评估程序;水准基点;水准点;水准标点
benchmark appraisal 基准点估价
benchmark at tidal station 验潮站水准标准
benchmark basin 参证流域;参考流域
benchmark built-in wall 墙上水准点
benchmark checking 校平
benchmark data 基本数据
benchmark description 水准点之记;水准点说明
benchmark elevation 基准点高程;水准点标高
benchmarking method and result 基准程序方法和结果

benchmark level(l)ing 基平;水准点高程测量(基平)
benchmark list 水准点高程成果表;水准点成果表
benchmark of shaft 井口水准基点
benchmark on rock 基岩水准标石
benchmark on wall 墙脚水准标志
benchmark problem 基准问题;基准题
benchmark program(me) 基准程序;标准检查程序
benchmark routine 水准程序
benchmark soil 水准点地面
benchmark station 参证站;参证水文基站;参照站【测】
benchmark statistic(al) data 综合统计数据
benchmark statistics 基本统计资料;标志性统计数据;综合统计(数字)
benchmark study 基准研究
benchmark task 基准题
benchmark test 基准测试
benchmark value 水准点标高
benchmark wastewater treatment plant 基准污水处理厂
bench method 分层开采法;台阶式挖土法;台阶式开挖法
bench micrometer 台用千分尺
bench milling machine 台式铣床
bench mining 台阶式开采
bench model 小型模型
bench mo(u)lder 短风锤
bench-mounted hardness tester 台式硬度计
bench of burners 炉组;喷燃器组
bench oil paint 油性底层涂料
bench photometer 光具座式光度计;台式光度计
bench placer 阶地砂矿
bench plane 木工台刨;台刨;手工短刨
bench preparation 露天矿阶段准备;采准
bench press 台式压榨机;台式压床
bench prestressing 台施预应力
bench rammer 台式砂桩
bench rebate plane 平刨;侧面刨
bench rebbet plane 平刨;侧面刨
bench roller 台辊
bench run 小试验;试验台试验
bench sander 台式砂轮;台式磨光机
bench saw 台锯
bench scale 小型;台阶式坡面
bench-scale analysis 实验室规模分析
bench-scale dissolver 小型(电解)溶解器
bench-scale experiment 实验室试验
bench-scale oxygen consumption colo(u)rimeter 小型氧消耗比色计
bench-scale test(ing) 小型试验;小试;小规模试验;台架试验
bench screw 木工台夹钳;台虎钳
bench section 横断面;横剖面
bench shaper 护道整形机;台式刨床
bench shaping machine 台式刨床
bench shears 台剪(机)
bench shooting 阶梯爆破
bench side 焦面
bench simulation test 台架模拟试验
bench slope 边坡倾斜角;台阶式边坡
bench snips 台剪(机)
bench stake 小铁钻
bench stop 木工台挡头;刨木台止推条
bench table 壁柱台座;石坐凳;墙台;墙基
bench tapping machine 台式攻丝机
bench terrace 阶梯式梯田;梯田;梯阶形梯田;梯田;台阶;台地;水平梯田
bench test 台架试验;工作台试验;小型试验;试验台试验
bench testing time 台架试验时间
bench the slopes 斜坡挖成台阶形
bench time 台架时间
bench toe 坡底线
bench tomb 有平台的坟墓
benchtop cave 半敞开小室
benchtop injection mo(u)lding 敞开式注模
bench torch 喷灯
bench trimmer 截锯机;台式修整机;台阶修整机;木工修整机;木工截切机
bench-type core blower 台式吹芯机
bench-type laboratory pultruder 台式实验用拉挤机
bench type spot-welder 台式点焊机
bench-type terracing 平坡梯田

bench vise 老虎钳;台用虎钳;台式老虎钳;台钳;台虎钳
bench vise for wood working 木工台钳
bench vise with anvil 带砧台虎钳
bench vise with anvil stationary base 固定式带砧台虎钳
bench wall 承拱墙
bench work 台上手工作业;案子工;作台工;钳工工作
bench worker 钳工;钳床工人
benchworking 钳工操作
bencyclane 苄环烷
bend 折弯;缚;绑结;曲隅部;弯道;使弯
bendability 可弯性
bendable pipe drilling 软管钻进;柔杆钻进
bendable plastic pipe 可弯曲塑管
bend a cable 系统钢索
bend allowance 钢管余量;钢管余量;钢筋弯曲公差;弯曲余量
bend alloy 易熔合金;弯管合金;弯管用易熔合金
bend angle 弯角
bend a sail 挂帆
benday 色彩加深;色彩加浓
Benday process 本戴底纹制版法
bend bar 元宝钢;挠钢;元宝筋;弯曲钢筋;弯起钢筋
bend capability 可弯性
bend capacity 可弯性
bend curvature 河弯曲率;弯段曲率
bend cutting 弯插
bend discharge machine 立弯式连铸机
bend ductility 弯曲延性
bended ore 条带状矿石
bended Pb-Zn ore 条带状铅锌矿石
bended strain ga(u)ge 粘贴式应变计
bender 挠曲机;折弯模;折弯机;弯曲;弯筋机;弯管器;弯管机
bender and cutter 挠曲折断两用机;折弯切割两用机;钢筋弯切两用机;弯曲截断两用机;弯切两用机(钢筋)
bender and cutter of steel bars 钢筋弯曲截断两用机
bender impression 弯曲模膛
Bender process 本德法
bend fitting 弯头配件
bend flattening 河曲变缓
bend flow 弯道水流
bend(flow) meter 弯管流量计
bend folding 横弯褶皱
bend glass 曲面玻璃
Bendien s test 本氏试验
bend improvement 裁弯取直;弯道治理;弯道整治
bendiness 转弯度
bend in exhaust pipe 排气管弯头
bending 卷刀口;挠曲作用;挠曲;拱曲作用;板曲作用;弯曲
bending action 弯曲作用(力)
bending allowance 弯曲容度;弯曲公差
bending amount 弯曲量;弯曲度量
bending analysis 弯曲分析
bending and denting clause (对桶装货的)弯曲和凹陷条款
bending and forming machine 折弯成形机
bending and quenching machine 成形淬火机;弹簧成形淬火机
bending and reinforcement assembly shop 弯扎钢筋车间
bending and straightening machine 弯曲整直两用机
bending and straightening press 弯曲矫正两用压力机;弯曲矫直两用机
bending and unbending test 反复弯曲试验;反复挠曲试验;曲折试验
bending angle 挠曲角;弯曲角度;弯角
bending apparatus 弯曲设备;弯曲器;弯钢筋机
bending area of fold axial trace 褶皱轴迹弯曲地段
bending area of large fault 大断裂转折地段
bending arm 受弯力臂;弯曲臂(扶手)
bending axis 弯曲轴
bending-back point 峰点
bending beam 曲梁;连系梁
bending bench 弯钢筋钢管的条凳
bending block 弯曲型砌块;蜂窝板;弯型块;弯材平台
bending bond failure 弯曲黏结力破坏

bending brake 板料折弯机;板料弯折机;板料弯曲机
bending capability 可弯性;弯曲能力
bending capacity 弯曲能力
bending chamber 弯板室
bending coefficient 挠曲系数;弯曲系数
bending compression 弯压
bending compression strength 弯压强度;弯曲抗压强度
bending compression zone 弯压区;弯曲应力区
bending connection 弯曲接头;弯曲连接
bending crack 弯曲裂纹;弯曲裂缝
bending creep 弯曲蠕变
bending creep test 弯曲蠕变试验
bending crew 弯管小队
bending curve 弯曲曲线
bending cycles 交变弯曲周期;交变弯曲循环
bending deflection 上挠度;挠曲偏斜;挠曲变位;挠偏转;挠度;垂度;弯曲变位;弯度
bending deflection of collars 钻具弯曲度
bending deformation 弯曲变形
bending diagram 挠矩图;弯矩图
bending die 弯曲模
bending dimension 弯曲尺度
bending displacement 弯曲位移
bending distribution 弯曲分布
bending disturbance 弯曲扰动
bending elasticity 挠曲弹性
bending endurance 弯曲耐久性;受弯耐力
bending energy 弯曲能量
bending failure 挠曲毁坏;弯曲破坏;弯曲破损;弯曲破坏;弯曲断裂;受弯坡坏
bending failure curve 弯曲破坏曲线;弯曲断裂曲线
bending failure stress 弯曲破坏应力;弯曲断裂应力
bending fatigue 挠曲疲劳;弯曲疲劳;受弯疲劳
bending fatigue strength 弯曲疲劳强度
bending fatigue test 弯曲疲劳试验
bending fiber 弯曲纤维
bending field 致偏场
bending flexure 弯曲
bending floor 蜂窝架;弯材平台
bending fold(ing) 隆曲褶皱;隆起褶皱
bending force 挠曲力;弯曲力;弯力
bending former 折弯设备
bending formula 弯曲公式
bending fracture 弯曲断裂
bending furnace 管坯炉;弯板炉
bending gang 弯小队
bending inertia 弯曲惯性
bending in two direction 双弧度
bending in two planes 在两个平面中受弯
bending iron 弯钢筋扳头;铅管调直器;弯铁;弯钢筋工具;弯钢管扳头
bending jackstay 吊帆绳或杆;天幕边索
bending joint 弯曲接头;弯曲连接
bending lehr 弯板退火炉
bending length 弯曲长度
bending lever 弯曲杠杆
bending limit 弯曲极限
bending line 弯度线
bending list 弯筋一览表;弯钢筋表
bending load 弯曲荷载
bending loading 横向荷载;弯曲荷载
bending loss 弯头损失
bending machine 卷板机;折弯机;屈挠试验机;弯折机;弯曲机;弯曲钢筋机;弯筋机;弯管机;弯板机
bending machine with fixed blocks 带定滑车的折弯机
bending mandrel 弯曲圆棒;弯管芯
bending member 弯曲构件;受弯构件
bending method 蝶形孔型轧制法
bending modulus 抗弯模数;抗弯模量;弯模数;弯曲模量
bending modulus of concrete 混凝土弯拉模量
bending moment 挠矩;弯曲力矩;弯矩
bending moment area 弯矩图面积
bending moment at a corner 转角处的弯矩
bending moment at mid-span 在跨中的弯矩
bending moment diagram 弯矩图;弯矩图
bending moment envelope 弯矩包络线
bending moment influence line 弯曲力矩影响线
bending moment-less 无弯矩
bending-moment ratio 弯矩比(率);弯曲比
bending-moment theory 弯矩理论
bending mo(u)ld 弯曲造型;弯曲模具;弯板模

bending of beams under transverse force 横梁弯曲
bending of flexure 弯矩
bending of light 光线弯曲
bending of stock rail 基本轨的弯曲
bending of strata 地层弯曲;底层弯曲
bending operation 弯曲操作
bending oscillation 弯曲振动;弯曲振荡
bending oscillation failure 弯曲振动破坏;弯曲振动断裂
bending oscillation strength 弯曲抗振强度
bending oscillation test 弯曲振动试验
bending party 弯管小队
bending pin 铅管调直器;调整管子工具;弯管钉
bending plane 弯曲平面;弯曲面
bending planer 曲面刨光机;曲面抛光机
bending plane slip 弯曲滑动
bending plate 弯板
bending pliers 弯管钳
bending point 挠曲点;弯曲点
bending power 弯曲能力;弯曲应力
bending press 压弯机;弯压机
bending pressure 弯曲应力
bending process 弯曲过程
bending property 可弯性
bending quenching 弯强化
bending radius 弯曲半径;挠曲半径
bending rail 弯轨
bending ratio 弯曲比
bending reinforcement 弯曲钢筋;加固钢筋
bending resistance 抗弯强度;抗弯性;抗弯刚;抗弯
bending-resistant 抵抗弯曲的
bending rigidity 抗弯刚性;抗弯刚度;弯曲刚度
bending rigidity of rail 钢轨抗弯刚度
bending roller 弯板机;滚板机;折页三角板夹纸辊
bending roll machine 滚管机
bending rolls 卷板机;辊子弯板机;辊子卷板机;卷管机;转向辊
bending rotation capacity 挠转能量
bending rupture 弯曲破坏;弯曲断裂
bending rupture curve 弯曲破坏曲线;弯曲断裂曲线
bending schedule 弯钢筋明细表;钢筋加工表;配筋表;弯钢筋表
bending shackle 联锚卸扣;联锚卸机;接锚卸扣;锚钩环【船】;锚端卸扣
bending shoe 弯管胎具
bending shore 接锚短锁
bending slab 弯材平台
bending slope 弯曲斜度
bending spring 弯曲铅管用弹簧;弯曲钢管用弹簧
bending stability 挠曲稳定性
bending steam 弯管小队
bending stiffness 抗弯劲度;抗弯刚性;抗弯刚度;弯曲刚度
bending strain 挠曲应变;弯应变;弯曲应变
bending strain energy 弯曲应变能
bending strength 抗弯强度;弯曲强度;抗折强度;抗弯能力;抗挠强度;弯拉强度
bending strength of wood 木材抗弯强度
bending strength tester 抗折强度仪
bending strength tester for corrugated sheet 波瓦抗弯试验机
bending strength under high temperature 高温下抗弯强度;高温抗折强度
bending stress 抗弯应力;弯应力;挠曲应力;弯曲应力
bending stress distribution 弯曲应力分布
bending stresses in shell 薄壳弯曲应力
bending stress formula 弯曲应力公式
bending table 弯曲工作台;弯钢筋工作台
bending temper(ing) 弯钢化;弯强化
bending-tensile strength 弯拉强度
bending-tensile test beam 弯拉试验梁
bending tension 弯曲拉力
bending tension failure 弯曲拉力破坏
bending tension strength 弯拉强度
bending test 弯曲试验;抗折试验;弯曲试验;挠曲试验;受弯试验
bending test beam 弯曲试验梁
bending tester 弯曲试验机;曲折试验器(涂膜的)
bending test for pole 电杆抗弯试验
bending test in tempered or quenched state 回火或淬火状态时弯曲试验
bending test machine 弯曲试验机
bending test specimen 弯曲试验试件
bending test specimen strength 弯曲试验试件强度
bending test under three-point loading 三次荷载弯曲试验
bending theory 弯矩理论
bending thick plates 厚板卷制
bending through 180 degrees 做180°弯折
bending tongs 弯管钳
bending tool 弯曲工具;弯管工具
bending torsion stability 抗弯扭稳定性
bending trestle 弯筋工作台;弯管工作台;弯板工作台
bending twisting coupling 弯曲扭转耦合;弯扭耦合
bending unit stress 单位弯曲应力
bending up 向上弯
bending up cables 向上弯曲缆索;钢缆弯曲
bending up method 蝶式孔型轧制法
bending up reinforcement 弯起钢筋
bending up technique 钢绞线绕技术
bending value 抗弯能力
bending vibration 挠曲振动;弯曲振动
bending vibration failure 弯曲振动破坏
bending vibration strength 弯曲抗振强度
bending vibration test 弯曲振动试验
bending wave 弯曲波
bending width 弯曲宽度
bending with small deflection 小挠度弯曲
bending wood 弯曲木(材)
bending yield point 抗弯屈服点
bend in river 河弯;河曲
Bendix gear 本迪克斯式齿轮
bend line 弯曲线
bend loading 弯曲负荷
bend loss 转向损失;弯头损失;弯水头损失;弯道损耗;弯段损失;弯段水头损失;弯道损失
Bendman arrester 本德曼避雷器
bend metal 弯管合金
bend meter 弯头水表;弯道流量计
bend muffler 消声弯头
bend neck 歪颈
bend of fault 断层弯折
bend of river 河曲
bend of road 道路弯曲;道路弯道;弯道(道路曲线段)
bend of vessel 容器的弯曲
Bendorf bridge 本道夫桥(德国)
bend outcrop 岩层露头
bend-over test 弯折试验
bend piece 弯管接头
bend pipe 挠管;弯管
bend pipe meter 弯管流量计;弯管水表
bend point 弯曲点
bendproof 抗弯曲的;抗弯的
bend pulley 转向轮;改向滚筒
bend radius 弯道曲率半径;弯道半径
bend radius of channel 航道弯曲半径
bend regulation 弯道治理;弯道整治
bend-rushing flow 扫弯水
bends 沉箱病;气压病;木船舷侧厚板;潜水员病;潜水病
bends and hitched 绳结
bend shoal 弯道浅滩
bend specimen 弯曲试样
bend summit 河弯顶点
bend test 挠曲试验
bend tester 弯曲试验机
bend test piece 受弯试件
bend tube 曲管
bend type expansion joint 弯管式伸缩接头;胀缩弯头
bend union 弯管活接头
bend up bar 弯起钢筋
bendway 两河弯间河段;浅滩;弯道
bend wheel 改向轮
bend whistle 弯航道信号
bend widening 弯道加宽
bend with double hub 双承弯管
bend with single hub 单承弯管
bend wood 弯曲材
beneaped 淤浅;搁浅的;大潮时搁浅;变浅
Benedictine abbey 本尼迪克丁修道院
Benedictine abbey church 本尼迪克丁修道院教堂

Benedictine choir 本尼迪克丁唱诗班
Benedictine church 本尼迪克丁教堂
Benedictine monastery 本尼迪克丁修道院
Benedictine quire 本尼迪克丁唱诗班
Benedict's solution 本尼迪克特溶液
Benedict's test 本尼迪克特试验
beneficial component 有益组分
beneficial cycle in reproduction 社会再生产的良性循环
beneficial enjoyment 财产享用权
beneficial estate 所有权尚未归属的房产
beneficial interest 可享用的利益;受益权益;受益利息
beneficial occupancy 实际使用权;受益使用权;空间利用;利用建筑空间
beneficial owner 受益所有者;受益所有人
beneficial owner of the royalties 特许使用费受益所有人
beneficial ownership 实际所有权
beneficial purpose calculation 兴利计算
beneficial right 受益权
beneficial spread 利益扩展
beneficial use 兴利;享用;有益利用;公益使用
beneficial use of dredged material 疏浚土的有效利用
beneficial use of water 水的有效利用;水的有益利用
beneficiary 享受保险赔偿者;受益者;受益人;受款人;收款人
beneficiary case 受益者角色
beneficiary endorsement 保险单上受益人背书
beneficiary of remittance 汇款收款人
beneficiary of transferable credit 可转让信用证的受益人
beneficiary pay principle 受惠者负担原则
beneficiary repayment 受惠者偿还原则
beneficiary's demand 受益人的要求
beneficiary's statement 受益人的声明
beneficiated burden material 精料
beneficiated iron core 铁精砂;铁精矿
beneficiated iron ore 精选铁矿砂
beneficiation 去除杂质改进材料的化学物理性质;精选;选矿;富集
beneficiation flow 选矿流程
beneficiation of aggregate 集料改善;骨料改善
beneficiation of low-grade gypsum 改进低等级石膏性质
beneficiation plant 选矿厂
beneficium excussionis 财产检索的权利
benefit 利益;利润;救济金;受益;得益
benefit assessment 效益评估
benefit building society 合作建筑协会
benefit cost 福利金
benefit-cost analysis 利益成本分析;效益费用分析;效益成本分析;效率费用分析;收益—成本分析
benefit-cost ratio 利润—成本比;收益投资比;利益成本比率;效益费用比(率);效益成本比率;效率—费用比;受益—成本比;收益—成本比
benefit-cost ratio method 效益成本比率法;受益—成本比法
benefit-cost relationship 效益成本关系
benefit-cost study 收益—成本分析
benefit country 受援国;受惠国
benefited group analysis 受益群体分析
benefited interest 受益方
benefited party 受益方
benefited-user charge 受益费
benefit effects 有利的影响
benefit-evaluation procedure 效益评价程序
benefit factor 受益率
benefit flow 利益流量;效益流量
benefit forecast 效益预测
benefit foregone 放弃的权益
benefit forgone 丧失的利益
benefit for the change of job 调职补助金
benefit function 效益函数
benefit fund reserve 职工福利基金准备;福利基金储备
benefit level 福利水平
benefit method of selling 强调顾客利益推销法
benefit of agreement 协议利益
benefit of division 分配利益
benefit of fall 降价利益
benefit of groundwater development and utilization 地下水开发利用效益
benefit of insurance clause 保险利益条款
benefit of invention 发明利益
benefit of inventory 存货利益
benefit of mankind 人类利益
benefit of the bargain 契约收益;合同收益
benefit paid 利用支付
benefit payable 应付养恤金
benefit payment 保险费的支出
benefit prediction 效益预测
benefit principle 受益原则
benefit principle of taxation 赋税受益原理
benefit program(me) 福利计划
benefit project 效益工程
benefit projection 效益预测
benefit ratio 受益率;受益比;收益投资比
benefits 福利
benefit spillover 界外受益
benefits scales 福利标准
benefit system 福利制度
benefit tariff 优惠税率
benefit theory of taxation 赋税受益理论;赋税利益理论
benefit-to-cost relationship 效益费用关系;效益成本关系
benefit trust 受益信托
benefit value 受益值
benefit year 保险赔偿年度
benemid 丙磺舒;本尼米德
benevolent association 慈善协会
benevolent interaction 有帮助的相互作用
benfix 预制的地板格栅
benfotiamine 苯磷硫胺
beng 大麻(印度)
bengala 铁丹
Bengal gelatime 琼脂
bengaline 罗缎
Benger unit 本格尔单位
Benghas gluta 马来漆树
Benguela current 本格拉海流
Benham Plateau 下哈姆高原
Benham top 贝哈姆转盘
benieflex (嵌入金属线的)防水油毛毡
benign circle 良性循环
benign cycle 良性循环
benihene 贝尼烯
Benin walnut 胡桃木(尼日利亚)
Benioff extensometer 贝尼奥夫伸长计
Benioff fault place 贝尼奥夫地带【地】
Benioff seismic zone 贝尼奥夫(地震)带
Benioff short-period seismograph 贝尼奥夫短周期地震仪
Benioff strain meter 贝尼奥夫应变计
benioff-type subduction zone 贝尼奥夫型俯冲带【地】
Benioff zone 贝尼奥夫带
benitier 圣水盆
benito 飞机导航装置
benitoite 硅酸钡钛矿
benizin 苯精
Benjamin 二苯乙醇树脂;香胶
Benkelman beam 杠杆挠沉仪;贝克曼弯沉仪;贝克曼梁;受沉仪(用于路面弯沉测量)
Benkelman beam deflection 贝克曼梁弯沉值;杠托弯沉(值)
Benkelman beam rebound 贝克曼梁回弹弯沉(值)
Benkelman quartz spectrophotometer 贝克曼石英分光光度计
benmoxin 苯莫辛
Bennet ion-mass spectrometer 贝纳特离子质谱仪
Bennhold's test 本霍尔德试验
ben oil 辣木油;贝昂油
Benoist's scale 本诺依计
benorilate 贝诺酯
Beno-Shilde drying machine 贝诺希尔德单层烘燥机
be not in accordance with the terms and conditions of credit 与信用证条款不符
Benoto boring plant 贝诺托钻孔压桩机
Benoto boring rig 贝诺托钻孔压桩机
Benoto grab 贝诺托式抓斗
Benoto hammer grab 贝诺托冲击式抓具
Benoto machine 贝诺托式(大口径)挖掘机
Benoto method 贝诺托(施工)法(大口径现浇混凝土桩)
Benoto method boring machine 全套管钻孔机
Benoto piling machine 冲抓式钻孔桩法;冲抓式钻机
Benoto process 贝诺托施工法(大口径现浇混凝土桩)
Benoto system 贝诺托法
benperidol 苯哌利多
benproperine 苯丙哌林
Benson boiler 直流锅炉;本生式锅炉;本生锅炉;本森锅炉
bent 排架
bent aerial 曲折天线
bent angle tee T 形弯折
bent antenna 曲折天线
bent apparatus 弯筋机
bent approach 弯道
bent axis type axial plunger pump 斜轴式轴向柱塞泵
bent axle 曲轴;弯臂轴
bent-back occlusion 后曲锢囚【气】
bent bar 起弯钢筋;挠曲钢筋;锚固钢筋;弯曲钢筋;弯筋
bent bar anchorage 弯筋锚固
bent beam 弯曲梁
bent blade 弯刀
bent blade snips 弯剪刀
bent cap 帽梁;框架盖梁;盖梁;排架盖梁
bent centering 排架结构鹰架
bent chisel 角凿;L 形凿;曲柄凿;弯凿
bent clamp 弓形夹具
bent coupling link 挠性联结
bent crystal spectrometer 弯晶体光谱仪
bent dent 弯形筘齿
bent draft tube 肘形尾水管;弯曲尾水管
bent element 曲杆;弯曲钢筋
bent face 弯曲面;弯曲工作面
bent finish 歪凸(玻璃制品缺陷)
bent frame (木船的)烘弯肋骨;横向排架;排架
bent glass 曲面玻璃;弯形玻璃
bent gouge 曲柄凿;曲柄弧口凿
bent grass 根茎草
bent gun 曲轴电子枪
benthal 底栖的;水底的
benthal demand 有机沉积需氧量
benthal deposit 河床堆积物;海底淀积(物);海底沉积(物);水底有机物沉积;水底沉积(物)
benthal oxygen demand 底栖需氧量;沉泥需氧量
benthal sediment 水底沉积(物)
bent handle single head wrench 弯柄单头扳手
bent-handle wrench 弯头扳手
benthic 底栖;水底的
benthic alage 底生藻类;底栖藻类
benthic animal 底栖动物
benthic biota 海底生物群
benthic biozone 底栖生物带
benthic community 底栖生物群落
benthic deposit 海底淀积(物);海底沉积(物)
benthic diatom 底栖硅藻
benthic division 海底区划
benthic ecology 海底生态学;底栖生态学
benthic ecosystem 海底生态系统;底栖生态系统
benthic environment 底栖环境
benthic fauna 底栖动物群;海底动物区系
benthic fishes 底栖鱼类
benthic flora 海底植物群;海底植物区系;底栖植物(区系)
benthic fluxes 海底通量
benthic infanau sampler 底栖动物取样器
benthic invertebrate 底栖无脊椎动物
benthic macroinvertebrate 底栖大型无脊椎动物;大型无底栖无脊椎动物
benthic micro-invertebrate 底栖微小无脊椎动物
benthic microorganism 底栖微生物;水底微生物
benthic mud 底泥
benthic mussels 底栖贝类
benthic nitrogenous oxygen consumtion rate 底栖含氮生物耗氧速率
benthic organic matter 水底有机物
benthic organism 海底生物
benthic oxygen demand 底栖生物需氧量
benthic oxygen uptake rate 底栖生物摄氧速率
benthic plant 底生植物;沉水植物

benthic population 底栖种群;底栖群落
benthic quality index 底水水质指数
benthic sample 水底样品
benthic sediment 海底淀积(物);海底沉积(物)
benthic zone 底栖区;底栖带
benthiocarb 杀草丹
benthithermoprobe 海底温度仪
benthograph 海深记录器;深水球形摄影仪;深海球形摄影仪
benthohypo-neuston 底栖水表漂浮生物
benthon 海底生物;水底生物
benth(on)ic 海底的;底栖的;海洋的
benthonic domain 底栖分区
benthonic fauna 底栖动物群
benthonic invertebrate 底栖无脊椎动物
benthonic microorganism 底栖微生物
benthonic organism 底栖生物
benthonic realm 底栖生境
benthonic region 海底区(域)
benthophyte 海底植物;底生植物;沉水植物
bentho-potamous 河流的
benthos 湖底动植物;海底生物;底栖生物;水底生物;深水底栖生物;深海底
benthos algae 底生藻类
benthos belt 底栖生物带
benthoscope 深海球形潜水器;深水球形摄影仪;深海球形摄影仪
benthos eater 食底栖生物者
benthos feeder 食底栖生物者
benthos sediment sampler 深海沉积(物)取样器
bent housing 弯管【岩】
bentiamine 苯甲硫胺
bentioite 蓝锥矿
bent iron 弯头捻缝凿
bent joint bar 弯夹板
bent latch needle 钢丝舌针
bent lever 弯杆;直角形杠杆;直角杠杆;曲杆
bent loss 弯段水头损失
bent member 曲杆
bent neck 歪颈
bent nose pliers 歪嘴钳
bentogene sediment 底栖生物沉积
Benton grease 本顿润滑脂
Benton group 本顿群
bentonite 菱碱土矿;有机皂土;皂土;斑脱岩;斑脱土;膨润土
bentonite bonded sand 膨润土砂
bentonite cement 膨润土水泥
bentonite cement pellets 膨润土水泥粒料
bentonite clay 胶质黏土;皂土;膨润黏土;膨润土
bentonite concrete 膨润土混凝土
bentonite consumption 膨润土消耗量(槽孔固壁)
bentonite deposit 膨润土矿床
bentonite diesel oil 膨润土柴油;胶质柴油
bentonite flocculation 膨润土絮凝(作用)
bentonite flocculation test 皂土絮凝试验
bentonite gel 膨润土凝胶体
bentonite grease 膨润土润滑脂
bentonite grout 斑脱土泥灌浆
bentonite grouting 胶质黏土灌浆;膨润土(灌)浆
bentonite mud 膨润土泥浆;膨润土胶泥;斑脱土泥浆
bentonite ore 膨润土矿石
bentonite pellet 膨润土球
bentonite powder 膨润土粉
bentonite process 膨润土施工法
bentonite shale 膨润土页岩
bentonite slurry 护壁泥浆;膨润土浆;斑脱土泥浆
bentonite stabilizing fluid 膨润土稳定液
bentonite suspension 膨润土悬浮液
bentonite treatment 膨润土处理
bentonite tunneling machine 泥水盾构
bentonite-type shield 膨润土盾构
bentonite-water solution 膨润土水溶液
bentonitic clay 膨润黏土;膨润土
bentonitic mud 膨润土;膨润土泥浆
bentorite 钙铬矾
bent pattern 弯板图式
bent pier 架式桥墩
bent pile flexible pier 排架桩柔性墩
bent pile pier 排架桩墩
bent pipe 弯头;弯管
bent plain tile 弯曲平瓦
bent plate 波形板;曲板;弯板

bent plywood 弯曲胶合板
bent pyramid 折线形棱锥(体);折线形金字塔
bentranil 草恶嗪
bent reinforcement 挠曲钢筋
bent reinforcement bar 起弯钢筋;弯曲钢筋
bent riffler 弯曲来福线锉
bent rod 曲杆;弯曲钻杆;弯曲钢筋
bent rod type stirrer 弯棒式搅拌器
bent shaft 曲轴
bent sheet metal 弯曲金属(钢)板
bent shoe 圆弧形弯的底座线脚
bent socket wrench 弯柄套筒扳手
bent spanner 弯头扳手
bent spar 弯曲翼梁
bent steel 挠曲钢筋;弯筋;弯曲钢筋
bent stem 倾斜茎
bent stem thermometer 曲管温度表
bent stock rail 弯曲基本轨
bent strip 镰刀形带材
bent strip test 弯条试验
bent structure 排架结构
bent sub 斜接头;弯接头
bent surface 曲面
bent-tail carrier 弯尾夹头
bent tap 弯柄丝锥;弯柄螺母丝锥
bent tempered glass 弯钢化玻璃
bent tile 槽瓦;曲瓦;曲瓦面
bent timber 弯曲木材
bent tongue depressor 角式压舌板
bent tool 弯头车刀
bent tree 马刀树(即醉林)
bent tube 弯头;弯管
bent-tube boiler 弯管式锅炉;弯管锅炉
bent-tube evapo(u)rator 弯管式蒸发器
bent tubular lever 曲管水平仪
bent up 弯曲的
bent-up bar 起弯钢筋;上弯钢筋;斜筋;向上弯曲钢筋;弯上钢筋;弯曲钢筋;弯起钢筋
bent-up end 弯起端
bent-up point 弯起点
bent-up portion 弯起部分
bent-up reinforcement 弯上钢筋;弯曲钢筋
Benturi Kascade tray 本图利阶梯式塔板
bentwood 弯曲木(材);弯曲材;(弯曲成型而非加工成型的)曲木
bent wood furniture 弯曲木家具
bent wood turning machine 弯式木工车床
bent wrench 弯头扳手
Bentzel tube 本特兹管;(测水流速度的)本茨尔管
Bentzel velocity tube 本茨尔低速测流管
Benxi epsilon structural system 本溪山字型构造体系
benzacridine 苯并吖啶
benzalaminophenol 亚苄基氨基苯
benzalaniline 亚苄基苯胺
benzalcohol 苄醇
benzaldehyde 苯甲醚;安息香醛
benzaldehyde acetal 苯甲醛缩二醇
benzaldehyde-carboxylic acid 苯醛酸
benzaldoxime 苯醛肟;苯甲醛肟
benzalin 苯胺黑
benzalkonium 苄烷铵;苯甲烃铵
benzalkonium bromide 苯扎溴铵
benzalkonium chloride 苯扎氯铵
benzamarone 苯苦杏酮
benzamide 苯酰胺;苯甲酰胺
benzamides 苯酰胺类
benzamidothiophenol 苯酰胺硫酚
benzamidoxime 苄胺肟
benzamine 苯扎明
benzaminic acid 氨基苯甲酸
benzanil black 亚苯基苯胺黑
benzanil blue 亚苯基苯胺蓝
benzanil colo(u)r 亚苯基苯胺染料
benzanilide 苯酰替苯胺
benzanthracene 苯并蒽
benzanthrone 苯绕蒽酮;苯并蒽酮
benzarone 苯扎隆
benzatropine 苯扎托品
benzazide 苯甲酰叠氮
benzazole 吲哚;氮茚
benzdiazole 吲唑
benzdioxan 苯并二恶烷
benzedrine 苯异丙胺

benzedrinum 苯丙胺
benzene 纯苯
benzene and its analogies determination 苯系物检测
benzene arsonate 苯胂酸盐
benzene arsonic acid 苯胂酸
benzene-azo-benzene 苯偶氮苯
benzeneazo cresol 苯偶氮甲基苯酚
benzeneazo-naphthylamine 苯偶氮萘胺
benzeneazo resorcinol 苯偶氮间苯二酚
benzene carbon amidine 苯基脒
benzene condenser 苯冷凝器
benzene dibromide 二溴化苯
benzene dicarboxylic acid 苯二甲酸
benzene dichloride 二氯化苯
benzene diiodide 二碘化苯
benzene dinitrile 苯二腈
benzene distilling apparatus 苯蒸馏器
benzene-disulfo-chloride 苯二磺酰氯
benzene disulfonate 苯二磺酸盐
benzene disulfonic acid 苯二磺酸
benzene-ethanol insoluble 不溶于苯—乙醇
benzene-ethanol soluble 能溶于苯—乙醇
benzene extract 苯抽出物
benzene formula 苯式
benzene hexacarbonic acid 苯六甲酸
benzene hexachloride 六六六
benzenehydrasinonaphthalene 苯肼基萘
benzene hydrocarbons 苯系烃
benzene insoluble 含杂质的苯;不溶于苯的
benzene metering tank 苯计量槽
benzene mixed 混合苯
benzene nucleus 苯核
benzene-pentacarboxylic acid 苯五甲酸
benzenephosphonic acid 苯膦酸
benzenephosphorus dichloride 二氯苯磷
benzene poisoning 苯中毒
benzene pollution 苯污染
benzene polycarbonic acid 苯多甲酸
benzene polycarboxylic acid 苯多甲酸
benzene ring 苯环
benzene series 苯系
benzene series content 苯系含量
benzene siliconic acid 苯硅酸
benzene soluble fraction 苯可溶分数
benzene still 苯蒸馏器
benzene storage 苯储[贮]槽
benzene sulfinate 苯亚磺酸盐
benzene sulfinic acid 苯亚磺酸
benzene sulfinic acid sodium salt 苯亚磺酸钠
benzenesulfinyl 苯亚磺酰
benzene sulfonate 苯磺酸盐
benzene sulfonic acid 苯磺酸
benzene sulfonic acid sodium salt 苯磺酸钠
benzene sulfonic amide 苯磺酰胺
benzenesulfonyl 苯磺酰基;苯磺酰
benzene sulfonyl anthranilic acid 苯磺酰邻氨基苯甲酸
benzene sulfonyl chloride 苯磺酰氯
benzenesulfonyl hydrazide 苯磺酰肼
benzene sulfonyl hydroxylamine 苯磺酰胺
benzene tetracarboxylic acid 苯四甲酸
benzenethiol 苯硫酚
benzene thiosulfonic acid 苯硫代磺酸
benzene-toluene-xylene 苯—甲苯—二甲苯
benzene tricarboxylic acid 苯三甲酸
benzene trisulfonic acid 苯三磺酸
benzenoid hydrocarbons 苯型烃类
benzenoid structure 苯型结构
benzflavone 苯并黄酮
benzfluorene 苯并芴
benzfluorenol 苯并芴醇
benzfuran 苯并呋喃
benzhydrazide 苯酰肼
benzhydrol 二苯甲醇
benzhydroxamic acid 苯氧肟酸
benzhydryl amine 二苯甲基胺
benzhydryl bromide 二苯甲基溴
benzhydrylcellulose 二苯甲基纤维素
benzidine 联苯胺;二氨基联苯
benzidine acetate 醋酸联苯胺
benzidine blue 联苯胺蓝
benzidine dye 联苯胺(结构)染料
benzidine orange 联苯胺橙

benzidine reaction 联苯胺反应
benzidine reagent 联苯胺试剂
benzidine rearrangement 二芳基肼重排作用
benzidine red 联苯胺红
benzidine tank vehicle 联苯胺槽车
benzidine test 联苯胺试验
benzidine yellow 联苯胺黄
benzil 苯偶酰
benzil dioxime 苯偶酰二肟
benzilic acid 二苯乙醇酸;二苯基乙醇酸
benzilonium bromide 苯咯溴铵
benzil osazone 苯偶酰脎
benzimidazole 苯并咪唑
benzimidazoline 苯并咪唑啉
benzimidazolone 苯并咪唑酮
benzimidazolyl- 苯并咪唑基
benzimide 苄基亚胺
benzindan 苯并二氢茚
benzindene 苯并茚
benzin(e) 挥发性油;挥发性油;苯精;轻质汽油;轻馏分油;汽油;石油挥发油
benzine-ligroin 轻汽油
benzine resisting hose 耐汽油软管
Benzinger metal lath(ing) 贝净格钢丝网
benziodarone 苯碘达隆
benzo 苯并
benzoate 苯甲酸酯;苯甲酸盐
benzoate fibre 苯甲酸酯纤维
benzo-azurine 苯并天青精
benzo-black-blue 苯并黑蓝
benzo blue 苯并蓝
benzo brilliant orange 苯并亮橙
benzocaine 对氨基苯甲酸乙酯;氨基苯甲酸乙酯
benzo chrome black 苯并铬黑
benzo chrome brown 苯并铬棕
benzochromone 苯色酮;苯并色酮
benzo-colo(u)rs 苯并染料
benzodiazine 苯并二嗪
benzodihydropyrone 苯并二氢吡喃酮
benzododecinium chloride 苯度氯铵
benzo fast blue 苯并坚牢蓝
benzo-fast-heliotrope 苯并坚牢淡紫
benzo-fast-orange 苯并坚牢橙
benzo fast pink 苯并坚牢桃红
benzo fast red 苯并坚牢红
benzo fast yellow 苯并坚牢黄
benzoffuanamine formaldehyde resin 苯代(基)三聚氰胺甲醛树脂
benzoflavine 苯并黄素
benzofuran 氧茚;苯并呋喃
benzofuran resin 苯并呋喃树脂
benzofuranyl 苯并呋喃基
benzoglycols 苯乙二醇
benzoguanamine 苯鸟粪胺;苯代三聚氰胺;苯并胍胺
benzoguanamine formaldehyde resin 苯鸟粪胺甲醛树脂;苯并胍胺醛树脂
benzoguanamme 苯代(基)三聚氰胺
benzo-hydroperoxide 过氧苯甲酸
benzohydroxamic acid 苯异羟肟酸
benzoic acid 苯甲酸;苯甲基;安息香酸
benzoic alcohol 苄醇
benzoic amide 苯甲酰胺
benzoic anhydride 苯酸酐;苯甲酐
benzoic conden-sation 安息香缩合
benzoic ether 苯甲酸酯
benzoic stripping 苯溶出
benzoid compound 苯型化合物
benzoid hydrocarbon 苯型烃
benzoin 二苯乙醇酮;苯偶姻;安息香树;安息香胶;安息香
benzoin acetate 乙酰苯偶姻;苯偶姻乙酸酯
benzoin blue 苯偶姻蓝
benzoin condensation 苯偶姻缩合
benzoin dark-green 苯偶姻暗绿
benzo-indigo-blue 苯并靛蓝
benzoin ethyl ether 苯偶姻乙醚
benzoin fast-red 苯偶姻坚牢红
benzoin gum 安息香胶
benzoin oxime 苯偶姻肟;安息香肟
benzoin test 安息香试验
benzoin tincture 安息香酊
benzois condensation 醇酮缩合作用
benzoisoquinoline 苯并异喹啉

benzol 粗苯;苯;安息油
benzol-acetone solvent 苯丙酮溶剂
benzol black 苯炭黑
benzol blends 苯混合物
benzole 粗苯
benzole mixture 混合粗苯
benzol equivalent 苯当量
benzol(e)scrubber 洗苯器
benzo light blue 苯并浅蓝
benzo light orange 苯并浅橙
benzo-light rubine 苯并亮玉红
benzoline 苯汽油;轻质汽油
benzolism 苯中毒
benzolized oil 苯化油
benzol removal 脱苯
benzonaphthalene 苯甲酸萘
benzonaphthalin 苯甲酸萘
benzonaphthol 苯甲酸萘酚
benzonaphtol 苯佐萘酚
benzonatate 苯佐那酯
benzonitrile 苯基腈
benzo orange 苯并橙黄
benzoperoxide 过氧化二苯甲酰;过氧苯甲酰
benzophenone 二苯甲酮;苯酮
benzophenone-anil 二苯酮苯胺
benzophenone carboxylic acid 二苯酮甲酸
benzophenone dicarboxylic acid 二苯甲酮二甲酸
benzophenone-oxime 二苯甲酮肟
benzophenone phenylhydrazone 二苯甲酮苯腙
benzo-phenone tetracarboxylic dianhydride 二苯酮四酸二酐
benzopiaselenole 苯并硒二唑
benzopinacol 苯频哪醇
benzopurpurin(e) 苯并红紫;苯紫;苯红紫
benzopyranyl 苯并吡喃基
benzopyrene 苯并芘
benzopyrene pollution 苯并芘污染
benzopyranone 苯并吡喃酮
benzopyrone series 苯并吡喃酮系
benzopyrrole 苯并吡咯
benzoquinaldine 苯并喹哪啶
benzoquinoline 苯并喹啉
benzoquinone 苯醌
benzo red 苯并红
benzo-red-blue 苯并红蓝
benzoresorcinol 苯酰间苯二酚
benzo-rhoduline red 苯并若杜林红;苯并碱红
benzo sky blue 苯并青
benzosulfimide 邻磺酰苯甲酰亚胺
benzosulfonazole 苯并磺酰唑
benzotetrazine 苯并四嗪
benzothiadiazole 苯并噻二唑
benzothiazole 苯并噻唑
benzothiazole polymer 苯并噻唑聚合物
benzothiazolyl 苯并噻唑基
benzothiophene 苯并噻吩
benzoxazine 氧氮杂萘;苯并恶嗪
benzoxazinyl- 苯并恶嗪基
benzoxazole 间氮杂氧茚;苯并恶唑
benzoxazolyl 苯并恶唑基
benzoxiquine 苯甲酰喹啉
benzoxonium chloride 苯佐氯铵
benzoxy 苯甲酸基
benzoxyperoxycarbonylation 苯酸过氧羰化作用
benzoyl 苯甲酰基;苯甲酰
benzoyl acetanilide 苯甲酰乙酰替苯胺
benzoyl acetic acid 苯甲酰乙酸
benzoyl acetone 苯甲酰丙酮
benzoyl acetonitrile 苯甲酰乙腈
benzoyl acid glycolic acid 苯甲酰乙醇酸
benzoyl acrylic acid 苯甲酰丙烯酸
benzoyl amide 苯甲酰胺
benzoylaminobenzoic acid 苯甲酰氨基苯酸
benzoylate 苯甲酰化物
benzoylated 苯甲酰化的
benzoylating agent 苯甲酰剂
benzoylation 苯甲酰化
benzoyl azide 苯甲酰叠氮
benzoylbenzoic acid 苯甲酰苯甲酸
benzoyl chloride 苯甲酰氯
benzoylcholine 苯甲酰胆碱
benzoyl fluoride 苯甲酰氟
benzoyl formic acid 苯甲酰甲酸
benzoylglycine 苯甲酰甘氨酸

benzoyl hydrazine 苯甲酰肼
benzoyl hydroperoxide 过氧苯甲酸
benzoyl isocyanate 异氰酸苯甲酰酯
benzoyl lactic acid 苯甲酰乳酸
benzoylnaphthol 苯甲酸萘酚
benzoyl oxide 苯甲酸酐
benzoyloxy 苯甲酰基
benzoyl peroxide 过氧化二苯甲酰;过氧化苯甲酰;过氧苯酰;过氧苯甲酰
benzoylphenyl-hydroxylamine 苯甲酰苯基羟胺
benzoyl phthalic acid 苯甲酰邻苯二甲酸
benzoyl ramanone 苯甲酸来门酮
benzoyl thiokinase 苯甲酰硫激酶
benzoyl thiourea 苯甲酰硫脲
benzoyl urea 苯甲酰脲
benzpinacol 苯频那醇
benzpinacolone 苯频哪酮
benzsulfamide 苯磺酰胺
benzyl 苄基;苯甲
benzyl abietate 松香酸苄酯
benzyl acetate 乙酸苄酯
benzyl acetone 苄基丙酮
benzyl acrylate 丙烯酸苄酯
benzyl alcohol 苄醇;苯甲醇
benzyl amine 苄胺
benzylamine carbonate 苄胺碳酸盐
benzylamine cinnamate 苄胺肉桂酸盐
benzyl aminophenol 苄氨基苯酚
benzyl aniline 苄基苯胺
benzylanilineazo-benzene 苄基苯胺偶氮苯
benzyl aniline resin 苄基苯胺树脂
benzylate 苄化
benzylating agent 苄化剂
benzylbenzene 二苯甲烷
benzyl benzoate 苄基苯甲酸酯;苄基苯甲酸盐;苯甲酸苄酯
benzyl benzoic acid 苄基苯甲酸
benzyl biphenyl 苄基联苯
benzyl boric acid 苄基硼酸
benzyl boron dihydroxide 苄基硼酸
benzyl bromide α-bromotoluene 苄基溴
benzyl butyl phthalate 邻苯二甲酸苄丁酯
benzyl butyrate 丁酸苄酯
benzyl carbamate 氨基甲酸苄酯
benzyl carbinol 苯甲醇
benzyl cellosolve 丁二醇单苄醚;苄基溶纤剂
benzylcellulose 苄基纤维素
benzyl chloride 苄基氯
benzyl cinnamate 肉桂酸苄脂
benzyl cyanide 苯乙腈
benzyl dimethylamine 苄基二甲胺
benzyl ether 二苄醚
benzyl ethyl ether 苄基乙基醚
benzyl fluoride 苄基氟
benzyl group 苯甲基
benzyl halide 苄基卤
benzyl hydrazine 苄基肼
benzylhydroperoxide 过苯甲酸
benzylidene 苯亚甲基
benzylidenebutyramide 苄烯丁胺
benzyl iodide 苄碘
benzyl-iso-thiourea hydrochloride 苄基异硫脲盐酸盐
benzyl mercaptan 苄硫醇
benzyl oxide 苄醚
benzylphenol 苄基苯酚
benzylphenyl carbamate 氨基甲酸苄基苯酯
benzylphenyl chloride 苄苯氯
benzyl resinate 松香酸苄酯;树脂酸苄酯
benzyl succinate 琥珀酸苄酯
benzylsulfamide 苄磺胺
benzyl sulfanilamide 苄基磺胺
benzyl sulfonic acid 苄磺酸
benzyltrim-ethy-lammonium bromide 苄基三甲基溴化铵
benzyltrim-ethy-lammonium chloride 苄基三甲基氯化铵
benzyltrim-ethy-lammonium hydroxide 苄基三甲基氢氧化铵
benzyltrim-ethy-lammonium iodide 苄基三甲基碘化铵
benzyltrimethyl ketone 苄基甲基酮
benzyne 苯炔;脱氢苯
benzyne intermediate 苯炔中间体

be obligated to pay tax 负有纳税义务
be off duty 下班
be open to traffic 交付运营
be operating under capacity 开工不足
beou oil 杷荏油
be out of debt 不欠债
be out of pocket 赔钱
bephenium 酚乙铵
be prepared against natural disasters 备荒
be proficient in 精通
be proficient in professional and technical work 精通业务和技术
be prolonged accordingly 依此法顺延
be prolonged in a similar manner 依此法顺延
be pulled out from the ground 从土地拔出
be purchased in small quantities 少量购买
be put in storage 入库(存货)
be qualified for 有资格
bequeath 遗赠动产
bequeather 遗赠人
bequest 遗赠；遗产
Beranek scale 白瑞纳克度标；白瑞纳克等级(噪声分类度标)
berangan 马来栲树
beraunite 簇磷铁矿
berboritae 水硼铍石
bercular corrosion 点状腐蚀
berdesinskiite 钛钒矿
Berea sandstone 贝瑞亚砂岩
Bereau Veritas 法国船级社
be recognized by their characteristic shape 根据其典型形状识别出来
Berek compensator 贝雷克补偿器
berel edge 倒角
berengelite 脂光沥青
berenil 重氮氨苯脒乙酰甘氨酸盐
beresite 黄铁绢英岩；倍利岩
beresite porphyry 黄铁细晶斑岩
beresitization 黄铁绢英岩石
beresovite 红铬铅矿
be responsible for losses and profits 统负盈亏
be responsible for one's own finance 财政包干
berezovskite 微镁铬铁矿
berg 大冰块；冰山
bergalite 黑云蓝方黄煌岩
bergapten 佛手内酯
Berg-Barrett method 反射法；X射线反射形貌法
bergblau 碳酸铜矿
berg deposit 冰山沉积
berge 停机坪闪光
bergenite 磷钡铀矿
Berger code 伯格码
bergere 高背扶手椅；围手椅
bergfall furrow 大块石下落沟槽；石沟
berg ice 冰山冰
Bergius process 伯吉尤斯法
Bergman generator 伯格曼发电机
Bergmann's law 伯格曼定律
Bergmann's rule 伯格曼规律；伯格曼法则；伯格曼定则
Bergman series 伯格曼系
Bergman's law 伯格曼定律
Bergman-Turner unit 伯格曼—吐纳单位
bergmeal 矽藻土；硅藻土
bergschrund 冰川裂缝；大冰隙(超过500米宽)；冰后隙；背隙窿
Bergschrunde 冰川边沿裂隙
Berg's diver method 潜沉法
bergslagite 羟砷钙铍石
berg till 浮冰碛；浮冰冰碛
bergy 冰山群
bergy bit 浮冰群；小冰川；中型
berhting vessel 靠泊船只；靠泊船舶
berillia 氧化铍
berillia ceramics 氧化铍陶瓷
berillite 白硅铍石
Bering Canyon 白令峡谷
Bering Sea 白令海
Berkefeld filter 伯克菲尔德(细菌)滤器；伯克菲尔德滤池
Berkeley clay 伯克里耐火高岭土
berkelium 锫
Berkovitvh indenter 伯氏压痕器
Berksonian line 伯克森线

berkyite 天蓝石
Berlese's organ 伯利斯器
Berlese's funnel 伯利斯漏斗
Berlin black 耐热漆；黑色沥青(耐热)漆；柏林黑；无光黑漆
Berlin black paint 无光黑漆
Berlin blue 亚铁氰化铁；普鲁士蓝；柏林蓝；深蓝色
berlin(e) 大四轮车
berliner 镶嵌大理石片的水磨石；大理石渣罩面水磨石
Berlin green 柏林绿
berlinite 块磷铝石；块磷铝矿；板磷铝矿
Berlin porcelain 柏林化学瓷
Berlin white 柏林白
Berl saddle 鞍形填充物；马鞍形填料；贝尔鞍形填料
Berl saddle packing 弧鞍形填料；贝尔鞍形填料；圆鞍形填料
bermanite 板磷锰矿；板磷镁锰矿
bermatypic coral 造礁珊瑚
Bermax bearing alloy 伯马克斯轴承合金
berm(e) 路肩；肩状阶地；马道；护坡道；护脚土台；护道；后滨阶地；小路；堆积岸堤；戗台；戗道；滩肩；台坎
berm(e) construction 护坡道构造
berm(e) crest 海滩阶坎前缘；滨后阶地脊部；滩肩脊
berm edge 马道边缘；海滩阶坎前缘；滨后阶地脊部
berm(e) ditch 边坡截水沟；护路排水沟；护道排水沟；护岸排水沟；傍山排水沟；戗道排水沟
berm(e) edge 滩肩外缘
berm(e) maintenance 马道养护；护坡道养护
berm(e) maintenance tool 护道养护机具
berm(e) of cofferdam 围堰戗道
berm(e) of ditch 沟岸小道
berm(e) scarp 滨后陡坎；滩肩崖；滩肩坎；滩壁
berm(e) shoulder 斜坡
berm(e) spillway 护岸溢洪道；护道排水沟；傍山排水沟；戗道排水沟
berm(e) stake 护道边桩
bermetic centrifuge 密封离心机
berm(e) trimming 戗道修整；马道修整
berm(e) type structure 戗台式建筑
berm for channel 渠道旁边的路
berming 设置护道
Bermuda 百慕大群岛
Bermuda asphalt 百慕大(石油)沥青
Bermuda high 百慕大高压
Bermuda platform 百慕大浅台
Bermudze 贝尔姆地沥青
Bernard's duct 伯纳尔管
Bernay's sponge 伯纳海绵
berndtite 三方硫锡矿
Berne Convention 伯恩公约
Berner's window 大钢窗；伯尔尼窗
Berne Union 伯恩联盟(信用及投资保险人国际联盟)
Bernhardt's formula 伯恩哈特公式
Bernheimer's fiber 伯恩海默纤维
Bernold lining method 贝尔纳衬砌法
Bernold sheet 拱形压花钢板
Bernoulle polynomial 伯努利多项式
Bernoulli distribution 二项分布
Bernoulli-Euler law 伯努利—欧拉定律
Bernoulli's analysis 伯努利分析
Bernoulli's assumption 伯努利假定
Bernoulli's constant 伯努利常数
Bernoulli's differential equation 伯努利微分方程
Bernoulli's distribution 伯努利分布
Bernoulli's effect 伯努利效应
Bernoulli's energy equation 伯努利能量公式
Bernoulli's equation 伯努利方程
Bernoulli's-Euler differential equation 伯努利—欧拉微分方程
Bernoulli's force 伯努利力
Bernoulli's formula 伯努利公式
Bernoulli's hypothesis 伯努利假定
Bernoulli's law 伯努利定律
Bernoulli's lemniscate 伯努利双纽线
Bernoulli's number 伯努利数
Bernoulli's number table 伯努利数表
Bernoulli's probability 伯努利概率
Bernoulli's process 伯努利过程
Bernoulli's series 伯努利级数
Bernoulli's spiral 伯努利螺线
Bernoulli's stochastic process 伯努利随机过程

Bernoulli's theorem 伯努利定律；伯努利定理
Bernoulli's theory 伯努利定理
Bernoulli's trial 伯努利试验
Bernoulli's variation 伯努利变异
Bernoulli's walk 伯努利游走
Bernstein's inequality 伯恩斯坦不等式
Bernstein's theorem 伯恩斯坦定理
berperacetylated rubber 过乙酸化橡胶
Berriasiangy 贝里亚斯岩
Berries graining 贝利漆画木纹工艺
berry 浆果
Berry asphalt 柏利地沥青
berryite 板铋铋铜铅矿
Berry machine 柏利砖坯成型机
Berry strain ga(u)ge 柏利式应变仪
Berry transformer 柏利变压器
berth 架床；锚位；船位；车位(港)；舱位；泊位；泊船处；铺位；停泊地；停泊处
berthage 码头税；码头靠泊费；码头吨税；系泊费；泊位费；停泊费
berthage space 停泊区
berth alongside 靠泊
berth alongside depth 泊位前沿水深
berth area 泊船区
berth availability 泊位可用率
berth bill of lading 码头提单
berth capacity 泊位通过能力；泊位能力
berth cargo 配载货物；外加货；填舱货
berth cargo rate 空航载货运费率；停埠船运费率
berth charges 码头靠泊费；停泊费
berth charter 船位包租
berth crane 码头起重机
berth deck 居住甲板；住舱甲板【船】
berth designation 指泊
berth dimension 泊位尺度
berth dues 码头税
berthed vessel 靠泊船只；锚泊船只；停靠在码头的船只
Berthele eyepiece 伯塞莱目镜
Berthelot's equation 贝特洛方程式
Berthelot's ralation 贝特洛关系
Berthelot's calorimeter 贝特洛量热器
Berthelot-Thomsen principle 贝特洛—汤姆逊原理
berth extension 码头扩建
berth fitting-out 船台舾装
berth for female 女舱室
berth for male 男舱室
berth geometry 泊位尺度
berthierine 磁绿泥石；铁铝蛇纹石
berthierite 辉锑铁矿；蓝铁矿
berthing 靠码头；靠泊；舷缘上列板；在泊位上；停泊(指靠码头作业)
berthing accommodation 靠船设备；靠泊设施；靠泊设备；锚泊设备；系泊设备
berthing aid system 靠岸测速仪
berthing alongside 靠码头
berthing area 碇泊区；停泊区
berthing barge 居住驳船；浮码头
berthing beam 靠船梁
berthing capacity 靠船能力；泊位容量
berthing clause 停泊条款
berthing compartment 住舱
berthing dock 船坞；船码头
berthing dolphin 靠船墩
berthing energy 靠船能(量)；靠泊能量
berthing face 靠船面；码头正面
berthing facility 靠泊设施；靠泊设备；系船设备；系泊设施；靠船设施；靠泊设备
berthing fender 码头防撞设施；码头防撞设备
berthing force 靠泊力；挤靠力【船】；泊力
berthing hawser 靠泊系缆
berthing head 突堤码头前缘靠泊位
berthing impact 靠船撞击力；靠船力；靠船边；锚泊冲击力；撞击力(船靠泊时)
berthing jetty 靠船栈桥
berthing line 码头线；泊位线
berthing load 靠船荷载；靠泊荷载
berthing maneuver 靠船作业；泊位操纵
berthing master 指泊员；泊位调度员
berthing member 靠船构件
berthing mode 靠泊模式
berthing note 停泊提示
berthing officer 指泊员；泊位调度员
berthing pier 靠船墩

berthing plan 停泊地平面图
berthing plant 靠船设备;锚泊设备
berthing platform 靠船平台
berthing pressure 靠挤压力
berthing reaction 靠泊反力
berthing room 泊位间隔;停泊周旋余地;停泊区
berthing room accommodation 住舱
berthing ship 靠泊船只;泊位船舶
berthing signal 停泊信号;靠泊信号
berthing space 泊位间隔;泊位;停泊周旋余地;停泊区
berthing speed 靠船速度;靠泊速度
berthing structure 靠船结构(物);靠船建筑物
berthing time 靠泊时间
berthing tug(boat) 停靠拖船;停靠拖轮;靠泊拖轮;靠离作业拖轮;靠泊用拖轮;移泊拖轮
berthing velocity 靠船速度;靠泊速度
berthing vessel 正在靠泊的船
berth jetty 栈桥码头
berth length 泊位长度
berth level 码头前高程
berth line 靠泊线;泊位线
berth liner 班轮
berth list 客运清单
Berthlot's calorimeter 贝司罗量热计
berth master 码头长
berth note 订舱单
berth number 泊位区号;铺位号码
berth occupancy 泊位占用率;泊位利用率
berth occupancy factor 泊位占用率;泊位利用率
Bertholide 伯托利化合物
Berthon dynamometer 伯松计;伯松测力计
berth order 指泊通知;指泊命令
berth orientation 泊位轴线方向
berth owner 班轮船主
berth plan 港泊图
berth rate 班轮费率
berth revenue 卧铺收入
berth's capacity of freight 泊位通过能力
berth schedule 指泊计划
berth shifting 移档;移泊
berth space 港泊区;停泊场
berth structure 锚泊设施
berth telephone 泊位电话
berth term 码头条件;船方不承担装卸费;泊位条件;班轮条款;泊拉条件
berth throughput 泊位吞吐量;泊位通过量;泊位生产量;泊位货运量
berth throughput capacity 泊位通过能力
berth ticket 卧铺票
berth time 停泊时间
bertossaite 磷钙钙锂石
Bertrand curve 伯特兰曲线
bertrandite 硅铍石;羟硅铍石
bertrandite ore 羟硅铍石铍矿石
Bertrand lens 伯特兰(透)镜;伯特兰镜头(偏光显微镜观察用)
Bertrand postulate 伯特兰假设
Bertrand qualifying equation 伯特兰验证方程式
Bertrand's Duopoly Model 伯特兰的双垄断模型
Bertrand's postulate 伯特兰公设
Bertrand's rule 伯特兰规则
beryl 绿柱石;绿宝石;海绿色
beryl-bearing metagranite deposit 含绿柱石交代蚀变花岗岩矿床
berylco alloy 铜合金
beryl concentrate 绿柱石
beryllia 氧化铍
beryllia ceramics 氧化铍陶瓷
beryllia porcelain 氧化铍瓷
beryllia refractory 氧化铍耐火材料
beryllia refractory product 氧化铍耐火材料制品
beryllia substrate 氧化铍衬底
beryllia valve seat 氧化铍阀座
berylliosis 铍肺
beryllite 水硅铍石
beryllium 铍
beryllium-bearing skarn deposit 含铍矽卡岩矿床
beryllium boride 硼化铍
beryllium bronze 铍青铜
beryllium-copper alloy 铜铍合金
beryllium hydroxide 氢氧化铍
beryllium method 铍法
beryllium nitride 氮化铍

beryllium ore 铍矿(石)
beryllium oxide 氧化铍
beryllium silicate 硅酸铍
beryllium window 铍窗
beryllium zinc silicate 硅酸锌铍
beryllon two 二型铍试剂
beryllonite 磷酸钠铍石;磷钠铍石
beryl ore 绿柱石铍矿石
beryl pegmatite deposit 绿柱石伟晶岩矿床
beryl-quartz vein deposit 绿柱石石英脉矿床
berzelianite 硒铜矿
berzeliite 黄砷榴石
Berzelius theory of valency 贝齐利阿斯价键理论
besabt 银币饰
besant 圆盘形装饰
bescot 弹簧门锁
besd sed of rim 鞍边
besel 监视孔;监视窗
be separated into other elements 分成其他元素
be separated into some plots to test 分成几个小区试验
beset with crisis 危机四伏
besiclometer 镜架宽度计
besides compost 除堆肥以外
beside-the-press grinder 压机旁研磨机
Besinc function 贝森克函数
besmirch 沾染
bespangle 使……灿烂发光;金银箔饰
bespeak 预订的货
bespoke building 单独设计的建筑物
Bessel ellipsoid of 1841 贝塞尔椭球体1841【测】
Besselian date 贝塞尔日期
Besselian day number 贝塞尔日数
Besselian elements 贝塞尔要素;贝塞尔根数
Besselian star constant 贝塞尔恒星常数
Besselian star number 贝塞尔星数
Besselian year 假年;虚构年;贝塞尔年
Bessel's date 贝塞尔日期
Bessel's day number 贝塞尔日数
Bessel's differential equation 贝塞尔微分方程
Bessel's ellipsoid 贝塞尔椭圆体;贝塞尔椭球
Bessel's equation 贝塞尔方程
Bessel's formula for solution of geodetic problem 贝塞尔大地问题解算公式
Bessel's function 贝塞尔函数
Bessel's horn 贝塞尔喇叭
Bessel's inequality 贝塞尔不等式
Bessel's interpolation formula 贝塞尔内插公式
Bessel's method 贝塞尔(图上定位)法
Bessel's point 贝塞尔支承点
Bessel's star number 贝塞尔星数
Bessel's transition coefficient 贝塞尔转移系数
Bessemer charge 酸性转炉炉料
Bessemer converter 贝氏转炉;贝塞麦转炉;酸性转炉
Bessemer furnace slag 转炉炉渣
Bessemer heat 酸性转炉熔炼
Bessemer iron 贝塞麦铁;酸性转炉铁;酸性转炉钢;酸性钢
bessemerizing 转炉吹冶炼法;酸性转炉吹炼(法)
Bessemer ladle 底注盛钢桶
Bessemer low carbon steel 酸性转炉低碳钢
Bessemer matte 转炉锍;吹炉锍
Bessemer mild steel 贝塞麦钢
Bessemer pig 转炉钢;酸性钢
Bessemer pig iron 贝塞麦铁
Bessemer process 贝塞麦方法;酸性转炉炼钢法
Bessemer steel 酸性钢;转炉钢;贝塞麦钢
Bessey unit 贝西单位
bessmertnovite 碲铜金矿
best 优越,最佳的
bestac (瓦楞石棉板打底或嵌填用的)混合剂
best anchor 备用船首大锚
best approximated 最近似的
best approximation 最优逼近;最佳估计;最近逼近
best available contamination control technology 可得的最佳控制污染技术
best available control technology 最佳有效控制技术
best available technology 最适用工艺;最佳可用技术;最佳可行技术
best available technology economically achievable 经济上可行的最佳技术
best available true heading 现实最佳航向

best bar 二次精练条铁
best bet 最好的措施
best bid 最高出价
best bower 右舷船首锚
best bower anchor 船首右大锚;备用船首大锚
best climbing angle 最好上升角
best coke grade tin plate 高级镀锡钢板
best cokes 最优薄锡层镀锡薄钢板;最优薄锡层
best compromise solution 最佳协调解
best conventional pollutant control technology 最常用污染物控制技术
best curve 最佳曲线
best decision 最佳方案
best depth range 最佳深度范围
best economy rating 最低燃料消耗率
best-effects selling 非包销发行
best efficiency 最佳效率
best efficiency point 最佳效率点
best efforts 最好承担(包工合同)
best efforts syndication 尽力联贷
best estimate 最佳估计
best estimator 最佳估计量
best-fir large circle 最佳拟合大圆
best-fir small circle 最佳拟合小圆
best fit 最佳配合
best fit approximation 最优近似
best fit equation 最佳拟合方程
best-fit girdle axis 最佳拟合环带轴
best-fit method 最优满足法
best-fit polynomial approximation 最优逼近多项式
best fitting 最佳拟合
best fitting curve 最佳拟合曲线;最符合曲线
best fitting polynomial 最佳拟合多项式
best focus 最佳焦点
best form lens 最佳形状透镜
best gold 磨光金
best hand-picked lime 人工精选上等石灰;手选优质石灰
best hand-picked quicklime 手精选的生石灰
best hydraulic cross-section 最佳水力断面;最佳水力断面;水力最佳断面
bestiary (中世纪教堂的)奇兽画或雕刻
best in quality 最佳质量;品质优良;品质优等
best invariant estimate of location parameter 局部参数的最佳不变量估计
best invariant test for location parameter 局部参数的最佳不变量试验
best iron 二次加工熟铁
best lending rate 最优惠贷款率
best linear fit 最优线性拟合;最佳线性拟合
best linear unbiased estimate 最佳线性无偏估计
best linear unbiased estimator 最优线性无偏估算子
best load 最大效率时的负荷
best management practice 最佳管理实践
best match file searching 查询最近似的文件
best match(ing) 最佳匹配;最佳配合
best obtainable price 最好可成交价格
best offer 最低报价
best operating point 最佳工作点
best order quantity 最合适的订购数
bestos wool 石棉绒
bestowal 赠品
bestow a prize 颁发奖金
best performance curve 最佳性能曲线
best plough steel wire 铅淬火高强度钢丝
best plough wire 高强钢丝
best pole of plane 平面的最佳极点
best possible merchandise 优良的商品
best possible value 最或然值
best power mixture 最大动力的混合气
best practible technology 最可行技术
best practicable control technology currently available 当前可获得的最佳实用控制工艺
best practicable means 最优实用手段;最佳实践方法
best practical technology 最佳实用技术
best practice 最优方法
best-practice labo(u)r productivity 先进企业的劳动生产率
best practices in human settlement improvement 改善人类居住环境最佳范例奖
best prevailing local conditions 当地最佳条件
best price 最优价格

best puddled bar 精制熟铁棒材
best quality 优质;优等品;最好的质量
best quality brick 优质砖
best reed 优质芦苇
best result 最好的效果
best route analysis 最佳线路分析
best season for planting trees 植树的最好季节
best selected copper 优质精铜
best seller 畅销品
best servo 最佳随动系统
best setting 最佳调节;最好调节
best solution 最佳解法
best speed 经济速率
best terms 最优惠条件
best unbiased test 最佳无偏检验
best use of economic resources 经济资源最优利用
best value 最佳值
best way 最佳方法
be subject to taxation 被抽税
be suitable for grow in tropical area 适用于热带地区种植
be suited to grow olive 适合种植油橄榄
be suited to local weather 适合当地的气温条件
be sure to 必定
bet 河漫滩;低泛滥平原
beta absorption ga(u)ge 按吸收原理工作的贝塔测量计
beta-alabandite 贝塔硫锰矿
beta-boracite 贝塔方硼石
beta chalcocite 辉铜矿
beta coefficient 贝塔系数
beta-cristobalite 贝塔方英石
beta decay 贝塔衰变
beta distribution 贝塔分布
beta fabric 贝塔织物
beta-fergusonite 褐铈铌矿;贝塔褐钇铌矿
betafite 贝塔石
beta function 贝塔函数;β函数
beta-gamma composite analysis 贝塔—伽马综合分析
beta-gamma documentary value 贝塔—伽马编录值
beta-gamma sampling 贝塔—伽马取样
beta-gamma survey under water 水下贝塔—伽马测量
beta interaction 弱相互作用
betalight 氚发光器件
beta-naumannite 贝塔硒银矿
beta particle 贝塔质子;乙种粒子;贝塔粒子
beta phase 贝塔相
beta plus gamma survey 贝塔—伽马测量
beta quartz 贝塔石英
beta radioactivity 贝塔放射性
beta radioactivity measuring 贝塔放射性测量
beta ray 乙种射线;贝塔射线
beta-ray spectrometry 贝塔射线谱法
beta site 临时场地
beta-split pin 开口销
beta-sulfur 贝塔自然硫
betatopic 差电子的;失电子
beta-tridymite 贝塔鳞石英
betatron 电子回旋加速器
betatron acceleration 电子回旋加速
betatron effect 电子回旋加速效应
betatron electromotive force image 电子感应电动势象
betatron(generator) 电子感应加速器
betatron mechanism 电子回旋加速机制
betatron process 电子回旋加速过程
beta-uranophane 贝塔硅钙铀矿
beta-ziesite 贝塔氧钒铜矿
bet effort 净作用力
betekhtinite 针硫铅铜矿
betel leaves oil 蒌叶油
betel nut 槟榔
be temporarily stiffened 临时加固
bethanise 钢丝电解镀锌法
bethanized steel fabric 电镀锌钢丝网
bethanized wire 电镀锌钢丝
bethanizing process 钢丝电解镀锌法
be the forfeit of 抵偿
Bethe-hole coupler 倍兹孔耦合器
bethel (非英国国教的)礼拜堂圣地
bethelizing 木材真空注油防腐;木材加压灌油防腐;木材注油
Bethel's method 贝塞尔法(木材防腐)
Bethel's process 贝塞尔防腐法;填满细胞法(木材防腐);木材加压灌油防腐法
Bethenod-Latour alternator 贝蒂诺德—拉图尔交流发电机
Bethe-Salpeter equation 贝蒂—萨尔彼尔彼方程
Bethe's salining method 贝蒂固定亚甲蓝染色法
Bethe-Weiz-sacker cycle 贝蒂—魏茨泽克循环
bet hit by a natural calamity 遭灾
bethlehem 圣餐房
Bethlehem beam 宽缘工字梁
Bethlehem section 宽缘工字截面
BET multilevel adsorption model BET 多层吸附模型
be to deal with on the spot 现场解决
beton 混凝土
betonac 混凝土表面硬化剂;金属混凝土(混凝土内含有细粒金属骨料)
beton arme 钢筋混凝土 T 形梁
beton brut 混凝土未加工表面;(拆模后的)混凝土粗表面
beton paint 混凝土用漆
beton project 喷射灌浆
beton translucide 玻璃集料混凝土;玻璃骨料混凝土
be too wet during the period of tillage 耕作季节过于潮湿
betpakdalite 砷钼铁钙矿
betrunked river 断尾河
betrunked stream 断尾河
betrunking of stream 河流干涸
Bettaskaf 钢脚手架
Bettendorff's method 白田道夫法
Bettendorff's reagent 白田道夫试剂
better assessment science in integrating point and non-point sources 综合点污染源和非点污染源最近评价技术
better bed fireclay 培特煤层耐火黏土
better business 业务改善
better distribution of investment 调整投资结构
better earth 良接地
better input 优质投入
betterment 扩建;修缮扩建;改善;改良投资;改建
betterment and extension 修缮和扩建
betterment and improvement of railway 铁路技术改造
betterment cost 修缮经费;固定资产改进费用;改善费
betterment expenses 改良费用
betterment investment 改善投资
betterment investment environment 改善投资环境
betterment levy 增值税;改善税
betterment of assets 资产改良
betterment of land 土地改良
betterment order 修缮通知单;添购自用资产通知单
betterments 修缮经费
betterment survey 改善测量
betterment works 扩建工程;修缮工程;改建工程;改善工程
better riding 良好行车
better-run enterprises 办得比较好的企业
better same order bias estimator 较优同阶有偏估计量
better than average 中上
better-than-average purity 高于一般纯度
better the price 让价
better the quality 较好的质量
Betterton-Kroll process 贝特顿—克洛耳法
better traffic facility 交通设施的改善
better water 良好的水
betting curve 博弈曲线
betting hall 赌场
Betti number 连通数
Betti's law 贝蒂定律
Betti's method 贝蒂法
betti's reaction 贝蒂反应
Betti's reciprocal theorem 贝蒂互易定律;贝蒂易定理
Betti's theorem 贝蒂定理
bettle 捣棒;木锤;木槌
Betts process 贝茨法
Betula alba 欧洲桦
betulinic acid 桦木酸
betulonic acid 桦木酮酸

between analyst bias 人员间偏倚
between batch bias 批次间偏倚
between centers 轴间距(离);轴间;中到中
between decks 夹层舱;中间甲板;中层甲板;主甲板下方的甲板
between-decks bulkhead 甲板间舱壁
between epoch difference 历元间求差【天】
between failures 二次故障之间
between frequency difference 频率间求差
between laboratory bias 实验室间偏倚
between-lens shutter 镜间快门
between method bias 方法间偏倚
between outside 外间距
between outside faces 在外表面之间
between perpendiculars 垂线距离;垂线间
between product 中间产品
between receiver difference 接收机间求差
between satellite difference 卫星间求差
between-season heating period 春秋季供暖时期
between-state variance 区间变异
between-the-lens shutter 中心快门
between-the-mean variation 平均值之间的变差
between tracks connection 轨道连接
between tree distance 株距
between two survey points 两测点间距
between wind and water 船体水线附近
betwixt land 中介地区;盾地
betwixt mountain 轴山;中界山脉;中间地块
Betz momentum theory 贝兹动量理论
beudantite 砷铅铁矾;砷菱铅矾
be under the hammer 被拍卖
be unequivocally established 明确证实
be up to sample 符合货物样品规格
be up to the neck in debt 负债累累
be used together 一起使用
beusite 磷铁锰矿
Beutlermethod 博伊特勒法
bevantolol 贝凡洛尔
bevatron 高功率质子回旋加速器;高能质子同步稳相加速器
bevel 开坡口;斜截;削面;制成斜面;割角;对切;割口;坡口;歪角曲尺
bevel angle 斜削角;斜面角;单边坡口角度;边缘斜角;板边角度;坡口面角度;(焊接的)坡口角(度)
bevel arm piece 斜三通管
bevel bar 斜型圆钢
bevel bearing plate 弧形支座
bevel board 斜角定线板;斜面定线板;斜角板;斜角木板;斜方板
bevel bonder 斜面砌墙石
bevel bondstone 斜面砌墙石
bevel brick 斜砖;斜角砖;斜边砖;拱脚砖
bevel chisel 斜眼凿(打刻榫用);斜边凿
bevel clutch 锥形离合器
bevel cocking 斜面槽形榫
bevel cogging 斜面槽形榫
bevel collar 铸铁承插管弯头
bevel cone friction clutch 锥形摩擦离合器
bevel corking 斜面槽形榫
bevel-corner end-grain wood block 斜角横切面铺地板木块
bevel(cut) 斜切;斜边切割
bevel cutter 坡口铣刀;伞齿铣刀
bevel-cutting 斜边切割
bevel design 斜面设计;斜角设计
bevel differential 锥形齿轮差动器
bevel door 斜边门
bevel drive 伞形齿轮传动;锥齿轮传动;伞齿轮传动
bevel drive pinion 主动小锥齿轮
bevel drive pinion ball bearing 主动小锥齿轮滚珠轴承
bevel edge 斜缘;斜削边;倒角边
bevel-edge mirror 磨边镜子
bevel end 坡口加工端;坡口端;坡管口
beveler 磨石板机
bevel etch 倾斜腐蚀
bevel face 斜角面
bevel frame 斜边肋骨
bevel friction gear 斜摩擦轮
bevel ga(u)ge 斜曲尺;斜角规;分度规;曲尺
bevel gear 盘形齿轮;斜面齿轮;锥形齿轮;锥齿轮;直齿伞齿轮;伞形齿轮;伞齿轮
bevel gear blank 锥齿轮坯料

bevel gear burnishing machine 伞齿轮研齿机
bevel gear cutter 锥齿轮铣刀
bevel gear cutting machine 锥齿轮切齿机床
bevel gear differential 锥齿轮差动装置;伞形齿轮差动器
bevel gear drive 斜面齿轮传动;伞形齿轮传动;斜齿轮传动;锥齿轮传动
bevel geared bulb turbine 斜齿轮传动灯泡式水轮机;锥齿轮传动灯泡式水轮机
bevel gear gate lifting device 斜齿轮闸门启闭机
bevel gear generator 伞齿轮刨齿机
bevel gear housing 锥齿轮传动箱
bevel gearing 斜面齿轮传动装置;锥齿轮装置
bevel gear jack 锥齿轮千斤顶
bevel gear main drive 锥齿轮主传动(机构);锥齿轮主传动机构;伞齿轮主传动(机构)
bevel gear mesh adjusting gasket 锥齿轮啮合调整衬垫
bevel gear operated valve 伞形齿轮操纵阀
bevel gear pair 锥齿轮副
bevel gear pinion 斜齿轮小齿轮
bevel gear ratio 伞齿轮速比
bevel gear reverse 锥齿轮换向法;伞形齿轮转换器
bevel gear system 锥齿轮系
bevel gear tester 锥齿轮检查仪
bevel gear thrust bronze plate 锥齿轮推力铜挡板
bevel gear wheel 圆锥齿轮
bevel glass 斜边玻璃
bevel grinding 斜磨;按角度研磨
bevel grinding attachment 斜度磨光装置
bevel halving 斜面半连接;斜面半搭接;斜面接头
bevel header 斜面露头石
bevel head rivet 锥头铆钉
bevel jack 斜锯架
bevel joint 斜削接合;斜面接头
bevel joint of sheet pile 板桩斜接头
bevel joint piston ring 斜开口活塞环
bevel lead 斜导程
bevel(l)ed 斜削的
bevel(l)ed-barrier bar trap 剖平—障壁砂坝圈闭
bevel(l)ed bead 斜角镶玻璃条
bevel(l)ed block 斜削木块
bevel(l)ed brick 斜面砖;斜面砖
bevel(l)ed brush 削角电刷
bevel(l)ed claw-cluth 斜齿离合器;锥形牙嵌离合器;锥形爪式离合器
bevel(l)ed closer 斜面封口砖;斜削砖;斜面层收屋砖;斜面顶砖;斩角砖;斜斩面封口砖;斜面砖层收尾砖
bevel(l)ed cogging 斜鸳鸯榫头
bevel(l)ed corner 斜角
bevel(l)ed door 斜边门
bevel(l)ed edge 斜缘;斜削边;斜条;坡口边;倒斜板边
bevel(l)ed-edge chisel 斜刃凿
bevel(l)ed gear 斜齿轮
bevel(l)ed gear jack 斜齿轮起重器
bevel(l)ed glass 斜边玻璃
bevel(led)glass door 斜边玻璃门
bevel(l)ed halving 斜削接;斜对接;斜接接头;斜面接头;斜面接合
bevel(l)ed hill top 平切丘顶
bevel(l)ed housing 斜槽插榫;斜槽眼
bevel(l)ed joint 斜削接合;斜面接头;斜削接头;斜削接缝
bevel(l)ed joist 斜角格栅;斜顶格栅
bevel(l)ed lip entrance 斜唇入口
bevel(l)ed lip of the pulley 滑轮的斜唇面
bevel(l)ed mirror glass 磨边镜玻璃
bevel(l)ed nailing strip 梯形截面受钉木条;梯形斜面受钉木条
bevel(l)ed pipe 斜头管
bevel(l)ed(plate)glass 带斜边的板玻璃;车边玻璃;带边的板玻璃
bevel(l)ed point of switch 尖轨斜削端
bevel(l)ed punch 斜刃凸模
bevel(l)ed-rabbeted window stool 斜嵌式窗台板
bevel(l)ed rectangular sleeper 斜面轨枕
bevel(l)ed rectangular tie 斜角枕木
bevel(l)ed shoulder 斜台阶;斜肩
bevel(l)ed sickle wheel 磨刀锥形砂轮
bevel(l)ed side 斜壁板
bevel(l)ed siding 斜面封檐板;斜面雨板板;斜面的外墙覆面木板;斜壁板

bevel(l)ed sleeper 楔形垫木
bevel(l)ed tie 不等厚枕木
bevel(l)ed washer 楔形垫圈;斜垫圈;楔形螺栓垫圈;斜面垫圈
bevel(l)ed wheel 斜棱滚轮
bevel(l)er 磨斜面机
bevel(l)ing 磨斜边;倒斜角;倒棱;磨斜棱;斜切;做成斜边;倒角;车边(将玻璃边缘磨出斜边)
bevel(l)ing cut of jack rafter 小椽的斜切面
bevel(l)ing machine 倒角机;倒斜边机;倒角机;刨边机
bevel(l)ing of the edge 坡口加工
bevel(l)ing plane 槽刨;榫槽刨
bevel(l)ing post 磨角架;磨角器
bevel(l)ing radius 弯曲半径
bevel material 斜切材料
bevelment 斜对切;斜切;削平
bevel mortise wheel 锥形嵌齿轮
bevel mo(u)lding 斜面线条;斜面线脚
bevel of compound bead 玻璃窗(镶板)滴水
beveloid 斜面轮
bevel on both surfaces 双面斜棱边
bevel pinion 斜面小齿轮;小锥齿轮;小伞齿轮;伞形齿轮
bevel pinion adjusting sleeve 锥齿轮校准套
bevel pinion differential gear 差动小锥齿轮;差动小伞齿轮
bevel pinion housing 斜面小齿轮罩
bevel pinion shaft 圆锥齿轮轴
bevel planetary gear drive 伞齿轮行星减速机
bevel planet gearing 行星锥齿装置
bevel plate heat exchanger 伞板式换热器
bevel point of sheet pile 板桩尖端;板桩斜角尖端
bevel polishing machine 斜边抛光机
bevel protractor 量角规;活动量角器;斜量角规;斜角规;万能角尺
bevel reaming shell 带内锥度扩孔器
bevel seat(ed)valve 斜面密封阀;斜形座阀;角阀;斜座阀
bevel shearing machine 斜剪机
bevel sheet 斜角板
bevel sheet pile 斜角板桩
bevel shell 锥套
bevel siding 互搭护墙板;互搭板壁;互搭板;披垒板壁;斜挡板;斜壁板;楔形挡板(板宽逐渐减少)
bevel spur gear drive 正伞齿轮减速机
bevel square 量角规;斜量尺;斜角规;分度规
bevel start hammer 斜角启动锤
bevel surface 斜面
bevel surface overflow 斜面式溢流堰
bevel tie 有斜面的系梁;有斜撑的系梁;梯形系杆
bevel tile 斜边砖
bevel tool 角切刀;锥形刀具
bevel trimmer 斜面切边机
bevel trimming cutter 斜面切边刀具;斜面修整机
bevel-type differential gear 锥式差动齿轮
bevel-type planetary gear 锥式行星齿轮
bevel-type planetary gearing 锥式行星传动装置
bevel wall 斜边墙
bevel(-wall)bit 带内锥度环形钻头
bevel wall core shell 带内锥度扩孔器
bevel washer 斜垫圈
bevelways 斜着
bevel weld 斜焊
bevel weld in a tee joint T形接头上斜面焊接
bevel weld in a T joint T形接头上斜面焊接
bevel welding 斜焊
bevel wheel 斜摩擦轮;锥齿轮;伞齿轮;伞形齿轮
bevel wheel change gear 锥齿轮变速装置;伞形齿轮变速装置
bevel wheel drill 锥形锪钻;角钻;角砧;斜齿轮
beverage 饮料
Beverage antenna 贝佛莱日天线
beverage bottle 饮料瓶
beverage dispenser 饮料配制器;自动饮料出售机
beverage industry wastewater 饮料工业废水
beverage room 饮料间
beverage store 饮料店
beveragfe distillery 酿酒厂
bevonium metilsulfate 贝弗宁甲硫酸盐
beware of falling sign 小心落物标
beware of fume 谨防漏气
beware of imitations 谨防假冒
beware of propellers 谨防车叶

bewel 挠曲;预留曲度
bexoid 尿素—甲醛塑料
beyerite 碳钙铋矿
Beyliss turbidimeter 贝利斯浊度计
beyond control 不能控制;无法控制的
beyond-critical Mach number 超临界马赫数
beyond local repair 无法就地修复
beyond national jurisdiction 国家管辖范围外
beyond number 无法计算的
beyond-Panamax 超巴拿马型
beyond repair 不能修理
beyond right 以远权
beyond-the-horizon communication 散射传播;超视距通信
beyond-the-horizon propagation 超越地平线传播;超视距传播
beyond-the-horizon transmission 超视距传输
beyond the mark 超出界限
beyond the means of ordinary consumer 超出一般消费者的经济能力
beyond tolerance 超限差
beyond unworked depth of ore body 矿体埋藏深
beyond visual range radar 超视距雷达
bezafibrate 苯扎贝特
Bezald Brucke phenomenon 贝措尔德-布鲁克现象
bezant 圆币饰;圆盘形装饰;银币饰
bezantee 金币花饰;圆盘花饰;列圆饰
bezel 聚光圈;小指示灯;镶框;座圈;遮光板;带槽框;玻璃框(仪器的);刃角;嵌玻璃沟缘;企口
bezoar 毛粪石
B-Ga-silicon dioxide forming film component 掺硼镓乳胶
B grade 乙级(指货品)
Bhaskara satellite 巴斯卡拉卫星
Bhattacharyya distance 巴特â里亚距离
Bhavani stilling basin 巴瓦尼消力池(印度巴瓦尼坝)
bhilawan nut shell oil 肉托果壳油
bhilawanol 肉托果壳酚
B-horizon 淀积层
bhur 小山
bhurland 小山
biabsorption 双吸收
bialamicol 卡马风
bialkali photocathode 双碱光电阴极
Bial's test 比亚耳试验
bialternant 双交替式
biamperometry 双安培滴定法
Bianchi cosmology 比安奇宇宙学
Bianchi identity 伯恒希等式;比安奇恒等式
bianchite 锌铁矾
biancola 人造大理石
biandan constant feeder 扁担式定量喂料机
biannual 一年两次的;半年一次的
bianode 双阳极的
bianry image 二值图像
biaryl synthesis 联芳基合成
bias 经济预测中的倾向性;斜痕;倾向;偏置;偏倚;偏压;偏性;偏见;偏航;偏度;偏磁;偏差
biasar 侧南蛇丘;冰砾堤
bias battery 偏压电池组
bias bell 偏动电铃
bias buff 斜裁布抛光轮
bias check 边缘校验;边缘检验
bias coefficient 偏度系数
bias control 偏置控制
bias correction 偏移修正值
bias crosstalk 偏磁串扰
bias current 偏流
bias cutter 斜切机;斜裁机
bias data 偏置数据;偏离数据
bias detection 偏流检波
bias detector 偏压检波器
bias distortion 偏置畸变;偏移失真;偏移畸变;偏离失真;偏畸变
bias drift 偏压漂移
biased 有偏的
biased blocking oscillator 负偏压间歇振荡器
biased critical region 偏倚临界域
biased data 分布不匀数据;分布不匀的数据
biased detector 偏置探测器
biased differential protective system 极化差动保护装置;比率差动保护装置;偏置差动保护装置
biased downward 向下偏倚

biased electrode 加偏压的电极
biased error 偏置误差；偏离误差
biased estimate 偏向估计
biased estimation 有偏估计
biased estimation of demand elasticity 需求弹性的倾向性估计
biased estimator 有偏估值；有偏估计量；偏置估计
biased exponent 偏置指数
biased forecast 有偏预测
biased low-strength field 偏低场
biased memory 双轴磁芯存储器
biased multivibrator 闭锁多谐振荡器
biased relay 带制动的继电器；百分率差动继电器
biased result 有偏结果
biased statistics 有偏估计量；有偏见统计
biased-strength field 偏高场
biased technological progress 不全面的技术进展
bias error 系统误差；固有误差；有偏误差；偏倚误差；偏移误差
bias factor 偏因
bias field 偏磁场
bias filling 纬疵
bias gear 偏动装置
biasing 加偏压；加偏磁
biasing characteristic 偏置特性
biasing circuit 偏置电路
biasing logic 偏置逻辑
biasing spring 偏动弹簧
biasing technique 偏置技术；偏磁技术
biasing voltage 偏压【电】
bias light 衬托光；本底光；背景光
bias light compensation 背景光补偿
bias lighting 跑光（照相）；基本照明；本地照明；背景光；衬底照明；背景照明；偏置光
bias meter 信号电压计；偏流计；偏流表；偏畸计
bias mute 偏置抑制杂波
bias off 加偏压截止；偏置截止
bias of spring 弹簧偏压
bias oscillator 偏置振荡器；偏磁震荡
bias-ply tire 斜交帘布轮胎
bias proportion 偏倚比例
bias pulse 偏置脉冲；偏压脉冲
bias rail box girder 偏轨箱形梁
bias rectifier 偏压整流器
bias resistor 偏压电阻
bias(s)ed estimator 偏估计量
bias(s)ed tolerance 偏斜容限（试验行车偏离指定横向地位的容许限度）
bias sputtering 偏置溅射；偏压溅射
bias test 拉偏试验
bias torque 偏转力矩
bias uncertainty 偏离不定度
bias winding 辅助磁化线圈；偏压线圈
biax element 双轴元件
biaxial 二轴的；双轴的
biaxial anisotropy 双轴各向异性
biaxial bending 双轴挠曲；双向弯曲
biaxial bending stress 双轴弯曲应力
biaxial compression 双轴抗压；双向压缩
biaxial compression strength 双轴抗压强度
biaxial compression type of strain ellipse 双轴缩短型应变椭圆
biaxial coordinates 双轴坐标
biaxial crystal 二轴晶；双轴晶体
biaxial deformation 双轴向变形
biaxial effect (of Poisson's ratio)（泊松比的）双轴向影响
biaxial extension type of strain ellipse 双轴伸长型应变椭圆
biaxial gyrostabilized platform 双轴陀螺稳定平台
biaxial indicatrix 双轴晶光率体
biaxial interference figure 二轴晶干涉图；双轴干涉图
biaxiality 双轴性
biaxiality factor 双向因子
biaxial lateral load 双轴向侧力
biaxial load 双轴荷载；双向荷载
biaxial loading 双轴向荷载；双轴向负荷
biaxial longitudinal strain 双轴伸缩应变
biaxiall-shaft mixer 双轴搅拌机
biaxial negative character 二轴负光性
biaxial optical symmetry 双轴光学对称
biaxial orientation 双轴取向；双轴定向
biaxial photoelastic ga(u)ge 双轴光弹仪；双轴光弹计；双向光弹应变计
biaxial positive character 二轴正光性
biaxial probe 双轴探头
biaxial shaking table 双轴振动台
biaxial state of stress 双轴应力状态；双向应力状态
biaxial strain 二轴应变
biaxial strength 双轴强度
biaxial stress 二轴应力；二向应力；平面应力；双轴(向)应力；双向应力
biaxial stress state 双轴应力状态
biaxial stress-strain relationship 双轴应力应变关系
biaxial stress system 二向应力系统；双轴向应力系统
biaxial stretch 双轴向拉伸
biaxial stretch-forming machine 双轴向拉伸成形机
biaxial system 双轴系统
biaxial tensile strain 双向拉伸应变
biaxial tensile test 双轴抗拉试验
biaxial tension 双轴向张力；双向张力；双向拉伸
biaxial test 双向试验
biaxial tiltmeter 双轴倾斜仪
biaxial viscosity 双轴向黏性
biaxial winding 双轴缠绕法
biax magnetic element 双轴磁芯元件
Biazzi process 毕阿兹法
bib 弯嘴旋塞
bibacke line 双贝克线
bib and spanner 用活动扳手操作的水龙头
bibasic 二代
bibasic amino acid 二氨基酸
bibbles 软含水层
bibbley rock 砾石岩；砾岩
bibbs 桅门加强板
Bibby coupling 毕培式联轴器；曲簧联轴器；蛇形弹簧联轴器
bib-cock 弯嘴旋塞；弯嘴龙头；弯管旋塞；水龙头；小水龙头；活塞；活门；旋塞
bibelot 框架上陈列的小珍品；小件古玩；装饰品
bibenzonium bromide 比苯溴铵
Biber-Donau interglacial stage 拜伯—多瑙间冰期
Biber glaciation 拜佰冰期
biberine 贝比碱
bibevel(l)ed 两边斜面的；双斜面的
bibiofilm 图书显微软片
biblio 书志目录
bibliographer 目录学家；书目编纂者
bibliographic(al) center 文献中心
bibliographic(al) database 文献数据库；文献目录数据库
bibliographic(al) information 文献信息
bibliographic structure 目录结构
bibliographic(al) term 文献目录
bibliographic(al) tool 目录工具书
bibliography 目录学；参考资料；参考书目（录）；文献学；文献目录；书目
bibliolite 纸状页岩；层片岩
bibliometer 吸水性能测定仪
bibliotheca 图书馆藏书；书库
bib nozzle 弯嘴水龙头；龙头嘴；旋塞嘴；小水龙头弯嘴；厨房洗涤盆喷头
Bibra alloy 比布拉铋锡铅合金
bib tap 弯嘴水龙头；弯嘴旋塞；弯嘴龙头；小水龙头
bibulous paper 滤纸；吸水纸
bi-cable aerial 双线架空索道
bi-cable aerial tramway 双线架空吊车索道
bi-cable ropeway 双缆索道；双索索道；双索架空索道
bi-cable system 双线式缆道
bi-cable tramway 双线索道
bicamera 双镜头照相机；双镜头摄影机
bicameral 二室的
bicapitate 二头的
bicarb 碳酸氢钠
bicarbonate 重碳酸盐；碳酸氢盐；酸性碳酸岩
bicarbonate alkalinity 重碳酸盐碱度；碳酸氢盐碱度
bicarbonate anion exchange column 重碳酸盐型离子交换柱
bicarbonate buffer 碳酸氢盐缓冲液
bicarbonate buffer system 重碳酸盐缓冲系
bicarbonate-carbonate system 碳酸氢盐-碳酸盐系统
bicarbonate concentration 重碳酸盐浓度
bicarbonate groundwater 重碳酸盐类水
bicarbonate hardness 酸式碳酸盐硬度；重碳酸盐硬度
bicarbonate ion 碳酸氢盐离子；重碳酸盐离子
bicarbonate ion concentration 重碳酸钙盐离子
bicarbonate mineral spring water 重碳酸盐矿泉水
bicarbonate radical 重碳酸根；碳酸氢根
bicarbonate resin 重碳酸盐树脂
bicarbonate-sodium type 重碳酸钠型
bicarbonate-sulphate system 重碳酸盐-硫酸盐体系
bicavitary 两腔的
bicchulite 羟铝黄长石
bice 灰蓝色（颜料）
bicellular sealant backing 双多孔密封膏背衬材料
bicentenary 二百年的
biceps 二头肌
bicharacteristic ray 双特征射线
bicharacteristics 双特征性
biche 打捞母锥
Bicheroux process 间歇压延法（平板玻璃）；比切鲁克斯法
bichloride of tin 二氯化锡
bichromate 重铬酸盐
bichromate cell 重铬酸盐电池；重铬酸电池
bichromated 重铬酸盐化
bichromated coating 重铬酸盐处理层
bichromate dipped finish 重铬酸盐浸渍处理
bichromate ion 重铬酸根离子
bichromate method 重铬酸盐法
bichromate process 重铬酸盐处理法
bichromate reduction 重铬酸盐还原
bichromate species 重铬酸盐物种
bichromate value 重铬酸盐化值
bichromatic analyzer 双色分析仪
bichromatic plate 双色版
bichrome 两色的
bichronmatic analyser 双色分析仪
bichu fiber 大荨麻纤维
bicipital rib 二头肋
bicircular 重圆
bicircular surface 四次圆纹曲面
bicirculating 双循环
bicirculation 双环流
bickern 砧角；丁字砧；双嘴砧
Bickford fuse 俾氏导火线；俾氏导火索
Bickford safety fuse 俾氏安全导火线；俾氏安全导火索
bick-iron 砧角；铁砧；砧
Bicknell sandstone 比克内尔砂岩
biclavate 两端成棒状的
biclothoid 双回旋曲线
bicoca 塔楼；看守塔
bicolo(u)rable graph 双色图
bicolo(u)r aerial film 双色航摄胶片
bicolo(u)red light 两色灯
bicolo(u)rimeter 双筒比色计；双色比色计；双层比色计
bicolo(u)rimetric 双色比色的
bicolo(u)rimetric method 双筒比色法
bicolo(u)rimetry 双筒比色法
bicolo(u)r system 双色系统
bicomponent composite fiber 双组分复合纤维
bicomponent extruded spinning head 双组分挤压纺丝头
bicomponent filament yarn 复合长丝
bicomponent glass fiber 双组分玻璃纤维
bicomponent nozzle 组合管口
bicomponent polyurethane coating 双组分聚氨酯涂料
bicomponent spinning device 双组份纤维纺丝装置
bi-compound pump 复合泵
biconcave 双凹面的；两面凹的；双凹形；双凹(的)；双面凹的
biconcave lens 双凹面透镜；双凹镜
biconditional gate 双条件门
biconditional proposition 等值命题
bicone lens 双凹透镜
biconic(al) 双圆锥的
biconical antenna 双锥形天线
biconic(al) connector 双圆锥活动连接器
biconic(al) gravity concrete mixer 双锥形自落式混凝土搅拌机
biconical reflector 双锥形反射镜
biconjugate 双共轭；双对
biconjugato-pinnate 重双羽状的

biconnected graph 双连通图
biconstituent fiber 双成分纤维
bicontinuous 双连续
bicontinuous mapping 双连续映像
biconvex 两面凸(的);双凸面(的);双凸(的);双面凸(的)
biconvex lens 双凸透镜;双凸面透镜
bicrofarad 毫微法拉
bicron 纳米;毫微米
bicrystal 双联晶;双晶体;双晶
bicubic 双三次
bicuiba 红棕色硬木(产于中美洲)
bi-current traction unit 交直流牵引装置
bicurve polarization figure 双曲线偏光图
bicurve section 两曲翼剖面
bicuspid 二尖的
bicuspidal 二尖的
bicuspidate 二尖的
bicuspid teeth 槽牙
bicycle 脚踏车
bicycle and pedestrian path 自行车人行道
bicycle dynamo 自行车发电机
bicycle ergometer 自行车功量计;自行车测力计;踏车功能试验器
bicycle gear 自行车式起落架
bicycle lane 自行车车道
bicycle park 自行车停放处
bicycle parking 自行车场
bicycle parking facility 自行车存放设施
bicycle path 自行车道
bicycle pedestrian path 自行车与人行道合用
bicycle plant 自行车厂
bicycle-racing arena 自行车比赛场
bicycle-racing track 自行车赛道;自行车比赛跑道
bicycle rack 自行车架;自行车存车架
bicycle room 自行车房
bicycle shed 自行车棚
bicycle shelter 自行车棚
bicycle stand 自行车停放架;自行车架
bicycle test 脚踏车试验
bicycle track 自行车专用道;自行车小路;自行车跑道;自行车路(径)
bicycle traffic 自行车交通
bicycle tube 自行车内胎
bicycle type bearing 自行车型轴承
bicycle undercarriage 双轮底盘
bicycle way 自行车道
bicycle wheel roof 车轮式屋盖;车轮式房屋;自行车轮式屋顶
bicyclic 二环的
bicyclic(al) compound 二环化合物
bicyclic(al) epoxide 二环氧化物
bicyclic(al) hydrocarbon 二环烃
bicyclic(al) sulfide 二环硫化物
bicycloalkane 双环烷烃
bicyclohexyll-ethane 双环己基乙烷
bicyclohexyll-methane 双环己基甲烷
bicyclononane 双环壬烷
bicyclooctane 二环辛烷;双环辛烷
bicylcle activities sphere 自行车活动范围
bicylinder 双柱面透镜
bicylinder lens 双柱面透镜
bicylindrical 双柱面;双圆柱
bicylindrical resonator 双圆柱共振器
bi-cylindro-conic(al) drum 双柱双锥式卷筒
bid 叫价;递盘;捕获信道;报价;投标书
bid abstract 分项标价综合单;标价总表;标价汇总表;标单提要综合单(投标商名称)
bid against a person 出价比别人高;与人竞标
bid against each other 竞相投标;投标相互竞争
bid a high price 出高价
bid a high price for 出高价买
bidalotite 直闪石
bid and asked 出价和讨价
bid and offered rates 出价和报价利率
bid and quotation 投标和估价;投标和报价
bid awarding 决标;合同授予
bid bond 押标金;押标保证;押标保证金;承包;保证金;投标押金;投标保证金;投标保证金;投标保函;投标保单;投标
bid bond and performance bond letter of credit 投标履约信用证
bid business 投标企业
bid call 招标

bid cock 弯嘴龙头
bid cost 投标费用
bid currencies 投标货币
bid date 招标期限;投标日期
bid deadline 投标(或承包)截止日期
bid deposit 投标保证金
bidder 竞买人;出价者;出价人;报价人;投标者;投标商;投标人
bidder's home country 投标者国籍;投标商国籍;投标人国籍;投标厂商国籍
bidders ring 围标
bidders sheet 投标商名单;投标人名单
bid design 发包设计
biddiblack 比地黑
bid(ding) 投标;出价;求职申请
bidding advertisement 招标广告
bidding advice 招标通知
bidding and inviting bids of projects 工程项目招标投标
bidding announcement 招标公告
bidding block 拍卖场
bidding book 投标书
bidding combination 投标联合体;承包联合
bidding condition 投标要求;投标条件
bidding contract 投标承包;投标契约;投标合同
bidding deadline 投标截止日期
bidding document 招标文件;(招标的)标书;投标文件说明书文件
bidding for project 工程投标
bidding group 投标小组;投标集团
bidding notification 招标通知;招标通告
bidding or negotiation phase 投标或议标阶段
bidding period 招标期;投标限期;投标期限(美国);投标阶段
bidding phase 投标阶段
bidding procedure 投标诉讼;投标手续;投标程序;投标裁决
bidding process 投标步骤
bidding quotation 标盘
bidding requirements 投标要求;投标须知
bidding results 投标结果
bidding rules 投标规则
bidding sheet 标单;标价单;投标单
bidding stage 投标阶段
bidding system 投标制
bidding volume 包工量
bid document 标书;投标文件
bideauxite 氯银铅矿
bidematron 毕代玛管
bident 两尖器;两叉矛
bidentate adsorbate 双配位吸附质
bidentate ligand 二合配位体
bidery metal 白合金
bid estimate 投标估算
bidet 净身室;净身盆;坐浴(浴)盆;妇女卫生盆;下身盆
bidet closet 净身盆间
bid evaluation 投标估价;评标;投标估计
bid examination 投标核查
bid form 标书;标书格式;投标格式;投标表格;投标表格格式(按表格填写)
bid format 投标形式
bid guarantee 承包担保;投标担保;投标保证书;投标保函
bidiagonal matrix 两对角线矩阵
bidimensional 两维
bi-dimensional coefficient 双向系数
bi-dimensional frame 二向框架
bi-dimensional ground motion 双向地面运动
bi-dimensional shaking motion 双向振动
bid indicator 捕获信道指示器
bid inquiry 询标
bid invitation 邀请投标(书);招标通知;招标
bid invitation for project 工程招标
bidirection 双向
bidirectional 双向的;双向作用
bidirectional bus 双向总线
bidirectional bus driver 双向总线驱动器
bidirectional cable 双向光缆
bidirectional chained list 双向链接列表
bidirectional clamping circuit 双向箝位电路
bidirectional clipping circuit 双向削波电路
bidirectional clutch 双向离合器
bidirectional combined type marshalling station 双向混合式编组站
bidirectional counter 双向计数器;双向耦合器
bidirectional filter 双向滤池
bidirectional flow 双向流程线;双向流
bidirectional heuristic 双向试探
bidirectional hydrophone 双向水听器
bidirectional laminate 双向增强夹层板
bidirectional lines 双向线路;双向线
bidirectional list 双向列表
bidirectional longitudinal type marshalling station 双向纵列编组站【铁】
bidirectional microphone 双向传声器
bidirectional modulator 双向调制器
bidirection(al) operation 双向操作
bidirectional pulse train 双向脉冲序列
bidirectional reflectance distribution function 双向反射分布函数;双反射分布函数
bidirectional relay 双向继电器
bidirectional replication 双向复制
bidirectional switch 双向开关
bidirectional test port 双向测试口
bidirectional traffic 双向运行;双向交通(量)
bidirectional transducer 双向换能器;双向传感器;双向变换器;双向转换器
bidirectional transistor 双向晶体管
bidirectional transversal type marshalling station 双向横列式编组站
bidirectional waveform 双向波形
bidirectional wire mesh 双向钢筋网
bid item 投标条款;投标项目
bid language 投标语言
bid letting 开标
bid market 出价市场
bid negotiation 议标
bid on 投标;出价;投标承包
bidonville 市郊贫民窟(北非)
bid open 开标
bid opening 开标
bid opening minutes 投标开标纪要
bid opening procedure 开标程序
bid or quotation 报价或估价
bid package 一揽子投标
bid preparation 投标准备;投标编写
bid price 出价;投标单;(投标的)标价;投标价格;投标价;投标报价
bid price quotation 投标报价单
bid procedure 招投标程序
bid proposal 承包建议书;投标建议(书)
bid quotation 报价估价
bid rate 买价;投标价;出价
bid-rent function 出价租金函数
bid rigging 操纵投标
bi-drum boiler 双汽包锅炉
bids and offers 买价与卖价
bid schedule of price 投标价格表
bid security 投标商合约保证金;投标担保;投标保证金
bid sheet 投标人名单;投标单
bid shopper 投标业务;投标商业务收购人;承揽商
bid's substantial responsiveness 投标的根本性反应
bid substance 投标实质内容
bid summary 分项标价综合单;投标商分项标价综合单;投标人分项标价综合单
bid tabulation 标价汇总表;投标商各项标价一览表;投标人各项标价一览表
bid time 招标期限;收标期限
bid up 竞出高价;哄抬(价钱)
biduplexed system 两双工系统
bid upon 投标;出价
bid validity 投标书有效期
bid wanted 征求报标;招标
bieberite 钴矾;赤矾
Biedermeierstil 毕德迈尔式(19世纪前半期德国、奥地利流行的家具风格)
bieletrolysis 双极电解
bielliptic(al) 双椭圆
biellipticity 双椭圆率
bielzite 脆块沥青
Bienayme-Chebycheff inequality 比安内梅—切比雪夫不等式
biennial 两年一次的;两年生的
biennial budget cycle 两年预算周期
biennial oscillation 两年振动

biennial plant 两年生植物
biennial programming 制定两年期计划
bier 棺材(架)
Bierbaum microcharacter 比尔鲍姆微压痕硬度计
Bierbaum scratch hardness 比尔鲍姆划痕硬度
Bierbaum scratch hardness equipment 比尔鲍姆刮刻硬度仪
Bierbaum scratch hardness test 比尔鲍姆划痕硬度试验
biethiol 比廷
biface 双界面
biface tool 双面工具
bifacial 双面的
bifacial tension 双界张力
bi-fibre 棉毛混纺地毯纱
bifid 叉形的
bifid rib 分叉肋
bifid tongue 分支舌
bifilar 双线；双股
bifilar choke 双绕扼流圈
bifilar current indicator 双线流向指示器
bifilar electromagnetic oscillograph 双线电磁示波器
bifilar electrometer 双线静电计
bifilar galvanometer 双线检流计
bifilar gravimeter 双丝重力仪
bifilar helix 双线螺旋；双螺线；双螺线；双股螺旋
bifilar helix slow-wave circuit 双螺线慢波电路
bifilar inductor 双线扼流圈
bifilar micrometer 复丝千分尺；双丝测微计；双螺旋线千分尺
bifilar oscillogram 回线示波器
bifilar oscilloscope 双线示波器
bifilar pendulum 双线摆
bifilar potentiometer 双线电位计
bifilar suspension 双张丝悬吊；双线悬置；双线悬挂
bifilar transformer 双线变压器；双线变压器
bifilar winding 双线绕组；双线绕法
biflaker 高速开卷机
biflow filter 双向滤池；双向流滤池
biflow regeneration 分流再生
bifluoride 氟氢化物
bifocal 两焦点；双焦点的
bifocal chord of a quadric 二次曲面的两焦点弦
bifocal eye glass 双焦眼镜；双光眼镜
bifocal lens 双焦透镜；双焦镜；双焦点透镜
bifocals 双焦点透镜；双筒望远镜玻璃；双焦镜片；双光眼镜
bifocal segment 双焦面；双焦点片
bifocal spectacle 双焦眼镜；双光眼镜
bifocal surface 双焦面
bifocal tube 双焦点管
bifocus 双焦点
bifolding door 双摺门；双折门
biform imaging system 双型成像系统
biforous spiracle 双室气门
bifrequency 双频率
bifrequency system 双频制
bifuel system 双燃料系统
bifumarate 富马酸氢盐
bifunctional 双官能团的
bifuracation angle 分叉角
bifurcate 两股；叉状的；分叉
bifurcated 分为二支
bifurcated channel 分汊水道；分汊航道
bifurcated chute 分叉斜槽；双股溜槽
bifurcated contact 分支接触体；分支触点；双叉接插件；双叉触点簧片
bifurcated discharger 叉形卸料机
bifurcated E layer 分叉E层
bifurcated gate 分水闸(门)
bifurcate differentiation 分叉分化
bifurcated launcher 分叉流槽
bifurcated line 分叉线路
bifurcated optic(al) fiber 分支光纤
bifurcated pipe 岔管；分叉管；三通管
bifurcated rivet 开口铆钉；分叉铆钉
bifurcated-rivet wire 开口铆钉用钢丝
bifurcated rudder 可张翼舵
bifurcated stair(case) 双分楼梯
bifurcated tube 分叉管
bifurcated type stair(case) 双分楼梯
bifurcate image 双像
bifurcate ligament 分歧韧带

bifurcate ripple 分枝波痕
bifurcating box 分路电缆套；双芯线终端套管；双叉分接盒
bifurcating feeder 叉式给料机
bifurcation 两歧状态【计】；河道分叉口；岔口；汊口；分枝；分支；分叉叉；分岔；分叉；二异状态；水道分汊(处)；双态；双叉口
bifurcational valve 分歧值
bifurcation buoy 航路分叉浮标；航道分叉浮标；左右通航浮标；洲头分汊浮标；中沙外端浮标
bifurcation connector 分支连接器；分叉连接器
bifurcation gate 分水闸(门)；分水门闸；沟渠或管道分流结构
bifurcation headgates of laterals 支渠分水闸；双支渠分水闸
bifurcation hypothesis 分歧假设
bifurcation index 分支指数；分叉指数
bifurcation launcher 分叉流槽
bifurcation mark 左右通航标；汉港标志
bifurcation of river 河道分叉点
bifurcation point 分歧点；河道分流口；中沙外端分流口；分汊点；歧点；水道分汊点(对进口船而言)；分叉点
bifurcation ratio 分支比；分流比；分汊比；分叉比
bifurcation signal 分汊航标
bifurcation structure 分水结构；分叉结构
bifurcation theory 分支理论；分歧理论
biga 两马两轮战车；由二马牵引的古代双轮战车(常用于雕塑)
bigalopolis 大都市(美国)
Big Apple 大苹果(美国纽约市的俚称)
big bad 不景气
big bang 巨大响声
big-bang cosmology 大爆炸宇宙论
big bang hypothesis 大爆炸假说
big-bang model 大爆炸模型
big-bang testing 大范围测试
big base plow 宽幅犁
big bath 巨额冲销
Big Bertha 大型客机；短柄大链钳
big blast 大爆炸；大爆破
big blaster 空气炮
Big Board 纽约证券交易所
Big Boffe cotton 大波菲棉(巴西)
big bounce 大反冲
big bowl 大碗
big bracket bush 大托架轴瓦衬
big brick chimney 扇形砖砌烟囱
big bridge 大桥
big bull 领工员；工长
big-bus 大客车
big business 大企业
big businessman 大企业家
big capacity of hump 大能力驼峰
bigcatkin willow 水杨
big cave 半洞
big chimney 大烟囱
big clear 冰湖的
big coal D 煤矿用D炸药
big concrete site 大型混凝土建筑现场；大型混凝土施工现场
big construction items 大型建设项目
big diameter borehole 大口径钻孔
big diameter drilling 大孔径钻进
big diameter welded tube 大口径焊缝管
Big Dipper 北斗七星；二苯胺杀虫剂
big dough 巨款
big earth current system 大接地电流系数
Big Eight 美国八大会计师事务所
bigeminal 二联的
big-end 大端；连杆曲拐头
big end bearing 连接杆支承；大端轴承
big-end bolts 连杆螺栓
big-end down 上小下大的
big end of connecting rod 连杆大头
big-end up 上大下小的
big-end-up mould 上大下小钢锭模
big floe 大浮冰块(宽500~2000米)；大冰盘
big-font files 大字体文件
big gap between income 分配悬殊
big gear wheel 大齿轮
Bigger Fool Theory 白痴理论(房地产用语)
bigger investments 投资高指标
bigger on the hoof 长的比较高大

biggin(g) 任何一种建筑物(英国)
bighead carp 花鲢；大头鱼
big hole 大眼井(直径10英寸以上套管的)(1英寸=0.0254米)
big hole drilling 大直径钻进
big hole perforator 大孔眼穿孔器
big hole rock drill 大孔岩石钻机
big hook 救援起重机
big house 官邸；住宅中生活社交活动区域；宅第
bight 开展海湾；开阔海湾；海湾弯曲部分；新月湾；小湾；大湾；冰绕湾；盘索；弯曲；索具套环；绳环
bight line 拉索
big inch 大直径管；大孔径的
big inch line 大直径管线
big inch pipe 大口径输油管
big internal gear 大内齿轮
bigleaf mahogany 大叶桃花心木
biglet 索环套头销
big log 大原木
big mill 开坯轧机；粗轧机
big minibus 大面包车
big M method 大M方法
bigness 大生意经
bigness scale 粗测
big one 千元美钞
big opening easy open lid production line 大开口易拉盖生产线
big open void 大开型孔隙
big open void ratio 大开空隙率
big panel form construction 大块模板施工法
big plaster board 大的熟石膏板；大灰胶纸柏板
big-pore adsorption resin 大孔吸附树脂
big power station 大型发电厂
big-power tractor 大功率拖拉机
big-production 大生产
bigraded group 双重分次群
bigraph 二部图(形)
big reed screen 粗苇帘
big repair 大检修
big restaurant 饭庄
big retailer 大规模的零售商
bigrid 双栅极
big rubbish 大型垃圾
big run sprinkler 大型喷洒器
big sand 大颗粒砂
big savage 领工员；工长
big-scale 大型的；大规模的
big-scale harbo(u)r 大型港口
big-scale port 大型港口
big-scale project 大型项目；大型工程项目
big-scale work 大型工程项目
big shareholder 大股东
big slip 插条
big spalled-off rock mass 大块剥落岩层
big-span floor 防火地板(其中之一)
big square 大方木(18英寸以上见方，长20英尺以上的方材)(1英寸=0.0254米,1英尺=0.3048米)
big steel 大垄断钢铁企业
big-stone bit 大粒金刚石钻头；粗粒金刚石钻头(粒度为1克拉8颗以上)
big store 大量销售店
Big Three 美国三大汽车公司
big-ticket 高价的
big ticket items 高价商品
big ticket leasing 巨额租赁
big tool box 大工具箱
big tractor 大型拖拉机
big trader 大贸易商
big tree 巨杉
big tree transplanting 大树移植
biguanide 缩二胍；双胍
big-volume resin 工业生产树脂
big ware kiln 大器窑
Big Windy 大风(美国芝加哥的俚称)
bigwoodite 钠长微斜正长石
big yield 大水量；大出水量
big zone of oil and gas accumulation 油气聚集大带
biharite 黄叶蜡石
biharmonic 双谐波；双调和的；双谐的
biharmonic equation 双调和方程
biharmonic function 双调和函数
biharmonic operator 重调和算子；双调和算子

Bi hollow cathode lamp 铋空心阴极灯
bi-ion active agent 两性离子表面活性剂
bijection 双向单射;双单射
bijective mapping 双向单射
bijou 宝石;珠宝
bijou house 很小的房屋
bijouterie 宝石类
Bijur type booster 毕朱尔式升压机
bijvoetite 羟碳钇铀石
bikarbit 弱性炸药(由消焰剂和少量敏化剂合成的)
bike and ride 自行车停车换乘
bike lane 自行车专用道;自行车车道
bike-linking 自行车道
bike way 自行车专用道
Bikini City 比基尼城(美国迈阿密的俚称)
bikitaite 硅锂铝石
bilaminar 二层的
biland 半岛
bilateral 两面的;两侧的;双向作用;双向(性);双侧;双边的
bilateral agreement 双边协议;双边协定
bilateral agreement on tax credit 双边税收抵免协定
bilateral aid 双边援助
bilateral amplifier 双向放大器;收发两用放大器
bilateral and multilateral economic cooperation 双边和多边经济合作
bilateral antenna 双向天线
bilateral arbitrage 双边套利
bilateral-area 双边面积
bilateral assistance 双边援助
bilateral bar 双向钢筋
bilateral centrifugal compressor 双面离心式压缩机
bilateral circuit 可逆电路;双向电路
bilateral clearing 双边清算
bilateral commitments of capital 双边性投资
bilateral contract 双边契约;双边合同
bilateral control 双向控制
bilateral convention 双边公约;双边协定
bilateral deflection 双向偏转
bilateral development loan 双边发展贷款
bilateral diffusion 双向扩散
bilateral diode 双向二极管
bilateral discharge 两侧卸料
bilateral duopoly 相互两家垄断
bilateral economic aid 双边经济援助
bilateral element 双向作用元件;双通(电路)元件
bilateral flow 双边对流
bilateral gear 双向行车装置
bilateral geosyncline 双面地槽
bilateral hinge 双向铰链
bilateral impedance 双向阻抗
bilateral import quotas 双边进口配额
bilateralism 两则对称
bilateral Laplace transform 双侧拉普拉斯变换
bilateral Laplace transformation 双边拉普拉斯变换
bilaterall-area track 双边调制声道
bilateral legal relations 双边法律关系
bilateral level(l)ing 对向水准测定;双转点水准测量;双向水准测定
bilateral lighting 两侧采光
bilateral loans 双边贷款
bilaterally adjustable 双向可调的
bilaterally harmonized standard 双边协调标准
bilaterally matched attenuator 双向匹配衰减器
bilaterally variable slit 双刀口狭缝
bilateral matching 双向匹配
bilateral midterms loan 双边中期贷款
bilateral monopoly 相互独家垄断;双边垄断
bilateral negotiations agreements 双边谈判和协定
bilateral network 双向网络
bilateral observation 双边观测
bilateral parking 两边停车;道路两侧停车;道路两侧存车
bilateral partners 双边合伙
bilateral payments agreement 双边支付协议;双边支付协定
bilateral portfolio investment 双边有价证券投资
bilateral quota 双边协定配额;协定配额;双边限额;双边配额
bilateral redeployment institution and fund 重新部署的双边机构和基金
bilateral rupture 双向破裂;双侧向破裂
bilateral scan(ning) 双向扫描

bilateral servo-mechanism 双向伺服机构
bilateral slit 双刀口狭缝
bilateral splitting 双向分岔
bilateral spotting 交会观测法
bilateral structure 双面结构
bilateral swap agreement 双边信贷互惠协定
bilateral switching 双向转换;双通开关
bilateral symmetry 两侧对称;左右对称
bilateral system 可逆系统;双向制;双向系统;双侧系统
bilateral tolerance 双向公差
bilateral trade 双边贸易
bilateral trade and payment agreement 双边贸易(及)支付协定
bilateral transducer 双向换能器;双向传感器;双向变换器
bilateral treaty 双边条约
bilayer 双分子层;双层
bilberry 乌板树属
Bilby steel tower 毕氏钢标
bilection 凸出平面的线脚;凸出嵌线
bilection mo(u)lding 凸出线脚;盖缝线脚
bilene 次甲基胆色素
bilevel 错落式住宅
bilevel operation 双电平工作
bilevel pantograph 两用绘图仪
bilge 隆起腹部;底舱;船的下水部分;舱底;桶腰;矢高
bilge alarm system 舱底进水警报系统
bilge altar 舭承凸座
bilge and ballast system 污水及压载管系
bilge and ballast water pipes 舱底污水和压载水管系
bilge arrangement 舱底水排水系统
bilge block 船腹边墩;侧盘木;侧垫块【船】;舱底垫块;边龙骨墩;舭墩(木);垫船块(造船用)
bilge block loading 船腹边墩荷载
bilge board 舷侧披水板;舱底水道盖板;水道盖板;污水沟盖板
bilge bracket 舭肘板;双层翼板
bilge ceiling 污水沟内底板
bilge chest 舱底水阀箱
bilge chock 消摆龙骨;舭龙骨
bilge circle 舷弧度;舭圆
bilge cribbing 舭墩(木)
bilged compartment 浸水舱;进水区间;底舱;水密区间;水浸区间
bilge delivery pipe 舱底导水管
bilge-diagonal intercept 舭对角线
bilge ejector 舱底喷射器;污水抽射器
bilge frame 船腹肋骨;舭肋材
bilge free 桶腰架空法
bilge hat 舱底污水阱;舭水井;污水阱;污水井
bilge hoop 桶腰箍
bilge injection 舭吸水管;污水管吸水装置
bilge injection pipe 船底喷水管
bilge injection valve 船底喷水阀
bilge injector 舱底喷射器
bilge inlet 船底进水口
bilge keel 减摇龙骨;船龙骨;船底龙骨;舭;鳍龙骨
bilge keel damping 舭内龙骨阻尼
bilge keelson 底内龙骨;舭内龙骨
bilge level 舱底水位
bilge limber board 污水沟盖板
bilge line 舱底吸水管;舭吸水管;污水抽水管系;滑道垫木;船台滑道垫木
bilge logs 上滑道
bilge main 舱底水总管;舭水总管
bilge main pipe 舱底水总管
bilge piece 减摇龙骨
bilge pipe 船底吸水管;舭吸水管;污水沟排水管系
bilge plank 舭外板
bilge plate 舭裂板;舭板
bilge plating 舭外板;舭部外板
bilge plug 舭泄水塞;舭排水栓;舭放水栓
bilge pump 船底(排)水泵;舱底污水泵;舱底(排)水泵
bilge pump accessories 舱底泵附件
bilge pump air vessel 舱底泵气腔
bilge pump cock 舱底泵塞门
bilge pump discharge pipe 舱底泵出水管
bilge pump gland 舱底泵压盖
bilge pump neck bush 舱底泵颈衬
bilge pump petcock 舱底泵小塞门

bilge pump plunger 舱底泵推水柱
bilge pump plunger rod 舱底泵推水柱杆
bilge pump rod 舱底泵挺杆
bilge pump stud 舱底泵枢键
bilge pump suction pipe 舱底泵吸水管
bilge pump valve 舱底泵阀
bilge radius 底边半径;舭部(曲度)半径
bilge saw 凸鼓形锯;桶形锯;桶板锯
bilge separator 舭油水分分离;舭部油水分离器
bilge shore 舭撑柱
bilge sludge and water 舱底油渣和水
bilge sounding pipe 污水测量管
bilge strainer 污水过滤器;污水过滤池
bilge strake 舭腹外板;船底板条;舭列板;舭部列板
bilge stringer 船底长桁;舭纵材
bilge strum box 舱底水滤盒;舱底管滤水器
bilge suction 船底吸水管;船底水吸口;舭吸水管
bilge sump 污水阱;污水井
bilge system 舱底水系统
bilge tank 污水柜
bilge timber 舭肋骨
bilge waste 舱底污水;舱底废物
bilge wastewater 船底污水;船舱废水;舱底污水
bilge water 舱底漏水;舭水;舭底水
bilge water alarm 漏水警报器;船漏警报器;船底漏水警报器
bilge water ga(u)ge 船底测漏用水位表;船底测漏水位表
bilge way 船台滑道;污水道;上滑道
bilge well 舭水井;污水阱;污水井
bilging 舱底破漏
Bilgram bevel gear shaper 彼尔格莱姆锥齿轮刨齿机
bilibinskite 碲铅铜金矿
bilinear 双直线;双一次性;双线性的
bilinear coefficient 双线性系数
bilinear diagram 普兰德耳双线应力图
bilinear expansion 双线性展开式
bilinear expression 双线性式
bilinear factor 双线性系数
bilinear form 双线性式
bilinear hysteresis-type spring 双线性滞变形弹簧
bilinear idealized hysteresis loop 双线性理想滞变曲线;双线性理想滞变回线
bilinear interpolation 双线性插值
bilinearity 双线性
bilinear law 双线性律
bilinear material 双线性材料
bilinear matrix form 双线性矩阵式
bilinear model 双线性模型
bilinear moment deflection relationship 二次弯矩挠度关系;双线性弯矩挠度关系
bilinear operator 双线性运算子;双线性运算号;双线性运算符;双线性运算符
bilinear programming 双线性程序设计
bilinear system 双线性系统
bilinear tariff 双率税则
bilinear transformation 双线性变换
bilinear transformation analysis 双线性变换分析
bilinear trial function 双线性试验函数
bilingual 两种文字对照的;两种语言的
bilingual legend 两种文字对照图例
bilingual map 两种语言对照图;有两文字的地图
bilinite 复铁矾;比林石
biliographical reference 参考书目(录)
bilithic filter 双片式滤波器
bill 壳扁片;锚爪尖部;议案;陡岬;单据;发票;备用现金;报单;清单;期票;票据
billabong 季节性河流;盲肠河汊;故河道;干涸河道;干涸道;死水洼地;死水潭
bill acceptance 票据承兑
bill accepted 票据承兑
bill accompanied by document 附单据的汇票
bill accompanied by warehouse receipts 附仓单汇票
bill account 票据账目;汇票清算
bill advice 支票到期通知
bill after date 开票后
bill and account payable 期付款项
bill and account receivable 期收款项
bill and accounts payable with terms 有条件期付款项
bill and charge system 开票记账法;发票记账法
bill as security 质押票据

bill at long sight 远期票据
bill at short sight 近期票据
bill at sight 即期票据;退期票据
bill at thirty day's sight 见票后三十天付款的期票
bill at usance 习惯期限的汇票
billboard 锚座;广告牌;广告栏;锚床;公告牌;置锚板;告示牌;告示板
billboard array 横列定向天线阵
bill book 支票簿;出纳簿;报单簿;票据簿
bill bought 购入票据
bill broker 证券经纪人;票据经纪人
bill business 票据交易
bill by negotiation 逆汇
bill center 押汇中心
bill clearing 票据清算
bill clerk 票据管理员
bill collection 票据托收
bill collector 收账员;收账人
bill department 开票部门;汇票业务部;发票部
bill deposited as collateral security 以汇票为抵押证券
bill discount 贴现汇票
bill discount deposit 票据贴现押金
bill discounted 贴现期票;已贴现票据;贴现票据
bill discounter 贴现业者
bill dishonored 拒付票据;拒付汇票
bill dishonored by nonacceptance 拒绝承兑汇票
bill drawn on a letter of credit 信用证汇票
bill drawn payable at a certain time after date 出票后若干日付款汇票;出票后…天后付款汇票
bill drawn payable at a certain time after sight 见票后若干天付款汇票
bill drawn payable at a fixed date 按规定日期付款汇票
bill drawn payable to bearer 无记名汇票
bill drawn to order 指示汇票;指定式汇票
bill draw payable at a certain time 出票远期承兑交单
bill duly protested 经制成拒绝证书汇票;制成拒绝证书的汇票
bill duplicate 票据副页
billed weight 提单上的重量
bill endorsed in bank 空白背书汇票
billet 木柴块;钢条;钢坯;钢垫板;短圆木;短木段;短厚木段;分配铺位;方钢坯;劈制木材;坯体;(金属的)坯段
billet and bar shears 钢坯型钢剪切机
billet bar 钢棒;条钢;型钢
billet bloom size 钢坯尺寸
billet car 低框架车
billet caster 方坯连铸机
billet chipper 钢坯清理装置
billeteer 钢坯剥皮机;粗加工机床
billet furnace 小钢坯加热炉;钢锭加热炉
Billet half lens 比累对切透镜
billeting capacity 开坯能力
billet mill 钢坯轧机;坯料轧机
billet mo(u)lding 错齿冶饰【建】
billet necking 钢坯切口
billet pusher 钢坯推出机;钢锭推出机
billet rod 钢棒;条钢
billet roll 钢坯轧机;钢坯轧辊
billet shears 剪铁条机;钢坯剪(切机);坯料剪切机
billet sizer 坯料回转定径机
Billet split lens 比耶对切透镜;比累对切透镜
billet steel 钢坯;短条钢;坯段钢
billet steel reinforcing bar 钢棒;条钢
billet steel reinforcing rod 钢棒;条钢
billet storage 坯料仓库
billet switch 坯料分配器
billet unloader 钢坯卸料机
billet unscrambler 钢坯自动堆坯机
billet-wood 圆材
bill exchanger 纸币兑换机
bill finance 短期债款;信贷
billfish 旗鱼
bill fold 单据夹;票据夹
bill for acceptance 认付汇票
bill for collection 托收期票;托收票据;托收汇票
bill for payment 付款票据
bill for premature delivery 提前交付的票据
bill for remittance 汇票
bill for term 兑换期票
bill goods 开列商品清单

bill head 空白单据
bill holder 持票人
bill-hook 钩镰
billi 千兆分之一
billiard-ball collision 弹性碰撞
billiard parlor 弹子房
billiard room 桌球房;弹子房
billiard saloon 弹子房
billiard table 弹子台;台球桌
billiary vessel 马氏管
billibit 千兆位;十亿位
billi capacitor 管状微调电容器
billi-condenser 管状微调电容器
billicycle 千兆周
billietite 黄钡铀矿
billing 列清单;开票;记账
billing clerk 开票员
billing cost 开单成本
billing demand 收费用水量
billing machine 开账机;开单机;记账计算机;发票登帐机;票据机;票据处理机;填票机;填表机
billing rate 开单价
Billingsellacea 伯灵贝类
billingsleyite 硫锑砷银矿
bill in process of clearing 交换中票据
bill insert 账单插入件
billion 垓(10^{12});万亿(英国);十亿(美国)
billion electron-volts 千兆电子伏特;十亿电子伏特(美国)
billion gallons per day 十亿加仑/天
billionth 十亿分之一(10^{-9})
billion years 十亿年
billisecond 毫微秒
bill loan 票据贷款
bill market 票据市场;汇兑市场
bill merchant 票据商
bill money 备用现金
Billnear method 真空养护法(混凝土);真空处理法(混凝土)
bill of approximate estimate 概算书
bill of clearance 出口报关单;出港申报表
bill of collection 托收凭单
bill of complaint 控诉状
bill of credit 取款单;银行信用证;取款凭证;取款凭单
bill of debt 债票
bill of entry 进口报关单;货物出入口报告书;报税通知单;报关单;入港呈报单;入港呈报表;入港报关单
bill of estimate 估价单
bill of exchange 交换单;汇票;银行汇票
bill of exchange in blank 空白汇票
bill of freight 货运单;运单
bill of goods 货品清单
bill of goods adventure 借贷抵押书
bill of health 健康证书;检疫证(书);免疫证书;船舶检疫证书
bill of lading 提货拼单;提货单;运货证书;运货单;提单
bill of lading clause 装载明细单;提单条款
bill of lading date 提单日期
bill-of-lading freight 提单运费
bill of lading minimum charges 提单单位运价
bill of lading ton 提单吨
bill of legal cost 诉讼费清单
bill of loading 运货单
bill of material processor 物品清单处理程序
bill of materials 材料清单;材料单;材料表;用料(清)单
bill of mortgage 抵押据
bill of parcels 发货单;货单;小货发单;包裹单;发票
bill of particulars 明细单
bill of payment 付款清单
bill of pratique 检疫证(书);无疫证书(船只)
bill of quantities 建筑工程清单;工程数量表;工程量清单;工程量表;数量清单;数量明细表;数量表
bill of quantity contract 工程量表合同
bill of sales 财产转让证书;卖契;卖据
bill of security 抵押票据
bill of sight 临时查验报关单;进口货临时报告书;(海关临时起岸的)报关单
bill of stores 海关免税单;再输入免税单;船上用品

免税单
bill of sufferance 免税装;免税单;海关落货许可单;卸货许可单
bill of the hook 大钩锁销
bill of transfer account 转账单
bill of works 工程报表
bill on demand 即期汇票
bill on deposit 以汇票为抵押证券;以汇票为担保的证券
bill on maturity 到期后付款汇票
billow 狂涛;巨浪;雪崩;大浪;大波;波涛;山崩
billow cloud 浪云;波状云
billowy 波涛汹涌的
bill passengers 填报乘客名单
bill payable 应付票据
bill payable at a definite time 预定时间付款汇票
bill payable at a fixed period after date 出票后定期汇票
bill payable at fixed date 定期付款票据
bill payable at fixed period after sight 见票后定期付款汇票
bill payable at long sight 见票后远期照付汇票
bill payable at sight 见票即付
bill payable at sight after a fixed period 定期见票后付票据
bill payable audit 应付票据审计
bill payable by installment 分期付汇票
bill payable by stated installments 分期付汇票
bill payable on demand 来取即付的汇票
bill purchased 出口押汇
bill quantity 产品、价目、数量单;产品数量表数量
bill receivable 应收票据
bill receivable audit 应收票据审计
bill register card 账单登记卡
bill rendered 开出票据
bills book 票据登记簿
bills discounted and remittance bought 贴现及买汇
bills discounted daily list 贴现票据日记账
bills discounted with collateral securities 抵押贴现票据;附担保的贴现票据
bills drawn under letter of credit 附有信用证的汇票
bills due 欠账
bills in process of clearing 票据交换
bills in three-parts 三联单
bills of payment 支付票据
bills of sale 卖据
bill sold on condition of repurchase 附有回购条件的售出票据
bills-only policy 专营国库券政策
bills payable 应付未付票据
bills payable account 库付票据账户
bill(s) receivable 应收未收票据
bills rediscounted 再贴现票据
bills retired 已清偿票据;赎回票据
bills sent for collection 托收款项
bills under letter of credit 附有信用证的汇票
bills under letter of instruction 附通知书汇票
bill to fall due 票据到期
bill to mature 票据到期
bill to order 记名票据;须签字照付的票据
bill to purchase 出口结汇
bill undue 期票;票期未到;未到期票据
bill without credit 非信用证签发的汇票;非凭信用证开发的汇票
billy 锅;石英岩层
bilobate 二叶的
bilobed wheel 双叶轮
bilocular 二格的
biloculate 二格的
bilp-scan radar 反射脉冲扫描雷达
bilux bulb 双灯丝灯泡
bimaceral 二重煤素质
bimaceral microlithotype 双组分显微煤岩类型
bimag 双态磁芯;双磁芯
bimagmatic 两代岩浆的;二元岩浆的;二代岩浆的
bimah (犹太教的)讲经台
bimaleate 马来酸氢酯;马来酸氢盐
bimalonate 丙二酸氢盐
bimanual examination 双手检查
bi-margin format 双边破图廓
Bimat film 载液胶片
bimatron 毕玛管

bimester 两月；一段为期两个月的时间
bimetal 复合金属板；双金属(片)
bimetal base apparatus 双金属基线测量器械
bimetal benchmark 双金属水准标
bimetal casting 双金属铸造
bimetal cover 双金属片盖
bimetal fuse 双金属熔丝
bimetal lamp 双金属灯
bimetallic 双金属的
bimetallic actinometer 双金属日射表；双金属感光计
bimetallic article 双金属制品
bimetallic balance 双金属摆轮
bimetallic bar 双金属杆件
bimetallic barrel 双金属挤出原料筒
bimetallic bearing 双金属轴承
bimetal(lic) blade 双金属片
bimetallic capacitor 双金属片电容器
bimetallic compensation strip 双金属补偿片
bimetallic condenser 双金属片电容器
bimetallic conductor 复合金属导线；双金属导线；双金属导体
bimetallic corrosion 双金属侵蚀；双金属腐蚀
bimetallic element 双金属元件
bimetallic fiber method 双金属丝法
bimetallic instrument 双金属仪表；双金属式仪表
bimetallic lined barrel 双金属衬筒
bimetallic material 双金属材料
bimetallic plate 双金属板
bimetallic regulator 双金属调节器
bimetallic rotor 双金属转子
bimetallic ruler 双金属尺
bimetallic saw blade for hack sawing machine 弓形锯床用双金属锯条
bimetallic spring 双金属弹簧
bimetallic(spring cone) seal 双金属密封
bimetallic standard 复本位
bimetallic strip 双金属带；双层金属带
bimetallic strip compensation 双金属片补偿
bimetallic strip ga(u)ge 双金属片传感器
bimetallic strip relay 双金属片继电器
bimetallic temperature regulator 双金属温度调节器
bimetallic thermocouple 双金属热电偶
bimetal(lic) thermometer 双金属温度计
bimetallic thermostat 双金属温度调节器；双金属恒温控制器
bimetallic tube 双层管
bimetallic wire 双金属线
bimetallism 金银二本位制；复本位制
bimetal plate 双层金属板
bimetal relay 双金属继电器
bimetal release 双金属片开关；双金属片断路器；双金属开关；双金属断路器
bimetal ruler method 双金属尺法
bimetal sheet 双金属板
bimetal strip 双金属条；双金属片；双金属组合条
bimetal timedelay relay 双金属延时继电器
bimetal tube 双金属管
bimetal type dial thermometer 双金属式度盘温度计
bimirror 双平面镜；双镜
bimodal 双模态的；双峰态的
bimodal cleavage 双型劈理
bimodal cross-bedding 双向交错层理
bimodal current rose 双向水流玫瑰图
bimodal curve 双峰型粒径曲线；双峰曲线
bimodal distribution 双型分布；双峰索道；双峰式(泥沙级配)分布
bimodal igneous activity 双模式火成活动
bimodality 双峰态性
bimodal lateral wave 双节横波
bimodal mica fabric 双峰云母组构
bimodal rift volcanism 双模式裂谷火山活动
bimodal size distribution 双众数粒度分布；双峰粒度分布
bimodal volcanic complex 双模式火山杂岩
bimodal volcanism 双峰火山作用
bimodel distribution 双峰分布
bimolecular collision 双分子碰撞
bimolecular film 双分子膜
bimolecular layer 双分子层
bimolecular termination 双分子终止(反应)
bi-moment 双力矩
bimonthly 两月一次；每月二次的；双月刊

bimorph 双压电晶片；双晶
bimorph cell 双压电晶片元件；双层晶片单元
bimorph crystal 耦联晶片；振荡互补偿晶体；双层晶体
bi-motor 双发动机
bin 料箱；料斗；料槽；料仓；接收器；集料台；储藏室；储藏谷物的场所
bina 坚硬黏土岩
binac 二进制计算机
bin activator 料斗抖动器；仓壁振动器
bin ag(e)ing 储[贮]存老化；胶料停放
bin and batcher plant 料仓和分批配料装置
bin and feeder firing 中间仓和喂煤机燃烧系统
bin and troll(e)y batcher plant 料仓和吊运配料装置
binangle 双角器
binarite 白铁矿
binary 二元的；二态的；二进位制；二成分；二叉的；二变量；物理双星；双体
binary accumulator 二进制累加器；二进位累加器
binary acid 二元酸
binary adder 二进制加法器；二进加法器
binary adding circuit 二进制加法电路；二进位加法线路
binary adding device 二进制加法器
binary addition 二进制加法
binary algebraic adder 二进位代数加法器
binary alloy 二元合金
binary alloy photoconductor 二元合金光电导体
binary alphabet 二进制字母表
binary amplitude 二元调幅
binary analog(ue) conversion 二进制模拟转化
binary baud rate 二进制波特率
binary code 二进制码；二进制代码；二进制编码；二进码；二进代码
binary-coded digit 二进制编码数字；二进制编码的数字
binary-coded Fraunhofer hologram 二进制码夫琅和费全息图
binary-coded interrupt vector 二进制编码的中断向量
binary-coded notation 二进制编码记数法；二进制编码表示法
binary-coded octal 二八进制；二进制编码的八进制
binary-coded output 二进制编码输出
binary code system 二进制编码系统
binary collision 二体碰撞
binary colo(u)r 合成色
binary column 二进制列
binary combination head 双重复合带头
binary commutation circuit 二进位转换线路
binary compound 二元化合物
binary computer 二进制计算机
binary condition 二进制状态；二进制条件；双值状态；双值条件
binary constant 二进制常数
binary converter 单级换流器
binary countering system 二进制计数系统
binary counting 二进制计数
binary cubic form 二元三次形式
binary cycle 双汽循环
binary cycle boiler 双工质锅炉
binary cycle geothermal power generator 双循环地热电站
binary cycle system 双循环系统
binary cycle two-phase biologic(al) process 双循环两相生物处理
binary cycly 双工质循环
binary data 二元数据；二进制数据
binary data item 二进制数据项
binary-decimal conversion 二进制转换；二十进制转换
binary-decimal number conversion 二进十进制数转换
binary decision 双择判决；双择判定
binary decision theory 二元判定理论
binary decoder 二进制解码器
binary dependent variable 二进制因变量
binary detection 双择检验；双择检测
binary device 二进制设备
binary diffusion 二元扩散
binary digit 二进制位；二进制数字；二进制数位；二进制数；二进位数；二进数位
binary directed tree 二元有向树；二叉有向树

binary divider 二进制除法器
binary divider stage 二进制分频级
binary division 二进制除法
binary divisive procedure 二分法
binary dump 二进制转储
binary dump program(me) 二进制转储程序
binary electrometer 双极静电计
binary element 二进制元素；二进制元件；二进制码；二进制单元；二进位单元；双态元件
binary eluant 二元洗提液
binary emulsion 二元乳状液
binary encoder 二进制编码器
binary encoding 二进制编码
binary engine 双元燃料发动机；双元发动机
binary equilibrium 二元平衡
binary equivalent 等效二进制位数；等效二进制数；等效二进位数
binary eutectic 二元共晶；二元低熔物；二元低共熔物
binary eutectic point 二元共结点
binary explosive 二元炸药
binary file 二进制文件
binary filter 二元滤波器
binary fission 二裂的；二分裂
binary fixed point constant 二进制定点常数
binary fixed point data 二进制定点数据
binary flip-flop 二进制触发器；二进双稳态触发器
binary floating-point constant 二进制浮点常数
binary-floating-point data 二进制浮点数据
binary floating-point resistor 二进制浮点电阻器
binary fluid 二元流体
binary fluid cycle 双流体循环
binary form 二元形式
binary format 二进制格式
binary forward counter 二进制正向计数器
binary fraction 二进制数小数部分；二进位分式
binary gain control 二进制增益控制
binary glass 二组分玻璃
binary granite 二云母花岗岩；二元花岗岩
binary half-adder 二进制半加器
binary hologram 二元全息图
binary image 二进制映象
binary incremental representation 二进制增量表示(法)
binary information 二进制信息
binary input 二进制输入
binary insertion 折半插入；二分插入法
binary instruction code 二进指令代码
binary instruction packet 二进制指令包
binary integer 二进制整数
binary internal number base 内部二进计数制
binary invariant 二元不变式
binary ionic crystal 二元离子晶体
binary item 二进位项目
binary level 二进制级
binary linear substitution 二元线性代换
binary link 两副杆
binary liquid mixtures 二元液体混合物
binary load 二进制装入
binary loader 二进制装入程序；二进制装配程序
binary loader tape 二进制装入程序带
binary logarithm 二进制对数
binary logic element 双值逻辑元件；双稳态逻辑元件；双态逻辑元件
binary magma 二元岩浆
binary mass balancing 二自由度系统的重力平衡
binary matching 二元匹配(法)
binary matrix 二元矩阵
binary matrix notation 二进制矩阵
binary matroid 二元拟阵
binary medium flow model 双重介质渗流模型
binary message 二进制信息；二进制消息
binary metal 二元合金
binary minimum 二元极小值
binary mixture 二元混合物；二元混合气体
binary mode 二进制方式；二进法
binary modulation 二元调制
binary multiplication 二进制乘法
binary multiplication rule 二进制乘法规则
binary multiplier 二进制乘法器；二进乘法器
binary node 二枝节点；二叉节点
binary non-algebraic adder 二进位非代数加法器
binary notation 二进制记数法；二进制计数法；二进位符号；二进位法；二进制
binary notation system 二进制记数制；二进制记

数系统
binary number 二进制数(字);二进数
binary number coding 二进信息码
binary(number) system 二进数制;二进制数系;二进位数制
binary numeral 二进制数值;二进制数码
binary numeration system 二进制数制;二进制计数系统;二进数制
binary object file 二进制目标文件
binary object format 二进制结果格式
binary octal 二八进制
binary octal decoder 二八进制译码器
binary of one's 二进制反码;二进制编码十进制
binary operation 二元运算;二元操作;二进制运算;二进制操作
binary operator 二元运算符;二目算符;二进制算符
binary orthogonal code 二元正交码
binary output 二进制输出
binary pattern 二进模式
binary phase diagram 二元相图
binary phase filter 二元相位滤波器
binary phase-shift keying 二相移键控;二进位相键移控法
binary picture 二态图
binary place 二进制数位
binary point 二进制小数点
binary polynomial 二进多项式
binary process 二元制处理
binary product 二进制乘积
binary product generator 二进制乘积发生器
binary pulsar 脉冲双星
binary pulse 二进制脉冲
binary pulse-code modulation 二进制脉码调制;双脉码调制;二元脉冲编码调制
binary punch 二进制穿孔
binary quadric form 二元二次形式
binary quantizer 二进制转换器;二进制数字转换器;二进制量化器
binary radix 二进制基数
binary random process 二元随机过程;二进制随机过程
binary raster 二元光栅
binary reflex-code 二进制反射码
binary relation 二元关系;二目关系
binary representation 二进制表示(法)
binary resolvent 二元消解式
binary ring 二进制计算环;二进位计算环
binary rock 二元岩
binary salt 二元盐
binary scale 二进制刻度;二进制记法;二进制标度;二进法
binary scaler 二进制计数器;二进制定标器;二进位换算电路;二进定标器
binary search 折半查找;对分检索;对分查找;对半搜索法;对半检索;二分法检索
binary search method 对半检索法
binary search tree 对分查找树;二进制搜索树;二叉检索树
binary sediment 二元沉积物;二相沉积物
binary selfdefining term 二进制自定义项
binary separation 二元分离
binary signal 二进信号
binary signal(l)ing 双态通信
binary silica optic(al)fibre 二元硅光学纤维
binary solid solution 二元固熔体
binary solution 二元溶液
binary solvus 二元溶线
binary sort 二分法分类
binary sorter 二元分类机
binary star 双星
binary-state variable 二值变量;二值变量;二态变量;二进制状态变量;双态变量;双态变量
binary steel 二元合金钢
binary storage cell 二进制存储单元
binary storage device 二进制存储单元
binary storage tube 二进制存储管
binary string 二进制串
binary subtracter 二进制减法器
binary subtraction 二进制减法
binary symmetric(al)channel 二进制对称通道
binary synchronous communication 二元同步通信规程;二进制同步通信;二进制同步通信;双同步通信
binary synchronous device data block 二元同步设备数据块
binary synchronous protocol 二进制同步协定
binary synchronous transmission 二元同步传输;二进制同步传输
binary system 二元制;二元(物)系;二进(位)制;二进记数制;二合体系
binary system piezo electric(al)ceramic 二元系压电陶瓷
binary tape assembler 二进制带汇编程序
binary time 二进制时间
binary-to analog(ue)converter 二进制—模拟变换器
binary-to decimal-conversion 二十进制转换;二十进制变换;二进制十进制变换
binary to decimal converter 二十进制转换器;二十进制转换程序
binary-to-hexadecimal conversion 二十六进制转换
binary-to-numeric conversion 二进数字变换
binary-to-octal conversion 二八进制转换
binary-to-octal converter 二八进制转换器
binary treatment 二进制处理
binary tree 二元树;二叉树
binary tree data structure 二进制树形数据结构;二叉树数据结构
binary tree search 二叉树检索;二叉树查找
binary tree traversal traversal of binary tree 二叉树的遍历
binary unit 二进制信息单位;二进制单位
binary unit distance code 二进制单位距离码
binary vapo(u)r cycle 二元蒸气循环;双气循环
binary vapo(u)r engine 双元蒸气发动机
binary variable 二状态变数;二元变数;二态变量;二进制变量
binary vector 二元向量
binary zero 二进制零
binary zone plate 二元波带片
binate 二分取样
binate leaves 双生叶
binaural 双耳的
binaural broadcasting 立体声广播
binaural effect 双取效应
binaural localization 双耳定位
bin board 散粮隔板;围囤木板
bin bottom 斗仓底部;料斗底
binburra 白山毛榉(产于澳大利亚)
bin card 储仓卡
bincoculars coil 双扎线圈
bin compartment 料仓分隔(间)
bin cure 胶料停放
bind 联结;捆轧;捆绑;捆;结合;胶泥;黏合;握裹;藤蔓;使受法律约束;使确定不变
bind a bargain 订买卖契约
bind catch basin 截水井
bind clip 接线夹
bind command 连接指令
binder 临时契约;连接averages;浸润剂;结合料;胶合物;胶合剂;基粘;黏粒;黏结粒;黏结剂;黏合料;黏合剂;系杆;装订机;装订工;柱内箍筋;保证金;保险单
binder aggregate mix(ture)黏合剂混合料
binder application 施加黏合料;黏合剂洒布
binder application section 黏结剂施加段
binder asphalt heating installation 黏合剂沥青加热装置
binder bar 拉结条;盖条
binder barrel 黏合剂桶
binder batcher 黏合剂配量器;黏合剂配料计量器
binder batching 黏合料按比例配料;黏合剂配料
binder batching pump 黏合剂配料泵
binder bolt 联结螺栓;连接螺栓
binder bridge 中间黏结剂
binder bulk storage 黏合料散装储[贮]藏
binder bulk storage and heating installation 沥青集料储[贮]存加热装置;沥青集料储[贮]存合加热装置
binder canvases 割捆机帆布输送带
binder-clay 胶黏土
binder coat 黏结层
binder content 黏合料成分;黏合剂含量
binder content determination 黏合剂含量测定
binder cooker 黏合料熔锅;熔化黏合料的锅;沥青熔锅
binder course 黏结层;丁砖层;联结层【道】;拉层;结合层
binder course mix(ture)黏结层混合料
binder curing oven 黏结剂固化炉
binder demand 基料需要量;漆基需要量
binder depositor 黏结剂施加装置
binder dispersion 黏合剂分散剂
binder distribution 结合料分布;结合料洒布
binder distributor 结合料洒布机
binder emulsion 黏合剂乳化液
binder-filler mix(ture)黏结料填充混合料
binder film 黏合料薄膜;黏合剂膜
binder for abrasive 磨具结合剂
binder for construction materials 建筑材料用的黏料
binder for porcelain paste 瓷件黏结剂
binder for pre-coating 预涂层的黏合料
binder for resin mortar 树脂砂浆胶合剂
binder for tractor draft with single canvas 单帆布带机引割捆机
binder gun 黏合剂喷枪;喷胶器
binder heater 黏合料熔锅;熔化黏合料的锅;沥青熔锅
binder heating 加热黏合料
binder heating installation 黏合料加热装置
binder heating trailer 加热黏合料的拖车
binder-hole card 多孔卡
binder injecting device 黏合料灌注器;黏合剂注入装置
binder injection 灌注黏合料;黏合剂注入
binder kettle 沥青熔锅
binderless sand 无黏结剂型砂
binder lever 系杆
binder material 胶结材料;黏滞黏质
binder matrix 结合混合料
binder measuring 黏合剂计量
binder measuring pump 黏合料配量泵;黏合料按比例混合用的泵;黏合剂配量泵
binder metering 黏合料按计量配料;黏合剂计量
binder metering pump 黏合料配料用的泵;黏合料计量泵;黏合剂计量泵
binder mix(ture)黏合料混合;黏合剂混合料
binder-moving lever 打捆装置移动杆
binder particle 黏合料颗粒
binder pipe 黏结料管
binder portion 黏合剂含量
binder portion determination 黏合剂含量确定
binder proportioning 黏结料按比例配料;黏合剂配料
binder proportioning pump 黏结料按比例配料用的泵;黏合剂配料泵
binder pump 黏结料配料用泵
binder ratio 基料比
binder recovery 黏合料回收
binder sickle 割捆机切割器割刀
binder soil 黏土;胶结土
binder spray 黏结剂喷涂
binder spray(ing)bar 地沥青喷射泵;黏结料喷射泵;沥青喷杆
binder spray(ing)machine 黏结料喷射机(器);沥青洒布机
binder spreading 撒铺黏结料;黏合剂洒布
binder stone 结合石
binder storage installation 黏结料储[贮]存装置
binder storage tank 结合料储[贮]槽
binder stud 结合柱螺栓;结合件双头螺栓;接合柱螺栓
binder suspension 黏合剂悬浊液
binder thermometer 地沥青温度计;黏结料温度计;沥青温度计
binder thread 滚边线
binder tube 沥青管道(沥青洒布机)
binder twine 细绳;单股捆包麻绳
binder weighing batcher 黏结料的按重量配料器;黏合剂称量配料器
binder weighing tank 结合料称量槽
binder with gathering arms 带集合臂的割捆机
bindery 装订所
binder yarn 绑扎纱
bin detection 货位检测器
bin detection device 双重货位检测装置
bindeton 结合黏土
bindheimite 水锑铅矿
bind image 连接映像
bind image table 连接映像表
binding 捆绑;拘束力;紧固;夹锯;黏结;黏合的;装

订;必须遵守的;包箍;包带;绑定;黏合
binding ability 结合性能;结合能力
binding admixture 黏结填加入物
binding agent 胶结料;黏结剂;结合剂;接合剂;黏合剂;载色剂
binding agglomerate 黏合集块岩
binding and layout of atlas 地图集装帧
binding apparatus 打捆装置
binding a program(me) 联编一个程序
binding attachment 打捆装置
binding base 结合基层
binding beam 连系梁;联系梁;联梁;联结梁
binding board 扎捆台
binding bolt 接合螺栓;联结螺栓;连接螺栓;固紧螺栓
binding character 约束性
binding clasp jaw 夹紧器
binding clause 拘束条款
binding clip 接合夹;线箍;钳头
binding coal 结块煤;黏结性煤;黏结煤
binding concrete course 混凝土黏结层
binding concrete layer 混凝土黏结层
binding constant 结合常数;接合常数
binding constraint 紧固约束
binding contract 有约束力合同
binding course 结合层;联系层;联结层【道】;拉结层;黏结层;黏合层
binding department 装订车间
binding edge 订口
binding effect 黏合效应
binding energy 结合能;束缚能
binding face 支承面;支撑面;贴合面
binding factor 结合因子
binding film 胶结膜;结合膜;黏结膜
binding force 结合力;黏着力;黏合力;内聚束力;约束力
binding force of elasticity film 弹性模约束力
binding form 结合态
binding gravel 掺黏结料的砾石
binding handle 结合手把;制动手把
binding head screw 接合螺钉(圆顶宽边);扁头结合螺钉
binding hours 限制时间
binding joist 联结格栅
binding layer 黏结层
binding lever 结合手把
binding log 封顶木
binding machine 捆扎机
binding material 胶结材料;胶凝材料;胶结物质;黏结材料;黏合材料
binding mechanism 打捆机构
binding mechanism for multilayer 层板包扎机械
binding medium 黏结料;胶浆;胶合剂;胶结介质;黏合剂
binding medium dispersion 胶结料弥散现象
binding medium emulsion 胶结料乳剂
binding medium suspension 胶结料悬浊液
binding metal 硬化锌合金;黏结金属;锌基合金;拜恩丁锌合金
binding nut 扣紧螺母;夹紧螺母;限动螺母
binding of lime 石灰黏合;石灰黏结作用
binding of metal 连接铁件
binding of stones 砌体结合;石砌体结合
binding pad 缠绕式垫片
binding piece 联结件;水平板条
binding pile 接线柱【电】
binding plate 角撑;角板
binding platform 打捆台
binding post 接线螺钉;接线柱【电】;接线端子
binding post nut 接线柱螺母
binding power 结合力;黏结力
binding press 装订机
binding primer 黏合底漆
binding process 联编过程
binding property 黏合性
binding quality 黏结性
binding rafter 承椽梁;联结椽;联椽;桁条;承椽木;檩
binding receipt 暂保收据;承保收据
binding reinforcement 连接钢筋;绑扎钢筋
binding ring 调整环;锁紧环
binding rivet 结合用铆钉;紧固螺钉;结合铆钉
binding rod 连接杆
binding screw 紧固螺丝;紧固螺钉;接线螺旋;接线螺钉;接合螺丝;固紧螺丝;固紧螺钉
binding site 结合部位
binding stone 拉结石
binding strake 联结箍条;联固板;加强外板
binding strength 结合强度;接合强度;黏结强度;黏合强度
binding tape 黏合带
binding thread 紧固螺纹;装订用线
binding turns 绑匝
binding unit 打捆装置
binding(up) 包扎
binding upon all parties 对双方面都有约束力;对有关各方面都有约束力
binding value under the action of water 水下作用的黏合值
binding varnish 黏合清漆
binding volcanic breccia 黏合火山角砾岩
binding wire 扎线;扎丝;绑扎用铁丝;绑扎用钢丝;绑扎线;绑线
binding wire coil 绑扎钢丝圈
bind in position 使定位
bin discharge 料仓卸货;料斗卸料
bin discharger 筒仓出料器
bindless briquetting 无黏结剂团矿法
bindle-stiff 流动工人
bind member 联杆;连接构件;连接杆件
bin door fill valve 料仓门上控制料流的阀;料仓装料活门
bin drainage system 自动进出料仓系统(材料)
bin drawing channel 料仓倾卸槽
bin drawing device 料仓出料装置
bind-seize 滞塞
bindstone 黏结岩
bind wire 绑缚线
bindwood 常春藤
bin effect 仓效应
Binet age 比奈年龄
Binet-Cauchy theorem 比奈一柯西定理
Binet's formula 比奈公式
bin feeder 料仓加料器;漏斗给料机;料斗给料机;斗式进料器;仓式喂料机;仓式进料器
bin feeding height 料斗加料高度;料斗送料高度
bin filling 料仓加料
bin flow device 料斗抖动装置;料斗抖动器
bin-flow test 仓流试验
binful 满仓;一满斗;一满仓;满一(料)斗;满斗
bing 矸石堆;垛;堆;材料堆;废石堆
bin gate 料仓闸门;料斗门;料仓门;仓门
bing brick 硬质砂岩
bing-cutter 锥形齿轮滚刀
Bingham body 宾汉塑性体
Bingham body plasticoviscous substance 宾汉体黏塑性体
Bingham flow 宾汉流动
Bingham fluid 宾汉流体
Bingham liquid 宾汉液体;塑性液体
Bingham model 宾汉模型
Bingham number 宾汉数
Bingham plastic 宾汉塑性体
Bingham plastometer 宾汉塑性计;宾汉可塑性测定仪
Bingham swelling 宾汉膨胀
Bingham viscous system 宾汉黏性体系
bing-hole 放矿溜口
bing ore 富铅矿
bin hang-up 料槽阻塞;料仓挂料;料斗阻塞
binifibrate 比尼贝特
bin indicator 料仓料位指示器
bin installation 料仓安装
binistor 四层半导体开关管
binit 二进制数位
bin level 料仓料位;料仓料面高度
bin level control housing 料仓料位控制箱
bin level detector 料仓高度探测器;料位指示器;料位计
bin level indicator 料位指示器;料位计;料仓料位指示器;储[贮]料斗存料指示器;仓内储[贮]量指示器
bin level limit switch 料仓料位限位开关
bin level measurement 料仓料位测量
bin level transmitter 料仓料位传感器
binnacle 罗盘座;罗盘箱;罗盘柜
binnacle ring 罗盘箱常平环
binnacle stand 罗盘柜座
Binnen container 内陆用集装箱
binning 装仓
binobal seiche 双节假潮
binocular 双眼单筒镜;双筒(镜);双目望远镜;双孔的
binocular accommodation 像散调节
binocular adjustment 双目调整
binocular body 双筒镜
binocular coil 双筒线圈;双孔线圈
binocular colo(u)r matching 双目配色
binocular colo(u)r mixing 双目混色
binocular contrast 双目衬比
binocular diplopia 双目双像
binocular field 双目视野
binocular field of view 双目视野
binocular fixation 双眼视轴交会
binocular hand level 双筒手提水准仪;双目手持式水准仪
binocular head 双眼观察头;双目镜头
binocular instrument 双眼仪器;双筒仪器
binocular integration 双眼集合
binocular loupe 双筒放大镜
binocular magnifier 双筒放大镜
binocular mat 双筒望远镜遮片
binocular microscope 双筒显微镜;双目显微镜;双目实体显微镜
binocular observation 双目观测
binocular parallax 双眼视差;双目视差
binocular photometry 双目光度学
binocular prism 双目望远镜棱镜
binocular prism telescope 双筒望远镜
binocular projector 双像投影器
binoculars 显微镜;望远镜;双筒望远镜;双目镜
binoculars box 望远镜盒
binocular sight 双镜瞄准具
binocular stereocamera 双镜立体摄影机
binocular telescope 双筒望远镜
binocular three dimensional display 双孔三维显示
binocular tube 双目镜管
binocular-type periscope 双筒潜望镜
binocular viewer 双镜观测镜
binocular viewing 双目观察
binocular viewing device 双目观察装置
binocular viewing head 双目观察头
binocular viewing system 双目观察系统
binocular vision 双目视觉;双目观察
binocular visual field 双目视场
binocular visual system 双目视系统
binoculus 双眼
binodal 双节
binodal curve 双节固溶曲线
binodal seiche 双节波动;水面双峰波
binode 双阳极管;双阳极
binode of a surface 二切面重点
binomen 双名
binomial 二项式
binomial antenna array 双正交天线阵
binomial array 双向阵
binomial array antenna 二项阵列天线
binomial coefficient 二项系数;二项式系数
binomial correlation 二项相关
binomial curve 二项式曲线
binomial density function 二项密度函数
binomial differential 二项式微分
binomial distribution 二项式分布;二项分布
binomial distribution function 二项分布函数
binomial distribution type 二项式分布型
binomial equation 二项方程式;二次方程(式)
binomial expansion 二项展开式;二项式展开
binomial experiment 二项试验
binomial expression 二项式
binomial factor 二项式系数
binomial formula 二项式公式
binomial frequency distribution 二项式频率分布
binomial index of dispersion 二项式离差指数
binomial law 二项定律
binomially distributed random number 二项分布随机数
binomial model 二项式模型
binomial nomenclature 二名法;双名法
binomial number system 二项数系
binomial population 二项式总体
binomial probability 二项式概率

binomial probability distribution 二项式概率分布
binomial probability paper 二项式概率坐标纸;二项概率纸
binomial series 二项级数
binomial surd 二项不尽根
binomial test 二项式检验;二项检验
binomial theorem 二项式原理;二项式定理;二项定理
binomial trial 二项试验
binomial trials model 二项试验模型
binomial two-sample model 二项双样本模型
binomial variable 二项式变量;二项变量
binomial variation 二项变异
binomial waiting time distribution 二项等待时间分布
binominal sheet 双编号图幅
binophthalmoscope 双眼检眼镜;双目检眼镜
binormal 次法线;副法线
bin outlet 料仓出口
bin plant 料仓的机械设备;料仓设备
bin-pressure test 仓压力试验
bin quality colo(u)r display 面元质量彩色显示
bin segregation 料斗内材料分层分布;料仓物料离析;料仓物料分层
bin shape 料仓形状
bin space 煤炭容量
bin stability 耐储[贮]性;储[贮]存稳定性
bin stopper 料仓闸门
bin storage 料仓储[贮]藏;料袋制动器;罐储[贮]藏;仓库储藏
bin stored material 罐存材料
B-instruction 变数指令
bin system 料仓系统
bin system of coal grinding and firing 煤粉集中粉磨经中间仓的间接燃烧系统
bin tag 库房标签
bintagnor 红色硬木(产于东印度)
binturong 熊狸
bin-type retaining wall 隔仓式挡土墙;仓式挡土墙
bin type retaining wall of steel 料仓式钢制挡土墙;仓式钢挡土墙
binuclear complex 双核络合物
bin unloader 料仓卸载装置
bin vibrator 料仓振捣器;料仓振动器
bin wall 料仓壁;隔仓式挡土墙;隔仓式桥台
bin weighing batcher scale 料仓按重量配料的秤盘;料斗定量秤
bio-accumulate 生物积累
bio-accumulated limestone 生物堆积灰岩
bio-accumulation 生物聚集;生物累积;生物堆积(作用);生物积累;生物积聚
bio-accumulation efficiency 生物积累效率
bio-accumulation element 生物积累元素
bio-accumulation factor 生物积累系数
bio-accumulation ratio 生物积累比
bioacoustics 生物声学
bioactivated carbon 生物活性炭
bioactivated carbon fiber 生物活性炭纤维
bioactivator 生物杀虫剂;生物活性剂
bioactive ceramics 生物活性陶瓷
bioactive compound 生物活性化合物
bioactive polymer 生物活性高分子
bioactivation 生物活化
bioactivity 生物活性;生物活度
bioactivity index 生物活性指标
bioadsorption of coefficient 生物吸附系数
bio-aeration 生物通气;活性通气;活性曝气法;生物曝气(作用)
bio-amine 生物胺
bio-amplification 生物放大(作用)
bioanalysis 生物分析(法)
bioarenite 生物砂屑岩
bio-argillaceous texture 生物泥质结构
bioassay method 生物测定法
bioassay 生物学鉴定法;生物鉴定;生物检验;生物检定;生物测试;生物测定
bioassay evaluation 生物测定评价
bioassay for water pollution 水污染生物测定
bioassay of pesticide 农药生物测定
bioassay of water quality 水质生物测定
bioassay parameter 生物检定参数;生物测定参数
bioassay response 生物测定特性曲线
bioassay system 生物检定系统;生物测定系统
bioassessment 生物评价

bioassessment of water pollution 水污染生物学评价
bioastrophysics 天体生物物理学
bio-atmospheric unit 生物大气单元
bioaugmentation 生物强化
bioaugmentation activated sludge reactor 生物强化活性污泥反应器
bioaugmentation fungus 生物强化真菌
bioaugmentation process 生物强化工艺
bioaugmentation process with liquid live microorganism 流动活微生物工艺
bioaugmentation technology 生物强化技术
bioaugmented sequencing batch membrane bioreactor 生物强化序批间歇式膜生物反应器
bioavailability 生物有效率;生物利用率;生物可利用性
bioavailability index 生物可利用指数
bioavailable phosphorus 生物可利用磷
biobalance 生物平衡
biobattery 生物电池
bioblooding 生物冲淋
bio-bomb 生物炸弹
biocalcarenite 生物碎屑灰岩;生物砂屑灰岩
biocalcilutite 生物泥屑灰岩
biocalcirudite 生物砾屑灰岩
biocarbon 生物炭
bio-catalyst 生物催化剂
biocell 生物电池
biocenological effect 生物群落效应
biocenose 生物群落
bio-ceramic filter 陶瓷生物滤池
bioceramics 生物陶瓷
biochemical 生物性药物;生物化学品;生物化学的
biochemical action 生物化学作用;生化作用
biochemical blood index 生化血指数
biochemical capacity of wastewater 生化污水容量
biochemical change 生物化学变化
biochemical characteristic 生化特征
biochemical coalification 生物化学煤化作用
biochemical consumption of oxygen 生化耗氧量
biochemical conversion 生化转化
biochemical cycle in ecosystem 生态系统中生物地球化学循环
biochemical degradation 生物化学降解作用;生化降解
biochemical deposit 生物化学沉积
biochemical deposition 生物化学沉积作用
biochemical ecology 生物化学生态学;生化生态学
biochemical effect 生化效应
biochemical effluent 生化出水
biochemical energy potential 生化势能
biochemical environment 生化环境
biochemical flocculation 生化絮凝
biochemical fuel cell 微生物化学电池;生化燃料电池
biochemical gelefication 生物化学凝胶化作用
biochemical hypothesis 生化假设
biochemical index 生化指数
biochemical indicator 生化指示物;生化指示剂
biochemical inhibition 生化抑制
biochemical interaction 生物化学的相互作用
biochemical lesion 生物化学损伤
biochemically treated municipal wastewater 生化处理城市污水
biochemical mechanism 生化机理
biochemical medium 生化培养基
biochemical methane potential 生化甲烷势
biochemical method 生物化学方法
biochemical mutant 生化突变(体)
biochemical mutation 生化突变
biochemical oxidation 生化氧化(作用)
biochemical oxygen demand 生物化学需氧量;生化需氧量(去除率);生化需氧量曲线;生化需氧量浓度;生化需氧量缓和剂
biochemical oxygen demand analyzer 生化需氧量测定仪
biochemical oxygen demand automatic recorder 生物化学需氧量自动记录仪;生化需氧量自记仪
biochemical oxygen demand determination 生化需氧量测定
biochemical oxygen demand-dissolved oxygen water quality model of river 河流生化需氧量—溶解氧水质模型
biochemical oxygen demand load(ing) 生化需氧量负荷
biochemical oxygen demand mass loading 生化需氧量总量负荷
biochemical oxygen demand meter 生化需氧量自记仪
biochemical oxygen demand meter for seawater 海水生化需氧量测定仪
biochemical oxygen demand releasing from organic bottom sludge 有机底泥释放的生物需氧量
biochemical oxygen demand settling and rising 生化需氧量沉浮
biochemical oxygen demand settling and rising coefficient 生化需氧量沉浮系数
biochemical oxygen demand sludge loading 生化需氧量污泥负荷
biochemical oxygen demand test 生化需氧量试验;生化需氧量测定
biochemical oxygen demand ultimate 最终生化需氧量
biochemical parameter 生化参数
biochemical pollution criteria 生化污染标准
biochemical pollution standard 生物化学污染标准
biochemical process 生物化学法;生化过程
biochemical purification 生化净化
biochemical reaction 生物化学反应;生化反应
biochemical reaction kinetics 生化反应动力学
biochemical reaction time 生化反应时间
biochemical reservoir 生物化学岩储集层
biochemical rock 生物化学岩
biochemicals 生物化学剂
biochemical-sedimentary ore deposit 生物化学沉积矿床
biochemical-sedimentary ore-forming process 生物化学沉积成矿作用
biochemical self-purification 生化自净化作用
biochemical sewage treatment 污水生化处理;生化污水处理
biochemical tank 生化池
biochemical technique 生物化学技术
biochemical treatment 生物化学处理;生化处理
biochemical treatment facility 生化处理装置
biochemical treatment of wastewater 废水生物化学处理法
biochemical variation 生化变异
biochemical warfare 生物化学战
biochemigenic rock 生物化学岩
biochemistry 生物化学
biochemistry-coagulation sedimentation method 生化—混凝沉淀法
biochemistry laboratory 生化试验室
biochemistry of ammonification 氨化生物化学
biochemistry zone of organic mater 有机质生物化学作用带
biochore 生境区
biochron 生物时
biochronology 生物年代学
biochronometer 生物钟
biochronostratigraphic(al) unit 生物年代地层单位
biocidal treatment 灭菌处理
biocide 生物毒剂;杀生物剂
bioclast 生物碎屑
bioclast-bearing micritic limestone 含生物屑微晶灰岩
bioclastic 生物碎屑的
bioclastic limestone 生物碎屑灰岩
bioclastic-micritic limestone 生物屑微晶灰岩
bioclastic rock 生物碎屑岩
bioclastics 生物碎屑岩
bioclastic texture 生物砍屑结构
bioclean 无菌的
bio-cleaning agent 生物净化剂
bio-clean room 生物洁净室
bioclimate 生物气候
bioclimatic change 生物气候变化
bioclimatic chart 生物气候图
bioclimatic frontier 生物气候界线
bioclimatograph 生物气候图解;生物气候图
bioclimatology 生物气候学
biocoated grain 生物包粒
biocoen 生物群落
biocoenology 生物群落学
biocoenosis 生物群落
biocoenotic connection 生物群落关联
biocolloid 生物胶体

biocommunity 生物群落
biocommunity method 生物群落法
biocompatibility 生物相容性;生物适应性
bio-concentrating 生物浓缩
bio-concentration 生物浓缩(作用)
bio-concentration factor 生物浓缩系数
bio-contact method 生物接触法
bio-contactor 生物接触器
bio-contact oxidation 生物接触氧化
bio-contact oxidation filter 生物接触氧化滤池
bio-contact oxidation pond 生物接触氧化塘
bio-contact oxidation process 生物接触氧化法
bio-contact reactor 生物接触反应器
bio-content 生物能含量
bio-control 生物电控制;生物控制
bioconversion 生物转化
biocriteria 生物基准
biocrystallography 生物结晶学
bi-octal system 二八进制记数法
biocurrent 生物电流
biocybemetics 生物控制论
biocycle 生物循环;生物环
bio-data 生物数据
biodatum 生物基准面
biodecolo(u)rization 生物脱色
biodegradability 可生物降解性;微生物分解性(能);生物可降解性;生物降解性;生物降解能力
biodegradability of chemicals 可生物降解化学物;生物降解化学物的能力
biodegradable 可生物降解的
biodegradable biosurfactant 可生物降解生物表面活性剂
biodegradable detergent 可生物降解洗涤剂
biodegradable dissolved organic carbon 可生物降解溶解有机碳
biodegradable dissolved organic content 可生物降解溶解有机碳含量
biodegradable fraction 可生物降解百分数
biodegradable glass 生物降解玻璃
biodegradable material 生物降解材料
biodegradable organic matter 可生物降解有机物质
biodegradable organic substance 生物降解有机物
biodegradable organism 可生物降解有机物
biodegradable plastic 可生物降解塑料
biodegradable polymer 可生物降解高分子
biodegradable substance 可生物降解物质;易生物降解物质
biodegradable surfactant 可生物降解表面活性剂
biodegradable tracer 可生物降解示踪剂
biodegradable volatile solids 可生物降解挥发性固体
biodegradable wastewater 可生物降解废水;生物可降解废水
biodegradation 生物降解(作用)
biodegradation kinetics 生物降解动力学
biodegradation module 生物降解模量
biodegradation of municipal refuse 城市垃圾的生物降解
biodegradation of organic chemicals 有机化学物生物降解
biodegradation of pollutant 污染物生物降解;污染物的生物降解
biodegradation of refuse 垃圾生物降解
biodegradation of solid wastes 固体废物的微生物降解
biodegradation pathway 生物降解途径
biodegradation potential 生物降解势
biodegradation process 生物降解法
biodegradation reaction 生物降解反应
biodegradation technique 生物降解技术
biodegradation wastewater 生物降解废水
biodeterioration 生物致劣;生物退化;生物腐蚀;生物变质
biodeterioratiorion 微生物老化
biodetoxification 生物解毒
biodetritus 生物碎屑
bio-diesel 生物柴油
biodisk filter 生物转盘滤池
bio-diversity 生物多样性
biodiversity criteria 生物多样性基准
biodynamic agriculture 生物动力学农业
bioecologic(al) potential 生态位能
bioecological system 生态系统
bioecology 生物生态学;生物群落学;生态学

bioeconomics 生物经济学
bio-effect 生物效应
bioeffluent 生物分泌物
bioelectric(al) control 生物电控制
bioelectric(al) current 生物电流
bioelectric(al) model 生物电模型
bioelectric(al) potential 生物电位;生物电势
bioelectricity 生物电
bioelectrochemical conversion 生物电化转换
bioelectrochemistry 生物电化学
bioelectrode 生物电极
bioelectrogenesis 生物电发生
bio-electroreaction tower 电生物反应塔
bioeleminable dissolved organic carbon 可生物消除溶解有机碳
bio-energizer 生物促进剂
bioenergy 生物能
bio-engineering 生物工程(学)
bioenhanced technology 生物强化技术
bioenhancement 生物强化
bioenhancement primary treatment method 一级生生物强化处理方法
bioenvironmental chemistry 生物环境化学
bioerosion 生物侵蚀(作用)
biofacial zones 生物相带
biofacies 生物相
biofacies-assemblage-zone 生物相组合带
biofacies map 生物相图
biofacies-paleotopographic(al) map 生物相古地理图
biofeedback 生物反馈
biofeedback control 生物反馈控制
bio-ferric process 生物铁工艺
bio-fertilizer 生物肥料
biofilm 生物膜
biofilm air-lift suspension reactor 气提式生物膜悬挂反应器
biofilm anaerobic fluidized bed 生物膜厌氧流化床
biofilm annular reactor 环状生物膜反应器
biofilm carrier 生物膜载体
biofilm colonization 生物挂膜
biofilm-controlled nitrifying trickling filter 受控生物膜硝化滤池
biofilm culturing 生物膜培养
biofilm density 生物膜密度
biofilm detachment mechanism 生物膜脱落机理
biofilm-electrode reactor 生物膜—电极反应器
biofilm formation 生物膜生成
biofilm growth 生物膜生长
biofilm loss rate 生物膜损耗率
biofilm-membrane bioreactor 生物膜—膜生物反应器
biofilm-membrane reactor 生物膜—膜反应器
biofilm monitor 生物膜监测器
biofilm organism 生物膜生物
biofilm reactor 生物膜反应器
biofilm sequencing batch reactor 序批式生物膜反应器
biofilm stability 生物膜稳定性
biofilm technique 生物膜技术
biofilm thickness 生物膜厚度
biofilm yield coefficient 生物膜产率系数
biofilter 细菌过滤器;生物滤器;生物滤池;生物过滤池
biofilter coupled process 生物滤池偶联工艺
biofilter denitrification system 生物滤池脱氮系统;生物滤池反硝化系统
biofilter loading 生物滤池负荷
biofiltration 生物过滤(法)
biofiltration method 生物过滤法
biofiltration process 生物过滤法
bioflocculant 生物絮凝剂
bioflocculant producing bacteria 生物絮凝剂产生菌
bioflocculation process 生物絮凝过程
biofloculation 生物絮凝(作用);生物混凝
biofluid mechanics 生物流体力学
biofog 生物雾
biofouling 生物污着;生物污染;生物附着
bioframework texture 生物骨架结构
biofunctionality 生物功能性
biogalvanic battery 生物电流电池
biogas 生物气体;沼气
biogas fermentation process 沼气发酵法

biogecoenology 生态地理群落学
biogenesis 生源说;生物起源说
biogenetic 生物成因的
biogenetic fractionation 生物成因分化作用
biogenetic gas zone 生物成因气带
biogenetic law 生物发生律
biogenetic limestone 生物石灰岩
biogenetic reworking 生物再造
biogenetic rock 生物岩
biogenetic texture 生物结构
biogenic 生物成因的
biogenic accumulation 生物聚集
biogenic anomaly 生物异常;生物成因异
biogenic carbonic acid 生物碳酸
biogenic chert 生物成因燧石
biogenic chlorine 源于生物的氯;生物产生的氯
biogenic coast 生物建造海岸
biogenic deposit 生物沉积(物)
biogenic dispersion 生物成因分散
biogenic gas 源于生物的气体;生物成因气
biogenic imprint 生物痕迹
biogenic migration 生物迁移
biogenic origin sand 生物成因泥沙
biogenic rock 生物岩
biogenic sediment 生物沉积(物)
biogenic structure 生物成因构造
biogenic substance concentration 生物剂浓度
biogenic sulfuric acid attack 生物硫酸侵蚀
biogenous rock 生物岩
biogenous sediment 生物沉淀
biogeochemical 生物地球化学的
biogeochemical anomaly 生物地球化学异常
biogeochemical barrier 生物地球学障
biogeochemical cycle 生物地球化学循环
biogeochemical cycle of pollutants 污染物的生物地球化学循环
biogeochemical disease 生物地球化学性疾病
biogeochemical ecology 生物地球化学生态学
biogeochemical enrichment 生物地球化学富集
biogeochemical modeling 生物地球化学模拟
biogeochemical nitrogen cycle 生物地球化氮循环
biogeochemical parameters 生物地球化学参数
biogeochemical process 生物地球化学过程
biogeochemical processes 生物地球化学作用
biogeochemical prospecting 生物地球化学探矿;生物地球化学探测
biogeochemical province 生物地球化学省
biogeochemical survey 生物地球化学测量
biogeochemistry 生物地质化学;生物地球化学
biogeochemistry cycle of pollutant 污染物的生物地球化学循环
biogeoclimate zone 生物地理气候区
biogeocoenosis 生物地理群落
biogeocoenosium 生物地理群落
biogeographic(al) province 生物地理区
biogeographic realm 生物地理分布
biogeography 生物地志;生物地理(学)
biogeology 生物地质状况;生物地质学;生物地质特征
biogeosphere 生物地理圈
bioglass 生物玻璃
biogliph 生物印痕;生物遗迹
bioglyph 生物印痕;生物遗迹
Biogon lens 比奥贡航摄镜头
biohazard 生物公害
bioherm 生物岩礁;生物丘;生物礁;珊瑚礁
bioherm limestone 生物礁灰岩
bioherm trap 生物礁圈闭
biohydrogen production 生物制氢
biohydrology 生物水文学
bioindex 生物指数
bioindication method 生物指示法
bioindication of polluted water 污染水生物指示法
bio-indicator 生物指示品种;生物指示剂;指示生物
bioinert ceramics 生物惰性陶瓷
bio-information 超感信息
biointensifying 生物强化
biokinetics 生物动力学
biokinetics constant 生物动力学常数
bioleaching 生物淋滤;生物沥滤;生物沥浸
bioleaching condition 生物淋滤情况
bioligcal monitoring of water pollution 水污染生物监测
biolimestone 生物灰岩

biolite 生物岩
biolith 生物岩
biolithite 生物灰岩
biologic(al) absorption coefficient 生物吸附系数
biologic(al) accumulation 生物蓄积；生物累积
biologic(al) accumulation efficiency 生物积累效率
biologic(al) accumulation factor 生物积累系数
biologic(al) accumulation ratio 生物积累比
biologic(al) action 生物作用
biologic(al) action spectrum 生物作用光谱
biologic(al) activated carbon 生物活性炭
biologic(al) activated carbon fiber 生物活性炭纤维
biologic(al) activated carbon filter 生物活性炭滤池
biologic(al) activated carbon filtration 生物活性炭过滤
biologic(al) activated carbon process 生物活性炭法
biologic(al) activated carbon sludge 生物活性炭污泥
biologic(al) activated filter 活性生物滤池
biologic(al) activated filtration 活性生物过滤
biologic(al) active filter 活性生物滤池
biologic(al) active rapid filter 快速活性生物滤池
biologic(al) activity 生物活性；生物活动
biologic(al) adsorption 生物吸附
biologic(al) aerated filter 曝气生物滤池
biologic(al) aerated filter effluent 曝气生物滤池出水
biologic(al) aerated filter filling 曝气生物滤池填料
biologic(al) aerated filter post-treatment process 曝气生物滤池后处理工艺
biologic(al) aerated flooded filter 溢流曝气生物滤池
biologic(al) aeration 生物曝气(作用)
biologic(al) aerobic treatment of wastewater 废水有氧生物处理；废水好氧生物处理；污水好氧生物处理
biologic(al) aerosol 生物性气溶胶
biologic(al) ag(e)ing 生物老化
biologic(al) agent 生物制剂；生物杀伤剂
biologic(al) amplification factor 生物放大因数
biologic(al) anaerobic treatment of sewage 污水厌氧生物处理
biologic(al) anaerobic treatment of wastewater 废水厌氧生物处理
biologic(al) analysis 生物检验；生物检测；生物分析(法)
biologic(al) analysis of sewage 污水生物分析
biologic(al) analysis of water quality 水质生物检验
biologic(al) and climate effects research 生物与气候效应研究
biologic(al) assay way 生物测定法
biologic(al) assay 生物学鉴定法；生物鉴定；生物检验；生物检定
biologic(al) assessment 生物评价
biologic(al) assessment of atmospheric pollution 大气污染生物学评价
biologic(al) assessment of pollutant 污染物的生物评价
biologic(al) assessment of river water pollution 河水污染生物评价；河流水污染生物学评价
biologic(al) assessment of water pollutant toxicity 水污染物毒性生物评价
biologic(al) association 生物组合
biologic(al) attack 生物侵蚀
biologic(al) availability 生物有效率
biologic(al) bacteria blocking 生物细菌堵塞
biologic(al) balance 生物学平衡；生物平衡
biologic(al) bench mark 生物基准(点)
biologic(al) carbon 生物炭
biologic(al) carbon tank 生物炭池
biologic(al) catalyst 生物催化剂
biologic(al) cell tissue 生物细胞组织
biologic(al) ceramsite filter 生物陶粒滤池
biologic(al) chain 生物链
biologic(al) characteristic 生物特征
biologic(al) characteristics of wastewater 废水生物特性
biologic(al) chemical treatment 生物化学处理
biologic(al) chemical weathering 生物化学风化
biologic(al) chemistry 生物化学
biologic(al) circulating fluidized bed 生物循环流化床

biologic(al) clarification 生物净化(法)
biologic(al) clarification plant 生物净化(处理)厂
biologic(al) classification 生物分类学
biologic(al) cleaning 生物净化(法)
biologic(al) clock 生物钟
biologic(al) coagulation 生物混凝
biologic(al) coast 生物海岸
biologic(al) community method 生物群落法
biologic(al) composite 生物医用复合材料
biologic(al) concentration 生物浓缩(作用)；生物浓集；生物富集
biologic(al) concentration factor 生物富集系数
biologic(al) constituents of wastewater 废水的生物组分
biologic(al) contact aeration equipment 接触曝气处理装置
biologic(al) contactor 生物接触器
biologic(al) contact oxidation 生物接触氧化
biologic(al) contact oxidation-flocculation sedimentation-filtration-disinfection process 生物接触氧化—絮凝沉淀—过滤—消毒工艺
biologic(al) contact oxidation pond 生物接触氧化塘
biologic(al) contact oxidation pond of elastic medium 弹性介质生物接触氧化塘
biologic(al) contact oxidation process 废水需氧生物处理；生物接触氧化法
biologic(al) contact oxidation reactor 生物接触反应器
biologic(al) contact oxidation unit 生物接触氧化装置
biologic(al) contamination 生物污染
biologic(al) control 生物控制；生物防治
biologic(al) control measure 生物防治措施
biologic(al) control of pests 虫害的生物控制
biologic(al) conversion 生物转化
biologic(al) corrosion 生物腐蚀
biologic(al) criteria 生物基准
biologic(al) cycle 生物循环
biologic(al) cycle of toxic substance 有毒物质生物循环
biologic(al) damage 生物危害；生物损伤
biologic(al) decay 生物衰变；生物腐败
biologic(al) decay constant 生物衰变常数
biologic(al) decomposition 生物分解(作用)
biologic(al) degradation 生物降解(作用)；生物分解(作用)
biologic(al) degradation technique 生物降解技术
biologic(al) denitrification 生物脱氮(作用)；生物的反硝化作用
biologic(al) denitrification system 生物脱氮系统；生物反硝化系统
biologic(al) denudation 生物剥蚀
biologic(al) depollution 生物法去污；生物除污；生物(学)去污
biologic(al) deposition 生物沉积作用
biologic(al) derived material 生物衍生材料
biologic(al) detection 生物检测
biologic(al) detection method 生物检测法
biologic(al) dike 生物堤
biologic(al) disc 生物转盘
biologic(al) discrimination 生物鉴别
biologic(al) disk treatment system 生物转盘处理系统
biologic(al) disturbance 生物失调
biologic(al) diversity 生物样性状态；生物多样性；生物多类状态；生物差异
biologic(al) diversity and protected area 生物多样性和保护区
biologic(al) divisions 生物区
biologic(al) dose 生物剂量
biologic(al) dosimeter 生物剂量指示剂
biologic(al) dosimetry 生物剂量学
biologic(al) drum 生物转筒
biologic(al) early warming 生物早起警告
biologic(al) effect 生物效应
biologic(al) effectiveness 生物有效性；生物有效率
biologic(al) effect of marine pollution 海洋污染生物效应
biologic(al) element 生物元素
biologic(al) emulsifier 生物乳化剂
biologic(al) endemic disease 生物性地方病
biologic(al) end point 生物学终点
biologic(al) energy source 生物能源

biologic(al) energy system 生物能系统
biologic(al) enhancement 生物强化
biologic(al) enrichment 生物富集
biologic(al) enrichment function 生物富集作用
biologic(al) environment 生物环境
biologic(al) environment pollutant 生物环境污染物
biologic(al) equilibrium 生物平衡
biologic(al) erosion 生物侵蚀
biologic(al) evaluation of river water pollution 河水污染生物评价
biologic(al) evolution 生物进化
biologic(al) examination 生物检验；生物检测
biologic(al) excess phosphorus removal 生物去除剩余磷
biologic(al) exposure 生物暴露量
biologic(al) exposure index 生物接触指数
biologic(al) exposure indicator 生物性暴露指示装置；生物暴露指标
biologic(al) factor 生物因子；生物因素
biologic(al) film 生物膜；生物薄膜
biologic(al) film-membrane bioreactor 生物膜—膜生物反应器
biologic(al) film treatment 生物膜处理
biologic(al) filter 生物滤器；生物滤池；生物过滤
biologic(al) filter appurtenances 生物滤池辅助措施
biologic(al) filter control system 生物滤池控制系统
biologic(al) filter flooding 生物滤池漫灌
biologic(al) filter floor 生物滤池底板
biologic(al) filter instrument 生物滤池仪表
biologic(al) filter load(ing) 生物滤池负荷
biologic(al) filter media 生物滤池填料；生物滤池滤料
biologic(al) filter media-plastic media 生物滤池滤料—塑料滤料
biologic(al) filter media plugging 生物滤池填料阻塞；生物滤池滤料阻塞
biologic(al) filter media-rock media 生物滤池滤料—碎石滤料
biologic(al) filter media-slag media 生物滤池滤料—炉渣滤料
biologic(al) filter ponding 生物滤池积水
biologic(al) filter process 生物滤池法
biologic(al) filtration 生物过滤
biologic(al) filtration bed 生物滤床
biologic(al) flo(a)tation 生物浮选
biologic(al) floc 生物絮凝体；生物絮凝物
biologic(al) flocculation 生物絮凝(作用)
biologic(al) flocculation and precipitation 生物絮凝和沉降(作用)；生物絮凝沉淀
biologic(al) fluidized bed 生物流化床
biologic(al) fluidized bed process 生物流化床法
biologic(al) fluidized bed reactor 生物流化床反应器
biologic(al) fluidized bed reactor of aerobic-anoxic integrated high-efficiency separation 好氧—缺氧一体化高效分离生物流化床反应器
biologic(al) fluidized bed technique 生物流化床技术
biologic(al) form 生物形态；生物小种
biologic(al) fouling 生物黏泥
biologic(al) fuel 生物燃料
biologic(al) fuel cell 生物燃料电池
biologic(al) gas 生物气；沼气
biologic(al) gradient 生物梯度
biologic(al) group 生物群；生物类群
biologic(al) growth 生物生长
biologic(al) habitat 生物生境
biologic(al) half-life 生物半衰期
biologic(al) hazard 生物危害
biologic(al) heterogeneity 生物多样性
biologic(al) hunbandry 有机肥耕作
biologic(al) husbandry 生物耕作
biologic(al) hybrid system 生物混合系统
biologic(al) improvement of soil 土壤的生物改良
biologic(al) index 生物指数
biologic(al) index assessment 生物指数评价
biologic(al) index of pollution 污染生物指数；生物污染指数
biologic(al) index of water-pollution 水的生物污染指数；水污染生物指数
biologic(al) indication 生物学标志

biologic(al) indication method 生物指示法
biologic(al) indication of water pollution 水污染生物指示法
biologic(al) indicator 指示生物;生物指示剂;生物指标;生物学指标;指示物
biologic(al) indicator method 生物指示法
biologic(al) indicator of pollution 污染生物指示器
biologic(al) indicator of water pollution 水污染生物指示器
biologic(al) indicator species 生物指示品种
biologic(al) induced phosphorus removal 生物诱导除磷
biologic(al) injury 生物性损伤
biologic(al) insect control 利用天敌防治昆虫;昆虫生物防治(法);生物治虫
biologic(al) integrity 生物完整性
biologic(al) integrity index 生物完整性指数
biologic(al) ion exchange process 生物离子交换法
biologic(al) iron and manganese removal filter 生物除铁除锰池
biologic(al) iron strengthening activated sludge process 生物铁强化活性污泥法
biologic(al) isolation 生物隔离
biologic(al) isotope effect 生物学同位素效应
biologic(al) land-farming 生物农耕
biologic(al) loading rate 生物负荷率
biologically accumulated element 生物积累元素
biologically activated filter 生物活性滤池
biologically active agent 生物活性剂
biologically active carbon 生物活性炭
biologically active floc 生物活性絮体
biologically active substance 生物活性物质
biologically active ultraviolet radiation 具有生物活性的紫外辐射
biologically available organic carbon 生物有效有机碳
biologically available phosphorus 生物可利用磷
biologically decomposable organic component 生物分解性的有机成分
biologically decomposable organic compound 可生物分解有机化合物
biologically diverse ecosystem 生物多样性生态系统
biologically effect ultraviolet radiation 具有生物效应的紫外辐射
biologically enhanced granular activated carbon 生物强化颗粒活性炭
biologically hard detergent 难生物降解洗涤剂
biologically treated sewage effluent 生物处理过的污水出水
biologic(al) magnification 生物浓集;生物密度扩大;生物富集;生物放大因数;生物放大(作用)
biologic(al) manganese removal 生物除锰
biologic(al) manganese removal filter 生物除锰滤池
biologic(al) material 生物医用材料;生物试样
biologic(al) materials 生物材料
biologic(al) mathematics 生物数学
biologic(al) measure 生物方法;生物测定
biologic(al) mechanism 生物机理
biologic(al) medium 生物培养基
biologic(al) membrane 生物膜
biologic(al) membrane method 生物膜法
biologic(al) membrane process 生物膜法
biologic(al) membrane technique 生物膜技术
biologic(al) metabolism 生物代谢(作用)
biologic(al) methane potential 生物甲基烷势
biologic(al) method 生物学方法;生物法
biologic(al) method for oxygen removal 生物除氧法
biologic(al) methylation 生物甲基化(作用)
biologic(al) metric 生物量度
biologic(al) microelectrolysis 生物微电解法
biologic(al) mineralization 生物矿化作用
biologic(al) minimum size 生物学的最小型
biologic(al) minimum temperature 生物学最低温度
biologic(al) mixing depth 生物混合深度
biologic(al) monitor 生物检验体
biologic(al) monitoring 生物监测
biologic(al) monitoring of atmospheric pollution 大气污染生物学监测
biologic(al) monitoring of pollution 污染生物监测;污染的生物监测

biologic(al) monitoring working party 生物监测工作组
biologic(al) nitrification denitrification 生物硝化脱氮作用
biologic(al) nitrogen fixation 生物固氮
biologic(al) noise 生物噪声
biologic(al) nutrient removal 生物去除营养物
biologic(al) nutrient removal plant 生物去除营养物装置
biologic(al) oceanography 生物海洋学
biologic(al) organic degradation 生物有机降解(作用)
biologic(al) organism 生物有机体
biologic(al) origin 生物起源
biologic(al) oxidation 生物氧化工艺;生物氧化(作用)
biologic(al) oxidation component 生物氧化分量
biologic(al) oxidation pond process 生物氧化塘工艺;生物氧化塘法
biologic(al) oxidation pre-treatment 生物氧化预处理
biologic(al) oxidation process 生物氧化
biologic(al) oxidation treatment 生物氧化处理
biologic(al) oxygen demand 生化需氧量;生物需氧量
biologic(al) oxygen demand concentration 生物需氧量浓度
biologic(al) oxygen demand dissolved 溶解生物需氧量
biologic(al) oxygen demand load 生物需氧量负荷
biologic(al) oxygen demand reaction kinetics 生化需氧量反应动力学
biologic(al) oxygen demand-releasing coefficient from bottom sludge 底泥释放生化需氧量系数
biologic(al) oxygen demand removal rate 生物需氧量去除率
biologic(al) oxygen demand test 生物需氧量测定
biologic(al) oxygen demand total 总生物需氧量
biologic(al) packing tower 生物填料塔
biologic(al) parameter 生物参数
biologic(al) parameters of water quality 水质生物参数
biologic(al) parameter uncertainty 生物参数不确定性
biologic(al) pathogen 生物病因
biologic(al) perturbation 生物扰动(作用)
biologic(al) pest control 生物病虫害防治
biologic(al) pesticide 生物农药
biologic(al) pharmaceutical wastewater 生物制药废水
biologic(al) phosphorus 生物磷
biologic(al) phosphorus removal 生物除磷
biologic(al) phosphorus removal process 生物除磷工艺
biologic(al) phosphorus removal system 生物除磷系统
biologic(al) physical weathering 生物物理风化
biologic(al) pollutant 生物污染物
biologic(al) pollution 生物性污染;生物污染
biologic(al) pollution assessment 生物污染评价
biologic(al) pollution of atmosphere 大气生物污染
biologic(al) pollution of soil 土壤生物污染
biologic(al) pollution of water body 水体生物污染
biologic(al) polymer 生物医用聚合物;生物聚合物
biologic(al) pond 生物塘
biologic(al) pond process 生物池氧化法
biologic(al) pool 生物塘;生物池塘
biologic(al) precipitation 生物沉淀(法)
biologic(al) preparation 生物制剂
biologic(al) pretreatment 生物前处理;生物预处理
biologic(al) process 生物过程;生物法;生物处理(法)
biologic(al) process of water body self-purification 水体自净生物工艺
biologic(al) productivity 生物生产力
biologic(al) protection 生物防治;生物防护;生物保护
biologic(al) purification 生物净化(作用);生物净化(法);生物沉淀法
biologic(al) purification in freshwater ecosystem 淡水生态系统生物净化
biologic(al) purification in ocean ecosystem 海洋生态系统生物净化

biologic(al) purification of atmospheric pollution 大气污染生物净化
biologic(al) purification of river 河水生物净化
biologic(al) purification of sewage 污水生物净化
biologic(al) purification of wastewater 废水的生物净化;废水生物净化
biologic(al) purification of water body 水体生物净化
biologic(al) purification process 生物净化工艺
biologic(al) purification tank 生物净化池
biologic(al) quality standard 生物质量标准
biologic(al) radioactive wastes 生物放射性废物
biologic(al) reaction 生物反应
biologic(al) reactor 生物反应器;生物反应池
biologic(al) reactor shield 生物对外来物质的保护
biologic(al) record center 生物记录中心
biologic(al) reduction 生物还原
biologic(al) regeneration 生物再生
biologic(al) remediation 生物修复
biologic(al) remediation of water body 水体生物修复
biologic(al) removal of iron and manganese 生物除铁除锰
biologic(al) repair 生物修复
biologic(al) reserve 生物保护区
biologic(al) residual toxin 生物残毒
biologic(al) residual toxin measurement 生物残毒测定
biologic(al) resistance 生物抗性
biologic(al) resource protection law 生物资源保护法
biologic(al) resources 生物资源
biologic(al) response 生物感应
biologic(al) response spectrum 生物反应谱
biologic(al) rotating disc 生物转盘
biologic(al) sample 生物样品;生物样本;生物试样;生物标本
biologic(al) sample analysis 生物样品分析;生物样本分析
biologic(al) sand filter 生物砂滤池(又称生物沙滤池)
biologic(al) sand filtration system 生物砂滤系统
biologic(al) scale 生物学标度;生物尺度;生物标度
biologic(al) screening test 生物筛选试验
biologic(al) selector 生物选择器
biologic(al) self-purification 生物自净(化)作用
biologic(al) sensitivity index 生物敏感性指数
biologic(al) settlement 生物沉降
biologic(al) sewage 生物污水
biologic(al) sewage disposal plant 污水生物处理厂
biologic(al) sewage disposal works 污水生物处理厂
biologic(al) sewage purification 废水的生物净化
biologic(al) sewage treatment 污水生物学处理;污水生物处理;生物污水净化处理
biologic(al) sewage treatment plant 污水生物处理厂
biologic(al) sewage treatment system 污水生物处理系统
biologic(al) shield 生物屏蔽
biologic(al) shielding 生物防治;生物防护;生物保护
biologic(al) shielding block 生物辐射防护装置
biologic(al) shielding concrete 生物辐射防护混凝土
biologic(al) shielding wall 生物辐射防护墙
biologic(al) simulation 生物模拟
biologic(al) slime 生物黏泥;生物黏膜;生物膜
biologic(al) slime content 生物黏泥量
biologic(al) slime process 生物膜法
biologic(al) sludge 生物渣滓;生物污泥
biologic(al) sludge solid 生物污泥固体
biologic(al) solid 生物固体
biologic(al) solid retention time 生物固体停留时间
biologic(al) sources of pollution 生物污染源
biologic(al) species 生物种;生物品种
biologic(al) specimen monitoring 生物材料监测
biologic(al) spectrum 生物型谱;生物景谱
biologic(al) spoilage of material 材料的生物性霉病
biologic(al) stability 生物稳定性
biologic(al) stabilization 生物稳定(作用);生物

稳定化
biologic(al) stabilization pond 生物稳定塘
biologic(al) stain 生物染色剂
biologic(al) standard 生物标准
biologic(al) standard substance 生物标准物质
biologic(al) state 生物状况
biologic(al) statistics 生物统计学
biologic(al) stream classification 生物流分类法
biologic(al) succession 生物演替
biologic(al) survey ship 生物调查船
biologic(al) synthesis 生物合成(作用)
biologic(al) system 生物系统
biologic(al) technique 生物技术
biologic(al) tertiary sewage treatment 污水生物三级处理
biologic(al) tertiary treatment 三级生物处理
biologic(al) test(ing) 生物试验;生物检测
biologic(al) threshold limit value 生物阈限值
biologic(al) time 生物时
biologic(al) time scale 生物地质年代表
biologic(al) tower 生物塔
biologic(al) tower filter 塔式生物滤池
biologic(al) toxicity 生物毒理学
biologic(al) toxic material 生物性有毒物质
biologic(al) transformation 生物转化(作用);生物变化
biologic(al) transport 生物转运;生物迁移
biologic(al) transport of organism 生物搬运
biologic(al) treatability 生物可处理性
biologic(al) treatment 生物处理(法)
biologic(al) treatment-catalytic iron inner cycle process 生物处理—催化铁内循环工艺
biologic(al) treatment kinetics of wastewater 污水生物处理动力学
biologic(al) treatment mechanism 生物处理机理
biologic(al) treatment method 生物处理法
biologic(al) treatment of wastewater by facultative microbe 废水兼性生物处理
biologic(al) treatment process 生物处理过程;生物处理工艺
biologic(al) treatment system 生物处理系统
biologic(al) trickling filter 生物滴滤池
biologic(al) trickling filtration 生物滴滤
biologic(al) unit process 生物单元处理法
biologic(al) universe 生物界
biologic(al) uptake 生物摄取
biologic(al) value 生物分类上的等级
biologic(al) waste treatment 生物废物处理
biologic(al) wastewater treatment 废水生物处理;污水生物处理
biologic(al) wastewater treatment system 污水生物处理系统
biologic(al) water 生物用水
biologic(al) water analysis 生物的水质分析
biologic(al) water quality indices 生物用水水质指标
biologic(al) water treatment process 生物用水处理工艺
biologic(al) weapons 生物武器
biologic(al) weathering 生物风化
biologic(al) weed control 生物治草;生物除草
biologic(al) zero point 生物致死温度;生物学致死温度
biologic photomicroscope 光学生物显微镜
biology 生物学
biology collection room 生物陈列室
biology department 生物系
biology filter bed 生物滤床
biology iron technology 生物铁技术
biology laboratory 生物试验室
biology of lake 湖泊生物学
biology of marine mammals 海洋哺乳动物生物学
biology of polluted water 污水生物学
biology of seaweeds 海藻生物学
biology of water and wastewater 水与废水的生物学
bioluminescence 生物发光(现象)
bioluminescent 生物发光的
bioluminescent bacterium 生物发光菌
bioluminescent reporter bacterium 生物发光指示菌
biolysis 生物分解(作用)
biolysis of sewage 污水生物分解
biolytic 破坏生物的

biolytic tank 生物处理池;生物分解池
biomacromolecule 生物大分子
biomagnetic effect 生物磁效应
biomagnetism 生物磁学;生物磁性
biomanipulation 生物操纵
biomarker 生物标志物
biomass 生物质;生物量
biomass accumulation efficient 生物质累积系数
biomass adsorbent 生物质吸附剂
biomass aggregation bioreactor 生物质群聚生物反应器
biomass biodegradation 生物质生物降解
biomass concentration 生物质密度
biomass density 生物质密度
biomass energy 生物质能
biomass filtration 生物质过滤
biomass gasification 生物质气化
biomass incineration power 生物质焚烧发电
biomass index transformation 生物量指标变换
biomass of lake algae 湖泊藻类生物量
biomass of lake aquatic organism 湖泊水生生物量
biomass of lake bacteria 湖泊细菌生物量
biomass of marine algae 海洋藻类生物量
biomass of marine plankton 海洋浮游生物的生物量
biomass physicochemical treatment 生物质物理化学处理
biomass pool 沼气池
biomass power generation 生物质能发电
biomass support particle 生物载体
biomass volatile solids concentration 挥发性生物固体浓度
biomass wastewater 生物质废水
biomaterial 生物材料
biomathematics 生物数学
biome 大生态区;生物群系;生物群落
biomechanics 工程心理学;生物力学
biomedical photogrammetry 生物医学摄影测量
biomembrane 生物膜
biomembrane-membrane bioreactor 生物膜—膜生物反应器
biomembrane process 生物膜法
biomere 生物层段
biometeorology 生物气象学
biometer sensor 生物仪传感器
biometrical method 生物统计学方法
biometrics 寿命测定;生物统计学;生物测量学;生物测定学
biometry 生物测定学
biome-type 生物群系型
biomicrite 生物微晶灰岩
biomicroscope 生物显微镜
biomicrosparite 生物微亮晶灰岩
biomicrudite 生物微晶砾屑灰岩
biomimic membrane 仿生膜
biomimic membrane bioreactor 仿生膜生物反应器
biomolecular reaction 生物分子反应
biomolecule 生物分子
biomonitor 生物监测器
biomonitoring 生物监测
biomotor 人工呼吸器
biomutation 生物变异
bionavigation 生物导航
bionic computer 仿生学计算机
bionics 仿生学;生体机械学
bionomic control 生态防治
bionomics 个体生态学;生态学
bionomy 生物群落学;生态学;生命规律学;生理学
biont 个体植物;生物个性
biont starting plane 生物初始面
biont terminal plane 生物终结面
biont transform plane 生物转换面
bioorganic chemistry 生物有机化学
bio-osmosis 生物渗透
bio-oxidation channel 生物氧化沟渠
bio-oxidation pond 生物氧化塘
bio-oxidation treatment 生物氧化处理
bio-oxidisable 可生物降解的;可生物氧化的
bio-oxygen stabilization 生物氧稳定化
biopak 生物舱
bioparticle deposition 生物微颗粒沉淀
biopelite 黑色页岩
biopelmicrite 生物球粒微晶灰岩
biopelsparite 生物球粒亮晶灰岩

biophile element 亲生物元素
biophotometer 光度适应计
biophysical chemistry 生物物理化学
biophysics 生物物理学
biophyte 寄生植物
biopisolite 生物豆状岩
bioplasm 活质;原生质
bioplex 生物复合体
biopoiesis 生物创建
biopolymer 生物聚合物;生物高分子
biopolymer flocculant 生物高分子絮凝剂
biopolymer hydration 生物聚合物水合作用;生物高分子水合作用
biopotential 生物潜能;生物电势
bioprecipitation 生物沉淀(法)
bio-precipitation process 生物沉降过程;生物沉降法
bio-pretreatment 生物预处理
bioprocess 生物过程;生物法
bioproductivity 生物生产力
biopsy 活组织检查
bioptix 电流式色温(度)计
Bioptix pyrometer 比色光学高温计
bioreactor 生物反应器
bio-regeneration 生物再生
bioremediation 生物修复
bioremediation technology 生物修复技术
biorhythm 生物节律
biorization 低温加压消毒
biorthogonal 双正交
biorthogonality relation 双正交关系
biorthogonal sets of functions 双正交函数集
biosafety 生物研究安全性;生物安全
biosatellite 载有生物的人造地球卫星
biosensor 生物传感器
bioseston 生物悬浮物
bio-simulation 生物模拟
bioslime 生物淤泥;生物污泥;生物软泥;生物黏泥
bioslime evaluation 生物污泥评价;生物黏泥评估
biosociology 生物社会学
biosolid 生物固体
biosorbent 生物吸附剂
biosorption 生物吸着(作用);生物吸附
biosorption aeration 生物吸附曝气
biosorption mechanism 生物吸附机理
biosorption of heavy metals 重金属生物吸附剂
biosorption process 吸附再生法;生物吸着法;生物吸附法
biosparging 生物注气;生物喷洒
biosparite 生物微亮晶灰岩;生物亮晶(石)灰岩
biosparitic calcirudite 生物亮晶砾屑灰岩
biosparrudite 生物亮晶砾屑灰岩
biospeleology 洞栖生物学
biosphere 生物圈;生物界;生物层;生命层
biosphere balance 生物圈平衡
biosphere centralism 生物圈中心主义
biosphere contamination 生物圈污染
biosphere evolution 生物圈演化
biosphere of planet earth 地球生物圈
biosphere protection area 生物圈保护区
biosphere reserve 生物圈保护区
biostability 生物稳定性
bio-stabilization 生物稳定化
bio-stabilization pond 生物稳定塘
biostabilizer 生物稳定剂
biostage 生物生长阶段
biostatic inhibition 生物静力抑制
biostatics 生物静力学
biostation 生物处理站
biostatistical area 生物统计区
biostatistical investigation 生物统计调查
biostatistics 生物统计学
biostereometrics 生物立体量测学
biostimulant 生物激活剂
biostimulation 生物刺激作用
biostorme trap 生物层闭
biostrata 生物层
biostratic unit 生物地层单位
biostratigraphic(al) classification 生物单位地层划分
biostratigraphy 生物地层学
biostratigrphic(al) correlation 生物地层对比
biostratigrphic(al) unit 生物地层单位
biostratigrphic(al) zone 生物地层区;生物地层带

biostratum 生物地层；[复]biostrata
biostromal limestone 壳灰岩；层状生物石灰岩；生物层灰岩
biostrome 层状生物礁；生物层
biostromic limestone 生物层灰岩
biostromic print 生物层面印痕
bio-supplement 生物添加剂
biosurfactant 生物表面活性剂
biosynthesis 生物合成(作用)
biosystem 生物系统
biosystematics 生物系统学；生物分类学
biosystematy 综合分类学
biot 毕奥特(圆偏振二向色性物质转动强度单位)
biota 动植物区(系)；地方生物志；生物群；生物区(系)
biota ecologic(al) feature 生物群生态特征
biotal distance 生物差距
Biotar lens 毕奥塔照相镜头；双高斯型镜头
biotar lens 双高斯透镜
biota transfer 生物群转移；生物群迁移
biotechnological issues 生物技术问题
biotechnological safety 生物技术安全
biotechnology 生物技术(学)；生物工艺学
biotelemetry 生物遥测(术)
biothion 双硫磷
biotic 生物的
biotic balance 生物平衡
biotic community 生物群落；生物界
biotic component 生物成分
biotic condition 生物条件
biotic condition index 生物条件指数
biotic control 生物控制
biotic ecotype 生物生态型
biotic energy 生物质能
biotic environment 生物环境
biotic environment factor 有机环境因素
biotic excreta 生物排泄物
biotic factor 生物因子；生物因素
biotic formation 生物群系
biotic index 生物污染指数
biotic index of organic pollution 有机污染生物指数
biotic index of pollution 污染生物指数
biotic index of water pollution 水污染生物指数
biotic influence 生物影响
biotic integrity 生物完整性
biotic integrity index 生物完整性指数
biotic interaction 生物交互作用
biotic level 生物层次
biotic metabolism 生物代谢(作用)
biotic nutrient 生物营养物
biotic oxygen demand 生物耗氧量
biotic pesticide 生物杀虫剂；生物农药
biotic pollution 生物污染
biotic pollution map 生物污染图
biotic province 生物区系省；生物地理区
biotic pyramid 生物金字塔
biotic resources 生物资源
biotic succession 生物演替；生物顺序
biotin 生物素
biotine 钙长石
biotinsulfone 生物素砜
biotite 黑云母
biotite andesite 黑云母安山岩
biotite beforsite 黑云母镁云碳酸岩
biotite dacite 黑云母英安岩
biotite diopside hornblende leptynite 黑云透辉角闪变粒岩
biotite diorite 黑云母闪长岩
biotite dolerite 黑云母粒玄岩
biotite-epidote gneiss 黑云绿帘片麻岩
biotite gabbro 黑云母辉长岩
biotite-garnet schist 黑云石榴片岩
biotite gneiss 黑云母片麻岩
biotite gneiss formation 黑云片麻岩建造
biotite gneiss-green schist formation 黑云片麻岩—绿片岩建造
biotite gneissic rock 黑云母片麻岩
biotite granite 黑云母花岗岩
biotite-granite gneiss 黑云花岗片麻岩
biotite granoblastite 黑云变粒岩
biotite granulite 黑云母麻粒岩
biotite hornbendite 黑云母角闪石闪长岩
biotite hornblende schist 黑云母角闪片岩

biotite hypersthene granulite 黑云紫苏辉石麻粒岩
biotite hypersthentite 黑云紫苏辉岩
biotite leptynite-leptite formation 黑云变粒岩—浅粒岩建
biotite leuco granoblastite 黑云浅粒岩
biotite monzogneiss 黑云二长片麻岩
biotite orthorhombic and monoclinic pyroxene granulite 黑云二辉麻粒岩
biotite pantellerite 黑云母碱流岩
biotite phlogopite alvikite 黑云母金云母方解石碳酸岩
biotite phyllite 黑云母千枚岩
biotite plagioclase gneiss 黑云斜长片麻岩
biotite quartzite 黑云母石英岩
biotite-quartz schist 黑云母石英片岩
biotite rauhaugite 黑云母白云(石)碳酸岩
biotite rhyolite 黑云母流纹岩
biotite schist 黑云母片岩
biotite sovite 黑云母黑云碳酸岩
biotite trachyte 黑云母粗面岩
biotitite 黑云母岩
biotitization 黑云母化
Biot law 毕奥定律
Biot modulus 毕奥模数
Biot number 毕奥系数
biotolerance 生物耐力
biotope 群落生境；生物小区；生物群落生境；生境；生活小区
biotopes and ecosystem 群落生境和生态系统
biotower 塔式生物滤池
biotoxication 生物中毒；生物致毒作用
biotoxicity 生物毒理学
biotoxin 生物毒素
biotransformation 生物转化
biotransformation rate 生物转化率
biotransformer 生物转化器
biotransmission 生物传播
biotransport 生物转运
biotreated domestic wastewater 生物处理过的生活污水
biotreated sludge 生物处理过的污泥
biotreatment 生物处理(法)
biotron 高跨导导孪生管
Biot-Savart law 毕奥—萨伐定律
Biot's consolidation theory 毕奥固结理论
bioturbate structure 生物扰动构造
bioturbate texture 生物扰动结构
bioturbation 生物扰动(作用)
bioturbite 生物扰动岩
biotype 纯系群；生物型
bioventing 生物通气；生物通风
biovolatilization 生物挥发
biowater 塔式生物滤池
bioxindol 异靛蓝
bioxirane 二氧化丁二烯；双环氧乙烷
biozeolite filter 沸石生物滤池
biozeolite reactor 沸石生物反应器
biozonation 生物分带性
biozone 生物带
bipack 彩色胶片；双重胶片
bipack film 二重胶片
bi-part door 双扇门
biparting 对开的
biparting door 双扇门；双向滑动门；对开门；同时开启的双拉门；双开拉门
biparting panel 对开的双闸门板
bi-parting sliding door 双极推拉门
bipartite 一式两份；双向的
bipartite church 由两部分构成的教堂
bipartite cubic 双枝三次曲线
bipartite graph 二部图(形)
bipartite matroid 二部拟阵
bipartition 对开；对分；分两部分
bipartition angle 对分角
bipass 二通管；双行(车道)；双通；双行道；双行车路
bipatch 双节距
biped 两足的
bipedal 两足的
biped robot 两足机器人
bipentene 双戊烯
biperforate 二穿孔的
biperiodical regime 双周期性状态
biphase 两相；二相的；双相
biphase current 二相电流

biphase equilibrium 两相平衡；双相平衡
biphase motor 两相电动机
biphase rectification 双相整流
biphase rectifier 全波整流器；双相整流器
biphasic 两相的；二双相性的
biphasic aerosol 二相气雾剂
biphasic biometanation 两相生物甲烷化作用
biphasic reaction 二相反应
biphasic toxicity 二相毒性
biphenyl 联二苯；二苯基
biphenyl acrylate 丙烯酸联苯酯
biphenylene terephthalate 对苯二甲酸联苯酯
biphenyl mercury 二苯汞
biphenyl mixture 联苯基混合物
biphenyl triphenol 联苯三酚
biphenylyl 联苯基
biphosphmmite 磷铵石
biphthalate 苯二甲酸氢酯；苯二甲酸氢盐
bipimelate 庚二酸盐
bipitch 双头；双螺旋的；双螺线；双节距
biplanar 二面；双平面
biplanar coupler 双平面耦合器
biplanar image tube 双平面显像管
biplane 双平面；双翼机；双翼
biplane butterfly valve 双翼式蝴蝶阀门
biplate 双板；双片
bipod 两脚架；二脚架；双脚架；双腿式起重机；安装用人字架
bipod crane 双腿式起重机
bipod mast 人字桅；双脚桅
bipolar 两极的；双极性的
bipolar amplifier 双极放大器
bipolar bit-slice microcomputer 双极位片微机
bipolar cell 双极元件；双极电解槽
bipolar circuit 双极电路
bipolar coding 双极性编码
bipolar coordinates 双极坐标
bipolar coordinate system 双极坐标系
bipolar current rose 双极水流玫瑰图
bipolar device 双极器件；双极性器件
bipolar distribution 双极性分布
bipolar dynamo 双极发电机
bipolar electric(al) motor 两极电机；双极电动机
bipolar electrode 双极性电极；双极电极
bipolar epiplankton 两极上层浮游生物
bipolar fabrication 双极器件制造
bipolar field 双极场
bipolar gate 双极性栅
bipolar generator 双极发电机
bipolar group 双极群
bipolar high voltage direct current system 双极高压直流输电系统
bipolar integration circuit 场效应集成电路；双极集成电路
bipolarity 双极性
bipolar machine 两极电机；双向电机；双磁极电机
bipolar magnetic driving unit 双极电磁驱动单元
bipolar mask bus 双极掩膜总线；双极屏蔽总线
bipolar memory 双极存储器
bipolar microcontroller 双极微控制器；双极微型控制器
bipolar modulation 双极性调制
bipolar power supply 双极电源
bipolar pressure 两端加压
bipolar processor 双极化处理器
bipolar pulse 双极性脉冲
bipolar rete 二极网
bipolar signal 双极性信号
bipolar socket 双向套筒
bipolar storage 双极存储器
bipolar sunspot 双极黑子
bipolar transistor 双载流子晶体管；双晶体管
bipolar transmission 双向传输
bipolar type 复极式
bipolar version 两极倒转术
bipolar violation 双极性破坏点
bipolar winding 双极绕组
bipolymer 二元共聚物；二聚物
bipositive 二正原子价的
bipost lamp 双接点聚光灯
bipotential 双电位
bipotential electrostatic lens 双电势静电透镜
bipotential lens 双电位透镜
bipp 铋泼糊

bippel 每个像素比特数
bipp paste 铋碘仿石蜡糊
biprism 双棱柱(体);双棱镜;双括棱镜;复柱
biprism interference 双棱镜干涉
bipropellant 双组元推进剂;双基火药
bipropenyl 联丙烯
bi-purposed bridge 公铁两用桥
bipyramid 双锥;双棱锥(体);双角锥(体)
bipyramidal 双锥体
biquadratic 四次幂;四次方程(式);四次方;四乘幂;双二次
biquadratic curve 四次曲线
biquadratic floating point gain control 四进制浮点增益控制
biquartz 双石英片
biquaternion 八元数
biquinary 二五混合进制的
biquinary code 二五混合码;二五混合进制码
biquinary-coded decimal number 二五混合码十进制数
biquinary notation 二五混合进制记数法;二五混合进位制计数法
biquinary number 二五混合进制数
biquinary number representation 二五混合进制数表示法
biquinary number system 二五混合进位制计数法
biquinary representation 二五混合进制表示法
biquinary scaler 二五混合制计数器;二五混合进制计数器
biquinary system 二五混合进制
biradial 双径向;双辐射
biradial symmetry 两侧辐射对称
biradical 双游离基;双基的
biramous 二支的
Biram's wind meter 毕拉姆式风速计
birch 桦木;桦;白桦
birch bark tar 桦木树皮焦油;桦皮焦油;白桦木树皮焦油
birch black 桦木黑
birch broom 桦木扫帚
birch camphor 桦木脑
birchen 桦树的;桦木的
birch faggot 桦木柴捆
birch tar oil 桦木焦油;桦焦油
birch veneer 桦木镶板;桦木饰面板;桦木夹板
birchwood 白桦木;桦木
birch wood tar 桦木焦油;白桦木焦油
bird 吊舱
bird and beast design 鸟兽纹
bird and beast pattern 鸟兽纹
bird applicator 刮板制膜器
bird banding 鸟类环志
birdbath 路面凹坑(雨后积水处);鸟池
birdcage 瓶内黏丝(玻璃制品缺陷);鸟笼(式);圆拱结构;玻璃瓶内搭丝
bird cage lightning 鸟笼形避雷器
bird caging 局部扭曲;钢丝打结
Bird centrifuge 伯德(型沉降式)离心脱水机
Bird coal filter 伯德离心脱水机
bird dog 猎取经济情报者
bird dogging 逐鸟把戏
bird droppings 鸟粪
bird ecology 鸟类生态学
bird foot delta 鸟爪状三角洲;多汊三角洲
bird-head bond 啄形接头
bird house 屋顶气窗;屋顶通气孔
bird impact 飞鸟撞击
bird impact resistance 抗鸟撞性
bird island 鸟岛
bird manure 鸟粪
bird migration 候鸟(定期)迁徙
birdnesting 团聚
bird of migration 候鸟
bird of passage 候鸟
bird pattern 鸟纹
bird peck 鸟啄斑纹;鸟啄纹
bird-receiving system 吊舱接收系统
bird sanctuary 鸟类保护区
bird's beak 鸟嘴式线脚
bird's beak mo(u)lding 鸟嘴式线脚;鸟嘴式线饰
bird's beak ornament 鸟嘴式装饰;鸟嘴式线饰
bird's beak ornamental 外窗台装饰
bird screen 防鸟网罩;防鸟筛孔栏板
bird's eye 俯视的;鸟瞰的;鸟眼(纹)

bird's eye coal 有鸟眼纹理的煤;眼球状煤
bird's eye figure 鸟眼纹理(木材)
bird's eye grain 鸟眼纹理
bird's eye gravel 细砾石
bird's eye in wood 鸟眼木纹
bird's eye lamp 白炽灯泡
bird's eye limestone 鸟眼灰岩
bird's eye maple 鸟眼槭树
bird's eye pattern 鸟眼纹理;鸟眼花纹
bird's eye perspective 概况;大纲;鸟瞰(图);鸟瞰透视图
bird's-eye photograph 鸟瞰照片
bird's-eye porosity 鸟眼孔隙
bird's eyes 鸟眼花纹(木材缺陷)
birdseye structure 鸟眼构造
bird's-eye survey 鸟瞰测量
bird's eye view 鸟瞰(图);俯瞰图
bird's eye wood 鸟眼纹材
bird's foot delta 鸟足状三角洲
bird-shaped cup 鸟形杯
bird-shaped jar with lid 鸟形盖罐
bird's-head mo(u)lding (中世纪的)鸟头线脚装饰
bird's-mouth 在椽子端再加一个角;承接口;凹角接;鸟嘴形;承接角口;凹角接
birdsmouth corner joint 斜向口转角缝
birdsmouthing 角口承接;鸟嘴形承接
birdsmouthing jointing 角口接合;斜面接合;带棒接合
birdsmouth joint 三角形企口接合;角口承接;承接角口;啮接
birdsmouth quoin 纯角承接屋角石块
birdsmouth quoin of masonry wall 圬工墙钝角
bird's nest 玻璃瓶内黏丝;桅上瞭望塔;桅斗
bird spike 惊鸟器
bird swing 瓶内黏丝
Bird unit 伯德单位
bird view 空中斜摄影;鸟瞰图
bireactant 双组份燃料;双元推进剂
birectangular 双直角形
birectangular quadrilateral 双直角四边形
birectangular spheric(al) triangle 两直角球面三角形
birectifier 双精馏器
bireflectance 双反射率
bireflectance of vitrinite 镜质体双反射率
bireflection 双反射
bireflectivity 双反射率
birefraction 双折射
birefringence 对折射;重折率;复折射;双折射率;双折射
birefringence anisotropy 双折射各向异性
birefringence compensating plate 双折射补偿板
birefringence crystal 双折射晶体
birefringence-dispersion method 重折率色散方法
birefringence filter 双折射滤光器
birefringencemeter 双折射检查仪
birefringence of flow 流动双折射
birefringence pattern 双折射图案
birefringence thermooptic(al) coefficient 双折射热光系数
birefringenct property 双折射性能
birefringent 双折射的
birefringent chain filter 双折射连锁滤光器
birefringent coating 光弹性贴片;双折射贴片
birefringent compensator 双折射补偿器
birefringent crystal 双折射晶体
birefringent demodulator 双折射解调器
birefringent filter 双折射滤光器
birefringent interference microscope 双折射干涉显微镜
birefringent interferometer 双折射干涉仪
birefringent medium 双折射媒质;双折射介质
birefringent monochrometer 双折射单色计
birefringent plate 双折射片;双折射板
birefringent polarizer 双折射偏振器
birefringent prism 双折射棱镜
birefringent retarder 双折射延迟器
birefringent waveguide 双折射波导
birefringent wedge 双折射光楔
biregular 双正则
Birge-Mecke rule 伯奇—梅克法则
biringuccite 比硼钠石
Birkeland-Eyde process 伯克兰—艾迪电弧法

Birkenhead creel 伯肯希德式粗纱架
Birkhill shales 贝克尔页岩
Birkhoff regularization 伯克霍夫正规化
Birkhoff's theorem 伯克霍夫定理
birkremite 紫苏花岗岩
Birmabright 伯马布赖特可锻铝镁合金;伯马布赖特合金
Birmal alloy 伯马尔压铸铝合金
Birmalite 伯马利特铸造铝基合金
Birmalite alloy 伯马利特(铸造铝基)合金
Birmasil 铸造铝合金;伯马西尔航空用铸造铝合金
Birmasil special alloy 伯马西尔硅铝特种合金
Birmastic 伯马斯蒂克耐热铸造铝合金
Birmetal 伯梅塔尔铝铜锌镁合金
Birmidium 伯米迪昂铝合金
Birmingham ga(u)ge 伯明翰规
Birmingham platina 伯明翰高锌黄铜
Birmingham platinum alloy 伯明翰铂合金
Birmingham wire ga(u)ge 伯明翰线径规;伯明翰(量)线规
Birnbaum-Sauders distribution 伯恩鲍姆—索德斯分布
Birnbaum-Tingey distribution 伯恩鲍姆—廷吉分布
birnessite 水钠锰矿
biro (可以吸墨水的)圆珠笔
birotary engine 双转子发动机
birotary turbine 双转子燃气轮机
bi-rotation 变旋光;双旋光;双异旋光
birotor 双转子
bi-rotor pump 双转子泵;双转轮泵
birr 冲量;机械转动噪声
birth and death process 增消过程
birth cohort analysis 出生群组分析
birthday stone 生辰石
birth-death ratio 生死比
birth function 出生函数
Birth House 漫密西神殿(埃及)
birth injury 产伤
birthmark 母斑
birth number 出生率
birth rate 出生率
birthrate fluctuation 出生率波动
birth rate of urban population 城市人口出生率
birth registration 出生登记
birth reporting 出生报告
birth to finishing housing 终生畜舍
birth trauma 产伤
birunite 硫碳硅钙石;硅碳石膏
bis-2-ethylhexyl phthalate 邻苯二甲酸双(2—乙基乙基)酯
bis-2-methylxyethyl phthalate 邻苯二甲酸双(2—甲氧基乙基)酯
bis-3-methyylbutyl peroxydicarbonate 过氧二碳酸双—3—甲基丁酯
bisacki 颗粒紫胶
bisacromial 二肩峰的
bisalt 酸式盐
bisamide wax 双酰胺蜡
bisatin 二乙酰酚靛红
bisaya 比萨耶麻
bisazo 双偶氮
bisazo benzil 双偶氮甲苯
bisazo compound 双偶氮化合物
bisazomethine 双甲亚胺
bisazo pigment 双偶氮颜料
bisazo yellow 双偶氮黄颜料
bisbenzimidazole 二苯并咪唑
bisbenzothiazole 二苯并噻唑
Biscay abyssal plain 比斯开深海平原
bischloromethyl ether 双氯烷基醚
bischofite 水氯镁石
Bischof process 毕肖夫法
Bischof's coefficient 毕肖夫系数;毕索夫系数
biscuit 灰褐色;制坯;饼坯;本色陶器;本色陶瓷;浅褐色;饼状模制品;盘料;未烧的干法琅层;外壳铸型;陶器素坯;素烧坯;素坯(瓷)
biscuit cutter 短岩芯管;短取芯器(钢丝绳冲击钻进时用);底端削尖的岩芯钻头
biscuit duck 粗厚帆布
biscuit fire 素烧;初次焙烧
biscuit firing 低温素烧;陶器素坯焙烧;素烧
biscuit for ceramics 瓷坯
biscuit furnace 坯炉;素烧窑
biscuiting 釉面烧成;素烧

biscuit kiln 素烧窑;坯窑;坯炉;陶器素坯窑
biscuit manufacture 饼干制造
biscuit metal 小块金属
biscuit mo(u)ld 素坯模具
biscuit oven 烧釉面砖的烘炉;陶器烤炉;陶器烘箱;陶器烘炉
biscuit painting 素坯彩涂
biscuit porcelain 素瓷
biscuit printing 素坯印花
biscuit throwing 旋坯
biscuit tile 烘烧瓷砖;素陶瓦;无釉地砖;陶瓷砖板
biscuit ware 充烧陶器;素烧器皿
bis-diazo compound 双重氮化合物
bisebacate 癸二酸氢盐
bisecant 二度割线
bisect 二等分(图);剖面线条;平分
bisecting compasses 比例两脚规;比例规
bisecting line 角的平分线;角的等分线;二等分线;二等分物
bisecting line of an angle 角的平分线;分角线
bisecting point 平分点
bisecting runway 交叉跑道
bisection 中剖;对分;对半开;二等分;平分
bisection error 平分误差
bisection line 平分线
bisection method 对分法;分半(方)法;二分法
bisection technique 分半技术
bisection theorem 中剖定理【计】;二等分定理
bisector 等分线;二等分线;二等分物;平分线
bisector of an angle 角的平分线;角的等分线
bisector of the strip map sheet 条图心中线
bisectrix 等分线;二等分线;二等分物
bisellium 荣誉座(古罗马)
biseptate 二分隔的
biserial 双列
biserial coefficient 双列系数
bi-serial coefficient of correlation 双列相关系数
biserial ratio of correlation 两双数列相关比率;双列相关比率
biseriate 二列的
biserrate 二重锯齿
bi-service 两用的
bisexual 两性体;两性的
bishomohopane 双升藿烷
bishop 手工夯具;基督教主教;人力夯具;手夯;手锤
Bishop's and Morgenstern slope stability analysis method 毕索与摩根斯坦(边坡稳定分析)法
bishop's cloth 粗厚方平织物
Bishop's graphic(al) construction 毕索普图解法
Bishop's method 毕索法
Bishop's miter 手夯斜切面
bishop's palace 主教的邸宅;主教官邸
Bishop's ring 毕索光环
Bishop's simplified method of slice 毕索简化条分法
bishydrazicarbonyl 环二脲
bis-hydroxy ethyl terephthalate 对苯二甲酸乙二醇酯
bi-signal zone 等强信号区;双信号区
bisilicate 偏硅酸盐类
bismaleimide resin composite 双马来亚胺树脂复合材料
bismanal 铋锰磁性合金
Bismarck brown 俾斯麦棕色
Bismarck Sea 俾斯麦海
bismite 铋华
bismoclite 氯铋矿
bismoid 铋悬液
bismuite 泡铋矿
bismuth active substance 铋活性剂
bismuth alloy 铋合金
bismuth amalgam 铋汞合金
bismuthate 铋酸盐
bismuthate glass 铋酸盐玻璃
bismuth base alloy 铋基合金
bismuth benzoate 苯甲酸铋
bismuth blende 闪铋矿
bismuth bronze 铋青铜(合金)
bismuth carbolate 苯酚二羟铋
bismuth carbonate 次碳酸铋
bismuth chromate 铬酸铋
bismuth-containing layer type structure 含铋层状结构
bismuth glance 辉铋矿
bismuth glass 铋玻璃
bismuth granule 铋粒
bismuth hydrate 氢氧化铋
bismuth hydroxide 氢氧化铋
bismuthic acid 铋酸
bismuthic deposit of penta element 五元素铋矿床
bismuthic veined sulfide deposit 含铋脉状硫化物矿床
bismuthic veined tungsten-tin deposit 含铋脉状钨锡矿床
bismuthide 铋化物
bismuth ingot 铋锭
bismuthinite 辉铋矿
bismuthinite-native bismuth-quartz vein formation 辉铋矿自然铋一矿石英脉建造
bismuthinte ore 辉铋矿矿石
bismuth iodide 碘化铋
bismuth iodoform paraffin paste 铋泼糊;铋碘仿石蜡糊
bismuth line 铋线
bismuth lustre 铋质光泽彩
bismuth magma 铋乳
bismuth meal 铋粉
bismuth molybdate 钼酸铋
bismuth naphtholate 萘酚铋
bismuth nitrate 硝酸铋;铋白
bismuth ocher 铋赭石;铋华
bismuth oleate 油酸铋
bismuth ore 铋砂;铋矿(石)
bismuth(ous) oxide 三氧化二铋
bismuth oxalate 草酸铋
bismuth oxide 氧化铋
bismuth oxycarbonate 次碳酸铋
bismuth pentoxide 五氧化二铋
bismuth phenate 苯酚铋
bismuth phenolate 苯酚铋
bismuth phenylate 苯酚铋
bismuth phosphate 磷酸铋
bismuth poisoning 铋中毒
bismuth potassium iodide 碘化铋钾
bismuth-silver-oxygen-cesium photocathode 铋银氧铯光电阴极
bismuth soap 铋皂
bismuth sodium triglycollamate 氨三乙酸铋钠
bismuth spiral 铋螺线
bismuth stannate 锡酸铋
bismuth strontium calcium copper oxide 铋锶钙铜氧化物
bismuth subcarbonate 碱式碳酸铋;次碳酸铋
bismuth subnitrate 碱式硝酸铋;次硝酸铋;铋白
bismuth sulfite agar 亚硫酸铋琼脂
bismuth test 铋试验
bismuth titanate 钛酸铋
bismuth titanate ceramics 钛酸铋陶瓷
bismuth trioxide 铋黄;三氧化二铋
bismuth white 碱式硝酸铋;铋白
bismuth yellow 铋黄;三氧化二铋
bismuthyl phenolate 苯酚二羟铋
bismutinite 辉铋矿
bismutoferrite 羟硅铋铁矿
bismutohauchecornite 硫双铋镍矿
bismutomicrolite 铋细晶石
bismutostibiconite 铋黄锑华
bismutotantalite 钽铋矿
bis-N-allyl pyromellitic imide 均苯四酰二烯丙(基)亚胺
bisolute system 双溶质体系
bisomus 分二室的石棺
bison 空心梁防火地板
bison floor 防火空心梁楼层;毕生楼板
bispectrum 双频谱
bispelmicrite 生物球粒微亮晶灰岩
bisphenoid 双楔
bisphenol A epoxy resin 双酚A环氧树脂
bisphenol A polyester resin 双酚A型聚酯树脂
bisphenol epoxide 双酚A环氧化物
bisphenol F epoxy resin 双酚F环氧树脂
bispheric lens 双球面透镜
bisque 本色陶器;本色陶瓷;素坯;陶瓷素坯;素瓷
bisque clause 修正条款
bisque firing 高温素烧;高温素爆
bisque sculpture 素雕
bisque tile 素陶瓦;无釉地砖
bissextile(year) 闰年
bistability 双稳性;双稳定性
bistable 双稳态式;双稳定的
bistable behaviour 双稳态性能
bistable circuit 双稳线路;双稳态电路;双稳电路
bistable device 双稳态器件
bistable light emitter 双稳态光发射体
bistable memory 双稳态存储器
bistable multioscillator 双稳态多谐振荡器
bistable multivibrator 双稳态多谐振动器;双稳多谐振荡器
bistable operation 双稳态工作
bistable optical device 双稳态光学装置
bistable optical element 双稳态光学元件
bistable optoelectronic element 双稳态光电子元件
bistable performance point 双重稳定性能点
bistable relay 双稳态继电器
bistable state 双稳态
bistable trigger 双稳触发器
bistable trigger circuit 双稳触发电路
bistable unit 双稳元件;双稳态部件
bistagite 透辉岩
bistagite of lime 石灰亚硫酸盐
bistatic radar 双站雷达;双基地雷达;双分雷达;收发分置雷达
bistatic reflectivity 复收反射性;双稳态反射率
bistatic sonar 收发分置声呐
bister 深褐色(颜料)
bistoury 细长刀
bistre 褐色水彩颜料;颜料褐;深褐色(颜料)
bistrique 磨头
bistro 小餐馆;夜总会
bisuccinate 丁二酸氢盐
bisufite 酸式亚硫酸盐
bisulfate 硫酸氢盐
bisulfide 二硫化物
bisulfite cooking 亚硫酸盐蒸煮
bisulfite of lime 石灰亚硫酸氢盐;石灰的亚硫酸盐
bisulphate 硫酸氢盐;亚硫酸盐;酸式硫酸盐
bisulphide 二硫化物
biswitch 双向硅对称开关
bisymmetric(al) 双对称的
bisymmetric(al) girder 两向对称大梁;双对称大梁
bisymmetric(al) mo(u)lding 双对称式修饰
bisymmetry 两对称
bisynchronous motor 双倍同步速度电动机
bisync protocol 双同步协议
bit 螺旋锥;钻头;凿子;刀具;刀片(头);二进制数位;比特;钎头;碎冰片
bit a cable 安装绳索
bit adapter 钻头接头
bit addressing 位定址
bitaleral constraint 不可解约束
bitangent 二重切;双切(线)
bitangent plane 二重切面
bit arbor 钻套
bit attachment 钻头连接(装置)
bit bailing 泥包卡钻;钻头泥包
bit bare 空位
bit blade 钻片;钻头刃
bit blank 钻坯
bit block 位块
bit block transfer 位块传送
bit body 钻头体
bit bounce 钻头跳动
bit brace 摇钻(柄);曲柄钻;手摇钻;钻孔器
bit brace tap 摇钻螺丝攻
bit breakage 钻头损坏
bit breaker 钻头装卸器;钻头拧卸器
bit break-in procedure 新钻头试验过程
Bit buffer unit 比特缓冲器
bit burnt out 烧钻
bitbus 位总线
bit-by-bit control 按位控制
bit-by-bit memory 按位存储器
bit-by-bit optical memory 按位光存储器
bit capacity 二进制数容量
bitch 弄污;打捞母锥;反向扒钉
bit change 更换钻头
bit changer 位变换器
bitch wood 热带硬木
bit clearance 孔壁与钻头的间隙;井壁与钻头的间隙;钻头与孔壁间隙;钻头出刃
bit-code 位码
bit combination 位组合

bit comparate 位比较
bit consumption 钻头消耗
bit contour 钻头轮廓;钻头唇部外形
bit core 导向头
bit cost 钻头价格;钻头费;钻头成本
bit count 钻头上金刚石数量
bit count appendage 附加按位计数
bit crown 钻头胎体
bit crown metal 钻头胎体金属
bit cutting angle 钻头切割边;钻头切割角;钻头磨角;钻头尖度;钻冠的尖度;钎子头刃角
bit cuttings 钻屑
bit deflection 钻头偏斜
bit density 二进制位密度;位密度
bit design 钻头设计
bit diameter 钻头直径;钎钎直径
bit die 钻头模具;锻钎模
bit drag 切削型钻头
bit dress 修钎机;锻钎杆
bit dresser 磨钎机,磨钎工;钻头修理器具;钻头修理工;锻钎机;修钎机
bit dressing 钻头修整
bit dressing crane 修理钻头用吊架
bit drop-in 信息混入
bit drop-out 信息丢失
bite 漏印;卡住;啮侵蚀;咬起;咬合;抓住(指锚抛稳);滚距;吃刀量;溶剂侵蚀;切削刀;脱印部分;酸腐蚀
bit edge 钻头刃
bite frame 咬合计录用架
bitegage 咬合测距尺
bite into 钻入;铲入
bite mark 咬痕
bit end 钻头连接端
bite of line 滑轮组绳索排距
biteplate 咬合板
bitermitron 双端管
biternary 双三进
bi-ter-nary system 二三进制
bite rot 根腐
bit error probability 误比特概率
bit error rate 比特误码率;比特误差率;误码率;误比特率;位出错率
bit error rate processor 比特误差率处理机
bitesize 一口体积
bite wing 咬合翼片
bite wound 咬伤
bit extension 钻头接杆;钻头接柄
bit face angle 钻头刃尖角
bit falling down hole 钻头落井
bit feed 钻头进尺速度;钻头给进
bit feeding mechanism 钻头给进机构
bit feet 钻头脚
bit footage 钻头长度;钻头进尺
bit forging machine 锻钎机
bit for underreaming 扩孔偏心钻头
bit frame 钻头体
bit frequency 比特频率
bit full down hole 孔内掉钻头
bit fuller 钎头平锤
bit gatherer 挑料工
bit ga(u)ge 钻头量规;钻头规;钻孔量规;对刀样板
bit grinder 钻头研磨机;磨机;钻头磨床
bit grinding 堵钻处理
bit grinding machine 磨钎机;锻钎机
bit head 钻头部
bit height 钻头高度
bithionol 硫双氯酚;比廷
bit holder 钻头夹持器;钻头打捞钩;钻套;打捞钩;钎夹
bit hook 凿头钩;钻头打捞钩
bit hustler 钻头送修工(美国)
bit hydraulic horsepower 钻头水马力
bit hydraulics 钻头水力特性
bitin 比廷
bit inclination 钻头倾角
biting 咬底;咬蚀;轧辊咬入轧件
biting angle 侵彻度
biting cold 严寒
biting force 咬力
biting force meter 咬力测测计
biting-in 腐蚀
biting wind 刺骨寒风
bit insert 钻头镶嵌块

bit inside diameter 钻头内径
bit-jitter 比特跳动
bit key 弹子锁钥匙;带齿钥匙
bit keyed lock 钥匙锁
bit leg 钻头爪;钻头巴掌
bit leg falling down hole 巴掌落井
bit length 钻头长度
bit life 钻头寿命;钻头使用期限
bit life mode 钻头寿命模式
Bitlis massif 比特利斯地块
bit literal 按位表示文字
bit load 钻压;(钻进时的)钻头压力;钻头负载
bit loading 钻头荷载
bit location 比特位置;位单元;数位位置
bit loss 位损
bit manipulation 位处理;位操作;二进制处理
bit manufacture 钻头出厂家
bit map 点阵图;位映像;位图;位变换
bit mask 位屏蔽
bit matrix 钻头胎体;比特矩阵
bit memory organization 位存储组织
bit meterage 钻头进尺
bit mo(u)ld 钻头模具
bit nozzle 钻头钻管;钻头水眼;钻头喷嘴
bit number 钻头编号;比特号码
bit of axe 斧刃
bit of drill 钻头;锥刃【岩】
bit of drill head 钻头割刀
bit of hatchet 小斧刃
bit of information 信息单位
bit of output signal 输出信号的位数
bit of the drill head 钎头切削刀
bit of the vice 老虎钳口;虎钳口
bitolterol 比托特罗
bitone ink 双色版油墨
bitoric lens 双曲面透镜;双复面透镜
bit orient 按位
bit-oriented 按位的
bit oriented memory 按位存取存储器
bit outside diameter 钻头外径
bit overfeed 钻头超速
bit parallel 位并行
bit parity 位奇偶
bit pattern 位组合格式
bit pattern generator 位模式发生器
bit penetration 机械钻速
bit performance 钻头工作性能(总进尺,单位进尺成本)
bit per inch 每英寸位数
bit pilot 钻头导向器(不取芯金刚石钻头中央凸出端部)
bit pin 钻头枢轴
bit plane 位平面
bit polishing 细粒金刚石钻头抛光(钻进极硬细粒岩石时);钻头磨光
bit pressure 钻压;(钻进时的)钻头压力
bit prong 活钻头;钻头牙轮齿;钻头尖;刮刀钻头翼片
bit puller 卸钻头器;钻头拧卸器
bit pulley and chain 链滑车(换钻头用)
bit rate 二进位信息传输率;比特率;位速率
bit reamer 扩孔钻;扩孔器;锥形铰刀;锥铰刀;整孔器
bit reaming shell 扩孔器(细粒金刚石)
bit reconditioning 钎头修整
bit recovering tap (金刚石的)钻头打捞公锥
bit reduction factor 比特减缩因子
bit register 位寄存器
bit resharpening 钎头修磨
bit Reynolds number 钻头雷诺数
bit ring 钻头(直径)量规
bitrochanteric 二转子的
bitrochanteric diameter 转子间径
bitrope 二点重切面
bitropic 两向性的
bit rotation speed 钻头旋转速度
bits 齿片
bit sample 钻头冲击取得的样品(钢丝绳冲击钻进时)
bit-seating arrangement (试验台上的)放置钎头的器具
bit seizure 钎头卡住
bit setter 钎头修理工
bit setting 钻头镶嵌;锻修钎头;嵌镶钻头
bit setting block 镶钻头夹具

bit setting ring 镶钻头环
bit setting tools 镶钻头工具
bit shank 钻杆;钻头丝扣部分;钻坯;钎尾;钎柄
bit sharpener 修钻头机;截齿修整机;修钎机;修整机;钻头磨床;锻钎机
bit sign 位符号
bit size 钻头尺寸
bit slice 位片
bit slice processor 位片处理机
bit slug 钻头镶嵌块
bit speed 钻头转速
bit speed coefficient 钻速系数
bit speed exponent 钻速指数
bits per second[bps] 比特每秒;每秒位数;位/秒
bit stealing 比特占用
bit stock 摇钻;摇柄钻;手摇钻;钻柄;钻头柄
bit stock brace 摆钻
bit stone 钵底垫粉;碎瓷粉
bit stop 钻头定程停止器;钻头钻进限制器
bit stream 比特流;位流
bit stream transmission 位流传输
bit string 位串
bit string constant 位串常数
bit string data 位串数据
bit string operation 位串运算
bit string operator 位串操作符
bit stub 钻头柄
bit stuffing aglorithm 位填充算法
bit style 钻头形式
bit switch 按位开关
bit symbol 钻头型号
bitt 缆柱;系缆柱;系缆柱;系船柱;带缆柱;带缆柱;挽在桩上;条船柱
bit table 位表
bit taper (放置岩芯提断器的)钻头内锥度;钎头磨刃角
bitt bracket 缆桩肘;系缆柱支承板
bit teeth 钻头齿刃
bitt end 索端
bitter 苦味的
Bitter coil 比特尔线圈
bitter earth 氧化镁;苦土;菱苦土
bitter-end 索端;船内锚链端
bitter fleabane 飞蓬
bitter lake 苦水湖;苦湖
bitter mineral water 苦矿质水
bittern 卤水;盐卤;天然盐水
bitterness 苦味
bittern liquor 苦水
bittern pan 卤水池
bitter orange 酸橙
Bitter pattern 比特粉纹图
bitter poplar 苦杨
bitter spar 纯晶白云石;菱镁矿;白云石
bitter spring 苦泉
bitter taste 苦味
bitter water 苦碱水
bitter well 苦水井
bitter wood 苦木
bitthead 系缆柱柱头
bit time 一位时间
bit time counter 比特时间计数器
bittiness 起块(粉刷)
bit tool 刀头;刀具
bit-tooth wear 牙齿磨损
bitt pin 系缆柱杆;柱栓
bit traffic 二进制信息通道
bit train 位列
bit transfer rate 比特传送速率
bitt standard 缆桩倒肘;系缆柱支柱
bitty 油漆表面裂纹;比蒂漆
bitty cream 凝结乳脂
bit type 钻头型号;钻头类型
bitubond 沥青混合物(用以浇筑洞孔)
bitulith 沥青混凝土(路面用)
bitulithic 沥青混凝土的
bitulithic pavement 沥青混凝土路面
bitumarin 含沥青及珐琅质的漆(用于铸件)
bitumastic 沥青砂胶;沥青玛琋脂的;脂沥青的;舱底沥青漆;石棉丝沥青胶
bitumastic adhesive 沥青砂胶粘合料
bitumastic compound 沥青覆面涂料
bitumastic enamel 沥青瓷漆

bitumastic lining 沥青衬里
bitumastic macadam 沥青碎石
bitumastic macadam mixing plant 沥青碎石拌和站;沥青碎石拌和车间
bitumastic paint 沥青漆;沥青涂料;沥青玛琋脂涂料
bitumastic pipe-line coating 沥青管道涂料
bitumastic roofing 沥青砂胶屋面
bitumastic sealer 沥青密封膏
bitumastic solution 水柏油
bitumen 沥青
bitumen addition 沥青添加剂
bitumen adhesion-preventing agent 防沥青黏合作用剂
bitumen adhesive 沥青黏结剂;沥青胶粘剂
bitumen adhesive composition 沥青黏结剂成分
bitumen adhesive compound 沥青粘结剂成分
bitumen-aggregate mixture 沥青集料混合物;沥青集料混合料;沥青骨料混合物;沥青骨料混合料
bitumen-aggregate ratio 油石比【道】
bitumen and tar(melting) kettle 沥青和柏油(溶解)烧锅
bitumen asbestos composition 沥青石棉合成(物)
bitumen asbestos compound 沥青石棉合成物
bitumen asbestos fibre cement 沥青石棉纤维水泥
bitumen asbestos mass 沥青石棉合成物
bitumen asbestos mastic 沥青石棉砂胶;沥青石棉胶合铺料
bitumen asbestos material 沥青石棉合成物
bitumen asbestos sealant 沥青石棉嵌缝膏
bitumen asbestos sheeting 沥青石棉层
bitumen base 沥青底子;沥青基
bitumen-based 含有沥青成分的
bitumen-based building mastic 含有沥青成分的房屋玛琋脂;含沥青建筑胶
bitumen-based mastic joint sealer 含有沥青的玛琋脂接缝密封料
bitumen-based paste 含有沥青成分的浆剂;含沥青胶;含沥青膏
bitumen-based protection paint 含沥青防锈漆(英国)
bitumen-based protective paint 含沥青防锈漆(英国)
bitumen-based rust protective paint 含沥青防锈漆(英国)
bitumen batcher 沥青配料器;沥青配料计量器
bitumen batching pump 沥青配料泵;沥青计量泵
bitumen binder 沥青胶合料
bitumen binder course 沥青黏合料层
bitumen binder(course) carpet coat 沥青黏结层
bitumen-bond flexible pavement 沥青黏结柔性路面
bitumen bonding adhesive 沥青黏结剂;沥青加固胶粘剂
bitumen bonding composition 沥青黏结剂成分
bitumen bound base 沥青结粒料基层;沥青结合料基层
bitumen briquet(te) 沥青试块;沥青模制(标准)试块
bitumen bucket 沥青料罐
bitumen building paper 沥青防潮纸;建筑沥青毡防水卷材
bitumen cable 沥青绝缘电缆
bitumen calcium silicate brick 沥青浸渍的灰砂砖;沥青硅酸盐砖
bitumen carpet 沥青地毯;沥青毡层
bitumen cement(ing agent) 沥青黏结剂;沥青胶结剂
bitumen cementing composition 沥青黏结料成分
bitumen coat 沥青表面涂层
bitumen-coated 涂沥青的
bitumen-coated chip(ping)s 沥青涂过的石屑;涂沥青片石;拌了沥青的石屑
bitumen-coated chip(ping)s carpet 沥青碎屑涂层地毯;沥青片石毡层
bitumen-coated gravel 涂沥青砾石
bitumen-coated material 涂沥青的材料;拌过沥青的材料;涂沥青材料
bitumen-coated pipe 涂沥青水管;涂沥青管(子)
bitumen-coated road metal 沥青涂过的筑路碎石
bitumen coating 沥青涂层
bitumen coating compound 沥青涂料混合物;沥青涂层混合料

bitumen coating material 沥青涂层材料
bitumen composition 沥青成分
bitumen composition roofing 沥青混合屋面料
bitumen concrete 沥青混凝土
bitumen concrete base 沥青混凝土打底
bitumen concrete carpet 沥青混凝土地毯;沥青混凝土覆盖层
bitumen concrete pavement 沥青混凝土路面;沥青混凝土铺面
bitumen concrete skeleton 沥青混凝土骨架
bitumen concrete surfacing 沥青混凝土表面层
bitumen content 沥青百分比;沥青含量
bitumen cooker 沥青熔化锅
bitumen cord filler 沥青麻绳填料
bitumen cutback 溶于石油馏出油的沥青
bitumen damp-proof course 沥青防潮层
bitumen deck surfacing 桥上沥青路面;沥青桥车行道路面
bitumen deposit 沥青矿
bitumen-dipped 浸沥青的
bitumen-dipped hemp 麻丝沥青
bitumen-dipped steel tube 沥青浸渍钢管;涂沥青钢管
bitumen dispersion 沥青分散作用
bitumen distributor 沥青喷洒器;沥青撒布机;沥青喷洒机;沥青喷布机
bitumen emulsion 沥青乳(浊)液;乳化沥青;沥青乳胶
bitumen emulsion cane fiber board 涂乳化沥青甘蔗板
bitumen emulsion injection 乳化沥青浇灌(法)
bitumen emulsion with colloidal emulsifier 用胶质乳化剂乳化沥青
bitumen enrichment 多加沥青
bitumen extractor 沥青混合料抽提机
bitumen exudate 沥青渗滤物
bitumen facing 沥青饰面
bitumen felt 沥青油毡;油毡
bitumen felt packing 油毡垫层
bitumen felt roof(ing) 沥青油毛毡屋面;沥青油毡屋面;油毡屋面;沥青卷材屋面
bitumen filler 沥青填料;沥青注入器
bitumen-filler mix(ture) 沥青填料混合物;沥青填充混合料;沥青填充拌和料
bitumen film 沥青薄膜
bitumen for building 建筑沥青
bitumen for dampproof coating 沥青防潮涂料;防潮层沥青
bitumen for road purposes 筑路用沥青
bitumen globule 沥青小珠;沥青球
bitumen grade 沥青等级
bitumen gravel 涂沥青的石子
bitumen gravel roofing 沥青砾石屋面;沥青砾石铺屋面料
bitumen grout 沥青灰浆
bitumen grouted macadam 沥青灌入的碎石;沥青灌浆碎石
bitumen grouting 沥青灌浆
bitumen gun 沥青喷枪
bitumen gunite 沥青压力喷浆
bitumen gunite equipment 压力喷射沥青的设备
bitumen gunite process 压力喷射沥青的程序
bitumen hardness(degree) 沥青硬度(数)
bitumen heater 沥青加热器
bitumen heating kettle 沥青加热炉
bitumen highway emulsion 用于公路的沥青乳化液;沥青道路乳液
bitumen hill 沥青丘
bitumeniferous 含沥青的
bitumen-impregnated 浸透沥青的;浸沥青的;用沥青浸渍的;沥青浸渍的
bitumen-impregnated calcium silicate brick 沥青浸渍的硅酸盐砖;沥青浸渍灰砂砖
bitumen-impregnated felt 沥青浸渍的油毛毡
bitumen-impregnated fiberboard 浸沥青纤维板;沥青浸渍纤维板
bitumen-impregnated insulating board 沥青浸渍绝热板
bitumen-impregnated lime-sand brick 沥青浸渍灰砂砖
bitumen-impregnated paper 沥青浸渍纸
bitumen-impregnated rock 沥青浸染岩石
bitumen-impregnated sand 沥青浸渍的砂
bitumen-impregnated sand lime brick 沥青浸渍的灰砂砖
bitumen-impregnated seal(ing) rope 沥青浸渍封塞绳子;沥青浸渍封口绳
bitumen-impregnated strip 沥青浸渍的板条
bitumen impregnating 沥青浸渍的
bitumen impregnation 沥青浸透;沥青浸渍
bitumen index 纯沥青指数
bitumen individual coat 沥青单独涂层
bitumen injection control valve 控制灌沥青的阀;沥青喷射控制阀
bitumen insulating coat 沥青绝缘层
bitumen insulating paste 沥青绝缘糊;沥青绝缘胶;沥青绝缘膏
bitumen insulating slab 沥青绝缘板
bitumenite 块煤
bitumenizing 涂沥青
bitumen joint 沥青接缝
bitumen jointing 沥青填缝;沥青封接
bitumen joint pouring compound 沥青接缝浇注混合料
bitumen joint runner 沥青接缝浇道;沥青接缝麻刀
bitumen Judaism 犹太沥青
bitumen kettle 熔化沥青的锅
bitumen-laminated 沥青叠层
bitumen-latex emulsion 沥青橡胶浆乳化液;沥青橡胶乳液
bitumen layer 沥青层;沥青铺洒机
bitumen limestone 沥青石灰岩;沥青灰岩
bitumen-lined 涂沥青的;沥青衬里的
bitumen liner 沥青衬里件;沥青衬垫
bitumen lining 沥青防水层;沥青衬里
bitumen macadam 沥青碎石(路);铺路用沥青碎石
bitumen macadam pavement 沥青碎石路面
bitumen macadam road 沥青碎石路
bitumen macadam surfacing 沥青碎石路面
bitumen macadam with gravel aggregate 沥青碎石混合砾石骨料;沥青碎石混合砾石集料
bitumen mastic 沥青砂胶
bitumen measuring pump 计量沥青的泵;沥青计量泵
bitumen melting boiler 熔化沥青的锅炉;沥青加热炉
bitumen melting kettle 熔化沥青的锅
bitumen melting tank 熔化沥青箱
bitumen membrane 沥青薄膜;沥青防水膜
bitumen metering pump 沥青计量泵
bitumen method 沥青法
bitumen mortar 沥青砂浆
bitumen of Judea 地沥青
bitumen paint 沥青漆
bitumen paper 沥青纸
bitumen pavement 沥青路面
bitumen pavement mix(ture) 沥青路面混合料
bitumen paver 沥青混合料摊铺机
bitumen penetration macadam 沥青浸透的碎石;沥青贯入式碎石路面
bitumen perforated sheet(ing) 沥青多孔挡板
bitumen plastics 沥青塑料
bitumen poisoning 沥青中毒
bitumen-polymer binder 沥青聚合物结合料
bitumen pouring compound 沥青灌缝混合料
bitumen pouring rope 沥青麻刀
bitumen preimpregnating 沥青预先浸渍;沥青预先浸透;预先浸渍
bitumen preimpregnation 预先浸渍
bitumen prepared roofing 预制沥青屋面覆盖料;沥青组成屋面料
bitumen pre-saturating 沥青预先饱和
bitumen prime coat 头道沥青;沥青透层;沥青结合层
bitumen primer 路面头道沥青
bitumen product 沥青产品
bitumen proportioning pump 沥青比量泵;沥青计量泵
bitumen protection coat 沥青保护层
bitumen protective coat 沥青保护层
bitumen protective coating 沥青保护涂层
bitumen protective felt 沥青保护毛毡;沥青保护油毡
bitumen pump 沥青泵
bitumen putty 沥青油灰;沥青腻子
bitumen rag-felt 预制沥青屋面油毛毡;沥青粗制油毡(屋面用)
bitumen rate 含油率

bitumen ratio 沥青比
bitumen ready roofing 预制沥青屋面材料
bitumen resin 沥青树脂
bitumen rich carpet 沥青过多的毡层
bitumen road emulsion 沥青路面乳剂;沥青道路乳液
bitumen road pavement 沥青路铺面
bitumen road surfacing 沥青路面层
bitumen rock 沥青岩
bitumen roll(ed-strip) roofing 沥青卷条屋面材料
bitumen roof cladding 沥青屋面覆盖;沥青屋面涂层
bitumen roof covering 沥青屋面覆盖;沥青油毡屋面覆盖层
bitumen roof(ing) 沥青铺屋面
bitumen roof(ing) cement 沥青屋面水泥;沥青屋面冷底子油膏
bitumen roof(ing) felt 沥青屋面油毡
bitumen roof(ing) sheet(ing) 沥青屋面护墙板
bitumen roof(ing) shingle 沥青屋面板
bitumen roof(ing) sheathing 沥青屋面覆盖层;沥青屋面望板
bitumen-rubber composition 沥青橡胶成分;沥青橡胶合成物
bitumen-rubber compound 沥青橡胶混合物;沥青橡胶合成物
bitumen-rubber mass 沥青橡胶合成物
bitumen-rubber material 沥青橡胶合成物
bitumen-rubber strip 沥青橡胶板条;沥青橡胶条
bitumen sand-lime brick 沥青浸渍的灰砂砖
bitumen-saturated cotton fabric 沥青浸渍棉织物
bitumen-saturated felt(ed fabric) 沥青饱和的毛毡织物
bitumen-saturated felt(ed fabric) mat 沥青饱和的毛毡坐垫
bitumen-saturated felted fabric pad 沥青饱和的毛毡织垫
bitumen-saturated(felted) mat 沥青浸渍毡衬垫
bitumen-saturated(felted) mat with cork 沥青浸油毡软木衬垫
bitumen-saturated(felted) pad with cork 沥青浸油毡软木衬垫
bitumen-saturated glass fabric 沥青浸渍玻璃布
bitumen-saturated paper 沥青饱和的纸
bitumen saturation 沥青饱和
bitumen screeding compound 沥青刮平层化混合物
bitumen seal(ing) 沥青封口(膏);沥青封缝
bitumen-seal(ing) compound 沥青止水料;沥青塞缝料;沥青填缝料
bitumen seal(ing) paste 沥青封口浆
bitumen seal(ing) sheet(ing) 沥青封口护板;沥青封口板
bitumen-sheathed 沥青绝缘的;沥青护面的
bitumen-sheathed paper cable 沥青绝缘电缆
bitumen sheet(ing) 沥青护墙板;沥青毯;沥青片材(材);油毛毡
bitumen sheet roofing 沥青毡屋面
bitumen shingle 沥青屋面板;沥青墙面板
bitumen ship 沥青运载机;沥青运输船
bitumen show 沥青显示
bitumen slip coating 沥青滑动层
bitumen slurry 沥青黏合板
bitumen slurry seal 沥青灰浆封层;沥青泥浆封层;沥青浆塞缝;沥青浆封缝
bitumen solidification 沥青固化
bitumen solution 沥青溶液
bitumen spray bar 沥青喷油棒;沥青喷油管
bitumen sprayer 沥青散播器;沥青撒布机;沥青喷洒机;沥青喷布机
bitumen spraying machine 沥青撒布机;沥青喷洒机;沥青喷布机
bitumen spreader 沥青洒布机
bitumen stabilization 沥青稳定土;沥青稳定(作用)
bitumen strip 沥青条
bitumen subseal(ing) 沥青封底;沥青基封底;沥青基层处理
bitumen surface coat(ing) 沥青路面涂层;沥青表面涂层
bitumen surfaced road 沥青路面
bitumen surfacing 沥青饰面
bitumen surfacing practice 沥青路面作法
bitumen survey 沥青测量方法

bitumen tank 沥青箱
bitumen tank car heater 沥青箱加热车
bitumen tape 沥青胶带;沥青绳
bitumen-tar binder 沥青焦油混合料
bitumen-tar blend 混合沥青;石油沥青—焦油沥青化合物
bitumen-tar mixture 石油沥青—焦油沥青化合物
bitumen test 沥青试验
bitumen top dressing 沥青浇面
bitumen trailer 沥青挂车
bitumen treated 浸透沥青的;沥青处治过的;沥青处理过的
bitumen-type wood fiberboard 沥青式木纤维板
bitumen varnish 沥青清漆
bitumen waterproofer coat 沥青防水层
bitumen waterproofer course 沥青防水层
bitumen wearing course 沥青磨耗层
bitumen weighing batcher 沥青按重量配料器
bitumen wood fibre board 沥青木纤维板
bitumen wool felt 预制沥青屋面料;沥青羊毛毡
bituminated filler 沥青水填料
bituminated hydrofiller 沥青水填料
bituminic 沥青质
bituminiferous 油沥青质的;含沥青的;沥青质的
bituminisation 沥青化
bituminised cement 沥青黏合剂;沥青膏
bituminite 烟煤;沥青质体;沥青煤
bituminite pneumoconiosis 油页岩尘肺
bituminization 沥青化;沥青固化;沥青处治;煤化作用;用沥青处理
bituminize 沥青化;沥青处理;使含沥青
bituminized 加沥青的
bituminized aggregate 涂沥青骨料;涂沥青的骨料;涂沥青集料
bituminized cement 沥青胶泥;加沥青水泥;沥青膏
bituminized chip(ping)s 沥青层碎屑;涂沥青片石
bituminized concrete 沥青混凝土;加沥青混凝土
bituminized cord 涂沥青的绳子;涂沥青绳
bituminized discrete aggregate 涂沥青松散骨料
bituminized fabric 沥青织物
bituminized fiber pipe 沥青纤维管;沥青处理纤维管
bituminized floor covering material 加沥青的楼面材料;涂沥青楼面覆盖材料
bituminized jute hessian cloth 浸沥青的粗麻布;预制沥青粗麻布面
bituminized mineral aggregate 涂沥青矿物集料;涂沥青矿物骨料;涂沥青的矿物骨料
bituminized mortar 加沥青砂浆
bituminized paper 涂沥青的纸;沥青纸;含沥青油纸
bituminized rope 涂沥青的绳子
bituminized stone 含沥青石
bituminizing 使与沥青混合;沥青处理
bituminosis 沥青末肺;沥青末沉着病;煤末沉着病
bituminous 沥青质的;沥青的;含油沥青的
bituminous adhesive composition 沥青的黏合成分;沥青胶结化合物
bituminous adhesive compound 沥青胶结化合物
bituminous agent 沥青物质
bituminous application 敷沥青
bituminous asbestos felt 沥青石棉油毡
bituminous asbestos mastic 沥青石棉砂胶;沥青石棉胶结料;沥青石棉玛琼脂
bituminous asbestos rope 沥青石棉绳
bituminous asphalt slab mattress 沥青块沉排
bituminous base(course) 沥青基层;沥青基底层;沥青底层;黑色基层;含沥青粗砂垫层
bituminous batcher 沥青配料器
bituminous batching pump 沥青配料泵;沥青配量泵
bituminous binder 沥青胶合剂;沥青结合料;沥青胶结料
bituminous binder course 沥青联结层;沥青结合层
bituminous binding material 沥青黏合料;沥青黏合剂
bituminous blend 混合沥青
bituminous board 沥青板
bituminous body coat 沥青主层
bituminous bond 沥青结合
bituminous bound 沥青结合
bituminous bound material 沥青结合料
bituminous broken-stone pavement 沥青碎石路面
bituminous brushable compound 沥青的可涂刷混合物;沥青涂刷混合物

bituminous building mastic 沥青房屋玛琼脂;沥青房屋砂胶
bituminous building material 沥青建筑材料
bituminous built-up roof(ing) 沥青组合屋面
bituminous carpet 沥青毡
bituminous carpet coat 沥青毡层
bituminous carpeting 沥青铺面;沥青碎石片面层
bituminous cement 膏体沥青;沥青胶泥;沥青胶结体;沥青胶结料;沥青胶结剂;沥青膏;加沥青水泥
bituminous cementing composition 沥青膏成分
bituminous clay 沥青黏土;沥青粉土
bituminous clay-lime pavement 油灰土路面
bituminous coal 沥青煤;烟煤
bituminous coat 沥青涂料面;沥青护面;沥青层
bituminous-coated chips 拌过沥青的石屑;黑色石屑;沥青石屑
bituminous-coated stone 沥青碎石
bituminous coating 沥青涂料面层;沥青涂层
bituminous coating composition 沥青涂料成分;沥青涂层混合料
bituminous coating compound 沥青涂层混合物
bituminous coating mass 沥青涂层混合料
bituminous coating material 沥青涂层材料
bituminous coating of aggregate 集料沥青包层
bituminous composition 沥青成分;沥青混合料
bituminous concrete 沥青混凝土
bituminous concrete facing membrane 沥青混凝土面层
bituminous concrete flooring 沥青混凝土楼地面
bituminous concrete mixture 沥青混凝土料
bituminous concrete pavement 沥青混凝土铺面;沥青混凝土路面
bituminous concrete paver 沥青混凝土摊铺机
bituminous concrete sidewalk 沥青混凝土人行道
bituminous cooker 沥青熔化锅;沥青烧锅
bituminous cover(ing) 沥青罩面;沥青盖层
bituminous damp-proofing agent 沥青防潮剂
bituminous damp-proofing and waterproofing 沥青的防潮和防水
bituminous damp-proofing water proofing 沥青防潮防水剂
bituminous decking 沥青薄面铺盖;铺薄沥青表层
bituminous dike 沥青质岩脉
bituminous dispersion 沥青分散液
bituminous disposal 沥青处治
bituminous distributor 沥青撒布机;沥青喷洒机;沥青喷布机
bituminous dressing 沥青敷面
bituminous emulsion 沥青乳(浊)液;乳化沥青
bituminous enamel 沥青釉质
bituminous epoxy paint 环氧沥青漆
bituminous epoxy resin 沥青环氧树脂
bituminous expansion joint 沥青伸缩缝
bituminous facing 沥青护面;沥青敷面
bituminous facing composition 沥青饰面料成分
bituminous felt 沥青油毡;沥青油毛毡;油毡;石油沥青毡
bituminous fiber filler 沥青纤维填(缝)料
bituminous fibre pipe 沥青纤维管
bituminous filler 沥青质填(缝)料;沥青填(缝)料
bituminous filter layer 沥青过滤层
bituminous finish composition 沥青涂层混合料
bituminous finisher machine 沥青整面机
bituminous flax 麻丝沥青
bituminous flax cord 沥青麻绳
bituminous flax felt 麻布油毡
bituminous floor cover(ing) 沥青楼面覆盖层;沥青楼层覆面
bituminous fuel 沥青燃料
bituminous glass fabric felt 玻璃纤维布油毡
bituminous grout 水沥青;含沥青溶液;沥青砂浆;沥青浆
bituminous grouting 沥青灌浆
bituminous heater 沥青加热熔化锅
bituminous heating installation 沥青加热设备
bituminous heating plant 沥青加热厂
bituminous heating trailer 沥青加热拖车
bituminous highway 沥青公路
bituminous highway construction 沥青公路建造
bituminous highway emulsion 沥青路面乳化液
bituminous highway mix(ture) 公路用沥青混合料
bituminous impervious element 沥青不透水层
bituminous impregnating composition 沥青浸

渍成分
bituminous impregnating mix(ture) 沥青浸渍混合物
bituminous impregnation 沥青浸渍
bituminous injection device 灌入沥青的设备；沥青灌入设备
bituminous insulating paint 沥青系防锈绝缘涂料；沥青绝缘涂料；沥青防锈绝缘漆
bituminous insulation(-grade) board 沥青绝缘(等级)板；沥青绝缘板
bituminous joint 沥青缝
bituminous joint filler 沥青填缝料
bituminous joints for pipes 管道沥青接缝
bituminous kettle 沥青烧锅
bituminous lacquer 沥青涂漆；沥青溶媒漆
bituminous level(l)ing course 沥青整平层；沥青找平层
bituminous lignite 沥青褐煤
bituminous limestone 沥青质石灰石；沥青石灰岩
bituminous limestone pavement 沥青石灰石路面
bituminous limestone powder 沥青石灰石粉
bituminous limestone quarry 沥青石灰石开采场
bituminous macadam 沥青碎石(路)
bituminous macadam and tar macadam mixing plant 沥青碎石和柏油碎石拌和装置
bituminous macadam base course 沥青碎石底层
bituminous macadam mixture 沥青碎石混合料
bituminous macadam pavement 沥青碎石路面
bituminous macadam road 沥青碎石路
bituminous marl 沥青灰泥
bituminous mass 沥青块
bituminous mastic 沥青砂胶；沥青玛琋脂
bituminous mastic concrete 沥青砂胶混凝土
bituminous mastic concrete paving 沥青砂胶混凝土路面
bituminous mastic concrete surfacing 沥青砂胶混凝土面层
bituminous mastic joint sealer 沥青砂浆接缝封层
bituminous mastic mixer 沥青砂胶拌和机
bituminous mat 沥青毡层；沥青垫层
bituminous material 沥青材料
bituminous matter 沥青质
bituminous mattress 沥青垫层
bituminous mattress revetment 沥青垫层护岸
bituminous measuring pump 沥青比例计量泵
bituminous(melting) kettle 沥青熔化锅
bituminous membrane 沥青薄膜；沥青防水膜
bituminous metering pump 沥青比例计量泵
bituminous mineral wool felt(quilt) 沥青矿棉毡
bituminous mixed pavement 沥青混合料路面
bituminous mixing plant 沥青拌和厂；沥青拌和装置
bituminous mix-in-place 路拌沥青混合料
bituminous mixture 沥青混合料；沥青拌和物
bituminous mortar 沥青砂浆
bituminous mortar flooring 沥青砂浆楼地面
bituminous mortar plaster 沥青砂浆粉刷
bituminous mulch top 沥青盖顶
bituminous neoprene paint 氯丁橡胶沥青漆
bituminous oakum 沥青麻筋
bituminous odo(u)r 沥青臭味
bituminous oil 沥青油
bituminous overlay 沥青罩面
bituminous paint 沥青涂料；沥青漆
bituminous painted surface 沥青涂刷面
bituminous patching 用沥青料修补；沥青补路
bituminous-patching crew 沥青修补小队；修补沥青路面小队
bituminous pavement 沥青面层；沥青铺面；沥青路面
bituminous paver 沥青混合料铺路机；沥青混合料摊铺机；沥青铺洒机；沥青铺路机
bituminous paver-finisher 沥青铺路护面整修机；沥青铺路面整机；沥青铺路及整修路面机
bituminous paving machine 沥青铺路机；沥青铺路护面整修机；沥青铺路护面整面机
bituminous peat 沥青泥煤；沥青泥炭；沥青质泥炭；沥青泥炭
bituminous penetrating macadam 沥青贯入式碎石路(面)
bituminous penetration 沥青渗透；沥青浇灌；沥青贯入(式)
bituminous penetration macadam 沥青贯入式碎石路面
bituminous penetration macadam base course 沥青贯入式碎石路基层
bituminous penetration macadam surface course 沥青贯入式碎石面层
bituminous penetration pavement 沥青贯入式路面
bituminous penetration road 沥青贯入式浇灌路；贯入式沥青路面；渗透式沥青路面
bituminous pipes 铺路沥青空隙
bituminous pitch 沥青柏油
bituminous plant mix 厂拌沥青混合料
bituminous plaster 沥青膏
bituminous plastic cement 沥青质塑性胶泥
bituminous plastics 沥青塑料
bituminous preservative for structures and building 结构和房屋的沥青防腐剂
bituminous prime coat(ing) 沥青打底层；沥青底涂层
bituminous primer 低黏性的沥青筑路材料；沥青冷底子油
bituminous priming solution 沥青冷底子油
bituminous product 沥青产品
bituminous proof coating 沥青防水膜层
bituminous proportioning pump 沥青比例分配泵；沥青配料泵
bituminous protective coat(ing) 沥青保护层
bituminous protective membrane 沥青保护薄膜；沥青防水薄膜
bituminous pump 沥青泵
bituminous putty 沥青膏；沥青油灰；沥青封泥
bituminous resin 沥青树脂
bituminous resin emulsion paint 沥青树脂漆
bituminous retreat 沥青重复处治；二次沥青处理
bituminous road 沥青路；黑色路
bituminous road construction 沥青公路施工
bituminous road material 沥青路面材料；沥青筑路材料
bituminous road-mixer 沥青混合料路拌机
bituminous road mix(ture) 公路使用的沥青混合料；路拌沥青混合料
bituminous road surface treatment 沥青路面处理
bituminous road surfacing finisher 沥青路面层修整机
bituminous rock 油沥岩；沥青岩(石)；沥青砂石
bituminous roof(ing) felt 沥青屋面油毡
bituminous roof(ing) membrane 沥青屋面膜
bituminous roof(ing) sheet(ing) 沥青屋面薄板；沥青屋面板
bituminous sand 沥青砂
bituminous sand mastic 沥青砂胶
bituminous sandstone 沥青砂石；沥青质砂岩
bituminous sandstone pavement 沥青砂石路面
bituminous saturant 沥青饱和料；沥青浸渍料；沥青浸渍材料；浸渍材料
bituminous saturating mix(ture) 沥青饱和混合料
bituminous saturation 沥青饱和
bituminous seal 沥青止水；沥青填缝；沥青封缝；沥青封闭
bituminous seal coat 沥青封闭层；沥青密闭层；黑色封层
bituminous sealed shoulder 沥青处治路肩
bituminous sealing compound 沥青填缝料；沥青封缝混合料
bituminous shale 沥青质页岩；沥青页岩；油页岩
bituminous sheet(ing) 沥青薄板；沥青护墙板
bituminous shield 沥青覆盖层；沥青挡水板；沥青护面
bituminous slurry 沥青稀浆
bituminous slurry seal 含沥青泥浆封闭
bituminous softening point test 沥青软化点试验
bituminous soil 沥青土
bituminous soil aggregate 沥青土集料；沥青土骨料
bituminous soil mixture 沥青土混合料
bituminous soil stabilization 沥青土稳定(法)
bituminous solution 沥青溶液
bituminous spray(ing) bar 沥青喷洒棒；沥青喷油管
bituminous spray(ing) machine 沥青撒布机；沥青喷洒机；沥青喷布机
bituminous spreading and finishing machine 沥青铺路整修机
bituminous stabilization 沥青稳定土；沥青稳定处理；沥青稳定(法)；沥青加固
bituminous stabilized sand pavement 沥青稳定砂路面
bituminous stone 浸沥青的花岗岩；浸沥青的石砂路面
bituminous storage and supply tank 沥青储[贮]罐
bituminous storage installation 沥青储存装置；沥青储[贮]存装置
bituminous storage tank 沥青储[贮]罐
bituminous street 柏油马路
bituminous structural material 沥青建筑材料；沥青筑路材料
bituminous substance 沥青物质
bituminous surface 沥青面层；沥青路面；黑色面层
bituminous surface course 沥青面层
bituminous surface covering 沥青罩面料；沥青罩面
bituminous surface disposal 沥青表面处治
bituminous surface treated road 沥青表面处治路(面)
bituminous surface treatment 沥青浇面；沥青表面处治；沥青表面处理
bituminous surfacing 沥青路面；沥青面层
bituminous surfacing composition 沥青面层成分
bituminous tack coat 沥青胶结层
bituminous texture 沥青结构
bituminous thermometer 沥青温度计
bituminous topping on set paving 小石块路面上沥青表面层
bituminous treated base 沥青处理的基层
bituminous treatment 沥青处理
bituminous tubes 铺路沥青间空隙
bituminous underseal 沥青底封层
bituminous varnish 沥青清漆
bituminous viscometer 沥青黏(滞)度仪；沥青黏(滞)度计
bituminous water-proofing coat(ing) 沥青防水层
bituminous water-proofing course 沥青防水层
bituminous water-proofing 沥青防水层；沥青防水措施
bituminous water-proofing membrane 沥青防水薄膜
bituminous wearing coat 沥青磨耗层
bituminous wearing course 沥青磨耗层
bituminous weigh(ing) batcher 沥青按重量配料器；沥青秤
bituminous wood 沥青木；具有木质外形的褐煤
bituminous work 沥青工种；沥青工程
bitumirous lump coal 烟煤块
bitumite 烟煤
bitumogence 沥青成因【地】
bituplastic 含沥青的塑料
bituseal 冷操作的沥青混合料(用于屋顶、地板等)
bitusol 含沥青油漆；溶胶沥青；天然沥青；固体分散胶溶沥青；天然地沥青
bituthene 橡胶(地)沥青
bituthene waterproofing membrane 橡胶地沥青防水膜
bit vector 位向量
bit wall 钻头壁(丝扣与胎体钻头的部分)
bit wall thickness 钻头壁厚
bit wear 钻头磨损
bit wear and tear 钻头磨耗
bit wear and tear per meter 每米钻头磨耗
bit weight 钻压；钻头负荷
bit weight exponent 钻压指数
bit weight per unit area 单位面积钻压；比钻压
bit wing 钻头翼片；钻头刃；钻头刀；凿刃
bit wing angle 钻头翼片角度
bit wing thickness 钻头翼片厚度
bitwise operation 逐位运算；逐位操作
bit with teeth 耙式钻头
bit with wings 有叶片的钻头；切削型钻头
bit wrench 钻头钳
bityite 锂铍脆云母
bit zone 标志位
biuret 缩二脲
biuret linkage 缩二脲键合
biuret reaction 双缩脲反应
bivalence 双化合价
bivalent 二价的
bivalent catio 二价阳离子
bivalent element 二价元素
bivalent metallic ion complex 二价金属络合物
bivalent radical 二价基
bivalent switch 两用开关
bivalve 双壳(类)；双瓣；蚌类

bivalve faunal province 双壳动物地理区
bivalve mollusks 蚌类
bivane 双向风向标;双风向标
bivariant 双变量的;双变(式)
bivariant state 双变状态
bivariant system 双变体系;双变量系统
bivariate 二变量;双变量;双变
bivariate analysis 双变量分析
bivariate binomial distribution 二元二项分布
bivariate Cauchy distribution 二元柯西分布
bivariate continuous distribution 二元连续分布
bivariate density function 二元密度函数
bivariate discrete distribution 二元离散分布
bivariate distribution 二元分布;二维分布;双变量分布
bivariate frequency table 二元频率表
bivariate generating function 二元母函数;双变量母函数
bivariate linear regression 二元直线回归
bivariate logarithmic distribution 二元对数分布
bivariate logarithmic series distribution 二元对数序列分布
bivariate negative binomial distribution 二维负二项分布
bivariate normal distribution 二元正态分布;双变量正态分布
bivariate normal integral 二元正态积分
bivariate normal probability density function 二元正态概率密度函数
bivariate normal surface 二元正态曲面;双变量正态曲面
bivariate normal target distribution 二维规范目标分布
bivariate Pareto distribution 二元帕累托分布
bivariate Pascal distribution 二元帕斯卡分布
bivariate Poisson distribution 二元泊松分布
bivariate Poisson population 二元泊松总体
bivariate polynomial 双变量多项式
bivariate population bar graph 二元总体条形图
bivariate probability distribution 二元概率分布
bivariate process 二元过程
bivariate regression model 二元回归模型
bivariate sample 二变量样本
bivariate scatter plot 二变量散点图
bivariate sign test 二元符号检验
bivariate stochastic process 二元随机过程;二维随机过程
bivariate table 二维表
bivariate type 2 distribution 二元二型分布
bivariate uniform distribution 二元均匀分布
bi-vault 双拱屋顶;双穹隆;双层穹顶
bivector 二维向量
bivectorial 双矢的
bivector multinomial distribution 二维向量多项式分布
bivicon 双视像管
bivinyl 丁二烯
bivinyl rubber 丁基橡胶;丁二烯橡胶
bivoltine 二化的
bivouac 露营
bivouac site 露营地
Biwa 1 reversed polarity subzone 拜韦一反向极性亚带
Biwa 2 reversed polarity subzone 拜韦二反向极性亚带
Biwa 3 reversed polarity subzone 拜韦三反向极性亚带
biweekly 两星期一次
biweekly mortgage 双周抵押
bixbite 红绿柱石
bixbyite 方铁锰矿
BJ₅ type container BJ₅型集装箱
bjarebyite 磷铝锰钡石
Bjerrim's relation 薄耶伦关系式
Blaauw mechanism 布劳机制
B-labeled door B级门(防火标准)
black 加密的;黑色(的);黑皮
black acre 黑地
black acrylic matt enamel 黑色丙烯酸消光瓷漆
black adobe soil 黑色冲积黏土
black afara 非洲伊地泡木
black after white 白洁黑;白后黑
black air drying paint 黑色风干漆
black alkali soil 黑碱土

black alum 黑矾
black amber 煤精;黑琥珀
black and brown pigments 黑色和褐色色素
black-and-white 黑白的
black-and-white checkboard 黑白格板
black-and-white chequered buoy 黑白方格浮标
black-and-white contrast 黑白对比
black-and-white edition 黑色印样
black-and-white film 黑白胶片
black-and-white hologram 黑白全息图
black-and-white horizontal stripes 黑白横纹
black-and-white iceberg 黑白冰山
black-and-white infrared airphoto 黑白红外航片
black-and-white map 黑白图
black-and-white negative film 黑白负片
black-and-white panchromatic airphotos 黑白全色航片
black-and-white photograph 黑白照片;黑白像片
black-and-white photograph process 黑白像片冲洗
black-and-white photography 黑白摄影
black-and-white picture 黑白照片;黑白像片;黑白图像;黑白画
black-and-white positive film 黑白正片
black-and-white print 黑版印样;黑白像片
black-and-white sensitized material 黑色感光材料
black-and-white spots 黑白斑
black-and-white vertical stripes 黑白直纹
black and white work 木石构造(木构架中填石和灰泥)
black and yellow 黄地黑花
black annealed wire 黑退火钢丝
black annealing 黑色退火;初退火
black anoxic layer 黑色缺氧层
black area 黑区
black ash 粗碱灰;黑灰;粗硫化钡
black ash ball 黑灰块
black ash cake 黑灰块
black ash liquor 黑灰液
black ash mortar 黑灰砂浆;石灰煤渣灰浆
black-ash revolver 黑炭旋转炉
black-ash waste 黑灰废液
black azimuth mirror 黑色方位镜
black background 黑色背景
black backing varnish 黑底漆
black baking varnish 黑色烘干漆
black ball 锚球
black bamboo 紫竹
blackband 煤铁矿;黑带;黑条纹;泥铁矿;黑矿层
blackband iron-ore 黑菱铁矿
black bar 市场上的钢材;小型轧材
black base(course) 黑色底层;黑色基层;沥青基层;沥青底层;黑底
black bean (产于澳大利亚的)暗棕色硬木;黑豆
black-belt 黑土带
black berg 黑色冰山
black bituminous baking enamel 黑沥青烘干瓷漆
black blasting powder 黑色火药
black blight 煤污病
black blind 无光百叶窗
black blizzard 黑尘暴
blackboard 黑板
blackboard bulletin 黑板报
blackboard enamel 黑板搪瓷
blackboard eraser 黑板擦
black board paint 黑板漆
black body 黑体
black body cavity 黑体空穴
black body coefficient 黑体系数
black body emission 黑体发射
black body locus 黑体轨迹
black body photocell 全吸收光电管
black body radiation 黑体辐射
black body radiometer 黑体辐射计
black body temperature 黑体温度
black bog 泥炭沼地;黑土沼;黑色沼泽;黑色沼泥;低沼
black bolt 粗制螺栓
black book 劳氏失踪船舶名册;劳埃德失踪船舶名册
black border 黑色边缘
black bottom 黑色洼地
black box 快速调换部分;黑箱;黑匣子;黑盒(子);黑方框
black-box model 黑盒模型;黑箱模型;暗匣模型

black-box simulation 暗匣模拟
black brick 黑砖;青砖
black bronze glaze 乌金釉
black brown 黑褐色
black building 无窗房屋
black-bulb temperature 黑球温度
black-bulb temperature method 黑球温度法
black-bulb thermometer 黑球温度计;黑泡温度计
black buoy 黑色浮标
black buran 黑色风暴
blackburn printer 布拉克本印花机
Blackburn Rivet Head machine 布拉克本铆绒地毯织机
black burst generator 黑场发生器
blackbutt 黑基木
black calcareous soil 碳酸盐黑土
black canker 黑水病
black carbon 碳黑
black carbon counter 炭黑计数器
black carp 青鱼
black chaff of rice 黑谷物
black chalk 黑垩
black cherry 美国黑果稠李
black Chien ware 黑建陶
black chuglum (产于印度的灰棕～橄榄棕色的)硬木
black cinder (水淬不完全的)黑渣硬渣
black circuit 传输加密信息的电路
black clay glaze pottery 泥釉黑陶
black cloth 黑布
black coal 黑煤
black coating 黑涂料;黑色涂料;黑色涂层;镀黑
black cobalt 钴土
black collar axle 带内凸肩车轴
black compilation 黑板编绘原图
black computer 黑色要素控制器
black concrete 黑色混凝土
black conic(al) shape 黑色锥形号型
black content 黑色量;黑色成分
black copper 黑铜;粗铜
black coral 黑珊瑚
black core 黑心;炭斑(黑心)
black cotton clay 黑棉土
black cotton soil 黑棉土
black countersunk head rivet 粗制沉头铆钉
black crystal 墨晶
black culvert 黑铁管涵洞
black current 黑潮
black cyanide 氰化钙
black dammar 黑达玛(树脂)
blackdamp 碳酸气;窒息性空气;含二氧化碳气体;窒息性气体;窒息毒气;炮烟
black death 黑死病;鼠疫
black deflection 强振幅
black diamond 黑色金刚石;黑金刚石
black discolo(u)ration 黑斑
Blackdown beds 布莱克当层
black drop 黑漆粒;黑滴
black durain 黑暗煤
black durite 黑色微暗煤
black dusters 黑色风暴
black dyes 黑色染料
black earth 黑土;黑钙土;褐煤
black-earthen mask 黑陶面具
black East India 黑达玛(树脂);东印度黑
black ebony 黑檀木;硬木(印度)
black-edged plate 黑边铁皮
black enamelled conduit 黑瓷漆铁管
blackened grains 变黑颗粒
blackened receiver 涂黑接收器
black engraved map 黑色印样
blackening 黑度;上黑涂料;黑化;涂碳粉;涂黑(法);变黑
blackening bath 黑化液
blackening line 黑度线
blackening measuring method 黑度测定法
blackening of glaze 釉面烟熏
blacker-than-black level 超黑电平
black factice 硫化油膏;墨油膏
black fallow 黑色休闲地;秋耕闲地
black fever 黑色热;黑热病
black-figure style 黑色图案
black-figure vase 黑人像瓶饰;黑色柱饰(科林恩或混合柱式柱头)

black filter 不可见光滤光镜
black finish adjust-able wrench 发黑处理可调扳手
black finish combination pliers 发黑处理钢丝钳
black finishing varnish 黑色罩漆
black flame 黑火头
black flange 黑色法兰;无孔凸缘管堵
black flat taper head rivet 粗制平锥头铆钉
black fog 黑雾
black forest soil 暗色森林土
black fracture 黑色断口
black friction (黑色的)绝缘胶布
black friction tape 黑胶布带;黑色绝缘胶布
black frost 厚冻雾;黑霜;黑冻;严霜
black fundamental matrix 黑色基质
black furnace 不加热炉
black fused alumina 黑刚玉
black gang 修理工作队;(土木工程承包的)设备维修小组
black gasoline 黑市汽油
Black Gate 尼日尔港
black glass 黑色玻璃;黑玻璃;中性滤光片;中性滤光镜
black glass for ultra-violet transmission 黑色透紫外玻璃
black glass reflectivity 黑玻璃反射率
black glaze 乌釉
black-globe temperature 黑球温度
black-globe temperature method 黑球温度法
black-globe thermometer 黑球温度计
black goggles 黑色护目镜
black granite 黑色花岗岩;黑花岗岩;闪长岩
black graphic(al) paint 黑色石墨涂料
black graphite paint 黑色石墨涂料
black green 黑绿色
black grey 黑灰色
blackground screen 黑底荧光屏
black gum 黑胶(美国)
black hard 半干状态
black hawthorn 黑山楂木
blackhead 粉刺
black heart 炭斑;(木材的)黑心
black heat 暗热
black heat-proof varnish 黑色耐热漆
black heat-resisting paint 黑色耐高温漆
black hematite 黑赤铁矿
black hill spruce 云杉
black hole 黑洞
black hole effect 黑洞效应
black hole neighbourhood 黑洞邻域
black hot 暗红热
black humus earth 黑色腐殖(质)土
black humus soil 黑色腐殖(质)土
black ice 黑冰;透明薄冰
black information 黑板要素
blacking 铸造用黑粉料;造型涂料;粉磨石墨
blacking bath 黑化液
blacking brush 涂料用毛刷
blacking hole 气孔;针孔;涂料气孔;石墨窝(铸件缺陷)
blacking mill 碳质涂料碾磨机
blacking mixer 黑色涂料搅拌机
blacking paint 黑色涂料
blacking up 泛黑
black ink 贷方
black insulating tape 黑色绝缘胶布
black iron 黑铁皮;黑钢板;平铁;黑铁;黑铁板
black iron conduit 黑铁导线管
black iron hexagon nut 黑铁六角螺丝帽
black iron ore 硬锰矿;磁铁矿
black iron oxide 氧化铁黑
black iron oxide pigment 黑氧化铁颜料
black iron pipe 黑铁管
black iron plate 黑铁板
black iron(sheet) 黑铁皮;黑铁板;未镀锌铁板
black iron washer 黑铁垫板
blackish 稍黑的;带黑色的;微黑的
blackish green 墨绿
blackish-green epoxy baking paint for blackboard 墨绿环氧烘干黑板漆
blackist ticket 黑名单票
blackjack 闪锌矿;粗黑焦油;烛黑
blackjack bearing series 含煤炭质黏土
black japan 沥青类涂料;黑色防腐漆;油性沥青漆;深黑漆;沥青漆

black japanned brass-plated iron hasp, staple 黑漆镀黄铜锁牌
black japanned iron hasp, staple 黑漆铁锁牌
black japanned iron hinge 黑漆铁铰链
black japanned iron wire hasp and staple 黑漆铁丝锁牌
black japanned screen door spring hinge 黑漆纱门弹簧铰链
black japanned spring hinge 黑漆双弹簧铰链;黑漆单弹簧铰链
black karuni 黑色拼花木地板材
black knot 黑节疤
black lacquer 黑色大漆
black lava glass 黑色熔岩玻璃
black layout 排样
black lead 黑铅;不纯石墨;笔铅;炭精;石墨
black-lead crucible 石墨坩埚
blackleading machine 涂石墨机
black lead lubrication 石墨润滑剂
black lead ore 石墨;黑色铅矿
black-lead paint 黑铅漆;石墨涂料;黑铅油漆
black lead powder 黑铅粉
black level 黑色电平;暗电平
black-level clipping 黑色电平切割;黑电平切割
black light 近紫外光;黑光;不可见光;暗光;荧光灯
black light crack detector 超紫外线或红外线探伤仪
black light lamp 黑光灯;不可见光灯
black lighting 黑光照明
black lightning 黑闪
black lignite 黑色褐煤;黑褐煤
black line 黑线(纹);黑体粗线
blackline print 黑色印图;黑版印样
black liquor 黑液;纸浆废液;醋酸铁液
black liquor oxidation process 黑液氧化法
black liquor oxidation tower 黑液氧化塔
black liquor recovery boiler 黑液回收锅炉
black liquor recovery unit 黑液回收装置
black liquor solid 黑液固体物
black loading 填加炭黑;炭黑填充量
black locust 洋槐;刺槐(木)
black lung 煤尘肺;煤肺病
black lung disease 黑肺病
black magnetic oxide 磁性氧化铁黑
black mahogany 赤红木
black maire 有黑色条纹的深褐色硬木(产于新西兰)
black malleable casting 美国韧性铸铁
Blackman window 布兰克曼窗口
black marble 黑大理石
black marble chippings 黑大理石渣
black marble in grains 黑大理石粒
black marble mantel clock 黑大理石座钟
black market 黑市;非法交易市场
black marketing 从事黑市交易
black market price 黑市价(格)
black market purchase 黑市采购
black market transaction 黑市交易
black mark-line 墨斗线
black masking 黑色蒙片
black meal process 黑生料法
black mercuric sulfide 黑色硫化汞
black metal 含煤黑页岩
black mica 黑云母
black mild steel carriage bolt and nut 黑铁马车螺丝闩
black mild steel fish bolt and nut 黑铁鱼尾螺丝闩
black mild steel flat head rivet 黑铁扁头铆钉
black mild steel flat head tinmen rivet 黑铁号头铆钉
black mild steel hexagonal nut 黑铁六角螺丝帽
black mild steel pan head rivet 黑铁锥头铆钉
black mild steel round head rivet 黑铁圆头铆钉
black mild steel square nut 黑铁四方螺丝帽
black mineral 黑色矿物
black mineral oil 重油;重矿物油
black mordant 醋酸铁液
black mortar 黑砂浆;黑灰砂浆;石灰煤渣灰浆
black mo(u)ld 黑土
black mud 黑泥
black mustard oil 黑芥子油
black negative 黑色为负
black negative or positive 黑负或黑正
blackness 毛面;黑度
blackness value 黑度值

black nut 粗制螺母
black ocher 锰土
black oil 黑油;黑机油;润滑重油;石油残渣
black-oil model 黑油模型
black on kingfisher blue 翠青地黑花
black-on-white scale 白底黑刻度
black onyx 黑宝石
blackout 熄灭灯光;封锁;黑视;消隐;一时性黑蒙;遮蔽;灯光管制
blackout amplifier 消隐放大器
blackout area 盲区
blackout blind 遮光(窗)帘;遮光挡板
blackout building 无窗建筑;无窗房屋
blackout dial 夜光表盘
blackout door 不透光门
blackout dosage of activated sludge 活性污泥的变黑剂量
blackout effect 遮蔽效应;关闭效应
blackout installation 不透光设备
blackout jalousie 不透光的固定百叶窗;不透光气窗
blackout lamp 防空灯
blackout light 防空灯
blackout louvers 不透光的天窗;不透光气窗
blackout paint 防空遮黑漆;遮光漆
blackout plant 无窗厂房
blackout pulse 消隐脉冲;熄灭脉冲
blackout slated blind 不透光的条板百叶窗
blackout switch 舞台灯光总开关
blackout voltage 截止电压
blackout window 不透光窗;无光窗
black overlay 着墨
black oxide of cobalt 黑氧化钴;锰钴土
black oxide of iron(pigment) 黑色氧化铁颜料
black paint 黑涂料;黑色漆;黑漆
black paint soil amendment and mulch 黑色涂料和覆盖
black panel temperature 黑板温度
black paper-making wastewater 造纸黑液
black paste 沥青膏
black patch 黑块;黑斑点;(钢材上的)未酸洗部分
black peak 黑峰
black peat 黑色泥炭
black persimmon wood 黑柿木
black petroleum products 石油重油
black phenolic ready-mixed paint 黑色酚醛调和漆
black phosphorous 黑色磷
black pickling 粗酸洗;初酸洗
black pigment 黑颜料;黑色颜料;炭黑
black pine 黑松;澳大利亚松
black pipe 无镀层管;非镀锌管;黑铁管;非涂锌管
black pitch 黑色松脂;黑沥青;黑焦油;船用黑焦油;柏油脂
black plate 黑钢板;黑铁皮;黑铁板
black plate for tinning 马口铁卷板
black porphyry 黑斑岩;暗粉岩
black positive 黑色为正
black pottery 黑陶(器)
Black Pottery Culture 黑陶文化
black pottery with clay glaze 泥釉黑陶
black powder 黑色炸药;黑火药
black powder fuse 黑火药导火线
black precipitate 黑色沉淀
black print 黑版
black process of production gas 生产气体炭黑过程
black product 黑色石油产品;石油重油
black pure phenolic resin primer 黑色纯酚醛底漆
black quartz 墨晶;黑石英
black radiator 黑辐射体
black rain 黑雨
black(raw) meal 黑生料
black red 黑红
black red heat 黑红色温度
black-reef 礁后
black ribbon 黑色带
black roofing adhesive 黑屋顶油毛毡胶粘剂;屋面油毡胶粘剂
black roofing felt 黑屋顶油毛毡;屋面油毡
blackroom 黑屋
black root 黑根
blackroot share 草根土角楔形犁铧
black rot 黑腐病
black rouge 黑铁丹;氧化铁黑;铁黑
black rough bolt 粗制螺栓

black rubber 炭黑橡胶
black rubber insertion 夹布胶片
black sample 黑试样
black sand 黑砂
black saturation 黑色饱和
black scale 黑色标度
black schist 黑色片岩
black scope 暗底显示管
black screen 中灰滤光屏;暗色滤光屏;暗色滤光镜
black screw 粗制螺丝
Black Sea 黑海
Black Sea-Arctic transgressive oscillation 黑海—北极式海侵颠动
Black Sea berth charter 黑海谷物班轮租船合同
Black Sea berth terms 黑海租船合同
black seal 黑色封层
black shaded 加黑斑补偿
black shadow 黑影
black shale 黑色页岩
black shape 黑色号型;手制陶器
black sheet 黑铁板;黑图;黑铁皮;黑钢板;薄钢板
black sheeting felt 中隔沥青毡层;沥青油毡(中)间层
black sheet iron 黑铁皮;马口铁皮;黑钢皮
black ship 黑船
black short 冷脆的;黑色裂口
black shortness 冷脆性
black sighting board 黑色测试板
black silicon carbide 黑碳化硅
black silicon carbide grain 黑碳化硅磨料
black silver 脆银矿
black skin 铸皮
black slag 黑渣
black slate 黑色板岩
Black slip stopper 布勒克滑钩链制动器
black slurry 黑料浆;炭黑水分散体;炭黑浆状液
black smear 黑色拖尾
blacksmith 锻工;铁匠;铁工;手锻工
blacksmithing 锻造;手工锻造
blacksmith's anvil 锻工铁砧
blacksmith's bellows 锻造风箱
blacksmith's chisel 锻工凿;锻工用凿子
blacksmith's chisel for hot iron 锻工热錾
blacksmith's flat(ter) hammer 锻工平锤
blacksmith's hammer 铁匠锤;铁匠锒头;锻工锤;锻锤
blacksmith's hardy 铁匠的方柄凿
blacksmith's leg vise 长脚虎钳
blacksmith's punch 锻工冲子
blacksmith's ruler 锻工尺
blacksmith's slag 锻渣
blacksmith's tongs 锻工钳
blacksmith's tool 锻工工具
blacksmith's welded joint 锻接接头
blacksmith's welding 锻接;锻焊;锻工焊接
blacksmith's(work)shop 锻工车间
blacksmith welded joint 煅接接头
blacksmith welding 锻接
black smoke 黑烟
black snap head rivet 粗制圆头铆钉;粗制半圆头铆钉
black softened 初退火
black soil 黑土
black speck 黑色斑点;黑斑点
black sphere 墨结球;黑球
black spot 黑点;事故发点;黑色斑;盲点;黑斑
black spot distortion 黑点失真
black spot interference 黑点干扰
black spotter 杂波抑制器
black spotting 黑斑
black spruce 黑云杉
black square 黑方格
black square head bolt 粗制方头螺栓
black square nut 粗制方螺母
black stain 黑色釉料
black stamping pad 纯黑印泥
black star sapphire 黑宝石
black steel 无镀层钢
black steel pipe 黑钢板
black steel sheet 黑钢皮
Black's test 勃拉克试验
black stix 装筒黑色炸药
black stock 黑浆;炭黑混合物;重油裂化原料
black stone powder 黑石粉

black stork 黑鹳
black storm 黑风暴(中亚东北风)
black storm incident 黑风暴灾害
black strap 黑色重润滑油;乳化黑色油
black stream 黑潮
black strip 黑带钢;热轧带钢
black stump chalk 黑炭粉
black-surface enclosure 黑表面包壳
black-surrounded tube 黑底管
black talc 黑滑石
black tape 黑色绝缘胶布;摩擦带;黑胶布;刹车带
black tarred tow 涂柏油的拖绳
black tea 红茶
black test 琼脂
black thorn 黑刺李
black tile 青瓦
black tin 二氧化锡
black tin cassiterite 锡石
black tin steel 黑钢板
blacktop 沥青路面层;黑色面层;黑色道路面层;黑色沥青材料
black-top finishing roller 黑色层修整滚筒;沥青道路表面修压路机;沥青道路表面修整路碾
black-top pavement 黑面层沥青路面;沥青路面
black-top paver 黑面层铺路机;沥青路面修整机;沥青路面洒布机;沥青路面铺料机
black-topped road 柏油路
blacktopping 铺筑黑色面层;用沥青铺路面;铺筑沥青路面层;铺筑青
black top road 沥青路面;黑色路(面)
black top soil 暗色表土;黑表土
black-top spreader without hopper 无料箱沥青路面铺料机
black top surface 黑色路面
black-to-white level 黑白电平
black-to-white transition 亮度跃迁
black Tura 黑侏罗纪【地】
black turf soil 黑色生草土;黑泥炭土
black vacuum 低真空
black varnish 黑色清漆;黑清漆;黑光漆
black varnished cambric cloth 黑漆布
black varnished cambric tape 黑漆带
black varnish finish 黑光漆面
black vista (产于美国加州的)黑色花岗岩
blackwall hitch 吊钩结
black walnut 胡桃楸木
black ware 黑色陶器
black wash 黑色薄涂层;造型涂料;黑色(造型)涂料;碳素涂料
black washer 毛垫圈
blackwash sprayer 涂料喷枪
black water 黑水
black-water estuary 黑水港湾
black-water fever 黑水热
black waxy soil 黑油土;深色毛
black-white control 亮度调节
black wire 黑铁丝
black wire rope 黑钢丝绳
blackwood 黑木金合欢;黑木
blackwook carving articles 红木雕刻品
blackwood furniture 红木家具
blackwood furniture with filigree inlay 红木嵌银丝家具
blackwood products 红木制品
blackwood rule 红木尺
blackwork 锻工物;锻工场;软钢(铁)锻件;锻件
black wrought iron 无镀层熟铁
blacolite 绢石蛇纹岩
blad cypress 落羽松
bladder 胶袋;囊状物;充气囊袋;铲板式平地机
bladder glue 鳔胶
bladder press 轮胎压床
bladder tank 储[贮]气袋
bladder wrack 球胆破坏
blade 型片;快门叶片;桨叶;桨片;尖刀;叶片;闸刀(盒);刀翼;刀身;刀片;草body;刃;托板;推土板;推进器叶片
blade accumulator cylinder 铲刀缓冲器
blade adapter 刀架
blade adjsutable fan 动叶可调风机
blade adjustable axial fan 叶片可调轴流风机
blade adjustable in still condition fan 静叶可调整风机
blade adjusting servomotor 桨叶接力器;叶片调节接力器

blade agitator 桨式搅动机;叶片式搅拌器
blade angle 刀片角度;桨叶角;叶片装置角;叶片角;叶片安装角;叶片安置角;叶片安放角;刃口角(度);托刃口角度
blade area 翼面积;叶片面积
blade arrangement 叶片配置
blade back and forth 往返铲刮(用平地机)
blade backfiller 推土机回填机;刮板回填机;填沟机平铲
blade beam 螺距测量器
blade bearer 梭形支撑;刀面式支承;刀口支撑
blade bearing 刃型支撑;刃形支承
blade bit 刮刀钻头
blade bowl 铲斗
blade bracket 平土铲托架
blade camber 叶片弯度
blade carrier 刀架
blade carrying axle 刀轴
blade cascade 叶栅
blade clearance 剪刀间隙;冲裁间隙
blade clearance cavitation 轮叶间隙空蚀
blade clip 刀形夹头
blade coater 刮刀涂布;刮板式涂布机
blade coating 刮涂;刮板涂布
blade coating machine 刮刀涂布
blade connection 刀形接触
blade contact 插口式插头;刀口式触点
blade control 卫护板控制;平铲控制
blade control lever 铲刀操纵杆
blade control wheel 平路机刀片操纵轮
blade coverer 铲式覆土器
blade crusher 叶片式破碎机
blade cultivator 平铲中耕机
blade curvature 叶片弯度
blade cutting angle 推土板切削角
blade cylinder 圆柱形叶片;平路机汽缸
bladed 叶片状;刃状的
bladed allowable speed 叶片的允许速度
blade damper 桨叶减震器
bladed crystal 叶片状晶体
blade developed area ratio 桨叶展开面积比
bladed habit 刃状习性
blade dozer 刮铲推土机;刮板推土机
blade drag 刀片式刮土机;刀片式刮路机;刀片刮土机;刃式刮路机;推土式平路机
blade drop below ground 地平线下深开挖
bladed rotor 装有叶片的转子
bladed rotor separator 装有叶轮的回转选粉机
bladed saw 板锯
bladed shoulder cavitation 叶扇空蚀
bladed shutter 叶片式快门
bladed structure 叶状构造;叶片状构造;刃状构造
bladed-surface aerator 叶片式曝气装置
bladed wheel 装有叶片的叶轮
blade edge 叶片缘
blade efficiency 桨片效率;叶片效率
blade equipment 平土机
blade erosion tester 叶片侵蚀测试计
blade extension 叶片伸长;平土铲加长件
blade face 叶片面
blade-face cavitation 叶面汽蚀;叶面空蚀
blade fan 叶片式风扇
blade feather 桨叶曲线
blade fin 导向滑板
blade for earth borer 地钻叶片
blade for plane 刨刀片
bladeful 一刀片之量
blade grader 刮板平土机;刀片式平地机;铲式平土机;铲式平路机;平铲式平地机;平铲式平地机;刃式平路机;刃式刮路机
blade grading 用平路机整平路道叶片式搅拌机路
blade grinder 磨刀石;磨刀砂轮
blade groove 叶片槽
blade harrow 叶形耙;刀耙
blade head 滑动刀架
blade heel 卫护板端部
blade height 叶片高(度);叶片长;推土板高度
blade holder 锯条夹;磨刀夹
blade incidence 叶片装置角;叶片冲角;叶片安装角;叶片安置角;叶片安放角
blade inclination 叶片倾斜度(水泵)
blade inlet angle 叶片入口角
blade into a single window 用推土机推成行;用

平路机将材料堆成单条长堆
blade latch 开关保险锁;闸刀保险销;断路器制动片
blade lattice 叶栅
blade length 桨叶长度
bladeless centrifugal pump 无叶片离心泵
blade letter 刀顶宽代号
blade lever 桨叶转动臂杆
blade lift arm 卫护板升臂;平铲提升臂
blade-lift coefficient 桨叶升力系数
blade lift control housing 叶片提升机械管理室;平铲提升控制箱
blade lift control pinion 叶片提升控制小齿轮;平铲提升控制齿轮
blade lift cylinders 铲刀升降油缸
blade lifting height 推土板提升高度
blade lifting time 推土板提升时间
blade lift link 叶片提升连杆
blade lift mechanism 叶片提升机械构造;平铲提升机械装置
blade lift post 平铲提升连杆
blade lift shaft 叶片提升轴
blade link 桨叶联杆
blade linkage 叶片联动装置
blade load 剪切力
blade loading 旋翼桨叶荷载
blade loading diagram 叶片荷载分布图
blade lock 叶片制动销;平铲闭锁装置
blade locking plunger and camshaft 叶片的制动冲杆及凸轮轴
blade loss 叶片损失
blade lowering depth 推土板下降深度
blade lowering time 推土板下降时间
blade machine 平路机;推土机
blade magnetic domain 刀片形磁畴
blade maintainer 单刃养路机
blademan 铲刮工;平地机驾驶员;平地机手
blade marker 叶片的标志圆环;叶片标号
blade milled from solid material 方钢铣制叶片
blade mixer 叶片拌和器;桨叶搅拌机;叶片式混砂机;刮板式搅拌机
blade mixer shaft 叶片拌和器轴
blade mixing 用刀片拌和;叶片拌和
blade of aviation engine 航空发动机叶片
blade of diaphragm 光阑叶片
blade of dredge pump 泥泵叶片
blade of grader 平路机铲刀;平路机
blade of rotor 动叶
blade of shovel 铲斗刃口;铲刀
blade of T-square 丁字尺身
blade of variable cross-section 变截面叶片
blade of water turbine 水轮机叶片
blade of water wheel 水轮机叶片
blade opening 叶片开口;叶片开度
blade outlet angle 叶片出口角(螺旋桨)
blade over the roadway 满堂摊铺(用平地机在整个车行道摊铺材料)
blade paddle mixer 桨叶式搅拌器;桨叶式搅拌机;桨叶式拌和机;桨式搅拌机;叶片式拌和机
blade-paddle mixer 桨叶式搅拌机
blade paddle stirrer 桨式搅拌机
blade paddle system 叶片搅拌系统
blade passage 叶片流道
blade pitch 卫护板开挖角度;叶片间距
blade point 叶片顶尖;刀刃;刀锋;刀宽度
blade pressure 叶片压力
blade profile 叶型;叶片轮廓
blade profile drag coefficient 桨叶剖面阻力系数
blade pump 叶片泵
blader 平路机;气囊;推土机
blade radius 刀尖圆角半径
blade rake 带齿活板
blade regulating valve 轮叶调节阀;轮叶调节器
blade reinforcement 卫护板加固;叶片加固
blade removal opening 桨叶检修孔;桨叶拆装孔;导叶接力器
blade retainer 叶片安装底架
blade reverse control 刀片水平回转操纵(平地机)
blader(grader) 平路机
blade rim 叶片轮缘
blade ring 叶片环
blade root (螺旋桨的)叶根
blades 扁叉
blade saw 片锯

blade saw frame 锯条架
blades cooling 叶片冷却
blade scraper 刮板式平地机
blade section 叶片截面
blade servomotor 叶片伺服马达;叶片接力器
blade setting 叶片装置;装定叶角
blade shaft 叶片搅拌轴;叶片式混砂机传动轴
blade shaft beam 叶片偏移光束
blade shape 叶片形式
blade-shaped diaphragm 叶片状隔板
blade sharpening machine 磨刀机
blade shield 叶片式盾构
blade shift beam 叶片转换杆
blade shoulder cavitation 叶肩汽蚀
blade shutter 叶片快门
blade side shift 叶片边移
blade side shift control 刀片侧向回向操纵(平地机)
blade smith 刀片工人
blade snow plough 犁式除雪机
blade spacing 叶片间距
blade spreader 刮板式(混合料)摊铺机;刮板式(混合料)分布机
blade spring 叶片弹簧;片簧
blade stabilizer 平土铲稳定器
blade stirrer 叶片搅拌机
blade stropper 磨刀片器
blade thrust 叶片推力
blade tilt 叶片倾角
blade tilt hydraulic system 桨叶倾斜的水力系统
blade tilt strap 桨叶倾斜部分
blade tip 叶梢;叶片尖端;推进器叶端
blade tip cavitation 叶尖汽蚀;叶尖空蚀
blade tip cylinder 平土铲翻转油缸
blade tip eddy 叶片顶端涡旋;叶顶漩涡
blade tip vortex 叶梢涡旋;叶尖涡流
blade toe 叶片尖端;刀刃;刀锋
blade trailing edge 叶片出汽边
blade twist 叶片扭转
blade type belt cleaner 叶片型皮带扫清器
blade type conveyer 刮板式输送机
blade-type cutter 叶片型刻刀
blade type cutter head 刮刀式钻头
blade type damper 叶片式挡板
blade-type jaw crusher 双肘杆颚式破碎机
blade type sensor 刀式传感器
blade type sharpening stone 刀形磨石
blade-type snow plough 叶片式扫雪机
blade vane 刀片
blade vortex 叶片涡流
blade wear indicator 叶片磨损指示器
blade wheel 叶轮
blade width 推土板宽度
blading 叶栅;叶片装置;叶片镶嵌;整平;铲平;平路
blading and dragging 用刮路机及拖斗车平整路面;铲刮整平
blading back 回推土;倒向运土
blading back of earth 回推土;倒向推土
blading compaction 整型压实;刮平压实
blading method of proportioning 配料平路法(用平地机或推土机)
blading operation 平路机工作;平路机操作(平路、刮路、整型、整平等工作)
blading work 叶片安装(工作);铲平作业;平路机工作
blae 劣质黏土页岩;灰青碳质页岩;灰青灰质页岩
blaenau 河源;[复]blaen
blaes brick 劣质黏土页岩砖
Blagden's law 布莱格登定律
Blaine air permeability apparatus 布莱恩透气性试验装置
Blaine air permeability method 布莱恩透气(性方)法
Blaine air permeability test 布莱恩透气性试验
Blaine apparatus 布莱恩比表面积测定仪
Blaine fineness 布莱恩细度
Blaine fineness sensor 布莱恩比表面积测定仪
Blaine fineness tester 布莱恩细(微粒)度测定仪
Blaine formation 布莱恩建造【地】
Blaine meter 布莱恩比表面测定仪
Blaine method 布莱恩透气法
Blaine number 布莱恩值;布莱恩渗透度法
Blaine permeability method 布莱恩渗透度法
Blaine specific surface 布莱恩比表面积

Blaine surface area 布莱恩表面积
Blaine test 布莱恩细度试验
Blaine test method 布莱恩比面测定法
Blaine value 布莱恩比面值
blair 平原
blairmorite 淡方钠岩
Blair oven 布累尔烘箱
blaize 硬砂岩
Blake bottle 布莱克培养瓶
Blake breaker 下动颚式破碎机;布莱克型颚式破碎机
Blake crusher 下动颚式破碎机;布莱克型颚式破碎机;勃雷克颚式破碎机
blakeite 红铁碲矿;红碲铁石
Blake jaw crusher 布莱克颚式压碎机
Blakely test 布莱克利试验法(坯釉适应性)
Blake number 布莱克数
Blake plateau 布莱克海台
Blake reversed polarity subchron 布莱克反向极性亚时
Blake reversed polarity subchronzone 布莱克反向极性亚时间带
Blake reversed polarity subzone 布莱克反向极性亚带
Blake's hydraulic radius 布莱克水力学半径
Blake type jaw crusher 下动颚式破碎机;布莱克型颚式破碎机;双肘杆颚式破碎机
Blancan 勃朗阶
Blancato type suspended gravity fender 布兰卡图悬挂重力式护舷
blanc fix(e) 硫酸钡粉;重晶石粉;沉淀硫酸钡
blanch 镀锡
Blanchard brush 布兰查德涂布刷
blanching reaction 转白试验;变白反应
blanching water 漂白水
blancmange mo(u)ld 带浮雕的模型
blancometer 白度计
Blanc rule 布朗规则
bland flange 管口盖凸缘
bland tracery 实心窗花格
blanfordite 钠锰辉石
blank 落料件;空格;空白式;空白片;空白盘;空白(表格);毛坯(件);预制棒;种板;钢坯;刚体【岩】;断路板;初型模;冲切;不显示;创成坯料;板坯;切断件;坯料;坯件;数字间间隔
blank acceptance 空白承兑汇票;不记名承兑票据
blank account book 空白账簿;空白账册
blank-and-cup die 落料模
blank and pierce die 剪料穿孔模
blank arbitration clause 空白仲裁条款
blank arcade 假拱廊;封闭拱廊;实心连拱(廊);实心拱廊
blank arch 假拱;装饰拱;拱形装饰;实心拱
blank assay 空白试验
blank back bill of lading 背面空白提单
blank bill 空白票据
blank bill of lading 空白提单
blank bit 空白钻头;圆钻头
blank bolt 非切制螺栓;无螺纹栓
blank bond 空白登据;不记名债券
blank book 空白簿
blank cap 毛坯盖;盲孔盖;盲盖板
blank carburizing 假渗碳;坯料渗碳
blank card 空插件;空白卡片
blank casing 无孔套管
blank-casing bit 空白套管钻头
blank ceiling mounting channel 安装毛坯天花板所用的槽铁
blank channel 坯斜槽
blank character 间隔符号;空白字符
blank chart 素图
blank check 空额支票;空白检验
blankcheck trust 空头信托
blank cheque 空白支票
blank code 空代码
blank coil 空心线圈;空卷;空白卷
blank column 空白专栏
blank command 空指令
blank common 空白公用(存储)区;空白公共区
blank common block 空公用块;空白公用块;无名公用区;无名公用块
blank common data block 空白公用数据块
blank correction 空白校正
blank cover 固定盖;封盖
blank credit 空额信用证;空白信用证;空白信用票

据;空白信用汇票;信用票据
blank crown 空白钻头
blank deleter 空白消除设备;空白删除器;空白符删除器;消隐脉冲消除器
blank determination 空白试验;对照试验
blank diameter 螺纹毛坯直径;毛坯直径
blank dimension 毛坯尺寸
blank door 封堵门;装饰门;假门;暗门
blank drawing 毛坯图
blanked element 空白要素
blanked lumber 粗制木材
blanked off 配有盖板的;配有堵头的
blanked off pipe 关闭管
blank endorsement 空白背书;不记名背书
blanker 冲切工;消隐装置;熄灭装置;关闭器;粗模膛;闪光;制坯工;下料工
blanker conversion ration 再生区转换比
blanker damp 橡皮布夹头
blanker wind 橡皮布紧轴
blanket 路面底层;溜矿槽衬底;矿席;胶印橡皮布;棉毡;毛毡;橡皮布;烧火装置;掩盖;毡;再生区;盖层;叠层坯料;垫层;敷层;全部接收;铺毡层;毯
blanket approval 全部同意;全部认可
blanket area 敷层面积;覆盖面积
blanket batch charger 毯式投料机
blanket bog 毡状酸沼;覆盖沼泽;覆被酸沼
blanket bond 总括保证保险
blanket charging 薄层加料
blanket chest 箱式家具(的一种)
blanket cleaning control 覆盖层清除控制
blanket cleaning device 覆盖层清除设备
blanket clearance 通用出入港许可
blanket clip 覆盖层夹具
blanket coat 黏层
blanket coater 垫带式刮涂机
blanket coating 垫带刮涂
blanket commitment request 一次总付要求
blanket condemnation 若干房屋、土地同时征用
blanket contract 一揽子合同
blanket conveyor belt 夹皮带输送机;覆盖带
blanket core tube 表土取芯管
blanket-count station 大面积观测站
blanket course 毡层;透水盖层
blanket currency 不兑换通货
blanket cylinder 胶印滚筒;橡皮滚筒;覆盖柱面;覆盖层柱面
blanket deed of trust 一揽子信托契据
blanket deposit 均厚沉积;席状沉积;平腹矿床;平伏矿床
blanket drain 铺盖排水
blanket drying machine 衬毯烘干机
blanketed with silt 粉砂覆盖
blanket encumbrance 揽总抵押
blanket export 综合出口
blanket feed 薄层加料;毯式投料
blanket feeder 毯式投料机
blanket fidelity bond 一揽子信誉保证书
blanket filter 毡滤器
blanket filtration 泥渣层过滤
blanket gas 保护气体
blanket grout 覆盖层灌浆
blanket grout hole 覆盖灌浆孔
blanket grouting 护面灌浆;固床覆盖灌浆;覆盖灌浆;铺盖灌浆
blanket heat exchanger 再生区热交换器;封闭式热交换机
blanket-height controller 泥面控制器
blanketing 毛毡;编织物选矿;强信号噪扰;铺盖;掩盖;覆盖;封上;包上
blanketing effect 覆盖效应
blanketing effect of dust 灰尘覆盖效应
blanketing fctor 覆盖因子
blanketing frequency 抑止频率
blanketing gas 填充气垫
blanketing material 铺盖材料
blanketing with gas 用气体盖覆
blanket insulation 绝缘毡;绝缘卷材;绝热毡;隔声毡;隔热毡
blanket insurance 总括保险;综合保险;统括保险;统保
blanket insurance policy 一揽子保险单
blanket license 总许可证;总括执照
blanketlike 毯状
blanket loom 毛毡织机

blanket method 毛毡干燥法
blanket mortgage 联合抵押;揽总抵押
blanket net 兜网
blanket of dry air 密封干燥空气层
blanket of glass wool 玻璃丝毡;玻璃棉毡
blanket of graded gravel 级配砾石铺盖
blanket of nitrogen 氮气层
blank(et) order 总括订(货)单;总订货单
blanket peat 毡状泥炭
blanket policy 一揽子保险单;总括保险单;总保单;统保单
blanket power 掩盖功率;再生区功率
blanket price 一揽子价格;共同价格;统括价格
blanket process 熄灭过程
blanket revetment 连片护岸;席状护岸;整体护岸;大面积护岸;成片护岸
blanket robe 车毡
blanket rule 适用于各种情况的规则;普通规则;普通适用的规则
blanket sand 冲积砂层;冲积砂;冲积沙;冲积覆盖砂层;层状砂;平伏砂层;砂席
blanket sandstone 平伏砂岩层
blanket socket 通用插座
blanket steam 毯层水蒸气
blanket stone 护面石
blanket strake 毡衬洗矿槽;绒衬洗矿槽
blanket trap 毡席圈闭
blanket trust deed 揽总抵押信托书
blanket-type cell 气孔底式浮选机
blanket-type insulant 毯状绝缘材料
blanket-type pneumatic machine 多孔底压气式浮选机;毯式压气浮选机
blanket vein 平腹脉;平伏矿脉
blanket washing machine 衬布洗涤机
blanket weed 覆盖杂草层
blanket with silt 粉砂覆盖
blank experiment 空转试验
blank fenestration 假的门窗布局;假门窗
blank field 空白字段
blank field descriptor 空白字段说明符;空白域说明符
blank fill 空白填充
blank flange 管口盖板;盲口缘;盲法兰;法兰盲板;法兰盘;无孔凸缘;无孔法兰(盘);死法兰
blank flat 毛坯板
blank flue 假烟道;掩蔽烟道
blank foil 纯箔
blank form 空白媒体;空白格式;空白表式;空白表格
blank form contract 空白合同
blank forms or schedules 空白表式
blank freezing 鼓风冻结
blank groove 哑纹;哑槽;未调纹
blank holding 坯料压紧
blank hole 未下套管井段
blank indorsement 不记名背书
blanking 落料;空白显示;截шки;过程;消隐;下料;熄灭;用盲板堵塞;遮没;断流;冲ljenog下斜;冲割;冲裁;切料
blanking amplifier 消隐脉冲放大器;熄灭脉冲放大器
blanking and embossing press 轧凹凸机
blanking bar 熄灭脉冲黑带
blanking circuit 熄灭电路
blanking clearance 下料间隙;冲裁间隙
blanking code 消除码
blanking control 消隐控制
blanking die 孔模;下料模;冲裁模;落料模
blanking disc 遮光盘;屏蔽盘
blanking gate 消隐门
blanking gate photocell 猝灭选通脉冲光电管;产生熄灭脉冲的光电管
blanking impulse 消隐脉冲
blanking impulse generator 消隐脉冲发生器
blanking input 截止输入
blanking instruction 间隔指令
blanking level 消隐电平;消隐信号电平;熄灭电平
blanking line 落料生产线;下料生产线;冲裁生产线
blanking machine 下料机;冲切机
blanking mixer 逆程消隐混合器;消隐混合器
blanking mixer tube 消隐脉冲混频管
blanking-off plug 浇口塞
blanking pedestal 消隐脉冲电平;熄灭脉冲电平
blanking plate 盲板

blanking press 压箔机;冲压机;冲割压力机
blanking pressure 下料力;冲裁力
blanking pulse 熄灭脉冲
blanking-pulse amplifier 熄灭脉冲放大器
blanking screen 遮光屏
blanking signal 消隐信号
blanking technology 消隐技术
blanking time 消隐时间;熄灭信号持续时间
blanking tolerance 下料公差
blanking voltage 截止电压
blanking wave 消隐波;熄灭波
blanking with opposed dies 对向凹模冲裁
blank instruction 空操作指令
blank jamb 盲边框
blank key 钥匙形片
blank label 空白标签
blank lacquer 空白胶片
blank layout 坯料排样;坯件布置
blank letter of credit 空白信用证
blank line 空白行
blank liner 无眼衬管;未穿孔的套管
blank map 空白图;空白地图;暗射(地)图;素图
blank masonry wall 不开洞的圬工墙;无开口圬工墙
blank material 种板材料
blank medium 空白媒体;空白介体;间隔介质;参考介质
blank mo(u)ld 初型模;初模;坯模
blank mounting 毛坯安装
blank nest 坯料定位窝
blank nitriding 空氮化
blank note 空白支票;空白汇票
blankoff 加盲板;消隐
blankoff flange 盲板;盲法兰
blankoff plate 盲底板;盲板
blankoff pressure 极限压强;极限低压强
blank-off volumetric(al) efficiency 零位容积效率
blank operation 一齐投入
blank order form 空白订单
blank out 使熄灭
blank-out sign 漏光式标记(雕空)
blank-out wall 难以通过的障碍物;无窗墙
blank panel 备用面板;空盘;空面板
blank paper tape coil 空白纸带圈;空白纸带卷
blank pipe 空管;无侧眼管子;无孔管
blank plate 盲板;底板;刻图片
blank plug 塞子
blank policy 不记名保单
blank power of attorney 不记名委托书
blank press 落料压力机
blank processing 毛坯加工
blank rate 总括保险费率
blank reaming shell 空白扩孔器
blank receipt 空白收据
blank rosette 假的圆花窗;实心圆花窗
blank run 空转;空机操作;无岩芯回次进尺
blank sample 零试样;空白试样
blank scatter 空白散射
blank sheet 空白图幅;素图
blank signal 空白信号
blank slug 毛坯条
blank specimen 空白对比试样
blank spot 漏耕地
blank stock 控制备料;不记名股票;调节备料
blank tape 空白纸带
blank tear 弯扭
blank test 空试车;空白试验;对照试验
blank through die 漏件式落料模
blank tracery 假的窗花格;实心窗板;无孔花格窗
blank transfer 空白过户凭证
blank-transmission test 空格传输测试;空传输测试
blank triforium 教堂拱门上面的假拱廊;教堂拱门上面的实心拱廊
blank vidicon retrace 消隐光导管回描
blank wall 暗墙;闷墙;无门窗墙;平壁;无门窗的墙;无窗墙;死墙;实墙
blank when zero 零时空格
blank window 封堵的窗;假窗;盲窗;暗窗
blank wire 裸线
blank with structure level numbers 结构层号中所具有的空格
blank zone 无衬区
blare 捻缝料;响声
blase furnace cast iron 高炉生铁
blasenschiefer (像扁桃样的)蔡希施坦白云石

Blashko effect 布拉什科效应
Blaslus theorem 白拉斯勒斯定理
blast 水分离;吹砂;分裂球丝;爆炸气浪;汽笛声;气浪;喷气器;送风
blastability 可爆性;抗爆力;招爆性
blastability of rock 岩石可爆性
blast aeration 鼓风曝气
blast aeration tank 鼓风曝气池
blast aerator 空气炮;鼓风曝气装置;鼓风曝气器
blast air 鼓入空气;鼓风;强制通风;喷射空气
blast airfan 鼓风机
blast apparatus 鼓风设备
blast area 爆炸区(域);爆破区
blast atomizer 喷射雾化器
blast attenuating process 气流喷吹工艺
blast blower 鼓风机
blast box 鼓风机;风箱
blast breakout 漏风
blast burner 高压燃烧器;喷灯
blast burner for soldering 焊接喷灯
blast cap 鼓风盖
blast capacity 鼓风能力;鼓风量
blast cap funnel cap 风帽
blast cap gasket 鼓风盖垫密片
blast cap screw 鼓风盖螺钉
blast chamber 起爆室;鼓风室;燃烧室
blast charging 鼓风装药
blast cleaning 喷丸清理;喷丸净化;喷射清理;喷砂清理(法);喷砂净化;喷砂除垢;喷吹清理;抛丸除锈
blast cock 放气旋塞
blast coil 鼓风盘管
blast cold 吹冷
blast conditioning 鼓风调剂
blast connection 风管接头
blast conveyer 气流式输送器
blast cooling device 炉水分离器
blast cupola 化铁炉
blast cupola furnace 鼓风化铁炉
blast damper 爆压减震器
blast deflector 喷焰偏转器
blast delay 爆炸延迟
blast densification 爆破增密;爆破振密
blast depth 爆破进尺
blast design 爆破设计
blast down 爆破
blast draft 爆炸压力气流;压力气流
blast drying box 鼓风干燥箱
blasted patch 喷砂斑
blast effect 喷气作用
blast engine 鼓风机
blast entity rock per circulation 每循环爆落实体岩石量
blaster 鼓风机;放炮器;放炮工;发爆器;爆破手;爆破器;爆破技术员;爆破机;起爆器;喷砂设备;喷砂机
blaster cap 起爆雷管
blaster fuse 导火线;引线;起爆引线
blast fan 鼓风机;风扇(叶轮);通风机
blast fence 涡轮机废气防护墙;气流挡板;射流折流栅
blast finishing 喷砂修整;喷丸处理;喷砂处理
blast flame spraying 火焰喷涂法
blast freezer 吹风速冻机;鼓风冷冻室;速冻冷库
blast fume 炮烟
blast furnace 鼓风炉;高炉
blast-furnace air 高炉鼓风
blast-furnace block 高炉砖;高炉大砖
blast-furnace blower 高炉鼓风机
blast-furnace bosh 鼓风炉炉腹
blast-furnace bottom 鼓风炉炉底
blast-furnace brick 高炉砖;高炉矿砖
blast-furnace brickwork 鼓风炉砌砖;高炉砌砖
blast-furnace bucket hoist 高炉料罐起重机
blast-furnace built-up roof(ing) 高炉矿渣组合铺屋面
blast-furnace burden 高炉配料;高炉配装料
blast-furnace burden scale 高炉配料秤
blast-furnace bustle pipe 高炉环状风管
blast-furnace car 高炉料罐车
blast-furnace casting 高炉出铁
blast-furnace cast iron 高炉铸铁
blast-furnace cement 炉水水泥;炉碴水泥;矿水水泥;高炉渣水泥;高炉水泥

blast-furnace cement concrete 矿渣水泥混凝土
blast-furnace charge 鼓风炉装料
blast-furnace cinder 高炉矿渣;鼓风炉渣;高炉熔渣
blast-furnace coal tar 高炉煤焦油
blast-furnace coal tar pitch 高炉煤焦油沥青
blast-furnace coke 高炉焦;鼓风炉焦炭;高炉用焦炭;高炉焦炭
blast-furnace crucibe 高炉炉缸
blast-furnace drainage 高炉排水
blast-furnace dust 高炉灰
blast-furnace ferromanganese 高炉锰铁
blast-furnace frame 高炉炉架
blast-furnace gas 鼓风炉煤气;高炉瓦斯;高炉煤气
blast-furnace gas burner 高炉煤气燃烧器
blast-furnace gas cleaner 鼓风炉炉气除尘器;高炉煤气净化器
blast-furnace gas engine 高炉煤气机
blast-furnace gas regulator 高炉煤气调节器
blast-furnace gas scrubber 高炉煤气洗涤塔
blast-furnace gas scrubbing wastewater 高炉煤气洗涤废水
blast-furnace gas scrubbing water 高炉煤气洗涤水
blast-furnace gas turbine 高炉煤气汽轮机
blast-furnace gas valve 高炉煤气阀
blast-furnace gate 高炉风阀
blast-furnace graphite 高炉石墨
blast-furnace gun 高炉泥炮
blast-furnace heater 鼓风加热器
blast-furnace hoist 高炉绞车
blast-furnace hot blast-stove 高炉热风炉
blast-furnace iron powder 高炉矿粉
blast-furnace lift 高炉升降机
blast-furnace line 高炉型ខ
blast-furnace lining 高炉衬;高炉衬砌
blast-furnace lump ore 高炉块矿
blast-furnace lump slag 高炉块渣;高炉块渣
blast-furnace man 高炉工;鼓风炉工
blast-furnace matte 鼓风炉锍
blast-furnace method 鼓风炉熔炼法
blast-furnace mud gun 高炉用泥炮
blast-furnace output 鼓风炉产量
blast-furnace pitch 高炉焦油沥青
blast-furnace plant 炼钢厂;高炉车间
blast-furnace platform 高炉工作台
blast-furnace Portland cement 矿渣硅酸盐水泥;矿渣波特兰水泥;高炉硅酸盐水泥
blast-furnace production campaign 高炉生产炉周期
blast-furnace profile 高炉内型
blast-furnace ring 高炉炉环梁
blast-furnace roasting 鼓风煅烧
blast-furnace shaft 高炉炉身
blast-furnace skip hoist 高速冷轧管机;高炉翻斗绞车
blast-furnace slag 矿渣;硬矿渣;高炉熔渣;高炉渣;高炉矿渣;冲天炉渣
blast-furnace slag aggregate 高炉矿渣集料;高炉矿渣骨料
blast-furnace slag cast stone 高炉渣铸石
blast-furnace slag cement 高炉矿渣水泥
blast-furnace slag cement flake board 高炉矿渣水泥木屑板
blast-furnace slag chip(ping)s 高炉炉渣碎屑
blast-furnace slag coarse aggregate 高炉炉渣粗集料;高炉炉渣粗骨料;高炉矿渣粗骨料
blast-furnace slag concrete 高炉矿渣混凝土
blast-furnace slag dust 高炉炉渣粉末
blast-furnace slag fiber 高炉炉渣纤维
blast-furnace slag filler 高炉炉渣填充料
blast-furnace slag fill(ing) 高炉炉渣填塞
blast-furnace slag(paving) sett 高炉矿渣铺路石;高炉炉渣铺路小方块
blast-furnace slag Portland cement 高炉矿渣波特兰水泥
blast-furnace slag sand 高炉矿渣扎石砂;高炉矿渣砂
blast-furnace slag sand concrete 高炉矿渣砂混凝土;高炉矿渣砂混凝土
blast-furnace stack 高炉烟囱
blast-furnace stack casing 高炉身外壳
blast-furnace steam blower 高炉蒸汽鼓风机
blast-furnace tapping 高炉出铁
blast-furnace tar 高炉焦油沥青
blast-furnace tear down apparatus 高炉旧炉衬拆除设备
blast-furnace throat 高炉炉喉
blast-furnace top bell 高炉料钟
blast-furnace top gas mud 高炉煤气尘泥
blast-furnace trass cement 高炉凝灰岩水泥;高炉火山灰水泥
blast-furnace well 鼓风炉炉缸;高炉炉缸
blast gas cloud 爆炸气体云
blast gate 风门;风阀;排气门
blast ga(u)ge 风压计
blast governing 鼓风调节
blast grit 喷丸
blast gun 吹砂枪
blast head 鼓风压头;风栅
blast head modules 鼓风机头模件
blast heater 空气预热器;热风炉;热炉
blast heating 热风;预热送风;鼓热风
blast-heating cupola 热风冲天炉
blast-hole 爆破眼;炮眼;爆破孔;风口
blast-hole arrangement 炮眼布置
blast-hole bit 爆破孔钻头;炮眼钻
blast-hole charge 炮眼装药
blast-hole charger 深孔装药器;炮眼装药器
blast-hole drill 爆孔钻;深孔凿岩机;爆破孔钻机;炮眼钻(机);炮孔钻
blast-hole drilling 爆破孔钻进
blast-hole machine 爆破孔钻机;炮眼钻机
blast-hole method 打眼法;深孔崩落开采法
blast-hole mining 深孔崩落开采
blast-hole pit 炮眼钻头
blast-hole rig 爆破洞装备
blast-hole ring 扇形炮眼
blast-hole springing 炮眼扩大;掏壶
blast hood 鼓风罩
blastic deformation 冲击变形
blastic texture 变晶结构
blast in borehole 井下爆破
blast indicator 鼓风指示器;测风计
blasting 过载失真;吹洗;爆破;喷丸清理;喷砂法;碎裂;鼓风
blasting accessory 爆破用具;爆破仪表
blasting accident 爆破事故
blasting action 裂开作用;鼓风作用;爆破作用
blasting agent 炸药;爆破剂
blasting agent storage 炸药库
blasting and sanding method 爆破灌砂法
blasting barrel 导火线套管
blasting battery 爆破机器;发爆器
blasting benches 梯段爆破
blasting board 放炮时的支撑保护板
blasting bridge 爆破电桥
blasting bulldozer 频爆式推土机
blasting cable 爆破母线
blasting cap 雷管(头);火雷管;引爆管;爆破雷管;起爆筒;起爆雷管;起爆管;普通雷管;引信
blasting cartridge 爆破炸药包;爆破筒;爆破管;起爆药筒
blasting chamber 爆破峒室;起爆室
blasting charge 炸药装置;炸药(包);爆破装药;爆破炸药
blasting circuit 爆破线路;爆破电路;起爆电路
blasting code 爆破规程
blasting coefficient 爆破系数
blasting compaction 爆炸挤密;爆炸压密;爆破压实
blasting compaction method 爆炸压密法;爆炸挤密法;爆破挤密法
blasting compound 爆炸成分
blasting concussion 爆破冲击作用;爆破震动
blasting cone 爆破漏斗
blasting consolidation 爆破加固
blasting contractor 爆破承包人
blasting crater 爆破漏斗
blasting crater hole 爆破漏斗孔
blasting crater test 爆破漏斗试验
blasting crew 爆破班组
blasting curtain 爆破防护帘;爆破挡帘
blasting density 爆炸密实度(按每立方厘米克计)
blasting depth 爆破深度;爆破深度
blasting device 放炮用具
blasting diaphragm 爆破隔膜
blasting displacement method 爆破位移法
blasting doubler 二次爆破
blasting effect 爆破影响;爆破效应;爆破效果

blasting efficiency 爆破效率
blasting energy 爆破能
blasting equipment 爆破用具;喷砂装置;喷砂设备
blasting experiment 爆破试验
blasting explosive 爆破炸药
blasting explosive density 爆破炸药密度
blasting force 爆力
blasting for internal compression(in rock) 内部压缩爆破(岩石中)
blasting for loosening rock 松动爆破
blasting formula 装药量计算公式;爆破药量公式
blasting for throwing rock 抛掷爆破
blasting for tunnel exploration 坑探爆破
blasting fragmentation 爆破块度
blasting fuse 雷管;引信;引爆导线;导火线;导火索;导爆线;导爆索;传爆线;传爆索;起爆引线
blasting gang 爆破班组
blasting gear 放炮用具;爆破设备
blasting gelatin(e) 胶棉炸药;炸胶;甘油凝胶;爆炸胶
blasting gelation 胶棉炸药
blasting hazards and limitations 爆破危害及其限制
blasting hole without stemming material 不用填塞材料进行爆破洞
blasting impulsive wave 爆破冲击波
blasting index 岩石爆破性指数;爆破指数
blasting kettle 喷砂壶
blasting layout 爆破点布置;爆破布置
blasting lead 爆破导线
blasting machine 起爆机;电爆机;引爆器;放炮器;发爆机;爆破器械;起爆器;喷砂机
blasting mat 爆破挡帘;防(爆)(钢)网;防爆垫;爆破垫
blasting match 放炮用点火线
blasting material 炸药;爆破器材;爆破材料
blasting method 矿山法(指隧道开挖);爆炸方法;爆破方法;爆破法
blasting monitor 爆破监测器
blasting-off the solid 原矿体爆破;不掏槽爆破
blasting of peat 爆炭泥炭土;泥炭爆破;沼泽地爆破
blasting of profiles 周边爆破;光面爆破
blasting ohmmeter 爆破欧姆表
blasting oil 硝化甘油;爆炸油
blasting operation 爆破作业;鼓风操作
blasting parameter 爆破参数
blasting party 爆破班组
blasting pattern 爆破顺序;爆破模式;爆破方式;爆破布置
blasting pellet 爆破雷管;球状炸药
blasting point 爆破点
blasting powder 黑色炸药;炸药;爆破炸药
blasting power 爆炸力
blasting practice 爆破操作;爆破技术
blasting pressure 爆破压力
blasting primer 引爆管
blasting procedure 爆破方法;爆破法;爆破程序
blasting propagation 传爆
blasting ratio 爆破比(单位耗药)
blasting regulation 爆破规程
blasting replacement method 爆破置换法;爆破排淤法
blasting resistance 爆破阻力
blasting round 爆破循环;炮眼组
blasting screen 爆破防护帘;爆破挡帘
blasting set 爆破用具;喷砂装置
blasting shock 爆破震动
blasting shop 喷砂车间
blasting shot 喷丸;喷的铁丸
blasting site 爆破场地;爆破地点
blasting slurry 浆状炸药
blasting supply 爆破物资;爆炸材料
blasting switch 爆破开关
blasting team 爆破班组
blasting technician 爆破技师;爆破工
blasting technique 爆破技术
blasting test 爆破试验
blasting timer 爆破定时器
blasting tools 爆破工具
blasting trenching 爆破开沟
blasting-tumbling machine 滚筒喷砂机
blasting unit 引爆器;电力放炮机;起爆器
blasting vibration 爆破震动;爆破振动;爆破引起的振动
blasting vibration observation 爆破地震观测
blasting wedge 爆破楔;爆破漏斗

blasting with angular grit 倾斜喷砂
blasting with round shot 旋转式喷砂
blasting work 爆破工作;爆破作业
blast injury 冲击伤;爆炸性损伤
blast inlet 鼓风进口
blast intensity 鼓风强度
blast lamp 喷灯
blast lamp for soldering 焊接喷灯
blast line 鼓风管;空气管
blast load 爆炸荷载;爆破荷载
blast loading 爆破装药
blast main 总风管;主空气管道
blast meter 测风计;爆破力测量计
blast mixer 车载干拌混凝土喷射装置
blast nozzle 风嘴;喷砂嘴;喷气嘴
blasto-amygdaloidal structure 变余杏仁构造
blastoaplitic texture 变余细晶结构
blastoash texture 变余凝灰结构
blastobreccia 变余角砾岩
blastobrecciatic texture 变余角砾状结构
blastocrystalloclastic texture 变余晶屑结构
blastodettritus texture 变余岩屑结构
blastodiabasic texture 变余辉绿结构
blast of air 气流
blasto-flow structure 变余流纹构造
blastofragmental texture 变余碎屑状结构
blast of steam 蒸汽冲击
blast of wind 阵风
blastogabbroic texture 变余辉长结构
blasto giant stratification structure 变余巨厚层层理构造
blastoglassyclastic texture 变余玻屑状结构
blastogranite texture 变余花岗状结构;变余花岗结构
blastogranitic 变余花岗状
blastogranitic rock 变余花岗岩
blasto heavy stratification structure 变余厚层层理构造
blasto-hemiclastite 变晶半碎裂岩
blasto-holoclastite 变晶全碎裂岩
blastoid extinction 海蕾绝灭
blasto medium heavy bedded stratification structure 变余中厚层层理构造
blastomere 分裂球
blastomylonite 变余糜棱岩
blastomylonitic texture 变余糜棱状结构
blasto-nodular structure 变余结核构造
blastopelitic texture 变余泥状结构
blastophitic 变余辉岩;变余辉绿状
blasto-pillow structure 变余枕状构造
blastopilotaxitic texture 变余交织结构
blastopore 原口
blastoporphyritic texture 变余斑状结构;变斑晶结构
blastoporphyritic 变余斑状
blastopsammite 变余沙岩
blastopsammitic 变余砂状
blastopsammitic texture 变余砂状结构
blastopsephitic 变余砾状
blastopsephitic texture 变余砾状结构
blastoresorption-crystal texture 变余熔蚀结构
blast-oriented 定向爆破
blast orifice 鼓风口
blasto-ripple mark structure 变余波痕构造
blastosilt texture 变余粉砂状结构
blasto-streaked structure 变余条带构造
blasto thin bedded stratification structure 变余薄层层理构造
blastotomy 分裂球分离
blast out 爆破
blasto-vesicular structure 变余气孔构造
blastovolcanic brecciatic texture 变余火山角砾结构
blast pipe 风管;乏气管;鼓风管;吹管;吹风管;废气管;放气管;排气管
blast pipe nozzle 吹管嘴
blast pipe with intermediate nozzles 带中间喷嘴的喷射管
blast pit 排气井
blast plate 防爆管;防烟板;喷气防护板
blast preheater 鼓风预热器
blast pressure 爆破波压力;鼓风压力;风压;爆炸压力;瞬间燃烧压力
blast pressure ga(u)ge 风压计
blast primer 预涂底漆

blast-produced 吹制的
blast-proof 防爆的
blast-proof design 防爆设计
blast protection 冲击波防护;强气流防护;防爆
blast purge 鼓风吹扫;气体清洗
blast rate 鼓风速率
blast regulation valve 鼓风调节阀
blast-resistant civil defence structures 防爆土木防护结构
blast-resistant construction 防爆结构
blast-resistant door 防爆门
blast rig 钻炮眼设备
blast roaster 鼓风焙烧炉
blast roasting 鼓风煅烧;鼓风焙烧法
blast room 鼓风室;喷砂室
blast-safe chamber 爆炸掩蔽室
blast sand 喷砂(用砂)
blast sanding 喷砂处理
blast sand with tumbling barrel 滚筒式喷砂机
blast shell 爆破弹
blast shelter 爆炸掩体;爆炸掩蔽室
blast shield 防爆屏蔽;反焰器;反焰板
blast shutter 爆炸快门
blast stone 爆破形成的毛石;石渣路
blast stower 风力充填机
blast supply 鼓风供应;供风
blast temperature 鼓风温度;爆炸温度
blast throwing 爆破抛掷
blast tube 喷嘴管;送风管
blast turbine 喷气发动机射流涡轮机
blast tuyere 风口
blast type loading 冲击型加载
blast vent 防爆孔
blast vibration 爆炸振动
blast volume 鼓风容积
blast volume controller 风量控制器
blast wall 防弹墙;防爆墙
blast wandering 风压波动
blast wave 激发波;冲击波;爆炸波;爆破冲击波
blast wave pressure 爆炸波压力
blatt(er) 横推断层;平推断层
Blatthaller loudspeaker 布拉特哈勒扬声器
blau gas 液化烃;蓝煤气;纯净水煤气
Blaut Lang process 布劳特兰(不锈钢)电解抛光法
Blavier's test 布莱维尔试验
Blaw Knox jet mill 四喷嘴对喷式气流粉碎机
blaze 刻记号(路标);刻标;火焰;测量标记;树号
blaze angle 闪耀角
blazed diffraction grating 闪耀衍射光栅
blazed grating 红外(线)光栅;炫耀光栅;定向光栅
blazed hologram 闪耀全息图
blazed pig iron 高硅生铁
blaze of grating 光栅闪耀
blazer 燃烧物
blaze the line 标出道路路线
blaze the rail 做路标
blaze the rail for 为……铺平道路
blaze wavelength 闪耀波长
blazing-off 回火;油中回火
blazoned glass 彩色玻璃
BL diazo fast bordeaux BL 重氮坚牢枣红
bleach 冲淡法;漂白
bleachability 漂泊性;漂泊能力
bleachable absorber 可漂白吸收体
bleachable ink 可漂白油墨
bleachalbe filter 可漂白滤色片
bleach-bath 漂白液
bleached bed 浅色层;漂白层
bleached beeswax 白蜡;白蜂蜡
bleached chemithermomechanical pulp 漂白化学热力学纸浆
bleached earth 漂洗土;漂白土
bleached fiber 漂白麻刀
bleached glue 漂白胶
bleached holographic(al) grating 漂洗全息摄影栅
bleached horizon 漂洗层;漂白层
bleached kraft-mill effluent 漂白牛皮纸厂污水
bleached kraft pulp 漂白牛皮纸浆
bleached oil 精制油;无色油;漂白油
bleached paper 漂白纸
bleached pulp 漂白纸浆
bleached pulp sewage 漂白纸浆污水
bleached sand 漂白砂
bleached(shel)lac 白虫胶;漂白紫胶;漂白虫胶

bleached soil 漂洗土
bleached sulfite pulp 漂白亚硫酸盐纸浆
bleached zone 褪色带
bleacher 露天看台;漂白器;漂白剂;漂白机;漂白工
bleacherite 露天看台的观众
bleachers 看台
bleacher seating 露天看台座位
bleachery 漂白间
bleachery effluent 漂白厂污水
bleachery waste(water) 漂白厂污水
bleach hollander 漂白机
bleaching 变白(漂白(的);褪色作用;涂面变色
bleaching action 漂白作用;漂白反应
bleaching agent 漂白剂
bleaching and dy(e)ing mill 漂染(工)厂
bleaching and dyeing wastewater 漂白染色污水
bleaching assistant 漂白辅助剂
bleaching bath 漂白液
bleaching clay 活性土;漂白黏土;漂白土
bleaching earth 活性土;漂白土
bleaching effect 漂白效应;褪色效应
bleaching effluent 漂白污水
bleaching energy sensitivity 褪色能量灵敏度
bleaching fluid 漂白粉液
bleaching in stage 分段漂白
bleaching intensity 漂白强度
bleaching liquid 漂白液
bleaching liquor 漂液
bleaching liquor settling tank 漂液澄清池
bleaching liquor storage tank 漂液储[贮]槽
bleaching lye 漂白液
bleaching machine 漂白机
bleaching-out 褪色(的)
bleaching plant 漂白车间
bleaching podzolisation 漂白灰化作用
bleaching powder 含氯石灰;漂粉;漂白粉
bleaching powder and liquor section 漂粉与漂液工段
bleaching powder machine 漂粉机
bleaching power 漂白能力
bleaching process 漂泊过程
bleaching signal 变透明信号
bleaching sludge 漂液残渣
bleaching soil 漂白土
bleaching solution 漂白液
bleaching tower system 塔式漂白装置
bleaching water 漂白水
bleaching works 漂白厂
bleach liquor 漂白液
bleach liquor settling tank 漂白液沉淀池
bleach oil 漂白油;无色滑油
bleach on the green 草地曝晒漂白
bleach out process 漂白法
bleach spot 漂白斑;白斑
bleancherite 露天看台的观众
bleary 模糊的
bleb 矿物包裹体;小包体;气泡;突起的小泡(陶瓷缺陷)
bleb ingot 有泡钢锭
bled steam 撤汽;废蒸汽
bled steam evapo(u)rator 抽气加热蒸发器
bled steam feed heater 抽气回热给水加热器
bled timber 泌脂原木
bleed 漏出;流失;冒油;泄放;排污
bleed air 放气
bleed air flow 抽气量
bleed cock 泄水孔塞
bleed connection 排气接头
bleeded feeder 出气帽口
bleeded head 出气帽口
bleeded riser 出气帽口
bleeder 降压电阻;泄流器;泄放器;泄出器;泄出器;分水口;放出管;放气阀;旁漏;疏水阀
bleeder circuit 泄放电路;分压器电路
bleeder cloth 吸胶布
bleeder cock 放水活门;泄水孔塞;放水龙头;放水旋塞
bleeder coil 排流线圈
bleeder condensing turbine 汽冷涡轮机
bleeder current 泄放电流;旁漏电流
bleeder heater 回热加热器;抽气加热器
bleeder hole 流放孔;抽气孔;气孔;排泄孔;通气孔;通风孔;出气孔
bleeder line of tile 排水瓦管

bleeder network 旁漏网络
bleeder nozzle 抽气管
bleeder pipe 集水管;泄水管;放水管;放气管;排泄管(道);排出管
bleeder plug 泄水孔塞;放油塞
bleeder resistance 泄漏电阻;分压电阻;分泄电阻;排泄阻力
bleeder resister 泄漏电阻器
bleeder resistor 分泄电阻器
bleeder screw 放水阀螺旋
bleeder steam 抽气;汽轮机抽气
bleeder tile 排气瓦(管);泄水瓦(管)
bleeder turbine 抽气透平;抽气式汽轮机;抽气轮机;放汽式汽轮机
bleeder type condenser 溢流式大气冷凝器
bleeder type condensing plant 滴水大气冷凝厂;溢流式大气冷凝装置
bleeder type steam engine 余热式蒸汽机
bleed(er) valve 放泄阀;吸出阀;出料阀;抽气阀;抽出阀;排泄阀;放气阀;泄放阀
bleeder well 减压井;排水井
bleed evacuation 放空水;放空气
bleed gas 放气
bleed heating 抽汽加热
bleed hole 通气洞;泄水眼;出铁口;排水孔;排出孔
bleeding 浸溃透过;泌水;泌水(指混凝土);泌浆(指混凝土);泅色;溢剂;固定相的流失;抽气;分级加热;放出;泛油(路面泛出多余沥青);泛水;倾倒(将袋装谷类倾入散装舱);排气操作;渗色;渗出
bleeding a project 放血(从一项工程项目中抽走资金,而忽略维护工作)
bleeding asphalt 沥青路面泛油
bleeding asphalt surface 泛油的沥青路面
bleeding capacity 砂浆泌水率;灰浆泌水率;泌水能力;泌水量;灰浆泌浆率;单位泌水量;单位泛水量
bleeding cement 泛浆(指混凝土表面泛出水泥浆);翻浆(混凝土表面);水泥浆沫(混凝土表面);水泥浮浆;水泥泛浆(混凝土表面);浮浆(指水泥)
bleeding channel 泌水通道
bleeding cock 泄放阀;排放旋塞
bleeding cycle 抽气循环
bleeding device 排放设备
bleeding edge 海图上延伸到图框外的图形边缘;破图廓
bleeding export 亏本的出口
bleeding fringe 扩散条纹
bleeding index 渗出性指数
bleeding iron 取样棍;取样工具
bleeding joint 石油沥青接缝
bleeding of a feeder 冒口回涨
bleeding of colo(u)r 颜料印流;染料印流
bleeding of concrete 混凝土泌浆;混凝土泌水现象;混凝土的泌水
bleeding of foam 泡沫泌水量
bleeding of the brake 制动器渗漏
bleeding of waste liquid 废液排除
bleeding pigment 渗色颜料
bleeding pipe 放水管
bleeding plug 放气旋塞;泄水孔塞
bleeding point 抽气口;收取点;收集点
bleeding pressure 根压;伤流压
bleeding rate 泌水速率;泌水率;泌浆率
bleeding ratio 泌水率;泌浆率
bleeding recovery 排气回收装置(制冷机冷媒)
bleeding resistance 抗渗色性
bleeding resistant grease 抗凝胶收缩润滑脂;抗流润滑脂
bleeding shutter (放气管的)节气门
bleeding steam 提取蒸汽
bleeding surface 渗色污染的表面
bleeding test 分油试验;渗色试验;渗出试验
bleeding turbine 抽气式汽轮机;放气式涡轮机;放气式透平
bleeding water 混凝土析水;混凝土沁水;混凝土泌浆
bleed line 溢流油管道;线条扩散
bleed-off 移去能力;溢流调节;放水口;放出过多液体
bleed-off belt 抽气环形室
bleed-off line 排出管线;排出管路
bleed-off passage 抽气通道
bleed-off pressure 排出压力
bleed-off point 油漆流淌

bleed opening 流出口;放出孔
bleed-out 放出;树脂渗出;漏钢;跑钢
bleed screw 带孔螺钉
bleed steam connection 抽气接管
bleed the tyre 减轻胎内压力
bleed-through 透背;渗胶;褪色;透胶
bleed valve 放气阀;泄水阀(门);泄放阀;排除阀;排出阀
bleed venting 排气通风
bleed water 分出的水
bleicherde 灰壤淋溶层
Bleininger's method 布莱宁格耐酸试验法
blemish 瑕疵;沾污;表面缺陷;缺陷;轻疵;污渍;污点
blemish-free surface 无瑕疵的表面;无缺陷表面
blemish surface 缺陷表面
blench 变白;退缩
blend 搅混;混和;混合;共混;掺和(物);掺和料;拌和
blendability 共混性
blendable 可混合的
blend composition 混合组分;共混组分
blende 褐色内光矿物;闪锌矿
blended aggregate 级配集料;级配骨料;混合集料;混合骨料
blended aggregate mixture 级配集料混合料;级配骨料混合料
blended asphalt 掺和(地)沥青;调配沥青
blended asphalt joint 填缝(石油)沥青掺和料
blended asphalt joint filler 填缝(石油)沥青掺和料;填缝沥青拌和料
blended cement 混合水泥;合成水泥;掺配水泥
blended coal 掺和煤
blended electrodynamic(al) brake 电空混合制动
blended electropneumatic brake 电空混合制动
blended fabric 混纺织物
blended filament yarn 长丝混合纱
blended frit 混合熔块
blended fuel 混合燃料
blended gasoline 混合汽油
blended lamp 自动镇流灯
blended liquid phase 掺杂液相
blended loan 混合息率贷款
blended material 混合材
blended meal silo 生料搅拌库
blended meal slurry 生料混合库
blended oil 混合油
blended Portland cement 混合硅酸盐水泥;混合波特兰水泥
blended product 混合产物
blended sand 混合砂
blended spun yarn 短纤维混纺纱
blended stock 混合原料
blended tar 掺和柏油;掺和焦油沥青
blended unconformity 混合不整合
blendent 配色
blender 混料师;混料机;混合器;獾毛软刷;捣碎器;掺和器;掺和机;拌和机;试验室用小型混合机
blender grinder 磨碎拌和机
blender-type drier 锥形混合式干燥器
blending 混料;釉变;倒圆;掺混;掺和(剂);掺拌;计量斗;润彩(涂装);法;配料;调合;松砂
blending abrasives 掺和的磨料
blending agent 调合剂
blending aggregate 掺和集料;掺和骨料
blending air 搅拌空气
blending basin 料浆搅拌池;搅拌池;掺和池
blending batch 掺和配合料;调整配方
blending bin 混料仓;搅拌仓;混合斗;掺料仓;拌和斗;配料舱;配料仓
blending bunker 混合斗仓;混合斗仓;混合仓;配合仓
blending chamber silo 混合室料仓
blending chart 混合图表
blending colo(u)r 配色
blending component 混合材
blending composition 掺和料组成
blending compound 掺和组分
blending cone silo 混合锥库
blending constituent 混合成分
blending container 混合容器
blending control 配料控制
blending conveyer 混合输送机
blending credit 协调信贷
blending effect 混合效果

blending fuel 掺和燃料
blending machine 搅拌机;掺和机
blending machinery 搅拌机械
blending material 混合材
blending mixer 混料机
blending mixture 搅拌混合物
blending octane number 混合辛烷值
blending of aviation gasoline components 航空汽油组成的调合
blending of fuel 燃料的掺和
blending of ink 调油墨
blending of sizes 粒度配比
blending pile 混合堆场
blending plant 混配设备
blending pond 料浆搅拌池
blending pool 料浆搅拌池
blending pump 混合泵
blending silo 均化仓;混料仓
blending system 混气设备
blending tank 搅拌池;混合槽;掺和槽
blending time 搅拌时间
blending valve 混流阀;混合阀
blend ores 掺和矿
blendous 含闪锌矿的
blend price 综合价格
blend ratio 掺和比
blend soil 混杂土
blend stop 挡条
Blenheim Palace 布雷牛姆宫
blenometer 弹力计;弹簧计;弹簧弹力测量仪
Blessum compensation 伯勒森补偿
blet aerator 皮带松砂机
blew out 沉箱放气
bliabergite 块云母
blibbing 表面气泡(石膏模缺陷)
blibe 细长气泡
Bligh's creep theory 布莱渗流理论
blight 受挫;枯萎(病)
blight and flight 衰落迁腐(城市)
blighted area 陋屋区;破落区;污染区;枯萎区(植物);荒芜杂草地区;荒芜地区;荒废地;荒地;阴影面积;遮蔽面积
blighted effect 遮蔽作用(指裂缝)
blighted house 陋宅
blight trial 不定样品试验
blight wood 盲枝
blimp 小型飞船;软式小飞艇;(可折叠的)软式汽艇;软式飞艇
blind 螺旋帽;盲目的;盲的;隐棚;遮帘;遮光物;障蔽之物;断流板;不着墨;不感光的;无出口的;素压
blind abut 隐蔽支座;隐蔽桥台;隐蔽扶壁
blind abutment 假装桥台;隐蔽支座;隐蔽桥台;隐蔽扶壁
blind advertisement 匿名广告
blindage 盲障;掩体;封堵材料
blind alley 死巷;尽头路;死胡同
blind anchor 隐蔽锚碇
blind anchorage 隐蔽锚地
blind and dumb asylum 盲哑院
blind angle 盲角;遮蔽角
blind anomaly 盲异常
blind approach 盲目着陆;盲目进场
blind approach beacon system(babs) 盲目进场信标系统
blind approach beam system 波束导航盲目进场系统
blind apron 卷帘百叶窗罩
blind arcade 盲拱廊;壁上拱廊;假拱廊;(不开洞的)装饰拱廊;封闭拱廊;无窗拱廊;实心连拱(廊);实心拱廊
blind arch 盲拱;实心拱;装饰拱;填塞的拱;假拱
blind area 封闭地块;雷达盲区;盲区;遮视区;房屋外墙遮盖;无信号区
blind assignment 随机分组
blind attic 紧接房屋顶下不装修而封闭的空间;未装修的阁楼;屋顶下封闭空间
blind axle 暗轴;静轴
blind bag 盲袋
blind baggage 铁皮货车(铁路)
blind balustrade 实心栏杆
blind bit 不取心钻头
blind blocking 无色凹凸印;素压印
blind bolt 封闭螺栓;暗插销
blind bond 砖砌体的一种砌合法

blind bore hole 盲孔
blind borehole technique 盲孔钻进技术
blind box 遮帘匣;窗帘内箱;百叶窗匣
blind buckler 锚链孔盖
blind car 行李车
blind casing 粗窗框;未加装修的毛窗匣
blind catch 关闭把手;百叶窗扣
blind catch basin(地下排水的)暗井;截留盲井;截流盲井;盲视式截留井;暗截流井;暗接井;暗集流井;渗水井;排水井
blind catchment basin 暗井
blind cave 盲洞
blind ceiling 中间层顶板
blind channel 盲水道
blind chintz 横条子窗帘布
blind closure 闭塞封闭器
blind coal 细薄干煤;无烟焦;天然焦;瘦煤
blind column 部分筑在墙内的柱
blind competition 盲目竞争
blind concave 闭式凹板
blind concrete 铺填混凝土
blind concrete beam 实心混凝土梁
blind controller system 盲目控制器系统
blind corner 碍视转角;碍视交叉口转角
blind cover 舷窗铁盖
blind creek 间歇河床;临时性小溪;间歇性小河;间歇性溪流;雨季河;干涸的小溪;干涸的小河;干河床
blind crossing 碍视交叉口;视线不良的交叉口
blind curtain 遮风幕;遮阳布篷
blind curve 阻视曲线;碍视曲线
blind curved road 阻碍视距的弯路;碍视弯路
blind cylinder 二端封闭筒
blind deposit 盲矿床;隐伏矿床
blind distance 阻视距离;碍视距离
blind ditch 盲沟;暗沟;排水盲沟
blind ditch design 盲沟设计
blind door 装饰门;假门;百叶门;暗门
blind dovetail 暗燕尾榫;暗楔榫
blind drain 盲沟;阴沟;暗渗(管)道;暗沟;排水暗沟;死排水沟;石砌排水沟;渗沟
blind drainage 内流水系;盲沟排水;阴沟排水;闭流水系;暗排水沟;暗流排水;无流水系
blind drainage area 闭流区
blind drain design 盲沟设计
blind drilling 无地质准备工作的钻孔;未取任何资料的钻进
blind driver 无翼缘主动轮
blind driving 盲目开车;关灯驾驶
blind driving wheel 无轮缘动轮
blind edge 暗缝边
blinded-pipe yield test 盲管流量测定
blind embossing 素压印
blind end 封闭端
blind entry 无说明分录;失实记录
blinder 舷窗铁盖
blind estuary 盲湾(河口泻湖湾)
blind fast 百叶窗固定器;百叶窗扣
blind fastener 隐蔽式扣件;单面紧固件;膨胀螺栓
blind fault 盲断层;隐伏断层
blind feeder 暗冒口
blind fence 半封闭篱笆;半透风篱笆
blind figure 失实的数字
blind film 色盲片
blind fish 盲鱼
blind flange 闷头法兰;盲法兰;盲板(集装箱);管口盖凸缘;管口盖板;管堵头法兰;盖板;堵头法兰;堵塞法兰;法兰盖;闭塞板;无孔凸缘;无孔法兰(盘)
blind floor 下层地板;桥面底层;毛地板;楼面底板
blind fret 盲纹
blind furrow 盲垄
blind gallery 无出口通道;死通道;死胡同;实心拱廊
blind green paint 暗绿色漆;暗绿漆
blind ground joint 磨口堵头
blindgut 盲管
blindgut wound 盲管伤
blind harbo(u)r 盲港
blind hatch 暗舱口
blind head 暗冒口
blind header 墙内丁砖;半砖;假顶砖;暗丁砖;暗丁石;半块砖
blind hinge 暗铰(链);暗合页
blind hoistway 封闭的升降机井道
blind hole 闷眼;盲眼;盲孔;非贯穿孔;不通孔

blind image 隐蔽像;不受墨图像
blind impression 平压印
blinding 细石屑;釉面失透;盖土;断流;堵塞;补路砂石;填隙用细石屑;填缝石屑;填封;填充道路面层孔的细石屑;石屑(填缝用);失透
blinding concrete 盖面混凝土;防水铺填混凝土;填充混凝土
blinding concrete course 找平的混凝土层;混凝土找平层
blinding course 封层
blinding glare 眩光
blinding layer 基础垫层;垫层
blinding layer of concrete 混凝土找平层
blinding material 黏合剂;包封材料;透水材料;填塞料;填塞材料;填充材料;透水填料
blinding of lithographic(al) plates 印刷板堵(油)量
blinding of screen 筛子阻塞;筛子堵塞;筛孔堵塞
blinding printing 空印
blinding sand 炫目的砂;填砂
blinding sandstone 嵌缝砂石
blinding stone 作填料层的石头;嵌缝石
blinding tile drain 暗瓦管;盖土瓦管;盖土排水瓦管
blind inlet 盲沟进水口;排水暗沟进水口
blind interception 盲目拦截
blind introduction 盲目引进
blind joint 盲节理;盲接头;隐节理;隐接头;封闭接头;暗缝;无间隙接头;无缝接头;瞎缝
blind keyboard 盲键盘
blind lake 冬季湖(夏天干枯);封闭湖(无源出口);死湖(无补给的湖)
blind landing 盲目着陆;仪表着陆
blind landing experiment 盲目着陆试验装置
Blind Landing Experiment Unit 盲目着陆实验所
blind lattice 假坚棱条
blind layer 隐蔽层
blind lead 单一出口水道;袋状海湾;冰川死水道
blind level 倒置虹管式排水巷道;水平排水暗管
blind-life 百叶窗拉手
blind lift 升降百叶窗执手;窗帘升落开关;百叶窗拉手;窗帘升降开关
blind lining 中间层壁板(冷藏车用)
blind lode 无露头矿脉
blind-loop 盲曲
blind machine 盲机
blind main 尽端暗管;尽端干管
blind monitoring 监控传声器;盲目控制;视场外监听
blind mortise 凹榫;暗榫眼;暗榫孔
blind-mortise-and-tenon joint 暗榫结合
blind-nailed 暗钉
blind nail(ing) 暗钉
blind navigation 盲目航行;仪表领航;仪表导航
blind navigation chart 仪表领航图
blindness 视觉缺失
blind niche 隐蔽壁龛
blind nipple 闷心短管
blind nut 螺帽;螺母;螺盖;盲孔螺母
blind off a line 关闭管路
blind operator 百叶窗片控制器
blind ore 盲矿
blind ore body 盲矿体
blind orebody beside borehole 井旁盲矿
blind orebody lower the bottom of borehole 井底盲矿
blind ore vein 无露头矿脉
blind pack 眼罩
blind panel 不开窗口的墙板
blind pass 死巷;尽头路;暗道
blind patch 孔口盖板;盖补板;堵孔板
blind perforator 盲穿孔机
blind pier 部分筑在墙内的墩子
blind pier capital 筑在墙内的部分墩顶;附墙柱头
blind piercement fold 隐刺穿褶皱
blind pipe 盲管;烟突;堵头管;无孔管
blind pipe cave 盲管洞
blind pit 暗坑;盲纹孔
blind plate 盲板
blind plug 空插头;绝缘插头;堵塞
blind pocket 暗匣
blind pool 盲式集资;盲式集团
blind power 电抗功率
blind production 盲目生产
blind prognosis 气候展望

blind pulley 三眼木饼
blind rams 全封闭闸板(防喷器)
blind rams blow(ing) out preventer 全封闭防喷器
blind riser 暗冒口
blind rivet 盲铆钉;埋头铆钉
blind roaster 套式炉
blind rock 暗礁
blind roller 长涌;暗浪
blind rope chopper 断绳器(事故处理工具)
blind rosette 实心圆花窗
blind row 剧院靠墙排座
blinds 百叶帘;百叶窗
blind sag 轨道暗坑;暗坑
blind sea 长涌
blind search 盲目搜索
blind seat 视线被挡的座椅
blind sector 盲区;荧光屏阴影区;扇形阴影
blind sector of the bow 船首盲区
blind sending 盲发送
blind(set) core 盲孔芯;盲芯
blind sewer 污水暗管
blind shaft 盲竖井;盲井;窗帘卷轴;暗竖井;暗井
blind sheet 空白页
blind shell 未炸的炮弹
blind shield 挤压式盾构;闭口盾构
blind shield method 闭胸盾构法
blind side 死角
blind siding 尽头线
blind slat 固定的百叶板;百叶的叶片;百叶窗帘板
blind slope 盲斜井
blind splice 捻接;拼接
blind spot 静区;盲点;哑点;死区;死角
blind stamping 拱花
blind stitch 暗线缝
blind stop 暗挡条
blind stop bar 窗帘止铁
blind stopper 窗帘止卡
blind stor(e)y 暗层;暗层;无窗楼层;无窗的层楼;(哥特式教堂大门上的)拱廊
blind street 袋形路
blind taper joint 锥柱堵头;磨口锥塞
blind tenon 暗榫头
blind test 盲试验;不加标识的商品测验
blind thrust 隐伏冲断层
blind tracery 无孔花格窗;实心窗(花)格
blind tree 盲树
blind trial 盲目试验
blind triforium 不点亮的教堂拱廊;实心拱廊
blind trust 绝对信任委托;保密委托
blind type shield 闭胸式盾构
blind tyre 无凸缘轮胎
blind unconformity 隐蔽不整合
blind up 填补
blind valley 盲谷
blind vault 封闭拱顶;封闭穹隆
blind vein 无露头矿脉
blind vent 暗通气
blind wall 无门窗的墙;无窗墙;实墙;闷墙
blind well 沙底水井
blind window 假窗;盲窗;隐窗口;百叶窗口;封堵的窗
blind window frame 百叶窗架
blind workings of the value principle 盲目地起作用
blind xenolith 隐伏捕虏体
blind zone 盲区;隐蔽层;物探屏蔽区
B-line 变址数寄存器;设计开挖线;计价线;允许超挖线;保持系
blink 勃林克;缩注
blink arch 假拱
blink comparator 瞬变比较镜;闪视镜;闪视比长仪
blink conduit 暗渠
blink drain 地下排水沟
blinker 闪光灯;移带叉;闪光警标
blinker light 闪光信号灯
blinker relay 闪光警戒继电器;闪灯继电器
blinkers 遮灰护目镜
blinker tube 闪光管
blink floor 毛地板
blinking 闪烁(现象);闪视(现象)
blinking characteristic 闪烁特性
blinking cursor 闪烁光标
blinking device 频内装置;闪视器
blinking method 闪视法;闪光法

blinking signal 闪光信号
blink microscope 瞬变显微比较镜;闪视显微镜;闪视比较镜;闪亮显微镜
blink wall 死墙
blip 理想红外线检测器;尖头信号;反射脉冲;标志;标记;报时信号
blip detector 标志信号检测器
blip facility 标志功能
blip-scan ratio 现扫比
blister 局部隆起;煤层圆形结核;浮泡;发泡剂;包皮;气泡;气孔;破图廓部分;水泡;生气泡;沙眼
blister after refining 再生气泡
blister bar 起泡熟铁条
blister box 起泡试验箱
blister cake 粗铜块;泡铜块
blister copper 粗铜;泡铜
blister copper ore 黄铜矿
blister corrosion 起泡腐蚀
blister cracking 起泡性开裂
blister design 凸皮
blistered 起泡的;多孔的
blistered casting 多孔铸件
blister figure 泡沫花纹;泡瘤状纹;(木饰面板上的)不均称疱状突起
blister furnace 泡铜熔炼炉
blister house 起泡加速试验箱
blister hypothesis 拱起假设
blistering 局部隆起;形成气孔;爆皮;起凸;起泡
blistering gas 糜烂性毒气
blistering in plaster 抹面起鼓
blistering of felt 油毡起鼓
blistering on the stone 石块爆破
blistering resistance 抗起泡性;耐起泡性
blister packaging 起泡包装
blister refining 粗铜精炼;泡铜精炼
blister repair 疱状突起修整
blister roasting 粗铜焙烧
blister sand 砂疤
blisters in plywood panel 胶合板鼓泡
blister steel 粗钢(韧性硬质钢);刨面钢;泡面钢;泡钢(由熟铁渗碳而成的钢);渗碳钢
blitz 空袭;闪电战
blitzed site clearance 闪电式清除场地
blixite 氯氧铅矿
blizzard 暴风雪;雪暴;大风雪;大吹雪;风暴
blizzard snow dune 暴风雪堆
blizzard wind 暴风雪;冷暴风
bloach 残印
bloat 膨胀
bloat agent 膨胀试剂
bloated brick 面包砖
bloated clay 膨胀黏土
bloated clay aggregate 膨胀黏土集料
bloated clay concrete 膨胀黏土混凝土
bloated clay concrete solid block 膨胀黏土混凝土砌块
bloated clay(concrete) wall slab 膨胀黏土混凝土墙板
bloated concrete aggregate 膨胀混凝土集料
bloated credit 信用膨胀
bloated pe(a)rlite 膨胀的珠光体;膨胀珍珠岩
bloated shale 膨胀页岩
bloated shale clay 膨胀页岩黏土
bloated shale concrete 膨胀页岩混凝土
bloated slate 膨胀板岩
bloated slate concrete 膨胀板岩混凝土
bloated slate factory 膨胀石板工厂;膨胀板岩工厂
bloated swollen aggregate 轻质集料;轻质骨料
bloater 膨胀剂;膨胀器
bloating 炉衬膨胀;鼓胀;膨胀
bloating agent 膨胀剂
bloating bulking 膨胀
bloating clay 胀性黏土;膨胀黏土
bloating clay aggregate 膨胀黏土集料
bloating clay concrete 膨胀黏土混凝土
bloating clay concrete solid block 胀膨黏土混凝土砌块
bloating clay concrete solid tile 膨胀黏土混凝土砖
bloating clay(concrete) wall slab 膨胀黏土混凝土墙板
bloating concrete aggregate 膨胀混凝土集料
bloating of the stone 膨胀黏土体
bloating pe(a)rlite 膨胀珍珠岩
bloating plant 膨胀设备

bloating shale 膨胀页岩
bloating shale coarse aggregate 膨胀页岩粗集料;膨胀页岩粗骨料
bloating shale concrete 膨胀页岩混凝土
bloating slate concrete 膨胀板岩混凝土
bloating slate factory 膨胀板岩工厂
bloating test 膨胀试验
blob 模糊点;小块;团斑;一滴;滴状;斑点
blob foundation 隔离的浅层底脚及基础;点状基础;独立式基础
blob of slag 火山渣块;渣饼
bloc 集图
blocage 毛石砌体
blocallocation 整笔拨款
bloc cast engine 整铸电动机
Bloch band 布洛赫带
Bloch colo(u)rimeter 布洛赫色度计
Bloch equations 布洛赫方程
Bloch function 布洛赫函数
Bloch-Siegert effect 布洛赫—西格特效应
Bloch-Siegert frequency 布洛赫—西格特频率
Bloch theorem 布洛赫定理
Bloch wall 布洛赫畴壁;布洛赫壁
Bloch wave 布洛赫波
bloc investment 集团投资
block 护面块体;块体;块料;均温块体;巨额证券;街区;街段;记录串;集团;小块;码块;料料;滑轮架座;滑轮;信息组;楔子;自保;铸造板坯;闸块;墩木;断块;短圆材;大料;粗料;传导阻滞;程序块;仓库运货车;采掘区;分程序;方块;闭塞物;闭塞区间;(程)区组;区域网【测】;区段;砌块;汽缸体;平台;盘;图贴;艇座架
block access 程序块访问;成组存取
block account 封存账户;被冻结的存款
block accumulator 条形极板蓄电池;条条极板蓄电池
block action 块体作用
block activation 分程序动用
block adaptable to extension 可以扩建的房子
block adaptation 程序块适配;部件适配
block add 块相加
block address 块地址
block addressing 程序块寻址
blockade 禁运;阻断;封闭;封锁;封闭
blockade and embargo 封锁与禁运;封锁禁运
blockade a port 封锁港口
blockade balance 封存结余
blockade deposits 冻结存款
blockade line 封锁线
blockade running 偷渡封锁线
blockade ship 封锁船
blockade zone 封锁区域;封锁地区
block adjustment 面积平差;分区平差;区域网平差
block adjustment by using bundle method 光束法区域网平差
block adjustment using independent model of method 独立模型法区域网平差
block aerial triangulation 区域网空中三角测量
blockage 结连堵塞;毛石砌体;堵塞(过程);封阻;封锁(交通);闭塞
blockage coefficient 断面系数
blockage effect 收缩效应
blockage factor 断面系数;断面比;堵塞系数
blockage in pumping 泵阻塞;泵的阻塞
blockage of borehole 钻孔堵塞
blockage of boring 钻孔堵塞
blockage stone 小方石的石料(美国)
blockage time of track 线路封锁时间
block a line 停用一条管路
block allocation 整笔拨款
block analysis 块分析;分程序分析
block analytic(al) aerial triangulation 区域网解析空中三角测量
block anchorage 块体锚固
block and block 滑车拉到头
block and cross bond 制造拳石或小方石的石料;丁顺砖交叉砌合
block and fall 滑轮组;神仙葫芦
block and pulley 滑轮组
block and start 长短砌合
block and tackle 滑轮组;滑车组
block angular system 方块角系统
block antenna 集合天线;共用天线
block apparatus 闭塞机;区截机

block applied method 闭塞代用法
block arch 砌块拱
block architecture 大砌块建筑；街坊建筑
block assembly 分组装配作业；分段装配
block assembly shop 分段装配车间；分段预装车间
block automation system 巨额交易自动报价制度
block basin 断块盆地
block beam 梁段；块体组合梁；分块拼装梁；预应力块梁
block bearing 止推轴承；支承轴承；推力轴承
block bell signal 闭塞铃信号
block beneath port neck 小炉脖底砖
block bill 宽斧；宽刃斧
block board 胶合夹芯板；木块芯细木工板；芯块胶合板；滑轮盘；细木板；夹芯板；板条芯胶合板
block body 分程序体
block bogie (农村用的)卡车；宽刃斧
block bond 块料砌合；丁砖与顺砖交叉或隔层砌合
block bonding 丁砖层与顺砖层交叉或隔层砌合；砌块砌合；多层啮合砌搭
block book 木刻版本
block booking 搭配定购
block-bottom bag 四角包底水泥袋；层叠底袋
block box 模块；密封组件
block brake 块闸；制动器；滑块式制动器；闸瓦制动(器)；闸瓦式制动器；瓦闸；瓦块式制动器
block brake unit 闸瓦制动装置
block brazing 块钎焊；加热块钎焊；块料钎焊
block breakwater 块石防波堤
block brick 大块砖
block bridging (小梁间加固的)支撑；横撑；格栅横撑
block brush 块状电刷；电刷块
block busing 以谣诼压价收买房地产；(房地产捐客的)房屋咬卖生意
block-by-block address range 街区到街区地址范围
block cabinet 闭塞机
block cancel character 信息组作废符号
block capital 块形柱头；罗马式柱头；方柱头；方块式柱头
block carrying truck 块体搬运车；方块搬运车
block cast 整铸；整浇
block casting 块铸
block cast motor 整铸电动机
block cathode 方块阴极
block caving 阶段崩落(开采)法；大块坍落；分块崩落(开采)法
block caving coal mining method 大块崩落采煤法
block-caving method 柱子坍陷法
block cement 堵漏水泥
block chain 块环链；滑轮链；滑轮导链；滑车链条
block chaining 分组连接
block chamber 机密室；闭塞的房间
block chart 块图；方框图；方块图
block check 块校验；块检验；棋盘格纹饰
block check character 块校验字符；块检验字符；信息组检查字符
block check procedure 块校验规程；码组校验过程
block check sequence 块检查序列
block chimney 砌块烟囱
block chopper 琢石工；割石机；凿石工；石块切割机；石工
block cipher 分组密码
block circuit 闭塞电路
block circulant approximation 分组循环近似法
block clay 泥石混杂物；混杂岩
block clear 闭塞区间开通
block cleared 解除闭塞
block clearing point 闭塞区间开通点
block clutch 闸瓦离合器；伸缩闸瓦离合器
block coal sample 块状煤样
block coast 块石海岸；断块海岸
block coat 过渡涂层；附着增进涂层
block code 块码；划区编码；信息组代码；程序块码；分组码；分批码
block code length 分组码长
block coding 定长编码
block coefficient 船(体方)型系数；充满系数；肥瘦系数；肥瘠系数；方形系数；填充系数
block column 房子柱子；建筑支柱
block complex 建筑集合体；建筑组合
block compression strength 石块受压强度；砌块抗压强度
block concreting 分块浇筑

block connection diagram 方块接线图
block consistency 固态稠度
block constant 表征数字组特性常数
block construction 积木式结构；砖块构造；单元结构；大型砌块结构；大型砌块建筑；大型砌块构造；部件结构；砌块构造
block construction method 大型砌块施工法；砌块构造(方)法
block construction wall 大块砌筑岸壁
block contact 联锁触头；联锁触点；闭塞触点
block control 块控制；闭塞控制
block copolymer 嵌段共聚物
block-cork insulation 软木块绝缘；软木块绝热
block correction 区域校正；区域改正
block count 块计数；程序块数；分程序块数
block count exit routine 块计数出口程序
block count field 块计数字段
block crane 钢锭起重机
block creeling 分段换筒
block curve 连续曲线；实线曲线
block-cutpoint graph 块割点图
block cutter 块料切割机；切板机
block cutting machine 块料切割机
block data 块数据；数据块
block data statement 数据块语句
block data subprogram(me) 数据块子程序
block datum plane 分区基准面
block density 颗粒密度
block depth 街坊深度
block descriptor word 分程序说明字
block design 木块设计测验；分组设计；分块结构设计；区划设计
block determinaion 分程序确定
block diagonal matrix 块对角矩阵；分块对角矩阵
block diagram 立体图；框图；块状图；阶梯式统计分布图；分区图；分块图；方框图；方块图；区划图；三相图；三维曲面图；草图
block diagram language 框图语言
block diagram of rock 岩石立体图
block diagram symbol 方块图符号
block diagram system 框图系统
block diameter 卷筒直径
block directed graph 块方向图
block distribution 集体分配法
block-dome 块岩丘
block down converting 分块下变换
block down valve 排污阀
block drill 印版钻孔机
block dumping 方块抛填
block economy 封锁经济
blocked 封闭的；闭塞了的
blocked access 成块存取
blocked account 冻结账户；封锁账户；封存账户
blocked acid catalyst 封闭的酸催化剂
blocked assets 冻结资产
blocked balance 封存结余
blocked calls cleared 阻塞呼叫清除
blocked capacity 起重臂提升能力
blocked capillary pore 封闭型毛细孔
blocked channel 淤塞河道；淤塞河槽
blocked check 冻结支票
blocked column 粗面块石柱
blocked course 块状砌层
blocked currency 不能兑换外汇的货币
blocked deposits 冻结存款
blocked doorway 安置崇拜雕像的壁龛；有雕塑的壁龛门道
blocked exchange 封锁外汇；封存外汇
blocked file 成块文件；封锁的文件
blocked fixed assets 封存的固定资产
blocked foreign exchange 封存外汇
blocked fund 冻结资金；冻结款项
blocked funds attestation 封存货款银行证明
blocked-grid keying 栅截止键控
blocked grid limiter 栅；栅截止限幅器
blocked heat 中止氧化
blocked impedance 约束阻抗；阻挡阻抗；固定阻抗；停塞阻抗
blocked income 冻结收入
blocked job 封锁的作业；分块作业
blocked joint 分段连接；分段接头
blocked level 闭锁电平
blocked level access 成组存取；成块的层次访问；封锁的层次访问

blocked line 封闭的线路
blocked oil 密封油
blocked operation 隔断操作
blockedout ore 可靠储量；块断储量
blocked polygon 封闭的多边形
blocked polyurethane paint 封闭型聚氨酯涂料
blocked process 封锁进程
blocked rand 封闭边缘
blocked record 成组记录；成块记录；封锁的记录；块式记录
blocked resistance 阻挡电阻；闭塞电阻
blocked road 封闭道路
blocked rotor 止转转子
blocked rotor current 止转转子电流
blocked rotor rest 制动转子试验
blocked rustication 分段粗琢；带凸块粗磨；粗面重块石砌筑
blocked section 区间阻塞
blocked stone 块石；块石
blocked stream 阻断河；堵塞河
blocked sulfoleate 封闭的磺化油酸盐
blocked system 闭塞系统
blocked to traffic 封锁交通
blocked track 封闭的线路；闭塞线路
blocked traffic 封锁交通
blocked up 过厚
blocked up valley 阻塞的山谷；阻塞河谷
block effect 块体效应；体效应
block elimination 越组约化；区域消元法
block encoding 分组编码；分块编码
block encryption 块加密
block end 块结束；程序块结尾；分程序结束
block entrance 房屋入口；分程序入口
block entrance door 建筑入口大门
block entry 块入口；分程序入口
block equipment 房屋设备；建筑设备；闭塞设备
blocker 预锻模膛；制坯模膛
block error rate 块出错率；码组差错率；信息组错误率；信息组差错率
block error rate test 块差错率检验
blockette 小组信息；小程序块；子信息块；分字组；分区块；分程序块；数字组；数据小区组
block event control 事件控制用信息组
block exit 分程序出口
block extension 房屋扩建；扩建
block face 沿街门面；沿街土地；块材的面；街区面积
block factory 建筑砌块厂
block failure 整体破坏
block failure strength of pile group 群桩的整体破坏强度
block fault 块断层；地块断层；块状断层
block-fault and rifted regions 块断区和裂谷区
block-fault folding 块断作用
block faulting 块状断层；断块作用
block field 滑块机构
block file 粗齿方锉；分程序文件
block filler 封闭漆；填孔漆
block filter glass 阻断滤片
block finish 混凝土表面磨光；混凝土表面修整；混凝土表面光洁度；表面磨光
block finishing 混凝土表面磨光
block fire 房屋着火
block floor 拼花地板
block flooring 块料地板；块材地板；分块地板；拼花地板
block folding 块断褶皱作用
block footing 块形大方脚；块形底脚
block for drawing 拉丝卷筒
block format 块格式；程序段格式；分程序格式；砌块规格；砌块大小；砌块尺寸
block for structural form 建筑部位砌块
block foundation 块式基础
block fracture texture 破碎块裂体结构
block frame 房屋构架；建筑框架；房屋框架
blockfront 街区的临街土地
block furnace 分段组合式电炉；方形炉
block gap 块间隙；块间隔；信息组间隙；程序块间隔
block ga(u)ge 块规
block Gauss-Seidel method 分块高斯—赛德尔法
block glacier 块状冰川；泥石流；石冰块
block glass 玻璃块
blockglide 块体滑坡；块体滑动；地块坍塌；地块滑坍
block graben 块垒地
block grant 一揽子赠款；分类财政补贴

block graph 框图;块状图;块图
block graphics 块图;图阵
block grease 块状润滑油;黄油块;润滑脂块
block group 块组
block hammer 汽锤;落锤
block handle 整边黏结杯柄
block handling 块体搬运吊装;方块搬运吊装
block harvesting 成片采伐
block head 块头;块首;程序块首部;分程序首部
block head cylinder 整体汽缸
block header 程序块首部;分程序首部
block heating 分区供暖;分片供热;分块供暖
block heel riser 叉尾垫块
block holder 块规夹持器
block hole 小炮眼;大石块上钻的炮眼;浅炮眼
blockhole blasting 岩块爆破;二次爆破
block holer 锤钻
blockholing 分块钻孔;巨石钻孔爆破;岩块爆破;爆破大石
blockholing method 分块钻孔法
block house 碉堡;工棚;木屋;线路所【铁】;砌块间;水泥掩体;砌块屋;钢筋混凝土半球形房屋;地堡;掩体;堡垒
block ice 冰流
block iceberg 陡峭冰山
block ice-movement 块状冰川运动
block identifier 分程序标识符
block ignore character 信息组作废符号
block in 堵塞;封锁;筹划;拟大纲;画草图;画略图
block-in-course 层砌方块;成层琢石砌体;层块体;方正块石密砌;方块砌筑;嵌入楔块层;层次整齐的琢石圬工
block-in-course block 成层琢石块体
block-in-course bond 砖楔块砌合;嵌入楔块层砌合
block-in-course masonry 锤琢石块的圬工;成层砌石圬工
block independence 分块独立(性)
block indicator 闭塞指示器
blocking 阻滞;联锁;块化;基团封闭;粘连;木填条;模块化;预锻;压箔;阻塞;阻断;中止氧化;中断振荡;止碳;滚沸(玻璃液);堵塞;堵口;冻结;单元化;大木夹胶法;粗型锻;粗模锻;船在干坞中撑垫材;传导阻滞;成组;成块;封锁;分块化;分段;并块;闭锁;闭塞(系统);背木;办理闭塞;区组化
blocking action 阻塞作用
blocking agent 基团封闭剂;阻滞剂;阻断剂;堵水剂;封堵剂
blocking and stuffing intake passage 填塞进水通道
blocking anti-body 封阻抗体
blocking anticyclone 阻塞高压;阻塞反气旋
blocking bias 截止偏压
blocking by tablet 路签闭塞
blocking cement 上胶盘
blocking chisel 宽口凿
blocking chromes 止碳硅铬铁
blocking circuit 间歇电路
blocking clement 闭锁元件
blocking coil 闭锁线圈
blocking condenser 隔直流电容器;隔流电容器
blocking construction method 分块施工法
blocking contact 断开触点
blocking contactor 间歇接触器
blocking coordinate system 整体坐标系
blocking course 檐头石;女儿墙;檐口墙;檐头墙;压顶
blocking degree of casing well 管井堵塞程度
blocking device 闭锁装置;区截装置
blocking effect 整体效应;封闭效应
blocking efficiency 整流效率
blocking element 联锁元件
blocking factor 块因子;封锁因素;封闭因子;分块系数
blocking filter 间歇滤波器;闭塞滤波器
blocking filtration 封闭过滤
blocking fixture 上盘定位夹具
blocking flow 阻塞流
blocking foil 金属箔
blocking gear 粗型锻传动机构;粗模锻传动机构
blocking group 保护基团;保护基
blocking high 阻塞高压(反气旋)
blocking hoe 分篼间苗锄
blocking impression 预锻模膛

blocking interlocking system 闭塞连锁系统
blocking joint 分段连接
blocking journal 止推轴颈
blocking layer 耗尽层;阻挡层;闭锁层
blocking layer photocell 阻挡层光电管
blocking level 阻挡层
blocking lever 联锁杆;闭塞杆;止动杆;停秒杠杆
blocking location 堵塞部位
blocking lock 联动锁
blocking machine (平板玻璃的)抛光机
blocking mechanism 闭锁机构
blocking nut 防松螺母;锁紧螺母;锁紧螺帽
blocking of concrete dam 混凝土坝体分缝分块
blocking of concrete placement 混凝土浇筑分缝分块
blocking of drainage 排水阻碍
blocking of ice 冰阻;冰塞
blocking of record 记录的分块
blocking operation 封锁行动
blocking order 闭锁指令
blocking oscillator 间歇振荡器;闭塞振荡器
blocking out 涂墨
blocking(partition) 分块
blocking passage of polluted water coming in 堵塞污水进入通道
blocking point 黏着温度;黏着点
blocking press 压印机
blocking process 分块过程
blocking property 黏结性能;黏合性能;黏附性能;整体性质
blocking relay 联锁继电器
blocking relay for closed circuit 合闸闭锁继电器
blocking resistance 抗黏着性;抗黏连性
blocking section 闭塞分区;封锁区间
blocking shaper 镜盘模
blocking signal 阻塞信号
blocking state 阻塞状态;阻断状态;闭(锁状)态
blocking temperature 间歇温度
blocking test 封闭试验;分块试验
blocking time 截止时间;闭锁时间
blocking tool 胶合模
blocking tube oscillator 电子管间歇振荡器
blocking up 泥沙填塞
blocking up of track 线路封锁
blocking up with silt 泥沙淤塞
blocking valve 闭锁阀;闭塞阀;锁气阀
blocking view 障景;抑景
blocking voltage 阻塞电压;闭塞电压
blocking water with grouting heavy curtain 帷幕灌浆堵水
blocking water with grouting of shaft and mining lane 井巷注浆堵水
block input 块输入;字组输入
block input/output 成组输入输出
block input/output instruction 成组输入输出指令
block instrument 闭塞器;闭塞机
block insulation 块状绝缘层;保温块
block insurance 船舶保险
block interface 块体接触面
block interlocking system 半自动闭塞系统
block in traffic 交通阻塞
block irrigation 畦灌
block irrigation system 分片灌溉制
block iterative method 块迭代法;成组迭代法
block Jacobi method 分块雅可比法
block joint 预留砖石孔;预留砖孔;预留混凝土孔;砌块预留孔
block justifying machine 整版机
block Kriging 块段克立格法
block lac 块状虫胶
block lava 块状熔岩;熔岩块
block lava flow 块状熔岩流
block-layer 砌块层
block laying 砌块砌墙;砌块砌筑
block layout 方块图
block layout planning 总体布置
block leaf 块料拼门;砌块门扉
block length 路段长度;块长(度);信息组长度;字组长度;字区长;分组长度;分程序长度;闭塞区段长度;区段长度
block length indicator 块长指示符
block letter 印刷体;大型字体;粗体字母
block level 框式水准器;钳位信号电平;气泡水准仪;气泡水准器;平放水准器

block lifting 方块起吊
block-lifting machine 起重机臂提升机;砌块提升车
blocklike 块状的
blocklike structure 块状构造;似块状结构
block limits 地段限额
block line 街区线;房屋建筑界线
block lining 混凝土板衬砌;砌块衬砌
block linkage 滑块联动装置
block lintel 房屋中砌块过梁;砌块过梁
block list 块表;分程序表
block load(ing) 基本负载;程序块装入;房屋荷载;巨额负载;巨额负荷;基本荷载
block lock state 块封锁状态
block logic(al) design 块逻辑设计
block machine 墙壁砌块制造机;砌块制造机
block making 制版;砌块制造
block-making machine 压块机;砌块成型机;制块机;制块材机
block-making plant 方块预制场;方块预制厂
blockman 砌石工人;路面铺砌工;板块工匠
block map 块状图;略图;框图;立体图;方块图
block mark 块标志;字组标志;字组标记;程序块符号;分程序符号;区标;卷筒条纹
block masonry wall 砌块圬工墙
block masonry(work) 房屋砌块圬工;砌块工程;砌块砌体
block mass 大块;岩块;古陆;地块;大砌块;团块
block matrix 块矩阵
block method 分组方法;分段统制账法
block mica 块云母;厚片云母
block mill 成组轧机
block mining 分区开采;分块开采;分块崩落(开采)法
block mining system 分区开采法
block mode 分块模式
block model 块体模式;实心木船模
block modillion 飞檐下无装饰块料的水平托架
block moraine 块砾碛
block motor 电动机部件
block mo(u)ld 块材模型;整体铸型;整体模型;整体模(具)
block mo(u)ld process 整体熔模铸造法
block-mound breakwater 抛筑块体防波堤
block mount 组合装配;组合安装
block mountain 块状山;断块山
block mountain coast 断块山海岸
block move 成块调动
block movement 块体运动;块体移动;块传送;断块运动;地块运动
block multiplexer channel 成组多路通道;数组多路信道
block multiplexer mode 成组多路转接方式;成组多路方式;成组多路传送方式
block multiplexing control bit 信息组多路控制位
block multiplex mode 成组多路方式
block multiplication 分块相连;分块相乘
block multiplication of matrices 矩阵分块乘法;矩阵的分块乘法
block name 块名;信息组名
blockness 块度
block number 成组传送号;分程序编号;区组数;批号
block nut 止动螺母;防松螺帽
block occupied 线路闭塞
block of dock 船坞墩木
block of dwelling houses 居住小区
block off 阻塞
block offer 成批出售
block office 线路所【铁】
block of flats 公寓楼区;公寓楼(房);套房楼
block-off valve 放气阀(门)
block of gas supply 供气区段
block of glass 玻璃棉板
block of graph 图块
block of groped shops 商业中心
block of houses 街坊;住房区段
block of housing 住宅区
block of ice 冰流
block of information 信息区;信息码组;信息块;信号字数
block of large blocks 大砌块房屋;大砌块住宅
block of large slabs 用大板构造的房屋;大板住宅
block of large tiles 大空心砖建筑
block of lense 透镜块

block of memory 存储块
block of offices 办公楼区;办公室大楼;办公大楼
block of photographs 像片组
block of rock 石块;岩块;大块石
block of shared memory 共享存储块
block of staggered design 错列式设计的房屋;错列式设计住宅
block of stone 石块;磐石
block of suspended arch 炉顶吊砖
block of uniform height 等厚块(体)
block of valve 阀锁
block openings 填塞孔洞;填塞空洞
block operator 线路所值班员
block-organized holographic memory 分块全息存储
block orientation 街区方位;建筑方位;建筑定位
block-oriented 面向块的
block-oriented associative processor 面向块的相联处理机
block-oriented memory 按块存储器;分方位存储
block-oriented random access memory 按区随机存储器
block-oriented random access storage 按区随机存储器
block oscillation 房屋振动;建筑振动
block out 预埋木块;勾画轮廓;孔槽;绘草图;画草图;预留孔堵块(浇混凝土后移去);勾轮廓规划;打草图;筹划
block out method 描绘制版法
block out mo(u)ld 钢板模板
block out ore reserves 作矿藏略图
block output 块输出;字组输出
block out type safety fence 阻挡式护栏
block paging 成组调页;成块调页
block panel system 分区短柱开采法
block parity 块奇偶性
block parquetry floor covering 镶木地板覆盖;镶木地板条块覆盖
block partition(wall) 砌块隔墙;块体隔墙
block pattern 街区型
block pavement 块状路面;块体铺面;块体路面;块石路面;块料路面;块材铺面;砌块铺砌路面;砌块路面层;砌块路面;铺石路面;石块铺砌;砌块路面
block paving 块料铺砌(法);块材铺面
block percolation 全渗透;全渗滤
block placement 分区填筑;分块浇筑
block plan 略图;建筑位置图;基地平面图;概略计划;地段规划;地段(平面)图;分区图;分区平面图;分区规划;区划图
block plane 横木纹的刨;短刨
block planning 建筑用地规划
block plant 砌筑车间;砌筑工厂;建筑砌块厂
block planting 群状栽植
block platform 组装石平台
block policy 统保单
block polymer 整体聚合物;成块聚合物;嵌段共聚物
block post 线路所【铁】;闭塞信号控制站;闭塞控制站
block post before hub 枢纽前方线路所
block post keeper 区号标志管理员
block post with distribution tracks 闸站【铁】
block power plant 单岸式水电站;河床式(水)电站;单岸式电站
block power station 单岸式电站
block prefix 信息组前缀
block prefix field 信息组前缀字段
block press 模压机;砌块模压制
block-press machine 压块机
block print 手工模板印花织物
block printing 木刻版印刷(术);木版印刷;人工印花
block prism 立方棱镜
block processing 块处理
block processor 块处理程序
block protection 块体护岸;块石护岸
block protector 块状保护装置
block pulley 滑轮组;滑车组;神仙葫芦
block push button 闭塞按钮
block quay wall 块料砌成的码头岸壁;码头挡土墙块
block quoin 屋角仿石体
block rack 磨削伤;屑伤(磨光玻璃缺陷)
block rake 划痕(平板玻璃表面缺陷);链状磨痕
block random mosaic 任意铺砌的马赛克
block ratio 荒料率

block record 块记录
block recursive system 分块递归方程组
block reference 分程序块引用
block relaxation 块松弛
block relay 闭锁继电器
block replacement 整块替换
block representating 方块图表示法
block representative 方块图表示法
block research 分块检索
block reservation 区保留
block reserve 断块储量
block resistance 抗粘连性
block ridge 断块海岭
block ring 阻流环
block road 块石路面
block(rock) 块石;块状岩
block run length 运行块长度
block sale 成批出售交易
block sample 块状试样;块体试样;区域抽样
block sampling 分块抽样
block schematic diagram 方框图
block scheme 方块图
block-schooled movement 大块运动
block screw 螺旋顶高器;千斤顶
block search 分组调查;分批搜索;分批试验(法);分块查找
block search technique 分批搜索法
block section 闭塞区间;闭塞区段
block separation 块分隔
block separator 块分隔符
block sequence 分段多层焊
block sequence indicator 信息组顺序指示符
block sequence welding 分段连续焊接;分段多层焊
block setter 分捆架导轨
block setting crane 钢块起重机
block-setting plant 块体吊装设备;方块吊装设备
block shears 剪切机
block shear test 块体剪切试验;断块剪切试验
block shellac 结块紫胶
block ship 囤船;堵江船;(港水道的)沉船;仓库船
block signal 交通区段志;铁路区截信号;阻塞信号;截止信号;分段信号;闭锁信号;闭塞信号;区段信号;通信信号机
block signal box 线路所【铁】
block signal(l)er 通过信号机
block signal machine 闭塞信号机
block signal system 闭塞信号系统
block silo 分区段的筒仓;混凝土砌块筒仓;混凝土砌块谷仓
block size 块体尺寸;街廓大小;程序块尺寸;分程序大小
block slide 块体滑坡;块体滑动;块滑;地块滑坡;大块滑动
block slope 块石坡
block-slump 地滑
block sort 信息分块;字组分类;字区分类;成组排序;分组;块分类
block span 防扭绳
block spar 块长石
block spirit level 平放水平仪
block splitter 块料分割器;块料切割机;砌块切割机
block splitting 块分离
block splitting machine 块料锯开机;切削机(石块加工)
block square 矩形角尺
block squeeze 管外加压注水泥
block stack 块体烟囱;砌块烟囱
block stand-by 备用数据块【计】
block station 线路所【铁】;闭塞站
block station track 封锁站线
block stock yard 立体分段堆场
block stone 加工石;块石;条石;大块石
block stone lining 块石衬砌
block stone pavement 块石铺砌的路面;块体铺砌路面;块石路面;铺砌路面
block-stone road 方块石路;铺方块石路
block storage 块体堆场;方块堆场
block stowage 码头装载
block stream 块状岩流;岩块流
block street 街区
block strength 块体强度;砌块强度
block stripe 块状条纹
block strop 滑车环索
block structure 块状结构;块体结构;块结构;结构块;断块山;断块构造;地块构造;程序块结构;成组结构;分程序结构
block-structured code 分程序结构码
block structure mechanism 块结构的机构;分程序结构原理;分程序结构机构
block subsection 闭塞分区
block supply 区域供电
block swage 粗型锻模
block switch 分组开关
block-switch technology 分组交换技术
blocks world-scene analysis 积木世界景象分析
block system 框图系统;分区切块开采法;分区开采法;分片灌溉制;分段造船法;分段式;方框图;闭锁系统;闭塞装置;闭塞制;闭塞;区划法;主体分段建造法
block system relay 闭塞系统继电器
block tackle 复滑车
block take-off 开挖设备
block technique 封闭术
block terminal 端子排;堵端;程序块终止;分线匣;分线盒
block terminal statement 程序块终端语句
block test 台上试验;台架试验
block testing machine 块料试验机(器);混凝土砌块试验机
block testing stand 发动机试验台
block texture 块状结构
block the fairway 阻塞航道
block theory 阻滞学说
block the passage of message 封锁消息
block thrust 块状冲断层
block tie 短枕
block tilting 地块翘起
block time 轮挡时间
block tin 锡块;纯锡块
block tipping 岩块倒落
block tire 实心轮胎
block-to-page mapping 从块到页面变换
block transaction 集团交易
block transfer 块传送;信息组传送;字组传送;整块转移;整块传送;成组传送
block transmission 块传送;块传输
block triangular matrix 分块三角矩阵
block triangulation 区域网法
block tridiagonal matrix 块三对角线矩阵
block truck 运货车;板车;手摇车;手推车(行李)
block type 建筑类型
block-type arrangement 组式布置
block-type boring cutter 插入式镗刀;插刀式镗刀
block-type connector 块式连接件
block type dock wall 砌块坞墙
block type foundation 块形基础
block-type insulant 块型绝缘材料
block-type insulation 块材保温
block type quay 砌块码头;砌块堤岸;砌块驳岸
block-type reamer 盘形铰刀
block type runoff-river power station 单岸式径流式电站
block type seismograph 组块地震仪
block-type tie 分块式轨枕
block type turbine 组装式涡轮机;组装式透平;组装式水轮机;组装式汽轮机
block type wharf 砌块码头
block up 垫高;停用;封闭;堵塞;封锁
block upheaval 地块隆起
block up water method in mining well 矿井堵水方法
block valve 普通阀;楔阀;重型阀
block variable 块变量
block vibrating machine 振动块料机器;混凝土砌块振动器
block vibration 砌块振动;建筑振动
block wall 块砌岸壁;巨砾砌;磁畴壁;方块岸壁;砌块隔墙;砌块隔断;砌块岸壁
block-walling 砌块墙工;块砌墙;砌块圬工
block water wall 组合式水冷壁
block weight 落锤
block welding 多段多焊缝;块焊接
block welding sequence 多层焊接法;分段焊接法;分段多层焊
block wheel 刀轮(装金刚石的切裁玻璃用)
block with external access galleries 外廊建筑
block with one wing 单边房住宅建筑
block with sight hole 看火孔砖

block with temperature measuring hole 测温孔砖
block wood 大木块;木块
block wood flooring 木地面
block wood pavement 木块铺面;木块砌路面
blockwork 砌块墙(体);块石作业;砌块工程;预制砌块;预制块砌体;预制大砌块;堵塞工程;方块砌体;块体工程;铁路闭塞工程
block working system 行车闭塞制;行车闭塞系统
block(work) partition 砌块隔墙;砌块隔断
blocky 大块的
blocky and seamy rock 破碎和裂隙岩石
block yard 混凝土构件场;砌块工场;块体预制场;混凝土砌块场;混凝土块厂;混凝土(方)块制造场;混凝土(方)块预制场;预制工场;方块预制场;预制混凝土构件场
blocky ferrite 不均质铁素体
blocky formation (留在岩芯管内的)卡塞岩块
blocky graphite 块状石墨
blocky shape 团块状
blocky soil structure 土壤块状结构
block(y) stones 大块石
block(y) structure 块状构造;块状结构
block zone 封锁区
bloc price 集团价格
blocrete 喷射混凝土
bloc trade 集团贸易
bloc voting 集团投资
bloedite 白钠镁矾
blo-holes 砂眼;气泡
Blom's method 布洛姆法
blomstrandine 钇易解石
Blondel 布朗德尔(光亮度单位)
Blondel diagram 相量图;布朗德尔曲线图
Blondel oscillograph 布朗德尔示波器
Blondel-Rey law 布朗德尔—雷定律
blond(e) oil 金黄油
blondin 悬索道(料场运石用);索道(式)起重机
blond in cable-crane 缆索式挖土机;缆索式挖掘机
blood adhesive 鲜血胶
blood albumin 血胶
blood albumin glue 血白蛋白胶水
blood bank 血库
blood examinating room 血液检验室
blood gas 窒息性毒气
blood glue 血胶
blood preserving bottle 储血瓶
blood rain 血雨;红雨
blood red 血红色
bloodstone 鸡血石;血滴石;赤铁矿
bloodstone burnisher 红石磨光器
blood waste 废血
blood wood 赤桉;红木
blood worms 红虫
bloom 水华;矿脉露头的风化;耐火土块;海绵铁球;油霜;砖墙风化;钢坯;钢块;钢锭;大量增殖;大钢(锻)坯;大方坯;初轧(方坯);粉化;表面起晕加工;起霜作用;图像浮散;铁块;水花区;散晕
bloom and slab piler 钢坯和板材卸垛区
bloomary 精炼炉床;土法熟铁吹炼炉
bloom base 支柱座
bloom base plate 支柱座板;支柱底板;生铁支柱座板
bloom-blank 大异型坯
bloom-block 大异型坯
bloom butts 方坯切头
bloom crop end 方坯切头
bloom disease 水中病
bloomed coating 无反射涂层
bloomed lens 镀膜透镜;敷霜透镜
bloomer 开坯机;初轧机
bloomery 熟铁吹炼炉
bloomery iron 熟铁块
bloomery process 熟铁吹炼法
bloom film 光学膜
bloom film glass 加膜玻璃
Bloom gelometer 布卢姆测胶计
blooming 起膜的油漆面;开坯;加膜;模糊现象;敷霜;起霜;无光油漆;图像发晕;白化;喷霜
blooming furnace 炼铁炉
blooming housing 初轧机外壳
blooming mill 轧钢车间;开坯机;钢坯延压机;初轧机;齿轮轧钢机
blooming mill housing 初轧机外壳
blooming(mill) stand 初轧机机座

blooming roll 方坯轧辊
blooming-slabbing mill 方板坯初轧机
blooming train 初轧机组;初轧机机组
bloom inhibitor 防起雾剂;防喷霜剂
bloomless oil 不起霜润滑油
bloom oil 起霜油
bloom out 起霜
bloom pass 开坯孔型;初轧轧槽;初轧机孔型;初轧道次
bloom pusher 初轧方坯推出机
bloom scale 大方坯磅秤
bloom shears 大钢坯剪切机;钢坯剪切机
bloom slab 扁钢坯
bloom steel 钢坯;初轧钢
bloom yard 初轧方坯仓库
blooper 接收机辐射信号;发杂音接收机
Bloor's test 布卢尔试验
blossom 煤华;风化煤层露头;铁帽
blossom cluster 花簇丛
blossom drop 落花现象;落花
blossom rock 落华石
blossom showers 梅雨
blot 瑕疵;污渍;污斑;涂脏;涂色
blotch 污点;污斑
blotchy 布满污渍的;斑渍;斑污
blot coat 浸入层;吸入层;表面渗入层
blot fat area 路面厚腻油斑
blotter 流水账;临时记录簿;吸墨纸;吸墨用具;吸墨水纸;草账
blotter aggregate 吸油集料
blotter coat 表面渗入层;吸油层;侵油层;吸油层;渗透层
blotter model 吸水纸模型
blotter press 压滤机
blotter treatment 吸油处理;污渍处理
blotting 抹去;涂去;吸干;吸去;浸透;沾吸作用
blotting book 流水账
blotting capacity 吸油能力
blotting material 吸油材料
blotting paper 吸墨纸;吸滤纸
blotting test 吸墨试验
blouse 工作服
bloused back 背囊
blout 块状石英
blow 断开电路;打击(锻压时);吹炼
blowability 吹成性
blow a line down 吹洗输气管道
blow-and-blow machine 吹—吹法成型机;吹制机
blow-and-blow process 吹—吹成型法
blow asphalt 氧化沥青;吹制(地)沥青
blow away 吹走
blow a well clean 喷射洗井(用清泥浆)
blowback 倒吹气;起跳压力与回复压力差;逆吹;后座;反吹;气体后泄;回爆;泵回
blowback gas 泵回气体
blowball 絮球
blow bar 打击棒
blow bending test 冲击弯曲试验
blow-by 窜漏;漏气;不密封
blow-by gas 压缩废气;排放漏气
blow-by of piston 活塞漏气
blow by the bilge 舱底水打到货物上
blow can 喷水壶
blow case 吹气扬酸箱;吹气箱
blow chamber 吹风室
blow cleaner 吹净器
blow cock 排放旋栓;排气栓
blow cold 低温吹炼
blow core 吹芯
blow count 标准贯入试验击数;锤击次数;冲击次数;打击次数;击数;贯入击数;贯击数;锤击数;锤击计数
blow count of light sounding test 轻便触探试验锤击数
blow count of SPT 标准贯(入)击数
blow count of standard penetration 标准贯入击数
blow cylinder process 人工吹管法
blow diagram 贯入度图
blowdown 放气;泄料;增压;吹除;放水;放空;破坏性风暴;排料;污染物排放;吹除;吹倒;吹净;吹落;吹扫;吹倒;吹落;(气压沉箱的)放气强迫下沉;(锅炉的)排污;吹净
blowdown apparatus 排放装置;排污装置
blowdown branch 排污支管;放空支管

blowdown condenser 放空冷凝器
blowdown connection 排污管接头;排污接管
blow down container 泄料罐
blowdown cooler 排污冷却器
blowdown drum 放空罐;排污罐
blowdown fan 增压风机
blowdown gas 吹气
blowdown line 卸料导管
blowdown period 减压时间(蒸压釜);降压历时
blowdown pipe 补气管;放空管;排泄管(道);排污管
blowdown piping 泄料管路;放空气管路
blow(down) pit 泄料池;放空池
blowdown plant 排污装置
blowdown presssure 排放压力;放泄压力
blowdown pressure of safety valve 安全阀关闭压力
blowdown recovery 降低压力采芯
blowdown return 排污回收
blow down separator 排放分离器
blowdown softening 排污水软化
blow(down) stack 放空烟囱;放空烟道;放空管;泄料烟道;排泄管(道)
blowdown system 泄料系统;放空系统
blowdown tank 快速排空罐;泄料箱;放空箱;排污箱
blowdown tube 放空管
blowdown tunnel 吹气风洞;放气式风洞
blowdown turbine 冲击式涡轮(机);排气透平机
blowdown valve 吹扫阀;排污阀;排泄阀;泄料阀;泄放活门;高压放气阀;放汽阀
blowdown water 排污水
blowdown water treatment 排污水处理
blowdown wind tunnel 放气式风洞
blowed ore supervision 放矿管理
blow efficiency 打击效率
blow energy 打击能量
blow engine 鼓风机
blower 压缩机;吹气机;增压器;增压机;鼓气机;吹气机;吹灰器;吹气器;吹风管;风扇;风机;玻璃灯工;玻璃吹制工;通风机调节器;通风机(调节风门);气винd吸粮机
blower aeration 鼓风曝气
blower base 鼓风机底座
blower blade 吹送机轮叶片
blower body 鼓风机体
blower bottom 鼓风机底部
blower casing 鼓风机壳(体)
blower control 鼓风机控制器
blower cooled diesel engine 鼓风冷却式柴油机
blower cooled engine 鼓风冷却式发动机;风冷式发动机
blower cooled project 风冷式投影机
blower cooler 风冷式冷凝器
blower cooling 鼓风冷却
blower door equipment 鼓风门测定仪
blower drain valve 鼓风机放泄阀
blower drive 鼓风机传动装置
blower drum 鼓风机外壳
blower duster 鼓风式喷粉机
blower elevator 鼓风抛送升运器
blower fan 鼓风机;风扇;风机;增压机;鼓风扇;吹风机
blower house 鼓风机房
blower kiln 鼓风干燥室
blower line 放汽管
blower motor 鼓风机用马达;鼓风电动机
blower nozzle 鼓风口
blower performance 通风机性能
blower pipe 吹送管;放水管;放气管
blower pipe union 吹风管连管节
blower pressure 吹风压力;吹风机压力
blower production line 鼓风机生产线
blower pump 增压泵
blower ratio 鼓风机增压比
blower regulation valve 风机调节阀
blower rod 鼓风机杆
blower room 鼓风机室
blower rotor 鼓风机转子
blower section 鼓风机部分
blower set 通风机组
blower shutoff 鼓风机关闭器
blower slinger 增压器分液环
blower sprayer 鼓风喷雾器
blower steam pipe 吹风进汽管

blower system 鼓风系统;吹风系统
blower turbine 鼓风机透平
blower-type snow plough 旋转式除雪机
blower-type specific gravity stoner 吹式比重法去石机
blower unit heater 吹风式暖风机
blower valve 鼓风机阀
blower wheel 鼓风机叶轮
blower wind machine 鼓风机
blow fan 鼓风机
blow film 吹塑薄膜
blow flexure test 冲击弯曲试验
blow form 衬砌模架
blow frequency 冲击频率
blow gun 吹嘴;喷雾器;喷枪;喷粉器
blow head 吹气头
blow-hole 窑顶排气孔;铸孔;铸件气泡;吹穴;吹砂孔;潮吹(指海蚀洞天井);(刚性路面的)喷泥机;风蚀穴;风浪穴;喷水孔;砂眼;气孔;喷呢孔(刚性路面上的)
blow-in 鼓风;开炉;涌入;吹入;跑气(气压盾构)
blowing 咽泥(刚性路面的喷泥现象);抹灰爆裂气;吹塑;吹气;放浆;表面起泡;表面爆裂;爆裂孔眼;气力吹开;喷气;吹吹;喷出;陶瓷表面起泡
blowing action (刚性路面的)喷泥作用
blowing activator 发泡活性剂
blowing agent 发泡剂;起泡剂;起爆剂;生气剂
blowing angle 吹炼角
blowing a well 使井喷出泥沙等物质
blowing boiler tube 锅炉吹灰管
blowing boundary-layer control 吹除式附面层控制
blowing burner 喷吹燃烧器
blowing cave 吹风洞
blowing chamber 集棉室;沉降室
blowing characteristics 发泡性
blowing chest 风栅
blowing current 熔断电流
blowing device 吹芯机
blowing down 锅炉在压力放气;停风;放水排污
blowing dune 内陆移动沙丘;移动沙丘
blowing dust 巨风尘;刮风沙;高吹沙;吹尘;尘暴;飞尘
blowing engine 鼓风机(械)
blowing equipment 鼓风设备
blowing fan 鼓风机;压风机
blowing from bosh 腰风
blowing gas 吹入气体;发泡气体
blowing glaze 吹釉
blowing hole 吹洗孔
blowing in 开始送风;送风
blowing-in burden(ing) 底炭;开炉底料
blowing-in shaft 送风竖井
blowing installation 吹管;玻璃吹管;鼓风装置;吹氧化装置;吹气装置
blowing intensity 鼓风强度
blowing in wild 敞喷;无阻井喷
blowing machine 鼓风机;增压器;吹风机;风箱
blowing making 吹制
blowing method 吹制法;吹放法
blowing method of ventilation 压入法通风
blowing mo(u)ld 吹模
blowing mo(u)lding 吹模法
blowing murmur 吹风样杂音
blowing nozzle 吹嘴
blowing off 吹散;吹脱;吹净
blowing-off effect 喷射效应
blowing off period 排污周期
blowing-off pond 吹脱池
blowing off time 排污时间
blowing-off tower 吹脱塔
blowing of plaster 抹灰爆裂;灰泥或墙粉爆孔
blowing of soil 土壤吹失
blowing on taphole 出铁口喷火
blowing on the monkey 渣口喷火
blowing opening 送风口;进风口;风口
blowing-out 停炉;吹洗炮眼;停风
blowing-out of lines 炉衬腐蚀
blowing out preventer 防喷器
blowing-over 工作面通风(爆破后)
blowing pipe 吹洗管
blowing piston 漏气活塞
blowing plant 压缩空气装置
blowing position 吹炼位置
blowing power 风力

blowing pressure 吹塑压力
blowing process 喷吹法
blowing promoter 发泡助剂
blowing rate 鼓风速率;风量
blowing respiration 吹气样呼吸音
blowing road 主进风平巷
blowing rock drill drifter 风动架式凿岩机
blowing sand 刮风沙;高吹沙;吹砂
blowing snow 驱傲雪;高吹雪;吹雪
blowing sound 吹音
blowing spray 浪花飞溅;高吹沫;水花
blowing still 沥青氧化釜;吹炼釜
blowing stone-removing machine 吹式去石机
blowing system of ventilation 压入式通风
blowing-tensile crack 鼓胀裂缝
blowing through 分支排气
blowing tower tray 吹液现象
blowing tube 排污管
blowing vent 送风口;进风口
blowing ventilaiton mode 吹入式方式
blowing ventilation 受压通风;压入式通风;鼓风通风;吹风换气
blowing ventilation pipe 压入式通风管
blowing well 吹风井;喷泉井
blowing wool 喷吹棉;吹吹玻璃棉
blow joint 吹接接头
Blow-Knox antenna 布洛—诺克斯天线
blow lamp 焊接灯;焊灯;吹管;喷燃器;喷灯
blow land 风蚀地
blow line 流道管;风道;放喷管(路);喷放管道
blow mo(u)ld 吹制模具
blow mo(u)lding 吹塑(成型)法;吹塑;吹气造型;吹喷成型
blow mo(u)lding hot stamping 吹制热压成型
blow mo(u)lding machine 吹塑机;吹模机
blow mo(u)lding machine for extrathin film 超薄薄膜吹塑机
blown 吹制(成的)
blown aeration 鼓风曝气
blown animal oil 吹制动物油
blown asphalt 氧化(地)沥青;吹制(地)沥青;吹气(地)沥青
blown asphaltic bitumen 吹制沥青;氧化沥青;吹制纯(地)沥青
blown Baltic oil 吹制波罗的海亚麻油
blown bitumen 氧化沥青;吹制沥青
blown Calcutta oil 吹制加尔各答亚麻油
blown casting 气孔铸件
blown castor oil 氧化蓖麻油
blown coal-tar pitch 氧化煤柏油脂;吹气硬煤沥青;吹气煤柏油脂;吹制硬沥青煤
blown cobalt-blue glaze 吹青釉
blown cylinder glass 吹制圆筒摊平玻璃;吹制筒状玻璃量器
blown-down silencer 排气管消音器
blown dust 飞尘
blown fatty oil 氧化脂肪油
blown fibre 吹制纤维
blown film 吹塑薄膜;多孔膜
blown film extruder 吹膜挤出机
blown fish oil 吹制鱼油
blown flat glass 吹制平板玻璃
blown fuse 保险丝熔断
blown fuse indicator 保险丝熔断指示器;熔线熔断指示器;熔断指示器
blown glass 吹制的玻璃制品;吹制玻璃
blown glass tumbler 吹制玻璃杯
blown glassware 吹制玻璃器皿
blown in burden 开炉配料
blown joint 吹接(头);封铅接合
blown linseed oil 亚麻厚油;吹制亚麻仁油
blown metal 吹炼金属
blown-off tank 吹泄箱
blown-off valve 吹泄阀
blown oil 氧化油;鼓气油;吹制油;气吹油
blown-out concrete 多孔混凝土;充气混凝土;泡沫混凝土
blown out hole 空泡;无效炮眼
blown-out land 风蚀地;风蚀区
blown-out shot 少量炸药;空炮;哑炮;瞎炮;废炮
blown out tyre 爆裂轮胎
blown oxidized asphalt 吹制沥青
blown nozzle 空气喷嘴
blown pattern 素乱散布

blown petroleum 吹制石油沥青;氧化石油沥青
blown pitch 氧化煤柏油脂;吹气硬煤沥青;吹气煤柏油脂;吹制硬沥青
blown plate oil 吹制阿根廷亚麻油
blown primer 轧飞雷管
blown red glaze 吹红釉
blown rubber 海绵胶
blown sand 漂沙;风积沙;风成砂;飞砂;飞沙
blown sheet 吹制薄片玻璃;吹制筒状玻璃量器;吹制的平板玻璃
blown smoke 吹烟
blow(n) soil 风积土(壤)
blown sponge 海绵胶
blown stand oil 氧化聚合油;吹制定油
blown staple fiber mat 吹制定长纤维毡
blown tip pile 爆底桩
blown tubing 吹塑模管
blown tumbler 吹制杯
blown tumbler with enamelled decorations 彩花吹制杯
blown tumbler with intaglio decoration 吹制车花杯
blow number 贯入击数
blown up 吹足;危险的堆货
blown-up halftone 放大半色调片
blown-up mosaic 放大镶嵌图
blown warm air hand dryer 热吹风干手器
blown wool 喷吹玻璃棉
blow of caisson 沉箱放气
blow of dynamic(al) sounding 动力触探锤击
blow-off 退火;吹离;吹除;刮走;吹出;放气;排污(口);排放;停炉;通风口;排污管;吹逸;吹泥;吹掉;喷出;停吹
blow-off barge 吹泥船
blow-off boiler 锅炉排污
blow-off branch 放水支管;排汽支管
blow off branch pipe 排气支管
blow-off branch with manhole 带人孔排泥分叉管
blow-off chamber 排气室;吹泄室;排泥井
blow-off check 放气阀(门)
blow-off cock 排泄栓;排泄龙头;排污旋塞;排气阀;吹除开关;排气栓
blow-off cock hole 放水塞孔
blow off collapse due to air blow 沉箱放气
blow-off connection 排污管接头
blow-off dredge(r) 吹泥船
blow-off gate 放泄闸门
blow-off hydrant 排泥放水龙头
blow-off line 排污管道;排汽管道
blow-off method 吹脱法
blow-off nozzle 断火嘴;吹嘴;排污短管
blow-off pipe 急泄管;吹泥管;吹除管;放泄管;放气孔;放气管;安全排气管;排污管;排气管;排泥管;排出管
blow-off plug cock 排气活塞栓
blow-off point 吹开点
blow-off pressure 放气压力;排气压力;停吹气压
blow-off tank 吹卸槽;排污池
blow-off tee 排污三通;排气三通
blow-off through valve 吹通阀门
blow-off valve 放水阀;放泄阀;排泄阀;排污阀;排料阀;排空阀;急泄阀
blow-off velocity 灭火速度
blow off ventilation 抽气通风;排气通风(法)
blow off water 排污水
blow of hammer 锤击
blow of penetration test 贯入试验锤击数
blow on 开炉
blowout 井喷;爆裂;暴发性火;熔断;喷出;砂涌;矿床风化露头;(堤坝的)决口;鼓风;吹蚀坑;吹出;风蚀洼地;风蚀坑;废炮;井喷;熔断;排灰;打穿;吹风;停吹
blowout box 蒸汽洗衣箱
blow out cock 气放塞门
blowout coil 消火花线圈;灭弧线圈;磁吹线圈
blowout contact 熄弧器触点
blowout current 熔断电流
blowout diaphragm 防爆膜;快速光闸;遮断膜片
blowout disc 保护隔膜
blowout door 防爆安全门
blowout door diaphragm 防爆门薄膜
blowout dune 吹蚀沙丘
blowout electricity 束缚电荷
blow out-fill trap 吹坑充填圈闭

blowoutfire 井喷失火
blowout hookup 防喷设备
blowout magnet 灭弧磁体;磁性灭弧用磁铁;熄弧磁铁;磁保熄弧磁铁
blowout of an engine 发动机灭火
blowout of compressed air 压缩空气瞬时喷出
blow out of tire 轮胎爆裂
blowout patch 补胎胶布;补胎胶片;轴圈
blowout plug 岩芯冲出器
blowout preventer 封井器;防喷装置;防喷器;石油防喷器
blowout preventer and spool 防喷器及其上端的法兰短节
blowout preventer of stuffing box type 盘根盒式防喷器
blowout preventer of the pressure packer type 压封式防喷器
blowout preventer of threestagepacker type 三段密封式防喷器
blowout preventer stack 防喷器组
blowout prevention 防喷
blowout prevention equipment 防喷设备
blowout process 吹动法
blowout roof 泄爆屋顶
blowout shot 瞎炮
blowouts in shaft kilns 立窑喷火
blow out switch 放气开关
blowout the furnace 停炉
blowout valve 排气管;排出阀
blowout water closet 冲击式大便器
blowout window 泄爆窗;爆裂窗;爆裂孔口(起安全阀作用)
blow-over 吹出料泡头(人工吹制玻璃制品)
blow over cooling 外部吹风冷却
blow-period 吹风期
blow pile 爆破桩
blow pipe 压缩空气输送管;吹气管;吹火筒;吹管;送风(支)管;焊枪;直吹管;喷气灯;喷焊器
blowpipe analysis 吹管分析;吹管反应分析
blowpipe assay 吹管鉴定;吹管分析
blowpipe flame 吹管焰
blowpipe head 气焊割炬嘴子
blowpipe igniter 吹管点火器
blowpipe lamp 焊接灯
blowpipe reaction 吹管反应
blowpipe reaction analysis 吹管反应分析
blow pipe stand 喷火管架
blowpipe test 吹管试验
blow pit 吹制坑;卸料槽
blow plate 吹芯板;吹砂板
blow plate bushing 吹砂板销套
blow purge 鼓风吹扫
blow rate 吹胀速度
blow-run 鼓风制气
blow sand 流沙(也称流砂);飞沙;飘沙;喷砂
blows foot 每条尺深锤击次数
blow-speed 打击速度
blows per layer 每层贯击数
blows per-minute (锻锤的)每分钟打击次数;每分钟击数
blow squeeze mo(u)lding machine 吹压式造型机
blow stress 冲应力;冲胁强;冲击应力
blow table 成形吹制台
blow-take off tower 吹脱塔
blow tank 泄料桶;出料槽;仓式泵;疏水箱;卸料槽;卸料桶
blow-test 吹气试验;撞击试验;冲击试验;爆破试验
blow the bilges 桶腹胀大
blow through 排污
blow through valve 吹除阀
blow torch 吹管;焊枪;焊灯;喷气灯;喷灯
blow torch blowing tube 吹管
blow-up 崩裂;(由冻胀引起的)隆起;鼓起;拱胀【道】;放漂;压曲破坏;炸毁;突起
blow-up failure(on the pavement) (路面的)冻胀破坏
blow-up mouth 上部吹扫口
blow-up of the charge 吹过金属层上面
blow-up patches (路面上的)冻裂修补处
blow-up performance(of pavement) (路面的)冻胀性状
blow-up ratio 吹胀比;模膜直径比
blow-up zone 充气区
blow valve 卸渣阀;通风阀;送风阀

blow vent 排气口;通气口;通风口;送风口
blow wash 压力吹洗;压水冲洗;压力冲洗;吹洗;喷洗
blow well 自喷井;自流井;涌泉
blow whistle mark 鸣笛标
blow wild 猛烈喷出
blub 起泡;灰泡;起皱;局部隆起;浇筑灰浆
blubber 鲸脂
blubber oil 鲸脂油
bludgeon 大头棒
blue 蓝色(的);上蓝色
blue agate 蓝玛瑙
blue-alert 台风警报
blue algae 蓝藻
blue alizarin(e) lake 蓝色茜素色淀
blue and green compilation 蓝绿版编绘原图
blue and green decorative painting 青绿彩画
blue and white 青花
blue and white porcelain 青花瓷器
blue and white with copper red colo(u)rs 青花釉里红;青花五彩莲池鸳鸯图碗
blue and white with overglaze colo(u)rs 青花斗彩
blue and white with pierced decoration 青花玲珑
blue and white with rice pattern 青花玲珑
blue annealed wire 发蓝钢丝
blue annealing 发蓝退火;软化退火;黑退火
blue annealing wire 发蓝钢丝
blue apex 蓝色点
blue appearance 蓝色印样
blue asbestos 蓝石棉;钠闪石石棉;青石棉
blue aura 蓝辉
Blue Azimuth Table 美国天体方位表(赤纬24°～70°)
blue band 蓝带
blue base 蓝图
blue beam 蓝色信号射束
blue Belgian limestone 比利时青石
blue bell 吊钟柳
blue bind 煤系中的页岩或泥岩;硬黏土
blue black 深蓝黑色;深蓝色的;蓝黑色
blue black level 蓝路黑电平
blue branch 蓝分支
blue brick 蓝砖;阴沟砖;青砖
blue brick paving 青砖铺面
blue brittleness 蓝热脆性;蓝冷脆(性);蓝脆(性)
blue-brown 蓝褐色
blue cap 蓝色焰晕
blue carbonate of copper 石青蓝;蓝铜矿(石青)
blue cell 蓝光电池
blue chalcocite 蓝辉铜矿
blue chip 热门证券;热门股票
blue chip firm 热门商号
blue chip shares 头等证券
blue chrome black 蓝色铬黑
blue circle 波特兰水泥(注册名称)
blue clay 青黏土;蓝黏土;青土
blue-collar worker 产业工人;蓝领工人;体力劳动者;体力工人
blue colo(u)r difference axis 蓝色差信号轴
blue colo(u)r difference matrix 蓝色差矩阵
blue colo(u)r difference modulator 蓝色差调制器
blue colo(u)r of sky 天蓝(色)
blue concrete 蓝色混凝土
blue connector 蓝色连接器
blue control grid 蓝色控制栅
blue copperas 硫酸铜;蓝矾;五水(合)硫酸铜
blue copper ore 蓝铜矿
blue copy 蓝图
blue dichroic mirror 蓝色二向色镜;蓝分色镜;蓝二向色反射镜
blue dip 蓝色汞浸液(镀银前)
blue drawing 描青
blued scissors 法兰剪
blue dwarf 蓝矮星
blue dye(stuff) 蓝色染料
blue earth 蓝地;角砾云榄岩;青土
blue edge 蓝界
blue elephant (置于井上的)立式直接作用泵
blue end 蓝磁极;磁针指南端
blue eye-protection glass 蓝色护目镜玻璃
blue filter 蓝色滤色片;蓝色滤光镜
blue finished 蓝色回火的

blue fish 海豚;深蓝色鱼类
blue flame burner 蓝焰喷灯
blue flash 绿闪;蓝闪
blue flashbulb 蓝色闪光灯
blue flax 暗黑色亚麻
blue-fluorescence 蓝色荧光
blue-form income tax return 蓝色所得税申报表
blue-form return 蓝色所得税申报表
blue-free filter 蓝色滤光镜
blue gas 蓝(焰)煤气;纯净水煤气;氰毒气;水煤气
blue-gas generator 水煤气发生炉
blue gas pipe 水煤气管
blue gas set 水煤气装置
blue gas tar 水煤气煤焦油;蓝煤气柏油
blue glass 蓝色玻璃
blue glaze 蓝釉;蓝(锌)粉
blue glazed porcelain 青釉瓷器
blue glow 蓝辉光
blue gold 蓝金;金铁合金
blue goods 具有蓝色污点的软木
bluegrass 早熟禾
blue-green 蓝绿色
blue-green algae 蓝丝藻;蓝绿藻
blue-green algal grwoth 蓝绿藻生长
blue-green flame 绿闪;蓝绿闪
blue-green region of spectrum 蓝绿光谱范围
blue-grey 蓝灰色
blue-grey clay 蓝灰色黏土
blue grindstone 青色磨石
blue ground 蓝土;蓝地
blue gum 蓝桉胶
blue heat 蓝热
blue-heat brittle 蓝脆(性)的
blue heating 低温加热
blue horizontal branch 蓝水平支
blue horizontal branch star 蓝水平支星
blue ice 纯洁冰;纯海冰
blue illuminating wave 蓝色照明波
blu(e)ing 发蓝;烧蓝
blu(e)ing treatment 发蓝处理
blue iron earth 蓝铁矿
blue jack 胆矾;五水(合)硫酸铜
blue jacket 水手
blue john 蓝块萤石;蓝萤石
blue Jun glaze 蓝钧釉
blue key 蓝图
blue lake 蓝色色淀
blue lamp 蓝灯
blue lateral convergence 蓝位校正;蓝位会聚;蓝色横向会聚
blue laws 蓝法
blue lead 蓝色金砂矿;方铅矿;碱水硫酸铅;蓝铅
blue lead-ore 方铅矿
blue-lead paint 蓝铅漆
Blue Lias 蓝里亚斯层【地】
blue lias lime (侏罗纪岩层石灰石中的)水硬石灰;蓝皮水硬石灰
blue-light source 蓝光源
blue lime 蓝色石灰
blue line 蓝线;伯顿氏线;蓝色印样;蓝色线条图
blueline board 蓝图板
blueline guide 蓝图
blue-line key 蓝图拼台纸
blueline print 蓝图
blueline process 晒蓝图
blueline reproduction 蓝线图
blue litmus paper 蓝石蕊试纸
blue lumber 青变材
blue lustre 蓝光泽彩
blue magnetism 蓝色南极磁性;蓝色磁性;蓝磁性
blue malachite 蓝铜矿
blue marl 青泥灰岩
blue metal 黏土质片岩;蓝锌粉;蓝铜锍;泥质页岩;铜硫层;碎青石
Blue Mosque at Istanbul 伊斯坦布尔的蓝清真寺
blue mo(u)ld 青霉
blue mud 青泥
blue nankeen 毛蓝土布
blue needle 蓝针
blueness 青蓝;蓝色;蓝度
blueness index 蓝度指数
blue of the sky 天空蓝度
blue oil 脱蜡后的油料;蓝色油;蓝油;地蜡矿蒸馏得到的重质油和蜡的混合物;粗柴油

blue organic pigment 蓝色有机颜料
blue-out paper 蓝晒纸
blue pearl 蓝珍珠
blue phosphor 蓝色磷光体
blue pigment 蓝色颜料;蓝色染料;青的颜料
blue pigment dyestuff 蓝色染料
blue pigment in oil 蓝颜料浆;蓝浆料
blue plan 蓝色计划
blue planished steel 发蓝薄钢板
blue pole 蓝磁极
blue powder 蓝(锌)粉
blue primary 蓝原色
blue primary information 蓝基色信息
blueprint 规划蓝图;图纸设计;晒蓝;蓝图;图纸(指蓝图)
blueprint apparatus 晒图设备;晒蓝机;晒蓝图设备;晒蓝图机
blueprint drawing 蓝图;蓝图清绘
blueprinter 晒图员;晒图机;晒蓝图机;晒蓝图工人
blueprinting 晒蓝图;晒图
blueprinting apparatus 晒图机
blueprinting machine 晒图机
blueprinting paper 晒图纸;蓝图纸
blueprinting room 晒图室
blueprint lamp 晒图灯
blueprint machine 晒图机;晒图设备
blueprint paper 蓝印纸;蓝图纸;晒图纸
blueprint solution 晒溶液
blue radiation 蓝色辐射
blue response 蓝光响应
blue restorer 蓝电平恢复器
blue ribbon 蓝带程序;精选的;最好的;最高荣誉的;头等的;第一流的
blue-ribbon connector 矩形接插件;矩形接头座
blue ribbon program(me) 蓝带程序;一次通过程序
blue roofing tile 小青瓦
blue rot (木材的)青腐
blue scaler 蓝色度盘
blue schist facies 蓝闪片岩相;蓝片岩相
blue sea 深海
blue seal 蓝色封缝砂浆
blue-sensitive 对蓝色感光的
blue-sensitive plate 蓝敏底片
blue sensitivity 蓝敏度
blue shade 蓝相;带蓝头
blue sheet 发蓝薄钢板;蓝钢皮
blue shift 蓝移
blue shortness 蓝冷脆(性);蓝脆(性)
blue sky 晴空;价值极微的
blue sky exploratory well 新区普查孔
blue-sky law 股票发行控制法;青空法;蓝天法
blue-sky market 户外市;露天市场
blue-sky scale 林克标【气】
blue spar 蓝铜矿;天蓝石
blue stain 霉斑;蓝霉斑;(木材的)蓝斑;青变
blue stain fungi 蓝霉菌
blue static convergence magnet 蓝静会聚磁铁
blue steel 蓝钢
blue stone 硫酸铜;蓝石;胆矾;青石(岩);天然硫酸铜;蓝砂岩;硬黏土;筑路用青石;蓝灰砂岩
blue strain 蓝变
blue striped heavy cees 蓝线平纹袋
blue-tinted flash bulb 蓝色闪光灯
blue tinted glass 蓝色玻璃
blue tone 蓝色调
blue tone process 晒蓝图
blue toning 吊篮谐
blue top 高度桩;坡面桩;坡度桩【测】;坡度标准桩
blue-top grade 旧路改建前的纵断面(美国)
blue tube 蓝辉光管
blue ultramarine 群青;佛青
blue verditer 蓝色铜盐颜料;铜蓝
blue-violet 蓝紫色
blue vitriol 硫酸铜;蓝矾;胆矾;五水(合)硫酸铜
blue wash 涂蓝色;涂蓝薄层
blue watch 第三班(船上值班表)
blue water 大海
blue water gas 蓝水煤气;蓝煤气;海底天然气
blue waters 公海
blue water vessel 钻孔船;水下钻孔船
blue whale 蓝鲸
blue wool scale 蓝色羊毛标准
bluff 岬陡岸;悬岩;悬崖;陡峭的;陡壁;肥型船首;不良流体陡坡;天然陡坡

bluff bank 陡岸
bluff body 非流线型体;扁平体
bluff body flame 障碍板稳定火焰
bluff bow 肥型船首;方形船头;方形船首
bluff erosion 陡壁冲刷;陡岸冲刷
bluff failure 陡壁损毁;陡岸坍毁;坍岸;陡壁坍塌
bluff falling 危崖落石
bluff formation 大孔性黄土层;粗黄土层
bluffing 褶边装置
bluff piece 爪件
bluff podzol 沼津灰土壤
bluff racking 空转
bluff work 边坡整平作业;陡坡整平工作;边坡整平工作;边坡修整作业
bluffy coast 陡岸
bluing 蓝化;加蓝去黄提白;着蓝;涂蓝;烧青砖法;上蓝
bluing of steel 钢加蓝
bluish 淡蓝色的;带青蓝色的;浅蓝色的
bluish-green 蓝绿色
bluish-green argon laser 蓝绿氩激光器
bluish-green spotted 青点
bluish-grey 蓝灰色
bluishing caused by thinner 由稀释剂(引起)的发白
bluish red 蓝红色;浅蓝红色的
bluish white 青白(色)
bluish white porcelain 青白瓷
Blumea oil 艾叶油
Blumenfield process 布卢门菲尔德水解法
Blum measure 波鲁姆测度
Blum reagent 波鲁姆试剂
Blum's compress theorem 波鲁姆压缩定理
Blum speed up theorem 波鲁姆加速定理
blunder 故障;粗差;误差
blunder against 冲撞
blunder detection 粗差检验
blunder free 无粗差的
blunge 用水搅拌;用水搅和;柔黏土;水掺和黏土
blunger 搅浆机;黏土和水搅拌器;黏土拌和器;黏土拌和机;圆筒形搅拌机;圆筒搅和机
blunge water 浆料
blunging 制成泥浆(陶瓷料);泥浆搅拌;湿混合
blunging of kaolin 高岭土捣浆
blunk 厚实印花布
blunt 钝的
blunt angle 钝角
blunt arch 垂拱(拱高度小于一半跨度者);(拱心在起拱线下的)平坦拱
blunt bit 钝钻头
blunt bolt 木螺栓
blunt cutting edge 钝切削刃
blunt drill 磨钝的钻头;磨钝的钎头
blunted cone 钝锥
blunt edge 钝缘
blunt-edged 去锐边
blunt edge stones 磨钝的金刚石
blunted spur 切平山嘴
blunt end 钝端
blunt ended body 钝尾物体
blunt-ended root 钝根
blunt-ended spur 钝端丁坝;方头丁坝
blunt end of pile 桩的钝头;桩的粗端
blunt file 直边锉;钝端锉;齐头平锉
blunt hook 钝钩
blunt mill file 平行细锉
blunt nail point 钉的钝端
bluntness 钝度;粗率
blunt-nosed 钝头的;平头的
blunt(-nosed) body 钝头物体
blunt nosed cone 钝头锥体
blunt nosed pier 平头墩
blunt pile 钝头桩
blunt point suture needle 锤头缝合计
blunt round file 齐头圆锉
blunt screw 圆头螺丝钉
blunt square file 平行方锉
blunt start 螺钉钝头
blunt stones 磨钝的金刚石
blunt trailing edge 钝后缘
blur 模糊不清;颜色不均匀涂层;污点;涂污
blur circle 模糊圈;弥散圆
blur factor 模糊因数
blur level 模糊级

blurred 模糊的
blurred contour 模糊轮廓
blurred image 模糊影像;模糊形象;模糊图像
blurred picture 模糊影像
blurred region 模糊区域
blurred signal 模糊信号
blurred spot 模糊点
blurred target 模糊目标
blurred vision 视力障碍;视力模糊
blur(ring) 污损;模糊
blurring computer 模糊计算机
blurring effect 模糊效应
blurring formula 模糊公式
blurring highlight test 耐酸试验后搪瓷表面侵蚀程度试验
blurring mapping 模糊变换
blurring transition function 模糊转变函数
blur spot 模糊斑点
blushing 变红;褪色;(油漆的)混浊膜;泛红色;发白;雾浊
blush proof for vinyl perchloside coating 过氯乙烯漆防潮剂
blush resistance 抗发白性
blush-resistance determination 抗发白测定
Blyth elutriator 布莱恩淘析器
Blyth's test 布莱恩氏试验
B-mineral 贝利特
bnnana oil 乙酸戊酯
Bnusen ice calorimeter 本生冰量热器
board 薄板;建筑用板;减速地点标;舷;仪器面板;仪表板;中薄板;登船;船舷;插件板;复合板材;板;条板;上车
board adhesive 木板板胶粘剂
board and batten 薄厚板镶接;板条芯板;板和条;板条盖缝的墙板
board and batten siding 宽窄板交替壁板
board and board 两舷相接;两舷平行
board and brace (木材的)镶接
board-and-brace work 镶接(木)板(作业)
board-and-lodging apartment 膳宿公寓
board as a rare commodity 奇货可居
board a ship 接舷攻击;登船
board-board-transshipment 船转船运输
board body type 板式体形
board butt joint 木板对接缝;喷射施工缝;喷射混凝土板式平接施工缝;喷混凝土的施工缝
board by board 两舷相接;两舷平行
boardcarton 石膏板护面纸
board ceiling 板条平顶
board chairman 董事会主席
board chronometer 天文钟
board clip 单板夹
board coal 纤维质煤;纤维褐煤
board code 插件板代码
board core 板心
board covering 覆盖板
board cutter 纸板裁切机
board cutting machine 锯板机
board deals 厚宽板材
board debugger 插件调试器
board deck 木板屋顶
board-down 下山
board drag 延板;刮板
board drop hammer 夹板落锤;夹板锤;木柄摩擦落锤;打桩锤
board drop stamp 夹板落锤捣碎机
board eaves 木檐;木板挑檐
boarded ceiling 板平顶;板顶棚;板吊顶
boarded door 木板门;直板门
boarded false ceiling 板条吊顶棚
boarded fence 木栅栏
boarded floor 木楼板地面
boarded gangway 铺板甬道
boarded leather 磨光皮革
boarded parquetry 木板嵌镶作业;木板镶嵌作业;木板镶嵌工作;木板嵌镶(工作)
boarded partition 木板隔墙
boarded up 钉上板子的
boarded web 木腹板
board facing 木面料;衬板
board fence 木栅(栏);木板栅栏;木板篱笆;板篱
board finish plaster 木板纹饰面粉刷;板面粉饰灰浆
board foot 板英尺;木料英尺;木料尺;板口尺;板材容积单位;板尺

board-foot measure 板英尺量度;木材等级;板英尺度量
board forms 板模;木模(板);木壁板
board formwork 木模(板);木壁板
board for stretching leather 钉皮板
board grapple 木板夹具
board guard 夹板防护罩
board guide 插件导轨
board hammer 重力式落锤
board heating 周边供暖
board hole 承接槽口
boarding 木装修板;地板;搓纹;起纹;铺木板;铺地板;铺板;围篱;贴上;上船
boarding book 上船检查登记簿
boarding car 宿营车
boarding gate 登记口
boarding house 膳宿公寓;供膳宿舍;招待所;供膳宿的私家住宅
boarding(in) 安装木板
boarding inspector 上船检验人员
boarding joist 裸(木)格栅;楼板格栅;板条托梁
boarding ladder 登船悬梯;登船梯
boarding-lodging house 供膳宿的住房;供膳宿住
boarding machine 定型机
boarding net 登船网
boarding of ship 接舷战
boarding pass 登记证
boarding passenger volume 上车客流量
boarding platform 候车站台(美国)
boarding press 热压定形机
boarding quantity on arrival 入港时剩余数量
boardings 验船条款
boarding school 寄宿学校
boarding station 候潮锚地
boarding time 上客时间
board insulation 木板隔热隔声;绝热板;热绝缘板;绝缘板
board joint 木板接合;平板接;平板接(缝)
board joint filler 板接缝防水填料
board knife 刮刀
board lath 石膏板(条);石膏条板;灰板条
board level 插件层次
board lifter 插件插拔器
board-like 板样的
board-like consis 板样硬度
board-like rigidity 板状强直
board machine 纸板机
board-marked 木纹板面;形成木纹表面
board-marked concrete 模板纹混凝土
board-marked concrete finish 木板纹混凝土饰面
board-marked finish 混凝土木质花纹饰面
board-marked texture 模板印纹
board measure 材积单位;材积;板英尺度量度;板积计(算);板积测量;板尺(计算);板材计量单位;按板尺计算
board measurement feet 板材量尺
board meter 板米(尺)
board mill 锯板厂;制板厂;纸板机;板材制板厂
board minutes 董事会会议(记)录
board-mounted controller 面板式控制器
Board of Adjustment 调节委员会
board of administration 理事会;董事会
board of advisors 顾问委员会
Board of Aldermen 市政委员会
board of arbitration 仲裁委员会
board of asbestos cement 石棉水泥板
board of audit(ors) 审计委员会
board of civil authority 地方民政当局
board of conciliation 调解委员会
board of consultants 顾问委员会
board of directors 理事会;董事会
board of estimate 财政监查委员会
Board of Exchequer 英国财政部
board of executive directors 执行董事会;常务董事会
board of experts 专家委员会
board of extraordinary directors 非常董事会
board of felt(ed fabric) 缩绒织物;油毡板
board officer 海关联络员
Board of Foreign Trade 国际贸易局
board of governors 理事会
board of health 卫生局
board of labo(u)r 劳动部门
board of management 理事会

board of paper 纸板
Board of Realtors 房地产同业公会
board of review 课税证价审查委员会
board of supervisors 监事会;县管理委员会(美国)
Board of Trade 贸易部(英国);商品交易所(美国);贸易促进会;英国贸易部;商业部;商务部
Board of Trade Unit 度(英国电量单位,即千瓦小时)
board of trustee 信任理事会
board of work 工程委员会
Board of Zoning Appeals 区划申诉委员会
board paper 纸板
board parquetry 木板镶嵌细作;镶嵌细作木板
board partition 木板隔墙;板墙
board platform 木板工作台
board roof 木板屋盖
board room 会议室;董事室;交易所的交易场所;矿房;董事会会议室
board rule 板尺;木板量尺;材积尺;板英尺;板尺
board runway 木板走道
boards 板纸;舞台面
boards and battens 垫板和薄板
board sawing machine 锯板机
board sawmill 制板厂
board scale 按板尺计算
board scraper 板式刮土机
board separator 分板机
board sheathing 木望板;望板
board shuttering 木板百叶窗;木模;木模(板);木壁板
board-side paint 船舷漆
board sling 吊货盘
board spacing 板距
board splitter 缝楔;分板楔
boardstone 界石
board tree 板材树
board type console 面板式控制台
board-type insulant 板状绝缘材料;板式绝热材料;板式绝缘材料
board-type insulation 板材保温
board up 用板遮住;用板围住;上山
board voltage 配电板电压
board wages 膳宿工资
board walk 海滩上步行;步行板;木板人行道
boardy feel 硬性手感
boar's nest 工房(俚语,施工场地的);工棚
boart 金刚石屑;金刚石砂;圆料金刚石;低质钻石
Boas-Oppler bacillus 嗜酸乳杆菌
boast 钢铁块;粗凿石
boasted ashlar 粗凿琢石;粗凿石圬工;粗凿方石;乱纹琢石饰
boasted finish of stone 粗凿石面
boasted joint surface 宽凿接缝槽面;刻平行槽石面
boasted surface 粗凿工作;粗琢表面;粗凿面
boasted work 刻槽琢石;刻槽作业;刻石作业;修凿;宽凿工;粗凿工作;粗凿工程
boaster 槽榫刨;粗堑凿;阔凿;宽凿
boaster chisel 片石粗琢
boaster fan 射流风机
Boas' test 博亚斯试验
boasting 粗凿石块;堑石;粗琢
boasting chisel 粗堑凿;阔堑;片石阔凿
boat 木材槽朴;小划子;船舶(尤指小型船舶);船;槽子砖;艇;挑料环
boatable 可航行的;可用小船运输的;可通小船的
boatage 小船运输;带缆小艇费
boat air tank 小艇空气箱;浮箱
boat anchor 小艇锚;艇锚;舢板锚
boat anchored by cableway 缆道吊船;过河索吊船
boatard ashler 毛石墙琢石;粗饰琢石
boat arrangement 救生艇布置图
boat awning 救生艇天幕
boat basin 锚地;小船泊池;小船泊地;港地;船坞;泊地;船渠
boat beam 承艇枕
boat block 吊艇滑车
boat body 船体
boat body design 船体设计
boat body stress 船体应力
boat boom 舷侧系艇杆;系艇杆;吊艇杆;撑艇杆
boat bottom type of deformation 船底式变形
boat box 小艇属具箱
boat bridge 舟桥;浮桥

boat builder 造船厂
boatbuilding 小船业;造船
boatbuilding design 造船设计
boatbuilding yard 造船厂
boat canal 船渠;浅水运河
boat car 运艇车
boat channel 小船航道
boat charges 带缆舢板费
boat chock 救生艇垫座
boat compass 救生艇罗经;航海罗盘(仪)
boat conformation 船式构型
boat construction 救生艇结构
boat course 赛船用划道
boat cover 艇罩
boat cover hitch 艇罩结
boat cradle 救生艇垫座
boat crane 吊艇起重机
boat crutch 桨叉
boat davit 吊艇杆;吊艇杆
boat deck 救生甲板;艇甲板
boat derrick 吊艇杆;船式起重机;艇式起重机;水上起重机
boat dock 小船码头;小船港池;带轮布景台;(舞台上迅速换景用的)车台
boat drain 艇底泄水孔
boat drill 操艇训练
boatel (集柱的)柱身;驾驶游客旅馆;游艇旅客馆;汽艇游客旅馆;水上旅馆;集柱中一个柱身
boat engine 救生艇发动机
boat equipment 救生艇用具;救生艇用具
boat fall 吊艇索;吊艇辘绳;吊艇滑车索;钩头篙
boat fall block 吊艇滑车
boat fall reel 吊艇辘绳卷筒;吊艇滚筒
boat fare 舢板租金
boat fast 艇缆
boat fender 艇护舷
boat ferry 渡河小船
boat fitting 艇舾装
boat form 船式
boat gear 吊艇装置
boat gong 艇用信号锣
boat gripe 稳艇带
boat guy 艇牵索
boat handling 救生艇起放;操艇
boat handling gear 吊艇装置
boat handling signals 救生艇收发指挥信号
boat handling winch 救生艇收放绞车
boat harbo(u)r 小船码头;小船港(池);小船避风港;船港;停船港
boat hire 小船运输;带缆小艇费;带缆舢板费
boat hoist 吊艇绞车
boat hook 吊艇钩;带钩撑稿;船拖钩;钩头篙;挽钩
boat hook stave 钩篙
boathouse 游艇停泊场;游艇之家;游艇旅社;船库;艇库
boat incline 船滑台
boating 划船;小艇驾驶术
boating facility 划船设施
boating pool 泊船池
boating spot 游船区
boat keel stay 龙骨支柱
boat label plate 艇的铭牌
boat lamp 救生艇灯
boat landing 小船码头
boat launch 船舶下水装置
boat lead 救生艇用测深锤
boat level 船形水平仪
boat lift 升艇机
boat lift railway 斜面上船道;小船升船滑道
boat line 系船缆;装卸油管线
boatman 桨手
boat(man) charges 带缆小艇费;小船运输费
boatmanship 划艇术;划船术
boat measurement 船测
boat-mounted acoustic(al) locating device 船用水声定位仪
boat-mounted streamflow measuring equipment 船用水文测验设备
boat nail 方钉;救生艇钉
boat nest 艇垫架
boat noise 船舶噪声
boat nomenclature 救生艇编号
boat note 交货记录;卸货授受单;船舶记录;配载清单

boat of refractory metal 高熔点金属舟(真空镀膜用)
boat painter 艇首缆
boat pattern 船型
boat pipeline 装卸油管线
boat plug 船底排水旋塞；艇底塞
boat-pole 篙
boat propeller gear 小艇推进器
boat pulling 荡桨
boat ramp 船用斜坡道；升船斜坡道
boat rations 救生艇备用食品
boat reactor 舟形反应器
boat repair gang 修船分队；修船班
boat repair group 修船小组
boat rising 艇侧纵条
boat rivet 艇铆钉
boat saddle 艇鞍架
boatsail drill 船帆布
boat sailing 救生艇驶风
boat sailing equipment 救生艇帆具
boat scaffold(ing) 船用活动滑台；悬挂(式)脚手架
boat's chart 救生艇海图
boat-shaped bushing 舟形漏板
boat shaped object 船形器
boat shaped vase 船形壶
boat shed 舢板棚
boat shop 小艇车间
boat-side paint 船舷涂料
boat skate 救生艇滑架
boat skid 护舷板；承舰枕
boat sling 救生艇吊索；船用吊索
boat spar 吊艇杆；撑艇杆
boat spike 船用方钉；船钉
boat's rig 小艇索具
boat's screw 救生艇艇员
boat stand 艇支柱
boat stations bill 救生部署表【船】
boat stations signal 救生演习信号
boat steerer 救生艇舵手
boat stretcher 脚蹬(舢板上)
boatswain 水手长
boatswain's chair 船工脚手板；高空操作座椅；吊椅(高空作业用的)；单人座板；施工吊椅；吊篮
boatswain's holder 帆缆工具箱
boatswain's locker 水手仓库
boatswain's mate 副水手长
boatswain's pipe 水手长口笛
boatswain's store 船具室；甲板部物料；舱面物料储〔贮〕藏室；帆缆材料室；水手长仓库
boatswain's storeman 甲板部物料间
boatswain's whistle 水手长口笛
boat tackle 吊艇绞辘
boat tackle fall 吊艇辘绳
boattail 船形尾部
boat-tailed 有流线型尾部的；船尾形的
boat tank 救生艇浮力箱
boat test 吊艇试验
boat thwart 桨手座
boat tiller 救生艇舵柄
boat tracker 纤工
boat tractor 机耕船
boat train (与船运衔接的)联运列车；水陆联运列车；水陆联用列车
boat truck 脚轮推车
boat-type aeroplane 船式飞机
boat varnish 船用漆；船用清漆
boat wharf 小艇码头
boat winch 吊艇绞车
boat work 小艇作业
boatwright 造船工
boatyard 小船厂；造艇工厂；造艇场；小船厂
boat yoke 横舵柄
bob 悬锤；振子坠；振子球；秤锤；测锤；擦光毡；暗冒口；铅锤
bobbed 形成串的
bobber 浮沉材；浮材
bobbin 胶卷轴；架圈；线圈；线圈；细绳；点火线圈；绕线管；拖网木滚；筒子；筒管
bobbin bit 槽形扁钻头
bobbin core 线圈核心；带磁芯
bobbin derrick 滚轮起重杆
bobbin disk 卷盘
bobbin feeder 绕线筒进给器；绕线架进给器；绕线管进给器

bobbin flow meter 电感应式流量计
bobbing 截短；标记干扰性移动
bobbing buoy 梭式浮标
bobbing machine 振动机
bobbinite 硫铵炸药；筒管(硫铵)炸药
bobbin latch 木闩；木门闩
bobbin oil 锭子油
bobbin peg 筒子插钉
bobbin seal 套管密封
bobbin sorter 筒管分径机
bobbin spinning machine 筒管纺纱机
bobbin stripper 清除筒脚车
bobbin winder tension bracket 倒线架小过线
bobby prop 顶梁短支柱；防爆木挡板
bob-clearing bit 非自洁式钻头
bobeche 烛台托盘；台灯柱环
bob ga(u)ge 浮标
Bobierre's metal 鲍氏黄铜
bobierrite 白磷镁矿
Bobillier's law 博比里尔定律
bob pendulum 连叉摆
bobrovkite 高镍铁矿
Bobrovska garnet 绿钙铁榴石
bob sled 大雪橇；二撬拖材车；雪橇
bobsledding 串联雪橇运输
bob sleigh 二撬拖材车
bobsleigh run 雪橇路
bobtail draw bridge 截尼仰开桥
bobtailing 沿地面集材
bobtail plant 瓦斯装置
bobtail pooling 拼凑的合伙经营
bobtail rig 轻便钻机
bobtail target 隐显目标
bobtail truss 截尼桁架
bobweight 铅锤；反重；砝码；平衡重；配重；平衡锤；配锤
bobwire 铅垂准线；带刺铁丝；刺铁丝
bobwire fence 刺铁丝围栏
boca 河口(西班牙)；海港入口；峡谷出口
bocage 林地；混交林区；支叶等的花纹(美术)
bocca 玻璃炉的炉口；熔岩口；喷口；喷火口；小锥体
boccaro 红陶器；宜兴陶瓷
boccaro ware 无釉红炻器(宜兴紫砂器皿)
bocconine 博落回碱
Bochner's theorem 博克纳定理
bock kiln 堆砌式烧砖容
Bock's three-component model 博克三分量模型
bod 黏土封口；泥塞；砂塞；塞子
Bodansky unit 博唐斯基单位
bodden 海湾(德国)
BOD-DO water quality model of river 河流水质BOD-DO 模型
Bodecker index 博德克指数
Bode diagram 波德图
bodega 酒窖(西班牙)
bodeneis 地下冰；底冰(德国)
bodenite 伯顿石
Bodenstein number 博登斯坦数
Bode's law 波德定则
bodger 木雕刻工；木车工；椅子车工；撬杆
Bodian staining method 博迪恩胶体银染色法
bodied bit 有体式钻头
bodied linseed oil 熟亚麻子油；干性亚麻子油；黏稠亚麻子油；聚合亚麻子油；稠亚麻子油；厚亚麻子油
bodied oil 聚合油；厚油
bodied rail 端墙内加镶条；端墙内镶条
bodiless 脱胎
bodiless bit 无体式钻头
bodiless chinaware 脱胎瓷
bodiless porcelain 脱胎瓷
bodiless ware 脱胎瓷
bodily injury 人身伤害
bodily injury insurance 人身保险；伤亡保险
bodily movement 整体运动；整体移动
bodily seismic wave 地震体波
bodily sliding 整体滑动
bodily tide 固body潮
bodiness 加厚；增稠
bodkin 锥孔器；锥子；粗针；穿孔锥
BOD 生化需氧量
BOD-releasing coefficient from bottom sludge 底泥释放 BOD 系数
body 质地；床身；本文；本体；全体；坯体；坯料；物体

body acid 主份酸
body anorthosite 岩体型斜长岩
body armo(u)r 防弹衣
body assembly 机架组件
body at the rear 后犁体
body axis 联系轴；机体轴(线)；物体轴线
body bloating 疙瘩(陶瓷缺陷)
body bolster 车身承梁；车架承梁
body bracket 车身托架
body break 辊身折断
body brick 炉墙砖；优质烧透砖；透烧砖
body brush 马刷(子)
body builder 车身制造厂
body burden 人体负荷；体含量
body burden of pollutant 生物体内污染物负荷量
body capacity 车身容量；人体电容
body case 外壳；壳体
body cavity 体腔
body-center 体心
body-centered 体心的
body-centedd cube 体心立方体
body-centered cubic(al) 体心立方的
body centered cubic(al) crystal 体心立方晶体
body-centered cubic(al) lattice 体心立方晶格；体心立方点阵
body centered cubic(al) packing 体心立方堆积
body centered cubic(al) structure 体心立方结构
body centered cubic(al) system 体心立方晶系
body centered lattice 体心晶格；体心格子；体心点阵
body centered tetragonal lattice 体心正方晶格
body charge 体电荷
body cleaner 车身清洗器
body coat 硬质油灰；中间涂层；质地涂层；底涂漆
body colo(u)r 主体色；体色(彩)；不透明色
body cone 体锥面
body construction 车身构造
body contact 接壳
body core 芯子；主芯
body corporate 法人团体
body cover 车身涂底
body crushing strength 土体抗碎强度
body crush injury 全身挤压伤
body current 水体流动
body damping 机身阻尼
body defect 坯体缺陷；体缺陷
body design 车身设计
body diagonal 体对角线
body disc 体盘
body distortion 本体畸变；本体变形
body drag 机体阻力
body dumping mechanism 车厢倾卸机构
body effect 人体效应
body end furring 车端木梁
body end plate 顶棚端梁；上端梁
body end rail 端墙内镶条
body failure 本体折断
body feed 主体加料
body-fixed coordinate system 地固坐标系(统)
body flange 法兰模板；凸缘体
body flex 人体屈曲度
body fluid 流体
body flux 坯用熔剂
body force 重力；质量力；惯性力；砌力；体积力
body force potential 体积力势能
body form 体状
body formation 坯体成型
body forming machine 罐身成型机
body frame 机体架；机架；灶体构造；车底架；灶体构架
body freedom 物体自由度
body girth 体围
body-glaze fit 坯釉适应性
body-glaze intermediate layer 坯釉中间层
body grip 伸长的拉手
body hardware 车身金属构件
body holddown bracket 车身托架
body image 体像
bodying 指涂增厚出亮；加厚；油漆基层工艺；增加体积；稠化；变稠
bodying agent 基础剂；增稠剂
bodying coat 下涂；底涂
bodying in 擦漆；全工序油漆法；涂罩光漆
bodying of oil 油的聚合；油引聚合
bodying speed 黏度增加速度

bodying temperature 热炼温度
bodying-up 擦漆;涂火酒漆;涂罩光漆
bodying velocity 热炼聚合速度
body interference 机身干涉
body jack 车身千斤顶
body label 罐身商标纸
body leakage 机壳漏泄;外壳漏电
body length 体长
bodyless porcelain 脱胎瓷
body light 车内灯
body line 车身装配线;车身外形线;车身轮廓线
body lines 线形图
body lining 车身材料
body lip 车身凸出部分
body longeron (飞机的)机身纵梁
body material 车体材料;釜体材质
body matter 正文;本文
body measurement 体尺
body measurement method 实体测定法
body model 车身形式
body mo(u)ld 模体;下模
body necking-in machine 罐身收颈机
body nodal point 机身重心
body of abutment 台身
body of air 空气
body of an instrument 文件文本
body of axle 轴身
body of ballast 道砟层
body of bolt 螺栓杆
body of bridge pier 桥墩墩身
body of car 车身
body of casting 铸体
body of dam 坝身
body of deed 契据本文
body of equipment 设备壳体;设备的壳体
body of freshwater 淡水水体
body of lathe bed 车床床身
body of masonry 道砟层;圬工主体;圬工体
body of niche 壁龛主体
body of normative document 标准文件的主体
body of oil 油基;润滑油底质
body of paint 油漆稠度
body of pump 泵体
body of railroad 铁路路基
body of regulator 调节器体
body of revolution 回转体
body of river 主河床
body of road 路基
body of rod 杆体
body of roll 辊身
body of salt water 盐水体
body of saw 锯架
body of screw 螺钉体
body of shaft 井身
body of ship 船身
body of the bit 钻头体
body of the deposit 矿体
body of the report 报告正文
body of the shield 盾构主体
body of the work 正文
body of upright breakwater 直立式防波堤堤身
body of upright wall 直墙堤堤身
body of valve 阀体
body of velocity 速度体
body of vertical breakwater 直立式防波堤堤身;直立堤堤身
body of vessel 船身
body of wall 墙身
body of water 储(贮)水池;水体储(贮)水池;水体;储水池
body of wheel 轮盘
body opening 体孔
body oriented coordinate system 固定于物体的坐标系
body paint 底彩
body paper 纸厚
body parts 镜箱部分
body peel off 坯爆
body pigment 打底颜料;体质颜料
body pitch 型体倾斜度
body pivot 翻斗枢轴
body plan 机身平面图;横剖面线图;横断面线图;正面图;弹体平面图;船体正面图
body-plate stiffener 车身加强板

body polish 车身抛光油
body polisher 车身抛光工具
body preparation 坯体制备;坯料制备
body products 生物产品
body proportion 体尺比例
body putty 车身腻子
body rail 机身纵梁(飞机);机身梁
body refuse 泥渣
body-revolving tower crane 回转塔式起重机
body rolling machine 罐身成圆机
body scab 气疤
body scrap 回笼泥
body screw 型体螺钉
body seam 壳体接缝
body section 机舱;床身
body shear bolt 车身保险螺栓
body shear bolt bracket 车身保险螺栓架
body shell 车身外壳
body shop 车身车间
body size 体格大小
body size hole 穿透孔;通孔
body sleeve 壳套
body slip 坯用泥浆
body slitting machine 罐身截断机
body solid 固体
body squeeze 全身挤压伤
body stab 刺断
body stain 坯用色料
body statement 本体语句
body stock 体砧
body stress 机身应力;内应力;自重应力;船体应力;体应力;体积力
body style 形体风格;体态
body support 支身架
body support rod 车身支杆
body surface 辊身表面;体表
body surface area 体表面积
body switchover (双向犁的)犁体翻转机构
body system 体系
body tender bolster 煤水车身承梁
body test of whole tank vehicle/trailer 整车车体检验
body tide 固体潮
body tide perturbation 固体潮摄动
body-tilting equipment 车体倾斜装置
body-tilting technique 车身倾斜技术
body-tinted glass 本体着色玻璃
body transom 底架横梁
body truss rod 车身骨架支柱
body truss rod bracket 车身内柱架
body type 体型;车身式样
body-type antenna 弹力天线
body-type graph 体型图
body up 变稠
body upholstery 车身衬里
body varnish 磨光清漆;磨漆;面层用清漆;车身涂料;长油性清漆;长油度清漆;外用清漆
body warp 地经纱
body waste 物质损耗
body wave 体波
body wave magnitude 体波震级
body wave radiation 体波辐射
body wave spectrum 体波谱
body weight 体重
body welding machine 罐身焊接机
body wetting before glazing 补水;上釉前补水
body width 体宽
bodywood 无节新材;主干材;无枝干材
bodywork 船体;机身制造(工艺);船体修造工作;船身制造(工艺);车身制造;车身保修作业
body wrinkles 拉延件侧壁皱纹
Boedeker's test 伯德克试验
Boehme hammer 勃姆锤
boehmite 薄水铝矿;一水软铝石;勃姆石;软水铝石
boehmite-diaspore ore 一水型铝土矿矿石
boehm lamellae 勃姆薄层
Boeing aircraft 波音飞机
Boeing Co. 波音公司(美国)
Boeing helicopter 波音直升机
Boeing Scientific Research Laboratory 波音科学研究所
Boeman 波音机器人
Boetius furnace 贝替斯窑;半直焰烧煤坩埚窑
Boeton asphalt 菩伊吞地沥青

Boettger's test 伯特格尔试验
boffin 科技人员
Boffle 背面开口扬声器箱
bog 泥塘;泥炭沼泽;沼泽;沼地;采空区井巷;酸沼
bog and marsh garden 沼泽园
Bogardus scale 博加德斯
bogaz 沼泽疏干;沼泽排水;深岩沟;[复]bogazi
bog bath 泥浴
bog blasting 泥沼爆炸;泥炭爆炸
bog canal 沼泽地渠道
bog coal 沼炭;土状褐煤
Bogda geodome series 博格达地穹列
Bogda-Halik tectonic zone 博格达—哈尔克构造带【地】
bogdanovite 碲铁铜金矿
Bogdanski screw pump 鲍格丹斯基型立式螺旋泵
Bogda tectonic segment 鲍格达构造段【地】
bog deposit 泥沼沼泽沉积
bog down 埋陷
bog drainage 湿地排水;沼泽地排水;沼泽疏干;沼泽排水
bog drill 泥炭钻机
bog earth 沼土(壤)
bogen structure 弧形组织;弧形结构;玻璃碎片构造
bogey 转向架架;非正式定额;台车
bog facies 泥炭沼泽相
bog flow 沼泽泥流
boggildite 氟磷钠铋石
bogginess 沼泽性
bogging 沼泽土化;沼泽化
bog(ging) down 陷入泥中;下陷
boggin line 应急舵链索
bog ground 沼泽地;沼泽湿地
boggy 多沼泽的
boggy country 泥沼地区;沼泽地区
boggy ground 卑湿地
boggy soil 沼泽性土(壤)
bog(gy) water 沼泽水
bog harrow 重型沼泽地缺口圆盘耙
boghead 烟煤
boghead cannel 沼油页岩;藻烛煤;藻油页岩
boghead cannel shale 沼油页岩
boghead(ite) coal 藻(烛)煤;泥煤;泥沼煤
boghead shale 沼油页岩;藻油页岩
bog hole 垃圾坑;泥沼坑;沼穴
bogie 矿车;活动窑底;行走机构;小车;支承轮;吊车转盘;船台小车;承轮梁;平衡装置;挖土机车架;台车;双后轴(轮架)
bogie adjusting gear 转向架调节装置
bogie against deflection rigidity 转向架扭曲刚度
bogie arm 地轮臂
bogie bath machine 摇动洗净机
bogie bolster 小车弹簧;转向架旁承;转向架承梁
bogie bracket 转向托架
bogie brake rigging 转向架基础制动装置
bogie car 转向车;架车
bogie center plate 转向架心盘
bogie-change track 换轮线
bogie-changing installation 转向架更换整备
bogie cleaning machine 转向架清洗机
bogie cut-out cock 转向架截断塞门
bogie diagonal 转向架对角线
bogie engine 转向机车【铁】
bogie exchange system 转向架交换方式
bogie frame 转向架构架;转向架架;转向架
bogie-frame member 转向架构件
bogie gudgeon 转向架耳轴
bogie hearth furnace 小车膛式炉;车底炉;活车炉
bogie kiln 活底窑;客车式间歇式窑;台车窑
bogie lamp 走行部灯
bogie landing gear 小车式起落架
bogie link 转向器连接架
bogie locomotive 转向架式机车;转向机车【铁】
bogie of car 车辆转向架
bogie pin 转向架转轴;转向架心轴
bogie pitch 转向架复原装置
bogie pivot 转向架旋转枢轴
bogie pivot pitch 转向架心盘中心距
bogie rocker arm 地轮摇臂
bogie side frame 侧架
bogie skirt 转向架脚板;转向架边界裙板
bogie slide 转向架导座
bogie solebar 转向架支承板
bogie steering mechanism 转向架转向机理

bogie stock 转向架小车；转向架库存
bogie storage rack 转向架安放台
bogie swing mechanism 台车回转装置
bogie truck 多轴车；矿车；转向架挂车；转向车；挂车；双轴车；转向车架
bogie-type furnace 车底式炉
bogie undercarriage 小车式起落架；多轮式起落架
bogie unit 中后轴平衡悬架（汽车）
bogie vehicle 转向架式车辆
bogie volute spring 转向架锥形弹簧
bogie wagon 转向架上铁路货车；有转向架的铁路车辆；转向车
bogie washing machine 摇动洗净机
bogie wheel 负重轮
bogie wheelbase 车轮固定轴距；转向架轴距
bogie wheel bracket 转向架轮托座
bogie wheel gudgeon pin 转向架轮耳轴销
bogie with false pivot 无心轴转向架
bogie with quadruple suspension 四系弹簧悬挂式转向架
bog iron ore 沼铁矿
bog lake 沼泽湖
boglet 小泥塘
boglime 沼泽石灰；沼灰土
boglime lake marl 沼泽淤泥
Bogli's reaction model 包格里反应模式
Boglolubov-Parasiuk theorem 巴哥罗伯夫—泊拉奇克定理
bog manganese 沼锰矿
bog margin material 沼泽边缘物质
bog marl 沼泽地泥灰岩；沼泽灰泥
bog material 沼泽物质
bog mine ore 沼铁矿；沼矿
bog moss 泽边苔藓
bog muck 泥炭；沼泽泥炭；沼泽腐殖土；沼泽腐泥
bog ore 沼铁矿；沼矿
bog peat 沼泽泥炭（土）
bog plant 沼泽植物
bog pool 泥潭；泥水塘；泥水矿
bog road 采空区内平巷
bog soil 沼泽土（壤）
bog taste 沼泽味
bog type 沼泽式（成土作用）
Bogue's formula 鲍格公式
bogus 伪造的；伪造物
bogus certificate 伪造的证件
bogus company 空头商行；虚设公司
bogus demand and supply system 虚假供需系统
bogus dividend 动用资本分红
bogus paper 仿制纸
bogus stock company 不合规章的股份公司
Bogu Studio 博古斋
bog water 泥沼水
bog waters 酸沼水域
bogy 转向车架
bogy undercarriage system 悬挂式行走机构
Bohai kiln 渤海窑
bohdanowiczite 硒铋银
boheic acid 茶酸
Bohemian crystal 波希米亚晶质玻璃
Bohemian crystal glass 波希米亚水晶玻璃；波希米亚晶质玻璃
Bohemian gemstone 波希米亚宝石
Bohemian gemstones 波希米亚宝石类
Bohemian glass 刻花玻璃；硬玻璃；波希米亚玻璃
Bohemian massif 波希米亚地块
Bohemian ruby 蔷薇石英
Bohemian topaz 黄晶
Bohemian vault 波希米亚拱顶；波希米亚拱形屋顶
bohler 银亮钢
Bohlin method 波林群法
Bohm diffusion 玻姆扩散
Bohna filter 平流式滤池（其中的一种）
Bohr atom 玻尔原子
Bohr-Coster diagram 玻尔—科斯特图
Bohr effect 玻尔效应
Bohr frequency 玻尔频率
Bohr frequency principle 玻尔频率定则
Bohr frequency rule 玻尔频率定则
Bohr magneton number 玻尔磁子数
Bohr magnetron 玻尔磁子
Bohr model 玻尔模型
Bohr orbit 玻尔轨道
Bohr postulates 玻尔假设

Bohr radius 玻尔半径
Bohr's correspondence principle 对应原理
Bohr-Sommerfield atom 玻尔—索末菲原子
Bohr-Sommerfield theory 玻尔—索末菲理论
Bohr theory 玻尔原理；玻尔学说；玻尔理论
Bohr transition 玻尔跃迁
Bohr-van Leeuwen theorem 玻尔—冯刘文定理
Bohr-Wheeler theory of fission 玻尔—惠勒裂变理论
boil 隐匿气泡；沸腾细气泡；沸腾声；泡水（指水文现象）
boil away 煮干；完全蒸发
boil combustion 沸腾燃烧
boilded water tank 沸水箱
boil down 缩短；蒸煮浓缩；精简
boil dry 煮干
boiled 煮沸的
boil-eddy-caused danger passage 泡漩险滩
boil-eddy flow 泡漩水流
boil-eddy rapids 泡漩险滩
boiled linseed oil 精炼亚麻子油；精炼熟胡麻油；亚麻清油；干亚麻油；熟亚麻仁油
boiled oil 聚合油；厚油；清油；熟油；熟亚麻仁油；熟桐油；熟炼油
boiled oil paint 干性油漆；熟炼油漆
boiled-out water 沸过的蒸馏水
boiled pat 蒸煮过的水泥扁饼
boiled perilla oil 熟苏籽油
boiled tar 脱水焦油；熟焦油
boiled tung oil 熟桐油
boiled urushi 熟大漆
boiled water 沸腾的水
boiled wood oil 熟桐油
boiler 炉膛；硝基纤维（素）漆（用）溶剂；煮锅；蒸煮器；蒸发器
boiler accessories 锅炉装配附件；锅炉附件
boiler air heater 锅炉空气加热器
boiler air preheater 锅炉空气预热器
boiler alarm 锅炉警报器；锅炉过压报警器；缺水报警器（锅炉）
boiler and burner 锅炉和燃烧器
boiler and furnace room 锅炉房
boiler and pressure vessel 锅炉及压力容器
boiler antiscaling composition 锅炉清洗剂；锅炉防锈剂；锅炉防垢剂
boiler anti-scaling compound 锅炉防垢剂
boiler area 锅炉面积
boiler arrangement of marine 舶用锅炉装置
boiler ash 炉灰；锅炉灰渣
boiler attendance 锅炉维护
boiler auxiliary 锅炉辅助设备；锅炉附件
boiler back end plate 锅炉后挡板
boiler back tube plate 锅炉烟管后板
boiler band 锅炉箍圈
boiler band joining bolt 锅炉带圈连接螺栓
boiler band joining eye bolt 锅炉带圈连接眼螺栓
boiler band joint piece 锅炉带圈接合板
boiler barrel 锅炉圆筒；汽包
boiler barrel third course 第三节炉筒
boiler bearer 锅炉座
boiler bearing 锅炉座
boiler bedding 锅炉座
boiler blasting 锅炉爆炸
boiler blow-down 锅炉排污
boiler blow-down water 锅炉泄水；锅炉排污水；锅炉冷凝水；锅炉回水
boiler blower 锅炉鼓风机
boiler blow-off 锅炉吹洗；锅炉排污；安全阀
boiler blow-off safety valve 锅炉排污安全阀
boiler blow-off valve 锅炉排污阀
boiler blow valve 锅炉排污阀
boiler body 锅炉体
boiler bottom 锅炉底
boiler bottom plate 锅炉底板
boiler brace 锅炉拉条
boiler bracket 锅炉托架
boiler brickwork 锅炉砖衬；锅炉的砖衬（用于绝热）
boiler bridge 炉坝；火坝
boiler capacity 锅炉蒸发量；锅炉容量；锅炉能量；锅炉出力
boiler case 锅炉围壁；锅炉套；锅炉外套；锅炉套箱
boiler casing 锅炉围壁；锅炉外罩；锅炉套；锅炉壳
boiler certificate renewal 锅炉年检

boiler certificating 锅炉认证
boiler check valve 锅炉止回阀
boiler circuit 锅炉循环回路；汽水系统
boiler circulation 锅炉环流
boiler circulation pump 锅炉循环泵
boiler cleaning 锅炉清洗；清炉
boiler clothing 炉衣；炉套；锅炉砖砌面层
boiler coaling installation 锅炉加煤装置
boiler coaling plant 锅炉加煤设备
boiler cock 锅炉旋塞；锅炉用旋塞
boiler code 锅炉规范
boiler code penetrometer 锅炉规范透度计
boiler compartment 锅炉舱
boiler composition 锅炉清洗剂；锅炉防锈剂；锅炉部件组成
boiler compound 锅炉防垢剂
boiler construction 锅炉构造
boiler control 锅炉控制；锅炉调节
boiler control box 锅炉控制箱
boiler control instrument 锅炉控制仪
boiler corrosion 锅炉腐蚀
boiler covering 锅炉绝热罩
boiler cradle 锅炉托架
boiler crown 锅炉顶
boiler dash plate 锅炉中隔板
boiler deposit 锅垢
boiler design 锅炉设计
boiler detergent 锅炉洗涤剂
boiler draft 锅炉通风力
boiler drill 锅炉钻孔
boiler drilling machine 锅炉钻机
boiler drum 锅炉外罩；锅炉壳；锅炉鼓；锅炉包；煮沸鼓
boiler dust 锅炉灰尘；锅炉粉尘
boiler economizer 锅炉省煤器
boiler efficiency 锅炉效率
boiler equipment 锅炉装置；锅炉设备
boiler erection 锅炉安装
boiler face 炉面
boiler fan turbine 锅炉风机透平
boiler feed and condensate 锅炉给水和冷凝水
boiler feeder 锅炉给水器
boiler feeding 锅炉加料；锅炉供水
boiler feed main 锅炉给水总管
boiler feed pipe 锅炉给水管
boiler feed pump 锅炉给水泵
boiler feed pump turbine 锅炉给水泵透平
boiler feed tank 锅炉给水箱
boiler feed valve 锅炉止回阀；锅炉给水阀
boiler feed(water) 锅炉补给水；锅炉给水
boiler feedwater heating 锅炉给水加热
boiler feed(water) pump 锅炉加水泵；锅炉给水泵
boiler feedwater regulation 锅炉给水调节
boiler feedwater treatment 锅炉水处理；锅炉给水处理
boiler ferrule 锅炉烟管套圈
boiler filling 锅炉注水器；锅炉上水
boiler fittings 锅炉装配附件；锅炉配件；锅炉附件
boiler fixture 锅炉附属设备
boiler flue 锅炉烟道
boiler fluid 防抗剂
boiler fly ash 锅炉飞灰
boiler foot 锅炉底座
boiler for domestic use 家用锅炉；生活用锅炉
boiler forge 锅炉锻工车间
boiler foundation 锅炉底座；锅炉基础
boiler frame 锅炉框架
boiler front plate 锅炉前挡板
boiler front tube plate 锅炉烟管前板
boiler fuel 锅炉燃料
boiler fullpower capacity 锅炉满功率容量（每小时产生的蒸气量）
boiler furnace 锅炉炉子；炉胆；锅炉炉膛；锅炉胆；锅炉火箱
boiler-furnace explosion 炉膛爆炸
boiler gas burner 燃气烧锅炉
boiler gasket 锅炉垫密片
boiler ga(u)ge 锅炉水位计
boiler grate 锅炉炉算
boiler heat balance 锅炉热平衡
boiler heating surface 锅炉受热面
boiler holder 锅炉支架
boiler hood 锅炉炉罩
boiler horsepower 锅炉马力（锅炉蒸发量单位，1

锅炉马力＝15.6 千克/时）
boiler house 锅炉房(间)
boiler house area 锅炉房面积
boiler house for power station 锅炉动力站
boiler house stack 锅炉房烟囱
boiler hydrostatic test 锅炉静压试验
boiler improvement 锅炉改进
boiler-incinerator composite 锅炉焚烧炉联合装置
boiler incrustant 锅炉水垢
boiler incrustation 锅炉水垢
boiler inspection 锅炉检查
boiler inspector 锅炉检查员;锅炉监察员
boiler installation 锅炉安装
boiler insulation 锅炉保温
boiler insurance 锅炉保险
boiler iron 锅炉铁板;锅炉钢
boiler jacket 锅炉外套;锅炉外壳
boiler kier 精炼锅
boiler lagging 炉衣;锅炉外套;锅炉外被;锅炉套箱;锅炉隔热外套;锅炉隔热套层
boiler lagging band 锅炉外壳箍
boiler lagging brass strap 锅炉套圈
boiler lagging plate 锅炉隔热套层板
boiler lagging plate band 锅炉外套板箍
boiler lagging plate band lug 锅炉外套板箍带头
boiler lagging plate joint piece 锅炉外套板接合板
boiler lagging plate set screw 锅炉套板止动螺钉
boiler lagging spacing piece 锅炉套间隔垫片
boiler load 锅炉负荷
boiler losses 锅炉损耗
boilermaker 锅炉制造商;锅炉制造厂;锅炉制造工
boilermaker's hammer 锅炉用锤;锅炉工(人)用锤
boiler maker's number 锅炉出厂号码
boiler maker's roll 锅炉工人用轧轮
boiler maker's shop 锅炉车间
boiler making 锅炉制造
boiler manufacturer 锅炉制造商;锅炉制造厂
boiler masonry(work) 锅炉衬砌工;锅炉圬工
boiler mounting block 锅炉架座
boiler oil 锅炉燃料(油);石油锅炉燃料
boiler output 锅炉输出能力;锅炉输出(功率);锅炉出力
boiler paint 锅炉漆
boiler patch bolt 锅炉补块拧紧螺栓
boiler performance 锅炉性能
boiler pick 锅垢锤
boiler pipe 锅炉管
boiler plant 锅炉房;锅炉装置;锅炉设备
boiler plate 锅炉钢板;锅炉板;平轧钢板;外壳
boiler plate bending machine 锅炉板弯曲机
boiler plate flanging machine 锅炉板摺边机
boiler plate ga(u)ge 锅炉钢板量尺
boiler plate planer 锅炉板削边刨床
boiler plug 锅炉塞
boiler plug alloy 易熔塞合金;易熔合金;易熔锅炉塞合金;锅炉安全塞合金
boiler pressure 锅炉压强;锅炉压力;锅炉气压
boiler pressure ga(u)ge 锅炉压力计
boiler proving pump 锅炉试验泵;锅炉检验泵
boiler pump 锅炉试验泵
boiler rating 锅炉额定功率;锅炉蒸发量;锅炉容量;锅炉能量;锅炉额定蒸发量;锅炉额定容量;锅炉出力
boiler rating output 锅炉额定出力
boiler regulator 锅炉调节器
boiler retarder 锅炉阻流板
boiler return 锅炉回路;锅炉回水
boiler return trap 锅炉回水存水箱;锅炉回水隔汽具
boiler riveting machine 锅炉铆接机
boiler room 锅炉间;锅炉房(间)
boiler room expenses 锅炉间费用
boiler room operation 锅炉间作业
boiler saddle 锅炉支座;锅炉托架
boiler safety 安全锅炉;锅炉保险
boiler safety valve 锅炉安全阀
boiler saline 锅炉盐水
boiler scale 炉垢;锅炉水垢;锅垢;锅垢
boiler scale and rust chipper 除锅炉中水垢及锈的凿子
boiler scale furring 锅炉内侧氧化皮
boiler scaling 成垢作用
boiler scaling appliance 水垢清除工具
boiler scaling hammer 锅垢锤
boiler scaling tool 锅炉去水锈工具;去水锈工具

boiler seam 锅炉接缝
boiler seat(ing) 锅炉座(架)
boiler sequence control 锅炉程序控制
boiler setting 锅炉基础;锅炉装置;锅炉构架与护板;锅炉安装
boiler shape 锅炉形式
boiler sheet 锅炉筒钢皮
boiler shell 锅炉壳体
boiler shell drilling machine 装锅炉外壳用钻机
boiler shell foot 锅炉筒角座
boiler-shell plate 锅炉包板;锅炉钢板
boiler-shell ring 锅炉外壳环箍;锅炉筒体卡箍
boiler shop 锅炉间;锅炉车间
boiler shovel 锅炉铲
boiler slag 锅炉渣
boiler sludge conditioning agent 锅炉泥渣调节剂
boiler space 锅炉房(间);锅炉舱
boiler stay 锅炉牵条;锅炉撑条
boiler stay-bolt 锅炉螺撑
boiler stay tap 锅炉牵条螺丝攻
boiler steam 锅炉蒸汽
boiler steam dome 锅炉气泡
boiler steam rate 锅炉出力
boiler steel 锅炉钢板;锅炉钢
boiler stool 锅炉座
boiler storage 锅炉储备器
boiler stud 锅炉双头螺栓
boiler suit 连衫裤工作服
boiler superheater 锅炉过热器
boiler supply water 锅炉用水
boiler support 锅炉支架;锅炉支撑架
boiler supporting structure 支承锅炉的结构;锅炉支架
boiler survey 锅炉检验
boiler suspender 锅炉悬挂架
boiler suspension 锅炉吊架
boiler system 供热系统
boiler tank 锅炉水柜
boiler test 锅炉试验;锅炉检验
boiler thermostat 锅炉恒温控制器
boiler tower 锅炉水塔
boiler trial 锅炉试车
boiler trim 锅炉附件连接管
boiler trimming 锅炉附件
boiler tube 炉管;锅炉管
boiler tube brush 锅炉管刷
boiler tube cleaner 锅炉清管器;锅炉管清洁器
boiler tube stopper 锅炉管塞子
boiler-tube washer 锅炉洗管机
boiler unit 锅炉组;锅炉机组
boiler wall 锅炉壁
boiler waste liquor 锅炉废水
boiler water 锅炉用水
boiler water circulating pump 锅炉水循环泵
boiler water concentration 炉水浓度;锅炉水含盐量
boiler water conditioning 锅炉水软化处理
boiler water deaeration 锅炉水除氧
boiler water ga(u)ge 锅炉水位表
boiler water level regulator 锅炉水位调节器
boiler water pump 锅炉给水泵
boiler water tank 锅炉水柜
boiler water treatment 锅炉水处理
boiler water tube 锅炉水管
boiler whistle 锅炉汽笛
boiler with large water space 多水量锅炉
boiler with single steam space 单腔蒸汽锅炉;单汽包锅炉
boiler working pressure 锅炉工作压力
boiler works 锅炉厂
boiling 冒水翻砂现象;滚沸;管涌现象;地基管涌现象;翻浆(冒泥);泡水(指水文现象);搪瓷面釉气泡;煮沸
boiling action loading 沸腾容器装载
boiling bed 流化层;沸腾床
boiling bed drier 沸腾床干燥器
boiling box 沸煮箱
boiling bulb 蒸馏锅;蒸馏釜
boiling bulk 蒸馏锅
boiling burner 灶面燃烧器
boiling chamber 沸腾室
boiling chip 沸石
boiling condition 沸腾条件
boiling constant 沸点升高常数
boiling cycle reactor 沸腾循环反应堆

boiling depth 沸腾深度
boiling down 湿蒸馏;煮浓
boiling down pan 蒸煮锅
boiling ebullition 沸腾
boiling fastness 耐煮性
boiling flask 长颈烧瓶;烧瓶
boiling geyser 激喷的间歇泉
boiling heat 沸点热力
boiling heat transfer coefficient 沸腾传热系数
boiling hole 煮生石灰坑
boiling hot 滚烫;滚热的
boiling house 沸腾室
boiling inclusion 沸腾包裹体
boiling kier 蒸煮锅;沸煮锅;漂煮锅
boiling kiln 沸煮炉
boiling lake 高温湖;沸湖
boiling mud 沸泥
boiling mud lake 沸泥湖
boiling mud pool 沸泥塘
boiling of cell 电解槽沸腾
boiling-off 煮练;蒸发;汽化;沸腾
boiling of sand 涌砂;管涌;涌沙
boiling of soil 土沸现象
boiling-on-grain 炒砂法;砂炒法
boiling over 沸溢
boiling period 石灰氧化期;沸腾期(转炉)
boiling point 沸点
boiling point analyser 沸点分析器
boiling point apparatus 沸点测定器
boiling point at depth 与深度对应的沸点
boiling point curve 沸点曲线
boiling point depression 沸点降低
boiling point diagram 沸点图
boiling point elevation 沸点升高
boiling point evapo(u)ration 沸点蒸发
boiling point gravity constant 沸点比值
boiling point gravity number 沸点比重值
boiling point index 沸点指数
boiling point lowering 沸点降低
boiling point of petroleum 石油的沸点
boiling point pressure 沸点压力
boiling point rising 沸点升高
boiling point temperature 沸点温度
boiling point test 沸点试验
boiling point thermometer 沸点温度计;沸点测高器
boiling point viscosity constant 沸点黏度常数
boiling pool 沸水塘
boiling pot 热带气旋中心
boiling proof 抗煮沸的
boiling range 管涌范围;沸腾范围;沸点范围;沸程
boiling river 沸水河
boiling sea 波涛汹涌的海面
boiling seasoning 煮沸干燥
boiling seawater spring 海水沸泉
boiling soil 土涌
boiling spread (石油馏分的)沸点范围
boiling spring 涌泉;高温泉;沸泉
boiling steel 沸腾钢
boiling sterilization 煮沸灭菌法;煮沸灭菌
boiling sterilizer 煮沸消毒器;煮沸灭菌器
boiling stone 沸石
boiling surface 沸腾表面
boiling temperature 沸点(温度)
boiling temperature for the reservoir pressure 与热储压力对应的沸点温度
boiling test 煮沸试验;(水泥的)蒸煮试验
boiling-through marks 泛沸法
boiling treatment (木材的)蒸煮处理
boiling tub 化灰池;熟化池(石灰)
boiling tube 蒸发管
boiling-up soil 冻胀土
boiling vessel 煮沸器
boiling water 汹涌的波浪;沸水;汽化水
boiling water column 沸腾水柱
boiling water cycling reactor 沸水循环反应堆
boiling waterfall 沸水瀑布
boiling water preparation 沸水制备
boiling water reaction 沸水反应
boiling water reactor 沸水反应堆;沸水反应器;沸水堆
boiling water resistance 耐沸水性
boiling water sealing 沸水封闭法
boiling water soundness test 煮沸安定性试验
boiling water sterilization 煮沸消毒

boiling water test 沸水试验;耐沸水试验
boiling waves 汹涌的波涛
boiling-without vacuum 常压油煮调湿法
boil mud 土涉;渗水和土上涌
boil notice 烧开水通知
boil-off 沸腾;汽化损耗
boil-off assistant 脱胶助剂
boil of sand 砂上升;管涌
boil over 沸溢
boil phenomenon 冒水翻砂现象
boil-proof 抗煮沸的;耐煮的
boil-proof bond 耐高温砌合
boil-proof-glued 耐沸腾胶的
boil resistant adhesive 抗沸腾胶粘剂;耐用胶粘剂
boils and mudsprouts 喷水冒砂
boils-vertical eddies 泡漩
boil up 滚沸
boil-vertical eddiesl boils vertical eddies 直涌漩涡
boil vortex 泡漩
boil-vortex flow 泡漩水流
boioler casing 锅炉外壳
boiserie 细木护壁板;镶板;嵌板细木工
Boise sandstone (产于美国爱达荷州的)波伊斯砂岩
boitel 半圆凸线脚
Boivin antigen 博伊文抗原
boke 小细(矿)脉
Bok globule 博克球状体
bokite 钒铝铁石
bolar 黏土的
bolbosehoenus maritimus palla 三棱草
bold 大块
bold bow 肥型船首
bold cliff 绝壁
bold coast 陡峭海岸;陡岸
bold corduroy 粗条灯芯绒
bold design 大胆设计;大胆的设计
boldenone 勃地酮
bold face 黑体字
bold-faced character 粗体字【计】
bold-faced word 粗体字【计】
bold face letter 黑体字;黑粗体字母;粗线体;粗体字【计】
bold figure 黑体字;黑体数字
bold hawse 高锚链孔
bolding ground 河缆储[贮]木场
Bolding's gulley 算盖可转动的雨水口
bold letter 黑粗体字母
bold line 黑体粗线;粗线;粗体字【计】
bold plan 大胆的设计
bold platform 陡台地
bold relief 粗糙地形
bold shore 陡海岸;陡岸;峭岸
bold to 陡深(指岸边或浅滩边缘突然变深)
bold-type letter 黑体字母
bold water 瀑布;急流
bole 胶块黏土;胶块土;胶结黏土;黏土;红玄武土;有色黏土;粗干材
bole area 干材断面积
bolection 镜框式线脚;凸平面线脚;凸出嵌线
bolection mo(u)ld(ing) 镜框式线脚;盖缝线脚;突线条;凸式线脚;凸嵌线;凸出嵌线饰
boleg oil 氧化矿物油
boleite 氯铜银铅矿;银铜氯铅矿
Bolen test 波伦试验
boletic acid 富马酸
bolexion mo(u)lding 凸出嵌线饰;凸出平面的线脚
Boli-Antu fault belt 勃利一安图断裂构造带【地】
Boli basin 勃利盆地
bolide 火流星
Boliden salts 卜立顿盐剂(一种木材防腐剂)
bolivar 博利瓦法兰绒
bolivarite 隐磷铝石
bollard 交通安全岛护柱;码头带缆柱;系绳柱;系缆柱;系缆柱(拉力);系船柱;带缆柱;带缆柱
bollard cleat 系柱;钩头缆柱
bollard eye 缆绳琵琶头(套系缆桩用)
bollard foundation 系缆柱基础
bollard head 码头系缆柱;系缆柱
bollard niche 系缆壁洞
bollard pull 系船柱拉力;系船柱拉力;系船力
bollard pull trial 系船柱拉力试验
bollard recess 系缆壁洞
bollard test 系柱试验(在岸边试验推进性能)

boll crushers 轧麻荚机
Bolley's gold purple 波利紫金色料
Bolley's green 波利绿
Bolling interstade 博林间冰阶
Bollmann extractor 波尔曼萃取器;波尔曼萃取机;立式吊篮萃取机
Bollmann truss 波尔曼式桁架
bollock 厚壳滑车
bollock block 厚壳滑车
boll rot 铃烂;铃腐
boll weevil 新钻工
boll-weevil lubricator 输气管道润滑器
boll-weevil tongs 结构简单的短重管钳
boll-weevil tubing head 结构简单的油管头
bolmantalate 勃金刚酯
bologna spar 重晶石
Bologna survey 博洛尼亚射电电源表
bologram 辐射热测量记录器;热辐射测量图
bolograph 测辐射热仪;测辐射记录器;辐射热测量记录器;辐射计;热辐射测量图
bolometer 电阻辐射热测定器;测辐射热器;测辐射计;辐射计量器;辐射热测量记录器;辐射热测量计;热辐射测量计
bolometer bridge 测辐射热计电桥;热辐射计电桥
bolometer detector 测辐射热计检验器
bolometric absolute magnitude 绝对热星等
bolometric amplitude 测辐射热计;热星等变幅;热变幅
bolometric correction 热星等改正
bolometric detector 辐射探测器
bolometric instrument 辐射热测试器
bolometric luminosity 热光度
bolometric magnitude 热星等
bolometric radiation 热辐射
bolometric standard irradiance 测辐射热标准辐照度
bolometric stellar magnitude 辐射热计测量体大小
bolometric wave detector 辐射热检波器
Bolomey's curve 鲍罗米(低渗混凝土级配曲线)
Bolomey's formula (计算混凝土强度或级配的)鲍罗米公式
boloscope 金属异物探测器
bolson 宽浅内陆盆地;季节性盐湖;荒芜盆地;河流汇交盆地;干湖地;封闭洼地;封闭盆地;沙漠盆地
Bolsover Moor stone 玻尔索瓦莫尔石
bolster 梁托;模垫;缓冲物;护板;摇枕;柱头横木;枕垫;垫枕;垫木;承枕;承托木;承垫车;车架承梁;长枕;软垫;托木;铁马凳;填石铁丝笼
bolster beam 摇枕吊杆
bolster chisel 套管凿
bolster connection 垫枕接合
bolster damping 转向架减振器
bolster hanger lug 摇枕吊耳
bolsterless bogie 无摇枕转向架
bolster naval hood 锚链枕垫
bolster plate 护承板;压床垫板;支承板;承梁板
bolster screw jack 车架螺旋千斤顶
bolster spring 垫枕弹簧;承梁弹簧
bolster spring cap 承梁弹簧帽
bolster spring device 摇枕弹簧装置
bolster spring plank 车架承簧板
bolster spring seat 承梁弹簧座
bolster stake 承载柱
bolster wagon 平板车
bolster-type shank 钎肩式钎尾
bolster work 承垫工;支撑工;隆腹状砌筑;支撑工程;枕木状砌筑
bolt 螺栓;锚杆;杆柱;插销;插销;栓销;锁舌;栓钉
bolt anchor 螺栓锚件
bolt anchorage 地脚螺丝固定
bolt anchor shackle 带销锚卡环
bolt and clip type fastening 螺栓及扣板式扣件
bolt and nail connection 螺栓及钉结合
bolt and nut 带螺帽螺栓
bolt and nut making machine 制螺栓螺帽机
bolt and nut split machine 螺栓和螺帽拼合机
bolted and welded steel bridge 栓焊钢桥
bolt bar 螺旋把手;螺栓杆;驻栓
bolt-bearing property 螺栓承压能力
bolt blank 无螺纹栓;螺栓毛坯
bolt cam 活动三角;螺栓闸门
bolt cap 螺母
bolt carton 螺栓套纸板匣;纸板螺栓套

bolt chisel 扁头錾;扁尖凿
bolt circle 螺栓分布圆
bolt circle diameter 螺栓圆直径
bolt clasp 螺栓夹扣
bolt clipper 螺栓切割器;断线钳
bolt clippers with pipe-type handle 管式断线钳
bolt clippers with tubular handle 管式断线钳
bolt-connected joint 用螺栓连接的节点;螺栓接合
bolt-connected tubular scaffold 螺栓连接钢管脚手架
bolt connection 螺栓接合;螺栓连接
bolt core 螺栓型心
bolt coupling 螺栓联轴节;螺栓管箍
bolt cropper 割螺栓器
bolt cutter 螺栓刀具;断线钳
bolt cutting lathe 螺栓车床
bolt die 板牙
bolt drill 螺栓孔钻
bolt driver 改锥;螺丝刀
bolted and welded steel girder or truss 栓焊钢梁
bolted bonnet 螺桩固定的罩;螺桩固定的盖;螺栓压盖;螺栓连接阀盖
bolted bridle joint 螺栓啮接
bolted cap 螺栓压帽
bolted closure 螺栓封闭器
bolted compression coupling 拴接压力
bolted connection 螺栓联结;螺栓连接;螺栓接合;螺栓结合
bolted connector 螺栓连接器
bolted element 螺栓零件
bolted fishplate 螺栓鱼尾
bolted fishplate splice 螺栓(鱼尾板)接合;螺栓夹板续接
bolted flange 螺栓连接法兰;螺栓结合凸缘;螺栓接合法兰盘
bolted joint 螺体连接;螺栓联轴节;螺栓结合;螺栓接头;螺栓接合;螺栓连接
bolted liner 螺栓衬板
bolted method 栓锁法
bolted moment-resisting steel joint 抗弯钢螺栓连接
bolted node connector 节点螺栓接头
bolted on connection 螺栓连接法
bolted pile 螺栓桩;螺栓接桩
bolted pump 螺栓连接泵
bolted rail crossing 钢轨拼装交叉
bolted rail joint 螺栓式钢轨接头
bolted rigid frog 组合辙叉;固定式拼装辙叉;钢轨组合辙叉
bolted sectional box type of floating dock 螺栓分节拼装的箱式浮船坞
bolted sectional dock 螺栓连接分段浮船坞;装配式分节浮船坞;拴接分部式浮(船)坞
bolted spherical node 螺栓球节点
bolted splice 螺栓铰接
bolted steel bridge 拴接钢桥
bolted support 螺栓支座
boltel 圆小杆;凸圆饰
bolt embedded welding 螺栓栽焊;电栓焊
bolt end 螺栓端
bolter 筛石机;细孔筛;分离筛;板筛;筛分机
bolter-up 外板装配工
bolt eye 螺栓眼
bolt fastening 螺栓联结;螺栓连接
bolt firing tool 螺栓枪
bolt flange 拴接法兰
bolt forcer 螺栓压入机;螺栓冲孔器
bolt forging machine 锻螺栓机
bolt for insulator steady 绝缘子固定螺栓
bolt former 螺栓锻造机
bolt for tool post 刀架螺栓
bolt-ga(u)ge 螺栓线规
bolt guide 闩头导槽
bolt handle 机柄;闩柄
bolt head 螺栓头
bolt header 螺栓头镦(锻)机;螺栓头锻(造)机;螺杆冷镦机
bolt heading machine 螺栓头镦压机;打螺钉头机
bolt head trimming 螺栓头冲压;冲压螺栓头
bolt head with feather 带销螺栓头
bolt hole 螺栓孔;联络小风巷;锚杆孔
bolthole borer 螺栓孔钻孔器
bolt hole chamfering 螺栓孔削角;螺栓孔刀棱
bolt hole circle 螺栓圆周;螺栓孔分布图

bolt hole crack 螺栓孔裂纹;螺栓孔开裂
bolt hole drilling 锚杆孔凿岩
bolt holes for nozzle flange 接管法兰的螺栓孔
bolt hole spacing 螺栓孔间距
bolt hole star crack 螺栓孔裂纹
bolthole vacuum cleaner 杆柱钻孔真空清扫器;干式除尘器
bolt hook 有螺杆和螺母可用作螺栓的钩子
bolt in double shear 双剪螺栓;受双剪螺栓
bolting 螺栓连接;螺栓固定;拧紧;锚固;锚杆(支护);振动筛筛分;栓位;拴接
bolting cloth 筛布
bolting down (机台的)螺栓固定
bolting iron 拴接榫凿(块石);锁舌眼凿子
bolting jumbo 安装锚杆台车
bolting lug 螺栓连接的凸缘
bolting machine 机械筛;锚杆安装机;杆柱安装机;施锚机械;筛分机
bolting mill 筛分机
bolting pull tester 锚杆拉拔器
bolting reel 转筒筛
bolting single shear 受单剪螺栓
bolting steel 螺栓钢
bolting support 锚杆支架
bolting technical parameter 锚杆支护工艺参数
bolt in single shear 单剪螺栓
bolt iron 螺栓铁
bolt joint 螺栓联结;螺栓连接
bolt length 锚杆长度
boltless 无螺栓
boltless fixing system 无螺栓固定系统
boltless liner 无螺栓衬板
boltless lining 无螺栓衬板
bolt lever 闩柄
bolt load 螺栓荷载
bolt load for gasket sealing 螺栓预紧荷载
bolt load for operating conditions 螺栓操作荷载
bolt load under pretension condition 螺栓预紧荷载
bolt-lock 连锁盒;螺栓保险(器);栓锁
bolt-making machine 螺栓机;制螺栓机
bolt neck 螺栓颈
bolt nut 螺母
bolt of blasting 爆固式锚杆
bolt of resinification 树脂式锚杆
bolt of sand grout 砂浆式锚杆
bolt-on 螺栓紧固
boltonite brick 镁橄榄石砖
bolt-on share 螺栓固定的犁铧
bolt pin 螺栓销;销子
bolt pitch 螺栓间距
bolt pitch-circle 螺栓轴心线圆
bolt point 螺栓尖
bolt press 螺栓冲压机
bolt pulling force 锚杆拉力
bolt rod air-pressing machine 螺杆式空气压缩机
bolt rope 帆缘索;帆边绳;天篷边绳
bolt rope needle 弯针
bolt rope stitch 帆边绳缝法
bolt row space 锚杆排距
bolts 短圆材
bolts and nuts for steel slotted 角钢螺丝
bolts and wire mesh 锚杆金属网
bolt shackle 螺栓钩环
bolt shank 螺栓杆
bolt shank extruding 螺栓挤压
bolt shoulder 螺栓肩
bolt sleeve 螺栓套(管)
bolt socket 螺栓插座;套筒膨胀螺栓
bolt space 锚杆间距
bolt spacing 锚杆间距
bolt spring 栓簧
bolt stay 撑螺栓
bolt stock 螺栓备料;螺栓存货;螺栓钢料
bolt stress 螺栓应力
bolt stud 柱螺栓;双头螺栓柱螺栓
bolt support 螺栓支撑;锚杆支护;杆柱
bolt tension 螺栓张力;螺栓拉力
bolt threader 螺栓割丝机;螺栓车床
bolt-threading machine 螺栓车绞机
bolt thread rolling machine 螺栓滚丝机
bolt tightener 螺栓紧固器;螺栓绞紧器
bolt timber 锯材;短圆木材;大块木材
bolt together 螺栓联结;螺栓连接

bolt torque 螺栓起动力矩
bolt-type parallel groove clamp 单螺栓并沟线夹
bolt-up 螺栓紧固
bolt up fishplate 用螺栓固定的接合板;用螺栓固定的鱼尾板
bolt-upright 直竖的
bolt up with handle 有柄插销;带柄插销
bolt washer 螺栓垫圈
bolt wire 螺栓钢条
bolt with feather 带滑键螺栓;螺鼻螺栓
bolt with nut 带榫螺栓
bolt with one end bent back 弯头螺栓;弯关螺栓
bolt with stop 制止螺栓
bolt with winged nut 翼形帽螺栓
boltwoodite 黄硅钾铀矿;硅钾铀矿
Boltzmann-Saha theory 波兹曼—沙哈理论
Boltzmann's constant 波尔兹曼常数
Boltzmann's distribution 波尔兹曼分布
Boltzmann's distribution law 波尔兹曼分布律
Boltzmann's Einstein system of equations 波尔兹曼—爱因斯坦方程组
Boltzmann's emission law 波尔兹曼发射定律
Boltzmann's entropy hypothesis 波尔兹曼熵假设
Boltzmann's equation 波尔兹曼方程(式)
Boltzmann's equation of state 波尔兹曼物态方程
Boltzmann's factor 波尔兹曼因子;波尔兹曼系数
Boltzmann's formula 波尔兹曼公式
Boltzmann function 波耳兹曼函数
Boltzmann's H-theorem 波尔兹曼 H 定理
Boltzmann's principle 波尔兹曼原理
Boltzmann's relation 波尔兹曼关系
Boltzmann's rheological model 波尔兹曼流变模型
Boltzmann's statistics 波尔兹曼统计
Boltzmann's transport equation 波尔兹曼输运方程
Boltz's development method 博尔茨扩展法
Boltz's method of substitution 博尔茨替代法
Boltz's reaction 博尔茨反应
Boltz's test 博尔茨反应
Boluohuoluo geodome series 博洛霍洛地穹列
bolus 胶块土;红玄武土;团块;填塞物;陶土
bolus abla 高岭土;白陶土
bolus of water 小水团
Bolzano's theorem 波尔察诺定理
Bolzano-Weierstrass theorem 波尔察诺—维尔斯特拉斯定理
Bolza problem in calculus of variation 变分法中的博尔赞问题
bomb 火山弹;高压液化气容器;储气瓶;瓶(形物)
bombard 轰击
bombarder 轰击器
bombardment 轰击;曝光
bombardment energy 冲击能(量)
bombax cotton 木棉
bombays 掺杂紧包黄麻
bombay twill 孟买斜纹布
bomb black wood 铁刀木
bomb breaking layer 爆破层;起爆层
bomb calorimeter 弹式测热器;弹式量热器;弹式量热计;爆炸量热器;量热弹;炸弹量热计
bomb calorimeter test 弹式量热试验
bomb calorimetry 弹式量热法
bomb complete round 完整炸弹
bomb crater 爆炸坑
bomb crucible 还原钢罐;还原弹坩埚
bomb dump 炸弹储[贮]藏处;野战炸弹库
bombed site 炸后废墟
bomb furnace 封管炉
bomb gas 钢瓶气体;瓶装气体
bombicesterol 蚕茧醇
bombing proof 防轰炸
bomb penetration 炸穿深度;炸弹炸穿深度
bombproof 防轰炸的;防炸的
bombproof basement 人防地下室
bombproof shelter 防空洞
bombproof vault 避弹窖;避弹窑
bomb-resistant 抗爆炸的
bomb sag 火山弹沉降
bomb shelter 防空弹体;防空洞;避弹掩壕
bomb shelter door 防空洞门
bomb shelter garage 防空洞车库;防空掩蔽车库
bomb store 炸弹储[贮]藏处;野战炸弹库
bomb store area 炸弹储[贮]藏面积
bombway 棕色硬木(产于安达曼群岛)
bometolol 波美洛尔

Bomi tectonic segment 波密构造段
Bommer type helical hinge 博默型螺旋铰链
bonaccordite 硼镍铁矿
bona fide 可靠的
bona fide bid 可信投标;有效投标;有效投书;可信标书
bona fide contract 善意合约
bona fide cost 真正成本;真实成本;实际成本
bona fide holder 真实持有人;善意执票人
bona fide possession 善意占用
bona fide purchaser 善意买方
bona fides creditor 真正债权人
bona fides taxpayer 正当纳税人
bona fide third party 善意第三方
bon ami powder 试车用粉
bona notabilia 贵重物品
bonanza 大矿囊;富矿窝;富矿脉;富矿带
bona transaction 公平交易
bonattite 蓝铜矾;三水胆矾
Bonawe granite 伯纳维花岗岩
bonchevite 斜方硫铋铅矿
bond 结合;胶结;黏结;化学键;证券;黏合层;债券;低阻电焊接头;导接线;浮动利率债券;保证金;保税(单);熔合区;契据
bondability 密着性;握裹力;(钢筋混凝土的)结合力
bondability index 黏结指数
bond account 债券账户
bondage 约束;束缚
bond agent 业务联系员;联系代表处;经理处
Bond albedo 邦德反照率
bond anchorage 黏结锚固
bond and mortgage record 债券及抵押款记录
bond and shares account 公债及股票账户
Bond and Wang crushing theory 邦德—王破碎学说
bond angle 键角
bond-angle deformation 键角变形
bond area 结合面积;黏结面(积);握裹面积
bond audit 债券审核
bond authorized 额定债券
bond beam 结合梁
bond-beam block 组梁砌块;组合砌块;预制混凝土结合梁块
bond beam tile 结合料面砖
bond behavio(u)r 黏合性能
bond between binder and aggregate 结合料与集料间的黏结(力);黏结剂黏合
bond between concrete and steel 混凝土与钢筋间的结合力;钢筋与混凝土的握裹力;钢筋混凝土的结合力;钢筋与混凝土的结合力;钢筋混凝土结合力
bond blister 黏结气泡;附着结疤
bond block 嵌接方块
bond breaker 隔黏剂;隔黏层;隔黏材料;分隔剂;防黏结材料;黏结分隔材料
bond breaker course 粘连间断层
bond-breaking 断开;黏结破坏
bond-breaking agent 断裂剂;分离剂(升板和倾板施工中,涂于板间以免黏结)
bond-breaking medium 隔离材料
bond broker 债券经纪人
bond certificate 债券
bond chlorine 结合氯
bond clay 胶结黏土;黏合黏土;砌合黏土
bond coat 黏合层;初层;增黏涂层;结合层;底涂层;打底层
bond collateral loan 公债担保贷款;债券抵押贷款
bond concrete 黏结混凝土
bond contact 键合接点;熔合接触
bond conversion 债券兑换
bond coupon 债券息票
bond course 拉结层;黏结层;砌砖层;(砖或石的)砌合层
bond creep 黏着徐变
bond dealer 债券自营商
bond department 债券经营部门;债券部
bond discount 债券折价;债务贴现;债券贴现
bond discount accumulation 债券折价累积;债券贴现累积
bond discount amortization 债券贴现摊销
bond discount and expenses account 债券贴现及费用账
bond discount and premium audit 债券折价和溢价审计
bond discount unamortized 未摊销债券贴现;未

摊销债券折价
bond dissociation energy 键裂解能
bond distance 键长【化】
bond dividend 债券股息;债务信息;债券股利
bonded 化合的
bonded abrasive products 固结磨具
bonded alumina-zirconia-silica 烧结锆刚玉
bonded arch 咬合拱;砌合拱
bonded area 保税区;保税地区
bonded block 结合部件;嵌砌方块;嵌接方块;砌合方块
bonded blockwork 嵌砌块体
bonded brick arch 磨砖密缝券;磨砖密缝拱
bonded brick work 咬合砌体
bonded cable 油轮接地电缆;接地电缆
bonded cargo 保税货物
bonded-carrier 键合势垒
bonded clay brick arch 规准黏土砖拱
bonded coating 胶粘涂料
bonded concrete 胶结混凝土
bonded concrete overlay 结合式混凝土加厚层
bonded construction 砌合结构
bonded debt 债券债务;债务负债
bonded fabric 黏结布;无纺织物;黏合衬;贴合布;不织布;编织结合纸
bonded fiber 黏结纤维
bonded fiber fabric 纤维黏合物;黏合织物;非织造织物
bonded fibers of nylon and polypropylene 尼龙和聚丙烯黏胶纤维
bonded flow 约束流动
bonded flux 陶质焊剂
bonded foctory 保税工厂
bonded glaze 黏釉
bonded godown 保税关栈;保税仓库
bonded goods 保税物品;保税货物
bonded goods yard 保税货场
bonded gravel roof 黏结的砾石屋顶
bonded gravel screen 胶结砾石滤筒
bonded joint 胶结紧密的接缝;黏结结合面;黏接接头;黏结接合;密接缝;黏合纤维网;黏合接头
bonded manufacturing warehouse 保税工厂;保税的制品仓库
bonded mat 黏结毡;黏结毡;黏合织物
bonded member 结合构件;黏结构件;黏合构件;预应力锚固筋;结牢杆件
bonded metal 黏合金属板;包层金属(板)
bonded-on 黏结住的
bonded overlay 黏结(路面)加铺层
bonded phase 键合相
bonded phase chromatography 键合相层析
bonded-phase coverage 键合相覆盖率
bonded phase packings 键合相填料;键合相填充物
bonded porous material 胶结多孔物料
bonded port 保税港口
bonded post-tensioning 后张预应力筋;黏结后张(法);后张灌浆工艺;后张法黏结
bonded price 保税价格
bonded reinforcement 黏着钢筋;黏结钢筋
bonded resistance 黏着抗力
bonded roof 窑炉顶部砌接;砌合屋顶;(厂商具保的)屋面材料
bonded rubber cushion(ing) 黏合橡胶(软)垫;黏合橡胶(软)垫;预制整体橡皮垫;条形橡皮垫
bonded shed 保税货棚;保税关栈;保税仓库
bonded(stationary) phase 化学键合相
bonded steel wire 黏着钢丝
bonded store 保税关栈;保税仓库;保税物料
bonded structure 砌合结构
bonded surface 砌合面层;砌合路面
bonded synthetic(al) fiber fabrics 合成粘胶纤维织物
bonded system 保税制度
bonded tendon 黏着力钢束;黏结钢筋束;黏结钢筋束;黏结(预应力)钢束;灌浆的(预应力混凝土中的)钢筋束或钢丝束;黏结的预应力筋
bonded terrazzo 黏结水磨石
bond(ed) test 黏结试验;结合力试验
bonded timber 墙结合木
bonded track 导通轨道
bonded transportation 保税运输
bonded type construction joint 黏结型施工缝
bonded value 关栈货值;保税价值
bonded value clause 保税价值条款
bonded vaults 保税地窖
bonded warehouse 海关保税仓库;关栈;保税关栈;保税仓库
bonded warehouse transaction 保税仓库交易
bonded wheel 结合砂轮
bonded wire 黏合漆包线
bonded wire strain ga(u)ge 黏合应变片;黏合应变计;粘贴电阻丝应变片;粘贴电阻丝应变计
bonded with asphaltic material 用沥青材料黏结;沥青黏结剂
bonded with synthetic(al) resinous material 用合成树脂黏结;合成树脂胶合
bonded wood 胶合木板
bonded wood construction 胶合木构造;胶合木结构
bond electron pair 成键电子对
bonder 拉结丁砖;黏结剂;锚石;丁砖;砌墙石;砌墙块石;束石
bonder brick 顶砖;丁头砖;黏结丁砖;砖缝砖;系石
bonder course 丁头砖层;系石层;丁砖层
bonder dam 砌石坝
bonder header 丁砖;横切砖
bonderisation 磷化处理
bonderite 磷酸盐处理层;磷酸盐(薄膜防锈)处理层;磷化剂
bonderization 磷酸盐处理
bonderize-ise 磷化
bonderized finish 磷酸盐处理表面;磷化处理表面
bonderized sheet iron 磷酸盐处理的耐蚀镀锌钢板
bonderizing 磷酸盐(表面)处理(法);磷化处理(法)
bonder mass foundation 毛石砌体基础
bonderolle 花杆;标杆
bonder wire 绑接钢丝(钢筋、弹簧等)
bond expenses 债券费用
bond external plaster 外部胶结粉刷
bond external rendering 外部胶结粉刷
bond face 黏结面;结合面
bond failure 结合损坏;黏着破坏;黏着力破坏;黏结破坏;黏结力破坏;握裹失效
bond fee 海关关栈仓租费
bond financing 发行公债筹措资金
bond finish 胶结粉刷饰面
bond fireclay 塑性耐火黏土
bond fission 键分裂
bond flo(a)tation 债券发行
bond flo(a)tation market 债券发行市场
bond flux 结合熔剂;结合焊剂;黏结焊剂;陶质焊剂
bond force 结合力;化合力;握裹力
bond force constant 键力常数
bond for objective function 目标函数的约束条件
bond funds 债券基金
bond graph 键合图
bond header 贯通石;丁头石;砌合头(石)咬合砌(石)法;系石;拉结丁砖
bond holder 债券持有者;债券持有人;持券人
bond indenture 债券契约;债券信托契约
bonding 接合;接地连接;胶合;键合;黏合;黏附;压焊;砖石墙丁头砌合;砖石砌体砌合;低电阻连接;存入关栈(保税仓库);保证单;砌合
bonding action 黏合作用;黏合强度
bonding additive 黏结促进剂
bonding adhesive 结合剂;黏结胶粘剂;黏结剂;黏合剂
bonding adhesive agent for concrete 混凝土砌合黏结剂;混凝土胶结剂
bonding adhesive based on coal tar 煤焦油脂黏结剂;煤沥青黏结剂
bonding adhesive for laying 浇筑用黏结剂;铺砌用黏结剂;敷设黏结剂
bonding admixture 接合剂
bonding agent 结合剂;键合剂;黏结剂;黏合剂;黏着剂;保税授信机构;业务联系人;联系代表处;经理处;分层黏结剂
bonding agent for concrete 混凝土黏结剂
bonding agent for laying 浇筑用黏结剂;铺砌用黏结剂
bonding agent for mortar 灰浆黏合剂
bonding alloy 黏结合金
bonding anchor 砌合锚;固定锚;黏结锚固
bonding area 黏结范围;黏结面(积)
bonding bar 闭锁器;连接杆;接触器
bonding brick 拉结砖;接合榫砖;束砖;空斗墙连接砖
bonding cable 接合电缆;屏蔽电缆
bonding capacity 胶结性能;胶结能力;胶合能力;黏力;担保能力
bonding cement 胶结水泥;黏结胶泥;黏结剂
bonding characteristic 黏结性;黏合性
bonding clay 结合黏土;造浆黏土
bonding coat 黏结涂层
bonding coating 胶粘涂料
bonding company 担保公司;业务代办公司;材料财务经理公司;部分业务经销处
bonding composition 胶结剂组成;胶结剂成分
bonding compound 胶结料;胶结剂成分;黏结混合物;黏结混合料;黏结(沥青)混合料;黏结剂;屋面沥青(冷铺或热铺)
bonding concrete 黏结混凝土;胶结混凝土
bonding conductor 接地导线;搭接片;跨接线;屏蔽接地(导)线
bonding course 结合层;黏合层;固结层;黏结层
bonding dispersion 黏结剂分散(作用)
bonding door gasket 箱门密封垫黏合
bonding electrons 成键电子
bonding emulsion 胶结乳液;黏结乳剂
bonding energy 结合能;黏结能
bonding equipment 焊线机
bonding exterior plaster 外部胶结抹灰;外部胶结粉刷
bonding exterior rendering 外部胶结抹灰;外部胶结粉刷
bonding failure 黏结破坏;黏结损坏;黏结失效
bonding fee 海关保税仓库费用;关栈费
bonding finish 胶结抹灰;胶结粉刷;砌合饰面
bonding fixture 黏合夹具
bonding force 键力;键合力;黏结力
bonding glue 合成树脂胶
bonding header 砌合丁(头)砖;露天砖;丁石;丁墙石
bonding header course 砌合丁头层;丁砖层
bonding insurance 保证保险;保税保险
bonding jumper 金属条;跨接线;金属片;搭接片;搭地线
bonding land 接合区
bonding layer 结合层;黏结层;黏合层;固结层;搭接层;水泥黏结层
bonding line 接合线
bonding machine 焊接机
bonding material 胶凝材料;胶结材料;胶合材料;黏结材料;黏合材料;黏结材料
bonding matrix 连接体;连接基岩
bonding mechanism 结合机理
bonding medium 黏结介质;黏合介质;黏合剂
bonding medium for concrete 混凝土黏结剂;混凝土黏结剂
bonding medium for laying 铺砌用黏结剂;铺砌黏结剂;砌筑胶结剂
bonding medium suspension 胶结悬浊液;胶结悬浮液
bonding metal 黏结金属
bonding method 砌合法
bonding mortar 黏结砂浆;砌筑砂浆
bonding of brickwork 砖的砌结;砖的黏结;砖的砌合
bonding of granules 粒料黏结;颗粒间黏着力
bonding of metal(s) 金属连接;金属连黏结;金属连焊接
bonding orbital 键轨函数;成键轨函;成键轨道
bonding pad 联装填料;连接区;结合区;结合片;焊接区;焊接点
bonding paper 胶粘纸
bonding paste 黏结浆糊;水泥浆
bonding paste coating 涂膏
bonding performance 黏结性能;黏合性能;黏附性能
bonding phase 黏结相
bonding plane 黏结面;砌结面
bonding plaster 黏结石膏;黏合墙粉;特制石膏灰浆;黏结性石膏灰浆
bonding pocket 砖层凸出;砖层凹齿
bonding point 结合点;接合点
bonding power 黏结能力;接合力;黏合力;成键能力
bonding primer 黏结性底漆
bonding process 黏结法
bonding property 黏结性能;黏合性能;黏附性能;黏结性质;胶接性能

bonding putty 黏结油灰;胶质水泥;黏结胶泥
bonding quality 黏结质量
bonding rendering 外部胶结粉刷
bonding resin 黏结树脂;胶接树脂;黏合树脂
bonding rubber 胶结橡胶
bonding scheme 键合形式
bonding shackle 连接卡环
bonding strength 黏结强度;黏合强度
bonding strength of glue joint 胶接黏合强度
bonding stress 黏结应力;握裹力
bonding strip 铅条
bonding stucco 胶结外部粉刷;黏结外部粉刷
bonding suspension 黏结悬浮度
bonding system 连接体系;黏合体系
bonding tape 胶带
bonding technique 黏结工艺技术;连接工艺技术;焊接技术
bonding technology 焊接工艺
bonding temperature 黏结温度;黏合温度
bonding test 黏结试验;结合力试验
bonding tie 连接拉杆
bonding timber (砖工中砌入薄砖墙内的)长梁
bonding tool 焊头
bonding treatment 黏结处理
bonding wire 焊线;等电位连接线
bond insurance 债券保险
bond interest 债券利息
bond interest expenses 债券利息支出
bond interest receivable 应收债券利息
bond interface 熔合面
bond investment 债券投资
bond investment audit 债券投资审计
bond investment trust 债券投资信托
bond issue 债券发行
bond issue cost 债券发行成本
bond issuing expenses 债券发行费用
bond issuing market 债券发行市场
bond layer 黏结层;(砖或石的)砌合层
bond length 键长【化】;锚着长度;锚固长度;钢筋锚着长度;钢筋搭接长度;握裹长度;紧固长度(钢筋)
Bondley process 邦德勒方法
bond life 债券期限;公债期限
bond limit 债券限额;公债限额
bond-limited subset 紧密结合的表现
bond line 胶层;黏结线;熔合线;黏合层;胶结层
bondline thickness 胶层厚度;黏结层厚度
bond log 水泥胶结测井图
bondman 担保人
bond market 债券市场
bond master 环氧树脂类黏结剂;环氧树脂类黏合剂
bond mechanism 黏结机理
bond metal 烧结金属
bond meter 胶接检验仪
bond moment 键矩
bond mortar 黏结砂浆
bond note 海关免税提货单;海关保税输出证;债券式票据;保税提货许可证;保税单
Bond number 邦德数;键数
bond of indemnity 赔偿保证书
bond open 耦合断开;焊接断开;焊缝裂开
bond order 键级
bond outstanding 未清偿债券
bond paper 高级书写纸;道林纸
bond parameter 键参数
bond performance 黏结性能
bond plaster 黏结灰泥;黏结灰膏;特制石膏灰浆(整体混凝土涂料)
bond plug 接线塞
bond-plus warrants 附有优先认购券的债券
bond premium 债务价;债券溢价
bond premium amortization 债券溢价摊销
bond premium unamortized 未摊销债券溢价
bond prevention 防止黏结;无黏结隔离
bond price 债券价格
bond pricing 债券计价
bond principle 债券原理
bond purchase audit 债券核算审计;债券购入审计
bond quality 黏结质量
bond rate 债券利率
bond rating 债券评级
bond redemption 债券偿还;债券兑回
bond refunding 债券换新
bond register 债券登记簿

bond rendering 胶结外部刷刷
bond repayment 债券偿还
bond resistance 抗黏合性;黏着强度;黏合抗力;黏着抗力
bond retirement fund 债券偿债基金
bond rope 连接绳
bonds and debentures 债券及信用债券
bonds and mortgage 债券及抵押款
bonds authorized 额定公司债;额定出债;批准发行的债券
bonds authorized and unissued 已批准发行但尚未实际发行的债券;已批准的但尚未发行债券
bond scission 键裂开
bond shifting 键位移
bond sinking fund 债券偿债基金
bonds in series 连续发行的债券
bonds issued 已发行债券
Bond's law 邦德定律
bond slip 黏结滑移
bond slippage 黏结滑动
bondsman 保证人;受契约束缚的人
bond soil 黏合土
bonds payable 应付债券
bonds payable audit 应付债券审计
bond split failure 握裹力劈裂破坏
bond splitting resistance 劈裂黏结强度
Bond's size reduction theory 邦德粉碎理论
bond stone 砌合石;系石;束石;连结石;拉结石;结合石
bondstone 系石;束石
bondstone course 系石层
bond strength 结合强度;键力;键合强度;黏着强度;黏结强度;黏合强度;砌合强度;握裹强度
bond stress 黏结应力;黏着应力;黏合应力;附着应力;握裹力
bond stress of mortar 砂浆握力
bond structure 黏着结构
bond study 黏结研究
bond subscription 债券认购(书)
bonds with warrants 权利债券
Bond's work index 邦德功指数
bond tables 债券表
bond tester 接头电阻测试器;胶接检验仪;黏结强度试验器
bond testing machine 黏结力试验机
bond tile 搭盖瓦
bond timber 系木;枕距护木;束木;墙结合木
bond transfer 砌合传递;结合传递
bond trustee 债券受托人
bond type 砌合类型;坏工砌合类型
bond type diode 键型二极管
bond underwriter 债券担保者
bond underwriting 债券包销
bond unit stress 单位黏结应力;单位握裹力
bond valuation 债券估价
bond value 债券价值;公债价值;黏着值;黏着度
bond vitrified 熔融黏结
bond washing 空头交易;债券清洗
bond waviness 接头波纹
bond welding 钢轨接头焊接
bond wheel 结合剂砂轮
bond with stock purchase warrant 带收买股票权的债券
bond with warrants 附有认股权债券
bond yield audit 债券收益审计
bond yields 债券利率;债券收益;债券市场利息率
bone 骨;船首白浪
Bone Age 骨器时代
bone article 骨器
bone ash 骨灰白
bone bed 骨层
bone black 兽炭黑;骨炭粉;骨炭(黑);动物黑;骨灰;骨黑
bone black pigment 骨炭颜料
bone breccia 骨角砾岩
bone brown 骨棕(色)
bone carving 骨刻
bone char 骨炭
bone charcoal 骨炭
bone china 骨瓷;轻质瓷器

bone coal 煤质页岩;骨炭;骨煤;高灰分煤
bone cutter 碎骨机
bone dry 完全无水的;绝对干燥(的);干透的;充分干燥(的);全干;完全干燥(的)
bone-dry aggregate 干透集料;干透骨料
bone dry bleached shellac 漂白紫胶
bone dry weight 绝干重量
bone-dry wood 全干木材;烘干木材
boned tile 脊瓦
bone dust 骨粉
bone fat 滑脂
bone formation 骨组织
bone gelatin 骨胶
bone glass 乳色玻璃;乳化玻璃;乳白色玻璃;乳白玻璃
bone glue 牛骨胶;骨胶
bone grain 骨粒
bone grease 骨润滑脂
bone grinder 碎骨机
bone grist 粗骨粉
bone-grubber 拣废者;拾荒人
bone hammer 骨锤
bone house 尸骨仓
bone knob 角柄
bone line 地性线
bone mallet 骨锤
bone meal 骨粉
bone needle 骨针;刺骨针
bone oil 骨油
bone pitch 骨沥青
bone plate 骨板
bone porous 多孔骨架的
bone punch 骨打孔凿
bone rasp 粗骨锉
bone retention 骨内滞留
bone screw and bolt outfit 骨螺钉螺栓器械包
bone sculpture 骨雕
boneset 穿心佩兰
bone-shaped column 骨状柱子
bone shears 骨剪
bone structure 骨结构
bone tar 骨柏油;骨焦油沥青
bone tar oil 骨焦油
bone-tar pitch 骨沥青;骨焦油沥青
bone trimmer 剔骨机
bone turquoise 齿绿松石
bone wax 骨蜡
bone white 骨灰白
bonfire 篝火;营火;水溅飞雾
bonification 居住与生产条件的改善;土地改良
boning 检验墙的垂直;测平法(用测杆测定开挖工程的标高);测平(检验墙身垂直度);施骨肥
boning board 测杆;视机【测】;测平尺;测平板;T形测平板
boning fall 用测平杆测得的管沟斜度
boning in 瞄准(检测二点间水平差);打测平桩;匀整坡度
boning out 定直线
boning peg 测平桩【测】
boning rod 测平杆;测杆;水平尺;水平测杆;整坡杆
boning stick 检验墙体垂直的木棒
boninite 玻安岩
bonitation 繁殖适度;发生适度;土地评价
bonito oil 鲣鱼油;松鱼油
Bonjean curve 邦金曲线
Bonjean's curves 邦金曲线图(船体横剖面面积与吃水关系图,用以计算各种浮态时排水量)
Bon maroon 邦紫红
Bonn Convention 邦恩公约
Bonnel type spring 邦内尔式弹簧锁
Bonner-Ball neutron detector 邦内尔—鲍尔中子探测器
Bonner durchmusterung 邦内尔星表
Bonner sphere spectrometer 邦纳球谱仪
Bonne's normal equivalent pseudo-conic(al) projection 邦内正轴等积伪圆锥投影
Bonne's projection 邦内尔投影
Bonne's pseudo-conic(al) equivalent projection 邦内等积伪圆锥投影
bonnet 炉罩;机罩;机器罩;火花避雷器;引擎罩;烟囱罩;遮风;阀门帽;阀帽;阀梗罩;阀盖;伐脊盖瓦;无边小圆软帽
bonnet bed 小帐顶床
bonnet body seal 阀帽体密封

bonnet bolt 盖螺栓;顶盖固定螺栓
bonnet gasket 阀帽体密封垫圈
bonnet headed door or window 带帽饰的门或窗;外抱斜削门或窗
bonnet hip 连接屋顶表面的角形瓦
bonnet hip tile 屋脊瓦;罩瓦;帽状脊瓦;屋脊弯瓦;屋脊盖瓦;弯面盖瓦
bonnet of an engine 发动机盖
bonnet rug 保护用毯子
bonnetted radiator 带罩散热器
bonnet tile 屋脊盖瓦;屋脊瓦;斜脊瓦盖;罩瓦;戗脊盖瓦;弯面盖瓦;盖瓦
bonnet type 凸头式
bonnet type cab 凸头式载重汽车
bonnet valve 帽状阀
bonney 矿囊;矿巢
Bonnybridge fire clay 邦尼桥耐火黏土
Bononian 邦诺阶【地】
bononian stone 重晶石
Bon pigment 邦系颜料
Bon red 邦红
bons 法国公债
bonsai 盆栽树木;盆栽植物;盆景;日本式盆景
bonshtedtite 本斯得石
bonstay 盲井;暗井
bont 提升装置(提升用的钢丝及其附件)
bonus 奖金;免费赠品;红利;额外津贴
bonus account 奖金科目;红利账户;奖金账
bonus and dividend fund 奖励分红基金
bonus and penalty clause 奖惩条例;奖罚条款
bonus and premium plan 奖金和津贴制
bonus bearing certificate 有奖储蓄
bonus bond 红利债券
bonus clause 奖励条款;奖金条例;奖金条款
bonus debenture 红利债券
bonus dividend 红利股息
bonuses issued on the basis of a general voicing 评奖
bonus for completion 竣工奖(励);竣工奖(金)
bonus for early completion of works 提前竣工奖金
bonus for topping quota 超额奖金
bonus incentives for development 开发奖励
bonus issue 作为红利发行的股票
bonus loading 附加保险费
bonus method 奖金法;分红法
bonus paid for extra-heavy work 超载工作奖金
bonus payable 应付奖金;应付红利
bonus payment reserve 奖金储备
bonus-penalty contract 奖罚合同;奖惩合同
bonus plan 奖金分配法;奖金制度
bonus pool 红利基金
bonus record 奖金记录
bonus scheme 津贴计划
bonus shares 分红股
bonus stock 赢利股
bonus system 奖励工资制;奖金制度;分红制
bonus system of wages 工资奖金制度;工资奖励制
bonus target date 预定结束日期奖金
bonus to employees 职工红利
bonus to officer 职员奖金
bonus to partner 合伙人奖金
bonus wage 奖金式工资
bony 干瘦
bony coal 页岩煤;骨煤
Bonyhard organization 博奈哈德结构
bony nodule 核骨
bony shell 骨包壳
booby hatch 便道小舱口;精神病院;活盖小舱口
booby trap 隐蔽爆炸装置;铅笔雷
boogie box 压浆泵;压力灌浆泵
boogie pump 低噪声泵;压浆泵;压力灌浆泵
boojie pump 压力灌浆机;压浆泵;气压灌浆泵
book 海存段;著作;注册;订货;登账;程序簿;册
bookable 可预购的
book account 往来账户
book amount 账面金额
book audit 账面审计
book backbone lining up machine 书脊黏纱布机
book back rounding machine 书脊扒圆机
book balance 账面差额
book binder's ink 封面油墨
book binder's varnish 书脊清漆

bookbinder wire 装钉钢丝
bookbindery 装订所
bookbinding 装订
bookbinding machine 装订机
book by double entry 复式账册;复式记账
book by simple entry 单式簿记
book capacitor 调角电容;书状电容器
bookcase 书柜
bookcase drier 书柜式干燥器
book clay 页状黏土;带状黏土;微薄层黏土
book closet 书橱
book cloth 封面布
book confirmation 确认定货
book conveyer 图书传送带
book copying 多份复制
book cost 资产购置成本;账面成本
book credit 账面信用;账面信贷;赊销金额
book date 记账日
book debt 账面债券;账面欠债
book depreciation 账面折扣
booked cargo 已订舱货物
booked mica 未分开的片页块云母
book entry 账面记录;账簿记录
book-entry security 记账证券
book fashion 按书写顺序摆放法(岩芯)
book holder 夹书架
bookhouse fabric 页状集束构造;书堆组构
bookhouse texture 片堆结构
booking 预订;订票;订船票;订舱;叠箱造型法
booking book 订舱簿
booking bookkeeping by card system 卡片式记账制度
booking cargo space 订舱
booking hall 售票处;预约处;登记处;售票厅
booking list 预订单;订舱清单;订舱单
booking menu 订舱单
booking note 订载单;订舱单;托运单
booking of cargo 承办货运
booking office 售票处;订舱营业所
booking office machine 票房售票机;售票机
booking period 预约承运期间
booking report 货运预订报告书
booking space 预订舱位
booking window 售票窗
booking yarn 打包纱绳
book inventory 账面盘存;账面存货
book inventory method 账面盘存法
book inventory record 账面盘存记
book inventory sheet 账面盘存表
bookkeeper 会计(指人员);簿记员;会计员;账房先生
bookkeeper's departure 账房
bookkeeping 内务操作;簿记
bookkeeping by card system 卡片式簿记(制)
book(keeping) by double entry 复式簿记
bookkeeping by Italian method 意大利式簿记;复式簿记记账法
bookkeeping by single entry 单式簿记
bookkeeping machine 簿记机
bookkeeping operation 程序加工操作;簿记操作
bookkeeping record 簿记录
bookkeeping typewriter 簿记打字机
book-kiosk 书亭
book land 注册土地
book(-leaf) clay 书页黏土
booklet 目录单;小册子
booklet of general drawing 总图图册
booklift 图书升降机
book loss 账面损失;账面亏损
book mark 书签
book matching 木纹拼花;正反配板法;书页式拼装法;书页式拼板
book message 多目标消息
book mica 书页云母
bookmobile 流动图书馆;活动图书馆
book mo(u)ld(ing) 叠箱铸型;分体模
book of abstract 摘要账
book of accounts 账簿
book of chronological entry 序时账簿
book of estimates 概算书
book of final entry 结账记录;终结账簿
book of original document for payments 支出凭证簿
book of original document for receipts 收入凭证簿

book of original entry 原始账簿
book of posting entry 过账记录
book of rate 税则
book of record 记录簿
book of reference 参考书
book of secondary entry 转账账簿
book of store balance 材料余额簿
book of tender 投标书
book on ceramics 陶雅
book out 冲销
book post 印刷品邮件
book price 账面价格
book profit 账面利润
book quantity 账存数量
bookrack 搁书架;书架
book rate 账面利率
book rate of return 账面收益率
book record 记事簿
book rental 租书处
book repository 藏书库;书库
bookrest 阅书架;图书架
books and reference materials 图书资料
bookseller's 书店
book sewing machine 锁线订书机
book shelf 搁口板;书架
book shop 书店
book slide 活动书架
book smashing machine 书籍压平机
books of planning task 计划(任务)书
bookstack 分层书架;书架
book stacker 堆书机
bookstack room 书库;固定书架室;固定书库
bookstall 书堆;书亭
bookstand 书亭;书架;书柜
bookstore 书页岩
bookstore 书店
book structure 页状构造;书页构造
book surplus 账面盈余
book system 账簿制度
book table 书架桌
book tally 记录理货(法);划号理货
book tax difference 账面税金差额
book test 弯合试验(厚板)
book tile (带凹凸边的)书形空心砖
book tower 多层藏书楼;多层图书馆
book value 账面价值
book value of capital stock 股本账面价值
book value of fixed assets 固定资产账面价值
book value per share(of stock) 每股股票的账面价值;每股账面价值;每股票面价值
book-value shares 账面价值股票
book vault 藏书室;书库
book with you 向你方买进
book your order for 卖给你方
Boolean 布尔(值);布尔符号;布尔变量
Boolean add 布尔加
Boolean algebra 逻辑代数;布尔代数
Boolean algebra of propositional logic 命题逻辑布尔代数
Boolean analyzer 布尔分析器
Boolean array 逻辑数组;布尔数组
Boolean calculation 逻辑代数运算;布尔运算
Boolean calculus 布尔演算;布尔计算法
Boolean character 布尔字符
Boolean choice model 布尔选择模型
Boolean choice pattern 布尔选择样式
Boolean coefficient 布尔系数
Boolean complementation 布尔求反;逻辑求反;布尔求否
Boolean computer 布尔计算机
Boolean condition 布尔条件
Boolean conjunction 布尔连接;布尔合取
Boolean connective 布尔演算符;布尔连接(符);布尔连接词
Boolean constant 布尔常数
Boolean data item 布尔数据项
Boolean data type 布尔数据类型
Boolean denotation 布尔标志
Boolean difference 布尔差分
Boolean difference method 布尔差分法
Boolean equation 布林方程;布尔方程
Boolean expression 布尔(表达)式
Boolean factor 布尔因式

Boolean field 布尔域
Boolean format 布尔格式
Boolean function 布尔函数
Boolean homomorphism 布尔同态
Boolean hypercube 布尔超正方体
Boolean literal 布尔常字
Boolean logic 二值逻辑;布尔逻辑
Boolean map 布尔图
Boolean marker 布尔标志符
Boolean matrix 布尔矩阵
Boolean method 布尔法
Boolean multiplication 布尔乘法
Boolean network 布尔网络
Boolean number 布尔型
Boolean operation 布尔运算
Boolean operation table 布尔运算表
Boolean operator 布尔运算符;布尔算符
Boolean pattern 布尔图样;布尔模式
Boolean position 布尔位置
Boolean primary 布尔一次量;布尔初等量
Boolean quantity 布尔量
Boolean recurrence function 布尔递归函数
Boolean recurrence solver 布尔递归求解器
Boolean ring 布尔环
Boolean secondary 布尔二次式;布尔二次量
Booleans to bits symbol 布尔转字位符号
Boolean symbol 布尔符号
Boolean term 布尔项
Boolean type 布尔(类)型
Boolean unit 布尔部件
Boolean value 布尔值
Boolean variable 布尔变数;布尔变量;逻辑变量
boom 栏木;拦船木栅;景气;木浮标;横江铁索;河中标柱;河绠;悬臂;钻臂;纵帆下桁【船】;张帆桁;构架;硬存木;浮栅;(木材港池的)浮式拦河埂;防浪浮架;帆脚杆;起重臂;嗡嗡声;水上航标;受电杆
boomage 护舷费
boom and bucket delivery 吊杆与戽斗递送法;吊杆和戽斗输送;吊戽运送法
boom and bust 经济繁荣与萧条的交替循环;一时性繁荣
boom and slump 经济繁荣与萧条的交替循环
boom angle 环状角铁;翼缘角钢;箍用角条;动臂倾角;吊机臂转角;吊臂角度;起重机吊杆的斜角度;伸臂角度
boom angle indicator 起重臂的角度显示器
boom attachment 轴头装置
boom-balancing system 臂平衡系统;吊臂平衡系统
boom belt conveyer 悬臂输送带
boom between pontoon and shore wall 趸船支撑
boom boring machine 桁架式钻杆
boom brace 吊臂斜杆;起重臂桁架;伸梁支架
boom bracing 起重臂桁架式结构;弦杆加劲
boom brake 起重臂桁架(式)结构
boom bucket 悬臂式铲斗
boom cable 流材索;绠缆
boom cat 悬臂起重机架;挖土机手;悬臂铺管机
boom chain 绠缆
boom chock 吊杆托架
boom chord 弦杆
boom city 新兴城市
boom concrete mixer 横杆混凝土搅拌机
boom control 起重臂调节
boom conveyer 悬臂输送机
boom cradle 吊杆托架
boom crane 桁梁起重机;悬臂起重机;吊杆机;臂式起重机;臂式吊车;伸臂式起重机;伸臂起重机
boom crowd ram 动臂推压油缸
boom crutch 吊杆托架
boom cylinder (装载机的)转臂油缸
boom defence vessel 基地布网船;布栅船
boom derrick 悬臂起重机;摇臂起重机架;桅杆式起重机杆;吊臂吊杆
boom derrick crane 支臂桅杆起重机
boom downpipe 下悬喷管
boom dragline 起重臂索;吊杆索斗铲;拉铲;索斗(式)挖土机;索斗(式)挖掘机
boom dredge(r) 悬臂挖泥船;边抛挖泥船
boom drilling cable 钢绳冲击钻进大绳
boom drum 吊架卷筒;起重杆卷筒
boom element 桁架构件;弦杆

Boomer 布麦尔震源;流动工人;临时工;(物探用的)轰鸣器;自动木栅;水下栅栏;声振器;自动水栅
boomerang 木三铰拱(框架)
boomerang arch 尖拱
boomerang effect 相反效果
boomerang sediment corer 自返式沉积取芯器
boom excavator 吊杆挖土机;臂式挖土机;壁式挖土机
boom extension 起重臂伸长;喷杆延长杆
boom extension ram 起重臂的附加桩锤
boom foot spool 起重臂底部的卷筒
boom for building construction 房屋用的起重臂
boom for concrete delivery 混凝土布料杆
boom harness 起重臂吊索
boom head 起重臂头部;吊臂上端
boom header 部分断面隧道掘进机
boom height 起重臂高度
boom hoist 旋臂起重机;悬臂起重机;悬臂起重机;吊臂式吊车;臂式起重机;臂式吊车;壁式起重机;伸臂起重机;吊杆;吊杆机
boom hoist drum 起重臂的卷盘起重;变幅卷筒
boom hoisting mechanism 悬臂升降机构;悬臂俯仰机构;前伸臂俯仰装置
boom hoisting speed 吊杆俯仰速度;俯仰速度
boom hoist limit switch 起重臂的限制起重量开关
boom hoist mechanism 吊臂升降机构
boom hoist rope 变幅钢丝绳
boom hydraulic lift cylinder 动臂液压缸
boom inclination 起重臂斜度
boom inclination angle 起重臂倾角
boominess 空腔谐振;箱谐振
boominess resonance 机箱共鸣;机盒共振;箱共鸣
booming 伸出;突出;高收入;闪冲砂矿法
booming dune 鸣沙沙丘
booming market 景气市场;繁荣的市场
booming sand 鸣砂
booming season 旺季
booming town 新兴城市
boom iron 吊杆端箍
boom jack 支承起重臂上的滑轮的重吊杆
boomkin 滑车伸出架
boom ladder hoist 多斗提升的起重臂
boom latch device 吊臂闭锁装置
boom length 起重臂长度;吊臂长度;起重机悬臂长度
boomless sprayer 无喷管式喷雾器
boomlet 小气景;短时景气
boom lift cable 起重臂提升缆索;起重机悬臂提升索
boom lift drum 起重臂提升圆筒
boom lift ram 起重臂提升桩锤
boom lift switch 起重臂提升限制开关
boom lift valve 起重臂提升阀
boom light 吊杆灯
boom line 起重臂缆索;悬臂绷绳
boom loader 转臂式装载机
boom lowering equipment 臂架放倒机构
boom lowering mechanism 吊臂放倒机构
boom man 装卸手;吊臂操作人
boom mounted drifter 凿岩台车;动臂凿岩机;架式风钻
boom-mounted plugger 动臂式凿岩机
boom-mounted wheel excavator 臂轮式挖土机
boom net 栅栏网
boom of an arched girder 弯曲大梁起重臂
boom of arch 拱环
boom of crane 起重机吊臂
boom of dredging 边抛疏浚
boom-out 最大伸距;最大伸臂距;最大伸幅;起重机最大伸距;起重机臂伸出极限长度;最大挖掘半径;伸臂长度
boom period 繁荣期;景气时期;繁荣时期
boom pin 起重臂底部枢轴
boom pitch 悬臂倾斜角
boom plan 景气计划
boom plane 桁架平面
boom plate 弦板;梁板;翼缘板
boom platform 吊杆台
boom point 起重臂端点;起重臂顶点;吊机臂端;伸臂末端
boom point pin 起重臂顶尖枢轴
boom point sheave 起重臂顶尖滑轮;臂端
boom point with fixed point 起重臂顶尖与锚固点
boom position 起重臂位置
boom prices 猛涨起来的物价

boom raising and lowering 起重臂提升及下降
boom raising and lowering cable 起重臂提升及下降缆索
boom raising and lowering gear 起重臂提升及下降齿轮;起重臂升降机构
boom raising and lowering motor 起重臂提升及下降电动机
boom raising and lowering speed 起重臂提升及下降速度
boom ram 起重臂撞头
boom reach 最大伸距;吊杆最大伸距;起重机伸臂长度;起重机臂伸出极限长度;最大臂距;伸臂长度
boom rest 吊杆托架
boom-revolving gear 吊杆旋转装置
boom rig 吊杆索具;臂式起吊设备;伸臂式机具
boom rod 吊杆;弦杆
boom rope 起重臂缆索;流材索
boom saddle 吊杆托架
boom scraper 臂式刮料机;臂式铲运机;臂式铲运车
boom section 起重臂截面;喷杆组(喷雾器的)
boom set 重吊杆;水栅
boom shaker 杆式抖动器
boom sheave 导向滑轮;臂架滑轮
boom sheet 吊杆端下支索
boom shipper shaft 起重臂移动轴
boom side sheave 起重臂靠边滑车轮
boom slew 转臂回转角度
boom slewing angle 吊杆旋回角
boom socket 吊杆承臼
boom sprayer 喷管式喷雾器
boom sprinkler 悬臂式喷灌机;长臂喷灌机;臂式喷洒机;悬臂杆式喷灌机
boom stacker 悬臂堆料机
boom stay 吊杆端舷侧支索
boom step bracket 轻型吊杆座
boom sticks 缏漂木
boom stiffening 加劲杆;桁架加固;弦杆加固
boom support 梁支架;弦杆支柱
boom support guy 挖土机臂索;机臂吊索;吊臂稳索
boom support post 吊杆支柱(俗称将军柱)
boom swing 吊杆回转;吊机臂转动
boom swing angle 起重臂转动角度;吊臂回转角
boom table 吊杆平台
boom tackle 吊杆索具;吊杆滑轮组
boom telescoping device 起重臂伸缩机构
boom tie 悬臂拉索
boom tooth 起重臂墙叉口(安装副吊臂用)
boom topping angle 吊杆仰角
boom town 新兴城市
boom type shovel 起重臂式铲土机;单斗式挖掘机
boom type trenching machine 起重臂式挖沟机;梯形挖沟机
boom well 构架槽
boom width 桁梁宽度
boom winch 起重臂上卷扬机
boom wire 吊杆拉索
boom with crab 起重臂上起重小车
boom year 景气年份
boondocks 荒野
boonies 远离城市的原野
boony fiber 木质纤维
Boord olefin synthesis 伯尔德烯烃合成法
Boord synthesis 伯尔德合成
boorga 布加风
boor topping 标色水线【船】
boost 加速;助爆;汽缸升压
boost capacitor 升压电容
boost charge 快速充电;加强充电;补充充电
boost competitive power 提高竞争能力
boost compressor 增压式压气机
boost control 增压控制;增压调节;升压控制
boost controller 增压调节器
boost dose 加强剂量
boosted boost circuit 升压增音电路
boosted circuit 附加电压电路
boosted voltage 附加电压;辅助电压;升高电压
boosted water supply 增压供水;给水;上水道
boost engine 助力发动机
booster 局部通风机;加压机;加强剂量;继爆器;运载工具;助推器;助推发动机;助力器;助爆器;增压装置;增压器;增压放大器;增能器;增幅剂;传爆剂;起动磁电机;排气辅助器;调压机;升压器;升压机;升压电阻;升压变压器

booster air heater 辅助加热炉
booster air pipe 增压空气管
booster air pump 启动气泵
booster air tube 增压空气管
booster alternator 增压交流发电机;增速交流发电机
booster amplifier 高频前级放大器;附加放大器;辅助放大器;升压放大器
booster battery 升压电池组
booster blower 升压鼓风机
booster brake 真空加力制动;增压制动器
booster cable 升压电缆
booster charge 助爆炸药;再充电;传爆装药;传爆药柱;升压充电
booster circuit 升压电路
booster coil 起动点火线圈
booster compressor 循环压缩机;增压压缩机;增压空压机;辅助压缩机;升压压缩机
booster converter 升压变流器
booster cylinder 助力液压缸
booster cylinder boot 助力汽缸罩
booster cylinder bracket 升压缸托架
booster device 升压设备
booster diffusion pump 增压扩散泵
booster diode 辅助二极管
booster dose 加强剂量;激发剂量;促升剂量
booster element 增益元件
boost(er) engine 加速机组;加速发动机;助推发动机
booster explosive 助爆破药
booster fan 加压风机;增压鼓风机;鼓风机;辅助扇风机;辅助鼓风机;升压风机
booster fan thrust 风机推力
booster flame 辅助火焰
booster gas turbine 加速燃气轮机;加力燃气轮机
booster heater 增温电热器;加热器;中间加热器;增热器;辅助加热器
booster hose clamp 助力器软管夹
booster hose clip 助力器软管箍
boostering pump 增压泵
booster injection 加强注射;激发注射
booster installation 增压站;升压站
booster locomotive 带增压器的机车;辅助机车
booster magneto 助力磁电机;启动磁力机
booster oil diffusion pump 增压油扩散真空泵
booster oil pump 增压油泵;升压油泵
booster plant 增压厂;增压站;升压站
booster platform 增压平台
boost(er) power plant 助推动力装置
boost(er) pressure 增压压力
booster pressurizer 增压器
booster pump 接力泵;加压泵;增压泵;前置泵;升压泵
booster pump boat 接力泵船
booster pumping station 升压水泵;增压泵站
booster pump station 中继泵站;加压站;增压泵站
booster relay 升压继电器
booster resistor 附加电阻器
booster response 加速响应
booster rocket 火箭助推器
booster(rod) 增益棒
booster set 增压机组
booster shot 加强注射
booster signal 提升信号
booster stage 助推级;增速级
booster station 接力电台;接力泵站;加压(泵)站;增压站;增压泵站;辅助中继电台;辅助电台;升压站;升压电台
booster supercharger 增压机
booster telephone circuit 电话增音电路
booster tension 提升电压
booster thermocouple 均衡热电偶;加温器热电偶
booster thrust 火箭助推器推力
booster-track gradient 补机坡度(美语)
booster transformer 吸流变压器;增压变压器;升压变压器
booster transformer feeding system 吸流变压器供电系统
booster transmitter 辅助发射机
booster tube 传爆管
booster-type diffusion pump 增压式扩散泵
booster-type master cylinder 增压式主油缸
booster valve 增压阀;辅助阀
booster voltage 辅增电压
booster water heater 附加热水器

booster well 传爆管孔
boost fan 鼓风扇
boost ga(u)ge 增压计;增压表;升压压力计;增压压力表
boost horsepower 加速功率
boosting 升压;增压;电压升(高);辅助加热
boosting anode 辅助阳极
boosting battery 快速充电蓄电池;加压电池;补充电池
boosting charge 升压充电
boosting impulse 助推器冲量
boosting of furnace 炉膛强化
boosting transformer 增压变压器;升压变压器
boosting transformer booster 吸上变压器
boosting voltage 补充电势
boosting wire 吸上线【电】
boost line 增压线;增压管道
boost melting 电热促熔
boost motor 加速器;助推器
boost output 提高产量
boost power plant 辅助动力装置
boost pressure 升压;增压(压力)
boost(pressure) ga(u)ge 增压压力表
boost price 提价
boost productivity 提高生产率
boost pump 推进泵
boost ratio 增压比
boost resistor 升压电阻
boost sales 推销
booststrap circuit 自益放大电路
boost the competitiveness 提高竞争能力
boost voltage 增高电压
boot 进料斗;缓冲柱;引出线罩;给料斗;长统靴;附得利益;保护罩;无底坩埚;枪套;枪裙;水落管槽
boot-beam plate 靴梁
boot boiler 靴形锅炉
boot cap 保护罩
boot clamp 密封套夹
boot compartment 行李舱
boot floor 行李舱底板
booth 临时售货棚;小室;小舍;小棚;小房(间);暗箱;喷漆橱;摊店;棚舍;公用电话间;工作台
Booth algorithm of binary multiplication 二进制乘法的布斯算法
booth front opening 放映室的前窗口
booth hood 柜形吸气罩
boothite 七水胆矾;铜水绿矾
booth microphone 暗箱式话筒;室内送话器
boot hook 长柄捻缝弯铲
Booth's algorithm 布斯算法
boot jack 带闩打捞器;钻井用捞钩
bootle brick 无底瓶形建筑装饰砖
bootleg 石柱;瞎眼炮眼;炮眼底;未爆炮眼
bootlegging 走私漏税;非法制造;非法运输;皮带打滑
bootleg hole 残炮眼
boot lintel 带凸边的门窗过梁;带凸边过梁;挑口过梁
boot loading elevator 隔式载重提升机
boot magnetron 长阳极磁控管
bootman (新浇混凝土工地的)着长靴工人;(用特别装备的卡车从公路涂油的)筑路工
boot money 在房屋修建中从事困难工作的酬金
boot of elevator 升降机底滑车箱
boot pulley 斗式提升机底部滑轮
boot-shaped 靴形
boot-shaped heart 靴形心
boot socket 带门打捞筒
bootstrap 自持系统;共益;辅助程序;仿真线;人工线路;自举作用;引导
bootstrap amplifier 阴极输出放大器;自益放大器;自举放大器
bootstrap block 引导块
bootstrap cathode follower 仿真线路阴极输出器
bootstrap circuit 自举电路
bootstrap driver 引导指令【计】;引导驱动器;阴极输出激励器;自举驱动器;自举激励器
bootstrap dynamics 靴袢动力学
bootstrap function 自举作用
bootstrap generator 仿真线路振荡器
bootstraping 引导指令【计】
bootstrap input program(me) 引导输入程序
bootstrap instructor technique 引导指令方法
bootstrap loader 引导装入程序;输入引导子程序
bootstrap loading routine 引导装入例行程序;引

导装入程序
bootstrap memory 辅助程序存储器
bootstrap process 自持过程
bootstrap program(me) 引导指令程序;引导程序【计】
bootstrap record 引导指令记录
bootstrap routine 辅助程序;引导例行程序;引导程序【计】;自举程序
bootstrap sawtooth generator 仿真线路锯齿波振荡器
bootstrap scheme 靴带方案
bootstrap system 自导程序系统;布氏空气冷却系统
bootstrap technique 引导指令方法
boot topping paint 水线涂漆;水线涂料;水线带漆;水线漆
boot truck 沥青路面喷射车;沥青路面喷射机
boot water-line paint 水线涂漆;水线涂料
bop control system 防喷器控制系统
bora 布拉风
boracic 含硼的
boracic acid 硼酸
boracic acid flint glass 硼酸火石玻璃
boracic frit 硼酸质熔块
boracic glaze 无铅硼釉
boracite 方硼石
boracite glass powder 方硼石玻璃粉
bora fog 布拉雾
boral 碳化铝
boralloy sprocket 硼钢制的链轮
borane 硼烷;硼氢化物
borasca 博拉斯科雷暴
borasque 博拉斯科雷暴
borate 熔融硼砂;硼酸酯;硼酸盐
borated 硼酸处理过的
borated concrete 含硼混凝土
borate deposit 硼酸岩矿床
borate flint(glass) 含硼火石玻璃;含硼电石
borate glass 硼酸盐玻璃
borate mineral 硼酸矿物
borate of lime 硼酸盐石灰
boratto 丝毛交织物
borax 月石;丹石;硼砂;天然硼砂;四硼酸钠
boraxal 氧化硼钙
borax anhydrous 无水硼砂
borax bead 硼砂珠;硼砂熔珠
borax bead test 硼砂珠试验
borax-carmine stain 硼砂卡红染剂
borax concrete 硼砂混凝土
borax glass 熔融硼砂;硼砂玻璃;硼玻璃
borax lake 硼砂湖
borax-lead technique 硼砂珠技术
borax paraffin collimator 硼—石蜡准直仪
borazine 环硼氮烷;硼嗪
borazole 硼嗪
Borazon 博雷崇(一种人造亚硝酸硼高硬度研磨材料);立方氮化硼
borazone 氮化硼半导体
borcarite 单斜碳镁钙硼石
Borda count 博达数
Borda loss 水管突扩大处流速水头损失
Borda's mouthpiece 博达管嘴;博达管口
Borda's pipe 博达管(以法国工程师命名的一种泄水管)
bord-down 下山巷道
bordeaux 枣红(色);酸性枣红
Bordeaux connection 波尔多式连接(钢丝绳与起重链的连接)
Bordeaux faience 波尔多锡釉陶器(法国)
bordeaux lake 酸性枣红色淀
Bordeaux mixture 波尔多液;波尔多混合液
Bordeaux turpentine 枣红松节油;波尔多松脂
bordello 妓院
Borden Inc. 博登公司(美国)
border 路缘;缘(饰);国境;滚边;埂(子);边缘(地区);边沿;边境;边界;契口键
border arc 边缘弧
border area 路边地带;边缘地区;边区;边疆
border bed 花圃
border break 破图廓部分
border check 畦田
border-check irrigation 畦灌
border contrast 边界衬度
border customs 边境海关
border data 图廓资料

border dike 边堤;畦底;围堤;田埂
border district 边缘地区;边地
bordereau 分保明细表;情况明细表
bordereaux 业务报表
bordered determinant 加边行列式;加边行列式
bordered matrix 加边矩阵
bordered pit 具缘纹孔;重纹孔;有边细孔
bordered pit vessel 具缘纹孔导管
bordered separation 底图
bordered symmetric(al) matrix 加边对称矩阵
border effect 周边效应;边行影响
border etching 边界蚀刻法
border evaluation of a matrix 矩阵的边缘估值
border facies 接触相;边缘相
border fault 边缘断层;边界断层
border figure 图廓数字注记
border flooding 畦灌
border growth 路及绿篱
border heating 周边供暖
border ice 固定冰;滨冰;边缘冰;岸冰;贴岸冰
border information 图廓注记
bordering 边界(标志)
bordering tool 折边工具
bordering zone 边缘区
border in headers 丁头镶边
border irrigation 沟灌;带状灌溉;边缘行业;边缘灌溉;畦灌
border joist 边梁;缘饰托梁
border lake 山麓湖
border land 边缘陆地;边缘古陆;边缘地;边界地带;边境区;边缘地带;边缘地方;交界地区
borderland slope 边缘地斜坡
borderless easel 无边界框
border light 边缘灯;舞台台口灯光;场界灯;顶光;舞台上部横挂的照明灯
border line 界线;国境线;国界;边线;边界线;图廓线
border line case 临界案情
borderline curve 轮廓线;边界线曲线
border line field 边缘(科学)领域
borderline hole 边界孔
borderline of expropriation land 征地红线
borderline of land 土地界线
border line of lot 建筑用地界线
borderline risk 难以确定的风险
borderline science frontier science 边缘科学
borderline well 边界井
border mask 边缘遮幅
border moraine 边碛
border note 图廓注记
border of a town 城镇的边缘
border of oval fossa 卵圆窝缘
border on 接近;毗连
border pattern 边纹图案;边花(纹)
border pen 边缘曲线用的绘图笔
border pile 边桩;边缘桩
border plane 边缘槽刨
border price 边境价格
border-punched card 边穿孔卡
border region 边区
border region currency 边币
border rim 反应缘;(变质作用的)反应边【地】
border ring 反应池;反应边;边界环
border ring texture 反应边结构
border river 边界河
border row 边行
border scale 图廓比例尺
border sea 边缘海
border sheet 底图
border spring 边井;边界泉
border station 国境站;国境车站
border stone 路缘石;界石;镶边石;边石
border stretcher 边拉伸器
border strip 边缘防护带;条形畦田
border strip flooding 分区漫灌
border tax 边境税
border tax arrangement 商定国境税
border tie 边纹穿吊
border tile 边线砖
border trade 边境贸易
border tree 行道树
border tropical zone 边缘热带
border wire 绑线钢丝;绑丝
border zone 边境免税贸易区;边区
Bordet's amboceptor 博代介体

Bordet's law 博代定律
Bordet's phenomenon 博代现象
Bordex 建筑板材(可防水和防火)
Bordini effect 包廷尼效应
bordorflex 防火轻质墙板
bordroom 煤房
bordroom-man 煤房杂工
bordroom width 煤房宽度
bord-up 上山巷道
bore 料孔;口径;激浪;怒潮;内腔;内径(数据库用);涌波;缸径;打眼;汽缸筒;炮眼;水力波;砂芯
boreability 可钻性;可钻度
boreal 北方的;北的
boreal ammonite realm 北方菊石地理区系
boreal-atlantic belemnite region 大西洋箭石地理
boreal belemnite realm 北方箭石地理区系
boreal brachiopod realm 北方腕足动物地理区系
boreal climate 北部气候
boreal coral realm 北方珊瑚地理区系
boreal forest 北方针叶林;北部林
Boreal life zone 北方生物带
boreal marine flora 寒温带海洋植物区系
Boreal period 北方期
boreal plankton 寒温带浮游生物
boreal species 寒温带种;北方种
bore and stroke 衡量人的能力;缸径与冲程(水泵)
bore-and-turning mill 旋转车床
bore area 孔径面积
Boreas 北风
bore bar 钻杆
bore bit 镗孔钻头;钻头(钻孔用)
bore brush 枪刷
bore casing 套管
bore check 精密小孔测定器
bore chip 镗屑
bore climb (海啸的)高潮冲击
bore core 岩芯
bore cross-section 钻孔断面
bored by percussion drill bucket 冲抓钻钻孔
bored cable tunnel 钻成的缆索隧道
bored cased pile 钻成带套的桩
bored casing 套管
bored casing pipe 钻成带套的管子
bored casing tube 钻成带套的管子
bored cast-in place pile 钻孔灌注桩;钻孔桩
bored cast-in-situ pile 钻孔(钻入)灌注桩
bored concrete pile 混凝土灌注桩
bored diaphragm 钻成排桩;钻横隔墙
bored heart timber 含心成材
bored-hole 钻孔
bored hole diameter measurement 钻孔直径检测
bore diameter 孔径;搪孔直径
bore diaphragm 螺旋挡墙
bored latch (装在门孔中的)弹簧锁
bored lock 圆柱锁
bored nut 钻孔螺帽
bored pile 螺旋(钻孔)桩;钻孔(灌注)桩;钻孔插入桩;挖孔(灌注)桩
bored pile embedded rock 嵌岩灌注桩
bored pile wall 钻孔桩墙
bored piling 钻孔桩
bored reinforced concrete pile 钢筋混凝土钻孔桩
bored roll 空心滚筒
bored screw nut 螺旋钻孔螺帽
bored spring 人工泉
bored tie 钻孔枕木
bored tube 钻井;套管
bored tube well 钻孔;钻井
bored tubular well
bored tunnel 钻探隧道;盾构隧道
bored under reamed pile 扩底钻孔灌注桩
bored well 螺旋钻井;井孔;钻孔;普通水井;钻孔灌注桩
bore face machine 镗孔锪端面加工机床
bore field 井田;井区;井场
bore frame 钻架
bore ga(u)ge 孔径规;缸径量规
bore generator 生波器(用于潮波模型试验)
bore hammer 凿岩落锤;凿岩机
bore height 怒潮高度
borehole 井眼;(木材的)虫眼;钻探孔;钻【岩】
borehole accelerometer 井下加速度计

borehole acoustic(al) televiewing 声波电视测井(曲线)
borehole acoustic(al) televiewing logger 声波电视测井仪
borehole anti-corrosion method 钻孔防腐法
borehole array 钻孔排列
borehole auger 钻洞机
borehole axial extensometer 钻孔轴向应变计
borehole axial strain indicator 钻孔轴向应变显示器;钻孔轴向应变计
borehole bit 不取心钻头
borehole blasting 钻孔爆破
borehole-borehole mode 井中—井中工作方式
borehole bottom 孔底
borehole cable 矿井电缆;钻孔电缆;测孔电缆
borehole calipers 钻孔卡规
borehole camera 井下摄影仪;钻孔照相机;钻孔摄影仪;钻孔壁摄影机
borehole camera inspection 钻孔摄影检查
borehole casing 钻孔外壳;钻孔套管
borehole cast-in-place(concrete) pile 钻孔灌注桩;混凝土钻孔灌注桩
borehole charge 钻孔装药;钻孔炸药;炮眼装药
borehole collar 钻孔口
borehole columnar section 钻孔柱状剖面图
borehole compensated acoustic(al) velocity log curve 井眼补偿声波速度测井曲线
borehole compensated acoustic(al) velocity logger 井眼补偿声速测井仪
borehole compensated sonic log 井眼补偿声波测井;补偿声波测井
borehole condition 井筒状况
bore hole construction profile 钻孔结构剖面图
borehole coordinates 钻孔坐标
borehole core 岩芯;钻孔岩芯
borehole deepening method for stress measurement 钻孔加深法
borehole deflectometer 钻孔弯度计
borehole deformation 钻孔变形
borehole deformation ga(u)ge 钻孔变形计
borehole deformation meter 钻孔变形仪;钻孔变形计
borehole deformation strain cell 钻孔变形应变计
borehole depth 炮井深度
borehole depth when logging 测井时钻孔深度
borehole deviation 钻孔偏差
borehole device 钻孔装置
borehole diameter 孔径【岩】;钻孔直径
borehole diameter correction 井径校正
borehole diametral strain indicator 钻孔径向应变显示器;钻孔径向应变计;钻孔横向应变计
borehole dilatometer 钻孔膨胀仪
borehole direct shear device 钻孔直剪仪
borehole drilling 旋转法钻进钻孔
borehole drilling machine 钻机
borehole electric(al) correlative method 井中电对比法
borehole electric(al) prospecting 井中电法
borehole electro-magnetic method 井中电磁法
borehole electromagnetic wave method crossing plot 井中电磁波法交会图
borehole electromagnetic wave method curve 井中电磁波法曲线
borehole equipment 钻孔设备
borehole equipment tripod 钻孔设备三脚架
borehole evidence 钻探结果
borehole expansion probe 钻孔膨胀仪
borehole extensometer 钻孔伸缩仪;钻孔伸长仪
borehole facility 钻孔设备
borehole flow instrument 钻孔流速仪
borehole for inspection 勘探孔
borehole geometric(al) factor 井眼的几何因子
borehole gravimeter 井中重力仪;井孔重力仪
borehole gravimetry 井中重力
borehole gravity measurement 井中重力测量
borehole hydraulic circulating system 钻孔水力循环系统
borehole inclination 井斜
borehole inclinometer 钻孔测斜仪
borehole inclusion stressmeter 井下包体式应力计
borehole inclusion stressmeter 钻孔包体式应力计;钻孔包裹式应力计
borehole initial geologic(al) logging 钻孔原始地质编录

borehole inspection 钻井勘探
borehole installation 钻孔设备
borehole jack 钻孔千斤顶
borehole jack method 钻孔塞孔法
borehole lead insulation 穿心引出线绝缘
borehole line 套管柱
borehole lining 钻孔套管;套管
borehole loading 炮眼装药;炮孔装药;炮孔炸药
borehole location 钻孔位置
borehole log 钻孔柱状图;钻孔岩性柱状图;钻孔记录;钻孔地质柱状图;钻井记录;测井图;测井曲线
borehole logging 钻孔编录
borehole low-frequency electro-magnetic method 井中低频电磁法
borehole measurement device 钻孔测试装置
borehole method 钻孔法;检层法(单孔上传法)【岩】
borehole mining 钻井采矿法
borehole mise-a-la-masse method 井中充电法
borehole mud 钻孔泥浆
borehole No. 钻孔编号
borehole number 孔号
borehole observation station 钻孔观测站
borehole of long-term observation 长期观察孔;长期观测孔
bore hole of water probing 探水钻孔
borehole pattern 钻孔布置形式
borehole penetrometer 钻孔贯入器
borehole periscope 钻孔潜望镜
borehole photography 钻孔摄影
borehole pick-up camera 钻孔摄影仪
bore(hole) plug 钻孔土样
borehole plugging 钻孔盖塞
borehole poisoning 钻孔毒杀
borehole position survey 钻孔位置测量
borehole pressure 钻孔压力
borehole pressuremeter 钻孔压力仪
borehole pressure recovery test 钻孔压力恢复试验
borehole producing by flow 水流产生的钻孔
borehole profile 钻孔断面图
borehole pump 钻探用泵;钻井泵;深井泵;钻孔泵
borehole record 钻孔记录
borehole rig 钻孔钻塔;钻孔设备
borehole sample 钻孔土样;钻孔样品;钻孔试样;岩芯试样
borehole sampling 钻孔取样
borehole scanner 钻孔扫描器
borehole sealer 封孔器
borehole seismic log(ging) 地震测井
borehole seismic prospecting 井中地震
borehole seismograph 井下地震仪
borehole seismometer 井下地震计
borehole self-potential method 井中自然电场法
borehole shear apparatus 钻孔剪切仪
borehole shear test 钻孔剪切试验
borehole sinking 钻孔【岩】
borehole site 孔位
borehole spacing 钻孔间距
borehole specimen 钻孔土样
borehole springing 炮孔扩底;掏壶;射孔
borehole strain ga(u)ge 钻孔应变计
borehole strainometer 钻孔应变计
borehole structure 钻孔结构
borehole-surface mode 井中—地面工作方式
borehole survey(ing) 钻测;钻探;井下勘探;井下测量;全测;钻孔勘测
borehole surveying instrument 地球物理钻孔探测仪
borehole televiewer 井中电视;井下电视;钻孔电视
borehole television 钻孔电视
borehole television camera 井电视摄影机;钻孔电视摄像机
borehole thermometer 井温仪
borehole throw 钻孔偏向;钻孔偏差
borehole transient electro-magnetic method 井中瞬变电磁法
borehole tube 钻孔套管;套管
borehole TV 钻孔电视
borehole type pump 钻孔式泵
borehole velocity instrument 钻孔流量仪
borehole wall 孔壁;井壁;钻壁
borehole wall consolidating mixture 孔壁固结剂;井壁固结剂
borehole water supply 钻孔给水

borehole well 钻井;深井
borehole yield 钻孔出水量
bore interferometer 孔径干涉仪
Borel covering theorem 波雷尔覆盖定理
Borel function 波雷尔函数
bore liner 井孔套管
Borell unit 波雷尔单位
Borel measurable 波雷尔可测
Borel measurable function 波雷尔可测函数
Borel measure 波雷尔测度
bore log 钻孔柱状图;钻井记录;柱状图(土层)
Borel set 波雷尔集
Borel set family 波雷尔集合族
Borel-Tanner distribution 波雷尔—坦纳分布
bore meal 钻屑;钻孔岩粉;钻粉
bore mud 钻泥;钻井泥浆
bore(of pipe) 管内径;管的内径
bore of water meter 水表口径
bore-out-of-round 孔不圆度
bore pipe 钻管
bore pit 钻探坑;浅井;探井
bore plug 土样;钻探岩芯;钻探土样
bore premature 膛炸
bore profile 钻井剖面
borer 钻心虫;钻孔器;钻机;钻工;蛀心虫;蛀虫;镗孔刀具;镗孔刀;镗床
bore record 钻孔记录
borer hole 蛀孔;虫孔
bore riding pin 膛内保险销
bore rod 钻杆
borer population 虫数
borer-proof plywood 防虫蛀胶合板
bore sample 岩芯
borescope 管道探测镜;校靶镜;内孔检视仪;钻孔检查显示器;光学孔径(测试)仪;管道镜
bore shaped pattern 怒潮形式
boresight 瞄准线;校靶(镜);瞄准点;视轴
boresight adjustment 瞄准线零位调整
boresight camera 瞄准摄影机;平行对准照相机
boresighted radiometer 觇孔辐射计
boresighting 轴线校正;轴线校准;膛内瞄准
boresighting error 瞄视误差
boresighting test 瞄准线检验
boresight slope 瞄准斜率
bore size 内径;孔径
bore spacing 汽缸中心距
bore specimen 钻探岩芯;钻探岩样;钻探土样;土样
bore stress 内径应力
bore stroke ratio 缸径冲程比
bore well 钻井;钻孔【岩】;自流井;凿井
bore with slurry 用泥浆钻进
Borg-Warner overdrive 博格-沃纳超速器
boric acid 硼酸
boric acid buffered potassium iodide 硼酸缓冲溶液
boric acid ester 硼酸酯
boric acid gas 硼酸气体
boric acid lotion 硼酸洗液
boric cement mortar 加硼水泥砂浆
boric fiber reinforced plastics 硼纤维增强塑料
boric filament 硼纤维
borickite 磷钙铁矿
boric ointment 硼酸油膏
boric oxide 氧化硼
boric(oxide) anomaly 硼反常;氧化硼的异常性
boric thermal water 含硼热水
boride 硼化物;硼化合物
boride base cerment 硼化物基金属陶瓷
boride ceramics 硼化物陶瓷
boride cermet 硼金属陶瓷
boride coating 硼化物涂层
borides 硼化物陶瓷材料
boriding 渗硼
borine radical 硼氢基
boring 铰孔;钻凿;钻穴;钻探;钻孔工作;钻孔【岩】;钻进;镗孔
boring accessories 钻探配件;钻探零件;钻探附件
boring accident 钻探事故
boring anchor 钻座
boring and blasting data 钻孔爆破资料;钻孔爆破数据
boring and claming hole trenching machine 钻抓斗式挖掘机
boring and drilling machine 镗钻两用机床

boring and facing machine 镗孔车面机床
boring and high pressure jet grouting pile 钻孔高压旋喷桩
boring and mortising machine 钻孔机;榫眼钻床
boring and turning machine 旋转车床;钻孔车削机床;镗车两用机床
boring and turning mill 镗车两用机床
boring-and-underreaming method 钻孔扩端法
boring apparatus 钻探设备;打眼设备;镗孔刀具
boring bar 钻杆;镗杆
boring barge 钻探平底船
boring bar tool 镗杆刀具
boring bit 活钻头;活钎头;钻头;钻探钻头
boring blade bit 钻探用桨叶式钻头
boring block 钻井装置;钻头(钻孔用)
boring block outfit 钻井台
boring break 钻孔停歇时间
boring breakers 钻屑
boring by percussion 冲击式凿岩(机)
boring by rotation 旋转式钻进
boring cable 钻孔用缆索;钻头索
boring camera 钻孔照相机
boring campaign 钻孔作业活动;钻孔作业周期
boring casing 钻探套管;钻孔套管;套管
boring chisel 钻探钻头;钻探用钻头
boring clam 蛀石(海)虫
boring clamshell 钻孔样品的夹具
boring cock 钻头(钻孔用)
boring condition 钻头情况;钻头条件
boring contract 钻孔合同;钻井合同
boring contractor 钻孔承包人;钻井承包商
boring control 钻孔管理;钻孔控制
boring core 钻探岩样
boring core bit 取岩芯钻头
boring crew 钻孔队员;钻井队
boring cut 镗锯法
boring cutter 钻进切削具;镗刀;实心钻头
boring deflection measurement 钻孔测斜
boring depth 钻孔深度;孔深【岩】;镗孔深度
boring device 钻井装置
boring direction 钻孔方向
boring dusts 钻屑
boring engine 钻探发动机
boring engineering 钻孔工程;钻孔技术
boring equipment 钻孔工具;钻井设备
boring field 钻孔工地;钻孔范围
boring finishing turning tool 镗孔光车刀
boring firm 钻探公司
boring fixture 镗孔夹具
boring for drainage 排水钻孔
boring for oil 石油钻探
boring frame 钻探架;钻架
boring gang 钻孔工作队;钻井队
boring ga(u)ge 钻深计;钻孔量测计;钻探深度计
boring grout 钻孔灌浆(沉井外围)
boring head 金刚石钻头的刀具;钻头;镗刀盘;镗床主轴箱
boring insect 蛀虫
boring installation 钻孔装置;钻探设备;钻机
boring island 钻孔平台;钻台
boring jig 镗孔夹具
boring journal 钻探报表;钻进日志;钻进记录
boring jumbo 钻孔大设备
boring lathe 镗床;镗车两用机床
boring length 钻孔长度
boring line 钻孔用缆索;钻头索
boring location 钻孔位置
boring location survey 钻孔位置测量
boring log 钻探柱状图;钻探框架图;钻探记录;钻孔柱状图;钻孔岩性柱状图;钻孔记录;钻孔地质分层图;钻井记录;钻进记录;测井图
boring machine 巷道掘进机;钻探机;钻孔机;钻机;镗床
boring machine for pile foundation 桩基础钻探机
boring machine operator 镗工
boring machinery 镗孔机械
boring mast 钻孔支架
boring master 钻探技工;钻井队长
boring method 钻孔方法;钻取法
boring mill 镗床
boring motor 钻探电动机
boring mud 钻孔泥浆
boring number 孔号
boring obligation 钻孔职责

boring obstacle 钻孔障碍
boring of cylinder block 缸体镗孔
boring of rail 钢轨钻孔
boring of shaft 井筒钻凿
boring of tunnel 隧道钻探
boring on the rake 在倾斜地上钻孔
boring operation 钻探操作
boring organism 钻孔生物
boring outfit 钻孔工具;钻井设备
boring party 钻孔工作队;钻井队;钻进组;钻进班
boring pattern 钻孔样板;钻孔图(式);钻孔排列形式;钻孔布置形式;炮眼组布置和放炮程序图;炮眼排列
boring performance diagram 钻孔进展图表;钻进率图表
boring pipe 钻管
boring plan 钻孔计划;钻探方案
boring platform 钻台;钻井平台
boring point 钻探地点;钻探工地
boring porosity 钻孔孔隙
boring position 钻孔位置;钻探位置
boring principle 钻进原理
boring profile 钻探剖面图;钻孔纵剖面(图);钻孔断面图
boring program(me) 钻孔程序;钻探计划;钻探方案;钻探程序
boring progress 钻孔进度;钻探进度
boring progress chart 钻孔进度(图)表;钻探进度(图)表
boring pump 钻探用泵
boring range 钻探范围
boring rate 钻孔速率;钻探进度
boring record 钻孔记录
boring record sheet 钻孔记录表
boring repair shop 钻机修理车间
boring requirement 钻机要求
boring resistance 钻孔阻力;钻探阻力
boring result 钻孔结果;钻探结果
boring rig 钻架;钻探设备;钻探机(具);钻孔机;钻井设备;钻机;钻车
boring rod 钻探杆;钻杆;镗杆
boring rod clamp (钻探用的)管钳;钻杆夹
boring rope 钻孔绳索;钻头索;钻孔钢绳
borings 金属切屑;钻屑;镗屑
boring sample 取岩样;岩芯样(品);岩芯试样;岩芯取样;钻探样品;钻探试样;钻探岩样;钻孔土样
boring sampling 钻孔取样
boring scheme 钻探图
boring shaft 钻井
boring shift 钻孔轮班
boring ship 钻探船
boring shot-holes in the face 正面钻炮眼
boring site 钻探场地;钻探工地
boring sludges 钻屑
boring spacing 钻孔间距
boring spindle 镗杆
boring sponge 穿孔海绵
boring stay 镗杆支柱
boring superintendent 钻孔监工员
boring system 钻孔方法
boring table 镗床工作台
boring taper 镗孔锥度
boring team 钻孔队;钻井队
boring template 枕木钻孔样板
boring test 钻孔试验;钻孔调查
boring time 钻孔时间;钻孔历时;钻进时间
boring time break 钻孔间歇时间
boring tool 钻探工具;镗孔车刀;镗刀
boring tool joint 钻具接头
boring tower 钻探塔架;钻塔;钻井架;钻塔架
boring tripod 钻井三脚架;钻探平台;钻架
boring tube 套管
boring turning tool 镗孔车刀
boring unit 钻井装置
boring vessel 钻孔船
boring well 钻深井;钻井
boring winch 钻机绞车
boring with line 钢绳冲击钻进;索钻
boring work 钻孔工作;镗孔工作
borishanskiite 铅砷钯矿
borism 硼中毒
borizing (金刚石的)镗孔
Borland type fish pass 鱼闸;闸式鱼道
Born approximation 玻恩近似

born city 新兴城市
bornelone 波尼酮
bornemanite 磷硅铌钠钡石
borne nut 齿形螺母
borneol 冰片
borneol flake 冰片
Born equation 玻恩方程式
borne trade 海上运输
Born-Haber cycle 玻恩—哈伯循环
bornhardt (干旱带的)岛山
bornhardtite 方硒钴矿
borning 用测平杆测定孔洞的标高;对某一表面测其水准面
bornite 斑铜矿
bornite ore 斑铜矿矿石
Born-Madelung model 玻恩—马德隆模型
born manager 天生的经理人员
Born-Mayer equation 玻恩—梅耶方程
Born-Oppen-heimer approximation 玻恩—奥本海默近似
Born-Oppenheimer method 玻恩—奥本海默方法
Born-von Karman theory 玻恩—冯卡曼理论
bornyl 茨醇基
bornyl acetate 醋酸冰片酯
bornylane 冰片烷
borocalcite 硼酸方解石;硼钙石
borocarbon resister 硼碳电阻器
boro-carbon resistor 碳硼电阻
boroferrite 鲍硼铁矿
borogen 硼酸乙酯
Borohoro anticlinorium belt 婆罗科努复背斜带
Borohoro tectonic segment 婆罗科努构造段
borol 硼硫酸钠
borolanite 霞榴正长岩
boroll 极地软土
Borolon 合成氧化铝
boron alloy 硼合金
boron-bearing skarn deposit 矽卡岩硼矿床
boron capsule diffusion 箱式硼扩散
boron carbide 碳化硼
boron cast-iron 含硼铸铁
boron chamber 硼电离室
boron-coated 涂硼的
boron containing aggregate 含硼集料;含硼骨料
boron containing cement 含硼水泥
boron diffusion 硼扩散
boron-doped diamond electrode 掺杂硼金刚石电极
boron doped silica 掺硼二氧化硅
boron-doping 掺硼
boron-dopped layer 硼掺杂层
boron-epoxy composite 硼纤维增强环氧复合材料
boron fertilizer 硼肥
boron fiber 硼纤维
boron fiber reinforced plastics 硼纤维增强塑料
boron filament 硼纤维
boron filament reinforced alumin(i)um 硼纤维增强铝
boron fluoride 氟化硼
boron-free glass 无硼玻璃
boron-free glaze 无硼釉
boron frit 硼玻璃料
boron glass 硼玻璃
boron group elements 硼族元素
boron hydride 氢化硼;硼氢化合物
boron hydride poisoning 硼氢化合物中毒
boron hydroxide 氢氧化硼
boron-iodine water 硼碘水
boronizing 渗硼
boron-lined 衬硼的
boron-lined counter 衬硼计数管
boron loaded 涂硼的
boron-loaded concrete 加硼密实混凝土;硼化密实混凝土;含硼集料混凝土;含硼(骨料)混凝土
boron nitride 氮化硼
boron nitride ceramics 氮化硼陶瓷
boron nitride coating 氮化硼涂层
boron nitride fiber 氮化硼纤维
boron nitride filament yarn 氮化硼长丝
boron nitride grease 氮化硼油脂
boron-nitrogen polymer 硼氮聚合物
boron ores 硼矿
boron oxide 氧化硼;二氧化硼

boron-phenolic resin 硼酚醛树脂
boron phenyl difluoride 苯硼化二氟
boron pollution 硼污染
boron polymer 硼聚合物
boron resin 硼树脂
boron silicate 硼硅酸盐
boron steel 硼钢
boron triethoxide 硼酸乙酯
boron trifluoride 氟化硼;三氟化硼
boron trifluoride etherate 醚合三氟化硼
boron water 含硼水;硼水
boron wool 硼棉
borophenylic acid 苯硼酸
borophosphate 硼磷酸盐
boroscope 内径表面检查仪
borosilicate 硅酸硼;硼硅酸盐
borosilicate crown 硼硅酸盐铬黄;硼硅铬黄
borosilicate crown glass 硼硅酸盐冕玻璃
borosilicate glass 硅酸硼(冕牌)玻璃;光学玻璃;硼硅酸盐玻璃;硼硅玻璃
borosiliconizing 渗硼渗硅处理;渗硼硅
borough 自治市;自治村镇(英国)
borovskite 亮碲锑钯矿
boroxane 硼氧烷
Borrmann anomalous-transmission technique 博曼反常透射技术
Borrmann effect 博曼效应
Borros pole 波罗杆
borrow 借用;借位数;贷;采料;取土;取料坑
borrow and loan contract 借贷合同
borrow area 采料场;取土面积;取土坑;取土场;取土区
borrow bank 挖土边坡;借土边坡
borrow by check 支票借款
borrow cut 借土方方;取土开挖
borrow digit 借位数
borrow ditch 借土沟
borrow earth 借土
borrowed capital 借入资金;借入资本;借贷资本
borrowed current fund 借入流动资金
borrowed fill 借土填方
borrowed fund 贷款资金;借入资金;借入资本
borrowed light 借光窗;间接采光(窗);间接采光(灯);室内窗
borrowed material 移用的材料;采料
borrowed money 借入资金;借入资本
borrowed money accounts 借入资金账户;借款账户
borrowed reserve 借入准备金
borrowed scenery 借景
borrowed security 借入有价证券
borrowed share 借入股票
borrowed stock 借入股票
borrowed view 借景
borrower 债户;借方;借款人;借款单位
borrower's bank 借贷银行;贷款银行
borrower's risk 借款人风险
borrow excavation 借土开挖;取土开挖
borrow exploration 料场勘探
borrow fill 借土填方
borrow fill material 借土
borrow generating device 借位发生器
borrow ideas from 借鉴
borrowing arrangements 借款安排
borrowing authority 借款授权
borrowing cost 借款费用;借款成本
borrowing demand 借款需求;借款限额
borrowing from affiliate 从联属企业借款
borrowing from the public 从社会借款
borrowing in advance 预借
borrowing-lending 借—贷款
borrowing limit 借款限额
borrowing needs 贷款要求
borrowing on non-concessional terms 非减让性借款
borrowing plan 贷款计划
borrowing potential (power) 借款能力
borrowing power 借款权限;举债权
borrowing rate 借贷利率
borrowing requirement 借款条件
borrowing short to lend 空头借入
borrowing space 借景
borrowing time 借用时间
borrow light 借光窗;间接采光窗

borrow material 借土土方；取用土料
borrow money 借债；借款
borrow(money) form 从……处借款
borrow money on 押借
borrow money on security 押款
borrow offer 借发价；借报价
borrow pit 采料场；采料坑；取土坑
borrow place 借位
borrow site 借土地点；借土场地；采料场；取土地点；取土场地
borrow soil 借土；取土；外来土
borrow survey 料场勘探
borrow without security 无抵押借款
borrow yardage 借土方数；取土方数
borsal 硼硅酸钠
borsing up 安装挑檐饰线
borsyl 硼硅酸钠
Bortala tectonic knot 博尔塔拉构造结
bortam 硅硼钛铝锰合金
borthiin 环硼硫酸
bort(z) 金刚石粉粒；金刚石粒；圆钻金刚石
bort(z) bit 低质金刚石钻头；金刚(钻)钻头；金刚石钻头
bort(z) powder 细粒金刚石粉
bort(z)-set bit 镶金刚石钻头；细粒金刚石钻头
borvarite 碳镁钙石
borway bit 齿状取芯钻头
boryslowite 硬地蜡
Bosanquet and Pearson's diffusion formula 博桑基特—皮尔桑扩散公式
Bosanquet's law 博桑基特定律
boscage 灌木丛；树丛
Bosch fuel injection pump 博施柴油喷射泵
Bosch oil lubricator 博施给油器
Bosch process 博施法
Bose-Einstein condensation 玻色—爱因斯坦凝聚
Bose-Einstein distribution 玻色—爱因斯坦分布
Bose-Einstein distribution law 玻色—爱因斯坦分布律
Bose-Einstein nuclei 玻色—爱因斯坦核
Bose-Einstein statistics 玻色—爱因斯坦统计(法)
Bose gas 玻色气体
Bose-Kishen square 玻色—基森方
bosh 炉腰；炉膛高温带；炉腹；冷却用水槽；冷却水槽；浸冷水槽；附着石英；水刷
bosh angle 炉腹角
Boshan ware 博山窑器
bosh area 炉腹区
bosh band 炉腹钢带
bosh break-out 炉腹破裂；炉腰烧穿
bosh brick 炉膛砖；高炉炉腹砖
bosh casing 炉腹外壳；炉腰外壳
bosh cooler 炉腹冷却装置
bosh cooling box 炉腹冷却箱；炉腹冷却器；炉腰冷却器
bosh cooling plate 炉腹冷却板
boshing 浸水冷却；浸水除鳞；浸冷
bosh jacket 炉腹外套；炉腹外壳
bosh line 炉腹水平线
boshplate 炉腹冷却板；炉腰冷却板
boshplate box 炉腹冷却箱
bosic chromatin 碱性染质
bosk(et) 矮林；小丛林；矮丛；树丛
bosom 角撑；角板；木肘材；中间；对缝连接角钢；衬托面；拖网中段
bosom bar 衬角钢
bosom knee 木肘材
bosom of angle bar 角铁内表面
bosom piece 衬角钢；衬角材；角撑；对缝连接角钢
boson 玻色子
boson fluctuation 玻色子波动
bosporus 博斯普鲁斯海峡
bosquet 滩地林；小丛林；树丛
boss 螺旋桨毂；经理；灰炭桶；岩瘤；岩窟；轴节；止挡；雕花垂饰；浮凸饰；凸台；凸饰；四角螺丝套
boss age 浮雕装饰；毛面浮雕；雕花垂饰
bossage without bevel 斜削浮雕装饰【建】
boss bearing of pulley 滑轮毂轴承
boss bolt 轮壳螺栓；轮壳螺栓
boss branch 三承四通；三承十字管
bossed 有浮雕装饰的；凸起的
bossed frame 包毂肋骨
bossed(on) both sides 双面浮雕饰；(有浮起装饰的)两侧凸起的
bossed(on) one side 单面浮雕饰；(有浮起装饰的)一侧凸起的
bossed ornament 凸饰
bosselated 有圆凸的
bosselation 小圆凸
boss end bracket 轴毂端支架
boss flange 凸法兰
boss hammer 大锤；碎石石工；碎石大锤；碎石锤；锻工锤
Bosshardt-Zeiss reducing tacheometer 博斯哈德—蔡思双影测速仪；双像视距仪
bossing 凸起部(分)；刻痕和堆焊；金属片加工；节疤；厚层砌底；轴包套；导流罩；锤碾金属；船尾管鼓出部；清除刷花痕迹；尾锚孔部
bossing mallet 金属片加工槌
bossing stick 加工木框；加工木槌
bossi work 大理石镶嵌技术
bosslike 穹隆状；穹状
bossmanship 企业精神
boss of quadrant 舵扇毂
boss of tiller 舵柄毂
boss plate 包毂板
boss ratio 内外径比；辐比
boss ring 轮毂箍；壳箍；毂箍
boss rod 主机轴
boss strike plate 门锁舌片
boss thickness 轮毂厚度
bossy 凸起的
boster 熟铁板
Bostock sedimentation balance 博斯托克沉降天平
Boston blue clay 波士顿蓝黏土
Boston caisson 凿井沉箱；管柱沉井施工法；波士顿沉箱
Boston hip 波士顿屋脊；波纹脊饰
Boston hip roof 波士顿式四坡屋顶
bostonite 淡歪细晶岩
Boston ivy 爬墙虎(地棉)；爬山虎【植】
Boston lap 波士顿式搭接
Boston level(l)ing rod 波士顿水准尺
Boston ridge 波纹脊饰；波士顿式屋脊；波士顿脊
Boston rocker 讲究的摇椅
Boston sash fast 波士顿窗扣锁
Boston stone 波士顿石磨
bostryx 螺状聚伞花序
bostwickite 多水硅钙锰石
bosum 座板升降结
bot 泥塞
botallackite 斜氯铜矿
Botallo's duct 博塔洛管
Botallo's foramen 卵圆孔；博塔洛孔
botanical garden 植物园
botanical insecticide 植物杀虫剂
botanical museum 植物博物馆
botanical name 植物名称
botanical park 植物园
Botanical Society of the British Isles 英国植物学会
botanical variety 形态变种；植物变种
botanical zone 植物带
botanist 植物学家
botanizing 植物调查
botanogeochemical sample 植物地球化学样品
botanogeochemistry 植物地球化学
botany 植物学
Botany Bay 波坦尼湾
botel 凸形圆饰；水上旅馆
both 双方(当事人)
both cauce 双层河(喀斯特)
both contracting parties 缔约双方
both days inclusive 首尾两天均包在内
both ends 两端；装到两港；起运港与目的港
both ends supported beam 两端支承梁
both end threaded 两端带螺纹的
both faces 两面
both inclined cross 二者斜交
both mature and young soils 成熟土和幼年土
both parties 两造；双方(当事人)
both perpendicular 相互垂直
both preference and ordinary shares 优先股和普通股
both principal and interest 本利合计
both regular and immediate command 常规和立即命令
bothridium 裂片
both sewer system 分流制下水道系统；分流下水道系统
both sideband 双边带
both side cargo handling 两舷装卸
both sides 两舷；两面；两侧；两边
both sides clear 借贷结清；银货两讫
both sides welding 双面焊(接)
both to blame collision clauses 双方有责任碰撞条款；双方过失碰撞条款；碰撞双方均有过失条款
both vendor and purchaser 买卖双方
both way 双向
both-way communication 双向通信
both way signal 双向信号
both-way trunk line 双向中继线；双向干线
bothy 独间小屋；棚屋
bot pick 钎子
botryite 赤铁矿
Botrylloides 拟菊海鞘(属)
Botryllus 菊海鞘属
Botryococus braunii 丛粒藻
botryogen 赤铁矾
botryoidal nodule 葡萄状结核
botryoidal structure 葡萄状结构
bott 泥塞；堵塞
botter 灌注机
Bottger ware 深红色炻器
botting 黏土泥塞；堵出铁口
botting clay 一种高塑性黏土；高塑性黏土
bottle 瓶子；瓶(形物)；漂瓶
bottle air 瓶装气体
bottle bank cullet 回收碎玻璃
bottle blowing machine 吹瓶机
bottle brick 单排圆孔空心砖
bottle brush 洗瓶刷
bottle cap 瓶盖
bottle capper 封瓶机；瓶盖机
bottle cellar 藏瓶地窖；瓶地窖
bottle cement 瓶装水泥
bottle chart 瓶测洋流海图
bottle coal 气煤
bottle conveyer 瓶子运输器
bottle cooler 瓶用冷却柜
bottle cork 软木塞
bottle cullet 瓶罐碎玻璃
bottled 瓶装的
bottled cement 压力瓶装水泥；压力罐装水泥
bottled gas 瓶装煤气；瓶装液化(石油)气
bottled water 瓶装水
bottle filling machine 装瓶机；灌瓶机
bottle filter 瓶式过滤器
bottle gas 瓶装液体煤气
bottle glass 制瓶玻璃；瓶罐玻璃
bottle gourd 葫芦
bottle gourd carving 葫芦雕刻品
bottle graft 瓶接
bottle grafting 插瓶靠接
bottle green 瓶绿色；深绿色
bottle holder 瓶式储气罐
bottle jack 螺旋举升器；瓶状千斤顶；瓶式千斤顶；瓶颈式千斤顶
bottle kiln 间歇操作立窑；瓶状竖窑；瓶形竖窑；瓶式竖窑
bottle(-making) machine 制瓶机
bottle method (测定液体和集料比重的)比重瓶法
bottle mixer 瓶形搅拌机
bottle-mouth 瓶口
bottleneck 瓶颈(口)；局部狭窄河段；局部束窄河段；涌塞；隘道；缩颈处
bottleneck analysis 隘路分析；瓶颈分析
bottleneck assignment problem 阻塞分派问题
bottleneck commodity 稀缺商品
bottleneck effect 瓶颈作用
bottle neck hitch 瓶口结
bottleneck inflation 瓶颈式通货膨胀
bottleneck of river 河道狭窄
bottleneck operation 生产薄弱环节
bottleneck path 隘径
bottleneck problem 薄弱环节问题(瓶颈问题)
bottleneck reach 卡口河段
bottleneck road 局部狭窄路段；狭路；瓶颈(式)道路
bottleneck section 咽喉区
bottleneck state 瓶颈状态
bottleneck traffic 交通易阻塞的狭口
bottleneck valley 瓶颈谷

bottle-nose curb 圆边踏步级；圆边滴水槽；铅皮屋顶沿边滴水槽
bottle-nose drip 瓶鼻状滴水槽；圆边滴水槽；屋檐（瓶口状）滴水
bottlenose oil 北极鲸油；槌鲸油
bottle-nose step 圆边踏步级
bottle-nose whale oil 槌鲸蜡油
bottle nosing 半突缘
bottle off 装瓶(在桶内)
bottle opener 启瓶器
bottle oven 瓶式窑
bottle oxygen 瓶装氧气
bottle packing paper 包瓶纸
bottle paper 漂瓶资料
bottle point experiment 瓶点试验
bottle point method 瓶点法
bottle post 漂流瓶
bottle pump 玻璃水银扩散泵
bottler 灌瓶器
bottle rinse water 洗瓶水
bottle rinsing machine 玻瓶冲洗机
bottle robbin winding machine 红木锭摇纱机
bottle sampler 瓶式采样器
bottle screw 螺旋起重器；螺旋起重机；松紧螺丝
bottle-shaped pillar 瓶形柱墩
bottle-shape filter 瓶式过滤器
bottle shop 灌瓶站；灌瓶车间；灌瓶厂
bottle silt sampler 瓶式淤泥取样器
bottle stone 暗绿玻璃
bottle tank 瓶罐玻璃池窑
bottle tight 密封的；束紧如瓶
bottle-tight joint 瓶密接头
bottle trap 直签式存水弯；直管存水弯；瓶形存水弯；瓶式曲管
bottle truck 瓶子卡车
bottle type suspension load sampler 瓶式悬移质取样器
bottle up 封锁
bottle valve 瓶头阀
bottle vendor 瓶用冷却售货柜
bottle washer 洗瓶机
bottle-washing machine 洗瓶机
bottle-washing plant 洗瓶机
bottle-washing waste 洗瓶废水
bottle-washing water 洗瓶水
bottle wrapping paper 包瓶纸
bottling 灌瓶；瓶装(的)；缩颈
bottling and soft drink plant 瓶装和软饮料厂
bottling area 装瓶区
bottling clay 堵瓶出铁口用黏土
Bottlinger diagram 玻特林格图
Bottlinger model 玻特林格模型
bottling hall 装瓶间
bottling installation 装瓶设备
bottling machine 装瓶机
bottom 煤层底部；下皮；下部；底音；底脚；底(床)；此端向下；船底
bottom accretion 河底淤高
bottom active stream sediment 河底活性沉积物
bottom adapter 底部插头
bottom anchor 下锚
bottom angle 齿夹角；齿根角
bottom anoxia 底层水缺氧
bottom anti-corrosive paint 船底防锈漆
bottom anvil 底砧
bottom area 底层区
bottom area of pit of water infiltration test 渗水试坑底面积
bottom arm 门系闭器连杆；下电极臂
bottom articulated joint 底部铰接缝
bottom ash 炉底灰；底灰
bottom ballast 底渣
bottom bank 浅滩；沙洲
bottom bar 下鼠笼条；底杆；底层钢筋
bottom beam 窑底托梁；窑底梁；底梁
bottom bed 底板岩层；下层；底层；台面
bottom bed load 推移质
bottom bell 底料钟
bottom belt 炉底生物带
bottom bench 下部台阶
bottom berth level 泊位底标高
bottom bidding price 招标底价
bottom biocenoses 底栖生物群丛
bottom blast forge 底吹锻铁炉

bottom block 带钩滑轮；窑底砖；池底大砖
bottom blowing 底风；底吹
bottom blown converter 底吹转炉
bottom blown oxygen process 底吹法
bottom blow valve 底部防水阀(锅炉的)
bottom blug 底塞
bottom board 活动底板；载型板；底板；船底护板；砂箱底板
bottom boiler connection 锅炉下接头
bottom bolt 门底闩；底插销；门底栓；门底插销
bottom boom 下弦
bottom boom bar 下弦杆；下弦杆钢条
bottom boom junction plate 下弦杆连接板；下弦节点连接板
bottom boom longitudinal bar 下弦纵向钢条；下弦纵杆
bottom boom member 起重臂的下弦杆；下弦杆
bottom boom rod 下弦杆
bottom border 下图廓；图廓下边
bottom bottle 底层浮标瓶
bottom bounce 海底反射
bottom-bounce sonar 海底反射声呐
bottom bracing 底撑
bottom bracket 下托架；底托架
bottom break 底部断裂
bottom brick 炉底砖；底砖
bottom bullnose 下部喷管；底部喷射口
bottom burden 底部抵抗线
bottom bush 底衬
bottom cam box assembly 下三角座箱
bottom cam box plate 下三角座底板
bottom cap 底盖
bottom car clearance 厢底净空(电梯)
bottom cargo 压舱物；压舱货(物)；垫底(货)；底货(载)；底舱货；舱底货物
bottom case 底箱；底座；底壳
bottom casing 底座箱
bottom casting 下铸(法)；底注(铸造)
bottom ceiling 船底衬板；舱底木板质铺板
bottom cement 船底混凝土
bottom center-fired pit 中心燃烧式均热炉
bottom chain 地链
bottom chain sprocket 下部链轮
bottom channel 底沟；隧道废沟
bottom characteristic 河床特征；河床特性；水底特性
bottom charge 钻孔最深处的炸药；底部装药；底部填药料；孔底装药
bottom-charting fathometer 海底地形测绘仪
bottom chart recorder 海底地貌记录仪
bottom check 底部裂缝
bottom chock 下轴承座
bottom chord 下弦；底弦
bottom chord bar 下弦钢条；下弦杆
bottom chord junction bar 下弦杆连接板
bottom chord longitudinal bar 下弦杆纵向钢条
bottom chord member 下弦杆
bottom chord stress 下弦应力
bottom chord wind bracing 下弦风撑
bottom clack 进油活瓣；进水活瓣
bottom clamping plate 底部压紧板
bottom clay 底部粘土；淤泥
bottom cleaning 河底清理；敲铲船底
bottom cleaning jet auger 喷水式清底螺旋土样钻
bottom cleaning plug 底部清除塞
bottom clearance 间隙(齿根与齿顶)；底部净距
bottom cloth 衬垫织物
bottom coat 底基层；底层；底涂层；底面涂层；船底涂层
bottom coating 底涂
bottom collector 塔底集液管
bottom community 水底群落
bottom compartment 底室
bottom composition 船底涂料
bottom composition B 二号船底漆
bottom concrete 船底混凝土
bottom condition 底部状况；底部条件
bottom configuration 海底地形；底形；底部形状；底形地形；水底地形
bottom construction 炉底结构
bottom contact platform 海底钻探平台；海底钻井平台
bottom contour 底部轮廓；湖底等高线；河底等高线；海底形状；海底地形；底部地形；海底等高线；等水深线；等深线；水底地形；水底等深线

bottom contour chart 海底地形图；海底地貌图；海底等深图
bottom contour map 海底地貌图
bottom contraction 堰底收缩；底部约束；底部缩窄；底部收缩
bottom cooling 局部冷却；窑底冷却
bottom coring 海底取岩心
bottom course 基层；底层
bottom course concrete 底层混凝土
bottom course structure 底层结构
bottom covering 底涂层；底面涂层
bottom crack inducer 底部引裂条(混凝土道路接缝处)
bottom cradle 下摇架
bottom-crawling submersible vehicle 海底履带爬行型潜水器
bottom-crawling vehicle 海底爬行潜水器；海底调车
bottom crest 海底峰
bottom culture 海底养殖；底层养殖
bottom cupola 化铁炉(底)
bottom current 底部流水；海底流；底层流；潜流
bottom current direction 底流方向
bottom current direction meter 底流方向仪；底流方向计
bottom current meter 底流仪；底层流速计
bottom cut 拉底；下部淘空；底槽；底部淘刷；掏槽；底板切割；下部掏槽
bottom cutter 根部切割器
bottom cutting 底部截槽；掏底槽
bottom damage 船底破损
bottom dead center 下死点；下止点；外死点
bottom dead-center indicator 下止点指示器
bottom dead point 下止点
bottom deck 底层甲板；底甲板
bottom definition 确定水底地形
bottom deposit 库底泥沙；库底沉积物；海底淀积(物)；海底沉积(物)；底部沉积
bottom deposit sampler 底部沉淀取样器
bottom depth 基底深度
bottom depth of observation tunnel 观测平硐底深
bottom depth of producing horizon 水泥塞底深度；生产层底深
bottom development road 底下的开发路面
bottom diameter 底径
bottom diameter of thread 螺纹底径
bottom die 下模具；底模
bottom discharge 底部排水；底部排泥；活底卸料；河底排放；下出料；底部卸载；底部卸料；底部排泄
bottom-discharge bit 底冲式钻头；唇面排水眼钻头；唇面排水孔钻头
bottom-discharge bit blank 底冲式原钻头；唇面排水眼原钻头；唇面排水孔原钻头
bottom-discharge bucket 底卸式漏斗；底卸式料斗；底卸式吊桶
bottom-discharge core barrel 连接底冲式钻头的岩芯管
bottom-discharge door 底卸活门；开启卸泥门；底卸门
bottom-discharge gate 底卸式闸门
bottom-discharge of sediment 底部排沙
bottom-discharge of water 底部排水
bottom-discharge pipe 底部流出的管子；泄水底管
bottom-discharge scow 底部倾卸的驳船；底部倾卸的平底船；底卸式泥驳
bottom-discharge semitrailer 底部倾卸的半挂车；底卸式半挂车
bottom-discharge skip 底卸式箕斗
bottom-discharge tractor 卸料车
bottom-discharge tractor-trailer 底部倾卸的拖挂车
bottom-discharge tractor-truck 底部倾卸的货车；底部倾卸的拖拉卡车
bottom-discharge tunnel 底部泄水隧洞
bottom-discharge type 底卸式；底冲式
bottom-discharge valve 底部冲洗阀；底部排料阀
bottom-discharge wagon 底部卸出的货车；活底卸料车
bottom dog 底钩
bottom dome 钟形底脚；穹顶底脚
bottom door 下门；清扫孔；底排水孔；泥门；底部排水孔
bottom door control device 泥门启闭装置
bottom door discharge 船底泥门排泥

bottom door rail 下冒头；下横挡
bottom door type of hopper barge 开底泥驳
bottom double seaming machine 罐底双重封缝机
bottom drag 河底曳引力；河底拖力；河底推移力；底部阻力；船底阻力
bottom drainage 底部排水
bottom drainage system 底部排水系统
bottom drained drydock 底排水式干船坞
bottom drain valve 底部排水阀
bottom draw cut 底部掏槽
bottom drier 底干料；底催干剂
bottom drift 底部横坑道；海底漂移；下导坑；下部导坑；下部半断面；底部导洞
bottom-drift and ring-cut method 下导坑核心支承开挖法
bottom-drift excavation method 漏斗棚架(式隧道开挖)法；下导坑先墙后拱法；下导坑漏斗棚架法
bottom-drift method 下导坑超前先拱后强法；下导洞推进法
bottom drive 井底驱动
bottom drive shaft 底部传动轴
bottom drop car 底开门车
bottom dropped out of 急剧跌价
bottom drying 底干
bottom-dump 底卸车；车底卸载，底卸式；底冲式
bottom-dump barge 底部倾卸的平底船；开底泥驳
bottom-dump bucket 活底卸料斗；底卸料斗；活底铲斗；底卸式料斗；底卸式铲斗；底部卸料斗
bottom-dump car 活底卸车；底卸车；底部卸料车
bottom-dump concrete bucket 底卸混凝土罐
bottom-dump door 底卸门
bottom dumper 底卸运土车；底卸式自卸汽车
bottom-dump haul(i)er 底卸式运输车；活底自动卸车；底卸式拖运机；底卸拖运车
bottom-dump hauling and spreading trailer 底部拖运和洒布挂车
bottom-dump hauling trailer 底部倾卸的拖运拖车
bottom-dumping 底部卸料；底部倾卸；车底卸载
bottom-dumping gismo 底卸式吉斯莫万能采掘机
bottom-dumping hopper body truck 底卸料斗式运货汽车
bottom-dumping muck-car 底部倾卸的弃土车；底卸式翻斗车
bottom-dump scow 底部倾卸的平底船；开底泥驳
bottom-dump semitrailer 底部倾卸的半挂车；底卸式半挂车
bottom-dump skip 底卸式箕斗
bottom-dump tipper 底部可拆开的自卸卡车
bottom-dump tractor-trailer 底卸拖拉机拖车
bottom-dump trailer 底部卸料拖车；底卸式自卸挂车；底部卸料拖车
bottom-dump truck 底卸式运货汽车；底卸车；底卸(式自卸卡)车；底部卸料车；开底车；底卸卡车
bottom-dump unit 底部倾卸装置
bottom-dump wagon 底部卸料车；活底卸车；活底料车；底卸车
bottom dweller 海底栖居者
bottom-dwelling organism 水底生物；底栖生物
bottom dyeing 底色染色；底色染色
bottom echo 底回波；反面回声
bottomed at N meters 钻到 N 米深
bottom edge 底边；南图廓；下缘
bottomed region 通导区
bottom effect 河底效应
bottom ejector plate 喷射底板
bottom elevation 底面标高；底标高
bottom elevation of borehole 孔底标高
bottom elevator car run-by 升降机轿厢底部超跑距离
bottom-emptying 底卸式；底冲式
bottom-emptying gallery 底部泄水坑道；底泄水廊道；闸门泄水管；闸底泄水管；底部泄水廊道
bottom-emptying skip 底卸式箕斗
bottom emptying wagon 活底料车；底卸式料车；底卸车；底开式料车
bottom enamel 底漆
bottom end 原木头；底端(桶)
bottom end cover 下端端
bottom end of stroke 行程的下死点
bottom end rail 下端梁(集装箱)
bottom end transverse member 下端梁(集装箱)
bottom entering backswept agitator 底伸式后掠形搅拌器
bottom entering mixers agitator 倒置式底面混合搅拌器
bottom entering type agitator 底伸式搅拌器
bottom entry 下端插入
bottom-environmental sensing system 水底环境传感系统；海底环境传感系统
bottom equipment 孔内设备(过滤器，深井泵等)；井内设备(过滤器，深井泵等)
bottom exhaust duct 底部废水管道；底部废气管道；地下排气管道
bottom face 孔底
bottom facing machine 光面机(底部转动)
bottom failure 海底失稳
bottom fastening bolt 底螺栓
bottom fauna 海底动物区系；底栖动物群；底栖动物区系；船底寄生物
bottom-fed evapo(u)rator 底部供液的蒸发器
bottom feed 底部喂料；底部加料
bottom feeder 食底泥动物
bottom feed waste oiler 底给棉纱头加油器
bottom feed wick oiler 底吸油芯加油器
bottom fermentation 底发酵；底层发酵
bottom fiber 底部纤维
bottom filling ewer 倒灌壶
bottom filling system 闸底灌水系统；底部充水系统(船闸)
bottom finder 测深仪
bottom fired heater 底烧炉
bottom fitting 底部配件
bottom flange 梁下翼(缘)；下翼缘
bottom flange bar 底部翼缘钢条
bottom flange junction plate 底部凸缘接合板；底部翼缘接合板
bottom flange plate 底翼缘板；凸缘底板
bottom flap 活底吊门；折叠底(板)
bottom flash 底飞边
bottom flood 河床底水流；底层水流
bottom floor 底板
bottom flow 回头浪；回卷流；回卷浪；底流；底层流
bottom flowing pressure 井底流动压力
bottom flue 小烟道；底烟道
bottom flush 底部齐平面
bottom flushing of sediment 底部排沙
bottom formwork 底模板
bottom frame 井底井框；船底肋骨
bottom framing 船底构架
bottom fresh air duct 底下的新鲜空气管道；地下净气管道
bottom friction 底摩擦；底部摩阻力
bottom friction value 底摩阻值；底摩擦值
bottom gate 底下的门；底注式浇注系统；底注式浇口；底孔闸门；底浇口；底部闸门
bottom gate leaf 底下护板
bottom gate skip 活底铲斗；底闸门活底铲斗
bottom gating 底注式浇注系统
bottom gear 一挡齿轮；低速排挡；低速挡；头挡
bottom girder 底梁；底板材；底大梁
bottom glade 河滩沼地；河谷；谷地；谷底；低谷
bottom glass liquid 池底玻璃液
bottom glazing flashing 天窗防水压条(瓦楞石棉板屋面)；底部玻璃泛水
bottom grab 啮合采泥器；海底挖式取土器；咬合采样器；蚌式采样器；水底挖泥抓斗；挖泥抓斗
bottom grade 河底坡度；河底比降
bottom grading 河底整平
bottom grass 底层草
bottom gravimeter 海底重力仪
bottom grazing 底层放牧
bottom grid 热绝缘底槽板
bottom ground 海底
bottom guide 底部导轨；底部导向装置
bottom guide bearing 下导轴承
bottom guide track 底部导向装置；底部导轨
bottom guide vane 底导板
bottom half 箱体底部；下半箱体
bottom head 底封头；底盘；底盖
bottom header pipe 闸首底部放水管；闸门底部放水管
bottom heading 下坑道；下部超前工作面；底层坑道(法)；底导洞(法)；底部导洞
bottom heading method 下导坑超前先拱后强法
bottom heading-over-head bench 全断面分块开挖法
bottom heat 坩埚底热量
bottom heating 底部预热
bottom heave 坑底隆起；基底隆起；隆起；底部隆起
bottom heavy 初稳心太高
bottom-hinged 底铰的；底部铰接的
bottom-hinged box type flap gate 卧式坞门
bottom-hinged inswinging window 底悬内开翻窗
bottom-hinged sash 底部铰接窗扇
bottom-hinged sash window 下部铰接的框格窗
bottom-hinged ventilator 底部铰接气窗
bottom-hinged window 下悬窗
bottom hold 底舱；船底层舱
bottom hold down 底部固定
bottom hole 孔底；井底；底孔
bottom-hole accident time 井下事故时间
bottom-hole choke 底孔管嘴；井底油嘴；井底阻流器
bottom-hole circulating pressure 井底循环压力
bottom-hole circulating temperature 井底循环温度
bottom-hole coverage 井底覆盖；钻头与孔底接触面积
bottom-hole drilling tools 潜孔钻具
bottom-hole flow bean 井底喷嘴
bottom-hole flow meter 井下流量计
bottom-hole orientation 孔底定向
bottom-hole packer 井底封隔器
bottom-hole partial jet reverse circulation core tool 喷射式孔底反循环取芯钻具
bottom-hole plug packer 塞式井底封隔器
bottom-hole pressure 井底压力；孔底压力
bottom-hole pressure bomb 孔底压力计
bottom-hole pressure indicator 井底压力指示器
bottom-hole pump 井下泵；井底沉没式泵；潜水泵
bottom-hole sample 孔底取样；井底样(品)
bottom-hole sampler 井下取样器；井底取样器；孔底取样器
bottom-hole sample taker 孔底取样器
bottom-hole sampling 取井底样
bottom-hole scavenging 井底清洗
bottom-hole separator 孔底分离器；井底油气分离器
bottom-hole spacing 井底距离；孔底间距
bottom-hole static pressure 井底静止压力
bottom-hole static temperature 井底静止温度
bottom-hole temperature 井底温度；孔底温度
bottom-hole temperature bomb 孔底温度计
bottom-hole temperature ga(u)ge 孔底压力计；孔底温度计
bottom-hole temperature recorder 井下温度记录仪
bottom-hole trouble 井底事故
bottom hopper barge 底卸式船；底开式驳船；开底泥驳
bottom horizontal departure 孔底水平位移；井底水平位移
bottom hung 下悬式(窗)；下悬；底旋窗
bottom hung projecting window 下铰的凸出窗
bottom hung sash window 底悬窗
bottom hung window 最低吊窗；下悬窗
bottom ice 锚冰【水文】；底层冰；底冰
bottoming 槽底找平；清底；铺底；碎石铺底；输入下限；输出下限；石料铺底；石块铺砌
bottoming bending die 校正弯曲模
bottoming bit 可卸钻头；活钻头
bottoming circulating funds 铺底流动资金
bottoming drill 平底钻
bottoming hand tap 手用平底丝锥
bottoming hole 加热孔
bottoming machine 封底机
bottoming reamer 精铰刀；平底铰刀
bottoming tap 盲孔丝锥；平底螺丝纹；三锥
bottoming type bit 孔底可卸式钻头；可卸式钎头；活动钻头
bottom initiation 孔底起爆
bottom installation 井底车场设备
bottom instrument 水下仪器
bottom intake rack 底孔拦污栅；底孔进水口拦污栅；进水底孔拦污栅
bottom inwall line 炉腰水平线
bottom joint 底节理【地】
bottom kerf 底槽
bottom knife coater 底刀式刮涂填孔机
bottom knockout 下脱模；下顶料
bottom land 河滩地；河漫滩；谷底；低沼泽(地)；齿槽底面；洼地；水泛地；盆地；滩地；低洼地；谷地；底部地；低地

bottom land forest 河谷林
bottom lateral 底部横向水平支撑杆;底部横向杆
bottom lateral bracing 下弦横向水平支撑;下平(纵)联;底横撑;底部横向支撑
bottom layer 下层;底层;贴地大气层
bottom-layer bridge 下承式公路桥
bottom-layer plate 底板;底梁
bottom lead type 底部引线形
bottom leaf 下部门页扇;下部窗页扇;底叶
bottom left-hand corner 图幅左下角
bottomless 无底
bottomless bucket 无底的吊桶
bottomless hole 穿透孔
bottom level 底标高;池底标高
bottom lever 手柄末端
bottom life 炉底寿命
bottom lift 底板升起;下端向上升
bottom lifter 底提升机构
bottom lifting spreader 兜底吊架
bottom lift sling 吊索顶升
bottom line 净收益栏;账本底线;过底绳
bottom line of teeth 齿根圆线
bottom liner 底衬
bottom lining 底衬
bottom lip surface 底唇面
bottom liquid 罗盘盆液
bottom living fish 底栖鱼
bottom living fishes 底栖鱼类
bottom living stage 底栖阶段
bottom load 底载质;底沙;底负载;底负荷
bottom loading 孔底荷载
bottom-loading conveyer 下部装载运输机
bottom load sampler 底沙采样器;推移质采样器
bottom lock 海底封闭
bottom log 船底计程仪
bottom longitudinal 底部纵骨;底部纵材;船底纵向构件;船底纵骨
bottom longitudinal distribution system (船闸的)闸底纵干管灌水系统
bottomman 井底把钩工
bottom margin 底线;下图廓;下白边;底边
bottom marker 底标志
bottom material 底质;底料;底积物;底部物质
bottom-mating bell 底部对接式潜水钟
bottom matter 底质
bottom mechanical float 底面抹光机;底面机动抹灰板;底部抹灰板
bottom mechanical trowel 底面机动修平刀;底部机动镘泥刀
bottom melt hypothesis 底冰融化假说
bottom member 船底构件;船底杆件
bottom mine 沉底雷
bottom moraine 底碛层;底碛
bottom-most 最下层
bottom mo(u)lding box 底箱;下砂箱
bottom-mounted array 海底基阵
bottom-mounted breakwater 混合式防波堤
bottom-mounted current meter 海底海流计
bottom-mounted mooring pylon 海底系泊塔
bottom-mounted sonar 海底声呐
bottom-mounted sonar network 海底声呐网
bottom-mounted tide recorder 海底潮汐记录仪
bottom-mounted wave direction meter 海底波浪方向仪
bottom mud 淤泥;底泥;底层泥浆
bottom note 注脚
bottom nourishment 淤滩护底;补沙护底
bottom of bed 河床(底);河道底部;渠道底部
bottom of blast furnace 熔铁炉底
bottom of boring 孔底
bottom of building pit 房屋坑底;基坑底
bottom of excavation 开挖地面
bottom of foundation 基础底板;基础底面;地基底板
bottom of hole 钻孔底
bottom of layer 层底
bottom of paintpot 漆桶中的硬沉淀物
bottom of process(ing) interval 处理井段最大深度
bottom of roof rock 盖层底界
bottom of sedimentary 沉积层底部
bottom of shaft 井底;竖井底部
bottom of side ditch in cutting 路堑侧沟底
bottom of slot 槽底

bottom of soil sample 土样底面
bottom of springs 弹簧失效
bottom of stack 栈底;堆栈的底部
bottom of the blade 叶片底叶
bottom of the excavation 槽坑底
bottom of the groove 槽底
bottom of the hole 孔底;井底;炮眼底
bottom of thread 螺纹底部
bottom of trench 沟底
bottom of valley 谷底
bottom oil 油脚;残油
bottom opening 底孔
bottom-opening hopper barge 开底泥驳
bottom opening of jaws 颚式破碎机的排料口
bottom-opening skip 开底式抓斗
bottom organism 海底生物;水底生物
bottom-oriented system 海底定向系统
bottom out 开始回升;钻到规定的深度
bottom outlet 底部出口;泄水孔;泄洪孔
bottom outlet diversion 底孔导流
bottom outlet door 底排水孔
bottom outlet gate 底部放水闸门
bottom outlet orifice 底部放水孔口
bottom outlet pipe 底部放水管
bottom packer 底部打捆器
bottom-packer method 下木塞(注水泥)法
bottom paint 船底漆
bottom pan 储[贮]漆槽
bottom panel 底面镶板;底截镶板;底部导流板;踢脚板;底墙
bottom pass 下轧槽
bottom pavement 铺砌窑底
bottom paving 池底铺砌;铺砌窑底
bottom paving brick 窑底砖;池底铺砖
bottom peat 低沼泽地的泥炭土;湖河泥炭
bottom photograph 海底照片
bottom photography 海底照相
bottom pinch roll 下夹送辊
bottom pintle 舵踵栓;枢(人字门)
bottom pitch 铧尖犁梁距离
bottom pivot 下支承座
bottom-pivoted ventilator 底部摇窗;底部旋气窗;底部旋窗
bottom plane 搜根刨
bottom-plankton sampler 海底浮游生物取样器;底层浮游生物取样器
bottom plate 枕木支承板;底梁;脚盘;主夹板;垫板;底板
bottom plate altitude of aquifer 含水层底板高程
bottom plate of column 分馏塔底塔盘;分馏塔底法
bottom plate screw 底板螺钉
bottom plate seam 模底线印
bottom plating 船底板
bottom plating of bulkhead 舱壁下列板
bottom plow 铧式犁
bottom plow with subsoiler 带心土铲的铧式犁
bottom plug 泄水孔塞;下胶塞;钻井注水泥下木塞
bottom poured 下注的
bottom pouring 下铸(法);下注;底注
bottom pour ladle 漏包;底注式浇包
bottom-pour mo(u)ld 底浇铸型
bottom pressure 底压力;底部压力
bottom pressure relief 底部减压
bottom price 最低价(格);底价
bottom priming 孔底起爆;底部点火
bottom printing 透明薄膜背面印刷
bottom probe 河底取样钻;底质样品;底质样(本);底部调查
bottom product 底部产物;底部产品;残液;残留产物;残留产品;塔釜残液;塔底产品
bottom profile 水下地形
bottom profiler 断面测量仪
bottom prop 炉底支柱
bottom protection 护底
bottom-pull method 沿底拖运法;锚碇(方)法
bottom punch 下模冲
bottom quality 劣等质量;底质量
bottom rack 底部拦污栅
bottom radiator bracket 散热器下部托座
bottom radius 槽底半径
bottom rail 下栏门窗;下横档;下冒头
bottom rail of door 门底边;下冒头
bottom ramming machine 炉底打结机
bottom-reference piston corer 海底固定活塞柱状取样器

bottom reflectance 海底反射率
bottom reflected coefficient 海底反射系数
bottom reflected pulse 海底反射脉冲
bottom reflected ray 海底反射声线
bottom reflection 海底声反射;海底反射
bottom reflection echo 海底反射波
bottom reflection loss 海底反射损失
bottom reflector 底部反射体;底部反射层
bottom region 底层区;水底区域
bottom register 沿踢脚板铺设的通风装置
bottom reinforcement bars 底部钢筋
bottom relief 海底地形;底面起伏;底部地形;水下地形;水底地形
bottom relief drawing 水底地形图
bottom resistance line 底部抵抗线
bottom reverberation 河底交混回响;海底混响
bottom ridge 底脊
bottom right-hand corner 图幅右下角
bottom ring 底圈;活塞环框圈;管靴承托环;底环
bottom-ripping shot 卧底爆破;抽底爆破
bottom ripple 河底波痕
bottom road bridge 下承式公路桥
bottom rock 基岩
bottom rock sample 井底岩样
bottom roll 下轧辊;下辊;底辊
bottom roller 河底回溜;河底回灌;下辊轮;下滚子;底辊
bottom rope 绳的下端
bottom rung 楼梯最下段
bottom rung of stair(case) 楼梯的最下段
bottom runner 底滚子
bottom running 底注法
bottomry 冒险借款;以船作抵押的借款;押船借款;舱舶抵押贷款
bottomry and respondentia 船舶和货物抵押借款
bottomry bond 冒险借款债券;押船契约;船舶抵押契约
bottomry lien 船舶抵押留置权
bottomry loan 船舶抵押借款
bottomry premium 航海利息
bottoms 底部产品;底部残留物;残留物;釜底残油;塔釜残油
bottom samping device 底质采样器
bottom sample 底质样品;底质样(本);底样
bottom sampler 河底取样器;海底采样器;底质采样器;底质采样器;底样采样器;底层取样器;底部泥沙取样器;采泥器;水底取土钻;水底取土器
bottom sampling 河底取样;海底采样;底质取样;底样采集;底面采样
bottom sampling device 水底取样设备;水底取样器;海底采样装置
bottom scanning sonar 海底扫描声呐
bottom scattering 海底散射
bottom scattering layer 海底散射层
bottom scavenging 孔底清洗
bottom scour 海底冲刷
bottom scuttle 船底孔洞
bottom seal 底部封闭;底止条;底部止水条封
bottom sealing by concreting 混凝土封底
bottom section 下部半断面;底断面;底部截面;底部断面
bottom sector gate 底部扇形闸门
bottom sediment 湖底泥沙;河底泥沙;河床质;河床底质;海底泥沙;海底淀积(物);海底沉积(物);罐底杂质;底沙;底沉积物;底部泥沙;底部沉积(物);沉积泥沙
bottom sediment and water 底部残渣
bottom sedimentation 底部沉积作用
bottom sediment distribution chart 底沙分布图
bottom sediment investigation 底质调查
bottom sediment(settling) and water 底部残渣与水
bottom-seeding method 底部晶种法
bottom set(bed) 底(碛)层;底积(层)
bottom set deposit 底层沉积
bottom setting net 底层定置网
bottom settlings 河床质;底脚;沉渣;水杂淀积;底沙;底部沉积(物)
bottom shaft 织机下地轴
bottom shaping 河底整平
bottom shear 底部剪力
bottom shear stress 底切应力;底部切应力

bottom shooting 海底发射
bottom shore 斜撑的贴墙横木;底撑;底部支柱
bottom shot 拉底炮眼;底眼
bottom shuttering 底模板
bottom(side) 底部
bottom side girder 船底纵桁(双层底);旁内龙骨
bottom side rail 下桁材;下侧梁(集装箱)
bottom sill 下槛;(车身的)底梁;底槛;潜坝
bottom silt 海底淤泥
bottom size 粒度下限;入粒下限
bottom sizing 底胶料;涂底
bottom skin 下表面蒙皮;底部蒙皮
bottom slab 底板
bottom slab slide 底部片滑
bottom slide 底部滑板
bottom slide of saddle 刀架底板
bottom slope 基底坡度;河底坡降;河底比降;河床坡度;底坡(度)
bottom slope of canal 渠槽底坡
bottom slot layer 下层绕组
bottom sludge and water 油渣和水;油脚和水
bottom sluice 泄水底孔;泄洪底孔;底部泄水闸
bottom sluice gate 泄水底孔;底部泄水闸门
bottom socket 下部承口;底接头
bottom soil 下层土;底质;底土
bottom soil sample 河底土样;底部取样
bottom sonar 海底地貌仪
bottom sounder 海底地貌探测仪
bottom spindle 下轴
bottom sprag 底部支撑
bottom spring 底弹簧
bottom sprocket(wheel) 链轮底部
bottom squeeze 底鼓
bottom squeeze mo(u)lding machine 底压式造型机
bottom stabilization 底板稳定(作用);护底;稳定河底;稳定海底
bottom stabilizing 护底
bottom stamp 底款(器皿底标记)
bottom stand 机座
bottom standing hydrostatic(al) corer 海底固定静水压柱状取样器
bottom station 底部踏步
bottom step 起步梯级;踏步首步
bottom stiffener 底部加强肋
bottom stitching machine 封底钉铆机
bottom stockade dam diversion 底拦栅坝引水
bottom stone 基石;底岩
bottom stowage 压载【船】
bottom strake 船底外列板
bottom stratum 底层
bottom stream 塔底流出物
bottom-suction self-jetting well point 底吸自射式井点
bottom sump 底部集水坑
bottom-supported platform 座底式平台;底撑式平台
bottom supporting drilling platform 座底式钻井平台
bottom surface 下表面;底面;低面
bottom surface dummy joint 下部表面假接缝;底面收缩缝
bottom surface error 底面误差
bottom survey 海底调查;船底检查
bottom swage 下陷型模;下凹模
bottom sweeping with towing rope 拖地扫测
bottom tap 下除渣口;平尾丝锥
bottom tapping 底部分流
bottom temperature 井底温度;底层温度
bottom tension wire 底张拉钢丝;底张拉钢筋
bottom terminal landing 底层终点站平台;底层电梯平台
bottom terrace 河底阶地
bottom timber 底木
bottom time 水底停留时间
bottom tin 粘锡;底部锡
bottom to be excavated 开挖的底部
bottom tool 下刀架;底锻模
bottom tooth thickness 齿根截面厚
bottom topography 底部地形;水底地形
bottom topography survey system 海底地形测量系统
bottom tow 底拖
bottom trace 底部记录线

bottom track 海底跟踪
bottom traction 底部牵引
bottom tractive drag 河底曳引力;河底拖力;河底推移力
bottom tractive force 河底曳引力;河底拖力;河底推移力
bottom transom 栏杆下档;下档(栏杆)
bottom transverse 底部横梁;船底横向构件;船底横骨;船底横材
bottom trash rack 底部拦污栅
bottom trawl 底拖网;水底拖网
bottom tumbler 下导轮
bottom turbidity 海底浊度
bottom unloader 底部卸载机
bottom unloading valve of cement silo 水泥库底卸料阀
bottom up 自底向上;倒置;船只倾覆;翻倒;确定孔深;倾覆【船】
bottom-up 底部朝上
bottom-up construction method 顺筑法
bottom-up design 自底向上设计
bottom-up effect 上行效应
bottom-up management 上行式的管理
bottom-up method 顺作法
bottom valve 底阀
bottom vane 底导流叶片;底导板
bottom velocity 河底速度;河底流速;流速;底流速;底流速
bottom ventilated bin 底部通风粮箱
bottom ventilation 池底通气
bottom view 下视图;仰视图;底视图
bottom wale piece(of the piling) 底部腰(护)板(打桩用)
bottom wall 下盘;底壁
bottom wall and anchor packer 孔壁及尾管封隔器;井壁及尾管封隔器;复合式井底封隔器
bottom wall of intrusion 岩体底盘
bottom walls 底墙
bottom water 深水;底水;底面水;底层水;底部水;低层水
bottom water ballast 双层底水压载
bottom water coning model of simple well 单井底水锥进模型
bottom water drive 底水驱动
bottom water of lake 湖底层
bottom water sampler 底水取样器;底层采水样器
bottom water screen 炉底水帘管;水筛炉底
bottom water sprays 下喷水器
bottom water velocity 底部水中波速
bottom wave 底波
bottom-wear chamfer 底倒角
bottom wedging 底部加楔作用
bottom well location 钻井的底点
bottom width 底宽;底部宽度
bottom wing 下翼
bottom withdrawal tube 底部排放管
bottom working roll 下工作辊
bottonm-facing machine 转底光面机
boucharde 凿毛锤(石面);凿毛锤
Bouchard's index 布夏尔指数
bouche 钻孔
boucherie 端部压注法
boucherie process 硫酸铜防腐法(木材)
boucherising 硫酸铜防腐法(木材)
boucherize 用蓝矾(硫酸铜)浸渍
boucherizing 落差式(硫酸铜)注入法(一种木材防腐剂);树液置代法
Boucherot circuit 波切洛特电路
Boucherot(squirrel-cage) motor 双鼠笼式电动机
Boucherot starting method 波切洛特起动法
bouchon 轴衬;点火机
bouchon wire 衬管丝材
boucle 结子线;珠皮呢织物
boudin(age) 石香肠
boudinage structure 香肠构造
Boudin continuous double-roll process 连续双辊压延法
boudin length 石香肠长度
boudin thickness 石香肠厚度
boudin width 石香肠宽度
boudle 布多(英国钢铁厂用重量单位,1 布多 = 56 磅)
boudle root 等根【数】
boudoir 女士的私室;闺房(法国)

Boudouard's reaction 鲍多尔反应
Bouffloux stoneware 鲍尔菲奥盐釉炻器
Bougainville trench 布干维尔海沟
bough 大枝(树的)
boughpot 大花瓶
bought 进货
bought book 进货簿
bought in components 外购件
bought invoice book 进货簿
bought ledger 进货总账;进货分类账;购买分类账
bought(-out) component 定型部件;定型配件
bought scrap 外来废铁
bought without any restriction 敞开供应
bougie 压力灌缝锅;灌缝锅;瓷制多孔滤筒;探条;陶质的多孔滤筒
Bouguer compass 袖珍罗盘
Bouguer-Lambert law 布格—朗伯定律
Bouguer plot 布格图
Bouguer's anomaly 布格异常
Bouguer's compass 布格罗盘
Bouguer's correction 布格校正(值)
Bouguer's gravity 布格重力值
Bouguer's gravity correlation 布格重力对比
Bouguer's halo 布格晕
Bouguer's law 布格定律;波格定律
Bouguer's reduction 布格校正;布格改正
Bouin's solution 布安溶液
boulangerite 硫锑铅矿;块硫锑铅矿
boulder 块石;加工石;巨砾;大卵石;大块石;漂石;漂砾
boulder and shingle foreshore 卵石及沙石海滩
boulder bank 巨砾滩地
boulder bar 卵石滩
boulder barricade 岸边砾石堤
boulder base 拳石基层;卵石基层;蛮石基础;蛮石地基;蛮石底基;圆石基层;大卵石基层;漂石基底
boulder beach 砾滩
boulder bed 粗砾河床;蛮石基层;巨砾层;蛮石底盘;大卵石层;粗砾地层;漂石基床;漂砾河床
boulder belt 漂砾带
boulder blaster 二次爆破工
boulder blasting 补充的爆破;岩石小爆破;大块石爆破;大块二次爆破;分块钻孔;爆破巨砾;爆破大石
boulder breccia 巨砾角砾岩
boulder buster 岩石破碎器;岩石破碎机
boulder channel 漂砾河槽
boulder clay 砾泥;泥砾(土);粗砾土;冰碛土;冰砾泥;漂石黏土;漂砾黏土;漂砾河床
boulder closure 大块石截流
boulder coast 群石海岸
boulder concrete 卵石混凝土
boulder cracker 岩石破碎器;岩石破碎机
boulder crusher 粗碎机
boulder dam 大块石坝;顽石坝;石坝;蛮石砌筑坝
boulder ditch 乱石盲沟;卵石沟;法式排水沟;盲沟
boulderet 小漂砾;中砾(石)
boulder fan 漂砾扇;冰砾列扇
boulder flint 卵石
boulder foundation 大块卵石填砂基础
boulder fragmentation 巨砾碎块;巨砾破块;巨砾解爆;巨砾爆破;大块碎裂(作用)
boulder gravel 巨砾(层);漂砾(层)
boulderhead 堤坝前防浪排桩;堤坝前桩
bouldering 拳石铺面;蛮石铺砌;大块石铺的路;大蛮石铺砌
boulder-like inclusions (松软岩石中的)石砾状杂层
boulder loam 冰碛土;冰砾壤土
boulder-massive grain grade 漂石—块石粒组
boulder mud 冰砾泥
boulder pavement 蛮石铺面;漂砾道面
boulder paving 砾石路面;大块石铺砌;漂石铺面
boulder prospecting 漂砾勘探
boulder quarry 蛮石开采;蛮石铺地
boulder river bed 粗砾河床;卵石河床
boulders concrete 砾石混凝土;卵石混凝土
boulder setter 蛮石铺砌层
boulder shingle 海滨巨大砂石;粗砾推移质;漂砾推移质
boulder shore 大卵石岸
boulder slope 大块废物堆
boulder stone 巨砾岩;杂石滩
boulder stream canyon 蛮石峡谷
boulder strip 条石

boulder studded 乱石嶙峋的
boulders with a trace of cobbles and gravels 混少量卵石和砾石的漂砾
boulders with some cobbles and gravels 混一些卵石和砾石的漂砾
boulder texture 巨砾结构
boulder train 漂砾列
boulder wall 蛮石墙;大卵石墙;大块石墙;漂石墙;漂砾墙;石墙
bouldery 卵石的,含巨砾的;大卵石的
boule 人造刚玉;刚玉;镶嵌木料;镶嵌装饰;通风木垛;台基;透气圆木
Boule dam 布雷坝
bouleuterion 集会所(古希腊);行政楼
boulevard 大街;林阴道;林阴大道;大道;干道;大马路
boulevard boultin 凸圆饰
boulevard light 干道路灯
boulevard restaurant 林阴大道饭店;林阴大道餐厅
boulevard stop 林阴大道停车处
Boule weir 布雷式堰
boulidou (喀斯特区的)间歇泉
boulle 镶嵌装饰
boulonite 重晶石
bo(u)ltel 凸圆线脚
boultin 馒头形饰
boultine 凸圆饰;凸圆线脚
Boulton formula 博尔顿公式
Boulton point 博尔顿点
Boulton process (木材的)真空压力防腐法;热槽处理;真空干燥(法)
Boulton reed count 博尔顿筘号
Boulton well function 博尔顿井函数
Boulwarism 博尔韦尔制度(用于劳资谈判)
Bouma cycle 鲍玛旋回
Bouma sequence 布马层序
bounce 支票拒付退还开票人;反跳;反冲;跳动;弹跳;弹回
bounce back 反喷;反射;反冲
bounce cast 突型脊【地】;弹跳铸型
bounce cylinder 缓冲汽缸
bounce dive 跳跃式潜水
bounce diving 短促潜水
bounced light 间接照明
bounce flash 间接闪光;反射闪光;跳闪
bounce frequency 弹跳频率
bounce mark 弹跳痕迹
bounce plate 反跳板
bouncer 弹跳装置
bounce resonance 反冲共振
bounce response 跳动响应
bounce table 冲击台;振动台
bouncing 图像跳动
bouncing against 跳跃式接触
bouncing characteristic 跳动特性
bouncing motion 图像跳动
bouncing pin 撞针;(仪表的)跳针;弹跳杆
bouncing-pin apparatus 跳针指示仪;弹针仪
bouncing putty 弹性油泥;弹性油灰
bouncing sum 巨大金额
bound 邻接界区;连接;界限(约束);弹跳;受约束;受限制;上下限;上下界
bound acid 结合酸
boundaries for reserve calculations 储量计算边界线
boundaries of discharge area 排放区边界
boundaries of districts 分区界线
boundary 境界;分界线;分层面;边界
boundary action 界面作用;边界作用
boundary address register 界地址寄存器
boundary adjustment 边界调整
boundary affect 边界影响
boundary alignment 对界;边界对准;边界定位
boundary angle 界面角;界边角钢;接触角;边界角
boundary approximation 边界逼近
boundary arch 端拱;边拱;边界拱
boundary band 境界线色带
boundary bank 边界堤
boundary bay 疆界海湾;边界海湾
boundary beam 边梁
boundary belt 保护带
boundary between classes of reserves 储量级别边界线
boundary between industrial types of ores 工业类型边界线
boundary between light and dark bars 亮线与暗线分界线
boundary between natural types of ores 自然类型边界线
boundary between technologic(al) classes of ores 工业晶级边界线
boundary breaking 界面破碎
boundary brittle fracture 界面脆性断裂
boundary bulkhead 隔舱壁
boundary cavitation 界面层空蚀;边界区汽蚀;边界区空蚀
boundary chain 边界链
boundary coding 界编码
boundary coefficient 界面系数
boundary concrete block 边缘混凝土砌块
boundary condition 极限条件;边界条件
boundary conduction layer 边导热层
boundary constraint 边界约束
boundary context 限界上下文
boundary contraction method 边界收缩(方)法;边界蚀刻法
boundary contrast 亮度差阈;边界对比度
boundary crack 界面裂纹;界面裂缝
boundary creep fracture 边界蠕变断裂
boundary current 边界流
boundary curve 界限曲线;相界曲线;边缘曲线;界曲线
boundary deformation condition 边界形变条件
boundary depth 极限深度
boundary depth of reservoir at bottom 储集层底界面深度
boundary depth of reservoir at top 储集层顶界面深度
boundary description 边界说明
boundary detection 边界检测
boundary determination 界线测定
boundary diffraction 边界衍射
boundary diffusion 界面扩散
boundary dimensions 边界尺寸
boundary displacement 边界位移
boundary distribution 判别边界
boundary disturbance 边界扰动
boundary drag coefficient 边界阻力系数
boundary echo 界回声;界面反射波;界回波;边界反射
boundary ecosystem 边界生态系统
boundary effect 附面层效应;边界效应;边际效应
boundary element 边缘构件;边缘部件;界面单元;边界单元;边界元(素)
boundary element method 边界(单)元法
boundary enhancement 边缘增强
boundary equation method 边界方程法
boundary error 边界错误
boundary face 界面;边界面
boundary fault 边界断层
boundary feature of microlithon 微劈石边界特征
boundary fence 边篱;边界栅栏
boundary file 边界文件
boundary film 界限膜;界膜;界面膜;边界膜
boundary flux 界面通量
boundary force 边限力;边界力
boundary fracture 边界断裂
boundary frame on crossing 道口限界架【道】
boundary frequency 边界频率
boundary friction 边界摩擦
boundary function 分界功能;界函数;边界功能
boundary gable 边缘山墙;边山墙
boundary geometry 边界形状;边界几何形状
boundary grid pattern 边界格水形
boundary grid point 边界格点
boundary indicator 边界指示器
boundary integral 定积分
boundary integral equation method 边界积分方程法
boundary integral method 边界积分法
boundary land-locked sea 疆界内海
boundary layer 界面层;附面层;边界导热层;边界层
boundary layer acceleration 边界层加速
boundary layer accumulation 附面层积厚
boundary layer blowing-off 边界层吹除
boundary layer casing treatment 边界层壁面处理
boundary layer control 边界层控制
boundary layer effect 界面层效应;边界层效应
boundary layer flow 边界层水流;边界层流动
boundary layer growth 边界层增厚
boundary layer meteorology 边界层气象学
boundary layer model 边界层模型
boundary layer motion 边界层流动
boundary layer noise 边界层噪声
boundary layer of laminar flow 层流边界层
boundary layer oscillation 边界层震动
boundary layer phenomenon 边界层现象
boundary layer photocell 边界层光电池;光生伏打电池
boundary layer probe 边界层探测管
boundary layer seam 附面层表面
boundary layer separation 边界层分离现象;边界层分离
boundary layer skin friction 附面层表面摩擦;边界表面摩擦
boundary layer suction 边界层吸除
boundary layer theory 附面层理论;薄膜理论(一种薄膜设计的理论);边界层学说;边界层理论
boundary layer thickness 边界层厚度
boundary layer transition 边界层转捩;边界层过渡(区)
boundary layer trip 边界层激流丝
boundary layer turbulence 附面层紊流(度);边界层湍流
boundary layer velocity profile 边界层速度剖面
boundary light 降落场界线灯;周边照明;边界标灯;边界灯
boundary line 境界线;禁区界线;界线;国境线;分界限线;边线;边界线;界限
boundary line of construction 建筑限界;建筑界限
boundary line of ore types and grades 矿石类型与品级边界线
boundary line of reserve grades 储量计算边界线;储量级别界线
boundary line of road construction 道路建筑限界
boundary line of street 道路红线
boundary line of zone lot 区域地块分界线
boundary line survey 境界线调查
boundary lipid 界面脂
boundary lubricant 边界润滑剂
boundary lubrication 边界润滑;界面润滑;边界润滑
boundary lubrication additive 边界润滑添加剂
boundary map 行政区划图
boundary marker 界桩;机场标志板;边界指点标;界标
boundary member 限制器;边缘构件;停车器
boundary module 边界模数
boundary module of orebody 矿体边界模数
boundary moment 边缘力矩
boundary monument 界桩;界标(石);界标(碑);界碑
boundary mount 界石
boundary network 境界线网
boundary network file 曲线网文件
boundary network node 分界网络节点;边界网络节点
boundary node 边界节点
boundary of a set 集合的界限
boundary of a set of points 点集的边界
boundary of basin 盆地的边界
boundary of carriage way 路缘标线
boundary of field 田边
boundary of frame on road 道路建筑界限
boundary offset 集合的界限
boundary of grout zone 灌浆区范围
boundary of known flow 已知流量边界
boundary of known water level 已知水位边界
boundary of open pit 露天采矿界
boundary of polygons 多边形周边
boundary of property 矿区地界
boundary of recharge area 补给区边界
boundary of refilling 回填边缘线
boundary of safe region 安全区域的边界
boundary of set 集合界限
boundary of the bright region 亮区边界
boundary of water body 水体边界
boundary or grain 晶粒界
boundary perturbation problem 边界扰动问题
boundary phenomenon 边界现象
boundary piezometer 界面流压计
boundary pillar 柱子的安全限度;分界矿柱;边界(安全)矿柱

boundary plane 分界面;边界面
boundary plank 木甲板边(缘)板;双层底边板
boundary planting 边界种植
boundary point 界址点;界限点;分界点;边界点
boundary point of graph 图的边界点
boundary position 极限位置
boundary post 界柱;界桩;界标
boundary potential 边界势
boundary problem 边界问题
boundary property 边界性质
boundary property line 道路边线
boundary property of basin 盆地边界性质
boundary raise 边界天井
boundary ray 极限射线
boundary reflectance 界面反射
boundary reflection 界面反射;边缘反射;边界反射
boundary region 边界区
boundary register 边界寄存器;上下限寄存器;上下限地址寄存器;上下界寄存器
boundary relation 边界关系
boundary resistance 边界阻力;边界电阻
boundary reverberation 边界混响
boundary river 界河;国境河流
boundary row 保护行
boundary scattering 界面散射;边界散射
boundary science 边缘科学
boundary settlement 勘界;划定界限
boundary shear 边界剪(切)力
boundary shear stress 边界上剪应力;边界剪(切)应力
boundary sign 界标
boundary similarity 边界相似
boundary solution method 边界解法
boundary space 边界空间
boundary spring 溢出泉;冲积坡泉;边界弹力
boundary stability 边界稳定性;边界稳定度
boundary state of stress 极限应力状态;应力边界条件
boundary stone 界石;边缘石;边石
boundary strait 边界海峡
boundary stratotype 界线层型;地层分界线;界限层型
boundary strength 界面强度
boundary stress 边缘应力;边界应力
boundary strip 边界地带;边角地带;田界
boundary stripping ratio 境界剥采比
boundary surface 界面;带边界曲面;分界面;边界面
boundary survey 地界测量;边界测量
boundary surveying 边界勘测
boundary symbol 境界符号
boundary tablet 界碑
boundary-tag allocation 边界标志定位
boundary tag method 边界标志法
boundary tag method of storage allocation 存储分配的边界标志方法
boundary temperature 边界温度
boundary tension 界面张力
boundary tension between gas water contact 气水界面张力
boundary tension between oil-gas contact 油气界面张力
boundary tension between oil-water contact 油水界面张力
boundary tensor 边界品位
boundary tissue 界限组织
boundary torque moment 边缘扭转力矩
boundary treatment 边界处理
boundary tree 界标树;边行树
boundary type of tectonic region 构造区边界类型
boundary value 界限值;边值;边界值
boundary value diagram 边界值图解;边界值图表
boundary value problem 极限值问题;边值问题;边界值问题
boundary value problem of potential theory 位理论边值问题
boundary vector 界标向量
boundary velocity 界面速度
boundary vicinity 边界附近
boundary wall (三顺一丁的)界墙(砌法);边界墙;围墙
boundary wall bond 三顺一丁砌合
boundary wall gutter 前后檐墙檐沟
boundary water 边境水域

boundary waters canoe area 边界水域舟状区
boundary waters canoe area wilderness 边界水域无边舟状区
boundary waters treaty 边界水域条约
boundary wave 界面波;边界波
boundary wavelength 边界波长
boundary well 边界井
boundary zone 边界范围;界面区;边缘带;边界区
boundary zone of capillary 毛细水作用带;毛细边带;毛细边界层;毛细管作用带;毛管作用带;毛管水边界层
boundary zone of capillary in soil 土中毛吸水作用带
bound auxin 束缚生长激素
bound a volume 装订成册
bound by ice 冰冰封的
bound charge 束缚电荷
bound component 界元素;界分量
bound control module 界限控制模块
bounded area 轮廓面积
bounded edge 包好的边
bounded fraction 有界函数
bounded function 囿函数;有界函数
bound edge 包边
bounded linear transformation 有界线性变换
bounded matrix 有界矩阵
bounded measure of probability 概率的有限测度
bounded medium 有限介质
boundedness 局限性
bounded operator 有界算子
bounded projection 有界投影
bounded region 有界域
bounded sequence 有界序列
bounded set 集合;囿集;有界集(合)
bounded set of numbers 有界数集
bounded set of points 有界点集
bounded test 界限检验
bounded variable 基本变量;有界变数;有界变量
bounded variation 有界变差
bound energy 跳动能;束缚能
bounder 矿山定界人;矿区丈量员
bound extracellular polymeric substance 固着性胞外聚合物
bound fiber 捆束纤维
bound flow 限约流
bound from 来自某地或某方向
bounding bar 界边角钢
bounding capacity 黏着力;黏合能力
bounding depth 极限深度
bounding dislocation 边界位错
bounding effect 行边效应
bounding erosion surface 边界侵蚀面
bounding hyperplane 边界超平面
bounding layer 边界导热层;边界层
bounding meridian 图廓经线
bounding mine 跳雷
bounding parallel 图廓纬线
bounding position in master structure 发育构造部位
bounding pulse 洪脉
bounding surface 分界面
bound lane 定向车道
boundless 无限的
bound method 界限法;限界法
bound mode 束缚模
bound module 界限模块
bound occurrence 界限出现
bound of aggregationn 聚集边界
bound of interconnection 关联界限
bound of parameter 参数的界
bound pair 界限对;界偶表;界对;限界对;对接
bound pair list 界偶表;界对表
bound pile 加筋混凝土桩;加箍混凝土桩;装有钢筋的钢筋混凝土桩;带铁箍桩
bound porosity 束缚孔隙度
bound rate 约束税率
bound rates of duty 受约束税率
bound register 界限寄存器
bound register for memory protection 供存储保护用的界限寄存器
bound reinforcement 束缚钢筋
bound scattering 束缚散射
bound setter 地形测绘者;测量员
bounds on error 误差界限

bound state 束缚态
bound stone 界石;黏结碳酸岩;黏结灰岩
bound surface 邻接面;边界面
bound symbol 连接符号;连接符;界限符
bound symbol sequence 连接符号顺序
bound task set 界限任务集
bound term 约束项
bound to failure probability 破坏概率限值
bound variable 界变量;约束变项;约束变量;受限变量
bound vector 束缚向量
bound volume 合订本
bound vortex 约束涡流;附着涡流;附体涡流;束缚涡流
bound water 结合水;限流水;吸附水;约束水;封存水;薄膜水;束缚水
bound water bond 结合水连接
bound water film 结合水膜
bound water inside the mineral 矿物内部结合水
bound water on mineral surface 土粒表面结合水
bound water-proof pavement 不透水路面
bound wave 合成波
bound with resinous material 用树脂材料黏结
bounty on exportation 出口津贴
bounty on import 进口津贴
bouquet 一揽子;顶端花饰
bouquet agreement 一揽子协议
Bourbon 布尔邦锡铝合金;带边绳
Bourbon metal 布尔邦锡铝合金
Bourdon ga(u)ge 布尔登(管式)压力计;布尔登量规;弹性金属曲管式压力计
Bourdon manometer 布尔登压力表
Bourdon pressure ga(u)ge 布尔登压力计;布尔登压力表
Bourdon pressure vacuum ga(u)ge 布尔登低压真空计
Bourdon's metallic barometer 布尔登金属气压计
Bourdon spring 布尔登弹簧;弹簧管
Bourdon's test 布尔登试验
Bourdon temperature ga(u)ge 布尔登温度计
Bourdon tube 布尔登管;布尔登压力计
Bourdon tube bottom hole pressure recorder 布尔登管式井下压力计
Bourdon tube ga(u)ge 布尔登管式压力计;布尔登管压力计
Bourdon-tube manometer 布尔登管压力表
Bourdon tube pressure ga(u)ge 布尔登管式压力计
Bourdon tube type manometer 布尔登管式压力计
Bourdon tube type pressure ga(u)ge 布尔登管式压力表
Bourdon tube type vacuum ga(u)ge 布尔登管式真空表
bourdony 布尔登呢
bourette 表面粗糙不平的织品;绵绸
bourette silk 绸丝
bourette silk card 绸丝梳绵机
bourette spinning machine 稠丝细纱机
bourette yarn 绸丝
bourg 村(镇);城镇
bourgeoisite 假硅灰石
Bourger's correction 布格校正
Bourges process 布尔格斯人工设色法
Bourg-La-Reine porcelain 鲍格—拉—莱恩软瓷(法国)
bourn(e) 范围;目的地;界限;边界;小河
Bournemouth beds 伯恩默斯层
bournonite 硫化锑铅铜矿;车轮矿
bourock 茅屋(苏格兰);大块石头;石堆
Bourquin-Sherman unit 布尔奎—谢尔曼单位
bourrelet 定心带
Bourry diagram 鲍里图样(坯样干燥收缩与时间关系曲线)
bourse 交易所;证券交易所
bourse tax 交易所税
bouse 用滑车吊起
Bousin's fixative 鲍辛固定液
Boussinesq's coefficient 布辛奈斯克系数
Boussinesq's elastic theory 布辛奈斯克弹性理论
Boussinesq's equation 布辛奈斯克方程
Boussinesq's formula 布辛奈斯克公式
Boussinesq's mathematics 布辛奈斯克数学
Boussinesq's problem 布辛奈斯克课题
Boussinesq's solution 布辛奈斯克解

Boussinesq's stress 布辛奈斯克应力
Boussinesq's theory 布辛奈斯克理论
boussingaultite 六水铵镁矾;铵铝矾
boussir 碎谷
boustay 盲井;暗井
bout 一次
bouteillenstein 暗绿玻璃
bouteillerie 备餐室;酒仓
boutell 环状半圆线圆装饰;四分之三倚柱
boutgate 井底通道;煤层间的通道;通向地面的人行道
boutique 镶嵌品;妇女用品商店;镶嵌珠宝的日用品;时装用品小店
bouton 棒头
bouton process (木材的)去湿法
Bouvealt-Blanc method 布维尔特—布朗法
Bouwers catadioptric system 鲍沃斯折反射系统
Bouwers telescope 鲍沃斯望远镜
Bouyoucos hydrometer 保氏比重计;(用于测定悬浮液固体颗粒大小)
Boveall Babbitt alloy 博维尔巴比特合金
Boveri's test 博韦里试验
Bovey bed 博韦特拉西层【地】
bovey coal 褐煤
Bovey Tracey bed 博韦特拉西层【地】
bow 锯框;弓状物;弓形弯;弓(形);弓弧形房屋;凹进翘曲;船头;弓头(方向)(左右45°范围内);船的前部;艏;舰首
bow action 弓形弯曲
bow anchor 船首锚;首锚
bow anchor line 船首主锚缆
bow and beam bearings 四点定位法
bow and buttock plane 首纵剖面
bow and string girder 弓弦式桁梁
bow angle 首舷角
bow array 艏基阵
bow asdic dome 艏声呐导流罩
bow assembly 分开造的船首
bow beam 弓形梁;弧形弓梁;弓形式梁
bow bearing 舷角;首舷方位
bow breast 前横梁
bow bulwark 首部舷墙
bow cal(l)ipers 弓形卡钳
bow cap 机头罩
bow chain saw 弓形链锯
bow charging crane 船头装货起重机
bow chock 船首遮浪板;船首导缆口;前喷口;前缆口
bow-chock mooring pipe 船首导缆孔;艏导缆孔
bow collision quay 船头触碰码头
bow compartment 船首舱;前舱
bow compasses 微调小圆规;量规;两脚规;小圆规;弹簧圆规;卡钳;测径规
bow construction 船首结构
bow contact 弓形接触
bow coupling system for pumping ashore 船首吹泥快速接头
bow crook 弧形弯曲;(木料的)翘曲
bow crossing range 船首穿越距离
bow crossing time 船首穿越时间
bow current collector 受电弓
bow curve 弓形曲线
bow cushion 首岸推
Bowden cable 鲍登线;鲍登安全电缆;钢闸绳;钢丝软轴
bowderling 大块石铺的路
Bowditch's rule 鲍狄奇(测量)平差规则;鲍狄奇规则
bow divider 弹簧分规
bow dome 艏声呐导流罩
bow door 首门【船】
bow draft 船首吃水
bow drill 弓钻;弓形钻;三叉钻
bowed 弓弯木材
bowed-branch layering 曲枝压条
bowed filling 弓纬
bowed height 矢高
bowed pastern 曲系
bowed roll 弓形辊
bowed roller 弧形辊;弯辊
Bowelism 翻肠倒肚式【建】
bowel urinal 大便池槽
Bowen fluorescence mechanism 鲍文荧光机制
bow-engined ship 艉机型船
bowenite 硬绿蛇纹石;鲍文玉

Bowen line 鲍文谱线
Bowen reaction dnes series 鲍文反应系列
Bowen's disease 鲍文病
Bowen's ratio 鲍文比值
Bowen's reaction series 鲍文反应系列
bower 主锚;船头锚;船头挂锚;船首锚;树荫处;首锚;艏锚
bower anchor 主锚;大锚;船头锚;船首锚;前锚;船首主锚;艏锚
Bower-Barff 鲍威尔—巴尔夫氧化铁水管防锈法
Bower-Barff process 鲍威尔—巴尔夫(钢件防锈)法
bower chain 首锚链
Bowers ridge 鲍威尔斯海岭
bower winch 船首主锚绞车
bow face 弓面
bow fairleader 艏导缆器
bow fast 头缆;艏系缆;艏缆;首缆
bow feather 船首碎浪
bow fender 船首护材
bow flare 船首外飘
bow flare slamming 船头猛倾
bow flood plain 弓形河漫滩
bow frame 集电弓支架;船首肋骨;受电弓框架;艏肋骨
bow gang-plank 船首跳板
bow girder 弓形大梁
bowgrace 船首防冰护具
bowguide 弓形导板
Bowie effect 鲍维效应
Bowie formula 鲍维公式
bowieite 硫铱铑矿
Bowie method of adjustment 鲍维平差法
bowing 回转窑弯曲变形;弧状弯曲;弓形翘曲;侧向偏移;突起变形
bowing in the length dimension 在长度方向卷边
bowing under load 负载弯曲
bow instrument 弹簧圆规
bow jet maneuvering system 船首喷射侧推装置
bow jet rudder 艏喷水侧向舵;首喷水侧向舵
bowk 吊桶;提升大吊桶
bow knot 缩帆结;活结;滑结
bowl 罗盘盆;料碗;木球;圆锥壳;转鼓;滚球;辊筒;大酒杯;浮槽;球形物;漂网上小浮标;碗状物;筒体
bow ladder 艏部桥架
bowl and slips 卡盘
bowl arrangement 碗状排列(构造);盆状排列(构造);碗状构造;碗形柱头装饰
bowl bottom 斗底
bowl brake valve 铲斗制动阀
bowl brick 托碗砖
bowl capital 碗形柱头装饰;碗形柱头
bowl carry-check valve 铲斗传送防逆阀
bowl centrifuge 无孔截头锥形筒离心机
bowl classifier 铲斗分选工;铲斗分选机;浮槽分级机;分极槽;分类槽
bowl clutch valve 铲斗离合阀(门)
bowl control cable 铲斗操纵缆
bowl control lift cable 铲斗提升索
bowl control valve 铲斗操纵阀(门);铲斗控制阀
bowl cover 杆盖;杯盖
bowl cover packing 杯盖密垫
bowl crown 杯状树冠
bowl depression land 碗状洼地
bowlder shore 大卵石岸
bowled floor 蝶形坡地板;滚球式地板;碗形地面
bowl end 驼峰编组场尾部
bowl-feed polish 滚筒抛光
bowl-feed polishing machine 滚筒进给抛光机
bowl-feed polishing process 滚筒抛光过程
bowl-feed technique 滚筒抛光技术
bowl grinding machine 滚球研磨机
bow light 旗杆灯;首灯
bow lighthouse 红绿灯板框
bow line 弓形线;弓形曲线;船首纵剖线;头缆;艏缆;张帆索;单套结【船】;帆脚索
bowline bridle 帆脚索
bow line chart 船首线形图
bowline cringle 帆脚索索环
bowline hitch 单套结【船】;帆脚索结
bowline knot 帆脚索结
bowline on a bight 双套结
bow line scow 船首方驳;艏缆方驳
bowling alley 地球场;保龄球场;地滚球场

bowling-alley test 电球强度试验
bowling-alley tester 滚球强度测试器
bowling center 保龄球中心;滚球中心;滚木球场
bowling centre hall 保龄球大厅
bowling green 地滚球场;木球草地;草地保龄球
bowling hall 保龄球房;滚球房
bowlingite 皂石
bowling stadium 滚木球场
bowl in the shape of a steamed bun 馒头形碗
bowl jack 铲斗传动装置;铲斗汽缸
bowl lamp 碗型灯
bowl level adjustment 铲斗高度调整
bowl lift 铲斗提升
bowl lift cable 铲斗提升用缆
bowl lift lever 铲斗提升用杠杆;铲斗提升吊杆
bowl line 铲斗缆;船首缆
bowl metal 粗铸锑
bowl mill 球磨机;碗磨;碗辊磨
bowl mixer 盘式搅拌机
bow loading 船首上货
bow loading and discharge arrangement 首部装卸设备
bow log 船首水压计程仪
bowl-rake classifier 浮槽耙式分级机
bowl ram 铲斗动力油缸
bowl scraper 斗式刮土机;斗式铲运机(车);铲运机
bowl-shaped magnet 碗形磁铁
bowl-shaped settlement 碗状沉降;碟形沉降
bowl-shell wall 离心机篮壁
bowl side 斗边
bowl sidewall 斗边侧翼;铲斗侧壁(刮土机、铲运机)
bowl side wall extension 铲斗侧翼伸延
bowl slip-sockets 抓钩;打捞器
bowl stop 铲斗碰挡
bowl temperature 滚筒温度
bowl track 驼峰峰顶股道
bowl travel(l)ing lock 铲斗运行阻碍
bowl-type centrifuge 碗式离心机
bowl-type classifier 分级槽;浮槽式分级机;分类槽
bowl-type cutter 弓形切坯器;切坯弓
bowl type mill 盘式辊磨机;碗形磨
bowl urinal 小便斗
bowl valve 铲斗阀
bowl valve group 铲斗阀组
bowl washer 洗碗机
bowl-washing machine 洗碗机
bowl-washing room 洗碗间
bowl with carved designs 雕花碗
bowl with fitted cover 盖碗(瓷器名)
bowl with sapphire blue glaze 宝石蓝釉碗(瓷器名)
Bowmaker's plasticity index 鲍麦克塑性指数
bowman 头桨手
Bowman's theory 鲍曼学说
bow member 弓形构件
bow mounted sonar 舰艏声呐
bow navigator 船首推进器
bowoar 头桨手
bow of a rudder 舵弓
bow of shackle 卸扣弯部
bow-on
bow-opening door 船首门
bow or head rope 船缆
bow painter 艇首缆
bow pen 划线笔
bow pencil 弓形铅笔圆规
bow pens 两脚规;小圆规
bow piece 木舵后边缘
bow-piece of corbel bracket 斗拱(的拱)
bow plating 船首外板
bow pointer 尖篾板
Bow porcelain 鲍氏软瓷(法国)
bow port 首货门
bow propeller 船首推进器;船首侧推器;艏推进器;艏螺旋桨
bow pump 船头泵
bow ramp 船首跳板门;船首跳板;船首门
bowranite (适用于多种特殊油漆的)防腐蚀保护层
bow room 弓形室
bow rudder 船首舵;艏舵
bow saw 弓形锯;弓形锯框;弓锯
bow saw frame 木工锯架
bow screen 弧形筛

bow seal 艏气封
bow seas 船首浪
bowse away 拉起
bow section 船首分段
bowser 加油船;加油车;加油泵;水柜车;水槽车;机场加油车
bowser boat 水柜艇;水槽艇;加油艇
bowse to a chain 用制链器制链
bow shackle 弓形钢丝绳夹头;弓形卸扣;弓形钩环
bow-shaped 弓形栏杆小柱;弓形(的)
bow-shaped member 弓形构件
bow shell door 艏门;首门【船】
bowshock 弓形激波
Bow's notation 鲍氏符号;鲍氏标记法
Bow's polygon 倒数力多边形;反数力多边形
bow spring 弓形片弹簧;船首倒缆;前倒缆
bowsprit 船首斜桁;船首第一斜桅
bowsprit gear 船首斜桅索具
bow stall 无栓分隔栏
bowstave 弓材
bow steering propeller 船首侧推器;艏转向推进器
bow stopper 掣链器
bow strap 艇首缆系环
bowstring 弓弦(式)
bowstring arch 悬梁拱;弓形拱;弓弦拱
bowstring arch bridge 系杆拱桥;有拉条拱桥;弓弦式拱桥;弓弦拱桥;弓弦式拱桥
bowstring arched girder 弓弦拱形大梁
bowstring beam 弓形梁;弓弦式梁;弓弦梁;弓弦拱梁
bowstring bridge 弓弦式桥;弓弦桥
bowstring girder 弓形横梁;弓弦式大梁;弓弦大梁
bowstring length between half-maximum point 半极值点弦长
bowstring length between half-maximum point of curve mid-gradient 中梯曲线半极值点弦长
bow string pattern 弦纹
bowstring roof 弧形屋顶;弓弦屋盖;弓弦式屋顶
bowstring roof truss 拱形屋架;弓弦式屋架
bowstring tangent distance 弦切距
bowstring tangent distance of combined profile curve 联剖曲线弦切距
bowstring tangent distance of mid-gradient curve 中梯曲线弦切距
bowstring truss 弓形桁架;弓弦式桁架;弓弦桁架
bowstring truss bridge 弓弦式桁桥
bowstring truss girder 弓弦式桁梁
bow strip 接触条;接触滑条
bow strop 艇首缆系环
bow superstructure 船首上层建筑
bow support 拱顶支座
bow supporter 拱顶支架
bow suspension 受电弓悬挂装置
bowtel(1) 四分之三倚柱;凸圆线脚;凸出圆饰;四分之三凸圆饰
bow thruster 船首转向推进器;船首推进器;船首横向推进器;船首侧推器;首推器
bow thrust machinery compartment 艏侧推器机械舱
bow thrust propeller 船首侧推器
bow thrust tunnel 船首侧推装置管隧
bow thrust unit 船首推进器;船首推进装置
bow-tie antenna 蝴蝶结天线
bow troley 触轮式电车
bow-twixter 弓形扭绞机
bow-type cutter 单弓切坯机
bow-type spring 叠板弹簧
bow-type track ga(u)ge 弓形道尺
bow unit 队首助航船
bow up 上挠
bow visor 船首门;首盔
bow warping 板材翘曲
bow wave 正激波;弓形波;顶头波;弹道波;船头浪;船头波;船首波;冲激波;头部激波;头波;艏波;激波;脱体波
bow weight 弓力
bow winch 艉锚绞车
bow window 凸肚窗;圆肚窗;弓形凸窗;弓形窗;凸窗
box 逻辑匣;框符;锯钮;小木箱;箱子;箱形物;箱匣;钻盒;程序中的逻辑单元;包厢;尾堵(舢板后段)
box abutment 箱形桥台;箱形拱座;箱式桥台
box-and-arrow notation 方盒箭头表示法

box and needle 罗盘;小罗经;手提罗经
box and pin 公母扣(接头)
box and pin type 公母接头型抽油杆
box annealed sheet 装箱退火的薄钢板
box annealing 密闭退火;装箱退火(板材);闭箱退火
box annealing furnace 箱形退火窑
box antenna 箱箱天线
box apron 箱式伸臂
box arch bridge 箱形拱桥
box baffle 扬声器助音箱;扬声器箱
box barge 方驳
box barrage 方形弹幕
box barrow 推车手(推)车
box base 箱座
box beam 箱形梁;匣形梁;方型梁
box beam floor 空心梁楼板;箱形梁楼板;箱形截面梁楼板
box bearing 空转轴承
box bed 凹凳床;有围栏床;有栏杆的床
box bell 母锥
box bit drill 梅花钻
box board 箱板;硬纸板;箱底板
box-body dump car 侧卸箱车;倾卸箱车
box bolt 盒式插锁;门锁插销;门插销
box bond 空斗砖
box bond brick wall 空斗墙
box bow 方形船头;方形船首
box bow and stern 方形艏艉
box breakwater 木笼防波堤;箱形防波堤
box bridge 箱形桥;匣式电桥;电阻箱电桥
box bubble 圆泡水准器;球形水准器
box building 箱状建筑物
box caisson 箱形沉箱;箱式沉箱;沉箱;方形沉箱;标准沉箱
box caisson foundation 沉箱基础
box canyon 箱状峡谷;箱形峡谷
boxcar 闷罐车;涵箱;箱式货车;有篷货车;棚车;厢车;料斗车;矩形波串;闷罐车;箱车;转换盒;带篷车;棚车
box carburizing 装箱固体渗碳;闭箱渗碳
boxcar function 矩形波函数;方脉冲函数
box casing 家具内衬;箱衬;衡重箱;箱形门框;衡重箱框;平衡重滑道框
box chamber 长方室
box channel 长方形槽
box chisel 起钉凿;起钉錾
box chuck 两爪卡盘;双爪卡盘
box circuit 警报装置电路
box closure 箱子闭锁;接合螺栓;系紧螺栓;套环螺栓;紧墙螺栓
box cofferdam 箱形围堰;围堰
box column 箱形柱;箱柱;箱形钢柱
box compass 罗盘仪;方框罗针
box compound 浇注电缆、套管的混合物
box conduit 箱形输水道
box connection 套管连接
box connector 接线盒;箱式连接器
box construction type 箱形构造类型;箱形结构形式
box container 集装箱
box cooler 箱式冷却器
box corer 箱形岩芯取样器
box cornice 空心跳檐;箱形飞檐;飞檐形窗帘盒
box cotton 标准样棉
box couch 内装储柜的床
box coupling 套管轴套;套管联结;接箍;内螺纹半锁接箍;箱形联轴节;箱形连接器;筒接头;套筒联轴节;套管接合;箱形联结器
Box-Cox transformation 博克斯—考克斯变换
box(cross-)section 方形截面;箱形横截面
box culvert 矩形涵箱;箱形涵洞;箱式涵洞;箱涵
box cut 开挖锄;箱形挖锄;箱形堑沟;桶形掏槽
box cutting method 箱形开挖法
box dam 箱形围堰;箱形坝;整体围堰;围堰;围坝
box data 框数据
box design 箱子设计;箱形设计
box diagram of exchange 交易的矩形图
box diagram of production 生产的矩形图
box diffusion 箱法扩散
box dock 箱形浮(船)坞;箱式浮坞;箱式船坞
box dovetail 盒式燕尾榫接;明箱燕尾形榫接
box drain 方形沟;箱形排水渠;匣形沟渠;方沟
box drier 箱式干燥器
box driver 套管螺丝起子

box dumper 箱式翻斗;倾箱机
box eaves 飞檐形窗帘盒
boxed 匣形的;箱形的
boxed air level 气泡水准仪
boxed anchor 双向锁定
boxed buttress wall 箱形墙垛
boxed cargo 箱装货(物);箱形货物
boxed cornice 空心挑檐
boxed dimension 轮廓尺寸;外形尺寸;最大尺寸;总尺寸;装箱总尺寸;全尺寸
boxed eaves 封闭檐口;封闭式屋檐
boxed frame 匣形窗框;箱形框架
boxed girder 箱形大梁
boxed gutter 箱形水槽;箱形檐沟
boxed heart 罩壳心材;函心材;去芯锯法(木材)
boxed-heart timber 带芯锯材
boxed mullion 门窗的匣形竖框;匣形竖框
boxed-off 隔成小间的
boxed pitch (木材的)含髓
boxed rod 方形连杆
boxed roof gutter 箱形屋顶水沟
boxed sheet piling 箱式板桩
boxed shutter 能折入箱中的百叶窗扇;能摺箱中的百叶窗扇
boxed stair(case) 箱式楼梯;箱式楼梯;箱式电梯
boxed steel column 箱形钢柱
boxed stringer 暗式楼梯梁
boxed tenon 匣式凸榫;匣式榫头
box end 母螺纹端;母扣端
box-end stacking 箱形堆垛法
box-end wrench 梅花扳手
boxer engine 汽缸对置发动机
box face 内螺纹面;母扣面;母扣端面
box falling gate 卧式坞门
box feeder 加料箱;箱式加料器;箱式给料机
box filter 箱式过滤器
box flange 轴箱凸缘
box flume 箱形渡槽
box fold 箱状褶皱
box footing 箱形基脚
box-form built steel column 箱式组合钢柱
box former 箱形梁模板
box for mirrors 镜箱
boxform lintel (金属门框上用的)钢过梁
box form(work) 箱式模板;箱形模板
box for weights 称重箱
box foundation 箱形基础;箱基
box frame 箱形肋骨;箱形构架;匣形构架;箱形框架;箱框;匣式窗框;空心窗框
box frame(d) construction 箱形框架结构;匣形结构
box frame in jack-in method 箱涵
box frame motor 框形电动机;箱形机座电动机
box frame of window 匣形窗框
box frame type reinforced concrete construction 箱形框架钢筋混凝土结构
box frame type shear wall 箱形框架式抗震墙
box freight car 箱式货运车辆;篷车;有车厢的货车;棚车
box furnace 箱式窑
box gas purifier 箱式气净化器
box gate 箱形船闸;卧式坞门
box ga(u)ge 盒式浮标水尺;箱式验潮仪;箱形潮位仪;箱式潮位计;箱式测潮表;浮子验潮仪;浮子潮水计
box girder 箱形截面梁;空腹梁;箱形桁;箱形(大)梁;匣形梁
box girder bridge 箱形梁桥;箱形(薄壁)梁桥;箱梁桥
box girder cantilever bridge 箱式大梁悬臂桥;箱形式悬臂桥
box girder crane 箱形截面梁式起重机
box girder with multiple webs 多腹板箱形梁
box girder with sloping exterior webs 斜腹板箱梁
box girder with two compartments 有两分隔间的箱式大梁
box grain 方格粒纹
box green 黄杨绿
box groove 箱形轧槽
box ground (产于英国威尔特郡的)奶油色石灰石
box guidance 轴箱定位装置
box gunwale 箱形舷缘
box gutter 箱形檐沟;箱形雨水槽;箱形水沟;箱形水槽;箱形路沟;箱形边沟;匣形水槽;槽形天沟

方形檐沟;平行雨水槽
box hardening 箱渗碳
box hat 钢锭帽;钢毂;钢壳
box header 盒式联箱
box header boiler 整屋(水管)锅炉;整联箱式锅炉
box-head window 箱头窗
box heart check 心裂
box heliotrope 匣式回照器
Box Hole 箱穴
box hooks 吊箱钩(吊装或桶装货用)
box-hopper spreader 箱式摊铺机;箱式分布机;箱斗式摊铺机;箱斗式分布机
box horn 喇叭形天线
box horse (手推车马道的)箱形支座
boxig 环焊
box-in 加边框式;加边框式;特异型制品装窑法
boxiness 呈四方状
boxing 钻盒;装箱;周围焊接;制箱木料;对口装窑;吊窗匣子;槽口嵌接材;复肋材;百叶窗匣;绕焊
boxing arena 拳击场
boxing field 拳击场
boxing footing 沉箱下部结构;沉箱基础结构;沉箱底层结构
boxing-in 浇注;砌入;装配;安装;箱形料架
boxing-in pocket setting 笼架装
boxing out 留洞(模板);留孔(模板)
boxing ring 拳击场
boxing shutter 百叶窗匣;(能藏在匣内的)软百叶窗;折叠式百叶;匣装百叶
boxing strip 封盒带;封箱带
boxing tenon 框榫
boxing the compass 连续罗经读数
boxing up 捣起道;做围圈;立板条;包装
boxing up gang 道砟整理工作班
box inlet 箱形进水口
box intake 箱形进水口
box iron 槽钢
Box-Jenkins control 博克斯—詹金斯控制
Box-Jenkins forecasting 博克斯—詹金斯预测
Box-Jenkins forecasting method 博克斯—詹金斯预测法
Box-Jenkins method 博克斯—詹金斯法
Box-Jenkins model 博克斯—詹金斯模型
Box-Jenkins seasonal model 博克斯—詹金斯季节模型
box jib 箱形挺杆
box jig 箱式夹具
box joint 内螺纹半锁接头;母接头
box keelson 箱形内龙骨
box key 梅花扳手;套筒扳手
box kiln 箱式干燥窑
box lamp 应急舱壁灯
box lattice(d) girder 箱形格构大梁
box launder 箱形流槽
boxless mo(u)lding 无箱造型
box level 圆水准器
box lewis 箱式起重爪
Box-like 箱状
box-like arm 盒形支臂
boxlike casting table 箱式铸铁工作台
box link 夹板滑环
box link motion 夹板滑环运动
box lock 包壳锁
box loop 环形天线;箱形天线
box lumber 箱材
box lumen meter 箱式流明计
box marking 装箱标志
box marking ink 号箱墨水
box member 内螺纹部分
box metal 减摩合金;轴承合金
box method 箱式配料法
box model 箱式模型
box mo(u)ld 箱式造型;箱式翻砂模;有箱造
box mo(u)lding 砂箱造型
box nail 平方钉;箱用钉
box nailing machine 打钉机
box nut 箱形螺母;外套螺母;盖螺母;方盔形螺母
box off 使船头离开风
box office 票房;售票室;售票处
box of shutters 百叶窗匣
box of tool joint 钻具连接母扣;钻杆母扣接头
box of tricks 差速机构
box opener 开箱器;开箱机
box out 空心箱;(模板的)凹出口;箱形出口;预留孔

模;安置箱型板梁;留孔(模板);留洞(模板)
box pallet 框盒底板;箱形托盘;箱底托盘;框盒托盘
box pass 矩形孔型;箱形孔型
box-pew 箱形靠背长椅
box photometer 盒形光度计;箱形光度计
box photometry 方格测光
Box-Pierce chi-square test 博克斯—皮尔斯卡方检验
box pile 箱形桩;箱形钢桩
box pin 箱销;砂箱定位销
box pipe 箱形管
box plane 槽刨
box planting 箱植法
box plate girder 箱形板梁
box pleating 胖裥;和合裥;箱式褶裥
box-plywood portal frame 箱形胶合板门式框架;箱形胶合板龙门架;箱形层压板门式钢架
box prestressed concrete structure 箱形预应力混凝土结构
box profile 箱形断面
box pump 箱形泵
box quarter 四瓣轴瓦的滑动轴承
box radiation calorimeter 盒式辐射热量计
box rainwater gutter 箱形雨水沟;箱形水槽;箱形路面排水沟
box rate 包柜费率
box reducer 箱形减速器;箱形减压器
box rib 盒形肋筋
box ribbed arch bridge 箱形肋拱桥;箱形拱桥
box rib profile 箱形肋断面
box roof gutter 箱式天沟
boxroom 箱子间;收藏室(皮箱、旅行箱等的,英国)
box scarf 箱式斜接;(木材的)直面搭接
box screen 箱形框纱窗
box sealing machine 封纸盒机
box search 方格检索
box seat 剧院包厢席;箱座;包厢座
box section 箱形隧洞;箱形断面;箱形截面;箱开截面;箱开断面;管道形式—箱形管
box section column 箱形截面柱
box section foundation 箱形断面基脚;箱形断面基础;箱形断面底脚
box section frame 箱形截面的框架;箱形框架
box section girder 箱形截面大梁;箱形截面梁
box section leader 匣形截面的水落管;匣形水落管
box section pile 箱式截面桩
box section sheet pile 有外壳式;箱式截面板桩
box section string 箱形断面索;箱式截面的纵梁
box section stringer 箱形截面楼梯斜梁
box section under frame 箱形底架
box segment 箱形管片
box separation device 分箱装置
box settle 木制有扶手的高背长靠椅
box sextant 袖珍六分仪
box shape 箱形
box shaped 箱形的
box-shaped bottom door 矩形泥门
box-shaped column 箱形柱
box-shaped hopper door 矩形泥门
box-shaped meander 箱形曲流
box-shaped module open on two side 两端开敞的箱式住宅预制单元
box shaped pillow 箱形枕
box-shaped pipe 箱形管
box-shaped prestressed concrete structure 箱形预应力混凝土结构
box-shaped ridge 箱形脊
box-shaped scraper bucket 箱形的刮土斗;(箱形的)刮土机斗
box share 盒形犁铧
box shear apparatus 盒式剪切仪;直剪仪
box shear test 盒式剪切试验;盒式剪力试验;箱式剪力试验
box sheeting 横板桩;平板支撑;水平支撑(法)
box shook 包装箱用材;箱板料
box-shrouded impeller 壳内叶轮
box shutter 匣内软肖叶;藏在匣内的软百叶窗
box shuttering 箱形模板
box sill 木框砖或混凝土潜台坎;楼板格栅端头钉板条;在木板框架中的砖或混凝土基础
box-sinking method 沉箱掘进法
box slab 箱形薄板;箱形(桥梁)
box sling 箱形吊索;箱货绳扣;吊箱索环

box slip 黄杨木护面条
box socket set 棘轮套筒扳手
box spanner 梅花扳手;道钉扳手;套筒扳手
box spreader 箱式(石料)摊铺机
box spring 席梦思床垫;弹簧软垫
box spring bed 箱式弹簧床
Box's rule 博克斯法则(用以计算排水量或排水管直径的供水问题)
box staff 箱尺;塔式标尺;塔尺【测】
box stair(case) 封闭式楼梯;箱式楼梯;箱型楼梯
box stall (厩的)分格栏;单畜房;单圈;兽栏;分隔房;方形畜舍栏
box stand 箱形座
box staple 门锁槽
box stapling conveyer 箱形钉固的输送器
box stapling machine 钉纸盒机
box steel 箱形钢
box steel sheet piling 箱形钢板桩;箱式钢板桩
box stool 带箱座凳
box stoop 箱形门阶
box store 箱形商店
box strike 锁舌罩
box strike plate 门锁舌片
box stringer 暗色楼梯侧梁(梁身与踏步榫接合)
box structure 箱状建筑物;箱形结构
box strut 箱形支撑梁;箱门支撑杆
box switch 箱形开关
box system 吊箱式;盒子体系;箱形体系;吊箱系统
box tap 打捞母锥
box tenon 匣形榫
box the paint 油漆调稠
box thread 母螺纹;母扣
box timber 箱板(木)材
box tire 轮带;滚圈
box to box 双母扣(接头)
box to box pipe 双母扣钻杆;双母扣管子
box tong space 搭管钳卡口
box tool 组合刀具
box-to-pin 公母扣(接头)
box trailer 箱形拖车
box truss 箱形桁架
box truss girder 箱形桁架梁
box tunnel 箱形隧道
box type abutment 箱形桥台
box type air filter 箱形空气过滤器
box type arm 盒形支臂
box type bit 内螺纹钻头;母扣型钻头
box type boiler 箱式锅炉
box-type boom 箱形吊杆
box-type brush holder 盒式电握夹
box-type clip 盒形卡簧
box type concrete spreader 箱形混凝土摊铺机
box-type construction 箱形构造;箱形结构
box type cooler 箱形冷凝器
box type cross brace 方箱形横梁
box-type cut 桶形掏槽
box-typed leg 箱形床脚
box-type enamelling furnace 箱式搪烧炉
box type frame 箱形架
box type furnace 箱式炉
box type heater 箱式加热炉
box type jib 箱形臂
box type jig 箱式夹具;箱式钻模
box type oven 箱式炉
box-type piston 箱形活塞
box-type regenerator 箱式蓄热室
box-type resistance furance 箱式电阻炉
box-type rheostat 箱式变阻器
box type sampler 匣式取样器
box-type scraper bucket (箱形的)刮土斗
box-type tipping wagon 翻斗车;倾卸矿车
box type truck wheel dump wagon 翻斗车
box type vertical drilling machine 方柱立式钻床
box ventilator 箱形通风管
box wagon 棚车;敞篷货车
box weight 容重
box weir 箱形围堰
box willow 打土机;威罗机
boxwood 黄杨木;箱板(木)材
boxwood bobbin 黄杨木蛋形球
boxwood calliper rule 黄杨木卡尺
boxwood carving 黄杨木雕
boxwood dresser 黄杨木修整器
boxwood folding rule 黄杨木折尺

boxwood tampin 细木管楦;黄杨木捣塞
boxwood T square 黄杨木丁字尺
boxwork 箱状构造;箱形构造;窗格构造;蜂窝状褐铁矿沉积;蜂窝状构造;网格构造
box wrench 梅花扳手;套筒扳手
boycott 联合抵制;拒绝交易
Boyd cone-bearing value 鲍特锥承值;保特锥承值
Boyden index 博伊登指数
Boyden radial outward flow turbine 博伊顿式辐向外流水轮机;博伊顿辐向外流水轮机;博伊登式辐向外流水轮机
Boyden technic 博伊登技术
Boyd winding machine 博伊德卷纬机;波埃德卷纬机
Boyer-Lindquist coordinates 博耶—林德奎斯特坐标
Boyes decision procedure 博伊斯决策法
Boylan automatic cone 波仑型自动调节圆锥分级机
Boyle-Charles' law 波义耳—查理定律
boyleite 四水锌矾
Boyle's bottle 波义耳瓶
Boyle's law 波义耳定律
Boyle's law suit 波义耳定律加压服
Boyle's machine 波义耳机
Boyle's scale 波义耳压力标
Boyle's temperature 波义耳温度
Boylis turbidimeter 波义理斯浊度计
Boys calorimeter 波伊斯量热器
Boys camera 波伊斯照相机
boys gymnasium 儿童体育馆;健身房;体育馆
Boys radio-micrometer 波伊斯辐射测微计
Bozeman's position 博斯曼位置
boziga 寓所;住所
bozo 底画
bozo line 提升方钻杆并放入鼠洞装置
Bozsin box 波琴箱
bozzetto 初步设计模型(意大利)
B-parameter 孔隙压力系数 B
B-pillar 中立柱
B-position 乙交换机
B-power(supply) 阳极电源;板极电源
Brabancon lace 布拉班康梭结花边
Brabander amylograph 布拉班德尔淀粉黏焙力测定仪
Brabant linen 布拉邦特亚麻布
Brabant loop 布拉邦特花边
Brabender plastograph 布拉班德塑性仪
bracciale 挑出的金属旗杆插座
brace 拉条;馈电线凡弦;海潮;纵撑;舵枢;吊带;大括弧;大括号;撑条;撑臂;托板;水平斜撑;手摇曲柄钻
brace and bit 手摇曲柄钻
brace angle 角钢撑架;支撑角钢;撑杆角铁;撑杆角钢
brace band 系带
brace bar 撑杆;拉杆
brace beam 支撑梁
brace bit 弓钻;摇钻;钻孔器;曲柄钻(孔器);手摇(曲柄)钻
brace bit with depth ga(u)ge 带深度规钻头
brace block 联结块;木键
brace bolt and nut 拉条螺栓及螺母
brace bracket 斜撑座
brace chuck 固紧夹盘
brace comb 支架窝
brace counter-sink 钝角菊花形尖头板钻
braced 加支撑的
braced abutment 空腹桥墩
braced and stayed surface 支持和支撑面;有拉撑和支撑的表面
braced arch 拱形桁架;桁架形拱;桁拱
braced arched bridge 拱形桁架桥;桁拱桥
braced arched girder bridge 有斜撑的拱式大梁桥
braced arched girder with parallel booms 有斜撑的平行弦杆拱式大梁
braced arch sprandrel 有斜撑的拱肩
braced beam 桁架式梁;斜撑梁
braced box frame 加支撑的箱形框架
braced chain (悬桥的)桁链
braced-chain bridge 悬链桥
braced-chain girder system 桁梁系统
braced-chain of suspension bridge 桁链悬索桥;悬桥桁链;悬链桥
braced chimney 有支撑的烟囱

braced cofferdam 撑架(式)围堰;撑架
braced core building 格架式筒体建筑
braced cut 支撑基坑
braced dolphin 联结靠船墩;加撑靠船墩
braced door 联结门;加对角撑门扇;直拼斜撑框门
braced excavation 支撑挖掘;支撑开挖
braced face 支撑面
braced frame 联结构架;加撑构架;斜撑构架;刚性构架;带支撑框架;撑系框架
braced frame construction 杆系框架结构;撑系框架结构
braced framing 联结构架;加撑构架;斜撑框架;支撑系结构;支撑构架;撑系框架
braced framing dam 支墩坝
braced girder 联结梁;桁梁;桁架梁;有刚性腹杆的梁式桁架;支撑系主梁;支撑系大梁
braced head 大管钳
braced key 大管钳
braced mo(u)lding 支撑式线脚(作大括弧形挑出)
braced panel 斜撑节段;支撑系节间
braced pier 联结式桥墩;联结墩;连接墩;空腹桥墩
braced pier system 联结墩系
braced purlin(e) 加支撑的檩条
braced quadrilateral 大地四边形
braced rail 支撑横杆
braced reinforced concrete flume 桁架式钢筋混凝土渡槽
braced rib arch 桁肋拱;桁架式肋拱;肋桁拱
braced sheeting 支撑挡板
braced skeleton 支撑骨架
braced structure 支撑结构
braced strut 联结支撑
braced web 斜撑腹板;斜撑梁腹
brace for lightning 避雷针拉铁
brace for lightning rod 避雷针支撑
brace frame 加固构架;斜撑构架;撑系框架;加斜撑的框架
brace framing 斜撑框架(施工法);支撑系结构
brace head 木柄手把
brace jaws 固紧钳;固紧夹盘
brace key 木柄手把
bracelet 护臂
brace member 联结杆
brace mo(u)ld 葱拱样板
brace mo(u)lding 龙骨线脚;葱花饰;盘木线脚;支撑式线脚(大括弧式挑出)
brace nut 拉条螺母
brace of a pile grating 桩台横撑
brace of roof truss 屋架外撑
brace pendant 转桁索
brace piece 联杆;联板;加劲杆
brace pile 支撑桩;斜桩;角撑桩
brace piles 叉桩
brace plate 撑板;缀合板;盖板
brace pole 支撑杆
Brace prism 布雷斯棱镜
bracer 捆扎工
brace rail 支撑横杆
brace rib arch 桁助拱
brace rod 拉杆;连接杆
brace root 支柱根
brace scale 斜撑长度刻度表
brace screw 撑柱螺丝
brace screw driver 撑臂起子
brace socket 接筋接连的套节;钻机立柱基座
braces of trench 脚手架水平杆件支撑
brace strut 斜撑;支撑(杆)
brace summer 支撑梁;双重梁
brace system 联结系
brace table 斜撑长度计算表;支撑桌;支撑台
brace tube 管状拉条
brace type 支撑式
Bracewell coater 喷雾涂布机
bracewellite 羟铬矿
brace wrench 曲柄头扳手;曲柄扳手
brachial fascia 臂筋膜
brachiform 臂形的
brachiocubital 臂肘的
brachiocyllosis 臂弯曲
brachiocyrtosis 臂弯曲
Brachiopoda 腕足(动物)门
brachiopod faunal province 腕足动物地理区

brachiopod limestone 腕足类灰岩
brachiostrophosis 臂扭转
brachistochrone 最速降线;最短时程
brachistochrone with bounded curvature 界曲率最速降线
brachy-anticlinal fold 短背斜褶皱【地】
brachy anticline 短轴背斜;短背斜
brachy anticline fold 短轴背斜褶皱
brachy anticline trap 短轴背斜圈闭
brachy anticline zone 短轴背斜带
brachy anticlinorium 短轴复背斜
brachy-axis 短轴
brachy-axis echelon type 短轴雁列式
brachy-axis juxtaposed type 短轴并列式
brachy-axis tandem type 短轴串列式
brachycephalic 圆头型
brachydactylia 短趾
brachydiagonal 短轴
brachydome 短轴穹隆;短轴坡面
brachygeosyncline 短地槽
brachy pinacoid 短轴面
brachy-prism 短轴柱
brachypterism 短翅
brachypterous 短翅的
brachy pyramid 短轴棱锥
brachyskelic 短腿型
brachy-synclinal fold 短向斜褶皱
brachy syncline 短(轴)向斜
brachy syncline fold 短向斜褶皱
brachy synclinorium 短轴复向斜
brachytherapy 短距离放射治疗
brachyural 短尾的
brachyurous 短尾的
braciale 挑出的金属旗杆插座
bracing 联结(系);连接系;交搭;加劲;加固;加撑;支撑;撑条;撑拉件
bracing angle 加劲角钢
bracing beam 联结条;联结梁;支撑梁
bracing boom 加劲杆;撑梁
bracing cable 拉索
bracing cage 支撑铁架;支撑铁笼
bracing channel 支撑槽钢
bracing column 支撑柱
bracing diagonal 加劲斜杆;连接斜杆;支撑斜杆
bracing diaphragm 支撑隔板;加劲隔板
bracing frame 加劲框架;加固构架;支撑架;带斜框架
bracing girder 加劲大梁;联结大梁;加劲梁
bracing guy stub 紧线桩
bracing in open cut 明挖支撑
bracing in tunnel support 支护顶撑;隧道支护顶撑
bracing iron 加劲铁;支撑铁
bracing member 加劲杆;支撑构件
bracing of crown 大碹支撑
bracing of regenerator 蓄热室支撑
bracing panel 支撑板
bracing piece 横梁;斜梁;斜撑;加撑杆;加劲杆;加劲板;支撑杆
bracing plane 联结面;支撑面
bracing plate 撑板
bracing pressure 支撑压力
bracing purlin(e) 支撑檩条
bracing reinforcement 联系钢筋
bracing rib 支撑肋;加强肋
bracing ring 支撑圈
bracing sheet 加强板
bracing slackness 放松紧度
bracing structure 支撑结构
bracing strut 斜杆;支杆;斜撑;支柱(撑杆)
bracing surface 支撑面
bracing system 支撑系统;杆件系统
bracing the bit 支撑钻头
bracing tied-arch 桁架式连杆拱
bracing to piers 支撑在墩上
bracing tube 撑管
bracing wall 加固墙;支撑墙
bracing wire 拉线;拉铁丝;紧固线;张线
bracing with verticals 竖杆联结系
brack 有缺陷的;分等;不合格的
brackebuschite 锰铁钒铅矿;水钒锰铅矿
bracker fan 台式风扇
bracket 括号;夹板;夹叉;牛腿【建】;悬臂;支托;支臂(架);斗拱;分档;标界消息;泵座;托臂
bracket angle 托架角铁

bracket arm 托柄扁担;单臂线担;托架臂;托臂
bracket baluster 挑出栏杆;踏步栏杆柱;踏步侧边栏杆;踏步边栏杆柱
bracket base 支架臂底座
bracket base of mid-anchor clamp for busbar 汇流排中心锚结下锚底座
bracket bearing 托架支承;托架轴承
bracket block 斗【建】;托架垫块
bracket board construction 悬挑板结构
bracket bracing 托座支撑
bracket capital (早期西班牙式的)牛腿柱冠;伸臂柱头
bracket clamp 托架夹
bracket column 托架柱
bracket column connection 梁柱托座结合
bracket communication 标界通信
bracket crab 壁装起重绞车;壁吊
bracket crane 悬臂式起重机;悬臂吊车;壁吊;伸臂吊车
bracket-drilling machine 托架钻床
bracketed arithmetic expression 括号算术表达式;带括号算术表达式
bracketed cornice 托座飞檐;悬臂式挑檐
bracketed logic(al) expression 括号逻辑表达式
bracketed stair(case) 悬臂式楼梯
bracketed step 悬臂式楼梯;悬臂楼梯
bracketed string 制动带;(贴有托座形装饰的)露明楼梯梁
bracket figure 括号内数字
bracket floor 空心肋板;组合肋板;构架肋板
bracket foot 托架脚
bracket for head-span wire 定位索底座
bracket frame 肘板框架肋骨
bracket fungus 多孔菌
bracket height 机架高度
bracket-hung wash basin 托架悬挂式脸盆
bracket-hung wash bowl 托架悬挂式脸盆
bracketing 底数托;承托底层;撑托
bracketing approach 插入法
bracketing method 夹叉试射法
bracketing method of mass measurement 质量测定插入法;插入法质量测定
bracketing technique 高低射标技术
bracket insulator 直脚绝缘子;卡口绝缘子
bracket joint 角接板;角板
bracket knee 梁肘
bracket lamp 壁灯;壁装式托架灯;托架灯
bracket leading jetty 托架式导堤
bracketless 无托架的
bracketless framing 无肘板结构式
bracket light 悬臂式壁灯;壁灯
bracket-like 撑架状
bracket-like column 托座式支柱;托臂式支柱;帽柱
bracket load 托座荷载;托臂荷载
bracket metal 托架轴承合金
bracket mount 托架;角撑架
bracket mount for discharge pipeline 排泥管架
bracket number 括号内数字
bracket of cantilever 悬臂肘板
bracket operation 括号运算
bracket pair 括号对
bracket panel 辅助盘;副盘
bracket pile 承台桩
bracket plaque 托架板
bracket plate 肘板
bracket pole 托臂支柱
bracket post 托座支柱;托臂支柱
bracket progression 分等递进征税
bracket protocol 分类协议;标界协议
bracket saw 框锯
bracket scaffold(ing) 附墙式脚手架;悬臂(式)脚手架;托架式脚手架;挑出式脚手架
bracket set 斗拱
bracket set on column 柱头铺作;柱头科
bracket set on corner 转角铺作;角科
bracket staging 挑出式脚手架
bracket state manager 分类状态管理程序;标界状态管理程序
bracket-step 挑出踏步
bracket stringer 挑出式楼梯边梁;挑出楼梯边梁;挑出的楼梯纵梁
bracket support 牛腿支座;牛腿支架;托臂支座;托臂支柱;托架托座
bracket suspension 横撑悬挂

bracket system 肘板式;托架装置
bracket table 托座工作台
Brackett continuum 布拉开连续区
Brackett limit 布拉开系极
bracket toe 肘板趾
bracket tray 托砖
Brackett series 布拉开系
Brackett spectral lines 布拉开谱线
bracket type quay wall 扶垛式岸壁
bracket-type retaining wall 托座式挡土墙
bracket-type sandslinger 单轨式抛砂机;壁行式抛砂机;壁行抛砂机
bracket winch 墙上绞车;托架绞车
brackish cooling water 半咸冷却水;微咸冷却水
brackish deposit 半咸水沉积
brackish facies 半咸水相
brackish lake 低盐湖;半咸水湖
brackish marsh 微咸水沼泽
brackishness 半咸性;微咸性;微碱性
brackish water 苦盐水;苦卤水;苦水;咸淡水;淡盐味水;淡海水;半咸水;微咸水;稍带咸味的水
brackish-water aquaculture 半咸水养殖
brackish water culture 咸淡水养殖
brackish water ecosystem 咸水生态系统
brackish water environment 半咸水环境
brackish water fauna 咸淡水动物区系;咸水动物区系
brackish-water lagoon 半咸水泻湖
brackish water lake 半咸湖;微咸湖
brackish water limestone 微咸水石灰石
brackish water plankton 半咸水浮游生物
brackish water resources 含盐水资源
brackish waters 微咸水域
brackish water zone 盐味水域区;咸淡水混合区;非饮用水区;盐淡水混合区
Bracklesham bed 布拉克尔歇姆层
bracteole 小苞片
bracttice 围板
brad 角钉;销钉;方钉;曲头钉;平头型钉;无头钉;土钉
bradawl 木工锥子;锥钻;打眼钻;小锥
Bradbury's formula for stresses due to thermal warping 布拉德佰里温度翘曲应力计算公式
bradding (牙轮轮齿的)挤压
braden gas (油层折)顶部天然气
bradenhead gas 井口瓦斯
bradenhead squeeze job 水泥挤入管柱作业
Bradford breaker 滚筒式碎选机;勃莱福破碎机
Bradford Clay 布莱德福特黏土
Bradford distribution 布莱德福分布
Bradford ga(u)ge 布莱德福式大容量雨量计;布莱德雨量斗
Bradfordian 布莱德福阶;布莱德福期
Bradford preferential separation process 选择性浮游分选法
Bradford system 勃拉福分级制
Bradford warping mill 布莱德福整经机
Bradley aberration 布拉德莱光行差
Bradley aberration method 布拉德莱光行差法
bradley fountain 圆形共用淋浴设备
bradleyite 磷碳镁钠石
brad nail 曲头钉;无头钉;角钉
brad punch 压钉器;钉具;钉钉器
brad pusher 曲头钉打钉器
brad set 钉具;钉钉器
brad setter 射钉枪;冲孔器;线脚钉夹钳;钉钉机
brady 活动百叶窗及房屋配件
bradye 布拉代染色法
bradypragia 动作缓慢
bradyscope 慢示器
bradyseism 缓震;海陆升降;地壳缓慢升降运动
brae 倾斜处;山坡;山腰;丘陵区;陡路
Bragaw tensile impact test 布拉格张力冲击试验
Bragg 布拉格
Bragg angle 布拉格角
Bragg-Brentano parafocusing system 布拉格—布伦塔诺仲聚焦系统
Bragg cell spectrometer 布拉格盒频谱仪
Bragg condition 布拉格条件
Bragg cone 布拉格衍射锥
Bragg coupling efficiency 布拉格耦合效率
Bragg curve 布拉格曲线
Bragg deflector 布拉格偏转器
Bragg diffraction 布拉格衍射;布拉格散射

Bragg diffraction spectrograph 布拉格光栅摄谱仪
Bragg diffractometer 布拉格衍射计
Bragg effect 布拉格效应
Bragg electro-striction 布拉格电致伸缩
Bragg equation 布拉格方程
bragger 牛腿;托架;挑出的承重砌体;挑台
Bragg focusing condition 布拉格聚焦条件
Bragg frequency 布拉格频率
Bragg grating sensor 布拉格光栅传感器
Bragg-Gray cavity principle 布拉格—格雷空腔原理
Bragg-Gray principle 布拉格—格雷原理
Bragg-Gray relation 布拉格—格雷关系
Bragg ionization spectrometer 布雷格电离分光计
braggite 硫镍钯铂矿;布拉格矿
Bragg-Kleeman rule 布拉格定则
Bragg law 布拉格定律
Bragg maximum 布拉格最大值
Bragg method 布拉格法
Bragg modulator 布拉格调制器
Bragg-Pierce law 布拉格—皮尔斯定律
Bragg plane 布拉格平面
Bragg reflection 布拉格散射;布拉格反射
Bragg reflection angle 布拉格反射角
Bragg reflector 布拉格反射器
Bragg rotating crystal method 布拉格转(动)晶体法
Bragg rule 布拉格定则
Bragg scattering 布拉格散射
Bragg's law 布拉格定律
Bragg spectrograph 布雷格摄谱仪
Bragg spectrometer 布雷格分光计;布拉格光谱计;布拉格分光计
Bragg-William's theory 布拉格—威廉姆理论
Bragg X-ray microscope 布拉格X射线显微镜
Brahman style 婆罗门式(建筑)
braid 条带;饰带;编带
braid bar 江心洲;江心滩;游荡沙坝;网状河沙坝
braided acetylene hose 编织乙炔胶管
braided asbestos cord covering 石棉绳编织的覆盖层
braided cable 编织电缆;编缆;编包电缆
braided cable jacket 编织电缆外套
braided channel 多汊水道;分汊河道;分汊河槽;辫状(河)道;辫状槽;网状水道;网状渠道
braided compressionresistant hose 编织耐压胶管
braided construction hose 编织胶管
braided course 多汊河段;分汊河流
braided distributary estuary 网状河口
braided door 栅栏门;编苇门;编竹门
braided drainage 辫状水系
braided electrode 编织电极
braided fabric 编织物;编苇织物
braided glass tube 编织玻璃纤维管
braided hose 加衬软管(编织物);编织物填衬软管
braided interchange 群桥交叉
braided intersection 多交叉点;多层立(体)交(叉);群桥交叉
braided mode 网状水系模式
braided nylon rope 编织尼龙绳
braided oxygen hose 编织氧气胶管
braided packing 编织填料
braided reach 游荡性河段;汊流河段;分汊河段;辫状河段;网状河段
braided river 乱流河道;多支汊河流;多汊河流;分汊河流;辫状河;网状河(流)
braided river course 分汊河道;辫状河道
braided river sedimentation model 辫状河沉积模式
braided rope 辫带式钢索;编织绳
braided sleeve 编织套管
braided slip surface 断层带网状(树枝状)错动面
braided spray hose 编织喷雾胶管
braided strand 编织线
braided stream 多支汊河流;多汊河流;汊河;分汊河流;辫状水道;辫状(河)道;网状河(流)
braided stream deposit 辫状河沉积
braided stream sand trap 辫状河道砂圈闭
braided stream sequence 辫状河层序
braided striation 辫状条痕
braided tape 编织带
braided tubing 编织套管
braided twill weaves 变化网形斜纹
braided type 多汊型
braided wire 编织电线;编织线

braided wire rope 编织钢丝网;编织钢丝绳
braider 编织机;编带机
braiding 河道分支;河道的分支;纤维编织保护层;编织(装饰);编带
braiding knot 渔网结
braiding machine 编织机;编线机
braid pattern 带形纹丝形饰;发辫形装饰
braid rug 编辫地毯
brail 卷帆索;角拉杆;斜撑梁;斜撑杆
brailer 圆锥网;抄网
brailing 抄鱼作业
brailing operation 抄鱼作业
Braille 布瑞勒通信系统
brail line 卷帆索
brail net 抄网
brain 电子脑;电子计算机
brain child 计划;作品;脑力劳动产物
brain damage 脑损伤
brain drain 知识外流;人才外流
brain industry 智力工业;知识产业
brain machine 自动计算机
brain power 能力;智能;智力;科学家;科学工作者
brain rafter 脑椽
Brain's reflex 布雷恩反射
brainstorming 研究讨论;发表创造性意见
brain trust 智囊团;专家顾问团
brainwash 推销
brain work 脑力劳动
braise 煤粉
brait 非研磨用的金刚石;未磨的粗金刚石
braithwaite 单框架结构物;预制房屋
braitschite 硼铈钙石
braize 焦屑;煤粉焦粉;煤尘焦粉
brake 唧筒柄;制动器闸;制动器;闸式测功器;大耙;碎土耙
brake accelerator 制动加速器
brake accumulator 制动闸压力积储器;制动蓄压器
brake action 制动作用
brake actuator 制动机构
brake adjuster 制动器调节器;闸调器
brake-adjusting device 闸调器
brake adjusting gear 制动调节机构
brake adjusting jack 制动调整的压力油缸
brake adjusting nut 制动调整的螺母
brake adjusting tool 闸调整工具
brake adjusting wedge 制动调整楔
brake adjustment 制动闸调整
brakeage 制动器的动作;制动力
brake air supply tank 制动空气供给箱
brake air valve 制动空气阀
brake anchor 闸瓦支持销
brake anchorage 制动块支座
brake anchor pin 制动块支撑
brake anchor pin bushing 锚碇螺杆衬套
brake anchor plate 制动块支撑板
brake and air compressor repairing workshop 制动和空压机检修车间
brake apparatus 制动装置;刹车装置
brake application 制动操作;制动作用;施行制动;施加制动
brake application time 制动作用时间
brake arm 制动杠(杆);制动臂
brake assembly 制动系统;刹车装置
brake assistor 制动增力装置
brake axle 制动轴;刹车轴
brake balancer 精确调整制动器
brake balancing arm 制动均衡梁
brake band 制动带;制动器带;制动带;闸带;刹车带
brake band anchor clip 制动带支座
brake band clevis 制动带系钎
brake band lining 制动片
brake band tightener 制动带张紧工具
brake bar 制动杠(杆);制动杆;闸杆
brake bead 拉延筋
brake beam 制动梁
brake beam adjuster 闸梁调整器
brake beams 制动器杠杆系统
brake beam safety chain eye 闸杆安全链眼
brake beam safety chain eye bar 闸梁安全链眼铁
brake beam strap link 闸梁系带环
brake beam tension member 制动器横杆受拉条
brake bell crank 制动器直角杠杆
brake bleeder 制动油路排气装置
brake block 制动瓦;制动铁鞋;制动块;闸瓦;制动片

brake block adjusting gear 制动器闸瓦调整装置
brake-block hanger 闸瓦吊
brake block holder 闸瓦压杆;闸瓦托;刹车片固定座
brake-block insert 闸瓦调整垫片
brake block key 闸瓦销
brake block pressure 闸瓦压力
brake block with removable liner 可更换衬的刹车片
brake booster 制动升压器;制动助力器;制动加力器
brake box 制动盒
brake bush 制动器摩擦衬片;制动衬带;闸衬片
brake cable 制动缆索;制动器拉索;制动拉索
brake cable wire 制动钢丝绳用钢丝
brake calliper 闸瓦卡规
brake cam 制动凸轮;闸凸轮
brake cam lever 闸凸轮杠杆
brake camshaft 制动凸轮轴
brake camshaft collar 刹车凸轮轴环
brake carrier 制动器挂架
brake chain 制动链
brake chamber 制动室;制动气室;闸盒
brake chamber push rod 闸盒推杆
brake chamber push rod guide 闸梁调整器导承
brake chamber push rod spring 闸盒推杆弹簧
brake changeover 制动转换
brake changeover weight 制动重量转换
brake check 闸颊板
brake circuit 制动电路
brake circuit governor 制动电路控制器
brake clamp 制动块
brake clearance 闸间隙
brake clevis 制动拉杆叉
brake clip 制动夹
brake clutch 制动离合器
brake coil 制动线圈
brake compressor 制动气泵
brake cone 制动锥(体)
brake connecting rod 闸联杆
brake control 制动器控制;制动控制;刹车控制
brake controller 闸控制器
brake control machanism 制动操纵机构
brake control shaft 制动控制轴
brake control unit 制动控制单元
brake control valve 转向制动器控制阀;制动控制阀
brake cooler 制动冷却器
brake cord 制动带
brake counterweight 制动器平衡锤;制动平衡重;平衡重
brake coupling 制动连接器;制动离合器;制动联轴器
brake coupling hose 制动连接器软管
brake covering 制动衬里
brake crank 制动手把;制动把
brake cross bar 制动梁
brake cross lever 横向制动杆
brake cross shaft 制动横杆
brake crusher 对肘板颚式破碎机;双肘板颚式破碎机
brake current 闸电流
brake current transducer 制动电流变换器
brake cylinder 制动圆筒;制动油缸;制动器缸;制动汽缸;制动缸;制动分泵;闸缸
brake cylinder clamps 制动器汽缸夹紧器
brake cylinder filling rate 制动缸充气时间
brake cylinder filling speed 制动缸充气速度
brake cylinder lever 制动缸杠杆
brake cylinder lever shaft arm 闸缸杠杆轴臂
brake cylinder pipe 闸缸管
brake cylinder piston 制动器汽缸活塞
brake cylinder piston cup 制动器汽缸活塞皮碗
brake cylinder pressure 制动风缸压力
brake deceleration 制动减速度
brake decelerometer 制动减速计
brake device 制动装置
braked idler 制动空转轮
brake die 折弯模
brake differential steering 制动差速转向装置
brake disc 制动圆板;制动片;制动盘;闸盘;制动轮
brake distance in breakwater entrance 口门制动距离【港】
braked mass 制动质量
brake doctor 制动摩擦片修磨机
brake drag 制动阻力
brake dressing 制动器润滑脂;刹车涂料

brake drum 制动鼓(筒);轮闸(鼓);制动圆鼓;闸轮;刹车鼓
brake drum anvil 制动鼓钻
brake drum drilling machine 闸轮式钻床
brake drum dust cover 制动鼓防尘罩
brake drum dust shield 闸轮防尘罩
brake drum lathe 闸鼓车床
brake drum liner 制动鼓板衬垫
brake drum oil deflector 制动轮导油器
brake drum retarder 刹车鼓缓速器
braked tie rod 制动系杆
brake dust collector 制动闸集尘器
brake dust shield 闸防尘罩
braked wagon 制动的货车;制动车辆
braked weight 制动的重力
braked weight of a train 列车中制动重量
braked weight percentage 制动重力的百分率
brake dynamometer 轮闸功率计;轮韧力计;制动测力计;制动测功计
brake eccentric 刹车偏心轮
brake eccentric wheel 制动偏心轮
brake effect 制动效应;制动效力;制动效果
brake effectiveness road 制动路
brake efficiency 制动效率
brake effort 制动力
brake electromagnet 制动电磁铁
brake engine 制动发动机
brake equalizer 制动平衡装置;制动平衡器
brake expander 制动器扩张器
brake facing 制动衬片;闸衬片
brake-fade 制动器热衰退
brake failure 制动毁坏;制动失效
brake-field triode 减速场管
brake-field tube 正栅管;二次电子抑制管
brake flange 闸凸缘
brake flap 减速板
brake fluid 制动液
brake fluid header 制动液箱
brake fluid system 液压制动系统
brake foot lever 脚踏闸杆
brake foot plate 闸踏板
brake force 制动力
brake(force) coefficient 制动力系数
brake force regulation 制动力调节
brake friction area 制动摩擦面积
brake gate 制动闸
brake-gear 制动装置;闸装置
brake-gear return spring 制动装置复位弹簧
brake-gear support 制动装置支架
brake governor 制动调节器
brake hand control lever 手控制动杆;手控闸杆
brake handle 制动手柄;制动器手柄
brake hand lever 制动的手拉杆
brake hanger bracket 闸瓦托吊座
brake head 闸瓦托
brake head hanger lug 闸瓦托吊耳
brake-holder block 制动器闸瓦;制动片
brake hoop 制动环
brake horse power 制动功率;刹车马力;制动马力
brake horsepower-hour 制动马力·小时
brake hose 制动软管
brake hose and coupling 闸软管及接头
brake hose clamp 制动软管夹
brake hose clip 制动软管夹
brake hose connecting nipple 制动软管螺纹接套
brake hose coupling head 制动软管接头
brake hose joint 制动软管连接
brake hose pipe 闸软管
brake housing 制动遮盖物;制动箱;制动器箱
brake hub 制动轮毂;刹车轮毂
brake hydrovac cylinder 闸瓦真空液压制动器缸
brake incident 制动事故
brake induction coil 制动感应线圈
brake inspection depot 制动检修所
brake interface unit 制动界面装置
brake intermediate shaft 制动器中间轴
brake iron 制动铁带
brake isolated 独立制动
brake jaw 制动爪
brake key 制动键
brake lag 制动延时;制动生效时间
brake lag distance 制动延时距离
brake lamp 制动灯

brake lamp switch 刹车灯开关
brake latch 制动爪;制动掣子
brake latch spoon 制动掣子柄
brake latch spring 制动掣子弹簧
brake lever 制动手把;制动杠(杆);闸杆;刹车手柄;刹车杆;刹车臂
brake leverage 制动杠杆臂长比
brake lever axle 制动杠杆轴
brake lever for rope drum 钢丝绳滚筒刹车(手)把
brake lever fulcrum 闸杆支架
brake lever latch 制动杆爪
brake lever pawl 制动杆爪
brake lever segment 制动杆扇形板
brake lever stop 闸杠杆止动器
brake lifting magnet 制动提升的磁铁
brake line 制动系统管路;闸气管
brake lining 制动内衬;制动器内衬;制动(摩擦衬)面;制动带摩擦片;制动衬面;制动衬带;闸瓦,闸片;闸衬片;闸衬(面);刹车面;刹车垫
brake lining disc 制动面摩擦片
brake lining for drill 钻机闸瓦
brake lining grinder 闸瓦衬带磨光器
brake lining pad 制动器衬垫
brake lining refacer 闸衬光面器
brake lining stret cher 制动摩擦片拉伸器
brake link 闸连杆
brake linkage 制动器连接杆;制动联杆;闸联动装置;刹车联动装置
brake load 刹车力;制动荷载;制动负荷
brake lock 制动闸
brake lock control lever 制动闸的控制杆杠
brake locking 抱闸
brake locomotive 制动机车
brake magnet 制动磁铁
brake main cylinder 主闸缸
brakeman 制动工人;制动员;司闸员
brakeman's box 制动工人小室
brakeman's cabin 制动员室【铁】
brakeman's caboose 司闸员守车
brake mast 闸杆
brake master cylinder assy 制动主缸总成
brake master cylinder cover 制动主缸盖
brake master cylinder filler plug 制动主缸加油器塞
brake mean effective pressure 制动有效平均压力;制动有效均压;制动平均有效压强;刹车平均有效压力
brake mean pressure 平均制动马力
brake mechanism 制动的机械构造;制动机构
brake metering valve 制动调节阀
brake method 制动测量法
brake moment 制动力矩
brake motor 制动电动机
brake nozzle 制动喷嘴
brake oil 制动油;制动液;刹车油
brake operating conditions 制动操作规则
brake operating lever 闸操作杆
brake operating pin 闸操作销
brake operating rod 闸操作杆
brake-operating spindle 制动器操作轴
brake operating system 制动操作系统
brake operating wedge 制动操作楔
brake operation device 制动操纵装置
brake pad 制动块
brake paddle 制动踏板;刹车踏板
brake parking lever 停车制动杆
brake pawl 制动器掣子
brake pedal 脚踏闸;制动器;刹车踏板;制动踏板;脚制动踏板
brake pedal arm 制动器踏板杠杆
brake pedal bushing 制动踏板轴衬
brake pedal shaft 制动器踏板轴
brake pedal valve 制动踏板阀
brake percentage 制动比;制动百分率
brake performance 制动器性能
brake phenomenon 制动现象
brake pipe 制动液管;制动系统管路;制动管;制动输气管
brake-pipe emptying accelerator 制动管排空加速器
brake pipe strainer 闸管滤气管
brake piping 制动的管线
brake piston 制动器汽缸活塞
brake plate 制动叠层板
brake plate anchor 闸瓦固定板

brake position 制动位
brake power 轴功率;制动力
brake-press 闸压床
brake pressure 制动压力;制动机压强;制动机压力
brake pressure ga(u)ge 制动压力机
brake pressure line 制动压力线
brake pressure valve 制动压力阀
brake propagation rate 制动波速
brake-protecting relay 制动保护继电器
brake proving indicator 制动证实指示灯
brake puck 制动圆盘
brake pull 制动曳力
brake pulley 制动轮;闸轮
brake pull-rod 拉闸杆
brake pulse 断路脉冲
brake pushrod 制动推杆
brake ratchet 制动爪;制动棘轮机构
brake ratchet wheel 制动轮;闸棘轮
brake ratio 制动比
brake reaction time 制动反应时间
brake ready for operation 准备制动
brake reel 制动绞盘
brake regulator 制动器调节器;制动调节器
brake relay 制动继电器
brake release 制动缓解;松开制动器
brake release curve 制动缓解曲线
brake release pedal 制动缓解踏板
brake release speed 制动缓解速度
brake release spring 制动器分离弹簧
brake release time 制动缓解时间
brake reliner 制动衬带更换机
brake reservoir 制动器罐;制动机的风缸
brake resistance 制动阻力;制动力;制动电阻
brake response time 制动反应时间
brake retainer 止轮器
brake retracting spring 制动拉簧
brake ribbon 刹车带
brake rigging 制动机试验台
brake-rigging efficiency 制动装置效率
brake-rigging support 制动装置支承
brake rim 制动盘;制动轮
brake ring 制动器止推环;制动器接合盘;制动环;闸板
brake rocker arm 制动摇臂
brake rod 制动拉杆;制动杠(杆);制动杆
brake-rod adjuster 制动拉杆调整器;制动调整器
brake-rod crevice 制动杆端叉形铁
brake-rod spring 制动杆弹簧
brake-rod yoke 制动拉杆叉
brake roll 制动辊
brake rope 制动绳
brake rubber 制动橡皮
brake rubber plate 制动橡皮带
brake scotch 制动瓦;制动三角木
brake screw 制动螺杆
brake screw handle 制动螺旋摇把
brake sealing cup 制动胀圈
brake selector(valve) 制动控制阀
brake service 脚踏闸;制动器操作
brake servo piston 制动器伺服活塞
brake set(ting) 制动调节
brake shaft 制动轴;制动器轴
brake shaft bearing 制动轴承
brake shaft bracket 制动轴架
brake shaft carrier 闸轴架
brake shaft lever 闸轴杆
brake-shoe 刹瓦;制动瓦;制动铁鞋;制动蹄;制动器闸瓦;制动块;闸瓦
brake-shoe adjuster 闸瓦调整器;闸瓦调节器
brake-shoe adjusting cam 闸瓦调整凸轮
brake-shoe anchor pin 闸瓦固定销
brake-shoe clearance 制动块间隙
brake-shoe expander 闸瓦扩张器
brake-shoe for train 火车闸瓦
brake-shoe ga(u)ge 闸瓦规
brake-shoe grinder 闸瓦磨床
brake-shoe guide 闸瓦导柱
brake-shoe guide pin 闸瓦导(定位)销
brake-shoe guide spring 闸瓦导簧
brake-shoe holder 闸瓦托
brake-shoe key 闸瓦销
brake-shoe lining 制动蹄摩擦面;闸瓦片带摩擦片
brake-shoe link pin 闸瓦联杆销
brake-shoe pressure 制动块压力;闸瓦压力

brake-shoe stop pin 闸瓦止动销
brake-shoe tester 闸瓦试验器
brake shuttle valve 制动空气阀
brake signal 制动信号
brake slipper 制动铁鞋;制动滑块
brakes man 制动驾驶员
brake spanner 制动扳手
brake spindle 制动轴
brake spring 制动弹簧;制动器弹簧
brake spring pliers 装制动器张紧弹簧用的夹子;安装制动弹簧用手钳
brake squeak 制动时的啸声
brake staff 制动系统
brake staple 制动卡子
brake steel plate 制动钢薄层板
brake step 制动踏板;闸踏板
brake stop 制动止点;制动器停机;制动块
brake-stopping distance 制动距离
brake supply reservoir 制动风缸
brake support 制动的支承;制动器支座
brake support shield 闸支板
brake surface 制动面;闸面
brake switch 制动开关
brake system 制动系统
brake test 制动试验
brake tester 制动试验台;制动器试验器
brake thermal efficiency 制动热效率;闸测热效率
brake thrust ring 制动器止推环
brake thrust screw 制动器止推调节螺钉
brake time 制动时间
brake toggle 制动肘节;制动凸轮
brake torque 制动扭矩;制动转矩;制动(力)矩
brake transmission speed 制动传送速度
brake trial 制动试验
brake truss 制动桁构
brake truss bar 构架式闸杆
brake tube 闸管
brake tube bracket 闸管托架
brake tube shutoff cock 闸管关断旋塞
brake tube tee 闸管丁字管节
brake turbine 制动涡轮;制动透平
brake unit 制动单元
brake vacuum booster cylinder 制动真空加力泵
brake value 制动值
brake valve 制动阀;闸阀;刹车阀
brake van 制动车;司闸车
brake van track 守车停留线
brake vehicle 机动车的刹车
brake way 滑距;惯性滑距
brake wedge 制动楔
brake weight 刹车重块
brake wheel 制动轮;闸轮
brake wire 制动拉索
brake with graduated release 有阶段缓解的制动机
braking 制动(闸);保持制动作用;刹车
braking ability 制动能力
braking absorption 阻尼吸收
braking action 刹车;制动作用
braking airscrew 反制螺桨
braking axle 制动轴
braking beam 制动梁
braking bench 制动试验台
braking bracing 制动联结系
braking characteristic 制动特性
braking chart 制动特性表
braking chopper 制动换能器
braking chopper module 制动调节模块
braking circuit 制动电路
braking clamp 制动夹(钳);制动块
braking clamp test bench 制动夹钳试验装置
braking club 闸辊
braking coefficient 制动力系数
braking contactor 制动接触器
braking current 制动电流
braking curve 制动曲线
braking deceleration 制动减速率;制动减速度
braking device 制动装置;制动设施;制动器;刹车装置
braking disc 制动盘
braking distance 制动距离;保留距离;刹车距离
braking effect 制动作用;制动效应;制动效能;制动效力;制动效果
braking efficiency 制动效率;刹车效率
braking effort 制动力

braking element 制动元件
braking ellipses 减速椭圆轨道
braking fluid 制动工作液
braking force 制动力
braking force curve 制动力曲线
braking force of train 列车制动力
braking friction 制动摩阻力
braking grade 制动等级
braking gradient 制动坡度
braking index 制动指数
braking inertia 制动惯性
braking jet 制动射流
braking length 制动距离
braking lever 制动手把；闸杆
braking load 制动荷载
braking loss 制动能高损耗
braking magnet 制动磁铁
braking main 制动主管
braking maneuver 制动操纵
braking mechanism 制动机构
braking method 制动方式
braking mode 制动方式
braking moment 制动转矩；制动（力）矩
braking of the load 重车制动作用
braking of the tare 空车制动作用
braking of winch 绞车制动
braking orbit 制动轨道
braking path 制动；制动距离
braking performance 制动性能
braking period 制动周期；制动时段
braking pier 制动墩
braking position 制动位置
braking power 制动力；制动功率
braking power at wheel rim 轮周制动功率
braking radiation 阻尼辐射
braking range 制动幅度
braking ratio 制动率
braking ratio deficiency 制动率不足
braking resistor 制动电阻器
braking resistor arrangement 制动电阻器风扇
braking rim 闸板
braking sheet 制动报表
braking signal switch 制动信号开关
braking skid mark 制动活动印迹
braking slip 制动滑移
braking slope 制动坡度
braking speed 制动速度
braking state of retarder 缓行器制动状态
braking stress 制动应力
braking surface 制动（作用）面
braking switch group 制动开关组
braking system 制动系统
braking test 制动试验
braking time 制动时间
braking to a stop 停车制动
braking torque 制动转矩
braking to suit line conditions 适应线路条件的制动
braking vane 制动板
braking way 制动距离；惯性滑距
braking wheel 制动轮；闸轮
brale 金刚石锥头；（洛氏硬度试验机的）金刚石圆锥压头；锥形金刚石压头
Bramah's lock 布拉默锁闩
Bramah's press 布拉默水压机
bramble 荆棘；悬钩子；复果植物
bramley fall 浅棕色硬砂岩（产于英国约克郡）
Bramley fall stone 布立姆雷佛尔石
brammallite sodium illite 钠伊利石
bran 粗硬麻布
bran bath 糠浴
brancart 效果照明装置
branch 支链；支管；港汊；分枝；分支机构；分行；分路；分公司；分店；分厂；分岔；部门；树枝；扇形拱支肋
branch abscission 枝脱离
branch account 分店账户；分公司账户；分支机构往来账；分类账；分行账户
branch accounting 分支机构会计；分店会计
branch address 转移地址
branch-admittance matrix 分支容许矩阵
branch air duct 支风道
branch and bound 分界界限法
branch and bound algorithm 分支定界法

branch-and-bound method 分支限界法；分支定界法
branch-and-bound model 分店模式
branch-and-bound search 分支界限搜索
branch-and-bound solution strategy 分支界限解法
branch-and-bound technique 分枝和限界法
branch angle 分叉角
branch angularity cage antenna 分支角笼型天线
branch arm 分支臂
branch balance book 分店总账余额本
branch balance sheet 分支机构资产负债表；分店资产负债表
branch bank 分行
branch beacon 航道支流的立标
branch bend 分支弯管；支弯管
branch-bound algorithm 分支界限算法
branch bounding method 分支界法
branch-bound method 分支估界法；分支限界法；分支界限法
branch-bound technique 分支限界技术
branch box 电线分线盒；分线匣；分线盒
branch box to a building 房屋上分线盒
branch business 分支机构；分店
branch cable 分支电缆；分路电缆
branch calling 转移调用
branch canal 分支渠道；分叉渠道；小火道；支通道；支渠
branch cash account 分支机构现金账户
branch cell 三通；支管接头
branch center 分区中心；小区中心
branch channel 支槽；汊道；分支河道
branch chuck 四爪平面卡盘
branch circuit 支路【电】；支回路；分支电路；分路；分流电路
branch clamp 分线线夹
branch conductor 分支导线
branch conduit 分支导管
branch connection 支线连接；分支连接；分路连接；分叉联结；分叉连接
branch consignments 分店承销品
branch control account 分支机构管制账户；分店控制账户
branch controller 支线控制器；分支控制器
branch current 支路电流；分支电流；分路电流
branch curve 分支曲线
branch cut 分支切割
branch-cut bridge 渡线
branch cutoff 分支齿墙
branch diameter 支管直径
branch distributing pipe 配水支管
branch distributor 分支分配器
branch drain 排水支管
branch duct 分支管道
branch earnings 分支机构收益
branched 分枝的
branched cable 分支光缆
branched cage antenna 分支笼形天线
branched candle holder 枝形烛架
branched chain 支链；分枝链
branched channel 分汊河道；分汊河槽；分汊水道；分汊航道
branched disintegration 分叉蜕变
branched distribution system 枝状配水管网；枝状管网
branched ear 分枝穗
branched endings 分枝状末梢
branched-guide coupler 短截线分支耦合器；分支波导耦合器
branched hair 分支毛
branched heating system 枝状热网
branched high polymer 支化高聚物；歧化高聚物
branched inflorescence 分枝花序
branched knot 分叉节；树节（木材缺陷）
branched light guide 分叉光导
branched line 支线
branched lode 支矿脉；分支矿脉
branched network 枝状管网
branched of material production 物质生产部门
branched optic(al) fiber bundle 多股光学纤维束
branched panicles 分枝圆锥花序
branched passage 分管
branched pipe 叉管；分支管
branched polyethylene 高压聚乙烯
branched polymer 支化聚合物；分支聚合物

branched road 歧路
branched root 分支根
branched root system 分枝根系
branched spark 分支火花
branched stack 集合烟囱
branched structure 分支结构
branched system 枝状管网
branched work 枝叶状装饰；枝叶状雕饰
branch elbow 支弯管
branch ell 分支弯管；支弯管
brancher 石匠；石工；石标
branches of non-material production 非物质生产部门
branches of shrubs 灌木枝条
branches of trade 行业；行息
branches of trees 树木枝条
branch establishment 分支机构
branch exchange 分局；交换分机
branch factory 分厂
branch fault 邻近断层；支断层；次要断层；副断层
branch feeding 分支供电；分路馈电；分路供电
branch financial statement 分支机构损益计算书；分支机构财务报表
branch fitting 支管装配件
branch flue 分支烟道；支烟道
branch form factor 枝材形数
branch forth 扩展分支机构（商店等）
branch gap 枝隙
branch gate 分支浇口
branch heading 错车巷道；分支巷道
branch highway 分支总线
branch hole 侧眼；分支钻孔
branch house 分店
branchion of molecule 分子文化
branch impedance 支路阻抗；分支阻抗
branching 扩大营业；支化；出岔；分枝合成；分支过程；分支（放射）；分接；分岔
branching action 文化作用；歧化（作用）
branching algorithm 转移算法；分支算法
branching and intersecting area of large fault 大断裂分支与交切地段
branching at an acute angle 锐角支管
branching at right angle 垂直支管
branching bay 河口湾；汊湾；分枝湾；支湾
branching block post 分歧线路所
branching cable 分支电缆
branching chain reaction 分支连锁反应
branching channel 夹江；枝状渠道；分汊水道；分汊河道；分汊河槽；分汊航道
branching coefficient 支化系数
branching composite variable 枝状组合变量
branching decay 分支衰变
branching diagram 分支图
branching duct 支管
branching factor 分子因子；分支因子
branching fault 分枝断层；分支断裂；分支断层
branching filter 分支滤波器；分路滤波器
branching fraction 分叉系数
branching habit 分枝习性
branching insertion 分枝着生
branching jack 分支塞孔
branching junction 分岔接点
branching Markov process 分支马尔可夫过程
branching network 枝状管网；分支网络
branching of a river 河的分叉
branching-off line 分歧线路
branching off point 分支点
branching operation 转移运算；转移操作
branching pattern 分枝类型
branching pipe 叉管；分支管；三通管；岔管；支管
branching point 分岔处；分支点；分歧点；分枝点
branching Poisson process 分枝泊松过程；分支泊松过程
branching probability 支化几率
branching process 分枝过程
branching program(me) 线路图
branching random walk 分支随机游走
branching ratio 分支放射比；分支比
branching reach 分汊河段
branching reaction 支化反应
branching renewal process 分支更新过程
branching renewal process of occurrence 断层分支复原过程
branching repeater 分支中继器

branching river 分汊河流
branching river-bed 叉河型河床
branching side of points 道岔支线侧
branching stream 分汊河流;分叉河流
branching structure 分枝状构造
branching switchboard 并联复式交换机
branching tube 支管
branching unit 支路装置;分向装置;分路器
branching variable 分支变量
branch institute 分院
branch instruction 分支指令;分岔指令;转移指令
branch interceptor 截流支管
branch intersection 分支交叉
branch interval 分流范围;分流区间;分类间距;支管间距
Branchiostoma 文昌鱼
branch island 河心岛;河间岛;分汊岛
branch joint 支管接头;分支套管;分支连接;分支接头;分线套管;分线接头;三通接头
branch knot 岔节;树节
branch ledger 分支机构(分类)账;分店总账
branch library 图书分馆
branch line 支线;断叉线;岔线【铁】;分支线(路);分支路;分支管道;分支管道;分流
branch line of turnout 道岔侧线
branch-line pumping system 分支管路泵送系统
branch line railway 支线铁路
branch line thermal substation 区域热力站
branch line to outlying terminus 尽头支线
branch linkage 分支连接
branch long-lived assets 分支机构永久性资产
branch loss 分枝损耗
branch main 分支干管;支干管
branch manager 分店经理
branch manifold 多歧管;分总管
branch mast 分支桅杆
branch method 分枝法;分支法
branch net income or loss 分店净损益
branch network 分支机构网;分支机构图
branch node 分支节点
branch number of an underground river 暗河分支数
branch of a curve 曲线的分支
branch off 分支;分叉
branch of fault 断层分支
branch office 分公司;分店;分支行;分支机构;分行;分社;分局
branch office account 支店账户
branch office current 分支机构往来;分店往来
branch office general account 分支机构往来账户;分店往来账户
branch office report 分支机构报告
branch office system 分支机构制度
branch of insurance 保险类别
branch of Kuroshio 黑潮支流
branch of mainspring 击针簧片
branch of tree 树干
branch of work 工作类别;工作部;工种
branch on 按……转移
branch on condition clear 清零条件转移
branch on count 计数转移
branch on equality 相等转移
branch on false 假条件转移
branch on index high 变址数大转移
branch-on indicator 转移指示器
branch-on-switch setting 按预置开关转移
branch on true 真条件转移
branch on zero 零转移
branch on zero instruction 零转移指令;零转移指令;按零转移指令
branch operation 分支操作;转移操作;分支动作;分路运算
branch order 转移指令;分支指令
branch out 分叉
branch piece 三通;歧管;套管;支管(接头);分支
branch pilot 区间引航员
branch pipe 支管道;岔管;歧管
branch pipe of inlet 雨水口支管
branch pipe strainer 支管过滤器
branch point 转移点;分歧点;文化点;分支点;分岔点;分叉点
branch pole 分支杆塔;分线杆
branch prediction and speculation execution 分支预测和推测执行

branch process 分枝过程
branch profit 分号利润;分店利润
branch profit and loss account 分店损益表
branch range 支脉
branch records on home office books 分店交易由总店记账
branch riser 支竖管
branch river 支流
branch road 支路;支道;岔路
branch root 侧根
branch sales office 销售分处
branch sewer 排水支管;支阴沟;分支阴沟;分支沟;污水支管
branch shaped timbering 枝叉型柱支撑;隧道枝叉型柱支撑
branch shipments 发往分店货物
branch sleeve 管节;连接套筒;支管套筒;筒形联轴器
branch socket 分支管筒;分支插口
branch split 枝条劈裂
branch statement 转移语句;分支机构报表
branch steam line 蒸汽支管
branch stem 侧枝
branch store 分店
branch strip 分支带
branch structure 分支结构
branch stub 支矩管
branch switch 分支开关;分路开关
branch switchboard 分路配电盘
branch system 枝状管网;分支系统;分路电话制
branch table 转移表
branch tee 三通接头;三通管
branch terminal 分支线端
branch terminal line 尽头支线;分支终线
branch thrust 分支冲断层
branch to a building 进户支线;用户引入线;进户支管
branch trace 枝迹
branch tracer 转移跟踪程序
branch tracery 枝形花格(窗);枝形窗格
branch track 支线;岔线【铁】
branch track of turnout 道岔侧线
branch transition 分支跃迁
branch transmittance 支路传递系数
branch traverse 支导线【测】
branch trust department 分行信托部
branch tube 支管道
branch turnout 分路道岔【铁】
branch type switch board 分立型配电盘
branch valley 支谷
branch value 分支值
branch valve 分支阀
branch vent 分支排泄口;分支排出管;通气支管;通风支管
branch water 由小河来的水
branch water supply pipe 分支给水管
branchwood 枝条材
branch work 分支网(地下水);分厂;分叉洞穴;树枝状洞穴;树枝状地下水系;辅助车间
branchy 多枝的
branchy stem 分枝
branchy wood 多节木(材);密枝木
brand 烙印;火印;印记;品种;品质;牌子;牌号
brand awareness 牌子意识;商标意识
brand barometer report 品牌晴雨报告
brand choice 牌号选择;商标选择
branded goods 牌名货;商标产品
branded oil 优质油;名牌油
branded product 有商标产品;名牌货
Brandenburg 勃兰登堡(德国)
Brandenburg Gate 勃兰登堡门
Brandenburg ice stage 勃兰登堡冰阶【地】
brander 钉板条;打商标机
brandering (垫木的)灰条木棒;钉抹灰板条用的横垫条;板条子操作;板条木榫;双层灰条
brand franchise 牌子影响范围
brand identification 商标识别
branding 打烙印;印号码;标记
branding and stamping of rails 钢轨打标记
branding iron 路面平压铁嵌;烙铁;刻压机
branding race 打号通道;打号栏
brand iron 烙铁
brandishing 城垛;穿孔女儿墙;透空花格;透空花纹的顶部装饰(哥特式)

brandisite 绿脆云母
brand leader 头号牌子
brand loyalty analysis 商品信赖度分析
brand manager 牌子产品经理
brand mark 品标;商标(符号)
brand name 牌号(名称);商标(名称)
brandname products 名牌产品
brand-new 全新的;最新出品的
brand-new idea 创见
brand of cement 水泥牌号
brand policy 厂牌政策;牌名政策
brand punch 烙印器
brandreth 栅栏;横杆围栏;井栏;铁架;三角架;支承木架;井口栏杆
brand share 牌子产品份额;商标(市场)占有率
brand switching 商标转换
brand-switching model 牌子改用模型
brandtite 砷锰钙石
Brandtzaeg's theory 勃兰特查各三维压力混凝土强度理论
branduster 除糠机
Brane's formula for flow (in slimy sewers) 巴恩斯水流公式(用于计算泥泞污水)
Branley coherer 布冉利金屑检波器
branner 磨机;磨光机;磨工;清净机;抛光机
brannerite 钛铀矿
branners 绒布磨轮
branning machine 钢板清净机
brannockite 锡锂大隅石
bran pellets 麸皮球团
Brans-Dicke cosmology 布兰斯—迪克宇宙论
Brans-Dicke theory 布兰斯—迪克理论
brash 风化碎石;崩解石块;瓦砾;碎冰群
brash chopper 枝梢切碎机
brash ice 碎冰群;碎冰块
brashiness 易碎性
brash wood 脆木
brashy wood 脆性木材;脆性木料;脆性的;崩解石块
brasier 火盆;烘篮
Brasilia 巴西利亚【建】
brasilic acid 巴西酸
brasilin 巴西灵
brasq(ue) 铸工捣实工具;炉衬;填料;耐火封口材料;衬料
brasqueing 坩埚衬碳
brass 黄铜;黄铁矿块;铜器
brass air cock 黄铜气塞
brass alloy 黄铜合金
brass and bronze 黄铜与青铜;有色金属
brass asbestos-filled gasket 黄铜石棉密封衬垫
brass asbestos-lined gasket 石棉衬里黄铜垫圈
brass ball 黄铜球
brass ball valve 黄铜球阀
brass band 黄铜带
brass bar 黄铜棒
brass basin 铜盆
brass bearing 黄铜轴承
brass bib cock with nose for hose 黄铜皮带水龙头
brass billet 黄铜环
brass bobbins 铜筒管
brass bolts and nuts with round head 黄铜圆头螺栓和螺母
brassbound 包黄铜的
brass brazing 铜焊
brass brazing alloy 黄铜焊接合金;铜焊焊料
brass bush(ing) 黄铜衬(套);黄铜轴衬;铜衬
brass button nickelled 镀镍铜纽扣
brass casting 黄铜铸件
brass chain 黄铜链
brass channel (section) 槽铜断面
brass coating 黄铜镀层
brass cock angle 黄铜弯柄旋塞;铜弯嘴旋塞
brass cock straight 铜直嘴旋塞
brass coiled sheet 黄铜卷片
brass coins 铜币
brass cotter pin 铜衬扁销
brass countersunk head wood screw 黄铜平头木螺丝
brass curtain rod 门窗帘铜棍
brass cylinder bushing 黄铜筒衬套
brass die casting 黄铜模铸
brass divider strip 黄铜分隔条
brass dividing strip 分格铜条
brass division strip 分格铜条

brass door handle 黄铜门拉手
brass dowel 黄铜销钉
brass drift 铜穿孔器;铜冲头
brass drum 黄铜鼓
brass electrode 黄铜电焊条
brasserie 啤酒店
brasses 黄铁矿块;铜瓦;铜板送经器
brass extrusion 挤压黄铜
brass file 铜锉
brass filled 黄铜填充的
brass finisher's lathe 加工有色合金的高速车床
brass fitting of steel window 钢窗铜配件
brass fittings 黄铜配件
brass flange 黄铜法兰
brass flat wire 黄铜扁丝
brass foil 黄铜箔
brass foundry 翻铜作;黄铜铸造;铸铜车间;铸铜场
brass fume 黄铜烟雾
brass furniture 黄铜家具
brass gas cock 黄铜煤气龙头
brass gate valve 黄铜门阀
brass glazing 黄铜上釉;黄铜玻璃窗
brass globe valve 黄铜球阀
brass grease cup 铜牛油杯
brass hammer 黄铜锤
brass hardware 黄铜小五金;黄铜硬件
brass hinge 镀铜铰链;黄铜合页;铜铰链
brass hot water tap 黄铜热水龙头
brassicasterol 菜籽甾醇
brassidic acid 巴西烯酸
brassil 黄铁矿;含黄铁矿的煤
brass industries waste 黄铜炼制工业废物
brassiness 黄铜质;黄铜色
brassing 镀黄铜;黄铜铸件
brass ingot 黄铜块
brassite 水砷镁石
brass jacket 黄铜套
brass lid 黄铜盖板
brass lining 黄铜衬(套)
brass links 黄铜链
brass lock 黄铜锁
brass lock cock 黄铜带锁扣水龙头
brass long tail cock 黄铜保暖水龙头
brass mill machine 轧铜机
brass mill sheet in coil 黄铜卷片
brass nail 黄铜钉
brass nipple 黄铜接头
brass ore 绿铅锌矿
brass parting strip 分格铜条
brass pattern 黄铜模
brass pipe 黄铜管
brass plate 板块镀黄铜;黄铜板;厚黄铜板
brass-plated heavy iron hinge 黄铜厚铁铰链
brass-plated iron hasp and staple 镀黄铜铁锁牌
brass-plated iron hooks and eye 镀黄铜铁窗钩
brass-plated iron square hinge 镀黄铜抽心铁方铰链
brass-plated light iron hinge 镀黄铜薄铁铰链
brass-plated L type iron tower bolt 镀铜L形铁插销
brass-plated mild steel round bead wood screw 镀黄铜圆头软钢木螺丝
brass-plated mild steel round head machine screw 镀黄铜圆头软钢机器螺丝
brass-plated steel wire 镀黄铜钢丝
brass plating 镀黄铜;镀合铜;电镀黄铜
brass polish 擦铜油
brass polish cream 擦铜膏
brass powder 黄铜粉
brass rag 擦铜布
brass right-angled valve 黄铜直角阀门
brass ring 黄铜圈
brass rod 黄铜条;黄铜棒材;黄铜棒
brass rolling mill 黄铜材轧机
brass scrap 废黄铜
brass screen 黄铜网
brass screw down bib cock 黄铜水龙头
brass screw plug 黄铜螺纹堵头
brass sheet 黄铜皮;黄铜板
brass shell 厚壁黄铜毛管
brass shouldered cup screw hook 黄铜螺丝弯钩
brass single jack chain 光身铜丝链
brass sleeve 黄铜箍;黄铜套管
brass smith 铜活工;铜工

brass smith shop 铜工车间
brass solder 铜钎焊;黄铜焊锡(料)
brass soldering 黄铜焊
brass spring 黄铜弹簧
brass spring coil 黄铜弹簧圈
brass square bar 方黄铜棒
brass square screw hook 黄铜螺丝曲尺钩
brass strip 黄铜带
brass swing check valve 黄铜横式逆止阀
brass tacks 黄铜平头钉;铜钉
brass tap 黄铜水龙头
brass thimble (防止木材裂开的)黄铜箍;黄铜套管
brass-toned button 镀铜扣
brass track (窗框滑槽用的)黄铜导轨
brass trap 黄铜凝气阀;黄铜存水弯
brass tube 黄铜管
brass turning tool 黄铜车刀
brass valve 黄铜阀
brass ware 黄铜器(皿)
brass-ware wire 黄铜器皿丝
brass washer 黄铜垫圈
brass watch case 黄铜表壳
brass welding rod 黄铜焊条
brass welding wire 黄铜焊条
brass wire 黄铜丝;割坯黄铜丝;铜线
brass wire brush 黄铜丝刷
brass wire gauze 黄铜丝网
brass wire mesh 黄铜丝网
brass wire traverse 铜丝排线器
brass wood screw 黄铜木螺丝
brass work 铜制品
brass work polish 擦铜剂
brassy 黄铜色的;似黄铜
brass yellow 铜黄色(的)
brassy-looking 黄铜色的
brastil 压铸黄铜
brat 原煤
brate force filter 平滑滤器
Brathay Flags 布拉雪大板层
bratisheen 硬衬毛毡
brat pan 炒炸盘;嫩煎锅
brattice 临时木建筑;临时隔墙;临时隔断;(坑通气用的)间壁矿;城堡中木塔;风墙
brattice cloth 黄麻或粗亚麻篷布;风障布
brattice drying machine 转帘烘干机
brattice road(way) 人工堆砌巷道(采空区内)
brattice sheet 风幕;风障
brattice wall 隔风墙
bratticing 墙头饰;脊饰;花格装饰;透空花格装饰
brattishing (围屏或檐头顶部的)镂空花饰;脊饰;透空花格
Braun curve 布朗曲线
braunite 褐锰矿;布氏体;布朗陨铁
Braun sample grinder 布朗试样研磨机;布朗样品研磨机
Braun's canal 布朗管
Braun's hook 断胎头钩
Braun's traction 布朗牵引
Braun's traction frame 布朗牵引架
Braun tube 电子束管;布朗阴极射线管;布朗管
Braun tube oscillograph 布朗示波管;布朗示波器
Bravais indices 布拉维指数;布拉菲指数
bravaisite 漂云母
Bravais lattice 布拉维空间格子;布拉维点阵
Bravais Miller indices 布拉维一密勒指数
Bravais plate 布拉维双片
Bravais symbol 布拉维符号
Bravais system 布拉维系统
Bravais unit cell 布拉维晶胞
brave the wind and waves 乘风破浪
bravoite 方硫铁镍矿
brawn drain 劳工(外流)
bray 碾碎;研碎;捣碎
brayer 手推墨辊
brayer roll 涂墨辊;刷色辊;手推油墨辊
Braystoke directional reading current flow meter 布雷斯托克多直读式海流计
Braystoke multi-parameter current flow meter 布雷斯托克多参数海流计
Braystoke self-aligning current meter 布雷斯托克多自调式海流计
bray stone 多孔砂岩;碎石
Brayton cycle 布雷顿循环;气涡轮机循环
brazan 巴西烷

braze 镶饰;钎焊铜焊
braze ability 钎焊性
brazed 钎焊的;铜焊接的
brazed carbide turning tool 焊接车刀
brazed connections for nozzles 接管的钎焊连接
brazed copper tube 黄铜管
braze fixture 铜焊件
brazed joint 钎焊接头;硬钎接合;硬钎焊连接;点焊接合;铜焊接合
braze metal 钎缝金属
brazed natural diamond tool 焊接式天然单晶金刚石刀具
brazed nipple 黄铜喷嘴;黄铜螺纹接口;黄铜接口(管)
brazed seam 铜焊缝
brazed shank tool 焊接刀具
brazed vessel 钎焊容器
brazed welding 钎焊接
brazen yellow 黄铜色
braze-on fitting 铜焊零件
braze over 镀黄铜
braze welding 硬焊;钎铜焊;钎接(焊);钎焊;铜焊
brazier 烘篮;火钵;火炉;火体;火盆;黄铜匠;制铜技工;铜匠;黄铜细工厂
brazier head rivet 扁头铆钉
brazier head screw 扁头螺钉
Brazil basin 巴西海盆
Brazil current 巴西海流
Brazilette 巴西木芯材
Brazilian aquamarines 巴西海蓝宝石
Brazilian balsam 巴西香脂
Brazilian chrysolite 电气石
Brazilian emerald 绿电气石;绿碧硒
Brazilian gem 巴西宝石
brazilianite 磷铝钠石
Brazilian kingswood 巴西王木
Brazilian mahogany 巴西桃花心木
Brazilian pebble 巴西石英;巴西卵石
Brazilian peridot 巴西橄榄石
Brazilian platform 巴西地台
Brazilian rosewood 巴西黑檀;巴西黄檀木;巴西玫瑰红木
Brazilian ruby 电气石;巴西红宝石
Brazilian sapphire 蓝碧硒;巴西蓝宝石
Brazilian scapolite 巴西方柱石
Brazilian shield 巴西地盾
Brazilian splitting test 巴西劈裂试验;间接拉力试验
Brazilian tensile test 巴西法抗拉试验;劈裂法抗拉试验
Brazilian tensiling method 劈裂法
Brazilian test 巴西试验;巴西式(径向)压缩试验;劈裂试验
Brazilian topaz 巴西黄玉;巴西黄宝石
Brazilian tourmaline 巴西碧硒
Brazilian twin 巴西双晶
Brazilian twin law 巴西双晶律
brazilic acid 巴西酸
brazilin 巴西苏木素
Brazilite 巴西石
Brazil law 巴西律
Brazillian test 巴西影试验(混凝土陶瓷的抗张强度试验)
Brazil nut 巴西果
Brazil old land 巴西古陆
Brazil-radioactive pollution incident 巴西放射性污染事故
Brazil rosewood 巴西玫瑰红木
Brazil split test 间接拉应力试验
Brazil twin 巴西双晶
Brazil twin law 巴西双晶律
Brazil wax 巴西棕榈蜡
Brazil wood 巴西苏木;巴西木;苏木
brazing 硬钎焊;铜焊
brazing alloy 硬焊合金;焊接用合金;硬钎料;钎料;钎焊合金
brazing apparatus 铜焊工具
brazing brass 焊料黄铜
brazing burner 铜焊喷嘴;铜焊枪
brazing clamp 焊接夹钳;铜焊钳
brazing filler metal 铜焊填充金属;硬钎料;钎料
brazing flap 钎焊片状板;铜焊挡板;钎焊挡板
brazing flux 钎焊剂
brazing furnace 焊接炉;钎焊炉
brazing in controlled atmosphere 保护气氛钎焊

brazing lamp 铜焊接灯；铜焊灯
brazing metal 金属焊料；锌铜合金；钎焊金属；钎焊合金
brazing pincers 铜焊钳；焊钳
brazing powder 粉状硬钎料；钎焊粉；铜焊粉
brazing rod 铜焊条
brazing seam 焊缝(铜)
brazing sheet 钎焊板
brazing socket 铜焊套筒；铜焊插座
brazing solder 焊铜；黄铜钎料；铜焊料；铜焊料
brazing spelter 黄铜钎料
brazing temperature 钎焊温度
brazing tongs 铜焊钳；焊钳；钎接用夹钳
brazing torch 钎炬；钎焊用喷灯；钎焊焊炬；气焊枪；喷灯；铜焊炬
brazing wire 丝状钎料
brazing with the unparallel clearance 不等间隙钎焊
brazinrg seam 钎缝
Brazi wood 胡核木
B.R.C. fabric 英制钢丝网
brea 沥青砂；焦油
brea bed 沥青砂土层
breach 裂口；溃决；(堤坝的)决口；激浪；不履行(契约)；缺口；破口
breach a contract 违约
breach before performance is due 到期前的违约
breach board 试验板
breach contract 违反合同
breached anticline 裂体背斜
breached cone 裂口火山锥；裂火山锥；缺裂锥
breached crater 裂火山口；马蹄形火山口；缺陷火山口
breacher 波浪冲击机
breach in dam 堤坝破坏
breach in dike 堤的破坏
breach of an express guarantee 违反明确的担保
breach of conditions 违反条件；违反合同条件
breach of confidence 泄露机密；破坏信用
breach of contract 违约；违反契约；违反合约；违反合同；违背契约
breach of contract before performance is due 履行期末到前违约
breach of contract conditions 违反合同条件
breach of contractor 承包者违约；承包商违约；承包人违约
breach of covenant 违背契约；违反契约；违反合约；违反合同
breach of discipline 破坏纪律
breach of duty 渎职；失职
breach of faith 失信
breach of law 违法
breach of obligation to pay an instalment 违反分期付款的义务
breach of privilege 滥用职权
breach of promise 毁约；违反诺言；违背诺言
breach of regulations 违章；违例；违反规则
breach of rules 违例
breach of the contract of carriage 违反运输契约
breach of trading warranty 违反航行范围的保证
breach of trust 背信；违反信托；违背信托
breach of warrant 违反担保
breach of warranty 违背担保；违反保证；违反特约条款（水险单上附加的条款）
breach of warranty of authority 违反有代理权的保证
breachway 堤坝决口冲路；连接的渠道
breadalbaneite 普通角闪石
bread-basket 产谷物地区
breadboard 模拟样板；线路板；试验电路板；试验板
breadboard area 模拟板区域
breadboard circuit 试验电路
breadboard design 模拟板设计；试验电路板设计；模拟板设计；电路模拟设计
breadboard model 模型板；试验板模型
bread board set up 模拟板装置
bread board socket 试验板座
bread-crust bomb 剥皮火山弹
bread-crusted bomb 层剥火山岩
bread-crusted boulder 层剥岩
bread-crusted structure 面包皮状构造；层剥构造
bread-in 磨合
breading cap 安全日
breadth 宽度；长宽高；幅面

breadth-and-depth-first search 宽度与深度优先搜索
breadth coefficient 分布系数；绕组系数
breadth depth ratio 宽深比
breadth draught ratio 船宽吃水比
breadth extreme 全宽(度)
breadth factor 分布系数
breadth-first 宽度优先
breadth-first generation 宽度优先生成
breadth-first procedure 宽度优先过程；横式生成过程
breadth-first search 宽度优先搜索
breadth length ratio 宽长比
breadth maximum 全宽(度)
breadth-mo(u)lded 型宽(度)；船型宽(度)
breadth of beam 梁宽(度)
breadth of deck 船宽
breadth of discharge opening 卸料开口宽度
breadth of earth-retaining structure 挡土结构的宽度
breadth of entrance 口门宽度；进港口门宽度
breadth of road 路幅
breadth of ship 船宽
breadth of tooth 齿宽
breadth of weld metal 焊缝宽度
breadths 幅边
breadth way 横向
breadth wise 横向地
break 溃决；间歇；减弱；煤层的不连续面；优选研磨粒度；(特性曲线的)转折；暂停；断入；断路(器)；断裂【地】；层缺
breakability 破碎性
breakable 易碎的；脆的
breakable glass seal 易碎玻璃密封
breakables 易碎品
break a contract 违反契约；违反合约；违反合同；违背合同
break address 中断点地址
break a deposit 预支未到期存款；提前支取未到期存款
breakage 堆货余隙；断裂；破损(赔偿额)；破裂(碎片)；破损片(材料)；损漏；损坏处；损耗量
breakage allowance 破损折扣
breakage clause 破损险条款；破碎险条款
breakage of packages 包装破损
breakage of packing 包装破裂险
breakage-proof 防破损
breakage rate of raw sheet 原板破损率
breakage risks 破碎险
breakage susceptibility 易碎性
break agreement 破坏协议
break a jam 拆垛
break alarm 停止信号
break a loan 借款中途解约；提前归还未到期借款
break an agreement 破坏协定；违约
break an appointment 背约
break angle 折角；错接角(焊)
break apart 崩裂
break arc 断路时的电弧
break a seal(ing) 破封
break auger 钻煤用螺旋钻
breakaway 断掉；断开(力)；拆毁；剥裂；气体分离；气流离体；气流分裂区域；脱流；跳闸
breakaway braked weight 减去制动重力
breakaway chain 防断安全链
breakaway connection 断销式离合器
breakaway connector 拉脱式电缆接头；易拆式管接头
breakaway corrosion 剧增腐蚀；脱皮腐蚀
breakaway coupling 断开式联轴节
breakaway device 安全脱钩装置；安全分离装置
breakaway force 起动力；起步阻力；起步力；脱离锯线的部位
breakaway friction 滑动摩擦
breakaway from conventions 打破常规
breakaway light pole 解体消能式灯杆；解体消能式标志柱
breakaway method 脱离镶嵌法
breakaway mount 解体消能基座；解体消能墩；缓冲柱座；缓冲标志柱
breakaway of detonation 爆炸中断
breakaway plow 带有脱钩装置的犁
breakaway plug 分离插头
breakaway point 起动点；启动点；脱离点

breakaway principle 解体消能原理
breakaway release 断开式装置
breakaway resistance 起步阻力
breakaway stress 破裂应力
breakaway torque (内燃机的)最小开机转矩；打滑转矩
breakaway velocity 分离速度
breakback 反击穿
breakback contact 静触点
breakback mechanism 碰返机构；复位安全机构
break barrow 碎土机；碎土车
break-before-make 先开后合
break-before-make contact 先开后合触点；先开后合接点
break bond 接缝破裂(层间)
breakbone fever 登革热
break bulk 开始卸货；件杂货；逐件装卸法
break-bulk berth 杂货泊位
break-bulk cargo 零担货(物)；散件货物；件杂货；杂类船货
break-bulk carrier 杂货船
break-bulk container ship 杂货集装箱船；杂货货柜
break-bulkhead 船楼端舱壁
break-bulk liner 杂货班轮
break-bulk operations 卸货作业
break-bulk point 卸货点
break-bulk ship 杂货船
break-bulk transport 杂货运输
break-bulk vessel 杂货船；件杂货船
break capacity 开断容量
break-circuit chronometer 断路时计
break circulation 恢复循环(钻孔内冲洗液)
break cistern 吸水水池
break coarse 粗压碎；粗碾
break communication 切断交通
break comparator 中断比较器
break connection 断路接头
break contact 静止接点；断开接点；断开触点
break contact spring 断开触点簧片
break contract 破坏合同
break counter 间歇计数器
break deck 船楼甲板
break delay 断开时间
break detector (钻眼内的)裂缝探测器；(钻眼内的)裂缝检查器
breakdown 分项数字；分解；离解；击穿；细目分类(法)；折断；故障；断损；断裂；分为细数；分类(细)账；分镜头；崩塌；崩溃；乳胶分水；气流离体；气流分离；剖料；破断；堤崩；损坏；事故停堆；事故；拆开；分类剖析；分布细目
breakdown agent (水力碎煤的)碎煤射流
breakdown avalanche 雪崩击穿
break down by grinding 碾碎；磨碎
breakdown cable 应急电缆；替换电缆
breakdown car 工程修理车
breakdown clause 故障条款；停租条款
breakdown crane 拆卸起重机；救援起重机；救险吊车；应急起重机；抢修起重机；(铁路的)事故起重机
breakdown crane wagon 事故起重车
breakdown current 击穿电流
breakdown die 开坯模
breakdown discount rate 损益平衡贴现率
breakdown drawing 分解图
breakdown drawing machine 粗拉伸丝机
break down economic boundaries 冲破经济界线
breakdown factor 故障系数
breakdown field strength 耐压强度；绝缘强度；击穿电场强度；击穿场强
breakdown fluid (水力压裂油层的)压裂液体
breakdown gang 救援队；抢修车；急救队(火场失事)
breakdown impedance 击穿阻抗
break down into simple compounds 分解成简单的化合物
breakdown law 全概率定律
breakdown light 故障指示灯；操纵失灵信号；操纵失灵灯；失控信号灯
breakdown lorry 故障救援车；救险汽车式起重机；救险起重机；起重车；工程救险车；抢修工程车；抢险起重机；抢险工程车
breakdown maintenance 故障维修；故障排除；损坏维护；事故维修
breakdown mechanism 破坏机制；破坏机理；破坏

机构
breakdown method 分解评估折旧法
breakdown mill 开坯机;碎研机
breakdown noise 击穿噪声
breakdown of amount of expenditures 支出款额剖析
breakdown of barge train 驳船队解队
breakdown of bid price 投标总价的细分
breakdown of cost 成本分析
breakdown of equipment 设备故障
breakdown of ink colloid 油墨胶体分解
breakdown of lump sum bid price 投标总价的细分
breakdown of lump sum items 包干项目的分项
breakdown of machinery 机器故障
breakdown of oil 油的澄清
breakdown of price 价格分析;价格分解
breakdown of roof 塌顶
breakdown of soil materials 土壤特质的碎裂
breakdown of tender prices 标价的分项表
breakdown of variation 变差分解
breakdown period 停工时间
breakdown point 破损强度;破损点;断裂点;击穿点;屈服点;破坏点;损益平衡点;损益两平点
breakdown potential 击穿电位;击穿电势
breakdown pressure 临界压强;临界压力;击穿压强;断裂压力;破碎压力;破裂压力;破坏压力
breakdown process 分解过程
breakdown region 击穿区;击穿范围
breakdown repair 故障修理;破损修理
breakdown reverse voltage 反向击穿电压
breakdown rolling 压缩碾压
breakdown service 修理站;抢修站(车辆);破损维修
breakdown signal 击穿信号;故障信号
breakdown slip 临界转差率;极限转差率;停转转差率
breakdown speed 破坏速度
breakdown stand 粗轧机座
breakdown strength 击穿强度;破裂强度;破坏强度
breakdown stress 破坏应力
breakdown switch 故障开关
breakdown tender 修理工程汽车
breakdown test 破损试验;断裂试验;击穿试验;耐压试验;折断试验;电击穿试验;漆膜击穿试验;破损试验;破坏性试验;破裂试验;稳定性试验
breakdown threshold 击穿阈值
breakdown time 击穿时间;破坏期;停工时间
breakdown torque 极限扭矩;破坏扭矩;停转转矩;停转力矩
breakdown torque speed 临界转速;停转转速
breakdown trailer 救援挂车
breakdown train 救援列车;抢修列车
breakdown van 救援车;急救车;修理车;抢修车
breakdown vehicle 救险起重车;抢修工程车
breakdown voltage 绝缘击穿电压;击穿电压;崩溃电压;破坏电压
breakdown wagon 破损货车
breakdown zone 断裂部分
breakdown locomotive 救援机车
breakdown pass 粗轧孔型
break earth 破土动工
breaker 裂碎机;开关闸;遮断器;轧碎机;隔热骨材;隔断器;断电器;淡水桶;揉布机;破土犁;破碎机;破碎波;破浪;破波;拍岸浪;碎浪机;碎波;水密容器
breaker arm 开关闸杆;断电臂
breaker arm adjusting screw 断电臂调整螺钉
breaker arm spring 断电臂弹簧
breaker ball 撞球(拆除建筑物用);拆房重锤
breaker bar 反击算条;反击棒
breaker beater 拌浆机;梳解机
breaker block 碎石机
breaker bolt 安全螺栓
breaker box 开关箱
breaker cam 断路器凸轮;断电器凸轮
breaker camshaft 断路器凸轮轴
breaker contact 开关触头
breaker contact point 断路器触点
breaker core 隔片型芯
breaker cover 断路器盖
breaker depth 碎波深度;(破波的)临界水深;碎波临界水深
breaker dike[dyke] 破浪堤

breaker distance 碎波距离
breaker drawing frame 头道并条机
breaker dust 碎石机粉尘
breaker failure protection 断路器失灵保护(装置)
breaker feeder 碎石机的加料器
breaker feed hopper 碎石机的加料漏斗;破碎送料斗
breaker height 临界波高;破波波高;碎波临界波高
breaker index 破波指数
breaker in failure 冲切剪切破坏
breaker lever 断路器杆
breaker lever axle 断路器杆轴
breaker line 破波线;碎波线
breaker lugs 钻头拧卸器提手
breaker machinery 破碎机械
breaker main contact 断路器主触头
breaker plate 断电器板;断路器板;冲击板;反击板;破碎机衬板;破碎板;碎石板
breaker point 断闭点;断闭点
breaker point wrench 电流开关扳手
breaker position 波浪卷碎位置;破碎位置
breaker prop row 破碎机的系列支撑柱
breaker props 破碎机的支撑柱;放顶密集支柱
breaker roll 轧碎机滚筒;破碎机轧辊
breaker roll type oil seed press 对辊式榨油机
breakers 浪花
breakers ahead 前面有暗礁
breakers along shore 岸边浪花
breaker scutcher 粗打麻机;揉麻梳麻机
breaker spring 断闭器弹簧
breaker strip 防断条;垫层
breaker terrace 破浪阶地
breaker timing 断电定时;定时断电;开关定时;断路器定时
breaker tool 破碎工具
breaker travel 碎波行程
breaker trip coil 自动断路器线圈;断路器脱扣线圈
breaker tyre cloth 缓冲帘子布
breaker unit 碎砂机单元
breaker wall 破浪墙
breaker zone 破浪带;破波带;拍岸浪带;碎波区;碎波带
break-even 均等的;盈亏平衡;收支相抵的
break-even analysis 收支平衡分析;保本分析;得失分析;盈亏分析;够本分析;保本点分析
break-even chart 保本分析;盈亏分界图表;收支平衡图;盈亏平衡图;盈亏分析图表;两平图表
break-even computation 保本计算
break-even cost 成本费用;无盈亏价格;损益平衡成本
break-even cut-off ore grade 无亏损最低品位;无亏损品位下限
break-even discount rate 损益平衡贴现率
break-even dollar sales 保本销售额
break-even economic grade of mineral commodity 矿产经济临界品位
break-even exchange rate 两平汇率;扯平汇率
break-even freight rate 保本运价
break-even graph 保本图
break-even investment 保本投资
break-even level of income 收支两平的收入水准
break-even load factor 保本积载因素
break-even method 保本计算法
break-even model 保本模式
break-even point 转折点;平均转效点;平滑转折点;损益平衡点;损益两平点;收支平衡点;盈亏相抵;盈亏平衡点;盈亏临界点;盈亏点;够本点;保本点
break-even point analysis 盈亏临界点分析
break-even point method 盈亏平衡点法
break-even point of composite economic grade 综合经济临界品位
break-even point of economic grade by mineral 伴生矿经济临界品位
break-even price 保本价格;不亏不盈价格
break-even pricing 收支平衡定价法
break-even probability 保本概率
break-even rate of operation 收支平衡营运率
break-even revenue 保本销售额
break-even sales level 不赔不赚的销售水平
break-even sales volume 保本销售量
break-even stripping ratio 经济合理剥采比
break-even traffic volume 保本运量

break-even units 保本销售量
break-even zone 无盈亏价格
break faith 丧失信用;背信;失信用
breakfast kitchen 早膳厨房
breakfast nook 早餐座
breakfast room 早餐室
break fault 间断断层
breakfinder 炮眼壁探缝器
breakfree oil 精制油
break frequency 折断频率(频率特征曲线的折断点);拐点频率
break friction 裂缝摩阻(力)
breakfront 凹凸面橱;凸肚型书柜;凸肚型橱柜
break grade 改变比降
break ground 挖地;破土;开垦;动工
breakhead 船首破冰板;船头破冰装置
break her sheer 锚链缠绞
break hiatus 间断
break impulse 切断脉冲
break-in 打断;插入;截断;木砖孔;(砖墙上用的)木落孔;变向
break-in a curve 曲线转折点;曲线拐点
break-in cross rates 套算汇率差距
break-in curve 曲线折断
break-in device 插入收发机构
break-induced current 断路感应电流
break-in facility 截断功能
break-in failure 冲切剪切破坏;切入式破坏
breaking 遮断;打开;破碎;破裂;破坏
breaking and expanding coefficient 碎胀系数
breaking angle of wave 波浪破碎角
breaking arc 掉闸电弧;切断电弧
breaking arrangement 断路装置
breaking ball 起破碎作用的球
breaking bending moment 断裂弯矩;极限弯距
breaking bulk 开舱(卸货)
breaking by curling 扭曲破碎;波浪卷跃破碎
breaking by falling 落下破碎
breaking capacity 遮断容量;遮断能力;断路容量;短路容量;分断能力;破裂能力;开断容量
breaking chain 间断整尺
breaking chamber 腔内破碎;破碎室
breaking circuit 断路【电】
breaking coil 跳闸线圈
breaking condition 破坏分析条件;断裂条件
breaking crack 断裂缝
breaking cross-section 破裂断面
breaking current 遮断电流;断路电流;断开电流;分断电流
breaking depth 碎波水深;碎波临界水深
breaking dissociation 离解
breaking down 圆木锯成板;出砂;开木料;分解;断裂;锤碾
breaking down mill 粗轧机
breaking down point 断裂点;破损点
breaking down process 断裂过程
breaking down roll 粗轧轧辊
breaking down stand 粗轧机座
breaking down strength 断裂强度;破裂强度
breaking down test 断裂试验;耐击穿试验
breaking down the pipe 拧卸钻杆
breaking-down tool 锻工凿;锻工錾
breaking-drop theory 水滴破碎理论
breaking edge (屋顶的)破损边缘;截断线;(顶板岩石的)崩落线
breaking elongation 致断延伸率;折断延伸率;总延伸率;断裂延伸率
breaking energy 破坏能(量)
breaking extension 致断延伸率;断裂延伸
breaking factor 破裂阻力系数;破断系数
breaking force 断裂力;崩落力;破断力
breaking forcebreaking strength 断裂强力
breaking groove 断裂槽
breaking ground 破土;地层爆破
breaking hammer 破碎锤
breaking head 破碎头;轧头
breaking-in 夯实;开始使用;开始生产;试车;开工;磨合;跑合;试运转
breaking-in bulking 压曲破坏;膨胀破坏
breaking index 断裂指数;碎波指数
breaking-in hole 掏槽孔
breaking-in period 下矿井时间;开动期;溶解期;试车时期
breaking-in process 试运转过程

breaking-in shot 掏槽爆破
breaking installation 破碎设备；破碎工厂
breaking joint 断缝；沉降缝；断裂节理；断裂缝；错（开接）缝；参差接缝
breaking knife 撞击刀；轧刀
breaking lag 起爆时滞
breaking length 裂断长度；断裂长度
breaking limit 断裂极限；断裂限度；破碎极限
breaking limit line 破裂极限线
breaking line （顶板岩石的）崩落线；碎波线
breaking link （水轮机的）脆性连杆；破断联杆（水轮机）
breaking load 极限荷载；最大负载；最大负荷；断裂荷载；断裂负载；破损荷载；破坏荷载；破断负载
breaking loading 破坏荷载
breaking machine 致断力试验机；切碎机；碎茎机
breaking moment 断裂力矩；切断瞬间；破坏力矩
breaking of arc 断弧
breaking of bulk 货物分卸
breaking of contact 断接；断路；触点断开
breaking of coupling 离合器分离
breaking of dike 堤防溃决；决堤
breaking of emulsion 乳浊液破坏；乳液离析；乳液分裂；乳胶体分层；乳乳（油）
breaking off 断管【岩】；剥蚀
breaking off of base 掉ек（玻璃制品缺陷）
breaking of ground 落岩；破土
breaking of monsoon 季风爆发
breaking of oil 油的澄清
breaking of pigs 破碎金属块
breaking of rail 钢轨折断
breaking of the drill stem 钻杆断裂
breaking of the meres 水体杂锦
breaking of waves 破浪；波浪破碎
breaking operation 破碎操作
breaking-out 喷火
breaking-out bulk 大量崩落；大块崩落
breaking-out cut 裁边碎边；辅助裁切
breaking-out of fire 发生火灾
breaking-out of ore 崩矿
breaking piece 保险零件；保险连接件；安全零件；安全连接器
breaking pin 剪断保险销；自断安全销；断裂安全销
breaking pin device 折断销装置
breaking plane 断裂面；破坏面
breaking plant 毁坏设备；破碎工厂；粉碎设备；破碎装置；破坏装置
breaking plow 开荒犁（装有近似螺旋型犁体）
breaking point 断点；击穿点；转换点；断裂点；强度极限；破损强度；破损点；破碎点；破乳点；碎波点
breaking point test 断裂点试验
breaking pressure 断裂压力；崩落压力；破损压力
breaking prop 复式支架；放顶密集支柱
breaking pulper 打浆机
break in grade 变坡点
breaking rate of diamond 金刚石破损率
breaking resistance 抗破坏强度；抗裂强度；抗断强度
breaking roof 破碎顶板
breaking scum 破碎浮渣
breaking sea 碎波（涛）；开花浪
breaking section 断裂截面；破坏截面
breaking short 低效崩落
breaking shot 掏槽炮眼；掏槽爆破
breaking staggering of joints 错列接缝
breaking strain 断裂应变；破坏应变
breaking strength 断裂强度；抗拉强度；抗断强度；极限破坏强度；击穿强度；断裂应力；破碎强度；破坏强度；破断强度
breaking strength test 断裂强度试验；破坏强度试验
breaking stress 极限强度；抗断应力；断裂应力；断裂应力；破坏应力
breaking stress circle 破坏应力圆
breaking stress condition 破坏应力条件
breaking surge 涌潮
breaking tape 分段丈量
breaking tenacity 断裂强度；扯断力
breaking tension 断裂张力
breaking test 破坏试验；断裂试验；破断试验
breaking through of water 突然涌水
breaking time test of Caroselli 卡罗萨利断裂时间试验
breaking time test of Weber and Bechler 韦伯和白撒勒断裂时间试验
breaking tooth （挖土机的）齿裂
breaking travel 碎波行程
breaking up 击碎；劈开；调漆；打匀；强力搅匀（稀释时）
breaking-up of boulders 大块的二次破碎
breaking-up plow 开荒犁（装有近似螺旋型犁体）
breaking-up pressure 放松压力
breaking-up ridge 间断式（分土）土埂
breaking velocity 断路速度
breaking water level 跌落水位；跌落水面
breaking wave 近破波；破碎波（涛）；断裂波；破波；碎波
breaking wave height 碎波波高；破波波高
breaking wave pressure 破波压力
breaking weight 断裂重量；断裂荷载；破坏荷重；破坏荷载
breaking zone 断裂区
break-in invert 槽底折断
break-in key 插话式键
break-in keying 插话式键控
break-in oil 磨合油；跑合用油
break-in operation 插话
break-in period 间断期；试运转期
break-in point 汇合点
break-in relay 插入继电器；强拆继电器
break-in slope 坡折（裂点）
break-in temperature 气温突变
break interrupt signal 中断请求信号
break-in the market 市价突然下跌
break into a face 交会测量；交会
break-in upon 打断；插入
break-in water main 供水干管破裂
break iron 修琢铁；刨刀护铁；刨顶压铁；铁夹头；铁挡块；碎花铁片
break jack 切断塞孔
break jamb 门窗框侧壁；门窗框边挺
break joint 断裂节理；断缝；间砌法；间缝；错缝（接合）；不连续接头
break key 断路键；打断键
break level 断路电平
break liberty 超过假期
break limit 断裂极限
break line 虚线；折断线；断开线；段末短行；破裂线；制图虚折线；部分图例
break line of spraying 喷涂分界线
breakling 对接线
break loose 迸发出来；挣脱出来
break make break contact 断合断触点
break-make contact 开合触点；断合触点
break make system 先断后通式
break of an earth bank 路堤滑坡
break of an engagement 毁约
break of conductor 导线断开
break of contact 触点断开
break of contract 违约；违反合同
break-off 折断；采折；暂停工作；中止
break of forecastle 首楼后端；船首楼后端
break-off point 折裂点
break-of-ga(u)ge station 过轨站
break of joint 节点错位；错缝
break of load 负荷的破坏
break of poop 船尾楼前端；尾楼前端
break of stratum 岩层破裂
break open 破裂；炸裂；裂开；拆开；撬开
breakout 裁板；炉衬破裂；漏钢；分开接头；分接；崩落；爆发；提升钻杆
breakout block （拧卸时把钻头固定在转盘上用的）方卡块；（拧卸时把钻头固定在转盘上用的）卡瓦块；转动卡块（扭断钻头用）
breakout break through 钢水冲击
breakout deuce 卸管手把
breakout effort 脱开力
breakout force 掘起力；爆发力；劈开力；突然涌出力；突然脱出力
breakout friction 静摩擦（力）
breakout gun 拧管机
breakout load 断开力
breakout man 管钳工
breakout material 易磨损材料
breakout of bottom bed 底板鼓起
breakout plate 劈开薄板；钻头拧卸器；钻头装卸器
breakout point 中断点；分叉点
breakoutput 断续输出
breakout table 拧管机
breakout tongs 大管钳
breakout torque 卸扣扭矩
breakover 穿通；圆脊；导通
breakover voltage 击穿电压；导通电压
break pattern 玻璃碎块形状；破裂形状
break pin 安全销（钉）
break plane 断裂面
break point 转折点；转效点；断流点；分割点；停止点；停车点；折点
break-point chlorination 断点加氯法；折点氯化（法）；氯化转效点（水净化）
break-point command 断点命令
break-point fault 断点故障
break-point halt 断点停止；断点停机
break-point information 断点信息；分割点信息
break-point insertion 断点插入
break-point instruction 断点指令
break-point of chlorination 氯化转效点
break-point operation 分割点操作
break-point order 断点指令；返回指令；射流返回指令；分割点指令
break-point switch 断点开关
break-point symbol 断点符（号）
break-point test 折点试验
break-point trap 断点自陷；断点俘获
break pressure cistern 减压水箱
break-pressure tank 减压箱；制压箱；截流箱
break-proof tap 防断丝锥
break pulse generator 断路脉冲发生器
break rate 破损率
break regulation 违反法规
break request 中断请求
break-resistance 抗断裂性
break-resistant 抗裂的；抗断裂的
break roller mill 辊式粉磨机
break rolls 对轨辊；带钢压紧辊
break rules and regulations 违章
break scraper 裂缝探测仪；裂缝探测器
break seal 拆封；破坏密封
break sequence 间断序列；间断顺序
breaksetting method of intermediate pile 防止中间桩下沉作业
break sheer 锚链缠绞
break shovel 松土铲
break sign 开始符号；分隔记号
breaks in overcast 阴天微隙
break spark 断路火花
breakstone 碎石；石渣
break test 断裂试验
break the bond 毁约
break the bulk 卸载
break the circuit 断开电路
break the joint 打开接头
break the leaf 折断叶子
break the normal procedure 打破常规
break the record 打破先例；打破记录
breakthrough 泄漏炉衬破裂；临界点；技术革新；转折点；重要发明；贯穿；缺口；打通；穿透；突破；通风联络小巷
breakthrough capacity 漏穿容量；泄漏容量；贯流容量；穿透容量；突透容量；突破量
breakthrough curve 穿透曲线
breakthrough in filter 滤池泄漏
breakthrough in science and technology 重大科学技术成就
breakthrough of turbidity 浊度穿透
breakthrough pitting 穿透点状腐蚀
breakthrough point 泄露点；临界点；漏过点；贯通点；穿透点
breakthrough survey 贯通测量
breakthrough sweep efficiency 水串时清扫效率
breakthrough time 保护作用时间
breakthrough volume 突破体积
break thrust 切穿逆断层；背斜上冲断层
break time 断开时间；间断时间；转效时间；切断时间；起爆时间
break time on fault 事故停产时间；事故断电时间
break tour （钻机搬迁后的）连续三班开钻
break transfer 断续传送
break-up 分离；粉碎；拆散；停止；天气变坏；分散；崩解；解体；崩裂；破裂
break-up age 分解年龄
break-up capacity of hump 驼峰解体能力

break-up clause 破损条款
break-up constant 断裂常数
break-up date 解冻日期
break-up energy 分裂能
break-up for colo(u)rs 制作分色版
break-up hard ground 松动硬土
break-up into 分裂成
break-up of emulsion 乳液分层
break-up of raft 散排
break-up of river 河流解冻
break-up of the ice in a river 开江
break-up of trains 解体调车
break-up operation 解体作业【铁】
break up the clods 打碎土块
break up to form a soil 粉碎而形成的土壤
break-up value 破产企业财产清理价值;破产清理价值;分析价值;清理价值
break valve 截止阀
breakwater 船头防波栏;港口外堤;挡浪板;防浪堤;防波设备;防波堤
breakwater alignment 防波堤的轴线布置
breakwater and sea-wall in shallow water 浅水防波堤和海堤
breakwater arrangement 防波堤布置
breakwater axis 防波堤轴线
breakwater berth 防波堤泊位
breakwater caisson 防波堤沉箱
breakwater capstone 防波堤压顶石
breakwater core 堤心
breakwater end 防波堤端
breakwater entrance 防波堤入口处;防波堤口门
breakwater gap 堤头口门;防波堤缺口;防波堤口门
breakwater gate 挡潮闸(门)
breakwater glacis 消能滩;消能坡;防波斜堤消能坡;防波斜堤堆石消能滩;防波堤码头;破坏浪石砌体;防波堤铺石面;破浪石砌体(桥墩式护岸)
breakwater head 堤头(防波堤)
breakwater in deep water 深水防波堤
breakwater layout 防波堤布置
breakwater light 防波堤灯标
breakwater of sea port 海港防波堤
breakwater of submerged reef type 潜礁式防波堤
breakwater pier 防波突堤;防波堤码头
breakwater pier head 防波堤头部;防波堤岸码头
breakwater quay 防波堤堤岸码头;防波堤堤岸
breakwater structure 防波堤建筑
breakwater superstructure 防波堤上部结构
breakwater tip 堤头(防波堤)
breakwater trunk 防波堤主体
breakwater wharf 防波堤码头
breakwater with concrete-block gravity wall 方块防波堤
breakwater with sloping faces 斜坡式防波堤
breakwater with vertical faces 直墙式防波堤
breakwind 挡风墙;挡风罩;防风设备;防风林;防风墙
break your pick 离职或解雇
bream 烘底船
bearer draft 不记名汇票
Brearley steel 布里阿雷高铬钢;布里阿雷不锈钢
breasing line 横缆(船中部)
breast 炉腰;炉胸;炉腹;梁底;煤房;窗腰;船的中段采矿工作面
breast abutment 胸式桥台;无斜翼桥台;无斜翼岸墩;胸墙式桥台
breast anchor 船舷锚
breast and pillar method 房柱采矿法
breast backstay 横后牵
breast beam 过梁;腰梁;船尾楼前端;船首楼后梁;船首横梁;船楼端梁;托墙梁
breast bench 锯房;台锯机
breast board 插板;背板;栏板;开挖面支护;挡土板
breast borer 曲柄钻;胸压手摇钻
breast brace 胸压钻孔器;胸压手摇钻
breast derrick 人字把杆起重机
breast drill 曲柄钻;脚摇钻;胸钻;胸压手摇钻;曲柄手钻;手持式钻机
breast dup 齐胸高的
breast element 窗下墙构件
breast fast 横缆(船中部)
breast height 窗下墙高度;胸墙高度
breast-high 齐胸高的
breast hole 腰眼;中央渣口;中部炮眼;清渣口;齐胸部炮眼
breast hook 尖蹼板;船首肘板;船首补强肘板;首顶肘板;加劲肘板(船)
breasting 宽巷道;锯齿形矫正;(水轮的)中部冲水式
breasting clustered piles 靠船簇桩
breasting dolphin 承冲护船桩;靠船柱;靠船架;靠船墩;靠船簇桩;承冲系船建筑物;承冲靠船架;承冲护船桩;承冲靠船桩(靠船架);岸壁系缆桩
breasting float 船与码头间的护岸材
breasting island 靠船墩;靠船地;靠船岛
breasting jack 胸腔千斤顶
breasting line 系船靠岸索;靠墙索;靠岸索;系船触岸索
breasting lining 胸高护墙面饰
breasting load 靠船力
breasting method 全面开挖法;全面开采法;全面掘进法;平掘开挖法
breasting platform 靠船平台
breasting wheel machine 掘进机;切削型掘进机
breast jack 挡土千斤顶
breast lead weight 胸部压重物
breast line 胸索;船中部横缆
breast-line dolphin 横缆墩
breast lining 窗肚封口板;窗台壁饰;窗腰板衬;窗台线;窗肚板;裙板;窗下护墙
breast machine 宽截盘截煤机
breast mo(u)lding 墙裙线脚;窗腰线脚;窗下墙线脚;窗盘线条
breast nogging 窗腰木砖;窗腰木栓
breast of a window 女儿窗凹;窗口墙;壁龛;凹进处
breast panel 窗下墙
breast pile 挡土桩;坡脚桩;坡脚挡土桩
breast plate 领盘【救】;挡风板;胸板;楼梯平台裙板
breast plate borer 胸压手摇钻
breast plate microphone 带式送话器
breast rail 腰栏;船楼端栏杆;(船侧、窗前的)栏杆
breast roll 胸辊;中心辊
breast roller 机架辊子
breast rope 横缆(船中部);安全索
breast shore 横撑木;船边撑木
breast-shot water wheel 中间轮;胸射式水轮机
breast stopping 全面采矿法
breast stroke 推桨
breast summer 横楣;过梁(木);大木;托窗梁;托墙梁
breast telephone 挂胸电话机
breast the sea 破浪前进
breast timber 斜撑支柱;沟槽脚手斜撑;横档;腰梁;斜撑支桩
breast wall 腰墙;山墙;齐胸高的女儿墙;护坡;码头墙;护岸;胸墙;挡土胸墙;挡土墙;侧墙;防浪墙
breast water wheel 中射式水轮;胸水轮;胸射式水轮机;腰部进水水轮;中间水轮
breast work 女儿墙;靠船设施;胸墙作业;胸墙;船楼端栏杆;壁炉胸墙;掩体
breathability 透热性
breath apparatus 吸气装置
breather 吸潮器;防毒面具;热带飑;通气装置;通气筒;通气设备;通气器
breather cap 排气器帽
breather filler cap 给油箱加油口盖
breather film 透气漆膜;透气薄膜
breather(hole) 通气孔;透气孔
breather line 通气管(道)
breather loss 呼吸耗损
breather mask 呼吸面具
breather membrane 透气薄膜;透气膜
breather panel 进气壁板
breather paper 透气防潮纸
breather pipe cap 通风管盖
breather plug 通气塞;制冷室通风管堵头
breather roof 涨缩屋顶;浮顶;透气屋顶;升降屋顶
breather roof tank 浮顶(油)罐
breather screen 排气器滤网
breather valve 呼吸阀;通气控制阀;通气阀
breather vane 通气控制阀
breather vent 通气口
breath-hold diving operation 屏气潜水作业
breath-hold diving skin diving 屏气潜水
breath holding 屏气
breathing 间歇缓变;通气
breathing ability 通气性能
breathing apparatus 呼吸器;氧气面罩;排气装置;通气装置;通气设备
breathing bag 呼吸袋
breathing cave 吹风洞
breathing coating 透气性涂层
breathing device 呼吸设备
breathing drier 放气干燥管
breathing equipment 呼吸装置;氧气设备
breathing hazard 呼吸危害性
breathing hole 呼吸孔;通气孔
breathing length 伸缩区长度
breathing line 通风线
breathing mash 防毒面具
breathing mask 呼吸面具
breathing panel 进气壁板
breathing pipe 通气管
breathing place 通气口;休息处
breathing property 透气性
breathing reserve 换气储备
breathing reserve ratio 换气储备比(值)
breathing resistance 呼吸阻力
breathing roof 浮顶
breathing snorkel 呼吸管
breathing space 休息处;膨胀区
breathing tank 浮顶(油)罐
breathing tolerance 空气中污染物的吸收容许量;空气中污染物的吸入容许量;容许吸入量
breathing tube 呼吸管
breathing valve 通气阀
breathing veneer dryer 呼吸式单板干燥机
breathing well 呼吸井;透气井;通气井
breathing zone 伸缩区
breath of discharge opening 排料口宽度
breath of skin 皮肤损伤
breccia 角砾(岩)
breccia-bearing conglomerate 含角砾砾岩
breccia dike 角砾岩岩脉
breccia-gravel filling 角砾碎石充填
breccia lava 角砾熔岩
breccia lava texture 角砾熔岩结构
breccia marble 角砾大理岩;角砾大理石
breccia pipe 角砾岩筒
breccia reservoir 角砾岩热储
brecciated 角砾的
brecciated Au ore 角砾状金矿石
brecciated coal 角砾煤
brecciated dolomite 角砾状白云岩
brecciated limestone 角砾状灰岩;角砾石灰岩
brecciated ore 角砾状矿石
brecciated Pb-Zn ore 角砾状铅锌矿石
brecciated structure 角砾状构造;角砾构造
brecciated texture 角砾结构
brecciated trap 角砾岩的圈闭
brecciation 角砾作用;角砾岩化
brecciation way 角砾(岩)化方式
breccioid 似角砾岩的
brecciola 角砾灰岩
Brechtel pressure pile 布雷赫载受压混凝土钻桩
breck 丘陵
bredigite 布列底格石;白硅钙石
Bredig's arc process 布雷德希电弧法
bred material 增生材料
bred uranium 增殖铀
breech 滑车尾
breechblock 闩体
breeches buoy 连裤救生圈;裤形救生吊笼;裤式救生圈
breeches chute 叉斜道
breeches fittings 叉形管配件
breeches pipe 滑轮组末端管;钢叉管;叉形管;Y形管
breech fitting 叉形配件
breeching 烟囱水平连接部;(帆船的)后支索;烟道;水平烟道;(烟囱、烟囱等的)水平连接处
breeching fitting Y形管接头;斜叉管接头
breech lock 闭锁卡铁;闩锁
breech-lock ring 镜头锁紧卡口环
breech opening 装药孔;炸药孔
breech operating mechanism 开闭机构
breech pressure 膛内压力
breech-sight 瞄准器
breed 品种
breed characteristic 品种特征
breeder 育种人员
breeder converter 增殖反应堆
breeder material 增殖材料

breeder plant 增殖堆电站
breeder reaction 增殖反应
breeder reactor 增殖反应堆
breeder's association 品种协会
breed fish sewage 养鱼污水
breeding 育种;增殖;繁殖
breeding activity 繁殖活动
breeding area 繁殖区
breeding fire 火灾征兆;自燃
breeding for special qualities 特殊品质的选择
breeding ground 育种场;繁殖场
breeding habit 繁殖习性
breeding herd 繁殖群
breeding nursery together 集中育种圃
breeding of aquatic products 水产养殖
breeding period 生殖期
breeding place 育种场;孳生地;繁殖场
breeding season 繁殖季节;繁育季节
breeding station 育种场
breed out 排除
breed spring 养殖泉
breed structure 品种结构
breeze 矿粉(煤渣);煤渣;焦屑;焦炭粉;焦末;煤粉
breeze aggregate 炉渣集料;炉渣骨料;炉碴集料;炉碴骨料;焦渣集料;焦渣骨料
breeze block 炉渣砖;炉渣砌块;焦渣石;焦渣砌块;焦渣块;焦渣水泥砖;煤渣砌块;煤渣块
breeze brick 炉渣砖;焦渣砖;煤渣砖
breeze cement brick 煤渣水泥砖
breeze concrete 炉渣混凝土;焦渣混凝土;煤渣混凝土;轻质煤渣混凝土
breeze fixing brick 焦渣水泥砖;焦渣固结砖;焦渣受钉砖;煤渣固结砖
breeze oven 焦末化铁炉
breezer 无篷汽车
breeze stone 焦渣石
breezeway 过道(房屋之间);连接廊;穿堂;房屋之间的走廊;通廊;走廊
breezing 不清晰
bregema 冠矢点
B-register 变址寄存器;加法寄存器;变数寄存器
Breguet hair spring 挑框游丝
Breguet spring 挑框游丝
breithauptite 红锑镍矿
Breit-Rabi formula 布赖特—拉比公式
Breit-Wigner equation 布赖特—维格纳方程
Breit-Wigner formula 布赖特—维格纳公式
Breit-Wigner theory 布赖特—维格纳理论
Brelave Bridge 勃拉凡桥(法国)
Bremen blue 蓝色碳酸铜;梅蓝;铜蓝
Bremen green 不来梅绿;铜绿
Bremen harbo(u)r 不来梅港(德国)
Bremer's test 布列默试验
Bremond porosimeter 布列蒙德孔径仪
bremsstrahlen 制动射流
bremsstrahlung 制动射流;韧致辐射
bremsstrahlung effect 韧致辐射效应
Bren carrier 履带式小型装甲车
brenkite 氟碳钙石
Brennan monorail car 布伦纳南单轨车
Brenner ga(u)ge 布伦纳量规;搪瓷厚度仪
Brennert ga(u)ge 布伦纳规
brennschluss 熄火;断火
brennschluss time 熄火时刻
brenstone 硫黄
brent 丘陵
Brentidae 三锥象虫科
Brereton scale 柏里列顿木材容积计算法
breriseptum 短隔壁;短隔板
bressummer 横楣;大木;托梁器;过梁
bretagne 米色或蓝色帆布
bretesse 防卫工事;望楼
bretisement 透空花格;透空花纹的顶部装饰(哥特式)
Bretonian 布锐东统【地】
Bretonian orogeny 布锐东造山运动【地】
Bretonian stratum 布锐东地层
Bretton Woods Agreement 布雷顿森林协定
Bretton Woods Agreement Act 布雷顿森林协定法
Bretton Woods conference 布雷顿森林会议
Bretton Woods Era 布雷顿森林时代
Bretton Woods Monetary System 布雷顿森林货币体系
Bretton Woods Regime 布雷顿森林货币体系

Bretton Woods System 布雷顿森林制度;布雷顿森林体系
Breuchaud pile 布鲁齐德桩;(在水中施工的)钢管套混凝土桩
breviseptum 短隔壁;短隔板
brevity 简洁;简短
brevity code 简码;简化码
brewage wastewater 酿造废水
Brewer anaerobic jar 布鲁尔厌氧罐
Brewer instrument 布鲁尔检测仪
Brewer-Mast electrochemical sonde 布鲁尔—马斯特电化学探空仪
brewer's pitch 啤酒桶涂料用沥青
Brewer's point 布鲁尔点
brewert waste 酒厂废水
brewery 酒厂;酿造厂;酿造厂;啤酒厂
brewery and distillery wastewater 酿造和蒸馏废水
brewery effluent 啤酒厂废水
brewery industry 酿酒工业
brewery process liquor 啤酒酿造废水
brewery residues 酿造渣;啤酒厂残渣
brewery waste 啤酒厂废水
brewery wastewater 啤酒厂废水
brewhouse 啤酒厂;酿酒厂
brewing industry 酿造业
brewing period of investment 投资酝酿阶段
brewing process wastewater 酿造废水;啤酒酿造废水
brew kettle 酿造锅
Brewster 布鲁儒斯特(光弹性单位)
Brewster angle 极化角;偏振角
brewsterite 锶沸石
Brewster's angle 布儒斯特角
Brewster's angled rod 布儒斯特角棒
Brewster's angle surface 布儒斯特角表面
Brewster's angle window 布儒斯特角窗
Brewster's band 布儒斯特干涉带
Brewster's coefficient 布儒斯特系数
Brewster's cut 布儒斯特切割
Brewster's fringe 布儒斯特条纹
Brewster's lamp 布儒斯特灯
Brewster's law 布儒斯特定律;(光的反射与折射定律)
Brewster's magnifier 布儒斯特式放大镜
Brewster's optics Kerr cell 布儒斯特角克尔盒
Brewster's point 布儒斯特(中性)点
Brewster's prism 布儒斯特棱镜
Brewster's process 布儒斯特法
Brewster's stereoscope 布儒斯特立体镜
Brewster's window 布儒斯特窗
brezinaite 硫铬矿
BrF₃ method 三氟化溴法
BrF₅ method 五氟化溴法
Brianchone's luster 布来考恩光泽彩(不需要还原气氛的光泽彩)
Brianchone's theorem 布来考恩定理
brianite 磷镁钙钠石
briar dress 拨料
Briar hill sandstone 玻利亚山砂岩(一种产于美国俄亥俄州比利尔山的砂岩)
briartite 灰锗矿
briar tooth (锯的)钩齿;偏锋齿
briar-toothed saw 横截锯;隙齿锯
bribe 贿赂;行贿
briber 行贿者;行贿人
bribery 行贿
bribery and kickbacks 贿赂和回扣
bribes 贿赂
bric-a-brac 小摆设;古玩
brick 砖形物;砖
brick aggregate 碎砖骨料;集砖料;碎砖集料
brick aggregate concrete 砖集料混凝土;砖骨料混凝土
brick anchor 砖锚杆;固定饰面的砖(预)锚件
brick-and-a-half wall 一砖半厚墙;半砖厚墙;三七墙
brick and brick 清水砖工;规准砖工;无灰浆砌筑
brick-and-concrete composite construction 砖混结构
brick-and-stone work 砖石工程
brick-and-stud 砖木墙壁;木架砖壁
brick-and-stud work 木架填砖墙;木龙骨间填砖作业;木架砖墙;砖木墙壁
brick and tile industry 砖瓦工业
brick-and-timber construction 砖木结构

brick apron 砖散水
brick arch 火箱砖拱;砖拱
brick arch bridge 砖拱桥
brick arch floor 耐拱楼盖;砖拱楼面;砖拱楼盖;砖拱楼板
brick architecture 砖砌建筑物
brick arch sewer 砖拱管
brick axe 双刃斧;尖尼锤;尖尼斧;劈砖斧;瓦工锤;瓦工槌
brick backing 砖衬背;砖背衬
brick baffle 砖隔墙;砖隔屏
brick ballast 砖碴
brick barrel-vault construction 砖筒拱结构
brick barricade 砖屏
brick base 砖砌基础;砖基;砖座
brick bat 砖头;砖块;半(块)砖;碎砖;砖片
brickbat drain 碎砖盲沟;碎砖排水(盲)沟
brick beam 配筋砖过梁;加筋砖过梁;砖过梁;钢筋砖过梁
brick bearing wall 砖承重墙;承重砖墙
brick bedding 砖铺垫层
brick bed joint 砖砌层水平接缝
brick block 砖砌墙
brick bond 砌砖体;砌砖法
brick bonder 丁砖
brick bonding 砌砖
brick bonding header 丁砖
brick bondstone 丁砖
brick boundary wall 界限砖墙;边界砖墙
brick breaker 碎砖器;碎砖机
brick buggy 装砖手推车
brick building 砖建筑(物)
brick built 砖砌体
brick burner 装砖工人
brick burning 烧砖;砖的烧制
brick buttress 砖垛
brick cage 装运砖的笼子;烧砖台
brick canal 砖渠
brick-cap brick face 砖顶压檐墙(砖压顶女儿墙)
brick-cap parapet 砖压顶女儿墙;砖顶压檐墙
brick carrier 运输砖的车子
brick cart 运砖的轻便车
brick carving 砖雕
brick cathedral 砖圆层顶
brick cavity wall 空心砖墙;砖砌空芯墙
brick cement 圬工防水水泥;砖砌防水水泥;砖筑水泥;砌筑水泥;烧结黏土水泥;防水水泥(砖石砌体用)
brick chimney 砖烟囱
brick chip(ping)s 砖碎片
brick chisel 修砖凿
brick clad building 镶砖房屋;砖镶房屋;镶砖建筑物
brick clamp 砌砖拉结件(用于砌空芯墙);砖堆;砖夹
brick clay 砖黏土;砖土;制砖黏土
brick clay deposit 制砖黏土矿床
brick clay ore 砖瓦黏土矿石
brick cleaner 砖的洗净剂
brick cleaning machine 砖的清洗机
brick coffering 衬砌井壁
brick column 砖柱
brick concrete 碎砖混凝土
brick-concrete structure 砖混结构
brick conduit 砖砌渠道
brick conduit-type sewer 砖砌下水道;砖管道式污水管
brick construction 砖石工程;砖石结构;砖砌结构;砖结构
brick coping 砖压顶
brick coping on gable 山墙砖压顶
brick corbel 砖叠涩
brick core 砖心;砖砌模
brick cornice 砖挑檐;砖砌挑檐
brick course 砖砌层;砖层
brick course joint 砖砌层水平接缝
brick crusher 碎砖器
brick cube 方砖块
brick cube pavement 方砖块路面
brick culvert(pipe) 砖涵洞;砖砌涵管;砖砌涵洞
brick cup 座砖
brick curb 砖道牙;砖缘石;砖缘边石;砖面缘石
brick cutter 砖砌机;切砖机
brick cutting 砍砖
brick cutting wire 切割砖的钢丝
brick dado wall 砖墙裙

brick dam 砖坝;砖砌坝
brick damp course 砖防潮层
brick defect 砖疵
brick die 砖模
brick dome 砖穹隆
brick dust 碎砖;砖屑;砖灰;砖粉;砂砖粉
brick dwelling 砖建住宅
brick earth 砖土;制砖土
bricked ladle 砖砌盛钢桶
brick efflorescence 砖面泛白
brick elevator 提升式运砖机
brick-encased 镶砖的;镶柱的;包砖的
brick-encased wall 包砖墙
brick enclosing wall 砖围墙
brick enclosure 砖围墙
brick engineering 砖石工程
brick fabric 砖结构
brick facade 砖正面;砖门面;砖立面
brick faced 砖饰面的
brick facing 砖饰面;砖贴面;砖砌面层;砌砖面层
brick fencing wall 砖围墙
brick fender 砖砌炉围;砖炉围
brickfield 制砖工场;砖厂
brickfielder 拨立克斐儿特风
brick fired with combustible additives 内燃砖
brick fireproofing 耐火砖
brick flat arch 砖平拱
brick floor(ing) 砖楼面;砖铺地(面);砖地(面)
brick flour 砖粉;砖屑
brick flue 砖烟道;砖砌烟道
brick footing 砖基础;砖大方脚
brick for arch(ed) roof 拱壳砖
brick for dampproof work 防潮用的砖
brick for foundation 基础砖
brick format 砖加灰缝的尺寸;砖的尺寸
brick for wedge use 楔形砖
brick foundation 砖基础
brick-framed wall 包框墙
brick fuel 炭砖
brick furnace 砖窑
brick gable 砖砌山墙
brick game 方块游戏机
brick ga(u)ge 皮尺数;砖层高度标准尺;砖标号
brick glaze 砖釉
brick grab 夹砖设备;抓砖器
brick grade 砖号
brick grease 块状润滑油
brick grill(e) 砖砌花格
brick grip 砖夹
brick gripping device 夹砖设备
brick hammer 尖尾手锤;瓦工锤;瓦工槌
brick hardcore 小块碎砖;碎砖骨料;碎砖集料
brick heating kiln 烤砖窑
brick holder 砖托
brick hollow wall 砖砌空芯墙
brick house 砖房
brick-in 砖衬的
brick inclusion 夹砖;夹入砖
brick industry 制砖业
bricking 砖瓦工程;砖衬;仿砖工作
bricking ring 砖壁座
bricking scaffold 砌砖吊盘
bricking-up 砖塞孔;砖砌;砖堵塞;砌体;用套塞孔
brick-in-mastic waterproofing 涂胶泥砖防水
brick joint 砖层接缝
brick kiln 砖窑;堆成拱形的砖坯
brick laid on edge 斗砖;侧砌砖;竖砌砖
brick laid on end 立砌砖
brick laid on flat 平砌砖;平铺砖
brick laid with mortar 砂浆砌砖
bricklayer 泥瓦工;砌砖工;泥水匠;泥水工;瓦工;砖层
bricklayer ladder 砌砖工的梯子
bricklayer line 砌砖工的拉线
bricklayer scaffold(ing) 砌砖工的脚手架
bricklayer's charge 泥瓦工工头
bricklayer's cleaver 瓦刀
bricklayer's hammer 尖尾手锤;砖工锤;瓦工锤;瓦工槌
bricklayer's hod 砖石工用砂浆桶
bricklayer's labo(u)rer 砌砖杂工
bricklayer's ladder 泥瓦工梯
bricklayer's line 泥工线
bricklayer's scaffold(ing) 砖工脚手架;砖工用的脚手架;瓦工脚手架
bricklayer's square scaffold 砖工用的以方木构成的脚手架;瓦工操作平台;瓦工方台脚手架
bricklayer's tool 砌砖工具;泥工工具
bricklayer's trowel 砖工镘(板);瓦工镘;灰镘;大铲
bricklayer's work 砖工工作;砖工工程;砌砖工的工作;砌砖工的工具
brick-laying 砖工;砌砖
brick-laying channel 砌筑地沟
brick-laying crew 砌砖队;砌砖工作队
brick-laying tool 砌砖工具
brick lift 砖提升机;砖的提升器
brick-lined 衬砖的
brick lined canal 砖衬砌渠道
brick-lined concrete slab 混凝土板边镶砖;镶砖混凝土板
brick-lined masonry 混合砖工;镶砖砖工;砖衬砖工
brick lining 砖内衬;砖衬(砌);砌砖内衬
brick lintel 砖砌过梁;过梁
brick made from sand-lime 用灰砂制的砖块
brickmaker 制砖工
brick making 制砖(法)
brick making equipment 制砖设备
brick making machine 制砖机
brick making machinery 制砖机械
brick making method 制砖法
brick making plant 制砖厂;砖厂
brick manhole 砖砌检查井
brick manufacture 制砖
brickmason 砌砖工;砖层;瓦工
brick masonry 砖工;砖石工;砖砌工;砖砌体;砖砌工程;砌砖体
brick masonry bearing wall 砖砌承重墙
brick masonry structure 砖石结构;砖工结构
brickmason's tool 砌砖工具
brick mass foundation 砖石砌体基础
brick mastaba 砖砌坟墓(古埃及)
brick mortar 砌砖砂浆;砌砖火泥;砌砖灰浆;砌筑砂浆
brick mo(u)ld 砖模;砖线脚;砖砌线脚
brick mo(u)lded sill 砖砌窗台;砖窗台
brick mo(u)lding 砖线脚;砖缝盖条;砖缝木嵌条
brick mo(u)lding machine 制砖机
brick mo(u)lding press 压砖机
brick niche 砖龛
brick nogged 砖填木架隔墙
brick nogged timber wall 砖填木构架墙
brick nogged wall 木架砖墙
brick nog(ging) 砖木墙壁;填充砖墙;砖填木架隔墙;木架砖壁;木构架填砖作业;木架填砖隔墙
brick nogging building 立贴式房屋;立贴房屋;木架砖墙房屋;木构填砖隔墙建筑
brick offset 退缩砖砌阶
brick oil 压砖机润滑油;砖油
brick on bed 平砌砖;卧砖
brick-on-edge 侧砌砖
brick-on-edge coping 侧砌砖压顶;侧砖压顶
brick-on-edge course 侧砌砖层
brick-on-edge inverted arch 倒拱侧砌砖
brick-on-edge pavement 侧砖铺砌路面
brick-on-edge paving 侧砖铺砌;侧砖铺面
brick-on-edge sill 侧砌砖窗台式门槛;侧砖窗台;侧砌砖门槛;侧砌砖窗台
brick-on-end 竖砖;立砌砖;竖砌砖
brick-on-end course 侧砌砖墙
brick-on-end soldier arch 竖砖拱
brick on header for border 丁砖镶边
brick on plate 平砌砖
brick ornamental string(course) 砖的装饰线列
brick pack 砖捆
brick package 砖头组装
brick pagoda 砖塔
brick pallet 运砖板
brick panel 预制砖板
brick panel(building) unit 预制砖板建筑单元
brick partition 砖隔墙
brick pattern 砌砖式码堆
brick-paved plinth 砖砌勒脚
brick-paved-red 红砖砌砖
brick pavement 砖铺面;砖铺地(面);砖路面;砖块路面;铺砖路面
brick pavement on mortar bed 砂浆垫层铺砖路面
brick pavement upon mortar bed 浆砌缸砖路面
brick paving 砖块铺砌;砖地(面);砖墁地;砖路面;铺砖
brick pier 砖墩;砖柱
brick pit 砖砌井坑;砖砌井筒;砖砌坑
brick plant 制砖厂
brick plinth 砖底座
brick pointed arch 砖砌二心同似挑尖拱
brick press 压砖机;制砖机
brick-pressing machine 压砖机;砖坯压制机
brick protective skin 砖保护层
brick quoin 砖饰墙隅
brick raking 刮砖
brick rattle 砖磨损试验
brick rattler 炼砖磨耗试验机;砖磨损试验机
brick recuperator 陶质换热器
brick-red 红砖色(的);橘红色;砖红色
brick reinforcement fabric 砖砌体钢筋(垫)网
brick reinforcement mat 砖砌体钢筋(垫)网
brick relieving arch 砌砖辅助拱
brick repress 砖坯再压机
brick reservoir 砖砌水库
brick retaining wall 砖挡土墙
brick rib 砖肋
brick road 砖砌道路;砖路
brick roll 砖式卷材(沥青浸涂毡材凸出似砖状)
brick rubbing 磨砖
brick rubbish 碎砖渣土
brick rubble 粗面砖块;碎砖
brick saw 锯砖机;砖锯;切砖锯
brick seat 砖支座;大放脚;墙合;横挡(墙或基础上);壁架
brick set 砍砖砧;砖凿;劈砖枕垫;劈刀;砍砖;砖衬
brick set on edge 侧砌砖
brick setting 砖工;镶砖;砖砌;砌砖体
brick sewer 砖砌阴沟;砖砌污水管;砖砌下水道;砖砌污水沟;砖砌沟管
bricks from sludge ash 污泥砖
brick shaft 砖砌柱身
brick shale 页岩砖
brick shape 砖形;异型砖
brick shaped 砖形的
brick shell 砖薄壳
brick sidewalk 砖砌人行道
brick sill 砖窗槛;砖窗台
brick size 砖的尺寸
brick-size external cladding klinker 外部覆盖面缸砖;精细烘焙的小砖
brick size tile 如砖大小的瓦
brick skin 砖表面;砖的表面;砖保护层
brick slip 砖滑道;仿砖贴面;面砖
brick soldier course 竖砖层
brick spread foundation 砖放大脚基础;砖砌大方脚
brick stacker 堆砖工
brick stair(case) 砖砌楼梯(间)
brick step 砖台阶;砖踏步
brick-stone masonry 砖石结构
brick stopping 砖砌隔墙
brick strength 砖强度;砖的强度
brick structure 砖建筑物;砖结构;砖石结构
brick supporter 砖托
brick(sur)facing 砖铺路面;砖砌面层
brick system building 砖体系建筑;预制装配式砖石建筑
brick tea 砖茶
brick test(ing) 砖的试验
brick testing machine 砖试验机
brick texture 砖体结构
brick tool 砌砖工具
brick tracery 砖砌窗花格
brick trimmer 砖托梁;砖面修整;磨砖工人
brick trowel 灰镘;砌砖镘刀;大铲
brick truck 运砖车
brick tunnel vault 砖砌筒形穹顶
brick type 砖型
brick underdrain 砖砌阴沟
brick vault 砖砌圆顶;砖砌穹顶;砖拱
brick-vault hall 无梁殿
brick vault roof 砖拱顶楼面
brick veneer 砖饰面;砖镶面;砖贴面;砖砌镶面层;砌砖面层;贴面砌砖
brick-veneered 砖镶面的;砖贴面的
brick-veneered construction 砖镶面构造
brick-veneer wall 砖面夹墙;砖贴面墙;砖镶面墙
brick volume per unit chamber space 格子体充填系数

brick wall 空斗墙;砖墙
brick wall arch 砖墙发券;砖砌实心拱
brick wall bearing construction 砖墙承重结构
brick wall expansion joint 砖墙伸缩缝
brick wall fence 砖围墙
brick walling 用砖筑墙;砖砌井壁;砖砌墙体
brick wall load bearing 砖墙承重
brick wall without plastering 清水砖墙
brick ware 陶砖;陶块
brick wetting 砖的浸湿
brick wheel window 砖砌轮形窗扇
brick whistle 泪孔;挡土墙缝滴水孔;流水孔
brick wicket 砖砌窑门
brick with grooves 带槽砖;大开条砖
brick with horizontal perforations 有长穿孔眼的砖
brick with lattice perforations 具有孔眼花格的砖;菱形孔空心砖
Brickwood 布里克伍德(商品名,一种可受钉的砌块)
brickwork 砖砌体;砖工;砖厂;砌砖工作;砌砖工程;炉墙;砖结构;砌体
brickwork arch 砖拱;砖砌拱
brickwork base 砖座;砖砌基座;砖砌基础
brickwork bond 砖砌体
brickwork casing 砖工刷灰;砖块镶面;砖工罩面;砖砌面层
brickwork castle 砖砌城堡
brickwork chaser 刻缝机
brickwork chimney 砖砌烟道
brickwork column 砖柱;砖砌柱
brickwork corbel 挑出砖牙
brickwork cube 立体砖砌体
brick worker 瓦工
brickwork fire place 砖砌壁炉
brickwork gable 砖砌山墙
brickwork individual base 砖砌独立基础
brickwork insert 砖工镶嵌
brickwork joint 砖缝;砖接缝
brickwork manhole 砖砌检查井
brickwork mortar 砖工灰浆;砖砌筑砂浆
brickwork mound 砖砌护堤
brickwork movement joint 砖砌体伸缩缝;砖墙变形缝
brickwork pier 砖墩
brickwork pointed arch 砖砌二内心挑尖拱
brickwork reinforcement 砖工加筋
brickworks 砖厂
brickwork system building 预制装配式砖石建筑
brickwork to top out 砖墙到顶
bricky 砖块状;砖的;砌砖的
brickyard 堆砖场地;砖瓦厂;砖厂;堆场场
bride 连接狭条
bride-door 双接门;连接门
bridestone 砂岩大圆石
bridge 连接梁,桥梁;结拱(塑);舰桥,驾驶室(又称驾驶台)(船);驾驶桥楼,弥合;航行驾驶台;过桥夹板;搭棚(炉内);船桥;桥楼(船);桥接线,桥接器;桥(梁);嵌锁;漆膜的搭接复盖;网桥
bridge abutment 桥台;桥墩
bridge acceptance loading test 桥梁验收荷载试验
bridge action 桥梁作用
bridge-adapter 桥式接合
bridge after bulkhead 驾驶室后端舱壁;桥楼后舱壁
bridge airdraft 高潮时桥梁架空高度
bridge amplifier 电桥用放大器;电桥式放大器;桥式放大器
bridge analysis simplified 桥梁简化分析法
bridge anchor 桥梁锚碇
bridge and ball (空气吹洗钻进时湿岩粉所形成的)泥包
bridge and tunnel maintenance team 桥隧维修工队
bridge annunciating device 网络通知设备
bridge annunciating signal 桥梁预示信号
bridge approach 引桥(工作);桥头引线;桥头引道;桥梁引路;桥梁引道
bridge approach alignment 桥头引道接线
bridge approach channel 桥区航道
bridge approach embankment 桥梁引道路堤;桥头路堤
bridge-approach fill 引桥路堤
bridge arch 桥拱;船尼拱架;尼框顶部
bridge architecture 桥梁美学
bridge arch span 拱桥跨

bridge arm 电桥臂;桥臂
bridge art 桥梁艺术
bridge aseismatic strengthening 桥梁抗震加固
bridge assembly equipment 桥架设用的工具;架桥设备
bridge axis 桥轴线
bridge axis location 桥梁轴线测定
bridge axis survey 桥轴线测量
bridge backwater 桥梁壅水
bridge balance 电桥平衡
bridge-balanced d-c amplifier 桥式平衡直流放大器
bridge balancing oscillator 电桥平衡振荡器
bridge ballasted floor 道砟桥面
bridge beam 桥主梁;桥式梁;桥梁大梁;桥的主梁;挑梁板
bridge beam run 便桥式脚手架
bridge bearing 桥梁支座;桥梁支承
bridge bearing pad 桥梁支座(衬垫)
bridge bearing plate 桥梁支座垫板
bridge block 大梁砖;砌砖
bridge board 斜梁;楼梯帮;楼梯梁;楼梯侧板;桥板;梯级搁板(木楼梯的斜帮,承载踏步);接梯侧板;短梯基
bridge board stair(case) 楼梯架的楼梯间
bridge boat 船搭浮桥;浮桥趸船;浮船;平底船
bridge body 桥体
bridge bond 桥键
bridge book 航海日志草本
bridge box 电桥箱
bridge branch 桥接旁路
bridge breaker 桥式断路器
bridge brick 桥砖
bridge builder 造桥人;调解人
bridge building 建造桥;造桥
bridge building engineer 建桥工程师
bridge building practice 建桥实践;桥施工技术
bridge cable 吊桥索;吊桥钢缆;桥索
bridge cable rake 桥式钢丝钯
bridge calculating circuit 桥式计算电路
bridge calibration 电桥校准
bridge camber 桥梁拱势;桥梁上拱度
bridge cap 桥墩顶座;桥墩帽
bridge carriageway 桥梁车行道
bridge carriageway pavement 桥上车行道路面;桥面
bridge carriageway surfacing 桥上车行道铺面;桥面
bridge checkup 桥梁鉴定
bridge circuit 电桥电路;桥式电路;桥路;桥接线路;桥接电路
bridge clearance 桥梁下净空;高潮时桥梁架空高度;桥下净空;桥梁净空;桥梁架空高度(高水位时)
bridge coating 桥梁漆
bridge column 桥墩
bridge compound 桥环化合物
bridge conductor (沿起重机的)桥架导线
bridge configuration 桥形结构
bridge connection 跨接;桥形结线;桥式连线;桥接
bridge connector 桥型连接件;桥式连接器;桥接条
bridge construction 桥梁施工;桥梁结构;桥梁建筑;桥梁建造;桥梁工程
bridge construction clearance 桥梁建筑限界
bridge construction practice 桥梁施工技术
bridge construction site 桥梁建筑工地;桥梁工地
bridge construction truck 架桥载重汽车
bridge constructor 建桥人
bridge consulting engineer 桥梁顾问工程师
bridge contact 分路触点;桥接式触点;桥接点;桥接触点
bridge control console 驾驶室控制台
bridge convertor connection 换流器桥式联结
bridge conveyer 高架桥式输送带
bridge corpuscle 桥粒
bridge/coupler switch 桥/联络开关
bridge crack 挤坏悬空开裂
bridge crane 行车;高架起重机;桥式起重机;龙门起重机;桥式行车;桥式吊(车);桥吊;行动吊车
bridge crane with hook 带钩桥式起重机
bridge crane with level luffing crane 装有起重杆升降设备的桥式起重机
bridge crossing 立交桥;跨线桥;桥水道;桥渡
bridge-crossing channel 桥区航道

bridge crossing of railway 铁路桥渡
bridge cross-section 桥梁的横断截面
bridge-culvert survey 桥涵洞测量
bridge current 桥式电流;桥路电流;桥接电流
bridge current indicator 电桥电流指示器
bridge cut-off 架桥取直;桥梁捷径;断桥;桥式断路
bridge-cut-off relay 断桥继电器;分隔继电器;桥式断路继电器
bridge-cut-off test 断桥试验
bridge dam 桥式坝;桥坝
bridge data 桥梁数据
bridge deck drainage 桥面排水
bridge deck formwork 桥面模板
bridge deck(ing) 驾驶室甲板;船桥甲板;桥面板;桥楼甲板
bridge decking system 桥面系
bridge deck pavement 桥面铺装
bridge deck renewal 桥面更换
bridge deck slab 桥面板
bridge-deck superstructure 桥楼甲板上层建筑
bridge dedication 献身于桥梁事业
bridge defect 桥梁病害
bridge design 桥梁设计
bridge design engineer 桥梁设计工程师
bridge designer 桥设计人;桥梁设计人
bridge design specification 桥梁设计规范
bridge detector 桥路检波器
bridged floor 带撑格栅路面;格栅地板
bridged graphite 桥接石墨
bridged gutter 渡槽;架空雨水槽
bridged hole 架桥孔;堵塞孔
bridge diagram 跨接图;电桥电路图;桥梁图式
bridge die 空心件挤压模;桥式孔形挤压模
bridge diplexer 桥接双工器
bridged joint 桥式连接;桥式接头
bridge dodger 桥楼挡雨幔
bridge drainage 桥的排水;桥梁排水
bridge drawing 绘桥图;桥梁图
bridge driver 桥式驱动器
bridged rotary 环形立体交叉
bridged rotary intersection 桥式环形立体交叉
bridged-T circuit 桥接T形电路
bridged-T filter 桥接T形滤波器
bridged-T network 桥接T形网络
bridged traffic route 桥上交通线路
bridged-T tap 桥接T形抽头;桥接T形水龙头
bridged-T trap 桥接T形陷波器
bridge duplex system 桥式双工系统;桥接双工系统
bridge end transition slab 桥头搭板
bridge engineer 桥梁工程师
bridge engineering 桥梁工程(学)
bridge erection 桥楼建筑安装;桥架设;桥梁加设
bridge erection without scaffolding 无支架施工
bridge erection by floating 浮运架桥法
bridge-erection crane 架桥机
bridge excavation 桥基开挖
bridge extension 桥楼伸展平台
bridge fabrication 桥梁拼装
bridge falsework 桥梁鹰架;桥梁脚手架
bridge fault 桥接故障
bridge field engineer 桥梁施工工程师
bridge financing 中间性筹款;过渡性筹款
bridge fittings (集装箱间的)横搭扣;桥式连接器
bridge floor 桥面;格栅地板;带撑格栅楼面
bridge floor expansion and contraction installation 桥面伸缩装置
bridge floor plank 桥面板
bridge floor system 桥面系
bridge floor without ballast 无砟桥面
bridge floor without ballast sleeper 无砟无轨桥面
bridge floor with telescopic device 桥面伸缩装置
bridge footing 桥梁底脚
bridge forecastle 驾驶室首楼
bridge foreman 桥梁工长;桥梁施工员;桥梁领工员
bridge for flow-ga(u)ging 测流便桥
bridge formation 粉料架桥现象
bridge for mula 桥梁照查实施要点
bridge form(work) 桥的模板工作
bridge for streamflow measurement 水文测桥
bridge for stream-ga(u)ging 测流桥;水文测桥
bridge for supporting pipe 管道栈桥
bridge foundation 桥梁基础;桥基
bridge frame 拱架
bridge framework 桥梁构架

bridge front bulkhead 桥楼前舱壁
bridge fuse 插接保险丝；桥接保险丝
bridge gallery 过街楼
bridge-gang 桥梁工区；桥梁工程队
bridge ga(u)ge 桥式量规；桥规
bridge girder 桥梁板梁；桥大梁
bridge girder erection equipment 架桥机
bridge girder erection machine 架桥机
bridge gore 桥头锥形护坡；桥头锥体护坡
bridge grab crane 桥式抓斗起重机；桥式抓斗吊（车）
bridge gradient 桥的坡度
bridge grafting 桥接法
bridge grating 桥上铁格子
bridge green 桥梁绿色
bridge guard house 护桥警卫室
bridge guard rail 桥梁护轨
bridge gutter 桥梁边沟；渡槽；架空雨水槽；架空水槽
bridge handrail 桥栏杆
bridge head 桥头（堡）；桥堡
bridgehead alignment 桥头定线
bridgehead construction 桥头建筑
bridgehead greening 桥头绿化
bridge heating 桥上路面受热
bridge hole 桥孔
bridge hose 过渡宿主
bridge host 中间宿主；桥梁寄主
bridge house 桥楼【船】；桥头警卫室
bridge improvement 桥梁加固
bridge infection 过渡侵染
bridge in furnace 炉坝；火坝
bridge inspection 桥梁检查
bridge inspection car 桥梁检修车
bridge inspection vehicle 桥梁检测车
bridge inspector 桥梁检查员
bridge insulation 桥的绝热
bridge islet 涨潮岛（退潮时是半岛）
bridge joint 架接；啮接；悬空接头；悬接；桥接
bridge key 桥路中电键
bridge ladder 驾驶室楼梯
bridge layer 架桥汽车
bridge-laying truck 架桥汽车
bridge layout 桥梁布置
bridge length 桥梁长度
bridge limiter 桥式限幅器
bridge line 连通铁路；联络线
bridge linkage 桥式联结
bridge load 桥梁荷载
bridge load spectrum 桥梁荷载谱
bridge location 桥位；桥址
bridge locking 活动桥锁闭
bridge loop 桥梁套线；桥梁（迂）回线；桥梁环行线
bridge machine 桥接计算机
bridge magnetic amplifier 串桥磁放大器
bridge maintenance 桥梁养护；桥梁维修
bridgeman 开合桥操作人员；架桥工人
bridge management system 桥梁管理系统
bridge manufacture 桥梁制造
bridge mark 桥涵标
bridge marker 桥梁标志
bridgemaster 桥梁管理（人）员
bridge measurement 电桥测量；桥式电路测定；桥测
bridge megger 桥式高阻表；桥式测高阻电表；电桥式兆欧计
bridge member 桥梁杆件；连接杆件；桥梁构件
bridge method 电桥法；桥路法；桥接法
bridge mixer 桥式可移动的混凝土搅拌机
bridge model 桥的模型
bridge model test 桥梁模型试验
bridge name plate 桥名牌
bridge network 桥形网络
bridge neutralizing 桥接抵消法
bridge observation 桥梁观测
bridge obstruction signal 桥梁遮断信号
bridge of boat 浮桥
bridge of boiler 火坝
bridge of circular plan form 曲线桥；弯桥
bridge of complex line 复线桥
bridge of fault 断层脊
bridge of the bit 钻头牙轮架；多牙轮钻头十字接头
bridge of the nose 鼻柱；鼻背
bridge of white marble 汉白玉桥
bridge on slope 坡桥
bridge open floor 明桥面

bridge opening （以墩分隔的）桥跨；桥孔；桥洞
bridge opening mark 桥涵标
bridge-oscillation 桥梁摆振
bridge oscillator 桥式振荡器
bridge over 跨越；桥接
bridge overgrade 铁路在下交叉
bridge overhead travel(l)ing crane 行车桥；（移动式高架重机的）吊车桥
bridge over railway 跨线桥
bridge pad 桥梁支座
bridge parts 架桥器材
bridge paver 桥式铺路机
bridge paving sett 桥面铺砌小方石
bridge piece 船尼柱框，（车床的）马鞍；（车床的）过桥；船尼拱架；桥梁片；尼框顶部
bridge pier 桥墩；墙垛
bridge pier body of ashlar stone face filled with rubble masonry 条石砌面中填毛石圬工桥墩身
bridge pier body of cellular reinforced concrete 空腹空格钢筋混凝土桥墩身
bridge pier body of rubble masonry 毛石圬工桥墩身
bridge pier light 桥柱灯
bridge pile 桥桩；桥脚
bridge pin 桥钉；桥枢
bridge pipe 桥接管
bridge plank 桥面板
bridge planning 桥梁规划
bridge plant 桥梁厂
bridge plate 架接板；装railing跳板；桥梁钢板
bridge plate girder 板梁桥；（桥的）板梁
bridge plug 可胀堵塞（用于孔内隔离生产层或灌水泥浆）；桥塞
bridge point 渡口
bridge polar duplex system 桥接双工制
bridge position 桥位
bridge post 桥梁标
bridge precamber 桥梁预拱度
bridge project 桥梁工程
bridge pylon 桥塔
bridge rail 活动短轨；桥形钢轨；桥式起重机轨道；桥式起重机钢轨；桥式护栏
bridge railing 桥梁栏杆；桥栏杆
bridge rail joint 桥式钢轨接头
bridge rail lock 活动桥钢轨连接锁闭器
bridge raising 桥的提升
bridge ramp 桥式坡道；跳板；上桥斜坡
bridge rating 桥梁检定
bridge rating test 桥梁检定试验
bridge ratio arm 电桥比率臂
bridge reamer 桥工铆钉扩孔器；桥工铆钉绞刀；长扩孔钻；桥工（铆钉）钻孔机；桥工铰刀
bridge rebuilding program(me) 桥梁的重建计划
bridge receiving room 桥楼无线电室
bridge reclaimer 桥式取料机
bridge reconstruction 桥梁改建
bridge rectifier 桥式整流器
bridge reinforcement 桥梁加固
bridge reinforcing 桥梁加固
bridge remote control 驾驶室遥控
bridge repair 桥梁维修
bridge ring 桥环
bridge road 联络线；桥上车道；桥面；驮道
bridge rope 桥缆（索）
bridge scale 地秤；地磅
bridge screw 夹板螺钉
bridge seal(ing) 桥式的防水
bridge seam weld 桥线焊；桥缝焊接
bridge seat 桥座；桥支座
bridge seat back wall 雉墙
bridge seating girder 桥座大梁
bridge secondary control station 驾驶室遥控站
bridge set 桥接装置
bridge set paving 桥上用小方石铺路面
bridge setting girder 桥座大梁
bridge shoe 桥靴
bridge shop 桥梁车间
bridge shoring 桥梁临时顶撑；桥梁脚手架
bridge shuttering 桥梁模板工作
bridge sign 桥梁标（志）
bridge site 桥址；桥位
bridge site stake 桥位桩
bridge site survey 桥址测量；桥位勘测
bridge slab 桥上薄板；桥面板；桥机板（混凝土）

bridge sleeper 桥枕；桥梁枕木
bridge socket 桥式承窝
bridge space 桥楼【船】
bridge span 桥梁跨度；（以墩分隔的）桥跨
bridge span in curved plan 曲梁
bridge spiral 桥梁螺旋线
bridge spot welding 单面搭板点焊；单面衬垫点焊；带接合板点焊
bridge-stabilized oscillator 桥式稳频振荡器
bridge stacking system 桥式堆料系统
bridge stair(case) 桥上楼梯；桥梁梯
bridge statue 桥梁雕像
bridge steel 桥梁钢
bridge steel plate 桥梁钢板
bridge stone 构筑桥梁的石块；盖板；水沟盖板；修桥梁石块；盖石；道口石；搭石
bridge strand 桥上绞合缆索；桥缆（索）
bridge structural test 桥梁结构试验
bridge structure 桥梁结构
bridge substructure 桥梁下部结构
bridge superstructure 桥梁上层建筑；桥梁上部结构；桥跨结构
bridge support 桥支座；桥式支架；桥梁支座
bridge support cap 桥墩帽
bridge survey 桥位测量；桥梁测量
bridge tabulation 编制换算表
bridge tamper 桥面夯实机器
bridge tap 分路抽头；桥接抽头
bridge team 架桥队
bridge technology 过渡性技术
bridge tender 桥梁管理（人）员
bridge test car 桥梁实验车
bridge test(ing) 桥梁试验
bridge testing train 桥梁试验列车
bridge the gap 弥合裂痕
bridge the hole 孔内架桥；钻孔下塞
bridge the price gap 解决差价
bridge thrust 拱桥水平推力；拱桥推力
bridge tie 桥枕；桥梁枕木
bridge tower 桥头堡；桥塔
bridge traffic 过境运输
bridge train 架桥队；架桥列车
bridge tramway 桥式吊（车）
bridge transformer 混合变压器；等差作用线圈；桥式差接变压器
bridge transition 桥式换接过程
bridge transporter 装卸桥；桥式运送机
bridge travel （起重机的）架桥行程
bridge truss 桥桁架；桥析架
bridge truss arm 电桥比率臂
bridge truss clearance 桥梁建筑限界
bridge-tunnel highway 桥梁隧道公路
bridge type 桥式
bridge type coal grab-unloader 桥式抓斗卸煤机
bridge type crane 桥架起重机
bridge type crane for boat lift 桥式垂直升船机
bridge type detector 桥式检测器
bridge type filter pipe 挤式滤水管
bridge type frequency meter 电桥式频率计
bridge type gantry crane 桥吊
bridge-type hammer 桥式锤
bridge-type photoelectric(al) amplifier 桥式光电放大器
bridge-type pontoon 桥式浮桥
bridge-type scraping reclaimer 桥式刮板取料机
bridge-type sludge scraper 桥式刮泥机
bridge-type stereoscope 桥式立体镜
bridge underpass 桥下孔道
bridge unit 桥梁预测构件
bridge up 砂桥卡钻；堵塞
bridge vibration 桥的振动；桥梁振动
bridge vibration mode analysis 桥梁振型分析
bridge voltage 电桥电压
bridge walkway 桥上人行道
bridge wall 浴炉隔墙；挡火墙；坝墙；熔炉隔墙（桥墙）；桥墙；管式炉的坝墙
bridge(wall) cover 桥墙盖板
bridge ward 守桥人
bridgeware 桥接件
bridge washer 阀桥衬垫
bridge washout 桥梁冲刷；桥梁冲毁
bridge waterway 桥区水道；桥梁过水孔径；桥梁河段
bridgeway 桥上道路；楼间架空通道
bridge weir 桥式堰
bridge welding 盖板焊；带接合板焊接；桥接焊

bridge wiper 并接弧刷
bridge wire 电桥标准导线
bridge wire resistance of electric(al) detonator 雷管桥线电阻
bridgework 桥托;桥梁工程;通楼天桥(带桥梁的上层建筑);桥梁厂
bridging 跨连;跨接;剪刀撑;架桥现象;架桥;加密;挂料;格栅撑;表面鼓泡(夹层玻璃缺陷);桥连;桥接;桥键;桥变;起拱;漆膜的搭接复盖;棚料
bridging action 联接作用;架桥作用
bridging agent 黏结剂
bridging amplifier 并联放大器;桥式放大器;桥接放大器
bridging beam 架接桥;横梁;过梁;渡梁
bridging board 格栅跨板;格栅撑板
bridging bond 跨接线
bridging coagulation 架桥凝聚;架桥絮凝
bridging coil 并联线圈;桥式线圈
bridging condenser 分接电容器;并联电容器
bridging crack 桥联裂纹
bridging crossing reach 桥渡河段
bridging distance of control points along strips 航向控制点跨度
bridging distance of control points cross strips 旁向控制点跨度
bridging effect 跨隙效应;架桥效应;架空作用;架空效应;架空现象;遮蔽作用(指裂缝)
bridging error 加密误差;衔接误差
bridging fiber 搭桥纤维
bridging floor 过街楼面;格栅楼面(仅由格栅承载的木楼板);格栅楼面(仅由格栅承载的木楼板);便桥式楼面;无梁格栅楼面;格栅撑楼面
bridging gallery 过街楼
bridging host 过渡寄主
bridging instrument 空中三角测量仪
bridging ion 桥连离子
bridging jack 并联塞孔;桥接塞孔
bridging joist 横格栅;桥式格栅;加撑的格栅;横梁;小梁;格栅加斜撑;渡梁
bridging line 共用线;桥接线
bridging loan 临时贷款;过渡性贷款
bridging loss 分流损耗;桥接损耗
bridging microscope 跨式放大镜
bridging model 架桥模型
bridging of models 模型连接
bridging order 连接指令;返回指令
bridging over 架设;敷设;跨越
bridging over bill 架桥票据(指临时过渡性票据)
bridging over flue 架空烟道;悬空烟道
bridging oxygen 桥氧
bridging part 连接器
bridging particles 堵漏剂
bridging piece 格栅横担(在格栅之间的一块木料,承载隔断墙);桥接片;挑梁板;挑板;加劲件;格栅撑
bridging plug 腰塞(下水泥用)
bridging pontoon 垫档趸船
bridging run 便桥式脚手架
bridging set 分机;并联电话机
bridging slab 桥头搭板
bridging theory 桥联理论
bridging wiper 桥接弧刷
bridging work 联结工作;联结工程
Bridgman anvil 布里奇曼钻砧;布里奇曼压砧
Bridgman effect 布里奇曼效应
Bridgman heat 布里奇曼热
Bridgman method 布里奇兹曼法(晶体生长法);布里奇曼坩埚下降法
Bridgman relation 布里奇曼关系式
Bridgman sampler 布里奇曼取样机
Bridgman-Stockbarger method 布里奇曼—斯托克巴杰坩埚移动法
bridgwater tile (英国制的)相互锁住的瓦片
bridle 龙须缆[船];跨线;紧船索;限动器;系带;系索;辊式张紧装置;短索;承接梁;板簧夹;八字形马鞍链;鞍桥;平衡链;驮运路;托梁;束带
bridle-and-pulley suspension 吊索滑落装置
bridle bar 系杆
bridle bridge 通马不通车的狭桥
bridle cable 锚固缆索;短索缆;直角锚固钢缆(与拉力成90°角)
bridle chain 马鞍链;悬挂罐笼的保险链;八字形马鞍链;平衡短链
bridle chord bridge 斜索桥;斜拉桥;牵索桥;撑架牵索桥

bridled anemometer 制动风速表
bridled pressure plate 制动风压板
bridle electrode 马笼头电极
bridle guiding device 吊索导向托座
bridle hitch 锚固系索(联结锚固缆索与滑轮的);鞍形连接装置
bridle iron 蹬筋;托架铁件;悬挂铁条;格栅托架;箍筋;格栅托座;钢支座;钢筋箍
bridle joint 斜榫接;双榫接;啮接
bridle of prepuce 包皮系带
bridle path 马大道;马道;大车路;大车道;跑马道;小路
bridle plate 轨距垫板
bridle pulley device 吊索滑落装置
bridle ring 吊线环
bridle road 马道;大车路;大车道;跑马道
bridle rod 保持铁路轨距的钢标尺;辙尖拉杆;拉杆
bridle rolls 活套张紧辊
bridle rope 缰绳;吊绳
bridle sling 起重索套
bridle suspension 吊索悬挂
bridle tower 导向塔;控制塔
bridle track 大车路;大车道
bridleway 大车路;大车道;马道
bridle wire 绝缘跨接线;跳线
bridle wire clamp 吊索线夹
bridle wire rope sling 束套式钢丝绳吊具
brief 概要;大纲
brief acceleration 短时加速度;瞬时有效加速度;瞬时加速度
brief analysis 简易分析
brief appraisal 简评
brief description 概述
brief description of the process 工艺流程简述
brief description on business 业务简介
brief dip-wash 短暂浸孔
brief flowchart 简要流程图;简单流程图
brief flowchart of soil classification 土分类简要流程图
briefing 简介
briefing room 新闻发布室
brief introduction 简介
brief introduction of project 工程简介
brief investigation 简易调查
briefly cloudy 少云
briefly frozen soil 暂冻土
brief minutes 纪要摘要
brief note 例条
brief(outline) 提要
brief stoppage 临时停工
brief sum-up 小结
brief survey 简要总结
brief test 简易试验
brief wash 短暂水洗
Brieger's test 布里格氏试验
brig 监狱;警卫室;方帆双桅船;布里格;双桅横帆船
brigade 工作组;工作队;组;队;班
brigantine 主桅纵帆
Briggean system of logarithm 常用对数;布对数;布里格对数
Brigg's clinophone 布里格测斜仪
Brigg's dispersion coefficient 布里格扩散系数
Briggs equalizer 布里格均衡器
Brigg's logarithm 布氏(常用)对数;常用对数;布氏对数;布里格常用对数
Brigg's standard 布里格标准(1862年制定的管材及螺纹标准);美国标准管螺纹
Brigg's standard pipe thread 布里格标准管螺纹;布氏标准管螺纹
Brigg's thread 布氏螺纹;布里格螺纹
bright 明亮的;光泽木材;光亮的;光亮的;高度光泽
bright adaptation 明适应
bright and black mild steel flat head tinmen rivet 号头铆钉
bright annealed wire 光亮退火钢丝
bright annealing 封闭退火;光亮退火;非氧化退火
bright-annealing line 光亮退火作业线
bright aqua 亮水色
bright band 亮带
bright-banded coal 亮煤;光亮条带煤
bright blast 出白级喷砂
bright board operating 亮屏运行
bright bolt 精制螺栓;光制螺栓;光螺栓

bright border 亮轮缘
bright cerulean blue 亮天蓝色
bright cherry-red 亮樱桃红色;樱桃红
bright coal 亮煤(质煤);优质沥青煤;光亮煤
bright colo(u)r 鲜艳色;鲜明色
bright contrast 亮反差;明反衬
bright coral red 亮珊瑚红
bright crystalline fracture 亮晶断口
bright-dark field condenser 亮暗场聚光镜
bright dip 浸渍抛光;浸渍磨光;亮漆剂
bright dipping 光亮浸渍
bright display 亮度显示
bright-drawing 光亮拉拔
bright-drawn bar 冷拉光条
bright-drawn steel 精拉钢;光亮拉拔钢材;光拔钢
bright dye 透明颜料
bright emitter 高能热离子管
bright emitter tube 白炽灯丝电子管
brighten 亮度;磨亮的;磨光;增亮
bright enamels 高级上光纸
bright enamel tile 明亮的琉璃瓦
brightened dot 亮点
brightener 增艳剂;增亮剂;增白剂;光亮剂;抛光剂
brightening 增亮;擦亮;发光;推光
brightening agent 增艳剂;增白剂
brightening impulse 亮冲击
brightening power 增亮力;增白力
brightening pulse 辉亮脉冲;照明脉冲
brightening towards the limb 临边增亮
bright eruption 喷焰
bright etching 光亮酸蚀刻;透明酸蚀刻
bright field 亮现场;亮场;明视场;明视野
bright field condenser 亮场聚光镜
bright-field illumination 亮场照明
bright-field image 亮场图像;明场图像;亮视像
bright-field imagery 亮场成像
bright-field microscope 亮现场显微镜
bright-field microscopy 亮现场显微术
bright field observation 亮场观察
bright field vertical illumination 亮场竖直照明
bright finish 磨光面;抛光(面);光亮
bright finished 镜面抛光的;磨光(的);抛光的
bright-finished steel 光亮精整钢
bright flame 白火焰
bright fracture 亮口(可锻铸铁件珠光体组织)
bright-frame finder 亮框取景器
bright fringe 明条纹;亮条纹
bright front (用冲洗过的泥土制成的)高质量砖块
bright glaze 亮釉;光泽釉
bright gold 亮金
bright green 鲜绿的
bright heat treatment 光量热处理;光亮热处理
bright image 亮像;清晰影像
bright iron flag hinge 光身旗铰链
bright iron square hinge 方形铁铰链
bright iron wire 光铁线
bright light 强光
bright light iron hinge 薄铁铰链
bright limb 亮边缘
bright line 亮线;明线
bright line spectrum 亮线光谱;明线光谱;明线光谱
bright liquid gold 亮金水
bright luster sheet 光亮钢板;镜面光亮薄板
bright metal 光亮金属(装饰)
bright mild steel cotter pin 光亮软钢开尾销
bright mild steel countersunk head wood screw 光亮软钢平头木螺丝
bright mild steel hexagonal nut 光亮软钢六角螺丝帽;六角螺帽;六角螺帽
bright mild steel hexagonal pressed nut 光亮软钢六角冲压螺丝帽
bright mild steel round head wood screw 光亮软钢圆头木螺丝
bright mottle 亮日芒
bright multifilament rayon yarn in skein 成绞有光人造复丝
brightness 明亮(度);明度;辉度;光泽;亮;从黑到白的无彩分度;视觉亮度;色觉亮度
brightness acuity 亮度敏锐性
brightness adaption 亮度适应(性)
brightness attribute 亮度属性;明度属性
brightness beat 亮度跳动;亮度差拍
brightness blink 亮度闪烁
brightness characteristic of aerial object 航摄景

物亮度特性
brightness coefficient 亮度系数
brightness colo(u)r image 色像亮度
brightness conservation 亮度守恒
brightness constancy 亮度恒定性
brightness contour 等亮度线
brightness contrast 亮度反差;亮度对比;明度对比
brightness contrast border 亮度对比界限
brightness contrast range 亮度对比范围
brightness control 亮度控制;亮度调节
brightness curve 亮度曲线
brightness decay 亮度衰减
brightness degree 亮度等级
brightness discriminaton 亮度鉴别(能力)
brightness distribution 亮度分布
brightness distribution curve 亮度分布曲线
brightness enhancing 亮度增强
brightness factor 亮度因子;亮度因数
brightness flicker 亮度闪烁
brightness flop 明度视角变异
brightness fluctuation 亮度扰动;亮度波动
brightness gain 亮度增益
brightness impression 亮度感
brightness in oil 吸油后白度
brightness matching 亮度匹配
brightness meter 亮度计;光度计
brightness modulation 亮度调制
brightness multiplication 亮度倍增
brightness noise 亮度噪声
brightness of a colo(u)r 颜色明度
brightness of daylight 日间亮度;日光亮度
brightness of flame 火焰亮度
brightness of image 影像亮度;成像清晰度;成像亮度
brightness of screen picture 图像亮度
brightness of sky 天空亮度
brightness of the spot 光点亮度
brightness of window surface 窗口亮度;采光口亮度
brightness pulse 亮度脉冲
brightness range 亮度范围;明度范围
brightness range of object 地物亮度范围
brightness ratio 亮度比
brightness regulator 亮度调器
brightness retention 亮度保持性
brightness reversion 回色
brightness scale 亮度分划;亮度比;明度梯级;明度标
brightness sensation 亮度感觉
brightness separation 亮度间隔
brightness signal 亮度信号
brightness signal detector 亮度信号检测器
brightness standard 亮度标准
brightness temperature 亮温度;亮光温度;亮度温度
brightness-temperature pyrometer 亮度温度高温计
brightness test 明度试验
brightness test chart 亮度测试图
brightness testing 亮度检查
brightness transfer characteristic 亮度传输特性
brightness transition 亮度跃迁
brightness value 亮度值
brightness-voltage characteristic 亮度电压特性
brightness zone 亮度区
bright normalizing 光亮正火
bright nut 光制螺母;精制螺母
bright ochre 亮赭石;亮赭色
brighton 蜂巢纹织物
bright pickling 光亮浸渍
bright plate 电解淀积物;光亮沉积
bright plating 光亮电镀
bright point 亮点
bright point section 亮点剖面
bright-polished sheet 磨光薄板
bright quenching 光亮淬火
Brightray 布赖特瑞镍铬合金
bright rays 亮纹
bright red 鲜红(色);朱红(色)
bright region 亮区
bright resistance 亮电阻
bright rim structure 亮环结构
bright rolled finish 光碾精加工
brightrope 光亮钢丝索

bright sap 净面树材;净面边材
bright sapwood 原色边材
bright satin anodize 阳极化处理缎光面
bright silver 亮银;银水;亮银色
bright soft wire 光亮软钢丝
bright spot 亮点;亮斑;辉点
bright spot technique 亮点技术
bright steel 光亮钢
bright steel bar 光亮型钢
bright steel products 光面钢制品
bright stock 精制润滑油料;高黏度油;浅色高黏度润滑油配料;透明油
bright sulfur 纯硫
bright-throwing power (对不规则工件的)均匀电镀能力;光亮电镀能力
bright tin plate 马口铁;白口铁;锡钢板
bright tonal anomaly 亮色调异常
bright trace 亮迹
bright valve 白炽灯
bright washer 精制垫圈;光垫圈
bright wire 光面线;光亮钢丝
bright wire rope 黑钢丝绳;光面钢丝绳
bright wisp 亮条
bright wood 优质木材
bright wool 浅色毛
brightwork 镀光五金家具;光亮零件;镀得发光的金属部分(车、船、机械上);五金器具;五金工具
bright yellow 嫩黄
Brij 玻雷吉(一种表面活性剂)
Brikollare system 布里科勒系统
bril 布里耳
brill 亮度;辉度
brillant blue 煌蓝
brilliance 亮度;明清度;鲜艳;耀度;光亮性;高频逼真度
brilliance control 亮度控制;亮度调节
brilliance modulation 亮度调制
brilliance modulation method 辉度调制法
brilliancy 辉度;耀度;多面型宝石光
brilliant 钻石;多面型
brilliant carmine 亮洋红;亮胭脂红
brilliant colo(u)r 亮色
brilliant cresyl blue 煌焦油蓝;灿烂甲酚蓝
brilliant cut 磨光刻花法
brilliant cutter 刻花机;磨光机
brilliant cutting 玻璃雕饰;磨光刻花法;多面型修饰;玻璃雕刻;玻璃片修饰
brilliant finder 镜式取景器;反转式检像镜
brilliant green 亮绿;碱性亮绿;煌绿
brilliant green tone 艳绿调色剂
brilliant image 清晰影像
brilliantly lit hall 灯火辉煌的大厅
brilliant orange 亮橘红;亮橙
brilliant polish 光亮抛光;抛光剂
brilliant scarlet 亮猩红
brilliant ultramarine 合成群青
brilliant varnish 发亮清漆
brilliant violet 亮紫
brilliant vital red 光辉活染红
brilliant white 亮白(色);炽白光;白炽光
brilliant yellow 亮黄;艳黄
Brillouin backscattering 布里渊反向散射
Brillouin diagram 布里渊图
Brillouin electronic efficiency 布里渊电子效率
Brillouin emission 布里渊发射
Brillouin fiber amplifier 布里渊光纤放大器
Brillouin field 布里渊场
Brillouin flow 布里渊流
Brillouin flux density 布里渊场强;布里渊磁通密度
Brillouin formula 布里渊公式
Brillouin function 布里渊函数
Brillouin laser 布里渊激光器
Brillouin light 布里渊光;布里渊辐射
Brillouin light amplifier 布里渊光放大器
Brillouin linewidth 布里渊线宽
Brillouin scattering 布里渊散射
Brillouin zone 布里渊区
Briouin scattering 布里渊衍射
brim 井栏;檐;水边
brimful capacity 满装容量
briming 海面磷光
brim of a well 井边沿
brimstone 硫黄色的;硫黄石;硫黄
brim-stretching machine 帽边扩张机

brindle 虎斑;变色的;杂色的;有斑纹的
brindled 有斑纹的
brindled brick 带斑的青砖;二等砖;斑纹砖(含铁质黏土);斑点砖;起皱纹的次砖;起条纹劣砖;起条纹次砖;(颜色差的)二级砖
brindles (颜色差的)二级砖
brindleyite 镍铝蛇纹石
brine 卤水;浓盐液;浓盐水溶液;浓咸水;海水;咸水;盐水处理
brine balance tank 盐水膨胀箱
brine calculation pipe 卤咸水测管
brine circulation 盐水环流
brine concentration 盐水浓度
brine cooler 海水冷却器;盐水冷却器;盐水冷却管
brine cooling 盐水冷却
brine cooling system 盐水冷却系统
brine corrosion 盐腐蚀
brine deposit 卤水矿(床)
brine dispenser 加盐水机
brine disposal 海上处理;盐水处理
brine drum 盐水桶
brine freezing system 盐水冻结装置;盐水冻结系统;盐水冻结方式
brine ga(u)ge 盐水比重计;盐浮计
brine hydrometer 海水比重计
brine-into-oil curve 水驱油曲线
brine lake 盐湖
brine lake deposit 盐湖沉积
brine leaching 盐溶液漫出
brinelling 测布氏硬度;布氏硬度试验
Brinell on delivery state 交货状态布氏硬度
Brinell's ball 布氏硬度试验球
Brinell's ball hardness 布氏球印硬度;布氏钢球压入硬度
Brinell's ball hardness testing machine 布氏球印硬度试验机;布氏钢球压入硬度试验机
Brinell's ball test 布氏球印硬度试验;布氏球印试验
Brinell's dent 布氏钢球压痕
Brinell's figure 布氏硬度值;布氏硬度数;布氏球印硬度值;布氏球印硬度数
Brinell's hardness 布氏硬度;布里涅尔硬度
Brinell's hardness number 布氏硬度值;布氏硬度数;布里涅尔硬度值
Brinell's hardness scale 布氏硬度刻度盘
Brinell's hardness test 布氏硬度试验
Brinell's hardness tester 布氏硬度试验仪;布氏硬度试验机;布氏硬度计
Brinell's hardness testing machine 布氏硬度试验机
Brinell impact hardness tester 锤击式布氏硬度试验机
Brinell impact test 冲击强度落球试验;球压试验
Brinell's instrument 布氏硬度仪
Brinell's machine 布氏硬度仪;布氏硬度试验机;布氏硬度机
Brinell's microscope 布里涅尔显微镜;布氏硬度测量显微镜
Brinell's number 布氏硬度值;布氏硬度数;布氏球印硬度值;布氏球印硬度数
Brinell's scale 布氏硬度标尺
Brinell's test 布氏(硬度)试验
Brinell's tester 布氏硬度计
brine mud 盐水泥浆
brine neutralizing tank 盐水中和槽
brine osmotic potential 盐渗析势
brine-pan 盐田;盐场
brine pipe 盐水管
brine pipe cooling 盐水循环冷却;盐水冷却管
brine pipe grid coil system 格子排盘管盐水冷却装置;格子排盘管盐水冷却系统;格子排盘管盐水冷却方式
brine pit 卤水坑;盐井
brine pollution from oil well 油井盐水污染
brine pool 卤水塘
brine preheater 盐水预热器
brine-proof paint 防盐水漆
brine pump 盐水泵
brine return tank 盐水回流箱
brine section 盐水工段
brine sludge 含盐污泥;盐泥
brine solution 盐溶液
brine solution tank 盐液箱
brine-spray method 喷洒盐水(养护)法
brine spray refrigerating system 盐水喷淋制冷装

置;盐水喷淋制冷系统;盐水喷淋制冷方式
brine spray system 盐水喷淋系统
brine spring 卤水泉;盐泉
brine system 卤水系统
brine tank 盐水箱
brine tube 盐水管
brine wastewater 含盐废水
brine water 卤水;海水;盐水
brine well 卤水井
bring about 导致
bring about profit 实现利润
bring a charge against sb. 控告某人
Bringal 康熙三彩
bring an action against somebody 起诉
bring a suit against somebody 起诉
bring back pillar 回采煤柱
bring bottom up 冲孔(钻孔)
bring down 浓缩;承转
bring down on account 账户结转
bring down price 降低价格;降价
bring forward 降低价格;重涂前修补处理;提前处理;套圈
bring home 恢复原状;证明;收紧锚(索)
bring home the anchor 绞锚
bring in a well 新井投产
bring in business 招揽生意
bring in focus 焦聚;使聚焦
bring in funds 吸收资金
bringing back 送回
bringing into service 投入运行
bringing-up 升温;加热
bringing-up section 加热段
bring in phase 使同相位
bring into action 使动作
bring into effect 实施
bring into line 使排列成列
bring into motion 开动
bring into operation 交付运用;投入运行;投入生产
bring into production 投入生产
bring into service 使运行;使工作
bring into step 整步;同步;使同步
bring more land under cultivation 扩大耕地面积
bring on stream 投入生产
bring recovery 休养生息
bring something under the hammer 把某物拿出拍卖
bring suit 起诉
bring the fire under control 控制火灾
bring to grade 定坡;填筑到设计线
bring to market 卖出;出售
bring to rest 停止运行;停车
bring up 船停住;船到终点
bring-up test 发出测试
bring up to date 使现代化
bring up with a round turn 急速停住(缆绳)
brining 用盐水处理;盐浸作用;盐浸处理
brinishness 含盐度
brink 边缘;边沿(河流等);边侧;岸边;峭河岸;沙丘滑面顶
Brink demister 布林克除雾器
brink depth 边缘水深
brink ice 边冰;岸冰
Brinkmann number 布林克曼数
brinkmanship 冒险政策;边缘政策
Brin's process 泊伦法
briny 海水的
briquettability 压制性;成型性
briquet(te) 八字形试块;模制试块;煤砖;制团;标准水泥试块;团块;水泥标准试件;试块;煤球;坯块;生料块
briquet(te)d coal 成型煤
briquet(te) press 压片机;压块机;标准水泥试块压制机;试块压制机;水泥试块成型机
briquetter 压块机;试样压制机
briquet(te) tension test 试块拉力试验
briquette(te) test specimen 模制(标准)试块(水泥)
briquetting 凝定;压块;制团;团矿;压制成块
briquetting asphalt 煤砖专用沥青
briquetting batch 压块料
briquetting die 压块用阴模
briquetting machine 压块机;压制机;压坯机;制团机
briquetting method 压制成块方法

briquetting of coal 压制煤砖
briquetting pitch 压硬沥青成块
briquetting plant 压坯厂
briquetting press 压块机;团压机;团矿机
briquetting pressing 制球机
briquetting pressure 成型压力
briquetting process 压块法
briquetting test 压制试验;标准试块试验
brisa 拨立柴风
brisa carabinera 卡拉伯恩风
brisance 猛度;炸药震力;炸药威力;爆炸威力;破坏效力
brisance index 炸药震力指数
brise-bise 半截式花边窗帘
brisement 裂断
brise soleil 百叶窗(用在热带地区)
brisk business 生意好
brisk commerce 商业兴旺
brisk demand 不断的需求
brisk market 繁荣的市场;市场活跃;市场繁荣
brisk sales 畅销;旺销
brisk trade 交易活跃;贸易繁荣
brisote 拨立沙脱风
Brissaud's reflex 布里索氏反射
bristle 硬毛;猪鬃;短粗纤维
bristle brush 鬃漆刷;硬毛漆刷
bristle roll 毛刷罗拉
bristle roller 硬毛辊
bristletail 蛀虫
Bristol alloy 布里斯托尔铜锌锡合金
Bristol board 细料纸板;光泽纸板;上等板纸(绘图用)
Bristol brass 布里斯托尔黄铜
Bristol diamonds 美丽石英
Bristol glaze 锌釉;陶釉;窑釉;生锌釉
Bristol joint 布里斯托布接头
Bristol paper 优质纸(绘图用);细料纸板;图案纸
bristol pennant 蓝灰色玄岩
Bristol porcelain 布里斯托布瓷器(英国)
Bristol steel belt lacing 布里斯托布钢带接头
bristol stone 细砂磨砖;美晶石英
Britannia 锡铜锑合金
Britannia cell 大不列颠型浮选机
Britannia joint 锡铜锡焊接;大不列颠焊接
Britannia metal 不列颠(锡铜铜)合金;锡锑铜合金
britholite 铈磷灰石;铈硅磷灰石
britholite-(Y) 钇硅磷灰石
British absolute system of units 英国绝对单位制
British Admiralty Chart 英版海图
British Aggregate Construction Materials Industries 英国集料建材工业公司
British Airways 英国航空公司
British-American schools 英美学派
British-American Tobacco 英美烟草公司
British and America system 英美制
British Architecture 英国建筑
British Association for the Advancement of Science 英国科学发展协会
British Association of Accountants and Auditors 英国会计师与审计师协会
British Association of Threads 英国螺纹学会
British Association ohm 英制欧姆
British Association Thread 英国螺纹协会
British association unit 英国标准单位
British Benchmark 英国标准水位点
British Broadcasting Corp 英国广播公司
British candle 英制烛光;英国坎德拉
British cast Iron Research Association 英国铸铁研究协会
British Corporation Register 英国船级社
British Council for the Promotion of International Trade 英国国际贸易促进会
British Engineering Standard Association 英国工程标准协会
British engineering units 英国工程单位制
British Export Board 英国出口局
British Geotechnical Society 英国岩土工程学会
British gum 淀粉胶;糊精
British gum glue 淀粉胶粘剂;糊精胶粘剂
British horse power 英制马力
British Hydromechanics Research Association 英国流体力学研究协会
British imperial gallon 英制加仑(1英制加仑≈1.2美国加仑,容量为277.42立方英寸)

British imperial pound 英国标准磅
British Institute of Electrical Engineers 英国电气工程师学会
British Insurance Association 英国保险协会
British Iron and Steel Research Association 英国钢铁研究协会
British Lloyd 英国劳氏船级社
British Military Grid 英国军用坐标网
British Museum 大不列颠博物馆(1823~1847年英国伦敦)
British National Maritime Board 英国海运总局
British Oil and Colo(u)r Chemists Association 英国油脂与颜料化学家协会
British Ordinance Datum 英国标准零点
British Overseas Airways Corporation 英国海外航空公司
British Overseas Trade Board 英国海外贸易局
British Paint Manufactures and Allied Trades Association 英国油漆制造商与相关行业协会
British Paint Research Association 英国涂料研究协会
British Patent 英国专利
British Petroleum Company 英国石油公司
British portable skid-resistance number 英国路面抗滑指数
British Portland cement Standard 英国普通水泥标准
British Ports Association 英国港口协会
British preferential tariff 英联邦特惠关税
British Rail 英国达比铁路中心
British Railways 英国铁路
British Rainfall Organization 英国降雨组织
British Ready Mixed Concrete Association 英国商品混凝土协会
British Road Federation 英国道路联合会;英国道路联合会
British Road Tar Association 英国路用柏油协会
British savings bonds 英国储蓄公债
British Ship Research Association 英国船舶研究会
British Society of Soil Science 英国土壤学会
British Standard[BS] 英国标准
British Standard Board 英国标准局
British standard candle 英国标准烛光
British Standard Code of Practice 英国标准开业守则
British standard dimension 英国度量标准
British Standard Fine Thread 英国细牙螺纹标准;英国标准细牙螺纹
British Standard Institution 英国标准学会;英国标准协会
British Standard Pipe Thread 英国管螺纹标准;英国标准管用螺纹
British Standards 英国工业标准
British standard scale 英国标准尺度
British standard screen scale 英国标准筛制
British standard section 英国标准型钢
British standard section iron 英国标准型钢
British Standard Sieve 英国标准网目;英国标准筛制
British standard specification 英国标准规格(书);英国标准规范;英国标准技术规范
British standard time 英国标准时
British Standard unit 英国标准单位
British Standard Whitworth Thread 英国惠氏标准螺纹;英国标准惠氏螺纹
British Standard Wire Ga(u)ge 英国标准线规
British steel wire 英制钢丝网
British Survey 英国测绘局
British system 英制
British system of units 英国单位制
British thermal unit 英制热单位;英国热(量)单位
British ton 英吨
British Transport Docks Board 英国运输码头局;英国货运码头局
British United press 英国合众社
British viscosity unit 英国黏度单位
British Waterways Board 英国河道局
British Waterworks Association 英国给水协会
British Waterworks Association Journal 英国给水协会会刊
British Welding Research Association 英国焊接研究会
British winter time 英国冬季时间
britonite 硝酸钾;硝酸甘油;脆通炸药;草酸铵炸药;波利通那炸药

brittle 脆性的;脆的
brittle alloy 脆性合金
brittle at blue 蓝脆(性)的
brittle at blue heat 蓝脆(性)的
brittle behavio(u)r 脆性
brittle clay 脆性黏土
brittle coating 应力(脆性)涂料法;脆性涂料
brittle coating method 脆性涂料法
brittle consistency 脆性结持度
brittle crack 脆性裂缝
brittle deformation 脆性变形
brittle-ductile fault 脆韧性断层
brittle-ductile transition 脆性延性转变;脆韧转变;脆韧性转变
brittle failure 脆性破坏;脆性疲劳;脆裂;脆断
brittle failure fashion 脆性破坏方式
brittle fault 脆性断层
brittle faulting 脆性断裂作用
brittle fiber 脆(性)纤维
brittle film 脆性薄膜
brittle fracture 脆性破裂;脆性破坏;脆性裂隙;脆性断裂;脆裂
brittle fractures of rail 钢轨折断
brittle fracture strength 脆性强度
brittle fracture stress 脆性断裂应力
brittle fracture surface 脆性断裂面
brittle glass 易碎玻璃;脆性玻璃
brittle heart 木材节疤;脆心材;树瘤纹
brittle iron 脆性铁;脆铁
brittle lacquer 脆性涂料;脆性漆
brittle-lacquer technique 裂纹漆工艺;脆性喷漆技术
brittle lattice 易碎晶格;脆弱晶格
brittle material 脆性材料
brittle materials tensile-strength tester 脆性材料抗拉强度测试仪
brittlement 脆性
brittle metal 脆性金属;脆金属
brittle mica 辉光云母;脆云母
brittle mineral 脆性矿物
brittle model 脆性模型
brittleness 易碎性;脆性;脆碎性;脆度
brittleness index 脆性指数
brittleness temperature 脆性温度
brittleness test 脆性试验
brittle pearl mica 辉光云母
brittle-plastic transition 脆性塑性转变
brittle point 冷脆点;脆折点;脆性点;脆裂点;脆化点;脆点
brittle point of asphalt 沥青冷脆点
brittle rachis 脆穗轴
brittle range 脆性区域
brittle resistance sulphur black 防脆硫化黑
brittle-ring test 脆性环状体试验
brittle rock 脆性岩石
brittle rupture 脆性破裂;脆性破裂;脆性断裂;脆裂
brittle shear failure 脆性剪切破坏
brittle silver ore 脆银矿
brittle solid 脆性固体
brittle strength 脆裂强度
brittle structure 脆性结构
brittle temperature 脆化温度;脆化点
brittle transition temperature 冷脆临界温度;脆性转折温度;脆性转变温度;变脆点
brittle-type structure 脆性结构
brittle varnish 脆性清漆
brittle wort 轮藻
brittle zone 脆性区
briviskop 布维硬度显示仪
Brix degree 布里克斯度
Brix hydrometer 布里克斯比重计
briza 拨立柴风
brize weld(ing) 硬焊(接)
Brnsted acid 布仑斯惕酸
Brnsted's relation 布仑斯惕关系
broach 尖头杆;尖塔;钢地;倒刺针;打眼;剥刀;拉削
broach back pilot 拉刀后导部
broach cargo 开箱偷货
broached 凿平的(墙面)方石
broached spire 有半角锥体撑住的塔尖
broached work 石工刻线板;石工初刻作业;斜纹缀面;在石面斜对角线墙;石面刻槽工作;石面曲线槽;刻对角线槽石面
broacher 拉床;铰孔机;剥刀机
broach for deep hole 深孔拉刀

broach for external broaching 外拉刀
broach grinder 拉磨床;拉刀磨床
broach grinding machine 拉刀磨床
broaching 扩孔;铰孔;拉孔;钻孔开采法;堆削;拉削
broaching bit 扩孔钻头;绞刀
broaching cutter 拉削刀具
broaching die 拉模;削孔模
broaching dovetail grooves 燕尾槽拉削
broaching fixture 拉削夹具;扩孔夹具;铰孔夹具
broaching machine 拉床;铰孔机;剥孔机
broach length 拉刀长度
broach out 穿孔
broach post 人字架中柱;桁架中柱;桁架中竖杆;桁架竖杆
broach reamer 锥铰刀
broach roof 尖塔屋顶
broach sharpener 拉刀磨床
broach sharpening machine 拉刀刃磨床
broach spire 尖塔顶;八角尖塔
broach squeezed pile 钻孔挤扩支盘桩
broach support 拉刀支架
broach taper 绞刀;钻;锥;铰刀
broach travel 拉刀行程
broad 扩孔刀具;广阔的;灯槽
Broad acre City 广阔一亩城市(美国等设想的城市规划方案)
broad aisle 宽阔走廊;宽阔过道
broad and shallow channel 宽浅水道
broad angle 宽角度的;钝角
broad azimuth trace 宽方位角线;大方位角线
broadband 宽波段;宽频带;谱带;宽带
broadband absorber 宽带吸收体
broadband absorption 宽带吸收
broadband amplifier 宽频带放大器;宽带放大器
broadband antenna 宽带天线;宽波段天线
broadband anti-reflection film 宽频带增透膜
broadband channel 宽带信道;宽带通道
broadband circulator 宽波带循环器
broadband coaxial cable 宽带同轴电缆
broadband communication network 宽带通信网(络)
broadband constant beam width sonar 宽带恒定束宽声呐
broadband converter 宽带变频器
broadband digital subscriber line 宽带数字用户线系统
broadband dissipative attenuator 宽带耗散衰减器
broadbanded salary structure 宽幅薪水结构
broadband electrooptic modulator 宽带光电调制器;宽带电光调制器
broadband equipment 宽带设备
broadband exactness 宽带精度
broadband exchange 宽频带交换;宽带交换
broadband exchange services 宽带交换业务
broadband filter 宽频带滤光片
broadband grating filter 宽带光栅滤波器
broadband hologram lens 宽波段全息透镜
broadband integrated services digital network 宽带综合业务数字网
broadband isolator 宽带隔离装置
broadband klystron 宽频带速调管
broadband light pump 宽带光泵
broadband matching 宽频带匹配
broadband metal-dielectric(al) interference filter 宽带金属—介质膜干涉滤光片
broadband noise 宽带噪声
broadband pass filter 宽带滤光片
broadband path 宽频带通路
broadband photoconductive detector 宽带光电导探测仪
broadband radiation 宽带辐射
broadband radio frequency cable network 宽带射频电缆网络
broadband random process 宽频带随机过程;宽带随机过程
broadband random vibration 宽频带随机振动
broadband receiver 宽频带接收机;宽带接收机
broadband relay network 宽频带中继网络
broadband shock-associated noise 冲击室宽频带噪声
broadband signal 宽带信号
broadband stationary noise 宽带稳态噪声

broadband stub 宽带调谐短截线
broadband system 宽带系统
broadband transformer 宽带变压器
broadband transmission 宽带传输
broadband transmission equipment 宽带雷达传输设备
broadband two-element system 宽波段双元件系统
broadband video detector 宽带视频探测器;宽带视频检波器
broadband wave plate 宽带波带片
broad base azimuth mirror 宽底方位镜
broad base terrace 宽广台地;宽基梯田;宽底阶地;宽底坡地;缓坡台阶式护岸(防止水土流失)
broad beam 宽缘梁;宽光束
γ-ray 宽束伽马射线
broad beam irradiation 宽束辐照
broad beam shielding 宽束屏
broad-body metro transit 鼓形地铁电动客车
broad-bottomed 平底的
broad bow 肥型船首
broad brush 粗线条的
broadcast address 广播地址
broadcast application 撒施法
broadcast band 广播波段
broadcast boom 宽幅悬臂
broadcast by television 电视广播
broadcast control desk 广播桌式播控台
broadcast data 广播数据
broadcast distribution 撒施
broadcasted distribution space 撒施距离
broadcast ephemeris 广播星历
broadcast equipment 广播设备
broadcaster 广播装置;广播员;散播机;撒播机
broadcaster model 广播设备模型
broadcast(ing) 广播;播送;撒施
broadcasting center 广播中心;播中中心
broadcasting control table 播音控制桌
broadcasting equipment 无线电广播设备
broadcasting equipment for the service line 运营线路广播系统
broadcasting equipment plant 广播器材厂
broadcasting house 播音室
broadcasting junctioning equipment 广播中继设备
broadcasting network 广播网
broadcasting on schedule 定时播发
broadcasting repeater 广播增音机
broadcasting room 广播室
broadcasting satellite service 广播卫星业务
broadcasting sodding 满铺草皮
broadcasting station 广播站
broadcasting studio 广播室;播音室
broadcasting tower 无线电发射塔
broadcasting transmission 无线电广播
broadcasting transmitter 广播发射机
broadcast message 广播消息;通知信息
broadcast method 撒施法
broadcast receiver 广播收音机
broadcast relaying 广播转播
broadcast satellite 广播卫星
broadcast seeder 散播机
broadcast sodding 满堂草皮;泛铺草皮
broadcast sower 撒播机
broadcast station 广播电台
broadcast studio 室内广播
broadcast television 广播电视
broadcast-tower antenna 铁塔广播天线
broadcast transmission 无线电广播广播
broadcast(transmission) station 广播电台
broadcast transmitter 广播发射机
broadcast transmitter intellect test instrument 广播发射机智能测试仪
broadcast transmitting station 广播电台;广播发射台
broadcast treatment 撒施处理
broadcast warning 广播警报
broad chisel 平凿
broad classification 粗分类法
broadcloth 绒面呢;各色细平布;阔幅布
broad concrete cross-tie 混凝土宽枕
broad concrete tie 轨枕板
broad cone nozzle 宽锥形喷嘴
broad crested measuring weir 宽顶量水堰
broad crested weir 阔顶堰;宽顶堰
broad daylight 全日(光)照

broaden business 扩大业务
broadened base period 扩张基期
broadened base system 扩张基期法;扩大基期法
broadened cross-section 加宽横截面
broad energy distribution source 宽能分布源
broad energy-spectrum source 宽能谱源
broaden(ing) 展宽;增宽(度);扩大;拓宽;加宽;放宽
broadening by damping 阻尼致宽
broadening factor 扩张因子;扩张因素
broadening of bridge 桥梁加宽工程
broadening of spectral line 光谱线增宽
broadening vision in management 开阔管理视野
broaden sales 扩大销路
broaden sources of income and reduce expenditures 开源节流
broaden the channels of circulation 扩大流通渠道
broader term 上界概念
broad evergreen forest 常绿阔叶树;常绿阔叶林
broad flange beam 阔翼梁;宽缘梁;宽缘工字梁;宽缘工字钢
broad-flanged 宽凸缘的;宽翼的
broad-flanged beam 宽缘工字钢;宽翼缘
broad-flanged girder 宽板'梁
broad-flanged I-section prop 宽缘 I 字截面支撑
broad flanged steel 宽缘工字钢
broad flange section 宽翼缘断面
broad footed rail 宽底轨(条);宽底钢轨
broad forming cut 粗成型切削
broad ga(u)ge 宽轨距的;宽轨(距)
broad-ga(u)ge railroad 宽轨铁路;宽轨铁道
broad-ga(u)ge railrway 宽轨铁路;宽轨铁道
broad glass 宽玻璃板;玻璃板
broad halo 延伸晕
broad hatchet 宽口小斧;小阔斧;短柄宽斧
broad heading 大项目;大类
broad image 模糊图像
broad irrigation 污水灌溉;漫灌;地面灌溉;大水漫灌;泛灌
broad irrigation of sewage 污水漫灌
broad knife 刮刀;铲刀
broad leaf 阔叶林
broadleaf evergreen 常绿阔叶树
broad leaf tree 阔叶树
broad-leaved evergreen 常绿阔叶树;常绿阔叶林
broad-leaved evergreen scrub 常绿阔叶丛
broad-leaved forest 阔叶林
broad-leaved plant 阔叶植物
broad-leaved tree 阔叶树;阔叶林
broad-leaved wood 有孔材
broad light 漫射光;散射光;散射灯
broad line strategy 宽产品系统战略
broadloom 磁控管可调频率干扰发射机;磁控管波段干扰发射机
broadloom carpet 宽幅地毯
broadly graded soils 宽级配土
broad market 大量交易的市场;交易活跃的市场;广阔的市场
broadness 宽度;广度
broadness of band 频带宽度
broad nosed finishing tool 阔头光车刀
broad nose pliers 阔嘴钳
broad ocean 公海;外洋
broad ocean area 公海区;外洋区
broad on/off the bow 左右舷角 45°
broad on the beam 左右舷角 90°
broad on the bow 前八字方向
broad on the quarter 后八字方向;左右舷角 135°
broad outline 轮廓;概要
Broad pass terrane 布罗特斯地体
broad planning 轮廓规划;大区规划;初步规划
broad pulse 宽脉冲
broad radiation pattern beam 宽辐射图光束
broad range regulator 广限调整器
broad reach 后舷风行驶
broad reeded 有沟槽的玻璃
broad river reach 宽河段
broad-sclerophyll vegetation 宽硬叶植被
broad seal 印鉴
broad search 域域搜索;广阔搜索;广泛搜索
broad search allocation 广阔搜索布局
broad search plan 广阔搜索方案
broad search sensor 广阔搜索探测设备
broad search time 广阔搜索时间

broad seas 公海
broad section 宽缘工字截面
broad sense 广义
broad shallow channel 宽浅水道
broad shallow reach 宽浅河段
broad share 宽铲
broadsheet 宽面纸
broadside 宽边方向;宽边;舷侧;船舷侧;船弦
broadside antenna 边射天线;同相天线
broadside array 宽边天线阵;垂射天线阵列
broadside array antenna 边射阵天线
broadside berthing 平移驶靠(船)
broadside directiona antenna 垂射天线
broadside diregate 垂射
broadside launch 横向下水
broadside layout 非纵观测系统
broadside line 非纵测线
broadside marine railway 横向滑道船闸
broadside mill 宽展机座(用于增大板坯宽度)
broadside on 靠舷侧;舷侧相向;船侧相向;船侧靠;侧对
broadside on berthing 平移靠码头
broadside paint 舷侧漆
broadside railway dry dock 轨道干(船)坞;修船厂
broadside rolling 宽展轧制
broadside sea(wave) 横波;横浪
broadside seismic reflection profiling 侧向地震反射剖面法
broadside slipway 横向船排滑道;侧边滑道
broadside spread 非纵排列
broadside stand 宽展机座(用于增大板坯宽度)
broadside wind 横风
broadsiding 侧移
broad spectral-range spectrograph 宽程摄谱仪
broad spectrum 广谱
broad spectrum noise 宽带噪声
broad spectrum pesticide 广谱性农药
broad spectrum preservative 广谱防腐剂
broadstep 楼梯平台;楼梯踏板
broad stone 琢石;铺面大石板;石板
broad strip mill 宽条轧钢机;宽带钢轧机
broad strip train 宽条街道;宽条公路
broad survey 普查的
broad survey method 普查(方)法
broad suturing gravity dam 宽缝重力坝
broad tie 宽轨枕
broad tool 宽凿;琢石工具;阔凿
broad tooled 阔粗凿石
broad tooling 阔粗凿石
broad T-rest 宽丁字刀托(木工车床)
broad tuning 宽调谐;粗调谐
broad valley 宽广谷地;宽谷;平底河谷
broad veining (宽排列的)脉状纹痕
broad walk 大道
broadway 圆路道路;横向;沿宽度方向
broad weather 泛天候
broadwise 横向地
broad wool 无弯曲毛;直毛
brob 楔形长钉;钩头钉;弯头道钉
brocade 锦缎;织锦
brocade glass 锦缎玻璃
brocade mill 织锦厂
Broca prism 布罗卡棱镜
Broca's area 布罗卡区
brocatel(le) 彩色大理石(黄色带深红条纹);凸花厚缎;花缎
broch 圆形石塔
brochantite 羟胆矾;水胆矾
broche 打眼;粗刨;锥形绞刀;尖塔;浮纹织物;提花织物
brochure 简介材料小册子;小册子;公司简介;单位简介
Brocken bow 反日晕
brocken-joint tile 参错接缝瓦
Brocken specter 布罗肯宝光
brockie oil 陶瓷业用润滑油;陶瓷工业用润滑油
brockite 水磷钙钍矿
Brock punch 布罗克打孔器
brockram 石膏灰岩角砾岩;砂泥石灰角砾岩
Brocot suspension 布罗珂脱悬吊摆
B-rod bit 不取心金刚石钻头
Brodhun photometer 布洛洪光度计
broeboe 布鲁贝风
brog 曲柄(手摇)钻

broggite 褐地沥青
brogue 半统工作靴
brogue hole 气孔
broguer 气孔焊工
broil 矿脉指示碎石
broiler 烤箱
broiler plant 童子鸡工厂
broke 回炉纸;废纸;碎片;碎块
broke beater 打浆机;废纸打浆机;废物磨粉机
broke disposal 废纸处置;废纸处理
broked stripe 鹿斑纹
brokee 破产人
broken 断开的
broken aggregate 轧碎集料;轧碎骨料;碎砖集料;碎砖骨料
broken and damaged cargo list 货物残损单
broken arch 装饰拱;中断式拱;断拱;缺口拱
broken ashlar 乱砌方石工程;杂纹方石工程;不规则琢石;不等形琢石块
broken ashlar masonry 错缝琢石砌体;不规则琢石圬工;不等形琢石圬工
broken back 横裂纹带
broken-back curve 断背曲线;断脊曲线;同向曲线
broken-back transit 折轴经纬仪
broken basalt 破碎的玄武岩
broken base 折形基线;轨底破裂
broken baseline 折基线
broken beading 脱节
broken-beam technique 截止束技术
broken belt 裂冰带(纯海水与密冰的过渡带);海冰过渡带;破冰区
broken-bladed conveyer 刮板螺旋输送机
broken blast furnace slag sand 高炉矿渣扎石砂
broken blister 开口小气泡;开口泡
broken blockness 破碎块度
broken bond 不规则砌缝;断续黏结
broken bow 彩光环
broken brick 碎砖
broken-brick base 碎砖基层
broken brick concrete 碎砖混凝土
broken brick concrete filler brick 碎砖混凝土填块;碎砖混凝土块;碎砖混凝土填充块体
broken brick hardcore 碎砖垫层
broken chainage 断链【测】
broken chaining peg 断链桩【测】
broken chain stake 断链桩【测】
broken chip(ping)s 碎砖屑
broken cinder sand 轧碎炉渣砂;高炉矿渣扎石砂
broken circle 虚线圆
broken circuit 开路;断路
broken clay brick 轧碎黏土砖
broken coast 曲折海岸
broken coastline 曲折海岸线;破碎海岸线
broken colo(u)r 多彩色;配合色(产品)
broken-colo(u)r work 配合色产品;古老漆法;仿古涂装
broken compensation 代偿机能不全
broken concrete 破碎的混凝土;混凝土碎块;碎混凝土块
broken concrete groin 碎混凝土块丁坝
broken concrete groyne 碎混凝土块丁坝
broken concrete wall slab 碎砖混凝土墙板
broken concreting sand 轧碎混凝土形成的砂;破碎混凝土用砂
broken condition 破裂性
broken corner 掉角;缺角;角裂
broken country 沟谷交错地带;丘陵地;崎岖不平地区
broken course 断序;断裂;断层;隔断层;十字砌合层
broken crest 折线堰顶
broken crow twill 四页破斜纹
broken curve 断背曲线;虚线曲线;虚曲线
broken direction observation 不完全方向观测
broken down 受损事故;损毁
broken edge 边裂;不规则边(缘)
broken emulsion 分层乳浊液;已破坏的乳化液
broken end 断头;缺径
broken expanded cinder 破碎膨胀性矿渣
broken eyepiece 折管目镜【测】
broken face 裂面
broken fine aggregate 轧碎细集料;轧碎细骨料
broken finish 瓶口开裂
broken-flight stair(case) 层间双折楼梯
broken foamed(blast furnace) slag 轧碎高炉泡

沫溶渣;破碎泡沫矿渣
broken formation 破碎地层
broken furrow 不整齐的垄条
broken gable roof 偏屋脊坡屋顶
broken glass 破碎玻璃
broken grain 纹裂
broken granite 轧碎花岗岩;花岗岩碎块;花岗石碎块;碎花岗岩
broken granulated cinder 轧碎粒状炉渣
broken gravel 碎砾(石)
broken gravel sand 碎砾石砂
broken ground 新垦地;新开垦地;耕翻地;断裂地层;不平地(面);丘陵地;崎岖地区;崎岖地带;破碎地层;破碎带;破裂地层;松软地层
broken hardening 分级淬火
broken hard rock 轧碎坚硬岩石;硬碎地层
broken height 断高【测】
broken herringbone 变化山形组织
broken-hooked fiber 断弯钩纤维
broken ice 破碎的冰;稀疏冰;碎冰块
broken in bit 磨合了的钻头
broken in stone 缺角的金刚石
broken-in surface 磨合表面
broken iron 废铁
broken joint 间砌法;相错式接头;错接;断缝;错列接头;错列接合;错缝接合
broken joint tile 简单搭接的瓦管;断缝瓦;错缝接合瓦
broken lava 轧碎熔岩
broken level(l)ing 断高【测】;短高
broken limestone 轧碎石灰石;碎石灰石
broken limestone sand 轧碎石灰石砂
broken line 虚线;羽状条纹;折线;断线;破折线
broken-line analysis 线段近似法;折线分析法
broken linear 折线状的
broken line crown 折线型路拱
broken line distortion 断线degree歪曲;断线扭曲
broken-line slope 折线型边坡
broken lot 散批货物
broken lump slag 轧碎块状矿渣
broken marble 轧碎大理石;碎大理石
broken-marble patterned flooring 碎拼大理石地面
broken material 轧碎材料;破碎材料;破碎物质
broken money 零钱
broken mosaic 断续镶嵌
broken number 分数;分式【数】
broken oil 澄清了的油
broken-out section 切面;断裂面;局部断面视图;切开断面;破断面
broken-out section view 破断面视图
broken parcel 破图廓地区;破图廓部分
broken pediment 裂山花;中断式山墙;中断三角楣饰;缺口三角楣饰
broken-period interest 截期利息
broken pick 断纬
broken pits 破碎坑
broken polygon 断开多边形
broken product 轧碎产品
broken rail 断轨
broken rail base 轨底崩裂
broken rail protection 断轨保障
broken rail state 断轨状态
broken-range ashlar 不分层琢石砌体;毛石砌体;不规则形琢石块
broken-range ashlar masonry 错列琢石圬工
broken range masonry 断层石圬工;断层石砌
broken range work 错列层砌石;断层石砌;断层工作;断层石砌(不同高度的石块砌筑)
broken ray 折方向线
broken resin 断树脂
broken ring 开口圈;缺口圈
broken rock 破碎岩石层;碎石
broken rock landslide 破碎岩石滑坡
broken-rock pavement 碎石路面;碎石铺砌层
broken-rock road 碎石路
broken sand 轧制砂;破碎砂
broken sea 碎浪
broken seam 间断缝
broken section 破碎剖面
broken section line 断开线
broken seed 开口泡
broken selvedge 裂边;破边
broken series 非连续数列
broken shoreline 破碎海岸线

broken sky 多云(天)
broken slag 矿渣碎石;轧碎矿渣
broken slag sand 破碎矿渣砂;高炉矿渣石砂
broken slate 轧碎板岩;破碎页岩
broken-sloped slipway 折线变坡滑道【船】
broken soffit 不平整的石楼梯的下部
broken space 亏耗货位;亏舱;舱位损失
broken stick 轧碎棒;轧碎辊
broken stone 轧碎集料;轧碎骨料;破碎石块;碎石;碎片
broken stone base 碎石打底;碎石基础
broken stone bed 碎石基床;碎石路基
broken stone chip 碎石渣
broken stone course 碎石层;碎石垫层
broken stone drainage layer 碎石排水层
broken-stone level(l)ing course 碎石找平层;碎石整平层
broken stone metaling 碎石道砟
broken stone packing 碎石基层
broken-stone pavement 碎石面层
broken-stone pile 碎石桩
broken-stone road 碎石路(面)
broken stone sand 碎石砂;混凝土细砂
broken stowage 亏耗货位;亏舱;舱内不规则装货
broken stripe 碎条(状花)纹
broken-strip(e) veneer 碎条形贴面板
broken surface 开裂的路面
broken tape detection 断带检测
broken tape measurement 破尺丈量
broken tape switch 电梯自动停驶开关
broken teeth (钻头的)大量坏齿
broken terrain 沟谷交错地带;断绝地;地势起伏地区;丘陵地;崎岖地区;崎岖地带
broken-tile foundation 瓦渣基础
broken time 零星时间;停工时间
broken top chord 多边形屋架;桁架折线上弦;折线上弦(梁)
broken transit 曲折经纬仪
broken transit instrument 折轴中星仪
broken twill 破斜纹
broken twill weave 斜纹断续的织物
broken type telescope 折轴望远镜
broken up 分化的;分隔的;分裂的
broken up compressed natural asphalt 分裂受压的天然沥青
broken up compressed rock asphalt 分裂受压的岩沥青
broken up material 分裂的材料
broken voyage 航程中断
broken water 浪花;激流;碎波
broken waters 有微波的水域
broken wave 远破浪;远堤破浪;已碎波;破碎波;碎浪;碎波
broken wave height 破碎后波高
broken weir crest 折线堰顶
broken white 复白;乳白色油漆;配合白
broken wire 断丝
broker 代理人;经纪人
brokerage 经纪业;回扣;佣金
brokerage business 委托销售企业
brokerage charges 经纪费;佣金(费用)
brokerage commission 经纪人手续费;经纪人佣金
brokerage expenses 佣金费用
brokerage lease 经纪租约
broker agent 佣人兼代理人;经纪代理人
brokerage office 经纪人事务所
broker and dealer loans 经理商及代理商贷款
broker associate 助理经纪
broker insurance 保险代理商;保险经纪人
broker's advice 经纪人通知书
broker's agency 经纪商
broker's board 交易所
broker's bond 经纪人债券
broker's call 选择通话
broker's contract note 经纪人契约
broker's cover note 保险经纪人暂保单
broker's lien 经纪人留置权
broker's loan 经纪人贷款
broker's order 船舶经纪人运货单
broker's return 经纪人装倒收据单;货物装船明细表
broke storage chest 废纸浆池
broking 经纪业;别口
bromak 气动刷子

bromal hydrate 水合三溴乙醛
bromate 溴酸盐
bromation 溴化(作用);溴处理
bromchlorphenol blue 溴氯酚蓝
bromellite 铍石
bromeosin 曙红
bromide 溴化物
bromide emulsion 溴化银乳剂
bromide of ammonium 溴化铵
bromide of sodium 溴化钠
bromide paper 溴素纸;溴化银像纸;放大纸
bromide print 溴化银像纸
bromide-rich lake water 富溴化物湖水
brominated rubber 溴化橡胶
bromination 溴处理
bromination volumetric(al) method 溴化容量法
bromine atom 溴原子
bromine-boron-lithium water 溴硼锂水
bromine-boron water 溴硼水
bromine-containing substance 含溴物质
bromine cyanide 溴化氰
bromine deposit 溴矿床
bromine index 溴指数
bromine-iodine ratio 溴碘比值系数
bromine-iodine-strontium water 溴碘锶水
bromine-iodine water 溴碘水
bromine-lithium water 溴锂水
bromine number 溴值
bromine oil 溴化油
bromine poisoning 溴中毒
bromine sulfide 二硫化二溴
bromine test 溴化试验
bromine trifluoride 三氟化溴
bromine tungsten filament lamp 溴钨丝灯
bromine water oxidation method 溴水氧化法
bromism 溴中毒
bromizating agent 溴化剂
bromization 溴处理
bromize 用溴处理
bromoacetic acid 溴乙酸
bromoacrylic acid 溴代丙烯酸
bromoallylene 烯丙基溴
bromoamine acid wastewater 溴氨酸废水
bromochloroacetate 溴氯乙酸盐
bromochloroacetonitrile 溴氯乙腈
bromochloromethane 溴氯甲烷
bromocresol blue 溴甲酚蓝
bromocresol green 溴甲酚绿
bromocresol purple 溴甲酚紫
bromoethane 溴乙烷
bromoethanesulfonic acid 溴乙烷磺酸
bromofluorocarbons 溴氟碳化合物
bromoform 三溴甲烷
bromometry 溴量法
bromophos 溴硫磷
bromotrifluorochloromethane 三氟三溴甲烷
bromotrifluoromethane 一溴三氟甲烷
Bromwich contour 布朗维奇围道
bronchial respiration 吹气样呼吸音
Brongniart's formula 布朗尼尔特公式(泥浆重与固体比重关系式)
Bronsil Shales 布龙西尔页岩
Bronsted acid 布朗斯台德酸
Bronsted acid base 布朗斯台德酸碱性
Bronsted base 布朗斯台德碱
Bronsted-Lowry concept of acid and base 布朗斯台德—劳瑞酸碱概念
bronteum 雷鸣发生器(古希腊、古罗马)
brontide(s) 湖吼;地声;湖鸣;轻微地震声
brontogram 雷雨自记曲线
brontograph 雷雨计;雷雨表;雷暴自记器
brontometer 雷雨表;雷暴计;雷暴表;雨爆计
bronze 古铜(色);青铜制品;青铜(色);铜器
Bronze Age 青铜(器)时代;铜器时代
bronze-age building 青铜(器)时代建筑
bronze-age palace 青铜时代宫殿
bronze age pottery 青铜(器)时代陶器
bronze alloy 青铜合金
bronze anchor 青铜锚(固)
bronze annulet 青铜圆箍线(枯环饰)
bronze-backed bearing 青铜背轴承
bronze-backed metal bearing 青铜基滑动轴承
bronze bearing 青铜轴承
bronze black 黑色颜料

bronze blue 深蓝色;金光铁蓝;青铜蓝;铜蓝色
bronze blue pigment 深蓝色颜料
bronze bolt 青铜螺栓
bronze builders furniture 青铜建筑设备
bronze bush(ing) 青铜衬套
bronze cage 青铜保持架
bronze casting 青铜铸造;青铜铸件
bronze coins 铜币
bronze colo(u)r 青铜色;古铜色
bronze compact 青铜坯块
bronze conductor 青铜导线
bronze connector 青铜连接件;青铜连接器
bronze curtain wall 青铜护墙;青铜幕墙
bronzed disease 青铜病
bronze door 青铜门
bronzed pottery 青铜色陶器
bronze dusting 泛铜光;泛金光
bronze extruded spindle 挤压黄铜轴
bronze filled polytetrafluoro ethylene 填充聚四氟乙烯的青铜
bronze fitting 青铜配件
bronze flagpole 青铜旗杆
bronze flake 青铜薄片
bronze foil 青铜箔
bronze glass 古铜色玻璃;青铜色玻璃
bronze-graphite contact material 青铜石墨电接触材
bronze grill(e) 青铜花格窗格栅;青铜花格
bronze guide 青铜导承;青铜导板
bronze hardware 青铜小五金;青铜硬件
bronze hinge 青铜铰链
bronze lacquer 金粉漆;青铜色漆
bronzeless blue 五金光铁蓝
bronze liner 青铜衬
bronze lustre 青铜光泽彩
bronze mirror 铜镜
bronze mo(u)ld 青铜脚;青铜模;青铜造型
bronze-on pipe fitting 镀铜管子配件
bronze paint 古铜色油漆;青铜色涂料;金粉漆
bronze paste 青铜胶;金属粉(色)浆;铜粉浆
bronze pendulum 青铜摆
bronze pigment 古铜色颜料;青铜色颜料
bronze-plated iron square hinge 镀青铜抽心铁方铰链
bronze plating 镀镍青铜
bronze pottery 古铜色陶瓷器
bronze powder 青铜粉
bronze printing 金粉印刷;印金
bronze printing ink 印金油墨;代金墨
bronze printing varnish 印金用清漆;凡立油
bronze profile 古铜色形象;青铜外形
bronzer 青铜匠
bronze red 金光红
bronze relief 古铜色浮雕;青铜浮雕
bronze ring 青铜环
bronze(roof) tile 青铜瓦;青铜屋面瓦
bronze scarlet 金光大红
bronze screw 青铜螺旋
bronze screw hook 古铜螺丝钩
bronze section 古铜色型材;青铜外形
bronze shaft-ring 青铜柱形线脚;青铜柱形饰
bronze shape 青铜外形
bronze sheathed 青铜披挡板
bronze sheen 青铜色泽
bronze sheet 青铜板
bronze sheet panel 青铜镶板;青铜护墙;青铜薄板;青铜金属护墙板
bronze skyscraper 古铜色摩天楼;青铜色摩天楼
bronze smoke 青铜烟灰色
bronze stain 混凝土表面青铜色污斑
bronze statue 青铜像;铜像
bronze steel 镀青铜钢
bronze surround 古铜色镶嵌;古铜色外包皮;青铜镶嵌;青铜包裹
bronze swing door 青铜转门;青铜双式弹簧门
bronze swivel with eye 青铜旋转双耳
bronze swivel with two trunnions 青铜旋转双耳
bronze table 铜标牌
bronze tablet 铜粉
bronze tape 铜卷尺
bronze thrust collar 青铜止推环;青铜推力轴环
bronze tie 青铜系材
bronze trim 青铜镶框;青铜装饰;青铜镶边
bronze tripod 铜鼎

bronze tube 青铜管
bronze unit 青铜构件;青铜外形
bronze vessels 青铜器
bronze ware 青铜器;铜器
bronze washer 青铜垫片
bronze wedge 青铜楔子
bronze welding 青铜焊
bronze window 青铜窗
bronze wire 青铜线;铜丝
bronze-working 青铜工作
bronze yellow 古铜黄色
bronzing 揩金;镀青铜;擦金;青铜色氧化;闪铜光
bronzing and dusting machine 擦金粉机
bronzing fluid 铜粉漆;调金漆料
bronzing lacquer 古铜色喷漆;青铜色腊克;青铜色喷漆;金粉漆
bronzing liquid 金属漆液;金胶;金粉浆;铜粉浆
bronzing machine 擦金粉机;烫金机
bronzing medium 调金漆料
bronzing varnish 金粉清漆
bronzite 古铜辉石;青铜辉石
bronzite chondrite 古铜辉石球粒陨石
bronzite peridotite 古铜辉石橄榄岩
bronzite picrite 古铜辉石苦橄岩
bronzitfels 古铜辉石岩
bronzitite 古铜辉岩;古铜辉石
bronzy 带青铜色的;青铜色的
brood 窝抱
brooder 暖棚育鸡房;暖房育鸡房;孵鸡器
brood fish 怀卵鱼;亲鱼
brood-grow-lay house 终生鸡舍
brood lac 种胶(紫胶)
brood tree 放虫树
brook 河沟;小河;溪(流)
brook bridge 跨溪桥;小河桥
brook-clearing 小河冲刷
brook culvert 小沟涵洞;溪流涵洞
Brookfield rotational viscometer 布鲁克菲尔德回转黏度计;布鲁克菲尔德旋转式黏度计
Brookfield viscometer 黏度计;布鲁克菲尔德黏度计
Brookfield viscosity 布鲁克尔德黏度
Brookfield yield value 布鲁克菲尔德致流值
Brookhill waffler 布鲁克希尔截装机
Brookings model 布鲁金斯模型
brookite 板钛矿
brooklet 小川;小溪;溪
brooklet stream 小溪流
Brooklyn bridge 布鲁克林桥(美国纽约)
Brooklyn suspension bridge 布鲁克林悬索桥(美国纽约)
Brookmire-Harvard method 布鲁克米亚—哈佛法
brook outlet 溪口
brook outlet rapids 溪口急滩
brook retention basin 流水保持池;小溪调节池;小河调节池
Brooks variable inductometer 布鲁克斯可变电感计
broom 桩顶开花;顶篷裂
broom box 清扫用具箱
broom brush 地板刷
broom clean 竣工清扫交付
broom closet 扫雪壁柜;清洁间;清洁工具柜(橱)
broom drag 刮路刷;带刷刮路器;拖挂扫路机
broomed pile 开花木桩
broomed pile head 蓬裂桩头
broomed wood pile 开花木桩
broom finish 路面扫毛;扫毛;表面扫毛(处理);刷清路面;扫面处理;扫面层;拉毛;刷面
broom-finish concrete 扫毛混凝土;拉毛混凝土
broom finishing 扫帚留存条痕;扫毛
broom-finish of concrete 混凝土刷面
broom head of pile 开花桩头;桩头开花;桩的蓬裂顶部;蓬裂桩头
brooming 桩顶发裂;木桩碎裂;抹印痕;桩头开花;刷涂(沥青);扫毛
broom-like veins 帚状矿脉
broom-pattern 帚状
broomstick 扫帚把;帚柄
broom texture 帚状结构
Broseley tile 黏土平瓦
brosoil 硼—硅—铁合金
brotch 茅草屋顶椽木
brothel 妓院
brother chain 吊链
brother node 兄弟节点;同级节点

brother of the brush 油漆工
brother task table 兄弟任务表;同级任务表
brotocrystal 隔蚀晶体;融蚀斑晶
Brotonne bridge 善鲁东奈桥(法国)
brougham 轿车;四轮车
brought down 接下页;减低
brought forward 结转;接下页;后页;转入下页;承前页
brought forward account 前期累结账目;前期滚结账目
brought forward from previous page 接前页
Broughton countersink 布劳顿锥坑钻
brought over 结转;转入下页
brought to course 随砌层
broussaisism 布罗塞学说
broussonetia papyrifera 构树
brouter 桥路器
Brouwer's fixed point theorem 布劳威尔不动点原理
Brouwer theorem 布劳威尔定理
brow 集材场;滑松跳板;檐板;遮水檐;窗檐;窗楣;池水槽;额
browing 抹灰垫层
browing coat 第二道抹灰
brow landing 跳板台(船舷上的木格板)
brow leakage 端部漏磁
brown 咖啡色;褐色;棕色的
brown acid 褐色酸;棕色酸
Brown agitator 布朗型搅拌器
brown algae 褐藻
brown alkali soil 棕色碱土
Brown and Sharpe ga(u)ge 美国线规;布朗—沙普锥度
Brown and Sharpe wire ga(u)ge 美制量线规;布朗沙普线规(美国线规)
Brown antenna 布朗天线
brown-black 褐黑色
brown black tone 棕黑色调
brown-blue 褐蓝色
brown body 褐色体
Brown-Boveri, Cie 勃朗—包维利股份公司(瑞士)
brown box 棕色指示器
brown calcareoul soil 棕色钙质土
brown cast 棕色调
brown china 褐斑釉陶瓷
brown clay 褐黏土;褐色黏土
brown clay ironstone 褐泥铁石
brown coal 褐煤;棕煤
brown coal ash 褐煤灰
brown coal briquet(te) 褐煤球;褐煤标准试块
brown coal briquetting plant 褐煤球厂
brown coal cableway excavator 褐煤索道开挖机
brown coal conveyer 褐煤输送机
brown coal deposit 褐煤矿床
brown coal furnace 褐煤炉子
brown coal getting 褐煤开采
brown coal grinding mill 褐煤碾磨厂
brown coal grit 褐煤砂粒;褐煤硬渣
brown coal international classification 褐煤国际分类
brown coal low temperature coke 褐煤的低热焦炭
brown coal mine 褐煤矿
brown coal opencast mine 露天开采的褐煤矿
brown coal seam 褐煤层;褐煤缝
brown coal shovel 开采褐煤的动力铲
brown coal tar 褐煤焦油(沥青)
brown coal(tar) pitch 褐煤沥青;褐色硬质沥青
brown coal underground mine 褐煤深矿
brown coal working 褐煤开采
brown coat 末道灰泥;褐色罩面;罩面基层灰;二道抹灰
brown colo(u)r glass 茶色玻璃
brown concrete 褐色混凝土
brown desert soil 褐色荒漠土;棕色荒漠土;棕漠土
brown dye 棕色染料;褐染料
brown earth 褐土;褐色土;棕壤(土);褐棕色土
brown-earth soil 棕壤土
brown earthy coal 土状褐煤
Browne correction 布隆尼改正
brown edge 铜色口沿;铁口
Browne heat test 桐油胶化试验
Brownell twister 布劳内尔捻线机
Browne terms 布朗项
brown forest soil 褐色森林土;棕色森林土

brown fume 棕色烟雾;棕色尘雾
brown gasoline 褐色汽油
brown glass 棕色玻璃
brown glaze 褐色釉;棕色釉;褐釉
brown-glazed brick 褐釉砖;盐釉砖;防潮砖
Brown gravity ga(u)ge 布朗重力仪
brown-green 褐绿色
brown-grey 褐灰色
brown grit 褐防潮砖煤砂
Brown gyrocompass 布朗型陀螺罗经
brown hard varnish 棕色硬质虫胶清漆
brown hay 褐色干草
brown haze 棕色轻雾
brown hematite 褐铁矿;褐赤铁矿
Brownian diffusion 布朗扩散
Brownian method 布朗方法
Brownian motion 布朗运动
Brownian motion process 布朗运动过程
Brownian movement 布朗运动
brownie 便携式雷达装置
brown induration 褐色硬结
browning 褐色氧化处理;褐变;变褐
browning agent 褐色试剂
browning coat 第二道抹灰(在三道抹灰中);抹第二层灰
browning paste 褐色糨糊
browning plaster 抹浆
Browning prism 布朗宁棱镜
browning reaction 褐变反应
browning salt 褐化盐
browning solution 褐化溶液
brown iron ore 褐铁矿
brown iron oxide 氧化铁棕
brown iron oxide pigment 褐色铁氧化物颜料
brownish 淡棕色;带褐色的
brown(ish) black 棕黑(的)
brownish-black colo(u)r 棕黑色
brownish red 棕红
brownish yellow(facing) brick 土黄色面砖
brownism 布朗学说
brown jersey gloves 褐色细毛绒手套
brown laterite soil and brown laterite area 砖红壤性土壤及砖红壤区
brown lignite 棕褐煤
brown lime 褐石灰;棕石灰;次石灰
brown limerized soil 灰棕色灰化土
brownline 褐色图纸;棕色线条图
brown loam 棕壤(土)
Brown lunar theory 布朗月球运动理论
brown madder 褐红(色)
brown madder lake 茜草棕色淀
brown matter 褐色物质
brown mechanical pulp board 褐色磨光的厚纸板
Brown metal 布朗黄铜
brown millerite 钙铁石
brown mixture 棕色合剂
Brown-Mood process 布朗穆德过程
brownness 褐色;棕色
brown ocher 褐铁矿
brown ore 褐色铁矿
brown-out 抹底灰;节电;限制用电;灯火管制
brown oxide 棕色氧化物
brown oxide of iron(pigment) 褐色铁氧化物颜料
brown packing paper 牛皮纸;包装纸
brown paint 褐色厚漆
brown patch 褐块
brown phosphorus 褐色磷
brown pigment 褐色颜料
brown pine 高罗汉松
brown pink 栎皮红棕色淀
brown podzolic 棕色灰化土
brown podzolic soil 褐色灰化土
brown powder 黑火药;有烟火药
brownprint 棕图;棕色图;晒棕图
brown purple 褐紫色
brown-red 褐红(色)
brown rice 极粗糙米
brown rice-grain 茶色米
brown-ring test 褐轮试法;棕色环试验;棕黄试验;棕环试验
brown rock 褐色岩石
brown root 褐根
brown rot 褐色腐朽;褐腐;棕色腐朽
brown salt glaze on pottery 陶器用褐色盐釉

Brown's correction value 布朗校正值
brown seaweed 褐藻
Brown-Sharpe type dividing head 布朗夏普分度头
Brown-Sharp Wire ga(u)ge 布朗夏普线规(美国线规)
brown sienna 赭色
brown size 褐色胶(料)
Brown's method 布朗方法
brown smoke 褐烟;棕色烟气
brown snow 褐色雪
brown soil 褐土;褐土(壤);棕色土;壤土(土);棕钙土
brown spar 铁菱镁矿;铁白云石
brown stain(ing) 棕褐色变(砌体灰缝);褐色斑;褐变
Brown's test 钢丝绳磨损试验
Brown's theory 布朗学说
brownstone 褐色砂石;褐色砂岩;褐石(镶面);褐砂岩;褐砂石墙的房屋
Brownstone Series 褐砂岩统
Brown taper 布朗锥度
brown tone 棕色调
Brown tongs 布朗钳
Brown-Twiss intensity interferometer 布朗—特威斯光强干涉仪
brown-violet 褐紫色
brown ware 褐色陶器;(褐色的)陶制品;棕色陶器
Brown water quality index 布朗水质指数
brown wood 木栓化部分
brown wood board 棕纸板
brown-yellow 褐黄色
brown ytterbium niobium concentrate 褐钇铌精矿
brow of the hill 山顶
brow piece 井口支架;门口梁;大块木材;门过梁
brow plate 带轮跳板
brow post 顶梁;横梁
browsability 索得率;随机索得率
browser 浏览器
browsing room 图书阅览室;浏览室
browumillerite 钙铁铝石
Bruce antenna 倒V形天线
brucellin test 波状热菌素试验
brucellosis 波状热
Bruce's bundle 角连束
Bruce Series 布鲁斯统【地】
Bruceton method 布鲁斯顿方法
Bruch's membrane 布鲁赫膜
brucite 含水氧化镁矿石;水镁石;水滑石
brucite asbestos 水镁石石棉
brucite marble 水镁石大理石;水镁石大理岩;水滑结晶灰岩
brucitite 水镁石岩
brueggenite 水碘钙石
brufen tablet 拔怒风片
brugnatellite 红磷铁镁矿;次碳酸镁铁矿
bruiachite 萤石
bruise 压扁;砸伤;挫伤;产生凹痕(木材和金属);擦伤痕;碰伤;伤痕
bruise check 碰撞裂纹
bruised place 擦伤痕
bruise mark 碰痕
bruiser 压碎机;压扁机;捣碎机
bruise resistance 捣击阻力
bruising 硬伤;压扁
bruising mill 碎矿机
Brulax system 布鲁勒氏系统
bruma 白罗马霾
brume 雾
brummer 工厂汽笛;防水木材填塞料
Brunauer Emmett and Teller 布鲁瑙厄·埃梅特·泰勒法
Brunauer Emmett and Teler equation 多分子层等温吸附式
Brunauer-Emmett and Tetter multilevel adsorption model 布鲁瑙厄·埃梅特-泰勒多层吸附模型
Brunhes normal polarity chron 布伦斯极性时
Brunhes normal polarity chronzone 布伦斯正向极性时间带
Brunhes normal polarity epoch 布伦斯正极性期
Brunhes normal polarity zone 布伦斯正向极性带
brunishing-in 跑合作业
brunisolic soil 棕壤(土)
brunizem 湿草原土

Brunk's test 布伦克检验
Brunner-Emmett-Teller method 比表面积测定法
Brunner's yellow 布鲁钠锑黄(色料)
brunofix 发黑氧化处理
brunogeierite 锗磁铁矿
brunonianism 布朗学说
brunorizing 特别常化法
Brunson jig transit 布朗桑工具经纬仪
Bruns pile 布隆斯桩
Bruns theorem 布隆斯定理
brunsvigite 铁镁绿泥石
Brunswick black 布伦斯维克黑(漆)
Brunswick blue 布伦斯维克蓝
Brunswick green 布伦斯维克绿;铅铬黄
Brunswick ratio 布伦斯维克比
Brunton compass 布隆顿(地质)罗盘
Brunton pocket transit 布隆顿罗盘
Brunt's formula 布隆特公式(晴天天空反射量计算公式之一)
brush 电刷;擦掉;热带密灌丛;涂料刷;刷子;刷大巷道;梢料
brushability 可刷性;耐刷能力;易刷性;涂刷性;刷涂性
brushable 可涂刷的
brushable consistence 可涂刷稠度
brushable consistency 可涂刷稠度;可刷稠度
brush aeration 转刷曝气
brush aeration system 转刷曝气系统;曝气刷系统
brush aeration tank 转刷曝气池
brush aerator 刷形充气器
brush and cable bank protection 梢料锚索护岸;梢捆护岸
brush and pile dam 梢料木桩坝
brush and pile dike 梢料木桩堤
brush and slotless motor 无电刷槽电动机
brush and stone barge 梢石料船;梢石料驳船
brush and wire envelop mattress 铁丝网柴(褥沉)排
brush angle 电刷倾斜角
brush application (油漆技术中的)涂刷方法;涂刷;涂抹
brush-applied 用刷涂敷的
brush-applied coat(ing) 刷敷涂层;电刷用涂层;刷涂(涂装);刷涂(涂)层
brush arbor seal 毛刷轴密封垫
brush arc 刷弧
brush arm 刷臂
brush assembly 碳刷组;炭刷组
brush back 后刷边
brush backed sander 自动旋转打磨机
brush backward lead 电刷反向超前
brush barge 梢料船;梢料坝
brush binder 漆刷刷毛黏合剂
brush box 红胶木;碳刷盒
brush bracket 刷架
brush breaker 灌木铲除机;灌木清除机;割灌机
brush broom 钢丝路刷
brush burn 擦伤
brush buster 灌木切除器
brush carbon 电刷碳
brush carriage 电刷支架
brush carrier 电刷架;刷握
brush cast 刷模
brush check dam 梢料谷坊
brush chipper 灌木铲除机
brush clamp 电刷夹持器
brush cleaner 清扫刷;漆刷清洗剂;刷式清选机
brush cleaning 毛刷清洁
brush clearing 扫清树枝;用刷子清除
brush coat 电刷涂料;敷层;涂料层;刷涂层;刷层
brush coater 刷涂机
brush coating 电刷涂布;刷敷涂层;抹涂;刷涂
brush coating method 刷涂法
brush collector 集流刷;集电刷
brush colter 灌木铲除器
brush-comb 毛刷梳棉
brush compare check 电刷比较检验
brush contact 电刷接点;电刷触点;刷形触点
brush contact drop 电刷接触电压降
brush contact reactor 电刷接触电抗器
brush contracting works 梢料束狭工程
brush cord 电刷软绳
brush corduroy 柴排加固土路;木排路;灯芯绒
brush corona 电晕放电;刷形电晕

brush cotton gin 刷式轧花机
brush country 丛林地区
brush coupling 刷形联轴节
brush covering factor 电刷覆盖系数
brush current 电刷电流
brush cutoff 刷式阻种器
brush cutter 灌木清除机；灌木铲除机
brush cutting 清除灌木
brush cylinder 刷式滚筒
brush dam 柳条坝；垃圾堤坝；枝条排坝；灌木堰；柴草坝；梢枝坝；梢料坝
brush-dewing machine 给湿刷毛机
brush dike 梢料堤
brush discharge 刷形放电
brush displacement 刷形夹角；刷位移
brush down 刷去
brush drag 拖刷；刷涂阻力
brush dust 刷灰；刷粉
brushed concrete 刷毛混凝土
brushed finish 粉刷；刷面处理；（混凝土路面的）刷毛面；刷饰面
brushed on strippable coating 在可剥离的涂层上涂刷
brushed plywood 刷光胶合板
brushed roofing membrane 涂刷的屋面薄膜
brushed surface 刷面层；（混凝土路面的）刷毛面
brushed surface finish 刷毛面饰
brush electrode 刷式电极
brush encoder 电刷编码器
brusher 机刷；机械刷；采煤工；清刷机；刷式清理机；刷鳞机
brushes of Ruffini's terminal cylinder 露菲尼终柱
brush etching 毛笔腐蚀
brush faults 寻状断层
brush feed 刷式排种器
brush feed mechaniam 刷式排种装置
brush fiber 制刷纤维
brush filler 刷涂用腻子
brush finish （混凝土的）刷毛；刷纹饰面；粉刷；刷子拉毛饰面；刷纹面层；刷面（处理）
brush fire 灌丛火
brush for aeroelectromachine 航空电刷
brush forward lead 电刷正向超前
brush friction loss 电刷摩擦损失
brush furnisher 毛刷给浆辊筒
brush gang 线路清理班组
brush ga(u)ge 电刷量具
brush gear 电刷装置
brush go-devil 除锈器
brush graining 刷饰木纹状饰面【建】；刷木纹状漆面；刷仿木纹
brush guard 护网
brush hand 有手艺的油漆工
brush(handle) wire 刷用钢丝
brush harrow 荆条拖耙；丛林耙
brush holder 电刷握臂；笔筒；刷握
brush holder insulating bushing 电刷柄绝缘套
brush holder pin 刷柄销
brush holder rod 刷握臂
brush-holder stud 刷握支柱；刷杆
brush holder yoke 刷握架
brush hook 截枝钩刀；剪灌弯刀
brush hopper 刷斗
brush housing door 毛刷仓盖
brush housing upper baffle 毛刷仓上挡板
brush hurdle 绿篱；树篱
brush inclination 电刷倾角
brushing 创面电灼术；清渣；涂刷；刷亮；刷尖放电；刷灰；刷光；刷布
brushing and steaming 刷蒸机
brushing compound 刷光涂料
brushing consistency 可涂刷稠度；可刷稠度
brushing decoration 刷花
brushing discharge 电刷放电
brushing down 刷下
brushing glazing 刷釉
brushing lacquer 刷漆；刷涂挥发性漆
brushing loss 刷损量（稳定土磨耗试验用）
brushing machine 刷光机
brushing machine for caul plate 刷垫板机
brushing off 擦掉；刷去
brushing of tunnel top 挑顶
brushing-on paint 刷涂用漆
brushing primer 刷涂底漆

brushing property 刷涂性；涂光性；刷光性
brushing quality 涂刷质量
brushing shot 清扫性放炮；挑顶刷帮炮眼；刷帮爆破
brushing test 刷损试验（试验稳定土的耐久性）
brushing wheel 刷轮
brushite 透磷钙石；透钙磷石
brush joints 寻状节理
brush latch opener 毛刷开针舌器
brush lead 电刷引线
brushless excitation 无刷励磁
brushless motor 无电刷电动机
brush line 平板玻璃；细丝通；细线道；细条纹
brush lipless guard 平顶短牙刃器
brush-lubricated 油刷润滑的
brush machine 擦光机；清道机；街道清扫机
brushman 刷子工
brush mark 刷状痕；刷印；刷痕
brush matters 梢料排
brush matting 柴排；柴排；梢料垫层
brush mattress 沉排；柴排；梢枝排；梢排垫；埽褥；梢扫沉排；梢褥
brush mattress protection 柴排护坡；柴排护岸
brush mattress revetment 柴排护坡
brush method 涂刷法
brush motor 带刷马达；整流式电动机；擦洗机
brush off 刷去；擦去；刷清
brush off blast 清扫级喷砂
brush of kernel 刷毛
brush-on 涂刷；刷涂
brush-on acid-resisting cement finish 耐酸水泥刷面
brush or spray coating 涂刷或喷洒（防水）层（在地面下基础外表）
brushout 除根；刷掉；小面积试涂；刷漆样板
brush out cards 遮盖力试验纸
brush over 用刷上色
brush painted decoration 点画花
brush painting 刷涂法
brush party 线路清理班组
brush paving 梢料护面
brush pencil 画笔
brush pick-up 刷式拴拾器
brush pig 除锈器
brush pilling tester 毛刷式起球试验仪
brush-plating 刷镀
brush plough 灌木清除机；铲掘机重型犁
brush plow 犁式除荆机
brush polish 刷光剂；上光剂
brush polishing 刷磨光
brush pot 笔洗
brush power chart 黑白格纸
brush pressure 电刷压力
brush rack 笔架
brush rake （安装在推土机前作清扫工作的）叉形推架；灌木丛铲除机
brush reader 电刷阅读器
brush resistance 电刷电阻
brush resistance loss 电刷接触电阻损耗；电刷电阻损耗
brush rest 笔架
brush retouching 毛刷修版
brush revetment 柴排护坡；柴排加固边坡；梢枝排护岸；梢排护岸
brush rigging 刷握；刷夹
brush ring 刷环
brush rocker 电刷摇移器；电刷摇杆
brush-rocker ring 电刷摇杆转环；移刷环
brush roll 辊筒刷；树枝梢捆
brush roller 毛刷罗拉
brush saw gin 毛刷式锯齿轧花机；刷式锯齿轧花机
brush scraper 钢刷刮管器；除锈器
brush scythe 割灌大镰刀
brush segment 清刷段
brush sensing 电刷读出
brush set 电刷组
brush shade 地貌晕渲
brush-shading(method) 晕渲法
brush shift 电刷位移
brush shifting device 电刷移动装置
brush-shifting motor 移刷型电动机
brush shifting speed 电刷移动速度
brush short-circuit current 电刷短路电流
brush shredder 灌木切除机

brush sickle 灌木切刀
brush sliding areas 电刷滑动面积
brush spark(ing) 电刷发火花
brush spreader 刷式涂布机
brush spring 电刷弹簧；刷握弹簧
brush spring adjuster 电刷簧调整器；刷握弹簧调整器
brush spring assembly 钢丝刷弹簧组
brush spring unit 钢丝刷弹簧件
brush stand 笔筒
brush station 电刷站；电刷读孔头；电刷测量点
brush stone separator 刷式石块分离机；刷式清石机
brush stop 钢丝刷止动器
brush stripping saw gin 毛刷式锯齿轧花机
brush stroke 刷毛行程（混凝土路面刷毛机的一次）；每一次行程（混凝土路面刷毛机）
brush stud 电刷柄
brush surface analyser 电刷表面分析器
brush sweeper 清扫毛刷
brush team 线路清理班组
brush-topped 柴排盖顶的；灌木遮盖的
brush-topped road 柴束路
brush trace 刷痕
brush track 刷迹圈
brush treatment 涂刷防腐剂；（木材的）涂刷防腐；涂刷处理
brush truck 刷洗车
brush-type aerator 刷式曝气器（污水处理）
brush-type folds 寻状褶皱
brush-type rotor 转刷（式转子）
brush-type stripper 刷式摘棉铃机
brush-type structure 寻状构造；刷型构造；寻型构造【地】
brush up 刷光；刷新
brush veins 寻状矿脉
brush voltage 电刷电压
brush-washer 笔洗
brush weather strip 毛绒防风条；刷形挡风雨条；刷形密封条
brush weir 枝条排堰
brush wheel 刷轮；磨轮
brush wicher-work 梢枝护岸设施；梢护岸设施
brush width 电刷宽度
brush wire 电刷线
brushwood 矮丛林；灌木林；灌木犁；灌木丛；灌木（材）；矮林；梢料
brushwood-and-stone dam 梢石坝
brushwood bundle 梢束；柴捆
brushwood checkdam 梢捆挡水坝；柴捆谷坊；柴捆挡水坝
brushwood chipper 灌木切割器
brushwood clearing 清除小树丛
brushwood cofferdam 梢料围堰
brushwood cutter 灌木切割器；灌木铲除机
brushwood dam 梢料坝
brushwood faggot 梢捆
brushwood faggot screen 灌木柴把屏障
brushwood fascine 柴束；梢捆
brushwood fence 杂木围墙；编柴墙
brushwood fender 柴排护舷
brushwood hurdle 灌木栽成的篱笆
brushwood layer 灌木层
brushwood mattress 梢埽沉排；梢蓐；柴排；埽褥
brushwood rake 灌木耙
brushwood revetment 梢捆护岸
brushwood road 柴束路
brushwood weir 梢枝堰；梢捆堰
brush work 毛笔画；涂刷；埽工；用刷子完成的工作；护岸设施
brushy 多灌木的
brush yoke 刷架
brusing mill 锤式碾碎机
Brussels carpet 布鲁塞尔地毯
Brussels carpet jacquard 布鲁塞尔地毯提花机
Brussels classification 布鲁塞尔十进分类法；布鲁塞尔分类法
Brussels Conference 布鲁塞尔会议
Brussels Convenntion 布鲁塞尔公约；（1924年布鲁塞尔统一提单的国际公约）
Brussels curtains 布鲁塞尔帘幔
Brussels nomenclature 布鲁塞尔命名法
Brussels Protocol 布鲁塞尔议定书
Brussels quilling 重针网眼
Brussels system 布鲁塞尔分类系统；通用十进制

系统
Brussels Tariff Nomenclature 布鲁塞尔关税商品分类
Brussels weather strip 布鲁塞尔式(毛圈)地毯
Brutalism 粗野主义【建】
brutalist architecture 粗野派建筑
brute force 强力
brute-force approach 粗略近似法；强制法
brute force feeder 强制喂料机
brute force focussing 强力聚焦
brute force radar 强力雷达
bruting 宝石互磨
brutonizing 钢丝热镀锌法
brutto-budget 总预算
Bruxellian 布鲁克塞阶
Brückner cycle 勃吕克纳循环
Bryant's traction 布赖恩特牵引
bryoflora 苔藓植物
bryophyte 苔藓植物
Bryozoa 苔藓虫纲
bryozoan faunal province 苔藓虫动物地理区
bryozoan limestone 苔藓虫灰岩
bryozoatum 海浮石
B-source 阳极电源；板极电源
BS speed 英国标准感光度
B-stage 乙阶段
B-stage resin 乙阶(段)树脂；半熔酚醛树脂
B-tectonite B 岩组构造岩；B 构造岩
Bttger's test 伯特格尔试验
B-type lineation B 线理
B type steel B 类钢
buaze fiber 蝉翼藤皮纤维
bubbing 析出气体
bubble 磁泡；发泡；前缘吸力式压力分布；气泡；气囊；气流离体区；泡状包裹体；水泡(音)；商业骗局
bubble absorber 泡沫吸收器
bubble aeration 鼓泡曝气；吹泡汽通；气泡曝气
bubble agitation tank 气泡搅拌池；气泡搅拌槽
bubble alumina 空心氧化铝球
bubble annihilation 磁泡消除
bubble annihilator 磁泡消除器
bubble attachment 气泡黏附
bubble axis 水准轴
bubble blockage 气泡栓塞
bubble bowl 球形玻璃缸
bubble breakwater 气囊防波堤
bubble bucket 采样桶（井口泥浆）
bubble bursting 气泡爆裂
bubble canopy 气泡式座舱罩
bubble cap 球状盖；球形盖；泡罩；泡帽
bubble cap fractionating column 泡罩分馏柱；泡罩分馏塔
bubble-cap plate 泡罩板
bubble-cap plate coloum 泡罩塔
bubble-cap plated tower 泡罩板式塔
bubble-cap plate tower 泡罩塔
bubblecaps 泡罩式分布器
bubble-cap tower 泡罩蒸馏塔
bubble cap tray 泡罩塔盘
bubble-cap tray tower 泡罩塔
bubble car 微型汽车
bubble cavitation 空泡气蚀
bubble cell 气泡式水准器；气泡浮选槽；泡沫浮选槽
bubble centre 气泡中心
bubble chamber 气泡室；泡沫室
bubble chamber photography 气泡室摄影
bubble chamber track 气泡室径迹
bubble chip 磁泡芯片
bubble chip control 磁泡片控制
bubble chip single-mask technology 磁泡芯片一次掩蔽工艺
bubble circuit 磁泡电路
bubble clinometer 气泡测斜仪
bubble coating 微泡涂层；微孔涂层
bubble collapse 空泡溃灭；气泡消失；气泡破灭
bubble collimating system 气泡水准器视准系统
bubble column 泡塔
bubble column chromatography 气泡柱色谱法
bubble-column flo(a)tation 浮选柱浮选；气升柱浮选
bubble-column machine 泡柱式浮选机
bubble company 虚设公司；皮包公司
bubble concept 气泡概念
bubble concrete 加气混凝土；泡沫混凝土

bubble convection 气泡对流
bubble correction 气泡校正
bubble counter 计泡器
bubble counting 气泡计数
bubble crack 汽泡崩裂
bubble data 磁泡数据
bubble data module 磁泡数据模块
bubble deck 起泡盘；泡罩塔盘
bubble delay device 磁泡延迟器件
bubble density 气泡密度
bubble device 磁泡器件
bubble device margin 磁泡器件容限
bubble diagram 气泡图
bubble disk latch 磁泡盘闩
bubble dispersant 气泡分散剂
bubble display 磁泡显示器
bubble distortion 气泡扭变
bubble distribution 气泡分布
bubble domain 磁泡畴；泡畴
bubble domain collapser 泡畴破裂器
bubble drive 磁泡驱动
bubble dust collector 泡沫除尘器
bubble dust scrubber 泡沫除尘器
bubble economy 泡沫经济
bubble eliminator 消除气泡器；气泡分离器
bubble extinguisher 泡沫灭火器；泡沫灭火机
bubble film 微泡膜
bubble flask 充气 U 形管
bubble floating force 气泡浮力
bubble flow 气泡状流动；气泡
bubble flow-steering switch 磁泡流导向开关
bubble formation 形成气泡；鼓泡；气泡形成
bubble former 起泡器
bubble frequency 气泡频率；气泡密度
bubble gap 磁泡隙
bubble gas scrubber 鼓泡式气体吸收器
bubble ga(u)ge 气泡水位计；气泡指示器；气泡式潮位计；气泡计
bubble generator 磁泡产生器
bubble glass 泡沫玻璃；气泡玻璃；泡沫玻璃
bubble guide channel 磁泡导向通道
bubble gun 射流枪
bubble hard error 磁泡硬错率
bubble heating 吹泡加热
bubble horizon 气泡地平
bubble idler 磁泡空转器
bubble impression 吹痕
bubble inclinometer 气泡倾斜仪；气泡测斜仪
bubble index 气泡指数
bubble internal energy 气泡内部能量
bubble in the interlayer 胶合层气泡
bubble ladder 磁泡梯形网络；泡梯
bubble ladder organization 磁泡梯结构
bubble lake 鼓泡湖
bubble lattice 磁泡晶格
bubble lattice device 磁泡点阵器件
bubble lattice file 磁泡点阵文件存储器
bubble layer 气泡层
bubbleless aeration 无泡曝气
bubble level 气泡水准仪；气泡水准器；水准器
bubble level(l)ing 气泡水准测量
bubble logic 磁泡逻辑
bubble machine 气泡式浮选机；起泡浮选机
bubble memory 磁泡存储器
bubble memory card 磁泡存储卡片
bubble memory device 磁泡存储器
bubble memory module 磁泡存储器模块
bubblement 起泡(状态)
bubblemeter 气泡检查仪
bubble method 气泡法
bubble method leak detection 气泡法探漏；泡沫法探漏
bubble mobility 磁泡迁移率
bubble mo(u)ld cooling 膜泡塑模冷却
bubble noise 气泡噪声
bubble of level 水准仪气泡
bubble on the wire 夹丝上的气泡
bubble oscillation 气泡振荡
bubble pattern 磁泡图
bubble pattern sensitivity testing 磁泡模式过敏测试
bubble physics 磁泡物理学
bubble plate tower 泡罩层蒸馏塔；泡罩层塔
bubble point 沸腾点；起泡点；泡点；始沸点

bubble point curve 泡点曲线
bubble point pressure 饱和压力；泡点压力
bubble point pump 饱和压力泵
bubble pressure 空泡压强；气泡压力
bubble pressure method 气压测孔法；气泡压力法
bubble proof 吹验法
bubble propagation 磁泡传播
bubble protractor 汽泡式量角器；气泡式分度规
bubble pulse 重复冲击；气泡脉冲
bubble push against airgun 气泡推斥式气枪
bubble quadrant 气泡水准测角仪
bubbler 扩散器；饮水口；鼓泡器；气泡器；起泡器；喷泉式饮水器；水浴瓶
bubble radius 气泡半径
bubble raft 泡筏
bubble reflector 水准管反射镜
bubble-release agent 释气剂
bubble replicator 磁泡复制器
bubbler fountain 喷水式饮水器；喷泉式饮水器
bubbler ga(u)ge 液量计
bubbler head 扩散式喷洒头
bubbler of level 水平仪气泡
bubbler system 气泡系统
bubbler tube 气泡管；水准仪水泡管；水平仪玻璃管
bubble sand 气泡砂
bubble scale value 水准仪格值
bubble scheme 空头计划
bubble screen 气泡屏
bubble scrubber 鼓泡洗涤塔
bubble sextant 气泡六分仪
bubble sextant with averaging gear 平均机气泡六分仪
bubble shift register 磁泡移位寄存器
bubble shutoff 不起泡关闭
bubble signal 磁泡信号
bubble size 磁泡尺寸；气泡大小
bubble soft error 磁泡软错
bubble sort 冒泡分类法；泡沫分类；上推排序法；上推分类法
bubble sort method 泡沫分类法
bubble spacing 气泡间隙；气泡间距
bubble spectrum 气泡谱；气泡分布
bubbles policy 泡沫政策
bubble stabilizer 气泡稳定剂；稳泡剂
bubble stage recorder 气泡水位计
bubblestone 泡沫水泥制品；轻水泥制品(分隔用)
bubble stretch 磁泡延伸
bubble stretcher 磁泡延伸器
bubble structure 釉泡结构；多孔构造；气泡状结构；气泡结构
bubble surface 气泡表面
bubble system 磁泡系统
bubblet 小气泡
bubble tank 鼓泡池
bubble test 鼓泡试验
bubble tester 水准器检验仪
bubble tight 不起泡的
bubble tower 鼓泡塔
bubble train 成串气泡；气泡串
bubble transfer switch 磁泡转移开关
bubble trap 磁泡陷阱
bubble tray 空气扩散板；鼓泡塔盘
bubble trier 气泡验准器
bubble tube 管状水准器；鼓泡管；气泡水平仪；水准管
bubble tube axis 水准管轴
bubble-type-flow counter 气泡型流量计数器
bubble type inclination 气泡倾斜仪
bubble-type sampler 气泡型采样器
bubble unit 发泡器
bubble viscometer 泡沫黏度计
bubble viscosity 泡沫黏度
bubble zirconia 空心氧化锆球
bubbling 连串起泡；冒泡；油漆起泡；鼓泡；飞溅；起泡；泼溅
bubbling carburet(t)or 气泡式汽化器
bubbling device 鼓泡装置
bubbling fluidized bed 鼓泡流化床
bubbling fluidized bed combustion 沸腾式流化床燃烧
bubbling hood 泡罩
bubbling point 起泡点
bubbling polymerization 气泡聚合
bubbling potential 起泡能力

bubbling spring 冒泡泉;含气涌泉;溢气泉;鼓泡泉;泡沸泉
bubbling stability 发气稳定性;起泡稳定性
bubbling tower 鼓泡塔
bubbling washer 煤气洗涤机
bubbling well 泡沸井
bubbly 多泡的
bubbly clay 起泡黏土
bubinga 青龙木;花梨木;西非黄檀木
Bubnov-Galerkin method 布巴诺夫—伽辽金方法
Bucaleni 布卡伦尼针织用聚酯卷曲丝
bucaramangite 淡黄树脂
Bucaroni 布卡鲁尼卷曲变形耐纶丝
buccaro 粗陶
Buchan trap 钟口 P 形存水弯
Bucherer reaction 布霍勒反应
Buchholtz protector 布克霍尔茨保护装置
Buchholz protective device 布克霍尔茨保护装置
Buchholz relay 布克霍尔茨(气体)继电器;瓦斯继电器
Buchi sand trap 比希截沙坑;比希沉沙池
buchite 玻化岩
Buchmann-Meyer effect 布赫曼-迈耶效应
Buchner filter 瓷平底漏斗;瓷漏斗
Buchner funnel 瓷漏斗;抽滤漏斗;布氏漏斗;平底漏斗
Buchner funnel test 布氏漏斗检验
buchonite 闪云灰玄岩
buchucamphor 布楚樟脑
buchwaldite 磷钠钙石
buck 锯台;锯开;碱水;大装配架;大模型架;搪烧支架;试样研磨;顶撞
buck anchor 够劲锚栓
buck-and-boost regulator 电压升降调节器;升降电压调节器
buck arm 转角横杆;补偿杆
buck basket 洗衣筐
buckboard 装弹簧座椅的四轮马车;四轮马车;踏脚板(在软土面上作业用);研磨板
buck-boost control signal 正反控制信号
buck dollar 美元
bucked 截成木材长度
bucker 碎木机;锯木架;碾压机;碾碎机;碎石锤;碎矿机;碎矿工;碎矿锤;破碎机
bucker excavator 斗式挖掘机
bucker method 吊桶法
bucker-up 铆钉工
bucket 料罐;泥斗;买空卖空;戽斗;消力坎;钻屑收集筒;抓斗;斗;吊桶;吊罐;存槽斗;称料斗;反弧面;斗(形电极);水桶(形电极);水桶;勺斗
bucket-and-belt elevator 斗带式提升机;斗带式升运器
bucket-and-chain dredge(r) 斗链式挖泥机;斗链式挖泥船;链斗式挖泥船
bucket and stopwatch technique 量桶停表法
bucket angle indicator 铲斗角度显示器
bucket apron 消力戽护坦;反弧段护坦
bucket apron conveyer 斗裙式输送机
bucket arm 斗柄;铲斗臂
bucket armchair 斗式扶手椅
bucket arm control valve 铲斗臂控制阀
bucket arm lowering time 铲斗臂下落的时间
bucket attachment 戽斗连接;水斗联结(装置)
bucket auger 勺钻;短管螺旋钻具;桶式钻头;勺形钻
bucket auto-leveler 铲斗自动放平装置
bucket backfiller 铲斗填土机;回填机铲斗;斗式回填机
bucket basin 戽斗式消力池
bucket blade 斗形搅拌叶片;铲斗刀片
bucket board 戽斗板
bucket body 斗体
bucket boom dredge(r) 多斗臂挖泥船
bucket boom excavator 多斗臂挖土机
bucket bottom gate 挖斗底卸式闸门;底卸式挖斗闸门
bucket brigade device 组练式器件;斗链器件
bucket brigade electronics 斗链电子学电路
bucket brigade switch 组练式开关
bucket can 凹罐
bucket capacity 料斗容积;泥斗容量;斗容(量)
bucket capacity at waterline 水线斗容
bucket capacity for various soils 不同土质泥斗容量
bucket car 戽车;斗式运输车;斗式运料车

bucket carrier 斗式转运机;斗式装载机;斗式装料机;斗式装料机;斗式转载机;斗式运送机;斗式运输器;斗式运料车;斗式输送机
bucket chain 斗链;铲斗链
bucket chain dredge(r) 斗链式挖泥机;斗链式挖泥船;链斗式挖泥船
bucket chain drive 斗链驱动机械
bucket chain excavator 斗链式挖掘机
bucket chain pitch 斗链节距
bucket chain reclaimer 链斗式取料机
bucket chain speed 斗链转速
bucket chain tension 斗链张紧
bucket closing cable socket 挖斗式卸货索插座
bucket content 斗内物料
bucket control 铲斗控制
bucket control lever 铲斗的控制杠杆
bucket control stop 铲斗控制停止
bucket control valve 铲斗控制阀
bucket conveyer 多斗输送机;多斗输送机;斗式提升机;链斗式输送机;斗式运输机;斗式提升机;斗式输送机
bucket crane 料罐起重机;抓斗起重机;斗式起重机;吊斗起重机
bucket current meter 叶片式流速计
bucket curve 挑流鼻坎曲线
bucket curve of spillway 溢流堰消能挑坎部分曲线;溢洪道消力戽曲线
bucket cutting edge 铲斗切割边;铲斗刃
bucket cutting teeth 斗齿
bucket cylinder 铲斗油缸
bucket discharge 用斗出料
bucket disc spring 斗盘弹簧
bucket ditcher 铲斗挖沟机
bucket door 铲斗挡板;斗底倒泥门;铲斗活底(门)
bucket drainage 吊桶排水
bucket dredge(r) 链斗式挖泥机;多斗挖泥机;多斗挖泥船;斗式挖泥船;斗式挖掘机;斗链挖泥船;斗链式挖掘船;斗式挖泥船
bucket dredge(r) well 链斗挖泥船桥档
bucket dredge(r) with elevated chute 高架溜槽链斗挖泥船
bucket dredge(r) with sand pump 泥泵链斗挖泥船
bucket drill 带钻粉筒螺旋钻机
bucket dump stop 铲斗倾斜停止件
bucket elevating method 斗式提升法
bucket elevation 斗式提升法
bucket elevator 多斗提升机;多斗式提升机;斗式提升机;斗式升运机;斗式升降机;斗式开运器;斗式升降机
bucket elevator belt 铲斗式升降机皮带
bucket elevator boom 铲斗式升降机吊杆
bucket elevator boot 斗式提升机底部料箱;斗链式提升机底部给料室
bucket elevator casing 斗式提升机壳体
bucket elevator chain 链斗式提升机;斗式提升机斗链
bucket elevator frame 斗式提升机机架
bucket elevator head 铲斗式升降机头
bucket elevator loader 铲斗式升降机荷载器
bucket elevator part 斗式升降机部件
bucket elevator pit 铲斗式升降机竖井;斗式提升机地坑
bucket elevator rubber belt 铲斗式升降机橡皮带
bucket elevator with central discharge 中间卸料的链斗式提升机
bucket elevator with centrifugal discharge 离心卸料斗式提升机
bucket elevator with double roller type chains 双辊链斗提升机
bucket elevator with return run 可反转斗式提升机
bucket excavator 多斗挖土机;斗链式挖掘机;斗链式挖掘机;斗臂挖掘机
bucket excavator chain 铲斗式挖土机链
bucket excavator for downward scraping 下挖式链斗挖土机;下挖式挖掘机;下挖式多斗挖土机;下挖式多斗挖掘机
bucket excavator for upward scraping 上挖式链斗挖土机;上挖式挖掘机
bucket excavator track 铲斗式挖土机轨道
bucket factor 满斗系数
bucket feed-discharge elevator 斗式装卸升运器
bucket feeder 斗式加料器
bucket ferris wheel 斗式料浆喂料机

bucket fill degree 满斗率
bucket fill factor 铲斗装满系数
bucket filling efficiency 斗内物料填充度
bucket flight 戽斗式斗架
bucket floor 挖土底板
bucket for demolition 拆除建筑物用铲斗
bucket for multipurpose 多功能铲斗
bucket for rock 装载石块铲斗
bucketful 满桶;一斗
bucket gate 底卸式斗闸门
bucket grab 抓石器;抓斗;斗式抓岩机
bucket guide 吊桶吊绳;吊桶导向装置
bucket handle 铲斗臂
bucket handle pointing 半圆凹槽形勾缝
bucket handling crab 铲斗移动起重小车
bucket hand-pump 铲斗式手泵;斗式手压泵
bucket hoist 吊桶提升机;单斗卷扬机
bucket hoist conveyer 斗式提升输送机
bucket hydraulic cylinder 铲斗液压缸(轮式铲运机)
bucket hydraulics 铲斗水力学
bucket inclined conveyer 斜斗输送机
bucket index 存储桶索引
bucket invert 戽斗反弧拱
bucket ladder 多斗挖掘机斗架;斗桥【疏】;斗架;挖土机斗架;挖泥船斗架
bucket ladder coal unloading device 链斗卸煤机
bucket-ladder dredge(r) 多斗式挖泥船;多斗式挖泥船;多斗式挖泥机;斗链式挖泥船
bucket-ladder excavator 多斗挖土机;链斗式挖掘机;开挖沟槽机;多斗臂挖掘机;斗式挖沟机;多斗式挖掘机
bucket ladder gantry 斗架吊架
bucket ladder winch 链斗绞车
bucket lanyard 吊桶绳
bucket latch cord 斗底活拴绳
bucket latrine 便坑厕所;斗式厕所;便桶
bucket-less jig 无提升机跳汰机
bucket lift 提斗提升(指加速)
bucket lift control lever 铲斗升降控制杠杆
bucket lifter 斗式提升机
bucket lift kick-out 铲斗升降开关
bucket lift lever 铲斗吊杆
bucket lift truck 斗式铲车
bucket line 斗链;铲斗链
bucket-line and hydraulic dredge 链斗式水力挖掘船
bucket line chain 铲斗升降链
bucket line dredge(r) 多斗式挖泥船;链斗式挖泥船
bucket line drive 铲斗链式传动
bucket line guide roller 斗链导辊
bucket link 链节板;斗链
bucket linkage 铲斗连接杆;铲斗悬挂铰链机构
bucket lip 溢洪道鼻坎;斗刃;斗口;铲斗刃口;铲斗前缘;铲斗刃;铲斗刃口;挑流鼻坎
bucket loader 铲斗式推土机;摄斗式挖土机;自动单斗装料机;链斗装载机;链斗装料机;斗式装载机;斗式装运机;斗式装料机;铲斗式装载机
bucket-loader excavator 吊桶装卸挖掘土机
bucket loader with conveyer 带运输机的斗式装载机
bucket locking piece 叶片锁块
bucket loss 叶片损失
bucket lower time 铲斗下降时间
bucket nose 桶嘴
bucket of a water wheel 水轮戽斗
bucket of dam 大坝反弧段
bucket of ogee 双弯曲线形铲斗
bucket on revolving wheel 转轮上的铲斗;斗轮
bucket operating cylinder 铲斗操纵油缸
bucket operating ram 铲斗操纵动力油缸
bucket outreach 铲斗伸出长度
bucket paternoster 斗链
bucket path 铲斗挖土机运行路线;戽运道路
bucket(payload) capacity 铲斗容量;铲斗有效负荷能力
bucket pin 泥斗销
bucket pitch 斗距
bucket plate apron conveyer 浅斗裙式输送机
bucket positioner 转斗定位器
bucket positioner assembly 铲斗定位器组(合)件
bucket positioner bellcrank 铲斗定位器曲柄
bucket positioner latch 铲斗定位器插销
bucket position indicator 铲斗位置显示器

bucket proofer 料斗检验台
bucket prover 间格斗式运输器
bucket pump 活塞式(汲油)泵;活塞式(抽水)泵;斗式提水机;斗式唧筒;斗式抽水机;手摇带阀活塞泵
bucket-pump dredge(r) 链吹挖泥船
bucket pump gun 戽斗式加油枪(加润滑油用)
bucket rabbling vat 斗式储浆池
bucket raise time 提升时间
bucket reach of dipper dredge(r) 戽斗挖泥船挖斗伸距
bucket rig 戽索;戽斗索具;吊罐索具
bucket ring 动叶环
bucket rock grab 铲斗式抓岩机
bucket rod 泵的活塞杆
bucket roller 戽内水辊;消力戽内(的)水辊
bucket running counter 泥斗运转计数器
bucket running speed 泥斗运转速度
bucket scale 自动戽斗定量秤
bucket scraper 将伐倒的树锯成圆木;斗式铲运机(车)
bucket seat 单人圆背低椅
bucket sheave 铲斗滑车轮;铲斗滑轮
bucket shop 投机交易所
bucket side 铲斗边板
bucket side edge 铲斗侧刃
bucket sink 落地水水槽
bucket size 斗容(量)
bucket spacing 斗距
bucket speed indicator 泥斗转速指示仪
bucket spindle 斗轴
bucket spindle stud and nut 斗轴柱螺栓及螺母
bucket steam trap 浮筒式回水阀;浮筒式汽水阀
bucket stick 铲斗臂
bucket stick sleeve 铲斗臂套筒
bucket suspension tackle 铲斗吊索滑车
bucket system 虹吸装置
bucket temperature 表面水温
bucket thermometer 吊杯式温度表;吊杯式水温表;水桶温度计
bucket tilt control lever 铲斗倾斜控制杠杆
bucket tilting force 斗子倾卸力
bucket-tipping car 翻斗式矿车
bucket tipping device 斗式提升机翻斗装置;(斗式提升机的)翻斗装置
bucket-tipping device 开斗装置;(挖土机的)翻斗装置
bucket-tipping machine (挖土机的)翻斗装置
bucket tooth 戽斗齿;斗齿;斗头齿
bucket transfer car 铲斗上移动小车;挖斗运输车
bucket transfer equipment 铲斗上传递装置;斗式运输机具
bucket trap 浮子式阻汽器;浮子式疏水器;浮子式凝汽阀
bucket trench digger 斗式挖沟机
bucket trencher 铲斗挖沟工人;斗式挖掘机;斗式挖沟机
bucket trenching machine 链斗式挖沟机;多斗挖沟机;斗式挖掘机;斗式挖沟机
bucket trip 铲斗倾斜装置;吊桶解扣
bucket trolley 抓斗起重小车
bucket type elevator 斗式提升机
bucket-type energy dissipator 戽斗式消能工
bucket-type grain elevator 斗式谷物升运器
bucket-type lifter 斗式提升机
bucket-type loader 铲斗式装载机
bucket-type privy 桶厕
bucket-type seat 铲斗式座位
bucket-type stationary elevator 斗式固定升运器
bucket-type wheel 水戽(斗叶);铲斗式透平
bucket unloader 斗式卸载机
bucket valve 活塞阀
bucket volume 斗容(量);挖斗容量
bucket wander 挖斗不正确的行程
bucket wheel 戽式链轮;斗轮;水斗轮轮;戽轮;杓轮
bucket wheel blower 叶轮鼓风机;鼓轮风箱
bucket wheel car loader 斗轮(式)装车机
bucket wheel cutter 斗轮铰刀
bucket wheel dredge(r) 斗轮(式)挖泥船;斗轮(式)挖掘机
bucket wheel excavator 戽头转轮挖掘机;戽头转轮挖土机;斗轮取料机;斗轮(式)挖掘机
bucket-wheel feeder 斗轮式喂料机
bucket wheel loader 斗轮加载器;斗轮链轮装货机;斗轮(式)装载机

bucket wheel machine 斗轮机
bucket wheel reclaimer 杓轮回收设备;轮斗式取料机;斗轮取料机
bucket wheel shovel 斗轮铲
bucket wheel stacker 斗轮堆料机
bucket wheel stacker-reclaimer 斗轮(式)堆取料机
bucket wheel suction dredge(r) 联斗式抽吸挖泥机;斗轮(式)挖泥船
bucket wheel trencher 斗轮(式)挖沟机
bucket wheel type agitator 多斗旋轮式搅拌机;联斗式搅拌机
bucket winding 吊桶提升
bucket wire 铲斗提升钢缆
bucket with side dump 侧倾式铲斗
buck-eye 翁格那木;橡树;七叶树
Buckeye sandstone 俄亥俄州砂岩(美国)
buck frame 粗木(门)框;芯框;装有凹槽的门框;背有凹槽的门框;有凹槽门框
buckhorn 仅留主枝的修剪
bucking 将伐倒的树锯成圆木;用船体反复冲击海水;冲冰航行;人工磨矿;顶撞
bucking bar 铆钉顶棒;打钉杆
bucking circuit 抵消电路
bucking coil 抵销线圈;反感应线圈;反感线圈;反磁线圈;补偿线圈;补偿绕组;去磁线圈
bucking coil loudspeaker 反作用线圈式扬声器;反线圈扬声器
bucking effect 抵消作用;反效应;反电动势效应
bucking electrodes 屏蔽电极
bucking failure 弯折破坏
bucking field 反向场
bucking hammer 地质锤;碎矿机
Buckingham's equation 布金汉方程
Buckingham's theorem 布金汉定理
bucking iron plate 研样板;碎矿板
bucking kier 煮
bucking ladder 梯形造材台
bucking-out system 抵消系统;补偿系统
bucking ram 冲击顶头修理工具
bucking saw 人工磨床锯;长椭圆形锯
bucking stress 弯折应力
bucking transformer 反作用变压器
bucking voltage 抵消电压;反作用电压;补偿电压
bucking winding 抵消绕组;补偿绕组;去磁绕组
bucklandite 黑帘石
buckle 螺丝扣;拉紧联结器;扣(板);锯扣;压瘪管(断气方法);折痕;带扣;边缘歪扭;起皱
buckle chain 活节链
buckle clamp (皮带等的)扣夹;卡夹
buckled 弯曲的
buckled frame member 弯曲的构件
buckled pipe 压瘪管(断气方法)
buckled plate 凹凸板;弯曲板;波纹板
buckled steel plate 压曲钢板
buckled track 胀轨;跑道【铁】
buckled zone 弯曲区
buckle fold 弯曲褶皱【地】;弯曲褶曲
buckle folding 纵弯褶皱;侧压褶;弯曲褶皱作用
buckle folding machine 栅栏式折页机
buckle insulator 茶台绝缘子
buckle latch 搭扣(安全钩)
buckle making machine 带扣制造机
buckle outward 向外翘曲
buckle pattern 翘曲图形;翘曲图式
buckeye plate 皱褶钢板;皱纹板;皱曲钢板;凹凸板;盆板
buckle plate press 扭曲板压机;拗曲板压机
buckle plate sheet piling 扭曲板桩;拗曲板桩
buckler 锚链孔盖;防水罩
buckler plate 锚链孔盖
buckle tear 穿孔撕裂
buckle-up 安全扣带
buckle wave length 皱折波长;翘曲波长
Buckley ga(u)ge 布克雷真空计
Buckley opener 布克雷式开棉机
buckling 压曲;纵向弯曲;纵弯曲;纹纹;皱曲;产生皱折;屈曲;翘曲;翘起;弯折;弯曲
buckling action 压屈作用;压曲应力;压曲作用
buckling amplitude 压曲幅度
buckling analysis 纵向弯曲强度计算;压曲强度分析
buckling and crushing tests 翘曲和压缩试验
buckling behavio(u)r 失稳状况;压曲特性
buckling breaking 压曲撕裂;压曲破坏
buckling case 压曲情况

buckling coefficient 压曲系数
buckling column fender 消能桩;曲柱式防撞桩
buckling condition 压曲条件;压曲状态
buckling configuration 压曲形状;压曲图形
buckling constant 曲率常数;弯折常数
buckling criterion 压曲标准
buckling deflection (受压弹簧的)挠折收缩量
buckling deformation 翘曲变形
buckling domain 屈曲域
buckling effect 压曲效应
buckling effective length 压曲计算长度
buckling factor 抗弯因数;扭曲因数;曲折因数;抗弯力系数;扭曲系数;压曲系数;曲折系数
buckling failure 压曲破坏
buckling force 弯曲力;压曲力;涨轨力
buckling formula 压曲公式
buckling height 压曲高度
buckling hypothesis 挠曲说
buckling index 失稳指数
buckling instability 压曲不稳定性;曲率不稳定性
buckling investigation 压曲试验
buckling length 抗弯长度;压曲长度
buckling limit 压曲限度;压曲强度极限
buckling load 临界纵向荷载;压曲临界荷载;压曲荷载;折断载荷;变形载荷;屈曲荷载
buckling loading 压曲临界荷载;压曲荷载
buckling load plate 压曲荷载板
buckling load strength 压曲荷载强度
buckling mode 压曲型;屈曲形式
buckling modulus 压曲模数;压曲模量
buckling of a vehicle 车辆下挠度
buckling of frame 机架弯曲
buckling of long column 长柱弯曲
buckling of plate 板的折曲;板的曲度
buckling of rail 涨轨
buckling of rod 杆的压曲;杆的屈曲
buckling of track 涨轨跑道
buckling of wall 墙身凹鼓
buckling pattern 翘曲图形
buckling point 压曲拐点;转折点;拐点
buckling pressure 压曲压力
buckling problem 压曲问题
buckling reinforcement 压曲铠装包装
buckling resistance 抗纵向弯曲力;抗弯阻力;抗弯性;抗弯能力;抗鼓出阻力;压曲力;压曲抗(阻)力;翘曲阻力;凸凹变形阻力
buckling risk 压曲危险
buckling rupture 压曲破裂;压曲破坏
buckling safety 压曲稳定性;压曲安全
buckling safety factor 稳定安全系数
buckling stability 压屈稳定(性);抗弯稳定性;压曲稳定性;屈服稳定性
buckling stiffness 压曲劲度
buckling strain 压曲变形;压曲应变;弯曲应变
buckling strength 抗纵向弯曲强度;翘曲强度;抗弯强度;抗弯能力;压曲强度;纵向弯曲强度;屈曲强度;弯屈强度
buckling stress 抗弯应力;扭曲应力;压曲应力;屈曲应力;压折应力;弯曲应力
buckling test 压曲试验;屈曲试验
buckling theory 压曲理论
buckling value 压曲临界值;压曲值
buckling vector 曲率矢量
Buckman apparauts 巴克曼仪器(量测溶液冰点和沸点)
buckmortar 推式研钵
buck opening 毛门洞
buck ore vessel 矿石船
buck-passer 推卸责任者
buck-passing 推卸责任
buck-plate 装配板;磨矿板;凹凸板
buck rack 集草耙
buckrake 搂草耙;集堆机;集草器
buckrake hay stacker 集草堆草机
buckram 硬衬布;厚麻布(装帧用)
bucksaw 框锯;架锯;木锯;弓锯;大木锯;架子锯
buck scraper 横板刮土机;刮土铲运机;弹板刮土机;铲形耙;斗式铲运机
buckshot 大钻粒;粗团块结构;熔岩粒
buckshot aggregate 大团粒
buckshot land 铁屑地
buckshot sand 粒砂;无棱角砂
buckshot soil 褐铁结核土
buckskin 鹿皮呢;被除去树皮的木头

buck stacker 堆垛机
buck stave 护炉钢架;夹炉板
buckstay 支柱;支撑;护炉钢架;拱边支柱;槽钢支架
buckstay heel 立柱柱脚
buckstone 不含金废石
buck-up 铆钉撑锤顶住铆钉头;拧紧(管接头)
buckwheat coal 大麦粒级无烟煤
buckwheat screen 无烟煤筛分机
bucky diaphragm 栅格光栏
bucladesine sodium 布拉地辛钠
buclosamide 丁氯柳胺
bucrane 牛头骨状(雕)饰(在古罗马爱奥尼和柯林斯柱式上的中楣内)
bucrane frieze 牛头骨状(雕)饰中楣;牛头骨状雕带
bucranium 牛头雕饰
bucranium frieze 牛头骨状雕带
bucranium mo(u)lding 牛头骨状(雕)饰
bucrilate 丁氰酯
bud 牙状物
Buda car 公铁两用车
Budan's theorem 布当定理
bud capital 柯林斯柱头
Budde effect 布德效应
Buddha 佛
Buddha pine 罗汉松
Buddha's hand citron 佛手壶
Buddha shrine 舍利(子)塔
buddha statue 佛像
Buddhism 佛教
Buddhist architecture 佛教建筑
Buddhist convent 庵
Buddhistic 佛教的
Buddhist monastery 佛寺
Buddhist nunnery 尼姑庵
Buddhist pagoda 佛塔
Buddhist scripture hall 经堂
Buddhist shrine 舍利塔
Buddhist statue 佛像
Buddhist temple 佛寺
budding ratio 发芽率
buddingtonite 水铵长石
buddle 淘汰盘;洗矿槽
buddle frame 固定淘汰盘
buddle jig 定筛跳汰机
buddle-work 淘选;淘汰盘选矿
buddling 淘选
Buddy 布迪式短壁截煤机
buddy diving 结伴潜水
buddy system for storage allocation 存储分配的伙伴系统
budger (油漆工的)绘波纹器具
budget 预算;支配
budget according to working drawing 施工图预算
budget account 预算账户;预算项目
budget activity 预算活动
budget allocation 预算拨款;预算分配;投资分摊;投资分配
budget allotment 预算分配额;分配预算
budget allowance 预备金
budget amendment 预算修正案
budget amount 预算总(金)额
budget and financial statement system 预算与决算制度
budget appropriation 预算拨款
budgetary 预算的
budgetary account 预算账户
budgetary accounting 预算会计
budgetary allowance 预算限额
budgetary balance sheet 预算平衡表
budgetary concept 预算概念
budget(ary) constraint 预算限制
budgetary control 预算控制(数)
budgetary cost 预算成本
budgetary cycle 预算循环
budgetary deficit 预算赤字
budgetary discipline 预算专业
budgetary effect 预算效果
budgetary engineer 预算工程师
budgetary equilibrium 预算平衡
budgetary estimate of capital construction 基本建设预算;基本建设概算
budgetary estimate of unit construction 单位工程概算
budgetary estimates and budget of a project 工程改善和预算
budgetary expenditures 预算支出
budgetary expenditures account 预算支出账户
budgetary forecast 预算预报
budgetary fund 预算内资金;预算资金
budgetary making 预算编制
budgetary outlay 预算支出
budgetary overhead 间接费用预算
budgetary payments 预算付款
budgetary performance 预算执行情况
budgetary planning 预算计划;预算编制
budgetary policy 预算政策
budgetary position 预算实况
budgetary price 预算价格
budgetary process 预算编制程序
budgetary receipts account 预算收入账户
budgetary reduction 预算削减
budgetary requirement 所需预算经费;预算需要
budgetary reserve 预算准备金
budgetary resource 预算资金;预算经费
budgetary revenue 预算收入
budgetary saving 预算结余;预算节余
budgetary security 预算审查
budgetary support 预算保证
budgetary surplus 预算结余
budgetary target 概算指标
budgetary value 预算值
budgetary view-point 预算观点
budget audit(ing) 预算核查;预算审查
budget authority 预算主管当局
budget authorization 预算权限
budget base 预算基数
budget bureau 预算局
budget calendar 预算一览表
budget chart 预算图表
budget classification 预算分类
budget committee 预算(编制)委员会
budget compilation 汇编预算
budget control 预算控制(数);预算管理
budget cycle 预算周期
budget data 预算数据
budget deferral 预算的延期
budget deficit 预算赤字
budget department 预算部门
budget director 预算主管人
budget dirigation 干预经济的预算政策
budget division 预算处
budget documents 预算文件;预算书
budgeted amount 预算金额
budgeted balance sheet 预算资产负债表
budgeted capacity 预算能力
budgeted capital 预算资本
budgeted cost 预算成本
budgeted expenses 预算内开支;预算费用
budgeted financial statement 预算财务报表
budgeted income statement 预计收益表
budgeted material price 材料预算价格
budgeted overhead 预算制造费用
budgeted performance 预计完成情况
budget enforcement 预算执行
budgeter 预算员;预算编制者;预算编制人
budget estimate 预算估计;概算
budget estimate for construction design 设计概算
budget estimate of cost 成本概算
budget estimates making 概算编制
budget estimates unit 概算单位
budget estimation cost 概算价值
budget execution 预算执行;预算实施
budget executive 预算执行部门
budget for annual expenditures 年度支出预算;岁出预算
budget for annual expenses 岁出预算
budget for annual receipts 岁入预算
budget for growth 发展预算
budget for pollution policy 污染治理(经费)预算
budget for revenues and expenditures 收支预算
budget gap 预算赤字
budget implementation 预算执行
budgeting 预计编制;编制预算
budgeting system 预算系统
budgeting technique 预算编制方法
budgeting to objective 目标预算法
budgeting to priorities 优先预算法
budget items 预算项目
budget law 预算法
budget layout 预算草案;预算编排;预算安排
budget ledger 预算分类账
budget level 预算标准
budget liberally and spend sparingly 宽打窄用
budget limitation 预算限度
budget making 编制预算
budget making for working design 施工图预算
budget management 预算管理
budget manual 预算指南
budget message 预算咨文;预算申请书
budget mortgage 预算抵押贷款(贷款人每月付税费和保险费)
budget needed for the completion of project 竣工决算
budget norm 预算定额
budget notice 预算通知
budget of central government 中央政府预算
budget of construction 工程建造预算
budget of construction drawing project 施工图预算
budget of construction drawing stage 施工图阶段预算
budget of expenditures 支出预算;费用预算
budget office 预算局
budget officer 预算员
budget of local government 地方预算
budget of production expenses 生产费用预算
budget of project 全部工程方案预算
budget of railway utilities unit 铁路事业单位预算
budget of undertaking expenditures 事业经费预算
budget of water demand and water supply 水量供需平衡
budget outlay 预算支出
budget performance 预算执行情况
budget period 预算期
budget plan 分期付款计划
budget planning 预算编制
budget practice 预算办法
budget presentation 预算编制格式
budget price 预算价格
budget procedure 预算汇总表;预算程序;预算编制程序
budget process 预算程序
budget program(me) 预算方案
budget proposal 预算建议书;预算草案;概算
budget provision 预算拨款
budget receipt 预算收入
budget report 预算报告
budget reserve 预算储备金
budget restraint 预算限制
budget restriction 预算限制
budget result table 决算表
budget retrench 预算紧缩
budget review 预算审查
budget revision 预算修正案
budget rollover 滚动预算
budget schedule 预算明细表
budget share 预算份额
budget sheet 预算表
budget source 预算资源
budget statement 预算说明书;预算书;预算申请书;预算表
budget summary 预算简表;预算汇总
budget surplus 预算盈余
budget system 预算制度
budget update 按当前情况修正预算
budget year 预算年度;估算年
budgust 截粉
Budleigh Salterton Beds 布德雷沙尔特顿层
buergerite 布铬电气石
Buerger's precession camera 伯格进动照相机
buetschliite 水碳钾钙石;三方碳钾钙石
bufeniode 丁苯碘胺
bufetolol 布非洛尔
bufexamac 丁苯羟酸
bufezolac 丁苯唑酸
buff 减振;米色;擦光轮;浅橘黄色;浅黄色;抛光轮
buffability 可磨面性
buffable 可磨面的
buffalo 船首舷墙;水陆两用拖拉机;水陆两用坦克
buffalo box 路缘阀门箱
buffalo grips 拉紧卡子(电线或绷绳)
buff brick 黄砖;浅黄砖

buff-burned pattern 抛光灼晕
buff-duff 高频无线电测向仪
buffedge 光边(板的)
buffed leather 磨面革
buffer 内缓冲器;缓冲物;缓冲器;缓冲剂;缓冲放大器;缓冲垫;驻退机;分配缓冲区;保险杆(汽车)
buffer action 缓冲作用;阻尼作用;隔离作用
buffer action of soil 土的缓冲作用
buffer address 缓冲器地址;缓冲地址
buffer address register 缓冲地址寄存器
buffer agent 缓冲剂
buffer allocation 缓冲区分配;缓冲器分配
buffer amplifier 缓冲放大器;隔离放大器
buffer area 缓冲区;缓冲地带
buffer assemblage 缓冲组合
buffer attribute 缓冲标志;缓冲特征
buffer bag 缓冲袋
buffer bar 缓冲杆
buffer battery 缓冲电池组
buffer beam 缓冲梁;缓冲杆
buffer belt 防护带
buffer bin 缓冲仓;中间储存仓
buffer blasting 缓冲爆破(法)
buffer block 缓冲块;车挡
buffer block deletion 缓冲区删除
buffer box 缓冲器座
buffer building 缓冲建筑物
buffer cap 减震垫圈
buffer capacitor 缓冲电容器
buffer capacity 缓冲容量;缓冲能力
buffer case 缓冲筒
buffer casing 缓冲器套筒
buffer cell 缓冲单元
buffer chain 缓冲链
buffer change valve 缓变阀
buffer chute 缓冲溜槽
buffer circuit 缓冲电路
buffer coke charge 层焦
buffer condenser 缓冲电容器
buffer control 缓冲控制
buffer control program(me) 缓冲控制程序
buffer control unit 缓冲控制装置;缓冲控制器;缓冲控制单元
buffer control word 缓冲控制字
buffer disc 缓冲盘
buffer drilling 减震钻进
buffer drum 缓冲磁鼓
buffer dynamo 减震发电机;缓冲发电机
buffered computer 缓冲存储计算机
buffered device 带缓冲的设备
bufferedge fiber 带缓冲层光纤
buffered input 带缓冲输入
buffered input/output 缓冲输入/输出;带缓冲器输入输出
buffered keyboard 开式键盘;缓冲键盘
buffered mode 缓冲存储方式
buffered solution 缓冲溶液;缓冲液
buffered station 带缓冲的站
buffered terminal 带缓冲的终端
buffer empty 缓冲区空
buffer error 缓冲错(误)
buffer flange 缓冲器座
buffer frame 缓冲架
buffer function 隔离作用
buffer fund 缓冲基金;平准资金;平准基金;调节资金
buffer gap 衬垫间隙
buffer gas 清管用气体
buffer green (工厂区和生活区的)防护绿地;隔离绿带
buffer green belt 缓冲绿带
buffer green(ground) 缓冲绿地
buffer green space 缓冲绿地
buffer height 减振器高度
buffer hopper 缓冲漏斗;中间仓
buffer index 缓冲指数
buffering 缓冲;中间转换
buffering action 缓冲作用
buffering capacity 缓冲容量;缓冲能力
buffering capacity in natural water bodies 天然水体缓冲能力
buffering control 缓冲控制
buffering curve 缓冲曲线
buffering device 缓冲装置

buffering effect 缓冲效应
buffering function 缓冲作用
buffering load 抖振荷载
buffering memory 缓冲存储器
buffering piston 缓冲器活塞
buffering pool 缓冲存储器
buffering power of soil 土壤缓冲能力
buffering solution 缓冲液
buffering storage 缓冲存储器
buffering store 缓冲存储器
buffering technique 缓冲技术;缓冲方法
buffering unit 缓冲装置
buffer input 缓冲输入
buffer input-output 缓冲输入输出
buffer intensity 缓冲强度
buffer intensity curve 缓冲强度曲线
buffer intensity of water 水的缓冲强度
buffer interface switch 缓冲器接口开关
buffer inverter 缓冲倒相器;隔离反相器
buffer layer 减震层;缓冲层
buffer length 缓冲区长度;缓冲器长度;缓冲长度
buffer material 缓冲材料
buffer module 缓冲组件;中间转换器
buffer of cable trolley wire 电缆小车钢丝绳缓冲装置
buffer offset 缓冲区位移量
buffer output 缓冲区输出
buffer pair 缓冲对偶
buffer plate 缓冲垫板;缓冲板;挡板
buffer plunger 缓冲柱塞;缓冲器活塞
buffer pool 缓冲池
buffer post 缓冲杆
buffer-precipitation test 缓冲沉淀试验
buffer rail 缓冲轨;缓冲窗;护墙栏(杆)
buffer ram 缓冲杆
buffer reagent 缓冲剂
buffer region 缓冲区;过渡区;过渡层
buffer register 缓冲寄存器
buffer resistance 缓冲电阻;阻尼电阻;放电电阻
buffer ring 阻尼器环
buffer rod 缓冲器杆;缓冲杆
buffer scheduling 缓冲区调度;缓冲调度
buffer section 缓冲段
buffer segment 缓冲区分段;缓冲段
buffer shell 缓冲器座
buffer shoe 减振器托板
buffer shooting 缓冲爆破(法)
buffer size constraint 缓冲区大小限制
buffer slip 缓冲滑片
buffer solution 缓冲溶液
buffer spindle 缓冲区活塞杆
buffer spring 阻尼弹簧;减震弹簧;缓冲器弹簧;缓冲弹簧
buffer stage 缓冲级;缓冲放大器
buffer station 缓冲工位
buffer stay 缓冲杆
buffer stem 缓冲柄
buffer stock 缓冲库存;缓冲存货;缓冲储存;缓冲储备;保险储备;安全储量;调节性库存储备;调节库存
buffer stock financing 调节性库存储备的资金供应
buffer stop 止冲器;车挡【铁】
buffer stop area 车挡长度
buffer-stop indicator 车挡标志;停车标志
buffer storage 缓冲存储(装置);缓冲仓库;中间存储(器);调节库存量;调节仓库
buffer storage buffer reservoir 缓冲库容
buffer storage computer 缓冲存储计算机
buffer storage unit 缓冲存储单元
buffer strip 缓冲地带;缓冲带
buffer stroke 缓冲器行程;缓冲器的行程
buffer support plate 缓冲器托板
buffer tackle 缓冲滑车组
buffer tank 缓冲池;缓冲油罐;缓冲罐;备用水箱
buffer tube 缓冲管
buffer unit 缓和装置;缓冲部件
buffer vessel 缓冲罐
buffer voltage 缓冲电压
buffer wagon 缓冲车
buffer washer 减震垫圈
buffer work area 缓冲工作区
buff(er) zone 缓冲地带
buffer zone 缓冲区;缓冲带;防护带
buffet 小卖部;小吃部;小餐室;圆角边缘;抖动;打击;餐具柜

buffet bar 打击杆;快餐酒吧
buffet car 餐车
buffeting 抖振翼尾颤振;颤振
buffeting of vane 叶片颤动
buffet kitchen 冷菜制作间
buffet lounge 快餐休息室
buffing 磨光;打光;擦光;软皮(布)抛光;抛光(屑)
buffing and draw gear 抛光拉伸装置
buffing attachment 抛光装置
buffing compound 磨光剂;抛光剂
buffing cone 抛光锥
buffing disc 抛光盘
buffing fabric 磨光织物
buffing gear 抛光装置
buffing head 抛光轮架
buffing lathe 磨光机;抛光机
buffing machine 磨光机;磨革机;软皮(布)抛光机;抛光机
buffing mark 磨痕
buffing of sinter metals 烧结金属抛光
buffing oil 磨光油;抛光油
buffing paper 磨光纸;磨光砂纸
buffing roll 擦光辊
buffing speed 抛光速度
buffing test 研磨试验
buffing wheel 摇摆的圆盘;抛光砂轮;抛光轮;弹性磨轮
buff leather 磨面绒革
buff on heading 墩头
buff-patched appearance 打磨不匀
buffs 磨光皮
buff standard brick 浅黄色标准砖
buff strap 对接贴板
buff titanium dioxide 浅橘黄色二氧化钛颜料
buff unit 抛光头;抛光动力头
buff wheel 擦光轮;磨光轮;软皮轮;软皮(布)抛光轮
buflomedil 丁咯地尔
buformin 丁双胍
bug 雷达位置测定器;错误;半自动发报键;(机器等的)缺陷;窃听器;双座小(型)汽车
bugas 瓶装液化(石油)气
bug check 调试检查
bug dust 岩粉(空气洗井钻进时);粉尘
bugduster 除粉器;除尘器;清除煤粉工
bugdusting 清除岩粉
bugeye 平底小船
bugeye lens 超广角镜头
buggy 料车;煤粉;混凝土手推车;小斗车;小车;二轮手推车;手推车
buggy away 用小车运走
buggy casting 小车浇铸
buggying 小货车;手推车
buggy ladle 钢包车;台车式泡包
buggy-loading hopper 装载小车的斗
buggy rock 多孔隙岩石
bug holes 晶穴;疵穴;小洞穴;麻面;蜂窝;气泡孔眼(混凝土)
bug holes of concrete 混凝土表面蜂窝
bugi hole (岩石中的)孔穴;晶洞
bugite 紫苏英闪岩
bug key 快速发报键;双向报键
bugle derrick (立根架在一侧的)钻塔
bugle factor 弯曲系数
bug monitor 错误检测程序
bug patch 勘误号;订正号;错误补块;错误标记码
Bugula neritina 总合草苔虫
buhl 镶嵌工艺品
buhl and counter 装饰细工
Buhler-Ming raw preheater 比勒—米亚格生料预热器
buhl saw 框架锯
buhl work 镶嵌饰物的家具
Buhrer kiln 毕列尔窑
buhrstone 磨盘;细砂质磨石;磨石(砂砾);油石;砂质多孔石灰岩
buhr(stone) mill 硅石磨盘;双盘石磨;盘磨机;石磨
buial hill zone 潜山带
buider 增洁剂
build 建筑;膜厚;造型;(圬工的)垂直缝;确立
buildability 可建性
buildable area 可建面积
buildable width 可建宽度
build a chapel 半边风掉抢

build address resolution table 内部地址分辨表
build a dike 筑堤
build and erection cost 建安工程造价
build and erection price 建安工程价格
build-down 减低;降落
builder 经营建筑业者;建筑工人;建造者;建造厂;建设者;修建人员;营造商;制造厂;次品砖;施工人员;施工单位
builder berth 滑道(船坞);造船台
builder bond 承包商保证单
builder cam 成形凸轮
builder commitment 建造单位承诺;建造者承担的义务
builder hand cart 施工用手推车
builder identification 建造商证明
builder iron supplier 建造五金器具
builder merchant 建材公司;建造材料商人
builder pick gear 成形撑牙
builder rough plank 建造毛板
builder rubbish 建造废料;建筑垃圾
builder's certificate 建造证书
builder's claw hammer head 建筑工羊角锤头
builder's diary 施工日志;施工日志;施工日记
builder's equipment 建筑施工机械设备;施工设备;施工机具
builder's finish(ing) hardware 建筑装修小五金;建筑五金
builder's fittings 建筑五金配件;建筑小五金
builder's fittings factory 建筑小五金工厂
builder's furniture 建筑器具;建筑五金配件;建筑施工装配器具;建筑施工五金配件
builder's glass fittings 建筑玻璃小五金配件
builder's hand cart 建筑用手推车
builder's handyman 零杂工;半熟练工;施工零杂工
builder's hardware 建筑五金;建筑小五金
builder's hoist 施工用卷扬机;施工卷扬机
builder's iron supplies 建筑五金器具
builder's jack 施工用挑架;施工千斤顶
builder's knot 绳索死结;死结;瓶结
builder's labo(u)r 建筑工人
builder's labo(u)rer 建筑杂工;小工(建筑施工)
builder's ladder 施工用踏步梯;木工梯;施工用扶梯
builder's level 气泡水准仪;定镜水准仪;简易水准仪;施工用水准仪;施工水准仪
builder's licence 建筑许可证;施工营业执照;施工许可证
builder's lift 建筑升降机;建筑工程升降机;施工升降机
builder's merchant 建筑材料商
builder's method 承包商法
builder's name plate 制造厂铭牌
builder's old measurement 旧吨位量法
builder's policy 造船保单
builder's risk insurance 工地财产保险;建筑工程财产保险;承建者风险保险;承包人财产安全保险
builder's risks 建造风险;船舶建造险
builder's road 施工人员便道;施工便道
builder's rough planks 建筑用毛板
builder's rubbish 建筑垃圾;建筑废料
builder's saw 建筑工人锯
builder's scaffold 用方木材构筑的脚手架;施工用脚手架
builder's sea trial 船厂试航
builder's square 施工角尺
builder's staging 顶升架;重型脚手架;施工台架;施工用脚手架
builder's tape 施工皮尺;施工用水准仪;施工卷尺
builders' temporary shed 工棚
builder's tonnage 原先吨位;设计吨位
builder's warranty 营造商的保证书
builder's winch 建筑用绞车;建筑绞车;施工卷扬机;施工绞车
builder's workshop 建筑工场;建筑车间
builder's yard 建筑场地;施工场地
builder treat 建筑处理
builder-upper 建立者
build-in 固有的;内装的;内部的;嵌入;嵌固
build in achievement indicator 规定完成的指标
buildiness 丰满度
building 建造;建设;营造(物);大厦;大楼;房屋(屋)
building, operation, maintenance and transfer 建设、运营、维修、移交
building accessory 建筑附件;附属构筑物;附属房

building account 建筑账(户)
building acoustic(al) measurement 建筑声学测量
building acoustics 建筑声学
building act 建筑法(规)
building activities and losses 建筑功能
building activity 建筑施工
building adaptable to extension 能延伸的建筑;能扩建的建筑
building adhesive 建筑胶粘剂;建筑用胶粘剂
building administration 建筑管理;建筑经营
building agreement 建筑协议;建筑协定
building alteration 建筑物改进;建筑物重建;建筑物改建;改建工程
building and construction industry 建筑与施工工业
building and contents 建筑物及其内部设施
building and equipment 房屋和设备
building and ground dimensions 建筑物及场地大小
building and loan agreement 建造和贷款协议(书)
building and loan association 建筑贷款公司;建宅互助金;建筑信用协会;房屋贷款协会
building and repair 建造
building annex 附设建筑物
building appliance 建筑设备;建筑用具;建筑机具
building application 建筑申请
building approval 建筑许可证;获得批准的建筑
building architecture 房屋建筑学
building area 基底面积;建筑基底面积;建筑占地面积;建筑面积
building area quota 面积定额;建筑面积定额
building article 建筑制品;建筑产品
building asbestos 建筑石棉
building aseismicity 建筑抗震
building asphalt 建筑石油沥青;建筑沥青
building authorities 建筑主管机关;建筑主管部门;建筑科;建筑部门
building authority 建筑权限;建筑主管人员;建筑工程局
building automatic system 车站设备控制系统
building automation 建筑自动化
building average zoning 建筑密度分区
building axis 房屋轴线
building axis survey 建筑轴线测设
building bar 建筑钢筋
building barracks (临时的)工棚施工
building basin 造船渠;造船船坞
building berth 滑道(船坞);造船台;船台;船架;造船位
building berth number 船台号码
building berth schedule 船台使用顺序计划
building bioclimate chart 建筑生物气候图
building blasting 房屋建筑爆破
building block 龙骨墩;结构单元;建筑砌块;积木式元件;积木式单元;积木块;装配用滑车;造船龙骨垫;构块;构件;船台墩垫;标准组件;墙壁砌块;砌块
building block counter 标准部件制成的计数器
building block design 模块化设计
building block factory 建筑砌块工厂;砌块工厂(混凝土);砌块厂(混凝土)
building block lintel 砌块过梁
building block masonry wall 砌块坯工墙
building block masonry(work) 砌块工程
building block module 砌块模数
building block module method 建筑砌块模数法;砌块模数法
building-block molecule 构件分子
building block plant 建筑砌块工厂;砌块工厂(混凝土);砌块厂(混凝土)
building block principle 积木构造原理;积木式原理;积木式结构原理
building block system 积木系统;积木式(系统);积木式结构方式
building block technique 模块技术
building block theory 积木理论
building block wall 建筑砌墙;砌块墙
building bloom 建筑全盛期
building board 建筑饰面板;建筑板材;墙板;匀质纤维板
building(board) facing 建筑衬板;建筑面板;建筑板饰面

building board for industrial construction 工业建筑用的建筑板
building(board) lining 建筑衬板;建筑板衬料
building bonding adhesive 建筑用黏结剂;建筑用黏合剂
building bonding agent 建筑用黏合剂
building boom 建筑吊杆
building boom cycles 建筑业景气周期
building brick 建筑砖;建筑用砖;黏土建筑砖;墙心砖;普通机砖
building brick clay 建筑砖用黏土
building buddle 累积式淘汰盘;斜面累积洗矿槽
building bulk 房屋容积;房屋体量;房屋的长、宽、高
building bulk zoning 建筑容积分区
building by-laws 建筑技术标准;建筑法规;建筑法规细则;建筑条例;建筑规则;建筑规程
building caisson 房屋基础沉箱;房屋基础沉井
building campaign 建筑竞争
building capital 建筑资本
building capitalization rate 建筑还原率
building carcass 建筑骨架
building cement 建筑水泥
building center 建筑中心
building certificate 施工凭证;建筑证(明)书
building characteristic 建筑特征
building chimney 房屋烟囱;生活用烟囱
building clearance 建筑限界
building climatology 建筑气候学
building clinker 建筑用烧结砖
building clock 大厦时钟
building club 建筑俱乐部
building code 建筑物规范;建筑规范;建筑法规;建筑法;房屋建筑规范
building code formula 建筑规范公式
building code requirements for reinforced concrete 钢筋混凝土建筑规范
building codes and standards 建筑法规与标准
building codes of practice 建筑执行法规
building coefficient 建筑系数
building coil 成形圈层
building column 建筑物支柱;建筑物支承;建筑物撑木
building combined drain 房屋合流下水道;房屋合流污水管(道);雨污水合流下水(管)道
building combined drain sewer 雨污水合流下水(管)道
building combined sewer 房屋合流下水管;房屋合流污水管(道);雨污水合流下水(管)道
building commissioner 建筑的行政管理人员
building company 建筑公司
building complex 建筑物综合体;建筑物组合件;建筑群;综合建筑群
building component (预制的)建筑单元;增效组分;建筑构件
building component design 建筑构件设计
building component manufactory 构件厂
building concerns 建筑业
building concession 建筑许可;建筑执照
building connection 建筑物连接管;通向建筑物的连接管
building construction 建筑施工;建筑结构;建筑构造(方式);房屋建筑(学);房屋构造
building construction activity 房屋建造业务
building construction administration 建筑施工管理
building construction and civil engineering 工民建及土木工程
building construction bureau 建筑工程局
building construction company 建筑工程公司
building construction completion certificate 房屋建筑竣工证明
building construction component 建筑工程构件
building construction contract 建筑工程承包合同
building construction cost 房屋建造造价
building construction department 建筑工程施工部门;建筑工程管理处;房屋建造部门
building construction drawing 房屋建造图纸
building construction engineer 房屋建筑工程师;建筑工程工程师
building construction expert 房屋建筑专家;建筑工程专家
building construction ground 建筑工程场地
building construction industry 建筑工程工业;房屋建造工业

building construction joint 建筑构造缝
building construction labo(u)r 房屋建造工人;建筑工程工人
building construction legislation 建筑工程立法
building construction lot 建筑工程基地;建筑工程用地;建筑工程地区
building construction material 房屋建造材料;建筑工程材料
building construction member 建筑工程构件;建筑工程(预制)构件
building construction method 建筑工程施工方法
building construction office 建筑工程事务所
building construction operation 房屋建造实施;建筑工程施工
building construction panel 房屋护墙板;建筑护墙板
building construction phase 建筑工程状态
building construction plan 房屋平面图;建筑平面图
building construction plastics 建筑工程塑料
building construction price 建筑工程价格
building construction price index 建筑工程价格指数
building construction procedure 建筑工程施工程序
building construction regulation 房屋建造章程;建筑工程规章;建筑工程法规;建筑工程条例
building construction research 建筑工程科学研究
building construction site 房屋建造场地;建筑工程现场;建筑工程工地
building construction site installation 房屋建造场地设施;建筑工程现场设置
building construction speed 建筑工程速度
building construction standard 建筑工程规范;建筑工程标准
building construction supervision 建筑工程施工检查;建筑工程监理
building construction system 建筑工程体系
building construction technician 房屋建造技术人员;建筑工程技术人员
building construction type 建筑工程类型
building construction with lifting method 升板施工法
building construction work 房屋建造工作;建筑工程工作
building construction worker 房屋建造工人;建筑工程工人
building contract 建筑合同;建筑承包合同
building contracting 建筑工程承包方式
building contractor 房屋营造商;建筑承包商;建造者;营造商;工程承包者;工程承包商;工程承包人;工程承包单位;承建人
building control 建造监理;建筑管理
building coordinate system 建筑坐标系(统)
building core 房屋核心;胶面轴心;混凝土心墙
building cost 建筑费用;建筑造价;建筑物造价;建造成本;造价;建筑总费用
building cost analysis 建筑造价分析
building cost calculator and valuation guide 建筑成本核算及估价指南
building cost estimate 建筑造价估价
building cost index 建筑费用指数;建筑造价指数
building cost of project 工程造价
building coverage 建筑系数;建筑面积率;建筑覆盖率;建筑覆盖度
building covered area 建筑占地面积
building cradle 支船架
building crane 房屋吊车;建筑用起重机;建筑起重机
building cycle 建筑周期
building damage 建筑物毁害
building dedication tablet 建筑物纪念提名匾
building deficiency 房屋短缺;建筑物缺陷
building demolition 建筑物拆除
building density 建筑物密度;建筑密度;建设用地密度;房屋密度
building department 房屋管理局;建筑工程局;建筑管理处;房屋管理处
building department cost 建筑部门费用;建筑部门成本
building depth 建筑进深
building design 建筑物设计;房屋设计
building design codes for seismic area 地震区建筑设计规范
building design competition 建筑方案
building destroying 建筑物破坏

building diagnosis 建筑诊所
building dimension 建筑结构尺寸
building disaster prevention 建筑防灾
building division 分户墙
building dock 造船坞;造船船坞
building document 建筑施工文件
building drain 室内排水管;建筑排水管;房屋排水管;排出管
building drainage 房屋排水
building drainage pipe 房屋排水管
building drainage system 建筑排水系统;房屋排水系统;室内排水系统
building drain system 房屋排水系统
building drawing 建筑制图;建筑图纸
building economic index 建筑经济指标
building economic norm 建筑经济定额
building economy 建筑经济
building efficiency 可出租面积比
building efficiency ratio 建筑有效面积比
building element 建筑物件;建筑构件;建筑单元;建筑部件
building elevation 建筑立面
building enclosure 建筑物围护结构;围护结构
building engineer 建筑工程师
building engineering 建筑工程(学)
building engineering school 建筑工程学校
building engineering survey 建筑工程测量
building enterprise 建筑企业
building entrance 建筑物入口
building entrance door 建筑物入口门
building entrance plastic door 建筑物入口塑料门
building envelope 建筑物维护结构;建筑工程区域地段;围护结构;建筑工地外围建筑
building equipment 建筑用设备;建筑设备;房屋设备;房屋建筑设备(指公用服务设施、机器设备等)
building erection system 房屋架设系统;建筑安装体系;建筑安装方法
building estate 建筑地产
building evacuation 建筑物拆除
building exit 房屋出口
building expansion 房屋扩建
building expenses 建筑费用;房屋费用
building expert 建筑专家
building expert system 建立专家系统
building extension 建筑扩展;建筑扩建
building extrusion product 房屋上突出物
building fabric 建筑结构
building face 房屋外貌
building facing tile 建筑面砖
building failure 房屋损坏;建筑破坏;建筑物损坏
building feature 建筑特征
building fiberboard 建筑纤维板
building field 施工现场;建筑工地
building finishing 建筑装修
building fire 装配用火;建筑火灾;建筑物火灾
building firm 建筑商行;建筑公司
building fit 房屋装配;建筑装配;建筑物装配
building floor space 建筑使用面积;建筑楼面面积
building footing 建筑物基础
building for equipment 设备用房
building for livestock 畜生(饲养)房
building for management 管理用房
building for port auxiliary operation 港口辅助生产建筑物
building foundation 房屋基础;建筑基础;建筑物基础;房基
building frame 建筑框架;采腊巢框;房屋框架;房屋构架
building fresh water barrier 建造淡水屏障
building fund reserve 建筑储备基金;建筑基金准备
building funds 建筑资金
building gable 建筑山墙;房屋山墙
building gas pipe 室内燃气管道
building geomorphology 构件地貌学
building glass 建筑玻璃
building goods 建筑制品;建筑产品
building grade 建筑物等级;房屋地面标高;设计标高
building granite 建筑用花岗岩
building graver 建筑物刻图器
building gravity drainage system 建筑物的重力排水系统;房屋重力排水系统
building ground 建筑基地;建筑工地
building ground elevation 地坪标高;室内地坪标高;室内地面标高
building group 建筑群
building guards 建筑物保护设施
building guide 建筑工程刻图规
building heating 房屋采暖;建筑物加热;建筑物供暖
building heating entry 热力入口
building heating installation 建筑物加热设备;建筑物供暖设备
building heating system 建筑物供热系统
building height 建筑高度
building height in stor(e)ys 建筑物的层数高度
building height zoning 建筑高度分区(规划)
building house drain 房屋雨水管;房屋废水总管
building house sewer 房屋污水管;房屋废水总管
building house storm sewer 房屋雨水总管
building house tax 建筑税
building implement 建筑设备;建筑用具;建筑用工具
building impost 建筑拱墩
building in 预埋构件;堵塞;添埋构件;施工期中的建筑设备;嵌固;固接
building industrialization 建筑工业化
building industry 建筑业;建筑工业
building-in fitting 预埋件
building information center 房屋信息中心;建筑信息中心
building-in furniture 镶壁家具
building-in-general 一般建筑物
building in landscape 园林建筑
building in series 连续生产
building inspection 建筑监理;建筑物验收;建筑检查;房屋检查
building inspector 建筑监理师;房屋监工员;建筑监察员;建筑检查员;房屋检查员
building installation works 建筑物安装工程
building insulant 建筑隔热隔声材料;建筑绝缘材料
building insulating article 建筑隔热制品;建筑隔声制品;建筑绝缘制品;绝缘产品
building insulating brick 建筑保温砖;建筑隔热砖;绝缘砖
building insulating felt 建筑保温毡
building insulating foil 建筑隔热薄片;建筑隔声薄片;建筑绝缘金属薄片
building insulating material 建筑绝缘材料
building insulating paper 建筑保温纸;绝缘纸
building insulating product 建筑隔热制品;建筑隔声制品;建筑绝缘制品;绝缘产品
building insulating sheet 房屋绝缘板;房屋隔热板;房屋隔声板;建筑绝缘板;建筑绝热板;建筑隔音板
building insulating slab 建筑隔声板;建筑绝缘板;建筑绝热板;建筑隔音板
building insulating unit 建筑隔热制品;建筑隔声制品;建筑绝缘制品
building insulating wool 建筑保温毛织品
building insulation 建筑隔热材料;建筑绝缘;建筑隔垫材料
building insulator 建筑隔离剂;建筑绝缘体
building insurance 建筑物保险;建筑保险;房屋保险
building interior 建筑物内部
building interior decoration 建筑物内部装饰
building investment 建筑投资
building iron 建筑钢
building jig 建筑夹具(用于砌砖的钢夹具)
building joinery 门与窗的建筑;建筑细木工
building labo(u)r(er) 建筑工人;建筑劳动力
building land 房屋地产;建设用地;建筑地
building land development 房地产开发
building land purchase 房地产购买
building law 建筑条例;建筑法(规)
building layout 建筑布局
building lease 建筑租约;建筑物租约;租地造屋合同
building ledger 建筑物分类账
building legislation 建筑立法
building lien 建筑扣押留置权
building lime 建筑用石灰;建筑石灰
building limit 建筑范围
building line 房屋的正面宽度;街区线;建筑线;建筑界线;建筑红线;红线;房屋界线;房屋建筑界线;房基线;施工线
building line platform 房屋建筑线;建筑平台线
building line setback 建筑沿街后退线
building line survey 建筑红线测量
building load 房屋荷载;建筑物荷载

building loan 建筑贷款
building loan agreement 房屋贷款协议；建筑贷款合同
building loan interest 建设贷款利息
building log 建筑材
building lot 地皮；建筑用地；建筑基地；宅地；房建用地
building lot coverage 建筑密度
building lots for sale 分段出售建筑用地
building low-rise 低层建筑
building machine 建筑机械；配套机
building machinery 建筑机械；土建机械
building machinery plant 建筑机械厂
building main 建筑总水管；引进建筑物的给水干管；房屋给水总管
building maintenance 建筑保养；房屋保养；房屋维修；建筑物维修
building maintenance expenses 建筑物维修费；建筑物维护费；房屋维修费
building maintenance project 楼房维修计划
building masonry wall 建筑圬工墙；建筑物墙体
building mass 建筑群；房屋总体积
building mastic 填缝料
building mat 建筑保护层
building material condition 建筑材料条件
building material consumption norm 建筑材料消耗定额
building material dealer 建筑材料商；建筑材料行
building material deposit 建筑材料存储
building material distribution 建筑材料运输
building material distributor 建筑材料销售人
building material engineer 建筑材料工程师
building material failure 建筑材料损坏
building material field area 料场面积
building material from solid wastes 固体废物建筑材料
building material industry 建筑材料工业；建材工业
building material industry wastewater 建筑材料工业废水
building material machine 建筑材料机械
building material manufacturer 建筑材料生产者；建筑材料生产厂（商）
building material market 建筑材料市场
building material plant 建筑材料厂
building material practice 建筑材料工艺；建筑材料业务
building material processing 建筑材料处理；建筑材料加工；建材加工；建材处理
building material producer 建筑材料生产者；建筑材料生产厂（商）
building material production 建筑材料制品；建筑材料制造；建筑材料生产
building material quality 建筑材料质量
building material quality control 建筑材料质量控制；建筑材料质量管理
building material requirement 建筑材料需要量；建筑材料要求规格
building materials 建筑材料；建材
building material saving 保存建筑材料；节约建筑材料；建筑材料节约
building material scale 建筑材料等级；建筑材料度量
building materials for civilian use 民用建筑材料
building materials from industrial solid wastes 工业废渣建筑材料
building material standards 建筑材料标准
building material store 建筑材料仓库；建筑材料储〔贮〕藏库；建筑材料商店
building material test(ing) 建筑材料试验；建筑材料检验
building material testing device 建筑材料试验设备
building material testing institute 建筑材料试验协会；建筑材料试验室
building material testing machine 建筑材料试验机
building member 建筑杆件；建筑构件；构件；房屋单元
building membrane 建筑用膜材
building metal 建筑金属材料
building method 建筑工方法；合成法
building model 建筑模型
building module 建筑模型；建筑模数；积木式模块
building modulus 建筑模数
building mortar 建筑砂浆；建筑灰浆
building motion 建筑物运动；成形装置；成形运动

building number 房屋号数；房屋门牌号码
building occupation permit 建筑占用许可；房屋使用许可证
building of barracks 营房建筑；临时工棚施工
building of civic 民用建筑
building office 建筑事务所
building official 建筑行政管理人员；建筑规范执行人员；工程检验人员；房屋检查员
building of high-rise 高层建筑
building of historic interests 有历史意义的建筑
building of industrial 工业建筑
building of large blocks 大型砌块建筑；大型砌块施工
building of large slabs 大型板材建筑；大板建筑
building of large tiles 大型砖块施工；大型砌块施工
building of poor materials and workmanship 低劣房屋
building of radial plan 辐射形平等建筑
building of refinery units 炼油厂建筑
building of reinforced concrete construction 钢筋混凝土结构
building of residential 居住建筑
building of skeleton construction 骨架构造建筑；骨架房屋；构架房屋
building of slag 造渣
building of staggered design 交错排列设计的房屋；交错排列设计的建筑
building of steel frame construction 钢框架房屋
building of wood(en) construction 木结构建筑
building operation 房屋管理；建筑作业；建筑物经营；营造；建筑施工
building ordinance 建筑条例；建筑法令
building orientation 建筑物定位方位；建筑物朝向；建筑朝向；朝向
building orientation coordinate network 建筑坐标网
building oscillation 建筑振动；建筑物振荡
building-out 补偿值
building out cable 补偿电缆
building out capacitor 附加电容器
building-out circuit 附加电路；匹配电路
building-out condenser 附加冷凝器；附加电容器
building outline 建筑物轮廓线；建筑轮廓线
building-out network 附加平衡网络；附加网络
building owner 房所有权人；房主；营造业主；业主
Building Owners and Managers Association International 国际房主与经理协会
Building Owners and Managers Association International standard 国际房主与经理协会标准
building paint 房屋油漆；建筑油漆；建筑物油漆；建筑物涂料
building panel 建筑物护墙板
building paper 油纸；油毛毡；隔音纸；隔声纸；防潮纸；保温纸
building part 建筑部分；建筑物细部；建筑物分部
building particulars and plans 建筑细目和平面图；建筑施工文件
building party 造标组
building pattern 构筑模式
building performance 建筑物性能
building perimeter 建筑周边线
building permit 建设许可证；施工执照；建筑许可证；建筑工程规划许可证
building permit application 建筑执照
building permit for construction 建筑施工执照
building photogrammetry 建筑摄影测量
building photography 建筑摄影
building physics 建筑物理（学）
building pipe 建筑管材
building pit 建筑物基坑
building pit blasting 基坑爆破
building pit closed by sheet pile 四周用板桩围住的基坑
building pit closed by sheet piling 房屋用板桩围住
building pit construction 房屋坑的建造；基坑施工
building pit drainage 基坑排水
building pit pump 房屋坑泵；基坑排水用泵
building pit sheeting work 建筑基坑挡板
building pit side 房屋坑边
building pit slope 房屋坑的斜坡
building pit water 房屋的坑水；基坑水
building plan 建筑平面图；建筑规划
building plane 建筑平面（图）

building planning 建筑规划；建筑施工计划
building planning regulations 建筑物设计条例
building plant scale 建厂规模
building plaster 建筑粉刷灰泥；建筑墁灰；建筑石膏（粉饰）
building plastic 房屋上用的塑料树脂；建筑塑料；建筑用塑料
building plastic film 建筑塑料薄膜
building plastic for external use 室外用建筑塑料
building plastic for internal use 室内用建筑塑料
building plastic sheeting 建筑塑料薄膜
building plot 建设地点；建筑用地；房屋分区图；房屋地区图
building plumbering 建筑水下水工程；建筑装置工程；装管工程
building plumbing system 建筑给水排水
building population 建筑物密度；建筑物内人口
building practice 建筑实践
building preservative 建筑物防蚀剂；建筑物保护料；建筑防腐剂
building pressure 建筑物压力
building price 建筑物价格
building price index 建筑物价格指数
building principal 主要建筑物
building principle 建筑原则；建筑法规
building process 建筑施工过程
building production 建筑生产
building products 建筑创作；建筑制品；建筑产品
building program(me) 建筑施工程序；建筑方案
building project 建筑工程计划；建筑工程（方案）；土建工程；建筑项目
building property 房产
building property business 房产经营
building property mortgage 房产抵押
building property registration 房产登记
building property right 房产权
building property title certificate 房产证书
building proposal 建筑计划；建筑申请；建筑投标
building quicklime 建筑用生石灰；建筑石灰
building quilt 被状隔热填料；隔板；垫板
building quota 施工定额
building rate 建筑费率
building reconstruction 房屋重建
building regulation 建筑规则；建筑规程
building reinforcing material 建筑用钢筋
building removal 房屋拆迁
building repair 建筑物修缮；建筑物修理；建筑物修补；建筑物维修
building repair cost 建筑物修理费
building research 建筑科学研究
building research academy 建筑科学研究院
building research establishment 建筑研究所
Building Research Institute 房屋研究所；建筑研究所；建筑研究院；房屋建筑研究所
building restriction 建筑限制；建筑限物条例
building restriction line 建筑限定线；建筑红线；建筑限制线；建筑范围
building ring 工作环；胎圈
building roof 屋面；屋顶
building roofer 建筑屋面工（人）；屋面工
building room 装配间
building rubber compound 橡胶组合
building rubbish 建筑垃圾；建筑废料
building rubble 建筑碎砖瓦砾；建筑垃圾
building ruins 建筑废墟
building rule 建筑条例
building safety inspection 房屋安全调查
building sand 建筑用砂；砌筑砂浆用砂
buildings and community systems 房屋及社区系统
building sand from natural sources 天然产地的建筑用砂
building sanitary drain 建筑生活污水管道；生活污水排泄
building sanitary sewer 房屋生活污水管道；房屋生活污水
building science 建筑科学
buildings collapse 建筑物倒塌
buildings cracking 建筑物开裂
buildings depreciation reserve 建筑折旧储备
building sealant 建筑用密封剂；建筑密封膏；建筑密封膏
building season 建筑施工季节
building section 建筑区段；施工区段

building sequence 建筑顺序;建筑次序
building service chute 服务性滑槽;房屋服务管道
building service department 房产服务部门
building service equipment 建筑服务设施
building service expenses 房屋服务费用
building services 建筑机械设备;房屋设备;建筑设施;建筑服务设施
building service system 建筑辅助系统
building setback line 建筑收进线;建筑后退线
building setback restriction 建筑后退线规定
building settlement 建筑物陷落;建筑物沉降
building settlement observation 建筑物沉降观测
building sewer 房屋污水管(房屋外墙一米处至公共水道的一般管道);室内污水管
buildings for port operation 港口生产建筑物
building shape 建筑形式;建筑形状
building sheet 建筑钢板;建筑板材
building sheet facing 建筑薄村板
building sheet for industrial construction 工业建筑用的建筑薄板
building sheet lining 建筑薄衬板;建筑衬板
building sheet surfacing 建筑薄衬板
building shell 建筑薄壳;建筑外壳
building side 建筑物的侧面
building siding 建筑物护墙板
building silhouette 建筑轮廓线(条)
buildings inclination 建筑物倾斜
building site 建筑基地面积;建筑用地;建筑物现场;建筑工地;建筑地址;建筑地点;建筑场址;工地
building site freight elevator 工地运货升降机
building site front 建筑基地沿街线
building site goods lift 工地运货升降机
building site installation 建筑场地设备;建筑工地装置;工地设备
building site latrine 建筑工地厕所;工地厕所
building site rent 建筑土地地租
building size 建筑体形
building skeleton 建筑骨骼骨架;房屋构架;房屋框架;建筑构架;建筑骨架;建筑框架
building slab facing 建筑衬板
building slab lining 建筑衬板
building slab(sur)facing 建筑衬板
building slip 造船台;造船滑台;船台
building slipway 建造船只;建筑施工滑道
building slip(way) crane 建筑船台吊车;码头起重机;岸壁起重机
building slip(way) winch 建筑船台卷扬机
building society 房产及信贷协会;住宅建筑合作社;建筑社团;建房互助协会;房屋建筑会
building space 房屋间距;建筑物空间;建筑空间
building spacing 房屋间距
building span 建筑跨度
building speed 建筑施工速度
building square 建筑物前集散广场;建筑广场
building square grid 建筑方格网
building square grid axle method 建筑方格网轴线
building square grid survey by method of control network 建筑方格网布网法
building stair(case) 房屋楼梯;建筑物楼梯
building standard law 建筑标准法
building standards 建筑标准
building status 建筑用地性质
building steel 建筑钢
building stock 建筑群
building stone 建筑石料;建筑石材;充填石料;石料
building-stone quarry 建筑材料开采场;建筑石料开采场
building storm drain 房屋暴雨排水管;建筑暴雨排水管;房屋雨水管(道)
building storm drainage 房屋雨水管(道)
building storm drain pipe 房屋的暴雨排水管
building storm sewer 房屋雨水沟管;房屋雨水管(道);室内雨水管
building(structural) system 建筑结构体系
building structure 大楼;房屋;构筑物;建筑物;建筑结构;房屋结构;房屋构造
building style 建筑风格
building subdrain(pipe) 房屋暗沟;房屋地下排水管
building sub-house drain 房屋地下排水系统
building subsidence 建筑物沉陷
building subsidence observation 建筑物沉降观测
building substructure 房屋地下结构;建筑地下结构;基础(基线以下结构)
building subsystem 建筑辅助系统;房屋附属设备

系统
building supervision 建筑管理
building supply 建筑物给水
building-supply water system 房屋供水系统
building support 房屋支柱
building survey(ing) 建筑测量
building surveyor 建筑勘测员;建筑检查员;房屋勘测员;房屋检查员
building symbol 建筑物符号
building system 房屋整体布置;建筑整体布置;工业化建筑;建筑系统;建筑体制;建筑体系
building tax 建筑税;房产税
building team 建筑施工队
building technical standard 建筑技术标准
building technician 房屋技术员;建筑技术员
building technique 建筑技术
building technology 建筑术语汇集;建筑工艺学;建筑技术
building the nation with science and technology 科技兴国
building tile 空心砖;屋瓦
building tile factory 建筑用瓦厂;建筑用制砖厂;制瓦厂;制砖厂;砌块厂(混凝土)
building tile lintel 建筑物砖过梁;砌块过梁
building tile(masonry) wall 砌块圬工墙;建筑圬工墙;建筑砖墙
building tile masonry(work) 建筑铺瓦工;砌块工程
building tile plant 建筑用瓦厂;建筑用制砖厂;制瓦厂;制砖厂;砌块厂(混凝土)
building timber 建筑用木料;建筑用木材;建筑木料;建筑木材
building timber grade 建筑用木材等级
building time 建设周期
building time on the stocks 船台周期
building tools 建筑器械;建筑用具
building tower 房屋塔楼
building trades 建筑业;建筑行业;建筑工种;建筑工程行业
building traffic 房屋场地交通;建筑场地交通;工地交通
building trap 房屋排水管防臭瓣;房屋存水弯
building type 建筑型式;建筑形式;建筑物类型;建筑类型
building under construction 建筑中的房屋
building unit 建筑构件;(预制装配的)建筑单元
building-up 建造;建立;装配;建成;安装
building-up curve 建起曲线;增长曲线
building-up process 合成过程
building-up property 提升性
building-up reaction 合成反应;叠合反应
building-up roofing 砂砾油毛毡屋面
building usable volume 建筑物可用空间
building use 建筑物用途
building user 使用单位
building using party 使用单位
building value 建筑造价
building vibration 房屋振动;建筑振动;建筑物的振动
building volume 房屋体积
building wall 房屋墙壁
building ware 建筑制品
building water-supply system 房屋供水系统
building width 面阔
building winch 房屋用卷扬机;建筑用卷扬机;建筑用绞车
building wiring system 建筑物布线系统
building with central space 集中式建筑;中间留空地的建筑
building with dwellings 设有住宅的建筑;公寓住宅
building with mechanical core 具有力学墙心的建筑
building with one wing 有一个侧厅的建筑
building with service core 设有服务中心的建筑
building worker 建筑工人
building works 房屋建造;建筑建造;建筑工程;建筑物
building yard 建筑场地;营造厂;造船厂
building zone 建筑用地;建筑范围
building zone map 建筑物区划图
build-in hear box 嵌入齿轮轮箱
build-in items in masonry 砖石工程中的嵌入件
build-in language 固有语言
build-in type suspended carrier 淹没式悬浮载体
build itself-up 自激(发)

build joint 构造缝;勾缝
build labo(u)rer 建筑工人
build lease 租地造物权
build number 建造号【船】
build-operate-transfer 建设、营运、转让
build-out resistor 匹配电阻
buildress 女施工员
build road-bed 铺道床
build the pressure 形成压力
build-up 连焊作用;加速蔓延;岩隆;累积作用;结圈;结皮;积累
build up a joint 连接起来
build up budget 编制预算
build up curve 压力恢复曲线
build-up effect 聚集效应;积累效应
build-up electrode 堆焊焊条
build-up factor 积累因子;复合系数
build-up flow 道口尾续车流;绿灯尾续车流
build-up in the(bore)hole 钻孔堵塞;埋钻
build-up lag 增长滞后
build-up member 装配(构)件;装配(部)件;拼装(构)件
build-up mica 人造云母
build-up of administration 建立管理机构
build-up of cash 现金滞留
build-up of dust 尘末聚积
build-up of fluid 液面升高;液面恢复
build-up of material 材料堆积;物料结块;物料积聚
build-up of pollutants 污染物的累积
build-up of pressure 建立(造成)压力;压力升(高);压力积聚;压力恢复;造成压力
build-up of pulse 脉冲增长
build-up of stress 应力增大
build-up of water production 出水量增加
build-up pressure 增大压力;恢复压力
build-up pressure survey 测恢复压力
build-up radiation 累积辐射
build-up sequence 熔敷顺序;焊道熔敷顺序
build-up the metal 熔接金属
build-up time of frequency stabilization 稳定频率建立时间
build-up tolerance 累积公差;装配公差
build-virtual-machine program(me) 编制虚拟计算机程序
built 装配的;搭架
built, operate and trasnfer 建造、运营、移交
built, subsidize, operate and transfer 建造、补贴、运营、移交
built arch 组合拱
built arch theory 固定拱圈理论
built area 建成区面积
built block 组合滑车
built channel 组合槽钢
built detergent 复配洗涤剂
built drainage system 建成的排水系统
built environment 建成环境;城市环境
built-for purpose tool 专用工具
built frame 组合构架
built H-column 组成工形柱
built-in 机内;内装式;装在机内的;固有的;安装在内部的;嵌入;嵌固的
built-in adapter 内部适配器
built-in airstair(case) 随机客梯
built-in antenna 机内天线
built-in antisplash 内装式防溅板;嵌入式防溅板
built-in appliance 嵌入式煤气灶
built-in arch 嵌入式发券;固端拱;嵌入式拱;嵌固拱
built-in arch theory 固定拱(圈)理论
built-in ashtray 内装式灰盆;嵌入式灰盆
built-in attribute 内部属性
built-in bathtub 镶入式浴盆;嵌入式浴盆
built-in battery 可充电电池;固定电池
built-in beam 木落砖固端梁;固端梁;嵌入梁;内部梁;固定梁;嵌固梁
built-in beam or slab 嵌入梁或板;固定梁或板
built-in bias 内建偏压
built-in bottom type gritter 装入式铺砂器
built-in breathing system 自带呼吸系统
built-in calibrator 机内校准器
built-in carbon dioxide fire extinguishing system 固定式二氧化碳灭火系统
built-in casing 嵌入式门套
built-in cavity 管内空腔谐振器
built-in center line 砌入路面中央分道线

built-in channel 内建管道
built-in check 内含校验;内部校验;内部检验;自动校验;固定校验
built-in checking 固定检查
built-in clutch 装入传动器
built-in colo(u)r changing system 内装颜色转换系统
built-in comfort 嵌入式生活设施;嵌入式生活设备
built-in comfort in building 建筑物的砌入式生活设备
built-in command 内务命令;内部指令
built-in computer 内装式计算机
built-in control 内部控制
built-in crutch 内装支柱;内装式支柱;内装式支承件;内装式支架(环)
built-in cupboard 壁橱;嵌入柜
built-in division logic 内部除法逻辑
built-in edge 固定边(缘);嵌入边(缘);嵌固边缘
built-in element 固定的部件
built-in end 固定端;嵌入端;嵌固端
built-in-end beam 砌端梁
built-in error 埋置误差
built-in error checking 内部错误检查
built-in error correction 内部误差校正;内部错误校正
built-in export department 附设出口部
built-in exposure meter 内装曝光表
built-in fail-safe behavio(u)r 内置故障防止效能
built-in field 内建场
built-in fitment 固定装设;固定装备;固定件
built-in fitting 埋设件;埋入件;镶入器件;预埋件(管子或电线等);固定件
built-in flexibility 内在伸缩性
built-in flow circuit 内装流路
built-in frame 砌入式窗框
built-in function 库函数;内在功能;内建功能;内部函数;内部功能;内部操作
built-in furniture 镶嵌家具;嵌入式家具;镶壁家具
built-in garage 楼房建有存车库;屋内车库
built-in gas fire 嵌墙式煤气炉
built-in gearbox 嵌入式齿轮箱
built-in girder 嵌入的大梁;固定端大梁(两端固定)
builting-up time 增长时间
built-in gutter 暗落水管;内落水管;暗水管
built-in hanger 预埋吊件
built-in homojunction 内均质结
built-in hotplate 嵌入式板式灶
built-in hydraulic jacks 装入的液压千斤顶
built-in items 固定装置设施;镶入设施
built-in jack 固定起重器
built-in journal 嵌入轴颈
built-in kitchen 内部厨房
built-in kitchen cabinet 嵌入式橱柜
built-in lamp 墙壁内灯;墙内灯;墙顶内灯;嵌墙灯;嵌入灯
built-in length 伸入长度
built-in light 固定照明
built-in lighting 嵌墙照明设施;隐蔽式照明;砌入照明;嵌入(式)照明设施;嵌入式照明(设备)
built-in lock 暗锁
built-in measuring mark 内装测标
built-in meter 内装曝光表
built-in microphone 机内话筒
built-in microscope 内装显微镜
built-in motor 机内电动机
built-in motor drive 单独内装电机传动;单独电动机驱动;单独电动机传动
built-in motor pocker 装入马达搅拌杆
built-in multiplication logic 内乘法逻辑
built-in nailing block 嵌墙木砖;木落砖
built-in obsolescence 内在陈旧性
built-in oscillation 固有振荡;固定振荡
built-in parts 镶入配件
built-in place pile 就地填筑桩
built-in problem 本身存在的问题;存在问题
built-in producer 附设发生炉
built-in pump 内装式泵
built-in radiator 墙内散热器;嵌墙散热器
built-in reactivity 剩余反应性
built-in reliability 结构可靠性
built-in sash 嵌固窗扇
built-in sections 由部件装配的;分段建造
built-in sharpener 嵌入式磨石装置;嵌入式的磨刃装置
built-in sheave 嵌装滑轮
built-in shower 内部小淋浴间
built-in shower stall 嵌入式小淋浴间
built-in software delay 内部软件延迟
built-in stabilization 内在稳定作用
built-in stabilizer 内在稳定器
built-in steps 建在河道中的梯级
built-in storage 内储存器
built-in strain 内部应变
built-in stress 残余应力
built-in subroutine 内部子程序
built-in support 嵌入端固定支座
built-in system 嵌入式系统
built-in test 机内测试
built-in test equipment 机内(自动)测试设备
built-in thermocouple 埋入式热电偶
built-in tool 机内工具
built-in tracing structure 内部跟踪结构
built-in tub 固定浴盆
built-in unit 生产线上设备;固定设备
built-in voltage 内建电压
built-in wall 嵌在墙上
built-in wardrobe 嵌入式衣橱
built-in welfare 固有福利
built-in wetting 内润湿剂
built-in wooden brick 预埋木砖
built kitchen cabinet 内部厨房
built mast 复接杆
built-on 加高的;添建的
built-on pump (与发动机安装在一起的)内置泵;固接式泵
built-on-the-job 现场制作装配
built pile 组合桩
built platform 堆积台地
built rib 组成肋材;组合肋
built shoe 组合座脚
built spar 组合焊柱;拼制木柱
built structure 建成结构
built terrace 浪积阶地;冲积阶地;冲击阶地;波成阶地
built to order 定制
built-to-scale 按比例制造的
built-up 组合(的);装配的
built-up arch 组合拱
built-up architrave 组合门窗框缘;组合框缘;组合门窗头线条板
built-up area 城市建成区;建筑完成面积;建筑区;建筑面积;建成区;建成街区;组合面积;吹填区;施工区
built-up asphalt ballast 铺装式沥青道床
built-up asphalt roof covering 组合沥青屋面覆盖层
built(-up) beam 组合梁;拼装梁;拼接木梁;叠合梁
built-up board 组合板(材)
built-up boat 组合船;分段建造的船舶;拼装式船舶
built-up bridge 组合体系桥;装配式桥(梁)
built-up broach 组合拉刀
built-up casing 组合式气箱
built-up channel 组合槽钢
built-up character 建成性质;建成特性
built-up coil 加重线圈;加重带卷;组合带卷
built-up column 组合柱
built-up common crossing with both rails curved 双曲轨组合式普通辙叉
built-up common crossing with both rails straight 双直轨组合式普通辙叉
built-up compression member 组合压力杆;组合压杆
built-up concrete 外表结皮的混凝土
built-up connection 组合连接
built-up construction 组合结构
built-up crank 组合式曲柄;组合曲柄
built-up crank axle 组合曲柄心棒
built-up crank shaft 组合曲柄轴
built-up crossing 组合辙叉;组合式辙叉;钢轨组合辙叉
built-up cross-section 组合横断面
built-up culvert block 组合涵洞砌块
built-up cutter 组合铣刀
built-up cutter suction dredge(r) 组合式绞吸挖泥船
built-up disk rotor 组合轮转子
built-up district 建成区
built-up dredge(r) 组装式挖泥船;组合式挖泥船
built-up edge 切屑瘤;切屑卷边;刀瘤
built-up flange 组合突缘
built-up flask 组合砂箱
built-up flat roof 卷材平屋面
built-up form 组合模板
built-up foundation 组装基础
built-up fraction 并排的分数式
built-up frame 组合肋骨;组合构架
built-up ga(u)ge 组合量规
built-up gear 组合齿轮
built-up girder 组合大梁
built-up glued beam 胶合梁
built-up gun 筒紧炮
built-up hatch board 组合舱盖板
built-up H-column 组合工字柱
built-up jig 组合夹具
built-up joint 组合接头
built-up laminated wood 胶合板
built-up land 吹填地;填筑区
built-up lattice 组合格构
built-up litter system 厚褥草法
built-up magnet 复合磁体
built-up mandrel 膨胀心轴
built-up mast 组合桅杆
built-up member 组合构件;组合杆(件);装配件
built-up membrane 组合薄膜;装配薄膜;复合油毡
built-up method 组合法
built-up mica 组合云母
built-up model 船舶组合模型
built-up mortar 外表结皮的砂浆
built-up mo(u)ld 组合压铸模
built-up mo(u)lding box 组合砂箱
built-up mullion 组合中梃
built-up obtuse crossing 组合式钝角辙叉
built-up pallet 组装式提升机
built-up pattern 组合空心模架
built-up pile 组合桩;组合柱
built-up pillar 组合柱
built-up piston 组合式活塞;组合活塞
built-up plate 组合模板;固定板;双面模板
built-up plate girder 组合板大梁;组合板梁
built-up pressure 升高的压力
built-up propeller 组合螺旋桨
built-up purlin(e) 组合檩条
built-up retaining wall 拼装式挡土墙
built-up rib 组合肋;叠合肋;木叠合肋
built-up rigid frame 组合刚性构架
built-up rim 组合轮缘
built-up roof covering 组合屋面层
built-up roof(ing) 铺组合屋面;组合屋面材料;组合屋面(层);柔性防水屋面;叠层油毡屋面
built-up roofing with coping 设有墙帽的组合屋顶
built-up rotor 圆盘电枢
built-up rudder frame 组合舵架
built-up section 组合截面;组合部分;组合式结构
built-up segmental arch 组合圆弧拱;组合弓形拱
built-up sequence 组合顺序;组合程序
built-up stair horse 组合楼梯斜梁
built-up steel section 组合型钢(截面)
built-up string 组合弧形木楼梯梁;组合楼梯斜梁
built-up terrace 堆积阶地
built-up timber 组合木构件;组合木材;拼接木料
built-up truss 组合桁架
built-up type scaffolding 装配式脚手架
built-up urban area 城市建成区
built-up voltage 增压电压
built-up welding 镶焊;堆焊;补焊
built-up wheel 组合车轮
built volume index 建筑体积指数
buke 小球;球管;球形物;简单房屋
Bukhara basin 布哈拉盆地
Bukipulp 布基浆粕
bukocite 硒铊铁铜矿
bukovkyite 羟种铁矾
bulachite 水羟砷铝石
bulb 抹灰凸包;小球;小电珠;电灯泡;灯泡;包壳;球状物;球缘;球形物;球形零件;球头;球茎;球管;气瓶;烧瓶
bulb angle 球头角钢;圆趾角钢;球缘角钢
bulb angle-bar 圆头角料;圆趾角钢;球缘角钢
bulb angle iron 圆头角铁;球缘角钢;球头角钢
bulb angle section 圆形角材;圆头角材
bulb angle steel 球头角钢
bulb bar 圆头铁条;圆肋钢筋;球缘扁铁;球缘扁钢

bulb beam 圆缘梁;球头工字钢梁
bulb bit 冲头
bulb-blowing machine 灯泡吹制机;吹球机;吹管泡机;吹玻壳机;玻壳吹制机
bulb bow 球形船首;球鼻船首
bulb-bowed ship 球鼻艏船
bulb edge 胖边;圆头形;圆头钢(平板玻璃磨成的)圆边;卷边子;厚圆边;边子;干板玻璃(其边缘为碾压成型)
bulb edge severing device 切边装置
bulbend cast in-place pile 护底桩
bulb flat steel 圆角扁钢
bulb for endoscope 内窥镜灯泡
bulb for spectacle glass 眼镜片玻璃球
bulb front plating 球鼻首前端板
bulb glacier 球鼻冰川;扇形冰川尾
bulb holder 灯头;灯座
bulb hydraulic generator 灯泡式水轮发电机
bulb iron 圆头铁条;球头角钢;圆趾铁;圆头铁;球缘铁
bulb neck crack-off machine 灯泡壳颈切割机
bulb-neck splicing machine 管颈连接机
bulb nose 扩散头部
bulboeapnine 波尔埃皮宁
bulb of percussion 打击形成半锥体
bulb of pressure 应力泡;压力泡;膨胀压力;土压力泡
bulb of the eye 眼球
bulbous 隆起的;球状的
bulb(ous) bow 球鼻艏;球形船头;球鼻船首
bulbous bow frame 球鼻首肋骨
bulbous dome 球形屋顶;球形穹顶
bulbous hull 球鼻(首船体)
bulbous pile 球形爆扩桩;球形状柱
bulb(ous) stem 球鼻艏;球形船首;球鼻船首;球形艇
bulb pile 爆扩桩;圆趾桩;扩端桩;葱头桩;爆破桩;球形扩脚桩;球基桩;球根桩
bulb planter 球根栽植器
bulb plate 球缘钢板;球头扁钢
bulb potential 玻壳电位
bulb pressure 泡状压力线;应力集中
bulb rail 圆头丁字钢;球头轨
bulb rail steel 圆头丁字钢;球头丁字钢
bulb rudder 整流罩舵
bulb sealing machine 封灯泡机
bulb section 圆形断面;圆头型材
bulb separation 分球
bulb-shaped 球状的
bulb-shaped base 扩展桩基;扩大底部的建筑物;基底扩大
bulb socket with Edison cap 螺口灯泡
bulb socket with swan cap 卡口灯座
bulb steel 圆头钢条;球扁钢
bulb stem 球缘首柱;球式首柱
bulb stern 球形船尾
bulb tee 圆头丁字铁;圆头丁字钢;圆筒丁字钢;球缘T形材;轨形截面
bulb tee patent glazing bar 圆头T形玻璃窗心条;圆头T形玻璃格条
bulb tee puttyless glazing bar 圆头T形不用油灰的玻璃格条
bulb thermomter 球管温度计
bulb trailer 泥浆挤出器
bulb tube 球形凸管;球管
bulb tubular turbine 灯泡型贯流式水轮机
bulb tubulating machine 接管机
bulb turbine 灯泡式水轮机
bulb-type charger 管式充电器
bulb type hydraulic generating set 灯泡式水力发电机组
bulb-type turbogenerator 灯泡型发电机
bulb unit 灯泡式机组
bulbus pili 毛球
bulb water turbine 灯泡式水轮机
bulb with bayonet cap 卡口灯泡
bulb with Edison cap 螺口灯泡
bulb with goliath screw cap 螺旋式灯泡
bulb with swan cap 卡口灯泡
buldozerite 拆建派;推倒重建派
Bulgarian architecture 保加利亚建筑
bulgarite 球状碱安岩
bulge 价格的急骤上涨;核球;胀形;鼓胀;鼓起;钢板凸曲;船腹;船侧凸出部分;隆起;凸起;凸片;凸出部分;桶腰
bulge arc 凸出弧
bulge clearance 膨胀间隙
bulge coefficient 胀形系数
bulged blade 凸出叶片
bulged finish 胀口;凸口
bulged in 压入的
bulged ride 器壁凸出
bulged side 器壁凸出(玻璃制品缺陷)
bulged tube 膨胀管;凸起管
bulge in 凸进
bulge nucleation 隆丘成核
bulge of the loop 回线凸出程度
bulge out 凸出
bulge stress 拱凸应力
bulge test 扩管试验;打压试验(焊接);胀管试验
bulge X-ray source 核球X射线源
bulging 胀形;鼓胀;鼓起;撑压内形法;凸起;凸出(部)
bulging and deflection 凸鼓变形
bulging die 胀形模;膨胀模;撑压模
bulging effect 起拱效应
bulging force 膨胀力
bulging in 使挤入;使压入
bulging in valley 谷底肿胀(冻土区)
bulging lathe 旋压车床
bulging of cell 钢板桩格型鼓胀;钢板桩格体鼓胀
bulging of tire 轮胎凸胀
bulging outward 外鼓
bulging shell 鼓出外板
bulging test 撑压内形试验
bulging wall 凸肚墙
bulgram 黑黏土夹层
bulk 基本部分;胀量;大块;大部分;船舱载货;松厚度;散装
bulk acoustic(al) wave delay line 体声波延迟线
bulk acoustic(al) wave transducer 体声换能器
bulk-active load bearing system 大体积活荷载体系
bulk additive 填充剂
bulkage 膨胀;涨方
bulk analysis 总分析;全分析;整体分析
bulk and general cargo loading 散装杂货运载
bulk arrival 大批到达;大量到达
bulk article 大量制品;大量生产的制品;大量生产产品
bulk asphaltic bitumen distributor 大容量沥青喷洒车
bulk assembly 大部件装配
bulk-bag unloading station 散装袋卸料机
bulk batching 按体积配料(法)
bulk bauxite carrier 散装矾土船
bulk-behavio(u)r region 大块性质区
bulk bin 集装箱;集装罐;散装储存箱;散装储存仓;散粒储[贮]存箱;散粒储存箱
bulk blasting 大爆破
bulk boat 散货船;散装船
bulk body 散粒物车厢
bulk buying 大量购买;大量采购
bulk capacity 散装容积;散装舱容;散货容积
bulk cargo 零担散货;大宗散货;大宗货物;散装物料;散货(物);散货仓库;散货
bulk cargo berth 散货码头
bulk cargo carrier 散装货船
bulk cargo car unloader 散货卸车机
bulk cargo clause 散装货条款
bulk cargo container 散装货集装箱
bulk cargo conveyer 散装货运送机
bulk cargo handling equipment 笨重货物装卸设备
bulk cargo hold 散货舱
bulk cargo loader 散货装车机
bulk cargo port 散装货物港;散货港(口)
bulk cargo ship 散装货船;散货船
bulk cargo ship unloader 散货卸船机
bulk cargo terminal 散货码头;散货货码头
bulk cargo transfer equipment 散货转载设备
bulk cargo transfer machine 散货转载机
bulk cargo transport 散货运输
bulk cargo unloader 散货卸载机
bulk cargo unloading crane 散货卸船起重机
bulk cargo unloading device 散货卸船设备
bulk cargo unloading facility 散货卸船设备
bulk cargo wharf 散装货物码头;散货码头
bulk cargo yard 散货堆场
bulk/car/ore ship 散货/汽车/矿石船
bulk carrier 散装运输装置;散装运输船;散装货运输工具;散料转运车;散料运输车;散货运输船;散货船
bulk carrier fleet 散货船队
bulk carrier ore strengthened 矿石加强式散货船
bulk carrier vehicle 散装载运汽车
bulk/car ship 散货/汽车运输船
bulk caving 大崩落
bulk cement 散装水泥
bulk cement barge 散装水泥驳船
bulk cement boat 散装水泥专用船
bulk cement car discharged by gravity 重力卸料式散装水泥车
bulk cement carrier 散装水泥输送装置
bulk cement loading station 散装水泥站
bulk cement lorry 散装水泥卡车;散装水泥货车
bulk cement silo 散装水泥库
bulk cement supply station 散装水泥中转站
bulk cement terminal 散装水泥中转站
bulk cement trailer unit 散装水泥拖车单元
bulk cement transporter 散装水泥运送机;散装水泥输送机
bulk cement truck 散装水泥卡车
bulk cement vehicle 散装水泥运载车
bulk ceramic fiber 散状陶瓷纤维
bulk charge 整体电荷
bulk charge transfer device 体电荷转移器件
bulk chartering 散装租船
bulkcheap 薄利多销
bulk checking 非开箱大宗验货
bulk chemical code 散装化学品规则;散装化学品法规
bulk chemical fertilizer 散装化肥
bulk chemical tanker 散装化学品船
bulk cleaner flo(a)tation 精矿全部再浮选
bulk client 主要用户
bulk coal carrier 散装煤船
bulk coefficient 容积系数;体积系数
bulk coining 体积精压
bulk column 球形扩脚柱
bulk commodity 大宗商品;大宗散货;大宗货物
bulk compliance 体积柔量
bulk composition 内部组成;总成分;体积组成;松散混合料成分
bulk compound 散装混合料;瓶装喷压密封膏
bulk compressibility 毛体积压缩系数;体积压缩性
bulk concentrate 整体精矿
bulk concentration 体积浓度
bulk concrete 大体积混凝土
bulk concrete structure 大体积混凝土建筑物;大体积混凝土结构
bulk conductivity 体积电导率;体电导率
bulk consumer 大量消费者
bulk container 散装容器;散装集装箱;散装货箱;散粒物容器;散货集装箱
bulk conveyer 散装输送
bulk cooler 散装物物冷却器
bulk core memory 大容量磁芯存储器
bulk cracking 块状裂化
bulk cross-section 总截面
bulk crushing 粗碎
bulk crystal 体晶
bulk crystallization 大量结晶
bulk cubic(al) yard (以立方码计的)爆破岩石松散体积;(以立方码计的)爆破松散体积;松散立方码
bulk cumoil carrier 散货和石油两用船
bulk cumore carrier 散货和矿石两用船
bulk data transfer 成批数据传送
bulk defect 体缺陷
bulk delivery 散装输送;散装发货
bulk density 毛体积密度;堆密度;堆积密度;堆积比重;单位体积重量;容重;容积密度;体密度;体积密度;松散密度;松散密度;散货密度(积载系数);散比重
bulk density of formation 地层体积密度
bulk density of grinding media 研磨体容重
bulk density of soil 土的容重
bulk density of wet clay 湿黏土的密度
bulk density test 毛体积密度试验
bulk deposition 大量沉积
bulk-differential flo(a)tation flowsheet 混合—优先浮选流程图
bulk diffusion 体积扩散

bulk distributor 松散材料撒布机；散货平舱机
bulk district 按建筑容积分区指定的地区
bulk dolomite carrier 散装白云石船
bulk doped infrared detector 体掺杂红外探测器
bulk drying curves 散装干燥曲线
bulk-dry specific gravity 松堆密度
bulk earthmoving 大量运土；大规模运土
bulked down 实体积的
bulked feed 体积大的饲料
bulked glass yarn 玻璃纤维膨体纱
bulked yarn 膨体纱；膨体变形纱
bulk effect 体效应；体效率；体积效应；体负阻效应
bulk effect device 体效应器件
bulk elasticity 体积弹性
bulk elastic modulus 体积弹性模量
bulk electroluminescence 体电致发光
bulk energy facility 大型能量设备
bulk envelope 限定外包轮廓；(建筑物的)可能体量
bulker 附有计量设备的水泥储箱；散装水泥运输车；吨位丈量员；舱货容量检查人；散货船；水泥堆料车
bulk eraser 整体消磁器
bulker discharge hatch 散货卸货门
bulker roof hatch 散货卸货门
bulk excavation 大面积挖土；大面积开挖；大量开挖；大规模挖土；大规模开挖
bulk factor 紧缩率；压缩因数；压缩率；容积因素；容积压缩因素；体积因子；体积因素；体积系数；体积比
bulk factoring 批量保付代理
bulk factor of mo(u)lding compound 模塑料压缩比
bulk feed 散装饲料
bulk fertilizer 散装化肥
bulk fill 松填土；松填方
bulk filler 增量剂
bulk filling 大量填土；大规模回填
bulk film 整装胶片
bulk filtration 外部过滤；散装过滤
bulk fishmeal carrier 散装鱼粉船
bulk flo(a)tation 混合浮选；全浮选
bulk flow 总体流动
bulk fluid phase 体液相
bulk fluid process 总体流动过程
bulk forming 整体成型；体积成形
bulk foundation 大块基础
bulk freight 粗松货物；散装货船；散货
bulk freighter 散货转运车；散货船
bulk freight port 散货港(口)
bulk freight transport 散货运输
bulk freight unloader 散货卸载机
bulk fuel oil carrier 散装燃油船；散装燃油运输船
bulk gasoline plant 汽油站
bulkgead 闷头
bulk getter 块状消气剂
bulk glass 灯泡玻璃
bulk goods 大量货物；大宗散货；大批货物；散装货物；散货(船)；散堆装货物
bulk goods freight station 大宗货运站
bulk goods port 散货港(口)
bulk grain 散装粮谷；散装谷物；散粮
bulk grain feeder 散装谷物灌补舱
bulk gritter body 大尺寸的铺砂机机身
bulk gypsum 散装石膏
bulk gypsum carrier 散装石膏船
bulk handling 大量货物搬运；大量货物装卸；散装装卸；散装运输；散装装卸；散装搬运
bulk handling equipment 散货装卸设备
bulk handling machine 松散材料输送机；散装物输送机
bulk handling machinery 散货装卸机械
bulk handling plant 散货装卸设备
bulk handling unit 散货装卸设备
bulk hauling 散装拖运
bulkhead 楼梯斜顶；楼梯间斜平顶；码头驳岸；护岸墙；拱顶锁砖；隔框；隔舱；隔壁；(浇灌混凝土时用的)堵头板；堵头；水闸门；堤岸；挡土板；挡墙；橱窗下玻璃下矮墙；舱壁；封端板；驳岸
bulkhead adapter 舱壁管道接头
bulkhead air grill(e) 栅状空气挡板；挡风格窗
bulkhead boundary bar 舱壁底边角材
bulkhead building 挡土墙；堤岸；岸壁
bulkhead clock 舱壁船钟

bulkhead coaming 舱壁围板
bulkhead connector 穿板式连接器
bulkhead deck 舱壁甲板【船】
bulkhead deck at side 船侧舱壁甲板
bulkhead dock 驳岸
bulkhead door 舱壁门
bulkhead draught 舱壁通风
bulkheaded 嵌入的
bulkhead flange 舱壁凸缘
bulkhead floor(plate) 舱壁肋板
bulkhead foundation 岸壁基础
bulkhead framing 舱壁构架
bulkhead gate 检修闸门；挡水闸门；平板闸门；堵水平板闸门
bulkhead lamp 舱壁灯
bulkhead line 港口护岸坡脚线；码头前沿线；堤岸线；挡水岸线；舱壁衬条；驳岸线；岸壁线；实体码头前沿线；实体岸壁建筑界线
bulkhead lock 隔舱闸(有两金属隔壁洞道的闸)；隔壁闸
bulkhead method 分层浇筑混凝土法；分层浇注混凝土法；分层浇灌混凝土法
bulkhead packing gland 舱壁压盖
bulkhead panel 舱壁镶板；舱壁板条
bulkhead piece 舱壁连接管
bulkhead plan 舱壁结构图
bulkhead plank of timber pile bent abutment 木桩排架桥台挡土板
bulkhead quay(wall) 堤岸码头；岸壁(式)码头
bulkhead recess 舱壁龛；舱壁凹入处
bulkhead slot 堵水闸门门槽
bulkhead sluice(door) 舱壁闸门；舱壁闸阀
bulkhead stiffener 舱壁防挠材
bulk head taxiway 一头不通的滑行道
bulkhead tee 长臂丁字尺
bulkhead valve 舱壁阀门
bulkhead wall 驳岸；码头驳岸；岸壁(码头)；挡土墙；挡水墙
bulkhead web 舱壁垂直桁
bulkhead wharf 堤岸式码头；堤岸码头；岸壁型码头；岸壁(式)码头；散货码头
bulk heat treatment 整体热处理
bulk hydrated lime 散装消石灰
bulk image 粗加工像片
bulk impregnated specific gravity 吸油后毛体积比重(集料)
bulk-in 散装(货)袋卸系统
bulk-in bag-out 散装(货)袋卸系统
bulk inclusion 大块夹杂物
bulk index 体积指数
bulkiness 膨松度；体容度
bulking 成块作用；按体积计量；体胀；湿胀
bulking agent 增量剂；膨胀剂；填充剂
bulking and foaming activated sludge plant 膨胀发泡活性污泥处理设备
bulking curve 湿胀曲线(砂的体积)；容积增大曲线；含水砂子体积变化曲线
bulking effect 膨胀效应；湿胀性；湿胀效应
bulking factor 搅松系数；湿胀系数；湿胀率；湿涨率；干砂与含水砂体积比系数
bulking figure 容重
bulking filler 疏松填料
bulking intensity 膨胀度
bulking liquor 补给液
bulking machine 膨化机
bulking of sand 砂的体胀；砂的湿胀
bulking phenomenon 湿胀现象
bulking pigment 填充性颜料
bulking power 膨体变形能力；膨松性
bulking property 松散性
bulking sludge 膨胀污泥
bulking thickness 堆积厚度
bulking value 重量体积；堆积值；膨化度值；蓬松度值
bulk installation 总体安装
bulk insulation 体积绝热
bulk intrinsic germanium 块状本征锗
bulk items list 散装货物清单
bulk joint 凸缝
bulk ladle 大铸勺
Bulkley pressure viscosimeter 包克莱压力黏度计(润滑脂用)
bulk lifetime 本体寿命
bulk lime spreader 松散石灰喷洒器；石灰撒布机
bulk limestone carrier 散装石灰石船

bulk line 吃水线；散装界线
bulk liquid 散装液体
bulk liquid cargo 液散货；散装液体货物
bulk liquid carrier 散装液体船
bulk liquid container 液体货物集装箱
bulkload 毛载；大体积货物
bulkload chassis 散装物运输底盘
bulkload compressor 装料用空气压缩机；松散材料压缩机
bulk loader 水泥散装设备；散装装料器；散装物料装载机；散粒物料装载机
bulk loading 毛载；散装(装车)；散装入船
bulk loading of cement 水泥散货散装；散装水泥
bulk loading station 散装装料站
bulk lorry 栏板卡车；散装物载重汽车
bulk mass 大体积；大块体
bulk material 粒状材料；基体材料；统装材料；松散材料；疏松材料；散装料；散装材料
bulk material absorption 块料吸收
bulk material grab 松散材料抓斗；松散材料抓扬机
bulk material in powder form 粉状大散材料
bulk material loader 松散材料搬运设备；松散材料装载机
bulk material loading plant 松散材料搬运机
bulk material scale 松散材料等级
bulk material scattering 块料散射
bulk material terminal 散装货码头
bulk material transfer equipment 松散材料转运设备
bulk material unloader 松散材料卸货器
bulk matter 大块物质
bulk memory 档案存储；大容量
bulk memory device 档案资料存储器；大容量存储器设备
bulkmeter 膨松度试验仪；容积计
bulk method 大量生产法；成品生产法；成批生产法；全巷法；批量生产法
bulk milk collection lorry 收集牛奶用奶罐车
bulk mixing equipment 散料混合设备
bulk modulus 体积模量；体积弹性模数
bulk modulus of compressibility 体积压缩模量
bulk modulus of elasticity 体积弹性模数；体积弹性模量
bulk mo(u)lding compound 块状模塑料
bulk movement 整体运动
bulk mud 粉装泥土
bulk nickel carrier 散装镍矿船
bulk noise 体噪声
bulk of building 房屋体积；房屋面积
bulk of molecule 分子的大小
bulk of reservoir rock 储油层厚度
bulk of the electrolyte 电解质主体
bulk oil 油库
bulk-oil breaker 富油开关
bulk-oil carrier 散货石油(运输)船
bulk-oil circuit-breaker 多油(式)断路器
bulk-oil flo(a)tation 油团浮选；多油浮选；全油浮选
bulk-oil storage 散装储油
bulk-oil terminal 运油车的终点站
bulk optic(al) plastics 块料光学塑料
bulk palletizer 大体积包装机
bulk parcel 零担散货；小批散货
bulk partition coefficient 总分配系数
bulk permeability 体磁导率
bulk phase 主体相
bulk phosphate carrier 散装磷酸盐船
bulk photoconductor 大块光电导体
bulk pile 扩底桩；大头桩
bulk piling 实积
bulk plant 油库；批发油库
bulk plastic flow 体积塑性流动
bulk polymer 本体聚合物
bulk polymerization 本体聚合(法)
bulk porosity 体积气孔率；体积孔隙率；松装孔隙率
bulk potential 体电势
bulk powder 散装炸药
bulk printer 大型印像机
bulk processed 成批处理的
bulk processing 大量处理；粗处理
bulk process(ing) imaginary 粗加工像片
bulk processing tank 大型冲洗池
bulk product 松散产品；散装材料
bulk production 大量生产
bulk property 厚层性质；体特性；主体性质；整体性

质;整体特征;大块性质;材料松散性能;体积性质
bulk purchase 大宗采购;成批采购
bulk queue 成批排队
bulk raw material 散装原料
bulk recombination 体复合
bulk reduction 小批装油
bulk refractory 不定型耐火材料
bulk refuse 大型垃圾
bulk regulation 体量管理条例;体量法规;建筑物体量管理条例;基地建筑容量规定
bulk resistance 体电阻
bulk resistivity 体电阻率
bulk resistor 体电阻
bulk rock 全岩
bulk rock analysis 全岩分析
bulk rockfill 抛填堆石(体);抛石体;抛石堆填
bulksale 大宗销售;大批出售
bulk sale contract 大宗销售合同
bulksales act 大批销售法
bulk salt carrier 散盐船
bulk sample 粗糙地取样;分散地取样;总样(品);总试样;整体样品;大样;大体积样品;大块样品;全样;主体样品
bulk sampling 总体采样;总取样;平均样品;平均样方
bulk selection 集团选择
bulk service 集团业务;大批服务;成批服务
bulk settlement 整体沉降
bulk settling 整体沉降
bulk shield 立体屏蔽(层);整体屏蔽
bulk shielding facility 整体屏蔽装置
bulk shieldling reactor 整体屏蔽堆
bulk ship 散装货船;散装船
bulk ship cargo 散装船货
bulk ship cargo certificate 散装船许可证
bulk shipment 散装运输;散装水运
bulk shipping of cement 散装运输水泥
bulk slurry carrier 散装泥浆船
bulk socket with swan 卡口灯泡
bulk soil sample 散装土样
bulk solid 块状固体;散装固体
bulk solid flow feeder 松散固体流量喂料机
bulk solid flowmeter 松散固体流量计
bulk solid loader and unloader 散料装卸机
bulk solution 本体溶液
bulk solution composition 本体溶液成分
bulk specific gravity 容重;假比重;毛体积密度;毛体积比重;毛比重;整体比重;堆积比重;体积比重;体比重;松散容重;散装比重;散容重
bulk specific volume 松比容
bulk specific weight 毛体积比重;散装比重;粒状物堆积比重
bulk spreader 松散材料撒布机;堆积物撒布器;土壤洒布密实机
bulk stability 容积稳(定)度
bulk stacking 密法法;实垛法
bulk station 配油站;散装油站
bulk stock 大量储备
bulk storage 后备存储器;大容量存储器;散装库存;散装储存;散装仓库
bulk storage building 散装仓库
bulk storage facility 散装货储[贮]藏设备
bulk-storage plant 储[贮]油场
bulk storage silo 散料储[贮]仓
bulk storage warehouse 散装货仓库
bulk strain 总应变;体积应变
bulk strength 块体强度;块强度;总体强度;体积力
bulk structural defect 体结构缺陷
bulk sugar carrier 散装糖船
bulk sulphur carrier 散装硫黄船
bulk supplies 批量供应
bulk supply 整体供电
bulk supply transformer 降压变压器;用户供电变压器
bulk tanker 散装水泥罐车
bulk technology 体效应技术
bulk temperature 整体温度;按体积计算的平均温度;群体温度
bulk terminal 批发油库
bulk test 总体试验
bulk tester 膨松度测试仪;膨化度测试器
bulk testing 整体试验
bulk timber carrier 散装木材船
bulk tipping container 散装体的倾倒容器

bulk trailer 散装货拖车;散装挂车
bulk train 散装货物列车
bulk transfer 大量转运;散装卸;散装输送;散装搬运
bulk transfer protocol 海量传输协议;大容量传输协议;成批传输协议
bulk transport 大宗运输;散装运输;散装输送;散装货运输
bulk transporter 散装物输送机
bulk trial 大量生产试验
bulk tube 球管
bulk type zinc oxide baristor 体型氧化锌压敏电阻器
bulk uniformity 膨松均匀度;膨化均匀性
bulk unitization charges 货载单位化费用
bulk unit weight 毛体积密度;毛容量;单位毛重;容度;体密重
bulk unloader 散装物卸载机
bulk unloading crane 散货卸船起重机;散货吊船起重机
bulk unloading device 散货卸船设备;散货吊船设备
bulk user 大宗用户
bulk variability 膨松不匀度;膨化不均度
bulk vehicle carrier 散装车辆船
bulk velocity 整体速度
bulk vessel 装载松散物体的船只;散货船
bulk viscoelasticity 体积黏弹性
bulk viscosity 本体黏度;体积黏性;体积黏度
bulk volume 松散体积;毛体积;毛容量;总体积;总容积;容积;松装体积;散装容积;散体积
bulk volume fraction of shale 泥质含量
bulk warehouse 散装仓库
bulk water 重力水
bulk wave filters 体波滤波器
bulk weigher 散装物秤
bulk weight 毛重;物料单位体积重量;松填重量;散重
bulk wood chip carrier 散装木片船
bulk wood-pulp/sulphuric acid carrier 散装木浆/硫酸船
bulk wool 原棉;散棉
bulky 粗大;松散的
bulky cargo 超大货物;笨重货;泡货(体大质轻)
bulky cargo charges 超大货物运费
bulky character 丰满特性
bulky colo(u)r 体色(彩)
bulky goods 笨重货物;笨大货物
bulky grain 大颗粒
bulky joint 粗缝;凸缝;凹凸缝
bulky refuse 松散垃圾
bulky waste 大宗废物;大量废物
bulky yarn 膨胀疏松纱;膨松纱
bulk zoning 容积分区;建筑容积分区
bull 利好;买空;强行实现;屋面用水泥(俚语)
bulla 大炮
bull account 买主账(户)
bull bay 洋玉兰
bull bit 一字形钻头;一字形钎头;凿钻头;凿形钻头
bull block 拉丝机;主索滑车;大型滑车;起模板
bull blocker 拉丝机
bullboat 牛皮艇;牛皮浅水船
bull brick 单圆角砖
bull capital 大型柱顶
bull chain 减速链;(木材拖拉机的)拖运链;传送链
bull clam 弯曲刮刀推土机
bull-clam shovel 铲运机;平土机
bull core structure 筒体结构
bull crack 裂纹;厚薄不均匀而裂
bull currency 看涨货币
bull deep-well pump 杆式深井泵
bull ditcher 大型开沟机
bulldog 狗头矛(打捞套管用)
bulldog casing spear 固定捞管器;套管打捞矛
bulldog clamp 脚踏钻杆夹持器
bulldog clip 颚形夹
bulldog connector 短鼻接合器;齿形板连接器
bulldog double slip spear 双卡瓦打捞器
bulldogged 卡住的
bulldog grip U形螺栓;钢丝绳夹;骑马螺栓;绳头夹;两端丝扣U形螺栓
bulldog pin socket 打捞母锥
bulldog plate 摩擦板(防止搭接木料移动);短鼻接合板;颚形板
bulldog single-slip spear 单卡瓦打捞器

bulldog slip 鳄形夹;(夹住板条或托起混凝土构件的)夹钳
bulldog slip socket 卡瓦式打捞筒
bulldog spear 补炉底材料的矛形尖;固定捞管器
bulldog wrench 鳄形扳手;颚式扳手
bulldong 补炉底材料
bull donkey 集材绞盘机
bulldose blasting 整平地面爆破
bulldozer 履带式推土机;压弯机;二次破碎工;弯钢机;推土机
bulldozer attachment 推土机附件
bulldozer attachment for jeeps 装在吉普车前面的推土机
bulldozer blade 推土机推(土)铲;推土机刮土铲;推土机刀片;推土机铲板;推土刀
bull dozer-equipped track-type tractor 履带式推土机
bulldozer fitted to wheel tractor 有橡皮轮胎的推土机
bulldozer lift cylinder 推土机提升油缸
bulldozer loader 推土机铲
bulldozer ripper 推土机松土器
bulldozer shovel 铲式推土机;推土机铲
bulldozer stabilizer 推土机稳定器;推铲支撑柱;推铲支撑架
bulldozer with angling blade 有转角翼的筑路机械;拖挂式筑路机械
bulldozer with removable equipment 带可移动装置的推土机
bulldozing 削平;推土;用推土机推土
bulldozing blade 挖板;刮板
bulldozing chamber 二次破碎室
bulldozing output 挖产量;刮产量
bulldust 粗粉渣;粗尘
bull earing 天幕角眼;上帆缩帆眼索
bull edger 多锯片板条锯
bullen 阔头钉
bull-engine 直接作用蒸汽机(与抽油泵相连的)
bullen nail 阔圆头钉;阔头钉
Bullen's earth model 布伦地球模型
Buller's ring 布勒(热收缩试验)环
bullet 油井爆破药筒;坠撞器;弹丸;半圆球壳;强调符;喷口整流锥
bullet amplifier 插塞式放大器
bullet-and-flange joint 插销带翼接合;插塞—凸缘接合
bullet block 蛋形滑车
bullet bolt 伸缩插销
bullet-bossed vibrating bar 球头式振捣棒
bullet casting 弹丸模铸法
bullet catch 弹簧门碰锁;碰珠门锁;阔头钉;门碰球;自动扣;弹子门扣
bullet chart 带首标符的图
bullet connection 插塞(式)连接
bullet connector 插塞接头
bullet defying 防弹的
bullet-die 弹丸模具
bullet-headed 圆筒的;子弹似的;圆头的
bulletin 新闻简报;公告;公报;通报
bulletin board 公告牌;布告牌;布告栏;布告板
bulletin board service 公告板服务
bulletin board system 简报板系统;公告牌系统;电子公告板系统
bulletin broad system 电子公告牌
bulletin colo(u)r 指定色
bulletin paint 广告涂料;公告牌用漆
bullet latch 弹子门扣
bullet loan 子弹头式贷款;一次还本贷款
bullet lubrication 弹丸润滑
bullet maturing 一次还本贷款
bullet maturity 一次还本贷款
bullet-nosed 子弹头似的;圆头的
bullet-nosed rod 圆头捣棒
bullet-nosed vibrator 球头式振捣器;球头式振捣棒
bullet-nosed median 弹丸式分隔带
bullet-nosed rod 圆头杆
bullet perforator 球形穿孔机;子弹式穿孔器;子弹式穿孔机;弹丸射孔器
bullet-proof 防弹的
bullet-proof car 防弹汽车
bullet-proof glass 防弹玻璃
bullet-proof material 防弹材料
bullet-proof sedan 防弹轿车
bullet-proof shelter 防弹掩蔽所

bullet-proof vest 防弹背心
bullet-resistant 防弹的
bullet-resistant glass 防弹玻璃
bullet-resistant glazing 防弹玻璃
bullet-resistant material 防弹材料
bullet-resisting 防弹的
bullet-resisting glass 防弹玻璃
bullet roofing tile 屋脊联结瓦
bullet train 子弹车
bullet transformer 一个超高频转换器;超高频转器
bullet type tank 圆筒形罐;卧式圆形储[贮]罐
bullet valve 球阀
bullet wave 弹射波
bullet wood 绿褐色硬木(产于西印度)
bulleye window 小侧窗
bull-fighting arena 大型斗技场;斗牛竞技场
bull float 大镘板;混凝土抹平器;混凝土刮平器;大型抹子;大抹子;长柄镘板
bullfloat finishing machine 长柄镘板整修机
bullfrog 隧道中推运泥车的小车;平衡重设备;上坡牵引
bull gang 辅助工班;铺路及敷设管道工作队;普通工作队
bull gear 从动齿轮;主齿轮;大齿轮
bull gear clamp 带背轮的塔轮接合销
bull gear drive 大齿轮转动
bull gear reducer 大齿轮减速机
bullgrader 侧铲推土机;大型平土机(械);大型平地机;平土机;平路机
bull head 大头T形管;堤岸;挡土墙;圆头方砖;平板箱形孔型;双头式(大圆头);露顶侧砖;小圆头
Bull-head bed 布尔赫德层【地】
bull-head pass 辊身平孔型
bull-head(ed) rail 圆头钢轨;小圆头栏杆;圆筒钢轨;工字钢轨;双头钢轨;双头钢轨
bull-head(ed) rail chair 双头轨座
bull head rivet 圆头铆钉
bull-head(ed) tee 大小头三通;异径三通;大头三通
bull header 露头侧砌丁砖;露顶侧砖;圆角丁砖;圆端丁砖;墙角突头砖;圆头丁砖
bull-holder 牛鼻钳
bull-horn 大功率定向扬声器;手提式扩音器
bulliat 煮沸
bulling 钻孔孔壁糊泥;钻入壁糊泥
bulling bar 插泥棒;钻杆
Bullington nomograph 布林登计算图
bullion 金银丝化;金银块;煤层结核;小圆窗;圆形金属饰;条形金属
bullion balance 金银秤
bullion bar 金锭;粗金属锭
bullion content 贵金属含量
bullion lead 生铅
bullion market 金银市场
bullion room 珠宝室
bullish 看涨
bullish factor 上涨因素
bullish market 利好市场
bullishness 景气现象
bull jig 粗料跳汰机
Bullky-ace 布尔克亚斯粘胶短纤维
bull ladle 大型浇包;大起重机式浇包;输送包
bull line 舱面拉货短索
bull market 多头市场;旺市
bull metal mould 大头的金属模子
bull-nlose 外围头盔;外圆角;圆柱砖;艉导缆钳
bull-nose bed 地毯托架
bull-nose bit 牛鼻形不աΠ芯钻头;楔形开孔钻头
bull-nose block 圆角砌块;圆角混凝土块体;外圆角形混凝土砌块
bull-nose brick 牛鼻形砖;圆角砖;单圆角砖;外圆角砖
bull-nose coping 圆角帽梁
bull-nosed plane 外圆角刨
bull-nosed step 楼梯起步的圆头踏板
bull-nose edge 抹圆角边
bull-nose guide 半球形导向头
bull-nose header 圆端丁砖
bull-nose plane 牛头刨;外圆角刨
bull-nose screen (混凝土摊铺机的)整平板
bull-nose step 圆角踏步;圆端踏步;楼梯起步的圆头踏板
bull-nose stretcher 圆角顺砖;圆端顺砖;圆边侧砖;立砌顺砖
bull-nose tool 拉荒车刀;通切刀

bull-nose trim 圆角边距;圆镶边;有圆边的装饰配件
bull-nose winder (楼梯的)圆角斜踏步
bullock 小公牛
bullock cart 牛车
bullock gear 辘轳;绞车
bullock shed 公牛棚
bull of the woods 工程主管人
bull operation 哄抬价格活动
bull pen 候审室;牛棚;宿营地
bull pin 圆头钉;大销钉;钢结构钻孔锥
bull pine 泡松
bull plug 管塞(子);大塞
bull plug nose 锥形塞头
bull point 角锥铲;大尖钻;(破碎岩石或砖砌石砌体的)钢钻;钢锥;大尖塔;十字铲;(破碎岩石或砖砌石砌体的)钢钻;钢凿
bull pole 公牛棍
bull press 型钢矫正压力机
bull prick 钢钎
bull pump 杆式泵;推力泵;双头泵
bull pup 无价值矿地
bull quartz 烟脂石英
bull rail 木码头护木;挡轮坎(码头前沿);挡车槛(码头前沿);码头护缘
bull ring 研磨圈;船首系缆圈;填料护盖;斗牛场
bull riveter 重型铆钉机;大型铆机
bull rod 钻杆;盘条
bull rope 牵引绳子;栏索;吊桠桅绳;吊艇杆前牵绳;粗绳;粗钢丝绳;传动索;舱内搬货安全索;防磨索;稳定索
bull screen 碎料筛
bull seller 空头
bullset 小壁石;小石壁
bull's eye 牛眼状(黄铁矿结核心);牛眼木饼;滑槽;旋风眼;小圆窗;圆天窗;圆玻璃窗;单眼滑车
bull's eye arch 圆拱;牛眼形拱
bull's eye cringle 帆梢木眼环
bulls'eye glass 牛眼玻璃
bull's eye lamp 牛眼灯
bull's eye lantern 圆形马灯
bull's eye lens 牛眼透镜
bull's eye level 圆水准器
bull's eye ring 牛眼环
bull's eye window 小圆窗;牛腿窗
bull shaker 摇筛;摇动筛;摇动溜槽
bull's liver 砂子与湿红黏土混合土壤
bull's nose 外圆角
Bull's ring 布尔窑
bull station 公牛站
bull stick 转杆
bull stock 公牛牵引杆
bull stretcher 路边侧桩;露边侧砖;露边侧砌斗砖;圆端顺砖;带外圆角的顺砖;圆角踏步板;侧砌的顺砖
bull switch 照明控制开关
bull the buoy 撞上浮标
bull the market 哄抬市场价格;使股票等的行情上涨
bull tongs 大管钳
bull tongue 窄松土铲
bull tongue shovel 重型窄锄铲
bull trawler 对拖网渔船;大拖网渔船
bull trowel 尖脊的镘刀;圆角泥刀;圆角抹子;圆角铰刀
bullule 小泡
bull wagon 套管车
bull weeks 缺勤少时期
bull wheel (钢缆冲击钻的)拉轮绳;牛轮;钻具滚筒;大转盘;大转轮;大轮;大卷筒;起重机水平转盘;(起重机的)水平转盘
bull wheel box 钻具滚筒端的轴颈套
bull wheel brake 钻具滚筒刹车
bull wheel crane 转盘起重机
bull wheel derrick 大转盘动臂起重机;转盘起重机;大转盘桅杆起重机
bull wheel drive 大齿轮转动
bull wheel driven binder 行走轮驱动的割捆机
bull wheel shaft clamps 钻具滚筒轴套
bull window 圆形或椭圆形窗
bully 井场辅助工
bulrush 芦苇;灯芯草;宽叶香蒲
bult 固定沙丘
bultfonteinite 氟硅钙石;水氟硅钙石
bulwark 甲板栏墙;舷墙;寨墙;船舷侧板;防浪堤;壁垒;寨墙堡垒;防御物

bulwark brace 舷墙斜撑
bulwark chock 舷墙导缆钩
bulwark freeing port 舷墙排水口
bulwark ladder 舷梯
bulwark line 舷墙线
bulwark netting 舷墙网
bulwark plating 舷墙板
bulwark port 舷墙排水口
bulwark rail 舷墙栏杆
bulwark stanchion 舷墙柱
bulwark stay 舷墙支柱
bulwark strake 舷墙纵材
Bulygen number 布莱根数
bum 质量低劣的
bum boat 装有交车的小船;小贩船
bumetanide 丁尿胺
bumf 手纸;便纸
bumicky 水泥石粉浆
buming glass edge 烧边
bummer 低轮集材车;低轮车
bump 岩石受压移动;颠簸(指车、飞机);低频噪声;冲撞;冲击地压;突出部;凸缘
bump at bridge-head 桥头跳车
bump bag 缓冲袋
bump check 冲击裂纹
bump clearance 震动间隙;弹簧振动间隙
bump contact 块形连接;大面积接触
bumpcut pavement 整平路面;整平混凝土路面
bumpcutter (machine) 整平机(整修水泥混凝土路用)
bumpcutting 隆起整平;混凝土路面整平
bumped head 冲压封头;凸形的底
bumper 缓冲器;橡胶碰门垫;震动台;挡板;冲击锤;车杠;防撞装置;震动器;保险杆(汽车);保险档;汽车保险杠
bumper arm 缓冲器臂
bumper bar 缓冲梁;缓冲杆;保险杆;减震梁
bumper beam 减震梁;缓冲梁
bumper block 缓冲块;防撞块;碰垫;弹性垫座
bumper frame 防撞架
bumper gate 放牧场栅栏门
bumper guard 缓冲保护器;保险杆护挡
bumper gusset 保险杆角撑
Bumper-Harvest Programme 丰收计划
bumper harvests for years running 连年丰收
bumper hook 门挡衣钩
bumper jack 保险杠起重器
bumper jar 震击器(事故处理用)
bumper mill 撞击弹回式分选器
bumper milling machine 锤式缩泥机
bumper pin 挡销
bumper post 缓冲柱;挡柱;车挡【铁】
bumper step 后踏板
bumper stone deflector 缓冲防石器
bumper stop 缓冲器行程止器
bumper stud 缓冲柱螺栓
bumper sub 钻探船钻杆滑动接头;打头(撞击被卡钻具用);伸缩钻杆
bumper support 缓冲器支架
bumper to bumper 一辆接一辆;车杠到车杠
bumper-to-bumper traffic 拥挤的交通;交通繁忙
bumper type screen 撞击筛;防撞击式网屏
bumper up 铆钉工的辅助工
bumpiness 颠簸性;气流变换不定
bumping 砖工中的撞击增压;空气冲击;以较高级职工代替低级职工;低频不稳定燃烧;崩沸
bumping bag 缓冲袋
bumping block 木结构中的防冲块;缓冲块;撞击块;防撞块;碰垫
bumping chute 缓冲溜槽
bumping collision 弹性碰撞
bumping conveyer 缓冲溜槽;冲震运输机
bumping hammer 开槽锤;大锤
bumping mallet 撞击槌
bumping mo(u)lding machine 振实造型机
bumping pile 缓冲桩
bumping post 缓冲柱;挡柱
bumping table 圆形振动台;碰撞式摇床
bumping test 冲震试验
bumping trough 缓冲溜槽;撞击式溜槽
bump integrator 撞击积算仪
bump joint 隆起接头;扩口接合;异径接头
bumpless changeover 无扰动转换
bump level(l)er 吸震器

bump method 隆起物法
bump shield 冲撞防护
bump storage 缓冲存储器
bump stress 撞击应力
bump stroke 撞击载荷
bump test 连续冲击试验；撞击试验；颠簸试验；冲撞试验；冲击试验；碰撞试验
bumpy air 颠簸空气
bumpy flow 乱流；涡流；紊流
bumpy road 崎岖道路
bumpy running 不稳定运转
buna 聚丁二烯人造橡胶
bunamiodyl 丁碘桂酸
Buna-N 丁氰橡胶
buna rubber 丁钠橡胶；丁苯橡胶；布纳橡胶
buna-S 丁苯橡胶
buna-S3 丁苯S3橡胶
bunch 集束；归堆；丛状造林；丛状的；草垛；群聚团
bunch blasting 明火放炮；成束导火线爆破
bunch builder 包脚纱成形装置
bunch density 凝块密度；群聚团密度
bunch discharge 束形放电
bunched cable 成束缆索；集束电缆；多股电缆；束状电缆
bunched charge 聚束电荷；群聚电荷
bunched conductor 成束导线
bunched cost 整批成本
bunched current 聚束电流
bunched frame alignment signal 集中式帧定位信号
bunched pair 线束对；成股线对
bunched up herding 密集放牧
bunched wire 绞合线；多绞线
buncher 聚束器；聚束极；聚群谐振器；集material机；堆垛机；搓捻机；群聚器；调制腔；调速电极；束线机
buncher attachment 集捆附加装置
buncher beam 群聚束
buncher cavity 聚束谐振腔；聚束腔
buncher coupling factor 聚束极耦合因素
buncher frequency 聚束频率
buncher gap 聚束隙
buncher grid 聚束栅
buncher resonator 聚束谐振腔
buncher space 聚束栅空间
buncher voltage 聚束电压
bunch grass 簇生草
bunching 聚束；聚群；串车；成组；成群；车群
bunching accelerator 聚束加速器
bunching admittance 聚束导纳
bunching angle 聚束角
bunching attachment 割草机的集捆附加装置
bunching cavity 聚束谐振腔；聚束腔
bunching effect of photons 光子聚束效应
bunching machine 搓捻机；束线机
bunching of broken rock 采落岩石堆
bunching parameter 群聚参数
bunching tube 聚束管
bunch light 聚束灯光
bunch machine 捆把机
bunch-map analysis 聚束图分析
bunch of cables 电缆束
bunch of electrons 电子束
bunch of fours 四钩吊货索
bunch of particles 粒子束
bunch of ships 船队
bunch planting 丛状栽植
bunch wire 多绞线
bunchy 成束(的)；穗状
bunchy blastic texture 束状变晶结构
bunchy fault-fold 断褶束
bunchy folds 褶皱束
bunchy texture 束状结构
bund 湖边；海滨路；沿江道路；堤岸
bunder 港口；港湾；驳船港；码头
bunding 岸堤；坝
bunding strengthening 岸堤加固
bundle 扎；捆(装)；捆扎；一捆；包裹；束(扎)
bundle adjustment 光束法平差
bundle aerotriangulation 光束空中三角测量
bundle block aerial triangulation 光束法区域中三角测量
bundle branch 束支
bundle buster 盘条挂送装置
bundle cable 束光缆
bundle cable assembly 束式光缆组件
bundle cable-stay bridge 束式斜拉桥
bundle cargo 捆扎货
bundle carrier 禾捆积运器
bundle conductor 导线束
bundled bar 钢筋束；成束钢筋；盘钢筋
bundled cables 成束敷设的电缆
bundled conductor 盘着的导线；导线束；分股导线
bundled software 捆绑软件
bundled steel wires 钢丝束
bundled tube 成束管；束筒；成束管组
bundled tube structure 成束筒结构
bundled tube system 束筒体系
bundle elevator 禾捆升运器
bundle factor 分裂系数
bundle finishing 束状纹
bundle floating 木排浮运
bundle iron 铁丝卷
bundle jacket 光纤束护套
bundle loader 禾捆装载机
bundle method 光束法
bundle of bars 钢筋束
bundle of capillary tubes 毛细管束
bundle of circles 圆把
bundle of columns 簇柱；群柱；柱群
bundle of electrons 电子束
bundle of folds 褶皱束；褶皱群
bundle of lath 板条捆
bundle of lines 直线把
bundle of piers 桩组；桩群
bundle of pillar 集柱；群柱
bundle of pipes 管束；管组；成束管组
bundle of quadrics 二次曲面把
bundle of rays 射线束
bundle of round reinforcing bars 圆钢筋束(混凝土)
bundle of services 服务群
bundle of steel wire 钢丝束
bundle of tube 管束
bundle pier 群桩形墙墩；群墩
bundle pillar 簇柱；群柱
bundler 捆束机；扎捆机；打包工人
bundle reinforcement 束筋
bundle resolving power 光纤束分辨率
bundles of fascine 梢枝捆
bundle spacing 分裂导线间距
bundle stock 成捆材
bundle stripper 禾捆抛掷机
bundle transfer function 光纤束传递函数
bundle triangulation 光线束法
bundle tube 成束管
bundle type cable-stayed bridge 伞形斜拉桥
bundle vibration 管束的振动
bundle wire carrier communication 分裂导线通信
bundline 堤岸线；岸线
bundling 打捆
bundling attachment 集草车
bundling cradle 管束框架
bundling effect 聚束效应
bundling machine 打捆机；包扎机
bundling press 小包机；打小包机；打包压榨机
bunds 堤岸工程
Bundsandstein 班特统
bund wall 堤墙；防液堤
bundyweld 双层蜡(焊)管法
bundyweld tube 铜接双层钢管；铜钎接双层钢管
Buneman instability 布尼曼不稳定性
bun foot 圆球脚
bung 甲板木塞；木塞；匣钵柱；反射炉盖；桶口；桶盖；塞头
bungaloid 平房式的
bungalow 平房庭院；高平房；(带回廊的)小住宅；(有凉台的)平房
bungalow court 平房庭院
bungalow siding 平房凉台挡板；平房壁板；安装封檐板
bungalow villas 平房别墅
bungee 橡皮筋；跳簧；松紧绳
bung gasket 塞盖垫
bunghole 桶口；桶侧孔；油桶口；桶孔
bunghole boiled linseed oil 桶装亚麻子油
bunghole oil 亚麻子油清漆；清油
bungle 粗制滥造
bungled(piece of) work 笨拙的工作；粗制滥造的工作
bung off 桶塞脱落
bung of setters 托架柱(容具)
bung stave 桶塞板条
bungum 新淤泥；(一种冲积的)粉砂；软泥；冲淤泥沙
bung up and bilge free 桶塞朝上
bung washer 塞盖垫圈
Buniakowski's inequality 柯西—许瓦尔兹不等式
buninoid 丘状的
B-unit 变址部件；变量部件
bunitrolol 布尼洛尔
bunk 链钩；框架床铺；积材车；座床；承枕；铺位；铁架帆布床
bunk apron 上铺挡板
bunk bed 叠架床；双层铺；双层床
bunk bed with ladder and guard rail 带床梯及护栏的双层床
bunk board 铺位舷侧木板
bunk car 宿营车
bunker 料斗(装料)；料仓漏斗；料仓；煤舱；煤斗；掩体；储[贮]槽；储[贮]仓；储[贮]冰室；斗仓；储料仓；储藏库(燃料)；燃料
bunker adjustment factor 燃料调整因素
bunkerage 储[贮]藏设备；燃料费
bunker barge 加油驳；船用燃料驳
bunker bay 煤仓间
bunker bin 煤箱
bunker boat 油槽
bunker capacity 燃料舱容量
bunker car 运煤车；仓车；石渣斗车
bunker clause 燃料条款
bunker coal 船用燃料煤；船用煤(炭)；船用燃煤
bunker content 仓内物料
bunker conveyer 料仓输送装置；煤炭运输机
bunker cushion 料斗缓冲垫
bunker door 煤舱门
bunker frame 煤舱肋骨
bunker fuel oil 船用重油；船用燃(料)油；船用锅炉燃料油
bunker gate 漏斗闸门；料斗闸门；矿仓闸门
bunker hold 燃料舱；下煤舱
bunkering 加装船用燃料；加燃料；装燃料；装仓；储[贮]燃料；燃料供应
bunkering boat 燃料供应船
bunkering capacity 煤仓容量；储仓容量
bunkering facility 加燃料设备；装燃料设备
bunkering jetty 船用燃料码头
bunkering machine 船用燃料装船机具
bunkering port 加油港；船用燃料补给港；船舶燃料供应港
bunkering schedule 燃料储存表
bunkering station 加油站；加油码头；加煤码头
bunkering tanker 加油船；供油船；燃油船
bunkering unit 船用燃料装船机具
bunkering wharf 加油码头
bunker in the ground 储[贮]存窖；材料储[贮]藏坑；材料储[贮]藏窖
bunker jetty 上燃料栈桥码头
bunker oil 船用油；船用燃(料)油；燃油
bunker on arrival 入港时燃料存量
bunker port 船用燃料补给港
bunker position indicator 煤斗装置指示器
bunker quantity on arrival 入港时燃料舱存油量；入港时燃料舱存煤量
bunker room 储槽；仓库
bunker scale 料仓秤
bunker silo 地面青储[贮]堆
bunker spectrophotometer 船用分光光度计
bunker station 燃料储[贮]藏站
bunker stay 煤舱壁横支撑
bunker surcharge 燃油附加费；燃料附加费
bunker train 槽式列车
bunker under track 轨下储[贮]仓
bunker vibrator 料斗振动器
bunker yard 储[贮]料场
bunket position-indicator 料斗装量指示器
bunk-feeder 给料器
bunk feeder with chain conveyer 输送链式饲料分送器
bunk feed wagon 饲料分选小车
bunkhouse 简易房屋；小平房；合宿处；工房；简易钻工房屋；简易工棚(建筑工地)；简陋小屋
bunkie station 泵送站
bunkload 压枕木；一批圆木

Bunlsandstein 斑砂岩统
bunning 木台架
bun ochra fibre 本奥克拉纤维
bunodont 丘齿
Bunsen beaker 平底烧瓶；平底烧杯；烧杯
Bunsen burner 本生燃烧器；本生喷灯；本生灯
Bunsen cell 本生电池
Bunsen combustion 部分预混(式)燃烧
Bunsen element 本生元素；本生电池
Bunsen flame 本生焰
Bunsen gas burner 本生煤气灯
bunsenite 绿镍矿
Bunsen-Kirchoff law 本生—基尔霍夫定律
Bunsen lamp 本生灯
Bunsen photometer 本生式光度计；本生光度计
Bunsen-Roscoe reciprocity law 本生—罗斯科互易律
Bunsen's extinction coefficient 本生消光系数
Bunsen voltameter 本生电量计
bunt 渔网的网身；半外筋斗
bunter 防撞器；撞头；触杆
Bunter 班特统
bunter plate 阻弹板
bunter sandstone 斑砂岩；杂色砂岩
bunt gasket 方帆的束帆索；十字系索
bunting 船旗；旗纱；街道装饰彩旗
bunting block 旗绳滑车
bunt line 拢风索
bunton 横梁；(井筒的)罐道梁；矩形罐梁；横撑；巨型井筒圈梁；轨道撑；防爬撑
bunton box 梁窝
Buntsandstein-chugwater polarity chron 奔特桑斯汀—楚怀特极性时
Buntsandstein-chugwater polarity chronzone 奔特桑斯汀—楚怀特极性时间带
Buntsandstein-chugwater polarity zone 奔特桑斯汀—楚怀特极性时器
Bunzlau ware 庞兹劳炻器
buon fresco 湿壁画
buoy 浮标；浮子；浮筒
buoyage 航标费；系船浮标使用税；装置浮标；浮筒系泊费；浮标装置；浮标使用费；水上航标制；浮标
buoyage fee 浮标使用税
buoyage system 浮标系统
buoyage system A 浮标系统 A
buoyage system B 浮标系统 B
buoyance 浮力
buoy anchor 浮筒锚；浮标锚
buoyancy 浮性；浮力(作用)
buoyancy aid 浮具
buoyancy bag 浮袋
buoyancy ball 浮球
buoyancy block 浮力软木块(救生艇上)
buoyancy calculation 浮力计算
buoyancy chamber 浮室；浮球；浮力室；浮力柜；浮力舱
buoyancy compartment 浮箱；浮力舱
buoyancy computation 浮力计算
buoyancy control system 浮力控制系统
buoyancy correction of weighing 浮力校正
buoyancy curve 浮力曲线
buoyancy density 浮力密度
buoyancy effect 浮力效应；弹性效应
buoyancy excess 浮超
buoyancy float 浮袋
buoyancy jet 浮射流
buoyancy level indicator 浮力液面指示器
buoyancy meter test 浮力仪试验
buoyancy moisture meter 浮力湿度计
buoyancy moment 浮力矩
buoyancy of water 水浮力；水的浮力
buoyancy parameter 浮力参数
buoyancy pontoon 浮筒
buoyancy pressure 浮力；向上压力
buoyancy pump 浮力泵
buoyancy raft 浮筏基础
buoyancy regulating rubber pocket 浮力调节橡皮囊
buoyancy regulating system 浮力调节系统
buoyancy system 浮标系统
buoyancy tank 空气箱；浮箱；浮力舱；浮舱
buoyancy test for life jacket 救生衣浮力试验
buoyancy transducer 浮力传感器
buoyancy trim 浮力纵倾调整

buoyancy tube 浮力管
buoyancy-type density transmitter 浮力式比重传感器
buoyant apparatus 救生浮具
buoyant apparatus tests 救生浮具试验
buoyant ascent 浮力上升法
buoyant belt 浮力救生带
buoyant center 浮心；浮力中心
buoyant chamber 浮室；浮力舱
buoyant compartment 浮力舱
buoyant deck chair 救生椅
buoyant density 浸没密度；浮力密度
buoyant density method 浮力密度法
buoyant effect 浮力效应
buoyant effluent 浮飘污水
buoyant fender moored to sea bed 系连海底的浮式护舷
buoyant force 浮托力；浮力
buoyant foundation 浮式基础；浮力基础；浮基；筏基
buoyant foundation dam 浮基坝
buoyant foundation on raft 浮筏基础
buoyant hinged gate 浮式铰链闸门
buoyant hollow box foundation 格箱式浮基
buoyant jet 浮射流
buoyant lift 浮升(力)
buoyant line and point source dispersion model 浮飘烟线与点污染源扩展模型
buoyant market 市场活跃
buoyant medium 浮力介质
buoyant monolith system 整体浮运法；浮运沉箱法(防波堤施工)
buoyant number 浮标参数
buoyant plume 浮烟；浮选烟
buoyant plume rise 浮选扬性缕烟上升；浮力烟羽上升
buoyant probe 浮飘探针
buoyant property 可浮性能
buoyant raft 浮筏
buoyant rescue quoit 救生索浮圈
buoyant roof 浮顶
buoyant smoke signal 可浮烟火信号
buoyant support of grain 颗粒浮托力
buoyant surface jet 表面浮射流
buoyant unit weight 浮重度；浮容重
buoyant vertical ship lift 浮筒式垂直升船机
buoyant weight (钻孔中充满冲洗液时的)钻具浮重
buoy-ball type standard pressure producer 浮珠式标准压力发生器
buoy bell 浮筒号钟
buoy boat 浮标船；航标敷设艇
buoy chain 浮标锚链
buoy craft 浮标敷设船
buoy davit 浮标吊柱
buoy derrick 浮标吊杆
buoy diving 浮标潜水
buoy dues 系船浮标使用税；浮筒捐；浮筒费；浮泊费；浮标税；浮标使用费
buoyed channel 设置航标的航道
buoyed hydrophone 浮标水听声
buoyed trap 浮陷阱网
buoy engineering 浮标工程
buoy float 浮标浮体
buoy hire 浮泊费
buoy hook 浮标系钩
buoying 浮筒系泊；设置浮筒；设浮标
buoying depth 设标水深
buoying device 浮具
buoy jumper 浮筒带缆工
buoy lantern 浮标灯
buoy-laying 安置浮标
buoy-laying vessel 安置浮标工作船
buoy light(ed) 灯浮标
buoy line 浮子网
buoy missing 浮标失踪
buoy mooring 系浮；浮筒系泊；浮筒锚碇装置；浮筒锚泊装置；浮泊；浮标系固；浮标系泊；浮标锚碇
buoy mooring facility 浮标系碇设备
buoy mooring mark 浮筒系泊标志
buoy motion package 浮标运动监测装置
buoy out of position 浮标离位
buoy repeater station 浮标(数据)中继站
buoy ring 救生圈
buoy rope 吊锚缆；浮子网；浮标索；浮标绳
buoy-satellite observation system 浮标—卫星观测系统

buoy sensor 浮标传感器
buoy shackle 浮筒卸扣；浮筒卡环
buoy sling 浮标索环；浮标吊索
buoy stone 锚碇重块；坠子
buoy superstructure 浮标灯架
buoy tender 航标工作船；航标敷设船；航标船；浮筒维修船；浮标供应船；浮标工作船；浮标敷设船；布标船；设标者
buoy-to-buoy time 轮挡时间
buoy-type seismic station 流动地震台
buoy up 使浮起
buoy whistle 浮标响哨
buoy with topmark 顶标浮标
buoy yard 浮标站
bur 圆头锉；磨石；凸珠；块燧石；草刺
buramate 布拉氨酯
buran 布冷风
burangaite 磷铝铁钠石
burbanite 黄碳锶钠石
bur-bark 伯巴克纤维
Burberry 柏柏丽雨衣
burble 扰流；气流分裂；涡流分离；漩涡
burble angle 泡流分离角；失速角
burble point 分离点；泡流分离角
burbling 流体起旋；气流分离
burbling cavitation 扰流空蚀
Burbon cotton 柏邦棉
Burchard-Liebermann test 伯查德—李伯曼试验
Burch grizzly 筒形格筛
Burch reflecting microscope 伯奇反射显微镜
burckhardtite 硅碲铁铅石
bur cleaner 芒刺清除机
bur clover 苜蓿草
burden 抗力线；泥沙荷载；荷载；载重吨位；载重吨；覆盖层；负担；爆破岩层；配料比
burden analysis 炉料分析
burden apportionment 负荷分配；负荷的分配
burden board 货物垫板
burden calculation 配料计算；炉料计算
burden charges 间接费用
burden charging carriage 配料车
burden depth 爆破纵深
burden distribution 负荷分配；分荷分配
burdened river 含沙河流；挟带泥沙的河流
burdened stream 含悬浮土粒的水流；含沙量大的河流；含沙河流；含泥沙河流；挟带泥沙的水流；挟带泥沙的河流
burdened vessel 义务船；让路船
burden error of power transformer 电源变压器负载误差
burden-fluxing sinter 间断烧结法；单体烧结
burden ga(u)ge 炮眼导向架；炮孔定向器
burden in explosion 爆破抵抗线
burdening 炉料；配料；载重；荷载
burdening the furnace 炉子配料
burden-in-process 在制造费用；在制品间接费用
burden line 最小崩矿层厚线
burden of contract 履行合同责任
burden of debt servicing 偿债负担
burden of paying tax 纳税负担
burden of persuasion 提供说明的责任
burden of proof 举证责任；提供证明的责任
burden of radio isotope 放射性同位素剂量；放射性同位素含量
burden of river drift 河流挟沙；河流挟砂；河流冲积层
burden of taxation 赋税分担；税负
burden rate 负荷率；装载量定额
burden rating 额定装载量；额定负载
burden removing 表土剥离
burden requirement 负荷要求；配料要求
burden share 间接费用分摊份额
burden sheet 配料表
burdensome 繁琐的
burden squeezer 炉料压紧器
burden statement 制造费用表
burden tax 赋税分担
burden test 负荷试验
Burdiehouse limestone 布代豪斯石灰岩
Burdigalian 波尔多阶
Burdilian 波尔迪阶
Burdorite 波尔多里特
Burdwood's sun's azimuth table 柏氏太阳方位表

bure（斐济群岛的）大型房屋
bureau 办公署；所
bureaucracy 官僚机构
bureaucratic form of organization 行政制度式组织
bureaucratic process 繁文缛节的手续
Bureau International del'Heure 国际时间局
Bbureau of Administration 管理局
Bureau of Audit 审计局
Bureau of Bridge Construction 桥梁工程局
Bureau of Budget 预算局
Bureau of Civil Administration 民政局
Bureau of Commerce and Industry 工商局
Bureau of Commercial Affairs 通商局
Bureau of Customs 海关总署；海关局
bureau of customs letter 海关内部通知书
Bureau of Electric(al) Power 电力局
Bureau of Enforcement 法令执行局
Bureau of Forestry 林业局
Bureau of Harbo(u)r Administration 港务管理局
Bureau of Highway Administration 公路管理局
Bureau of Industry 工业局
Bureau of Labo(u)r Insurance 劳工保险局
Bureau of Labo(u)r Statistics 劳工统计局；劳动统计局
Bureau of Land Management 土地管理局
Bureau of Lighthouse 灯塔管理处
Bureau of Mines 矿务局
Bureau of Public Roads 公路管理局；公路局
Bureau of Public works 市政工程局
Bureau of Reclamation 垦务局
Bureau of Research and Development Center 研究与发展中心局
bureau of standards 度量局；标准局
Bureau of Technical Standard 技术标准局(美国)
Bureau of Weather Reports 气象局
Bureau of Yards and Docks 美国海军厂署
Bureau Public Road Classification System 美国公路局土的分类系统
bureau rate 公定保险费率
bureautique 办公自动化
buret(te) 量管；滴定管
buret(te) clamp 滴定管夹
buret(te) float 滴定管浮标
buret(te) holder 滴定管支架
buret(te) meniscus reader 滴定管液面读镜
buret(te) stand 滴定管架
buret(te) visco(si)meter 滴定管黏度计
buret(te) with rubber pipe 带橡皮管的滴定管
buret(te) with straight stopcock 带直路活塞的滴定管
buret(te) with three-way stopcock 带三路活塞的滴定管
burg 市(美国)
burga 布加风；城镇
burgee 磨光废砂；燕尾旗；研磨平板玻璃的废砂；研磨废砂
Burger-Beer's law 布格-比尔定律
Burger's dislocation 布格错位；布格位错
Burger's model 布格模型
Burger's plate 布格板
Burger's vector 布格矢量
burgh 自治市
burgher economy 民有经济
burglar alarm 防盗器；防盗警铃；防盗警报器；防盗报警器
burglar alarm system 防盗警报系统
burglar bar 防护铁栅
burglar-proof 防窃的
burglar-proof glazing 防盗玻璃
burglar-resisting glazing 防盗玻璃
burglary 盗窃行为；偷盗保险
burglary accident 被盗事故
burglary alarm 防窃警报器
burglary alarm system 防窃警报装置
burglary insurance 盗窃保险
burglary protection 防窃；防盗窃保护
burglary resistance 抗盗设备
burglary-resisting door 防盗门
burglary-resisting installation 防盗设备
burglary-resisting window 防盗窗
burglary resistive 抗盗设备
Burgos lustre 伯戈氏红色光泽彩
Burgund guidance system 勃艮制导系统

Burgundian church 勃艮地教堂
Burgundian Gothic(style) 勃艮地哥特式风格
Burgundian portal 勃艮地大门
Burgundian style of sculpture 勃艮地雕塑形式
Burgundian vault 勃艮地穹隆
burgundy pitch 胡树脂；白树脂
burgy 细粉；研磨废砂
bur holder 持钻头器
burial 埋藏
burial at sea 海葬
burial chamber 墓室；墓地
burial-chamber hypogeum 地下墓室
burial depth 埋藏深度
burial dome oil-gas field 潜伏隆起油气田
burial ground 埋弃场；埋藏场；坟地；墓地
burial ground of radioactive wastes 放射性废物埋藏场
burial hill 潜山
burial hill oil-gas field 潜山油气田
burial hill pool 潜山油(气)藏
burial hill pool trend of basement rock 基岩潜山油气藏趋向带
burial hill pool trend of carbonate rock 碳酸盐岩潜山油气藏趋向带
burial hill pool trend of clastic rock 碎屑岩潜山油气藏趋向带
burial layer 埋层
burial metamorphism 埋藏变质作用
burial mound 古墓
burial of sewage screenings 污水筛渣掩埋
burial of sludge 污泥填埋
burial pit 缓冷坑
burial place 墓地
burial tank 埋弃储[贮]槽；埋弃储槽；埋藏槽；储[贮]埋槽(放射性废物)
burial trial 土埋试验(纺织品耐腐性试验)
burial vault 穹形墓穴；穹隆形墓穴
buried 埋入地下的
buried abutment 埋置式桥台；埋式桥台；埋入式岩墩；埋入式桥台；埋入式桥墩；埋入式岸墩；埋墩式桥台
buried ancient planation surface 埋藏古夷平面
buried anomaly 埋藏异常
buried antenna 埋地天线；地下天线
buried arc 埋弧；潜弧
buried blasting 埋藏爆破
buried cable 埋置电缆；埋式电缆；埋地电缆；埋层电缆；置埋电缆；地下电缆
buried cage 深鼠笼
buried channel 埋置沟道；埋藏河道；埋藏河槽；古河槽；地下河道；地下沟渠
buried-channel charge coupled device 埋沟电荷耦合器件
buried circular conduit 埋入式圆形涵管
buried concrete 埋入混凝土；地下混凝土；地埋混凝土
buried condition of aquifer 含水层埋藏条件
buried conduit 地下埋管
buried control centre 地下控制中心；地下指挥所
buried convex area 潜伏凸起
buried conveyer 埋入式运输机
buried copper 槽内铜导线
buried crack 内埋裂纹
buried crestal trap 潜脊圈闭
buried culvert 暗涵
buried depth 埋置深度；埋设深度；埋入深度；埋藏深度；入土深度(指桩)
buried depth map of phreatic water 潜水埋藏深度图
buried depth of anomaly body 异常体埋深
buried depth of an underground river 暗河埋藏深度
buried depth of apical bed of aquifer 含水层顶板埋藏深度图
buried depth of aquifer 含水层埋深
buried depth of basement 基底的埋深
buried depth of bedrock 基底埋深
buried depth of breaking zone 破碎带埋深
buried depth of catchment pipe 集水管埋深
buried depth of groundwater 地下水位埋深
buried depth of moisture measuring probe 湿度测量探头埋设深度
buried depth of motherboard 底板底埋深
buried depth of ore-body 矿体埋深

buried depth of polarized body 极化体埋深
buried depth of rail top 轨面埋深
buried depth of reservoir top 储集层顶界埋深
buried depth of rock body 岩体的埋深
buried depth of route 线路埋深
buried depth of salt-fresh water interface 咸淡水界面埋深；咸淡水界面的埋深
buried depth of soft layer 软弱层埋藏深度
buried depth of sole-plate bottom 底板底埋深
buried depth of spillpoint 溢出点埋深
buried depth of station 车站埋深
buried depth of top of effective source rock 有效母岩顶面埋深
buried depth of trap top 圈闭顶埋深
buried depth of water level at mining funnel 开采漏斗中心水位埋深
buried depth of water table 地下水埋藏深度
buried dip-slope trap 潜山倾向坡圈闭
buried drain 埋入地下的排水管；埋式排水管；埋藏式排水；隐式排水
buried dump 埋场土场；填入土
buried erosion surface 埋藏侵蚀面；埋藏冲刷面；地下冲蚀面
buried escarpment trap 鼻丘圈闭
buried explosion 地下爆炸
buried explosive 地下爆炸物
buried explosive source 地下爆炸源
buried fire hydrant 埋置消火栓
buried flaw 内埋缺陷
buried flexible pressure conduit 埋入式柔性压力涵管
buried focus 地下焦贴(地震勘探)
buried fracture 埋藏裂隙
buried glacier 埋藏冰川
buried halo 埋藏晕
buried-heated pipeline 埋地热油管道
buried heat source 隐伏热源
buried hill 埋丘；埋藏山；埋藏丘；古潜山；潜山；潜丘
buried hill trap 潜山圈闭
buried-hill zone 潜山带
buried ice 埋藏冰
buried installation 地下敷设
buried joint 埋式接缝
buried karst 埋藏型岩溶；埋藏型喀斯特
buried landform 埋藏地形
buried layer 埋置层；埋层；埋藏层
buried laying 埋设
buried length 埋置长度
buried lifeline 埋入式生命线
buried line 埋线路；埋藏线；地下线路；地下管线
buried manganese nodule 埋藏锰结核
buried membrane 埋置膜；埋置沥青膜
buried oil pipeline 地下油管
buried optic(al) cable 埋式光缆
buried ore body 埋藏矿体
buried outcrop 埋没露头；隐伏露头
buried penstock 埋入式压力水管；埋入式压力管道；埋入式压力钢管；埋管；埋藏式压力水管
buried pier 埋入式桥墩
buried pile cap 低桩承台
buried pipe 埋置管；地下管
buried pipe and culvert 地下埋设物
buried pipe line 地下管道；掩埋管道
buried piping 地下管道
buried placer 埋藏砂矿
buried planation surface 埋藏夷平面
buried pluton 埋藏深成岩体
buried pluvial fan 掩埋式洪积扇
buried power station 地下电站
buried pressure pipeline 地下压力管道
buried relationship 埋藏关系
buried ridge 埋藏山脊
buried river 埋藏河；地下暗河
buried scraper coal-feeder 埋置刮板给煤机
buried shelter 地下掩蔽部；地下掩蔽所；地下防空洞；防空洞
buried soil 埋藏土(壤)；下层土；古土壤；腐土
buried soil horizon 埋藏土层
buried storage 地下储库
buried storage tank 埋藏储油罐
buried stratum 埋藏层
buried stream 埋没河流；埋藏河
buried structure 潜伏构造；埋入式结构物；隐蔽结构物；掩体建筑物；地下结构(物)；地下建筑物；

浅埋结构
buried submarine cable 埋式海底光缆
buried surface 隐匿面
buried suture 埋藏缝合
buried system 地下系统
buried tank 地下储[贮]罐;地下储油罐;地下储槽
buried terrace 地下阶地;埋藏阶地
buried treasure 宝藏
buried tubular conduit 埋入的管道;埋设的设备
buried underground 埋地敷设
buried uplift 潜伏隆起
buried valley 埋藏谷;掩埋谷
buried valley trap 谷地圈闭
buried valve with hand wheel 液下手轮阀
buried valve with key 液下控制阀
buried wire 埋线线;暗线
buried wiring 隐蔽布线;暗线
burin 雕刻风格;錾刀;雕刻刀
burkeite 碳酸钠矾;碳钠矾
Burkholder approximation 布克霍尔德逼近
Burkina Faso 布基纳法索
Burkli-ziegler formula for runoff 伯克利—齐格拉径流公式
burl 粒结;树瘤(纹);树疤
Burlana 布尔拉纳聚丙烯腈系纤维
burlap 麻布;打包粗麻布;粗麻布;粗帆布
burlap bag 麻袋;粗麻袋
burlap canvas mat 麻袋帆布村垫;麻袋帆布铺盖
burlap cloth 打包麻布
burlap drag 粗麻布的阻力;麻布拉平(水泥混凝土路面施工用);刷面刷具
burlap-drag finish 底部覆盖以粗麻布;麻布拉平(水泥混凝土路面施工用)
burlap finish 麻布饰面;麻布拉毛;布面装饰
burlap mat 粗麻布底垫
burlapping 粗麻布袋装
burlap sack 粗麻袋
Burleigh (需要两人抱住的)重型凿岩机
burley clay 蠕状耐火黏土
burling 修补;修呢;去除毛结;修布
burling-irons 修布钳
Burlington slate 布林顿板岩(英国)
burl veneer 涡形饰纹镶面板
Burma architecture 缅甸建筑
Burma jade 缅甸玉
Burma lancewood 缅甸枪木
Burma mahogany 缅甸桃花心木
Burma-Malaya geosyncline 缅甸—马来亚地槽【地】
Burma tulipwood 缅甸红色黄檀;缅甸郁金香木
Burmese jade 缅甸玉
Burmese lacquer 缅甸天然漆
Burmese scapolite 缅甸方柱石
Burmese style 缅甸式(建筑)
burmished paper 高光泽纸
Burmister's theory 波米斯特理论【道】
burmite 缅甸硬琥珀
burn 灼烧;灼伤;焚烧;烧绝;燃烧
burnability 可燃性;易倾性
burnability factor 易烧性系数
burnability index 易烧性指数
burnability scale 易烧性指标
burn a bit 烧钻
burnable 可燃物;易燃的;可烧绝的
burnable liquid 可燃液体
burnable poison 可燃吸收体;可燃毒物
burnable quantity 烧失量
burnable refuse 可燃垃圾;可燃废物
burn-back 炉衬烧损
burn clean fuel 使用清洁能源
burn cut 直眼掏槽;平行钻孔爆裂;平行掏槽
burn cut along tunnel roof 沿顶板陶槽爆破
burn-cut near top of working face 上部陶槽爆破
burned 燃烧的
burned ballast 烧结黏土道砟
burned brick 炼砖;烧结砖;烧结黏土砖
burned clay 煅烧黏土;熟黏土;烧结黏土
burned clay brick 烧结黏土砖
burned clay curved roofing tile 烧制黏土曲瓦
burned clay curved tile roof 烧结黏土曲瓦屋面
burned clay hip tile 烧结黏土屋脊盖瓦
burned clay light(weight) aggregate 煅烧黏土轻集料;烧绝黏土轻骨料
burned clay pavement 烧结黏土路面
burned clay product 烧结黏土制品
burned clay ridge tile 烧结黏土脊瓦;烧结黏土盖瓦
burned cut 空眼掏槽
burned cut blasting 空眼掏槽爆破;直线掏槽爆破
burned degree 烧毁程度
burned diamond 烧毁金刚石
burned dust 烧结黏土泥料粉
burned filler 熟料;烧结黏土泥料粉
burned finish 木纹突起处理
burned fireclay 耐火土熟料
burned gas 燃烧过的气体;排出的气体;废气;燃气;已燃气
burned impregnated basic brick 烧成油浸碱性砖
burned-in image 烧附图像
burned iron 脆铁
burned lead joint 铅熔(焊)接
burned lime 生石灰;烧石灰
burned ocher 烧赭石
burned oil-shale aggregate 油页岩烧制集料
burned product 烧制产品
burned refuse 烧掉的垃圾
burned sand 焦砂
burned shale 煅烧页岩;烧页岩
burned shale product 烧页岩制品
burned sienna 暗红橙色的;煅黄土;暗橙色
burned steel 过烧钢
burned umber 煅棕土;烧成赭色;烧褐土
burned ware 烧制品
burned whiteness 焙烧白度
burned wood 木炭
burner 看火工;焊接喷灯;烧烧嘴;燃烧器;汽焊工;气焊工;气割工;喷燃器;喷火嘴;喷灯
burner block 小炉砖;喷嘴砖;喷火口砖;烧嘴砖
burner blower 燃烧器的鼓风机;烧嘴风机(热风炉)
burner blowing process 火焰喷吹法
burner blowing wool-forming aggregate 火焰喷棉机组
burner box 喷燃器壳
burner capacity 燃烧器最大热负荷;炉膛热功率;燃烧器能力;燃烧炉容量
burner characteristic 喷燃器特性
burner-condenser 燃烧冷凝器
burner cup 喷嘴头
burner efficiency 燃烧室效率
burner equipment 气割机
burner fire-brick 烧嘴耐火砖
burner firing block 喷燃器耐火块
burner fixing plate 烧嘴固定板
burner flame 燃烧器火焰
burner fuel oil 灯用燃料油
burner gas 炉气
burner glasses 看火镜
burner grate 锅架
burner guard 护焰罩
burner head 喷灯射口;燃烧器喷嘴
burner hearth 炉床
burner hose 燃烧器软管
burner housing 喷燃器壳体
burner inner liner 火焰筒
burner insert 燃烧芯
burner jet 燃烧器喷嘴;喷灯射口
burner level 喷嘴标高
burner liner 火焰管
burner lining 炉衬
burner loading 喷燃器热能量
burner man 看火工;烧火工
burner manifold 燃烧器燃料管;喷燃器燃料管
burner muffle block 燃烧器的减声装置
burner nozzle 喷灯射口;燃烧喷嘴;喷灯嘴;喷灯口;烧嘴
burner oil 燃烧油;喷射燃料
burner opening 燃烧器喷嘴;喷燃器喷嘴
burner pipe 喷煤(油)管
burner piping 烧嘴管线
burner pliers 烧嘴钳
burner port 火箱嘴;火孔;燃烧室(口);烧嘴孔
burner reactor 燃烧堆
burner ring 燃烧口砖;喷灯环
burner setting 炉子砌体;燃烧室砌体
burner's platform 看火平台
burner system 燃烧器具
burner tap with T handle 丁字柄灯头螺丝攻
burner tile 耐火瓦;炉瓦;燃烧器耐火砖;烧嘴砖;喷嘴砖
burner tip 燃烧器喷尖;喷煤嘴
burner tube 喷烧管
burner tunnel 燃烧器火道
burner turndown factor 燃烧器降燃因数
burner unit 燃烧器装置
burner vaporizer 蒸发式燃烧器
burner wind box 喷燃器风箱
burner with stand 托炉
Burnes act 伯恩斯法
Burnett effect 伯内特旋转转化效应
burnettizing 氯化锌浸渍(木材);氯化锌防腐法(护材作业法)
Burnett method 伯内特法
Burnett's axiom 伯内特原则
Burnett's disinfecting fluid 伯内特氏消毒液
Burnett's fluid 伯内特防腐剂
Burnett's process 氯化锌全吸收法;伯内特木材氧化锌防腐处理法
Burnett's solution 伯内特氏溶液
burn hole 空炮眼
burn hole drilling machine 中孔凿岩机
burn-in 烙上;老化;焊上;烧上;考机;局部加厚处理;预烧;烧焊
burning 金属氧化;焙烧;烧灼;烧成;燃烧
burning action 燃烧作用
burning and collision 火灾和碰撞
burning appliance 燃烧设备
burning appliance room 燃烧设备室
burning area 燃烧区;燃烧面积;焚化场
burning bar of lead 铅焊条
burning behavio(u)r 燃烧特性
burning-brand test 灼烙试验(屋面材料)
burning by catching fire 火灾蔓延烧绝
burning capacity 燃烧能力;燃烧量
burning chamber 燃烧室
burning characteristic 燃烧特性
burning colo(u)r 燃烧的颜色
burning constant 燃烧常数
burning cost 超额赔款分保费;赔款成本
burning crack 燃烧裂缝;温度裂缝
burning defect 燃烧损伤
burning degree 燃烧程度
burning diagram 燃烧曲线
burning duration 燃烧持续时间
burning effect 燃烧作用;燃烧效果
burning expansion 燃烧膨胀;耐火材料锻膨胀
burning feet 烧灼样足
burning furnace 焚化垃圾炉;燃烧炉
burning gas 燃烧气体
burning gate 栅形炉算
burning glass 点火镜;取火镜;凸面镜
burning grate 炉算
burning hearth 炉膛
burning heat sensation 灼热感
burning-in 金属渗入砂型;钢包砂;熔接;熔挂法(一般指滑动轴承上减摩合金的浇挂法);漆膜热修补
burning index 烧除指标
burning-in kiln 焙烧窑
burning-in period 老化时间
burning installation 加热设备;燃烧装置
burning into sand 夹砂
burning kerosene 燃烧煤油
burning kiln 燃烧炉;燃烧窑;煅烧窑;焙烧窑
burning limestone 煅烧石灰
burning line 燃气管线
burning loss 烧损
burning machine 焙烧机;气割机
burning method 焚烧法
burning mixture 可燃混合气体
burning mountain 火山
burning of bearing 轴承预热
burning of bit 烧钻
burning of clay 黏土烧成
burning of contact 触点烧坏
burning of contact material 电接触器材烧绝
burning off 加热(火焰)除漆;燃除;烧脱;烧除
burning off and edge-melting machine 爆口烘口机
burning off paint 加热法清除油漆;油漆燃除
burning of lime sludge 白泥煅烧
burning of limestone 煅烧石灰石
burning oil 煤油馏出物;灯油

burning-on 金属熔补;熔补
burning-on method 熔接法;熔补法
burning-out 烧净;烧熔;烧尽;烧坏;烧除
burning period 盛燃时间
burning plant 燃烧设备;燃烧装置
burning point 燃烧点;燃点
burning point of asphalt 沥青燃点
burning preventer 防燃器
burning process 燃烧过程;燃烧工艺
burning process with common meal 白生料烧成法
burning process with ordinary raw meal 白生料烧成法
burning process with secondary nodulization 二次成球烧成法;包壳球烧成法
burning-quality index 燃烧质量指数
burning range 燃烧范围
burning rate 燃烧速率;燃烧率
burning-rate accelerator 助燃剂
burning-rate constant 燃烧速率系数
burning ratio 焚烧率
burning recording 燃烧记录法
burning reinforcement 切割钢筋(用氧气乙炔切割炬);气割钢筋
burning resistance 抗烧性;耐烧性;耐烧性;耐燃烧性;阻燃性
burning rope 毒品
burning sand 烧结砂块
burning section 火区
burning shrinkage 煅烧收缩;烧成收缩;烧结收缩;烧成收缩
burning speed 燃烧率
burning speed of fuse 导火索燃速
burning stone 硫磺石;硫磺
burning surface 燃烧表面
burning system 燃烧系统
burning technique of coal water mixture 煤水浆燃烧技术
burning temperature 燃烧温度
burning test 煅烧试验;燃烧试验;试烧
burning through 烧尽;烧穿
burning to(cement) clinker 烧结水泥熟料;烧成熔块
burning torch 点火枪;气割
burning velocity 燃烧速度
burning well 井在燃烧
burning zone 燃烧段;燃烧区;燃烧带;燃烧层
burning zone of kiln 窑燃烧区;窑的烧成区
burn-in screen 老化筛选;高温功率老化筛选
burnish 抛光;磨削;磨亮的;磨合;制坯;涂光
burnish broach 挤压推刀
burnish broaching 精拉削;挤压推削;挤光拉削
burnished bolt 精制螺栓
burnished edge 磨光书边
burnished gold 亮金黄色;磨光金;抛光金
burnished metal 磨光金属
burnished silver 磨光银;抛光银
burnished straw 亮麦秆色
burnished tincoated paper 高光泽锡粉涂布纸
burnisher 挤光器;磨光器;磨棒;摩擦光器;辊光器;抛光器
burnish gilding 涂光
burnishing 挤光;摩擦抛光;打光;擦亮;擦光;磨光
burnishing action 磨光作用
burnishing barrel 抛光滚筒
burnishing brush 磨光辊;抛光辊
burnishing die 挤光模
burnishing gold 磨光金
burnishing-in 挤光;辊光
burnishing lathe 抛光车床
burnishing machine 磨光机;抛光机
burnishing of gear teeth 轮齿研磨
burnishing oil 抛光油
burnishing powder 磨光粉;抛光粉
burnishing roll 磨光辊;抛光辊
burnishing shell 抛光壳体
burnishing silver 磨光银
burnishing stick 磨光条;抛光条
burnishing surface 磨光面
burnishing wheel 磨光轮;摩擦抛光轮;抛光轮
burnish machine 打磨机
burnish resistance 抗磨光性
burnish sand 研磨砂
burnish silver 磨光银

burn mark 火刺
burn-off 雾消;烧化;烧割
burn-off edge 焦边
burn-off length (摩擦焊的)摩擦变形量
burn off paint 烧除油漆
burn-off rate 焊耗速率;燃尽率;摩擦变形速度
burn-on 黏砂
burn-out 烧毁;熔断;全部曝光;停止燃烧;烧光;烧断;烧坏;烧完
burn out check 断线探测
burn out condition 耗尽状态
burn-out flux 烧毁热通量
burn-out heat flux 临界热负荷;破坏性热负载;烧毁热通量
burn-out indicator 烧毁指示器
burn-out life 灯丝寿命;失效寿命;烧毁寿命
burn-out pipe 通气孔
burn-out plasma 烧毁等离子体
burn-out poison 可燃毒物;烧尽毒物
burn-out proof 耐高温的
burn-out relay 烧断继电器
burn-out resistance 烧毁电阻
burn-out time 熄火时间
burn pit 燃烧废油坑
burn's 多种铸铁卫生设备
Burnside boring machine 伯恩赛德钻孔机
Burnside's lemma 伯恩赛德引理
Burns-McDonnell activated sludge treatment system 伯恩斯—麦克唐奈活性污泥处理系统
burnstone 硫黄石
burnt 烧伤
burnt aggregate 烧结集料;烧结骨料
burnt and ground lime 磨细生石灰
burnt ballast 烧结黏土道砟
burnt bearing 磨损的支承
burnt bit 烧钻;烧毁金刚石钻头
burnt brick 烧透砖;烧结砖
burnt chalk 煅烧白垩
burnt clay 烧结黏土
burnt clay article 烧结黏土制品
burnt clay ballast 烧结黏土道砟
burnt clay brick 烧结黏土砖
burnt clay cement 烧结黏土水泥
burnt clay curved roof(ing) tile 烧结黏土屋面弧形瓦;烧结黏土屋面曲瓦
burnt clay curved tile roof 烧结黏土弧形瓦屋面;烧结黏土曲瓦屋面
burnt clay hip tile 烧结黏土屋面脊盖瓦
burnt clay light(weight) aggregate 烧结黏土轻集料;烧结黏土轻骨料
burnt clay masonry(work) 烧结黏土圬工(砖建筑)
burnt clay pavement 烧结黏土路面
burnt clay product 烧结黏土制品
burnt clay ridge tile 烧结黏土脊瓦;烧结黏土盖瓦
burnt clay tile 烧结黏土瓦
burnt coal 天然焦(炭);烧变煤
burnt colo(u)r 烧后颜色
burnt dolomite 煅烧白云石
burnt down 烧成平地;烧光;烧毁;全毁
burnt down by fire 焚毁
burnt dust 烧制成的灰;烧结黏土泥料粉
burnt edge 焦边
burnt filler 烧制填料
burnt gaize 烧生物蛋白石
burnt gault 煅烧白垩;烧黏土
burnt gypsum 烧石膏
burnt hair 发灰
burn-through 焊穿;烧过头;烧透;烧毁
burn through melt down 烧透
burnt-in sand 机械黏砂;包砂
burnt iron 过烧铁铸铁;过烧钢
burnt kiln 煅烧窑
burnt lime 氧化钙;煅烧石灰;生石灰;煅石灰
burnt material 烧过的材料
burnt metal 过烧金属
burnt ocher 烧赭石
burnt oil 烧炼油
burnt-on sand 黏砂;起隔子
burnt paper sunshine recorder 焦纸日照计
burnt pipe 煅烧管
burnt plaster 煅石膏;烧石膏
burnt plate oil 烧成板油
burnt potash 氧化钾

burnt product 煅烧制品
burnt pyrite 煅烧黄铁矿
burnt refractory 烧成耐火材料
burnt rivet 热处理铆钉
burnt sand 焦砂
burnt sand casting 半永久型砂型铸造
burnt shale 页岩残渣;烧页岩
burnt shale product 烧页岩制品
burnt sienna 富铁煅黄土;焦茶;煅土黄;煅黄土;深褐色;烧赭土;烧过的富铁黄土
burnt slate 烧结石板瓦
burnt spot 焦斑
burnt state 烧结状态
burnt steel 过烧钢
burnt stone 烧变岩
burnt structure 过烧组织
burnt temperature 烧成温度
burnt through 烧穿
burnt umber 煅棕土(一种红棕色颜料);煅土黄;烧棕土;烧赭土;烧褐铁矿
burnt umber brown 烧褐色赭土
burnt ware 烧成器皿;烧成品
burnt weld 氧焊
burnt wood 烧损木;烧过的木材;木炭
burnt zone 热带
burnu 低角
burn-up 燃耗;烧尽;烧毁
burn-up analysis 燃耗分析
burn-up determination of nuclear fuel 核燃料烧耗测定
burn-up fraction 燃耗份额;燃耗比
bur oil 芒壳油
burr 磨目石;毛边(毛刺);脉结;硬矿块;铸模合缝;杆环;草刺;板金坯件;树瘤纹;三角凿刀;毛口;毛刺;圆头锉
burr bit 倒角钻头
burr-drill 圆头锉;钻锥
burred edge 毛边
burr edge 粗边
burred image 模糊印样
Burrel apparatus 柏瑞气体分析器
Burrel-Orsat apparatus 柏瑞—奥赛特气体分析器
burried arc furnace 埋弧炉;埋弧电炉
burried layer process 埋层法
burring 毛口磨光;去毛口;去毛刺;去毛边
burring chisel 清除毛刺用凿
burring grinder 磨盘式磨碎机;磨盘式磨粉机
burring machine 修整机;去翅机
burring mill 磨盘式磨碎机;磨盘式磨粉机
burring of taper prop 地压支柱尖端的压裂
burring reamer 锥形去毛刺手铰刀;去毛刺整孔钻
burring toothing 去毛刺滚齿
Burri's method 布里法
burr mill 磨盘粉磨机;双盘石磨;双盘磨机;石磨
burrock 河中小堰;堤堰
Burroughs data access method 宝来公司数据存取方法
Burroughs network architecture 宝来公司网络体系结构
burrow 掘穴;洞;虫孔;废石堆;挖洞
burrower 穴居者
burrowing animal 掘穴动物;挖洞动物
burrowing organism 掘穴生物
burrow porosity 生物潜穴孔隙
burr pan 草刺盘
burr picker 除草籽机
burr pump 舱底泵
burr reamer 毛头铰刀
burr removing machine 毛口磨光机(器);去毛刺机
burrs 过火砖
Burr's distribution 伯尔分布
burrstone 磨石
burrstone mill 石磨机
burr wire 锯齿钢丝;钢刺条
burr wood 有树瘤的木材;草籽毛
Bursa 布尔萨聚酯纤维
bursa 中世纪大学的学生宿舍
burse 证券交易所
Bursk polarity superchron 布尔斯克极性超时
Bursk polarity superchronzone 布尔斯克极性超时间带
Bursk polarity superzone 布尔斯克极性超带
burst 漏缺;溃决;猛烈爆发;脉冲穿发;脉冲串;信号序列;纤维爆裂;冲破;冲决;冲击;分张;分片;二进制

位组;爆破;突发;瞬时脉冲群;色同步信号
burst amplifier 闪光信号放大器
burst apart 崩裂
burst blanking pulse 彩色同步信号消隐脉冲
burst-can detector 燃料元件破裂检测器
burst controlled oscillator 短脉冲串控制振荡器;猝发振荡器;色同步控制振荡器
burst-correcting code 纠正突发错误码
burst-correction 突发纠正
burst data 成组数据
burst detector 元件破损检测器
burst device 成组设备
burst diaphragm 安全隔膜
burst disk 安全隔板
burst-distance indicator 爆炸距离指示器
burst edge 裂边
burst-energy-recovery efficiency 色同步再生效率
burster 炸药;分纸器;分离器;爆炸剂;爆炸管;爆炸故障;起爆药
burster charge 爆炸装药
burst error 猝发误差;爆发错误;猝发差错;成组错误;突发错误;突发差错
burst-error-correction 猝发误差校正
burster-trimmer-stacker 分纸器—剪切器—堆垛器
burst event 分歧事件
burst failure criterion 爆破失效准则
burst flag 取样脉冲
burst former 色同步形成器
burst gate tube 猝发选通管;猝发放大控制管;闪光控制管
burst generator 短时脉冲群发生器;短脉冲串发生器;瞬时脉冲群发生器
burst height 爆炸高度
burst hose protection 防爆裂软管
bursting 爆裂;炸裂;波峰尖;爆碎
bursting break 裂口;决口
bursting by compressed air 压缩空气爆破
bursting chamber 炸药室
bursting charge 炸药
bursting crack 胀裂
bursting diaphragm 放空门;防暴门;爆破膜
bursting disk 爆破隔膜;安全隔板
bursting disk of heat-insulating jacketed wall 隔热夹套爆破膜
bursting expansion 爆裂膨胀
bursting explosive 炸药
bursting force 爆破力
bursting fracture 爆破裂隙
bursting layer 竖固掩盖(防空洞上);抗爆层;爆炸层
bursting mud in mining tunnel 井巷突泥
bursting of a tube 管子爆裂
bursting of bags 货袋破裂
bursting of boiler 锅炉爆炸
bursting of dam 溃坝;垮坝
bursting of embankment 路堤破坏
bursting of monsoon 季风爆发
bursting of nodules 料球爆裂
bursting of pellets 料球爆裂
bursting panel 防爆阀
bursting point 爆破点;爆发点
bursting pressure 爆破压强;爆破压力;破裂压力;胀破压力;爆裂压力;破损压力;突发压力
bursting pressure of tank shell 罐身爆破压力
bursting pressure test 破裂压力试验;抗冲压试验
bursting range 爆裂距离
bursting reinforcement 抗裂张钢筋;横向张力钢筋;防爆钢筋
bursting space 爆裂空间
bursting strain 耐爆应变
bursting strength 抗破裂强度;抗断裂强度;抗爆强度;胀裂强度;脆裂强度;爆裂强度;破裂强度;爆破强度
bursting strength tester 抗断裂试验机;爆炸强度试验机
bursting stress 爆裂应力
bursting tension 裂张;破裂性张拉力
bursting test 爆破试验
bursting test machine 爆破试验机
bursting time 起爆时间
bursting transmission 快速传输
bursting tube (变压器的)安全管
bursting water 定水
bursting water at stope 采场突水

bursting water from apical bed 顶板突水
bursting water from bottom bed 底板突水
bursting water from goaf 老窑突水
bursting water in mine 矿井突水
bursting water in mining tunnel 巷道突水
bursting water of fault 断裂突水
bursting wave 爆破波
bursting with biz 买卖兴旺
burst into flame 着火;点火
burst isochronous 等时脉冲串;成组传送
burst isochronous signal 段等时信号
burst key 连锁键;爆炸键
burst length 脉冲时间;脉冲串长度;突发长度
burst load 极限负荷
burst mode 脉冲串式;脉冲串方式;猝发模式;猝发方式;成组方式
burst mode refresh 集中更新
burst noise 脉冲噪声;猝发噪声;突发噪声
burst of light 光猝发
burst of rail base 轨底崩裂
burst of rail bottom 轨底崩裂
burst-oriented data transmission 猝发式数据传输;成组式数据传输
burst out 猝发;爆发
burst period 猝发周期;爆炸时间;爆发期间
burst phase 彩色同步信号副载波相位
burst phenomenon 猝发现象
burst pipe 爆管
burst pressure 爆裂压力
burst pulse 短脉冲;色同步脉冲
burst rate 猝发速率
burst separator 彩色同步信号分离器;色同步分离
burst signal 脉冲串信号;彩色同步信号
burst signal generator 猝发信号发生器
burst size 释放量
burst slug 破损燃料
burst slug detector 释热元件损伤探测器
burst strength 抗爆裂强度;胀破强度
burst tearing test 胀破试验
burst test 耐破度试验;爆裂试验
burst time 猝发时间
burst transmission 脉冲串方式传输;脉冲串传输;猝发传输;成组传送
burst type mud-stone flow 溃块型泥石流
burst waveform 脉冲群波形
burtite 羟锡钙石
burton 滑车组;钢脚手架;辘轳;复滑车
burton boom 舷外斜桁;吊杆;伸出舷外的栏杆
burton fall 吊索
burton gear 船舶吊具
burtoning 联杆吊货法;装卸货固定伸臂
burtoning boom rig 联杆索具
burtoning system 双索单钩吊货系统;双吊杆联合作业法
burton man 联杆作业吊货时的接钩手
burton pendant 重货吊绳
Burton's line 伯顿口线
burton system 联杆吊货法
burton tackle 联杆吊货法
Burundi 布隆迪
bury 埋藏;地穴;软黏土;桅杆的甲板下部分
bury by sand 砂埋;沙埋
burying barge 埋管驳船
burying depth of pool base 油气藏底界埋深
burying depth of pool top 油气藏顶界埋藏深度
burying depth of soil cave 土洞埋深
burying garbage 埋垃圾;埋厨房垃圾
burying ground 坟地
burying method 填埋法
burying place 埋弃场
burying storage 埋藏法
bury length 埋置长度
bury of a mast 甲板下的桅杆
bury sludge 填埋污泥

bus 母线;汇线;汇流线;信息转移通路;总线;公共汽车
bus access conflict 总线访问争用;总线访问冲突
bus access module 总线存取模块
bus acknowledge 总线应答
bus adapter 总线衔接器
bus admittance matrix 节点导纳矩阵;母线导纳矩阵
bus allocator 总线分配器
bus and truck map 公路图

bus available 总线可用
bus available pulse 总线可用脉冲
bus available signal 总线可用信号
busbar 母线;汇流排(管);导电条;汇流线;汇流条;分段断路器
busbar arrangement 母线布置
busbar chamber 母线箱;母线室
busbar channel 母线通道;母线管道;槽形母线
busbar clamp 母线夹
busbar connection 母线联结
busbar connection arrangement 主线布置
busbar corridor 母线检查廊道
busbar coupler 母线连接器;母线联络开关
busbar coupler circuit breaker 母线联络断路器
busbar coupler switch panel 母线联络开关柜
busbar coupling 母线耦合;母线接合
busbar cross section 母线截面
busbar current transformer 母线电流变压器
busbar disconnecting switch 分段母线隔离开关
busbar duct 母线进线管道;母线导管
busbar expansion joint 母线伸缩接头
busbar fault 母线故障
busbar frame 母线架
busbar gallery 母线廊道
busbar grounding 母线接地
busbar insulator 母线绝缘子;母线绝缘器
busbar interconnection switch 母联
busbar partition 母线间隔;母线隔板
busbar potential transformer 母线电压互感器
busbar protection 母线保护
busbar protective devices 母线保护装置
busbar sectionalizing switch 母线分段开关
busbar section circuit breaker 母线分段断路器
busbar supporting insulator 母线支持绝缘子
busbar system 母线系统;汇流排系统
busbar terminal 母线接线端子
busbar tie 母联
busbar tie circuit breaker 母联断路器
busbar tie circuit breaker coupler 母线联络断路器
busbar tie cubicle 母联开关柜
busbar tie switch 母线联络开关
busbar tunnel 母线廊道
busbar unit 母线单元
busbar voltage 母线电压
busbar wall bushing 母线式穿墙套管
busbar wire 汇流条;方条铜线;扁条铜线;汇流排
bus bay 公共汽车停车带;公交停靠站;公共汽车港湾式停车站;港湾式停靠站;港湾式公共汽车站
bus-bite breaker 母联断路器
bus cable 总线电缆
bus category 总线类别
bus chamber 母线盒
bus chassis 公共汽车底盘
bus circuit 总线电路
bus clock 总线计时器
bus communication equipment 母线通信设备
bus compartment 母线室;汇流条隔离室
bus compatible 总线兼容的
bus compensator 母线伸缩接头
bus connector 总线插接器
bus controller 总线控制器
bus control logic 总线控制逻辑
bus control unit 总线控制部件
bus coupler 母线耦合器
bus-coupling 母线耦合;母线接合
bus cycles 总线周期
bus depot 公共汽车库;巴士库;公共汽车总站;公共汽车终点站;公共汽车站;公共汽车停车场
bus driver 母线驱动器;总线驱动器;公共汽车司机
bus duct 母线进线管道;母线管道;母线槽;汇流管道;插接母线
bus duct work 母线管道工程
Busemann apple-curve 布泽曼苹果曲线
Busemann solution for conical flow 布泽曼锥型流解
bus fleet 公共汽车队
bus gate 公共汽车闸门
bush 林丛;灌木(丛);丛枝灌木;丛生灌木;矮树丛;矮灌木;热带密灌丛;嵌轮心套
bush bearing 轴瓦
bush bleaching 草地曝晒漂麻法
bush breaker 灌木犁
bush chain 滚子链
bush chipper 灌木切碎机;灌木铲除机

bush cleaning 灌木清除
bush clearing 清除矮树丛
bush clover 胡枝子;灌木胡枝子
bush crusher 灌木丛压碎器
bush cutter 灌木铲除机
bushed 装了套筒的
bushed bearing 轴承管
bushel 浦耳(英制容积单位)
bushel iron 碎铁
bushelled iron 熟铁
bush-faced 凿毛面
bush-faced masonry 凿面块石砌体;凿面块石圬工
bush fallow 燎荒期;燎荒灌丛;灌丛;丛林
bush fallow system 森林休耕作业法
bush fire 丛林火
bush for rear bearing lower half 后轴承下半部轴衬
bush fruit 灌木性果树
bush hammer 修整锤;鳞齿锤;花锤;汽动凿毛机;凿石锤;凿毛锤;气动凿毛机
bush-hammered concrete 花锤饰面混凝土;锤凿混凝土
bush-hammered 用凿石锤粗修的
bush-hammered dressing 剁斧琢面;剁斧石做法;石锤修琢;石锤修饰
bush-hammered face 石锤琢面
bush-hammered finish (face) 剁斧石面;花锤饰面;锤琢饰面
bush-hammered plaster 凿石锤整修的灰泥涂层;凿石锤整修的墁灰;凿锤石膏
bush-hammering 凿毛(混凝土);打毛(混凝土);锤凿石面;堑凿加工;凿石锤;剁斧石做法
bush harrow 轻型覆土耙
bush hook 长柄大镰刀;钩刀
bushing 拉丝坩埚;内外螺母;轴衬(套);高压套管;电熔铂坩埚;除灌;补心;缩接管;绝缘套;衬套
bushing assembling 漏板装配
bushing assembly 漏板式坩埚装置;漏板装置
bushing baseplate 漏板底板
bushing bearing 径向滑动轴承;滑动轴承;套筒式轴承;轴瓦
bushing block 漏板砖;衬砖
bushing blower 喷吹设备(玻璃纤维);漏板吹风器
bushing bracket 漏板托梁
bushing chain 柱平环链;套筒链
bushing clamp 漏板夹头;电极套;电极夹头
bushing cooling coil 漏板蛇形冷却管
bushing current transformer 环形电流互感器;套管式电流互感器
bushing current-transformer 套管式变流器
bushing driver 套装拆卸器
bushing ear 漏板耳朵
bushing electric 漏板电器;坩埚电器
bushing electrode 漏板电极
bushing extractor 衬套拔出器
bushing feed hole 坩埚加球孔
bushing flow block 漏板供玻璃液砖
bushing frame 坩埚框架
bushing gasket 漏板垫片
bushing holder 套管握持器
bushing insulator 绝缘管;电瓷套管;套管绝缘子
bushing lid 坩埚盖
bushing lock 连杆轴瓦锁
bushing lug 坩埚耳朵
bushing needle 漏板调节针
bushing packing 漏板衬
bushing partition 炉位隔板
bushing plate 钻模板
bushing position 坩埚位置
bushing press 衬套压入机;套管压入机
bushing puller 轴衬拆卸器;套筒拆卸器
bushing reamer 套筒扩孔器
bushing renewing tool 更换套筒工具
bushing replacer 套管置换设备
bushing ring 衬圈;衬环
bushing screen 套筛
bushing shell for transformer 变压器套管
bushing split 拼合衬套
bushing supporting beam 漏板托梁
bushing terminal 漏板耳朵;漏板端子;坩埚耳朵;套筒式端头
bushing test tap 套管试验抽头
bushing throughout 漏板流量
bushing tip 漏板喷丝孔;漏嘴

bushing tool 衬套装卸工具;衬套拆卸工具;套筒装卸工具
bushing transformer 漏板变压器
bushing-type condenser 套管式电容器;穿心电容器
bushing type current transformer 套筒式电流互感器
bush kauri 鲜贝壳松脂
bush knife 刮轴衬用的刀;割灌刀
bushland 荒野;灌木地;灌灌丛;矮灌木地
bush log (护岸用的)灌树原木
bush metal 轴承合金;衬套金属;衬套合金
bush nursery 临时苗圃;林间苗圃
bush of jack 塞孔衬套
bush puller 灌木掘根机
bush rake 灌木耙
bush roller chain 轴衬滚轮链;变径接箍;套筒滚子链
bush-rope 藤
Bush's analogy 布什比拟法
bush set screw 衬套制动螺钉
bush sickness 灌木病
bush swamp 灌木林沼泽
bush sweep 灌木扫除机;灌木清除
bush-topped 柴排盖顶的
bush tree 矮干树
bush type 灌木型
bus hub 总线插座
bushveld 丛林地带;丛林草原(南非)
Bushveld chromium deposit 布什维尔德铬矿床
bush veldt 灌木韦尔德草原
Bushveld type of ultramafic rock body 布什维尔德型超镁铁岩体
bushwash 油罐底残渣
bushwood 灌丛
bushy crown 灌木型树冠
bushy flame 素流火焰
bus idle machine cycle 总线空闲机器工作周期
bus in 总线输入
business 交易;买卖;业务;职责;事务;生意;商务
business ability 营业能力
business account 企业往来账户
business accountability 独立核算制
business accountancy 商业会计
business accounting 经营核算;经济核算;商业簿记
business accounting at different levels 分级核算
business accounting system 经济核算制
business accounting unit 经济核算单位
business activity 业务活动;企业活动
business adjustment 商业调整
business affairs 经营事务;营业事务
business agencies 业务部门
business agency 营业机构
business agent 工会代表;业务代表;行业代理机关
business aircraft 商用飞机
business analysis 经营分析;商情分析
business analyst 经营分析人员;商业分析员
business application 商业应用;商业上的应用
business area 商业区
business arrangement 商业调度
business assessment 营业评估
business assets 营业资产
business association 工商协会
business audit 业务审计
business automation 经营自动化;营业自动化;业务自动化;事务自动化;商务自动化
business background 企业背景
business bank 商业银行
business barometer 营业晴雨表;商情指标;商情晴雨表
business behavio(u)r 营业行为
business block 商业区;商业建筑
business books 商业账簿
business boom 商业景气;商业繁荣
business boom led by investment 投资主导型经济繁荣
business borrowing 工商业贷款
business boundaries 经营疆域
business building 商业房屋;办公楼;商业性建筑物;商业建筑
business calculation 业务核算;商业计算
business capacity 经营实力
business capital 商业资本
business capital formation 企业资本积累
business car 公务用车

business card 名片;业务名片;商业名片
business career 实业界
business cash flow 营业现金流动
business center 商业中心
business chance 业务机会
business check-up 商务检查
business circle 企业界;商业界;商界
business circumstance 经营情况
business climate 商业气候
business closed 停业
business code 业务法规
business combination 企业合并
business communication 商业通信
business communication system 公务通信系统
business community 实业界
business compiler 商业编译程序
business complex 商业综合体
business computer 经营用计算机;商用计算机
business condition digest 商情摘要
business condition fluctuation 商情波动
business condition research 商情研究
business condition 商情
business conference 业务会议
business confidence 经营信心
business conglomerate 大联合企业;企业集团
business connection 门路;业务联系
business consideration 业务报酬
business consultation 商业磋商
business contact 业务往来
business contract 经济合同;贸易合同;业务合同
business control 业务控制
business cooperation 业务合作
business corporation 法人企业;商业公司
business correspondence 商业通信;商业书信
business cost 企业成本
business credit 商业信用;商业信贷
business crisis 营业危机;经营危机;商业危机
business curve 商情曲线
business customs 商业习惯
business cycle 经济周期;景气变动;经济循环;业务周期;商业循环周期;商业循环
business cycle analysis 经济周期分析;经济循环分析;商业循环分析
business cycle fluctuation 商业循环波动
business cycle indicators 商业循环指示数字
business cycle marketing 市场景况
business cycle model 经济周期模型;商业循环模型
business data processing 业务资料处理;业务数据处理;事务数据处理;商业资料处理;商业数据处理;商务数据处理
business day 营业日
business dealing 经营交易;商业行为;商业交易
business decision 商业决策
business department 营业部;业务部
business department objective 业务部门目标
business deposit 营业存款;商业存款
business depression 经营萧条;经济萧条
business development zone 经济发展区;商业发展区
business diagnosis 业务分析;企业诊断
business diary 营业日记;业务日志
business dip 营业下降
business directory 厂商指南
business dispute 商业争端
business district 营业区;商业区
business diversification 经营多样化
business division 营业部;业务部
business document 业务凭证
business done 成交
business downturn 营业下降
business earnings 企业盈余
business economics 经营经济学
business economists 经营经济学家
business economy 企业经济;商业经济
business education 经营管理教育;企业教育;商业教育
business efficiency 经营效率
business encouragement policy 鼓励企业政策
business end of tool 刀具工作端
business enterprise 经营企业;商业单位
business entertainment 业务招待费;企业招待费
business entertainment expenses 交际接待费用
business entity 营业实体;企业实体
business entity convention 财务报表按营业单位

编制的惯例
business environment 经营环境;企业环境;商业环境
business environmental management 企业环境管理
Business Environmental Risk Index 贸易环境风险指数
business environment risk information 营业环境风险信息
business equipment 营业设备;企业设备
businesses economics 商业经济学
business establishment 商业机构;营业处
business estimate 营业评估
business ethics 营业道德;经营道德;商业道德
business evaluation 经济评价
business events 经营业务;企业业务
business expansion 业务扩张
business expenditures 业务费支出
business expenses 经营费用;营业费(用);业务支出
business expenses of insurer 保险事业营业费
business failure 企业破产;企业倒闭
business field 业务范围
business finance 企业财务
business firm 商号
business fluctuation 商业波动
business fluctuation theory 经济周期波动论
business forecast(ing) 经营情况预测;商情预测
business form 商业报表格式
business formation 工商业组成
business frontage 商业区街面;商业区街道;商业街面
business function 企业机能
business fund 营业资金;经营资金;营业基金;企业基金
business fund account 业务资金账户;业务基金账户
business game 经营管理策略;业务竞争;业务策略;商业竞赛
business gift expenses 业务礼品费用
business graphics 商业图表;业务图表
business growth 企业成长
business guide 经营指导;营业指南;业务指南
business guild 同业组织
business guild trade organization 同业公会
business hall 营业厅;营业大厅
business hotel 商业用旅馆
business hours 营业时间;工作时间
business house 商店
business in brief 商业简报;商情简报
business income 营业收入;业务收入
business income tax 经营所得税;工商所得税
business index 商业指数;商情指数
business indicator 经营指标;商情指标
business industry 商办工业
business information system 商业情报系统
business insolvency 企业破产
business institution 企业机构
business interest rate 经营利率
business interests 企业界
business interruption insurance 营业中断保险;业务中断保险
business investor 企业投资者
business items 营业项目
business knowledge 商业知识
business law 行政法规;企业法;商事法;商法
business law and regulation 企业法规
business leader 企业领导
business league 同业组织
business licence 营业执照;营业证
business life insurance trust 企业寿命保险信托
businesslike 商业化
business lines 专用线;业务种类
business liquidity 企业资金的流动性
business-living room 商住楼
business loan 营业贷款;工商业贷款
business local street 商业区街道
business location 企业场所
business logistics 企业物资物流通体系;营业后勤
business loss 营业亏损
business machine 事务管理机;事务用计算机;事务处理机;商用机器
business machine clocking 事务用计算机时钟;事务处理机定时

business machinery 业务机械;商业机器
businessman 实业家;商人;工商业者
business management 经营管理;业务管理;企业管理;商业管理
business management consulting 企业管理咨询
business management contract 企业管理合同
business management rights 经营管理权
business management system 企业管理体系
business manager 营业主任;企业经理;业务经理
business manpower 企业人力
businessman's investment 工商业者投资;商业性投资
business material 商业资料
business measurement 经济观测
business meeting 业务会议
business merger 企业合并
business monopoly 商业垄断
business mortality 营业道德;企业倒闭率
business motive 经营动机
business name 业务用名;商业名称
business negotiation 交易磋商;洽谈业务
business occupancy 商业使用
business of dealer 自营推销业务
business of distributor 推销业务
business office 事务处;商业事务所
business of good standing 信誉良好的企业
business of insurance 保险业
business of producing 生产业务
business of vacation 休业
business on exclusive agency 独家代理业务
business operating with foreign capital 外资企业
business operations 经营活动
business opportunity 商业企业的资产
business or activity not for profit 非盈利性经营活动
business organization 商业机构;企业组织
business outlook 经济展望
business owner 业主;企业主
business panics 经济恐慌
business paper 营业票据
business parade 商业区街道;商业广场
business park 商业园
business patent 营业专利权;营业特许
business performance 经营成绩
business period 营业期
business personnel 业务人员
business philosophy 经营哲学
business pick-up 经济恢复
business picture 经济情况;营业状况
business place 营业处
business planning 业务计划;业务规划;企业计划
business policy 经营政策;业务方针;商业政策
business policy-making 经营决策
business practice 经营惯例;业务实习
business premises 办公室;事务所;商业建筑
business presence 在场业务
business pressure 生意忙
business principle 经营原则
business process 商务处理
business profit 营业利润;商业利润
business profit tax 营业收益税;盈利税
business profit tax return 营业利润税退税;盈利税申报书
business programming 商业程序设计
business property 企业财产;商业地产
business proprietor 业主
business prosperity 经济繁荣
business quarter 营业区;商业区
business rationalization 经营合理化
business receipt 营业收入
business recession 经济衰退
business recovery 经济复苏
business reference 业务参考;商业备咨
business registration 商业登记
business registration certificate 营业登记证;商业登记证
business registration fee 营业登记费
business regulations 企业规程;商业规程
business relations 业务关系
business remises 商业事务所
business reply cards 商业回信用明信片
business report 营业报告;业务报告;商情报告
business report of condition 业务情况报告
business reputation 营业信誉

business research 企业研究
business resources 企业资源
business restraining policy 经济紧缩政策
business result 经营成果
business result analysis 经营成果分析
business revival 经济复兴;经济复苏
business risk 经营风险
business risk and insurance 企业风险和保险
business rivalry 商业竞争
business rivalship 商业竞争
business route marker 商业区路线指示标;商业区路标
business routine of export trade 出口贸易的日常业务
business rules 业务法规
business sales 流转额总数;营业总额
business sales tax 工商营业税
business savings 企业积蓄金
business school 商业学院;商业学校;商学院;商学校
business science 经营学;经营科学
business scope 经营范围;业务范围
business secret 业务秘密;商业秘密
business segments 核算单位;业务部门
business separation 业务分割
business setback 经济衰退
business share 业务份额
business simulation 商务仿真
business situation 经济状况
business slack 经济停滞
business slump 营业不振
business solvency 具有业务偿债能力;企业偿债能力;企业偿付能力
business speculation 商业投机
business stagnation 经济停滞
business standing 企业信信情况;商业信用状况
business statistics 企业统计;经济状况统计;业务统计;商业统计
business status 营业状况
business store 营业仓库
business strategy 经营战略;经营策略;企业策略;商业策略
business strategy plan 企业战略计划
business street 商业区街道;商业街(道)
business structure 企业结构;企业家
business studies 经营研究
business survey 景气动向调查;业务调查;商业考查;商情调查
business survey index 经济动向调查指数
business suspended 停业
business system 商业系统;商务系统
business-system analyst 商业系统分析员
business talks 商务会谈
business taxes 工商业税;营业税
business tax rate 工商税率
business title 商号
business tout 经纪人
business traffic 商业运输
business transaction 经营贸易;交易往来;业务交易;商业交易
business transfer payment 营业转让支付费
business trend 经营趋势;商情趋势
business trend and cycle analysis 景气趋势及循环分析;商情趋势周期分析
business trip 业务出行;公务出行
business trust 营业信托;业务信托机构;企业信托;商业信托
business-type budget 业务性预算
business unit 商号
business upswing 经济回升
business uptrend 经济回升
business upturn 经济回升
business valuation 经营评价
business venture 商业冒险
business volume 营业额;商业性交通量
business vouchers 企业证券;业务凭证;商业单证
business warning indicator 经济警告指标
business with a small capital 小本经营
business world 企业界
business year 会计年度;营业年度;财政年度
business zone 贸易区;商业区;营业区
bus interface 总线接口
bus interface emulation 总线接口仿真
bus interface level 总线接口电平

bus lane 公交(专用)车道;公共汽车(车)道
bus line 汇流线;汇流条;方条铜线;总母线
bus-line coupler 汇流线联络开关
bus-loading-bay 公共汽车停车港;公共汽车停车站
bus master 总线控制者;总线控制器
bus monitor 总线监控器
bus multiplexing 总线多路传送
bus-only lane 公共汽车专用(车)道
bus only street 公共汽车专用街道
bus organization 总线式结构
bus organized structures 总线结构
bus out 总线输出
bus-out check 总线输出校验
bus overhaul works 公共汽车检修厂
buspirone 丁螺环酮
bus plane 电源层
bus platform 公共汽车站台
bus pocket 袋形停车处
buspool 公共汽车(车)场;合乘公共汽车
bus preemption signal 公共汽车优先信号
bus priority 公共汽车优先
bus priority control 公共汽车优先控制
bus priority in 总线优先输入
bus priority lane 公共汽车优先通行道;公共汽车优先车道
bus priority out 总线优先输出
bus priority signal 公共汽车优先通行信号
bus priority structure 总线优先级结构
bus priority system 公共汽车优先通行系统
bus queue barrier 公共汽车排队候车用栅栏;公共汽车候车栏
bus rapid transit 公共汽车快速交通
bus register 总线寄存器
bus request 总线请求
bus request cycle 总线请求周期
bus ride 乘公共汽车旅行
bus ring 集电环
bus rod 汇流母线;圆条母线;圆汇流线
bus room 母线室
bus route 公共汽车路线
bus schedule 公共汽车时刻表
bus section reactor 分段电抗器
bus section switch 分段母线开关
bus separator 总线分离器
bus shelter 公共汽车站罩棚;公共汽车候车亭
bussing 高压线与汇流排的连接;高压线汇流排连接
bus slave 总线受控者
Bussman-Simetag press 巴塞式上下双动水压机
bus standard 总线标准
bus station 公交车站;公共汽车站
bus stop 公交车站;公共汽车站;公共汽车停靠站;公共汽车停车站;车站
bus-stop area 公共汽车停车场
bus structure 母线支架;总线结构
bus system 总线系统
bust 经营失败;胸像;操作错;暴降;半身像;失误
Bustamente furnace 布斯塔蒙特式竖炉
bustamite 锰硅灰石;钙蔷薇辉石
busted convertible 非更换性
bustee (印度的)简陋住屋群
buster 落煤机;铆钉铲;整坯模;钢楔子;钉头切断机;爆破筒;无火焰发爆器
buster hole 空炮眼
bus terminal 公共汽车总站;公共汽车终点站
buster plough 双壁起垄犁
buster plow 双壁开沟犁
buster-slab 防弹墙
bus-tie-in 汇电板
bustife 顽火无球粒陨石
bustle pipe 环形风管;环管;促通管
bustling 繁华
bus topology 总线型拓扑
bust pier 上半部桥墩
bus traffic network study tool 公共汽车优先网络研究法
bus trailer 公共汽车拖车
bus transit 公共汽车运输
bus transportation 公共汽车运输
bust this 传输失败
bus turnout 公共汽车驶出用分支车道
bus turnover rate 客车周转率
bus-type current transformer 母线式电流互感器
bus voltage regulator 母线电压调整器
busway 公共汽车专用(车)道;配电通道;母线通道;母线管道;汇流通道
bus wire 总线;母线;汇流线
busy 操作中;繁华
busy back 忙回信号
busy-back jack 忙音塞孔
busy-back signal 占线信号
busy-back tone 占线音
busy bit 忙碌位
busy-buzz 忙蜂音
busy channel 繁忙通道
busy condition 忙状态;占用状态
busy contact 忙接点
busy crossing 繁忙的交叉口
busy freight line 繁忙货运线路
busy hour 忙时
busy lamp 占线指示灯
busy line 忙线;繁忙线路
busy mainline 繁忙干线
busy mainline railway 繁忙干线铁路
busy period 忙期;忙碌期间;高峰时间;繁忙时期;繁忙期;全占用时间;旺季
busy period distribution 忙期分布
busy queuing/callback 繁忙排队/自动回叫
busy road 交通繁重的道路
busy season 旺季
busy signal 忙音信号;忙碌信号
busy street 闹市
busy test 忙碌状态测试;忙碌测试;满载试验;占线测试
busy time 繁忙时间
busy time of line 线路繁忙时;道路繁忙时间
busy tone 忙音
busy tone signal 忙音信号
busy tone trunk 忙音中继线
busy traffic line 繁忙线路
busy trunk line 繁忙干线
busy verification 占线证实
busy waiting 忙碌等待;忙等待
but 房子的外部房间(苏格兰)
butachlor 丁草胺
butadiene 丁二烯
butadiene-acrylonitrile copolymer 丁二烯丙烯腈共聚物
butadiene-acrylonitrile rubber 丁腈橡胶;丁二烯丙烯腈橡胶
butadiene dimer 丁二烯二聚物
butadiene dioxide 二氧化丁二烯;双环氧乙烷
butadiene epoxy resin 丁二环氧树脂
butadiene latex 丁二烯胶乳
butadiene polymer 丁二烯聚合物
butadiene rubber 聚丁烯橡胶;丁二烯橡胶;丁二苯乙烯橡胶
butadiene rubber plate 聚丁胶板
butadiene-styrene copolymer 丁苯共聚物
butadiene-styrene latex 丁二烯苯乙烯胶乳;丁苯乳胶
butadiene styrene rubber 丁苯橡胶
butadiene styrene synthetic rubber 丁二烯—苯乙烯合成橡胶
butadiene-styrene-vinyl pyridine rubber 丁苯吡橡胶
butadien sulfone 环丁烯砜
butadiyne 丁二炔
butamben 氨苯丁酯
Butamer process 丁烷异构法
butanamine 丁氨
butane 丁烷
butane-air mixture 丁烷—空气混合气
butane-air plant 丁烷—空气掺混装置
butane blowlamp 丁烷喷灯
butane burning engine 丁烷发动机
butane-butylene fraction 丁烷—丁烯馏分
butane cryophorus 冰凝器
butane dehydrogenation 丁烷脱氢
butane diacid 丁二酸
butanedial 丁二醛
butane diamine 丁二胺
butane dicarboxylic acid 己二酸
butanedinitrile 丁二腈
butanedioic acid 琥珀酸;丁二酸
butanediol 丁二醇
butanedione 丁二酮
butanedioxime 丁二肟
butane flame methanometer 丁烷火焰瓦斯检定器
butane fuel gas tube 丁烷气管
butane storage 丁烷储[贮]藏
butanethiol 丁硫醇;丁硫醇
butanetriol 丁三醇
butane vapo(u)rphase isomerization 丁烷气相异构化
butanoic acid 丁酸
butanol 丁醇
butanone 丁酮
butantetraol 丁四醇
butazolidin 苯丁唑酮;苯丁唑啉
butazolidine 苯基丁氮酮
but buttress 支持物;肋墩;支墩;扶壁;前扶垛
butcher block 厨用砧板
butcher lienen 粗粘胶纤维仿亚麻布;平纹厚亚麻布
butcher rayon 平纹人造丝织物
Butcher's Hall at Haarlem 荷兰哈勒姆肉商行会厅
butcher's knife 屠刀
butchery 屠宰场
butch flax 亚麻芥
but down derrick 放井架
butenal 丁烯醛
Butenandt unit 布特南脱单位
butene 丁烯
butene-1 丁烯—1
butene dioic acid 丁烯二酸
butene diol 丁烯二醇
butenoic acid 丁烯酸
butenoic aldehyde 丁烯醛
butenol 丁烯醇
butenolide 丁烯酸内酯
butenylidyne 次丁烯基
butethal 布特萨
Butex 布特克斯流程
but-for income 特殊原因的收益
Butha block 布特哈地块
buthalital sodium 丁硫妥钠
butler 司膳人
butler finish (板材表面的)无光精整
butlerite 基铁矾
Butler oscillator 巴特勒振荡器
butler's pantry 备餐室;膳务室
butler's table 备餐矮桌
butler's tray 腰圆形木盘
butler's window 送菜窗口
butlery 配餐室;膳务室
butline hitch 帆根结
butment 邻接;拱座;桥台
butment cheek 榫孔周围与榫肩对的颊面
butment masonry 拱座圬工;墩座圬工;桥墩圬工
butment pier 靠岸桥墩;岸墩
butment wall 桥台;扶壁;桥台壁
Buton asphalt 布通沥青
butonate 丁酯磷;布托酯
Buton resin 布通树脂
butopyrammonium iodide 布托碘铵
butoxide 增效醚
butoxy benzene 苯丁醚
butoxyethanol ester 丁氧基乙醇酯
butoxyethyl laurate 月桂酸丁氧基乙酯
butoxy resin 丁氧基树脂;丁醇醚化树脂
butt 截土;桩头;铸块;干基;对接;底部;大头;大酒桶;粗端;背皮;暴露煤面;平接(合)
butt and break 对头错开;对头错缝
butt and butt 两头对接;一头接一头;边接;头对头接合
butt-and-butt joint 两头对接
butt-and-collar joint 套筒接口;套筒接合;环对头接合
butt and collar joint for sewer 管道套筒接口
butt and lap joint 互搭接头
butt-and-miter joint 上半部对接而下半部斜接的木工接缝
butt and strap hinge 丁字铰链
butt and strapped joint 搭板对接
butt angle 平接角钢
butt block 对接贴板
butt bolt 铰链筒
butt box 大块氧化皮收集箱
butt cap 悬臂梁
butt-carriage 转材车
butt casement hinge 平开窗铰链
butt chain 对接链
butt chisel 平头凿;平头铲刀

butt chock 角贴片(木船肋材补角材)
butt cleat 端割理;次级劈理;次级割理
buttcock line 尾线
butt connection 对接接头;端接
butt connector 对接接缝
butt contact 对接触点;半圆形触点
butt-contact switch 对接触点开关
butt contraction joint 明收缩缝
butt coupling 套筒联轴节
butt cover plate 鱼尾板;拼接板;对接板
butt cracks 端头裂纹
butt diameter 根端直径
butt dowel 边接榫钉
butte 小尖山;小孤山;孤山;孤峰;地垛
butted 顶部连接;对接的
butted bridging joist 顶棚格栅
butted frame 平接门框;平装门框
butted joint 对接
butted tight 撞紧的
butted tube 粗端管
butt end 木材横截面;根端;大头;平头端;平头部;平端头
butt-end 粗端;大头(木桩)
butt-end bolt 平头螺钉;缓冲螺栓
butt-end joint 对接缝;平端接缝
butt-end treatment 接触端(防腐)处理;木尾防腐
butt end watermark 端头水印
butter 黄油;涂抹;焊膏
butter adjuster 齐根器
butter adjuster lever 撞击器调节杆
butter churn 黄油搅拌器
butter concrete 高坍落度混凝土
butter control unit 缓冲控制元件
buttercover plate 对接鱼尾板;对接板
buttercup yellow 锌黄
butter deflector 齐根器
butter dish 黄油碟
butter dish with cover 带盖黄油碟
buttered 沿边将砖墙铺平在灰浆上
buttered joint 薄浆缝
buttered masonry 挤浆砌筑;抹浆砌筑
butter finish 半光泽抛光;无光精整(板材表面)
butterfly 列车飞条;节流门;节流挡板;蝴蝶;旋转挡板;纸蝶;蝶式棉块
butterfly agitator 蝶式搅拌器
butterfly bamper 蝶阀
butterfly bolt 蝶形螺栓;双叶螺栓
butterfly capacitor 蝶形电容器
butterfly circuit 蝶形电路
butterfly circuit type frequency meter 蝶形电路式频率计
butterfly clack 蝶形活门;蝶形瓣
butterfly cock 蝶形龙头;蝶形旋塞
butterfly dam 蝶闸坝;蝶形闸门
butterfly damper 节气门;蝶形气流调节器;蝶形阀板(门);蝶形阀板;调节门;调风门
butterfly diagram 蝴蝶图
butterfly fastener 蝶形扣件
butterfly flower 蝶粉花
butterfly gate 蝴蝶闸门;蝶形闸门;蝶阀
butterfly hinge 蝴蝶铰链;蝶形铰链;碟形铰链;蝶形合页;蝶式铰链;碟形铰链;长翼铰链
butterfly map projection 蝶状地图投影
butterfly method 蝶形孔型轧制法
butterfly mixer 蝶形搅拌器;蝶形混合器
butterfly nut 蝴蝶螺母;蝶形螺母;双叶螺母
butterfly oscillator 蝶形振荡器
butterfly reinforcement 蝶形加固件
butterfly resonator 蝶式谐振器
butterfly roof 蝶形屋顶;双货棚屋顶
butterfly screw 元宝螺钉;蝶形螺(丝)钉
butterfly seam 蝶形接缝
butterfly self-close damper 蝶形自闭门
butterfly self-closing damper 蝶形自闭门
butterfly self-closing gate 蝶形自闭闸门
butterfly shaped roof 蝶形屋顶
butterfly spindle 蝶形纱锭
butterfly spring 蝶形弹簧
butterfly table 折面桌;折叶圆桌;蝶形折叠桌
butterfly tail 蝶形尾部
butterfly throttle 蝶形节流阀
butterfly throttle valve 加速门;油门;风门;蝶形节流阀
butterfly throttling valve 蝶形节流阀

butterfly tile 蝴蝶瓦
butterfly tuner 蝶式调谐器
butterfly tuner type frequency meter 蝶形调谐式频率计
butterfly twin 蝶形双晶;蝶形孪晶
butterfly-type frame 蝴蝶架
butterfly type of lock valve 蝶形船闸进水阀;闸门蝶形进水阀
butterfly valve 混合气门;蝶形阀(门);碟式阀;衬胶蝶阀;风门;双叶阀门
butterfly valve body 蝶形阀罩子
butterfly valve of lock 船闸蝶形阀
butterfly valve with electric(al) actuator 电动蝶阀
butterfly valve with lockable actuator 手动蝶阀
butterfly valve with pneumatic actuator 气动蝶阀
butterfly wall tie 蝶形系墙铁;碟形墙拉结筋
butterfly wedge 燕尾榫搭角接;蝶形木楔;蝶形楔
butter homogenizer 黄油均化机
buttering 预堆边焊;用镘涂灰浆;隔离层堆焊;涂灰浆;涂盖;抹灰
buttering of mixer 搅拌机的预涂
buttering technique 堆叠焊法
buttering trowel 灰刀;桃铲;大铲;铺灰镘;涂砂浆镘刀;涂灰镘(板)
butter knife 黄油刀
butter making machine 黄油制作机;黄油提制机
butter milk 脱脂乳
butter mixing machine 黄油搅拌机
butternut 胡桃;白胡桃木
butter of tin 四氯化锡
butter powder 黄油粉
butter rock 铁明矾
butter shaping and packing machine 黄油形成和包装机
butter with mastic 用油灰黏牢;用油灰填平;打腻子
butter worker 黄油搅拌器
butterworth cleaner 货油舱洗舱器
Butterworth filter 勃特沃斯滤波器
Butterworth head 水龙带转动头
Butterworth tank cleaning system 油轮洗舱清洁法
buttery 伙食房;食品小卖部;食品储藏室;配膳室
buttery concrete 塑性混凝土;高坍落度混凝土
butter yellow 奶油黄;奶油黄色;对二甲氨基偶氮苯
buttery hatch 备餐室与大厅的隔断;配膳室与食堂之间的窗口;食堂发售食物窗口
butter zone 防护地带
butte temoin 外露层
butt-fused joint 对熔接头
butt fusion 对头熔接;熔接;熔融对接
butt ga(u)ge 木作尺;铰链规;木工画线盘;木工画线规
buttgenbachite 毛青铜矿
butt glazing 对接镶玻璃
butt heat fusion polyethylene plastic fitting 对接热熔聚乙烯塑料管件
butt hinge 合页;铰链;明合页;平接铰链;平铰接
butt hinge pin 铰链销
butt hook 连接钩
butt-hung door 铰链门
butting 扎接;平接;对接
butting angle 平接角钢
butting collision of trains 列车互撞
butting optic(al) fiber 对接式光纤
butt ingot 钢锭切头
butting plow 搅土器;双板犁
butting technique 拼接技术
butt iron 捻缝尖刀
butt joining 对头接合
butt joint 对焊缝;丁字焊缝;丁字连接;丁字对接;对头连接;对接(头);对接焊接头;对焊接头;端接;顶头连接;平头焊;平头接;平接
butt joint corner 对接角
butt jointed 平接的;对抵相接;对接的;裱糊的
butt-jointed seam 对接焊缝
butt-jointed seam with strap 夹板对接焊缝
butt-jointed shell course 对接锅炉身
butt-jointed shell ring 对接炉身环箍
butt joint for sewer 管道对接口
butt joint with cover 带盖板的对接接头
butt joint with double straps 双搭板对接
butt joint with double traps 双夹板对接
butt joint with single strap 用单搭板对接
butt junction 对抵连接;对接(头);对抵接头;平接头

butt lagging 平接壁板
butt lap 板端接迭法
butt laying 平接壁板
butt leather 底革
butt log 根端原木;树基材
butt mill 立铣刀;端面铣刀
butt muff 对接套筒
butt-muff coupling 刚性联轴器
buttock 煤面拐角;工作面端部;船尼;艄型;船艄(凸面)
buttock face 台阶式工作面
buttock getter 机窝采煤机
buttock line 船尼纵剖线;船尼型线;船尼线形;船体纵剖线
buttock machine 立槽割煤机
butt-off 补捣
butt of pile 桩头;桩的根端;桩的钝头;桩的粗端
button 路钮;金属小珠;没有价值的小东西;焊缝试样;旋钮;小块;压坑;电钮
button attaching machine 钉扣机
button ball 美国梧桐
button bit 球齿形钎头;碳化钨钻头
buttonboard 钮子板
button bottom 圆头桩基
button bottom pile 牛形底桩
button breaking machine 凸码揉包机
button capacitor 小型电容器
button catch 门扣
button cell 扣式电池;纽扣型电池
button control 按钮控制;按钮操纵
button device 按钮型制图输入设备;按钮型设备
button die retainer 嵌入式圆形凹模固定板
button dies 可调圆扳手;纽扣式冲模;模具镶套
button drainage 钮式引流法;串珠水系;串珠河
button ga(u)ge 中心量柱
button-head 圆形端头;圆头(式张拉锚具系统)
button-head anchorage 徽头锚;圆头锚碇
button-head bolt 圆头螺栓
button-head cap screw 圆头帽钉;半圆头紧固螺钉
button-headed bolt 半圆头螺栓
button-headed screw 圆筒螺钉;半圆头螺钉
button-head rivet 圆头铆钉;半圆头铆钉
button-head screw 半圆头螺钉;圆头螺丝;圆头螺钉;圆头钉
button-head spike 圆头道钉;圆头长钉
button-head spring 钮头弹簧
buttonhole maker 锁眼工
buttonhook contact 钮钩触点
button interphone 按钮式对讲机
button lac 钮状紫胶
button-like insert 小球状镶嵌物
button microphone 纽扣型传声器
button mounting 小型晶体座
button panel 按钮型控制板
button plate 凸点钢板;凸点钢板
button pulling 按钮拔出
button punching 冲边搭接
button rivet 圆头铆钉
button safety-cap 按钮保险罩
button-screen pipe 钮扣状过滤管
button set 铆钉墩座;按钮组;铆钉圆头成型器
button-sewing machine 钉钮机;钉扣机
button sizer 纽扣自选机
button sleeker 球面慢切
button socket 按钮灯口
button spot weld 按电钮点焊
button stem 芯柱;推拉用小把手
button stem machine 芯柱机
button switch 圆头开关;按钮开关
button test (钢丝的)自身缠绕试验;熔球试验
button up 扣住;扣紧;盖上
button weights 试金砝码
buttonwood 美洲梧桐;美洲悬铃木
butt pin 铰链销
butt plate 掩蔽贴比;端接板
butt plate splice 钢板对头拼接
butt-prop 立柱撑杆;对接杆叉
butt rammer 平头捣锤;平头锤
butt ramming 对头击;对捣
butt resistance welder 电阻对接焊机;电阻对焊机
butt resistance welding 对头电阻焊接;对接电阻焊;电阻对接焊;电阻对焊
buttress 肋墩;撑墙;扶垛;扶壁;边坡支护;板根;坝

垛;墙垛;前扶垛;水管支墩
buttress and frame system 支柱和框架系统
buttress blade 不对称梯形齿锯带
buttress brace 垛间支撑梁;加强梁;加劲梁;支撑刚性梁;垛间支撑
buttress bracing 垛间支撑
buttress bracing strut 有扶垛支撑的支柱;横撑;加劲梁;刚性梁
buttress cap 扶壁压顶
buttress casing coupling thread tensile strength 梯形套管接箍丝扣抗拉强度
buttress casing thread tensile strength 梯形套管丝扣抗拉强度
buttress centers 坝垛中心(间)距;支墩中心距;支墩间距
buttress contraction joint 支墩收缩缝
buttress dam 扶壁式坝;肋墩坝;支墩坝;垛坝;扶垛式坝
buttress drain 扶墩式排水(沟)
buttressed abutment 扶壁式桥台;扶垛式桥台
buttressed arch 支墩拱;扶壁拱
buttressed column 扶壁柱
buttressed dam 扶壁式坝;大头坝
buttressed pier 扶柱
buttressed quay-wall 扶壁式码头岸壁
buttressed reinforcement 支墩钢筋
buttressed retaining wall 扶垛式挡土墙;扶壁式挡土墙
buttressed type retain wall 外支式挡土墙
buttressed wall 扶壁码头;扶壁结构;前抹墙;前扶墙;外支墙;扶垛墙
buttressed wing 扶壁侧翼
buttressed zone 扶垛突起带
buttress(flying) 连拱扶垛
buttress head 支墩头;垛头
buttressing 支撑
buttressing pier 支墩;拱座
buttressing pier or wall 扶垛或扶壁墙
buttress lift 用混凝土层加固的柱子
buttress niche 扶壁壁龛
buttress of pier 扶壁墩子
buttress pack 超前充填带
buttress pier 支柱防波堤;支墩;扶壁墩
buttress pile 细桩
buttress power station 墩内式水电站
buttress reinforcement 加固柱子的钢筋;扶垛钢筋;扶壁钢筋
buttress root 板状根
buttress rubber 异丁橡胶
buttress sand 支撑砂层
buttress sand trap 支撑砂岩圈闭
buttress screw thread 锯齿形螺纹
buttress shaft 支撑桩;细柱
buttress spacing 柱轴线间距离;支墩间距;坝垛间距
buttress stiffener 柱子支撑;扶壁加劲肋
buttress strip 扶壁衬板;扶壁垫板
buttress strut 加劲撑
buttress thread 锯齿形螺纹;梯形丝扣
buttress thread casing 偏梯形扣套管;梯形扣套管
buttress thread coupling 偏梯形扣接箍
buttress thread screw 斜梯形纹螺钉
buttress tower 支墩塔;拱门两旁的塔楼
buttress type dam 扶垛式坝
buttress wall 支壁墙;扶壁墙;扶垛墙
buttress weir 支墩堰;垛堰
buttress wing 扶壁翼
butt rigging 连接索具
butt riveted joint 铆钉对接
butt riveting 对头铆接;对接铆
butt rivet joint 平头铆接;夹板铆接;对头铆接;对抵铆接;平铆接头
butt roller 平缝辊筒
butt rot 根腐
butts 合页铰链;短硬黄麻根纤维
butts and bonds 地界(地界的宽窄长短)
butt saw 截木锯
butt-scarf joint 铰链嵌接;对嵌连接;对嵌接
butt seal 对头封接
butt seam 对头接缝
butt seam inspection method 对接焊探伤法
butt seam welder 对接滚焊机
butt seam welding 滚对焊;对接滚焊
butt seam welding machine 对接缝焊机
butt severing device 根部切除装置

butt-sintering 对接烧结
butt sling 单环吊索
butt splice 对缝接头;对头拼接;对拼接(夹板);对抵拼接;平接接头
butt stile 门窗铰接梃
butt strap 平接盖板;对接搭板;对接衬板;对接垫板;对接板;边接贴板;平接板;平搭接盖板
butt-strap(ped) joint 贴板边接法;搭板对接(合)
butt strip 鱼尾板;对接衬条;平接片;拼接板
butt-swell 根端膨大的
butt-swelling 干脚;脚材;杆脚
butt together 螺纹两端面互相顶紧
butt tool 接缝工具
butt trap 对抵拼接夹板
butt treatment 根端处理
butt turning 顶岸掉头
butt-type door frame 对撞门架
butt type expansion installation 对接式伸缩装置
butt-type heat insulating material 厚板状隔热材料
butt veneer 卷曲花纹贴面板;曲纹贴面板;树根薄板;根部单板;粗纹饰面
butt weld 平头焊接;对接焊(缝);平焊;对头焊接;对缝焊接;对焊;平式焊接
butt-welded 对焊的
butt-welded body seam 平焊桶体接缝
butt-welded drill 对头焊接钻头
butt-welded joint 对焊接头;对接焊头
butt-welded pipe 对缝焊管;对焊焊管;对头焊管;焊缝管
butt-welded rail ends 对焊轨端
butt-welded seam 对接焊缝;对焊缝
butt-welded tube 对口焊接钢管;对缝焊管;卷焊钢管
butt-welded with chamfered ends 斜头平接焊;斜头对接焊
butt-welded with square ends 方头平焊接;方头对接焊
butt weld ends 对头焊接端
butt welder 对接焊机;对焊机;碰焊机
butt-welding 接触对焊;对接焊;对接搭板;对焊;对抵焊接;平对焊;碰焊
butt-welding die 对接模子;对焊模子
butt-welding joint 对焊接头
butt-welding machine 对焊机;对接焊机
butt-welding process 对头焊法;对焊法;对接焊法
butt-welding with chamfered ends 斜头平接焊;斜头对接焊
butt-weld in the downhand position 对接平焊
butt weld joint 对接焊缝;对缝焊接;对头焊接
butt-weld pipe mill 对焊管轧机
buttwood 还孔材;树桩木
butt work 对接工作
butty boat 对槽船;顶推驳(船)
butty gang 零包工人;合伙工作队
butuminic 沥青质
buty cycle 忙闲度
butyl acetate 乙酸丁酯;醋酸丁酯
butylacetic acid 己酸
butyl acetoacetate 乙酰乙酸丁酯
butyl acrylate 丙烯酸丁酯
butyl aetate 醋酸丁酯
butyl alcohol 丁醇
butyl alcohol bacteria 丁醇细菌
butylaldoxime 丁醛肟
butylamine 丁胺
butylated melamine 丁醇醚化三聚氰胺
butylated melamine-formaldehyde resin 丁醇醚化三聚氰胺甲醛树脂
butylated methylol melamine 丁醇醚化羟甲基三聚氰胺
butylated resin 丁醇醚化树脂
butylated urea resin 丁醇醚化脲醛树脂
butylbenzene 丁基苯;丁苯
butyl benzoate 苯甲酸丁酯
butyl benzyl phthalate 邻苯二甲酸丁基月桂酯;邻苯二甲酸丁苄酯
butyl borate 硼酸三丁酯
butyl butyrate 丁酸丁酯
butyl carbamate 氨基甲酸丁酯
butyl carbitol 丁基卡必醇
butyl carbitol formal 二甘醇二乙醚甲醛
butyl cellosolve 乙二醇单丁醚;丁基溶纤剂
butyl cellulose 丁基纤维素

butylcitrate 柠檬酸丁酯
butyl coated fabric 丁基橡胶涂层织物
butyl cyclohexane 丁基环己烷
butyl cyclopentane 丁基环戊烷
butyl diglycol carbonate 丁基二甘醇碳酸酯
butylene 亚丁基;丁烯
butylenealdehyde 丁烯醛
butylene glycol 丁二醇
butylene oxide 氧化丁烯;丁撑氧
butylene plastics 丁烯塑料
butyl etherified melamine resin 丁醇醚化三聚氰胺树脂
butyl ethyl malonate 丙二酸丁乙酯
butyl flufenamate 氟灭酸丁酯
butyl glycol 乙二醇单丁醚;丁基乙二醇
butyl group 丁基
butyl hydrogen sulphate 丁硫酸
butylidyne 次丁基
butyl lactate 乳酸丁酯
butyl mercaptan 丁硫醇
butyl mercury 丁基汞
butyl methacrylate 甲基丙烯酸丁酯
butyl octadecanoate 硬脂酸丁酯
butyl oleate 油酸丁酯
butyl oxalate 草酸二丁酯
butyl oxamate 草酰胺酸丁酯;草氨酸丁酯
butyl oxide 二丁醚
butyl para-hydroxybenzoate 对羟基苯甲酸丁酯
butylphen 对叔丁基苯酚
butylphenamide 丁苯柳胺
butyl phenol 丁基苯酚
butyl phenylacetate 苯乙酸丁酯
butylphenyl anisate 大茴香酸对丁基苯酯
butyl phenylhydrazine 丁基苯肼
butyl phthalate 邻苯二甲酸丁酯
butyl propionate 丙酸丁酯
butyl ricinoleate 蓖酸丁酯;蓖麻酸丁酯
butyl rubber 异丁(烯)橡胶
butyl rubber base 丁基橡胶底座;异丁橡胶基层底板
butyl rubber beading 异丁烯橡胶墙角护条
butyl rubber(building) mastic 异丁橡皮胶;异丁基橡皮胶
butyl rubber elastomer 丁基橡胶弹性体
butyl rubber foam 丁基橡胶泡沫;异丁橡胶泡沫
butyl rubber sealant 异丁橡胶嵌缝膏;异丁烯橡胶嵌缝膏;丁基橡胶密封膏
butyl rubber tape 异丁烯橡胶带
butyl sealing gasket 丁基密封垫物
butyl silicon triisocyanate 丁基甲硅烷三异氰酸酯
butyl stearate 丁基硬脂酸盐;防潮剂(混凝土用无色无臭);硬脂酸丁酯
butyl synthetic rubber sheeting 丁基合成橡胶板
butyl titanate 钛酸丁酯
butylvaleriate 戊酸丁酯
butynediol 丁炔二醇
butyraceous 含油
butyral 缩丁醛
butyraldehyde 丁醛
butyraldehyd oxime 丁醛肟
butyral resin 丁缩醛树脂;丁醛树脂;缩丁醛树脂
butyranilide 丁酰苯胺
butyrate 丁酸酯;丁酸盐
butyrate resin 丁酸酯树脂
Butyribacterium 丁酸杆菌属
butyric 酪酸的
butyric acid 丁酸
butyric anhydride 丁酸酐
butyric ester 丁酸酯
butyric ether 丁乙酯
butyric fermentation 丁酸发酵
Butyrivibrio 丁酸弧菌属
butyrk alcohol 正丁醇
butyrolactone 丁内酯
butyronitrile 丁腈
buxine 黄杨碱
Buxu microphylla var 黄杨
buyable spare part 可外购备件
buy and change gold 收兑金银外币
buy and sell at reasonable price 买卖公平
buy-and-sell shop 旧货店
buy a share 入股
buy as required 现用现买
buy at a bargain price 议价购买
buy at a good bargain 买得便宜

buy at retail 零买
buy back 回购;产品返销
buy back crude 回购原油
buy back deal 回购贸易
buy back price 回购价格
buy boat 渔贩船
buy-build-sell-lease transaction 购—建—销—租业务
buy cheap 贱买
buy cheap and sell dear 贱买贵卖
buy commission 买货佣金
buydown (1~5年内的)通过预付款降低贷款初期利率;折扣售房
buyer 进货员;技术引进方;买主;买方;购买人;采购员
buyer and supplier contract 买方和卖方直接合同
buyer in good faith 善意买方
buyer's broker 买方经纪人
buyer's buying intention 买主意图
buyer's credit 买方信贷;买方贷款
buyer's credit guarantee 买方信贷担保
buyer's entrance 顾客入口处
buyer's exit 顾客出口处
buyer's hedging 买方套购保证
Buyer's inspection to be final 以买方检验为最后依据
buyer's interest 买主权益
buyer's market 买方市场
buyer's monopoly 买方垄断
buyer's opportunity 买方有利时机
buyer's option 买主选择权;买货人选定
buyers over 买主过多
buyer's price 对买主有利的价格
buyer's risk 买方风险
buyer's sample 买方提供的样品
buyer's surplus 买主剩余
buyer's usance letter of credit 买方远期信用证
buy for cash 用现金买
buy foreign securities 购买外国公司证券
buy for future delivery 预购
buy for ready money 现款采购
buy forward 购买期货
buy goods at the sale 买廉价货
buy goods on arrival 买新到货
buy goods on tick 赊购货物
buying acquisition 请购单
buying agent 购买代理人;采购代理商;采购代理人
buying and selling 买卖
buying and selling channels 购销渠道
buying and selling on commission 代客买卖
buying and selling operation of securities 证券买卖业务
buying by institutional investors 投资机构购入
buying by tender 通过招标方式购买
buying capacity 购买力
buying clerk 采购员
buying commission 采购手续费;买方佣金;购货佣金
buying contract 买卖契约
buying cost 购入原价;采购成本;购入成本
buying department 采购部
buying disposition 买方安排
buying exchange 买汇
buying expences 进货费用;采购费用
buying for both spot and forward delivery 现货和期货购进
buying for business use 业务采购
buying for current needs 现用采购
buying for resale 销售采购
buying for ultimate consumption 消费采购
buying hedge 买期保值
buying inclination 购买倾向
buying in retail business 零售商购货
buying intention 购买意向
buying limit 购买限度
buying long 多头购入;投机购买
buying off 行贿收买
buying offer 卖方发盘;买方出价
buying on a yield basis 为获利而购进
buying on margin 交押金买入;边际购买
buying on scale 按一定规模采购
buying operation 购进业务
buying operation under repurchase agreement 按回购协定进行的购买业务

buying order 采购订货;定购单
buying over 行贿收买
buying party 买方
buying power 购买力
buying price 买入价(格);买价;购买价格
buying quota 采购限额;购买分配额
buying rate 买入价;买入汇率
buying rate for time bill 期票买入
buying rate for usance bill 远期汇票买入价
buying requisition 采购申请单
buying sight rate 按现货价格买进
buying tender 投标购货;标购
buying title 购入产权
buying up 全买
buying wrong ticket 误购
buy in stock 买进股票
buy insurance 买保险
buy into 买进……的股票
buy long 买空
buy national commodity 买本国货
buy-now-later 先买后付
buy off 收买
buy on a large scale 采办
buy on credit 赊购
buy one's ticket after the normal time 补票
buy on installment 分期付款购买
buy on margin 押金购买
buy on scale 零买
buy on tally 赊购
buy on the installment plan 用分期付款办法购买
buy on the nod 赊购
buy on tick 赊购
buy on trust 赊购
buy out 买下……的全部产权
buy out a business 买招牌
buy outright 不附带条件的购买
buy over 贿买;收买
Buys-Ballot's law 白贝罗定律
buy secondary-hand 买旧货
buy-sell contract 买卖合同
buy-sell offer 或买或卖出价
buy sth. on the never-never 分期付款购买某物
buy the refusal 买得优先权
buy the refusal of 优先购买权;付定钱
buy up 囤积
buy up the market 囤积货物
buy up wholesale 批发采购
buy wholesale and sell retail 整买零卖
Buzaglo's stain 布扎格洛染剂
Buzin roughness coefficient 巴金粗糙度系数;巴金糖率系数
buzz 噪声;高频颤动;蜂音;蜂鸣音;不稳态间跃迁;嗡鸣(进气);汽笛
Buzzard's reflex 巴扎德反射
buzzer 工地电话;蜂音器;蜂鸣器;蜂铃;轻型凿岩机;轻型掘岩机;轻型穿孔机;汽笛
buzzer coil 蜂鸣器线圈
buzzer exciter 蜂音激磁机
buzzer frequency meter 蜂鸣器频率计
buzzer generator 蜂音发生器
buzzer oscillator 蜂音振荡器
buzzerphone 轻便电话机;军用轻便电话机;蜂音信号;蜂鸣器
buzzer relay 蜂音继电器;蜂鸣继电器
buzzer stop 蜂鸣制动器
buzzer stop/confirmation 蜂鸣器响/确认
buzzer-type tester 蜂鸣式测试仪
buzzer wave meter 蜂鸣器式波长计
buzz plane 圆刨(床)
buzz planer 刨床
buzz saw 圆盘锯;圆锯机;圆锯
buzz stick 蜂音绝缘子测试棒
buzz stick method 绝缘棒探测法
buzz tester 蜂音试验机
buzzy 伸缩式凿岩机;伸缩式风钻
B watertight door 乙种水密门
B/W pan airphotos 黑白全色航片
BX cable 软电缆;铠装电缆
by account 计算的
by agreement 经同意
by-alley 侧巷
by-altar 附属圣坛
by and between 双方(当事人)
by and large 大体上

by appointment 预先约定
byat 井筒横梁
byatt 挖槽水平跳板;水平木;横木
B-Y axis 蓝色差信号轴
by-bidding 抬价出卖
by boat 用船装运
by-branch 支行
by causing organic material to rot 通过腐烂有机物
by-channel 旧河床;溢水道;溢洪道;支汊;支渠;侧渠;侧流渠;侧流管;副航道;分流河槽
by coast, insurance and freight 以到岸价计
by contract 承包;按照合同;按合同
by degrees 逐渐地
by dozen 以打计
bydrostatic balance 比重秤
by dutch auction 降价拍卖
bye channel 支渠;旁侧溢洪道;溢水道
by-effect 副作用;副作业;附加作用
bye-hole 侧孔
by-end 附带目的
bye-pass 支管路;旁通路
bye-pass gallery 分流走廊
bye-path 旁通路
byerite 黏结沥青煤
byerlite 氧化石油沥青
byerlyte 炼焦煤,(贝尔勒法生产的)氧化石油沥青
Byer's process 拜尔(方)法
by-estimate 据估计
by estimation 按照估计
bye-wash 分水槽;溢水道
bye-work 副业
by fair means foul 不择手段
byflow filter 双向流滤池
by free on board 以离岸价格计
by friction of liquids damping 液体摩擦减震
bygone 以往的
bygone water 旧河道
Bygrave slide rule 拜格赖夫计算尺
by gravity 自坠(落);在重力作用下
by hand 用手
by-heads 断续式的;脉动式的;间歇自喷
by hook or by crook 不择手段
byland 半岛
by-lane 僻巷;小巷(道);支巷;私道
by-law 附则;地方法则;地方法规;细则;章程
by-law of corporation 公司章程
by lead 溢洪道
by-level 辅助平巷
by-level broaching machine 双层拉床
by-lime putty 碳化钙石灰膏
by-line 平行线;支线
by means of compensation lines 通过补偿系试验
by measure 按尺寸
by-mineral cost 副产品综合费用
by mutual consent 双方同意
by-name 假名
by negotiations 经过谈判
by notation 注明
byon 含宝石的黄褐色土
by or on behalf of the ship's owner 由或代表船东签发
by-pass 分路;近路;间道;夹管;支管路;侧道;分路迂回;绕行路;绕行道(环形公路);绕流管;绕过;旁通
by-passable 可改道的
by-passable traffic 分支交通(量)
by pass accumulator 浮充蓄电池(组)
by-passage 便道;旁路
by-pass air duct 旁通空气导管
by-pass arrester 旁路避雷器
by-pass baffles 旁通挡板
by-pass bar screen 超越道粗筛
by-pass battery 缓冲电池组
by-pass block 旁路信息组
by-pass canal 路沟;旁沟;溢流道;绕道的渠;间渠;旁渠;(水库的)溢水渠;支运河;支渠;分水渠
by-pass capacitor 旁路电容器
by-pass channel 分水槽;分水渠;分流渠;分流河槽;并联信道;旁通渠;旁路(信道)
by-pass cock 旁路旋塞;调整旋塞;调节开关
by-pass coil 旁路线圈
by-pass collar 回流装置;旁通装置
by-pass condenser 分流电容器;旁路电容器
by-pass conductor 迂回导线

by-pass conduit 溢流道；旁通管道；旁通管；旁通道；旁通导管；溢水槽
by-pass connection 旁通连接
by-pass control 旁通控制；旁道控制
by-pass control valve 旁通控制阀
by-pass convention to promote someone to leading post 破格提拔
by-pass culvert 旁通涵洞
by-pass cyclone 旁通式旋轮吸尘器
by-pass damper 跨越闸门；旁通风门；旁路挡板；溢流闸门
by-pass decoupling 旁路去耦
by-pass diversion structure 岔道分水构筑物
by-pass drainage 迂回排水道
by-pass dredging 旁通施工【疏】
by-pass duct 旁路管道
by passed 加分路的
by passed oil 残留石油
by-passed stone 未过筛孔石料；回笼石料
by-pass engine 双路式涡轮喷气发动机
by-pass factor 旁路系数；旁道系数(旁道空气量与总空气量之比)
by-pass feeder 旁路供电线
by pass filter 分路滤清器
by-pass flow 侧流；旁通流
by-pass flue 分支烟道；旁通烟道
by-pass gallery 分流走廊；侧路渠
by-pass gate 旁通闸门
by-pass governing 旁通控制；旁通调节
by-pass highway 旁路；间路；分支公路；支路；公路支道
by-passing 分路作用；分流；旁泄；旁导；分路
by-passing door 旁通门；边门；旁门
by-passing loop 迂回线
by-passing of gas 气流转移
by-passing of river 河道支流
by passing plant 泥沙旁通设施；排沙装置；排沙设备；(绕道的)输沙设备
by-passing track 迂回线
by-passing yard 通过车场【铁】
by-pass left turn 绕行左转
by-pass line 支管线；旁通线；旁路线；旁路管线
by-pass marker 绕道间道指示标
by-pass method 旁通法
by-pass monochrome image 平行单色图像
by-pass motorway 绕道行驶的机动车道；汽车道(路)
by-pass of flood 分洪道
by-pass of pumping house 泵房岔道
by-pass oil filter 旁通油过滤器
by-pass on 合闸
by-pass opening 旁通孔
by-pass operation 捷径术；旁路转流术
by-pass outlet 回流出口
by-pass panel 旁路开关柜
by-pass penstocks 岔道闸门
by-pass pipe 溢流管；短旁通管；旁通管
by-pass pipeline 旁通管道；放风管道
by-pass pipeline for equalizing pressure 平衡压力溢流支管道
by-pass plug 旁泄塞；旁路塞
by-pass port 旁通孔
by-pass procedure 旁路过程；旁路处理
by-pass product 副产品
by-pass rate 旁通管流量
by-pass ratio 旁通比；旁路比
by-pass rectifier 分流整流阀；旁路整流器
by-pass relay 旁路中继
by-pass relief channel 分洪道
by-pass relief valve 旁通减压阀；旁通安全阀
by-pass resistor 分路电阻器
by-pass road 支路；间路；分支路
by-pass route 支线；侧线；间路；迂回线；支路；绕行路
by-pass screen 旁通筛道

by-pass seepage 绕渗
by-pass set 旁路接续器
by-pass stage 旁路级
by-pass stop valve 备修旁通阀
by-pass strainer 旁路过滤器
by-pass street 绕道可走的道路；小街
by-pass superheater 旁通式过热器
by-pass switch 旁通开关；旁路开关
by-pass system 旁通系统；旁路系统；旁路放风系统
by-pass tee 旁路三通
by-pass temperature control 旁路调温
by-pass to ground 旁通到地；旁路接地
by-pass to waste 排泄(指水、气等)
by-pass track 迂回线
by-pass tube 分路管；旁通管
by-pass tunnel 分支隧道；异流涵洞；旁通隧洞；旁通渠
by-pass-type cotton fan 侧吸式输棉风扇
by-pass valve 分流阀；旁通阀；旁路放风阀；旁路阀
by-pass valve body 旁通阀体
by-pass valve escape pipe 旁通阀泄管
by-pass valve piston rod 旁通阀活塞杆
by-pass vent 分支排气(孔)；支管排气孔；排气支管
by-path 便道；小路；间道；分路；僻径；旁通；旁路；支路；支道；侧管
by-path meter 旁道水表
by-path principle 分路原理
by-path system 旁路系统
by-path valve 旁通阀
by-pipe 支管
by pipeline 旁通管道
by-pit 副井；风井
byplace 穷乡僻壤；偏僻地方；偏僻处
by point 偏点
by procuration 以代理人之身份
by-product 副产物；副产物；附(加)产品
by-product aggregate 副产品骨料；副产品集料
by-product and scrap 副产品及废料
by-product anhydrite 人造硬石膏；合成硬石膏
by-product anhydrite screen 合成硬石膏抹灰面层
by-product coke 副产焦炭
by-product coke oven 副产焦炉
by-product coke-oven gas 炼焦炉副产煤气
by-product coke process 有副产品的炼焦法
by-product cost 副产品成本
by-product costing 副产品成本计算
by-product inventory 库存副产品
by-product light(weight) aggregate 副产品轻集料；副产品轻骨料
by-product lime (电石法产生的)副产石灰；炭化钙石灰
by-product lime putty (电石法产生的)副产石灰油灰
by-product material 副产物
by-product method 副产品法
by-product plant 副产品回收装置
by-product power 副功率；副电力
by-product production 副产品生产
by-product recording 附带记录
by-product recoveries 副产品成本补偿
by-product recovery 副产物回收；副产品回收
by-product recovery unit 副产品回收设备
by-product sales 副产品销售
by-product sludge 副产物污泥
by-products recoveries 扣除副产品
by-product steam 副产水蒸气
by public tender 通过公开投标
by purpose 按用途(划分)
by rail 用铁路运输
Byrd-Dew method 伯德—迪杜法
byre 畜栏；畜棚；牛栏
by Renminbi 以人民币计
by return of post 请即回示
by-road 间路；岔路；支路；小路

by-road subsidiary road 副林道
by roll 支滚
by-room 侧室
by rough estimate 据粗略估计
by sample 按样品
by sea 经由海路；经海路运输
by selfting or backcrossing 通过自交或回交
by separate mail 另邮
by series 逐次
by sight 目测；肉眼观察
bysma 填ృ物；塞子
bysmalith 岩柱；岩栓；垂直岩体
by snatches 断断续续地
by sounding 用测深法
by spells 轮流；轮班
byssinosis 棉屑沉着病；棉尘沉着病
byssolite 绿石棉；纤闪石
byssus 海丝
by stages 分阶段的
by-street 旁街；支巷；支街
bystromite 锑镁矿
byte 信息组；字节【计】；二进位组
byte-addressable storage 按字节编址存储器
byte addressing 按字节寻址
byte address machine 按字节寻址计算机
byte data 字节数据
byte disinterleaving 字节消间插
byte error detecting code 字节错误检测码
byte instruction 字节指令
byte interleaving 字节间插
byte manipulation 字组处理
byte memory 二进位组存储器
byte mode 二进位组式
by-tender 以投标形式；投标形式
byte oriented operand 按字节操作数
byte-oriented operation 按字节操作
bytes per inch 每英寸字节数
byte storage 信息组存储器
by-the-book 精心设计
by the day 按日
by the head 船首纵倾；首倾【船】
by the hour 按小时计算
by the job 论件；计件(的)；包做；按件计(工资)
by the month 按月
by the piece 按件计算
by the piece of work 计件
by the run 按进尺(计算)
by the stern 后倾【船】；船尾纵倾
by the way 顺便
bytownite 倍长石；培长石
by-turn bidding 轮流投标
by-turning 间路
by usage 按照常规
by volume 以体积计；按体积
by-walk 私路；行人小道；僻径
by-wash 河岸排水道；溢洪道；排水沟；排水支管沟；侧流排水道
by-wash channel 排水槽；排水沟
by-water 牛轭湖；由水路；旧水路；旧河床；废河道
by-water lake 牛轭湖
byway 小径；次要方面；次要部分；僻径；小路
by weight 以重量计；按重量计算；按重量
by-work 副业
by-workman 短工；临时工(人)
byzant 列圆节；圆盘形装饰；银币饰；系列圆盘线脚(诺曼底式建筑)
Byzantine architecture 拜占庭式建筑
Byzantine art 拜占庭式艺术
Byzantine building 拜占庭式建筑
Byzantine capital 拜占庭式柱头；梯形柱头
Byzantine church 拜占庭式教堂
Byzantine column 拜占庭式柱
Byzantine dome 拜占庭式圆穹顶
Byzantine house 拜占庭式住宅
Byzantine style 拜占庭式

C

Ca activation log 钙活化测井
Caada (麦加大清真寺院中的)穆斯林圣堂
caatinga 卡丁加群落
cabal glass 钙硼铝玻璃
cab alongside engine type 侧驾驶室型【船】
cabana 帐篷屋;小屋;棚屋;简易浴室
cabane 顶架;翼柱
cabane radiator 鞍形散热器;屋顶散热器
cabane strut 翼支柱
cab apron 渡板;驾驶室盖板;驾驶室渡板
cab apron hinge 渡板铰链
cabaret 音乐餐厅;夜总会;双人用成套咖啡具
cabature 求积法
Cabba 小屋;(麦加大清真寺中的)小石屋
cabbage flea beetle 曲条跳甲
cabbaging press 包装压榨机
cab behind engine type 凸头式
cab body 舱身
cab control 座舱控制
cab control cable 驾驶室控制电缆
cab equipment 驾驶室设备
cab fault indication panel 驾驶室故障指示板
cab-forward type vehicle 平头型卡车
cab front 驾驶室前窗
cab guard 驾驶室护板
cab(in) 操作室;驾驶室;客舱;小室;机舱;小屋;座舱;载客吊笼;工作间;船舱;舱室;舱
cabin altimeter 航空测高仪
cabin altitude 座舱高度
cabin blower 座舱增压器
cabin bulkhead 房舱舱壁
cabin cableway 吊舱索道
cabin camp 小屋野营;分隔帐篷
cabin car 守车
cabin class 三等舱
cabin compass 倒挂罗经
cabin court 公路旅馆;汽车旅馆
cabin cruiser 摩托艇;娱乐游艇(有娱乐及生活、住宿等设备)
cabin de luxe 特等舱
cabin de suite 特等客舱套间
cabin differential pressure 座舱压差
cabin distribution box 舱室分电箱
cabin enamel 船舱漆
cabin en suite 特等客舱套间
cabinet 机壳;机柜;小室;小裁;柜子;柜橱;柜;橱;塑料机壳
cabinet air conditioner 柜式空调器
cabinet air purifier 柜式净空气器;空气过滤箱
cabinet base 底柜;床脚;台座
cabinet bonds 分柜债券
cabinet bulb 阵列橱窗
cabinet convector 罩式对流器
cabinet door 柜门;橱门
cabinet door hook 船舱门钩;橱门钩
cabinet drawer kicker 抽屉拉出防坠装置
cabinet drawer runner 抽屉滑条
cabinet drawer stop 抽屉推入挡块
cabinet drawing 斜二轴测图
cabinet dryer 干燥橱;干燥室;室式干燥器
cabinet file 半圆锉;小锉;细木工锉
cabinet filler 橱墙间垫块;小室隔断
cabinet finish 细木工装饰;精细装修;细木工涂装;细木壁饰;壁板装饰(室内);室内壁板装饰
cabinet for dry washing 干洗间;干洗柜;洗衣干燥柜(住宅中的)
cabinet for television set 电视机柜
cabinet ga(u)ge control unit 真空计线路箱
cabinet hardware 家具小五金
cabinet heater 柜式取暖器;柜式供暖器;柜式采暖器;橱式散热器
cabinet heating 柜式取暖;柜式供暖
cabinet hinge 家具铰链
cabinet hinged door 铰链柜橱门
cabinet jamb 饰面门框
cabinet latch 柜橱碰锁;水平推拉栓锁;小五金门锁
cabinet lavatory 间隔盥洗室;分间式盥洗室
cabinet leg 机床底脚;床脚;台座

cabinet lock 弹簧插销;弹簧锁
cabinet maker 细木工;家具(木)工
cabinet making 家具制造;细木工艺
cabinet mirror 柜镜
cabinet of curiosities 美术品陈列室
cabinet oven 柜式烤炉
cabinet panel 配电盘;配电板
cabinet pile 积木式堆装法
cabinet projection 斜角立体投影(工程画)
cabinet rake 机箱架
cabinet rasp 细木锉
cabinet resonance 机箱共鸣
cabinet respirator 箱式呼吸器
cabinet scraper 箱形刮泥器;木工(小)括刀;刮刨;平刨(橱柜用);精细刮刨;刮削刨
cabinets for documents 文件柜
cabinet sliding door 柜橱拉门
cabinet speaker 箱式扬声器
cabinet switch 开关箱
cabinet tray drier 橱式托盘干燥箱
cabinet-type air conditioner 箱式空调器
cabinet-type electric(al) furnace 箱式电炉
cabinet-type governor 箱式调速器
cabinet-type shower 分间(式)淋浴室;分级淋浴间
cabinet-type switch 柜式开关
cabinet urinal 间隔小便池;分式小便器
cabinet water-proof switch 柜式防水开关
cabinet window 陈列橱窗
cabinet with mirror door 带镜立柜
cabinet wood 细木工用材;细木工用木料
cabinet work 细木家具工;细木(工)作
cabin heater 室内吊车;分隔间取暖器;座舱加温器;室内取暖器
cabin hook 窗钩;门(窗)钩;风钩
cabin lighting 室内照明
cabin lightning 座舱照明
cabin passenger 房舱旅客
cabin plan 舱室面置图
cabin port 舱室舷窗
cabin ration feeding machine 仓式定量给料机
cabin ropeway 吊舱索道
cabin store 舱室用品库;舱室用品
cabin supercharger 座舱增压器
cabin-tank lifecraft 带座舱救生筏
cabin top 半显舱室
cabin transport 舱式运输机
cabin trunk 扁式硬皮箱;半显舱室
cabin type crane 座舱式起重机
cabin ventilator 舱室通风机
cabin wall ventilator 舱壁通风筒
cable 缆(索);(直径在10英寸以上的)巨缆;海缆电报;钢索;钢丝束;电缆;电报;索缆
cable action 缆索操作
cable-actuated excavator 缆控挖土机
cable address 电报挂号
cable anchor 缆索锚(栓);锚索;钢缆锚固
cable anchorage 缆索锚具;缆绳锚固;钢索锚碇;钢丝绳锚具;钢缆锚固;锚固装置
cable anchorage pier 缆索锚桩;电缆夹钢丝绳锚固支柱;电缆夹钢绳锚固支墩;电缆夹钢丝绳锚固台;钢绳锚固支墩;钢丝绳锚固台
cable and accessory 电缆及附件
cable and pipe trench 电缆和管沟;电缆管道沟
cable-and-trunk schematic 电缆和干线简图
cable an offer 电报报盘
cable area 海底电缆区
cable armour cutter 电缆钢带铠装剥切刀
cable armo(u)r(ing) 电缆铠装
cable assembly 光缆组件;钢索组件;电缆组件
cable band 缆(索)箍;缆绳卡箍;钢丝卡箍;索箍
cable barrier 缆式栅栏
cable-based local area network 基于电缆连接的局域网
cable basement 电缆基础层;地下电缆室
cable bay 电缆电线间隔
cable beacon 电缆标杆
cable bearer 电缆支架
cable belt conveyer 缆索传送带式运输机;夹钢丝芯胶带输送机;钢绳胶带输送机;绳带(式)输送机;索道输送机

cable bend 链索结
cable bent 缆索垂度;悬索曲度
cable bent tower 缆索塔架;索塔
cable bit 钢丝冲击钻头;冲击钻头
cable boat 放缆船
cable bond 电缆连接器;缆索联结;缆索接头;电缆接头
cable boring 冲击钻孔;冲击钻机
cable boring tool 电缆表面穿孔器
cable box 有线电视机;电缆箱;电缆套管;电缆盒
cable brace 拉绳
cable bracket 电缆(托)架
cable braiding machine 编绳索机
cable brake 索闸
cable branch(ing) 电缆分线;电缆分支;缆索分支
cable branching terminal box 分向电缆盒
cable breakdown 电缆击穿
cable breaking 电缆断裂
cable bridge 缆索桥;索桥;悬索桥
cable buoy 锚索浮标;锚标;海底电缆浮标;电缆浮标;导缆浮标(港内没有系泊浮筒时);水底电缆浮标
cable burying vehicle 电缆埋设潜水器
cable calibration amplifier 电缆校正放大器
cable cantilever bridge 悬索桥
cable cantilevering roof 悬索悬挑屋顶
cable capacitance 电缆电容
cable car 缆车;悬空缆车;载客吊笼;吊笼;电缆车;索车
cable car crane 缆车起重机
cable car pumping station 缆车式泵站
cable carriage 缆车;架空缆车;悬空缆车
cable cart 电缆盘拖车
cablecast 电缆广播
cable certificate 锚链强度证书
cable chain 缆索链;锚链
cable chamber 电缆室
cable change-over box 电缆转换箱
cable channel 电缆沟;缆索槽;电缆管道;电缆槽
cable check 电缆检查
cable checking 校缆
cable chock 导缆器;导缆孔
cable chute 缆索滑(运)道;电缆沟;电缆槽
cable clamp 缆索夹;线夹;钢丝绳夹;电缆夹;电缆挂钩;索卡;索夹;绳夹
cable clamp collar 电缆连接卡头
cable clamshell 缆索抓斗挖土机;索式抓斗;绳式蛤壳形抓斗
cable clamshell excavator 缆索抓斗挖土机
cable cleat 线夹;电缆夹具
cable clench 系链扣座
cable clevis 钢绳活环
cable clinch 系链扣座
cable clip 钢索夹头;电缆夹头;电缆线卡;电缆卡钉;电缆夹;电缆挂钩
cable coil 电缆盘
cable collector drum 卷缆筒;集缆筒;电缆筒
cable column 卷缆柱
cable communication 有线通信;有线传输;电缆通信
cable compass 电缆罗盘
cable compensation 电缆矫正
cable complement 电缆对群
cable compound 钢绳油;钢缆油;电缆油;电缆绝缘胶;电缆膏
cable compressor 锚链掣
cable conduct 电线套管
cable conductor 电缆线
cable conduit 电缆套管;电缆(通)道;电缆管(道);电缆沟;电缆导管;电缆槽;电缆暗沟
cable confirmation 电报确认
cable connecting 接线
cable connecting plug 电缆连接插头
cable connection 缆索接头;悬索接头;电缆连接
cable connection panel 电缆接线板
cable connector 电缆连接盒;电缆连接器;电缆接

头;电缆插销
cable connector cable wax 电缆连接装置
cable continuity test 电缆连续性测试
cable control 钢索控制;钢索操纵
cable control device 缆索控制装置
cable-controlled 缆索控制的
cable-controlled brake 钢索操纵制动
cable-controlled scraper 索控式铲运机
cable-controlled underwater research 缆挖式调查潜水器
cable control system 缆索控制系统
cable control unit 拖拉推土机;索控装置;缆索操作机械;缆索控制装置
cable conveyer 缆索输送机;钢丝网吊运机;钢索传送机;缆索运输机;缆索输送机;缆道输送机;吊篮输送机;索道输送机;绳索输送机;缆车
cable core 缆芯;电缆芯
cable core tube assembly 绳索取芯装置
cable correction 电缆校正
cable correction unit 电缆校正装置
cable coupler 电缆连接器
cable coupling box 电缆接线盒;电缆接线盒
cable cover(ing) 电缆沟盖;电缆包皮;电缆护套;电缆(内)套
cable covering machine 电缆包编机
cable-covering press 电缆包皮压力机
cable cover lead cutter 电缆破铅皮刀
cable crane 缆索运送机;缆索式起重机;缆索吊机;缆式起重机;索起重设备;索道(式)起重机;索道起重设备
cable-crane winch 缆索吊车的绞盘
cable crane with swinging leg 摇摆式缆索起重机
cable credit 电报信用证
cable crossing 过河电缆
cable culvert 电缆管道
cable current transformer 电缆用电流互感器
cable cutter 钢索剪断器;钢绳割刀;电缆剪
cable cylinder 钢索缸
cabled 用缆索系固
cable dancing 电缆摇荡
cable dead end 缆索锚固
cable delay 电缆延迟
cable depth indicator 电缆深度指示器
cable detecting device 电缆检测设备
cable detector 探测缆索器;电缆检验器
cable deviation equalizer 电缆偏差均衡器
cabled fluting 肋形柱槽;凹弧与平肋相间的装饰
cabled hyperbolic paraboloidal roof 悬索双曲抛物面屋顶
cable diameter 钢绳直径
cable diameter ga(u)ge 缆径测量仪
cable differential protection 电缆差动保护
cable distribution box 交接箱;电缆交接箱;电缆分线箱;电缆分线盒
cable distribution head 电缆分线箱;电缆分线盒
cable distribution point 电缆交接点
cable distributor 电缆配线架
cable draft 电汇票
cable drag scraper 缆索拖铲;缆式铲运机;索曳铲运机;索铲
cable-drawn scoop 钢索传动斗
cable draw-off gear 抽出缆索的工具
cable draw pit 缆索滑运道
cable dredging machine 钢索式挖掘机;缆索式挖泥机
cable drill(er) 缆索钻机;钢索冲击(式)钻机;钢丝绳冲击(式)钻机;钢绳冲击(式)钻机
cable drilling 缆索钻井;缆索钻进;缆索打井法;钢索冲击式钻进;钢绳冲击钻进;冲击钻进;冲击钻;索钻
cable drilling bit 缆索式钻头
cable drilling machine 钢丝绳冲击(式)钻机
cable drilling system 钢丝绳冲击式钻眼法
cable drilling tool 钢绳冲击钻具;钢绳冲击钻钻具;索钻钻具
cable drill rig line up 钢绳冲击钻机组合
cable drive 缆索传动
cable-driven car 钢索牵引车
cable-driven excavator 机械传动挖掘机
cable-driven ingot buggy 钢丝绳牵引的送锭车;钢索牵引送锭车
cable dropper 缆索吊架;电缆吊架
cable drum 缆绳滚筒;钢索卷筒;钢丝绳卷筒;钢缆卷筒;钢索鼓筒;电缆盘;电缆滚筒

cable drum carriage 电缆放线车
cable drum handling equipment 绞盘式装卸设备
cable drum jack 电缆盘架
cable drum table 卷放电缆支撑架;电缆转盘台
cable drum trailer 缆索卷筒运输拖车;电缆卷筒拖车
cabled staple fiber yarn 定长纤合股纱
cable duct 缆索管道;缆线道;缆索钩;公用隧管;钢丝束孔道;电缆管道;电缆道;电缆暗沟
cable duct making machine 制造缆索通道的机器
cable duct tube 电缆导管
cable dummy section 电缆交接箱
cable-dump truck 钢索自卸卡车;索式自卸卡车
cable eccentricity ga(u)ge 钢索偏心度测量仪
cable echo 电缆回波
cable-electrical 电缆
cable encasing-tube 缆索套管;钢索套管
cable end 缆索终端
cable end connecter 电缆端接盒
cable engineering and fittings 电缆工程及配套材料
cable entry 电缆入口;电缆进线口
cable equalizer 电缆均衡器
cable excavator 塔式缆索挖掘机;吊铲;缆索式挖土机;缆索(式)挖掘机;缆索挖掘机;索挖式铲运机;铲挖掘机
cable excavator tower 缆索开挖的塔
cable eye 钢丝绳套环;绳眼
cable factory 电缆厂
cable fastenings 电缆紧固件
cable fault 漏电;电缆损伤;电缆故障
cable fault detector 电缆故障探测器;电缆故障检测器
cable fault finder 电缆故障测定仪
cable fault indicator 电缆故障指示器
cable fault locator 电缆故障定位仪
cable fault search set 电缆事故检查器
cable ferry 缆索摆渡;缆索渡口;缆索牵引渡口;缆索牵引渡船;跨河缆道;牵引渡船
cable fill 电缆占用率
cable filler yarn 电缆填充线
cable filling compound 电缆填料
cable film 电缆电影
cable fishing tool 钢绳冲击钻进(用)打捞工具
cable fitting 电缆配件
cable flat roof 索索平屋顶
cable flaw detector 电缆探伤仪
cable flit-box 可拆卸式电缆接线盒
cable for anti-wind vibration 防风震缆索
cable for lighting 照明用电线;照明电缆
cable form 电缆模板
cable-former 成缆机
cable for prestressed concrete 预应力混凝土用钢索
cable for ship 船用电缆
cable-free 不需接外电源的
cable gallery 电缆廊道
cable gate hoist 卷扬式启闭机
cable gear 敷设电缆的机械;敷电缆机
cable geometry 缆索几何形状
cable-grade talc 电缆级滑石
cablegram 海缆电报;海底电报;水线电报
cable grease 缆索润滑脂;电缆脂
cable grip 钢丝绳夹;电缆轴摇把;电缆钳;电缆扣;电缆夹
cable gripper 电缆夹
cable groove 钢索槽
cable guard 缆索的外护层;钢丝绳防护装置
cable guard rail 缆索护栏;链索护栏
cable guard railing 钢索护栏
cable guidance 有线制导
cable guide 缆索引出管;电缆(引)导管;导缆器;绳罐道
cable gum 电缆胶
cable handling 敷设电缆
cable handling equipment 布缆设备
cable hanger 缆索吊杆;缆索吊架;电缆挂钩;电缆吊架
cable hanging unit 电缆悬挂装置
cable haulage 缆索运输
cable haulage machine 缆索运输机器
cable-hauled bucket 刮土机;索式耙斗
cable hauling gear 绞缆装置
cable hay stack 绞盘式干草堆垛机
cable head 电缆终端接头;电缆接头;电缆分线盒;电缆终端箱;电缆终端盒;电缆头

cable head plug 电缆插头
cable head receptacle 电缆插头
cable hoist 缆索绞车;卷扬机;钢丝绳吊车;钢绳卷扬机;电缆绞盘
cable hoist conveyer 索道输送机
cable hoisting pulley 钢缆起重滑车
cable holder 锚(机持)链轮;电极夹;持链轮
cable hook 锚链钩;电缆(挂)钩;电缆箍圈
cable hose clamp bracket 电缆软管夹托架
cable hull fittings 电缆填料涵(通过船体外板和隔板处的装置)
cable hut 电缆分线箱;电缆分线盒
cable in code 简码电报
cable in cotton 棉纱包电缆
cable inlet 电缆入口
cable input plug 电缆输入插头
cable inspection 锚链检验
cable installation 电缆线路敷设
cable installation work 电缆安装工程
cable in stock 备用电缆
cable insulating paper 电缆绝缘纸
cable insulation 电缆绝缘
cable interlayer 电缆夹层
cable iron 钢索铁;钢绳铁
cable isolating paper 电缆绝缘纸
cable isolator 电缆隔离器
cable jack 钢绳千斤顶;电缆盘千斤顶;电缆卷轴架;电缆卷筒
cable jacket 光缆护套;电缆外壳
cable jack test 锚索千斤顶试验
cable joint 缆索接头;悬索接头;电缆接头;绳接头
cable joint box 电缆接线箱;电缆套;电缆交接箱
cable jumper 连接电缆
cable kilometer 电缆敷设长度(以公里计)
cable knife 缆索刀;电缆刀
cable lac 电缆胶漆
cable lacquer 电缆漆
cable ladder 梯级式架桥
cable-laid rope 绳股捻成的缆索;拧索;左捻钢(丝)绳;多花绳;多股钢丝绳
cable lamp 行灯
cable lay 电缆绞距
cable layer 缆索架设工;海底电缆铺设船;电缆敷设船;电缆敷设机;电缆敷设船;布缆船
cable layer ship 电缆敷设船
cable laying 电缆铺设;电缆敷设
cable laying equipment 布缆设备
cable laying machine 敷设缆索机;缆索埋管机
cable laying plant 敷设缆索的设备
cable-laying ship 电缆敷设船;海底电缆铺设船;海底电缆敷设船;放缆船;布缆船
cable-laying truck 电缆敷设车
cable laying winch 敷设缆索的绞车
cable layout 电缆敷设图
cable lay wire rope 缆式钢丝绳
cable lead 电缆端套管
cable length 链(海上距离长度单位);链长;一链长
cable length switch 电缆长度转换开关
cableless linking of traffic signals 无电缆交通信号
cableless submersible 无缆式潜水器
cable level 缆索高度
cable lift 卷扬机;缆索提升机;缆索式起落机构;钢索起重机
cable lift bascule bridge 缆索提升仰开桥;缆索提升开启桥;索升式倾开桥
cable lifter 锚机持链轮
cable line 钢缆;电缆线路;绳索
cable line tester 电缆路径测试仪
cable load 电缆负载
cable locator 探测缆索器
cable lock 钢丝绳锁定器;钢丝绳保险锁
cable locker 锚链舱
cable locking system 绳索锁紧装置
cable log haul 缆索拉木机
cable loss 电缆损耗
cable lost 掉电缆
cable lower station 缆索下端站
cable lubricant 钢缆润滑剂
cable lug 连接线夹;电缆终端;电缆接头头
cable lug with two connecting holes 双孔连接线夹
cable machinery 布缆机械
cable manufacture 电缆制造
cable mark detector 电缆记号探测器

cable mark(er) 电缆标(志);电缆标(记);管线标
cable marking paper 电缆标记用纸
cable mat heating system (埋于楼板内的)网式供热系统;席式供暖系统
cable message 海外电报
cable messenger 悬缆线;吊线(电工);悬缆索
cable modem 电缆调制解调器
cable monitor 电缆监控器
cable mooring swivel 锚链转环
cable mo(u)lding 螺旋形线脚;卷绳线脚;卷缆花饰
cable negotiation 电报商议
cable net 大网眼窗纱
cable net system 索网体系
cable net with opposite curvature 反向曲面缆网
cable network 电缆网络;悬索网;电缆网
cable net(work) structure 钢索网(络)结构
cable nippers 电缆钳子;电缆拔钳
cable noise 电缆噪声;电缆干扰
cable of 48 seven mm wires 48根7毫米直径的钢丝索;48根7毫米直径的钢丝绳
cable oil 钢绳油;电缆油
cable-operated 缆索传动的
cable-operated brake 钢索操作制动器
cable-operated crawler excavator 缆索带动的履带式挖土机
cable-operated excavator 用缆索操纵的挖掘机;机械挖掘机
cable-operated face shovel 钢缆操纵正铲
cable-operated mucker 抓斗式装岩机
cable-operated observation and rescue device 缆挖式观察与救生装置
cable-operated shovel 钢缆操纵铲
cable optic(al) fiber 成缆光纤
cable order 电报订货
cable outlet 电缆出口
cable pairs 电缆芯线对
cable paper 电缆(绝缘)纸
cable path 钢丝束走向
cable percussion 钢绳冲击钻
cable percussion rig 钢绳冲击式钻架
cable phone 有线电话
cable picking and laying machine 卷缆机;电缆收放机;布缆机
cable pipe 电缆管道;电缆保护套管
cable piping 电缆管道
cable piping excavator 电缆管道挖沟机
cable pit 电缆沟;电缆槽
cable placing 电缆敷设
cable plane 索(平)面
cable plant 电缆厂
cable plough 钢索牵引犁;绳索牵引犁
cable plow 钢索牵引犁
cable plow unit 电缆敷设机
cable post 缆柱
cable pot-head 电缆端套
cable-powered 缆索带动的
cable power factor 电缆功率因数
cable protection pipe 电缆保护管
cable protection system 电缆保护系统
cable pulley 索轮;绳轮
cable pulley cradle 缆索滑轮托架
cable pull force 钢索张力
cable-pullout unloading method 绳拉卸车法
cable puncture 电缆击穿
cable pylon 电缆铁塔
cable quadrant 拉线盘
cable rack 电缆(托)架
cable railroad 缆车铁道;悬索铁路;悬索铁道
cable railway 电缆铁道;缆道;缆车铁道;悬索铁路;悬索铁道
cable railway terminal 缆车码头
cable ranging 排链待检
cable rate 电报汇率
cable reach (钢绳冲击钻进过程中的)钢绳弹性拉伸
cable record 电缆说明书
cable reel 缆索绞盘;缆索卷筒;卷索轴;(绳)卷筒;电缆盘;电缆轮;电缆卷筒;电缆绞车
cable reel capsule 钢绳卷筒筒封壳
cable reel jack 电缆卷轴架
cable reel locomotive 电缆式电机车
cable reel truck 缆索绞盘卡车
cable reinforced rubber belt conveyer 钢芯胶带运输机
cable relay 电缆继电器

cable release 快门线;线条快门开关;电缆释放装置
cable releaser 弃链器
cable release socket 快门线插口
cable reliever 锚链解脱器
cable repair ship 海底电缆维修船
cable resplicing 电缆重编接
cable-restrained air-supported structure 缆束充气建筑
cable restraint 电缆限制器
cable retention 光缆保持力
cable rig 缆索钻(探)架;顿钻钻机;索钻架;绳式钻具
cable ring 电缆圈
cable riser 电缆立管
cable road 缆道;索道
cable roadcable picking and laying machine 缆道
cable roller 电缆导轮
cable roof 悬索屋盖;悬吊屋顶
cable roof with saddle shape 鞍形悬索屋顶;悬吊鞍形屋顶
cable rope 缆索;钢丝缆
cable run 电缆敷设(线路);缆道线路;光缆主干;电缆线路;索道线路
cable running 电缆敷设
cable running list 电缆布线表
cable saddle 电缆鞍;索鞍;(悬索桥塔上的)绳索支承鞍座
cable sag 缆索垂度;绳索挠度
cable sag measurement 缆索垂度测量
cable scraper 缆索刮土机;缆索挖土机;塔式挖掘机;索(挖)式铲运机
cable semicontinuous casting machine 钢丝绳半连续铸造机
cable sender 电缆传送机
cables gutter 电缆槽
cable shackle 锚链卸扣
cable shaft 缆索驱干;电缆(竖)井
cable shaker 绳索式抖动器
cable shears 电缆剪刀
cable sheath 光缆皮;电缆外皮;电缆(护)套;电缆护皮;电缆(的保护)包皮
cable sheathing 电缆包皮
cable sheathing press 缆索铠装机
cable sheave 电缆绞轮
cable sheave bracket 钢丝绳滑轮支架
cable shelf 电缆支架
cable shield 电缆护套;电缆的输入套管;电缆的输入管道
cable ship 海底电缆铺设船;海底电缆敷设船;电缆敷设船;布缆船
cable shoe (吊桥上的)缆索脚座;电缆终端套管;电缆终端;电缆套管
cable shopping 缆线购物
cable shroud 电缆套
cables in double plane 双面索
cables in simple plane 单面索
cable skidder 索道集材机
cable skidding 索道集材
cable slackening switch 钢丝绳张弛调节器;电缆垂度调节开关
cable sleeve 电缆连接套管;电缆(接头)套管;电缆盒
cable's length 链(相当于0.1英里,1英里=1609.344米)
cable slide 绳索滑座
cable sling 缆式吊索;钢索吊钩;电缆吊钩
cable sock 钢丝保护套
cable socket 钢丝管套;电缆端头;电缆承口;电缆插口
cable spacer 钢丝定位件
cable speed 缆车速率;钢绳卷绕速度;缠绳速度
cable-spinning equipment 缆线绞车
cable splice 电缆拼接;锚索插接;电缆接头
cable splicing 钢丝绳接头
cable spreading room 电缆分布室
cable stack 拉线式堆垛机
cable stacker 缆索升降摄像机台
cable stand 电缆架
cable stay 缆索支撑;(吊桥上的)缆索支墩
cable-stayed bridge 斜拉桥;张拉桥
cable-stayed bridge with inclined cable plane 斜索面斜拉桥
cable-stayed bridge with mixed deck 混合桥面斜拉桥
cable-stayed bridge with single cable plane 单索面斜拉桥
cable-stayed bridge with single pylon 独塔斜拉桥
cable-stayed construction 斜拉结构
cable-stayed girder bridge 斜拉(板)梁桥
cable-stayed structure 斜拉结构
cable-stayed system 拉索体系
cable steel 缆索钢材
cable stopper 螺(旋锚)链掣;锚链制动器
cable strand 电缆线股
cable strength 锚链强度
cable stress detector 钢索测力仪
cable structure 悬索结构
cable stud 链环挡
cable subway 电缆管道;电缆地下管道;电缆隧道
cable supply 电缆式供电
cable support 电缆支架;缆索支承架
cable-supported 缆索支持的
cable-supported bridge 悬索桥;缆索支撑桥
cable support plate 钢丝绳支架金属板
cable support rack 电缆引入架
cable support tower 索塔
cable-suspended cantilever roof 缆索悬挂屋盖;悬索悬挂屋顶;索悬挂屋盖
cable-suspended current meter 悬挂式流速仪
cable-suspended feeder 钢索悬吊式给料机
cable-suspended hydrometric(al) cableway 悬索式水文缆道
cable-suspended point-integrating sampler 悬挂式积点采样器;悬挂式点积采样器
cable-suspended screen 悬索筛
cable-suspended structure 悬索结构
cable suspender 缆索吊桥;电缆挂钩;电缆吊架
cable suspension bridge 悬索桥;缆索悬(索)桥;缆式悬桥;钢悬索桥;钢索吊桥;索吊桥
cable suspension current meter 吊挂式流速仪
cable-suspension idler 钢绳托滚
cable suspension wire 接触线
cable switching box 电缆配件箱
cable swivel collar 钢丝绳回转环
cable system 缆索体系;钢绳冲击钻进(方)法;电缆系统;电缆网
cable system drill 钢绳冲击(式)钻机;顿钻
cablet 小缆;缆索;多股反搓缆绳
cable tail 电缆引线;电缆尾
cable take-up 钢绳弹性收缩(钻头冲击井底后)
cable take-up system 钢丝绳收紧装置
cable tank 电缆箱;电缆槽
cable tapping box 电缆分接箱
cable television 有线电流
cable television system 有线电视系统
cable telpher 高架索道;电动缆车
cable tensiometer 电缆张力计
cable tensioner 钢丝绳张紧轮
cable tension indicator 钢绳张力指示器;钢绳引力计
cable terminal 电缆(终)端;电缆接头
cable terminal box 电缆终端箱;电缆(终端)盒
cable terminal box on a post 杆上电缆盒
cable terminal rack 电缆转接架
cable terminating equipment 电缆终端设备
cable termination 电缆封端
cable termination box 终端电缆盒
cable tester 电缆测试器
cable testing 电缆试验
cable testing bridge 电缆故障测验桥
cable testing set 电缆测试器
cable threading 缆索卷成螺纹;钢丝绳张紧
cable tile 电缆瓦管;电缆套
cable tire 电缆胎
cable tool 索钻钻具;绳索钻钻具;钢丝绳冲击钻具
cable tool bit 钢绳冲击钻头
cable tool core 钢绳冲击钻进的岩芯
cable tool core barrel 钢绳冲击钻进的岩芯管
cable tool cuttings 钢绳冲击钻进岩屑
cable tool drill 钢绳冲击(式)钻机;钢绳冲击钻
cable tool drilling 钢绳钻从钻探;钢(丝)绳冲击钻进(法);索式冲击钻进法
cable tool drilling equipment 钢绳冲击钻井设备
cable tool drilling outfit 钢绳冲击钻进设备和工具
cable tool drilling system 钢绳冲击钻进(方)法
cable tool drilling unit 钢绳冲击钻进装置
cable tool jack 钢绳冲击钻千斤顶
cable tool joint 钢绳冲击钻接箍
cable-tool method 索式冲击(钻探)法;绳索钻具

法【岩】
cable tool outfit 钢绳冲击(式)钻机
cable-tool rig 顿井机组;钢绳冲市(式)钻机;缆索凿井机组
cable-tool well 凿井;钢缆绳冲击钻井
cable tool wrench 冲击钻扳手
cable-towed machine 钢索牵引机
cable towing traction 缆索牵引力;索式拖运
cable tracer 电缆检验器
cable tracing set 缆索探测器
cable track 索道
cable traction 缆索牵引
cable trailer 电缆(盘)拖车
cable tramcar 缆道电车
cable tramway 缆索电车道;缆车道;架空索道;索道电车
cable-tramway of single rope 单线索道
cable transfer 缆索运输;电汇
cable transporting truck 电缆运输车
cable transporting vehicle 电缆运输车
cable tray 电缆托架;电缆槽;托盘式架桥
cable trench 缆索沟;电缆通道;电缆沟;电缆槽
cable trench with chequer 带方形盖板的电缆沟槽
cable trolley 电缆小车
cable trough 电缆走线架;电缆(走线)槽;电缆暗渠;电缆通道;电缆沟
cable trunk 电缆管道
cable trunking 电缆匣式护套
cable truss 索桁架;悬索桁架
cable tunnel 电缆隧道;电缆廊道
cable type 电缆类型
cable type bulldozer 钢索操纵推土机
cable type crane 缆索起重机
cable type hitch 钢索联结装置
cable type shovel 钢绳传动单斗挖掘机
cable type suspension bridge 钢缆悬索桥
cable tyre 电缆胎;橡皮绝缘的
cable unit 电缆装置
cable varnish 电缆(清)漆
cable vault 电缆(地下)室;地下电缆检修孔
cable vessel 电缆船
cable vulcanizer 电缆硫化器
cable wander 缆索偏移
cable wax 电缆蜡
cableway 索道;架空索道;钢绳吊索;缆索起重机;缆(索)道;缆车索道;架空索道;架空缆道;钢索道;电缆通道;电缆管道;电缆沟;电缆槽
cableway bucket 缆索起重机吊斗;缆索起重机吊运的料罐;缆式吊灌;缆道运载斗;吊斗;索道吊罐;索道吊斗
cableway control house 缆索车道操纵室
cableway crane 架线起重机
cableway erecting equipment 缆索吊装设备
cable(way) excavator 缆索开挖机
cableway measurement 缆道测量
cableway power scraper 塔式挖掘机
cableway skidder 架空集材机
cable(way) tower 缆道塔;索道塔
cableway transporter 缆索运输机;缆道运输机;索道运输机
cable wheel 卷筒;锚机持链轮
cable winch 电缆卷扬机;绞(线)车;钢索绞车;钢绳绞车;钢绳绞车;电缆绞车
cable winder 卷缆机;绞缆机;绕线机
cable winding drum 缆索卷筒
cable wire 钢丝绳
cable work 电缆工程;敷设电缆
cable working vehicle 电缆工程车
cable works 钢绳绞车;电缆厂
cable wrapping machine 绕缆索机
cable wrapping paper 电缆包装纸
cable yarn 钢索股缆;电缆纱线
cabling 卷绳饰;结构化布线;架设电缆;电缆线路;电缆铺设;电缆合线;电缆并捻;打海底电报;敷设电缆;并捻;拖索除灌;卷绳状雕饰;卷缆饰
cabling diagram 电缆线路图
cabling machine 搓绳机
cabnite 砷硼钙石
cabo 海角
cabochon 圆形;馒头形;金刚石磨法;弧面型;顶部磨成圆形的宝石
cabonic ether 碳酸盐乙酯
cabook 砖红壤(土)
caboose 列车后守望车;舱面小厨房;守车

caboose parking track 守车线
caboose track 守车停留线
caboose working system 随乘制【铁】
cab-operated 驾驶室操纵的
cab-operated crane 驾驶室操纵的起重机
cab-operated overhead crane 驾驶室操纵的桥式起重机
caboret 有歌舞厅餐厅
Cab-O-sil 硅石粉
cabosil 气相法二氧化硅;气相法白炭黑
cabotage 沿海航行(权);沿岸交易;国内交通运输权;沿海贸易
cabotage right 沿海航运权
cabouche 舱面小厨房
cab over engine truck 平头型货车
cab over engine type 平头式载货汽车
cab rank 出租汽车停车处;出租汽车站
cabreuva 胡桃木(产于美国的棕色硬木)
cabriole leg 弯腿(S形曲线的家具腿)
cabriolet 活顶小轿车;篷式汽车
cabritte 锡铜钯矿
cab roof 驾驶室顶(板)
cab service 马车业
cab signal 车内信号;驾驶室信号
cab signal box 驾驶室信号箱
cab signal inductor location 机车信号作用点
cab signal testing section 机车信号测试区段
cab singal(l)ing 机车信号
cab singal(l)ing equipment 机车信号设备
cabstand 马车出租站;出租汽车站;出租汽车停车处
cabtire 有厚橡胶套的
cabtire cable 原橡胶绝缘电缆;厚橡胶软电缆
cabtire cord 橡皮绝缘软线
cabtire sheath(ing) 硬橡胶套管;硬橡胶护套
cabtire wire 厚橡胶线
cab turret 驾驶室分汽阀
cabtype cable 软电缆;厚橡皮绝缘软电缆;橡皮绝缘电缆;软管电缆
cabtyre 原橡胶绝缘电缆
cabtyre connector 橡皮绝缘插座
cabtyre cord 橡皮绝缘软线;橡皮绝缘软线
cabtyre rubber 橡皮绝缘软性电缆
cabtyre sheath(ing) 硬橡胶护套;硬橡皮电缆护套
cacaerometer 空气污染检查器;空气纯度测定器;空气纯度测定管
cacahuananche oil 柯莞油
cacciatore 水银地震计
cache 隐含存储器;储[贮]藏处;储[贮]藏场所;高速缓存;超高速缓(冲)存(储器)
cache buffer memory 超高速缓(冲)存(储器)
cache card 高速缓冲存储卡
cache contour 存储等高线
cachectic fever 黑热病
cache memory 高速缓冲存储器;超高速缓存
cache memory system 高速缓冲存储系统;超高速缓冲存储器系统
cachepot 花盆箱
cache storage 高速缓冲存(器);超高速缓(冲)存(储器)
cache storage hit 超高速缓冲存储器命中
cache storage miss 超高速缓冲存储器告缺
cache system 超高速存储器系统
cacholong 美蛋白石
cacimbo 加新坡雾【气】
cacinotron 返波管
cackling 防霉卷
CaCl₂ mud 氯化钙泥浆
cacocenite 黄磷铁矿
cacoclasite 钙铝黄长石
C/A code 来往账编码
cacodylate 二甲胂酸盐
cacoethic 不良的
cacogeusia 恶味
cacosmia 恶臭味
cacotopia 令人厌恶的社会;极不理想的居住地
cactus 仙人球;仙人掌
cactus grab 仙人掌式抓斗;多爪抓石器;多瓣抓斗
cactus type grab 多爪抓岩机
cadaster 地政局;地籍图;地籍册;地籍簿;房地产估算簿
cadastral control 地籍测量控制
cadastral district 地籍区
cadastral file 地籍测量文件

cadastral information system 地籍信息系统
cadastral investigation 地籍调查
cadastral list 地籍表册
cadastral map 地籍图
cadastral mapping 地籍测图
cadastral map series 地籍图系列;地籍图辑
cadastral plan 地籍(平面)图
cadastral sheet 地籍图
cadastral survey 地籍测量;地籍册
cadastral surveying and mapping 地籍测绘
cadastral survey manual 地籍测量细则
cadastral survey system 地籍测量系统
cadastration 地籍测量
cadastre 河流志;地籍图;地籍册;地籍簿;水册
CAD 计算机辅助设计
CAD automatic editing tool 计算机辅助设计自动编辑工具
cadaver 尸体
caddis fly 石蛾
caddy 零件搬运车;茶叶罐;茶罐;搬运工具
cade oil 杜松油
cadet engineer 轮机实习员;实习轮机员
cadet officer 驾驶实习员
cadger 小型注油器;小油壶
cadmium chlorate 氯酸镉
cadinene 杜松萜烯
cadion 镉试剂
cadmia 锌壳
cadmic compound 正镉化合物
cadmiferous 含镉的
cadminate 丁二酸镉(杀菌剂)
cadmiosis 镉尘肺
cadmium 自然镉
cadmium absorption 镉吸收
cadmium acetate 乙酸镉
cadmium amalgam 镉汞合金
cadmium antimonide 锑化镉
cadmium bearing alloy 镉轴承合金
cadmium bromide 溴化镉
cadmium cell 镉电池
cadmium chloride 氯化镉
cadmium coat 镀镉层
cadmium coating 镀镉
cadmium colo(u)r 镉颜料
cadmium column method 镉柱法
cadmium column reduction 镉柱还原法
cadmium compound 镉化合物
cadmium copper 镉铜
cadmium copper stranded conductor 镉铜绞线
cadmium copper wire 镉铜线
cadmium crimson 镉大红颜料
cadmium cycle 镉循环
cadmium deep orange 深镉橙颜料
cadmium deep red 深镉红颜料
cadmium deposit 镉矿床
cadmium dihydrogen phosphate 磷酸二氢镉
cadmium dithionate 连二硫酸镉
cadmium eliminator column 除镉塔
cadmium filter 镉过滤器
cadmium finish 镀镉层
cadmium fluoride 氟化镉
cadmium fluoride pho-ton spectrometer 氟化镉光子谱仪
cadmium-germanium detector 镉锗探测器
cadmium gold 镉金装饰合金
cadmium green 镉绿(色)
cadmium hydride 氢化镉
cadmium hydroxide 氢氧化镉
cadmium ingot 镉锭
cadmium iodide 碘化镉
cadmium ion 镉离子
cadmium ions removal 除镉离子
cadmium lamp 镉灯
cadmium lemon yellow 镉柠檬黄颜料
cadmium light orange 浅镉橙颜料
cadmium line 镉谱线
cadmium lithopone 镉黄;镉钡颜料
cadmium lustre 镉光泽彩
cadmium magnesium alloy 镉镁合金
cadmium maroon 镉紫红颜料;镉枣红颜料
cadmium mercury 镉汞
cadmium mercury lithopone 镉汞钡颜料
cadmium mercury red 镉汞红颜料
cadmium mercury sulphide 镉汞橙颜料

cadmium metal 镉合金
cadmium metallurgy 镉冶金
cadmium metasilicate 硅酸镉
cadmium neutron 镉中子
cadmium-nickel accumulator 镉镍蓄电池
cadmium nickel alloy 镉镁合金
cadmium-nickel storage battery 镉镍蓄电池
cadmium-nickel storage cell 镍镉蓄电池;镉镍蓄电池
cadmium niobate 铌酸镉
cadmium oil poisoning symptom 镉油中毒症状
cadmium orange 镉橘黄;镉橙
cadmium ore 镉矿
cadmium orthosilicate 原硅酸镉
cadmium oxalate 草酸镉
cadmium oxide 氧化镉
cadmium oxide photovoltaic cell 氧化镉光生伏打电池
cadmium photocell 镉光电池
cadmium pigment 镉颜料
cadmium plasma 镉等离子
cadmium plated 镀镉的
cadmium plating 电镀镉;涂镉;镉镀层;镀镉
cadmium poisoning 镉中毒
cadmium pollutant 镉污染物
cadmium polluted fertilizer 镉污染肥料
cadmium pollution 镉污染
cadmium primrose 镉草黄颜料
cadmium purple 镉紫颜料;镉紫色
cadmium ratio 镉比
cadmium red 镉红;法国朱红色
cadmium red pigment 镉红颜料
cadmium salt 镉盐
cadmium scarlet 镉猩红颜料
cadmium selenide 硒化镉颜料;硒化镉
cadmium selenide cell 硒化镉光电管
cadmium-selenium red 镉硒红
cadmium silver 镉银合金
cadmium-silver oxide cell 镉银氧化物电池;镉氧化银电池
cadmium solder 镉焊料
cadmium standard cell 镉标准电池
cadmium sulfate 硫酸镉
cadmium sulfide 硫化镉
cadmium sulfide cell 硫化镉光电管
cadmium sulfide exposure meter 硫化镉曝光表
cadmium sulfide photocell 硫化镉光电管
cadmium sulphide 硫化镉
cadmium sulphoselenide 硫硒化镉颜料
cadmium telluride 碲化镉
cadmium test 镉棒测试
cadmium test voltmeter 镉棒测试专用伏特计
cadmium tungstate 钨酸镉荧光颜料
cadmium vapo(u)r arc lamp 镉汽弧灯
cadmium vermil(l)ion 朱红色的;镉汞红颜料
cadmium wastewater 含镉污水;含镉废水
cadmium yellow 镉黄颜料;镉黄
cadmium yellow pigment 镉黄(颜料)
cadmium-zinc eutectic alloy 镉锌共晶合金
cadmium zirconate 锆酸镉
cadmopone 镉黄
cadmoselite 硒镉矿
cadmous compound 亚镉化合物
Cadomian orogeny 卡多米造山运动
Ca'd'Oro 黄金府邸(15世纪意大利威尼斯)
caducei symbol 医疗单位标志;双蛇杖标志;神使杖标志
caducity 脱落性
caducous 早落的;脱落的
cadux HS 光亮镀锡
cadwaladerite 氯羟铝石
caecum 盲端
caementicius 古罗马粗毛石建筑
caenis 小蜉蝣
Caen-stone 波纹奶色石灰石;褐色海生石灰岩;亮褐色海生石灰岩(英国)
caeoma 裸锈子器
caesious 青灰色的
caesium chromate 铬酸铯
caesium iodide 碘化铯
caesium-oxygen cell 充气铯光电管
caespiticolous 草栖的
caespitoso-graminosa 生草禾本石草原

CaF₂ film 二氟化钙薄膜
cafarsite 砷钛铁钙石
cafe 露天餐馆
cafe chantant 音乐餐厅
cafe curtains 餐馆帘;半截帘(遮住门窗下半部的帘子);褶裥窗帘
cafeteria 小吃部;自助食堂;自助餐馆
cafetite 钙钛钛矿;钙钛铁矿
cafetorium 礼堂;食堂— 礼堂;自助食堂两用大厅(学校或大楼内)
cafe-terrace 露天咖啡馆
Caffali process 卡发利上蜡法
caffeine 咖啡因
cafferata 隔墙石板;隔墙平板
cage 笼;护圈;栅笼;罐笼;隔离罩;电梯厢;保持架;(竖井内的)升降车
cage adapter 外壳连接器
cage antenna 笼形天线;鼠笼天线
cage assembly 升降装置;升降台
cage bag filter 栅式布袋过滤器
cage bar 栅笼壁条
cage beam 罐道梁
cage chair 罐座
cage circuit 笼形电路
cage clearance 罐笼间隙
cage compartment 罐笼隔间
cage construction 笼式结构;框架(式)结构;骨架(式)结构;骨架(式)构造
cage control 笼内控制;车厢内控制
cage culture 笼养(法);网箱养殖
cage cylinder 栅条式滚筒
cage disintegrator 笼式磨碎机;笼式粉碎机
caged ladder 后边有安全装置的梯子
cage drum 笼式滚筒
cage effect 笼蔽效应
cage elevator 笼式升降机
cage failure 保持架损坏
cage fish culture 网箱养鱼
cage goniometer 笼形侧向器
cage grid 笼形栅极
cage guide 罐道;电梯导轨;升降机笼导轨
cage hanger 罐笼悬挂装置
cage hoist 升降机筒;笼式升降机;罐笼提升机
cagekep 罐托;罐座
cage knob 锁钮
cageless anti-friction bearing 无夹圈滚动轴承
cageless ball bearing 无保持架的滚珠轴承
cageless rolling bearing 无保持架轴承
cage lifter 笼式升降机
cagelike 笼形的
cage mast 笼形桅;格子桅
cage mill 笼形磨机;笼式磨碎机;笼式粉碎机
cage mill disintegrator 高速转运破碎机
cage-mill dryer 笼式研磨干燥机
cage mixer 笼式搅拌机;笼式混料器
cage motor 鼠笼式电动机
cage of reinforcement 钢筋组架;钢筋笼;钢筋骨架
cage oil press 笼式榨油机
cage platform 装罐台
cager 井口信号工;把钩工
cage rack 笼形拦污栅;笼式拦污栅
cage reaction 笼闭反应
cage rearing 笼养(法)
cage relay 笼形继电器
cage ring 隔离圈;鼠笼端环
cage roller 笼形镇压器
cage rotor 笼形转子;笼式转子;转笼;短路式转子;鼠笼转子;鼠笼式转子
cage-rotor aeration 转筒曝气系统
cager rocker shaft 装罐用推车机的推杆
cage screen 笼形筛网;笼式筛网;笼式拦污栅;笼式格栅;笼筛;铁网笼
cage sheave 罐笼提升天轮
cage shoe 笼架底座;罐爪;罐耳
cage structure 笼形结构
cage switch control 锁定转换控制
cage-tainer 框架集装箱
cage type recuperator 笼形换热器
cage type valve 笼形阀
cage wheel elevator 笼轮式升运器
cage winding 罐笼提升;鼠笼式绕组
cage winding plant 罐笼提升设备
cage work 通花制品

cage zone melting 笼式区域熔化
cage zone refining 笼式区域提纯
caging 吸持;装罐
caging device 限位装置
caging machine 装罐机
caging mechanism 锁定机构
caging section 锁紧机构
caging system 锁紧系统
caging unit 装罐设备
cahot 坑洞;涡穴
cahtode glow 阴极辉光电
Caillet 凯列特过程(气体由高压到低压自由膨胀过程)
cailloutis 卵石;砾石
Caiman trench 凯曼海沟
Cainozoic era 新生代【地】
Cainozoic group 新生界
Cain's formula 凯恩公式
Ca-ion exchange capacity 钙离子交换容量
cairn 圆锥形石堆;堆石路标;堆石(觇)标;石砌陆标;石堆
cairngorm 烟色的石英;烟水晶
cairn point 堆石觇标点
cairn's lodge 最佳石灰石
caission disease 潘函病
caisson 藻井;凿井活箱;隔离空舱;弹药箱;打捞浮筒;船坞闸门;沉箱;浮箱式坞闸;浮动坞门;潜箱(修理船底用);屋顶曝晒箱;屋顶曝晒架
caisson base ring 沉箱圈梁
caisson box 封底式沉箱
caisson breakwater 沉箱(式)防波堤
caisson breakwater having a vertical wall penetrated by holes 开孔箱防波堤
caisson built on artificial island on shallow water 筑岛沉箱;筑岛沉井
caisson built on bank and floated to the site 浮式沉井
caisson ceiling 花格平顶【建】;藻井天花板;藻井平顶;古建筑中的藻井
caisson chamber 船坞闸门槽室;沉箱室
caisson completion system 沉箱式完井系统
caisson cutting edge 沉箱的刃口;沉井刃脚
caisson cutting shoe 沉箱的刃脚
caisson decompression 沉箱减压
caisson design 沉箱设计
caisson disease 沉箱病;潜水员病;潜涵病;气压沉箱病
caisson dock 沉箱坞
caisson drill 沉井桩孔钻孔
caisson flight 沉箱踏步
caisson floor plate 空腹楼板
caisson floor slab 空腹楼板
caisson foundation 沉箱基础
caisson foundation structure 沉箱基础结构
caisson gate 箱式坞门;沉箱闸门;沉箱式坞闸;浮坞门
caisson hatch covering 沉箱封舱
caisson launching 沉箱下水
caisson method 沉箱法
caisson mole 沉箱堤
caisson monolith construction 整体式沉箱结构;整体沉箱结构
caisson or pier foundation 沉箱或墩台基础
caisson pattern 藻井形态
caisson perdu(e) system 隐蔽沉箱系统;隐蔽沉箱结构
caisson pier 沉箱式墩;沉箱墩
caisson pile 凿井沉箱;沉管灌注桩;现场灌注桩;灌注桩;管桩;管柱桩;管柱沉井施工法;沉箱桩
caisson placing 沉箱安放
caisson platform 沉箱(平)台
caisson precasting yard 沉箱预制场
caisson precast platform 沉箱预制平台
caisson prefabricating yard 沉箱预制厂
caisson quaywall 沉箱码头;沉箱岸壁
caissons 井式平顶;井式顶棚
caisson separator 沉箱分离室;沉箱分离器
caisson-set 沉箱套;沉箱结构
caisson shell 沉箱外壳
caisson sinking 沉箱下沉;沉箱法凿井
caisson slab 空腹板
caisson slipway 沉箱滑道
caisson soffit 沉箱底面;空腹拱
caisson stall 沉箱发生故障;沉箱截流

caisson storage 沉箱储[贮]存
caisson structure 沉箱结构
caisson-supported pier 沉箱支承的墩
caisson system 沉箱系统
caisson towing 沉箱拖运
caisson type 沉箱类型
caisson-type pile 沉箱式桩
caisson wall 沉箱挡土墙
caisson wharf 沉箱码头
caisson with digging wells 有横墙的开口沉箱
caisson work 沉箱施工法
caisson worker 沉箱工人
caisson working chamber 沉箱工作小室
caisson working chamber pressure 沉箱工作室压力
caisson works 沉箱工程
caisson yard 沉箱预制场；沉箱预制厂
Caithness 凯思内斯暗灰色砂岩
Caithness flag 凯思内斯板层
cajeputene 白千层萜
cajeputol 桉油精
cajuelite 金红石（矿）
cajuput oil 白千层油
Caka tectonic knot 茶卡构造结【地】
cake 块状物；泥饼；钻尾皮；团块
cake asphalt 沥青饼
cake breaker 滤饼破碎机
cake compressibility 压缩指数
cake conveyer 滤饼运输带
cake core 型芯片；饼状岩芯；扁平型芯
caked clinker 块状熟料；熟料结块；熟料大块
cake discharge 滤饼卸出装置；料饼卸出装置
caked mass 结块；烧结体；烧结块
caked-on ink 结块油墨
cakefactory 取暖室
cake filtration 滤饼过滤
cake fork 点心叉
cake friction factor 泥饼摩擦系数
cake ice 小冰块；板冰
cake lac 饼状虫胶
cake mass 结块体；烧结体
cake mill 碎饼机
cake mixer 拌泥机式搅拌机
cake of alum 明矾块
cake of filter-press 压滤机滤饼
cake of metal 金属锭
cake press 颜料滤饼
cake scraper 滤饼刮板
cakes of cement 小僵块
cake sticking 泥饼黏附卡钻
cake tea 茶砖
cake thickness 泥饼厚度
cake wax 蜡盘
cak(e)y 成（了）块的
caking 油漆块渣；结焦性；块结；结块；加热黏接；埋版；油漆硬结；堆版；烧结
caking capacity 黏结能力；固结能力
caking cementation 黏结
caking coal 黏结性煤；黏结煤；煤饼
caking index 黏结指数
caking index G graduation of coal G 指数分级
caking index of coal G 指数
caking of crystals 晶体胶结；晶体的块结
caking of oil 油的黏结
caking power 黏结能力；烧结能力
caking property 黏结性
caking test 结焦性试验
caky 成了块的
cakzarlite 氟铝钠钙石
calabash 葫芦（做淘沙盘用）；排管式
calaboose 监狱；拘留所
Calabrian 卡拉布里亚阶【地】；阿斯蒂阶
Calacas-type iron deposit 卡腊贾斯型铁矿床
calacata 灰纹理大理石
calaite 绿松石
Calal 卡拉尔钙铝合金
calamander 柿木（产于东印度）
calamansanay（菲律宾产的黄红色~玫瑰色的）硬木
calamendiol 菖蒲二醇
calamina 炉甘石
calamina praeparata 精制炉甘石
Calamine 卡拉敏锌铅锡合金；炉甘石；菱锌矿；锌铅锡合金；异极石
calamine cream 锌膏

calamine lotion 炉甘石洗液；炉甘石洗剂
calamitous mudflow 灾害性泥石流
calamito-vitrite 芦木微镜煤
calamity 灾难；灾荒；灾害
calamity area 受灾害地区
calamity danger district 灾害危险区
calamity ecology 灾害生态学
calamity foreknowledge 灾害预测
calamity precaution planning 城市防灾规划
calamity prevention 防灾
calamus oil 菖蒲油
calandria 加热体；加热管群；中央循环管式蒸发器；蒸发设备；多效真空蒸发器；排管体；排管式（加热器）；排管
calandria evapo(u)rator 短管立式蒸发器
calanque 狭海湾
calathus （柱头花饰的）花篮形内心；钟形柱头；考林辛式柱头
calaverite 碲金矿
calawberite 骨石磁铁岩
calbe elliptic(al) roof 椭圆形缆索屋顶
calbe making machinery 制造电缆的机械
calc-alkali 碱性的
calc-alkali basalt 钙碱性玄武岩
calc-alkalic coefficient of Peacock 皮科克钙碱系数
calc-alkali diabase 钙碱性辉绿岩
calc-alkali gabbro 钙碱性辉长岩
calc-alkali granite 钙碱性花岗岩
calc-alkali lamprophyre 钙碱性煌斑岩
calc-alkali microsyenite 钙碱性微晶正长岩
calc-alkali nasaltic glass 钙碱性玄武玻璃
calc-alkaline series 钙碱系列
calc-alkali rock 石灰碱石头
calc-alkali syenite 钙碱性正长岩
calc-alkali trachyte 钙碱性粗面岩
calcaneal region 跟区
calcaneus branches 跟支
calcaneus rete 跟网
calcaphanite 钙质隐晶岩
calcar 熔窑；熔玻璃窑；熔炉；煅烧炉
calcarenaceous sandstone 灰屑砂岩；钙屑砂岩
calcarenite 灰屑岩；钙质岩；钙屑灰岩；砂屑灰岩
calcarenyte 钙屑岩
calcareous 含钙的；钙质（的）；石灰质的；石灰性
calcareous aeolianite 钙质风成岩
calcareous aggregate 钙质集料；钙质骨料；石灰质集料；石灰（质）骨料
calcareous alabaster 石灰质石膏；钙质石膏
calcareous alga 石灰（质）藻
calcareous algae 钙（质）藻；钙性藻类；石灰藻
calcareous alluvial soil 钙质冲积土；石灰性冲积土
calcareous anhydrite-gypsum ore 碳酸盐质硬石膏—石膏矿石
calcareous anhydrite ore 碳酸盐质硬石膏矿石
calcareous aquifer 钙质含水层
calcareous-bearing mudstone 含钙质泥岩
calcareous binding 钙质胶结物
calcareous breccia 钙质角砾岩
calcareous brick 浮石砖；钙质砖
calcareous cement 含钙水泥；钙质水泥；钙质胶结物；水硬水泥；水硬灰灰；石灰质合料；石灰类黏结料
calcareous cementing material 钙质黏结材料
calcareous chert 钙质燧石
calcareous clay 钙质黏土；白垩黏土；石灰质黏土；石灰性黏土
calcareous clay slate 石灰质黏板岩；钙质黏板岩
calcareous concretion 钙质结核
calcareous conglomerate 钙质砾岩
calcareous crust 钙质壳；钙积层
calcareous deposit 钙质沉积
calcareous dolostone 钙质白云岩
calcareous earth 石灰性钙；含钙土；钙质土；石灰质土
calcareous earthenware 石灰质陶器
calcareous earthenware type material 石灰质陶器材料
calcareous environment 钙质环境
calcareous eolianite 钙质风成岩
calcareous facies 石灰质外形；钙质相；石灰相
calcareous glaze 石灰釉
calcareous gravel 石灰质砾石
calcareous grit 石灰质粗砂岩；钙质粗砂岩；钙质粗砂岩

calcareous gypsum 简单石膏矿石；碳酸盐质石膏
calcareous gypsum-anhydrite ore 碳酸盐质石膏—硬石膏矿石
calcareous iron-stone 石灰质铁石
calcareous karst 石灰岩岩溶
calcareous layer 钙质层
calcareous-magnesian phosphorite ore 钙镁质磷块岩矿石
calcareous marl 钙质泥灰岩；石灰质泥灰岩
calcareous material 含钙材料；含钙物质；石灰质材料
calcareous mud 钙质（烂）泥；石灰黏泥
calcareous mudstone 钙质泥岩
calcareous ooze 钙质软泥；石灰质淤泥
calcareous pan 石灰盘
calcareous peat 钙质泥炭
calcareous petrosilex 石灰硅酸盐石头
calcareous quartz sandstone 钙质石英砂岩
calcareous ratchel 钙质料姜石
calcareous rock 含钙（石灰）岩；钙质岩；石灰岩
calcareous sand 钙质砂；石灰质砂
calcareous sandstone 灰质砂岩；钙（质）砂岩；石灰质砂岩
calcareous sediment 钙质沉积（物）
calcareous shale 石灰质页岩；灰质页岩；钙质页岩
calcareous sharn 钙质矽卡岩
calcareous shell 钙质介壳
calcareous silex 钙质石英；石灰质燧石；石灰质石英
calcareous siltstone 钙质粉砂岩
calcareous sinter 钙华；石灰华
calcareous-sinter cave 灰华洞
calcareous slack 钙屑
calcareous slag 石灰质矿渣；石灰炉渣
calcareous slate 钙质石板；钙质板岩
calcareous soil 石灰性土；含钙土；钙质土；石灰质土；石灰（性）土壤
calcareous spar 方解石
calcareous spring 钙酸泉；碳酸泉
calcareous stone 石灰质岩石
calcareous tufa 钙质凝灰岩；钙华；石灰华
calcareous tuff 石灰质沉凝灰岩；灰质沉凝灰岩；钙质凝灰岩；钙华
calcareous tuffite 钙质沉凝灰岩
calcareous water 钙质水
calcareous whiteware 石灰质白坯陶瓷；石灰质精陶
calcarious 含钙的
calcarious film 钙质薄膜
calcarium 耐洗的水漆
calc bentonite ore 钙质膨润土矿石
calc-chlorite schist 钙质绿泥片岩
calcedony 玉髓
calcflinta 钙质燧石
calc gneiss 钙质片麻岩
calc granite 钙质花岗岩
calc-hornblende schist 钙质角闪片岩
calcia 氧化钙
calcia-chondrodite 粒状硅钙石
calcia clinker 氧化钙熟料
calcic 含钙；钙质的；石灰的
calcic brown soil 棕钙土
calcic carbonatite lava 钙质碳酸岩熔岩
calcic granite 钙质花岗岩
calcic horizon 钙积层
calcic horizon distribution 钙质层分布
calcic magnesium carbonatite lava 钙镁质碳酸岩熔岩
calcicoater 石灰涂料
calcicole 钙生植物
calcicopiapite 钙叶绿矾
calcicosis 钙尘肺
calcic-phyllite 钙质千枚岩
calcicrete 结结钙
calcic rock series 钙质岩系
calcic silicate glass 硅酸钙玻璃
calcic-silicate hornfels 钙硅角岩
calcic slate 钙质板岩
calcierrite 钙磷铁矿
calciferous 含石灰质的；含（碳酸）钙的
calciferous petrosilex 石灰角页岩
calciferous sandstone 含钙砂岩
calcific 化成石灰的；钙化的
calcification 钙化（作用）；沉积作用；石灰化作用
calcifuge 嫌钙植物；非钙土植物；避钙植物
calcify(ing) 石灰化；钙化

calcigenous 石灰造成的
calcilization 方解石化(作用)
calcilutite 泥屑灰岩；灰泥岩；钙质泥岩；石灰泥岩
calcilutyte 泥屑灰岩；灰质碎屑岩；灰泥岩
calcimangite 锰方解石
calcimeter 碳酸计；碳酸测定仪；石灰(含量)测定器
calcimine 老粉；可赛银粉；墙粉；水浆涂料；刷墙水粉；刷墙石灰浆
calcimonzonite 钙质二长岩
calcimorphic soil 钙成土
calcimurite 钙氯岩
calcinate 煅烧产物；焙烧；脱水物
calcinated soda 无水苏打
calcinating 焙烧
calcination 煅烧；焙烧；焙解；灼烧
calcination furnace 灰化炉；煅烧炉；焙烧炉
calcination in dumps 堆垒煅烧
calcination in heaps 堆摊煅烧
calcination loss 灼烧损失；灼烧减量
calcination plant 煅烧(石灰)厂；煅烧设备；烧成车间
calcination progress 煅烧法
calcinations ratio 煅烧产率
calcination temperature 煅烧温度；分解温度
calcination zone 煅烧带
calcinator 料浆蒸发机；煅烧窑；煅(烧)炉；焙烧炉
calcinatory 煅烧的
calcine 焙烧
calcine car 焙砂小车
calcine cooler 焙砂冷却器
calcined alum 煅烧明矾
calcined alumina 煅烧氧化铝
calcined baryta 氧化钡；煅烧重晶石
calcined bauxite 煅烧铝矾土
calcined bauxite in powder 矾土粉
calcined calcium carbonate 煅烧碳酸钙
calcined clay 黏土熟料；煅烧高岭土；煅烧的黏土；煅烧瓷土；焙烧黏土
calcined colemanite 煅烧硼钙石
calcined diatomite 煅烧硅藻土
calcined dolomite 煅烧白云石；烧白云石
calcined dolomite in powder 白云石灰
calcined expanded aggregate 煅烧膨胀集料；煅烧膨胀骨料
calcined flint chips 煅(烧)燧石屑
calcined gypsum 建筑石膏；煅(烧)石膏；熟石膏；烧石膏
calcined kalinite 煅烧高岭石
calcined kaolin 煅烧高岭土；煅烧瓷土
calcined lime 生石膏
calcined limestone 煅烧石灰石
calcined magnesia 煅烧氧化镁；轻烧镁砂
calcined magnesite 煅烧菱苦土；煅烧镁氧土；煅烧菱镁矿
calcined magnesium carbonate 煅烧碳酸镁
calcined pigment 煅烧颜料
calcined plaster 熟石膏；建筑石膏；煅石膏；煅烧石膏
calcined product 煅烧产物
calcined raw meal 煅烧的生料
calcined soda 纯碱；苏打灰
calcine leaching 焙砂浸出
calciner 煅烧炉；分解炉；焙烧窑
calciner cyclone 分解炉旋风筒
calcines leaching circuit 焙砂浸出系统
calcining 生石灰；烧成物；煅烧
calcining chamber 煅烧室
calcining combustor 分解炉
calcining compartment 煅烧分隔间
calcining furnace 煅烧炉；分解炉；焙烧炉
calcining heat 煅烧热
calcining installation 煅烧装置
calcining kiln 煅烧窑；带卧式料浆蒸发机的窑；焙烧窑；焙烧炉
calcining oven 煅(烧)炉
calcining plant 焙烧装置；煅烧设备；煅烧厂
calcining rate 分解率
calcining region 煅烧带
calcining shaft 立筒式分解炉
calcining zone 煅烧带；分解带(碳酸盐)
calcinm oxalate calculus 草酸钙结石
calcioancylite 碳钙铈矿
calciobetafite 钙贝塔石
calcioborite 斜硼钙石
calciocarnotite 钙钒铀矿

calcio-chondrodite 钙质硅镁石
calcioferrite 钙磷铁矿；水磷钙铁石
calciogadolinite 钙硅铍钇矿
calciotalc 镁钙珍珠云母
calciotantite 钙钽石
calciovolborthite 钙钒铜矿
calciovollerthite 钒钙铜矿
calcipelite 泥屑石灰岩
calcipenia 钙质减少
calcipete 适钙植物
calcipexis 钙固定
calciphilous plant 钙土植物
calciphobe 避钙植物
calciphobous plant 避钙植物
calciphyre 斑纹大理石；斑花大理岩
calciphyte 钙土植物
calcipulverite 钙质细屑岩
calcirudite 砾屑灰岩；钙质砾岩；钙质砾屑灰岩；钙结砾岩
calcirudyte 砾屑灰岩
calcisiltite 钙质粉屑岩；粉砂屑石灰岩；粉砂屑灰岩；石灰粉砂岩
calcisphere 钙球
calcite 寒水石；方解石类矿物；方解石
calcite aegiapite 方解霓磷灰石
calcite ankerite magnesite beforsite 方解石铁白云岩菱镁矿镁云碳酸岩
calcite ankerite magnesite rauhaugite 方解石铁白云石菱镁矿白云(石)碳酸岩
calcite-bearing argillaceous dolomite 含方解石泥质白云岩
calcite-bearing dolomite mudstone 含方解石白云质泥岩
calcite beforsite 方解石镁云碳酸岩
calcite birefringence 方解石双折射
calcite cement 方解石胶结物
calcite compensation depth 钙的补偿深度；方解石补偿深度
calcite concrete 方解石混凝土
calcite crystal prism 方解石晶体棱镜
calcite deposit 方解石矿床
calcite diopside grossular hornfels 方解石透辉石钙铝榴石角岩
calcite dolomite 灰质白云岩
calcite dolomite-bearing mudstone 含方解白云石泥岩
calcite ettringite 方解钙矾石
calcite flour 石灰石粉
calcite-fluorite ore 方解石—萤石矿石
calcite grain 方解石粒
calcite in grain 方解石粒
calcite interferometer 方解石干涉仪
calcite limestone 方解石灰岩
calcite lining 方解石衬砌
calcite lysocline 方解石溶跃层
calcite marble 方解大理石
calcite rauhaugite 方解石白云(石)碳酸岩
calcite rhombohedron 方解石菱形体
calcite saturation index 碳酸盐饱和指数
calcite saturation(level) 碳酸钙饱和度
calcite sedimentation method 碳酸盐沉淀法
calcite streaks 方解石纹理；方解石夹层
calcite syenite 方解正长岩
calcite tremolite epidote hornfels 方解石透闪石绿帘石角岩
calcitic 石灰质的
calcitic dolomite 方解白云石
calcitic lime 方解石灰
calcitic limestone 方解石灰岩；高钙石灰岩
calcitic marble 方解大理石；石灰质大理石
calcitization 方解石化作用
calcitrant 耐火的
calcituff 灰质凝灰岩
calcium acetate 乙酸钙；醋酸钙
calcium acetylide 碳化钙
calcium acrylate 丙烯酸钙
calcium acrylate treatment 丙烯酸钙处理(化学灌浆)
calcium activation 钙活化反应
calcium age 钙龄
calcium alginate 藻酸钙
calcium alkalinity 钙碱度
calcium alkylbenzenesulfonate 合成磺酸钙
calcium alloy 钙合金

calcium aluminate 铝酸钙；钙矾土
calcium aluminate carbonate hydrate 水化碳铝酸钙
calcium aluminate cement 高铝水泥；矾土水泥；铝酸钙水泥；氧化铝水泥；高矾土水泥
calcium aluminate chloride hydrate 水化氯铝酸钙
calcium aluminate ferrite hydrate 水化铁铝酸钙
calcium aluminate glass 铝酸钙玻璃
calcium aluminate hydrate 铝酸钙水化物；水化铝酸钙
calcium aluminate silicate hydrate 水化硅铝酸钙
calcium aluminate sulphate hydrate 水化硫铝酸钙
calcium-alumin(i)um borosilicate glass 钙铝硼硅酸盐玻璃
calcium-alumin(i)um-silicon alloy 钙铝硅合金
calcium aluminoferrite 铝铁酸钙；铁铝酸钙
calcium aluminoferrite cement 铁铝酸钙水泥
calcium aluminosilicate 铝硅酸钙
calcium alumo-ferrite hydrate 水化铁铝酸钙
calcium and magnesium acetate 乙酸钙镁
calcium and magnesium oxide content 氯化钙氯化镁总含量
calcium antimonate 锑酸钙
calcium arsenate 砷酸钙
calcium arsonite 亚砷酸钙
calcium base grease 钙基润滑脂
calcium base lubricating grease 钙基润滑脂
calcium base titanox 钛完钙白颜料
calcium benzamidosalicylate 苯沙酸钙；苯甲酰胺水杨酸钙
calcium bicarbonate 重碳酸钙；碳酸氢钙
calcium bicarbonate spring 重碳酸钙泉
calcium bicarbonate thermal water 重碳酸钙型热水
calcium bicarbonate water 重碳酸钙水
calcium bioxalate 草酸氢钙
calcium bisulfite 亚硫酸氢钙
calcium bleach 漂白粉
calcium borate 硼酸钙
calcium boride 钙硼合金；硼化钙
calcium brine 钙卤水
calcium butyrate 丁酸钙
calcium carbide 电石；碳化钙
calcium-carbide furnace transformer 电石炉变压器
calcium carbide powder 碳酸钙粉末；电石粉
calcium carbide reflectivity 碳化钙反射率
calcium carboaluminate hydrate 水化碳铝酸钙
calcium carbonate 重质碳酸钙；白垩；碳酸钙
calcium carbonate activated 活性炭酸钙
calcium carbonate cement 碳酸钙胶结物
calcium carbonate compensation depth 碳酸钙补偿深度
calcium carbonate crystal 碳酸钙结晶
calcium carbonate equivalent 碳酸钙当量
calcium carbonate floc 碳酸钙絮凝物
calcium carbonate layer 碳酸钙层
calcium carbonate precipitation potential 碳酸钙沉淀势
calcium carbonate protective scale 碳酸钙保护层
calcium carbonate super-fine powder 超细碳酸钙
calcium cataplelite 钙锆石
calcium chloride 氯化钙
calcium chloride compounding tank 氯化钙配制槽
calcium chloride desiccator 氯化钙干燥器
calcium chloride injection 氯化钙针剂
calcium chloride powder 粉状氯化钙
calcium chloride solid fused 固体熔融氯化钙
calcium chloride treatment 氯化钙处理
calcium chloride tube 氯化钙管
calcium chloride water 氯化钙型水
calcium chloroacetate 氯乙酸钙
calcium chloroferrite hydrate 水化氯铁酸钙
calcium chromate 铬酸钙
calcium citrate 柠檬酸钙
calcium clay 钙质黏土；钙质黏粒
calcium cloud 钙云
calcium coefficient 钙质系数
calcium compound 钙混合物
calcium content 钙含量
calcium cuprate 铜酸钙
calcium cyanamide 氨氰化钙；氰氨(化)钙
calcium cyanide 氰化钙
calcium cycle 钙循环
calcium dialuminate 二铝酸钙；二铝酸一钙

calcium drier 钙干料;钙催干剂
calcium earth 钙质土
calcium electric rapid determination method 钙电极快速测定法
calcium electrode 钙电极
calcium extended titanium dioxide 钛钙白颜料
calcium ferrite 铁酸钙
calcium ferrite hydrate 水化铁酸钙
calcium ferrite sulphate hydrate 水化硫铝酸钙
calcium flocculus 钙谱斑
calcium fluoride 萤石;氟化钙
calcium fluoride infrared transmitting ceramics 氟化钙透红外陶瓷
calcium fluoride laser 氟化钙激光器
calcium fluoride structure 氟化钙结构
calcium fluoride substrate 氟化钙衬底
calcium fluoride type ceramics 氟化钙型陶瓷
calcium fluorite film 氟化钙薄膜
calcium fluorophosphate 氟磷酸钙
calcium fluosilicate 氟硅酸钙
calcium formate 甲酸钙
calcium grease 钙皂脂;钙基脂
calcium hardness 钙(质)硬度
calcium hardness of water quality 水质钙硬度
calcium hexa-aluminate 六铝酸钙
calcium humate 腐殖酸钙
calcium hureaulite 钙红磷锰矿
calcium hydrate 熟石灰
calcium hydride 氢化钙
calcium hydrogen carbonate 碳酸氢钙
calcium hydrosilicate 水化硅酸钙
calcium hydroxide 消石灰;氢氧化钙;熟石灰
calcium hydroxide crystal 氢氧化钙晶体
calcium hydroxide paste 氢氧化钙糊剂
calcium hypochlorite 漂白粉;次氯酸钙
calcium hypochlorite lime 漂白粉
calcium hypochlorite tablet 漂粉精片
calcium hypophosphate 连二磷酸钙
calcium hypophosphite 次磷酸钙
calcium iodate 碘酸钙
calcium iodide 碘化钙
calcium ion 钙离子
calcium-ionized clay 钙离子化黏土
calcium iron silicate 铁硅酸钙
calcium laevulinate 乙酰丙酸钙
calcium larsenite 钙铅锌矿
calcium lead 钙铅合金
calcium light 磷钙救生浮灯
calcium lignin(e) sulphonate 木质素磺酸盐;木素磺酸钙
calcium lignosulfonate 木质磺酸钙
calcium lignosulphonate 钙磺化木质素;木质(素)磺酸钙
calcium lime 未消石灰;生石灰
calcium line 钙线
calcium lithol red 钙立索红
calcium lithol toner 钙立索色原
calcium-magnesium alloy 钙镁合金
calcium magnesium aluminate cement 钙镁铝酸盐水泥
calcium-magnesium bicarbonate thermal water 重碳酸钙镁型热水
calcium magnesium carbonate 碳酸钙镁
calcium magnesium metasilicate 偏硅酸镁钙
calcium magnesium phosphate fertilizer 钙镁磷肥
calcium magnesium silicate 硅酸镁钙;硅酸钙镁
calcium maleate 马来酸钙
calcium manganate 锰酸钙
calcium-manganese-silicon alloy 钙锰硅合金;硅锰钙合金
calcium mesoxalate 丙酮二酸钙
calcium metal 金属钙;含钙铅基轴承合金
calcium metalolism 钙代谢(作用)
calcium metasilicate 偏硅酸钙
calcium metatitanate 偏钛酸钙
calcium mica 钙云母
calcium molybdate 钼酸钙
calcium monohydrogen phosphate 磷酸氢钙
calcium naphthenate 环烷酸钙
calcium network 钙网络
calcium niobate 铌酸钙
calcium nitride 二氮化三钙
calcium noleate 亚油酸钙
calcium oleate 钙皂;钙油酸脂;油酸钙

calcium orthoplumbate 原高铅酸钙
calcium orthosilicate 原硅酸钙
calcium oxalate 草酸钙
calcium oxalate crystal 草酸钙结晶
calcium oxide 生石灰;氧化钙;石灰
calcium oxide burn 氧化钙烧伤
calcium-oxygen isotope correlation 钙-氧同位素相关性
calcium pantothenate 泛酸钙
calcium parallax 钙吸收视差
calcium perchlorate 高氯酸钙
calcium permanganate 高锰酸钙
calcium peroxide 过氧化钙
calcium phenate 苯酚钙
calcium phosphate 磷酸钙
calcium phosphate gel 磷酸钙凝胶
calcium phosphide 磷化钙
calcium plage 钙谱斑
calcium plaster 石膏浆;石灰(砂)浆
calcium-plastic material 钙塑材料
calcium plumbate 高铅酸钙
calcium plumbate anti-rust paint 高铅酸钙防锈漆
calcium plumbate-coated pigment 高铅酸钙包膜颜料
calcium plumbate-coated silica 包核高铅酸钙颜料;高铅酸钙包膜二氧化硅
calcium plumbite 铅酸钙
calcium polyacrylate 聚丙烯酸钙
calcium polyphosphate 聚磷酸钙
calcium polysulfides 石硫合剂
calcium polysulphide 多硫化钙
calcium precipitating organism 沉淀钙的生物
calcium propionate 丙酸钙
calcium quicklime 生石灰
calcium rate 钙率
calcium reduction route 钙还原法
calcium resinate 钙松脂酸盐;松香酸钙;树脂酸钙;石灰处理的松香
calcium-rich coal 富钙煤
calcium-rich peat 富钙泥炭
calcium salt 钙盐
calcium/silica pigment 钙离子交换二氧化硅防锈颜料
calcium silicate 硅酸盐产品;硅酸钙
calcium silicate block 硅酸钙大砖
calcium silicate board 硅酸钙板
calcium silicate brick 硅酸钙砖(过梁);硅酸盐砖
calcium silicate concrete 硅酸盐混凝土;硅酸钙混凝土
calcium silicate face brick 硅酸盐面砖;硅酸钙面砖
calcium silicate facing brick 硅酸钙面砖;石灰砂砌面砖
calcium silicate hydrate 硅酸钙水化物;水化硅酸钙
calcium silicate insulation 硅酸钙绝缘材料;硅酸盐绝缘材料
calcium silicate thermal insulation material 硅酸钙绝缘材料
calcium silicide 硅化钙
calcium silicofluoride 硅氟化钙
calcium silicoluminate hydrate 水化硅铝酸钙
calcium silicon 硅钙合金
calcium soap 钙皂
calcium-soap grease 钙皂脂;钙皂(基)润滑脂;钙基皂润滑脂
calcium soil 钙质土
calcium stannate 锡酸钙
calcium stannate ceramics 锡酸钙陶瓷
calcium star 钙星
calcium stearate 硬脂酸钙
calcium-strontium ratio 钙锶比值系数
calcium sulfate 硫酸钙;石膏
calcium sulfate cement 石膏胶凝材料;硫酸钙水泥素
calcium sulfate hemi-hydrate 半水石膏
calcium sulfate hydrate 石膏铝钙水化物
calcium sulfate plaster 石膏灰浆
calcium sulfate spring 硫酸钙泉
calcium sulfate water 硫酸钙水
calcium sulfide 硫化钙
calcium sulfite 亚硫酸钙
calcium sulfoaluminate 钙硫化铝酸盐;硫化铝酸钙
calcium sulphate 石膏;硫酸钙
calcium sulphate anhydrate 无水石膏
calcium sulphate dihydrate 二水石膏
calcium sulphate hemihydrate 烧石膏;半水硫酸钙;半水合硫酸钙;煅石膏;熟石膏;碳酸钙的半水化合物
calcium sulphate incrustant 硫酸钙积垢;石膏装饰板
calcium sulphate incrustation 石膏装防火隔声板
calcium sulphate plaster 石膏墙粉;硫酸钙墙分;·石膏(抹面)灰浆;石膏抹灰面
calcium sulphate plaster screen 石膏墙粉整平板;硫酸钙墙粉整平板
calcium sulphate retarder 石膏缓凝剂
calcium sulphide 硫化钙
calcium sulphoaluminate 硫铝酸钙(水化物)
calcium superoxide 过氧化钙
calcium superphosphate 过磷酸钙
calcium supply 钙的供应
calcium tallate 松浆油酸钙
calcium tartarate 酒石酸钙
calcium thiosulfate 硫代硫酸钙
calcium titanate 钛酸钙
calcium titanate porcelain 钛酸钙瓷
calcium titanium silicate 硅酸钙钛
calcium tolerance 钙容忍度
calcium treated mud 钙处理泥浆
calcium tungstate 钨酸钙
calcium wolframate 钨酸钙
calcium yellow vinyl primer 钙黄乙烯底漆
calcium-yttrium silicate oxyapatite laser 硅酸盐钙—钇氧磷灰石激光器
calcium zirconate 锆酸钙
calcium zirconate ceramics 锆酸钙陶瓷
calcium zirconate clinker 锆钙砂
calclacite 醋酸氯钙石;醋氯钙石
calclithite 碱屑灰岩;灰屑岩;钙质石屑岩;钙岩屑砂屑岩;碎屑灰岩
calc-mica schist 钙质云母片岩
calcnrsilite 钙硅铀矿
calcomalachite 钙孔雀石
calcon 钙试剂
calcon-carboxylic acid 钙指示剂
calco-sodium glass 钙冕玻璃
calcothar 铁丹
calcrete 灰质结砾岩;钙(质)结砾岩;钙质角砾岩
calc-schist 钙质片岩
calc-silicate hornfels 钙质硅酸盐角页岩
calc-silicate marble 钙质硅酸盐大理岩
calc-silicate rock 钙硅质岩
calc-sinter 多孔石灰岩;石灰华
calc-sinter fan 钙华扇
calc-slate 钙板
calc-spar 方解石;双折射透明方解石;钙质晶石;冰洲石
calc-sparite 亮方解石
calc-spar lining 方解石衬砌
calc-spessartite 钙斜煌岩
calc-spot 钙质花纹
calc-stripe 钙质条带
calc-tufa 石灰华
calculability 可计算性
calculable 可计算的;能计算的
calculagraph 记录时器;计时仪;计器
calculary 石的
calculate 核计;运算
calculate by inspection 观察计算
calculated according to constant price 按不变价格计算
calculated according to current price 按现行价格计算
calculated address 计算地址;形成地址;生成地址
calculated altitude 计算高度
calculated area 计算面积
calculated assay 计算矿物的有用成分
calculated azimuth 计算方位(角)
calculated braking distance 计算制动距离
calculated brickwork 结构计算确定的砖砌体
calculated capacity 计算容量;计划容量
calculated circulation circuit 计算流转环路
calculated coordinates 推算坐标
calculated curve 计算曲线
calculated data 计算数据
calculated density 计算比重
calculated diameter of fiber 纤维计算直径
calculated dose 理论剂量;计算剂量;设计剂量
calculated feed(ing) 计量的送料量;计算入料;重组给料

calculated filling 设计充满度
calculated flame 火焰传播计算速度
calculated flow rate 计算流率;计算流量
calculated gradient 计算坡度
calculated growth 推算出的生长率
calculated head loss 计算水头损失
calculated horsepower 计算马力
calculated interest 估算利息
calculated length of railway vehicles 车辆计算长度【铁】
calculated live load 计算活荷载
calculated load 计算荷载;设计负载
calculated man-days in attendance 应出勤人日数;应出勤工日数
calculated method of heat reserve 热储量计算方法
calculated octane number 计算辛烷值
calculated on gross profit 依毛利计算
calculated on the basis of their original price 按原价计算
calculated on the comparable basis 按可比口径计算
calculated over-depth 计算超深
calculated over-width 计算超宽
calculated oxygen demand 计算需氧量
calculated performance 计算性能
calculated period 计算期
calculated power 计算功率
calculated rise 计算矢高
calculated rise of arch 计算拱矢高
calculated risk 计算风险;计划风险;预计风险
calculated screen cut 计算的遮板切割大小;计算断面
calculated self-weight collapse 计算自重湿陷量
calculated settlement 计算的沉降;计算的沉陷
calculated settlement of pile foundation under maximum loading 桩基受最大荷载的计算沉降量
calculated span 计算跨度
calculated speed 计算速度
calculated strength 计算强度
calculated strength of subgrade 地基计算强度
calculated target point 计算停车点【铁】
calculated tractive effort 计算牵引力
calculated value 计算值
calculated velocity 计算速度
calculated velocity of locomotive 机车计算速度
calculated weight 计算重量
calculated weight of locomotive 机车计算重量
calculate heat accumulation coefficient 材料计算蓄热系数
calculate point of hump height 峰高计算点
calculate reckoning 推算
calculating 计算的
calculating area 计算面积
calculating basis 计算根据;计算基本原理
calculating board 计算台
calculating center 计算中心
calculating chart 计算用表;计算图(表)
calculating coordinate system 计算坐标系统
calculating device 计算装置
calculating error 计算误差;计算错误
calculating hourly rainfall 计数小时降雨量;计算时降雨量
calculating inspection 计算检查
calculating instrument 计数仪
calculating machine 计算机(器)
calculating method 计算方法
calculating method of influence radius 影响半径计算方法
calculating method of water yield parameter 涌水量参数计算方法
calculating methods of model ages 模式年龄计算方法
calculating office work 内业计算工作
calculating of heat exchange 热量交换计算
calculating parameter 计算参数
calculating parameter of corrosion equation 溶蚀方程计算参数
calculating plan of outer stripping amount of orebody 矿体外剥离量计算平面图
calculating plan of outer stripping amount of ore deposit 矿床外剥离量计算平面图
calculating procedure 计算程序
calculating roughly 粗略计算
calculating rule 计算尺

calculating runoff 计算径流量
calculating scale 计算尺(度)
calculating sorting-machine 分类计算机
calculating table 计算(用)表
calculating time 计算时间
calculating velocity 计算流速
calculation 计算;运算;推算
calculation accuracy 计算精度
calculation and analysis fo foundation engineering 基础工程计算与分析
calculation and propagation of reliability 可信度计算和传播
calculation boundary 计算边界
calculation by iteration 迭代法计算
calculation condition 运算条件
calculation error 计算误差;计算错误
calculation for bolt force of metal flat gasket 金属平垫密封的螺栓力计算
calculation for bolt force of midseparate closure 中分面密封的螺栓力计算
calculation for bolt load of double cone seal 双锥密封的螺栓力计算
calculation forms in commercial enterprises 商业企业核算形式
calculation for normally consolidation clay 正常固结黏土沉降计算
calculation interference flow 计算干扰流量
calculation interval 计算时间
calculation load for wedge gasket 楔形垫密封的计算荷载
calculation location mode 计算型定位方式
calculation map of reserves 储量计算图
calculation means 计算方法
calculation of assessment 评估计算
calculation of charges 费用计算
calculation of coal ash absorption 煤灰吸收量计算
calculation of cutting and filling 挖方和填方计算;土方挖填计算;土方量计算
calculation of development indicator 开发指标计算
calculation of earth volume 土方计算
calculation of formation curvature 岩层曲率计算
calculation of groundwater resources 地下水资源计算
calculation of heat insulation 绝热计算;隔热计算
calculation of heat losses 热损失计算
calculation of loading 荷载计算
calculation of mud barge required 泥驳需要量计算
calculation of net cost 成本计算
calculation of offset 偏移计算
calculation of protection setting 保护装置的整定计算
calculation of raw mix proportions 生料配料计算
calculation of regulation guarantee 调节保证计算
calculation of reserves 估算;储量计算;储量估算
calculation of reservoir capacity 水库容量计算
calculation of residues 留数计算
calculation of riverbed deformation 河床变形计算
calculation of shear 剪力计算
calculation of skirt support 裙座计算
calculation of stability 整体稳定性计算;稳定计算
calculation of strength 强度计算
calculation of the buckling strength 压曲强度计算
calculation of the effect of expense incurred 费用支出效果核算
calculation of tractive effort 牵引力计算
calculation of transit period 运输期限的计算
calculation of variation 变分法
calculation of wall thicknesses 墙厚度计算
calculation on dry basis 折干计算
calculation plan of overburden ratio 剥离量计算平面图
calculation price 估算价(格)
calculation procedure 计算机方法;计算程序;计算步骤
calculation report of floodability 破舱稳性计算书
calculation result 计算结果
calculation sheet 计算书;计算单;计算表格
calculations method of reserves 储量计算方法
calculation specification 运算说明
calculation stretch 计算段
calculative cost 计算价格;造价
calculator 解算装置;解算机;计算装置;计算员;计算器

calculator chip 单片计算器
calculator with algebratic logic 代数逻辑计算器
calculator without addressable storage 不带可编址存储器的计算器
calculator without programmability 程序不变的计算器
calculeus 多石的
calculi 建筑砖
calculiform 卵圆形的
calculifragous 碎石的
Calculon 凯尔库仑(1972年推广的一种砖的尺寸);小砖
calculous 杂有石块的;似砂石的
calculous soil 砾质土;坚隔土
calculus 演算;微积分(学)
calculus of corrosion velocity 溶蚀速度计算法
calculus of differences 差分计算;差分法
calculus of enlargement 差分演算
calculus of finite difference 有限差分法;差分演算;差分学;差分法
calculus of fluxion 微积分
calculus of observation 观测演算
calculus of probability 概率计算
calculus of probability integral 概率积分计算
calculus of proposition 命题演算
calculus of residues 残数计算;残值计算
calculus of tensors 张量演算
calculus of variations 变分学;变分计算;变分法
calculus of variations in the large 宏观变分学
calcurmolite 钼钙铀矿
Calcutta hemp 黄麻
Calcutta tonnage scale 加尔各答运费吨(以50立方英尺为1吨)
calcybeborosillite 硅硼铍钇钙石
caldarium 高温浴室(古罗马澡堂);热浴(室)
caldera 火山(喷)口;破火山口
caldera lake 火山口湖
caldera subsidence 火山崩坍构造
Calder-Fox scrubber 卡尔德—福克斯洗涤器
Calder-Fox separator 卡尔德—福克斯除尘器
calderite 锰铁榴石
Calderon charger 卡尔德郎废钢装料机
Calderwood cement 卡尔德伍兹水泥(一种天然水泥)
Caldivia deep 瓦尔迪维亚海渊
caldron 火山口;大锅;海盆;煮皂釜;敞口锅;釜
caldron basin 锅状洼地;锅状盆地
caldron-shaped crown 锅形树冠;杯形树冠
caleareous tufa 钙华
Caledon 加里东【地】
Caledonian-Appalachian geosyncline system 加里东—阿帕拉契地槽【地】
Caledonian basement 加里东期基底
Caledonian brown 天然铁棕颜料;苏格兰棕颜料
Caledonian curtain 加里东幕【地】
Caledonian epoch geosyncline 加里东期地槽【地】
Caledonian geosyncline 加里东地槽【地】
Caledonian marine trough 加里东海槽【地】
Caledonian movement 加里东运动【地】
Caledonian old land mountains 加里东山地【地】
Caledonian orogency 加里东造山运动【地】
Caledonian orogeny period 加里东期【地】
Caledonides 加里东造山带【地】;加里东山系【地】
caledonite 铅绿矾
calefaction 热污染;发暖作用
calefactor 发暖器;温暖器
calefactory 寺院内的取暖室;寺院内的起居室;寺院内的暖房
caleium zinc cyclotetraphosphate pigment 环四磷酸钙锌(防锈)颜料
calendar 历书;历法;日历;日历表;全年日程表
calendar clock 历钟
calendar cut 折腾
calendar date 历日期
calendar date line 国际日期变更线
calendar day 历书日;历日;民用日;日历日
calendar day and date watch 双历手表
calendar file 计时文件
calendar hours 日历时间
calendar induced variations 日历诱导变动
calendar inspection 定期检查
calendar line 日界线

calendar loading arrangement 日历装车安排
calendar machine-hours of cargo handling machinery 装卸机械日历台时
calendar machine-time 日历台日数
calendar man-days 日历计算人日数;日历工日数
calendar month 历月;历日月;民用月;日历月
calendar of conference 会议日期表
calendar plan 日历计划
calendar progress chart 计划进度表;计划进度安排;每日工作进度表;工作计划进度表;日进度表;日程表
calendar spread 日历差幅
calendar standard 日历标准
calendar stone 历石
calendar subroutine 计时子程序
calendar time 日历时间
calendar unlocking yoke 双历瞬跳杆
calendar variation 日历变动
calendar watch 日历手表
calendar week 日历周
calendar workdays 日历工日数
calendar year 历年;民用年;回归年;自然年;分至年;日历年(度)
calender 轮压机;碾光机;研光机;研光棍;压延机;压光机;辗光机
calender blackening 辊压黑道
calender bowl 研光滚
calender coater 辊压涂布机;涂胶研光机
calender coating 辊压涂布;辊压机涂漆
calender colo(u)ring 研光上色
calender crush finish 压纹装饰
calendered cloth 电绝缘用研光布
calendered coating 辊压机涂漆;封面纸罩光涂膜
calendered paper 蜡光纸
calendered sheet 辊压板材
calender finish 研光装饰
calender ga(u)ge 辊压用测厚仪
calender grain 研光效应;压延效应
calender grease 研光机脂
calendering 碾光;研光;压延
calendering process 压延法
calender mark(ing) 辊压条痕
calender method 日程法
calender press 辊压机
calender roll 研光辊
calender run 研光
calender scabs 辊压疤点
calender spot 辊压斑点
calenderstack 研光机
calender take-off system 轮压机启动系统
calender train 研光机
calendic acid 金盏酸
calendulin 金盏花黄
calescence 变热
calescene 渐增热
calf 小浮冰块
calf and maternity quarters 小牛栏及产房
calf-dozer 推扫机;小型推土机;小型散货平舱机
calf-house 犊舍
cal-fibre brick 凸条砖
calf kneed 凹膝
calf line 包装线;滑车钢绳(钢绳冲击钻进起下套管用)
calf pen 犊子栏
calf reel 钢绳滚筒
calf's tongue mo(u)lding 牛舌线脚;牛舌饰
calf wheel 拔管绞盘;钢绳滚筒
calf wheel box 钢绳滚筒轴承箱(钢绳冲击钻机上)
calf wheel tug rim 拔管绞盘的传动滑轮
calgon 六聚偏磷酸钠
Calgon method 卡尔康法(提高钻井出水量的方法)
caliber 口径;卡规测径器;轧辊型缝;管内径;管径;尺寸
caliber compasses 微调小圆规;量规;弯脚圆规
caliber ga(u)ge 测径规
caliber number 机芯序数;机芯编号
caliber radius head 弹头蛋形部曲率半径
caliber size 圆柱径;品质;口径尺寸;口径;管径大小;管径
caliber square 测径尺
calibrate 校准;定分度;定标;使合标准;使标准化
calibrated 有刻度的

calibrated absorber 校准用吸收体
calibrated air speed 仪表修整速度(飞机);修正空速
calibrated altitude 校准高度;观测高度;标定高度
calibrated block 标准块
calibrated chain 标准链
calibrated coefficient 率定系数
calibrated curve 率定曲线
calibrated dial 校准度盘;分划盘;分划尺;标准度盘
calibrated disc 刻度盘
calibrated error 标定误差
calibrated feeder 校准计量给料机;定量供料器
calibrated flask 校正细颈瓶
calibrated flow meter 校准的流量计
calibrated flux 校准通量;校准流量
calibrated focal length 校准焦距;标定焦距
calibrated gate 校准量水闸门(水工模型用)
calibrated manometer 校准压力计
calibrated meter 已校(准)仪表
calibrated orifice 校准孔;标定孔口
calibrated pipet(te) 校准用吸移管
calibrated radiation source 标准辐射源
calibrated reseau photography 已校正格网摄影法
calibrated resistance 已校准的电阻
calibrated sand method 校准堆砂法;标定堆砂法;(测定密实度用的)砂充法
calibrated scale 分度尺
calibrated sluice 标准闸门(水工模型用)
calibrated solar cell 标定太阳能电池
calibrated step wedge 刻度级变
calibrated test weight 校准的检查砝码
calibrated thrust engine 标定推力发动机
calibrate for error 误差校准
calibrate model 校准模型
calibrate practice 校准技术
calibrater 测厚仪;校准器;校径规
calibrate unit 校准设备
calibrate solution 校准溶液
calibrate test 校准试验
calibrating 校准
calibrating accuracy 校准精度;标定精度
calibrating apparatus 校准用器;检定装置
calibrating arm 校准臂
calibrating burette 校准用滴定管
calibrating circuit 校准电路;校正电路
calibrating device 率定装置;校准设备;率定设备
calibrating frame 承片框
calibrating gas 标准气体
calibrating oscillator 刻度校正振荡器
calibrating pipette 校准用吸移管
calibrating plot 校准曲线
calibrating point 标定点
calibrating pressure 校正压力
calibrating procedure 校准程序
calibrating receiver 校准接收机
calibrating resistor 校准电阻(器)
calibrating run 检验螺纹头
calibrating set for resistance attenuator 电阻衰减器校正装置
calibrating signal 校准信号
calibrating source 校准噪声发生器
calibrating spot 校准点
calibrating stand 校准架子
calibrating stem 校准杆
calibrating strap (拉模的)定径带
calibrating tank 校准箱
calibrating terminal 测试接线柱;测试端子
calibrating weight 校准重量
calibration 率定;量口径;校准;检定;标定
calibration accuracy 率定精度;校准准确度;校准精(确)度;校正精度
calibration and verification of water quality model 水质模型标定和验证
calibration bar 卡钳;测径规;指标杆
calibration battery 校准电池
calibration beacon 校正导航标
calibration bin 校正仓
calibration block 校准试块
calibration capacitor 校准电容器
calibration certificate 校验证书;率定证书;刻度合格证书;校准证书;检验证书
calibration chart 校准图;校准表
calibration check compound 校核化合物
calibration circle 刻度员;校准圆
calibration coefficient 率定系数;校准系数;标定

系数
calibration compensator 校准用补偿器;校正仪
calibration constant 校准常数
calibration curve 率定曲线;刻度曲线;校准曲线;校正曲线;定标曲线;标准曲线;标定曲线
calibration data 检定数据
calibration depot for ultrared hot box detector 红外(线)轴温探测设备检测所【铁】
calibration device 校准装置
calibration error 刻度误差;校准误差;校正误差;定标误差
calibration factor 率定系数;校准因子;校准系数;校正因素;标定因素;标定系数
calibration figure 校准数值
calibration filter 校准滤光片
calibration function 标定函数
calibration gas 校准气
calibration gas generating system 校正用气体配气装置
calibration gear 校准装置
calibration grid plate 检核用坐表格网板
calibration instrument 校准仪表
calibration lamp 校准灯
calibration leak 校准漏孔
calibration line 校准线;校正线(路);标定线
calibration loop of caliper 井径刻度环
calibration loop of induction log 感应测井刻度环
calibration mark 校准标志;校准标记
calibration marker 校准指示器;标准标示器
calibration measurement 校准量测
calibration method 校准法
calibration mixture 校准混合物
calibration negative 检定底片
calibration notch 校准槽
calibration of a measuring instrument 测量仪器的标定
calibration of gravimeter constant 重力仪常数标定
calibration of measuring equipment 监测仪器的标定;测量仪器的标定
calibration of power 功率校准
calibration of reference fuels 参比燃料之校准
calibration of test ga(u)ge 试验压力计的校核
calibration of testing instrument 试验仪器的校准
calibration of testing machine 校准试验机
calibration of viscosity 黏度计标定
calibration of wavelength scale 波长标度校准
calibration period 测定期;检定期
calibration plate 校准板
calibration point 校准点;标定点
calibration power 检定功率;标准功率
calibration practice 校准技术
calibration principal-distance 检定主距
calibration pulse 校准脉冲;标准脉冲
calibration radio beacon 校准无线电标;供校准用的无线电信标
calibration regulator 校验调节器
calibration resistor 校准电阻(器);标准电阻器
calibration result 校准结果
calibration ring 校准环;校准压力环
calibration sample 校准样品;校准试样;校核试样;标定样品
calibration scale 标定表
calibration screw 校准螺旋;校准螺钉
calibration sheet 校准图表;检定表
calibration source 参考源
calibration specimen 校核试样
calibration stability 校准稳定性
calibration stand 校准状态
calibration standard 校准用标准
calibration star 定标星
calibration station 校准站;校准台
calibration system 校准系统;检定装置
calibration table 校准(数值)表;检定表
calibration tails 刻度校验记录;刻度线;(附在电测曲线前后的)校正曲线;校验记录
calibration tank 率定池;率定槽;校正料桶;校正料罐;检定池;检定槽;标定罐
calibration technique 校准技术;标定技术
calibration temperature 校准温度;检定温度
calibration template 检定模片
calibration templet 校正样板;检定模板
calibration test 校准试验;校正试验;标定试验
calibration test equipment 校准检验装置
calibration tolerance 校准限差

calibration trace 校准曲线
calibration wheel 探测轮
calibration zone 标定区
calibrator 校准者；校准仪；校准器；校正器；校径规；管径测量仪；定标器；测径器；测厚仪
calibrator for compressive testing machine 压力试验机校准仪
calibrator for transducer-calibrating confining pressures 围压率定机
calibrator(unit) 校准设备
calibre 卡规测径器；圆柱径；轧辊型缝
calibre ga(u)ge 测径规
calibre size 口径尺寸
calibre square 测径尺
caliche 泥灰石；卡利许【地】；硝石(层)；钙结层；钙积层；生硝
caliche crust 钙质壳
caliche nodule 钙质结核
calicheras 小水洼
calichified limestone 钙结石灰岩
calico 印花布
calico bandage 白布绷带
calico marble 印花大理岩
calico-printing 印花
calico rock 印花岩
Calido 卡里多镍铬铁合金
caliduct 暖水管；暖气管；热气管
califont 浴水快热器；浴室内热水
California bearing capacity 加利福尼亚州承载力
California Bearing Ratio [CBR] 加利福尼亚州承载比；加利福尼亚承载比；加利福尼亚州承载量比
California Bearing ratio method 加利福尼亚州承载比设计法；加利福尼亚州承载比方法
California Bearing Ratio method of flexible pavement design 加利福尼亚州承载比柔性路面设计法
California Bearing Ratio test 加利福尼亚州承载比试验
California bearing reaction 加利福尼亚州承载力
California current 加利福尼亚洋流；加利福尼亚海流
California derrick 中层平台有护栅的钻塔
California design method 加利福尼亚州设计法
California drag-head 加利福尼亚州耙头；平靴形耙头
California fog 加利福尼亚雾
California-Japan route 加利福尼亚—日本航线
California kerosene equivalent 加利福尼亚州煤油恒等式
California method of mixture design 加利福尼亚州(沥青)混合料配合比设计法
California Nebula 加利福尼亚星云
californian onyx 淡褐霰石
Californian stamp 加利福尼亚型捣矿机
California pattern bit 工字形钻头(冲击钻用)
California polymerization 加利福尼亚聚合作用
California ranch architecture 加利福尼亚农场式建筑
California sampler 活塞取样器
California stabilometer 加利福尼亚州稳定度仪
California stabilometer method 加利福尼亚州稳定度仪设计法(柔性路面厚度设计)
California stabilometer method for pavement design 加利福尼亚州稳定度仪路面设计法
California storepipe drilling 加利福尼亚烟囱式钻井法
California switch 浮放道岔
California-type dredge 加利福尼亚型采金砂船
California type ideal fishtail bit 标准工字形鱼尾钻头(转盘钻进用)
californite 玉符山石
californium 锎
calina 加林那霾
caliper rule 卡尺
caliper square 滑动卡尺
Calite 卡利特铁镍铝铬合金；铁镍铝合金
calkinsite 水碳镧铈石
call 要求支付分保摊赔金额；请求；通话；调入；提前赎回证券；赎回
callable 可随时支取的；可赎回的；应付的
callable bond 可通知赎还的债券；可提前兑付债券；可提前偿还的债券；活期债券；通知赎还债券；可收回债券
callable capital 请求即付资金；通知即缴的股本；通知即付资本
callable loan 可赎回贷款；通知贷款；通知拆放

callable preferred stock 可提前回收优先股；可收回优先股
callable subprogram(me) 可调用子程序
callable subroutine 可调用子程序
callable trust bond 通知信托债券
call a bond 招人领取公债券金额；通知公债券还本；收兑债券
call accepted 接受呼叫
call accepted message 接受呼叫信息
call-accepted signal 接受呼叫信号
call address 引入地址；传呼地址；调用地址；调入地址
callaghanite 水碳铜镁石
calla green 水芋绿色
callainite 绿磷铝石
call-a-matic telephone 存拨电话
call analyzer 呼叫分析仪
call and signal(l)ing system 呼叫与信号系统
call announcer 呼叫指示器
Callao painter 不洁雾
callao rope 货驳靠把
call assignment switch 呼叫分配器
call at 停靠
call at a port 停泊(指锚泊作业)
call attempt 冲击呼叫
call-back 重复呼叫；复查；回调；唤回；回叫；收回待修产品
call-back pay 加班工资
call barred 封锁呼叫
call bell 信号铃；呼叫铃；电铃
call bell indicator 电铃号码牌
call-board 公告板
call book 呼号簿；催缴簿
call box 公用电话亭；公用电话间；电话亭；电话室；电话间
call broker 通知借款经纪人；通知贷款经纪人
call button 呼人或报警铃的按钮；呼叫按钮
call by descriptor 按描述符调用
call by location 地址调用；按位置调用；按位调用；按地址调用
call by name 代入名；按名调用
call by passing 传递调用
call by reference 引用调用
call by result 结果调用
call by value 代入值；赋值调用；按值调用
callche 钙结岩
callche fine conglomerate 钙结细砾岩
callche hardened shell conglomerate 钙结硬壳砾岩
call circuit 呼叫电路；业务线
call-circuit button 呼叫按钮
call completing rate 接通率
call congestion ratio 呼损率
call controller 呼叫控制器
call control procedure 呼叫控制规程
call control signal 呼叫控制信号
call data save 调入数据保存
call date 提前偿还日期
call deposits 即付存款；通知存款
call directing code 调入指向码
call direction code 指定收信人住址代码
call display position 简式交换机；号码指示位置
call divider 呼叫分配器
call down 满筒
call down rate 满筒率
called bond 已通知还本的公债；通知偿还债券
called forward 提支
called line identification facility 识别被呼叫线路设备
called line identification signal 被呼线路识别信号；识别被呼线路信号
called number 受话号码
called party 被呼叫用户；被呼方；受叫方
called procedure 被调过程
called program(me) 被调用程序；被调程序
called station 被叫台；被呼站；被调用位置；被调入位置；通信台
called subscriber answer 被叫用户应答
called-up capital audit 投入资本审计
called-up capital authenticity audit 投入资本真实性审计
called-up capital legality audit 投入资本合法性审计
callee 电话受话人；被访问者；受话人

Callendar's equation 卡伦德方程
Callendar's formula 卡伦德公式
Callendar's system 土壤的防潮处理；卡伦德系统
Callendar's thermometer 铂电阻温度计
callendrite 纯沥青卷材(用于防潮)
callenia 叠层
call entry 引入呼叫
call error 调用错误
call establishment 调用建立
call executive 调用执行
calley stone 泥质硬砂岩
call for an agreement 征求同意
call for bids 需价；要求开价；招标通告；招标
call for bids on 就……进行招标
call for capital 招股；筹集资本
call for funds 集资
call for offers 询价
call for share capital 筹集股份资本
call for tenders 招标；公开招标
call frame 调用栈帧
call hold 永久性保留；通话保留
call hour 小时呼
Callier Q Coefficient 卡里尔 Q 系数
Callier quotient 卡里尔商
Callier's Q factor 卡里尔 Q 系数
calligraphy 书法
call in 调入
call in a loan 收回贷款
call indicator 呼叫指示器
calling 呼叫；调用
calling argument 调用变元
calling branch 调用转移
calling card 名片
calling detector 车辆到达检测器
calling device 呼叫装置；呼叫设备
calling equipment 调度设备；呼叫装置
calling for order 待命靠泊；卸货港序
calling for proposals 索取承包商估价书
calling for tenders 招标
calling frequency 呼号频率
calling indicator 引入指示；引入指令
calling-in-point 呼叫点
calling lamp 呼叫灯
calling line identification facility 识别呼叫线路设备
calling line identification signal 识别呼叫线路信号
calling list 呼叫表
calling machine 铃流发生器
calling-magneto 振铃手摇发电机
calling-on signal 引导信号【铁】
calling order 发送程序
calling party 呼叫用户
calling party's category signal 主叫方类别信号
calling population 要求服务的总体
calling probability 呼通概率
calling processor 调用处理程序
calling program(me) 调用程序
calling rate 呼叫率；访问顾客率
calling relay 呼叫继电器
calling remote monitoring system 呼叫遥控监测系统
calling segment 调用程序段
calling sequence 呼叫序列；引入序列
calling station 呼叫局
calling trace 调用跟踪
call instruction 调用指令；调用命令
call-in-time 呼入时间
cal(l)iper 纸板厚度；轮尺
cal(l)iper calibration value 井径仪刻度值
cal(l)iper detector 厚度探测器
cal(l)iper diameter 井径测斜仪
cal(l)iper ga(u)ge 蹄形卡规；卡规；井径仪；井径规；测径规
cal(l)ipers hook 两脚规
cal(l)iper log 井径剖面；井径测井(记录)；钻孔直径记录图；测径记录；孔径钻剖面；钻孔柱状剖面
cal(l)iper logging 井径测井记录图；孔径量测；测井记录；井径测量
cal(l)iper machine 找平机；定厚机
cal(l)iper measure 测径；轮尺量度；轮尺量材法
cal(l)iper measure method 测径法
cal(l)iper pig 管道内壁检测器
cal(l)iper profiler 厚度调节器
cal(l)iper rule 卡尺
cal(l)ipers 两脚规；卡钳；卡尺；卡规；测径器；测径

cal(l)iper scale 计;测径规;双脚规
cal(l)iper scale 轮尺检杖
cal(l)iper splint 两角规形夹;双脚规形夹
cal(l)iper square 游标规
cal(l)iper stage 有左右侧台的舞台
cal(l)iper survey 井径测量
Callippic cycle 卡利普循环;卡利普周期
callisthenics hall 健美体操馆
call lamp 信号灯;呼叫灯
call letter 催款信;推销商品的信件;船名呼号;通信呼号
call library 调用程序库
call loan 短期同行拆借;活期贷款;通知放款;通知拆放
call loan rate 通知放款利率
call loan secured 活期抵押放款
call lose 呼叫丢失
call lost system 明显损失制
call macro 调用宏
call management 通信管理
call margin 追加押金;追加保证金
call market 通知放款市场
call memory 拆款
call meter 计数寄存器;呼叫计数器
call minute 分钟呼叫
call money 通知放款;通知拆放
call not accepted signal 不接受呼叫信号
call number 呼号;引入数;调用数字;调用号;调入数字;调入数;图幅号;书架号码
call-off system 通知送货制
call on 访问;停靠
call on shareholder 要求交付股款
call option 购买选择权
callout 图纸上着重注意点的标注
Callovian 卡洛夫阶
callow 黏土矿床上的废土与废石;低沼泽(地);低沼;浮盖层
Callow cone 卡路锥
Callow cone hydroseparator 卡路锥形水力分级机
Callow flo(a)tation cell 卡路浮选槽
Callow flo(a)tation process 卡路浮选法
callow rock 水化石灰;生石灰
Callow screen 卡路筛
C-alloy 钙铝合金;铜镍硅合金
call pay 通知工资
call persons 召集人员
call pickup 来话代答
call premium 卖回溢价;需要溢价;提前赎债溢额;赎回溢价
call premium on bonds 债券赎回溢价;提前偿还债券贴水;赎回债券溢价
call price 赎债价格;通知偿还价格;提前偿还债券价格;赎回价格
call privilege 提前偿还权
call provision 提前赎债条款;提前兑回条款;提前偿付条款
call provision on hand 提前兑回债券的条款
call rate 活期贷款利率;通知贷款利率
call receipt 催款收据
call release 调用释放;调用解除
call report 财政决算
call request 调用请求;电话挂号
call request signal 呼叫请求信号
call rerouting 重呼叫路由
call return met 通知借款余额
call routing 呼叫路由选择
call secured loan 通知抵押放款
call sequencer 呼叫排序器
call sign 呼叫信号;呼号;船名呼号;通信呼号
call signal 呼叫信号;呼号;识别信号
call signal apparatus 呼号机
call sign placement number 呼号排列号码
calls in advance 预收催缴资本
calls in arrears 延付催缴股款
call slip 领料单;请拨单
call stack 调用栈
call statement 调入语句
call subroutine 调用子程序
call switch 呼叫开关
call system 呼叫系统
call telephony 通话
call through test 接通试验;综合试验
call transaction 期价交易
call transaction made 通知借款余额
call transfer 呼叫转移
call turnover made 通知借款周转余额
call-up capital 催缴资本;已缴股本
call up on 对……的催付
callus 假皮;木疵
call waiting 呼叫等待;调用等待
call wire 联络线;记录线;传号线
call word 调用字;调入字
callys 片岩;板岩
calm 零级风;道格拉斯零级浪;蒲福零级风;无风(无浪)镶嵌铅条;淡色页岩
calmagite 钙镁试剂
calmalloy 热磁合金;铜镍铁合金
calm and tranquil seas 风平浪静
calmative 镇静剂
calm belt 静风带;无风带
calm control eye 无风眼
calm day 平静日
Calmet 卡尔梅特铬镍铝奥氏体;铬铬铝奥氏体耐热钢
calmet burner 垂直燃烧器
Calmette's tuberculin 沉淀结核菌素
calm-glassy sea 无浪
calming section (柱的)减震部分
calm inversion 无风逆温
calm inversion pollution 宁静逆温污染;无风逆温污染
calm layer 无风层;稳静层
calm of Cancer 北回归线无风带
calm of Capricorn 南回归线无风带
calm sea 零级浪;海平如镜;平静海面;无浪
calm sedimentation 静沉淀
calm smog 宁静烟雾;稳静烟雾
calms of Cancer 副热带无风带;北回归线无风带;北半球副热带无风带
calmus oil 菖蒲油
calm waters 静水区;风平浪静的水域
calm water zone 静水区
calm weather 无风天气
calm zone 宁静区;无风区;无风带
calm zone of Capricorn 南半球副热带无风带
calnitro 钙硝肥料
calobiosis 同栖共生
caloing 裂冰
calomel 汞膏;甘汞
calomel cell 甘汞电池
calomel electrode 甘汞电极
calomelene 汞膏
calomel half cell 甘汞半电池
calomelite 汞膏
calomel poisoning 轻度中毒
Calomic alloy 卡劳密克镍铬铁合金
caloradiance 热辐射线
caloreceptor 热能感受器
calorescence 炽热;热光
calorex 阻光或滤热玻璃
calor gas 瓶装液体煤气
caloric 热质;热量的
calorically perfect gas 发热的理想气体
caloric balance 热量平衡
caloric disease 高温病
caloricity (发)热量;热容量
caloric meter 热量计
caloric of heat 热量平衡
caloric power 发热量;热值
caloric quotient 热量商(数)
caloric receptivity 热容量
caloric requirement 需热量;热能需求量;热量需要
caloric test 变温试验
caloric unit 热量单位;卡
caloric value 热值;热价
calorie 卡路里;热量单位;热卡
calorie conversion factor 热量转换系数
calorie intake 热量摄取
calorie meter 卡计;量热器;量热计
calorie unit 卡路里
calorie value 热量值;热值
calorifacient 生热的
calorific 热量的
calorification 发热
calorific balance 热平衡
calorific capacity 热值;卡值;发热量;热容量
calorific conduction 热传导
calorific effect 热效应
calorific effect of furnace 炉子热效率
calorific efficiency 热值;卡值;发热量;热效率
calorific equivalent 发热当量
calorific intensity 热强度;热卡强度
calorific power 卡值;发热量;热量(功率);热值
calorific power of fuel 燃烧值;燃料热值
calorific radiation 辐射热;热辐射
calorific receptivity 感热性;热容量
calorific requirement 需热量
calorifics 热学
calorific unit 热(量)单位
calorific value 卡值;发热值;发热量;热值;热价;热当量
calorific value at constant pressure 恒压发热量
calorific value at constant volume 恒容发热量
calorific value determination in a steel bomb 铁弹发热量
calorific value of coal 煤的发热量
calorific value of fuel 燃烧值;燃料(发)热值
calorific value of gas 煤气热值
calorifier 加热器;换热式热水器;热水器;热风炉;热风机
calorigenic 生热
calorimeter 量热器;量热计;卡路里计;测热计;热量计
calorimeter assembly 量热装置
calorimeter boiler 量热计蒸发器
calorimetering method 量热法
calorimeter instrument 量热器
calorimeter of openflow system 放流式量热器
calorimeter thermometer 量热用温度计
calorimeter vessel 量热器筒;量热计筒
calorimeter without heat transmission 不传热的量热计
calorimetric 量热;测热;热量的
calorimetric analysis 热量分析
calorimetric balance 热量平衡
calorimetric bomb 量热弹;热量弹;测热弹
calorimetric data 热量数据
calorimetric detector 热量探测器
calorimetric measurement 热量测量;热量测定
calorimetric method 量热法
calorimetric power meter 量热式功率计
calorimetric pyrometer 量热高温计
calorimetric test 热量试验;量热试验;测热试验
calorimetric test for impurities 检验有机杂质的热量试验
calorimetric titration 温度滴定
calorimetry 量热学;量热术;量热法;测热法;发热量测定;热量法;热量测定
calor innatus 本体热
Caloris basin 卡洛里斯盆地
caloriscope 热量器
calorised steel 渗铝钢
Calorite 卡洛利特镍铁合金;耐热合金
calorization 热化
calorizator 热法浸提器
calorize 热镀铝;铝化(处理);渗铝处理
calorized 铝化(处理)的
calorized steel 涂铝钢;渗化钢
calorizing 铝化(处理);防腐蚀渗铝(法)
calorizing process 渗铝
calorizing steel 铝化钢
calorstat 恒温器;恒温箱
Calot's triangle 卡洛氏三角
calotte 极帽;帽状物;帽罩;回缩盘;圆顶
calotype 光力摄影法;碘化银纸照相法
calp 灰蓝灰岩
calp box 钙塑瓦楞箱
calpis 乳浊液
Cal-Red 钙指示剂
calsibar 钙硅钡合金
calsomine 铝碳(水)粉
calthrop 棘刺
calumetite 蓝水氯铜矿
calutron 铀同位素分离器;电磁(型)同位素分离器
Calvary 耶稣受难像
Calvert lighting 卡尔浮照明系统
calves' tongue mo(u)lding 牛舌线脚;牛舌饰
calving 裂冰作用;冰裂作用
calx 矿灰;金属灰
calxiouranoite 水钙铀矿
calyculate 蔓形
calyx 钻屑收集筒;取粉管
calyx bit 颗状钻头

calyx borehole 颚状钻孔
calyx core drill 萼状岩芯钻
calyx drill 回转式岩芯管钻粒钻机;钻粒钻机;大口径萼状(取芯)钻;颚状取心钻;萼状钻;萼状取芯钻
calyx drill boring 萼状钻钻探
calyx drill hole 颚状钻孔
calyx drilling 花萼钻凿井;花萼钻钻探;钻粒钻进
calyx rod 钻粒钻进用钻杆
calyx tube 萼筒
Calzecchi-Onestic effect 卡尔齐启—奥乃斯梯效应
calzirtite 钙锆钛矿
calzone 运河地带
cam 凸轮
camacite 梁状铁
cam-actuated 凸轮驱动的
cam actuated clamp 偏心压板
cam adapter 凸轮联轴器
cam adjusting gear 凸轮调整装置
camanchaca 浓湿雾
cam and lever mechanism 凸轮杠杆机构
cam and piston pump 凸轮活塞式(水)泵
cam and ratchet drive 凸轮和棘轮传动装置
cam angle 凸轮角;凸轮分度角;凸轮包角
camarin 公共建筑密室
cam band brake 凸轮带式制动器
cam bearing 凸梁;弯梁
cam bearing clearance 凸轮支承的空隙;弯梁的净空
camber 梁拱;弧拱;小船坞;向上曲曲;中凸形;中拱;挠度;车轮外倾;反壳度;反挠度;反拱;坝顶余幅;坝顶超填量;曲面;曲率;曲度;桥拱;起拱度;起拱;弯度;凸弯度;凸曲;凸度;上挠度;上拱(度)
camber angle 中心线弯曲角;外倾角
camber angle adjustment 外倾角调整
camber arch 平拱;外平内弯拱;弯拱
camber(back) truss 弓形桁架
camber bar 壁炉变梁
camber barge board(gable) 弓形山墙封檐板
camber beam 弯背梁;弓形梁;弓背梁;反挠梁;曲线梁;上挠梁;弯梁;桁架水平拉杆
camber block 拱型垫块,反挠垫块;桥梁膺架支撑垫块;砌拱垫块
camber board 弓背板;路拱板;弓形(模)板;起拱板
camber control 曲面控制
camber correction tool 曲面修整工具
camber curvature 中弧线弯度
camber curve 路拱曲线;梁拱曲线;拱度曲线
camber diagram 起拱图
cambered 弧形的;拱形的;弓形的
cambered arch 平圆三心拱;弯拱
cambered axle 曲轴;弯轴
cambered blade 曲面桨叶;弯曲叶片
cambered blade section 弯曲桨叶型
cambered board 弓形板
cambered ceiling 弓形平顶;弓形顶棚;平拱形天花板;平拱形顶棚
cambered deck 拱形甲板
cambered flatter 起拱形校正锤
cambered girder 起拱大梁
cambered model 正向弯曲模型
cambered plate 弓形板
cambered road 拱形路;弓形路面
cambered roll 凸面轧辊
cambered slopeway 弓形滑道
cambered surface 弧面
cambered truss 弓形桁架;带上曲弦桁架
camber grinding 中凹度磨削;曲线磨削
cambering 向上弯曲;中凹度磨削;翘曲;弧高
cambering gear 凹面加工装置;凸面加工装置
cambering machine 压型机;钢梁矫直机;钢轨矫直机;钢板钢梁矫直机;弯曲机
cambering mechanism 辊型磨制自动装置(轧辊磨床上)
cambering roll 预弯辊
camber jack 压弯机
camber-keeled ship 弧底船
camber line 倾斜线;挠线;弧线
camber member 弓形件
camber of a ship 船中拱
camber of beam 梁拱
camber of bearing 支承反挠度
camber of bridge span 桥梁上拱度
camber of ceiling 预拱度

camber of paving 路面拱度
camber of sheet 板材的翘曲;钢板翘曲;钢板的翘曲
camber of truss 桁架拱度;桁架高度;曲线形钻架
camber piece (支砖拱用小弯度的)起拱板;向上弯曲的构件;拱材;砌拱垫块;凸形构件
camber ratio 弧度比
camber rod 桁架式腹杆;桁架梁拉杆;拱势拉杆;拱式拉杆
camber roof 弓形屋顶
camber slip (砌平拱的)微起拱模架;弧形板;砌拱模架;砌拱垫块
camber templet 路拱板
camber test 翘曲试验;平面弯曲试验(板材)
camber window 弧形窗
cambia 束柱带
cambial zone 形成层带
cambic horizon 过渡层
cambisol 始成土
cambist 外汇经营者
cam bit 凸轮块
cambium 形成层;新生层
cambium layer 青色纤维层;木材青色纤维层;青面纤维层
cam block 三角滑块
Cambodian architecture 柬埔寨(式)建筑
camboge 穿孔混凝土块
cambogia 藤黄
Cambo-Ordovician era 寒武—奥陶纪
cambose 舱面小厨房
cam bowl 凸轮滚子
cam box 凸轮箱;三角箱
cam brake 凸轮制动器
cam breakout 凸轮启动装置
Cambrian epoch 寒武世
Cambrian era 寒武纪
Cambrian glacial stage 寒武纪冰期
Cambrian period 寒武期
Cambrian system 寒武系
cambric 麻纱;黄蜡布;白漆布;细麻布;细薄布
cambric cable 细麻索缆;细麻绳纱
cambric insulation 黄蜡布绝缘;细麻布绝缘
cambric paper 葛纹纸;布纹纸
cambric shade 葛布卷帘
cambric tape 葛布带
Cambridge model 剑桥模型
cambridge ring 楔形环
Cambridge system 剑桥系统;(一种二维矩阵的)剑桥存储系统
Cambridge torque fender 剑桥式扭变护舷
cam carrier 凸轮推杆
cam catch 凸轮挡
cam chuck 凸轮卡盘
cam clearance 凸轮间隙
cam contactor 凸轮接触器
cam contour 凸轮廓
cam control 凸轮控制
cam control gear 凸轮控制装置;凸轮控制机构
cam cutter 凸轮机床
cam diagram 凸轮图
cam disc 偏心盘;凸轮(圆)盘
cam dog 凸轮挡块
cam drive 凸轮驱动;凸轮传动
cam-driven knockout 凸轮驱动出坯
cam drive rotary head 凸轮驱动旋转喷洒头
cam driving gear 凸轮传动装置
cam drum 偏心开槽式圆筒;凸轮盘;凸轮鼓
came 镶嵌铅条;有槽铅条;嵌窗玻璃铅条;铅棂条
cam ejector 凸轮顶出装置
camel 码头边的木靠把;护岸浮木排;吸粮机橡皮管头上的嘴;打捞浮筒;充气浮筒;浮桥船;浮垫;起重浮箱
camel back 桁架拱形上弦
camel-back bridge 罗锅桥;驼峰式桥
camel-back curve 驼峰曲线
camel-back top chord 桁拱弦上梁;拱形上弦;(桁架的)驼背上弦杆
camel-back truss 上弯桁架;折线形桁架;驼背式桁架;驼峰桁架;弓形桁架
camel-back-type crusher 驼峰形锤式破碎机
cameleon 织物闪色效应;织物闪光效应;变色蜥蜴
camel hair 驼毛刷
camel-hair brush 毛刷(涂料用)
camel-hair mop 毛鬃油漆刷
Camelia alloy 卡梅利亚合金

camellia 山茶花
camellia carbon 山茶木炭
camellia oil 山茶油
camel thorn 骆驼刺
cameo 建筑装饰浮雕;浮雕玉石贝壳;浮雕雕刻;浮雕宝石;饱和度从低到适中
cameo blue 玉石蓝色
cameo brown 玉石棕色
cameo cut 宝石雕刻
cameo glass 浮雕玻璃;宝石玻璃(多层套色并具有浮雕的玻璃器皿)
cameo green 玉石绿色
cameoid relief 凸纹墙纸
cameo incrustation 宝石镶嵌
cameo paper 花面涂布纸
cameo pink 玉石粉红色
cameo printing 浮凸印刷;凸版印刷;凸纹转筒印花
cameo ware 有浮雕陶制小装饰品;浮雕器皿
cameo yellow 玉石黄色
camera 小箱;照相机;暗箱;暗室;摄影机;摄像机
camera aerial number 航摄仪编号
camera alignment 射束校正
camera altitude 摄影机姿态
camera angle 照相机头视角;摄影角度
camera aperture 取景框;曝光门
camera array 摄影机系统
camera axis 镜箱轴【测】;光轴
camera back 镜箱后盖;照相机背盖
camera bag 暗袋
camera base 摄像机底座
camera bellows 照相机皮腔;暗箱
camera body 照相机机身;摄影机主体
camera calibration 摄影机检定
camera calibration field 摄影机检定场
camera calibrator 摄影机检定器
camera car 摄影车
camera cardan 投影器方向节
camera constant 航摄仪常数
camera container 照相机外套
camera crane 照相机三脚架
camera device 照相机室
camera display 记录照相显示
camera extension 摄影镜头延伸
camera eye 摄影机取景孔
camera for deep-sea 深海摄影机
camera for measuring cloud 测云照相机
camera for single photograph 单像幅摄影机;单镜摄影机
camera function 镜箱性能
camera geometry 摄影机几何参数
camera gun 照相机镜头;摄影枪
camera head 摄像装置
camera hood 漏光防护罩
camera housing 照相机外套
camera in space 空间照相机
camera installation 照相设备;摄影机暗箱
camera lens 照相机镜头;暗箱透镜;暗箱镜头
cameralistics 财政学
camera lucida 描像器;显微描绘器;显画器;明箱;像片转绘仪;显微描绘仪;转绘仪
cameraman 摄影师
camera marker 摄影机的标志灯
camera microfilm 摄影显微胶片
camera monitor 摄像机监控器
camera-mount 照相机架
camera mounting 摄影机架
camera obscure 暗箱;投影器;像片转绘仪
camera observation 照相观测
camera of deep-sea 深海照相机
camera of projection 投影器
camera operator 航空摄影员
camera port 摄影窗口
camera processor 摄像处理机
camera-read theodolite 摄影读数经纬仪
camera recycle rate 摄影机曝光
camera remote control unit 摄像机遥控装置
camera-scan(ning) pattern 摄像管扫描图;照相机扫描图形
camera screen 摄像屏
camera section 照相机机构
camera shaft 摄影机轴
camera shutter 照相机快门;摄影机快门
camera stability 航摄仪稳定性
camera-stand 复照仪底座

camera station 空中摄影站；摄(影)站
camerated 隔成小室的
camera test 摄影机检验
camera tilt 摄影机俯仰运动
camera tilt down 摄影机俯摄
camera timer 摄影机计时器
camera transit 摄影经纬仪
camera tripod 三脚照相架
camera tube 摄像管
camera tube lag meter 摄像管余像测量仪
camera tube with high velocity scanning beam 高速电子扫描摄像管
camera window 摄影窗口；摄影窗孔
camera with eyepiece 目镜照相机
camera with ground glass focusing 毛玻璃聚焦摄影机
camera work 摄影技巧
camettes tile 半透明上釉的瓷砖
cam face 凸轮表面
cam finger 偏心爪
cam follower 凸轮随动轮；凸轮随动件；凸轮跟随器；凸轮从动轮
cam follower guide 凸轮随动件导承
cam follower lever 凸轮从动杆
cam follower needle roller bearing 滚轮滚针轴承
cam follower pin 凸轮从动滚轮销
cam follower roller 凸轮推杆滚柱
cam forming and profiling machine 凸轮仿形机床
cam gear 偏心盘；凸轮机构；凸轮(传动)装置
cam governor 凸轮同速器；凸轮调速器
Ca-Mg ratio 钙镁比(率)
cam grinder 凸轮磨床
cam grinding 磨成凸轮形；凸轮磨削
cam handle 转动把手
cam holder 凸轮架
caminoids 卡米伊德
cam inversor 曲线模板控制器
camion 军用货车；军用汽车；载货汽车；载货车
Camkometer 剑桥旁压仪
camlet 防雨驼毛织物
cam lever 凸轮杆
cam lever shaft 凸轮杆轴
cam-lift 凸轮升程；凸轮升降器；凸轮升度
cam lining 扬料砖衬
cam lobe 凸轮(的)凸角
camloc 管式脚手架连接器
cam-lock 凸轮锁；偏心轧头；偏心夹；偏心闩；凸轮锁紧
cam lock bracket 锁杆轮座(集装箱)
cam lock holding mechanism 凸轮锁紧机构
cam lock spindle nose 凸轮锁紧轴端
cam loom 踏盘织机
Camloy 卡姆镍铬铁合金；镍铬铁耐热合金
commander 指挥者
cam mechanism 凸轮机构；凸轮装置
cam member 凸轮件
Cammett table 肯梅特摇床
cam milling attachment 凸轮铣切装置
cam milling machine 凸轮铣床
camming 凸轮系统
cam motion 凸轮运动
cam nose 凸轮尖
cam nut 止动螺母；止动螺帽
cam of camshaft 凸轮轴凸轮
Camoform 卡马风
cam of variable lift 变升程凸轮
camomatic grinder 全自动凸轮磨床
cam-operated 凸轮推动的；凸轮操纵的
cam-operated switch 凸轮开关
cam-operated(tamping) block machine 凸轮夯实机
camopy manipulation 林冠修整
camouflage 伪装(色彩)
camouflage coat 伪装外罩；迷彩；伪装涂层；伪装色
camouflage coat of paint 油漆的伪装层
camouflage colo(u)ring 伪装(着)色
camouflaged building 伪装建筑物
camouflage detection 伪装侦察
camouflage detection film 伪装探测胶片
camouflage lacquer 伪装油漆
camouflage paint 保护色油漆；保护色涂料；伪装涂料；伪装色彩；伪装涂料；伪装用涂料
camouflage painting 伪装涂漆
camouflaging 掩蔽

camouflet 爆破孔眼；地下空洞
camp 营地；帐篷；工人营地；保留呼叫
campaign 炉役；炉期；工业窑炉炉龄；衬料使用寿命
campaign fire 大火
campaign for the conservation of the water's edge 保护河海边缘运动
campaign length 炉龄
campaign life 炉期；炉龄
Campamuria chinesis 中国钟螳【地】
campana 排钟
campaniform 钟形
campanile 钟塔；钟楼(车站)；(与主楼分离的)独立钟楼
Campanion 坎帕阶
campanulate 钟状的
campanulate corolla 钟状花冠
campas grass (产于南美，有丝光，可用于装饰的)南美草
cam path 凸轮槽
cam pawl 凸轮爪
camp bed 行军床；帆布床
Campbell's formula 坎贝尔公式
Campbell's ga(u)ge 坎贝尔气压计；坎贝尔量规
Campbell's theorem 坎贝尔定理
Campbell-Stokes sunshine recorder 坎贝尔—斯托克斯日照计
Campbell-stokes' heliograph 坎贝尔—斯托克斯式日照光度计
camp building 工地宿舍；工地房屋；施工营地
camp car 工地住宿车；宿营车；施工宿车
camp ceiling (斜的或凸圆的)天花板；坡形顶棚；(斜的或凸圆的)顶棚；闷顶楼板；帐篷式天花；帐篷式顶棚
camp chair 轻便折椅；帆布椅
camp construction 临时房屋建设
campeachy wood 苏木
camp equipment 工地营房设备；外业装备；外业设备
camper 野营车
camper trailer 野营拖挂车
camper winding 阻尼绕组
campesterol 菜籽甾醇
camp fire 营火
camp ground 露营场地；野营场地
camphane 樟烷
camphene 莰烯
camphoid 樟脑火棉浆
camphor 樟脑
camphor glass 樟脑玻璃
camphor laurel 樟树
camphor oil 樟脑油
camphor oil white 白樟油
camphoronic acid 分解樟脑酸
camphor tablet synthetic 合成樟脑块
camphor tree 樟树
camphor wood 樟木
camp hospital 野营医院
campigliaite 坎锰图矾
campimeter 平面视野计
campimetry 平面视野计检查法
cam pin 偏心销(钉)
camp infirmary 兵站医务所
camping area 野营场地
camping bag 露营包
camping cylinder 旅游用气瓶
camping screw 紧固螺钉
camping site 野营场地
camping tent 露营帐篷
cam plastometer 凸轮塑性计
cam plate 平板形凸轮；凸轮盘；轮盘
cam plate type axial piston pump 斜盘式轴向活塞(油)泵
Camp-Meidell condition 坎普—迈德耳条件
Camp-Meidell inequality 坎普—迈德尔不等式
campoboard 树脂胶合板
camp-on 预占；暂等
camporee 野营会
camp sheathing 板桩护岸；河岸挡土岸壁；挡土岸壁；板桩排(用于河岸或松软土壤)；软泥基中板排桩；棚屋护板
camp sheathing column 轻型板桩墙
campshed 排桩岸壁(两排桩中填土)；插板板桩墙(岸壁)；排柱中填土；排桩护岸
camp shedding 河岸挡土岸壁；挡土岸壁；棚屋护板

camp sheeting 河岸挡土岸壁；挡土岸壁；板桩排(用于河岸或松软土壤)；板桩挡土岸壁；插板板桩墙(岸壁)
camp sheltering 插板板桩墙(岸壁)
camp shop 野外修理站
campshot 堤防铺木板；堤身堆土石；插板板桩墙(岸壁)
camp site 露营地；营地；工地；施工现场；野营区
campsite laboratory 驻地实验室
campstool 轻便折凳
camp structure 帐篷构造
camptonite 棕闪斜煌岩；闪煌岩
campus 校园；大学校园
campus building 校园建筑
campus plan 校园平面布置图(大学)
campus planning 校园规划
campus town 大学城
campylotropal 弯曲的
cam-ram machine 凸轮夯实机
cam reduction gear 凸轮减速装置
cam relationship (导叶开度与桨叶角度的)协联关系；组合关系
cam return spring 凸轮回位弹簧
cam ring 凸轮环
cam-ring chuck 三爪卡盘
cam rise 凸轮升程
cam roll 凸缘滚辗
cam scale 凸盘秤
camshaft 凸轮轴；曲轮轴；偏心轮轴；桃轮轴
camshaft bushing 凸轮轴衬套
camshaft chain and sprocket drive 偏心轴链条传动
camshaft control 凸轮轴控制
camshaft cover 凸轮轴盖
camshaft drive 凸轮轴传动
camshaft gear 凸轮轴(上)齿轮
camshaft gear drive 偏心轮齿轮传动
camshaft housing 分配轴箱
camshaft packing gland 凸轮轴填密压盖
camshaft phasing gear 凸轮轴上控制盘
camshaft sprocket 凸轮轴链轮
camshaft textolite gear 凸轮轴夹布胶木齿轮
camshaft thrust bearing 凸轮轴止推轴承
camshaft thrust plunger 凸轮轴止推塞
camshaft time gear 凸轮轴定时齿轮
camshaft timing gear 凸轮轴定时齿轮
camshaft timing gear hub 凸轮轴定时齿轮壳
camshaft turning lathe 凸轮轴车床
camshaft wedge 凸轮楔
cam-shaped piston 卵形活塞；靠模加工活塞；仿型加工活塞
cam shaper cutter 齿轮插刀
cam sleeve 凸轮联轴节；三角套筒
cam slide 凸轮滑化
cam slot 曲线槽；凸轮槽
cam spindle 凸轮轴
cam stroke 凸轮动程
cam surface 凸轮面
cam switch 凸轮(操纵)开关
cam template 凸轮样板
cam template for idle stroke 空程凸轮样板
cam throttle 凸轮节流阀；凸轮调节阀
cam-throttle type screen 冲击筛
cam throw 凸轮动程
cam-turning 靠模车削
cam-type axial piston motor 斜盘式轴向柱塞电动机
cam type screen 凸轮式遮板；防撞遮板
camus 装配式建筑施工法
Camus system 卡莫斯式体系建筑(法)
cam wheel 凸轮
camwood 非洲红木
can 密封外壳；马口铁罐；罐形容器；罐头；容器
Canada balsam 冷杉树脂胶；加拿大香树脂；加拿大香胶；加拿大枞树香脂；枞香脂
Canada bolt 加大插销
Canada hatchet 加拿大短柄小斧
Canada hemlock 加拿大铁杉
Canada pitch 加拿大松香；铁杉脂
Canada total environmental quality index 加拿大总环境质量指数
Canada turpentine 加拿大松节油；枞香脂
Canadian architecture 加拿大(式)建筑
Canadian asbestos 温石棉；加拿大石棉

Canadian ax(e) 加拿大短柄小斧
Canadian balsam 加拿大香树脂;枞香脂
Canadian dollar 加拿大元
Canadian Engineering Standards Association 加拿大工程标准协会
Canadian Environmental Protection Act 加拿大环境保护法
Canadian Federal Survey Bureum 加拿大联邦测绘局
Canadian Good Roads Association 加拿大大好路协会
Canadian grid 加拿大格网
Canadian Hydrographic Office 加拿大水道测量局
Canadian latch 门窗插销;加拿大锁;压开锁
Canadian life zone 加拿大生物带
Canadian Institute of Surveying 加拿大测量学会
Canadian method of flexible pavement design 加拿大柔性路面设计法
Canadian oil of turpentine 加拿大松节油
Canadian pattern trowel 加拿大式泥刀
Canadian red pine 加拿大红松
Canadian roll-over mortgage 加拿大式滚动低压贷款
Canadian shield 加拿大地盾
Canadian spruce 加拿大云杉木
Canadian Standard Association 加拿大标准协会
Canadian Standard Freeness 加拿大标准游离度
Canadian window 加拿大式窗
canadite 钠霞正长岩;钠霞正长石
canadite pegmatite 钠霞正长伟晶岩
canadol 坎那油
canaigre ink 销迹墨水
canal 运河;管;沟;渠道;渠;水渠;水沟
canal advantage rate 土地增值税;水渠收益税
canal alignment 运河定线
canal appurtenance 运河的从属权利;渠道附属设施;渠道附属建筑物
canal aqueduct 运河桥;运河立交渠;跨水管;跨渠管道;跨渠渡槽;渡槽;输水渡槽;输水渠道
canal arch 通路碹
canal bank 运河堤岸;运河岸(坡);运河护坡;运河堤
canal bank protection 运河护岸
canal barge 运河驳船;平底船
canal base width 运河底宽
canal basin 渠槽
canal bed 运河(河)床;渠底
canal berm 运河戗道
canal black 槽法炭黑
canal boat 运河船
canal bottom 渠床(底);渠底
canal bottom width 运河底宽;渠底宽度
canal branch 沟渠连接支管;迂回支流
canal bridge 跨梁桥;运河桥;渡槽;渠道桥
canal building machinery 筑运河的机械
canal-built 运河专用的
canal capacity 运河运输能力;运河通过能力;渠道输水能力;渠道容量;渠道排灌能力;渠道流量;渠道过水能力
canal capacity at the headwork 渠首处渠道过水能力;渠道口流量
canal channel 运河河槽
canal charges 运河收费
canal check 运河水位控制设备;渠道节制阀;渠道节制闸
canal cleaning 渠道疏浚
canal closure 水渠关闭
canal concrete paver 筑运河用移动混凝土模板;渠道混凝土浇筑机
canal construction 运河建造;运河工程;渠道建筑;渠道施工;渠道开挖
canal crossing 运河渡口;渠道交叉建筑物;渠道渡槽
canal cross-section 运河的横断面;运河断面;渠道剖面;渠道截面;渠道断面
canal curve 运河曲线;河渠弯道;运河弯段;渠道弯曲;运河曲线段
canal cut 截水沟;截流渠道
canal cut on sloping ground 筑在斜地面上的运河
canal desilting 渠道清淤
canal distribution system 渠道配水系统
canal ditch 泥浆槽
canal dock 运河码头;渠式干船坞;双端干船坞;运河船坞
canal drainage 明渠排水

canal dredge(r) 运河挖泥机;航道挖泥船;运河挖泥船
canal dryer 管道干燥器;隧道(式)干燥机;隧道干燥窑;隧道干燥器
canal drop 渠道跌水
canal dues 运河通行费;运河税
canal embankment 运河堤
canal engineering 运河工程;渠工学;渠道工程
canal entrance 运河入口
canalette blind 滑槽卷帘;花条痕状百叶窗
canal fee 运河税
canal feeding 运河水量补给
canal feeding by pumping 用泵补偿运河需水
canal flush 沟形水槽
canal for goods-carrying barges 装货驳船用的运河
canal for rafting wood 浮运木料的运河;运筏渠
canal freeboard 渠岸出水高度
canal garden 运河庭园(文艺复兴时期的一种几何式花园)
canal grade line 渠道纵坡线
canal gradient 运河比降;渠道比降
canal harbo(u)r 运河港
canal head 运河起点建筑物;渠首;运河起点构筑物
canal head gate 渠首闸门
canal headwork 渠首建筑物;渠首部设施;渠道口工程;渠首工程
canal hydraulics 运河水力学;渠道水力特性
canaliculus 三槽板上的小沟;三槽板上的小槽
canal in a cut 堑槽运河;挖筑运河;挖方渠道
canal in a cutting 堑槽运河
canal incline 运河斜坡道
canal incline lift 运河斜面升船机
canal in cutting 挖凿运河;挖方运河;挖方渠道
canal inland lock 运河(内陆)船闸
canal inlet 渠道进水口
canal intake 渠首;渠道进水口
canal irrigation 渠灌
canalis 凹圆嵌条;爱奥尼亚式螺旋饰;管
canalis pudendalis 阿尔克氏管
canalis semicircularis lateralis 外半规管
canalization 河网化;沟渠化;穿通;渠(道)系统;渠(道)化
canalization dam 通航用坝(用坝将河道分段)
canalization lock 通航船闸;通航闸(用闸将河道分段)
canalization of river 河道渠化
canalization project 河道梯级开发工程;河道渠化工程;渠化工程;梯级开发工程;河道梯级化工程
canalization river 梯级化河道;通航河道;渠化河道;运河化河流
canalization section 渠化剖面;开挖运河的剖面
canalization works 渠化工程
canalize 开挖水道;渠化
canalized channel 渠化河道;渠化航道
canalized engineering 渠化工程
canalized river 开浚河道;人工开挖河道;渠化河流
canalized river section 渠化河段
canalized river stretch 渠化河段;通航河段
canalized stream 人工开挖河道
canalized transition 渠道渐变段
canalized wastage 渠道损失
canalized waterway 开挖水道;渠化水道;渠化航道;通航水道
canal junction 运河(交汇)口
canalled dock 运河式船坞
canal(l)er 运河船;运河船员
canal lift 运河升船机;运河水级
canal lining 运河边坡衬砌;渠道衬砌;运河衬砌
canal lock 运河闸;运河船闸;渠闸
canal mouth 运河口
canal mud 河泥
canal navigable stage control 运河通航水位控制
canal navigation 运河上航运;运河航运
canal navigation lock 运河船闸
canal of Corti 螺旋器隧道
canal off-let 渠道斗门;渠道出水口;渠道放水门;渠道放水口
canal on embankment 堤槽运河;填方渠道;填堤运河
canal permissible scouring velocity 渠道容许冲刷流速
canal pilot 运河引航员
canal plugger 根管充填器

canal pond 船闸间河段;渠段;渠首前池;渠道前池
canal pool 渠段
canal port 运河上港口;运河港
canal prism (棱柱形的)运河河槽
canal pumping station 运河泵站
canal ramps 运河梯级;梯级运河
canal rapids 陡槽;湍流渠
canal rating 率定池;率定槽;检定槽
canal ray 极隧射线;正阳线
canal ray tube 极隧射线管
canal reach 运河段;渠段
canal regime 运河情况
canal rudder 运河舵
canal scouring sluice 渠道泄水闸
canal section 渠道剖面;渠道截面;渠道断面
canal sedimentation 渠道淤积
canal seepage 渠道渗漏
canal seepage loss 渠道渗漏损失;渠道渗流损失
canal seepage prevention 运河防渗
canal side 运河边
canal side slope 运河边坡;渠道边坡
canal silt clearance 渠道清淤
canal silting 渠道淤积;渠道淤积
canal slipform 渠道滑模;渠道滑动模板
canal slope 运河坡度;渠道岸坡
canal slope concrete paver 用混凝土铺料机建筑运河坡
canal slope protection 运河坡保护;运河坡;运河护岸;渠道护坡;渠岸护坡
canal slope trimmer 渠道边坡修整机
canal slope trimming machine 整修运河坡的机器;渠道边坡修整机;渠岸修整机
canal steamer 运河轮船
canal structure 渠系建筑物
canal surcharge 绕闸附加费
canal surface 管道曲面
canal survey 渠道测量
canal system 渠系;渠道系统
canal system power station 引水渠式发电站
canal theory 海洋分渠理论(一种潮汐运动)
canal tolls 运河通行税
canal tonnage 运河(计费)吨位;通过运河吨位
canal transition 渠道渐变段
canal transit time 运河通过时间
canal transport 运河运输
canal trimmer 渠道修整机
canal tunnel 运河(隧道);运河通航隧洞;渠道隧洞
canal tunnel(l)ed in rock 在石头中开挖的运河
canal wall 运河边墙;渠道边墙
canal wastage 运河水量损失;渠道水量损失
canal water 运河的水;渠道水
canal water depth 运河水深
canal widening 渠道加宽
canal with locks 有船闸的运河;设闸运河;设闸渠道;街道泻水沟
canal works 渠道工程
canal zone 运河区;运河地区
Canal Zone Code 运河区法规
can and container making machinery 罐头和容器制造机
cananga oil 衣兰油
can-annular combustion chamber 环管式燃烧室
canaopy (圣坛、神龛上的)罩盖;掩蔽顶篷
canaopy lip (圣坛、神龛上的)罩盖披檐;掩蔽顶篷边缘
canard 鸭式水翼简图
canard configuration 鸭式图形;鸭式布局
canard layout 鸭式设计方案;鸭式构图
canaria 金丝雀色;鲜黄色
canaries 高频噪声
canary 金丝雀色;鲜黄色
Canary basin 加那利海盆
Canary current 加那利海流
canary glass 鲜黄色玻璃;黄色玻璃
canary lamp 充气黄色灯泡
canary sassafras 黄樟(昆士兰州产)
canary stained wool 污黄毛
canary whitewood 白木(美国产)
canary wood 加那利木
canary yellow 嫩黄;金丝雀黄
canasite 硅碱钙石
Canastotan 卡纳斯托特阶【地】
canavesite 硼碳镁石
can bleeding tester 罐头渗漏检验机

can buoy 立式圆筒浮标;罐形浮标;筒形浮标;桶浮标;浮筒
can burner 单管燃烧室;罐式燃烧器
canbyite 硅铁石
Cancani absolute intensity scale 坎坎尼绝对烈度表
can-cap dryer 罐盖烘干机
cancel 消除;相约;作废;组成网格状;注销;撤销;取消;删去;删除字
cancelable lease 可消租约
cancel a block 取消间塞
cancel a cheque 注销支票
cancel a contract 取消合同
cancel a debt 销账
cancel after verification 核销
cancel an offer 撤销发盘
cancel an order 取消订货
cancel arch 胜利拱门;凯旋门;古罗马凯旋门
cancel a route 取消进路
cancel a signature 签字作废
cancel a tenancy 退佃
cancel character 作废键;作废(字)符;取消(字)符;删去(字)符
cancel circuit 消除电路
cancel closedown 异常停机
cancel code 作废码
cancel debt 销账
canceled project 停建项目
canceled structure 格构
canceler 补偿设备
cancel from an account 销账
canceling machine 消除机
cancel key 消除键;符号取消键;清除键;切断按钮
cancellate 蜂窝状;网格状
cancelled 组成格构式的
cancellation 桁架腹杆体系;消去;相约;作废;注销;抵消;取消;退保注销
cancellation amount 注销金额
cancellation amplifier 补偿放大器
cancellation and refunding of development credit 开发信贷的撤销与偿还
cancellation charges 注销费
cancellation clause 销约条款;撤销条款;取消条款
cancellation cost 撤销费用
cancellation date 注销日
cancellation interlace 取消交织
cancellation law 可约律;消去律;相消律
cancellation mark 次品标志;次品标记;删除符
cancellation money 解约金
cancellation network 抵消网络
cancellation notice 注销通知
cancellation of a contact 撤销合同
cancellation of a lease 撤销租约
cancellation of a lost instrument 注销遗失的票据
cancellation of authorization to use 撤销委任使用
cancellation of building licence 取消营造执照;建筑禁止期;建筑封锁期;收回或取消建筑许可证
cancellation of contract 撤销合同;取消合同
cancellation of debt 债务的取消
cancellation of future performance 撤销未到期的合同
cancellation of indebtedness 免去债务;注销债务;取消债务
cancellation of intensities 振动强度的抵消
cancellation of project 取消项目
cancellation of purchase contract 取消订货合同
cancellation of registration 取销登记
cancellation of reinsurance contract 注销分保合同
cancellation of special drawing right 取消特别提款权
cancellation of the order 取消订货
cancellation of treaty 注销契约;注销合同
cancellation of warning signal 解除风暴警报
cancelled 已解约
cancelled bond 已注销债券
cancelled check 已注销支票;已付支票;注销支票;注销的支票;废票;付讫支票
cancelled in error 误被注销
cancelled leaf 不合格的印张;取消的叶
cancelled structure 格组构件;空腹结构;格形构造;格构(式)结构;透空式结构
canceller 消除器
canceller output voltage 补偿设备输出电压
cancelli 教堂中国屏;挂幕;隔断;栏杆铁棍支杆;小柱;屏障小柱
cancelling 解约;解除合同;取消出口
cancelling circuit 补偿电路
cancelling clause 解约条款;解除合同条款;注销条款
cancelling date 解约日期(船租合同);解除合同日期;相约取消日期;撤销合同日期
cancelling returns 注销退款;注销保险单退费;退还保险费
cancelling returns only 只退还保险金
cancelling signal 取消信号
cancelling stamp 作废图章
cancelling stocks 退库
cancello 花格围屏;花格围栏;教堂围屏
cancel mark switch 取消符号开关
cancel message 作废信息;作废信号;撤销信号
cancel of reserves 储量注销;储量报销
cancel order 注销订货单
cancel statement 撤销语句
cancel the order 退货
cancel transmission 取消传送
cancer by radiation 辐射致病
cancer due to occupation 职业癌
canch 挑顶卧底;斜壁沟;狭底水沟
canchizzarite 卡辉铋铅矿
can coating 罐头涂料;罐头涂层
can coding machine 罐头代号打印机
can coding set 罐头代号打印机
can coiler 圈条器
can combustor 筒形燃烧室
can cooler 罐头冷却器
can corrosion 罐头腐蚀
can corrosion inhibitor 罐头防蚀剂
can-corrosion tester 罐腐蚀检验机
can count 管壳计数
cancrinite 钙霞石
cancrinite canadite 钙霞钠霞正长岩
cancrinite litchfieldite 钙霞钠霞云霞正长岩
cancrinite nepheline syenite 二霞正长岩
cancrinite nordsjoite 钙霞霞正长岩
cancrinite syenite 钙霞正长岩
cancrinite tinguaite 钙霞丁古岩
candela 新烛光;新国际烛光(单位);新国际光单位;烛光
candela(ed) 坎德拉(发光强度单位)
candelabra 枝状大烛台或灯台
candelabra base 小形灯座
candelabrum 华表;分枝烛架;华柱;烛台架;枝形烛台
candelabrum base 蜡台形灯座;小型灯座;小形灯座;小型蜡台形灯座
candela per square meter 坎德拉每平方米
candela per unit area 单位面积新烛光
candelilla wax 小烛树蜡
candescence 炽热;白热
candescent 炽热;白热(化)的
candidate for tendering 投标候选人
candidate material 备用材料
candidate plant 备选电站
candidate volume 候补卷宗
candle 蜡烛;烛枝;烛光;陶瓷过滤芯
candle balance 光度计;照度平衡
candle beam 悬灯梁;教堂插烛横杆
candle berry tree 石栗
candle black 蜡烛烟黑
candle coal 长焰煤;烛煤
candle filter 烛式过滤器
candle filtration 烛式过滤
candle-foot 烛(光英)尺
candle holder 烛台
candle-hour 烛光·小时
candle ice 针状冰
candle lamp 烛焰形装饰灯;烛光灯;烛形灯
candle lantern 点烛灯笼
candle light(ing) 烛光(照明);人造光;人工柔光;烛光力
candle machine 蜡烛机
candle-meter 烛米(照度单位);烛光·米
candlenut 石栗
candlenut oil 烛果油;石栗子油
candle pitch 硬脂酸沥青;硬脂酸焦油;脂硬脂沥青;柏脂
candle power 以烛光表示的发光强度;烛光
candle power brilliance 烛光亮度
candle power distribution 配光
candle power hour 烛光·小时
candles per unit area 单位面积上的烛光
candlestand 烛台架
candlestand table 落地烛台
candlestick 蜡烛台;蜡台;烛台
candle tar 烛脂柏油;烛脂焦油沥青
candlewick 烛芯纱盘花簇绒织物;灯芯绒
candle wood 烛木;树脂木
candle works 蜡烛工厂
candling egg 照光检查
candlot's salt 硫代铝酸钙;坎德拉盐;水泥杆菌;硬化水泥石中的硫铝酸钙结晶
can draw press 罐头开合机
can dump 倒冰架;脱冰机
candy bucket 糖果吊桶(俚语);工地料斗
candy pink 淡红糖色;砂糖粉红色
candy store 糖果店
candy stripe 条纹图案
candy wagon 糖果车(俚语);工地运输车
cane 蚕丝经线;藤料
cane-and-white ware 藤色釉白坯陶瓷
cane bolt 插销;竹栓
canebrake 竹丛;藤丛
canebreak 植物碎屑(灌浆用填料)
cane chair 藤椅
cane crusher 甘蔗榨汁机;甘蔗压榨机
cane easy chair 大藤椅
cane fender 管制防冲垫;藤条碰垫;藤料防冲器
cane fiber board 甘蔗纤维板;蔗渣板;甘蔗板
cane fiber insulation(cane fibre insulation) 甘蔗板隔热隔声
cane filler 植物纤维填缝材料
cane glass 玻璃棒
cane glaze 浅黄釉
can egualizer 奶桶压力补偿器
cane-juice squeezer 甘蔗榨汁机
cane lifter 甘蔗提升机
cane-lifting machine 甘蔗提升机
cane loader 甘蔗装载机
cane masher 甘蔗压榨机;干蔗压榨机
cane mat 藤席;藤科编织物
cane mattress 藤垫;柴排;竹排;竹茎沉排
can-end press 罐头封顶压力机;罐盖冲床
canephora 女郎头顶祭盘的装饰品;古希腊少女雕像
canephorae 女雕像柱
canephoros 顶篮女像柱
cane press 甘蔗压榨机
canes 镶玻璃铅条;藤科
canescent 变灰白的;变白的;被灰白毛的
cane-sugar 蔗糖
canete 单丝绢丝法
cane trash 甘蔗废料
cane trash board 甘蔗废料板
cane trash sheet 甘蔗废料薄板
cane trash tar 甘蔗废料焦油
cane type shift 细长杆式变速;直接操纵式变速
cane work 藤编加工;藤制工艺晶
can extrusion 筒形件挤压
caney 藤孔状的
Canfield anaerobic ammonia oxidation 肯弗尔德厌氧氨氧化
canfieldite 硫银锡矿;硫锡银矿
Canfield's reagent 肯弗尔德试剂
can filler 装罐机;注水器
can filling 装听;装罐;桶装
can flanging machine 罐身翻边机
canga 铁角砾岩
cangiante lustre 彩虹光泽
can hook 吊桶钩;带钩撑杆
can hooks sling 吊桶钩绳环
can ice 大块冰
canicular days 三伏天
canidrome club 跑狗场
Canif mixer 卡尼夫砂浆拌和机
canine teeth 尖牙
canister 金属容器;罐形容器;钢筒;防护面罩滤器;滤毒罐;咖啡罐;茶叶罐
canister elbow (铁烟道的)F形接头
canister float 水下双浮标;深水浮标
canister purge line 过滤器净化管
canker growth of wood 木瘤
canker of tea tree 茶叶枝
can labeller 罐头贴标机

can lacquer 罐头涂料
can liner 罐头内壁涂漆机
can lining enamel 罐头内壁涂料
can lining machine 罐头涂漆机
can mixer 罐头混
can mo(u)lding 瓶罐模制
Cannabis sativa 大麻
Canna indica 美人蕉
cannal cut 截水沟
canned 装罐的;全封闭的
canned-candle filter 烛式过滤机
canned computer program(me) 编好的计算机程序
canned cycle 存取循环;存储循环
canned data 已有数据;已存数据;存储信息
canned food 罐头食品
canned format 固定格式
canned fruit industry 罐头蔬菜工业
canned motor 密封(式)发动机;封闭电动机
canned motor pump 密封(式)电动泵;屏蔽(式)潜水电)泵
canned paragraph 固定段
canned powder extrution 粉末包套挤压
canned pump 密封(的)轻便泵;密封泵;封闭泵;屏蔽泵
canned record 盘声记录
canned-rotor pump 密封转子泵
canned routine 固定例行程序;定型例行程序
canned software 固定软件
cannel 烛煤
cannel bass 含煤页岩
cannel boghead 烛藻煤
cannel coal 烛煤;长焰;残烛煤
cannelite 烛煤
canneloid 类烛煤
cannel shale 烛煤页岩
cannelure 细纹;纵向槽;(物体表面的)沟槽;纵向切口;槽线
cannery 罐头食品厂;罐头厂;监狱
cannery shop 罐头品车间
cannette 单丝绢丝法
cannibalise 零件拆用
cannibalism 同类相食;同类相残
cannibalization 利用旧件;同型装配
cannibalize 利用旧件;拆用零(配)件;拼修;同型装配
cannikin 小罐;木桶
canning 分装;包套
canning factory 罐头食品制造厂
canning factory ship 鱼类罐头加工船
canning glass jar 包装用玻璃罐
canning industry waste 罐头食品厂废料
canning industry waste water 罐头食品厂废水
canning plant 罐头食品厂
canning plant ship 鱼类罐头加工船
canning waste 罐头食品厂废料
canning waste(water) 罐头食品厂废水
can(n)ister 空气过滤器;金属罐;箱式真空吸尘器;炸药罐
can(n)ister float 水下双浮标;深水浮标
Cannizzaro reaction 坎尼扎罗反应
cannned food production wastewater 罐头生产废水
cannon 空心轴;空心圆筒;加农高速钢;火炮
cannonball 炮弹;快车
cannonball bed (四角短柱各饰大圆球的)球柱床
cannon connector 加农插头与插座
cannon pinion 分轮管
cannon plug 加农插头;圆柱形插头
cannon-proof 防弹的
cannon tube shield 筒形屏蔽罩
cannon-type projector 短管式喷射器
can not be closed 不能成交
cannula 插管;套管
cannular burner 环管燃烧室;环管形燃烧室
cannular combustion chamber 联筒燃烧室;筒形燃烧室;筒形燃烧室
cannular combustor 环管燃烧室
cannulation 套管插入术
canny 罐头食品厂废料
canny waste water 罐头食品厂废水
canoe 开敞式小艇;划艇;独木舟
canoe automobile 水陆两用(汽)车
canoe birch 纸皮桦

canoe fold 舟状向斜;舟形褶皱
canoe-shaped valley 皮舟式山谷;马蹄形山谷
canoe valley 船形峡谷;马蹄形山谷
can of paint 漆桶
canography 加农摄影法
canoic acid 硬脂酸
canonic(al) 正则的;正测的;典型的
canonic(al) answer 规范回答
canonic(al) assembly 正则系统
canonic(al) base 典范基(底);标准基(底)
canonic(al) block 正则块
canonic(al) change 正则变化
canonic(al) conjugate 典型共轭量;正则共轭
canonic(al) conjugate variable 正则共轭变量
canonic(al) conjunctive form 逻辑积标准形
canonic(al) constant 正则常数
canonic(al) convex programming 标准凸规则
canonic(al) coordinates 典型坐标;典范坐标;正则坐标
canonic(al) correlation 典型相关;典范相关
canonic(al) correlation analysis 典型相关分析;典范相关分析
canonic(al) correlation coefficient 典型相关系数;正则相关系数
canonic(al) correspondence analysis 典型对应分析
canonic(al) decomposition 标准分解
canonic(al) decomposition of matrix 矩阵的典型分解
canonic(al) disjunctive form 逻辑和标准形
canonic(al) distribution 正则分布
canonic(al) element 正则根数
canonic(al) ensemble 正则系统;正则集
canonic(al) equation 典型方程;标准方程(式);正则方程
canonic(al) equation of state 正则状态方程式
canonic(al) extension 正则扩充
canonic(al) form 规范型;规范形式;典型形式;典型格式;典范式;典范格式;范式;标准形(式);标准格式;正则(形式)
canonic(al) form theorem 标准形定理
canonic(al) function 标准函数
canonic(al) generation 规范生成
canonic(al) implicant 标准内项
canonically conjugate variable 正则共轭变量
canonically labelled tree 典型标记树
canonic(al) mapping 标准映射
canonic(al) matrix 典型矩阵;标准矩阵;正则矩阵
canonic(al) measure 标准测量;标准测度
canonic(al) momentum 正则动量
canonic(al) non-thermal source 典型非热源
canonic(al) parameter 典型参数
canonic(al) parse 规范语法分析
canonic(al) path 正则路线
canonic(al) plotting 标准曲线图
canonic(al) product 标准积
canonic(al) profile 标准剖面
canonic(al) regression analysis 典型回归分析
canonic(al) representation 标准形式
canonic(al) representation of separable game 可分对策的典式
canonic(al) scale 标准尺(度)
canonic(al) sentential form 规范句型
canonic(al) species 典型种
canonic(al) statistics 标准统计(数字)
canonic(al) structure 典范结构
canonic(al) sum 标准和
canonic(al) system 规范系统;典型系统;正则系统
canonic(al) tanalysis 正则分析
canonic(al) time unit 正则时间单位
canonic(al) transformation 典型变换;典范形式;标准变换;正则变换
canonic(al) transition matrix 规范转移矩阵;标准转移矩阵
canonic(al) trend surface analysis 典型趋势面分析
canonic(al) trend variation 规范的趋势变化
canonic(al) variable 典型变量;标准变量;正则变量
canonic(al) variable value of marine crude oil 海相原油的典型变量值
canonic(al) variable value of terrigenous crude oil 陆相原油的典型变量值
canonic(al) variate 标准变量;典范变量
canonization vectogram 规范化向量图

canon of economy 经济法规
canon of marine transgression 海进规程
canon of transgressions and regressions 海水进退规程
canons of art 艺术标准
canons of descent 不动产继承顺序
canons of taxation 课税准则;课税原则;征税准则;赋税准则
can opener 罐头刀
canopied 遮有天篷的
canopy 华盖;雨罩;雨篷;雨棚;座舱盖;遮篷;盖篷;顶盖;挑棚;天幕;天盖;舰板天幕
canopy actuating cylinder 座舱盖开启作动筒
canopy brace 遮棚撑梁
canopy ceiling 顶篷装饰;防雨天盖
canopy class 林冠级;林冠层
canopy cover 覆盖;地面覆盖;林冠覆盖面
canopy density 林冠郁闭度;郁闭度;植冠密度;森林郁闭度
canopy door 上悬挑门;上悬活动挑门
canopy frame 天篷肋骨
canopy guard 护页
canopy hood 雨篷罩;保温炉排风罩;伞形罩;伞形吸气罩
canopy interception 树冠截留
canopy jettison initiator 舱盖起爆器
canopy layer 林冠覆盖
canopy lip 雨棚边缘;坡屋边缘
canopy manual operating handle 座舱盖手控柄
canopy-protected 遮盖防护
canopy sidewalk 骑楼式人行道;挑棚式人行道
canopy switch 顶盖开关;电车顶盖开关;天棚开关
canopy top 天篷;天盖式车顶
canopy tree 树冠茂密的树木
canopy tubes 管棚
canopy window 上旋窗;挑窗
can overfilling 灌装过满
can pack 灌装
can packer 条筒撅压器
can packing machine 罐装机
can rinser 洗桶机;洗罐机;罐头洗涤机
canroy machine 卷刷机
can sealer 罐头密封器;封罐机
can seamer 罐(头)封口机
can slightly leaking 罐微漏
can sling 吊桶索套;吊桶绳环
can stability 有效期限;储[贮]存稳定性;罐存稳定性;罐藏稳定性
cant 曲线超高;面角造型;毛方(木);斜面;撑斜面;超高;屋隅;外径超高(的坡度);四角木材
cant angle 交角
cantango 顺价
cant bay 三边形突肚窗;斜砌涵洞;多边形凸窗;缺口
cant-bay window 多边形凸窗;三边形突肚窗
cant beam 斜梁;转动梁
cant board 带冷板;带棱板;侧立板;天沟侧板;斜面板
cant body 斜肋骨部
cant bonder 斜面砌缝砖石;斜切砌缝
cant bondstone 斜面砌缝石
cant brick 斜面砖
cant chisel 栓钉錾
cant column 多角柱;菱形柱
cant course 倾斜走向
cant deficiency 欠超高
cant dog 木杆钩;钩杆;带钩撬棒(俚语);短把平头搬钩
canted 切去棱角的
canted antiferromagnetism 斜交反铁磁性
canted bearing plate 斜垫板;削角边板
canted column 多角柱;切角柱
canted corner 切角
canted folding 斜面折叠;倾斜折叠
canted hook 活动铁钩
canted mo(u)lding 倾斜线脚
canted nozzle 扭曲喷管
canted plate 斜坡垫板
canted radiator tube 倾斜的散热管;斜置冷却管
canted rail 轨头外倾轨;超高钢轨;设置轨底坡的钢轨
canted shot 倾斜镜头拍摄
canted tie plate 斜面垫板;有轨斜坡的垫板
canted wall 角交墙;互支角墙;互交角墙;斜交墙;转弯墙;曲角墙
canted window 多角窗

canteen 流动食堂;工人食堂;餐具箱;休息室;小卖部;公共食堂;工地食堂;炊事箱;饭厅
canteen building 饮食店建筑
canteen counter 饮食店柜台
canteen ship 伙食船
canteen store 饮食店
cantelcas 升降式悬臂栈桥
cant excess 过超高
cant fall 斜绞辘
cant file 斜面锉;扁三角锉
cant firmer chisel 杆凿
cant floor 斜肋板
cant frame 斜肋骨(框架);斜筋骨;斜架
cantharene 二氢邻二甲苯
cantharides lustre 银黄色光泽彩
cantharus (早期教堂建筑的)门廊瓶饰;教堂前院中的喷水池
cant header 斜面丁砖;斜列丁砖
cant hook 木杆钩;钩杆;吊桶钩绳;带钩撬杠;(带钩的)搬运木杠;平头搬钩
cant hook sling 吊桶钩绳
cant(ick) quoin 桶堆楔块
cantihook 转杆器;传杆器
cantilever 交叉支架;纸条盘;腕臂;伸臂
cantilever action 悬臂作用
cantilever anchorage 悬臂锚碇
cantilever arch 悬臂拱
cantilever arch bridge 悬臂拱桥
cantilever arched bridge 悬臂式拱桥;悬臂拱形桥;伸悬臂拱式大梁;悬臂拱形大梁;悬臂拱桥;悬臂弓形架
cantilever arched girder 悬臂拱式大梁
cantilever arch truss 悬臂(拱形)桁架;拱形桁架
cantilever arch truss bridge 悬臂拱形桁架桥
cantilever arm 悬臂(部分);悬臂距;挑出臂;伸出臂;桥悬臂部分
cantilever bar 悬臂(顶)梁
cantilever bar screen 悬臂条屏架;悬臂铁栅筛
cantilever beam 肱梁;悬(臂)梁;挑梁
cantilever beam bridge 悬臂梁桥
cantilever block 悬臂块;横支柱;小牛腿;支柱石
cantilever boom 起重悬臂;水平悬臂梁
cantilever braced truss 悬臂支承桁架
cantilever bracket 悬臂支托;悬臂托座;悬臂托架;悬臂牛腿
cantilever bridge 悬臂桥;单端固定桥;伸臂梁桥
cantilever bridging 悬臂式加密
cantilever buttress dam 悬臂式支墩坝
cantilever caging device 悬臂限位装置
cantilever casting 悬臂浇筑
cantilever caves rafter 飞椽(古建筑)
cantilever column 悬臂柱
cantilever construction 悬臂施工(法)
cantilever construction carriage 悬臂施工架;悬臂天平
cantilever course 出挑砖层
cantilever crane 悬臂(式)起重机;悬臂吊车
cantilever cupola 悬臂式圆屋顶
cantilever-deck bridge 上弦承重悬臂桥
cantilever-deck dam 悬臂(式)平板坝
cantilever derrick 悬臂式井架;折叠式井架
cantilever descending anti-overspeed device 悬臂下降超速限制器
cantilever diaphragm 悬臂式膜窗
cantilever dome 悬臂式穹顶
cantilever driving plant 悬臂打桩设备;悬臂式打桩机
cantilever eaves rafter 飞檐;翘飞椽
cantilevered balcony 外挑阳台;挑阳台
cantilevered beam impact tester 悬臂梁冲击试验机
cantilevered corner beam 角梁
cantilevered element 悬臂式构件
cantilevered end 悬臂端
cantilevered foot path 悬臂式人行道
cantilevered footway 悬臂式人行道
cantilevered frame 悬臂式构架
cantilevered gantry 悬臂式脚手架走道;悬臂起重机台架;单脚高架起重机
cantilevered landing 悬臂楼梯平台
cantilevered platform 外挑平台
cantilevered quay wall 悬臂式岸壁
cantilevered return wall 桥台耳墙

cantilevered sheet wall structure 悬壁板墙结构
cantilevered stairway 悬挑楼梯
cantilevered steps 半悬臂级;悬臂式楼梯;半悬踏步
cantilevered stor(e)y 悬挑楼层
cantilevered wall 悬臂墙
cantilevered wing wall 耳墙
cantilever effect 悬臂作用
cantilever erection 悬臂架设(法)
cantilever erection construction 悬臂架设施工法
cantilever erection method 悬臂架设法;悬臂拼装法
cantilever extension 悬臂延伸段;悬臂引伸
cantilever flange 悬臂桥面板
cantilever flap gate 悬臂式卧倒门
cantilever flying strip 支航线
cantilever folded plate roof 悬臂折板屋面
cantilever foot bridge 悬臂式人行桥
cantilever footing 悬臂(式)基础(挡土墙);悬臂基脚;伸臂底座
cantilever foot path 桥的悬臂人行道
cantilever footway 悬臂式人行道
cantilever for footway 悬臂式人行道
cantilever form 滑模;悬臂式模板
cantilever formwork 悬臂模板
cantilever for pin insulator 针瓷肩架
cantilever for post insulator 瓷横担肩架
cantilever foundation 悬臂(式)基础;挑梁式基础
cantilever frame 悬臂肋骨;悬臂(框)架
cantilever framing 悬臂式构架
cantilever gantry 悬臂起重(机台)架;单脚高架起重机;带悬臂的装卸桥
cantilever gantry crane 悬臂式门吊;悬臂式龙门起重机
cantilever girder 伸臂大梁;悬臂架梁;悬臂大梁
cantilever girder bridge 悬臂梁桥
cantilever grizzly 悬臂条屏;悬臂条筛;悬臂格栅;外伸式格筛
cantilever hipped-plate roof 悬臂斜脊屋顶
cantilever hood 悬臂回转式排风罩
cantilever hydrofoil 悬臂水翼
cantilevering 悬臂;悬挑
cantilevering arched girder 悬臂拱梁
cantilevering assemble 悬臂拼装法
cantilevering beam 悬臂梁
cantilevering brick 出挑砖
cantilevering component 悬臂构件
cantilevering conoid 挑出的圆锥体
cantilevering crane(runway) girder 悬臂行车梁
cantilevering curved girder 悬臂弓形梁
cantilevering disk 挑出的圆盘
cantilevering dome 挑出的圆顶
cantilevering end 悬臂末端
cantilevering floor 悬挑楼板
cantilevering floor slab 悬挑楼面板
cantilevering folded plate roof 悬挑折板屋顶
cantilevering folded slab roof 悬挑折板屋顶
cantilevering girder 悬臂大梁
cantilevering hipped-plate roof 悬挑复折屋顶
cantilevering landing 悬挑楼梯平台
cantilevering length 悬臂长度
cantilevering load 悬臂荷载
cantilevering masonry 悬臂圬工
cantilevering member 悬挑构件
cantilevering method 悬臂法
cantilevering plate 悬挑板
cantilevering roof 悬挑屋顶
cantilevering roof slab 悬挑屋顶板
cantilevering segment 悬挑段
cantilevering sheet 悬挑薄板
cantilevering shell 悬挑壳体
cantilevering slab 悬挑板
cantilevering stair(case) 悬挑楼梯
cantilevering steps 悬挑台阶;悬挑踏步
cantilevering system 悬挑体系
cantilevering terrace 悬挑平台;悬挑露台
cantilevering tilted-slab roof 悬挑倾斜板屋面
cantilevering unit 悬臂单元
cantilever jib 悬臂吊杆
cantilever lattice girder bridge 悬臂式花格大梁桥;悬臂式桁架桥
cantilever leg 张臂式支柱
cantilever length 悬臂长度
cantilever load 悬臂荷载
cantilever loading 悬臂负荷

cantilever lock wall 悬臂式闸墙
cantilever masonry(work) 悬臂砌筑
cantilever mast 腕臂柱
cantilever method 悬臂方法
cantilever method of design 悬臂设计法
cantilever method of(wind stress) analysis 悬臂(风应力)分析法
cantilever moment 悬臂力矩
cantilever out 悬臂伸出
cantilever over-hanging footway 悬臂式人行道
cantilever pedestrian bridge 悬臂人行桥
cantilever pile 悬臂桩
cantilever pile driving plant 悬臂打桩设备
cantilever platform 悬臂平台;挑台
cantilever platform roof 悬臂的车站雨棚
cantilever portable drill rig 悬臂式轻便钻架
cantilever portion 悬臂部分
cantilever prismatic shell roof 悬臂折板薄壳屋顶
cantilever reach 腕臂活动
cantilever reinforced concrete slab 悬挑钢筋混凝土板
cantilever retaining wall 悬臂(梁)式挡土墙;无锚碇挡土墙
cantilever roof 翅棚;悬臂(式)屋顶
cantilever screen of bars 悬臂条子铁栅筛
cantilever section 悬臂梁截面
cantilever segment 悬臂段;(桥梁结构上的)悬臂扇形体;悬臂部分
cantilever segmental concreting 分段悬臂浇注(混凝土)
cantilever segmental erection 分段悬臂拼装
cantilever shear wall 悬臂剪力墙
cantilever sheet pile 悬臂(式)板桩;无锚碇板桩;无拉杆板桩
cantilever sheet pile(ed) quaywall 悬臂板桩岸壁;无锚碇板桩岸壁
cantilever sheet piling 悬臂(式)板桩;无拉杆板桩
cantilever sheet piling wall 悬臂式板桩墙
cantilever shuttering 悬臂模板
cantilever single-wall 悬臂单排板墙
cantilever single-wall sheet pile structure 无锚单排板桩墙结构
cantilever single-wall structure 悬臂单排板墙结构
cantilever slab 挑出的平板;悬臂板;挑板
cantilever spacing 悬臂距离
cantilever span 悬臂跨度
cantilever spring 悬臂弹簧
cantilever stair(case) 挑出的楼梯;悬臂式楼梯
cantilever steel truss 悬臂式钢桁架
cantilever steps 悬挂的楼梯(防火用);悬臂踏步;悬臂梯级;悬臂踏步;半悬踏步
cantilever stone on eave 挑檐石
cantilever strip triangulation 悬臂航线三角测量
cantilever structure 悬臂结构
cantilever strutted roof truss 悬臂加撑杆的屋顶桁架
cantilever support 腕臂支撑
cantilever swivel bracket 腕臂底座
cantilever tank 舷侧水柜
cantilever-through bridge 下弦承重悬臂桥
cantilever timber beam bridge 伸臂梁桥
cantilever timbering 前探支架
cantilever truss 悬臂式桁架;悬臂桁架
cantilever truss bridge 悬臂桁架桥
cantilever truss with tension rod 加撑杆的悬臂桁架
cantilever-type hyperbolic paraboloidal roof 悬臂双曲线抛物面屋顶
cantilever-type retaining wall 悬臂式挡土墙;无锚碇挡土墙
cantilever type traffic sign 悬臂式交通标志
cantilever vault 悬挑拱顶;横支柱;小牛腿;支柱石
cantilever vibration 悬梁式振动
cantilever walking crane 行走式悬臂吊车;壁装移动式悬臂起重机
cantilever wall 悬臂式挡土墙;钢筋混凝土悬臂式挡土墙;钢板桩墙(堤坝、码头、护岸等)
cantina 小酒吧
canting 撞翻;弯道外缘的超高;弄斜;压成倾斜
canting piece 助钩角
canting saw table 斜置锯子桌;倾斜锯子桌
canting strip 泻水倾斜面板;披水条(外墙角);承雨线脚;墙壁突出底座;披水面;挑水倾斜面板;排水槽铁面板

canting table 锯木台;倾斜锯台
cantline 桶间凹隙;绳索绞纹
cantling 分隔砖层;分隔部分
cant mo(u)lding 斜(面)状(装)饰;斜面线条;斜面线脚
cant of a wall 墙上洞口的斜面
cant of curve 曲线超高
cant of rail 钢轨内倾角
cant of the jet 喷管倾斜
cant of the super-elevation 超高度
cant of track super-elevation 超高顺坡【铁】
cant of vehicle 车辆内倾
canton 驻军营房;外角;分区;凸隅角石
cantonal banks 瑞士的州银行
Canton blue 广东蓝(紫蓝色陶瓷色料)
Canton china 广东瓷
canton crane 轻便落地吊车
cantoned building 隅角建筑物
cantoned column 隅角柱
cantoned pier 隅角墩柱;隅角墩;哥特式复式柱
Canton enamel 广州搪瓷
Canton glue 广胶
cantonment 居民点;兵站;临时营房
can top 油壶盖
can top crimp test 罐头卷边试验
cantoria 教堂唱诗班座台
cant purchase 斜绞辘
cant quoin 堆装楔块
cantraflexure 反弯曲
cantraflexure stretch 反弯曲段
cant scraper 三棱刮刀;三角刮刀
cant stretcher 斜面顺砖;斜刃砖;斜面砌砖
cant strip 檐板;三角条;倾斜嵌条;披水条;封檐板;镶边压缝条;镶边板条;垫瓦条
cant timber 斜肋板
cant wall 斜交墙
cant window 三边形突肚窗;多边形凸窗;多角窗
can-type 罐装式;罐形的
can-type(combustion) chamber 罐式燃烧室;罐形燃烧室
can-type combustor 罐式燃烧室;筒形燃烧室
can-type generators 筒式烟雾发生器
can-type precipitation ga(u)ge 罐式雨量计
can vacuum tester 罐头真空检测仪
canvas 粗帆布
canvas air conduit 帆布送风管;帆布输气管
canvas and rope work 帆缆作业
canvas apron 帆布输送带
canvas-apron unloader 输送带式卸载机
canvas awning 帆布凉篷
canvas bag 帆布袋
canvas bed 帆布床
canvas belt(ing) 帆布带传动装置;帆布(皮)带
canvas binder 帆布带式割捆机
canvas board 帆布纹纸板
canvas boat 帆布艇
canvas boot 帆布罩
canvas bucket 帆布桶
canvas buckle 帆布输送器扣
canvas canoe 帆布小舟;帆布划艇
canvas canopy 帆布罩棚
canvas climber 上桅水手
canvas cloth 帆布
canvas connecting strip 帆布输送带两端连接杆
canvas connection 帆布连接
canvas container 集装袋;帆布套
canvas conveyer 帆布运输机;帆布输送机;帆布带运输机;帆布带输送机;帆布传送带
canvas cot 帆布床
canvas cover 帆布篷;帆布罩;帆布套
canvas dam 帆布闸;帆布坝
canvas decking 防水帆布屋面
canvas deflector 帆布挡帘
canvas dodger 防波幕
canvas draper 帆布输送器;帆布带输送机
canvas duct 帆布管道;帆布软管
canvased light van 轻型帆布篷车
canvas extension 帆布延长板
canvas fabric airslide 帆布制的输送空气管道
canvas fender 帆布碰垫
canvas filter 帆布过滤器
canvas gate 帆布门帘;帆布挡板
canvas gears 帆布船具
canvas grommet 帆布保护垫圈

canvas guide 帆布输送带两边导板
canvas hood 帆布罩
canvas hose 帆布水带;帆布水龙带;帆布软管
canvas joint 帆布接头
canvas lining 帆布衬
canvas loom 帆布织机
canvas needle 缝帆针
canvas painting 油画
canvas pallet 大浮子
canvas paste paint 帆布厚漆
canvas reticulation 绘图格网
canvas roller 帆布输送带辊轴
canvas roof 帆布篷
canvas screen 帆布遮蔽物;帆布筛
canvass customer opinions 用户调访
canvas seaming tools 缝帆工具
canvasser 推销员
canvass for cargo 揽货
canvass for insurance 兜揽保险
canvass for orders 兜揽订货
canvas shroud 帆布罩
canvassing note 揽货通知书
canvassing report 揽货报告单
canvas slat 帆布输送器条板
canvas slat clip 帆布带板条夹箍
canvas slat conveyer 帆布条板式输送器
canvas slide 帆布输送带导向板
canvas sling 帆布吊兜(吊货用)
canvas sling conveyer 帆布传送带;袋式输送器;袋式输送机
canvas sling elevator 帆布带升降机
canvas suitcase and bag 帆布箱包
canvas tack 帆布输送带钉卡箍
canvas tarpaulin 油帆布
canvas thread 帆线
canvas tube 帆布空气管道
canvas-type pickup 帆布条板式捡拾器
canvas-type platform 帆布输送带式收割台
canvas ventilation pipe 帆布风筒
canvas ventilator 帆布通风筒;帆布风斗
canvas wall 粗帆布壁;粗帆布罩
canvas wall-surface 帆布墙面
canvas work 缝帆工艺
can vendor 罐用冷却售货柜
canvert the capital into a product 资本转化为产品
Canvet 考维特式建筑(一种法国建筑体系)
can weigher 罐头称重机
can weight machine 罐头称重机
can with thumb button 带阀油壶
canyon 狭谷式热室;峡谷
canyon bench 峡谷阶地
canyon current 峡谷流
canyon delta 谷口三角洲
canyoned river 深谷河流
canyoned river reach 峡谷河段
canyoned stream 峡谷河流;深谷河流
canyon effect(of high buildings) (高层建筑的)峡谷效应
canyon-fan system 海底谷扇系
canyon fill 峡谷淤积
canyon-fill trap 峡谷填充圈闭
canyon shape factor 河谷形状系数
canyon side 峡谷陡壁
canyon street 峡谷街道
canyon-type pump 屏蔽泵
canyon wall 峡谷陡壁;峡谷壁
canyon wind 下降风;下吹风;重力风;山风
Canzler 坎兹雷尔铜合金
caoutchouc 天然橡胶;弹性橡皮;生橡胶
cap 金属罩;集件帽;门窗上梁;帽梁;烟囱内罩帽;柱头垫木;盖罩;电灯泡的灯头;灯口;大礶;采板;棚架顶梁;桅尖
capability 可能性;能力;本领;权能;权力
capability computing system 权力计算系统
capability correction 功率校正
capability curve 可能输出曲线
capability diagram 出力图
capability manager 权力管理程序
capability margin 能力裕度
capability of ion(ic) exchange 离子交换能力
capability of land 土地生产能力;地力
capability of load transfer 传荷能力
capability of settlement 沉陷的可能性
capability of swelling 可膨胀性

capability of welding vertically upwards 直上焊接能力
capability vector machine 权力向量机
capable assistant 得力助手
capable fault 可能活动断层
capable of bearing wheel loads 能够承受的车辆荷载
capable person 有行为能力的人
capacidin 能杀菌素
capacious 广阔的
capacitance 机械容量;电容式;电容量;电容
capacitance altimeter 电容式高度计;电容式测高计
capacitance box 电容箱
capacitance bridge 电容电桥
capacitance bushing 电容式进线套管
capacitance coefficient 电容系数
capacitance comparator 电容比较仪
capacitance compensation 电容补偿
capacitance connecting three point type oscillator 电容三点式振荡器
capacitance constancy 电容恒定度
capacitance coupled flip-flop 电容耦合触发器
capacitance coupling 电容耦合
capacitance current 电容(性)电流
capacitance error selector 电容误差分选仪
capacitance-frequency method 电容频率法
capacitance grading 电容分级
capacitance integrator 电容积分器
capacitance level indicator 电容式水平指示器
capacitance level transducer 电容式液位传感器
capacitance measuring instrument 电容测量仪
capacitance measuring tester 电容测试仪
capacitance meter 电容计;电容测试器;电容表;法拉计
capacitance method 电容法
capacitance micrometer 电容测微器;电容测微计
capacitance of condenser 电容器
capacitance oscillator 电容反馈振荡器
capacitance per unit length 单位长度电容
capacitance potential device 电容分压器;电容电位计
capacitance potential transformer 电容式电压互感器
capacitance potentiometer 电容电位器;电容电位计
capacitance pressure transmitter 电容式压力变送器
capacitance probe 电容探针;电容探头;电容探测器
capacitance probe type wave recorder 电容(探针)式波高计
capacitance ratio 电容比;容量比
capacitance reaction 电容(性)反应
capacitance recorder 电容记录器
capacitance relay 电容式继电器;电容操纵继电器
capacitance-resistance 电容电阻的
capacitance resistance coupling 电容电阻耦合
capacitance-resistance filter 阻容滤波器
capacitance section 电容节
capacitance strain ga(u)ge 电容应变计
capacitance strain transducer 电容应变传感器
capacitance test 电容试验
capacitance tester 电容试验器
capacitance transducer 电容换能器;电容传感器
capacitance tube 电容管;电容电抗管
capacitance tuning 电容调谐
capacitance-type electroscope 电容式验电器
capacitance-type level-meter 电容式料位计
capacitance-type regularity tester 电容式均匀度试验机;电容式规律性试验机
capacitance-type resistor 电容电阻器
capacitance-type sensor 电容式传感器
capacitance-type strain ga(u)ge 电容式应变仪;电容式应变计
capacitance-variable transistor 变容晶体管
capacitance-voltage method 电容电压法
capacitance water-level detector 电容式水位检测器
capacitance wire ga(u)ge 电容线规
capacitated transportation 限量运输
capacitatory factor 获能因子
capacitive approach switch 电容式接近开关
capacitive character 电容性
capacitive charging capacity 电容充电容量
capacitive circuit 电容电路;容性电路
capacitive commutator 电容换向器
capacitive component 电容成分

capacitive cooling 热容式冷却
capacitive coupling 电容耦合；容性耦合
capacitive coupling amplifier 电容耦合放大器
capacitive coupling meter 电容耦合测试器
capacitive current 电容性电流
capacitive discharge ignition 电容性放电点火
capacitive discharge pilot light 电容式放电领示灯
capacitive divider 电容(式)分压器
capacitive edge 容量边
capacitive electromotive force 电容电动势
capacitive feedback 电容性反馈
capacitive feedback circuit 电容回授电路
capacitive filter 电容滤波器
capacitive impedance 电容性阻抗
capacitive iris 电容性膜片
capacitive joint 电容性连接
capacitive lag 电容惰性
capacitive leakage current 电容泄漏电流
capacitive level indicator 电容式料位指示器
capacitive level switch 电容式料位开关
capacitive load(ing) 电容性负载；容荷
capacitively loaded antenna 加容天线
capacitive micrometer 电容(式)测微计；电容(式)测微器
capacitive moisture sensor 电容式湿敏元件
capacitive motor 电容电动机
capacitive pick-up 容性传感器
capacitive position transducer 电容式位置传感器
capacitive pressure transducer 电容压力传感器
capacitive probe 电容探测器
capacitive reactance 电容(性)电抗；容抗
capacitive reading 电容读数
capacitive resistance tube 电容性电抗管
capacitive salinometer 电容式盐度仪
capacitive saw-tooth generator 电容式锯齿波发生器
capacitive storage circuit 电容存储[贮]电路
capacitive transducer 电容式传感器
capacitive tuning 电容调谐
capacitive voltage divider 电容分压器
capacitive voltage transformer 电容式电压互感器
capacitive window 电容性窗口
capacitivity 电容率；电容量
capacitometer 电容测量仪
capacitoplethysmograph 电容脉波计；调频式电容脉波计
capacitor 凝气器；小腔室；电容器
capacitor analyzer 电容器分析仪；电容器试验器
capacitor antenna 电容器天线
capacitor bank 电容器组(合)
capacitor box 电容器箱
capacitor bushing 电容式套管
capacitor ceramics 电容器陶瓷
capacitor charge 电容器充电
capacitor colo(u)r code 电容器色标
capacitor-controlled oscillator 电容控制振荡器
capacitor-coupled logic 电容耦合逻辑
capacitor coupling 电容耦合
capacitor diode storage 电容器二极管存储器
capacitor discharge spot welder 电容放电式点焊机
capacitor-discharge welding 电容放电焊接
capacitor filter 电容滤波器
capacitor filtering 电容滤波
capacitor firing furnace 电容器焙烧炉
capacitor fluviograph 电容式水位计
capacitor glass 电容器玻璃
capacitor hydrophone 电容水听器；电容式听水器
capacitor induction motor 电容电动机
capacitor-input filter 电容输入滤波器
capacitor in series and parallel 电容器的串联和并联
capacitor loudspeaker 电容器式扬声器；静电扬声器
capacitor material level indicator 电容式料位指示器
capacitor memory 电容器存储器
capacitor microphone 电容式传感器；电容传声器
capacitor modulator 电容调制器
capacitor moisture measurer 电容式水分测定仪
capacitor motor 电容电动机；电容式单相电动机；电容器起动电动机；电容起动电动机
capacitor paper 电容器纸
capacitor phase shifter 电容器移相器
capacitor pickup 电容拾音器
capacitor plate 电容器极板

capacitor-resistor unit 封装阻容
capacitor shaft 电容器轴
capacitor split-phase motor 电容分相式电动机
capacitor start 电容器起动
capacitor start and run motor 电容起动行驶式电动机；电容式启动和运转的电动机
capacitor starter 电容式起动机；电容起动器
capacitor start induction motor 电容起动感应电动机
capacitor start motor 电容起动电动机；电容式启动电动机；电容器起动电动机
capacitor start-run motor 固定分相电容器式电动机
capacitor storage 电容器存储器；电容存储器
capacitor store 电容器存储器
capacitor tachometer 电容式转速计
capacitor tester 电容器试验器；电容器测试器
capacitor tissue 电容器纸
capacitor trigger 电容触发器
capacitor voltage divider 电容分压器
capacitor voltage transformer 电容式电压互感器
capacitron 电容汞弧管
capacity 通行能力；计算效率；能量；能力；扬量；载量；负载量；负载额量；本领；容量；容积；生产额
capacity altimeter 电容式测高计
capacity antenna 电容性天线
capacity as a subject of law 法律主体资格
capacity attenuator 电容衰减器
capacity balance 电容平衡
capacity battery 高能干电池
capacity booster 电容性升压电机
capacity bottleneck 生产能力薄弱环节
capacity bridge 电容电桥；电容的电桥
capacity-bridge method 电容桥法
capacity building 扩大能力
capacity cell 电容元件
capacity certification 容量证明
capacity charge 电容充电
capacity checker 电容器检验器；电容检验器
capacity coefficient 容量系数
capacity commutator 电容转换器
capacity constant 容量常数；设备利用常数
capacity control 功率调节
capacity conversion 容量交换
capacity conversion for safety valve 安全阀的容量换算
capacity correction 容积校正
capacity cost 最高生产能量成本；生产能力成本
capacity coupled 电容耦合的
capacity coupler 电容耦合器
capacity coupling measuring set 电容耦合测量仪
capacity coupling meter 电容耦合测试仪
capacity current 电容性电流
capacity curve 库容曲线；容量曲线
capacity decision 能量决策
capacity-demand analysis 通行能力—交通需求分析(法)
capacity dimension 容量维数
capacity divider 电容分压器
capacity earth 地网
capacity elevation 毛细上升
capacity-elevation curve 高程—库容(关系)曲线；水位—容积曲线
capacity-elevation relation 库面高程—库容关系
capacity enhancing on existing line 既有线能力加强
capacity estimate 容量估计
capacity estimate of passage 通过能力估计
capacity exceeding number 超位数
capacity expansion 容量膨胀
capacity expansion investment 扩充生产能力的投资；能量扩充投资
capacity expenses 生产能力费用
capacity factor (水库的)库容系数；能量因素；能力(利用)系数；装载容量利用系数；装机容量利用系数；发电利用率；容量因子；容量因数；容量系数；设备容量利用率
capacity factor of a blast furnace 高炉利用系数
capacity factor of open-hearth furnace 平炉利用系数
capacity fall-off 电容量减退
capacity for heat transmission 传热能力
capacity formula 渠道流量公式
capacity for on and off 乘降量
capacity for picking up coating 挂油皮能力

capacity for private rights 民事权利能力
capacity for reaction 反应能力
capacity for responsibility of juristic person 法人的责任能力
capacity for rights 权利能力
capacity for work 作功能力
capacity goniometer 电容测向器
capacity gravimeter 容积重力仪
capacity head curve 容量压差曲线
capacity in bales 包装货舱容量
capacity-input filter 电容输入滤波器
capacity insulation 隔热；热容绝缘；热容；绝热能力；隔声能力；储热能力
capacity-intake ratio 容量一进水比
capacity in tons per hour 每小时生产吨数
capacity investment 生产能力投资
capacity level 蓄水池中可能的最高水位；最大蓄水高度；生产能力水平
capacity level indicator 电容式料位指示器
capacity limitation 容量限制；容量极限
capacity load 满载；满负荷；充分荷载；额定载荷
capacity loss 容量损失
capacity material level indicator 电容式料位指示器
capacity measure 容量测定
capacity measuring set 电容测试器
capacity meter 电容计；电容测试器
capacity method of quadratic programming 二次规划容量法
capacity moment 极限弯矩
capacity multiplier 电容倍增器
capacity needs 扩大生产能力的需要
capacity of a gallery 坑道的蓄水能力
capacity of a hydroelectric(al) plant 水力发电厂容量(正常水头满流情况下，电厂最大发电量)
capacity of artesian well 自流井出水量
capacity of berth 泊位通过能力
capacity of body 车体负载
capacity of boiler 锅炉蒸发量；锅炉容量；锅炉能量；锅炉出力
capacity of bridge 桥梁(交通)容量
capacity of capillarity 毛管容量
capacity of cargo hold 货舱容积
capacity of carriage 车载量
capacity of coal bunker 煤仓容量
capacity of condenser 电容器的电容量
capacity of corporate 法人资格
capacity of coverage 遮盖力
capacity of crank press 曲柄压力机能力
capacity of culvert 涵洞容量
capacity of decomposition 分解能力
capacity of driven pile 打入桩的承载量；沉桩的承载能力
capacity of drum 滚筒容量
capacity of drydock 干船坞的能力
capacity of electric(al) production 年发电量
capacity of equipment 设备能力
capacity of execution 施工能力
capacity of facility 设备能力
capacity of frost heaving 冻胀量
capacity of furnace 炉能量
capacity of geologic(al) environment 地质环境容量
capacity of groundwater reservoir regulation 地下水库调蓄能力
capacity of heat 热容(量)
capacity of heat exchanger 热交换器容量
capacity of heat transmission 导热能力；传热量；热气传送量；热传导能力
capacity of highway 公路通行能力
capacity of holding deposit 容纳淤积能力
capacity of hump yard 驼峰编组站作业能力；驼峰编组作业能力
capacity of hydroelectric(al) plant 水力发电厂容量
capacity of initial set(ting) 初凝能力
capacity of installing power equipment 安装动力设备能力
capacity of interchange 互通式立体交叉通行能力
capacity of intersection 交叉口通行能力；交叉口通过能力
capacity of ion(ic) exchange 离子交换能力
capacity of lighterage 过泊通过能力
capacity of lock 船闸容量
capacity of mud pump 泥浆泵流量

capacity of navigation lock 船闸通过能力
capacity of network 路网(通过)容量;路网通过能力
capacity of noise reduction 降噪量
capacity of party 当事人能力
capacity of passenger transportation 客运能力
capacity of penetration 贯入能力
capacity of port 港口通过能力
capacity of pump 抽水机出水率;泵的生产率;泵的排水量;泵出水率;水泵出水率;水泵额定流量
capacity of pumping equipment 泵设备容量
capacity of reinjection 回灌能力
capacity of replenishment 补给能力
capacity of reservoir 库容;水库容积
capacity of river 河流过水能力;河道流量
capacity of road 道路通行能力;道路容量
capacity of road net 路网容量
capacity of saturation 饱和容量
capacity of scale 秤的容量
capacity of set(ting) 凝结能力
capacity of shaft kiln for unit cross-sectional area 立窑单位断面积产量
capacity of shaft kiln for unit volume 立窑单位容积生产能力
capacity of shiplift 升船机的能力
capacity of short-circuit 短路容量
capacity of spillway 泄洪能力;溢洪能力
capacity of storage 储存能量;储存能力
capacity of storage battery 蓄电池容量
capacity of stream 河流过水能力
capacity of synchronous condenser 同步调相机容量;调相容量
capacity of tank car 油槽车容量
capacity of telephone exchange 电话局容量
capacity of the market 市场容量
capacity of the wind 风沙负荷量
capacity of track 线路通过能力【铁】;通过能力【铁】
capacity of transport tunnel 隧道输送量
capacity of unloading 卸车能力
capacity of vehicle 车辆装载量
capacity of vessel 船负荷量
capacity of wagons 货车载重量;货车载重能力;货车标记载重(量)
capacity of water body self-purification 水体自净能力
capacity of water pump 抽水机出水量
capacity of water shed 汇水流域内的蓄水量
capacity of water supply equipment 给水设备能力
capacity of well 井的出水量;井的出水率
capacity of wharf 码头通过能力
capacity of wind 风的挟带力
capacity operating rate 开工率;设备利用率
capacity operation 满载量操作;满载操作;全容量操作
capacity output 产出能力
capacity packing 包装能力;容量包装
capacity paper tape reader 电容式纸带输入机
capacity payload 容量有效载荷
capacity plan 舱容图;容积图
capacity point 容载限点
capacity power 输出功率
capacity probe 容量探示器
capacity production 生产能力;正常生产量;标定产量
capacity profile 容量纵剖面
capacity range 试验机测程
capacity rating 额定生产能力;额定生产率;额定容量;容量计算;生产率
capacity ratio 容量比;生产能量比率;生产能力比率
capacity ratio of shothole 炮眼利用率
capacity reaction 电容反馈
capacity record 容量记录
capacity regenerative circuit 电容再生电路
capacity regulating device 容量调节装置
capacity regulating valve 容量调节阀
capacity regulation 容量调节
capacity regulator 功率调节器;电容调节器;出力调节器
capacity relay 电容式继电器
capacity required 要求容量
capacity reserve 容量储备
capacity seismograph 电容地震仪
capacity seismometer 电容式地震检波器;电容地震计;容积式地震计

capacity sensitive circuit 电容敏感电路
capacity shunt ammeter 电容分路安培计
capacity standard 电容标准
capacity starting 电容式起动
capacity strengthening on existing line 既有线能力加强
capacity study 能力研究;通行能力调查
capacity study of passage 通过能力研究
capacity tensiometer 电容式张力计
capacity test 能率试验;容量试验;排量试验
capacity to borrow 借款能力
capacity to combat natural adversities 抗御自然灾害的能力
capacity to contract 订约资格;订约能力
capacity ton/hour 负荷吨/时
capacity tonnage 载重量;标记载重
capacity to sound 隔声性
capacity to stand 持久性;稳定性
capacity to submit to arbitration 提请仲裁的资格
capacity unbalance 电容不平衡
capacity usage ratio 能量利用率
capacity utilization 利用能力
capacity utilization factor 容量利用系数
capacity utilization rate 生产能力利用率
capacity utilization variance 能量利用差异
capacity value 荷载量;电容值;功率;容量
capacity variable range 容量变化范围
capacity variance 能量差异;能力差异;生产能力差异
capacity volume variance 容量差异
capacity waste 田间流失
Capaco process 凯帕科石膏型铸造法
capadyne 电致伸缩继电器
Capa Horn Current 合恩角海流
cap and fuse blasting 明火放炮
cap and fuse firing 火炮爆破;传爆线起爆(法)
cap-and-pin insulator 帽销式绝缘子;帽盖装脚式绝缘子;球形连接盘式绝缘子
capanus 船蛆
capaswitch 双电致伸缩继电器
capatcity discharge reading 电容放电法读出
cap badge 帽徽
cap bar finger 指形棒
cap block 大斗;护斗;桩垫;柱头垫木;替打(垫块);桩帽;锤垫
cap bolt 盖螺栓;倒角螺栓
cap cable 短钢索;顶盖锚缆(预应力混凝土联结用的);负弯钢筋;钢丝束(安置在预应力混凝土梁负弯矩区域)
cap chamber 底火室
cap charge 雷管中装料
cap clamp 盖夹
cap cloud 驻云【气】;山帽云【气】
cap collar gasket 螺母垫圈;螺帽垫圈
cap concrete 帽盖混凝土
cap contact 灯头触点
cap copper 帽铜;带状黄铜
cap cover furnace 罩式炉
cap crimper 雷管帽卷边机;雷管拉索
cape 岬(角);海(岸)角;斗篷;低角
cape asbestos 青石石棉;兰石棉;南非石棉
Cape basin 开普海盆
cape blue asbestos 南非石棉;青石棉
cape bolt 圆头螺栓;固定螺栓
cape chisel 岬錾;扁尖凿;削凿(刀);拐角凿子
cape cloud 山顶白云
Cape Cod house 科德角式房屋;低矮的海滨别墅
cape diamond 黄金刚石;南非金刚石;淡黄色金刚石
cape doctor 道格特角风(指南非沿海一种强东南风)
cape foot 南非尺
cape ga(u)ge 扁尖量规
Cape geosyncline 开普地槽面
Cape gum 好望角胶
cape hood 顶盖
capel 脉壁黏土带;铁索眼环接头;石英质岩石
capelitic limestone 沥青质灰岩
Capell fan 卡贝尔扇风机
Cape mountain tectonic belt 开普山纬向构造带
cape of Good Hope 好望角
cape of Good Hope deviation additional 好望角绕航附加费
cape overcoat 斗篷
cape ruby 红榴石;南非红宝石

cape top 篷车顶
Cape Verde 佛得角群岛
Cape Verde basin 佛得角海盆
Cape Verde fracture zone 佛得角破裂带
cape walnut 臭木
cap flange 螺母垫圈;螺帽垫圈
cap flashing 层搭防雨板;金属盖片(防漏用);帽盖泛水;盖泛水
cap form (制造桩帽用的)桩帽模型
cap formation 帽形成
cap for plugging oil cut-out valve 给油总阀盖
cap gasket 盖垫密片
cap grouting 顶盖灌浆法
cap height 大写字母高度
capillarimeter 毛细液度器;毛细管仪;毛细管计;毛细管测液器;毛细管检液器;毛细管测液器
capillariomotor 毛细管运动的
capillarioscopy 毛细管显微镜检查
capillarity 毛细作用;毛细性;毛细现象;毛细作用;毛细管学;毛细管性;毛细管现象;毛管作用
capillarity absorption 毛细管吸收作用
capillarity constant 毛细常数
capillarity correction 毛细校正
capillarity of concrete 混凝土的毛细管作用
capillarity of fiber 纤维毛细作用
capillarity phenomenon 毛管现象
capillarity suction head 毛细管吸升水头
capillaroscope 毛细管显微镜
capillaroscopy 毛细管显微镜检查
capillary absorbed water 毛细(管)吸附水;毛管吸附水
capillary absorption 毛细吸湿(状态);毛细(管)吸收(作用)
capillary action 毛细(管)作用;毛细管现象;毛细管湿作用;毛细管定形生长;毛管作用
capillary action shaping technique 毛细管定型生长技术
capillary active compound 毛细活性化合物
capillary activity 毛细(管)活性
capillary adjustment 毛管调节
capillary adsorbed water 毛细吸附水
capillary adsorption 毛细吸附(作用);毛细吸附力
capillary air washer 毛细管净气器
capillary analysis 毛细管分析(法);毛细分析(法)
capillary analysis method 毛细管分析法
capillary analytical method 毛细分析法
capillary apparatus 毛细管作用仪器
capillary arc ion source 毛细弧光离子源
capillary array 毛细管整列
capillary ascension 毛细(管)上升;毛管上升
capillary ascent 毛细(管)上升;毛管水上升高度
capillary attraction 毛细(吸)引(力);毛细管作用;毛细管引力;毛细管现象;毛细管吸引(力);毛细管吸力;毛管吸引力
capillary behavio(u)r 毛细管特性
capillary bond 毛细黏结;毛细水连接;毛细结合
capillary bore 毛细管内径;毛细孔
capillary brazing 紧隙铜焊法
capillary break 毛细管中断;毛细管破坏;毛细管堵塞;防水槽;毛细现象隔层
capillary break-up 毛细管破裂
capillary capacity 毛细容量;毛(细)管吸湿量;毛细(管)容量;毛细管含水量;毛管持水量;毛细(管)容(水)量
capillary cavity 毛细管空腔
capillary cell washer 毛细管喷淋洗涤机
capillary channel 毛细孔道;毛细管道;毛细孔通道
capillary chemistry 毛细(管)化学
capillary chromatographic column 毛细管色谱柱
capillary cohesion 毛细(管)内聚力
capillary collector 毛细收集器
capillary column 开管柱;毛细(管)柱;戈雷氏柱
capillary column chromatography 毛细柱色谱法
capillary condensation 毛细冷凝;毛细(管)凝聚;毛细管凝结;毛管凝缩作用
capillary condensation theory 毛细管凝结理论
capillary condensed water 毛(细)管凝结水
capillary conductivity 毛细管传导率;毛细管传导性;毛细传导率;毛细传导度;毛管传导度
capillary constant 毛细管常数;毛细常数
capillary control 毛细管控制作用
capillary copper 铜毛
capillary correction 毛细管改正;弧形液面修正
capillary correction chart 弧形液面修正图

capillary correction graph 弧形液面修正图
capillary correction table 弧形液面修正表
capillary crack 毛细裂纹;毛细裂缝;发状裂纹;发状裂缝;发细裂纹;发丝裂纹
capillary cross section effect 毛细管截面效应
capillary cut-off 毛细水隔离层;毛细隔离层
capillary depression 毛细水下降;毛细(管)下降;毛管水下降
capillary detector 毛细管检测器
capillary dewatering 毛细脱水
capillary dewatering sludge 毛细脱水污泥
capillary dewatering unit 毛细管脱水装置
capillary diffusion 毛细水扩散
capillary dip 毛细管浸渗
capillary drainage 毛细管引流法;毛细管导液法
capillary drying 毛细(管)干燥
capillary effect 毛细管的灯芯作用;毛细效应;毛细(管)效应
capillary ejecta 火山毛
capillary electrode 毛细电极
capillary electrolysis 交界面电解;毛细管渗透电解;毛细电解
capillary electrometer 毛细(管)静电计
capillary electrophoresis 毛细管电泳
capillary elevation 毛细上升作用;毛细(管)上升
capillary embolism 毛细管栓塞
capillary embolus 毛细管栓子
capillary energy 毛细(管)能
capillary-entrance-pressure drop 毛细管入口压力降
capillary equilibrium 毛细平衡
capillary equilibrium height 毛细平衡高度
capillary equilibrium level 毛细平衡液位
capillary feed 毛细管给油
capillary filter 毛细滤器(湿式集尘装置用)
capillary filtering 毛细管过滤
capillary filtration coefficient 毛细管过滤系数
capillary fissure 毛细龟裂纹
capillary fitting 毛细管接头
capillary flask 毛细瓶
capillary flow 毛细流动;毛细管流(动);毛管水流;毛管流(动)
capillary flow measuring method 毛细管黏度测定法
capillary flow meter 毛细流量计;毛细管流量计
capillary force 毛细作用力;毛细吸力;毛细力;毛细管力;毛管力
capillary fragility 毛细管脆性
capillary fringe 毛细作用带;毛细水边缘;毛细带;毛细管层;毛细管边缘;毛管作用带;毛管水边缘;毛管上限;毛管边缘;水位以上毛细水边缘
capillary-fringe belt 毛细管水边缘带
capillary fringe of groundwater 地下水毛细饱和带
capillary fringe zone 毛细水区
capillary front 毛细水顶;毛管湿锋
capillary gas chromatography 毛细管气相色谱(法);毛细管气相层析
capillary geometry 毛细管几何形
capillary glass electrode 毛细管玻璃电极
capillary glass tube 玻璃毛细管
capillary groove 防水槽;防阻毛细管水沟槽
capillary ground water 毛细地下水
capillary head 毛细水头;毛细管水头;毛管水头
capillary heat pipe 毛细热管
capillary height 毛细管水上升高度;毛细水高度;毛细上升高度;毛细管高度
capillary-height method 毛细管高度法;毛细高度法
capillary hydrodynamics 毛细管流体动力学
capillary hydrostatic pressure 毛细水压力
capillary hysteresis 毛细迟滞性;毛细滞后现象
capillary imhibition 毛细管渗透作用
capillary inlet 毛细管入口
capillary intersection 毛细管断面
capillary interstice 毛细间隙;毛细管孔隙;毛细间隙;毛管孔隙;毛管间隙
capillary ion source 毛细管离子源
capillary jet 毛细管射流
capillary joint 毛细管接合(在承插式接合中利用毛细管作用吸入熔化焊料而连接的方法)
capillary layer 毛管水层;毛细管水层
capillary leak 毛细管漏孔;毛细泄漏;毛细管泄漏;毛细管漏孔;毛管渗漏
capillary lift 毛细上升;毛细管上升;毛管水上升高度;毛管上升

capillary lift test 毛细管上升高度试验
capillary limit 毛细限
capillary lubricant 毛细管润滑剂
capillary lubrication 毛细管润滑
capillary manometer 毛细管压力计
capillary manometric method 毛细管压力法
capillary melting point 毛细管法熔点
capillary membrane 毛细管膜
capillary meniscus 毛细管弯液面
capillary mercury lamp 毛细管水银灯
capillary method 毛细管法
capillary microcell 毛细管微池
capillary microscope 毛细显微镜
capillary microscopy 毛细显微术
capillary migration 毛细移动;毛细水移动;毛细迁移;毛细管移动;毛细管水移动;毛管移动;毛管水迁移
capillary moisture 毛细水;毛细管水分;毛管水分
capillary moisture capacity 毛细管容量;毛细管含水率;毛管容量;毛细含水率;毛细管持水量
capillary motion 毛细运动;毛细管运动
capillary movement 地下水毛细管流;毛细移动;毛细(管)运动;毛管(水)运动
capillary network 毛细管网
capillary number 毛细管数
capillary oiler 毛细管给油器
capillary opening 毛细孔;毛细管孔(径)
capillary oscillation 毛细面振动
capillary osmosis 毛细渗透作用
capillary outlet 毛细管出口
capillary packed column 毛细管填料柱
capillary path 毛细途径
capillary pen 毛细管绘图笔
capillary penetration 毛细渗透;毛细孔穿透;毛细管透过作用;毛细管渗透(作用);毛管渗透
capillary percolation 毛细(管)渗透;毛管渗透
capillary permeability 毛细管渗透性
capillary phenomenon 毛细(管)现象
capillary pipette 毛细吸管
capillary pore 毛(细)管孔隙;毛细管孔
capillary pore volume 毛细孔体积
capillary porosity 毛细孔隙率;毛细孔隙(度);毛细管孔隙度;毛细管空隙度;毛细管孔隙率;毛管孔隙度
capillary porous body 毛细多孔体
capillary potential 毛细水位势;毛细潜能;毛细位能;毛细管水势差;毛(细)管势;毛细管潜能;毛管水位能
capillary potential energy 毛管潜能
capillary potential gradient 毛细势差;毛细势梯度;毛管势(能)梯度;毛管势力差
capillary power 毛细管吸力
capillary pressure 毛细张力;毛细压强;毛细(水)压力;毛细管(水)压力;毛管水压力
capillary pressure at saturation medium value 饱和度中值的毛细管压力
capillary pressure head 毛(细)管压头
capillary pull 毛细水张力
capillary pulling power 毛管张力
capillary pump 毛细泵
capillary pumping head 毛细唧送压头
capillary pumping parameter 毛细唧送参量
capillary pyrite 针镍矿;毛发状黄铁矿
capillary quartz cell 毛细石英盒
capillary radius 毛细管半径
capillary resistance measure apparatus 毛细管阻抗测定器
capillary rheometer 毛细管流变仪;毛细管流变计
capillary ripple 毛细波痕;毛细波
capillary rise 毛细升高;毛细上升高度;毛细上升;毛细管上升;毛管水上升高度;毛管上升;毛细水高度;毛管上升
capillary rise correction 毛细管上升校正
capillary rise fringe 毛管水上升边缘
capillary rise height 毛细提升高度;毛细上升高度
capillary rise method 毛细管上升法
capillary rise rate 毛细管上升(速)率
capillary rise test 毛细管上升高度试验
capillary rise velocity 毛细(管)上升速度
capillary rising fringe 毛细管上升边缘
capillary rising height 毛细(管上)升高度
capillary rising height test 毛细管上升高度试验
capillary rising rate 毛细管上升速度
capillary saturation 毛细(管)水饱和;毛细饱和;毛细饱和
capillary seal 毛细管封闭
capillary seepage 毛细渗流;毛(细)管渗漏;毛(细)管渗流
capillary seepage trench 毛细管渗水槽
capillary-shear-induced orientation 毛细管切变取向
capillary sheath 毛细管鞘
capillary siphoning 毛(细)管虹吸作用
capillary soil moisture 土壤毛细水
capillary soil water 土壤毛细水
capillary space 水泥砂浆中夹入气泡;毛细管空间;毛细管间隙
capillary splitter 毛细管分流器
capillary stage 毛管水位
capillary stopcock 毛细活栓
capillary structure 毛细孔结构;毛细(管)结构
capillary suction 毛细吸力
capillary suction head 毛管吸水头;毛管吸升高度
capillary suction time 毛细管吸力时间;毛细管附时间;毛细抽吸时间
capillary suction time test 毛细抽吸时间试验
capillary superconductivity 毛细管超导性
capillary support 毛细管载体
capillary surface 毛细表面
capillary syringe 毛细吸管
capillary technique 毛细管技术
capillary tension 毛细水张力;毛细孔张力;毛细(管)张力
capillary tension theory 毛细管张力理论
capillary test 毛细试验
capillary theory 毛细作用理论
capillary thermometer 压力式温度计
capillary transport 毛细管输运;毛细管输送
capillary tube 微管;毛细管作用速率;毛(细)管
capillary tube agglutination test 毛细管凝集试验
capillary tube method 毛细管法
capillary-tube type of viscometer 毛细管黏度计
capillary tubing 毛细管
capillary vessel 毛细导管
capillary visco(si)meter 毛细(管)黏度计;毛管黏度计
capillary visco(si)metry 毛细管测黏(度)法
capillary void 毛细管空隙
capillary wall 毛细管壁
capillary water 毛细(管)水;毛细水
capillary water absorption test 毛细管吸水试验
capillary watering 毛管灌水
capillary water test 毛细管水试验
capillary water zone 毛细水带
capillary wattmeter 毛细管式功率表
capillary wave 界面波;毛细波;毛管波;张力波;表面张力波
capillary wetting method 毛(细)管湿润法
capillary zone 毛细管区;毛细管带;毛管带
capillator 毛细(管)比色计
capillometer 毛细(管)试验仪
capillomotor 毛细管运动的
cap iron 护铁(木工刨刀)
capital 资本;柱顶;首都
capital account 资本账(目);资本账户
capital account balance sheet 资本账户平衡表
capital account data 资本账目资料
capital account items 资本账户科目
capital account outflow 资本账户外流
capital accumulation 资本积累
capital additions 增资
capital adjustment account 资本调整账户
capital allowance 投资限额;资本减让;投资优惠扣除
capital amount 资金额;资本金额
capital analysis 资金分析
capital and interest 本利和
capital and labo(u)r 劳资
capital and liability ratio 资产负债比(率)
capital and running cost 资本费用和经营费用
capital appreciation 资本增值(税)
capital appropriation 资本拨款;投资拨款
capital arising from re-capitalization 由调整资本而产生的资本
capital arrangement 资金安排
capital array 基本台阵
capital asset pricing model 资本资产评价模型
capital assets 基本资产;资本资产;固定资产

capital assets fund 资本资产基金
capital-attracting 对投资有吸引力的
capital-attracting standard 吸引资本标准
capital authority 投资权
capital-authorized 法定资本;额定资本
capital availability 可得到的资本
capital balance 资本差额
capital base 资本基础
capital bonus 资本红利
capital borrowed 借入资本
capital budget 资金预算;投资预算;基建预算;基本建设预算;资本预算
capital budgeting 资本预算编制;编制资本预算;投资方案选择
capital carving 柱头雕塑
capital charges 资本支出;造价;利息费用;基建投资
capital check 资金稽查
capital coefficient 资本系数
capital combination 资本联合
capital commitment 资本承诺
capital common stock 普通股本
capital composition 资本构成
capital construction 基建;基本建设
capital construction accountant 基本建设会计
capital construction analysis 基本建设投资分析
capital construction budget 基建预算;基本建设预算
capital construction cost 基本建设费(用)
capital construction expenditures 基本建设开支;基本建设支出
capital construction financial plan audit 基本建设财务计划的审计
capital construction investment 基建投资;基本建设投资
capital construction investment plan audit 基本建设投资计划的审计
capital construction item 基本建设项目
capital construction loan 基建贷款
capital construction management for highway engineering 公路工程基本建设管理办法
capital construction plan audit 基本建设计划的审计
capital construction plan of current year 本年基本建设计划
capital construction procedure 基本建设程序
capital construction project 基本建设项目;基本建设工程
capital construction project design estimation audit 基本建设项目设计概算的审计
capital construction project plan audit 基本建设项目计划的审计
capital construction statistics 基本建设统计
capital construction work for the earlier stage 基本建设前期工作
capital construction works 基建工程
capital consumption 资本消耗
capital consumption allowance 基本消耗的限定;资本消耗扣除;资本消耗补偿
capital contributed in excess of par value 超票面值缴入资本
capital contribution 资本摊缴;认缴资本
capital control 资金管理;资本控制
capital cost 基建投资费(用);基本投资;基本建设费(用);基本费用;资金成本;资本值;资本成本;主要费用;动态投资;投资性成本;投资费(用);投资
capital cost analysis 资本费用分析
capital cost of reactor plant 反应堆装置投资
capital credit 资本信用
capital debenture 资本债券
capital debt 基建借款
capital decrease 减资
capital deepening 资本增密
capital deficit 资本亏损
capital demanded 资本需要量
capital depreciation 基本折旧;固定资产折旧
capital destruction 资金耗损;资本耗损
capital development fund 资本发展基金
capital dividend 资本股利
capital dredging 基建性挖泥;基建性疏浚
capital duty 必须缴纳的印花税
capital efficiency 资本效率
capital employed 使用资本额
capital endowment 捐赠资金

capital equipment 基本工具;大型设备;大型工具;基建设备;基本设备;资本设备;主要设备;固定设备;生产资料
capital-excess account 超额资本账户
capital exodus 资本流出
capital expenditures 基建投资;基建费(用);基本建设投资;基本建设费(用);基本费用;资本支出;置产费用;固定资本投资;投资支出
capital expenditure analysis 资本支出分析
capital expenditure and receipt 资本支出与收入
capital expenditure and revenue expenditure 资本支出与收益支出
capital expenditure control 资本支出控制
capital expenditure decision 资本支出决策
capital expenditure expansion 扩大基本建设支出
capital expenditure to sales ratio 资本支出对销售的比率
capital expenses 集资费用
capital export 资本输出
capital export country 资本输出国
capital facility 固定资产;不动产
capital financing 资金筹集;集资;资本理财;筹集资本;提供资本
capital flight 资本外逃
capital flow 资本流量
capital formation 资本形成;投资
capital formation in residential building construction 居民住宅建筑投资
capital formative discharge 主要造床流量
capital for trading purpose 商业资本
capital from appraisal adjustment 由资产重估所得资本
capital from treasury stock transactions 库存股本交易所得资本;由库存股本交易所得资本
capital fund 资本基金;投资基金
capital gain free of taxation 免税资本利得
capital gain option 可列作资本利得
capital gains 资本收益;资本升值收入
capital gains and losses 资本的利得与损失
capital gains or loss 资本损益
capital gains tax 资本增值税;资本收益税;资本收益分配
capital gap 资本差额
capital gearing 业主权益与债务资本比率;资本构成
capital goods 资本货物;生产资料
capital grant 支助拨款;公共工程投资项目
capital grant progress payments 分期拨款的资金
capital immobility case 资本不流动情况
capital improvement budget 改进生产设备费用预算
capital improvement loan 改进生产设备贷款
capital improvement program(me) 资本刷新计划
capital in budget 预算内资金
capital in cash 资本金
capital income 资本收益;资本收入
capital incorporation 投入资本
capital increase 增加资本;增资
capital in excess of stated value 超设定值缴入股本
capital inflow 资本流入
capital influx 资本流入
capital infusion 扩充资本;资本引入
capital in general 一般资本
capital input 资本投入;投入资本
capital insurance 资本保险
capital intensity 资本密集程度;投资额;资本密集度
capital intensity of production 生产资本密集程度
capital-intensive 资金密集的;资本密集的
capital intensive enterprise 资本密集企业
capital intensive goods 资本密集货物
capital intensive industry 资本密集工业
capital intensive investment scheme 资本密集投资计划
capital intensive project 资金密集项目;资本密集型项目
capital interest 资本利息
capital interest tax 资本利息税
capital invested 投资
capital investment 基金投资;基建投资;基本投资;基本建设投资;资本投资;投资支出;投资
capital investment analysis 基金投资分析
capital investment and expansion 资本投入与扩展
capital investment appraisal 资本投资评价
capital investment decision 资本投资决策;投资决策

capital investment in machine or plant decision 投资于设备或厂房的决策;投资于机械或厂房的决策
capital investment loan 基建借款
capitalism market 资本主义市场
capital issue 股票发行;发行股票
capitalist cycle 资本主义周期
capitalistic production 资本主义生产
capitalistic system 资本主义制度
capitalist market economy 资本主义市场经济
capitalist world monetary system 资本主义世界货币体系
capital item 资本项目
capitalization 资本总额;资本折算;资本还原过程;资本化;转作资本
capitalization approach 收益法
capitalization issue 资本化证券;债券股票等的发行
capitalization of earning power 盈利能力的资本化
capitalization of interests 滚利作本;以利息化为资本
capitalization rate 资本还原率;资本化率;资本化利率;投资收入率
capitalization recapture 资本回收
capitalize 列为长期投资项目;计算现值;变换成生产资金;投资
capitalized 大修
capitalized annual cost method 资本化年费用法
capitalized cost 资本值;核定投资额;资化成本;资本化费用;资本化成本;成本;本金化费用;投资成本
capitalized expenses 资本化费用;费用的资本化值
capitalized interest 资本化利息;滚利作本;本金化的利息
capitalized interest rate 资本化利率
capitalized profit 资本化利润
capitalized ratio 资本化还原比率
capitalized syndication fee 开办费的资本化值
capitalized total cost 核定投资总额
capitalized value 核定资本值;资本化价值;本金化价值
capitalizing rate 资本核算率
capitalizing the front-end fee 启用费资本化
capital-labo(u)r ratio 资本—劳力比
capital lease 资本租赁;财务租赁;融资租赁
capital lease compared with financing lease 资本租赁与财务租赁比较
capital leasing 融资租赁
capital letter 大写字母
capital leverage 资本的杠杆作用;财务杠杆作用
capital levy 资本税;资本课税;财产税
capital liabilities 固定负债;资产负债;资本负债
capital liability ratio 资本负债比率
capital loan 资本借贷
capital loss 资本损失
capital management information system 资金管理信息系统
capital manager 长期资本经营者
capital market 资本市场
capital market control system 资本市场管理制度
capital market reporting system 资本市场报告系统
capital markets department commitment fee 资本市场承诺费
capital mobility case 资本流动情况
capital monopoly 资本垄断;资本独占
capital movement 资本流动
capital of account 会计科目名称
capital of Byzantine 拜占庭式柱头
capital of circulation 流通资本
capital of composite (column) 混合式柱头
capital of construction 基本建设资金
capital of Corinthian column 考林辛柱头
capital of Doric column 陶立克柱头
capital of Gothic column 哥特式柱头
capital of Lonic column 爱奥尼克柱头
capital of Moorish column 摩洛哥柱头
capital of Romanesque column 罗马式式柱头
capital of Tuscan column 塔斯康柱头
capital of Tuscan order 塔斯康774柱头
capital on call 待收资本
capital operation 资本业务
capital optimum 资本限额
capital ornament 柱头(装)饰
capital outflow 资本外流

capital outlay 基建投资;基建费(用);基本投资;基本建设费(用);资本支出
capital out of budget 预算外资金
capital-output 资本产出
capital-output coefficient of technology 技术的资本产出系数
capital-output ratio 投资产出比;资本与产量比率;资本产值比(率);资本产出(比)率
capital overhaul 技术大修
capital owned 自有资本
capital ownership 资本所有权
capital paid-in 交入资本;实缴股本;实收资本
capital payment 资本支付
capital pay-off 投资收回期;资金偿还;投资收回;投资回收
capital pay-off time 投资回收期
capital personal account 股东个人账户
capital planning 投资计划
capital plant 主要工厂;大型工厂
capital pool 资本总额
capital premium 资本溢价
capital producing 生产资本
capital productivity 资本生产率
capital profit 资本利润
capital program(me) 基本建设计划
capital project 基建项目;基本建设项目
capital project funds 基建项目基金;基本建设项目基金;投资项目基金
capital proposal 投资建议
capital raised under share incentive scheme 以职员优先股计划筹集资金
capital raising 集资;筹集资金
capital rating 资本定级
capital ratio 资本比率
capital rationing 资金限额;一定期限内的投资限额;资本配额
capital receipts 资本收益;资本收入
capital recipient country 资本接受国
capital reconciliation statement 资本调节表
capital recovery 资本回收系数;投资收回;投资还本;收回资本
capital recovery cost 还本付价;资本还原成本;投资还原成本;回收投资费;回收基建费
capital recovery factor 还本因素;资本因素;资本回收系数;资本恢复因子;投资收益率;投资回收因数
capital recovery period 还本周期;投资回收周期
capital-recovery-plus-interest 资本回收加利息
capital redemption 资本赎回
capital redemption reserve fund 偿还资本准备基金
capital repair 资本补偿;大修理;大修;全面拆修
capital repair brigade 大修队
capital repair of pavement 路面大修
capital repatriation 资本调回
capital requirement 资金需要(量)
capital reserve 资本储备金;资本准备;资本公积金;资本储备
capital resources 资本财力
capital-re-switching 资本再转移
capital return 资本收益率
capital returned to stockholders in dividends 以股息方式摊还资本;摊还股本
capital revenue 资本收入;资本收益
capital saving 资本节省
capital-saving invention 节约资本发明
capital-saving technical progress 节约资本的技术进步
capital sharing system 资本分配制度
capital ship 主力舰
capital sources 资金来源
capital spending 投资;资本支出;资本花费
capital standard 资本标准
capital statement 资本变动表
capital stock 资金;股金总额;股本;公司资本额
capital stock assessment 股本的估价
capital stock authorized 法定股本;额定股本
capital stock issued 已发行股本;已发股本
capital stock of subsidiary and affiliated company 附属公司和联营公司股本
capital stock outstanding 净发行股本;已发在外流通股本
capital stock paid in 实缴股本
capital stock paid-up 已缴股本
capital stock preferred 优先股本

capital stock premium 股本溢价
capital stock registered 已注册股本
capital(stock) subscription 认缴股本
capital stock subscriptions receivable 应收认缴股款
capital stock unpaid 未缴股本
capital stone 柱顶石
capital structure 资本结构
capital structure decision 资本构成决策
capital structure ratio 资本结构比率
capital subscription 认购股本;出资额
capital subscription items 认缴股本项目
capital sum 资本总额;本金
capital supply 资本来源
capital supply schedule 资金供应时间表
capital surplus 资本盈余;资本盈利;超面值缴入资本
capital surplus reserve 资本盈余准备
capital surplus statement 资本盈余表
capital tax 资本税
capital territory 首都地区
capital tie-up 不能自由运用的资本
capital transaction 资本交易
capital transaction account 资本交易账户
capital transfer 资本转移;资本转让;资本流动
capital transfer tax 资本转让税;资本过户税
capital turnover 资本周转
capital turnover period 资本周转期
capital turnover rate 资本周转率
capital turnover ratio 资本周转率
capital uncalled 未收资本;未收回资本
capital value 资本价;土地期望价
capital value of premises 楼宇资本价值;房地产资本价值
capital wastage cost curve 资本损耗成本曲线
capital wave 基波
capital widening 资本扩大
capital works 基本(建设)工程;主要工作
capitate 头状的;槌形的
capitation 筑路壅水;按人收费(人头税);按人计算;人口税
capitation grant 按人计算补助费
capitation tax 人头税;人口税
capitive shop 内部商店
capitol 议会大厦;立法机构大厦;立法机构行政会议厅
cap jet 伞形喷口;副喷口;辅助喷射口
cap jib 斜桅三角帆
cap key 闭口扳手;套筒扳手
cap lamp 帽灯;头灯
caplastometer 黏度计
caple 脉壁黏土带;石英质岩石
caple glass 板玻璃
cap lever 护铁压杆
cap line 顶线
cap liner 封口片
cap log 挡轮坎(码头前沿);挡车槛(码头前沿)
cap model 帽檐模型;盖帽模型
cap mo(u)ld(ing) 压顶线脚【建】;门窗顶线脚
capnometer 二氧化碳检测计
cap nut 螺母;螺帽;盖形螺母
capoc cloth 木棉布
cap of bulb 电灯泡的灯头
cap of column 塔泡罩
cap of end wall of portal 洞门墙顶顶帽
cap of lock 锁帽;锁盖
cap of pier 桥墩台顶帽
cap of pile 桩台;桩帽;盖梁
cap of screw 螺(钉)帽
capok cloth 木棉布
caponier 阴沟隔断
capped anchorage 压顶锚具;排张拉锚具
capped bullet 被帽弹丸
capped butt 有盖铰链
capped column 冠柱晶;带箍柱
capped end 加帽端
capped fuse 引火管;带雷管的导火线
capped ingot 钢锭;封顶钢锭
capped nonionics 封端的非离子表面活性剂
capped nut 螺帽;盖螺母
capped pile 带箍桩;安上桩冒的桩
capped piling 带箍桩
capped pipe 排水管的帽管;加盖管;帽管(加盖管);封头管
capped post 带箍柱

capped quartz 截短的石英;冠状石英
capped reactor 加盖反应器
capped screw 镀铬圆头螺钉
capped steel 加盖钢;沸腾钢;半镇静钢;脱氧钢
capped stone 顶盖石
capped tube 封头管
capped yield model 盖帽屈服模型
cappelenite 硼硅钡钇矿
capper 压盖机;采板工;封口机;切裁工
cap piece 喷头;帽罩;压瓦板;柱帽子;帽材
cap piece of keel block 盘木顶块
cap piece of sheet pile 板桩联系帽梁(船坞用);板桩联系帽条
capping 矿帽;聚合物端基封闭;加盖;压顶;盖面;顶板;槽盖;封端;封闭剂;封堵器;铜盖顶;铁帽;台盖梁
capping agent 封端剂
capping beam 帽梁;压檐梁;压顶梁;顶梁
capping beam of sheet-pile 板桩帽梁
capping block 护面块体;盖顶块体;防波堤护面块体
capping brick 帽桩;檐砖;压檐砖;压顶砖
capping cables 锚夹预应力负弯矩钢筋
capping cover layer 压顶防护层;盖顶防护罩;防波堤上部防护面层;(防波堤的)上部防护面层
capping crane 加盖起重机
capping machine 加盖机
capping mass 糊炮(封)泥;压顶物;盖顶物;盖层物;覆盖岩层;浮土;剥土;表土
capping materials 封顶材料
capping metal 浇灌金属
capping of specimen 试件封顶
capping phenomenon 成帽现象
capping piece 帽材;压檐木;压檐梁;压顶梁;压顶板;帽木
capping plane 圆弧刨;圆角木护手(栏杆上);托板;盖板
cap(ping) plug 盖塞
capping rock 盖顶块石;盖层岩;防波堤护面块石
capping shutter 叠合快门
capping slab 压顶板;盖板
capping spring 覆盖泉
capping stone 压顶石;护面(大)块石;盖顶石
capping tile 压檐瓦;牝瓦
capping wall 码头胸墙
cap plate 柱头板;压顶板;帽木;盖板;托板
cap plate of stanchion 柱顶托板
capraldehyde 癸醛
capramide 癸酰胺
caprate 癸酸酯;癸酸盐
capreolate 卷须状的;卷曲状的
capreolus 古木屋顶支撑;古木屋顶拉杆
Capri blue 卡布水蓝色;卡布里蓝染料
capric acid 癸酸
capric acid chloride 癸酰氯
capric aldehyde 癸醛
capric amide 癸酰胺
capric anhydride 癸酸酐
capricious value 任意价格;不可靠价格
capric nitrile 癸腈
capricornoid 犀角线
cap ridge 帽缘
caprifig 突变无花果
caprin 癸酸甘油酯
caprinoyl 癸酰
capristor 封装阻容
caproaldehyole 己醛
caproate 己酸酯;己酸盐
cap rock 岩顶;盖岩;覆岩;岩盘;冠岩;盖层;护面块石;盖顶块石;顶盖岩
caprock-type geothermal aquifer 有盖层的地热含水层
caproic aicd 己酸
caprolactone 羟基己酸内酯
caproleic acid 癸烯酸
Caprone 卡普隆(商品名);聚己内酰胺纤维
Caprotti valve-gear 卡普罗提式阀动装置
caproyl 己酰(基)
caproylamine 己胺
caproyl chloride 己氯
capryl alcohol 辛醇
capryl compounds 辛基化合物
caprylic acid 辛酸
caprylic alcohol 辛醇
caprylyl acetate 醋酸正辛酯

capsanthin 辣椒红色
cap screw 有头螺栓；螺纹盖；有头螺钉；带头螺栓；内六角螺栓；封口螺钉
cap scuttle 有盖小舱口
cap seal 帽形密封；封帽
cap seat 瓶盖毡
cap sheet 面层油毡；最上层屋面板；压顶层；屋面露明卷材（多层油毡屋面面层）；盖板
cap-sill 盖木
cap size 大写字母尺寸；倾覆；翻到
capsizing angle 倾覆角
capsizing moment 倾覆力矩
cap sleeve 帽套
capsomere 壳粒；壳微体
capstan 立轴绞车；卷筒；卷缆绞盘；绞盘；绞船柱；绞车；主导轮；刀盘；索式卷扬机
capstan axle 主导轴
capstan bar 绞盘杆；绞盘棒
capstan barrel 绞盘筒
capstan bolt 绞盘螺栓；带销螺栓
capstan box 绞盘箱
capstan canvas 绞盘帆布罩
capstan cover 绞盘帆布罩
capstan crab 起锚机
capstan drum 绞盘卷筒
capstan engine 绞盘机；起锚机；卷扬机
capstan error 主动轮误差
capstan feed 主动轮传动；主导轴输送机构
capstan handle 十字手柄
capstan handwheel 绞盘手轮
capstan head 六角刀架；绞轮；猫头轮；刀具盘
capstan headed screw 绞盘螺钉
capstan head slide 六角头滑板；转塔刀架
capstan idler 主导轮空转轮
capstan lathe 六角车床
capstan motor 主动轮电动机；主导电动机
capstan nut 有孔螺母；带孔螺母；带孔螺帽；槽形螺母
capstan oscillator 主导振荡器
capstan partner 绞盘垫
capstan rest 六角刀架转塔
capstan roller 竖滚柱；输带辊
capstan saucer 绞盘承座
capstan screw 绞盘旋；转塔丝杠
capstan servo 绞盘伺服机构
capstan shaft 主导轴
capstan turret 六角刀架转塔
capstan winch 绞盘
capstan windlass 立式锚机；绞盘式卷扬机；起锚绞盘
cap starter brick 拱脚砖
cap stay 桅间支索；桅顶旗索
capstone 帽石；顶层石；盖顶石；顶层石；拱顶石
cap stopper 帽塞
cap strip 帽材
Capstrum filtering 复塞谱滤波
capsular jack 扁千斤顶；千斤顶
capsular-spring ga(u)ge 测量压力的弹簧起爆仪
capsulate(d) 内部安装的；装入雷管的
capsulated ga(u)ge 屏罩应变计；屏罩应变片
capsulation 封装；密封
capsule 胶囊；膜片；密封舱；小盒；舱；瓶帽；瓶盖
capsule aneroid 无液气压机
capsule aneroid barometer 空盒气压计
capsule cabin 座舱
capsule form 胶囊形式
capsule ga(u)ge 膜盒式压力表
capsule pressure ga(u)ge 膜盒式压力表
capsuler 套瓶帽机
capsule stack 膜盒组件；膜盒组
capsule-type injector 胶囊式注射器；灯泡式喷头
capsule-type manometer 膜盒式压力计
capsule-type pressure 囊式压力计
capsule(-type vacuum) ga(u)ge 膜盒（式）真空计
captain 指挥者
captain's bed 船长床（下设抽屉柜）
captain's bill of lading 船长提单；随船提单
captain's chair 船长椅；矮靠背鞍座椅
captain's copy 随船提单抄本
captain's copy of a bill of lading 提单的船长用抄本
captain's log 航海日志草本
captain's mail 随船送达邮件
captain's protest 海事声明；海难抗议书

captain's public room 船长办公室
captain's room 船长室
captain's walk 屋顶便道
captance 容抗
captation 截水；集水装置；筑坝壅水
cap tendon 盖顶预应力钢筋；顶部预应力筋
cap the climax 超高限度
captimber for pile bent 桩排架盖木
cap tin 柱帽锡
caption 插图说明；插图标题；标题
caption blanking 标题消隐
caption foundry 铸造车间
caption of account 会计科目
caption of a drawing 图签
caption test 工作台试验
caption title 书名标题
captivator 夹具
captive air bubble 捕获的气泡
captive balloon 系留气球
captive balloon survey 系留观察气球；系留气球勘测
captive barge 码头趸船；系留趸船；定线驳船
captive breasting barge 码头趸船
captive company 附属公司
captive container 限定集装箱
captive eddy region 固定涡流区
captive fastener 系留紧固件
captive float 系留趸船；系定浮筏
captive foundry 铸造车间；铸工车间
captive insurance company 自保保险公司
captive item 自产自用项目
captive key 滑键；弹性销；弹性键
captive market 垄断市场；固定销路
captive nut 扣紧螺母；栓住螺母
captive pin 安全针；安全枢轴；安全钉；安全销；合缝钉
captive pontoon 系留趸船
captive producer 附庸生产者
captive production 自用生产
captive screw 系紧螺钉；自动保险螺钉；外加螺钉
captive shop 附属工厂
captive tape 缓动传送带
captive terminal 货主码头；专用码头
captive test 静态试验；工作台试验；捕获试验；台架试验
captive use 垄断使用
captor river 袭夺河
cap trim 门窗顶贴脸；门窗顶线脚；盖线
capture 俘获；捕获
capture and recapture experiment 标识放流
capture angle 俘获角
capture area 吸收面
capture area of antenna 天线有效面积
capture cross section 俘获截面
captured air bubble craft 侧壁式气垫船；封闭气泡式气垫船；双体气垫船
captured river 截断河；断头河；裁短河；被袭夺河；（流）
captured rotation 俘获自转
captured stream 断头河；被夺河（流）
captured vessel 捕获船
capture effect 遮蔽效应；俘获效应
capture efficiency 俘获效率
capture event 俘获事件
capture hypothesis 俘获假说；捕获假说
capture market 争夺市场；争取市场；占领市场
capture of underground water 地下水收集
capture process 俘获过程
capture radiation 俘获辐射；伴随俘获的辐射
capture radius 捕获半径
capture range 同步范围
capture rate 销售占有率；捕获效率
capture ratio 俘获率；俘获比
capture region 捕获区域
Capture Theory 捕获理论
capture time 捕获时间
capture velocity 控制风速；集气速度；吸取速度；捕捉速度
capture zone 捕获带
capturing 找准
capturing hood 外部吸气罩
capturing river 袭夺河；夺流河
capturing stream 袭夺河；夺流河
capucine lake 旱金莲胭脂红色

Capuron's cardinal point 卡普隆氏点
Capuron's point 卡普隆氏点
caput articularis 关节头
caput mortuum 残渣
caput reflexum 反折头
cap vault 拱形顶
cap wale 桩墩盖梁
cap weld 最后焊层；盖面焊缝
cap wire 雷管导线
capwise 直交方向；横梁方向
cap wrench valve 扳手盖阀
car 轿船；吊舱；车辆；搬运人；梯厢
Carabine 卡拉伯恩风
car accelerator 车辆加速器
car accounting 车辆核算
caracol(e)曲径；螺旋（式）楼梯；旋梯
caracolite 氯铅芒硝
caradle vault 筒形屋顶；筒形穹顶
carafe 玻璃水瓶；饮料瓶；球形窄颈瓶
carajura 秋海棠红色
caralogable permanent file 可编目永久性文件
caramel 焦糖色；酱色；表面陷坑
caramel brown 焦糖棕色
caranalyzer 定碳仪
caranda 卡兰达棕；巴西扇棕
caranda wax 卡兰达蜡
caranday 巴西扇棕
car annunciator 轿厢信号显示器；电梯楼层显示盘
carapace 上壳翼【地】
carapa oil 烛果油；西印度红木
caraphoresis 电泳
car arrester 阻车器
carat 克拉（宝石重量单位，1 克拉等于 200 毫克）
carat balance 克拉天平
carat count 克拉计数
Caratheodory outer measure 卡拉西奥道里外测度【数】
caratloss 金刚石消耗（钻进时）；金刚石耗损（钻进时）
car attendant 列车员
carat weight 克拉重量
caravan 大篷车；商队；篷车拖车和活动房屋；旅行队；搬运车
caravan city 汽车拖挂式房聚居地
caravan of vehicles 一序列车辆
caravan park 旅行车停车场（英国）
caravan route 商队路
caravansary（中、西亚供商队住宿的）客店；旅馆；客店；篷车旅店
caravanserai 旅馆
car axle 车轴
carbacidometer 大气碳酸计；碳酸计
carballoy 卡波碳化硬质合金；碳化钨硬质合金
carbamamidine 胍
carbamate herbicides 氨基甲酸酯类除草剂
carbamate insecticides 氨基甲酸酯类杀虫剂
carbamate pesticide poisoning 氨基甲酸酯农药中毒
carbamate pesticides 氨基甲酸酯类农药
carbamic acid 氨基甲酸
carbamic acid ester 氨基甲酸酯
carbamide 碳酰二胺
carbamide formaldehyde resin 脲甲醛树脂
carbamide peroxide 过氧化脲
carbamide resin 聚脲树脂；碳酰胺树脂
carbamide resin adhesive 尿素树脂黏结剂；尿素树脂胶粘剂
carbamidine 胍
carbam(in)ate 氨基甲酸酯；氨基甲酸盐
carbamino alanine 氨甲酰丙氨酸
carbamonitrile 氨基化氰
carbamoyl 氨基甲酰基；氨基甲酰
carbamyl 氨基甲酰
carbamyl bromide 氨基甲酰溴
carbamyl chloride 氨基甲酰氯
carbamyl fluoride 氨基甲酰氟
carbamyl phosphate 氨甲酰磷酸；氨基甲酰磷酸
carbamylphosphonate 氨甲酰膦酸酯
carbanalyzer 定碳仪
carbanilic acid 苯氨基甲酸
carbanilino 苯氨基甲酰基
carbaniloyl 苯氨基甲酰；苯氨基甲酰
carbanion 阴碳离子
carbanion addition 阴碳离子加成作用

carbankerite 铁白云石质煤
carbargilite 泥质煤
car barn 电车库
carbarsus 纱布;麻布
carbaryl 胺甲萘;胺甲苯
car battery 车用蓄电池
carbazole 卡唑
carbazole dye 咔唑(结构)染料
car beam straightening equipment 调梁设备
car beam straightening siding 调梁线【铁】
carbel table 挑檐
carbendazol wettable powder 多菌灵可湿性粉剂
carbene 卡宾;聚炔;亚碳;碳烯
carbene chemistry 碳烯化学
carbenes 碳质沥青
carbenium ion 碳正离子
carbetamide 草长灭
carbide 电石;碳化物;碳化钙
carbide alloy 硬质合金;碳化物合金
carbide annealing 球化退火;碳化物退火
carbide ash 碳化物灰
carbide base(d) cemet 碳化物基金属陶瓷
carbide bit 硬质合金钻头;碳化物刀头
carbide black 碳化黑
carbide blade 硬质合金托板;硬质合金刀片
carbide brick 碳化砖;碳硅砖
carbide bur 碳化物牙钻
carbide carbon 化合碳
carbide casting plant 碳化物铸造厂
carbide ceramics 碳化物陶瓷
carbide cermet 碳化物金属陶瓷
carbide chamber 电石篮
carbide chip 硬质合金刀片
carbide cross bit cutting 硬质合金十字钻头切割边
carbide cutter 硬质合金刀具;碳化物刀具
carbide cutting element 硬质合金切削具
carbide cylindric(al) surface cutter 硬质合金圆柱平面铣刀
carbide die 硬质合金拉模
carbide draw ring 硬质合金拉延环
carbide drill 硬质合金钻(头)
carbide drill bit cutting edge 硬质合金钻头切割边
carbide drum 电石储[贮]罐
carbide end mill 硬质合金端铣刀
carbide-feed generator 乙炔发生器;气体发生炉
carbide furnace 碳化炉;炭精电极炉
carbide insert 硬质合金镶块
carbide-insert core bit 硬质合金取芯钻头
carbide lamellartiy 碳化物带状组织
carbide lamp 矿用电石灯;电石灯
carbide lime 碳化钙石灰油灰;碳化钙石灰;电石石灰
carbide method 碳化钙测型砂水分法
carbide miner 遥控自动化采煤机
carbide mud residue 电石渣
carbide network 网状碳化物
carbide nuclear fuel 碳化核燃料
carbide of calcium 碳化钙;电石
carbide of silicon 金刚砂;碳化硅
carbide of silicon brick 碳化硅砖
carbide of silicon mortar 碳化硅灰浆
carbide of silicon slab 碳化硅板
carbide percussion drilling 硬质合金钻具的风动冲击钻进
carbide press die 硬质合金冲压模
carbide promoter 碳化物促进剂
carbide punch 硬质合金凸模;硬质合金冲模
carbide reduction process 碳化物热还原法
carbide refractory 碳化物耐火材料
carbide refractory product 碳化物耐火制品
carbide reservoir 电石槽
carbide scraper 硬质合金刮刀
carbide segregation 碳化物偏析
carbide side cutter 硬质合金三面刃铣刀;硬质合金侧铣
carbide slag 碳化物渣;电石渣
carbide slag slurry wastewater 电石渣浆废水
carbide stabilizer 碳化物稳定剂
carbide tip 硬质合金片;硬质合金刀头;精磨刀片
carbide-tipped 硬质合金的
carbide-tipped bit 硬质合金(镶)的钻头
carbide-tipped center 硬质合金顶尖
carbide-tipped core drill 硬质合金空心钻
carbide-tipped cutter bit 硬质合金刀刃
carbide-tipped drill 硬质合金钻头
carbide-tipped milling cutter 硬质合金铣刀
carbide-tipped reamer 硬质合金铰刀
carbide-tipped saw blade 硬质合金锯条
carbide-tipped slip 硬质合金镶卡瓦
carbide-tipped steel 硬质合金镶钢钎
carbide-tipped tool 硬质合金刀具
carbide-tip tool grinding machine 硬质合金工具磨床
carbide tool 硬质合金工具
carbide to water acetylene generator 电石入水式乙炔发生器
carbide to water generator 电石入水式乙炔发生器;投入式乙炔发生器
carbide tungsten 硬质合金;碳化钨
carbide type bit 硬质合金块式钻头
carbide type cement 碳化物型金属陶瓷
carbimide 异氰酸;碳酰亚胺
carbine 弹簧钩
carbinol 伯醇
car bit 杠钻
Carbite 卡拜特炸药
carbitol solvent 卡必醇溶剂
Carbium 卡毕阿姆铝铜合金
car block 网具装卸滑车
carboalkoxy 烷氧羰基
carboatomic ring 碳环
carboborite 水碳硼石;水碳钙镁石
carbocation 碳阳离子
carbocer 稀土沥青
carboceric acid 二十七酸
carbocernaite 碳铈钠石
carbo-charger 混气器
carbocholine 碳胆碱
carbocoal 半焦碳;半焦
carbocoal tar 半焦柏油;低温焦油;半焦油
carbocoal yield 半焦收率
carbo-corundum 碳刚玉
carbocycle 碳环
carbocyclic(al) compound 碳环化合物
carbodiimide 碳化二胺
car body 底座;车厢;车体;车身
car body coupler 车钩
car-body dropping 落车
car body energy absorption 车体能量吸收
car body industry 车体制造业
car body interior decoration 车体内装修
car-body jacking 架车【铁】
car body lettering 车厢写字
car body mounting 车体附件
car body painting 车体上漆
car body repair shed 车体整修库
car body rest frames after lifting jack released 铁马
car body shell 车体构架
car body supporting base 车体支承台
carbofrax 金刚硅砖料;碳化硅
carbogel 煤胶体
carbohm 电阻定碳仪;电阻测碳仪
carbohydrate 糖类;碳水化合物
carbohydrate allowance 碳水化合物供给量
carboid 油焦质
carboids 炭青质;似碳物(沥青组分)
carboirite 羟锗铁铝石
carbol-alcohol 石炭酸酒精
carbolate 酚盐;石炭酸盐
carbolfuchsin 碳酸复红;酚品红液
carbolfuchsin solution 石炭酸品红溶液
carbolfuchsin stain 石炭酸品红染剂
carbolic acid 苯酚;石炭酸
carbolic carbolineum 焦油护木剂(一种木材防腐材料)
carbolic oil 酚油
carbolic soap 酚皂
carbolineum 焦油护木剂(一种木材防腐材料);焦油防腐剂(一种木材防腐混合材料);蒽油防腐剂;蒽油
carbolism(us) 石炭酸中毒
carbolite 卡包塑料;卡包立(酚醛塑料)
carbolon 卡包纶(碳化硅的商品名);碳化硅
carboloy metal 碳化钨硬质合金
carboloy nozzle 硬质合金喷嘴
carboloy-set bit 硬质合金胎体的钻头
carbometer 空气碳酸计;定碳仪;二氧化碳计;二氧化碳测定计
carbometry 二氧化碳定量法;碳酸定量法
carbominerite 矿化煤
carbon 碳精;炭精
carbon-13 magnetic resonance 碳—13 核磁共振
carbon-14 碳—14
carbon-14 dating 放射性碳测定年龄;碳—14 年代测定法
carbon absorption 炭吸附
carbon absorption column 炭吸附柱
carbon absorption method 碳吸附法
carbonaceous 碳质的
carbonaceous adsorbent 含碳吸附剂
carbonaceous atmosphere 碳质气氛
carbonaceous biochemical oxygen demand 含碳生化需氧量
carbonaceous biochemic(al) oxygen demand removal 含碳生化需氧去除
carbonaceous biological oxygen demand 含碳生物需氧量
carbonaceous brasque 碳质耐火堵泥
carbonaceous chondrite 碳质球粒陨石
carbonaceous clay 碳质黏土
carbonaceous coal 柴煤;半无烟煤
carbonaceous compound 碳质化合物
carbonaceous deposit 碳质沉积(物)
carbonaceous fuel 含碳燃料;碳质燃料
carbonaceous gas 含碳气体
carbonaceous ironstone 碳质铁石
carbonaceous limestone 碳质石灰岩;碳质灰岩
carbonaceous material 含碳物质;碳素材料
carbonaceous matter 含碳物质
carbonaceous meteorite 碳质陨石
carbonaceous mudstone 含炭泥岩;炭质泥岩
carbonaceous mudstone inclusion 炭质泥岩包裹体
carbonaceous organic material 含碳有机物
carbonaceous oxidation 碳质氧化
carbonaceous parting 碳质脱膜剂
carbonaceous paste 电极糊
carbonaceous pneumoconiosis 碳素尘肺
carbonaceous reducing agent 碳质还原剂
carbonaceous refractory 碳素耐火材料
carbonaceous refuse 含碳废弃物;碳质废弃物
carbonaceous residue 碳质残渣
carbonaceous rock 碳质岩
carbonaceous sediment 碳质沉积(物)
carbonaceous shale 碳质页岩;炭质页岩
carbonaceous shale waste 碳质页岩废物
carbonaceous slate 碳质板岩
carbonaceous zeolite 碳质沸石
carbonado 墨金刚石;黑金刚石
carbonado bit 黑金刚石钻头
carbon alcohol extract 碳醇萃取物
carbon along with other materials 碳及其他物质
carbon amber glass 琥珀玻璃;有色玻璃
carbon analyser 定碳仪
carbon and charcoal 碳和木炭
carbon and low-alloy steel vessels 碳钢及低合金钢容器
carbon anode 碳正极
carbon arc 碳极电弧;碳弧;炭精电弧;炭弧
carbon arc air gouging 碳弧气刨
carbon-arc-air process 电弧气刨法
carbon arc brazing 碳弧硬钎焊
carbon arc cutting 碳极弧割;碳弧切割;炭极弧割;炭弧切割
carbon arc electrode 碳弧电极
carbon arc gouging 电弧气刨清理
carbon arc lamp 碳弧聚光灯;碳精电弧灯;碳棒灯;炭棒灯
carbon arc spotlight 碳弧聚光灯
carbon arc torch 弓形碳精喷灯
carbon arc weld(ing) 炭极弧焊;碳极弧焊;碳弧焊(接)
carbon arrester 炭质避雷器;炭质放电器;炭精避雷器
carbon assimalation 碳素同化(作用)
carbonatation 二氧化碳饱和
carbonate 碳酸盐
carbonate acid/base group 碳酸盐酸/碱基
carbonate activity coefficient 碳酸盐活度系数
carbonate aegirine-pyroxenite 碳酸霓辉岩
carbonate aggregate 碳酸盐集料;碳酸盐骨料

carbonate alkalinity 碳酸盐碱度
carbonate allochems 碳酸盐异化粒
carbonate analysis logging map 碳酸盐分析测井图
carbonate-apatite 碳酸磷灰石
carbonate attack 碳酸盐侵蚀
carbonate balance 石炭与二氧化碳的平衡
carbonate bank facies 碳酸盐岩滩相
carbonate barrier 碳酸盐障
carbonate basin 碳酸盐盆地
carbonate-bearing aegirine-pyroxenite 含碳酸霓辉岩
carbonate-bearing ijolite 含碳酸霓霞岩
carbonate-bearing rock group 含碳酸岩类
carbonate-bearing siltstone 含碳质粉砂岩
carbonate buffer system 碳酸盐缓冲系；碳酸缓冲系
carbonate buildup facies 碳酸盐岩隆相
carbonate build-up trap 碳酸盐岩隆圈闭
carbonate calculus 碳酸盐结石
carbonate carbon dioxide 碳酸盐二氧化碳
carbonate cement 碳酸盐胶结物
carbonate compensation depth 碳酸盐补偿深度；碳酸岩补偿深度
carbonate compensation level 碳酸盐补偿界面
carbonate concentration 碳酸盐浓度
carbonate concretion 碳酸盐结核
carbonate-containing species 含碳酸盐物种
carbonate continental shelf 碳酸盐陆棚
carbonate-cyanotrichite 碳绒铜矾
carbonate cycle 碳酸盐循环
carbonated beverage 充气饮料
carbonate deposit period 碳酸盐沉积周期
carbonated hardness of water 水的碳酸盐硬度
carbonate distribution 碳酸盐分布
carbonated lime brick 碳化砖
carbonated lime concrete 碳化石灰混凝土
carbonated lime foam concrete 碳化泡沫石灰混凝土
carbonated lime hollow-core slab 碳化石灰空心板
carbonated lime sand brick 碳化石灰砂砖
carbonated rock 碳酸盐地层
carbonated rock rubbish brick 碳化石屑砖
carbonated spring 碳酸泉
carbonated thermal water 碳酸热水
carbonated water 苏打水
carbonate equilibrium 碳酸盐平衡
carbonate facies 碳酸盐相
carbonate formation 碳酸盐建造
carbonate-free lime 纯石灰
carbonate group 碳酸盐基
carbonate hardness 水的碳酸盐硬度；碳酸盐硬度
carbonate-hydroxylaptite 碳羟磷灰石
carbonate ijolite 碳酸霓霞岩
carbonate interbed 碳酸盐夹层
carbonate lake 碳酸盐湖
carbonate leach 碳酸盐浸出
carbonate magnesia 碳酸镁
carbonate medium 碳酸盐介质
carbonate method 碳酸盐法
carbonate mineral 碳酸盐矿物；碳质矿物
carbonate mineral spring water 碳酸盐矿泉水
carbonate mound trap 碳酸盐丘圈闭
carbonate nodular forms 碳酸盐团块状
carbonate of iron 蓝石英
carbonate of lead 碳酸铅
carbonate of lime 碳酸钙；石灰石
carbonate of potash 碳酸钾
carbonate of silver 碳酸银
carbonate of soda 碳酸钠
carbonate of zinc 碳酸锌
carbonate ore 碳酸盐矿石
carbonate platform 碳酸盐台地
carbonate pump 碳酸盐泵
carbonate radical 碳酸根
carbonate reduction zone 碳酸盐还原带
carbonate reef trap 碳酸盐礁圈闭
carbonate reservoir 碳酸盐岩油储
carbonate rock 碳酸盐岩(石)
carbonate rock deposit 碳酸盐岩矿床
carbonate rocks formation 碳酸盐岩建造
carbonate rock reservoir 碳酸盐储集层
carbonate-saline soil 碳酸盐渍土
carbonate salinized soil 碳酸盐盐渍土
carbonate sediment 碳酸盐沉积物

carbonate shallow shoal 碳酸盐浅滩
carbonate slope 碳酸盐斜坡
carbonate-sulfate salinized soil 碳酸盐—硫酸盐土
carbonate system 碳酸盐系
carbonate tidal flat 碳酸盐潮坪
carbonate trap 碳酸盐岩圈闭
carbonate water 碳酸水
carbonate weathering crust 碳酸盐风化壳
carbonating chamber 碳化室
carbonation 充碳酸气；碳酸盐法；碳酸化(作用)；碳酸饱和；碳化作用
carbonation crazing 碳化裂缝
carbonation gas 碳酸饱充用气
carbonation juice pump 充气汁泵
carbonation of fresh concrete 新拌混凝土的碳化
carbonation of hydrated lime 消石灰的碳化
carbonation treatment 碳酸盐法处理；碳酸饱和处理；碳化作用处理
carbonatite 碳酸盐岩；碳酸岩
carbonatite-carbon dioxide-water system 碳酸盐岩—二氧化碳—水体系
carbonatite deposit 碳酸岩矿床
carbonatite lava 碳酸岩熔岩
carbonatite lava group 碳酸岩熔岩类
carbonatization 碳酸盐化作用；碳酸盐化；碳酸化(作用)；碳酸饱和作用
carbon atom net 碳原子网
carbonator 碳酸化器
carbon backplate 碳素护板
carbon balance 碳平衡
carbon bar furnace 碳棒电炉；石墨电极炉
carbon-base coating 碳质涂料
carbon battery 碳电池
carbon beads concrete 碳珠混凝土
carbon-bearing 含碳的；带碳
carbon-bearing dust 含炭粉尘
carbon-bearing mudstone 含炭泥岩
carbon biologic(al) oxygen demand 含氮生物需氧量
carbon bisulfide 二硫化碳
carbon bisulfide insoluble 不溶于二硫化碳
carbon bisulfide soluble 能溶于二硫化碳
carbon black 碳烟灰；炭黑
carbon black dispersion 炭黑分散体
carbon black factory 炭黑厂
carbon black gel complex 炭黑凝胶复合体
carbon black industry 炭黑工业
carbon black pigment 炭黑颜料
carbon black pneumoconiosis 炭黑尘肺
carbon black process 炭黑生产过程
carbon black undertone 炭黑法底色
carbon blaster 积炭喷净装置
carbon block 碳块体；炭块体
carbon blow 吹碳期；碳气化期
carbon body 电刷
carbon boil 脱碳沸腾；碳沸腾期
carbon bolometer 碳质测辐射热计
carbon bonded refractory 碳结合耐火材料
carbon bottom 碳块炉底
carbon branch 碳分支
carbon breaker 碳极断路器；碳断路器
carbon break switch 碳触点开关
carbon brick 碳砖；碳质砖；碳质耐火砖；碳素耐火砖
carbon bronze 碳青铜
carbon brush 碳刷；炭刷
carbon brush for electric(al) machine 电机碳刷
carbon build-up 积炭
carbon burning 烧炭；碳燃烧
carbon button 碳精按钮
carbon capsule 炭精盒
carbon-carbon composite 碳—碳复合材料
carbon carburizing steel 渗碳(碳素)钢
carbon carrier gas 碳的载体气体
carbon cell 碳极电池；炭极电池
carbon cement 碳黏泥；碳素黏结剂；碳胶(脱水剂)；石墨胶合剂
carbon ceramics 碳陶瓷
carbon chain isomer 碳链异构
carbon chains 碳链
carbon chamber 炭精盒
carbon chloroform extract 碳氯仿萃取液
carbon chloroform extraction 碳氯仿萃取(法)
carbon clearance test 碳粒清除试验

carbon clip 炭精夹
carbon cloth 碳(纤维)布
carbon-coated 被碳沉积盖覆的
carbon coating 涂碳层；碳涂层
carbon coherer 炭屑检波器
carbon collector 集炭器
carbon colo(u)r test 比色定碳试验
carbon column 碳柱
carbon comparison meter 定碳比较仪
carbon component for mechanical engineering 碳素机械零件
carbon construction(al) quality steel 优质碳素结构钢
carbon construction(al) steel 建筑用碳钢；碳素结构钢
carbon construction(al) steel round 碳素结构圆钢
carbon consumption rate 碳耗率
carbon contact 碳触点；炭质接点
carbon contact pick-up 碳粒拾音器
carbon-containing alloy 含碳合金
carbon-containing alloy steel 含碳合金钢
carbon-containing compound 含碳化合物
carbon content 含碳量；碳含量
carbon content of steel 钢含碳量
carbon content of weld materials 焊接材料的碳含量
carbon copy 副本；复写本
carbon credit 炭额度
carbon crucible 石墨坩埚
carbon cycle 碳(素)循环
carbon dating 碳定年
carbon-deoxidized metal 碳脱氧的金属
carbon deposit 煤烟附着；碳沉积
carbon determinator 定碳仪
carbon diamond 黑金刚石
carbon diaphragm 碳膜；炭精振动膜片
carbon dichloride 二氯化碳
carbon dioxide 二氧化碳；碳酸气
carbon dioxide absorbent 二氧化碳吸收剂
carbon dioxide absorber 二氧化碳吸收器
carbon dioxide absorbing canister 二氧化碳吸收罐
carbon dioxide absorption cartridge 二氧化碳吸收筒
carbon dioxide absorption tube 二氧化碳吸收管
carbon dioxide acidosis 二氧化碳性酸中毒
carbon dioxide analyser 二氧化碳分析仪；二氧化碳分析器
carbon dioxide analysis in flue gas 废气二氧化碳分析
carbon dioxide arc welding 二氧化碳(气体)保护焊
carbon dioxide band 二氧化碳吸收带
carbon dioxide beam fiber drawing machine 二氧化碳激光束拉丝机
carbon dioxide bottle 二氧化碳筒
carbon dioxide capacity 二氧化碳容量
carbon dioxide carbon dioxide extinguishing system 二氧化碳灭火系统
carbon dioxide-combination curve 二氧化碳结合曲线
carbon dioxide combining power 二氧化碳结合力
carbon-dioxide complex 干冰生产全套设备
carbon dioxide concentration method 二氧化碳浓度法
carbon dioxide content 二氧化碳含量
carbon dioxide cycle 二氧化碳循环
carbon dioxide cylinder 二氧化碳瓶
carbon dioxide cylinder station 二氧化碳站
carbon dioxide density method 二氧化碳密度法
carbon dioxide dissociation 二氧化碳解离量
carbon dioxide dissociation curve 二氧化碳解离曲线
carbon dioxide effect 二氧化碳效应
carbon dioxide electrode 二氧化碳电极
carbon dioxide equivalent 二氧化碳当量
carbon dioxide exchange 二氧化碳交换
carbon dioxide exchange rate 二氧化碳交换率
carbon dioxide extinguisher 二氧化碳灭火器；二氧化碳灭火机
carbon dioxide fire extinguisher 二氧化碳灭火器
carbon dioxide fixation 固二氧化碳作用
carbon dioxide flushing 二氧化碳洗井
carbon dioxide free distilled water 无二氧化碳蒸馏水
carbon dioxide gas 含二氧化碳天然气；二氧化碳气

carbon dioxide gas incubator 二氧化碳气体培养箱
carbon dioxide gas miscible flooding 二氧化碳混合驱动
carbon dioxide gas shielded arc welding 二氧化碳气体保护电弧焊
carbon dioxide generator 二氧化碳发生器
carbon dioxide inclusion 二氧化碳包裹体
carbon dioxide indicator 二氧化碳检测器
carbon dioxide induced changes of climate 二氧化碳引起的气候变化
carbon-dioxide inflatable rubber boat 充气橡皮救生艇
carbon dioxide insufflation apparatus 二氧化碳吹入仪
carbon dioxide lamp 二氧化碳灯
carbon dioxide laser 二氧化碳激光器；二氧化碳激光
carbon dioxide laser auto-copy engraver 二氧化碳激光仿形雕刻机
carbon dioxide laser cutting and carving machine tool 二氧化碳激光切割及雕刻机床
carbon dioxide laser energy detector 二氧化碳激光能量探测器
carbon dioxide laser interferometer 二氧化碳激光干涉仪
carbon dioxide laser perforator 二氧化碳激光穿孔机
carbon dioxide laser processing machine tool 二氧化碳激光加工机床
carbon dioxide laser range finder 二氧化碳激光测距仪
carbon dioxide laser spectrum analyzer 二氧化碳激光谱线分析仪
carbon dioxide mo(u)lding 二氧化碳砂造型
carbon dioxide multi-jet 二氧化碳防火复式喷头
carbon dioxide-nitrogen zone 二氧化碳氮气带
carbon dioxide output 二氧化碳放出量
carbon dioxide piping 二氧化碳管路
carbon dioxide poisoning 二氧化碳中毒
carbon dioxide pollution 二氧化碳污染
carbon dioxide process 二氧化碳硬化砂法；二氧化碳法；二氧化碳处理法
carbon dioxide receptor 二氧化碳受体
carbon dioxide recorder 二氧化碳气体自动记录仪；二氧化碳记录仪；二氧化碳记录器
carbon dioxide reduction rate 二氧化碳还原率
carbon dioxide refrigerating machine 二氧化碳制冷机
carbon dioxide refrigerator 二氧化碳冷冻机
carbon dioxide released 二氧化碳释放
carbon dioxide removal system 二氧化碳净化系统
carbon dioxide retention 二氧化碳储留
carbon dioxide scrubber 二氧化碳洗涤器
carbon dioxide sensor 二氧化碳传感器
carbon dioxide silicate mo(u)ld 二氧化碳水玻璃砂型
carbon dioxide sodium silicate process 二氧化碳水玻璃硬化砂法
carbon dioxide sodium silicate sand 二氧化碳水玻璃砂
carbon dioxide solid refrigerator 二氧化碳固体制冷器
carbon dioxide specific volume method 二氧化碳比容法
carbon dioxide survey 二氧化碳法
carbon dioxide system 二氧化碳系
carbon dioxide tension 二氧化碳张力
carbon dioxide trap 二氧化碳冷阱
carbon dioxide water mixer 二氧化碳水混合器
carbon dioxidoice 固体二氧化碳(俗称干冰)
carbon discharge 碳排
carbon diselenide 二硒化碳
carbon disk electrode 碳盘电极
carbon disk microphone 碳(盘)传声器
carbon disulfide 二硫化碳
carbon disulfide poisoning 二硫化碳中毒
carbon disulphide 二硫化碳
carbon dust 碳尘；炭屑；尘
carbon dust resistance 碳粉电阻
carbon dust resistor 炭末电阻
carbon electric(al) contact 碳触点
carbon electrode 炭(素)电极；炭精电极；炭棒
carbon element for electrovacuum technique 电真空石墨元件

carbon elimination 去碳
carboneous 炭质
carbon epoxy composite 碳纤维增强环氧复合材料
carbon equivalent 碳当量
carbon equivalent meter 碳当量仪
carbon exchange rate 碳交换率
carbon family 碳族化合物
carbon-felt 炭毡
carbon fiber 碳纤维
carbon fiber ceramics 碳纤维陶瓷
carbon fiber composite 碳纤维复合材料
carbon fiber concrete 碳纤维混凝土
carbon fiber cone 碳纤维锥
carbon fiber packing 碳纤维盘根
carbon fiber reinforced cement 碳纤维增强水泥；石墨纤维增强水泥
carbon fiber reinforced composite 碳纤维增强复合材料
carbon fiber reinforced concrete 碳纤维增强混凝土
carbon fiber reinforced metal 碳纤维增强金属
carbon fiber reinforced plastics 碳纤维增强塑料
carbon fiber reinforced polymer 碳纤维增强聚合物
carbon fiber tape 碳纤维带
carbon filament 碳丝
carbon filament atomizer 碳丝原子化器
carbon filament atom reservoir 石墨灯丝原子储存器
carbon-filament lamp 碳丝灯
carbon filled hollow fibre 充碳中空纤维
carbon filler ring 碳素填料环
carbon film 炭膜
carbon-film potentiometer 碳膜电位器
carbon-film resistor 碳膜电阻器
carbon filter 炭滤池；活性炭滤池；活性炭过滤器；碳(过)滤器
carbon fin 散热片
carbon flash 碳闪
carbon flow controller 活性炭流量调节器
carbon-fluoride 碳氟化物
carbon foam 泡状碳
carbon for filter 过滤用碳
carbon-free 不含碳的
carbon-free non-machinable steel 不可加工的无碳钢
carbon-free stainless steel 无碳不锈钢
carbon-free steel 无碳钢
carbon freezing 用二氧化碳冷冻
carbon from dissolved carbonate 溶解碳酸盐碳
carbon fund 炭基金
carbon furnace 炭炉
carbon gas 碳化气
carbon gland 碳环压盖
carbon gradient 含碳梯度
carbon granular agglutination 碳粒凝集反应
carbon granule 碳粒；炭粒
carbon graphite fibre 碳石墨纤维
carbon group elements 碳族元素
carbon hearth bottom 碳质打结炉底
carbon heating element 碳素发热材料
carbon holder 炭刷握
carbon hydrogen nitrogen analyzer 碳氢氮元素分析仪
carbon-hydrogen-oxygen diagram 碳氢氧相图
carbon-hydrogen-oxygen ratio 碳氢氧比例
carbon hydrogen ratio 碳氢比
carbonic 二氧化碳的；碳的
carbonic acid 碳酸
carbonic acid attack 碳酸侵蚀
carbonic acide method 碳酸法
carbonic acid erosion index 碳酸侵蚀指标
carbonic acid gas 碳酸气(体)
carbonic acid in air 大气中的碳酸
carbonic acid inclusion 碳酸包裹体
carbonic anhydride 碳(酸)酐
carbonic ester 碳酸酯
Carbonic limestone 石炭纪灰岩【地】；碳质石灰岩
carbonic oxide 一氧化碳
carbonic oxide cell 一氧化碳电池
Carbonic period 石炭纪
carbonic rock 煤层；含碳岩石
carbonic sandstone 碳质砂岩
carbonic snow 干冰
carbonide 碳化物

carbonide spring 碳酸泉
carboniferous 含碳的
carboniferous ground 含碳的岩层
Carboniferous limestone 石炭纪灰岩
carboniferous material 含碳物质
Carboniferous period 石炭纪
carboniferous-Permian glacial stage 石炭二叠纪冰期
carboniferous sandstone 碳质砂岩
Carboniferous system 石炭系；泥炭系
carbonific 生碳化合物
carbonification 煤化(作用)；碳化作用
carbonify 碳化
carbon impregnated vycor glass 掺碳高硼硅酸耐热玻璃
carbon ink 炭墨；炭黑
carbon in materials for welding 焊接用材料中的碳
carbon ion 负碳离子
carbon iron 碳素铁
carbon iron balance 铁碳平衡
carbon isotope 碳同位素
carbon isotope geothermometer 碳同位素地质温度计
carbon isotope ratio 碳同位素比
carbon isotopic geothermometer 碳同位素地热温标
carbonite 锯屑炸药；硝酸甘油；不溶沥青；天然焦；碳质炸药
carbonitride 碳氮化物；碳氮化合物
carbonitride coating 碳氮化物涂层
carbonitrided steel 碳氮化钢；碳氮共渗钢；氰化钢
carbonitride segregation 碳氮化合物夹杂
carbonitriding 氰化；碳氮共渗
carbonium 带正电的有机离子
carbonium ion polymerization 碳离子聚合(作用)
carbonization 焦化作用；焦化；增碳；干馏；碳化作用；碳化过程；碳化；炭化；渗碳处理
carbonization at low temperature 半焦化(作用)
carbonization chamber 碳化室
carbonization effluent 焦化流出水；碳化污水；碳化出水
carbonization gas 干馏煤气
carbonization index of oil 油的碳化指数
carbonization(method) 碳化法
carbonization of wood 木材炭化
carbonization plant 碳化装置；碳化厂
carbonization sludge adsorbent 碳化污泥吸收剂
carbonization test in alumin(i)um retort 铝甑干馏试验
carbonization zone 干馏区
carbonize 碳化
carbonized bones 碳化骨；骨炭
carbonized brick 渗碳耐火砖
carbonized cathode 碳化(物)阴极
carbonized cellulose 碳化纤维素
carbonized clay 渗碳黏土
carbonized cloth 碳化布
carbonized cork 碳化软木
carbonized depth 碳化深度
carbonized felt 碳纤维毡
carbonized fibre 碳化纤维
carbonized fuel 碳化燃料
carbonized gypsum slab 碳化石膏板
carbonized peat 木炭化泥炭
carbonized pottery 黑陶
carbonized sand 碳化砂
carbonized test 碳化试验
carbonized thoriated tungsten cathode 碳化钍钨阴极
carbonizer 碳化塔；碳化器
carbonizing 碳素印刷；碳化
carbonizing chamber 碳化窑
carbonizing flame 还原焰；碳化焰
carbonizing of gas 煤气渗碳
carbonizing period 增碳期；碳化周期
carbonizing plant 碳化装置
carbonizing programme 推焦程序
carbonizing steel 渗碳钢
carbon knock 积炭敲击声；积炭爆震
carbon lamp 碳弧光灯；碳棒灯；碳丝灯
carbon lattice 碳素晶格
carbon laydown 碳沉积；炭沉积
carbon line 碳线；炭砖炉衬

carbon-lined mold 衬碳模
carbon lining 碳素涂底;碳质内衬
carbon log 碳测井
carbon log curve 碳测井曲线
carbon loss 碳损
carbon magnetic resonance 碳核磁共振
carbon mark 碳迹(玻璃制品缺陷)
carbon material 碳材料
carbon matrix 碳基
carbon membrane-aerated biofilm reactor 炭膜曝气生物膜反应器
carbon mesophase microbead 碳质中间相小球体
carbon metal bond 碳金属键
carbon modifier 碳修饰剂
carbon molecular sieve 碳分子筛
carbon molybdenum steel 钼碳钢
carbon monoxide 煤气;一氧化碳控制器;一氧化碳
carbon monoxide analyser 一氧化碳分析仪
carbon monoxide canister 一氧化碳消毒罐
carbon monoxide controller 一氧化碳控制器
carbon monoxide converter 一氧化碳转化器
carbon monoxide dectector 一氧化碳检测器
carbon monoxide diffusion capacity 一氧化碳弥散量
carbon monoxide disintegration 一氧化碳分解
carbon monoxide emission rate 一氧化碳排放率
carbon monoxide filter 一氧化碳过滤器
carbon monoxide index 一氧化碳指数
carbon monoxide meter 一氧化碳监测仪
carbon monoxide monitor 一氧化碳监测仪
carbon monoxide poisoning 煤气中毒;一氧化碳中毒;氧化碳中毒
carbon monoxide pollution 一氧化碳污染
carbon monoxide recorder 一氧化碳记录仪;一氧化碳记录器
carbon monoxide resistance 抗一氧化碳
carbon monoxide self-rescuer 一氧化碳自救器
carbon mo(u)ld 碳质模型
carbon nitrogen bond 碳氮键
carbon-nitrogen cycle 碳氮循环
carbon-nitrogen index 碳氮指数
carbon-nitrogen-oxygen group 碳氮氧基
carbon-nitrogen ratio 碳氮磷比;碳—氮比(率)
carbon noise 碳质电阻噪声;碳粒噪声;炭质电阻噪声
carbon non-activated 非活性炭
carbon nozzle insert 石墨喷管衬套
carbon nutrition 碳素营养
carbon of expulsive oil and gas 排出油气中碳
carbon of kerogen in source rock 母岩中干酪根碳
carbon of residual bitumen in source rock 残留母岩中的沥青碳
carbon of residual hydrocarbon in source rock 残留母岩中的烃碳
carbonolite 碳质岩
carbonometer 二氧化碳计;碳酸计
carbonometry 二氧化碳定量法;碳酸定量法
carbonous 碳的
carbon-oxygen isotope correlation 碳氧同位素相关性
carbon/oxygen log 碳氧比测井
carbon/oxygen log curve 碳氧比测井曲线
carbon oxygen sulphide 氧硫化碳
carbon oxysulfide 硫化碳
carbon-packed gland 碳精填密涵盖
carbon packing 碳质填料;碳素填料;碳素垫料;碳素衬垫
carbon paint 碳质涂料
carbon paper 复写纸;碳素印相纸
carbon paper ink 复写纸用墨
carbon paper type recorder 碳素纸式记录器
carbon particle 碳末颗粒;碳粒子
carbon paste 碳胶(脱水剂);碳糊;碳膏
carbon period 石炭纪
carbon pick-up 碳化;渗碳;吸收碳;增碳
carbon piece 碳棒
carbon pigment printing 碳素印相法
carbon pile 稳压用碳柱;碳堆
carbon pile pressure transducer 碳堆压力传感器
carbon pile regulator 碳堆稳压器;碳堆稳压管
carbon pile voltage transformer 碳堆变压器
carbon piston ring 炭质活塞环
carbon plate 炭片
carbon-plates chamber 碳板火花室

carbon plate shelf 碳板架
carbon-point curve 碳线
carbon-point rod 碳极棒
carbon pole 碳素电极
carbon-polytetrafluoroethylene O_2-fed cathode 碳聚乙烯充氧阴极
carbon potential 碳势
carbon potential meter 碳势计
carbon powder 碳精粉;炭末
carbon preference index 碳优势指数
carbon pressure recording 碳压记录法
carbon product 碳素制品
carbon raiser 增碳剂
carbon ratio 定碳比;碳比
carbon ratio theory 碳比理论
carbon reducing method 碳还原法
carbon reduction 碳还原
carbon refractory 碳质耐火材料;碳素耐火材料;炭质耐火材料
carbon refractory block 碳素耐火砖
carbon remover 除积炭器;碳尘消除剂
carbon removerand cleaner 除积炭清洗剂
carbon replica 碳复型;炭复制器
carbon replica method 碳复型法
carbon replication 碳模制
carbon residue 残碳值;残碳
carbon residue test 碳渣(值)试验
carbon resistance 碳电阻
carbon resistance film 碳膜电阻
carbon resistance furnace 小碳粒电阻炉;碳阻电炉
carbon resistance rod 碳电极
carbon resistance thermometer 碳电阻温度计;炭阻电温度计
carbon resistor 碳质电阻器;碳电阻
carbon resistor block 碳素电阻体
carbon resistor bolometer 碳阻测辐射热计
carbon resistor furnace 石墨电阻炉
carbon resistor rod 碳阻棒;碳电极
carbon restoration 复碳法
carbon rheostat 碳质变阻器;碳变阻器;炭质变阻器
carbon ribbon apply indicator 色带使用程度指示器
carbon rich waste 富碳有机废物
carbon ring 碳圈密封;碳精环;炭环
carbon ring gland 炭精气封圈;炭环压盖
carbon rivet steel 铆钉碳钢
carbon rock 碳质岩
carbon rod 炭(精)棒;炭棒
carbon rod atomizer 碳棒原子化器
carbon roller 炭质滚轮
carbonrundum 碳化硅;人造刚玉;金刚砂
carbonrundum brick 碳化硅砖;金刚砂砖
carbonrundum cloth or paper 金刚砂布或纸
carbonrundum disc 金刚砂圆锯
carbonrundum dust 金刚砂粉
carbons 碳素材料
carbon sand 碳素砂
carbon-sand filter 炭砂滤池
carbon scraper 积炭刮除器;刮碳器;刮煤机
carbon scraping 残碳
carbon screen 炭屏
carbon seal 石墨密封环
carbon seal ring 石墨密封圈
carbon sequence 碳序
carbon-set bit 黑金刚石钻头
carbon settling chamber 碳粒沉积室
carbon shale 碳质页岩
carbon shirt-circuiting furnace 碳管炉
carbon shoe 碳刷;石墨密封条
carbon silicide brick 碳化硅砖
carbon silicide mortar 碳化硅灰浆
carbon silicide refractory material 碳化硅耐火材料
carbon silicide slab 碳化硅板
carbon slab 炭板
carbon sorbent 炭吸附剂
carbon source 碳源
carbon spot (金刚石晶体中的)石墨杂质
carbon steel 碳(素)钢
carbon steel bit 碳素钢钻头
carbon steel covered electrode 碳钢焊条
carbon steel curtain wall 碳钢幕墙
carbon steel pipe 碳钢管
carbon steel plate 碳素钢板(和型材)
carbon steel rail 碳素钢轨
carbon steels for machine structural purposes 机械结构用碳素钢

carbon steel tubing 碳素钢管
carbon stick 炭精棒
carbon-stick microphone 炭棒传声器
carbon strip 碳滑条
carbon structural quality steel 优质碳素结构钢
carbon structural steel 结构用碳钢;碳素结构钢
carbon suboxide 二氧化三碳
carbon subsulfide 二硫化三碳
carbon switch contact 炭质开关接点
carbon test 测定含碳量
carbon tetrachloride 四氯化碳
carbon tetrachloride activity 四氯化碳活度
carbon tetrachloride fire extinguisher 四氯化碳灭火机
carbon tetrachloride fuse 四氯化碳熔丝;四氯化碳熔断器
carbon tetrachloride poisoning 四氯化碳中毒
carbon theory 定碳比理论
carbon thermal reduction 碳热还原
carbon thrust ring 石墨推承环
carbon tissue 碳素印相纸;碳素相纸;碳素(相)纸
carbon tissue transfer machine 碳素纸过版机
carbon-to-carbon linkage 碳—碳键合
carbon to nitrogen ratio 碳—氮比(率)
carbon tool steel 碳(素)工具钢
carbon tracer 示踪碳
carbon transducer 碳粒传感器;碳粉传感器;炭精式换能器
carbon transfer recording 碳粒转移记录
carbon transformation 碳转化率
carbon trichloride 六氯乙烷
carbon tube furnace 碳管炉
carbon tube membrane-aerated biofilm reactor 炭管膜曝气生物反应器
carbon-type seal 炭密封
carbon value 碳值
carbon-white 合成轻质二氧化硅;白炭黑
carbonylation 羰基化作用;羰基化
carbonyl chloride 光气;碳酰氯
carbonyl compact 羰基法粉末坯块
carbonyl compound 羰基化合物
carbonyl core 羰基铁芯
carbonyl dithiocarbonic acid 二硫代碳酸
carbonyl fluoride 碳酰氟
carbonyl group 羰基
carbonyl iron 高纯度铁;羰基铁
carbonyl iron dust core 羰基铁粉磁芯
carbonyl iron powder 羰基铁粉
carbonyl platinous chloride 羰基亚铂氯
carbonyl powder 羰基粉末
carbonyl sulfide 硫化碳酰;氧硫化碳;碳酰硫
carbonyl value 羰基值
carbon zinc 碳化锌
carbon-zinc battery 碳锌电池
carbophenothion 三硫磷
carbopol 聚羧乙烯
carbopolyminerite 多矿物质煤
carbopyrite 黄铁矿质煤
carboradiant kiln 金刚砂电炉
carborane 碳硼烷
carborane methyl silicone 碳硼烷甲基硅酮
car-borne 汽车转运
car-borne equipment 车载设备
car-borne radioactivity survey 自动放射性测量
car-borne unit 车载单元
carborundum 金刚砂;人造金刚砂;碳化硅;碳硅砂
carborundum brick 碳化硅砖
carborundum circular saw 金刚砂圆锯
carborundum cloth 金刚砂布
carborundum cutting wheel 刻花砂轮;金刚石切割砂轮
carborundum detector 金刚砂检波器;碳化硅检波器
carborundum disc 金刚砂片;金刚砂盘
carborundum fillet 金刚砂带
carborundum grinding wheel 金刚砂磨轮;金刚砂轮;砂轮
carborundum grit slide-proof strip 金刚砂防滑条
carborundum paper 金刚砂纸;碳化硅砂纸
carborundum paste 金刚砂研膏
carborundum powder 碳化硅粉
carborundum refractory 碳化硅耐火材料
carborundum saw 金刚砂锯;碳化硅砂锯;砂锯

carborundum tile 金刚砂砖
carborundum tube 碳化硅管
carborundum wheel 金刚砂轮
carbosand 碳化砂
carboscope 光学测烟仪
Carboseal 卡波夕耳(收集灰尘用润滑剂);集尘用润滑剂
carbosilicite 硅质煤
carbothermic method 碳热还原法
carbothermic process 碳热还原法
carbothermic smelting of aluminum 碳热法炼铝
car-bottom batch-type furnace 车底式批生产加热炉
car bottom electric(al) furnace 车底式电阻炉
car-bottom furnace 活底炉子;车底式加热炉
car-bottom hearth furnace 台车式炉
car-bottom kiln 台车窑
carbowax 碳蜡;水溶性有机润滑剂
carboxamide 羧基酰胺
carboxyhemoglobin 碳氧血红蛋白
carboxylate 羧酸酯;羧酸盐;羧化物
carboxylated styrene butadiene rubber 羧基化丁苯橡胶
carboxylation 羧化(作用)
carboxylic acid 羧酸
carboxylic acid polymer 羧酸聚合物
carboxylic acid resin 羧酸树脂
carboxylic resin 羧基树脂
carboxyl sulfide 硫化碳
carboxy methyl cellulose 羧甲基纤维素(泥浆添加剂)
carboxymethyl ether 羧甲基醚;烃氧基乙酸
carboxy methyl hydroxyethyl cellulose 羧甲基羟乙基纤维素(泥浆添加剂)
carboxy reactivity 煤对二氧化碳反应性
carboxyvinyl polymer 聚羧乙烯
carboy 护套大玻璃瓶;小口大玻璃瓶;用柳条或竹笼装的细口瓶;钢瓶;气筒;坛(子);酸坛
car buffer 梯厢减振器
car builder 车辆制造厂
car bumper 车挡
carbuncle 红榴玉;红榴石红色;红榴石;红宝石
carbunk(运木材往炉内烘干用的)手推车
carburant 增碳剂;渗碳剂
carburated case depth 渗碳硬化层深度
carburated gas 干馏煤气
carburated water gas 增碳水煤气
carburation 渗碳(作用)
carburator 渗碳器;汽化器
carburet 增碳;碳化物
carburetant 增碳剂;碳化剂
carburet(t)ed air 增炭空气;汽化空气;渗汽空气
carburet(t)ed engine 汽油机
carburet(t)ed iron 碳化铁
carburet(t)ed spring 碳酸泉
carburet(t)ed water gas installation 水煤气发生装置
carburet(t)ed water gas scrubbing wastewater 加热水煤气洗涤废水
carburet(t)ed water gas tar 与水煤气碳化合的焦油
carburet(t)er 化油器;增碳器
carburet(t)er jet 汽化器喷口
carburet(t)er overhaul shop 汽车修理工场
carburet(t)ing pilot 空气煤气混合式引燃器;气化式引燃器
carburet(t)ion 增碳作用
carburet(t)or 化油器;增碳器;内燃机化油器;汽化器(内燃机)
carburet(t)or air lever 汽化器进气阀杆
carburet(t)or air scoop 汽化器进气口
carburet(t)or anti-icer 汽化器的防冰器
carburet(t)or backfire arrester 汽化器回火制止器
carburet(t)or choke 汽化器阻风门
carburet(t)or engine 汽油机
carburet(t)or flow bench 汽化器试验台
carburet(t)or jet 汽化器喷嘴
carburet(t)or main jet 汽化器主射口
carburet(t)or manifold 汽化器歧管
carburet(t)or mixing chamber 汽化器混合器
carburet(t)or muff 汽化器扩散管
carburet(t)or primer 汲油按钮;汽化器起动注油器
carburet(t)or strainer 汽化器滤网
carburet(t)or throttling plate 汽化器节流板

carburet(t)or tickler 汽化器打油泵
carburet(t)or engine 柴油机化油器
carburising 硬化处理;渗透处理;表面渗碳
carburization 渗碳作用;渗碳
carburization of gas 煤气渗碳
carburization zone 渗碳层
carburize 碳化;渗碳
carburized case 渗碳表面层
carburized(case) depth 渗碳层深度
carburized layer 渗碳层
carburized metal 渗碳金属
carburized structure 渗碳组织
carburizer 碳化剂;渗碳剂
carburizing 硬化处理;渗透处理;表面渗碳
carburizing apparatus 碳化装置
carburizing atmosphere 渗碳气氛
carburizing bath 加碳液池
carburizing by solid matters 固体渗碳
carburizing cycle 渗碳期
carburizing depth 渗碳深度
carburizing flame 碳化火焰;渗碳火焰
carburizing furnace 渗碳炉
carburizing period 强渗期
carburizing process 渗碳过程
carburizing reagent 增碳剂;碳化反应
carburizing steel 碳钢;渗碳钢
carburizing temperature 碳化温度;渗碳温度
carbusintering 渗碳烧结
carbutamide 氨磺丁脲
carbylamine 异氰化物
car cable 车辆用电缆
car cableway 架空索道斗车
car-carrier 载车船;汽车运载船;装载汽车的货车;汽车装运车
car-carrier train 装运汽车的列车
car-carrier wagon 汽车专用货车
carcase 轮胎胎壳;框架;壳体;基本结构;骨架;车架;房屋骨架
carcase roof(ing) 屋顶骨架;毛屋顶
carcase saw 大鸠尾锯
carcase work 预埋管线工程;预埋工程(管子或电线等);骨架工程
carcasing 骨架制作
carcasing timber 构架材
carcass 加强层;基本结构;主体骨架;骨架;构架木料
carcass flooring 毛地板;楼层骨架
carcassing 预埋管线工程;运架设;骨架制作
carcassing lumber 结构木材
carcassing timber 结构木材;木构架构件
carcassing work 主体工程
carcass roofing 毛屋顶;屋顶骨架
carcass-saw 大鸠尾锯
car casting 车铸法
Carcel 卡索,灯光度单位(1卡索=6.9英国烛光单位)
carcenet 卡塞内特风
carcer 圆剧场兽室;赛车起点栏;监狱
carchedonius 石榴石
carchedony 石榴石
carcinogen(e) 致癌物
carcinogen health risk 致癌健康风险
carcinogenic compound 致癌化合物;致癌物质
carcinogenic contaminant 致癌污染物
carcinogenic environment 致癌环境
carcinogenic industrial chemicals 致癌工业化学品
carcinogenicity 致癌性
carcinogenicity of asbestos 石棉致癌性
carcinogenic pollutant 致癌污染物
carcinogensis 致癌作用
carcinotron 回波管
carclazyte 高岭土;白土;全白土;陶土
carclazyte catalyst 全白土催化剂
car cleaner 车辆洗净剂;车辆清洗剂
car cleaning plant 洗车库
car cleaning track 货车洗刷线
car container 汽车集装箱
car conveyer 小车式输送机;车台式输送机
car counterweight 吊车缆车平衡重;吊车缆车平衡锤
car coupler 车钩
car cushioning 车辆减震
card 罗盘面;罗盘卡;卡板;卡片;程序单;插件板;插件;纹板;梳刷
card access 卡片存取

card accounting 卡片式会计
card address 插件地址
cardan 平衡环;平浮环
cardan axis 卡登轴;万向节轴;万向关节轴
cardan axle 万向轴
cardan bracket 万向关节托架
cardan drive 万向接头传动
cardan fulcrum 万向接矢支轴
cardan gear 万向接头传动装置;万向接头传动
cardanic mirror 万向反射镜
cardanic suspension 万向悬挂架
cardan joint 万向联轴器;万向联轴节;万向节;万向接头;卡登节
cardan link 万向关节
Cardan motion 卡登运动
Cardan mount 卡登架
cardan mounting 万向悬挂架
cardanol 卡丹酚;腰果壳酚
cardan ring 万向环
cardan shaft (汽车的)中轴;万向轴;万向节轴;推进轴
Cardan's suspension 卡丹式悬架
card back 卡片背面
card base 卡片库
card base system 卡片库方式
card bed 卡片座
card bin 卡片箱
cardbisphenol 卡双酚
cardboard 卡纸板;厚纸板;硬纸板;薄纸板;卡片
cardboard bolt sleeve 纸板螺栓套
cardboard box 纸板箱
cardboard bushing 硬纸套管
card-board computer 插件板式计算机
cardboard drain 纸板排水;排水纸板
cardboard drain method 排水板法
card-board flange 纸板法兰垫
cardboard mo(u)ld 纸板模型
cardboard roof 纸板屋顶
cardboard sleeve 硬纸套筒
cardboard space former 填充纸板
cardboard tube 硬纸管;纸站
cardboard wall 纸板护墙
cardboard wax 纸板石蜡
cardboard wick 硬纸板灯芯
card box 卡片盒;卡片柜
card bus 插件总线
card cabinet 卡片柜
card cage 插件箱;插件框架;插件机架;插件盒
card capacity 卡片容量
cardcase 卡片盒;名片盒
card catalogue 卡片式目录;卡片目录;目录卡
card catalogue cabinet 卡片目录柜
card catalogue hall 卡片目录室
card chase 小型版框
card chassis 插件底板
card checking 卡片检验
card circuit 直插式电路
card clothing 钢丝布;梳理机针布
card code 卡片码;卡片代码;卡片编码
card collector 卡片整理机;卡片校对机
card column 卡片上的一列孔;卡片列孔;卡片列
card compass 平板罗盘
card connector 插件插头座
card control 看板管理
card counter 卡片计数器
card cycle 卡片周期
card data 卡片数据
card data converter 卡片数据转换器
card deck 卡片组;卡片叠;一组卡片;一叠卡片
card deck input 卡片组输入
card design 插件设计
card dialer 卡片拨号机
car deck 装车架;车辆甲板
car deck pontoon 车辆甲板箱体
carded cotton waste 粗棉废棉
card edge connecter 直接连接器;卡片边沿连接器
card edge type 直接插入式
card edge type connector 插件边缘型插头座
card editor 卡片编辑程序
card ejector 插件拔出器
car depot 车辆段
carder 刷毛机
car derailment 车辆脱轨
car detained for repair 扣车【铁】

car detention time 车辆途中延误时间;车辆停留时间
car detention time under accumulation 集结时间
card face 卡片正面;卡片使用面
card feed 卡片馈送装置;卡片馈送部件;卡片传送(机构);卡片传递;送卡
card feed delay 卡片馈送延迟
card feed device 卡片馈送机构
card feeder 穿孔卡输入机
card field 卡片字段;卡片信息组;卡片栏;穿卡区
card file 卡片储[贮]存器;卡片文件;卡片目录;卡片柜;卡片存储器;目录卡
card filing cabinet 卡片文件箱;卡片档案箱
card fillet 钢丝针布
card form 卡片形式
card format 卡片格式
card(-form) ledger 卡片式分类账
card frame 名片框(门上)
card frame cage 插件机架
card ga(u)ge 卡片量规;卡片测量器
cardguide 插件导轨
card-holder 卡片支架;卡片夹;插名片的框(门上)
card hole 卡片孔
card hopper 卡片传送斗;储卡箱;储卡机;送卡箱;输入卡箱
card-house fabric 片架(状)组构
card-house structure 片架(状)结构;排架结构;纸房状构造
cardilate 赤藓醇四硝酸酯
cardiloid 赤藓醇四硝酸酯
card image 卡片映象;卡片信息;卡片图像;穿孔卡片(信息)图(式)
card image mode 卡片图式
car dimensions 车厢尺寸;梯厢尺寸
cardinal 主要的;蝶铰
cardinal altitude 主要飞行高度
cardinal direction marker 基向指示器
cardinal distance 基距
cardinal effect 基本效应
cardinal heading 正方位艏向
cardinality of a fuzzy set 模糊集的基数
cardinality of a set 集合的基数
cardinality of fuzzy sets 模糊集基数
cardinal ligament 主韧带
cardinal line 主线
cardinal margin 主缘
cardinal mark 方位标志;方位标(记)
cardinal mark control 方位标志控制
cardinal marking system 方位制航标系统
cardinal marks of buoyage system A 浮标系统A方位标志
cardinal number 基数;纯数
cardinal numerals 基数词
cardinal plane 主平面
cardinal point 罗盘基点;基点方位;基点;主点;方位基点
cardinal point effect 方位基点效应
cardinal points of ecliptic 黄道四基点(春分、夏至、秋分、冬至)
cardinal power 基数幂
cardinal principle 基本原理;根本原则;根本原理
cardinal product 基数积
cardinal red 深红色的;深红色
cardinal sign 黄道带主宫
cardinal spline 基数样条
cardinal stimuli 主刺激
cardinal sum 基数和
cardinal symptom 主要症状
cardinal system 基本方位浮标设置系统;方位制系统
cardinal system of buoyage 方位浮标制
cardinal utility 计数效用;基数效用
cardinal utility function 基数效用函数
cardinal versus ordinal utility 基数作为序数的效用;基数对序数效用
cardinal wind 主导风向;主要风向
card index 卡片(式)索引;卡片目录;目录卡;档案盒
carding 梳理
carding bottom 梳莱凹板
carding cylinder 梳摘滚筒
carding engine 刷毛机
carding machine 梳理机;梳粉机
carding tool 梳刀
carding willow 开松机;扯麻机
card input 卡片输入

card input editor 卡片输入编辑机
card input machine 卡片输入机
card input magazine 卡片输入箱
card insertion and extraction tool 插件插拔工具
card interpreter 卡片解释程序
cardioid 心脏形(曲)线;心形曲线
cardioid microphone 单向传声器
cardioid pattern 心形图
cardiotonic tablet 定心丹
card jam 卡片阻塞;卡片堵塞
card lacing 纸板串连带
card lacing machine 纸板串联机
card layout 卡片格式设计
card leading edge 卡片前沿
card level module 插件级模件
card loader 卡片装入程序;卡片引导程序
card loan 卡片贷款
card machine 卡片机器
card matching 卡片对比;卡片的核对
card middle 衬纸
card mode 卡片状态
card model 卡片模型
card module 插件模件
card nipper 纹板冲孔机
card No. of the user 订货部门代号
cardo 轴节
card of account 分类账目录;分类账目簿
card of ledger 卡片式总账;卡片式分类账
card of patterns 装有几个模型的型板
card of work order 施工说明卡片
cardol 腰果(间)二酚
car door 车门
car door contact 门碰联动开关
car-door electric(al) contact 笼门触点
car door mechanism testing stand 门机试验台
car door operating handle 车门把手
car door power closer 车门的电动开关
Cardox method 卡多思(火烧)方法
card pack 卡片堆
card path 卡片通道;卡片导轨
cardphone 磁卡电话
card pile-drawing machine 钢针拉毛机
card plate (门上的)名片框
card pocket 卡片袋;名片袋
card power supply 插件电源
card printer 名片机
Card process 卡氏蒸炼木材法;氧化锌—杂酚油防腐法
card processing machine 卡片处理机
card processor 卡片处理机
card programing 穿孔卡片程序设计
card program(me) 卡片程序
card-program(me) calculator 卡片程序计算器
card program(me) computer 卡片程序式计算机
card programmed calculation 卡片程序计算
card-programmed calculator 卡片计算机;卡片程序计算器
card programmed computer 穿孔卡片程序(控制)计算机
card programming 卡片程序设计
card programming control 卡片程序控制
card proof punch 卡片验证机
card puller 插件板插拔器;插件拔插器;拔插件把手
card punch buffer 卡片穿孔缓冲器
card punch code 卡片穿孔码
card punch control unit 卡片穿孔控制装置
card punch(er) 卡片穿孔(机);纹板冲孔机;穿卡机
card punching 卡片穿孔
card punching machine 卡片穿孔机
card punch reproduce check unit 卡片穿孔校对机
card rack 插件板导轨
card raising machine 钢丝起绒机
card reader 卡片阅读器;卡片阅读机;卡片输入机;卡片读出机;读卡器;读卡机;穿孔卡片阅读器
card reader control unit 读卡机控制装置
card reader punch 卡片输入穿孔机
card read punch 卡片阅读穿孔机;读卡穿孔机
card read-punch unit 卡片读出—穿孔机
card receiver 卡片接收器;接卡箱
card reconditioner 卡片修整机;卡片调整机
card recorder 卡片记录器
card register 卡片寄存器
card reproducer 卡片复制机;穿孔卡片复制机
card reproducing puncher 卡片复制穿孔机

car driver (小型)汽车驾驶员
card room 打牌室;娱乐室(打纸牌用)
card row 卡片行孔;卡片行;卡片穿孔段
card row punch 卡片行穿孔机
card run 卡片运行
cards 连续卡片
card saver 卡片修改纸带
card sensor 卡片读出器
card sorter 卡片分类机
card sorting 卡片分类
card sorting machine 卡片分类机
cards per minute 每分钟卡片数
card stacker 叠卡器;叠卡机
card-stacking method 卡片分类法
card stock 卡片材料;穿孔卡片原料
card storage 磁卡片存储器
card stripper 抄针装置
card stripping 抄针
card stripping roller 抄钢丝辊
card system 看板制度;卡片制;卡片式账簿(制);卡片式记账法;信用卡记账法
card system cabinet 卡片保存器
card system of accounting 卡片式会计
card table 折叠桌;牌桌
card tailing edge 卡片后沿
card track 卡片轨道;卡片道;卡片导轨;卡片传送导轨
card transcriber 卡片转录器;卡片读数器;卡片抄录器
card translator 卡片译码器
card type indicator 卡片式指示器;图表式指示器
car dump 卸货车;车库;翻车卸载装置;停车场
car dumper 货车倾卸机;自动倾卸车;翻车机;倾卸货车;倾倒卸货车;倾卸式开底车;汽车倾卸机
car dumping crane 提升式(汽车)翻车机
car dumping facility 翻车设备
card unit speed 卡片机速度
card verifier 卡片校核机;验卡机
card weight 卡片压重
card wire 针布钢丝
car dynamometer 测力小车;车式测功仪
care 小心
care, custody and control insurance 管理、监护、支配等意外保险
care and maintenance 保养与维修
care and maintenance regulations 保管条例
care and regeneration of forest 森林抚育更新
careass flooring 楼层骨架
careenage 修船处
careen grid 船架
careening 使船倾斜
careening basin 修船港池
careening grid 格子板翻船架;格子船台;格框船台;铁格子
careening wharf 倾斜修船码头
careen site 倾船场
careen tackle 防倾绞辘(船在干坞用)
career apparel 职员工作服
career counseling 职业前程辅导
career customer 关键顾客
career development training 职业发展训练
career employee 常期雇员
career-long employment 终身雇佣制
career management 职业前程发展管理
career-oriented women 职业妇女
career plan 企业管理制度
career planning 职业规划
career prospect 职业前途
career service 长期服务
career training 就业训练;专职训练;专业训练;职业训练
career woman 职业妇女
careful distillation 精馏
careful investigation 精密调查
careful reading 精读
careful treatment 精心处理
care home 康复(医)院
care in applications 小心使用
care label 使用须知标签
careless 毛糙的;草率;不注意的;不用心的
careless application of irrigation water 粗放灌水
car elevator 轿厢式升降机;装车输送机;汽车(用)电梯
care mark 小心标志;注重标志;注意标志;注意标记

car enclosure 轿厢顶盖和四壁
car end number 1 车厢的一位端
car end number 2 车厢的二位端
carene 龙骨状突起
car entry speed 车辆进入速度
care of 转交
care of field 田间管理
care of grain binder 割捆机的维护
care of instrument 维护工具
care of works 工程的照管
caret 插入记号
caretaker 看守工人；看门人；清洁人员；看守者；看守人；看管者；看管人
caretaker's flat 管理员住房
caretakers' room 值班员室；管理员室
caretaker ticket rate 押运人票价
caretonite 碳硅碱钙石
carex marsh 莎草沼泽
carex peat 莎草泥炭
carex swamp 苔草沼泽
Carey-Foster bridge 交流电桥
car fare 交通费
carfax (四条以上马路的)交叉路口
car ferry boat 汽车渡轮
car ferry(crossing) 汽车渡口；火车轮渡；火车渡轮；火车渡口；车辆轮渡；车辆渡船；车渡；汽车轮渡；汽车渡船
car ferry hovercraft 载车气垫轮渡
car ferrying 汽车轮渡；汽车渡口；火车轮渡；火车渡口
car ferry-steamer 汽车轮渡
car-ferry terminal 汽车渡站
car filler 装车
car finish 车身抛光
car float 火车渡船；车辆轮渡船；车辆渡船；非自航式火车轮渡；汽车浮box
car-floor contact 电梯底板接触器
car flow 车流
car flow routing 车流经路
car-following theory 跟车理论；车辆跟随理论
car for traffic use 运用车
car frame 轿车骨架；轿车构架；车架
car-free mall 步行商场；步行区
car furnace 台车式加热炉
car gasoline 车用汽油
car gate 轿厢闸门(升降机)
car gate power closer 轿厢闸门开门器；大门电动开关
car ga(u)ge 车辆轮距
car-ga(u)ge clearance 车辆界限
cargo 货物；船装货
cargo accommodation 货运设备；装货设备
cargo accounting advice 货物账款通知单
cargo afloat 业经装船之货；水运中的货物
cargo airliner 货运班机
cargo air tariff 货物空运费率表
cargo allocation plan 货物配载计划图
cargo alternative 替换货物
cargo and train ferry 货物和列车轮渡
cargo arrival operation income 货物到达作业收入
cargo attendant 货物押运员
cargo barge 货驳；装货驳船
cargo batten 货舱壁护条；舱壁护条
cargo bay 货位；装货单位；装货车位
cargo beam 杆式吊架；吊梁
cargo berth 货船停泊位
cargo billing system 货物单证制
cargo block 货栈；装卸货滑车；吊货滑车
cargo boat 货船
cargo boat note 卸货记录
cargo body 货舱
cargo book 货物清单；货物登记册；装货记录簿
cargo boom 吊(货)杆；船上吊杆
cargo-box barge 平底船
cargo bridle 四钩吊货索
cargo building 货物建筑；仓库
cargo cage 货笼
cargo capacity 货物容量；货舱容积；装货容量；装货能力；载重能力；载货容量；载货容积；船货容量；车辆载重(量)
cargo carried together with bill 货单同行；单货同行
cargo carrier 货运工具；货船；运输机具；运货工具
cargo carrying ability 载货能力

cargo carrying barge 货驳
cargo carrying capacity 载货量；额定装货量；货运量；装货能力；载货吨位
cargo certificate 货单
cargo chain 链扣；链斗钩；吊货短链
cargo characteristics 货物特性
cargo checker 点货员
cargo chute 溜槽；溜板；货物装卸滑槽；滑板
cargo claim 货物索赔
cargo clause 货物保险条款
cargo clearance 货物结关
cargo clip 吊货夹
cargo cluster light 货舱移动丛灯
cargo compartment 货舱
cargo compartment hatchway 货舱口
cargo conservator 货舱管理员
cargo consolidation 集运；货物集中
cargo consolidation of port 港口集运
cargo container 行李箱；行李袋；集装箱；货箱；货物集装箱
cargo containment area 货物围护区域
cargo cooling 货物冷却
cargo crane 船上(货物)起重机；甲板起重机；码头起重机；船货起重机
cargo damage prevention 货损预防
cargo damage report 货损报告
cargo damage survey 货损检验
cargo deadweight 净载重量；载货重量
cargo deadweight ton 载货吨
cargo deadweight tonnage 实载货吨位
cargo deck 载货甲板
cargo dehumidification system 货物减湿装置
cargo delivery notice 提货通知(单)
cargo delivery receipt 交货收据
cargo delivery system 交货系统；货运系统
cargo density 货物密度；船舶单位长度载货量
cargo derrick 吊货杆；船舶吊杆
cargo derrick guy pendant 吊货杆牵索
cargo derrick guy rope 吊货杆稳索
cargo destined 指定货物
cargo documents 货运单据
cargo door 装货门
cargo dues 货物港务费
cargoes imported and exported 进出口货物
cargoes loaded and unloaded 货物装卸量
cargoes proportion of general average 共同海损货物分摊额
cargo examination 验货
cargo expenses 装卸费(用)
cargo factor 积载因素；积载因数；货物装载因子；货物积载因数；货物积载系数
cargo fall 吊货索
cargo flood light 装卸用大功率照明灯
cargo flow 货流
cargo fork 货叉
cargo gear 货物装卸设备；货物装卸机具设备；装卸设备；船上吊货设备
cargo gear arrangement 装卸设备布置图
cargo gear certificate 装卸工具证书
cargo gear hire 装卸工具租金
cargo gear register 装卸设备合格证
cargo gin block 单饼铁滑车
cargo grating 货物格栅
cargo hand hook 搭货手钩
cargo-handled by owner 自理装卸
cargo handling 货物装卸滑车；货物装卸机；货物装卸；货物装饰；货物的搬运和操作；船货装卸；驳船装卸
cargo-handling anchorage 水上装卸锚地
cargo-handling and transport system 货物装卸机和运输系统
cargo-handling appliance 货物装卸设备；港口货物装卸机械
cargo-handling bay 装卸港池；装卸车位
cargo-handling berth 货运码头；装卸车位
cargo handling building 货物装卸建筑(物)
cargo-handling by conveyer 输送机装卸
cargo-handling by deck crane 起重机装卸
cargo-handling by derrick 吊杆装卸
cargo-handling by lighter 驳船装卸货物
cargo-handling capacity 货物装卸能力；吞吐量；装卸能力
cargo-handling capacity of seaports 港口吞吐量
cargo-handling cargo in out 货物装卸情况

cargo-handling control board 货油装卸控制台
cargo-handling district of port 港口作业区
cargo-handling equipment 货物装卸设备；装卸设备；装卸机械
cargo-handling gear 货物装卸设备；装卸设备；装卸机械；船货搬运装置
cargo-handling machine 货物装卸机械；装卸机械
cargo-handling machinery 装卸机械
cargo-handling norms 装卸定额
cargo-handling on sea 海上装卸
cargo-handling on water 水上装卸
cargo-handling operation 货物装卸作业；装卸作业
cargo-handling operation line 装卸作业线
cargo-handling operation of port 港口装卸作业
cargo-handling operation with fixed machinery and moving ship 定机移船装卸作业
cargo-handling operation with movable machinery and fixing ship 定船移机装卸作业
cargo-handling output per man-hour 装卸工时产量
cargo-handling plan 货物装卸计划；装卸计划
cargo-handling plant 货物装卸设备；装卸机械；装卸设备
cargo-handling process 装卸过程；装卸工艺
cargo-handling productivity 货物装卸能力
cargo-handling ramp 装卸货跳板
cargo-handling system 装卸工艺系统
cargo-handling system in container terminal 集装箱码头装卸方式
cargo-handling system of anchorage 锚地作业工艺系统
cargo-handling technology 装卸工艺
cargo-handling technology system 装卸工艺系统
cargo-handling terminal 货站；货运终点站；货运码头
cargo-handling unit operation 装卸工序
cargo-handling volume 货物装卸量
cargo-handling winch 装卸货物用的起重绞车
cargo-handling wire rope 货物起吊钢丝绳；起重钢丝绳(装卸货物用)
cargo-handling with double boom 双杆装卸法
cargo-handling with married fall 双杆装卸法
cargo harbo(u)r 货运港(口)
cargo hatch pontoon cover 箱形货舱口盖
cargo hatch side coaming 纵向货舱口围板
cargo hatch(way) 货舱(舱)口
cargo heater 货物加热器
cargo heating system 货物加热系统
cargo hoist 货物提升装置；货物升降机；吊货设备；吊货辘绳；吊货绞车
cargo hoisting block 吊货滑车
cargo hold 货舱
cargo hold bulkhead 货舱舱壁
cargo hook 装卸手钩；吊货钩；起货钩
cargo hook assembly 吊钩装置
cargo hose 装卸软管
cargo hose compartment 液货软管舱
cargo icebreaker 破冰型货船
cargo in bag 袋装货
cargo in bale 袋装货
cargo in bond 存入关栈的货物
cargo in boxes 箱装货(物)
cargo in bulk 散装货仓库；散装货；散装
cargo in bundles 捆装货
cargo in cans 罐或坛装货
cargo in cartons 箱装货(物)
cargo in case 箱装货(物)
cargo in cask 桶装货
cargo in chests 箱装货(物)
cargo in coil 卷装货；成卷货物
cargo incompatible with each other in storage 忌装
cargo in crates 箱装货(物)
cargo in drum 桶装货
cargo inherent vice 货物内缺陷
cargo in hold 混载货物
cargo in jars 罐或坛装货
cargo in large amount 大宗货
cargo in packages 件货
cargo in roll 卷装货；成卷货物
cargo in sack 袋装货
cargo insurance 货物保险
cargo insurance rate 货物保险费率

cargo interest 货方(利益);货主部门
cargo in transit 过境货(物)
cargo jack 货物起重器;压包机;撬货千斤顶
cargo jetty 货运栈桥;货运栈桥;防波堤
cargo joiner 加固货物连接件
cargo lamp 货物装卸照明灯
cargo lashing chain 货物捆绑链
cargo lien 货物留置(权)
cargo lift 货物升降机;船用升降机
cargo light 货物装卸照明灯;货舱灯;装卸作业灯
cargo liner 货运班轮;货运班机;运货班轮;装货班轮;定期货运班轮;定期货轮;定期货船;大型货(运飞)机
cargo list 货物清单;货物计数单;小舱单
cargo load 堆货荷载
cargo loaded 已装货
cargo loading control console 装卸货油操纵台(油轮)
cargo loading/discharge crossover head 货物装卸转换联箱
cargo-load team yard 整车货场
cargo-lot mark 发货符号
cargo manifest 货物舱单;舱单
cargo marine insurance 货物海上保险;海运货物保险
cargo mark(ing) 货物标志
cargo marks 货种标志
cargo mast 吊货桅杆;吊货高杆;起重杆;货物起重墙杆(在岸上货棚墙上设置的起重杆)
cargo mat 货垫;隔货席子;垫席
cargo measurement 货物丈量
cargo mill 船运材制材厂
cargo mixing 货物混杂
cargo movement 货物搬运;货流
cargo navicert 货物准运证;货物航运执照;货物海运执照
cargo net 货网;卸货网;装卸网兜;吊货网(兜)
cargo net sling 吊货索网兜
cargo notation 货种标志
cargo of contradictory nature 性能相抵触货
cargo oil 货油;船运油(料)
cargo oil control system 货油控制系统
cargo oil head 货油压头
cargo oil heating system 货油加热系统
cargo oil hose 输油胶管
cargo oil pump 船舶装油泵;输油泵;装油泵
cargo oil remote control system 货油遥控系统;装油遥控系统
cargo(oil) tank 货油舱
cargo on deck 甲板货
cargo owner 货主
cargo-owner's wharf 货主码头
cargo package 货物包装
cargo packing 货物包装
cargo papers 货单
cargo parachute harness 投物伞吊带
cargo passage 货物航程
cargo permeability 货物渗透性
cargo pilfering 货物失窃
cargo pin 单饼铁滑车
cargo pipe line 输油管系(统)
cargo plan 积载图;配载图
cargo plane 货运飞机
cargo planning 货物配载计划;装卸计划;配载计划;配船计划
cargo policy 货物保险单
cargo pooling agreement 货载分配协定
cargo port 货运港(口);舷侧装货门;装卸货的侧舷门;驳门
Cargo Preference Act 货物特惠法
cargo premium 货物保险费
cargo pump 货油泵;装卸油泵
cargo pump room 货油泵舱
cargo purchase 货卸索具
cargo quantity 货量
cargo receipt 货运收据;陆运收据;交货收据;收货单
cargo receiver 收货人
cargo reconditioning expense 货物整理费
cargo record book 货物记录簿
cargo reflector 货舱移动丛灯
cargo refrigerated space 冷藏货舱
cargo rig(ging) 吊货索具
cargo room 货物存放室
cargo rope 吊货索

cargo runner 吊货索;吊货辘绳
cargo saving net 舷边安全网
cargo sawmill 船运材制材厂
cargo separate net 隔飘票网
cargo sharing 货载分摊
cargo shed 货棚
cargo sheet 货物计数单;收货单
cargo sheet clerk 装卸交货总管理员;船舶货运员
cargo shifting expenses 倒舱费;翻舱费
cargo ship 货轮;货船
cargo ship dock 货船码头;货船船坞
cargo shipped in bulk 散货船货
cargo shipped to and fro 往返货运
cargo shipping contract 货运合同
cargo shipping order 货运单;货物装船指示;货物装船单;发货单
cargo ship safety construction certificate 货船构造安全证书
cargo ship safety equipment approval record 货船设备安全认可记录
cargo ship safety equipment certificate 货船设备安全证书
cargo ship safety radio approval record 货船无线电安全认可记录
cargo ship safety radio certificate 货船无线电安全证书
cargo ship safety radio station certificate 货船无线电台安全证书
cargo ship safety radiotelegraphy certificate 货船无线电报/电话安全证书
cargo sized container 中型货物集装箱
cargo skid 滑货木板箱
cargo sling 吊具;吊货套索;吊货套索;吊货钢丝索环
cargo sling chain 吊货链环
cargo space 载货容量;载货容积;载货舱位;船货空间;舱位
cargo sparring 货舱壁护条;舱壁护条
cargo stack 货垛
cargo stage 装卸台
cargo steamer 货轮
cargo stevedore 货物装卸工人
cargo storage 堆货
cargo stowage coefficient 船货堆装系数
cargo stowage factor 积载因素;积载因数
cargo stowage plan 货物配载计划图;船舶积载图
cargo stowing 装载
cargo stripping pump 清舱泵;扫舱泵
cargo suction pipe 货油吸管;液货吸管
cargo superintendent 货物管理员
cargo surveyor 货物检验师;货物检验人员;商检人员
cargo sweat 货载汗漏;货物汗湿;潮腐蚀
cargo sweeping(residue) 地脚货
cargo sweep(ning) 扫舱货
cargo tally corporation 理货公司
cargo tallyman 货物理货员
cargo tank 货舱;油船;载油舱
cargo tanker 货轮
cargo tank ga(u)ge 货油舱液面指示器
cargo tank valve 货油阀
cargo terminal 货运终点站;货运码头
cargo throughput 货物吞吐量
cargo throughput of port 港口货物吞吐量
cargo ton 计费吨
cargo tonnage 货物吨位;货物吨(总称);载货吨位
cargo tractor 运输牵引车
cargo trade 船货贸易
cargo traffic 货物交通;货运转点;货运量;货运;货物转运
cargo transportation density among stations 站间货物运输密度
cargo transportation insurance 货物运输保险
cargo-transport plane 运输机
cargo tray 吊货托盘
cargo trimmer 平舱工人
cargo truck 运货汽车;运货车;载重汽车;载货车
cargo truck management information system 货车管理信息系统
cargo tunnel 货车隧道
cargo turnover 货运周转量;货物交接
cargo type 货物类型
cargo underwriter 货物承运人;货物保险承保人
cargo unit 货物组
cargo unloaded 已卸货物

cargo unprepared 货未备妥
cargo valuation form 货价表
cargo valve 货油阀
cargo vehicle 载货车;运货汽车
cargo vessel 货轮;货船
cargo wharf 货运码头
cargo whip 吊货索;吊货辘绳
cargo winch 货物绞车;吊货卷扬机;船货卷扬机;起货绞车;起货机
cargo wire 吊货索;吊货钢丝
cargo wire block 吊货钢索滑车
cargo wire net 钢绳兜
cargo wire rope 吊货钢丝
cargo wire runner 吊货钢丝
cargo without bill of lading 没有提单的货物
cargo work 货物装卸;装卸作业
cargo-worthy 适货
car grease 车辆润滑油
car harmonic current interference 车辆谐波干扰
car haul 矿车牵引;钢索矿车运输
car hauler 调度绞车
car hearth furnace 活底炉
car heater 车用暖风设备
car hopper 车斗
Caribbean area 加勒比海地区
Caribbean bivalve province 加勒比海双壳类地理区
Caribbean Community 加勒比海共同体
Caribbean Conservation Association 加勒比养护协会
Caribbean current 加勒比海流;加勒比海洋流
Caribbean Environment Program(me) 加勒比环境方案
Caribbean plate 加勒比海板块
Caribbean Sea 加勒比海
Caribbean Sea shipping line 加勒比海航线
Caribbean Sea whirl structure 加勒比海漩涡状构造
Caribbean waters 加勒比海海域
caricaturist 漫画家
carillon 钟组
carina 龙骨(状)突(起);峰板;[复]carinae
carinal canal 脊下道
carinate 龙骨形的;肋状;脊状;船骨状的
carinate anticline 脊状背斜;肋状背斜
carinate fold 脊状褶皱
car indexer 调车定位装置
car industry 汽车工业
car inspecting track 检车线(路);车辆检修线
car inspection 车辆检查
car inspection and repair depot 列检所
car inspection and repair line 检修线
car inspection track 车辆检修线
car insurance 汽车保险
car insurance policy 车辆保险单
Carinthian furnace 卡林塞反射炉
carinthine 亚蓝闪石
Carius method 卡留斯法
car jack 车辆起重器;汽车千斤顶
car-jack type sounding apparatus 汽车式触探器
car-kilometer 车辆公里(计程表)
car ladder 汽车舷梯
car lane 车行道
car lashing 车辆绑缚
Carle valve 卡尔阀
car-leveling device 轿厢平层装置;轿厢平层调平机构
carley float 救生浮具
Carley raft 卡利式救生筏
carlfriesite 碲钙石
carlhintzeite 水氟铝钙石
car license 车辆执照
car lift 车辆提升机;升车机
car lighting 车灯
car lightning arrester 车辆避雷器
carline 拱条;拱梁;短纵梁;电车线路
carline box 梁间储[贮]藏室
carline knee 短纵梁肘
carling 短纵梁
carling main beam longitudinal 舱口纵桁
carling sole 短纵梁垫板;短纵梁底基
carlinite 辉铊矿
Carlin-type gold deposit 卡林型金矿床
Carlit explosive 过氯酸盐炸药
Carlmahr's cylinder ga(u)ge 卡尔玛斯缸径规
carload 装在车上用地磅过秤;整车;车辆负载;车

辆负荷;一整车;一节货车满载量;整车载荷;车辆载荷;汽车荷载;铁路货车每辆积载量
carload area 货位
carload delivery 车辆输送
carloader 装车机
car load factor 车辆负荷系数
carload(freight) 整车货(物)
carload freight unloaded at two or more stations 整车分卸
car load(ing) 车辆载重;装车;货车运输量
car loading by directions 各线路装车数
car loading by groups 成组装车
car-loading charges 汽车装卸费
car loading facility 装车设备
car loading list 货车装载清单
car loading 汽车装载
car loadings 装车数;铁路货运积载量
car-loading system 装车工艺系统
car-loading technology system 装车工艺系统
car load lean yard 整车货场
carload lot 整车货(物)
carload rate 整车运价
carlot wholesaler 分配批发商
Carlsbad twin 卡斯巴双晶
Carlsbad twin law 卡斯巴双晶律
carlsbergite 氮铬矿;氮络矿
Carlson compass 卡尔逊罗盘
Carlson instrument(ation) 卡尔逊仪表;差动电阻仪
Carlson meter 卡尔逊仪表;差动电阻仪
Carlson resistance ga(u)ge 卡尔逊电阻计
Carlson rotating auger 卡尔逊旋(转)钻(探装置)
Carlson sensor 卡尔逊传感器
Carlson strain meter 卡尔逊应变计
Carlson stress meter 卡尔逊应力计
Carlson stress meter for concrete 卡尔逊混凝土应力计
Carlson type strain ga(u)ge 卡尔逊式应变仪
Carlton joint 卡尔顿管接头
carmaker 汽车制造者
carman 车务人员
Carman equation (测定水泥比表面积的)卡门公式
Car Manufacturers Institute 汽车制造商协会
car manufacturing factory 车辆制造厂
car manufacturing waste 汽车制造废水
carmatron 卡玛管
Carmazine 代森锰锌杀菌剂
Carmelite church 圣衣会教堂;加尔默罗会教堂
carmetta 洋红色
car mile 车英里
car mileage 行车旅程
carmin(e) 卡红;胭脂红色;涂(染)成胭脂红色;洋红色;洋红
carmine lake 洋红色;胭脂红色淀
carminette 洋红色
carmine vermil(l)ion 硫化汞颜料
carminite 砷铅铁石;砷铅铁矿
carmoisine 红色酸性染料
car monitoring system 车辆控制系统
car-mount radio telephone 车载无线电话机
carnallite 光卤石
carnallitolite 光卤石岩
Carnarvon arch 卡那文式拱;平半圆混合拱
carnary (公墓小教堂里的)尸骨存放所
carnation 杂色麝香石竹;淡红色;肉色
carnationed 成红色的
carnation red 石竹红色
carnation rose 石竹淡红色
carnauba 巴西扇棕
carnaubanol 巴西棕榈醇
carnauba wax 巴西棕榈蜡;巴西蜡蜡;加洛巴蜡;蜡棕榈
carnaubic acid 巴西棕榈酸
carnaubyl alcohol 巴西棕榈醇
carnegieite 三斜霞石
Carnegie section 卡尼奇型钢
carnel 城垛的斜面洞口
carnelian 红玉髓;红玛瑙;光玉髓
carnelion red 光玉髓红色
carne(o)se 肉色
carnet 海关文件;通关卡
carnificina 古罗马地下囚牢
carniole 多孔石云岩;多孔白云岩
Carnish single-flue boiler 科尼斯单烟道锅炉
carnivore 食肉性生物;食肉动物;食虫植物

carnivorous animal 食肉动物
car not for traffic use 非运用车
carnotite 卡诺石;钒钾铀矿
Carnot's cycle 卡诺循环
Carnot's efficiency 卡诺效率
Carnot's engine 卡诺发动机
Carnot's law 卡诺定律
Carnot's principle 卡诺原理
Carnot's reagent 卡诺试剂
Carnot's refrigerating cycle 卡诺制冷循环
Carnot's refrigeration cycle 卡诺制冷循环
Carnot's test 卡诺试验
Carnot's theorem 卡诺定理
car number 车号
carobbiite 方氟钾石
caro bronze 磷青铜(锡7.5%~9%,磷0.11%~0.4%);卡洛磷青酮
car occupancy 汽车占用率
car of particular class 特种车
caroid 心形的
carol 阅览隔间;凸窗座
Carolina Bays 卡罗来纳湾洼地
caroline 奶油色大理石
Carolingian architecture (公元8世纪~公元10世纪法兰西时期的)加洛林王朝建筑;先罗马式建筑形式
Carolingian basilica(n) (公元8世纪~公元10世纪法兰西时期的)加洛林王朝长方形的教堂建筑
caroll 凸窗座
carol(le) 小书斋(修道院)
caronato 含碳酸盐的
caronium ion 阳碳离子
carousel 圆盘传送器;圆盘传送带;转盘传递带;四轮运坯车
carousel rod changer 轮盘式移摆管装置
carousel type mo(u)lding lay-out 转盘式造型生产布置
car owner 车主
car ownership 车辆拥有量
car ownership rate 私人占有汽车的平均数
car ownership ratio 汽车拥有率
car-owning household 有车家庭
car park 停车处;停车场
car park basement 地下室停车场;地下车库
car park building 停车房
car parking 汽车停放
car parking roof 屋顶停车场;停车屋顶;停车棚
car passer 错车装置;会车装置;横向移车机
Carpathian old land 喀尔巴阡古陆
carpathite 黄蜡烯
carpenter 木匠;木工工人;木工
carpenter ant 木蚁
carpenter art 木工技艺
Carpenter brake 双箱制动器
Carpenter centrifuge 卡氏离心脱水器;竖式离心脱水器
Carpenter Gothic 尖拱式木建筑;哥特式木建筑
carpentering 木工业;木工操作
carpenter machinery 木工机械
carpenter's and joiner's tools 粗细木工具;粗纹木工具
carpenter's apprentice 木工学徒
carpenter's art 木工工艺
carpenter's auger 木(工)钻
carpenter's axe 木工斧
carpenter's band saw 木工带锯
carpenter's bench 木工(工作)台
carpenter's bench knife 木匠木凳刀
carpenter's boast 屋顶桁架的柱环与椽子的接榫
carpenter's brace 木工钻
carpenter's bracket scaffold 木材金属托架平台脚手架;木材金属平台鹰架;木工托臂脚手架;木工挑出式脚手架
carpenter's certificate 木匠执照;木匠证书
carpenter's chipping-off chisel 木工平錾
carpenter's clamp 木工夹
carpenter's drawing compass 木工绘图圆规
carpenter's finish 细木(工)装修;木(工)装修
carpenter's flat chisel 木工凿
carpenter's flat plane 木工平刨
carpenter's ga(u)ge 木工平面规;木工划线盘
carpenter's glue 牛皮胶;木工胶
carpenter's hammer 鱼尾锤;羊角锤;木工锤
carpenter shop 木工车间

carpenter's ink box 木工墨斗
carpenter's ink box and line 墨斗线
carpenter's ink number 墨斗
carpenter's jack plane 木工粗刨
carpenter's joint 榫头
carpenter's level 木匠水平尺;木工水平仪;木工水平尺
carpenter's level rule 木工水平尺
carpenter's line 木工墨线;墨线
carpenter's mallet 木匠用的木槌
carpenter's mate 木匠助手
carpenter's mattock 鹤嘴斧;鹤嘴锄
carpenter's nail 木工(用)钉
carpenter's overalls 木匠外套
carpenter's pincers 木工(用)胡桃钳;胡桃钳
carpenter's plane 木工刨
carpenter's planer 木工刨床
carpenter's punch 木工钉冲
carpenter's roofing 木屋盖
carpenter's room 木工间
carpenter's rule 木工尺
carpenter's saw 木工锯
carpenter's sawing machine 木工锯机;木工锯床
carpenter's scene 幕间小节目
carpenter's scraper 木工刮刀
carpenter's screw clamp 台钳
carpenter's shop 木工车间;木工厂
carpenter's smooth plane 木工细刨
carpenter's square 矩尺;角尺;木工(角)尺;曲尺
carpenter's square hammer 木工方锤
carpenter's steel square 木工钢角尺
carpenter's store 木匠储[贮]藏室
carpenter's tool 木工工具
carpenter's tool kit 木工工具箱
carpenter stopper 缆索掣
carpenter's trade 木工职业
carpenter's trying plane 木工粗刨
carpenter's undertaking 木作企业
carpenter's water level rule 木工水平尺
carpenter's work 木作
carpenter's workshop 木工厂
carpenter's yard 木工厂
carpenter trade 木工行业;木工手艺
carpenter undertaking 木工企业;木工操作
carpentry 木作;木工业;木工工作;木工(指工作);木钉;大木作;木器;木结构
carpentry joint 木工接缝;木工接点;建筑的连接;木作接头
carpentry shop 木工厂
carpentry technology 木工工艺学
carpentry tongue 木工凿;木工斧;木雄榫
carpentry work 大木作
carpet 磨耗层;地毯;毯
carpet adhesive 毡层黏结剂;地毯黏合剂;地毯黏结剂
carpet backing 地毯基层;地毯垫面
carpet base 地毯垫层
carpet beater 地毯拍打器
carpet beating machine 地毯拍打机
carpet bed(ding) 地毯式花坛;地毯花坛;模样花坛
carpet bonding agent 地毯结合剂;地毯黏结剂
carpet brocade 地毯织棉
carpet cleaner 地毯清洁剂
carpet coat 磨耗层;毡层;道路面层
carpet cushion 地毯垫层;地毯衬垫
carpet cut 地毯卡槽
carpet density 地毯密度
carpet edging cover 地毯边盖
carpet edging strip 地毯边条
carpet factory waste 地毯厂废物
carpet factory wastewater 地毯厂废水
carpet fiber 地毯纤维
carpet filament 地毯用长丝
carpet float 毡贴面木抹子
carpet herb 地皮草;铺地草本植物
carpet holder 地毯压条
carpeting fiber 地毯用纤维
carpeting work 铺筑毡层
carpeting yarn 地毯用粗细绒
carpet jacquard 地毯提花机
carpet latexing machine 地毯上乳胶机
carpet lining paper 地毯纸
carpet loom 地毯织机
carpet manufacturing waste water 地毯厂生产废水

carpet overedge machine 地毯包边机
carpet-overlock machine 地毯拷边机
carpet pick 地毯纬密
carpet pile 地毯绒毛
carpet pile height 地毯绒毛厚度；地毯绒头高度
carpet pin 地毯钉
carpet pitch 地毯经密
carpet rod 压地毯条；地毯棍
carpet rolling-and measuring machine 地毯量卷机
carpet runner 通道地毯
carpet slitting and trimming machine 地毯割绒修整机
carpet soap 地毯皂
carpet strip 门槛；地毯挡条；踢脚压条；条形地毯；踢脚板底缝压条（与地板之间的线脚）；地毯扣边；地毯板条
carpet stuffer 地毯背底增厚线料
carpet sweeper 地毯吸尘器；地毯清扫机
carpet tester 射频脉冲发生器；射频脉冲发生机试验器
carpet testing 地毯检验
carpet tile 拼合地毯；小方地毯
carpet treatment 表面处治；铺筑毡层
carpet underlayment 地毯垫层；地毯衬垫
carpet veneer 毯层；毡层；表面处治
carpet warp 地毯经线
carpet wear tester 地毯磨损试验机
carpet weft 地毯纬线
carpet with Persian knots 有波斯结的地毯
carpet with Turkish knots 有土耳其结的地毯
carpet wool 地毯毛
carpet yarn 地毯纱线
carpet yarn precision winder 地毯纱精密络筒机
carpholite 纤锰柱石；纤锰闪石
carphosiderite 草黄铁矾
car piler 车式堆垛机
car pincher 推车装车工
car place 停汽车车位
car planned requisition list 要车计划表
car platform 轿厢平台；电梯厢平台
car polish 车体抛光；汽车抛光蜡
carpool 汽车合乘；汽车公用组织；合用汽车；合乘汽车
car pooling 合用小汽车；合乘车
carpool lane 合用汽车车道；合用车专用车道
carpool priority control 合用汽车优先控制
carpophyte 显花植物
carport 简易车库；存车港湾；车库；汽车棚；停车棚；多层停车棚；汽车间
car puller 拖车绞盘；推车机
car pusher 推车器；推车机
carr 运载工具；托架；承重构件；卡尔群落
carracing rental service 汽车出租处
carracing retarder 行轨式制动闸
carracing track 赛车跑道；汽车跑道
car radiator 车用散热器
car(r)ag(h)een 鹿角菜；卡拉胶；鹿角菜精宁；角叉藻；角叉(藻)胶
car(r)ag(h)eenin 鹿角菜精宁
car ramp 汽车匝道；汽车用坡道；汽车坡道
Carrara glass 卡拉拉玻璃
Carrara marble 卡拉拉大理岩；雕像用白大理石
carrboydite 镍铝矾
carreau 格子花纹；方形或菱形玻璃或彩瓦；玻璃上釉瓷砖块；装饰窗玻璃
carrefour 交叉口；十字路口；广场
car registration 登记车数；注册车数
carrelage 瓷砖；普通砖定型面积；空心砖
carrel(l) 阅览隔间；书库内小阅览凹室；单人阅览室；带书架阅览桌
carrel(l) cubic(al) 阅览隔间
carrel(l) stall 阅览隔间
car rental service 租车机构
car repair depot 车辆修理所
car repair(ing) shed 修车棚；修车库；(车的)修理棚
car repair shop 车辆修理厂
car repair track 修车线【铁】；车辆检修线
car replacer 车辆复轨机；复轨器
carrequip 载波设备
car resistance 车辆阻力
car retarder 车辆减速器【铁】；矿车减速器；缓行器；车辆缓行器【铁】
carriage 楼梯踏步梁；楼梯格栅；机器滑动部分；机床拖板；滑鞍；运输；运费；印刷版台；支撑框支撑架；载运；载货台【机】；工作台车；大刀架；承船斜车架；车厢；车辆；车架；叉座；拖板；托架；铁路客车

carriage, insurance paid to 运费和保险费付至
carriage and wagon works 客货车工厂
carriage apron 车床拖板箱
carriage axle 车轴
carriage body construction 车厢结构
carriage bolt 方颈螺栓；螺栓；车箱螺栓；车身螺栓
carriage by air 空运；航空运送
carriage by land 陆运；陆路运费
carriage by road 公路运输
carriage by sea 海运；海洋运送；海上运输
carriage cam 走车运动凸轮；车床靠模
carriage character 反转符
carriage clamp 木工用夹紧装置；刀架固定手柄
carriage clock 车钟
carriage collector 集水车
carriage contract 运输合同
carriage control character 托架控制字符
carriage control tape 传动控制带；托架控制带
carriage coordinate system 车架坐标系【测】
carriage draw spring 滚轮架拉力弹簧
carriage elevator 悬吊式升降机；缆索式升降机
carriage entrance 车辆入口；车辆入口通道
carriage expenses 搬运费(用)
carriage forward 交货时付运费；运费未收；运费未付；运费交货照付；运费交货时付
carriage free 免付运费；免费运送；运费付；运费已付
carriage freight 运费
carriage gate 车辆出入门
carriage guide 刀架导槽
carriage guideways 托架导板
carriage hand wheel 拖板手轮
carriage hoist 举车机
carriage house 车房
carriage insurance and freight (CIF) 保险运输费
carriage jack 车辆千斤顶；车辆起重器
carriage kilometer 客车公里
carriage lock screw 拖板锁紧螺钉
carriage motorway 客运快速干道
carriage mounting 装在车上的；台车钻架
carriage nut 车架螺母
carriage of automatic welding machine 自动焊机走架
carriage of container 集装箱运输
carriage of contraband 载运违禁品
carriage of dangerous goods 危险品运输；危险货物运输
carriage of goods by inland river 内河货物运输
Carriage of Goods by Sea Act 海运货物法例；海上货物运输法
carriage-out 出车
carriage outward 销货运费
carriage paid 免付运费；运费已付；运费付讫
carriage paid to 运费付至……
carriage paid to the named point of destination 运费付至目的交货地
carriage piece 楼段加劲杆；车架构件；楼梯斜梁；楼梯格栅；楼梯踏步梁
carriage porch 车廊；车道门廊；车厢门廊；停车门廊；停车廊（入口处）
carriage positioning 托架定位
carriage prepaid 预付运费
carriage rail 回车导轨；托架导轨
carriage return 回车；刀架复位
carriage return button 回车按钮；复原按扭
carriage return character 回车字符；托架回车符号；托架返回符号
carriage road 大车路；大车道
carriage rope 牵引索；牵引绳；起重机小车上提升用的钢丝绳
carriage saddle 刀架座
carriage shed 存车棚；车棚；车内作业
carriage slide track and slide frame 大车滑轨和滑架
carriage space key 托架空推键
carriage sprayer 果园动力喷雾机
carriage spring 车架弹簧
carriage turn table 转车台
carriage-type switchgear 车架式开关装置
carriage underframe 车子底盘
carriage water 运水车

carriageway 行车道；车行道路；车行道；车道
carriageway cover 井道顶棚；车行道面层；井盖（行车道上）
carriageway edge marking 车行道边缘标志
carriageway expansion joint 行车道伸缩缝；行车道连接缝；行车道变形缝
carriageway for turning traffic 转弯车辆用车道；转弯车道
carriageway foundation 车行道路基；公路路基
carriageway line 车行道线
carriageway maintenance 车行道保养；道路维修；道路保养
carriageway marking 车行道标线；行车道标志线；车行道划线；车行道标线
carriageway pavement 车行道路面
carriageway slab 车行道板
carriageway width 车道宽度
carriage wheel 车轮
carriage with semi-cushioned berthes 硬席卧车
carriage with semi-cushioned seats 硬席座车
carriage with two axles 两轴客车
carriage works 车辆厂
Carribel explosive 卡里别尔炸药
carrick bend 大绳接结；单花大绳接头
carrick bitts 支承座立柱；起锚机柱；起锚机系柱；支撑起锚机的缆柱
carrick heads 起锚机系柱
car rider 守车员；随车制动工
car rider's box 守车员车厢；守车员室
carried forward 接后；转入次页；转次页；上期转来
carried interest 附带权益
carrier 记录媒体；携带者；运载体；运载工具；运输者；运输人；运输器；运输汽车；运输公司；运输船；载体；载流子；载波通信系统；底盘框架；带菌体；传递体；承运人；舱口梁座；搬运人；托板架；托板；胎体；数据媒体
carrier accumulation 载流子累积
carrier-actuated relay 载频驱动继电器
carrier addition method 载体添加法
carrier air tube system 由压缩空气操作的管道（输送设备）
carrier amplitude 载波振幅
carrier amplitude regulation 载波幅度调整；载波幅度变动率
carrier-and-stacker 堆垛车
carrier angle 支托角钢
carrier assay 运载装备
carrier assembly 运载装备
carrier balance 载频平衡
carrier bar 托梁；支托钢条；承载梁
carrier-based aircraft 舰载飞机
carrier bearer 承载器架
carrier beat 载波拍频
carrier beat phase 载波拍频相位
carrier bed 过渡层；传导层
carrier block 承载块
carrier-borne aircraft 舰载飞机
carrier-break pushbutton 载波切断按钮
carrier buffer 载频缓冲器
carrier cable 吊索；吊力索；缆绳钢索；载重索；承载索
carrier capacity 载波功率
carrier chain 输送链
carrier channel 载波信道
Carrier chart 卡里尔(湿空气)曲线图
carrier-chrominance signal 色度信号；载波色度信号
carrier communication 载波通信
carrier compound 载体；负荷体
carrier compression 载波振幅压缩
carrier concentration 载体浓度；载流子浓度
carrier control modulation 载波控制调制
carrier current 载波电流
carrier-current blocking 高频闭锁
carrier-current phase differential protection 相差高频保护
carrier-current phase differential protection device 相差载波保护装置
carrier-current protection 载波电流保护(装置)；高频保护(装置)
carrier-current relaying 载频中继
carrier density 载流子密度
carrier detection 载波检波
carrier deviation 载频偏移
carrier displacement 载体置换(法)

carrier distillation 载体蒸馏;载体分馏
carrier drain 降压排水管;输水渠
carrier drift transistor 载流子漂移型晶体管
carrier drop 载波跌落
carrier effect 载体效应
carrier equipment 载波设备
carrier film 承载膜
carrier filter 载波滤波器
carrier flooding 载流子充满
carrier flutter 载波颤动
carrier for dark plate 暗板托架
carrier frame 托架;运输车架;底盘框架;汽车底盘框架
carrier-free tracer 无载体指示剂
carrier frequency 载(波)频(率)
carrier frequency amplification 载频放大
carrier frequency amplifier 载频放大器
carrier frequency hologram 载频全息图
carrier-frequency offset 载波频率偏置;载频偏置
carrier gas 载运气体;载气;载流气体
carrier gas inversion 载气逆流
carrier generation 载体产生
carrier generator 载波发生器
carrier hay loader 传送带式装干草机
carrier hole 导孔
carrier hole modility 载流子空穴迁移率
carrier ilder 输送带托辊;承载轮
carrier injection 载流子注入
carrier interleaving 载波交错
carrier isolating choke coil 载波隔离扼流圈;隔载波扼流线圈
carrier leak 载波漏泄
carrier lever 载波电平
carrier line 载波线路;输油线
carrier liquid 载液
carrier load 载波负载
carrier load control 载波功率控制
carrier loader 运载车
carrier loading 载波加载
carrier lorry 运营货车
carrier material 底基
carrier-mediated transport 载体性转运
carrier medium 载体(介质)
carrier metal 载体金属
carrier method 载流法
carrier mobility 载流子迁移率
carrier network 货运网
carrier noise 载波噪声
carrier-noise ratio 载波噪声比
carrier of a game 对策的施行
carrier of pathogens 病原体的携带者
carrier pack 携带包装
carrier-pellet 催化剂载体片
carrier phase 载波相位
carrier phasor 载波矢量
carrier-pilot relzy system 载波遥控中继方式
carrier pilot system 载波导频系统
carrier plate 托板;炉腰环架;顶板;承板
carrier position 载波的分布位置
carrier power 载波功率
carrier power-output rating 载波额定输出功率
carrier precision offset 载频精确偏置
carrier radioisotope 载体同位素
carrier recovery 载波恢复
carrier reduced signal side band 减载波单边带
carrier reduction 载波降低度
carrier regeneration 载波再生
carrier registration expenses 运输工具注册费
carrier reinsertion 载频重置
carrier relaying 载波中继制
carrier return 载波恢复
carrier ring 承环
carrier ripple 载频脉动
carrier rocket 运载火箭
carrier rod 顶杆
carrier roller 履带托轮;链规托轮;导辊;承载辊子;托辊
carrier roller of ejector 推卸器支重滚轮
carrier round braider 支锤编带机
carrier's agent 承运人代理
carrier's allowance 承运人津贴
carrier sense multiple access 载波检测多址
carrier sense multiple access with collision detection 载波监听多路访问/冲突检测

carrier separator 载体分离器
carrier serving area 载波服务区
carrier's haulage 承运人接运
carrier shift 载波偏移;载波偏移
carrier side 传送带的载货面;皮带运输机的上(面传送)带
carrier signal 载波信号
carrier signal(l)ing 载波信令;载波发信
carrier's immunity and responsibility 承运人权责
carrier's insurance 承运人保险
carrier size 载体大小
carrier's liability 承运人责任
carrier's liability clause 承运人责任条款
carrier's liability insurance 承运人责任保险
carrier's liability to cargo 承运人对货物的责任
carrier's lien 承运人留置权
carrier's manifest 承运人舱单
carrier's maximum liability 承运人的最高限度责任
carrier's negligence 承运人的疏忽
carrier's note 取货证;取货通知;取货单;提货证;提货单
carrier source 载体振荡器;载波振荡器
carrier's pack 承运人装箱
carrier's risks 承运险
carrier's statement 承运记录
carrier storage 载波积聚
carrier's trade 运输行业;雇工运输行业
carrier suppression 载波抑制
carrier suppressor system 载波抑制方式
carrier swing 载频摆幅;载波频移;载波摆值
carrier system 载波制;载波式;载波(通信)系统;输导系统
carrier system traffic 载波制通信
carrier telephone system 载波电话系统
carrier telephone terminal 载波电话终端
carrier terminal 载波终端机
carrier-to interference ratio 载波干涉比
carrier-to-noise ratio 载波噪声比;载噪比
carrier track 高架轨道(卸货用)
carrier transfer device 载流子传输器件
carrier transfer filter 载波传输滤波器
carrier transmission 载波传输
carrier transmission system 载波传输系统
carrier transport 载运
carrier transport phenomenon 载流子输运现象
carrier transposition 载波交叉
carrier truck 运营货车
carrier vehicle 运输汽车;运载体;运输车辆;搬运汽车;运输车
carrier voltage 载波电压
carrier wave 载波
carrier wheel 移动齿轮;托带轮
carrier wire 载波线;载波电缆
carring cost 存储费用
carriole 雪橇;小篷车;单人马车
car rod aerial bent 汽车天线杆弯曲
car rollability 车辆溜放特性
car rolling resistance 车辆溜放阻力
car rolling resistance line 车辆基本阻力能高线
car rolling resistance lost 车辆基本阻力能损失
car rolling speed 车辆溜放速度
carrollite 硫铜钴矿
car roll-overdevice 车辆翻转装置
car roof 车顶
carrotless charge 不填塞装药
carroty 胡萝卜色
carrousel 回转架
car row 车列
car-row length 车列长度
Carr-Price unit 卡尔—普赖斯单位
Carruthers current meter 卡卢德斯测速仪;卡卢德斯流速仪
Carruthers residual current meter 卡卢德斯余(海)流计
Carruthers vertical log 卡卢德斯测流垂直旋杆;卡卢德斯测流重量旋杆
carry 支承;搬运
carryable belt conveyer 移动式皮带传送带;可移动式皮带传送带;可移动式皮带传送机
carryable track 可移动轨道;工地轻便的临时铁道
carry-account 亏损账户
carry a contract 履行合同
carry a dry hole 钻干孔
carryall 军用大客车;运料车;刮泥机;大型载客汽车;大型客车;铲运机;牵引式刮土机
carryall loading hopper 轮斗式装运机
carryall scraper 连续式整装机;连续式铲运机;轮式铲运机;轮式铲土机;通用铲运机
carryall tractor 万能拖拉机
carry a load of debt 亏欠;负债
carry-around oxygen cylinder 携带式氧气瓶
carry away 卷走
carry a wet hole 钻湿孔
carry back 运回;移回扣减
carry-back of losses 扭转亏损;亏损转回
carry bit 进位
carry chain 进位链
carry charges 仓储摊销
carry circuit 转移电路;进位电路
carry clear signal 进位清除信号;允许进位信号
carry complete 进位完毕
carry-complete signal 进位完成信号;进位结束信号
carry completion 进位完成
carry control 进位控制
carry-delay 进位延迟;进位迟延
carry delay time 进位延迟时间
carry digit 进位位;进位数(字);移位数字;移位数
carry down 结转下;结转
carry equation 进位方程
carry failure 进位失败
carry flag 进位标志
carry flip-flop 进位触发器;移位触发器
carry forward 结转;上期结转
carry forward of azimuth 方位角传递
carry forward of bearing 方位角传递
carry gate 进位门;进位阀
carry gate signal 进位门信号
carry generate 进位发生
carry generator 进位发生器
carry heavy debt 负担重债
carry-in (退火窑的)装料
carry in bit 输入位【计】
carry-in boy 装容工(玻璃退火时装送制品的人员)
carry indicator 进位指示符
carry-in fork 装容用叉子
carrying 承重;承载
carrying agent 携带介质;承载介质
carrying angle 臂外偏角
carrying area 承压面积
carrying a rudder 压舵
carrying axle 支承轴
carrying bar 支架顶梁;顶梁;承梁
carrying bogie 承载转向架
carrying bolt 支承螺栓
carrying bridge wall 支承式挡火墙
carrying by means of barrels 用桶装运
carrying cable 装载索;载运索;运载索;承载索;承载缆索;承力钢索
carrying capacity 挟运能力;运载能力;运输能力;允许载荷量;支承能力;载重;载运量;载畜量;载流量;载货量;承载能力;承载量;承压力;负荷试验;负荷能力;容许负荷量;容纳量;水能力;输送能力
carrying capacity in the section computed on the basis of non-parallel train working diagram 非平行运行图的区间通过能力
carrying capacity in the section computed on the basis of parallel train working diagram 平行运行图的区间通过能力
carrying capacity limiting section 限制区间
carrying capacity of a section 区间通过能力
carrying capacity of a station 车站通过能力
carrying capacity of canal 运河运输能力;运河通过能力;渠道输水能力;渠道排灌能力;渠道过水能力
carrying capacity of chassis 底盘负载量
carrying capacity of crane 吊车起重能力
carrying capacity of environment 环境负荷量;环境负担能力
carrying capacity of groundwater environment 地下水环境容量
carrying capacity of liferaft 救生筏乘员定额
carrying capacity of line 单线输送能力;铁路线路通过能力
carrying capacity of line network 线网负荷强度
carrying capacity of locomotive facility 机务设备通过能力
carrying capacity of pile 桩的承载力

carrying capacity of pipe 管道输水能力;管道过水能力;管道过流能力
carrying capacity of railway 铁路通过能力
carrying capacity of receiving-departure track 到发线通过能力;到达线通过能力
carrying capacity of station 铁路车站通过能力
carrying capacity of station throat 车站咽喉通过能力
carrying capacity of the block section 区间通过能力
carrying capacity of the environment 环境的负担能力
carrying capacity of throat points 咽喉道岔通过能力;咽喉区道岔通过能力
carrying capacity of track 线路通过能力【铁】;通过能力【铁】
carrying capacity of vehicle 车辆装载能力;车辆输送量
carrying capacity of wagon tagged 货车标记载重(量)
carrying capacity of water 过水能力
carrying capacity of water environment 水环境容量
carrying case 携运箱;携带式仪器箱
carrying channel 槽形金属条;进位通道;载流通道;承重槽钢;承载槽;主龙骨;槽形金属;U 形承载槽
carrying charges 维护费;资产持有费用;持用财产的费用;财产维持费;分期付款购货的附加价格;保管费;土地持有费;加价格
carrying contour 合成等高线
carrying cost 运输成本;储囤成本(指存货);财产维持费
carrying cost of capital 资本的附加费用
carrying cupola 承重穹顶;承载的钟形屋顶
carrying current 极限电流;容许负载电流;容纳流;瞬时极限电流
carrying cylinder 支承筒;承重筒
carrying freezer 低温冷藏室
carrying girth 承重皮带
carrying handle 提携手柄;承提把手;吊吊把;提把
carrying heavy debt 负重债
carrying idler 空载;托辊
carrying implement 搬运机
carrying-in expenses 搬入费
carrying intensity of line network 线网负荷强度
carrying ladle 运输桶;载流桶
carrying load 运送负载;承压货重
carrying no floor load 不支承楼板荷载
carrying-on-activity 在进行中的生产活动
carrying over 飞料损失
carrying plane 生力面;支承面;承压面
carrying plate 炉腰环梁
carrying point 支承点
carrying pole 扁担
carrying power 携带岩屑能力(泥浆);携带岩粉能力(泥浆);挟运能力;挟带能力;挟持能力;载重量
carrying power of pile 桩的承载力
carrying productivity 运输生产率
carrying pushcart 搬运车
carrying rail 承压轨;承载轨道
carrying rate 载畜量
carrying ratio 实载率
carrying rod of ceiling 吊顶拉杆上的横筋
carrying roll(er) 支承辊;承压滚轴;托轮
carrying rope 吊重用;承重索;承载轴;承载索;承力钢索;受力绳;受力索
carrying scraper 铲运机
carrying shaft 支承轴;承载轴
carrying ship 载货船只
carrying skip 铲运斗;单轨运输道上的运输箱
carrying steel 受力筋
carrying strand 吊重绳;承力钢索
carrying strap 承重盖板
carrying surface 外力面
carrying times per shift 车班载客次数
carrying tongs 运载钳
carrying traction rope 承载牵引索
carrying trade 货物运输业;运输业;转口贸易
carrying trade network economy characteristic 运输业网络经济特性
carrying traffic 承担交通量
carrying up 砌墙到规定标高;砌墙
carrying value 置存价值;现存价值;账面结存价值

carrying value of a bond issued 已发行公司债的现行价值
carrying vessel 载货船只
carrying wall 支承墙;承重墙;承压墙
carrying way with engine stopped 尚航
carrying wheel 承载车轮
carrying wire rope 承载钢丝绳
carrying wire rope for aerial tramways 架空索道用承载钢丝绳
carry initiating signal 进位起始信号
carry-in kiln 装窑
carry input 进位输入;移位输入端
carry-in terminal 输入端
carry into execution 实行;实施
carry into practice 实施
carry line 进位线
carryload scraper 铲运机
carry log 高轮拖木车
carry logic 进位逻辑
carry lookahead 先行进位(法);超前进位法
carry look ahead adder 先行进位加法器;超前进位加法器
carry number 进位数;移位数
carry off 带走;搬开
carry-off heat 排热
carry of spray 喷雾射程
carry operation 进位操作
carryout 履行;实行;外卖食物餐馆
carry out a contract 执行合同
carry out a test 进行一次试验
carry-out bit 输出位
carry out flag 进位输出标记
carry-out restaurant 外卖食物餐馆
carry-out terminal 输出端
carry out the duties 履行职责
carry out to the letter 不折不扣执行
carry out without any reservation 不折不扣地执行
carry-over 结转;带出(粉尘);携带;移后扣减;滞后【水文】;带出物
carry-over bar 动齿条
carry-over bed 冷床
carry-over clause 转交条款
carry-over coefficient 承载系数
carry-over effect 滞后效应
carry-over efficiency 传递效率
carry-over facility 转记装置,进位装置
carry-over factor 传递系数;传递因子
carry-over file 结转的档案
carry-over flow 前期降雨径流;前期流量
carry-over indicator 转移指示器
carry-over loss 带出损失;级间余速损失
carry-over moment 传递弯矩;带过力矩;传递力矩
carry-over of boiler water 炉水携带
carry-over of catalyst fines 催化剂粉末带出
carry-over of losses 亏损结转
carry-over pinch roll 拉辊;牵引辊
carry-over point 传递点
carry-over rate 转期利率
carry-over reservoir 多年调节水库
carry-over soil moisture 携带土壤水分
carry-over sound 传播声音;传递声
carry-over stock 结转库存量
carry-over storage 多年蓄水;多年调节水库;多年调节库容;转期调节
carry-over storage capacity 多年调节库容量
carry-over storage plant 多年调节电站
carry-over storage station 多年调节电站
carry-over to the next term 结转下期
carry pipe right to the bottom 将钻杆下到孔底
carry portion 进位部分
carry position 运位位置
carry propagate 进位传输
carry propagation 进位传送
carry pulse 进位脉冲
carry rates 运输费率
carry register 进位寄存器
carry reset 进位清除;进位复位
carry ripple 进位脉冲
carry save adder 进位保留加法器;保留进位加法器
carry scraper 刮土机;刮具;铲运机
carry separation 进位分离
carry set 进位组
carry shift 主动机构位移
carry signal 进位信号

carry signal terminal 进位信号终端
carry skip 跳跃进位
carry skip method 跳跃进位法
carry storage 进位储存
carry the bearing forward 方位角传递
carry time 进位时间
carry to two decimal places 计算到小数点后两位
carry-under 水中带汽
carry up 墙砌到规定高度
carry water across field 引水入大田
car safety 轿厢安全装置
car sampling 槽车取样
cars checking 车辆清查
cars collection and delivery charges 取送车费及其他
carse 沿河低冲积地;低湿地;冲积平原;(肥沃的)冲积河滩
car seal(ing) 货车施封
Carsel lamp 卡塞尔测光灯
car service 铁路车辆服务
car service station 汽车服务站
car shakeout 振动卸车装置;摇摆卸车机;车皮振动器
car shaker 车辆振动器
car shall 停车场上划分好的汽车停放位置
car shed 客车库;车棚;车内作业;停车棚
car-shed hanger 电车库吊架
car shelter 车棚
car shifting platform 移车台
car shredder 汽车破碎机
car sickness 晕车
car sickness and sea sickness 晕车晕船
car signal 机车信号
car silo 地下车库
car-sleeper train 卧铺列车
car sling 小汽车吊具;汽车吊架
cars of particular class dispatching plan 特种货车挂运计划
cars on hand 现在车数
cars open to traffic 运用车
car speed 车速
cars per cut 钩车(车组)
car spotting hoist 调度绞车
car spring 车弹簧
cars sector expenses 车辆部门支出
cars statistics of railway 铁路车辆统计
carst 喀斯特;岩溶
car stall 汽车停车车位;汽车车位
carstone 砂铁岩;铁砂石层
car stop 车挡;停车处
car stop indicator 车挡标志
car stopper 挡车器;车辆固定索具
car storage park 存车场
car storage track 存车线【铁】
car stowage 汽车装载
carst river 岩溶河(流)
carst spring 岩溶泉
carst stream 岩溶河(流)
cars under repair 检修车
cars under repair controller 车辆检修调度员
car suspension cable 吊舱缆
cars waiting for discharging 待卸车
car-switch control 笼内开关控制;车厢开关控制
car switching equipment 空重车交换装置
car-switch operation 轿厢启停操作
cart 运货马车;独轮手推车;大车;轻便车
cartage 马车运输;货车运输费;货车运费;运费;短途运输(市内);车运费;车运费;搬运费(用)
cartage book 搬运登记簿
cartage department 搬运处
cartage expenses 搬运费(用)
cartage-in 进货搬运费
cartage note 搬运许可证
cartage-out 销货车费;销货搬运费
cartage port 搬运港
cartage service 搬运业务
cartage undertaking 搬运业
cart and barrow hoist 混凝土手推车提升机
cartap(hydrochloride) 巴丹
cart away 运走
cart balance 车秤
carte blanche 空白委托书;全权委任
cartel 卡特尔
car telephone 车载电话

cartel-like international agreement 卡特尔式国际协定
cartel price 卡特尔价格
cartel ship 交换俘虏船
cartel tariff 卡特尔关税(保护关税的一种)
carter 搬运工
Carter chart 卡特图
Carter lead 卡特法铅白
Carter's coefficient 卡特系数;卡氏系数;气隙系数
carter's cradle (悬挂在导轨上的)移动式起重吊架
Carter-Wolf source 卡特一沃耳夫光源
Cartesian axis 直角坐标轴;笛卡儿轴
Cartesian coordinates 直角坐标;笛卡儿坐标
Cartesian coordinate system 笛卡儿坐标系
Cartesian diver 浮沉子
Cartesian diver manostat 浮沉子式恒压器
Cartesian geometry 解析几何(学);笛卡儿几何
Cartesian hydrometer 笛卡儿比重计
Cartesian method of control 笛卡儿坐标控制法
Cartesian ovals 笛卡儿卵形线
Cartesian product 笛卡儿乘积
Cartesian reference system 笛卡儿坐标系
Cartesian surface 笛卡儿曲面
Cartesian tensor 笛卡儿张量
Cartesian vector 笛卡儿矢量
cart grease 车辆润滑油;车滑脂
carthamus red 红花红色
Carthusian monastery 天主教卡尔特修道院
Cartier hydrometer 卡蒂尔比重计
car tilting device 车辆倾斜装置
carting 马车运输;运出;手推车运输
car tipper 矿车翻笼;自动前端重力翻车机;翻车机
car tipping plant 翻车机设备
car tippler 横倾翻斗车
cart-ladder (车上的)货架
cartload 运货车;手推车装载量
cartobiliography 地图目录
cartogram 图解;统计图
cartograph 地图
cartographer 制图员;地图制图员;编测人员
cartographic(al) accuracy 制图精度
cartographic(al) analysis 地图分析
cartographic(al) and graphic(al) arts type setting system 制图美术式记录系统
cartographic(al) annotation 地图注记
cartographic(al) association 地图制图学会
cartographic(al) base 底图
cartographic(al) camera 复照仪
cartographic(al) classification 制图分类
cartographic(al) communication 地图传输
cartographic(al) compilation 地图编绘
cartographic(al) craft 制图技巧
cartographic(al) data 地图资料;地图数据
cartographic(al) data bank 地图数据库
cartographic(al) data base management system 地图数据库管理系统
cartographic(al) digital model maintenance 地图数字模型更新
cartographic(al) digitizing data bank 地图数字数据库
cartographic(al) digitizing plotter system 数控绘图系统
cartographic(al) documents 制图资料;地图资料
cartographic(al) draftsman 绘图员
cartographic(al) draughtsman 绘图员;制图员
cartographic(al) editing 地图编辑
cartographic(al) editing system 地图编辑系统
cartographic(al) engineer 制图工程师
cartographic(al) engineering 制图工程(学)
cartographic(al) error 制图误差
cartographic(al) evaluation 地图评价
cartographic(al) exaggeration 地图夸大
cartographic(al) expert system 制图专家系统
cartographic(al) feature 地物图像;地图要素
cartographic(al) feature data bank 地图要素数据库
cartographic(al) field 制图学领域
cartographic(al) file 制图文件
cartographic(al) film 绘图胶片
cartographic(al) function key 制图功能键盘
cartographic(al) generalization 制图综合;地图概括;综合制图(法)
cartographic(al) generalization of relief 地貌综合
cartographic(al) generation 制图综合

cartographic(al) grid 地图格网;制图网格;制图格网
cartographic(al) hierarchy 制图分级
cartographic(al) history 地图学史
cartographic(al) house 制图室
cartographic(al) information 地图资料;地图信息
cartographic(al) information bank 地图信息库
cartographic(al) information retrieval system 地图资料检索系统
cartographic(al) information storage 制图信息库
cartographic(al) installation 制图设备
cartographic(al) language 地图语言
cartographic(al) lettering 地图注记
cartographic(al) manipulation 制图操作
cartographic(al) mask 制图蒙片
cartographic(al) model 地图模型
cartographic(al) monitor software 制图监控软件
cartographic(al) operation 制图工作
cartographic(al) organization 地图内容结构
cartographic(al) planning 制图规划
cartographic(al) probability 制图概率
cartographic(al) product 地图成果
cartographic(al) program(me) bank 制图程序库
cartographic(al) projection 地图投影;制图学投影
cartographic(al) projection equation 地图投影方程
cartographic(al) representation 现图法
cartographic(al) reproduction 地图制印
cartographic(al) reproduction equipment 地图制印设备
cartographic(al) reproduction technique 地图制印技术
cartographic(al) reproduction technologist 地图制印技术员
cartographic(al) room 制图室
cartographic(al) satellite 测图卫星;测绘卫星
cartographic(al) scanner 地图数字化扫描器;地图扫描仪;地图扫描器
cartographic(al) selection 制图选取
cartographic(al) simplification 制图简化
cartographic(al) sketching 手制草图
cartographic(al) specifications 编绘规范
cartographic(al) survey(ing) 地形测量;制图测量(地图);测图
cartographic(al) symbol 地形符号;图式符号;地图图式
cartographic(al) symbolism 图式符号
cartographic(al) symbols of topographic(al) maps 地形图图式
cartographic(al) technique 制图技术
cartographic(al) terminology 制图学术语
cartographic(al) topography 地形测图摄影
cartographic(al) training 制图培训
cartographic(al) unit 制图单位;编测单位
cartographic font 制图字体
cartography 绘图学;制图学;制图法;地形测绘;地图制图学;地图学
cartography course 地图制图学课程
cartography department 地图制图科
cartography of soil 土壤制图学
cartology 海图学;地图学
cartometric aid 量图工具
cartometry 量图学;地图量算;图上量算
carton 卡片纸;纸箱;纸盒;纸板箱;纸板盒
carton board bulged 箱板凸起
carton board patched 箱板修补
carton broken at side 箱旁破裂
carton chafed 箱擦花
carton contents rattling 箱内容物有破碎声
carton deformed 箱变形
carton flap open 条盖松开
cartoning machine 装盒机
cartoning sealing machine 封箱机
carton-pierre 箱型纸
carton-pierre and gravel roofing 混凝纸板砾石屋面
carton pipe 厚纸管
carton plank dented 箱板瘪
carton staple 钉箱钉
carton stapler 钉纸盒机
carton stitching machine 钉纸箱机
carton torn, dented 箱撕破压扁
carton wrapper torn 箱皮撕裂
cartoon 卡通片;足尺图案;动画片;动画

cartooning 装纸箱
car top 客车的台面
cartophoto 分区相片镶嵌图
cartoscope 图像显示屏
cartouch(e) 涡形装饰;图廓花边;涡卷饰;涡形装置;漩涡花饰
cartouch(e) release mechanism 喷射筒排放装置
cart paper 地图纸
car track 电车轨道
car-track lane 有轨电车道
car-track line 电车轨道;有轨电车路线
car trailer 拖车
car transfer time 车辆中转时间
car transformer 车用变压器
car transportation 汽车运输
cartridge 卡盘;药包;管壳筒;灯座;熔丝盒;枪筒;拾音器心座
cartridge actuated initiator 火药驱动起始器
cartridge antenna 盒形天线
cartridge brass 黄铜
cartridge clip 套夹
cartridge container 尾管
cartridge cover 滤筒顶盖
cartridged water gel explosive 弹药筒水胶爆药
cartridge dynamite 炸药包;炸药筒
cartridge element 放热元件
cartridge explosive 药卷;炸药筒
cartridge filter 滤芯过滤器;纸筒过滤器;过滤筒;筒式过滤器
cartridge filtration 滤芯过滤
cartridge fuse 管形断熔器;管形熔断器;保险丝管;筒形熔断器;熔丝管;熔断器;保险丝
cartridge heater 加热筒;加热管;圆筒加热器;筒式加热器
cartridge holder 滤芯座
cartridge housing 轴承壳
cartridge igniter 装药点火管
cartridge ignition 火药点火;药包点火
cartridge inserted valve 筒式插装式阀
cartridge interface 盒式接口
cartridge loader 装药工(人)
cartridge paper 绘图纸;转筒纸;图画纸
cartridge pleats 没熨过的窄褶;管形褶裥
cartridge powered tool 锚枪
cartridge receiver 机匣
cartridge replacement 更换滤芯
cartridge seal 盒式密封
cartridge silk 火药袋绸
cartridge stop 弹挡
cartridge tube fuse out 筒状保险装置
cartridge type fuse 熔丝管;保险丝管
cartridge-type pump 芯式泵
cartridge valve 插装式阀;筒形插装式阀
cartridging 包装药卷
car trip 小客车出行
cart road 马车路;马车道;货运道路;乡村(道)路;大车路;大车道
car truck 车辆转向架;车辆转动台
cart rut 车辙
cart track 马车路;乡村道路;大车路;大车道;车道
car tunnel kiln 窑车式隧道窑;车辆装运隧道窑;隧道式窑
car turnaround time 车辆中转时间
car turnround time 货车周转时间
cart way 马车路;马车道;乡村道路;大车路;大车道
cart way panel 脚手架的小车道板
cartwheel 车轮
cart wright 造车工匠;车辆制造者
car-type airless shot blasting machine 车式无空气抛丸清理机
car type conveyer 车型输送机
car-type cooling conveyer 车式冷却输送机
car type furnace 车式电炉;台车式炉
car-type heat treating furnace 台车式热处理炉
car type mo(u)ld conveyer 小车式铸型输送机
car type oven 台车式烘炉
car unloader 卸车机
car unloading operation 卸车作业
car unloading platform 卸车台
car unloadings 卸车数
car unloading technology system 卸车工艺系统
carvan 高碳钒铁
carved altar 雕花祭台;雕刻祭坛
carved and filled in 雕填

carved and inlaid ware 雕镶器
carved bamboo ware 翻簧竹刻
carved boat 木雕船
carved brick facing 雕砖饰面
carved brickwork 砖浮雕；雕刻砖砌体
carved camphorwood chest 雕刻樟木箱
carved capital 雕饰柱帽；雕刻饰柱帽；雕花柱头
carved concrete 雕塑混凝土
carved decorating feature 雕刻装饰特征
carved decoration 刻花；雕刻装饰
carved decorative finish 雕刻装饰面
carved design 刻花
carved embellishment 雕刻装饰
carved enrichment 雕刻装饰
carved fancy blackwood article 雕刻红木小件
carved figure 雕像；雕刻形象；雕刻图样
carved foliage 雕刻叶饰；雕刻的树叶或花形饰物
carved furniture 雕刻家具；雕花家具
carved gilded decoration 雕金
carved ink-stone 雕花墨砚
carved ivory ball 雕刻象牙球
carved ivory vase 牙雕素瓶
carved lacquer cabinet 雕漆柜
carved lacquer folding screen 雕漆围屏
carved lacquer jewellery 雕漆首饰
carved lacquer necklace 雕漆项链
carved lacquer pendant 雕漆垂饰
carved lacquer pendant with tassel 雕漆项坠
carved lacquer screen 雕漆屏风
carved lacquer screen on black ground 黑地雕漆屏风
carved lacquer stool 雕漆凳
carved lacquer wardrobe 雕漆衣橱
carved lacquer ware 雕漆器皿；雕漆
carved motif 雕刻主题；雕饰图式
carved ornamental 雕刻装饰
carved ornamental finish 雕塑装饰
carved pattern 雕刻图案；雕花
carved porcelain 雕瓷
carved pulpit 雕刻布道坛；雕饰的教堂讲台
carved red lacquer 剔红（朱红雕漆）
carved relief 浮雕
carved rug 剪花地毯
carved stone 石雕
carved style 雕塑风格
carved ware 雕刻品；雕花器皿
carved waterspout 雕花水落管；滴水嘴装饰（奇形怪状雕塑像）
carved wood(en) animal 木刻动物
carved wood(en) bracelet 木雕镯子
carved wood(en) cabinet 木雕橱
carved wood(en) chair 木雕椅
carved wood(en) chest 木雕箱
carved wood(en) coffee table 木雕咖啡桌
carved wooden door 雕刻木门
carved wood(en) fancy chest 木雕小套箱
carved wood(en) figure shaped pendant 木雕人物坠
carved wood(en) fire bench 木雕火炉凳
carved wood(en) fruit plate 木雕果盘
carved wood(en) furniture 木雕家具
carved wood(en) jewellery 木雕首饰
carved wood(en) jewellery box 木雕首饰盒
carved wood(en) lamp 木雕灯
carved wood(en) low table 木雕炕几
carved wood(en) necklace 木雕项链
carved wood(en) pendant 木雕垂饰
carved wood(en) screen 木雕屏风
carved wood(en) stool 木雕凳
carved wooden wall hanging 雕刻木挂屏
carved wood(en) wall panel(l)ing 木材雕刻的墙面装饰
carved wood(en) writing desk 木雕写字台
carved work 砖石雕刻；雕刻作业；雕刻工作
carvel built 平镶；平铺
carvel built boat 平铺式舢板
carvel joint 平镶接合；平接缝
car velocity head 车辆速度能高
carvel planking 平铺船壳板；平铺（式）船壳板
carvel shackle 埋头螺钉卸扣
carvel system 平铺式
carve of logging record 测井记录曲线
carver 雕匠；雕刻者；雕刻师；雕工
carve up 压碎；划分；分割

carveyor 有座位的活动人行道
carving 镂刻；雕刻物；雕刻术；雕刻品
carving-and-engraving decoration 雕饰
carving-and-scraping 雕削
carving carbon 切割碳棒
carving chisel 雕刻凿（刀）
carving-grade talc 雕刻级滑石
carving groove 刻槽
carving instrument 雕刻工具
carving knife 雕刻刀
carving machine 刻模机；雕刻机
carving on wood 木雕
carving ornamentation 雕饰
carving stone 雕刻石
carving tool 雕刻工具
carving up the market 分割市场
carving wood 锯料；锯材；木刻材料
car visor 车用遮阳板
carvonone 香芹酮
car wash 洗车房；洗车处；汽车清洗处
car wash booth 洗车厢
car washer 洗车机
car washing 汽车冲洗
car washing canal 洗车沟
car washing line 洗车线
car washing room 洗车房
car washing stand 洗车台
car washing track 洗车线
car-washing wastewater 洗车废水
car wash yard 洗车场
car wax 汽车蜡
car weigher 车辆衡器
car weight 空车重量；车辆重量
car wheel boring machine 车轮镗床
car wheel drilling machine 车轮钻床
car wheel lathe 车轮车床
car wheel repair shop 车轮修理厂；车轮厂
car width 车辆宽度；车辆控制宽度
car windscreen 汽车挡风玻璃
car working record 车辆履历表
caryatid 墩构件；女像柱；女雕像柱
caryatid porch 女像柱门廊
caryinite 砷锰铅矿；砷锰钙矿
caryogram 核型图
caryolysis 核溶解
Caryophyllaceae 石竹科
caryophyllaceous type 直轴式
caryophyllia 佛手珊瑚
caryophyllus oil 丁香油
caryopilite 肾硅锰矿
carystine 石棉
casa 房屋（西班牙的一种房屋）
casale closure with external screw 外螺纹卡扎里密封
cascade 阶梯；阶式布置；急湍；急滩；小瀑布；险滩；挂滑坡；拱滑；格状物；波状花边；梯级
cascade accelerator 级联加速器
cascade aeration 多阶段曝气（法）
cascade aerator 阶式曝气器
cascade amplification 级联放大
cascade amplifier 级联放大器；多级放大器；串级放大器
cascade amplifier klystron 级联放大速调管
cascade analysis 筛分分析
cascade arrangement of tanks 电解槽的阶梯式排列
cascade battery 级联电池组
cascade blending system 串级搅拌系统
cascade breakers 顺序启闭断路器组
cascade buncher 级联聚束器
cascade calc-sinter 瀑水钙华
cascade carry 按位进位
cascade center-peter 串联式中心收集器
cascade centripetal sampler 阶式向心取样器
cascade chromatography 级联色谱法；级联层析
cascade chute 分段式溜子；分段式溜槽
cascade compensation 级联补偿；串联校正；串联补偿
cascade-compensation network 串联校正网络
cascade concentration plant 阶式蒸浓装置
cascade concentrator 阶式蒸浓器
cascade condenser 阶式冷凝器；级联冷凝器
cascade configuration 叶栅外形
cascade connected 级联的；串联的

cascade connection 级联
cascade connection in series 串联
cascade connection tandem 串级
cascade control 级联控制；级联调节；串联控制；串联调速；串级控制；串级调速；分段控制
cascade control system 级联控制系统；串级控制系统；串级调节系统
cascade converter 级联变换器；级联变换机
cascade cooler 级联冷却器；阶式冷却器
cascade coupling 级联耦合；串联耦合
cascade culvert 阶梯式排水道；阶梯式涵洞
cascade current transformer 级联电流互感器
cascaded 级联的
cascade dams 梯级坝
cascaded analog(ue) to circulation code converter 级联模拟循环码转换器
cascaded analog(ue) to pure binary converter 级联模拟纯二进制转换器
cascaded carry 逐位进位；逐级进位；串联进位
cascaded delay system 串行延迟系统
cascade decoder 级联译码器
cascade dense-medium cyclone 梯流式重介质旋流器
cascade development 级联显色；干粉显影；梯级开发
cascaded feed-back canceler 级联反馈补偿器
cascade digital 级联数字
cascaded image converter 级联图像转换器
cascaded laser 级联激发器
cascade drier 多段式干燥器
cascade drop 连续跌水；多级跌水；梯级跌水
cascade dyeing 连续冲流染色
cascade effect 级联效应；叶栅效应
cascade emission 级联发射
cascade entry 级联入口；串接表目
cascade-evaporation 串级一蒸发
cascade evapo(u)rator 阶式蒸发器
cascade exciter 级联激发器；级联励磁机
cascade fall 多级跌水；梯级跌水
cascade firing 级联激发
cascade flow 叶栅流动
cascade fluorescent screen 积层荧光屏
cascade fold 滑曲褶皱；滑落褶皱；重力弯曲褶皱；拱滑褶皱；叠褶（重力滑曲褶皱）；迭褶
cascade gamma emission 级联伽马发射
cascade generator 级联加速器；级联发电机；串级发电机
cascade grain drier 阶梯式谷物干燥机
cascade graphs 级联图
cascade grinding mill 湍流式磨坊；急流磨坊；级联式磨坊
cascade heater 串联加热器
cascade heat exchanger 阶梯式热换器
cascade homogenization 串级均化
cascade hydroelectric(al) power plant 梯级水电站
cascade hydroelectric(al) station 梯级水电站
cascade hydropower stations 多级水电站
cascade hypothesis 瀑布假设
cascade image converter 级联变像管
cascade image intensifier 级联（式）图像增强器
cascade image tube 级联摄像管
cascade imaging system 级联成像系统
cascade impact electrostatic precipitator 多段冲击式静电除尘器
cascade impacter 阶式低速碰撞采样器
cascade impactor 级联无砧座锻锤；阶式撞击（采样）器；阶式碰撞采样器；阶式冲击采样器；级联冲击（取样）器
cascade input 级联输入
cascade jet impactor 阶式喷嘴冲击器；阶式喷嘴碰撞器；阶式喷嘴碰撞采样器
cascade klystron 级联速调管
cascade laser 级联激光器；级联激光器
cascade liquefaction 级联液化
cascade list 级联表；串接表
cascade lubrication 帘状润滑
cascade luminescence 级联发光
cascade machine 高落式浮选机；串联机
cascade mass spectrometer 级联质谱计
cascade mechanism 级联机理
cascade merging 级联归并
cascade method 级联法；逐级测量法；串级连接法
cascade mill 级联式坊；多级粉磨；气唐磨
cascade mixer-settler 级联混合澄清槽

cascade motor 级联电动机
cascade motors 级联电动机组
cascade network 阶梯形网络;级联网络
cascade noise 串级噪声
cascade of mixers 级联混合器
cascade of settlers 级联沉降器
cascade of straightline profile 平板型叶栅
cascade oiling 环给油;油杯润滑;梯流润滑
cascade outage 串级停电事故;串级式断电
cascade pitchchord ratio 叶栅节弦比
cascade portion 急流段;险滩段
cascade process 阶式法;级联过程;串联过程;串联法
cascade process of refrigeration 制冷级联过程
cascade protection 分级保护
cascade pulverizer 多段粉碎机;串联粉碎机;梯流粉碎机
cascade pump 级泵;串联自吸式泵
cascade pumping stations 多级泵站
cascader 扬料器;扬料板
cascade refrigerating system 串级式制冷系统;串级制冷系统
cascade regulating system 串级调节系统
cascade regulation 串级调节
cascade relay 级联继电器
cascade reservoirs 梯级水库
cascade reuse 阶式再用
cascade reuse with pretreatment 有预处理的阶式再用
cascade rotary pump 级联泵
cascade screen 级屏蔽;级联屏蔽;串级屏蔽
cascade sequence 重叠焊层沉积顺序;串列顺序;串级叠置法
cascade set 级联式机组;串级机组
cascade shower 级联簇射
cascade solidity 叶栅稠度
cascade sort 级联分类
cascade spacing 栅距
cascade spillway 阶梯式消能溢洪道
cascade starter 级联起动机
cascade synchrotron 级联同步加速器
cascade synthesis 链接法
cascade system 级联系统;级联方式;串联系统;复迭式系统
cascade tank 阶式水箱
cascade tax 阶式税
cascade test 叶栅试验
cascade thermoelectric(al) cooling system 级联热电冷却系统
cascade thermoelectric(al) refrigerating unit 覆叠式热电制冷器
cascade transformer 级联变压器;级间变压器
cascade transformer type accelerator 级联变压器式加速器
cascade transistor circuit 串联晶体管电路
cascade transition detector 级联跃迁探测器
cascade tripping 逐级跳闸;串级跳闸
cascade tube 级联管
cascade type 阶式
cascade type drier 阶式干燥器;井筒式干燥机;多段式干燥器
cascade unit 辐射长度
cascade voltage doubler 级联倍压器
cascade washer 阶式洗涤器;级联洗涤器;冲流式洗涤机;冲碱装置
cascade washing 阶式水洗;梯流水洗
cascade welding 阶梯式焊;山形多层焊
cascade welding sequence 串级叠置法
cascade wind tunnel 叶栅风洞
Cascadian orogeny 卡斯卡底(造山)运动
cascading 级联作用;级联效应;级联;瀑球瀑落;梯流
cascading channel of grinding media 研磨体提升槽
cascading flow 梯级跌水;梯级跌流;瀑布式梯级跌流
cascading glacier 瀑布冰川;湍降冰川
cascading of grinding media 研磨体脱落
cascading water 泄流;下沉水
cascadite 橄榄云煌岩
cascandite 羟硅钙钪石
casco 蛋白质胶合剂(防潮用)
cascode amplifier 共射共基放大器
casco insurance 海损保险;船壳保险
cascophen 酚醛树脂黏合剂

case 壳体;壳;箱子;箱体;箱盒;主体;案情;案例;案件;情况;情景;外罩;外型(石膏模);外壳;外表面;套;数据组;实例
case 2 firing 第二瞄准法
case 2 pointing 第二瞄准法
case adapter 套管接箍
case analysis 实例分析
case at law 法律案件
case band 箱箍
case bay 明间;开间;(跨搁在大梁上的)格栅;桁间;楼地板梁空挡;梁间距;梁间隔
case bay part 椽梁间距部分
case bolt 圆头螺栓
casebook 专题资料集
case broken 箱破损
case butt coupling 套筒式联轴器
case-by-case 逐项给予
case-by-case basis 逐例进行
case by case screening 逐个审查
case carbon 表面含碳量
case-carbonizing 表面渗碳
case-carburizing 表面渗碳
case-chilled 表面冷激的;表面冷硬的
case chip 封装片
case closed 结案
case column 匣形柱
case contact 接地片
case cord 包芯纱缆线
case crush 齿面剥落
case crushed 箱摔碎
case crushing 表面碎裂
case cube reckoning 货箱立方计算
cased 外加套管的
cased beam 箱形梁;匣形梁
cased book 封壳书
cased bore hole 下套管的井;套管钻孔;下套管钻孔;有套管钻孔;装有套管的钻孔
cased column 匣形柱;空心柱;加套柱;箱形柱
cased concrete pile 带套管的混凝土桩;套管混凝土桩
case death ratio 疾病死亡(比)率
case depth 表面深度
cased frame 箱形框;箱式窗架;匣形窗框;上下拉窗的空心框架
cased girder 箱形大梁
cased glass 镶色玻璃;套色玻璃;叠层装饰玻璃;套覆玻璃
cased glass cutting 套色刻花玻璃
cased hole 下套管的钻孔;套管钻孔
cased hole completion 套管钻孔;带孔的预制件
cased-hole-test 套管钻孔试验
cased hollow ware 套料玻璃器皿
cased impeller 箱形叶轮;装在外壳内的叶轮
cased-in blower 盒形鼓风机
cased-in pile 带套桩;带套管的钻孔灌注桩
cased-muff coupling 刚性联轴器
cased off the(bore)hole 孔内套管隔离的
cased opening 饰边门洞;有贴脸的门洞
cased pile 带套桩;带套管的钻孔灌注桩;套管桩;钢壳混凝土桩
cased pile Aba-Lorenz 阿巴洛伦兹(带)套管桩
cased post 外包支柱;有套的桩
cased pressure pile 带套管压力桩
case drain line 壳体泄油管;壳排泄管
cased sash frame 匣形窗框
cased seal 带壳密封
cased structure 箱形结构
cased through well 全部下套管井
cased type fan 有护网风扇;有保护网的风扇
cased wall 被服壁(日本式);橱壁
cased well 管井;套管深井;套管井
cased window 匣形窗架
cased window frame 箱形窗框;匣形窗框;箱式窗框;匣形窗框
cased worm gear 闭式蜗轮传动
case empty 箱空
case fatality 病死率
case for changeover braking device 制动转换器箱体
case for gyro 陀螺房
case fragile 箱子脆弱
case frail 箱脆弱
case front 橱的正面
case glass 叠层装饰玻璃

case goods 箱装货(物)
case half empty 箱半空
case handle 装入凹槽的门拉手
case harden 热处理;木材干曲;表面硬化
case-hardenability 表面可淬硬性
case-hardened 表面硬化的;(玻璃的)钢化;表面干裂的(木材)
case-hardened glass 表面硬化玻璃;钢化玻璃;表面钢化硬化
case-hardened steel 表面(渗碳)硬化钢;渗碳钢
case-hardening (无损试验的)表面硬化法渗;表面渗碳硬化;表面淬火;天然石中含水量使表面硬化;钢的表面渗碳硬化;表面干燥的木材;表皮硬化;表面硬化成壳;表层硬化;外层硬化;表面硬化
case hardening bath 表面硬化池
case-hardening carburizer 表面硬化渗碳剂
case-hardening furnace 表面硬化炉;箱式渗碳炉
case hardening mixture 表面硬化混合物
case-hardening stee 渗碳钢
case-hardening steel 表面硬化钢
case hardness 表面硬度
case head 套管头
case heating-treatment 局部热处理
case histories 事例
case history 专题资料;专题档案;工程事例;工程实录;工程实例;典型例证;病历;事故历史;实例记载;施工记录
case history file 工程史料档案
case history package 实例历史包
case hook 箱装货吊钩
case in 下套管
casein 酪朊
casein adhesive 酪朊黏合剂
caseinate 酪朊酸盐
casein cement 酪素黏合剂
casein emulsion 酪素乳液
casein finish 揩光浆;乳酪涂饰剂
casein formaldehyde resin 酪素甲醛树脂
casein glue 酪素胶;酪朊胶;酪蛋白胶;干酪胶
casein medium 酪素介质
caseinogen 酪朊
casein paint 干酪漆;酪素涂料;酪素漆;酪朊漆;酪蛋白涂料;酪蛋白漆
casein pigment 酪素颜料膏
casein plastics 酪蛋白塑料;酪素塑料;酪朊塑料
case in point 例证
casein powder 酪蛋白粉;粉状酪素
casein-powder paint 酪朊粉末涂料
casein resin 酪素树脂
casein size 酪素白胶水
casein sizing material 酪蛋白胶凝材料
case investigation 个案调查
casein wash 可赛银粉刷
casein water paint 酪素水溶性涂料;酪朊质水溶性涂料
case irreducibility 事件不可简化
case-knife 餐刀
case label 分情形标号
case law 判例法
case leak 管壳漏泄
case-lid box 套盖箱
Casella automatic microscope 卡塞拉自动显微镜
case-load 事故例数
casemate 空心造型;防炸弹掩蔽部;防弹掩蔽体;避弹窖;壁弹堡垒;掩蔽部;暗炮台
casemate wall 城堡墙上的掩蔽体;城堡墙上的掩蔽体
casement 孔模;空型;空心造型;窗扇;窗扉;玻璃窗扇;玻璃窗扉;平开窗扇
casement adjuster 窗扇撑杆;窗风撑
casement bolt 玻璃窗插销
casement cloth 薄窗帘布;薄织物
casement door 长窗;玻璃门扉;玻璃门
casement fastener 窗扇固定撑件;钢窗扳手;窗销;窗插销
casement frame 竖铰链窗框架;窗扇框;窗扇边框
casement hinge 窗扇铰链
casement light 有枢轴窗;有铰链窗;竖铰链气窗
casement opening in 内开扇;内开(式)窗扇
casement opening out 外开窗扇;外开窗扉
casement pintle 玻璃窗枢轴;玻璃窗铰链
casement repp 厚棱窗帘布
casement sash 竖铰链窗扇
casement screen 平开纱窗;平开隔帘

casement section 窗框钢;窗框型钢;窗框钢材
casement section steel 窗框钢
casement stay 窗撑头;支窗棍;窗扇支撑;窗风钩;窗风撑;窗撑杆
casement ventilator 平开(窗)通风门
casement wall 侧墙
casement window 双扇窗;门式窗;玻璃窗;平开窗;竖铰链窗
casement window bolt 玻璃窗插销;平开窗插销
casement with sliding upper-sash 上扇滑动式窗扉
case method 案例研究法
case mo(u)ld 模盒;壳模;(石膏的)框架模型;母模;二代种模;石膏壳模
case mo(u)ld plaster 母模用石膏
case-mounted controller 箱装式调节器
case nearly empty 几乎空箱
case noise 壳体噪声
case number 箱号
case of avertence 偏转式
case of bending 弯曲状态
case of degeneracy 退化情况
case of differential 差速器箱;分速器箱
case of divergence 离向式
case of drawing instrument 盒装绘图仪
case of enforcement of taking up land unlawfully 非法占用土地强制执行案
case of final drive 最终转动箱
case of horizontal swing 平面旋转式
case of loading 荷载方式
case of lock 锁壳
case of longitudinal tilt 等偏式
case of pump 泵壳
case of stability 稳定状态
case oil 箱装油;桶装油
case opener 开箱器;开箱机
caseous tubercle 黄色结核结节
case package 管壳封装
case pallet 箱盒运送板架
case pile 钢壳混凝土桩
case pipe 套管
caser 装箱机
Case Rancho roofing tile 坎撒·兰奇欧屋面瓦
case record 专题资料
case repaired 箱子经修补
case ripper 套管割刀
casern 小型临时工棚;要塞内兵营
case rotating shutter 盒形旋转快门
cases bottoms off 箱底脱落
cases broken at bottom 箱底破裂
cases broken at corner 箱角破裂
cases cracked and renailed 箱子摔破重钉
case sealed 箱被封妥
case seal off 箱封条脱落
case selector 分情形选择子;分情形选择器
case shell 表面层次;汽缸外罩
cases jammed at the side 箱边压坏
case smashed 箱撞碎
case spring 闭锁弹簧
case statement 选择语句
case steel 表面硬化合金钢
case study 情况研究;专题研究;工程实例;个例研究;案例研究;事例研究;实例研究;实例分析
case study house 实验性住宅
case study method 方案研究法
cases with bands broken 箱箍断脱
cases with bands off 箍条脱落
cases with boards and battens broken 箱板条板破裂
cases with boards staved in 箱板撞破
cases with end battens split 箱末端板条裂开
cases with ends split 箱端裂开
cases with nail holes 箱板有钉眼
cases with panels split 箱镶板裂开
cases with side batten broken 箱旁板条破裂
cases with side boards broken 箱旁板破裂
case unsufficient packed 包装不固
case way wiring system 箱式布线法
casework 装修木工作业;细木工;调查工作;案例调查
cash 现款;现金;现钞(尤指新钞);备用现金
cash ability 变现能力
cash a bill 根据期票领取现款
cash account 现款交易;现金账目;现金账户
cash accounting 现金会计
cash a check 支票兑现;兑支票

cash administration 现金管理
cash advance 预付现金
cash against delivery 交货付款;货到付款
cash against documents 见单据付款;凭单据付款;凭单付款;押汇证;付现交单;凭单证付款
cash against documents at port of shipment 在启运港凭单据付现
cash against payment 付款交单
cash against(shipping) documents 交单付款
cash agent 现金代理人
cash allowance 现金折让;现金津贴;备用费项
cash and bank balance 现金及银行结存
cash and bank book 现金和银行出纳账
cash-and-carry 现款现货;现购自运;现金交易;仓库现价交货
cash and receivable to current liabilities ratio 现金及应收款对流动负债的比率
cash assets 现金资产
cash audit 现金审计;现金审核
cash balance 现金结存;现款余额;现金余额
cash balance on hand 库存现金
cash basis 现收现付制;收付实现基础
cash basis accounting 收付实现制会计
cash basis of accounting 现金往来账户
cash basis of revenue recognition 营业收入的收付实现确定
cash before delivery 货到前付款;交货前付款;付现金交货;付现后提货;付款后提货
cash before shipment 装船前付现;装船前付款
cash blotter 现金出纳备查账
cash bonus 现金红利;现金分红
cash book 现金账簿;现金账目;现金账户;现金记录本;现金出纳簿;现金簿
cashbox 钱柜;钱箱
cash break-even point 现金收支平衡点
cash budget 现金预算
cash-buoy 桶形浮标
cash by return mail 回邮付现;回程邮递付现
cash by return steamer 回航付现
cash call 摊付分保现金赔款
cash center 现金出纳中心
cash check 现金支票
cash circulation 现金流通
cash claim 现金索赔
cash collateral invested 投入的现金抵押
cash collection schedule 现金回收一览表
cash compensation 现金赔偿
cash concept 现金观念
cash consumption 现金消费
cash contribution 现金捐助
cash control 现金管理
cash count 现金点查
cash counterpart 对应现金
cash covered options 现金抵补期权
cash credit 活期信用放款;活期贷款;现金借方;现金贷方;暂付贷款;保证放款
cash credit slip 现金支出传票;现金信贷传票;现金付出传票
cash crop 经济作物;商品作物
cash currency 现金货币;现金通货
cash customer 以现金交易的顾客
cash cycle 现金循环
cash day 付款日
cash debit slip 现金收入传票
cash deficit 现金亏绌
cash delivery 现款交割
cash deposit 现金存款;保证金
cash deposit as collateral 交现金保证;押金;保证金
cash deposited for bond interest 备付债券利息存款
cash deposit in advance 预付现金
cash deposit in sinking fund 偿债基金银行存款;现金存款比率
cash-deposit ratio 现金存款比率
cash disbursement 现金付出;现金支出
cash disbursement book 现金支出簿
cash disbursement register 现金支出登记簿
cash disbursement schedule 现金支出一览表
cash disbursements journal 现金支出日记账
cash disbursement voucher 现金支出凭证
cash discount 现金折扣;付现折扣
cash distribution plan 现金分配计划
cash dividend 现金股息;现金股利;现金分红
cash dividend paid 已付现金股息
cash down 即付现款;即期付款

cash drain 现金消耗;现金外流;现金枯竭
cash drawing 提取现金
cash earnings 现金收益
cash earnings per share 每股现金收益
cashed check 已兑现支票
cash equivalency 现金等值法
cash equivalent value 现金等值
casher box 铁筐(手工成型时盛放吹制玻璃瓶)
casher's department 账房
cashew 腰果(实)
cashew lake 金棕色
cashew nut aldehyde plastic 漆酚醛塑料
cashew nut oil 腰果油
cashew nut shell liquid resin 腰果壳液树脂
cashew nut shell oil aldehyde resin 腰果壳油酚醛树脂
cashew resin 腰果油树脂;漆酚树脂
cashew resin enamel 腰果树脂瓷漆
cashew resin paint 腰果树脂涂料
cashew resin putty 腰果树脂油灰
cashew water coat 腰果油水性防锈底漆
cash financing program(me) 现金筹借计划;现金筹措方案
cash flow 资金周转;现金流转;现金流量;现金流动;资金流动
cash flow adequacy ratio 现金流量充足程度比率
cash flow analysis 现金流通分析;现金流量分析;资金流动分析
cash flow before tax 税前现金流量
cash flow budget 现金流预算;现金流循环
cash-flow diagram 货币流向图
cash flow effect 内部资金效应
cash flow estimate 现金流量估计
cash flow from operation 营业现金流动
cash flowing 现金流通
cash flow statement 现金收支一览表;现金流程报表;收支款项期报
cash flow stream 现金流
cash flow table 现金流量表;财务现金流量表
cash forecast 现金预测
cash forward contract 现金交货合同
cash fund 保证金;现款资金;现金基金
cash generated shares 每股所获得现金额
cash-grain farm 粮食农场
cash guarantee 现金保证;保证金
cash held in foreign countries 存国外现金
cash holding 库存现金
cash holding audit 库存现金审计
cashier 出纳员
cashier's check 银行本票
cashier's counter 出纳柜台
cashier's desk 出纳台
cashier's funds 出纳备用金
cashier's office 出纳处
cashier's order 本票
cash in 结算;兑现
cash in advance 现金预付;先付;预付货款
cash income 现金收入
cash income and cost 现金收益和成本
cash in deposit 存银行现金
cash inflow 现金流入
cash inflow analysis 现金流入分析
cashing check 兑现支票
cashing dividend 股利兑现
cashing in a prize 现金发奖
cashing interest 领取利息;利息兑现
cash in hand 库存现金
cash in hand of agent 存代理人现金
cash in order 现款订货;订货付现(金)
cash in transit 在途现金
cash in transit insurance 现金运输保险
cash in transit policy 现金运输保险单;现金保险单
cash in treasury 库存现金
cash in trust account 现金信托账户
cash invoice 现购发票
cash item 现金科目
cash items in process of collection 托收中款项
cash joint 平齐接缝
cash journal book 现金日记簿
cash journal method 现金分录法
cash keeper 现金保管员
cash liability 流动负债
cash liquidity 现金流动性
cash losses 现金赔款;保险金的现金赔款

cash management 现金管理
cash management account 现金管理账户
cash management bills 现金管理票据
cash management model 现金管理模式
cash management service 现金管理服务
cash market 现金交易市场;现货市场;付现市场
cash nexus 现金交易关系
cash of letter of credit 现金信用证
cashomat 自动提款机
cash on arrival 货到付款
cash-on-cash 税前现金除以投资股本
cash-on-cash return 现金投资现金收益(率)
cash on delivery 交货付现;交货付款;货到收款;货到付现款;货到付款;现款交货;现金交货;银货两讫;到货付款;到货收款
cash on delivery sales 交货收款销售
cash on documents 见证付款;付现交单
cash on hand 留存现金
cash on hand account 库存现金账户
cash on hand in bank account 库存现金及银行存款账户
cash on hand statement 可即时支付的现金声明
cash on receipt of merchandise 收货时付款
cash on shipment 装货付款;装船付款
cash on the bank 现款
cash on the mail 邮寄现款
cash operating statement 现金收支执行情况表
cash order 现金订(货)单;现付票
cash out 现金支出;现金流出;现金不足
cash outflow 现金外流;现金流出
cash over and short 现金余缺
cash over and short account 现金超过和不足账户
cash paid book 现金支出账;现金支出簿;现金付款登记簿
cash payment 交付现金;现金支出;现金付款;现付;付现钱;付现金
cash payment book 现金支出簿
cash payment journal 现金支出日记账
cash payment register 现金支出登记簿
cash payment slip 现金支出传票
cash plow 现金周转
cash policy loan 以保险单作抵押的贷款
cash position 现金头寸
cash position sheet 现金部位表
cash price 现款售价;现金售价;现金付款的最低价格;付现金价格;付现价格
cash provided by operations 运营现金收入;经营现金收入
cash purchase 现款购买;现金购买;现购
cash rate 现钞汇率
cash ratio 现金比率
cash realizable income 变价收入
cash receipts 现金收入
cash receipts and expense audit 现金收支业务审计
cash receipts book 现金收入簿
cash receipts journal 现金收入日记账
cash receipts register 现金收入登记簿
cash-receipt voucher 现金收入凭证
cash received 收入现金
cash received on behalf of third party 代收款
cash record 现金记录
cash redemption 现金偿还
cash register 现金出纳机
cash remainder 现金余额
cash remittance 汇款;汇汇
cash remittance note 解款单
cash rent 现金地租
cash report 现金收支报告
cash requirement 现金需要量;现金需要
cash reserve 流动储备;现金余额;现金储备
cash reserve equation 现金储备方程式
cash reserve ratio 现金储备率
cash resource 现金资源
cash reward 现款奖金
cash room 点钞室
cash sale 现销;现金销售;现货销售
cash sale book 现金销售簿
cash sale invoice 门市发票;现销发票;现金销售发票
cash sale price 现销价格
cash sale slip 现金销售传票
cash sedimenatation 现金沉淀
cash settlement 现金结算;现汇结算

cash share 现金股票
cash slip 现金传票
cash source 现金来源
cash spending approach 现金耗用渠道
cash statement 库存表;现金(报)表
cash status 现金状况
cash status prediction 现金状况预测
cash stave broken 桶板破裂
cash store 现金兑换店
cash summary book 现金汇总簿
cash surplus 现金结余;现金盈余
cash surrender value 保险积存金;人寿保险退保解约金值;退保金额
cash surrender value of life insurance 人身保险退保现值
cash term 付现条件
cash throw-off 税前现金流动
cash tickets 门市发票;现销票
cash-to-revenue ratio 现金对销货收入比率
cash totalizer 现金累计器
cash-to-total assets ratio 现金对总资产比率
cash train 运款列车
cash transaction 现金买卖;现金交易;现金
cash transfer 现金调拨
cash turnover 现金周转
cash value 现金价值
cash vault 现款保险库
cash verification 现金核实;现金核查
cash verification certificate 现金核查证(书)
cash vs. accrual 应收与实收的现金
cash with agent for dividend 发放股利专款
cash with order 现款订货;下定付款;订货时付款;随订单支付现金;随订单付款
cash without discount 付现无折扣
Casimir operator 卡西姆算子
casing 壳体;壳;机壳;装箱;柱套管;遮板;罩布;包装;汽缸;汽车外胎;围壁;外套;外壳;套色
casing adapter 管柱接头;套管异径接头;套管接箍
casing anchor packer 套管堵塞器
casing anchor point 汽缸固定点
casing appliances 下套管用工具
casing barrel 套管筒
casing barrel reaming shell 回转套管法钻进用扩孔器
casing bead 手头装饰条;门窗洞口珠饰;嵌条;分隔条;石膏边缘
casing-bead door-frame 带嵌条的门框
casing bit 套管钻孔用钻头;套管钻头
casing block 下套管滑车;起下套管游动滑车
casing boring platform 套管钻孔平台
casing bowl 套管打捞筒
casing breakdown type 套管破坏分类
casing buckling critical length 套管弯曲临界长度
casing buoyed weight 套管浮重
casing cap 螺旋管塞;螺栓管塞
casing catcher 捞管器;钻孔套管防坠器
casing cementing job 套管注水泥作业
casing cement plug 套管水泥塞
casing centralizer 套管定中器;套管找中器;套管扶正器
casing clamp 套管卡;套管夹板
casing clearance 套管环状间隙
casing collapse depth 套管抗挤下入深度
casing collar 套管接箍
casing collar kick 磁力钻孔测深器
casing collar locator 套管接箍定位器
casing column 套管柱
casing connections 套管接箍
casing cost 套管费;套管成本
casing coupling 套接管(俗称缩节);套管(连箍);套管接箍;缩节(即套接管)
casing cutter 套管割刀
casing cutter jar 套管割刀震击器
casing cutter sinker 套管割刀加重杆
casing cutter wedge 套管割刀楔
casing data 套管数据
casing depth 套管深度
casing depth of producing zone 生产层套管深
casing design 套管设计
casing diameter-thickness ratio 套管径厚比
casing dog 捞管器;套管打捞矛
casing drift ga(u)ge 套管内径规
casing drill-in 套管送下程序
casing drilling 套管钻进

casing drive head 套管打入头
casing drive shoe 套管打入靴
casing elastic transition zone unstable breakdown 套管弹性过渡区不稳定破坏
casing elastic unstable breakdown 套管弹性不稳定破坏
casing element 外壳件
casing elevator 箱形升降机;套管吊机;管式升降机;管式电梯;套管提引器;套管吊卡
casing fin tube 套片式翅片管
casing fittings 套管配件
casing flange 套管突缘;套管法兰(盘);管子法兰(盘)
casing float 套管浮阀
casing float collar 套管浮箍
casing float shoe 套管浮靴
casing force type 套管受力分类
casing form 套管类型
casing for pendulum clock 摆钟外壳
casing for spark adjustment 火花调整箱
casing ga(u)ge 套管螺纹规;套管规
casing glass 镶色玻璃;套色玻璃
casing grab 套管打捞抓
casing grades 套管钢级
casing grip 套管夹
casing grouting 套管灌浆
casing guide 套筒滑板;套管引靴;套管导正装置;套管导向帽
casing guide of pump 泵体导槽
casing gun 套管冲孔枪;弹筒打孔机;套管射孔枪
casing hanger 套管挂;套管吊卡
casing hanger slips 套管悬挂卡瓦
casing head 螺旋管塞;套管头
casing-head gas 井口气(体);天然气;套管头气体;油井气
casing-head gasoline 天然汽油;井口气分离出来的油
casing head plant 天然气汽油厂
casing head stuffing box 井口密封装置;套管头填料函
casinghead tank 套管头储罐
casing head top 套管头上端
casing hole-boring method 套管护壁钻孔法
casing hook 套管吊钩
casing housing 套管头
casing hydraulic tongs 套管液压吊钳
casing injection 套管注浆
casing inside diameter 套管内径
casing installation 下套管
casing jack 套管千斤顶
casing joint 套管连接;套管接头;套管接口;套管接合;套管接箍
casing knife 修饰刀;切管器;切管刀;墙纸切边刀;套管割刀
casing leak 机壳漏泄;外壳漏泄;套管漏泄
casing leakage tester 套管测漏器
casing less completion 无套管成井
casing line 起落滑车绳;套管钢丝绳
casing line pulley 套管钢丝绳滑轮
casing liner 装箱衬垫
casing line sheave 套管缆绳滑轮
casing liquid level detecting 测套管液面
casingl jacking 套料
casing log 套管测井
casing lug 汽缸猫爪
casing machine 钻井设备;钻井机械;钻孔加固机
casing mandrel 套管整形器
casing mark 套管标记
casing method 箱移法
casing method of cementing 套管注水泥法
casing mount 泵座
casing-mounted drill rig 套管安装钻孔机
casing nail 饰钉;小头钉;包装用图钉;包装钉
casing nipple 套管公接头
casing nominal diameter 套管公称直径
casing of column 柱子护面
casing off 套管隔离
casing off hole 套管固孔
casing of pile 打桩架;填筑柱箍
casing of wells 井栏
casing or driving pile 套管或击入管
casing O-ring 泵壳 O 形环
casing oscillator 箱式振动器
casing outside diameter 套管外径

casing packer 套管封隔器
casing paper 衬筒用纸
casing part 套管部件
casing perforator 套管冲孔器;套管射孔器
casing pipe 钻管;井壁管;套管
casing pipe for slipping form 滑模套管
casing pipe shield method 套管护壁法
casing plastic unstable breakdown 套管塑性失稳破坏
casing platform 钻井平台;套管平台
casing ply 骨架层
casing point 钻管下入深度;下套管深度;下套管部位
casing pressure 套管压力
casing program(me) 井身结构;钻孔加固设计;入井套管层数
casing protector 钻杆橡胶护箍;套管护圈;钻杆橡胶护圈;套管护箍;套管
casing pulley 起下套管滑车
casing record 下套管记录
casing ripper 套管932割刀;套管割刀
casing roller 胀管器;套管修正器;套管修整器
casing rope percussion-grab drill 有套管绳索冲抓钻机
casing-rotating device 回转套管装置
casing screw head 套管护丝(母扣)
casing screw protector 套管护丝(公扣)
casing seat 套管座;套管承托环
casing section mill 套管铣刀
casing severing 切割套管
casing shoe 套管靴;套管头;套管管靴
casing shoe bit 套管靴钻头
casing shoe depth 套管靴深度
casing size 套管尺寸
casing slips 套管卡瓦
casing snubber 套管压入头
casing spear 套管(打捞)矛;套管(深孔技术)
casing spider 套管卡盘
casing splitter 套管(纵向)割刀;套管劈裂器
casing spool 套管钢丝绳滚筒
casing spreader 外(胎)扩展器
casing spring 框簧
casing squib 套管炸药包
casing starter (套管柱中的)下部套管
casing steel grade coefficient 套管钢级系数
casing steel minimum limit strength 套管钢材最小极限强度
casing steel minimum yield strength 套管钢材最小屈服强皮
casing strength 套管强度
casing string 套管柱
casing string design 下套管程序设计
casing sub(stitute) 套管转接器;套管异径接头;套管大小头
casing suspender 套管悬挂器
casing swage 胀管器;套管修正器;套管修整器
casing swedge 挤管器
casing swivel 套管(水)龙头
casing tap 套管打捞公锥
casing tester 套管测漏器;套管试验器;套管测试器
casing test leading line 套装测试引线
casing test pressure 泵壳试验压力
casing thread data 套管丝扣数据
casing threads 套管螺纹
casing thread strength 套管丝扣强度
casing tongs 套筒扳子;套管扳手;套管吊钳;套管扳钳
casing tool 套管工具
casing top half 上半汽缸
casing treatment 机匣处理
casing trim 门窗的额楣
casing trouble 套管事故
casing tube 钻管;井壁管
casing unit weight 套管单位重量
casing unstable breakdown 套管失稳破坏
casing valve 套管浮阀
casing wagon 运套管车;套管车
casing wall 缸壁;侧墙;汽缸壁
casing wall thickness 汽缸壁厚
casing water swivel 套管水龙头;套管进水转座连接器;套管进水回转接头
casing weight 套管重量
casing weight elongation 套管自重伸长
casing well section 套管井段
casing wire nail 包装钉

casing with bell-mounted collars 承插式套管
casing wrench 套筒扳手
casing yield breakdown 套管屈服破坏
Casino 卡西诺高速钢
casino (公园等的)凉棚;娱乐场;赌场
casino folie 娱乐场
casino pink 茜草玫瑰红色
casiterite-quartz vein formation 锡石—石英脉建造
cask 木桶;中木桶
cask beer store 桶装啤酒店
casket 小箱;手箱;吊斗;吊桶;容器
cask filling machine 装桶机
cask hook off 桶钩
cask leaking through seam 桶板缝渗漏
cask packing machine 装桶机
cask-re-coopered 桶重修理
cask sling 吊桶索套;吊桶绳环;桶钩
cask stave off 桶板脱落
cask wagon 酒类容器车
Casle's seal with bolts 螺栓卡扎里密封
caslox 合成树脂结合剂磁铁
Casparian dot 凯氏点
Caspian Sea 里海
caspidite 枪晶石
Cassadagan 凯瑟达格统【地】
Cassagrande dot 卡萨格兰德孔压测头
Cassagrande hydrometer 卡萨格兰德比重计
Cassagrande liquid limit apparatus 卡萨格兰德液限仪;卡氏液限仪
Cassagrande liquid limit machine 卡萨格兰德液限测定机
Cassagrande plasticity chart 卡萨格兰德塑性图
Cassagrande shear test apparatus 卡萨格兰德液限剪力试验机
Cassagrande soil classification 卡萨格兰德土分类
Cassagrande soil classification chart 卡萨格兰德土分类图
Cassagrande system of classification of soil for airport projects 卡萨格兰德飞机场跑道土分类法
Cassagrande type of consolidometer for combined compression 卡萨格兰德液式复合受压渗透(固结)仪
cassava adhesive 木薯粉胶结剂
cassava alcohol wastewater 木薯酒精液
cassava glue 木薯粉胶;木薯粉糨糊
cassava starch 木薯粉
Cassegrain(ian) antenna 卡塞格林天线
Cassegrain(ian) camera 卡塞格林照相机
Cassegrain(ian) focus 卡塞格林焦点
Cassegrain(ian) focus corrector 卡塞格林校正器
Cassegrain(ian) horn antenna 卡塞格林喇叭形天线
Cassegrain(ian) mirror 卡塞格林镜
Cassegrain(ian) objective 卡塞格林物镜
Cassegrain(ian) radiometer 卡塞格林辐射计
Cassegrain(ian) reflecting telescope 卡塞格林反射望远镜
Cassegrain(ian) secondary 卡塞格林副镜
Cassegrain(ian) system 卡塞格林系统
Cassegrain(ian) telescope 卡塞格林望远镜;双折射望远镜
Cassel brown 卡塞尔棕;深棕色
Cassel earth 卡塞尔棕
Cassel green 卡塞尔绿;锰绿
Cassella's fibre test 卡塞尔纤维鉴定法
Casselmann green 卡塞尔曼绿色颜料
Cassel's yellow 碱性氯化铅黄色染料
Cassel tester 卡塞尔试验机
Cassel yellow 卡塞尔黄;氯氧化铅黄色颜料;氯化铅黄
casserole 有柄蒸发皿;有柄瓷皿;有柄瓷埚;勺皿;砂锅
casseroles brown coal 褐煤式颜料
cassette 胶卷的卷盒;暗盒;片匣
cassette ceiling 藻井顶棚;格形顶棚
cassette data recorder 盒式数据记录器
cassette design 格子形设计
cassette floor 双向密肋楼板
cassette holder 盒座
cassette pattern 格子形图案;格子形模式
cassette photoelastic indicator 便携式光弹仪
cassette recorder 盒式记录器
cassette slab 格子板
cassette soffit 格状拱腹;格状拱顶

cassia 桂皮树
cassia lignea 樟木板
Cassiar orogeny 卡西阿尔造山运动
cassia scrapped 桂芯
cassia scrapped broken 桂芯碎
cassia scrapped whole 桂芯条
cassidyite 磷钙镍石
cassie 包损纸
cassimere twill 两上两下斜纹
Cassini formula of gravity 卡辛尼重力公式
Cassini's projection 卡西尼投影
cassinite 钡正长石
cassiterite 锡石含量;锡石
cassiterite ore 锡石矿石
cassiterite ore associated by sulfide mineral 硫化物型锡石矿石
cassiterite ore of chlorite type 绿泥石型锡石矿石
cassiterite ore of greisen type 云英岩型锡石矿石
cassiterite ore of tourmaline type 电气石型锡石矿石
cassiterite-quartz vein deposit 锡石石英脉矿床
cassiterite-scheelite-quartz vein formation 锡石—白钨—石英脉建造
cassiterite-sulfide deposit 锡石硫化物矿床
cassius gold purple 洒金紫
Cassius purple 金锡紫
Cassler yellow 卡斯勒黄颜料;碱式氯化铅黄
cassonade 粗糖
Cassone equation 卡松式
cassoon 沉箱;沉井;凹镶顶棚;顶棚镶板
cast 模型;铸型品;铸型化石;管型;锤测深;抛掷;投掷
castability 可铸性;铸造性
castability test 铸造性试验
castable 可铸的;可塑的;可浇铸材料;浇注料;浇注成型的
castable alumina lining 可浇注的刚玉衬里
castable layer 浇注料层
castable period 浇注时期
castable refractory 可铸耐火材料;浇注耐火材料;浇灌耐火材料;耐火泥;耐火浇注料;耐火浇灌料;铸制耐火材料;熔铸耐火材料
castable refractory concrete 耐火混凝土
cast account 算账
cast aerated concrete 多孔混凝土
cast aerated concrete wand panel 多孔混凝土墙板
cast aerated concrete wand wall 多孔混凝土墙板
castaingite 硫钼铜矿
cast alloy 铸合金;铸造合金
cast alloy iron 合金铸铁;合金铸件
cast alloy steel piston 合金铸钢活塞
cast alloy tool 铸造合金工具
cast alumin(i)um 生铝;铸铝
cast alumin(i)um alloy 浇铸的铝合金;铸铝合金
cast alumin(i)um letter 铸铝字母
castana 巴西千果树
cast analysis 铸造分析;铸件分析
cast anchor 锚定重块;沉锤;抛锚
castanea 栗树;栗色
castanha 巴西千果树
castanha de cotia kernel oil 巴西千果树坚果油
cast anode 铸造阳极
cast architectural component 浇制的建筑部件
cast articles 浇筑的制品;浇制的制品
cast ashore 搁岸
castaway 抛弃;遇难船
cast babbit metal bearing 铸造巴比合金轴承
cast balcony 浇筑的阳台
cast basalt 铸玄武岩;熔融玄武岩
cast baseplate 铁垫板
cast bay 浇筑的底板
cast beam 浇制的梁
cast beam and filler floor 浇筑的密肋梁楼板
cast beam floor 浇筑的梁式楼板
cast beam lintel block 浇制的梁式过梁块
cast bearing 铸造轴承
cast bit 铸镶金刚石钻头
cast blade 铸造叶片;铸镶叶片;铸铁叶片
cast brass 铸造黄铜;铸铜;生黄铜
cast bronze 铸造青铜;铸(青)铜
cast building 浇筑的房屋
cast building member 浇注的建筑构件
cast building stones 浇注建筑石料;浇注建筑材料
cast building unit 浇筑的建筑单元

cast carbide 铸态硬质合金
cast carbon steel 铸钢
cast case meter 铸壳煤气表;硬壳煤气表
cast chaplet 铸造的芯撑
cast charge 浇注火药柱;发射剂
cast cladding slab 浇制镶面板;浇制覆盖板
cast-coated paper 高光泽印刷纸;可塑涂布纸
cast coater 流延涂布机
cast coating 流延涂布法
cast cold 低温浇注
cast concrete 浇注混凝土;浇灌混凝土;浇捣混凝土;模筑混凝土
cast concrete admix(ture) 混凝土混合料
cast concrete anchor 混凝土坠石
cast concrete beam floor 预制混凝土梁式楼板
cast concrete block 预制混凝土实心砌块
cast concrete bomb shelter 预制混凝土防空洞
cast concrete capping 混凝土压顶
cast concrete cill 预制混凝土窗台;预制混凝土窗槛
cast concrete cladding 预制混凝土镶面;预制混凝土表面处理
cast concrete factory 混凝土浇制工厂
cast concrete floor rib 预制混凝土楼板肋条
cast concrete green tile 预制混凝土砖坯;预制混凝土新砖
cast concrete gutter 预制混凝土檐槽;预制混凝土天沟
cast concrete infill(ing) slab 预制混凝土填板
cast concrete lining component 浇制混凝土内衬构件
cast concrete pilaster tile 预制混凝土壁柱砖
cast concrete producer 浇制混凝土生产者
cast concrete products 浇注混凝土制品
cast concrete profile(d) panel 预制混凝土异形板;预制混凝土成形板
cast concrete septic tank 浇制混凝土化粪池
cast concrete shelter 预制混凝土防空掩蔽所
cast concrete silo 预制混凝土筒仓
cast concrete sinker 混凝土坠石
cast concrete solid tile 预制混凝土实心砖
cast concrete stor(e)y height wall panel 预制混凝土楼层高墙板
cast concrete symposium 预制混凝土学术讨论会
cast concrete tread 预制混凝土楼梯踏板
cast concrete umbrella 预制混凝土烟囱顶罩
cast concrete valley beam 预制混凝土天沟槽形梁
cast concrete valley girder 预制混凝土屋面槽形大梁
cast concrete valley gutter 预制混凝土屋面天沟
cast concrete vent(ilation) duct 预制混凝土通风管道
cast concrete ware factory 浇制混凝土制品工厂
cast concrete worker 浇制混凝土的生产工人
cast construction 浇制建筑分隔设施;浇制建筑
cast construction method 浇制建筑施工方法
cast copper 铸铜
cast crossing 铸铁辙叉
cast cupola 浇制穹顶
cast curb of white granite aggregate 白色花岗岩集料浇制(混凝土)路缘石
cast cylinder 整铸缸体
cast detachable chain 铸造钩头链
cast die 铸模
cast dome 浇制穹隆
cast door frame 浇制门框
cast dwelling tower 浇制住宅塔楼
caste 等级制度
cast eaves gutter 浇制檐沟
cast eaves trough 浇制檐槽
cast edge beam 浇制边梁
cast electrolytic iron 铸造电解铁
castelet 小城堡
castellan 城堡主;寨主
castellate 建造城堡
castellated 蝶形的;齿形的;城堡式的
castellated beam 空腹梁;腹板带孔的梁;空腹钢梁;蝶形梁
castellated bit 铸钢齿状钻头;槽沿钻头
castellated bridge parapet 蝶形桥梁护栏
castellated house 城堡形建筑
castellated nut 开花螺帽;开花螺母;槽形螺母;槽顶螺母;槽型螺帽
castellated parapet 蝶墙;齿状矮墙;城垛
castellated(screw) nut 蝶形螺母;槽顶螺帽

castellated shaft 花键轴
castellation 雉蝶墙;城堡状建筑物;城堡形建筑
castellum 城堡;经建筑处理的蓄水池
cast end-block 浇制端部砌块
caster 脚轮;浇板机;小脚轮;自位轮;铸工;主销倾角;翻砂工(人);撬棍
caster angle 主销后倾角
castered floor container 带轮大箱
caster effect 转向轮后倾效果
caster tyre 滑轮胎
caster wheel 脚轮
caster with rubber wheel 脚轮
cast expanded concrete building component 浇制膨胀性混凝土建筑构件
cast expanded concrete building unit 浇制膨胀性混凝土建筑构件
cast explosive 浇铸装药
cast exposed aggregate panel 浇制外露集料嵌板;浇制外露集料护墙板
cast exposed aggregate slab 浇制外露集料板
cast facade 浇制正面
cast face 浇注面
cast filler(block) 浇制填块
cast filler slab 浇制填料板
cast film 流延薄膜;平挤薄膜
cast film extrusion 冷辊式挤出;平挤薄膜挤塑;辊冷式挤塑
cast flange 浇铸突缘;铸造法兰;铸成凸缘;固定凸缘;无孔凸缘;无孔法兰(盘)
cast flange adapter 铸造法兰接头
cast flight of stair(case) 浇制楼梯段
cast flooring construction 浇制楼板建筑
cast flooring system 浇制楼板体系
cast floor member 浇制楼板构件
cast floor rib 建筑楼板肋条
cast floor slab 浇制楼面板
cast flue 浇制烟道
cast form 铸造成形;铸型
cast frame 浇制框架;浇制构架
cast from the bottom 底注
cast from the top 顶注;上注
cast-fused refractory 熔铸耐火材料
cast gable 浇制山墙
cast gable beam 预制人字梁
cast garage 浇制汽车库
cast garden building unit 浇制花园建筑构件
cast gas concrete compound unit 浇制加气混凝土组合构件
cast gate 流道;浇口;铸口
cast gear 铸造齿轮
cast girder 浇制大梁
cast glass 浇铸型玻璃;压铸玻璃;铸(造)玻璃
cast glass balcony parapet 压铸玻璃阳台护墙
cast glass alorex 压铸滤热玻璃
cast glass canopy 压铸的玻璃天篷
cast glass door 压铸玻璃门
cast glass partition(wall) 压铸玻璃隔墙
cast glass roof overhang 压铸的玻璃挑出屋顶
cast glass tile 压铸玻璃砖;压铸玻璃瓦
cast glazed tile 压铸玻璃板
cast goods 浇制制品
cast grain 铸粒
cast grandstand 浇制正面看台
cast green block 浇制新砌块
cast green tile 浇制新砖
cast gripping block 浇铸的握固块;铸钢锚碇块;铸钢夹板
cast gutter 浇制排水沟
cast gypsum product 浇制石膏制品
cast hardening 注浆硬化
cast high-rise block of flats 浇制高层套房砌块
cast hole 浇孔
cast home construction 浇制住宅建筑
cast house 建筑住宅;浇注场;铸造浇注场;出铁场
cast house crane 炉前吊车;出铁场吊车
cast house crew 出铁场工人
cast housing construction 浇制住宅建筑
Castigliano's principle 卡斯底葛吕诺定理
Castigliano's theorem 卡斯底葛吕恩诺定理;卡斯的龙定理;卡氏最小功定理;卡氏定理
Castile soap 橄榄皂
cast-in 浇入
cast in a narrow channel 狭水道掉头
cast-in blade 铸入式叶片

cast-in bored pile 钻孔灌注桩
cast in cement 浇筑水泥
cast in chilling 冷硬铸造
cast-in-concrete reactor 混凝土芯反应堆;混凝土芯电抗器
cast-in diaphragm 铸造隔板
cast-in-drilled hole 钻孔灌注
cast-in drilled pile 钻孔灌注桩
cast industry 浇制工业
cast(in) filler panel 浇制填料板
cast infilling slab 浇制填缝板
casting 流延;浇铸;浇注;现浇的;铸造法;铸造;铸塑品;浇铸;注浆成型;刮涂膜;翻砂;熔铸;铸塑;铸件
casting aisle 浇锭跨;铸锭工段
casting alloy 铸造用合金;铸造合金
casting alloy crown 铸造合金冠
casting analyser 铸件成品检查仪
casting and blading of materials 材料的堆筑与整平
casting and forging factory 铸造锻铁工厂;锻铸工厂
casting and laying 浇注和铺设
casting area 浇铸工段;浇注场地;浇注面
casting base 铸座
casting bay 铸件场;预制场;铸造车间;铸造间;铸造厂
casting bed 预制场地;铸造台;浇灌台座;浇灌台;预制构件场地;预制床;铸场;铸床
casting blank 铸坯
casting bogey 铸锭车
casting bogie 浇铸小车;铸锭车
casting box 型箱;砂箱;模板;翻砂箱;砂型箱
casting box for curved stereos 曲面铅版浇铸机
casting brass 铸造黄铜
casting bronze 铸造青铜
casting carriage 铸桶车
casting character 铸造性能
casting characteristic 铸造性能;铸造特性
casting clasp 铸造卡环
casting cleaning 铸件清理
casting cleaning machine 铸件清理机
casting clean-up 铸件清理
casting coke 铸造用焦(炭)
casting composition 浇铸组成;浇制构造
casting compound 浇铸化合物;流延用混合料;铸塑用混合料
casting concrete 浇筑混凝土
casting concrete ware 浇制混凝土制品
casting cooling system 铸件冷却装置
casting copper 铸铜;铸造用铜
casting copper alloys 铸铜合金
casting crack 铸造裂纹
casting crown 铸造冠
casting crown post 铸造冠桩
casting cycle 浇注周期;混凝土浇铸周期
casting defect 铸造缺陷;铸件缺陷;铸疵
casting department 铸工车间
casting design 铸造设计
casting die 压模;铸造模;铸模
casting equipment 铸造设备
casting factory 浇制厂;铸造厂
casting fin 铸件周缘翅片;铸件披缝;铸件飞边
casting finish 铸件精整
casting flaw 铸造缺陷
casting-forging method 模压铸造
casting form 铸模;铸件模样
casting furnace 铸造用炉
casting glass 压铸玻璃
casting gold 铸造金
casting groove 铸成槽
casting harness 保定带
casting head 冒口;注浆料入口
casting hopper 注浆料斗
casting-in 浇合
casting-in open 开放型浇注;明浇;无盖箱铸造;敞开式浇造
casting-in rising stream 底注;底铸(法)
casting iron pan 铸铁锅
casting iron pipe 铸铁管
casting ladle 浇注勺;浇铸桶;浇(铸)包
casting lap 铸件皱纹
casting layout machine 铸件设计机;铸件测绘缩放仪

casting lead model 浇铅模具
casting leakage 铸件渗漏
casting machine 流延机;压铸机;铸注机;压匹机;铸造设备;铸造机;铸塑机;注浆机;涂布机
casting machine for non-ferrous metals 非铁金属的铸造设备
casting mass 浇制大型块件
casting material 浇铸材料;浇筑材料
casting metal 铸造金属
casting method 流延法;铸造方法;铸塑法
casting modulus 铸件模数
casting mortar 浇筑灰浆
casting mo(u)ld 浇模;铸型;注浆模
casting nozzle 铸口
casting nut 槽顶螺母;槽顶螺帽
casting of commerce 商业铸件
casting of lining 浇注衬砌
casting of piston ring pots 活塞环的筒体铸造
casting of rocks 岩石搬运
casting oil 铸芯油;铸件油
casting-on 浇补;补铸
cast ingot 铸锭
casting-out-nine check 除九检验
casting-out nines 舍九法;除九校验
casting-out-nines check 除九校验
casting pattern 铸造模(型)
casting pig 铸铁(块)
casting pig slag 铸造生铁炉渣
casting pit 铸(造)坑;铸锭坑
casting-pit brick 铸锭砖
casting-pit refractory 铸制耐火材料;铸钢用耐火材料
casting plan 铸造设计
casting plant 浇制厂
casting plaster 铸塑石膏;浇铸(石膏)灰浆;铸模石膏
casting plaster board 浇注石膏板
casting process 注浆过程(陶瓷);浇铸过程;浇铸方法;铸造方法;注浆成型法
casting production line 注浆生产线
casting property 铸造性能
casting rate 注浆速率
casting refractory 铸制耐火材料
casting resin 浇注树脂;浇塑树脂;流延树脂;流性树脂;铸模树脂;充填树脂;充填料
casting ring 浇注的圆环;铸型圈
casting roller 压延辊(平板玻璃)
casting-rolling process 浇注辊压成型法
casting runner 流铁沟;流铁槽
casting sand 型砂;铸造用砂;造型砂
casting scar 浇注斑疤
casting schedule 出铁时间表;出铁程序
casting scrap 余浆;铸造废钢
casting scull 铸造披缝
casting sealer 铸件渗补剂
casting seam 铸缝;注件模缝
casting section thickness 铸件壁厚
castings examination 铸件检验
casting shop 铸造车间;铸工车间
casting shovel 铸工铲
casting shrinkage 铸铸收缩;铸造收缩
casting site 浇注场地
casting skin 黑皮;铸皮
casting slip 浇铸浆;浇注用泥浆;铸型滑泥;注浆用泥浆
Casting-Slodzian mass analyzer 直接成像质谱仪
casting speed 铸造速度
casting spot 注浆斑点
casting sprue former 铸道形成针
casting station 铸造部
casting strain 浇铸应变;铸造应变;铸缩应变;注浆产生的应变
casting stress 浇铸应力;铸造应力;注浆产生的应力
casting surface 铸件表面;铸造表面
casting surface area 铸件表面积
casting surface defect 铸件表面缺陷
casting syrups 铸塑浆
casting system 浇制体系
casting table 浇注台
casting technique 浇铸技术;铸造工艺
casting temperature 压铸温度;浇注温度;铸造温度
casting under low-pressure 低压铸造
casting unit 铸锭设备
casting-up 铸型浇注

casting volume 铸造批量
casting wax 铸造蜡
casting wheel 浇铸轮
casting with bell mounted collars 承接式套管
casting yard 制造块体场地(防波堤用);浇铸场;浇注场(地);预制构件场;预制场;制作块体场地
cast-in heater 铸入加热器
cast-in iron band 预埋铁箍
cast-in member 现浇浇制构件;现浇构件
cast-in metal 浇铸轴承合金
cast-in oil lead 镶入油管;附铸油管
cast-in-pile 打入灌注桩
cast-in-place 就地浇铸;就地浇筑(的);就地灌筑;就地灌注;现场浇铸;现场浇筑;现场浇注;现场浇灌;现场灌筑;现浇(的)
cast-in-place balancing cantilever method 平衡悬浇法
cast-in-place cantilever method 悬臂浇筑法
cast-in-place cased concrete pile 就地浇注的套管混凝土桩
cast-in-place cased pile 现场浇注的箱桩;现场浇制钢壳混凝土桩
cast-in-place concrete 就地浇筑混凝土;模筑混凝土;现浇混凝土;现场浇筑混凝土;现场浇注混凝土;现场浇灌混凝土
cast-in-place concrete balcony 就地浇筑混凝土阳台
cast-in-place concrete cable duct 就地浇筑混凝土电缆管道
cast-in-place concrete ceiling 现浇混凝土顶板
cast-in-place concrete floor 就地浇筑混凝土楼板
cast-in-place concrete pile 钻孔灌注的混凝土桩;就地浇筑混凝土桩;就地浇捣混凝土桩;就地灌注混凝土桩;现浇混凝土桩;挖孔灌注桩
cast-in-place concrete pile rig 灌注桩钻孔机
cast-in-place concrete rib(bed) floor 就地浇筑混凝土加肋地板
cast-in-place concrete shell 就地浇筑混凝土壳体
cast-in-place concrete stair(case) 就地浇筑混凝土楼梯
cast-in-place concrete structure 就地浇筑混凝土结构
cast-in-place diaphragm wall 就地灌注地下连续墙
cast-in-place floor 现浇混凝土楼板
cast-in-place light(weight) concrete 就地浇捣轻质混凝土
cast-in-place lining 整体式衬砌
cast-in-place pile 钻孔灌注(的混凝土)桩;就地浇注的混凝土桩;就地浇注桩;就地浇注桩;现(场)浇(筑)桩;现场浇制桩;现场浇灌桩;灌注桩;打入灌注桩
cast-in-place pile underpinning 灌注桩托换
cast-in-place reinforced concrete structure 现浇钢筋混凝土结构
cast-in-place shelless pile 就地浇注无壳桩;就地灌注无壳桩;就地浇注无壳桩
cast-in-place shell pile 现场浇制带壳桩
cast-in-place tension pile 就地浇灌拉桩
cast-in-place uncased pile 就地浇注无壳桩
cast-insert bit 铸嵌钻头
cast-in-site 现(场)浇(捣)的;就地浇(的);现场浇筑(的)
cast-in-situ 就地灌筑;现浇;现场浇注;现浇的
cast-in-situ aerated concrete 现浇加气混凝土;现场浇捣加气混凝土
cast-in-situ architectural concrete 现浇建筑混凝土
cast-in-situ breast wall 现浇胸墙
cast-in-situ capping 现浇帽梁
cast-in-situ concrete 模筑混凝土;现浇混凝土;现场浇注混凝土;原地混凝土
cast-in-situ concrete base course 现浇混凝土垫层
cast-in-situ concrete parapet wall 现场浇混凝土胸墙
cast-in-situ concrete pavement 现浇混凝土铺面;现浇混凝土路面
cast-in-situ concrete pile 现场浇注的混凝土桩;就地灌注混凝土桩;现浇混凝土桩;钻孔灌注桩
cast-in-situ concrete piled wall 现浇混凝土桩墙
cast-in-situ concrete pipe 现浇混凝土管
cast-in-situ concrete slab 现浇混凝土板
cast-in-situ construction 现浇混凝土施工
cast-in-situ floor 现浇地坪
cast-in-situ mortar 现浇灰浆;现浇筑灰浆
cast-in-situ of underwater concrete pile 水下混

凝土桩灌注
cast-in-situ pier 现浇柱;现浇墩
cast-in-situ pile 灌筑桩;灌注桩(打入式);就地灌注桩
cast-in-situ piling machine 现场浇制桩打桩机;钻孔灌注桩打桩机
cast-in-situ reinforced concrete 现场浇捣钢筋混凝土;现浇钢筋混凝土
cast-in-situ reinforced concrete structure 现浇钢筋混凝土结构
cast-in-situ reinforced concrete wall 现浇钢筋混凝土墙
cast in situs 现场浇筑
cast-in-situ shipbuilding 整体造船法
cast-in-situ tension pile 就地浇注的拉力桩
cast-in-situ terrazzo 现浇水磨石
cast-in-situ terrazzo 现浇筑水磨石
cast-in-situ uncased pile 无套管现浇(混凝土)桩
cast-in socket 预留锚固孔;铸嵌插孔
cast inspection chamber 浇制检查井
cast insulating transformer 浇注绝缘变压器
cast integral 整体浇筑的
cast integrally runner 整铸转轮
cast integral test bar 主体铸造试棒
cast-iron 生铁;铸造生铁;铸铁
cast-iron adapter 生铁连接管
cast-iron alloy 特种铸铁
cast-iron apron 铸铁铺板
cast-iron area grating 铸铁场地栅栏
cast-iron ballast weight (浮标的)铸铁压重
cast-iron(bath) tub 铸铁浴盆
cast-iron bedplate 铸铁床板
cast-iron bell and spigot pipe 承插铸铁管
cast-iron belt pulley 铸铁皮带轮
cast-iron bend 铸铁弯管;铸铁弯头
cast-iron biological shielding wall 铸铁生物屏蔽墙
cast-iron block 铸铁铺地砖;生铁块
cast-iron boiler 铸铁锅炉
cast-iron boot 铸铁开沟器
cast-iron box 铸铁方垫箱
cast-iron box gutter 铸铁箱形排水沟
cast-iron box rainwater gutter 铸铁箱形雨水排水沟
cast-iron bracket 巴掌铁;铸铁托架;铸铁架
cast-iron bridge 铸铁桥梁
cast-iron chemical equipment 铸铁化工设备
cast-iron circular dished heads 铸铁圆形碟形封头
cast-iron closet bend 铸铁马桶弯头
cast-iron column 铸铁柱
cast-iron column for street lighting 街道照明用铸铁灯柱
cast-iron construction 铸铁构造
cast-iron corrosion 铸铁腐蚀
cast-iron cover 铸铁盖
cast-iron cover(ing) plate 铸铁盖板
cast-iron cylinder 铸铁筒柱
cast-iron discharge pipe 铸铁排水管
cast-iron drain 铸铁排水管
cast-iron drive ferrule 生铁套圈
cast-iron electrode 铸铁焊条
cast-iron electrode with steel core 钢芯铸铁焊条
cast-iron enamel 铸铁搪瓷
cast-iron engine part 铸铁发动机部件
cast-iron facade 铸铁正面
cast-iron fish-bellied girder 铸铁鱼腹式大梁;鱼腹式铸铁大梁
cast-iron fittings 铸铁管件;铸铁(接头)配件
cast-iron flange 铸铁法兰;铸铁凸缘管;铸铁凸缘管
cast-iron flange(d) pipe 铸铁法兰管
cast-iron flange(pipe) fittings 铸铁法兰管件;铸铁凸缘管件
cast-iron flat gate 铸铁平板闸门
cast-iron floor plate 铸铁地面板
cast-iron flushing cistern 铸铁(厕所用)冲洗水箱
cast-iron flushing tank 铸铁(厕所用)冲洗水柜
cast-iron for enameling 搪瓷用铸铁
cast-iron front 铸铁立面;(由预制件组成的)承重的建筑正面
cast-iron gate valve 铸铁闸阀
cast-iron girder 铸铁大梁
cast-iron grating 铸铁栅;铸铁箅子
cast-iron grill 灰口铸铁笼
cast-iron guide shoe 铸铁引靴
cast-iron header 生铁集管

cast-iron lateral 生铁分支管
cast-iron lining 铸铁衬砌；铸铁衬里
cast-iron lining segment 铸铁衬段
cast-iron manhole cover and frame 检查井铸铁盖座
cast-iron manhole ring 铸铁井圈
cast-iron mast 铸铁支柱
cast-iron offset 生铁迂回管
cast-iron paving 铸铁块铺砌
cast-iron perforated drain pipe 铸铁花管
cast-iron pipe 铸铁管；生铁管
cast-iron(pipe) fitting 铸铁管件
cast-iron pipe with bell and spigot 承插式铸铁管
cast-iron plate 铸铁板
cast-iron pot 铸铁罐；铸铁坩埚
cast-iron pressure cooker 高压铸铁锅
cast-iron pressure pipe 铸铁压力管
cast-iron products for sewerage systems 排水系统用的铸铁制品
cast-iron pull(handle)铸铁拉手
cast-iron radiation shielding wall 铸铁防辐射屏蔽墙
cast-iron radiator 铸铁散热器
cast-iron rail 铸铁护栏
cast-iron rainwater goods 铸铁雨水(排水)制品
cast-iron rainwater outlet 铸铁雨水斗
cast-iron rainwater pipe 铸铁水落管
cast-iron rainwater product 铸铁雨水(管件)制品
cast-iron register 壁炉铁框
cast-iron ribbed washer 铸铁加肋垫圈
cast-iron riser and cover 铸铁井升和盖
cast-iron road paving 铸铁铺路块
cast-iron(road) paving block 铺路铸铁块
cast-iron roof gutter 铸铁箱形屋顶天沟
cast-iron scrap 旧铸件；铸铁屑；铸铁废料；废铸铁
cast-iron screwed fittings 铸铁螺口接件
cast-iron screwed pipe 螺口铸铁管
cast-iron screw pipe 铸铁螺纹管
cast-iron sectional boiler 铸铁片式锅炉
cast-iron sectional tank 铸铁标准水箱
cast-iron segment 铸铁管段；铸铁管片
cast-iron sewer pipe 铸铁排水管
cast-iron shield 铸铁保护板
cast-iron shielding wall 铸铁屏蔽墙
cast-iron socket 铸铁套管
cast-iron socket tee 铸铁承口三通管
cast-iron soil pipe 铸铁粪便污水管；铸铁污水管和配件；铸铁污水管
cast-iron soil pipe support 铸铁污水管支架
cast-iron spigot and socket discharge pipe 铸铁窝接式排水管接头
cast-iron spigot and socket draining-pipe 铸铁窝接式排水管接头
cast-iron spigot and socket waste pipe 铸铁窝接式污水管接头；铸铁窝接式废水管接头
cast-iron standard parts 铸铁标准部件
cast-iron stave 铸铁立式冷壁
cast-iron step 铸铁踏步
cast-iron strainer 铸铁箅子
cast-iron structural pipe 铸铁结构管子；铸铁结构管材
cast-iron structural tube 铸铁结构管子；铸铁结构管材
cast-iron tee 铸铁 T 字管节
cast-iron timber connector 铸铁木材连接器；铸铁木材结合环
cast-iron tracery 铸铁窗花格
cast-iron trap 铸铁存水弯
cast-iron tub 铸铁盆
cast-iron tube 铸铁管
cast-iron tubing 铸铁管
cast-iron tubular column 铸铁管柱
cast-iron valve 铸铁阀门
cast-iron vessel 铸铁容器
cast-iron volute casing 铸铁蜗形机壳
cast-iron washer 铸铁垫圈；铸铁衬圈
cast-iron waste pipe 铸铁污水管；铸铁废水管；铸铁(给)水管
cast-iron water pipe and fittings 铸铁水管和配件
cast-iron water waste preventer 铸铁水中污物预防器
cast-iron window 铸铁窗
cast jacket 整铸套箱
cast joint 浇铸连接；浇注连接；铸焊接

cast kerb of white granite aggregate 白色花岗岩集料浇制路缘石
cast landing 浇制(楼梯)平台
castle 船楼；城堡
cast lead 铸铅
cast lead trap 铸铅存水弯
castle architecture 城堡建筑；宫殿建筑艺术
castle building 城堡建筑；宫殿建筑
castle church 城堡教堂
castle circular nut 六角圆顶螺母
castle-court 城堡大院
castled nut 带槽螺母
castle garden 城堡花园
castle-gate 城堡大门
castle-like town 城堡式市镇
castle manipulator 高架式机械手；双柱窝式机械手
castle nut 开花螺帽；开花螺母；蝶形螺母；槽形螺母；槽顶螺母；槽顶螺帽
castle ruin 废古城堡
castle set in artificial lake 位于人工湖上的城堡
Castle's intrinsic(al) factor 内源因素
castle sited on low-lying ground 位于低地上的城堡
castle site on high-lying ground 位于高地上的城堡
castle-tower 城堡塔楼
castle-wall 城堡墙
castle-wheel system 塔型轮系
cast light concrete box construction type 浇制轻质混凝土箱式构造类型
cast lighting column 浇制照明灯柱
cast lighting mast 浇制照明灯柱
cast lightweight component 浇制轻质构件
cast lightweight factory 建筑轻质(混凝土)工厂
cast lightweight factory member 建筑轻质工厂构件
cast lightweight member 建筑轻质构件
cast lightweight unit 建筑轻质构件
cast lining 浇筑容衬；浇筑衬料
cast lintel 浇制过梁；预制混凝土过梁
cast lintel block 浇制过梁砌块
cast(load) bearing skeleton 浇制承重框架；浇制承重骨架；浇制承重构件
cast loading 熔注装药
cast manganese steel frog 整铸锰钢辙叉
cast manganese steel pan 锰钢链板
cast manhole 浇制检查井
cast manway 建筑人行道；建筑井筒梯子格
cast marble tile 建筑大理石砖
cast mast 浇制支柱
cast material 铸造材料；模型材料
cast member 铸造构件；预制混凝土构件
cast metal 铸造金属；铸金属
cast metal base 铸造基托
cast-metal bit 铸造钻头
cast metal letter 金属铸造字母
cast metal matrix 铸造胎体
cast mo(u)lding 浇注成型(法)；注塑成型；浇铸的装饰件；浇铸成型
cast nail 铸制钉
Castner cell 卡斯纳电解池
Castner process 卡斯纳法
Castner's process 卡斯纳过程
cast No. 浇注号
cast non-housing construction 浇制非住宅式建筑
cast non-residential building 浇制非住宅式建筑
cast off 开船；摆脱；起锚；版面计算
cast off buoy 离浮筒
cast off the sail 展帆
cast off wharf 离开码头
castolin 铸铁焊料合金
castomatic method 钎料棒自动铸造性
cast-on 熔补
cast on scaffolding method 鹰架浇法；就地浇筑法
cast-on test bar 铸造试棒
cast-on test piece 主体铸造试棒
castor 自位轮；(家具下面装的)小脚轮；脚轮；胶脚轮
castor bean 蓖麻子
castor bean poisoning 蓖麻子中毒
castor-bean sheller 蓖麻脱粒机
castor-bean stripper 蓖麻籽摘取机
castor cocoon 蓖麻茧
Castor fiber 河狸

castor grey 河狸灰色；海狸灰色
castor leaf 蓖麻叶
castor machine oil 蓖麻机器油
castor meal 蓖麻粉
castor oil 蓖麻油
castor oil acid 蓖麻油酸
castor oil alkyd 蓖麻油醇酸树脂
castor oil capacitor 蓖麻油电容器
castor oil fatty acid 蓖麻油脂肪酸
castor oil isocyanate adduct 蓖麻油异氰酸酯加成物
castor oil plant 蓖麻
castor pomace 蓖麻油渣
castor root 蓖麻根
castorseed cake 蓖麻籽饼
castorseed oil 蓖麻油
castor seeds 蓖麻籽
cast panel 浇制板
cast panel fence 浇制嵌板栅栏
cast panel wall 浇筑板墙
cast panel with window opening 有窗孔的浇制板
cast parachute 浇制防坠器
cast parapet 浇制女儿墙；浇制护栏
cast partition block 浇制隔墙砌块
cast parts 铸件
cast perimeter frame 浇铸边框底架
cast permanent lightweight concrete shuttering 浇制永久性轻质混凝土模壳；浇制永久性轻质混凝土模板
cast permanent lightweight forms 浇制永久性轻质混凝土模壳；浇制永久性轻质混凝土模板
cast phenolic resin 铸塑酚醛树脂；注塑酚醛树脂
cast pilaster block 浇制壁柱砌块
cast pilaster tile 浇制壁柱砖
cast pile 浇制桩
cast pipe 铸管
cast pipe fittings 铸铁管配件
cast plank 浇制板
cast plastics 铸塑塑料
cast plate 整铸双面型板
cast plough 铸铁犁
cast pointed arch 浇筑尖券；浇筑尖拱
cast portal(frame) 浇制桥门式框架；浇制桥门式构架
cast prestressed component 浇制预应力构件
cast prestressed compound unit 浇制预应力组合构件
cast prestressed floor 浇制预应力楼板
cast prestressed floor member 浇制预应力楼板构件
cast prestressed lintel 浇制预应力过梁
cast prestressed member 浇制预应力构件
cast prestressed panel 浇筑预应力板
cast prestressed string 浇筑预应力楼梯斜梁
cast prestressed structural system 浇制预应力结构体系
cast prestressed system construction 浇筑预应力体系建筑
cast prestressed tee-section 浇筑预应力 T 形断面构件
cast prestressed wall 浇制预应力墙
cast prestressed wall slab 浇制预应力墙板
cast product 铸件(生产)
cast product maker 浇筑制品制造者
cast product manufacturer 浇制制品制造商
cast profile(d) panel 浇制异形板；浇制定型板
cast pumice component 浇制浮石构件
cast pumice concrete 浇筑浮石混凝土
cast pumice unit 浇筑浮石构件
cast purlin(e)浇制的檩条
cast ranking strut 浇制支柱；浇制撑条；浇制撑杆
cast refractory 浇铸成型耐火材料；整体耐火材料；熔注成型耐火材料
cast reinforced beam 钢筋混凝土梁
cast reinforced beam floor 钢筋混凝土梁式楼板
cast reinforced compound unit 钢筋混凝土构件
cast reinforced concrete rib 钢筋混凝土肋条
cast reinforced construction 钢筋混凝土建筑
cast reinforced fair-faced manufacturing yard 钢筋混凝土磨光盖面品生产场地
cast reinforced fair-faced panel 钢筋混凝土磨光盖面板
cast reinforced fair-faced pile 钢筋混凝土磨光盖面桩
cast reinforced floor 钢筋混凝土地板

cast reinforced frame 钢筋混凝土构架
cast reinforced member 钢筋混凝土构件
cast reinforced system construction 钢筋混凝土体系构造
cast reinforced unit 钢筋混凝土单元
cast reinforced wall 钢筋混凝土墙
cast reinforced wall slab 钢筋混凝土墙板
cast resin 浇注树脂;铸造(用)树脂;铸型树脂;注塑树脂
cast rib 浇制肋
cast rib and filler floor 混凝土空心砖密肋楼板
cast rib(bed) floor 混凝土带肋楼板
cast rib slab 浇制肋板
cast ring 铸铁环
cast rink 铸铁环
cast roof 浇制屋顶
cast(-set) bit 铸造钻头
cast-set coring bit 铸镶取芯钻头
cast-set diamond bit 金刚石钻头;铸镶金刚石钻头
cast setting 机械镶嵌;铸造法镶嵌
cast-setting material 铸镶材料
cast shadow 投射阴影;投影
cast sheet glass 浇注的薄板玻璃
cast sheet lead 铸铅片
cast shell 浇制薄壳
cast shell process 毛管铸造法
cast shop 浇制车间
cast sill 浇制的窗槛
cast slab 浇筑板;扁钢锭
cast slab floor 浇筑地板
cast slag block 浇注矿渣砌块
cast slag copper 渣铜浇铸
cast slag sett 浇铸矿渣小方块
cast slag tile 浇注矿渣砖
cast soldering 浇铸连接;滴焊连接
cast solid block 浇制实心砌块
cast solid tile 浇注实心砖
cast spiral case 铸铁蜗壳;铸钢蜗壳
cast stainless steel 铸造不锈钢
cast stand 铸钢座
cast steel 铸钢(件)
cast-steel 2-wing bit 双翼铸钢钻头
cast-steel 3-wing pilot bit 中心突出的三翼铸钢钻头
cast-steel anchor block 铸钢锚碇块
cast-steel bucket 铸钢泥斗
cast-steel case 铸钢机壳
cast-steel chain cable 铸钢锚链
cast-steel crossing 铸钢辙叉
cast-steel flight 铸钢楔扬料器;铸钢楔扬料板
cast-steel gate 铸钢闸门
cast-steel grade 铸钢品种
cast-steel leaf 铸钢叶片
cast-steel lifter 铸钢楔扬料器;铸钢楔扬料板
cast-steel lining 铸铁管片
cast-steel member 铸钢杆件
cast-steel pipe 铸钢管
cast-steel pipe flange 铸钢管法兰;铸钢管凸缘
cast-steel plate 铸钢板
cast-steel rocker 铸钢能摇动底座;铸钢滚轴支座;铸钢摇动底座
cast-steel separator 铸钢横梁;铸钢隔板
cast-steel shot 钢砂
cast-steel support 铸钢支座
cast-steel wedge 铸钢楔块
cast-steel work 铸钢;铸钢构件
cast-steel yoke 铸钢轭
cast stone 人造石材;铸石;再造石;人造石块;人造石
cast stone finish 人造石饰面
cast stone floor cover(ing) 铸石地坪覆面
cast stone floor(ing) 铸石地坪
cast stone machine 铸石(块)机
cast stone paving flag 铸石铺路板
cast stone shop 铸石车间
cast stone skin 铸石表面
cast stone stair(case) 铸石楼梯
cast stone tile 铸石砖
cast stone tile floor cover(ing) 铸石砖地坪覆面
cast stone waterproofer 铸石防水剂
cast stone work 铸石工
cast string 浇筑楼梯斜梁
cast structural concrete 浇筑的结构混凝土
cast structural skeleton 浇筑的结构构架
cast structure 浇筑结构;铸造结构;铸态组织
cast system 浇筑体系

cast system construction 浇筑体系构造
cast tablet 铸造匾牌
cast terrazzo 浇制水磨石
cast test block 铸造试块
cast the lead 打水砣(用手砣测深);抛水砣
cast threshold 浇制的门槛
cast to shape 不再进行机械加工的铸造
cast tow 短麻
cast trim 浇铸装饰件
cast trimming 模型修整
cast tube 铸造管;铸管
cast umbrella 混凝土遮阳;混凝土顶盖
cast unit 浇制单元;整件铸件
cast valley gutter 浇筑的斜沟槽
cast vault 浇筑的拱顶
cast wall 浇筑墙
cast wall panel 浇制墙板
cast ware 浇铸制品;浇注制品;注浆陶瓷
cast-weld construction 铸焊结构
cast welded rail joint 铸焊钢轨接头
cast welding 铸焊
cast wheel 铸造车轮;铸轮
cast window frame 浇制的窗框
cast window glass 浇制窗玻璃
cast wood concrete block 浇制的木纤维混凝土砌块
casual 临时的
casual acceleration 随机加速度
casual audit 临时审计;不定期审计
casual bar 临时性沙洲;不固定沙洲
casual business 偶然交易
casual expenses 临时费用
casual income 临时收入
casual inspection 临时检查;抽检;不定期检查
casual labo(u)rer 零工;临时工;短工
casual labo(u)r office 零工就业服务所
casual observation 随机观测
casual profit 偶然利益;偶得利润
casual repair 临修【铁】
casual sands 不规则砂层
casualty 意外事故;变故;伤亡事故
casualty date 事故日期
casualty dressing and examination room 急救室
casualty insurance 意外灾害保险;意外保险;灾害保险;伤害保险
casualty loss 伤亡事故损失
casualty of non-railway man 路外人员伤亡
casualty of railway man 路内人员伤亡
casualty person 伤亡人员
casual ward 临时收容所;临时救济所
casual worker 临时工;散工
caswellsilverite 阴硫钠铬矿
cat 硬耐火土;(填塞灰板条缝的)草泥浆
catabaltic wind 冷烈风
catabasis 希腊正教堂祭台下放圣骨和圣物处
cable ripeness 食用成熟度
catabolin 异化产物
catabolism 异化作用;分解代谢;破坏变质
catabolite 异化产物;分解代谢产物;分解产物
catabolite repression 分解物阻遏
catacaustic 反射焦线
catachosis 破碎变质;碎裂变质
cataclasis way 碎裂作用方式
cataclasite 破裂岩;碎裂岩
cataclasite series 碎裂岩系列
cataclasitic coal 碎裂煤
cataclasm 碎断
cataclastic 碎裂的
cataclastic breccia 碎裂角砾岩
cataclastic conglomerate 碎裂砾石
cataclastic flow 碎裂流动
cataclastic granulation way 碎粒化方式
cataclastic metamorphism 碎裂变质作用;碎裂变质
cataclastic rock 碎裂岩
cataclastic series 碎裂岩系
cataclastic structure 压碎构造;碎裂构造
cataclastic texture 碎裂结构
cataclastic zone 破碎带
cataclinal river 顺斜河;顺向河
cataclinal stream 顺向河
catacline 下倾型;顺向的
cataclysm 洪水泛滥;骤变(地壳);灾变;特大洪水
cataclysmic event 激变事件
cataclysm theory 灾变理论
catacomb 地下墓窖;陵墓;酒窖;地下墓穴

catacomb grave 地下墓穴
cata-condensed hydrocarbon 背缩烃类
catacorolla 副花冠
catadecamethyloctosiloxane 十八甲基八硅氧烷
catadioptric 反折射的
catadioptric apparatus 反折射器
catadioptric objective 折反射物镜;反折射物镜;反折射物镜
catadioptrics 反折射学
catadioptric system 反射折射系统;反折射系统;反折射光系;反折射复合系统
catadioptric telescope 折反射望远镜;反折射望远镜
catadromous migration 降海洄游
catadromy 降海繁殖;降海产卵
catadupe 急流;瀑布
catafalco 灵台;灵柩车;灵柩台
catafalque 灵柩台
catafront 下滑锋【气】
catagenesis 后生作用;后成作用;退化
catagenetic stage 后生退化作用阶段;深成作用阶段
catagraph 凝块石
cataiite 氯铟碱矾
Catalan forge 卡塔兰熟铁炉
Catalan process 卡塔兰法
catalase 过氧化酶
catalectic systole 不全收缩
catalimetric titration 催化滴定(法)
catalin 丙烯酸类聚合物;铸塑酚醛塑料
cataloged data set 目录式数据集;编目数据集
cataloged file 编目文件
cataloger 编目程序
cataloging and compilation section 测绘资料科
cataloging code 编目规则
cataloging of objects 藏品编目
cataloging syytem 编目系统
catalog(ue) 目录(册);目录表;一览表数据;商品目录;一览表;产品样本;产品目录;编目(录)
catalog(ue) appliance showrooms 目录商品展室
catalog(ue) cards 目录卡片
catalog(ue) case 卡片柜
catalogued 编目的
catalog(ue) data base 目录数据库
catalogued data set 编目数据组
catalogued file 编目档案
catalog(ue) directory 编目目录;目录字典;目录索引
catalogued procedure 编目手续;编目过程
catalog(ue) file 带目录文件;目录文件
catalog(ue) function 目录函数;目录功能
catalog(ue) listing 目录单;目录表
catalog(ue) memory 目录存储器
catalog(ue) No 分类顺序
catalog(ue) number 目录号;星表号数;样本号;产品样本号
catalog(ue) of charts and other navigation publications 航海图图书目录
catalog(ue) of geodetic stars 测地星表;大地测量星表
catalog(ue) of necessary commodity item 商品必备目录
catalog(ue) of property 财产目录
catalog(ue) of railway material supply 铁路物资供应目录
catalog(ue) of typified drawing 定型图图纸目录
catalog(ue) price 目录价(格);商品目录价格
catalog(ue) raisonne 分类目录
catalog(ue) record 编目记录;目录记录
catalog(ue) recovery area 目录恢复区
catalog(ue) retrieval system 目录检索系统
catalog(ue) search 目录检索
catalog(ue) system 编目系统
cataloguing of file 档案编目
Catalonian architecture 加泰隆尼亚建筑(11世纪西班牙)
catalpa oil 梓油
catalpic acid 梓油酸
catalysant 被催化物
catalysate 催化产物
catalysed chemical oxidation 催化化学氧化
catalysis 催化作用;催化;触媒(作用)
catalysis technology 催化技术
catalysis wastewater 催化作用废水
catalyst 接触剂;催化剂;触媒剂;水合催化剂

catalyst accelerator 催化促进剂
catalyst activity 催化剂活性
catalyst agent 催化剂;媒介剂
catalyst attritor 催化剂磨碎器
catalyst basket 催化剂筐
catalyst bulk density 催化剂堆密度
catalyst carrier 催化剂载体
catalyst case 催化剂室;反应室;反应器
catalyst chamber 触媒室
catalyst cracker 催化裂化设备
catalyst damage 催化剂中毒
catalyst deactivation 催化剂失活
catalyst-equipped passenger car 装有催化净化废气系统的小客车
catalytic hydrogenation of coal tar 煤焦油催化加氢
catalyst injection system 催化剂注射系统
catalyst oxidation 催化剂氧化
catalyst poisoning 催化剂中毒
catalyst promoter 助催化剂
catalyst reactivation 催化剂再生
catalyst recovery 催化剂再生
catalyst reducer 催化剂还原器
catalyst regeneration 催化剂再生
catalyst scrubber column 催化剂洗涤塔
catalyst selectivity 催化剂选择性;催化剂选择度
catalyst settler 催化剂沉降器
catalyst spraying 催化喷涂;双口喷枪喷涂
catalyst stripping 催化剂汽脱
catalyst surface 催化剂表面
catalyst susceptibility 催化剂感受性
catalyst-to-crude ratio 催化剂对原料比
catalyst-to-oil ratio 催化剂对油料比
catalyst tube 催化剂管
catalyst wax 催化剂表面沉积蜡
catalytic 催化的
catalytic action 接触作用;催化作用;催化触媒作用;触媒作用
catalytic activity 催化活力;催化活度
catalytic agent 催化剂;触媒剂
catalytically 催化地
catalytically blown asphalt 催化氧化(地)沥青;催化吹制(地)沥青
catalytically-blown asphalt 催化吹制沥青
catalytically cracked gas 催化裂解气
catalytically cracked gasoline 催化裂化汽油
catalytically oxidized asphalt 催化氧化沥青
catalytic amorphous glass fiber 催化用玻璃纤维
catalytic apparatus 催化仪器
catalytic asphalt 氧化石油沥青;催化沥青
catalytic carrier 催化载体
catalytic chromatography 催化色谱(法)
catalytic cleaning 催化净化
catalytic CO detector 催化一氧化碳探测仪
catalytic colo(u)rimetry 催化比色
catalytic combustion 催化燃烧
catalytic combustion system 触媒燃烧装置
catalytic combustion type oxygen analyzer 催化燃烧式氧分析仪
catalytic converter 催化转化器
catalytic cracker 催化裂化器
catalytic cracking 经催化的裂化;催化裂化法;催化裂化
catalytic cracking unit 催化裂化装置
catalytic crystallization 催化析晶
catalytic cure 催化熟化;催化固化
catalytic cycle oil 催化循环油
catalytic decomposition 催化分解
catalytic dehydration 催化脱水作用
catalytic desulfurization 催化脱硫
catalytic detector 催化检测器
catalytic exchange reactor 催化交换反应器
catalytic exhaust purifier 废气催化净化器
catalytic filter 催化(过)滤器
catalytic fluorimetry 催化荧光法
catalytic gas oil 催化瓦斯油
catalytic gasoline 催化裂化汽油
catalytic gas reforming 催化气体转化
catalytic heater 催化加热器
catalytic hydrogenation 催化氢化
catalytic hydrogenation apparatus 催化氢化器
catalytic hydrogenolysis 催化氢解
catalytic ignition 催化剂点火
catalytic ionization detector 催化电离检测器

catalytic iron internal electrolysis 催化铁内电解法
catalytic isomerisation 催化异构化
catalytic metal 催化金属
catalytic method 催化法
catalytic microwave oxidation process 微波催化氧化法
catalytic muffler 催化式消声器
catalytic odo(u)r treatment 臭气催化处理
catalytic oscillopolarography 催化示波极谱法
catalytic oxidation 催化氧化(作用)
catalytic oxidation process 催化氧化法
catalytic ozonation 催化臭氧氧化
catalytic ozonation biologic(al) aerated filter process 催化臭氧氧化曝气生物滤池工艺
catalytic ozonation process 催化臭氧氧化法
catalytic ozone 催化臭氧
catalytic ozone converter 催化臭氧转换器
catalytic ozone destruction cycle 臭氧的催化分解循环
catalytic ozone oxidation 催化臭氧氧化
catalytic poison 催化毒物
catalytic polarography 催化极谱(法)
catalytic polyforming 催化聚合重整
catalytic polymerization 催化聚合
catalytic purification 催化净化
catalytic reaction 催化反应;触媒反应催化反应
catalytic reactor 催化反应器
catalytic reduction 催化还原
catalytic reformer 催化重整装置;催化重整炉
catalytic reforming 催化重整
catalytic reforming furnace 催化重整炉
catalytic reforming unit 催化重整装置
catalytic residue 催化残基
catalytic reverse shift reaction 催化逆移反应
catalytic site 催化部位
catalytic space heater 催化燃烧采暖炉
catalytic spectrophotometry 催化分光光度法
catalytic subunit 催化亚单位
catalytic technology 催化技术
catalytic thermal treatment 催化热处理
catalytic titration 催化滴定(法)
catalytic vapo(u)rizer 催化汽化器
catalytic wave 催化波
catalytic wet air hydrogen peroxide oxidation process 催化湿空气过氧化氢氧化工艺
catalytic wet air oxidation 催化湿空气氧化
catalytic wet air oxidation technology 催化湿空气氧化技术
catalyzed aggregate 经催化的集料
catalyzed coating 催化型涂料
catalyzed iron internal electrolysis 催化铁内电解法
catalyzed lacquer 催化固化型漆
catalyzed oil 催化(聚合)油
catalyzed ozonation 催化臭氧氧化
catalyzed urethane resin 催化氨基甲酸(乙)酯树脂
catalyzer 催化剂;触媒剂
catalyze redox reaction 催化氧化还原反应
catamaran 打捞筏;长方筏;捕鱼竹排筏;捕鱼木排筏;双体筏;双体船
catamaran buoy tender 航标敷设双体船
catamaran driller 双体钻探船
catamaran drilling vessel 双体钻探船
catamaran fender 双体浮座护舷
catamaran platform 双体平台
catamaran-stern hull 双尾船体
catamaran towboat 双体推轮
catamaran trawler 双体拖网渔船
catamaran tug 双体推轮
catamorphism 风化变质(作用);碎裂变质作用;碎裂变质
cat and can 履带式拖拉机牵引的铲运机;履带式铲运拖拉机拖引刮土平整机
cat-and-clay chimney 稻草泥烟囱
catapepsis 完全消化
cataphalanx 冷锋面
cataphoresis 阳离子电泳;电泳现象
cataphoresis effect 电泳效应
cataphoresis laser 电泳激光器
cataphoretic coating 电泳涂覆法
cataphoretic migration speed 电泳迁移速度
cataphoretic mobility 电泳迁移率
cataphoretic velocity 阳离子电泳速度
cataphorite 红钠闪石;褐钠闪石
cataplasm 泥敷剂

catapleiite 钠锆石
catapult 导弹发射器;抛送机;抛射;弹射器;弹射
catapult launching 起飞喷射装置;弹射器发射
catapult mechanism 弹射机制
cataract 矿山制动装置;急瀑布;急流;跌水;低速度运转(球磨机);低临界速度运转(球磨机);大瀑布;水力制动器;水力制动机
cataract action 急流作用
cataracting 瀑泻;冲击研磨
cataracting angle 瀑落角
cataracting of grinding media 磨球瀑落
cataracting point 瀑落角
catarinite 镍铁陨星
catarobia 清水生物
cataspilite 斑点云母
catastrophe 灾难;灾害;灾变;大灾害;大事故;突然的大变动
catastrophe hazard 巨灾(危)险
catastrophe loss 巨灾损失;严重全漏失
catastrophe reinsurance cover premium 巨灾损失分保保险费
catastrophe risk 巨灾(风)险;特殊危险
catastrophic cancellation 巨量消失
catastrophic change 灾害性变化
catastrophic cloudburst 灾难性暴雨;灾害性暴雨;特大暴雨
catastrophic collapse 灾害性倒塌;灾害性倒毁
catastrophic collision 灾难性碰撞
catastrophic damage 灾害性破坏
catastrophic degradation 突然退化
catastrophic earthquake 灾害性地震
catastrophic event 灾难性事件;灾变事件;灾变
catastrophic failure 严重损坏;严重破坏;严重故障;灾害性破坏;突然失效;突然故障
catastrophic flood 灾害性洪水;非常洪水;特大洪水
catastrophic flood level 非常洪水位
catastrophic fluctuation 严重波动
catastrophic hypothesis 灾变假说
catastrophic landslide 灾难性山崩
catastrophic mortality 灾难性死亡
catastrophic mudflow 灾害性泥石流
catastrophic phenomenon 灾难现象
catastrophic variation 灾害性变化
catastrophic vibration 突变振动
catastrophism 灾变说;古生灾变说【地】
catastrphe 灾变物
catathermal period 降温气候期
catathermometer 冷却温度计;干湿球温度计
cat back 锚钩索
cat bar 木插销
cat bar bolt 木插销
cat block 吊锚滑车;大型起锚滑车(船舶)
cat-boat 独桅艇
catch 轮挡;内燃机凸轮挡;门拉手;门后扣钩;门后挂钩;门挡;挂钩;蓄水点;抓爪;抓住;制动片;窗钩;风铃;捕获物;捕获量
catchable population 可捕种群
cat chain 吊锚链
catch-all 总受器;装杂物的容器;总收集器;杂物容器;提包;垃圾箱;截流器;截液器;分沫器
catch-all exceptions 一揽子免责
catch-all steam separator 分沫器
catch attention 引起注意
catch bar 推动杆
catch basin 收集盘;滤污器;流域;汇水盆地;解泥井;截留井;截流井;集水井;集盘;雨水井;储[贮]水池;沉沙井;沉泥井
catch basin and manhole block 集水井和检查井砌块
catch bolt 自动螺栓;止动螺栓;弹簧门锁
catch boom 流材挡栅;截污栅;(木材港池的)浮式拦河埭
catch button 挡钮
catch cable 挡缆
catch crop 农闲作用;填闲作物
catch-drain 集水沟;截水(暗)沟;泄水沟;盲沟;排水沟
catch drainage 截水沟;排水沟
catcher 抓爪;抓器;接受器;接钢工;获能腔;制动装置;捕捉者;捕捉器(载货器自由下落停止器);捕集器;捕获器;收集器;收板装置
catcher boat 渔猎船;捕鲸船
catcher current 捕集器电流
catcher foil 捕集箔

catcher gap 捕获隙;补获隙
catcher grid 集流栅;集电栅;捕获栅
catcher mark 夹痕
catcher space 收注栅空间
catcher's side 接钢边
catches 捕获数
catch eye effect 醒目效应;注目效果
catch feeder 灌溉小沟;灌溉农渠;灌溉沟渠;灌溉沟
catch fire 着火
catch fire temperature 着火温度
catch for sliding door 滑门锁扣
catch frame 格栅;格框;栅栏;(雨水井口的)格孔板;截流栅;挡泥板
catch gallery 集水廊道
catch gear 棘轮;罐笼(断绳)防坠装置;闭锁装置;锁紧装置;锁紧机构
catch-handle interlocking 键柄连锁
catch holder 保险器;保险丝盒;配电变压器保险器
catch hook 掣子爪;回转爪
catch hoom 截木栅
catching 拦截;捕收;捕捉
catching bargain 期待物权合同
catching bell 打捞筒【岩】
catching bowl 打捞筒【岩】
catching device 紧固装置;捕捉器(载货台自由下落停止器);卡夹装置
catching fire 起火
catching groove 打捞工具的槽;卡夹切口
catching of toothed wheels 齿轮啮合
catching range 捕捉范围
catching regulation of fishing and catching 捕鱼和获量管理规则
catch inlet 集水井
catch in number 渔获数
catch knob 带动锁钮
catch lever 掣子杆
catch light 捕获光
catch limit 渔获限量
catch member 抓挡件
catchment 汇水
catchment area 吸引地区;流域面积;流域(截水区);集水区(域);集水面积;汇水区;汇水面积;汇流面积;含水层补给区;水源区;受雨区;受水面积;项目服务面积
catchment area of lake 湖泊汇水面积
catchment area survey 汇水面积测量
catchment aspect ratio 流域长宽比
catchment basin 流域;汇水面积;集污池;集水盆地;集水流域;汇水盆(地);储(贮)水池
catchment basin sewerage 流域排水系统
catchment board 流域管理处;水利局
catchment boundary 流域分界线
catchment channel 截水渠
catchment fault 汇水断层
catchment glacier 吹雪冰川
catchment-intercepting wall 截水墙
catchment lake 集污池
catchment lake receiving waters 集污池纳污水体
catchment lake volume 集污池容积
catchment management 流域管理
catchment of spring 泉水汇集
catchment of water 集水;汇水
catchment population 服务对象人口
catchment quality control 流域水质控制
catchment rock vein 汇水岩脉
catchment slope 流域坡降;流域坡度
catchment structure 汇水构造
catchment tray 集水槽;汇水槽
catchment (water) basin 流域;集水池
catchment water channel 集水槽
catchment water drain 截水沟
catchment water drainage 汇水区排水
catchment yield 流域(出)水量;汇水径流量
catch net 保护网;圈网;捕鱼网
catch of used oil 废油收集
catch on wheel 挡轮器
catch pan 人孔里盖;承接盘
catch pan of spreader 涂布机集料器
catch pawl 抓子钩;挡爪
catch pin 挡杆;带动销
catch pit 沉砂井;截留井;截流井;集水坑;集水井;泥浆沉淀池;汇水洼地;汇水盆地;泄水井;排水井
catchpit 汇水盆地
catch plate 导夹盘;拨盘;制动板
catch platform 施工防护平台;坠台;截挡板
catch plug 挡住螺栓
catch point 支撑点;制动道岔;安全(线)道岔
catchpole 法警
catch pot 凝结水井;油气分离罐;泉坑
catch prop 临时立柱;辅助支柱
catch quota 捕鱼限额
catch ratline 加强桅梯绳
catch reversing gear 棘轮回动装置
catch siding 安全线【铁】
catch space 集尘能力
catch spring 挡簧
catch switch 安全道岔
catch tank 凝气管;预滤器;顶滤器;捕集槽
catch tray 收集盘;收集板
catch-up 追尾;非图文部分着墨
catch-up program(me) 赶上形势需求;赶上形势需要的规划
catchwater 截水管;灌溉渠水;分汽器;汽水分离器;截水;集水;集排水沟
catch-water basin 集水区;集水池;汇水盆地
catch-water chamber 集水槽
catch-water channel 集水槽
catch-water ditch 拦水沟
catch-water drain 排水沟;集水(排水)沟;截水沟;截流管
catch-water yield 集水量
catch weight coil 非定重线捆
catchwork 灌溉(水)渠;集水排水沟;集水工程
catchwork combined channel 集水沟
catchwork irrigation 漫灌;蓄水罐;蓄水池
cat coal 富黄铁矿煤
cat cracker 催化裂化器;催化裂解装置
cat-cracking 催化裂解
cat davit 吊锚柱;吊锚(吊)杆;起锚柱
catechol 邻苯二酚
catecholamine 邻苯二酚胺
catechol tannin 茶邻二酚
catechol violet 邻苯二酚紫
catechu mordant 儿茶媒染剂【化】
cated design radiator 带有斜管的散热器;带有斜置管的冷却器;带有倾斜冷却管的冷却器
categorical data 分类资料;分类数据;非数值数据
categorical distribution 分类分布
categorical grant 分类财政补贴
categorical grants aid 分类用途的财政补贴
categorical measure 类别量数;分类测度;分类变量;类型变量
categorical proposition 分类命题
categorical series 类别数列
categorical standard 绝对标准
categories of geomagnetic measurements 地磁测量类别
categories of natural conservations 自然保护区的类型
categories of reservations 自然保护区的类型
categories of reserve 储量类别
categories or classes of reserves 储量的分类和分级
categorization 分门别类;编目方法
categorization of geotechnical projects 岩土工程分级
categorized accounts summary 科目汇总表
categorized average margin rate method 分类平均差价率法
categorizer 分类器
category 类别;模式范畴;种类;分类单位;范畴;编组;品类
Category 2 goods 二类物资
category analysis 分类分析法
category message 分类信息
category method 分类比较法
category of automaton 自动机的范畴
category of behavio(u)r 行动的范畴
category of circular features 环形体种类
category of coal seam 煤层型别
category of damage 破坏类别
category of earthquake damage 震害类别
category of fuzzy sets 模糊集范畴
category of lineaments 线性体种类
category of machine 机器范畴
category of market 市场的类型
category of model 模型分类;模型范畴
category of out-of-ga(u)ge 超限种类
category of remote sensing applications for engineering geologic(al) survey 遥感工程地质调查应用种类
category of remote sensing applications for environmental geologic(al) survey 遥感环境地质调查应用种类
category of results of remote sensing applications for geology 遥感地质应用成果种类
category of roads 道路种类
category of roughness 糙率分级
category of set 集的范畴
category of site 场地分类
category of site soil 场地土分类
category of soil classification 土壤分类单元
category of taxes 税种;税目
category of traction 牵引种类【铁】
category of vector space 向量空间的范畴
category of vessel 容器类别
catelectrotonus 阴极电紧张
catena 链(条);耦合;串链;土链
catenarian 悬索线;悬垂线
catenarian arch 悬链拱
catenary 悬垂线
catenary action 悬链(线)作用
catenary anchor leg mooring 悬链腿式锚泊装置
catenary arch 悬链状拱;悬链(线)拱;悬链曲线拱;悬垂拱;反悬链式拱
catenary arch bridge 悬链线拱桥
catenary bridge 悬链线桥
catenary construction 悬链线结构;悬链构造
catenary control 弛度控制
catenary correction 垂度校正;悬链线校正;垂曲校正;垂曲改正
catenary curve 悬链(曲)线;垂曲线
catenary installation car 架线作业车;安装作业车
catenary kiln 悬拱窑
catenary lighting 街道悬垂式照明;悬垂式照明
catenary line 链型悬挂;垂曲线;悬链线
catenary mast 接触网支柱
catenary of equal resistance 等阻悬链线
catenary overhead contact line 链线架空接触线
catenary poise 链状砝码
catenary quick twister 接触网快速扭弯器
catenary ripple 链状波痕;悬链状波痕
catenary sag 承力索弛度
catenary suspension 悬链(吊架)
catenary system 接触网
catenary system tower 接触网支柱
catenary taping 悬尺测距
catenary unloader 悬链斗式卸机
catenary voltage 网压
catenary wire 悬链钢丝绳;承力索
catenary wire clip 承力索线夹
catenary wire clip support 承力索线夹支持座
catenary wire support clamp 接触网支撑装置线夹;承力索双线支撑线夹
catenary wire suspension clamp with hook 双耳鞍子
catenary work car 架线作业车
catenate 链接;串珠状的
catenated 链接的;链痕的;以链形花纹装饰的
catenated string 连接串
catenation 链接;链化作用;链;级联;耦合;并置
catenation operator 连接算符
catenoid 悬链面;悬链回转面;垂曲面;悬线链曲面;悬链线挠度;悬线(链)垂度;悬链(曲)面;悬链面【数】
catenulate 链形的;串珠状的;成链状的
catenulate sink hole 串珠状穴;串珠状溶坑;串珠落水洞;串珠状灰岩坑
catenuliform 串珠状的
cater angle 后倾角
cater-corner 对角线
cater-cornered 对角线的
caterer problem 包伙者问题
catering 大型厨房
catering period 灌溉水周期
caterpillar 履带式拖拉机;履带式的;履带车辆;环状车道;爬行车;毛虫
caterpillar asphalt concrete paver 履带式沥青混凝土摊铺机
caterpillar asphalt finisher 履带式沥青路面机;履带式沥青路面整修机
caterpillar asphalt paver 履带式沥青路面铺料机

caterpillar band 履带
caterpillar bituminous paving machine 履带式沥青铺路机
caterpillar black-top spreader 履带式黑路面喷洒机
caterpillar bulldozer 履带式(沥青路面)推土机;履带式开土机
caterpillar chain 履带(链)
caterpillar crane 履带(式)起重机;履带吊车
caterpillar crest gate 履带式堰顶闸门
caterpillar cut-off 履带式闸门
caterpillar diesel-driven grab 柴油履带式抓斗起重机
caterpillar drive 履带运行;履带(式)传动;履带(驱动)
caterpillar elevator 履带式提升机
caterpillar equipment 履带式机械
caterpillar excavating machine 履带式挖土机;履带式挖掘机
caterpillar excavation machine 履带式挖掘机
caterpillar excavator 履带挖土机;履带式挖掘机
caterpillar finisher 履带式路面机
caterpillar flip-over bucket loader 履带式翻转装载机
caterpillar gate 履带式闸门;链轮闸门
caterpillar grinder 连续式碎木机
caterpillar guide 履带导向装置;履带导承
caterpillar hydraulic loader 履带式液压操纵装载机;履带式液压装载机
caterpillar light 履带式照明
caterpillar loader 履带式装载机;履带式装运机;履带式装料机;推土装载机
caterpillar loading shovel 履带式铲土车;履带式铲土机
caterpillar logging arch 履带式拱钩;超长货物用的履带式拱钩
caterpillar machinery 履带式机械
caterpillar-mounted 安装履带的
caterpillar(-mounted) excavator 履带(式)挖土机
caterpillar overhead loader 履带式高位装载机;履带式高架装载机
caterpillar over-loader 履带式高位装载机;履带式高架装载机
caterpillar overshot loader 履带式高位装载机;履带式高架装载机
caterpillar paver 履带式铺路机
caterpillar rocker shovel 履带式翻铲装载机
caterpillar running mechanism 履带行走机构
caterpillar scraper 履带式刮土机
caterpillar shovel 履带式电铲
caterpillar track 履带;链轨
caterpillar track section 环形截面
caterpillar traction 履带牵引;用履带式拖拉机牵引
caterpillar tractor 履带式拖拉机;履带式牵引车;履带牵引机
caterpillar tread 履带轮底;履带
caterpillar tread wagon 履带车
caterpillar type oxidation bleaching range 履带氧漂机
cat eye (玻璃内的)细长气泡;木材节疤;小木节;反光路标
catface 猫脸(抹面瑕疵);麻点裂纹;油漆裂隙缺陷;抹面产生瑕疵污点;粗糙凹坑
cat fall 吊锚索;吊锚缆
catforming 催化转化法;催化重整
cat gold 金色云母;云母;白云母
cathamplifier 阴极放大器
catharometer 热导计;气体分析仪;气体分析器
Cathaysia 华夏古陆
Cathaysian 华夏系【地】;华夏式【地】
Cathaysian costal plain 华夏沿海平原
Cathaysian floral realm 华夏植物地理区系
Cathaysian inshore basin 华夏沿海盆地
Cathaysian island sea 华夏岛海
Cathaysian mountains 华夏古陆山地
Cathaysian old land 华夏古陆
Cathaysian structural system 华夏式构造体系
Cathaysian tectonic region 华夏构造区
Cathaysian tectonic system 华夏(式)构造体系
cathead 绞车;锚栓;转换开关凸轮;吊锚架;吊锚杆;起重机臂;铁矿结核;升降机支柱;螺栓上的定位螺母
cathead a derrick up 用安全锚头竖钻塔
cathead anvil 打头
cat head chuck 带螺钉套筒夹头

cathead clutch 起锚机离合器
cathead installation 起锚装置;拔管装置
cathead job 起锚工作
cathead line 起重索;起锚索;锚头绳
cathead line sheave 起重滑轮;起锚滑轮
cathead man 绞盘驾驶员;副司钻
cathead shaft 锚头轴;起重杆;起锚杆
cathedra 主教座
cathedral 教堂;大教堂
cathedral angle 下反角
cathedral choir 大教堂唱诗班
cathedral dome 大教堂穹顶;教堂半球形屋顶
cathedral glass 仿古窗玻璃;大教堂玻璃;教堂内拼花玻璃;教堂窗玻璃;压花玻璃;仿古玻璃;拼花玻璃
cathedral glass sheet 拼花玻璃
cathedral Gothic 歌特大教堂式
cathedral library 大教堂藏经室;大教堂藏经阁
Cathedral of Notre Dome 巴黎圣母院
cathedral quire 大教堂唱诗班
cathedral sculpture 大教堂雕塑
catheometer 高差表
Catherine 车轮形窗
Catherine wheel 圆花窗
Catherine wheel window 轮形圆窗;车轮形窗;玫瑰形窗
catheter 导(液)管
catheter holding forceps 夹套管用钳
catheterising telescope 测高望远镜
catheterization 导管插入术;插入导管术
catheter mount 导管接合器
catheter puller 导管拆卸器
cathetometer 精密高差仪;高差仪;高差计;垂高计;测高仪;测高计
cathetron 汞气整流器
cathetus 中直线
Cathie unite 凯塞单位
cathodal 负极的
cathode 阴极;负(电)极
cathode activation 阴极激活
cathode active coefficient 阴极有效系数
cathode activity 阴极活性
cathode analysis 阴极分析
cathode/anode ratio 阴阳极比
cathode aperture 阴极(插入)孔
cathode arrester 阴极放电管
cathode assembly 阴极组
cathode back-bombardment 阴极回袭
cathode bar 阴极棒
cathode base 阴极芯
cathode-base amplifier 阴极接地放大器
cathode base metal 阴极基金属
cathode beam 阴极射线
cathode bias resistor 阴极偏压电阻器
cathode block 阴极碳块;阴极块
cathode breakage 阴极断裂
cathode breakdown 阴极烧坏;阴极击穿
cathode busbar 阴极导电母线
cathode cadmium 阴极镉
cathode capacitance 阴极电容
cathode carbonization 阴极碳化
cathode-casting machine 阴极铸造机
cathode cavity 阴极空腔谐振器
cathode cell 阴极罩
cathode chamber 阴极空间
cathode circuit 阴极线路;阴极电路
cathode cleaning 阴极清洗
cathode coating 阴极涂层
cathode cold end effect 阴极冷端效应
cathode collector bar 阴极导电棒
cathode column 阴极区
cathode compartment 阴极箱;阴极室
cathode compensation 阴极补偿
cathode contamination 阴极污染
cathode copper 阴极铜;电解铜
cathode corrosion 阴极腐蚀
cathode-coupled amplifier 阴极耦合放大器
cathode coupled circuit 阴极耦合电路
cathode-coupled clipper 阴极耦合限幅器
cathode coupled oscillator 阴极耦合式振荡器
cathode coupled stage 阴极耦合器
cathode coupler 阴极耦合器
cathode coupling 阴极耦合
cathode current 阴极电流

cathode current intensity 阴极电流强度
cathode dark current 阴极暗电流
cathode dark space 阴极暗区
cathode degeneration 阴极负反馈
cathode degeneration resistance 阴极负反馈电阻
cathode degeneration resistor 阴极负反馈电阻器
cathode density 阴极密度
cathode deposit 阴极淀积;阴极沉积(物)
cathode disintegration 阴极崩解
cathode drive 阴极激励
cathode drop 阴极(电)压降
cathode effect 阴极效应
cathode efficiency 阴极效率;阴极电流效率
cathode emission 阴极发射
cathode emission current density 阴极发射电流密度
cathode emission efficiency 阴极发射效率
cathode emissivity 阴极发射率
cathode emitter 阴极发射体
cathode end 阴极引出端
cathode evaporation 阴极蒸发
cathode fall 阴极势降;阴极(电)压降
cathode feedback amplifier 阴极反馈放大器
cathode filament 阴极丝
cathode film 阴极膜
cathode flame 阴极气流
cathode fluorescence image 阴极荧光像
cathode follower 阴极输出器;阴极耦合器;阴极跟随器
cathode follower amplifier 阴极输出放大器
cathode follower detector 阴极输出检波器
cathode follower stage 阴极输出级
cathode glow 阴极辉
cathode glow space 阴极辉光区
cathode grid 阴极栅;抑制栅
cathode grid lens 抑制栅透镜
cathode heater 阴极加热器
cathode inductance 阴极电感
cathode injection 阴极注入;阴极输入
cathode-input amplifier 共栅放大器;阴极输入放大器
cathode interface impedance 阴极界面阻抗
cathode interlayer 阴极介层
cathode keying 阴极键控法
cathode lamp 阴极灯
cathode layer arc technique 阴极层电弧技术
cathode layer effect 阴极层效应
cathode layer enrichment 阴极区富集法;阴极层富集
cathode layer method 阴极层弧光法
cathode lead inductance 阴极引线电感
cathode leg 阴极引线;阴极臂
cathode lens 阴极透镜
cathode life 阴极寿命
cathode light 阴极发光
cathode liquor 阴极电解液
cathode load 阴极负载
cathode-loaded 阴极负载的
cathode loop 阴极回路;阴极环
cathode luminescence 阴极射线致发光;阴极(激)发光;阴极辉光
cathode luminescence image 阴极发光图像
cathode luminous sensitivity 阴极光照灵敏度
cathode material 阴极材料
cathode modulation 阴极调制
cathode non-corrosive method 阴极防蚀(法)
cathode parameter 阴极参量
cathode particle 阴极涂层粒子
cathode peaking 阴极(高频)峰化;阴极高频补偿
cathode phase inverter 阴极导向器
cathode photocurrent 阴极光电流
cathode plate 阴极板
cathode potential 阴极电位;阴极电势
cathode potential regulator 阴极电位调节器
cathode power connection 阴极电源接头
cathode power lead 阴极电源引线
cathode preheating time 阴极预热时间
cathode product receiver 阴极产物接收器
cathode protection 阴极保护
cathode protection method 阴极保护法
cathode quantum efficiency 阴极量子效率
cathode radiant sensitivity 阴极辐射灵敏度
cathode ray 阴极射线
cathode-ray accelerator 阴极射线加速器

cathode-ray apparatus 阴极射线仪器
cathode-ray beam 阴极射线束;阴极电子束;电子束
cathode-ray camera 阴极射线摄影机
cathode-ray curve tracer 阴极射线特性曲线描记器;阴极射线曲线图示仪
cathode-ray direction finder 阴极射线显示器测向仪
cathode-ray furnace 阴极射线炉
cathode-ray gun 电子枪
cathode-ray instrument 阴极射线仪器
cathode-ray luminescent method 阴极射线发光法
cathode-ray memory tube 阴极射线储存器
cathode-ray oscillograph 阴极射线示波器;阴极射线记录仪;阴极射线录波器
cathode-ray oscillography 阴极射线示波术
cathode-ray oscilloscope 阴极射线示波器;阴极射线示波管
cathode-ray output 阴极射线输出
cathode-ray pencil 阴极射线束;电子束
cathode-ray picture tube 阴极射线显像管
cathode-ray polarograph 阴极射线极谱仪
cathode-ray polarography 阴极射线极谱法
cathode-ray receiving tube 阴极射线接收管
cathode-ray scan display 阴极射线扫描显示(器)
cathode-ray spectroradiometer 阴极射线光谱辐射计
cathode-ray storage tube 电子束存储管
cathode-ray treatment 阴极射线处理
cathode-ray tube 阴极射线管;电子束管
cathode-ray-tube adder 配置阴极射线管的加法器
cathode-ray tube controller 阴极射线管显示控制器;阴极射线管控制器
cathode-ray tube display 阴极射线管显示(器)
cathode-ray tube graphics terminal 阴极射线管图形显示终端
cathode-ray tube plotter 阴极射线管绘图仪
cathode-ray tube presentation 阴极射线管显示
cathode-ray tube proof 显影屏打样法
cathode-ray tube record 阴极射线管记录
cathode-ray tube refraction seismograph 阴极射线管折射地震仪
cathode-ray tube sealing machine 阴极射线管封接机
cathode-ray tube spot scanner 阴极射线管光点扫描器
cathode-ray(tube) terminal 阴极射线管终端
cathode-ray tube type powermeter 阴极射线管型功率计
cathode-ray voltmeter 阴极射线伏特计
cathode reduction 阴极还原
cathode regulation 阴极稳压
cathode resistor 阴极电阻
cathode sensitivity 阴极灵敏度
cathode shield 阴极罩;阴极护罩
cathode space 阴极区
cathode sparking 阴极打火
cathode spot 阴极斑点
cathode-spray machine 阴极材料喷涂机
cathode sputtering 阴极溅射;阴极溅镀
cathode sputtering process 阴极溅射镀膜法
cathode terminal 阴极端子
cathode transformer 阴极变压器
cathode trap 阴极陷波电路
cathode voltage 阴极电压
cathodevoltage stabilized camera tube 低速电子束摄像管
cathodic 阴极的;负极的
cathodic area 阴极区
cathodic bias 阴极偏压
cathodic cell 阴极电池
cathodic cleaning 电解清洗
cathodic coating 阴极涂覆法;阴极镀层
cathodic control 阴极控制
cathodic corrosion 阴极腐蚀;电化学腐蚀
cathodic corrosion protection 阴极防蚀(法);阴极防腐法
cathodic current 阴极电流
cathodic current intensity 阴极电流强度
cathodic deposit(ion) 阴极沉积(物)
cathodic disbonding 阴极脱粘
cathodic electrodeposition 阴极电泳;阴极电沉积
cathodic etching 阴极蚀刻
cathodic ground 阴极接地
cathodic inhibitor 阴极抑制剂;阴极缓蚀剂

cathodic leakage 阴极漏泄
cathodic light 阴极辉光
cathodic luminescence 阴极射线发光
cathodic membrane 阴极隔膜
cathodic peeling 阴极剥落
cathodic pickling 电解酸洗
cathodic polarization 阴极极化
cathodic protection 阴极保护;阴极防蚀
cathodic protection coating 阴极保护涂层;阴极保护膜
cathodic protection corrosion 阴极保护腐蚀
cathodic protection equipment 阴极保护设备;防电化学腐蚀装置
cathodic protection of pipeline 管道的阴极保护
cathodic protection of steel 钢材阴极保护法
cathodic protection of steel pipe 钢管阴极保护
cathodic protection parasites 妨碍阴极保护的物质
cathodic protective primer 阴极保护底漆
cathodic-ray display 阴极显示仪
cathodic reaction 阴极反应
cathodic reduction 阴极还原
cathodic resistor 阴极电阻器
cathodic sensitivity of photomultiplier tube 光电倍增管的阴极灵敏度
cathodic spattering 阴极喷镀;阴极溅射(镀膜)
cathodic spattering method 阴极溅射法
cathodic stripping 阴极析出
cathodic stripping voltammetry 阴极溶出伏安法
cathodic wave 阴极波
cathodochromic 阴极射线致色的
cathodochromic dark-trace tube 阴极射线致色暗迹管
cathodochromism 阴极射线致色;电子致色
cathodofluorescence 阴极射线荧光
cathodogram 阴极射线示波图
cathodograph 电子衍射照相机
cathodoluminescence 阴极射线(致)发光
cathodophone 阴极送话器
cathodophospho-rescence 阴极射线磷光
cathole 尾缆孔;猫洞(门、墙上);出入孔
catholyte 阴极电解质;阴极(电解)液
cat-hook 吊锚钩
cathophorite 磷钙钍石
cat ice 薄冰壳
cating coin 桶堆楔块
cation 阳离子;正离子
cation acid 阳离子酸
cation activity 阳离子活度
cation adsorption 阳离子吸附
cation analysis 阳离子的分析
cation-anion radius ratio 阴阳离子半径比值
cation balance 阳离子平衡
cation composition 阳离子组成
cation coordination 阳离子配位作用
cation defect 阳离子缺陷
cation emulsion 阳离子乳化剂;酸乳剂
cation equivalents 阳离子吸收当量
cation exchange 阳离子交换
cation exchange activity 阳离子交换活动性
cation exchange and adsorption 阳离子交替吸附作用
cation exchange bed 阳离子交换床
cation exchange capacity 阳离子交换容量;阳离子交换能力;阳离子交换量
cation exchange capacity measurement 阳离子交换测定
cation exchange capacity test 阳离子交换容量试验
cation exchange chromatography 阳离子交换层析
cation exchange filter 阳离子交换过滤器
cation exchange membrane 阳离子交换膜
cation exchange method of water 阳离子交换软水法
cation exchange property 阳离子交换特性
cation exchanger 阳离子交换器;阳离子交换剂
cation exchange resin 阳离子(型)交换树脂
cation exchange softener 阳离子交换软水器
cation exchanging compound 阳离子交换剂
cation exclusion 阳离子排斥(作用)
cationic activation 阳离子活化
cationic activity 阳离子活度
cationic additive 阳离子添加剂;阳离子掺和剂;阳离子附加剂
cationic adhesion agent 阳离子黏着剂
cationic analysis 阳离子分析

cationic asphalt emulsion 阳离子(地)沥青乳液
cationic binder 阳离子黏合剂
cationic charge 阳离子电荷
cationic chemiluminescent polymer 阳离子型化学发光聚合物
cationic chemiluniescent monomer 阳离子型化学发光单体
cationic clarificant 阳离子净水剂
cationic complex 阳离子络合物
cationic conductor 阳离子导体
cationic curing 阳离子固化
cationic detergent 阳离子洗涤剂;阳离子型去污剂
cationic emulsified bitumen 阳离子乳化沥青
cationic emulsifier 阳离子乳液;阳离子乳化剂
cationic emulsion 阳离子乳(化)液;酸性乳剂
cationic emulsion slurry seal 酸性沥青乳液密封剂
cationic exchange filter 阳离子交换滤水器;阳离子交换过滤器;阳离子交换反应器
cationic fixation 阳离子固定
cationic-free curing 无阳离子固化
cationic germicide 阳离子杀菌剂
cationic grouping 阳离子分组
cationic head 阳离子头
cationic inorganic microparticle 阳离子型无机微粒
cationic latex 阳离子乳胶
cationic layer 阳离子层
cationic membrane 阳离子隔膜
cationic membrane cell 阳离子膜电池
cationic modification 阳离子改性
cationic modified starch 阳离子改性淀粉
cationic polyacrylamide 阳离子型聚丙烯酰胺
cationic polyelectrolyte 阳离子电解质
cationic polymeric flocculant 阳离子高分子絮凝剂
cationic polymerization 阳离子聚合
cationic reagent 阳离子试剂
cationic resin 阳离子树脂
cationic slurry 酸性稀浆;阳离子(地)沥青乳液
cationic soap 阳离子皂
cationic species 阳离子物种
cationic stabilization 阳离子稳定
cation(ic) surface active agent 阳离子表面活化剂;阳离子型表面活性剂
cationic surface active agent micelle 阳离子型表面活性剂胶团
cationic surfactant 阳离子性表面活性剂;阳离子型表面活性剂;阳离子表面活化剂
cationic surfactant micelle 阳离子型表面活性剂胶团
cationic wetting agent 阳离子润湿剂
cation interchange 阳离子交换
cationite 阳离子交换剂
cationoid 类阳离子
cationoid reagent 类阳离子试剂
cation penetration 阳离子渗入
cation polyelectrolyte 阳离子型聚电解质
cation replacement 阳离子置换
cation resin 阳离子交换树脂
cation resin exchange column 阳离子树脂交换柱
cation surface-active disinfectant 阳离子表面活性消毒剂
cation tower 阳离子交换塔
cationtrophy 阳离子移变
cation vacancy 阳离子空位
cat-jockey 履带式车驾驶员
cat ladder 吊梯;垂直吊梯;便梯;墙上竖梯;爬梯
catladder 直爬梯;墙上竖梯
catlin 两刃切断刀
catline 起重索;起锚索
catline guard 锚头绳导向轮
catline hook 锚头绳钩
catline sheave 锚头绳滑车
catling 两刃切断刀
catloft 谷仓阁楼
catmon 红棕色硬木(东印度)
cat-mouse station 航向指挥台
catoctin 古准平原残丘
catogenic metamorphism 热液变质
catonic starch 阳离子淀粉
cat operator 履带式拖拉机驾驶员;履带式拖拉机传动轮
catoptric 反射的
catoptric apparatus 反射器
catoptric arrangement 反射结构
catoptric imaging 反射成像

catoptric lens 反射式物镜
catoptric light 反射光
catoptric lighthouse 反射光灯塔
catoptric micro objective 反射式显微物镜
catoptric objective 反射物镜
catoptrics 反射光学
catoptric system 反射光系统;反射系列;反射光组;反射光系统;反光系统;反光镜系统
catoptric telescope 反射望远镜
catoptrite 黑硅锑锰矿
catoptroscope 反射物镜;反射检验物镜;反光检查器
catplant 催化剂装置
cat reformer 催化重整器
cats and dogs 价值低的股票;投机性股票
cat sapphire 黑绿蓝宝石
cat's ass 缆索上的纽结
cat scratch 抓痕
cat's eye 猫眼石;猫眼;(玻璃中的)细长气泡
cat's eye diaphragm 猫眼光栏
cat's eye reflector 猫眼反射镜
cat's eye retroreflector 猫眼后向反射器
cat's eye waveform 猫眼式波形
cat's head 悬臂提升机;套管;鸟头饰;猫头石;吊锚架
catshead 兽头形装饰
cat's head mo(u)lding 鸟头饰线脚(中世纪);猫头饰线脚(古世纪)
cat silver 银云母
Catskill beds 喀士基层(晚泥盆世);
catskinner 履带式拖拉机驾驶员(俚语)
catslide 低斜坡屋顶
cat's-paw knot 钢索栓钩结
catstep 滑坡阶地;梯形山墙顶;踏步形山墙
cat tackle 吊锚绞辘;吊锚复滑车
Catt concept 卡特论点
catted chimney 稻草泥烟囱
catter 冰脚
cattierite 方硫钴矿
catting shackle 吊锚卸扣
cattle 家畜;牛;牲畜
cattle carrier 牲畜装运船;牲畜(运输)船;牲畜车;牲口船
cattle-carrying ship 牲畜运输船
cattle-container ship 牲畜集装箱船
cattle farm 畜牧场;养牛场
cattle fitting 牲口运输装置;牲畜栏
cattle grid 防畜栏;牲畜防护网栏
cattle guard 防畜(护)栏;牲畜护栏
cattle hair felt 牛毛毡
cattle house 家禽舍
cattle loading ramp 牲畜装车斜台
cattle manifest 牲畜舱单
cattle manure 家畜粪肥
cattle pass 畜力车道;公路下(立体交叉)牲畜通道;牲畜小道;畜力车路
cattlepass marker 兽力车过道路标
cattle pen 牛栏
cattle pier 牲畜码头
cattle sling 牲畜吊具
cattle stall 牲畜栏
cattle transport track 家畜搬运车
cattle wagon 牲畜车
cattle waste 牲口排泄物
cattle wharf 牲畜码头
catty 市斤;斤
cat wagon 履带式小车
catwalk 马道;狭窄小道;照明天桥;高架狭窄人行道;步桥;人行便道;轻便梯;桥形通道;桥上人行道;跳板;天桥;施工步道
cat walk bridge 步桥
catwhisker 针电极
cat whisker set 晶须装置
Caucasia carpet 高加索(式)地毯
Caucasus geosyncline 高加索地槽
cauce 河床
Cauchy-Riemann equation 柯西—黎曼方程
Cauchy's boundary condition 柯西边界条件
Cauchy-Schwarz inequality 柯西—施瓦茨不等式
Cauchy's condensation test 柯西并项判别法
Cauchy's dispersion formula 柯西色散公式
Cauchy's distribution 柯西分布
Cauchy's equation of motion 柯西运动方程
Cauchy's inequality 柯西不等式

Cauchy's integral formula 柯西积分公式
Cauchy's integral theorem 柯西积分定理
Cauchy's law 柯西定律
Cauchy's law of elasticity 弹性相似定律
Cauchy's law of similarity 柯西相似(定)律
Cauchy's matrix 柯西矩阵
Cauchy's mean 柯西中值
Cauchy's mean value theorem 柯西中值定理
Cauchy's number 柯西数
Cauchy's principal value 柯西主值
Cauchy's principle of the argument 柯西幅角原理
Cauchy's problem 柯西问题
Cauchy's product 柯西乘积
Cauchy's quadric 柯西二次曲面
Cauchy's radical test 柯西根式判别法
Cauchy's ratio test 柯西比率判别法
Cauchy's residue theorem 柯西残数定理
Cauchy's sequence 柯西序列
Cauchy's similarity criterion 柯西相似准则
Cauchy's test for convergence 柯西收敛准则
Cauchy's theorem 柯西定理
Cauchy's transcendental equation 柯西超越方程
Cauchy's type extreme value distribution 柯西型极值分布
Cauchy-type distribution 柯西型分布
cauda 震尾
caudad 尾向
caudad acceleration 向尾加速度
caudal 尾部的
caudal end 尾端
caudal horn 尾角
caudal region 尾部
caudal saddle 尾鞍
caudal style 尾干
Cauer formula 考尔公式
cauf 蓄鱼箱;罐;吊桶;吊筐
caught on 卡于孔内的;卡于井内的
caught on a filter 滤出来的
cauk 白垩岩;白垩纪
caul 均衡压力用覆盖板;薄板曲压机;填块
caulcole (考林辛式柱头的)叶梗饰;叶丛卷茎饰柱头
ca(u)ldron 大锅;火山陷落区;火山口洼地;陷落火山口;锅
ca(u)ldron basin 锅状洼地;锅状盆地
ca(u)ldron shaped Ding 釜形鼎
ca(u)ldron subsidence 火(山)口沉陷
ca(u)ldron wagon 凹底平车
Cauliculus 考林辛式柱顶板下纤细柱身
caulis 圆装饰线脚;叶饰中段柱身
ca(u)lk 敛缝;油膏;凿缝(封);嵌塞;嵌缝料;填密
ca(u)lk adhesion 油膏黏结性
ca(u)lkage 捻缝填料
ca(u)lked 堵缝的;嵌实的
ca(u)lked edge 填缝边;装饰线脚
ca(u)lked end 开尾螺栓
ca(u)lked joint 填缝接头;打嵌头(管道);敛密缝;接口接头;凿密缝;嵌实缝;嵌密缝;嵌缝
ca(u)lked rivet 錾紧铆钉;敛缝铆钉
ca(u)lked seam 敛密缝;凿密缝;嵌实缝
ca(u)lked socket joint 嵌缝承插接头
ca(u)lker 敛缝机;敛缝工人;敛缝锤;捻缝工;堵塞工具;平锤;填隙料;填缝工具
ca(u)lker box 捻缝工具箱
ca(u)lker material 填隙料
ca(u)lker's drift 敛缝冲头
ca(u)lker's oakum 敛缝麻絮;堵缝麻絮
ca(u)lking 敛缝堵头;捻口;捻缝(口);管道填缝;堵缝;嵌缝;填隙;密缝凿
ca(u)lking butt 齐端堵缝
ca(u)lking cartridge 嵌缝填料筒;筒装嵌缝胶
ca(u)lking chisel 长circledot敛缝凿;填隙凿;捻缝凿;管道填缝凿;凿密凿;凿固凿;嵌固凿;嵌缝凿(刀);填缝凿
ca(u)lking compound 敛缝料;捻料;腻子胶;管道填缝混合料;堵缝化合物;填缝料;嵌缝料;嵌缝胶;嵌缝化合物;嵌缝膏;嵌缝材料;填缝混合料;填缝油膏
ca(u)lking cotton 捻缝棉条
ca(u)lking ferrule 填缝套圈
ca(u)lking filter 嵌缝工具
ca(u)lking groove 捻缝隙;嵌缝;嵌槽
ca(u)lking gun 胶合喷枪;挤枪;灌缝喷枪;堵缝枪;堵缝喷枪;捻缝枪;嵌缝胶枪;嵌缝胶注射器
ca(u)lking gun loader 嵌缝枪加料器

ca(u)lking hammer 接缝手锤;敛缝锤;捻缝锤;铆锤;凿密锤;管道填缝锤;堵缝锤;嵌缝锤;填隙锤;填缝锤
ca(u)lking hards 填缝麻屑;堵缝亚麻粗纤维
ca(u)lking iron 捻缝凿;密缝凿;密缝鏨;堵缝工具;封口铁;填隙器;凿缝凿;填隙凿
ca(u)lking joint 塞接口;捻口接头;接口填缝;堵缝;打口;嵌缝
ca(u)lking lead 填缝铅
ca(u)lking mallet 堵塞木槌;敛缝锤;捻缝木槌;捻缝槌;堵缝锤;填隙锤
ca(u)lking mastic 嵌缝膏
ca(u)lking material 堵缝材料;嵌缝材料
ca(u)lking metal 敛缝合金;堵缝合金;填隙合金
ca(u)lking nail 嵌缝钉;填隙钉;鞋钉;开尾钉
ca(u)lking nut 自锁螺母
ca(u)lking oakum 堵缝麻絮
ca(u)lking of segment 砌块堵填
ca(u)lking of tubes 堵塞管缝
ca(u)lking piece 敛缝片
ca(u)lking pocket 填塞穴;填塞腔;填塞槽;嵌槽;捻缝槽
ca(u)lking recess 承插口凹缝
ca(u)lking ring 加劲圈;加固圈;密封圈;涨圈;隔环;垫圈
ca(u)lking segment 填密片
ca(u)lking set 敛缝工具;凿密具;填缝工具
ca(u)lking side 捻缝边
ca(u)lking socket joint 嵌缝泵插接头
ca(u)lking space 插口缝
ca(u)lking strip 敛缝软钢条;敛缝软钢带;捻缝条;嵌缝条;塞缝条
ca(u)lking stuff 捻缝填料
ca(u)lking tool 压力灌注玛琋脂(接缝中防水);敛缝工具;捻缝工具;密缝凿;密缝鏨;凿密工具;管道填缝工具;填隙器;凿缝凿;填缝工具;凿密;嵌缝工具
ca(u)lking tow 填缝麻屑;堵缝亚麻粗纤维
ca(u)lking up hole 填塞洞眼
ca(u)lking weld 塞焊焊缝;密实焊缝;密封焊接;缝焊;填缝焊
ca(u)lking work 填缝工作
ca(u)lk joint 填缝
ca(u)lk piece 填密片
ca(u)lk seam 捻密缝;填实缝
ca(u)lk stone 软绿砂岩
ca(u)lk up hole 堵塞洞眼
ca(u)lk volatility 油膏挥发率
ca(u)lk walk bridge 纵向天桥
ca(u)lk weld 塞焊(焊缝)
ca(u)lk workability 油膏施工性
ca(u)nch 斜壁沟
causa 直接原因
causa donatio 遗赠
causal and effect 因果
causal association 因果联系
causal biostratigraphy 因果生物地层学
causal chain 因果链
causal circle 因果循环
causal factor 病原
causal forecasting 因果预测
causal forecasting method 因果分析预报法
causal hypothesis 因果假设
causality 因果性;因果(规)律
causal(ity) analysis 因果分析
causality condition 因果(律)条件
causality fault 发震断裂;发震断层
causal mechanism 发震机制
causal method 因果分析法
causal network 因果性网络
causal organism 病原生物
causal relation 因果关系
causal relationship 因果关系
causal repair 临时修理
causal repair of vehicle detached from train 摘车轴检
causal sequence 因果顺序
causal system 因果系统
causa mortis 遗赠
causans 直接原因
causa sine qua non 必要的原因
caustic lime mud 灰质淤泥
causative action 要因行为

causative action of agreement 要约
causative agent 病原体
causative contract 要因合同
causative fault 发震断裂;发震断层
cause 事业
cause an accident 造成事故
cause and effect 因果
cause and effect analysis chart 要因分析图;特性要因分析图
cause and effect chain diagram 因果链图
cause and effect diagram 因果图
cause and effect forecast 因果预报;确定性预报
cause-and-effect relationship 因果关系
cause and effect system diagram 因果系统图
cause-consequence chart 因果图
cause directly leading to death 直接死因
caused repair at marshalling yard 边线修【铁】
cause of accident 事故原因
cause of action 案由
cause of anchorage 锚泊原因
cause of bankruptcy 破产原因
cause of damage 损害原因;致损原因
cause of death 死因;死亡原因
cause of depreciation 折旧原因;贬值原因
cause of discarded mining 弃采原因
cause of earthquake 地震成因
cause of effect diagram 特性图解
cause of eustatic movement 海平面变化原因
cause of fire 火灾原因
cause of formation 构造成因;成因
cause of formation of beach 海滩成因
cause of formation of nodule 结核成因;锰结核成因
cause of groundwater continuous drawdown 地下水位持续下降原因
cause of infection 传染原因
cause of membrane fouling 膜污着成因
cause reason 原因
causes and effects chain system diagram 因果链系统图
causeway 海堤;路堤;堤道;长堤;湿地;高架人行道
causewaying (用石块或花岗岩石铺砌的)堤道
causeway section 长堤段
causey 堤道
causing organic material to rot 使有机物变腐烂
caustic 苛性的;焦散的;散焦点
caustic alkali 苛性碱
caustic alkalinity 苛性碱液;苛性碱度
caustic alumina 苛性氧化铝
caustic ammonla 苛性氨
caustic and acidic cleaning solution 酸碱洗液
caustic attack 苛性腐蚀
caustic baryta 氢氧化钡
caustic brittleness 苛性脆化;碱性脆化
caustic contact tower 碱接触塔
caustic corrosion 苛性腐蚀;碱腐蚀
caustic cracking 苛性裂纹;苛性脆化;苛性脆裂;碱蚀开裂;碱裂
caustic curve 焦散曲线
caustic dip 苛性浸渍;浸碱;碱液清洗
caustic dolomite 苛性白云石
caustic embrittlement 苛性脆裂;苛性脆化;碱蚀致脆;碱脆;腐蚀性脆化
caustic embrittleness 苛性脆化
caustic etch 浸蚀饰面;雪花饰面
caustic extraction 碱萃取法;碱抽提
caustic flaking machine 苛性钠刨片机
caustic fusion 碱熔法;碱熔
caustic holder 腐蚀剂点棒
caustic hydride process 苛化氢化法(一种除垢法)
caustic in flakes 片状烧碱
causticity 苛性;腐蚀性
causticization 苛性化;苛化作用
causticizer 苛化器;苛化剂
causticizing agent 苛化剂
caustic lime 氧化钙;苛性石灰;熟石灰
caustic lime mud 灰质软泥
caustic lye 苛性钠液
caustic lye of soda 烧碱液
caustic magnesia 氢氧化镁;苛性镁;轻烧镁砂
caustic magnesite 苛性姜石土;氢氧化镁;轻烧镁砂
caustic metamorphism 烘烤变质作用
caustic mud 石灰泥浆;烧碱盐泥

caustic neutralizer column 碱中和塔
caustic point 焦散点
caustic potash 苛性钾;钾碱;氢氧化钾
caustic pretreating 碱预处理
caustic product pump 成品碱泵
caustic product storage tank 成品碱储[贮]槽
caustic quenching 碱液淬火
caustics 焦散线
caustic salt 苛性盐
caustic scrubber 碱洗气器
caustic scrubbing 碱洗
caustic slag 碱性炉渣
caustic sludge 碱渣
caustic soda 苛性苏打;苛性钠;氢氧化钠;烧碱
caustic soda ash 苛性苏打灰
caustic soda concentration unit 烧碱浓缩装置
caustic soda process 苛性苏打法
caustic soda production 烧碱生产
caustic soda softening 苛性苏打软水法
caustic soda solid fused 固体熔融烧碱
caustic solution 苛性碱溶液
caustic stain 碱性锈蚀
caustic steaming of cargo tanks 苛性蒸汽洗油舱法
caustic surface 焦散面
caustic trauma 腐蚀伤
caustic treatment tower 碱处理塔
caustic wash 碱洗
caustic washing 碱洗
caustification 苛性作用
caustobilolithe 可燃性生物岩
caustobiolith 可燃(性)有机岩;可燃性生物岩
caustolite 可燃岩
caustophytolith 可燃(性)植物岩
caustozoolith 可燃性动物岩;可燃动物岩
Causul metal 镍铬铜合金铸铁
cauter 烙器;烧灼器
cauterantia 烧灼剂
cauterization 烙术;灼法;灼烙;烧灼术
cautery 烙术;烙器;烧灼术
cautery knife 烙刀
cautery needle 烙针
cautery therapy with heated needle 针烙
cautery unit 电烙装置
cautionary characteristic of light 警告性灯质
caution(ary) mark 小心标志;注意标志;警戒标志
cautionary note 禁区注记
caution board 警告牌
caution light 警告灯
caution money 保证金
caution money paid on contract 按合约支付保证金
caution position 注意位置
caution security 保证金
caution sign 警告标志
caution signal 警告信号;缓行信号;注意信号
caution speed 警告速度
cautionwarning sign 告示牌
caution zone 注意区
cautio pro expensis 费用担保
cautious running 小心运行
Cauvet 考维特式建筑(一种类似卡莫斯式的体系结构)
cavaedium 中庭;罗马式住房内院
cavalcade 船队;车队
cavalier (城堡中的)高土台
cavalier drawing 斜等轴测图
cavalier projection 斜投影
cavalry carrier 装甲运输车
cavansite 水硅钒钙石
cave 洞(穴);地下窑洞;屏蔽室;山洞
cavea 罗马半圆形露天剧场的阶级座位;演讲厅
cave accumulation soil 洞穴堆积土
cave animal 穴居动物
cave archaeology and paleontology items 洞穴考古和洞穴古生物
cave architecture 窑洞建筑
cave art 洞穴艺术;洞窟艺术
caveat 防止误解的说明;停止支付通知;停止诉讼申请
caveat emptor 货物出门概不退换
caveat vendor 卖方负责;买方负责
cave biogenic deposit 洞穴生物堆积物
cave breakdown 洞穴崩塌堆积(物);洞穴崩塌
cave breccia 洞穴碎屑物;洞穴角砾岩

cave ceiling karren 洞顶溶沟
cave chaitya hall 岩洞神祠
cave church 洞穴教堂
cave collapse 洞穴坍陷
cave collapse deposit 洞穴崩塌堆积(物)
cave coral 石珊瑚
cave deposit 洞穴沉积
cave deposition rate 洞穴沉积速度
cave depth below the ground surface 洞穴垂向深度
caved goaf 崩落的采空区;岩石崩落带
caved ground 坍陷地面
caved material 崩塌材料;崩塌料;坍落体
cave dwelling 穴居;窑洞住宅;窑洞;洞穴居所
cave earth 洞穴土
cave environment 洞穴环境
cave fauna 洞穴动物群
cave flag 石旗
cave floor karren 洞底溶沟
cave flower 石花
cave formation time 洞穴形成时期
cave hall 洞穴大厅
cave height 洞穴高度
cave-house 窑洞
cave hydrologic(al) elements 洞穴水文要素
cave ice 洞穴冰;洞冰
cave-in 冒顶;下陷;岩石崩塌;坍塌;沉陷;崩陷;凹陷的;坍塌;坍方;塌陷;(修筑上的)塌方
cave-in lake 解冻湖;冰川湖;融陷湖;融冻湖
cave-in of roof 顶板陷落
cave-in risk 坍落危险
cave in rock 岩洞
cave in soil 土洞
cavel 石工锤
cave length 洞穴长度
cave lotus leaf 石荷叶
cave-man 洞穴人;炉前工;穴居人
cave meteorology and climate zones 洞穴气象气候变化带
cave meteorology elements 洞穴气象要素
cave meteorology factors 洞穴气象因素
cave monastery 岩洞修道院
caven 洞窟
cavendish 板烟
Cavendish balance 卡文迪什天平
Cavendish experiment 卡文迪什实验
cavenger jig 扫选跳汰机
cave-obstructed borehole 坍塌岩石堵塞的孔眼;坍塌岩石堵塞的井眼
cave of debouchure 出水洞(穴)
cave organism 洞穴生物
cave painting 洞穴壁画;洞窟壁画
cave pearl 穴珠
cave period 穴居时代
cave position 洞穴位置
caver 卡浮风
cave residues 洞穴残积物
cave river 洞穴河流
cavern 空洞;岩洞;洞穴;洞室;洞;地下窑洞;大岩洞;大洞穴;石窟
cavern breccia 洞穴角砾岩
cavern filling 洞穴充填
cavern flow 洞穴水流;洞穴流;溶洞水
cavern limestone 多孔石灰岩
cavern of mountain 山洞
cavernous 海绵状的;洞穴状的;洞穴的
cavernous body 海绵体
cavernous dolomite 孔状白云岩
cavernous formation rock 多孔岩层
cavernous height 洞体高度
cavernous limestone 孔状灰岩
cavernous porosity 洞穴状孔隙率
cavernous rock 洞穴状岩石;多孔岩(石)
cavernous rock formation 多孔岩层
cavernous structure 孔状构造
cavernous texture 孔状结构
cavernous weathering 孔状风化;孔状风化
cavern rock 多孔岩(石)
cavern spring 洞穴泉
caverns supported by rock bolts 岩石锚杆加固的洞室
cavern water 岩洞水;洞穴水;溶洞水;岩溶水
cave sanctuary 洞穴圣所;洞穴庇护所
cave sepulcher 洞穴安息所

cave shields(palettes) 石盾
cave shooting 山洞爆炸
cave sinkhole 洞穴落水洞
cave skylight 洞穴天窗
cave span 洞穴跨度
cave spring types 洞穴泉的类型
cave stream 洞穴溪流
cave temple 石窟
cave tomb 岩洞坟墓
cave top height 洞顶高度
cave to the surface 地面坍陷
cave trend 洞穴走向
cavetto 修圆；削圆角；打圆；凹环形线脚；凹弧饰
cavetto cornice 凹弧形屋檐
cavetto mo(u)lding 凹圆线脚；凹弧饰；四分之一凹圆线脚
cavetto vault 中部(为)平顶的反水槽式穹顶；正方形拱顶；凹形拱顶
cave volume 洞穴体积
cave volume for flood preservation 洞穴蓄洪体积
cave wall karren 洞壁溶沟
cave water fall 洞穴瀑布
cave width 洞穴宽度
cave without wind 无风洞
cave with wind 有风洞
cavey formation 易坍塌地层
cavil 尖锤
cavil axe 小尖锤；尖斧锤
caviling days 矿工换区日
caving 冒顶；崩落开采法；淘空；坍落；塌落
caving angle 崩落角
caving area 陷落区；崩落带
caving bank 崩塌河岸；崩岸；坍塌堤岸；受冲岸
caving bin 落顶开采法漏斗
caving brick 空心砖
caving formation 易坍塌岩层；易坍塌地层；陷落地层；坍塌岩层
caving ground 陷落地层；易坍塌地层
caving hole 易坍塌孔；易坍塌井
caving-in 坍落；崩坍；顶板坍落；冒顶
caving in of ground 地面塌陷
caving line 冒落线；坍落线
caving method roof control 垮落法控制顶板
caving nature 塌陷性
cavings 坍落的岩石碎块(孔内)
cavings filler 清理孔内岩石碎块的钻具(活塞抽筒状)
caving shale 易坍塌页岩
caving stopping method 崩落采矿法
caving wall 空芯墙
cavitary 腔的
cavitas 空洞
cavitating flow 空腔水流
cavitation 空穴现象；空隙现象；空蚀；空腔形成；空泡现象；空化作用；空化；空穴形成；穴蚀；成穴；成洞；汽蚀；气穴；气蚀现象
cavitation aeration system 空蚀曝气系统
cavitation air flotation 涡凹气浮
cavitation attack 空蚀破坏作用；汽蚀破坏作用；气穴侵蚀
cavitation bank 气蚀水箱
cavitation bubble 空化气泡；气蚀气泡
cavitation bubble collapse 空泡溃灭；气泡溃灭
cavitation characteristic (泵的)空隙特性空蚀特性(曲线)；空化特性(曲线)；汽蚀特性(曲线)；气蚀特性
cavitation coefficient 空蚀系数；空化系数；气蚀系数
cavitation core 空蚀中心；空化中心；气蚀中心
cavitation corrosion 空蚀；空腔腐蚀；气蚀（腐蚀）
cavitation criterion 空蚀准则；空化标准
cavitation damage 涡浊；气蚀麻面；内爆；空穴缺陷；空蚀(破坏)；气蚀损坏；混凝土表面气穴损害；气蚀损伤；气蚀损坏；气蚀破坏
cavitation degradation 空化降解
cavitation degree 空蚀度；空化度；气蚀度
cavitation effect 空化效应
cavitation erosion 空腔侵蚀；空隙腐蚀；空；空腔侵蚀；空化腐蚀；麻坑；凹蚀；汽蚀腐蚀；气蚀腐蚀
cavitation flow 空蚀水流；空化水流；气蚀水流
cavitation fracture 空洞断裂
cavitation free flow 无气蚀水流
cavitation-free operation 空蚀运行；气蚀运行；无空化运行
cavitation-free performance 空蚀性能；气蚀性能；无空化性能
cavitation guarantee 空化保证；气蚀保证
cavitation hysteresis 空化滞后；气蚀滞后
cavitation in centrifugal pump 离心泵气蚀
cavitation inception 空化初生
cavitation index 空穴数；空泡系数；空化指数；气蚀指数
cavitation indicator 空泡指示器
cavitation in pump 水泵气蚀
cavitation intensity 空泡强度；空化强度；气蚀强度
cavitation level 空蚀级别；空蚀程度
cavitation limit 空穴度；空蚀度；气蚀限度；涡凹限度
cavitation mechanics 成穴力学
cavitation mechanism 空化机理
cavitation meter 空化仪；气蚀仪
cavitation noise 空化噪声；气蚀噪声
cavitation number 空蚀数；空泡系数；空化数；气蚀数
cavitation of concrete 混凝土气眼
cavitation of pump 水泵气蚀
cavitation of pump impellers 水泵叶轮汽蚀
cavitation of tube 管的空蚀(作用)；管道气蚀
cavitation of tubing 管的空蚀(作用)
cavitation parameter 空化参数；气蚀参数
cavitation pattern 空蚀类型
cavitation phenomenon 气蚀现象
cavitation pitting 空蚀坑；空蚀痕迹；空蚀剥损；点气蚀；点空蚀；气蚀坑；气蚀痕迹；气蚀剥损
cavitation pocket 空蚀穴；气蚀穴；涡凹
cavitation range 气蚀区域
cavitation reactor 涡凹反应器
cavitation resistance 抗气蚀性；抗空蚀；抗空化
cavitation resistant alloy 抗气蚀合金；抗空蚀合金；抗空化合金；抗气蚀合金
cavitation scale 空蚀程度；气蚀等级
cavitation sensor 空蚀传感器；空化传感器；气蚀传感器
cavitation sigma 空蚀系数；空化系数；气蚀系数
cavitation test 空隙现象试验；空蚀试验；空腔试验；空泡试验；空化试验；气蚀试验
cavitation tunnel 风道；空泡试验筒；风洞
cavitation wear 穴蚀磨损
cavitator 气穴曝气池
cavitron 手提式超声波焊机
cavity 孔穴；孔洞；空穴；空腔；空气层；空洞；模槽；铸件气孔；洞穴；凹处；天然洞穴
cavity and pocket 空洞
cavity antenna 谐振腔天线
cavity-backed radiator antenna 空腔反射式天线；背腔式天线
cavity barrier 空心屏障；空腔隔板；空腔火障
cavity base 窑洞垫基
cavity batten 墙腔板条；墙腔木板条
cavity block 空心砌块；凹模；空心块(体)；阴膜；阴模块；阴模；空心混凝土砌体
cavity block floor 空心砖楼板
cavity block for walls 墙用空心砌块
cavity block making machine 空心砌块制造机
cavity block masonry wall 空心块墙
cavity block mold 空心砌块模
cavity block roof 空心砖屋顶
cavity block step 空心砖台阶
cavity block wall 空心砌块墙
cavity brick 空心砖
cavity brick masonry(work) 空心砖圬工
cavity brick wall 空心砖墙
cavity brickwork 空心砖工
cavity brick work wall 空心砖工墙
cavity charge 锥孔装药
cavity closure block 封口空心砌块
cavity concrete block 空心砌块；空心混凝土砌块
cavity concrete block for walls 墙用空心混凝土砌块
cavity concrete tile for walls 墙用空心混凝土砖
cavity concrete wall 空心混凝土墙；混凝土空芯墙
cavity configuration 腔结构
cavity construction 空心构造
cavity corrosion 空腔腐蚀
cavity coupling 空腔耦合
cavity cross-wall 空心横隔墙
cavity dam 空心坝；框格坝；空心填土坝
cavity detachment 空腔分离；游离型空腔
cavity die 型腔模；阴模；凹模
cavity diffusion length 共振腔扩散长度
cavity disinfection 窑洞消毒
cavity drop subsidence 火山沉陷(区)
cavity-dumped laser 倾腔激光器
cavity dumper 腔倒空器
cavity dumping 倾腔；腔倒空
cavity effect 空腔效应
cavity-embossing rubber die 橡皮压印模
cavity emissivity 空腔发射(能力)；容积辐射率；气体容积辐射
cavity expansion theory 扩孔理论；空穴膨胀理论；空间膨胀理论
cavity external masonry wall 空心外墙
cavity fill 空腔填充料；型腔的充填
cavity filling 空心填料；充填作用；窑洞充填
cavity filling phase 充填型腔的阶段
cavity filter 空腔滤波器
cavity fixing 自挤压螺旋铆件(薄墙壁用)
cavity flashing 空腔挡水板；空芯墙披水；空芯墙泛水
cavity flow 空心流；空腔水流；空泡流
cavity flow theory 空泡流理论
cavity flushing 冲洗洞壁
cavity frequency meter 谐振腔频率计；共振腔频率计；空腔频率计
cavity grouting 空腔灌浆；回填灌浆；盾尾灌浆；盾尾灌浆
cavity insulation 孔穴绝缘；空心保温
cavity in the concrete 混凝土中孔穴；空洞；(由于施工不良而造成的)混凝土里蜂窝
cavity lath (筑空芯墙时用以挡住灰浆滴落的)长条压条；(筑空芯墙时用以挡住灰浆滴落的)长条木板
cavityless casting 实型铸造；实模铸造法
cavity lining 洞衬料
cavity location 洞穴方位
cavity magnetron 空腔磁控管；多控磁控管
cavity masonry(work) 空心圬工墙；空芯墙
cavity mode 空腔谐振模；共振腔模式
cavity-molding 模孔铸型
cavity panel 空心板
cavity panel wall 空心板墙
cavity partition(wall) tile 空心隔墙砖
cavity party wall 空心共用墙；防火空芯墙
cavity piston 凹头活塞
cavity pocket 空腔；空洞(气泡)；气泡
cavity preparation 制洞；窑洞制备
cavity radiation 空腔辐射
cavity radiator 空腔辐射体；空腔辐射器
cavity reactor 腔式反应堆
cavity reflecting element 共振腔反射元件
cavity reflector 共振反射体
cavity reflector alignment 谐振腔反射镜准直；共振腔反射镜调准
cavity resonance 空腔谐振；空腔共振
cavity resonator 空腔谐振器；空腔共振器
cavity resonator method 空腔共振器方法
cavity resonator wavemeter 谐振腔波长计
cavity restoration 窑洞修复
cavity ring 空心环
cavity shaft wall 空心风道墙
cavity shell 中空壳；空心壳
cavity-stabilized oscillator 空腔稳频振荡器
cavity structure 孔洞构造
cavity sub frame 空芯墙上的框架用以固定钢窗
cavity temporary sealant 窑洞暂封剂
cavity theory 空穴理论
cavity tile 空心砖
cavity-tile floor 空心砖楼板
cavity-tile making machine 空心砖制造机
cavity-tile mo(u)ld 空心砖模
cavity-tile wall 空心砖墙
cavity transfer mixer 阴模压铸混合机
cavity tray 孔穴泛水；空芯墙披水；空腔挡水板；空穴泛水；空腔泛水
cavity tuning 空腔调谐
cavity venting 空隙通风
cavity wall 空斗墙；空芯墙；隔墙；窑洞壁；双层墙；双层壁
cavity wall insulation 空芯墙绝热材料；空芯墙隔热材料
cavity wall reinforcing(ladder type) 空芯墙梯式加固
cavity wall tie 空芯墙系铁；空芯墙连系件；空芯墙拉杆
cavity waterproofing 空腔防水

cavity wavemeter 空腔波长计;共振腔波长计
cavity well 空穴井;洞穴井
cavity with rarefied air 有稀薄空气的孔穴
cavity zone 空腔区
cavo-relievo 凹浮雕;平面浮雕;凹雕
cawk 氧化钡;重晶石
cawk stone 不纯重晶石(英国)
caxton floor 加筋空心耐火砖地板(用于防火)
cay 小礁岛;(海上低潮露出的)沙洲;沙礁
Cayley-Hamilton theorem 凯莱一哈密顿定理
Cayley inverse matrix 凯莱逆矩阵
Cayley-Klein parameter 凯莱一克莱因参数
cay rock 礁岛岩
cay sandstone 礁砂岩;礁岛砂岩
caysichite 碳硅钙钇石
Cayugan 卡尤加统
Cazenovian 卡泽诺维阶
cazin 共晶合金;低熔合金
cbale-equalizing amplifier 电缆均衡放大器
C-battery 电池组;丙电池(组)
CBR tester 加利福尼亚州承载比测定仪
CBR value 加利福尼亚州承载比值
C-clamp 弓形堵漏夹;C 形夹钳
C-class 丙类
C class division 丙级分隔
cctahydroestrone 八氢雌甾酮
cctocyclic(al) compound 八环化合物
cctohydroxylated acid 八羟基酸
Cd-Ni button cell 镉镍扣式电池
CD-R 可刻录光盘
CD-ROM 光驱【计】
CD-RW 可擦写光盘
cease 中止;停止
cease and desist order 产业停售通知;结束和停止命令
ceased and deferred projects 停建缓建项目
ceased project 停建项目
cease-fire line 停火线
cease operation 停止作业
cease payment 停止支付
cease to be available 失效
cease to be in effect 失效
cease to have effect 停止生效
cease to hold office 终止任职
cease to run(of limitation period)时效终止
cease tracking 停止跟踪
cease work 停工;收工
ceasing stick 摺缝刀
cebaite 氟碳铈钡矿
cebollite 纤维石
Cebu hemp 马尼拉麻
cecal 盲的
cechite 施钒铅铁石
cecilite 黄长白榴岩
cecocentral 中心盲点的
ceco-drop hammer 气动落锤
cecos tamp 不规则件压纹压印机
cecum 盲端
cedant 分出保险人;分保分出人;保险让与人
cedant company 分出公司
cedar 灰红色;雪松;柏木
cedar chest 香柏木箱
cedar elm 榆木;厚叶榆
cedar gum 雪松胶
cedar leaves oil 雪松叶油
cedar-nut oil 雪松坚果油
cedar of Lebanon 西南亚雪杉
cedar oil 雪松油;香柏油
cedar panelled door 杉木镶板门
cedar seal 树皮纤维堵漏剂
cedar slatted ceiling 西洋杉木板条顶棚
cedartex 屋面卷材面上凸起的花纹;压花雕饰(颗粒屋面板上)
cedar tree 雪松
cedar-tree laccolith 雪松树岩盖
cedarwood 杉木
cedar wood oil 柏木油;雪松(木)油
ceded line 分出保险部分
ceded share 分出股份
cede insurance 分(出)保险
ceding company 分出公司;分保分出公司
cedramber 柏木中醚
cedrol 雪松醇;柏木脑
Cedros trench 塞德罗斯海沟

cedrus 雪松
cedryl acetate 结晶乙酸柏木酯
cefferdam 空隔舱
Ceiba fibre 木棉
ceil 装顶棚
ceiler 壳板装配工
ceiling 云幕;云底高(度);云底;最高限度;顶篷;顶棚;舱底铺板;舱底垫板;舱壁护板;风渠隔板;天花;天顶
ceiling adhesive 顶棚胶黏剂
ceiling air diffuser 天花板空气散流器;顶棚空气散流器
ceiling altitude 升限
ceiling amount 最高(限)额
ceiling and floor 最高和最低(限度)
ceiling and visibility unlimited 能见度极好
ceiling area lighting 顶棚照明
ceiling balloon 云幂气球;测云气球
ceiling beam 平顶梁;顶棚大梁;顶梁;舱底板垫梁;天花梁
ceiling binder 平顶格栅中间支撑
ceiling block 顶棚(隔)块;灯线盒;顶棚花叶圆饰
ceiling board 平顶板;花板条天棚;盖板;顶(棚)板;舱底板;天花板
ceiling boarding 平顶盖板;顶棚覆板;天花板
ceiling box 平顶箱
ceiling brick 顶窑砖
ceiling button 天棚按钮;顶部按钮
ceiling caisson 藻井;平顶藻井;沉箱
ceiling capacity 最大能量
ceiling cavity 吊顶空间
ceiling channel 顶板沟道
ceiling classification 云幂分类
ceiling coat 顶棚面涂层
ceiling coil 顶盘管;灯线蛇形管
ceiling coil for heating 顶棚供暖盘管
ceiling collapse 洞顶坍塌
ceiling concentration 极限浓度
ceiling conduit 顶棚内管道
ceiling convector 顶棚对流加热器
ceiling cooling 顶棚冷却
ceiling cornice 顶棚(周围檐口式)线脚;吊顶线脚;平顶线脚;平顶飞檐式线脚
ceiling crab 吊架起重绞车;吊架起重机
ceiling design 天花板图案
ceiling diffuser 顶棚散流器;顶棚空气扩散器;天花板漫射灯具
ceiling distance 风机—隧洞顶距
ceiling dome light 顶棚顶灯;圆屋顶采光
ceiling drilling machine 悬装钻机
ceiling duct 天花板管道;顶棚通风管;平顶通道
ceiling effect 天花板效应;平顶效应
ceiling element 平顶部件
ceiling enrichment 顶棚装饰;楼板装饰
ceiling exchange rate 最高汇率
ceiling excitation 极限励磁
ceiling facing 顶棚饰面
ceiling fan 顶板通风器;吊(式)风扇;风扇;平顶(内装)风扇;天花板吊扇
ceiling-fan motor 吊扇电机
ceiling filling 附顶加灯
ceiling finish 顶棚装修
ceiling finishes 顶板表层
ceiling fitting 顶棚装修;顶棚灯具配件;天棚照明设备;天棚灯;平顶装配
ceiling flange 顶棚孔盖
ceiling floor 顶棚顶板;吊顶板(材);平顶结构;顶棚吊架
ceiling flourish 顶篷花饰
ceiling framing 顶棚结构
ceiling free of trussing 无梁天顶;无梁平顶
ceiling fresco 顶棚画
ceiling girder 顶梁
ceiling grid 顶棚格子;顶管
ceiling grille 顶棚花格
ceiling grinder 顶棚面磨光机
ceiling groove 平顶开槽
ceiling guard 天花板防护条;顶棚防护条
ceiling hanger 天花板悬挂物;天花板吊杆;顶棚悬挂物;顶棚吊杆;平顶吊筋
ceiling hatch 货舱底开口
ceiling heating panel 供暖顶板
ceiling height 室内净高;升限;云幕高度;云幂高度;顶棚高度;房间净高度;板高(度);天花板离楼地面高度
ceiling height indicator 云幕高度探照灯;升限指示器
ceiling hook 房顶吊钩
ceiling illumination 顶棚照明
ceiling incorporation services 内装设施的顶棚
ceiling inspection hole 顶棚检查孔
ceiling installation 天花板安装
ceiling insulation 顶棚保温
ceiling jack 顶棚灯;平顶千斤顶
ceiling joint 顶棚接缝;顶板接头
ceiling joist 天花板小梁;吊平顶龙骨;吊(平)格栅;吊顶(木)龙骨;平顶龙骨;平顶格栅
ceiling lamp 悬灯;吸顶灯;云幕灯;吊灯;舱顶灯;天棚灯;顶棚灯
ceiling lamp fitting 顶灯附件
ceiling lamp fixture 天棚照明配件
ceiling lath 顶棚抹灰板条
ceiling lattice work 天花板格子;顶棚格子;顶管
ceiling light 云幕灯;顶棚灯;顶棚采光;顶灯;吊灯;车厢顶灯;舱顶灯;平顶照明
ceiling light cupola 浅穹隆顶棚
ceiling light diffuser 顶灯散光罩
ceiling light fitting 顶棚灯具
ceiling lighting 顶棚照明
ceiling lighting fitting 平顶照明装置
ceiling light project 顶棚投光器
ceiling limitation 最高限额
ceiling lining 顶棚衬
ceiling load 最大载重量;顶棚荷载;平顶重;平顶荷载
ceiling luminaire(fixture)顶棚照明设备
ceiling manhole 顶棚检查孔
ceiling material 天花板材料;顶棚材料
ceiling mo(u)lding 平顶线脚;顶棚线脚
ceiling mounted air diffuser 顶棚散流器
ceiling mounted air outlet 顶棚送风口
ceiling mounted fan 吊扇
ceiling mounted fixture 装于顶棚的设备
ceiling mounted heating coil 顶棚供暖盘管
ceiling-mounted lamp 吸顶灯
ceiling mounting 倒置安装
ceiling mounting channel 顶棚安装电路;顶棚安装槽
ceiling night light 顶棚夜间灯
ceiling of asbestos cement sheets 石棉水泥板平顶;石棉水泥板顶棚
ceiling of convection 对流高度
ceiling on loan 贷款限额
ceiling on remittance of royalties 使用费汇款的最高限额
ceiling on rent 房租的最高标准
ceiling on wage 最高工资
ceiling ornament 顶棚装饰
ceiling oscillating fan 摇转吊扇
ceiling outlet 顶棚引出线(灯头用);顶棚(喷口型)送风口;顶棚电线头
ceiling paint 顶棚涂料
ceiling panel 顶棚镶板;平顶镶板;天花板
ceiling panel heating 顶棚辐射采暖;顶棚(板面)供暖;顶板供暖
ceiling panel(l)ing 天花镶面板
ceiling panel strip 古建筑的天花板枝条
ceiling paper 顶棚纸板
ceiling pattern 天花彩画
ceiling picture 顶棚绘画
ceiling plan 顶棚平面图;平顶布置图
ceiling plank 舱底木板
ceiling plaster 平顶涂层;楼板粉饰
ceiling plastering 顶棚抹灰
ceiling plaster lath(ing) 顶棚抹灰板条
ceiling plate 顶棚嵌板;吊线板;舱内复板;顶棚镶板
ceiling plenum 顶棚气隙
ceiling point 顶棚引出线(灯头用)
ceiling price 最高限价;最高价(格);最大限价;上限价格
ceiling profile 顶棚形状
ceiling projector 云(幂)灯
ceiling protection 顶棚防火层;天棚防火层
ceiling pull switch 顶棚拉线开关
ceiling rafter 顶棚格椽;天棚梁
ceiling rail 顶棚纵向梁;顶棚周边的压条
ceiling rate 最高价格;限价

ceiling reed 平顶弹簧片
ceiling reflection factor 顶棚反射系数
ceiling reflector 天花板反射板
ceiling rent 最高租价;租金限价
ceiling return inlet 顶棚回风口
ceiling rose 吊线盒;挂线盒;顶棚灯线孔盖
ceiling rosette 顶棚圆花饰
ceiling runner track 顶棚上的隔墙龙骨
ceiling sander 平顶打磨器;平顶打磨机
ceiling sanding machine 平顶打磨器;平顶打磨机
ceiling saucer dome 浅拱顶棚
ceiling screen 挡烟垂壁
ceiling sheet metal 薄钢板制天花板
ceiling shower 平顶淋浴器
ceiling sill 顶棚纵向梁
ceiling slab 隧道顶板
ceiling soffit 顶棚下皮
ceiling sound transmission 顶棚传声;吊顶传声
ceiling Sound Transmission Class[STC] 吊顶传声等级
ceiling speed 极限速度
ceiling spotlight 面光
ceiling strap 平顶板条;顶棚板条;楼板格栅木条（钉吊顶棚用）
ceiling strip 顶棚板条
ceiling strut 顶棚支撑
ceiling suspension 顶棚悬吊
ceiling suspension system 顶棚悬吊体系;吊顶体系
ceiling switch 顶棚拉线开关;天棚拉线开关
ceiling system 顶棚体系
ceiling temperature 规定最高温度;顶棚温度;上限温度
ceiling tile 吊顶板（材）
ceiling track 顶棚板墙筋;隔墙上槛
ceiling trim 修饰;蒙顶板
ceiling trimming 平顶饰;平顶镶边
ceiling-type direct-fired unit heaters 顶棚直燃式采暖机组
ceiling-type fan 吊扇
ceiling-type unit heater 平顶式单元采暖器
ceiling value 最高（限）值
ceiling visibility 云底能见度
ceiling void 吊顶空间
ceiling voltage 最高电压;峰值电压
ceiling wiring conduit 顶棚导线管
ceiling with exposed beam 露梁平顶;露梁顶棚
ceiling without trussing 无梁天顶;无梁平顶
ceil(l)ometer 云高计
ceilograph 云幕仪
ceilometer 云高计;云幕仪;云高指示器
ceilure 十字架上面的花格挑顶;祭台
ceinmicroscopy 电影显微术
celadon 雾青色;灰绿;艾青;青瓷（一种中国著名的瓷器）
celadon ewer with single handle 青瓷单柄壶
celadon flak 青瓷扁壶
celadon glaze 青瓷釉
celadonite 绿鳞石
celadon lotus shaped cap 青瓷莲花形盅
celadon spitton 青瓷垂壶
celadon ware 青瓷（一种中国著名的瓷器）
celadon with dark mottes 青釉褐斑
celadon with flyspots 飞青
Celanese 西乐尼丝（商品名）
celanna 醋酯里子布
celature 金属表面浮雕;金属上雕刻花饰
celcure 铜铬木材防腐剂;铜铬防腐剂;铬酸铜（木材防腐剂）
celedon 艾青
cele-navigation 天文航海法
celerity 运动速度
celerity of flood wave 洪水（波）波速
celerity of wave 波速
celery 浅盘
celery top pine 澳大利亚松树
celescope 天体镜
celeste blue 天青色料
celeste glaze 天青釉
celestial atlas 天体图集
celestial axis 天轴
celestial blue pigment 宇宙蓝颜料;天青蓝颜料
celestial body 天体
celestial body azimuth 天体方位
celestial body diurnal apparent motion 天体周日视运动
celestial body's 天体的
celestial body true motion 天体真运动
celestial cartography 天体制图学
celestial chart 星图;天体图
celestial clock 天体钟
celestial compass 天文罗盘;天体罗盘
celestial concave 天穹
celestial coordinate comparator 天体坐标量测仪
celestial coordinates 天体坐标;天球坐标
celestial coordinate system 天球坐标系
celestial ephemeris pole 天球历书极
celestial equator 天体赤道;天球赤道;天赤道
celestial equator system of coordinates 赤道系统;天赤道坐标系（统）
celestial fix 天文定位;天文船位
celestial fixing 测天对时;测天定位
celestial fixing position by astrocalculation 天文定位
celestial geodesy 天文大地测量学
celestial globe 星球仪;天体仪;天球仪
celestial guidance 天文导航
celestial horizon 真地平线;真地平（圈）;天文地平（线）;天球地平（圈）;真正地平
celestial-inertial guidance 天文惯性导航
celestial inertial navigation aids 天文惯性导航设备
celestial latitude 黄纬
celestial line of position 天文位置线
celestial longitude 黄经;天文经度
celestial map 天文图
celestial mechanics 天体力学
celestial meridian 赤经圈;天球子午线;天球子午圈
celestial navigation 天文领航;天文航海（学）;天文航法;天文导航;天体导航
celestial navigation instrument 天文航海仪表
celestial observation 天文观测;天体测量
celestial orientation 天体定向
celestial origin 天体成因
celestial parallel 黄纬圈;赤纬圈;赤纬平行圈;天球纬圈
celestial photography 天体照相学
celestial photometry 天体光度学
celestial planisphere 平面星图
celestial polar distance 天体（极）距
celestial pole 天极
celestial refraction 天文蒙气差
celestial sphere 天体;天球
celestial stem 天干
celestial system 天文导航系统
celestial triangle 天文三角形;天球三角形
celestial ultraviolet telescope 紫外天文望远镜
celestine 天青石含量
celestite 天青石
celestite cement 天青石胶结物
celestite concretion 天青石结核
celestite ore 天青石矿石
C eliminator 代丙电池
celite 硅藻土;次乙酰塑料;C水泥
cell 镜框;信元;小室;小房(间);小电池;匣铁垛;隔室;电池;单元;单人牢房;存储单元;翅室;浮选槽;前置炉;炮孔
cella 内殿;古希腊或罗马庙宇里的内殿
cellactite （石棉保护的）金属屋顶材料
cella die 挤泥机模口
cella door 内殿门
cella facade 内殿内墙立面
cell alternative 单组选择元
cellar 酒窖;堆栈;地下室;地窖
cellarage 地窖容积
cellar air shaft 地下室通风（采光竖）井
cellar area 单元面积
cellar bolster 用品箱车垫木
cellar control 钻孔上的闭塞机构
cellar door 地下室门
cellar door lock 地下室门锁
cellar drain(age) 地下室排水
cellar drainage sump pump 地下室排水沟泵
cellar dwelling 地下室住房
cell area 面积元（素）
cellar entrance 地下室入口
cellar excavation 地窖挖掘
cellar floor 地下室地面
cellar for branches 楼体地下连接通道
cellar frame 地下室框架
cellar fungus 木材干枯;木材干腐
cellar garage 地下车库
cellar grating 地下室格栅
cellar gulley 地下室集水沟
cellar hole 地下坑洞
cellarino （托斯卡纳或陶立克式柱头的馒形下的）柱颈
cellarless 无地下室
cellar light well 地下室采光天井
cellar masonry wall 地下室（砖石）墙
cellar masonry work 地下室墙体工程
cellar moistening 地窖湿润
cellar nesting storage 后进先出存储器
cellar niche 地下室壁龛
cellar pipe 地下室灭火管
cell array 单元阵列
cellar recess 地下室壁凹
cellar room 地下室房间
cellar rot 地下腐烂;湿朽
cellar rubber 地下绝缘
cellar shelter 地下室掩蔽处
cellar stair(case) 地下室楼梯
cellar structure 地下室结构
cellar under a street 街道下地下室
cellar vault 地下室拱顶
cellar wall 地下室墙
cellarway 地下室通道
cellar window 地下室窗;地窖气窗
cellastic tire 纤弹轮胎
cell atmosphere 电解槽气氛
cell attribute 单元属性;单元表征
cella wall 内殿墙
cella window 内殿窗
cell bath 电解槽电解液
cell battery 太阳能电池
cell belt 隔板式输送带
cell block 监狱分区;单元块
cell body 池体
cell bottom 槽底
cell box 电池箱
cell box girder 多孔箱形梁
cell building 单元建筑
cell busbar 电解槽导电母线
cell call 单元引入;单元调用
cell capacity 电池容量
cell case 电瓶壳
cell chart 方格图
cell classification 标准分挡
cell coalescence 晶胞聚结
cell complex 单元复合体
cell concrete 多孔混凝土
cell configuration 单元形状
cell connector 联级板;电池连接板
cell constant 电池常数
cell construction 电解槽结构
cell container 电解槽容器
cell cover 电瓶盖
cell cube 单元体;单元立方体
cell cup 蜡碗
cell current 槽电流
cell data 单元数据;方格数据
cell delay variation 信元延时变化
cell delineation 信元定界
cell design 牢房设计
cell dimensions 电槽外形
cell drier 室式干燥器;室式干燥机
cell enamel 填彩搪瓷;景泰蓝
cell error ratio 信元差错比
cellestron silk 醋酯丝
cell feed material 电解槽供料
cell fender 鼓形护弦;鼓形（空心）橡胶碰垫
cellfibre 单纤维
cell field model 蜂窝场模型
cell fill 格体填料
cell frequency 方格频数
cell furnace 电熔窑
cell gas 电解槽气体
cell gate 单元门
cell geometry 电解槽的几何形状
cell guide 箱格导柱;格导;格槽导引装置;格舱导轨
cell header 信头
cell hopper 分选料斗
cell house 教养院;监狱;单间监狱
cell-in-cell-out method 池入—池出法

cell in hollow tile 空心砖孔
cell inspection lamp 电池检验灯
cellinsulate brick 绝热砖
cell interconnection mask 单元互连掩模
cell letters 呼号
cell library 单元库
cell life 电解槽寿命;单元槽寿命
cell limit 标准档次范围
cell line 拱肋旁的腹线;电解槽系列
cell lining 电解槽内衬
cell liquor head tank 电解液高位槽
cell liquor pump 电解液泵
cell loss ratio 信元丢失比
cell maintenance 电解槽维护
cell matrix 单元矩阵
cell mean 方格平均值
cell membrane 细胞膜
cell method 分格法
cell model 电解槽模型
cell motor 电池电动机
cell mount 镜框架
cell name 单组名;单元名
cell number 标准档次编号
cellodion section 火棉胶切片
cello foils 乙烯箔
celloidin 棉胶;火棉(液)
celloidin paper 棉胶纸
celloline 玻璃纸
cellon lacquer 硝化纤维清漆
cellophane 玻璃纸;透明纸;塞珞分
cellophane cover 透明玻璃纸面
cellophane ink 玻璃纸印刷油墨
cellophane paper 玻璃纸;透明纸
cellophane web 玻璃纸卷材
cell orifice 单元孔
cellosolve 溶纤剂
cellosolve acetate 乙酸溶纤剂;乙二醇一乙醚乙酸酯
celloyarn 玻璃纸条;玻璃带纤维
cellpacking 管壳
cell pair 室对
cell pair resistance 室对电阻
cell parameter 晶胞参数
cell plate 窝眼盘
cell pressure 盒压式压力盒;压力盒压力;压力盒压;(旁压仪的)腔压;室压(三轴试验)
cell quartz 多孔石英
cell quay wall 格式岸壁;格体式码头岸壁
cell rate decoupling 信元速率解耦
cell reaction 电解反应
cell resistance 槽电阻
cell room 牢房;电解车间;单人牢房
cell sampling 格子采样
cell segregation 粒状偏析
cell sidewall 电解槽侧壁
cell size 单元大小;蜂窝格子尺寸;孔度
cell-surface hydrophobicity 细胞表面疏水性
cell switch 电池转换开关
cell terminal 电池接柱;电池接线端
cell test 荷兰式三轴试验;压力盒试验
cell tester 电池测试器
cell thickness 单元厚度(垃圾掩埋场);堆层厚度
cell transfer delay 信元传递延时
cell type 单元类型;程控;槽型
cell-type buffer 鼓形护弦本体
cell-type container ship 格栅型集装箱船
cell-type corrosion 电池型腐蚀
cell-type electrocardiograph 电池式心电图仪
cell-type fender 鼓形(空心)橡胶碰垫
cell-type heater 单元式加热炉;管式加热炉;槽型加热器
cell-type incinerator 格算式焚化炉
cell-type soaking pit 分隔式均热炉
cell-type sound absorber 小格型消声器
cell-type switch tube 盒式开关管
cell-type tube 电池式电子管
celluiose 钎维素
cellula 小房(间);(古罗马小神庙中的)小圣堂
cellulae 含气小房
cellular 格胞式;格形的;格体式的;格式式的;多孔状、蜂窝状的;分格式的
cellular abutment 格间式桥台;框格桥台;框格式边墩;空心桥座;空心桥墩;空心边墩;隔仓式桥台;格形桥台;格式桥台;多孔桥台
cellular adhesive 多孔胶黏剂;泡沫胶黏剂

cellular aggregate 蜂窝状集料;蜂窝状骨料
cellular air filter 分隔型空气过滤器
cellular array 单元(的)阵列
cellular array processor 单元阵列处理机
cellular automation 分区自动化
cellular beam 格型梁;格形梁
cellular bin 分格式储[贮]料箱
cellular block 空心砖混凝土块;孔心砌块;空心砌块;格孔砌块;格孔方块;多孔砖;多孔砌块
cellular board 隔舱板;多孔板;蜂窝夹层板
cellular bond 花墙砌法;格形砌合
cellular box structure 分格箱形结构
cellular breakwater 格构式防波堤
cellular brick 多孔砖
cellular building 格笼式建筑;箱式建筑;分格式建筑
cellular bulkhead 格形堤岸;格形岸壁;格体岸壁
cellular buttress 空心支墩;格型支墩;格形支墩
cellular buttress gravity dam 格形扶壁式重力坝
cellular caisson 框格式沉箱;组格式沉箱;格(体)式沉箱;格墙沉箱;格孔式沉箱;多室沉箱;分格沉箱
cellular cell 格体
cellular chain 单元(式)链
cellular circulation 环型环流;涡胞环流
cellular cloud 细胞状云
cellular cofferdam 框格式围堰;空心围堰;格型围堰;格形围堰;格体围堰;格式围堰;格孔式围堰
cellular concrete 多孔混凝土;泡沫混凝土;轻质多孔混凝土;加气混凝土
cellular concrete block 格形空心混凝土;格形混凝土块体;多孔混凝土砌块;泡沫混凝土砌块
cellular concrete slab 加气混凝土板;泡沫混凝土板
cellular conductor 空心导线;穿管导线
cellular configuration 格舱形状
cellular construction 多孔构造;多孔(混凝土)施工方法;框格式建筑;细胞状结构;格型建筑(物);格型构造;格形构造;格体结构;格式结构;格式构造;单元结构;蜂房式构造;分格式建筑(法)
cellular construction method 框格式建筑法
cellular container 格舱式集装箱
cellular container ship 格体式集装箱船;格舱式集装箱船
cellular container vessel 组格式货柜(集装箱货船)
cellular convection 环形对流
cellular cooler 蜂房式烧结矿冷却机
cellular core door 蜂窝状芯门
cellular core wall 格箱芯墙;框格芯墙;孔格芯墙;箱格形芯墙;格体式芯墙
cellular dam 格箱(式)坝;框格坝;箱式坝;重力支墩坝;格体式坝
cellular deck 分格式桥面
cellular double bottom 框格双层底;组格式双层底
cellular ebonite 蜂窝硬质胶
cellular elastomeric gasket 多孔弹性体密封垫
cellular-expanded concrete 加气混凝土;蜂窝膨胀混凝土;多孔混凝土
cellular-expanded concrete(building) block 膨胀多孔混凝土砌块
cellular-expanded concrete slab 加气膨胀混凝土板
cellular-expanded gypsum 膨胀网ేపు泡沫石膏
cellular-expanded material 泡沫膨胀材料
cellular-expanded mortar 微孔砂浆;泡沫砂浆;加气砂浆
cellular filter 蜂窝状滤器;蜂窝(过)滤器
cellular flap gate 分隔卧式坞门;分格卧倒门
cellular floor 格形楼板;格型楼板
cellular floor raceway 蜂窝结构地板电缆管道
cellular foam 多孔状泡沫体
cellular foamed polystyrene 聚苯乙烯泡沫塑料
cellular frame 格形构架
cellular framing 箱形框架;蜂窝框架构筑法;网格式构架
cellular girder 空腹梁;箱形型梁;格(形)大梁;分格大梁;空心梁
cellular glass 多孔玻璃;泡沫玻璃
cellular glass block 多孔玻璃块
cellular glass ceramics 蜂窝玻璃陶瓷
cellular gravity dam 箱格式重力坝;格箱式重力坝;格体式重力坝
cellular growth 细胞形增长;分格生长
cellular gypsum 多孔石膏
cellular gypsum plaster board 多孔石膏建筑板材

cellular heat exchanger 格子式热交换器
cellular height 格舱高度
cellular hold 箱格式货舱
cellular horn 多格喇叭;分格号角
cellular inserts 格子式热交换装置
cellular insulant 泡沫保温材料;多孔绝缘材料
cellular insulant material 多孔绝缘材料
cellular insulating material 多孔绝缘材料
cellular insulation 泡沫绝缘
cellular inverted list 单元式倒排表
cellularity 多孔性;网架型
cellular lava 多孔熔岩
cellular leather cloth 微孔人造革
cellular limestone 蜂窝状石灰岩
cellular list 单元细目表
cellular logic 单元逻辑
cellular material 泡沫材料;多孔材料;蜂窝状材料
cellular metal 蜂窝形金属
cellular metal floor raceway 蜂窝结构金属地板电缆管道
cellular method 分格法
cellular mobile tele-communications 无线移动电话系统
cellular moisture 毛细水分
cellular mortar 泡沫砂浆
cellular multilist 多表单元
cellular phone 移动电话
cellular pier 空心桥墩;格型墩
cellular plastics 蜂窝塑料;多孔塑料;泡沫塑料;微孔塑料
cellular polyethylene 泡沫聚乙烯
cellular polyhedron 分格多面体
cellular polyisocyanate 聚异氰酸酯泡沫
cellular polyisocyanate thermal insulation board 聚异氰酸酯泡沫保温板
cellular production 单元生产
cellular pyrite 白铁矿
cellular quartz 多孔石英
cellular quaywall 格体式码头岸壁
cellular raceway 格型线槽;分格导线槽
cellular radiator 蜂窝(式)散热器;蜂窝式放热器
cellular raft 格型板式基础
cellular reproduction 细胞繁殖
cellular resin 泡沫树脂
cellular retaining wall 框格型挡土墙;空箱式挡土墙;箱格式挡土墙;格形挡土墙;格间式挡土墙
cellular rubber 多孔橡胶;海绵橡胶;橡胶海绵;泡沫皮;泡沫橡胶
cellular rubble 泡沫循环
cellular section 蜂窝状断面
cellular shape 蜂窝形状
cellular shear 格状剪切
cellular sheet pile 格形板桩
cellular sheet pile breakwater 格形板桩围堰防波堤;箱格钢板桩防波堤;格形板桩防波堤
cellular sheet pile bulkhead 板桩格体岸壁
cellular sheet pile dolphin 组格式板系船墩;格形板桩墩
cellular sheet pile(d) structure 格形板桩结构
cellular sheet pile jetty 格形板桩码头
cellular sheet pile piper 格形板桩码头
cellular sheet pile quay 板桩格体码头
cellular sheet pile wharf 组格式板桩码头;格形板桩码头;板桩格体码头
cellular ship 隔舱式货轮
cellular silencer 泡沫消声器;多孔状消声器
cellular silica 多孔硅石;多孔二氧化硅
cellular soil 多孔状土壤;多边形土;蜂窝状土壤
cellular splitting 单元分裂
cellular steel deck 多孔蜂窝状钢甲板(屋面)
cellular steel floor 折叠形钢楼板;格型钢楼板
cellular steel sheet pile 格形钢板桩
cellular steel sheet pile breakwater 格形钢板桩防波堤
cellular steel sheet pile bulkhead 钢板桩格型岸壁;钢板桩格体岸壁
cellular steel sheet pile cofferdam 格形钢板桩围堰
cellular steel sheet pile jetty 格形钢板桩码头
cellular steel sheet pile pier 格形钢板桩码头
cellular steel sheet pile quaywall 格形钢板桩岸壁
cellular steel sheet pile wharf 格形钢板桩码头
cellular striation 多孔塑料条纹缺陷;蜂窝状夹层构造
cellular structure 蜂窝式结构;箱格型结构;细胞状

结构;格形结构物;格形结构;多箱结构;多孔构造;单元式构造;蜂窝状构造;蜂房式构造;薄壁蜂窝结构;网格结构
cellular substructure 格型下部结构
cellular switchboard 分区开关板;分格式配电盘
cellular system 组格式;单元系(统)
cellular telephone 手机
cellular texture 蜂窝状组织;蜂房式结构;网状组织;网格结构
cellular thermal insulation 泡沫绝热材料
cellular tile 多孔砖
cellular tubing 分格式配管;分格型配管
cellular-type 多箱式
cellular-type block 分格式房屋;单元式房屋
cellular-type building 单元组合型建筑;单元组合型房屋;蜂窝状建筑
cellular-type cofferdam 蜂窝式围堰;蜂窝式潜水箱
cellular-type core 蜂窝型心
cellular-type dam 蜂箱式坝
cellular-type of floor 密肋式楼面;密肋式楼盖;密肋式楼板
cellular-type radiator 多孔式散热器
cellular unit 多孔单元
cellular unmanned production system 成组无人生产系统
cellular vessel 导轨式集装箱船
cellular vortex 环型漩涡;环形漩涡
cellular wave 格形波;微波
cellular-wedge core wall Amburser dam 平板支墩蜂窝式芯墙坝
cellular wheel sluice 分格轮(式)闸门
cellulase 纤维素酶
cellulated ceramics 多孔轻型陶瓷;多孔陶瓷制品
cellulated glass 多孔玻璃
cellulation 泡沫结构
cellulith 赛璐里斯
celluloid 假象牙;明胶;赛璐珞
celluloid lacquer 赛璐珞漆
celluloid paint 透明油漆;透明漆
celluloid scrap 赛璐珞废料
celluloid sheet 赛璐珞片
celluloid template 赛璐珞模板
cellulolytic 分解纤维的
cellulolytic bacterium 分解纤维素菌
cellulos-bearing wastewater 含纤维废水
cellulose 纤维素;纸浆
cellulose acetate 乙酸纤维素;醋酸纤维素
cellulose acetate butyrate 乙酸丁酸纤维素;醋酸—丁酸纤维素
cellulose acetate butyrate lacquer 乙酸丁酸纤维素喷漆
cellulose acetate butyrate plastics 乙酸丁酸纤维素塑料
cellulose acetate dye 醋纤浆料
cellulose acetate fibre 醋酸纤维素纤维
cellulose acetate film 醋酸纤维素薄膜
cellulose acetate lacquer 醋酸纤维素清漆
cellulose acetate membrane 醋酸纤维薄膜
cellulose-acetate membrane electrophoresis 醋酸纤维膜电泳
cellulose acetate mo(u)lding powder 醋酸纤维素软性塑胶粉
cellulose acetate paint 醋酸纤维素漆
cellulose acetate-phthalate 邻苯二甲酸醋酸纤维素
cellulose acetate plastics 醋酸纤维素塑料
cellulose acetate propionate 醋酸丙酸纤维素;乙酸丙酸纤维素
cellulose acetate sheet 醋酸纤维素片材;醋酸纤维素薄板
cellulose acetobutyrate 乙酸丁酸纤维素
cellulose acetopropionate 乙酸丙酸纤维素
cellulose adhesive 硝酸纤维素黏结剂;纤维素黏合剂;纤维素胶黏剂
cellulose-asbestos 石棉纤维素
cellulose bacteria 纤维细菌
cellulose cement 纤维素胶合剂
cellulose-coated electrode 纤维素型焊条
cellulose coating 纤维素涂料
cellulose decomposing 纤维素分解细菌
cellulose decomposing capacity 纤维素分解能力
cellulose decomposition 纤维素分解(作用)
cellulose derivative 纤维素衍生物
cellulose diacetate 二乙酸纤维素
cellulose diacetate film 二醋酸纤维素薄膜

cellulose enamel 硝酸纤维素;快干瓷漆;快干瓷釉;硝化纤维素漆
cellulose enamel paint 快干瓷漆
cellulose ester 纤维素酯
cellulose ether 纤维素醚
cellulose fermenting bacteria 纤维素发酵菌
cellulose fiber tile 纤维板;木丝板
cellulose filler 纤维素填料
cellulose film 纤维素膜;纤维素胶片
cellulose filter 纤维素过滤器
cellulose flooring 硝化纤维素地板
cellulose foil 纤维素薄片
cellulose glue 纤维素胶
cellulose gum 纤维素(树)胶
cellulose hydrates 水化纤维素
cellulose lacquer 纤维素(喷)漆
cellulose mat 纤维素毡
cellulose medium 纤维素介质
cellulose membrane 纤维素膜
cellulose membrane filtration 纤维素膜过滤
cellulose modified acrylic lacquer 纤维素改性丙烯酸漆
cellulose nitrate 硝酸纤维素;硝化棉
cellulose nitrate lacquer 纤维素漆
cellulose nitrate paint 硝酸纤维素漆
cellulose nitrate plastics 硝酸纤维素塑料;硝化纤维素塑料
cellulose nitrate stopper 硝酸纤维素填料
cellulose nitroacetate 硝酸乙酸纤维素
cellulose oxalate 草酸纤维素
cellulose paint 纤维素漆
cellulose paper 纤维素纸
cellulose powder 纤维素粉
cellulose propionate 丙酸纤维素
cellulose seal 纤维素封口
cellulose sheet 纤维素薄板
cellulose solvent 纤维素溶剂
cellulose stopper 纤维素填充料
cellulose support 纤维素片基
cellulose tape 纤维素胶带;透明胶带
cellulose thickener 纤维素增稠剂
cellulose thinner 纤维素稀释剂
cellulose triacetate 三醋酸纤维素
cellulose varnish 纤维素清漆;纤维素凡立水
cellulose vehicle 纤维素媒介
cellulose wall 纤维素壁
cellulose wool 人造羊毛;合成羊毛;化学纤维织物
cellulose xanthate 黄原酸纤维素
cellulose zymolytic wastewater 纤维素发酵废水
cellulosic 纤维(素)质
cellulosic coating 纤维质涂料
cellulosic exchanger 纤维素类交换剂
cellulosic fiber 纤维素纤维;木纤维素纤维
cellulosic insulation material 纤维素绝热材料;纤维素保温材料
cellulosic ion exchanger 纤维素离子交换剂
cellulosic paint 纤维质涂料
cellulosic plastics 纤维素塑料
cellulosic resin 纤维素树脂
cellulosic sizing agents 纤维质填料;纤维素填料
cellulosic varnish 纤维质清漆;纤维素涂料;纤维素漆;纤维质涂料
cellulosic waste 纤维质废物;纤维素废品;纤维素废(旧)料
cellulosis 纤维分解
cell vault 格型拱顶
cell voltage 电池电压;槽电压
cell wall 槽壁
cell without transference 无迁移电池
cell with packets of block anodes 块状阳极簇电解槽
cell with prebaked multiple anodes 多预焙阳极电解槽
cell with split anodes 对开阳极电解槽
cell zinc tube extruder 电池锌筒挤压机
cell zinc tube pinch-trimming machine 电池锌筒切口机
celo 赛洛
celonavigation 天文航海(学);天文航法;天文导航
Celor lens system 塞洛尔透镜系统
celotex board (木质纤维毡压制的)纤维板;(木质纤维毡压制的)绝缘板;隔声材料;(木质纤维毡压制的)隔热板;吸声隔板;隔声纤维板;甘蔗吸声板;色罗提隔声板

celsian 钡长石
celsian ceramics 钡长石(陶)瓷
celsian feldspar 钡长石
celsian porcelain 钡长石瓷
Celsius 摄氏
Celsius scale 摄氏温标;百分温度标
Celsius temperature 摄氏温度
Celsius thermometer 百分温度计;摄氏温度计;摄氏温度表
Celsius thermometric scale 摄氏温(度)标
Celtic architecture 凯尔特式建筑
Celtic cross 凯尔特十字柱头;凯尔特十字架(英国)
Celtic ornament 凯尔特式纹样(英国)
Celtic ornamentation 凯尔特式装饰
celure 筒形屋顶的镶板装饰(教堂十字架或圣坛上);十字架上面的花格挑顶
celvacenie grease 油脂
cembra(n) pine 瑞士松
cemedin(e) 结合剂;胶合剂;黏合剂
cement 结合剂;胶粘剂;胶黏材料;胶结物;胶;黏固粉;黏质;水泥井;水泥
cement acceleator 水泥促凝剂
cement additive 水泥外加剂;水泥添加剂
cement admix(ture) 水泥混合料;水泥外加剂;水泥掺和剂
cement agent 水泥添加剂
cement-aggregate bond 水泥—骨料间粉结;水泥—集料间黏结
cement-aggregate compatibility 灰集比的正确选择;水泥集料间的相互适应性;水泥骨料间的相互适应性
cement-aggregate ratio 灰集比(指水泥—骨料比);水泥骨料比(率);混凝土中水泥—骨料比率;水泥集料比(率)
cement-aggregate reaction 水泥骨料间反应;水泥集料间反应
cement-and-grouted steel bolt 钢筋砂浆锚杆
cement and sand cushion 水泥砂浆垫层
cement article 水泥制品
cement artificial marble 水泥人造大理石
cement asbestos 石棉水泥
cement asbestos article 水泥石棉制品
cement asbestos board 水泥石棉板;石棉水泥板
cement asbestos box roof gutter 石棉水泥方形屋顶天沟
cement asbestos building member 水泥石棉建筑构件
cement asbestos coating 水泥石棉涂层
cement asbestos conductor 水泥石棉水落管
cement asbestos corrugated board 水泥石棉瓦楞板
cement asbestos discharge pipe 水泥石棉排水管
cement asbestos distance piece 水泥石棉定距块
cement asbestos face 水泥石棉面
cement asbestos fence 水泥石棉栅栏
cement asbestos fittings 水泥石棉配件
cement asbestos floor(ing) 水泥石棉地板
cement asbestos flue 水泥石棉烟道
cement asbestos gutter 水泥石棉天沟;水泥石棉边沟
cement asbestos mortar 石棉水泥砂浆
cement asbestos pipe 水泥石棉管;石棉水泥管
cement asbestos rainwater article 水泥石棉雨水排水制品
cement asbestos refuse water pipe 水泥石棉废水管
cement asbestos ridge capping tile 水泥石棉(屋)脊瓦
cement asbestos roof covering 水泥石棉瓦
cement asbestos roof gutter 水泥石棉天沟
cement asbestos roofing board 水泥石棉屋面板
cement asbestos roofing shingle 水泥石棉瓦
cement asbestos roofing slate 水泥石棉屋面瓦
cement asbestos rubber tile 橡胶面石棉水泥面砖
cement asbestos separator 水泥石棉定距块
cement asbestos sewage pipe 水泥石棉污水管
cement asbestos sheet 水泥石棉薄板
cement asbestos shingle 水泥石棉屋面板
cement asbestos siding 水泥石棉墙面板
cement asbestos slate 水泥石棉板
cement asbestos solid board 水泥石棉实心板
cement asbestos spacer 水泥石棉定距块
cement asbestos tile 石棉水泥瓦

cement asbestos tube 水泥石棉管;石棉水泥管
cement asbestos wall panel 水泥石棉墙板
cement asbestos ware 水泥石棉制品;水泥石棉器皿
cement-asphalt 水泥沥青
cementation 晶体并合;胶结(作用);黏结作用;黏结(性);置换沉淀;水泥喷浆;水泥胶结;水泥灌浆;水泥固化;水泥封闭;渗透处理
cementation agent 渗碳剂
cementation bond 胶结连接
cementation by gases 气体沉定置换法
cementation condition 胶结情况
cementation effect 胶结效应
cementation factor 胶结系数
cementation furnace 渗碳炉
cementation index 黏结(性)指数;胶结(性)指数;硬化指数;硬化率;水泥硬化指数;水泥硬化率
cementation material 黏结剂
cementation of fissures 裂缝灌浆;灌缝;裂隙灌浆
cementation of lenses 透镜胶合
cementation power 胶结能力
cementation process 灌浆过程;水泥灌浆硬化法;水泥灌浆固结法(软地基);水泥灌浆工艺;水泥灌浆法;渗碳过程;渗碳法;压力灌浆法
cementation pumping equipment 注水泥设备
cementation purification 置换净化
cementation sinking 灌浆法凿井
cementation steel 渗碳钢
cementation test 黏结性试验;黏结试验
cementation zone 胶结带;渗碳层
cementatory 胶结的;黏结的;水泥的
cement bacillus(e) 水泥杆菌
cement baffle collar 水泥塞挡圈
cement bag 水泥袋
cement bagging machine 水泥装袋机
cement bag opening machine 拆水泥袋机
cement balls 水泥团
cement barge 运水泥专用船;水泥专用船
cement base 水泥基础
cement-based adhesive 水泥黏浆;水泥基黏合剂;水泥基黏结剂
cement-based glazed finish 水泥粉光面
cement-based on coal tar 含煤沥青水泥
cement-based product 水泥基制品;含水泥制品
cement-based waterproof coating 水泥基防水涂层
cement-base paint 水泥(基)涂料
cement-base waterproof coating 水泥基防水涂料;水泥基防水面层
cement basket 水泥胀圈;水泥料筒;水泥篮
cement batcher 水泥送料计量器;水泥配料器;水泥搅拌机;水泥配料计量器
cement batching 水泥计量
cement batching bin 水泥配料仓
cement batching plant 水泥配料装置;水泥称量装置
cement batching scale 水泥配料秤
cement batching screw 水泥配料螺旋输送器
cement bed 水泥层
cement bentonite milk 膨润土掺水泥乳液(沉井外围防水灌浆用);膨润土水泥乳浆;水泥膨润土乳浆
cement bentonite pellets 水泥膨润土团矿
cement bin 水泥称装斗;水泥储[贮]藏室;水泥库;水泥(储[贮])仓
cement block 预制水泥型块;混凝土块;混凝土(方)块体;水泥预制块;水泥预制块;水泥石棉管;水泥砌块;水泥(型)块
cement blower 水泥喷枪
cement board 水泥板
cement board roof covering 水泥板盖屋顶
cement-bond 水泥封闭;水泥胶结
cement-bonded mo(u)lding 水泥砂造型
cement-bonded particleboard 水泥刨花板
cement-bonded sand 混合料;水泥砂
cement-bond log 水泥胶结测井
cement-bond log curve 水泥胶结测井曲线
cement-bond sonic log 固井声波测试图
cement-bound 水泥结合的
cement-bound base 水泥结(粒料)基层
cement-bound excelsior building slab 水泥木屑建筑板;水泥木丝建筑板
cement-bound granular materials 水泥稳定粒料;水泥结颗粒材料
cement-bound macadam 水泥(胶结)碎石路

cement-bound macadam pavement 水泥结碎石路面
cement-bound material 水泥结合料;水泥胶结料;水泥胶结材料
cement-bound road 水泥砖铺地;水泥砖铺(路)面;水泥(结)碎石路面;水泥结粒料路面;水泥结砾石路面
cement-bound surface 水泥结路面
cement-bound surfacing 水泥结路面
cement box 水泥堵漏箱
cement box method 水泥堵漏法
cement brand 水泥标号;水泥牌号
cement brick 水泥砖
cement briquet(te) 水泥试块
cement(bucket) elevator 水泥提升机;水泥升降机;斗式水泥提升机
cement bulk transporter 散装水泥运输工具
cement bunker 水泥储[贮]藏室;水泥库;水泥(储[贮])仓
cement burning 水泥煅烧
cement burning process 水泥煅烧过程;水泥煅烧工艺;水泥煅烧方法
cement burnt 混凝土灼伤
cement cableway 水泥输送索道
cement cake 水泥试饼
cement carbon 渗碳
cement carrier 水泥运输船;散装水泥船
cement casing 注水泥套管
cement casing head 水泥套管头
cement casing shoe 水泥注套管头;水泥套管靴
cement caulked joint 水泥嵌缝
cement cementite 水泥碳化铁体
cement channeling 水泥窜槽
cement channeling proofing 验串
cement chemistry 水泥化学
cement chipwood 水泥碎木板;水泥刨花板
cement chuck 平面卡盘
cement cinder 水泥焦渣
cement cinder block 水泥焦渣预制块
cement clay grouting 水泥黏土灌浆
cement clay mortar 水泥黏土砂浆
cement clay ore 水泥黏土矿石
cement clinker 水泥熟料;水泥烧块;水泥熔渣;水泥熔碴
cement clinker chemistry 水泥熟料化学
cement clinker composition 水泥熟料成分
cement clinker grain 水泥熟料颗粒
cement clinker grinding 水泥熟料研磨
cement clinker phase 水泥熟料相
cement-coated nail 水泥涂面钉;涂胶钉
cement coat(ing) 水泥抹面;水泥涂层;水泥护面(层)
cement coefficient 水泥系数
cement colo(u)rant 水泥着色剂
cement colo(u)rs 水泥颜色;彩色水泥;颜色水泥
cement compression strength 水泥抗压强度
cement concrete 水泥混凝土
cement concrete aggregate 水泥混凝土骨料
cement concrete finisher 水泥混凝土整面机
cement concrete gunite machine 水泥混凝土喷枪
cement concrete lining 水泥混凝土衬砌
cement concrete mark 水泥混凝土标号
cement concrete mixing plant 水泥混凝土混合料拌和设备;水泥混凝土拌和设备
cement concrete mixture 水泥混凝土混合料
cement concrete mixture pav(i)er 水泥混凝土混合料摊铺机
cement concrete mixture pump 水泥混凝土混合料泵
cement concrete pavement 水泥混凝土铺面;水泥混凝土路面
cement concrete pavement cutter 水泥混凝土路面切割机
cement concrete pump 水泥混凝土泵
cement concrete road 水泥混凝土路;水泥混凝土道路
cement concrete runway 水泥混凝土跑道
cement concrete spreading machine 水泥混凝土摊铺机
cement consistency 水泥浆稠度
cement constituent 水泥组成物;水泥成分
cement consumption 耗灰量;水泥消耗量
cement container 散装水泥罐
cement container car 散装水泥车

cement content 水泥用量;水泥含量
cement content test 水泥含量试验
cement content titration test 水泥含量滴定试验
cement contraction ga(u)ge 水泥收缩仪
cement contractor 水泥承包商
cement conveyer 水泥运输机;水泥输送机
cement conveying pipe 水泥输送管
cement cooler 水泥冷却剂;水泥冷却器
cement cooling plant 水泥冷却设备
cement copper 沉淀铜;沉积铜
cement core 水泥核
cement cost 水泥费
cement covering 水泥罩面;水泥抹面;水泥护面(层);水泥盖面
cement curtain 水泥帷幕
cement cushion block 水泥垫块
cement cutt(ing) 水泥侵蚀
cement decorated floor tile 水泥花砖
cement de fondu 矾土水泥
cement delivery truck 水泥运输车;散装水泥车
cement density ga(u)ge 水泥容重计
cement deposit 胶结矿床;胶结沉积;天然水泥产地
cement dispersing agent 水泥弥散剂;水泥分散剂
cement dispersion admixture 水泥磨细(分散)附加剂
cement distribution terminal 水泥发运中转站
cement distributor 水泥分配机;水泥摊铺机;水泥分布机
cement dosing system 水泥称量系统
cement down 将水泥黏牢
cement dressing 水泥罩面;水泥饰面;水泥敷面;水泥粉刷;粉刷
cement dump 卸水泥机;水泥垛;水泥堆;水泥罐;水泥抽筒
cement dust 水泥粉尘
cement economiser 水泥节约剂
cemented 胶结的;灌水泥的
cemented ballaste 道砟板结
cemented belt joint 皮带胶结
cemented carbide 胶结碳化物;硬质合金;烧结硬质合金;烧结碳化物
cemented carbide bit 硬质合金钻头
cemented carbide matrix 胶结碳化物胎体
cemented carbide powder 硬质合金粉末
cemented carbite 硬质合金;烧结碳化物
cemented casing 已注水泥的套管柱
cemented catalyst 胶结催化器
cemented chip board 水泥刨花板
cemented component 胶合元件
cemented doublet 双胶透镜
cemented excelsior board 水泥刨花板;水泥木丝板
cemented gravel 砾岩;胶结砾石
cemented iron 渗铜铁;烧结铁
cemented joint 胶结(接)缝;胶合接头;水泥结合;水泥(接)缝
cemented metal 烧结金属;渗碳金属
cemented multicarbide 多元碳化物硬质合金
cemented multiple carbide 烧结多元碳化物
cemented objective 胶合物镜;黏合物镜
cemented oxide 胶结氧化物
cemented prism 胶合棱镜
cemented quartzite 胶结石英岩
cemented rock 胶结岩石
cemented roof-bolt 钢筋混凝土锚杆
cemented seam can 边缝黏合器
cemented socket joint 水泥管接头
cemented soil 胶结土
cemented steel 表面(收缩)硬化钢;渗碳钢
cemented structure 胶结结构
cemented surface 胶合面
cemented templet steel 渗碳样坯钢
cemented tube 接合管材;黏结管材
cemented tungsten carbide 黏合碳化钨;碳化钨硬质合金;烧结碳化钨
cemented wood fibre board 木纤维水泥板
cement elevator 水泥升降机
cement emulsion mixes 水泥乳浊液混合物
cement equipment 注水泥装置
cementer 注水泥装置
cement evaluation log 固井评价测井
cement extender 填充性水泥混合材料
cement external plaster 水泥外粉剂
cement facing 水泥罩面;水泥抹面;水泥护面(层);水泥盖面

cement factor 水泥用量；水泥系数；水泥含量
cement factory 水泥厂
cement family coating material 水泥系涂料
cement fastness 水泥黏固性
cement feeder 水泥送料器
cement fibrolite plate 纤维水泥板；水泥纤维板
cement fibrous plate 水泥纤维板
cement filler 水泥填(缝)料
cement filler grout 水泥填(缝)料浆；水泥稀浆
cement filler slurry 水泥填(缝)稀浆
cement fillet 水泥压缝条；水泥砂浆贴条；水泥圆角线；水泥压线条；水泥填角；水泥八字角(也称水泥填角)；水泥圆线脚
cement filling method 胶结充填法
cement film 水泥薄膜
cement fineness 水泥细度
cement fine ratio 水泥与细骨料比值
cement finish 水泥罩面；水泥修面；水泥饰面；水泥抹面；水泥面层；水泥护面(层)
cement finisher 水泥整平抹光机；水泥平整器
cement finishing process 水泥粉磨工艺
cement fixing method 黏结法
cement flag 水泥板
cement flag pavement 水泥板(铺砌)路面
cement flag paving 水泥板铺面
cement flashing 水泥散水坂；水泥泻水；水泥泛水
cement flo(a)tation 水泥抹面
cement floating 水泥抹面；水泥面层镘平
cement floor(ing) 混凝土楼面；混凝土地坪；混凝土地面；水泥地坪；水泥地面；水泥地板
cement floor tile 水泥地面砖
cement flour 水泥粉
cement flue dust 水泥窑灰
cement flyash gravel pile 水泥粉煤灰碎石桩
cement fondu 快硬水泥；矾土水泥
cement for crown bridge and inlay 冠桥及嵌体用黏固粉
cement for grouting 灌浆水泥
cement for injection 灌浆水泥
cement for joint filling 填缝水泥
cement for joints 填缝水泥；浇注水泥；水泥接口；水泥接缝
cement for road-use 道路水泥
cement free castable 无水泥浇注料
cement gel 水泥黏结剂；水泥凝胶；水泥胶体；水泥胶(结料)
cement glass plate 黏固粉玻璃板
cement glazed coat(ing) 水泥光泽涂层
cement glue 水泥黏结剂；水泥凝胶；水泥胶(结料)
cement grade 水泥等级；水泥标号
cement grain 水泥晶粒；水泥颗粒
cement gravel 胶结砾石；胶结卵石；黏结砾石
cement gray 水泥(灰)色
cement grinding 水泥研磨
cement grinding control 水泥粉磨控制
cement grinding mill 水泥研磨机；水泥磨碎机
cement grinding station 水泥粉磨站
cement grit 粒状水泥熟料；粗磨水泥；粗(粒)水泥；水泥烧粒
cement grout 稀水泥浆；薄水泥浆；水泥注浆；水泥压力灌浆；水泥浆；水泥浆
cement grout anchors 水泥浆锚固
cement grout discharge hose 水泥浆输送管
cement-grouted 水泥灌固的；灌水泥浆的
cement grouted bolt 水泥砂浆锚栓；灌浆锚杆
cement grouter 水泥灌浆机
cement grout filler 水泥砂浆填料；水泥浆填料；水泥灌浆填料；水泥灌浆填缝
cement grouting 水泥喷浆；水泥灌浆(法)；水泥灌浆材料；灌浆水泥浆
cement gun 水泥(喷)枪；水泥浆喷枪
cement gun concrete 喷射水泥混凝土；水泥枪喷涂混凝土
cement gun shooting 水泥浆喷射法
cement gun work 水泥喷浇工作
cement handling 水泥运输；水泥装卸
cement handling installation 水泥输送设备；水泥装卸设备
cement hangar 水泥(仓)库
cement hardener 水泥速凝剂；水泥(粉刷)硬化剂
cement hardener rendering 水泥增硬剂粉刷
cement haulage unit 水泥运输工具
cement head 水泥头

cement high in iron oxide 高铁水泥
cement high in lime 高钙水泥
cement high in lime and iron oxide 高铁高钙水泥
cement hog 刮板式水泥装卸机
cement hollow block 水泥空心砌块
cement hollow slab 水泥多孔板
cement-hydrate 水泥水化物
cement hydration 水泥水化(作用)
cement hydration product 水泥水合产品
cement hydration test 水泥水化试验
cement imitation marble 水泥仿大理石
cementin 黏合质
cement in barrels 桶装水泥
cement in bulk 散装水泥
cement industry 水泥工业
cementing 晶体并合；胶结；胶合；注水泥(浆)；固井
cementing action 胶凝作用；胶结作用
cementing agent 沥青黏结剂；结合剂；胶结剂；黏结剂；黏合剂；黏结介质
cementing agent based on coal tar 煤沥青基黏结剂
cementing basket 灌水泥浆吊桶
cementing between two moving plugs 双柱法注水泥
cementing bond 黏结力
cementing capacity 黏结力
cementing casing head 注水泥套管头
cementing composition 胶结混合物
cementing compound 胶黏混合物；胶黏剂
cementing condition of fault 断层胶结情况
cementing cost 固井费用
cementing design 注水泥设计
cementing equipment 注水泥设备；灌浆设备
cementing evaluation and station 固井评价和岗位
cementing float collar 注水泥浮箍
cementing float shoe 注水泥浮靴
cementing force 胶结力
cementing head 水泥灌浆压力头
cementing hole 灌浆孔；水泥灌浆孔
cementing hose 灌浆软管
cementing index 黏结指数
cementing injector 水泥浆注入器
cementing jig 胶合夹具
cementing job 灌浆工作；注水泥
cementing layer 黏胶层；胶合层
cementing line 黏结线
cementing machine 擦胶机
cementing material 黏结料；黏结材料；胶凝材料；胶结料；胶合材料；胶结材料
cementing matrix 水泥基
cementing matter 胶结物
cementing medium 胶结介质；黏结介质
cementing metal 黏结剂金属
cementing method 胶结法
cementing mix 胶凝性混合物
cementing mortar 胶结灰浆
cementing off 水泥固井
cementing of well 注水泥固井
cementing outfit 水泥灌浆设备；注水泥设备；灌浆装置；灌浆设备
cementing paper 胶水纸
cementing paste 水泥浆；稠水泥浆
cementing plant 渗碳装置
cementing plug 灌浆胶结塞；注水泥用塞
cementing point 注水泥井段
cementing position 胶结部位
cementing pot 渗碳箱
cementing power 胶结能力；黏结(能)力
cementing practice 注水泥技术
cementing process 渗碳硬化法；渗碳过程
cementing property 胶粘性；胶结性能；黏结性
cementing pump 水泥灌浆泵；固井泵
cementing quality 黏结性；胶结性；黏结质量
cementing rubber 橡皮胶水
cementing shoe 灌浆筒座；灌浆管座；注水泥套管靴
cementing strength 黏结强度
cementing stress 胶合应力
cementing structure 胶结结构
cementing surface 胶合面
cementing system 胶结体系
cementing through 通过穿孔套管注水泥
cementing time 注水泥时间；固井时间
cementing tool 水泥灌浆工具
cementing truck 水泥灌浆车

cementing under pressure 高压注水泥
cementing value 胶结力；黏结值
cementing zone 黏结带
cement injection 压力灌射水泥浆；灌水泥浆；水泥注浆；水泥喷浆；水泥灌注；水泥灌浆
cement injector 水泥撒布机；水泥喷枪；水泥灌浆机
cement iron 水泥铁
cementite 胶铁；碳化铁(体)；渗碳体
cementitious 胶凝性的
cementitious agent 结合剂；黏结剂
cementitious brick 浮石水泥砖；泡沫水泥砖；水泥黏合砖
cementitious material 胶凝材料；胶结(材)料；黏结(材)料
cementitious matter 胶结物；胶结料
cementitiousness 胶结能力；黏结能力
cementitious sheet 石棉水泥毡；水泥板；石棉水泥板
cementitious value 黏结度；黏结值
cement jet 水泥喷枪
cement joggle 水泥砂浆嵌(齿合)缝；水泥啮合；水泥接榫
cement joggle joint 水泥榫接合
cement joint 水泥接合；水泥接缝；水泥接口
cement joint for pipe 管子水泥接口
cement jointless floor(ing) 无缝水泥地坪
cement joint of pipes 管子的胶接头
cement kiln 水泥窑
cement kiln dust 水泥窑灰尘
cement kiln feed 水泥窑喂料
cement lab(oratory) 水泥试验室
cement-laid masonry 浆砌圬工
cement latex 水泥乳液混凝土
cement less concrete (不用水泥而用高分子材料作为胶凝材料的)塑料混凝土
cementless insulator 无胶绝缘子
cement level(l)ing-culvert 水泥平衡涵洞
cement level(l)ing-culvert intake 水泥平衡涵洞进水口
cement level(l)ing-culvert outlet 水泥平衡涵洞出水口
cement-like 水泥状
cement like mixture 水泥状化合物
cement lime concrete 石灰水泥混凝土
cement lime mortar 混合砂浆；水泥石灰浆
cement-lime sand mix(ture)水泥、石灰、砂混合料
cement lime sand mortar 水泥石灰砂浆
cement lime sand mortar plaster 水泥石灰砂浆粉刷
cement-lined 水泥衬砌的
cement-lined casing 水泥衬里套管
cement-lined ditch 水泥砂浆衬砌沟渠
cement-lined pipe 水泥衬砌管(道)；水泥衬里管(道)
cement lining 水泥炉衬；水泥衬砌；水泥衬里
cement lining pipe 水泥砂浆内衬管
cement liquid 黏固粉液
cement lithin paint 彩色水泥(树脂)涂料
cement log 水泥测井
cement lute 水泥封涂
cement macadam 水泥碎石路(面)
cement making plant 水泥生产设备
cement man-made marble 人造水泥大理石
cement manufacture 水泥制品
cement manufactured marble 水泥制造的大理石
cement marble 水泥大理石
cement mark 水泥牌号；水泥标号
cement mastic 水泥砂胶；水泥胶合料
cement material 胶结物
cement matrix 水泥结合料；水泥胶结料；水泥基料
cement measuring plant 水泥计量设备
cement measuring screw 水泥计量螺旋输送机
cement milk 水泥乳浆
cement mill 水泥研磨机；水泥磨机；水泥磨；水泥厂
cement mineral 水泥矿物
cement mix 水泥混合料
cement mixer 混凝土搅拌容器；混凝土搅拌机；水泥搅拌机；水泥浆搅拌机；水泥拌和机；砂浆拌和机
cement material 胶结材料
cement mixing test 水泥配料试验；拌水泥试验
cement modified aggregate 水泥稳定集料；水泥稳定骨料；水泥处治的集料；水泥处治的骨料
cement modified asphalt emulsion mixture 掺水泥的沥青乳液混合料；掺水泥的沥青乳液配合料
cement modified soil 水泥改善土(壤)；水泥改良

土(壤);水泥处理土壤
cement modified subgrade 水泥改善路基;水泥改良路基
cement mortar 水泥(砂)浆
cement mortar anchor 砂浆锚固
cement mortar base course 水泥砂浆底层
cement mortar bedding cushion 水泥砂浆垫层
cement mortar covering 水泥砂浆罩面
cement mortar finish 水泥砂浆罩面
cement mortar flooring 水泥地面;水泥地板
cement mortar grouting 水泥灰浆灌注;水泥砂浆灌浆
cement mortar injection 水泥灰浆注射
cement mortar joint 水泥砂浆接头;水泥砂浆接口;水泥砂浆接缝
cement mortar lining 水泥砂浆衬里
cement mortar mark 水泥砂浆标号
cement mortar mixer 水泥砂浆拌和机
cement mortar plaster(ing) 水泥砂浆粉刷;水泥砂浆抹面;砂浆抹平
cement mortar plaster sectioned 水泥砂浆粉刷分格
cement mortar pointing 水泥砂浆勾缝
cement mortar protective course 水泥砂浆保护层
cement mortar rendering 水泥砂浆粉面
cement mortar screed 水泥砂浆找平层
cement mortar screeding 水泥砂浆找平
cement mortar screeding to falls 水泥砂浆找坡
cement mortar surface 水泥砂浆地面
cement mortar waterproofing system 水泥砂浆防水做法
cement mortar whitewashing 水泥砂浆喷浆
cement mortar with waterproof additive 水泥砂浆加防水剂
cement mortar with waterproof compound 水泥砂浆加防水剂
cement mo(u)ld 水泥型
cement nail 水泥钉
cement needle (测定水泥凝结时间用的)水泥针;水泥硬固检验针;水泥凝固时间测定针;水泥(凝固)测针
cement of low index 低标号水泥
cement of rapid initial set(ting) 快速初凝水泥
cement out 置换出来;沉淀析出
cement overlay 水泥罩面
cement packer 水泥包装机
cement packing machine 水泥包装机
cement packing plant 水泥包装设备;水泥包装车间
cement paint 水泥用漆;耐碱漆;水泥涂料;水泥涂层;水泥饰面涂料;水泥漆;水泥浆涂料;涂饰饰面水泥
cement paint brush 水泥刷
cement particle 水泥颗粒
cement particleboard 水泥刨花板
cement paste 水泥净浆;水泥浆体;水泥浆剂
cement pat 水泥(试)饼;水泥扁饼
Cement Pavilion 水泥馆(1939年苏黎世博览会马耶设计的薄壳建筑)
cement paving floor 水泥结地面
cement penetration 水泥胶结碎石(路面)
cement penetration method 水泥贯透法
cement phase 水泥相
cement physics 水泥物理
cement pigment 水泥颜料
cement pile 水泥桩
cement pipe 水泥管
cement pipe fittings 水泥管异形件
cement plain(roofing) tile 水泥平瓦
cement plant 水泥厂
cement plaster 水泥石膏灰泥;水泥砂浆(抹面);水泥抹面灰泥;水泥灰泥;水泥粉饰;石膏水泥灰泥;石膏泥;石膏粉
cement plaster-coat waterproof(ing) 水泥抹面防水
cement plaster finish 水泥抹面;水泥抹灰
cement plastering 水泥抹面;水泥粉刷;水泥抹灰
cement-plastic cold glaze 水泥塑料冷涂釉料
cement-plastic cold-glazed wall coat(ing) 水泥塑料冷涂釉层墙面
cement-plastic vitreous surfacing 水泥塑料透明涂层
cement plate 水泥板
cement plugger 黏固粉充填器
cement plug length 水泥塞长度

cement pneumatic conveyer 水泥风动输送机
cement pneumoconiosis 水泥尘肺
cement-polyvinyl acetate emulsion 水泥聚乙烯酸酯乳液浆料
cement-polyvinyl acetate emulsion concrete 水泥聚乙烯酸酯乳液混凝土
cement powder paint 粉状水泥涂料
cement powder spraying pile 粉喷桩
cement pressing (在石膏模中模制的)水泥装饰品
cement product 水泥制品
cement proportioning plant 水泥级配装置
cement proportioning scale 水泥级配磅秤
cement proportioning screw 水泥级配螺旋输送机
cement pump 水泥泵
cement putty 纯水泥灰泥
cement quality class 水泥质量等级
cement quantity 混凝土量
cement quay 水泥码头
cement ratio (混凝土中的)骨料水泥比
cement raw material 水泥原料
cement raw material commodities 水泥原料矿产
cement raw meal 水泥生磨
cement receiving hopper 水泥受料斗
cement rendering 水泥刷面;水泥抹面;水泥粉刷
cement replacement 水泥代用(品)
cement replacement material 水泥代用材料
cement requirement 水泥要求
cement retainer 水泥承托器
cement retarder 水泥缓凝剂
cement-rich concrete 富水泥混凝土
cement-rich roller compaction concrete mix 多灰碾压混凝土拌和料
cement-rich Rolling Compaction Concrete mix 富灰碾压混凝土混合料
cement rock 水泥(用灰)岩;水泥生产用(石)灰岩;水泥灰岩
cement roof(ing) tile 水泥(屋)瓦;水泥屋面瓦;水泥屋顶瓦
cement roofing tile machine 水泥屋顶瓦制造机
cement rotary kiln 水泥回转窑
cement rotary kiln lining 水泥回转窑窑衬
cement rubber latex 水泥橡胶胶乳;柔质乳胶
cement rubber latex flooring 水泥柔性地面
cement-rubble 水泥毛石的
cement rubble masonry 水泥毛石圬工;水泥毛石砌体;水泥浆砌毛石圬工
cement rubble retaining wall 水泥毛石挡土墙
cement sample 水泥样品
cement sampler 水泥取样器
cement sand bed 水泥砂浆层
cement sand grout 灌水泥砂浆;水泥砂浆层
cement-sand grout mixer 水泥砂浆拌和机
cement-sand mix(ture) 水泥一沙混合料
cement sand mortar 水泥砂灰浆
cement suspension 水泥悬浮液;水泥稀浆
cement-sand ratio 灰砂比
cement-sand ratio method 灰砂比法
cement sand screed 水泥砂整平板
cement scouring 水泥侵蚀
cement screed 水泥灰饼;水泥砂浆找平层;水泥砂浆面层;水泥砂浆地面;水泥找中筋
cement screed-coat 水泥找平层
cement screeding 水泥浆(面)找平;水泥浆操平
cement screed material 水泥找平材料
cement screw(conveyor) 水泥螺旋输送机
cement screw(conveyor) for steep conveying 水泥螺旋的陡坡输送机
cement screw feeder 螺旋式水泥进料器;水泥螺旋喂料机
cement seal 水泥封闭;水泥密封
cement sealing hole 水泥封孔
cement setting 水泥凝结;水泥凝固
cement setting test 水泥凝结试验
cement shaft kiln 水泥立窑
cement shed 水泥储[贮]存站;水泥储棚;水泥仓库
cement sheet 水泥薄板
cement sheet roof cladding 水泥薄屋面板
cement sheet roofing 水泥薄板屋面
cement shoe 水泥靴
cement silo 水泥储[贮]藏室;水泥筒仓;水泥库;水泥(储[贮])仓;散装水泥筒舱
cement silo aerator 水泥筒仓风动设备
cement skin 水泥薄膜;水泥表(皮)层
cement slab revetment 混凝土板护坡;水泥板护岸
cement slag 制水泥用矿渣;水泥矿渣

cement slide 水泥运输滑道;水泥滑道
cement slurry 水泥浆;水泥灌浆料;水泥稀浆;水泥(泥)浆;水泥粉刷料
cement slurry fluid property 水泥浆流动性能
cement slurry mixer 水泥浆拌和机
cement slurry pump 灌浆泵;水泥浆泵
cements of clastic rocks 碎屑岩胶结物
cement-soil base 水泥土基层
cement soil stabilization 水泥稳定土壤;水泥加固土;水泥土加固
cement solidification 水泥固化
cement-solidified 水泥固化的
cement soundness 水泥安定性
cement soundness test 水泥安定性试验
cement spacer 水泥砂浆垫块
cement-space ratio 水泥与集料孔隙比(率);水泥与骨料孔隙比(率);水泥量与集料孔隙比;水泥量与骨料孔隙比;灰隙比
cement spreader 水泥分配器
cement squeeze 水泥挤压
cement squeezing technology 挤水泥工艺
cement stabilization 水泥灌浆加固;水泥稳定(法);水泥加固(作用);水泥固定处理法(软地基)
cement stabilization of soil 土的水泥加固法
cement stabilization soil 水泥稳定土(壤)
cement stabilized 水泥加固的;浆砌的;水泥稳定的;浆结的
cement stabilized block 浆砌块石
cement stabilized gravel 水泥固结卵石;水泥固结砾石
cement stabilized pavement 浆砌路面;水泥稳定的路面
cement stabilized road 水泥稳定的土路
cement stabilized sand pavement 水泥稳定积砂路面
cement stabilized soil 水泥稳定土(壤);水泥稳定砂土;水泥加固土
cement standard screen 水泥标准筛
cement standard specification 水泥标准规范
cement stone 水泥用石灰岩;水泥石;水泥灰岩
cement stone-dust mixture 水泥石屑拌合料
cement storage 水泥仓库
cement storage silo 水泥筒仓;水泥储[贮]藏室;水泥库;水泥储库;水泥(储[贮])仓
cement storage tank 水泥储槽
cement store 水泥仓库
cement stucco 水泥拉毛粉刷;水泥拉毛抹灰;室内装饰抹灰
cement stuff 水泥浆料;水泥抹面砂浆
cement substitute 水泥代用品
cement superplasticizer 水泥超塑化剂
cement surface 水泥(路)面
cement suspension 水泥悬浮液;水泥稀浆
cement sweep(ning)s 扫库水泥;扫地水泥
cement system 水泥体系
cement tank 水泥罐
cement tanker 水泥罐车;散装水泥罐车;散装水泥船
cement technology 水泥工艺
cement temper 水泥调和剂;水泥增强剂;水泥石灰灰浆
cement terminal 水泥中转站;水泥码头
cement test(ing) 水泥试验;水泥检验
cement test(ing) machine 水泥试验机
cement(throwing) jet 水泥喷枪;喷浆机;水泥喷浆
cement tile 水泥瓦;水泥瓦片;水泥板;水泥砖;水泥花砖
cement tile pavement 水泥砖铺地;水泥砖铺(路)面;水泥板(铺砌)路面
cement topping 水泥面层
cement trailer unit 水泥拖车
cement transport in bulk transporters 散装运输水泥
cement treated 水泥处治的;水泥处理的
cement treated base 水泥处理基层;低标号水泥混凝土基础;水泥加固的路面;水泥加固的承载层;水泥处理基底
cement treated crushed rock 水泥胶结的破碎岩石
cement treated grout 水泥调浆
cement treated road 水泥处理(土)路
cement treated sand-gravel base 水泥稳定砂砾底层
cement treated soil 水泥加固土
cement treated soil grout(ing) 水泥土浆
cement treated soil slurry 水泥土稀(砂)浆;水泥

土泥浆
cement treated subgrade 水泥加固的基础
cement treated supporting layer 水泥处治的承重层;水泥处理的承重层
cement trowel 水泥镘刀
cement tube 水泥管
cement type 水泥类型
cement unloading equipment 卸水泥设备;水泥卸载设备
cement user 水泥用户
cement value 黏力;胶黏能力;胶结能力
cement valve 注浆阀;水泥(止回)阀
cement veneer 水泥饰面
cement vermiculite 水泥蛭石
cement vertical kiln 水泥立窑
cement vessel 水泥容器;散装水泥罐
cement void ratio theory 水泥空隙比理论
cement volume soundness 水泥体积安定性
cement V washer V 形水泥砂浆垫块
cement wash 水泥涂料;水泥浆涂刷;水泥浆刷面;刷水泥浆
cement washer 水泥砂浆垫块
cement-water factor 灰水比
cement-water glass grout 水泥—水玻璃双浆;水泥砂浆砌片石
cement-water grout 稀水泥浆;水泥薄浆
cement-water mixture 水泥净浆
cement-water paint 水泥浆涂料;水性水泥涂料
cement waterproofer 水泥止水材料;水泥防水材料
cement waterproofing coating 水泥防水涂层
cement waterproofing powder 水泥防水粉
cement-water ratio 灰水比;水灰比(例)
cement weigh-batching device 水泥称量装置
cement weigh-batching unit 水泥称量装置
cement weigher 水泥秤
cement weighing 水泥称量
cement weighing batcher 水泥重量投配料斗
cement weighing hopper 水泥重量投配器;水泥称重给料器;加灰料车
cement weighing machine 水泥计量机;水泥称量机
cement weighing scale 水泥重量秤
cement-welding 金属陶瓷焊接
cement wharf 水泥码头
cement with excessive gypsum content 含过量石膏的水泥
cement with high sulphate resistance 高抗硫酸盐水泥
cement with low hydration heat 低水化热水泥
cement with moderate sulphate resistance 中等抗硫酸盐水泥
cement without clinker 无熟料水泥
cement wood 混凝土集料用的碎木块
cement-wood floor 木块集料混凝土地面;木块骨料混凝土地面;水泥木屑地面;水泥木屑楼板
cement-wood flooring 水泥木料地面处理
cement worker's pneumoconiosis 水泥工尘肺
cement works 水泥厂
cemestos (一种预制好的)石棉水泥房屋
cemetery 陵园;墓地;公墓;坟场
cemetery church 墓地教堂
cemetery site 公墓用地
cemixene 预制水泥混合物(加水即可使用)
cemprover 水泥改性剂(一种改善水泥制品性质的液体)
cemseal (水泥或混凝土的)防水胶质
cenaculum 小餐室(古罗马);食堂(古罗马住宅)
Cenco pump 森科高真空回转油泵
C end frame C 形端架
cenesthesia 存在感觉
cenogenesis 后生变态
cenology 地面地质学
Cenomanian 森诺曼阶;【地】
cenophyte 新生代植物
cenosphere 煤胞;空心微球
cenotaph 墓碑;(无遗体埋葬的)纪念碑;纪念塔;墓;衣冠冢
cenote 岩洞陷前;洞状陷穴;天然井;(石灰岩溶蚀塌陷形成的)天然井;竖井
cenotypal rock 新相火山岩
Cenozoic era 新生代【地】
Cenozoic erathem 新生界
Cenozoic fold 新生代褶皱
Cenozoic group 新生界
Cenozoic ocean 新生代海洋

Cenozoic system 新生界
cenozone 群集带
cenral mileage 中心里程
censer 香炉
censer mechanism 摇动机制
censored observation 终检值
censored sample 截尾抽样;已校对样本
censorship 保密检查
census 人口普查
census analysis 人口调查分析
census block 人口普查街区
census computer 人口调查计算机
census data 调查资料;调查数据
census enumeration district 人口统计区
census form 普查表
census map 人机调查图
census method 种群调查法;国情调查法;普查法
census methodology 普查方法论
census of business 企业调查
census of communication 交通运输普查
census of distribution of services 服务分布情况的普查
census office 普查办公室
census of housing 住房普查;定期住宅普查
census of industry 工业调查
census of manufacturing 工业调查
census of production 生产普查
census of shareholder 股东情况调查
census of the sea 海洋普查
census of trade 贸易普查
census of transportation 交通运输调查
census paper 人口调查表
census registration station 人口登记站
census report 普查报告
census schedule 采访表;普查表
census statistic method 人口调查统计法
census tract 人口普查区段;普查地段
cent 分(货币单位)
centafold door 中悬折叠门
centage 百分率
cental 百磅(重)
centare 一平方米
centaurus 半人半马像;人首马身像
centec 袖珍电动刻纹器;袖珍电动槽刨机
centenary 百年纪念
Centenary Hall at Breslau 布雷斯百年纪念
centennial variation 百年变化
center 中心(点);找中心;顶尖;枢
center adjustment 调整中心
center aisle coach 中央通道客车
center anchor 船首中锚
center angle 圆心角
center arch 中间拱;中心拱
center arch span 中跨
center bar 江心洲;江心滩
center barrier 中部栅栏;路中挡册;高速公路狭长绿草地边上的护栏
center batch processing 集中成批处理
center bay 中跨;中心跨距;中心跨度
center beaded 珠饰环中心
center bearing 中心支承
center bearing swing bridge 中心支承平旋桥
center bearing type 中心支承式
center bit 摇柄钻;矛式钻头;转柄钻;中心钻(头);三叉钻头
centerbit of claw type 切齿中心钻
center block 齐心斗;槽升子
center-block type joint 中心滑块式连接
center board 中央拔水板;中心板材;中心板;防摇(垂)板;中插板【船】
centerbody 中心体
center bolt 中心螺栓
center bore 钻头内径;中心孔
center-bound tie 中心联杆
center brake 中央制动闸;中央制动器
center buckle (板材的)中心凸起
center buff 中心孔抛光轮
center buffing 抛光中心线
center bulb 建筑上半圆花边止水条
center burner 中心喷管
center business district 中心商务区
center by-pass (阀的)中立旁通
center cal(l)ipers 测径规
center casting 铸坑

center collector ring 中心集流环;中心集电器;中心集电环;中心滑环
center column 中柱
center computer 主计算机
center conductor 中心(导)线
center cone 中心锥
center contact rail 中间接触轨
center control room 中控室
center convergence 中心收敛;中心会聚
center coupler 车钩对心
center crank 中心曲柄
center-crank arrangement 中间曲柄装置
center-cut method 中心开挖法
center cut(ting) 中心开挖;中央陶槽;中心掏槽;锥形掏槽
center depth regulator 中央深浅调节器
center depth to the magnetic body 磁性体中心埋深
center diecasting 离心铸造
center differential 中央差动机构;桥间差速器
center-diffusion tube gas burner 中心管进气喷燃器
center discharge 中心卸料
center discharge rod mill 中心排料(式)棒磨机
center disk cutters 中央滚动
center distance 轴间距(离);中心距(离);顶尖距
center distance modification 轴向位移(指光)
center distance modification coefficient 轴向位移系数(指光)
center distance of riser 升气管中心距
center drag 舯耙
center drain 中心排水管;中心排水沟
center drawer guide 抽提中心导轨
center drift 中央导坑;中心偏移
center drift method 中央导坑法;中央导洞法
center drift tunneling 中心导孔隧洞掘进法;中心导孔隧道掘进法;中心导坑隧洞掘进法;中心导坑隧道掘进法
center drilling 中心钻削
center drilling(bore)hole 中心钻孔
center drilling tool 打中心孔工具
center drive 中心传动
center-driven antenna 中心激励天线
center dump car 底部卸料车
center dump trailer 底部卸料拖车
centered difference 中心差分
centered rectangular lattice 面心长方点阵
centerest sink faucets 冷热水(混合)龙头
center facies zone 中心相带
center failure storage unit 中央故障存储单元
center-fed aerial 中点馈电天线
center-fed track circuit 中央馈电式轨道电路
center feed 中央加料;中央馈电;中心馈电
center filling 中心装料
center fire 中心发火;中心点火
center fired hot top 闭式灶面板
center firing 中心点火法
center flower 中心花饰
center folding 对折
center for biology of natural systems 自然科学生物中心
Center for Environment(al) and Sanitary Engineering 环境和卫生中心
Center for International Research of Environment and Development 国际环境研究和发展中心
center form 形心;中模;内模
Center for Surface and Coating Research of American 美国表面与涂装研究中心
center fraise 中心孔铣刀
center frame 中心架
center frequency 中心频率
center frogs 中心辙岔
center gangway coach 中心通道客车
center gap 中板离缝
center gate 中心浇口
center-gated mo(u)ld 中心流道模
center gate stop 中心止门装置;中心门挡
center ga(u)ge 定心规;中心规
center gudgeon 中心枢轴
center gutter 斜沟槽;屋面天沟
center half 中间二分之一
center head 顶心头;求心规
center heading 中央导坑;隧道的中央导洞

center height 中心高度
center height of gravity for car loaded 重车重心高
center height of upper tumbler 上导轮中心高
center hinge 门窗枢轴
center hinge pivot 中支枢轴
center hole 顶尖孔;中心孔
center-hole bit 中心水口式钻头;带中心水眼的钻头
center hole drill 中心钻
center hole jack 穿心式千斤顶
center hole lapping machine 中心孔研磨机
center hole reamer 中心孔铰刀
center hopper 中舱
center-hung door 中旋门;中悬门
center-hung pivot 中悬转轴
center-hung sash 中旋式(气)窗框;中旋窗框
center-hung sash window 中悬窗
center-hung swivel window 中悬窗
centering 建筑中的拱架;中心校正;置中;找正;拱圈模型架;对中;定中心;定圆心;定心;打中心孔;找中心;对准中心
centering adjustment 合轴调整;对准中心;对中调整;定心调整;中心调整;对中
centering amplifier 中心调节放大器
centering angle 弧心角;中心角
centering apparatus 定心器
centering bracket 垂球定中夹
centering by clamped planks in arch shape 夹板拱架
centering chain 对中链【船】
centering control 居中调整;中心调整;居中调节;中心调节
centering core 泥芯
centering device 置中器;对中装置;对中器;对心装置;定心装置
centering drill 中心钻;钻孔定心冲头
centering element 归心元素
centering error 对心误差;偏心差
centering guide 定心器;扶正器
centering hinge 中心铰
centering jaw 中心凸轮
centering machine 中心孔加工机床;定心机;顶尖孔机床;拱架定心机
centering magnet 中心位置调整磁铁
centering mark 合轴标记;定心标
centering of arch 拱顶鹰架;拱顶脚手架
centering of bubble 气泡置中;气泡居中
centering of instrument 仪器对心
centering of level(l)ing bubble 水准气泡对中
centering of pencil 铅笔对中
centering of photograph 像片归心
centering piece 对中件;定心装置
centering pin 置中销;定心销
centering plate 对中板
centering point 对中杆尖端
centering potentiometer 定心电位器
centering projection 归心投影
centering ring 裂口圈;定心环
centering rod 对中杆;定心杆
centering Rogla 罗格拉鹰架;移动钢桁架式鹰架
centering rope 定中心的缆索
centering scaffold 鹰架
centering shoulder 对中肩
centering slips 楔形小木条
centering spring seat 中心弹簧座
centering tongs 定心卡具
centering tool 定心工具
centering tube 中心调整管
centering under point 点下对中
centering unloading 拱架卸落
center iron 中心铁
center-island 中心岛
center jar socket 钢绳钻进震击器打捞筒
center jet drill bit 中心喷射式钻头
center joint 中心缝(路面的纵向缝)
center journal 转轴;中心立轴
center jump 中心跳跃
center keel 重板龙骨;中央拔水板;中竖龙骨
center keelson 中心(内)龙骨
center key 中心键;拆锥套楔
center laboratory 中心实验室
center lane 中央车道;中间车道
center-lane traffic 中央车道交通
center-latch elevator 中心插销吊卡(堆开式)
center lathe 顶针车床;顶尖车床;普通车床

centerless 无中心的
centerless grinder 无心磨床
centerless grinding 无心研磨;无心磨削
center light 中央灯具;中间窗;中间采光
center limb variation 中心边缘变化
center line 轴线;中(心)线;笔画中线;几何中线
center line average 中线平均值;算术平均值
center line average method 算术平均法
center line bulkhead 中纵舱壁
center line concentration 中心线浓度
center line dilution 中心线稀释度
center line erection 中线安装
center line grade 中线坡度
center line grade elevation 中线红色标高;道路中线坡度标高
center line ground elevation 中线黑色标高;中线地面标高
center line keelson 中心内龙骨
center line knot 中心线节
center line marker 中线标划机
center line marking 标划中线
center line marking of road 标划道路中线
center line method 中线法
center line mount 中线安装
center line of bridge 桥梁中线
center line of main channel 主航道中心线
center line of plume 烟羽中心线
center line of road 道路中心线
center line of support 悬挂中心线
center line of survey 路线测量中心线
center line of track 线路中(心)线
center line of tunnel 隧道中(心)线
center line of ventilation adit 风道中线
center line of wave 波浪中心线
center line peg 中线木桩;中线木钉
center line platform 中线平台
center line radius 中心线半径
center line rudder 中舵
center line runway lights 跑道中线灯
center line shrinkage 轴线缩孔
center line shrinkage cavity 轴线缩孔;中心线收缩
center line slope 中线坡度
center line stake 中线桩;中心桩【测】
center line stake level(l)ing 中线桩高程测量(中平)【测】
center line survey(ing) 中线测量;线路中线测量
center line through plate 中纵通板
center line track 主线轨道
center line tunnel 主线隧道
center line velocity 中心线流速
center lining 标画路中线;画出中线【道】
center lining of no-passing zone 在禁止超车区画出中线
center loading 跨中荷载
center-lock 中心锁定
center mall 道路分隔带;路中林荫带
center management 计算中心管理
center mark 中心标记;中线符号;起90点;中心标志;定中心点
center matched 中线相配;中心榫接;中心合榫材
center mileage 中心里程
center mixer 混凝土中心搅拌站;混凝土中心搅拌机
center mix(ing) 集中拌和
center mount 中心装
center mount support 中间支承;中间支撑
center nailing 中间钉法(屋面石板瓦);对头顶屋面板
center nib 板簧中心楔
center of a circle 圆心
center of action 作用中心
center of a lattice 格的中心
center of anallatism 准距中心【测】
center of analysor 分析器中枢
center of arch 拱心
center of area 面心;面积中心;形心
center of a rope 绳芯
center of atmospheric action 大气活动中心
center of attraction 引力中心
center of borehole 井筒中心线;钻孔轴线
center of buoyancy 浮力中心
center of buoyancy curve 浮心曲线
center of business district 商业中心区
center of circle 圆心
center of crown 路拱中心

center of cult 礼拜场所
center of curvature 曲率中心
center of curve 曲线中心
center of curve of curved bridge 弯桥曲线中心
center of discharge area 排放区中心
center of distortion 畸变中心
center of distribution 分布中心;配电中心
center of disturbance 扰动中心
center of drag 拉力中心;阻力中心
center of effort 作用中心
center of equilibrium 平衡中心
center of figure 面心;体积中心
center of flexure 挠曲中心;弯曲中心
center of flo(a)tation 船舶水线断面形心;浮心;浮体重心;浮体水面质量中心;漂心
center of force 力心
center of forecasting and analysis 分析测试中心
center of form 形心
center-off polarized relay 中位断开式极化继电器
center of friction 摩擦中心
center of gravity 重心
center of gravity for car loaded 重车重心
center of gravity line 重心线
center of gravity member section 杆件断面重心
center of gravity path 重心运动轨迹
center of gyration 回转中心;回转半径;旋心
center of housing estate 小区中心
center of ice dispersal 浮冰分布中心
center of immersed bulk 浸水体中心
center of impact 命中中心;撞击中心;冲击中心
center of inertia 惯性心
center of inertia system 惯性中心系统
center of information feedback 信息反馈中心
center of instrument 仪器中心
center of inversion 倒心;倒反中心;反演中心;反伸中心
center of jet exit 喷口截面中心
center of lift 升力中心
center of light 光心
center of location 定位中心
center of mass 质心;质量中心
center of mass coordinate 质心坐标
center of mass coordinate system 质心坐标系
center of mass frame 质心系
center of mean distance 均距中心
center of moment 力矩中心;矩心
center of momentum coordinate system 质心坐标系
center of motion 运动中心
center of movement 运动中心
center of origin 震源;起源
center of oscillation 摆动中心
center of percussion 撞击中心;振心;打击中心
center of pressure 压强中心;压力中心
center of pressure coefficient 压力中心系数
center of pressure shift 压力中心移动
center of projection 投影中心
center of radiation 辐射中心
center of range 全矩中心
center of regular polygon 正多边形中心
center of resistance 阻力中心
center of reticule 十字线中心;十字丝中心
center of rigidity 刚度中心
center of rise 升压中心
center of rotating movement 旋转中心
center of rotation 旋转中心;转动中心
center of section 截面中心
center of sensitization 感光中心
center of similarity 相似中心
center of similitude 相似中心
center of sphere 球心
center of springing 起拱中心
center of stress 应力中心
center of submerged lateral area 中线面浸水部分中心
center of supply 供电中心
center of surface 曲面中心
center of switch 转辙器中心
center of symmetry 对称中心
center of the trace 轨道中心
center of top 拱顶中央
center of torsion 扭转中心
center of tracking gate 波门中心
center of tropic(al) cyclone 热带气旋中心

center of turnout 道岔中心
center of twist 扭转中心;扭曲中心
center of typhoon 台风中心
center of vibration 振动中心
center of view 目视中心
center of water drag 水阻力中心
center of weir crest circle 堰顶圆弧中心
center og buoyancy 浮心
center opening 中央开口;中跨;明间;中心孔
center-opening door panel 中分双扇门
center operator 中央操作员
center-opposition development program(me) 中央对角式开拓系统
center-outlet bunker 中央出口式储[贮]仓
center owncoding routine 中央扩充工作码程序
center-parallel development program(me) 中央并列式开拓系统
center parking 路中停车;街心停车地带;街心停车处
center pedestal 中心基座
center peg 中心掏槽爆破;中心小木桩
center peripheral discharge 中间周边卸料
center piece 中央花饰(块);(顶棚的)中央部位之装饰品;中心块;中心件;中心花饰
center pier 中墩
center pile 中间桩
center pillar 中立柱
center pin 中枢;中心轴;中心销
center pin bearing 中央滚柱轴承;中枢轴支承
center pin bolster 转向架摇枕
center pin steering 中枢轴式转向
center pin support 中心销支承
center pivot 中心枢轴;心销
center-pivoted door 中旋门;中心支承旋转门
center pivoted irrigation 自转喷灌
center pivoted pin 中心枢轴销
center pivoted sprinkler 时针式喷灌机;自转式喷灌机;中心枢轴式喷灌机
center pivoted window 中心支承旋转窗;中旋窗;中枢支承旋转窗
center pivot steer(ing) 中枢轴式转向
centerplank 中心板(取材圆木中部)
center plate 心盘;中心板;中线板
center plate loading force 心盘荷载
center plate rudder 单板舵
center plug 中心镶嵌塞(金刚石钻头);中心塞钻头
center point 中(心)点
center-point galvanometer 中心零位电流计
center-point load 集中荷载;集中负载;中点荷载
center point of resistance 反力中心
center point steering 中心点转向(车辆)
center pole 路中式电车杆;中心支柱
center porosity 中心缩松;中心疏松
center position 中心位置
centerpot 商业中心
center punch 中心冲孔器;定中心;中心冲头;冲心錾
center punched grouser shoe 中心冲装履带抓地齿;中心冲装的轮爪
center punching 中心冲凿;中心冲孔
center quad 中心四绕组
center rail 中冒头
center ramp (立体交叉的)中央匝道;中央接坡;中心匝道
center resolution 中心分辨率
center rest 中心架;顶尖架
center routine 中央程序
centers 中到中
center safety island 路中安全岛
center sample recovery 反循环连续取芯钻进
center scraper 中心括板
center second 中心秒针
center section 中心剖面;中间部分
center shaft 转门立轴;转门中轴;中轴
center shed jacquard 中开口提花机
center shift 水平推力装置
center shift drive shaft 中心位移驱动轴
center shift link 中心位移连杆
center shift pinion 中心位移齿轮
center shift rack 中心位移支架
center shop 中心店
center shot 中心掏槽爆破;中心掏槽炮眼
center sill (车的)中间梁
center sleeve 顶尖套
center span 中孔;中跨;桥梁中央跨
center spinning 离心铸造法

center-spiral 定中螺旋箍
center split pipe 沿中心线切断管子;对开管
center split supporting ring 半圆支承环
center split tube 对开管
center spot room (设在剧场后墙中心的)后墙投光室
center square 中心角尺
center stake 中桩【测】;中心标桩;中线桩
center station 中心桩号
center stringer 楼梯中间斜梁
center strip 路中分隔带
center strut 中支柱
center support 中心支撑;中心支承
center support system 中心支撑系统
center surveying 中线测量
center system 中央系统
center tank 中间舱
center tap 中心抽头;中间抽头
centertap keying 中点键控
center tester 中心试验器
center test station 中心试验站
center thickness 中心厚度
center-to-center 中心距;中(心)到(中)心;轴间距(离)
center-to-center distance 中心距;中至中距离;顶尖间距
center-to-center method 心对式心法;中心连接法
center-to-center of rivets 铆钉中(心)距
center-to-center spacing 中心距
center-to-center window 中间窗
center-to-cente distance 中距
center-to-edge 中心至边缘
center-to-end 中心到端面
center-to-face 中心到表面
center toolbar 中部悬挂机具架
center-to-side baffles 双缺圆折流板
center transformer 中心变电所
center trigger 中央触发器
center type 中心式
center-type cylindric(al) grinding 中心外圆磨削
center valve 中心阀
center vent 中心放气孔
center vertical keel 船底中纵桁材
center washer 离心洗涤器
center weight 中心锤
center weight governor 加重控制器;加重调速器
center well 中心井;船井(海洋钻探船)
center wheel 中心轮
centerx 集中式交换;市话分局交换
centerx central office 集中式交换中心局
center-zero ga(u)ge 中心零位压力计
center zero instrument 中心零位仪表
center zero meter 中心指零式测量仪表
center zero relay 中间零位继电器
centesimal 百进位的;百分制的;百分之一(的)
centesimal balance 百分天平
centesimal(circle) graduation 百分分度;圆圈百分分度;四百度制度盘分划
centesimal degree 百分制的度
centesimal notation 百进位符号
centesimal scale 百分制标
centesimal system 百进(位)制;百分制;百分度制
cent(i)are 平方米
centibar 中心杆;厘巴(1厘巴=1/100巴)【物】
centicell unit 多管式除尘器
centi-degree 摄氏度
centigrade 百分温标;百分(刻)度的;百分度制;百分度(数);摄氏温度
centigrade degree 百分温度;百分度(数);摄氏度数
centigrade heat unit 百分温标热单位;百(分)度热量单位
centigrade scale 百分温度标;百分刻度;百分标;百度表;摄氏温(度)标
centigrade temperature 百分温度
centigrade thermal unit 百分温度热单位;摄氏热量单位
centigrade thermometer 百分温度计;百分温度表;百分寒暑表;摄氏温度计;摄氏温度表
centigram(me) 公毫;厘克
centigram method 厘克法
centilane 分巷
centile 百分位数
centiliter 厘升;百分之一升
centimeter 厘米;公分

centimeter-excess object 厘米超天体
centimeter-gram-second electromagnetic system 绝对电磁单位制
centimeter-gram-second system 厘米克秒制;厘米克秒单位制;米制
centimeter-gram-second system of units CGS 单位制
centimeter-gram-second system unit 厘米克秒制单位
centimeter-gram-second unit 厘米克秒单位
centimeter-gram-second unit system 厘米克秒单位制
centimeter height finder 厘米波测高计
centimeter per second 每秒厘米(数)
centimeter per second per second 每秒每秒厘米(加速度单位)
centimeter square 平方厘米
centimeter wave(centimetre wave) 厘米波
centimeter wavelength 厘米波长
centimetric emission 厘米波辐射
centimorgan 厘摩;分摩
centinormal 厘规;百分之一当量的
centi-octave 百分之一八度音程
centipede 百脚
centipoises 厘泊(动力黏度单位)
centisthene 厘斯(坦)(力的单位)
centistoke 厘斯托;厘泡(动力黏度单位)
centi-tone 百分之一音程
centival 厘克当量;百分单位
centner 50千克;50公斤;生奈尔(重量单位)
cent per cent 百分之百的;百分之百;无例外
centrad 厘弧度
central 中央的;中心的;中枢的;电话总机;电话接线员
centralab 中心实验室
central address memory 集中编址存储器
central-adjustable handle-adjustable wrench 双开活动扳手
central administrative expenses 中枢管理费用
central aerial television 共用天线电视
Central-African copper deposit 中非铜矿床
Central-African Customs and Economic Union 中非关税经济同盟
central air blow at higher location 高中心风
central air conditioner 集中式空调;集中空调系统;中央空气调节器
central air conditioning 集中空调
central air conditioning equipment 集中式空调装置
central air conditioning installation 中央空气装置
central air conditioning system 集中空调系统;中央空气调节系统
central air-handling unit 集中空气处理机组
central air heating 集中空气加热
central air supply 中央送风
central altar 主祭坛
Central America 中美洲
Central America architecture 中美洲建筑
Central America Common Market 中美洲共同市场
Central America mahogany 中美洲红木
Central American Economic Cooperation Committee 中美洲经济合作委员会
Central American Trade Area 中美洲自由贸易区
central angle 圆心角;中心角
central angle of arch 拱圈中心角
central angle of curved bridge 弯桥中心角
central anode photocell 中心阳极光电管
central anticlinal belt 中央背斜带
central aperture 中心孔径
central apparatus room 中央设备室
central archway 中央拱廊
central area 中心区
central area park 市中心停车场
central Asia Mongolian geosyncline system 中亚—蒙古地槽系
central asphalt mixing plant 集中式地沥青拌和厂
central assessment 集中估价
central avenue 中央大道;中央林荫道
central axial filament 中央轴丝
central axis 轴;央轴;中轴线;中央轴线;中心轴
central band 中央光带
central bank 中央银行
Central Bank of Argentina Republic 阿根廷共和

国中央银行
Central Bank of Iraq 伊拉克中央银行
Central Bank of Trinidad, Tobago 特立尼达和多巴哥中央银行
central bar 江心洲;江心滩;心滩
central basin fault 中央断块断层
central batching plant system 集中投配方式;集中投配(操作方)法
central bath 中心浴场
central battery 中央电池组
central battery signalling unit 共电制电话信号装置
central bay 明间;中心间(古建筑)
central bearing 中心轴承
central bearing swing bridge 中心支承平旋桥
central bearing type 中心支承式
central biologic(al) system 集中式生物系统
central bit 中心钻头
central block 主楼;中心街区;中心块
central blowing at higher position 高中心风
central body 中心体
central boiler house 中心锅炉房
central boiler installation 中心锅炉装置
central brake 中央制动闸;中央制动器
central broadcasting controller microphone call 中央广播控制站
central buffer 中心缓冲器
central buffer coupling 车钩缓冲装置连接
central bundle 中央输导束
central buoy 中泓浮标
Central Bureau of Meteorology 中央气象局
central burner 中心喷管
central burning appliance 中心燃烧设备
central burst 中心开裂
central business area 商业中心区
central business district 中心商业区;商业中心区;中心商务区
central buying 集中采购
central cable plane 中央索面
central caisson 中心集水井
central calm 中心无风区;中心无风带
central canal 中央管
central cantilever truss 中心悬臂桁架
central cataract 中央白内障
central chain angle 链条中心角
central chamber 中央镜箱;中心室
central chamber blending silo 中心室搅拌库
central Chile chess-board structure 智利中部棋盘格式构造
central China semiarid to arid subtropical zone 中国中部半干旱—干旱亚热带
central circulating tube 中央循环管
central city 核心城(市);中心城市
central clamping mechanism 中央夹紧机构
central clamping screw 中央固定螺旋
central cofferdam 混凝土芯墙围堰
central collecting compartment 中央骨料室
central collision 迎面碰撞;直接碰撞;正碰
central commercial district 中心商业区
Central Committee for Contacts 中央合同委员会
central communication line 中央通信线路
central compartment 中央室
central complex 中央处理器群;中心杂岩
central compressed column 中心受压柱
central compression 中心压缩
central computer 主计算机;中央计算机
central computer input-output 中央计算机输入输出
central concave 中央凹的
central concentrated load 中心负荷
central concentration 中心密集度
central concrete core 混凝土心墙
central concrete core wall cofferdam 混凝土芯墙围堰
central concrete membrane 混凝土芯墙
central concrete mixing plant 混凝土集中搅拌设备;集中搅拌装置(混凝土)
central concrete wall 混凝土芯墙
central concrete wall type dam 混凝土芯墙式坝
central condensation 中心凝聚物
central conditioning installation 中心空调装置
central conditioning plant 中心空调设备
central conductor 中间导线
central conductor (magnetizing) method 贯通法磁粉探伤

central conduit 中心管
central cone 料仓中心锥体
central cone silo 以中心轴分室的多室料仓;中间锥形圆库
central confidence interval 中心置信区间
central configuration 中心构形
central conic 中心圆锥线;中心二次曲线
central control 集中控制;中心控制;中央控制
central control board 中央控制台;中心控制台
central control computer 控制中心计算机;中央控制计算机;中央操纵计算机
central control console 中央操纵台;中央控制台
central control cycle 集中控制循环
central control deck 中央控制台
central control desk 中央控制台
central control device 集中控制装置
central control equipment 集中控制设备
central controller 中央控制台;中央控制器
central controlling organization 中央控制机构
central control post(aboard) 操纵中心【船】
central control room 中央控制室;中心控制室
central control station 集中控制站;中央控制站
central control system 中央控制系统
central control unit 中央控制器
central cooling 集中供冷
central cooling plant 集中冷却装置
central coordinate of area element 面元中心点坐标
central core 中间核心;中心核
central core of strength 强度核心
central core wall 中心筒墙
central corporate expenses 公司总部费用
central corridor 中央走廊;内廊;内走廊;中间过道
central-corridor residential building 内廊式住宅建筑;内走廊房屋
central course 中央航道
central court(yard) 中间庭院;内庭
central crawlway 中央履带车道
central cross-section 中心截面
central cusp deformity 畸形中央尖
central cyclone 中央气旋
central cylindric(al) projection 圆柱中心投影
central data bank 中央资料库;中心数据库
central database 集中数据库
central data collection system 中央数据收集系统
central data conversion equipment 中央数据变换设备
central data performance monitor 中央数据性能监控器
central data processing 集中数据处理
central data processing computer 中央数据处理计算机
central data processing system 中央数据处理系统
central data processor 中央数据处理机
central data station 中央数据站
central data terminal 中央数据终端
central date 中央日期
central definition 中心清晰度;场中心清晰度
central dehumidification system 集中空气干燥系统(舱中)
central dense granule 中心致密颗粒
central difference 中心差分
central difference interpolation 中心差分插值(法)
central discharge 中心卸料
central dislocation 中心脱位
central dispatching room 中心调度室
central dispatching station 中心调度所
central display unit 中央显示装置;中央显示器
central distance 中心距(离)
central distribution and marshalling station 集配站
central distribution board 集中配电盘
central district 中心区
central district line net density 中心区线网密度
central dogma 中心法则
central domal highland 穹隆中央高地
central dome 中央圆屋顶
central door 中门
central down-comer 中央循环管
central downcomer evapo(u)rator 中央循环管蒸发器;中央降液管蒸发器
central drain 中央排水渠;中央排水管;中心排水管;中心排水沟

central drainage 集中排水;中央排水
central drainage system 中央排水系统
central drainage tunnel 中心排水隧洞
central driven log frame 中心驱动排锯机
central drive peripheral discharge type mill 中心传动和边缘卸料的多仓磨
central drop duct 中央落料管
central dry mix 混凝土干料
central ducting system 集中式通风管道系统
central dumping tower 中心进档塔
central dust collecting system 集中集尘系统;集中集尘方式
central earthquake 中心震
Central East China Sea-Diaoyudao uplift region 东海中部—钓鱼岛隆起地带
central eclipse 中心食;中心蚀
central eclipse curve 中心蚀曲线
central electric(al) power plant 中心发电厂
central electric(al) power station 中心发电站
central electronic command 集中电子操纵
central elevation 中心标高
central end 中枢端;工作端
central engine 中置发动机
central entrance 中间入口
central equatorial Pacific 赤道中太平洋
central eruption 中心式喷发;中心喷发
central European karst 中欧型岩溶
central examination station 中心检查站
central excavating process 中心开挖法
central exchange 电话总机;中心交易所
central exchange rate 中心汇率
central-excitation system 中心励磁式
central facies 中心相
central facility 中心设施
central facility area 设备区
central factorial moment 中心阶乘矩
central factory 中心工厂
central failure storage unit 中央故障存储单元
central fan heating 激动式热风供暖;热风集中供暖
central fan system 集中通风系统
central feature 主要特点
central feed 中央加料;中心供给
central field of vision 中心视野
central figure 中心构形
central file on-line 主文件联机
central file system 中央文件系统
central fill 坝体芯墙
central fire 中央燃烧设备
central fire detecting equipment 中心火警探测器
central fixation point 中心注视点
central float 中泓浮标
central flower 中央花饰
central force 心力;中心力
central force field 中心力场
central force orbit 有心力轨道
central fore-and-aft vertical plane (船体的)中纵剖面
central forecasting office 预报中心
central foreman 总监工
central fovea 中心凹
central frame 中梁式车架
central frequency 中心频率
central fringe 中央条纹
Central fume 二四混剂(杀虫剂)
central furnace 中心高炉;集中供热炉
central gallows for counterpoise 地网中心架
central gear 中心齿轮;太阳轮;太阳齿轮
central government 中央政府
central granite 中心花岗岩
central granule 中心粒
central gravity field 有心重力场
central grinding plant 集中粉磨车间;粉磨站
central grinding system 集中粉磨系统
central group 中央群
central hair (光学仪器的)中心丝
central hall 中央大厅
central heating 集中加热;集中供热;集中供暖法;集中供暖;集中采暖(法);暖气
central heating area 集中采暖区
central heating boiler 集中采暖用的锅炉
central heating chimney 集中供热烟囱
central heating for region 分区集中供热
central heating gas appliance 集中供热煤气设备
central heating installation 集中供暖装置

central heating plant 供暖总站;供热总站;集中供热设备;集中采暖设备
central heating supply 集中供热
central heating system 集中热水供应系统;集中供热系统;集中供暖系统;中心供热系统
central heating zone 集中采暖区
central heat supply 集中供热
central hole 中心孔
central-hole nozzle 中心孔式喷嘴
central hole of the axle 车轴中心孔
central hopper 泥船;中舱
central hot water preparation plant 集中热水制备装置
central humidifier 集中增湿器
central hydrothermal eruption 中心式水热喷发
central igniter 中心引爆装置;中心点火器
central image 中心图像
central immersed longitudinal 中纵浸水平面
central impact 对心碰撞;对头相碰
central impervious core dam 防渗芯墙坝
central incineration 垃圾中焚烧;集中焚烧
central industry 中央工业
central initiation zone 中央起源区
central intermediate grey matter 中央中间灰质
central interrupt register 中央中断寄存器
central interrupt system 中央中断制
central involution 中心封合
central ion 中心离子
centralised control panel 中央控制面板
centralised monitoring 集中监控
centralised traffic control 交通集中控制;交通集中管制
centraliser 定中心装置
central island 街心行人安全岛;江心岛;中心环岛;中心安全岛;叉口环形中心岛
centrality 集中性;向心性;中心性
centralization 集中式;集中(化);集于中心;集权体制;向心化;中央集合;中心化
centralization of authority 集权
centralization of authority and responsibility 权责集中
centralization of capital 资本集中
centralization of control 集中控制
centralization of goods transport 货运业务集中化
centralization of operations and maintenance 操作与维修集中化
centralization system 中央集中制;排水系统布置—集中式系统
centralize 集中
centralized accounting 集中核算
centralized accounting system 集中核算制
centralized AC system 集中空调系统
centralized adaptive routing 集中式自适应路由选择
centralized air-conditioner system 集中空调系统
centralized algorithm 集中化算法
centralized arbiter 集中判优器
centralized attendant service 集中式话务员服务
centralized authority 统一管理机构
centralized automatic message accounting 集中自动计费;集中式自动信息计算;中央自动信息记录
centralized boiler room 集中锅炉房
centralized branch accounting system 分支机构会计集中制
centralized bridge control 驾驶室遥控;船桥集中操纵
centralized building 中心枢纽建筑物
centralized command 集中指挥
centralized communication structure 集中通信结构
centralized communication system 集中通信系统
centralized computer 集中式计算机
centralized computer-aided supervisory control 集中化计算机辅助监控
centralized computer facility 集中式计算机设施
centralized computer network 集中式计算机网络
central(ized) concrete mixing 混凝土集中搅拌;混凝土集中拌制
centralized concrete mixing plant 混凝土中心搅拌厂
centralized configuration 集中式配置
centralized control 集中控制;集中操纵;中央控制;中心控制

centralized control desk 集中控制台;中央控制台
centralized control device 集控装置;中央控制装置
centralized control input, output 集中式控制输入输出
centralized control mechanism 中央控制机构
centralized control on floor 地面集中控制
centralized control panel 中央控制面板
centralized-control point 集中控制点
centralized control room 集中控制室;中央控制室
centralized control structure 集中控制结构;集中管理结构
centralized control system 集中控制系统;集中控制方式;集中管理制
centralized control unit 中心控制单元
centralized data processing 集中式数据处理;数据集中处理
centralized design 集中设计
centralized dewatering facility 集中脱水设施
centralized dictation system 集中式指令系统
centralized direction 集中指挥
centralized dispatching 集中支配;集中调度
centralized dispatching control 调度集中
centralized distant control 集中遥控
centralized dust collecting system 集中式除尘系统
centralized electronic control 集中电子控制
centralized engine control panel 轮机集中操纵台
centralized engine room control 机舱集中控制
centralized equipment remote control 集中遥控装置
centralized factory 集中化工厂;固定工厂
centralized force-feed lubrication 集中压力润滑
centralized grease automatic system 自动干油集中润滑系统
centralized grease lubrication 集中油脂润滑;集中润滑装置;集中润滑系统;集中干油润滑
centralized heat supply 集中供热
centralized industrial wastewater treatment 工业废水集中处理
centralized inspection 集中检验
centralized installation of welding machine 多站焊接
centralized instrument panel 集中仪表板
centralized intercept bureau 集中式监听台
centralized interlocking 集中联锁
centralized labor allocation system 统包统配的劳动制度
centralized laboratory 集中实验室
centralized lubricating system 集中润滑系统;集中供润滑油系统
centralized lubrication 集中润滑(法);中心润滑法
centralized lubrication system 集中润滑系统
centralized maintenance 集中维修
centralized management 集中管理;一元化管理
centralized management of wastewater 集中式污水管理
centralized marketing organization 集中运销组织
centralized mixing 集中搅拌;集中拌制(法)
centralized monitoring 中心监视;集中监视;集中监控
centralized monitoring and controlling equipment 集中监控装置
centralized monitoring system 集中监控系统;集中监测系统
centralized network 集中性网络;集中式网络
centralized plan 集中化平面图
centralized planning 集中规划
centralized power source 集中电源
centralized power supply 集中供电;集中电源
centralized pressure-grease system 干油集中压力润滑系统
centralized processing 集中式处理;集中处理
centralized regulation 中央调整
centralized remote control 集中遥控
centralized repair 集中修理
centralized reserve 集中准备
centralized resource management 集中式资源管理
centralized servicing 服务的集中管理
centralized sewerage system 集中式下水道系统
centralized supervisory and control equipment 集中管理控制设备
centralized system 中心化系统
centralized telecontrol 中心遥控
centralized tool grinding 集中磨刀
centralized traffic control 交通集中控制;交通集

中管制;集中行车管制;集中行车调度;集中交通控制;调度集中
centralized traffic control area 调度集中区域
centralized traffic control at terminal 枢纽遥控
centralized traffic control branch 调度集中遥控分机
centralized traffic control device 调度集中设备
centralized traffic control system 集中式交通控制系统
centralized traffic supply 集中供电
centralized treatment plant 集中处理厂
centralized vacuum-cleaning plant 集中式真空除尘设备
centralized ventilation 集中式通风
centralized ventilation system 中央通风系统
centralized wastewater collection and treatment system 集中式污水收集处理系统
centralized wastewater treatment 集中式污水处理
centralizer 定心(扶正)装置;导中器;换位矩阵;中心化子;对冲器;定心装置;定心器;定心夹具;扶正器
centralizing switch 集中开关
central joint 中线缝
central junction box 中心接线盒
central kitchen 中心厨房
central kitchen heating 中心厨房加热
central lab 中心实验室
central laboratory 中心实验室;中心检验室;中央实验室
centrallasite 白沸钙石
central lateral plane (船体的)中纵剖面;中心侧向平面
central laundry 中心洗衣房
central layer 芯层;中心层;中层
central leader training 中轴式整枝
central lens plane 透镜中心面
central level 中心级;中心标高
central library 中心图书馆
central-lift 中管提升
central lift agitator 中管提升搅拌器
central limit theorem 中心极限定理
central line 中心(线)
central line load 中心荷载
central line of carriageway 车行道中心线
central line of channel 航道中心线
central line of lateral overlap 旁向重叠中线
central line of plume 烟羽中心线
central line surveying 中线测量
central link 中心环节
centrallized-control box 集中控制箱
central loading 集中荷载;中心荷载;中心负载
central lobe 中心瓣
central log 中央记录
central longitudinal bulkhead 中纵舱壁
central longitudinal plane 中纵线面
central low-pressure boiler station 中心低压锅炉房
central lubricating system 集中供冷却液系统
central lubrication pump 集中润滑用泵
centrally administered municipality 直辖市
centrally compressed column 中心受压柱
centrally compressed member 中心受压构件
centrally controlled train lighting 集中控制列车照明
centrally-heated 中心加热
centrally housed materials 集中放置的材料
centrally mixed concrete 厂拌混凝土;集中拌制混凝土
centrally mounted 中间悬挂式
centrally notched specimen 中心缺口试样
centrally operated switch 集中操纵的道岔
centrally operated turnout 集中道岔【铁】
centrally pivoted vertical window 立轴中悬窗
centrally planed 中心放射型(城市)规划
centrally planned economy (城市)计划经济;中央计划经济
centrally ported 中心配流
central machine room 中心机房
central main computer 中央主计算机
central mall 路中林荫带;中央林荫分隔带
central managerial function 大城市中枢管理功能
central manifold 中心干管
central market 中心市场
central master key 中央控制钥匙;中央控制键
central master-keyed lock 中央控制锁

central median 中央分隔带
central memory 中央存储器
central meridian 中央子午线;日心子午线
central meridianal plane 中央子午面
central meridian distance 日心距[离]
central meridian of gnomonic chart 心射图中央子午线
central meridian path 日心径
central meridian transit 中央子午线中天
Central Meteorological Bureau 中央气象局
Central Meteorological Observatory 中央气象台
central microtubule 中央微管
central mix 集中拌和;集中搅拌混凝土
central mixed concrete 厂拌混凝土;集中搅拌混凝土;集中拌和混凝土
central mixer 中央搅拌站;中心搅拌机;集中搅拌站;集中搅拌机;固定搅拌机
central mixing 集中搅拌;集中拌制(法)
central mixing concrete 中央搅拌混凝土
central mixing method 集中搅拌法;混凝土集中拌和法;混凝土集中搅拌法
central mixing of concrete 混凝土集中混合
central mixing plant 集中搅拌设备;集中搅拌机;集中搅拌厂;集中搅拌厂;混凝土集中搅拌站;混凝土集中搅拌厂;混凝土集中拌和站;混凝土集中拌和厂;混凝土拌和中心;搅拌中心;中心拌合厂
central moment 中心力矩
central moment of inertia 中心惯性矩
Central Mongolian deep fracture zone 中蒙古深断裂系
central monitor 中心监护器
central motif 中心主题
central mountain 中央峰
central mounting 中央悬挂
central nave 中厅
central net system unit 中心网络管理系统单元
central office 中央(电话)局
central office exchange 中心交换局
central office line 用户线
central office termination 局端
central oil distribution system 集中配油系统
central oiling 中心润滑
central opening 集中孔洞
central operation system 中央操作系统
central overlap technique 中心重叠法
central pallor 中心染色过浅
central panel 集中配电盘;集中控制台
central parallel circle 中央平行圈
central park 中央公园
central parking 路中央停车(处);路中停车;街心停车地带;街心停车处
central parking district 市中心停车区
central part 中心部分
central partition plate 中心隔板
central passage 中心通道
central peak 中央峰
central peak of crater 坑底中央突起
central peg 中心小木桩
central perforation 中央穿孔
central perspective 中心透视
central-perspective correlation 中心透视相关
central-perspective image formation 中心透视成像
central piece 芯板;中心块
central pier 中心墩
central pillar 中柱;中墩
central pilot 引火管小火
central pilot tunnel(l)ing method 中央导洞法
central pipe 中心缩孔;中心缩管;中心管
central pivot 中央枢轴;中心轴枢;中心枢轴;中枢
central pivoted window 中心支承旋转窗
central-pivot irrigation system 中心支枢喷灌系统
central place 中心位置;中心场所
central place theory 中心地学理论;中心地(带)理论
central plan 总计划;中心平面图
central plane 中心面;中心面
central plane of objective 物镜主平面
central plant 总动力厂;总厂
central plant mixing 集中搅拌法;集中厂拌(法)
central-plant refrigeration system 集中制冷系统
central plant system 集中机房系统
central platform 中心工作平台
Central platform uplift 中部台隆[地]

central plaza 中心市场
central point 中心点
central point figure 中点多边形
central point load 中点荷载
central polygon system 中点多边形系
central porosity 中心疏松;中部多孔性;中心缩孔;中心孔隙
central portion 中央部
central portion of painted beam 枋心
central position 中心位置
central post 中心柱(木结构佛塔);中心支柱
central post slewing ring 柱式回转支承(环)
central post slewing supporting device 柱式旋转支承装置
central power 中心电站;中心泵站
central power plant 总动力厂;中心发电厂
central power station 中心发电站
central pressure 中心气压[气]
central pressure index 台风中心气压指数
central pressure regulator 中心压力调节器
central price 中心价格
central principal axis 重心惯性主轴;中心主轴
central prism of a rangefinder 测距仪中心棱镜
central processing 中央处理
central processing system 中央处理系统
central processing unit 中央行车程序处理设备;集中运算器;主机板;中央处理装置;中心处理装置;中心处理机;中央处理机
central processing unit loop 中央处理机主循环;中央处理机主程序
central processor 集中处理装置;中央处理机
central processor organization 中央处理机结构
central process unit 中央处理装置;中央处理器
central projection 心射投影;中心投影
central projection method 中心投影法
central projection of sphere 球面心射投影
central proportioning plant 集中配料装置;集中配料厂;中央配料车间;中央配料厂
central punch 中心冲头
central punching 中心冲孔
central purlin(e) 中檩条
central radio laboratory 中央无线电实验室
central rail 中冒头
central(railway) station 中央火车站;火车总站
central ramp 中央接坡
central rate 中心汇率
central ray 中央射线;中心射线;中心光线
central reactor flux 反应堆中心通量
central receiving aerial 共用接收天线;共用天线
central recorder 中心记录器
central recording accelerograph 中心记录加速度仪
central recording station 中央记录站
central recording system 中心记录系统
central reducer unit 中央减速器
central refrigerating 集中制冷
central refrigerating plant 中央制冷装置;中央制冷设备;集中制冷设备;冷冻总厂;集中制冷装置;制冷总厂
central refrigerating system 集中制冷系统
central refrigeration 集中制冷
central refrigeration plant 中央制冷装置;中心制冷站
central refuge 街心安全岛;路中安全岛;街心行人安全岛
central register 中央寄存器
central repair shop 集中修理厂;总修理厂;中心修理厂
central repository 中心馆
central representation 中心表现
central reprocessing plant 中心后处理厂
central reservation 中央分车带;路中预留地带;路中预留分车带;路中央保留用地
central reverse 分隔带
central rod 中央柱
central roll 中辊
central rotating joint 中心回转接头
central screw 中心螺旋;中心螺钉;定心螺钉
central screw extruder 中心螺杆式挤出机
Central sea basin uplift of South China Sea 南海中央海盆隆起地带
central second moment 中心惯性矩
central second moment of area 面积中心二次矩
central section 中间部分;中央断面
central sedimentation basin 中心进水沉淀池

central separate line 中心隔离线
central separator 中央分隔带
central shaft 中心轴
central(shaft) drive 中心传动
central shaft drive ball mill 中心传动球磨机
central sheet 中央管板;中央薄板
central shrine 主神龛
central shutter 中心快门
central site 中心站
central skid chassis 中橇架
central space 中央空间;中心地位
central spacing effect 中心间隔效应
central span 中(心)跨距;中心跨度;中(间)跨
central spigot 中心插口
central spot market 中央现货市场
central square 中央广场;中心广场
central stage 中心式舞台
central standard time 中央标准时;中部标准时(间)
central star 中央星;中心星
central station 总(发电)站;总厂;中央台;中心(工作)站;中心电站;总电站
central station control 集中式控制
central-station cycle 区域电站热力系统
central steam heating 中央蒸汽供热
central stock 中央岩株
central storage 中心库
central store 中心商店
central stratum 中央层
central street refuge 道路中央安全地带
central string(er) 中央楼梯侧板
central strip 中央分隔带
central sug system 集中垃圾收集系统
central Sumatra basin 中苏门达腊盆地
central support 中间支承
central surface velocity 中泓水面流速
central-suspended lighting 街心悬挂式照明;路中央悬挂式照明;路中悬挂式照明
central-symmetric(al) 中心对称的
central symmetric degree analysis of lineament 线性体中心对称度分析
central symmetry 中心对称(性)
central system 集中系统;中央系统;中心制;中心系统
central tank 中央储[贮]槽
central tax 国税
Central Telecommunication Office 中央电信局
central telephone exchange 电话中心局
central telephone office 电话中心局
central temperature 中心温度
central tendency 集中趋势;中心倾向
central tension bolt 中心拉紧螺栓
central terminal 中央终端
central terminal station 总枢纽站;中转枢纽;中央终点站
central terminal unit 中央终端装置
central thread 中丝
central timing equipment 中央计时设备
central timing system 统一计时制
central timing unit 中央计时装置;中央计时设备
central tower 主塔
central town 中心城镇
central track 中央航道
central traffic control 调度集中
central traffic district 市中心交通区
central train control 中央列车控制
central transmission equipment 中心传输设备
central treasury 中央金库
central treatment 集中处理
central treatment facility 集中处理设施
central trenching method 中央挖槽施工法
central trunk 集群总机
central tube 总管;中心管道;中心管
central tube frame 单管梁式框架
central tube method 中心管方法
central tube of sedimentation tank 沉淀池中心管
central tube tool carrier 单管梁式自动底盘
central type 中央型
central uplift 中央隆起
central vacuole 中央液泡
central vacuum system 中央抽真空系统
central valley 裂谷
central value 中心值;代表值
central value of distribution 分布(的)中心值

central ventilating station 供风中心
central ventilation 中央式通风
central vision 中心视觉
central vision mixer 主视频混合器
central warehouse 中转仓库;中心仓库
central warm-air heating system 集中热风供暖系统;集中热风采暖系统
central waste water treatment 集中(式)废水处理;污水集中处理
central water 中央水
central water heating 中心热水站
central water planning unit 中央给水规划单位
central water supply 集中供水
central water treatment 水的集中处理
central weight 中心锤
central well 中心进料孔
central well down take evapo(u)rator 中央循环管蒸发器;中央降液管蒸发器
central wholesale market 中心批发市场
central wire 中丝
central zone 中央带;中心区;中心带;芯墙(土坝)
centra steel 球状石墨铸钢
centrate 离心滤液;离心分离液
centrate pump 离心液泵
centrate wet well 离心液(湿)井
centration 共轴性;对中
centration error 共轴性误差
centraxonial 中轴的
centre = center
centre adjustment 置中
centre aisle coach 中央通道客车
centre align 对中心
centre anchor 船首备用锚
centre angle 圆心角;中心角
centre arbor 中心心轴
centre arch 中心拱
centre bank 江心洲;江心滩
centre bar keel 方形龙骨
centre barrier 高速公路中间狭长绿草边上的护栏
centre bearing 中(心)轴承;中心支承
centre bearing companion yoke nut 中心轴承连轭螺母
centre bearing housing 中心轴承壳
centre bearing housing bracket 中心轴承托架
centre bearing housing cap 中心轴承壳盖
centre bearing housing carrier 中心轴承托架
centre bearing housing insulator 中心轴承壳软垫
centre bearing mud slinger 中心轴承除泥罩
centre-bearing swing bridge 中心平旋桥
centre bearing type 中心支承式
centre bit 矛式钻头;转柄钻;中心钻(头);三叉钻头;打眼锥
centre-blade frame saw 中心刀片框架锯
centre board 心板材;重板龙骨;防摇(垂)板;去心板材
centrebody 中心体
centre bore 钻头内径;中心孔
centre-bound tie 中心联杆
centre brake 中央制动机;中央制动器
centre brake drum 主制动轮
centre brake flange 主制动凸缘
centre bridge 定心孔塞
centre bubble 气泡置中
centre buff 抛光轮轮毂
centre buffing 抛光中心线
centre bulb 中心胀大
centre burst 中心裂纹
centre bushing 中心杆衬套
centre bypass valve 中间(位置)旁通(的)换向阀
centre casting 铸坑
centre casting crane 铸坑起重机
centre castle 船中部上层建筑
centre chamber 中央镜箱
centre charging 中心装料
centre chord 中心弦
centre cleavage 中心裂缝
centre collector ring 中心集流环;中心集电器;中心集电环;中心滑环
centre column drilling machine 中心床身式钻床
centre column transfer machine 中心床身式连续自动工作机床
centre combustion stove 中心燃烧式热风炉
centre conductor 中心导线
centre conductor rail 中间接触轨

centre conic 中心圆锥线
centre-contact holder 中心接触的灯座
centre control room 中控室
centre core wall 中心筒墙
centre corridor 中央走廊;内廊
centre coupler 中心连接器;车钩对心
centre crank 中心曲柄
centre crank shaft 中心曲轴
centre cross member 中央横梁
centre-cut method 中心开挖法
centre cut(ting) 中心掏槽
centred 同轴的
centre depth regulator 中央深浅调节器
centre differential 轴间差速器
centre discharge 中心卸料
centre distance 中心距(离);顶尖距
centre distance between outrigger 支腿中心间距
centred lattice 定心栅格
centred n-sided polygon 中心N边形
centred optic(al) system 合轴光学系统;共轴光学系统
centre drain 中心排水管;中心排水沟;马路中线下排水管
centre drawer guide 抽屉中心导轨
centre drawer hinge bracket 公铰链
centre drift method 中央导坑法;中央导洞法
centre drift tunneling 中心导孔隧洞掘进法;中心导孔隧道掘进法;中心导坑隧洞掘进法;中心导坑隧道掘进法
centre drill 中心钻;中心孔钻(头)
centre drilling 中心钻削
centre drilling(bore)hole 中心钻孔
centre-drilling lathe 中心孔钻床
centre drilling tool 打中心孔工具
centre drive lathe 中心传动车床
centre drive sludge collector 中心驱动集泥器
centre drvie motor mower 中央驱动动力割草机
centre dump car 底卸料车
centre earthquake 中心震
centre fed 中心馈电的;对称供电的
centre feed 中央馈电;中央加料;中心供电
centre-feed circular sedimentation basin 中心进水圆形沉淀池
centre feed hole 中导孔
centre filling 中心装料
centre fixture 定心装置
centre flower 中央花饰
centre focus 中心聚焦
centre folding 对折
centre force 中心力
Centre for Disease Control 疾病控制中心
centre for geologic(al) remote sensing applications/ministry of nuclear industry 核工业部地质遥感应用中心
Centre for International environment-al Information 国际环境信息中心
centre form 内模
centre for scientific research 科学研究中心
Centre for Shipping Information and Advisory Services 海运情报及咨询服务中心
centre frequency 中心频率
centre frogs 中心辙岔
centre gangway coach 中心通道客车
centre gate 中间控制级
centre ga(u)ge 中心规;中央量油口
centre girder 中龙骨立板;船底中纵桁材
centre girder of ship bottom 船底中桁材
centre grinder 有心磨床;中心磨床
centre grinder attachment 中心磨床附件
centre half 中间二分之一
centre head 顶尖架
centre heading 中央导坑
centre height 中心高度
centre-height ga(u)ge 中心高度规
centre hinge 门窗枢轴
centre hole 中心孔;顶针孔;顶尖孔
centre-hole bit 中心水口式钻头
centre-hole grinder 中心孔磨床
centre hole jack 穿心式千斤顶
centre-hole lap 顶尖研具
centre-hole machine 中心孔机床
centre-hole of lathe 车床顶尖孔
centre hole planting 穴中栽植
centre(-hole) reamer 中心孔铰刀

centre-hung 中旋;中悬
centre-hung door 中旋门
centre-hung horizontally pivoting window 横轴中悬窗
centre-hung sash 中旋式(气)窗框;在枢轴上转动的窗框
centre-hung swivel window 中悬窗
centre-hung window 中悬翻窗;中悬窗扇
centre-island 中心岛
centre jam 江心木材捆塞
centre jet drill bit 中心喷射式钻头
centre joint 中央纵缝(混凝土路面的);中缝;中心枢轴;中心缝
centre junction 中心结
centre keel 中竖龙骨;中龙骨;(单底的)竖龙骨
centre lane 中间车道;中央车道
centre-lane traffic 中央车道交通
centre lapping machine 顶尖精研机
centre lathe 顶针车床;顶尖车床;普通车床
centreless 无中心的
centreless grinder 无心磨床
centreless grinding machine 无心磨床
centreless grinding machine external 无心外圆磨床
centreless grinding machine internal 无心内圆磨床
centreless grinding machine principle 无心磨削原理
centreless honing machine 无心珩床
centreless internal grinder 无心内圆磨
centreless lapping machine 无心精研机
centre line 中心线;中线
centre line adjustment 中心线校正
centre line average 中心线平均值;算术平均法
centreline average height 中线平均高度;平均高度
centre line average method 平均高度法
centreline camber 中线弯度
centre line concentration 中心线浓度
centre line dilution 中心线稀释度
centre line grade elevation 中线红色标高
centre line ground elevation 中线黑色标高;道路中线地面标高
centre(-line) joint 中缝;中线接缝
centre line keelson 中内龙骨
centre line marking machine 中线标划机
centre line of a ship 船舶中线
centre line of boiler 锅炉轴线
centre line of bridge 桥梁中线
centre line of crane rail 起重机轨道中心线
centre line of inertia 惯性中线
centre line of mine-swept route 扫雷航道中心线
centre line of road 道路中心线
centre line of support 悬挂中心线
centre line of track 铁路中心线;轨道轴线
centre line of tunnel 隧道中线
centre line of ventilation adit 风道中线
centre line of wall 墙中心线
centre line peg 中心桩【测】
centre line point 中线点
centre line shrinkage 轴线缩孔
centre line stake 中线桩
centre line stake level(l)ing 中线桩高程测量(简称中平)【测】
centre line strake 中心线板条
centre line survey(ing) 中线测量
centre line velocity 中心线流速
centre line witness mark 中线参考标志
centre loading 跨中荷载
centre lubrication 集中润滑(法)
centre mall 路中林荫带;道路分隔带
centre mark 中心标记;定中心点
centre mix(ing) 集中拌和;中央搅拌
centre mixing plant 集中拌和厂
centre mounted type 中悬挂式
centre mount support 中间支承;中间支撑
centre nailing 中间钉(瓦)法
centre nib 板簧中心楔
centre notes 栏间注释
centre nozzle 中部水口
centre of a conic(al) 二次曲线的心
centre of action 作用中心
centre of a magnetic well 磁阱中心
centre of anallatism 准距中心【测】
centre of an involution 对合中心

centre of apex 顶尖中心
centre of arch 拱心
centre of area 面积中心
centre of atmospheric action 大气活动中心
centre of beacon 觇标中心
centre of borehole 井眼中心线;钻孔轴线
centre of buoyancy 浮心;浮力中心
centre of buoyancy above top of keel 浮心在龙骨上高度
centre of circle 圆心
centre of collineation 共线中心
centre of compression 压力中心
centre of conic(al) section 二次曲线心
centre of crest circle 坝顶曲线中心点
centre of crown 路拱中心
centre of cult 礼拜场所
centre of curvature 曲线中心;曲率中心
centre of cyclone 气旋中心
centre of dispersion 散布中心
centre of displacement 排水中心
centre of disturbance 扰动中心
centre of draft 阻力中心
centre of drag 拉力中心
centre of effort 作用力中心
centre of energy release 能量释放中心
centre of equilibrium 平衡中心
centre of figure 形心
centre of flo(a)tation 船舶水线断面形心;浮心;浮面中心;水线面中心
centre of force 力心
centre of forecasting and analysis 分析测试中心
centre of friction 摩擦中心
centre of girder 龙骨立板
centre of gravity 重心
centre-of-gravity disturbance 重心扰动
centre of gravity for compartments 船舱重心
centre of gravity member section 杆件断面重心
centre-of-gravity motion 重心运动
centre-of-gravity mounting system 重心支承系统
centre of gravity of geoid 大地水准面重心
centre of gravity of ship 船舶重心
centre-of-gravity position 重心位置
centre-of-gravity range 重心范围
centre-of-gravity shift 重心偏移
centre-of-gravity suspension 重心悬置法
centre of gyration 回转中心;旋心
centre of homology 透射中心
centre of inertia 惯性中心
centre of inertia system 惯性中心系统
centre of infection 传染中心
centre of instrument 仪器中心
centre of lateral pressure 侧压中心
centre of lead curve 导偏中心
centre of lift 升力中心
centre of light 灯光中心
centre of location 定位中心
centre of maritime studies 海事研究中心
centre of moment 力矩中心;矩心
centre-of-momentum system 动量中心系统
centre of motion 运动中心
centre of movement 运动中心
centre of nodal circle 波节圆中心
centre of origin 震源
centre of oscillation 摇摆中心;摆动中心
centre of parallel forces 平行力系中心
centre of percussion 打击中心
centre of perspectivity 透视中心
centre of pressure 压强中心;压力中心
centre of pressure travel 压力移动中心
centre of projection 射影中心
centre of punch 冲心
centre of quadric 二次曲面心
centre of radiation 辐射中心
centre of regular polygon 正多边形中心
centre of resistance 阻力中心
centre of rigidity 刚度中心
centre of rotation 旋转中心
centre of science and technology 科技中心
centre of sensitization 感光中心
centre of sheet 图幅中心
centre of similitude 相似中心
centre of source line 源线中心
centre of span 跨距中点;跨度中心
centre of sphere 球心

centre of springing 起拱中心
centre of storm 风暴眼
centre of stress 应力中心
centre of supply 供电中心
centre of suspension 悬心;支点
centre of switch 转辙器中心
centre of symmetry 对称中心
centre of syncline 向斜中心
centre of the earth 地心
centre of the Ionic volute 爱奥尼亚式涡卷眼
centre of the trace 轨道中心
centre of tropic(al) cyclone 热带气旋中心
centre of turnout 道岔中心
centre of twist 扭转中心;扭曲中心
centre of vertex 顶点中心
centre of vibration 振动中心
centre of water plane 水线面浸水部分中心
centre of water resistance 水阻力中心
centre of weir crest circle 堰顶圆弧中心
centre of wind pressure 风压中心
centre opening 中心孔
centre operation 顶尖操作
centre parking 路中停车;街心停车地带;街心停车处
centre pedestal 中央基座
centre peg 中心桩【测】
centre piece 中间部件;中心块;十字轴
centre pier 中心桥墩;中心墩;中墩
centre pile 中心桩;中间桩
centre pilot 中心导向
centre pin 中心销;球端心轴
centre pin guide 中枢导承
centre pinion 中心齿轴
centre pin socket 中心销承窝
centre pivot 心销
centre pivoted window 中旋窗;中心支承旋转窗
centre-pivot irrigation 旋臂灌溉
centre-pivot irrigation system 中枢灌溉系统
centre pivot pin 中心枢轴销
centre plank 树心板;中心板;材心板;去心厚板
centre plate 心盘;中线板
centre plate loading force 心盘载荷
centre plate oak block 中心板橡木(垫)块
centre-point earthing 中点接地
centre-point load 中点荷载;集中荷载;集中负载
centre-point loading 中心载荷
centre-point steering 中心点转向
centre pole 街心电杆;路中式电车杆;中间杆
centre portion 中部
centre position 中心位置;中间位置
centre post 中线标桩
centre prism 中央棱镜
centre punch 中心冲头
centre punching 中心冲孔
centre quad 中心器绕组
centre rail 中央轨
centre ramp (立体交叉的)中央匝道
centre recording station 中心记录站
centre region 中心区
centre rest 中心架
centre ridger body 中央作垄型体
centre roll of calender 压延机中辊
centre routine 中央程序
centre scale 中心标度
centrescope 定点放大镜
centre section 中心剖面
centre shaft 转门中轴;中轴
centre shift 转盘中心移动装置(平地机)
centre shop 中心店
centre shot 中心掏槽爆破;中心掏槽炮眼
centre sill 中梁;(车的)中间梁
centre site 中心站
centre-slot system 中心槽制
centre spacing 中距
centre span 中心跨距;中心跨度;桥梁的中央跨
centre spigot 中心插口
centre split pipe 对开管
centre split supporting ring 半圆支承环
centre spool valve 中心柱形阀
centre square 中心角尺
centre staff 中心小轴杆
centre stake 中线桩;中桩【测】;中心桩【测】
centre station 中心桩号
centre-stitching 中央缝缀
centre stringer 中间楼梯梁

centre strip 路中分隔带
centre strut 中心支柱
centre support system 中心支持系统
centre suspension cord 中悬索
centre system 中央系统
centre take-off water 中心出水
centre tank 中间舱
centre tap 中心抽头
centre-tapped secondary 中心抽头次级线圈
centre tester 中心试验器
centre test station 中心试验站
centre thickness 中心厚度
centre through plate 中内龙骨
centre-to-centre 中(心)到中(心)
centre-to-centre distance 中心距(离);中到中距离;顶尖间距
centre-to-centre method 中心连接法
centre-to-centre of rivets 铆钉中(心)距
centre-to-centre spacing 中心距(离)
centre to edge 中心至边缘
centre triangulation 中性辐射三角测量
centre tube 中心调整管
centre-type cylindric(al) grinder 中心外圆磨床
centre-type cylindric(al) grinding 中心处圆磨削
centre valve 中心阀
centre vent 中心放气孔
centre washer 中圈
centre weight 中心锤
centre well 中心井;船井(海洋钻探船)
centre-zero instrument 中心零位(式)仪表
centre zero relay 三位辅助继电器
centre zero scale 双向刻度
centre zone profile 中心区分布
centric 中心的;中枢的
centric(al) impact 中心冲击;对心碰撞
centric cleaner 锥形除渣器
centric diatom 环纹硅藻;中心硅藻
centri-chromatograph 离心色谱(法)
centri-cleaner 锥形离心式除渣器
centric load 轴心荷载;中心载荷;中心负载
centriclone 锥形除渣器;中心旋流器;涡流旋流器;水力旋流器
centric pattern 向心型
centric texture 向心结构
centrifiner 离心精炼炉
centrifugal 离心沉淀法
centrifugal absorber 离心吸收器
centrifugal acceleration 离心加速度
centrifugal acceleration of equator 赤道上离心加速度
centrifugal action 离心作用
centrifugal aerator 离心曝气器
centrifugal air classifier 离心式选粉机;离心空气分级器;离心式空气分级机
centrifugal air compressor 离心式空(气)压(缩)机;离心空气压缩机
centrifugal air separation 离心空气分离
centrifugal air separator 离心式选粉机;离心空气分离器
centrifugal analysis 离心分析
centrifugal apparatus 离心机
centrifugal aspirator 离式吸气器;离心式吸尘器
centrifugal atomization 离心式喷雾;离心式弥雾;离心雾化;离心喷射
centrifugal atomizer 离心雾化器;离心(式)喷雾器
centrifugal autoclaved concrete pile 离心蒸压混凝土桩
centrifugal ball mill 离心式球磨机
centrifugal barrel finishing 离心转鼓涂漆
centrifugal barrier 离心势垒
centrifugal basket 离心机转筒;离心机筛篮;离心吊篮;离心过滤转鼓
centrifugal blender 离心式搅拌机
centrifugal blower 离心式吹风机;离心通风机;离心式增压器;离心式鼓风机;离心式风机;离心鼓风机
centrifugal bolting mill 离心机转筒筛
centrifugal booster pump 离心式增压泵
centrifugal bowl 离心机转筒
centrifugal brake 离心制动器;离心式制动器;离心式制动机
centrifugal breaker 离心式破碎机;离心式断路器
centrifugal breather 离心式通风器
centrifugal bucket elevator 离心式斗式提升机

centrifugal casting 离心铸造;离心铸管法;离心式浇注;离心浇铸;离心浇注(成型)
centrifugal casting concrete 离心式浇注混凝土
centrifugal casting concrete pipe 离心浇注混凝土管
centrifugal casting machine 离心铸造机;离心式浇注机;离心浇铸机
centrifugal casting mo(u)ld 离心(铸造)铸型
centrifugal casting of concrete 离心浇注混凝土
centrifugal cast iron pipe 离心法铸铁管
centrifugal chiller 离心式冷水机
centrifugal chuck 离心卡盘
centrifugal clarification 离心净化
centrifugal clarifier 离心澄清器;离心澄清机;离心澄清池
centrifugal classification 离心选粉;离心分选;离心分类;离心分级
centrifugal classifier 离心分级器;离心分级机
centrifugal classifying 离心分选;离心分级
centrifugal cleaner 离心清洁器
centrifugal clutch 离心(式)离合器
centrifugal coating 离心涂覆
centrifugal collector 离心(捕集)器;漩涡除尘器;(吸集气体悬浮微粒用的)离心吸集器
centrifugal compacting 离心塑型
centrifugal compacting process 离心成型工艺
centrifugal compressor 离心空气压缩机;离心压气机;离心式压缩机
centrifugal concentration method 离心浓缩法
centrifugal concentrator 离心浓缩机
centrifugal concrete 离心(法浇注)混凝土
centrifugal concrete pile 离心(法浇制)混凝土桩;离心(法)浇注混凝土桩
centrifugal concrete pipe 离心制混凝土管;离心混凝土管
centrifugal concretion 离心结核
centrifugal coolant clarifier 离心式冷却液过滤器
centrifugal couple 离心力偶
centrifugal cream separator 离心乳脂分离器
centrifugal crusher 离心式破碎机
centrifugal deduster 离心除尘器
centrifugal deep-well pump 深井离心泵;离心式深井泵
centrifugal dehydrator 离心脱水机
centrifugal development 离心展开
centrifugal dewatering 离心脱水(法)
centrifugal dewatering of sludge 污泥离心脱水
centrifugal dirt collector 离心集尘器
centrifugal discharge 离心卸料
centrifugal discharge bucket elevator 离心链斗式提升机;离心出料斗式机
centrifugal discharge elevator 离心卸料提升机
centrifugal disc sprayer dryer 离心盘式喷雾干燥器
centrifugal disk atomizer 离心导流转盘式雾化器;离心导流转盘式喷雾器
centrifugal disk microfiltration 离心隔膜微滤
centrifugal distortion 离心畸变;离心变形
centrifugal distribution 离心分布
centrifugal distributor 离心式撒分器
centrifugal drainage 离心水系;辐射状水系
centrifugal drainage pattern 离心状水系;辐射水系型
centrifugal drawing 离心拉丝
centrifugal dredge pump 离心式泥泵
centrifugal dryer 离心式干燥器;离心式干燥机;离心干燥器;离心干燥机
centrifugal dry dust collector 离心旋风干式集尘器
centrifugal drying 离心干燥
centrifugal ductile iron pipe 高压韧性离心球墨铸铁管
centrifugal dust arrester 旋压式除尘器
centrifugal dust collector 离心式集尘器;离心式吸尘器;离心式吸尘器;离心集尘器;离心除尘器
centrifugal duster 离心除尘器
centrifugal dust remover 旋压式除尘器
centrifugal dust separator 离心除尘器
centrifugal effect 离心作用
centrifugal effort 离心力
centrifugal emulsor 离心乳化剂
centrifugal enamelling 离心涂漆
centrifugal error 离心误差
centrifugal exhauster 离心排气机
centrifugal extraction analysis 离心提取分析
centrifugal extraction method 离心抽提法

centrifugal extractor 离心提取器;离心萃取器
centrifugal factor 离心因子
centrifugal fan 离心(式)通风机;离心式风扇;离心(式)风机
centrifugal fan system 离心风机系统
centrifugal fan type wet scrubber 离心扇形湿式洗涤器
centrifugal fast analyzer 离心式快速分析仪
centrifugal fault 离心断层
centrifugal feed 离心式排种器
centrifugal-film evapo(u)rator 离心薄膜式蒸发器
centrifugal filter 离心(式)过滤器;离心滤器;离心过滤器
centrifugal filtration 离心过滤
centrifugal filtration process 离心过滤法
centrifugal finishing 离心涂覆;甩涂
centrifugal fire pump 离心式灭火水泵
centrifugal-flow compressor 离心式压气机
centrifugal force 离心力
centrifugal force control 离心力控制
centrifugal force field 离心力场
centrifugal force mill 离心式磨机
centrifugal force of train 列车离心力
centrifugal force pelletizer 离心力成粒器
centrifugal force tachometer 离心式转速计;离心式转数计
centrifugal gas cleaner 离心涤气机
centrifugal gas compressor 离心式气体压缩机
centrifugal gas washer 离心气洗机
centrifugal governor 离心(式)调速器;离心式调节器
centrifugal grease interceptor 离心集油器
centrifugal grinder 离心研磨机
centrifugal hammer crushing machine 离心锤式粉碎机
centrifugal head 离心水头
centrifugal humidifier 离心增湿器;离心湿润器
centrifugal hydroextractor 离心液体分离机;离心脱水器;离心脱水机
centrifugal immersion process 离心浸涂(合金)法
centrifugal impact mill 离心冲击研磨机
centrifugal impeller 离心(式)叶轮
centrifugal impeller mixer 离心叶轮混合器;离心式叶轮搅拌机;离心式叶轮混合机;离心式叶轮拌和机
centrifugal impeller web 离心叶轮连接板
centrifugal inertial force 离心惯性力
centrifugal inertial separator 离心式惯性除尘器;离心惯性分离器
centrifugal integrator 离心积分器
centrifugal interceptor 离心式分离器
centrifugalization 离心过程;离心(分离)作用;离心分离;远心沉淀
centrifugalize 离心分离
centrifugal jet pump 离心式射流泵
centrifugal jig 离心跳汰机
centrifugal kerosene equivalent 离心煤油当量
centrifugal laboratory test 离心实验室试验
centrifugal lift 离心升力
centrifugal load(ing) 离心荷重;离心荷载
centrifugal lubrication 离心注油
centrifugal lubricator 离心(式)润滑器
centrifugally cast cluster 离心串注件
centrifugally cast column 离心法浇制柱子
centrifugally cast concrete 离心法浇注混凝土;离心成型混凝土
centrifugally cast concrete drain pipe 离心法浇制混凝土排水管
centrifugally cast concrete drive pile 离心浇注混凝土打入桩
centrifugally cast(ed) 离心浇铸的
centrifugally casted pipe 离心铸造管
centrifugally cast glass-fiber-reinforced thermosetting resin pipe 离心成型玻纤增强热固性树脂管
centrifugally casting concrete 离心浇注混凝土
centrifugally casting concrete pipe 离心浇注混凝土管
centrifugally cast-iron pipe 离心铸铁管
centrifugally cast pipe 离心铸铁管
centrifugally lubricator 离心注油器
centrifugally prefabricated 用离心浇注法预制的
centrifugally spun 离心制的;离心浇注的
centrifugally spun mo(u)lding 离心成型

centrifugally spun concrete conduit 离心式旋制混凝土管
centrifugally spun concrete pipe 离心制混凝土管;离心式旋制混凝土管;离心法制混凝土管
centrifugally spun concrete tube 离心式旋制混凝土管
centrifugal machine 离心机
centrifugal method 离心法
centrifugal microscope 离心显微镜
centrifugal migration 离心迁徙
centrifugal mill 离心磨矿机;离心式研磨机;离心磨石机
centrifugal mixed flow compressor 离心轴流混合压气机
centrifugal model test 离心模型试验
centrifugal moisture equivalent 离心含水当量
centrifugal moisture of soil 土壤离心含水率
centrifugal moment 离心力矩;离心力矩
centrifugal moment of inertia 离心惯矩
centrifugal mo(u)lding 离心造型;离心模制;离心法成型
centrifugal mo(u)lding for pipe 离心制管
centrifugal mud machine 离心式泥浆净化器
centrifugal muller 快速混砂机;摆轮式混砂机;离心式混砂机
centrifugal nozzle 离心(式)喷嘴
centrifugal oiler 离心加油器
centrifugal oil filter 离心滤油器
centrifugal oil pump 离心(式)油泵
centrifugal oil purifier 离心式滤油器
centrifugal pendulum 离心(力)调速器;离心摆
centrifugal phase-barrier recompression distillation 离心相障再压蒸发法
centrifugal photometric analyzer 离心光度分析仪;离心光度分析器
centrifugal pouring 离心浇注
centrifugal precipitation 离心力除尘器
centrifugal precipitation mechanism 离心沉淀机理
centrifugal precipitator 离心沉淀器
centrifugal pressure casting 离心铸(造)法
centrifugal prestressed concrete pile 离心预应力混凝土桩
centrifugal process 离心工艺;离心法
centrifugal pull 离心力
centrifugal pulp cleaner 离心除渣器
centrifugal pump 离心水泵;离心式增压泵;离心抽水机;离心泵
centrifugal pump drainage 离心泵排水
centrifugal pump laws 离心泵定则
centrifugal pump of lining teflon 衬里离心泵
centrifugal pump of multistage type 多级(式)离心泵
centrifugal pump of single-stage type 单级离心泵
centrifugal pump of turbine type 涡轮式离心泵;水轮(机)式离心泵
centrifugal pump with bladeless impellers 叶轮无叶片的离心泵
centrifugal pump with horizontal axis 卧式离心泵
centrifugal pump with lining Teflon 聚四氟乙烯塑料衬里离心泵
centrifugal pump with vertical axis 立式离心泵
centrifugal purification 离心净化
centrifugal purifier 离心净化机
centrifugal radial drainage pattern 放射状水系形式
centrifugal refrigerating machine 离心(式)制冷机
centrifugal refrigeration machine 离心制冷机
centrifugal regulator 离心调节器
centrifugal reinforced concrete pile 离心(浇注)钢筋混凝土桩
centrifugal reinforced concrete pipe 离心(浇注)钢筋混凝土管
centrifugal relay 离心继电器
centrifugal ring-ball mill 离心环球式磨机
centrifugal rolling process 离心辊压法
centrifugal roll mill 离心辊磨
centrifugal roof ventilator 离心式屋顶通风器
centrifugal sand mixer 离心搅砂机
centrifugal sand-pump 离心砂泵
centrifugal screen 离心式筛分机;离心筛;离心滤网
centrifugal screw pump 离心螺旋泵
centrifugal scrubber 离心洗涤器;离心涤气器
centrifugal sedimentation 离心沉降
centrifugal separation 离心分离
centrifugal separation method 离心分离法

centrifugal separator 离心选粉机;离心(式)分离器;离心(式)分离机;离心式除尘器;离心澄清机
centrifugal settler 离心沉降器
centrifugal settling 离心沉降
centrifugal sewage pump 离心(式)污水泵
centrifugal sheller 离心剥壳机
centrifugal shield 离心套管
centrifugal sieve 离心筛
centrifugal simulator 离心模拟器
centrifugal sizing 离心分粒
centrifugal speed 离心速度
centrifugal speed regulator 离心调速器
centrifugal spinning disc 离心式撒布器;离心式撒布盘
centrifugal spray 离心喷雾;离心喷涂
centrifugal sprayer 离心式喷雾器
centrifugal spraying humidifier 离心式喷雾加湿器
centrifugal spray tower 离心喷雾塔
centrifugal spreader 离心式撒布器;离心式撒布机
centrifugal sprinkle head 离心式喷头
centrifugal stability 离心稳定器
centrifugal starter 离心式起动器
centrifugal starting switch 离心式起动开关
centrifugal steel 离心铸钢
centrifugal still 离心蒸馏器
centrifugal stirrer 离心搅动器
centrifugal stop bolt 离心式止动螺栓
centrifugal stower 抛掷充填机
centrifugal stress 离心应力
centrifugal stretching 离心拉伸
centrifugal subsidence 离心沉降
centrifugal subsider 沉降式离心机
centrifugal supercharger 离心式增压器
centrifugal super-fractionator 离心超分馏器
centrifugal swing 离心摆动
centrifugal switch 离心式开关
centrifugal switching 离心切换
centrifugal symmetry 离心对称
centrifugal tar extractor 离心焦油提取器;离心焦油萃取器
centrifugal tendency 离心倾向
centrifugal tension 离心张力
centrifugal thickener 离心浓缩器;离心浓缩机
centrifugal thickening 离心加厚
centrifugal tube 离心管
centrifugal turbine compressor 离心式涡轮压缩机
centrifugal type 柴油机驱动离心水泵
centrifugal-type automatic clutch 离心式自动离合器
centrifugal type injection valve 离心式喷射阀
centrifugal type of speedometer 离心式速率计
centrifugal type spray drier (离心雾化型)喷雾干燥器
centrifugal ultrafiltration 离心超滤
centrifugal ventilator 离心式通风机
centrifugal vibrating process 离心振动法
centrifugal wastewater concentrator 离心式污水浓缩机
centrifugal water chiller 离心式冷水机组
centrifugal water chilling unit 离心式水冷却置;离心式水冷却器
centrifugal water-packed gland 离心水封
centrifugal water pump 离心水泵;离心式水泵
centrifugal wet collector 离心式湿法集尘器
centrifugal wet scrubber 离心式湿式涤尘器
centrifugal wheel 离心轮
centrifugal whirling of shaft 轴的离心旋动
centrifugal with surrounding blowers 离心喷棉机
centrifugate 离心(液)
centrifugation 离心分离作用;离心(沉淀);分离(作用)
centrifugation potentials 离心势
centrifugation process 离心工艺
centrifugation separation 离心分离
centrifuge 离心脱水(作用);离心机
centrifuge action 离心作用
centrifuge adhesion 离心法附着力;抗甩性
centrifuge basket 离心机盘
centrifuge brake drum 离心制动轮
centrifuge casting 离心铸造
centrifuge contactor 离心接触器
centrifuged apparatus 离心设备
centrifuged compact 离心沉降物
centrifuged laboratory experiment 离心实验

centrifuge drum 离心浇铸制动鼓
centrifuge(d) steel 离心铸造钢
centrifuge effect 离心分离效应
centrifuge kerosene equivalent 离心煤油当量(一种测定细集料表面吸油量指标)
centrifugel dehumidifier 离心脱水器
centrifuge method 离心机法;离心测量法(含砂量测定)
centrifuge microscopy 离心显微镜检查
centrifuge mill 离心式磨机
centrifuge modeling 离心机模拟
centrifuge moisture equivalent 离心湿度当量;离心含水当量;离心持水当量;湿度当量
centrifuge of sludge 污泥离心机
centrifuge reclaiming 离心机法(废润滑油)再生
centrifuge reclaiming waste oil 离心法再生废油
centrifuge refining 离心精炼
centrifuge rotor 离心机转子
centrifuge simulation 离心器模拟
centrifuge stock 离心处理燃料
centrifuge test 离心试验;离心机试验
centrifuge test system 离心试验系统
centrifuge treating 离心机处理
centrifuging 离心脱水(法);离心浓缩污泥法;离心串注;离心处理;离心作用
centrifuging process 离心法
centrigrade system 百分度制
centring 中心校正;光栅对准中心;对中;定中心;定(圆)心
centring bar 定中心棒
centring cam 对中心凸轮
centring chisel 定中心凿
centring chuck 定中心夹头
centring circuit 中心位置调节电路
centring cone 定心圆锥
centring control 居中调节
centring control current 中心调节电流
centring device 定心装置;定心机构
centring element 对中元件
centring jaw 定心凸轮
centring machine 对中心机;定心机
centring mo(u)lding machine 定心造型机
centring of diaphragm 隔板定中心
centring of origin 震源
centring potentiometer 定心电位器
centring projection 归心投影
centring ring 定心环
centring roll(er) 对中辊
centring screw 定中心螺钉
centring sleeve 定心套
centring spigot 定心凸出物
centring spool 对中套筒
centring spring 对心弹簧
centring spring case 自对中心弹簧盒
centring spring rod 中心弹簧杆
centring spring seat 复原弹簧座
centring support 中分面支持;汽缸中分面支持
centring system 中心调整系统
centring tongs 定心卡具
centring wedge 定中心楔
centriole 中心粒
centriole pinwheel 中心粒小轮
centripetal 利用向心力的;向中的;求中的
centripetal acceleration 向心加速度
centripetal development 向心展开
centripetal drain 向心排水
centripetal drainage 向心水系
centripetal fault 向心断层;正断层
centripetal force 进合力;向心力
centripetal migration 向心迁徙
centripetal movement 向心移动;向心交通
centripetal pattern 向心型
centripetal pump 向心(式水)泵
centripetal replacement 向心交代【地】
centripetal tendency 向心倾斜
centripetal turbine 内流式涡轮机;内流式透平;内流式水轮机
centripetal urbanization 离心型城市化
centrix 瓦片切割器
centrobaric 重心的
centrocecal 中心盲点的
centroclinal dip 向心倾斜;向心倾角
centroclinal dip projection method 归心投影法
centroclinal fold 向心褶皱

centroclinal selection 向心选择
centrode 瞬心轨迹
centroid 矩心;面心;形心曲线;形心;心迹线;重心;质心;质量中心;出行中心
centroidal 质心的;穿过重心
centroidal axis 矩心轴;质量中心轴线
centroidal compressive stress 重心压应力
centroidal distance 形心距离;重心距离
centroidal frequency 形心频率
centroidal moment 质心矩
centroidal principal axis 向心主轴
centroid aspect ratio 中心方向比
centroid center of volume 体积中心
centroid label 质心标记
centroid method 重心波长法
centroid of area 面积形心;面积矩心;面的矩心
centroid of buoyancy 浮心
centroid of drainage basin 流域形心
centroid of storm rainfall 暴雨重心;暴雨中心
centroids of areas and lines 面积和线的形心
centron 原子核
centronics 并行接口标准
centronucleus 中心核
centrosome 中心体
centrosphere 重ітак;中心球;地心圈;地核
centrostigma 集中点
centrosymmetric(al) crystal 中心对称晶体
centrosymmetry 中心对称(性)
centrotaxis 趋中性
centrum (地震的)震源;中心(体)
cents-less accounting 整元会计
centum clause 保费基数条款
centum weight 英担
centuple 百倍的
centurial year 世纪年
century 百年;世纪
century schoolbook font 古教书字体
cenuglomerate 泥流砾岩;泥流角砾岩
cenuglomerite 泥流角砾岩
ceoefficient of spread 展宽系数
cephalad acceleration 向首加速度
cephalic module 头尺寸中数
cephalodium 地衣瘿
cephalograph 头描记器
cephalometer 头测量器
cephalometry 头测量法
cephalophorous 托住头的;主大梁;主梁
Cephalopoda 头足纲【地】
Cephalopoda limestone 头足类灰岩
cepstrum 对数逆谱;对数倒频谱;倒频谱
cepstrum vocoder 对数倒频谱声码器
ceptor 介体
cera 蜂蜡
cera alba 白蜂蜡
ceraceous 蜡状的;蜡色;似蜡的
ceracircuit 瓷衬底印刷电路
cera flava 黄蜂蜡
ceralumin 铝铸造合金
Ceralumin alloy 塞拉卢明合金
ceram 陶瓷器;陶器;陶瓷
ceramagnet 钡恒磁;陶瓷磁体
ceramal 金属陶瓷;金属合金陶瓷;合金陶瓷;陶瓷金属;陶瓷合金;烧结金属学
ceramet 陶瓷金属;烧结金属学
cerametallics 金属陶瓷;烧结金属学
ceramet bit 金属陶瓷刀头;金属陶瓷车刀
ceramet resistance 涂釉电阻
ceramic 陶瓷的;陶瓷的
ceramic adhesive 陶瓷基黏结剂;陶瓷胶黏剂
ceramic aggregate 陶瓷团粒;陶瓷集料;陶瓷骨料
ceramic amplifier 压电陶瓷放大器;陶瓷放大器
ceramic and refractory industry 陶瓷及耐火工业
ceramic armo(u)r 陶瓷铠甲
ceramic arrester 陶瓷避雷器
ceramic article 陶瓷制品
ceramic attenuator element 陶瓷衰减器元件
ceramic ball 瓷球
ceramic bearing 陶瓷轴承
ceramic binder 陶瓷黏合剂
ceramic bit 陶瓷钻头;陶瓷刀头
ceramic blade 陶瓷叶片
ceramic block 陶瓷砌块
ceramic body 陶瓷体;陶瓷坯体
ceramic body composition 陶瓷配料;陶瓷坯体组分

ceramic bond 黏土黏结料;黏土黏合剂;陶瓷黏结剂;陶瓷黏合剂;陶瓷结合剂
ceramic bonded Alumina Zirconia Silica brick 烧结锆刚玉砖
ceramic bonded refractory 烧结耐火材料
ceramic bonding 陶瓷结合
ceramic brake 陶瓷制动片
ceramic building material 陶瓷建筑材料
ceramic building member 陶瓷建筑构件
ceramic building unit 陶瓷建筑单元
ceramic burner 陶瓷燃烧器喷嘴
ceramic candle filter 陶瓷烛式过滤器
ceramic cap 陶瓷帽
ceramic capacitor 陶瓷电容器
ceramic cartridge 陶瓷拾音头;陶瓷拾音器芯座;陶瓷换能头
ceramic ceiling tile 陶瓷顶棚面砖
ceramic cellular block 陶瓷多孔砖
ceramic chamber lining 陶瓷衬套
ceramic chip 瓷片;陶瓷片
ceramic clay 陶瓷(用)黏土;陶瓷土
ceramic cleat 陶瓷夹板
ceramic-coated 敷有陶瓷的
ceramic coated cutting tool 陶瓷涂层切削工具
ceramic coated metal cutting tool 陶瓷涂层金属切削工具
ceramic coat(ing) 陶瓷敷层;陶质涂层;陶瓷涂层;陶瓷表层
ceramic cobalt blue 钴蓝颜料;瓷蓝(颜料)
ceramic colo(u)r 瓷器用颜料;陶瓷颜料;陶瓷色釉;陶瓷色料
ceramic colo(u)rant 陶瓷着色剂;陶瓷色材
ceramic colo(u)r glaze 陶(瓷)釉;陶瓷彩(色)釉(面)
ceramic column plate 陶瓷搭板
ceramic component 陶瓷构件
ceramic composite material 陶瓷复合材料
ceramic constituent 陶瓷成分
ceramic constitutional structure 陶瓷组织结构
ceramic core 陶瓷芯
ceramic cover with high shock degree 高冲击度用陶瓷罩
ceramic crayon 陶瓷蜡笔
ceramic crossflow microfiltration 陶瓷交叉流微滤
ceramic crucible 陶瓷坩埚
ceramic cup 陶瓷杯
ceramic cutting tools 陶瓷刀具
ceramic cylinder 陶瓷汽缸
ceramic delay line 陶瓷延迟线
ceramic die 陶瓷模
ceramic dielectric 陶瓷介质
ceramic dielectric(al) material 陶瓷介电材料
ceramic discharge pipe 陶瓷排水管
ceramic discriminator 陶瓷鉴频器
ceramic door knob 陶瓷门执手;陶瓷门把手
ceramic drain 陶瓷排水管
ceramic draining pipe 陶瓷排水管
ceramic dual in-line package 陶瓷双列直插式封装
ceramic encapsulation 陶瓷封装
ceramic engine 陶瓷引擎;陶瓷发动机
ceramic extrusion machine 陶瓷压挤机
ceramic-faced glass 陶瓷面玻璃
ceramic facing 陶瓷饰面
ceramic factory 陶瓷厂
ceramic fiber 陶瓷纤维;硅酸盐纤维
ceramic fiber blanket 陶瓷纤维毯
ceramic fiber board 陶瓷纤维板
ceramic fiber castable 陶瓷纤维浇注料
ceramic fiber felt 陶瓷纤维毡
ceramic fiber insulation 陶瓷纤维绝热层
ceramic fiber insulation material 陶瓷纤维绝热材料
ceramic fiber mat 陶瓷纤维编织物
ceramic fiber paper 陶瓷纤维纸
ceramic fiber rope 陶瓷纤维绳
ceramic fiber strip 陶瓷纤维带
ceramic fiber tape 陶瓷纤维带
ceramic filler 陶瓷填料
ceramic film 陶瓷薄膜
ceramic filter 陶瓷滤头;陶瓷滤器;陶瓷过滤器
ceramic filter dust separator 陶瓷过滤除尘器
ceramic filter media 陶瓷过滤介质
ceramic flat head 陶瓷扁平磁头
ceramic floor 陶瓷地面

ceramic floor cover(ing) 陶瓷地面覆盖层
ceramic flooring tile 陶瓷铺地砖;陶瓷地面砖
ceramic flux 陶质焊剂
ceramic foam 陶瓷泡沫
ceramic foam filter 泡沫陶瓷过滤器
ceramic former 瓷模
ceramic forming 陶瓷成型
ceramic fracture power 陶瓷断裂功
ceramic fracture toughness 陶瓷断裂韧性
ceramic fuel 陶瓷燃料
ceramic fuel element 陶瓷燃料元件
ceramic glass enamel 陶瓷玻璃釉
ceramic glaze 陶瓷釉;陶釉;陶瓷釉
ceramic glazed coat(ing) 陶瓷釉面
ceramic glazed sewer pipe 陶釉污水管;上釉下水管;上釉的下水管;上釉的水管
ceramic glazed sewer tube 上釉的水管
ceramic glazing 陶瓷涂釉
ceramic glossary 陶瓷词汇
ceramic grade talc 陶瓷级滑石
ceramic granular facing material 陶瓷粒状饰面材料
ceramic granules 陶瓷粒料
ceramic grinding ball 陶瓷磨球
ceramic grinding media 陶瓷研磨体
ceramic hastelloy 陶瓷耐蚀耐高温镍基合金
ceramic heat exchanger 陶瓷热交换器;陶瓷换热器
ceramic history 陶瓷史
ceramic hollow slab 陶瓷多孔板
ceramic honeycomb 蜂窝陶瓷;陶瓷蜂窝结构
ceramic ignitor 陶瓷点火装置
ceramic industry 窑业;陶瓷业;陶瓷工业
ceramic infrared emitter 陶瓷红外线发热器
ceramic infrared heater 陶瓷红外线加热器
ceramic ink 瓷贴花印刷油墨;陶瓷墨水(含陶瓷色料的墨水)
ceramic-insulated 陶瓷绝缘的
ceramic-insulated band heater 陶瓷绝缘带式加热器
ceramic-insulated coil 陶瓷绝缘线圈
ceramic insulating material 陶瓷绝缘材料;陶瓷绝热材料
ceramic insulation 陶瓷绝缘
ceramic insulation material 陶瓷绝缘材料;陶瓷绝热材料
ceramic insulator 陶瓷绝缘子
ceramic isolating material 陶瓷绝缘材料
ceramicite 陶土岩
ceramic kiln 陶瓷窑炉
ceramic knob 陶瓷执手;陶瓷把手
ceramic landscape 陶瓷盆景
ceramic lavatory 陶瓷洗手盆
ceramic lavatory basin 陶瓷盥洗室盆
ceramic lay-in panel 陶瓷嵌板
ceramic lifter 陶瓷扬料板
ceramic lighter 陶瓷点火器
ceramic-like 像陶瓷的;似陶瓷的
ceramic-lined chamber 陶瓷衬里的燃烧室;陶瓷燃烧室
ceramic-lined thrust chamber 陶瓷衬垫推力室
ceramic liner 瓷衬
ceramic lining 陶瓷衬料
ceramic loudspeaker 陶瓷扬声器
ceramic magnet 氧化物磁铁;铁磁陶瓷;陶磁体;陶瓷磁体;烧结磁铁
ceramic material 陶质材料;陶瓷原料;陶瓷材料
ceramic material molecular engineering 陶瓷材料分子工程
ceramic matrix composite 陶瓷基复合材料
ceramic membrane 陶瓷膜
ceramic membrane-biochemical process 陶瓷膜—生化工艺
ceramic membrane-bioreactor 陶瓷膜—生物反应器
ceramic membrane filtration 陶瓷膜过滤
ceramic membrane microfiltration 陶瓷膜微滤
ceramic membrane treatment 陶瓷膜处理
ceramic membrane ultrafiltration 陶瓷膜超滤
ceramic metal 金属陶瓷;陶瓷金属
ceramic-metal adhesive combination 金属陶瓷黏合制品
ceramic-metal composite 金属陶瓷复合材料
ceramic metallization 陶瓷金属化
ceramic metallizing 陶瓷金属化

ceramic metal system 陶瓷—金属系统
ceramic microfiltration membrane 陶瓷微滤膜
ceramic microphone 陶瓷扬声器;陶瓷传声器
ceramic microstructure 陶瓷显微结构
ceramic mosaic 马赛克;陶瓷砖;陶瓷马赛克;陶瓷锦砖
ceramic mosaic tile 陶瓷锦砖;马赛克
ceramic mo(u)ld 陶瓷铸型
ceramic mo(u)ld casting 陶瓷模注法
ceramic mo(u)lding material 制陶瓷型材料
ceramic mo(u)lding plyboard 陶瓷模制层压板
ceramic nuclear fuel 陶瓷核燃料
ceramic oil 陶瓷用油
ceramic package 陶瓷外壳;陶瓷封装
ceramic packing 陶瓷填料;陶瓷垫圈;陶瓷垫片
ceramic paint 陶瓷颜料
ceramic particle filtration 陶粒过滤
ceramic particle medium 陶粒介质
ceramic particles-manganese sand particles filter 陶粒—锰砂粒滤池
ceramic particulate reinforcing agent 陶瓷颗粒增强剂
ceramic paste 陶瓷坯泥
ceramic phase 陶瓷相
ceramic photography 陶瓷器摄照像术
ceramic pickup 陶瓷拾声器;钛酸钡陶瓷传感器;压电陶瓷拾声器
ceramic piezoelectric(al) microphone 压电陶瓷微声器
ceramic pigment 陶瓷色釉
ceramic pin grid array 陶瓷针型栅格阵列
ceramic pipe 缸瓦管;陶(土)管
ceramic plasma 陶瓷等离子体
ceramic plate 陶瓷片;陶瓷板
ceramic porous tube 陶瓷多孔管
ceramic post sintering 陶瓷烧结
ceramic pottery plaster 陶瓷用熟石膏
ceramic powder granulation 陶瓷粉体的造粒
ceramic preparation science 陶瓷制备科学
ceramic printing 陶瓷印花
ceramic process 陶瓷制造工艺
ceramic process pump 陶瓷化工泵
ceramic product 陶瓷产品;陶瓷制品
ceramic pump 陶瓷泵
ceramic raw material 陶瓷原料
ceramic raw material commodities 陶瓷原料矿产
ceramic reactor 陶质反应堆;陶瓷元件反应堆;陶瓷燃料堆;陶瓷反应器;陶瓷反应堆;陶瓷材料反应堆
ceramic receiver 陶瓷受话器
ceramic recuperator 陶质换热器;陶瓷换热器
ceramic resistor 陶瓷电阻器
ceramic resonant device 陶瓷谐振器件
ceramic ring 陶瓷环
ceramic rod flame spraying 陶瓷棒火焰喷涂法
ceramic rotor blade 陶瓷动叶片
ceramics 窑业;制陶术;陶管;陶瓷(制品);陶瓷业;陶瓷学;瓷瓦工艺
ceramic screen brick 陶瓷筛砖;栏杆砖
ceramics crucible 陶瓷坩埚
ceramic sculpture 陶瓷雕塑
ceramic seal 陶瓷密封;陶瓷封接
ceramic sensor 陶瓷传感器
ceramic sensor element 陶瓷传感器
ceramics flux 非熔炼焊剂
ceramics for daily use 日用陶瓷
ceramic shell mo(u)ld 陶瓷壳型
ceramic shower tray 陶瓷淋浴(浅)盆
ceramic sintering technology 陶瓷烧结术
ceramics magnet 烧结氧化物磁铁
ceramic solidification 陶瓷固化
ceramic sound-control brick 陶瓷声控砖
ceramic spray coating 陶瓷喷敷层
ceramic spray shower tray 陶瓷喷淋浴浅盆
ceramic stain 陶瓷色素;陶瓷色剂
ceramics-to-metal seal by Mo-Mn process 钼—锰法陶瓷金属封接
ceramic strip heater 陶瓷带加热器
ceramic structural material 陶瓷结构材料
ceramic structural member 陶瓷结构构件
ceramic substrate 陶瓷基片;陶瓷衬底
ceramic substrate glaze 陶瓷衬底釉
ceramic superconductor 陶瓷超导体
ceramic superplasticity 陶瓷超塑性

ceramic surfacing 陶瓷铺面
ceramics with low shrinkage 低收缩陶瓷;微收缩陶瓷
ceramics with super-plasticity 超塑性陶瓷
ceramic technology 陶瓷工艺学
ceramic tessera 镶嵌陶瓷细片
ceramic thermal protection system tile 陶瓷防热瓦
ceramic thick film 陶瓷厚膜
ceramic thin film 陶瓷薄膜
ceramic tile 瓷砖;陶瓷(面)砖
ceramic tile adhesive 陶瓷砖胶结剂
ceramic tile block 陶瓷砖块
ceramic tile facing 瓷砖贴面
ceramic tile finish 瓷砖面
ceramic tile floor 瓷砖地面
ceramic tile flooring 瓷砖(楼)地面
ceramic tile panel 陶瓷面砖
ceramic tip 金属陶瓷刀片;陶瓷测头
ceramic to metal seal 陶瓷金属封接
ceramic-to-metal seal by active metal process 活性金属法陶瓷—金属封接
ceramic tool 金属陶瓷刀具;金属陶瓷车刀
ceramic transducer 电致伸缩换能器;陶质换能器;陶瓷换能器
ceramic transfer picture 陶瓷贴花
ceramic transformer 陶瓷变压器
ceramic trimmer 微调瓷介电容器
ceramic tube 陶瓷管
ceramic type insulation 瓷绝缘
ceramic unit 陶瓷元件
ceramic vacuum tube 陶瓷真空管
ceramic varistor 陶瓷变阻器
ceramic veneer 陶瓷镶面;陶瓷饰面;陶瓷面砖;陶瓷板镶面;陶瓷板
ceramic vertical tiling 贴陶瓷面砖
ceramic vessel 陶瓷容器
ceramic wafer 陶片;陶瓷饰面板;陶瓷片;陶瓷基片
ceramic wall 陶瓷墙
ceramic wall tile 陶瓷墙面砖
ceramic ware 陶瓷制品
ceramic wash basin 陶瓷脸盆;陶瓷盥洗盆
ceramic washbowl 陶瓷脸盆
ceramic waste pipe 陶质污水管
ceramic whisker 陶瓷晶须
ceramic whisker composite 陶瓷晶须复合材料
ceramic whiteware 白瓷
ceramic window sill 砖窗槛
ceraminator 倍音检波元件;陶瓷压电元件
ceramin-metal tube 金属陶瓷管
ceramisito 陶粒
ceramisite clay 陶粒用黏土
ceramisite concrete 陶粒混凝土
ceramisite shale 陶粒页岩
ceramist 窑业家;陶瓷工作者;陶瓷工人
ceramodontia 瓷牙学
ceramography 瓷相学;陶瓷组织学;陶瓷结构学
ceramohalite 毛矾石
ceramoplastic 陶瓷塑料
ceramsite concrete wall panel 陶粒混凝土墙板
ceramsite pavement 陶粒路面
cerap 伴音中频陷波元件;陶瓷压电元件
cerargyrite 角银矿
cerarryrite 氯化银
cerasin(e) red 角铅矿红颜料
cerasus 樱桃树
cerate 铈酸盐;蜡剂;蜡膏
cerated 涂蜡的
ceraunograph 雷电计
ceravital 高硅钙生物玻璃系
cercis (希腊剧场阶梯过道间的)梯形座位群
cerdip 陶瓷浸渍
cere 蜡膜;黄蜡;涂蜡;上蜡
cereal 谷物;谷类的
cereal binder 面粉黏结剂
cereal crops 禾谷类作物;粮食作物
cereal expanding machine 谷物膨化机
cereal production 粮食生产;谷物生产
cereals 禾谷类作物
cereals and oil-seeds processing machinery 粮油加工机械
cereal testing machine 谷物试验机
cerebrose 半乳糖
cerebrovascular disease 脑血管疾病
cerein 蜡样菌素

ceremonial courtyard 礼仪庭院
ceremony 仪式;典礼
Cerenkov resonant effect 波粒共振效应
Cerenkov's detector 切伦科夫探测器
Cerenkov's effect 切伦科夫效应
Cerenkov's glass 切伦科夫玻璃
Cerenkov's radiation 切伦科夫辐射
cerepidote 褐帘石
Ceresan universal trockenbeize 灭菌硅
ceresin(e) 硬脆蜡;纯地蜡;微晶蜡;氯化乙汞
ceresin(e) wax 无定形蜡;防水剂;纯地蜡;防水粉;石蜡
ceria 氧化铈;铈石
cerianite 方铈石
ceric 高铈的
ceric hydroxide 氢氧化高铈
ceric sulfate 硫酸高铈
ceride 蜡脂;蜡类脂
cerificate of quantity 数量证(明)书
cerin(e) 蜡素
cerinite 杂白钙沸石
ceriopyrochlore 铈烧绿石
cerise 鲜红色;樱桃色;粉红色
cerise glass 宝石红玻璃
cerite 硅铈石;铈硅石
cerium-ankerite 铈铁白云石
cerium boride 硼化铈
cerium carbide 碳化铈
cerium dioxide 二氧化铈
cerium glass 铈玻璃
cerium hydroxide 氢氧化铈
cerium-lanthanum naphthenate 环烷酸铈镧
cerium magneside 镁化铈
cerium mischmetall 铈烯土
cerium naphthenate 环烷酸铈
cerium nitride 氮化铈
cerium ores 铈矿
cerium oxalate 草酸铈
cerium oxide 氧化铈
cerium rouge 氧化铈抛光粉
cerium stearate 硬脂酸铈
cerium-thermic reducer 铈热还原剂
cerium-thorium getter 铈钍吸气剂
cerium vanadate 钒酸铈
cermet 金属陶瓷(合金);合金陶瓷;硬质合金;陶瓷合金
cermet bit 硬质合金刀头
cermet button 金属陶瓷钮扣式摩擦块
cermet coat(ing) 金属陶瓷外层;金属陶瓷涂层
cermet contactor 金属陶瓷接触器
cermet facing 金属陶瓷饰面
cermet fibre 金属陶瓷纤维
cermet finish 金属陶瓷饰面
cermet resistance 金属陶瓷电阻
cermet resistor 金属陶瓷电阻器
cernada(h) 游廊
cernuous 俯垂的
cernyite 铜镉黄锡矿
cerography 蜡刻法;蜡版术
ceroid 蜡样质
cerolipoid 蜡脂质
cerolite 蜡蛇纹石
ceroma 浴室的人体涂油室(希腊、罗马)
ceromel 蜜蜡
ceroplastic 蜡塑的
cerorthite 铈褐帘石
cerosin(e) 地蜡;精制地蜡;蔗蜡
cerotate 蜡酸盐
cerotic acid 蜡酸
cerotin 蜡精;蜡醇
cerotinic acid 蜡酸
cerotungstite 铈钨华
cerous fluoride 三氟化铈
cerous hydroxide 氢氧化铈
cerous hypophosphate 次磷酸铈
cerous oxalate 草酸铈
cerphosphorhuttonite 铈磷硅钍石;磷硅铈钍石
cerreal 谷类货
Cerro 塞罗铋基低熔合金
cerrobase alloy 塞罗贝斯铋铅合金;铋铅合金;低熔点铅合金
Cerrobend alloy 塞罗本德合金
cerrogreen 角须绿色
Cerromatris alloy 塞罗马特里克斯合金

Cerruti problem 塞露蒂课题
cers 塞尔斯风
certain 必然的
certain annuity 确定年金;确定固定年限的年金
certain event 必然事件
certain micro-organisms 某些微生物
certain plants 某些植物
certainty 必定的事情;必然性;确定性;可靠性
certainty and valuewise independence 确定性与值态的独立性
certainty decision 确定性决策
certainty equivalence 确定性等价;确定当量
certainty equivalents 确定等值
certainty of measurement 测量精(确)度
certificate 检定书;检定;许可证;执照技术合格证;执照;证书检验证;证书;证明书;证明;证件
certificate and list of measurement and (or) weight 容量重量证明书
certificate approval of works 工程核准证明书
certificate backed mortgage 存款单担保抵押贷款
certificate book 股票簿
certificate deposit 证券存款
certificated lifeboatman 有证艇员
certificated master 合格船长
certificated of reasonable value 合理价值证明书
certificated quality level 质量保证书
certificated stock 已检验的存货
certificate fee 签证费
certificate fo analysis 化验证(明)书
certificate for alteration of export permit 出口许可证修改证明书
certificate for cargo gear 起货设备证书
certificate for completion 竣工证书;竣工证明
certificate for export 出口检验合格证书;输出合格证
certificate for fire extinguishing and detecting apparatus 消防及火警设备证书
certificate for import inspection 进口检查合格证
certificate for insurance 保险单
certificate for payment 付款凭证;付款证明(书)
certificate for refrigerating apparatus 冷藏设备证书
certificate issue voucher 商品出口证明书;商品产地证明书
certificate market 结汇证券市场
certificate of acceptance 接纳证(明)书;验收证(明)书;验收合格证(书);验收单
certificate of agent's authority 代理权证书
certificate of allocated exchange 结汇证
certificate of amount owing 欠款金额证明书
certificate of analysis 分析证明书
certificate of analyzing 化验证(明)书
certificate of appointment 委托证书;委任状
certificate of appraisal 估价证书
certificate of approval 批准证书;设备合格证
certificate of approval of the marking 载重线证书
certificate of auditing 审计证明书
certificate of authority 许可证;准许证书;授权书
certificate of authorization 许可证;准许证书;授权证书;授权书
certificate of authorization of code symbol stamp 规范符号标志的认可证书
certificate of average 海损理算书
certificate of balance 存款凭单
certificate of balance sheet 资产负债表证明书
certificate of bank balance 银行存款的证明
certificate of birth 出生证
certificate of breakage 货物毁损证书
certificate of business 营业执照;营业证
certificate of chemical analysis 化学分析证明书
certificate of clearance 结关证(书);输油证
certificate of competency 能力证书;合格执照;合格证(书);资格证(明)书
certificate of completion 竣工证书;竣工证明;竣工合格证书;完工证书
certificate of completion by stages 分期竣工证书
certificate of completion of handling over 交接证明书
certificate of completion of works 工程竣工证书
certificate of compliance 合格证(书);一致的证明;符合证明(书)
certificate of conformity 合格证明书;证明商品符合合同中的规格的证书
certificate of corporation 法人认可证
certificate of credit standing issued by bank 银

行出具的资信证明书
certificate of damage 货损证明;残损证明(书);损失证明书
certificate of date of sailing 开航证明书
certificate of death 死亡证
certificate of deduction of interest tax 扣除利息税证明书
certificate of delivery 交货证(明)书;交货运单;交货单;交付证书;交船证书
certificate of delivery and redelivery 交船与还船凭证(租船)
certificate of departure port 离港证明书
certificate of deposit 存款证书;存款单;押金收据;存单
certificate of discharge 离职证书
certificate of disinfection 消毒证书
certificate of domestic container 国产集装箱证书
certificate of eligibility 合格证书
certificate of entrustment 委托书
certificate of environment(al) impact assessment 环境影响评价证书
certificate of estoppel 禁止翻供证书(贷款人签署有关欠款数额及利率的法律文件)
certificate of exchange settlement for import 进口结汇证书
certificate of expenditures 支出证明书
certificate of export inspection 出口检验合格证明书
certificate of financial standing 金融状况资信证明
certificate of fitness 适装证书
certificate of freeboard 干舷证书
certificate of fumigation 熏舱证书
certificate of health 健康证书;无疫通行证
certificate of historic(al) buildings 历史性建筑保护证明
certificate of identification 身份证
certificate of import 进口证明;进口许可证
certificate of import licence 特许进口证明书
certificate of incorporation 公司注册证;公司登记执照;登记证明书;法人认可证
certificate of indebtedness 借据
certificate of indebtness 一年期的公债
certificate of independent public accountant 独立会计师查账报告
certificate of inspection 检验合格证;检验证(明)书;合格证(书)
certificate of insurance 保险证(明)书;保险凭证;保险单;保险保证书
certificate of interest 息票
certificate of inward clearance 进口检查证;海关检验单
certificate of license and enrollment 航业执照
certificate of lifeboatman 救生艇艇员合格证书
certificate of load line 载重线证书
certificate of loss 损失证明书
certificate of loss or damage 残损证明(书);损失或损坏证明
certificate of manufacture 制造许可证;制造商证书;工厂证明书;出厂证明书
certificate of manufacturer 制造厂证明书;制造厂家证明书
certificate of measurement 木材尺寸检查证(美国);吨位丈量证书
certificate of merchandise 出厂证(书)
certificate of mortgage 抵押证书
certificate of nationality 国籍证书
certificate of necessity 需要证书
certificate of occupancy 建筑使用证明书;占有许可证;符合使用证;使用证书;居住证
certificate of origin 产品来源证明;产地证(明)书;商品出口证明书;商品产地证明书
certificate of originals 原产地证明书
certificate of ownership 所有权证(明)书
certificate of patent 专利执照;专利证(书)
certificate of payment 支付证书;领款证明书;领款凭证;支付凭证;付款证明书;付款凭证
certificate of port of registry 船籍港证书
certificate of possession of land 土地所有权证书
certificate of pratique 检疫证(书);(检疫后发给的)入港证书;无疫通行证
certificate of pressure test 压力试验证(明)书
certificate of proof 试验证(明)书
certificate of protection 保护证书(美国)
certificate of protest 拒绝证书;拒付证书;海事报告

certificate of public necessity 公需证书
certificate of purchase (房屋、房地产等的)购买证书
certificate of purchase A or B 采购证明书 A 式或 B 式
certificate of purity 纯度证明书
certificate of quality 质量证(明)书;品质证(明)书
certificate of quality or quantity 品质或数量证明书
certificate of quality test 质量检验证书;品质试验证明书
certificate of receipt of export prepayment 出口预收货款证明书
certificate of redemption 购买证书
certificate of registration 注册执照;注册证(书)
certificate of registry 登记证书;船籍证书;船舶登记证
certificate of reinsurance 分保单
certificate of sale 销售证书
certificate of service 送达证明书
certificate of settlement of import drafts 进口结汇证书
certificate of share 股票
certificate of shipment 装船证书;出口许可证;发运证书
certificate of ship's inspection 船舶检验证书
certificate of soundness 合格证书
certificate of specification 规格证书
certificate of survey 检验(明)书
certificate of tare weight 皮重证明书
certificate of tax payment 纳税证明书;完税证
certificate of test 试验证书;检验证书
certificate of testing 检验证书
certificate of the auditor 审查报告
certificate of title 所有权证;产权证书;物权证书;所有权凭证
certificate of valuation 估价签定书
certificate of value 货价证书
certificate of weight 重量证(明)书;重量检定证书
certificate of welder's quality 焊工合格证书
certificate on damaged cargo 验残证书
certificate on hold 验舱证书
certificate on progress 进度证明书
certificate on tank 验舱证书
certificate test 检定试验
certificate to practice license 执业证书
certificate(trade) mark scheme 商标图式证(明)书
certificate vaccination 检疫证(书)
certificate value 认可值
certification 检验;凭单的发给;书面证明
certification activity 认证活动
certification and Watchkeeping for Seafarers 国际船员训练、发证和值班标准公约
certification body 认证机构
certification by the competent authority 主管当局的证明书
certification by vessel manufacturer 容器制造厂的出厂合格证明
certification fee 认证费
certification flight test 动力装置飞行试验
certification mark 商品标志
certification marking of product quality 产品质量认证标志
certification of authorization 核准证书;委任证明
certification of bargaining agent 谈判代理人合格证
certification of carriage 运输证明书
certification of completion 竣工合格证书
certification of fitness 合格证(书);质量合格证
certification of historic buildings 历史性建筑保护证明
certification of inspection 检验证书;产品合格证
certification of materials and products 材料与制品鉴定
certification of non-destructive personnel 无损检验人员证明书
certification of product 产品证明书
certification of proof 检验证书
certification of service 服务证明书
certification on tax payment 纳税证明书
certification procedure of product quality 产品质量认证程序
certification quality 合格证(书)
certification requirement 认证要求
certification scheme 认证计划

certification system 认证体系
certification system of product quality 产品质量认证制度
certification test 鉴定试验;检验试验;合格检验;合格检查;验证试验;发照考试
certified account 已查证账目;已查核账目;经会计师证明的账目
certified accountant 注册会计师;审定会计师
certified aircraft 有执照的飞行器
certified assessment evaluator 持证评估员
certified balance sheet 经会计师证明的资产负债表;已查证资产负债表;已查核资产负债表
certified ballast 检定合格的镇流器
certified board resolution 有保证的委员会决议
certified burette 检定滴定管
certified check 保付支票;认付支票;签证支票
certified cheque 保付支票
certified colo(u)rs for food 合法食用色素
certified concentration 合格浓度
certified copy 经核证的抄本;经过证明的副本;抄件
certified copy documents 文件的正式副本
certified correct 证明无误
certified crew list 经签证的船员名册
certified documents 公证文件
certified drainage ditch 合格排水沟
certified emission reductions 核证的排放消减量
certified emulsion 合格乳剂
certified engineer 合格轮机员
certified financial analyst 合格财政分析家
certified financial statement 经会计师证明的财务报表
certified fire lane 合格防火线
certified fumigant 合成熏蒸剂
certified helitanker 合格灭火直升飞机
certified horticultural tractor 合格园艺拖拉机
certified internal auditor 内部审计师
certified international property specialist 注册国际财产专家
certified invoice 证实发票;签证发票
certified management accountant 公证执照管理会计师
certified marcottage 合格压条法
certified officer 合格驾驶员
certified pollution control facility 合法污染控制设备
certified products 正品
certified protest 拒绝证书
certified public accountant 会计师;有证会计师;执业会计师;公证会计师;特许会计师
certified reagent 检定试剂
certified seed 合格种子
certified signature 经验证的签名
certified smoothing harrow 合格平地耙
certified sod 合格草皮;优质赛马场草皮
certified soil particle 合格土粒
certified soil protection wood 合格林地保护树
certified standard 合格基准
certified statement 经审定的财务报表
certified submersible sump 合格地下泵
certified test block 合格试验区
certified tool list 检定合格的工具单
certified transfer 经签证的转让;经核证转让
certified two-wheeled double share plough 合格多铧犁
certified union 特许工会
certified value 认可值
certified volumetric(al) flask 检定量瓶
certified water bombing 合格飞机投水弹
certified weights 检定砝码
certifier 证明者;证明人;保证人
certifying bank 保付银行
certifying officer 核证人
certiorari 复审令
certosa 卡尔特教会的修道院
certosec 抗凝结的漆
centre-driven antenna 中点馈电天线
certus glue 酪蛋白黏结剂
cerulean 天蓝色;蔚蓝色
cerulean blue 青天蓝;锡酸钴蓝;钴天蓝;天蓝(色)
cerulean pale 淡青
ceruleite 块砷铝铜石
ceruleolactite 微晶磷铝石
cerumen 巢质

ceruranopyrochlore 铈烧绿石
ceruse 铅白(碳酸);碳酸铅
cerussite 白铅矿;天然碳酸铅
cervantite 黄锑矿
cervinus 红灰色
ceryl alcohol 蜡醇
cesanite 钙钠矾
Cesaro convergent series 蔡查罗收敛数列
cesarolite 泡锰铅矿
Cesaro sum(mation) 蔡查罗(求)和
cesbronite 羟碲铜矿
cesium bichromate 重铬酸铯
cesium chromate 铬酸铯
cesium clock 铯钟
cesium deposit 铯矿床
cesium hydrate 氢氧化铯
cesium hydroxide 氢氧化铯
cesium iodide 碘化铯
cesium-kupletskite 铯锰星叶石
cesium monoxide 一氧化铯
cesium ore of pegmatite type 伟晶岩型铯矿石
cesium ores 铯矿
cesium oxide 氧化铯
cesium-oxygen-filled cell 充气铯电池
cesium perchlorate 过氯酸铯;高氯酸铯
cesium salt 铯盐
cesium source 铯源
cesium vapo(u)r magnetometer 铯蒸汽磁力仪
cespitose 丛生的
cess 多孔排水管
cessation 终止
cessation of business 停业
cessation of infringement 停止侵害
cessation of liability 终止责任
cessation of works 工程停建
cessation reaction 链终止反应;终止反应
cesse clause 责任终止条款
cesser 权利终止期
cessing 油漆表皮的小孔;油漆表皮的污点
cessio bonorum 以货抵债;放弃财产
cessio legis 法定让与
cession 财产权利等的转让;分出项目(保险)
cessionary 财产权利等的受让人
cessionary bankrupt 以产业供债权人分配者
cesspipe 粪管;污水管
cesspit 雨水斗;污水渗井;落水斗;粪坑;粪池;污水储[贮]池;污水坑
cesspool 粪坑;粪池;污水储[贮]池;污水渗井;污水坑;污水池;渗井
cesspoolage truck 污水车
cestibtantitie 铯锑钽矿
cestui que trust 信托资产收益受益人
Cetacea 鲸目
cetacean 鲸目动物(的);鲸类
Cetal 赛达铝锌合金(6.5%硅,3%铜,10%锌,余量为铝)
cetane 鲸蜡烷;十六烷
cetane number(of diesel oil) 十六烷值
cetanol 鲸蜡醇
cetene 鲸蜡烯
cetene number 十六烯值
ceti 缅甸寺院的塔
cetificate of monthly payment 月支付证书
cetoleic acid 鲸蜡烯酸
cetuplicate 百倍
cetyl acetate 鲸蜡醇醋酸酯;醋酸十六烷酯
cetyl alcohol 鲸蜡醇
cetylic acid 棕榈酸;软脂酸
cevadine 瑟瓦定
ceylan cedar 南岭楝树;楝木
Ceylon ebony 斯里兰卡乌木
Ceylon graphite 斯里兰卡天然石墨
Ceylon ironwood 斯里兰卡铁力木
ceylonite 镁铁尖晶石
Ceylon kino 斯里兰卡吉纳树胶
Ceylon satinwood 东印度椴木
Ceylon zircon 斯里兰卡锆石
C-factor C 因素
C-frame hydraulic press 单臂式液压机
c.g. datum 重心基准点
c.g. limits 重心极限
CG quartz-spring gravimeter CG 石英弹簧重力仪
chabasite 菱沸石
chabka 古干水道

chabourneite 硫砷锑铅铊矿
chace air meter 袖珍含气量测定仪
chad 冲屑;穿孔纸屑;石砾;砂砾
chadacryst 客晶;捕虏晶;捕获晶【地】
chadded tape 穿孔带;半孔屑纸带;无孔屑纸带
chadless paper tape 半穿孔纸带
chadless perforation 半穿孔
chadless tape 无孔屑纸带;部分穿孔纸带;半孔屑纸带;半穿孔带
chadrac (电化钢处理过的)防火玻璃
chaetoplankton 硬毛浮游生物
chafe 摩擦
chafed yarn 擦伤纱
chafe mark 擦伤痕
chafer 胎圈包布
chafer house 酒店(英旧称)
chaff (干扰雷达用的)金属带;敷金属纸条;箔条;箔片
chaff cutter 铡草机;切草机
chaff cutter thrower 茎叶切碎抛送机
chaff device 雷达干扰装置
Chaffee spark-gap 查菲放电器
chaffer 讲价钱;讨价还价
chafferer 议价人
chaff peat 植屑泥炭
chaff slicer 铡草机
chaffy wool 草屑毛
chafing batten 防擦木板
chafing block 防擦块
chafing board 防擦木板
chafing chain 拖索防擦段(一段铁链)
chafing check 擦损检查
chafing corrosion 摩擦腐蚀
chafing dish 火锅;暖锅
chafing fatigue 磨蚀疲劳
chafing gear 防摩擦装置;防擦装置
chafing mat 防擦席;防擦垫
chafing net 防擦网
chafing path 摩擦挡布
chafing piece 防擦网
chafing plate 防擦板;包角铁板
chafing plate at well side 桥挡护木
chafing strip 防磨条
Chagos trough 查戈斯海槽
chagrenate-corrugate 粗皱缩
chain 链状;链条;链式连接;链;连锁;线性有序集;工程测链;台组;锁链
chain across ship's head 链过船头
chain addition program(me) 链加程序
chain addressing 链式寻址;链式访问
chain adjuster 链(条)调整器
chain adjusting screw 链调节螺钉
chain adjustment 网锁平差
chainage 测链数
chain aggregation 链聚集
chain anchorage 链锚固
chain and backet elevator 环链式斗式提升机
chain and barrel 链条与圆筒;安全门链
chain-and-bucket elevator 链斗式升运器
chain(-and-)bucket excavator 多斗挖掘机
chain and chisel mortise machine 链凿式榫眼机
chain-and-drag 链条牵引
chain-and-ducking dog mechanism 拖运机
chain-and-flight conveyer 链刮板输送机;链板运输机
chain-and-flight type stationary elevator 链叉式固定升运器
chain-and-salt unloader 链板式卸载输送机
chain-and-scraper conveyer 链式铲运机
chain-and-segment linkage 分段传动装置
chain-and-slat type elevator 链耙式升运器
chain and sprocket 链和链轮
chain-and-sprocket drive 链条齿轮传动;链和链轮传动
chain-and-tackle 链滑轮
chain-and-tackle block 链滑轮组;链条起重滑车
chain average 连锁平均数
chain axle 链轴
chain balance 链码天平
chain banking 连锁银行
chain barker 链式剥皮机
chain barrel 链筒;链卷筒
chain barrier 链栅栏

chain bar spreader 链式粗梳机
chain belt 链带;传动链
chain-belt conveyer veneer drier 链带式单板干燥机
chain billet switch 链式坯料分配器
chain bin 锚链舱
chain block 起重葫芦;吊链;链条滑车;链条葫芦;链式差动滑轮;链式差动滑车;链滑轮组;链滑车;链动滑轮;拉链起重器;倒链;手拉葫芦;神仙葫芦
chain block and tackle 链式滑车
chain block crane 链滑车起重机
chain boat 港内运送长期系泊设备的小船;港口运送长期系泊设备的小船
chain bolt 链螺栓;带链销;带链插销
chain bond 链式砌合(法);链式搭接;砌合夹钳;埋铁条砌体
chain book 野外记录簿;野外记录本
chain box 链箱
chain brake 链闸;链条制动机;链犁
chain branching 链支化;连锁分枝
chain break 链断裂
chain breakage 链锁中断;链断裂
chain breast coalcutter 链式短壁截煤机
chain breast machine 短工作面链式截煤机
chain bridge 链式吊桥;吊桥;链(悬)桥;链索吊桥;链式悬桥
chain bridle 马鞍链
chain broaching machine 链形拉刀拉床
chain broadcasting 联播
chain bucket 链斗;提料斗
chain bucket car unloader 链斗卸车机
chain bucket dredger 链斗式挖泥机;链斗式挖泥船
chain bucket elevator 链斗式提升机;链带式升运机;链式升降机
chain bucket excavator 链斗式挖土机;链斗式挖泥机;链斗式挖掘机;链斗式开沟机
chain bucket loader 链斗(式)装载机;链斗装料机;链斗式装载机
chain bucket sampler 链斗式采样器
chain bucket ship unloader 链斗卸船机
chain bucket trencher 链斗式挖沟机
chain cable 链索
chain cable certificate 锚链证书
chain cable common link 锚链普通链环
chain cable compressor 制动链;掣链器
chain cable controller 制动链;掣链器
chain cable end link 锚链端环
chain cable fairlead 导链滚轮
chain cable fairleader 锚链导向器
chain cable mark 锚链标记
chain cable proving house 锚链检验室
chain cable shackle 锚链卸扣;锚链连接卸扣
chain carrier 链锁载体;链锁进位;传链子;反应活性中心
chain carry 链锁进位;循环进位
chain case 链传动箱
chain catenary 锚链悬链线
chain-cessation 链终止(作用)
chain chair (悬索桥的)吊链锚座
chain claw conveyer 链爪输送机
chain clip 链卡子
chainclosure 链锁合
chain coal cutter 链式截煤机
chain code 链式(编)码;循环码
chain colllector 链式刮泥器;链式刮泥机
chain combination 链结合
chain complex 链复形【数】
chain compound 链条复合组;链化合物
chain compressor 掣链器
chain configuration 链构型
chain connection 链连接;联结;串级连接
chain contact 链动接点
chain continuous broaching machine 链形拉刀连续拉床
chain contract 连锁合同
chain conveyer 链式输送器;链式运输机;链条搬动机;链条输送机;链式运送器;链式传送机;链传送带;链板式输送机
chain-conveyer for basket 链式吊篮输送机
chain-conveyer furnace 链式炉
chain conveyer type furnace 传运链式炉
chain conveyor for basket 链式吊篮输送机
chain coupling 链形连接(器);链式联轴节;链耦合

chain course 夹钳砌合的露头石层;钳砌石层;拉结层
chain cover 链(条)罩
chain crab 链起重绞车
chain creeper 链条输送器
chain crowd(ing) 链板推进
chain curtain 链条帘幕;粗料定量控制器;链式闸门;链幕
chain curtain feeder 粗料定量送料器
chain cut-off saw 链截锯
chain cutter 链条拆卸器;链条拔销器;链式切削器;链式切碎机;链式切割机;割管器
chain cutter bar 链式截盘
chain cutter mo(u)lding machine 链刀造型机;链刀起槽机
chain cutting machine 链式截煤机
chain data 链式数据;数据链
chain data flag 链式数据标志;数据链标记
chain debts 三角债
chain deep 链海渊
chain-deformation 链形变
chain delivery 链式传送
chain delivery mechanism 链式输送机构;链式送纸机构
chain density 链条密度
chain detacher 链条拆卸器
chain device 链形装置
chain dimensioning 链式尺寸标注
chain direction 锚链方向
chain discount 连锁折扣;连环折扣
chain dog 链条扳手;链钩;钳子;拔钉钳
chain door fastener 链式门扣件;门的扣链;安全门链
chain-dot 链形网点
chain-dot line 点划虚线
chain dot screen 链形网屏
chain-dotted line 点划线
chain drag 链式刮器
chain draw bridge 链式活动桥
chain dredge(r) 链式挖泥船;链式挖泥机
chain dryer 链式干燥窑
chain drive 链条传动;链齿轮传动
chain-drive lubricant 传动链润滑剂
chain-drive mechanism 链传动机构
chain-drive motorcycle 链动机器脚踏车
chain drive(n) 链传动
chain-driven elevator 链传动升降机
chain-driven planter 链条传动式栽植机
chain-driven rotary table 链动转台
chain-drive rotary 链条传动的转盘
chain-drive section 机头部
chain-drive table 链传动的转盘
chain-drop hammer 链式落锤
chain drum 链条卷筒;卷链筒
chain dryer 链式干燥器
chained command 链接命令
chained command flag 链接命令标志
chained file 链式文件;链接文件
chained library 链式程序库;链接库
chained list 链(式)表;链接表;连接表;系列性表格
chained program(me) 链接程序
chained program(me) access method 串行程序的访问方法;串行程序存取法
chained record 链式数据;链式记录;链接记录;连锁记录
chained scheduling 链接调度
chained sector 链接区段
chained segment buffer 链式段缓冲器
chained sequential operation 链锁程序操作
chain effect 连锁效用;连锁效应
chain element 链节
chain elevating 链索提升
chain elevator 链式升运机;链式升降机;链式提升机
chain equipment 链式设备
chainer 方形隅石块
chain extender 扩链剂;增链剂
chain extension 链延长
chain fall 链绳;链式差动滑轮;链式差动滑车;倒链
chain feature 链式特征
chain feeder 连续加料器;链式喂料机;链式料斗;链式进料机;链式加料器;链式给料机;悬挂式喂料机
chain feed(ing) 链式加料;连续加料;链条给进;链带送料

chain fender 链式防撞装置;防撞链
chain fendering 链式防撞装置
chain ferry 缆索牵引渡口;缆索牵引渡船;拉链渡船;索链渡船
chain figure 链条形图案
chain filter 链型滤波器;链形滤波器;多节滤波器
chain fission yield 裂变链产额
chain fittings 锚链配件;锚链附件;链装配件
chain flail stirrer 链式连枷状搅拌器
chain flight feeder 链板给料器
chain follower 后面的持链人;(用链测量长度时的)后链
chain for fender 护木挂索
chain fracture zone 链破裂带
chain free radical reaction 链自由基反应
chain gate feeder 链幕喂料机
chain ga(u)ge 链绳式水位计;测绳式水位计;链式水(标)尺;测深链
chain gear 链传动;链齿轮传动装置;链齿轮;传动链
chain gear for drum shaft 转鼓轴链轮
chain gearing 链条传动装置;链传动装置;链齿轮传动装置
chain grab 起重葫芦;吊链;锚机持链轮
chain grab bucket 链式抓桶;链式抓具
chain grate 链条炉排;链条炉算;链式炉排;链式炉算;链炉算
chain grate stoker 链栅加煤机;链条炉排加煤机;链式炉排加煤器;链式炉算加煤机
chain grease 链条润滑脂
chain grip jockey 矿车的无极绳抓链
chain grit 链式筛
chain grizzly 链筛
chain group 链群
chain growth polymerization 链增长聚合(作用)
chain guard 链罩;护链槽
chain guide 链条导板
chain hanger 悬链吊架;链式吊钩
chain harrow 链式耙;链耙
chain hoist 链式起重机;神仙葫芦;链式起重滑车;链式启闭机;链式滑轮组;链式差动滑轮;链式差动滑车;倒链;手动葫芦
chain home beamed 制导系统
chain homomorphism 链同态【数】
chain hook 链(斗)钩;锚链钩
chain hung type screen 悬链(式)筛
chain hydrometer 链式比重计
chain image 循环图像
chain-in 链通道输入
chain index 链索引;连续指数;连锁指数;环比指数
chain index number 连锁指数
chaining 链锁;链接;链环执行;链测;用链丈量;丈量;给车轮装链;成捆拖集;长度丈量;拖链除灌
chaining address 链锁地址;链式地址
chaining arrow 测针【测】
chaining buck 拉力架
chaining check 链锁检验;链接检查
chaining error 丈量误差
chaining overflow 链式溢出
chaining pin 测钎
chaining research 链式检索
chaining search 链式检索;链接检索;循环检索
chain initiation 链引发;链起始作用;连锁开始
chain-initiation reaction 链引发反应
chain initiator 链引发剂
chain insulation 链状绝缘子
chain insulator 绝缘子串
chain intermittent fillet weld 链式断续贴角焊;链型断续贴角焊;链式分段角焊;并列间断贴角焊;并列间断角焊缝
chain intermittent fillet welding 链式断续角焊;并列断续角焊缝
chain intermittent weld 并列焊接
chain interruption 链断裂
chain iron 链环
chain islands 列岛
chain jack 链式千斤顶;链式起重器;链式起重机(俗称神仙葫芦);链式顶重器
chain jib 链式截盘
chain job 链式作业
chain joint 链连接;链接
chain kep 链式罐笼座
chain knot 链节
chain-leader 前链员【测】
chain leading 锚链方向

chain leading aft 锚链朝后
chain leading forward 锚链朝前
chain legs 双链钩
chain length 链长
chain lifter 链式升运机
chain lightning 链状闪电
chain like structure 链状构造
chain like texture 链状结构
chain-like veins 锁链状矿脉
chain line 点划线
chain link 链索;链节;链接链;链环;锚链环
chain linkage 链索
chain link door 钢丝网门
chain-linked conveyer 链式输送机
chain link fence 钢丝网围栏;链状栅栏
chain link fencing 链接围栏
chain link record 链式连接记录
chain loader 链式装载机
chain loading station 输送链装料站
chain lock 链锁;闭合链节
chain lockage 锚链舱
chain locker 锚链舱
chain locker pipe 锚链管
chain lubrication 链注油;链润滑
chain macromolecule 链状大分子
chain making machine 制链机
chainman 测量员;测链员;司链员
chain-mapping 链映像
chain mark 链带痕;钢丝外露
chain measure 链测
chain measuring 链测
chain mechanism 链锁机构
chain mesh of roof-bolt 护顶锚杆金属网
chain method 链锁法;链法
chain migration 链锁迁移;链式迁移;连锁性迁移
chain mobility 链移动性
chain mortiser 链式开榫机
chain mortising machine 链式开榫机
chain motif(pattern) 链状基型
chain mo(u)lding 链饰(线脚);链形线脚
chain network 链形网络;链式网络
chain of bucket 斗链
chain of bursts 爆发链
chain of causation 因果链
chain of command 控制面;控制跨度;命令系统
chain of cones 链状火山锥群
chain of egg insulators 蛋形绝缘子串
chain-off thread back-tacking 辫线夹子
chain of hinges 铰接链
chain of islands 列岛;群岛
chain of lakes 成串湖泊;湖系;湖泊群
chain of locks 多层闸室;多层船闸;梯级(式)船闸
chain of lozenge 菱形锁;用于辐射三角测量的菱形锁
chain of mountains 山系;山脉;山峦
chain of one regular 单调节环
chain of pot 斗链
chain of power plants 梯级电站
chain of rings 连环状的
chain of several regulator 多调节环
chain of simplices 单形链
chain of stations 台链
chain of stirred tanks 串联搅拌釜
chain of triangles 三角锁
chain of triangulation 三角网系;三角测量锁
chain-oiled bearing 油链润滑轴瓦
chain oiler 链条加油器
chainomatic balance 链条天平;链码天平
chain on the beam 锚链在正横
chain on the bow 锚链在首舷方向
chain on the quarter 锚链在尾舷方向
chain-operated ferry 拉链渡船
chain operation system 连锁运作体系
chain orientation 链取向;链定向
chain oven 链式加热炉
chain pendant 吊灯;灯具吊链;拖缆短链;链条吊灯
chain pendant lamp 吊链灯
chain pendant type 链吊式
chain pin 链销;链测标杆;测钎
chain pipe 锚链管
chain pipe spanner 链条管扳手
chain pipe tongs 链钳
chain pipe vice 链台钳;链式台钳
chain pipe wrench 链钳;链条管(子)钳;链条管

（子）扳手；链式管子扳手
chain pitch 链节距
chain plank conveyer 链板式输送机
chain plate 桅侧支索牵条
chain-plate conveyer 链板式喂料机
chain plot 测链样方
chain polymer 链形聚合物
chain procedure 联锁分类法
chain program(me) 链接程序
chain propagation reaction 链增长反应
chain proved 锚链检验合格
chain pull 链条张紧力
chain pulldown feed 链条强制给进
chain pulley 链式滑轮组；链式滑车；链（滑）轮
chain pulley block 链滑轮组
chain pulley collar 链轮环
chain-pull switch 拉线开关
chain pump 链式泵；链斗式水车；链斗式提水器；链泵
chain radar beacon 多雷达信标
chain rail 链栏
chain railway 链索轨道
chain-reacting pile 链式反应堆；原子反应堆
chain reaction 链锁反应；链式反应；链反应；连锁反应
chain-reaction bankruptcy 连(锁)破产；破产连锁反应
chain reaction pile 链式反应堆
chain reflex 链(锁)反射；连锁反射
chain relative 连锁比例
chain relay 连锁继电器；串动继电器
chain repercussion 连锁影响；连锁银行
chain return cylinder 带链传动的液压缸
chain ring 锚链环
chain rivet 并列铆钉；链铆钉
chain-riveting 并列铆(接)；并列铆钉；排钉；平行铆接；平行铆钉；链型铆接；链式铆钉
chain-rivet joint 并列铆钉；并列铆钉接合
chain rivet remover 链条拔销器；拆链器
chain roll 链辊
chain rule 链锁法；链式法则；连锁法；测链（长 66 英尺）（1 英尺＝0.3048 米）
chain rupture 链锁中断；断链
chains 联营企业
chain saddle （悬索桥的）吊链支座
chain saw 叠锯；链（式）锯
chain saw file 链锯锉刀
chain scale 链尺
chain scheduling 链式调度
chain scission 断链(作用)
chain scope 出链长度
chain scraper 链式刮泥器；链式刮泥机
chain screen 链屏
chain seaming 链缝
chain search 链式检索；链接检索
chain section 链条区
chain segment 链段
chain sequence 链顺序
chain shackle 链环
chain shaker 抖动链式升运器
chains hanging in festoon 花环链
chain sheave 链卷筒；链滑轮
chain shop 锚链车间
chain shotblasting machine 履带式喷丸机
chain sieve 链式筛
chain silicate 链硅酸盐
chain slark 链条垂度；链条弛度
chain sling 悬链；链式吊索；链环；链钩；吊梁；吊链
chain sling with two ears 双扣吊链
chain sounding line 测深链
chain splice 链条绞接
chain sprocket 铰链；扁环节链；链轮
chain stitch machine 锁缝机
chain stoker 链式加煤机；链条炉排
chain stopped alkyd 断链醇酸树脂；止链醇酸
chain stopper 锚链制动器；制链器；掣链器
chain store 联营商店；联号商店；连锁店；连锁商店
chain structure 链式结构；链式结构；网形结构
chain survey(ing) 距离测量；链测；测链丈量；测链测量
chain suspension bridge 链悬吊桥
chain swivel 锚链旋转接头；链锚旋转环
chain system 链型系统；连锁制；链型坞系统（船坞）

chain system installation in cement kiln 水泥窑的链条串联系统设施
chain system of docks 串联式港池
chain taker pipe 锚链管
chain tape 测量链尺；链尺；测深链；测量链；测链尺；测链（长 66 英尺）；钢卷尺
chain taut 锚链紧了
chain tension 链条张力
chain tensioner 张链器；拉链器
chain tensioning 紧链装置
chain termination 链终止(作用)
chain testing machine 链条试验机
chain texture 链状影纹
chain tiller 链条舵柄；链控舵柄
chain timbers 木圈梁；系木
chain tongs 链条钳；链条管子钳；链条管(子)扳手；链式管钳；链式大管钳
chain towing traction 链条牵引力
chain-track tractor 履带式拖拉机；履带式牵引车；一套表格；表格系列
chain traction 链索牵引
chain transfer case 链条传动箱
chain transmission 链传动
chain transmission device 链传动装置
chain traverse 测链导线
chain tread 链带踏级；履带
chain trolley 空中吊运仓库
chain trolley conveyer 吊挂式链输送机
chain truck （悬索桥的）吊链支座
chain tube cutter 链管切削刀具
chain type bucket elevator 链斗式提升机
chain type cooling bed 链式冷床
chain type cutter 链式切坯机
chain type grate 链排式炉栅
chain type hay loader 链耙式装干草机
chain type horizontal stripper 水平甩链式除茎叶器
chain type pusher for transfer car 链式传送顶车机
chain type rear-mounted side-rake 后悬挂链轨式侧向搂草机
chain type side rake 链式侧向搂草机
chain type side-rake for motormower 动力割草机的链指式侧向搂草器
chain type spaced bucket elevator 链斗式提升机
chain type stone saw 链式锯石机
chain type structure 链式结构
chain vice 链式老虎钳，链条虎钳
chain warp 链经
chain welding 并列焊接
chain well 锚链舱
chain wheel 链轮
chainwheel valve 链轮阀
chainwheel with crank 带曲柄链轮
chain winch 链(式)绞盘；链式绞车
chain winding 链形绕组
chain wire 链条钢丝
chain wrench 链条管(子)钳；链式扳手
chain zone 链条带
chair 椅子；辙枕；轨座；罐托
chair-bed 坐卧两用椅
chair car 有活动座椅的客车【铁】；活动坐位客车
chaircolling （绘壁画前准备的）灰泥表面涂层
chair form 椅式
chair hitch 座板升降结
chair lift 单座架空滑车；有座椅的缆车；座椅电梯；单座架空滑车
chairman 董事长
chair of kiln tyre 旋窑轮带支座
chair plate 座板；垫板
chair rail 靠椅护墙板；靠椅扶手；护墙栏；护墙板；靠椅栏
chair rail cap 护墙靠椅栏顶边线脚
chairs for reinforcing steel 钢筋支架座
chair tie 椅带
chair work 人工成型；手工玻璃成型；自由吹制成型
chaitya 岩窟寺
chaitya cave 岩窟寺院；石窟寺院
chaitya hall 佛堂
Chai ware 柴窑器
chalaza 合点
chalcanthite 蓝矾；胆矾
chalcedony 玉髓；石髓
chalcedony aggregate 玉髓骨料
chalcedony cement 玉髓胶结物
chalcedony-opal 玉髓蛋白石

chalcedonyx 带纹玉髓
chalchewete 绿松石；硬玉；翡翠
chalcidicum 教堂附加的柱廊；基督教的门厅；古罗马司法行政房屋
chalcoalumite 铜明矾；铜矾石
chalcoamprite 烧绿石
chalcocite 辉铜矿
chalcocite ore 辉铜矿矿石
chalcocitization 辉铜矿化
chalcocyanite 铜靛石；铜靛矾
chalcogen 硫属元素；硫属
chalcogenide 硫属化物
chalcogenide glass 硫(属)化物玻璃
chalcography 雕铜术
chalcolite 铜铀云母
chalcomenite 兰硒铜矿
chalcomorphite 硅铝钙石
chalconatronite 蓝铜钠石
chalcone 查耳酮
chalcone derivative crystal 查尔酮衍生物晶体
chalcophanite 黑锌锰矿
chalcophile element 亲铜元素
chalcophyllite 云母铜矿；叶硫砷铜石
chalcopyrite 辉铜矿；黄铜矿
chalcopyrite ore 黄铜矿矿石
chalcopyrrhotite 铜磁黄铁矿
chalcosiderite 磷铜铁矿；铁绿松石
chalcosine 辉铜矿
chalcostibite 硫铜锑矿；硫锑铊铜矿
chalcothallite 硫铊铜矿
chalcotrichite 毛铜矿（一种赤铜矿）；毛赤铜矿
chaldron 查尔特隆
chalet （瑞士山中倾斜屋顶的）木屋；农舍式房屋；木造农舍；公共厕所
chalice 高脚酒杯；杯木造状花
chalicosis 石末肺；石工肺
chalk 胡粉；蛤灰；粉化产物；粉笔；垩化；垩；贝壳灰；白尘粉；天然碳酸钙
chalk-based paint 白垩基油漆；白垩基涂料
chalk black 黑色碳质黏土
chalkboard 黑板；粉笔书写板
chalkboard paint 黑板漆
chalkboard trim 带小五金的黑板框
chalk brick 掺白垩黏土砖
chalk deposit 白垩矿床
chalk drawing 粉笔画
chalked 粉化的
chalk fade 粉化褪色
chalk flint 白垩燧石；矿质白垩
chalk grade titanium dioxide 抗粉化型二氧化钛
chalk humus soil 白垩腐殖土
chalkiness 光泽暗淡发白
chalking 灰化；油漆打粉底；打粉底（油漆）；粉化；白垩处理；起霜；起垩；涂层粉化
chalking compound 隔离剂
chalking paint 粉化型白漆
chalking pigment 易粉化型颜料
chalking resistance 耐粉化性；耐起粉性
chalking resistant marine paint 不褪色船用瓷漆；不粉化船用油漆
chalking tester 粉化试验剂
chalk kauri 白粉状贝壳杉脂
chalk lime 白垩石灰
chalk line 划白垩；粉绳（弹线用）；粉线；白粉笔画线；施工粉笔画线
chalk line marking 石灰标线
chalk marl 白垩泥灰岩
chalk masking 粉化变色
chalk mill 白粉磨
chalk overlay paper 白垩纸
chalk paper 白垩纸
chalk pit 白垩取土坑；白垩矿场
chalk powder 白垩粉
chalk putty 白垩土油灰
chalk quarrying 白垩开采
chalk rating 粉化等级
chalk reservoir 白垩储集层
chalk resistance 耐粉化；抗粉化性
chalk rock 白垩石
chalk slate 白垩板岩粉
chalk slurry 白垩浆
chalk soil 白垩土
chalkstone 白垩(石)；石灰岩；石灰石
chalk test 白垩试验；渗透探伤试验

chalk up a new record 创纪录
chalk whiting 白垩粉
chalky 白垩的
chalky clay 白垩(质)黏土;泥灰岩;白垩土
chalky deposit 白垩沉积物
chalky enamel 无光搪瓷
chalky limestone 白垩质石灰石
chalky soil 白垩质土(壤);白垩土
challantite 黄水铁矾
challenge 顶板上光剂(其中之一种);攻击
Challenge disk feeder 加兰齐型转盘给料机
challenge organism analysis 可凝有机物分析
challenger 询问器
Challenger fracture zone 挑战者破裂带
challenge switch 呼叫开关
challenge water 可凝性水
challenging unit 攻击单位
challie 毛料
Challiho 查里荷风
challis 毛料
chalmersite 方黄铜矿
chalybeate 含铁的;钢蓝色;铁盐渗透的;铁泉;似铁
chalybeate spring 含铁矿泉;铁泉;铁盐矿泉
chalybeate water 含铁水;铁质水
chalybeatus 钢蓝色
chalybe(o)us 钢蓝色
chalybite 蓝石英;陨铁
chamaephyte 地上芽植物
chamber 箱盒;药室;闸室(船闸);峒室;船舱;便携式加压舱
chamber acid 含硝硫酸;铅室酸;铅室法硫酸
chamber bench excavation 坑道台阶式挖掘
chamber blasting 洞室爆破;峒室爆破;大爆破
chamber box 水闸箱
chamber brick 火泥砖
chamber burette 球滴定管
chamber capacity 燃烧室容积
chamber charge 洞室装药
chamber collector 室式收集器;室内收集器
chamber condition 燃烧室条件
chamber coolant system 燃烧室冷却系统
chamber cooling 燃烧室冷却
chamber cross form 洞室断面形状
chamber crystal 铅室晶体
chamber depth 闸室水深
chamber detritus 沉沙池颗粒
chamber dive 舱内模拟潜水
chamber diving 舱内模拟潜水
chamber dock 箱式船坞
chamber dryer 箱式烘干器;分箱干燥器;分室烘干机;分室干燥器;室式干燥器;室式干燥机
chamber drying oven 房式干燥炉;房干燥炉
chambered 分隔的
chambered corridor 两旁有房间的走廊
chambered edge 倒棱
chambered kiln 阶梯窑;房式窑
chambered level tube 有气室水准器;附有空气的水准管;气室水准器;气泡水准管(如木工水平尺)
chambered spirit 带气室水准器
chambered vein 囊状矿body
chamber filling conduit 闸室充水廊道;闸室充水管道;闸式充水管道
chamber filter press 箱式压滤机
chamber floor 闸室底板;(船闸的)闸底
chamber front end 燃烧室前端
chamber furnace 箱式炉;分室炉
chamber gate 箱形船闸;闸室(闸)门
chamber head 燃烧室头
chambering 扩孔爆破;扩孔;内腔加工;炮眼掏壶;炮眼扩孔
chamber interceptor 窨井前存水弯;竖井闸门
chamber interior materials and supplies 舱内器材和用具
chamber kiln 闭腔窑;室式窑
Chamberlain and Hookman 水银旋转计数器
chamberlet 小房;小室;小层
chamber method 洞室试验法(隧洞水压试验)
chamber mist 喷雾洞室
chamber(navigation) lock 箱式船闸;闸室式船闸
chamber of commerce 商会
chamber of commerce and industry 工商会
Chamber of Commerce and Industry 日本工商会
chamber of commerce noting fee 商会证明书费

chamber of dry dock 干船(坞)坞室
chamber of manhole 人孔室
chamber of ore 矿囊;矿瘤
chamber of shipping 航运公会
chamber of spring 弹簧室
chamber operator 操舱员【救】
chamber passage 燃烧室通道
chamber performance 燃烧室性能
chamber pot 便壶
chamber pot colo(u)r 溺壶色
chamber pressure 燃烧室压力
chamber pressure record 燃烧室压力记录
chamber regulator 调节室
chamber saturation 展开槽饱和;槽饱和
chambersite 锰方硼石
chamber slab 闸室墙板;闸室底板
chamber stor(e)y 寝室层
chamber support 峒室支撑
chamber temperature 燃烧室温度
chamber test 建筑材料防火试验;容器静态试验
chamber tomb (由巨石长道引向用土盖起的)墓室
chamber type 洞室类型
chamber volume 燃烧室容积
chamber wall 炉墙;闸室墙;闸墙;燃烧室壁
chamber-wall stress 燃烧室壁应力
chamber white lead 室法铅白
chambranle 框饰;门窗框饰;窗口饰
chambray 钱布雷绸;钱布雷布;条纹布
chameaecyparis pisifera 花柏
chameanite 砷硒铜矿
chameleon 变色蜥蜴
chameleon fibre 光敏性变色纤维;变色纤维
chameleon glass block 变色眼镜片
chameleon paint 变色漆;热变颜料;热变漆;示温涂料
chameleon segment 可变程序段
chameleon thermometer 变色示温温度计;示温片
Chamet bronze 贾梅特青铜;锡黄铜
chamfer 走线架;割角;凹线;切面;倒棱
chamfer angle 倒棱角
chamfer bit 扩孔钻;铰刀
chamfer cutter 倒角铣刀;倒角车刀
chamfered angle 斜削角;斜切角
chamfered board 削边板
chamfered column 削角柱
chamfered corner length 过渡刃长度
chamfered edge 斜削边;斜切边;削(角)边;倒棱缘
chamfered edge brushing machine 倒棱角刷光机
chamfered entablature 雕饰柱顶盘
chamfered gravity breakwater 削角防波堤
chamfered groove 角槽;三角形断面槽
chamfered joint 斜削接头;斜削接缝;斜接
chamfered panel 削边镶板
chamfered rustic work 削边粗石工
chamfered section 导切部分
chamfered shoulder 倒棱肩
chamfered square bar 斜面方钢;刻槽方钢条
chamfered step 削边踏步
chamfered teeth 倒角齿
chamfered valve 削角阀
chamfered washer 倒角垫圈
chamfer head screw 沉头螺钉
chamfering 斜切;截角;坡口加工;削角;刻槽;倒角
chamfering bit 斜角铣床;埋头钻
chamfering gear shaper cutter 倒角插齿刀
chamfering hob 倒角滚刀;齿轮倒角滚刀;齿轮
chamfering machine 磨边磨角开槽机;磨边倒角机;倒棱机;倒角机
chamfering mechanism 倒棱机构
chamfering of rail 钢轨斜切
chamfering plane 削角刨
chamfering symbol 倒棱符号
chamfering tool 刻槽工具;倒棱机;倒棱工具;倒棱刀具;倒角刀具;成型切刀
chamfering unit 倒棱清理机床
chamfer needle 槽式移圈针
chamfer plane 倒棱刨;倒角刨;能调整角度的刨
chamfer shape 斜切形;倒角形
chamfer stop 倒棱(挡);倒角形砖
chamfer strip 镶边压缝条;镶边板条;垫瓦条;斜边板条;倒棱(板)条;嵌缝压条;施工缝小木条
chamfer-tongued jointing 斜企口接缝
chamfer type transfer needle 槽式移圈针
chamois 羚羊皮;麂皮(色)

chamois leather 麂皮;油鞣革
chamois lustre 红中带黄光泽彩
chamois skin 麂皮色
chamosite 鲕绿泥石
chamosite mudstone 鲕绿泥石泥岩
chamosite rock 鲕绿泥石泥岩
chamosite-siderite mudstone 鲕绿泥岩—菱铁矿泥岩
chamositic cement 鲕状绿泥石胶结物
chamositic oolite 鲕绿泥石鲕状岩
chamot(te) 火泥;熟耐火黏土;焦宝石;黏土砖;黏土熟料;耐火黏土(熟料);耐高温材料
chamot(te) brick 耐火砖
chamot(te) burning 火泥烧制
chamot(te) ceramics 耐火黏土陶瓷
chamot(te) concrete 耐火泥混凝土
chamot(te) facing 火泥盖面
chamot(te) flour 火泥粉末
chamot(te) lining 火泥衬砌
chamot(te) mortar 火泥灰浆
chamot(te) mo(u)ld 泥型
chamot(te) pipe 火泥管
chamot(te) point 热塑变形点
chamot(te) product 火泥制品
chamot(te) rotary kiln 熟料转窑
chamot(te) sand 烧结砂;熟料砂
chamot(te) vent pipe 火泥通风管
chamot(te) ware 火泥制品
champ 平坦的表面;场地
champagne geyser 冷间歇泉
champaign 平原
champing jaw 夹钳
champion data 优异数据
champleve 凹凸珐琅
champleve enamel 镶嵌式陶瓷
champleve enamel ware 珐琅器
chamshell 抓斗
chamshell excavator 抓铲挖掘机
chanalyst 故障探寻仪
Chanar 阿根廷刺木
Chanar steppe 加纳干草原
chance 机遇;机会;采运难易度
chance cause 机遇因素
chance coincidence 偶然重合;偶然一致
Chance cone agitator 浅司型重介质锥形分选机的搅拌器;漏斗式搅拌器
chance cone silt skimmer 漏斗式钻粉取样器
chance correlation 偶然相关
chance distribution 随机分布
chance error 机误;概率误差
chance event 偶然事件;不确定结点
chance factor 机遇因素
chance failure 偶然故障;意外事故
chancel 高坛;圣坛
chancel aisle 圣坛侧廊;通高坛耳堂
chancel arch 高坛拱;圣坛拱顶;高坛拱顶
chancel flask method 气体比重精密测定法
chancellery 官邸;大使馆邸
chancel rail 圣坛围栏
chancel screen 圣坛屏饰
chance machine 概率计算机
chance mutation 随机突变
chance of acceptance 接受机会
chance of occurrence 出现概
chance rate 概率
chancery 档案馆
chance sample 任意取样;随机样品;随机样本
chance sampling 随机取样;随机抽样;随机采样
chance seedling 机油实生苗
chance variable 随机变数;随机变量
chance variation 机会变差
chandelier 花灯;枝形灯架;枝形(大)吊灯
chandi Prambanan 九世纪爪哇岛的普朗巴南寺
chandler 杂货商;杂货店;烛台
Chandler period 张德勒周期
chandlery 灯具库;灯具店
chandry 灯具库;灯具店
chanduy 仓裴风
changbaiite 长白石;长白矿
change 偶然性;换岗;作业区;转换;转变;找零钱;更换;改动;改变;变迁;变化;变更;变革;变动
changeability 可换性
change a bill 换成零钱
changeable 可变的

changeable boring bar 可调正镗杆
changeable colo(u)r revolving lamp for electric(al) fan 电风扇自动变色旋转灯
changeable cost 变动成本
changeable message sign 可变信息标志
changeable optics 可置换光学装置
changeable phase object 可变相位物体
changeable pitch propeller 可变螺距推器；可变螺距螺旋桨
changeable route signal 可变线路信号
changeable sign 可换式墙上标志
changeable storage 可换存储器；可换储存器；内容可更换的存储器
changeable wall sign 可换式墙上标志
changeable weather 多变天气
change accumulation 改变累积
change agent 变革促进者
changeant 闪光效应
change a way 改道
change bit 更换位；改变位；变更位
change can mixer 换罐式混合机
change character 改变字符；变更字符
change chart 变化图
change-coil instrument 变换线圈仪表
change colo(u)r 改变颜色
change condition 条件变更条款
change condition of water level in rivers and canals 河渠水位变化条件
change control board 变速控制板
change course 转向
change cutting 交变切削法
change day 换班
changed condition 已变更的条件
changed diameter paraffin scraper 变径刮蜡器
change-detection images 变化检测图像
change diameter 换径；变径
change directive 更改指令
change down 降速
change drive 变速传动
change dump 信息更换；信息(改变区)转储；变更转储
changed wall thickness pipeline 变壁厚管道
change face 正倒镜【测】
change factor 偏差系数
change file 变换文件；变更文件
changeful-gear 交换齿轮
change fund 变动基金；找零备用金
change furnace 变换炉
change gear 交换齿轮；换挡齿轮；转换齿轮；挂轮；变向齿轮；变速传动的；变速齿轮；变速齿轮
change gear box 交换齿轮箱；变速(齿轮)箱；变速器
change gear case 变速齿轮箱
change gear device 变速齿轮装置
change-gear drive 变速齿轮传动
change gear lever 变速齿轮杆
change gear plate 挂轮架
change gear ratio 交换齿轮速比
change gear set 交换齿轮组；变速齿轮组
change gear shaft 变速齿轮轴
change gear shaft sleeve 变速齿轮轴套
change gear stud 变速齿轮双端螺柱
change gear train 交换齿轮系；变速(齿)轮系
change hand 换主人；易主
change house 更衣室
change in business inventory 企业库存的变动
change in capacitance 电容变化
change in capacity 电容变化
change in channel form 河槽形状变化
change in channel shape 河槽形状变化
change in channel state 河槽状况变化
change in demand 需求变动
change in design 修改设计；设计修改；设计(的)更改
change in design clause 变换商品设计条款
change in determinants of general equilibrium 一般均衡决定因素的变动
change in direction 改变方向
change in discharge 流量变化
change in distortion 畸变变化
change in exchange rate 汇率变动
change in factor demand 需求要素变动
change in ga(u)ge 直径变化
change in gradient 变坡

change in length of stroke 柱塞冲程变化
change in load 负荷变化
change in natural mortality with age 随年龄变化的自然死亡率
change in nature of risk 危险性质变更
change in offer 修订报价
change in par value 变更票面额
change in quality 质变
change in quantity demanded 需求量变动
change in quantity supplied 需求量变动
change in scour and fill 冲淤变化
change in sea level 海平面变化
change in sensitivity 灵敏度变化
change in size 尺寸变化
change in standard free energy 标准自由能变化
change in surface water quantity 地表水体的变化量
change in the work 工程变更
change into 变成
change in value 价值变化；数值变化；数量变化
change inversely 反方向变化
change in water quantity of aeration zone 包气带水量的变化量
change in weather pattern 天气型的变化；天气模式变化
change in weight 重量变化
change key 单用钥匙
change level 变更级
change lever 变速杠杆
change management 变革管理
changement 换向机构；转换设备
change of address 改址；地址更改
change of air 换气
change of base 基底变换
change of cant 改变超高；超高改变
change of chromatic adaptation 色适应变化
change of class 改变等级
change of colo(u)r 变色
change of commission rate 变更佣金率
change of contract 合同的更改
change of control 功能控制改变
change of course 河道改道；改变路线；改变航向；改变航线；改变方向
change of current direction 流向变化
change of curvature 曲率变化
change of declivity 坡度变化
change of direction 改变方向；方向变化
change of displacement per inch of trim 每英寸纵倾排水量变化
change of diurnal temperature 昼夜温度变化
change of energy flow 能流改向
change of engine 更换机车
change off 交替；换班
change of fragments 碎屑物的变化
change of ga(u)ge 改轨
change of geologic(al) condition 地质情况变化
change of gradient 变坡；梯度变化
change of hole diameter 换径钻进
change of image area 像场变化
change of layer 层次变化
change of levels 高程变化
change of line 改线
change-of-linkage law 磁链变化定律
change of loading 荷载变化
change of management 变更管理权【船】
change of mineral 矿物变化
change-of-mode facility 公交换乘设施
change of momentum 动量的变化
change of name 更名
change of oil 换油
change of oil and gas in the reservoir 储层中油气变化
change of order 命令变更
change of origin 原点改变
change of ownership 变更所有权
change of plan 计划变更；变更计划
change of polarity 极性变更
change of population 人口变动
change of pore diameter 孔隙直径变化
change of position 换位
change of registration 变更登记
change of reservoir storage 水库蓄水变化
change of residence 住址变动；迁居；居住地变更；搬家

change of risk 风险变更
change of riverbed 河床变化
change of river course 河流改道；河道改向
change of river pattern 河型变化
change of river route 河道演变
change of running direction 改变运行方向
change of scale 变换比例尺；变更尺度
change of shift 换班
change of sign 符号改变
change of slope 坡度变化
change of sovereignty 领土主权改变
change of states 状态变化；物态变化
change of style 形式变化；风格转变
change of super-elevation 改变超高
change of thickness in sedimentary beds 沉积层的厚度变化
change of tide 转潮；潮位变化
change of tools 钻具更换
change of variable 更换变量；变数的更换；变量变换
change of variable technique 变量更换方法
change of variable theorem 变量更换定理；变量变换定理
change of variation 磁差变量
change of voltage 电压变化
change of voyage 航线变更
change of water and sediment condition 水沙条件的改变
change of water quality environment 水质环境变化
change of wave packet shape 波包形状变化
change of weather 气象变化
change oil 换油
change on gas 送气期
change on one method 不归零法
change order 变更更改指令；合约条款变动；修改通知书；修改通知单；工程变更通知(单)；更改通知单；更改令；变更指示；变更通知；变更命令
change-over 换火；换向；转接设备；转换；改变；倒转；重调
change-over arrangement 转换装置
change-over box for disable people's elevator 残疾人电梯切换箱
change-over box for smoke curtain 挡烟垂幕电源切换箱
change-over box for waste water pump 废水泵电源切换箱
change-over box of power 电源切换箱
change-over button 换向按钮
change-over circuit 转换电路
change-over clock (仪表的)转接时钟
change-over cock 转换开关；转换旋塞
change-over contact 转向接点；转换接点
change-over device 转换装置
change-over door 翻泥门；分泥门
change-over for guiding sign 导向牌切换箱
change-over gear 逆转装置；转换装置
change-over key 换向电键；换向按钮；转换键
change-over lever 操纵杆
change-over mechanism 转接机构
change-over module 换向模块
change-over of appliance 燃具改装
change-over period 更换周期；转换时间
change-over plug 转换插座
change-over point 转换点
change-over portion 过渡段
change-over relay 切换继电器
change-over speed gear 变速箱
change-over spray head 转换型喷洒头
change-over sub 异径接头
change-over switch 换向开关；双向开关；转接开关；转换开关
change-over switch controller 转换开关控制器
change-over system 转换系统
change-over to manufacture on imported models 仿造外国机器
change-over valve 变换阀；活页阀；换向阀；转换阀
change-panel signs 可换片的广告牌
change point 转车(地)点；改变点；变异点；变换点；变化点；转换点
change point of gradient 坡度变化点
change pole 换极；变极
change-pole moter 变极电动机
change positively 正方向变化
change presser device 变换压板装置

change pressure 充气气压
change pulley 变速轮
changer 变量器;变换装置;变换器
change rate 变化率
change rate of solute flowing-out volume element 流出元内溶质量变化速率
change rate of spring discharge 泉水流量的变率
change record 变更记录
change report 更改报告;变更报告
changer lever 切换杆
change room 更衣室
change set 齿轮变速机
change shifts 换班
change side of double line 换侧;测】【铁】
change sign 改变数的符号;改变符号;变更符号
changes in personnel 职员的更换;人员的更换
changes in the water table 地下水位的变化
changes in the work 高程变动;工程变更;工程项目更动
change size 换径;变径
changes of acidity 酸度变化
changes of fluorescence 荧光量的变化
changes of water table 地下水位变化
change-speed gear 齿轮速度变换;换向齿轮;换排挡;变速齿轮
change-speed gear box 变速齿轮箱
change-speed lever 变速杆
change-speed motor 变速电动机;分级调速式电动机;变速马达
change spur gear 变速轮
change stand 可换机座
change-stress creep test 变应力蠕变试验
change-switch 转换开关
change tack 换向迎风行驶【船】;换枪
change-tank mixer 换罐式混合机
change tape 记(录修)改带;修改带;变更带
change temperature hardening 变温锻炼
change the line of production 转产
change the order of crops 变化轮作次序
change tide 朔日潮
change trial 变换试验
change-tune switch 波段(转换)开关;调谐开关;变换调谐开关
change valve 换向阀;转换阀
change wastes into valuables 变废为宝
change wheel 换向轮;变速装置;变速(齿)轮;变换齿轮;配换齿轮
changing atmosphere 大气变化
changing attachment 变换装置
changing bag 换胶片袋
changing barges without a break in production 换驳不停产
changing cabin 转换舱;化妆车;更衣车;更衣室
changing channel 不稳定河床;变动河床;变动河槽
changing colo(u)r enamel 变色釉
changing colo(u)r enamel manufacture 变色釉生产工艺
changing composition of the atmosphere 大气成分的变化
changing crew at midway system 中途换班制【铁】
changing cubicle 更衣小室
changing debiteuse 换槽子砖
changing discharge method 变流量方式
changing down 降速;换底挡速;变慢
changing drill string 倒换钻具
changing environment 变化着的环境
changing field size 可变像场大小
changing into nitrate 变成硝酸盐
changing load 可变荷载;交变荷载;交变负载;活荷载;负荷改变;变换荷载
changing loading 交变荷载
changing magazine 换暗盒
changing of bit 调换钻头
changing order point 变序点
changing-over stove 热风转换炉
changing pattern 变化模式
changing point in side slope 边坡变化点
changing pressure 变化压力
changing rate of hole azimuth angle deviation 钻孔方位角偏斜率
changing rate of hole zenith angle deviation 钻孔顶角偏斜率
changing rate of separation zone 分离带变化率

changing rig 换辊装置
changing room 更衣室;更衣隔间;生活间
changing shifts 倒班
changing side 换辊侧
changing stage 变动水位
changing station (测量站的)中间位置
changing stress 交变应力
changing switch 转接开关
changing the direction of a stream of water 改变水流方向
changing the distribution of investment 调整投资结构
changing the formation 变换队形
changing the foundation soil for sand 地基的挖土换砂
changing the run direction of through passenger train 变更运行方向通过旅客列车
changing the tug 换拖【船】
changing time 供料时间;更换时间
changing up 换高速挡;增加速率
changing weight 可变权数;变动权数
changing winder 换筒拉丝机
Changjiang delta 长江三角洲
Changjiang estuary 长江口
Changjiang River 长江
Changlin diamond 常林钻石
Chang's seismograph 地动仪
channel 航道;联箱;孔道;海峡;信道;管箱;沟渠;沟道;沟;道路砖;道路砌块;船闸引航道;槽;波道;渠道;渠;频道
channel accretion 河床淤高;河床加积(作用);河床冲积;河槽淤积;航道淤积;渠道淤积
channel adapter 通道适配器
channel address 分路地址
channel address field 通信地址区
channel address word 分路地址代码;通道地址字
channel aggradation 河床淤高;河槽淤积;航道淤高
channel alignment 河道取直;航道定线;渠道定线
channel alinement 河道线向
channel allocation 信道分配;信道地址;波道分配
channel alteration 改槽
channel amplifier 信道放大器
channel amplitude characteristic 通路振幅特性
channel anchor 龙架锚固;槽钢锚固
channel arc 航道弯段
channel arm of bascule bridge 仰开桥桥翼
channel armoring below dam 坝下河槽粗化
channel assignment 信道指配;信道分配
channel assignments table 波道分配计划表
channel associated signal(l)ing 随路信令
channel avulsion 河槽改道
channel axis 河道中线;河槽(轴)线;海峡轴
channel bandwidth 信道带宽
channel bank 河岸;信道组;信道架;信道处理单元;运河边坡;渠岸
channel-bank slumping 河岸滑塌作用
channel bar 河心滩;河道沙洲;河道沙坝;河道浅滩;河铁;槽钢(顶梁);槽杆
channel bar concave 槽杆式凹板
channel-bar deposit 心滩沉积
channel bar facies 心滩相
channel beam 槽形梁;槽开梁;槽钢
channel bed 河底;河床;渠底
channel bed elevation 航道底高程
channel bed level 航道底高程
channel bed of river 河床
channel bend 管道弯头
channel bend radius 航道曲率半径
channel bifurcation 河道分叉;水道分汊(处)
channel black 槽法炭黑
channel block 航道堵塞;槽形砌块;槽块;通路砖;水道堵塞
channel boat 海峡轮船
channel border 风洞壁
channel boundaries 河槽界线
channel boundary 河道边界
channel boundary condition 河槽边界条件
channel braided 河槽游荡
channel breadth 航道宽度
channel buffer storage 通道缓冲存储器
channel buoy 航道浮标;水道浮标
channel busy 信道占用
channel capacity 河槽蓄水能力;河槽容量;河槽过水能力;航道阻塞;航道通过能力;信息通道容量;信道容量;渠道输水能力;渠道排灌能力;渠道过水能力;频道容量;通路容量;通道频带宽度;通道传输能力
channel carrier frequency 波道载波频率;通信电路载频
channel cast 水道铸型;水道模
channel centerline 河槽中线
channel change 河床演变;河道演变;河道变迁;河道变化;河槽变化;水道改道;水道改变
channel characteristic 河道特性;河床特征;河床特性;河槽特征(值);河槽特性
channel chart 航道图
channel check (沟内防止冲刷的)挡水小堰;渠道节制闸
channel circuit 分岔电路
channel classes 航道等级
channel cleanout 河槽清淤;河渠清淤;航道清淤;渠道清淤;清槽
channel clip 槽形夹子
channel coding 信道编码
channel column 槽形柱;槽钢柱
channel command 信道命令
channel compression 信道压缩;波道压缩
channel condition 航道条件
channel conductance 沟道电导
channel connector 信道连接器
channel constraint condition 渠道约束条件
channel constriction 河道压缩;河道束窄;河槽束窄;河槽收缩
channel constriction work 河道束窄工程
channel contraction 河道缩窄段;河道束窄;河槽压缩;河槽束窄;河槽收缩
channel contraction work 河道束窄工程
channel control 河槽控制;航道控制;信道控制;渠道控制
channel controller 通道控制器
channel control unit 传送控制装置
channel cooling arrangement 槽式冷却装置
channel correction 河槽整治
channel cover 泥浆槽盖;管箱盖
channel creek 小港道
channel cross-section 河槽横断面;航道(横)断面;槽钢截面
channel current 沟道电流
channel curve 渠道弯曲
channel cutoff 河道裁弯
channel cutting 河槽下切;河槽刷深;河槽冲刷;渠道冲刷
channel decoder 信道译码器
channel definition format 频道定义格式
channel degradation 河道刷深;河槽刷深;航道刷深;渠道刷深
channel demarcation 航道设标;航道标界;航槽界;航槽边界;港道标界
channel demultiplexing 信道分用
channel density 河网密度
channel depletion mode junction 沟道耗尽型结
channel deposit 河道淤积;河道泥沙;河道沉积物;河道沉积;航道淤积
channel deposition 航道淤积
channel deposits 河槽沉积物;河槽沉积泥沙
channel deposit silt 河道沉积泥沙
channel depth 河底深度;河槽深度;航道水深;渠道深度;计量安装玻璃尺寸长高;瓷釉工艺计量
channel depth clearance 航道富余水深
channel design 管道设计
channel designator 信道序号
channel desilting 河道清淤;航道清淤
channel detention 河槽滞蓄量
channel deterioration 河道恶化
channel development 航道开发
channel device 沟道器件
channel digging 渠道开挖
channel dimensions 航道尺度
channel discount 特殊折扣;特别折扣
channel distortion 信道失真
channel distortion coefficient 通路非线性失真系数
channel diversion 明渠分水;明渠导流
channel documentation 探槽编录
channel downcutting 河槽下切
channel dredging 河道疏浚;航道疏浚;疏浚水道;河渠疏浚(工作)
channel dredging engineering 河道疏浚工程
channel dredging works 河渠疏浚工程

channel earthenware pipe 槽形陶管
channel edge depth 航道边缘水深
channel edge marking 航道边缘标志
channeled oil ring 开槽油环
channel effect 沟渠效应;沟道效应
channel efficiency 信道速率
channel electron multiplier 渠道式电子倍增器
channel end bunt-on 两端撑紧槽钢
channel entrance 航道入口;水道入口
channel equilibrium 河槽平衡
channel erosion 河道冲刷;河床冲刷;河槽侵蚀;河槽冲刷;河槽冲蚀;航道冲刷;沟蚀;槽形侵蚀;槽蚀;渠道冲刷
channel etching 沟道刻蚀
channel excavation 沟通开挖;河槽开挖
channel expansion 河道扩展;河槽扩大(段);航道加宽;渠道扩大(段)
channel exploration 航道开发
channel ferry 海峡渡轮
channel fever 陆地热;河床热
channel-fill cross-bedding 水道充填交错层理
channel fill deposits 河槽沉积物;河槽沉积泥沙
channel fill(ing) 河槽淤积;航道淤积
channel fills 河槽沉积物;河槽沉积泥沙
channel fill sand trap 河道砂体圈闭
channel-fill swamp 河道填积沼泽
channel filter 波道滤波器
channel firth 海门;峡道口;港道口
channel fish culture 渠道养鱼
channel fixation 河槽固定
channel floor lag 河底滞留相
channel flow 螺流(量);明渠流;河道流量;河槽水流;渠(道水)流;明槽流
channel flow accretion 河槽水流增加
channel flow aircushion vehicle 横流式气垫船
channel flow depletion 河槽水流亏耗;河槽水流减损
channel flow time 河槽汇流时间
channel flow velocity 河道流速
channel form 河道类型;河槽形状
channel formed by breakers 破浪冲蚀沟
channel forming die 槽形成型模
channel forming discharge 造床流量
channel-forming process 河床演变;造床过程
channel for oiling 加油槽;油路
channel for working 施工用航道
channel frame 槽形构件框架;槽钢框架
channel frequency 河流频数;河道频数;河道密度;通道频率;水流密度频率
channel frequency converter 分各变频器
channel frequency response 波道频率响应
channel friction head loss 河槽摩擦水头损失
channeled frith 峡道口
channel galley 带沟捕捉槽
channel gasket 槽型密封圈
channel gate 信道门
channel geometry 航道断面形状;渠道几何条件
channel girder 槽形大梁
channel glazing 槽嵌玻璃法;槽内贴面板;安装玻璃工艺
channel grade 渠道比降;信道等级
channel gradient 河坡降;河道比降;河床坡度
channel gradient ratio 河流比降比(率)
channel grant high 高优先通道授予
channel groove (物体表面的)沟槽
channel group 信道组
channel group separating and combining equipment 分路并路机
channel guide 槽钢罐道
channel gull(e)y 排水沟;集污槽进水井
channel habitat 河床生镜
channel head 槽钢顶梁
channel hydraulic characteristic 河槽水力特性
channel hydraulics 航道水力学;明渠水力学;河槽水力学
channel idle noise 通路固有噪声
channel improvement 河道整治;河道整治;航道治理;航道整治
channel improvement works 河道整治工程
channel in 沟流
channel incision 河道下切;河槽刷深;航道刷深
channel indicating lights 引航道灯
channel inflow 河道入流;河道来水量;河槽入流(量);渠道进水量

channel in reservoir region 库区航道
channel intake 渠道进水口
channel interception 河面降水
channel interface 通道接口
channel interval 道间距
channel iron 槽铁;槽钢;凹形铁
channel iron frame 槽铁框架
channel island 江心洲;江心滩;河间岛;汊河岛
channel islands 海峡群岛
channelization 导流;交通渠化;河网化;管道化;渠道化
channelization island 导流岛【道】
channelization project 河道渠化工程
channelize 交通渠化;信道划分;使成沟渠
channelizing intersection 渠化路口;渠化交通交叉口
channelizing island 分车岛;渠化交通岛;路口分车岛;道路分路台
channelizing line 交通渠化线
channelizing traffic 渠化交通
channel jam 航道堵塞
channel jamb 槽钢门框
channel lag 河底滞留沉积;河槽滞(流)时(间);河槽汇流滞后时间
channel lag deposit 河底滞留沉积物;河床残留沉积物;河床残余堆积物
channel lane 航道
channel lead 航道
channel(l)ed linear space 沟槽式直线型空间
channel(l)ed plate 皱纹板;条纹板;菱形网纹钢板;开槽板
channel(l)ed quoin 细琢方块石(上边缘企口接合)
channel(l)ed rail 槽形导轨
channel(l)ed runoff 河槽径流
channel(l)ed spectrum 沟槽光谱
channel(l)ed-steel wheel rim 槽钢制轮辋
channel(l)ed substrate 开槽底板
channel(l)ed-substrate-planar structure 沟道衬底平面结构
channel(l)ed upland 槽蚀高地
channel length 航道长度
channel(l)er 开渠机;开渠机;挖沟机;挖槽机;凿沟机;海峡隧道建造机
channel light 航道灯标;水道指示灯
channel line 航界界线;河道边线;河流中泓;河道主流线;河道中泓线;河道深泓线;航道整治线;航道线;主槽线;中泓线;深泓线
channel(l)ing 沟流的形成(玻璃液中);窜槽;开渠;开沟;多路传输;凿沟;槽道现象
channel(l)ing aharacteristics 沟流特性;成沟特性
channel(l)ing cutter 铣槽刀;槽铣刀
channel(l)ing effect 沟道效应
channel(l)ing effect factor 隧道效应系数
channel(l)ing in column 塔中形成沟流
channel(l)ing machine 凿沟机;滚槽机
channel(l)ing method 刻线法
channel(l)ing radiation 波道辐射
channel lining 河床铺砌
channel lining machine 河床铺砌机;河槽粉面机
channel(l)ized-bulb intersection 灯泡形渠化交叉口
channel(l)ized erosion 集中冲蚀;沟状冲刷;冲沟型冲刷
channel(l)ized intersection 导流交叉口;分道转弯式交叉【道】;渠化交叉口
channel(l)ized island 道路分路岛
channel(l)ized lane 渠化车道
channel(l)ized layout 渠化布置
channel(l)ized load 渠化交通荷载
channel(l)ized loading in pavement and deck 路面和码头面的分带荷载
channel(l)ized marking 渠化标线
channel(l)ized river 渠化河流
channel(l)ized river section 渠化河段
channel(l)ized stream 渠化河流
channel(l)ized stream section 渠化河段
channel(l)ized stretch 渠化河段
channel(l)ized strip 渠化路带
channel(l)ized time dividing 时分多路通信
channel(l)ized traffic 渠化交通
channel(l)ized transmitter 信道发射机
channel(l)ized Y intersection 分路式 Y 形交叉
channel(l)ized zone 导流带
channel loss 河槽水量损失;水道损失;渠道渗流

损失
channel maintenance 航道维护
channel margin 河道边界
channel marker 航道标(志);航道标志;水道标示器
channel mark(ing) 航道标(志);水标;航道设标
channel-marking buoy 航道浮标
channel material 槽型材料;槽形材料
channel meandering 河道弯曲;河流弯曲
channel meander model 河槽蜿蜒试验模型
channel member 槽形构件
channel metamorphosis 河槽变形
channel migration 河槽迁移
channel miles 信道英里
channel mill 环磨机
channel mopping 条黏;条涂
channel morphology 河相学;河相关系;河床形态;河槽形态
channel morphology dimensions 河槽形态尺度
channel mo(u)ld 下水道盖板;槽模
channel mo(u)lding 槽饰
channel-mouth bar 拦门沙;河口沙洲
channel multiplexor 通道多路转换器;通道多路复用器;通道多路调制器
channel navigation 狭水道航行
channel navigation information service 航道导航信息业务
channel net 河道网;水系
channel net loss 通路净衰耗
channel net loss stability 通路净损失持恒度;通路净衰耗持恒度
channel noise 沟道噪声
channel nozzle 管箱接管
channel number 信道编号;电路编号;频道数;通道号
channel nut 槽形螺帽
channel of approach 引(水)渠;引水渠;引槽
channel of communication 交流渠道;信息交流渠道
channel of disbursement 支付渠道
channel of distribution 销售渠道;分配渠道;分配流通渠道
channel of facsimile 传真通路
channel of leakage 渗漏通道
channel of main river 主河道
channel of main stream 主河道;主河槽;主槽
channel of promotion 推销途径;特销渠道
channel of river 河槽
channel of the fluid 流体通道
channel order 河流等级;支流级
channel packet 海舱邮船
channel packing 信道合并
channel pan 河塘
channel pass(age) 通航水道
channel patch 槽形补缀
channel pattern 河网类型;河道类型;河槽形态;沟道图案;水系类型;水道类型
channel-phase runoff 渠相径流
channel phasing 信道相位调整
channel pilot 海峡引航员
channel pin 槽楔钉
channel pin bond 导电轨销钉接合器
channel pipe 槽形管;半圆明槽排水沟;梯田台地沟堑;半圆排水管
channel piracy 河道袭夺
channel plan 频道设计;频道方案
channel plate 槽形板;通道板
channel point 成沟点
channel pond 河塘;渠段
channel pool 河塘
channel precipitation 河面降水;河道降水;河床沉淀(作用);河槽降水;水面降水;水道降水
channel process 槽法
channel profile 槽形型材;槽形剖面
channel program(me) 通道程序
channel program(me) translation 通道程序转换
channel promotion 晋升渠道
channel property 河床特性;河槽特性
channel protection 河维护;航道维护;水道保护
channel protection system 波道备用制;波道备用方式
channel pulse 信道脉冲
channel purlin(e) 槽钢檩条
channel rail 槽形轨条;沟形轨条;槽形栏杆
channel reach 河段;渠段

channel reactivation 河道复活
channel realignment 河流改道；航道改线；航道改道；渠道改线
channel rectification 河道整治；河槽整治；河槽整直；航道整治
channel regime(n) 河道特征；河道流态；航道流态；河流状况；河流情态；河流情势；河流情况；河道水情；水道状况
channel regulation 河道整治；河槽治理；河槽整治；航道整治
channel regulation project 河槽整治工程
channel regulation works 河道整治工程；水道整治工程；河槽整治工作；河槽整治工程
channel reliability 信道可靠性
channel relocation 航道改线；航道改道
channel replacement furnace 代槽炉黑
channel restriction 河道缩窄段；河道束窄；河槽束窄；河槽收缩
channel revetment 河槽护坡；河道护岸；渠道护岸
channel rib 槽肋
channel ring 槽形钢圈
channel roof slab 槽形屋面板
channel roughness 河床粗糙度；河床糙度；河槽糙率；河槽不平整度；沟床糙度；渠道糙率；河道糙率
channel routing 河道洪水演算；河道洪水演进；河槽演算；河槽洪水演算；河槽洪水演进
channel routing of solute 河道溶质量演算
channel rubber 橡皮衬里；橡胶夹层
channel runner 槽形导板；主龙骨
channel runoff 河槽径流；渠道径流
channel sample 掘槽采样
channel sampling 刻槽取样；刻槽采样
channel sand 河床砂；深槽砂
channel sand pool trend 河道砂油气藏趋向带
channel scour 河床冲蚀；河床冲刷；河道冲刷；床侵蚀；航道冲刷；渠道冲刷
channel scouring 渠道冲刷
channel scraper 河床挖泥机
channel scraper and elevator 下水道清理设备
channel section 河槽断面；航道断面；槽形截面；槽形断面；槽钢截面；槽钢；渠道截面
channel section axle 河床断面轴线；槽钢截面轴线；槽钢轴
channel-section-bar guard 安全号槽
channel(-section) glass 槽形玻璃
channel-section iron 槽铁
channel-section spar 槽形梁
channel-section steel 槽形钢
channel sediment 河道沉积(物)；河道淤积物；河槽泥沙
channel sedimentation 河道淤积；河槽淤积；航道淤积；渠道淤积
channel sediments 河槽沉积物；河道沉积泥沙
channel sediment silt 河道沉积泥沙
channel seepage 渠道渗漏
channel seepage loss 渠道渗漏损失；渠道渗流损失
channel segment 河流分支；渠道分支
channel selector 信道选择器；波道转换开关；波道选择器；波道选择开关；频道转换开关；通路开关
channel separation 信道间距；信道间距
channel sequence number 信道序号
channels for purchasing goods wholesale 批发进货渠道
channel shape 河槽形状
channel-shaped 槽形的
channel-shaped floor slab 槽形楼板
channel shaped wall section 槽形墙截面
channel share 槽形挖掘铲；槽形犁铧
channel shear connector 抗剪槽钢连接件
channel shifter 信道移动器；波道搬移电器
channel shortening 河槽缩短；航道缩短；缩短航道；截弯取直
channel side 防波堤之临河侧
channel silting 河道淤积；航道淤积；渠道淤积
channel size 频道宽度
channel slab 槽形板
channel slope 河床比降；河床坡降；河槽坡降；河槽坡度；航道边坡；运河边坡；渠道坡度；河道坡度
channel snow plough 挖槽除雪机
channel snow plow 挖槽除雪机
channels of distribution 交货途径
channel spacing 信道间隔；频道间隔
channel span 跨越水道的桥跨；河道跨(度)；航道宽度；渠道宽度

channel speed 信道速率
channel spillway 河床式溢洪道
channel spin 道自旋
channel splay (河渠的)扩展淤积；河滩淤积
channel spread 信道扩展
channel spring 河谷泉；沿河泉
channel stability 河床稳定性；河槽稳定(性)；航道稳定性；信道稳定性；渠道稳定性；通路稳定度；通道稳定性
channel stabilization 河槽稳定化；渠道稳定化；渠道抗冲措施；稳定河槽形态
channel status 通道状态
channel status word 通道状态字
channel steamer 海峡轮船
channel steel 槽钢；U形钢
channel steel bar 小型槽钢
channel steel beam 大型槽钢
channel stone 排雨槽
channel stopper 沟道抑制环；沟道截断环
channel storage 河槽储[贮](水)量；河槽蓄水(量)；河槽储水量；渠道蓄水量；槽蓄
channel storage capacity 河槽蓄水容积；槽蓄(容)量；河槽调蓄能力；河槽调节能力
channel storage equation 槽蓄方程
channel storage flood routing 河槽蓄洪演算
channel storage routing 槽蓄演算
channel straightening 河道整直；河槽(裁弯)取直
channel string 槽钢；U形楼梯斜梁
channel stringer 槽钢楼梯梁
channel strip 信道带
channel strut U形铁支柱；槽形支柱；槽钢支柱；槽钢撑杆
channel stud 槽钢立筋；U形龙骨
channel stud partition 槽钢立筋隔断
channel-subdivider 信道分路器
channel-substrate breakdown 沟道衬底击穿
channel sump 集污槽
channel supervisor 河道监督员
channel survey 航道测量
channel surveying reference level 航道绘图水位
channel switch 波道(转换)开关
channel switcher 频道转换开关；频道转换电路
channel-switching equipment 波道倒换机
channels within reservoir region 水库库区航道
channel synchronizer 信道同步器
channel system 水系；渠道系统
channel table 通道表
channel terminal 信道终端
channel terminal bay 电路终端架
channel terrace 槽形阶地；等高台地水渠
channel test 沟试验
channel thread 河道主流
channel tie 槽钢拉杆
channel tile 槽形瓦；槽瓦；(意大利、西班牙式的)方砖基底
channel time delay 河道滞(流)时(间)
channel time slot 信道时隙
channel toothing 槽形齿轮啮合
channel training 河道整治
channel training works 河道整治建筑物；河道整治构筑物；河槽调治工程；水道整治工程；水道导流工程
channel transition 渠槽渐变段
channel trap 信道陷波电路
channel travel 河槽流程；河槽汇流
channel travel isochrone 河槽汇流等时线；等流时线
channel travel isochrone chart 河槽汇流等时线图
channel travel time 河槽汇流时间；河槽传播时间
channel tray 槽形托盘
channel trigger circuit 信道触发电路
channeltron 渠道倍增器
channel twisting 摆动河道；摆动河床
channel-type carbon black 槽法炭黑
channel-type electron multiplier 槽式电子倍增器
channel-type erosion 冲沟型冲刷
channel-type head race 引水渠；前渠
channel-type idler 槽形托辊
channel-type jig 槽式钻模
channel-type race 明渠式进水渠道
channel-type spillway 槽形溢洪道
channel-type tail race 槽形尾水渠
channel-type terrace 槽式梯田；槽形阶地

channel unit 传送装置
channel unit address 传送装置地址
channel utility 航道利用
channel variation 河槽变化
channel vecoder 谱带式声码器
channel viscosimeter U形黏度计
channel washing 航道冲刷
channel water 渠道水
channel wave 槽波；通道波；声道波
channel waveguide 波导管
channel way 河床；河槽；通道
channel weed 河床杂草；渠床杂草
channel-wheel 双缘轮
channel wheel separator 槽轮式选粉机
channel widening 河道展宽；河槽展宽；航道加宽；渠道展宽
channel width 频道宽度；槽宽；河道宽度；河槽宽度；航道宽度；信道宽度；波道宽度；计量安装玻璃尺度长宽
channel-width variation 波道宽度变化
channel wing 槽形机翼
channel with cover 盖板渠
channel within lake 湖泊航道
channery 碎石块；碎石
Chanoine wicket dam 查诺尼式门坝；查诺尼式门堤
chantalite 钱羟硅铝钙石
chantilly lace 尚蒂利细花花边
chantlate (连接在公共椽子上的)短椽；檐口滴水条
chantry (捐献的)小教堂；附属教堂；祈唱堂
chantry-altar 奉献祭坛
chantry chapel 附属教堂
Chanty-Mansijsk massif 汉特一曼西斯克地块
chaoite 赵石墨
chaos 巨屑混杂岩；巨屑混杂堆积；混杂角砾岩；混杂；混乱
chaos of conception 概念混乱
chaotic boulder clay 混杂漂砾黏土
chaotic deposit 混杂沉积
chaotic geological body 地质混杂体
chaotic melange 混杂堆积；混杂沉积；混乱堆积
chaotic motion 不规则运动
chaotic seismic reflection configuration 不规则地震反射结构
chaotic structure 不规则结构；混杂构造；混乱构造
chaotte 火泥
chaoyang basin 朝阳盆地
chap 龟裂；变粗糙；裂纹；裂缝
chapada 高地；台地；上升高原
chapapote 裂沥青
chaparral 矮树林；灌木群落(常绿阔叶)
chapbook 小册子
chape 线头焊片；搭扣；搭口；扒钉
chapeau de mangan 锰帽
chapeau de plumb 铅帽
chapeirao 孤礁
chapel 礼拜室；小教堂；附属教堂
chapel arcade 礼拜堂拱廊
chapelet 链斗(式)提水机；链斗式提升机；链斗式疏浚机；链式泵；链斗传送器；斗式提升机
chapel-in-the-round 圆形礼拜堂
chapelling a ship 操舵转风
Chapel of Charlemgne at Aachen 亚琛的查理曼礼拜堂
chapel of ease 地段小教堂
Chapel of the Resurrection 耶稣复活礼拜堂
chapel royal 宫内教堂
Chaperon resistor 查佩龙电阻器
Chaperon winding 查佩龙线圈
chapiter 大斗(柱上的)；柱头
chaplash 硬木(印度产的黄色～金棕色树)
chaplet 串珠花饰；花环；型芯撑；串珠饰【建】；撑子
chaplet area 芯撑支撑面积
chaplet nail 型芯撑钉
chapman 小贩
Chapman equation 查普曼方程
chapmanite 硅锑铁矿；羟硅锑铁矿
Chapmanizing 查普曼氰化法
Chapman region 查普曼区
Chapman's test 查普曼氏试验
chapman test 聚乙烯薄膜印墨黏附试验
chappy 龟裂的
chapterhall 大主厅；大厅
chapter house 俱乐部；(会社的)分会所；牧师会集堂
chapter-room 团体聚会室

chapters and sections 章节
chaptrel 哥特式(建筑的圆)柱顶;拱基;拱墩
char 低温焦炭;散工;烧炭工;女杂工;木炭
char absorption 炭吸附(作用)
charachtron 显字示波管
character 性质,符号
character array 字符数组
character attribute 字符属性
character book 征信书
character buffer 字符缓冲器
character check 字符校验;符号检验;数字检验
character code 记号码;信息码;字(母)码
character coefficient 特性系数
character controlled generator 符号控制工作程序
character curve of well 井的特性曲线
character curve-window based supervised classification 特征曲线—窗口法监督分类
character data 字符数据
character deletion character 删去字符
character density 符号密度
character design 字符设计;符号设计
character diagram 特性图
character display 行车字符显示器;信息显示;数字字母显示器
character display tube 显字管
charactered material reserve 特准储备物资
character expansion factor 字符扩展因子
character fill 字符填充
character font 字体
character format 字符格式
character generator 符号多项式;数字字母发生器
character group 特征标群
character handling 字符处理
character height 字符高度
character index 特征指数
characteristic 性能;特征;特性的;特性;特色;特点
characteristic absorption peak 特征吸收峰;特性吸收峰
characteristic adaptive control 特性自适应控制
characteristic age 特征年龄
characteristic analysis 特征分析
characteristic analysis method 特征分析法
characteristic analysis technique 特征分析技术
characteristic and continuous fluorescence correction 特征和连续荧光修正
characteristic and mantissa of logarithm 对数的首数与尾数
characteristic band 特征谱带
characteristic beauty 特征美
characteristic boundary condition 特征边界条件
characteristic bowl-shaped curve 特征圆锥曲线
characteristic chamber length 燃烧室的换算长度;燃烧室特征长度
characteristic class 特征曲线类型
characteristic code 特征码
characteristic code list 特征码表
characteristic coding feature coding 特征编码
characteristic coefficient 特征系数
characteristic coefficient of member 构件特征系数
characteristic component 特征组分;特性组分
characteristic concentration 特征浓度
characteristic concentration curve 特征浓度曲线
characteristic concrete cube strength 混凝土立方块特征强度
characteristic condition 特征条件
characteristic cone 特征锥
characteristic conoid 特征锥
characteristic constant 特征常数
characteristic contaminant 特性污染物
characteristic coordinates 特征坐标
characteristic cube strength 混凝土特种立方体强度
characteristic curve 性能曲线;感光特性曲线;特征曲线;特性曲线
characteristic curve method 特征曲线法
characteristic curve method of water drive 水驱特征曲线法
characteristic curve of discharge pipeline 排泥管特征曲线
characteristic curve of film 胶片特性曲线
characteristic curve of sand pump 泥泵特性曲线
characteristic curve slope 特性曲线陡度
characteristic data 特征数据
characteristic data of equipment 仪器性能数据
characteristic data of instrument 仪器性能数据

characteristic date 指定日期
characteristic definition card 特征码定义卡
characteristic delay time 特性滞后时间
characteristic depth of water 特性水深
characteristic design 特性设计
characteristic determinant 特征行列式
characteristic diagram 特征线图;特性要图;特性图解;特性(曲线)图
characteristic diameter 公称直径;特征直径
characteristic difference hybrid method 特征线—差分混合法
characteristic dimension 基准尺寸;特征尺寸;特性尺寸
characteristic discharge 特征流量;特性流量
characteristic dislocation 特征位错
characteristic distortion 特性失真;特性畸变
characteristic divergence 特性分歧
characteristic doublet peak 特征双峰
characteristic dyadic 特征并矢
characteristic element 特征元素;特性要素
characteristic emulsion curve 乳剂特性曲线
characteristic energy 特征能量
characteristic equation 特征方程;特性方程
characteristic equation of a matrix 矩阵的特征方程
characteristic error 特征误差;特性误差
characteristic exhaust velocity 特征排气速度;特性排气速度
characteristic exponent 特征指数;特性指数【数】
characteristic extraction program(me) 特征抽出程序
characteristic factor 特性因素;特性因数
characteristic family 特性曲线族
characteristic feature of an invention 发明的特点
characteristic form 特征形式
characteristic fossil 标准化石;特征化石
characteristic frequency 特征频率;特性频率
characteristic frequency spectrum 特征频谱
characteristic function 本征函数;特征函数
characteristic function of a set 集(合)的特征函数
characteristic function of probability distribution 概率分布特征函数
characteristic function of the binomial law 二项分布律的特征函数
characteristic head 特征水头
characteristic hydraulic feature 特征水力性能
characteristic hydraulic parameter 特征水力参数
characteristic hydrograph 指标过程线;特征过程线;特性水文曲线
characteristic hyperplane 特征超平面
characteristic identification 特征识别
characteristic impedance 特征阻抗;特性阻抗
characteristic impedance of explosive 炸药特性阻抗
characteristic-impedance termination 特性阻抗终端负载
characteristic index 特征指数;特征指标;示性指示
characteristic information 特征信息
characteristic infrared group frequency 特征红外基团频率
characteristic initial value problem 特征初值问题
characteristic instant 特征瞬间
characteristic ion 特征离子
characteristic isochromatic spectroscopy 特征单色谱
characteristic length 特征长度;特性长度
characteristic length of acoustic(al) vibration 声振动特征长度
characteristic letter 特征字
characteristic line 特征线;特性线
characteristic line method 特征线法
characteristic line spectrum 标识线状光谱
characteristic loss spectroscopy 特征损失能谱学
characteristic manifold 特征流形
characteristic mantissa 尾数
characteristic mark 特征记号;特征标志;识别
characteristic mass 特征质量
characteristic matrix 本征矩阵;特征矩阵
characteristic method 特征法;特征曲线法
characteristic mineral 特征矿物
characteristic net 特征曲线网
characteristic number 特征值;特征数

characteristic number of logarithm 对数的特征数
characteristic odo(u)r 特性恶臭;特性气味
characteristic overflow 阶码溢出;特征上溢【计】
characteristic parameter 特征参数;特性参数
characteristic pattern 特征图
characteristic peak 特征峰
characteristic point 特征点;特性点
characteristic pollution graph 特性污染图
characteristic polynomial 特征多项式
characteristic probability 特征概率
characteristic product 特征产品
characteristic proper 特征向量
characteristic quadratic form 特征二次型
characteristic quality 特性质量
characteristic quantity 特征量;特性数
characteristic radiation 标识辐射;特性辐射
characteristic radius 特性半径
characteristic rate 特征速率
characteristic rate of decay 特征衰变率
characteristic ratio 特征比
characteristic ray 标识射线;特征射线
characteristic relation 特性关系曲线
characteristic residual current 特性残余流
characteristic river reach 特征河段
characteristic Roentgen ray 标识X辐射
characteristic root 特征值;特征根
characteristic root of a matrix 矩阵特征根
characteristic root test 特征根检验
characteristics 规格参数表
characteristic scalar 特征标量
characteristic series 特征数列
characteristic shape 特征形状;特性形式
characteristic sheet 图式
characteristics identification 标志
characteristic signal 特征信号
characteristic site period 场(地)特征周期
characteristic size curve 粒度特性曲线
characteristics of activated sludge 活性污泥特性
characteristics of a logarithm 对数的首数
characteristics of atmospheric transmission 大气传输特性
characteristics of a water-course 流域特征
characteristics of basin 流域特征(值)
characteristics of borehole test section 钻孔试投特征
characteristics of canal 运河水力特性
characteristics of centrifugal pump 离心泵特征曲线
characteristics of channel 河槽特征(值);河槽特性;通路特性
characteristics of char residue 焦渣特性
characteristics of coke 焦块特性
characteristics of correspondence 对应的特征
characteristics of crucible non-volatile residue 焦渣特征
characteristics of demand 需求特点
characteristics of digester gas 消化池气(体)特性
characteristics of drainage 水系特征
characteristics of earthquake 地震特征
characteristics of earthquake rupture propagation 地震破裂扩展特征
characteristics of electrode consumption 电极消耗特性
characteristics of environment 环境特征
characteristics of fluvial morphology 河流地貌特征
characteristics of fluvial process 河床演变特性
characteristics of hood 集气罩性能
characteristics of internal combustion engine 内燃机特性
characteristics of landfill leachates 填埋滤液特性
characteristics of light 灯光性质;灯光特征
characteristics of line 管道特性
characteristics of logarithm 首数【数】
characteristics of material 材料特性
characteristics of motion 运动规律
characteristics of motor vehicle 汽车性能
characteristics of nuclear energy 核能特性
characteristics of odo(u)r and taste 臭味特征;气味和味道特征
characteristics of particulate matter 颗粒物特征
characteristics of pipe-net 管网特性
characteristics of pump 水泵特性
characteristics of radioactive tracer 放射性示踪

剂特性
characteristics of returned value 返回数值的特性
characteristics of river geomorphology 河流地貌特征
characteristics of scale 标度尺特性
characteristics of separation 分离特性
characteristics of seawater 海水特性
characteristics of sewage 污水特性;污水特性
characteristics of shingles 屋面板特性;对屋面板的特殊要求
characteristics of surface water body 地表水体特征
characteristics of wagon flow 车流性质
characteristics of waste 废料特性
characteristics of water 水的特性
characteristics of wave 波浪特征;波浪特性
characteristic solution 特征解;特性解
characteristic species 特征种
characteristic spectrum 标识光谱;特征(光)谱
characteristic speed 特征速度;特性转速;特性速度
characteristic stage 特征水位
characteristic strength 特征强度;特性强度
characteristics under load 负载特性
characteristic surface 示性曲面
characteristic temperature 特征温度;特性温度
characteristic test 性能试验;特性试验;特性测试;特性测定
characteristic theory 特征理论
characteristic time 特征时间;特性时间;时间常量
characteristic time response 特性时间反应
characteristic time scale 特征时间尺度
characteristic triangle method 特征三角形法
characteristic type 特征型号;特性型号
characteristic underflow 阶码下溢
characteristic underload 欠载特性曲线
characteristic value 本征值;特征值;特性值;特定价值
characteristic value for strength of materials 材料强度标准值
characteristic value of action 作用标准值
characteristic value load 荷载标准值
characteristic value of the vertical bearing capacity of a pile 单桩竖向承载力特征值
characteristic value of watershed 流域特征值
characteristic value of wave 波浪特征值
characteristic variable 特征变量
characteristic vector 特征向量;特征矢量
characteristic vegetation 特征植被
characteristic velocity 特征(排气)速度;特征流速;特性流速
characteristic vibration 特征振动;固有振动
characteristic water level 特征水位
characteristic wave height 特征波高
characteristic white 特征白色
characteristic width 特征宽度
characteristic X-radiation 标识X辐射;特征X辐射;特性伦琴辐射;特征X辐射
characteristic X-ray 标识X辐射;特性X-射线
characteristic X-ray absorption coefficient 特征X射线吸收系数
characteristic X-ray area scanning image 特征X射线面扫描象
characteristic X-ray image 特征X射线图像
characteristic X-ray intensity 特征X射线强度
characteristic X-ray point scanning image 特征X射线点扫描象
characteristic X-ray systems 特征X射线系
characterization 性能描写;表征;品质鉴定;特征化;特性说明;特性描述;特性鉴定;特性化
characterization analysis 表征分析
characterization factor 特征因数;特性因素;特性因数
characterization of adsorption 吸附特性
characterization of sorption 吸附特性
characterization of water body 水体特性
characterization technique 表征技术
characterizing factor 特性系数;特性因数
characterizing mirror 标识反射器
characterizing non-point pollution 特性非点源污染
character level 字符级
character line generator 特征发生器
character loan 信用借款;信用贷款
character manipulation 字符处理;字符操作
character mode 字符方式

character of a light 灯质
character of bed 河床特征;河床特性
character of classification 船级符号;入级符号
character of double refraction 重屈折性;重折射性
character of folded beds 褶皱岩层性质
character of karst shape 岩溶形态特征
character of operation 操作性质
character of outcrop 露头特征
character of pal(a)eobasin 古盆地特征
character of pal(a)eoclimate 古气候特征
character of pal(a)eoland 古陆特征
character of regional geochemistry survey 区域地球化学调查性质
character of regional stress 区域应力性质
character of reservoir induced earthquake 水库诱发地震特点
character of river 河性;河流特性
character of service 使用特性;工作制(度)
character of surface 表面特征;表面特性
character of surrounding rock 围岩特性
character of surveyed region 工作区性质
character of swell 涌浪级别
character of traffic 交通性质;交通特性;运量特征
character of variation 变化性质
character pattern 字形
character per inch 每英寸字符数【计】
character picture specification 字符形象说明
character pitch 字符间隔
character plotting 字符绘图
character processing 字符处理
character rate 字符传输率
character reader 读符器;符号读出器
character recognition 字体识别;字符识别;符号识别;文体识别;特征辨认
character sensing strip 读出字段
character sequence 字符顺序
character set(field) 符号组;字符组;字符集
character signal 字符信号
characters incised in the paste 暗款
character spacing 字符间隔
character spacing reference line 字符定位线
character species 典型种
characters per frame 每帧字符数
characters per second 每秒字符数
character string 字符串序列;字符串行
character stroke 符号笔画
character structure 字符结构
character subset 字符子集
character timer 字母定时器
character up vector 字符上矢量
charactery 记号法;征象法
charactron 显字管
charactron gun 数码管枪
chara limestone 轮藻灰岩
char bucket conveyer 木炭斗式输送机
charco 平坦谷地;水洼
charcoal 木炭;暗紫灰色;炭
charcoal adsorption process 木炭吸附过程;木炭吸附法;活性炭吸附过程;炭吸附法
charcoal adsorption test method 木炭吸附测定法
charcoal agglutination 炭凝试验
charcoal agglutination test 炭粒凝集试验
charcoal and firewood energy 木炭和木材能
charcoal bed filtration 木炭床过滤
charcoal black 木炭黑
charcoal blacking 碳黑涂料;炭黑涂料
charcoal blast fuhmace 木炭高炉
charcoal burner 木炭炉
charcoal cartridge 木炭加热筒
charcoal consumption 木炭消耗
charcoal crayon 炭笔
charcoal diffuser bed 木炭扩散层
charcoal drawing 木炭画;炭画;炭笔画
charcoal dust cake stove 木炭粉饼炉
charcoal electrode 炭电极
charcoal filter 木炭滤池;木炭过滤器;活性炭过滤剂;炭滤器;炭滤池
charcoal filtration 木炭过滤法
charcoal finishing 木炭抛光
charcoal ink factory 炭素厂
charcoal iron 木炭熨斗;木炭铁;木炭生铁
charcoal kiln 炭窑
charcoal knobbled iron 木炭精炼铁
charcoal liquid oxygen explosive 木炭液氧炸药

charcoal non-activated 非活性碳
charcoal paper 木炭纸
charcoal pencil 炭精铅笔;炭笔
charcoal pig iron 木炭生铁
charcoal rot 木炭病;炭腐病
charcoal stick 木炭笔
charcoal test 炭吸附测定;试验木炭(用于测定天然气中的含油量)
charcoal test method 木炭吸附测定法;活性炭吸附测定法;炭吸附测定法
charcoal tin plate 厚锡层镀锡薄钢板
charcoal trap 炭阱
charcoal wire 超低碳钢丝
chare 杂务;零活
Charfered Surveyor 工料估算员(英国政府注册)
charge 借记;加料;带电;进料;磨内物料与钢球总量;要价;装药;装炉;装料量;支出;负责;负载;费用;索价;收费
chargeability 电荷率;极化率
chargeable duration 通话计费时间
chargeable heat 消耗热
chargeable income 应纳税的入息
chargeable labo(u)r 可记在某人账上的劳务费
chargeable time 计费时间
chargeable time counter 收费计时器
chargeable time lamp 报时灯
chargeable weight 收费重量
charge account 零售账;客户购账;赊购账
charge against capital 资本支出;冲底资本
charge against gross national balance 国民总支出
charge against revenue 费用支出;收益支出
charge against the general reserve 一般储备的借项
charge amount 装料量
charge amplifier 电荷放大器
charge and discharge board 充放电配电盘
charge and discharge key 充放电钥;充放电键
charge and discharge rubber(lined) hose 装卸软管
charge and discharge statement 信托财产的经管及支出情况
charge and discharge valve 装卸阀(门)
charge assembly 装药组合
charge atom 带电原子
charge balance 电荷平衡
charge balance analysis 电荷平衡分析
charge balance equation 电荷平衡方程
charge bank 料坡
charge bar 料棒
charge book 作业记录;装料记录
charge bridge 装料桥
charge button 充电按钮;电容充电按钮
charge by adding or subtracting a ratio 运价加、减成计费
charge by meter 按表收费
charge cable 充气电缆
charge calculation 配料计算
charge can 装料罐
charge capacity 炉料容量;药室容量;装载力;装载(容)量;充电容量;充电额;(电池的)蓄电量
charge card 赊账卡
charge carriage 装料机
charge carrier 电荷载体;载流子
charge carrier recombination dynamics 荷载复合动力学
charge-carrying belt 充电带;输电带
charge cask 装料容器
charge characteristic 充电特性
charge chute 装料槽
charge clause 费用条款
charge coefficient 满度系数;装载系数;装药系数;填充系数
charge coil 充电线圈
charge collect 费用向收货人索取;费用到付;收到付款
charge collectable 收费项目
charge compensation 电荷补偿
charge composition 炉料组成;炉料配比;炉料成分
charge computer 收费计算机
charge concentration 装药密度
charge control agent 电荷控制剂
charge control room 加料控制室
charge-coupled device 电荷耦合装置;电荷耦合器
charge current 充电电流

charged 带电的;充电的
charged adsorbate 带电吸附质
charge d'affaires ad interim 临时代办
charged against the budget 从预算下支付
charged body 带电体
charged call 收费电话
charged capillary membrane 带电毛细管膜
charged corpuscle 带电微粒
charged current interaction 带电流相互作用
charge deficient 电荷不足
charge density 电荷密度
charge density wave 电荷密度波
charged extra 额外收费
charge(d) hole 装药孔;装药炮孔;装料口
charge diameter 药包直径
charged in account 记账
charged in full 全部费用
charged interface 带电界面
charge distribution 电荷分布
charged latent image 带电潜影
charged main 运行中管道
charged media electronic air filter 滤材感电式空气过滤器
charged membrane 带电膜
charged meson 带电介子
charged microporous membrane 带电微细孔膜
charged particle 带电粒子;带电颗粒
charged particle accelerator 带电粒子加速器;带电粒子加速计
charged particle activation analysis 带电粒子活化分析
charged particle detector 带电粒子探测器;带电粒子检测器
charged particle energy analyser 带电粒子能量分析仪
charged particle energy spectrometer 带电粒子能谱仪
charged particle scintillation spectrometer 带电粒子闪烁能谱仪
charged particle spectrograph 带电粒子谱仪
charged pressure 充入压力;充气压力
charged profit and loss method 损益收费法
charged reverse osmosis membrane 带电反渗透膜
charge-driven piston mechanism 分离机构
charged rollers 进料辊
charged species 带电物种
charged state 充电状态
charged surface 带电表面
charged wall 带电壁
charged weight 计费重量;货物计费重量
charge efficiency 充电效率;给料效率
charge establishment 装药量确定
charge exchange 电荷交换
charges for collection 代收票据费
charges for conditional water service 条件供水收费
charges for detention at station 停留费
charges for disposing pollutants 排污费
charges for goods keeping temporarily 货物暂存费
charges for licensing of import permit 进口签证费
charges for loading and unloading 路外、路内装卸费
charges for negotiation of export bills 出口押汇费
charges for returning of private carriage or special passenger train stock 空驶费
charges for sorting out cargo mark 分标记费
charges for the depreciation of fixed assets 固定资产折旧费
charges for the use of circulating funds 流动资金占用费
charges for trouble 手续费
charges for tug's service 拖轮费
charges for use 使用费
charges forward 买方收货后付款;货到收款人付费;费用先付;费用后付;货到后收款人自付运费
charges for water 给水率;自来水费;水费
charge hand 装卸工;组长;领班
charge hoist 加料起重机;装卸料吊车
charge-in 进料
charge index 收费指数
charge indicator 充电指示器;炉料牌
charge kiln 装窑

charge leakage 充电漏泄
charge length 装药长度
charge level 料线;料面;投料料位
charge level indicator 料位计
charge level measurement 料面测量
charge lien 监管的留置权
charge limit 极限装药量;装药限度
charge line 加料管线
charge list 赊欠清单
charge magnetism 充磁
charge make-up area 备料场
chargeman 竣炮工;充电工
charge-mass ratio 荷质比
charge material 载质;炉料
charge mechanism 装料机构
charge method 充电法
charge mixing machine 炉料混合机
charge mixture 配料
charge modified filter media 电荷改性过滤介质
charge motor 充电用电动机;充电马达
charge negotiable 协商费(用)
charge neutrality 电中和
charge number 炉料号;账号;荷载数;负载量;批号
charge of an explosive 炸药装填量
charge of briquettes 团块炉料
charge of confined water elastic storage 承压水弹性储量的变化量
charge of cullet 加碎玻璃料
charge of dynamite 黄色炸药装药量;黄色炸药装填
charge off 列销;注销;出账;剔除
charge of oil 装油
charge of rupture 破坏荷载;坡坏载荷
charge of safety 安全荷载
charge of specifically adsorbed ions 特性吸附离子电荷
charge of surety 安全荷载;许可荷载;容许荷载
charge of the electron 电子电荷
charge on assets 资产置留权;财产税
charge on corporation 公司捐税
charge on money 个人捐税
charge on real estate 征收地产税
charge on water pollution 水污染费
charge-overvalve 超负荷阀
charge paid 杂费已付;各项费用付讫;费用已付
charge pattern 充电曲线
charge period 加料期
charge pipe 加料管
charge plate 赊货率
charge-potential relationship 电荷—电位关系
charge power supply 充电电源
charge prepaid 预付费用;费用预付(讫);费用先付
charge pressure indicator 充气压力指示器
charge proportion 进料比例;混合料配比;喂料比例
charge pump 给料泵;供给泵;进料泵;加料泵
charger 控告者;加载装置;加料器;加料机;蓄电池充电器;装料设备;装料人;充电人;装料工;装料装置;充电设备;充电器;充电机;充电工;委托者;委托人
charger-air cooler 增压空气冷却器
charge rate 充电率;进料速率;进料量
charge ratio 充电比;炉料组成;满载系数;满载比;装载系数;装药系数;装填系数;装填比;充气比;配料比
charger-reader 充电读数装置
charger resistance furnace 电阻炉
charger shape 锭料形状
charger sheet 配料单
charger unit 电晕器
charge sales 赊售
charge send 发货;收费送货
charge set 充电装置
charges forward 货到后收货人自付运费
charge shaft 垂直火道
charge sheet 案件记录;记料单
charge side 贷方
charge-space symmetry 荷空对称性
charge stock 进料
charge surface 磨内装球表面积
charge switch 充电开关
charge temperature 着火温度
charge ticket 赊购发票
charge time 充电时间;装料时间
charge time constant 充电时间常数
charge titration 电荷滴定

charge to income 营业支出
charge-to-mass ratio 荷质比
charge transfer complex 电荷转移复合物
charge transfer spectrum 传荷光谱
charge type of swamp water 沼泽水的补给类型
charge unit 充电装置;计价单位;计费单位
charge unit cleaner 充电装置清洗器
charge up circuit 充电电路
charge value 电荷值;充气阀
charge volume 装炸药体积;负载率
charge weight 装填重量;装药量
charge well 垃圾箱
charging 加料;加料装料;装炸药;装炉;装料;注油;充风;投料
charging accessories 装药用具
charging air into pontoon 浮筒充气
charging aisle 炉子跨
charging alternator 交流发电机
charging and discharging equipment for accumulator 蓄电池充放电装置
charging and drawing crane 装卸吊车
charging and service jumbo 多种作业台车
charging aperture 加料孔
charging apparatus 进料装置;装料装置
charging appliance 进料设备;装载设备;料设备
charging area 装料场;反应堆活性区
charging bag 装填袋
charging barrel 料筒
charging basket 料桶;料筐;加料桶
charging bay 装料跨
charging bell 加料钟罩;加料钟
charging bell gear 加料钟装置
charging belt 充电带;输电带;上料皮带
charging board 充电盘
charging box 加料箱;加料槽;铸模
charging bucket 装料斗;料斗;加料桶;料料罐;投料斗
charging by batch 间歇进料;分批进料
charging by batches 分批装料
charging by conduction 传导充电
charging cable 充电电缆
charging capacitor 充电电容器
charging capacity 电池蓄电量;装载量;充电(容)量
charging car 加料斗车;装炸药车;装料车
charging carriage 装料机;炉用推料机;装载小车;装料小车
charging chain 充电链;输电链
charging chamber 装料室
charging change-over 充气转换
charging choke 充电扼流圈
charging chute 进料溜子;进料溜槽;加料斜槽
charging coefficient 充电系数
charging compressor 充气压气机
charging conduit 落料槽
charging connection 充电连接电路
charging connector 充气嘴;充气接头
charging construction 装药结构
charging contactor 充电接触器
charging conveyer 装料运输机
charging crane 加料(起重)机;加料吊车;装卸机;装料起重机;装料吊车
charging current 电容电流;充电电流
charging current of condenser 电容器充电电流
charging curve 充电曲线
charging deck 加料(平)台;装料平台
charging degree of grinding media 研磨体填充率
charging density 装载密度;装药密度;电荷密度
charging device 进料装置;加料装置;装料装置;装料设备;充电设备
charging door 加料口;加料门;装料门
charging dynamo 充电发电机
charging efficiency 充气效率;充电效率
charging elevator 加料提升机
charging emission 进料排气
charging end 加料端;装料端;充电发电机
charging equipment 加料设备;装货设备;充电设备;上料设备
charging face 装料面
charging facility 装料设备
charging feeder 送料斗
charging floor 炉台;装置台;装料平台
charging flue 进气管
charging formula 填充公式

charging frequency 装料次数
charging function 计费功能
charging funnel 装料斗
charging fuse 充电熔丝
charging gantry crane 高架装料吊车
charging ga(u)ge 充电指示器
charging gear 装料机构
charging generator 充电发动机；充电发电机
charging grate 上料格栅
charging hatch 装料口（集装箱）
charging hoist 上料卷扬机
charging hole 装料口；装料孔
charging hole seat 装煤孔座
charging hopper 料斗；加料斗；装料漏斗；装料斗；装料仓；供料斗
charging in alternate layers 分层布料
charging indication 充电指示
charging indicator 充电指示器；充电指示灯
charging inductance 充电电感
charging in groups 成组充电
charging installation 上料设施；加载设施；装料设备
charging interval 充电时间；充电间隔
charging ladle 装料桶
charging level 充电电平
charging line 加液管道；供水管道；供料线；充气管道；供电线路；充液管路；充气管路；送料管道；进料线；供水管线；供给线
charging liquid 充电液
charging load 充电负载
charging lorry 加料斗车；装料斗车；装料车
charging machine 炉用推料机；装料器；装料机；充电机
charging magazine 装料台
charging manning 充分船员配备
charging material 炉料；批料
charging method 装料方式
charging nozzle 带电荷喷嘴
charging of mix 加混合物料；混合加料；充填配合剂
charging of the cell 电解槽加料
charging on a thermal basis 按燃气热值收费
charging opening 加料口；装料门；装料孔
charging operation 充装作业
charging order 负担责任令
charging panel 充电控制板
charging peel 装料机推杆
charging period 充电时间
charging pipe 装料管；供给管；充水管
charging plate 装料板
charging platform 送料平台；装料台；喂料平台
charging point 填充点
charging port 装料孔
charging position 装料位置
charging pump 进料泵；装料泵；灌注泵；供料泵；供给泵；给料泵；充液泵；充水泵；送料泵；上水泵
charging rack 充电架
charging ram 推料杆
charging rate 进料速度；进料率；进料量；加料速度；充气率；充电率
charging ratio of fire-extinguishing agent 灭火剂装充率
charging reactor 充电扼流圈
charging rectifier 整流器充电；充电整流器
charging resistor 充电电阻器
charging room 充电室；充电间
charging scaffold 加料平台
charging scale 加料秤；装料磅秤
charging schedule 装料程序
charging scoop 装料铲
charging sequence 加料顺序
charging set 增压装置；充电装置；充电器；充电机（组）
charging sheet 配料单
charging side 加料侧
charging skid 装料台架
charging skip 加料桶；加料(箕)斗；装料(翻)斗；进料小车
charging spout 料斗；加料槽
charging stage 进气阶段；进料阶段；充电阶段
charging state 充电状态
charging station 充电站
charging stroke 进气冲程；充气冲程
charging subsystem 计费子系统
charging system 加料系统；装料机构
charging-tank 供应罐

charging temperature 装炉温度
charging time 装料时间；充电时间
charging time constant 荷电时间常数
charging tray 装料盘
charging tube 注入管；充电管
charging turbine 透平增压装置
charging-turbine set 增压涡轮机组
charging unit 充电机
charging-up 加注；加添
charging valve 加液阀；加料阀；充注阀；充水阀；充气阀；充电管；送料阀
charging voltage 充电电压
charging volt-ampere 充电伏安容量
charging wagon 上料车；上料车
charging weight 充装重量
charging well 上料井
chargino mass statex 超荷子质量态
charginos 超带电子
chargometer 充电计
charhing unit 充电单元
chariot 柱形摛纵机构板状部分
charity bazar 义卖市场
charity hospital 慈善医院；慈善病院
charity institution 慈善机关
charity sale 义卖
charity school 慈善学校
chark 焦炭；烧炭；焦化
Charles Bridge at Prague 布拉格卡尔斯桥
charlesite 水硼铝钙矾
Charles' law 查理定律
Charles University at Prague 布拉格卡尔斯大学
Charles' wain 北斗七星
Charley forest 莱斯特郡石（英国）
charley paddock 弓形锯（俚语）；钢锯；大锯；粗锯
Charlie-Gibbs fracture zone 查尔利－吉布斯断裂带
Charlie Noble 厨房烟囱罩
Charlier check 查利尔验算法
charlock 田芥菜
Charlotte colloid mill 查洛特胶体磨
Charlton blanching test 查尔顿氏褪色试验
Charlton photoceramic process 查尔顿陶瓷感光法
charmeuse 查米尤斯绸级
charm(ing) price 有吸引力的价格
Charmouthian 察尔毛茨阶【地】
charnel 藏骸所
charnel house 藏骸所；藏骨屋；尸骨存放处
Charnian series 查尼统
charnockite 紫苏花岗岩
Charnoid direction 查恩诺德方向
char-oil energy development process 木炭石油能源开发过程
charoite 紫硅碱钙石
Charonian steps 希腊剧院舞台中央通乐队的踏步；希腊剧院舞台中央通乐队的扶梯
charophyta 轮藻类
char-pit method 火烧清除树根法
Charpy impact machine 摆锤冲击试验机；单梁式冲击试验机；查贝卜梁式冲击试验机
Charpy impact strength 查贝冲击强度；摆锤冲击强度
Charpy impact test 单梁式冲击试验；查贝(式)冲击试验；摆锤(式)冲击试验
Charpy impact tester 查贝冲击试验机；摆式冲击试验机；摆锤(式)冲击试验机
Charpy key hole specimen 钥匙孔形缺口冲击试样
Charpy pendulum 摆式冲击试验机；摆锤(式)冲击试验机
Charpy's alloy 贾皮锡锑铜合金
Charpy test 查贝试验；单梁(式)冲击试验
Charpy toughness test 查贝韧度试验
Charpy-type test 查贝冲击试验
charpy unnotched impact value 查贝无缺口冲击值
Charpy V-notch 查贝 V 形切口
Charpy V-notch test 查贝 V 形缺口试验
charr 烘底船
charred 烧成的
charred bones 炭化骨；骨炭
charred coal 焦炭；焦煤
charred peat 焦化泥炭；石炭化泥炭
charred pile 焦头桩
charred spot 炭化斑
charred stone 斜凿石
charring 焦化；碳化；炭化（防腐）
charring ablative material 碳化烧蚀材料；炭化烧蚀材料
charring ablator 炭化烧蚀体
charring layer 炭化层
charring rate 炭化速率
charring spot 炭化斑
charry 炭化；似碳的
chart 航海图；程序框图；图表
chart abbreviation 海图上缩写
chart amendment patch 区域改正
chart atlas 海图集；海图册
chart block 海图改正贴图
chart board 图夹；图板
chart border 海图图廓；海图图框
chart card 海图卡片
chart catalog(ue) 海图目录
chart comparison unit 雷达标图板；海图比较器；雷达测绘板
chart compass 罗盘花；向位圈
chart compilation 海图编制
chart conditional sign 海图符号
chart constant 制图常数
chart correction 海图改正
chart correction service 海图修改业务
chart cutter 切换录纸刀
chart datum 海图深度基准面；海图零点；海图基准(面)；水深基准点
chart depot 海图库；海图供应站
chart desk 海图桌；图板
chart distortion 海图变形
chart division 表格刻度
chart drawing pen 鸭嘴直线笔
chart drawing set 绘图仪器
chart drive mechanism 记录纸传动装置；拖带机构
Charte d' Athenes 雅典宪章（1933 年确定城市规划原则的，法国）
charted coast 图注海岸线
charted depth 海图水深；图注水深
charted position 图注位置；图上位置
charted screen 坐标投影屏
charted visibility 海图视界
charted visibility of light 图注灯标射程
chart equipment 海图设备
charter 租船契约；租船合同；执照；公司执照；契约；特许证；特许权；特许成立公司
charterage 租船费
charter back 卖出旧船再租回
charter bare 空船租船契约
charter base 租船合同租金计算标准
charter by time 租船按时计费；期租
charter by voyage 租船按航次计费
charter commission 租船佣金
charter contract 包租合同
chartered accountant 注册会计师；会计师；持有皇家特许证的会计师；特许会计师
chartered back 返租
chartered bank 特许银行
chartered building surveyor 房屋估算员；估算员；特许房屋勘测员
chartered company 特许公司
chartered concession 特许证；特许权
chartered engineer 注册工程师；特许工程师
chartered freight 租船运价；租船运费
chartered hire 船舶租金
chartered institute of transport 特许运输学会
Chartered Institution of Building Service 特许建筑服务设施协会
chartered owner 租船主
chartered plane 包机
chartered right 特许权
chartered ship 租用船；被租船
chartered surveyor （英国政府注册的）工料估算员；（英国政府注册的）工程检查员；特许测量员
chartered vehicle 包车
chartered voyage 租船航次
charterer 租船人
charter(er) certificate 租船证书
charterer pays dues 租船人支付费用；承租人负责税金
charter fee 包租费
charter flight 包租的班机
charter foreign vessels 租用外轮
charter hire 租费
charterhouse 卡尔特教会修道院
Charterhouse at Granada 格兰特纳达教堂议事

厅;格兰特纳达教堂圣器室
chartering 租船
chartering agent 租船代理(商);租船代理(人)
chartering broker 租船经纪人
chartering company 租船公司
chartering market 租船市场;包租市场
chartering of tonnage 租船
chartering order 租船证;租船单
charter-land 特许保留地
charter loan 信用借款
charter member 发起人;创始成员
charter money 租(船)费
charter of cargo 货物包租
charter of company 公司章程
charter of concession 特许证
charter of freight 包租运费
charter of natural conservation 自然保护宪章
charter of the corporate body 法人章程
Charter of the United Nations 联合国宪章
Charter on Groundwater Management 地下水管理章程
charter order 租船委托书
charter party 空船租船契约;租船契约;租船合同
charter party bill of lading 租船合同提单
charter party by demise 光船租船合同
charter party forms 租船合同标准格式
charter period 租船期
charter rate 租船费率
charter's operator 租船承运商
chart feed 供图
chart folio 海图夹;海图集
chart for break even analysis 保本分析图
chart for super-elevation 超高图表
chart gallery 图箱;图库
chart holder 图架
chart house 海图室
chart index 接图表;图幅接合表;图表索引
charting 制图;制表;图示行为法;图表研究;填图
charting base level 绘图水位
charting datum 绘图基面
charting photography 地形测图摄影
chart integrator 图形积分器
chart lamp 海图灯
chart legend 海图标题栏
chartlet 小海图;区域改正;贴图
chartlet for correction 贴图改正
chart level 绘图水位
chart light 海图灯
chart magazine 记录纸盒
chartmaking 作图
chart making institution 制图部门
chart making unit 制图室
chart matching 图形匹配;图形符合
chart matching device 图重合仪
chart name 海图图名
chart number(ing) 海图编号;海图号
chart of accounts 会计科目表
chart of background value of soil 土壤背景值图
chart of business 商业图表
chart of distribution of base rocks 基岩分布图
chart of equal magnetic force 等磁力线图
chart of equal variation 等磁差曲线图
chart of goods intercourse 货物交流表
chart of lines of equal horizontal magnetic force 等水平磁力曲线图
chart of lines of equal magnetic dip 等磁倾角曲线图
chart of lines of equal magnetic variation 等磁差曲线图
chart of magnetic field affected by direction variation 改航对磁场影响图
chart of marine gravity anomaly 海洋重力异常图
chart of mixture ratio 混合比图表
chart of percentage comparison 百分率比较图
chart of possible fire 火力配制图
chart of soil background value 土壤背景值图
chart of standardization 标准化图表
chart of the planet Mars 火星图
Chart of the World General Surface Current Circulation 世界表层环流图
Chart of Time Zones of the World 世界时区图
chartographer 制图员
chartographer drawer 制图员
chartography 制图法;制图学

chartometer 测图器
chart on issue 发行的地图
chartophylacium 档案室
chart paper 海图纸;图(表)纸
chart plate 海图印刷板
chart plotter 填图员
chart portfolio 海图夹
chart position indicating apparatus 图上定位装置
chart projection 海图投影(法)
chart quadrat 图解样方;图记样方
chart rack 海图架;图架
chart reader 卡片阅读仪;海图放大镜;图表阅读者
chart reading 海图解释
chart reading glass 海图放大镜
chart recorder 绘图记录仪;载图记录仪;图表记录器;图案记录器
chart-recording instrument 绘图记录仪;曲线记录仪;图记录式仪表
chart relationship 接图表
chart representation 图示法
chartreuse 黄绿色;浅黄绿色(的);卡尔特教会修道院(法国);淡黄绿色
chartreuse colo(u)r cloth 草黄色布
chart rolling mechanism 记录纸滚动机构
chart room 航海室;海图室;图表室
chart scale 海图比例尺;地图缩尺
chart scale length 记录纸标尺宽度
charts coverage 海图分幅
chart screen 图形投影屏
chart series 地图系列
charts for reference 备考图
chart sheet 地图图幅;图页;图幅
chart size 海图图幅
chart speed 记录纸速度;图移速率;图表记录速度
chart-speed shifter 记录纸变速器
Chart Standardization Committee 海图标准委员会
chart symbol 海图图式;海图符号;海图代号
chart table 海图桌;图表台
chart title 海图标题
chart weight 压图块;图镇
chart work 海图作业;制图工作
chase 模套;用梳刀刻螺纹;管子槽;管线槽;沟;版框;凹沟;暗线槽;切螺纹;排版架
chase bonding 竖槽接桩;凹槽结合(法)
chased helicoid 法向梯形齿廓螺旋面
chase form 成槽木模;凹槽模
chase hole 管子槽;楼面洞口(运料用)
chase indicator 跟踪指示器
chase leaks 检漏
chase lintel 有浮雕的金屋小梁;槽楣
chase mortise 槽榫;榫眼槽
chase mortising 凿榫眼;开材榫;开凿木榫;开木榫;开槽榫(法)
Chase National Bank 大通银行
chase partition 管槽隔墙;竖槽隔墙
chaser 石棉碾;揉泥碾;螺纹梳刀;轮碾机;雕刻工;螺纹梳刀;螺纹刀;螺压机;边碾机;送桩器;梳刀盘
chaser die head 螺纹梳刀头;螺纹梳刀盘
chaser grinder 螺纹梳刀;螺纹切削板牙磨床
chaser mill 螺纹梳刀机;轮碾机;辊磨机;干式辊碾机;干碾机
chase the pipe 清洗管子
chase the threads 清洗螺纹
chase wedge 楔形槽
chashou 叉手
chasing 雕刻工作;螺旋板;金属锤;铸件最后抛光;雕镂
chasing bar 攻丝装置;切制螺纹装置
chasing calender 雕镂压延机
chasing control hodograph 追逐时距曲线系统
chasing dial 螺纹指示盘
chasing hammer 雕镂锤;修整铸件锤
chasing lathe 螺纹车床
chasing tool 螺纹刀具;螺丝刀具
chasm 断层;裂口;陷坑;大陆裂缝;深渊
chasma 呵欠
chasorrite 黏土
chasse 圣徒遗物存放箱
chassignite 纯橄无球粒陨石
chassis 框架;机箱;机壳;机具台架;底盘车;底盘;车底架
chassis assembly 整底盘;底盘组装;底盘总成;底

架总成
chassis base 底板
chassis black 底盘防锈用黑色涂料
chassis cross member 底盘横档
chassis dynamometer 框架测力计
chassis earth 底盘接地
chassis fittings 底盘配件
chassis for yard 场地用底盘车
chassis frame 底(盘车)架;起落架
chassis grease 底脂
chassis ground 框架接地线;底盘接地
chassis height 机身高度(飞机);底盘高度
chassis lubricant 底架润滑剂
chassis mainframe 底盘
chassis-mounted construction 底盘式结构
chassis-mounted gritter 底盘式铺砂机;底盘式铺砂车
chassis number 底盘号
chassis spring 底盘弹簧
chassis system 底盘车方式;底盘(车队)系统
chassis train 底盘车队
chassis underframe 底架
chassis wiring cable 底盘电线
chassis wiring terminal sleeve 底盘接线柱管
Chastetree 木姜;牡荆属
chastity 精洁;高雅
chat 打包粗麻布;燧石(砾岩);碎石;碎片
chatean grey 霜灰色
chateau 别墅;庄园;城堡
chatelet 小宫殿;小型城堡
chatkalite 硫锡铁铜矿
chatki 粉碎紫胶用的石磨
chatoyance 猫眼效应;变彩;闪光现象
chatoyant 金绿宝石;猫眼石;闪光石(又称猫眼石)
chat-roller 焙烧矿石辊碎机
chat room 聊天区
chats 矿山废石;选矿中间产物
chat-sawed finish 粗切锯面
chat-sawn finish 燧石屑锯面
chattel (除房地产以外的)财产
chattel charging order 动产扣押令
chattel mortgage 动产抵押;财产抵押
chattel mortgage bond 动产抵押债券
chattel mortgage financing 动产抵押融通
chattel personal 个人的动产
chattel real 准不动产(指租借地权等)
chattels 动产;杂用品
chattels personal 人的准动产
chattels real 物的准动产
chatter 撞击震颤声;震颤;振动声;刀跳
chatter-bar 颤动路标
chatter bump 路面不平凸起;砾石路面隆起;震凸变形
chatter decoration 并纹装饰法
chattering 间歇电震;震颤(波)纹;震颤;跳跃现象;跳刀
chattering dither 颤振
chattering drive 震颤行车
chattering of gears 齿轮颤动
chattering of the valve 活门跳动
chattering prevention block relay 防跳闭锁继电器
chatter mark 振痕;震颤(波)纹;振纹;颤痕;颤动擦痕
Chatterton's compound 查特顿化合物
Chattock ga(u)ge 微压计
Chattonella marina 海洋卡盾藻
chat wood 矮林;灌木
chauffage 烘烙法
chauffer 小炉子
chauffeur (小型的)汽车驾驶员;驾驶员(小汽车)
chauffeurette (汽车的)女驾驶员
chauffeurs' room 汽车驾驶员室;驾驶员室
chauffeuse (汽车的)女驾驶员
chaulmoogra oil 大风子油;晁模油
chaulmoogrene 晁模烯
chaulmoogric acid 晁模醇
chaulmoogroyl 晁模酰
chaulmoogryl 晁模基
chaulmoogryl alcohol 晁模醇
chaulmugra 晁模子
Chaumitien series 炒米店统【地】
chauntry 祈唱堂;附属教堂;教堂中的小纪念堂
Chauvel safety glass 夏飞尔安全玻璃
chawls 多层寄宿舍(印度)

Chayu block 察隅地块
Chayu Tengcnong seismotectonic zone 察隅腾冲地震构造带
Chazyan 夏西统【地】
chcesc ingot 八方钢锭
cheanser 清铲工
cheap article 廉价货
cheap credit 低利息贷款
cheap dollar 已贬值的美元
cheapen 跌价
cheapener 降价填料;廉价添加料
cheap goods 廉价品
cheap goods bargain 廉价商品;廉价交易
cheapie 廉价商店(美国)
cheap John 卖廉价商品的商人
cheap labo(u)r 廉价劳动
cheap labo(u)r force 廉价劳动力
cheap money 廉价货币;低息借款;低息贷款
cheap-money policy 低息政策
cheap price 低价
cheap restaurant 经济餐馆
cheater 加力管;加长管截
cheat in work and cut down material 偷工减料
Chebyshev approximation 极小极大法;切比雪夫逼近法
Chebyshev adjustment 切比雪夫平差
Chebyshev polynomial 切比雪夫多项式
Chebyshev's differential equation 切比雪夫微分方程
Chebysher's equalizer 切比雪夫均衡器
Chebyshev's filter 切比雪夫滤波器
Chebyshe's smoothing 切比雪夫平滑法
Chebyshev's inequality 切比雪夫不等式
Chebyshev's norm 切比雪夫模
Chebyshev's polynomials 切比雪夫多项式
Chebyshev's series representation 切比雪夫级数展开
Chebyshev's set 切比雪夫集合
Chechotta formula 切乔特公式
check 节制阀;校验;校对;检查;检测;支票;账单;对照;对联;微裂纹陶瓷器底部边缘
check 4 mm opening 四毫米锁闭
check-acceptance survey of open pit 露天采场验收测量
check account 支票账户;查账;盘账
check action (门窗的)制动作用;检验行动
check alteration 涂改的支票
check analysis 校核分析;检验分析;验证分析;对照分析;成品分析
check and accept 验收;点收
check and accept suggestion 成果验收意见
check and accept the completed capital construction project 基本建设项目竣工验收
check and approve 储量审批
check and balance 制约与平衡;制衡原则
check and decide 审定
check and drop 压降检验;落差检验
check and hand over 点交
check and receive 点收
check and shake 裂纹
check and store 校验与寄存
check angle 校核角
check apron 挡板;挡帘
check as cash 视同现金支票
check authorization record 检验特许记录;检验核准记录
check back signal 后验信号
check ball 止逆球阀
check band 制动绳;制动带
check bar 校验棒;校核杆;试验杆
check base(line) 校核根据;检核基线;验核基线;校对基线;校核基线
check basin 格田
check-basin irrigation 分块灌溉;方块灌溉
check beam 导航射线;导航波束
check bearing 校核方位角;校核方位
check beater 反转逐稿轮
check benchmark 校核水准点
check bit 校验位【计】;校验比特;校验位;监督位
check biting 咬颊
check bit sum 校验位和
check board 检验板;挡板;边模(板);逐稿器挡帘
checkboard door 棋盘门
checkboard load 棋盘式装载

checkboard system 棋盘式街道系统;棋盘式街道体系
check bolster 防松垫木;防松承梁
check bolt 制止螺杆;防松螺栓
check book 理货簿;支票簿
check-book money 支票货币
check-boring 控制钻进;验证钻进
check bound street system 棋盘式街区
check box 校验盒
check bus 校验总线;检验总线
check button 尺度索结扣
check by sampling 抽查
check by sight 目视检核;肉眼检查;视力检查
check cable 控制索;拦阻索;安全索
check calculation 核(对计)算
check card 校验卡;检验卡片
check chain 限位链;保安链
check chain clevis 保安链 U 形夹
check chain eye 安全链眼
check character 校验字符;校验符号
check clearing 票据交换
check clock 校核钟;穿孔钟
check clone 对照品系
check-cloth suitcase 格布箱
check code 校验码
check collection 支票托收;支票兑现;支票兑取
check collector 支票兑取人
check column 校验列
check committee 查账委员会
check computation 校验计算;核算
check course 防隔层
check crack 收缩裂缝;细裂缝;网裂;收缩裂纹
check crossed specially 特别划线支票
check cross-section 对照断面
check curtain 挡帘
check dam 拦水坝;拦沙坝;护底坝;谷坊坝;谷坊;挡水坝
check dam for building farmland 拦沙造田坝;淤地坝;造田坝
check damp 烟囱闸板
check damper 风挡
check dam with willow 插柳谷坊
check deposit 支票存款
check design 方格纹
check digit 校验数位;检验数字;检验数位;校验数码;检查数字;核对数字
check dike 拦沙堤
check disposition 棋盘式排列
checkdraft 烟道挡风板;烟道洞口;烟囱门洞
check drawer 支票出票人;出票人
check-drilling 控制钻进;验证钻进
check drop 节制跌水闸
check drum 逐稿轮
checked and adjusted capacity 查定能力
checked bottom joint 底缝开裂
checked by 校核;复核
checked crepe 点格绉
checked dam 格坝
checked direction 检查方向
checked discharge 检算流量;验算流量
checked edge 裂边
checked flood(water-)level 校核洪水位
checked gauze 格子绉
checked ground 企口底板
checked mosaic 有控制点镶嵌图
checked spot elevation at site 现场校核高程
checked surface 格纹面;龟裂(表)面;布满细裂纹的表面
checked-up lake 堰塞湖
checker 理货员;检验员;校核者;检验装置;检验员;检验设备;检验器;量员;检查者;检查器;格子体;点货员;查对者;方格子;方格图案(美国)
checker and accepter 验收者
checkerboard 万格盘
checkerboard array 棋盘阵
checkerboard brushout method 黑白格测缝盖灰方法
checkerboard door 棋盘门
checkerboard drilling 方格式钻进;棋盘式钻进
checkerboard effect 检测板效应
checkerboard frequency 棋盘频率
checkerboard image 棋盘图像
checkerboarding 棋盘法
checkerboard loading 棋盘式荷载布置

checkerboard(masonry) bond 格子形砌合墙体
checkerboard of colo(u)r filters 嵌镶式滤色器
checkerboard pattern 棋盘状图案;棋盘式图案
checkerboard pattern test signal 棋盘(测试)信号
checkerboard regenerator 砖格蓄热室
checkerboard reticle blade 试验板调制盘叶片
checkerboard reticule 棋盘形调制盘
checkerboard roof 格子炉顶
checkerboard street system 方格式街道网;棋盘式街道系统;棋盘式街道体系;方格式街道系统
checker brick 格子砖;方格砖
checker brickwork 格子砖砌体
checker chamber 蓄热室
checkered 交错的;成格子状的;方格的
checkered buoy 花格浮标
checkered iron 花纹钢;格纹铁;网纹铁板
checkered pattern 芦席纹图形;方格图案(美国)
checkered plate 网纹钢板;花铁板;方格板;网(格)纹板;防滑钢板
checkered shear wall 格纹剪力墙
checkered sheet 网纹钢板;划格片材;划格板材;网状钢板
checkered steel plate 花纹钢板;网纹钢板;防滑钢板
checkered straw mat sand-break 草方格沙障
checkered surface 方格式块料路面
checkered type brick paver 方格铺地砖
checkered(work) pavement 方格式路面
checker fire brick 火口砖;格子耐火砖;方格耐火砖
checker flue 砖格气道;砖格烟道
checkering cradle 滚花托架
checkering line count 刻花直线密度
checkering machine 刻花机
checkering test 格孔检查
checkering tool 滚花工具
checker parquet flooring 拼花硬木地板
checker passage 砖格孔道
checker pattern 钢化应力斑;风嘴印
checker paving brick 方格铺地砖
checker plate 格子花板(用于厨房炉灶下面通风);菱形网纹钢板;网纹钢板
checker plow 筑埂犁
checkers 砖格;方格式铺砌(美国)
checker setting 格子砖码法
checker supporting column 格砖支柱
checkerwork 方格式铺砌(工作)(美国);方格花纹;格式装置;方格式砌体;棋盘形细工;格子砖砌体
checker work grillage 砖格(子)
checker work pavement 直角交替式路面;方格式路面;棋盘式铺面
check experiment 检查试验;核对试验;对照试验;对照检验;对比试验
check face 沟深
check facility for snow slide 雪崩防止设备
check feature 检验性能
check feed valve 给水止回阀
check figure 校核数;核对数字
check figures in posting 核对过账数
check-fillet 挡水嵌条;屋面沥青天沟;边饰
checkfire 防火复合门
check firebrick 格子火砖
check flight 技术检查飞行
check flood 校核洪水
check flooding 过高壅水;分畦淹灌;分畦漫灌
check-flooding irrigation 分区漫灌;畦灌;分畦漫灌
check for collection account 催收支账户
check for errors 检查错误
check for interference fringes 色环检查
check formula 验算公式
check for transfer 转账支票
check for zero 零点校核;校验零点
check gate 节制闸(门);斗门;辅助门;配水闸(门)
check ga(u)ge 校验仪表;校验量规;校核水尺
check gauging 验证计量
check goods 查点货物
check harrow 分块耙
checkhead 尺度索扣拨器
check high water level 校核高水位
check holder 支票执票人;支票持有人
check hole 检查口
check-in 进港登记;登记
check-in detection 检入检测器
check indicator 校验指示器;检查指示灯

checking 核对；校核；检验；减风；木材裂开；细裂；釉裂；复核；浅裂纹；起裂纹；坯裂；微裂纹；涂膜浅裂；涂层表面发裂
checking account 核对账目；支票存款；对账
checking amplifier 监听用放大器
checking and recovery error 校验和恢复错误；差错检出及排除
check in gate 进港登记闸门
checking bit 奇偶校验位
checking bollard 挡车柱；控制系缆柱；码头带缆柱；挂缆柱；防松式系船柱
checking brake 减速制动器
checking braking 减速制动
checking by resubstitution 置换检查；重新代入检验
checking calculation 验算
checking certificate 检验证书
checking circuit 检查电路；校验线路；电路检查
checking colo(u)ration 校对色；标准色
checking computation 检验计算；验算
checking cylinder 制动缸
checking dam 拦洞坝；砂坝；防洪坝；拦砂坝；节制坝；护坝；谷坊；防冲坝；挡水坝
checking deposit 以支票提取的活期存款
checking device 校正装置；检验装置；尺度索扣拨器；锁定装置
checking draught 校验吃水深度
checking drill string time 钻具检查时间
checking experiment 检查实验
checking feature 校验性能；校验能力；检验特性
checking floor hinge 地垄；控制式落地门枢
checking for 零位调整
checking for zero 零位调整
checking ga(u)ge 检查规；检查仪
checking information 检验信息
checking key 校正键
checking law 票据法
checking line 检核线
checking list 清点材料
checking load 验算荷载
checking loop 检验循环；检验回线
checking measurements 检验测量
checking motion 自动多梭箱档；防冲装置
checking number 检验数；检查数
checking of building line 验线
checking of dimensions 核对尺寸；尺寸的核对
checking of drawings 审图
checking of estimate 预算书的检查
checking off symbol 查讫符号
checking of invoice 核对发票
checking of materials 清点材料
checking of operations 操作检查
checking of quality 品质复验
checking of routing 路线检查
checking of weight 重量核对
checking point 检核点；检查点；校验点
checking positive 检核正片
checking posting 核对过账
checking procedure 校验过程；检验步骤；检查步骤；核对手续
checking profile 校核剖面；校核断面
checking program(me) 检验程序；检查程序
checking resistance 抗细裂性；抗龟裂性；耐龟裂性
checking routine 检验程序；检查程序；校验例程序
checking sample 检查抽样；检验样品
checking scale 检验比例尺
checking screening 检验筛分
checking section 校核断面
checking sequence 检查序列
checking sheet 检查单
checking subroutine 检验子程序
checking symbol 校验符号
checking table 检查台
checking the telegraph 对车钟
checking the terrain at the actual ground 现地对照地形
checking tickets system 自动检票系统
checking traverse 检核导线
checking up on the projects under construction 清理再建项目
checking weight by draft 水尺计重
checking weight by volume 容量计重
check ink 支票油墨
check investment items 审定投资项目
check in wood 辐裂；木材辐裂

check irrigation 方格灌溉；围畦灌溉；畦灌；格田灌溉
check it 截住
check jump 突阻水跌
check key 校验键；校验关键字；检查键；止动监听按钮
check knot 尺度索结扣
checkless banking system 不用支票的银行制度
checkless society 不用支票的社会
check levee 畦埂
check level 校核水准
check lever 止回杆
check light 校验指示灯
checkline 校核线；靠岸缆索
check line of sounding 检查测深线
check-line survey 查核线调查
check list 校验表；校对表；检查项目表；检查单；检查表；核查一览表；项目核对法；一览表；验证单；核对清单；核对表；清单（目录）
check list method 表格检验法
check list system 校验表格系统；检验表格系统
check load 校核载荷；验算荷载
check loading 验算荷载
check loading of bridge 桥梁验算荷载
check lock 门锁防松装置；保险（附）锁；保险螺帽；门锁螺栓防松器
check locking 照查锁闭【铁】
check lock lever 锁紧杆
check low water level 校核低水位
checkman of train 检车员
check mark 校验标记；检验记号；检验标志；检查记号；核对符号
check master 裂纹检验器
check matrix 校验矩阵；复核表
check measurement 控制尺寸；校核测量；检测
checkmeter 校验仪（表）；检验仪；计量表
check method of irrigation 分畦灌溉法
check mix 校正配合比
check mode 检核方法
check nozzle 自动关闭喷嘴
check number 校验数
checknut 校核螺母；止动螺母；防松螺母；防松螺帽；锁紧螺母；锁紧螺帽；保险螺母；保险螺帽
check observation 检验观测
check observation line number 检查观测线号
check observation point number 检查观测点号
check observation value 检查观测值
check of calculating and collecting petty expenses in transport 运杂费计算和核收的审查
check of drawing 校图
check of equipment 设备的检查
check off 顶底板会合；查讫；验证性试验；验讫
check of foundation subsoil 验槽
check of procedure 程序的复核
check of sealed sections 检查封锁区段
check of test equipment 试验设备的检查
check of wellbore technical state 井内技术情况检查
check of wood 木材裂纹
checkon base 基线检核
check on connection correction 接线正确性检查
check only for account 只能转账的支票
check operator 校验员
checkout 付账后离去；检测；检验；检出；检查；查出；测试；验算；校正；结账；调整
checkout console 检验台；检测板；测试操纵台
checkout counter 结账柜台
checkout detector 检出池检测器
checkout environment area 结账区；检查环境区
checkout gear 测试设备
checkout operation manual 检验操作手册
checkout routine 校验程序；校验例行程序
checkout system 检出系统
checkout test set 检测设备
checkout time 清除时间；扫清时间
check out valve 止回阀
check over 清查
check pawl 止回棘爪
check pawl pivot 止回爪枢
check payable in account 以记账支付的支票
check payable to bearer 来人即付的支票；无记名支票
check piece 制速器；支柱；支架；自动停车器
check pin 防松销钉
check pipe 检查口；检查管

check plant 对照植株
check plate 挡板；制动板；止动板
check plot 对照样地；对照（小）区
check plus minus 正负校验
checkpoint 检验点；检查站；检查台；检测点；核对校准点；核对（基准）点；抽点检验；校验点；查录；测试点；边境检查站
checkpoint and restart procedure 检验点和再启动过程
checkpoint code 抽点检验电码
checkpoint data set 检查点数据集；检验点数据集
checkpoint dump 检验点（信息）转储；检验点清除
checkpoint file 检查点文件
checkpoint record 检验点记录
checkpoint request record 检查点请求记录
checkpoint restart 检查点再启动；检查点再启动
checkpoint restart service facility 检验点再启动性能；检查点再启动服务程序
checkpoint routine 检查点子程序；校验点例程；检查点例行程序
checkpoint sorting 校验点分类
checkpoint start 检验点启动；检查点启动
checkpoint subprogram(me) 抽点检验子程序
checkpoint subroutine 检查点子程序；抽点检验子程序
check post 检查哨所
check pot 称量桶
check practice 实施的核查
check price 限制价格；限价
check problem 校验问题；校验题；检验问题
check profile 校核剖面
check program(me) 校正程序；校验程序；测试程序；试验程序
check protect 检验保护
check protection 检验保护
check rail (双悬窗的)中横挡；榴板条；挡条；会挡；护轮轨；护轨；拔水条；碰头挡
check rail sash 垂直滑移窗扇
check rail track 有护轨的线路
check rate 即期汇兑率；支票利率
check reading station 读数检核台
check receiver 监控接收机
check register 检验寄存器；支票登记簿
check relay 核对断电器
check replacement 支票挂失
check requisition 照查条件
check reset key 校验复位键
check result 校核结果；检验结果；检查结果；核对结果；对照结果
check returned 退回支票
check ring 垫圈；止动环；挡圈；弹簧挡圈；锁紧环；锁环
check rod 检验棒；抑止杆
check roller 扣拨器滚轮
check rolling 复验碾压
check room 小件物品寄存处；衣帽间；更衣室；存物室；行李寄存处；行李寄放处
check rope 防松索
check routine 校验例行程序；校验例程
check routine test 校验程序测试
check routing symbol 支票清算路线标识
check row 纵横行植；棋盘式种植；校验行；带形物
checkrow boot 方形穴播开沟器
checkrow bunch planter 方形穴播机
checkrow clutch 方形穴播同步离合器
checkrower 方形穴播装置
checkrowing device 尺度索扣拨器
checkrow planter 方形穴播机；方形点播机
checkrow planting 方形点播
checks 斧颊；发裂
check sample 控制试件；校正试样；校核试样；校对样品；检查用试样；检查样品；对照样品；对照样本；对照试样
check sampling 检验取样
check screw 压紧螺丝；压紧螺钉；止动螺钉；固定螺丝；固定螺钉
check sheet 对账单；报账单；设备及工具选择专用卡片
check signing machine 支票签字机
check sluice 逆止水阀；逆止水闸
check sorter 检查分类器
check specimen 核验样品；检查用试样
check spring 止动弹簧；调整弹簧
check stand 验货收款台

check station 检查站
check stop 扯启窗止;检查停止;阻挡压条;双悬窗框底边线沟槽
check stopper 止索
check strip 分隔条
check structure 节制建筑物;拦水建筑物
check stub 支票存根
check-sum 检验总数;检查和;检验和
check sum code 检查和码
check sum error-detecting system 检和式误差测定系统
check surface 表面龟裂;龟裂面
check survey 检查测量
check symbol 检验符号
check system 检验系统
check tag 检验标识
check test 控制试验;校核试验;鉴定试验;检查试验;核对试验;核对检验;对照检验;测试
check the market 查对市场行情
check the quality 检查品质
check the stock 核对存货
check throat(ing) 窗槛止水槽;窗台滴水;滴水线;滴水槽
check ticket 附支票单
check time 超记时间
check to bearer 不记名支票
check to order 记名支票;认人支票;抬头支票
check total 检验总数;检查和
check track 校验道;检验道
check trading 支票交易
check transit number 支票传送号码
check truncation 支票电托收
check trunk 校验总线;检验总线
check-type boot 方形穴播开沟器
check type calculation of heat exchanger 换热器的核算型计算
check-up 检查;核对
check-up account 对账
check-up lake 堰塞湖
check-up of drawings 审图
check-up of warehouses 清仓查库
check up on direction 方向检核
check-up on the financial work 财务检查
check-up o production capacity 查定生产能力
check-up system 考核制度
check-up thoroughly 彻底检查
check-up through statistic(al) means 统计检查
check valve 检验开关;检验阀;检查阀;逆止阀;单向活门;单向阀;防逆(止)阀;止回阀
check-valve ball 止回阀球阀;单向活门球阀;止回阀珠
check-valve ball seat 止回阀球座
check-valve ball spring 止回阀球弹簧
check-valve body 止回阀体;单向阀壳体
check-valve body cap 止回阀体盖
check-valve cap 止回阀盖
check-valve case 止回阀箱
check-valve case gasket 止回阀箱密封垫
check-valve clamp 止回阀夹
check valve for hot water inlet 温水入口止逆阀
check valve for washing water 洗涤水止逆阀
check-valve guidance 止回阀导座
check-valve hole liner 止回阀孔衬圈
check-valve lever 止回阀杆
check-valve lever spring 止回阀杆弹簧
check-valve nozzle 止回阀喷嘴
check-valve plunger 止回阀柱塞
check-valve plunger guide 止回阀柱塞导座
check-valve pump 止回阀配流泵
check-valve spring 止回阀弹簧
check-valve to manifold pipe 止回阀至歧管管
check variety 对照品种
check verification record 检验核实记录
check warehouse stocks 清仓查库
check washer 止回垫圈;防松垫圈;防松垫片
checkweigher 检验衡器
check weight 数量复验
check wheel 棘壁
check-wire guide 尺度索导向器
check-wire reel 尺度索卷绕轮
check with construction 复核建筑
check word 校验字
checkwork 直角交缝式;方格花纹;方格式铺砌工作;砌砖格;棋盘形细(木)工;格子体
check writing machine 支票填写机

chee 红色硬木(印度支那产)
Cheecol 轻量混凝土(其中之一种)
cheek 滑车壳面板;小炉侧墙;中型箱;成对部件;边框【建】;榫眼侧
cheek block 单面滑车;桅顶滑车
cheek board 边模板;侧模(板)
cheek clutch 颚形离合器
cheek course 避湿层
cheek cut 椽子端部斜切口;斜锯面
cheek damp course 避湿层
cheek flask 中(砂)箱
cheek knee 首柱破浪材肘板
cheek listing method of environment(al) assessment 环境评价列表清单法
cheek nailing 对钉法
cheek of block 块料侧面
cheek of crank 曲柄臂
cheek pattern 中型箱铸模
cheek pieces 侧壁;边框;成对侧板【建】
cheek plate 桅肩;碎矿机颊板
cheeks 成对侧板【建】
cheek tile 老虎窗侧挡板
cheek wall 侧壁;边墙
cheerful room 阳光充足房间
chees block 垫块
cheese 筒子纱
cheese aerial 盒形天线
cheese antenna 盒形天线;饼形天线
cheese cloth 稀纱布;粗滤布
cheesecloth 粗平布
cheesecloth filter paper 粗孔滤纸
cheese hard 半干状态
cheese-hard clay 硬乳酪状黏土
cheese head 高平头;平头;凸圆头;丝饼头
cheese head bolt 高平头螺栓
cheese head rivet 高平头铆钉
cheese head screw 圆头螺丝;圆头螺钉;高平头螺丝
cheese industry 干酪工业
cheese room 干酪室
cheese screw 圆头螺丝;圆头螺钉
cheesewring 蘑菇石(美国科尼什产的一种浅灰色花岗岩)
cheesiness 酪皮;乳酪状膜;乳酪状干燥
cheesy 漆膜疲软
cheesy pus 稠脓
chef 厨师
cheiroscope 手导镜
Cheirotonus 彩臂金龟
chela 夹子;钳爪
chelant 螯合试剂
chelate 螯形的;螯合(物)
chelate agent 螯合剂
chelate complex 螯合物
chelate compound 螯(形化)合物
chelate effect 螯合效应
chelate fibre 螯合纤维
chelate formation 螯合物形成
chelate group 螯合基
chelate laser 螯合(物)激光器
chelate polymer 螯合聚合物
chelate resin 螯合树脂
chelate ring 螯(形)环;螯合环
chelate solution 螯合溶液
chelate sorbent 螯合吸附剂
chelate structure 螯合结构
chelating agent 螯合剂
chelating agent-loaded resin separation 螯合剂负载树脂分离法
chelating capacity 螯合能力
chelating coordination compound 螯合配位化合物
chelating effect 螯合效应
chelating extract method 螯合萃取法
chelating ion-exchanger 螯合(型)离子交换剂
chelating ligand 螯合配位体
chelating polymer 螯形聚合物
chelating property 螯合性能
chelating reagent 螯合试剂
chelating resin 螯合树脂
chelation 螯合(作用)
chelatometric estimation 螯合估测
chelatometric indicator 螯合指示剂
chelatometric titration 螯合滴定法;螯合滴定
chelatometry 络合滴定(法)
chelator 螯合剂

Chele formula 恰尔爱公式
cheliceral sheath 螯肢鞘
chelic polymer 螯形聚合物
cheliform 螯器
cheliped 螯足
chelkarite 水氯硼钙镁石
Chellean Age 旧石器时代
Chelmsford gray 灰色花岗岩(美国马萨诸塞州产)
chelmsfordite 中柱石;方柱石
chelomery 螯合测定法
chelometric titration 螯合滴定
chelometry 螯合滴定法
chelura 蛀木(海)虫
chelura spengler 木蠹
chelura terebrans 钻木昆虫
cheluviation 螯合淋溶作用【地】
Chelyosoma siboja Oka 龟海鞘
chemasthenia 化学过程减弱
chemboard 硬质纤维板;硬化纤维板
chemcia environmental pollution 化学环境污染
chem-crete 化学白垩
chemiadsorption 化学吸附
chemibarotrophic bacteria 化能自养细菌;化学气体营养细菌
chemical 化学制剂
chemical 化学的
chemical absorbent 化学吸收剂
chemical absorber 化学吸收剂
chemical absorption 化学吸收
chemical abstracts service 化学文摘服务
chemical acceleration 化学加速
chemical accelerator 化学加速剂;化学促凝剂
chemical achromation 化学消色
chemical action 化学作用
chemical activation 化学活化
chemical activation barrier 化学活化能
chemical activity 化学活(动)性
chemical activity coefficient 化学活度系数
chemical addition agent 化学添加剂
chemical additive 化学添加剂
chemical adhesion 化学黏合
chemical adhesive 化学黏合剂
chemical admixture 化合物外加剂;添加剂
chemical adsorption 化学吸附
chemical affinity 化学亲和性;化学亲合力
chemical afterburning 化学复燃
chemical aftereffect 化学滞后效应
chemical agent 化验剂;化学药剂;化学试剂;化学剂
chemical air conditioning 化学空气调节
chemical amelioration 化学改良
chemical amendment 化学改良法
chemical amplification 化学放大
chemical analysis 化学分析法;化学分析
chemical analysis for external examination 外检样品化学分析
chemical analysis for internal examination 内检样品化学分坼
chemical analysis instrument 化学分析仪器
chemical analysis method 化学分析方法
chemical analysis of ashed plant 灰化植物化学分析
chemical analysis of gas 气体化学分析
chemical analysis of ore 矿石化学分析
chemical analysis of rock 岩石化学分析
chemical analysis of sewage 污水化学分析
chemical analysis of single-phase mineral 单矿物化学分析
chemical analysis of soil 土壤化学分析
chemical analysis of vapo(u)r 水汽化学分析
chemcia analysis of water 水化学分析
chemical analysis sample 化学分析试样
chemical anchor 化学锚固
chemical and biologic(al) warfare 化学战和生物战
chemical anhydrite 化学硬石膏
chemical anti-corrosion 化工防腐
chemical anti-dote 化学性解毒剂;化学解毒剂
chemical anti-rust paint 化学性防锈漆
chemical apparatus 化学仪器
chemical applicator 化肥撒肥机
chemical assay 化学测定
chemical atmosphere 化学大气
chemical atomic weight 化学原子量
chemical attack 化学作用;化学性侵蚀;化学侵蚀

chemical attractant 化学引诱剂
chemical attraction 化学吸引;化学吸力;亲合力
chemical augmentation 化学强化
chemical availability 化学有效性
chemical bactericide 化学杀菌剂
chemical balance 化学天平;分析天平
chemical barge 化学品驳船
chemical barking 化学剥皮法
chemical barrier 化学屏蔽
chemical beam epitary 化学束外延
chemical behavior 化学行为
chemical binder 化学结合剂
chemical binding component 化学接合分量
chemical binding effect 化学结合效应
chemical biologic(al) flocculation 化学生物絮凝
chemical biologic(al) flocculation effluent 化学生物絮凝污水
chemical biologic(al) flocculation process 化学生物絮凝工艺
chemical biologic(al) method 生物化学方法
chemical biologic(al) warfare agent 化学细菌战争媒介
chemical blocking 化学阻塞
chemical blowing agent 化学起泡剂
chemical board drain 化学板排水;排水化学板
chemical bomb 化学炸弹
chemical bond 化学粘合;化学结合;化学键
chemical bonded alumina brick 化学结合高铝砖
chemical bonded brick 化学结合砖
chemical bonded ceramics 化学结合陶瓷
chemical bonded magnesite brick 化学结合镁砖
chemical bonded phase chromatography 化学键合(固定)相色谱(法)
chemical bonded refractory cement 化学结合耐火水泥
chemical bonding 化学键接
chemical bonding resin 化学键合型树脂
chemical bound water 化学结合水
chemical breakdown 化学分解;化学损坏
chemical brick 耐酸砖
chemical brown stain 化学褐变
chemical buffer 化学缓冲剂
chemical building material 化学建筑材料
chemical burn 化学灼伤;化学性烧伤
chemical calculated 化学计算的
chemical calculation 化学计算
chemical calculation method 化学计算法
chemical capacitor 电解质电容器
chemical capacity 化学性质;化学制品生产量
chemical carcinogenesis 化学致癌(作用);化学物致癌作用
chemical carrying ship 化学品运输船
chemical cartridge respirator 化学药筒防毒面具
chemical catalyst 化学催化剂
chemical cell 化学电池
chemical ceramic bonding monolithic refractory 化学陶瓷结合整体耐火材料
chemical ceramics 化学陶瓷
chemical chalk 碳酸钙
chemical change 化学变化
chemical character 化学特征;化学特性
chemical characteristic 化学特征;化学特性
chemical characteristics of natural water 天然水化学特征
chemical characteristics of wastewater 废水化学特性
chemical characerics of water 水化学特征
chemical characterization 化学表征
chemical churning pile 化学旋喷桩;旋喷桩(即喷射桩);喷射桩
chemical churning process 旋喷(法)
chemical circles 化工界
chemical circulation cleaning 化学循环清洗;化学品循环清洗
chemical clarification 化学澄清
chemical clarifier 化学澄清器;化学澄清剂;化学澄清池
chemical clarifying agent 化学澄清剂
chemical clarifying tank 化学澄清池
chemical cleaner 化学清洁剂;化学除锈剂;化学除草剂
chemical cleaning 化学洗涤;化学脱垢;化学清洗;化学除锈;化学除草
chemical cleaning cycle 化学清洗周期

chemical cleaning method 化学清洗法
chemical cleaning of coal 煤的化学净化
chemical cleaning of well 化学洗井
chemical closet 化学制品橱;化学大便器;化学处理厕所;化学厕座;化学厕所;消毒厕座
chemical cloud 毒气团
chemical coagulation 化学凝结;化学混凝
chemical coagulation-iron-carbon microelectrolysis-electrodialysis process 化学混凝—铁碳微电解—电渗析法
chemical coagulation method 化学混凝法
chemical coagulation process 化学凝聚处理
chemical coagulation treatment 化学混凝处理
chemical coalification process 化学煤化作用
chemical coating 化学涂料;化学涂层;化学覆盖层
chemical colo(u)ring 化学着色
chemical colo(u)ring process 化学着色法;化学染色法
chemical combination 化合作用;化合
chemical combine 化学化合
chemical combustion 化学燃烧
chemical combustion chamber 化学燃料燃烧室
chemical compatibility 化学相容性;化学混容性
chemical complex 化合络合物
chemical component 化学组分;化学组成;化学成分
chemical component of nodule 结核化学成分
chemical component of ore 矿石化学成分
chemical composition 化学组分;化学组成;化学成分
chemical composition bitumen 沥青的化学组成
chemical composition of aerosol 气溶胶的化学组成
chemical composition of coal 煤的化学组成
chemical composition of corrosion solution 溶蚀液化学成分
chemical composition of minerals 矿物化学组成
chemical composition of oil field water 油田水化学组成
chemical composition of ore 矿石的化学组成
chemical composition of organism 有机体的化学组成
chemical composition of precipitation 降水的化学组成;降水化学组成
chemical composition of samples 试样化学成分
chemical composition of soil 土壤化学成分
chemical composition of soil and rock 岩土的化学成分
chemical composition of water 水的化学成分
chemical composition quality of rock and soil 岩石土壤化学成分分析
chemical compound 化合物
chemical concentration 化学选矿法;化学浓缩;化学计量浓度
chemical concrete technology 混凝土化学
chemical condenser 化学冷凝器;电解质电容器
chemical conditioner 化学调理剂;化学调理池
chemical conditioning 化学调理
chemical conditioning of sludge 污泥化学调理;污泥的化学调理
chemical consolidation 化学凝固;化学加固(法);化学固结
chemical consolidation of soil 土的化学加固
chemical constant 化学常数
chemical constituent 化学组分;化学组成
chemical constituents of wastewater 废水化学组分
chemical constitution 化学成分
chemical construction 化学结构
chemical construction corporation 化学工程公司
chemical consumption 化学药品消耗(量);再生剂耗量
chemical contaminant 化学污染物
chemical contamination 化学污染
chemical contamination of water body 水体化学污染
chemical content of water 水的化学成分含量
chemical control 化学控制;化学防治
chemical control system 化学药剂控制系统
chemical conversion 化学致癌物
chemical conversion coating 化学转化膜;化学处理膜
chemical conversion film 化学转化膜;化学处理膜
chemical conversion treatment 化学转化处理;化学法(金属)表面处理

chemical coolant 化学冷却剂
chemical cooling 化学冷却
Chemical Coordinating Center 化学问题协调中心
chemical coprecipitation process 化学共沉淀方法
chemical corrosion 化学溶蚀(作用);化学腐蚀;化学防腐
chemical coupling 化学偶合
chemical crystallography 化学结晶学
chemical cure 化学硫化;化学固化
chemical dampcourse 化学防潮层
chemical damproofing 化学防潮法
chemical dating 化学法测定年龄
chemical decay 化学分解;化学性腐烂;化学变性
chemical decladding 化学去壳
chemical decolo(u)rization 化学脱色
chemical decomposition 化学分解
chemical decontamination 化学去污;化学净化
chemical defecation 化学澄清法
chemical defect 化学缺陷
chemical defoliant 化学脱叶剂
chemical degradation 化学性破碎;化学(性)降解;化学剥蚀
chemical degradation of waste 废物的化学降解
chemical degreasing 化学脱脂
chemical dehydrator 化学脱水器;化学脱水剂;化学干燥剂
chemical deliming agent 化学脱灰剂
chemical demagnetization 化学退滋
chemical demagnetizing solution 化学退磁溶液
chemical demagnetizing temperature 化学退磁温度
chemical demagnetizing time 化学退磁时间
chemical demineralization 化学脱矿质;化学水软化
chemical demulsifier 化学破乳剂
chemical denudation 化学剥蚀作用;化学剥蚀
chemical deposit(ion) 化学镀层;化学沉积(物);化学沉淀物;化学沉淀
chemical deposits inside cave 洞穴化学沉积物
chemical derivatization 化学衍生(作用)
chemical derusting 化学除锈
chemical desalting 化学脱盐;化学除盐
chemical desalting method 化学脱盐法
chemical design criterion 化学设计判据
Chemical Design Institute 化工设计院
chemical desulfurization 化学脱硫
chemical detergent 化学洗涤剂
chemical deterioration 化学腐蚀;化学变质
chemical determination of inclusion 包裹体成分分析法
chemical determinator 化学测定仪
chemical development 化学显影;直接显影
chemical dewatering 化学脱水
chemical dewatering of slurry 料浆化学脱水
chemical differentiation 化学分化
chemical differentiation way 化学分异方式
chemical diffusion 化学扩散
chemical digestion 化学(性)消化
chemical dip brazing 化学浸液钎焊
chemical dip coating 化学浸渍覆盖层
chemical disinfectant 化学消毒剂
chemical disinfection 化学消毒(法)
chemical dispersant 化学分散剂
chemical dissolving box 化学药剂浴解箱;药剂溶解箱
chemical dose 化学剂量
chemical dosimeter 化学剂量仪
chemical dosimetry 化学剂量学
chemical dosing 化学剂量测定;投药
chemical drier 化学干燥器
chemical drill 化学腐蚀成孔钻机
chemical drilling 化学钻进
chemical drip tube 化学滴管
chemical dry feeder 化学的干加料器
chemical dump 化学废料
chemical durability 化学稳定性;化学耐久性
chemical durability of layers 膜层的化学稳定性
chemical ecology 化学生态学
chemical effect 化学效应
chemical efficiency 化学效率
chemical effluent 化学排出物
chemical electrovalence 化学电价;地球化学电价
chemical element 化学元素
chemical element balance 化学元素平衡
chemical element balance method 化学元素平

衡法
chemical eluviation 化学淋溶作用;化学淋溶
chemical emergency 化学紧急事故
chemical emergency response information system 化学紧急情况回答信息系统
chemical enamel equipment 化工搪瓷设备
chemical endemic disease 化学性地方病
chemical energy 化学能
chemical energy potential 化学能势
chemical engineering 化学工程(学);化工
chemical engineering abstracts 化工文摘
chemical engineering and construction division 化学工程建设部
chemical engineering design 化工设计
chemical engineering machinery 化工机械
chemical engineering plant 化工厂
chemical engineering science 化学工程学
chemical engineering unit operation 化工单位操作
chemical engine hose 化工软管
chemical engraving 化学刻图
chemical enhanced coagulation 化学强化混凝
chemical enterprise 化工企业
chemical environment 化学环境
chemical environmental factor 化学环境因子
chemical-environmental monitoring of freshwater ecosystem 淡水生态系统化学—环境监测
chemical equation 化学(反应)方程式
chemical equilibrium 化学平衡
chemical equilibrium condition 化学平衡条件
chemical equilibrium constant 化学平衡常数
chemical equilibrium software 化学平衡软件
chemical equipment 化工设备
chemical equipment in sets 化工成套设备
chemical equipment manufacture 化工设备制造
chemical equipment manufacturer 化工设备制造厂
chemical equipment parts 化工设备零部件
chemical equivalent 化学当量
chemical erosion 化学(性)侵蚀
chemical etching 化学侵蚀;化学刻蚀;化学浸蚀;化学腐蚀
chemical evaluation of environment 环境的化学演化
chemical evolution 化学演化;化学进化
chemical examination 化验;化学检验
chemical examination of water 水化学检测
chemical experimental equipment 化学实验设备
chemical exploration 化学探索
chemical extraction 化学萃取法;化学抽提
chemical factor 化学因子
chemical fade 化学淡入
chemical fallout 化学粉尘
chemical fastness 化学牢固性;化学抗性
chemical feeder 加药器;化学药剂投加器;化学物进料器;化学品加料机
chemical feeding 化学药剂投加
chemical feed machine 投药机
chemical feed pump 加药泵
chemical feed room 投药间
chemical fertilizer 化学(肥料)
chemical fertilizer industrial wastewater 化肥工业废水
chemical fertilizer industry 化肥工业
chemical fertilizer plant 化肥装置;化肥厂
chemical fertilizer pollution 化肥污染
chemical fiber 化(学)纤(维)
chemical fiber industry 化学纤维工业
chemical fiber paper 化学纤维纸
chemical fiber plant 化纤厂
chemical fiber webbing 化纤带
chemical fibre factory 化纤厂
chemical fibre plant 化学纤维厂
chemical fibre pump 化纤泵
chemical fibre wastewater 含化学纤维废水
chemical filling 滤毒剂
chemical film 化学薄膜
chemical filter 化学过滤器
chemical filter respirator 化学过滤(式)呼吸器
chemical finish 化学涂层;化学涂料
chemical(fire) extinguisher 化学灭火器
chemical firm 化学公司
chemical fixation 化学固定法
chemical flashing 化学被膜
chemical floc 化学絮凝物;化学絮凝剂

chemical flocculant 化学絮凝剂
chemical flocculation 化学絮凝(法)
chemical floccules 化学絮凝物
chemical flow 化学流动;化学径流;溶质径流
chemical flushing 化学挤水法
chemical flushing of well 化学洗井
chemical flux 化学溶剂
chemical flux cutting 氧熔切割;氧熔剂切割
chemical foam 化学泡沫
chemical foam extinguisher 化学泡沫灭火器
chemical focus 光化焦点
chemical fog 化学灰雾;显影蒙翳
chemical food poisoning 化学性食物中毒
chemical force 化学力
chemical form 化学形态
chemical formation 化学结构
chemical formed rock 化学导成岩
chemical formula 化学(公)式
chemical formula of minerals 矿物化学式
chemical for the separation of ore 分选矿石用化学制品
chemical fractionation 化学分部分离
chemical fragmentation of rock 岩石化学破碎
chemical fuel 化学燃料
chemical fungicide 化学杀菌剂;化学杀虫剂
chemical furnaces 化工炉类
chemical gas feeder 气体加药器
chemical gas generator 化学气体发生器;化学气体储压器
chemical ga(u)ging (of flow) 化学测流(法);化学方法测定;化学法测流;化学测定
chemical geography 化学地理
chemical geothermometer 化学地热温标
chemical gilding 化学镀金
chemical glass 化学(仪器)玻璃
chemical glue 化学胶水
chemical ground pulp waste 化学细磨纸浆废料
chemical ground pulp wastewater 化学细磨纸浆废水
chemical group composition 化学族组成
chemical grout 化学注浆
chemical grouting 化学灌浆;注浆施工;灌注化学品加固
chemical grouting agent 化学灌浆材料
chemical grouting method 化学灌浆法
chemical grouting process 化学灌浆加固法
chemical grouting stabilization 化学灌浆加固
chemical grouting system 化学(剂)灌浆系统
chemical gypsum 化学石膏;工业废料石膏
chemical hardening 化学硬化
chemical hazard 化学性危害;化学危害物
chemical hazardous substance 化学危险物质
chemical hazards response information system 化学危险品响应信息系统
chemical heat production 化学性产热
chemical heat storage 化学热蓄积
chemical heat treatment 化学热处理
chemical herbicide 化学除莠剂;化学除草剂;除莠剂;除草剂
chemical homogeneity 化学均匀性
chemical hydrology 化学水文(学);水文化学
chemical hydrometer 化学比重计
chemical hygrometer 化学湿度计
chemical ice 添加化学保藏剂的冰
chemical identification 化学鉴定
chemical improvement of soil 土壤的化学改良
chemical incompatibility 化学互克性
chemical index 化学指标
chemical industrial organic refractory wastewater 难降解化工有机废水
chemical industrial organic wastewater 化工有机废水;化工厂废水
chemical industrial products 化工制品
chemical industry 化学工业;化工
chemical industry furnace 化学工业炉
chemical inertness 化学惰性
chemical inhibitor 抗氧剂;化学抑制剂
chemical injection 化学(药液)注浆;化学灌浆;药剂喷射
chemical injection process 化学注液法;化学药剂注入法(用于稳定土);化学药剂灌注法;化学灌浆加固法;灌注化学品加固法
chemical injury 化学性损伤
chemical insecticide 化学杀虫剂

chemical instability 化学不稳定性
chemical insulation 化学隔离
chemical interaction 化学相互作用
chemical interference 化学干扰
chemical intermediate 化学中间体;化学半成品
chemical inventory 化学品存货
chemical investigation 化学研究;化学试验;化学检验
chemical ionization 化学电离
chemical ionization mass spectrometry 化学电离质谱法
chemical ionization source 化学离子源
chemical irritant 化学刺激物
chemical irritation 化学性刺激
chemical isomer 化学异构物
chemical kinetics 化学动力学
chemical laboratory 化验室;化学试验室
chemical laser 化学激光器
chemical leaching 化学淋溶;化学淋滤
chemical lead 化学纯铅
chemical leather 合成革
chemical lesion 化学性损害
chemical lime 化学石灰
chemical limestone 化学石灰岩
chemical lines 化工行业
chemical liquid deposition 化学液相沉积
chemical load 溶解质
chemical luminescence 化学发光
chemically active 化学活性的
chemically active fluidized bed 化学活性流化床
chemically active pigment 化学活性颜料
chemically active plasticizer 化学活性增塑剂
chemically active substance 化学活性物质
chemically active surface 化学作用(表)面
chemically adsorbed adsorbate 化学吸附的吸附质
chemically bonded brick 化学砌筑砖;化学结合砖;非砂浆砌筑砖;不烧砖
chemically bonded moiety 化学键合部分
chemically bonded phase 化学键合相
chemically bonded phase packing 化学键合相填充剂
chemically bound water 化合水
chemically clean surface 化学净化表面
chemically coagulated sludge 化学混凝的污泥;化学法絮凝的污泥;化学法凝聚的污泥;化学(法)凝结(的)污泥
chemically combined 化学结合的
chemically combined hydrogen 化学结合氢
chemically combined water 化学结合体;化学结合水
chemically cured finish 化学固化型涂层;化学反应性涂料;催化固化型涂层
chemically curing sealant 化学固化密封膏;化学反应型密封膏
chemically deposited sedimentary rock 化学沉积岩
chemically disintegrated powder 化学粉碎的粉末
chemically dissolved amount 化学溶解量
chemically enhanced primary treatment method 一级化学强化处理方法
chemically fixed energy 化学固定能
chemically flocculated and clarified raw sewage 化学絮凝净化的原污水
chemically flocculated wastewater 化学絮凝污水
chemically foamed plastics 化学泡沫塑料
chemically formed rock 化学成因岩石
chemically inert 化学惰性(的);不起化学作用的
chemically mechanical polishing 化学机械抛光
chemically mechanical regeneration system 化学机械更新装置
chemically mechanical welding 化学机械焊(接)
chemically modified fiber 化学变性纤维
chemically oxidizable organic matter 化学可氧化有机物质
chemically oxidizing atmosphere 化学氧化性大气;化性大气
chemically precipitated sludge 化学法沉淀的污泥;化学沉淀污泥
chemically prestressed concrete 化学预应力混凝土;自应力混凝土
chemically prestressing cement 化学预应力水泥;自应力水泥
chemically pure 化学纯
chemically pure chrome yellow 化学纯铬黄

chemically pure water 化学纯水
chemically pure zinc 化学纯锌
chemically reactive dyes 反应性染料
chemically resistant cement 防化学腐蚀水泥
chemically resistant glass fiber 耐化学玻璃纤维
chemically reversed photomask 化学法做掩模板
chemically sedimentated sludge 化学沉淀的淤渣
chemically stabilized earth lining 化学加固的土质衬砌
chemically stable 化学稳定的
chemically strengthened glass 化学钢化玻璃
chemically structured kaolin 化学改性高岭
chemically toughened glass 化学钢化玻璃
chemically treated timber 化学处理的木材
chemical machine 化工机器
chemical machinery manufacture 化工机械制造
Chemical Machinery Research Institute 化工机械研究院
chemical machining 化学加工
chemical manure 化肥
chemical manure fertilizing machine 化肥撒肥机
chemical mass 化学质量
chemical mass balance 化学质量平衡
chemical materials 化工材料
chemical measurement 化学测量法；再生剂计量
chemical mechanical tests 化学物理试验
chemical mechanism 化学机制；化学机理
chemical mediator 化学介质
chemical medicine 化学药品
chemical medium 化工介质
chemical metal cleaning wastewater 化学金属清洗污水
chemical meteorology 化学气象
chemical metering 化学计量(法)
chemical metering pump 化学计量泵
chemical method 化学法
chemical methods of wastewater treatment 污水化学处理法
chemical milling 化学蚀刻；化学加工
chemical mineral composition 化学矿物组成
chemical mineral mater 化学成因矿物质
chemical mineralogy 化学矿物学
chemical mining 化学采矿
chemical mixer 化学药品混合器
chemical mixing 化学混合
chemical mixing unit 化工搅拌装置
chemical mobility 化学活性
chemical modification 化学修饰；化学改性
chemical monitoring 化学监测
chemical mutagen 化学诱变物；化学诱变剂
chemical mutagenesis 化学物诱变作用
chemical name 化学名称
chemical nature 化学性质
chemical neutral 化学中性
chemical nickel plating wastewater 化学镀镍废水
chemical nomenclature 化学命名
chemical oceanography 化学海洋学；海洋化学
chemical overlay 化学被覆(物)
chemical oxidation 化学氧化
chemical oxidation-coagulation process 化学氧化—混凝工艺
chemical oxidation polymerization 化学氧化聚合
chemical oxidation pretreatment 化学氧化预处理
chemical oxidation process 化学氧化(处理)法
chemical oxidation reduction 化学氧化还原
chemical oxygen consumption 化学需氧量；化学耗氧量
chemical oxygen demand 化学需氧量
chemical oxygen demand analyser 化学需氧量测定仪
chemical oxygen demand concentration 化学需氧量浓度
chemical oxygen demand determination 化学需氧量测定
chemical oxygen demand dissolved 溶解化学需氧量
chemical oxygen demand loading 化学需氧量负荷
chemical oxygen demand removal 除化学需氧量
chemical oxygen demand removal efficiency 化学需氧量去除效率；除化学需氧量效率
chemical oxygen demand test 化学需氧量测定
chemical oxygen demand total 总化学需氧量
chemical paper 化学纸；防潮纸
chemical parameter 化学参数
chemical parcel tanker 零担液体化学品专用船；小批化学液体专用船
chemical passivity 化学钝性
chemical pattern recognition 化学模式识别
chemical pesticide 化学农药
chemical petrologic classification 岩石化学分类
chemical petrology 化学岩石学
chemical pharmaceutical factory 化学制药厂
chemical phase 化学相
chemical phenomenon 化学现象
chemical phosphorus removal 化学除磷
chemical physics 化学物理学
chemical pigment 化学颜料
chemical pinching agent 化学整枝剂
chemical pipeline 化学管线
chemical piping 化学剂管路；化学管涌
chemical plant 化(学)工厂
chemical plant installations 化工装置
chemical plaster 化学石膏灰；干硬性石浆
chemical plating 化学镀(敷)；化学电镀
chemical plating agent 化学镀剂
chemical poisoning 化学品中毒
chemical polarization 化学极化
chemical polish 化学抛光剂
chemical polishing 化学抛光；化学磨光；化学擦光
chemical polishing agent 化学抛光剂
chemical pollutant 化学污染物
chemical pollution 化学污染
chemical pollution index 化学污染指数
chemical polymer 化学聚合体
chemical porcelain 化学瓷(器)
chemical potential 化学位；化学势
chemical potential energy 化学位能；化学势能
chemical potential gradient 化学位梯度；化学势能梯度
chemical potentiometer 化学势计
chemical powder 化学能源
chemical powder extinguisher 干粉灭火器；干粉灭火机
chemical powder mixture 化学干粉混合剂
chemical precipitate 化学沉淀物
chemical precipitate blocking 化学沉淀物堵塞
chemical precipitation 化学沉积；化学沉淀作用；化学沉淀
chemical precipitation agent 化学沉淀剂
chemical precipitation method 化学沉淀法
chemical preservation 化学防污；化学防腐(作用)
chemical preservative 化学防腐剂
chemical prestressing 化学预(加)应力法；化学法预(加)应力
chemical pretreatment 化学预处理
chemical priming 化学预处理
chemical principle 化学原理
chemical process 化学加工；化学过程；化工工艺
chemical process analysis 化学过程分析
chemical process equipment 化学加工设备
chemical processes zoning of phreatic water 潜水化学作用分带
chemical process industry 化学加工工业
chemical processing plant 化学加工厂
chemical processing reactor 化工辐照用反应堆
chemical process pump 化工工艺用泵
chemical products 焦化类；化学产品
chemical products from coal 煤化学制品
chemical promoter 化学促进剂
chemical-proof 耐化学药剂性
chemical proof floor 防腐地面
chemical properties of water 水的化学性质；水的化学性状
chemical property 化学性质；化学性能；化学特性
chemical property of solid bitumen 固体沥青的化学性质
chemical proportioner 比例加药器
chemical protection 化学防护
chemical pruning 化学修枝
chemical pulp 化学纸浆；化学砂浆
chemical pulping 化学制浆
chemical pulverization 化学粉碎
chemical pump 化工(用)泵
chemical purification 化学净化(法)；化学法提纯
chemical purification plant 化学净化装置；化学净化设备；化学净化厂
chemical quality 化学性质
chemical race 化学品种
chemical radiation effect 化学辐射效应
chemical raw material 化工原料
chemical ray 化学射线
chemical reaction 化学反应
chemical reaction detector 化学反应检测器
chemical reaction equipment 化学反应设备
chemical reaction method 化学反应法
chemical reaction rate 化学反应速率
chemical reactive coating 化学反应包层
chemical reactive furance 化工炉类
chemical reactivity 化学反应性
chemical reactor 化学反应器；化学反应堆；化工反应器
chemical ready for use 备用化学药品
chemical reagent 化学试剂
chemical reagent for scale removal 化学清垢剂；化学除垢剂
chemical recovery 化学品回收
chemical recovery coke oven 化学回收炼焦炉
chemical recovery furnace 化学回收炉
chemical rectifier 电解整流器
chemical reducing 化学还原
chemical reduction 化学还原作用；化学还原
chemical refining 化学精炼
chemical refining process 化学精制过程
chemical reflex 化学反射
chemical refrigeration 化学制冷；化学冷却
chemical refrigeration process 化学冷冻法
chemical refuse 化学废弃物；化工垃圾
chemical regeneration 化学再生
chemical release agent 化学脱模剂
chemical remanent magnetization 化学剩余磁化强度
chemical remediation 化学修复
chemical removal 化学分离
chemical reprocessing liquid waste 化学再处理废液
Chemical Research Institute 化工研究院
chemical reservoir 化学岩储集层；化学储层
chemical residence time 化学品停留时间；化学品存在时间
chemical residue 化学残渣；化学残留物
chemical residue analyzer 化工残渣分析仪
chemical resistance 抗斑污(楼地面)；抗化学侵蚀性；抗腐蚀；耐化学(药品)性；化学阻力
chemical resistance masonry unit 耐化学侵蚀的砌体单元
chemical resistant 化学抗力；耐化学侵蚀性
chemical-resistant ceramics 化学稳定陶瓷
chemical-resistant coating 耐化学涂层；耐化学药品涂层
chemical resistant masonry 耐化学品圬工
chemical-resistant paint 耐化学涂料；耐化学漆；耐化学药品涂料；防化学漆
chemical resisting 抗化学作用
chemical retardation 化学阻滞
chemical retting 化学浸解
chemical rising technique 化学清洗法；槽内处理法
chemical rock 化学岩
chemical rocket 化学燃料火箭
chemical rust removing 化学处理法除锈
chemicals 化学制品；化学制剂；化学药品；化学药剂；化学物；化学品；化合物
chemical safety 化学安全
chemical sampling 化学取样
chemical sanitary 化学卫生的
chemicals closet 药橱；化学制品橱柜
chemical's concentration gradient 化学物浓度梯度
chemical screening 化学除皮
chemical scrubbing 化学清洗
chemical sealing 化学密封
chemical seasoning 化学干燥
chemical sediment 化学沉积物；化学沉淀物
chemical sedimentary deposit 化学沉积床
chemical-sedimentary deposit by true solution 真溶液化学沉积矿床
chemical sedimentary differentiation 化学沉积分异作用
chemical sedimentary rock 化学沉积岩
chemical sedimentation 化学沉淀法
chemical self-purification 化学自净(化)作用
chemical sense 化学感觉
chemical sensibiligen 化学致敏原

chemical sensitization 化学增感;化学敏化
chemical sensitizer 化学增感剂;化学敏化剂
chemical sensor 化学传感器
chemical separation 化学法选矿
chemical separation of minerals 矿物化学分离
chemical setting 化学凝固
chemical sewage sludge 化学污水污泥
chemical shift 化学位移
chemical shift value 化学位移值
chemical shim 化学控制剂;化学补偿(物)
chemical shrinkage 内在收缩;化学收缩
chemical similitude 化学相似性模拟法
chemical simulation 化学模拟
chemical sludge 化学污泥
chemical sludge floc density 化学污泥絮凝物密度
chemical slurry 化学稀浆;化学灌浆
chemical smoke 化学烟雾;化学烟气
chemical softener 化学软水剂
chemical soil stabilization 化学药物土壤稳定(作用);化学稳定土;土壤化学稳定法;化学剂稳定土壤
chemical solidification 化学固结
chemical solution 化学溶液
chemical solution feeder 化学溶液加料器
chemical solution tank 化学溶液池;化学溶液槽;药品溶液池
chemical solvents 化学溶剂
chemical sorting equipment 化学分选装置
chemical specialities 化学特制品
chemical species 化学物种;化学物类
chemical specks 化学异点
chemical spectroscopy 化学光谱学
chemical spill 化学溢漏
chemical spreader 化学品撒布机;除冰剂撒布机
chemical sprinkler system 化学喷淋系统
chemicals resistant clayware 耐化学药品的陶土制品
Chemicals Specilty of Manufacturers Association 美国特种化学品制造商联合会
chemical stability 化学稳定性
chemical stability of fungicide 杀菌剂化学稳定性
chemical stabilization 化学稳定化;化学稳定(作用);化学稳定(法);化学加固
chemical stabilization agent 化学稳定剂
chemical stabilization process 化学加固法
chemical stabilizer 化学稳定剂
chemical stain 化学污斑;化学色彩
chemical staining 化学着色处理(木材);化学着色
chemical state 化学状态
chemical sterilization 化学药物灭菌法;化学杀菌
chemicals that are useful 有效的化学品
chemical stimulus 化学刺激物
chemical stoichiometry 化学计量法
chemical stoneware 耐酸陶器;耐酸炻器;化学陶器;化工陶瓷;化工炻器
chemical storage 化学药剂储[贮]存;化学(药剂)储[贮]藏
chemical stratification 化学分层
chemical strengthened glass 化学增强玻璃;化学强化玻璃
chemical strengthening 化学增强
chemical stress 化学应力
chemical stress relaxation 化学应力松弛
chemical stripper 化学涂层剥除剂
chemical structural formula 化学结构式
chemical structure 化学结构
chemical substance disbalance method 化学物质不平衡法
chemical substance index 化学物质索引
chemical substances 化学物质
chemical substitute 化学替代物质;化学代用品
chemical subtraction 化学去污
chemical surface hardening 化学表面硬化法
chemical surface treatment 化学表面处理
chemical symbol 化学符号
chemical synthesis 化学合成(法)
chemical synthesized pharmaceutical wastewater 化学合成制药废水
chemical system 化学系统
chemical tank 化学槽;化工槽罐;药剂槽
chemical tanker 化学品运输船
chemical technology 化学工艺(学)
chemical technology of wood 木材化学工艺
chemical tempering 化学强化

chemical tempering process 化学钢化法
chemical tendering 化学软化;化学脆化
chemical tension 化学电位
chemical tensioning method 化学强拉法
chemical test 化学试验
chemical themodynamics 化学热力学
chemical theory 化学学说
chemical toilet 化学掩臭剂;化学化妆品;化学(处理)厕所;化学便桶
chemical toning 化学调色
chemical tool 陶瓷工具
chemical toughness 化学韧性
chemical towers 化工塔类
chemical toxicant 化学毒素;化学毒剂
chemical toxicant monitoring 化学毒物监测
chemical trace constituent 化学痕量成分
chemical tracer 化学示踪剂
chemical tracing 化学示踪
chemical transfer process 化学转印法
chemical transformation 化学转化
chemical transmission 化学(性)传递
chemical transmission theory 化学传递说
chemical transmitter 化学递质
chemical transport 化学迁移
chemical transportation 化学搬运
chemical transporting amount 化学搬运量
chemical transport model 化学传输模式
chemical transport reaction 化学运输反应
chemical trauma 化学伤
chemical treatment 化学药剂处理;化学法处理;化学处理
chemical treatment method 化学处理法
chemical treatment of mud 泥浆的化学处理
chemical treatment of sewage 污水化学处理法
chemical treatment of sludge 污泥化学处理
chemical treatment of soil 土壤化学处理
chemical treatment of waste 废物化学处理
chemical treatment of wastewater 废水化学处理法
chemical treatment process 化学处理法
chemical treatment program(me) 化学处理计划;化学处理方案
chemical type fire extinguishing system 化学型灭火系统
chemical type of groundwater type of groundwater 地下水化学类型
chemical type of ultra-mafic rock 超镁铁岩化学类型
chemical type plasticizer 化学型增塑剂
chemical unconformity 化学不整合
chemical uncoupler 化学解偶联剂
chemical unit 化学单位
chemical unit process 化学单元处理法
chemical unstable mud 化学性质不稳定泥浆
chemical valence 化合价
chemical vapo(u)r deposition 化学蒸气沉积;化学气相沉积
chemical vapo(u)r infiltration 化学气相渗透
chemical vapo(u)r transport 化学气相输送
chemical variation 化学变化
chemical variation diagram 化学变化图解
chemical warfare agent 化学战剂
chemical warfare protective equipment 防化设备
chemical waste 化学废物
chemical waste water 化工废水
chemical wastewater treatment 化学废水处理(厂)
chemical water analysis 水(的)化学分析
chemical water pollution 化学水污染
chemical water purification 水的化学净化;化学净水
chemical water quality index 水质化学指标
chemical water quality parameter 水质化学参数
chemical water-softening plant 化学软水设备
chemical water treatment 化学水处理;水化学处理
chemical weapons 化学武器
chemical weathering 化学性风化;化学气候;化学风化(作用)
chemical weathering rate 化学风化速率
chemical weed control 化学除莠;化学除草
chemical weed control practices 化学除草法
chemical weeding 化学除草
chemical weed killer 化学除莠剂;化学除草剂;除莠剂
chemical weed killing 化学除莠;化学除草

chemical wood preservation 木材化学防腐法
chemical wood-pulp paper 化学木浆纸
chemical works 化(学)工厂
chemical works waste 化工厂废物
chemical yarn, fabric 化纤纱布
chemichromatography 化学反应色谱法
chemico 化学工程公司
chemicobiology 化学生物学
chemico-mechanical pulp 化学机械纸浆
chemico-mechanical welding 化学机械焊(接)
chemicophysics 化学物理学
chemico-thermal treatment 化学热处理
chemiground pulp waste 化学细磨纸浆废料
chemiground pulp wastewater 化学细磨纸浆废水
chemi-ground wood 化学机械木浆
chemi-ground wood pulp 化学磨木浆
chemi-ground wood pulp wastewater 化学磨木纸浆废水
chemigum 丁腈橡胶
chemihydrometry 化学水文测验;化学水文测量;化学测流(法)
chemi-ionization 化学电离
chemi-ionization detector 化学电离检测器
chemi-ionization process 化学电离过程
chemi-ionization reaction 化学电离反应
chem I-ject construction method 铝酸钠施工法
chemiluminescence 冷焰光;化学发光;低温发光
chemiluminescence analysis 化学发光分析
chemiluminescence detector 化学发光检测器
chemiluminescence light source 化学发光光源
chemiluminescence nitrogen oxides monitor 化学发光氮氧化物监测仪
chemiluminescence ozone monitor 化学发光臭氧监测仪
chemiluminescence reaction 化学发光反应
chemiluminescent indicator 化学发光指示剂
chemiluminescent material 化学发光材料
chemiluminescent NOx analyzer 化学发光式氮氧化物分析仪
chemimechanical pulp 化学机械纸浆
chemin-de-ronde 城墙后通道
cheminduction 化学诱导
cheminosis 化学物质病
chemio-pile method 生石灰桩施工法
chemiosmotic hypothesis 化学渗透假说
chemiosterilization 化学消毒(法)
chemise 衬墙;复面层;土堤岸护墙;土堤岸(护面)
chemism 化学性质;化学历程;化学机制;化学机理;化学关系
chemisorb 化学吸附
chemisorbed adsorbate 化学吸附的吸附质
chemisorbed water 化学吸附水
chemisorption 化学吸着;化学吸附
chemisorption method 化学吸附法
chemisorptive fibre 化学吸着性纤维
chemist 鲕绿灰岩
chemistry 化学
chemistry and composition of atmosphere 大气化学及其组成
chemistry coal 化工用煤
chemistry laboratory 化学试验室
chemistry of air pollution 大气污染化学
chemistry of atmosphere 大气化学
chemistry of atmospheric pollution 大气污染化学
chemistry of cement 水泥化学
chemistry of coal utilization 煤炭利用化学
chemistry of corrosion control 化学腐蚀控制
chemistry of estuary 港湾化学
chemistry of process 过程之化学
chemistry of radioactive substance 放射性物质化学
chemistry of seawaters 海水化学
chemistry softening 化学软化
chemist's shop 药房;药店
chemitype 化学蚀刻凸版;化学制版
chem-mill 化学蚀刻成型
chemoautotroph 化学自养生物;化能自养生物
chemoautotrophic bacteria 化能自养细菌
chemoautotrophy 化能自养
chemoautoytophic bacteria 化学营养细菌
Chem-O-Bam 代森铵
chemo-barotropic(al) bacteria 化能自养细菌
chemobiotic zone 化学生物带
chemoceptor 化学性感受器

chemocline 化跃面;化跃层;化学斜层;化变层
chemocoagulation 化学凝固法
chemodifferentiation 化学分化
chemodynamics 化学动态(学)
chemoelectrical transducer 化电换能器
chemofacies 化学相
chemofaciescherokee rose 金樱子
chemogenic 化学成因的
chemogenic sediment 化学沉积物;化学沉淀物
chemogenic structure 化学成因构造
chemoheterotroph 化能生物
chemoinduction 化学感应
chemokinesis 化学动态(学)
chemolithoautotroph 化能无机自养生物
chemolithoautotrophy 化能无机自养
chemolithotrophic bacteria 化能无机营养菌
chemolithotrophy 化能无机营养
chemoluminescence 化学发光
chemolysis 化学溶蚀(作用);化学分析;化学分解
chemomanalytic 化学分析的
chemo-mechanical regeneration system 化学机械更新装置
chemometrics 化学统计学;化学计量学
chemomorphosis 化学诱变;化学性变态
chemomotive force 化学动力
chemonite 开莫耐特溶液;亚砷酸铜氨液木材防腐剂
chemoorganoheterotroph 化能有机异养生物
chemoorganotrophic bacteria 化能有机营养菌
chemoorganotrophy 化能有机营养
chemopause 臭氧顶层;臭氧层顶
chemoplast 化学质体
chemoprophylactic drug 化学预防药
chemoprophylaxis 化学预防;化学药物预防;化学品预防
chemoreception 化学感应
chemoreceptor 化学性感受器;化学受体;化学接受体
chemoreflex 化学反射
chemoresistance 抗化学性;耐化学性;化学抗性
chemorheology 化学流变学
chemosensitivity 化学敏感性
chemosetting 化学固化
chemosmosis 化学渗透作用;化学渗透
chemosorbent 化学吸收剂;化学吸附剂
chemosphere 光化层;臭氧层
chemostat 化学环境恒定器;恒化器
chemostat dynamics 化学稳定器动力学
chemosterilant 化学消毒剂
chemosterilization 化学杀菌作用
chemosynthesis 化学合成(法);化能合成(作用)
chemosynthetic(al) autotroph 化能合成自养菌
chemosynthetic(al) microorganisms 化学合成微生物
chemosynthetic bacteria 化学自养细菌;化能自养细菌
chemotactism 化学向性
chemotaxin 化学吸引素
chemotaxis 趋药性
chemotaxonomy 化学分类学(动植物)
chemotec 环氧树脂类黏结剂;环氧树脂类黏合剂
chemotoxic zone 化学毒性带
chemotrophy 化学营养
Chemung group 舍蒙群
Chemungian 舍蒙统
Chemungian stage 舍蒙阶【地】
chemurgy 农艺化学;农业化工
cheneau 装饰上檐口;瓦檐饰
chenevixite 绿砷铁铜矿;砷铁铜石
chengal 棕黄色硬木(马来西亚)
Chenghua ware 成化窑
chenier 海沼沙脊;沿海沙脊;沼泽沙丘
Chenier 千尼尔
Chenier deposit 千尼尔沉积
chenier plain 接岸平原;海沼沙脊平原
chenier structure (码头的)接岸结构
chenille 绳绒线;雪尼尔线织物;绳绒织物
chenille axminster carpet 雪尼尔地毯
chenille carpet 雪尼尔地毯;绳绒地毯;长毛绒地毯
Chenorveth pile 切诺韦思柱
cheque 支票
cheque account 支票账户
cheque board scan 棋盘格扫描
cheque-book 支票簿
cheque-book stubs 支票簿存根

cheque card 支票卡
cheque drawer 支票出票人
cheque holder 支票持票人
cheque ink 支票油墨
cheque lost 支票挂失
cheque paper 支票纸
cheque plate 棋格板
chequer 砖格;方格;格子窑;格子图案;方格图案(英国)
cheque rate 支票利率
chequer board construction 隔仓施工法
chequerboard screen 模拟形网屏
chequer brick 方格砖
chequered buoy 方格纹浮标
chequered lost head nail 棋盘格的切头柱
chequered material 花格子材料
chequered order 棋盘式排列
chequered pattern 方格图案(英国);棋盘式图案
chequered plate 网纹钢板;花钢板;格形盖板;格子板;防滑板
chequer work 棋盘形铺设;方格式砌砖(英国)
cheque to bearer 不记名支票
cheque trading 支票交易
cheque work 棋盘形细工
cheque writing machine 支票打字机
cheralite 磷钙钍矿
chernovite 砷钇石
chernozem 黑土(带);黑钙土
chernozem(ic) soil 黑(钙)土
chernozem-like soil 黑钙土状土壤
chernykhite 钡钒云母
cherry 鲜红色的;樱桃木;樱桃;樱木;樱红色
cherry birch 矮桦;山桦
cherry blossom 樱桃花红色
cherry coal 结焦煤;樱煤;暗煤
cherry-hard brick 缸砖
cherry mahogany 樱红色桃花心木;樱桃红木
cherry picker 移动升降台;移车器;格构斜臂式起重机;动臂装载机;动臂装卸机;调车工;车载升降台;车载起重机;叉车吊高器;万能装卸机;车载式吊车
cherry processing 樱桃加工
cherry red 樱桃红(色)
cherry-red heat 樱桃红热
cherry rose 樱桃蔷薇色
cherry stone 樱桃核
cherry tree 樱桃树
cherry wine 深红色
chersonese 半岛
chert 角岩;黑燧石;浅燧石
chert aggregate 硅质岩集料;硅岩岩骨料;燧石集料
chert arenite 燧石砂屑岩
chert-bearing 含燧石的
chert bit 燧石钻头
chert breccia 燧石角砾岩
chert flint 燧石
chert gravel 燧砾石
chertification 燧石化(作用)
chert limestone 燧石灰岩
chert nodule 燧石结核
chert nodule-bearing limestone 含燧石结核灰岩
chert shale 燧石页岩
cherty flint 燧石
cherty limestone 含燧石的石灰岩;硅质石灰岩;高硅质石灰石
cherty soil 石英质土
Cherub rotor 丘勒勃计程仪转子
chervetite 斜钒铅矿
Cheshu lacquer 生漆;漆树漆
chessboard design 棋盘式设计
chess-board structure 棋盘格式构造
chess-board type 棋盘格式
chesserite 奇斯克石
chessy copper 石青
chessylite 蓝铜矿;石青
chessylite blue 石青色(灰绿蓝色)
chest 金库;浆槽;箱子;柜;水手(用)箱
chest bellows 箱形锻造风箱
chest coverset of screw and nut 成套螺钉及螺帽箱
chest drying machine 多层干燥机
chesterfield 低背长椅;睡椅
Chesterfield's process 带钢淬火法
chesterlite 白微斜长石

chester sampling 整群抽样
chest freezer 冰柜;冷冻柜;家用冰箱
chestnut 栗子;栗木;褐栗色
chestnut-chestnut 橙色
chestnut coal 栗级无烟煤;小块煤;栗状煤
chestnut colo(u)r 栗色
chestnut colo(u)red soil 栗钙土
chestnut oak 锥栗木
chestnut paling 栗林篱笆
chestnut roan 红砂栗
chestnut shell 板栗壳
chestnut soil 栗色土;栗钙土
chestnut tree 栗树
chestnut(wood) shingle 栗木片瓦;栗木盖屋板
chest of drawers 五斗柜;衣柜
chest-on-chest 两节柜;叠(式立)柜
chest rope 系艇缆;牵引索;牵引绳;拖船安定绳
chest saw 小手锯
chest type refrigerator 顶开门式冰箱
cheval-defrise 路障;柜马;墙壁尖刺
chevalet 架柱
cheval-glass 穿衣镜
cheval mirror (可转动的)穿衣镜
chevee 凹雕宝石
chev(e)ron 锯齿饰【建】;锯齿形花饰;人字(形)纹(样);人字形饰;人字形断口
chev(e)ron baffle 人字形障板;迷宫形障板;百叶障板;人字形挡板
chev(e)ron bar tread 人字条纹
chev(e)ron crack 中心裂纹
chev(e)ron cross bedding 人字形交错层理
chev(e)ron cross-bedding structure 人字形交错层理构造
chev(e)ron design metal roof 曲折缝金属屋面
chev(e)ron drain 人字(形)排水沟
chev(e)ron drainage 人字形排水管系
chev(e)ron dune V形沙丘
chev(e)ron fin 人字形散热片
chev(e)ron fold 角棱褶皱;尖顶褶皱
chev(e)ron mark 锯齿痕;山峰纹
chev(e)ron method 人字形堆料法
chev(e)ron mo(u)lding 波形饰;锯齿形线脚;曲折线脚;波浪饰
chev(e)ron notch V形切口;人字槽;山型缺口
chev(e)ron pattern 加工成人字形虚线;人字形花样
chev(e)ron ring 人字密封圈
chev(e)ron ring packing 带槽环轴衬
chev(e)ron seal 迷宫式密封;人字形密封
chev(e)ron seam metal roof 曲折缝金属屋面
chev(e)ron slat 人字形木板条;人字形金属板条;百叶板条
chev(e)ron style fold 尖顶式褶皱
chev(e)ron tube settler 人字形管式沉降器;人字形管式沉降计
chev(e)ron type 锯齿形;波浪形;人字形;人字斜纹;山形
chevet 教堂的内室;圆室
chevilled silk 光丝
chevkinite 硅钛铈铁矿;硅钛铈矿
chewing movement 咬合运动
chews 中等大小的煤
chew wax 嚼用蜡
cheyne-stokes breathing 潮式呼吸
cheyne-stokes respiration 潮式呼吸
Chezy coefficient 谢氏系数
Chezy equation 谢氏方程
Chezy formula 谢氏公式
Chezy velocity factor 谢氏速度系数
chi[a unit of length] 市尺
Chiadni's figures 查特尼图
chian 暗色大理岩;沥青;柏油
chian varnish 醇溶性松香清漆;松香钙脂清漆
chia oil 墨西哥油
chiaro(s)curo 黑白色画;明暗对比法;明暗配合;图画影光;图画明暗法
chiasmate frequency 交叉频率
chiastolite 空晶石
chiavennite 水硅锰钙铍石
chibinite 希宾岩
Chicago boom derrick 芝加哥式悬臂起重机
Chicago caisson 小型墩台沉井;芝加哥(闸墩)式沉箱;桥墩小型基础围堰
Chicago school 芝加哥学派
Chicago well 小型墩台沉井

Chicargo blue 芝加哥蓝
Chicargo boom derrick 芝加哥式悬臂起重机
Chicargo window 芝加哥式大窗
chick (印度东南亚的)竹帘
chicken coop 鸡舍
chicken farm 养鸡场
chicken grit 大理石渣;大理石屑;石灰石屑
chicken ilex wood carving 冬青木雕
chicken ladder 木板梯;爬行板
chicken pot 天鸡壶(瓷器名)
chicken run 养鸡场
chicken's-claw-like terrain 鸡爪形地带
chicken wire 六角形网格钢丝网;细号钢丝网;铁丝织网;铁丝网
chicken-wire cracking 龟裂;网状裂纹;网状裂缝
chicken wire fabric 网格组构
chicken wire structure 鸡皱构造
chickrasey 浅棕色有光泽的硬木(印度及缅甸产)
chick sale 施工现场活动厕所
chicle 粗地胶
chicot 半截站杆
Chideruan 希德鲁阶【地】
chief 主要的;主任;负责;取样器;首要;首领
chief accountant 总会计师
chief accountant room 总会计师室
chief adviser 首席顾问
chief architect 总建筑师;主任建筑师
chief auditor 总审计师;审计长
chief cashier 出纳主任
chief checker 理货长
chief civil engineer 土木总工程师;土木主任工程师
chief complaint 主诉
chief component 主要组成(部分)
chief comptroller 审计长
chief constructor 造船总工程师
chief container 服务组长
chief controller 主计长
chief crop 主要作物
chief delegate 首席代表
chief design 总设计
chief design engineer 设计总工程师
chief designer 总设计者;总设计师;设计总负责人
chief despatcher 调度长
chief dispatcher 主调;调度长
chief draftsman 绘图室主任
chief electric(al) and mechanical engineer 机电总工程师
chief engineer 轮机长;总工程师
chief engineering maintenance of works 工务总工程师
chief engineer of the project 设计总负责人
chief engineer's log 轮机长日志
chief engineer's room 轮机长室
chief engineman 轮机员
chief engineman room 轮机员室
chief executive 最高层管理者
chief executive officer 总经理;首席执行官;业务最高负责人
chief finance officer 财务长
chief fitter 装配工长
chief foreman 总领班
chief frame 主肋骨;主(框)架
chief inspector 总检查员
chiefly 主要地
chief maintenance 工务段长
chief manager 总经理;经理
chief mate 大副【船】
chief mechanic 总机械师;主任机械师
chief mechanical engineer 机械总工程师
chief motive power engineer 动力总工程师
chief of customs 海关关长
chief of division on duty 值班科长
chief of engineer 工程兵司令
chief officer 大副【船】
chief of gang 班长
chief of goods dispatchers 主任货物调度员
chief of inspector 检验主任
chief of locomotive dispatchers 主任机车调度员
chief of party 队长;班长
chief of passenger train controller 主任客运调度员
chief of section 工段长
chief of team 队长;班长
chief of transportation 运输主任
chief operating officer 业务总裁

chief operator 操作班长
chief operators desk 值班台长
chief permanent way engineer 线路总工程师【铁】
chief pilot 领港主任;首席领航员
chief production control and costing engineer 生产与成本管理主任工程师
chief program(me) team 主程序组
chief purser 业务主任【船】
chief quality engineer 总检查师
chief ray 主光线
chief representative 总代表;首席代表
chief resident architect 驻工地主任建筑师;主任建筑师;驻场主任建筑师
chief resident engineer 驻工地总工程师;驻段工程师;驻地主任工程师;驻地(盘)总工程师;驻场主任工程师;工段主任工程师
chief scientist 首席科学家
chief series 主列
chief source of water for animals 动物的主要水源
chief species 主林木
chief steel erector 钢建筑工长
chief stevedore 装卸组长;装卸长
chief steward 管事;大台
chief stoker 司炉
chief tally clerk 理货长
chief tallyman 理货长
chief town planner 城市总设计师
chief value 主要价值
chief wall 主墙
Chien kiln 建窑(中国古名窑)
chiffonier 化妆台;五斗柜;碗碟柜
Chihsia limestone 栖霞灰岩(早二叠世)
chi-hung 祭红
chika 糙斑
chiklite 铈钠闪石
Chiksan loading arm 奇克山式动臂油管塔
Chiksan marine loading arm 奇克山式铰接装油臂
chilblain 冻疮
childbirth 生产
child boarding group 育龄组;生育年龄组
child boarding home 儿童寄宿处
child care center 托儿中心;托儿所
child care home 托儿所
child care institution 托儿所
child day-care 儿童日托
child day care home 日托托儿所
child fender 滚筒碰垫;滚动碰垫
child health 儿童保健
child labor(er) 童工
child labor laws 童工保护法
Child-Langmuier equation 蔡尔德-朗缪尔公式;二分之三次方定律
child mortality 儿童死亡率
child night time home 夜托托儿所
child of school age 学龄儿童组
children hostel 儿童寄宿舍
childrenite 磷铝铁矿
children library 儿童图书馆
children maze 儿童迷宫
children park 儿童公园
children playing space 儿童活动区
children pool 儿童游泳池
children's allowance 子女津贴
children's benefit 子女补助金
children's center 少年之家
children's club 少年之家
children's hall 少年宫
children's hospital 儿童医院
children's house 少年之家
children's library 儿童图书馆
children's nursery 托儿所
children's paint 儿童物品用漆
children's palace 少年宫
children's play area 儿童游戏场
children's playground 儿童乐园;儿童活动中心
children's playground for traffic education 儿童交通公园
children's playroom 儿童游戏室
children's reading room 儿童阅览室
children's room 儿童室
children's shop 儿童用品商店
children's slide 滑梯
Child's law 蔡尔德定律

child ticket 儿童票
chile bar 含硫粗铜
Chile niter 智利硝石
Chile pine 南美杉
Chile slatpeter 智利硝石
chili 奇利风
chiliad 一千年
chilisaltpeter 钠硝石
chill 冷硬;冷灼;冷激层;冷冻;冷藏;急冷皱痕;激冷铸型;寒冷;白口层
chillagite 钼钨铅矿
chill area 激冷面积
chill-back 降温材料;激冷剂
chill bar 激冷试棒;导热衬垫
chill block 三角试块;冷铁;激冷试块
chill box 冷藏箱;冷藏室
chill car 冷藏车
chill cargo 低温货
chill cast(ing) 冷铸;冷硬铸件
chill cast ingot 金属模锭
chill cast share 冷硬铸铁犁铧
chill check 表面(微)裂纹;表面裂缝
chill coating 冷铁涂料
chill coil 螺旋形内冷铁;激冷圈
chill control 白口检验
chill crack 激冷裂纹;(热轧钢材表面上的)辊裂印痕;急冷裂纹
chill crystal 激冷晶体
chill depth 激冷深度;白口深度
chilldown 冷却
chill-down duct 冷却导管
chilled bottom 冷硬铸铁犁铧体
chilled car 冷藏车
chilled cargo 冷货(冷藏船装运保持华氏 30~40 货物);冷藏货;轻度冷藏货
chilled cast(ing) 冷硬铸件;冷硬铸造
chilled cast-iron 冷淬铸铁;冷硬铸铁
chilled cast iron mo(u)ld board 冷硬铸铁犁壁
chilled cast iron roll 冷硬铸铁轧辊
chilled cast iron shot 冷冷铸铁珠;激冷铸铁球
chilled cast iron shot concrete 冷硬铸铁砂混凝土
chilled cast steel 冷硬铸钢
chilled contact 冷凝接触
chilled cooling water return 冷却水回路
chilled depth 冷硬深度
chilled food storage room 冷却食物储藏室
chilled glass 淬火玻璃;钢化玻璃
chilled goods 冷藏货
chilled hardened steel 淬火钢
chilled hold 轻度冷藏舱
chilled iron 冷硬铁;冷淬铁;激冷铁
chilled iron roll 冷硬铸铁轧辊
chilled iron shot 冷硬丸粒;冷淬铁珠;冷硬铸铁丸;冷硬弹头
chilled iron wheel 冷硬铸铁车轮
chilled meat 冷冻肉
chilled meat ship 冷藏肉船;冷藏加工船
chilled roll 冷硬轧辊
chilled shot 冷硬丸粒;激冷钻粒
chilled shot-bit 钻粒钻头
chilled shot drill 钻粒钻机
chilled shot drilling 钻粒钻进
chilled shot system 钻粒钻进法
chilled skim 冷硬铸铁小前犁
chilled spot 冷硬点;激冷铸块
chilled spring wire 淬火弹簧钢丝
chilled steel 冷淬钢(铁);硬化钢;淬硬钢;淬火钢
chilled steel shot 钢砂;钢粒
chilled sump 冷却坑
chilled transport 低温运输
chilled water 冷制;冷(却)水;冷冻水;急冷水;制冷水
chilled water cooler 冷水冷却器
chilled water inlet 冷冻水入口
chilled water outlet 冷冻水出口
chilled water refrigerating system 冷却水制冷装置;冷却水制冷系统;冷却水制冷方式;冷冻水制冷装置;冷冻水制冷系统;冷冻水制冷方式
chilled water return 冷却水回路
chilled water storage 冷却水储存库
chilled water supply 冷却水供应
chilled water system 冷冻水装置;冷冻水系统;冷水式系统
chilled water system with primary pumps 一次

泵冷水系统
chilled water system with secondary pumps 二次泵冷水系统
chilled water unit 冷却水装置
chilled work 冷硬加工
chiller 冷却器;冷凝器;冷冻装置;冷冻(机)器;激冷器;致冷装置
chiller crystallizer 冷却结晶器
chillers and cooling towers 致冷和冷却塔
chill hardening 冷硬法
chilling 冷却;冷凝;冷淬;云状花纹;致冷;激冷
chilling action 激冷作用
chilling chamber 冷冻室;冷藏室
chilling effect 激冷效应
chilling injury 冷却损伤;冷害;寒害
chilling pipe 冻结管
chilling press 冷压机
chilling rate 冷冻速率
chilling resistance 急冷抵抗性;耐急冷能力
chilling room 冷却间
chilling sensitive 寒害敏感
chilling sump 冷却坑
chilling technique 冷冻技术
chilling temperature 冷冻温度
chilling tendency 白口倾向
chilling unit 冷却设备;致冷装置;致冷设备
chill in the hearth 高炉冷炉缸
chill mark 表面皱纹;表面冷斑
chill material 激冷材料
chill metal instrument 冷金属露点仪
chill mirror 冷镜面法
chill mo(u)ld 冷(硬铸)模
chill nail 激冷钉
chill oil 激冷油
chill-pass roll 激冷槽轧辊;冷硬轧槽(铸造)轧辊;冷硬孔型轧辊
chill plate 激冷板
chill point 凝固点;冻结点;冰冻点
chill-pressing 冷压;低温压制
chill removing annealing 消除白口退火
chill rod 棒状内冷铁
chill roll 冷硬轧辊;冷硬(铸造)轧辊;冷轧辊;骤冷辊
chill-roll extrusion 冷辊式挤出;辊冷式挤塑
chill room 冷冻间;冷藏室;食品冷冻间
chill test 冷却试验;激冷试验;楔形试验;白口层深度试验
chill time 激冷时间
chill wash 冷铁涂料
chill wind factor 寒风指数
chilly 寒冷的
chilmark 浅棕色石灰岩(英国威尔特郡产)
chilns 云斑
chimb 木甲板上的边沟;油漆桶凹边
chimborazite 霰石;文石
chime 木甲板上的边沟;油漆桶凹边;编钟;桶底凸缘
chime hoop 凸边箍
chimera 狮头羊身蛇尾饰
chime reinforcement 凸边加强件
chimerism 嵌合性;嵌合现象
chime unit 钟声装置
chime whistle 和声汽笛
chimney 木垛通气道;烟囱罩;管状矿脉;冰种瓯穴;冰川瓯穴;通风筒
chimney above roof 出屋顶烟囱
chimney apron 烟囱(群板)突裙;烟囱群板泛水
chimney arch 烟道拱;烟囱碹
chimney aspirator 烟囱风帽;烟囱抽气罩;烟囱抽气器;抽风器
chimney attenuator 烟囱状衰减器
chimney back 壁炉后墙;烟道调节板;烟囱衬壁;壁炉背衬
chimney back plate 烟囱调节板
chimney bar 壁炉条;壁炉铁杆
chimney base 烟囱座基
chimney block 混凝土砌块;烟囱砌块;筒形砖
chimney board 烟囱调节板
chimney bond 顺砖砌合;烟囱砌合
chimney breast 壁炉突角;壁炉台;壁炉架;壁炉腔
chimney brick 烟囱砖;烟囱砖块
chimney can 烟囱管帽;烟囱筒帽;烟囱顶管
chimney cap 烟囱顶罩;烟囱帽
chimney capital 烟囱帽顶;烟囱顶(部)
chimney casing 烟囱外壳
chimney cheek 烟囱侧壁

chimney chest 室内凸出烟囱
chimney cloud 烟云;烟囱式云
chimney concrete 烟囱混凝土
chimney connector 烟囱连接管(道)
chimney construction hoist 烟囱施工卷扬机
chimney construction scaffold(ing) 烟囱施工脚手架
chimney cooler 冷却塔;烟囱冷却器;管式冷却器
chimney cooling tower 风筒式冷却塔
chimney corner 壁炉墙角;炉角
chimney coul 烟囱通风帽(烟囱用的金属旋转通风器)
chimney cowl 通风顶管;烟囱通风帽;烟囱通风罩
chimney crane 壁炉吊臂
chimney cricket (烟囱后面排水用的)斜沟小屋顶
chimney crook 壁炉挂钩
chimney current 烟囱式气流
chimney damper 烟囱风门;烟囱烟挡;烟囱调节器
chimney draft 烟囱拔风;烟囱通风;烟囱气流;烟囱抽力;抽风作用
chimney drain 竖向截水体
chimney drain pipe 垂直排水管
chimney drain system 竖向排水系统;(土坝的)垂直排水系统
chimney draught 烟囱通风;烟囱气流;烟囱抽风
chimney draught regulator 烟囱气流调节器;烟囱拔风调节器
chimney effect 烟囱效应;自然上升效应;自然抽风效应;抽吸作用
chimney emission 烟囱排放
chimney fan 烟囱鼓风机
chimney flap 烟道门;烟囱闸板;烟囱铰链板
chimney flashing 烟囱防雨板;烟囱泛水
chimney flue 烟囱管(道);烟囱道
chimney fondu 烟囱管
chimney foot 烟囱座
chimney for locomotive 火车头烟囱
chimney gas chamber 烟气室
chimney gutter 烟囱雨水槽;烟囱排水槽
chimney head 烟囱帽;烟囱头;烟囱上部;烟囱顶(部)
chimney height 烟囱高度
chimney hood 烟囱帽
chimney hook 壁炉挂钩
chimneying 中心管道现象
chimney intake 烟道进口
chimney intake at base 烟囱底部烟道进口
chimney jack 旋转式烟囱帽;烟囱旋帽
chimney jambs 烟囱侧壁;壁炉侧墙
chimney junction 烟囱连接点
chimney ladder 烟囱爬梯
chimney lid 烟囱盖头;烟囱调节板
chimney liner 烟囱内衬;烟囱内套
chimney lining 烟囱内衬;烟囱衬壁;衬砌料
chimney loss 烟囱热耗
chimney mantle 烟囱盖顶
chimney neck 烟道
chimney netting 烟囱隔网
chimney of hollow clay tile 空心黏土砖烟囱
chimney orebody 柱状矿体
chimney or vent 烟囱或烟管
chimney outlet 烟囱出口
chimney piece 壁炉饰面;壁炉架;壁炉台
chimney pipe 烟道;烟囱管(道)
chimney plume behavio(u)r 烟囱烟羽状况
chimney pot 烟囱衬块;烟囱顶管;烟囱管帽;烟囱顶盖
chimney raft 烟囱筏基
chimney register 烟道节气门;烟囱节气门;烟囱调节板
chimney rock 浪蚀石笋;柱状石
chimney shaft 烟筒;烟囱;烟囱(筒)身
chimney shaft attached to wall 附墙烟囱
chimney shell 烟囱外壳
chimney shell wall 烟囱外壳墙
chimney slide 烟囱滑道;烟囱导板
chimney sliding damper 烟道滑挡
chimney soot 油烟;烟垢;烟囱烟灰
chimney spot 烟垢;烟囱地点烟垢
chimney stack 屋顶烟囱;集合烟囱;壁炉烟道;烟囱体;总烟囱;工厂高烟囱;高烟囱;多烟道烟囱;丛烟囱
chimney stacker 旋转式烟囱帽
chimney stalk 工厂高烟囱;高(大)烟囱;丛烟囱

chimney superelevation 烟囱超高
chimney sweep 烟囱扫除
chimney sweeper's carcinoma 扫烟囱工人癌
chimney sweeping 烟囱扫除;通烟囱
chimney taper 烟囱收分
chimney terminal 烟筒;烟囱帽
chimney throat 壁炉咽喉;壁炉喉口;炉壁烟道口
chimney throating 烟道咽喉
chimney top 烟囱顶(部);烟囱帽
chimney tun 一排烟囱;高烟囱
chimney valve 烟道节气门
chimney ventilation 烟囱通风;烟囱抽风;自然通风
chimney waist 炉壁烟道口;烟道咽喉
chimney wall 烟囱壁
chimney wing 烟囱收缩侧壁;壁炉烟道口翼墙
chimopelagic 冬季表层的
chin 刃
china 瓷器;瓷;白坯瓷器
China Armand pine 华山松
China Association for Engineering Construction Standardization 中国工程建设标准化协会
China Association for Science and Technology 中国科学技术学会
China Association for the Management of Construction cost 中国建设工程造价管理协会
Chinabells 拟赤杨【植】
chinaberry 楝树;紫花树
chinaberry tree 楝树
China blue 中国蓝;青瓷色;酸性水溶青
China Building Industrial Corporation 中国建筑工业公司
china cabinet 陈列柜;瓷器室
China carpet 中国地毯
China Classification Society 中国船级社
china clay 高岭土;瓷土;白瓷土
china clay chamot(te) 瓷土熟料
china clay in powder 瓷土粉
china clay method 制瓷方法;陶瓷法
china clay standard solution 陶土标准溶液
china clay waste 瓷器厂废料
china closet 瓷质大便器;瓷器室;瓷马桶;瓷便桶
China commodity inspection Bureau 中国商品检验局
China Communications, Transportation Association 中国交通运输协会
China Communications Planning and Design Institute for water Transportation 中交水运规划设计院
China Construction Engineering Corporation 中国建筑工程公司
china-cypress 水松属
China Digital Data Network 中国数字数据网
China Education and Research Network 中国教育科研网
chinafir 杉属
China flexible pavement design method 中国柔性路面设计法
China Gas Society 中国煤气学会
chinagraph pencil 特种铅笔
china grass 苎麻
China grass cloth 芦席帘;草席
China green 孔雀石绿;中国绿
chinagreen 广东万年青;亮丝草
China Harbo(u)r Engineering Company 中国港湾建设总公司
China Highway and transportation Society 中国公路运输学会
China-hospital 陶器修理所
China ink 中国墨水;墨
china-ink painting 墨彩(陶瓷装饰法)
China insurance clause 中国保险条款
China Interactive Network Association 中国交互网络协会
China International Contractors Association 中国对外承包工程商会
china-making pencil 特种铅笔
China Maritime Search and Rescue Center 中国海上搜索救助中心
China Merchants Steam Navigation Co., LTD 中国招商局轮船股份有限公司
China Merchants Steamship Company 招商局
China Merchants Steamship Navigation Company 招商局(中国)轮船公司
chinampa 墨西哥式人造地坪;浮园耕作法

China National Construction Machinery company 中国建筑工程机械公司
China National Foreign Trade Transportation Corp. 中国对外贸易运输(集团)公司
China National Technical Import & Export Corporation 中国技术进出口总公司
China network 中国网
China Ocean Shipping Agency 中国外轮代理公司
China Ocean Shipping (Group) Company 中国远洋运输(集团)总公司
China Ocean Shipping Tally Company 中国外轮理货公司
China Official System of Units 中国计量单位制
china orange 橙色
china painting 陶器面图画
China Petrochemical Corporation 中国石化总公司
China pink 石竹
China Ports 中国港口(杂志名)
china process 陶器生产方法；瓷器制造法
chinar 法国梧桐；藤悬木
China railway industry 中国铁路工业
China rose 月季【植】
China rug 中国地毯
chinarump 硅化木
china sanitary wares 卫生陶瓷制品；陶瓷卫生洁具
china shop 瓷器店
China silver alloy 中国银合金
China Society of Civil Engineering 中国土木工程学会
China Society of Geology 中国地质学会
China Society of Mechanical Engineering 中国机械工程学会
China Society of Seism 中国地震学会
China squash 南瓜
China Standardization BS yearbook 中国标准化年鉴
china stone 高岭土化花岗岩；瓷(土)石
china tea-pot 瓷茶壶
Chinatown 中国城；唐人街
china varnish 柏油清漆
china vase 瓷花瓶
chinaware 瓷器
China white 锌白；纯铅白；白颜料
China wood oil 桐油
chinbeak mo(u)lding 波纹形线饰；S形凹凸线饰
chin-chin hardening 冷模；激冷
chinckle 纽结
chine 木甲板上的边沟；舷线(舭线)；舭缘线；刨式水平镗—钻—铣机床；山脊
chine boat 尖底船
chine cupboard 碗碟架
chine line 舭部折角线
Chinese Academy of Medical Sciences 中国医学科学院
Chinese Academy of Sciences 中国科学院
Chinese alligator 扬子鳄
Chinese and Foreign joint-venture vessel 中外合营船舶
Chinese arbor-vitae 侧柏
Chinese Architecture 中国建筑(杂志名)；中国式建筑
Chinese architecture decoration 中国建筑装饰
Chinese art metal 中国工艺品用铅锡黄铜
Chinese ash 枫杨；白蜡树
Chinese Astronomical Society 中国天文学会
Chinese azalea 八厘麻
Chinese bamboo water bucket 中国竹斗岸水车
Chinese binary 中国式二进制；直列二进制；竖式二进制
Chinese black pine 罗汉松
Chinese blue 中国蓝；华蓝；优质普鲁土蓝
Chinese bronze 中国青铜
Chinese capstan 双径绞滚筒绞盘
Chinese character 汉字【计】
Chinese character library 汉字库
Chinese chestnut 板栗
Chinese chrysanthemum 菊花
Chinese classical garden 中国古典园林
Chinese coir palm 棕榈
Chinese convex and concave tile 蝴蝶瓦
Chinese coptis 黄连
Chinese decorative art 中国式装饰艺术
Chinese denglas fir 黄杉；华帝杉
Chinese dolphin 白鳍豚

Chinese Douglas fir 黄杉
Chinese dragon pump 龙骨水车；中国式水车
Chinese drill stock 弓形钻；辘轳钻
Chinese elm 榔榆
Chinese equivalent 折合市制
Chinese evergreen chinquapin 甜槠栲；甜橙
Chinese falsepistache 银鹊树
Chinese fan-palm 蒲葵
Chinese filbert 华榛；山白果
Chinese finless porpoise 江豚
Chinese fir 杉木
Chinese flowering apple 海棠花
Chinese font store 汉字模库
Chinese food 中餐馆
Chinese-foreign joint ventures 中外合资企业
Chinese gabled roof 硬山屋顶
Chinese gallotannic acid 五倍子丹宁酸
Chinese green 中国绿；绿胶；绿膏
Chinese hazel 榛树
Chinese hemlock-spruce 铁杉
Chinese herring 曹白鱼
Chinese hipped roof 庑殿屋顶
Chinese honeylocust 皂荚
Chinese Hydraulic Engineering Society 中国水利学会
Chinese ilex 冬青
Chinese Industrial Standards 中国工业标准
Chinese information processing 汉字信息处理
Chinese ink 黑墨(汁)
Chinese ink-and-wash painting 中国水墨画
Chinese insect wax 中国(虫)蜡；虫白蜡
Chinese juniper 桧
Chinese lacquer 毛生漆；中国生漆；中国大漆；大漆；生漆
Chinese lake 胭脂红色
Chinese lake dolphin 洞泊白鳍豚
Chinese lantern 中国灯笼；彩色折纸灯笼
Chinese larch 红杉【植】
Chinese littleleaf box wood carving 黄杨木刻
Chinese locust 槐木
Chinese lug 肋杆纵帆
Chinese Medical Team 中国医疗队
Chinese mountain and water garden 中国山水园
Chinese National Committee on Irrigation and Drainage 中国灌溉排水委员会
Chinese No.7 Signal(l)ing 中国七号信令【铁】
Chinese oil 桐油
Chinese orange 金橘色；茜素橙色
Chinese overhung gable-end roof 悬山屋顶
Chinese painting 中国画
Chinese parasol 梧桐
Chinese pasque-flower 白头翁
Chinese phonetic script 汉字拼音字母
Chinese pine 油松
Chinese platform 中国地台
Chinese pump 差动式泵
Chinese red 朱红色；橘红色；中国红
Chinese red pine 马尾松
Chinese remainder theorem 中国剩余定理
Chinese restaurant 中餐馆
Chinese river dolphin 江河白鳍豚
Chinese rockery 假山园
Chinese rose 月季
Chinese round ridge roof 卷棚屋顶
Chinese sapium 乌桕
Chinese screen 中国屏风
Chinese seismic degree 中国地震烈度
Chinese Shaku-do bronze 中国谢库多铜合金
Chinese silvergrass 芒草
Chinese Society of Agriculture 中国农学会
Chinese Society of Coeanology and Limnology 中国海洋湖沼学会
Chinese Society of Environmental Sciences 中国环境学会
Chinese Society of Hydraulic Engineering 中国水利学会
Chinese Society of Oceanology 中国海洋学会
Chinese Society of Surveying and Mapping 中国测绘学会
Chinese Standard 国标
Chinese stewartia 紫茎
Chinese sturgeon 中华鲟
Chinese style 中国式
Chinese sucker 胭脂鱼

Chinese sumac 盐肤木
Chinese sweet gum 枫香树
Chinese sweetleaf 华山矾
Chinese system of weights and measures 市制
Chinese tallow tree 乌桕
Chinese-traditional paired stelae 华表
Chinese translation 中译
Chinese tree wax 中国白蜡
Chinese tuliptree 鹅常楸
Chinese varnish 中国漆
Chinese varnish tree 石栗
Chinese(vegetable) tallow 乌桕(皮)油；桕脂；皮油
Chinese vermil(l)ion 中国朱砂
Chinese wax 中国蜡；虫蜡；白蜡
Chinese weeping cypress 柏木
Chinese wheat straw picture 麦秸画
Chinese wheel 水车
Chinese white 锌白；氧化锌；中国白；白色颜料
Chinese white poplar 毛白杨
Chinese white stand oil enamel 锌白瓷漆
Chinese windlass 差动绞筒；差动绞盘
Chinese Wire Ga(u)ge 中国线规
Chinese wistaria 紫藤
Chinese wood oil 桐油
Chinese yellow 皇衣黄色；中国黄；雌黄黄色
Chinese yellowwood 小花香槐
Chinesization 汉化
Chiness speculum metal 中国镜铜
chine to chine 舷对舷
Ching Te Chen kiln 景德镇窑
chink 裂缝；开裂；隙缝；龟裂；缝隙；透过缝隙的一束可见光
chinker 填隙物
chinking 填缝料；油灰；腻子；灰泥
chinky 有裂缝的；多孔隙的
chinley coal 高级烟煤；块煤
Chinlung limestone 青龙灰岩(早三迭世)
Chinoiserie 仿东风建筑；中国式装饰艺术；中国艺术风格
chinometer 水平计
chinook 钦诺克风
chinsing iron 填塞凿
chintz 像印花棉布一样的墙纸；印花棉布；轧光印花织物
chintz finish 摩擦轧光整理
chiolite 锥冰晶石
chionathin 流苏树脂
chioniolphobous plant 避雪植物
chionophobous plant 嫌雪植物
chionophytia 雪地植物群落
chip 石屑；料屑；孔屑；集成电路芯片；集成电路块；基片；木屑；木片；码子；芯片；薄片剥裂；玻璃制品缺陷；爆边；微型组件；碎落；石碴
chip area 基片面积；切屑通道；排屑槽
chip ax(e) 劈斧；剁斧(头)；琢石斧
chip ballast 碎石渣；石碴
chip basket 切屑篮
chip bit 碎片钻头
chip black 黑漆片
chip blasting 浅孔爆破；真底岩爆破
chip-blower 气枪
chipboard 木屑板；再生纸板；刨花板；碎木胶合板；废纸制成的纸板；粗纸板；刨光板；碎纸胶合板
chip board press 碎料板压机
chip bonding 片接合；片焊接
chip box 孔屑箱
chip-breaker 压碎机(木片、石片)；分屑沟；破屑机；断屑台；木片破碎机；断屑器；断屑槽
chip-breaker distance 断屑台到切削刃的距离
chip-breaker grinder 断屑台磨床；刀具断屑台磨床
chip-breaker grinding machine 断屑台磨床
chip breaker groove 断屑槽
chip-breaker groove radius 断屑槽底半径
chip-breaker groove width 断屑槽宽
chip-breaker height 断屑台高度
chip-breaker wedge angle 断屑台楔角
chip-breaking flute 断屑沟
chip bucket 切屑桶；松散料铲斗
chip cap 木刨压板
chip capacitor 片状电容器
chip capacity 切削能力
chip carving 雕饰；雕刻品
chip catcher 捞粉器；岩粉收集器
chip chisel 碎石选分设备

chip cleaner 木片清洁机;排屑器
chip clearance 切屑间隙
chip component 片状元件
chip compression ratio 切屑收缩比
chipcore 碎料板芯
chip crack 碎裂纹;微裂纹;发裂
chip crusher 木片压碎机;木片破碎机
chip curl 切屑螺旋
chip deformation 切屑变形
chip diamond 碎片状金刚石
chip dispersion method 压片分散法
chip distributor 石屑摊铺机;石屑撒布机
chip dividing groove 分屑槽
chip flow 切屑流
chip flow type 易排屑型
chip formation 切屑形成
chip former 卷屑器
chip grinder 削片打磨机
chip guard 切屑防护器
chip handler 晶片分布控制器
chip hold down effect 岩片抑制作用
chip injector 铁屑喷射器
chip input protection 集成电路输入保护
chip junction 片结
chip layout 片设计
chipless machining 无屑加工
chipless turning 无屑车削
chipless working 无屑加工
chip level 片级
chip-load 切屑抗力;切削荷载
chip-load per tooth 每齿切屑荷载
chip log 肩板计程仪;(测程仪;拖板计程仪)
chip marking 削片压痕
chip marks 碎片压痕(木材);毛刺沟痕;削片压痕;碎屑压痕
chip marks on integrally forged vessel 整体锻造容器上的缺口标志
chip method 拣块法
chip microprocessor 芯片微处理器;单片微处理机;单片式微处理器
chip microprocessor process 单片式微处理机工艺
chip-off 削去
chip of stone 碎石片
chip packer 木片装料机
chip packing 切屑填塞
chippage 面砖剥落;面砖脱皮;剥落(火灾后果)
chip pan 承屑盘
chipped gear tooth 剥蚀的轮齿
chipped glass 冰花玻璃;斑纹玻璃
chipped grain 刨光木纹;沟痕;刨痕;抛光木纹;刀痕;切削沟纹
chipped marble 爆皮玻璃球
chipped marble finish 干黏石
chipped stone 料石;琢石
chipped stone surface 石屑铺面琢石面
chipped surface 琢石面;破碎面;石屑铺面
chipped wave 削平波
chipped wood 木材削片
chipped wood concrete 碎木混凝土
Chippendale 朱色的;(英国18世纪的)希宾式轻巧家具
Chippendale mahogang 高级硬红木
chipper 机械化清理装置;削片机;凿子;錾子;錾凿;缺陷修整工;缺陷清理工;切片机
chipper with dust extractor 带收尘器的碎片机
chipping 机械脱漆;琢毛;凿平;凿毛;錾平;剁錾;铲修;铲平;剥釉;刨削;清铲;切割;切成碎片;无缝地板装修;涂膜碎屑;涂膜片落
chip(ping) axe 剁斧
chipping bed 铲凿清理台
chipping carpet 石屑毡层
chipping chisel 剁凿;剁錾;铲凿;平凿;平头铲刀;碎石凿;石錾;石凿;石錾;平头錾
chipping-crushing action 剪切压碎作用
chipping degree 碎片度
chipping edger 切边机;双削齐边锯
chipping goggles 敲锈眼镜
chipping hammer 尖锤;錾平锤;敲渣锤;敲锈锤;敲水垢锤;碎石锤;琢石锤
chipping heading 制材削片联合机
chipping knife 切刀
chipping machine 混凝土打毛机;切片机
chipping off 削去;削掉
chipping-out 热洗炉;清炉壁

chipping plane 边刨
chipping qualities 切片质量
chipping rating standards 涂膜片落等级标准
chip(ping) resistance 抗片落性;耐碎片性;耐崩裂性
chippings 细碎石;填缝骨料;碎屑;碎石填缝骨(细粒);碎片;石片;筛屑
chippings aggregate 石屑骨料;石屑集料
chippings breaker 轧碎机;轧石机
chippings carpet 石屑(垫)层
chippings compound 石屑混合物
chippings concrete 石屑混凝土;碎石集料混凝土;碎石骨料混凝土
chippings crusher 轧碎机;轧石机;刨屑压碎机
chippings from brick ruins 碎砖
chippings from masonry ruins 砖石墙;碎砖石;碎块
chippings machine 铺碎石机
chippings precoated with tar 煤沥青预拌石屑
chippings spreader 铺碎石机
chippings stone course 琢石层
chippings surfacing 石屑铺面
chipping test 片落试验
chipping tool 平口钳
chippy 研碎的;粉碎的;碎片的
chippy cage 副井罐笼
chippy hoist 副井提升机
chip rejector 碎石选分设备;石屑选分器
chip removal 排屑;排出岩屑(炮眼中)
chip resistance primer 抗片落底漆
chip resistor 片形电阻器
chip room 排屑槽
chiprupter 碎屑器
chips 金属屑;干颜片料;切屑;树脂片
chip sample 岩屑试样
chip sampling 采收岩屑试样
chips breaker 破碎机
chips concrete 石屑混凝土
chip scratch 屑痕
chips crusher 破碎机
chip seal 石屑罩面
chip separation 切屑分离
chipset 芯片组
chip shield 保护(垫)圈
chip silicon gate 片状硅栅
chip size 集成电路片尺寸;电路片尺寸
chip slice status 单片状态
chip soap 皂片
chips of glass 玻璃屑
chip space 排屑槽
chip spreader 石屑撒布机;碎石摊铺机
chip stone 细碎石;碎石片;石片
chipswringer 片屑绞段机
chip technology 芯片工艺
chip trawl system 刀具送进系统
chip trough 切屑槽
chip washer 保护(垫)圈
chip way 排粉槽
chipway space 排屑间隙(钻头)
chip weight 屑重
chip wringer 碎片榨干器;碎片榨干离心机
chirality 空间螺旋特性;手征
Chireix-Mesny antenna 锯齿形天线
Chireix modulation 异相调制
chirograph 亲笔字据;骑缝证书
chirography 书法;笔迹
chiroscope 手导镜
chirp 线性调频脉冲
chirping 鸣叫;啁啾声
chirp radar 脉冲压缩雷达;线性调频雷达
chisel 刻刀;凿子;凿式凿;錾子;錾凿;钢凿
chisel bar 钎杆;凿钎;撬棍
chisel bar lever 撬杆
chisel bit 凿形钻头;凿头;单刃钻冠;冲击式钻头;取样钻头
chisel bit tool 钢绳冲击钻头
chisel bond 凿形焊接
chisel-bond cracking 凿焊龟裂
chisel breaker 碎石锤
chisel breaker vessel 碎石凿船
chiseled ashlar 粗凿石砌体;精凿石;粗琢石坏工;粗凿石坏工
chisel edge 横刃;凿尖;切削刃(钻头);凿锋;刀刃;(钻头的)刀刃尖

chisel edge angle 横刃斜角;凿尖角
chisel edge thinning 修磨横刃
chiseled slate 凿面石板
chiseled tooth saw 凿齿锯
chisel finish 锤凿饰面
chisel for cutting iron 割铁凿
chisel groove 凿槽
chisel impact pressure chamber 钻头动力作用试验仪
chiseling 凿平;凿开;凿边
chiseling off 凿掉
chisel jumper 长凿
chisel knife 铲刀;扁铲;凿刀
chiselled ashlar 粗凿石
chiselled stone face 錾光石面
chisel-like 錾形的
chisel(l)ing 凿开(路面);凿缝;凿边
chisel(l)ing of tunnel surface 隧道补挖
chiselling point 凿点
chiselly 多砂砾的
chiselly soil 砾质土
chisel nail point 扁钉尖;楔形钉尖
chisel planting 沟种;沟播
chisel point 凿尖
chisel-pointed nail 凿钉尖;凿钉
chisel rod 一字形整体钎杆
chisel set 成套凿子
chisel-shaped 錾形的
chisel-shaped bit 錾形钻头;冲击式一字形钻头
chisel steel 凿子钢;凿钢
chisel teeth (牙轮钻头的)光齿
chisel-tipped steel 钎字杆
chisel tong 凿钳
chisel tool steel 工具钢
chisel-tooth shovel 凿齿松土铲
chisel-type bit cutting edge 凿头刃口
chisle 砾石滩
chisley soil 石质土
chi-square 卡方
chi-square confidence interval χ^2置信区间
chi-square criterion χ^2准则
chi-square distribution χ^2分布;χ平方分布
chi-square goodness of fit test 卡方拟合优度检验
chi-square test 卡方检验;概平方测检;χ^2检验;χ平方检验
chit 小额欠款单据;单据;便条
chitin 壳质;甲壳质;几丁质
chitinization 壳质化
chitinous 壳质的
chitinous layer 壳质层
chitinous wall 几丁质壁
chitose 壳糖;甲壳糖
chittering 微裂纹陶瓷器底部边缘;微裂纹
chitty 便条;小额欠款单据
chkalovite 硅铍钠石
chladnite 顽辉石陨星物质
chloanthite 复砷镍矿
chloasma 褐黄斑
chlophenacemide 苯氯乙酰脲
chlor 氯绿色
chloracetone 氯丙酮
chloracid 含氯酸
chloral 氯醛
chloral hydrate 水合三氯乙醛;水合氯醛
chloralkali plant 氯碱工厂
chloralkane 氯烷烃;氯代烷烃
chloralum 氯化铝合剂
chloraluminite 氯矾石
chloraluminte 氯铝石
chlorambucil 苯丁酸氮芥
chloramination 氯胺化
chloramine 氯胺
chloramines compound 氯胺化合物
chloramine solution 氯胺溶液
chloranil electrode 氯醌电极
chloranthus bouquet 百花珠兰
chlorapatite 氯磷灰石
chlorargyrite 氯银矿
chlorate 氯酸盐
chlorate(blasting) explosive 氯酸盐炸药
chlorating agent 氯化剂
chloration 氯化作用;加氯作用
chlorauric acid 氯金酸
chlorazide 叠氮化氯

chlorazotic acid 王水
chlordane 氯丹(杀虫剂);八氯六亚基萘
Chlorella 小球藻属
chlorellin 绿藻素;小球藻素
chlorendate 氯菌酸酯;六氯降冰片烯二酸酯
chlorendic acid 六氯降冰片烯二酸
chlorendic anhydride 氯菌酸酐
chlorendix acid 氯菌酸
chlorhydric acid 盐酸;氢氯酸
chlorhydrocarbon 氯化烃
chlorhydrogenation 氢氯化作用
chloric acid 氯酸
chloric water 氯水
chloride 氯泥石;氯化物;漂白剂
chloride-bromine ratio 氯溴比值系数
chloride calcium 氯化钙
chloride-calcium type 氯化钙型
chloride colo(u)r 氯化法二氧化钛色相
chloride concentration 氯离子浓度;氯化物浓度
chloride content 氯化物含量;含氯量
chloride extraction 氯化物萃取
chloride feedstock 氯化法二氧化钛原料
chloride flux 氯化物通量
chloride free 无氯盐
chloride index 氯化指数
chloride ion 氯离子
chloride ion shift 氯离子转移
chloride lake 氯化物湖
chloride-magnesium type 氯化镁型
chloride mass balance 氯化物质平衡
chloride mineral spring water 氯化物矿泉水
chloride of lime 氯化石灰;漂白粉
chloride of methyl 甲基氯化物
chloride of water 水的氯化物
chloride ore 氯化物矿石
chloride paper 氯化银象纸
chloride pollution 氯化物污染
chloride pre-scrubber 氯化物预洗涤器
chloride process titanium dioxide 氯化法二氧化钛
chloride rubber paint 氯化橡胶涂料;氯化橡胶漆
chloride salinized soil 氯化物盐渍土
chlorides bromides iodides 氯化物溴化物碘化物
chloride spring 氯化物型泉
chloride stabilization of soil 氯盐土壤稳定法;土的氯盐稳定法
chloride stress corrosion cracking 氯化物腐蚀裂纹
chloride-sulfate salinized soil 氯化物—硫酸盐盐土
chloride tar 氯化焦油
chloride technology 氯化法工艺
chloride thermal water 氯化物型热水
chloride trifluoride 三氟化氯
chloride water 氯化物水
chloride water zone 氯化物水带
chloridization 氯化处理;氯化(作用)
chloridizing roaster 氯化焙烧炉
chloridizing roasting 氯化焙烧
chloridometer 氯量计
chlorimet 耐热镍基合金;耐蚀金属
chlorimetry 氯量滴定法
chlorinatable titanium slag 可氯化钛渣
chlorinate 氯化产物
chlorinated acetone 氯丙酮
chlorinated additive 氯化添加剂
chlorinated alkane 氯化烷烃
chlorinated aromatic compound 氯化芳香化合物
chlorinated asphalt 氯化地沥青;氯化沥青
chlorinated benzene 氯苯
chlorinated biphenyl 氯化联苯
chlorinated butadiene 氯化丁二烯
chlorinated copperas 氯化绿矾
chlorinated copper phthalocyanine 氯代铜酞菁;酞菁绿
chlorinated effluent 氯化污水;氯化废水
chlorinated ethylene-propylene-diene copolymer 氯化三元乙丙橡胶
chlorinated ethylene reaction 氯化乙炔反应
chlorinated hydrocarbon 氯化烃;氯化碳氢化合物;氯代烃类
chlorinated hydrocarbon insecticides 氯化烃类杀虫剂
chlorinated hydrocarbon refrigerant 氯化碳氢化合物制冷剂
chlorinated ice 加氯冰
chlorinated insecticide 氯化杀虫剂

chlorinated isocyanurates 氯化异氰尿酸盐
chlorinated lime 氯化石灰;含氯石灰;漂白粉
chlorinated lubricant 氯化润滑剂
chlorinated methane 氯化甲烷
chlorinated naphthalene 氯化萘
chlorinated organic compound 氯化有机化合物
chlorinated organic pesticide 氯化有机杀虫剂;氯化有机农药
chlorinated organic substance 氯化有机物
chlorinated paraffin 氯化链烷烃
chlorinated paraffin wax 氯化石蜡
chlorinated para nitraniline red 氯化对位红
chlorinated para red 氯化对位红;永固银朱
chlorinated phenol 氯化(苯)酚
chlorinated phenolic compound 氯化—酚混合剂
chlorinated phenoxyacetic acid 氯化苯氧基乙酸
chlorinated phosphate 氯化磷酸酯
chlorinated polyether 氯化聚醚
chlorinated polyether pipe 氯化聚乙醚管
chlorinated polyethylene 氯化聚乙烯
chlorinated polypropylene 氯化聚丙烯
chlorinated polyvinyl chloride 氯化聚氯乙烯;过氯乙烯
chlorinated polyvinyl chloride fibre 过氯乙烯纤维
chlorinated polyvinyl chloride paint 氯化聚氯乙烯涂料
chlorinated polyvinyl chloride pipe 过氯乙烯管
chlorinated polyvinyl chloride plastic fitting 氯化聚氯乙烯塑料管件
chlorinated polyvinyl chloride plastics 氯化聚氯乙烯塑料;过氯乙烯塑料
chlorinated polyvinyl chloride resin 过氯乙烯树脂
chlorinated polyvinyl chloride solution adhesive 过氯乙烯胶液
chlorinated polyvinyl coated fabric 氯化聚乙烯涂层织物
chlorinated polyvinyl pipe 氯化聚乙烯管
chlorinated resin 氯化树脂
chlorinated rubber 氯化橡胶
chlorinated rubber additive 氯化橡胶胶黏剂
chlorinated rubber base 氯化橡胶基
chlorinated rubber coat 氯化橡胶涂层
chlorinated rubber curing compound 氯化橡胶养护混合物
chlorinated rubber enamel 氯化橡胶瓷漆
chlorinated rubber glue 氯化橡胶黏结剂
chlorinated rubber lacquer 氯化橡胶清漆
chlorinated rubber paint 氯化橡胶涂料;氯化橡胶漆
chlorinated rubber pigmented varnish 氯化橡胶加色罩光漆
chlorinated rubber polymer 氯化橡胶聚合物
chlorinated rubber priming paint 氯化橡胶底漆
chlorinated rubber varnish 氯化橡胶清漆
chlorinated solvent 氯化(了的)溶剂
chlorinated water 氯化水;加氯水
chlorinating light tube 氯化灯管
chlorination 氯化处理;氯化(作用);加氯法;加氯处理;吹氯除气精炼法
chlorination breakdown 氯化分解
chlorination breakpoint 氯化转效点;氯化折点;折点加氯
chlorination catalyst 氯催化剂
chlorination chamber 氯化池;加氯室;加氯间
chlorination compound 氯化化合物
chlorination contact tank 氯化接触槽;加氯接触池
chlorination disinfection 氯化消毒;加氯消毒
chlorination disinfection by-products 氯化消毒副产物
chlorination disinfection of water 水的氯化消毒法
chlorination effect 氯化效应
chlorination furnace 氯化炉
chlorination installation 加氯设备
chlorination in trade-waste treatment 工业废水的加氯处理
chlorination of cooling water 冷却水氯化
chlorination of drinking water 饮用水氯化处理;饮用水加氯消毒
chlorination of potable water 饮用水加氯消毒
chlorination of sewage 污水加氯;污水氯化
chlorination of water 水的氯化处理;水的加氯消毒(法)
chlorination of well 井水加氯消毒
chlorination plant 氯化设施;氯气消毒装置;加氯杀菌装置

chlorination process 氯化过程;氯化法
chlorination reaction 氯化反应
chlorination route 氯化法
chlorination table 加氯处理表
chlorination tower 氯化塔
chlorinator 加氯器;加氯机
chlorinator installation 加氯设备
chlorine 氯(气)
chlorine absorptive property 吸氯特性
chlorine-ammonia method 氯—氨法
chlorine-ammonia process 氯—氨法
chlorine-ammonia treatment 氯—氨处理
chlorine and derivative disinnfectants 氯系消毒剂
chlorine-bearing reservoir 氯吸收库
chlorine-bearing substance 含氯物料
chlorine bleaches 含氯漂白剂
chlorine breakpoint 折点加氯
chlorine catalyst 氯的催化作用
chlorine centrifugal compressor 离心式氯气压缩机
chlorine chemistry 氯化学
chlorine concentration 含氯浓度
chlorine contact chamber 氯接触池;氯气消毒室
chlorine-containing plastics 含氯塑料
chlorine-containing substance 含氯塑料
chlorine content 含氯量
chlorine content after intrusion 入侵后氯离子含量
chlorine content before intrusion 入侵前氯离子含量
chlorine corrosion 氯气腐蚀
chlorine demand 需氯量
chlorine demand of sewage 污水需氯量;污水净化需氯量;污水耗氯量
chlorine dioxide 二氧化氯
chlorine dioxide disinfectant generator 二氧化氯消毒剂发生器
chlorine dioxide disinfection 二氧化氯消毒
chlorine dioxide foam separation process 二氧化氯泡沫分离法
chlorine dioxide generator 二氧化氯发生器
chlorine dioxide oxidation method 二氧化氯氧化法
chlorine disinfection 氯消毒
chlorine dosage 加氯量;用氯量
chlorine feeder 加氯器
chlorine feed pump 投氯泵;加氯泵
chlorine gas monitor 氯气监测仪
chlorine gas poisoning 氯气中毒
chlorine hydrate 水合氯
chlorine ice 氯冰
chlorine index 氯化物指数
chlorine-induced oxidant 氯致氧化剂
chlorine log 氯测井;含氯量录井
chlorine monoxide 氧化氯
chlorine oxide 氯的氧化物
chlorine peroxide 过氧化氯
chlorinator poisoning 氯中毒
chlorine pollution 氯污染
chlorine products 氯产品
chlorine recorder 含氯量记录器
chlorine required 需氯量;需氯量
chlorine requirement 需氯量
chlorine reservoir 氯吸收库
chlorine residual 余氯
chlorine residual measurement 余氯测定
chlorine residue 余氯
chlorine room 加氯室;加氯间
chlorine saline soil 氯盐渍土;亚氯盐渍土
chlorine selectivity 氯的选择性
chlorine sink 氯吸收槽
chlorine tanker 氯气运输船
chlorine valve 氯阀
chlorine war gas 氯毒气
chlorinity 氯含量;氯度;氯当量;含氯量
chlorinity of seawater 海水氯度
chlorinolog 含氯量录井
chlorinolysis 氯化水解法
chloriodide 氯碘化物
chlorion 氯离子
chlorisoin-dolinone orange 氯化异吲哚酮橙颜料
chlorisoin-dolinone red 氯化异吲哚酮红颜料
chlorite 绿泥石;亚氯酸盐
chlorite albite muscovite quartz hornfels 绿泥纳长白云母石英角岩
chlorite-albite muscovite-schist 绿泥纳长白云母

片岩
chlorite amphibolite 绿泥石角闪岩
chlorite cement 绿泥石胶结物
chlorite cummingtonite schist 绿泥石镁铁闪石片芒
chlorite emulsion 氯化银乳剂
chlorite gneiss 绿泥石片麻岩
chlorite hornblende schist 绿泥石角闪片岩
chlorite-oolite 绿泥石鲕状岩
chlorite phyllite 绿泥石千枚岩
chlorite rock 绿泥石岩
chlorite schist 绿泥(石)片岩
Chlorite-sericite schist 绿泥绢云母片岩
chlorite serpentine schist 绿泥蛇纹片岩
chlorite shale 绿泥页岩
chlorite slate 绿泥板岩;绿泥石板
chlorite spar 硬绿泥石
chlorite tremolite schist 绿泥透闪片岩
chloritic 绿泥石质沉凝灰岩
Chloritic marl 绿泥泥灰岩层
chloritic mineral 绿泥矿物质
chloritization 绿泥石化
chloritoid 硬绿泥石
chloritoid phyllite 硬绿泥石千枚岩
chloritoid schist 硬绿泥(石)片岩
chlorizate 氯化产物
chlorization 绿泥石化;氯化作用;加氯作用
chlor kalk 漂白粉
chlormanganokalite 钾锰盐
chlormethine 氮芥
chlornitrofen 草枯醚
chloro 氯代
chloroacetaldehyde 氯乙醛
chloroacetate 氯乙酸
chloroacetic acid 氯醋酸
chloroacetic chloride 氯乙酰氯
chloroacetone 氯丙酮
chloroacetophenone 氯苯乙酮;氯乙酰苯
chloroacrolein 氯丙烯醛
chloroalkane 氯化烷烃
chloroalkane oxidation 氯烷氧化
chloroalkene 氯化烯烃
chloroaniline 氯苯胺
chlorobenzene 氯苯
chlorobenzene poisoning 氯苯中毒
chlorobenzoic acid 氯苯甲酸
chlorobenzol 氯苯
chlorobiphenyl 氯化联苯
chlorobutanol 氯丁醇
chlorobutyl 氯丁基
chlorocalcite 氯钾钙石
chlorocarbon 氯碳化合物
chlorocarbon refrigerant 氯烃制冷剂
chloro-carbon solvent 氯烃溶剂
chlorochrous 近似绿色的
chlorocobalamine 氯钴胺
chlorocosanes 氯化石蜡油
chlorocyclohexane 氯代环己烷
chlorodibromide 二溴化氯
chlorodioxide 二氧化氯
chloroepoxy propane 环氧氯丙烷
chloroethane 氯乙烷
chloroethene 氯乙烯
chlorofluorocarbon 氯氟碳化合物
chlorofluorocarbon oil 氟氯油
chlorofluorocarbon plastics 聚氯氟烃塑料;含氯氟烃塑料
chlorofluorocarbon refrigerant 氯氟碳化合物制冷剂
chlorofluorocarbon refrigeration fluid 氯氟碳化合物制冷液
chlorofluorocarbons 氯氟烃
chlorofluoroethane 氟氯乙烷
chlorofluorohydrocarbon plastics 氯氟烃塑料
chloro-fluoro-hydrocarbons 氯氟烃
chlorofluoromethane 氯氟甲烷
chloroform 氯仿;三氯甲烷
chloroform extract 氯仿萃取
chloroform extract method 氯仿萃取法
chloroform poisoning 氯仿中毒
chloroform water 氯仿水
chloroformyl chloride 光气
chloroheptanoic acid 氯代庚酸
chlorohydrin(e)乙氯醇;氯代醇
chlorohydrocarbon 氯代烃

chlorohydroquinone 氯对苯二酚;氯代氢醌
chloroiodomethane 氯碘甲烷
chloromagnesite 氯镁石
chloromethane 甲基氯
chloromethyl dimethyl chlorosilane 氯甲基二甲基氯硅烷
chloromethyl silane 氯甲基硅烷
chlorometry 氯定量法
chloronitrous acid 王水
chloronorgutta 聚氯丁(二)烯;氯丁橡胶
chloro-organic compound 有机氯化物;含氯有机化合物
chloro-organics 氯代有机物
chloropal 氯脱石
chloroparaffine 氯化石蜡
chloropentafliorocthane 五氟氯乙烷王水
chlorophaeite 褐绿泥石
chlorophenate 氯苯酚盐
chlorophenesic acid 二氯苯酚
chlorophenol 氯(苯)酚
chlorophenol decomposition 氯苯酚分解作用
chlorophenol red 氯酚红
chlorophenol taste 氯酚味
chlorophenoxy propionic acid 氯苯氧基丙酸
chlorophenylmethylketone 氯苯乙酮
chlorophoenicite 绿砷锌锰石;绿砷锌锰矿
Chlorophyceae 绿藻纲
chlorophyll 绿素石;叶绿素
chlorophyllinite 叶绿素体
chlorophyllite 叶绿石
Chlorophyta 绿藻门
chloropicrin 氯化苦
chloroplast 叶绿体
chloroplastic pigment 叶绿色颜料
chloroplatinate 氯铂酸盐
chloroplatinic acid 氯铂酸
chloropolyether 氯化聚醚
Chloropon 茅滴混剂
chloroprene 氯丁二烯
chloroprene gum rubber 氯丁二烯橡胶
chloroprene latex 氯丁二烯胶乳
chloroprene poisoning 氯丁二烯中毒
chloroprene polymer 氯丁二烯聚合物
chloroprene resin 氯丁二烯树脂
chloroprene rubber 氯丁(二烯)橡胶
chloroprene rubber adhesive agent 氯丁胶粘剂
chloropropane 氯丙烷
chloropropene 氯丙烯
chloropropylene oxide 表氯醇
chloropropyl functional silane 氯丙官能基硅烷
chloroproyl trimethoxysilane 氯丙基三甲氧基硅烷
chlororganic compound 氯有机化合物
chlorosapphire 绿宝石
chlorosilane 氯硅烷
chlorosis 褪绿;失绿病
chlorosity 体积氯度
chlorospinel 绿尖晶石
chlorostyrene 氯苯乙烯
chlorosulfonated polyester rubber 氯磺化聚酯橡胶
chlorosulfonated polyethylene sealant 氯磺化聚乙烯密封膏
chlorosulphonate 氯磺酸盐
chlorosulphonation 氯磺(酰)化
chlorosulphonic acid 氯磺酸
chlorosuphonated polyethene 氯磺化聚乙烯
chlorothiionite 氯钾胆矾
chlorotile 绿砷铜石
chlorotrifluoroethane 氯三氟甲烷
chlorotrifluoroethylene polymer 氯三氟乙烯聚合物
chlorotrifuoroethylene resin 三氟乙烯树脂
chlorowax 氯化石蜡
chloroxiphite 绿铜铅矿
chloroxylenol 氯二甲苯酚
chlorpheniramine maleate 变应素
chlrphenol 一氯酚
chlrphentermine 对氯苯丁胺
chlorphonium 氯化磷;矮形磷
chlorthion 氯硫磷
chlotoform bitumen A content 氯仿沥青A含量
chmeleon paint 变色涂料
CHM type rubber fender 充气帽型橡胶护舷
chock 轮挡;木片楔子;木垛支架;楔形垫块;止动楔块;止动器;止动;垫舱;导缆钳;导缆口;导缆钩;稳定架;塞块;塞舱

chock a block off 满舱
chock beam 堵塞架
chock blasting 岩柱爆破
chock block 止动楔
chock bolt 制动螺栓
chocked layer 填层
chocked throat 壅塞喉道
chocker 挂木素
chocker hook 挂木索钩
chock feed 填塞进料;填满喂料
chock full 塞满
chock ga(u)ge 塞规
chock holes (自行式钻机车轮下面的)浅坑
chocking 塞紧
chocking cavitation 阻塞气蚀
chocking section 堵塞截面
chocking-up 淤填;栓紧;楔住
chocking-up degree 堵塞程度
chock length 节流长度
chock off 舱内塞紧(防止货物松动)
chock pile 填塞桩
chock plug 塞头
chock release 气门塞降压;扼流释放;减压梁;减压垫;卸载梁
chock release valve 气门塞释放阀;扼流安全阀
chock slide 轧辊轴承座导板
chock-to-chock time 轮挡时间
chock vibrator 塞块振动器
chocolate 黑褐色;细云母片岩;深褐色
chocolate-brown 棕褐色(的);深棕色
chocolate lead 栗色含铅颜料;巧克力色含铅颜料
chocolate mousse 水油乳化液
chocolate-tape boudinage 巧克力方盘吞型构造
chocolate varnish 巧克力色饰油
choice 精选;选择;备选者;备选品
choice box 选择框
choice function 选择函数
choice goods 精选品;精选货(品)
choice mode 选择方式
choice of alternative 比较方案选择
choice of(bore)hole diameter 钻孔直径的选择
choice of bridge site 桥位选择
choice of cross-section 断面选择
choice of dam type 坝型选择
choice of forum clause 选择法庭条款
choice of importing equipment 进口设备选择
choice of law clause 选择(性)法律条款
choice of level of risks 风险水平的选择
choice of occupation 职业选择
choice of profile 断面选择
choice of ruling grade 限坡选定
choice of site 场地选择
choice of structure 结构(的)选择
choice phase 选择阶段
choice point 选择点
choice reaction 选择反应
choice rule 选择规则
choice sources of supply 选择供货单位
choice-start 开选
choicest quality 精选品质
choice trip 最佳的出行;选择的出行
choice variable 选择变量
choice wood 贵重木材
choir 教堂中唱诗班的席位;唱诗班
choir aisle 唱诗座旁的走道
choir arcade 唱诗席席位供廊
choir arch 唱诗席席位拱门
choir architecture 唱诗班席位建筑
choir bay 唱诗班席位
choir chapel 唱诗班礼拜堂
choir limb 唱诗班突伸席位
choir loft 楼台唱诗席;围唱诗座的栏杆
choir rail 围唱诗座的栏杆
choir screen 唱诗班屏栏;围隔唱诗班的屏障或围栏(教堂中)
choir side aisle 唱诗班席位走道
choir stall 唱诗班席位
choir tower 唱诗班塔楼
choir wall (拱廊下分隔唱诗座与走道的)柱间墙;拱廊下分隔唱诗座与走道的柱间墙
chokage 堵塞
choke 节汽门;节流(口);淤塞;阻气(塞)门;阻风;扼流(器);扼波器;气阻;瓶颈过窄;填塞的;缩口部;缩颈

choke aggregate 填塞集料；填塞骨料；拱顶石
choke and kill line 压井放喷管线
choke area 节流口面积
choke block 节流板
choke button 阻气阀操作按钮
choke cable 阻力拉索
choke chamber 阻气室
choke circuit 抗流电路；扼流电路
choke coil 抗流圈；节流圈；节流盘管；阻流圈；扼流（线）圈
choke control 阻气门控制；扼流控制
choke control knob 阻塞控制的旋钮
choke-coupled-amplifier 抗流圈耦合放大器
choke coupling 扼流耦合；波导管阻波凸缘
choked 阻塞的
choke damp 二氧化碳气；碳酸气；窒息性气体；窒息（毒）气
choked bore 瓶颈过窄
choked flange 抗流凸缘；节流接头；节流法兰；扼流接头
choked flow 阻流
choked-flow turbine 阻流式涡轮机；阻流式透平；超临界压降涡轮机
choked galley 阻塞的水沟
choked jet 阻塞喷注；超音速喷流
choked lake 堰塞湖
choked layer of sand 砂填层
choked neck 节流颈；瓶颈过窄
choked reflex 不动反射
choke-fed 滞塞给料的
choke feeding（颚式碎石机的）进料阻塞；滞塞给料；过饱进料；堆挤装料
choke filter 展平滤波器；扼流圈滤器
choke flange 节流孔板；节流法兰；阻波凸缘；凸缘
choke-flange joint 扼流凸缘接头
choke flow 扼流
choke for oil delivery pipe 送油管堵塞
choke joint 扼流接头；波导接头
choke lever 吸气阀杆
choke line 阻塞线；放喷管（路）
choke manifold 油嘴集管；油嘴管汇
choke material 填塞（材）料
choke modulation 抗流圈调制；扼流圈调制
choke modulator 扼流圈调制器
choke nozzle 阻气喷嘴
choke piston 扼流活塞
choke plug 闷头；塞头
choke plunger 扼流活塞；波导管阻波突缘
choke-point 滞塞点；阻塞点
choke protection 抗流圈保护
choker 捆柴排扎；夹具，阻气门；堵缝材料；填缝料
choker aggregate 嵌缝集料；嵌缝扎料；填隙集料；填隙骨料；填缝扎料
choker check-valve 阻气单向阀
choker course of aggregate 集料填塞层；骨料填塞层
choker hitch 钳式索结
choker hole 穿索肉
choker hook 捆木索钩；圆钩
choke ring 阻塞环
chokerman 捆木工
choker rod 风门拉杆
choker transformer 抗流变压器
choke stone 中心（石）；拱顶石；堵缝石；嵌片（基层）；嵌缝石；填缝料
choke suppress 扼止
choke transformer 扼流圈变压器；扼流变压器
choke transformer coupling 抗流圈变压器耦合
choke tube 阻流管
choke valve 节流阀；蝴蝶阀；阻气阀；阻流阀；阻风阀门
choke valve body 阻塞阀体
choke valve bracket 阻气阀托架
choke valve shaft 阻气阀轴
choke valve spring 阻气阀弹簧
choking 堵塞
choking action 阻塞作用
choking cavitation 阻塞性汽蚀；阻塞性空蚀
choking coil 阻流圈
choking device 阻塞装置
choking effect 节流作用；壅塞效应（风洞）；扼流作用；扼流效应
choking factor 扼流系数
choking field 反作用场

choking flow 闭塞流
choking gas 窒息性毒气
choking limit 阻塞极限
choking of filter 滤池阻塞；滤池堵塞
choking of the ignition 发火管阻塞
choking phenomenon（风洞的）阻塞现象
choking resistance 阻塞阻力；扼流电阻
choking section 临界截面
choking setting 卡装
choking winding 抗流线圈；扼流线圈
chokon 高频隔直流电容（器）
choky 拘留所；税收站
cholamine 氨基乙醇
cholera 霍乱
cholera epidemic 霍乱流行
Cholesky decomposition 乔里斯基分解（法）
cholestyramine 除胆树脂
choline 胆碱
choline iodide 碘化胆碱
choline theophyllinate 胆茶碱；茶碱胆碱
choloalite 丘碲铅铜石
choltry 篷车旅馆；（印度集会用的）大型农村厅堂或场地
chomecephalic 扁头的
chomophyte 岩隙（石生）植物；它旁杂草
chomophytes 石生植物
chomper 劈石机
chondrite 球粒陨石
chondrite-normalized abundance 球粒陨石标准化丰度
chondritic meteorite 球粒陨石
chondrodite 粒硅镁石
chondrule 粒状体；陨石球粒
chondrus 角叉藻属
chondus 鹿角莱；鹿角菜精宁；角叉藻
Chongsheng trough basin 冲绳海槽盆地
Chongyang 冲阳
chonolith 岩铸体
chooner 二桅帆船
choose 选购；挑选
choose sources of supply 选择供货单位
choose the best alternatives 选择最优方案
choosing star 选星
choosing the location 选址
chop 裂口；斫开；短峰波；风浪突变；切断；劈；图章
chop ahead 捣碎
chopass 高频隔直流电容（器）
chop feeder 箱式计量给料器
chop hammer 劈锤；砍锤
C-H-O phase diagram 碳氢氧相图
chop-hoop winding 缠绕—喷射结合工艺
chop-house 小饭店
chophytol 朝鲜蓟蓟
chop mat 短纤维毡片
chop off 切开
choppability 短切性
chopped beam 间歇束；间断束；切割束
chopped beam method 断续波束法
chopped fiber pellet 短切纤维粒料
chopped fibre 短纤维
chopped ice machine 碎冰制造机
chopped light 斩切光
chopped light power meter 断续光功率计
chopped photocurrent 截光电流
chopped pulse 削顶脉冲
chopped radiation 调制辐射
chopped reinforcing fiber 短切增强纤维
chopped strand 短切原丝
chopped strand machine 短切原丝机
chopped strand mat 短切原丝（薄）毡；短切纤维毡
chopped strand mat machine 短切原丝薄毡机组
chopped wave 截尾波；斩波
chopper 锯掇工；交流变换器；削波器；限制器；振动换流器；振动变换器；斩光器；断波器；断续装置；断续器；断路器；短切机；斧头；切碎器；切碎机；切光器；切割机；替续器
chopper amplifier 振簧式放大器；斩波放大器；断续放大作用；断续放大器
chopper bar 给进弓形杆
chopper-bar recorder 点（划）线记录器
chopper barrel 短切机筒体
chopper-blower 切碎吹送机
chopper body 短切机筒体
chopper control 削波控制；限幅控制；斩波控制

chopper control equipment 斩波控制设备
chopper frequency 间歇频率；间断频率；遮光频率
chopper knife 切碎装置刀片；切碎器刀片
chopper mill 切碎碾磨机
chopper motor 斩波器供电电动机；断路电动机
chopper spectrometer 斩波分光计
chopper speed control 脉冲调速
chopper stabilization system 换流器移动补偿制
chopper supply 斩波器电源
chopper switch 闸刀开关；断续开关；刀形开关
chopper thresher 切碎脱粒机
chopper transistor 断续器晶体管
chopper wheel 调制盘
chopping 砍；截短；消除危石；斩波；剁碎；断续；风向无定的；风向常变的
chopping action 调制作用
chopping bit 凿尖；剁碎钻头；顿钻钻头；冲击（式）钻头；扁凿钻头；扁铲钻头
chopping blade 切断刀片；砍刀
chopping block 切碎木块；砍断石块；砧板
chopping bucket 冲击抓斗
chopping characteristic 短切性能
chopping circuit 削波电路
chopping drill 冲击钻机
chopping drum 切碎滚筒
chopping flow 有浪水流
chopping frequency 削波频率
chopping jump 波状水跃
chopping knife 切断刀片；砍刀
chopping light 截光
chopping load 短切阻力
chopping machine 短切机；切片机
chopping oscillator 断续振荡器；断续动作发生器
chopping phase 调制相
chopping primary mirror 斩波调制主镜
chopping room 短切室
chopping roving 短切无捻粗纱
chopping sea 波涛汹涌的海面；三角浪；三角波
chopping shutter 断口快门
chopping speed 切断速度
chopping wheel 切割转盘
chopping wind 风向不定的阵风；方向不定风
choppy 裂缝多的；有皱纹的；多裂缝的；波涛汹涌的
choppy cross-bedding structure 锯齿状交错层理构造
choppy grade 锯齿形（纵）断面；波浪形纵断面
choppy sea 海面细浪；细浪；波涛汹涌（的海面）；三角浪；三角波
choppy street gutter 锯齿形街沟
chops 港湾入口；钳口
chop-suey 杂碎
chop-thresher plant 切碎脱粒装置
chop-type feeder 断续式给料器
chop up 钻碎
choragic monument 文艺纪念碑
choragium（古希腊、罗马剧院舞台后的）未被占用的建筑空间
C-H-O ratio 碳氢氧比例
choraula 教堂唱诗队练唱室
chord 弧弦
chordal addendum 弦齿高；测量齿高
chordal height 弦齿高
chordal pitch 弦节距
chordal running 弦接运行
chordal thickness 弦齿厚（度）；固定弦齿厚
chord angle 翼缘角钢；翼弦安放角
chord-angle method 弦角法
chorda oblique 斜索
chord area 弦杆面积
chord axis 联系轴；桁弦轴；弦轴；翼弦轴
chord bar 弦杆
chord boring machine 弦杆式钻孔机
chord bracing 桁弦交叉支撑
chord contact method 弦切法
chord deflection 弦偏距
chord deflection angle 弦（线）偏（向）角
chord deflection offset 弦线支距；弦线偏距
chorded winding 弦绕组
chord element 弦杆构件
chord force 弦向分力；平行于基准线的力
chord head 弦杆端孔眼（弦杆端头加大便于插销钉）
chord length 弦长
chord line 弦线；翼弦线

chord measurement 弦测法
chord member 弦构件;弦杆
chord member of truss 桁架弦杆
chord method 在设计组合梁时只考虑翼缘抗弯矩的方法
chord modulus 弦向模量;弦(杆)模量;弹性模数;弹性模量
chord of airfoil 翼弦
chord of an arc 弧弦
chord of a truss 桁架的弦杆
chord of bowstring arch 拱弦
chord of contact 切点弦
chord of curvature 曲率弦
chord offset method 弦线支距法
chord of varying sections 变截面弦杆
chord plane 弦杆平面
chord plate 弦板;翼缘板
chord rod 弦杆
chords of a truss 桁架的弦杆
chord spacing ratio 叶栅稠密度
chord splice 桁弦接合板;弦杆拼接板;弦杆接合板
chord stationary method 弦线桩定法
chord stiffening 桁弦加劲
chord stress 桁弦应力
chord-tangent angle 弦切角
chord width 翼缘宽度
chord width of barrel vault 筒拱弦宽
chord winding 分距绕组
chord wire 弦线
chordwise 弦向;沿翼弦方向
chordwise component 弦向分量
chordwise distance 弦向距离
chord(wise) force 弦向分力;弹体坐标系切向力
chorisogram 等值线图
chorisogram method 分区统计法
chorisopleth 等值区域线
chorisopleth map 等值区域线图
C-horizon 母质层;底土层
chorochromatic map 分区着色图
chorofluorocarbon 含氯氟烃;含氯氟甲烷
chorofluorocarbon polymer 含氯氟烃聚合物
chorogram 等值图;分区统计图
chorogram method 等值区域法;分区统计图法
chorograph 位置测定器;地图绘制法;地区地图
chorographic(al) map 一览图
chorography 地志学;地区地图;地方图编制学;区域地图
chorohydrocarbon 含氯烃
choroisogram 分区等值线图
choro-isopleth diagram 等值线图
choro-isotherm 等温线;空间等温线;地区等温线;分区等温线图
chorology 分布学;生物分布学;生物地理学
chorometry 土地测量(学)
choromorphographic(al) map 地貌形态图
choropleth 等值区域线
choropleth map 等值区域线图;分区地图
choropleth technique 分区着色图法
chose 精选的
chosed-cell closure strip 波形屋面板;闭合网络封条;肋形开口镶板
chosen latitude 假设纬度;假定纬度;选择纬度
chosen longitude 假设经度;假定经度;选择经度
chosen position 假设船位;假定位置;假定船位;选择位置;选择船位
choses in action 可经诉讼取得的财产权;诉讼财产;权利财产
choses in possession 可实际占有的物;实际占有的动产;占有动产
chosure of pores 毛孔闭塞
choudrite 管枝迹
choultry 大型农村厅堂或场地(印度集会用);篷车旅馆
chow chow 杂碎
chow water 短浪激流
chram st. vita 圣维特大教堂(10世纪初捷克布拉格)
chresard 可用水量;有效水分
Chretien system 克里蒂安反射镜系统
Chretien telescope 克里蒂安望远镜
chrismatory 圣油瓶龛
chrismon (希腊字母组成的)救世主符号
Christ-cross 十字形记号
Christensen model 克里斯蒂森模型
Christensen unit 克里斯蒂森单位
Christian architecture 基督教建筑
Christian basilica 基督教巴西利卡;长方形基督教堂
Christian bema 基督教讲坛
Christian church architecture 基督教教堂建筑
Christian era 公元
christianite 钙长石
Christianity 基督教
Christiansen effect 克里斯琴森效应
Christiansen filter 克里斯琴森滤光器
Christiansen interferometer 克里斯琴森干涉仪
Christiansen wavelength 克里斯琴森波长
christite 斜硫碲汞铊矿
Christmas 圣诞节
Christmas tree 圣诞树;井口采油装置;采油树;三缆定位装置【疏】
Christmas tree antenna 枞树形天线
christmas-tree formation 形成支晶
christmas tree ga(u)ge 采油井口压力表
Christmas tree pattern 反光图案
christobalite 方英石
Christogram (希腊字母组成的)救世主符号
christophite 铁闪锌矿
chrochtron 摆线管
chroismite 混合岩
chroisotherm 时间等温线
chroma 彩度;色品;色彩浓度;色饱和度
chroma amplifier 彩色信号放大器
chroma amplifier stage 彩色信号放大级
chroma bandpass amplifier 彩色同步脉冲放大器
chroma channel 彩色信号通道
chroma circuit 彩色信号电路
chromacoder 色度编码器;信号变换装置
chromacoder system 彩色编码制
chroma control 彩度调整;色品控制;色品调整;色度控制;彩色饱和调整;色度调整
chroma demodulator 彩色解调器;色度信号解调器
chromador 铬铜建筑钢
chromaffin 嗜铬染色
chromaffin body 副节
chroma inverter 色度倒相器
chromaite 铬钙石
chroma key 色品键控;色度键
chroma-key generator 色度键控信号发生器
chromaking 铬化
chromalize 镀铬
chroma luminance 色品亮度;色度亮度;色彩亮度;颜色明度;色度明度
chromammine 氨络铬
chroma modulation 彩色信号调制
chroma module 解码板(色度通道板)
Chroman alloys 克罗曼铁镍铬基合金
chroman(e) 色满
Chromang 克罗曼格不锈钢
chromanin 电阻合金
chromansil 铬锰合金;结构用合金钢;铬锰硅钢
chromanyl 苯并二氢吡喃基
chroma oscillator 彩色(信号)振荡器
chromarod 色谱棒
chroma saturation 色饱和度
chroma scale 彩度标
chroma separation 亮边分离
chroma separator 色度分离器
chromasia 颜色
chroma signal 彩色信号;色饱和度信号
chromatape 色带
chromate 用铬酸盐处理;铬酸盐
chromate coating 铬酸盐涂料;铬化处理涂层
chromate compound 铬酸盐化合物
chromate contaminated wastewater 含铬废水
chromated copper arsenate 加铬砷酸铜;铬化砷酸铜
chromate duplicate copy 铬胶翻版【测】
chromated zinc arsenite 加铬砷酸锌
chromated zinc chloride 加铬氯化锌;铬锌合剂
chromate gelatin 铬胶
chromate phosphate process 磷酸铬酸(盐表面)处理(法)
chromate pigment 铬酸盐颜料
chromate red 铬红
chromate treatment 铬酸盐处理;镀铬
chromate wastewater 含铬废水;铬酸盐废水
chromatic 有色的;彩色
chromatic aberration 色(像)差;色散
chromatic aberration of foci 焦点色差
chromatic aberration of lens 物镜色差
chromatic aberration of position 位置色像差
chromatic adaptation 彩色适应;色适应
chromatic coherent 色相干
chromatic colo(u)r 彩色
chromatic component 色彩组成部分
chromatic constancy 色度恒定性
chromatic contrast 色度对比
chromatic correction 色差校正
chromatic curve 色差曲线
chromatic defect 色差
chromatic diagram 色度图
chromatic difference of focal length 焦距的色差
chromatic difference of magnification 放大率色差
chromatic discrimination power 辨色能力
chromatic dispersion 色散(现象);色散射
chromatic distortion 色失真;色畸变
chromatic effect 色效应
chromatic equidensity 色的均等密度
chromatic fidelity 色彩逼真度
chromatic halo 色晕
chromatic image 色差像
chromatic index 色指数
chromatic information 色度信息
chromaticity 染色性;色品;色度;色彩质量
chromaticity aberration 色品像差
chromaticity bandwidth 色度信号带宽
chromaticity coefficient 色度系数;色品系数
chromaticity component 色品分量
chromaticity coordinates 色品坐标;色度坐标
chromaticity demodulator 彩色反调制器
chromaticity diagram 配色三角形;色品图;色度图;色度表;三色图
chromaticity difference 色品差
chromaticity difference diagram 色品差图;色度差图
chromaticity flicker 彩色闪烁;色品闪烁
chromaticity modulator 彩色调制器;色品信号调制器
chromaticity printing 彩色印刷;套色印刷
chromaticity sensor 色度传感器
chromaticity signal 色度信号
chromaticity signal bandwidth 色度信号带宽
chromaticity subcarrier sideband 色度副载波边带
chromaticity system 色品系统
chromaticity test 色度试验
chromaticity transducer 色度传感器
chromaticity value 色值
chromatic lens 色透镜
chromaticness 色质;色品度;色度感
chromatic number 色数
chromatic overcorrection 色的过度校正
chromatic paper 彩色磨光纸
chromatic parallax 色视差
chromatic pigment 彩色颜料
chromatic polarization 色偏振
chromatic polynomial 色多项式
chromatic printing 多色印刷
chromatic printing machine 彩色印刷机
chromatic refraction 色折射
chromatic rendition 彩色再现
chromatic resolution 彩色清晰度;彩色分辨力
chromatic resolving 色彩分辨率
chromatic resolving power 颜色分辨力;色分辨能力;色分辨率;色分辨本领
chromatic resolving power prism 色分辨率棱镜
chromatics 颜色学;色学;色彩学
chromatic scale 半音音阶
chromatic sensation 色觉
chromatic sensitivity 感色性;感色度;色差敏感度
chromatic spectrum 彩色光谱
chromatic subcarrier 彩色副载波频率
chromatic threshold 色阈
chromatic transference scale 彩色沾色样卡
chromatic value 色品值;色度值
chromatic vision 色觉
chromatid 染色半体
chromating 铬酸处理;铬酸盐处理;钝化处理
chromatism 异常着色;彩色学;色(像)差;色幻觉
chromatism of foci 焦点色差
chromatism of magnification 放大率色差
chromatizing 铬酸盐钝化处理

chromatobar 色谱棒;色带
chromato-diffusion 色谱扩散
chromatodisk 色谱盘
chromatofuge 离心色谱(法)
chromatogram 层析谱;色谱(图);色层分析图;色层(分离)谱
chromatogram analyzer 色谱分析器
chromatogram calculation 色谱图计算
chromatogram-gas logging 色谱气测井
chromatogram scan(ning) 色谱图扫描
chromatograph 用套色法印刷;彩印件;套色版;色谱仪;色谱分离;色层谱仪;色层分析仪;色层分析谱;色层分离谱
chromatographable organic halogen compound 可色层分离有机卤素化合物
chromatographable trace level organics 可色层分离微量级有机物
chromatograph chamber 色谱分离室
chromatographer 色谱工作人员
chromatographia 色谱学
chromatographic 色谱的
chromatographic absorption 色层吸收
chromatographic adsorption 色谱吸附
chromatographic analysis 分离分析;色谱分析;色层分析
chromatographic analysis reagent 层析试剂
chromatographic apparatus 色谱仪器
chromatographic column 色谱(吸附)柱;色层柱
chromatographic detection 色谱测定法
chromatographic detector 色谱仪检测器
chromatographic effect 色谱效应
chromatographic fractionation 色谱分离;色谱(法)分级;色层分离
chromatographic function 色谱函数;色谱功能
chromatographic grade 色谱级
chromatographic method 色谱法;层析法
chromatographic paper 色谱纸
chromatographic process 色谱过程;色谱方法
chromatographic science 色谱学
chromatographic sheet 薄层色谱板
chromatographic solution 色谱溶液
chromatographic solvent 色谱溶剂
chromatographic species 色谱分析类型
chromatographic technique 色谱技术;色层技术
chromatographic tube 色谱管
chromatograph operator 色谱操作人员
chromatography 茨维特法;层析(法);彩色学;套色法;色谱术;色谱法;色层分析法;色层(分)离(法)
chromatographymass spectrography 色谱分析质谱测定法
chromatography mass spectrometry 色谱—质(谱)联用
chromatology 色彩学
chromatolysis 铬盐分解
chromatomap 色谱图
chromatometer 色觉计;色度计;色觉仪;色度图表
chromatometry 色度学;色度法
chromatopack 色谱纸束
chromatopencil 色带;色谱棒
chromatophore 载色体;色素细胞
chromatophorous 载色体的
chromatophotometer 彩色光度计
chromatopile 色谱堆;色层分离堆
chromatoplate 薄层色谱板;色谱板
chromatopolarograph 色谱极谱
chromatoptometry 色觉检查
chromatoroll 色谱圆筒
chromatoscope 彩光折射率计;反射望远镜;配色镜
chromatosheet 薄层色谱板;色谱板
chromatostack 色谱堆
chromato-stick 色谱丝
chromatostrip 色谱条
chromatothermography 热色谱(法)
chromatotube 色谱管
chroma tracking circuit 色度跟踪电路
chromatron 彩色电视显像管;色标管
chromatrope 成双的彩色旋转幻灯片
chromatype 铬盐相片;彩色相片
chroma-under 彩色下置
Chromax bronze 克罗马克青铜
chromazol yellow 铬唑黄
chromazurine 铬天青
chromdravite 路镁电气石
chrome 增加色彩;铬黄颜料;铬;镀铬物;涂铬

chrome acid 铬酸
chrome acid treatment 铬酸处理
chrome acid waste treatment 铬酸废物处理
chrome alloy 铬合金
chrome alum 铬明(矾)
chrome-alumina brick 铬铝砖
chrome alumina pink 铬铝桃红
chrome alumina red 铬铝红
chrome-alumina refractory 铬铝(质)耐火材料
chrome-alumin(i)um wire 铬铝丝
chrome avanturine 铬(质)金星玻璃
chrome azo complex dye 铬偶氮配合染料
chrome azurol 铬天青
chrome-base refractory 铬基耐火材料
chrome bed 镀铬床
chrome black 铬黑
chrome board 彩色石印纸板
chrome brass 铬黄铜
chrome brick 含铬耐火砖;铬(矿)砖
chrome carbide 碳化铬
chrome-carbon steel 铬碳钢
chrome citron 铬柠檬色
chrome-cobalt alloy solder 钴铬合金焊
chrome collagen fiber 含铬的胶质纤维
chrome colo(u)r 铬质颜料;铬色
chrome complex 铬络合物
chrome complex dye 含铬染料
chrome-containing solid waste 含铬废屑
chrome-containing waste 含铬盐废水
chrome copper 铬铜合金
chrome-corundum brick 铬刚玉砖
chrome-corundum refractory 铬刚玉耐火材料
chrome-dolomite brick 铬白云石砖
chrome-dolomite refractory 铬白云质耐火材料
chrome-dolomite refractory product 含铬白云石耐火制品
chrome dye 铬媒染料
chrome dy(e)ing waste 铬染废水
chrome-faced 表面镀铬
chrome-faced piston ring 表面镀铬(的)活塞环
chrome fast cyanine 铬坚牢花青
chrome fast green 铬坚牢绿
chrome fast orange 铬坚牢橙
chrome fast yellow 铬坚牢黄
chrome film 彩色反转片
chrome-finished 镀铬的
chrome firebrick 铬耐火砖
chrome-free brick 无铬砖
chrome green 铬绿(色);铅铬绿;三氧化二铬
chrome green glass 铬绿色玻璃
chrome green pigment 铬绿颜料
chrome hardened steel 铬硬化钢
chrome inter diameter micrometer 镀铬内径千分尺
chrome iron 铬铁
chrome iron ore 铬铁矿石
chrome kid 铬鞣羔革
Chromel 克罗梅尔镍铬耐热合金
chromel 镍铬合金;铬镍合金
chromel-alumel couple 铬铝热电偶
chromel-alumel thermocouple 铬镍铝镍热电偶
chromel-alumel thermometer 铬镍铝镍温度计
chromel-constantan thermocouple 镍铬—康铜热电偶;铬镍康铜热电偶
chromel-copel thermocouple 铬镍铜镍热电偶;铬镍康铜热电偶
chrome leather 铬革
chrome leather sleeve 铬革套
chrome lemon 铬柠檬色
chrome lemon yellow 柠檬铬黄颜料
chromel-filament 镍铬丝
chrome lignin(e) 木质素铬盐
chrome lignin(e) sulphonate 木素磺酸铬(泥浆添加剂)
chrome-lignite 木质素铬
chrome lignosulfunate 木素磺酸铬(泥浆添加剂)
chrome-magnesia refractory 铬镁(质)耐火材料
chrome-magnesite brick 铬镁砖
chrome-magnesite firebrick 铬镁(质)耐火砖
chrome-magnesite refractory 铬镁(质)耐火材料
chrome-manganese nitrogen steel 铬锰氮钢
chrome-manganese-silicon alloy steel 铬锰硅合金钢
chrome manganese steel 铬锰钢

chrome mask 铬掩模
chrome molybdenum 铬钼
chrome-molybdenum steel 铬钼钢
chrome mordant 铬媒染剂
chrome mud 铬泥浆
chrome nicked steel 铬镍钢
chrome-nickel 铬镍合金
chrome-nickel alloy 铬镍合金
chrome-nickel wire 铬镍线
chrome ochre 铬赭石;铬绿;铬华
chrome orange 桔铬黄;铬橙(色);铅铬橙
chrome orange pigment 铬橙颜料;铬黄颜料
chrome ore 铬矿石;铬矿砂
chrome oxide 氧化铬
chrome oxide green 氧化铬绿
chrome paper 铜版纸
chrome peacock green 铬孔雀绿
chrome permalloy 铬透磁钢;铬波莫合金
chrome pickle 铬酸清洗
chrome pigment 铬颜料
chrome plate 镀铬压(平板)
chrome plated 镀铬的
chrome plated adjustable wrench 镀铬活络扳手
chrome plated automatic wire stripper 镀铬自动剥线器
chrome plated brass flexible hose 镀铬铜软管
chrome plated castor 镀铬脚轮
chrome plated claw hammer 镀铬拔钉锤
chrome plated dome 镀铬圆头螺钉
chrome plated finish 镀铬饰面
chrome plated grating with filter 带滤网的镀铬滤栅
chrome plated handle 镀铬拉手
chrome plated house-hold scissors 镀铬家用剪刀
chrome plated inner tube 镀铬内管
chrome plated iron tube 镀铬铁管
chrome plated narrow blade scissors 镀铬柳叶剪刀
chrome plated part 镀铬零件
chrome plated piston ring 镀铬活塞环
chrome plated PVC flexible hose 镀铬聚氯乙烯软管
chrome plated steel nozzle 镀铬钢喷管
chrome plated universal bevel protractor 镀铬万能角度尺
chrome plated vernier calliper 镀铬游标卡尺
chrome plate-sides ring 多面镀铬环
chrome plating 镀铬
chrome plating bath 镀铬槽
chrome plating plant 镀铬车间
chrome potash alum 铬明矾
chrome powder 粉状铬鞣剂
chrome primrose 铬淡黄色
chrome printing yellow 铬印染黄
chrome recovery 铬鞣液回收
chrome red 橘铬;铬红;铅铬红
chrome red pigment 铬铅红颜料
chrome refractory 铬质耐火材料
chrome refractory brick 铬耐火砖
chrome rutile yellow 钛铬黄
chrome scarlet 铬猩红
chrome silica brick 铬硅砖
chrome-silica refractory 铬刚玉质耐火材料
chrome-silicon 铬硅
chrome-silicon steel 铬硅钢
chromesillimanite refractory 铬硅线石耐火材料
chrome-spinel brick 铬尖晶石砖
chrome spinel(le) 铬尖晶石
chrome steel 铬钢
chromesthesia 色联觉;色幻觉
chromet 铝硅合金
chrome tanned leather 铬鞣革
chrome tin pink 铬锡红
chrome-tin red 铬锡红
chrome treatment 铬处理
chrome-tungsten steel 铬钨钢
chrome-vanadium steel 铬钒钢
chrome vermil(l)ion 钼铬橙;铬朱红
chrome yellow 莱比锡黄;铬黄;浅铬黄
chrome yellow orange 深铬黄
chrome yellow paint 铬黄漆
chrome yellow pigment 铬黄颜料
chromhalloysite 铬埃洛石
chromic acid 铬酸;铬酐

chromic acid battery 铬酸电池
chromic acid cell 铬酸电池
chromic acid mixture 铬酸混合液
chromic acid oxidation coating 铬酸(阳极)氧化膜
chromic alum 铬(明)矾
chromic anhydride 铬酐
chromic colo(u)r 铬染料
chromic compound 铬化合物
chromic fluoride 氟化铬
chromic hydrate 氢氧化铬
chromic hydroxide 氢氧化铬
chromic image 彩色像片
chromic iron 铬铁
chromic iron ore 铬铁矿
chromic lanthanum ceramics 铬酸镧陶瓷
chromic orange 铬橙(色)
chromic oxide 氧化铬;三氧化二铬
chromic oxide coating sheet 镀铬(氧化物)薄板
chromic oxide pigment 氧化铬颜料
chromic oxide polishing composition 绿色抛光块
chromic potassium oxalate 草酸铬钾
chromic potassium sulfate 硫酸铬钾
chromic potassium sulfur 铬钾矾
chromic salt 铬盐
chromic slag 铬渣
chromic sulfate 硫酸铬
chromic wastewater 含铬废水
chromiferous dye 含铬络合染料
chrominace lock 色度锁相
chrominance 彩色信号;彩矢量;色品;色度
chrominance adjustment 色品调整;色度调节
chrominance amplifier 色度放大器
chrominance axis 色度轴
chrominance cancellation 色度消隐
chrominance carrier 彩色载波;色度信号载波
chrominance-carrier reference 彩色载波标准;色度载波基准(频率)
chrominance channel 色度信道;色(度)通道
chrominance circuit 彩色信号电路
chrominance delay line 色品延时线
chrominance demodulation 色度解调
chrominance demodulator 彩色信号解调器
chrominance frequency 色度频率
chrominance gain control 色度增益调节
chrominancelluminance gain inequality 色度与亮度增益差
chrominance-lock receiver 彩色锁相接收机
chrominance modulator 彩色信号调制器
chrominance-noise measurement 色度杂波测量
chrominance phase 色度相位
chrominance primary 基色刺激;色度基色
chrominance pulse 彩色脉冲
chrominance resolution 彩色清晰度;彩色分辨力;色度清晰度
chrominance response 比色反应
chrominance signal 色度信号
chrominance signal carrier 色度信号载波
chrominance signal vector diagram 色度信号矢量图
chrominance subcarrier 彩色副载波频率
chrominance-subcarrier demodulator 彩色信号解调器
chrominance-subcarrier modulator 彩色信号调制器
chrominance-subcarrier oscillator 彩色信号振荡器
chrominance-subcarrier reference 色度载波基准频率
chrominance-subcarrier regenerator 色度副载波再生器
chrominance transcoding 色度代码转换;色度变码
chrominance unit 色度单元
chrominance vector 色度矢量
chrominance video signal 色度视频信号
chroming 镀铬
chroming machine 镀铬机
chrominium 科铬铅矿
chromising 铬化
chromite 亚铬酸盐;铬铁矿
chromite brick 铬(铁)砖
chromite ilmenite magnetite apatite-rich ultramafic rock group 富铬铁矿钛铁矿磁铁矿
chromite ilmenite magnetite-rich rock 富铬铁矿钛铁矿磁铁矿岩
chromite ore 铬铁矿石

chromite refractory 铬铁矿耐火材料
chromite-rich rock 富铬铁矿岩
chromitite 铬铁岩;铁矿岩
chromium 自然铬;铬
chromium alloy runner 铬合金(钢)转轮
chromium ammino complex method 铬氨络合物法
chromium analyzer 铬分析仪
chromium base alloy 铬基合金
chromium boride 硼化铬
chromium-bronze 铬青铜
chromium carbide 碳化铬
chromium carbide ceramics 碳化铬陶瓷
chromium cermet coating 铬金属陶瓷镀层
chromium chloride 氯化铬
chromium-chlorite 铬绿泥石
chromium coating 镀铬层
chromium compound 六价铬化合物;铬化合物
chromium-containing electroplating wastewater 含铬电镀废水
chromium-containing wastewater 含六价铬废水
chromium-copper 铬铜合金
chromium-copper wire 铬铜线
chromium cycle 铬循环
chromium dioxide 二氧化铬
chromium ethylenediamine tetraacetic acid 乙二铵四乙酸铬
chromium finish 铬表面处理
chromium firebrick 铬耐火砖
chromium fog inhibitor 铬雾抑制剂
chromium glass 铬玻璃
chromium green 氧化铬绿颜料
chromium group element 铬族元素
chromium humic acid 铬基腐植酸
chromium hydrate 氢氧化铬
chromium hydrate green 氢氧化铬绿
chromium hydroxide 氢氧化铬
chromium implements 渗铬法
chromium impregnation 铬化;渗铬
chromium-in-contact mask 铬接触掩模
chromium ion 铬离子
chromium iron 铬铁
chromium irons 铬铁合金
chromium lined barrel 镀铬枪管
chromium-magnesium brick 铬镁砖
chromium-manganese steel 铬锰钢
chromium mineral 铬矿物
chromium-molybdenum alloy 复合冷硬铬钼合金
chromium-molybdenum stainless steel 铬钼不锈钢
chromium-molybdenum steel 铬钼钢
chromium mordants 铬媒染剂
chromiumnickel austenite stainless steel 铬镍奥氏体不锈钢
chromium-nickel electrode 铬镍焊条
chromium-nickel stainless steel 铬镍不锈钢
chromium-nickel steel 高强度耐蚀铬镍钢;铬镍钢
chromium nitride 氮化铬
chromium ore 铬矿石;铬矿(砂)
chromium oxide 铬氧化物
chromium oxide green 氧化铬绿
chromium oxide grinding grease 氧化铬研磨膏
chromium oxide pigment 氧化铬颜料
chromium pentoxide 三氧化二铬
chromium phosphate 磷酸铬
chromium plate 铬板
chromium-plated blades 镀了铬的叶片
chromium-plated door pull 铬镀门拉手
chromium-plated four-purpose vernier calipers 镀铬四用游标卡尺
chromium plated handle 镀铬拉手
chromium-plated inside micrometer calipers 镀铬内径千分尺
chromium plated iron pipe 镀铬铁管
chromium plated iron tube 镀铬铁管
chromium-plated iron window catch 镀铬铁窗插销
chromium plated material 镀铬材料
chromium-plated mild steel animal chain 镀铬软钢动物链条
chromium-plated part 镀铬件
chromium-plated roller 镀铬滚筒
chromium-plated steel 镀铬钢
chromium-plated steeltube folding chair 镀铬钢管折椅
chromium-plated thin-walled steel cylinder liner 钢质薄壁镀铬汽缸衬套
chromium-plated tooth thickness calipers 镀铬镶合金齿厚卡尺
chromium-plated tube 镀铬管
chromium-plated universal bevel protractor 镀铬万能角规
chromium-plating 镀铬
chromium poisoning 铬中毒
chromium polish 铬抛光剂
chromium pollutant 铬污染物
chromium powder 铬粉
chromium salt 铬盐
chromium silicide 硅化铬
chromium-silicon coating 铬硅镀层
chromium slag 铬矿渣
chromium sludge 含铬污泥
chromium soap 铬皂
chromium stainless steel 不锈铬钢;铬不锈钢
chromium star 铬星
chromium stearate 硬脂酸铬
chromium steel 铬钢
chromium tetr(a)oxide 四氧化铬
chromium-titanium-antimony oxide composite anode 铬—钛—锑氧化物复合阳极
chromium titanium iron phosphate-rich glasses 富玻璃的铬钛铁磷酸盐
chromium trichloride 三氯化铬
chromium trifluoride 三氟化铬
chromium triiodide 三碘化铬
chromium trioxide 三氧化铬
chromium-tungstenvanadium steel 铬钨钒钢
chromium-zirconium-boride coating 铬硼化锆镀层
chromized coating 渗铬镀层
chromized steel 镀铬钢;渗铬钢
chromizing 扩散镀铬;铬化;渗铬
chromizing process 铬化处理
chromo 彩色石印(版)
chromo-chalcography 彩色凹版
chromo-citronine 柠檬铬黄
chromocollotype 彩色珂罗版
chromocyclite 彩鱼眼石
chromodiangnosis 色泽诊断
chromodynamics 色动力学
chromogen 铬精;发色体;色原
chromogen bonded polymer 彩色聚合物
Chromogene dye 铬精染料
chromogenesis 颜色形成;色素形成
chromogenetic 产色的
chromogenic 产色的
chromogen isomerism 异色异构(现象)
chromogenous 产色的
chromogram 立体彩图;彩色图
chromograph 胶版复制器;化学元素快速半定量分析仪;彩色印件;彩色平版
chromo-isomer 异色异构
chromo-isomerism 异色异构(现象)
chromolithograph 彩色印件;彩色石画画;彩色石印版;彩色平版;彩色石印图
chromolithographer 彩色石印师
chromolithography 彩色石印术;彩色平版印刷术;多色石板印刷术
chromomere 染色粒
chromometer 比色计;色度计
chromometry 色觉检查法
chromo-optometer 色视力计
chromopaper 彩色平印纸;彩印纸
chromophobe 难着色;嫌色(的);憎色(性)
chromophobia 嫌色性
chromophobic 嫌色的;憎色的
chromophore 发色团;生色团;色基
chromophoric 发色的
chromophoric colo(u)r 发色团颜色
chromophoric dissolved organic matter 有色溶解有机物
chromophoric electrons 发色电子
chromophoric group 发色基团
chromophoric property 发色性质
chromophoric theory 发色团学
chromophotograph 彩色照相
chromophotometer 比色计
chromophyll 叶色素
chromoplast 有色体;有色粒(体)

chromoplastid 有色体;有色粒(体)
chromoscan 彩色扫描
chromoscan densimeter 彩色扫描密度计
chromoscope 显色管;验色器;彩色摄像管;表色管验色管;色觉检查器;色度镜
chromoscopy 染色检查;色觉检查
chromosensitometer 彩色感光仪
chromosome 染色体
Chromosorb 红色硅藻土色谱载体
chromosphere 色球(层)
chromosphere-corona transition region 色球日冕过渡区
chromosphere telescope 色球望远镜
chromospheric bubble 色球泡
chromospheric ejection 色球抛射
chromospheric emission spectrum 色球层的发射光谱
chromospheric eruption 色球爆发
chromospheric evapo(u)ration 色球蒸发
chromospheric facula 色球光斑
chromospheric fine structure 色球精细结构
chromospheric flare 色球耀斑
chromospheric flocculus 色球谱斑
chromospheric knot 色球结
chromospheric material 色球物质
chromospheric mottling 色球日芒
chromospheric network 色球网络
chromospheric spectrum 色球光谱
chromospheric spicule 色球针状物
chromospheric spike 色球针状物
chromospheric temperature 色球温度
chromospheric whirl 色球漩涡
chromotronic microjet printing machine 电控微喷射印花机
chromotrope 铬变素
chromotropic acid 铬变酸;变色酸
chromotropic acid method 变色酸法
chromotropic acid spectrophotometry 变色酸分光光度法
chromotropism 向色性;异色异构颜色变化;异色异构(现象)
chromotropy 向色性;异色异构颜色变化;异色异构(现象)
chromotype 彩色印刷版;彩色印版;彩色石印图;套色石印图;彩色照相
chromotypography 彩色凸版印术;彩色凸版印术
chromotypy 彩色凸版印术
chromous acid 亚铬酸
chromous oxalate 草酸亚铬
chromous oxide 一氧化铬;氧化亚铬
chromous sulfate 硫酸亚铬
chromow 铬钼钨裂
chromowulfenite 铬钼铅矿
chromoxylograph 彩色木刻版印件
chromoxylography 彩色木刻版印刷术
chrompicotite 硬铬尖晶石
chromyl chloride 二氯二氧化铬
chron 年代
chronaxie 时值
chronaxie meter 时值计
chronaximeter 记时计
chronaximetry 时值测量
chronaxy 时值
chronel 铬镍合金
chronic adaptation 慢速适应
chronic arsenic poisoning 慢性砷中毒
chronic atrophic lichenoid dermatitis 白苔癣
chronic bombardment 慢性轰击
chronic bromism 慢性溴中毒
chronic bronchitis 慢性支气管炎
chronic cadium poisoning 慢性镉中毒
chronic contamination 长期污染
chronic damage 慢性伤害
chronic-disease hospital 慢性病(医)院
chronic effect 慢性效应
chronic experiment 慢性实验
chronic exposure 慢性照射;慢性暴露;持续照射;持续辐照;持续暴露
chronic fluorine poisoning 慢性氟中毒
chronic hazard 慢性危害
chronic imbalances 迁延性比例失调
chronic immobilization test 慢性活动抑制试验
chronic intake 慢性摄入
chronic irradiation 慢性辐照

chronicle 大事记
chronic mercury poisoning 慢性汞中毒
chronic mountain sickness 慢性高山病
chronic(oxygen) poisoning 慢性中毒
chronic pollution 慢(性)污染;长期污染
chronic pollution source 慢性污染源
chronic radiation 慢性辐射
chronic radiation disease 慢性放射病
chronic radiation effect 慢性辐射效应
chronic radiation hazard 慢性辐射危害
chronic radiation injury 慢性放射损伤
chronic slow growth of economy 经济长期发展缓慢
chronic threshold concentration 慢性阈浓度
chronic threshold dose 慢性阈剂量
chronic toxic effect zone 慢性毒作用带
chronic toxicity 慢性毒性;长效毒性
chronic toxicity test 慢性毒性试验
chronic toxicity test of aquatic organism 水生生物慢性毒性试验
chronic trade deficit 长期贸易逆差
chronic unemployment 经常性失业
chronistor 长时计
chronite 铬镍特种合金
chronoamperometry 计时安培分析法
chronoanemoisothermal diagram 时间—方向—温度图
chronocomparator 时间比较仪
chronocoulometry 计时库仑分析法
chronogeochemistry 地质年代化学
chronografic recorder 校表仪
chronogram 计时图;记时仪的记录;记时图
chronograph 录时器;精密计时计;记时仪;记时器;记录仪;秒表;时间记录器
chronographic(al) tape 记时纸带
chronograph reader 记录器读数模板
chronograph sheet 记录时仪纸带
chronograph stop watch 记时停表
chronograph watch 精密记时表
chronograph wheel finger 计时器轮拨销
chronography 时间记录法
chrono-interferometer 记时干涉仪;记录干涉仪
chronoisotherm 记时等温线;时间等温线
chronolith 年代地层单位
chronolithologic unit 年代地层单位
chronological 年代学的;按年月日顺序的
chronological age 编年龄
chronological average 序时平均数;时序平均数
chronological books 序时账簿
chronological chart 过程线
chronological order 年月日次序
chronological ordering 按年代顺序排列
chronological record 序时记录
chronological scale 地史年代表;地质年代表;编年表
chronological series 地质年代表;编年序列;时间系列
chronological succession 地层序列表;地层时序表
chronological table 年代时序表;年(代)表;地质年代表
chronological time scale 年代时序表;地层时间表
chronologize 按年代排列
chronology 纪年法;年代学;年表
chronology of faulting 断层形成时代
chronology of folding 褶皱地质年代
chronometer 精确时计;精密记时仪;精密测时计;经线仪;记时器;计时器;计时器;测时器;天文钟;时计
chronometer circuit 时计电路
chronometer clock 记时计;精密计时
chronometer correction 表差;天文钟差;时计校正;时计差
chronometer error 时计误差
chronometer house 天文钟室
chronometer lock 定时锁
chronometer rate 计时表日差;表速;天文钟差率
chronometer room 天文钟室
chronometer time 精密时间
chronometer watch 测天表;天文表;天文比对表
chronometric(al) 记时式
chronometric data 计时数据
chronometric difference 天文钟比差
chronometric distance 时距
chronometric encoder 记时编码器

chronometric invariant 计时不变量
chronometric observatory 测时天文台
chronometric radiosonde 计时无线电探空仪
chronometric tach(e)ometer 计时测速计;计时式转速计;钟表式转速计;比伦转速器;时序式发电机
chronometric tachymeter 计时测速计
chronometric thermometer 精密计时温度计
chronometry 记时法;计时学;计时电流法;测时术;比色法;时间测定法
chronon 双时元
chronopher 自动报时器;电控报时器;电动报时器;报时器
chronophotograph 连续照相片;连续照相
chronophotography 记录摄影
chronopotentiometry 计时电位(分析)法;计时电势(分析)法;计时电力分析法
chrono-release 计时器;计时仪(用于重要度调整)
chronoscope 精密计时计;记时器;记时计;计时镜;千分秒表;千分表;瞬时计;瞬时测时器
chronosphygmograph 记时脉搏描记器
chronostratigraphic(al) correlation chart 年代地层对比表
chronostratigraphical facies 年代地层相
chronostratigraphic(al) unit 年代地层单位
chronostratigraphic(al) zone 年代带;地层带
chronostratigraphy 年代地层学
chronotherm 时控恒温计;计时恒温调节器
chronothermometer 计时温度计;平均温度表
chronotron 毫微秒计时器;延时器;摆线管
chronotropic 变时性的
chronotropic(al) action 变速作用;变时作用
chronotropic(al) effect 变速性作用
chronotropic(al) response 变时性反应
chronotropic deceleration 时限减速
chronozone 年代带;地层带
chronzone 时带
chrysalic oil 金蛹油
chrysanthemum stone 菊化石
chryselephantine 金和象牙制器
chrysin 柯因
chrysoberyl 金绿宝石;金绿玉
chrysochalk alloy 铜锌铅装饰合金
chrysochlorous 带金黄绿色的
chrysocolla 孔雀石绿色;硅孔雀石
chrysol 菊稔
chrysolite 纤蛇纹石;贵橄石
chrysolit green 黄橄榄石绿色
chrysopal 金绿宝石
chrysophoron 琥珀
chrysoprase 绿玉髓;英卡石
chrysoprase green 玉髓绿色
chrysotile 纤蛇纹石
chrysotile asbestos 加拿大石棉;温石棉;蛇纹石棉
chrysotile-asbestos deposit 蛇纹石石棉矿床
chrysotilite 纤蛇纹石;水合硅酸盐石棉;温石棉
chuanposhi 穿破石
Chubb method 交流波峰值测量法
chuck 卡盘;夹头;修坯承座;钻轧头;导缆口;轻叩
chuck arbor 夹具柄
chuck bare 凿岩机导向套筒
chuck barrel 锤钻轴套
chuck block 吸盘用工作垫块;刀头箱;刀轮架
chuck bolt 卡盘螺栓;卡盘顶丝
chuck bucket elevator 链斗式提升机;链斗升料机
chuck bushing 钻头卡头;钎尾套筒
chuck capacity 卡盘最大装夹直径
chuck collet 卡盘钳夹
chuck drill 冲击钻机
chuck drilling tools 冲击钻具
chucked work 夹在卡盘上的工件
chucker 六角车床;卡盘车床
chucker two-axis slant bed machine 双轴斜床身卡盘机床
chuck face-plate 带爪花盘;带爪卡盘
chuck flange 卡盘凸缘
chuck handle 卡盘扳手
chuck holder 卡盘架
chuck holes 坑洞(路面上);坑洼
chucking 夹具
chucking automatic 自动夹紧
chucking capacity 夹具容量;夹紧能力;(卡盘的)最大装夹直径
chucking capstan and turret lathe 夹盘转塔车床

chucking eccentric 夹持偏心
chucking effect 夹紧力
chucking fixture 夹紧装置
chucking grinder 卡盘磨床
chucking grinding machine external 夹盘外圆磨床
chucking lug 定位耳
chucking machine 卡盘式转塔车床;卡盘式机床
chucking method 夹持法
chucking power 夹持力
chucking reamer 机用铰刀
chucking work 卡盘工作
chuck jaw 卡盘爪;夹头爪;卡盘卡瓦
chuck key 夹头钥匙
chuck lathe 卡盘车床
chuck lever 卡盘把手
chuck master 卡盘扳子
chuck nut 夹头螺帽;钎尾套筒
chuck piece 卡盘卡瓦
chuck ring 夹圈
chuck set screw 卡盘固定螺钉
chuck type 卡盘类型
chuck up 夹紧机上钻杆
chuck valve 逆止阀;单向阀
chuck-wagon 流动炊事车
chuck with holdfast 爪卡盘;抓卡盘
chuck with three-jaws 三爪卡盘
chuck work 卡盘作业
chuck wrench 卡盘扳手
chuco 芒硝层(钙积层的上部)
chudobaite 砷镁锌石
chuff 欠火砖;裂缝砖;炸纹废砖;废品砖;爆裂砖;纹裂砖;黑头砖
chuff brick 过火砖;低质红砖;欠火砖
chuffing 低频不稳定燃烧
chuffy brick 翘砖;欠火砖;疙瘩砖;黑头砖
Chugaev reaction 楚加耶夫反应
chugging 功率振荡;低频不稳定燃烧
chuglum 印度硬木
chukhrovite 水氟钙铝矾
chukhrovite Ce 水氟钙铈矾
Chulitna terrane 朱利特纳地体
chum 修坯承座
chump 石块;木片;块;木块;粗短木头
chunk (木头等的)大块;零块玻璃;碎玻璃块
chunk breaker 大块破碎机
chunk glass 玻璃毛坯;碎玻璃(片)
chunk method 土块试验法(测土壤密度)
chunk out 清扫便道(俚语)
chunk samping 大量抽样
chunk sample 块状试样
chunnel 海峡隧道
chur 河心沙洲
churada 楚勒打暴雨
church 教堂
church architecture 教堂建筑
Church at Qalb Louzeh 盖尔卜·洛格占教堂
church bulletin board 教堂布告板
church center 教堂中心
church fittings 教堂家具
church form 教堂形式
church hall 教区集会厅
church hall block 教区会堂建筑
church house 教区非宗教性活动房
churchill 工厂预制的钢板房屋
church-in-the-round 圆形教堂
churchite 水磷钇矿
church method 教堂法
church monument 教堂纪念物
church nave 教堂中殿;本堂
church oak varnish 教堂栎木清漆
church of Corpus Christi 基督圣体教堂
church of the Apostles at Cologne 科隆的使徒教堂
church of the Apostles at Constantinople 君士坦丁堡使徒教堂
church of the Holy Cross 圣十字教堂
church of the Holy Sepulchre at Jerusalem 耶路撒冷的圣墓教堂
church of the Holy Spirit 圣灵教堂
church of the(Holy) Trinity 圣三一教堂
church of the Magdalen 马格德林教堂
church of the Miraculous Virgin 圣母玛利亚教堂
church of the Nativity at Bethlehem 伯利恒的耶稣降生教堂

church of the Resurrection 耶稣复活教堂
church's baryta process 石头防腐处理(一种处理石头方法)
church steeple 教堂尖塔
church stile 教堂讲堂
church-text 黑体字
churchwarden's pew 教会执事座位
church window condenser 尖顶管束式凝汽器
church with central space 有中央空间的教堂
churchyard 教堂墓地,教堂院
churchyard cross 教堂庭院十字架
churlish 粗糙的
churn 流线式生产;剧烈搅动;搅拌器;摇转搅拌筒;摇转拌和筒;转动搅拌筒;钢丝绳冲击(式)钻机;冲击式钻机;手钻炮眼
churn bit 冲钻
churn drill bit 钢绳冲击钻头
churn drill blasting 冲击钻进爆破眼;冲击钻进爆破孔
churn drill equipment 冲击钻进设备
churn drill(er) 钢绳冲击(式)钻机;冲击钻进司钻;旋转冲钻;顿钻;吊绳冲击机;舂钻机;冲击钻
churn drill hole 冲击钻孔
churn drilling 钢丝绳冲击钻进;顿钻钻井;舂钻;冲击钻(探);冲击钻进
churn drill outfit 冲击钻机
churn drill rig 冲击钻机
churner 手摇式长钻
churn flow 翻腾流
churn flow grit chamber 涡流式沉砂池
churn hole 锅穴
churning 搅拌;反复买卖
churning loss 搅动损失;漩涡损失
churning motion 冲击运动
churning pile 旋喷桩(即喷射桩);喷射桩(即旋喷桩)
churning screw 搅拌螺旋
churning stone 混凝土石碴
churning time 搅拌时间;搅拌时间
churn lorry 奶桶卡车
churn mo(u)lding 齿形饰;波浪式装饰线条;诺尔曼式曲折线脚;之字线脚
churn of wave 猛冲海岸(指波浪)
churn shot drill 顿砂钻
churn shot drill rig 钢绳冲击式钻机
churn type percussion drill 钢绳冲击(式)钻机
churrah(gum) 切拉树脂
Churrigueresque style 过分华丽的建筑风格
Churrigueresque architecture 大量装饰的巴洛克式建筑
chussums 废丝
chutable 可用溜槽输送的
chutable concrete 可槽运混凝土;流态混凝土
chute 流料槽;流槽;溜管;溜道;决流槽;急流槽;滑槽;斜槽;舷侧出灰口;走线架;直管;直沟;直道;陡槽;冲流槽
chute and funnel 滑槽斗,滑槽斗
chute and funnel device 槽斗联合装置
chute and pool structure 槽泊构造
chute apron 溜槽底
chute bar 流槽沙坝
chute-barrel spooler 槽筒式络纱机
chute blades 滑片
chute block 陡槽式消能墩;陡槽(式)消力墩
chute board 滑(道)板;斜槽板
chute channel 泄槽;陡槽段
chute closure dike 支汊锁坝;串沟锁坝;串汊锁坝
chute concrete 溜槽浇注的混凝土
chute conveyer 往复摇动式运输机
chute cut-off 串沟裁弯;撇河切滩;流槽裁弯;切滩裁弯
chuted concrete 溜槽浇混凝土
chute delivery 滑槽输送
chute deposit 串沟沉积
chute direction control 滑槽的方向操纵杆
chute discharge 斜槽出料
chute door 溜槽闸门;垃圾门
chute drop 急流跌水
chute facies 串沟相
chute-fed incinerator 滑槽进料焚化炉;斜槽进料焚化炉
chute feed 斜槽进料装置
chute feeder 溜槽给料器;斜槽进料器;槽式给矿机;喂料槽

chute flap 溜泥槽分泥门
chute gantry 溜泥槽吊架
chute grate 链炉箅;斜槽炉排
chute jaw 漏口侧板
chute pitman 滑槽枢轴
chute raft 斜槽式筏道;浮运水槽
chute rail 滑运轨道
chute riffler 槽式取粉器
chute sill 溜槽底
chute spillway 斜槽式溢洪道;斜槽式溢流道;陡槽式溢洪道;槽式溢洪道
chute swivel joint 管道铰接头
chute-type milking bail 槽式挤奶台
chute unloading 溜槽卸料
chute winch 溜泥槽绞车
chuting 溜槽运送;滑槽运输;斜槽运输;斜槽输送
chuting concrete 用滑槽运送混凝土;溜槽浇注混凝土;斜槽溜混凝土;用斜(滑)槽运输混凝土
chuting device 溜槽装置
chuting equipment 溜槽装置
chuting installation 滑槽设备
chuting of concrete 混凝土浇筑
chuting plane 溜槽装置
chuting plant 溜槽(运送)设备;溜装置;溜运料设备;混凝土浇灌设备;斜槽运输设备;斜槽运料设备
chuting system 溜槽(输送)系统;斜槽装料系统;斜槽运输系统;斜槽运料系统
chuting unit 溜槽装置
chylification 糜化
ciborium 祭台上之天盖;龛室;祭坛华盖
ciborium altar 顶盖祭坛
cicerone 向导;导游人
cicoil 快干植物油
cicrotoic acid 环丁烯酸
cienega 地表地下水;沼泽
Ciereszko unit 西埃列兹柯单位
CIE standard overcast sky CIE 标准晴天阴天空
CIF[coast, insurance and freight] port of entry 目的港到岸价
cifax 密码传真
C.I.F.C5 包括百分之五佣金的到岸价
cigarette burn 香烟烧痕;端部燃烧
cigarette burning 端部燃烧;端部起燃
cigarette drain 卷烟式引流条;卷烟式引流管
cigarette factory wastewater 卷烟厂废水
cigarette ignition resistance 耐烟头点燃性
cigarette industry 卷烟工业
cigarette in packet 包装卷烟
cigarette lighter 打火机
cigarette making machine 卷烟机
cigarette packer 卷烟包装机
cigarette paper 卷烟纸
cigarette smoking 香烟烟熏
cigar lighter 火星塞;点烟器;打火机
cigar-shaped vallvey 雪茄烟型河谷;铁蹄型河谷
cigar store 香烟店
ciliary hyperemia 抱轮红赤
cilice 马毛
cill =sill
cill block 窗槛条块
cill cover 窗槛板
cill(l)ery 柱头卷叶饰;柱头的叶形雕饰
cill height 窗槛高度
cill tile 垫砖;窗台砖;窗槛面砖
cillus anaerobicus 厌氧杆菌
cima-inversal 上凸下凹双曲线脚
cima recta 枭混线【建】
cima reversa 反枭混线【建】
cima reversal 上凸下凹反曲线;袅混线脚
cimatium 反曲线盖板
cimbia (柱身上的)环箍饰
cimbination mill 组合式轧机
cimborio 穹隆状采光顶楼(屋顶上);灯笼式天窗;(教堂十字塔上的)采光塔
cimeliarch 教堂珍藏室
Ciment fondu 矾土速凝水泥(一种快硬水泥的商品名)
cimex 杀虫用熏剂
cimex lectularius 温带臭虫
ciminite 橄辉粗面岩
cimita 硅铝石
Cimmerian orogeny 细末里运动
cimolite 漂砾黏土;泥砾土;水磨土

cimolite formation 水合物生成
cimpressor work 压气机功
cinarcaf 朝鲜蓟酸
cincfoil 五叶梅花饰;五叶形饰
cinch anchor 加固锚栓
cinch bolt 膨胀螺栓
cinch mark 卷痕
cinchocaine 二丁卡因
Cincinnati-Blue epsilon structural system 辛辛那提一篮岭山字型构造体系
cincture (围绕圆柱的)线脚条;(柱头及柱座处的)环柱饰;环带装饰;柱环带;柱带
cincubative stage 潜伏期
Cindal 铝基合金
cinder 矿渣;焦渣;煤渣;煤屑;煤灰;灰渣;粗灰
cinder aggregate 矿渣骨料;焦渣集料;焦渣骨料;煤渣集料;煤渣骨料
cinder ballast 焦渣道砟;(轻便铁道的)煤渣道砟
cinder bed 炉渣通气孔;焦渣垫层;煤渣床
cinder block 炉渣砖;矿渣(混凝土砌)块;焦渣砌块;煤渣砖;煤渣砌块;煤渣混凝土块;堵渣口块
cinder block partition 焦渣砖隔断
cinder bogie 清渣车
cinder box 灰箱
cinder brick 炉碴砖;矿渣砖;焦渣砖;煤渣砖
cinder brick partition 焦渣砖隔墙
cinder catcher 集尘器;接渣器;集渣器;集灰器
cinder chip(ping)s 炉渣屑
cinder chute 煤渣漏斗
cinder cleaning hole 清渣孔
cinder coal 劣焦煤;极劣焦煤;天然焦
cinder coarse aggregate 炉渣粗集料;煤渣粗骨料
cinder concrete 煤渣混凝土;炉碴混凝土;矿渣混凝土;焦渣混凝土;煤渣混凝土
cinder concrete block 焦渣混凝土砌块;煤渣混凝土砌块
cinder concrete brick 煤渣混凝土砖
cinder concrete insulating course 焦渣混凝土保温层
cinder concrete lintel 焦渣混凝土过梁
cinder concrete roof(ing) slab 炉渣混凝土屋面盖板
cinder concrete wall slab 炉渣混凝土墙板
cinder cone 火山渣锥;岩渣锥;碎屑锥
cinder cooler 渣口冷却套
cinder dump 煤灰堆场
cinder dust 炉渣粉末;矿渣填充材料
cinder fall 煤渣坑;渣坑
cinder fiber 炉渣纤维
cinder fill 焦渣填实;煤渣填土;煤渣回填;煤渣垫实
cinder filler 炉渣填料
cinder filling 煤渣填实
cinder floor 煤渣地面;炉渣地面;焦渣地面
cinder heat 熔渣热度
cinder hole 渣孔
cinder hollow block 焦渣空心砖
cinder inclusion 夹渣;包渣
cindering work 摊铺煤渣
cinder ladle 渣罐
cinder loading 含灰量
cinder loss 大渣或飞灰热损失
cinder mill 炉渣研磨机;煤渣研磨机;煤渣粉碎机;碎渣机
cinder notch 出口渣
cinder path 炉渣小路;焦渣跑道;煤渣(小)路;煤渣步行道
cinder paving sett 煤渣铺路小方块
cinder pig 炉渣生铁;夹渣生铁
cinder pit 煤渣坑;渣坑
cinder pocket 沉渣室
cinder pot 炉渣空心砌块;渣桶
cinder removal plant 除碴设备
cinder road 煤渣路
cinder(road) surface 煤渣路面
cinder roofing 焦渣屋面
cinders 炉灰;煤渣跑道;火山岩烬
cinder sand 炉渣砂
cinder sand block 炉渣砂砌块
cinder sand concrete 炉渣砂混凝土
cinder sand cored block 炉渣砂空心砌块
cinder sand tile 炉渣砂面砖
cinder separator 倾斜式运输带分离机
cinder sett 炉渣小方块
cinder(side)walk 煤渣步行道

cinder sifter 煤渣筛
cinder slab 炉渣板
cindersnapper 渣口工
cinder spots (管材的)刻痕
cinder spout 流渣槽;渣沟流嘴
cinder tile 炉渣砖
cinder track 焦渣跑道;煤渣跑道
cinder trap 集灰器
cinder valve 卸灰阀
cinder wool 火山毛;矿渣绒;矿(渣)棉
cindery rock 火山渣岩
cine camera 电影摄影机
cinecolo(u)r 彩色电影
cinederivometer 计时计(用于重叠度调整);计时器
cine editor 编辑代
cinefaction 煅灰法;灰化
cine film 电影胶片
cinefilm studio 电影胶片厂
cinefluorography 荧光电影摄影
cineholographic(al) method 全息电影照相术
cineholomicroscopy 显微全息(电影)照相术
cinekodak 小型电影摄影机
cinema 电影院
cinema auditorium 电影院观众席
cinema coach 影视客车
cinema equipment 电影院设备
cinema house 电影院
cinema sign 电影广告
cinema theatre 电影院
cinema-theatre building 影剧院建筑
cinematheque 小电影馆;电影馆
cinematograph 电影摄影机
cinematographing department 摄影车间
cinematography 电影制片术
cinenic acid 桉烯酸
cineol 桉油醇
cineole 桉油精;桉叶素;桉树脑
cineolic acid 桉树脑酸
cinepanoramic house 全景宽银幕电影院
cineradiography 射线活动摄影术
cinerama 宽银幕电影
cinerarium 藏骨灰所;骨灰存放处
cinerary 灰的
cinerary urn 骨灰坛;骨灰罐
cinerary urn pit 骨灰坛墓穴
cineration 焚化;煅灰法
cinerator 垃圾焚化炉;火葬场;焚化炉
cinerator room 焚化间
cinereour 像灰一样的;由灰构成的
cinereous 似灰的;灰白色的
cinerite 火山渣沉积;火山渣岩
cineritious 烬灰色的;灰色的
cine-sextant tracking telescope 电影六分仪跟踪望远镜
cinesimeter 运动测量器
cinestrobe 频闪闪光装置
cinetheodolite 电影经纬仪;高精度光学跟踪仪
cinetheodolite record 电影经纬仪胶卷记录
cine transit 电影经纬仪
cinfiguration interaction 构型作用
cinlined plank settling tank of upflow 升流斜板沉淀池
cinnabar 朱红色;朱砂;晨砂
cinnabaric 含朱砂的;朱砂的;朱红的
cinnabarine 含朱砂的;朱砂的;朱红的
cinnabarite 银砂
cinnabar ore 辰砂矿石
cinnamene 肉桂烯
cinnamenyl 苯乙烯基
cinnamenyl angelic acid 苯乙烯基当归酸
cinnamomum camphora 樟树
cinnamon 桂皮树;桂皮(色);樟属植物;黄棕色
cinnamon brown 肉桂棕色
cinnamon soil 褐色土
cinnamon soil and cinnamon grey soil area 褐土及褐灰土区
cinnamon stone 桂榴石
cinnoline 邻二氮(杂)萘
cinqfoil 五叶形饰
cinquecento (16世纪的)意大利艺术
Cinquecento architecture 意大利文艺复兴时期建筑
cinquefoil 五联拱;梅花饰;五叶形饰;五瓣花饰
cinquefoil arch 五心连拱;五瓣(形)拱
ciodrin 丁烯磷

cipher 密押;暗号
cipher code 代码;暗号;密码;数字码
cipher component 密码元件
cipher device 密码装置
ciphering 译成密码
ciphering equipment 密码装备
cipher key 密码索引;暗号注解
cipher machine 密码机
cipher officer 译电员
cipher telegram 密码电报
cipher text 密码正文
cipher tunnel 假烟囱
cipherware 密码器件
ciphony 密码电话机
ciphony equipment 密码电话机
ciplyte 磷硅钙石
cipola dust collection 炉渣集料器;炉灰骨料器
cipolin(o) 云母大理岩;云母大理石;意大利产的白绿花纹的大理石
Cippoletti weir 西波列蒂堰;梯形堰
cippus 界碑
C. I. P. W. system of rock classification 四氏岩石分类法
circa 在周围;大约
circadian rhythm 昼夜节律;日周期节律
circalittoral community 潮周带群落
Circassian walnut 锡卡西亚胡桃木(欧洲南部)
circatidal rhythm 潮汐周期节律
circinate 环形的
circle 楼厅;节圈;马戏场;环状物;环(行);圆周;圆形;圆盘;二楼楼厅;施业区
circle adjustment 成圆调整
circle-arc tooth 圆弧齿
circle average deviation 圆上平均偏差
circle average dispersion 圆上平均散度
circle belt 环形构造带
circle bench 环形补偿器
circle bend 环形膨胀接头;圆曲管;圆弯管
circle bisect line method 圆圈等分法
circle brace 圆拉条;支撑环
circle brick 环砖;弧形砖;拱圈砖
circle brick on edge 环形侧砌砖;弧形砖;边部环砖
circle chains 圆环链
circle chart 圆(形)图
circle clearance 圆周间隙(转车盘)
circle communication network 圈式信息交流网络
circle coordinates 圆坐标
circle cutter 环刀;圆孔切割器;划圆刀
circle-cutting attachment 圆周切割装置
circle-cutting guide 圆形切割导向器
circle-cutting shears 圆周剪切机
circle cylindric(al) polar coordinate plot 圆柱面极坐标图
circle diagram 圆(向量)图
circle distribution 环形配水
circle distribution system 环形配水系统;环式配水系统
circle division 度盘划分;度盘分划
circle draft frame ball socket 圆牵引加万向接头;活用接头
circle drawbar 转盘牵引架(平地机)
circle drawing device 绘图器
circle drive 度盘微动螺旋
circle drive motor 转盘驱动马达
circle drive pinion 转盘驱动齿轮
circle end 半圆形起始踏步
circle fluorescent lamp 环形荧光灯
circle glass cutter 圆玻璃切刀
circle graduation 度盘刻度
circle graph 圆形图
circle-head window 半椭圆饰窗;半卵形窗亮子(通常安装在门口上方)
circle-hyperbolic system 圆一双曲线系统
circle illumination 度盘照明
circle-in 外光圈打开
circle jack (拧紧钻具螺纹用的)棘轮扳手
circle left 正镜【测】;盘左
circle levee 环堤;月堤;围堤
circle method of pipe-grid analysis 圆圈法管网分析法
circle microscope 度盘放大镜
circle network 环网络
circle number of grouting hole 注浆孔圈数
circle of altitude 地平纬圈

circle of a motor grader 机动平地机转台
circle of celestial latitude 黄纬圈
circle of celestial longitude 黄经圈
circle of confidence 置信圆
circle of confusion 模糊圆;散射圆盘;散光圆
circle of contact 切圆
circle of convergence 收敛圆;收敛图
circle of correction 校正圆;基圆
circle of curvature 曲率圆
circle of declination 赤纬圈;赤纬度盘;时圈
circle of diffusion 漫射图;漫射圈
circle of equal altitudes 等高(度)圈;等地平纬度圈;等位圈
circle of equal declination 赤纬圈;赤纬平行圈
circle of equal probability 圆径概率误差;等概率圈
circle of error 误差圆
circle of failure 破坏圈
circle of freeze-thaw 冻融循环
circle of good definition 清晰图;清晰区
circle of higher order 高阶圆
circle of hues 色彩圈
circle of inertia 惯性圆
circle of influence 影响圆;影响函数;影响范围
circle of influence of well 井的影响圈
circle of inversion 反演圈
circle of latitude 黄纬圈;黄经圈;纬度圈
circle of least confusion 模糊圈;明晰圈
circle of literature and art 文艺界
circle of longitude 经度圈;黄经圈
circle of Mohr 莫尔圆
circle of perpetual apparition 恒星圈
circle of perpetual occultation 恒隐圈
circle of position 船位圆;船位圈;位(置)圆;位置圈
circle of probable error 圆径概率误差
circle of reference 参考圆
circle of reflexion 反射圆
circle of right ascension 赤经圈
circle of rupture 破裂圈;破坏圆
circle of simultaneity 同步圆
circle of sphere 球面圆
circle of spontaneity 同步圆
circle of stress 应力圆
circle of uncertainty 误差圆
circle of vegetation 植被圈;群落环
circle of visibility 视程圈
circle of wall plates 承梁板凸缘
circle-on-circle 双曲工(立面和平面都是曲线)
circle-on-circle face 蛋面面
circle orientation 度盘定向
circle orienting knob 度盘定向钮
circle-out 外光圈关闭
circle pattern 圆周形
circle position 度盘位置
circle range 圆上极差;度盘标度
circle reading 度盘读数
circle reverse (平地机刮刀的)顺逆转角;(平地机刮刀的)调角装置;回转角
circle reverse control housing (平地机刮刀的)顺逆转驱动轴
circle reverse drive shaft (平地机刮刀的)顺逆转换向控制箱
circle right 倒镜;盘右
circle ring gear 转盘齿圈
circle road system 环式道路系统
circle round 回路
circle scanner 度盘扫描器
circle search for a mobile hider 对机动隐藏者的圆形搜索
circle setting 度盘位置;度盘换置;度盘定位;度盘调置
circle setting mechanism 度盘安装机构
circle setting screw 度盘安置螺旋;水平度盘安置螺旋
circle shears 圆盘剪(切机)
circle sheet 三点曲线图
circle sprinkler 环形喷灌机;摇臂式喷灌器;环动喷灌器
circle standard error 圆的标准偏差
circle structure 环形结构
circle support 圆形支柱
circle system 环形道路系统;环行系统;环式系统
circlet 小环;锁扣
circle template 画图样板
circle test 循环试验

circle theorem of hydrodynamics 流体动力学的圆柱绕流定理
circle throat clearance 圆形喉通道
circle throw vibrating screen 旋转振动筛
circle trowel 圆泥刀;曲面抹子
circle variance 圆上方差
circle vibration 圆周振动
circle vision cab(in) 全视野舱(车厢)
circlewise 成环状
circline 环形
circline lamp 环形灯
circling approach area 盘旋进场区
circling attachment 圆规
circling guidance light 盘旋引导灯(机场)
circling lamp 环形灯
circling mark 旋转式标志
circling radius 回转半径
circlip (开口的)簧环;定位环;卡簧;弹性挡圈
circlip for hole 孔用弹性挡圈
circlip pliers 卡簧手钳
circlips for shaft 轴的弹性挡圈
circuit 回路;环行;环路【电】;线路【电】;电路载流量;电路;水准闭合环
circuital current 电路电流
circuital forces (平衡力系中矢量相加所形成的)闭合回路
circuit algebra 线路代数
circuital law 环流定律
circuit allocation 电路分配
circuital magnetization 螺线管磁化;环形磁化
circuit analysis 线路分析;电路分析
circuit analyzer 电路分析器;伏欧毫安表
circuit analyzer and tester 电路分析测试仪
circuit analyzing test set 电路分析测试设备
circuit angle 变流器相位角
circuit aperture 圆孔
circuit arrangement 电路布置
circuit balancing 电路平衡
circuit-base plate 回路底板
circuit bin compartment 格子式料仓
circuit block 回路中间油路板
circuit board 电路板
circuit board drill 电路板钻头
circuit board plant wastewater 线路板厂废水
circuit bonding jumper 电路跨接线
circuit break 断路
circuit breaker 开关;断路继电器;线路断路器;断路闸;断路器;断路开关;断流器;电路断路器
circuit-breaker oil 开关油
circuit-breaker oil-storage tank 断路器油箱
circuit-breaker plate 断路器板
circuit breaker plug 断路插头
circuit-breaker unit 断路器单元
circuit buffer 电路缓冲存储器
circuit capacity 线路容量;电路容量;电路能力
circuit card 电路插件
circuit changer 转接器;电路开关;开关
circuit changing switch 电路转换开关
circuit cheater 模拟电路
circuit closer 开关;接电器;电路闭合器;闭路器;通路器
circuit closer and breaker 通路器与断路器
circuit closing 电路接通
circuit-closing connection 闭路接法
circuit-closing contact 闭路接点;闭合接点
circuit closure 环线闭合差
circuit code 闭路码
circuit common 电路公共端
circuit component 电路元件
circuit conduit 线路导管
circuit configuration 电路排布
circuit connection 电路接线
circuit connector (电路板的)插头座
circuit constant 电路常数
circuit controller 电路控制器
circuit coupling 电路耦合
circuit description table 电路描述表
circuit design 电路设计
circuit diagram 线路图;电路图
circuit drill 电钻
circuited track 装有轨道电路的股道
circuit efficiency 回路效率;电路效率
circuit equation 电路方程
circuit error 环线闭合差

circuit feed 回路给料;循环给料
circuit grade 电路级别
circuit grouting 中心圆管灌浆
circuit identification code 电路识别码
circuit image 电路图像
circuit impedance 电路阻抗
circuit interlocking 电路闭锁
circuit interrupter 电路断流器
circuit interrupting device 断路装置
circuit layout 线路布置
circuit layout card 电路布线图
circuit layout record 电路记录卡片
circuit layout record card 电路记录卡
circuit letter 通知书
circuit load 电路负载
circuit loading 电路荷载
circuit logic 电路逻辑
circuit loss 电路损失
circuit main 环形总管
circuit manifold 回路底板
circuit matrix 回路矩阵
circuit media 电路媒质;电路媒介
circuit misclosure 环线闭合差
circuit model 线路模型
circuit module 电路组件
circuit net loss 线路净损耗;电路净损耗
circuit network 网路
circuit noise 线路噪声
circuit noise level 线路噪声电平;电路噪声电平
circuit noise meter 电路噪声测试器
circuit number 电路号码
circuit of the cells 槽路
circuit-opening connection 开路接法
circuit-opening contact 开路接点
circuit opening relay 断路继电器
circuitous pattern 电路图形
circuitous philosophy 电路原理
circuitous resistance 电路电阻
circuitous resonance curve 电路谐振曲线
circuitous road 迂回路
circuitous traffic 迂回运输
circuit parameters 电路参数
circuit path 电路通道
circuit period 电路周期
circuit power factor 电路功率因数
circuit pressure 环路压力;循环液压力
circuit protection 电路保护
circuit protector 电路保护器
circuit quality monitor 电路质量监控器
circuit railroad 环形铁路;环形铁道;环行铁路;环行铁道
circuit railway 环行铁路;环行铁道
circuit rank 回路秩
circuit recloser 电路自动重合闸
circuit reliability 电路可靠度
circuit resistance 环路阻力
circuit rider approach 循环制导法
circuit rider concept 循环制导原理
circuit road 环线;环路【道】
circuitron 电路管
circuitry 整机电路;电路系统
circuitry module 电路模件
circuit schematic diagram 电路原理图
circuit section 电路区段
circuit shift 循环移位
circuit simulator 电路模拟器
circuit substrate 电路衬底
circuit supervision 电路监控
circuit switch 电路转接;电路开关
circuit switched network 线路交换网
circuit switching 线路转接;线路交换;电路接转;电路接转;电路交换
circuit switching concentration 电路交换集中
circuit switching connection 电路开关连接
circuit symbol 电路符号
circuit technology 电路工艺
circuit terminal 电路接线端
circuit test 断路检查试验;电路测试
circuit tester 线路测试器;万用电表;电路试验器;电路测试器;放炮检流仪;放炮检流计;万用表
circuit testing 电路测试
circuit testing screwdriver 测电起子
circuit theory 电路理论
circuit time recorder 电路计时器

circuit to hold a route for shunting 非进路调车电路
circuit tracing machine 电路绘图机;电路跟踪描图机
circuit track 环形线(路)
circuit tuning 回路调谐
circuit unit 电路组件;电路单元
circuit vent 环路通气口;环路通气管;环路排气管
circuit vent pipe 环路通气管;环路排气管
circuit voltage 线电压;导线间电压
circuit voltage stabilizer 电路稳压器
circuit winding 单循环缠绕
circuit with distributed constants 分布常数电路
circuity 流程;连接法;布线;罗盘;环行
circuity of action 票据的流回
circulant design 循环行列式设计
circulant determinant 循环行列式
circulant matrix 轮换矩阵;循环矩阵
circular 盘状;通函;通告;通报
circular abacus 圆形柱顶板
circular acceleration 圆周加速度
circular accelerator 回旋加速器
circular adsorption chromatogram 圆形吸附色谱
circular adsorption chromatography 圆形吸附色谱法
circular antenna 环状天线;圆形天线
circular arc 圆弧
circular arc analysis 圆弧分析法
circular arch 弧拱;圆(弧)拱
circular arch bridge 圆弧拱桥
circular arched girder 圆弧拱梁
circular arch or shell 圆弧拱或壳体
circular arc method 圆弧法
circular arc profile 圆弧叶型
circular-arc ruler 曲线板
circular arc type servomotor 环形接力器
circular area method 圆面积计算法
circular array 圆形天线阵
circular auditorium 圆形大厅
circular balcony 圆形阳台
circular bar 圆钢筋;圆棒
circular barrel siphon 圆筒形虹吸管
circular barrel vault 圆筒形拱顶;圆筒形穹顶
circular bead 台肩;圆角
circular beam 圆梁
circular beam connection 圈梁结合
circular bearing plate 圆形支承板
circular bell mouth entrance 圆喇叭形进水口
circular bench saw 圆形台锯
circular birefringence 圆双折射
circular bit 旋转钻孔器;旋转钻孔机;旋形钻孔器;圆钻;圆车刀
circular blade type rail sawing machine 圆片式锯轨机
circular bottom parking 环形井底车场
circular bowl 圆形斗;圆碗状物
circular box beam 圆形空心梁
circular bracket 圆括弧
circular bubble 圆水准器
circular building 圆形建筑(物)
circular butt welding 圆周焊接;圆周对接焊
circular buying 循环采购
circular caisson 圆形沉箱
circular cam 凸轮盘
circular campanile 圆形钟楼;圆形钟楼
circular canal tunne 圆形输水隧洞;圆形输水隧道
circular canal tunnel 圆形截面隧洞
circular capital 圆柱头;圆柱顶
circular carriageway 环行车行道
circular carved panel 圆雕饰板
circular casing 环形泵壳
circular cell 圆形单元;圆形格箱板桩墙;圆形格型结构;圆形格体
circular cella 圆形小室;圆形内殿
circular cellular cofferdam 圆形格箱板桩围堰
circular chamber 圆形房间
circular chart 极坐标记录纸
circular chart recorder 流量计
circular chaser 圆形螺纹梳刀
circular chimney 圆形烟囱
circular chimney bond 圆烟囱砌合
circular chisel 圆凿(子)
circular chromatography 环形色谱法
circular-circular face 蛋圆面

circular-circular sunk 双重曲度的凹陷部分
circular-circular sunk face 凹蛋面
circular clarifier 圆形澄清池
circular coal 眼状煤;圆(形)煤(基)
circular cofferdam 环形围堰;圆形围堰
circular cold saw 圆形冷锯
circular column 环形柱;圆柱
circular conchoid 圆蚌线
circular conduit 圆形管道
circular conduit-type sewer 圆管(式)下水道;圆形断面排水管
circular cone 圆锥(体)
circular conoid 圆锥体
circular constant-thickness arch 等厚圆形拱
circular contact method 环状接触法
circular control weir 圆形控制堰
circular conveyer 弧环输送机
circular cooler 环式冷却机
circular core 圆形芯体
circular correlation 循环相关
circular correlation function 循环相关函数
circular corridor 环形走廊;回廊
circular counter 盘点式计数器;表盘式计数器
circular cover 圆形顶盖
circular crack 环状裂隙
circular crevasse 环状裂纹
circular cross-cut saw 横割圆锯;横切圆锯
circular cross section 圆(形横)断面;圆截面
circular culvert 圆形涵洞
circular curve 圆曲线
circular curved beam 圆截面弯梁
circular curve location 圆曲线测设
circular curve with transition 有缓和段的圆曲线
circular cut 环形开挖
circular-cut file 圆铣刀;圆锉刀;弧形锉刀
circular-cut file with curved teeth 有曲齿的弧形锉刀
circular-cut file with straight teeth 有水平齿的弧形锉刀
circular cutter holder 回转刀架
circular cutting disc 回转刀盘;圆盘刀
circular cutting path 环形刀路
circular cylinder 圆柱(形);圆柱体
circular cylindric(al) coordinate 圆柱坐标
circular cylindric(al) shell 圆柱壳体;圆柱筒壳;圆形筒壳;圆筒壳
circular degree 圆度
circular departure 环线发车
circular diagram aerial 无方向天线;全向图天线
circular dial 圆度盘
circular diall-type scale 数字显示天平;数字显示秤
circular dichroism 圆振二向色性;圆二色性;圆二色散
circular die 圆板牙
circular diminutive tower 圆形小塔楼
circular disk feeder 圆盘给料机
circular disk grounding device 圆盘接地器
circular distribution 圆形分布
circular distributor ring 集电环
circular dividing table 分度台
circular division 圆分度
circular dome 圆(形)屋顶;圆穹顶
circular domical vault 环状圆顶;环状穹顶
circular donjon 圆形城堡主塔
circular down pipe 圆形水落管
circular drainage 环状水系
circular drawing chamber 圆形作业室
circular drying 循环干燥
circular dungeon 圆形地牢;圆形城堡主塔
circular dust collector 圆形集尘器
circular earthquake 环形分布地震;圆圈地震
circular economy 循环经济
circular edge 环形边
circular edge guide 圆形吸边器;圆形边导器
circular electric(al) mode magnetron 同轴磁控管
circular electrode 接触焊圆盘状电极;圆盘式电极;盘状电极
circular enclosure 圆形边围
circular engine house 圆形机车库【铁】
circular error 圆周误差;周期误差
circular error of probability 径向公算误差
circular error probability 概率误差圆
circular face 圆凸面;凸圆面
circular failure analysis 圆弧破坏分析

circular fault 环形断层
circular features of bioherms 生物礁环形体
circular features of dome structures 穹隆构造环形体
circular features of faulted blocks 断块环形体
circular features of intrusive bodies 侵入岩体环形体
circular features of salt domes and diapirs 盐丘及底辟环形体
circular features of structural basins 构造盆地环形体
circular features of volcanic mechanism 火山机构环形体
circular feeder 圆盘式喂料机;喂料盘
circular feed(ing) 回转进给
circular field 旋转磁场;圆磁场
circular file 圆锉
circular filling machine 转盘灌装机
circular flap 环形瓣
circular flat-plate 圆板
circular flow 环形射流;环流;循环流动;周转过程
circular flow method 圆形流法
circular flow of economic system 经济体系周流;经济体系的循环流动
circular flow of economy 经济循环
circular flue 圆形烟道;圆火道
circular folded slab roof 圆形折板屋顶
circular footing 圆形地脚;圆形基脚;圆基础
circular for confirmation 查证通函
circular form 循环型
circular forming cutter 成型圆铣刀
circular foundation 圆形基础;圆基础
circular foundation pier 圆形基础柱
circular frequency 角频率;圆周(频)率;圆(弧)频率
circular frequency of vibration 振动圆频率
circular function 圆函数;三角函数
circular gallery 圆形廊
circular galvanometer 圆形电流计
circular geometrical stair(case) 圆盘旋(楼)梯;圈盘旋梯
circular girder 圆形大梁
circular-glass photo stress ga(u)ge 玻璃圆柱式光应力计
circular graphical chart 圆图表
circular grate 圆形格栅;圆炉算
circular grate pelletizing machine 环式焙烧机
circular grinder 外圆磨床
circular grinding machine 外圆磨床
circular groove 环钩;圆槽
circular groove crack test 圆槽抗裂试验
circular guidance 弧形导承
circular guide 弧形导承
circular guideway 圆形导轨
circular hall 圆形大厅
circular hammer 圆夯
circular head 圆形封头
circular headed skylight 圆顶形天窗
circular helix 圆柱螺旋线;普通螺旋线
circular highway 环形公路;环路
circular hipped-plate roof 圆形折板屋顶
circular hole 圆孔
circular hole sieve 圆孔筛
circular hoop 圆箍
circular horn 圆喇叭
circular hot sawing machine 圆形热态锯切机
circular hut 圆形棚屋
circular inch 圆英寸
circular in circular 叠套环状的
circular index 循环指数;圆分度头
circular indexing table 圆形分度工作台
circular initial disturbance 环向初始扰动;环间初始扰动
circular integration 非竞争性企业联合
circular interference fringe 干涉光边缘
circular interference ring 干涉光环
circular interlocking pipe 圆形锁链管
circular interpolation 圆弧内插法
circular iris 圆形光圈
circular isobar 等压线圈
circularity 圆(形)度
circularity ratio 圆度比
circularization 通函询证
circular jack 径向千斤顶;环形千斤顶
circular jet 圆流

circular jib 圆截盘
circular jig 圆形跳汰机
circular joint 环向接缝;环接
circular keep 圆形囚牢
circular kiln 轮窑;圆窑;圆筒式干燥炉
circular lance 圆弧切口
circular layer 圆形层;环(状)层
circular letter 通知;通函
circular letter of credit 流通信用证
circular level 圆水准仪;圆水平仪;(圆)(盒)水器器
circular level adjustment screw 圆水准校正螺旋
circular life-belt 救生圈
circular linear load 圆形线荷载
circular line for loading 装车用环形线
circular lining 圆形衬砌
circular lip 圆法兰
circular list 环表
circular loaded area 圆形荷载面积
circular load(ing) 圆形荷载;循环加荷;循环荷载
circular load plate 圆形加载板
circular luminaire 圆形照明器
circularly polarized light 环状偏振光;圆偏振光;圆形偏极光
circularly polarized wave 圆偏振波
circular magnetic field 旋转磁场
circular magnetization 圆形磁化
circular main 环形总管;环形干线;环形干管
circular main motion 主回转运动
circular main movement 主回转运动
circular manhole 圆形检查车
circular matrix 循环矩阵
circular mean deviation 三角均值离差
circular mean difference 三角均值差
circular measure 弧度法
circular measurement 弧度测定
circular membrane 圆形薄膜
circular mil 密耳圆;圆密耳
circular milling 圆铣
circular milling attachment 圆形铣切装置
circular mitre 圆直面接合
circular mobility 循环流动性
circular motion 环形运动;圆(周)运动
circular mo(u)lding 圆形线脚
circular mound altar 环丘坛;天坛圜丘坛
circular multi-compartment bin 圆形多室料仓
circular multi-compartment silo 圆形多窖料仓
circular newel stair(case) 旋梯;环柱旋梯
circular nick 循环裂点
circular nomogram 圆算图(列线图)
circular nomograph 圆形诺模图
circular normal distribution 圆正态分布;正态分布圆
circular note 通告照会
circular notice of ship's movement 船位通报
circular nozzle 环形喷嘴
circular nut 圆螺母
circular nut with lock 带锁紧槽圆螺母
circular of stock-taking 盘存点货通知
circular openings 圆形洞室
circular openings not requiring reinforcement 不要求补强的圆形开孔
circular operation 循环作业
circular operation quota 循环作业定额
circular optimization 循环优化
circular optimization method 循环优化法
circular orbit 环形轨道;圆(形)轨道
circular orbital velocity 圆形轨道速度
circular order 循环次序;旋风次序
circular orifice 圆洞口;圆孔(口);圆隔片;圆隔板
circular orthomorphic projection 圆正形投影
circular oven 循环炉
circular pallet 等臂式叉瓦;等圆弧擒纵叉
circular pallets escapement 圆形擒纵机构
circular pan mixer 立式搅拌机;圆盘(式)搅拌机
circular paper chromatography 圆型纸色谱法
circular-patch crack(ing) test 圆形镶块抗裂试验;环形镶块抗裂试验
circular path 环形路线
circular pavilion 圆亭
circular paving 圆形铺面
circular pediment 弧形山墙
circular perimeter flow grit settling tank 圆形周边流沉沙池
circular perimeter temple 圆形列柱式神殿;圆形列柱式庙宇
circular peristyle 圆列柱廊
circular permutation 循环排列
circular permutation code 循环排列码
circular pierhead 圆形码头
circular pile 圆形桩;圆形堆场
circular pile terminal 圆形桩码头
circular pipe 圆管;循环管
circular pit 圆形均热炉
circular pitch 圆周节距;周节
circular pitch error 周节误差
circular plane 曲面刨;圆刨(床);圆弧刨
circular planing 圆形刨削
circular planing attachment 圆柱面刨削装置
circular planting 环植
circular plastic spacer 环形塑料定位件
circular plate 圆(形)板
circular pleochorism 圆形多色性
circular point 虚圆点;圆点
circular point group 循环点群
circular point on a surface 曲面上的虚圆点
circular polariscope 圆偏振光仪;圆偏光器
circular polarization 圆偏振
circular polarized loop vee 圆偏振环形 V 天线
circular polarized wave 圆(形)极化波
circular pond 圆池
circular preblend stockpile 圆形预均化堆场
circular prestress 圆向预应力
circular prestressing 环形预应力;圆周预加应力(法)
circular probable error 圆(径)概率误差
circular profile 圆截面型材
circular protractor 圆分度器
circular pump 环流(水)泵
circular pumping 循环抽取
circular purl machine 圆型回复机
circular(radio) beacon 环射无线电指向标;无定向无线电航信号台
circular(radio) beacon station 环射无线电指向标台;全向无线电指向标台
circular railway 环形铁路;环形铁道
circular raised table 圆形柱头托板;圆形提升台(水泥试验用)
circular rammer 硪;圆夯;飞硪
circular ramp 环形匝道;环形坡道;螺旋式坡道
circular reaction 连锁反应;循环反应
circular receiving 环线接车
circular recess 圆形槽
circular recorder 圆盘记录器
circular reinforcement bar 圆形钢筋
circular reinforcing cage 圆形加筋笼;圆形钢筋笼
circular repolarization 再生圆偏振
circular reservoir 圆形蓄水池
circular response 连锁反应
circular rib machine 圆型罗纹机
circular ring 圆环
circular ring girder 圆形圈梁;环形梁
circular ring(shaped) plate 圆环状平板
circular rip saw 纵切圆锯;圆形粗齿锯
circular rise 循环上涨
circular road 环路;环形路线
circular road system 环形道路系统
circular rod 圆钻杆
circular rod level 标尺圆水准器
circular rod web 钢筋腹杆
circular roller path 压路机圆形碾压路线;环形轨道
circular roll with knife 带刀圆辊
circular runout 径向振摆跳;径向跳动
circular rupture model 圆形破坏模式(梅耶霍夫关于混凝土路面破坏形式的一种设想)
circular saw 轮锯;圆盘锯;圆锯(机);盘锯
circular saw bench 圆锯架
circular saw blade 锯片;圆锯片
circular saw blade for hot metal cut 金属热切圆锯片
circular saw friction cutting 圆锯摩擦切割
circular saw sharpener 圆锯刃磨机
circular scale 环形标度
circular scan 圆形扫描
circular scanning 圆周扫描;圆形扫描
circular screen 圆形网目板;圆孔筛板;圆(孔)筛
circular screen feeder 筛式圆盘喂料机
circular screwing die 圆板牙
circular seam 环缝
circular seam welder 环缝对焊机;滚焊机;圆周线焊机
circular seam welding 环缝对焊;圆周焊接
circular section 圆形截面;圆形断面;圆切面
circular section of strain ellipsoid 应变椭球圆截面
circular section tunnel 圆形(断面)隧洞
circular sedimentation tank 圆形沉淀池
circular segment 圆弓形
circular sewer 圆形下水道;圆形污水管
circular shaft 圆杆;圆轴;圆(形)竖井
circular shake 木材环裂;环裂
circular shape 圆形
circular shaped 圆形的
circular shaped shield 圆形盾构
circular shears 回转剪切机;开圆盘剪切机;圆盘剪床
circular shear stress ratio 循环剪应力比
circular shear test 循环剪切试验
circular shed 圆形披屋
circular shift 循环移位
circular shit 循环位移
circular sieve 圆筛
circular sight glass 圆形视镜
circular silo 圆形储[贮]仓
circular silo compartment 圆库中间仓
circular simple shear test 循环单剪试验
circular skip 圆筒形吊斗
circular slide 圆弧滑动
circular-slide analysis 圆形滑坡分析
circular slide rule 圆计算尺
circular sliding plane 圆形滑动面
circular sliding plane method 圆形滑动面法
circular slip 圆弧滑动
circular slip surface 圆弧滑动面
circular sluice 圆形泄水孔
circular snips 圆剪
circular soil cutter 切土环刀
circular source 圆形源
circular spacer 支撑环;隔离环;定距环
circular spectrometer 回旋质谱仪
circular speed 圆周速度
circular spider type joint 十字叉连接
circular spike 齿形键;齿盘(金属制的木材连接件)
circular spillway crest 环形溢流堰顶
circular spirit level 圆形水平仪
circular spray sprinkler 远射程人工降雨机
circular sprinkler 环形喷灌机;圆形喷灌机;全圆周喷头
circular stair(case) 螺旋(楼)梯;圆形楼梯;盘旋楼梯
circular standard deviation 圆形标准差
circular stationary buddle 固定凸圆形淘汰盘
circular stone 磨盘
circular storage 循环储存器;循环储存
circular stratum 环层
circular street 环状街道;环行街道
circular stressing head 预应力混凝土中的锚头
circular striation 环状横纹
circular string border machine 大身圆型针织机
circular sunk face 凹圆面
circular system of locomotive running 循环运转制[铁]
circular tank 圆池
circular telegram 通电
circular temple 圆形神庙
circular test 循环试验;循环检验
circular texture 环状结构
circular thickener 圆形浓缩机;圆形厚浆池
circular thickness 弧齿厚度
circular-tipping concrete skip 四向翻倾混凝土运载车;混凝土运输车
circular titanium claded copper bar 圆形钛铜复合棒
circular tooth and cylindric(al) worm reducer 圆弧齿圆柱蜗杆减速机
circular tooth gear 圆弧齿轮
circular tooth thickness 弧齿厚
circular torsional shear test 循环扭剪试验
circular tower 圆塔
circular trace indicator 圆形扫描指示器;圆形扫描显示器
circular traceried window 圆形花格窗
circular track 循环
circular track test 环道试验【道】
circular transmission line chart 传输线圆图

circular trench 环钩
circular triaxial test 循环三轴试验
circular tube 圆管
circular tunnel 圆形隧道
circular tunnel vault 圆形隧道拱顶
circular turbulent jet 圆形紊动射流
circular turning basin 圆形回转试验水池
circular turret 圆形塔楼
circular type cellular cofferdam 圆环格状围堰;圆环格形围堰;圆格型围堰;环状格形围堰
circular type concrete hinge-bearing 圆形混凝土铰支座
circular valve 环阀
circular variable filter 环形可变滤光片
circular vault 圆拱顶;圆顶穹隆
circular velocity 环绕速度;圆(周)速度;周转速度;周速(度)
circular vent 环形通气管
circular vertical-flow settling tank 圆形竖流式沉淀池
circular vibration 圆振动
circular voided concrete slab 圆孔空心混凝土板
circular vortex 圆涡旋
circular wagon vault 圆筒状顶;圆筒形穹顶
circular washer 圆垫圈
circular wash fountain 圆形喷水式洗手器
circular wave 圆形波;同心波
circular waveguide taper 圆波导过渡器
circular web 圆形针织物
circular wedge 环形楔
circular weight 圆坠陀
circular weir 环堰;圆形堰
circular wetting 循环加湿
circular wetting and drying 循环干湿法
circular winding stair(case) 螺旋楼梯
circular window 圆(形)窗
circular wire brush 圆形钢丝绳
circular work 圆木作;圆形的红木工件
circulary-scanning reticle 圆周扫描调制盘
circulate 流通;流传;循环
circulated aeration tank 循环曝气池
circulated flow 循环水流
circulated rate 循环速率
circulated water analysis 循环水分析
circulating aerial ropeway 循环式架空索道
circulating air 循环空气;循环风
circulating air separator 循环空气(式)选粉机
circulating air supply 循环送风
circulating alkali 循环碱
circulating antibody 循环抗体
circulating assets 流动资产
circulating ball gear (转向装置的)滚珠式蜗杆机构
circulating bottom current 河底环流
circulating capital 流动资金;流动资本;营运资金;周转资金
circulating capital goods 流动资本货物
circulating chamber 循环室
circulating check 流动支票
circulating cheque 流动支票
circulating circle 循环圆
circulating conduit 循环油管;循环水管
circulating cooling 循环冷却
circulating cooling discharged water 循环冷却排水
circulating cooling system 循环式冷却装置;循环式冷却系统;循环式冷却方式
circulating cooling water 循环冷却水
circulating current 环流;循环液流
circulating current protection 环流保护
circulating-current protective system 环流保护系统
circulating decimal 循环小数
circulating device 循环装置
circulating direction 循环方向
circulating discharged wastewater 循环排污水
circulating disturbance 循环干扰
circulating ditch 循环沟;循环槽
circulating door 旋转门;循环门
circulating dust 循环飞灰
circulating electro-Fenton reactor 循环电芬顿反应器
circulating elevator 循环提升机
circulating equipment 流通设备;环流设备;循环设备

circulating factor of depot repairing 断修循环系数
circulating fan 回风扇;循环风机;摇头式风扇;通风扇
circulating filtration 循环过滤
circulating fireman 满火红
circulating floating bed reactor 循环浮床反应器
circulating flow 环流;循环水流
circulating flow method 环流法
circulating fluid 循环液
circulating fluidized bed combustor 循环流化燃烧室
circulating force 循环力
circulating fraction 循环分数
circulating fuel reactor 循环燃料反应堆
circulating fund 流动资金;流动资本;流动基金
circulating fund for every hundred Yuan of output value 百元产值占用流动资金
circulating gas preheater 循环气预热器
circulating head 循环压力;循环压头;循环(水)头
circulating heater 循环加热器
circulating heating system 循环加热系统
circulating hole 泥浆出口
circulating hole of the bit 钻头水眼
circulating liability 流动负债
circulating line 环流管道
circulating load 循环负荷
circulating load ratio 循环负荷率
circulating loss 环流损失
circulating lubrication 循环润滑
circulating market 流通市场
circulating medium 通货;流通媒介;循环介质;冲洗介质
circulating memory 回转存储器;循环存储[贮]器;循环存储;循环储存器;动态存储器
circulating mixer 循环混合机;带循环装置的搅拌机
circulating oil 循环油
circulating oiling 循环加油
circulating oil strainer 循环油滤器
circulating oil system 循环润滑系统
circulating openings 钻头水眼;冲洗眼
circulating over-shot 冲洗打捞器;循环打捞筒
circulating pipe 循环油管;循环水管;循环管路
circulating pipe for cooling water 冷却水环流管
circulating polarized light 圆形偏极光
circulating preblending silo 圆形预均化库
circulating pressure 循环水头;循环压力
circulating pump 环流(水)泵;循环(水)泵
circulating quantity 回流量;循环量
circulating rate 循环速率
circulating ratio 冷却倍率;循环倍率
circulating real capital 流通资产;动产
circulating reflux 循环回流
circulating register 循环寄存器
circulating scrap 循环废钢铁
circulating sedimentation tank 循环沉淀池
circulating shower 循环淋浴
circulating sludge 回流污泥;循环污泥
circulating-solution reactor 循环溶液反应堆
circulating speed 循环速度
circulating spray bar 循环喷雾器的喷杆;循环喷油管
circulating storage 循环存储[贮]器
circulating sub 环流阀瓣;环流阀
circulating system 循环系统;冲洗系统
circulating tank 循环水箱;环流箱槽;循环水舱
circulating time 冲孔时间
circulating tube 循环油管;循环(水)管
circulating turbocompressor 透平循环压缩机
circulating type room heater 循环式室内加热器
circulating velocity 循环速度
circulating velocity of deposit money 存款货币流通速度
circulating velocity of money 货币流通速度
circulating ventilation 循环通风
circulating warehouse receipt 可流通仓单
circulating water 冷却水;活水;循环水;散热水
circulating water bath 循环式水浴锅
circulating water head 循环水柱
circulating water reuse 循环水回用
circulating water system 循环水系统
circulating zone of active replacement 积极交替循环带
circulating zone of slow replacement 缓慢交替循环带

circulation 流转;流通;循环;闭回线积分;速度环量
circulation analog(ue)/digital converter 循环模数转换器
circulation anomaly 环流异常
circulation area 来往交通面积;流通面积;交通面积;人过流面积
circulation around circuit 封闭环流
circulation boiler 循环(式)水管锅炉
circulation boring method 正循环钻孔法
circulation by circulation 泵循环
circulation by induction degassing 磁感循环真空脱气法
circulation capital 流动资金;流动资本
circulation capital reductive rate 流动资金降低率
circulation cascade 阶式循环;级联循环
circulation centre 环流中心
circulation channel 流通渠道
circulation circuit 环路【电】;循环电路
circulation control 环流控制
circulation control system 流通管理系统
circulation cost 流动费用
circulation current in river bend 弯道环流
circulation degassing process 真空循环脱氧法
circulation design 正循环设计
circulation detection 巡回检测
circulation discharge 循环流量
circulation distribution 环流分布
circulation disturbance 循环干扰
circulation drill 正循环钻机
circulation factor 循环倍率
circulation feed lubrication 环流润滑
circulation feed water heater 循环(补)给水加热器
circulation flow 环流;循环流(动);循环总量
circulation fluid 循环液;循环流体
circulation flux 环流通量
circulation forecast 环流预报
circulation-free 无环流的
circulation fund 周转基金
circulation funds at fixed quota 定额流动资金
circulation heating 循环供热
circulation heating installation 循环供热装置
circulation implied list 隐循环表
circulation index 环流指数
circulation indicator 机油显示器
circulation integral 圆周积分
circulation layer 环流层
circulation line 循环管线;循环管路
circulation loss 流通损耗;循环中断;循环液漏失;冲洗液漏失
circulation loss of stabilizing fluid 稳定液逸散;散逸稳定液
circulation lubrication 循环润滑
circulation manager 营业主任;营业经理
circulation map 路线图;交通图
circulation market 流通市场
circulation of a bank 银行纸币发行额
circulation of acceleration 加速度环量
circulation of atmosphere 大气环流;大气环境
circulation of a vector 矢量的环流量
circulation of commodity 商品流通
circulation of electrolyte 电解液的循环
circulation of flushing medium 冲洗介质循环
circulation of goods 货物流转
circulation of heat 热循环
circulation of lakes and reservoirs 湖泊和水库的环流
circulation of materials 物料循环
circulation of money 货币流通
circulation of mud (旋钻技术中的)泥浆循环
circulation of traffic 交通运转
circulation of washing medium 冲洗介质循环
circulation of water 水文循环;水(的)循环
circulation of water vapo(u)r 水汽环流
circulation oiling 循环加油润滑
circulation on the earth 在地球上循环
circulation oven 循环(热气)炉
circulation passage(ways) 循环通道;冲洗液通道
circulation pattern 环流型;环流式;循环型;流通模式
circulation pattern of the atmosphere 大气环流模式
circulation period 循环期
circulation pipe 环流管;循环管
circulation pipework 环流管线工程

circulation prediction 环流预报
circulation pressure 循环压力
circulation pressure loss 循环压降
circulation principle 环流原理
circulation process 流通过程;循环流程
circulation pump 循环泵
circulation quantity 循环量
circulation rate 循环速率;循环倍率
circulation ratio 循环系数
circulation respirator 呼吸设备
circulation return 循环回水
circulation return pipe 循环回水管
circulation road 环行车道
circulation route 室内交通线
circulation section 循环段
circulation sedimentation tank 循环沉淀池
circulation space 来往交通空间
circulation supply 循环给水
circulation supply pipe 循环给水管
circulation supply system 循环供暖系统
circulation system 环流系统;循环(水)系统
circulation system lubrication 环流润滑系统
circulation theorem 环流定理
circulation time 循环时间
circulation time determination 循环时间测定
circulation tower 来往交通塔楼;楼房过人竖井
circulation tube 循环管
circulation tubing 循环管
circulation valve 环流阀;循环阀
circulation volume 循环容量;循环流量
circulation water 循环水
circulation water inlet valve 循环水入口阀
circulation zone 循环区
circulator 回转器;回流器;环行器;环流管;循环式热水器;循环器;循环电路
circulator clarifier 水力循环澄清池
circulator maser 循环器脉塞
circulator pressure type clarifier 水力循环澄清池
circulatory 循环的
circulatory arrest 循环骤停
circulatory failure 循环衰竭
circulatory flow 循环流动
circulatory function 循环功能
circulatory function test 循环机能检查
circulatory integral 围路积分
circulatory motion 循环运动
circulatory pool 循环池
circulatory stasis 循环停滞
circulatory system polygraph 循环系统多导记录仪
circulus venosus 脉环
circum 圆周
circum-Arctic molluscan realm 环北极软体动物地理区系
circum-Atlantic brachiopod region 环大西洋腕足动物地理大区
circumaviation 环球飞行
circumcenter 外心;外接圆心
circumcenter of a triangle 三角形的外心
circumcircle 外接圆
circum-continental geosyncline 环陆地槽【地】
circumdenudation 环状侵蚀;环状剥蚀(作用);环形剥蚀;环蚀;周围削磨
circumduction 环行运动
circumerosion 环形剥蚀;环蚀
circumference 圆周(长);周缘;周线;周长
circumference force 圆周力
circumference ga(u)ge 圆周规
circumference in equal parts 等分圆周
circumference of cannon bone 管围
circumference pressure 圆周压力
circumference ratio 圆周率
circumference slide calipers 圆周滑动卡尺
circumference stress 环向应力;圆周应力
circumference velocity 圆周速度
circumferenter 矿山罗盘仪
circumferential 环向的;圆周的;周缘的
circumferential angle 圆周角
circumferential butt 环状盖板
circumferential butt joint ring 周边对接圈
circumferential clearance 周刃隙角;刀刃后角
circumferential compressive force 切向压应力
circumferential corrugation 圆周波纹
circumferential distortion 周向畸变
circumferential drainage system 周边下水系统;围绕下水系统;环形排水系统
circumferential driven thickener 周边驱动型的浓缩池
circumferential efficiency 周边效率
circumferential expressway 快行环路
circumferential fault 环周断层;环形断层;周边断层
circumferential flow 环流;圆周流;切线水流
circumferential force 环向力;圆周力;切向力;切线力
circumferential freeway 环形高速公路
circumferential highway 环形道路;环行公路
circumferential joint 环形接头;环向连接;环缝;圆周接缝;周向连接
circumferential lamellae 侧蜡板
circumferential line load 圆周切线荷载
circumferential load 切向载荷;切向负荷
circumferential make up 周向拧紧力
circumferential mode 环向振型
circumferential notch 圆周槽口;圆形缺口
circumferential pitch 圆周节距;周节
circumferential pressure 环向压力;圆周压力;周边压力;切线压力
circumferential prestressing 环向预(加)应力
circumferential register screw 圆周定位螺丝
circumferential reinforcement 周向钢筋;环向钢筋;横向钢筋
circumferential rib tread 环周肋纹
circumferential ring gear 外圈大齿轮(搅拌机)
circumferential road 环行路
circumferential route 环形路线
circumferential scaffolding 周围脚手架;环形悬臂脚手架
circumferential seam 环形焊缝;圆周接缝
circumferential speed 圆周速度;周缘速度
circumferential strain 圆周应变
circumferential street 环形街道
circumferential stress 切向应力;环向应力;圆周应力;周向应力;箍应力
circumferential taxiway 环形滑行道
circumferential tendon 周向钢筋束;环向张拉
circumferential tensile stress 圆周拉应力;周边拉应力
circumferential velocity 圆周速度
circumferential weld 环焊缝(整周焊缝)
circumferential welding 圆周焊接
circumferential winding 环向缠绕;周向缠绕
circumferential wire 周围线
circumferentor 圆周罗盘;地质罗盘;测周器
circumflex branch 支
circumflexion 曲率;弯曲度;弯成圆形
circumfluence 环流;周流;绕流
circumfluent 环流的;绕流的
circumfluential 绕流的
circumfluential motion 绕流
circumfluous 绕流的
circumgyrate 陀螺运动;旋转
circumgyration 回转
circumhorizontal arc 日承
circumjacent 环绕的;周围的
circumlittoral 海滨的;沿海岸的
circummeridian altitude 拱子午线高度
circumnavigate 环球飞行
circumnavigation 环球航行
circumnutation 回旋转头运动
circumoceanic 环洋的
circumocean(ic) andesite 环洋安山岩
circumocean(ic) basalt 环洋玄武岩
circum-Pacific 环太平洋的
circum-Pacific belt 环太平洋带
circum-Pacific geosyncline system 环太平洋地槽系
circum-Pacific island arc 环太平洋岛弧
circum-Pacific orogenic zone 环太平洋造山带
circum-Pacific province 环太平洋区;太平洋套
circum-Pacific seismic belt 环太平洋地震带
circum-Pacific seismic zone 环太平洋地震带
circum-Pacific seismotectonic zone 环太平洋地震构造带
circum-Pacific tectonic active belt 环太平洋构造活动带
circum-Pacific tectonic region 环太平洋构造域
circum-Pacific tectonic zone 环太平洋构造内带
circum-Pacific trough 环太平洋海槽
circum-Pacific volcanic belt 环太平洋火山带
circum-Pacific zone regression 环太平洋带海退
circumplanetary space 环行星空间
circumpolar 环极(的)
circumpolar constellation 拱极星座
circumpolar current 环极流;绕极海流
circumpolar distribution 环极分布
circumpolar map 环极地图
circumpolar region 拱极区
circumpolar star 拱极星;绕极星
circumpolar vortex 环极涡旋;绕极漩涡
circumpolar water 环极水域;绕极水
circumpolar westerlies 环极西风(带)
circumpolar whirl 极地涡旋
circumpolar zone 拱极区
circumradius 外接圆半径
circum-sail 环球航行
circumscissile 周裂的
circumscribe 划边界线
circumscribed about circle 外切四边形
circumscribed circle 外切圆;外接圆
circumscribed cone 外切圆锥
circumscribed cylinder 外切圆柱
circumscribed figure 外切形;外切图形
circumscribed halo 外切晕
circumscribed massif granite 外围花岗岩体
circumscribed polygon 外切多边形
circumscribed polyhedron 外切多面体
circumscribed prism 外切棱柱
circumscribed pyramid 外切棱锥体
circumscribed sphere 外切球(面)
circumscribed triangle 外切三角形
circumscribing 限制;划界
circumscription 花边;界限;外接;界线;范围
circumstance 环境(情况);情况;事件
circumstance of mining 开采情况
circumstances of survey region 工作区概况
circumstantial evidence 间接证据;参考证据;情况证据;旁证
circumstantial report 详细报告;情况报告
circumterrestrial 绕地球的;环球的
circumvallate 用墙围起
circumvallation 壁垒;城堡;防御墙
circumvolution 卷缠(饰);同轴旋转
circumzenithal 拱天顶的
circumzenithal arc 幻日弧光;环天顶弧;日增
circus 马戏场;环形广场;圆形竞技场;圆形广场;杂技场
circus'big top 马戏场主帐篷
circusearth orbit 地球卫星的环行轨道
circus loading 现场装载
circus movement 环行运动;转圈运动
circus movement hypothesis 环行运动学说
cire-perdue method 失蜡铸造法
cire-perdue process 失蜡铸造法
cireumferenter 圆周罗盘
cirouablli 绿心硬木(产于圭亚那)
cirque 环形物;圆形(山)谷;圆形场;圆环;冰坑;冰斗;盆地谷
cirque cutting 冰斗削蚀;冰斗切刻
cirque erosion 冰斗侵蚀
cirque floor 环形楼层
cirque glacier 斗冰川;冰斗冰川
cirque moraine 冰斗冰碛
cirque platform 冰斗台
cirque step 冰斗阶地
cirque terrace 冰斗阶地
cirrhosis 硬变
cirrie 冰斗
cirrocumulus 卷积云
cirrocumulus lacunosus 网状卷积云
cirrolite 黄磷铝钙石
cirrostratus 卷云层;卷积云;卷层云
cirrus 卷云;细干卷
cirrus cloud 卷云
cirrus densus 密卷云
cirrus glass (有特殊光照效果的)压延玻璃
cirrus intortus 乱卷云
cirrus radiatus 辐辏状卷云
cirrus spissatus 密卷云
cirrus uncinus 钩卷云
cirrus ventosus 风卷云
cirscal meter 大转角动圈式电表
ciruclar bead 圆角
cis 顺式

cis-crientation 顺纹作用
cis-dimethyleterahydronaphthalene 顺二甲基四氢化萘
cislunar space 地月轨道空间
cis-martian space 火地空间
cismatan 埃及决明子
cis-orientation 顺向定位;在这一方向
cis-polybutadiene 顺式聚丁二烯
cis-polybutadiene rubber 顺式聚丁二烯橡胶
cis-position 顺位
cis-rich polybutadiene rubber 高顺式聚丁二烯橡胶
cissing 回缩;轻微的涂膜疙瘩;轻微的表面涂布不均;起纹;跑漆;缩边(漆病);缩孔;轻度龟裂
cissoid 蔓叶线(一种几何曲线)
cissoidal curve 蔓叶类曲线
cistactic polymer 顺式有规聚合物
Cistercian abbey 西妥僧侣修道院
Cistercian abbey-house 西妥僧侣修道院会堂
Cistercian church 西妥僧侣修道会教堂
Cistercian Gothic(style) 西妥僧侣修道会哥特式建筑
Cistercian monastery 西妥僧侣修道院
cistern 蓄水池;小储[贮]水池;油槽车;储[贮]液杯;储[贮]水器;池;容器;人工池(塘);水柜;水池
cisterna ambiens 环池
cisterna chiasmatis 交叉池
cisternal 池状的
cistern barometer 槽式气压计;杯状水银气压计;杯式气压计;水银槽气压计
cistern car 槽车;水罐车
cistern head 天沟水斗;水落斗
cistern hopper 水落斗
cistern manometer 液槽压力计
cistern of chiasma 交叉池
cistern valve 冲水阀
cistern wagon 液柜车
cistern with cover and pull chain 带盖及拉链水箱
cistern with cover and switch handle 带盖及开关拉手水箱
cistvaen 凯尔特人的石墓室;史前之石棺或石墓
citadel 要塞;多暗堡堡垒;城堡;避难所
citamci fiber 低绞西沙尔麻(墨西哥)
citation 条文
citation index 引证索引
citation matrix 引用矩阵
cities and towns 城镇
citify 城市化
citizen 城市居民
citizen hall 市民会堂
citizen organization 市民团体
citizen participation 公民参与;市民参与
citizens' band 民用频带
citizens band channels 民用电台频道
citizens band radio 民用波段无线电通信
citizenship 国籍
citizens' radio band 民用无线电频带;民用频段
citizens radio service 民用无线电业务
citizen's rights and duties for environment 公民的环境权力和义务
citizen's square 公共集会广场;市民广场
citraconic(al) acid 甲基顺丁烯二酸;柠康酸
citraconic(al) anhydride 甲基顺丁烯二酸酐;柠康酸酐
citrate 柠檬酸盐
citrate acid 柠檬酸
citrated copper salt 柠檬酸铜盐
citrate plasticizer 柠檬酸酯增塑剂
citrate soluble phosphate 柠檬酸盐溶性磷
citrate species 柠檬酸盐物种
citrate utilization test 柠檬酸盐利用试验
citraurin 橙色素
citreous 枸橼色
citric acid 柠檬酸
citric acid cycle 柠檬酸循环
citric acid ester 柠檬酸酯
citric acid gypsum 柠檬酸石膏
citric acid industrial wastewater 柠檬酸工业废水
citric acid wastewater 柠檬酸废水
citric acid wastewater sludge 柠檬酸废水污泥
citriculture 柑橘业
citrine 柠檬色;柠檬黄(色);黄水晶(色);黄晶;茶晶
citrite 茶色水晶
citron 柠檬色;柠檬黄(色);黄晶;香橼香橼色(的);香橼树;香橼色;香橼(黄绿色);枸橼

citronella 枫茅
citron green 枸橼绿色
citron grey 枸橼灰色
citron yellow 柠檬铬黄;锌黄颜料;枸橼黄色
citrus fruit 柑橘果
citrus grove 柑橘林
Citrus limon 柠檬
city administration 城市管理
city aesthetics 市容
city agglomeration 城镇群;城市群
city aggregate 城市群
city air blanket 城市热气空气层
city and region systems 城市和区域系统
city and town land use tax 城镇土地使用税
city appearance 城市景观;城市风景;市容
city archetype 城市原型
city architecture 城市建筑
city area 市区
city art 都市艺术
city as an economic center 经济中心城市
city beautiful movement 城市美化运动
city blight 都市衰落
city bond 市债
city-bound traffic 入市交通;辐射形交通
city building 城市房屋
city cable 城市电缆;市内电缆;市话电缆
city center 市中心
city church 教区教堂;(伦敦城)
city classification 城市分类
city climate 城市气候
city club 城市俱乐部
city collection 同城托收
city construction 城市建设
city contamination 城市污染
city council 市议会;市政厅
city culture 城市文化
city development 城市开发;城市发展
city development pattern 城市发展模式
city development planning 城市发展规划
city distribution 城市配电
city district 市区
city district planning 城市分区规划
city drainage 城市排水
city dweller 城市居民
city enterprise 城市企业
city estate 市有地(美国)
city extension zone 城市发展区
city farm 城市农庄
city flower 市花
city fog 城市烟雾
city forestry 都市林业
city for sight-seeing 游览城市;观光城市
city fuel economy 城市(车辆)燃料经济(试验)
city function division 城市功能分区
city gas 城市煤气
city gas dispatch 城市煤气调度
city gas gate station 城市煤气门站
city gas supply 城市煤气供应
city gate 城市入口;城门
city gate station 煤气降压站;煤气城边站;煤气城边点;城市门站
city gate tower 城楼
city general planning 城市总体规划
city green activity agreement 绿化协定
city green-area 城市绿化
city greening-area quota 城市绿地定额
city group 城市组园
city hall 市政厅
city hall complex 市政厅建筑群
city health 城市卫生
city health administration 城市卫生行政管理
city heat supply engineering 城市供热工程
city hinterland 城市腹地
city improvement 城市改建
city inset map 城市插图
city landscape 城市景观
city landscaping 城市景观
city land use planning 城市土地利用规划
city layout 城市规划;城市布局
city lighting 城市照明
city limits 市区范围;市区界
city line 市界
city loan company 城市放款公司
city-lot drilling 城区钻探;城区钻进

city man 金融家
city management 城市管理
city manager 市行政官
city map 城市图
city mass transit planning 城市公共交通规划
city metro 城市地铁
city moat 城市濠;护城河
city network 城市网络;城市网架;城区电力系统;城区电(力)网
city noise 城市噪声
city of standard urban structure 标准城市结构
city opening 开放城市
city operation 城市管理
city-oriented transport system 城市定向运输系统
city-owned undertakings 市办公共事业
city-owned utilities 城市公用事业;市办公共设施
city panorama 城市全景(图)
city park 城市公园
city parking lot 市内停车场
city pattern 市形式;城市格局;城市布局;城楼格局
city plan 城市(平面)图;城市规划图
city planner 城市规划师
city planning 都市规划;城市规划
city planning administration 城市规划事业管理
City Planning and Administration Bureau 城市规划管理局
city planning area 城市规划区(域);城市规划面积
city planning commission 城市规划委员会
city planning law 城市规划条例;城市规划法(规)
city planning map 城市规划图
city planning office 规划机构
city planting system 城市绿化系统
city pollution 城市污染
city population 城市人口
city proper 城(市)区;市内
city railway 城市铁路;市区铁路
city refuse 城市垃圾
city rehabilitation 城市改建
city ring road 绕城公路
city road 城市道路
city road network 城市道路系统;城市道路网
city road system 城市道路系统;城市道路网
city sanitation 城市(环境)卫生
city sanitation measures 城市环境卫生措施
city satellite 城市卫星
city scape 城市风貌;城楼景象;城市景观;城市风景
city scenery 城市景观
city's environment 城市环境
city's environmental decay 城市环境恶化
city's environmental pollutant 城市环境污染物
city's environmental pollution 城市环境污染
city sewer 城市下水道
city sight 城市景观
city siting 城市选址
city size 城市规模
city skyline 城市天际线
city square 城市广场
city state 城市国家
city street 市街;城市街道;市区道路;市内街道
city style and features 城市风貌
city subway 城市地铁
city survey 城市测量
city system 城市系统;城市道路网
city terminal 城市港口
city terminal service 市内运送
city thoroughfare 城市干道
city thoughfare 城市干道
city tower 城楼
city traffic 城市交通
city traffic survey 城市交通观测;城市交通调查
city transit 城市公共交通系统
city transmission network 城市输电网
city transport(ation) 城市运输
city transportation planning 城市交通规划
city traverse survey 城市导线测量
city triangulation 城市三角测量
city wall 城墙;城廓
city wall brick 城(墙)砖
city waste 城市污水
city water 自来水;城市给水
city water distribution net(work) 城市配水网
city water supply 城市自来水供应;城市供水;城市给水;自来水
city water tank 自来水箱

city-wide settlement of accounts 同城结算
citywood (产于圣多明各最佳质量的) 桃花心木
city zoning 城市分区
city zoning ordinance 城市土地区划管制条例
civery 祭台上装饰性天篷; 穹顶分隔间 (哥特式建筑穹隆顶棚的开间)
civette green 灵猫绿色
civic architecture 城市建筑
civic area 城市用地
civic art 城市 (建筑) 艺术
civic auditorium 城市大会堂; 礼堂
civic axis 城市轴线
civic basilica 世俗巴利卡; 长方形大会堂 (古罗马)
civic building 城市房屋; 市政建筑 (物)
civic center 公共中心; 市 (政) 中心
civic central area 城市中心区
civic centre planning 市中心区规划
civic crown 槲叶饰; 花帽箍; 叶环饰
civic design 建筑群体设计; 城市设计; 城市建筑艺术
civic Gothic(style) 世俗哥特式建筑
civic landscape 城市景观; 城市风景
civics 市政学
civic survey 城市调查
civic survey map 城市调查图
civil 民用的; 民间的
civil action 民事诉讼
civil action jurisprudence 民事诉讼法学
civil adjudication 民事审判庭
civil administration 民政机关; 民政 (管理); 市政
civil administration department 民政部门
civil aerodrome 民用机场
Civil Aeronautic Administration Methods 美国民航管理局法
civil aeronautic(al) 民用航空的
Civil Aeronautics Board 民用航空局 (美国)
civil aeroplane 民航机
civil affair public health 地方公共卫生
civil affairs 民政事务; 民政
civil aircraft 民用飞机; 民航机
civil air defence 人防
civil air-defence planning 城市人防规划
civil air defence works 人防工程
civil air defense 人防工程
civil air ensign 民航机标记
civil airfield 民用机场
civil airfield construction 民用 (飞) 机场
civil airport 民用机场
civil air regulations 民航条例
Civil and Structure Engineer Review 土木工程与结构工程师评论 (期刊)
civil appeal 民事上诉
civil architect 民用建筑师
civil architecture 民用建筑; 土木建筑
civil authority clause 消防损失条款
civil aviation 民用航空; 民航
civil aviation lines 民用航空线
civil aviation organization 民用航空组织
civil bail 民事保释
civil boundary 行政区划界
civil building 民用建筑 (物)
civil CAD 土木工程计算机辅助设计
civil calendar 民用 (日) 历
civil case 民事诉讼; 民事案件
civil charge 民事指控
civil claim 民事权利请求
civil code 民法法典
civil commotion 内乱; 民众骚乱; 民变险; 动乱
civil commotion and strikes 暴动、内乱和罢工
civil compensation 民事赔偿
civil complaint 民事投诉
civil conspiracy 民事方面的共谋
civil construction 土 (木) 建筑; 土建工程
civil construction facility 土木建筑设施
civil contingency fund 民用储备金
civil contract 民间契约
civil contractor 土建承包商
civil cost 土建费用
civil court 民事审判庭; 民事法庭
civil court of sessions 民事法庭
civil damages 民事损害赔偿
civil date 民用日期
civil day 民用日
civil debt 民事债务
civil defence 民防; 人民防空

civil defence construction 人民防空工程
civil defence planning 城市人防规划
civil defence sign 民防标志
civil defendant 民事被告
civil defense 人民防卫; 城市人防
civil defense against air raids 人民防空; 民用防空
civil defense construction 民防构筑; 防空建筑; 人民防空工程; 人防建筑
civil defense planning 城市人防规划
civil defense shelter 人防隐蔽所; 防空洞; 人防建筑物
civil defense shelter door 民用防空洞
civil defense shelter ventilation 民用防空洞通风装置
civil defense structure 民用防空建筑; 民防建筑
civil defense structures for radiation protection 防辐射用的民防建筑物; 防辐射结构
civil design 土木工程设计; 土建设计
civil dispute 民事纠纷
civil district 文教区
civil division 民事法庭
civil economics 民航经济学
civil economics of capital construction 民用基本建设经济学
civil embargo 内部禁运
civil engineer 土木工程师
civil engineering 民用工程; 土木工程学; 土木工程; 土建工程
civil engineering and building construction 土木建筑工程
Civil Engineering and Public Works Review 土木工程与公共建筑综论 (英国期刊名)
civil engineering contractor 土木工程承包人; 土建承包商
civil engineering cost 土建费用; 土木工程费用
civil engineering department 土木工程系
civil engineering drilling 工程施工钻探
civil engineering fabric 土工织物
civil engineering firm 土木工程公司; 建筑公司
civil engineering hydraulics equipment 水工机械
civil engineering inspector 土木工程检查员
civil engineering machinery 土木工程机械
civil engineering material 土建材料; 建筑工程材料; 土木建筑材料
civil engineering project 土木工程师项目
Civil Engineering Research Association 土木工程研究协会
civil engineering site 土木工程施工场地; 建筑工地
civil engineering standard method of measurement 土木工程定额测算方法
civil engineering stress 建筑结构应力
civil engineering structure 土木工程构筑物
civil engineering work 土木工程建筑
civil enterprise 民用企业
civil evidence 民事案件证据
civil expenditures 民用支出
civil feature 土建设施
civil forfeiture 民事 (上的) 没收; 民事 (上的) 罚款
civil hospital 地方医院
civil housing 民用住房
civilian 民间的; 公职人员; 公务员; 平民; 文职人员
civilian air photography 民用航空摄影学
civilian application 民用
civilian authority 民政当局
civilian construction 民用住房; 民用建筑
civilian goods 民用物资; 民用品
civilian industry 民用工业
civilian industry product 民用产品
civilian laborers working on public project 民工建勤
civilian labor force 民用劳动力
civilian map 民用地图
civilian production 民用生产
civilian pyrotechnics 民用烟火
civil(ian) structure 民用建筑物
civilian supplies 民用补给品
civilian technology 民用技术
civilian visit 民间访问
civilian worker 民工
civil industry 民用工业
civil injury 民事损害
civilization 文化
civilized river 驯化河流
civilized stream 驯化河流

civil judgment 民事判决
civil judicial statistics 民事 (司法) 统计
civil jurisdiction 民事管辖 (权); 民事裁判权
civil juristic act 民事法律行为
civil justice 民事审判
civil law 民法
civil lawsuit 民事诉讼
civil legal relationship 民事法律关系
civil legislation 民事立法
civil liability 民事责任
civil liability insurance 民事责任 (保) 险
civil libel 民事诽谤罪
civilll-law-rule for water rights 水权的民用条例
civil mapping 城镇制图
civil mean time 民用时
civil mediation 民事调解
civil minimum 城市生活设施最低水平
civil negligence 民事过失
civil object 民事客体
civil obligation 民事债务; 法定债务; 民事义务
civil parish index 行政区划示意图
civil penalty 民事罚款; 民事处罚
civil plaintiff 民事原告
civil plan 城镇平面图
civil prisoner 民事犯
civil procedure 民事诉讼 (程序)
Civil Procedure Act 民事诉讼 (程序) 法 (例)
civil procedure law 民事诉讼法
civil proceedings 民事诉讼程序
civil property relations 民事财产关系
civil protection 民用间防护
civil relation 民事关系
civil remedy 民事责任赔偿; 民事补救 (办) 法; 民事补偿
civil responsibility 民事责任
civil responsibility insurance 民事责任保险
civil right 民事权利
civil salvage 民间救助
civil sanction 民事制裁; 民事处分
civil service 文职公务人员; 行政事务; 文职人员
civil service system 公务员制
civil ship 民用船
civil speed 安全速度
civil strife 国内冲突
civil subject 民事主体
civil suit 民事诉讼
civil sunrise and sunset 民用日出没
civil technician 土木技术员
civil time 民用时; 平时
civil town 行政区
civil township line 城镇界
civil trespass 民事侵权行为
civil trial 民事审判
civil truck 民用载重车
civil twilight 民用曙暮光; 民用晨昏蒙影
civil twilight time 民用曙光时
civil-use civilian use 民用; 民事诉讼
civil version 民用型
civil war 内战
civil ware 民窑
civil works 民用工程; 公共工程; 人工建筑物; 土木工程设施; 土木工程; 土建工程; 土建
civil works investigation 土木工程勘测
civil works office 土建室
civil works on site 现场土木工程
civil wrong behavior 民事不法行为
civil year 历年; 民用年; 回归年; 自然年; 分至年
claborate-style painting 工笔画
clac-alkalic rock series 钙碱质岩系
claciner 煅烧炉
clack box 阀箱; 舌阀; 瓣阀箱
clack for shell 阀壳
clack mill 棘轮钻机; 手搬钻机
clack seat 阀座
clack valve 瓣阀; 闪动阀; 吸入阀; 逆止阀; 单向阀
clad 镀过的; 衬里; 包覆层; 阿尔克莱德包铝
clad alloy 包覆合金
clad alumin(i) um 复合铝板
clad brazing sheet 复合铜焊片
cladded bridge 石板桥
cladded fiber 包层光纤
cladded fiber optical waveguide 包层光学纤维波导
cladded optical fibre 包层光纤
cladded plate 复合板

cladded slab dielectric(al) waveguide 包层条形介质波导
cladded strip 覆盖带条;复合钢板
cladded yarn 包芯纱
cladding 金属包层;骨架外墙;骨架充填墙;镀覆;镀层;复合钢板法;复层法;被覆;包层;喷镀;外墙挂板;外墙;外包层;涂覆;贴面
cladding by rolling 轧制包层
cladding center 包层中心
cladding concrete 混凝土包壳
cladding diameter 包层直径
cladding eccentricity 包层偏心率
cladding element 墙面覆盖材料
cladding glass 幕墙玻璃;镀金属玻璃;包层玻璃
cladding glaze 盖底釉
cladding-guided mode 包层传导模
cladding index 包层折射率
cladding loss 包层损失
cladding material 镀层;覆盖材料
cladding metal sheet 复合金属板
cladding mode 包层模
cladding mode stripper 包层模消除器
cladding of fibre 光纤包层
cladding of roof overhang 檐口吊顶
cladding panel 覆面镶板;外挂板
cladding plate 包覆板
cladding profile 覆盖轮廓;覆盖断面
cladding sheet 覆盖薄板;骨架填充板材;覆盖板;外墙镶面板
cladding sheet steel 复合钢板
cladding slab 墙面板
cladding steel 包层钢(板);包覆钢;复合钢
cladding surface diameter deviation 包层表面直径偏差
cladding technique 包层技术
cladding thickness 包覆层厚度
cladding wall 围护墙
clad fiber 涂层纤维
clad laminate 叠压板;敷箔板
cladless HIP 无包套热等静压
clad material 镀过金属的材料;覆层材料;包覆材料
clad metal 包层钢板;层压金属;复合金属;包层金属;包层金属(板)
cladogenesis 分枝发生学
clad pipe 复合管(道)
clad plate 装饰板;复合(钢)板;包装板;涂层板
clad plate with corrosion resistant liners 带有腐蚀衬里层的钢板
clad sheet 包覆板;包覆纸
clad sheet steel 包覆层薄钢板;复合钢板
clad silica fiber 包层石英光纤
clad steel 包层钢(板);多层钢;复层钢;双金属钢;双层钢;复合钢
clad surface 覆层表面
clad tube sheet 复合管板
clad vessel 衬里容器;覆层容器;复合容器
clagging 涂料粘附
claim 开采权;要求权;债权;采矿权申请书;索赔金额;损害赔偿要求
claim against carrier 向承运人索赔
claim against damages 要求索赔损失
claim against insurance company 向保险公司索赔
claim against underwriter 向保险人索赔
claim amount 赔偿金额
claimant 原告;申请人;提出要求人;索赔(债权)人
claim assessor 估损人
claim attention 值得注意
claim barred by reason of limitation 受时效限制丧失索赔权
claim based on physical loss or damage 对实质性破坏的索赔
claim based on tort 对侵权行为的索赔
claim board 索赔委员会
claim clause 索赔条款
claim compensation 要求赔偿
claim compensation for losses 要求赔偿损失
claim compensation from a third party 向第三方索赔
claim documents 索赔单据
claimed accuracy 规定的精确度;要求精(确)度;要求的精确度
claimee 被索赔人;索赔债务人
claimer 提出要求人;索赔人
claim expenses 索赔费用

claim for additional security 要求增加担保
claim for compensation 要求赔偿;索赔
claim for compensation of damages 赔偿要求
claim for damages 赔偿损害要求;要求赔偿损失;赔偿损失的要求;索赔损失;索赔
claim for extension of time 延长工期索赔
claim for interest 索取利息
claim for loss and expenses 开支亏损索赔
claim for payment 要求付款
claim for proceeds 应得价款的请求;要求取得应收价款;索取应得价款
claim for reduction 要求减价
claim for refund 要求退款
claim for reimbursement 要求偿还
claim for restitution 要求归还原物
claim for salvage 要求支付救助费;要求付给救助费
claim for tax refund 申请退税
claim for trade dispute 商务纠纷索赔
claim in a patent application 请求保护专利范围
claim indemnity 索赔;要求赔偿
claiming bank 求偿银行
claim letter 要求赔偿书;索赔书;索赔函
claim note 保险索赔清单
claim notice 索赔通知(书)
claim of damages 货物索赔;要求赔偿损失
claim of deferring time limit of a project 工期索赔
claim of right 权利要求
claim payment 要求付款
claim reimbursement 索汇;要求付款
claim rejected 拒赔
claim report 索赔报告
claim reserve 索赔准备金
claim right insurance 赔偿权保险
claims board 索赔委员会
claims commission 索赔(审理)委员会
claims documents 索赔证件;索赔文件
claims documents of pollution damage 污染损害索赔(报告)书
claim section 索赔科
claim secured by mortgage 有抵押权担保的债权
claims estimated 赔偿预算;赔偿估算
claim settling agent 理赔代理人;赔偿代理人
claim settling clerk 理赔员
claim settling fee 理赔代理费
claims excepted from limitation 非限制性债权
claims expenses 理赔费(用)
claims-made policy 索赔保险单
claims paid 赔偿付讫
claims payable abroad 国外可付赔偿
claims priority 已申请(专利)
claims set-off against each other 相互抵消的债权
claims settlement 理赔;清理赔偿
claims settling fee 理赔代理费
claims subject to limitation 限制性债权
claims surveying agent 理赔检验代理人
claim statement 索赔清单;损失清单
claim tax relief 减免税款的提出
claim to order 记名债权;指示式债权
Clairaut's differential equation 克莱劳微分方程
Clairaut's equation 克莱劳方程
Clairaut's formula 克莱劳公式
clairecolle 细白垩胶;粘贴金箔用的透明涂料;明胶底漆;打底明胶;打底胶泥;白铝胶;白垩胶
clairite 硫砷铜矿
Claisen distillation flask 克莱森蒸馏瓶
Claisen flask 克莱森烧瓶
clam 蛤蜊
clam bank sandstone 蚌壳滩砂岩
clamber 攀登
clam bucket 抓斗
clam dredge 采蛤器
Clamer's alloy 克莱默合金
clam gun 锹;铲
clammer 夹子;夹钳
clamming 封窑门;封闭墙(窑炉)
clamp 卡箍;接线端子;夹子;夹住;夹线板;夹具;夹紧;夹管;夹钩;夹头;合模;箱夹;线箍;单面滑车;副承梁材;钳位;卡钉;土窑厂垛;夹持器;箱位
clamp bar 版夹
clamp bias 固定偏压;钳位偏左;钳位偏压
clamp bias potential 钳位偏置电平
clamp bit 装夹式车刀
clamp bolt 夹紧螺栓;紧固螺栓;箍紧螺栓
clamp bolt hole 夹紧螺栓孔

clamp bracket 夹紧托架
clamp bucket 夹斗;蛤斗
clamp-burned brick 土窑砖
clamp-burnt brick 土窑砖
clamp changer 麻电变换器
clamp cheek 夹颊板
clamp clip(terminal) 夹子;线夹;接线柱
clamp coupling 夹紧联轴器;壳形联轴器;纵向夹紧联轴器;套筒连轴节
clamp cover 堆藏覆土机
clamp covering machine 盖堆机
clamp current 钳位电流
clamp cutting edge 夹持刃口;夹持车刀;固定刃口
clamp device 紧固装置;钳位装置;夹紧装置
clamp dog 夹块;制块
clamp drill 夹钻
clamped amplifier 钳位放大器
clamped beam 梁端固定梁;固支梁;夹紧梁
clamped edge 夹紧边(缘);夹固边;固定端;固定边(缘)
clamped edge plate 固端板;周边夹紧平板;固边板
clamped edges plate 周边夹紧的平板
clamped impedance 夹紧阻抗;钳位阻抗
clamped insert-fitting joint 夹固嵌入管件接头
clamped joint seal 夹持型接头密封
clamped plate 边缘固定板
clamped terminal 夹子接线端
clamped tool 夹固刀具
clamper 卡箍;箱位器;钳位器;钳位电路;接线板;夹具;工具锚具;夹持器
clamper amplifier 信号电平固定放大器
clamp error 夹紧误差
clamper tube 箝位管;钳压管;钳位管
clamp excavator 抓斗式挖掘机
clamp for anchor rope 锚接绳固定线夹
clamp force 夹紧力;合模力
clamp for headspan cross messenger wire 横撑力索线夹
clamp forming machine 集堆机;垛堆机
clamp for upper cover 上盖夹紧装置
clamp frame 夹钳
clamp-free beam 悬臂梁
clamp furnace 围窑
clamp handle 紧固手把
clamp hanger 夹紧提升器;夹钳吊架
clamp holder 夹持器;夹柄;持夹器
clamp hook 双抱钩
clamp hose connection 用夹子连接软管
clamping 电平钳位;电平固定;箱位
clamping action 箱位作用;钳位作用
clamping apparatus 卡具;夹具;夹固件;工夹具
clamping arrangement 夹紧装置
clamping bar 夹杆
clamping belt 夹袋输送胶带
clamping block 可调节压块
clamping bolt 夹紧螺栓;扣压螺栓;固紧螺栓
clamping cap 模具压板;压紧盖
clamping chuck 卡盘;夹头;夹盘
clamping circuit 钳位电路
clamping claw 夹钳
clamping collar 夹圈
clamping coupling 夹壳联轴器
clamping cylinder 夹紧缸
clamping device 夹紧装置;固定装置;夹持机构;模具装夹机构;制动装置
clamping die 夹紧模
clamping disc 夹紧盘;离合器盘
clamping error 制动误差
clamping flange 固定用凸缘
clamping force 扣(件扣)压力;夹紧力
clamping frame 夹架
clamping handle 制动柄
clamping hoop 卡箍
clamping interval 钳位时间
clamping jaws 夹爪;夹(紧)钳(子)
clamping length (夹具的)夹紧长度
clamping level 钳位电平
clamping lever 夹紧杆;紧固柄;夹紧手柄
clamping means 夹紧方法
clamping mechanism (对焊机的)夹钳机构
clamping moment 固端变矩;固端弯矩
clamping nut 锁环;紧箍;夹紧螺帽;夹紧螺母
clamping nut slot 夹紧螺母槽
clamping of jaws 减速器夹钳

clamping of the grinding wheel 磨轮夹紧法
clamping on 夹紧
clamping paste 紧固用油膏;箱位油膏;钳位油膏
clamping plate 夹紧板;夹(固)板;夹骨齿夹откровень,压铁;压板;(电机的)齿压板
clamping pressure 夹持压力;合模压力
clamping pulse 箱位脉冲
clamping ring 夹(固)圈;压圈;锁紧环
clamping ring stop 止动夹环
clamping screw 制动螺旋;固定螺丝;固定螺钉;夹紧螺丝
clamping shoe 夹紧瓦;制动瓦
clamping slab 夹紧板
clamping sleeve 夹紧连接轴套
clamping slot 螺栓连接槽;夹钳槽
clamping spring 夹片弹簧
clamping stagnation 卡滞
clamping stirrup 夹紧卡箍
clamping stud 夹紧销
clamping support splenial teeth 夹固支架
clamping surface 夹紧面
clamping table 夹紧工作台
clamping time 夹固时间;加压时间;胶合时间
clamping tool 夹紧工具
clamping washer 夹紧垫圈
clamping yokes 夹紧轭杆
clamp iron 铁夹
clamp kiln 简易烧砖窑
clamp level 钳位电平
clamp lever 夹紧把手
clamp lock point 勾锁锁闭道岔
clamp lug 压耳
clampman 接管夹扶工
clamp moment 固端弯矩
clamp nail 夹钉
clamp nut 紧固螺母;夹固螺母;花螺母;紧固螺帽
clamp-off 掉砂;冲砂
clamp-on 挤入;保留呼叫;钳位
clamp-on ammeter 钳式安培计
clamp on amperemeter 钳形电流表
clamp-on pipe vise 管子台虎钳
clamp-on tool 夹紧刀具
clamp-on type ultrasonic flowmeter 夹管式超声波流量计
clamp-on vibrator 附着式外振捣器
clamp-on vice 台虎钳
clamp opening 夹钳开度
clamp output voltage 钳位输出电压
clamp pin 夹销;制动把;千分尺制动把;锁紧销
clamp pinion 夹紧小齿轮
clamp plate 压板;夹板
clamp pulse 钳位脉冲
clamp-pulse generator 控制脉冲发生器
clamp rail 固端板
clamp ring 夹紧环;压紧环;锁紧圈
clamps coupling 管夹接头
clamp screw 紧固螺栓;紧固螺钉;夹紧螺丝;夹紧螺钉;压紧螺丝;制动螺丝;制动螺钉
clamp screw sextant 夹紧螺丝型六分仪;制动螺钉六分仪
clamps for pipe hanger 吊架管卡
clamp shaft 夹紧轴
clampshell digging 抓斗挖泥
clampshell hopper 抓斗加料器
clampshell jaw 抓斗
clamp silage 堆藏青储[贮]料
clamp sleeve 套管夹
clamp splice 夹块
clamp stand 固定支座;固定支架
clamp strap 夹(紧)带
clamp stroke 合模行程
clamp sub 夹箍;夹头
clamp system 夹紧机构
clamp timber 木撑木;木夹板;木夹件
clamp time 模压时间
clamp type end fitting 对壳式软管终端接头
clamp type entry guide 入口夹板
clamp upset 弯压铁
clam rake 蛤耙
clam seal 蚌式密封
clamsheel excavator 抓斗挖土机
clamshell 蛤壳;抓斗;两瓣抓斗;蛤壳式抓斗;抓石机;抓斗;蛤壳型抓斗;蚌式挖斗;合瓣式抓斗
clamshell brake 钳夹制动器

clamshell bucket 两瓣抓斗;夹斗;合瓣抓斗;合瓣铲斗;蛤壳斗;蛤斗;抓斗;蛤壳式抓斗;蚌壳式抓斗;双颚式抓斗
clamshell(bucket) dredge(r) 抓斗式蒸汽挖泥船;蛤斗式(蒸汽)挖泥机;合瓣式挖泥船
clamshell car 自卸吊车
clamshell crane 合瓣式起重机;抓斗式起重机;抓斗式吊车
clamshell dredge(r) 抓斗式挖泥机;抓斗式疏浚船;蛤壳式挖泥船
clamshell dredging 用抓斗式挖泥船挖泥
clamshell-equipped crane 抓斗吊车;抓斗式起重机
clamshell excavator 抓铲挖掘机;合瓣式挖土机;合瓣式挖掘机;合瓣式开挖机;抓斗式挖土机;抓斗式挖掘机
clamshell furnace 蛤壳式炉
clamshell gate bucket 双瓣扇形斗门
clamshell grab 合瓣式挖土机;合瓣式挖泥机;合瓣式开挖机;蛤斗式抓斗;蛤壳式抓岩机;蛤壳式抓斗;蛤斗式抓岩机;蚌壳式抓斗;双颚式抓斗;双瓣(式)抓斗;合瓣式抓斗
clamshell grabbing crane 抓斗起重机;蛤壳式起重机
clamshell grab crane 抓斗式起重机
clamshell hydraulic backhoe 合瓣式液压反铲
clamshell loading 抓岩机装岩
clamshell mucker 抓岩机
clamshell nozzle 蚌壳式可调喷管;双活门可调喷管;双调节片可调喷管
clamshell quench 合瓣式淬冷
clamshell sampler 蚌式采样器
clamshell scoop 双瓣料斗
clamshell shovel 抓斗式挖土机;抓斗铲土机
clamshell type bucket 两瓣抓斗
clamshell type door 扇形门
clamshell type dump bucket 抓斗式料桶
clamshell type gate 扇形(闸)门
clam type loader 抓斗式装载机
clan 岩类;集群
clandestine sale 秘密交易
clandestine trade 秘密贸易
clang association 谐音联想
clanification system 净化系统
claosed water body 闭合水区
clap 轻拍;拍击
clapboard 隔板;木楔形墙板;护墙楔形板;护壁楔形板;楔形板;墙面板;桶板
clapboard ga(u)ge 护檐板规;墙面板卡规;护墙板卡规
clap collar 夹紧轴环
Clapeyron-Clausius equation 克拉珀龙方程
Clapeyron's equation 克拉珀龙方程
Clapeyron's theorem 克拉珀龙定理;三弯矩定理
Clapeyron's theorem of three moment 克拉珀龙三弯矩定理
clapotement 击水音
clapotis 立波;驻波;定波
clapped valve 单向阀
clapper 铃舌;铃锤;警钟锤;钻机取杆器舌门;止回阀(瓣);阀;取杆器舌门(钻机);拍板;抬刀装置;枢轴衔铁;舌板
clapper block 抬刀滑块
clapper box 摆动刀架;拍板座;抬刀座
clapper bridge 石桥
clapper die spotting press 合模机
clapper door 分泥门
clapper pin (牛头刨床的)摆动刀架轴销
clapper relay 铃锤式继电器;衔铁吸入式断电器;拍合式继电器
clapper switch 铃锤式开关
clapper-type armature 拍板式衔铁
clappet valve 单向阀;止回阀;瓣阀
clap(ping) sill 止水横挡;止水压条;消力槛;闸槛;人字门槛
Clapp oscillator 克林泼振荡器;电容反馈改进型振荡器
clap post 橱柜中梃
Clapton's line 克拉普顿氏线
clap valve 瓣阀
claracet red 酒红颜料
clarain 亮煤
claraite 克水碳锌铜矿
claratex 描图布
clarendon 中等线体字
Clarendonian 克拉里登组【地】

claret 紫红色;枣红(色)
claret brown 紫红色
claret red 酒红颜色;酒红色
claret wine 红葡萄酒色
Clari cone clarifier 克拉里锥式澄清池
clarificant 澄清剂
clarificate 净化;澄清
clarification 净化作用;净化;澄清作用;澄清报告;澄清;浅化
clarification basin 净化池;沉淀池
clarification chamber 沉淀室
clarification compartment 沉淀室
clarification drawing (工程修改补充的)说明图纸;说明图;增补详图
clarification equipment 澄清设备;净水构筑物
clarification of objective 客观的特征;目标的特征
clarification of sewage 污水净化(法);污水澄清
clarification of tendering documents 招标文件的澄清
clarification plant 水厂;净水厂;净化厂;澄清装置;澄清车间
clarification preparation 澄清制剂
clarification structure 净水构筑物
clarification tank 净化池;澄清池
clarification unit 澄清设备
clarification well 澄清井
clarification zone 澄清区
clarificator 澄清器
clarified effluent 净化污水
clarified liquor 澄清液
clarified oil 澄清油
clarified sewage 净化的污水;澄清的污水
clarified thickener 沉降增稠剂
clarified wastewater 净化的废水;澄清废水
clarified water 净(化)水;澄清水
clarified water channel 净水渠
clarified water pump 清水泵
clarifier 沉淀池明晰器;澄清器;澄清剂;澄清池;透明剂
clarifier basin 澄清池
clarifier hydrodynamics 沉淀池水动力学
clarifier tank 澄清池
clarifixator 离心均质机
clariflocculation 澄清絮凝
clariflocculator 絮凝澄清池;澄清絮凝器
clarify 澄清;阐明
clarifying 澄清的
clarifying agent 澄清剂
clarifying basin 澄清池
clarifying bath 澄清液
clarifying centrifuge 净化离心机
clarifying chamber 澄清室
clarifying efficiency 澄清效率
clarifying filter 滤净器;澄清过滤器
clarifying filtration 澄清过滤
clarifying tank 净化池;澄清池;沉淀池
clarifying tower 净化塔
clarifyling centrifuge 澄清离心器
clarify one's position 表态
clarinet 单簧管
claringbullite 水羟氯铜矿
clarinite 亮煤体;亮煤素质
clarite 微亮煤
clarite E 壳质微亮煤
clarite V 镜质微亮煤
clarithickener 澄清浓缩池
clarity 澄清度;清澈度
clarity of detail 碎石清晰性
Clark alignment curve 克拉克定线曲线
Clark beam 木组合梁
Clark cell 克拉克标准电池
Clark cycle 克拉克循环;二冲程循环
Clark degree 英制硬度
clarke 克拉克
Clarke beam 组合式木梁
Clarke ellipsoid of 1866 克拉克地球椭圆体(1866年)
clarkeite 水钠铀矿
Clarke of concentration 浓度克拉克值
Clarke's formula 克拉克公式
Clarke's perspective azimuthal projection 克拉克透视方位投影
Clarke spheroid of 1866 克拉克地球椭圆体(1866年)

Clarke's projection 克拉克投影
Clarke's value 克拉克值
Clark number 克拉克数
Clark process 克拉克法
clark's alloy 克拉克合金
clark's bit 支架上扩大的钻头
clark's process 硬水软化法
clarocollite 微亮煤质无结构镜煤
clarodurain 亮暗煤
clarodurite 微亮暗煤
clarofusain 亮丝炭
clarofusite 微亮质丝炭
clarotelite 微亮煤质结构镜煤
clarovitrain 亮镜煤
clarovitrinite 富氢镜质体
clarovitrite 微亮质镜煤
clash 对撞
clash and breakage risks 碰损破碎险
clash gear 滑动齿轮
clash gearbox 滑动齿轮传动箱
clash gear transmission 滑动齿轮传动(装置)
clashing of gear 齿轮的噪声
clashing risks 碰损险
clash risks 碰损险
clasmoschist 硬砂岩;片状碎屑岩;片状砂屑岩
clasolite 岩屑岩;碎屑岩墙;碎屑岩(脉);碎屑岩;碎屑石
clasp 扣紧物;扣紧接合法;扣钩;卡环;钩紧;弹簧钩
clasp brake 夹紧制动器
clasp braking 双侧制动
clasper 抓握器;抱(握)器
clasp handle 键柄
clasp-headed nail 钩头钉
clasp hook 弯脚钩
clasp joint 搭扣接合;钩接
clasp knife 折(叠)刀(式)
clasp lock 自动弹簧锁
clasp nail 扣钉;抓钉;扁钉;扒钉
clasp nut 开合螺母
clasp nut engaging lever 对开螺母开关杆
class 类程;类别;级别;种类;等级;分类;品类
class A 甲类;甲级
class A amplification 甲类放大
class action 集体诉讼
class A design institute 甲级设计院
class A evapo(u)ration pan A 型蒸发器
class A fire 甲类火
class A highway 甲级道路【道】
class A insulation A 级绝缘
class A investigation and design institute 甲级勘察设计院
class A investigation and design institute for harbo(u)r and waterway engineering 甲级水运勘察设计院
class A land evapo(u)ration pan A 型陆上蒸发器
class A land pan A 型陆上蒸发器
class annual survey 船级岁检
class A power amplifier 甲类功率放大器
class A ship 甲类船
class A signal area 强电视信号区
Class B amplification 乙类放大
class B inspection 乙级检验
class boat 入级(船舶)
class boundary 分组略线;组界;分组界限
class bowl 驼峰峰底
class C amplification 丙类放大
class C circuit 丙类电路
class certificate 船级证(明)书
class clause 分类子句
class code 类别符号;分类符
class condition 类别条件;分类条件
class-conscious dies 分级模具
class cost system 分类成本制度;分等成本计算法
class C power amplifier 丙类功率放大器
class D power amplifier 丁类功率放大器
classed of mainly exposed rock 主要露头岩石
classer 选粒机;分级机
classer's description 分级标志
classes for cracking control 裂缝控制等级
classes of differentiated intrusive 侵入体分异分类
classes of geologic(al) observation point 地质点类型
classes of igneous rock 火成岩大类
classes of pesticide toxicity 农药毒性等级

classes of pollution 污染等级
classes of reconnaissance survey 踏勘分类
classes of reserve 储量级别
classes of rise and fall 升降坡度分级
classes of tectonic layer 构造层类型
classes of unconformity 不整合类型
classfication of coal 煤分类法
class finding 分类归并
class group 等级组
class hardness 分级硬度
class heading 分类标题
classic 标准著作;古典的;经典的;典型的;传统的;标准的
classic(al) approach 古典方法
classic(al) approximation 经典近似
classic(al) architecture 古典(式)建筑;古建筑(物)
classic(al) art 古典艺术
classic(al) assumption 经典假设
classic(al) astronomy 经典天文学
classic(al) atlas 古代历史地图集
classic(al) Baroque 古典巴洛克式(建筑)
classic(al) biologic(al) filter 经典生物滤池
classic(al) building 古典(式)建筑
classic(al) calculation method 古典计算方法
classic(al) cascade 阶式蒸发器
classic(al) ceramics 传统陶瓷
classic(al) Chinese garden 中国古典园林
classic(al) conception 古典概念
classic(al) conception of probability 概率论的古典概念
classic(al) conditioning 经典性条件反射
classic(al) confidence interval 经典置信区间
classic(al) construction method 传统施工方法
classic(al) control 典型控制;经典控制;古典控制
classic(al) cooperative game 古典合作对策
classic(al) economic theory 经典经济理论
classic(al) electrodiagnosis apparatus 古典法电诊断仪
classic(al) electrodynamics 经典电动力学
classic(al) equation 经典方程
classic(al) error theory 经典误差理论
classic(al) fluid mechanics 经典流体力学;古典流体力学
classic(al) form 古典形式
classic(al) formalism 经典形式;经典体系
classic(al) gravitative model 经典重力学模型
classic(al) ground 古迹
classic(al) hydrodynamics 古典流体动力学
classic(al) hydromechanics 古典流体力学
classic(al) integral 经典积分
classic(al) landscape 古典风景
classic(al) mapping technique 分类制图法
classic(al) mathematical analysis 经典数学分析
classic(al) mechanics 经典力学
classic(al) melting method 古典熔融法
classic(al) method 经典方法;传统方法
classic(al) motif 古典式主题
classic(al) network theory 经典网络理论
classic(al) normal linear regression model 经典正态线性回归模型
classic(al) optics 经典光学
classic(al) optimization 经典优选法
classic(al) orders 古典柱型;古典柱式;古典式柱型
classic(al) oscillator 经典振荡器
classic(al) painting 古典画
classic(al) pathway 经典途径
classic(al) probability model 经典概度模型
classic(al) problem 经典问题
classic(al) purity 纯真的古典
classic(al) rectification 常规纠正
classic(al) regression model 经典回归模型
classic(al) restraint 经典的约束
classic(al) result 经典结果
classic(al) revival 古典复兴式
classic(al) revival architecture 古典复兴建筑
classic(al) revival form of decoration 复古式装饰
classic(al) rheobodies 经典流变体
classic(al) scattering 经典散射
classic(al) severity 经典的严格性;古典式朴素
classic(al) silk 次优级生丝
classic(al) smog 经典式烟雾
classic(al) statistical mechanics 经典统计力学
classic(al) statistical theory 经典统计理论

classic(al) statistics 经典统计学;经典统计法
classic(al) structure 古典式建筑物
classic(al) style 古典式
classic(al) temple 古典式神殿;古典式庙宇
classic(al) theory 经典理论
classic(al) thermodynamics 经典热动力学
classicism 古典主义
Class I consumer 一级用户
classifcation of investment 投资分类
classifcation screen 分极筛
classification 类别;科目;分组;分类;分级;分等;入级
classification analysis of ordered samples 有序样分类分析
classification and coding system 分类编码系统
classification and designation 分类与命名
classification and determination 分类鉴定
classification and division of dangerous goods 危险货物类项
classification and evaluation of local traps 局部圈闭分类和评价
classification and imputation of transaction 交易的分类和归属
classification and inspection of vessel 船舶定级和检验
classification and name of rocks 岩石的分类和名称
classification and regionalism of exploration intensity 勘探程度类型和分区
classification bowl 驼峰峰底
classification brief flowchart for soil 土的统一分类简要流程图
classification by construction location 按建设地区分类
classification by destination 按目的地编组【铁】;按到站编组【铁】
classification by dichotomy 二重分类;二分法分类
classification by function 按功能分类
classifcation by kind 按性质分类
classification by product 按产品分类
classification by region 按地区划分
classification by special function 按专业分类
classification by specific gravity 按比重分级
classification by the process of production 按生产过程分类
classification by the type of customers 按顾客类型分类
classification by type of construction activity 按建筑活动分类
classification category 密级
classification certificate 船级证(明)书
classification characteristic 分类特性
classification chart 分类图;分类表
classification clause 船级条款
classification committee 船级委员会
classification control 分类控制
classification copy 编图资料分类
classification count 分类计数
classifcation criterion 分类判据
classification declaration 分类说明
classification-departure yard 编发场
classification design method 分类计算法
classification effectiveness 分类效率
classification efficiency 选粉效率;分级效率;筛分效率
classification error 分类错误
classification example 分类例子
classification facility 调车设备
classification for break-up of train 解体调车
classification for winter annual 冬性一年生植物分类
classification ga(u)ge 分类规
classification grade 分类等级
classification group 工作级别
classification group of mechanism 机构工作级别
classification group of S/R machine 有轨巷道堆垛工作级别
classification item 明细科目
classification level 分类等级
classification list 分类表
classification list conveyor system 票据传送设备
classification make-up of trains 编组调车
classification manual 科目分类细则;分类说明

classification map of geochemistry work 地球化学工作分类图
classification map of geophysics work 地球物理学工作分类图
classification map of lithology 岩性分类图
classification map of morphology 地貌分类图
classification map showing grades of prospective mineralization area 成矿远景区级别分类图
classification map showing prospective mineralization area 远景矿产分类图
classification mark 分类号印
classification of …… 的分类
classification of abrasive 磨料分级
classification of accounts 会计科目表；分类会计科目表；账户分类
classification of acute toxicity 急性毒性分级
classification of asbestos fiber 石棉纤维分级
classification of balance calculation area 均衡计算区等级
classification of bridge and culvert according to span length 桥涵按跨径分类
classification of bridges 桥梁分类
classification of buildings 房屋分类
classification of bursting water 突水分类
classification of carbonatites 碳酸岩的分类图
classification of cartographic(al) projection 地图投影分类
classification of chemical composition of volcanic rock 火山岩化学成分分类图
classification of climate 气候分类
classification of cluster analysis 聚类分析分组
classification of coal mining roof 煤矿顶板分级
classification of coasts 岸线分类
classification of component in sedimentary rock 沉积岩组分类型
classification of consumption expenditures 消费支出分类
classification of control point 控制点类别
classification of cost 成本分类；费用分类
classification of crystal symmetry 晶体对称分类
classification of defects 故障类别；缺陷分类
classification of dredging soil 疏浚土分类
classification of earthquakes 地震类别；地震分类
classification of economic grouping 经济类型的划分
classification of electric(al) source 电源类别
classification of energy source 能源类别
classification of environment(al) hazard 环境危害类别
classification of field book 手簿类别
classification of fine sand 细砂分类
classification of foundation soil at seismic region 地震区地基土类型
classification of freight 货物分类
classification of freight rate 货物运价等级
classification of frost heaving of foundation soil 地基土冻胀性分级
classification of frozen soil 冻土分类
classification of fuzzy functions 模糊函数(的)分类
classification of gabbroic intrusive rocks 辉长质侵入岩分类图
classification of gassy mines 瓦斯煤矿分级(中国)
classification of geochemical anomaly 地球化学异常类型
classification of geomorphic(al) age 地貌年龄分类
classification of goods 商品分类
classification of gravity anomalies 重力异常分类
classification of gravity corrections 重力改正分类
classification of gravity survey 重力测量分类
classification of groundwater chemical composition 地下水化学成分分类
classification of groundwater quality 地下水水质分类
classification of heaving property of frozen soil 冻土的冻胀性分类
classification of highway 公路分类
classification of highways and roads 道路分级
classification of hydrogeochemical environment 水文地球化学环境分类
classification of igneous rocks in the double triangle Q-A-P-F 火成岩在双三角形 Q-A-P-F 中的分类图

classification of ijolites 霓霞岩的分类图
classification of industrial wastewater 工业废水分类
classification of industry 工业分类
classification of inland aids to navigation 内河航标等级
classification of items 项目分类
classification of kimberlites 金伯利岩的分类图
classification of lamprophyres in the Q-A-P-F 煌斑岩在 Q-A-P-F 中的分类图
classification of land 土地分类
classification of land use 用地分类
classification of layer sequence 层序划分
classification of line 线路分类
classification of local structural 局部构造类型
classification of loess by geologic(al) ear 黄土按地质时代分类
classification of loess in highway engineering 公路土分类中黄土分类；公路工程中黄土的分类
classification of luggage and parcel accident 行包事故种类
classification of magnitude of earthquake 震级分类
classification of map projection 地图投影分类
classification of market 市场分类
classification of melilitites 黄长石岩分类图
classification of melt-settlement of frozen soil 冻土沉融性分级
classification of meteorites 陨石矿物
classification of mine planning 建设类别
classification of mineral commodities 矿产分类
classification of mineral deposit-hydrogeology 矿床水文地质分类
classification of mineralogy 矿物学分类
classification of mineral resources 矿产资源分类
classification of mountain height 山地测高分类
classification of natural gas according to its composition 天然气按成分分类
classification of oil and gas field 油气田分类
classification of oil field water according its salinity 油田水按盐度分类
classification of ore reserve 储量级别
classification of out-of ga(u)ge freight 超限货物等级
classification of petroleum 石油分类
classification of petrology 岩石学分类
classification of photogrammetry 摄影测量分类
classification of plate tectonic units 板块构造单元分类
classification of pollutants 污染物分类
classification of pool 油气藏分类
classification of prestressed concrete 预应力混凝土的分类
classification of pumping house 泵房分类
classification of pyroclastic rocks 火山碎屑岩的分类图
classification of quantitative system 定性系统分类
classification of rail(way) 铁路等级
classification of railway lines 铁路线等级
classification of refractory raw material 耐火原料分级
classification of remote sensing 遥感分类
classification of reservoir 水库分类
classification of result list 成果表类别
classification of risk 危险分极
classification of river 河流分类；河道分类
classification of road 道路分类
classification of rock abrasiveness 岩石研磨性分级
classification of rock brittle-plasticity 岩石脆塑性分级
classification of rock hardness 岩石硬度分级
classification of rock mass 岩体分类；岩体分级
classification of rock stability 岩石稳定性分级
classification of rock strength 岩石坚固性分级
classification of sales 销售额分类
classification of saline soil 盐渍土分类
classification of salty soil 盐渍土分类
classification of salty soil in highway engineering 公路土分类中盐渍土分类；公路工程中盐渍土分类
classification of sedimentation 沉降分级
classification of seismic fault 地震断裂的分类
classification of seismic wave 地震波类型
classification of service 用户分类

classification of shale 页岩分类
classification of ship 船舶入级；船舶分类；船舶登记
classification of singular surfaces 奇异曲面的分类
classification of site 场地分类
classification of site soil 场地土分类
classification of soils 土壤分类(法)
classification of solid mineral reserves 固体矿产储量分类
classification of statistic(al) prediction of deposits 矿床统计预测分类
classification of stratigraphy 地层学分类
classification of stream 河流分类；河道分类
classification of structure 构造类别
classification of surrounding rock 围岩分类(法)
classification of surrounding rock of tunnel 洞室围岩分类
classification of swell 涌浪级别
classification of swelling property 膨胀性分类
classification of tariff 税则分类
classification of the crustal waves 地壳波浪分类
classification of the qualitative system 定性分类法
classification of the quantitative system 定量分类法
classification of tracks 轨道类型
classification of traffic 交通分类
classification of treatment 处理(法)分类
classification of triangulation 三角测量等级
classification of ultra-fine particles by air 超细颗粒空气筛分
classification of ultra-mafic rocks 超镁铁岩分类图
classification of ultramafic rocks with hornblende 含普通角闪石的超镁铁岩分类图
classification of urban road 城市道路分类
classification of vessel 船级；船舶入级；船舶分类；船舶登记
classification of volcanoes 火山分类
classification of wastes 废物分类
classification of wastewater 废水分类；污水分类
classification of wastewater treatment process 废水处理方法分类
classification of water corrosivity 环境水侵蚀性分类
classification of water resources 水资源分类
classification of waterway 航道分级
classification of waterway maintenance 航道维护分类
classification of well completion 完井类型
classification of well testing 测试分类
classification of zones of groundwater quality suitable 可适用地下水水质区分类
classification on the age of neotectonism 新构造作用的时间分类
classification on the influence 影响地域分类
classification on the movement character 运动特点分类
classification on the movement direction 运动方向分类
classification on the movement intensity 运动强度分类
classification on the movement mode 运动方式分类
classification on the movement result 运动结果分类
classification on the movement velocity 运动速度分类
classification on the occurring location 发生位置分类
classification operation 解体作业【铁】
classification problem 分类问题
classification process 分类过程
classification property 分类特性
classification rating 运费等级；分等运价
classification repair 入级修理
classification scheme 分类表图
classification screen 分级筛
classification sedimentation 分级沉降
classification service 分类服务
classification society 验船协会；船级社
classification society rule 船级社规定
classification society's surveyor 船级社验船师
classifications of tide 潮汐分类
classification sonar 目标识别声呐；目标鉴别声呐；分类声呐
classification stage 分类阶段

classification statement 分类表
classification station operation income 编组站作业收入
classification statistic 分类统计(量)
classification survey 野外调绘;船级检验;船舶入级检查;入级检验
classification surveyor 验船师;船级社验船师
classification symbol 分类号码;分类符
classification system 分类制度;分类系统;分类法
classification table 分类表
classification test 分类试验;分级试验
classification theorem 分类定理
classification track 调车线【铁】;编组线【铁】
classification used in China 国内分类
classification used in USA 美国分类
classification yard 调车场;编组站【铁】;编组车场;编组场【铁】
classificator 粒选机;精选机;分级器
classificatory grade of mineral deposit-hydrogeology 矿床水文地质分类等级
classificatory plan 分类方案
classificatory scale 分类尺度
classified advertisement 分类广告
classified application 分类应用
classified balance sheet 分类资产负债表
classified bonds 分类债券
classified catalog 分类目录
classified channel 等级航道
classified cost system 分类成本制度
classified counting 分组计数
classified data 分组资料
classified depreciation 分类折旧
classified depreciation rate of fixed assets 固定资产分类折旧率
classified documents 保密文件
classified facility 分类设施
classified feed 分级入选;分级给料
classified financial statement 分类财务报表
classified highway 等级公路
classified image 分类图像
classified income tax 分类所得税
classified index 分类索引
classified information 分类报告;分类信息
classified loan 秘密贷款;低于标准的贷款
classified or scheduled tax 分类所得税
classified population 职业人口
classified road 列入公路等级的道路;分级的道路;等级公路
classified road network 分级公路网
classified rubble fill 毛石选块填筑;用分级碎石填方
classified searching files 分类检索文档
classified security 分类证券
classified shelf arrangement 分类排列法
classified statistics 分类统计
classified stock 分类股票
classified tax 分类税
classified telephone directory 分类电话号码簿
classified trial balance 分类试算平衡表;分类试算表
classified variable 分类变量
classified water course 等级航路
classified waterway 等级航道
classifier 料粒分选器;粒选机;选分机;澄清器;分选器;分选工;分粒器;分选器;分类机;分类符;分级器;分级机
classifier air separator 选粉机
classifier cyclone 分级旋流器
classifier for fine separation 细分分级器
classifier gate 分选器挡板
classifier mill 分级式粉碎机
classifier overflow 分级机溢流
classifier pocket 分级容器;分级料箱
classifier rake 筛分耙
classifier separator 分级选矿机;分级机式分选机
classifier sieve 颗粒分级筛
classifier using screw principle 螺旋原理分级器
classify 分粒;分类;分(等)级
classify in a figurative sense 形象化分类
classifying accounting 分级核算
classifying bunker 分级料仓
classifying by equal falling 等降分级
classifying chamber 分级室
classifying cone 锥形颗粒分级器
classifying crystallizer 分级结晶器
classifying cyclone 分级旋流器

classifying drum 圆筒分级机;鼓形分级机
classifying effect 分级效应
classifying liner 分级衬板
classifying lining plate 分级衬板
classifying rules 分类规则
classifying screen 选分筛;分类筛;分级筛
classifying the items of work 分类工作项目
classifying washer 分级选矿机
classify in mechanics 机械学分类
classify of geothermal field 地热田分类
classify of geyser 间歇泉分类
classify of mine reserves 矿产储量分类
class index number 分类指数
class initial mark 船级缩写记号
class interval 组距;分组间隔;分类区间;分类间隔;标度分类间隔
class interval series 组据数列
class limit 组限;组界
class lock 类别锁
class mark 类中值;类代表值
class median 组中值
class mid value 类中值;组中值
class name 类别名
class notation 船级符号
class number 组数;层号;分类号码
class of accuracy 精密度分级
class of assessment 评议等级
class of benchmark 水准点等级
class of buildings 建筑等级
class of business 业务种类;业务范围
class of channel 信道种类
class of commodity 商品等级
class of concrete 混凝土等级
class of construction 工程等级;建筑等级
class of earthquake of Japan 日本地震分级
class of explosive-proof 防爆等级
class of fit 精度配合的分级;配合类别
class of gas 含气级别
class of goods 货物等级
class of hardness 硬度等级
class of heaviness of planning mine 设计项目等级
class of highway 道路等级;公路等级
class of insulation 绝缘等级
class of liabilities 负债类别
class of load 荷载等级
class of loading 荷载类别;荷载等级
class of luggage and parcel accident 行包事故等级
class of map sheet 图件类别
class of mortar 灰浆等级
class of oil 含油级别
class of options 同种期权
class of port engineering structure 港口水工建筑物;港口工程建筑物等级
class of precision 精度等级
class of procedures 程序分类
class of railway 铁路等级
class of resistance to acids 抗酸等级
class of resistance to alkalis 抗碱等级
class of restriction 限制等级
class of risk 危险类别
class of service 用户类别;业务类别;服务类别;服务级别;服务等级
class of service gas 用气分类
class of service table 服务类别表;服务级别表
class of site soil 场地土类别
class of soil particles 土壤颗粒分级
class of station 车站等级
class of taxpayer 纳税人分级
class of track 线路等级
class of traffic 运输等级指慢;运输等级指快
class of train 列车等级
class of triangulation 三角测量等级
class of triangulation point 三角点等级
Class one highway 一级公路
Class one Motor Carrier 一级汽车运输公司
Class one railroad 一级铁路
classons 经典子
class price 等级价格;分类价格
class probability 类别概率
class production 分(等)级生产
class-product production 等级品生产;分级产品生产
class range 组段

class rate 分级保险费率;分等运价
class rates 等级运价率
class rating 分类等级;级别;额定等级
classroom 课堂;教室
classroom building 课室建筑
classroom information system 课堂信息系统
classroom unit 课室单元
classroom window 教室窗(子)
class settling 分级沉降
class special survey 船级特检
class stamp tax 分等印花税
class survey 入级检验
class symbol 分组符号;分类符(号)
class test 分类试验
class two transformer 第二类变压器
class unionism 分类工会制度
class value 分级价值
class vector 分类向量
class withdrawal 取消船级
class with respect to a module 关于一模的类
clast 岩石碎屑;碎屑
clastation 碎裂作用
clastic 碎屑状(的);碎屑的;碎块状
clastic anomaly 碎屑异常
clastic association 碎屑组合
clastic breccia 剥蚀角砾岩;碎屑角砾岩
clastic constituents 碎屑成分
clastic deformation 破碎变形;碎屑变形
clastic deposit 底部沉积(物);碎屑沉积(物)
clastic dike 沉积岩墙;碎屑岩墙;碎屑岩脉;碎屑比
clastic dispersion 碎屑分散
clastic eruptive rock 火成岩碎屑
clastic fabric 碎屑组构
clastic grain 碎屑颗粒
clastic in subcontract 分包合同条款
clasticity index 碎屑度指数
clastic limestone 碎屑石灰岩
clastic lime tuff 石灰质凝灰岩碎屑
clastic loosen texture 碎块散体结构
clastic marl 碎屑泥灰岩
clastic mechanics 碎块(体)力学
clastic model 碎块体模型
clastic phenocryst 碎斑(晶)
clastic ratio 碎屑比
clastic reservoir 碎屑油储
clastic rock 碎屑岩
clastic rocks 碎屑岩类
clastic rock type 碎屑岩类型
clastics 碎屑物
clastic sediment 底部沉积(物)
clastic sedimentary rock 碎屑沉积岩
clastic shoe-zone system 碎屑滨岸带体系
clastic texture 碎屑织构;碎屑结构;碎裂结构
clastic tidal facies 潮汐碎屑相
clastic tuff 凝灰岩碎屑
clastic wedge 碎屑岩楔;碎屑楔状体;碎屑楔形层
clasto-crystalline 碎屑结晶质
clastogen 断裂剂
clastomorphic deformation 碎屑侵蚀变形
clastoporphyritic 碎裂斑状(结构)的
clasustrophobia 幽闭恐惧症
clathrate 格子状;窗格形的;网状的;笼形的
clathrate compound 包合物
clathri 铁栅
Clauberg unit 克劳伯格单位
Claude process 克劳德过程;克劳德法
claudetite 白砷石
claugh 山口
Claus blue 克劳斯蓝
clause 科目;条项;条款
claused bill 附有条件的票据;附条款票据
claused bill of lading 附有条件的提单;附条款提单
claused bill of exchange 附条款汇票
clause for cargoes shipped on deck 甲板货物条款
clause letter of credit 附条款信用证
clause limiting insurer liability 限定保险人责任的条款
clause rider 附加条款;补充条款
clauses 附加税
clause stamped 贴了印花的正式文件
Clausius-Clapeyron equation 克劳斯修斯—克拉珀龙方程式
Clausius equation 克劳修斯方程
Clausius law 克劳修斯定律

Clausius-Mosotti equation 克劳修斯—莫索蒂方程
Clausius number 克劳修斯数
Clausius range 克劳修斯范围
Clausius statement 克劳修斯说法
Clausius theorem 克劳修斯定理;克劳修斯不等式
Clausius unit 克劳修斯单位
Clausius virial theorem 克劳修斯维里定理
Claus kiln 克劳斯窑
clauster 禁室;堰;峡谷
clausthalite 硒铅矿
claustra 漏空石墙
claustral 附属于回廊的
claustrum 带回廊的修道院
clausum fregit 不动产占有侵犯
clava 棒状体;棒节
claval 棒状体的
claval suture 爪片的
clavate 棒状体的;棒状的
clavated antenna 棒状触角
Clavaxinellida 棒轴目
clavel 楔石;拱顶石
clavicle 锁骨
clavicorn 锤角
clavier 键盘
claviform 棒状的
clavis 拱顶石;楔石
claw 爪(钩)钳;爪状钩;卡爪;起锭器;起钉器
claw back 加税弥补
claw bar 木工夹具;爪杆;道渣耙;拔道撬棍;撬棍
claw beam 钳杆
claw belt fastener 皮带连接锁扣
claw bolt 爪栓
claw chisel 爪凿;齿刃凿
claw chuck 爪式卡盘
claw clutch 爪形离合器;爪式离合器
claw coupling 牙嵌离合器;爪形联轴器;爪形联轴节;爪形联轴器;爪形离合器;爪式联轴器;爪式联轴节;爪式联轴器;爪式离合器
claw crane 爪式起重机;钳式起重机
clawel 楔石
claw foot 弓形足;弓形趾
claw-foot crack 爪形皱裂
claw for lumbering 运木夹具
claw for timbering 运木夹具
claw hammer 鱼尾锤;羊漆郎头;羊角锤;拔钉锤;撬棍;起钉锤;钉锤
claw hammer black enamelled 黑漆羊角锤
claw hammer with octagon neck 八棱脖羊角锤
claw hand 爪形手
claw hatchet 爪斧;拔钉斧;起钉斧
clawker 棘轮撑头
claw-like hook 爪形钩
claw magnet 爪形磁铁
claw mounts 瞄准镜插座
claw plate 夹板;棘齿板;爪板;齿爪板
claw shell brake 专用双闸瓦制动器
claw stop 止爪
claw suspension gear 爪悬架
claw tipping mechanism 料耙倾翻(卸)机构
claw toe 爪状趾
claw tool 石工齿凿
claw wedge 钳口板
claw weeder 爪式除草器
claw wrench 钩形扳手
clay 黏土;泥土
clay absorption 黏土吸附
clay acid 黏粒酸
clay aggregate 陶粒
clay aggregate concrete 黏土集料混凝土;黏土骨料混凝土
clay anchor 黏土锚碇
clay and straw 滑秸泥;柴泥
clay and straw plaster 柴泥抹面;草泥抹面
clay appliqué 堆贴
clay article (烧制的)黏土制品;陶瓷制品;烧土制品
clay atmometer 陶瓷蒸发计
clay auger 泥土(钻)铲;开口的钻孔匙;黏土钻头
clay backfill 泥土回填;黏土回填
clay ball 土团;黏土团;黏土球
clay band 泥铁石;黏土夹层
clay bank 黏土堆棕色;黏土堤
clay bank sand-break 黏土砂障
clay barrel 带衬管的双管
clay barrier 黏土隔层

clay base 黏土基;黏土基座
clay base mud 含黏土泥浆
clay-bearing 含黏土的;含泥的
clay-bearing dolomite 含黏土白云岩
clay-bearing limestone 含黏土质灰岩
clay-bearing soil 黏性土;含黏土的土
clay bed 黏土路床;黏土基床;黏土地基;黏土层
clay belt 黏土带
clay binder 黏土黏合料;黏土结合剂;黏土胶结物;黏土胶结料
clay bit 黏土钻头
clay blanket 黏土帷幕;黏土铺盖;黏土护层;黏土覆盖层;黏土封层;黏土层;垫层
clay block 陶质砌块;黏土块
clay body 土体;烧制陶器
clay body brick 土坯砖
clay body crushing strength 土体抗碎强度
clay bond 黏土剂;造型黏土;黏土黏合
clay bonded castable refractory 黏土结合浇注耐火材料
clay bonded silica brick 黏土结合的硅砖
clay book tile 黏土屋脊瓦;空心舌槽砖;空心舌槽瓦
clay boulder 泥砾层;冰砾泥
clay-bound 泥结的
clay bound macadam 泥结碎石路面
clay-bound macadam/gravel 泥结碎石/砾石
clay-bound macadam pavement 泥结碎石铺面;泥结碎石路面
clay-bound macadam road 泥结碎石路
clay brick 黏土砖;砖坯
clay brick aggregate 黏土砖集料
clay brick aggregate concrete 黏土砖集料混凝土
clay brick aggregate concrete block 黏土砖集料混凝土砌块
clay brick beam 黏土砖梁
clay brick-built garage 黏土砖建汽车库
clay brick cathedral 黏土砖建大教堂
clay brick cavity wall 砖砌空斗墙;黏土砖砌空芯墙
clay brick construction 黏土砖结构
clay brick corbel 黏土砖梁托
clay brick cross-wall 黏土砖隔墙
clay brick dust 黏土砖粉尘
clay brick exterior wall 黏土砖外墙
clay brick floor 黏土砖地面
clay brick foundation 黏土砖基础
clay brick grille 黏土砖格子窗
clay brick hardcore 碎黏土砖
clay brick insulation 耐火砖绝缘
clay brick plinth 黏土砖柱基
clay brick setting 黏土砖砌筑
clay brick shaft 黏土砖竖井
clay brick step 黏土砖砌台阶
clay brick tracery 黏土砖窗花格
clay brick tunnel vault 黏土砖砌隧道拱顶
clay brick wall arch 黏土砖发券
clay brick wheel window 黏土砖砌轮式窗
clay brickwork fireplace 黏土砖砌体壁炉
clay bubble 黏土气泡
clay bucket 黏土挖掘铲斗
clay building brick 房屋用黏土砖
clay building material 陶质建筑材料
clay building member 陶质建筑构件
clay building unit 陶质建筑构件
clay burner 黏土煅烧窑
clay burning 黏土煅制;黏土煅烧
clay burning curve 黏土煅烧曲线
clay burning unit 黏土煅烧窑
clay buster 圆盘翻土机
clay cable cover 电缆陶套壳
clay castable 黏土质耐火浇注料
clay catalyst 黏土催化剂
clay ceiling tile 陶质顶棚面砖
clay cement 黏土胶结物;黏土胶结料
clay-cement grouting 黏土水泥灌浆
clay-cement injection 黏土水泥灌浆
clay-cement mortar 黏土水泥砂浆
clay-chemical grout 加有化学药剂的泥浆
clay chimney pot 陶质烟囱顶管
clay chip(ping)s 土块;黏土屑
clay chunks 黏土块
clay cleaner 黏土净化器;净土机
clay coat 黏土涂层
clay-coated finish 滑泞涂料;黏土涂面
clay coating 黏土涂覆;黏土盖层;黏土包壳;黏粒胶膜;泥皮;黏土涂层
clay coating machine 黏土涂铺机
clay cofferdam 黏土围堰
clay colloid 黏粒胶体
clay column 黏土条;泥条
clay concrete 黏土混凝土
clay concrete blanket 黏土混凝土铺盖层
clay containing concrete 黏土水泥混凝土
clay containing sea silt 含海沙的混凝土
clay content 黏土含量;黏粒含量
clay core 黏土(防渗)芯墙
clay core type embankment 黏土夹心式防波堤
clay core wall 黏土芯墙
clay coring barrel 取黏土岩芯管
clay course 黏土脉壁;黏土(夹)层
clay court 红土网球场
clay creep 黏土蠕变
clay cross 黏土小球架
clay crucible 黏土坩埚
clay crucible process 陶土坩埚法
clay crusher 碎土机;黏土破碎机
clay cupola 黏土穹顶
clay curved roof(ing) tile 曲面陶瓦
clay curved tile roof 曲面陶瓦屋顶
clay cut-off wall 黏土截水墙
clay cutter 黏土切割机;黏土铰刀;黏土(砖)切割器;混土(机)器;切土筒;切土机;碎土机
clay cutter dredge(r) 切土式挖泥机;切土式挖泥船
clay cutter suction dredge(r) 切土式吸泥船
clay cutting 黏土切割
clay cutting machine 黏土切割机;黏土切碎机
clay deposit 黏土沉淀;黏土矿床;黏土沉积
clay deposit for cement 水泥配料有黏土矿床
clay desert 黏土荒漠
clay digger 掘土铲;黏土铲;挖土铲;手持黏土铲
clay digging 挖土
clay dike 黏土岩脉
clay disintegrator 笼式黏土破碎机;黏土破碎机
clay dome 黏土穹顶
clay drab 黏土黄褐色
clay drain 瓦管排水
clay drainage 瓦管排水
clay drainage tile 黏土管瓦;瓦管
clay dryer 黏土干燥机
clay drying 黏土烘干
clay earthenware 黏土质陶器
clay emulsion 黏土乳液;稀泥浆
clay engineering brick 工程用黏土砖
clay excavator 掘土机
clayey 黏土质的;黏土(状)的
clayey aquitard 黏土隔水层
clayey cement 泥质胶结物
clayey colo(u)r 黏土色
clayey conglomerate 含泥砾岩
clayey dolomite 黏质白云岩
clayey filling 泥质充填
clayey fine sand 黏质细砂(土)
clayey gravel 含土砾石;黏土(质)砾石;泥(质)砾;带黏土砾石
clayey gypsum 含土石膏
clayey limestone 黏质灰岩
clayey loam 黏(质)壤土;黏质垆姆;亚砂土
clayey marl 黏质泥灰岩
clayey mud 黏质淤泥
clayey sand 黏质砂土;黏土质砂(土);亚砂土;黏砂土;砂壤;砂混黏土
clayey sandstone 含土砂岩;黏土砂岩
clayey sediment 黏质底泥;黏土质泥沙;黏土质沉积物
clay(ey) shale 黏土页岩
clayey silt 黏质粉土;黏土质粉砂;泥质粉砂;轻亚黏土
clayey soil 亚黏土;黏性土;黏土质(壤)
clayey stone 黏土质风化岩
clayey stratum 黏土层
clayey vessel 黏土器皿
clay facing tile 特制陶瓦(美国);饰面瓷砖
clay fault 黏土断层
clay fertilizer 黏土肥料
clay figure 泥塑像
clay figure modelling 泥塑
clay-filled cutoff 黏土填筑的截水墙
clay-filled inclined wall 黏土斜墙
clay-filler 黏土填料

clay fill(ing) 回填黏土;泥土填充;黏土填筑体;黏土填料
clay film 黏土涂层;黏土膜
clay flocculation 黏粒絮凝
clay floor 黏土地面
clay flushing 泥浆洗井
clay for brick-tile 砖瓦用黏土
clay for cement burden 水泥配料用黏土
clay for foundry 铸型用黏土
clay formation 黏土形成
clay foundation 黏土地基
clay fraction 黏土粒级;黏土成分;黏土部分;黏粒成分;黏粒(粒)组;黏粒粒级
clay-free drilling fluid 无黏土钻进冲洗液
clay-free rock 纯岩石
clay gall 黏土球粒;黏土片
clay getting 黏土加工
clay glaze 泥釉;土釉
clay gouge 黏土夹层;断层泥;耳巴泥(断层泥)
clay gouged intercalation 黏土化夹层;泥化夹层
clay grain 黏土颗粒
clay grain grade 黏粒组
clay-graphite crucible 黏土石墨坩埚
clay-graphite mixture 黏土石墨搪料
clay grease 黏土润滑脂;黏土润滑剂
clay grinding machine 磨泥机
clay-grinding pan 磨泥盘
clay-grit 黏土砂砾
clay grout 泥浆;黏土浆
clay grouting 黏土灌浆;黏粒灌浆;灌泥土浆
clay gun 泥炮
clay gypsum 土石膏
clay hardpan 黏土硬磐
clay hydration 黏土水化(作用)
clay idol 泥偶像
clay industry 黏土工业;砖瓦工业
claying 抹泥;煅烧黏土
claying bar 大槌;撞槌;土槌
claying knife 抹泥刀
clay-in-thick beds 厚粘土层;厚层黏土
clay intrusion 泥土混入
clay-iron floc 黏土—铁絮凝物
clay ironstone 黏土铁质岩;泥铁岩
clayite 高岭石
clayization 黏土化
clay key 黏土(小)团(球)
clay kneading machine 黏土搅拌机;捏泥机
clay-laden liquid 含黏土泥浆
clay laminae 黏土叠层
clay lath 泥(板)条;黏土条;泥板岩
clay lathing 黏土板条;泥条
clay layer 黏土层
clay lay-in panel 陶瓷镶板
clay lens 黏土透镜体
clay lenses 黏土扁平体
clay-lime mixture 黏土石灰混合料
clay lining 黏土衬砌;黏土衬层
clay loader 装土机
clay loam 黏质壤土;黏质炉ców;黏土炉姆;黏磐土;亚黏土
clay lump 黏土团(块);土团;土块
clayly 黏土(状)的
clay-magnesium floc 黏土—镁絮凝物
clay marl 黏质泥灰岩;黏土(质)泥灰岩;泥灰岩
clay mask 黏土面具
clay masonry unit 黏土坯;黏土砌块
clay mass 黏土体;黏土块
clay member 陶土构件
clay membrane 黏土防渗墙;黏土防渗层
clay mill 碾土机;黏土拌和器;黏土拌和机;陶土厂;陶瓷厂;碎土机
clay mineral 黏土矿物;黏土矿物;泥质矿物
clay mineral composition 黏土矿物组合
clay mineral composition analysis 黏土矿物成分分析
clay mineral dehydration 黏土矿物脱水
clay mineral materials 黏土矿物材料
clay mineralogy 黏土矿物学
clay mineral particle 黏土矿物颗粒
clay mineral property 黏土矿物性质
clay minerals and soil 黏土矿物和土壤
clay mineral structure 黏土矿物结构
clay mining 黏土采矿
clay mixer 黏土拌和器;黏土拌和机;混土(机)器

clay mixing consolidation 深层搅拌
clay model 泥塑模型
clay modeling 黏土制模
clay mortar 黏土砂浆;黏土泥浆;黏土灰浆;泥浆
clay mortar mix (磨细的)黏土灰浆混合料;黏土掺料;泥浆拌和
clay mould 陶器模
clay mo(u)lding 黏土模型
clay movement 黏粒移动
clay mud 黏土泥(浆)
clay mud rock 黏土泥岩石
clay object 黏土质物体
clay of high plasticity 高塑性黏土
clay onion-skin bond 黏皮葱皮黏结
clay ore 黏土矿石
clay outwash plain 黏土沉积平原
clay overburden 黏土覆盖层
clay packing 黏土膏浆填塞;黏土压实;黏土铺盖
clay pan 黏土硬层;黏土盘;隔水黏土层;隔水盘;黏磐
clay pan chernozem 黏磐黑钙土
clay pantile 黏土双曲瓦
clay particle 黏土颗粒;黏粒
clay parting 黏土夹层
clay pellet 黏土(小)团(球)
clay pick 土镐
clay pillar 黏土岩柱
clay pipe 瓦管;陶土管;陶管
clay pipe drainage 陶土排水管
clay pipe line 陶土排水管道
clay pit 黏土矿;黏土坑
clay plain 黏土平原
clay plant 陶土制品工厂
clay platelet 黏土片晶
clay plate mo(u)ld 制瓦模
clay plate press 制瓦机
clay plug 黏土栓;黏土(堵)塞;泥塞;堵口黏土体
clay pocket 黏土窝;黏土穴
clay pot 烟囱顶陶管;陶罐;陶瓷坩埚;黏土坩埚
clay pot floor 空心黏土砖地面
clay pot for tiled roofs 屋顶用空心黏土砖
clay pottery 泥质陶
clay-pottery building material 陶质建筑材料
clay pounding 舂土
clay powder 黏土粉末
clay preparation 黏土制备;黏土调料
clay processing 黏土加工处理
clay product 陶土制品;黏土质材料
clay puddle 黏土(砂)浆;黏土膏;捣实黏土;捣烂黏土
clay puddle core wall 黏土膏芯墙;捣实黏土芯墙
clay puddle lining 黏土膏衬层
clay puddle seal 黏土膏浆填封;捣实黏土防漏
clay purifier 黏土净化器;黏土提纯器
clay quarry 黏土开采;黏土开采场
clay range 黏土粒度范围
clay reactivation 黏土再活化
clay refining 黏土精炼
clay-rich 重黏土的;富黏土的
clay road 黏土路
clay rock 黏土岩;泥岩
clay rock type 黏土岩类型
clay roll 黏土滚筒
clay roofing tile 黏土屋面瓦
clay room 耐火材料制品间
clay sampler 黏土取样器
clay sand 黏质砂土;砂混黏土
clay-sand-gravel mix(ture) 黏土、砂、砾石混合料
clay sandstone 黏土砂岩
clay-sandy siltstone 黏土砂质粉砂岩
clay sculpture 泥塑
clay sealing 黏土密封;引黏土止水;黏土填缝
clay sealing hole 黏土封孔
clay seam 黏土夹层
clay settlement 黏土沉降
clay sewer pipe 排水瓦管;排水陶土管
clay sewer pipe and fittings 陶土污水管配件
clay shale 黏质页岩;黏土质页岩;泥页岩
clay shingle 黏土瓦;陶瓦
clay shower tray 陶瓷(淋)浴盆
clay shredder 碎土机
clay silicate 硅酸白(粉)
clay silo 黏土筒仓
clay silt 黏质粉土

clay silty sandstone 黏土粉砂质砂岩
clay sintering process 黏土烧结方法
clay size 黏土粒径
clay slab 黏土板
clay slate 泥板岩;黏土板(岩);黏板岩
clay slate ground 黏土板地面;黏质板岩地面
clay slide 黏土滑坡;黏土滑动
clay slip 黏土糊;泥浆;浮渣;黏土浆;黏土滑坡;黏土滑动;黏土断层
clay slope 黏土斜坡
clay slope revetment spreader 黏土斜坡铺布器
clay sloping core 黏土斜墙
clay slurry 黏土浆;黏泥浆;白土泥浆
clay slurry jacket 泥浆润滑套
clay slurry preparator 淘泥机
clay slush 泥浆
clay soil 黏土土壤;黏土;黏性土(壤)
clay soil containing coarse grains 含粗粒黏质土
clay soil containing lightly coarse grains 微含粗粒黏质土
clay solution 黏土胶体溶液
clay-sorting machine 分泥机
clay spade 黏土铲;风铲
Clay spar 石状黏土(商品名)
clay spray shower tray 陶瓷喷淋浴盆
clay stabilization 黏土稳定
clay-stabilized 黏土稳定的
clay-stabilized material 黏土稳定砂
claystone 黏土质风化岩;黏土岩;黏土石;含黏土岩;变朽黏土岩
claystone porphyry 变朽黏土斑岩
clay storage 黏土储藏
clay stratification 黏土层
clay stratum 黏土层;[复]strata
clay strength anisotropy 黏土强度各向异性
clay strength isotropy 黏土强度各向同性
clay strip 泥板条
clay-strip building method 泥条筑成法
clay-strip forming technique 泥条盘筑成型法
clay structural material 陶质结构材料
clay structure 黏土结构
clay substance 标准黏土;黏土质物体;黏土物质
clay surface 黏土路面;黏土表面
clay suspended water 黏土悬浮液
clay suspension 黏土悬浮液;黏土悬浮体;黏土浆;黏泥悬浮质
clay swell(ing) 黏土膨胀
clay tamping 黏土塞料;炮泥
clay tempering 捣黏土
clay terrace 黏土平台
clay tile 黏土瓦;布瓦;瓦管;陶瓦;陶土瓦;陶土瓷砖
clay tile bonding agent 瓷砖胶合剂
clay tile factory 黏土瓦工厂
clay tile filler 黏土空心砖填充物
clay tile floor cover(ing) 黏土砖铺地面
clay tile joist floor 混凝土格栅间的陶土地板砖
clay tile line 陶土瓦管管道
clay tile roof 黏土瓦屋顶
clay tile roof cladding 黏土瓦屋面
clay tile roofing 陶瓦屋面
clay tile valley 陶瓦屋谷
clay till 漂砾泥;冰碛土
clay treated 白土精制的;白土处理的
clay tube 料筒
clay unit 陶土构件
clay vein 黏土脉
clay vent pipe 陶土通风管
clay vessel 黏土容器
clay visible under-face 露地面的砖瓦
clay wall 黏土墙;土塘;土墙
clay wall tile 烧土砖墙
clayware 陶土制品;黏土制品
clay wash 黏土洗涤剂;黏土淘洗;白泥浆
clay-water micelle 黏性胶粒
clay-water mix(ture) 黏土泥浆
clay-water relationship 土—水关系
clay winning 黏土采掘
clay with considerable gravel 含大量砾石的黏土
clay with flints 含燧石黏土
clay with long spacing 长间距黏土
clay work 黏土细工
clay working 黏土加工;黏土开采
clay-working machine 黏土加工机
clay yield 净土收获率;黏土处理收率

cleading (开挖竖井的)支撑;隔热板;包皮;木闸门;木沉箱板;护罩;护墙板;护壁板(隧道);保热套;套板
clean 洁净砂石;洁净;洁化;光洁;干净;船尾呈尖形的(指水下部分);擦拭;清澈的
cleanability 易洁性;除尘度;漆膜的易洁性
clean acceptance 不附保留条件的承兑;无条件接受
clean a firebox 火箱清洗
clean aggregate 洁净骨料;洁净集料;清洁集料;清洁骨料
clean air 洁净空气;纯洁空气;清洁空气
Clean Air Act 空气净化法;空气清洁法
clean air manifold 净化空气总管
clean air outlet 净化空气出口
clean air package 罐装清洁空气
clean air policy 空气清洁政策
clean anchorage 清爽锚地;无障碍物锚地
clean annealing 光亮退火
clean ash 清炉除灰
clean a ship 清舱
clean atmosphere 洁净大气
clean back 光洁石面;拉结石外露端
clean ballast 洁净道砟
clean ballast pump 净压载水泵
clean ballast tank 清洁压载舱
clean ballast water 清洁压载水
clean bench 净化台;净化操作台;除尘台;清洁台
clean bill 信用票据;光票;不随货押汇票;清洁票据;清单
clean bill for collection 托收光票
clean bill of exchange 普通汇票
clean bill of health 无疫证书(船只);无疫健康证明书
clean bill of lading 洁净提单;单纯提单;不附带条件的提单;清洁提(货)单
clean blast 全部爆破
clean-bole 无枝干材
clean bond 清洁公债
clean bottom 干净海底(没有危险物、礁石等)
clean breach 浪扫全舱面
clean break 无火花断路
clean burn 全烧
clean by laminar flow 层流净化
clean cargo 净货;好货;清洁货物;完整货物
clean charter 清洁租船合同
clean circuit 净化电路
clean circulation 干净循环
clean/clear 清澈透明
clean coast 安全海岸(指没有沉船或暗礁等)
clean collection 光票托收;托收光票
clean colo(u)r 纯洁色
clean core 净堆芯
clean corridor 洁净走廊;清洁走廊
clean credit 纯信用证;清洁信用;无条款信用书
clean-cut 净切削;光洁(的)裁切;加工光洁的;轮廓鲜明的;光洁的;体型美观的
clean-cut separation 确切分离
clean-cut timber 无疵(病)木材;光洁木材;光洁木板
clean cutting formation 孔底清洁的地层;不黏钻头的地层
clean cuttings 清除岩粉
clean-cut wood 光洁木材
clean data output 清机数据输出
clean-deal 光洁(的)木板;光洁木材
clean development mechanism 清洁发展机制
clean discontinuities 无充填物的软弱结构面;无充填物的不连续面
clean documents 清洁单据
clean draft 普通汇票
clean drilling 正常泥浆钻进
clean-dry sand 洁净干砂
clean-dumped rockfill 纯石块填石;抛石;纯石堆填;水工上的抛石;纯抛石充填
cleaned coal 净煤;精煤
cleaned coal moisture 精煤水分
cleaned gas 净化气体
cleaned gas dust content 净化气体含尘量
cleaned map 洁化图
cleaned off rough stone 清除粗石
cleaned off rubble 清除毛石
clean energy 清洁能源
cleaner 滤清器;洗涤剂;洗涤液;刮油器;纯化剂;除垢器;去污剂;清洗剂;清洗剂;清理设备;清洁器;清洁工(人);清除器;提匀;提纯器

cleaner air 净化空气
cleaner bar 刮土板
cleaner cell 精选槽;浮游精选机
cleaner chain 分离升运链;清选输送器链
cleaner-elevator 清理升运器;清选升运器
cleaner flo(a)tation 精浮选;再浮选
cleaner for air conditioning equipment 空调装置清洗设备
cleaner for cast and natural stone 铸石和天然块石清洗剂
cleaner for cooking use 炊用清洁剂;炊用除垢器
cleaner-grader 清选分级机
cleaner insert part 滤清器芯子
cleaner-loader 清理装载机
cleaner production and environmental protection industry 清洁生产与环保产业
cleaner production audit 清洁生产审核
cleaner production option 清洁生产方案
cleaner production promotion law 清洁生产促进法
cleaner rod 除草杆;清洁器杆
cleaners and dyers 洗染店
cleaner-separator 清选分离机
cleaner's naphtha 清洗用石脑油
cleaner's sink 落地污水盆
cleaner's solvent 洗涤剂
cleaner tailings 精选尾矿
cleaner technology 净化技术;清洁技术
clean face cutting 净方面锯切法
clean fallow 无草休闲地
clean fire 清炉除灰
clean-fleece weight 净毛重量
clean float 自由浮动(汇率);纯净浮动
clean floating 清洁浮动
clean floating exchange rate 清洁浮动汇率
clean formation 含泥很少的地层;纯地层
clean fuel oil tank 清洁燃油舱
clean full 尽量吃风;展全帆;全满
clean gap graded aggregate 净选集配集料
clean gas 洁净煤气;净化煤气;纯净气体;无尘煤气
clean gas outlet 净化气出口
clean grind 均匀的研磨浆;研磨均匀
clean harbo(u)r 无碍航物的港
clean hole 清洁钻孔
clean hole punching 干净穿孔
cleaning 滤清;洗净;清洗;清砂;清洁(处理)
cleaning action 清洗作用
cleaning agent 清洁剂;洗涤剂;清洗剂
cleaning and degreasing 洗净和脱脂
cleaning and disposal procedure 清洗及处理程序
cleaning and lining process 清管和衬管过程;清管衬法
cleaning and priming machine 除锈和涂底漆机
cleaning and sweeping holds 洗扫舱
cleaning apparatus 除尘装置
cleaning area 清洗场
cleaning balcony 擦窗阳台
cleaning bee 清洁蜂
cleaning before welding 焊接前的清理
cleaning berm 清扫平盘
cleaning between the rolls 辊缝
cleaning blower 皮老虎;吹风净化器;喷砂器;喷气净化器
cleaning brush 清洁刷
cleaning bucket 清理铲斗
cleaning by shaking 振打清灰
cleaning cage 滚净筒;清洗滚筒;清理滚筒
cleaning cartridge 净化器
cleaning cell 净化器;精选槽
cleaning certificate 清管合格证
cleaning chamber 清洁室;排水管通沟室
cleaning cloth 抹布;除尘器用布
cleaning cock 排污龙头
cleaning compartments for wreck to keep afloat positively 清舱浮定
cleaning compound 洗涤剂
cleaning cone 清选锥体
cleaning cover 通沟室盖;清洁罩
cleaning cradle 清洗窗墙吊架
cleaning cutting 除伐
cleaning cylinder 清选滚筒
cleaning device 清洗装置;清扫装置;清灰装置
cleaning ditch 壕沟清理;清理沟渠
cleaning doctor 清洁刮刀;涂料刮刀

cleaning door 修炉口;工作门;清渣门;清扫门;清理门;清灰口
cleaning dozer 拔根机
cleaning drum 清砂滚筒
cleaning effect 净化效果
cleaning effect by rain 雨水清洁作用;雨水净化作用
cleaning efficiency 净化效率
cleaning emulsion 清洗乳液
cleaning equipment 净化设备;洗净装置;清洗设备;清理设备
cleaning equipment closet 清洗工具橱
cleaning equipment room 清洗工具室
cleaning eye 清扫口;清理孔;清理机;清除孔
cleaning flap 清扫板
cleaning flo(a)tation 再浮选
cleaning fluid 清洗液
cleaning for electroplating 电镀前的清洗
cleaning frequency 清洗周期
cleaning go-devil 清管机
cleaning guide 清管规定
cleaning gun 清洗枪(压缩空气);喷洗枪
cleaning hinge 长翼窗铰链;阔隙窗铰链;间隙窗铰链
cleaning hole 出砂孔;清扫孔;清除孔
cleaning hole cover latch 清洗孔盖闩
cleaning hose 冲洗管
cleaning line 清洗作业线
cleaning liquid 清洗液
cleaning liquor 洗涤液
cleaning machine 清洗机
cleaning machine for bottom sockets 底部承口清洗机
cleaning machine for motor 电机清洗机
cleaning means 修整工具;清洗工具
cleaning mill 滚净筒
cleaning mixture 洗涤剂;洗涤(混合)液
cleaning of bottom 清底
cleaning of brazed surfaces 钎焊表面清理
cleaning of catch basin 截泥井清泥
cleaning of coal energy 煤炭能量净化
cleaning of construction site 清理施工现场;清基
cleaning of element 元件的清洗
cleaning-off 清理
cleaning of reservoir 水库清理
cleaning of sand 洗砂
cleaning of sediment 清淤
cleaning of welding deposits 焊接沉积的清理
cleaning oil 轻油;挥发性油
cleaning opening 清洗孔
cleaning operation 清洗操作;清洁作业
cleaning paste 清洁膏
cleaning pig 清管机
cleaning pipe 清管
cleaning pit 清洗坑
cleaning plant 清理装置
cleaning plant equipment 清洗工场设备
cleaning plant reject 清洗厂废物
cleaning plate 清洗板
cleaning plug 清除塞;清除堵头
cleaning port 清除口
cleaning powder 去污粉;洗涤粉
cleaning process 洁化过程;带材的净化脱脂过程;清洗方法
cleaning rock surface 清理岩面
cleaning rod 长柄清洁杆
cleaning rust by sand blasting 喷砂除锈
cleanings 垃圾
cleaning sash (一端固定的)擦窗时可打开的窗扇
cleaning scraper 清管机
cleaning screen 清洗筛
cleaning screw 清洗螺钉;放水螺旋
cleaning shoe 清粮架;清粮筛
cleaning shop 净化车间
cleaning siding 洗刷线
cleaning sieve 草籽筛
cleaning solution 洗(涤)液;洗涤剂;洗涤混合液;澄清液;清洗液;清洁液
cleaning solvent 清洗溶剂;清洗剂
cleaning station 洗舱站
cleaning strainer 滤池;滤净器;过滤器;粗滤清器
cleaning table 选矿台;清理台
cleaning tank 冲洗池
cleaning tank charges 洗舱费
cleaning track 洗刷线

cleaning truck 清扫车
cleaning tube 清洗管
cleaning unit 选矿设备;选矿单位;钢材清理机组;清粮室;清粮设备
cleaning up 细木工的最后加工;收起(起落架);清扫;清除;清理
cleaning up information pollution 清除信息污染
cleaning vacuum 真空吸空装置
cleaning vacuum device 真空吸尘装置
cleaning vacuum plant 真空吸尘装置;真空吸尘设备
cleaning vacuum system 真空吸尘系统
cleaning waste 擦拭用回丝
cleaning web 清洗丝网;清洗卷筒
cleaning web indicator 清洗(丝)网指示器
cleaning work 场地清除工作
clean-in-place 就地清洗
clean lattice 净栅
clean letter of credit 光票信用证;不跟单信用证;清洁信用证
clean lines 流线型船
cleanliness 良流线性;良好绕流性;净度;光洁度;清洁度
cleanliness of site 场地清理
cleanliness of surface 表面光洁度
cleanliness standard 清洁标准
clean loan 无抵押借款
clean map 无误差的地图数据
clean material 净料
clean mining 高回收率开采法
clean negotiable bill of lading 清洁可转让提单
clean off balance outstanding 付清所欠余款
clean oil 白油;轻质油;轻质石油;透明油
clean oils 轻质石油产品
clean on board 洁净装船
clean on board bill of lading 已装船清洁提单;清洁装运提单
clean opinion 无保留意见
clean-out 清淤;清扫口;掏堵;管道清扫孔;清理完毕;清洁口
clean-out auger 清孔钻;抽汲筒(钻探工具);清孔钻头
clean-out bailer 抽砂筒(清孔用)
clean-out bit 清底钻具;清孔钻头
clean-out box 液体容器侧面清理方孔
clean-out cap 清洁孔盖板
clean-out chamber 清扫井;清通井
clean-out clearance 清除口净距;清洁孔净距
clean-out cock 净塞
clean-out cover 人孔盖;清扫口
clean-out crew 修井队
clean-out disc 清洗盘
clean-out door 出渣门;出灰门;出灰口;出粮口;清出口;清扫门
clean-out eye 清除孔
clean-out fitting 清除装置
clean-out frequency 清垢频率
clean-out hatch 出渣口;清扫口【船】;人孔
clean-out hole 清扫口
clean-out jet auger 抽汲筒;射流式清理井眼螺钻
clean-out job 修井工作
clean-out machine(ry) 修井机械
clean-out of hole 洗孔
clean-out of wall cavities 空芯墙清除口
clean-out of well 洗井
clean-out opening 清扫孔;清扫口;人孔
clean-out pit 清除坑
clean-out plate 人孔盖板;清扫口盖板
clean-out plug 清洁口堵头;疏水旋塞
clean-out port 清除口
clean-out screw 清除螺钉
clean-out string 修井钻柱;修井钻杆
clean-out timber 光洁木材;光洁木板
clean-out tools 清(洗)孔机具
clean-out tube 清除管
clean payment credit 光票付款信用证;全部预支信用证
clean payment letter of credit 全部预支信用证
clean petroleum product 轻质成品油
clean plastic 水晶胶
clean polyester 水晶胶
clean producer gas 净化发生炉煤气
clean proof 付印样;清样
clean rain 干净的雨水;未污染雨水

clean reactivity 净堆反应性;新堆反应性
clean remittances 不跟单汇兑
clean remittances of commercial paper 不跟单汇兑
clean risk of liquidation 破产风险
clean river 清水河(流);清洁河流
clean rock 纯岩石
clean room 绝尘室;净化室;(供实验用的)经消毒的房间;洁净室;清洁室
clean run 流线型船尾
clean sand 净砂;洁净砂;纯砂层;清洁砂
clean sandstone 纯砂岩
cleanse 净化;纯化;擦拭
cleanser 滤水器;洗涤剂;凿断工;擦亮剂;清洁器;清洁剂;清洁工(人);提绞器
cleanser drum 净化剂储[贮]罐
cleanser for masonry 砖石墙面清洗剂
cleanser injector 净化剂注入器
cleanser mill 清整转筒
cleanser powder 去污粉
clean ship 空船;清洁船舶;轻质油船
cleans hip 没载货的船
clean shipping documents 清洁装船单证
cleansing 提纯
cleansing agent 洗涤剂
cleansing agent tank 净化剂储[贮]罐
cleansing blower 吹净器;喷气净化器;喷砂器
cleansing doctor 刮浆刀
cleansing emulsion 清洗乳液
cleansing filter element 滤器净化元件
cleansing paste 清洁膏
cleansing powder 去污粉
cleansing solution 清洗液;洗涤液
cleansing tissue 擦拭纸
clean slide 干净载玻片
clean soil 洁净土壤
cleansol 软化剂
clean state principle 重新开始原则
clean steel 洁净钢;纯(净)钢
clean stmopshere 天然大气
clean stream 清水河(流);清洁河流
clean stub 新导电棒
clean stuff 优质木材(无节疤等)
clean summer fallow 夏季绝对休闲;夏耕休闲
clean superconductor 纯净超导体
clean surplus concept 净盈余观念
clean tank 轻油柜
clean tanker 轻油油轮
clean thread 光洁螺纹
clean tillage 清耕法
clean timber 无疵木材;精木料;无缺陷木材(又称无节疤陷的木材)
clean title 清洁物权
clean transport documents 清洁运输单据
clean-type ventilation 清洁式通风
clean-up 混凝土表面光洁验讫;清除;清舱;扫除(工地渣土);建筑工程完成验讫;光面验讫(木工作业);洗涤;清扫;清理现场
clean-up activity 净化活动;清洁活动
clean-up and move out 竣工清扫;现场清理;清扫并运出工地;清理;清除
clean-up barrel 清理滚筒
clean-up chute 精选溜槽;清洗溜槽
clean-up cost 净化费用
clean-up device 清舱装置
clean-up effect 净化作用;清除效应
clean-up job 清孔工作
clean-up method 净化法
clean-up of cofferdam foundation 围堰基础清理
clean up of oil spills 漂油清除
clean-up operation 清理作业
clean-up past due bill 清理逾期票据
clean-up process 净化工艺
clean-up pump 吸收泵;清除泵
clean-up reactor 气体吸收反应器
clean-up scraper 刮土板
clean up solvent 清洗溶剂
clean-up trip 冲孔回次
clean-up work 清理工作;清洗工作
clean urban river environment 净化城市河流环境
clean use of coal 煤的清洁用法
clean water 净水;洁净水
Clean Water Act 水(体)清洁法;清洁水法令;流水保护条例

Clean Water Act Amendments 水清洁法修正案
clean water algae 清水藻;清洁藻
clean water basin 清水池
clean water gallery 清水隧洞
clean-water lake 清水湖
clean water performance 清水性能
clean water reservoir 清水池
Clean Water Restoration Act 清洁水恢复法案;净水修复法令
clean water strategy 净水对策
clean-water supply 净水供应
clean water tank 清水箱;清水池
clean well 无垢井孔
clean wool 洗净毛
clean yellow 纯黄(色)
clean yield 净得
clear 解开;计数(器)归零;明晰的;没有障碍物;卸空;付清;晴朗、清晰的;清零;清机;清除;清澈的;无枝的
clear acceptance 单纯承兑
clear account 清账
clearage 裂解;清理;清除;开垦地
clearage with grinding wheel 砂轮清理
clear a hawse 清链
clear a hold 清舱
clear air 纯净空气;晴空
clear-air turbulence 晴空湍流
clear-all heart 全心木材
clear amber 透明琥珀
clear amount 净值
clear an account 结清账户
clear an aircraft to pass 允许飞机通过
clearance 廓清;净空;结关证(书);间隙;海关清税;芯头间隙;限界;余隙;轴配合间隙;纯益;出/入港证;全伐;去除率;清算;清理;清关;票据交换;通关
clearance above bridge deck 桥面净空
clearance above bridge floor 桥面净空
clearance account 结算账户
clearance adjuster 间隙调整器
clearance adjustment 间隙调整;孔隙调整
clearance advice 出港通知(书)
clearance angle 间隙角;后角;隙角
clearance area 余隙面积;应拆除住宅区;拆迁区;拆迁地区;不卫生地区
clearance at berth face 泊位面的富裕净空
clearance at parking 停车余隙
clearance between bubble cap and tray 泡罩与塔板间隙
clearance between ga(u)ge line of guard rail and frog center 护轨与心轨查照间隔
clearance between ga(u)ge line of guard rail and wing rail 护轨与翼轨查照间隔
clearance between lines 管线间距
clearance between rolls 辊间间隙
clearance between surfaces to be brazed 钎焊表面间隙
clearance between tie rods 支杠间间隙
clearance between wheel flange, ga(u)ge line 轮轨游间
clearance car 限界检测车
clearance cavitation 空隙气蚀;间隙气蚀
clearance certificate 结关证书;出港证书;结关单;验关证
clearance channel 缝隙槽
clearance chart 余隙指示器
clearance compensation 缝隙损失
clearance cost (土地上住户的)动迁费用
clearance curve 清除曲线
clearance date 结关日
clearance depot 结关货场;集散站(指货物);验关站;验关仓库;清关货场
clearance detector 限界检查器
clearance diagram 净空图;限界图;外形轮廓断面图
clearance diagram for rolling stock 机车车辆限界图
clearance diagram for structure 建筑接近限界图
clearance diameter 留隙直径;净空直径;隙径
clearance distance 间隙距离
clearance drawing 余隙指示图
clearance envelop 限界包络
clearance factor 余隙系数
clearance fee 结关手续费;出港手续费;结关费
clearance fit 间隙配合;转动配合;动座配合;松动

配合
clearance for expansion 胀隙;膨胀间隙
clearance formality 结关手续;报关验关手续
clearance for oversize commodities 宽大货物限界
clearance for pulling condenser tubes 凝汽器抽管距离
clearance for railway 铁路建筑限界
clearance for traffic of tunnel 隧道通行限界
clearance gate 量载规;界限门
clearance ga(u)ge 量隙规;建筑接近限界;间隙规;测隙规;塞尺
clearance ga(u)ge for pantographs 集电弓隔离限界
clearance goods 结关货场
clearance grinding 间隙磨削
clearance groove 空刀槽
clearance headway 净高(桥下)
clearance height 净高;间隙高度;限界高度;净空高(度)
clearance height of bridge 桥梁净空高度
clearance hole 出砂孔;排屑孔径;排沙孔
clearance indicator 净空显示器;净空显示牌
clearance inspection 间隙检查
clearance instruction 清除指令
clearance interval (交叉口的)清车时间
clearance inward 进口结关证明
clearance label 结关单;出港证
clearance lamp 净空灯;嵌合灯
clearance lamp side light 轮廓灯
clearance leakage 间隙漏水;间隙泄漏;间隙泄漏;缝隙漏水;不紧密
clearance light 轮廓灯;净空灯
clearance limit 净空界限;净空界线;限界(框);余隙限度;测隙极限
clearance limitation 净空限制
clearance limit for freight with exceptional dimension 宽大货物限界
clearance limit measurement 净空界限测量
clearance limit of bridge 桥梁净界限
clearance limit survey 净空区测量;净空界限测量
clearance line 净空线;限界线
clearance loading ga(u)ge 载货限定外形尺寸
clearance loss 净室损失;净空损失;间隙损失;余隙损失
clearance marker 净空标志
clearance note 进口结关证明
clearance notice 净空(界)限;进口通知;余隙限度;出港通知(书)
clearance nut 余隙螺母
clearance of a retarder 减速器接近限界
clearance of bridge 桥上净空
clearance of bridge opening 桥梁出水净高
clearance of cargo 货物报关
clearance of certificate 输油证
clearance of clearing dust 排粉间隙
clearance of expansion 胀隙
clearance of fouling of sea valve 清理污物的海底阀
clearance of goods 货物报关;报关
clearance of navigation obstructions 清航
clearance of pipeline blockage 清洗管路堵塞
clearance of railway accounts 铁路间账目清算
clearance of reservoir bottom 库盘清理
clearance of residual mud in a wreck 扫舱(救捞)
clearance of retarder 缓行器接近限界
clearance of rolling 机车车辆限界
clearance of ship 航迹间富裕宽度
clearance of side portal 侧门架净空
clearance of site on completion 竣工(现场)清理;工地完工清理
clearance of span 跨度净空;净跨(度);桥下净空;桥跨限界;桥跨净空
clearance of traffic in tunnel 隧道限界
clearanceometer 测隙计
clearance operation 清拆作业
clearance order 清除指令;断路指令
clearance outward 出口结关证明
clearance paper 结关文件;许可证(指船出海用);准单;出入港证;出港证;许可(证);出港单;报关许可证
clearance paper for port 船舶出港许可证
clearance period 清车时间(交叉口)
clearance permit 许可证(指船出海用);出港证;出港许可(证)
clearance pocket 补充余隙;补充容积
clearance point 计算停车点【铁】;卸货点;安全限界点
clearance pressure 间隙压力
clearance print 推动芯头
clearance profile 限界检查框
clearance program(me) 排除计划
clearance radius 净空半径
clearance rate 清除率
clearance ratio 余隙比
clearance restriction 净空限制;建筑接近限制
clearance sale 减价出售存货;销售底货;清出存货;放盘;清底销售;清仓拍卖
clearance sector 偏航指示区
clearances for highway tunnels 公路隧道界限
clearance site 清除现场;清除地段
clearance size 净空尺寸
clearance slippage 缝隙损失
clearance slum area 贫民窟清除地区;陋巷清除地区
clearance space 货堆空隙;余隙空间
clearance terminal 装卸口内结关(集装箱)
clearance test 廊清试验;清除试验
clearance testing car 净空测定车
clearance through customs 结关
clearance time 清车时间(交叉口)
clearance time at crossing 车辆通过交叉口时间
clearance time of signal 信号清尾时间
clearance tolerance 间隙公差
clearance treadle 限界检查器
clearance under bridge superstructure 桥下净空
clearance under valley tiles 天沟两侧瓦下的间隙
clearance volume 余隙容积;容积余隙
clearance widening 限界加宽
clearance width 净空宽度
clear anchor 锚清爽
clear anchorage 宽敞锚地;清爽锚地;无碍航物锚地
clear anodized alumin(i)um 本色氧化铝
clear a port 结关;出港
clear arch span 拱的净孔
clear area 空白区;净面积;有效面积;有效截面;干净区;畅通区;安全道;清晰区域;清零区;清洁区;未存储区;应拆(除的)居住区;清除区
clear area of screen 筛孔净面积;筛的有效面积
clear a ship 结关;卸清船上货物;船舶办理出口手续结关
Clear astern 船尾清爽
clear away 清除
clear back 反向拆线
clear band 空白区;空白段;干净区;清洁区
clear base 清洁基
clear bearing 避险方位
clear binder 透明黏合剂
clear boiled soap 抛光皂
clear boled 主干无枝的
clear breach 无碍的冲刷浪
clear bulb 普通灯泡;透明玻璃(灯)泡
clear buttress spacing 墩间净距;支墩净间距;坝垛净间距
clear ceiling height 楼层净高
clear ceramic glaze 透明釉;透明陶瓷光釉面
clear ceramic glaze tile 透明釉面砖
clear channel 开敞信道;专用信道;无干扰频道
clear character 清除字符
clear chill roll 纯白口冷硬轧辊
clear circuit 清零电路
clear clay 纯泥;纯净黏土
clear coagulation 明显聚沉
clear coast 无碍航物的海岸
clear coat 清漆涂层;透明涂层
clear coating 透明涂层;透明涂膜;透明涂面
clearcole 细白亚胶;白铅胶;油漆垫底;黏贴金箔用的透明涂料;明胶底漆;打底明胶;打底胶泥;白垩胶
clear colo(u)r 清色
clear confirmation 清除确认
clear control 清除控制
clear control logic 清除控制逻辑
clear cover 保护(覆盖)层;覆盖层
clear cross-section 净横断面
clear cryptographic(al) key 清除密码键
clear crystal 无色晶体
clear-cut 轮廓鲜明的;轮廓清晰;轮廓清楚的
clear-cut division of labo(u)r 明确分工
clear-cut margin 边缘锐利;边缘清楚
clear-cut timber 无疵(病)木材
clear cutting 净切边;皆伐(作业);全伐
clear data 空白数据;纯数据
clear day 晴天;晴日
clear days 净日数;整天天数;整天工数
clear debt 清理债务
clear diamond 纯净金刚石
clear dimension 内尺寸;净尺寸
clear directive variation 明显方向变化
clear display 清晰显示;清除显示
clear displayed 显示明显
clear dissolution 明显溶蚀
clear distance 净空;净距(离)
clear distance between tracks 股道间距
clear distance betwoen uprights 立柱净空距
clear dope 透明涂料;透明蒙皮漆
clear down 准许下降(高度)
cleared 已结关
cleared area 清除的场地;清理过的场地
cleared as planned 按计划清除
cleared condition 清零状态;清除条件
cleared export declaration 结关出口申请
cleared-out 净出
cleared site 清理的现场
cleared slope 光秃山坡
cleared slum area 贫民窟清除地区;陋巷清除地区
cleared to land 准许落地(飞机)
clear effluent 清洁水
clear entry 清除表目
clear entry key 清除输入键
clearer 澄清剂;清洁辊;清除器;排除器
clearer roller 绒辊
clearer spring 绒辊弹簧
clearer's solvent 清洁用溶剂
clear etching 玻璃刻蚀;透明酸蚀刻
clear etching bath 透明酸蚀刻槽
clear evidence 明显(的)证据
clear expense(s) 收支相抵
clear face 净面
clear felling 皆伐;透明漆膜
clear felling and stump-grubbing 伐木除根
clear film 透明薄膜
clear filtrate 清滤液
clear filtrate outlet 滤液出口
clear finish 光洁整理;清漆罩面;抛光处理;透明面漆
clear flag 清除标志
clear float glass 透明浮法玻璃
clear floor sealer 透明地面密封剂
clear foil 透明片
clear for action 备战
clear forward signal 正向话终信号
clear foul rope 理清绳缠
clear frit 透明熔块
clear fuel 清洁燃料
clear gas 无焦油煤气
clear glass 净片玻璃;透明玻璃
clear glass filter 透明滤光片
clear glaze 透明釉
clear goods 海关放行货物;已纳税货物
clear goods from customs 货物结关
clear hawse slip 滑钩链式止动器
clear headroom 净高;净空
clear headway 桥下净空;净车间时距;净高;净空高(度);净空
clear headway of bridge 桥下净空;桥下净高
clear height 净高;余高
clear height of bridge 桥下净高
clear hold 空净货舱;开敞货舱
clear ice 明冰;纯净冰
clear in 办理海关进口手续
clear indicator 清除指示符
clearing 拆线;非现金结算;清算;清扫;清理场地
clearing account 清理账目;暂记待结转账户;清算账户;清结账户
clearing agent 交换代理;澄清剂;清算代理人;清结代理;清洁剂
clearing agreement 清算协定
clearing agreement trade 记账贸易
clearing a hawse 清解锚链
clearing and controlling center 结算和监督中心
clearing and degreasing compound 净化去油剂
clearing and grubbing 树木挖根清出场地;清除地面障碍物

clearing and stripping 清理拆模;清基
clearing automation 结算业务自动化
clearing away 清除;除去
clearing balance 结算差额;票据交换差额
clearing bank 参加票据交换的银行;清算银行
clearing bath 除斑液;澄清溶槽
clearing book 交换(票据)登记簿
clearing business 票据交换业务
clearing center 集散中心;货运中心
Clearing Center for Intra-Asia 亚洲内部清算中心
clearing checks 交换中支票
clearing clerk 货运业务员
clearing contract 清算交易;期货交易
clearing corporation 清算公司
clearing dangerous rock 清除危石
clearing debits 清算债务数
clearing depth 清扫水深
clearing dollar 清算美元
clearing exchange slip 结算外汇单
clearing expenses 清理费用
clearing factor 清除因子;清除因素
clearing fee 出港费
clearing firm 清算商行
clearing form 结算方式
clearing form alongside 安全离开码头
clearing from alongside by after spring 后倒缆离码头
clearing from steep shore 离陡岸
clearing fund 清算基金
clearing glade 冰湖的
clearing ground 清理地面;清除地面;清场地
clearing hole 略大于螺栓的孔;钉孔;正常直径的孔;通过孔
clearing hospital 前方医院
clearing house 情报交易所;理算行;交换站;交换所;技术情报交换室;清算机构;票据交货所;票据交换所
clearing house association 票据交换协会
Clearing House Automated Transfer System 自动结算转账系统
clearing house balance 票据交换差额
clearing house credit ticket 票据交换贷方传票
clearing house for trade information 贸易情报交换所
clearing house funds 清算所基金;票据交换所资金
clearing house item 待清算项目
clearing house loan certificate 票据交换所贷款证券
clearing-house mechanism 情报交换所机制
clearing house proof 票据交换所决算表
clearing house settlement sheet 票据交换所清算表
clearing house statement 票据交换所清单;票据交换所报表
clearing index 净化指数;净化指标
clearing indicator 拆线指示器
clearing instruction 结算通知书;结算通知单;结关通知书
clearing interrupt 清除中断
clearing iron 刮缝凿;刮缝刀
clearing item 交换项目;清算项目;交换物件
clearing key 清除键
clearing label 出港证;出港许可
clearing labor 票据交换工
clearing lamp 指示信号灯
clearing line 安全导航线
clearing machine 绒毛净化器
clearing marks 导航标;安全导航标
clearing member 清算所成员;票据交换所会员银行
clearing notice 出港通知(书)
clearing of account 非现金结算
clearing of a fault 排除故障
clearing of a river bed 清理河床
clearing of bills 票据交换;票据
clearing of checks 票据交换
clearing off 清除
clearing of reservoir 放空水库;清洗水库
clearing of site 清理场地;现场清理;场地清理;场地清扫;清理现场
clearing of wire rope 清缆
clearing operation 清除工作;清算业务
clearing out 清除
clearing outward report 出口报告书
clearing period 无封冰期
clearing plug 泄水栓;疏水旋塞

clearing port 结关
clearing price 结算价格;清算价格
clearing rake 清扫把
clearing range marks 导航叠标
clearing return 交易清单
clearing revenue 清算收入
clearing sands with converged flow 束水攻沙
clearing section 净断面
clearing section size 净断面尺寸
clearing services 票据交换业务
clearing sheet 交易清单;交换清单;票据交换所贷借决算表
clearing signal 开放信号
clearing signalling 发出话终信号
clearing site 清场地
clearing station 医疗救护站
clearing stock 出清存货
clearing stock sale 清货减价
clearing system 清算制度
clearing the line 出清线路
clearing the market 出清市场
clearing the track 出清股道
clearing time 澄清时间;清澈时间
clearing time of fault 故障消除时间
clearing title 所有权清理
clearing union 清算同盟
clearing up 清扫;场地竣工后清理
clearing up oil spill 清理漏油
clearing view disc 离心式扫雨器
clearing water basin 清水池
clearing water reservoir 清水库;清水池
clearing water tank 清水池
clearing wheel 清除轮;疏渠管用的清除轮
clearing width 净宽(度);清除宽度
clearing wood(en) floor 木底板去污
clearing work 清除工作
clear input 归零输入
clear instruction 清除指令;结关通知书
clear interrupt 清除中断
clear inward 办清进口手续
clear key 清除键
clear key switch 清机键盘开关
clear lacquer 亮漆;硝基清漆;罩漆;清喷漆;透明漆;赛璐漆;透明亮漆
clear lamp 透明灯(泡)
clear launch 正确发射
clear length 净长度
clear length of bridge opening 桥孔净长
clear line 空(闲)线(路)
clear-line image 明线图像
clear liquid 清液
clear liquid preparation 清洁白液体制品
clear list 清除排列表;清除表
clear log 锯材原木
clear lumber 净材;无疵锯材;上等(锯)材;无疵材;精木材
clearly marked prices 明码实价
clearly definable natural point 明显地物点;明视地物点
clearly defined 轮廓分明的
clear medium 透明黏合媒质
clear memory 清除储器
clear mesh 网孔净径;净空;内径
clearness 清晰度
clearness number of sky 天空清晰度
clearness of active faults displayed on image 活动断层影象显示清晰程度
clearness of field 视场的明净
clear objective aperture 物镜通光孔径
clear octane number 不加铅辛烷值
clear off 偿债;清清;清偿;清除
clear off an account 付清账
clear of the waterline 在水线以上
clear of water 在水面以上
clear opening 净宽;净孔;(管道的)净断面;有效面积;有效截面(涵管等)
clear opening of bridge 桥梁净空
clear operation 清除操作
clear order 清除指令
clear out 卖光;安全出港;清除
clear out all the holding 出清全部存货;全面清仓
clear outward 办清出口手续
clear overflow 明流溢水
clear overflow weir 溢流堰;明流堰

clear pale yellow 净苍黄色
clear panel key 清面板键;清除面板键
clear plastic sheet 透明塑料膜;透明塑料板
clear point 明朗点;回零点
clear profit 净利(润);纯利(润)
clear reader 清除读出器
clear request 断开请求;清除请求
clear rib interval 肋的净间距
clear rib spacing 肋的净距
clear river 清水河(流)
clear roadway 路面净宽
clears 无疵材;精木料
clear scan radar 清晰显示雷达
clear shaper 复位脉冲形成器
clear sheet 净片玻璃
clear sheet glass 透明薄玻璃
clear-side roller 净边路碾;净边压路机
clear sight distance 明晰视距
clear sight triangle 视距三角线
clear signal 开通信号;拆线信号
clear signal indication 开放信号表示;信号开放表示
clear sky 晴空;无云天空
clear solution 清液
clear space 空货位;空地;净空空间;净空间;净空;洁净区
clear space between bars 钢筋间净距
clear spacing 净距(离);净间距
clear spacing of counterforts 扶垛净间距
clear span 净翼展;净宽;净跨(度);净孔;净空跨度;桥涵净跨径
clear span building 单跨建筑物;无内柱建筑物
clear span of bridge 桥梁净跨径;桥梁净跨
clear span of girder 大梁净跨
clear span tent 净跨帐篷
clear spring 回零弹簧
clear staff 精木料
clear stipple embossed 无色点饰浮雕;透明点状压花
clear stock 无刻痕板材;清货
clear storage 清除储器
clear store 清除存储指令
clear story 纵向气楼;阁楼天窗;气楼;高窗;开窗假楼;纵向天窗或气楼
clearstory lighting 高侧窗采光
clear stream 清水河(流)
clear stuff 无疵材;精木料;无节疤木材;无疵(病)木材;无缺陷料
clear supernatant 上层清液
clear switch 清除开关
clear synthetic(al) baking varnish 合成清烘漆
clear system 贸易清算制
clear terrain 开阔地
clear text 明文;明码通信;直读文本
clear the circuit 切断电路
clear the debt 清偿债务
clear the port 安全出港
clear through vision 清晰景象;辨明景象
clear timber 无疵材;精木料;无节疤木材
clear title 完整的产权
clear top(coating) 罩光清漆;面漆
clear topographic(al) feature 明置地物;明显地物
clear to send 待发;清除发送;清除待发
clear-to-send circuit 清除发送线路
clear track 空闲线路
clear up 清理
clear-up ability 清偿能力
clear-up debt 清理债务
clear up sludge 清淤
clear valve diameter 阀孔径
clear varnish 透明清漆
clear varnish finish 透明清漆罩面
clear varnish medium 透明清漆介质
clear varnish system 透明清漆体系
clear varnish vehicle 透明清漆介质
clear view 清晰视野
clear vision distance 明晰视距
clear water 净水;安全水域;清水;清洁水;无碍航物水域
clear water basin 净水池
clear water circulation 通畅水区航行
clear water drilling 清水钻进
clear water fixed-bed model 清水定床模型
clear-water lake 清水湖
clear water model 清水模型

clear water pump 清水泵
clear water repellent coating 透明憎水涂料
clear water reservoir 净水池;给水水库;清水库
clear water scour 清水冲刷
clear water type swivel 清水钻进水龙头
clear waterway 有效水道宽度
clear water zone 清水带
clear way 高速公路(采用立体交叉,确保不间断交通);扫清道路
clearway 快行道;净空道(机场跑道端外);高速道路;超速道路;超(高)速公路
clearway valve 全开阀
clear weather 晴天;天气晴朗
clear well 清水井;清洁井
clear white 纯白
clear white lead 净白铅
clear width 净宽(度)
clear width of carriageway 行车道净宽
clear width of deck 桥面净宽
clear window glass 透明窗玻璃
clear wire glass 透明嵌丝玻璃;透明嵌金属钢玻璃
clear wood 无疵材;无节疤木材;无疵(病)木材
clear woollen finish 毛呢光面整理
clear working days 十足工作天数
clear working place 安全工作地点
clear zone 清亮区
cleat 夹条;加强角片;内生裂隙;内生裂缝;楔子;系索耳;系绳角铁;系绳板条;系缆墩;羊角系缆柱;固着楔;割理;导缆钳;瓷夹板;瓷夹;防裂楔;陶瓷夹板
cleat-cut division of work 明确分工
cleated belt conveyer 条式输送带
cleated conveyer 条式输送带
cleated-wheel tractor 条轮拖拉机
cleating 夹板
cleating device 舱盖板压紧装置
cleating of reservoir 放空水库
cleat insulator 绝缘夹板;夹壳绝缘子;瓷夹
cleat plane 楔子劈入而
cleat spear 富铁白云石
cleat stanchion 梯柱
cleat strength 楔子强度
cleat tire chain 轮胎防滑(铁)链
cleat tyre chain 轮胎防滑(铁)链
cleat wiring 绝缘夹支承电缆;用瓷夹板固定电线;瓷夹布线
cleavability 可劈性;可裂性;可解理性;可节理性;解理性;劈裂性
cleavability of asbestos 石棉劈分性
cleavability of wood 木材可裂性
cleavable wood 可劈木材
cleavage 裂纹;裂开;开裂;晶体节理;解理性;解理;节理;层压制品层间开裂损坏;劈裂;劈理【地】
cleavage banding 劈理夹层
cleavage belt 劈理带
cleavage block 裂块
cleavage brittleness 晶间脆裂;解理脆性
cleavage crack 解理断裂;解理裂纹;劈裂
cleavage cracking 劈裂缝
cleavage crystal 解理晶体
cleavage domain 劈理域
cleavage energy 解理能
cleavage face 解理面
cleavage failure 劈裂破坏
cleavage fan 劈理扇;扇状劈理
cleavage fold 剪褶皱;劈理褶皱
cleavage fracture 解理破碎;可裂性破坏;解理断裂
cleavage grade 解理等级
cleavage in trace 劈理痕迹
cleavage method 解理法
cleavage-mullion 劈理式窗棂
cleavage peel-strength 劈裂剥离强度
cleavage plane 岩石天然裂面;裂开面;解理面;劈裂面;劈理面
cleavage product 分裂产物
cleavage reaction 分裂反应
cleavage splitting test 劈裂实验
cleavage strength 抗劈强度;抗劈力;解理强度;破坏强度
cleavage structure 劈理构造
cleavage surface 解理面
cleavage test 抗劈试验;抗剪试验;劈裂试验
cleavage velocity 解理速度

cleave 劈(开)
cleaved glass 黏合玻璃;夹层玻璃
cleaved wood 剖开的木材;劈开的木材
cleavelandite 叶钠长石
cleaver 瓦刀
cleaving 劈开
cleaving hammer 劈锤
cleaving saw 大锯
cleaving stone 页岩;板岩;片状岩
cleaving tile 组合地砖;片状瓦;裂纹地砖
cleaving timber 锯料;锯材;制材
Clebsch-Gordan coefficient 矢量耦合系数
cleck rod 牵条
cledge 漂白土上层
cleet 梁柱间的楔子
clef-non 碳酸钙
cleft 裂痕;裂缝;劈开的;深裂的
cleft-chestnut fencing 顺纹劈制栗木篱笆
cleft-chestnut paling 顺纹劈制栗木栅栏
cleft cutting 割插
cleft girdle 裂开环带
cleft graft 劈接法
cleft grafting 劈接
cleft-grafting stub 劈接
cleftiness 裂隙;节理
cleft timber 劈制材;顺纹劈开的木材;顺锯木材
cleft weld 裂口焊
cleft welding 裂口焊接
cleft wood 深裂木材;裂口木材
C-leg C形支柱;C形支腿;C形支架;C形支杆;C形支承
cleiophane 纯闪锌矿
cleithral 早期希腊建筑(有一个全遮蔽的屋顶)
Clemen counter vacuum flo(a)tation cell 克莱门式逆流真空浮选机
Clemen hardness apparatus 克莱门式刻痕硬度仪
Clemen hardness tester 克莱门式硬度计
Clemen scratch(ing) tester 克莱门式划痕硬度试验仪
Clement's driver 平衡拨盘
Clement's shuttering 克莱门活动窗板;克莱门活动百叶窗
Clemen vacuum flo(a)tation cell 克莱门式真空浮选机
clench 钉牢;钉住;敲弯;钳住
clench bolt 弯头螺栓
clencher 紧钳;夹子
clenching 砸弯;敲弯
clench nail 弯钉钉合用钉;弯头钉
clench nailing 敲弯;敲平钉子;钉牢;弯脚钉合;弯钉钉合
clench planking 盖瓦式叠板
clepsydra (古代计时用的)漏壶
clerestor(e)y 纵向天窗或楼;高窗;气楼
clerestor(e)y lighting 天窗采光
clerestor(e)y purlin 纵向天窗或楼楼条
clerestor(e)y roof 阁楼屋顶;天窗气楼屋顶;天窗顶板
clerestor(e)y window 阁楼天窗;顶部气窗;高侧窗
clerical accuracy 抄账无误
clerical cost 事务费用
clerical error 记录错误;笔误;书写错误
clerical machine 会计计算机
clerical staff 事务员
clerical type 书写体
clerical typographical error 笔误
clerical work 抄写工作;文书工作;文本工作
clerk 职员;办事(人)员
clerk general 一般工作人员
clerk of county court 县法院书记官
clerk of the works 监工(员);工程现场监工;工程管理员
clerk of work 工程管理员;现场监工员(英国)
clerk's office 秘书室;办事员室;事务室
cleuch 沟谷;拗沟
cleugh 峡谷;沟谷
cleve 悬崖;悬岩
cleveite 钇铀矿
Cleveland condensing humidity cabinet 克利弗兰冷凝型潮湿室
Cleveland flame point 克利弗兰法闪点
Cleveland flash tester 克利弗兰开杯闪烁式试验器
Cleveland high humidity cabinet 克利弗兰高湿试验箱

Cleveland open cup 克利弗兰开杯;克利弗兰敞口杯
Cleveland open cup flash point tester 克利弗兰开杯式闪点试验器
Cleveland open cup test 克利弗兰开杯闪烁式试验
Cleveland open cup tester 克利弗兰开杯(闪烁式)试验器
Cleveland tester 克利弗兰开杯试验器
Cleve's acid 克列氏酸
clevice (铁丝绳端的)吊环;U形夹子;马蹄钩;叉头
clevis 马蹄形钩;马蹄夹;马蹄钩;叉头;V形块;U形钩;弹簧安全钩
clevis and tongue coupling 槽型连接
clevis-base bracket for drop tube 立柱压管底座
clevis bolt 插销螺栓;套环螺栓
clevis drawbar 牵引环
clevis end clamp 双耳线夹
clevis end fitting for tube 管双耳接头
clevis end holder 套管双耳
clevis end wedge-type clamp 双耳楔型线夹
clevis eye 马蹄形钩孔;U形钩眼圈
clevis for phone rack 挂板
clevis joint 脚架接头;拖钩
clevis pin 马蹄钳栓;U形夹销;牵引钩联结销
clew iron 帆耳环;三角环
clew line 帆耳索
clew line block 隐轮滑车
chloride ion 氯根离子
cliachite 含铁铝土矿
clichder built ceiling boarding 互搭接钉天花板
cliche pattern plate 镶嵌型板;拼合型板
click 棘(轮)爪;单击【计】;掣子;掣手;插锁
click banking stop 棘爪限位片;止爪片
click bore 曲柄钻
clicker 冲裁机
clicker die 带刀切割模具
clicker press 带刀切割压力机
clicker press machine 冲裁机
clicket 阀
click filter 电键喀呖声消除滤波器
click motion 棘轮运动机构
click pulley 棘轮
click spring 棘轮弹簧;定位锁弹簧
click stop 止动爪;锁定光圈
click test 碰响试验(估计水泥土等相对硬度用)
click track 信号声带
clickwork 闸轮机构
client 建设单位;客户;交易人;业主;顾客;当事人;发包人;委托人;委托单位;事主
clientage 委托人;委托方;买方;挂号用户;委托关系
client concerned 有关当事人
clientele 事主;委托人;顾客
client full responsibility 买方负全责
client ledger 委托人总分类账;顾客总账;客户分户账;委托人分账户
client network 雇主网
clients' accounts 客户账户
client's adjusting entries 客户调整分录
client-server 客户机—服务器;主从式
client-server relationshop 雇主—服务者关系
client's personnel 业主职员
client's property 业主财产
client's representative 业主代表
client's written representations 客户签署意见
client terminal system 用户终端系统;顾客终端系统
cliff 悬崖(峭壁);断崖;陡壁;陡岸
cliff brick 煤矸石砖;歪斜的黏土砖
cliff channel 槽式
cliff coast 峭壁海岸
cliff debris 岩堆;坠积物;坡积物;山麓堆积
cliff drawing 地貌描绘
cliff dwelling(settlement) 峭壁居所;窑洞住宅;岩屋
cliffed coast 悬崖海岸;陡岸
cliffed headland 陡崖岬角
cliff fall 海蚀崖崩
cliff glacier 雪坑冰川;悬崖冰川
cliff in coral 珊瑚暗礁
cliff ladder 登崖坎梯
cliff line 陡岸线
cliff-maker 造崖层
cliff of displacement 断(层)崖
cliffordite 克碲铀矿
cliff protection 悬崖防护
cliffside 岩边;悬岩壁
cliff spring 悬岩泉;悬崖泉

cliffstone 崖石
cliffstone Paris white 响岩巴黎白
cliff work 地貌描绘
cliffy coastline 多峭壁海岸线
clift 硬泥岩
Cliftonian 克利夫顿期【地】
cliftonite 方晶石墨
cliket 节气阀;节气门;阀
climafrost 季节冻土;气候冻土层间融冻层
climagram 气候图解
climagraph 气候图表
climate 风土;气候
climate accident 气候偶变
climate alert 气候警报
climate analysis 气候分析
climate and soil conditions 气候条件和土壤条件
climate and weather 气候和天气
climate anomaly 气候异常;气候距平;气候反常
climate application 气候应用
climate application referral system 气候应用检索系统
climate belt 气候带
climate cabinet 气候条件模拟箱
climate cell 人工气候试验室;气候试验室
climate chamber 人工气候室
climate change 气候变迁;气候变化;气候变动
climate change cycle 气候变化周期
climate change detection project 气候变化检测计划
climate classification 气候分类
climate comfort 调节气温
climate computing 气候计算技术、方法
climate conditioning 调节气温
climate cycle 气候循环;气候(变化)周期
climate data 气候数据
climate divide 气候分界
climate during brief spring and summer 短暂的春夏两季气候
climate fluctuation 气候波动;气候变迁;气候变化;气候变动
climate forcing 气候作用力
climate forecast 气候预测;气候预报
climate hazard 气候危害
climate indicator 气候指示物
climate information 气候资料
climate issues 气候问题
climate long range investigation mapping prediction study 气候长期研究制图计划
climate model 气候模型
climate modification 人工影响气候
climate monitoring 气候监测
climate of coniferous forest on the subfrigid zone 亚寒带针叶林气候
climate of subtropic(al) forest 副热带森林气候
climate of Taiga 泰加林气候;原始森林气候;亚寒带气候;副极地气候
climate of tropic(al) table-land 热带高地气候
climate of tundra 冻土地带气候;苔原气候
climate optimum 气候适宜
climate parameter 气候参数
climate plant formation 气候植物区系
climate prediction 气候预测;气候预报
climate process 气候过程
climate record 气候记录
climate region 气候区域
climate region of building 建筑气候分区
climate science 气候科学
climate-sensitive activity 影响气候的活动;对气候敏感的活动
climate simulation 气候模拟
climate snow line 气候雪线
climate-specific cost 特定气候费用
climate stratigraphic(al) unit 气候地层单位
climate study 气候研究
climate system 气候系统
climate system monitoring 气候系统监测
climate trend 气候趋势
climate upheaval 气候激烈变化
climate variation 气候变迁;气候变化;气候变动
climate warming 气温升高;气候变暖
climate watch 气候监视
climate weathering cabinet 气候模拟室
climate/weather modification 气候/天气改变
climate zonation 气候地带性
climate zone 气候区(域);气候带
climatic adaptation 气候适应
climatic agencies 气候机构;测候机构
climatic amelioration 气候改善
climatic anomaly 气候异常;气候距平;气候反常
climatic arkose 气候长石砂岩
climatic belt 气候带
climatic booth 合片室
climatic cabin 合片室
climatic chamber 人工气候室;人工环境室
climatic chart 气候图
climatic climax 气候(演替)顶极
climatic climax vegetation 气候顶极植被
climatic comfort 气候舒适
climatic condition 气候条件
climatic conditions clause 气候情况条款
climatic control 气温控制;气候控制
climatic cycle 气候循环
climatic data 气候资料;气候数据;气候记录
climatic data center 气候资料中心
climatic data management 气候资料管理
climatic degeneration 气候恶化
climatic diagram 气候图解
climatic divide 气候分界
climatic division of humidification 气候的湿润性分区
climatic effect 气候效应
climatic element 气候因素;气候要素
climatic environment 气候环境
climatic environmental date 气候环境资料
climatic event 气候事件
climatic facility 气候设施
climatic factor 气候因子;气候因素;气候成因
climatic fluctuation 气候振动;气候变迁;气候变化;气候变动
climatic forecast 气候展望
climatic front 气候锋
climatic gasoline 适于气候的汽油
climatic graph 气候图表
climatic hygiene 气候卫生
climatic indicator 气候显示仪;气候标志
climatic information 气候资料
climatic laboratory 人工气候室;气候试验室
climatic life form 气候生活型
climatic map 气候图
climatic melioration 气候改良
climatic modification 气候控制;气候改善
climatic observation 气候观测
climatic observation station 气候观测站
climatic optimum 气候最优期;气候最适条件;气候最适度;气候适宜期
climatic oscillation 气候变动
climatic pal(a)eozone 古气候带
climatic pathology 气候病
climatic phenomenon 气候现象
climatic province 气候区
climatic record 气候记录
climatic region 气候区
climatic resources 气候资源
climatic rhythm 气候韵律
climatic scourge 气候灾害
climatic sensitivity 气候感应性
climatic shift 气候改变
climatic snow line 气候雪线
climatic soil formation 气候性成土作用;气候土壤形成
climatic soil types 气候性土类
climatic strain 气候应变
climatic stress 气候应力
climatic subdivision 气候付区
climatic symbols 气候符号
climatic terrace 气候阶地
climatic test(ing) 气候试验
climatic treatment 气候疗法
climatic trend 气候趋势
climatic type 气候型
climatic variability 气候的可变性
climatic variation 气候变迁;气候变化;气候变动
climatic year 气候年;水文年(度)
climatic zone 气候区域;气候带
climatic zone and climatic type 气候带和气候类型
climatic zoning 气候分带
climatic zoning for highway 公路自然区划
climatization 气候适应过程
climatized cabin 空气调节舱;气温调节舱
climatizer 气候实验室
climatogram 气候图(表)
climatograph 气候图(表)
climatography 气候志
climatolith 气候层
climatological atlas 气候图集;气候图册
climatological chart 气候图
climatological condition 气候情况;气候条件
climatological data 气候资料;气候数据
climatological day 气候日
climatological diagram 气候图表
climatological effect 气候效应
climatological forcast 以气候学方法预报
climatological normals 气候正常值;气候多年平均值
climatological normal value 气候正常值
climatological observation 气候观测
climatological parameter 气候参数
climatological prediction 气候预测;气候预报
climatological region 气候地区
climatological station 气候记录站
climatological statistics 气候统计学
climatological table 气候表
climatological zoning for buildings 建筑气候区划
climatologist 气候学家
climatology 风土学;气候学
climatology and meteorology 气候学和气象学
climatoloqical station 气象站
climatron 人工气候室
climax 极期;极点;最高峰(值);最高点;高阻镍钢;高峰期(间);高电阻铁镍合金;高潮;顶极;顶峰;顶点
climax alloy 铁镍整磁合金
climax avalanche 最大雪崩;强烈崩坍
climax biotic community 顶极生物群落
climax community 演替顶极
climax complex 顶极复合体
climax ecology 高峰生态
climax episode 高潮幕
climax forest 顶极森林
climax forest type 顶极森林类型
climax-life form 顶极生活型
climax of succession 演替顶极
climax phase of fault activity 最强烈活动时期
climax prairie 顶极高草原
climax soil 顶极土壤
climax species 顶极种
climax vegetation 顶极植被
climb 攀登;爬高
climbable gradient 能爬升的坡度(汽车)
climb and fall 上下坡
climb-cut grinding 同向磨削
climb cutting 同向铣削;顺铣
climber 攀缘植物;爬生器
climber crane 爬升起重机
climbers 脚刺
climb hobbing 同向滚削;顺向滚铣
climb indicator 升速指示器
climbing 攀移
climbing ability 车辆爬坡能力;爬坡能力
climbing ability test 爬坡性能试验
climbing aperture 提升窗口
climbing apparatus 攀登架
climbing bog 爬升沼泽地
climbing capacity 车辆爬坡能力;攀缘能力;爬坡能力;升高能力
climbing corridor 爬山廊
climbing crane 自升式塔吊;自升式起重机;攀缘式起重机;攀移式起重机;爬升式起重机(爬塔)
climbing cross lamellar structure 上攀沙纹交错纹理构造
climbing entada 过岗龙
climbing equipment 爬杆器
climbing fiber 爬行纤维
climbing-film evapo(u)rator 升膜蒸发器;按薄膜上升原理工作的蒸发器
climbing form 爬模;滑模;滑升模板;滑动上升模板;爬升模板;提升(式)模板;滑升模板
climbing formwork 提升式模板;爬升模板
climbing frame 攀登架;爬升套架
climbing iron 爬杆脚扣;金属踏步;脚扣;埋入金属踏步
climbing kiln 阶梯窑

climbing lane 加宽回车道;慢车坡道;爬坡车道
climbing mechanism 顶升机构;爬升机构
climbing nightshade 白英【植】
climbing number 上升系数
climbing of wheel flange onto rail 轮缘顺轨爬升
climbing plant 攀缘植物
climbing pole 爬杆
climbing power 爬坡能力;上升力;上坡牵引力
climbing ripple 爬升波痕
climbing ripple lamination 爬升波痕纹层
climbing rod 攀登杆
climbing rope 架索安全绳
climbing rose 攀缘蔷薇(植物)
climbing scaffold(ing) 爬升式脚手架;滑升脚手架
climbing sequence 攀升式层序
climbing shaft 攀登柱
climbing shrub 攀缘灌木;藤本植物
climbing shutter(ing) 提升模板;滑升模板
climbing space 攀登空间
climbing speed 爬升速度
climbing switch 攀登道岔;爬坡道岔
climbing therophyte 攀缘一年生植物
climbing tidal-cyclic(al) sequence 攀升式潮后周期层序
climbing tower crane 爬升式塔吊;爬升塔式起重机
climb milling 同向铣削;顺铣
climb motion 攀移运动
climb of dislocation 位错攀移
climb process 攀移过程
climb very steep slopes 爬陡坡
clime 气候;风土;地方
climograph 气象图;气候图表
climosequence 气候序列
clinch 钉住;待处理状态;打弯钉脚使固定;半结;敲弯;钳住
clinch a deal 成交
clinch bolt 夹紧螺栓;弯头螺栓
clincher 紧钳;铆钉;钳入式轮胎;扒钉
clincher boarding 鱼鳞板
clincher built ceiling boarding 搭钉天花板
clincher iron 密缝凿;填隙器
clincher rim 紧钳轮辋;钳入式轮辋
clincher tool 铆接工具
clincher tyre 紧嵌式轮胎;嵌合轮胎;箱入式轮胎;钳入式轮胎
clincher work 搭接工程;互搭接工程
clinching 砸弯(钉尖);敲弯
clinch joint 搭接接头
clinch nail 弯头钉;搭马【救】;抱钉;敲弯钉
clinch nailing 弯脚钉合
cline 渐变群;单向渐变群;梯度种;生态群
cline stratum 倾斜(地)层
cling 黏着;黏住;卷住
clingage 油垢(黏着在油箱壁的)附着油
clinging nappe 贴附水舌
clinic 门诊所;门诊部;医务所;诊疗所;卫生所;卫生室
clinical laboratory 临床化验室
clinical thermometer 医用温度计;体温计;体温表
clinicar 流动医疗车
clinic delegate 学术讨论会参加者
clink 裂口;(挖掘砾石或路面用的)尖棒;钢凿;劈楔
clinked ingot 开裂钢锭
clinker 炼渣;煤渣;灰渣;渣块;叠接列板;熔渣;熔块渣;熔块;熔渣;渣块;头等货;熟料带走热;熟料
clinker activity curve 熟料活性曲线
clinker adhesion 渣瘤
clinker aggregate 熔渣集料;熔渣骨料;熔块集料;熔块骨料;熟料;烧成集料;烧成骨料
clinker aggregate concrete 熔渣集料混凝土;熔渣骨料混凝土
clinker aggregate for concrete 混凝土熔渣骨料
clinker asphalt 溶渣地沥青
clinker asphalt mixture 溶渣地沥青混合料
clinker-bearing slag cement 矿渣硅酸盐水泥;含熟料矿渣硅酸盐水泥
clinker bed 熔渣块层;熟料层
clinker bleaching 熟料漂白
clinker block 矿渣混凝土砌块;熔渣砌块
clinker breaker 熟料破碎机
clinker brick 过烧砖;炼砖;缸化砖;熔渣砖;熔结砖;渣砖;熔渣缸砖
clinker built 盖瓦式叠板;叠接式构造法;搭接式(壳板);鳞状搭接的

clinker bulk density 熟料松密度
clinker burning 熟料烧成;熟料煅烧
clinker cement 矿渣水泥;熟料水泥
clinker cement brick 熟料水泥砖
clinker chemistry 水泥熟料化学
clinker chute 熟料溜槽
clinker clew 熔块
clinker clue 熔块
clinker coating 窑皮;熔渣盖层
clinker composition 水泥熟料组成;水泥熟料成分
clinker concrete 矿渣(集料)混凝土;矿渣骨料混凝土;熔渣混凝土
clinker constituent 水泥熟料成分
clinker constitute 熟料组分
clinker conveyer 熟料输送机
clinker cooler 熟料冷却器;熟料冷却机
clinker cooler efficiency 熟料冷却机效率
clinker cover 缸砖覆盖
clinker crusher 熟料破碎机
clinker crusher pit 熟料破碎机坑
clinker discharge gate 熟料卸料闸门
clinker discharge loss 熟料卸出带走的热损失
clinker dolomite 白云石熟料
clinker drag chain conveyer 熟料拉链机
clinker dust 熟料粉尘
clinkered body 烧结体
clinker field 块熔岩区
clinker floor 熔渣地面;矿渣砖地面;矿渣式混凝土;缸砖楼地面;缸砖地面
clinker formation 水泥熟料形成;熔渣形成;熟料形成
clinker-forming zone 熟料形成带
clinker for road construction purpose 道路建筑炼砖
clinker-free cement 无熟料水泥
clinker glass 炉渣玻璃
clinker grain 水泥熟料颗粒;熟料粒
clinker grate 倾倒炉算
clinker grinding 水泥熟料研磨;熟料粉磨
clinker grinding mill 水泥熟料磨;熟料粉磨机;水泥磨
clinker grinding test 熟料粉磨试验
clinker handling conveyer 熟料输送机
clinker hard-burned brick 炼砖
clinker hole 熔渣孔
clinker hopper 熟料斗;熟料斗
clinkering 炉条烧结;熔结;排渣;烧炼;烧渣;烧结
clinkering coal 易结渣煤
clinkering contraction 烧结收缩
clinkering crack 烧结裂缝
clinkering expansion 烧结膨胀
clinkering iron 烧结铁
clinkering point 成熟点(指熟料);熔结温度;熔结点;烧结点
clinkering property 结渣性
clinkering property curve 结渣性曲线
clinkering range 烧结范围
clinkering rate 结渣率
clinkering strain 烧结变形
clinkering zone 熔结带;烧结区
clinkerization diagram 烧结相图
clinkerless cement 无熟料水泥
clinker liquid 熟料液相
clinker masonry 缸砖砌体;熔渣(块)圬工
clinker material 水泥熟料矿物质;熔渣材料
clinker matrix 熟料基体
clinker mill 熟料磨;水泥磨
clinker mineral 熟料矿物
clinker mineral of cement 水泥熟料矿物
clinker nodules 熟料球
clinker of alumina oxide 铝氧熟料
clinker outlet 熟料出口
clinker output 熟料产量
clinker pavement 矿渣铺面;缸砖路面;熔渣路面
clinker paving 缸砖铺地
clinker phase 水泥熟料状态;熔渣状态;熟料相
clinker pile 熟料堆场
clinker pit 熟料储[贮]库;熟料坑
clinker planking 盖瓦式叠板
clinker plating 缸砖覆盖;盖瓦叠板式
clinker porosity 熟料孔隙度
clinker pot 空心熔渣砌块
clinker produced under reduced condition 还原状态下生成的熟料

clinker recirculating process 熟料循环过程
clinker reduction 熟料破碎
clinker retarding lip 挡料圈
clinker ring 水泥熟料结圈;熟料(结)圈
clinker road 熔渣路;炼砖路;缸砖路
clinker sausaging 熟料结大块
clinker scale 熟料秤
clinker screen 滤烟网
clinker screw 熟料螺旋输送机
clinker sintering 熟料烧成
clinker slab 熔渣混凝土板
clinker slag cement 熔渣水泥;炉渣水泥
clinker spreader 熟料散布器
clinker stockpiling facility 熟料堆存设施
clinker stockpiling plant 熟料堆场
clinker storage(building) 熟料储房;熟料储[贮]库
clinker storage hall 熟料棚;熟料库
clinker storage shed 熟料堆棚
clinker store 水泥熟料库
clinker strake 缸砖砌筑长槽
clinker tile 熔渣混凝土砖;烧结釉面砖
clinker-tile tube 缸瓦管
clinker tongs 拔火钳
clinker waste heat 熟料余热
clinker with brown core 黄心料
clinkery 熔渣的;烧结的
clinkery brick 缸砖
clinking 内裂缝
clink paving brick 金砖;特硬方砖
clinks 响裂
clinkstone 响岩
clino 海底斜坡
clino-amphibole 单斜闪石
clino-axis 斜轴
clino-axis circular current 斜轴环流
clinobisvanite 斜钒铋矿
clinochalcomenite 斜兰硒铜矿
clinochlore 斜绿泥石
clino-chrysotile 单斜纤维蛇纹石
clinoclase 斜节理;光线石;光线矿
clinoclasite 斜轻砷铜矿;光线矿;光线石
clino-diagonal 斜轴;斜径
clinodome 斜轴坡面
clinoenstatite 斜顽辉石
clinoenstatite porcelain 斜顽辉石瓷
clinoferrosilite 斜铁辉石
clinoform 陆坡地形;斜坡地形;斜坡沉积;倾斜型
clinograph 孔倾计(用以测量竖坑的倾斜度);绘图平行板;斜线尺;钻孔测斜仪;倾斜记录仪;平行板画图器;平行板
clinographic(al) curve 坡度曲线
clinographic(al) projection 斜射投影法
clinohedral 斜面体
clinohedrite 斜晶石
clinoholmquistite 斜锂闪石
clinohypesthene 单斜紫苏辉石
clinoid 鞍突
clinojimthompsonite 斜镁川石
clinokurchatovite 斜硼镁钙石
clinometer 量坡仪;测斜仪;测倾仪;测角器;倾角仪;倾角计;倾侧计
clinometer case 测斜仪外壳
clinometer rule 测斜仪量角器
clinometer shell 测斜仪外壳
clinometric height 由测斜仪测定的高程
clinophone 测斜仪
clinophosinaite 斜磷硅钙钠石
clinopinacoid 斜轴面
clinoprism 斜轴柱
clinopyramid 斜轴锥
clinopyroxene peridotite 单斜辉石橄榄岩
clinopyroxenite 单斜辉石岩
clinopyrrhotite 斜磁黄铁矿
clinopyxene 单斜辉石
clinorhoboidal system 三斜晶系
clinorhombic system 单斜晶系
clinorhomboidal 三斜
clinosafflorite 斜砷钴矿
clinoscope 侧滑指示器;倾斜仪;水平孔测斜仪
clinosol 坡积土
clinostat 回转器
clinothem 斜坡岩层;斜坡沉积
clinotyolite 斜铜泡石
clino-unconformity 斜交不整合

clinoungemachite 斜碱铁矾
clinozoisite 斜黝帘石
clinquant 金光闪闪的;银光闪闪的
clint 岩沟
Clintonian 克林顿阶【地】
Clinton iron deposit 克林顿铁矿床
clintonite 绿脆云母
Clinton Limestones 克林顿石灰岩
Clinton shales 克林顿石灰岩
clip 扣钩;剪断;夹(子);夹住;夹头;夹片;蚂蟥钉;回形针;小夹;钢夹;钢轨扣件;切砖;钳牢;卡子
clip and shave 铲修
clip angle 扣角钢;耳状角铁
clip applying and removing forceps 创缘夹联合钳
clip band 夹条节;夹条带
clip bar 夹杆
clip board 夹纸(垫板);记录板夹;弹簧夹板
clip bolt 夹箱螺栓;夹紧螺栓;夹箍螺栓
clip bond 斜切式砌合法;钳制砌合;顺砖内角斜切砌合体
clip brake 夹紧制动器
clip connector 夹(子)连(接)器;夹子接头;夹式接头
clip cord 夹子软线
clip(er) hooks 双抱钩
clip fastener 封管机
clip fastening 扣板式扣件
clip fit 滑配合
clip for frog legs 蛙腿夹
clip for rainwater downpipe 弹夹(用于水落管的)
clip ga(u)ge 夹持式变形计
clip-holding forceps 持夹镊
clip hook 抱钩
clip joint 钳制接头;加厚灰缝;修整缝;砖砌体钳制接头;砖砌厚缝
clip joints of rubble masonry 毛石圬工钳制接头
clip level 削波电平;限幅电瓶;限幅电平
clip nut 夹紧螺母
clip-on ammeter 钳形电流表
clip-on ceiling panel 卡装式吊顶板
clip-on cooling unit 可装配式冷却器
clip-on-refrigerating machine 集装箱制冷却
clip-on skirting 卡装式踢脚线
clip-on unit 夹箍式冷冻装置;夹接部件
clip-on voltammeter 钳形伏安表
clip ped bond 钳制砌合
clipped circuit 限幅电路
clipped eaves 短檐;截头檐
clipped gable 四坡屋顶;截山尖(斜坡屋顶);歇山屋面;半山头【建】
clipped gable roof 折山头屋顶;嵌制砌合的人字屋顶
clipped header 假丁砖;修饰丁(头)砖;截丁砖;半头砖
clipped hedge 整剪绿篱
clipped lintel 修饰过梁;截过梁
clipped noise 消波噪声
clipped trace 限幅迹道
clip ped tree 剪形树
clipped wave 限幅波;被削波
clipped wire 金属粒
clippe joint 钳连接
clipper 快速帆船;剪板机;削波器;限制器;斩波器;裁切机;人力切草机;钳位器
clipper amplifier 限幅放大器
clipper bottom 草地用熟地型犁体
clipper bow 飞剪型船首;飞剪式船首;曲船首
clipper circuit 削波电路
clipper-limiter 双向限幅器
clippers 剪取器;剪刀;尖口钳子;尖嘴钳;夹口钳;大剪刀;钳子
clipper seal 钳压密封
clipper service 快速服务;快捷服务
clipper stem 飞剪型船首;飞剪式船首;曲线船首
Clipperton fracture zone 克利帕顿破裂带
Clipperton latitudinal tectonic belt 克利帕顿纬向构造带
clipper transport 快速运输
clipper tube 削波管
clippie 石板瓦工的脚手架
clipping 剪下物;剪辑;剪裁;剪截;限幅
clipping amplifier 削波放大器
clipping and sculpturing technique 剪片工艺

clipping circuit 削波电路;限幅电路
clipping constant 削波器时间常数
clipping distortion 削波畸变
clipping indicator 剪取指示符
clipping machine 剪断机;修器器;切棒机
clipping machine for brick and tile 砖瓦切割机
clipping pasture 打草场
clipping path 剪取路径
clipping plane 剪取平面;裁剪平面
clipping press 切边压力机
clipping room 安装间
clip plate 夹轨底垫板;夹板;压板
clip protection 空转保护
clip ring 开口环;弹簧挡圈
clips 铰剪;大剪刀
clip screw 调置螺旋
Clipsham stone 克里普山姆石;(英国产的)浅黄色石灰岩
clip spot loom 浮点纹织机
clip-spring switch 弹簧开关
clip stretcher 链式展幅机
clip terminal 线夹端子
clip test 敲裂试验
clip tile 工字钢底砖
clip tingle 压板铁片
clip type connection 活动安装连接
clip with wood(en) board 木板夹
clip wood 薄木片
clique 集团型;操纵股票市场小集团;阀
clitellum 环带
clithral 早期希腊建筑(有一全遮蔽的屋顶)
cliver 吊环(吊钻杆用);吊钩(吊钻杆用)
clivis 山坡
clivus 斜坡
clivus multrum toilet 生物处理厕所(其中的一种)
clivvy 弹簧安全钩
cllipsoidal harmonics 谐和函数
Cll-tyretskite 三斜氯羟硼钙石
cloaca 暗渠;便所;下水道;阴沟
cloaca maximum 最大排水沟(古罗马)
cloak 斗篷;覆盖物(防水用)
cloak hook 挂衣钩
cloak rail 挂衣板;挂衣板
cloakroom 寄物处;行李存放处;衣帽间
cloakroom locker 衣帽间锁柜
cloak screen 衣帽间围屏;遮幕
cloak stand 衣帽架
clob 泥炭田
clobber 乱码
clobbered ware 加彩瓷器
clock 节拍;同步脉冲;时钟
clock amplifier 时钟脉冲放大器;时钟放大器
clock bit 时钟位
clock buffer 节拍缓冲器
clock card 考勤卡
clock chain 时钟链
clock chamber 钟室;教堂钟塔上的钟室
clock circuit 时钟(脉冲)电路
clock coder 时间编码
clock comparator 时钟比较器
clock comparison 时钟对比
clock control 钟控;时钟控制
clock controller 定时控制器
clock control system 钟控系统;时钟控制系统
clock correction 时差;时钟校准
clock counter 时钟计数器
clock cycle 同步脉冲周期;时钟脉冲周期
clock diagram 相量圆图;矢量圆图
clock dial micrometer 钟盘测微计
clock driver 时钟(脉冲)驱动器
clocked flip-flop 时标触发器
clock edge 时钟脉冲边沿
clocked inverter 拍频倒相器
clocked logic 时钟逻辑
clocker 计时员;钟楼
clock error 钟差;时钟校准
clock fork 钟摆纵叉
clock frequency 节拍频率;钟频;时钟(脉冲)频率;时钟节拍频率
clock gate 时钟脉冲门
clock ga(u)ge 千分表
clock generator 时钟生成器;时钟(脉冲)发生器
clock generator and driver 时钟发生驱动器
clock glass 表面皿;表面玻璃

clock-hour figure 钟时序数
clock-house 记录时间室;守卫室
clock in 时钟脉冲输入
clock in and clock out 上下班计时
clock indication 钟面读数;时钟读数
clocking 定时;产生时钟信号
clock input 时钟(脉冲)输入;时标输入
clock installation 时钟装置
clock interrupt 时钟中断
clockless 无节拍
clock mechanism 钟表机械装置;时钟机构
clock meter 钟表式计数器;钟式计时器;钟表机构计数器
clock module 时钟组件
clock motor 计时电动机;电钟用电动机
clock multivibrator 时间间隔发送器
clock oscillator 时钟振荡器
clock paradox 时钟佯谬
clock pen 时钟笔尖
clock pendulum 钟摆
clock period 时钟周期;时钟脉冲周期
clock-phase diagram 直角坐标矢量图
clock plot 时标图
clock pulse 时钟节拍脉冲;计时脉冲;同步脉冲;时钟脉动;时钟脉冲
clock pulse frequency 时钟脉冲频率
clock pulse generator 时钟脉冲发生器
clock pulse source 计时脉冲源
clock-pulse width 时钟脉冲宽度
clock qualifier 时钟脉冲限制器
clock radio 时钟收音机
clock rate 钟速;钟频;同步脉冲重复频率;时钟速率;时钟日差;时标速率
clock reading 针盘读数
clock recorder 钟表式计时器
clock recovery 时钟恢复
clock repetition rate 时钟脉冲重复频率
clock room 钟房
clock run-in sequence 时钟脉冲进入顺序
clock set 时钟校准
clock shift register 时钟移位寄存器
clock signal 时钟信号
clock signal frequency 时钟信号频率
clock signal generator 时钟信号发生器
clock skew 时钟歪斜;时钟脉冲相位差
clock source 时钟脉冲源
clock star 对时星;测时星
clock switch 时间开关
clock synchronization 时间同步
clock synchronizer 母钟;中心电钟;时钟同步器
clock system 钟控制系统;子母钟系统;时钟系统;时钟脉冲系统
clock terminal 时钟脉冲终端
clock test 时钟信号测试
clock time 钟面时;时钟时间
clock time of the middle time signal 中央时刻
clock time of transit 中天面时
clock-time scheduling 时钟调度
clock tower 钟楼;钟楼
clock track 时钟(脉冲)道;时标道
clock transaction 时钟更新
clock transportation 搬钟
clock turret 钟楼
clock-type dial 圆形刻度盘
clock type speedometer 钟式速度计
clock-unit 时钟单位
clock valve 单向铰链阀
clock weight 钟锤
clockwise 顺时针方向的
clockwise angle 顺时针角
clockwise cycling 顺时针循环
clockwise direction 顺时针方向
clockwise drift 顺时针方向偏移
clockwise drive 右旋传动
clockwise helical polarization 顺时针螺旋偏振
clockwise motion 顺时针方向运动
clockwise polarized electromagnetic wave 顺时针偏振电磁波
clockwise polygon 顺时针方向多边形
clockwise rotation 顺时针(方向)转动;顺时针(方向)旋转
clockwise vortex 顺时针漩涡
clockwork 发条装置;时钟机构
clockwork anemometer 钟制风速表

clockwork fuse 定时引爆
clockwork indicator 钟表装置指示器
clockwork motor 时间电动机
clockwork-motor camera 钟表机构控制摄影机
clockwork-triggering primer charge 带定时装置起爆药
clod 泥块;煤层软泥土顶板;煤层顶部底板页岩;岩块;成块;软质顶板;土块
clod-air pipe 冷气管
clod breaking roller 碎土镇压器
clod buster 土块破碎机
clod clearer 土块清除器
clod disintegrator 土块粉碎器
cloddy pulverescent structure 碎块状构造
cloddy pulverescent texture 碎块状结构
cloddy structure 块状构造;团块状构造
cloddy texture 团块状结构
clod eliminator 土块分离器
clod patch 冷法修补
clod settler 低温澄清池
clod smasher 土块破碎机
clod structure 碎块状构造
clod test 土块实验
clod texture 碎块状结构
clod top 黏土覆盖层;页岩覆盖层
clog 止轮器;止动器
clogged 阻塞的
clogged filter 阻塞了的滤池;阻塞滤池
clogged lines 阻塞了的管线
clogged sand filter 堵塞(了的)砂层
clogged sand layer 堵塞(的)砂层
clogged screen 阻塞了的筛网
clogged tube 管子阻塞;闭塞管
clogged up altitude of free water elevation 地下水壅高值
clogging 淤塞;阻塞;堵眼;堵塞现象;堵塞;闭合
clogging capacity 堵塞容量;容尘量
clogging front 阻塞锋;堵塞锋
clogging machine 堵眼机
clogging of a trickling filter 滴滤池污塞
clogging of cutterhead 铰刀头堵塞
clogging of filter element 滤清器元件堵塞
clogging of oil screen 滑油滤网堵塞
clogging of sewers 管道堵塞
clogging of the burner 烧嘴堵塞
cloggy 易堵塞的
clogproof 防堵塞的
clog-proof crawler track 防阻塞履带
clog seize 黏附
clog snow 黏雪
clog up 堵塞
cloisonne 景泰蓝
cloisonne enamel 景泰蓝
cloisonne pigment 景泰蓝彩颜料
cloister 回廊;寺院;走廊;拱廊
cloister arch 回廊拱
cloister cemetery 修道院墓地
cloister garth 回廊中庭;回廊庭院
cloister vault 回廊穹隆;回廊穹顶
cloister walk 走廊式人行道;游廊步道
cloistral 寺院的
clomiphene 哥罗米酚
clone 纯系;无性繁殖系(植物)
clone test 系比试验
clone trial 系比试验
Cloquet's canal 玻璃体管
closable 可关闭的
close 紧密的;截止;结束;结清;密贴;密集的;密的;弥合;闷热的;合闸;闭市
close a bargain 成交
close aboard 接近船舶
close a business 歇业;停业
close a case 结案
close account 结账;结算;结清户头;清账
close a deal 达成交易
close aggregation contact 近凝聚
close agreement 吻合
close alley 小胡同
close a mountain pass 封山
close an account 结算;结清账户
close an account with 停止交易
close analysis 精密分析;周密分析
close annealing 箱中退火
close approximation 相当准确的近似值

close at hand 迫近
close a well 封闭井眼
close ban 壁橱墙
close bed 紧密层;密实层
close bend test 对折弯曲试验
close bevel 锐角
close bid price 出价收盘
close binder(course) 密实结合层;结合层;密结合料;密级配结合层;密合层
close blade pug mill 密闭式叶片搅拌器
close block 关闭程序
close board 屋顶锚固瓦条;封板;围栏木板
close-boarded 密封衬板;整片铺设木板;密合铺板;封板(围栏)
close-boarded(battened) roofing 密铺板屋面
close-boarded fencing 密合挡板围栏;密排木版围篱;鱼鳞板围篱
close-boarded hoarding 临时篱板
close-boarded roof 密铺木板屋顶;整片铺设屋面板
close-boarded screen 密排模板隔墙
close-boarding fencing 鱼鳞板围篱;鱼鳞板围栏
close-boiling mixture 窄沸点混合物
closeburn (产于弗立斯的)红色砂岩
close-burning 成焦;熔结
close burning coal 炼焦煤
close business sale 停业削价出售
close button 电路接通按钮
close buyer 专买便宜货的人
close canopy 密集林冠
close cap flash point 闭阀闪点
close ceiling 密铺舱底板
close cell 闭合环流
close centered training 闭心式整枝
close chain 闭链
close check 严格检验;严格检查
close circuit battery 常流电池组
close circuit TV equipment for station 车站闭路电视设备
close-circuit wind tunnel 回路式风道
close classification 细目分类(法);细分类法
close clearance 紧公差;密合间隙;小间隙
close coating 牢固镀层;致密涂层;致密镀层
close coiled helical spring 密卷螺簧
close-coiled spring 闭式盘簧;密绕螺旋弹簧
close collision 近距离碰撞;小冲击参数碰撞
close-confinement 封闭限制
close confinement junction 紧限制结
close-confinement laser 封闭限制式激光器
close-connected bucket chain 连续斗链
close-connected bucket dredge(r) 连续斗链挖泥船
close connection 紧密联系
close connections of trains at districts and marshalling stations 列车紧密衔接
close contact 紧密接触
close-contact adhesive 紧密触压胶粘剂;贴合黏结剂;贴合胶粘剂;梳状黏合胶粘剂
close-contact glue 密缝胶
close continuous mill 多机轧制的连续式轧机
close control 仔细测试;精确检查;精确测试;近距离控制;紧密控制;接近引导
close corporation 内股公司(股票不对外公开)
close correlation 密切关系
close couple 紧密合;顶对椽子;紧密联结;密耦;人字屋架上弦的紧密联结
close-coupled 密背的;强耦合
close-coupled closet 双连厕所;双连壁橱
close-coupled integral water closet 带水箱的坐式大便器
close-coupled mounting 近联式悬挂
close-coupled pump 双吸泵;密耦泵;封闭式耦合泵;双级泵;紧偶合泵
close-coupled roof 带拉条的人字形屋面;简易人字屋顶
close-coupled solar water heater 紧凑式太阳热水器
close-coupled tank and bowl 密耦水箱坐式大便器
close-coupled wash down closet 紧密装配冲落式坐便器
close-coupled water closet 紧密装配式抽水马桶
close-couple roof 三脚架屋顶;有拉杆的人字屋
close coupling 紧耦合;密耦合;强耦合
close critical control 精密临界控制
close-cropped 矮茬的
close cut 截短;间道;近路

close-cut hip 准确切割戗脊瓦;密合屋脊;闭合戗脊;屋顶戗脊
close-cut valley 闭合屋谷;屋顶斜沟;准确切割斜沟瓦
close-cycle control 闭路控制
close cycle system 闭式循环系统;全回风系统
closed 封禁的;封闭的;闭合的;办公时间已过
closed abutment 封闭式桥台
closed access 封闭式出入口
closed account 结清账户;已结清账户;已结平账款
closed accounting 已结清账户
closed aerial 闭路天线
closed aerodrome 关闭航空站
closed aggregate 闭集
closed agitator 密闭搅拌器
closed-air aspirator 空气封闭循环式吸气器
closed air-cooling system 封闭式气冷系统
closed air cycle 闭式空气循环
closed and self-holding 闭合并自锁
closed angle 尖角;锐角
closed annealing 闷罐退火;密闭退火
closed antenna 闭路天线
closed anticline 闭合背斜【地】
closed area 禁区;禁航区;闭合面积
closed armo(u)ring 迭盖铠装电缆;封闭式铠装
closed array 封闭数组;闭数组;闭(合)阵列;闭合数组
closed assembly time (黏结的)叠装时间;预定装配时间;堆积(受压)时间;(黏结的)叠接时间
close date 截止收货日期;交运截止日
closed azimuth angle 闭合方位角
closed backwater 闭式回水
closed ball mill 闭路球磨机
closed barge carrier system 封闭式驳船货运系统
closed basin 内陆渠;内流流域;有闸港低;封口港池;封闭式港池;封闭盆地;封闭流域;闭合流域;闭合港池
closed bay 隐蔽海湾;封闭海湾;闭合海湾
closed bead 封闭的筋
closed bearing 闭式轴承
closed bell 闭式潜水钟
closed-belt conveyer 封闭型带式输送机;闭式传送带
closed bevel 锐斜角
closed bid 秘密出价投标;限员投标;有限投标;不公开招标;暗标兑投
closed bidding 非公开招标
closed bids 密封递价
closed bin 封闭式料仓
closed binder 密结合层
closed bloc 封闭集团
closed block 闭锁部件
closed body 轿式车身
closed booth 大容积密闭罩
closed box container 封闭式集装箱
closed box groove 矩形孔型;闭口式箱形孔型
closed box pass 闭口箱形孔型
closed breeding 闭锁育种
closed bridge 遮蔽式桥梁
closed brine system 封闭式盐水系统
closed buffer case 闭式缓冲筒
closed buffer cycle 闭式缓冲筒
closed buffer shell 闭式缓冲筒
closed building 封闭的建筑;有围护的房屋
closed building block module 紧密建筑的区段模式;紧密建筑的街区模式
closed butt gas pressure welding 闭式加压气焊
closed butt joint 紧密连接;紧密对接
closed cage 封闭型阀门罩(深井泵)
closed-capsule diffusion 闭管扩散
closed car 轿车
closed cargo 列入表定运费的货物;非自由货物
closed ceiling 封闭式天棚
closed-cell 密封槽;密闭槽;独立气泡;闭孔气泡
closed-cell cellular material 闭孔泡沫材料;闭孔式多孔材料
closed-cell foam 密封微孔泡沫;闭孔泡沫;闭孔发泡料
closed-cell foam plastics 闭孔泡沫塑料
closed-cell foam seal 闭孔气泡密封剂
closed-cell sealant backing 闭孔密封膏背衬材料
closed-cell structure 闭孔结构
closed center (阀的)中立关闭
closed center hydraulic system 闭心式液压系统

closed center valve 中间位置封闭的阀
closed centrifugal pump 封闭式离心泵
closed chain 固定链系
closed chain compound 闭链化合物
closed chain hydrocarbon 闭链烃;闭链烃
closed chain series 闭链系
closed chamber 封闭舱室
closed chamber fusion weld 闭室熔焊
closed channel 闭流路;闭槽路
closed chimney packing 烟囱式码法
closed chock 导缆孔;闭式导缆钳
closed chute 封闭溜槽
closed cipher system 封闭的密码体制
closed circuit 闭路循环;闭路(电路);闭合环路;闭合电路;圈流
closed circuit air cooling 闭环空气冷却
closed circuit air turbine 闭式循环空气涡轮机;闭式循环空气透平
closed circuit ball mill 闭路球磨机
closed circuit battery 闭路电池组;持续作用电池组
closed circuit breather 闭式循环呼吸器
closed circuit breathing apparatus 封闭式潜水呼吸器
closed circuit breathing gear 闭式循环呼吸装置
closed circuit breathing rig 闭式循环呼吸装置
closed circuit breathing system 闭式循环呼吸系统
closed circuit cell 常流电池
closed circuit closed circuit crushing system 闭路破碎系统
closed circuit communications system 闭路通信系统
closed circuit connection 闭路接线法
closed circuit cooler 闭式环流冷却器
closed circuit cooling 闭式环流冷却;环路冷却;二次循环冷却
closed circuit cooling system 闭路冷却系统
closed circuit coupling 闭路式连接
closed circuit crusher 闭路循环破碎机
closed circuit crushing 闭路破碎
closed circuit cyclone separator 带旋风筒的循环空气选粉机
closed circuit deduster 闭路除尘器
closed circuit diver's breathing system 闭式回路潜水员呼吸系统
closed circuit diving apparatus 密闭式循环潜水装置;闭锁回路式潜水装具
closed circuit fisling system 闭路加注系统
closed circuit grinder 闭路磨矿机
closed circuit grinding 闭路磨碎;闭路磨矿;闭路粉磨;圈闭流磨;密闭系统磨细
closed circuit grinding in combination with air separation 带选粉的闭路粉磨
closed circuit grinding system 闭路粉磨系统;圈流粉磨系统
closed circuit grouting 环式式灌浆;闭路(压力)注浆(法);循环式闭路灌浆(法);闭合压力灌浆
closed circuit hot water diving apparatus 闭式回路热水潜水装置
closed circuit hot water diving gear 闭式回路热水潜水装置
closed circuit hot water diving suit 闭式回路热水潜水服
closed circuit laser 闭路激光器
closed circuit lubrication 闭式环流润滑;闭路润滑
closed circuit method 闭路法;闭路粉磨法;闭式测定法
closed circuit mill 闭路磨
closed circuit mill with bucket elevator 带斗式提升机的闭路磨
closed circuit nozzle 闭循环式喷嘴
closed circuit oiling 闭式环流润滑
closed circuit oil system 闭路循环油系统
closed circuit operation 闭路循环作业;闭路操作
closed circuit oxygen equipment 闭路循环氧气设备
closed circuit pipe system 闭路(式)管(道)系(统)
closed circuit pressured lubrication 闭合加压润滑
closed circuit pulverizing 闭路研磨法
closed circuit radiator 封闭式采暖炉
closed circuit recording 闭路录像
closed circuit refrigeration 闭合循环致冷
closed circuit screen 闭路筛
closed circuit scuba 闭路自携式水下呼吸器

closed circuit self-contained underwater breathing apparatus 闭式回路自携式水下呼吸器
closed circuit signal(l)ing 闭路信号发送;闭路通信
closed circuit steam heating 封闭式蒸汽加热
closed circuit system 闭式循环系统;闭路制;闭路式;闭电路制
closed circuit telegraph system 闭路电报系统
closed circuit television 有线电视;闭路式电视;闭路电视
closed circuit television camera 闭路电视摄像机
closed circuit television common antenna system 闭路电视共用天线系统
closed circuit television control system 闭路电视控制系统
closed circuit television set 闭路电视接收机
closed circuit torque testing 闭路力矩试验
closed circuit transmission 闭路传输
closed circuit TV camera 闭路式电视摄像机
closed circuit underwater breathing apparatus 闭路式水下呼吸器
closed circuit ventilation 闭路式通风
closed circuit voltage 闭路电压
closed circuit water cooling 封闭式水冷却
closed circuit water system 闭路循环用水系统
closed circuit wet grinding system 闭路湿法粉磨系统;圈流湿法粉磨系统
closed circuit wet raw grinding 闭路湿法原料粉磨;圈流湿法原料粉磨
closed circuit working 闭路式工作
closed circulation 闭型循环
closed circulation of ash-sluicing water 冲灰水闭路循环
closed circulation system 闭锁循环系统
closed clamp 闭合夹
closed clause 闭子句
closed coat 紧密饰层;紧密涂层;密上胶层;密砂纸;密砂布
closed coil 通路线圈
closed coil armature 闭圈式电枢;闭圈电枢
closed coil spring 密圈弹簧
closed coil winding 闭圈绕组
closed combinatorial manifold 闭组合流形
closed community 密生群丛;密闭群落;郁闭群落;闭生群落
closed complex 闭复合形
closed condenser 闭式冷凝器
closed conduit 暗沟;密闭管道;封闭(式)管道;封闭管渠;暗渠;暗管
closed conduit drop 涵洞坡降;暗沟落差
closed-conduit flow 封闭式管道流;暗渠流
closed-conduit hydraulics 暗渠水力学
closed conduit wiring 暗管线
closed conference 闭门同盟
closed configuration 闭合磁位形
closed construction 封闭式构造;封闭结构;闭合多边形刚架实体结构;实体式结构;隐蔽结构
closed contact 闭合接点
closed container 密闭容器
closed contour 闭合等高线
closed convex hull 闭凸包
closed convex set 闭凸集
closed cooling box 密闭式冷却箱
closed cooling system 封闭冷却系统;闭式冷却系统
closed cooling tower 闭式冷却塔
closed cooling tower system 闭式冷却塔系统
closed core 闭口铁芯;闭合铁芯
closed core transformer 闭口铁芯变压器;闭磁路变压器
closed core type 闭合铁芯式
closed core type reactor 闭合铁芯型电抗器
closed core type transformer 闭合铁芯式变压器
closed cornice 匣形飞檐;窗间盒;封闭式挑檐
closed corporation 封闭式公司
closed correspondence 闭对应
closed country 荫蔽地区;断绝地;丘陵地带
closed-coupled pump 固定连接泵
closed coupling 永久接合;刚性联轴节;死接合
closed courtyard 封闭式庭院
closed crankcase compressor 闭式曲轴箱压缩机
closed crankcase ventilating system 闭式曲轴箱通风装置
closed crosshead 封闭十字头
closed cross section 闭合截面

closed cup flash point test 闭杯闪点试验
closed cup flash point tester 闭杯闪点试验器
closed cup flash test 闭皿闪点试验
closed cup flash tester 闭杯闪点测定器
closed cup test 闭皿闪点试验;闭口杯试验
closed cup tester 闭杯试验器
closed curve 封闭曲线;闭曲线;闭合曲线
closed cutter 闭式绞刀
closed cutter head 闭式绞刀头
closed cycle 封闭(式)循环;闭式循环;闭路循环;闭合循环
closed-cycle cascade helium refrigerator 闭合循环级联式氦致冷器
closed cycle control 闭环调节;闭合循环调节
closed-cycle control system 闭环控制系统
closed cycle cooler 闭环致冷器
closed-cycle cooling 封闭循环循环冷却;闭式冷却(法)
closed-cycle cooling maser 闭(合)环冷却微波激射器
closed-cycle cryogenic refrigerator 闭合循环致冷器
closed-cycle fuel cell 闭合循环燃料电池
closed-cycle gas turbine 闭循环燃气轮机
closed cycle geothermal power generation 闭式循环地热发电
closed-cycle helium turbine 闭式循环氦气透平
closed-cycle hydraulic piston pump 闭式循环水力活塞泵
closed cycle iodine laser 闭合通路碘激光器;闭合回路碘激光器
closed cycle magnetohydrodynamic(al) generation 闭式循环磁流体发电
closed cycle Oceanic Thermal Energy Coversion [OTEC] system 闭式循环海洋热能转换系统
closed-cycle reactor 闭循环反应堆
closed-cycle system 闭循环系统;闭式循环系统
closed-cycle turbine 闭式循环燃气轮机
closed-cycle underwater system 闭合回路式潜水呼吸系统
closed delivery boot 闭式输种开沟器
closed delivery drill 闭式输种条播机
closed design 闭锁型设计
closed diaphragm 封闭式膜片
closed die 模锻模;封闭模;闭合模
closed-die forging 模型锻造;闭式模锻;闭模锻造;紧模锻压;有飞边模锻
closed die steel 闭锻模钢
closed dike 锁坝
closed discharge filter press 闭卸式压滤机
closed diving bell 密闭式潜水钟
closed diving suit 密闭式潜水服
closed dock 有闸港池;船坞;封闭式港池;通潮闸坞;湿坞
closed domain 闭域;闭合磁畴
closed-door policy 闭关自守政策;闭关政策
closed dowmwarped extracontinental basin 封闭性陆外下陷盆地
closed drain 排水暗管
closed drainage 内陆水系;内流水系;关闭引流法;闭合式排水;闭式引流法;闭水系;暗管排水;暗沟排水
closed drainage basin 内流流域;封闭流域;闭流域
closed drum concrete mixer with reverse chute discharge 反向出料式混凝土搅拌机
closed drum concrete mixer with tipping chute discharge 翻斗出料式混凝土搅拌机
closed duct 闭合管道
closed eaves 匣式檐口;封闭檐口;封闭式挑檐
closed ecological life support system 闭合生态生活支援系统
closed ecological system 密闭生态系统;封闭(的)生态系(统);闭陷生态学系统;闭合生态系统
closed economy 封锁经济;封闭性经济;封闭的经济;闭关(自守)经济
closed economy model 封闭经济模式
closed-edge gutter stabilizer 封缘涡蜡稳定器
closed effuser 闭合式喷管
closed electric(al) furnace 封闭电炉
closed electrolysil circuit 闭路电解系统
closed enclosure 密闭匣
closed end 闭端
closed-end appropriation 限额拨款

closed end cap 封闭端盖
closed-end driveway 环行汽车道
closed-ended question 闭合问题
closed-end fund 限额基金
closed-end investment company 股份固定投资公司；封闭型投资公司
closed end investment fund 定额投资基金
closed-end mortgage 货款额固定的抵押；定额抵押；限额型抵押；闭口抵押
closed-end mutual fund 封闭型共同基金
closed-end needle 闭端(管状)针
closed end pipe pile 封底管桩
closed-end reel 闭端式拨禾轮
closed-end resonator 闭端式谐振控
closed-end spanner 闭口扳手
closed-end spark plug 闭端火花塞
closed-end type 闭端式
closed-end type investment trust 闭锁投资信托
closed-end wrench 闭口扳手
closed environment 密闭环境；封闭环境
close depression 封闭洼地
closed estuary 封闭式河口湾
closed expansion tank 密闭式膨胀箱；封闭式膨胀水箱
closed extruding shield 挤压闭胸盾构
closed eye bolt 环孔被锁住的螺栓
closed face 密闭型
closed fair-leader 导缆环；甲板导缆孔
closed fault 闭合断层【地】
closed-feed system 闭式馈给系统；闭式给水系统
closed-feed water 密闭式给水
closed-feed water heater 密封给水加热器；封闭给水加热器；闭式给水加热器
closed fermentation method 密闭发酵法
closed figure 闭合图形
closed figure item 闭合图形项
closed file 关闭文件
closed filter 密式滤池；封闭式滤池
closed filter press 密闭压滤机
closed fire-proof lifeboat 封闭式耐火救生艇
closed fireroom 密闭式锅炉
closed fireroom system 闭式火室系统
closed fire-season 防火封禁季节
closed fishing season 禁止捕鱼季节
closed fix package 封闭式固定程序包；闭的固定程序包
closed flash point 闭口闪点；闭杯闪点；阿贝尔闪点
closed flume 封闭式渡槽
closed-flux device 闭磁路装置
closed fold 闭合褶皱【地】；闭褶皱
closed foliation structure 闭叶理构造【地】
closed-force polygon 闭合力多边形
closed forest 封山育林；密林；封禁林
closed forging die 闭式锻模
closed formed solution 闭合式解
closed-form solution 闭合解
closed formula 闭合公式
closed fracture density 闭裂缝强度
closed frame 密封棚子；闭合刚构
closed front seat 前圆形便器座
closed front tower 正面封闭(型)钻塔
closed full flow return 比式满管回水
closed furnace 暗炉
closed fusing 闷炉
closed fuzzy information 闭合模糊信息
closed game 关闭博弈；闭式博弈
closed game-season 禁猎季节
closed geodesic line 闭合大地线
closed geothermal bore 封闭的地热井
closed grain 细纹理；细纹板材
closed-grained 密实的；细粒的；密纹的
closed graph theorem 闭图定理
closed groove 闭口式轧槽
closed half line 闭半直线
closed half space 闭半平面
closed harbo(u)r 闭合海港；隐蔽港湾；封闭港；非开放港(口)；不开放港口；不开放港埠；不对外国船只开放的港口；闭口港
closed heater 封闭加热器
closed heating system 封闭式供暖系统
closed high 闭合高压
closed hole 密封孔
closed hood 密闭集气罩
closed hopper 封闭式泥舱

closed hot water heating system 封闭式热水(供)暖系统
closed hydraulic circuit 闭式回路
closed ideal minor ring 闭理想子环
close difference 弥补分歧
close differential relay 近差继电器
close dill planting 窄行播种
closed impeller 封闭式叶轮
closed impeller type dredge pump 闭式叶轮泥泵
closed in bottom hole pressure 关井时井底压力
closed indent 限定(进口)订单
closed inequality 闭型不等式
closed in erection 封闭的上层建筑
closed injury 闭合性损伤
closed in pressure valve 关闭阀；封闭压力阀门
closed in production 关井时产量
closed in space 封闭舱室
closed installation 封闭设施
closed instruction loop 闭合指令循环回路
closed integration formula 闭型积分公式
closed interval 闭区间
closed in trunk 封闭围阱
closed ionization chamber 密闭电离室
closed-iron transformer 闭铁芯式变压器
closed jet 闭合流
closed-jet wind tunnel 闭口式风洞
closed job-shop 内部加工车间；封闭车间
closed joint 接头瞎缝；密合接头；密缝接头；密缝接合；密封接头；瞎缝【铁】；封闭接头；闭式连接；闭合接头；无(间)隙接头
closed joint chain 闭式连接链
closed kernel 闭核
closed lake 内陆湖；闭口湖泊；无出流湖
closed language 闭语言
closed level 封闭圈
closed level circle 水准闭合环
closed level circuit 闭合水准网；闭合水准环
closed level(l)ing line 闭合水准路线
closed lifeboat 封闭式救生艇
closed line 最后确定的保额；闭锁系；闭合线
closed linear manifold 闭线性流形
closed linear model of production 封闭式的线性生产模型
closed link 闭式链节
closed linkage 紧密耦合
closed link motion 闭合式滑环运动
closed list of bidders 密封的投标人名单；审定的投标人名单
closed lock 闭合闸门；闭合船闸
closed loop 封闭系统；闭循环；闭路循环；闭环；闭合循环回路；闭合回路；闭合回线；闭合环(路)；封闭环
closed loop adapter 闭环转接器；闭环适配器
closed loop adaptive optical system 闭环自适应光学系统
closed-loop automated water treatment plant 闭环自动化水处理厂
closed loop characteristic 闭路特性
closed-loop circuit 闭合回路；闭环电路
closed-loop computer control system 闭环计算机控制系统
closed-loop control 闭环线路控制系统；反馈控制；闭回路控制；闭合回线控制；闭合反馈回路控制；闭环控制
closed-loop control of fuel metering 燃料计量的闭环控制
closed-loop control system 闭环控制系统；封闭式环形控制系统；反馈控制系统
closed-loop domestic water heater 民用闭环太阳能热水器
closed-looped estimation 闭环估计法
closed-loop frequency response 闭环频率响应
closed-loop gain 闭合回路增益；闭环增益
closed-loop instruction 闭合环指令
closed-loop layout 封闭式设置
closed-loop phase lock system 闭环锁相系统
closed-loop policy 闭环方案
closed-loop process control 闭环过程控制
closed-loop program(me) 闭环程序
closed-loop recycling system 闭环循环系统
closed-loop regulation 闭环调节
closed-loop resolvent matrix 闭环分解矩阵
closed-loop servo-control system 闭环伺服控制系统

closed-loop servo system 闭环伺服系统；反馈随动系统
closed-loop system 闭循环系统；闭(合)环系统
closed-loop system matrix 闭环系统矩阵
closed-loop telemetry system 闭合回路遥测系统
closed-loop testing 闭合回路试验
closed-loop tracking system 闭合回路跟踪系统
closed-loop travel(l)ing grate cooler 回转算式冷却机
closed-loop valtage gain 闭环电压增益
closed-loop wastewater system 闭环废水系统
closed-loop water system 闭式环路水系统
closed low 闭合低压
closed macro 闭式宏程序
closed magnetic circuit 闭合磁路
closed magnetic configuration 闭合磁位形
closed magnetic-vibrating feeder 封闭型电磁振动给料机
closed manometer 闭管压力计
closed mapping 详细测图；闭映射
closed market 封闭市场
closed-mesh screen 细筛
closed method 闭算法；闭式方法
closed mix 密(实)混合料；密级配混合料
closed mixolimnion 闭合交换层
closed model 封闭性模型；封闭式模型；闭模型
closed moraine 封闭型冰丘
closed mortgage 限额抵押
closed mortgage bond 限额抵押债券
closed mo(u)ld 封闭式模具；闭式(压)模
closed mo(u)lding 闭模成型技术
closed mo(u)ld mo(u)lding 闭合模塑
closed neighborhood 闭邻域
closed nervuration 闭式脉序
closed network 闭环网络
closed network control 闭路网络控制；闭合网络控制
closed newel 砖砌旋梯的中柱；封闭式中芯墙(盘旋楼梯)
closed node 关闭节点
closed nozzle 闭式喷油嘴
closed-on-itself traverse 闭合导线【测】
close door 歇业
close door operation 闭门操作
closed-open 接通一断开
closed operation 闭运算；闭操作
closed operator 闭算子
closed orbit distortion 闭合轨道畸变
closed order 限定订货
closed orthonormal set 完全正交规范集
closed oscillation circuit 闭合振荡电路
closed oscillator 封闭式振子；闭路振子；闭路振荡器；闭合振荡器
closed oven 暗炉
closedown 关闭；停炉；关机；停歇；停机
close-down case 停业状况
closed-packed hexagonal structure 密堆积六角结构
closed-packed hexagonal system 密集六方晶系
closed packing 密合装填
closed pair 闭合对
closed pass 闭口式孔型；闭口孔型
closed path 闭路；闭合回路；闭合电流；通路
closed pavement structure 封闭式路面结构
closed pipe 一端封闭的管子
closed pipe system 封闭管道系统
closed pitching 密间隔
closed plane 闭平面
closed plane figure 封闭平面图形
closed planer 龙门刨床
closed plastic refining machine 密封式炼塑机
closed point rail 密贴尖轨
closed polygon 封闭的多边形；闭合环(路)；闭合多边形
closed polygonal rigid frame 闭合多边形刚架
closed polygon of force 力的闭合多边形；闭合力多边形
closed polyline 闭合折线
closed population 封闭性人口；封闭式人口；封闭人口；闭锁群
closed pore 封闭孔；闭气孔；闭口气孔；闭口孔
closed pore volume 闭孔体积
closed porosity 隐孔隙率；隐孔隙度；闭口气孔率
closed port 上游港(在河的上游)；封港；封闭港；

closed portion 非开放港(口);不开放港口;不开放港埠;不对外国船开放的港口
closed portion 关闭部分
closed position 关闭位置;封闭状态;闭合状态;停止位置
closed pot 有盖坩埚
closed pressure 封闭井的最大压力
closed process 闭式作业法;闭路工艺
closed ratio 密接变速比
closed reach 有闸河段
closed recirculating cooling system 密闭式循环冷却系统
closed recirculating cooling water system 密闭式循环冷却水系统
closed recirculating diving apparatus 密闭式循环潜水装置
closed recirculating system 密闭循环系统;密闭式循环系统
closed recirculation 闭路循环
closed recirculation system 封闭循环系统;封闭式循环系统;闭合再循环方式
closed refrigeration circuit 封闭式制冷回路
closed region 封闭区(域);闭区域
closed regional cooperation 限某区域的合作;封闭型区域合作;排他性区域合作
closed reservoir 封闭油灌;封闭油储;闭合储油层
closed resonator 闭合谐振腔
closed respiratory gas system 密闭呼吸气体系统;封闭式呼吸气体系统
close drift ice 密集流冰群
close drilling 密网钻进
close drill sowing 窄行播种
closed ring 闭环
closed ring dowel 闭口环榫
closed-ring hydrocarbons 闭环烃
closed ring reef 封闭环状礁;闭锁环礁
closed riser 暗冒口
closed roller chock 闭式滚轮导缆钳
closed roll mill 密闭式轧机
closed rotative gas lift 闭式气举
closed routine 闭型例行程序;闭型程序;闭列程;闭合程序
close drying 接近干
closed sand 密实砂
closed sandwich type panel 封边的组合板;闭合式夹心幕墙;封边夹层板;闭合式夹心护墙;闭合嵌板
closed screen 封闭式挡泥屏
closed sea 封闭(式)海;领海;内海;在狭窄海峡内的海区;闭(塞)海;受岬角包围的海区
closed sea canal 有闸通海运河
closed search trajectory 封闭搜索轨迹
closed season 禁渔期;禁猎期
closed section 闭口截面
closed security zone 保密区
closed segment 闭线段
closed semantic tree 封闭语义树
closed-sequence control 闭合程序控制
closed sequential scheme 闭合连续方案
closed-series combination method of measurement 闭环组合测量法
closed set 闭集
closed shackle 闭式钩环
closed shaft 有顶天井;封闭竖井
closed sheathing 连续式挖方支撑;隐式板撑
closed shed 闭合梭口
closed sheeting 围合板桩;连续式挖方支撑;隐蔽木板支撑
closed shell 满充壳层;闭合壳层
closed shell and tube condenser 闭式壳管式冷凝器;卧式可管式冷凝器
closed shell and tube evapo(u)rator 卧式壳管式蒸发器
closed-shell atom 闭壳层原子
closed-shell configuration 闭壳层组态
closed shelter decker 封闭式遮蔽甲板船
closed shelter-deck ship 闭式遮蔽甲板船
closed shelves 闭架
closed shelving 隐蔽式壁架
closed shield 密封屏蔽;封闭式结构;闭胸式盾构
closed shop 应用程序站;只雇佣某工会会员的工厂;封闭性工厂;封闭式机房;不开放计算站;闭式工厂
closed-side setting 破碎机排料口闭合时的间隙宽度

closed slide 闭合滑板;平行导轨
closed slot 闭口槽;闭合槽
closed-slot armature 闭槽电枢
closed sludge barrel 封闭式取粉管
closed socket 闭合锚杯
closed space 闭合空间;稠密地区
closed space vapo(u)r deposition 密间隔气相沉积
closed spacing 密间隔
closed specification 施工说明书(严格做法和材料的);详细说明书;详细规范;指令性说明;特殊产品规范说明
closed spiral array 闭合螺线阵列
closed spiral auger 闭合式螺旋管钻
closed spiral conveyer 封闭式螺旋输送机
closed squeezing shield 挤压闭胸盾构
closed stack system 闭架式
closed stair(case) 箱式楼梯;箱形楼梯;封闭(式)楼梯
closed stair string 封闭式楼梯斜梁
closed stairway 闭合楼梯
closed stand 闭式机座
closed state 闭态
closed steam 间接蒸汽;封闭蒸汽
closed sterilizer 密闭式杀菌器
closed stirrup 封闭(式)箍筋;封闭钢筋;闭口箍(筋)
closed stock 封闭式货物;配套商品
closed stokehold 密闭式锅炉舱
closed storage 封闭油灌
closed string(er) 封闭式楼梯斜梁;全包的楼梯梁;遮盖的楼梯梁
closed string stair(case) 斜梁有侧板的楼梯;封闭式斜楼梯梁;封闭式踏板楼梯;遮盖的楼梯梁
closed structure 闭式组织结构
closed subprogram(me) 闭型子程序
closed subroutine 封闭式子程序;闭型子程序;闭合(子)程序
closed subset 闭合子集
closed superstructure 封闭式上层建筑;封闭船楼
closed supply circuit 闭式供应系统
closed surface 密实面层;密闭式表层;闭曲面
closed surge tank 封闭式调压井;闭式调压塔;闭合调压塔
closed system 建筑定形式;密闭系统;限额制;专用建筑体系;封闭系统;封闭体系;闭系;闭锁体系;闭式系统;闭塞系统;闭路(循环)系统;闭合系统;闭合体系;局限体系
closed system model 封闭系统模式
closed-system of cooling 闭式冷却系统
closed system of heating 封闭式供暖系统;闭式供暖系统
closed system of product 专用产品生产系统
closed system of ventilation 封闭式通风系统;闭式通风系统
closed-system separator 闭路循环选矿机
closed-system test (土体冻胀试验的)闭式试验法
closed-system water line 闭式系统(指水循环)
closed tank 密闭储[贮]罐
closed tappered bore 闭式锥形孔径
closed tee joint 紧密T形接合
closed temperature measuring device 封闭式温度测量装置
closed tender 指定竞争投标
closed termination 封闭端
closed tester 闭杯闪点测定器
closed-throat with tunnel 闭锁形风洞
closed tie 箍筋
closed T joint 紧密T形接合
closed to car 禁止汽车通行
closed-top 顶部封闭的
closed-top duplex lantern 闭顶双灯(一种航标灯)
closed topology 闭合拓扑
closed-top single lantern 闭式单航标灯
closed to traffic 禁止车辆通行;禁止通行;禁止交通
closed towing chock 闭式拖缆钳
closed track circuit 正常闭合轨道电路
closed trade 结束交易;完结交易
closed transducer 密封换能器
closed traverse 闭合导线【测】
closed treaty 不开放条约
closed triangle 闭合三角形
closed triangle of force 力的闭合三角形

closed trickling filter 封闭式生物滤池;封闭式滴滤池
closed tube 连通管;联通管;闭管
closed-tube method 闭管法
closed type 封闭式;密封型;密闭式;闭式
closed-type air slide 闭式空气输送斜槽
closed-type bearing 密闭形轴承
closed-type bucket 闭式铲头
closed-type concave 闭式凹板
closed-type die 闭式模
closed-type double action crank press 闭式双动曲柄压力机
closed-type fuel valve 密闭形燃油阀
closed-type heater 闭式加热器
closed type hotplace 封闭式灶面板
closed-type hot-water heat-supply network 闭式热水热网
closed type layout 封闭式布局
closed-type manlift 封闭式载人电梯
closed-type screen 封闭式筛
closed-type single action crank press 闭式单动曲柄压力机
closed-type single point press 闭式单点压力机
closed-type square 封闭(式)广场
closed-type toilet enclosure 带门厕所间隔间
closed-type track circuit 闭路式轨道电路
closed ullage system 闭式液位测量装置
closed union 闭式工会;入会条件严格的工会
closed unit 封闭式装置
closed universe 闭宇宙
closed-up 闭路
closed user 封闭的用户
closed user group 封闭型用户组;封闭(式)用户组
closed valley 闭合屋面天沟;封闭式排水斜沟;暗沟
closed valve gear 封闭式阀动装置
closed van type container 封闭式集装箱
closed vapo(u)r cycle 闭式蒸汽循环
closed venation 闭锁脉序
closed ventilated container 封闭式通风集装箱
closed ventilation system 闭合通风系统;封闭式通风系统
closed vessel 密闭容器
closed vessel furnace 密闭式炉
closed void 封闭孔隙;闭型孔隙;闭口孔隙;闭合孔隙
closed volume 闭合体积
closed vortex line 封闭涡线
closed walk 闭途径
closed water body 闭合水域
closed water circuit 封闭式水循环
closed water feed 闭路供水
closed water heater 密闭热水器
closed water piping 闭式水管路
closed water-piping system 封闭式水管系统
closed water recycling 闭式水循环
closed water tank 比式水箱
closed waveguide 闭合波导
closed weir 闭合堰
closed weld 底边无缝焊;无间隙焊缝
closed well 封闭井
closed world assumption 闭型世界假定
closed worm conveyer 封闭式蜗杆输送机;封闭式螺旋输送机
closed wound 闭合性损伤
closed yoke 闭合轭
close earthquake 近震
close earth satellite 近地卫星
close encounter 紧密交会
close end socket 闭口绳头套环
close epiphytotic disease 密集病害流行
close etching 精蚀
close examination 精密检测;仔细检查
close feedbox 密集饲料箱
close fishing area 禁渔区
close fit 紧配合;密配合;密合丝扣
close-fitting cover 密封盖板
close-fitting metal cap 密接金属套
close flash point tester 封闭式闪点测定器
close folding grazing 超集约轮换放牧
close for cargo 停止收货
close fuel injector 闭式喷油器
close-graded 密级配(的)
close-graded aggregate 密级配集料;密级配骨料
close-graded aggregate asphalt(ic) concrete 密级配集料沥青混凝土

close-graded bituminous concrete 密级配沥青混凝土
close-graded bituminous surfacing 密级配沥青面层
close-graded pavement 密级配路面
close grain 木材язык纹; 密纹理; 密(木)纹; 细纹(理); 细晶粒
close-grained 木纹细密的; 细粒的
close-grained iron 密实铁
close-grained structure 密实级配结构; 细密结构; 细粒结构
close-grained timber 密纹木(材)
close-grained wheel 细砂轮
close-grained wood 密纹木(材); 细纹木(材)
close-grain rule 细纹标准
close-hauled ply 迎风航驶
close headway 密集车头时距
close herding 集中放牧; 密集放牧
close hydrant 密集龙头
close ice 密集流水; 密集流冰群; 密集冰
close impeller 闭合式叶轮
close in 迫近; 封闭; 包围
close in data 近场数据
close indent 闭口订货单
close in fault 近区故障; 近距离故障
close inspection 严格检验; 严格检查
close investigation 详细调查
close joint 密缝(接头); 密缝接合; 密封接头
close-joint cleavage 密集节理式劈理; 闭节劈理
close junction 关闭连接
close knit surface 密实面层; 密实表面
close level 闭合水准路线
close lid 密合的盖子
close limit 窄极限
close linkage 强耦合
close-link chain 核桃链; 短环链
close loop control circuit 闭环控制回路
close loop interferometric fiber optic(al) sensor 闭环干涉型光纤传感器
closely coiled arm 紧卷旋臂
closely coincide 精确符合
closely coupled 紧密耦合的
closely graded 密级配的
closely graded product 粒度均齐的产品
closely graded sand 密级配黄砂
closely graded soil 粒度均一土; 细密级配土(壤)
closely held company 股东人数有限公司; 封闭型控股公司
closely knit surface 密实面层; 密实表面
closely linked 一环扣一环
closely packed snow 密实雪
closely pitched 细牙
closely ringed timber 细纹木材
closely spaced 紧密间距的; 密集的; 邻接近的
closely spaced array 密置天线阵; 密排天线阵
closely spaced frame 密排框架
closely spaced lattice 紧密晶格; 紧密点阵
closely spaced spiral 密纹螺线
close mail 封邮
close mapping 详测; 详细测图
close market 歇市
close-mechanical type 密闭机械式
close medium shot 中景镜头
close-mesh 密网筛
close-meshed 密网眼的; 细网眼的; 细筛孔的
close-meshed rectangular grid 小方格网
close-meshed screen 密孔网
close-meshed sieve 细眼筛
close mesh screen 细孔筛
close-mesh street network 密网街道网; 密格街道网
close mo(u)ld 封闭铸型
close mowing 低割
closeness 紧密度; 密集度; 密闭; 封闭性
closeness of contours 等高线密集; 等高线闭合
closeness of fissure 裂隙密集度
closeness of packing 填充紧密度
closeness of winding 绕组紧密度
close nipple 螺纹接口(管); 螺纹接管; 螺纹管接口; 螺纹接套; 无隙内接头
close of business 停止营业
close off 结账
close-off rating 全闭额定压力差
close of horizon 全圆方向观测法
close of polygon 导线闭合

close of the season 季末
close of the year 年底
close on 临近
close-open control switch 合分控制开关
close out 卖完; 卖底货; 抛售
close out sale 放盘出售
close-over 砂胎
close-over hexagonal lattice 密集六角晶格
close-over lattice 紧装晶格; 密排晶格; 密集晶格; 密集点阵
close pack 紧密排列; 密集流水; 密集冰
close package 紧凑包装; 密装
close-packed 密集; 密堆积的
close packed code 紧充码
close-packed crystal 密堆积晶体
close-packed cubic(al) structure 立方密堆积结构
close-packed hexagonal 密排六方
close-packed hexagonal lattice 密排六方晶格
close-packed hexagonal structure 六角密集堆积结构; 密排六方结构
close-packed lattice 密集晶格; 密集点阵; 密堆积点阵
close-packed pumping system 闭合泵浦系统
close-packed structure 密堆积结构
close-packed traffic 高密度交通
close pack ice 密集浮冰群
close packing 紧密装填; 紧密包装; 密排; 密集; 密堆积
close-pass 闭口式轧槽
close pattern return bend 闭合式180度弯头; 闭合式回转弯头
close perspective recording 近摄; 近拍
close pile 紧排桩; 密排桩
close-piled 实体堆积的
close planting 密植; 密播; 窄行条播
close plating 紧密涂敷
close point 闭合点; 闭点
close port 内陆商港(英国); 河港; 不开放(的海)港; 未开放港口
close position 通路位置
close program(me) 闭型程序
close push button 闭合按钮
close quarter 狭隘的住所; 近距(离); 狭窄的居住空间; 拥挤处; 狭窄处
close quarter piston air drill 角隅风钻
close quarter riveting hammer 角隅铆钉锤
close-quarter work 狭窄场所作业
closer 铆钉模; 镶墙(边)的砖石; 插缝砖; 封口砖; 封口板桩; 闭路器; 闭合器; 填塞物
close rafter 有系梁椽; 带系杆人字木; 封闭椽
close range 近距离; 近景
close-range fading area 邻近衰落区
close range instrumentation 近景测量仪
close-range interval 短侧距
close range photogrammetry 近景摄影测量
close range photograph 近景像片
close range shooting 近距离射击
close-ratio gearbox 传动比接近的变速箱
close ratio two-speed motor 近比率双速电动机
closer brick 接合砖; 插头; 封口砖
close region 闭域
close register 精确配准
close regulation 细致调整; 准确调节
close restraint 严格限制
close return bend 闭合式U形弯头; 回弯头; 闭合回转弯头
close riveting 密铆
closer mo(u)ld 封层(砖)样板
close roaster 闭式烤炉
closer object 近距离物体
closer reinforcing sleeve 关门套加固器
closer row-spacing 较密行距
close running fit 紧转配合; 紧动配合
closes 死胡同
close sample 小间隔取样; 小间隔样品
close sampling 间隔不大的取样; 密集取样; 小间隔取样
close sand 致密砂层; 密级配砂; 填густ砂; 密实砂
close satellite orbit 近地球卫星轨道
close satellite theory 近点卫星理论
close scanning 细扫描
close seeding 密植; 密播
close selector 精密选波器
close service 不开放服务

close-set rolls 闭置对辊机
close-settled soil 密实土(壤); 密实地壤
close shearing 重剪
close-sheeted roofing 密铺屋面板
close sheeting 挡土排板; 密排板桩; 密封衬板; 封板(围栏)
close shift counter 闭合移位计数器
close shot 近摄; 近景摄影
close silage aftermath 密集再生草
close sliding fit 特小间隙滑动配合
close sonar contact 直接声呐接触
close-spaced method 密排法; 窄间隔法
close spaced structure 紧限位结构
close-spaced technique 密排技术
close-spaced wells 密网布置钻井
close spacing 密植距; 密植
close stand 密苗
closest approach 最近点; 最接近状态; 最接近点; 最接近道路
closest approach point 人卫近站点
close statement 闭合语句
closest discharge point 泉眼密集点
closest instant 最接近瞬间
closest offer 最近初发盘
close-stool 马桶; 有盖马桶; 室内大便器
close stowing anchor 马丁式锚; 有杆转爪锚
closest-packed crystal plane 密集的晶面
closest packing 最密装填; 最密堆积; 最紧密堆积
closest point 最近点
closest point of approach 会遇最近点
closest range 最近距离
closest rational approximation 最近有理逼近; 最近分数式
close string 带护板楼梯斜梁; 封闭式楼梯斜梁
close string housed string 闷扶梯基
close studding 密肋抹灰构造; 密排板墙筋
close study 仔细研究
close supervision 精细管理; 严密监督; 严格管理
closet 炉室; 蒸馏炉炉室; 橱; 壁柜
closet approach point 近站点
closet bar 衣橱挂衣钩; 壁橱挂衣钩
closet basin 盥洗室水盆
closet bend 厕所弯管
closet bolt 厕所用螺栓
closet bowl 马桶; 便桶; 便器陶瓷部件
closet cubicle 厕所小隔间; 分隔厕所间
closet door 盥洗室门; 壁橱门
close texture 细颗粒织构; 致密质地
close textured 密实组织的
closet facility 盥洗室设施
closet fan 盥洗室通风扇
closet floor flange 大便器地面连接法兰
closet flushing 盥洗室冲洗
closet flushing cistern 盥洗室冲洗水箱
closet flushing pipe 厕所冲洗水管
closet flushing water 厕所冲洗水
closet flush pipe 大便器冲洗管
closet hanging device 壁柜吊挂装置
close the border 封锁边境
close the country to international intercourse 闭关自守
close the feet 两脚并拢
close the gap 截流合龙; 缩短(弥合)差距
close the land 接近陆地
close the plant 关闭工厂
closet horn 马桶排出口; 大便器排水口; 便桶排出口
close timbering 密撑; 木板墙; 密闭支撑
close time 合闸时间; 下班时间
closet knob 衣橱门执手; 壁橱门执手
closet lining 壁橱内衬板
close to 接近
close to corn field 挨近玉米地
close to finish forging 精密锻件
close tolerance 紧公差; 闭合公差
close tolerance casting 精密铸件
close tolerance forging 精密锻造; 紧公差锻造
close to market 邻近市场
close-toothed 小齿距齿
closet pan 大便器
closet pole 壁橱挂衣杆
close translation 准确翻译
close trim 精密切边
closet rod 壁橱挂衣杆

closet room 盥洗室
closet screw 厕所用螺钉
closet seat 便桶座(圈)
closet seat lid 马桶盖
closet spindle 衣橱门执手心轴;壁橱门执手心轴
closet stop valve 厕所用断流阀
closet suite 盥洗室全套设备;成套卫生瓷
closet tank 马桶水箱;大便器水箱;厕座水箱;便桶水箱
closet valve 大便器冲洗阀;厕所水箱门阀
closet wastewater 厕所废水
closet wastewater pipe 厕所污水管
close type 封闭型
close type crushing plant 闭式联合碎石机组
close-type evapo(u)rator 闭式蒸发器
close-type injector 闭式喷油器
close type manual shield 密闭人工开挖式盾构
close up 紧排;接通;关闭;围护
close up camera 近景摄影机
close-up casement hinge 平开窗密合式铰链
close up fault 近距离故障
close up focusing 近距摄影调焦
close up lens 近摄镜头
close up mark 接岸标
close up photogrammetry 近景摄影测量
close up photography 特写摄影
close up shot 特写镜头
close up the gap 弥合差距
close up the rivet 铆平铆钉
close up view 局部放大图;近视图;近摄;特景;近观图;全貌图;特写图;特写镜头
close void ratio 闭空隙率
closeweight cargo 高比重货物
close well spacing 密井网;小井距
close winded 接近顶风
closewise tube 闭塞管
close with an offer 接受报盘
close with brick wall 砌墙封住
close woodland 密集林地
close work 狭窄场工作
close working fit 紧滑配合
closing 合闸;合型;合门;封闭;不动产交易的交割;闭合
closing account 结账;决算
closing a commitment 结束约定
closing action 关闭动作;闭合作用
closing adjustment 结账整理;决算整理
closing a file or device 关闭文件或设备
closing a flow 截止流通;断流
closing agreement 结案书(美国所得税)
closing and checking action 关闭、制动联合动作
closing and checking mechanism for doors 关门及制动装置
closing a position 结束部位
closing apparatus 闭合设备
closing asked price 要价收盘
closing a terminal 关闭终端
closing balance 期终余额;期末余额
closing balance account 结账余额
closing balance in account 账户结清余额
closing bid and asked price 收盘出价和要求
closing board 挡板;封板
closing branches for strengthening main channel 塞支强干
closing brick 锁砖
closing business 倒闭营业;闭业
closing cable 关闭拉索
closing call 收盘叫价
closing capacity 遮断容量;遮断能力;闭路电流容量
closing ceremony 闭幕式
closing check relay 合闸鉴定继电器
closing chute 封闭溜槽
closing clamp 缝合夹
closing cock 封闭塞;闭锁旋塞;闭锁活门
closing coil 接通线圈;合闸线圈;闭合线圈;闭合螺管
closing component 锁合
closing contact 闭合触点
closing cost 房地产成交手续费;决算费用;结算费用;成交价
closing current 闭合电流
closing cylinder 合模缸;闭模汽缸;闭合模汽缸
closing dam 堵口坝;围堰
closing date 截止(收货)日期;结账日期;结算日期;交运截止日

closing date of offer 报价截止日
closing day 决算日;结账日;结束日;停业日
closing debit balance in account 账户结清借方余额
closing dental space 关闭牙间隙
closing device 关门装置;关闭设备;闭合装置;自动关门装置;防火门自动开关装置
closing dike 截流堤;合拢堤;堵口堤;封闭堤
closing dike rockfill 堵口堤抛石;截流戗堤抛石
closing direction 闭合方向;闭合方向
closing down 停机
closing effect 接近效应
closing entries 结转分录;结账分录;结清分录
closing error 闭塞差;闭合误差;闭合差【测】
closing error in coordinate increment 坐标增量闭合差
closing error in department 横距闭合差
closing error in latitude 纵距闭合差
closing error of gravimetry 重力闭合差
closing file 关闭文件
closing flag 关闭标记
closing force 合型力;闭合力
closing handle 紧固手轮
closing high 高价收盘;收盘上涨
closing horizon 补角观测(指全圆观测法)
closing hour 停止营业时间
closing in 在新建筑配玻璃的挂门;关井;关闭;(空心铸件或管件的)收口
closing inventory 最后存量;期末库存量;期末存货
closing inventory valuation 期末存货估计
closing jaw pressure 颚压力;闭口压力
closing law 闭合规律
closing levee 截流堤;闭合堤
closing line 挖掘线;接合线;封口钱;封闭线;闭线;闭合线
closing low 收盘降价
closing machine 合绳机;封口机;刻槽机;合箱机;压盖机;封口压盖机
closing mechanism 关闭机械装置;闭合机构
closing meeting 末次会议;闭幕会议
closing member 关闭件
closing membrane 封闭膜
closing motor 密闭电动机
closing movement 闭合动作;关闭动作
closing noise 闭合噪声
closing of account 结清账户
closing of a file 关闭文件
closing of circuit 电路闭合
closing of contact 触点闭合
closing off 封闭交通
closing of point 道岔闭合
closing of polygon 导线闭合
closing of port 封港
closing order 闭合顺序
closing period 停止时间;结算期;闭合时间
closing piece 锁块
closing pile 结尾板桩;封(口板)桩;末桩(指在一排中最后打入的桩)
closing pin 合箱销
closing plant of drydock 干船坞闭坞装置
closing plate 闭板
closing plug 密封塞;合箱柱塞;关闭塞
closing point 闭合点
closing polyline 封闭折线
closing pontoon 箱式坞门
closing pressure 闭合压力;关闭压力;闭锁压力
closing pressure surge 闭合时的压力冲击
closing price 收市价;收盘价(格)
closing procedure 结账手续;结账程序
closing purchase 结清购买
closing quotation 收市行情;收市报价;收盘行市;收盘
closing range 闭合范围
closing rate 收市汇率;收盘汇率
closing ratio 配量比
closing reading 闭合读数
closing relay 接合继电器;合闸继电器
closing ring 止动环;关闭环;卡圈
closing roller 封闭式辊筒
closing rope 闭斗绳(抓岩机)
closing sale 结清出售
closing screw 螺栓塞;螺纹规
closing section of vaulting 拱顶接合段
closing segment 关闭图段

closing session 闭幕会议
closing shock 闭合时的压力冲击
closing signal 关闭信号
closing sleeve 夹紧套筒
closing slip 保险承保条款
closing solenoid 合闸线圈;合闸螺线管
closing speed 关闭速度;接近速度;闭合速度
closing spring 合闸弹簧
closing stile 闭合侧门窗梃;装锁边梃
closing stile for paired casements 成对窗扇的闭合梃
closing stock 期末存货
closing stroke 闭合行程
closing structure 闭合结构;闸门建筑物
closing sum 闭合和
closing switch 闸刀开关
closing system 关闭体系
closing the account 结清银行账户
closing the bank account 结清银行账户
closing the door to the outside world 闭关自守
closing the ledger 结算总账户
closing the revenue 结清收益
closing the section 封锁区间
closing the trustee's books 信托账户
closing time 下班时间;停止工作时间;截止时间;合闸时间;封闸时间;闭合时间
closing time of governor 调速器关闭时间
closing transport account 运输结算
closing traverse 闭合导线【测】
closing trial balance 结账试算表
closing unit 结算单位;封土器
closing up 铆钉铆合
closing up of the arch ring 合拢拱圈;拱圈合拢
closing value 闭合值
closing valve 隔离阀;节制阀;停汽阀
closing volume 闭合气量
closure 截止;合龙(指堤坝);线形闭合;终结;关闭;封赛;封闭器;封闭;闭合度;闭合;闭包;圈闭;停业;停歇;填塞砖;塞孔砖
closure against grazing 禁止放牧
closure ages 封闭年龄
closure algebra 闭包代数
closure assumption 闭包假设
closure axiom 闭包公理
closure bar 盖缝条
closure borrow area 截流取料场
closure brick 填塞砖
closure by jacking and sealing-off crown 千斤顶法封顶
closure by wedging-in crown 刹尖;尖拱法封顶
closure channel 封闭沟槽;有盖沟槽
closure condition 闭包条件
closure construction equipment 截流施工设备
closure construction schedule 截流施工进度
closure correlation of dense point 密点闭合对比
closure dam 截流坝;合龙坝;堵口坝;锁坝
closure dike 截流堤;锁坝;合龙堤
closure discrepancy 闭塞差;闭合差【测】
closure domain 闭域;闭合畴
closure door 保护盖
closure drop 截流落差
closure embankment 截流堤(岸)
closure error 闭合差【测】
closure error azimuths 方位角闭合差
closure error of azimuths 方位闭合差
closure error of closed figure 闭合图形闭合差
closure error of connecting line 附合路线闭合差
closure error of coordinate 坐标闭合差
closure error of elevation 高程闭合差
closure error of horizon 圆周闭合差;水平角闭合差
closure error of level circuit 水准环闭合差
closure error of loop 路线或环线闭合差
closure error of traverse 导线闭合差
closure error of triangle 三角形闭合(误)差;三角闭合差
closure fill 合拢填方
closure-finite complex 边缘有限复合形;闭合有限复合形
closure-finite complexes 边缘有限的复合形
closure flange 封闭凸缘
closure flow velocity 截流流速
closure gap 截流龙口;龙口
closure grouting 封闭灌浆
closure head 截流水头;闷头;封盖;外壳顶盖

closure member 关闭件
closure model test 截流模型试验
closure of a crevasse 堵口
closure of arch ring 拱圈封顶
closure of a set 集的闭包
closure of fault 断层闭合度
closure of fold 褶皱闭合度
closure of force polygon 力多边形闭合
closure of horizontal angle 水平角全测
closure of interdental space 关闭牙间隙
closure of liquidation 清算完结
closure of port 封港;闭港
closure of rift 裂谷的闭合
closure of track 线路封闭
closure of trap 圈闭闭合度
closure of triangle 三角闭合(差)
closure operation 合拢作业;闭包运算
closure phase 闭合相位
closure pile 闭合桩
closure position 闭合位置;闸门位置
closure project 截流工程
closure property 封闭性质;封闭特性
closure quantity 截流工程量
closure rail 合拢轨
closure rail of special length 短合拢轨
closure relation 闭包关系
closure-sealing embossed machine 密封压花机
closure section 闭合块;闭合段
closure slot 封闭槽;接缝灌浆槽
closure strip 堵缝板(条);肋形开口镶板;油毡收口压条;密封条
closure strip of felt 油毡收口压条
closure structure 闭合结构
closure system 封闭体系
closure temperature 接缝灌浆温度
closure test 弹簧压缩试验
closure theorem 闭合定理
closure time 黏着时间;黏合时间
closure tolerance 封闭公差;闭合公差
closure to prevent reverse flow 禁止反向交通流;防逆流;闭合构造
closure using special wedge seal 特种楔形密封
closure valve 闭合阀门
closure with multiplewedge gasket 组合式楔密封
closure works 截流设施;截流工程;合龙工程
clot 混块;凝块;荒坯;粗坯
cloth 布织物
cloth-backed paper 衬布纸
cloth base 织物底层;布底
cloth beam 布轴
cloth binding 滤布堵塞
cloth breaking machine 柔布机
cloth buff 布抛光轮
cloth-cap 劳动阶层
cloth-coating 织物涂料
cloth count 织物经纬密度
cloth discharge device 出布装置
cloth disk 布抛光轮
cloth doubler 层布贴合机
cloth dust collector 布袋集尘器
cloth-edge speaker 布边扬声器
cloth edition 布面装订板
cloth emerizing machine 金刚砂辊擦布机
cloth envelop collector 布袋过滤器
clothes chute 脏衣溜槽;洗衣槽
clothes closet 立柜;衣橱
clothes dryer 衣服烘干机;干衣机
clothes drying cabinet 衣服干燥室
clothes-hook 挂衣钩
clothes horse 晒衣架
clothes line 晒衣绳
clothes line pole 晒衣绳杆
clothes locker 存衣架
clothes-locker room 存衣间
clothes pole 晒衣绳柱
clothes shop 服装商店
clothes stand 衣帽架
clothes tree 衣帽架;柱式衣架
cloth fast black 布帛坚牢黑
cloth fast blue 布帛坚牢蓝
cloth fast yellow 布帛坚牢黄
cloth filler 布屑填料;布帛填料
cloth filter 织物吸尘器;布(过)滤器;布袋过滤器
cloth finishing mop 抹光布轮;抛光布轮

cloth flashing 织物泛水
cloth hanging 墙面油漆前糊布
cloth hanging mechanism 布袋悬挂机构
clothing 服装
clothing and textile industry 服装和纺织工业
clothing department 服装部
clothing expenses 衣着费
clothing factory 服装厂;被服厂
clothing industry 服装工业
clothing insulation 服装保温
clothing issue room 储物舱
clothing lined board 衬布纸板
clothing monitor 衣服沾染监测器
clothing store 服装商店
clothing store room 被服储藏室
cloth insert(ing) 织物嵌填
cloth inspecting machine 验布机
cloth joint 布接缝
cloth lath(ing) 织物板条;布条
cloth-lined paper 衬布纸
cloth-lining paper 布裹纸
cloth map 布质地图
cloth mark 布痕(玻璃缺陷)
clothmeasure 布尺
cloth measuring and cutting machine 白呢量裁机
cloth mellowing machine 叠层轧光机
clothoid 回旋曲线
clothoid biparties curve 两部分构成的缓和线
clothoid curve 回旋曲线
clothoid-shaped bend 回旋形弯道
cloth pasting machine 包裱辊机
cloth pinion (麻布帆布等制成的)布齿轮
cloth plaiting machine 折布机
cloth plaster 胶布膏
cloth ply 织物层
cloth polishing 毛毡抛光
cloth press 打布包机
cloth print 织物印刷;织物印花
cloth-reinforced 织物加强的
cloth-reinforced wire 纱包线
cloth rewinder 卷布机
cloth rewinder roll 法兰绒辊
cloth ring 布环
cloth roller 卷布辊
cloth rolling machine 卷布机
cloth roll stands 布卷架
cloth roof 帆布屋顶
cloth screen 布筛;布袋收尘器
cloth slitting machine 切绷带机
cloth strainer frame 布滤架
cloth tape 布卷尺
cloth tester 织物引张试验机
cloth turning machine 翻布机
cloth vacuum extractor 织物真空吸水机
cloth wearing tester 织物磨损试验机
cloth-wear testing machine 织物磨损试验机
cloth wheel 布抛光轮
clotting (液体变为固体的)凝结;凝固
clotting time 凝固时间
cloucrack 浮云
cloud 发混
cloud absorption 云吸收
cloudage 云量
cloud albedo 云的反照率
cloud amount 云量
cloud and collision warning 烟云及碰撞警报
cloud and collision warning system 防云及碰撞警告系统
cloud atlas 云图
cloud attenuation 云层衰减
cloud band 云带
cloud bank 云堤
cloud banner 旗状云
cloud bar 云带
cloud base 云底
cloud base ceiling 云幂底面
cloud-base recorder 云底记录仪
cloud basin 云盆
cloud behavior 云雾状态
cloud blue 云蓝色
cloud box 云虹
cloud break 云的破裂
cloudburst 大暴雨;豪雨;喷丸;喷铁砂
cloud burst flood 暴雨洪水

cloudburst hardness test 喷丸硬度试验
cloudburst process 喷丸处理法;钢球喷射法;喷丸硬化法
cloudburst test 喷丸硬化处理试验;喷丸试验
cloud burst treatment 表面倾淋处理;喷丸处理;喷丸硬化处理
cloud cap 帽状云;山帽云【气】
cloud ceiling 云幂
cloud chamber 气雾室;云雾室;云室
cloud chart 云图
cloud classification 云的分类
cloud climatology 云气候学
cloud cluster 云团
cloud column 云柱;烟云柱
cloud condensation nuclei 云凝聚核;云凝结核
cloud cover 云量;云覆盖;云层
cloud cover factor 云量系数
cloud crest 盔云;云冠
cloud deck 云盖
cloud depth 云深
cloud-detection radar 测云雷达;探云雷达
cloud discharge 云中放电
cloud-dragon 云龙
cloud drift 浮云
cloud drip 雾滴
cloud droplet 小云滴
cloud droplet spectrum 云滴谱
cloud-drop sampler 云滴取样器
cloud dust 云状粉尘
cloud echo 云的回波
clouded glass 毛玻璃;云纹玻璃
clouded jade 碧犀
clouded leopard 云豹
clouded ribbon 波纹条带
clouded ware 斑纹器皿
cloud environment system 云周围环境系统
cloud field 云区
cloud flash 云中放电
cloud forest 温带雨林
cloud form 云状
cloud formation 云的形成
cloud form sludge 云状污泥
cloud-free area 无云区
cloud gray 紫灰色
cloud height 云高
cloud height indicator 云高指示器
cloudier 人造云
cloudiness 混浊性;云翳;云量;云度;浊度;发蒙;发浑
clouding 混浊的;云状花纹;云纹工艺;发浑;污斑;塑料发白;发混
clouding point 浊点;云斑(图像上);始凝点
cloud-kissing 摩天的;高耸入云的
cloud layer 云层
cloudless 晴朗
cloudlessness 无云
cloudless sky 晴空
cloudlet 小云朵
cloud level 云高
cloudmachine 舞台幻灯机
cloud mass 云块
cloud meter 测云区
cloud metre 测云仪
cloud mirror 测云镜
cloud modification 云的人工影响
cloud negative 有云影底片
cloudness 混浊
cloud nuclei 云核
cloud of electrons 电子云
cloud of smoke 烟云
cloud on title 产权的凝点
cloud particle 云粒;云滴
cloud-phase chart 云相图
cloud physics 云物理学
cloud picture 云图
Cloud pillar 天安门
cloud pink 云粉红色
cloud point 混浊点;云斑;浊点;雾点
cloud print 云斑印疵;印刷云斑;印花云斑
cloud process 云形成过程;成云过程
cloud-projector 测云器
Cloud Props project "空中支架"设计(1924年艾尔里斯特兹克与马特斯坦所设计的,建立在巨大支架上跨越城市通道的办公大楼)

cloud pulse 电子云脉冲
cloudrack 浮云
cloud-radiation feedback 云辐射相互作用
cloud resulting form industry 工业污染云
clouds 多云天
cloud scale 云级
cloud seeding 云的催化；云层播雨；造雨；播云；人工降雨
cloud seeding agent 云催化剂
cloud shadow 云影
cloud shield 云幛
clouds terrace 长城云台
cloud street 云街
cloud superseeding 过量加入云的催化物
cloud symbol 云符号
cloud system 云系
cloud system of a depression 低压云系
cloud temperature 混浊温度；混浊点
cloud test 浊点试验
cloud theodolite 测云经纬仪
cloud thickness 云厚
cloud-to-cloud discharge 云间放电
cloud-to-ground discharge 云地间放电
cloud top 云顶
cloud-top height 云顶高度
cloud-top temperature 云顶温度
cloud track 云雾径迹
cloud veil 云盖
cloud velocity gauging 浊液测流法；浊度测流
cloud with vertical development 直展云
cloud wreath 黑云环
cloudy 不匀滑油漆面；多云(的)
cloudy crystal-ball model 混浊晶球模型
cloudy day 阴天
cloudy day factor 阴天率
cloudy film 雾状薄膜；混浊膜
cloudy flow 云流
cloudy patch 云点；云斑
cloudy sky 多云天气；多云天空
cloudy state 混浊状态
cloudy stream 云流
cloudy surface 暗黑表面；无光泽表面
cloudy water 浊水
cloudy weather 多云天气
cloudy wool 褪色毛
clough 深洞；涵洞闸门；拗沟；水闸门；山口
clound burst treatment forming 喷丸成型
clound meter 测云仪
cloup 落水洞；灰岩坑；石灰坑
clouring 清理现场多余石料；刻痕结合(墙饰面)；墙面刻痕
clour-printing paper 彩色印相纸
cloustonite 发气沥青
clout 大头钉；破布；碎布；流量孔板；防磨铁板；垫片
clout nail 平头钉；大帽钉；扁头钉；油毡钉；毡钉；大头钉
clove 小鳞茎；峡谷；丁香花；丁香；深峡
clove brown 丁香树棕色
clove bud oil 丁香花蕾油
clove flavoured olive 丁香橄榄
clove hook 双抱钩
clove leaf pattern 心型图案
cloven board 削边板
clove oil 丁香油
clove oil cement 丁香油黏固粉
clove plum 丁香李
clover 苜蓿
cloverleaf 苜蓿式；四叶形立体交叉；四叶式交叉
cloverleaf antenna 多瓣形特性天线；苜蓿形天线
cloverleaf body 三座车身
cloverleaf cam 三星凸轮
cloverleaf cell 花瓣形格体
cloverleaf coupon 梅花试棒
cloverleaf crossing 苜蓿叶形互通式交叉；苜蓿叶式(立体)交叉；四叶式交通布置；四叶式立体交叉；四环式立体交叉
clover-leafed cam 三星凸轮
cloverleaf(flyover) junction 苜蓿叶形立交
cloverleaf grade separation 苜蓿叶形互通式交叉；苜蓿叶立体交叉；蝶形立交
cloverleaf interchange 苜蓿叶立体交叉；苜蓿叶形(互通式)交叉；苜蓿叶形互通式交叉；四环式立体交叉
cloverleaf intersection 蝶形立交；四叶式立体交叉；四叶式立体交叉
cloverleaf junction 四叶式立体交叉；四环式立体交叉
cloverleaf layout 苜蓿叶式交叉布置；四叶式布置；三叉式布置；四叶式交叉布置(道路立体交叉)
cloverleaf loop 苜蓿叶式匝道；四叶式环形道
cloverleaf loop ramp 四叶式环形坡道
cloverleaf pattern 三叶草形
clover leaf plate 十字形板【船】
cloverleaf type plane 拉丁十字形平面(教室)
cloverotor pump 三叶转子泵
cloverpink 车轴草花粉红色
clover sod 三叶草栽培地
clove trio-pack olive 丁香橄榄球
clowring 整理现场多余石料
club 俱乐部；棒节
clubbing 杵变；漂流；脱锚；拖锚飘流；拖锚
club building 俱乐部建筑
club chair 低背安乐椅
club dolly 顶铆具
club foot 象角船首
club-footed pile 扩底桩
club-foot roller 弯脚羊足碾；羊角碾
club-foot sheep's foot roller 弯脚羊蹄压路机
club-fraternity-lodge 小旅馆；联谊会；俱乐部
club fungi 珊瑚菌
club hammer 大锤；墙工槌；砌砖用的手锤；手锤；石匠锤
clubhouse 俱乐部会所；运动员更衣室；会所
club line 段末短行
club link 联锚链环；锚环
club moss 石松
clubmosses 石松
Club of Rome 罗马俱乐部
club premises 房产事务所
club resource 筹集资金
clubroom 交谊室；俱乐部聚会室；集会室
club shackle 联锚卸扣；联锚卸机
club-shaped 球棍状的；棍棒状的
club-shaped structure 杆状构造
club skew 斜座石；歪斜的托梁
club small means together 聚拢微薄的资力
club terms 保险协会条款
club tool 圆慢杆
clubtooth escape wheel 叉瓦式擒纵轮
clubtooth escepement 叉瓦式擒纵机构
clue 线索
cluffs 炸纹废砖
clumb 严重冒顶
clump 丛生秆；铅条
clump block 厚壳滑车；粗笨滑车；强厚滑车
clumping 聚丛
clumping techniques 聚丛技术
clump of buildings 建筑群
clump of houses 住宅群
clump of piles 集桩；簇桩；群桩
clumpy 成块的
clumpy conglomerate 块状砾岩
clumsy ship 难操纵船
clunch 炉黏土；耐火黏土；煤(层)底黏土；细粒泥质岩；硬(质)白垩；硬(化)黏土；铁黏土
Clunise 克鲁尼斯铜镍锌合金
clunk 沉闷的金属声
Cluny 克吕尼修道院(10 至 12 世纪法兰西)
cluse 横谷
clusia rosea 藤黄
Clusius column 克鲁斯热扩散柱
Clusius-Dickel column 热扩散柱
cluster 集中站；集丛；模组(熔模铸造用)；货舱移动丛灯；花簇；一簇；一丛(群)；组群；组；族；住宅群；丛灯；分组；群束；群集器；集聚
cluster analysis 聚类分析；聚集分析；聚丛分析；集群分析；点群分析；成簇分析法；群组分析；群分析
cluster analysis model 聚类分析模型
cluster bedding 花丛式坛植
cluster bent 群柱排架
cluster bins 料仓群
cluster burner 聚口灯头
cluster casting 叠箱铸造；层串铸法
cluster centre 团中心
cluster city 簇群城市
cluster college (综合性大学内的)专科学院
cluster compound 原子簇化合物
cluster cones 簇状火山群
cluster control 群控
cluster controller 组合控制器；群(集)控(制)器
cluster controller node 群控节点；群集控制器节点
cluster control unit 簇控制器；群控器；群集控制装置；集群控制器
cluster coupling 多头管接头
cluster crystal 簇形结晶；簇晶
cluster cup 密集杯
cluster data 分类数据
cluster development 组团式建设；分组改进设计
cluster drill-hole 丛式钻孔
cluster drilling 多孔底粉钻进；丛式钻进
cluster economy 聚集经济
clustered carbide 小颗粒硬质合金(3～4毫米)
clustered chimneys 集束烟囱
clustered columns 组合柱；束柱；集柱；簇柱；群柱
clustered deploy(ment) 集群部署
clustered piers 群柱；集束桥墩；簇柱；集墩；群墩
clustered piles 桩束；集桩；簇桩
clustered pillar 集柱；集墩
cluster engine 发动机组
cluster entry 簇目录表；簇表目
cluster expansion 集团展开式
cluster factor 群集系数
cluster fig 丛生榕
cluster fitting 多头管接头
cluster ga(u)ge 仪表组；组合仪表
cluster gear 联体齿轮；连身齿轮；齿轮组；齿轮块；塔式齿轮；塔齿轮
cluster hardening 分散硬化
cluster head 多轴头
cluster houses 住宅群
cluster housing 住宅群；组团式住宅
clustering 聚类；集群法；成组；成团；成束；成簇；分组；分类归并；群聚
clustering algorithm 集群算法
clustering file 聚类文件
clustering method 聚类分析方法
clustering of disease in family 家庭聚集性
clustering of earthquake 群震
clustering of seismic events 群震
clustering technique 群集技术
clustering wool machine 簇绒机
cluster joint 交聚接头；丛接头；束状接头
cluster knot 群集节
cluster lamp 多灯照明器；丛灯；货舱移动丛灯
cluster lattice 群集栅格
cluster mill 多辊轧机；多辊式轧(钢)机
cluster model 群集模型
cluster of aromatic sheet 芳香核
cluster of columns 集柱；柱群
cluster of cones 火山锥族；群火山群
cluster of dendrites 树枝状晶体簇
cluster of domains 磁畴丝；畴丛
cluster of electric(al) cables 电缆索；电缆束
cluster of engines 发动机簇
cluster of fender piles 护桩群；防撞簇桩
cluster of flocculates 絮凝团
cluster of fuel elements 释热组件
cluster of grains 颗粒团
cluster of keys 关键码串
cluster of magnetization 磁化线束
cluster of needles 针状体簇
cluster of particle 颗粒团
cluster of piles 桩组；桩群；簇桩；群桩
cluster of reflectors 反射器组
cluster piles for fendering 防撞簇桩；防冲簇桩
cluster plan 群集式平面布置
cluster point 聚点
cluster random sampling 分群抽样法
cluster rock 簇礁(石)
cluster roll 多辊式轧(钢)机
cluster sampling 属性取样；丛块采样；集中采样；密集样本；整群取样；整群抽样；成群取样；成群抽样；成串采样；分组取样；分组抽样；分群抽样法；群组抽样；群集抽样；群体抽样
cluster shape 聚类形态
clusters of crystal 晶簇
cluster spring 组合弹簧
cluster structure 簇柱结构；簇集结构；葡萄状结构；团粒构造
cluster switch 组(件)开关
cluster system of gear 齿轮组

cluster terminal 群集终端
cluster texture 团粒结构
cluster theory 簇理论
cluster treatment 多户处理
cluster-type diamond dressing 簇状金刚石修整工具
cluster-type fuel element 棒束型燃料元件
cluster weld 丛聚焊缝
cluster zoning 组团式区划;住宅群体分区规划;规定住房密度的区划
clustration 集团化;成组团
clustre ＝cluster
clustre(bore)hole 丛式钻孔
clutch 离合器;咬合;钩形企口;扼
clutch adjusting arm 离合器调节臂
clutch adjusting cover 离合器调整盖
clutch adjusting nut 离合器调整螺帽
clutch adjusting screw 离合器调整螺钉
clutch adjustment 离合器调整
clutch air tank 离合器空气罐
clutch alignment 离合器对中
clutch and pinion operating lever 离合器齿轮操作杆
clutch arm 离合器弯爪
clutch back plate 离合器盖;离合器挡板
clutch block 离合器摩擦块
clutch body 离合器体
clutch bolt 离合器螺栓
clutch bowl 离合器筒;离合器壳(体)
clutch box 离合器箱;离合器壳(体)
clutch bracket 离合器托架
clutch cam 离合器凸轮
clutch capacity 离合器容量
clutch carrier 离合器支承
clutch case 离合器盖
clutch collar 离合器套环;连接器;接合套
clutch control 离合器操纵;离合器控制
clutch control booster 离合器操纵增强器
clutch control lever 离合器控制杆
clutch control rod 离合器控制杆;离合器操纵杆
clutch coupling 离合器;离合联轴节;牙嵌联轴节
clutch cushion 离合器盘;离合器缓冲装置
clutch diaphragm spring 离合器隔膜弹簧
clutch disc 离合器圆盘;离合器盘;离合器摩擦片
clutch disc facing 离合器盘衬片
clutch disc pressure 离合器盘压力
clutch disc torsion spring 离合器盘扭转弹簧
clutch disengaging axle 离合器脱离轴
clutch drive 离合器驱动
clutch driven compressor 离合器传动压缩机
clutch driven plate 离合器从动盘
clutch driving shaft flange 离合器传动轴凸缘
clutch driving strap 离合器传动片
clutch engagement screw 离合器接合螺丝
clutch engaging 离合器接合
clutcher 司钻
clutch facing 离合器摩擦衬面;离合器垫片;离合器衬片
clutch facing rivets 离合器面片铆钉
clutch foot lever bracket 离合器(脚)踏杆座
clutch foot lever bushing 离合器脚踏杆衬套
clutch fork 离合器叉
clutch friction spring 离合器摩擦弹簧
clutch gear 离合器齿轮
clutch gold 铜箔
clutch handle 离合器手柄
clutch head 防拆卸螺钉头
clutch housing 离合器壳(体)
clutch housing cover 离合器箱盖
clutch housing dust plate 离合器壳护板
clutch housing pan 离合器下壳
clutch housing under pan 离合器底壳
clutch jaw 卡盘
clutch judder 离合器颤动
clutch knob 离合器钮;离合钮
clutchless gear change 不分离离合器变速
clutch lever 离合器操纵杆
clutch lift 棘轮式自动起落机构
clutch liner 离合器衬片
clutch lining 离合器摩擦片;离合器摩擦衬面
clutch linkage 离合器联动装置;离合器分离机构;离合器操纵联杆
clutch lock 离合器保险(装置)
clutch locking 离合器接合

clutch magnet 离合器电磁铁;啮合电磁铁
clutch mechanism 离合器机构
clutch motor 带离合器的电动机
clutch of sheet pile 板桩导梁
clutch of shift 换挡离合器
clutch on-off 离合器离合
clutch operating device 离合器操纵机构
clutch operating lever 离合器操纵杆
clutch operating pin 离合器压盘销
clutch operating rod 离合器控制杆
clutch operating shaft 离合器操纵轴
clutch pedal 离合器踏板
clutch pedal bushing 离合器踏板轴衬套
clutch pedal clearance 离合器踏板间隙
clutch pedal connecting link 离合器踏板连杆
clutch pedal pad 离合器踏板垫
clutch pedal play 离合器踏板自由行程;离合器踏板间隙
clutch pedal retracting spring 离合器踏板拉簧
clutch pedal return spring 离合器踏板复位簧
clutch pedal rod 离合器踏板杆
clutch pedal shaft 离合器踏板轴
clutch pedal shaft bushing 离合器踏板轴衬套
clutch pedal shaft lever 离合器踏板轴杆
clutch pilot bearing 离合器导向轴承
clutch plate 离合器盘;离合器摩擦片;离合片
clutch point 离合点
clutch pressure 离合器压力
clutch pressure assembly 离合器片压紧机构
clutch pressure lever 离合器压杆
clutch pressure modulating valve 离合器调压阀
clutch pressure operating pin 离合器压销
clutch pressure plate 离合器压盘;离合器压板
clutch pressure plate pin 离合器压板销
clutch pressure plate screw 离合器压板螺钉
clutch pressure ring 离合器压环
clutch prongs and slots 离合器拨叉和凹槽
clutch pulley 离合皮带轮
clutch pull rod 离合器拉杆
clutch pully 离合器皮带轮
clutch push rod 离合器推杆
clutch release bearing 离合器分离轴承
clutch release bearing carrier 离合器分离轴承座
clutch release bearing sleeve 离合器放松轴承套
clutch release finger 离合器压盘分离杆
clutch release fork 离合器分离叉
clutch release fork lever 离合器分离叉杆
clutch release fork sleeve 离合器放松叉套
clutch release lever pin 离合器放松杆销
clutch release lever spring 离合器放松杆弹簧
clutch release rod 离合器分离拉杆
clutch release sleeve 离合器分离套
clutch rod 离合器结合杆
clutch roller bearing 离合器滚柱轴承
clutch shaft 离合器轴
clutch shifter 离合器结合杆;离合器分离叉;离合器拨叉
clutch shifter collar 离合器滑动环箍
clutch shoe 离合器瓦
clutch slippage 离合器滑转(量)
clutch(slipper) yoke 离合器拨叉
clutch spider 离合器从动盘
clutch spider hub 离合器从动盘毂
clutch spindle 变速箱第一轴
clutch spreader 离合器摩擦圈
clutch spring 离合器弹簧
clutch spring cap 离合器压紧弹簧座
clutch spring collar 离合器弹性环
clutch spring ring 离合器弹簧圈
clutch sprocket 离合器链轮
clutch sprocket wheel 离合器链轮
clutch steering 离合器式转向装置
clutch stop 离合器小制动器;止动凸爪
clutch tap holder 丝锥铰杠;丝锥角扣
clutch throwout bearing 离合器分离轴承
clutch throwout lever 离合器推杆
clutch throwout yoke 离合器分离叉
clutch thrust bearing 离合器止推轴承;离合器分离推力轴承
clutch type transmission 离合器传动
clutch yoke 离合器分离器;离合器拨叉;离合杆
clutch yoke bolt 离合器分离叉调整栓
clutter 地面反射干扰;混杂信号;杂乱回波;杂波;地物回波;散射干扰

clutter filter 防散射干扰滤波器;静噪滤波器;杂波滤波器;反干扰滤波器
clutter gating 避杂波选通
clutter lock 抗地面干扰
clutter noise 杂波噪声
clutter rejection 消除本机干扰;杂波抑制
clutter suppression 杂波抑制
clutter suppressor 杂波抑制器
Clwia sp 美螅属
Clyburn spanner 活(络)扳手
clypeus 盾状体
CML system 集装箱海上运输公司方式
C-M pattern of traction current 牵引流的 CM 图像
C-M pattern of turbidity current 浊流的 CM 图像
CN cycle 碳氮循环
CN index 碳氮指数
CO_2 acidosis 二氧化碳酸中毒
CO_2 exchange 二氧化碳交换
CO_2-fire extinguisher 二氧化碳灭火器;二氧化碳灭火机
CO_2 fire-extinguishing system 二氧化碳灭火系统
CO_2-induced changes of climate 二氧化碳引起的气候变化
CO_2 tension 二氧化碳张力
coabsorption 共吸收(作用);共吸附(作用)
co-acceptor 共同承兑人
coacervate 凝聚体;凝聚层
coacervate course 凝聚层
coacervated phase 凝聚相
coacervated system 凝聚体系
coacervation 凝聚(作用)
coach 教练员;大型客车;车厢;长途客车;长途(公共)汽车
coach axle-kilometre 客车轴公里
coach bolt 方头螺栓;方颈螺栓
coachbuilder 车身制造厂
coach bus 长途汽车;长途客车
coach cleaning yard 客车洗车场;客车清洗场
coach clip washer 方头垫圈
coach fitted with audio-visual equipment 带视听设备客车
coach house 车马房
coaching of supervisors 辅导管理员
coaching stock 客车车库
coaching traffic 客车运输
coach joint 弯边接头
coachmaker 车身制造厂
coach maker's bevel edge chisel 车匠斜边凿
coach paint 车辆漆
coach park 旅游车停车场;客车场
coach road 汽车路
coach roof 半显露舱室
coach screw 六角螺钉;方头木螺钉;方头螺钉;板头尖端木螺钉
coach servicing shed 客车整备库(棚)
coach spring 客车弹簧;平簧;弹簧片
coach washing machine 洗车机
coach whip 长三角旗
coach whipping 花样缠扎绳头
coach with automatic coupling 装自动联结器的客车
coach with center gangway 中央通道客车
coach with children's play area 儿童游戏车
coach with easy access for handicapped people 残疾人客车
coach with integral body 整体式客车
coach with open aisle 开敞型通道客车
coach with open gangway 非封闭通道客车
coach with tilting body 可倾斜式客车
coach with wooden body 木质车体客车
coach with wood-framed body with metal panels 钢木混合车体客车
coach wrench 双开活络扳手
coach yard 客场;客车场;车场
coa consumption rate 煤耗率
co-acting force 互相作用力
co-acting pulling rollers 复动式拉丝对轮
co-acting roller process 对辊法
co-acting signal 复示信号
coaction 相互作用;共同作用
coactivation 共激活作用;共活化作用
coactivator 共激活剂;共活化剂
coactive pattern 相互作用型
co-adaptation 互相适应;互相迁就

Coade stone 柯德石料
coadjacent 邻接的
coadjutor 助手;协作者
coadjutress 女助手
co-administrator 合作遗产管理人
coagel 凝聚胶
coagent 伙伴;合作者
coagent case 辅助主动者角色
coagglutination 协同凝集
coaguin 凝固素
coagulability 凝聚性;凝聚能力;凝结性;凝结力;凝固性;混凝能力
coagulant 聚沉剂;凝聚剂;凝结剂;凝固剂;促凝物质;促凝剂;促凝的;混凝剂
coagulant agent 促凝剂;混凝剂
coagulant aids 助凝剂;辅助凝聚剂
coagulant demand 混凝剂需量
coagulant dosage 混凝剂量
coagulant dosing apparatus 混凝剂投加设备
coagulant feeder 加矾设备
coagulant sedimentation 混凝沉淀法
coagulate 凝固;凝结物;混凝
coagulate anomaly 凝聚异常
coagulated aerosol 凝聚型气溶胶
coagulated matter 凝聚物
coagulated silt 凝结淤泥
coagulating basin 凝聚池
coagulating bath 凝结槽
coagulating chamber 凝聚室
coagulating effect 混凝效果
coagulating equipment 混凝设备
coagulating mechanism 混凝机理
coagulating point 凝(结)点
coagulating power 凝聚力;凝结力
coagulating property 混凝性能
coagulating(re)agent 凝结剂;凝雾剂;凝固剂;混凝剂;聚沉剂
coagulating sedimentation 凝结沉降(作用);混凝沉淀(法)
coagulating sedimentation method 凝结沉淀法
coagulating tank 凝聚池
coagulating value 凝聚值
coagulation 聚沉;胶凝;凝聚(作用);凝聚增长;(液体变为固体的)凝结;凝固;凝结;絮凝
coagulation action 凝聚作用
coagulation air flo(a)tation 混凝气浮法
coagulation and flocculation process 混凝土沉淀法
coagulation and flocculation treatment of wastewater 废水凝聚及絮凝处理
coagulation basin 凝聚池;混凝池;混凝池;沉淀池
coagulation-bio-ferric process 混凝—生物铁法
coagulation box 凝聚箱
coagulation chamber 凝聚室;混凝室;凝结室
coagulation-chlorine dioxide process 混凝—二氧化氯法
coagulation deposit 混凝沉淀
coagulation disorder 凝固障碍
coagulation effect 混凝效果
coagulation factor 稳定因子
coagulation factor deficiency 凝固因子缺乏
coagulation factors 凝固因子
coagulation filter 混凝滤池
coagulation-flo(a)tation reactor 混凝气浮反应器
coagulation-hydrolysis-oxidation process 混凝—水解—氧化工艺
coagulation kientics 混凝动力学
coagulation-magnetic separation process 混凝磁分离工艺
coagulation mechanism 混凝机理
coagulation-microbubble air flo(a)tation process 混凝—微气泡气浮法
coagulation of colloid 胶体凝聚
coagulation of water 水的凝聚;水的混凝
coagulation performance 混凝性能
coagulation point 凝结点
coagulation-precipitation-compatible anaerobic and oxygenic biomembrane oxidation ditch-biologic(al) contact oxidation treatment technology 混凝沉淀—沉淀—兼性厌氧和含氧生物膜氧化沟—生物接触氧化处理工艺
coagulation pretreatment 混凝预处理
coagulation reaction 凝结反应;混凝反应
coagulation sediment 凝聚沉淀

coagulation sedimentation 凝聚沉淀;混凝沉淀(法)
coagulation sedimentation-cavitation air flo(a)tation-V type filter process 混凝沉淀—涡凹气浮—V形滤池工艺
coagulation-sedimentation filtration 絮凝沉淀处理;混凝沉淀过滤
coagulation sedimentation method 混凝沉淀法
coagulation sedimentation rapid gravity filtration 混凝沉淀快速重力过滤
coagulation sedimentation tank 混凝沉淀池
coagulation stability 聚结稳定性
coagulation structure 凝结构造
coagulation tank 凝聚箱;混凝池;混凝反应池
coagulation test 混凝试验
coagulation time 凝固时间;混凝时间
coagulation treatment 凝结处理;混凝处理;絮凝处理
coagulation treatment wastewater 废水混凝处理法
coagulation-two-phase anaerobic-anoxic process 混凝—两相厌氧—缺氧—好氧工艺
coagulation value 聚沉值;凝聚值;凝结值;混凝值;絮凝值
coagulative 凝结的;促凝固的
coagulative device 凝结设备
coagulative precipitation 凝聚沉淀
coagulative precipitation tank 混凝沉淀池
coagulative process 凝聚过程
coagulative sedimentation 混凝沉淀(法)
coagulative sedimentation method 混凝沉淀法
coagulative tank 凝聚沉淀池
coagulator 聚沉剂;凝聚器;凝聚剂;凝结器;絮凝剂;凝集剂;凝固剂;促凝剂;沉淀凝集装置
coaguloreaction 凝固反应
coagulum 乳凝;粒子附聚体;凝块;凝聚物;凝结物;凝结块;凝固物;凝固乳胶
coai-tar product 煤焦油产品
coak 榫接头;栓销;木销钉;(木轮中的)金属下去物;柄接
coak scarf joint 雄榫斜口接合;雄榫嵌接
coal accumulated basin analysis 聚煤盆地分析
coal accumulation 成煤作用
coal accumulational area 聚煤区
coal accumulational center 聚煤中心
coal accumulational zone 聚煤带
coal accumulation basin 聚煤盆地
Coal Act 煤炭法(英国)
coal-alkali reagent 煤碱剂
coal alkylation 煤烃化作用
coal analysis 煤(质)分析
coal analysis result table 煤质分析试验成果表
coal apple 煤结核
coal ash 煤灰;煤的灰分;灰分
coal ash absorption 煤灰吸收
coal ash analysis 煤灰成分分析
coal ash and slag 煤灰渣
coal ash ceramisite 粉煤灰陶粒
coal ash fusibility 煤灰熔融性
coal ash fusibility graduation 煤灰熔融性分级
coal a ship 给船上煤
coal ash monitor 煤灰监测仪
coal ash ring 煤灰圈
coal ash viscosity 煤灰黏度
coal auger 煤钻;煤螺旋钻
coal ball 煤结核;煤层石球
coal ball apple core 煤核
coal ball concretion 煤中结核
coal band 煤条带
coal barge 煤驳;供煤驳
coal barge jetty 煤驳码头
coal barrow 运煤手车
coal base 煤炭基地
coal-based power generation 煤基发电
coal-based synthetic(al) fuel 煤合成燃料
coal basin 煤田;煤炭田;煤盆地;含煤盆地
coal basset 煤层露头
coal bauxite-bearing ferrous formation 含煤铝土矿铁质建造
coal-bearing 含煤的
coal-bearing coefficient 含煤率
coal-bearing density 含煤密度
coal-bearing formation 煤系;含煤系;含煤建造
coal-bearing parameter 含煤性
coal-bearing strata 含煤岩系;含煤地层

coal bed 煤层
coal-bed firing combustor 层燃炉
coal-bed gas pressure 煤层瓦斯压力
coal bed methane 煤层甲烷
coal bed pitch 煤层倾角
coal bed profile 煤层剖面
coal bed texture 煤炭结构
coal berth 煤炭泊位;煤码头
coal bin 煤箱;煤仓
coal bit 钻煤钻头
coal black 煤黑色;漆黑的;墨黑的
coal blending 配煤
coal blending rate 配煤比
coal block 煤柱
coal board slide 煤滑板
coal borer 煤钻
coal brass 煤中黄铁矿结核;黄铜矿;富碳黄铁矿
coal breaker 煤舱;碎煤机
coal breccia 煤角砾岩
coal brick 煤砖
coal bridge 运煤桥
coal briquet(te) 煤砖;煤渣砖;煤饼
coal briquet(te) for civillan use 民用型煤
coal bucket 煤斗
coal bump 煤条带;煤炭突出
coal bunker 煤舱;煤仓;储煤仓;燃煤舱
coal bunkering 加装船用煤
coal bunkering wharf 供煤码头
coal burner 煤粉喷燃器;粉煤燃烧嘴;喷煤嘴
coal burning 烧煤
coal burning boiler 燃煤锅炉;烧煤锅炉
coal burning gas turbine 燃煤的燃气轮机
coal burning installation 炉膛设备;燃煤装置
coal burning power station 燃煤发电厂
coal burning vessel 烧煤船
coalburster 水压爆煤筒
coal by-product 煤副产品;煤炭干馏副产品
coal capacity 装煤量
coal carbonization 煤炭化;煤干馏
coal carbonization by-product 煤炭干馏副产品
coal carbonization coking 焦化
coal carbonization industry 煤干馏工业
coal carbonization wastes 炼焦厂废水;煤的炭化废水
coal car-dumper 煤炭翻车机
coal carrier 煤船;运煤船
coal cassification 煤分类法
coal cell 煤窖
coal cellar 地下煤库;煤窖
coal characteristic 煤特性
coal charger 加煤车;加煤机
coal chemical industry 煤化(学)工(业)
coal chemicals 煤化学制品
coal chemical waste water 煤化学废水
coal chemistry 煤化学
coal chute 输煤管;加煤槽;煤溜槽;煤槽;装煤槽;输煤溜槽
coal chute crane 提升倾卸式煤炭翻车装船机
coal cinder 煤渣;煤块
coal cinder road 煤渣路
coal classification 煤的分类
coal classification system 煤炭分类法;煤分类系统
coal clay 耐火黏土;底黏土
coal cleaning leachate 洗煤渗沥液
coal cleaning plant 洗煤厂;煤清洗厂
coal cleaning plant building 洗煤厂建设
coal cleaning waste leachate 洗煤废弃物渗沥液
coal coke 煤焦(炭)
coal-coking process 炼焦法
coal collector 集焦器
coal combustibility 煤可燃性
coal combustibles 煤可燃成分
coal compartment boat 运煤方驳
coal complex 煤合成物
coal conduit 落煤槽
coal constitution 煤组构
coal consumption 煤炭消耗;煤耗;耗煤量
coal consumption rate 燃煤率
coal content instrument 煤含量计
coal conversion wastewater 煤转化中产生的污水
coal conveyer 煤炭输送机;运煤机
coal conveyer tunnel 运煤坑道
coal(conveyor) trestle 运煤栈桥
coal core length 煤芯长度

coal core sample 煤芯煤样
coal correlation 煤层对比
coal cracker 碎煤机;煤压碎机
coal cutter 截煤机;采煤机
coal-cutter-loader 联合截煤机
coal cutting 采煤;掘煤机;截煤
coal-cutting machine 截煤机
coalcutting machinery 采煤机械
coal cylindrical dryer 筒式煤炭干燥箱
coal damper 落煤控制板
coal delivered energy 煤释放能
coal delivery system 煤输送系统;煤的输送系统
coal deposit 煤田;煤床;煤沉积;煤层
coal deposits 煤藏
coal depot 煤栈;煤炭堆场;煤场
coal desulfurization 煤脱硫
coal desulfurization and denitrification 煤脱硫脱氮
coal distributor 煤炭撒布机;煤炭平舱机;布煤器
coal district name 煤矿区名
coal dock 煤码头
coal door 装煤舷门
coal dozer 推煤机
coal dressing 选煤
coal-dressing plant 选煤厂
coal-dressing wastewater 洗煤废水;选煤废水
coal drift 煤巷
coal drill 煤钻;打煤机
coal-drilling 打煤钻进
coal drop 煤溜槽;卸煤机
coal dry cleaning process 煤干洗方法
coal dryer 煤干燥器
coal-drying drum 煤烘干机
coal dump 煤堆
coal dust 煤屑;煤炭粉末;煤末;煤粉(尘);煤尘
coal dust bin 煤粉仓
coal dust brick 煤灰砖
coal dust burner pipe 喷煤(油)管
coal dust cementing agent 煤尘黏合剂
coal dust classifier 煤粉分离器
coal dust engine 煤粉发动机
coal dust explosion 煤粉爆炸;煤尘爆炸
coal dust index 煤尘指数
coal dust ring 煤灰圈
coal dust stained yarn 煤灰纱
coal dust storage hopper 煤粉仓;烧煤仓
coal dust suppressant 煤粉层抑制剂
coal dust suppression 煤尘排除
coal economizer 节煤器;省煤器
coal elevator 煤提升机
coal energy cleaning 煤炭能量净化
coal equivalent 煤当量;当量煤
coaler 煤炭专用线;煤商;煤船;运煤铁路;运煤船;船供应商人;船供应商船;运煤车
coalesced copper 聚结铜
coalesced glaze 聚釉
coalesced pigment 包核型复合颜料
coalescence 聚结
coalescence efficiency 并合率
coalescence of waves 波浪合成
coalescence process 并合过程
coalescent 聚结剂;成膜助剂
coalescent debris cone 联合泥石锥;聚合泥石堆;复合碎石锥
coalescent filter 凝聚过滤器
coalescent knots 重叠结点
coalescent pack 聚结填料
coalescer 聚结器;凝聚过滤器
coalescing agent 聚结剂;成膜助剂
coalescing alluvial fan 联合冲积扇;结合冲积扇;接合冲积扇
coalescing element 聚结元件
coalescing particles 聚结粒子
coalescing pediment 接合麓原
coalescing-separating principle 聚结分离原理
coalescing solvent 聚结溶剂
coalescing temperature 聚结温度
coalescive neomorphism 聚结新生变形作用
coaleum 煤烃
coal exploration 煤田勘探
coal extract 煤萃;煤馏出物
coal face 煤层截面;采煤工作面
coal face machinery 采煤工作面机械
coal facies 煤相
coal factor 煤系数;煤商

coal fall 煤崩落
coal feed 加煤装置
coal feeder 煤浆进料机;供煤机(加煤用);给煤机;饲煤机
coal feeding system 喂煤系统
coal field 煤田;煤炭田;产煤区
coal field area 煤田区
coalfield geology 煤田地质
coalfield name 煤田名称
coal field prediction 煤田预测
coalfield self-combustion 煤田自燃
coalfield structure 煤田构造;藻田构造
coal fines (0~0.5毫米)煤粉;煤炭粉末
coal-fir boiler 烧煤锅炉
coal-fired boiler 燃煤锅炉
coal-fired central heating 烧煤集中采暖(法)
coal-fired furnace 燃煤(加热)炉
coal-fired gas turbine 燃煤的燃气轮机
coal fired kiln 烧煤的窑
coal-fired power plant 煤炭发电厂;燃煤电厂
coal-fired power station 燃煤火电站;燃煤发电机;燃煤发电厂;火电厂
coal-fired range 燃煤炉灶
coal-fired stove 燃煤火炉
coal-fired unit 燃煤锅炉机组
coal firing 烧煤
coal firing hot-air generator 燃煤热风炉
coal flo(a)tation 煤浮选
coal for humic acids 腐殖酸用煤
coal formation 含煤岩系
coal-forming material 成煤物质
coal-forming model 聚煤模式
coal-forming period 聚煤期
coal-forming process type 成煤作用类型
coal-forming stage 成煤阶段
coal for refining oil 炼油用煤
coal fuel enrichment 煤燃料富集
coal fuel ratio 煤燃料比
coal gangue 煤矸石
coal gangue wool 煤矸石绵
coal gas 制煤气;煤气
coal gas and coke manufacture 煤气和焦炭制造业
coal-gas burst 煤与瓦斯突出
coal gas generator 煤气发生器;煤气发生炉
coal gasification 煤炭气化;煤气化;煤的气化
coal gasification in situ 煤就地气化
coal gasification process 煤气化法
coal gasification wastewater 煤气化废水
coal gasification wastewater treatment 煤气化废水处理
coal-gas outburst 煤与瓦斯突出
coal-gas outburst mine 煤和瓦斯突出矿井
coal-gas producer 煤气发生炉
coal gas recovery valve 煤气回收阀
coal-gas washing wastewater 煤气洗涤废水
coal-gas waste 煤气废水
coal-gas waste-water 煤气残液
coal gear 装煤装置
coal geography 煤炭地理学
coal geologic(al) hole 煤田地质钻孔
coal geology 煤田地质学;煤炭地质学;煤地质学
coal getter 采煤工
coal grab 抓煤机
coal grab bing bucket 煤吊篮;煤吊斗
coal grading 煤分级
coal grinding 煤研磨
coal grinding mill 煤磨;磨煤机
coal grinding tube-type mill 管式煤磨机
coal hammer 碎煤锤
coal handling 煤的装卸
coal handling facility 装卸煤设施
coal handling plant 煤炭装卸设备;输煤设备
coal harbo(u)r 煤港
coal haulage 煤炭运输
coal heap 煤堆场;储[贮]煤场
coal-heated brooder 煤炉加热式育雏器
coal heating 用煤加热;用煤供暖;燃煤供暖
coal heaver 煤炭装卸工(人)
coal hoist 扬煤机;吊煤机;提煤机
coal hoist craft 加煤船
coal hold 煤舱
coal hole 煤舱孔
coal hopper 煤漏斗;煤斗;煤舱;煤仓
coal horizon 煤层

coal house 煤库
coal hulk 储[贮]煤船
coal hydrogasification 煤氢气化
coal hydrogenation 煤氢化;煤的氢化;伯吉尤斯法
coal hydrogenation unit 煤氢化装置
coalification 煤化(作用)
coalification break 煤化间断
coalification gradient 煤化梯度
coalification jump 煤化跃变
coalification pattern 煤化型式
coalification series 煤化系列
coalification stage 煤化作用阶段
coalification track 煤化轨迹
coalified wood 煤化木
coal ignitability 煤自燃倾向性;煤可燃性
co-alignment 调准装置;匹配装置
coal-in-air suspension 悬浮煤粉尘
coal inclusion 煤包裹体
coal industry 煤炭工业
Coal Industry National Consultative Council 煤炭工业全国咨询委员会
Coal Industry Nationalization Act 煤炭工业国有化法(英国)
Coal Industry Society 煤炭工业协会
coal industry wastewater 煤炭工业废水
coaling 加煤
coaling bridge 装煤桥;加煤桥
coaling capacity 煤容量;载煤量
coaling crane 运煤吊车;装煤起重机
coaling day 装煤日
coaling depot 加煤站
coaling equipment 加煤装置
coaling gear 装煤工具
coaling hatch 煤舱
coaling installation 装煤设备;上煤设备
coalingite 片碳镁石
coaling pier 装煤码头
coaling plant 加煤装置;加煤设备;煤厂
coaling port 装煤港(口)
coaling road 上煤线
coaling ship 加煤船;供煤船
coaling stage 燃煤装卸台
coaling station 加煤站;装煤港(口);上煤站;装煤站
coaling track 上煤线
coaling trunk 进煤井道;装煤导筒
coaling vessel 装煤船
coaling wharf 加煤码头
coaling whip 装煤单绞机
coal-in-on suspension 煤在油中悬浮液
coal in pile 煤堆
coal in solid 整块煤
coal intergrinding 生料中掺煤共同粉磨
coalite 焦炭砖;低温焦油;低温焦炭;半焦炭
coalite tar 半焦油
coalition 联盟
coalition analysis 联合分析
coalition change 联盟变更
coalition structure 联盟结构
coal layer 煤层
coall chute 输煤管
coal leads 断层煤线
coal lift 煤田提升机
coal-lifting pump 煤水泵
coal lighter 煤驳;运煤驳(船)
coal liquefaction 煤炭液化;煤的液化
coal liquefaction rate 煤液化速率
coal liquid 煤浆
coal liquification 煤液化
coal liquor 煤气水
coal loader 装煤机
coal-loading apparatus 装煤器具
coal loading berth 煤炭装船泊位
coal loading bucket 装煤吊桶
coal loading crane 装煤起重机
coal loading dock 煤炭装船码头
coal loading facility 煤炭装船设施
coal loading terminal 煤炭装船码头
coal loading wharf 煤炭装船码头
coal lock 加煤锁斗
coal machinery 采煤机械
coal master 煤矿主
coal meal 煤粉
coal meal silo 煤粉仓
coal measure clay 夹煤层黏土

coal measure gas 煤系气
coal measures 煤系;煤层
coal-measures shale 煤系页岩
coal-measures strata 煤系地层
coal-measures unit 煤系旋回;煤系单位
coal metamorphism 煤变质作用
coal meter 煤计量器
coal microbiology 煤微生物学
coal mine 煤矿
coal mine drainage 煤矿排水
coal mine geology 煤矿地质(学)
coal mine name 煤矿名
coal miner 煤矿工人;采煤工
coalminer's lung 煤矿工人尘肺
coal mine wastewater 煤矿矿井废水
coal mine water 煤矿水
coal mine worker 煤矿工人
coal mine working-out section 煤矿采空区
coal mining 煤矿开采
coal mining and processing wastes 煤矿开采和加工废水
coal-mining emulsifying explosive 煤矿乳化炸药
coal mining machine 采煤机
coal mining methane 煤层气
coal mining methane drainage 煤层气抽放
coal mining methane emission 煤层气排放
coal mining methane utilization 煤层气利用
coal mining method 采煤方法
coal mining region 采煤区
coal mining waste 采煤废水
coal-mining water gel explosive 煤矿水胶炸药
coal mixing plant 配煤厂;配煤设备
coal mixing rate on viscosity 煤黏度混合比
coal mud 煤泥
coal-natural gas basin 含煤—天然气盆地
coal oil 煤(馏)油;煤焦油
coal-oil fuel 煤油混合燃料
coal-oil mixing liquefaction method 煤油混炼液化法
coal-oil mixture fuel 煤焦油混合燃料
coal-oil water fuel 煤油水混合燃料
coal opening pit mining 露天采煤矿
coal ore 煤矿石
coal-ore-carrier 运煤/矿石船
coal-ore trade 煤矿贸易
coal outbreak 煤层露头
coal oven gas 煤气
coal paleobotany 煤古植物学
coal parting 煤层夹矸
coal passer 装煤工
coal paste 煤浆
coal pebble 煤卵石
coal petrography 煤(岩)相学;煤岩石学
coal petrography classification of kerogen 干酪根煤岩学分类
coal petrology 煤岩(石)学
coal pick 采煤镐;刨煤镐
coal picker 选煤机
coal pier 装煤栈桥;突堤式煤码头
coal pile drainage 煤堆排水
coal pile trimmer 煤堆修整器
coal pillar 煤柱
coal pillar thrust 煤柱压碎
coal pinch out 煤层尖灭
coal pit 矿井;煤矿;煤坑
coal pitch 煤沥青;煤焦油脂
coal planer 刨煤机;刨煤机
coalplex 煤(的)综合利用
coal plough 刨煤机
coal pocket 煤仓
coal porosity 煤孔隙率
coal port 煤港;装煤舷门
coal powder 煤粉
coal powder fuel 粉煤燃料
coal powder injection equipment 高炉喷煤粉设备
coal powder line marking 弹粉煤线;弹粉线
coal preparation 选煤;选煤
coal preparation machine 选煤机
coal preparation plant 选煤车间;煤制备厂;煤制备装置
coal preparation process 煤炭洗选工艺
coal processing 煤炭加工
coal production 煤产量
coal production under unified central planning 统配煤
coal proportioning system 配煤系统
coal province 含煤区;产煤区
coal pulp 煤浆
coal pulverization 煤的粉碎
coal pulverizer 碎煤机;煤炭粉碎机;煤磨
coal pulverizing installation 碎煤装置
coal pulverizing mill 煤磨机;煤磨
coal pumping 煤泵送
coal puncher 冲击式截煤机;冲击式采煤机
coal pyrite 碳质黄铁矿
coal pyrolysis 煤(的)热分解
coal quality 煤质
coal quality analysis plot 煤质分析图
coal quality control 煤质管理
coal quality log 煤质测井
coal railway 运煤铁路
coal rake seam 煤层
coal range 煤炉
coal rank 煤品级;煤级
coal rank semifusinite 煤化半丝质体
coal receiving track 进煤线
coal recovery 采煤回收率
coal recovery drill 采煤螺旋钻
coal refuse 煤渣
coal reserves 煤炭储量;煤储量;煤藏
coal resources 煤炭资源
coal resources data base 煤炭资源数据库
coal resources exploration 煤炭资源勘探
coal-rich center 富煤中心
coal-rich zone 富煤带
coal road 煤巷;煤炭专用线
coal roadway 煤巷
coal room 煤间;储(贮)煤室
coal rotary dryer 煤炭旋转干燥器
coalsack 煤袋
coal samling 煤炭采样
coal sample 煤样
coal sample division 煤样的缩分
coal sample for checking production 生产煤样
coal sample for determination of apparent specific gravity 视比重煤样
coal sample for determination of dust 煤尘煤样
coal sample for determination of gas 瓦斯煤样
coal sample for determination of oxidized zone 氧化带煤样
coal sample for determination of weight of unit volume 块比重煤样
coal sample for petrographic(al) analysis 煤岩煤样
coal sample number 煤样编号
coal sample preparation 煤样设备;煤样的制备
coal sample total gravimetry before sieve 筛分前煤样总重量
coal sampling 采煤样
coal-sand dual media filter 煤砂双层滤料滤池
coal scale 煤磅
coal scarce area 缺煤地区
coal scatter 分煤设备
coal science 煤炭科学
coal screening plant 煤田筛选设备;煤田筛选长
coal screenings 煤屑
coal scuttle 煤桶;煤斗
coal seam 煤层
coal seam boundary 煤层边界
coal seam correlation 煤层对比
coal seam dip angle 煤层倾角
coal seam false floor 煤层伪底
coal seam false roof 煤层伪顶
coal seam floor 煤层底板
coal seam gas bearing capacity 煤层瓦斯含量
coal-seam gas pressure 煤层瓦斯压力
coal seam inclusion 煤层包体
coal seam infusion 煤层注水
coal seam methane content 煤层瓦斯含量
coal seam pitch 煤层倾角
coal seam profile 煤层剖面
coal seam roof 煤层顶板
coal seam sample 煤层煤样
coal seam sketch 煤层素描图
coal seam structure 煤层构造
coal seam texture 煤层结构
coal seam thickness 煤层厚度
coal-seam uncovering 煤层剥露
coal sear 底黏土
coal seismic prospecting 煤田地震勘探
coal-sensing probe 底板留煤的核探测器
coal-separating plant 选煤厂
coal separation station 选煤站
coal series 煤系
coal shed 煤库;煤仓
coal shoot 装煤槽;放煤槽
coal shovel 掘煤电铲;煤铲
coal skip-winding plant 煤田翻斗提升设备
coal slack 煤屑;煤粉
coal slag 煤渣
coal slaking 煤炭粉化
coal sludge 煤淤泥
coal slurry pipeline 煤泥(输送)管道;煤浆管道
coal smoke 煤烟
coal smoke pollution 煤烟污染
coal split 夹层煤;分裂煤层
coal spoil 煤矸石
coal spoil brick 煤矸石砖
coal spontaneous combustion 煤自燃;煤的自燃
coal spontaneous combustion tendency 煤自燃倾向性
coal sprinkler 煤洒水器
coal stone 块状无烟煤;煤石;煤矸石;油页岩;硬煤;无烟煤
coal storage 煤栈;煤仓;煤储(贮)库;煤储库;储煤
coal storage and bagging plant 煤炭储藏与装袋厂
coal storage ground 煤场
coal storage system 煤库;储煤仓库(不包括煤场)
coal storage tower 储(贮)煤塔
coal storage yard 堆煤场;煤炭堆场;煤场;储(贮)煤场
coal store 煤库;储煤仓;煤栈
coal stratingraphy 煤地层学
coal streak 煤线
coal stripping 煤炭露天开采
coal supply 给煤
coal supplying facility 供煤设备
coal supply kilometerage 供煤里程
coal supply tonnage 供煤量
coal swamp 煤沼木本沼泽
coal-tar 焦油沥青;煤焦油沥青;煤焦油;柏油
coal-tar asphalt 煤焦油沥青
coal-tar-base enamel 煤沥青瓷漆
coal-tar chemistry 煤焦油化学
coal-tar colo(u)r 煤焦油染料
coal-tar creosote 煤沥青木材防腐剂;煤焦(油)杂酚油
coal-tar creosote oil 煤沥青杂酚油
coal-tar creosote solution 木材防腐油
coal-tar dye 煤焦油染料
coal-tar enamel 煤沥青瓷漆;煤焦油瓷漆
coal-tar epoxide paint 环氧煤沥青涂料;环氧煤沥青漆;煤焦油环氧涂料;环氧煤焦油漆
coal-tar epoxy build paint 厚浆型环氧沥青漆
coal-tar epoxy coating 煤焦油环氧涂层;环氧煤焦油涂层
coal-tar epoxy paint 环氧煤沥青涂料;环氧煤沥青漆;煤焦油环氧涂料
coal-tar finish 煤沥青罩面
coal-tar fuel 煤焦油燃料
coal-tar gas 煤焦油煤气
coal-tar glass felt 煤焦油玻纤油毡
coal-tar heavy oil 煤焦重油;重焦油;高温煤焦油
coal-tar hydrocarbon 煤焦油烃
coal-tar hydrogenation 煤焦油氢化
coal-tar-impregnated 煤沥青浸透的
coal-tar industry 煤焦油工业
coal-tar light 轻焦油
coal-tar light oil 煤焦轻油
coal-tar middle oil 中(温煤)焦油
coal-tar oil 煤焦油
coal-tar ointment 煤焦油软膏
coal-tar paint 煤焦油涂料
coal-tar pitch 煤焦油脂;煤焦(油)沥青;煤沥青;硬煤沥青
coal-tar pitch cement 煤沥青胶泥
coal-tar pitch dispersion 硬煤沥青悬浮液
coal-tar pitch emulsion 煤沥青脂乳液
coal-tar pitch solution 煤沥青溶液
coal-tar primer 煤焦油底漆
coal-tar product 煤焦沥青产品

coal-tar resin 煤焦油树脂;苯并呋喃—茚树脂
coal-tar resin enamel 煤焦油树脂瓷漆
coal-tar-saturated 煤沥青饱和的
coal-tar saturated organic felt 煤焦油油纸
coal-tar-saturated roofing felt 煤沥青屋面油毡
coal-tar solvent 煤焦油溶剂
coal-tar thinner 煤焦稀释剂
coal tarurethane coating 煤沥青聚氨酯涂料
coal-tar urethane paint 煤沥青聚氨酯漆;煤焦油聚氨酯涂料
coal tempering 煤加湿
coal tender 供煤船
coal terminal 煤码头
coal testing 煤质试验
coal thermal stability K graduation 煤的热稳定性K分级
coal tip 煤渣场;煤炭倾卸装置;煤堆场;卸煤器
coal tipper 煤车翻车机
co-altitude 同高度;天(体)顶距
coal-to-coke replacement 煤置换比
coal-to-coke replacement ratio 煤焦炭更换比
coal tongs 火钳
coal tower scale 煤塔秤
coal tows 运煤驳船队
coal track 上煤线
coal trademark 煤炭牌号
coal transport(ation) 煤炭运输
coal transporter 推煤机;输煤机
coal trans-shipment 煤的装卸
coal trimmer 煤炭平舱机
coal trimming 加煤
coal type 煤型
coal unit 装煤装置
coal unloader 卸煤机
coal unloading berth 煤炭卸船泊位
coal unloading dock 煤炭卸船码头;卸煤码头
coal unloading facility 煤炭卸船设施
coal unloading terminal 煤炭卸船码头;卸煤码头
coal unloading wharf 煤炭卸船码头;卸煤码头
Coal Utilization Council 煤炭利用理事会
Coal Utilization Research Advisory Committee 煤炭利用研究咨询委员会
coal utilizaton rate 煤炭利用率
coal valuation 煤质评价
coal wall 长(壁)工作面
coal washability 煤(的)可选性
coal washer 洗煤机
coal washery 选煤厂;洗煤厂
coal washing 洗煤
coal washing and dressing 煤炭洗选加工
coal-washing plant 洗煤设备;洗煤厂
coal-washing wastewater 洗煤废水
coal waste 煤渣;煤焦油废水;煤矸石
coal water gas 干馏水煤气
coal water mixture 煤水混合物
coal-water-oil emulsion 煤水油乳胶
coal-water pump 煤水泵
coal-water ratio 煤水比
coal wedge 煤柱
coal weigher 煤磅
coal wharf 煤码头
coal whipper 卸煤机;卸煤工人
coal winch 煤炭起货机
coal-winning machine 采煤机
coal wiper 刮煤器
coal wood 煤化木
coal-worker's pneumoconiosis 煤矿工人尘肺;煤工尘肺
coaly 含煤的;似煤的
coal yard 煤炭堆场;煤场
coal yard management 煤场管理
coal yield 煤产量
coaly lignite 炭质褐煤
coaly polymeric 煤质聚合物
coaly rashings 软页岩
coaly type 煤质型
coaming 天窗人孔围坎;进口挡水围槛;挡水围墙;边衬;孔口栏板;洞口围坎(天窗或屋顶洞口处的围栏突边);挡水缘围;舱口围板【船】;舱口拦板;舱底栏板;凸起天窗
coaming angle 围板角钢
coaming bar 围板角钢
coaming chock 木围框导缆口
coaming cleat 舱口楔耳

coaming stay 舱口围撑板
coaming stiffener 舱口围撑板
Coanda effect 康达效应
Coanda effect amplifier 附壁型放大器
coarctation 狭窄;缩小;拧紧;紧压
coarray 并合台阵
coarse 近似的;毛糙的;粗粒的;粗粉;粗(糙)的;不精确(的)
coarse abrasive 粗磨料
coarse/acquisition code 粗码
coarse adjusting 粗调(整)
coarse adjusting rheostat 粗调变阻器
coarse adjustment 粗调整;粗调节;粗调
coarse adjustment knob 载物台粗调螺旋
coarse adjustment pinion 粗调小齿轮
coarse admixture 粗粒杂质
coarse aggregate 最大颗粒不超过1/4英寸(1英寸=0.0254米)的聚集体;粗粒料;粗集合体;粗聚集体;粗骨料;粗粒料;粗(大聚)集料
coarse(aggregate) analysis 粗集料筛分析;粗骨料筛分析
coarse aggregate concrete 粗集料混凝土;粗骨料混凝土
coarse aggregate factor 粗集料系数;粗骨料系数
coarse(air) bubble aeration 大气泡曝气
coarse and fine aggregate of expanded clay 黏土陶粒和陶砂
coarse and fine aggregate of expanded shale 页岩陶粒和陶砂
coarse and fine aggregate of sintered pulverized fuel ash 粉煤灰陶粒和陶砂
coarse and fine grinding system 粗细粉磨系统
coarse and fine shearing and punching machine 粗精剪冲机
coarse angular error sensor 粗略角误差传感器
coarse asphalt 原生沥青;粗沥青;粗级配沥青混凝土
coarse asphalt concrete 粗集料沥青混凝土;粗骨料沥青混凝土
coarse asphaltic concrete 粗骨料沥青混凝土
coarse balance 零点粗调;粗略调平;粗调
coarse bar rack 粗格栅
coarse(bar) screen 粗格栅
coarse bed 粗填料基层;粗集料层;粗骨料层;厚岩层;厚地层
coarse bed load 粗底沙(质)
coarse bed material 粗底沙(质)
coarse benthic organic matter 水底粗粒有机物
coarse bentonite 粗粒膨润土
coarse black 粗粒炭黑
coarse blast 粗粒喷砂
coarse blend(ed) aggregate 粗粒混合集料;粗混合骨料
coarse breaker 粗破碎机;粗碎机
coarse breaking 粗碎
coarse-breaking zone 粗破碎区
coarse break-up 粗粒分散体;粗分散;破碎成为粗
coarse broken stone 粗碎石;粗粒碎石
coarse bubble 大气泡
coarse bubble aeration system 大气泡曝气系统
coarse bubble diffuser 大气泡扩散器
coarse bubble system 大气泡系统
coarse carpenter file 粗木工锉
coarse cement 粗水泥
coarse cement flour 粗磨水泥粉
coarse chamber 粗测度破碎腔室
coarse channel 粗测通道
coarse chip(ping)s 粗石屑;粗粒切屑
coarse chrominance primary 窄带色度信号
coarse clarification zone 粗分类区
coarse classifier 粗选机
coarse clastic soil 粗碎屑土
coarse clastic texture 粗碎屑结构
coarse clay 粗黏土
coarse cleaner 粗滤清器
coarse clearance fit 松转配合
coarse cloth 粗织物;粗布
coarse coal 粗煤
coarse coir fibre 粗椰壳纤维
coarse comminution 粗破碎
coarse compensation 粗补偿
coarse component 粗粒部分
coarse concentrate 粗粒精矿
coarse concrete 粗集料混凝土;粗骨料混凝土
coarse content 粗颗粒含量

coarse control 粗调节;粗控;粗调控制
coarse control rod 粗操纵杆
coarse crumb 粗屑粒
coarse crushed 粗碎的
coarse crushed sand 粗轧碎砂
coarse crushed stone 粗碎石
coarse crusher 一级破碎机;粗碎机
coarse crushing 粗轧;粗碎;粗破碎;粗粉碎
coarse crushing chamber (轧石机的)粗轧室;(破碎机的)粗碎室
coarse-crushing zone 粗碎区
coarse crystal 粗晶体;粗晶
coarse crystalline 粗晶(质);粗晶(质)的
coarse crystalline dolomite 粗晶白云岩
coarse crystalline limestone 粗晶灰岩
coarse crystalline material 粗晶状物质
coarse crystalline texture 粗晶结构
coarse crystallization 粗结晶
coarse crystallized grain 粗晶粒
coarse crystallized grained 粗晶粒的
coarse crystalloblastic texture 粗粒变晶结构
coarse cut 粗锉纹
coarse cut file 粗纹锉
coarse cutter 粗切机
coarsed ashlar 层砌条石;层砌方石
coarse delay 近似延时;粗延迟
coarse delay dial 粗延时度盘
coarse deposit 粗粒沉积
coarse diamond knurl 粗菱形滚花
coarse diamond-point knurling roll 粗金刚钻压花滚刀
coarse disperse 粗分散
coarse dispersion 粗粒弥散系;粗分散(体)
coarse display 粗显示器
coarsed masonry 粗层砌体
coarse drainage layer 粗排水层
coarse drainage texture 粗疏水系结构
coarse drill 粗斜纹
coarse drive 粗动
coarse dust 粗粉粒;粉剂
coarse earthenware 粗陶器
coarse elutriation 粗淘
coarse emulsion 粗(滴)乳液
coarse end 粗经
coarse face 粗糙面
coarse feed stylus 粗线刻刀
coarse fiber 粗纤维
coarse-fibered 粗纤维的
coarse file 毛锉;中粗锉;粗齿锉
coarse filler 粗粒填料
coarse filter 大网眼过滤器;粗滤器;粗滤池;粗反滤层;初滤器
coarse filter layer 粗过滤层
coarse fine control 粗细调节
coarse fine sand with some clay 粗细砂混黏土
coarse fine sand with some silt 粗细砂混粉土
coarse firn 粗粒雪
coarse fit 粗配合
coarse flat file 粗扁锉
coarse focusing knob 粗调焦旋钮
coarse food grain 粗粮
coarse fraction 粗粒百分率;粗料(粗(粒)组分;粗粒级;粗粒含量;粗粒部分
coarse fractional 粗组分的
coarse fraction analysis 粗粒级分析;粗粒部分分析
coarse fraction content 粗粒组含量
coarse fragment 粗碎片;粗碎屑;(直径大于2毫米的)粗碎块
coarse-fragment ground 大块碎屑类土
coarse-fragment soil 大块碎屑类土
coarse glass paper 木砂皮纸
coarse glue 胶液;粗胶
coarse goods 粗制品
coarse gradation 粗级配
coarse-graded 粗粒级的;粗级配的
coarse-graded aggregate 粗级配集料;粗级配骨料;粗组集料;粗组配骨料;粗粒级集料;粗粒级骨料
coarse-graded aggregate asphaltic concrete 粗级配集料沥青混凝土
coarse-graded asphaltic concrete 粗级配沥青混凝土
coarse-graded bituminous concrete 粗级配沥青混凝土

coarse-graded bituminous concrete pavement 粗粒式沥青混凝土路面
coarse grading 粗粒径筛分;粗粒级分等;粗粒分级;粗级配
coarse grain 粗纹理;粗粒;粗颗粒;粗晶粒;粗晶
coarse-grainded sandstone 粗粒砂岩
coarse-grained 粗纹的;粗粒(状)的;粗粒级的;粗晶粒的;纹理粗疏的
coarse-grained abrasive 粗磨料
coarse-grained aggregate 粗颗粒集料;粗粒粒骨料
coarse-grained clastics 粗粒碎屑
coarse-grained clay 粗粒黏土
coarse-grained colo(u)red intrusive rock 粗粒浅色侵入岩
coarse-grained distribution 粗粒分布
coarse-grained filler 粗颗粒填料
coarse-grained filter 粗料滤池
coarse-grained fracture 粗晶断口;粗纹裂面
coarse-grained granite 粗粒花岗岩
coarse-grained gravel 粗颗粒砾石
coarse-grained ingot 粗晶锭
coarse-grained iron 粗晶生铁
coarse-grained limestone 粗粒灰岩
coarse-grained material 粗粒材料;粗粒物料
coarse-grained meander belt deposit 粗粒曲流带沉积
coarse-grained mortar 粗颗粒砂浆
coarse-grained paving 粗粒石铺
coarse-grained sand 粗粒砂
coarse-grained soil 粗颗粒土;粗(颗)粒土(壤)
coarse-grained steel 粗晶粒钢
coarse-grained stone 粗磨石
coarse-grained texture 粗(颗)粒结构;粗晶结构
coarse-grained timber 粗纹木材
coarse-grained wood 粗纹木
coarse grain film 粗颗粒胶片
coarse grain reservoir 粗粒储集层
coarse grain zone 粗晶区
coarse granitoid texture 粗粒花岗状结构
coarse granular 粗颗粒的;粗团粒
coarse granular cast 粗颗粒管型
coarse granular psamitic texture 粗粒砂状结构
coarse granular texture 粗粒结构
coarse granule 粗(大)颗粒
coarse grass cloth 粗夏布
coarse gravel 粗砂砾;粗卵石;粗(粒)砾石;粗砾
coarse gravel aggregate 粗砾石集料
coarse-gravel filling 回填粗砾砂
coarse gravel for concrete 混凝土用粗砾石
coarse gravel including cobbles 混一些卵石的粗砾石
coarse gravel-surfaced 粗砾石面层的
coarse grinding 粗磨
coarse grinding chamber 粗粉磨室;粗粉磨仓
coarse(grinding) mill 粗磨机
coarse grit 粗砂岩
coarse grit scythe 粗硬大镰刀
coarse ground cement 粗研水泥;粗磨水泥
coarse gyrasphere crusher 球旋型粗碎机
coarse half round file 粗半圆锉
coarse impact crusher 反击式粗碎机
coarse indicator 粗测指示器
coarse in feed(to screen)进筛粗料
coarse jaw breaker 粗碎颚式破碎机
coarse jaw crusher 粗碎颚式破碎机
coarse jig 粗粒跳汰机
coarse jigging 粗粒(跳)汰选
coarse joining method 毛接法
coarse knife file 粗刀锉
coarse laid wire rope 硬绳
coarse lamellar pearlite 厚层状珠光体
coarse layers of sediment 粗砂层
coarse level(l)ing 粗略调平
coarse limestone 粗粒石灰石
coarse line 粗线
coarse linearity control 直线性粗调
coarse loam 粗壤土
coarsely crystalline 粗晶质
coarsely dressed 粗糙修琢的
coarsely graded 粗粒级的
coarsely graded aggregate 高粗(颗)粒的级配集料;高粗(颗)粒的级配骨料;粗级配集料;粗级配骨料
coarsely grained 粗纹理的

coarsely grained coal 粗纹理煤
coarsely grained fracture 粗纹裂面
coarsely grained wood 粗纹木(材);粗纹理木材
coarsely granularcast 粗粒性管型
coarsely granular particle size 粗颗粒粒度;粗粉粒度
coarsely ground cement 粗磨水泥
coarsely ground cork 粗软木屑
coarsely ground foundry sand 红砂粉
coarsely ground talc 粗磨滑石粉
coarsely ringed timber 宽纹木材;宽年轮木材;粗纹木(材)
coarsely shaped 粗糙整形的
coarsely textured aggregate 粗纹理集料;粗纹理骨料
coarsely textured timber 粗纹理木材
coarsely thread screw 粗纹螺钉
coarse macadam surface 粗碎石面层
coarse manufactured sand 粗制砂
coarse material 粗粒物质
coarse material retained on screen 筛上粗物料
coarse material retained on sieve 筛上粗物料
coarse mesh 粗(眼)网目;粗筛孔;粗孔筛
coarse mesh filter 粗网过滤器
coarse mesh screen 粗孔滤网
coarse metal 粗晶硅酸铁
coarse middling 中粗的
coarse mineral granules 粗矿物粒料
coarse mix(ture) 浓水泥浆;粗(粒)混合料;粗混合物
coarse mode 粗粒模
coarse mortar 粗(颗粒)砂浆
coarse motion 粗动
coarse mull 粗腐熟腐殖质
coarseness 粒度;粗度;粗糙(度)
coarseness factor 粗度因数
coarseness of coal 煤的粒度
coarseness of grading 级配(颗)粒(粗)度;粒度组成;颗粒粒度
coarseness of grouping 分组的粗略性
coarseness ratio 粗糙度比
coarsening 粒度增大;粗化
coarsening of particles 颗粒粗化
coarsening of riverbed 河床粗化
coarsening rate 长大速度
coarsening-upward sequence 向上变粗层序
coarse of a screen 筛余粗粒
coarse oil filter 机油粗滤清器
coarse oil filter handle rod spring 机油粗滤清器手柄拉杆弹簧
coarse oil screen 粗油滤器
coarse paper 粗质纸;草纸
coarse para 粗橡胶
coarse particales 粗分子
coarse particle 粗(颗)粒
coarse particle size 粗粒度
coarse particulate organic carbon 粗颗粒有机碳
coarse particulate organic matter 粗颗粒有机物
coarse pearlite 粗珠光体;粗(晶)粒珠光体
coarse pebble 粗中砾
coarse pick 粗纬
coarse pitch 大螺距;大节距;粗节距
coarse-pitch cutter 大螺距齿铣刀
coarse-pitch drill tools 粗径钻具
coarse-pitch gear 粗径节齿轮;大径节齿轮
coarse-pitch involute system 粗径节渐开线制;大径节渐开线制
coarse-pitch screw 大螺距螺钉
coarse pitch thread 粗螺纹
coarse plain emery wheel 粗砂轮
coarse plaster 粗(料)粉刷
coarse plate 厚(钢)板
coarse porcelain 粗瓷
coarse pore 粗孔(指混凝土中粒径为 0.1～1 微米的孔隙)
coarse pored 大孔状;大孔隙的
coarse pore filtration activated sludge process 大孔隙过滤活性污泥法
coarse pore filtration activated sludge system 大孔隙过滤活性污泥系统
coarse porosity 粗孔率;粗孔隙性;粗孔隙度;粗大气孔率
coarse positioning 粗调定位
coarse powder 粗糙粉末;粗粉体;粗粉(末)

coarse premix 粗预混合
coarse product 粗粒成品;粗粒产物
coarse pugging 粗练
coarse rack 粗格拦污栅;粗栅;疏拦污栅
coarse radio location 无线电粗略定位
coarse-range scope 距离粗测器
coarse reading 粗读(数)
coarse reduction 粗粉碎
coarse reduction gyratory 粗碎圆锥破碎机
coarser mesh 粗网目;粗孔
coarse rock 粗岩;粗石
coarse roll 粗碾滚筒;粗碾辊
coarse rope 粗(麻)绳
coarse round file 粗圆锉
coarse salt 粗盐
coarse sampling period 粗采样时段
coarse sand 粗砂砾;粗质砂;粗砂
coarse sand beach 粗砂海滩;粗砂滨
coarse sand bond 粗砂砌体
coarse sand bottom 粗砂底(底质)
coarse sand filling 回填粗砂
coarse sand filter 粗砂滤池
coarse sand mortar 粗砂砂浆
coarse sandstone 粗粒砂岩;粗砂砂石
coarse sandy clast 粗砂屑
coarse sandy loam 粗亚砂土;粗砂壤土
coarse sand(y) ware 粗砂陶
coarse scale modulator 粗刻度调制器
coarse scan(ning) 疏扫描;粗扫描
coarse screen(ing) 粗筛选;粗筛(屑);拦污栅;大孔(格)筛;粗滤器;粗筛;粗矿筛;粗孔筛
coarse screen well 粗格栅井
coarse screw tap 粗丝锥
coarse scrubbed concrete finish 粗糙擦洗处理的混凝土表面
coarse scrubbed concrete surface 粗糙擦洗处理的混凝土表面
coarse sediment 粗粒泥沙;粗粒沉积物
coarse seed 粗晶种
coarse selsyn 粗示自动同步机;粗调自动同步机
coarse sensor 粗略传感器
coarse set(ting) 粗凝固;粗调整;粗调;初步瞄准
coarse sewage screen 下水道粗格栅
coarse sieve 粗筛
coarse sighting telescope 概略照准望远镜
coarse silk 粗罗;粗绢筛
coarse silt 粗泥砂;粗粉土;粗粉砂
coarse siltstone 粗粉砂岩
coarse silt texture 粗粉砂结构
coarse silty clast 粗粉砂屑
coarse size grading 粗颗粒级配
coarse size reduction 粗碎
coarse slag chip(ping)s 粗矿渣屑
coarse slurry 粗料浆
coarse soil 粗质土(壤)
coarse solder 粗焊接;铅焊料
coarse solid 粗固体
coarse square file 粗方锉
coarse steel 粗晶粒钢
coarse stipple finish 粗拉毛饰面
coarse stoneware 粗炻器
coarse stopper 粗填料
coarse stopping 粗填料
coarse strainer 粗滤网;粗滤器
coarse structure 粗视结构;粗糙结构
coarse stuff 刮草打底;打底料;粗填料;粗涂抹料;粗(涂)料;草筋灰;抹灰打底材料(一般使用灰泥)
coarse suspended dediment 粗粒悬沙
coarse suspension 粗粒悬浮体
coarse sweeting aggregate 粗嵌隙料
coarse system of bearing 方位粗测系统
coarse tail grading 粗级向上变细现象
coarse tar concrete 粗柏油混凝土
coarse texture 大颗粒结构;粗质地;粗纹理;粗结构
coarse-textured 粗粒结构的;粗结构的;纹理松的;粗(木)纹的
coarse textured drainage 粗稀结构水系
coarse textured soil 粗质土(壤);粗粒结构土(壤)
coarse textured timber 纹理粗的木材
coarse textured topography 粗切地形
coarse thread 粗牙螺纹;粗(螺)纹
coarse threading attachment 大螺距螺纹切削装置

coarse thread joint 粗牙螺纹接头
coarse thread screw 粗螺纹螺钉
coarse thread series 粗螺纹系(列)
coarse thread tap 粗螺纹丝锥
coarse time delay 粗延时
coarse-to-fine-aggregate ratio 粗细集料比;粗细骨料比;石砂比
coarse-to-fine filter 多层滤料滤池
coarse to-fine filtration 粗到细滤料过滤
coarse toothed cutter 粗齿切割刀;粗齿铣刀
coarse toothed saw 粗齿锯
coarse topography 粗切地形
coarse transition 粗粒反滤层;粗粒的过渡层
coarse trash rack 粗格拦污栅
coarse tremor 慢震颤;粗大震颤
coarse triangular file 粗三角锉
coarse tuff 粗粒的凝灰岩
coarse tuning 粗调谐
coarse vacuum 低真空;粗真空
coarse vibrated concrete 低振动混凝土
coarse waste 粗岩屑
coarse waste stone 粗废石
coarse wheel 粗砂轮
coarse wire 粗拔钢丝
coarse wire rope 粗铜丝绳;粗(丝)钢丝绳
coarse yarn 粗支纱
coast 海岸;惯性滑行;惯性跟踪;岸边
coast, insurance and freight 到岸价(格)
coast aggradation 海岸堆积
coastal 海岸的;沿(海)岸的
coastal accretion 海岸加积
coastal acoustical wave ga(u)ge 岸用声学测波仪
coastal aggradation 海岸加积
coastal aids 沿海航标
coastal aquifer 滨海含水层
coastal area 海岸区;沿海区;沿海地区;沿海地带;沿海岸水域;沿海带
coastal area development 海岸区开发
coastal area management 海岸区管理
coastal bar 河口潜洲;海口砂洲;沿海沙洲;沿海沙滩;沿海潜洲;沿海砂礁;沙洲;海滨河
coastal barrier 海岸沙堤
coastal barrier project 海岸堤堰工程
coastal basin 海岸盆地
coastal bays 沿海海湾
coastal beach 海滨;沿海沙洲;沿海沙滩
coastal belt 海岸带;海岸地带;沿海地带
coastal berm 后滨阶地
coastal berth 沿海泊位
coastal bevel 海岸削平;海岸对切
coastal boundary 领海边界
coastal breaker 沿海防波堤
coastal breakwater 沿海防波堤
coastal canal 沿海运河
coastal cargo-liner 近海货轮
coastal cargo vessel 沿海货船
coastal cave 海蚀(岩)穴;海蚀洞
coastal chain 海岸山链
coastal change 海岸演变
coastal chart 海岸图;沿海海图;沿海地图
coastal circulation 近岸环流;沿岸环流
coastal city 沿海城市
coastal city opening 沿海开放城市
coastal cliff 海岸峭壁;海崖
coastal cliff trap 海岸崖圈闭
coastal climate 海岸气候;沿海气候
coastal climate resistance 耐近海气候性
coastal coean pollution assessment 海岸污染评估
Coastal Commission 海岸委员会
coastal conservation 海岸保护
coastal contamination 海岸带污染物
coastal contour line 海岸地形线
coastal convergence zone 海岸集流区
coastal crude oil 海湾原油
coastal current 近岸流;海岸流;沿(洋)流;沿岸海流
coastal curvature 海岸弯度
coastal debris 海滨碎石
coastal defence work 海岸保护工事
coastal deposit 海岸沉积(物);岸积物
coastal depot 海岸油库
coastal desert 海岸沙漠
coastal desert climate 海岸沙漠气候
coastal destroyer 边海驱逐舰

coastal development 海岸发育;海岸带开发;沿海开发;滨海地区的发展
coastal development and protection 海滨地区开发与保护
coastal dike 防潮堤;海堤
coastal discharge 海岸排放
coastal drainage 海岸排放
coastal dredging 沿海疏浚;滨海挖泥作业
coastal drifting 海岸迁移
coastal drift(sand) 海岸漂沙
coastal dune 海岸沙丘;滨海沙丘;沿海沙丘;沿岸沙丘
coastal dune field 海岸沙丘原
coastal ecology 海(洋沿)岸生态学;沿海生态学
coastal economic open zone 沿海经济开放区
coastal ecosystem 海滨生态系统;海岸生态系统;沿海生态系统
coastal effect 海岸效应
coastal embayment 滨岸海湾
coastal encroachment 海岸进侵
coastal engineering 海塘工程;海岸工程(学)
Coastal Engineering Conference 海岸工程会议
coastal engineering hydraulic model 海岸工程水力模型
coastal engineering research center 海岸工程研究中心
coastal engineering ship 沿海工程船
coastal environment 海岸环境;海岸环境
coastal environment noise 海岸环境噪声
coastal erosion 海岸侵蚀;海岸冲刷;海岸冲蚀;沿岸侵蚀
coastal erosion management 海岸侵蚀管理
coastal error 海岸折射误差
coastal evolution 海岸演变
coastal facility 海岸设施
coastal feature 海岸特征;海岸地貌
coastal feeder service ship 近海支线货船
coastal fishery 沿海渔业
coastal fishery harbour 沿海渔港
coastal fishery state 沿海渔业国
coastal fishing ground 沿海渔场
coastal fleet 沿海船队
coastal flood 潮汐河口潮洪
coastal flooding 沿海洪水
coastal flow 近岸流
coastal fog 滨海雾;岸雾
coastal friction 海岸摩擦
coastal geomorphology 海岸地形学;海岸地貌学
coastal geomorphy 海岸地貌
coastal geothermal system 滨海地热系统
coastal groin 海岸防波堤;海岸丁坝
coastal groundwater 海岸地下水
coastal groyne 海岸防波堤;海岸丁坝
coastal guard activity 海岸防护活动
coastal harbo(u)r 海岸港;海港;沿海港口;沿海港
coastal hazard 海岸灾害
coastal highway 滨海公路
coastal ice 固定冰;岸冰
coastal ice blink 海岸冰崖
coastal industrial area 滨海工业区
coastal industrial region 沿海工业区
coastal industrial zone 沿海工业区
coastal industry 海岸工业
coastal inlet 海口;沿海河口湾
coastal jurisdiction 沿海管辖权
coastal lagoon 河口礁湖;海岸泻湖;海岸礁湖
coastal lake 沿岸湖;滨海湖
coastal land 滨海土地;沿海地区
coastal landfill 滨海填英
coastal landform 海蚀地形;海滨地貌;海岸地形;海岸地貌
coastal landslide 海岸滑坡;海岸滑动
coastal lead 海岸冰沟
coastal levee 防潮堤;海塘;海(岸)堤(防)
coastal light 海岸灯(光);灯标
coastal lighting 海岸照明;海岸灯标
coastal line 海岸线
coastal liner 沿海航线班轮
coastal management 海岸管理
coastal management plan 海岸管理计划
coastal management problem 海岸管理问题
coastal management unit 海岸管理单位
coastal map 海岸图
coastal mapping 海岸测绘

coastal marine environment 沿海海洋环境
coastal marsh 海滨沼泽;海岸沼泽;海岸沼地
coastal material 海岸泥沙
coastal monitoring 海岸监测;沿岸监测
coastal morphology 海岸形态学;海岸地貌学
coastal motor boat 海岸汽艇;海岸汽船
coastal mountain 沿海山丘
coastal mudflat 海岸泥坪
coastal navigation 沿海航行
coastal nip 海岸低悬崖
coastal observation 海岸观测;海滨观测
coastal ocean 近岸海洋;海洋边缘
coastal ocean monitor satellite 沿海海域监视卫星
coastal ocean monitor satellite system 沿岸海域监视卫星系统
coastal oceanography 近岸海洋学;沿岸海洋学
coastal onlap 海岸上超
coastal open city 沿海开放城市
coastal outline 海岸线
coastal park 海滨公园
coastal patrol interdiction craft 沿海巡逻截阻艇
coastal pilot 沿海航线引航员
coastal plain 海岸平原;海岸平原;沿海滩地;沿海平原;滨外平原;滨海平原;滩地
coastal plain estuary 海岸平原河口
coastal plain soils 海岸平原土;沿海平原土壤
coastal plain swamp 海岸平原沼泽;沿海沼泽
coastal planning 海岸规划
coastal platform 海岸台地
coastal pollution 沿岸污染;近岸污染;海滨污染;海岸污染
coastal pollution research 沿海污染研究
coastal pond 海岸池沼
coastal populant 海岸带污染物
coastal port 海岸港;沿海港口
coastal price 海边价格
coastal process 海岸演变过程
coastal progradation 海岸进积作用
coastal progress 海岸进展
coastal protected area 沿海保护区
coastal protection 海岸维护;海岸防护;海岸保护
coastal protection embankment 海岸防护堤
coastal protection structure 海岸防护建筑物
coastal protection works 海岸维护工程;海岸防护设施;海岸防护工程
coastal provinces 沿海省份;沿海各省
coastal radar 海岸雷达
coastal radar chain 海岸雷达链
coastal radar television 海岸雷达电视
coastal radio station 海岸(无线)电台
coastal range 沿海山脉
coastal reclamation 海涂围垦
coastal reflection 海岸反射
coastal refraction 海岸折射
coastal refraction error 海岸折射误差
coastal region 沿海区域;海岸区;沿海地区;沿岸区;滨海区
coastal region markets 沿海市场
coastal regulation 海岸整治
coastal resort 海边休养地
coastal resources 沿海(水产)资源;近岸资源;海涂资源;海岸资源;沿海资源
coastal resources management 海岸资源管理
coastal river 沿海河流;滨海河流
coastal road 滨海道路
coastal salt marshes and mangrove swamps 海滨沼泽
coastal salt water intrusion 沿海盐水侵入
coastal science 海滨科学
coastal sea 近海海洋;海洋边缘
coastal sea-bed economic area 沿海海床经济区域
coastal seas 沿海海域
coastal sediment 海岸沉积(物);岸积物;海岸泥沙
coastal service 沿海航线
coastal shelf 沿海大陆架
coastal shipping 沿海航运
coastal spit 海岸沙嘴
coastal stabilization 海岸稳定
coastal state 沿海州;沿海国(家);沿岸国(家)
coastal state priority zone 沿海国家优先区
coastal state resource jurisdiction 沿海国家资源管辖权
Coastal States Gas 沿海各州天然气公司(美国)
coastal station 海岸站;海岸台站;防潮堤

coastal station service 海岸电台业务
coastal stream 沿岸流
coastal street 岸边街道
coastal strip 海岸带;沿海地带
coastal structure 海岸构筑物;海岸结构;海岸建筑物;海岸防护建筑物
coastal submarine domain 沿海海底领土
coastal subsidence 海岸坍陷
coastal succession 海滨演替
coastal surveillance 沿岸监视
coastal surveillance system 沿岸监视系统
coastal survey 海岸测量
coastal surveying ship 沿海测量船
coastal swallow hole 海岸落水洞
coastal swamp 海滨沼泽;海岸沼泽
coastal tanker 沿海油轮
coastal terrace 海岸阶地;海岸台地;沿海台地
coastal terrain 海岸地貌
coastal tide 近岸潮汐
coastal toplap 海岸顶起
coastal topography 海岸地形;沿海地形;沿岸地形
coastal towage 沿海拖运
coastal town 沿海城市;滨海城市
coastal trade 沿岸交易
coastal traffic 沿海运输
coastal transportation 沿海运输
coastal type 海岸型
coastal upwelling 沿岸上升流
coastal utilization 海岸利用
coastal vegetation 海岸植被
coastal vessel 沿海(航行)船
coastal waiter 海关稽查员
coastal warning 沿海警告
coastal water 近岸水(体)
coastal water area 近海水域
coastal water environment 海岸水环境
coastal water pollution 沿海水污染
coastal waters 沿海水域;近海水域;近岸水域;海岸水域;沿岸水域;岸边水域;水体边缘
coastal waters pollution 海岸水域污染;岸边水域污染
coastal waterway 沿海航道
coastal waterway route 沿海航路
coastal wave 近岸浪;近岸波;击岸波
coastal wetland 沿海湿地
coastal wind 近岸风
coastal work 海岸工程
coast(al) zone 海岸带;海区;沿海地区
coastal zone colo(u)r scanner 海岸带彩色扫描器;沿岸水色扫描仪
coastal zone contamination 海岸带污染
coastal zone management 海岸带管理
Coastal Zone Management Act 沿海区管理法;海岸带管理法
coastal zone management law 海岸带管理法
coastal zone oil spill model 海岸带溢油模型
coastal zone pollution 岸带污染
coastal zone population 海岸带污染
coastal zone research vessel 沿海海域科学考察船
coast and geodetic magnetic observatory 沿岸地磁观测站
coast and geodetic survey 沿岸大地测量
Coast and Geodetic Survey Board 海岸大地测量局(美国)
coast and river-bank protection works 河海岸防护工程
coast beacon 海岸信标;沿海航标
coast change 海岸变迁
coast change in trend 海岸变迁趋向
coast charges 海岸费
coast chart 沿岸海图
coast confluence zone 沿海汇流区
coast coral reef 珊瑚礁岸
coast current 岸边流
coast defence radar 海防雷达
coast defence ship 海岸防卫舰;岸防舰
coast defence system 海岸防御系统
coast defense work 沿海防卫工作
coast delta 滨海三角洲
coast deposit(ion) 海岸沉积;滨海沉积(物)
coast depot 海运终点
coast dike 防波堤
coast discharge 岸边排放
coast down 减(退)

coast drainage 岸边排放
coast earth station 海岸地球站;海岸地面站
coast elevation 海岸上升
coaster 近海船;沿海巡船;沿海航船;沿海船舶;沿海(航行)船;单向联轴节
coaster brake 脚刹车;倒轮制动;倒轮式刹车
coaster gate 定轮闸门
coast erosion 海岸侵蚀;海岸冲蚀
coast evolution 海岸演化
coast features 海岸地形
coast fog 沿海雾
coast for orders 指定海岸待命
coast geomorphology of marine accumulation 海积海岸地貌
coast geomorphology of marine erosion 海蚀海岸地貌
coast guard 海岸防护;海防守卫队;海防哨兵;海岸警卫队
coast guard boat 海上警卫艇
coast guard ship 海上警卫艇;海岸护卫船;海防卫舰
coast guard station 边防站
coast imperfectly known 未精测海岸线;未精测滨线
coasting 滑行;海岸线;沿海航行;沿岸航行;惯性运动;惰行;惰力运转
coasting ability 滑行能力
coasting area 沿海地区
coasting boat 沿海(航行)船
coasting body 惯性体
coasting chart 沿海海图
coasting craft 沿海船舶
coasting drifting 惰行
coasting flight 惯性飞行
coasting grade 滑行坡度;惰性坡度;惰行坡度
coasting gradient 滑行坡道
coasting harbo(u)r 沿海港口
coasting lead 沿海用测深锤(30~50磅重)(1磅=1.44822牛)
coasting lights 沿海灯标
coasting line 沿海航线
coasting manifest 沿海货运清单
coasting mode 滑行工况
coasting navigation 沿海航行
coasting of temperature 温度惰性上升
coasting operation 惰走
coasting pilot 沿海航路指南
coasting port 沿海港口
coasting resistance 惰行阻力
coasting sailing 沿海岸航行
coasting service 沿海航运;沿海航线
coasting shipping line 沿海航线
coasting speed 滑行速度;惰行速度
coasting tanker 沿海油轮
coasting test 溜坡试验;滑行试验
coasting time for one way 旅行时间
coasting trade 沿海贸易
coasting trade area 沿海贸易区
coasting trade vessel 沿海贸易船舶
coasting trajectory 惯性运动轨迹
coasting vessel 沿海船舶
coasting vessel quay 沿海船舶码头
coast investigation 海岸调查
coastland 沿海陆地;沿海地区;沿海地带
coast leveled by marine erosion 海蚀夷平海岸
coast leveled of marine accumulation 海积夷平海岸
coast-lighting 海岸灯标
coastline 海岸线;海岸线;海岸地形;沿海航线
coastline denudation 海岸剥蚀
coastline effect 沿岸效应;海岸效应
coast line error 海岸误差
coastline feature 海岸线特征
coastline mapping 海岸线测图
coastline measurement 海岸线测量
coastline of Atlantic type 大西洋型(海)岸线
coastline of Dalmatian type 达玛西亚型岸线;达尔马提亚型海岸线
coastline of Pacific type 太平洋型海岸线
coastline profile 岸线纵断面
coast liner 沿海(海)班轮
coastline survey 海岸线测量
coast lining 海岸线测绘;海岸线勘测
coast mud 海岸泥

coast noise 海岸噪声
coast of elevation 上升海岸
coast of emergence 隆起海岸;上升(海)岸
coast of mobile region 变动(地域)海岸
coast of retrogression 后退海岸
coast of stable region 稳定海岸
coast of submergence 溺岸;下沉(海)岸;沉陷海岸
coast of subsidence 下沉岸
coast(or channel) for orders 沿海(或航道)待命
coast patrol 海军巡逻队
coast pilot 航路图志;沿海(航线)引航员;沿海航路指南
coast piloting 近岸引航
coast plain 海岸平原
coast pollution 岸边污染
coast protection 海岸防护
coast-protection dam 海岸防护坝
coast-protection structure 海岸防护构筑物
coast-protection works 海岸防护工程
coast protective forest 海岸防护林
coast radar station 海岸线雷达站;海岸雷达站
coast radio-telephone station 海滨无线电话站
coast railway 海岸铁路;沿海岸铁路
coast range marine trough 海岸山系海槽
coast recession 海岸后退
coast reef 海岸礁;岸礁
coast rift 海岸礁
coast sand 海岸砂
coast scour 海岸冲刷;海岸冲蚀
coast shelf 近岸大陆架;海岸水底平原;海岸大陆架;沉没海岸平原
coast shelter forest 海岸防护林
coast shower 海岸阵雨
coast side (齿轮的)不工作齿侧
coast side of tooth 齿的非工作面
coast signal 沿海航标
coast state 沿海国(家)
coast station 海洋观测站;海岸电台
coast surveillance center 海岸监视中心
coast survey 海岸调查
coast terrace 海岸阶地
coast trade 沿海贸易;沿海贸易
coast trade duties 沿海贸易税
coast trend 海岸走向
coast trunk road 沿海干道
coast vicissitude 海岸变迁
coastward 向海岸的;朝着海岸
coast watcher 海岸瞭望哨
coast watching unit 近岸观测设备
coast waters 沿海海域
coast wave current 海岸浪流
coastwise 沿海运输的
coastwise carriage 沿海运输
coastwise channel 沿海航道
coastwise craft 沿海船只
coastwise line container ship 沿海航线集装箱船
coastwise master 沿海船长
coastwise navigation 沿海航行
coastwise route 沿海航线
coastwise shipping 沿海航行
coastwise survey 沿岸测量
coastwise trade 沿海交易;沿海贸易
coast zone and mudflat survey 海岸带与海涂调查
coast zone geomorphologic(al) map 海岸带地貌图
coat 涂敷;涂底漆;涂布层;包覆;涂层;涂布;涂层
coatability 可涂性;结皮性能;挂窑皮能力;涂布性能
coat and hat hook 衣帽钩
coat closet 衣帽壁橱;存衣壁柜
coat colo(u)r 毛色
coat defect 涂层缺陷
coated 涂铺的
coated abrasive 外涂磨料;砂纸;被覆磨料;涂敷磨料;砂布
coated abrasive grinder 砂带磨床
coated abrasive machining 砂纸加工;砂带加工;砂布加工
coated alumin(i)um zinc steel stranded wire 铝锌合金镀层钢绞线
coated area 涂敷面积;涂布面积
coated asbestos cement pipe 涂层石棉水泥管
coated base felt 沥青底层油毡;涂沥青的底层毡
coated base sheet 沥青浸渍防潮油毡
coated board 涂饰板

coated both sides 双面要涂层的;双面涂布
coated calcium carbonate 带涂层的碳酸钙;包膜碳酸钙
coated carbide 涂层硬质合金
coated card 被覆卡
coated cast-iron pipe 涂层铸铁管
coated cathode 敷料阴极
coated charge 铠装火药柱
coated chip 涂层刀片
coated chippings 预拌石屑;拌沥青石屑;拌沥青砂粒
coated chipping spreader 拌沥青石屑撒布机
coated chips 拌沥青石屑;拌沥青砂粒
coated cloth 漆布
coated cold rolled flat products 冷轧涂层钢板
coated composite powder 涂层复合粉
coated duplex board with grey back 灰底白板纸
coated duplex board with white back 白底白板纸
coated electrode 敷料电极;包剂焊条;涂药电焊条;涂料焊条;涂剂焊条;涂覆电极
coated fabric 涂层织物;人造革;漆布(一种代用革);涂布织物
coated face 涂层面
coated fiber 涂覆光纤;被覆光纤;涂覆光纤
coated filament 直热式覆氧化物阴极;敷料灯丝
coated film 涂膜
coated film disintegration 涂膜蜕变破裂
coated filter 镀膜滤色镜
coated foil 涂层片
coated glass 镀膜玻璃;玻璃幕墙
coated glass fabric 玻璃纤维涂层织物
coated grade whiting 包膜级白垩粉
coated grain 包粒;包壳颗粒;外包颗粒
coated grain dolomite 包粒白云岩
coated grain limestone 包粒(石)灰岩
coated grain phosphorite 包粒磷块岩
coated gravel 拌沥青砾石
coated gravel plant 沥青砾石拌和设备
coated grit 拌沥青石屑;拌沥青砂粒
coated hot rolled flat product 热轧涂层钢板;热轧涂层扁钢制品
coated kraft paper 涂覆牛皮纸
coated lens 加膜物镜;镀膜透镜;镀膜镜头
coated limestone 包粒(石)灰岩
coated macadam 沥青覆盖层碎石路;预拌碎石;有沥青盖层的碎石路;表面处理的碎石路;拌有沥青的碎石路;柏油碎石路
coated macadam plant 黑色碎石拌制设备
coated material 带涂层材料;拌沥青材料;涂布物;涂布材料
coated mesh fabric 涂层网状织物
coated metal 镀层金属;被覆金属;包镀金属
coated metallic electrode 涂层金属电极
coated metal powder 被覆金属粉末;包膜金属粉;包镀金属粉末
coated nail 涂面钉
coated optic 涂层镜片
coated optics 加膜光学;涂敷光学
coated paper 涂料纸;涂布纸;铜版纸
coated particle 涂层颗粒;包覆粉粒;涂敷粉粒
coated phosphor particles 包膜磷光粉
coated photograph 裱板像片
coated pigment 包膜颜料;包核颜料
coated pile 涂面桩
coated pipe 涂过的管道;涂层钢管
coated plastics 镀层塑料;涂膜塑料
coated powder 涂层粉末
coated prior to arrival on site 到达工地前预先涂绝缘层
coated product 带涂层制品
coated rod 涂料焊条
coated(roofing)felt 涂沥青油毡
coated sheet 涂层薄板;覆层薄板;包层薄板;镀层薄板
coated side 纸张上光面;涂层面;涂布面
coated spun iron pipe 旋制涂层铁管
coated steel 涂面钢材;镀层钢板;涂层钢材;涂布钢材;贴面钢材
coated steel sheet 涂层钢板
coated surface 涂膜面
coated tape 涂磁粉带
coated-tips 涂层刀片
coated tool 涂层工具
coated tube 涂过的管子
coated type aggregate 轻质面层集料;轻质面层骨料
coated weight 涂覆量
coated wire 涂层钢丝;被覆线
coated with baked enamel 涂搪瓷的
coated with stucco 用灰泥粉刷
coated yarn 涂层纱
coater 涂膜机;涂镀设备;涂层器;敷涂器;涂刷器;涂漆机;涂料器;涂镀装置;涂布机;涂布工
Coatest graph 科茨图
coat film 薄膜涂层
coat film repair agent 涂膜修补剂
coat for radon 防氡覆盖层
coat glass 覆层玻璃
coat hanger 衣架
coat hanger die 衣架模
coat hook 衣(帽)钩
coating 蒙皮;一度漆;罩面;镀层;表面处治层;被覆(层);包膜;包覆(层);喷涂饰面;涂装;(焊条的)涂药;涂漆;涂料;涂覆;涂层;涂(布)
coating ability 结composability能力;遮盖力
coating action 涂布操作方式;盖覆作用
coating additive 涂料助剂
coating adhering 挂窑皮
coating adhesive 涂料添加剂
coating agent 涂层剂;颜料表面包膜剂;涂饰剂
coating aids 涂布助剂
coating and lining 涂层和衬层
coating and warping 绝缘和包缠
coating and warping crew[gang/party/team] 绝缘包缠班
coating application 涂料施工
coating applicator 涂料器;涂布机
coating asphalt 涂屋面的沥青;沥青护层
coating attenuator 涂敷衰减器
coating bar 涂层钢筋
coating base 漆基;涂膜片基
coating bath 镀槽
coating blister 涂层气泡
coating body stock 涂布原纸
coating bond 涂层结合强度;涂层附着力
coating brilliance 涂层明度
coating by inorganic cementitious bonding 无机胶凝黏结层
coating by vaporization 蒸涂
coating calender 涂布压光机;涂布辊压机;贴胶压延机
coating clay 黏土涂料
coating colo(u)r spot 涂层色斑
coating composition 涂料组分;涂料组成;涂料成分
coating composition for optic(al)fibre 光导纤维涂料
coating composition for prevention of icing 防结冰涂料
coating composition to prevent dirt 防积尘涂料
coating compound 涂料
coating content 涂料量;涂布量
coating damage 涂层损坏
coating density ratio 涂层相对密度
coating device 涂布装置
coating discontinuities 漏涂处
coating discontinuity 涂层不连续性
coating effect 外包效应
coating efficiency 涂渍效率
coating effluent 涂料排出物
coating equipment 涂布设备
coating fabric 涂层织物;涂层基布
coating facet 涂装范畴
coating factory 涂料厂
coating failure 垮窑皮
coating film 涂膜;涂布薄膜
coating film lap 涂膜搭接
coating fluid 涂层液体
coating for foots 胶鞋罩光清漆
coating for internal surface of can 罐内面涂料
coating for light metal 轻金属涂料
coating formation 挂窑皮
coating-forming capacity 形成窑皮能力
coating formula(e) 涂料配方
coating formulation 涂料配方
coating for preventing dew condensation 防结露涂料
coating for wound 铂涂层
coating fusing oven 涂装塑化炉
coating getter 消气剂涂层
coating glasses 镀膜眼镜
coating grade clay 涂布级瓷土
coating heat 涂布热量
coating hood 涂层雾罩
coating immersion 涂层渍浸
coating-in 涂油漆层
coating industry 涂料工业
coating industry wastewater 涂料工业废水
coating industry wastewater treatment 涂料工业废水处理
coating inspector 涂层探伤仪
coating knife 涂料刀;涂布刀
coating layer 涂渍层
coating layer thickness 涂层厚度
coating line 镀膜线
coating lump 涂料块
coating machine 涂铺机;镀膜机;涂漆机;涂覆机械;涂敷机;涂布机
coating mass container 涂油槽
coating material 包膜物;涂(盖)料;涂层材料;饰面材料;膜料;盖层
coating material for optic(al)fibre 光导纤维涂料
coating medium 涂料;涂布料
coating metal 电镀用金属
coating method 涂装方法;涂层法;涂布方法
coating of cement 水泥涂层;水泥结壳
coating of grinding media 糊球
coating of many colo(u)rs 多彩涂料
coating of mud 泥的包壳作用
coating of reinforcement 钢筋涂层
coating of rolls 辊轴涂面
coating on fruit fare 水果表面涂料
coating operation 涂渍操作
coating optic(al) fibre 涂层光纤
coating pan of spreader 涂布机涂料盘
coating penetration 涂渗
coating performance 涂覆性能;涂层性能
coating pipe 喷镀管道
coating pistol 涂料喷枪;涂胶用喷枪
coating plant 涂铺设备
coating plastics 涂面塑料
coating powder 涂面粉末;包覆粉
coating practice 涂覆业务;涂覆技术
coating process 黏层成球法;涂刮法;涂层(方)法
coating property 涂覆性能
coating protection 涂层保护
coating pulling up 涂料拉起
coating quality 涂面质量
coating region 结窑皮区
coating relative density 涂层相对密度
coating removal 脱漆
coating resin 涂覆树脂;涂面树脂;涂料用树脂;涂布用树脂
coating roll 涂刮辊;涂布辊;多辊涂布机
coating room 涂装室;涂漆间
coating scale of copper pipe 铜管涂膜
coatings for greenhouse 温室用涂料
coating solution 感光液;涂渍溶液;涂料溶液;涂层液;涂布液
coating space 收缩缝
coatings pick-up 涂料涂覆量
coating steel pipe 镀锌钢管
coating stock 涂布原纸
coating streak 涂层条痕
coating substrate 涂布基层;被涂底材
coating surface 喷涂面;涂漆面;涂布面
coating system 涂料方法
coating tank 涂油池
coating technology 涂层技术
coating temperature 涂布温度
coating test 挂窑皮试验
coating thickness 涂层厚度;涂布厚度
coating-thickness ga(u)ge 涂层测厚仪
coating thickness meter 镀层厚度测量仪
coating thickness testing instrument 涂层测厚仪
coating treatment 包膜处理;包覆处理
coating type 涂料类型;涂层形式
coating varnish 罩光清漆
coating vehicle 漆料;涂料展色剂
coating weight 膜重测定;涂漆量;涂覆量;涂布量
coating weight test 镀层重量试验
coating with a knife 刮刀涂布
coating with grease 涂润滑脂

coating with plastics 涂(以)塑(料)
coating with soft 新感觉涂层
coating with soft hand 手感柔软涂层
coating with zapon lacquer 涂硝化纤维板清漆
coating yarn 涂层纱
coat nail 挂衣钉；家具用圆钉
coat of asphalt 石油沥青涂层
coat of colo(u)r 颜料层；着色层
coat of glue(bound) water paint 胶性涂料涂层
coat of metal 金属镀层
coat of paint 漆皮；涂漆层；涂料；色漆涂层
coat of synthetic(al) resin 合成树脂涂层
coat of tar 煤焦油涂层；煤沥青涂层；涂漆层
coat of wax 蜡涂层
coat rack 挂衣架
coat room 衣帽间
coat skin 涂层薄膜
coat space 衣物间
coat surface 涂层表面
coat susceptor 钝化的衬托器
coat system 涂层系统
coat the lining with fused material 熔融物料刮窑皮
coat thickening 涂层加厚
coat thickness 涂层厚度
co-attitude 天极距
coat-top antenna 平顶天线
coat with paint 涂色；涂覆色漆
coat with rubber 涂橡胶
coat with varnish 涂(覆)清漆
coax 同轴电缆
coaxality 共同轴性；同轴性
coaxial 共轴的；同轴的
coaxial antenna 同轴天线
coaxial attenuator 同轴衰减器
coaxial beam spectrometer 共轴光束光谱仪
coaxial bolometer 同轴辐射热计；同轴测辐射热计
coaxial bolometer unit 同轴测辐射热组合
coaxial cable 共轴电缆；同轴电缆
coaxial cable carrier telephone system 同轴电缆载波电话系统
coaxial cable connector 同轴电缆连接器
coaxial cable equalizer 同轴电缆均衡器
coaxial cable information system 同轴电缆信息系统
coaxial cavity 同轴空腔(谐振器)
coaxial cavity magnetron 同轴谐振腔磁控管
coaxial circles 共轴圆；同轴圆
coaxial circuit 同轴电路
coaxial coil 同轴线圈
coaxial concentrator 同轴集光器
coaxial cone 共轴圆锥
coaxial configuration 同轴结构
coaxial connection 同轴线接头
coaxial connector 同轴线接插件；同轴(电缆)连接器
coaxial contrary rotating propeller 同轴对转螺旋桨
coaxial cord 同轴软线
coaxial correlation 合轴相关；共轴相关；同轴相关性
coaxial cylinders 共轴圆柱
coaxial diagram 共轴曲线图
coaxial dry load 同轴功率吸收器
coaxial extruded pellet 同轴挤出料粒
coaxial extrusion 同轴挤出
coaxial fabric 共轴的组构
coaxial feeder 同轴电缆；同轴馈(电)线
coaxial fitting 同轴电缆接头
coaxial fixed attenuator 同轴固定衰减器
coaxial fixed direction-coupler 同轴固定定向耦合器
coaxial flashlamp 共轴闪光灯
coaxial glass fibre 同轴玻璃纤维
coaxial hybrid 同轴混合器
coaxial inner conductor 同轴电缆心线
coaxial isolator 同轴隔离器
coaxiality 共轴性；同轴度
coaxial jack 同轴插孔
coaxial jets 同轴射流
coaxial laser pumping 同轴激光抽运
coaxial line 共轴线；同轴线(路)；同心导线电缆
coaxial-line isolator 同轴线去耦装置；同轴电缆去耦装置
coaxial line oscillator 同轴线振荡器
coaxial line termination 同轴线终端负载
coaxial line tube 同轴线式管
coaxial line tuner 同轴线调谐器
coaxial liquid coating 同轴液相镀技术；同轴液态涂覆(法)
coaxial magnetron 同轴磁控管
coaxial of calorimeter 同轴射频量热计
coaxial output circuit 同轴线输出器
coaxial package 同轴型组件
coaxial pad 同轴衰减器
coaxial pad insert 同轴衰减插头
coaxial peak pulse powermeter 同轴峰值脉冲功率计
coaxial phase shifter 同轴移相器
coaxial plasma engine 同轴等离子体发动机
coaxial plug 同轴插头
coaxial power cable feeding system 同轴电力电缆供电方式
coaxial power gear 同轴动力转向装置
coaxial probe 同轴探头
coaxial progressive deformation 共轴递进变形
coaxial pump laser 同轴抽运激光器
coaxial rainfall-runoff relation 降雨径流共轴关系曲线
coaxial relation 同轴关系
coaxial resonant cavity 同轴谐振腔
coaxial resonator 同轴谐振器
coaxial rod gap 同轴电极间火花隙
coaxial rotary joint 同轴线旋转接头
coaxial rotary viscosimeter 同轴回转黏度计
coaxial section 同轴电缆段
coaxial separator 同轴分离器
coaxial silo 同轴双筒库
coaxial spark-gap 共轴火花隙
coaxial spheric(al) system 共轴球面系统
coaxial spheric(al) wave 共轴球面波
coaxial spiral 轴向螺旋；定中心螺旋(孔内套管)
coaxial standing-wave detector 同轴驻波检测器
coaxial-streams injector 同轴流束喷射器
coaxial stub 同轴短线
coaxial superposed folds 共轴叠加褶皱
coaxial supply line 同轴馈(电)线
coaxial switch 同轴开关
coaxial system 共轴系统
coaxial-torus condenser 共轴环形electric容器
coaxial-to-waveguide transducer 同轴线—波导管匹配换能器
coaxial-to-waveguide transformer 同轴线—波导管耦合变压器
coaxial-to-waveguide transition 同轴线—波导管过渡
coaxial transmission line 同轴电缆；同轴传输线
coaxial trochotron 同轴电子开关
coaxial-waveguide output device 同轴波导型输出器
coaxial wavemeter 同轴波长计
coaxing 疲劳限渐增现象；人工提高金属疲劳强度(法)
co-axis 共轴
coaxitron 同轴管
coaxswitch 同轴(电路转换)开关
cobalt acetate 醋酸钴
cobalt alloy 钴合金
cobalt aluminate 铝酸钴；钴铝尖晶石；钴蓝颜料
cobaltamine complex 钴胺络合物
cobaltammine 氨合钴；氨络钴
cobalt arsenate 砷酸钴
cobalt-base alloys 钴基合金
cobalt benzoate 苯甲酸钴
cobalt black 一氧化钴；钴黑
cobalt bloom 八水合砷酸钴；钴华
cobalt blue 氧化钴；钴青；钴蓝；瓷蓝；豌青
cobalt blue glass 钴蓝玻璃
cobalt blue glaze 钴蓝釉
cobalt blue pigment 豌青颜料；钴青颜料
cobalt bomb 钴弹
cobalt-bonded tungsten carbide 钴作黏结金属的碳化钨粉末
cobalt-bonder 钴胶结剂
cobalt bronze 磷酸铵钴；钴青铜
cobalt brown 钴棕颜料
cobalt cemented titanium carbide 钴钛硬质合金
cobalt cemented tungsten carbide 钴钨硬质合金
cobalt-cementing phase 钴粘结相
cobalt chloride 氯化钴
cobalt-chromate 铬酸钴
cobalt-chromium alloy 钴铬(铸造)合金
cobalt-chromium alloy for high frequency casting 高频铸造用钴铬合金
cobalt-chromium steel 钴铬钢
cobalt-chromium-tungsten alloy 钴铬钨合金
cobalt complex dye 含钴复合染料
cobalt concentrate 钴精矿
cobalt deposit 钴矿床
cobalt disulfide 二硫化钴
cobalt dryer 钴干料；钴催干剂
cobalt edetate 依地酸钴
cobalt effect 钴效应
cobalt ethylenediamine tetraacetic acid 乙二铵四乙酸钴
cobalt ferrite 钴铁氧体
cobalt Fisher-Tropsch catalyst 费托水煤气合成钴催化剂
cobalt glance 辉砷钴矿
cobalt glass 钴玻璃
cobalt green 锌酸钴；钴绿(色)
cobalt high-speed steel 钴高速钢
cobalt hydrate 氢氧化钴
cobaltic chloride 氯化高钴
cobaltic dichlorotetrammine salt 二氯四氨合高钴盐
cobaltic dinitrotetrammine salt 二硝四氨合高钴盐
cobaltic flavo salt 二硝四氨合高钴盐
cobaltic fluoride 三氟化钴
cobaltic hexammine salt 六氨络高钴盐
cobaltic hydroxide 氢氧化高钴
cobaltic oxide 氧化高钴
cobaltic sulfate 硫酸高钴
cobaltic violeo salt 二氯四氯合高钴盐
cobalticyanide paper 氰化高钴试纸；氰高钴酸盐试纸
cobaltiferous wad 钴锰土
cobalt ion 钴离子
cobaltite 辉砷钴矿
cobaltkoritnigite 水砷钴石
cobalt lead crystal glass 钴铅晶质玻璃
cobalt linoleate 亚油酸钴
cobalt lustre 钴光泽彩
cobalt magnesia red 钴化钴镁红；钴桃红
cobalt-manganese drier 钴锰催干剂
cobalt metasilicate 偏硅酸钴
cobalt molybdate 钼酸钴
cobalt naphthenate 环烷酸钴
cobalt-nickel 钴镍合金
cobalt-nickel alloy 钴镍合金
cobalt-nickel cast iron 钴镍铸铁
cobaltocalcite 钴方解石
cobalt ocher 钴土；钴华
cobaltomenite 硒钴矿；水硒钴石
cobalt ore 钴矿(石)；钴矿(砂)
cobaltosis 钴尘肺
cobaltous 二价钴的
cobaltous acetate 醋酸钴
cobaltous chloride 氯化钴
cobaltous dichloride 二氯化钴
cobaltous fluorosilicate 氟硅酸钴
cobaltous hydroxide 氢氧化钴
cobaltous oxalate 草酸钴
cobaltous oxide 氧化钴
cobaltous perchlorate 高氯酸钴
cobalt oxalate 草酸钴
cobalt pellet 含钴颗粒料
cobalt pentlandite 钴镍黄铁矿
cobalt phosphide 一磷化二钴
cobalt pink 钴桃红色
cobalt poisoning 钴中毒
cobalt potassium nitrite 亚硝酸钴钾
cobalt powder 钴粉
cobalt rare earth magnet 稀土钴磁铁
cobalt recovery circuit 钴回收系统
cobalt red 钴红颜料；钴红色
cobalt resinate 树脂酸钴
cobalt salt 钴盐
Cobalt series 科博尔特统【地】
cobalt siccative 钴干燥剂
cobalt smaltite 钴砷矿
cobalt soap 钴皂
cobalt spinel 钴尖晶石
cobalt steel 钴钢

cobalt tallate 松浆油酸钴
cobalt thiocyanate 硫氰酸钴
cobalt trifluoride 三氟化钴
cobalt ultramarine 钴蓝(颜料)
cobalt unit 钴源装置
cobalt violet 钴紫(色)
cobalt violet pigment 钴紫颜料
cobalt yellow 亚硝酸钴高钴酸钾;钴黄(色)
cobalt zincate 锌酸钴;钴绿颜料
cobalt zippeite 水钴铀矾
Cobanic 可巴尼克镍钴铁合金
Cobb-Douglas production function 科布—道格拉斯生产函数
cobber 磁选机
cobbing 锤击选矿;人工敲碎;人工破碎;人工击碎;手选矿石块
cobbing separator 选矿分离机
cobble 卵石级煤;卵石;圆块煤;中卵石;中砾;大卵石;粗砾大卵石;粗砾;鹅卵石;半轧废品;铺路石;碎石
cobble bar 鹅卵石滩
cobble-bed river 卵石河床河流
cobble boulder 中砾石;圆石;大卵石;鹅卵石;铺路用卵石
cobble breccia 粗砾角砾岩
cobble conglomerate 粗砾砾岩
cobbled 用卵石铺设的
cobbled canal 卵石(衬护的)渠道
cobbled road 石子路
cobble foundation 卵石基础
cobble gravel 中砾(石)
cobble gutter 铺圆石水沟
cobble mix 中块石混凝土
cobble pavement 卵石路面;圆石路面;大卵石路面
cobbler's bench 鞋匠板凳;矮茶几
cobblestone 大圆卵石;圆石;中砾石;大卵石;粗砾岩;鹅卵石
cobblestone foundation 卵石基础
cobblestone masonry 大卵石砌筑
cobblestone pavement 大卵石路面;圆石路面;鹅卵(铺砌)的路面
cobble stone paving 大卵石路面
cobblestone road 大卵石路
cobblestone street 石子路
cobble texture 粗砾结构
cobble up 粗制滥造
cobbling 卵石铺砌;圆石铺砌
cobbly 用大卵石铺的
cobbly soil 粗砾质土
cob brick 草泥砖;土砖
Cobb sizing test 卡卜施胶度试验法
cobcoal 圆煤块;成团煤炭
cob construction 黏土夯实
cobelligerent 共同交战国
cobenium 恒弹模数钢
cob(house) 夯土建筑;土窑
coblat oxide treatment 钴氧化处理
coble 平底渔船
Coblenzian 柯布兰兹阶【地】
coble oar 扁阔型短桨
cob mill 黏性料块粉碎机
cob mo(u)lding 凹圆形线脚
co-borrower 共同借款人
coboundary 上边界
coboundary complex 上边缘复(合)形
coboundary operator 上边缘算子
co-broker 合作经纪人
cobs 钟形失真
cobstone 圆石;大卵石;粗砾岩
coburn fitting 重型拉门滑动装置
cob wall 土墙;土坯墙
cob walling 土坯墙;生土墙;夯土墙;板筑墙;空心方块式(一种防波堤块体);黏土墙;滑秸泥;糊墙纸;小圆块;柴泥;草泥;手锤选矿
cobweb 蜘蛛网状物
cobwebbing 蛛网(状);喷拉丝涂覆法;网纹涂覆法;蛛网状涂层;网纹喷胶法;喷出网状漆丝
cobwebbing during spraying 喷涂时形成的网纹;喷涂时蛛网
cobweb coating 拉丝涂料;网纹涂料
cobweb pattern 蛛网状图案
cobweb random rubble masonry 蛛网形缝乱石砌工
cobweb rubble masonry 蛛网形缝毛石砌工;蛛网

缝毛石砌体;蛛网形缝毛石坊
cobwork 圆木建筑;糊墙泥
cocap 伴帽
cocatalyst 共催化剂;辅催化剂;辅触媒
cocave cover 凹形盖
cocci 球菌
coccode 球状小粒
coccoid 粒状体
coccolith 颗石(藻)
coccolithglobigerina ooze 球菌抱球虫软泥
coccolith ooze 颗石(藻)软泥;球石软泥
Coccolithophorida 球石藻类
coccolithus 颗石(藻)类
coccus 球菌
coceine 鲜红(色)
cochain complex 上链复形
co-channel 同信道;同波道
co-channel interference 同波道干扰;同信道干扰
co-channel separation 同波道分离
cochineal 胭脂(虫)红
cochineal insect 胭脂虫
cochineal redcarmine 胭脂红
Cochin oil 科钦油;精制椰子油
Cochiti event 扣齐地事件
Cochiti normal polarity subchron 扣齐地正向极性亚时
Cochiti normal polarity subchronzone 扣齐地正向极性亚时间带
Cochiti normal polarity subzone 扣齐地正向极性亚带
cochlea 螺旋形楼梯(间);蜗牛壳状
cochlear 蜗形的
cochlear branch 蜗支
cochleare 匙
cochleare parvum 茶匙
cochlear model 蜗壳形模型
cochleary 蜗牛壳状
cochleary cochleated 螺旋形
cochleated 螺旋形;蜗牛壳状
cochleoid 蜗牛线
cochleopalpebral reflex 戈尔特氏反射
Cochran boiler 立式横烟管锅炉;立式多横火管锅炉;柯克兰锅炉
Cochran-Orcutt interative method 柯克兰—奥克特迭代法
Cochran-Orcutt method 柯克兰—奥克特法
Cochran-Orcutt two-step technique 柯克兰—奥克特两步法
Cochran's criterion 柯克兰判定准则
Cochran's rule 柯克兰准则
Cochran's test 柯克兰检验
cochromite 钴铬铁矿
cocinerite 杂硫银铜矿
cocircuit 共电路
co-citation indexing 共同引文索引
cock 考克;节气门;活栓;旋塞开关;旋塞(阀);液流开关;风信鸡;方向标
cock-a-bill 吊锚
cock bead 凸缘;(木作的)凸出边缘;隆珠
cockbill 吊锚
cockboat 小船(大船的供应船);小型救生艇或供应艇;船上小舢板;小艇;小舢板
cock body 旋塞体;塞体
cock carrying platform 集草车
cock-casing 活栓套管
cock comb pot 鸡冠壶
Cockcroft-Walton accelerator 高压倍增器;高压倍加器
Cockcroft-Walton generator 考克罗夫—瓦耳顿发电机
Cockcroft-Walton type accelerator 考克罗夫—瓦耳顿式加速器
cocked bead 凸出边缘
cocked finish 瓶口歪斜
cocked hat 船位误差三角形
cocked position 扳起位置
cocked switch 待动开关
cocker 采掘面上的支柱
cockering 人字形支架
cocket 海关印;海关放行证;小舢板
cocket handle 考克扳手;水龙头把手
cockeye grass 鸡眼草
cock handle 水龙头把手
cock head 凸出边缘

cock hole cover 旋塞孔洞盖
cocking 嵌齿扣合
cocking button 竖起钮
cocking-down 下垂
cocking lever 扳机
cocking piece 短人字木;飞檐椽;扒钉;翘檐板;扣链齿;钉在椽头顶部的楔形木块
cocking-up 上翘
cock key 旋塞式开关;旋塞扳手
cockle (玻璃中的)不均匀皱纹;小挠艇;皱皮;(薄板边缘的)皱裂(纹);白浪(花);外围挡板
cockle boat 小舢板
cockle cuts 皱裂
cockle cylinder 麦郎选除筒
cockled surface 表面鼓起
cockle finish 皱纹面漆;粗装饰;起皱加工
cockle separator 麦郎选除器
cockle stair(case) 螺旋梯;螺旋(式)楼梯;盘旋楼梯
cocklifter 运煤机;移堆机
cockling 扭转;缠绕;起皱
cockling sea 短促浪;三角浪
cockloft 阁楼;顶楼;顶层(指楼房)
cock meter 水表
cockpit 岩溶盆地;船尾舵手座;灰岩盆地;漏斗状石灰坑;漏斗状渗水井;灰岩盆地;座舱;斗鸡场;船尾座位;艇尾舱
cock-plug head 旋塞头;旋塞柄
cockroach killer 灭蟑螂药
cock saddle red 红尖毛
cock saddle white 白尖毛
cock saw 钢丝锯
cockscomb 鸡冠花;梳状刮刀;鸡冠饰
cock spanner 旋塞扳手
cockspur 三足支架
cockspur fastener 鸡爪扣;窗扉扣件;玻璃窗销扣
cock stop 旋塞龙头;开关龙头;旋塞触止
cocktail 鸡尾酒会;鸡尾手锯
cocktail belt 市郊高级住宅区
cocktail lounge 旅馆的鸡尾酒吧间;酒座;酒厅;鸡尾酒吧
cock tap 旋塞;龙头
cock wheel 棘轮;中间轮
cock wrench 龙头扳手
coclad 金属包层钢板
cocoa brown 可可棕色
co-coagulation flo(a)tation-biocontact oxidation process 共混凝气浮—生物接触氧化工艺
cocoa mat 椰壳纤维席纹布
coco(a)nut 椰子(色);椰子壳
coco(a)nut brown 椰子棕色
coco(a) nut butter 椰子奶油
coco(a)nut charcoal 椰壳炭
coco(a)nut fatty acid 椰子油脂肪酸
coco(a)nut fiber[fibre] 椰壳纤维
coco(a)nut mat(ting) 椰棕席子;椰棕地毯
coco(a)nut oil 椰子油;椰子脂
coco(a)nut palm 椰子树
cocobolo 毒黄檀木
cocombustion 合并焚烧
co-condensation 共缩合(作用)
co-condensation polymer 共缩聚物
coconinoite 硫磷铝铁铀矿
co-consultant 共同咨询人;共同协商者
cocontraction 协同收缩
cocontractor 共同签约人;共同承揽人;联合承包商
coconut capacitor 大型真空电容器
coconut matting 椰棕地毯
coconut oil 椰子油
coconut oil fatty acid 椰子油脂肪酸
cocoon 防护性喷涂;防护层
cocoon agitator 索绪器
cocoon beater 索绪装置
cocoonery 蚕室
cocoon fiber 蚕丝
cocooning 塑料披盖;茧状包覆;防护喷涂;保护措施;速率披盖
cocoonization 作茧喷涂法
cocoon packing 被覆包装
cocoon spraying 喷涂时喷丝
cocopan 小型矿车
Cocos plate 可可斯板块
coco wood 椰子木
co-crystallization 共结晶;共同析晶
cocrystallized Al-Si alloy 共晶铝硅合金

coctile 烧制品(建筑用)
coctolabile 不耐煮沸的
coctoprecipitation 煮沸沉淀
coctoprecipitinogen 煮沸沉淀原
coctostabile 耐煮沸的
co-cultivation 共生培养
co-curing 共固化
cocurrent 平行电流;同时发生的
cocurrent air and water backwash 并流气水反洗
cocurrent air-water backwash 并流气水反洗
cocurrent air-water-flow 气水汇流
cocurrent air-water scour 并流气水洗
cocurrent contact 顺流接触
cocurrent cooling 顺流冷却
cocurrent extraction 顺流萃取
cocurrent flow 合流;并流;同流
cocurrent flow gasification 并流气化
cocurrent leaching 并流浸出
cocurrent line 等潮时线;同时潮流线;同潮流时线
cocurrent observation 同时观察;同时观测
cocurrent observation method 同时观测法
cocurrent operation 并流操作
cocurrent regeneration 顺流再生
cocurrent solid bowl centrifuge 并流式无孔转鼓离心机
cocurrent water operation 并流气水操作
cocurriculum 辅助课程
cocusing of headlamps 头灯聚焦
cocuswood 椰子木;黄棕色硬木(产于西印度及缅甸);美洲乌木
cocycle 闭上链
cod 大西洋鳕;台风路线转折点;湿砂型芯
Codabar code 码条代码
codan lamp 接收指示灯
codastre 地政局
coda wave 震尾波
coda wave amplitude 震尾波振幅
coda wave spectrum 震尾波谱
Codazzi condition 科达齐条件式
codazzite 铈铁白云石
codding 烟囱侧壁基础;烟囱侧壁底脚
Coddington lens 科丁顿透镜
Coddington magnifier 科丁顿放大镜
coddled iron 混风生铁(高炉用冷热混合风冶炼的生铁)
code 密编码;码;一级代码;规约;规范;电码;代码;代号;法规;法典;法案;标准化代号
code address 电报挂号
code alphabet 码符号集
code and answering pendant 回答旗
code and standard 规范
code aperture 编码孔径
code area 编码区
code bar 码条
code bar drive magnet 码条驱动磁铁;编码条驱动磁铁
code base 码基数;编码基数
code beacon 码信标;电码指向标;闪光灯标
code better 形状编号(钢筋等)
code bit 码位;代码单位
code block 码组
code book 译码本;电码本;代码簿
co-debtor 共同债务人
codec 编解码器
code call 选码振铃;编码呼叫
code call indicator 编码呼叫指示器
code capacity 电码容量;编码容量;编码能力
code cases 规范案例
code character 码组
code check 码率;码检验;代码检验;代码检查
code checking time 代码检验时间
code circuit 编码电路
code classification 代码分类
codeclination 极距;同轴磁偏角;天体极距;余赤纬
code combination 码组合;代码组合
code compare 代码比较
code comparison 类号对比;代码对比
code compatible 兼容性代码
code control system 编码控制系统
code conversion 码变换;码变换;电码变换;转换;代码变换
code converter 码变换器;代码转换器;代码变换器;变型变换器;变码器;编码变换器
code current 编码电流

code current control 编码电流控制
code cycle 电码周期
coded 编码的
coded abstract 编码式文摘
coded address 编码地址
coded arithmetic data 编码(的)算术数据
code data 编码数据
code data conversion 编码数据转换
code data matrix 编译数据矩阵
coded automatic gain control 编码自动增益控制
coded character 编码字符
coded character set 编码字符集
coded circle 编码度盘
coded combination 编码组合
coded command 编码指令;编码命令
coded continuous wave 编码连续波
coded conversion 编码转换
coded data 编码(的)数据
coded data decoder 编码数据译码器
coded data pattern 编码数据模式
coded decimal 编码(的)十进制;十进编码数字
coded decimal adder 编码十进制加法器;十进代码加法器
coded decimal calculating machine 编码十进计算机
coded decimal digit 编码的十进制数字
coded decimal machine 编码的十进制计算机
coded decimal notation 以(二进制)编码表示的十进制记数法;编码(的)十进制(记数法);十进制编码
coded decimal number 编码十进制数;编码的十进制数
coded decimal presentation 编码十进制表示法
coded-decimal system 编码的十进制系统
coded diffraction grating 编码衍射光栅
coded-digital modulation 编码数字调制
coded disc 编码盘;码盘【计】
coded diversity 编码分集
code delay 码延迟;信号延迟;电码延迟;电码积压;编码延迟
code-dependent system 相关码体系
code design 代码设计
code designation 规定牌号
code device 编码装置;编码器
coded extension character 编码的扩展字符
coded form 编码形式
coded format 编码格式
coded Fourier-transform hologram 编码傅立叶变换全息图
coded graphics 编码制图;编码图形
code dialling 代码拨号
coded identification 译码表示法;编码识别;编码符号
coded image 编码图像
coded image space 编码图像空间
coded impulse 编码脉冲
coded information 编码信息;编码信息
coded initial 缩写词
coded input 编码输入
coded interrogator 编码询问器
code direct 绝对代码
code disc 码盘
code discriminator 鉴码器
code distance 码距;代码距离;代码间距
code distinguishability 码分辨率;编码分辨率
code distortion 码失真
coded items 条目检索号
code division 码分;代码分隔
code division multiple access 码分多址(方式)
code division multiple access cellular system 码分多址蜂窝系统
coded laser beam 编码激光束
coded-light identification 编码光识别;灯码识别
coded lock 密码锁
coded message 编码信息;编码消息
coded number 编码数;编码号
coded order 编码指令
coded parameter 代码参数
coded passive reflector 编码无源反射器
coded passive reflector antenna 码控无源反射器天线;编码无源反射器天线
coded program(me) 编码程序;代码程序;上机程序
coded program(me) system 编码程序系统
coded radio beam 编码无线电波束

coded reference hologram 编码参考全息图
coded reference wave hologram 编码参考波全息图
coded representation 编码表示法
code drum 代码鼓
coded scale 编码尺
coded scanning 编码扫描
coded scanning technique 编码扫描技术
coded-scan tomogram 编码扫描层析 X 射线照片
coded selector 编码选择开关
coded sequence 编码序列;编码程序
coded stop 程序停机;编码停机
coded string 编码字符串
coded system 编码系统;码码方式
coded tape valve 数码带射流阀
coded track circuit 编码轨道电路
coded word 编码字;编成代码的字
code efficiency 码效率;编码效率
code element 码元;代码单位;编码元素;代码表示
code element unit 编码元素单元;代码单元装置
code enforcement 执行法规
code error 代码错误;编码错误
code error rate 误码率
code extension character 代码扩充字符
code fetch 取(出)码
code fetch cycle 取码周期
code flag 信号旗
code for architectural and structural design 建筑及结构设计规范
code for architectural design 建筑设计规范
code for design of railway line 铁路线路设计规范
code for electric(al) design and installation 电气设计安装规范
code for erection and inspection 安装验收规范
code for fire protection 防火规范
code for information interchange 信息交换用代码
code form 电码型式
code format 代码格式
code for reinforced concrete structure 钢筋混凝土结构规范
code for structural design 结构设计规范
code for technology of railway engineering 铁路工程技术规范
code for water supply and sewerage 给排水规范
code generation 代码生成
code generator 编码发生器
code group 码群
code hole 电码孔;代码孔
code-independent data communication 独立于代码的数据通信
code independent transmission procedure 独立编码传送过程
code information bit 代码信息单位
code inhibit 代码禁止
code invariance 码恒定性
code keyer 代码键控器
code length 码长(度)
codeless input system 无编码输入系统
code letter 码字;代码字母
code level 代码结构级别;代码级
code light 信号灯(光);符号灯
code line 记码区;编码行
code line index 编码索引区
code machine 译码机;编码机
code mark 代码符号
code memory 代码存储器
code message 密码通信
code missing 漏码
code modulation 编码调制
code name 代名;代号
code not allocated 尚未分配的代码
code notation scheme 代码标志表
code notches 标记缺口
code number 密押;码数;序号;代号
code of civil procedure 民事诉讼法典
code of conduct 经营法典;行为规则;行为守则
Code of Conduct on Transnational Corporations 跨国公司行动守则
Code of Conducts for Scientific Diving 科研潜水规范
code of construction 建筑规则;建筑规范;建筑规程
code of fold 褶皱代号
code of international conduct 国际行为准则;国际行为标准

code of marketing in order 销售量所居位数
code of mineral commodity 产品牌号
code of photography area 摄影地区代号
code of practice 混凝土施工规程;业务守则;业务规章;业务法规;规定;定额;操作规范;实用规范;实施规程;施工规范;建筑规则;建筑规程
code of procedures 工艺规程;操作规程
code of professional ethics 职业道德准则
code of recommended practice 推荐业务条例
code of regime observation line 动态观测线编号
code of reserve-report 储量报告编号
code of stratigraphic(al) nomenclature 地层命名法则
code of survey area 测区代号
code of surveys 测量规范
code optimization 代码最优化;编码(最)优化
code optimize 代码优化
code order 码序
code oscillator 编码信号发生器
code pattern 代码格式;编码图
code phone 保密电话
code-piece 井架节段木制连接板
code plus decimal point 代码加小数点
code-point 码点
code polynomial 码多项式;代码多项式
code position 共积作用;穿孔位置;共(沉)积;同时沉积;同期沉积
code power 电码化电源
code preserving permutation 保码排列
coder 记发器;脉冲编码装置;发码器;编码员;编码器;编码机
code rate 码率;码比率;编码速率;编码率
coder-decoder 编码译码器
code reader 代码读出器
code receiver 收码器
code redundance 码剩余度
code reference wave 编码参考波
code regeneration 码再生
code register 代码寄存器
code repertoire 指令代码;指令表
code repertory 指令系统;指令代码;指令表
code requirement 采用规范;适用标准
code revision 修订规范
code rewriting 码再生;代码重写;代码再生
code ringing 编码信号振铃
coder operator 译码算子
code rule 编码规则
codes and standards 规范(和)标准
code scanner 代码扫描器
code scheme 编码方案
code segment 代码段
code selection 电码选择
code selector 选码器
code sender 发码器
code sending 发码
codesending radiosonde 发送电码无线电探空仪
code sequence 码序(列)
code sequence generator 码序发生器
code set 代码集
code sign 码符号;合作设计
code signal 码信号;编码信号;暗号
codes issued by Ministry 部颁规范
code sonar 编码声呐
code specification 规范
code stamp 规范印章;安装后规范印章的可及性
code standards 法规标准
code storage 代码存储装置
code storage circuit 编码存储线路
code structure 编码结构
code subroutine 编码子程序
code switch 码条式接线器;码开关;代码开关
code symbol 码符号;规范符号
code table 指令表;代码表
code telegram 明码电报;密码电报
codeterminants of price 价格决定的共同因素
codeterministic system 共确定系统
code test 标准试验
code theodolite 编码经纬仪
code track 代码道
code track circuit 电码轨道电路
code translator 解码器;译码装置;译码器;电码转换器;编码器
code transmitter 发报机
code-transparent system 明码系统

code tree 码树;编码树
code type 码条式
code-type radiosonde 发送电码无线电探空仪
code validity test 编码有效性测试
code vector 码矢
code weight 码重
code wheel 编码轮
code wire 隔离导线;标准线号导线
code word 密码命令;码字;信号码;电码字;代码字
code word-locator polynomial 码字定位多项式
Codex Alimentarius 食品法典委员会
co-diagonalization 相互对角比
codicil 遗嘱修改附录;遗嘱更改
codification 译成电码;编码
codified procedure 自动设计程序
codimer 共二聚体
coding 译成电码;编制程序
coding amplifier 编码放大器
coding and transmission selection 编码与传输选择
coding cam 编码凸轮
coding capacity 编码容量
coding card 编码卡
coding check 编码检查
coding clerk 程序员
coding collar 编码环
coding convention 编码约定
coding delay 信号延迟;电码延迟;编码延迟
coding design 编码设计
coding disk 编码盘
coding efficiency 编码效率
coding error 编码错误
coding form 编码形式;编码方式
coding gain 编码增益
coding gate 编码选通电路
coding image 编码影像;编码图像
coding impulse gate 编码脉冲门
coding legend 编码图例;符号解释;图例
coding line 指令行;编码行;编码线
coding manual 编码手册
coding mask 编码盘
coding method 编码方法
coding network 编码网络
coding noise 编码噪声
coding of continuous track circuit 轨道(电路)电码化
coding office 密码室;译电室
coding of railway statistics 铁路统计编码
coding pulse multiple 脉冲编码系统
coding relay 编码继电器
coding room 译电室
coding sheet 程序纸;编码纸
coding specification 编码分类
coding standard 编码标准
coding system 编码制;编码系统
coding theory 编码理论
coding translation data system 译码数据系统
coding tree 编码树
coding tube 编码管
coding videotelephone 编码电视电话
coding violation 编码违例
coding wave 编码波
co-disposal 联合处置;联合处理
codistor 静噪稳压管;静噪调压管
cod-liver oil 鳕鱼肝油
cod oil 鳕鱼油;松香油
codominance 共显性
co-dominant 次优势木;次优势的;共显性的
codominant gene pair 等优势基因对
codon 密码子;码字
codope 共掺杂;双掺杂
cod projection 吊砂
codrawer 共同出票人;共同发票人
codress 编码地址
co-driver 副驾驶员
co-driver's desk 副驾驶台
coed transit 编码经纬仪
coefficient 系数;因数
coefficient affecting fundamental period 基本周期影响系数
coefficient between layers 管涌比;层间系数(即管涌比)
coefficient checker work structure 格子体构筑系数
coefficient concerning the form of member 构件形状系数
coefficient concerning the surface effect 表面效应系数
coefficient conversion 按系数换算
coefficient distribution 系数分布
coefficient domain 系数域
coefficient during peak season 峰期系数
coefficient for branch loss 支管压力损失系数
coefficient for combination value of actions 作用组合值系数
coefficient for gradual enlargement 逐步扩大管压力损失系数
coefficient for orifice 孔口系数
coefficient for importance of structure 结构重要性系数
coefficient frost heaving 冻胀率
coefficient heat transmission 传递系数
coefficient in radiation shield 辐射防护系数
coefficient matrix 系数矩阵
coefficient multiplier 系数乘法器;比例乘法器
coefficient of abandoned fixed assets 固定资产拆除系数
coefficient of absolute roughness of pipe 管子的绝缘粗糙系数
coefficient of absolute viscosity 绝对黏(滞)度系数
coefficient of absorption 吸收系数;吸附系数
coefficient of absorption of light 光的吸收系数
coefficient of absorption of radiant heat 辐射热吸收系数
coefficient of acceleration 加速度系数
coefficient of accumulation 堆积系数
coefficient of accumulation of heat 蓄热系数
coefficient of accuracy 精度系数
coefficient of acidity 酸度系数
coefficient of acoustic(al) transmission 传声系数
coefficient of acoustics 吸声系数;声响系数
coefficient of action effect 作用效应系数
coefficient of active earth pressure 主动土压力系数
coefficient of adhesion 黏着系数;附着系数
coefficient of adiabatic compressibility 绝热压缩系数
coefficient of adjustment 校正系数
coefficient of admission 装满系数;充满系数
coefficient of adsorption 吸收系数
coefficient of adsorption velocity 吸附速度系数
coefficient of aggregate correction 骨料修正系数
coefficient of agreement 一致性系数
coefficient of air excess 空气过剩系数
coefficient of air friction 空气阻力系数
coefficient of air permeability 透气系数
coefficient of air resistance 空气阻力系数
coefficient of alienment 相疏系数
coefficient of alienation 离异系数;非相关系数;不相关系数
coefficient of amplification 放大系数
coefficient of apparent diffusion 表观扩散系数
coefficient of apparent expansion 表观膨胀系数;视膨胀系数
coefficient of aqueous migration 水迁移系数
coefficient of aquifer 含水层系数
coefficient of area 面积系数
coefficient of assimilation 同化系数
coefficient of association 相联系数;相伴系数;伴联系数
coefficient of atmospheric transmission 大气透明度系数
coefficient of at rest earth pressure 静止土压力系数
coefficient of attenuation 衰减系数;衰变系数
coefficient of audibility 可听度系数
coefficient of autocorrelation 自相关系数
coefficient of average accumulation efficiency 平均积累效果系数
coefficient of average deviation 平均偏差系数
coefficient of axial deformation 轴向变形系数
coefficient of axial displacement 轴向位移系数
coefficient of balance 平衡系数
coefficient of basin shape 流域形状系数
coefficient of basin slope 流域坡度系数
coefficient of beam utilization 光束利用系数
coefficient of bearing capacity 承载系数;承载能力系数;承载力系数
coefficient of bending stiffness 抗弯刚度系数

coefficient of bit pressure drop 钻头压降系数
coefficient of bottom friction 底摩阻系数;底摩擦系数
coefficient of braking 制动系数
coefficient of brightness 亮度系数
coefficient of brittleness 脆性系数
coefficient of bucket filling 泥斗充泥系数
coefficient of bulking 胀大系数
coefficient of capacitance 电容系数
coefficient of cargo-handling 操作系数(指装卸)
coefficient of charge 装料系数;负荷系数;装药系数;占空系数(线圈);填充系数
coefficient of charge weight 重量充装系数
coefficient of chemical resistance 耐蚀系数
coefficient of chemical resistant 耐化学侵蚀系数
coefficient of Chezy formula 谢氏系数
coefficient of circularity 圆度系数
coefficient of circulating pressure drop 循环系统压降系数
coefficient of climate 气候系数
coefficient of coal cutting resistance 截割阻力系数
coefficient of coal firmness 坚固性系数
coefficient of coal thickness variation 煤层厚度变异系数
coefficient of cohesion 黏聚系数;内聚(力)系数
coefficient of collapsibility 湿陷系数
coefficient of community 群落系数
coefficient of compaction 压实系数
coefficient of complementary 补偿系数
coefficient of compressibility 压缩系数
coefficient of compressive deformation of loess 黄土压缩变形系数
coefficient of compressive stiffness 抗压刚度系数
coefficient of concentration 集中系数;凝聚系数
coefficient of concordance 协同系数;一致性系数
coefficient of condensation 冷凝系数
coefficient of conductivity 导电系数;传导系数
coefficient of confidence 可靠系数
coefficient of configuration 装置系数
coefficient of confinement pressure 侧压力系数
coefficient of conformity 相似系数;适合度系数
coefficient of connection 联络系数
coefficient of consistency 稠度系数
coefficient of consolidation 压实系数;固结系数;渗压系数
coefficient of consolidation for radial flow 径向固结系数
coefficient of construction mechanization 施工机械化系数
coefficient of container load 集装箱荷载折减系数
coefficient of contingency 或然系数;相依系数
coefficient of continuity 连续性系数
coefficient of contraction 缩窄系数;收缩系数
coefficient of contraction for orifice 节流孔收缩系数
coefficient of convection 对流系数
coefficient of convective heat transfer 对流放热系数
coefficient of conversion 折算系数
coefficient of correction 校正系数;修正系数;改正系数
coefficient of correlation 相关系数;对比系数
coefficient of corrosion 腐蚀系数
coefficient of coupling 耦合系数
coefficient of creep 徐变系数
coefficient of cross-correlation 互相关系数
coefficient of cross elasticity 交叉弹性系数
coefficient of cubic(al) elasticity 体积弹性系数
coefficient of cubic(al) expansion 容量膨胀系数;体(积膨)胀系数
coefficient of current 流速系数
coefficient of current pressure 水流压力系数
coefficient of curvature 曲率系数
coefficient of curvity 曲率系数
coefficient of daily necessities 日常生活必需品系数
coefficient of damping 阻尼系数
coefficient of damping of critical 临界阻尼系数
coefficient of dead weight 自重系数
coefficient of decay 衰减系数;衰变系数
coefficient of deceleration 减速系数
coefficient of deduction 扣除系数
coefficient of deformation 形变系数;变形系数
coefficient of deformation of ground 地基振动变形系数
coefficient of depreciation 贬值系数
coefficient of derailment 脱轨系数
coefficient of destruction 破坏系数
coefficient of determination 可决系数;决定系数;测定系数;判定系数
coefficient of developed line 展线系数
coefficient of deviation 离差系数;自差系数;偏差系数
coefficient of deviation for magnetic 电磁罗盘偏差系数
coefficient of dielectric(al) loss 介质损耗系数
coefficient of differentiation 分化系数;变异系数
coefficient of diffuse(d) reflection 漫反射系数
coefficient of diffusion 扩散系数;散布系数
coefficient of diffusion for gas 气体扩散系数
coefficient of dilatation 膨胀系数;体胀系数
coefficient of dilution 稀释系数
coefficient of direction 方向系数
coefficient of discharge 流量系数;卸料系数;放电系数
coefficient of dispersion 离匀系数;离散系数;离差系数;扩散系数;分散系数;色散系数
coefficient of displacement 置换系数
coefficient of distribution 分配系数
coefficient of disturbance 扰动系数
coefficient of divergence 断裂带离散系数;分散系数;发散系数
coefficient of divergency 发散系数
coefficient of domestic sewage production 生活污水产生指数
coefficient of drag 曳力系数;阻力系数;水流力系数
coefficient of drying sensitivity 干燥敏感系数
coefficient of dry shrinkage 干燥收缩率
coefficient of dry volume shrinkage 干燥体积收缩率
coefficient of durability 耐久性系数
coefficient of dust removal 除尘率
coefficient of dynamic(al) force 动力系数
coefficient of dynamic(al) subgrade reaction 动基床反力系数;地基反力动力系数
coefficient of dynamic(al) viscosity 动黏滞(度)系数;动力黏滞(度)系数;动力黏滞度系数
coefficient of earth pressure 土压力系数
coefficient of earth pressure at rest 静止土压力系数
coefficient of economic results 经济效果系数
coefficient of eddy diffusion 涡流扩散系数;涡动扩散系数
coefficient of eddy thermal diffusion 涡流热扩散系数
coefficient of eddy viscosity 涡流黏滞(度)系数;涡动黏滞系数
coefficient of effective action 有效作用系数
coefficient of effective aperture 有效孔径系数
coefficient of efficiency 利用系数;效率;有效(作用)系数;有效率系数
coefficient of elasticity 杨氏系数;弹性系数
coefficient of elasticity test in situ 现场弹性系数测定
coefficient of elastic non-uniform compression 弹性非均匀压缩系数
coefficient of elastic of ground soil 地基土弹性系数
coefficient of elastic recovery 回弹系数;弹性恢复系数
coefficient of elastic shear 弹性剪切系数
coefficient of elastic uniform compression 弹性均匀压缩系数
coefficient of electron coupling 电子耦合系数
coefficient of elongation 拉伸系数;伸长系数;伸长率
coefficient of end restraint 端(部)约束系数
coefficient of energy 能量系数
coefficient of energy absorb 能量吸收系数
coefficient of energy loss 能量损失系数
coefficient of enlargement reckon 扩算系数
coefficient of equal triangle 图形强度系数
coefficient of equivalence 当量系数
coefficient of erosion 冲刷系数
coefficient of evapo(u)ration 蒸发率
coefficient of even lighting 照度均匀系数
coefficient of excess 超出系数;峰态系数;峰度系数
coefficient of excess air 空气过剩系数
coefficient of expansion 膨胀系数
coefficient of expansion and contraction 膨胀与收缩系数
coefficient of expansion by heat 热膨胀系数
coefficient of expansion of petroleum 石油的膨胀系数
coefficient of expectation 期待系数
coefficient of extension 延伸系数;延伸率;伸长系数
coefficient of extension line 展线系数
coefficient of extinction 消光系数;削弱系数
coefficient of fatigue 疲劳系数
coefficient of feeding 供给系数
coefficient of fineness 精度系数;细度系数;光洁度;肥瘦系数;肥瘠系数
coefficient of fineness delta 方形系数
coefficient of firing linear shrinkage 烧成线性收缩率
coefficient of firing shrinkage 烧成收缩率
coefficient of firing volume shrinkage 烧成体积收缩率
coefficient of fissuration 裂隙系数
coefficient of fissure 裂隙系数
coefficient of fixation 稳定系数
coefficient of float 浮子修正系数
coefficient of flood prediction 洪水水位推算系数
coefficient of flood recession 洪水水位下降系数;洪水降落系数
coefficient of flow 流量系数
coefficient of flow rate 排量系数
coefficient of flow resistance 水流阻力系数
coefficient of fluctuation 波动系数
coefficient of fluctuation in passenger traffic 客运波动系数
coefficient of fluidity 流度(系数);流动性系数
coefficient of foaming 发泡倍数
coefficient of forestation 流域森林面积系数;森林面积系数
coefficient of form 船型系数
coefficient of formation 编队系数
coefficient of foundation ditch's rebound 基坑回弹系数
coefficient of foundation soil deformation 地基变形系数
coefficient of freight(traffic) unbalance 运输不平衡系数
coefficient of frequency modulation 调频系数
coefficient of friction 内摩擦系数;摩擦系数
coefficient of friction(al) resistance 摩阻系数
coefficient of frost resistance 抗冻系数
coefficient of fuel feeding 供油系数
coefficient of fullness 满蓄率;装满系数;充满系数
coefficient of gabarite efficiency 尺寸利用系数
coefficient of gas transfer 气体转移系数
coefficient of goods train full-load 货物列车满轴系数
coefficient of goods train net-load 货物列车净载系数
coefficient of gradation 级配系数
coefficient of grain size 粒度系数
coefficient of gravel 砾石系数
coefficient of ground reaction 地基反力系数
coefficient of hard filth 硬垢系数
coefficient of hardness 硬度系数
coefficient of haze 烟雾(强)度系数;雾系数
coefficient of heat accumulation 蓄热系数
coefficient of heat conductivity 导热系数;导热率;热传导系数
coefficient of heat convection 对流给热系数;热对流系数
coefficient of heat efficiency 热利用系数
coefficient of heat emission 放热系数;散热系数
coefficient of heat exchange efficiency 热交换效率系数
coefficient of heat expansion test 自由膨胀率试验
coefficient of heat insulation 绝热系数;隔热系数;保温系数
coefficient of heat passage 导热系数
coefficient of heat perception 感热系数;受热系数
coefficient of heat storage 蓄热系数
coefficient of heat transfer 换热系数;热传输系数;热传导系数;传热系数
coefficient of heat transference 线热胀系数
coefficient of heat transmission 导热系数;传热系数;热传导系数;热传导率
coefficient of higher order 较高阶系数

coefficient of high order deviation 高阶自差系数
coefficient of homogeneity 均质系数;均匀系数;同质性系数;同一性系数
coefficient of horizontal consolidation 水平固结系数
coefficient of horizontal interference 水平干扰系数
coefficient of horizontal pile reaction 水平桩反力系数
coefficient of horizontal piling reaction 桩的水平反力系数
coefficient of horizontal pressure 侧压系数
coefficient of horizon(tal) soil reaction 地基水平反力系数
coefficient of horizon(tal) subgrade reaction 地基水平反力系数
coefficient of hydration 水化系数
coefficient of hydraulic conductivity 导水率;导水系数;渗透系数
coefficient of hydraulic flow 湍急系数
coefficient of hysteresis 磁滞损失系数
coefficient of impact 动力系数;冲击系数
coefficient of impedance 阻抗系数
coefficient of imperviousness 不透水系数;不渗透系数
coefficient of income elasticity of demand 需求收入弹性系数
coefficient of incompressibility 不可压缩性系数
coefficient of increase and decrease 增减率
coefficient of increase of mass of a train 列车回转质量系数
coefficient of induced deviation 感应自差系数
coefficient of inductive coupling 感应耦合系数
coefficient of infiltration 渗透系数
coefficient of injection 喷射系数
coefficient of injury 危害系数;损害系数
coefficient of insulation 绝缘系数
coefficient of intensity decrement 烈度衰减系数
coefficient of interclass correlation 组合相关系数
coefficient of internal friction 内摩擦系数
coefficient of internal friction of soil 土的内摩擦系数
coefficient of intraclass correlation 组内相关系数
coefficient of investment recovery 投资回收系数
coefficient of investment yields 投资效果系数
coefficient of irregularity 不平整系数;不规则系数
coefficient of joint sets 节理组数系数
coefficient of kinematic viscosity 运动黏(滞)度系数;动(力)黏滞系数
coefficient of kinetic eddy viscosity 涡流动力黏滞系数
coefficient of kinetic friction 动摩擦系数
coefficient of kinetic viscosity 动黏滞(度)系数
coefficient of kurtosis 峰度系数
coefficient of land used for buildings 建筑占地系数
coefficient of lateral pressure 侧压力系数
coefficient of leach deformation of loess 黄土溶滤变形系数
coefficient of leakage 漏损系数
coefficient of leaking 漏损系数;泄漏系数
coefficient of leeway 风压差系数
coefficient of lift 升力系数
coefficient of light diffusion 光扩散系数;散光系数
coefficient of lighting 采光系数
coefficient of light loss 光损失系数
coefficient of limit friction 极限摩擦系数
coefficient of limiting friction 极限摩擦系数
coefficient of linear continuity 线连续性系数
coefficient of linear expansion 线胀系数;线(性)膨胀系数;直线膨胀系数
coefficient of linear thermal expansion 线(性)热(膨)胀系数
coefficient of line development 展线系数
coefficient of liquefaction 液化系数
coefficient of load 装载系数;荷载系数
coefficient of load applied 荷载作用系数
coefficient of loading 装载系数;舱容系数
coefficient of loading characteristics 荷载特征系数
coefficient of load transference 荷载传递系数
coefficient of load transmission 荷载传递系数
coefficient of local resistance 局部阻力系数;局部阻抗系数;局部强度系数;局部电阻系数
coefficient of log error 计程仪误差改正系数
coefficient of losses 损失系数;损耗系数;亏损系数;耗损系数
coefficient of losses leakage 磁漏系数
coefficient of magnification 放大系数
coefficient of magnification of mapping 测图放大系数
coefficient of mass 质量系数
coefficient of mass absorption 质量吸收系数
coefficient of mass transfer 质量传递系数;传质系数
coefficient of mass utilization 质量利用系数
coefficient of mean deviation 均差系数;不匀率;平均差不匀率
coefficient of mean square contingency 均方列联系数
coefficient of mechanical efficiency 机械效率系数
coefficient of melt-settlement 融沉系数
coefficient of mineablity 可采系数
coefficient of mineralization 含矿系数
coefficient of mining 采动系数
coefficient of mixing 混合系数
coefficient of mobility 运动流度
coefficient of moisture adsorption 吸水系数;吸湿系数
coefficient of moisture transition 变湿系数
coefficient of molecular diffusion 分子扩散系数
coefficient of molecule diffusion 分子扩散系数
coefficient of moment 力矩系数
coefficient of moment of inertia 转动惯量系数
coefficient of momentum transfer 动量传递系数
coefficient of mortality 死亡系数
coefficient of movement 移动式系数
coefficient of mud pump 泥浆泵参数
coefficient of multiple determination 复测定系数
coefficient of multiple regression 复回归系数
coefficient of multiplication 增殖率
coefficient of mutual inductance 互感系数
coefficient of mutual induction 互感系数
coefficient of natural illumination 自然照度系数
coefficient of natural increase 自然增长系数
coefficient of natural lighting 自然采光系数
coefficient of natural mortality 自然死亡系数
coefficient of non-determination 不可测定;非可决系数;不可决系数
coefficient of non-sphericity 非球形系数
coefficient of non-uniformity 不均匀系数
coefficient of nozzle discharge 管嘴流量系数
coefficient of number of locomotive required 机车需要系数
coefficient of number of trains in the predominant traffic direction to that in non-prdominant traffic direction 运行图不成对系数
coefficient of occupation of seats 座席利用系数
coefficient of offset 移距系数
coefficient of opacity 不透明系数
coefficient of orifice 孔口流量系数
coefficient of oscillation 振荡系数
coefficient of overall heat transmission 传热系数;总传热系数
coefficient of overflow 溢流系数
coefficient of overlap 叠覆系数
coefficient of over-load 超载系数
coefficient of oxidation 氧化系数
coefficient of oxygen 氧利用系数
coefficient of parasite drag 寄生阻力系数
coefficient of partial correlation 偏相关系数
coefficient of partial determination 偏测定系数
coefficient of partial expansiveness 局部的扩充性系数
coefficient of passengers to inhabitants 居民乘车系数
coefficient of passive earth pressure 被动土压力系数
coefficient of pavement friction 路面摩擦系数
coefficient of percolation 渗透系数;渗透率;渗滤系数;渗漏系数;渗流系数
coefficient of performance 性能系数;效率;特性系数;使用系数
coefficient of performance of cross-section 截面效率系数
coefficient of permeability 渗透系数;渗透率
coefficient of permeability to water 透水性系数
coefficient of pipe friction 导管摩擦系数
coefficient of pirce elasticity of demand 需求价格弹性系数
coefficient of pitch of building 建筑物间距系数
coefficient of planar continuity 面连续性系数
coefficient of plasticity 塑性系数
coefficient of plunge 断裂带倾伏系数
coefficient of pore pressure 孔压系数
coefficient of potential 电位系数
coefficient of pressure 压力系数
coefficient of pressure conductivity 压力传导系数
coefficient of pressure distribution 压力分布系数
coefficient of probability 系综密度
coefficient of probe demarcation 探头标定系数
coefficient of production 生产系数
coefficient of product-moment correlation 积距相关系数
coefficient of products 货物运输系数
coefficient of proportionality 比例系数
coefficient of propulsion 推进系数
coefficient of pump pressure 泵压系数
coefficient of purification 净化系数
coefficient of radiation 辐射系数
coefficient of rainfall 降雨系数
coefficient of range 全距系数
coefficient of reaction 抗力系数;反力系数
coefficient of readiness of cargo handling machinery 装卸机械完好率
coefficient of reaeration 复氧系数
coefficient of realization of cargo-handling norms 装卸定额完成率
coefficient of rebound 回弹系数
coefficient of recovery 回收系数;回收率;复原系数
coefficient of redistribution of stress 应力重分布系数
coefficient of reduction 换算因子;换算因数;换算系数;折减系数
coefficient of redundancy 超静定系数
coefficient of reflection 反射系数;反差系数
coefficient of refraction 折射系数
coefficient of regime(n) 状态系数;洪枯流量比;河性系数
coefficient of region 区域系数
coefficient of regression 回归系数
coefficient of relative permeability 相对渗透系数
coefficient of relative roughness 相对粗糙度系数;相对糙率系数
coefficient of relative settlement 相对下沉系数;相对沉降系数
coefficient of relative strength 相对强度系数
coefficient of reliability 可靠系数
coefficient of reproduction 再生产系数
coefficient of reserve capacity 能力储备系数
coefficient of resilience 回弹系数
coefficient of resistance 阻抗系数;阻尼系数;阻力系数
coefficient of restitution 抗冲系数;回弹系数;恢复系数;还原系数;偿还系数
coefficient of retardation 延缓系数;阻滞系数;滞留系数;扼流系数
coefficient of rigidity 刚性系数;刚度系数
coefficient of rippling 波纹系数
coefficient of river regime(n) 河性系数
coefficient of road adhesion 路面附着系数
coefficient of rock abrasiveness 岩石研磨性系数
coefficient of rock decay 岩石风化程度系数
coefficient of rock drillability 岩石可钻性系数
coefficient of rock permeability 岩石的渗透系数
coefficient of rock resistance 岩石抗力系数;岩石抵抗系数
coefficient of rock stiffness 岩石坚固系数
coefficient of rolling friction 滚动摩擦系数
coefficient of rolling resistance 滚动阻力系数
coefficient of rotation 旋转系数
coefficient of roughness 粗糙系数
coefficient of rugosity 粗糙度系数
coefficient of runoff 径流系数;流出系数
coefficient of rupture 破裂系数
coefficient of safety 安全系数
coefficient of safety against cracking 抗裂安全系数
coefficient of safety against rupture 抗断裂安全系数
coefficient of scattering 散射系数
coefficient of scour(ing) 冲刷系数
coefficient of scouring resistance 岩石抗冲刷系数
coefficient of seasonal fluctuation in freight

traffic 货运(季节)波动系数
coefficient of seasonal fluctuation in goods traffic 货运季节波动系数
coefficient of seasonal fluctuation in passenger traffic 客运季度波动系数
coefficient of secondary compression 次压缩系数;二次压缩系数
coefficient of secondary consolidation 次固结系数
coefficient of secondary electron emission 二次电子发射系数
coefficient of section 断面系数
coefficient of section(al) form 断面(形状)系数
coefficient of seepage 渗漏系数;渗流系数
coefficient of seismic effect 地震影响系数
coefficient of selectivity 选择性系数
coefficient of self-induction 自感(应)系数
coefficient of self-oscillation 自振动系数
coefficient of self-weight collapse 自重湿陷系数
coefficient of self-weight collapsibility 自重塌陷系数
coefficient of sensitivity 灵敏度系数
coefficient of sensitivity of price 价格敏感系数
coefficient of separate determination 分别测定系数
coefficient of settlement 下沉系数
coefficient of shear 剪力系数;切变系数
coefficient of shearing resistance 抗剪系数
coefficient of shearing stiffness 抗剪刚度系数
coefficient of ship weight 船自重系数
coefficient of short tube 短管(流量)系数
coefficient of shrinkage 干缩系数;收缩系数
coefficient of side contraction 侧收缩系数
coefficient of silt transfer 泥沙输送系数
coefficient of similarity 相似系数
coefficient of sinuosity 弯曲系数
coefficient of skew 偏态系数;弯曲系数
coefficient of skewness 不对称系数;偏斜系数;偏态系数;偏度系数;偏差系数
coefficient of sliding 滑动系数
coefficient of sliding friction 抗滑动系数;滑动摩擦系数
coefficient of sliding friction of cover plate 路面板滑动摩擦系数
coefficient of sliding resistance 抗滑系数;抗滑动系数
coefficient of softness 软化系数
coefficient of softening 软化系数
coefficient of softening of rock 岩石软化性系数
coefficient of soil permeability 土的渗透系数
coefficient of soil preconsolidation 土的前期固结系数
coefficient of soil pressure 土压力系数
coefficient of soil reaction 路基(反力)系数;地基土反力系数;土(壤)反力系数
coefficient of solubility 溶度系数
coefficient of sorting 分选系数
coefficient of sound adsorption 吸声系数;吸音系数
coefficient of sound degree 完整性系数
coefficient of sound insulation 隔音系数
coefficient of source efficiency 光源效率
coefficient of spectral transmission 光谱传播系数
coefficient of spring depletion 泉水消耗系数
coefficient of sprinkling 喷洒系数
coefficient of stability 稳定性系数;稳定率
coefficient of standard deviation 标准偏差系数;标准离差系数;标准(变)差系数
coefficient of starting friction 起动摩擦系数
coefficient of static earth pressure 静止土压力系数
coefficient of static friction 静摩擦系数
coefficient of static friction of pavement 路面静摩擦系数
coefficient of stiffness 劲度系数;硬度系数;刚性系数;刚度系数
coefficient of storage 蓄量系数;储水系数;蓄水系数
coefficient of strain-hardening 应变硬化系数
coefficient of stream regime(n) 河性系数
coefficient of strength 强度系数
coefficient of stress concentration 应力集中系数
coefficient of stress redistribution 应力重分布系数
coefficient of structural response 结构反应系数
coefficient of subgrade 基床系数
coefficient of subgrade friction 路基摩擦系数
coefficient of subgrade reaction 路基反力系数;基底反力系数;基床刚度系数;基床反力系数;地基(土)反力系数;地基抗力系数
coefficient of subgrade resistance 路基阻力系数
coefficient of subgrade stiffness 路基劲度系数;路基刚度系数
coefficient of sudden contraction 骤缩系数
coefficient of surface configuration 表面形状系数
coefficient of surface polarization 面极化系数
coefficient of surface resistance 表面阻力系数
coefficient of surface runoff 地面径流系数
coefficient of suspense system 悬吊系统参数
coefficient of swelling 膨胀系数;湿胀系数
coefficient of temperature conductivity 导温系数;温度传导系数
coefficient of temperature correction 温度修正系数
coefficient of tension 拉力系数;张力系数
coefficient of thaw 融化系数
coefficient of thaw compression 融化压缩系数
coefficient of thaw subsidence 融陷系数;融沉系数
coefficient of thermal conductivity 导热系数;传热系数;热导性系数;热导率
coefficient of thermal conductivity of soil 土的导热系数
coefficient of thermal efficiency 热效率系数
coefficient of thermal expansion 热膨胀系数
coefficient of thermal movement 热移动系数
coefficient of thermal resistance 热阻(力)系数
coefficient of thermal transmission 传热系数;热传导系数
coefficient of thermometric conductivity 导温系数
coefficient of thermometric conductivity of soil 土的导温系数
coefficient of thermoosmotic transmission 热渗系数
coefficient of the shape of the basin 流域形状系数
coefficient of thixotropy 触变性系数
coefficient of thrust 推力系数
coefficient of tide 潮汐系数
coefficient of time lost 时间损失系数
coefficient of topographic(al) feature 地貌系数
coefficient of torsion 扭转系数;扭曲系数
coefficient of torsion(al) stiffness 抗扭刚度系数
coefficient of total correlation 总相关系数;复相关系数
coefficient of track efficiency 履带行走效率
coefficient of traction 牵引系数
coefficient of tractive resistance 牵引阻力系数
coefficient of train removal 列车扣除系数
coefficient of transmissibility 可透性系数;可渗透系数;导水系数;传导系数
coefficient of transmission 导水系数;传递系数;传导系数;传播系数
coefficient of transparency 透气系数;透明系数;透明度
coefficient of travel(l)ing to technical speed of passenger trains 旅客列车旅行速度系数
coefficient of travel time 行程时间系数
coefficient of turbidity 浊度系数
coefficient of turbulence 紊流系数;湍流系数;湍急系数
coefficient of turbulent transport 紊流输送系数
coefficient of twisting 扭转系数
coefficient of ultilization of wagon volume 货车容积利用率
coefficient of unbalance 不平衡系数
coefficient of unbalanced counter flow 反方向交通不平衡系数
coefficient of unbalanced freight traffic 货物运输(上下行)不均衡系数
coefficient of unevenness 不平整系数
coefficient of unevenness of goods traffic in two direction 货运方向不平衡数
coefficient of unfullness 不充满系数
coefficient of uniformity 均质系数;均匀系数;匀质系数
coefficient of unserviceable rolling stock 非运用车系数【铁】
coefficient of ununiformity 不均匀系数
coefficient of urban land-use 城市土地利用系数
coefficient of useful effect 功能系数;效率
coefficient of use-up 耗用系数
coefficient of utility 使用系数
coefficient of utilization 利用系数;利用率;使用系数
coefficient of utilization for car loading capacity 货车载重力利用系数
coefficient of utilization of container's capacity 箱载(重)利用率
coefficient of utilization of the loading capacity of wagon 货车载重力利用系数
coefficient of vapo(u)r permeation 蒸气渗透系数
coefficient of variability 变异系数
coefficient of variance 方差系数
coefficient of variation 变化系数;离中系数;离差系数;差异系数;差量系数;波动系数;变异系数;变动系数;变差系数
coefficient of variation in speed 变速系数
coefficient of variation of annual discharge 年流量变动率
coefficient of variation of thickness 厚度变化系数
coefficient of variation unevenness 不匀率变异系数
coefficient of variation with depth 随深度变化系数
coefficient of velocity 流速系数;速度系数
coefficient of vertical consolidation 竖向固结系数
coefficient of vertical contraction 垂直收缩系数
coefficient of vertical eddy diffusion 垂直涡流扩散系数
coefficient of vertical interference 垂直干扰系数
coefficient of vertical pile reaction 桩的垂直反力系数
coefficient of vessel volume 船舶容积系数
coefficient of viscosity 黏滞系数;黏性系数;黏度(系数)
coefficient of viscosity of serous material 浆液黏滞性系数
coefficient of viscous damping 黏性阻尼系数
coefficient of viscous traction 黏性牵伸系数
coefficient of visible light transparency 可见光透过系数
coefficient of volume compressibility 体积压缩系数
coefficient of volume decrease 体减系数
coefficient of volume expansion 容积膨胀系数;体膨胀系数
coefficient of volume of compressibility of soil 土的体积压缩系数
coefficient of volume polarization 体极化系数
coefficient of volumetric expansion 松散系数
coefficient of wagon loading capacity utilized 货车载重利用系数
coefficient of waste 浪费系数;磨损系数;废物系数
coefficient of water adsorption 吸水率
coefficient of water decrease 水量减小系数
coefficient of water-level 水位传导系数
coefficient of water-level conductivity 水位传导系数
coefficient of water-lines 浸水系数;浸润系数
coefficient of(water) permeability 透水系数
coefficient of water plane 水线面积系数
coefficient of water saturation 饱和吸水率
coefficient of wave shape 滤光片的波形系数
coefficient of wear 磨损系数
coefficient of wear efficiency 重量利用系数
coefficient of weathering for rock 岩石风化程度系数
coefficient of weight 加权系数
coefficient of weighted correlation 加权相关系数
coefficient of weight loss 失重系数
coefficient of wind force 风载系数;风力系数
coefficient of working condition 工作条件系数
coefficient of work per shunting locomotive 调车机车工作系数
coefficient potentiometer 系数(式)电位计;系数电位器
coefficient specific resistance 迫位抗力系数
coefficient tape 系数带
coefficient unit 系数组;系数部件;常数部件
coefficient value 系数值
coefficient value of amplitude recover 振幅恢复系数值
coefficient vector 系数向量
coek hopper 焦炭斗
coelanaglyphic relief 平浮雕
Coelenterata 腔肠动物
Coelinvar 柯艾林伐磁性合金
coelosphere 坐标仪;天球

coelostat 定向仪;定天镜
coemulsifier 共乳化剂;辅助乳化剂
c(o)enaculum 古罗马小餐室;(古罗马房屋中的)餐室
c(o)enatio 古罗马家庭正式餐厅
coenatio (古罗马房屋中的)正式餐厅
coenecium 共室
coenenchyma 共质轴
co-energy 共能量
coenesthesia 存在感觉
coenobiology 群落生物学
coenobiosis 群落生活
coenocline 群落生态群;群落渐变群
coenoecium 共室
coenology 群落学
coenosarc 共体
coenosis 生物群落
coenosium type 群落的类型
coenotype 群落型;群落类型
coenozone 群落带
coenvelope 上包络
coequal 同等的
coequalizer 余等геометр射
coercibility 可压缩性;可压凝性;可凝性;压缩性
coercible gas 可压缩气体
coercimeter 矫顽磁力计
coercion policy 高压政策
coercive circulation 强制性流通
coercive comparison 强制比较
coercive control 强制性控制
coercive correspondence 强制对应
coercive field 矫顽磁场;退磁磁场
coercive force 矫顽力;矫顽磁力;强制力;强迫力;抗磁力
coercive force meter 矫顽磁力计
coercive function 强制函数
coercivemeter 矫顽磁性测量仪;矫顽磁力计
coerciveness 强制性
coercive regulations 强制性规定
coercivity 矫顽磁性
coeruleoactite 钙绿松石
coesite 柯石英;高压人造石英
co-ester 共酯;复酯
coes wrench 活动扳手
coeval 同时代的
co-executor 合作遗嘱执行人
coexist 同时存在
coexistence border 双相分界线
co-existence element 共存元素
coexistence region 双相区
coexistence temperature 两态共存温度
coexisting organics 共存有机物
coexisting phases 共生相;共存相;共沉相
co-extraction 共(同)萃取
co-extrude 共挤出
coextruded film 复合薄膜
coextruded polyvinyl chloride plastic pipe 共挤聚乙烯塑料管
coextrusion 挤压法;共挤;双金属挤压
coextrusion casting 共挤流延
coextrusion coater 共挤涂机
coextrusion coating 共挤涂覆;共挤贴合
cofactor 合作因素;余因子;余因数;辅助因素;辅因(子)
cofactor matrix 余因数矩阵
coffee grounds 咖啡渣
coffee house 咖啡馆
coffee maker 咖啡壶
coffee palace 咖啡馆
coffee roasting 咖啡焙烘
coffee roasting industry 咖啡焙炒工业
coffee room 咖啡室
coffee shop 小餐馆;咖啡店;咖啡室
coffee table 咖啡茶几
coffee-tavern 禁卖酒旅馆;咖啡小餐室;小饭店
coffee urn 咖啡壶
coffee waste 咖啡废料
coffer 藻井;顶棚嵌板;沉箱;浮船坞;保险箱;潜水箱;平顶镶板;围堰
coffer ceiling 藻井平顶;井格天花板
cofferdam 隔离舱;防水堰;潜水箱;围堰(墙)
cofferdam bracing 围堰支撑
cofferdam construction 围堰修建;围堰施工
cofferdam construction method 围堰施工方法

cofferdam construction sequence 围堰施工程序
cofferdam demolition 围堰拆除
cofferdam foundation 围堰基底;围堰基础
cofferdam foundation cleanup 围堰基础清理
cofferdam method 围堰法
cofferdamming 修筑围堰;筑围堰
cofferdam outer rockfill dike 围堰迎水面堆石体;围堰迎水面堆石堤
cofferdam piling 围堰板桩
cofferdam quarry 围堰石料场
cofferdam removal 围堰拆除
cofferdam riprap protection 围堰抛石护坡
cofferdam sheeting 围堰板桩墙
cofferdam stockpile 围堰储料场
cofferdam stockyard 围堰储料场
cofferdam support 围堰支撑
coffer design 围堰设计;沉箱设计
coffered block 镶入砌体;镶入砌块
coffered bridge floor 格式桥面
coffered ceiling 井格天花板;花格平顶【建】;藻井天花板;藻井平顶;格子顶棚;格子顶棚;方格天花板
coffered floor 格式楼面;格式桥面
coffered foundation 箱式基础;箱式基础;沉箱基础;围堰底座
coffered light panel 吸顶灯
coffered wall 围堰墙;镶嵌砌块墙;方格墙
coffer foundation 箱形基础
coffering 藻井天花板;藻井平顶;方格天花板;方格顶棚;作围堰;衬砌空心筒柱(通常水压下不透水)
coffer lock 箱形船闸
coffers 井式平顶;井式顶棚;金库;国库
coffer-wall 围(堰)墙
coffer work 砌片石墙;沉箱工程
Coffey-still 科菲蒸馏器
cofficient of variation 变热系数
coffin 装运罐;棺材;屏蔽罐
coffin annealing (板材的)装箱退火
coffin chamber 墓室
coffin corner 危角
coffin crane 匣形起重机
coffin hoist 匣形起重机;匣升降机;小型链式起重机;小链轮起重机
coffinite 铀石;水硅铀矿
coffin plate 轴包板
coffret 传输接口
cofinal 共尾
cofinance 共同筹资
cofinancing 共筹资金;联合融资;联合贷款
cofiring 共加热;共焙烧
coflocculant aid 共絮凝助剂
coforest 余树林
co-founder 共同创立者
cofree 余自由
CO from aircraft 飞机排放的一氧化碳
CO from motor vehicles 汽车排放的一氧化碳
co-fumed leaded zinc oxide 炉法含铅氧化锌
cofunction 余函数
cog 轮牙;木垛;雄榫;小渔船;小舢板;小快艇;初轧;嵌齿;榫头
cog belt 楔形齿皮带;齿形带
cogelled 共凝胶的
cogeneration 联合发电;利用工业余热发电;余热发电;废热发电;热电联产
cogeneration facility 共发电设施
cogeneration plant 共发电设施
cogeneration power plant 热电厂
cogeneration turbine 供热式汽轮机
cogeoid 补偿大地水准面
cogeoid of the condensation reduction 压缩改正的调整大地水准面
coggea 榫头连接;嵌齿连接
cog gear 嵌齿轮
cogged belt 三角皮带
cogged bit 冲击式凿岩机;齿形钻头
cogged bloom 初轧方坯
cogged flywheel 有齿飞轮
cogged ingot 初轧坯;初轧钢锭
cogged joint 舌榫接合;对开十字接头;雄榫接合;插榫接合;榫接合;榫接;榫齿接合
cogged rail 齿轨
cogging 开坯;雄榫装入;雄榫嵌装;插榫接合;切削齿
cogging ca(u)lking 开跨接

cogging down 开坯
cogging-down pass 开坯孔型
cogging-down roll 开坯机轧辊
cogging effect 嵌齿效应
cogging engine 开坯原动机
cogging hammer 开坯锻锤
cogging joint 齿节;榫接;榫(齿)接(合);齿轮接合
cogging mill 开坯机;初轧机;齿轮扎钢机
cogging roll 粗轧辊
cogging shears 初轧坯剪切机
cognac oil 庚酸乙酯
cognate ejecta 同源喷出物
cognate inclusion 同源包体
cognate xenolith 同源捕虏体
cognitive limits 认识上的制约
cognitive mapping 认知制图
cognitive model 认识模型
cognitive process 认识过程
cognitive science 理性科学
cogongrass 茅草;白茅
cog railroad 齿轨铁路;齿轨铁道
cog railway 齿轨铁路;齿轨铁道;绳索铁道;缆车道
cographical matroid 余图拟阵
cogs 枕木块
cog scarf joint 雄榫(斜口)接合
cog timbering 垛式支架
cog-tooth 轮齿
co-guarantee 共同保证人
co-guarantor 共同担保人
cogwheel 钝齿轮;嵌齿轮
cog-wheel casing 嵌齿轮套
cog-wheel coupling 齿形联轴器;离合联轴节
cog-wheel gearing 齿轮传动装置
cog-wheel grease 嵌齿轮润滑脂
cogwheel ore 车轮矿
cogwheel phenomenon 齿轮现象
cog-wheel pump 齿轮泵
cogwheel rigidity 齿轮状强直
co-hade 断层倾角;补伸角【地】
cohenite 陨碳铁
cohere 相干;连贯
cohered video 相关视频信号
coherence 连贯(性);凝聚性;凝聚力;黏聚性;黏结(性);内聚力;相干性;咬合;共格
coherence array 相干组合
coherence effect 相干效应
coherence emphasis 相干加强
coherence filtering 相干滤波
coherence function 相干函数
coherence length 黏着长度
coherence matrix 相干矩阵
coherence modulation 相干调制
coherence noise 相干噪声
coherence noise filtering 相干噪声滤波
coherence of image formation 影像相干性
coherence of light wave 光波相干性
coherence range 相干范围
coherence technique 相干技术
coherence time 相干时间
coherency 连贯性;凝聚性;相干性;共格性
coherency function 凝聚函数
coherency strain 相干应变
coherent 相关的;相干;相参;共格的
coherent aperture 合成孔径
coherent background 相干背景
coherent beam 相干光束
coherent-brittle stage 脆性粘结阶段
coherent bundle 相关光纤束
coherent carrier 相干载波
coherent carrier system 相干载波系统
coherent coating 附着涂层;附着包膜层
coherent condition 聚集状态;黏附状态
coherent degree 相干度
coherent detection 相干检波
coherent detector 相干检波器
coherent echo 相干回波
coherent emission 相干发射
coherent enhancement 相干加强
coherent enhancement section 相干加强剖面
coherent fiber bundle 传像束
coherent flexible material 黏结柔性材料
coherent flow 黏流
coherent frequency 相干频率
coherent holography 相干光全息术

coherent illumination 相干照明
coherent image 相干成像
coherent imaging system 相干成像系统
coherent-impulse method 相干脉冲法
coherent infrared radar 红外（线）激光雷达；相干红外雷达
coherent interrupted waves 相干断续波
coherent length 相干长度
coherent light 相干光；相参光
coherent light communication 相干光通信
coherent light radar 相干光雷达
coherent light source 相干光源
coherent limit 黏着界限
coherently scattering region 相干散射区
coherent material 黏结剂
coherent microdensitometer 相干式微密度计
coherent monochromatic wave 相干单色波
coherent operation 相干运转
coherent optical information processing 相干光信息处理
coherent optical locator 相干光定位器
coherent optical mapping 相干光学测图
coherent optical parallax processing 相干光平行处理
coherent optical processing 相干光处理
coherent optical processing system 相干光学处理系统
coherent optical radar 相干光雷达
coherent optical radar system 相干光雷达系统
coherent optics 相干光学
coherent oscillator 相干震荡器；相干发生器
coherent phase shift keying 相干相移键控制
coherent precipitate 联结沉淀
coherent projection 相干投影
coherent-pulse radar 相关脉冲雷达；相干脉冲雷达
coherent pulses 相关脉冲
coherent radar 相干雷达
coherent radar photograph 相干雷达像片
coherent radiation 相干辐射
coherent Raman spectrometry 相干喇曼光谱法
coherent range 相干范围
coherent receiver 相干接收器
coherent reception 相干接收
coherent reference 相干参考
coherent reflecting 相干反射
coherent rock 黏合岩
coherent rotation 一致转动
coherent scattering 相干散射；相参散射
coherent secondary wave 相干次波
coherent separation 黏附选矿法
coherent side looking radar 相干侧视雷达
coherent soil 黏结性土壤
coherent source 相干（光）源
coherent spectrometer 相干分光计
coherent spectrum 相干光谱
coherent structure 内聚结构；相干结构
coherent system 相干系统；单调关联系统
coherent system of units 一贯单位制
coherent transformation function 相干传递函数
coherent transmitter 相干发射机
coherent transponder 相干转发器
coherent twin boundary 相干孪晶界面
coherent units 相关单位
coherent video 相干影像；相干视频信号
coherent wave 相干波；相参波
coherent wave train 相干波列
coherer detector 粉末检波器
coherer effect 检波器效应
cohesible 能黏聚的；能黏结的
cohesiometer 黏聚力仪；黏聚力计；黏度计；黏力计；附着仪
cohesion 结合（力）；凝聚性；黏聚性；黏聚力；黏附；内聚性；内聚力；黏结（力）
cohesional coefficient 黏聚系数；内聚力系数
cohesional resistance 凝聚阻力；黏聚抗力
cohesional work 内聚功
cohesion failure 内聚破坏
cohesion-friction-strain test 内聚力—摩擦力—应变试验
cohesion height 黏着高度
cohesion intercept 黏聚力截矩
cohesionless 非黏结性的；无黏性
cohesionless material 非黏性材料；无黏性材料；无黏聚力材料；松散材料

cohesionless sediment 无黏性泥沙；无黏性沉积物
cohesionless soil 无黏性土；无黏聚力土壤
cohesionless soil sample 无黏性土样
cohesion limit 凝集极限
cohesion mechanism 内聚机制
cohesion moment 黏聚力矩
cohesion of rock 岩石的黏结
cohesion of soil 土壤黏性；土壤黏聚力；土壤黏结力；土壤黏合性；土的黏聚性；土的内聚力
cohesion pressure 内聚压力
cohesion stress 抗滑应力
cohesion test 黏聚试验
cohesion type 内聚力种类
cohesive action 黏结作用；黏聚作用
cohesive affinity 凝聚力
cohesive air filter 黏滞式空气滤池
cohesive backfill 黏性填土
cohesive bed 黏性河床；黏性土层
cohesive bed sediment 黏性河床底泥
cohesive bonding 内聚黏结
cohesive clay 黏性土层
cohesive compressible soil stratum 黏性土压缩层
cohesive concrete 黏聚混凝土；不离析混凝土
cohesive-debris-flow deposit 黏性泥石流沉积
cohesive deposit 黏性泥沙；黏性沉积物
cohesive effect 黏性效应
cohesive end 黏性末端
cohesive energy 内聚能
cohesive energy density 凝集能密度；内聚能密度
cohesive failure 黏结失效；内（黏）聚（力）破坏；接触表面黏结破坏
cohesive force 凝聚力；黏聚力；黏结力；内聚力
cohesive fracture 内聚力断裂
cohesive index 黏结性指数
cohesiveless backfill 无黏性填土
cohesiveless material 无黏性材料
cohesiveless sediment 无黏性泥沙
cohesiveless soil 无黏性土
cohesive limit 黏性限
cohesive material 黏结材料；黏性泥沙；黏聚性材料；黏结剂
cohesive matter 黏聚性物质
cohesiveness 聚结性；凝聚性；黏着性；黏结性；内聚性
cohesiveness of concrete 混凝土的黏聚性
cohesiveness test 黏结性试验
cohesive pressure 结合压力
cohesive property 凝聚性
cohesive property of soil 土的黏性
cohesive resistance 黏聚强度；黏结阻力
cohesive river bed 黏性河床
cohesive river bottom 黏性河底
cohesive sediment 黏性泥沙；黏性沉积物
cohesive soil 黏性土（壤）；黏性土壤
cohesive soil landslide 黏性土滑坡
cohesive stratum 黏性土层
cohesive strength 黏聚力；黏结强度；内聚强度
cohesive ultimate 黏性限
cohesive work 内聚功
cohistory register 并列经历寄存器
Cohn fractionation 科恩分部分离法
coho 相干震荡器；相参振荡器
cohobation 回流蒸馏
cohomology theory 上同调理论
cohort 人群；群体；队列
cohort analysis 队列分析；群组分析
cohort coding system 分类编码系统
cohort mortality 群组死亡率
cohort observation 群组观察
cohort study 队列研究；定群研究；群组研究
cohort survival 群体幸存
cohydrol 石墨胶态溶液
cohydrolysis 共水解（作用）
coign 隅石；墙角石；角石
coign brick 外角砖；墙角砖
Coignet 凯哥涅式建筑
Coignet pile 隅桩；楔桩；凯哥涅柱
Coignet system 凯哥涅式建筑体系
coil 卷绕；卷片筒；卷material；一盘纸带；匝；绕组；屏线圈；盘绕；盘圈；盘曲管
coil adjuster 线圈调节器
coil all rope with the sun 顺盘绳法
coil and gravity starter 电流线圈重力式起动机
coil annealing furnace 带卷退火炉

coil antenna 框形天线；环形天线
coil a rope against the sun 反盘绳法
coil arrangement 线圈布置
coil assembling apparatus 下线装置；嵌线机
coil assembly 线圈组
coil axis 线圈轴线
coil banding machine 带卷捆扎机
coil base 炉台
coil boiler 盘管锅炉
coil box 线圈盒
coil brake 盘簧制动器
coil break 板卷折纹；卷裂
coil buckling 扭丝；（线材或轧件的）拧绞；边部浪（薄板缺陷）；（钢丝绳的）死扣
coil buggy 带卷自动装卸车；带卷装卸车；升降运输车
coil build-up line 带卷端头焊接作业线；拼卷作业线
coil bukling 拧绞
coil bumper 带卷防撞器
coil capacity 线圈电容
coil car 卷材移动台车
coil cargo 盘卷货
coil clamp 钢卷夹钳
coil clutch 螺旋弹簧离合器
coil-coated metal 涂金属卷板
coil coater 卷材涂装机
coil coating 盘管涂刷；盘管喷涂；卷材涂料；卷材连续涂覆；线圈涂料
coil coating paint systems 卷材涂料体系
coil column 盘旋柱
coil comparator 线圈测试器；线圈比较器
coil condenser 盘管冷凝器；蛇管冷凝器
coil constant 线圈质量因数；线圈常数
coil conveyer 带卷运输机；卷材输送机
coil cooler 盘管冷却器
coil cord 弹簧型软线
coil covering 线圈绝缘层；线圈包扎层
coil cradle 开卷机的带券座；卷料架；卷料进给装置
coil(curl) 线圈
coil current 线圈电流
coil cut-up line 带卷横剪作业线
coil deck 盘管盖板
coilded capped fuse 装有雷管的导火线
coil delivery ramp 带卷输出台
coil depth 盘管深度
coil dialyzer 管型透析器
coil diameter detector 带卷直径检测器
coil down 盘卷
coil downending machine 翻卷机
coil drag 螺旋捞矛
coiled alumin(i)um alloy sheet 铝合金卷片
coiled bar 成盘钢筋；成盘条钢
coiled coaxial electric(al) heater 盘旋共轴电热器
coiled coil 螺线形灯丝；复绕灯丝
coiled-coil filament 螺线式灯丝；双螺旋（线）灯丝
coiled-coil heater 双螺线加热器；复绕加热器；盘香状热子
coiled-coil lamp 螺线式灯丝灯泡
coiled condenser 盘管冷凝器；旋管冷却器；旋管冷凝器；蛇形冷凝管
coiled cooling pipe 双管冷却管（道）；复绕冷却管（道）
coiled expansion loop 环形膨胀节
coiled expansion pipe 盘旋膨胀管
coil edge 卷材边缘
coiled hemp rope 成卷麻绳
coiled key 旋簧键
coiled lamp 卷丝灯
coiled material 卷材
coiled oval 椭圆形旋管；椭圆形蛇形管
coiled pipe 蛇管；盘管
coiled pipe cooler 盘管冷却器
coiled pure aluminium sheet 纯铝卷片
coiled radiator 螺旋管（式）散热器；盘管散热器
coiled roofing sheet 屋面卷板；屋面卷材
coiled sheet 卷板
coiled spring 盘簧；螺旋弹簧
coiled steel 成卷带钢
coiled stock 带卷
coiled waveguide 卷绕式波导管；旋管式波导管；涡旋式波导管
coil-ejector 钢皮卷推出器
coil enamel 线圈瓷漆；绕组瓷漆
coil end 卷材端头

coil-end leakage 线圈端漏磁
coil end leakage reactance 线圈端漏电抗
coil entry ramp 带卷输入储存台
coiler 卷机机;旋管;缠绕机;缠卷装置;盘管机
coiler for trimmings 废边卷取机
coiler furnace 卷取炉
coiler gear 辊式卷取机
coiler kickoff 拨卷机;推卷机
coiler mandrel 卷取机卷筒;卷取机卷筒
coiler pulpit 卷取机操纵台
coiler roll 成型辊
coiler stand 卷取机支座
coiler tension rolling mill 张力冷轧带钢机
coiler wheel 圈条齿轮
coil evapo(u)rator 旋管蒸发器;蛇管蒸发器
coil face area 盘管迎风面积
coil feed line 开卷线
coil filter 螺旋过滤器;旋管过滤器;盘管过滤器;蛇管过滤器
coil for heating 加热螺管
coil form 线圈管
coil for preheating air 预热空气用蛇管
coil friction clutch 螺旋弹簧摩擦离合器
coil galvanometer 线圈电流计
coil/globule transition 聚合物链构形的圈/球转变
coil grab 钢卷吊钩;圈形物件抓具
coil grading 线圈分段
coil grinding 卷带磨光
coil group 线圈组
coil handling apparatus 送卷装置
coil handling crane 卷材装卸起重机
coil heater 盘管式加热器
coil heat exchanger 螺旋换热器;盘管的热器;蛇管换热器
coil heating 盘管加热;盘管供热;盘管供暖
coil height 盘管高度
coil holder 带线圈支持器
coil impregnating varnish 线圈浸渍清漆;绕组浸渍漆
coil in 进线;盘管进入接头
coil in box 箱内旋管;盘香管
coil in box cooler 箱内旋管式冷却器
coil inductance 线圈电感
coiling 螺旋;成卷;绕制线圈;绕圈;绕线;绕成螺旋;盘筑泥条;盘泥条成型;上卷筒
coiling apparatus 卷取装置
coiling direction 旋壳旋转方向
coiling geometry 旋壳几何形态
coiling length of drum 鼓轮绕索长度
coiling machine 卷料机;绕机机;盘绳机;弹簧机;盘管机;弹簧机
coiling partitions 卷绕隔墙(可卷绕箱中)
coiling reel 带材卷取机
coiling soup 卷花
coiling speed 卷取速度
coil inserting apparatus 下线装置;嵌线机
coil insulation 线圈绝缘
coil insulator tester 线圈绝缘测试器
coil jack 带卷升降车
coil jacket 蛇管夹套
coil joint 线圈接头
coil lashing 线圈端部绑扎
coil length 带卷长度;盘管长度
coil lift-and-turn unit 带卷升降回转台
coil lifter 带卷升降台
coil loaded cable 加感电缆
coil loaded circuit 加感电路
coil loading 加感
coillon 隅石;楔形石
coil magnetizing method 磁化绕组法
coil magnetometer 线圈磁力计;感应式磁强计
coil method 盘绕法
coil neutralization 线圈中和;感应中和
coil of cable 电缆卷
coil of strip 带卷
coil of wire 线匝;线卷
coilonychia 匙状形甲
coil opening machine 松卷机
coil out 盘管引出接头
coil pack 线圈组件;盘管组件
coil packing 盘(条)形填料
coil painting 卷材涂装
coil panel 线圈盘
coil pick-up 感应传感器

coil pipe 旋管;盘(曲)管;蛇管
coil pitch 线圈节距;盘管节距
coil planet centrifuge 螺管行星式离心机
coil polymer 螺旋状聚合物
coil positioner 卷材固定装置
coil preparation unit 带卷预整装置
coil pulser 线圈脉冲发生器
coil ramp 开卷机装料台
coil rod 成盘钢筋;盘条
coil roller 螺旋形镇压器
coil rolling 成卷轧制
coil rope 盘卷绳索
coil rotating rig 带卷回转装置
coil scales 卷材磅秤
coil segment 线圈段
coil serving 被覆物
coil sheet 成卷薄板;薄板卷
coil-side 线圈边
coil-skin pass mill 卷材平整机
coil sleeve 绕组套筒
coil space factor 线圈占空系数
coil span 线圈节距;线圈间距
coil span factor 线圈节距系数
coil spreading machine 线圈拉型机
coil spring 螺旋(形)弹簧;螺旋弹簧;卷圈弹簧;盘簧
coil spring dampener 弹簧缓冲器
coil spring drain 圆弹簧排水管
coil-spring retainer 盘簧底圈;盘簧承座
coil-spring shank 带弹簧螺圈的弹性柄
coil spring shock absorber 螺旋减震器
coil-spring standard 圈簧支柱
coil spring switch 螺簧接线器
coil-spring type vibration isolator 盘簧隔振器
coil steel 盘圆钢
coil stock 卷材;卷料
coil stocking bay 带卷储[贮]存跨
coil straightener 卷材矫直机
coil strip 成卷带材
coil stripper 卸线钩;剥线圈绝缘器
coil tap 线圈抽头;盘管抽头
coil temper-mill 卷材平整机
coil tie 线圈连接(线)
coil-to-coil breakdown 线圈间击穿
coil transfer buggy 带卷运送(装卸)车
coil transfer car 带卷移送车
coil transfer lifter 带卷移送升降机
coil transformer 线圈变压器
coil tube heater 蛇形管加热器
coil tubing unite 可卷油管车
coil type accumulator 旋管式储液器
coil type auger 弹簧式推动螺旋
coil type preheater 螺管式预热器
coil type vacuum filter 旋管真空滤器
coil upender 翻卷机
coil upender attachment 带卷翻转装置
coil varnish 线圈清漆;绕组清漆
coil weighing machine 带卷秤
coil weld 卷板对接焊;板卷焊
coil width 线圈宽度
coil winder 卷丝机;线材拉拔机;绕线机
coil winding 线圈绕组;圈线
coil winding machine 打盘机;绕线圈机
coil with sliding contact 滑触线圈
coil wound rotor 线绕转子
coil wrench mark 板卷折印
coil zipper 环扣拉链
coimage 余像
coin 压花铸币;钱币
coinage beneath face value 不足价货币
coinage gold 金币合金
coinage office 宝源局
coin(box) telephone 投币式公用电话
coin certificate 兑换券
coincide 重合;符合
coincidence 重合;符合;吻合
coincidence adder 重合加法器;符合加法器
coincidence adjustment 焦点距离调整;符合调整
coincidence amplifier 重合放大器
coincidence analyzer 重合分析仪;重合分析器;符合分析器
coincidence apparatus 符合器
coincidence arrangement 重合装置
coincidence array 重合阵列;重合电路列
coincidence-blocking oscillator 重合间歇振荡器

coincidence boundary 共格边界
coincidence circuit 重合电路
coincidence correction 重合校正;符合校正;同频校正;死时间改正
coincidence counter 重合计数器;重合电路;符合计数器;同步计数器
coincidence counting 符合计数
coincidence counting technique 重合计数技术
coincidence decoding 重合译码;符合译码
coincidence detector 重合检测器
coincidence effect 叠合效应;重合效应;吻合效应
coincidence error 重合性误差;重合误差
coincidence factor 同时使用系数;同时工作系数
coincidence focometer 重合式焦距仪
coincidence formula 重合公式;符合公式
coincidence frequency 相干频率;重合频率;吻合频率
coincidence gate 一致门;重合电路;符合门
coincidence indicator 一致指标
coincidence interval 符合间隔
coincidence lattice 符合点阵
coincidence level 符合水准仪;符合水准器
coincidence loss 符合误差
coincidence magnet 符合磁铁
coincidence method 重合法;符合法
coincidence method of measurement 符合测量法
coincidence micrometer 重合测微器
coincidence microscope 符合显微镜
coincidence number 重合数
coincidence of a correspondence 对应的叠合
coincidence of bubble 气泡重合;气泡相合
coincidence of flood peak 洪峰遭遇
coincidence of need 需要的一致
coincidence optimeter 重合光电比色计
coincidence pattern 重合图
coincidence pendulum 重合摆
coincidence prism 符合棱镜
coincidence proportional counter 比例重合计数器;比例符合计算器
coincidence range finder 复合焦点测距仪;符合测距仪
coincidence rate 合格率
coincidence-reading device 重合读数装置
coincidence resolving power 符合分辨能力
coincidence scaler 符合定标器
coincidence scintillation spectrometer 符合闪烁分光仪
coincidence selection 重合选择性
coincidence selection system 重合法选择系统
coincidence sensitivity distribution 符合灵敏度分布
coincidence sensor 重合检测器;符合传感器
coincidence setting 重合调节;符合对准
coincidence sorter 符合分离器
coincidence spectrometer 符合光谱仪
coincidence spectroscopy 重合式光谱术;符合光谱学
coincidence spectrum 重合光谱
coincidence system 重合方式;符合光学系统
coincidence transponder 重合脉冲转发器
coincidence tube 符合管
coincidence tuning 重合调谐
coincidence-type adder 重合型加法器;重合式加法器
coincidence unit 重合单元;重合单位;符合单元
coincidental correlation 叠合相关
coincidental indicator 同步指标
coincidental section 重复剖面
coincidental starter 风门起动器
coincidental starting 风门起动
coincident basin 重合盆地
coincident configuration 重合形
coincident current memory 电流重合法存储器
coincident-current selection 电流重合(选择)法;重合电流选择
coincident-current storage 电流重合法存储器
coincident demand 同时需气量
coincident demand effieiency 同时最大需用效率
coincident demand power 同时最大需用功率
coincident draft 火灾时同时取水
coincident economic indicators 同步经济指示数字
coincident estimation 重合估计
coincident factor 同时率
coincident-flux magnetic storage 磁通重合磁存

储器
coincident indicator 一致指标;同步指标
coincident indicators 重合经济指标
coincident-range-zone 一致延限带
coincident reaction 同时反应
coincident spectral density 共谱密度
coincident wave 重复波
coincide selection 符合选择
co-incineration 共焚化
coin circulation 流通硬币
coin collector 集币箱
coined gasket seal 矩形垫圈密封
coined part 精压零件
co-information 共信息
coin gold 币金
co-inhibitor 阻抑剂
coinidence unit 重合装置
coining 精压;印压;压花;冲制;模压
coining compact 精压坯块
coining die 压印模
coining dimpling die 内孔翻边镦粗模
coining machine 冲压机
coining mill 压花机
coining press 压印压力机
coin-in-the slot machine 自动售货机
coinjection type structural foam machine 共注型结构泡沫成型机
coin marking 金属(硬币)划痕
coin money 获暴利
coin-operated dryer[drier] 投硬币启动的干燥机
coin refund 退币口
coin scratch test 硬币划痕试验
coin silver 银币合金
coin-silver ring 银质接触环
coin slot 投币口;投币孔
co-insurance 共保;联合保险;共同担保;共同保险
co-insurance clause 共同保险条款
co-insurance clause in property insurance 财产保险中的共同保险条款
co-insured 联名被保险
coinsurer 共同保险人
cointegrate 共合体
coin telephone set 投币电话机
coin telephone station 投币电话站
coin test 硬币试验
coin-up 自动洗衣机(投币机器自行开动)
coin validator 硬币识别器
co-invariant 协不变量
coion 伴离子;同离子
coir board lath 棕丝板条
coir building mat 低密度建筑用棕丝毯
coir door mat 棕丝门毯
coir insulating sheet(ing) 棕丝隔热板
coir insulation strip 棕丝隔热板条
coir mat 棕席;棕垫
coir material 棕丝材料
coir rope 椰棕绳;棕索;棕绳;棕缆(绳)
coir-type gypsum plank 棕丝型石膏板
coir wallboard 棕丝墙板
coke 焦炭
cokeability 结焦性;焦化性;粘结性;成焦性
coke adsorption 焦炭吸附
coke and gas waste 炼焦煤气废水
coke bed 焦床;底焦
coke bin 焦炭斗
coke blacking 焦炭粉;涂焦炭粉
coke boiler 焦炭锅炉
coke breaker 轧焦机;碎焦机
coke breeze 粉焦灰;焦炭屑
coke breeze concrete 焦渣混凝土;焦炭屑混凝土;煤渣混凝土;粉煤灰混凝土
coke breeze concrete brick 煤渣混凝土砖
coke briquette 团块焦;炭砖
coke burden(ing) 焦炭配料
coke burner 炼焦炉
coke button 焦块
coke byproduct 炼焦副产品
coke charge 层焦;批焦
coke colo(u)r 焦块色泽
coke crack edge 焦块绽边
coke discharging ram 卸焦炭推杆
coke-drawing machine 耙焦机
coke drum 焦炭鼓
coke dust 焦炭粉

coke expenses ratio 焦比
coke extractor 焦炭提取机
coke filter 焦炭过滤器;填焦过滤器
coke fired furnace 焦炭炉
coke fissure 焦块缝隙
coke fork 焦炭叉斗
coke-forming hydrocarbon 成焦烃
coke furnace 炼焦炉
coke fused condition 焦块熔合情况
coke grate 壁炉(炉)栅
coke guide 拦焦车;导焦槽
coke-hole 坩埚炉
coke hopper 储[贮]焦斗
coke industry 焦炭工业
cokeite 天然焦
coke knocker 除焦机
coke-like sludge 类焦炭污泥;焦状污泥
coke making 炼焦
coke mill 焦炭球磨机
coke moisture meter for blast furnace 高炉焦炭水分计
coke number 焦值
coke-oven 炼焦炉
coke-oven acid firebrick 炼焦炉酸性耐火砖
coke-oven battery 炼焦炉组
coke-oven brick 焦炉砖
coke-oven brown coal tar 焦炉棕色煤焦油
coke-oven coal tar 煤焦沥青;焦炉煤焦油
coke-oven effluent 焦炉废水
coke-oven emission 焦炉排气
coke-oven gan(n)ister brick 炼焦炉耐火砖
coke-oven gas 焦炉(煤)气;炼焦煤气
coke-oven gas booster 焦炉气升压机
coke-oven gas compressor 焦炉煤气压缩机
coke-oven gas engine 焦炉煤气发动机
coke-oven industry 炼焦工业
coke-oven plant 焦化厂;炼焦厂
coke-oven plant for town gas production 生产城市煤气的炼焦厂
coke-oven regenerator 焦炉热再生器
coke-oven silica brick 炼焦炉硅砖
coke-oven system 焦炉系统
coke-oven tar 焦炉焦油;焦炉柏油
coke-oven wastewater 焦炉废水
coke-over gas engine 焦炭煤气发动机
coke(per) charge 焦批
coke percolator 焦炭渗滤器
coke plant waste 焦炭厂废水
coke plant wastewater 炼焦厂废水
coke plant wastewater treatment system 炼焦厂废水处理系统
coke plate 镀锡钢板
coke pore 焦块孔隙
coke powder 焦粉
coke producer 炼焦煤气发生炉
coke pusher 推焦机
coke pusher shoe 推焦杆头
coke quenching 淬焦
coke quenching car 淬焦车
coke quenching effluent 熄焦废水
coke quenching machine 熄焦机
coke quenching water 熄焦水
coker 炼焦设备;焦化设备
coke ratio 焦比
coke residue 焦炭渣
coker gasoline 焦化汽油
cokery 炼焦厂
cokes 薄锡层镀锡薄钢板
coke scrubber 焦炭洗涤器;填焦洗涤器
coke slurry 焦泥浆
coke split 隔离焦
coke spongy body 焦块海绵体
coke strength index 焦炭强度指数
coke test 焦渣试验;成焦试验
coke tinplate 薄锡层镀锡薄钢板
coke tray aerator 焦(炭)盘(式)曝气器;浅盘型焦炭曝气器
coke velocity 焦炭燃烧速度
coke-washing wastewater 洗焦废水
coke yield 焦炭产率
coking 炼焦;结渣性;成焦
coking capacity 结焦性
coking chamber 焦化室
coking coal 炼焦煤;焦(性)煤;成焦煤

coking industry 炼焦工业
coking plant 炼焦厂;焦化厂
coking plant waste 焦化厂废物
coking plant wastewater 炼焦厂废水
coking process 焦化过程
coking property 成焦性
coking refinery 炼焦厂
coking residue 炼焦残渣
coking steam coal 结焦锅炉用煤
coking stoker 焦化炉排;焦化加煤机
coking tar 炼焦焦油;高温焦油
coking tendency 结焦倾向
coking unit 焦化装置
coking value 焦化值
col 鞍形区;鞍形低压;鞍形;气压谷;山峡;山口;山脊口;山坳
colalloy 考拉洛铝镁合金
colamine 胆胺
colander 滤器
colarim (古典圆柱的)柱颈花边饰
colas 沥青乳胶;沥青乳化液;沥青乳(浊)液;沥青浮化液
colasmix 沥青砂石混合物;沥青铺料混合物;沥青和铺路材料混合物
colateral dipole 并列偶极子
colation 过滤
colatitude 余纬(度)
colature 滤液;粗滤产物
Colburn method 科尔伯恩法;玻璃板制造法
Colburn process 平拉法(指玻璃)
Colburn sheet process 科尔伯恩法
Colby's bars 考尔巴杆尺
colcothar 煅烧铁红;铁丹;褐红色铁氧化物
colcrete 胶体混凝土;预填骨料(灌浆)混凝土
colcrete constructed in sandwich process 分层预置骨料灌浆混凝土
colcrete method 预填骨料灌浆法
cold 冷寒色;低色度的;不通电的;无光(彩)的
cold acclimatization 冷适应;寒冷气候适应
cold adaptation 寒冷适应
cold adhesion method 冷黏法
cold adhesion test 低温黏结试验
cold adhesive 冷黏结剂
cold adhesive composition 冷用黏结剂组成
cold adhesive compound 冷用黏结剂复合物
cold advection 冷平流
cold agglutination 冷凝集反应
cold agglutination phenomenon 冷凝集反应现象
cold agglutination test 冷凝集试验
cold agglutinin 冷凝集素
cold agglutinin titer 冷凝集素效价
cold aggregate 冷集料;冷骨料
cold aggregate conveyer 冷料运输机
cold aggregate feeding unit 冷集料供给装置;冷骨料供给装置
cold aggregate feed system 冷集料进料系统;冷骨料进料系统
cold air 冷(空)气
cold air blast 冷风
cold air blower 冷空气吹送机
cold air circulating system 冷风循环装置;冷风循环系统;冷风循环方式
cold air curtain door 冷空气屏蔽门
cold air damper 冷空气调节器;冷风风门
cold air duct 冷空气管道
cold air flow 冷气流
cold air inlet duct 冷空气入口管
cold air intake 冷空气吸入口
cold air intrusion 冷空气侵入
cold air machine 冷空气机;冷风机;空气制冷机
cold air mass 冷气团
cold air pipe 冷风管;冷气管
cold air refrigerating machine 冷空气冷冻机;空气式冷冻机
cold air region 冷气团区
cold air return 冷(空)气回流
cold air sink 冷气穴
cold alkali purification 冷碱净化
cold allergy 寒冷变应性
cold and hot brittleness 冷脆热性
cold anticyclone 冷高压;冷性反气旋
cold appearance 表面冷斑
cold application 冷施工;冷灌筑;冷用;冷铺
cold application asbestos felt 冷用石棉油毡

cold application composition 冷用填缝合成物；冷用密封合成物
cold application creosote 冷用杂酚油；常温液态杂酚油
cold application sealer 冷用填缝材料；冷用密封材料
cold-application tar 冷铺柏油；冷用焦油沥青
cold applied 冷用的
cold-applied coall-tar coating 室温施工煤焦沥青涂料
cold-applied coating 冷涂（涂料）；室温施工涂料
cold applied malthoid 冷贴油毡
cold applied polyethylene sheet 冷铺（聚）乙烯薄膜
cold applied tape 冷用黏带
cold asphalt 冷（铺地）沥青
cold asphalt-base adhesive 冷沥青基黏结剂
cold asphalt concrete 冷沥青混凝土
cold asphalt emulsion 冷用沥青乳液
cold asphaltic-bitumen emulsion 冷用沥青乳液
cold asphaltic-bitumen mortar 冷用拉砂浆
cold asphaltic concrete pavement 冷铺地沥青混凝土路面
cold asphalt macadam 冷铺地沥青碎石
cold asphalt mix(ture) 冷拌沥青混合料
cold atmospheric leaching 常温常压浸出
cold atomic absorption method 冷原子吸收法
cold atomic absorption spectrophotometry 冷原子吸收分光光度法
cold atomic fluorescence spectrometry mercury analyzer 冷原子荧光测录仪
cold atomic fluorometric method 冷原子荧光法
cold-banded steel pipe 冷箍钢管
cold bath 冷水浴
cold bay 低温段
cold bed 冷床
cold belt 冷带；寒带
cold belt environment 寒带环境
cold bend(ing) 冷弯（曲）；冷矫直；冷弯弯头
cold bend(ing) grade 冷弯度
cold bend(ing) test 冷弯（曲）试验
cold bend machine 冷弯机
cold bitumen 冷铺沥青
cold bitumen-based adhesive 冷沥青基黏结剂
cold bitumen emulsion 沥青乳(浊)液；乳化沥青
cold bitumen grouting method 冷沥青灌浆法
cold bitumen mortar 冷沥青砂浆
cold bituminous road 冷铺沥青路
cold blast 冷鼓风；冷吹（风）
cold blast air 冷风
cold blast cupola 冷风化铁炉
cold blast engine 冷风机
cold blast main 冷风总管
cold blast pig iron 冷风生铁
cold blast slide valve 冷风滑动阀
cold blending 冷混；冷拌和
cold blocking 冷压印
cold blooded animal 冷血动物
cold blowing 冷吹
cold boiler 冷却沸腾器；真空蒸发器
cold bonding 冷黏结
cold bonding agent 冷黏结剂
cold bonding composition 冷黏结料
cold bonding compound 冷黏结复合物
cold bonding medium 冷黏结介质
cold boot 冷启动
cold bootstrap 冷自举；冷引导
cold boundary temperature 冷却面温度；冷却边界温度
cold box 低温试验箱
cold box core 冷芯盒砂芯
cold box core blower 冷芯盒射芯机
cold box core making machine 冷芯盒制芯机
cold box process 冷芯盒法
cold break 冷却残渣；冷淀物
cold breakdown 冷滚
cold breaker 土块击碎器
cold bridge 冷桥；防寒冷桥
cold bridge effect 冷桥效果
cold brittleness 冷脆性；低温脆性
cold burden 冷料
cold butt-type 冷对接式
cold calender 冷轧机
cold can 冷管壳

cold cap 寒带
cold capacity 冷态容量
cold cast 冷型浇注
cold casualty 冷冻伤亡
cold catch pot 低温载液罐
cold-application tar 冷铺柏油；冷用焦油沥青
cold cathode 冷阴极
cold cathode canalray tube 冷阴极极隧射线管
cold cathode counter tube 冷阴极计数管
cold-cathode counting tube 冷阴极计数管；计数放电管
cold-cathode discharge 辉光放电
cold cathode emission 冷阴极发射
cold cathode ga(u)ge 冷阴极真空计
cold cathode indicator tube 冷阴极指示管
cold cathode ionization ga(u)ge 冷阴极电离真空计
cold cathode ion pump 冷阴极离子泵
cold cathode ion source 冷阴极离子源
cold cathode lamp 冷阴极灯；荧光灯管
cold cathode magnetron ga(u)ge 冷阴极磁控管真空计
cold cathode rectifier 冷阴极整流管
cold cathode tube 冷阴极管
cold ca(u)lking 冷填缝
cold cave 冷洞
cold cellar 冷窖
cold cementing composition 冷胶凝材料
cold cementing compound 冷胶凝复合物
cold chain 冷（藏）链；冷藏环节
cold chamber 冷冻房；常温容器；低温室
cold chamber die casting 冷压铸造
cold chamber die-casting machine 冷室压铸机；冷式压铸机
cold chamber machine 冷室压铸机
cold charge 冷装（料）
cold charged ingot 冷装炉（钢）锭
cold check(ing) 冷裂纹；低温裂纹；冷爆；低温细裂
cold check resistance 抗冷裂性；抗冷脆性
cold check test 冷脆试验；低温细裂试验
cold chimney 冷烟囱
cold chisel 钢筋凿；钢筋錾；冷凿；金属削凿刀；凿子；錾子
cold circuit 迟延系统
cold circular saw 冷圆锯
cold clean criticality 冷态；冷净临界
cold clearance 冷时间间隙
cold climate 冷气候；寒冷气候
cold climate cell 常温气候元件
cold climate with moist winter 冬干寒冷气候
cold coat 冷涂层
cold coating 树脂涂料
cold coil 冷却盘管
cold coiling 冷卷
cold coining 冷精压
cold colo(u)r 冷色；寒色
cold-colo(u)r printing 瓷面印刷；玻璃印刷
cold colo(u)rs 冷色料；未烧油彩
cold compress 冷压缩
cold compression method 冷压法
cold compressive strength 常温耐压强度
cold compressor 制冷压缩机
cold condition 冷态
cold cone closure 冷锥密封结构
cold confinement 冷圈密集养牛
cold constant 冷态常数
cold container method 冷坩埚法
cold core box 冷芯盒
cold core cyclone 冷低压【气】
cold core high 冷高压
cold core low 冷低压【气】
cold crack(ing) 低温开裂；冷凝（开裂）；冷裂缝
cold crack resistance 抗低温细裂性；抗低温开裂性
cold crack stability 冷裂稳定性
cold crack temperature 冷裂温度
cold cream 冷霜
cold cross-section 冷截面
cold-crucible 水冷坩埚
cold crucible technique 冷坩埚法
cold crushing strength 冷碎强度；低温压碎强度；常温耐压强度
cold cure 冷硫化；冷处理；冷补；常温固化
cold-cured paint 冷养护涂料；冷养护漆
cold-cured resin 冷养护树脂
cold curing 冷固；环境温度固化；自干；低温固化

cold-curing material 冷养护材料
cold curing resin 冷固型树脂
cold current 寒流；寒潮
cold cut 冷溶；冷拼法；凿切割；冷作切割
cold cutter 冷切刃具；冷作业；钳工錾
cold cutting system 冷切割系统
cold cut varnish 冷溶清漆；常温溶解清漆
cold cycle 冷循环；低温循环
cold cyclone 冷气旋；冷低压【气】
cold damage 冷害
cold deformation 冷变形
cold deformation strengthening 冷变形强化
cold-deformed 冷变形的
cold depression 冷低压【气】
cold desert 冷荒漠；寒漠土壤；寒漠；寒地荒原
cold dew 寒露
cold differential test pressure 冷差试验压力
cold digestion 低温消化；不加温消化
cold dip coating 冷浸涂（料）
cold dipping 冷浸涂（料）
cold distillation 低温蒸馏
cold dome 冷堆
cold draft 冷送风；冷(吹)风
cold draw 冷抽；冷拉
cold draw equipment 冷拔设备
cold drawing 冷拉（伸）；冷拉拔；冷冲压；冷拔
cold drawing bench 冷拉床
cold drawing die 冷拉模
cold drawing lubricant 冷拔用润滑剂
cold-drawing 冷拉；冷拔
cold-drawn appearance 冷拔加工状态
cold-drawn bar 冷拉钢筋；冷拔条钢；冷拔钢筋
cold-drawn copper 冷拉铜
cold-drawn low carbon wire 冷拔低碳钢丝
cold-drawn pipe 冷拉管；冷拔管
cold-drawn round steel 冷拉圆钢
cold-drawn steel 冷拉钢；冷拔钢
cold-drawn steel bad 冷轧钢丝
cold-drawn steel bar 冷拔钢筋
cold-drawn steel pipe 冷拉钢管
cold-drawn steel tube 冷拉钢管
cold-drawn steel wire 冷拉钢丝；冷拔钢丝
cold-drawn-stress relief 冷拔应力消除
cold-drawn tube 冷拔管
cold-drawn wire 冷拔金丝；冷拉线
cold-drawn wire reinforcement 冷拉钢丝配筋
cold drink and snack counter 冷食部
cold-driven rivet 冷压铆钉；冷铆铆钉
cold drying 低温干燥
cold-dry type 干冷式
cold-dry type of refuse channel 冷干式垃圾管道；低温干式垃圾渠道
cold ductility 可冷锻性
cold effluent 冷排出物
cold elbow 冷弯弯头
cold elbowing with core 有芯冷弯管
cold elevator 冷料提升机
cold embossing 冷压（凹凸）印
cold emission 冷阴极放射；冷发射；场致发射
cold emission ion source 冷发射离子源
cold emulsifiable cleaning 室温乳化清洗
cold emulsion 冷乳化（沥青）；冷乳液
cold emulsion hand sprayer 冷乳液手提喷洒器
cold emulsion sprayer 冷黏合料喷布器
cold enamel 冷搪瓷
cold end 冷接点（热电偶）；冷端；冷电位端
cold end coating 冷端镀膜
cold end compensating unit 冷端补偿装置
cold end corrosion 冷端腐蚀；低温腐蚀
cold end corrosion and deposition 冷端腐蚀和沉积
cold-end element 低温元件
cold-end junction 冷接点（热电偶）
cold end operation 冷端作业
cold-end specks 冷端斑点（浮法玻璃缺陷）
cold endurance 耐寒性；耐寒力
cold energy 冷能
cold environment 冷环境
cold epoch 寒冷期
cold epoch 1 寒冷期一
cold expanding 冷（膨）胀；冷扩张
cold extractable copper 可冷提取的铜
cold extractable heavy metal 可冷提取重金属
cold extraction 冷提取；冷抽提

cold extraction method 冷提炼方法
cold extraction technique 冷提取技术
cold extruded 冷挤;冷拔的
cold extruded thread anchorage 轧丝锚具
cold extrusion 冷挤压;冲挤
cold extrusion of gears 齿轮的冷挤压
cold extrusion of steel 钢的冷挤压
cold face (耐火材料不面向热源的)冷面
cold fault 冷机故障
cold feed 冷料输送;冷加料
cold feed bin 冷料斗
cold feed extruder 冷进料挤出
cold feed pipe 冷水进水管
cold feed rubber machine 冷喂料橡胶挤出机
cold fine asphalt 冷铺细沥青(混合料)
cold finger 冷凝管;冷指;冷测厚规;指形冷冻器
cold finger collector 冷凝管捕集器
cold finish 冷矫直;冷(加工)精整
cold-finished bar 冷加工钢筋
cold-finished extruded shape 冷加工精制钢条
cold-finished product 冷加工产品
cold-finished shaft 冷光法制轴
cold-finished steel 冷加工精整钢;冷加工钢材
cold finisher 冷轧(机);精轧机(座)
cold flame 低温火焰
cold flat 无取暖设备公寓
cold flat rolling 冷轧扁
cold flattened steel 冷轧钢筋
cold flex 冷挠曲
cold flexibility 冷挠曲性
cold flex temperature 冷挠曲温度
cold flex tester 冷挠曲试验机
cold flexure test 冷弯试验
cold flow 室温蠕变;冷变形;冷塑加工;冷流;冷吹;低温流动
cold flow harden 冷变形硬化
cold flow test 冷流试验
cold fluid 冷却液
cold forest zone 寒带林带
cold forging 冷镦;冷锻;冷冲压
cold forging die 冷锻模
cold forging drawing 冷锻件图
cold forging machine 冷锻机
cold forging steel 冷锻钢
cold form 冷成型
cold formed product 冷成型制品
cold formed section 冷成型型材
cold-formed steel 冷弯型钢
cold formed steel shape 冷成型钢
cold-formed unit 低温成型件
cold former 冷锻机
cold forming 冷成型;冷作;冷加工;冷变形
cold fracture 冷裂
cold frame 冷床;种植玻璃房;不供暖的种植玻璃房;植物玻璃温室
cold frequency 冷态频率
cold freshwater 冷淡水
cold front 冷锋
cold front-like sea breeze 冷锋状海风
cold front rain 冷锋雨
cold front storm 冷锋(风暴)
cold front thunderstorm 冷锋雷雨
cold gagging 冷矫直;冷弯(曲)
cold galvanising 电镀
cold galvanising paint 冷镀锌涂料;冷镀锌漆;电镀锌涂料;电镀锌漆
cold galvanizing 电镀;冷镀锌;电镀锌(法)
cold galvanizing paint 冷镀锌涂料;冷镀锌漆;电镀锌涂料;电镀锌漆
cold gas efficiency 冷煤气效率
cold-gas reheater 低温级再热器
cold glacier 极地冰川
cold glaze 冷釉
cold glue 冷黏结剂;冷胶
cold gluing 冷(法)胶粘;(在室温下进行的)冷胶黏结
cold gum 低温(聚合)橡胶;低温(聚合)丁苯橡胶
cold hammer 冷锤
cold-hammered iron 冷锻铁
cold hammering 冷压;冷锻
cold hammering house 冷锻车间
cold harden(ing) 冷作硬化;冷(加工)硬化
cold hardiness 耐寒力
cold header 冷锻机

cold heading 冷镦;冷锻
cold-heading rod 冷镦条材
cold hearth melting 冷床熔炼法
cold heat 冷熔
cold heat stage 冷热台
cold high 冷高压
cold hobbing press 冷剂制模压力机
cold house 冷室
cold impedance 冷阻抗
cold infusion 冷浸剂
colding crack 冷裂纹
cold injury 冷害;冷(冻损)伤;寒冷损伤;冻伤;冻害
cold inspection 冷检验
cold insulant 冷绝缘材料
cold insulation 保冷
cold insulation mastic 冷绝缘胶合铺料;冷绝缘防潮胶合料
cold insulator 保冷材料
cold iron 冷铁
cold isostatic compaction 冷等静压成型
cold isostatic press 冷等静压
cold joint 冷(缩)缝;(新旧混凝土之间的)冷接缝;冷接;虚焊;施工缝
cold joint line 冷接缝线
cold junction 冷结;冷接合;冷端
cold junction compensation 冷端温度补偿;热电偶冷端补偿
cold junction temperature 冷端温度
cold kiln 冷窑;低温窑
cold laboratory 冷实验室(无放射性实验室)
cold-laid 冷铺的
cold-laid asphalt(ic) concrete 冷铺沥青混凝土
cold-laid asphalt surface 冷铺沥青面层
cold-laid bituminous concrete 冷铺沥青混凝土
cold-laid bituminous pavement 冷铺(地)沥青路面
cold-laid coarse tar concrete 冷铺粗煤沥青焦油混凝土
cold-laid concrete 冷混凝土
cold-laid liquid coating material 冷铺液体涂层材料
cold-laid mastic 冷铺玛琋脂
cold-laid method 冷(却)铺法
cold-laid mixture 冷铺混合料;冷铺拌和料
cold-laid moist bending 由湿冷引起的弯曲
cold-laid pavement 冷铺(沥青)路面
cold-laid pilgered pipe 冷铺皮尔格(无缝钢)管
cold-laid plant mixture 冷铺厂拌(沥青)混合料
cold lap 冷压折;冷隔(未焊);冷搭;重皮
cold lapping 未焊缝
cold laying 冷铺
cold-leach 常温浸出
cold level(l)ing 冷矫直
cold lift 冷料提升机
cold light 冷光
cold light illumination 冷光照明
cold light illuminator 冷光源
cold light lamp 冷光灯
cold light mirror 冷光镜
cold light reflector 冷光反射镜
cold-light source 冷光源
cold lime neutralization 冷石灰中和(作用)
cold lime-soda process 冷石灰苏打法
cold lime softening 冷法石灰软化
cold link 冷链接
cold liquid 冷却液
cold liquid metal 低于浇温金属液;低于浇温的金属液
cold locomotive 无火机车【铁】
cold loess 冷黄土
cold loop 冷回路
cold loss 冷损耗
cold low 冷低压【气】
cold maceration 冷浸法
cold manure 冷性肥料
cold mastic wax 冷用接蜡
cold medium 冷媒
cold melt 低温出钢
cold melt process 低温熔炼法
cold metal(drag) 粗筋
cold metal process 冷料法
cold metal saw 冷金属锯
cold-metal work 白铁工程
cold-metal worker 白铁工
cold method 冷冻法

cold-mill 冷轧机
cold milling 冷磨
cold milling equipment 冷磨设备;冷磨机具
cold milling machine 冷磨机
cold mineral spring 冷矿泉
cold mirror 冷光镜;冷反光膜
cold mirror reflector 冷镜反射器
cold-mix asphalt 冷沥青混合料;冷拌沥青
cold-mix asphalt plant 冷拌沥青设备
cold-mix bituminous macadam 冷拌沥青碎石
cold-mix bituminous macadam road 冷拌沥青碎石路
cold-mix bituminous pavement 冷拌沥青路面
cold-mix mixed asphaltic concrete 冷拌沥青混凝土
cold mixed concrete 冷拌混凝土
cold-mixed grease 低温混合润滑脂
cold mixing 冷混合;冷拌
cold mixing method 冷拌法
cold-mix plant 冷拌设备;冷拌和厂
cold-mix recycle 冷拌再生利用
cold-mix surface 冷拌(沥青)路面
cold mix(ture) 冷拌料;冷拌料
cold-mix type 冷拌冷铺
cold-mix type pavement 冷抖式路面;冷拌(沥青)路面
cold model 冷态模化
cold model test 冷态模型试验
cold-molding 造型冷压
cold mortality 冻死率
cold mo(u)ld 表面冷斑
cold mo(u)ld arc furnace 冷模电弧炉
cold mo(u)ld arc melting 水冷坩埚电弧熔炼
cold-mo(u)lded plastics 冷模塑塑料
cold-mo(u)ld furnace 自耗电极真空电弧炉
cold mo(u)lding 冷压模塑;冷塑;冷模压;冷模成型;常温压制
cold neck grease 冷轧辊颈润滑脂
coldness index 寒度指数
cold nodule 冷结节
cold nosing 盲目钻探
cold occlusion depression 冷性锢囚低压【气】
cold occlusion front 冷锢囚峰【气】
cold oil 冷却油;印油
cold oil return 冷却油回流
cold operation 冷运行;低温运行;低放射性物质的或非放射性的工作
cold operation area 冷作业区
cold pack 冷包裹法
cold painting 冷涂饰
cold patch 冷法补坑(用沥青混合料补坑);冷补
cold patching 冷法补坑;冷补工作(用沥青混合料补坑)
cold path mixture 冷补混合剂
cold path work 冷补工作(用沥青混合料补坑)
cold pat test 冷饼试验
cold penetration 冷灌;冷贯入
cold penetration bituminous 冷灌沥青
cold penetration bituminous macadam 冷灌沥青碎石(路)
cold penetration construction 冷灌沥青路面施工
cold period 冷期
cold pigiron 冷生铁
cold pile 剩余灰浆堆;冷馅
cold pinch 冷夹
cold plant-mixing 冷料厂拌的
cold plant-mix type 冷式厂拌(沥青混合料)
cold plasma technology 低温等离子技术
cold plastic anti-fouling paint 冷塑性(船底)防污漆
cold plastic deformation 冷塑性变形
cold plasticity 冷塑性
cold plastic matrix 冷塑性基料
cold plastic paint 冷塑性(船底)漆
cold plastics 低温塑料
cold plate 冷(凝)板;冷却板
coldplate-mounted 装在冷却板上的
cold plate refrigerated wagon 冷板冷藏车
cold plug 冷室火花塞
cold point 冷点
cold pole 冷极;寒极
cold polymerization 冷聚合
cold position 冷态(位置)
cold post potential 冷阴极电位
cold pour 低温流动性
cold pourable 冷可浇注性

cold poured 冷浇注的
cold poured material 冷灌材料
cold poured material for joint 接缝用的冷灌材料
cold pouring 低温浇注
cold press 冷压(机);常温压制
cold-pressed 冷压的
cold-pressed compact 冷压坯块
cold-pressed nut 冷压螺母
cold-pressed plywood 冷压胶合板
cold-pressed process 冷压方法
cold-pressed sheet steel 冷压薄钢板
cold-pressed steel 冷压钢
cold-pressed steel plate 冷压钢板
cold pressing 冷压(制);冷挤压;冷冲压
cold-pressing property 冷压性质
cold-pressing quality plate 冷压优质钢板
cold-pressing quality steel 冷弯钢
cold press method 冷压法
cold press mo(u)lding 冷压成型
cold pressor test 冷水升压试验;冷加压试验;寒冷升压试验
cold press ram 冷压冲杆
cold press resin 冷压树脂
cold pressured 冷压的
cold pressure welding 冷压焊
cold prevention transport 防寒运输
cold primer 冷底涂料
cold primer-oil 冷底子油
cold probe 低温探针
cold process 冷铺法(铺筑沥青路面);冷(加工)法
cold process asphalt 冷(浇)沥青;冷敷沥青
cold process cement 矿渣水泥
cold processing 常温加工
cold process roofing 冷法施工屋面材料;冷作业屋面做法;冷施工屋面;冷铺屋面
cold process sedimentation 冷法沉淀
cold process slag cement 冷法矿渣水泥
cold producing medium 制冷剂
cold production 冷制产品
cold proof 耐冷的
cold proof construction 防寒构造
cold proof dwelling house 防寒住宅
cold proof rabbet (门窗上的)防寒错口
cold proof rebate 防寒企口
cold pulling of steel 钢筋冷拔
cold pump 低温泵
cold punched nut 冷冲螺母
cold punching mo(u)ld 冷冲模
cold quartz mercury vapo(u)r lamp 太阳灯
cold quenching 零下淬火;冷介质淬火;冷淬;冰冷处理
cold recovery plant 冷量回收装置
cold rectifying hard-rolled 冷轧
cold reduced 冷轧(的);冷减径
cold-reduced grain-oriented silicon steel core 冷还原取向性硅钢片铁芯
cold reduced steel 马口铁卷板
cold reduced(steel) sheet 冷轧薄(钢)板
cold reducing department 冷轧车间
cold reducing machine 冷缩口机
cold reduction 冷碾压;冷(碾)压缩;减厚冷轧
cold reduction mill 冷轧机
cold reduction mill wastewater 低温还原厂厂废水
cold reduction of tubes 管材冷轧
cold reflux 冷回流
cold reheat 低温再热
cold reheater 低温级再热器
cold repair 冷态检修;冷(法)修(补)
cold requirement 冷法要求
cold reserve 冷储备
cold reserving board 保冷板
cold reserving material 保冷材料
cold reserving work 保冷工程
cold resistance 冷(态)电阻;抗冷性;抗寒性;耐溶性;耐冷性(能);耐冷度;耐寒性;耐冻性;耐冬性
cold resistant plant 抗寒植物
cold resisting property 抗冻性;耐冻性
cold restart 冷(态)再启动
cold ring 冷却环
cold rivet(t)ing 冷铆(接)
cold roll 冷轧机;冷轧辊;常温滚压
cold-rolled 冷轧的
cold-rolled band 冷轧带材
cold-rolled beam 冷轧钢梁

cold-rolled carbon spring steel 冷轧弹簧碳钢
cold-rolled carbon steel 冷轴碳钢
cold-rolled channel 冷轧槽钢
cold-rolled deformation 冷轧变形钢筋
cold-rolled deformed bar 冷轧变形钢筋
cold-rolled drawing quality sheet 冲压用优质冷轧薄板
cold-rolled drawing sheet 冲压用冷轧薄板
cold-rolled formed shape 冷弯型钢
cold-rolled forming section 冷轧型钢
cold-rolled hoop 冷轧箍钢
cold-rolled metal 冷轧金属
cold-rolled narrow strip 冷轧窄钢条
cold-rolled products 冷轧产品
cold-rolled ribbed steel bar 冷轧竹节钢筋
cold-rolled section 冷轧型材
cold-rolled sheet metal 冷轧金属(钢)板
cold-rolled sheet steel 冷轧薄钢板
cold-rolled silicon iron 冷轧硅钢板
cold-rolled steel 冷轧钢
cold-rolled steel bar 冷轧钢筋
cold-rolled steel pipe 冷轧钢管
cold-rolled(steel) plate 冷轧钢板
cold-rolled steel section 冷轧型钢
cold-rolled steel strip 冷轧带钢
cold-rolled strip 冷轧钢条
cold-rolled strip in cut lengths 定长冷扎钢条
cold-rolled strip steel 冷轧带钢
cold-rolled thread 冷轧螺纹;冷轧钢丝
cold-rolled uncoated flat product 无涂层冷扎扁钢制品;无涂层冷扎扁钢板
cold roll forming 冷轧成型;冷滚成型
cold-roll forming machine 辊式冷弯(成型)机
cold-rolling 冷轧;冷碾
cold-rolling mill 冷轧机;冷轧钢材厂;钢筋冷轧机
cold-rolling oil 冷轧润滑油
cold-rolling practice 冷轧法
cold-rolling property 冷轧性
cold-rolling reduction 冷轧(压缩)
cold-rolling wastewater 冷轧钢废水;冷轧(厂)废水
cold roof 冷房;冷屋面;冷平台式屋顶
cold room 冷房;冷藏间
cold room test 冷室内试验
cold roughing mill 冷粗轧机
cold rubber 冷(聚合)橡胶;低温橡胶;低温(硫化)橡胶;低温聚合橡胶
cold runner mo(u)ld 冷流道模
cold running 冷滑
cold rupture 冷裂
cold saw 冷锯
cold sawing machine 冷锯机
cold scrap process 冷装废钢法
cold scuffing 低温磨损
cold seal pressure vessel 冷封式高压釜
cold season 冷季;寒冷季节;寒季
cold-season plant 寒季植物
cold sector 冷区
cold separation unit 冷冻分离装置
cold set 冷作用具;冷凝固;冷变定;钢筋凿;钢筋錾;低温固化;冷塑化;冷固化;常温固化
cold set grease 冷压煮剂润滑脂
cold set ink 冷固着油墨
cold-set printing 冷凝印刷;用冷凝油墨印刷
cold-set resin 冷硬树脂;常温固化树脂
cold-setting 冷凝固;自硬化;冷硬化;冷胶合的;冷变定;常温凝固;室温硬化;冷塑化;冷固化;常温固化
cold-setting adhesive 常温固化黏合剂;冷凝黏合剂;冷凝固胶粘剂;冷固化黏合剂(20℃以下变定);冷固化胶粘剂(20℃以下变定);冷变定黏合剂;低温硬化胶着剂
cold-setting binder 自硬黏结剂
cold-setting glue 常温固化胶;冷固化黏合剂(20℃以下变定);冷固化胶粘剂(20℃以下变定)
cold-setting ink 冷凝油墨
cold-setting lacquer 常温固化(挥发性)漆
cold-setting process 自硬法
cold-setting resin 冷固化树脂;冷固型树脂
cold-setting sand 自硬砂
cold-settled cylinder oil 低温沉降汽缸油
cold settler 低温澄清器;低温澄清池
cold settling 冷淀积
cold shaping steel 冷变形钢
cold shearing 冷剪切

cold shears 冷剪
cold sheet mill 薄板冷轧机
cold sheet rolling mill 薄板冷轧机
cold shield angle 冷屏角
cold shield efficiency 冷屏效率
cold short 冷缩裂缝;冷脆的
cold short iron 冷脆钢
cold short material 冷脆材料
cold shortness 冷淬;冷脆(性);低温脆性;常温冷脆;常温脆性
cold short steel 冷脆钢
cold-shot 冷珠;铁豆
cold shower 冷水淋浴
cold shut 冷结;冷接;(铸件的)冷隔;冷疤
cold side 背面;冷面;阴面;玻璃板背面
cold sink 冷却散热片
cold site 浸满水的地面
cold sky 无源天区
cold slap 常温撞击声
cold slug 注塑冷料
cold sluggish lubrication 低温下流动迟缓的润滑油
cold slug well 冷料阱
cold soaking 冷浸
cold softening process 冷软水法
cold-solder 冷结合料(金属)
cold soldering 冷焊接
cold solder joint 冷焊料接头;虚焊
cold source 冷源
cold spark plug 常温火花塞
cold spell 冷汛;寒潮
cold spell in spring 春寒(期)
cold sperm 精冷
cold spinning 冷旋压
cold spot 冷点
cold spotty appearance 冷斑
cold spray 冷法喷洒
cold spring 冷泉
cold spruing 冷打浇冒口
cold stage 冷台
cold stamping 冷冲压;用油墨冷模压
cold stamping die 冷压模
cold start 冷启动;冷态起动;冷起动
cold start emission 冷起动时排出的废气
cold-starting 常温起动;低温起动;冷起动
cold-starting ability 低温起动能力
cold-starting device 低温起动装置
cold-starting heavy-oil engine 低温起动柴油发动机
cold-starting performance 冷起动性能
cold-starting test 低温起动试验
cold start injector 冷起动喷油器
cold start lamp 冷启动灯
cold start-up 冷启动
cold state temperature 冷态温度
cold steam 寒冷蒸气
cold steel 冷器
cold sterilization 冷消毒;冷灭菌;冷法灭菌
cold stock 冷装(料)
cold-storage 冷库;冷冻储[贮]存;冷藏装置;冷藏器;冷藏(库)
cold-storage boat 冷藏(运输)船
cold-storage cooler 冷藏室
cold-storage door 冷(藏)库门
cold-storage hold 冷藏舱
cold-storage locker 冷库;储冰槽
cold-storage locker plant 冷藏柜装置
cold-storage plant 冷冻厂;冷藏装置;冷藏厂设备
cold-storage room 冷藏室;冷藏间
cold-storage wall 冷库墙壁
cold-storage warehouse 冷藏(仓)库
cold store 冷(藏)库;冷藏仓库
cold-store door 冷藏室门
cold store insulation 冷库保温
cold stowage 冷藏装货法
cold straightening 冷直;冷矫正;冷法拉直;冷(法)矫直
cold strain 冷应变;冷变形
cold-strained 冷拉紧的;冷变形的
cold stratification 低温层积法
cold stream process 冷流(粉碎)法
cold strength 冷态强度;耐火混凝土的热前强度
cold stress 冷应力
cold stretch 冷拉
cold stretching 冷拉
cold stretched steel bar 冷拉钢筋

cold strip 冷轧粉带;冷轧带材
cold strip mill 冷轧带钢机;带钢冷轧机;带材冷轧机
cold strip reel 冷轧带钢卷取机
cold strip steel 冷轧带钢
cold-strip steel rolling mill 冷轧带钢机;带钢冷轧机
cold stroke 中寒
cold sum 负温度总和
cold support scratch 冷架印
cold surface treating road tar 冷法表面处理道路
cold surface treatment 冷法表面处理;表面冷处理
cold syndrome of middle-jiao 中寒
cold tandem mill 连续冷轧机
cold tap 冷水龙头
cold tar 冷浇柏油;冷用焦油沥青
cold tar concrete 冷柏油混凝土
cold target 冷目标
cold tarring 冷浇柏油
cold tear 冷裂
cold-technical 纯技术的
cold-temperature needle-leaf forest 寒温性针叶林
cold tensile strength 冷拉强度
cold tensioning of steel 钢筋冷拉
cold test 冷态试验;冷凉试验;冷冻试验;空车试验;耐低温试验;低温试验;常温试验;不通电试验
cold testing 冷试验
cold test oil 耐冷油
cold tin 冷镀锡
cold tolerance 耐寒性;寒冷耐性
cold tongue 冷舌
cold top 冷顶
cold trap 冷却捕集(器);冷凝气阀;冷井;冷槽;冷捕集器
cold treatment 冷处理;低温处理
cold trim(ming) 冷切边;冷修整;冷精整
cold trough 冷空气低压槽;冷槽
cold twisted 冷扭的
cold-twisted special reinforcing bar 冷扭特种钢筋
cold twisted steel bar 冷扭钢筋
cold twisting 冷扭
cold type composition 冷排
cold-type floor mastic 冷用地板胶;冷用地板玛琋脂
cold upsetting 冷锻粗;冷顶锻
cold vapo(u)r atomic absorption method 低温蒸气式原子吸收法
cold vapo(u)r atomic fluorescence spectrometry mercury analyzer 冷蒸气原子荧光测录仪
cold vapo(u)r machine 低温蒸气式冷冻机
cold vapo(u)r method 冷蒸气法
cold vapo(u)r refrigerator 低温蒸气式冷冻机
cold varnish 冷漆
cold vibration method 冷捣法
cold vulcanization 冷硫化
cold wall method 冷壁法
cold wall vacuum chamber 冷壁真空容器
cold waste 低效放射性废物
cold water 冷水
cold water apartment unit 供冷水的公寓单元
cold water circuit 冷水循环
cold water cistern 冷水槽
cold water consumption 冷水耗用量
cold water current 寒流
cold water detergent 低温洗涤剂
cold water feed 冷水供应
cold water feed tank 冷水供应箱
cold water fish 冷水性鱼
cold water flat 无热水供应的公寓
cold water high low pressure washer 高低压冷水清洗机
cold water indicator 冷水性指示种;寒流性指示种
cold water line 冷水(输送)管线
cold water main 冷水总管
cold water meter 自来水表;冷水表
cold water paint 冷水涂料;冷水(色)漆;常温水性涂料
cold water pipe 冷水管
cold water pressure pipe 冷水压力管
cold water pumping house 冷水泵房
cold water rinse 冷水清洗
cold water service 冷水供应设施
cold-water sphere 大洋冷水圈
cold water storage cistern 冷水箱
cold water supply 冷水供应;冷水供给

cold water system 冷水系统
cold water tank 凉水槽;冷水箱;冷水池
cold water tap 冷水龙头
cold water thermometer 冷水温度表
cold wave 寒流;寒潮
cold wave current 寒潮
cold wave warning 寒潮警报
cold-weather concrete 冬季施工的混凝土
cold-weather concreting 冬季浇筑混凝土;冬季浇注混凝土;冬季浇灌混凝土
cold-weather construction 冬季施工
cold-weather curing 冬季浇注混凝土的养护;冬季浇灌混凝土的养护
cold-weather engine primer pump 冬季发动机启动泵
cold-weather lubrication 冬季润滑
cold-weather protection 冷天保护
cold-weather starting 冷天机械启动
cold-weather test 低温试验
cold weld 冷压接;冷焊
cold welding 冷压焊;冷焊(接)
cold welding pressure 冷压焊
cold welding seed 冷焊粗粒
cold wet compress 冷湿敷
cold wind 寒风
cold winter survival suit 防寒救生衣;防寒救生套装
cold workability 冷加工性
cold-workable 可冷加工的
cold-worked 冷加工的
cold-worked bar 冷处理钢筋;冷加工钢筋
cold-worked deformed steel bar 冷加工异形钢筋
cold-worked material 冷加工材料
cold-worked steel 冷作钢;冷加工钢材
cold-worked steel reinforcement 冷拔钢筋
cold-worked twisted special renforcing bar 冷处理特殊旋纹钢筋
cold work hardening 冷作硬化
cold work(ing) 冷作;冷处理;冷加工;金属冷作
cold working hardening 冷加工硬化
cold working steel 冷作钢
cold work of glass 玻璃冷加工
cold zone 冷却带;寒带
cold zone ecosystem 寒带生态系统
coleassee 共同承租人
Colebrook equation 科尔布鲁克关系式
Colebrook's pipe friction formula 科尔布鲁克管道摩擦公式
Cole-Cole model 柯尔—柯尔模型
colection perennial herb 多年生植物收集
coled-rolled forming section 冷弯型钢
colemanite 硬硼钙石;硼钙石
Cole Pitometer 克尔皮托管流速表头
Cole Pitometer method 克尔皮托管流速计法(测流速用)
co-level 上位集
Colgnet system 柯尔涅工业化混凝土构造体系(法国)
colgrout 胶状水泥浆;胶体灰浆;预填骨料专用砂浆;预填骨料灌砂浆
coli-aerogenes bacteria 大肠产气细菌
colibacillus 大肠杆菌
colicin(e) 大肠杆菌素
colidar 相干光雷达;光雷达
coli determination 大肠杆菌测定
coliform bacteria 大肠菌类细菌;大肠杆菌
coliform bacteris value 大肠菌值
coliform biomass 大肠菌生物量
coliform concentration 大肠菌浓度
coliform contamination 大肠菌污染
coliform density 大肠菌密度
coliform determination 大肠菌测定
coliform-group bacteria 大肠菌群
coliform group test 大肠杆菌群检验
coliform index 大肠菌指数
coli-ground index 大肠菌群指数
coli-group 大肠杆菌群
coli-group index of water 水的大肠菌指数
coli-group titre of water 水的大肠菌群值
coli-index 大肠(杆)菌指数
colina 小丘
co-linear 共线的;同线
colinearity 同线性
colineated line of sight 瞄准重合线
coliseum 圆形露天剧场(古罗马);圆形大运动场

colitap 大肠菌快速验纸片
coli test 大肠杆菌试验
colititre 大肠(杆)菌值
colk 锅穴
colla 可拉风;静止气旋;缓和气旋
collaborating style 协作型作风
collaboration 合作;协作;共同研究
collaboration of steel and concrete 钢筋与混凝土的共同(受力)作用
collaboration organization 协作部门
collaborator 协作者;协作单位;共同研究者
collacin 胶质素
collada 可拉大风
collage 抽象派拼贴画;(装饰工艺之一的)拼贴技术;拼贴画;拼贴
collagen 胶原;生胶质
collagenolytic 溶胶原的
collain 无结构镜煤
collapse 毁灭;陷落;折叠;倒塌;崩陷;坍坏;崩溃;暴跌;坍陷;坍塌;坍缩;塌陷;(修筑上的)塌方;湿陷
collapse acceleration 倒塌时加速度
collapse after excavation 开挖后崩坍
collapse angle 陷落角
collapse arch 坍落拱
collapse area extent 坍陷区范围
collapse blasting 抛坍爆破
collapse breccia 崩塌角砾岩
collapse caldera 塌陷破火口
collapse condition 坍塌条件;破坏条件;倒塌条件
collapse crater 塌陷火山口
collapsed concrete 流态混凝土
collapsed doline 塌陷溶陷坑;塌陷落水洞;塌陷灰岩坑
collapse deformation 坍陷变形
collapse design 破坏阶段设计;极限状态设计
collapse design method 破坏阶段设计法;极限状态设计法;极限强度设计(方)法;破坏阶段设计法
collapsed face 工作面坍垮;工作面切顶;冒顶(隧洞)
collapsed mandrel (卷取机的)收缩状态的卷筒
collapse dolina 塌陷坑
collapsed portion of the bank 坍岸
collapsed slump 塌陷型坍落度
collapsed spring 压扁状态下弹簧
collapsed storage tank 可折叠油罐;收缩油罐;坍坏油池(地下储油箱)
collapse earthquake 陷落地震
collapse field of magnetic bubble 磁泡缩灭场
collapse fissure 倒塌裂缝;坍陷裂缝;塌陷裂缝
collapse height 拆卸后最小高度
collapse in tunnel construction 隧道施工坍方
collapse land 坍陷地
collapse layer 坍陷深度
collapse limit state 倒塌极限状态
collapse load 临界纵向荷载;极限荷载;倒塌荷载;破坏荷载;破坏性负载
collapse load factor 破坏荷载系数
collapse mechanism 倒塌机制;崩塌机理;破损机构;破坏机制;破坏机理;破坏机构;坍陷机理
collapse method of structural design 结构的破坏阶段设计法;破坏性结构设计法;破坏法结构设计
collapse of apical bed 顶板冒落
collapse of casing 外挤力;套管的压坏
collapse of dike body 坝体坍塌
collapse of gallery roof 顶板崩落
collapse of loess type 黄土湿陷类型
collapse of piston skirt 活塞裙故障
collapse of roof 坍顶(隧洞)
collapse of setting 倒垮
collapse of tunnel 隧道坍塌;隧道崩塌
collapse phase 挤压节拍
collapse phenomenon 扁塌现象;坍陷现象
collapse point 崩溃点
collapse pressure 破坏(性)压力
collapse prevention 倒塌预防
collapse protection 崩塌防治
collapse radius of magnetic bubble 磁泡缩灭半径
collapse resistance 抗陷落
collapse resistant 抗倒塌
collapse resistant coefficient 抗倒塌系数
collapse rubbish 塌崩碎屑
collapse settlement 湿陷量

collapse sink 塌陷落水洞
collapse sinkhole 塌陷落水洞
collapse slump 坍塌形坍落度
collapse soil 颓积土
collapse sticking 坍塌卡钻
collapse strength 倒塌强度;破坏强度
collapse structure 塌陷构造
collapse threshold earthquake intensity 倒塌阀限地震强度
collapse to take up their natural angle of repose 坍到其自然休止角
collapse treatment 塌方处理
collapse type 坍塌型
collapse velocity of bank 坍岸速度
collapse volume of bank 坍岸量
collapse zone 冒落带
collapsibility 崩溃性;退让性;湿陷性
collapsibility grading index 分级湿陷量
collapsibility of loess 黄土湿陷性;黄土的湿陷性
collapsible 可折叠;集装箱;活动的;折叠式的
collapsible bag sampler 皮囊取样器
collapsible beam 可拆卸梁;可拆式梁;装配可拆式梁
collapsible bit 可卸钻头;活头钻头;活动钻头
collapsible boat 折叠艇;折叠式小艇
collapsible cladding 紧裹包壳
collapsible container 可折叠式容器;可折叠式集装箱;可折叠货柜;可收缩容器;可拆装集装箱;可拆装的集装箱;折叠式集装箱
collapsible cooperation 折叠式公司(一种投机性质的房地产公司)
collapsible core 可拆型心
collapsible core box 可拆式芯盒
collapsible davit 可放倒式吊艇柱
collapsible derrick 折叠式钻塔
collapsible die 可拆模;组合模(具)
collapsible drilling mast 折叠式钻杆
collapsible element 可压溃元件
collapsible filter 自滲填缝反滤层
collapsible flash board 可拆卸闸板;自溃式闸板
collapsible form(work) 拼装式模板;活动模板
collapsible gate 活动闸门;可拆卸闸门;活栅门;折叠(大)门
collapsible grade of loess foundation 黄土地基湿陷等级
collapsible grading of loess 黄土湿陷性分级
collapsible life boat 橡皮帆布艇;折叠式救生艇
collapsible loess 湿陷性黄土
collapsible mandrel 可折叠管芯
collapsible mast 可放倒式桅杆;可拆卸桅杆
collapsible metal core 分辨金属芯;分瓣金属芯
collapsible mo(u)ld 分室模;分片模
collapsible needle weir 可折叠式针堰;活动叠梁堰
collapsible pan 单眼木模
collapsible prop 活动支座
collapsible rubber dam 可缩坍橡胶坝;可伸缩的橡胶坝;可拆除的橡胶坝;充水式橡胶坝;充气式橡胶坝
collapsible seat 折椅;可折座椅
collapsible soil 湿陷性土
collapsible spinning block 拼成的旋压模
collapsible sprue 弹性直浇口;弹簧直浇口
collapsible steel fender 拼装式钢护舷
collapsible steel form 工具式钢模板;拼装式钢模板;活动钢模板
collapsible steel shuttering 活动钢模板;装配式钢模;工具式钢模板;拼装式(活动)钢模板
collapsible storage tank 可折叠油罐
collapsible stripping spider 伸缩式线盘卸料器
collapsible tank 可拆卸叠箱;可拆卸油罐
collapsible tap 伸缩丝锥
collapsible tube 可拆卸软管;软锡管;收缩管;软管
collapsible tube paint 软管涂料
collapsible tube printing 软管印刷
collapsible-whip antenna 可折叠鞭状天线
collapsible whipstock 伸缩式造斜器
collapsing 压毁;压坏;断裂
collapsing breaker 回涌碎浪;滚奔破浪
collapsing breaker wave 坍滚破波
collapsing car gate 扭曲的轿厢闸门
collapsing cavity 溃灭空穴
collapsing force 破坏力
collapsing liner 塌缩隔离器
collapsing load 临界负载;临界负荷;屈压荷载;破坏性负载;破坏荷载;失稳荷载
collapsing mast 折叠式轻便钻塔
collapsing motion 落绞装置
collapsing pressure 崩塌压强;破坏(性)压力;外压强;外压力
collapsing pulse 陷落脉
collapsing soil 崩坍土;湿陷性土
collapsing strength 抗挤强度;破坏强度
collapsing stress 破坏应力
collar 卡箍;井口;挤泥机头;抹带;眼环;柱环;轴环;支索端眼【船】;轧辊环;管子卡环;根颈;底梁;卡圈;凸缘;套圈;套环
collar a hole 开钻
collar-and-clamp 环夹式铰链(船闸闸门铰链的一种,否则称为锚环式)
collar-and-tie roof 系梁三角屋架
collar annular return velocity 钻铤环空返速
collar band 压板,钳;夹具;弹夹
collar beam 系梁;拉梁;系圈;系梁;系杆;地脚梁
collar beam roof 系梁(人字)屋面;系梁(人字)屋顶
collar beam roof truss 系梁人字木屋架;有系梁的人字木屋架
collar beam roof with jamb walls 有边框墙的系梁人字木屋架
collar bearing 环形(推力)轴承;环行轴承;环式止推轴承
collar bolt 环螺栓
collar bound pipe 接箍连接的管子
collar brace 环撑;柱环支撑
collar bracing 系杆
collar bush 主轴套;轴环套
collar bushing (深井泵的)内接头
collar buster 割管器【岩】
collar clamp 夹板;管箍
collar connection 井口装置
collar control gate 井口调节闸
collar crushing forceps 压环钳
collar diameter 冲击钻杆颈直径
collared drill pipe 带环箍的钻管
collared hole 钻头定位孔
collared joint 加环接头;加箍接头
collared scops owl 领角鸮
collared sucker rod 加箍的泵轴
collar end 轴环端
collar end bearing 轴环端轴承
collar end of the casing 套管的接箍端
collar extension 外口;领口;环口
collar flange 环状凸缘
collar flow velocity 钻铤内流速
collar for a horse 套包
collar for flange connection 凸缘连接用颈圈
collar grab 打捞爪;(螺纹破坏时的)打捞卡套
collar-head screw 环头螺钉
collar heart 单眼木饼
collar hole 锁口钻孔
collar hook up 井口装置
collar ice 贴岸冰
collar in a hole 打炮眼
collaring (古典圆柱的)柱颈花边饰;用套管扶正钻杆;加轭;灰浆砌檐瓦;灰浆砌檐瓦;作凸缘,轧件缠辊;打眼;缠辊现象;标定炮眼位置;收颈
collaring a bit 开孔钻头
collaring machine 曲边机
collar injury 套包伤
collarino 柱下部圆柱体;系梁;维梁
collar jar 打箍器
collar joint (圬工墙的)水平接缝;系砖节点(空斗墙);系杆节点(人字木屋架);(圬工墙的)垂直接缝;轴环接头;轴环接合;人字木屋架系梁杆节点;竖砖缝
collar joint casing 接箍连接的套管
collar journal 止推轴颈
collar knot 脆顶结
collarless 无箍的
collar locator 接箍定位器
collar log 接箍测井
collar marks 辊环痕
collar nut 接头螺帽;环形螺帽;接头螺母;圆缘螺母;圆缘螺帽;凸缘螺帽;凸边螺帽
collar of anchorage device 锚口
collar of thrust bearing 止推轴承环
collar oiled bearing 轴环润滑轴承
collar oiler 加油环
collar oiling 轴环注油
collar piping 井口管;表层套管
collar plate 圆盘;垫圈;垫板;钢板管环
collar pounher 管线工
collar rafter 系梁;系杆;系椽
collar rafter roof 系梁人字木屋顶;有系梁的人字木屋(顶)
collar receiver 轴环支承
collar rim 轮圈;突出轮缘
collar roof 人字屋顶;系梁人字屋面;系梁人字屋顶;带系杆屋盖;三角屋架
collars 辊环
collar saddle 接头鞍座
collar scissors 冠颈钳
collar screw 头部带凸缘的螺钉;带环螺钉
collar set 锁口盘
collar shank 带凸肩的钎尾
collar socket 接箍打捞筒
collar stop 环形挡块
collar strap 颈圈;环箍
collar stretcher at tunnel springing 隧道起拱处上挑环
collar stud 环状柱头
collar thrust 止推环定炮眼位置;止推环
collar thrust bearing 环形张力轴承;环行止推轴承
collar tie 屋架拉条
collar(tie) beam 圈梁
collar tie roof 系梁三角屋架
collar tie roof with jamb walls 有边框墙的系梁三角屋架
collar type elevator 套筒性升降机
collar vortex 涡环
collar work 艰巨的工作
collastin 胶质素
colla taurina 明胶
collate 排序
collated correct 校对无误
collated fibre 成束纤维
colla tempestade 可拉风
collate program(me) 整理程序;排序程序;校对程序
collateral 间接的;担保品;侧的;附属的;附属担保品;附属担保物;并行;平行
collateral acceptance 附担保承兑
collateral action 并行作用;并行动作
collateral agreement 附属协议;附属协定;补充协议;补充协定
collateral arrangement 内外排列
collateral assignment 间接转让;附属转让;担保物的转让
collateral bond 有担保的债券;质押公司债;动产抵押公司债
collateral branch 侧支;侧副支
collateral circulation 侧支循环
collateral clause 并行子句
collateral condition 附带条件
collateral contact 并联触点;双触点
collateral contract 附属合约;附属合同
collateral damage 附带损失;附带损害
collateral declaration 并行说明
collateral elaboration 并行制作;并行加工;并行阐述
collateral evidence 证据;旁证
collateral heir 旁系继承人
collateral information 辅助编图资料
collateral in raising loan 举债的抵押
collateralize 提供抵押
collateralized junior mortgage 将已得贷款作另一贷款的抵押品
collateralized loan 附属抵押品贷款
collateral lines 平行线
collateral loan 抵押贷款
collateral note 有担保的期票;附有抵押的本票;附带期票;附带本票
collateral pledge agreement 存款抵押保证书
collateral presumption 附属论据
collateral radiation 次要辐射;伴随辐射
collateral security 证券抵押;质押品;附属抵押品;附属担保品;附加担保物
collateral series 支系;平行系;旁系
collateral sulcus 侧副沟
collateral surety 副保证人
collateral term 附属条款
collateral trigone 侧副三角
collateral trust bond 担保信托债券
collateral vessel 侧副管

collateral warranty 附带保证
collating 依序整理
collating sequence 核对顺序;排序序列
collating unit 整理装置;排序装置
collation 校对;排序
collation map 分级统计地图
collation of all geologic(al) data 整理所有地质资料
collation of information 整理情报
collation pass 整理通过
collator 校验机;校对机;整理者;整理机;对照者;分类器;分类机;配页机;排序装置;排序机;排序程序
colleague 同事
collect 捡;征收;采集;收集;收藏
collect account 收账
collect a draft 托收期票;托收票据
collect and process 采制
collect before breaking bulk 在开舱卸货支付
collect call 对方付款电话
collect data 资料搜集;收集资料
collect duties 征税
collected and delivered 货款两清;运费货物收讫交毕
collected current 集极电流
collected-current ratio 二次发射有效系数;收集电流比
collected drawings 图集;设计图集
collected dust 捕集粉尘;收下的粉尘
collected edition 统一版
collected liability 托收负债
collected mineral 采集的矿物
collected stack 集合(式)烟囱
collected works 全集;文集
collect fees arbitrarily 乱收费用
collect freight 到付运费
collectibility 可收回性;可收回的程度
collectible 可收账款
collectible goods 可收集商品
collecting 捕收
collecting agent 追债代理人;代收人;促集剂;捕收剂;捕集剂;收款代理商
collecting and dispatching capacity of port for cargoes and passengers 港口客货集疏运能力;港口集疏运能力
collecting and dispatching mode 集疏运方式
collecting and distributing point of passenger flow 客流集散点
collecting anode 集电(阳)极
collecting aperture 聚光孔径;集能孔径;集光孔径
collecting aqueduct 集水道
collecting area 集水区;接收面积;集水面积;集流区;集合区域;地面集水区
collecting bar 汇流排
collecting basin 集水箱;集水区;集水池;集流区;地面集水区;补给区;水池
collecting basket 采掘篮
collecting belt conveyer 集料带式传送机
collecting bill 收款汇票
collecting board 收集板
collecting bow 弓形集电器
collecting box 集水箱;收集箱
collecting brush 集流刷;集电刷;汇流刷;刮泡刷
collecting bucket 电铲勺斗
collecting canal 集水渠(道)
collecting cargo of port 港口集运
collecting cell 静电集尘单元
collecting chamber 集污坑;集水室
collecting channel 干管;集水沟;集水槽;总渠
collecting circus 集流区;集流漏斗
collecting comb 集电梳;梳集板
collecting commission 托收手续费
collecting company 集资公司;代收公司;追债公司
collecting conduit 集水管;集水暗沟
collecting conveyer 集矿运输机;集棉输送机;沉降输送机
collecting cylinder 收纸滚筒
collecting device 集电设备;收集设备
collecting direction 收集方向
collecting dirt 污物堆积
collecting ditch 集水沟
collecting drain 回流排水沟;集水排水沟
collecting drum 收集辊;集棉辊
collecting duct 集合导管

collecting electrode 集电极
collecting electrode pipe 集尘电极管
collecting electrode vibrator 集尘极振荡器
collecting field 集电极场;收集场
collecting flue 总烟道
collecting funnel 集流区;集流漏斗;汇集漏斗(指圆形山谷)
collecting gallery 集水廊道
collecting gutter 集水沟;集水槽;收集槽
collecting hopper 集料布料斗
collecting land use fees on different levels 分层次征收地产使用费
collecting layout 集水布局
collecting lens 聚场透境;集光镜
collecting line 汇集管线;集水管线
collecting lorry 垃圾车
collecting main 回流条;回流排;集水干线;集水干渠;集水干管;集流管;母线;总管(气、水等)
collecting main pressure 集气管压力
collecting manhole 集污窖井;有沉砂室的竖井
collecting mirror 聚光镜;聚场镜
collecting note 收据票据
collecting objective 聚光物镜
collecting of electrons 电子收集
collecting of lint 纸粉聚积;纸粉堆积
collecting pan 集油盘;受电弓滑板
collecting part 集尘段
collecting passage 集水沟;截流沟
collecting pipe 聚积管;集水管;集汽管;集气管;集尘管;汇流管
collecting plate 集尘板;沉降电极;沉淀极
collecting pocket 集水窝
collecting post 战俘收容所;伤病员收容所;代办收款的邮局;收容所
collecting potential 收集电位;收集电势
collecting pump 集水泵
collecting reagent in flo(a)tation 浮选中级收集剂
collecting reel 捡拾滚筒;滚筒式捡拾器
collecting ring 集水环;集汽环;集流环;集电环;汇流环
collecting sewer 污水支管;收集管【给】
collecting site 收集地点
collecting smoke funnel 集烟尘漏斗
collecting snow fence 扫雪栅栏
collecting soil solutions 收集土壤溶液
collecting station 难民收容所;废物收集站;伤病员收容所
collecting structure 集水建筑物
collecting sump 集水坑;集料仓;收集坑
collecting surface 集能面
collecting sweeper 扫路垃圾车
collecting system 集水系统;收集系统;排水系统
collecting system of port 港口集运系统
collecting tank 集储[贮]罐;集液池;集水池;储[贮]罐;储[贮]槽
collecting tray 集线器;汇流槽
collecting trough 集水槽
collecting vehicle 收集车;垃圾车
collecting vessel 集水容器
collecting water supply system 集中式给水
collecting well 集水井;集油阱
collecting work 集水工程;引水工程
collecting yield 集水量
collection 集合;汇集;选样;一批;一堆;代收;采收;取款;托收;搜集;收纸;收集物;收集
collection and acceptance 托收承付
collection angle 收集角
collection arrangement 托收安排
collection basis 征收基础
collection bill 托收票据;托收汇票
collection book for bills 票据回收账簿
collection box 汇水箱;采集箱
collection center 废品站;收集站
collection chamber 集气室;集合室
collection channel 集水槽
collection charges 代收手续费;取款手续费;托收费
collection clerk 收账员;收账人
collection commission 托收手续费
collection cost 成本代收;托收成本
collection department 收账部门
collection depot 回收站
collection efficiency 除尘效率;收集效率
collection efficient of particulate 集尘效率

collection electrode 集尘电极
collection expenses 托收费用;收账费用
collection fee 代收票据费;托收(票据)费
collection float 收账流动值
collection frequency 收集频率
collection frequency of refuse 垃圾收集次数
collection hopper 集料斗;灰斗
collection items 托收项目
collection ledger 托收账
collection letter 托收委托书
collection line 用户污水管;街道沟管;集水管线;汇水管线
collection memo 托收单
collection note 收款本票
collection of a bill 托收
collection of cash 托收现款
collection of classified refuse 垃圾分类收集
collection of combined sewerage 合流污水收集
collection of compulsory contributions 派款
collection of customs duty 关税征收
collection of data 资料收集;整理资料;收集资料
collection of debt 托收债款
collection of deposits 取得存款
collection of digester gas 消化池气(体)收集
collection of documentary bill 托收跟单汇票
collection of drawings 图集
collection of foreign business information 海外商情收集
collection of gases 气体收集
collection of household refuse 家庭垃圾的收集
collection of national railway statistic data 全国铁路统计资料汇编
collection of non-classified refuse 混合垃圾收集;不分类的垃圾收集
collection of payment 兑现;收款
collection of refuse at every door 逐户收集垃圾
collection of refuse at gathered places 垃圾定点收集
collection of sample 样品采集;样本采集
collection of sets 集族
collection of sewage 污水收集
collection of solid wastes 固体废物收集
collection of specimens 标本采集
collection of stormwater 雨水收集
collection of tax 税款的征收
collection of trade charges 代收货款
collection of unclassified refuse 不分类的垃圾收集
collection of waste oil 废油收集
collection of wastes 污水收集
collection of wastewater 废水收集
collection of water 集水
collection of water sample 水样本采集
collection on delivery 交货收款;代收货款
collection on documents 跟单托收
collection order 委托书;托收委托书;托收单
collection period 托收期;收账期
collection permit 托收证书
collection plan 装配平面图
collection policy 收账政策
collection pressure 接收压力
collection procedure 收集程序
collection processing 集中处理
collection rate 账款回收率;收款率(指账款回收率)
collection ratio 收款比率
collection receivable for customer 应收未收客户款
collection report 托收报告书;收账报告书
collection schedule 收款清单
collection service 托收服务
collections of papers 论文集
collections trusted 代收各种款项
collection sump 收集坑
collection system 集水系统;收集系统
collection tank 集水槽
collection through the bank 银行托收
collection time 集水时间;汇流时间;汇集时间;收集时间
collection track 待取列车停留线
collection trap 收集井
collection trusted 代收款项
collection tube 接收管
collection voucher 托收凭单;收入传票
collection well 集水井
collection with immediate settlement 立即清算

债务的收款
collection without acceptance 托收无承付;托收不承付
collective 集合的;天气集合
collective accelerator 集团加速器
collective account 统驭账户
collective accounting 集中核算
collective accumulation 集体积累
collective advertising 联合广告
collective agreement 集体协议;集体合同
collective antenna 共用天线
collective bargaining 集体协议
collective broadcast receiving station 通播接收台
collective broadcast sending station 通播发射台
collective call letters 集合呼号【无】
collective contract 集体合同
collective control 集中控制
collective control elevator 集中控制电梯
collective crystallization way 集合结晶作用方式
collective custody 集中保管
collective deposit 集体存款
collective design drawings 设计图集
collective diagram 综合图
collective distribution 共同分配
collective distribution method 共同分配法
collective dose equivalent 集体剂量当量
collective drawings 图集;装配图(样)
collective dwellings 集体住宅
collective ecological security 集体生态安全
collective economy 集体经济
collective expenses 集中用费
collective farm 集体农庄
collective flow curve 汇流曲线
collective forward error correction 集群型前向纠错
collective freight controller 零担货物调度员
collective fund 集体基金
collective goods 集体财产;共同财产;公共设施(公园、道路等)
collective goods controller 零担货物调度员
collective grains 集结颗粒
collective idea 集体观念
collective income 集体收入
collective lens 聚光透镜;会聚镜头
collective lighting 集中供电照明
collectively screened cable 集中屏蔽电缆
collective management 集体经营
collective medium 通用培养基
collective model 综合模型
collective motion 集体运动
collective nursery 搜集圃
collective ownership 集体所有(制)
collective phenomenon 集体现象
collective pitch control 集体变距操纵
collective plasma 集积等离子体
collective property 集体财产
collective recrystallization 聚集再结晶
collective regulating and stopping valve 集流调节闭塞阀
collective regulating cock 集流调节龙头
collective sampling unit 成组抽样单位
collective savings 集体储蓄
collective security 集体安全保障
collective settlement 集中结算
collective shelter 公共防空洞;公共隐蔽处
collective sump 集水坑
collective system 集敛系统
collective tank 集水池
collective transition 集体跃迁
collective welfare institution 集体福利设施
collective welfare project 集体福利事业
collect on delivery 交货收款;货到收款;货到付现金;钱货两讫;收货付款
collect on delivery sales 交货付款销售
collector 集束器;集流器;集合器;集管;集电器;换向器;整流子;采集器;捕集剂;编译模块;收税员;收税人;收款员;收集者;集流极
collector agent 浮选捕集剂
collector array 集热器陈列
collector barrier 集电极位垒
collector-base 集电极—基极
collector-base bias circuit 集电极基极偏置电路
collector beam lead 集电极梁引线
collector body 集电极本体
collector box 集流管

collector brush 集流刷;集电刷
collector cone 锥形骨料斗
collector contact 集电极接点(集电极接触)
collector coupling 集电极耦合
collector cover plate 集热器盖板
collector current 集电极电流
collector current cut-off 集电极电流截止
collector current runaway 集电极电流击穿
collector cut-off 集电极截止
collector cut-off current 集电极截止电流
collector cylinder 收集极圆筒
collector detection 集电极检波
collector dissipation 集电极损耗
collector-distributor 集散厅;集散干道
collector-distributor lane 集散车道
collector-distributor road 集散道路
collector-distributor square 集散广场
collector-distributor street 集散街道
collector ditch 集水沟
collector drain 排水管;集水(暗)沟
collector drainage 集水管
collector duct 集合管沟
collector efficiency 集热器效率;集电极效率;太阳集热器效率
collector efficient factor 集热器效率因子
collector electrode 集电极;捕集电极;收集电极
collector electrode pipe 集尘电极管
collector film 集电器膜;集电极膜
collector filter 集尘过滤器
collector flow factor 集热器流动因子
collector grid 集电栅;捕获栅
collector groove 汇集槽
collector gutter 集水沟
collector hauler 垃圾工(人)
collector heat removal factor 集热管转移因子
collector junction 集电(极)结
collector lead 集电极引线
collector lens 聚场透镜
collector manifold 集水支管;集水管线
collector mesh 收集栅网
collector modulation 集电极调制;集电极调幅
collector node 集电极节
collector orifice 集水孔
collector pipe 集水管;集汽管
collector plate 集电片
collector rail 集电轨
collector-reflector 集电极—反射极
collector region 集电极区域
collector resistance 集电极电阻
collector return resistor 集电极回路电阻
collector ring 集水环;集流环;集电环;汇流环;汇电环;整流环
collector ring rotor 集电环转子;回流环转子
collector road 支路
collector saturated voltage 集电极饱和电压
collector shoe 集电靴
collector-shoe gear 汇流装置
collector slipper 集电器滑板
collector street 辅助道路;汇集(交通)街道;辅助街道;(连接干道与地方道路的)辅助干道
collector subsystem 集热器子系统
collector supply voltage 集电极电源电压
collector system network 集水管系通
collector terminal 集电极引线(端);集电极端
collector tube 集液管;集丝管
collector unit 收集单元
collector voltage 集电极电压
collector well 集水井
collector work 集水支管
collect outstanding account 收回旧欠
collect specimen 采集标本;搜集标本
collect-to-traverse speed ratio 成型速比
college 学院;专科学校
college for professional training 专科学校
college grounds 大学校园
college housing 大专院校住房
college of applied art 应用艺术学院
college of architecture 建筑学院
college of design 设计制图学院
college of fine arts 美术学院
college of music 音乐学院
college of science and technology 理工学院
college of traditional Chinese medicine 中医学院
colleges and universities 高等院校;高等学校

collegiate architecture 学院风格建筑
collegiate chapel 牧师会组织的小教堂
collegiate Gothic 大学哥特式建筑
collegiate mosque 社团的清真寺
coller for motor 电动机冷却器
collet 开口内桩;夹头;有缝夹套;根颈;底托;废玻璃;套爪;弹性夹;绝缘块;筒夹
collet attachment 筒夹装置
collet cam 夹簧控制凸轮
collet chuck 套爪卡盘;弹簧夹;弹簧夹头;筒夹
collet clamping head 弹簧夹头
collet closing and broaching tool 弹簧夹头钳
collet closing tool 弹簧夹头闭合器
collet head 套筒夹
collet holder 套爪夹
collet index fixture 套筒转位夹具
collet segment 机头滑块
collet system 楔形系统(锚具或夹具)
collet type packet 夹套式封隔器
collide 碰撞
collided ship 被撞船
collided vessel 被碰船
collider 碰撞机
colliding ship 碰撞他船的船
colliding station 争用站;争用网;碰头站
colliding vessel 碰撞船
collier 矿工;煤矿工人;煤船;运煤船船员;运煤船
colliery 煤矿;煤井;采煤场
colliery arches 采煤用钢拱
colliery effluent 矿井污水
colliery engineering 煤炭技术;煤矿工程
colliery guarantee 煤矿保证条款
colliery products 矿产品
colliery screened 精选的煤
colliery siding 煤矿专用线
colliery spoils 煤矸石;煤矿废料
colliery waste 煤矿废物;煤矿废水;煤矸石
colliery waste tip 煤矿废物堆
colliery wire rope 煤矿用钢丝绳
colliery working day 矿井工作日;煤业工作日
colligation 连接;绑扎
colligative effect 依数效应
colligative property 依数性质;依数性
collimate 瞄准;准直;照准;对准
collimated beam 准直束;平行束
collimated image converter 平行光变像器
collimated lamp 准直灯
collimated light 准直光;照准光线;平行光
collimated light beam 准直光束
collimated monochromatic light 准直单色光
collimated output beam 平行输出光束
collimated principal point 照准主点
collimated roving 平绕粗纱
collimated γ-ray scintillation spectrometer 准直伽马射线闪烁分光计
collimating 准直
collimating aperture 准直孔径
collimating cone 准直锤体
collimating device 准直器;平行光管
collimating eyepiece 准直目镜
collimating fault 瞄准误差;平行校正误差
collimating grid 准直栅格
collimating level 准直水平仪
collimating line 校准轴线;准直轴;视准线
collimating mark 准标
collimating mirror 准直镜;平行光镜
collimating optics 平行光学
collimating point 框标点;准直点
collimating ray 视准线
collimating rod 视准杆
collimating sight 瞄准具;平行瞄准具
collimating slit 准直缝
collimating staff 视准标杆
collimating telescope 准直望远镜
collimation 准直;照准视线;照准;平行性;平行校正;视准
collimation accuracy 视准精度
collimation adjustment 视准校正;视准法
collimation axis 准直轴;照准轴;照准线;视准轴
collimation band 准直带
collimation cone 视锥
collimation constant 视准常数
collimation converter 准直仪
collimation correction 视准校正;视准改正

collimation error 瞄准误差;准直(误)差;照准差;透视差;视准(误)差;视觉误差
collimation error correction 视准差校正
collimation lens 准直透镜
collimation level 准直水平
collimation line 准直线;视(准)线
collimation line method 视准线法
collimation mark 框标;准直标(志);视准标志
collimation method 视准法
collimation plane 准直面;视准面
collimation system 用视线高度测水平法;准直系统;视准系统(大地测量)
collimation tower 瞄准塔
collimator 准直仪;准直十字线;准直器;准直管;照准仪;平行光管;视准仪;视准管
collimator field glass 准直望远镜
collimator-finder 视准检景器
collimator prism 准直仪棱镜;视准管棱镜
collimator set 准直仪
collinear 共线的
collinear acousto-optic(al) interaction 共线声光相互作用
collinear acousto-optic(al) tunable filter 共线声光可调滤波器
collinear array 直线天线阵;直排天线阵
collinear drainage 共线性水系
collinear forces 共线力;共力线
collinear heterodyning 共线外差
collinear holography 共线全息术
collinear image formation 共线成像
collinearity 共线性
collinearity condition 共线条件
collinearity condition equation 共线条件方程
collinearity equation 共线(性)方程
collinearity method 共线法
collinear load(ing) 线性荷载
collinear pattern 共线性
collinear phasematching 共线相位匹配
collinear planes 共线面
collinear points 共线点
collinear ray 共线光线
collinear second harmonic generation 共线二次谐波振荡
collinear solution 共线解
collinear vectors 共线向量
collineation 直射变换【数】;直接变换;照准视线;共线(性);测量仪视线
collineatory transformation 直射变换【数】
collinite 凝胶煤素质;无结构镜质体;无结构腐殖体
Collins bubble viscometer 柯林斯气泡黏度计
Collins capital 柯林斯柱头
collinsite 磷钙镁石;淡磷钙铁矿
colliodal form structure 胶状结构
colliquefaction 溶合
collision 抵碰;冲突;冲击;碰撞
collision accident 碰撞事故
collisional broadening 碰撞致宽
collisional damping 碰撞阻尼
collisional deactivation 碰撞减活作用
collisional heating 加热
collisional narrowing 碰撞致窄
collisional process 碰撞过程
collisional radio source 碰撞射电源
collisional turbulence 碰撞湍动
collision analysis 碰撞事故分析
collision angle 碰撞角
collision at sea 海上碰撞
collision avoidance 防撞;防止碰撞;避碰
collision avoidance grid 避碰网络
collision avoidance radar 防撞雷达
collision avoidance radar aids 避碰装置
collision avoidance radar system 避碰雷达系统
collision avoidance radar trainer 防撞雷达训练器
collision avoidance system 防撞装置;防撞系统;避碰系统;冲突回避系统
collision basin 碰撞谷盆地
collision bearing 碰撞方位
collision belt 碰撞带
collision between seagoing vessels 海船碰撞
collision blasting 碰撞爆破
collision blip 碰撞标志
collision boundary 碰撞边界
collision bulkhead 防撞舱壁
collision bumper 碰撞缓冲器

collision clause 碰撞条款
collision coefficient 碰撞系数
collision compartment 艏尖舱
collision course 相撞航线;冲突路线;冲突航向;碰撞航向;碰头航向
collision-course homing 迎击航向引导
collision course indication 航向防撞指示;避碰航向信号
collision cross-section 碰撞截面
collision damage 碰撞损坏
collision detection 冲突检测
collision detection circuit 冲突检测电路
collision diagram 事故分析图;碰撞事故分析图
collision drill 堵漏演习
collision due to vis major 不可抗力的碰撞
collision efficiency 碰撞效率;碰撞率
collision efficiency factor 碰撞效率因素
collision excitation 碰撞激发
collision factor 碰撞因子
collision force 碰撞力
collision-free bow shock 无碰撞弓形激波
collision frequency 碰撞频率
collision graph 碰撞事故分析图;交通事故分析图
collision hypothesis 碰撞假说
collision induced instability 碰撞感生不稳定性
collision induced spectrum 碰撞感生谱
collision injury 撞击伤
collision insurance 碰撞(保)险
collisionless Boltzmann equation 符拉索夫方程
collisionless Boltzmann equations 无碰撞玻耳兹曼方程
collisionless damping 无碰撞阻尼
collisionless shock wave 无碰撞激波;无碰撞冲击波
collisionless tearing instability 无碰撞撕裂不稳定性
collision liability 碰撞责任
collision load 撞击力(船靠泊时);撞击荷载
collision loss 碰撞损失
collision mat 堵漏毯(海工工程);堵漏毡;堵漏网垫;堵漏垫;防撞毡;防漏毡;防漏垫;防撞栅网;防撞垫层
collision matrix 散射矩阵
collision-mat(tress) 防撞柴排
collision method 撞击法
collision of plates 板块碰撞
collision of the first kind 第一种碰撞;第一类碰撞
collision of the second kind 第二种碰撞;第二类碰撞
collision on land 陆地上碰撞
collision orbit 碰撞轨道
collision parameter 碰撞参数
collision party 堵漏小组
collision post 防撞柱;防护柱
collision prevention 防止碰撞;避碰
collision prevention system 防撞系统
collision probability 碰撞概率
collision protection 列车防撞系统
collision quarter 堵漏部署
collision radiation 碰撞辐射
collision-radiative recombination 碰撞辐射复合
collision rate 碰撞速率;碰撞率
collision regulations 避碰规则
collision safety coefficient 交会安全系数
collision strength 碰撞强度
collision strut 防撞支杆;防冲支撑;事故分析图
collision suture 碰撞缝合
collision table 冲突分解;冲突表
collision test 撞击试验
collision theory 碰撞理论
collision theory of reaction rate 反应速率碰撞理论
collision time 碰撞时间
collision transfer of energy 碰撞能量传递
collision-type orogeny 碰撞型造山作用
collision warning 碰撞警告;碰撞警报
collision warning equipment 雷达碰撞警报装置
collision warning indicator 碰撞警报指示器
collision warning system 碰撞警报系统
collite 微无结构镜煤
collocation 搭配;配置
collocation point 配置点
collochemistry 胶体化学
collochore 配对区
colloclarite 微无结构镜质亮煤

collodion 罗甸;胶棉;火棉胶;火棉;醇溶(低氮)硝化棉
collodion cotton 大胶棉;低氮硝化纤维素
collodion dry plate 胶棉干版;棉胶干板
collodion elastique 弹性火棉胶
collodion filter 火棉胶滤器
collodion membrane 火棉胶膜
collodion of sulfonated bitumen 磺化沥青·火棉胶
collodion process 棉胶湿片法;火棉胶法
collodion varnish 含铝清漆(仅用于内部作业)
colloform 胶状结构;胶体(结构);胶粒结构
colloform ore 胶状矿石
colloform structure 胶状构造
colloid 胶质;胶体;胶态(化)
colloidal admixture 胶体混合物
colloidal agglutination 胶体凝结
colloidal aging 胶体陈化
colloidal albumin solution 胶质蛋白溶液
colloidal alumina 胶体氧化铝
colloidal alumin(i)um 胶态铝
colloidal amphoion 胶体两性离子;胶态两性离子
colloidal antimony pentoxide 胶态五氧化锑
colloidal bearer 胶体溶液载体;胶态载体
colloidal capacity of bituminous substance 沥青形成胶态能力;沥青材料形成胶体能力
colloidal carbonyl powder 超细羰基法粉末
colloidal carrier 胶态载体
colloidal carrier of enzyme 胶态载媒体
colloidal cement 胶质水泥
colloidal cement grout 胶体水泥浆
colloidal cement grout mixer 胶体水泥浆搅拌器;胶体水泥浆搅拌机
colloidal cement mortar 胶状水泥砂浆
colloidal chemistry 胶体化学;胶态化学
colloidal clay 胶质黏土;胶性黏土;胶体黏粒;胶态黏土;胶态瓷土;胶态白土
colloidal cluster 胶团
colloidal cluster structure 胶团结构
colloidal colo(u)r 胶体颜料;胶体色料
colloidal complex 胶状复合体;胶质络合物;胶质复合物;胶质复合体
colloidal composition and classification 胶体的组成和分类
colloidal compound 胶体化合物
colloidal concrete 胶质混凝土;胶体混凝土;预填骨料混凝土
colloidal condition 胶体状态
colloidal content 胶质含量
colloidal degeneration 胶样变性
colloidal dispersion 胶体弥散;胶状分散质;胶体分散;胶态分散(体)
colloidal dispersion system 胶体分散系
colloidal dye 胶体染料
colloidal electrolyte 胶体电解质
colloidal emulsifier 胶态乳化剂
colloidal emulsion 胶体溶液
colloidal emulsion stabilizer 胶体溶液稳定剂
colloidal entrapment 胶体截留
colloidal film 胶质膜
colloidal flo(a)tation 胶体浮选
colloidal form 胶体态
colloidal fouling 胶态污着
colloidal fuel 胶质燃料;煤在油中悬浮液
colloidal gel 胶质凝胶;胶态凝胶
colloida gel system 胶体凝胶体系
colloidal gold 胶体金
colloidal gold test 胶体金试验
colloidal grain 胶体颗粒
colloidal graphite 胶体石墨;胶态石墨
colloidal graphite for cathode tube inner coating 电子束管内导电石墨乳
colloidal graphite for fibre glass 玻纤涂膜石墨乳
colloidal graphite for press forging 锻压石墨乳
colloidal graphite mixed with water 水基石墨润滑剂
colloidal group 胶团
colloidal grout 胶质灰浆;胶质溶液灌浆;胶质浆体;胶体溶液灌浆;胶体浆液;胶体灰浆;胶体灌浆
colloidal grouting 胶体灌浆
colloidal index 胶质价
colloidality 胶性;胶度
colloidal load 胶质挟带量(河流)
colloidal load of stream 河流胶质携量
colloidal loads 胶态悬移质

colloidal lubricant 胶态润滑剂
colloidal mask 胶体遮蔽作用
colloidal material 胶质材料;胶体物质
colloidal matter 胶质;胶体物质;胶体;胶态物质
colloidal medium 胶体介质;胶态介质
colloidal medium test 胶介质试验
colloidal metal hydroxide particle 胶态金属氢氧化物颗粒
colloidal microparticle 胶态微颗粒
colloidal mill 胶体研磨器;胶体磨碎机;胶体磨;胶态磨
colloidal mineral 胶体矿物
colloidal mixer 胶浆搅拌机;胶体搅拌机
colloidal mortar 胶态砂浆;胶质砂浆;胶体砂浆
colloidal mortar visco(si)meter 胶体砂浆黏度计
colloidal movement 胶态运动
colloidal mud 胶体泥浆
colloidal nature 胶体性质
colloidal nucleus 胶核
colloidal organic carbon 胶体有机碳
colloidal organic material 胶态有机物质
colloidal osmotic pressure 胶体渗透压
colloidal particle 胶体微粒;胶(体)粒(子);胶体颗粒;胶态粒子;胶质颗粒
colloidal pollutant 胶体污染物
colloidal powder 超细粉末
colloidal precipitation 胶体沉淀
colloidal properties of clay 黏土胶性
colloidal property 胶体性质
colloidal quality 胶体质
colloidal rate 胶体率
colloidal resin 胶体树脂
colloidal rock 胶状岩
colloidal sediment 胶质泥沙;胶体沉淀物;胶态沉积物
colloidal silica 胶状硅石体;胶质硅石;胶体硅石;胶体(二氧)化硅;胶态(氧化)硅;胶态二氧化硅
colloidal silica coating 胶态二氧化硅类涂料
colloidal silver 胶态银
colloidal size 胶粒粒度
colloidal sol 溶胶
colloidal solid 胶质固体
colloidal solid matter 胶质固体物
colloidal solid substance 胶质固体物质
colloidal solution 胶体溶液;胶态溶液
colloidal stability 胶质稳定性;胶体稳定性;胶体安定性;胶态稳定性
colloidal stain 胶体着色剂;胶体颜料
colloidal state 胶(体)状态
colloidal substance 胶状物质;胶样物
colloidal suspension 胶性悬浮;胶状悬浮;胶悬(体);胶体悬液;胶体悬浮;胶体悬浮(物);胶态悬浮体
colloidal suspension particles 胶体悬浮颗粒
colloidal system 胶体体系;胶态系统
colloidal test 胶体试验
colloidal texture 胶状结构
colloidal theory 胶体学说
colloidal trace metal 胶质痕量金属
colloidal transfer 胶体转印
colloidal turbidity 胶体浓度
colloidal valence 胶质价
colloidal viscosity 可见黏性
colloidal water 胶体水
colloidal water treatment 胶性水处理
colloidal zeolite 胶态沸石
colloid chemical deposition 液体化学沉积作用
colloid complex 胶质复合物
colloid content 胶质含量;胶粒含量
colloided fuel 胶体燃料
colloid filter 胶棉过滤器
colloid fraction 胶粒(粒)组;胶质部分;胶分
colloid grain grade 胶粒组
colloidin 胶体素
colloidization 胶态化作用
colloidizing 胶化
colloidizing agent 胶化剂
colloid mill 竖式转锥磨机
colloidogen 胶态原
colloidor 胶质助凝器;胶体助凝器
colloid particle 胶态微粒;胶质微粒
colloid plate 胶板
colloid properties of clay 黏土胶性
collopake 胶质混凝剂(一种用于墙壁着色及防水的物质)
collophane 胶磷矿
collophore 黏管
colloquia 论文集
colloquium 座谈会;学术讨论(会);学术报告;讨论会(学术方面)
colloquy 正式会谈
colloresinite 镜质树脂体
collose 木质胶
collosol 溶胶
collotype 珂罗版
collotype ink 珂罗版印刷油墨
colloxylin 火棉
collusion 串通舞弊;串谋;勾结;共谋
collusive tender 欺诈性投标
collusive tendering 串通投标
colluvial clay 崩积黏土;塌积黏土
colluvial deposit 重力堆积物;崩流沉积;崩积物;崩积泥沙;崩积沉积;塌积物;塌积沉积
colluvial mantle rock 塌积盖层石
colluvial period 崩积期
colluvial soil 崩塌堆积土;崩积土;塌积土
colluviarium 进人孔(管道维修及通风用);人孔;水渠中通道
colluviation 崩积作用
colluvium 坠积物;崩积物;崩积层;塌积物
colluvium anomaly 塌积物异常
collyrite 微光高岭土
colly Weston slate 片石(用于铺盖屋顶)
Colman method 蒸汽预热法(防腐用)
colmascope 珂罗玛镜;胁变观察器
colmatage 淤灌(法);放淤(沉沙);放灌
colmation 淤灌;放淤
colmation works 放淤工程
Colm formula 柯尔姆公式(用于长期经济预测)
Colmol miner 柯尔木尔型连续采煤机
Colmonoy 科尔莫诺伊合金
colocthar 红铁粉
cologarithm 余对数;反对数
cologenide 硫硒碲化合物
cologne brown 科隆棕;煅棕土(一种红棕色颜料);煅烧天然棕土
Cologne cathedral church 科隆主教堂(德国西部一个城市)
cologne earth 科隆土;煅棕土(一种红棕色颜料);煅烧天然棕土
cologne spirit 科隆酒精
cologne yellow 科隆黄
colombage 大方料(木)结构
Colombia 哥伦比亚
colombier 对开图纸
colombo root 可坡根
colomite ore 钒云母矿石
Colomony 铜镍硼合金
colon classification 冒号分类法
colonette 小型柱
Colonial 科尼尔耐蚀铬镍合金钢
colonial 群体的
Colonial architecture 殖民地建筑
colonial architecture of America 美国初期建筑
colonial bill of credit 美国最早的货币
colonial blue 群青色
colonial casing 美国初期门窗饰边
colonial city 殖民城市
colonial coral 群体珊瑚
colonial house 殖民地住房建筑;美国初期住房建筑
colonial life 迁移性的生活
colonial nodule 群体结核
colonial organism 群体生物
colonial panel door 美国初期方格嵌板门
colonial revival 殖民地建筑风格重现
colonial siding 殖民地时期外墙板(美国)
colonial style 美国初期建筑风格;殖民地式
colonial yellow 群黄色
colonist 移位植物
colonization 占据
colonizing species 群居物种
colonnade 列柱;柱状节理下段;柱廊
colonnaded 有柱廊的
colonnaded avenue 有柱廊的道路;连拱廊通道
colonnaded building 柱廊式建筑
colonnaded court 有柱廊的庭院
colonnaded street (罗马建筑艺术中的)室内街道
colonnade-enriched piazza 柱廊围成的广场
colonnade foundation 管柱基础;柱群基础;柱列基础;管桩基础
colonnade foundation process 管柱钻孔法;管桩基础施工方法;管柱基础施工法
colonnette 小柱;装饰性小圆柱
colony 菌落;晶团;集群;殖民地;多种集群;蜂群;群体
colony counter 菌落计群器(自来水细菌检验);计群器
colony-forming unit 集落形成单位;群落形成单位;菌落形成单位
colony microstructure 晶团显微组织
colony of bacteria 菌群
colony unit 聚居单元
colophon 版权页;版权记录
colophonic acid 松香酸
colophonium 松脂;松香;树脂
colophony 松香;树脂
colophony soldering paste 松香焊膏
colophony soldering wire 松香焊锡丝
Coloradoan stage 科罗拉多阶【地】
Colorado low 科罗拉多低压
Colorado ruby 科罗拉多红宝石
Colorado topaz 科罗拉多黄玉
colorbitu 卡乐比丘(一种含沥青的水溶性颜料)
colossal monument 巨大纪念碑
colossal order 高巨柱型;巨柱(式);高柱式
colossal pilaster 高巨柱型壁柱
colossal profits 暴利
colossal statue 高巨柱型雕象
colossal statue of Ramses 拉姆萨斯高巨柱式
colossal tanker 超巨型油轮
colossal temple 高巨柱式神庙
Colosseum 古罗马的圆形大剧场;罗马圆形剧场;角斗场
Colosseum at Rome 罗马的大型建筑;弗立文圆形露天剧场
Colossi of Memnon 孟依巨像(古埃及底比斯)
colossus 庞然大物;名人巨像;巨人;巨型雕塑
colossus of Rhodes 罗得希旦人像
colo(u)r 颜色;彩色保真度;色彩
colo(u)r aberration 色像差;色差
colo(u)rability 可着色性;着色力
colo(u)rable graph 可着色图
colo(u)r absorber 滤光器;滤光片;色吸收体
colo(u)r absorption 颜色的吸收
colo(u)r acceptance 着色性
colo(u)r adaptation 彩色适应
colo(u)r adapter 彩色附加器
colo(u)r adder 彩色混合器
colo(u)r addition 彩色叠加
colo(u)r additive viewer 彩色合成仪
colo(u)r adjusting 墨色调整;颜色调整
colo(u)r admixture 染色掺料;染色掺和
colo(u)radoite 碲汞矿
colo(u)r aerial photograph 彩色航空照片
colo(u)r aerial photography 彩色航空摄影学
colo(u)r after firing 烧后颜色
colo(u)r agnosia 失辨色能
colo(u)r airphoto 真彩色航片
colo(u)r alteration index 色变指数
colo(u)rama 彩色光
colo(u)rama lighting 彩色照明;调色照明
colo(u)r amplitude code 彩色振幅编码
colo(u)r analyser 分色器
colo(u)r analysing filter 析色器;彩色分析滤光器
colo(u)r analysis 颜色分析;彩色分解;色分析
colo(u)r analysis display computer 彩色分析显示计算机
colo(u)r analyzer 颜色分析仪;颜色分析器;彩色分析仪;彩色分析计
colo(u)r and lustre 色泽
colo(u)r and plant 色彩与植物
colo(u)r and turbidity in water 水的色度和与浊度
colo(u)r anodizing 彩色阳极氧化
colo(u)r anomaly 彩色异常;色觉异常
colo(u)rant 颜色膏;着色剂;染料;(塑料的)色料
colo(u)rant dispenser 着色剂分配器
colo(u)rant dispersion 颜料分散体;着色剂分散体;着色剂的分散
colo(u)rant double coating 颜料复合包膜
colo(u)rant fiber 颜料纤维
colo(u)rant match 着色剂配色
colo(u)rant mixture 混合着色剂

colo(u)rant mixture computer 配色计算机
colo(u)rant slurry 色浆
colo(u)r appearance 色表(面)
colo(u)r appraisal 评色
colo(u)r aptitude 颜色适应性
colo(u)r aptitude test 颜色适应性试验
colo(u)r asphalt 着色沥青
colo(u)r Association of United States 美国色材学会
colo(u)r assortment 颜色搭配
colo(u)ration 显色;颜色;着色(作用);染色
colo(u)ration by colloid(al) particles 胶体粒子着色
colo(u)ration by radiation 辐射着色
colo(u)ration formation 颜色生成
colo(u)ration removal 去色
colo(u)r atlas 彩色图谱;色样本;色谱(集);色卡图册
colo(u)r autoradiography 彩色放射自显影
colo(u)r axis 色轴
colo(u)r balance 颜色调谐;彩色平衡;色(彩)平衡
colo(u)r balance adjustment 彩谐调整
colo(u)r band method 色带(测流)法
colo(u)r bar 彩(色)条
colo(u)r bar code 彩带码
colo(u)r bar dot crosshatch generator 彩色条点交叉图案信号发生器
colo(u)r bar generator 彩条信号发生器;色条信号发生器
colo(u)r bar pattern 色带图
colo(u)r bar pedestal 彩色消隐脉冲电平
colo(u)r bar signal 彩条信号
colo(u)r bar test pattern 彩条测视图
colo(u)r bar vector 色带矢量
colo(u)r bar Y buffer 彩条亮度缓冲器
colo(u)r bar Y level 彩条亮度电平
colo(u)r base 底色;底漆;彩色中间涂层;彩色底漆;发色母体;染色基料;调色基料;色基
colo(u)r base control 色底控制
colo(u)r batch 色母料
colo(u)r bias 色偏差
colo(u)r black 着色用炭黑;炭黑颜料
colo(u)r bleeding 泗色
colo(u)r bleeding resistance 抗渗色性;抗汜色性;抗混色性
colo(u)r blend 混色(料);调色
colo(u)r blindness 色盲
colo(u)r blindness film 盲色胶片
colo(u)r bloom 色起霜
colo(u)r boost 色升高
colo(u)r box 彩色盒
colo(u)r breakup 色乱;色分解;颜色分层
colo(u)r brightness 颜色明度;颜色亮度;彩色明亮度;色明度;色(彩)亮度
colo(u)r-brightness tester 色亮度测试仪
colo(u)r buffing 镜面抛光;消色
colo(u)r burn-out 油墨热致变色
colo(u)r burst 彩色同步信号脉冲;彩色副载波群;彩色定向信号脉冲;色同步(信号)
colo(u)r burst amplifier 彩色同步脉冲放大器
colo(u)r burst amplifier stage 彩色同步脉冲放大级
colo(u)r burst flag 彩色同步键控信号
colo(u)r burst interval 彩色闪光时间
colo(u)r burst pedestal 彩色基准波群台阶
colo(u)r camera 彩色摄影机;彩色摄像机
colo(u)r camera registration control 彩色摄像机配准控制
colo(u)r candle 彩色蜡烛
colo(u)r canning printing 分色样图
colo(u)r carbon ink 彩色复写墨
colo(u)r card 颜色样本
colo(u)r carrier 载色剂;彩色信号载波
colo(u)r carrier modifier 彩色载波修正器;彩色载波调节器
colo(u)r-carrier reference 色度载波基准频率
colo(u)r cast 偏色;色偏;彩色设计
colo(u)r cast paper 高光泽色纸
colo(u)r catalog 彩色样本
colo(u)r cell 彩色元件;彩色单元
colo(u)r cement 加色水泥;有色水泥
colo(u)r center 彩色中心;色心
colo(u)r center 色心
colo(u)r center absorption 色光中心吸收

colo(u)r center crystal 色心晶体
colo(u)r center laser 色心激光器
colo(u)r chalking 色料粉化
colo(u)r chalking of plastics 塑料色料起霜
colo(u)r change 换色;色调变化
colo(u)r change in flower 花色改变
colo(u)r change interval 变色区域
colo(u)r-changing agent 变色剂
colo(u)r-changing temperature indicating paint 示温涂料
colo(u)r channel 彩色信号通道
colo(u)r character display 彩色字符显示器
colo(u)r chart 颜色样卡;彩色图表;彩色测试图;比色图表;色彩图表;色表
colo(u)r check 比色检验;色度鉴定
colo(u)r chip(ping)s 颜色样本;色母片;色卡
colo(u)r circle 颜色环;色(相)环;色谱圈
colo(u)r circuit 彩色信号电路
colo(u)r class 色组
colo(u)r coat/clear coat 彩色涂层/透明涂层系统
colo(u)red concrete aggregate 涂色的混凝土集料;涂色的混凝土骨料
colo(u)r-coated steel 涂色钢材
colo(u)r coated steel sandwich board 彩钢复合板
colo(u)r-coat(ing) 彩色涂层;加色层
colo(u)r coating glass 彩色玻璃;涂色玻璃
colo(u)r code 涂色作标记;涂色规定;色码;色标
colo(u)r-coded cable 色码电缆
colo(u)r coded computer memory 彩色编码计算机存储器
colo(u)r-coded graph 彩色编码图
colo(u)r-coded light beam 彩色编码光束;色编码光束
colo(u)r-coded map 彩色编码(地)图
colo(u)r coded thermal infrared imagery device 彩色编码红外线成像装置
colo(u)r code number 色码数;色标数
colo(u)r coder 彩色(信号)编码器;比色仪
colo(u)r coding 彩色编码
colo(u)r coding paint 彩色编号涂料
colo(u)r coding system 彩色编码系统
colo(u)r coefficient 彩色系数
colo(u)r collotype 彩色珂罗版
colo(u)r colo(u)r chart 色卡
colo(u)r-colo(u)r diagram 两色图
colo(u)r-colo(u)r plot 两色图
colo(u)r combination 彩色合成;配色;色彩调合
colo(u)r command 彩色指令
colo(u)r comparator 比色仪;比色器;比色计
colo(u)r comparator pyrometer 比色高温计
colo(u)r comparimetry 比色法
colo(u)r comparison 比色
colo(u)r comparison tube 比色管
colo(u)r compensating 彩色补偿
colo(u)r compensating filter 彩色补偿滤光镜;补偿滤色片;补偿滤色镜;彩色补偿滤色镜
colo(u)r compensation 光补偿;色彩补偿
colo(u)r compensation technique 色补偿技术
colo(u)r component 彩色分量;彩色成分
colo(u)r component oscilloscope 彩色分量示波器
colo(u)r composite 彩色合成
colo(u)r composite image 彩色合成影像;彩色合成像片
colo(u)r compositing technique 彩色合成技术
colo(u)r composition 彩色组成;彩色合成;彩色成分
colo(u)r compound 着色混合料
colo(u)r computer 配色计算机
colo(u)r computer graphic(al) system 彩色计算机图像系统
colo(u)r concentrate 浓颜料制备物;浓色母料;颜料提浓物
colo(u)r concentration 颜色鲜艳度;颜色浓度
colo(u)r concentration correction 颜色浓度校正
colo(u)r conditioning 色彩设计;彩色调配;彩色调节
colo(u)r cone 色锥
colo(u)r consistency 颜色一致性;色浆稠度
colo(u)r constancy 颜色守恒;色觉恒常;色恒定性;色感一致性
colo(u)r constant 色恒量;色度常数
colo(u)r contamination 串色;彩色污染;彩色混杂;色(彩)混(杂)
colo(u)r content 纯色含量;色调范围;色彩成分

colo(u)r contour 彩色等值线图
colo(u)r contrast 颜色反衬;颜色对比;彩色对比度;色反差;色对比(度);色调对比;色衬度;色彩对比(度)
colo(u)r control 彩色调节;色度调整;色的控制
colo(u)r controller 颜色控制器
colo(u)r-control monitor 彩色图像监控器
colo(u)r control scale 色标
colo(u)r control strip 彩色控制条
colo(u)r control system 颜色控制系统
colo(u)r control unit 彩色控制装置
colo(u)r convention 用色惯例
colo(u)r conversion filter 彩色转换滤光镜;色转换滤色镜
colo(u)r converter 彩色图像变换器
colo(u)r coordinates 色坐标
colo(u)r coordinate system 彩色坐标系
colo(u)r coordinate transformation 色坐标变换
colo(u)r coordination 色彩协调
colo(u)r copy 彩色复印
colo(u)r cord 彩色软线
colo(u)r-cord colo(u)rimeter 色带比色计
colo(u)r core 彩色软线
colo(u)r corrected lens 消色差透镜
colo(u)r-corrected mercury lamp 颜色校正过的水银灯
colo(u)r correction 色修正;色改正;色差校正
colo(u)r correction factor 色修正系数
colo(u)r correction filter 彩色校正滤光片;彩色校正滤光镜
colo(u)r correction mark 彩色校正遮片
colo(u)r correction mask 彩色校正膜片
colo(u)r corrector 彩色校正器;色彩校正器
colo(u)r coupler 彩色成色剂
colo(u)r coupling component 成色剂
colo(u)rcrete 彩色快硬水泥;彩色混凝土
colo(u)r-critical graph 色临界图
colo(u)r curve 颜色曲线
colo(u)r cycle 流行色重复出现;色相环;色轮
colo(u)r darkening 颜色变暗
colo(u)r data 彩色信号数据;彩色信号参数
colo(u)r data display 彩色数据显示器
colo(u)r decoder 彩色信号译码电路;彩色解码器
colo(u)r decoder module 彩色解码组件
colo(u)r decoding 彩色解码
colo(u)r definition 颜色鲜明度;颜色清晰度;颜色测试
colo(u)r definition chart 彩色测试卡
colo(u)r demodulator 彩色信号反调制器;彩色像解调器
colo(u)r densitometer 彩色密度计
colo(u)r densitometry 彩色密度测定
colo(u)r density 颜色密度;彩色密度;色(彩)密度
colo(u)r density ratio 色调强度比;色彩密度比
colo(u)r density slicing 彩色密度分割
colo(u)r dependent error 色相关误差
colo(u)r depth 颜色浓度
colo(u)r design 彩色设计;色彩设计
colo(u)r designation 颜色标志
colo(u)r deterioration 退色
colo(u)r developer 彩色显影液;彩色显影剂
colo(u)r developing agent 彩色显影触媒剂;显示剂
colo(u)r developing solution 彩色显影液;彩色显影剂
colo(u)r development 彩色显影;显色
colo(u)r deviation 彩色失真;色偏差
colo(u)r diagram 比色图;色图
colo(u)r-difference 色差
colo(u)r difference acuity 色差视力;色差视觉;色差灵敏度
colo(u)r-difference amplifier 色差放大器
colo(u)r difference computer 色差计算机
colo(u)r difference equation 色差方程
colo(u)r difference formula 色差公式
colo(u)r difference meter 色差计
colo(u)r difference sensitivity 色差灵敏度
colo(u)r difference signal 色差信号
colo(u)r difference voltage 色差信号电压
colo(u)r diffusion 色扩散
colo(u)r dilution 色冲淡;色饱和度降低
colo(u)r dilution method 染料溶液法
colo(u)r dimension 色维
colo(u)r dimensions 颜色属性

colo(u)r disc 滤色片;色盘;色板
colo(u)r discrimination 颜色识别;颜色辨别;辨色能力;色鉴别
colo(u)r displacement 色位移
colo(u)r display 彩色显示器;彩色显示;色彩显示
colo(u)r display monitor 彩色显示监视器
colo(u)r distortion 彩色失真;色失真;色畸变
colo(u)r distribution 颜色分布
colo(u)r-division multiplexing 色分复用
colo(u)r doctor 色浆刮刀
colo(u)r dot 彩色斑点;色点
colo(u)r dressing 彩色整饰
colo(u)r drift 颜色发飘
colo(u)r drum 彩色镜鼓;彩色滚筒;上色滚筒
colo(u)r dub 彩色复制
colo(u)r duplication 彩色复制
colo(u)r dynamics 色彩浓淡法;色彩调配
colo(u)red 着色的;彩色
colo(u)red after image 彩色余像
colo(u)red aggregate 彩色集料;彩色骨料
colo(u)red alumin(i)um 着色的铝
colo(u)red anodic finish 彩色阳极保护(饰面)层
colo(u)red art pen 彩色水笔
colo(u)red asbestos-cement tile 彩色石棉水泥瓦
colo(u)red asphalt 着色沥青
colo(u)red background coating 有色衬底涂料
colo(u)red body 色坯
colo(u)red body china 色胎瓷
colo(u)red bulb 彩色灯泡
colo(u)red cement 有色水泥;彩色水泥
colo(u)red cement concrete 彩色水泥混凝土
colo(u)red cement face brick 彩色水泥(外)砖(墙)
colo(u)red cement spraying 彩色水泥喷涂
colo(u)red cement tile 彩色水泥瓦
colo(u)red chalk 彩色粉笔
colo(u)red clay 带色黏土
colo(u)red compound 有色化合物
colo(u)red concrete 着色混凝土;彩色混凝土
colo(u)red concrete road 有色混凝土路(面)
colo(u)red cones 导向色灯
colo(u)red contour map 彩色等量线图
colo(u)red covernote notebock book 彩色皮笔记本
colo(u)red cover paper yellow 黄色书面纸
colo(u)red crepe paper 彩色皱纹纸
colo(u)red crystal 有色晶体;增色晶体
colo(u)red defect 带色毛
colo(u)red display of data 数据彩色表示方法
colo(u)red drawing 彩画
colo(u)red electric(al) bulb 彩色(电)灯泡
colo(u)red enamel(glaze) 彩色瓷釉
colo(u)red felt-tipped pen 彩色毡尖笔
colo(u)red filter 滤色器;滤色片;滤色镜
colo(u)red finish 有色的装修;彩色罩面;彩色饰面
colo(u)red flux 染色助熔剂
colo(u)red foam glass 彩色泡沫玻璃
colo(u)red fog 有色雾翳;有色灰雾
colo(u)red form 着色样张
colo(u)red fringe 彩色条纹
colo(u)red frit 带色熔块
colo(u)red galvanized iron 彩色镀锌铁皮
colo(u)red garden tile 彩色花园砖
colo(u)red gelatin filter 明胶滤色片
colo(u)r edging 彩色镶边;彩色边缘
colo(u)red glass 有色玻璃;着色玻璃;彩色玻璃;色玻璃
colo(u)red glass beads 着色玻璃细珠
colo(u)red glass block 有色玻璃砖块
colo(u)red glass fibre 着色玻璃纤维
colo(u)red glass filter 有色玻璃滤光片;彩色玻璃滤光片
colo(u)red glaze 琉璃;彩釉;色釉
colo(u)red glazed coat(ing) 有色釉料涂层
colo(u)red glazed finish 彩色装修
colo(u)red glazed tile 彩色瓷砖
colo(u)red glazed wall tile 彩色釉面墙砖
colo(u)red-glaze flower pot 彩釉花盆
colo(u)red glazing 有色玻璃装配;上有色釉
colo(u)red glazing tile 琉璃瓦;彩釉瓦
colo(u)red hair 彩色发
colo(u)red ink 有色墨水;彩色油墨
colo(u)red ion 有色离子
colo(u)r edition 彩色版

colo(u)red knops 彩色结子纱
colo(u)red knot 带色节疤
colo(u)red laminated glass 彩色叠层玻璃
colo(u)red lamp 彩色灯
colo(u)red layer 色层
colo(u)red lead light 多色彩小方格玻璃
colo(u)red light 有色(灯)光
colo(u)red lustre ware 彩釉制品
colo(u)red map 彩色地图
colo(u)red marble 彩色大理石
colo(u)red masaic 彩色马赛克
colo(u)red mastic asphalt 有色沥青玛琋脂
colo(u)red match 彩色火柴
colo(u)red mineral aggregate 彩色矿物集料;彩色矿物骨料
colo(u)red mixed plaster 彩色灰浆
colo(u)red mortar 彩色砂浆;彩色灰浆
colo(u)red mortar finish 彩色砂浆罩面;彩色灰浆饰面
colo(u)red mortar spraying 彩色砂浆喷涂;彩色灰浆喷涂
colo(u)redness 有色性
colo(u)red newsprint 彩色新闻纸
colo(u)red noise 有色噪声
colo(u)red oiled crayon 彩色蜡笔
colo(u)red ophthalmic glass 有色眼镜片;有色眼镜玻璃
colo(u)red optic(al) glass 有色光学玻璃
colo(u)red optic(al) lens 光学彩色镜片
colo(u)red organics 有色有机物
colo(u)red original 着色原图
colo(u)red paint 有色涂料;彩艺术油漆;彩色油漆;彩色涂料;色漆
colo(u)red paper 彩色刻纸
colo(u)red pavement 加色路面;彩色路面
colo(u)red perspective 彩色透视
colo(u)red pigment 着色颜料
colo(u)red pigment for concrete 混凝土用着色颜料;混凝土用颜料
colo(u)red plaster 彩色石膏;彩色粉刷
colo(u)red plate glass 彩色平板玻璃;彩色板玻璃
colo(u)red Portland cement 彩色硅酸盐水泥;彩色波特兰水泥
colo(u)red rendering 彩色渲染
colo(u)red resin reagent 有色树脂试剂
colo(u)red ring 色环
colo(u)red road surfacing 加色路面
colo(u)red rubber-covered barbell 彩色包胶杠铃
colo(u)red safety lens 有色防护眼镜
colo(u)red-sand 染色砂
colo(u)red-sand experiment 染色砂试验
colo(u)red scenery picture 彩色风景画
colo(u)red sculpture 彩塑;彩雕
colo(u)red sensitized material 彩色感光材料
colo(u)red shade 色镜
colo(u)red shadow 有色阴影
colo(u)red sheet glass 彩色平板玻璃;彩色窗玻璃
colo(u)red silica glass 彩色石英玻璃
colo(u)red smoke 有色烟幕
colo(u)red speck 色斑
colo(u)red spectacle 互补色眼镜
colo(u)red stainless steel 彩色不锈钢
colo(u)red suspended matter 有色悬浮物;带色悬浮物
colo(u)red tea set with white band 彩色拉丝茶具
colo(u)red terrazzo 彩色水磨石
colo(u)red thick glass 彩色厚玻璃
colo(u)red thread 彩色线
colo(u)red tile 彩色砖瓦
colo(u)red tracer 着色示踪物
colo(u)red varnish 着色清漆;透明色漆
colo(u)red vase 彩色花瓶
colo(u)red viewing system 互补色观察系统
colo(u)red wastewater 有色废水
colo(u)red water 有色水;带色水
colo(u)red woodcut 套色木刻
colo(u)r effect 颜色效应
colo(u)r electrostatic printer 彩色静电复印机
colo(u)r element 彩色像素;色素
colo(u)r-emitting phosphor dot 彩色荧光屏发光点
colo(u)r enamels 珐琅彩
colo(u)r encoded focused image hologram 彩色编码聚集图像全息图
colo(u)r encoding files 彩色编码文件
colo(u)r encoding unit 彩色编码器
colo(u)r enhancement 彩色增强
colo(u)r enlarger 彩色扩印机;彩色放大机
colo(u)r equalizer 彩色均衡器
colo(u)r equation 颜色方程;彩色方程;色(谱)方程
colo(u)r equivalent 色当量
colo(u)rer 油漆工;彩画工
colo(u)r error 颜色误差;色差
colo(u)r evaluation 色评定
colo(u)r evaluation of weight 色彩的重量感
colo(u)r excess 彩色过剩;色余
colo(u)r-extinction pyrometer 吸色高温计
colo(u)r eye 分色睛
colo(u)r facility 颜色设备
colo(u)r facsimile 彩色传真
colo(u)r facsimile apparatus 彩色传真机
colo(u)r-fast 不褪色的;颜色坚牢的
colo(u)r fastness 颜色坚牢度;不褪色性;色牢度
colo(u)r fastness daylight exposure 耐日晒色牢度
colo(u)r fastness grading 色牢度评级
colo(u)r fastness rating 色牢度评级
colo(u)r fastness to light 耐光色牢度;颜色耐日晒牢度
colo(u)r fatigue 色觉疲劳
colo(u)r fidelity 彩色逼真(度);色(彩)保真度
colo(u)r field 彩色场;色(视)场
colo(u)r field corrector 色场校正器
colo(u)r figure 色图
colo(u)r film 彩色影片;彩色片;彩色(胶)片
colo(u)r film analyzer 彩色胶片分析仪
colo(u)r film layers-separate sensitivity 彩色片分层感光度
colo(u)r filter 滤色器;滤色片;滤色镜;滤光片;颜色滤光片;彩色转盘;色滤光片
colo(u)r filter frame 滤色片架
colo(u)r filtering 滤色
colo(u)r finish 着色饰面
colo(u)r finishing 彩色装饰
colo(u)r fixing 媒染剂;颜固定;定色(料)
colo(u)r flash 闪色
colo(u)r flexer 彩色编码器
colo(u)r flicker 彩色闪烁
colo(u)r floating 浮色;发花
colo(u)r floating glaze 颜色浮面釉
colo(u)r floc 有色絮凝体
colo(u)r-flop 随角异色;随观察角度变化的颜色
colo(u)r-flop coating 颜色随观察角度而变化的涂层;随角异色涂层
colo(u)r fluctuation 颜色波动;彩色起伏;色度波动
colo(u)r flying spot instrument 彩色飞点扫描设备
colo(u)r formation 颜色生成
colo(u)r forme 彩色印版
colo(u)r former 成色剂
colo(u)r former agent 彩色成色剂
colo(u)r frame 滤光片;色帧
colo(u)r-frame frequency 色帧频
colo(u)r fringe 彩色边纹
colo(u)r fringing 彩色镶边;彩色边纹现象
colo(u)rful projective luminaire for large screen 大屏幕彩色投影灯具
colo(u)r fusion 色融合
colo(u)r gain, delay test set 彩色增益与时延测试仪
colo(u)r gamut 彩色范围;全色图;色域
colo(u)r gate 彩色选通电路
colo(u)r gelatine filter 彩色明胶滤色镜
colo(u)r generation 呈色
colo(u)r generator lock 彩色同步锁相
colo(u)r geometry 色几何学
colo(u)r glass 颜色玻璃
colo(u)r glass bulb 彩色玻璃灯泡
colo(u)r glass filter 有色玻璃滤光片;玻璃滤色镜
colo(u)r glaze 颜色釉
colo(u)r glaze brick 彩色釉面砖
colo(u)r-glazed sand paint for building decoration 建筑装饰彩釉砂涂料
colo(u)r glaze porcelain 彩釉瓷器
colo(u)r gloss 色彩光泽
colo(u)r gradation 色等级
colo(u)r-gradation technique 叠晕染法
colo(u)r grade 比色卡;色谱;色度;色标

colo(u)r grader 色度测量计;色度测定仪
colo(u)r gradient 分层设色【测】
colo(u)r grading 色度分级
colo(u)r graduation 色阶
colo(u)r graph 彩图;色图;电子分色计
colo(u)r graphic 彩色图示
colo(u)r graphic monitor 彩色图像监视器
colo(u)r graphic network software 彩色网络图形软件
colo(u)r graphic plotter 彩色图形绘图机
colo(u)r graphic printer 彩色图形打印机
colo(u)r graphic software 彩色图形软件
colo(u)r graphic terminal 彩色图形终端机
colo(u)r grating 色彩分级
colo(u)r gravure 彩色照像凹版;彩色凹印
colo(u)r grid 色栅
colo(u)r grinder 颜料研磨机;色料研磨机
colo(u)r grinding machine 颜料研磨机;色料研磨机
colo(u)r group 色群
colo(u)r guide 彩色控制条;分色样图
colo(u)r harmonic theory 色彩调色理论
colo(u)r harmony 颜色调和;彩色和谐;色彩协调;色彩调和
colo(u)r harmony manual 颜色调和手册
colo(u)r hearing 色联觉;色幻觉
colo(u)r hexagon 色六角
colo(u)r hold circuit 彩色信号同步电路
colo(u)r hologram 彩色全息照相;彩色全息图
colo(u)r holographic(al) map 彩色全息地图
colo(u)r holography 彩色全息术
colo(u)r hue 色相;色调
colo(u)r identification 显色鉴定;颜色鉴别
colo(u)r identification of pipe lines 管线颜色标准化
colo(u)r illuminating 彩色照明
colo(u)r image 彩色图像;彩色图文;色像
colo(u)r image combination device 彩色图像合成仪
colo(u)r image display system 彩色图像显示系统
colo(u)r image dissector 彩色析像管
colo(u)r image map 彩色影像地图
colo(u)r image separation 彩色图像分离
colo(u)rimeter 比色器;比色表;测色计;比色计;色度计;色彩计
colo(u)rimeter density 测色浓度
colo(u)rimeter oxygen detector 比色计测氧仪
colo(u)rimeter with lamp 带灯色度计
colo(u)rimetric 比色的
colo(u)rimetric analysis 比色分析(法);色度分析
colo(u)rimetric analyzer 比色分析仪
colo(u)rimetric application 比色应用
colo(u)rimetric assay 比色定量
colo(u)rimetric card 比色卡
colo(u)rimetric comparator 比色器
colo(u)rimetric cylinder 比色管
colo(u)rimetric determination 比色测定
colo(u)rimetric determination method 比色测定法
colo(u)rimetric disc 比色盘
colo(u)rimetric dosimeter 比色剂量计
colo(u)rimetric estimate 比色测定
colo(u)rimetric measurement 比色测定
colo(u)rimetric method 比色法;色度法
colo(u)rimetric oxygen detector 比色测氧仪
colo(u)rimetric parameter 杯色参数;色调参数
colo(u)rimetric pH measurement 比色法pH测定
colo(u)rimetric photometer 比色光度计;滤色光度计
colo(u)rimetric purity 明度纯度;比色纯度;色度纯度;色纯(度)
colo(u)rimetric pyrometer 比色高温计
colo(u)rimetric quality 色品度
colo(u)rimetric radiation detector 比色辐射探测器;色度测定辐射探测器
colo(u)rimetric reaction 比色反应
colo(u)rimetric reagent 比色试剂
colo(u)rimetric relay 光波波长继电器
colo(u)rimetric scale 比色计刻度
colo(u)rimetric shelf 比色架
colo(u)rimetric shift 比色偏移;色(度)位移
colo(u)rimetric solution 比色溶液
colo(u)rimetric stand 比色架
colo(u)rimetric standard 色度标准

colo(u)rimetric standard illuminant 彩色标准光源
colo(u)rimetric system 色度体系
colo(u)rimetric test 比色测定;比色法检验;比色试验
colo(u)rimetric titration(method) 比色滴定法
colo(u)rimetric tube 比色管
colo(u)rimetric value 集料中不纯成分;比色(分析)值;色度值
colo(u)rimetrist 色度学家
colo(u)rimetry 测色学;测色(法);比色分析(述);比色法;色觉检查;色度学;色度术;色度测量
colo(u)rimetry measurement loci 色度测定域
colo(u)rimetry measure method of reflection colo(u)r 反射色的色度测量法
colo(u)rimetry with platinum-cobalt 铂—钴比色法
colo(u)r improver 颜色改良剂;颜色改进剂
colo(u)r index 现色指数;颜色指数;颜色指标;彩色指数;染色指数;染料索引;色指数;色相指数;色率;色标;色儿
colo(u)r-index error 彩色测定误差
colo(u)r-indexing circuit 彩色定相电路
colo(u)r indexing pulse 彩色指示脉冲
colo(u)r index name 染料索引名
colo(u)r index number 染料索引号
colo(u)r index pulse separator 彩色指示脉冲分离器
colo(u)r indicator 彩色指示器;色指示器
colo(u)r induction 色感应
colo(u)r industrial TV camera 彩色工业电视摄像机
colo(u)r information 彩色信息
colo(u)r infrared airphoto 彩色红外航片
colo(u)r infrared film 彩色红外(胶)片
colo(u)r infrared photograph 彩色红外像片
colo(u)r infrared photography 彩色红外摄影(术)
colo(u)ring 颜色法;轧色浆;染色;上色
colo(u)ring admixture 加色剂;着色外加剂
colo(u)ring agent 有色物(质);颜料;着色剂;染色剂;染料;色素
colo(u)ring brush 颜料刷
colo(u)ring by tempering 回火着色
colo(u)ring concrete 有色混凝土
colo(u)ring earth 矿质颜料;矿物染料
colo(u)ring material 着色剂;着色材料;色料
colo(u)ring matter 颜料;产色物质;染料
colo(u)ring metal 金属涂色
colo(u)ring of asphalt 沥青着色
colo(u)ring of crystal 晶体着色
colo(u)ring of glass fiber 玻璃纤维着色
colo(u)ring pigment 彩色颜料;油漆颜料;着色颜料;着色剂;色料
colo(u)ring power 染色能力;着色强度;着色力
colo(u)ring substance 着色物体;着色材料
colo(u)ring triangle 原色三角形
colo(u)ring value 着色值
colo(u)ring varnish 漆色浆
colo(u)r ink 彩色颜料
colo(u)r in oil 油性色浆;颜料油性分散体;调色漆;颜料油漆
colo(u)r integrity 颜色完整性
colo(u)r intensified photomap 彩色增强影像地图
colo(u)r intensity 着色强度;颜色强度;着色力;彩色信号强度;彩色明度;彩色亮度;白色颜料消色力
colo(u)r international standard depth 颜色国际标准深度
colo(u)r in varnish 颜料油性分散体
colo(u)risically 在色彩上
colo(u)rist 着色师;善于应用色彩的设计师;善于应用色彩的画家;配色师;配色人员
colo(u)ristic property 颜色性能
colo(u)rity 颜色;色度
colo(u)rity determination 色度测定
colo(u)r jet printer 彩色喷射打印机
colo(u)r jewel 彩色宝石
colo(u)r-key 以颜色分类
colo(u)r key sheet 分涂参考图
colo(u)r killer 消色器;彩色抑制电路
colo(u)r killer circuit 消色电路;彩色信号切断电路
colo(u)r killer stage 彩色信号抑制级
colo(u)r killer tube 彩色抑制管
colo(u)r kinescope 彩色显像管

colo(u)r kinescope recording 彩色显像管记录;彩色屏幕录像
colo(u)r knife 色浆刮刀
colo(u)r label on a bottle 彩色瓶贴
colo(u)r lacking uniformity 颜色不均
colo(u)r lacquer 彩色亮漆
colo(u)r lake 色淀
colo(u)r lamp 色泡
colo(u)r lens 有色透镜
colo(u)rless 无色的
colo(u)rless base 无色底釉
colo(u)rless calcium aluminate glass 无色钙铝酸盐玻璃
colo(u)rless coupler 无色成色剂
colo(u)rless flux 无色熔剂
colo(u)rless glass 无色玻璃
colo(u)rless glaze 无色釉
colo(u)rless hollow ware 无色空心玻璃制品
colo(u)rless lens 无色透镜
colo(u)rlessness 无色性
colo(u)rless optic(al) glass 无色光学玻璃
colo(u)rless optic(al) glass classification 无色光学玻璃分类法
colo(u)rless powder 无色的粉剂
colo(u)rless spectacle glass 无色眼镜玻璃
colo(u)r level 彩色信号电平
colo(u)r light 彩色光
colo(u)r-light analogue 色光模拟
colo(u)r light band 彩色光带
colo(u)r light glass 彩光玻璃
colo(u)r lighting 彩色照明
colo(u)r-light sequence 色灯顺序
colo(u)r light signal 彩光信号机;色光信号;色灯信号
colo(u)r limitation 着色限度
colo(u)r line screen 彩色线式荧光屏
colo(u)rlithograph 彩色平版
colo(u)rlithographic(al) printing 彩色平版印刷
colo(u)r lock receiver 彩色锁相接收机
colo(u)r lock transmitter 彩色锁相发射机
colo(u)r magnetic board 彩色磁板
colo(u)r magnification error 色放大率误差
colo(u)rman 颜料商;染色师
colo(u)r marble chips 色石渣
colo(u)r mark 颜色标志;颜色标记
colo(u)r marker 色指示器;色标识器
colo(u)r marking 有色标记;彩色标志;色标志
colo(u)r mask 滤光片;彩色障板;彩色蒙片
colo(u)r masking 彩色遮蔽;彩色修饰;彩色蒙版法
colo(u)r masking computer 色彩校正计算机
colo(u)r master differential colo(u)rimeter 主色微差比色计
colo(u)r match 颜色一致性;色匹配;配色;颜色匹配
colo(u)r matcher 配色人员
colo(u)r matching 配色;颜色调配;等色;调色;色匹配;颜色匹配
colo(u)r matching aptitude test 配色性能试验
colo(u)r matching computer 配色计算机
colo(u)r matching filter 彩色匹配滤光片
colo(u)r matching function 配色函数
colo(u)r matching method 配色法
colo(u)r matching program(me) 调色程序;配色程序
colo(u)r-match(ing) unit 着色匹配单位
colo(u)r material 彩色感光材料
colo(u)r matrix 彩色矩阵
colo(u)r matrix unit 彩色矩阵电路
colo(u)r measurement 颜色测量;颜色测定;测色;彩色测量
colo(u)r measurement by spectrophotometry 分光测色法
colo(u)r measurement system 测色装置
colo(u)r measuring 颜色测量
colo(u)r measuring instrument 色度仪;测色仪器;比色测量仪器
colo(u)r measuring system 色彩测量装
colo(u)r medium 有色介质
colo(u)r metallograph 彩色金相照片
colo(u)r meter 测色计
colo(u)r method of measuring velocity 颜色测流(速)法;色水测流法
colo(u)r microfilm 彩色缩微胶片
colo(u)r microphotography 彩色缩微摄影(术)

colo(u)r migration 串色;色移;色迁移
colo(u)r mill 颜料磨碎机;色料研磨机
colo(u)r-mirror reflector 二向色反光镜
colo(u)r misconvergence 色失聚
colo(u)r mix 色料混合
colo(u)r mixer 混色器;颜色混合器;调色机
colo(u)r mixing 混合颜色;调和颜色;混色;颜色混合;调色
colo(u)r mixing room 配色间
colo(u)r mixing tray 调色碟
colo(u)r mixture 混合色料;颜色混合
colo(u)r mixture computer 颜色混合计算机
colo(u)r mixture curve 颜色混合曲线
colo(u)r mixture data 混色数据
colo(u)r mixture diagram 混色图;色度图
colo(u)r model 分涂参考模型
colo(u)r modelling 彩塑
colo(u)r mode selector 彩色制式选择器
colo(u)r modifier 颜色调整剂
colo(u)r modulation 调色
colo(u)r modulator 彩色调制器
colo(u)r moment 色矩
colo(u)r monitor 颜色控制仪;彩色(图像)监控器;彩色监视器
colo(u)r monitoring 彩色图像监控
colo(u)r motif 颜色基调;色彩基调
colo(u)r multipicture separator 彩色多画面分割器
colo(u)r multiplexing 彩色信号多路传输(系统);彩色副载频正交调制
colo(u)r name 颜色名称;色名
colo(u)r name charts of ISCC-NBS 美国色彩联络协会和国家标准局制订的色名表示方法
colo(u)r natural-scene analysis 彩色自然景物分析
colo(u)r negative(film) 彩色负片;彩色底片;彩色照相胶卷;彩色负像
colo(u)r negative film developer 彩色负片显影剂
colo(u)r negative film process(ing) solution 彩色负片冲洗液
colo(u)r net system 彩色电视网系统
colo(u)r noise 彩色信号杂波
colo(u)r notation 颜色数标法;颜色表示法
colo(u)r notation system 表色系统
colo(u)r note 色标
colo(u)r number 染色值;色值;色数
colo(u)r of clinker 熟料颜色
colo(u)r of fault rock 断层岩颜色
colo(u)r of ground 底色
colo(u)r of inclusion 包裹体颜色
colo(u)r of kerogen 干酪根颜色
colo(u)r of petroleum 石油的溯色
colo(u)r of reservoir 储集层颜色
colo(u)r of right 表面上的权利
colo(u)r of rocks 岩石的颜色
colo(u)r of seawater 海水色度
colo(u)r of sedimentary rock 沉积岩颜色
colo(u)r of title 冒名
colo(u)r of wash liquid 冲洗液颜色
colo(u)r of water 水的色度
colo(u)r oil 调色油
colo(u)r on colo(u)r 重色配色法
colo(u)r optics 彩色光学
colo(u)r order system 表色系统
colo(u)r oscillator 彩色(信号)振荡器
colo(u)roto 彩色轮转凹印
colo(u)r oven 铅丹炉
colo(u)r overlapping 搭色
colo(u)r overload 色过载;色过量;色过饱和
colo(u)r oxide 着色氧化物
colo(u)r pack camera 背包式彩色摄像机
colo(u)r painter 彩色画
colo(u)r painting 彩色画
colo(u)r pan 调色桶
colo(u)r paper 彩色相纸
colo(u)r paper-developer 彩色相纸显影剂
colo(u)r particle 颜料粒子;色料粒子
colo(u)r particle coating 颜料粒子包膜层
colo(u)r particle power 颜料粒子间的黏结力
colo(u)r paste 色浆;染料糊
colo(u)r patch 小彩色胶片;色渍
colo(u)r-patch map 分区着色图
colo(u)r pattern 彩色整饰;彩色样图;色型
colo(u)r pattern generator 彩色测试图发生器
colo(u)rpavet 彩色沥青瓦
colo(u)r pellet 色料片

colo(u)r pen 彩笔
colo(u)r pencil 测温(色)笔;彩色铅笔;笔型(显色)测温计;笔型温度计
colo(u)r pencil drawing 彩色铅笔画
colo(u)r pen plotter 彩色绘图仪
colo(u)r perception 色知觉
colo(u)rphalt 彩色沥青
colo(u)r phase 彩色相位
colo(u)r phase alternation 彩色(信号)相位交变
colo(u)r phase alternation multivibrator 彩色相位变换多谐振荡器
colo(u)r phase chart 彩色副载波矢量图
colo(u)r phase constrat microscopy 彩色相衬显微术
colo(u)r phase detector 彩色相位检测器;彩色相位检比器
colo(u)r phase setter 彩色相位给定器
colo(u)r phase stabilizer 彩色相位稳定器
colo(u)r phasing amplifier 彩色信号定相放大器
colo(u)r phosphor 彩色磷光粉
colo(u)r photo facsimile transmitter 彩色照片传真发送机
colo(u)r photograph 彩色照片;彩色像片
colo(u)r photographic(al) paper 彩色像纸;彩色(照)相纸
colo(u)r photographic(al) processing 彩色摄影处理
colo(u)r photography 彩色照相术;彩色摄影(术)
colo(u)r photometer 测色仪
colo(u)r photomicrograph 彩色显微像片;彩色显微照相版
colo(u)r photomultiplier 彩色光电倍增管
colo(u)r photo paper 彩色照相纸
colo(u)r pickup 着色量
colo(u)r pick-up tube 彩色摄像管
colo(u)r picture 彩色图像;彩色图片
colo(u)r pictured pottery 彩绘陶
colo(u)r picture processing system 彩色图像处理系统
colo(u)r-picture rate 彩色影像频率;色像频率
colo(u)r picture signal 彩色图像信号
colo(u)r picture tube 彩色显像管
colo(u)r picture tube bulb outline measurer 彩色显像管玻壳外形测试设备
colo(u)r pigment 色彩颜料;彩色颜料
colo(u)r plane 彩色(平)面
colo(u)r planning 色彩设计
colo(u)r plate 显色板;颜色样板;彩色版;比色板
colo(u)r plate cylinder 彩色印版滚筒
colo(u)rplexer 彩色形成器;三基色信号形成设备
colo(u)r plotter 彩色绘图仪
colo(u)r popularity 流行色
colo(u)r Portland cement 彩色水泥
colo(u)r positive 彩色正片
colo(u)r positive film 彩色正片
colo(u)r positive film developer 彩色正片显影液
colo(u)r potential 浅色
colo(u)r press 彩(色)印(刷)机
colo(u)r press block 彩色凸版
colo(u)r previewer 彩色预检器
colo(u)r prime white 原白色
colo(u)r print 彩印件;彩色印样;彩色印刷品
colo(u)r printed map 彩色地图
colo(u)r printer 彩色打印机
colo(u)r print film 彩色正片
colo(u)r printing 彩(色)印(刷)
colo(u)r-printing laminated package bag 彩印复合包装袋
colo(u)r prints 彩色像片
colo(u)r process 彩色加工;彩色传输过程
colo(u)r processing 彩色(信号)处理
colo(u)r produced by spinodal decomposition 不稳定分解呈色
colo(u)r produced in glazes 釉的呈色
colo(u)r-producing reaction 显色反应
colo(u)r-producing reagent 显色试剂
colo(u)r projection system 彩色投影系统
colo(u)r projection TVreceiver 彩色投影电视机
colo(u)r proof 彩色印样
colo(u)r proof of map 地图彩色校样
colo(u)r proof press 彩色打样机
colo(u)r property 色彩特性
colo(u)r protection 色保护

colo(u)r purge 色料清洗
colo(u)r purifier 纯色器
colo(u)r purity 颜色纯度;彩色纯度;色纯(度)
colo(u)r pyramid 色棱锥
colo(u)r pyrometer 颜色高温计;比色(光学)高温计;色测高温计
colo(u)r quad image processor 彩色四分割图像处理机
colo(u)r quality 彩色品质;色质
colo(u)r quantization 颜色量化
colo(u)r quartzite 彩色石英岩
colo(u)r radioisotope scanner 彩色放射性同位素扫描仪
colo(u)r range 颜色范围;彩色范围;色域
colo(u)r range of indicator 指示剂变色范围
colo(u)r ratio 颜色比(例);色彩比值;色比
colo(u)r ratio pyrometer 比色高温计
colo(u)r reaction 显色反应;比色反应;色反映
colo(u)r reaction plate 显色反应板
colo(u)r reagent 显色(试)剂
colo(u)r receiver 彩色接收机
colo(u)r recombined instrument 彩色合成仪
colo(u)r record 分色片
colo(u)r recording 彩色录像
colo(u)r record negative 分色阴图片
colo(u)r reduction 消色
colo(u)r reference 彩色基准
colo(u)r reference signal 彩色基准信号
colo(u)r reflection hologram 彩色反射全息图
colo(u)r register 套色(装置)
colo(u)r registration 彩色重合;彩色图像混合;彩色配准;彩色会聚;彩色对准;套色
colo(u)r registration guide 彩色套合规矩线
colo(u)r registration proof 彩色叠合打样法
colo(u)r relief 彩色浮雕
colo(u)r removal 除色;脱色
colo(u)r removing mask 减色蒙片
colo(u)r-removing preparation 除色制剂
colo(u)r rendered plan 彩色渲染图
colo(u)r rendering 彩色重现;现色性;显色性;颜色呈现;颜色呈现;传色;彩色再现;色彩再现;色彩还原
colo(u)r rendering index 显色指数;彩色再现指数
colo(u)r rendering of light source 光源显色性
colo(u)r rendering property 彩色再现特性
colo(u)r rendition 彩色再现;彩色还原
colo(u)r representation 彩色表示
colo(u)r reproduction 彩色复制;彩色复印;彩再现;色复现
colo(u)r resolution 彩色分辨率
colo(u)r resolution pattern 彩色清晰度测试图;彩色分辨力测试图
colo(u)r response 彩色响应;色响应;色感应
colo(u)r response curve 光谱感应灵敏度曲线;彩色响应曲线;光谱特性曲线;色感应曲线
colo(u)r retention 颜色维持能力;保色性;色彩保持
colo(u)r reversal 彩色反转法
colo(u)r reversal film 彩色反转(胶)片
colo(u)r reversal material 彩色反转照相材料
colo(u)r reversal photographic(al) material 彩色翻转片
colo(u)r reversible electrochemical filter 彩色可逆电化学滤光器
colo(u)r reversible film 彩色反转片
colo(u)r reversible roll film 彩色反转片胶卷
colo(u)r rotation 彩色旋转
colo(u)r saddle 彩色鞍座
colo(u)r sample 颜色样本;色样;色彩样品
colo(u)r sampling device 彩色取样装置
colo(u)r saturation 颜色饱和度;色饱和度
colo(u)r-saturation control 饱度调整;饱和色度调整;色度调整
colo(u)r scale 火色温度计;彩色温标;彩色梯尺;比色刻度尺;色度标准;色标(度)
colo(u)r scanner 电子分色机;彩色扫描器;分色扫描机
colo(u)r scanning image 彩色扫描图像
colo(u)r scheme 色调;色别法;颜色组合;颜色配合;配色法;色彩设计;色标
colo(u)r schlieren 彩色纹影
colo(u)r schlieren system 彩色条纹系统;彩色纹影系统
colo(u)r screen 彩色荧光屏;彩色显示屏;彩色网屏;彩色网目版;彩色滤光屏

colo(u)r seismic display 彩色地震显示
colo(u)r-selecting-electrode system 选色电极系统
colo(u)r-selecting-electrode transmission 选色电极系统透过率
colo(u)r selection 颜色选择
colo(u)r-selection technique 色选法
colo(u)r selective mirror 分色镜
colo(u)r-selector deflection system 选色用偏转系统
colo(u)r sensation 彩色感觉;色觉;色感(觉)
colo(u)r sense 色觉
colo(u)r sense test 色觉检查
colo(u)r-sensitive 感色的
colo(u)r sensitivity 光谱灵敏度;颜色灵敏度;感色性;感色(灵敏)度;彩色灵敏度
colo(u)r sensitivity curve 彩色感光度曲线
colo(u)r sensitization 色增感
colo(u)r sensitizer 感色剂;彩色增感剂;彩色感光剂;色增感剂
colo(u)r sensitometer 彩色感光计
colo(u)r sensitometry 彩色感光学
colo(u)r separate 分色
colo(u)r separated copy 分色原图
colo(u)r separation 分色(参考图)
colo(u)r separation drafting 分色清绘
colo(u)r separation drawing 分色清绘
colo(u)r separation enlarger 分色放大机
colo(u)r separation film 分色感光片
colo(u)r separation filter 分色滤光镜
colo(u)r separation masters 分色原版
colo(u)r separation method 分色法
colo(u)r separation negative 分色度片
colo(u)r separation of photograph 照片分色
colo(u)r separation photographic(al) material 分色片
colo(u)r-separation plate 分色板
colo(u)r separation procedure 分色法
colo(u)r separation process 分色制版法
colo(u)r separation system 色分离系统
colo(u)r sequence 显示顺序(色灯信号);印刷套色顺序;灯色顺序;彩色发送序列
colo(u)r-sequence rate 色序率
colo(u)r sequence with memory 彩色顺序储备
colo(u)r sexing 毛色鉴定性别法
colo(u)r shade 色相
colo(u)r shading 彩色黑点
colo(u)r shading control 底色均匀度调整;色明暗度调整
colo(u)r shift 色移;色偏移
colo(u)r shifting circuit 彩色移动电路
colo(u)r shop 配色间
colo(u)r sideband 彩色信号边带
colo(u)r signal 彩色信号
colo(u)r signal source 彩色讯号源
colo(u)r simulation device 彩色模拟显示装置
colo(u)r slab 色板
colo(u)r slide 彩色幻灯片
colo(u)r-slide scanner 彩色飞点幻灯机
colo(u)rs of mineral 矿物的颜色
colo(u)rs of title right 假托
colo(u)r soiling 油墨污垢
colo(u)r solid 颜色立体模型;彩色立体;色立体图;色立体(用三度空间表示色彩之间关系)
colo(u)r solution bottle 染液瓶
colo(u)r sorting machine 辨色分选机
colo(u)r space 色隙;色立方
colo(u)r spacing 色空间
colo(u)r specification 颜色规格;色量;色度标志;色表示
colo(u)r specimen 墨色样本
colo(u)r spectrum 色谱
colo(u)r splitting 彩色分离;色乱;分色分离
colo(u)r splitting system 分色系统
colo(u)r spot 色斑
colo(u)r spray coating 彩色喷涂
colo(u)r spraying 彩色喷涂
colo(u)r stability 彩色材料的稳定性;颜色稳定性;色安定性;呈色稳定性
colo(u)r stability index 颜色稳定性指数
colo(u)r stabilizer 颜色稳定剂
colo(u)r stabilizing amplifier 彩色稳定放大器
colo(u)r stack 彩色方箱
colo(u)r staining 染色;色渍;色污染

colo(u)r stainless steel 彩色不锈钢
colo(u)r standard 颜色标准;标准色;色谱;色标;比色标准
colo(u)r standard of geologic(al) map 地质图色标
colo(u)r standard solution 比色标准液;色度标准液
colo(u)r stimulus 颜色刺激;色刺激
colo(u)r stimulus function 色刺激函数
colo(u)r stimulus specification 色品;色规格;色刺激值
colo(u)r stone 彩石
colo(u)r streak 色线
colo(u)r strength 颜色强度;颜色浓度;着色力;白色颜料消色力
colo(u)r strip 色带
colo(u)r stripe 色线;色条;色斑
colo(u)r stripe gauze 彩条绡
colo(u)r stripe signal 彩条信号
colo(u)r strip generator 色带信号发生器
colo(u)r stripper 脱色剂
colo(u)r style map 彩色地图
colo(u)r subcarrier 彩色副载波频率
colo(u)r subcarrier generator 彩色副载波发生器
colo(u)r subcarrier oscillator 彩色信号振荡器
colo(u)r superimposition 颜色重叠;彩色重合;彩色配准
colo(u)r swatch 颜色样本;色样
colo(u)r switching 颜色变换;彩色图像开关
colo(u)r symmetry 色对称
colo(u)r-sync channel 彩色同步信道
colo(u)r sync generator 彩色同步机
colo(u)r synchronism 彩色同步
colo(u)r sync process(ing) channel 彩色同步信号形成信道
colo(u)r synthesis 彩色合成
colo(u)r synthetic(al) jewel 彩色人造宝石
colo(u)r system 信号系统;配色表;色系;色灯系统
colo(u)r technology 彩色技术
colo(u)r television 彩色电视
colo(u)r television picture tube 彩色显像管
colo(u)r television set 彩色电视机
colo(u)r television studio 彩色电视播放室
colo(u)r television system 彩色电视系统
colo(u)r temperature 颜色温度;彩色温度;比色温度;示色温度;色温(度);色度温度;色测温度
colo(u)r temperature changing glass 色温变换玻璃
colo(u)r temperature measurement 色温测定
colo(u)r temperature meter 色温计
colo(u)r terminal display production line 彩色终端显示器生产线
colo(u)r test 显色试验;比色试验;色试验;色度鉴定
colo(u)r test cards 色觉检查表
colo(u)r testing machine 色彩测试机
colo(u)r theory 彩色理论;色原学说
colo(u)r thermograph 彩色热谱图
colo(u)r threshold 辨色阈值;色差阈
colo(u)r tile 彩色瓷砖
colo(u)r tintage 着色
colo(u)r tolerance 容许色差;色宽限;色宽容度;色差容限
colo(u)r tolerance unit 色容限单位
colo(u)r tone 色调
colo(u)r top 颜色混合器;色陀螺
colo(u)r trace tube 彩色显像管;色迹管
colo(u)r track 彩色径迹;彩色跟踪
colo(u)r transfer 彩色传送
colo(u)r transfer paper 彩色转印纸
colo(u)r transformation 彩色变换;色变换
colo(u)r transition 色过渡;色变换
colo(u)r transition point 指示剂颜色转变点;色变点
colo(u)r translating microscope 变换颜色显微镜
colo(u)r transmission 彩色(图像)传输
colo(u)r transparency 彩色透明片;彩色幻灯片
colo(u)r travel 漆膜颜色随角变化
colo(u)r triad 三色组合
colo(u)r triangle 原色三角形;配色三角形;色三角;色品图
colo(u)r triangle original 原色三角形
colo(u)r tube 测深色管
colo(u)r tube camera 比色管暗箱
colo(u)rtuft machine 多色地毯簇绒机

colo(u)r tutanaga 有色镀锌板
colo(u)r TV production line 彩色电视机生产线
colo(u)r TV set 彩色电视机
colo(u)r-TV-telephone set 彩色电视电话机
colo(u)r type 彩色印件;彩色版(印刷)法
colo(u)r ultra-sonograph 彩色超声描记器
colo(u)r-under 彩色下置
colo(u)r under gilt 金边
colo(u)r-under technique 彩色下置技术
colo(u)r undertone 彩色浅淡
colo(u)r uniformity 色调均匀性
colo(u)r unit 颜色单位
colo(u)r value 给色量;色(标)值
colo(u)r variation 颜色发飘;颜色不一致;颜色变化
colo(u)r variation method 色变法
colo(u)r vector graphics system 彩色矢量绘图系统
colo(u)r-velocity ga(u)ging 色水测流法;颜色测流(速)法
colo(u)r-velocity method 颜色测流(速)法
colo(u)r versatile copier 彩色多用途复印机
colo(u)r video amplifier 彩色视频信号放大器
colo(u)r video camera 彩色摄像机
colo(u)r video display unit 彩色图形显示器
colo(u)r vision 颜色视觉;色(视)觉;色彩视觉
colo(u)r vision test 色视觉测验
colo(u)r vivid 颜色鲜明的
colo(u)r wall protection plate 彩色护墙板
colo(u)r wash 易脱(性)彩色涂料;易脱(性)彩粉浆;彩色涂料;涂刷易脱(性)彩粉浆;刷色;上彩色涂料
colo(u)r washing machine 彩色洗涤机
colo(u)r water 富色水
colo(u)r way 颜色的组合;配色色位;色彩设计
colo(u)r weakness 色弱
colo(u)r wheel 颜色轮
colo(u)r wire-lock system 彩色有线锁相系统
colo(u)r with red shade 颜色带红头的
colo(u)r wood carving 彩绘木雕
colo(u)r working 调色
colo(u)ry 色泽优良的
colo(u)r yield 给色量
colo(u)r zone 色区
Colpitts oscillator 科耳波兹振荡器
colprovia 冷铺沥青混合料
colpus median 沟中线
colquiriite 氟铝钙锂石
colset 多色防水沥青混合剂
colt 自调通风机;烟囱风雨帽
colter 开沟器;切割器;犁刀
colter arm 刀柄
colter boot 开沟器
colter clamp 犁头卡夹
colter dip 犁头卡夹
colter disk 圆盘刀
colter lift 开沟器起落机构
colter raising mechanism 开沟器起落机构
colter seeder 带开沟器的播种机
colter shank 犁刀柄
colt's tail 卷云
columbarium 鸽房;壁龛;墙中凹梁座;尸骨安置所
columbary 鸽棚
Columbia blue 哥伦比亚蓝色
Columbian pine 花旗松
Columbia unit 哥伦比亚单位
columbite 铌铁矿
columbite-tantalite 铌钽铁矿(含量)
columbium carbide 一碳化铌
Columbus type davit 倒式吊艇架;倒竿式吊艇架
columella 小柱;小型柱
column 栏(目);圆柱体;钻车上钻机悬臂;柱(子);柱形物;柱形浮标;支架基座;床身;报刊的栏位;塔器;塔;竖筒
column abundance 柱丰度
column action 长柱(受压)作用
column-adding routine 按列相加程序
column address 列地址
column adsorption chromatography 柱吸附色谱(法)
column aggregate 柱状团聚体
column analogy method 柱比(拟)法;似柱法
column analogy method of design 柱比法(设计)
column analysis 分栏解析

column analysis sheet 分栏式分析表
column anchorage 柱的锚碇
column and knee type milling machine 升降台式铣床
column-and-panel wall 镶式板柱墙;柱夹镶板式墙;柱间镶板墙;柱板式墙
column and tie construction 穿斗式构架
columnar 柱状的
columnar accounting 多栏式账户
columnar account method 多栏式分类账法
columnar aggregate 柱状团聚体
columnar alkali soil 柱状碱土
columnar architecture 列柱(式)建筑;柱式建筑
columnar basalt 柱状玄武岩
columnar basilica 列柱式长方形会堂
columnar book 多栏式账簿
columnar budget 多栏式预算表
columnar building 列柱建筑
columnar buttress dam 组柱支墩坝;柱墩坝
columnar cash book 多栏式现金日记账
columnar cash payment journal 多栏式现金支出日记账
column architecture 多柱建筑
columnar cleavage 柱状劈理
columnar coal 柱状节理煤
columnar coal sample 柱状煤样
columnar contrast chart in ore layers 矿层柱状对比图
columnar crystal 柱状晶(体)
columnar decoration 列柱式装饰
columnar deflection 柱的(纵)向弯曲;柱变位;柱(纵向)挠曲
columnar ferrite 柱状铁素体
columnar figure 柱状图
columnar finger motion 柱指运动
columnar form budget 多栏式预算表
columnar fracture 柱状断口
columnar geyser 柱式间歇泉
columnar grain 柱状晶粒
columnar graph 条形图
columnar interior 室内列柱
columnar joint(ing) 柱状节理
columnar journal 多栏式日记账
columnar lateralis 侧柱
column arm 悬臂;钻架杆臂
columnar order 柱型;柱形
columnar ore body 柱状矿体
columnar ornament 柱子装饰
columnar pier 柱式墩
columnar pile 柱桩;端承柱
columnar pit sinking 柱形沉井
columnar portal 柱式门架
columnar posterior medullae spinalis 后灰柱
column arrangement 柱网排列;柱网布置
columnar record 多栏式记录
columnar resistance 柱电阻
columnar rostrata 船头形饰纪念柱【建】
columnar sampling 柱状采样
columnar section 柱状图(土层);柱状剖面(图)
columnar section of drilling 钻孔柱状图
columnar section of electric(al) property 电性柱状图
columnar section of quaternary system 第四系柱状图
columnar section of stratigraphic(al) magnetism 地层磁性柱状图
columnar section program(me) 柱状剖面图程序
columnar speaker 组合扬声器
columnar statistics table 柱统计表
columnar stroked dressing 槽纹修琢
columnar structure 柱状构造
columnar style 柱的式样
columnar texture 柱状结构
columnar type oil hydraulic press 柱式油压机
columnar window 柱状窗
columnar zone 柱状结晶区
columnated window stair(case) 柱支楼梯;柱支撑梯(便于开窗采光)
column autoclave 柱式高压釜
column axis 柱轴
column back-pressure 柱反压力
column bar 柱中钢筋;柱筋;竖条
column base 柱座;柱脚;柱基(础);柱底;柱础
column base block (凿岩机的)支架基础

column base plate 柱座板;柱底支承板
column basilica 列柱式长方形会堂
column beam 柱型梁;柱形梁
column bearing 柱(的)支承
column bent 柱子排架
column bent pier 柱墩
column binary 直列二进制;竖式二进制(数)
column binary card 栏二进制卡
column binary code 列二进码;竖式二进制代码
column-binary mode 竖二进制方式
column binder 柱内箍筋
column blasting 柱状装药爆破
column bleed(ing) 柱流失;柱损失
column blockwork 柱状块体
column body type 柱式体形
column boiler 蒸馏锅
column bottoms pump 塔底泵
column box 柱外壳;柱表贴件;柱形模板
column bracing 柱间支撑
column bracket 雀替【建】
column bracket[que-ti] 雀木(中国古典木建筑中,梁柱交接处的托座)
column bracket structure 柱架式结构
column buckling 柱状压屈
column cage 柱用预制钢筋骨架;柱用钢筋笼
column cap 立柱罩壳;柱头;柱帽;柱尖;柱顶
column capacity 柱容量
column capital 柱帽;柱头;柱尖
column capital form 柱头形式
column cartridge 柱状炸药包
column casing 柱壳;钢柱(防火)外壳;柱筒;柱套
column center line 柱中线
column charge 柱状装(炸)药;柱状炸药包;柱式装饰
column chart 柱状图
column chromatograph 柱色层
column chromatographic method 柱色谱法;柱层析法
column chromatography 柱型色层分离法;柱色谱法;柱色层法;柱层析法
column clamp 柱模板箍铁;柱模(板);柱夹;三角屋顶
column classification 柱分类
column cluster 柱群
column coating 柱涂层
column collimator 圆柱准直器
column combination 柱组合
column concrete 柱子混凝土
column conditioning 柱老化;柱调节
column connection 柱的连接
column constraint 列约束
column content of ozone 大气气柱中的臭氧含量
column control system 杆式控制系统
column count 列计数;列的计算
column counter 列计算器
column crane 塔式起重机;柱式起重机;管形塔式起重机
column creep 柱的徐变;柱的蠕变
column cross-section 柱(的)截面;柱剖面图
column crystal 柱晶
column crystallization 柱状结晶过程
column curve 柱的弯曲线
column deflection 柱纵向弯曲;柱变位
column density 柱密度
column design 柱(的)设计
column detailing 柱子细部结构
column development chromatography 柱展开色谱(法)
column diagram 方条图
column diagram of chemical composition of the sediments 沉积物中化学成分柱状图
column diagram of superficial sediments cores 表层沉积物岩芯柱状图
column diameter 柱径
column dimension 柱的尺寸
column distance 列距
column distiller 蒸馏柱
column dominance 列支配
column dried 竖筒式干燥
column drier 竖筒式干燥机
column drill 立式钻床;架式风钻;柱状岩芯钻;柱状取心钻
columned hall 列柱式厅堂
column effect 列效应

column effective length 柱的有效长度
column efficiency 柱效能;柱效率
column electrophoresis 柱电泳法
column elevator 柱式提升机
column elution program(me) 柱洗提程序
column end moment 柱端力矩
column engaged to the wall 附墙半身柱;半柱;贴墙柱
column evapo(u)ration 蒸发柱
column evapo(u)rator 浓缩柱
column extractor 萃取塔;柱式萃取器;萃取柱
column face 柱面
column facing 柱面的面饰;柱的覆面
column-figure 柱面花饰
column-figure portal 有花饰的面柱
column filling 柱填充(物);填柱物
column filter 柱状滤池;柱状滤器;柱式过滤器
column fireproofing 防火柱
column flange 柱的凸缘
column flotator 浮选柱
column flow rate 柱流量
column foot 柱脚;支架基础;塔基
column footing 柱脚;立柱基础;柱基(础);柱底脚
column form 柱的模板;柱模(板)
column format 列格式
column forming 柱体立模
column formula 压柱公式
column foundation 柱(形)基础
column foundation block 柱基础块
column fractionation 柱上分级
column-free 无列;无柱的
column free hanger 无柱头飞机库
column gate 列选通器
column gear shift 杆式变速
column generation 列生成
column generation method 列生成法
column Greek temple 列柱式希腊神庙
column grid 柱网
column grid line 柱网线
column grid pattern 柱面格子型花饰
column grinder 立柱式打磨机
column guard 柱保护钢板;柱保护钢板;墩保护钢板;柱角护
column-guided holder 直升式储气罐
column head 柱帽;柱头;柱顶
column-head cater 柱头梁座
column-head entasis treatment 柱头卷杀
column height 柱高
column hinge 柱铰
column hoist 柱架式起重机
column hold-up 填塔液;柱滞留
columniation 列柱式;列柱(法);群柱排列法
columniform 圆柱形的
column inch 栏英寸
column index 列指数
column indicator 列指示器
columning 棉塞支托法
column instability 柱的不稳定性
column interval 柱的间距
columnization 棉塞支托法
column jack 架式千斤顶;圆筒式千斤顶;柱式千斤顶;柱架式千斤顶
column jacket 柱管;柱的外壳;外柱
column jack-up 圆柱自升式平台
column jib crane 塔式起重机;塔式悬臂起重机
column jump 列转移
column lavatory basin 圆柱状排列的盥洗盆
column layout 柱的布置;柱网布置
column ledger 分栏式分类账
column length 柱长
column life 柱寿命
column-like pier 柱形桥墩
column line 柱子中心线
column lining 柱的衬砌
column load 柱状火药包;柱状装(炸)药;柱状炸药包;柱载重;柱荷载;柱负荷
column loading 柱载
column lots 容许修建柱基的土地
column material 柱材料
column matrix 列矩阵;直列矩阵
column median 列中位数
column method of execution 柱状施工法;分段施工法(混凝土坝)
column milling machine 立柱式铣床

column mixer 混合柱;塔式拌料箱
column moment 柱内力矩
column move and delete 列移动和删除
column movement 柱的移动
column name 列名称
column number 栏数;柱号
column of air 柱空气
column of angles 角钢柱
column of angle steel 角钢柱
column of atmosphere 大气气柱
column of built channels laced 槽钢缀合柱
column of channel box section 箱形截面槽钢柱
column of clay 黏土柱
column of concrete filled tube 混凝土填(塞)管柱
column offset 墙柱砌口;砌口
column of gravel 砾井
column of hollow section 空心截面柱
column of hydrometallurgy and pyrometallurgy 水冶和火冶联合法
column of Jupiter 朱庇特性
column of liquid 液柱;水柱
column of Marcus Aurelius at Rome 罗马马库斯奥里利斯柱
column of mercury 水银柱
column of mud 泥浆柱
column of radiator 暖气管道;暖气管柱;柱式汽炉片
column of sand 砂井
column of solid section 实心截面柱
column of tape 数控带的信道
column of tapered sheet construction 锥形薄板构成的柱
column of Trajan 图拉真纪念柱(古罗马)
column of trays 层板;板式塔
column of triumph 胜利柱(古罗马);凯旋柱(古罗马)
column of two channels laced 二槽钢缀合柱
column of variable sections 变截面柱
column of water 水柱
column of welded sheet construction 钢板焊成的柱
column opening 柱上的开孔
column operations 列运算
column order 纵向按序排列;柱式
column overloading 柱过载
column-packing 柱填充物
column-packing agent 柱填充剂
column-packing funnel 柱填充漏斗
column-packing material 柱填充材料
column-packing method 柱填充法
column-packing procedure 柱填充程序
column-packing structure 柱填充结构
column packing technique 装柱技术
column pad 柱基(础)
column pair 双柱;对柱
column parameter 柱参数
column pedestal 柱脚;柱的台座
column performance 柱性能
column permeability 柱渗透性
column pier 柱(式桥)墩
column pile 柱桩;支承桩;端承桩
column pipe 柱状排水管
column pitch 列间距
column plate 塔板【化】
column-plate retaining wall 柱板式挡土墙
column plinth 柱础
column precipitator 置换柱;沉淀柱
column precision 列精度
column preconditioning program(me) 柱预老化程序
column press 柱式压(力)机
column profile 柱状剖面图
column program(me) 柱状图程序
column-programming 变换柱
column radiator 立式散热器;柱形散热器;柱式散热器
column radius 柱半径
column rank 列秩【数】
column records 分栏记录
column reinforcement 柱筋
column resistance 柱阻力
column rib 柱肋
column rigidity 柱的刚度
column scan 列扫描
column schedule 柱施工一览表

column section 柱型材;柱的一节;柱的一段;柱截面;柱断面
column settlement 柱沉陷
column shaft 柱身
column shape 柱形
column-shift unit 移列部件
column shuttering 柱的模壳;柱体模板
column side 柱身模板;柱模侧板
column side board 柱身模板
column sideway mechanism 柱侧倾机理
columns in series 串联柱
column size 柱的尺寸
column sleeve 柱套
column socle 雕像座;柱基座
columns of area element mesh 面元网的列数
column space 列空间
column spacing 柱距;柱间距离
column spiral 柱用螺旋钢筋;螺旋箍筋
column splice 柱式接头
column split 列分组;列分隔;列分割;列分辨;分位机构
column split hub 分列插孔
column stability 柱稳定性
column stabilized drilling unit 支柱稳定式钻探船;支柱稳定式钻井船
column steel 柱钢
column steel hooping 配有箍筋的柱
column stiffness 柱的劲度;柱的刚度
column still 蒸馏塔
column strength 柱的强度
column stress 柱的应力
column strip 柱列带;柱上板带;柱顶条板
column structure 柱式结构
column support 柱支座;立柱架;柱载体
column supported 柱支承的
column supported gallery 柱承楼座
column supported pier 墩座栈桥
column supported trestle 柱墩式栈桥
column sweep algorithm 列扫描算法
column system 柱系统
column test 柱试验
column-throw stand 柱式扳道座
column tie 柱系杆;柱钢筋籀;籀筋
column-to-column joint 柱与柱的接合
column-to-floor connection 柱与地板的连接
column-to-footing connection 柱与柱脚的连接
column top 柱顶
column-top corbel-bracket set 柱头科川拱
column-to-slab connection 柱与板的连接
column totals 金额栏总计
column tray 塔盘;塔板【化】
column tube 转向轴管;柱形管
column tubing 柱管
column type 柱型
column type biofilter 塔式生物滤池
column-type diaphragm wall 柱列式地下连续墙
column type radiator 柱式散热器;管柱式散热器
column type scrubber 柱石洗气器
column upright drill machine 圆柱立式钻床
column valve 柱阀
column vector 列向量;列矢量;纵矢量
column vertical 柱竖筋
column vibrator 振柱器
column void volume 柱空体积
column wall 柱内壁
column warp 列整理
column wash basin 圆柱状排列的洗脸盆;圆柱状排列的清洗盆
column wash bowl 圆柱状排列的水洗槽
column washer 洗涤柱
column web 柱的腹部
column wheel 导柱轮
column width 柱的宽度
column with bracket 有牛腿的柱
column with clamped ends 固定端柱
column with constant cross-section 等截面柱
column with cosmetic work 经过装饰的柱子
column with cranked head 弓形床架
column with hinged ends 端部铰支柱
column with lateral reinforcement 横向钢筋柱;纵向配筋柱
column with lateral tie reinforcement 配有箍筋的柱
column with services 安装了水电等管线的柱子

column with spiral hooping 螺旋箍筋柱;螺旋钢筋柱;配有螺旋筋的柱
column with steel hooping 螺旋箍筋柱
column with two brackets 有两个托臂的柱子;有两个牛腿的柱子
column with uniform cross-section 等截面柱
column with variable cross-section 变断面柱;变截面柱
column zone 柱子地带;柱丛区
colura 讲坛雨棚;坟墓雨棚
colure 分至圈
Colusa sandstone 科鲁萨砂岩(产于美国加州的一种砂岩)
colusite 硫钒锡铜矿
coluvial slope 崩积斜坡
colvent inlet 溶剂入口
colymbethra 希腊教堂洗礼室;希腊教堂洗礼池
colza oil 菜籽油
coma 毛撮
coma aberration 彗形象差
comagmatic 同源岩浆的
comagmatic assemblage 同源岩浆岩组
comagmatic region 同源岩浆岩区
comagmatic rocks 同浆岩浆
co-maker 担保联署者
comalong 备焊机具
co-management 共同管理
co-manager 联合经理人
Comanchean 科曼齐系【地】
comancheite 卤汞石
Comanchian 科曼齐系【地】
comaparative mortality index 标准化死亡指数
comb 浪峰;蜂窝;蜂房;脱圈沉降片;梳状蜂窝;梳状刀片;梳轮;梳机
comb antenna 梳形天线
Combarloy 康巴高导电铜
comb arrester 梳形避雷器
combat 格斗
combatant craft 战斗艇
combat-car 轻型装甲车
combat carrier 装甲运输汽车
combat chart 作战海图
combat map 作战图
combat simulation on ground 地面空战模拟器
combat store ship 军需品补给船
combatting ag(e)ing 反老化
comb-back chair 梳形温莎椅
comb board 脊板
comb-chieselling 算式錾凿
comb chiselled finish 密线凿石面
comb cleat 梳形系索耳
comb collector 梳形集电极
comb cut 竖截面(椽子顶端与脊檩的接合面)
comb dent 梳齿
comb die 切边模
comb diversion 梳齿导流
comb dresser 梳刷转筒;梳齿转筒
combe 狭谷;峡谷;冲沟;深海谷
combed fascine raft 梳理的粗杂材
combed finish 梳刷面饰
combed-finish tile 齿面瓦;表面带槽纹的面砖
combed glass 带条纹的装饰玻璃;条纹装饰玻璃
combed joint 燕尾榫;齿结合;角榫接;马牙榫接【建】;插榫交接
combed plywood 蜂窝胶合板
combed stucco 齿形镘涂;蜂窝状墁涂;蜂窝状拉毛粉饰
combed ware 带条纹器皿;条纹器皿
combed work 钢齿琢石
combeite 菱硅钙石;菱硅钙钠石
comber 卷浪;精梳机;滚浪;长卷浪;分纸器;分纸滚轮;白头浪;深水白头浪
comber cylinder 梳理滚筒
comber lap machine 精梳成卷机
combescure transformation 梳状变换
comb filter 梳状滤波器;梳形滤波器;梳齿滤波器
comb foundation 巢础
comb function 梳状函数;梳齿函数
comb ga(u)ge 齿形检查量规
comb generator 梳状波发生器
comb grain 径向纹理;斜径木纹
comb-grained 斜纹的
comb-grained wood 密纹木;心木;心材;算梳纹理木材;梳形纹木材

comb grain timber 四开木材
comb guide 梳式导丝器
comb hammer 砌砖工的锤
combi-cooler 复式冷却机
combinableness 可结合性;可化合性
combinated oil and gas accumulation zone 复含油气聚集带
combinated sample 组合样品
combinated seal 联合封闭
combinated string of casing 复合套管
combinate form 聚形
combination 联合;化合;合并;组合;搭配
combination actuator 复合执行机构
combinational burial hill oil-gas field 复合潜山油气田
combinational circuit 组合电路
combinational logical element 组合逻辑元件
combinational network 组合网
combinational oil-gas field 复合油气田
combinational reef type oil-gas field 复合礁型油气田
combinational selection 组合选择
combinational switch circuit 组合开关电路
combinational switching network 复合开关网络
combination angle depth ga(u)ge 组合角度深度尺
combination angle ga(u)ge 组合测角器
combination arc 联合弧
combination automatic controller 组合自动控制器;复合自动控制器
combination band 组频谱带;组合谱带;综合谱带
combination-bar operating spring 组合条作动弹簧
combination beam 混成梁;组合梁;合成梁
combination beam bridge with slight curve slab 微弯板组合梁桥
combination bearing 组合轴承
combination berth 多用途泊位
combination bevel 组合量角规;联合斜面;组合斜角规;通用斜角规
combination bit 综合钻头
combination bit and mud socket (美国制造的)钻掏泥砂泵;带钻头的捞砂筒
combination board 合成纸板
combination body 活顶车(车顶可打开)
combination boiler 复式锅炉;分节锅炉
combination brake of air and electricity 空电混合制动
combination bridge 组合式桥梁;组合结构桥
combination bucket-wheel stacker-reclaimer 斗轮式堆取料机
combination budget 联合预算;混合预算
combination buffer and draft gear 缓冲车钩
combination building 混合管纱成型
combination buoy 组合浮标;灯光音响浮标;声响灯浮标
combination burn cut 混合掏槽
combination burner 联合燃烧器;多种燃料燃烧器
combination burner tool 组合式喷烧工具
combination cable 组合电缆
combination cal(l)ipers 内外卡钳;复式内外长卡钳
combination car 组合车辆
combination cargo 杂货
combination carrier 联合货船;油一散货混装船;通用货船
combination casing column 复式套管
combination center drill 组合中心钻
combination chamber vessel 组合室容器
combination characteristic curve 组合特性曲线
combination chuck 单动联动两用卡盘;复式卡盘;复动夹头
combination circulation 混合循环
combination coagulation 组合式混凝沉淀池
combination coagulation system 联合混凝系统
combination coaster 沿海客货船
combination code 组合编码
combination coefficient 组合系数
combination collar 异径接箍
combination colo(u)r 合成色
combination column 组合(式)柱(子);混凝土混成柱;混成柱;型钢混成柱;复合式柱子
combination compartment 淋浴分隔间
combination computer aiding method of environmental impact assessment 组合计算机辅助环境评价法

combination concrete curing and sealing compound 混凝土养护与密封两用复合物
combination connector 万能连接器;混合连接器
combination construction 混合结构;通用建筑
combination control 联合控制;综合控制
combination cooler 复式冷却机
combination cracking 联合裂化
combination crossing 组合交叉
combination current 组合电流
combination curve 协联特性曲线
combination cut pliers 剪钳
combination cutter 联合裁切机
combination cutting and loading machine 采矿联合机
combination cutting and twisting pliers 万能钳(子)
combination cutting tool 复合刀具
combination cycle 混合循环
combination cylinder 复合式筒仓
combination deal 联合经营
combination-deal trade 进出口结合贸易
combination density and moisture meter 密度湿度综合探测计
combination depth-and-angle ga(u)ge 组合深度角度规
combination diagonal pin type die set 对角导柱模架
combination dial pressure ga(u)ge 复式刻度压力计
combination die 组合式模;组合式凹模;组合模(具);多腔模
combination door 双扇门;组合(式)门;双层门
combination draw-redraw die 拉延再拉延组合模
combination draw-reverse draw die 拉延反拉延组合模
combination drill 组合钻;综合钻机;冲击回旋两用钻机;复合钻机
combination drilling 冲击回转钻进;复式钻机
combination drilling equipment 综合钻进设备;冲击回转钻进设备
combination drilling machine 组合钻床
combination drilling outfit 综合钻进设备;冲击回转钻进设备
combination drive 混合驱动
combination dump and platform body 堆卸平台
combination electrode 组合电极
combination engine 组合发动机
combination equipment item 组合设备项目
combination excavator and hauler 铲土运输机械
combination factor trap 复合因素圈闭
combination fan scrubber 联合扇形洗涤器;复式扇形洗涤器
combination faucet 冷热水(混合)龙头;混合水龙
combination feeder 联合给料机
combination ferryboat-trailer ship 拖拽两用船
combination firing burner 油气混用燃烧器
combination fishing vessel 混合式渔船
combination fixture 组合洗涤设备;组合卫生设备;复合洗涤设备;成套洗涤设备;成套卫生设备;组合装置;整套卫生设备;复合卫生设备
combination flow regulator 组合式流量调节阀
combination forbidden 禁止组合;非法组合
combination for action effect 作用效应组合
combination for long-term action effects 长期效应组合
combination for short-term action effects 短期效应组合
combination frame 混合构架;组合框架
combination frequency 组合频率;复式频率;并合频率
combination furniture 组合家具;卫生设备套件
combination fuse 复合式引信
combination gas 富天然气
combination gas and oil burner 油气联合燃烧器
combination gas-lift 混合气举
combination gas pliers 气管剪钳
combination ga(u)ge 组合量规
combination grain 混合粒度
combination grate 供热口
combination head 两用头;组合头
combination hot or cold mix plant 冷热两用拌和设备
combination hub 接插座
combination hydrant and fountain 给水喷泉两用栓;给水喷泉两用龙头

combination infrared laser tracker-ranger 红外(线)激光跟踪测距装置
combination in restraint of trade 贸易限制中的联合
combination interference 混合干扰
combination lathe 万能机床
combination law 并合律
combination level 复合能级
combination level-board 水平轨距尺
combination level-ga(u)ge 水平轨距尺
combination lever 联合杆
combination line 组合谱线
combination link 合并杆
combination load 组合荷载
combination loader 混合型装药器
combination lock 字码锁;暗码锁
combination locomotive 双能源机车
combination long roundnosed pliers 长嘴钳
combination machine 联合机
combination machinery 联合机械
combination man 具有多种手艺或不同行业会员证工人;技术工人
combination mat 复合毡
combination method system (of traffic signal) 交通信号组合法系统
combination microphone 复合传声器
combination mill 联合轧钢机;联合式轧机;联(合)磨;多仓磨(机);串列一横列联合式轧机
combination misalignment 复合误差
combination muffler 阻抗复合消声器
combination of actions 作用组合
combination of blowing and exhaust system of ventilation 混合式通风
combination of different materials 不同材料的组合
combination of drill tool 钻具组合
combination of dynamic(al) load 动荷载组合
combination of earthquake effect 地震作用效应组合
combination of exploring-drilling 坑钻结合
combination of free blocks 空闲块的组合
combination of instruments 仪表组合
combination of lens 透镜组合
combination of load 荷载组合
combination of load effects 荷载效应组合
combination of materials 组合材料;材料组合
combination of outdoor pipelines 室外管线综合图
combination of pipelines 管道综合
combination of point-bearing and friction pile 支承和摩擦混合桩
combination of pollutant 污染物混合
combination of pulley 滑车组合
combination of pumps 泵组
combination of risks 风险组合;风险的结合
combination of sectional charges 分段运费的汇总
combination of sentence 复合命题
combination of share 联合股份
combination of significance tests 显著性检验组合
combination of slip turnout and scissors crossing 道岔组合
combination of stress 应力组合
combination of triangulation and trilateration 边角测量
combination of vehicles 带半拖车的牵引车
combination of zones 复合带
combination oil seal 组合油封
combination operating condition 协联工况;组合工况
combination operation 组合操作
combination outfit 综合钻机
combination packaging 组合包装
combination packer 综合封隔
combination padlock 组合锁
combination painting 涂混合漆
combination pile 混成桩
combination pilot, drilling and reaming bit 导向、钻进、扩孔多用钻头
combination plane 万能刨;组合刨
combination plant 混合装置
combination plate 全要素板
combination pliers 足迹;接管器;剪钳;多用钳;鲤鱼钳;钢丝钳;万能钳(子)
combination pliers insulated handle 胶柄钢丝钳
combination pliers slip joint insulated handle

胶柄鲤鱼钳
combination pliers with side cutting jaws 花腮钳
combination pool 复合油(气)藏
combination pressure and velocity stage turbine 复合压力—速度级涡轮机
combination pressure/temperature transducer 复合压力/温度传感器
combination price 批发价(格)
combination principle 化合原理;并合原理
combination printing 混合印刷
combination probe 组合探针
combination property 综合性能
combination pump 组合泵;复合泵
combination quota 联合配额
combination railing 组合栏杆;组合扶手
combination range 组合行列;组合系列;组合炉灶
combination rate 联运(运)费率;合并运价
combination reaction 化合反应
combination reinforcement 组合增强
combination reinforcing mat 复合增强毡
combination relay 组合继电器
combination residual chlorine 结合性余氯
combination restrictive and by pass flow regulator 节流旁通组合式流量调节阀
combination rig 综合钻机;冲击旋转联合钻机
combination roller 组合式压路机
combination safety relief valve 复合安全泄压阀
combination sale 批发
combination sampler 综合取样器
combination saw 组合锯
combination saw blade 多用锯齿片
combination scale 组合刻度
combination scattering 复合散射
combination screw driver set with tester 带测电笔的套装螺丝刀
combination screw-screed spreader 螺旋式整平摊铺机(混凝土);螺旋式摊铺机(混凝土)
combination set 万能测角器;组合角尺
combination ship and barge dock 载驳货轮及子驳共用码头
combination ship yard 修造船厂
combination silo 组合均化库
combination sink cabinet and cooker 组合式水柜与炉灶单元
combination sleeve 组合联结器;组合联轴器
combination snips 带四联杆的剪刀
combination socket 复式打捞筒;组合套筒
combination space divide 组合(式)隔断
combination spanner 多用扳钳;花扳手
combination spreader 综合撒布机
combination spudding bit 组合开孔钻头
combination square 组合矩尺;万能工具;多功能量具;组合曲尺;组合角尺;什锦角尺
combination stack 综合式堆垛机
combination stacker-reclaimer 堆取料机
combination stair(case) 复合楼梯;组合式楼梯
combination standard ga(u)ge 万能塞规
combination standpipe 组合式立管(供消防和生活用水);组合立管(供消防和生活用水)
combination starter 混合启动器;综合启动器
combination static mixer/heat exchanger 复合静态混合机/热交换机
combination station 客货运站【铁】
combination stopper 组合式制止器
combination striation 聚形纹
combination string 复式套管;不同直径或不同壁厚组成的钻杆
combination structure 混合构造
combination suction and flushing line 抽吸冲洗两用管
combination sweep rake and stack 综合式堆垛机
combination switch 复合开关
combination switch board 组合开关箱
combination system 混合钻进方法
combination tail and stop lamp 车尾及停车组合灯
combination tank 整套冷热水箱
combination tanker 石油/散货两用船
combination tap assembly 组合接线装置
combination terminal 两用码头
combination theory 组合理论【数】
combination thermonet-sonde 测温网位仪
combination toilet tissue cabinet 组合式手纸盒
combination tool block 组合刀架
combination topping and cracking plant 拔顶裂化联合装置
combination train 组合列车
combination trap 组合圈闭;复合圈闭
combination traveling block (与起重钩一体的)联合游动滑车
combination tread 混合纹
combination treatment 联合处理;混合处理
combination treatment method 混合处理法
combination tube mill 联合磨管机
combination tumbler 字码锁机构;转字锁转轮
combination turbine 联合涡轮机;联合透平;联合水轮机;联合汽轮机;盘鼓型汽轮机
combination turret lathe 组合塔式六角车床
combination type 复合型
combination type(arterial) highway (同时有立交和平交的)混合的公路
combination type dam 混合式堤坝;砌筑和填筑混合坝
combination-type expressway 混合式高速公路
combination-type freeway 混合式高速公路
combination type gear inspection instrument 齿轮综合检测仪
combination-type miller 组合铣床
combination type of industry 联合式制造业
combination-type press 组合压机
combination-type road system 混合式道路系统
combination unit 联合装置;组合设备
combination valve 组合阀;复合阀
combination value of actions 作用组合值
combination variable 共生变化
combination vehicle 组合车;拖挂式汽车;拖挂列车
combination ventilation system 混合式通风
combination Venturi scrubber 联合文丘里洗涤器
combination vessel 客货船;混合型船舶;多用途船(舶);客货轮
combination vibration 组合振动
combination wall and anchor packer 两用封隔器(悬吊和锚碇的)
combination waste-and-vent system 排气和通气联合系统;排水及通气合用系统
combination water 结合水
combination well 联合井;连通井
combination wet scrubber 联合湿式洗涤器;复式湿式洗涤器
combination wheel 组合轮
combination whirler flood and guide shoe 带浮阀旋回阀的管锥
combination window 两用窗(冬、夏);混合窗;组合窗;双层窗;双扇窗
combination without repetition 非重复组合
combination with repetition 重复组合
combination wrench 多用扳手;组合扳手;两用扳手
combination yarn 混合纱
combinative resonator 复合共振腔
combinative scattering 联合散射;并合散射
combinative table 组合表;复合表
combinator 配合操纵器(水力透平机)
combinatorial 组合的
combinatorial analysis 组合分析
combinatorial chemistry 组合化学
combinatorial optimization 组合(最)优化
combinatorial probability 组合概率
combinatorial problem 组合问题
combinatorial property 组合性质
combinatorial statistic(al) data 综合统计资料
combinatorial sum 组合和
combinatorial theory 组合理论【数】
combinatorial topology 组合拓扑
combinatorics 组合学
combinatory analysis 组合分析
combinatory patterns of earthquake 地震断裂组合形式
combinatus 联合制
combine 联合机械;结合
combine apron 联合收割机滑板
combine baler 联合压捆机
combine body 联合收割机壳体
combine column footing 联合柱座
combined 化合的;组合的
combined account 综合账户
combined action 联合作用;组合作用;综合作用;共同作用
combined action of accelerators 加速器的联合作用
combined adhesion and rack railway 黏着和齿轨合用铁路
combined adjustment 联合平差
combined adjustment method 综合平差法
combined aerobic treatment process 联合好氧处理工艺
combined aggregate 混合集料;混合骨料;组合骨料
combined aggregate grading 集料混合级配;骨料混合级配;混合集料级配;混合骨料级配
combined air and water curing 联合养护(混凝土用空气和水)
combined air force 空气动力合力
combined air handling units 组合式空(气)调(节)器
combined air-water backwash 混合气—水反洗
combined analog(ue)-digital computer 模拟数字混合计算机
combined antenna 共用天线
combined appliance 联合燃具;多用途燃具
combined austempering 组合等温淬火
combined available chlorine 结合性有效氯;化合性有效氯
combined available residual chlorine 结合性有效残余氯;化合性有效余氯;组合有效余氯
combined axial and radial flow pump 轴流辐流泵
combined balance sheet 合并资产负债表
combined bauxite ore 复合型铝土矿矿石
combined bending and axial loading 弯曲和轴向组合荷载
combined biologic(al) carbon 化合生物碳
combined biologic(al) effect 混合生物效应;综合生物效应
combined bit 双用钻头
combined bolting and shotcrete 喷锚支护
combined bridge 两用(公路)桥;混合式夹板;组合夹板;公路铁路两用桥;铁路公路两用桥
combined broach 组合拉刀
combined bucket wheel loader 斗轮式堆取料机
combined bud grafting 复式芽接
combined building drain 建筑综合排水
combined building materials 组合建筑材料
combined building sewer 建筑合流污水管;房屋合流污水管(道);房屋的合流污水管
combined cable system 组合式缆索体系
combined cage and skip 罐笼箕斗混合提升容器
combined capacity 综合能力
combined capping slab 联合盖板
combined carbon 结合碳;化合(的)碳
combined car flow 混合车流
combined cargo carrier 两用货轮
combined carrier 通用货船
combined carrier system 复合输导系统
combined casing column 混合套管
combined casting 综合注浆;复合注浆
combined cement 混合水泥
combined center drill 复合中心钻
combined centrifuging and gas attenuating process 离心喷吹法
combined characteristic 组合特性;综合特性
combined characteristic curve 总特性曲线;综合特性曲线
combined checker work packing 混合式格子砖码法
combined chlorine 化合氯
combined chlorine residual 化合性余氯
combined churn 联合黄油搅拌器
combined circulation boiler 复合循环锅炉
combined city 综合城市;联合城市;群集城市
combined clamp 组合夹具
combined closing and check action 联合关闭和锁紧动作
combined coefficient 合成效率
combined collection of refuse 垃圾混合收集
combined column 复合柱
combined combined 联合燃烧器
combined compressed-air and vacuum brakes 空气真空两用制动机
combined concrete aggregate 组合混凝土骨料;混合混凝土骨料
combined condition collision diagram 路况事故(分析)图
combined construction 混合结构;混合建筑
combined contra-flow uniflow drying drum 逆流与单流两用烘缸;逆流与单流两用干燥筒
combined core box 组合芯盒

combined correction factor 综合修正系数
combined cost 综合成本;联产品成本
combined cost and general ledger 联合成本及普通分类账
combined course 结合层
combined creep 综合蠕变
combined critical speed 轴系临界转速
combined cropping system 乔林混农作业
combined curb-and-gutter 联合式街沟;整体式侧平石
combined curve 总曲线;复合曲线
combined curve-and-side planing machine 组合曲面侧面刨床
combined cutter loader 联合采煤机;采煤联合机
combined cutting 复合切削
combined cutting and welding blow-pipe 焊割两用炬
combined cutting and welding torch 焊割两用炬
combined cyanide 化合氰
combined cycle 联合循环
combined cycle gasification process 煤气化联合循环工艺
combined damper 组合风阀
combined data base 组合数据库
combined deflection 总挠度
combined degeneration 混合变性
combined depreciation and upkeep method 折旧与保养合并法
combined development 联合开拓
combined development method 联合开拓法
combined dewatering 联合疏干
combined diagram 合并图
combined diagram level 合图水平仪
combined die 组合式模;组合式凹模
combined digester 复合消化池
combined discharge 混合排放
combined discharge area 混合排放区
combined disk clutch 组合盘形离合器
combined dispersion effect 联合散射效应;并合散射效应
combined distortion 混合畸变
combined distress alerting network 组合遇险报警网络
combined distribution fire-extinguishing system 组合分配灭火系统
combined distribution frame 组合配线架
combined ditch 合流排水沟渠
combined documents 联合单据
combined drainage area 混合排放区
combined drainage system 合流制排水系统
combined draw and buffer gear 缓冲车沟
combined draw and buffing gear 复合拉丝抛光装置
combined dredge(r) 有排泥设备的挖泥船
combined drill 组合钻头;双用钻头
combined drill and countersink 组合中心钻;带护锥复合中心钻
combined drill and mill machine 铣钻联合机床
combined drill and reamer 钻铰两用刀
combined drying and crushing 烘干兼破碎
combined drying and grinding 同时烘干和粉磨
combined drying and grinding mill 烘干磨;研磨烘干机
combined dustpan dredge(r) 多用吸盘挖泥船
combined dwelling 商店(兼用)住宅
combined dwelling house 住宅商店组合建筑
combined earthquake coefficient 组合地震系数
combined earthquake response spectrum 组合地震反应谱
combined echo ranging echo sounding 组合回声测距系统
combined effect 混合效应
combined effect leverage 联合杠杆效应
combined effects meter 复合应流量仪表
combined effect of chemicals 化学物质联合作用
combined efficiency 合成效率;总效率;综合效率
combined electric(al) dust collector 组合电集尘器
combined electrofilter 组电集尘器
combined entry 复合分录;复合分类
combined error 混合差率;总误差
combined estimates of correlation 相关的合并估计
combined excutcheon plate (门执手和钥匙孔的)合一盖板

combined extensometer 联合伸缩仪
combined extract and impact system 抽气送气混合系统
combined extract and input system 抽气送气混合系统
combined extrusion 复合挤压
combined failure 复合破坏
combined fatty acid 化合脂肪酸
combined fertilizer 复合肥料
combined fertilizer-and-seed drill 施肥条播机
combined file 组合文件;输入输出共同文件
combined filter 组合式过滤器;混合式过滤器
combined filtering layer 混合式过滤层
combined financial statement 合并财务报表
combined firing 多种燃料混合燃烧
combined flat bending and compression 平弯曲和压缩的综合作用
combined flexure 联合弯曲
combined flocculant 复合絮凝剂
combined flow 合流水量
combined flow field 复流场
combined flow turbine 混流式涡轮机;混流式透平;混流式水轮机;混流式汽轮机
combined flushing and ramming 水冲锤击(混合打桩)法
combined focal length 组合焦距
combined footing 联结底盘;联合柱基;联合基础;联合底座;联合底脚;多柱基础;复式基础;双柱底脚;复合基柱
combined force-feed and splash lubrication 压力飞溅合用润滑
combined forging and heading 重复联合镦锻
combined form 化合态
combined foundation 联合基础;多柱基础;复式基础
combined frame 组合门框;带窗门框
combined frame antenna 复合框形天线
combined front-and-side loading 前端侧方联合装载
combined gas chromatography mass spectrometry 色谱质谱联用法
combined generalized stiffness 联合广义刚度;组合广义刚度;复合广义刚度
combined glass break detector 复合式玻璃破碎检测器
combined governing 综合调节
combined gradation 混合级配
combined gradation of aggregate 骨料混合级配;骨料的组合粒度
combined grading 混合级配
combined grinding 研磨;混合粉碎;联合粉磨
combined gutter 集水沟
combined halftone and line 网点线条混合版
combined harmonic 复合谐波
combined harmonic antenna 复合谐波天线
combined head 读写组合头
combined heat and power 热电联供
combined heat and power generation 供热并发电
combined heating and refrigerating system 供热供冷联合系统
combined heat supply 联合供热;联合采暖
combined H-Na ion exchange 混合氢钠离子交换(法)
combined hopper dredge(r) 有挖泥和排泥设备的(自航式)挖泥船
combined humic acid 结合腐殖酸
combined hybrid computer 组合式混合计算机
combined hydraulic torque converter 综合液力变矩器
combined hydrogen 化合氢
combined impulse and reaction turbine 冲动—反动混合式汽轮机
combined-impulse turbine 组合冲击式涡轮机
combined inclinometer 联合测斜仪
combined income and surplus statement 合并收益及盈余表
combined index 总指数
combined industrial and commercial tax 工商业统一税
combined industrial effluent 混合工业污水
combined inlets chamber 汇流井
combined intersection 联合交会
combined iron shaving fluidized bed pretreatment-catalytic oxidation-coagulating sedimentation process 铁屑流化床预处理—催化氧化—混凝沉淀组合工艺

combined isothermal treatment 组合等温处理
combined joint 铆焊混合接头
combined journal and ledger 日记总账
combined key 两用键
combined kitchen equipment 组合的厨房设备
combined land/sea 水陆联运
combined latent image 复合隐像;复合潜像
combined lathe 组合车床
combined leading-syn-chronous type 超前同步混合型
combined leveller and shears 矫直剪切联合机组
combined levitation guidance system 磁悬浮列车提升导向系统
combined lighting 混合采光;综合照明
combined lime 混合石灰;结合石灰
combined lime stone 带纹石灰石
combined line and recording operation 长途连接制通信
combined liquid plastic limit device 液塑限联合测定仪
combined loaded and empty flow 混合车流
combined load fatigue testing machine 复合疲劳试验机
combined load(ing) 组合荷载;混合载重;混合荷重;混合荷载;荷载组合;综合荷载
combined loading by loader and conveyer 装岩机与转载机联合装岩
combined longitudinal projection of orebody 矿体复合纵投影图
combined lumpsum and unit contract 总包单价付款合同
combined magnet system 复合磁铁系制品
combined main and exhaust steam turbine 双汽源汽轮机
combined main and intermediate distributing frame 联合配线盘
combined maintaining plate 双历压片
combined management 联合经营
combined maritime-riverine shipment 海江河联运
combined mark 联合标记
combined market and planning economy 市场和计划经济组合
combined mass 组合质量
combined material 组合材料
combined mean 合并均数
combined mechanical and electric(al) strength 机电综合强度
combined mechanical and electric(al) strength test 机电联合强度试验
combined member 组合构件
combined metal oxide semiconductor 复合金属氧化物半导体
combined method 组合方法;综合方法
combined method of triangulation and trilateration 边角同测法
combined microfiltration and biologic(al) process 微滤和生物组合工艺
combined mill effluent 混合工厂污水
combined milling and mixing plant 粉碎和混合联合装置
combined mining and dressing 采选综合能力
combined mining technology 综合开采工艺
combined mixing and paving machine 搅拌铺路联合机
combined moisture 结合水;化合水分
combined mooring and warping bollard 码头带缆桩;系泊牵引两用桩
combined offer 联合提供的标书;联合发盘;联合发价;联合出价;联合报价
combined oil-processing 石油联合加工
combined-oxide formula 复合氧化物化学式
combined-oxide pigment 复合氧化物颜料
combined packing 组合型填料
combined parlour and sitting room 客厅起居两用室
combined paver 联合铺路机
combined pedestrian-vehicle phase 人车合用信号显示;人车通行信号显示
combined pen brush 排笔
combined permanent deformation 复合永久变形
combined pile footing 联合桩基
combined pile foundation 联合桩基

combined pipe tap and drill 管用钻孔攻丝复合刀具；管丝锥钻头复合刀具
combined pipe transmission 合管输送
combined piping loss 合流水头损失
combined pitot-static 全压静压探测管
combined pitotstatic probe 总静压联合探头
combined pitot static tube 皮托静压组合管
combined plannar bending and compression 平弯曲和压缩的综合作用
combined plan of pipelines 管道综合平面图
combined plasticizer and air entraining agent 增塑加气剂
combined pneumatic conveyer 混合式气力输送机
combined pollutant index 联合污染物指数；复合污染物指数；污染物综合指数
combined pollution 混合污染；复合污染
combined pollution index 联合污染指数；复合污染指数
combined polymerization 共聚合(作用)
combined pool 联营企业
combined portal-to-shaft-to-portal ventilation 吹入式通风系统
combined potential 总电势
combined power consumption 综合电耗
combined prehomogenization 混合预均化
combined press brake and plate shears 板料折弯剪切机
combined pressure ga(u)ge 复式压力计
combined primer/top coat coating 底面合一涂料
combined probability 组合概率
combined process 联合生产法；联合处理工艺
combined production process of cement 水泥综合生产
combined profiling curve 联合剖面曲线
combined profiling method 联合剖面法
combined profit and loss statement 合并损益表
combined projection 组合投影
combined punching and shearing machine 冲剪两用机
combined Rabitz type wire cloth and reed lath(ing) 组合的拉比兹钢丝网布和芦苇板条
combined radial and axial bearing 径向推力联合轴承
combined radial and axial flow turbine 辐流—轴流式涡轮机；辐流—轴流式透平
combined rail and road transport 铁路公路联合运输
combined rail and road vehicle 铁路公路两用车；铁轨与道路两用车
combined rate 联合运价(率)
combined recording depth and temperature meter 深度温度联合记录仪
combined record-playback head 复合磁头
combined red and green lantern 合座红绿灯舷灯
combined reducer 混合减弱剂；混合减薄液
combined regulation 联合管理；综合调节
combined residual chloride 化合性余氯；结合性余氯
combined resistance connection 组合电阻联结法
combined resonator 复合式谐振器
combined reverberation 混合反射；复合交混回响
combined ribtubular knit 带螺纹的圆筒形针织物
combined river 合流河
combined riveter and punch 冲孔铆接联合机
combined roller 振动静压复合压路机
combined rooms 多用途房间；组合房间
combined rotor 组合转子
combined rupture 复合破裂
combined safety and air valve 安全与进气组合阀
combined scheme 混合式布置；综合布置
combined sea/land 水陆两用车
combined service(d) ceiling 有公用事业综合设施的顶棚
combined service drawing 综合管线图
combined sewage 合流制排水系统；混合污水；合流污水
combined sewage flow 混合污水流
combined sewage pumping house 合流泵房
combined sewage system 雨污水合流制；合流制排水系统；沟渠合流制；合流下水道系统；雨污合流下水道系统
combined sewer 合流下水(管)道；合流污水管；合流水道；合流沟渠；合流沟道；雨水合流下水(管)道；综合污水渠

combined sewerage 合流排水道
combined sewerage regulator 合流排水管调节装置
combined sewerage system 合流制排污(水)系统；合流污水系统；合流排水系统；合流制；雨污水合流系统
combined sewer area 混合下水道区
combined sewer flow 合流水量；混合污水流
combined sewer network 混合污水管网
combined sewer outfall 合流污水排口
combined sewer overflow 混合下水道溢流排放；混合下水道溢流纳污水体监测；混合下水道溢流(槽)
combined sewer regulator 合流排水管调节装置
combined sewers 合流式下水道
combined sewer system 混合下水道系统
combined sewer wastewater treatment system 混合下水道污水处理系统
combined sheet 套色印样
combined shipment 混合装船
combined sidelights 合座红绿灯舷灯
combined sieve and sedimentation test 筛分和沉淀综合试验
combined silica 结合硅石
combined sinewave signal 复合正弦波信号
combined sketch 拼接草图
combined spectral line 复合光谱线
combined speed 合速度
combined spinning trimming and curling machine 平皱剪卷联合机
combined sponge and soap holder 海绵和肥皂合用盛器
combined spreader and softener 软麻延展机；软麻粗梳机
combined stacker-reclaimer 堆取料机；斗轮式堆取料机
combined standard deviation 合并标准差
combined statement 联合决算表；联合报表；合并报表
combined station 复合站
combined steel and concrete column 钢筋混凝土柱；钢筋混凝土柱
combined stiffness matrix 组合刚度矩阵
combined stop and emergency plant 快速关闭主汽门
combined storage 联合储备
combined storage hall 联合储库
combined storage structure 联合储水构造
combined store 联合堆场
combined store door lock 仓库门组合锁
combined storm and sanitary 混合暴雨与生活污水管道
combined storm and sewage outlet 混合暴雨与生活污水排口
combined stoves and hot-water heater 热风及热水混合供热器；热风(和)热水混合供暖器
combined strainer and check-valve 组合滤气止回阀
combined stream 合流河
combined strength 复合强度
combined stress 合成应力；组合应力；综合应力；复合应力
combined stress fatigue tester 复合应力疲劳试验机
combined stress strength 复合应力强度
combined string of casing 组合套管柱
combined structural sections 联合构造剖面
combined study 联结考察
combined style 组合式
combined suction and force pump 联合真空压力泵；吸压两用泵；吸压力；吸压泵
combined suction and pressure conveying system 吸压两用传送系统
combined sulfur 化合硫；复合硫黄
combined supervision and control 监控
combined supply air fan and exhaust fan system 吹入式通风系统
combined support 混合支架
combined support with bolting and shotcreting 喷锚联合支护
combined surcharge preloading consolidation 复合预压加固
combined surface 并合面
combined surface pressure 复合表面应力
combined switch 组合开关

combined symbol system 联合标号法
combined system 合流(制排水)系统；混合(体)系；合流制；组合体系；复合系统
combined system arch bridge 组合体系拱桥
combined system intercepting sewer 合流系统中的截流污水管
combined system of floating dock and repairing berth 浮坞船台联合系统
combined system of framing 混合骨架式
combined system of girder and beam 大小梁组合体系
combined system of sewerage 合流下水道系统
combined system of sewers 混合排水系统
combined system of ventilation 联合通风系统
combined system sewer 合流制下水道
combined tedder-and-side rake 干草摊晒和侧向搂草联合作业机
combined tension 合拉力
combined-test run 联合试车
combined thermal analysis 综合热分析法
combined thermit welding 加压铸焊
combined thermomechanical treatment 组合形变热处理
combined thickness 综合厚度
combined three-electrode sounding 联合三极测深
combined through transpor 直达联运
combined through transport 多种方式联运；水陆联运
combined thrust and radial bearing 组合推力径向轴承
combined tool 组合刀具
combined tower crane 综合塔式吊车
combined toxicity of pesticide 农药联合毒性
combined track ga(u)ge and level 万能道尺
combined trade restrictions 限制竞争的协议
combined train 组合列车
combined transport 联动单据；联运；综合运输
combined transportation 联合运输
combined transport bill of lading 联运提单；联运运输提单
combined transport document 联合运输单证；联(合)运(输)单据
combined transport operator 联运经营人；联合运输经营人
combined treatment 联合处理
combined turbine 联合涡轮机；联合透平；联合水轮机；联合汽轮机
combined twills 复合斜纹
combined type 组合式；复合式
combined type double tube core barrel 复合式双管取芯钻具
combined type junction terminal 混合型枢纽
combined type of crane 复式起重机
combined type of quay crane 复式门吊；复式码头起重机
combined-type passenger station 混合式客运站
combined type wharf crane (码头用的)母子起重机；混合型岸吊
combined unit 联合单元
combined universal machine tool 组合万能机床
combined use district 混合作用区(工业、商业、居住等各种用途的建筑)
combined vacuum electroosmotic surcharge preloading method 真空预压法
combined ventilation system 联合通风系统
combined version 内外倒转术
combined vertical and horizontal milling machine 立卧铣组合机床
combined vibrating roller 复合振动压路机
combined voltage and current transformer 电压电流两用变压器
combined voltage current transformer 电压电流两用互感器
combined-voltage generator 交直流电压发电机
combined vortex 复合涡旋
combined waste pipe 合流废水管
combined wastewater 混合污水；混合废水
combined wastewater overflow 混合污水溢流
combined wastewater quality 混合污水水质
combined wastewater-stormwater collection system 混合污水—暴雨水收集系统
combined wastewater treatment 联合污水处理
combined water 结合水；混合水；化合水；束缚水
combined water and land air base 水陆机场；水

陆航空港
combined water-and-land transport 水陆联运
combined water sealing 联合止水
combined water source 联合水源
combined water treatment 混合水处理
combined weighing and mixing machine 称重拌和联合机;自动称量搅拌机;过秤拌和两用机
combined well cleaning 联合洗井
combined whistle system 联合号笛系统
combined wind box 复合进风废气箱
combined wire 复合焊丝
combined W ore 复合型钨矿石
combined working 联合开采
combined yard 联合(调)车场
combine functional test 联合机能试验
combine-harvester 联合收割机
combine indigenous and foreign method 土洋结合
combine machinery 联合式机械
combine plotter-digitizer 测图仪数字化器联合装置
combiner 混合器;合成器;合并器;组合器
combiner circuit 组合电路
combine rope 组合缆
combiner tree 组合器
combine system 综合制图
combine tillage operation 综合耕作
combine transport 联合运输
combine unit line connection 联合单元结线
combing 算纹;梳制花纹;梳刷
combing ability 配合力
combing beater 梳刷轮
combing-coaming 舱口围板【船】
combing machine 精梳机
combing of brick face 砖面修刮
combing roller 分梳辊
combing wheel 分纸滚轮
combining 联合收割;化合的
combining ability 组合力
combining affinity 化合亲和势
combining capacity 结合能力
combining circuit 组合电路
combining cone 联合喷嘴
combining effect of chemicals 化学物综合作用
combining estimates of variance 方差联合估计;方差的合并估计
combining filter 双工滤波器
combining flow 合流
combining heat 化合热
combining leverage 联合杠杆
combining machine 复合机;贴合机
combining network 汇接网络
combining power 结合力;化合力
combining proportion 化合比数;化合比例
combining running-in fee 联合试运转费
combining site 结合簇;结合部位
combining sound level 混合声级
combining strength criterion 联合强度理论
combining volume 化合体积
combining-volumes principle 化合体积原理
combining weight 化合量
combining zone 化合区
combinite 沥青屋架
comb in micrometer 测微器梳尺
combi-rope 麻钢混捻钢丝绳
comb joint 榫接接合
comblainite 羟碳钴镍石
comb lightning arrester 梳形避雷器
comb-like 梳状
comb-like structure 梳状构造
comb-like texture 梳状结构
comb nephoscope 梳状测云器
comb pattern 算纹
comb pitot 梳状皮托管
comb plaster 梳状粉刷;拉毛粉刷
comb plate 算形防滑板;算齿板;梳形板
comb poles 梳形电极
comb polymer 梳形聚合物
comb pottery 算纹陶
comb rack 梳形支架(窑具)
comb rendering 梳状抹灰;梳状打底
comb ridge 锯状山脊;锯齿状山脊;刃脊
comb roof 三角屋顶;两坡屋顶;双坡屋顶;人字屋顶
comb scale 梳尺
combscomb ewer 天鸡壶(瓷器名)

comb-shaped electrode 梳状电极
comb-shaped fold 梳状褶皱
comb-shaped ridge 梳状山脊
comb-shaped transverse spreading adder 梳状横向扩展全加器
comb structure 梳状结构;蜂窝状构造
comb system 梳式布置(港池)
comb type circuit 梳形电路
comb-type double beam spectrometer 梳形双光束光谱仪
comb type slipway 梳式滑道
comburant 助燃物;助燃料
combustibility 可燃性;易燃性;燃烧性能
combustibility grading period 燃烧性按时间分级
combustibility rating 可燃性等级
combustibility test 燃烧性试验
combustible 可燃的
combustible basis 可燃基
combustible blind fence 可燃实体围篱
combustible building material 可燃建筑材料
combustible case 可燃壳体
combustible charge 燃料
combustible component 可燃成分;燃烧体
combustible constituent 可燃组分;可燃部分
combustible construction 易燃结构;易燃构造
combustible decoration 可燃装饰
combustible dust 可燃粉尘;易燃粉尘
combustible fence 易燃栅栏;易燃围篱
combustible frame construction 易燃框架结构
combustible fume detection system 可燃气体探测系统
combustible gas 可燃气(体);易燃瓦斯;燃气
combustible gas alarm 自燃报警器
combustible gas indicator 自燃指示器;可燃气体指示仪;可燃气体指示器
combustible in refuse 垃圾中可燃物
combustible limit 着火极限浓度;爆炸极限;着色浓度极限
combustible liquid 可燃液体
combustible loss 可燃物损失
combustible material 可燃物质;可燃材料;易燃品;易燃材料
combustible matter 可燃物质
combustible mixture 可燃混合物
combustible peat 可燃泥煤
combustible refuse 可燃(性)垃圾
combustible rubbish 可燃烧废物
combustibles 易燃物
combustible schist 可燃片岩
combustible screen fence 可燃性围篱
combustible shale 可燃的油页岩;可燃的泥板岩;油页岩;塔斯曼油页岩
combustible storage building 油库建筑;易燃物仓库建筑
combustible sulfur 可燃硫
combustible vapo(u)r 可燃蒸气
combustible waste 可燃(烧)废物
combustion 发火;爆震燃烧;燃烧;燃尽;起烧;燃燃
combustion adjuvant 助燃剂;燃烧助剂
combustion aerosol 燃烧气溶胶
combustion air 燃烧空气
combustion air blower 助燃风机
combustion air inlet 助燃空气入口;供给空气于燃烧区
combustion analysis 燃烧分析
combustion analyzer 燃烧分析器
combustion arch 燃烧拱
combustion area 燃烧区
combustion automatic control 燃烧的自动控制
combustion blow-by 爆发时漏气
combustion boat 燃烧舟
combustion box 燃烧室
combustion by explosion 爆炸燃烧
combustion calculation 燃烧计算
combustion casting process 爆炸铸造法
combustion cell 燃烧池
combustion chamber 火箱;火室;炉膛;窑头灶;反应室;燃烧室
combustion chamber deposit 燃烧室积炭
combustion chamber draft 炉膛负压;燃烧室负压
combustion chamber hopper 冷灰斗;沉渣斗
combustion chamber shape 燃烧室形状
combustion chamber superheater 炉膛过热器;燃烧室过热器

combustion chamber turbine 燃气室涡轮机;燃气室透平
combustion characteristic 燃烧特性
combustion control 燃烧(过程)控制;燃烧调节
combustion control instrument 燃烧控制仪
combustion controller 燃烧控制器
combustion control system 燃烧控制系统
combustion curve 燃烧曲线
combustion deposit 燃烧室积灰
combustion device 燃烧设备
combustion duration 耐烧时间
combustion dust 燃烧尘埃
combustion efficiency 燃烧效率
combustion emission 燃烧废气
combustion engine 内燃机;热机;燃烧(发动)机
combustion engineering 燃烧工程
Combustion Engineering Inc. 内燃机公司
combustion equipment 燃烧设备
combustion expansion ratio 燃烧膨胀比
combustion fan 燃烧用送风机
combustion flame 燃烧火焰
combustion flue 炉道;回火道;烟道
combustion front 燃烧面
combustion fuel gas apparatus 燃烧排气测定器
combustion furnace 焚烧炉;燃烧炉
combustion gas 炉气;烟道气;燃气体;排出废气
combustion gas analyzer 炉气分析仪
combustion-gas stream 燃气流
combustion gas turbine 燃气(涡)轮机;燃气透平
combustion header 燃烧室集气管;燃气收集器
combustion heat 燃烧热
combustion heater 燃气加热器
combustion improver 助燃剂
combustion indicator 燃烧指示器
combustion-in-humidity articles 遇湿易燃物品
combustion instability 燃烧不稳定性
combustion installation for kiln 窑用燃烧设备
combustion intensity 炉膛热强度;燃烧强度
combustion in the regeneration furnace 再生炉中燃烧
combustion kinetics 燃烧动力学
combustion knock 发动机爆震
combustion liner 燃烧室耐火层;燃烧室衬套
combustion liquid 可燃液体
combustion loss 燃烧损失
combustion mechanism 燃烧机理
combustion method 燃烧(方)法
combustion mode 燃烧模式
combustion motor 内燃机
combustion noise 燃烧噪声
combustion nuclei 燃烧产生的核
combustion nucleus 燃烧核
combustion of gas and vapo(u)r 气体燃烧
combustion(of oil)insitu 就地燃烧
combustion of sludge 污泥燃烧
combustion parameter 燃烧参数
combustion performance 燃烧性能
combustion pipe 燃烧管
combustion plant 燃烧设施;燃烧车间
combustion plasm(a) 燃烧等离子体
combustion pot 燃气发生器
combustion potential 燃烧势
combustion process 燃烧过程
combustion product 燃烧产物
combustion rate 炉膛热负荷;燃烧(速)率
combustion ratio 燃烧比率
combustion reaction 燃烧反应
combustion recorder 记录式燃烧室气体分析仪;燃烧记录器
combustion residue 燃烧残渣;燃烧残余;燃烧残留
combustion safety controller 燃烧安全控制器
combustion shock 燃震
combustion source 燃烧源
combustion space 炉膛容积;燃烧室;燃烧空间
combustion spiral 燃烧旋管
combustion stability 燃烧稳定性
combustion stage 燃烧阶段
combustion stroke 燃烧冲程
combustion-supporting 助燃的
combustion-supporting air 助燃空气
combustion-supporting gas 助燃气体
combustion surface 燃烧表面
combustion synthesis 燃烧合成
combustion system 燃烧系统

combustion temperature 燃烧温度
combustion test 燃烧试验
combustion time 工作时间;燃烧时间
combustion-titrimetric apparatus 灼烧滴定法(测定水泥中硫酸根含量)
combustion train 燃烧装置
combustion tube 燃烧管
combustion tube cooling 燃烧管冷却
combustion tube furnace 燃烧管炉
combustion turbine 燃气涡轮;燃气轮机
combustion turbine starter 燃气涡轮起动装置
combustion unit 点火设备
combustion value 热值;燃烧值
combustion velocity 燃烧速度
combustion wave 暴燃波;燃烧波
combustion zone 燃烧区;燃烧带
combustion zone temperature 燃烧区温度
combustor 燃烧室;燃烧器
combwebbing of spray gun 喷枪拉丝
comby 梳状的;蜂窝状的;蜂窝似的;蜂蜂状的
comb zero 齿板零划分
come about 掉抢
come-along 一件设备(俚语);一件工具;万能螺帽扳手;摊铺混凝土铲;伸线器;紧线器;吊具;紧线夹;紧绳夹;混凝土摊平工具;自由钳;管子扳手
come alongside 靠舷侧
come-and-go 来回;先收敛再发散;收缩膨胀
come-back bottle washing machine 旋转式单向洗瓶机
comedo 粉刺
come down 陷落;下跌;崩落
come from oversea 来自海外的
come high 售价高
come into effect 开始生效;生效
come into force 开始生效;生效
come into operation 开始运转;开始工作;开工;实施;生效
come into the market 流入市场
come into vogue 开始流行
comendite 钠闪碱流岩
comentropy 信息熵
come off work 下班
come-on system 临时插入系统
come on water 开始出水
come out of a well 因钻孔缩径而不能继续钻进
come out of the red 转亏为盈
comercial message 商用消息
come round 船头转向更接近风向
comer strip 压角条
comes (用以嵌合窗上玻璃的)铅条
come short of the mark 不合格
co-metabolism 共代谢作用
cometamorphic 早期变质的
cometamorphism 早期变质(作用)
come to 船已被锚拉住
come to a climax 达到顶点
come to an agreement 商定
come to an end 期满
come to an understanding 达成协议
come to deficit 弥补亏损;造成亏损
come to hand 收到
come to terms 议定;订约;达成协议;商妥
come under the hammer 被拍卖
come up system 按期处理方式
come up to standard 合乎标准
come up to the standard 达到标准
comfimeter 空气冷却力计
comfinement 约束
comforming article 合格品
comformity 同形度
comfortability 舒适性
comfortable degree 舒适度
comfortable feeling 舒适感
comfortableness of environment at station and on train 站车环境舒适度
comfort air conditioning 舒适空(气)调(节);舒适性空(气)调(节)
comfort chart 舒适焓湿图;舒适(度)图
comfort condition 舒适条件
comfort cooling 舒适冷却;舒适降温;适度冷却
comfort cooling system 舒适冷气系统;舒适冷却系统
comfort cooling unit 舒适冷气装置
comfort curve 柔和曲线;舒适(度)曲线

comforter batt 被胎
comfort fiber 舒适纤维(一种平织用的高张力聚酯纤维)
comfort heating system 舒适采暖系统
comfort humidity 舒适湿度
comfort index 舒适指标;温湿指数;舒适(性)指数
comfort line 舒适带;舒适(温度)区;快感线
comfort requirement 舒适要求
comfort room 公共厕所
comfort standard 舒适范围
comfort station 公共卫生间;公共厕所
comfort temperature 舒适温度
comfort zone 舒适区(域);舒适范围;舒适带
comfort zone chart 快感区图
comfort zone with air conditioning 空气调节舒适区
COM graph plotter 计算机缩微胶片图像输出仪
comichem 组合化学
comic ink 彩色报纸印刷油墨
comic mode 连环图式
comic-oriented image 连环式定向画面
coming crisis 面临的危机
coming generation 下一代
coming out 新发行的股票
coming pressure 来压
comitium 市民集会所
com junction 角式接缝
Comleyan 科姆累阶【地】
Comley sandstone 科姆累砂岩
comma bacillus 弧菌
comma free code 免逗点码
command 控制力;命令;指令;指挥权;指挥(部);司令部
command abbreviation 命令缩写
command activated sonobuoy system 指控声浮呐标系统
command active sonobuoy system 主动式指令声呐浮标系统
command address 命令地址;指令地址
command and control 指挥控制
command and control device 指挥控制装置
command and control information system 指令与控制信息系统
command and control system 指令(与)控制系(统);指挥(与)控制系统
command and data acquisition station 指令与数据收集站;指令数据汇集台;指令和数据接收站
command and data handling 指令与数据处理
command and data handling console 指令与数据处理控制台
command and telemetry data handling 指令与遥测数据处理
commandant 指挥官
command antenna 指令天线
command area 命令区
command authority 命令授权;命令权限
command bit 命令位
command button 指令按钮
command car 指挥车
command channel 命令通道
command character 指令符号
command code 命令码;命令编码;指令码;操作码
command code group 操作编码组
command code register 操作码寄存器
command coding scheme 指令编码方案
command confirmation 命令批准
command control 命令控制
command control centre 指令控制中心
command control panel 指令控制板
command control program(me) 命令控制程序
command data processor 命令数据处理机
command destruct 破坏指令
command destruction 命令破坏;破坏指令
command dialogue 指令对话
command-directed economy 控制经济
command echelon 指挥组;指挥系统
command economy 控制经济
commanded land 引水自流灌溉地
command encoder 命令编码器;命令编码程序
commander 指挥员;指挥官;大木槌;手夯
commander equipment 指挥设备
commander receiver of remote control 遥控指令接收器
commander sender 指令发送器

command(er) ship 指挥舰
command facility 命令设施;命令设备
command field 命令信息组
command file 命令文件;指令文件
command file processor 命令文件处理机
command format 命令格式
command frame 命令帧
command function 功能命令
command fuze 指令引信
command gate 命令门
command generator 指令发生器
command group 命令组
command guidance 命令制导;指令导引
command guidance system 指令导引系统
command information 命令信息
commanding apparatus 指令设备;操纵设备
commanding communication 指挥用通信
commanding elevation 制高点
commanding height 制高点
commanding heights of economy 经济制高点
commanding house 指挥所
commanding impulse 指令脉冲
commanding officer 指挥员;指挥官;值班驾驶员
commanding point 制高点;调度站
command input buffer 命令输入缓冲
command instruction 命令指令
command instruction format 命令指令格式
command interpretation 命令解释
command interrupt mode 命令中断方式
command job 命令作业
command language 命令语言;指令语言
command line 命令总线;命令行;指令传送线;指挥线
command link 主令线路;传令线路
command list 命令(列)表
command logic 命令逻辑;指令逻辑
command memory 指令存储器
command mode 命令(状)态;命令模式;命令方式
command module 指令组件;指令舱;指令轮;指挥舱
command name 命令名称
command net 指令网
command of the sea 制海权
command operation 命令操作
commando sales team 突击销售队
commando ship 登陆艇
commando vessel 登陆艇
command phase 命令状态;命令阶段
command pointer 命令指针
command post equipment 指挥站设备
command privilege class 命令优先类;命令特权级
command processing 命令加工;指令处理
command processor 命令处理程序;指令处理程序
command profile 命令简要表;命令分布
command program(me) 命令程序
command prompt 命令提示符
command prompt area 命令提示符区
command pulse 命令脉冲
command quantity 命令量;指令变量
command reader 命令阅读程序
command register 指令寄存器
command reject 命令拒绝
command response 命令响应
command retrieval system 指令检索系统
command retry 命令重执行;命令复执;指令再试
command rollback 命令返回
command scan(ning) 命令扫描
command search 命令检索
command sequence 命令序列;指令序列
command set 指令组;指挥台
command set instruction 命令置位指令
command sheet 作业命令单;命令单
command signal 命令信号;指令信号;指挥信号
command state 命令状态
command stream 命令流
command structure 指令结构
command subsystem 指令子系统
command switch 命令开关
command system 命令系统
command table 命令表
command terminal 指令终端
command tower 指挥塔
command type 指令形式
command vector 命令向量;命令矢量

comma-shaped cuttings 逗号形切屑
commaterial 同一种材料;同物质
commeasurability 可公度性;可比性
commemoration hall 纪念堂
commemorative 纪念的
commemorative arch 纪念性拱形建筑
commemorative architecture 纪念性建筑
commemorative basilica 纪念性长方形厅堂
commemorative chapel 纪念性小教堂
commemorative church 纪念性教堂
commemorative column 纪念柱
commemorative figure 纪念性图形;纪念性塑像
commemorative hall 纪念性大厅
commemorative monument 纪念塔;纪念牌(坊)
commemorative stone 纪念石碑
commemorative structure 纪念性结构
commemorative ticket 纪念票
commence 开始;开工(尤指土木工程方面)
commencement 开端
commencement and termination of cover 保险责任的起讫(日期)
commencement date 开工日期;工程开工日期
commencement date of the work 开工日期
commencement of business 开始营业
commencement of commercial operation 正式运行
commencement of construction work 施工开始;建筑开工典礼
commencement of cover 保险责任开始
commencement of drying 干燥开始
commencement of fuel supply 供油始点
commencement of initial setting 初凝(的开始)
commencement of injection 供油始点
commencement of operation 运行开始
commencement of raft towing 木排起拖
commencement of risks 风险(的)开始;保险责任开始
commencement of the limitation period 时效期开始(日期)
commencement of towing 始拖
commencement of works 开工(尤指土木工程方面);工程开工
commencement report 开工报告
commencement time 开工时间
commencement time of the work 开工时间
commence search pattern 开始搜寻方式
commencing speed 开始速度;初速
commendation and penalization 奖惩
commensal 共栖者;共栖体;共栖生物
commensalism 共生(现象);共栖(现象)
commensalisms 偏利共栖
commensurability 通约性;可比性;可通约性;公度;同量;通约
commensurable 有同量的;有公度的;能成比例的;可通约的;可比的
commensurable motion 通约运动
commensurable number 可同度量数;可比数值;比例数
commensurable quantity 可公度量
commensurate 同量的;同单位(的);适应的
commensurate orbit 通约轨道
comment 注解;评论
commentary 解说词
commentary channel 旁示信道
commentator's booth 实况广播间
comment card 注解卡片
comment code 注解码
comment convention 注解约定
comment entry 注释项;注解项
comment field 注释栏;注解栏
comment line 注释行;注解行
comment on revision 修改意见
comments on a report 批示
comments on the contracts 签署意见
commercail compiler 商用编译程序
commerce clause 通商条款
commerce-destroyer 通商破坏舰
commerce maritime 海洋贸易;海上贸易
commerce power 商业电力
commercial 商业楼宇;办公室
commercial account 厂商价款;商业往来账户;商务账;商业账簿
commercial accounting 商业会计;商业核算
commercial acid 工业酸

commercial acre 商业用地;可供租的土地
commercial act 买卖
commercial action 商业行为
commercial activity 商务活动
commercial advertising 商业广告
commercial affair 商业事务
commercial age 商业时代
commercial agency 商业代表
commercial agent 贸易官员;代理商(行);商务代表;商务代办
commercial aggregates plant 商业骨料厂
commercial agreement 贸易协定;商务契约
commercial aircraft 商用飞机
commercial alloy 商品合金
commercial alumina 工业氧化铝
commercial aluminum 民用铝;商业铝(纯度大于99%)
commercial analysis 商业分析
commercial and administrative centre 商业及行政中心
commercial and industrial circles 工商界
commercial and industrial noise 商业和工业噪声
commercial annealing 中间退火
commercial application 商业应用
commercial arbitration 商业仲裁;商务仲裁
commercial architecture 商(业)用(的)建筑物
commercial area 商业(地)区
commercial art 商业美术
commercial article 商业新闻;商品
commercial asbestos 商品石棉;出售的石棉
commercial at 单价记号
commercial attaché 商务参赞;商务专员
commercial availability 市场购买可能性
commercial aviation 商业航空业
commercial awning 商业遮篷
commercial bank 商业银行;商人银行
commercial bank credit 商业银行信贷
commercial banking 商业银行业
commercial bank loan 商业银行贷款
commercial base 商业基地;贸易基地
commercial bed 有开采价值的矿层
commercial benzene 工业苯
commercial bill 贸易票据;商业票据
commercial bill market 商业汇票市场
commercial bill of exchange 商业汇票
commercial bill of lading 商业提货单
commercial blanket bond 商业诚实保证书;商业总担保
commercial blast 美国国家腐蚀工程师协会规定的金属表面喷砂处理的第三级
commercial blast cleaning 工业喷砂清理(表面)
commercial block 商业区段;商业街坊
commercial blockade 商务封锁
commercial body 商用车身
commercial book-keeping 商业簿记
commercial breed 商用品种
commercial brokerage 商业佣金
commercial bronze 工业青铜;商用青铜合金
commercial building 商业建筑(物);商业房屋
commercial burner oil 工业用燃烧油
commercial business 商业经营;贸易经营
commercial canopy 展示盖篷;商业盖缝
commercial capital 商业资本;贸易资本
commercial capital construction 商业基本建设
commercial carrying 商业性运输
commercial case 商业诉讼;商业案件
commercial casting 商品铸件
commercial catch 渔业捕获量
commercial cement 商品水泥
commercial center 闹市区;商业中心(区);商业区
commercial character 商用字符;商用符号;商业道德
commercial circle 商(业)界
commercial circulating funds 商业流动资金
commercial city 商业城市
commercial class 商务舱
commercial clear cutting 商业性采伐
commercial clearing account 贸易清算账户
commercial club 商会
commercial code 商务条例
commercial college 商业学校
commercial colo(u)r match 商业配色
commercial company 贸易公司;商业公司
commercial competition 贸易竞争;商业竞争;商业竞赛

commercial complex 综合性商业建筑
commercial computer 商用计算机
commercial concern 商店
commercial concrete 商品混凝土
commercial contract 贸易合同
commercial control 商业管制
commercial cooling 商品冷藏
commercial copper 商品铜
commercial cork 商品软木
commercial corporation 贸易公司
commercial correspondence 商业通信;商务函件
commercial cost 商业成本
commercial counsellor 商务专员;商务参赞
commercial counsellor's office 商务(参赞)处
commercial court 商业法庭;商事法院
commercial credit 商业信用;商业信贷;商业文件
commercial credit agency 商业征信所
commercial credit agreement 信用证约定书
commercial credit bureau 商业征信所
commercial credit company 商业信托公司
commercial credit corporation 贸易信贷公司
commercial credit documents 商业信用单据
commercial credit insurance 贸易信用保险;商业信用保险
commercial crisis 贸易危机;商业危机
commercial criterion 商业标准
commercial custom of trade 贸易常规
commercial customs 商业习惯
commercial data management system 商用数据管理系统
commercial data recorder 大量生产数据自动记录器
commercial deal 商业交易
commercial delegation 商务代表团
commercial department 商业部门;贸易部门
commercial depression 商业萧条
commercial designation 商品标志
commercial disappearance 商业销售
commercial discount 贸易折扣;商业贴现;商业折扣
commercial disignation 商品代号
commercial distemper 商品色粉涂料
commercial distribution 商业流通
commercial distributive network 商业网
commercial district 商业区
commercial diving 商业潜水
commercial dock 商用码头;民用码头;商用船坞
commercial documents 商务文件;商业单证
commercial draft 商业票据;贸易票据;商业汇票
commercial economy 商业经济
commercial efficiency 贸易效益;经济效益
commercial efficiency analysis 经济效果分析
commercial elastic limit 技术弹性限度
commercial electricity frequency 市电频率
commercial embargo 贸易禁运;商业禁运
commercial enterprise 商业企业
commercial environment 民用环境
commercial establishment 商业场所;各种店铺(包括服务业)
commercial exchange 贸易汇兑;商业往来;商业汇兑
commercial exchange rate 商业汇率
commercial exploitation 生产性开发
commercial explosive 民用炸药
commercial factor for coal 煤炭工业指标
commercial farm 商品农场
commercial fast breeder reactor 商用快中子增殖反应堆
commercial fertilizer 商品肥料
commercial fiber 商品纤维
commercial field 工业地热田
commercial finance 贸易金融;商业资金
commercial finance company 商业金融公司
commercial financial company 商业金融公司
commercial financial corporation 商业金融公司
commercial financial institution 商业金融机构
commercial financial plan 商业财务计划
commercial financing 提供商业资金;商业措施
commercial firm 厂商公司;商行
commercial fishery 商品渔业
commercial flat glass 商品平板玻璃
commercial forest 材用林地;经济林;用材林
commercial forest land 商业性林地
commercial forestry 商业性林业

commercial form 商业形式;商品形式
commercial formulation 商品化的加工制剂
commercial foundry 铸造厂
commercial franchised barge 营业性驳船
commercial free port 贸易自由港;商业自由港
commercial free zone 商业免税区
commercial frequency 市电频率;工业用电频率;商用频率
commercial frequency meter 工业用电频率计
commercial frontage 商业街面
commercial fuel 商品燃料
commercial garage 营业性停车库;商业停车库
commercial gas appliance 商业燃(气用)具
commercial goods imported 进口商品;外来商品
commercial grade 商业等级;工业级;商品(等)级
commercial grade steel 商品等级钢
commercial granite 商品花岗岩
commercial grower 商品栽培者
commercial growing 大面积栽培
commercial guild 商业公会
commercial harbo(u)r 商港
commercial heavy weight diving dress 商业重潜水服
commercial heavy weight diving suit 商业重潜水服
commercial height 商品材高度
commercial herd 商品畜群
commercial high-rise block 商业高层建筑区段;商业高层建筑街坊
commercial holding 商品农场
commercial hose diving 商业软管供气潜水
commercial hotel 营业性旅店;营业性饭店;商业旅馆
commercial house 交易委托行
commercial humic acid 工业腐殖酸
commercial incinerator 商业焚化炉
commercial indirect expenses 商业间接费用
commercial industrial building 商品化工业化房屋
commercial installation 工业设备;商业装置
commercial instruction set 商用指令系统;商用指令集
commercial instrument 营业证书;商用仪器
commercial insurance 商业保险
commercial intercourse 贸易往来;通商;商业联系;商业交往;商务往来
commercial inventory 商业库存
Commercial Investment Real Estate Institute 商业投标不动产研究所
commercial investment trust company 贸易投资信托公司;商业投资信托公司;商业投资公司
commercial invoice 商业发票
commercial iron 通用型铁;商品铁
commercialist 商业家
commercial item 商品
commercialization 投放市场;使商品化;商业化;商品化
commercialization of reseach findings 科技成果转化
commercialization of technology 技术商业化
commercialize 使供应市场
commercialized process 商品化过程
commercialized residence 商品化住宅
commercialized society 商业化社会
commercialize invention 发明商品化
commercial kennel 家畜训练服务店;家畜商店
commercial kitchen 商业厨房
commercial laboratory 商业试验室
commercial land development 商业土地开发
commercial language 商业用语
commercial law 贸易法;商业法(规);商法
commercial lead 工业铅
commercial lead time 商业交付周期
commercial lease 商业租契
commercial length 商用长度
commercial letter of credit 商业信用证
commercial leverage 商业调节(手段)
commercial lime 商品石灰;出售的石灰
commercial loan 商业资金;商业信贷;商业贷款
commercial loan rate 贸易贷款利率;商业贷款利率
commercial loan theory 商业贷款理论
commercial lubricating oil 商品润滑油
commercially available 市场上可买到的
commercially available substitute 市场上可以买到的代用品

commercially dry 吹干
commercially dry sludge 吹干污泥;商业干污泥;商品干污泥
commercially pivoted window 翻窗(商业、工业建筑用)
commercially pure 工业纯的;商业纯
commercially pure alumin(i)um 工业纯铝
commercial magnesium 工业镁
commercial management 商业管理;贸易管理
commercial manufacture 工业制造;商品生产
commercial marble 工业用大理石;商品大理石
commercial marine 商船
commercial measurement 工业测量
commercial message 商用报文;商业文件;商业广告;商情
commercial metal 商品金属
commercial mission 商业代表团
commercial mixed feed 商品混合饲料
commercial morality 商业道德
commercial motor 商用电动机
commercial museum 商品展览馆;商品陈列馆
commercial name 商(品)名(称)
commercial navigation 商业航行
commercial network 商业网点;商业网
commercial network planning 商业网点规划
commercial nickel 工业镍
commercial noise 商业噪声
commercial occupancy 商业使用
commercial offer 商业报价
commercial oil 工业用滑油
commercial operation 商业行为
commercial order 商业订货;商业订单
commercial organization 商业机构;贸易机构;商业组织
commercial packaging 商业包装
commercial paint 普通油漆;商品油漆;商品涂料
commercial paper 商业票据;商业本票
commercial paper exchange 商业票据交易所
commercial paper house 商业证券经纪行;商业票据承销公司
commercial paper market 商业票据市场
commercial par 商业平价
commercial parking area 收费停车场
commercial parking facility 商业停车场;收费停车场
commercial parking lot 收费停车场
commercial parking space 收费停车处;商业停车处
commercial partnership 商业合股
commercial pedestrian street 商业步行街
commercial pipe 商品管子;商品管(材)
commercial plan 商业规划
commercial plant 工业设备
commercial planting 大规模栽植
commercial pledge 商业上的抵押
commercial plywood 商业胶合板
commercial policy 商业保险单
commercial port 商港;贸易港;商埠
commercial portfolio 商业证券
commercial poultry production 商品养禽业
commercial power 电网电源;市电;商业电力
commercial practice 商业惯例;商业惯常做法
commercial premises 事务所
commercial prestressing member 商用预应力构件
commercial process 工业化生产(过程)
commercial procurement 贸易采购;商业采购
commercial production 商品性生产
commercial production machinery 商业机械
commercial production of sugar 生产商品食糖
commercial products 工业品
commercial profit 贸易利润;商业利润
commercial profitability 商业盈利率
commercial projected window 翻窗(商业、工业建筑使用)
commercial property 商业财产
commercial proposal 带报价的建议书
commercial quality 商品质量
commercial rate 贸易汇率
commercial reactor 商业反应堆
commercial rebate 商业回扣
commercial receiver 民用接收机
commercial reciprocity 互惠贸易;商业互惠
commercial recreation 营业性游乐场
commercial red brass 商品红黄铜
commercial refrigerating plant 商用致冷装置

commercial refrigerating system 商用制冷系统
commercial refrigeration unit 商业制冷机组
commercial refuse 商业垃圾
commercial register 商务注册
commercial registration 商业注册;贸易注册;商业登记
commercial relation 商业关系;贸易关系
commercial reliability 民用设备的可靠性
commercial representative 商务代表
commercial reprocessing 商用回收处理
commercial reputation 商业信誉
commercial requirement 商品要求
commercial reservoir 工业储量
commercial residential buildings 商品住宅
commercial risk 商业风险
commercial rock gas 天然石油气
commercial rolled section 商用轧制型钢
commercial room 行商展售室
commercial rules 商务规章
commercial run 工业生产
commercial rustless steel 商用不锈钢
Commercial Sale Rooms 商业交易所
commercial salt 粗制盐
commercial sample 商品试样
Commercial Samples Carnet 样品免税通关卡
commercial satellite 商用卫星
commercial-scale 商业规模的;大规模的
commercial scale process 大规模生产法
commercial scale production 大规模生产
commercial school 商业学校
commercial secretary 商务秘书
commercial service 商业供气
commercial service system 商业服务系统
commercial sheet 商品钢板
commercial shellac 商品虫胶
commercial sintered bronze product 工业烧结青铜制品
commercial sintered product 工业烧结制品
commercial size 商用规格;工业规模(的)
commercial size unit 工业规模设备
commercial society 商业界
commercial solid waste 商业垃圾
commercial sort 商业分类;商品品质
commercial space 商业面积
commercial specification 产品说明书;商品规格
commercial speed 旅行速度【铁】;满载正常航速;额定车速
commercial stable 商业性牛棚;马厩
commercial standard 商用标准;商业标准
commercial standard plywood 商业标准胶合板
commercial standing 商业信用;商业地位
commercial station 商用电台
commercial statistics 商业统计
commercial steel 型钢;条钢;商品钢材
commercial steel pipe 商品钢管
commercial stock length 成品轧材长度;成品轧材的标准长度
commercial stone-crushing plant 商业性轧石机
commercial strip 商业街
commercial structural steel 商用结构钢
commercial sulphuric acid 工业用硫酸
commercial survey organization 商业测量部门
commercial symbols 商业符号
commercial tall building 商业用高层建筑
commercial telegrams 商业电报
commercial television 商业电视
commercial terms 商业用语;商业条件
commercial territory 销售区
commercial test(ing) 工业(性)试验;全部产品试验;委托试验;大量生产时的全部产品试验
commercial thinning 商业性疏伐
commercial time-sharing 商业分配
commercial tin 工业锡
commercial titanium 工业钛
commercial tolerance 普通公差
commercial town 商业城市
commercial traffic 商业运输
commercial traffic bulletin 货运交易公报
commercial transaction 商业贸易;买卖;商业交易
commercial transport 商业运输
commercial travel(l)er 商业旅行者;行商
commercial treaty 贸易条约;通商条约;商约;商业条约
commercial trip 商务旅行;乘车采购

commercial triplex 商品三层玻璃
commercial tube 民用管
commercial-type appliance 生产类用具;工业用具;工业仪表
commercial types of mineral deposits 矿床商业类型
commercial undertakings 商业企业
commercial unit 工业设备;商业单位
commercial usage 商业惯例
commercial use 商业应用
commercial utilization 商业用气
commercial U-tube manometer 工业 U 形管压力计
commercial value 工业价值;商业价值
commercial variety 栽种品种;商品种子
commercial vehicle 民用车辆;货运汽车;载货卡车;商业运输工具
commercial vehicle rate 载重车混入率
commercial vehicles interview survey 营业车访问调查
commercial veneer 表面薄板
commercial vessel 商船
commercial washing compound 工业洗涤剂
commercial waste management 工业污水管理
commercial wastes 商业废物
commercial wastewater 商业废水
commercial water course 商业性水道
commercial waterway 商业性水路;商业性水道
commercial weight 正量;纯重;商用重量
commercial well 有工业价值的油井
commercial whaling 商业性捕鲸
commercial window 商业用窗;商业型窗
commercial world 商(业)界
commercial wrought iron pipe 商品熟铁管
commercial xylene 混合二甲苯;工业二甲苯
commercial year 商业年度
commercial zinc 工业锌
commercial zone 商业区;商业地带
commercium 商业;[复]commercia
commingled fiber 混合纤维
commingled fund 综合储备
commingled roving 混合无捻粗纱
commingler 混合器;搅拌器
commingling 混合资金
comminuted fibres 粉碎的纤维
comminuted polymer 聚合物粉末
comminuted powder 粉碎粉末
comminuted solid 粉碎固体
comminuted steel shot 钢粉
comminuter 磨粉机;粉碎器;捣碎器;粉碎机
comminuting machine 粉碎机
comminuting screen 粉碎筛
comminuting sludge 粉碎污泥
comminution 细碎;粉碎过程;粉碎(作用);粉刷;粉磨;破碎
comminution by gas stream 气流粉碎法
comminution by impact reduction 撞击粉碎;舂碎;舂粉
comminution equipment 粉碎设备
comminution of screenings 筛余碾碎物
comminution ratio 粉碎比
comminutions 小碎粒
comminution step 粉碎步骤
comminution theory 粉碎理论
comminutor 切碎机;磨碎机;造粒机;粉碎机
comminutor desintegrator 粉碎机
commision paid on discounted account 客账贴现佣金
commissary 委托人;食品商店
commission 经纪费;代理费;酬金;任命;委员会;委托;手续费
commission account 佣金账目;酬劳费账目;手续费账目
commission agency 经纪业;代理贸易
commission agent 贸易代理人;代销商;经纪人;信托商店;中间商;购销代理人;代理商;代理人
commissionaire 穿制服的看门人
commissionaire's room 门房
commission a ship 配备船舶人员
commission broker 捐客;证券介绍人;经纪人
commission business 委托贸易
commission charges 佣金;手续费
commission contract 代理贸易合同;商业代理合同
commissioned 定岗位
commissioned ship 现役船

commissioner 局长;海关关长;专员;地方长官;委员
commissioner of customs 海关税司
commissioner's deed 经官方签署的契约
commission fee 佣金;代理贸易费;代理费
commission fees and other charges 手续费及其他费用
Commission for Aeronautical Meteorology 航空气象学委员会
Commission for Atmospheric Sciences 大气科学委员会
Commission for Basic System 基本系统委员会
Commission for Climatology 气候学委员会
Commission for Special Applications of meteorology and Climatology 气象学和气候专门应用委员会
Commission for Synoptic(al) Meteorology 天气学委员会
Commission for Synoptic(al) Weather Information 天气情报委员会
commission house 信托商店;代办行;委托商行;委托商店
commissioning 开工;交付使用;带料生产;投入运行;投产;调试;试运转;试运行
commissioning application 启用申请
commissioning application forms 启用申请表格
commissioning date 投产日期;运行日期;启用日期;投入运行日期
commissioning expenditures 佣金支出
commissioning of individual equipment 单机调试
commissioning test 交接试验;船舶地球站启用实验;启用试验;投产(交换)试验;试车;试运行
commissioning test run 投料试生产
commissioning test run of pipeline 管道试运行
commissioning time 试运行时间
commission manufacture 委托加工制造
commission merchant 贸易代理商;中间商;代售商;代办商
commission mortgage loan 佣金抵押借款
Commission of Agricultural Meteorology 农业气象委员会
commission of audit 审计官
commission of authority 授权书
commission of conciliation 调解委员会
commission of inquiry (以询问方式为主并有权获得证据的)调查委员会
commission of investigation (深入持久进行审慎而系统调查的)调查委员会
commission of legal affairs 法制委员会
Commission of Space Research 国际宇宙研究委员会
Commission of the European Communities 欧洲共同体委员会
commission on a sliding scale 递加佣金
commission on drafts sold 出售外汇汇票佣金;出口外汇汇票佣金
Commission on Natural Resources 自然资源委员会
Commission on the Carriage of Dangerous Goods 危险货物运输委员会
commission order 代理委托书
commission over-ride 超额佣金
commission paid on discounted 客户贴现佣金
commission paid on discounted account receivable 应收账款折现佣金支出
commission paid on discounted accounts 客账贴现佣金支出;应收账款贴现佣金支出
commission past us 过手佣金
commission payable 应付手续费
commission receivable 应收手续费
commission sale 经售;委托销售;经销;寄售;代销
commission shop 信托商店;委托商店
commission trade 代理贸易
commissure 焊接处;焊缝;接合点;合缝处;(石砌体的)缝口;石层缝;结合点
commit 调拨
commit bribery 行贿
commitment 债务承诺;付款承诺;承租承诺;交托;承诺;承担义务;委托
commitment and involvement 信托与介入
commitment authority 承诺权
commitment charges 义务承担费
commitment control 工作区间管理
commitment fee 手续费;承诺费;承诺费
commitment letter 贷款许可;贷款承诺书;承诺信

commitment of fund 资金承诺;承付资金
commitment period 贷款有效期;承诺期;保证期
commitment unit 单位工作区间
commit no nuisance 禁止倾倒垃圾
committal in civil proceedings 民事诉讼中的拘押
committed 待发
committed capacity cost 折旧费
committed cost 约束成本;已承诺费用;不可取消的成本
committed dose equivalent 约定剂量当量;待积剂量当量
committed fixed cost 拘束性固定成本;约束固定成本
committed step 关键步骤
committee 委员会
Committee for Climate Changes and the Ocean 气候变化和海洋委员会
Committee for Development Planning 发展规划委员会
Committee for Environment(al) Conservation 环境保护委员会
Committee for Environmental protection 环境保护委员会
Committee for European Construction Equipment 欧洲建筑设备委员会
Committee for Industrial Development 工业发展委员会
Committee for Marine Meteorology 海洋气象学委员会
Committee for Oceanic Meteorology 海洋气象学委员会
committeeman 委员
committee member 委员
Committee of Communication Member Countries 通信成员国委员会
Committee of Safety of Nuclear powered Ship 核动力船安全委员会
Committee on International Geodesy 国际大地测量学委员会
Committee on International Ocean Affairs 国际海洋事务委员会
Committee on Marine Technology 海洋技术委员会
committee on medical appraisal of labor fitness 医务劳动鉴定委员会
Committee on Trade and Development 贸易发展委员会
Committee on Water Research 国际水文协会水研究委员会
Committee on Water Resources 水资源委员会
committee room 会议室;委员会办公室
commnopath reference wave 共光程参考波
commode 马桶;洗脸台;五斗柜
commode step 圆角踏步;宽踏步;弧形踏步;凸面踏脚板踏步;踏步凸面竖板
commode-type closet 坐式便桶
commode-type toilet 有便溺器的盥洗室;坐式便器
commodious 适宜的;宽敞的;方便的
commodiousness 宽敞
commodities 矿种
commodities at par 平价商品
commodities exempt from taxes 免税货物
commodities fair 商品展览会
commodities for energy source 能源矿产
commodities for fair 商品展销会
commodities for industrial material 工业原料矿产
commodities for the home market 内销商品
commodities in short supply 紧缺商品
commodities output value 商品产值
commodities purchased by choice 选购商品
commodities used to be rationed 凭票证供应的商品
commodity 货物;商品
commodity agreement 商品协议;控制商品产销协定
commodity allocation plan 商品调拨计划
commodity boom 商品繁荣
commodity broker 商品经纪人
commodity bundle 商品包
commodity capital 商品资本
commodity catalog(ue) 商品目录
commodity circulation 商品流通
commodity classification 商品分类
commodity code 商品编码

commodity composition 商品结构;商品构成
commodity composition of import and export 进出口商品结构
commodity concrete 商品混凝土
commodity considered as a use-value 使用物品的商品
commodity coverage 商品范围
commodity credit corporation 商品信用公司
commodity crop 商品作物
commodity custody 商品保管
commodity-dealing capital 商品经营资本
commodity demand volume 商品需求量
commodity differentiation 商品差别化
commodity draft 商品提单
commodity economy 商品经济
commodity-equivalents 商品等价物
commodity exchange 期货交易(农产品等);商品交易所;商品交易会
commodity export 商品输出
commodity fair 商品展览会
commodity flow analysis 商品流动分析
commodity freight 分货种运费
commodity grain 商品粮
commodity information 商品信息
commodity in short supply 短缺商品
commodity inspection 商品检验
commodity inspection and testing bureau 商(品)检(验)局
commodity inspection bureau 商品检验局
commodity inspection law 商品检验法规
commodity loan 商品贷款
commodity map 产品分布图
commodity market 商品市场
commodity money 实物货币;商品货币
commodity-money relationship 商品货币关系
commodity output 商品产量;商品产出
commodity output matrix 商品产出矩阵
commodity package 包装物的回收复用
commodity packaging 商品包装
commodity paper 押汇票据;商品票据(指银行本票和跟票提单)
commodity price 初级商品价格;商品价格
commodity product 商品产品
commodity projection 货品估计
commodity purchased 外购商品
commodity rate 货物运转率;个别商品运费率;个别货物特别运价
commodity's domestic value 商品的国内价值
commodity shunting 倒流运输;往复运输
commodity's international value 商品的国际价值
commodity sold overseas 销售国外的商品
commodity space 商品空间
commodity standard 商品本位制
commodity standardization 商品标准化
commodity stocks 库存商品
commodity supply 货源;商品供给;商品储蓄
commodity survey 商品概览
commodity talc 商品滑石
commodity tax 货物税;商品税
commodity transaction 商品交易
commodity turnover 商品流通
commodity wastage 商品损耗
commom fissure 普通龟裂(纹)
common 寻常的;共用的;共享的;共同的;公共的;普通的;通用的
common account 共同账户
common accumulation 公共积累
common accumulation fund 公积金
common activated sludge method 普通活性污泥法
common address space section 公用地址空间段
common adventure 海上共同事业
commonage 土地共有;共有;公用权;公地
common agent 通用肋剂
common aggregate 普通集料;普通骨料
common ailment 常见病
common air chamber 空气收集器;集气室;通气总管
common air mass 公有空气团
common air traffic control system 军民通用航空管理系统
common aisle 公共通道
commonality 通用性
common anchor 有杆锚

common-anode amplifier 共板极放大器
common antenna 共用天线;公用天线
common antenna television 公用天线电视
common antenna television system 共用天线电视系统
common apex 公共锥顶
common arch 粗拱;普通拱
common area 公用部分;共用面积;公用区域;公用区;公用(场)地
common area charges 公用区收费(房屋)
common area control 公用区控制
common area item 公用区项
common-area maintenance 公用部分(的)维护
common area of memory 存储器公用区
common ashlar 普通方石
common assembler 公用汇编程序
common association 公用结合
common auger 普通采样钻
common augite 普通辉石
common axis 共轴;公共轴线
common axis of rotation 共同的旋转轴
common axis of symmetry 常对称轴
common back-up area 公共备用堆置场
common bald cypress 落羽松
common bank 公用存储体
common bar iron 普通铁条
common base 共用基座;共用底座;共同基座;共同基础;共基极
common-base amplifier 共基极放大器
common-base characteristic 共基极特性
common bathing pool 浴池
common battery 中央电池组;共用电池(组);共电式电池
common battery central office 公用电源中心站;公用电源供电局;共电制电话总局
common battery exchange 共电式交换机
common battery station 中央蓄电池站
common-battery switchboard 共电制交换台
common-battery system 共电制
common beam (两楼板之间的)共用的梁;普通梁
common bed 公用底座;底座
common bevel gear 直齿伞齿轮
common bit 带尖钻;普通钻头
common block 公用块;公用区
common block list 公用块表
common bond 普通砌筑式;美式砌法;普通砌砖式;普通砌砖法;普通砌合
common boundary 共同边界
common bower 普通有杆首锚
common box tree 千年矮;黄杨
common branch 共分支
common brass 铜锌合金;标准黄铜
common brick 红砖;普通(建筑)砖
common buffer 公用缓冲区
common bus 总线;总汇流条;公用总线
common bus multiprocessor 公用总线多处理机
common bus system 共母线制;公用总线系统
common camellia 山茶花
common capital stock 普通股
common carport 公共停车库;公共停车场
common carriage 公用运输(工具);公共滑架
common carrier 公用事业公司;运输公司;公用载波;公共运业者;公共承运人;电信公司;运输行;公共运输(工具)
common carrier by water 海上托运人
common carrier telecommunication 共载波远距离通信
common cause 共同的目标
common cause failure 一般原因故障
common cement mortar 通用水泥砂浆
common centering 通用拱架;普通拱架
common center of gravity 公共重心
common centre of curvature 共同的曲率中心
common chamotte brick 普通耐火砖
common-channel interference 同波道干扰;同路干扰
common channel interoffice signalling 公共信道局间信号传输
common channel signal(l)ing 公共信道信号(方式);共路信令
common channel signal(l)ing subsystem 公共信道信号子系统
common channel terminal 公共信道终端
common chord 公弦

common classification of mineral resources 矿产资源常用分类
common clay brick 普通黏土砖
common cock 共用水龙头;公用水龙头
common code 公共代码
common collateral 共同担保
common-collector 共集电极
common-collector amplifer 共集电极放大器
common cololapse 普通塌陷
common communication adapter 公用通信适配器;公共通信适配器;通用通信适配器
common computer software 公用计算机软件
common concrete 普通混凝土
common consumer system 普通消费者系统
common control 一般控制
common controlling equipment 公共控制设备
common control section 公用控制节
common control switching arrangement 公用控制转换设备;公用控制交换网;公共控制开关装置
common control system 集中控制系统;集中控制方式
common control unit 公用控制部件;公共控制器;公共控制单元
common coordinate system 共同坐标系
common cordgrass 大米草
common core 主存储器公用区
common corridor 公共走廊
common corundum 普通刚玉
common cost 共同成本;公共费用;无差别成本
common crack 普通龟裂(纹)
common crane 灰鹤
common curvature 共曲率
common customs tariff 共同海关税则;共同关税(率)
common cut nail 普通方钉
common cycle time 公用周期时间
common data area 公用数据区
common data base 公用数据库
common data network 公用数据网
common davit 旋转式吊艇柱
common declaration 公共说明
common declaration statement 公用说明语句
common denominator 公共说明;公分母
common dense timber 普通密纹木材
common depth point 共深度点;共点深度
common-depth-point gather section 共深度点集合剖面
common-depth-point number 共深度点数
common depth point shooting 共深点爆炸
common depth-point stack 共深度点叠加
common description 普通描述
common difference 公差
common digitizer 通用数字转换器;通用数字读出机
common discharge 普通流量
common disease 常见病
common display logic 通用显示逻辑
common divisor 公约数;公益使用
common dog spike 普通狗头道钉
common dollar 共同货币;等值货币
common dollar financial statement 不变币值财务报表
common dovetail 普通燕尾榫;普通鸠尾榫
common drain 共漏极
common draw size 终拔前尺寸;钢丝终拔前尺寸
common drying room 集中干燥室;常用干燥室
common duckweed 浮萍(属)
common element 普通元素
common element structure 共用元件结构
common elm 大叶榆【植】
common emitter 共发射极
common emitter amplifier 共发射极放大器
common-emitter connection 发射极接地连接
common emitter junction phototransistor 共发射极结型光电晶体管
common envelope 共有包层
common equipment of crop drying 普通干燥设备
common equity 普通股产权
common establishment 潮候时差;标准潮汛;平均朔望高潮间隙;朔望高潮间隙
common event block 公用事件块
common event flag cluster 公用事件标志束
common excavating method 普通掘进法
common excavation (不含大块石的)一般挖方;普通挖方(指不含岩石的挖方);普通开挖

common excretory duct 总导管
common exit point 共用出口点
common expenses 共同费用
common expression 公用表达式;公共表达式
common facility 公共设施
common factor 共同因素;公因数;公(共)因子
common factor variance 公因子方差
common fastener 普通紧固件
common feldspar 普通长石
common fertilizer 普通肥料
common field 共用区;公用域;公用区;公用场;公共信息组
common financing fund 共同资金
common flax 亚麻
common flue 共用烟道
common focal plane 共焦点平面
common form 普通形式
common form bill of lading 通用格式提单
common fraction 普通分数
common frequency curve 一般频率曲线
common fund 共同基金
common fuse 普通导火线
common gain control 公共增益控制
common garage 普通汽车间;公用汽车间
common gas outflow 普通涌出
common gate 共栅极
common-gate connection 共栅连接
common gateway interface 公共网关接口
common genera 普通种属
common geomorphologic(al) map 普通地貌图
common glassine bag 普通玻璃纸袋
common goal 共同的目标
common good 公共利益
common grade 普通等级
common grate 普通栅极;共用格栅;普通格栅
common ground (墙面墁灰用的)灰板条;公地;普通木砖;木嵌条
common guideline 共同准则;通则
common gypsum 普通石膏
common hallway 公共过道;公共门厅
common hardware 公用的硬设备
common header 共用集装箱;共用集液箱;共用集水箱;共用集气箱
common high brass 通用高铜合金
common hornblende 普通角闪石
common house 供住宿的客店
common hyacinth 风信子
common hydrant 普通龙头
common illumination 普通照明;公用照明
common imperial scale 常用英制
common independent tank 普通独立舱
common information model 公共信息模型
common-inpedance coupling 公共阻抗耦合
common integrated processor 公共综合处理器
common interest 共同权益;双方的利益
common interface 共同接触面;公共界面
common interface language 公用接口语言
common ion 共用离子;共同离子;常见离子
common ion effect 共(同)离子效应;同离子效应
common iron 捻缝凿;普通生铁;普通钢材
common joist 地板格栅;共用格栅;楼板格栅;地板龙骨(英国);普通格栅;普通格梁
common juniper 欧洲刺柏;刺柏
common kitchen 普通厨房;公用厨房
common label 公用标号
common labo(u)r 杂工;非熟练工人
common labo(u)rer hour 普通工工时
common land roller 普通压地滚
common language 共同语言;公用语言;公共语言
common lap 普通搭接
common lathe 普通车床
common laundry 普通洗衣间;公用洗衣间
common law 习惯法(规);不成文法;普通法
common-law exchange 普通法上的兑换
common-law liability 习惯法责任
common law lien 普通法留置权
common law mortgage 习惯法抵押;普通抵押
common-law trust 习惯法信托机构;商业信托
common layerage 普通压条法
common lead 普通铅
common lead method 普通铅法
common lever 普通杠杆
common licence 普通许可证
common light 普通光

common lighting 普通灯光;公用照明
common lime 普通石灰
common line 主管线;共用线;公用线
common line system 共线系统
common link 一般链环;普通链环
common liquid application 普通液施法
common local optimization 公用局部优化;公用地区优化
common lock seam 普通卷边接缝
common lodging house 简易旅社
common log 扇板测速仪;扇板测程仪
common logarithm 常用对数;普通对数;十进对数
common logic address 公用逻辑地址
common logical address space 公用逻辑地址空间
common lounge 公用躺椅;公共休息室
common low alloy steel 普通低合金钢
common low water full and change 平均朔望低潮间隙
commonly adopted 通常采用的
commonly encountered disease 常见病
commonly used method 常用方法
common machine language 公用计算机语言;公用机器语言;通用机器语言
common main 总水管;总管(气、水等)
common management information protocol 公用管理信息协议
common management information service 公用管理信息服务
common manifold 均压复式接头
common mapping 普通制图
common market 共同市场
common mask 公用掩码;公用屏蔽
common material 普通材料
common measure 公约数;公用量度;公测度
common memory 共用存储器;公用存储器
common memory information transfer 公用存储器信息传送
common metal 普通金属(元素)
common method 普通方法
common mica 普通云母;白云母
common-midpoint number 共中心点数
common midpoint stack 共中心点叠加
common mineral fertilizer 普通矿质肥
common mixed fertilizer 普通混合肥
common mode 共模
common-mode choke 共模扼流圈;共式扼流圈
common-mode conversion 共态变换【电】
common-mode disturbance 共模干扰
common-mode error 共态误差
common-mode failure 共态失效
common-mode gain 共态增益;同模增益
common-mode input 同模输入
common-mode interference 纵向干扰
common-mode noise 共态噪声;共模噪声
common-mode operation 共模操作
common-mode performance 共态性能
common-mode rejection 共态抑制
common-mode rejection ratio 共模抑制比;同相抑制比;共态抑制比
common-mode voltage 共模电压
common monopoly 普通专利
common mountain-forest 普通山地森林土
common multiple 公倍数
common nail 普通钉
common name 通用名称
common natural resources 公有自然资源
common needs of society 社会共同需要
common network 公用网(络)
common neutral 共用中线;公共中性线
common nomenclature 习惯命名法
common normal 公共(法)线
common object request broker architecture 公共对象请求中介结构体系
common obligation 共同债务
common of estovers 木材采伐权
common offset gather section 共炮检距集合剖面
common of piscary 渔场共用权
common oil well cement 普通油井水泥
common olive 油橄榄
common open space 公用敞地;公用旷地
common ownership of land 土地公有制
common page 公用页
common paint 普通油漆
common parking area 公共停车场

common partition 共用隔墙;普通间壁;(不承重的)木间壁
commonpath interferometer 公光程干涉仪
common peripheral channel 公用外围通道
common peripheral interface 公用外围接口
common perpendicular 公有垂线
common physical parameter 普通物理参数
common pile driver 人工打桩机
common pine 樟木
common pitch 普通沥青;普通坡度
commonplace 普通现象
common plane of symmetry 常对称面
common plaster 普通粉刷
common point 公共点
common pollution control facility 公害防治设施
common port 公共端
common practice 常例;常规
common pricing 共同定价
common product of intermediate metabolism 共同中间代谢产物
common program(me) 公用程序
common programming language 普通程序设计语言;通用程序设计语言
common property 共有物;共有财产;共性;共同财产
common property line 公共地界线
common property resources 公共财产资源
common pumping station 普通排灌站
common pumpl motor base 泵与电动机的共用底座
common purpose 共同目的
common purse 共同资金
common pyrite 黄铁矿
common pyroclastic rock 普通火山碎屑岩
common quota 共同配额;共同现额
common raceway (鼓风炉风嘴处焦炭的)公共燃烧空窝
common rafter 中椽木;大椽木;大瓦木;普通椽木;共用椽木
common rail injection system 共轨喷射制
common ram 手锤;手夯
common-range-gather stack 同距选排叠加
common ratio 公率;公比
common reactance 互感
common-receiver-point gather section 共接收点道集剖面
common reed 芦苇
common-reflected-point channels number 共反射点道集数
common-reflected-point number 共反射点数
common reflection fin dividing 共反射面元划分
common-reflection-point stack 共反射点叠加
common register system 集中记发方式
common regularity 共同规律性
common relocation 公用再定位;公用浮动
common requirement clause 通用要求条款
common reserve fund 公积金
common reservoir 普通蓄水库
common return 共同回路;公共回线;公共回路
common ridge planting 普遍垄作
common rights 共同权利
common roofing 普通屋面
common room 公共活动室;教员休息室;公共休息室
common root 公根【数】
commons 公地;统材
common safety 共同安全
common-sallow 黄花柳【植】
common salt 氯化钠;食盐
common salt spring 普通食盐泉
common satisfaction of needs 满足共同需要部分
common scaffolding 普通脚手架
common scale of notation 常用记数法
common-scale strip 相同比例尺地图系列
common school 公立学校
common score card 普通鉴定表
common seal 普通密封
common section 共用节;共同段;公用段
common seed viability 普通活力
common segment 共用程序段;公用(程序)段
common segment bit 公用段位
common seizing 普通绑礼
common sense 经验常识
common sense method 常识法

common sense reasoning 常识性推理;常识理解
common sense rule 常识性规定
common service area 公用服务区
common services 共同事务;公益服务
common sewer 总下水道;普通下水道;共用下水道;公用下水道;公用污水管(道)
common shipworm 船蛆
common short stay bolt 普通短螺撑
common-shot depth migration 共炮点深度偏移
common-shot-point gather section 共炮点道集剖面
common-shot ray tracing migration 共炮点射线追踪偏移
common single turnout 普通单开道岔
common-size analysis 同型分析
common-size balance sheet 只有百分比的资产负债表
common-size income statement 只有百分比的收益表;同型损益表
common-size statement 百分比报表
common slipway 普通滑道
common software 公用软件;通用软件
common solder 标准软焊料
common source 共源极;共用(资)源;公用源
common-source connection 共源连接
common sources epidemic 同源流行
common space 共用空间;共用面积;公用空间
common spraying 普遍喷粉
common stability 基本稳定
common staff cost 一般人事费
common stake 共同利害关系
common state 公用状态
common steel tube 一般钢管
common stock dividend 普通股利
common stock fund 普通信托投资基金;普通股票投资资金;普通股票基金
common stock ratio 普通股比率
common stock[shares] 普通股
common stock subscribed 已认购的普通股
common storage 共用库容;共用存储器;公用存储;公共存储器;普通储[贮]藏
common storage area 共用存储区;公用存储器区;公共存储器区
common straight carbon steel 普通碳素钢
common strand wire rope 普通扭纹钢丝索
common strike 总走向【地】
common strontium 普通锶
common structural form 常见结构形状
common stucco 普通涂抹;普通墁涂;普通粉刷
common subexpression 公共子表达式
common subexpression elimination 公共子表达式消去
common subroutine 公用子程序;公用的子程序;通用子程序
common subroutine area 公用子程序区
common substrate 共衬底
common survey(ing) 普通测量学
common surveying instrument 普通测量仪器
common system 共用系统(环境卫生工程)
common tangent 公切线
common tangent point 公切点
common tap 共用水龙头;公用水龙头
common tariff 共同关税(率)
common tariff nomenclature 通用海关用语
common terminal 共接头
common thread 普通螺纹
common tile 普通瓦
common time signal 普通时号
common timing system 中心计时系统;共用计时装置
common tire 普通轮胎
common to both alternatives 两项方案都可通过
common tractor 通用拖拉机
common traffic 公共交通
common traffic facility 公共交通设施
common traffic service 公共交通公司
common transit facility 公共交通设施
common transport(ation) 公共交通;公用交通
common transport service 公共交通公司
common trap 共用疏水器;公共水封;共用存水弯
common trend 总走向【地】;总趋向
common trust 共同信托基金
common turpentine 普通松节油
common unit 公用部件

common unknown variance 共同的未知方差
common use 多畜共牧
common user berth 公共泊位
common-user channel 共用信道
common user circuit 公共用户电路
common user network 用户公用网络;公共用户网
common user of locomotive 机车轮乘制
common user service 公共用户服务
common user terminal 公用码头;公用码头
common-use size 公用尺寸
common vent 公用排泄口;共用风管;公共通气管
common venting system 共同排气系统
common volume-weight 普通容重
common wall (住宅的)分户隔墙;共用(隔)墙;用墙;公共墙
common walnut 胡桃
common washroom 公共盥洗室
common waste 公共污水
common waste pipe 共用的废水管道
common water 常水
common watercourse 普通水道
Commonwealth Agricultural Bureaux 英国联邦农业科技情报研究所
Commonwealth Preference System 英联邦特惠制
Commonwealth Secretariat 联邦秘书处
Commonwealth Telecommunications Organization 联邦电信组织
common weld 共用焊缝
common well 普通水井
common whipping 绳头缠扎
common wire 中性线;公共导线
common wire nail 普通钢丝钉
common worker 一般工人
common year 平年
common zero point 公共零点
common zeros 公共零点
common zone 公共区
commotion 混乱;电震
commotion in price 价格紊乱
communal 共有的;群居的
communal access 合建式出入口
communal building 公共建筑;合建建筑;公用建筑
communal carport 公共停车场;公共停车场
communal consumer system 社会消费者系统
communal drying room 公共干燥室
communal facility 公共设施
communal forest 公有林
communal garage 公用车库
communal garden (公寓式住宅的)公用花园;公共庭园
communal heating facility 公共取暖设备
communal illumination 公共照明
communalism 地方自治主义
communality 公因子方差
communality square error 公共因素方差
communal kitchen 公用厨房
communal land 公用地
communal lighting 公共照明
communal lounge 公共休息室
communal room 公用房间
communal services 公共服务
communal space 共享空间
communal television aerial 共用电视天线
communal washroom 公共盥洗室
Communaute Financiere du Pacifique franc 太平洋金融共同体法郎
commune 市区;社团
commune/brigade run enterprise 社队企业
commune/brigade run industry 社队工业
commune hospital 卫生院
commune-run enterprise 社办企业
communicability 传染性
communicable disease 传染病
communicate 通信;交通;传递;传达
communicating branch 交通支
communicating bridge 通道天桥
communicating canal 联系运河;联系渠;运河;通航渠道
communicating door 套间门;共用门;联络门
communicating door lock 套间门锁;联通房间门锁;共用门锁;联络门锁
communicating frame 互通门框
communicating lock 联系船闸
communicating operator 交换算子

communicating pipe 用户接管;连系管;连通管;通话管
communicating rooms 连通房间;互通房间;有门互通房间
communicating signal 联络信号
communicating the system 系统协调
communicating tube 连通管
communicating vessel 连通器
communication 联络;连络;交通机关;交际;协调服务;通信
communication adapter 通信转接器;通信线路连接器;通信适配器
communication and navigation facility 通信导航设施
communication and signal(l)ing equipment depreciation expenses 通信信号设备折旧费
communication and tacking system 通信与跟踪系统
communication and tracking subsystem 通信与跟踪子系统
communication and transportation 交通运输
communication apparatus 通信设备
communication apparatus of port 港区通讯设备
communication area 通信区域
communication auxiliary equipment 通信辅助设备
communication band 通信频带;通信波段
communication based system 联机通信系统;基于通信的系统
communication based train control 列车通信控制
communication boat 通信船
communication buffer 通信缓冲区;通信缓冲器
communication building 交通建筑;通信建筑
communication bus 通信总线;通信信息转移电路
communication cable 电信电缆;通信电缆
communication campaign concept 信息交流运动概念
communication capacity 通信容量
communication cell 通信用单元
communication center 交通中枢;通信中心;通信枢纽
communication center inter administration 局间通信枢纽
communication center of railway administration 局部通信枢纽【铁】
communication center of railway branch administration 分通信枢纽【铁】
communication channel 消息交流渠道;通信信道
communication circuit 通信线路;通信电路
communication clearing time 通信断开时间
communication code 通信密码
communication common carrier 公用通信载波(公司)
communication computer 通信计算机
communication condenser 换向电容器
communication conduit 地下电讯电缆管道
communication console 通信控制台
communication control center 通信控制中心
communication control character 通信控制字符
communication controller node 通信控制点
communication control package 通信控制程序包
communication control processor 通信控制处理装置;通信控制处理机
communication control system 通信控制系统
communication control unit 通信控制设备;通信控制器;通信控制单元
communication countermeasure 通信对抗
communication data processing 通信数据处理装置
communication data processor 通信数据处理机
communication data system 通信数据系统
communication design 通信系统设计
communication distance 交通运输距离;通信距离
communication emulator 通信仿真系统
communication engineering 电信工程(学);通信工程
communication equipment 通信设备
communication equipment room 通信设备室
communication executive 通信执行程序
communication expenses 通信费(用)
communication facility 交通设施;交通工具;通信设施;通信设备
communication flows 通信流量
communication function 信息交流功能
communication gallery 交通廊道

communication in port area 港口地区通信
communication interface 通信接口
communication interruption rate 通信中断率
communication jammer 通信干扰器
communication junction 交通枢纽
communication line 交通线;通讯线;通信线路
communication line depreciation expenses 通信线路折旧费
communication line insulator 通信线路绝缘子
communication line terminal 通信线路终端
communication link 通信线路;通信链(路);通信连接装置;数据自动传输装置
communication linkage controller 通信连接控制器
communication link handler 通信线路处理机
communication log 通信记录
communication map 交通图
communication matrix 通信矩阵
communication medium 传播介质
communication mix 信息交流组合
communication mode control 通信方式控制
communication module 通信模块
communication multiplex 通信多路传输机
communication network 交通网;信息交流网络;通信网(络);通信网
communication network control program(me) 通信网控制程序
communication network of water transportation 水运通信网
communication network processor 通信网络处理机
communication nodal point 交通枢纽站;交通枢纽点
communication of city 城市交通
communication operating zone identification 通信作业区标志
communication optic(al) cable 通信光缆
communication package 通信程序包
communication path 通信路径
communication pipe 传送管;连接管道
communication port 通信口
communication process 信息交流程序
communication processing 传输数据处理
communication processing system 通信处理系统
communication processor 通信处理器
communication program(me) 通信程序
communication protocol 通信协议
communication receiver 通信接收机
communication region 联系区;交流区
communication register unit 通信寄存装置
communication relay ship 通信中继船
communication relay(station) 通信中继站
communication reliability 通信可靠度
communication report center 通信中心站
communication resources 通信资源
communications 交通
communications and data-link system 通信与数据传输系统
communications and data system integration 通信和数据组合系统
communications and transport 交通运输
communication satellite 通信卫星
communication satellite coverage 通信卫星覆盖范围
communication-satellite earth station 通信卫星地面站
communication satellite orbit 通信卫星轨道
communication-satellite space station 通信卫星太空电台
communication satellite transponder 通信卫星转发器
communication security 通信安全
communication server 通信服务器
communication service 通信业务
communication serviceability facility 通信维修功能
communication ship 通信船
communications intelligence 电信侦察
communication software 通信软件
communication sonar 通信用声呐
communications processing center 通信处理中心
Communications Satellite Corporation 通信卫星公司(美国)
communications security 交通安全
communication station 通信站
communication structure 交通建筑(物)

communication subsystem 通信子系统
communication switching 通信转接
communication switching unit 通信转接器;通信转接部件
communication system 通信系统
communication system to provide two-way conversation 双向对话通信系统
communication task 通信作业;通信任务
communication technical satellite 通信技术卫星
communication technology 通信技术
communication terminal 通信终端
communication terminal equipment 通信终端设备
communication theory 信息(理)论;通信理论
communication tie-station 通信中心站
communication tools 通信工具
communication traffic 通信量
communication transmitter 通信发射机
communication trunk 通信中继线
communication user program(me) 通信用户程序
communication valve 联络阀
communication vector table 通信向量表
communication workstation 通信工作站
communication zone 通信区
communication zone indicator 通信区域指示器
communicativity 可交换性
communicator 联络员;磨碎机;发信机;通信器;通信装置;通信员;通话装置
communicator for sinking-kibble 凿井吊桶之间的联系
communion table 圣餐台
community 居民;共同体;公众;群落;团体;社区
community administration building 地区行政大楼
community air-borne waste 居民区上空飘的废气
community allotment 凭证放牧地
community amentities 社区康乐设施
community and ecosystem model 群落与生态系统模型
community antenna 共用天线;公共天线
community antenna television 共用天线电视;公用天线电视;公用电视天线;电缆电视
community antenna television line extender 公用天线电视线路扩展器
community antenna television main-line amplifier 公用天线电视主线放大器
community antenna television system 共用天线电视系统
community apartment project 社团公寓工程;社区共有的公寓项目
community assimilation number 群落同化度
community association 社区协会
community atmosphere 城市大气
community automatic exchange 公用自动交换机;公用自动电话交换;区内自动电话局
community biomass 群落生物量
community center 公共会堂;社区中心
community classification 群落分类
community cohesion 居民点聚集
community college 社区大学
community complex 群落复合体
community component 群落成分
community composition 群落成分
community consumption 生活用水;社会消费
community daycare housing 社区托儿所
community degradation index 群落消退指数
community design 社区设计
community development 地区共同体开发;社区开发(工程);社区发展;社区建设
community development corporation 社区开发公司
community development model 社区建设样板
community development plan 社区发展计划
community development program(me) 社区开发计划
community dial office 社团自动电话局;公用自动电话局
community diversity 群落多样性
community domestic wastewater treatment 小区生活污水处理
community dynamics 群落动态
community ecology 群落生态
community ecosystem 群落生态系统
community facility 社区设施
community facility plan 乡镇公用设施规划图;社区设施规划

community filtration rate 群落滤水率
community forest 公有林
community fund 公共资金
community ground 公众场地
community gymnasium 公众健身房
community habitat 群落生境
community health 公众卫生;公共卫生;社会保健
community kitchen 共用厨房
community level 社会水平
community line 共用线路
community living 社区生活
community management 社区管理
community metabolism 群落代谢
community migration 群落迁移
community noise 外界噪声;公共(场所)噪声;公众噪声;城市噪声
community noise aspect 居住区噪声问题
community of goods 财产共有
community of interest 共同利益;财务共管组织
community of land 土地的公共所有
community of private siding 私营线共用
community of stock 联股
community ownership 集体所有(制)
community participation 社区参与
community periodicity 群落周期性
community plan 社区规划
community planning 居住区规划;居民区规划;乡镇规划;城市规划;社区规划
community pollution 居民区污染
community pollution value 群落污染量
community property 共同财产;团体财产
community receiving system 集体接收制;集体接收系统
community recreational facility 社区游憩设施
community relations administrator 公共关系管理员
community residence 精神病院
community-run workshop 街道工厂
community section of line 共线区段
community service 公共服务;社区设施
community shopping center 社区购物中心;联营商业中心
community stability 群落的稳定性
community stock 普通股
community structure 群落结构;社区构成;社会结构
community structure and function 群落结构与功能
community structure regulation 群落结构调整
community succession 群落演替
community survey 地区调查;社会调查
community trust 社会信托;公益信托
community type 群落(的)类型
community view 共同意向
community wastewater disposal 公共废水处置;公共废水处理
community wastewater treatment 公共废水处理
community water system 公用给水系统;公共供水系统
community welfare 社会福利
community workshop 街道工厂
community zone 群落区
commutable 可整流;可换算;可抵偿
commutant 换位矩阵
commutate 交换;整流
commutate ammeter 多档安培计
commutating brush 换向电刷
commutating capacitor 换向电容器;整流电容器
commutating capacity 换向电容器
commutating current 整流电流
commutating device 换向装置;整流装置
commutating electromotive force 整流电动势
commutating field 换向(磁)场;换相磁场;整流磁场
commutating flux 换向磁通
commutating group 换相组
commutating machine 整流(子式)电机
commutating magnetic pole 换向磁极
commutating number 换相数
commutating optic(al) beam 换转光束
commutating pole 间极;换向极;整流极;辅助极
commutating-pole converter 换向磁极变流机
commutating pole generator 换向极发电机
commutating pole motor 换向极电动机
commutating pole winding 换向(磁极)绕组
commutating ratio 整流比

commutating reactance 整流电抗
commutating reactor 换向电抗器;整流电抗器
commutating segment 整流片
commutating tooth 换向齿;正齿;整齿
commutating voltage 整流电压
commutating winding 极间绕组;换向线圈;换向绕组;整流线圈;整流极绕组
commutating zone 整流带
commutation 交换;换向;转换;整流
commutation angle 换向角;安全角
commutation arc welding machine 整流弧焊机
commutation capacitor 换向电容器;整流电容器
commutation changeover 整流换向
commutation circuit 换向电路;切换电路
commutation coil 换向极线圈
commutation condenser 换向电容器
commutation cycle 整流周期
commutation diode 换向二极管
commutation factor 整流因数
commutation failure 换向失灵;整流故障;颠覆
commutation fare 市郊定期车票票价
commutation matrix 可换矩阵
commutation of annuity 年金折换
commutation rejection 对易关系
commutation relation 交换关系
commutation rule 交换法则;交换定则;对易定则
commutation table 换算表;折算表
commutation ticket 长期车票;月票
commutative 可交换的
commutative image 可换像
commutative law 可换律;交换律;互换律;对易律
commutative law of addition 加法交换律
commutative law of vector 向量交换律
commutative matrices 可换矩阵
commutative matrix 可交换矩阵;可换矩阵
commutative operator 交换算子
commutative production system 可交换的产生式系统
commutative rejection 可换关系
commutative relation 交换关系
commutative ring 可换环;交换环
commutativity 可换性
commutator 交换台;换向器;换位子;转接器;转换器;转换开关;整流子;对易子;分配器;切换开关;替续器
commutator armature 带换向器电枢
commutator bar 换向器杆;换向片;整流(子)片;整流条
commutator bar assembly 整流片组
commutator circuit 环形计数器;分配器线路
commutator-controlled weld 换向控制焊接
commutator end felt wick 整流器端毡心
commutator end frame 整流器尾架
commutator frequency changer 整流子频率变频器
commutator frequency converter 整流式变频
commutator generator 整流子式发电机
commutator grinder 整流子磨光机
commutator hub 整流子毂
commutator insulating segment 整流子绝缘隔片
commutatorless machine 无换向器电机
commutator lug 整流子接线片
commutator machine 换向器电机;整流子式电机
commutator modulator 换向调制器
commutator motor 整流(子)式电动机;电动机换向器
commutator pulse 定时脉冲
commutator rectifier 换向整流器;机械整流器
commutator ring 整流子(夹)环
commutator ripple 整流波纹
commutator riser 整流子竖片;整流子接线片;整流子接线叉
commutator segment 换向片;整流器扇形片;整流器片
commutator shaft 整流器轴
commutator shell 整流子套;整流子毂
commutator shrink ring 整流子绑箍
commutator sleeve 换向器套筒;整流子套筒
commutator switch 换向开关;按序切换开关
commutator tag 整流子接线片
commutator-type meter 整流式计数器
commuted refuse 粉碎垃圾
commuter 月票居民区;月票乘客;整流子;通勤者
commuter airliner 长期票乘客班机
commuter belt 居住地带(乘公共汽车上下班的人)
commuter bus 上下班公共汽车
commuterization (往返城市和郊区住所的)生活方式
commuter journey 月票乘客行程
commuter land 月票乘客居住区
commuter movement 经常客流;月票客流
commuter railroad 市郊铁路
commuter route 上下班路线
commuter time 上下班时间
commuter traffic 上下班交通
commuter train 市郊(旅客)列车
commuter transit 通勤铁路;上下班公共交通
commuter trip 经常性往返出行;通勤出行
commuter zone 月票乘客集中地区;通勤旅客区
commuting 通勤
commuting case 交换箱
commuting distance 通勤交通距离
commuting journey to work 上下班行程
commuting lever 交换杆
commuting movement 上下班客流
commuting operator 可换算符
commuting population 上下学人口;上下班人口
commuting ratio 月票乘客率;通勤交通率
commuting sphere 通勤交通圈
commuting traffic 通勤交通;市郊运输;上下班交通
Co-mo 考莫钴钼高速钢
como-cast process 浸入铸造法
Comol 考莫尔钴钼永磁合金
comonomer 共聚(用)单体
comonomer ratio 共聚(用)单体配比
Comoros 科摩罗
co-mortgagor 联合抵押人
comoving coordinates 共动坐标
comoving coordinate system 地球固定坐标系统;共动坐标系
compacost advantage 比较成本利益
compact 紧凑的;结实;密实图;密实的;合约;袖珍的;小型汽车;致密的;捣实;稠密的;契约
compactability 紧实性;密集性;易密性;致密性;可紧致性
compactable soil 可压缩土(壤)
compact battery 紧装电池;简装干电池组;小型电池
compact bulk density 捣实容重
compact cake 坚实泥皮
compact camera 袖珍摄像头
compact car 小型汽车;紧凑式汽车
compact city 小型城市
compact coarse sand 压实粗砂
compact coating 致密涂层;致密包膜层
compact code 紧致码
compact coding technique 紧凑编码法
compact conformation 体型结构紧凑
compact construction planning 紧凑建筑计划
compact crop 密植
compact crystalline graphite 致密结晶状石墨
compact density 压块密度
compact design 紧凑设计
compact diamond 聚晶金刚石
compact disk-read-only memory [CD-ROM] 只读光盘存储器
compact disk rewritable [CD-RW] 可重写型光驱
compact district 密集区
compacted 夯实的
compacted backfill 夯实回填土;压实(回)填土;捣实回填(土)
compacted cinder fill 压实焦渣填层
compacted clay 压实黏土;捣密黏土
compacted coarse sand 压实粗砂
compacted cohesive soil 压实黏性土
compacted column 挤密桩
compacted concrete 夯实混凝土;压实混凝土;捣实混凝土
compacted density 夯实密度;压实密度;振实密度
compacted depth 夯实深度;夯实厚度;压实深度;压实厚度
compacted earth 压实土;拍实土
compacted earth cofferdam 夯土堆筑围堰
compacted earth fill 压实填土
compacted earth in layers 分层拍实素土
compacted earth lining 压实底土衬砌;夯土衬砌
compacted earth pile 压土桩挤密
compacted earth rockfill 夯实的填土
compacted equipment 压实设备
compacted expanded base concrete pile 夯实扩底混凝土桩
compacted fill 夯实填土;压实填土
compacted fill density 夯实填土密度
compacted fine sand 压实细砂
compacted force 压实力
compact edge 压实边缘
compacted ice lee edge 密集冰的下风边缘
compacted impervious fill 压实防渗填料
compacted layer 坚实层;压实层
compacted lift 夯实分层厚度;压实分层厚度;压实层厚度;压实层
compacted lime earth 夯实灰土
compacted method 压实方法
compactedness 稠密;填空度;填充度
compacted pervious fill 压实的透水性填土
compacted pile 挤密桩
compacted pile method 挤密桩法
compacted rockfill 压实石块填方;碾压的堆石;压实填石;压实堆石;压实的堆石
compacted sand 压实砂土
compacted sands and gravels 压实砂和砾
compacted snow 压积的雪
compacted snow formation 压积的雪堆
compacted soil 紧实土(壤);坚实土;夯实土(壤);压实土
compacted soil condition 压实土状态
compacted stone 夯实的堆石
compacted subgrade 压实底层
compacted subsoil base 压实底土基层
compacted surface 碾实面层;夯实面层
compacted surface layer 密实面层
compacted thickness 压实深度;碾实厚度;夯实厚度;压实厚度
compacted volume 压实体积;实方体积
compacted weathered sand 压实风化砂
compacted yard 石方量;压实土方量
compact embankment 密实路堤
compact engine 小型发动机
compacter = compactor
compact extraction 出坯
compact fill 密填
compact fine sand 压实细砂
compact flare 致密耀斑
compact form 紧式;紧密结合形式
compact formation 坚硬岩层;致密岩层
compact grain 紧密颗粒;密实颗粒;致密晶粒
compact grained 细晶粒的;组织紧密的
compact-grain structure 密纹组织;致密(晶粒)组织
compact gravel 致密砾石
compact ground 板结地
compact gypsum 雪花石膏;致密石膏;纯白石膏
compact heat exchanger 紧凑式热交换器
compatibility 紧密性;可压实性;可夯实性;紧密度;密实性;密实度;压塑性;压实性;成型性
compatibility of a powder 粉末压塑性
compatibility of cohesiveless soil 无黏性土压密度
compatibility of soil 土的击实性;土壤的密实性
compact ice 密实冰;密集冰;冻冰块
compactification 紧化
compacting 压制;压实(工作);压坯;压紧
compacting action 压实作用
compacting and finishing machine 压实修整机
compacting by electric(al) impulse 电冲击成型
compacting by extrusion 挤压压塑
compacting by hand 手工捣实
compacting by high explosive 高爆炸力成型
compacting by rolling 轧压成型
compacting concrete by surface vibrator 用表面振捣器捣实混凝土
compacting concrete by vibration 振动捣实混凝土
compacting (cover) plate 压紧盖板
compacting effect 密实效果;压实效果
compacting effort 压实功
compacting equipment 压实设备
compacting factor 击实系数;密实因数;密实系数;夯实系数;压实因数;压实系数;压紧系数;致密系数;捣实因素;捣实因数;压实系数试验
compacting factor apparatus 压实系数试验仪;捣实因数试验仪
compacting hammer barge 夯平船
compacting head 紧压成型机头
compacting machine 压实机;镇压器;轻便电机;紧凑型电机

compacting machinery 压实机械
compacting mechanism 成型机理
compacting memory 紧缩内存;紧凑内存
compacting of fill 填土压实
compacting of foundation 地基夯实
compacting pass 压实通道;紧凑的通道
compacting pile 挤密桩;挤密砂桩;压实桩
compacting plant 碾压设备;夯碾设备
compacting press 成型压机
compacting pressure 紧密压力
compacting process 压制(成型)过程
compacting ratio 压实系数;松实比
compacting settling 稠密沉积
compacting test 压实试验
compacting tool set 压制模具
compacting width measurement 压实宽度测定
compacting work 压实工作
compact inorganic sand and silt mixture 压实无机质砂与粉砂混合物
compaction 精简;紧凑;简缩;击实;密集;夯实;压制工具;压缩凝结;压实;压紧;致密作用;粉料挤粒;频率分析精简法;拍实
compaction and undercompaction 五实和欠压实
compaction anticline 压实背斜
compaction-burial machine 埋压机
compaction by compression 压实
compaction by double-action 双效压塑;双向压制;双面加压
compaction by explosion 爆炸密实法;爆炸挤实;爆炸挤密;爆破密实
compaction by jolting 振动(密实)
compaction by layers 分层碾压;分层(填土)夯实
compaction by rolling 碾压;碾实;用碾压实;滚碾压实
compaction by single-acting 单效压塑
compaction by tamping 夯实;重锤夯实法
compaction by vibrating roller 振动碾压(法)
compaction by vibration (and compression) 振动压实;振实;振动压实(法);振动密实(法)
compaction by watering 注水压实(法);注水密实(法)
compaction coefficient 压实系数
compaction control 密实度控制
compaction control method 压实度控制法
compaction curve 击实曲线;密实度曲线;压实曲线
compaction deformation stage 压密变形阶段
compaction depth 压实深度
compaction device 击实仪;压实仪
compaction effect 压实作用;压实效果
compaction effort 击实功
compaction energy 压实能
compaction equipment 密实设备;碾压设备;夯实设备;夯碾设备;压实设备;振捣设备
compaction factor 密实系数;压实系数;捣实因素;捣实因数
compaction factor test 捣实因数试验
compaction fold 压实褶皱;致密褶皱
compaction grouting 压密注浆;挤密灌浆
compaction index 密实指数
compaction in layers 逐层密实;分层压实;分层夯实
compaction interval 间夯时间
compaction limit 压密极限
compaction machine 压实机械
compaction machine type 压实机械种类
compaction method 密实方法
compaction moisture 密实湿度
compaction mo(u)ld 击实筒;击实模子
compaction of earth dam 土坝压实
compaction of expansive clay 膨胀黏土的压实
compaction of fill 填土压实
compaction of landfill 填土压实
compaction of municipal 城市垃圾压实
compaction of municipal refuse 城市垃圾压缩处理
compaction of refuse 垃圾压实处理
compaction of soil 土壤压密
compaction of solid wastes 固体废物压缩;固体废物高压成型
compaction of underwater bedding 水下基床夯实
compaction pass 压实遍数
compaction pile 挤压桩;挤密桩;密实桩;压实桩;砂桩(普通砂井)
compaction pile method 挤密桩法;压实桩施工法;锤击桩施工(法)
compaction plane 压实面

compaction plant 压实厂房设施;碾压设备
compaction rate 密实程度;压实率;压实程度;压密率
compaction roll 压实轧轮;密实轧机
compaction roller 压路机;压路辊;压实辊;滚压碾
compaction sample process 压样法
compaction sand pile 挤密砂桩
compaction space 夯击点间距
compaction stage of argillaceous sediment 泥沉积物压实阶段
compaction table 压缩表
compaction technology 压实技术
compaction temperature 压实温度
compaction test 击实试验;压实(度)试验;冲击试验
compaction test apparatus 击实仪
compaction theory 密实理论
compaction type and zone 压实类型和分带
compaction unit weight 压实容重
compaction vibration 捣实
compaction weight 击实容重;压实容重
compaction work 压实工作;压实工程
compactive effort 压实作用;压实效应;压实效果;压实力;压实功
compactive energy 压实能(量)
compactive force 压实力
compact layer 致密层
compact limestone 密实的石灰石;坚实的石灰石;致密灰岩
compact machine 紧凑型电机
compact mass 硬块
compact material 密实材料
compact method 简捷(方)法
compactness 击实度;紧凑性;紧密性;紧密度;紧凑性;紧密度;坚实性;坚实度;密实性;密实度;压实度;致密性;填充度
compactness factor 紧密因素
compactness meter 紧实度计
compactness of crystal lattice 晶格紧密性
compactness of sand 砂土的密实(程)度
compactness of soil 土壤密度
compactness property 紧致特性
compactness theorem 紧致定理
compact object 致密天体
compact operator 紧算子
compactor 夯实工具;夯具;夯击机;压实器;压实机(械);压实工具;压路机;压路器;镇压器;捣固锤;废物压实机
compactor collection vehicle 压实式垃圾车
compactor pass 压实遍数;压实次数
compactor tips 碾压轮碾脚尖
compactor truck 垃圾收集车
compact pack ice 密集堆冰
compact part 密实部件;密部
compact planning 密集(建筑)规划;紧凑建筑计划
compact planting 密植
compact pole 紧致极点
compact polishable limestone 坚实可磨光的石灰石
compact powder 夯实粉末;压型粉末
compact radio source 致密射电源
compact reactor 紧密堆
compact reclaimer 简易旧砂再生装置
compact region 紧致域;紧域
compact retrieval device 精简式数据检索装置
compact rock 密实岩石;坚岩;致密岩石
compact sand 密(实)砂
compact sandstone 微密砂岩
compact section 坚实截面;坚实断面
compact set 紧致集;紧凑集
compact settlement 密集居住区
compact slag 致密渣;致密熔渣
compact snow 密实雪
compact soil 坚实土;密实土(壤)
compact space 紧(致)空间
compact spinning unit 短程纺丝机
compact storage 紧密存储
compact-stranded wire 压缩绞线
compact structure 密实结构;致密构造
compact substance 密质
compact tension specimen 紧凑拉伸(试件)
compact texture 致密结构
compact texture of single particle 紧密单粒结构
compact the soil 镇压地
compact tissue 致密组织

compact towage 堆装整齐
compact type 袖珍型;小型
compact unit 密实部件;装配紧密的部件
compact urban traffic control system 小型高效城市交通控制系统
compact voice communication 简化语言通信
compact volume 振实体积
compages 综合结构;结构;骨架
compander 压(缩)扩(张)器
companding 压缩扩展
compandor 压(缩)扩(展)器;展缩器
compandor compressor-expandor 压伸器
companion 成对物件之一;升降口围罩;升降口盖
companion action 伴随作用
companion blasting 联合爆破
companion blind flange 配对法兰盖
companion crop 间作作物;混作作物;伴生作物
companion fault 副断层
companion flange 联合凸缘;结合法兰;接合法兰;对应法兰;对接凸缘;成对法兰;配对法兰
companion hatch(way) 舱室升降口
companion keyboard 伴随键盘
companion ladder 舱梯;升降(口)梯
companion lode 副矿脉
companion matrix 伴随矩阵
companion paper 室内装修墙纸
companion project 姐妹项目
companions 伴生种
companion specimen 同类试件;同组试件;同类样本
companion star 伴星
companion store backup 伴随存储备份
companion to the cycloid 伴旋轮线
companion tree 伴生树
companion way 座舱走道;水下梯道;升降口;舱梯;升降(口)扶梯
compansatory financing schemes 补偿筹资办法
company 伙伴;公司
company account 公司核算
company contract 公司契约;公司合同
company credit 公司信贷
company evaluation 公司评价
company flag 公司旗
company fuse 组合导火线;组合保险丝
company housing 公司(提供的)住房
company image 公司商誉
company incorporated(Inc.) 注册公司
company law 公司法
company limited 有限公司
company limited by guarantee 担保有限公司
company limited by shares 股份有限公司
company name 厂商
company object 公司目标
company office 公司办公室
company of unlimited liability 无限公司
company policy 公司政策
company promoter 公司发起人
company property 公司财产
company's administration building 公司的行政大楼
company's articles of association 公司组织章程
company's risk 损失由公司负责
company standard 出厂标准;企业标准
company store 公司仓库
company's tower 公司大楼
company strategy 公司策略
company suspending transactions 停止交易的企业
company town 公司城;企业城(市)
company train 专用循环直达列车
company union 公司工会
company-used gas 工厂自用气
company with limited liability 有限责任公司
comparability 可比(较)性;比较
comparability graph 可比较图
comparability test 比较试验(法)
comparable aggregate 可比集
comparable analysis 可比分析
comparable data 参照数据
comparable form 可比形式
comparable function 可比函数
comparable income 可比收益
comparable life 可比年限
comparable machine and equipment 可比机器

设备
comparable measure 可比度量
comparable method 可比方法
comparable period 比较期
comparable population 比较人口
comparable price 可比价格
comparable product 可比产品
comparable regulations 可比规范
comparables 可比项目;可比价
comparable sales approach 可比销售法
comparable standard 可比标准
comparable terms 可比项
comparable uncontrolled price method 可比非控制价格法
comparand 比较(用)字(符);比较数;被比较字
comparand register 比较字寄存器
comparascope 显微比较镜
comparation potentiometer 标准电位计
comparative 比较的
comparative acceptable river pollution 相对容许河流污染
comparative advantage 相对有利;相对优势;相对利益;比较优势;比较利益
comparative advantage of a region 区域相对优势
comparative advantage theory 比较利益论
comparative advertising 比较广告法
comparative analysis 对比分析;比较分析
comparative analysis method 比较分析法
comparative analysis method of structure 结构对比分析法
comparative and predictive information 比较性和预测性情报
comparative balance sheet 比较资产负债表
comparative budget and actual income sheet 预算与实际损益比较表
comparative cartography 比较地图学
comparative characteristic 比较特性
comparative characters 比较特性
comparative concrete 对比混凝土
comparative cost 比较造价;比较成本;比价
comparative cost advance 比较成本优势(法)
comparative cost advantage 比较成本利益
comparative cost concepts 比较成本概念
comparative cost difference 比较成本差异
comparative cost planning 比较成本分析法
comparative cost sheet 比较成本表
comparative cost studies 比较成本研究;比较成本调查
comparative cover 重复摄影资料
comparative data 对比数据
comparative density index 人口密度比较指标
comparative design 比较设计;比较方案
comparative design estimate 设计比较估算
comparative difference in cost 成本比较
comparative disadvantage 比较劣势
comparative dynamics 比较动态学
comparative earnings standard 比较利益标准
comparative economy 经济比较
comparative effectiveness 比较效果
comparative estimating method 比较估计法
comparative experiments 比较实验
comparative factory site 比较厂址
comparative figures 比较数字
comparative financial statement 比较财务报表
comparative harbo(u)r site 比较港址
comparative income account 比较损益计算书;比较收益账户;比较收入表
comparative income statement 比较损益表
comparative indicator 差动电流计
comparative karyon type analysis 比较核型分析
comparative level 比较电平
comparative lifetime 相对寿命;比较寿命
comparative list 价格条件比较表;比较单
comparatively 比较地
comparatively relaxed economic environment 比较宽松的经济环境
comparatively well-off social development planning 小康社会发展规划
comparative map of coal seam 煤层对比图
comparative map of coal seam and strata 煤岩层对比图
comparative map of exploration and exploitation data 探采对比图
comparative map of logging curve 测井曲线对比图
comparative marketing 比较营销学
comparative measurement 比较测量;比较测定(法)
comparative method 比较法
comparative mineral refractive indices 对比矿物折射率
comparative mortality figure 比较死亡指数
comparative obscure displayed 显示较隐晦
comparative observation 比较观测
comparative oceanography 比较海洋学
comparative operating statement 比较营业表
comparative performance 比较效果
comparative plan 比较方案
comparative planetology 比较行星学
comparative polarography 比例极谱法;比较极谱法
comparative port site 比较港址
comparative principle of satisfying needs 满足需要可比原理;满足需要比较原理
comparative profit and loss statement 盈损比较表;比较损益表;比较收入表
comparative rate of profit 比较利益率
comparative reactivity 相对反应性
comparative readout 比较读出
comparative returns on investments 比较投资效果
comparative risk 相对危险(性);相对危险(度)
comparative route 比较路线
comparative salary ratio 薪金比率
comparative scale 比较计
comparative size 对比尺寸
comparative standard 比较的标准
comparative statement 比较表
comparative statement of cost per productive hour 生产小时成本比较表
comparative statement of operation cost 经营成本比较表;作业成本法
comparative statement of order cost 分批成本比较表
comparative statement of product cost 产品成本比较表
comparative static analysis 比较静态分析
comparative statistics 比较统计
comparative steady 较稳定的
comparative strength 对比强度
comparative structural studies 比较结构研究
comparative study 方案比较;比较研究
comparative table 对照表
comparative test 对比试验;比较试验(法)
comparative test series 对比试验组;对比试验系列
comparative trial balances 比较试算表
comparative unit method 单位比较法
comparative utility 比较效用
comparative value 比较数值
comparative valueless 比较无价值
comparator 校验器;检定器;比较仪;比较器;比较电路;比长仪;比长器;比长计;比测器
comparator base 比长基线
comparator block 比色座;比色匣;比色块
comparator box 比色盒
comparator check 比较器校验
comparator circuit 比较电路
comparator coordinates 坐标仪量测坐标
comparator-densitometer 比较测微光度计
comparator Hellige system of comparator 黑利格系比色计
comparator hysteresis 比较器迟滞
comparator micrometer 钟表千分尺;比较千分尺
comparator microscope 比较显微镜
comparator plate 千分表平板
compare 比较
compare and print 比较和打印
compare byte 比较字节
compare check brush 检验刷
compared with last year 与上年相比
compare facility 比较能力
compare instruction 比较指令
compare logical register 比较逻辑寄存器
compare operation 比较操作
comparer 比较器
compare test 比较测试
compare the time 对钟;对时
compare word 比较字
comparing 比较
comparing brushes 比较刷
comparing check 比较检验
comparing control change 控制改变
comparing data 比较数据
comparing element 比较装置;比较元件;比较电路;比较单元;比较部分
comparing indicator 比较指示器
comparing of colo(u)rs 颜色的比较
comparing rule 比例尺
comparing unit 比较装置;比较器;比较部件
comparing watch 比较表(指仪表)
comparison 对照;对比物;对比;比较
comparison among groups 组间比较
comparison and analogy method 比较类推法
comparison auditing 比较审计学
comparison base 对照基线
comparison baseline 比较基线
comparison basis 比较基础
comparison beam 比较束
comparison bridge 比较电桥
comparison bydirect deflection method 直接偏转比较法
comparison circuit 比较电路
comparison coder 比较编码器
comparison colo(u)rimeter 比色计
comparison column 参比柱
comparison cost 比较成本
comparison curve 比较曲线
comparison definition 比较定义
comparison error 比较误差
comparison expression 比较表达式;关系式
comparison eyepiece 比较目镜
comparison frequency 比较频率
comparison function 比较函数
comparison goniometer 比较测角仪
comparison index 比较指数
comparison indicator 比较指示器
comparison in the apparent resistivity 视电阻率拟合
comparison in the resistivity transform function T 转换函数 T 拟合
comparison lamp 比较灯
comparison line 检核基线;比较线(路)
comparison measurement 对比测量;比较测定(法)
comparison measurer 比值器
comparison measuring 比较测定(法)
comparison method 比较法
comparison microscope 比较用显微镜;比较显微镜
comparison of alternative 方案比较;比较方案
comparison of alternative projects 方案比选
comparison of alternative schemes 方案比选;方案比较
comparison of bid 比标价;投标出价的比较
comparison of chronometers 天文钟比对
comparison of matched pairs 配对比较
comparison of material characteristics 材料特性比较
comparison of mean direction 平均方向的比较
comparison-of-pair sorting 成对比较分类
comparison of precision 精度的比较
comparison of prices 比价
comparison of projects 方案比较
comparison of time-sequential maps 按时间顺序的图幅比较
comparison operation 比较运算;比较操作
comparison operator 比较运算符;比较操作(符)
comparison order 比较指令
comparison oscillator 比较振荡器
comparison oscilloscope 比较示波器
comparison postmortem 比较检错(程序)
comparison principle 比较原理
comparison prism 比谱棱镜;比较棱镜
comparison shopping 采购条件的比较调查
comparison size 对比尺寸
comparison slenderness 比较细长度
comparison solution 比较溶液
comparison sort 比较分类
comparison sorter 比较分类器
comparison specimen 对比试样
comparison spectroscope 比谱分光镜
comparison spectrum 比较光谱
comparison standard 参比标准
comparison strength 比较强度
comparison stress 比较应力

comparison study 比较研究
comparison table 比较表
comparison test 比较试验法;比较检验(法);比较测试
comparison tester 比较试验机
comparison test series 对比试验组;对比试验系列
comparison with adjacent chart 邻图拼接对比
comparison year 比较年
comparoscope 比较仪
compart 舱;分隔
compartition 分开
compartment 林班;客车包房;罐包装(集装箱等);隔室;隔间;隔舱;舱;分隔;区划
compartmentalized cabin 间隔密封舱
compartmentalized vehicle 分格收集垃圾车
compartment analysis 分离分析;区划分析
compartmentation 间隔化;水密分舱区划
compartment bin 分隔料斗;分隔料仓;分格式储[贮]料箱
compartment boat 分节式方驳
compartment boundary 分区界;数据分区界
compartment ceiling 分隔顶棚;分格棚顶;井口天花;格子平顶
compartment coach 包房客车
compartment designation 舱室名称
compartment division masonry wall 分隔房间的坞工墙
compartment dryer 分格干燥器;间隔烘干机;间隔干燥器;分室烘干机;分室干燥器
compartmented agitator 隔室式混合器
compartment floor 舱内地板;隔火楼板;隔舱楼板
compartment for detrainment ramp winder 紧急落放下车斜坡道绞盘室
compartment furnace 房式炉;多室炉
compartment history 林班记录
compartment hot well 分段热水井
compartment kiln 间隔窑;周期干燥窑;分室窑
compartment line 林班线
compartment loaded in combination 共同装载舱
compartment mill 多室式粉碎机;多仓磨(机);复式磨(机);分格磨
compartment mode 分室方式
compartment model 隔室模型;分室模型
compartment reserved for mail 邮政包间
compartment roofing 分隔屋面
compartment shield 分隔屏蔽
compartment shielding 分段型屏蔽
compartment silo 分隔式筒仓;分隔料仓
compartment steamer 分格蒸箱;分格蒸笼;分格蒸锅
compartment stoker 格子加煤机;分段加煤机
compartment storage hopper 分格储斗;分格储料仓
compartment system 分布系统;区分系统;伞代更新
compartment test 舱室试验
compartment tube ball mill 多仓管式球磨机
compartment type filter press 间格式压滤机
compartment wall 分隔墙分格储料斗;隔火墙;隔断墙
compass 罗盘(又称罗经);指南针;指北针
compass adjuster 罗盘校正师
compass adjustment 罗盘校正;罗盘调整
compass adjustment buoy 罗盘校正浮标
compass amplifier 罗盘放大镜
compass amplitude 罗针摆幅;罗盘幅度
compass angle 罗盘角
compass axis 罗盘轴
compass azimuth 罗盘方位(角)
compass base 罗盘校正台
compass bearing 罗盘象限角;罗盘方位
compass binnacle 罗盘盒;罗盘柜
compass binnacle top 罗盘柜盖
compass block 弧形木制品
compass bonding header 弧形丁砖
compass bowl 罗盘碗;罗盘盆
compass box 罗盘盒
compass brick 弧形砖;楔形砖;拱砖
compass bridge 罗盘观测台
compass buoy 罗盘校正浮标
compass calibration 校正罗经(又称校正罗盘);摆动法罗盘校正
compass cal(l)iper 弯脚卡钳
compass card 罗盘玫瑰图;罗盘刻度盘;罗盘卡;罗盘(方位)盘;罗盘标度板;平板罗盘
compass card axis 罗盘卡基线
compass chamber 罗盘柜上半部
compass circle 罗盘(分度)圈;罗盘度盘;编码度盘
compass compensation 罗盘自差校正;罗盘自差补偿
compass-controlled gyroscope 磁罗盘修正陀螺仪
compass correction 罗盘校正;罗盘改正量
compass correction card 罗盘修正表
compass corrector 罗盘校正器;罗经校正器
compass corrector unit 罗盘校准器
compass course 罗盘航向;罗盘航线
compass cover 罗盘盖
compass deck 罗盘甲板;罗盘观测台
compass declination 磁针偏角
compass declinometer 罗盘磁偏计
compass deflection 罗盘偏角
compass determination 罗盘速向
compass deviation 罗盘自差;罗盘偏差
compass deviation chart 罗盘自差图
compass diagram 罗盘自差图
compass dial 罗盘日晷;罗盘刻度盘;罗盘卡;罗盘标度盘;罗经面板
compass direction 罗盘方向
compass door 罗盘柜门
compassed by the sea 四周环水
compassed the earth 环绕地球
compass engineering brick 弧形工程砖
compass error 罗盘(误)差;罗经误差
compass error by celo-observation 天测罗经差
compasses 两脚规;圆规
compasses beam dividers 分规
compasses key 圆规扳子
compasses of proportion 比例规
compass flat 罗盘观测台
compass float 罗盘卡浮室
compass for drawing 绘图用圆规
compass format 弧形格式
compass graduation 罗盘刻度
compass hangs 罗盘卡转动不灵
compass hard brick 圆弧形硬砖
compass-headed arch 半圆拱;半圆拱
compass heading 罗盘船首向
compass hood 罗盘帽;罗盘柜帽
compass inclinometer 罗盘测斜仪
compass index error 罗盘指示差
compassing timber 弯木(料)
compass integrator 罗盘积分器
compass is wild 罗盘卡摆动不定
compass journal 罗盘日志
compass lengthening bar 圆规延伸杆;圆规腿
compass light 罗盘照明灯
compass liquid 罗盘液体
compass locator 罗盘定位器
compass lubber-line 罗盘标线
compass magnet 罗盘磁铁
compass march 按罗盘行进
compass meridian 罗盘子午线
compass motion 罗盘运动误差
compass needle 罗针;罗盘针
compass needle clamp 罗针制动
compass north 罗盘北(向)
compass of proportion 比例规
compass pedestal 罗盘座
compass pivot 罗盘轴针
compass plane 双用刨;圆刨;凹刨;曲面刨
compass plant 磁石植物
compass platform 罗盘观测台
compass point 罗盘上的刻点;临界点;罗盘点;罗经点
compass police 罗盘校正
compass prime vertical 罗盘的东西圈
compass rafter 轮椽;弯曲椽
compass reading 罗盘读数
compass reading glass 罗盘放大镜
compass rebate plane 曲槽刨
compass repeater 罗盘转发器;罗盘复示器;分罗经
compass roof 跨形屋顶;弧形屋顶;圆弧形屋顶;半圆(形)屋顶;曲线形屋盖
compass rose 罗盘(仪)刻度盘;罗盘面;罗盘玫瑰图;罗盘花;罗盘;罗经刻度盘;向位圈;磁向图
compass rule 罗盘仪法则;闭合差边长配赋法
compass saw 细木锯;截圆锯;斜形狭圆锯;圆锯;曲线锯
compass saw with wood(en) handle 木柄鸡尾锯
compass scale 罗盘标度
compass sensibility 罗针灵敏度
compass setting 用两脚规截取距离
compass sketch 罗盘仪测量草图
compass solid block 弧形实心砌块
compass solid brick 圆弧形实心砖
compass solid tile 弧形实心瓦
compass spindle 罗盘轴针
compass stand 罗盘柜座
compass station 罗盘仪测站
compass stretcher 弧形顺砌砖
compass survey 罗盘测量;罗盘仪测量
compass swing 校正罗经(又称校正罗盘)
compass swinging rose 罗盘旋转刻度盘
compass tach(e)ometer 罗盘视距仪
compass tachymeter 罗盘视距仪
compass theodolite 罗盘经纬仪
compass tile 弯瓦;弧形砖
compass timber 弯曲木(材);弯木(料)
compass torque motor 罗盘矫正电动机
compass track 罗盘轨迹方向
compass transmitter 罗盘发送器;方位传送器
compass traverse 罗盘仪导线
compass tripod 罗盘仪三脚架
compass type restricting side arms 罗盘形单面横臂
compass variation 罗盘变差;罗盘磁差
compass variation error 罗盘变动误差
compass well-burned brick 弧形透烧砖
compass window 凸肚窗;弧形窗;圆肚窗;弓形窗;半圆形凸窗
compass with diopter 照准仪罗盘
compass work 弧形木制品
compatibility 两用性;可兼容性;兼容性;混溶性;混溶性;互换性;协调性;相容性;一致性;共存三角形;亲合性;适应性;适合性
compatibility approach 兼容性法
compatibility box 兼容框
compatibility condition 协调条件;相容条件;相容情况
compatibility equation 协调方程;相容方程(式)
compatibility feature 兼容性特性;兼容性功能;兼容特性
compatibility law 相容性法则
compatibility method 相容性法;(结构力学上的)内力协调方法
compatibility mode 兼容模式;兼容方式
compatibility mode exception 兼容方式异常
compatibility objective 兼容性目标
compatibility of binders 拌和料间的相容性
compatibility of cement aggregate 水泥与集料的相容性;水泥与骨料的相容性
compatibility of deformation 变形协调
compatibility of goals 目标的一致性
compatibility relation 兼容关系;相容关系
compatibility test 兼容性检验;兼容性测试;相容性试验;配合试验;适应性试验;适合性检验
compatibility test unit 适配性试验设备
compatibility torsion 协调扭转
compatible 可兼容的;兼容的;相容;亲和的
compatible chart 相容图
compatible circuit 兼容电路
compatible colo(u)r system 兼容彩色系统;彩色兼容制
compatible data 兼容数据
compatible equation 相容方程(式)
compatible event 相容事件
compatible hardware 兼容硬件
compatible hybrid integrated circuit 兼容混合集成电路
compatible information system 兼容信息系统
compatible interface 兼容式接口
compatible laser system 兼容激光系统
compatible layer 兼性层
compatible mode 相容模型
compatibleness 协调性
compatible observation 相容观察
compatible operation system 兼容操作系统
compatible peripheral device 可兼容外围设备
compatible route 平行进路【铁】
compatible single sideband 兼容单边带
compatible single sideband system 兼容单边带系统

compatible software 兼容软件
compatible state of deformation 变形(的)协调状态
compatible storage 相容存储
compatible support 兼容支援
compatible table 对照表
compatible time-sharing system 兼容分时系统
compatible transmission 兼容制传输
compatible treatment 适应性处理
compel enforcement 强制性执行
compelled coagulation sedimentation equipment 强制凝结沉降装置
compelled current drainage 强制排流
compelled deformation 强迫变形
compelled excitation increasing device 强行励磁装置
compelled signalling 强制发信号
compelling force 强制力
compelling relay 闭锁继电器
compementer 补助器
compendency 连接因素；凝集性
compendium 简述；摘要；提纲
compendium of classification opinions 分类意见提要
compendium on low-and-nonwaste technologies 低废和无废技术概要
compens 补偿控制线保护系统
compensability 可补偿性
compensate control 补偿控制
compensated 补偿的
compensated acceleration method 补偿加速度法
compensated air thermometer 补偿空气温度计
compensated amplifier 补偿放大器
compensated aneroid 补偿气压表
compensated attenuator 补偿式衰减器
compensated bar 补偿杆尺
compensated basin 补偿盆地
compensated bogie 平衡架
compensated bridge 补偿式电桥
compensated carbureter 补整汽化器
compensated cavity 补偿空腔共振器
compensated cavity resonator 补偿空腔谐振器
compensated chamber 补偿室
compensated commutator motor 补偿整流电动机
compensated compass 补偿电罗经
compensated current 补偿流
compensated current transformer 补偿电流变感器；补偿变流器
compensated curve 补偿曲线
compensated demand curves 补偿需求曲线
compensated density log 补偿密度测井
compensated dewatering method 补偿疏干法
compensated dollar 美元补偿；补偿美元
compensated dollar plan 美元补偿计划
compensated dynamo 补偿发电机
compensated error 补偿误差
compensated flow 补偿流
compensated flow control valve 带压力补偿流量控制阀；补偿流量控制阀；调速阀
compensated formation densilog 补偿地层密度测井
compensated formation densilog curve 补偿地层密度测井曲线
compensated formation density log 补充地层密度测井
compensated foundation 补偿性基础；补偿式基础
compensated gamma-ray chamber 补偿型伽马射线箱
compensated generator 补偿式发电机
compensated impurity 补偿杂质
compensated-impurity resistor 补偿杂质电阻器
compensated induction generator 补偿感应发电机
compensated induction motor 补偿式感应电动机；补偿感应(式)电动机
compensated instrument transformer 补偿测量变压器
compensated intrinsic(al) material 补偿本征材料
compensated ionization chamber 补偿式电离箱；补偿电离室
compensated joint 补强接头；补偿接头
compensated level 补偿水准(仪)；补偿水准(器)
compensated line 补偿线路
compensated log 补偿测井
compensated loop 补偿环

compensated-loop direction finder 补偿(式)环形(天线)测向器
compensated microphotometer 补偿测微光度计
compensated model 补偿模型
compensated motor 补偿电动机
compensated neutron log 补偿中子测井
compensated neutron log curve 补偿中子测井曲线
compensated operational amplifier 补偿运算放大器
compensated optical fiber 补偿光纤
compensated pendulum 补偿摆
compensated planimeter 补偿求积仪
compensated receiver 补偿接收机
compensated reflector 补偿反射器
compensated regulator 补偿稳压器；补偿调节器
compensated relief valve 平衡式溢流阀
compensated repulsion motor 补偿推拆(式)电动机；补偿式推斥电动机；补偿感应推斥电动机
compensated retaining wall 平衡重式挡土墙
compensated scan(ning) 补偿扫描；展开(式)扫描
compensated semi-conductor 补偿型半导体
compensated sense winding 补偿(的)盘绕线
compensated series motor 补偿串励电动机；补偿串激式电动机
compensated shunt box 补偿式分流器箱
compensated sight 补偿视准器
compensated spectrum 补偿光谱
compensated spring scale 补偿弹簧秤
compensated topography 补偿地形
compensated value 补偿价值
compensated variometer 补偿磁变仪
compensated voltmeter 补偿式伏特计；补偿伏特表
compensated volume control 补偿音量控制
compensated wattmeter 补偿瓦特计
compensated wave 补偿波
compensate for loss 赔偿损失
compensate for wear 磨损补偿
compensate payment for secured value 保价赔偿款
compensate semi-conductor 补偿半导体
compensate transport expense liquidated revenue 运输支出补偿的清算收入
compensate-type micropressure meter 补偿式微压计
compensating 补助；补偿的
compensating ablancer 补偿摆轮
compensating accumulator 补偿累加器
compensating action 补偿作用
compensating air valve 补偿空气阀
compensating amplifier 补偿放大器
compensating apparatus 补偿装置
compensating arrangement 补偿装置
compensating axle-load cylinder 补偿承载轴液压缸
compensating balance 最低应存款；补偿存款金额
compensating bar 等制器；补偿杆；均力杆；校正铁；平衡杆
compensating base-bar 补偿基线杆尺
compensating base-line measuring apparatus 补偿式基线测量器械
compensating basin 补偿区
compensating beam 补偿梁；平衡梁；平衡杆
compensating brush 补偿电刷
compensating buffer 补偿缓冲器
compensating cable 均电缆；补偿电缆
compensating capacitor 补偿电容器
compensating cariation 补偿变差
compensating cell 补偿试池
compensating chain 补偿用的链条
compensating circuit 校正电路；补偿线路；补偿电路
compensating coil 补偿线圈
compensating collar 补偿轴环
compensating computation 平差计算
compensating condenser 补偿电容器
compensating conductor 补偿导线；补偿导体
compensating controller 互补调节器
compensating control point 补偿控制点
compensating criterion 补偿准则；补偿判据
compensating current 补偿电流
compensating curve 补偿曲线
compensating cylinder 补偿油缸；平衡缸
compensating delay 补偿延迟
compensating delay line 补偿延迟线
compensating depth 补偿深度

compensating developer 补偿式显影液
compensating device 补偿装置；平衡装置
compensating diaphragm (测量望远镜中的)测距补偿装置；补偿十字线片；补偿膜片；补偿光阑
compensating digit 补偿数字
compensating disc 补偿盘
compensating donor 补偿施主
compensating drive 补偿驱动(装置)
compensating duty 补偿税
compensating effect 补偿作用；补偿效应
compensating electric(al) field 补偿电场
compensating emission 补偿发射
compensating equalization 补偿均衡
compensating error 抵消(性)误差；冲销差错；补偿误差
compensating evapo(u)rator 补偿蒸发器
compensating excitation 补偿激发
compensating eyepiece 补偿目镜
compensating factor 补偿因子
compensating feedback 补偿反馈
compensating feedback loop 补偿反馈环
compensating feedforward 补偿前馈
compensating feed stoker 连续给煤层燃炉
compensating fee for electricity supply works 供电工程贴费
compensating field 补偿磁场；补偿场
compensating filter 补偿式滤色镜；补偿滤色镜；补偿滤光片；补偿滤波器
compensating fire detector 补偿式火灾检测器
compensating flow 补偿水流
compensating fluctuation 高低相抵
compensating foundation 补偿性基础
compensating ga(u)ge 补偿片；补偿计
compensating gear 差速齿轮；差动滑车；差动齿轮；差别齿轮；补偿装置
compensating glass 补偿镜
compensating grade 折减坡度
compensating groove 平衡槽
compensating heater 补偿加热器
compensating heating 补偿供暖
compensating hemming foot 补偿卷边压脚
compensating hub 补偿锁壳
compensating illumination 补偿照明
compensating impedance 补偿阻抗
compensating inductance 补偿电感
compensating ionization chamber 补偿电离室
compensating jet 补偿喷嘴；补偿量孔
compensating lattice 补偿帘子
compensating lead wire 补偿导线；补偿引线
compensating lens 补偿透镜
compensating line 补偿线(路)
compensating magnet 校正磁棒；补偿磁铁；调整磁铁
compensating magnetic field 补偿磁场
compensating mask 补偿蒙片
compensating measuring eyepiece 补偿测量目镜
compensating mirror 补偿镜
compensating network 补偿网络
compensating nozzle 补偿油嘴；补偿喷嘴
compensating of error 误差校正
compensating of load 负载力补偿
compensating osmometer 补偿渗透计
compensating parameter 补偿参数
compensating period 补偿期
compensating phase shift 补偿相移
compensating pipe 补整管；补偿管；平衡管；膨胀管；调整管；伸缩管
compensating piston 补偿活塞
compensating planimeter 补偿求积仪
compensating plate 补偿片；补偿板
compensating plate lock-up 装版补偿器
compensating point 补偿点
compensating pole 换向极；附加极；补偿极
compensating pressure transducer 补偿压力传感器
compensating prism 补偿棱镜
compensating procedure 补偿程序
compensating pulley 均衡滑轮；平衡滑轮
compensating push rod 补整推杆
compensating reservoir 补偿(调节)水库；调节水库；平衡水库
compensating resistance 补偿电阻
compensating resistor 补偿电阻器
compensating roller 补偿辊

compensating rope 补充用的钢丝绳
compensating self-recording instrument 补偿自动记录仪
compensating shaft 补偿轴
compensating sheave 补偿滑轮
compensating sheave tower 补偿滑轮塔
compensating shunt 补偿(式)分流器
compensating sight 补偿瞄准器
compensating spring 平衡弹簧
compensating squeeze head 弹性压头
compensating stop 补偿光阑;补整停止器
compensating stub 补偿短截线
compensating stuffing box 调整填料函
compensating surface 辅助面;补偿面
compensating surface current 补偿表层流
compensating tariff 补偿(性)关税(税则)
compensating tongue 补偿舌
compensating torque 补偿转矩
compensating trace test 补偿跟踪试验
compensating tube 补偿管
compensating unit 补偿装置;补偿器;补偿部件
compensating universal shunt 补偿(式)万用分流器;补偿式通用分流器
compensating valve 平衡阀;补偿阀
compensating ventilation 补偿通风
compensating vertical index 补偿竖直(度盘)指标
compensating voltage 补偿电压
compensating water 补偿水
compensating weight 补偿重物
compensating winding 补偿绕组
compensation 校正;抵偿;代偿(作用);代偿功能;补整;补贴;补偿(费);报酬;赔偿;赔偿金
compensation according to cost 照价赔偿
compensation action 补偿作用
compensation adjustment 补偿度的裁定;补偿调整
compensation adjustment system 补偿调整系统
compensation agreement 补偿协定
compensational regeneration 补偿再生
compensation ampereturns 补偿安匝
compensation analysis 补偿分析
compensation apparatus 补偿(仪)器
compensation arrangement 补偿安排
compensation authority 补偿权
compensation award 补偿赔款
compensation balance 补整平衡;补偿平衡;补偿摆轮;平衡摆
compensation bar 补偿杆
compensation brake rigging 平衡闸装置
compensation by result 酬赏成果化
compensation calorimeter 补偿量热器
compensation capacitor 补偿电容器
compensation capacity 补偿能力
compensation circuit 补偿电路
compensation claim 赔偿要求
compensation coefficient 补偿系数
compensation colo(u)rimeter 补偿色度计
compensation correction 补偿校正;补偿改正
compensation cost 补偿费用
compensation criterion 补偿准则;补偿判据
compensation current 补偿气源;补偿(气)流;补偿海流;补偿电流
compensation curve 补偿曲线
compensation degree 补偿度
compensation depth 补偿深度
compensation device 补偿装置;补偿器
compensation duty 补充关税;抵制关税
compensation equipment 补偿设备
compensation factor 修正系数;补偿系数
compensation fee 补偿费
compensation fee for land 土地补偿费
compensation fees for acquisition of land 征用土地补偿费
compensation filter 补偿滤色镜;补偿滤光器;补偿滤光片;补偿滤波器;调整滤波器
compensation financing facility 出口变动补偿性贷款办法
compensation financing of export fluctuation 出口变动补偿性贷款办法
compensation flow 补偿水流;补偿流(量)
compensation for a loss 赔偿损失
compensation for building removal 拆迁补偿费
compensation for cancellation of contract 解除合同赔偿费
compensation for damage 损失赔偿;损失补偿;损害赔偿;损害补偿
compensation for delay 交货逾期赔偿
compensation for demolition 拆迁补偿费
compensation for injury 工伤补偿
compensation for labor object 补偿劳动对象
compensation for land 土地补偿费
compensation for living labour 补偿劳动报酬
compensation for loss or damage 补偿损失
compensation for non-fulfillment of contract 不履行合同的补偿
compensation for non-payment 未付款补偿
compensation for removal 迁移赔偿费;解职金;拆迁费(用);拆除费(用);搬迁费;遣送费;遣散费;迁移费(用)
compensation for temporary land occupation 临时用地补偿
compensation function 代偿机能
compensation fund 补偿基金
compensation gear 补偿装置
compensation geoid 补偿大地水准面
compensation grade 补整坡度;折减坡度
compensation insurance 补偿保险;赔偿保险
compensation item 补偿项(目)
compensation joint 补强接头;补强接缝;调整缝
compensation layer 补偿层
compensation level 补偿水准(仪);补偿水准(器);补偿水平线;补偿水平(面)
compensation line 校正线(路)
compensation loan for export fluctuations 出口变动补偿性贷款办法
compensation magnetometer 补偿磁力计
compensation method 对消法;补偿法;配赋法
compensation micromanometer 补偿式微压计
compensation micro-pressure meter 补偿微压计
compensation money 赔偿金
compensation ocular 补偿目镜
compensation of compass 罗盘校正
compensation of curve 曲线坡度折减
compensation of deviation 自差消除
compensation of employees 受雇人员酬金
compensation of end-effect 端部效应补偿
compensation of errors 平差;误差调整
compensation of grades 纵坡折减
compensation of gradient 坡度折减
compensation of gradient in tunnel 隧道坡度折减;隧道坡道折减
compensation of gradient on curve section 曲线坡度折减
compensation of non-linear distortion 非直线性畸变补偿
compensation of power factor 功率因数的补偿
compensation of sharp curve 小半径曲线坡度折减
compensation of solution 溶液补充
compensation of thermal expansion 热补偿
compensation of undulation 起伏补偿
compensation of value 价值补偿
compensation on grade curve 曲线折减
compensation payment 补偿报酬
compensation pendulum 补偿摆
compensation pin 补偿引线
compensation planimeter 补整面积仪;补偿求积仪
compensation planting 补偿造林;补偿林
compensation plate 补偿板
compensation point 补偿点
compensation pond 补偿水塘
compensation principle 补偿原则
compensation process 补偿方法
compensation products 补偿产品
compensation projection 平衡投影
compensation pyrheliometer 补偿直接日射强度计;补偿式日温计
compensation rate 补偿量;补偿汇率
compensation reactor 补偿电抗器
compensation regulation 补偿调节
compensation reservoir 补偿水库
compensation resistor 补偿电阻器
compensation return 软反馈装置;弹性回复
compensation ring 补偿环
compensation rod 补偿杆
compensation sac 平衡水袋
compensation side pond 补偿侧水塘
compensation signal 补偿信号
compensation space 补偿空间
compensation spring 调整弹簧
compensation strand 补偿束
compensation surface 补偿面
compensation system 进出口差损补偿制度;补偿装置
compensation tank 补偿振荡槽路;补偿水箱;调整水舱
compensation technique 校正方法;补偿技术
compensation temperature 抵消温度;补偿温度
compensation test 补偿检验
compensation theorem 补偿定理
compensation to living labor 补偿活劳动报酬
compensation trade 补偿贸易
compensation trade project 补偿贸易项目
compensation transaction 补偿贸易
compensation transmitter 补偿发射机
compensation value 补偿值
compensation valve 平衡阀
compensation voltmeter 补偿式伏特计
compensation water 补充水;补偿水
compensation water outlet works 补偿水出水口工程
compensation water turbine 补偿水流涡轮机;补偿水流水轮机
compensation wave 空号波
compensation wheel 平衡摆
compensation winding 补偿绕组
compensative-birefringence method 补偿式双折射法
compensative network 补偿网络
compensator 校正铁;胀缩件;补助器;补偿器;补偿(棱)镜;补偿电位计;补偿薄板;平衡罐;膨胀节;赔偿人;稳压罐;调整器;调压罐;调相器;伸缩调整器
compensator alloy 补偿线合金
compensator-amplifier unit 补偿放大器
compensator balancer 补偿平衡器
compensator control 补偿调速
compensator for thermal expansion 补偿器
compensator frame 自耦变压器框架
compensator jet 补整器喷口
compensator knob 补偿器旋钮
compensator level 自动安平水准仪;补偿水准(仪);补偿水准(器)
compensator level(l)ing instrument 自动安平水准仪
compensator piece 膨胀补偿节
compensator plate 补偿板
compensator pretension 补偿器预拉伸
compensator receiver 补偿器式检波器
compensator reservoir 补偿水库
compensator setting 补偿(器)调整
compensator spring 平衡弹簧
compensator stand 补偿器支架
compensator starter 自耦变压器式起动器;补偿器启动器
compensator starting 自耦变压器启动
compensator system 补偿器系统
compensator tank 调节罐;缓冲罐
compensator transformer 补偿变压器
compensator valve 补偿阀
compensator weight 补偿锤
compensator winding 补偿(器)绕组
compensator with concentrated series capacitance 集中式串联电容补偿装置
compensator with movable cross hair 活动十字丝补偿器
compensator with movable objective 活动物镜补偿器
compensator with scattered series capacitance 分散式串联电容补偿装置
compensator with suspended lens 悬挂棱镜补偿器
compensatory 赔偿的;代偿的
compensatory adjustment 代偿性调节
compensatory allowance 津贴;补助
compensatory amount 补偿金
compensatory approach 补偿办法
compensatory balance 最低应存款
compensatory budget policy 弥补预算政策;补偿预算政策
compensatory circulation 侧支循环
compensatory concession 补偿性减让
compensatory damages 补偿损害赔偿金
compensatory deposit 补偿性存款

compensatory display 补偿式显示
compensatory error 冲销差错
compensatory expansion of demand 补偿性需求扩张
compensatory financing 补偿性资金供应;补偿贷款
compensatory financing facility 补偿贷款
compensatory fiscal policy 补偿性财政政策
compensatory gains 补偿增重
compensatory goods 补偿货物
compensatory growth 补偿生长
compensatory hypertrophy 代偿性增生
compensatory investment 补偿投资
compensatory leave 补偿假
compensatory lens 补偿透镜
compensatory mechanism 代偿机理;补偿办法
compensatory michanism 补偿机理
compensatory payment 补偿金;补偿支付
compensatory pressure 补偿压力
compensatory reflex 补偿反射
compensatory replacement of demolished housing 原拆原建
compensatory shear 补偿剪切
compensatory spending policy 补偿性支出政策
compensatory tariff 补偿性关税税则
compensatory tax 补偿税
compensatory theory 报酬的理论
compensatory time 加班时间
compensatory trade 补偿贸易
compensatory wage increase 补偿性的增加工资
compensatrix 平衡水袋
compentory payment 赔偿金
compete enumeration 全数调查
competence 机械搬运力;能力;资格;反应能力;权限
competence of auditor 审计权限
competence of river 河流挟沙能力;河流输沙能力
competence of rock 坚固度
competence of stream 水流强度;河流输沙能力
competence of wind 风的挟带力
competency 资格;权限
competency of non-destructive test operator 非破坏性试验操作工的资格
competent 能胜任的;有法定资格的
competent assistant 得力助手
competent authorities 主管部门;主管当局
competent authorities at higher level 上级主管机关
competent bed 强岩层;强层
competent condition 容许状态
competent court 主管法庭
competent department 主管部门
competent department for forestry 林业主管部门
competent department in charge of grassland 草原主管部门
competent discharge 适应流量
competent fold 强(性)褶皱
competent folding 强褶皱
competent formation 不易坍塌地层;稳定地层
competent meander 强蛇曲;强曲流
competent party 法定一方
competent person 胜任人员;合格人;能胜任的人;环境卫生检验员;合格人员;安全检查员
competent river 夹泥沙河流;挟沙河流;有挟沙能力的河流
competent rock 非塑性岩;硬质岩石;强岩(石)
competent rock bed 强岩层
competent rock stratum 强岩层
competent stream 挟沙河流
competent surveyor 合格检验人
competent velocity 起动流速
competent velocity scale 起动流速比例
competent worker 能胜任的工人
compete on an equal basis 平等竞争
compete on an equal footing 平等竞争
competing adsorbate 竞争吸附质
competing goods 竞争品
competing product 竞争(产)品
competition 竞争;竞赛
Competition and Credit Control 竞争及信贷控制条例
competition curve 竞争曲线
competition design 竞赛设计
competition entry 竞赛项目;竞赛登记
competition item 竞赛项目
competition mechanism 竞争机制

competition model 竞争模型
competition policy 竞争政策
competition process 竞争过程
competition resource use type 资源利用竞争型
competition studies in 竞争性能的研究
competitive 竞争性(的)
competitive adsorption 竞争性吸附
competitive advantage in the export of 出口有利性
competitive advertising 竞争性广告
competitive assumption 竞争性假说
competitive auction 竞争性标卖
competitive bid 竞争(性)投标;公开投标;竞投;有竞争性投标;公开招标的;比价的;投标竞争;投标
competitive bid contract 竞争性投标合同
competitive bidding 竞争性招标;竞争性递价;竞争性报价;竞争出价
competitive bidding system 比价制;竞争性招标制;招标制;投标制
competitive business environment 竞争的商业环境
competitive capacity 竞争能力
competitive cementation 竞争胶结(作用)
competitive combination 竞争性结合
competitive commodities 竞争品
competitive comparison method 竞争性对比法
competitive compexation 竞争性络合
competitive conjugation 竞争性结合
competitive contract 投标承包
competitive contractor 投标承包者;投标承包商;投标承包人
competitive cooperation 竞争的合作
competitive coordination 竞争性配位;竞争性络合
competitive decay 竞争衰变
competitive demand 竞争需求
competitive depreciation 竞争性贬值
competitive design 竞赛设计;竞争设计;比较设计
competitive devaluation 竞争性贬值
competitive economy 竞争(性)经济
competitive edge 竞争优势;竞争界限
competitive effect 竞争作用;竞争效应
competitive enterprise 竞争企业
competitive equilibrium 竞争(性)平衡;竞争(性)均衡
competitive equilibrium curve 竞争性平衡曲线
competitive equipment 可竞争设备
competitive exchange depreciation 竞争性外汇贬值
competitive exclusion principle 竞争排斥原理(即高斯原理)
competitive experiment 竞争实验;竞赛实验
competitive factors 竞争因素
competitive fringe 竞争边缘
competitive funding 竞争性拨款
competitive import 竞争性输入;竞争性进口
competitive information 竞争信息
competitive inhibition 竞争性抑制
competitive interconnection 竞争性关联
competitive intermediate market 竞争性中间产品市场
competitive international bidding 国际公开招标
competitive investment 竞争性投资;竞争投资
competitive ion 竞争性离子
competitive Langmuir adsorption 竞争性朗格缪尔吸附
competitive leasing 竞租
competitive list 比较单
competitive market 竞争(性)市场
competitive market analysis 竞争市场分析
competitive mechanism 竞争机制
competitive model 竞争模型
competitive needs limitations 竞争需要限制规定
competitiveness 竞争能力
competitiveness of export 出口竞争能力
competitive non-revaluation 竞争性的非升值
competitive offer 竞争性报价
competitive operator 竞争的经营者
competitive-oriented pricing 竞争导向的价格政策
competitive position 竞争状态;竞争地位
competitive power 竞争(能)力
competitive power of export 出口竞争能力
competitive price 竞争(性)价格;有竞争能力的价格;投标价(格)

competitive process 竞争过程
competitive rate 竞争费率
competitive rates 竞争性费率;有竞争能力的费率
competitive reaction 竞争反应
competitive redio-assay 竞争性放射性测定
competitive relation 竞争关系
competitive response 竞争反应
competitive sale 竞争性销售
competitive sealed bid 竞争性密封标单;密封标单
competitive situation 竞争性的局面
competitive stage 竞争阶段
competitive strains 竞争品系
competitive strategy 竞争性战略;竞争策略
competitive subsidization 竞争性补贴
competitive system 竞争制度
competitive tariff 竞争性运价表
competitive tender 竞争(性)投标;公开投标
competitive tendering 公开招标
competitive tender(ing) action 投标程序
competitive trades 竞争性的行业
competitive trial 比较性试验
competitive use of resources 资源的竞争使用
competitive wage 竞争性工资
competitor 竞争者;竞争对手;对手;替代电站
compexation of metal 金属络合作用
compexitities of adsorption reaction 复杂吸附反应
compexometric agent 络合滴定剂
compexometric indicator 络合指示剂
comphrehensive development of agriculture 农业综合发展
comphrehensive evaluation of enterprise performance 企业效绩综合评价
comphrehensive parameter measurement apparatus for silicon elements 硅元件综合系数调试仪
compilation 编纂;编制;编译;编辑;编绘
compilation base 编绘底图
compilation camera 光学编绘仪
compilation data 整编资料
compilation editing 编绘原图审校
compilation history 图历表
compilation instruction 编绘细则
compilation manuscript 作者原图
compilation method 编图法
compilation note 编图说明
compilation of budget 编制预算
compilation of geologic(al) investigation report based on remote sensing 遥感地质调查报告编写
compilation of hydrologic(al) data 水文资料整理;水文资料整编
compilation of plan 计划编制
compilation of statistics 编制统计
compilation of terms 技术用语汇编
compilation opinion 编辑意见
compilation organizer 编图单位
compilation phase 编译阶段;编译方面
compilation plot 编绘原图
compilation procedure 编图程序
compilation process 编译过程;编图过程
compilation program(me) 编图程序
compilation scale 编图比例尺;编绘图比例尺
compilation specification 编制规格书;编制规范
compilation technique 编译技术
compilation time 编译时间
compilation together with drafting 连编带绘
compilation tool 编译工具
compilation unit 编译单元;编译单位;编图室
compile 编纂;编译;编写
compile a budget 编制预算
compile and go 编译及执行;编译并执行;编译并立即执行
compile and run time 编译和运行时间
compile budget 编制预算
compile cost 编译成本
compiled code modelling technique 编译代码模型化法
compiled date 编图时间
compiled information 编图资料
compiled map 编图原图
compiled number of angle domain 角域编号
compiled original 编绘原图
compiled plot 编绘原图
compiled program(me) 编成程序;编译程序

compiled sheet 编绘原图
compiled signalling 编表信号方式
compiled unit 编图单位
compiled unit in chief 主编单位
compile duration 编译期间
compile error 编译错误
compile link 编译连接
compile link and go 编译联结并立即执行;编译连接并执行
compile list 编译表
compile phase 编译阶段
compiler 编者;编译器;编辑者;编绘员
compiler algorithm 编译算法
compiler automation 编译自动化
compiler base 编译程序库
compiler binding 编译器装帧;编译程序结合
compiler building system 编译设计系统
compiler call 编译程序调入
compiler communication table 编译程序通信表;编译程序(联)系表
compiler-compiler 编译程序的编译程序
compiler construction 编译程序构造
compiler cost 编译代价
compiler description 编译程序描述
compiler design 编译程序设计
compiler directing sentence 编译程序指示句;编译程序规定句
compiler directing statement 编译指示语句;编译程序引导语句
compiler directive 编译指示语句;编译命令;编译程序指示
compiler directive statement 编译指示语句
compiler driven simulation 编译驱动模拟;编译程序控制模拟
compiler generator 编译程序用编制程序;编译程序(的)生成程序
compiler interface 编译程序的接口
compiler interpreter system 编译程序解释系统
compiler language 编译(程序)语言
compiler level 编译级
compiler level language 编译(程序级)语言
compiler limit 编译程序限制
compiler option 编译程序(可)选项
compiler organization problem 编译程序组织问题
compiler program(me) 偏译程序【计】
compiler program(me) checking 编译程序检查
compiler rate 编译速率
compiler routine 编制器程序;编译例行程序
compiler routine method 编译程序法
compiler source program(me) 编译程序源程序
compiler source program(me) library 编译程序的源程序库
compiler source routine library 编译程序的源程序库
compiler statement 编译程序语句
compiler structure 编译程序结构
compiler subroutine library 编译程序的子程序库
compiler switch 编译程序开关
compiler system 编译程序系统
compiler theory 编译理论
compile shape 编辑型
compile step 编译步骤
compile-time activity 编译时动作;编译时的作用;编译时的活动
compile-time error 编译时错误
compile-time facilitate 编译时的简化;编译时的方便
compile-time facility 编译时功能
compile-time intrinsic(al) function 编译时间内在函数
compile-time message 编译时信息
compile-time stack 编译时栈
compile-time statement 编译时(间)语句
compile time table 翻译时间表
compile-time table or array 编译时间表或数组;编译时的表格或数组
compile type simulation 编译类型模拟
compiling 编译;编排;编辑
compiling algorithm 编译算法
compiling and linking 编译和连接
compiling automation 编译自动化
compiling basis 编制依据
compiling computer 编译计算机
compiling data 整编资料
compiling date 编写日期

compiling duration 程序编译持续时间;编译周期
compiling economy 编表经济体
compiling explanation 编制说明
compiling method 编译法;编码法
compiling phase 编译阶段
compiling program(me) 编译程序
compiling routine 编译程序
compiling scope 编制范围
compiling sheet of geologic(al) map 编测图幅【地】
compiling stream flow data 整理测流资料
compiling survey 编测
compiling system 编译系统
compiling technique 编译技术
compiling time 编译时间
compitable elements 相容元素
compital 叉处生的
complain 申诉
complainant 控诉人;起诉人;申诉人
complaint 申诉
complaint investigation 意见调查;陈诉调查;危害性调查
complaints and claims 抗议与索赔;申诉与索赔;控告与索赔;投诉与索赔
complanation 变成平面;平面化
complement 整套;补元;补图;补体;补数;补色;补码;补充;配套
complement address 补码地址
complemental 互补的;补足的;补充的
complemental air 补吸气
complemental code 补码
complemental space 补充隙
complement and carry add circuit 补码移位加法电路
complementarity 互余性;互补性;并协性
complementarity law 互余律
complementarity principle 互余原理;互补原理;附加性原理
complementary 互补(的);补充的;补偿的
complementary acceleration 补充加速度
complementary account 补充账户;补充科目
complementary action 互补作用
complementary after image 补色残像;互补余留影像
complementary angle 余角;互余角
complementary area 补偿面
complementary assay 互补试验
complementary bay construction 补仓施工
complementary beam 补充光束
complementary benefit 补助救济金
complementary binary 码二进制
complementary cargo 补充货
complementary cell 填充细胞
complementary chromatic adaptation 补色适应性
complementary chromaucity 互补色度
complementary circuit 补码电路
complementary clock 互补时钟
complementary code 补码
complementary colo(u)r 互补色;余色;补色;辅助色
complementary colo(u)r pairs 补色配对
complementary colo(u)r stimulus 加色法互补色
complementary condition 互补条件
complementary control point of analysis mapping 图根解析补点
complementary current wave 补余电流波
complementary data 互补数据
complementary differential amplifier 互补差动放大器
complementary dike 余脉
complementary dipole 互补偶极子
complementary diversity 互补分集法
complementary dominant wavelength 补色主波长
complementary economic structure 相辅形成的经济结构
complementary emitter follower 互补发射极输出器
complementary energy 余能;补余能量
complementary equation 补余方程
complementary equipment 补充设备
complementary error 互补误差
complementary error function 误差函数的补函数
complementary event 互补事件;相补事件;对立事件;补事件

complementary factor 互补因子;辅因(子)
complementary filter 互补滤波器
complementary financial facility 追加贷款
complementary frequency response 互补频率响应
complementary function 互补函数;余函数
complementary Gaussian distribution function 余高斯分布函数
complementary geometric(al) programming 几何规划(法)
complementary goods 互补商品;相关货物;辅助货物;补充品
complementary graph 补图
complementary hue 互补色调
complementary image 互补色图像
complementary induction 补充诱导;补偿感应
complementary influence line 互补影响线
complementary inverter 互补倒相器
complementary ion 互补离子;补偿离子
complementary load 互补荷载
complementary logic 互补逻辑
complementary market 补助性市场;补充市场
complementary mask 补色蒙片
complementary minor 余子式
complementary modul 余模
complementary modulus 补模
complementary network 附加网络;互补网络
complementary notation 补记记数法
complementary offset binary 补偿码二进制
complementary operation 补运算;补码操作;求反操作
complementary operator 补(数)算子;补码算子;补充算子;求反运算符
complementary output 双相输出
complementary partitions 余划分
complementary phenomenon 补色现象
complementary probability 互补概率
complementary product 附件商品;补充产品
complementary programming 补规划
complementary pulse circuit 互补脉冲电路
complementary related industries 相关产业助推
complementary scheme 补色配合
complementary series 互补序列
complementary set 补集
complementary shearing stress 余剪应力
complementary ship cargo 补充船货
complementary society 补充群居
complementary space 余空间
complementary stain 补充染色
complementary standard 补充性标准
complementary stress 互补应力
complementary structure 互补结构;互补构造
complementary subgraph 余子图
complementary submatrix 余子矩阵
complementary supply 补充供给
complementary surface 全曲面
complementary symmetric(al) amplifier 互补对称放大器
complementary symmetry 互补对称
complementary tax 附征税
complementary test(ing) 补充试验
complementary ticket 招待券
complementary tissue 补充组织
complementary tracking 补充跟踪
complementary tractor 辅助拖拉机
complementary treaty 补充条约
complementary unit 辅助机构
complementary use of resources 资源的辅助性使用
complementary wave 互补波;余波;副波;补偿波
complementary wavelength 互补(光)波长;补色波长
complementary work 辅助工作;相补功
complementation 互补(作用);补数法;补码法
complementation agreement 补充协议
complementation distribution 互补分布
complementation law 互补律
complementation of a fuzzy set 模糊集的补集
complementation test 补充试验
complement base 补码底数;补码(的)基数
complement bypass 补体旁路
complement circuit 补码电路
complement consumption test 补体消耗试验
complement defect 补体缺陷
complement deviation 补体偏差

complemented 补体致活的
complemented element 补元素
complemented error function 补余产差函数
complemented lattice 有补格;补格
complemented representation 补码表示
complemented subspace 补充子空间
complementer 补助器;补数器;补码器;取补装置;求反器
complement exploratory report 补充勘探报告;补充勘察报告
complement flip flop 互补双稳态触发器
complement form 补码形式
complement function 余函数
complementing flip-flop 状态求反触发器;求反触发器
complement inhibitor 补体抑制物
complement instruction 补码指令
complement integrated navigation 互补式组合导航
complement number 补数
complement number system 补数系
complement of a set 集(合)的补集
complement of geologic(al) data 地质数据的补齐
complement of personnel 满员编制
complement of ten's 十进制补码
complementoid 类补体
complement-on-nine 九的补数
complement-on-one 一的补数
complement-on-ten 十进制补码
complementor 补偿器
complement paper 补体纸
complement prospecting report 补充普查报告
complement pulse 补码脉冲
complement report 补充报告
complement representation 补码表示
complement representation of negative number 负数补码表示法
complement rule 补码规则
complement system 补体系统
complement unit 补体单位
complement vector 余矢量;补矢量
complemetary demand 相辅需求
complete absorption 完全吸收
complete a business transaction 交割
complete acid/base composition 完全酸碱成分
complete a contract 合同执行完毕
complete activated sludge oxidation process 活性污泥完全氧化法
complete air changes 全室换气次数
complete air exchange 全部换气
complete alternation 整周期;整人循环;全循环
complete analysis 整体分析;全(量)分析
complete appliance 整机
complete a project 完工
complete aquifuge 隔水层完整
complete assemblage of element 整体单元组分
complete assembly 完成装配
complete athodyd 冲压式空气喷气发动机
complete attenuation 总衰减
complete audit 全面审计;全部审计
complete austenitizing 完全奥氏体化
complete automatic device 全自动装置
complete automatization 全自动化
complete balance sheet 综合资产负债表
complete bath (设有浴盆,洗脸器,便器等设备的)整套浴室
complete Baumbach corona 完全鲍姆巴赫日冕
complete beta function 完全贝塔函数
complete black body 绝对黑体
complete black meal 全黑生料
complete block fracture texture 完整块裂体结构
complete Bouguer reduction 完全布格改正
complete budgetary control 全面预算控制
complete built unit 整套制成单元
complete built-up 整台
complete call 完成调用
complete carcinogen 完全致癌物
complete carry 全进位
complete catalogue 完整星表
complete cemented 完全胶结的
complete certificate 竣工说明书
complete chassis underframe 总成的底盘架
complete checking 详细核对
complete chemical analysis 化学全分析;完全化学分析
complete chill 全白口
complete circuit 闭合电路;全电路
complete closure 全闭合;完全关闭
complete cold end line 冷端全套生产线
complete colo(u)r information 彩色全信息
complete combination 完全组合
complete combustion 烧尽;完全燃烧
complete compensatory pause 完全代偿间歇
complete component analysis 整套部件分析
complete compression 完全压缩
complete condenser 全凝器
complete conductivity 全导电性
complete consumption coefficient 完全消耗系数
complete consumption coefficient of environment(al) resources 环境资源完全消耗系数
complete consumption coefficient of water resources 水资源完全消耗系数
complete contraction 全约束;完全收缩
complete contraction orifice 完全收缩孔(口)
complete control 完全控制
complete coolant equipment 全套冷却设备
complete correlation 完全相关
complete cost 总成本;全部成本
complete cotton and blend spinning equipment 成套棉与混纺纺纱设备
complete crown 全冠
complete curing 完全硬化;完全固化
complete customs union 完全关税同盟
complete cutoff 完全截水墙
complete cycle time 全循环时间
completed amount 完成工程量
complete date 完成日期
completed construction 完工建筑
completed contract method 全部完工毛利计算法
completed contract method of revenue recognition 营业收入的履约确定法
complete decomposition 完全分解
complete defoliation 全部脱叶;完全脱叶
complete demand coefficient 完全需要系数
complete detailed audit 全部详细审核
complete determinate 完全确定
completed gasification gas 全气化煤气
complete diamagnetism 全抗磁性
complete differential 全微分
complete diffusion 全漫射光
complete directed diagram 完全有向图
complete-directed graph 完全有向图
complete disability 完全丧失劳动力
complete discharge 全部卸载
complete dislocation 全脱位;完全脱位
complete dissociation 完全离解
completed item 建成(投产)项目;建成投入生产项目
complete diversion 完全分水
completed jobs journal 已完工批号日记账
completed layout 整套的工厂布置图
completed operation insurance 完满运行保险;完成施工保险
completed order 完成订单;完成订货
complete double circulation 完全双循环
completed project 完成项目;竣工工程;竣工项目;全部竣工项目
completed quantities 完成工程量
complete drainage spring 全排泄型泉
complete drilling crew 成建制钻机组;成建制机台【岩】
complete drilling date 完工日期
completed ship 完工船
completed transaction 已完成会计事项
complete dual monotonic 完全对偶单调
complete dull date 完钻日期
complete duration series 全部延时序列(洪水频率计算用)
completed well 完成的井
completed work amount 已完成工作量
completed work order 竣工工作通知
complete eclipse 完整食
complete(end) restraint 完全约束(的梁端);完全固定(的梁端)
complete enumeration 全面调查
complete equilibrium 不可逆平衡;完全平衡;完全平衡
complete equipment 整套装置;整套设备;成套设备;全套设备;成套装置
complete equipment for cane sugar factory 甘蔗糖厂成套设备
complete equipment for making bricks 成套制砖设备
complete equipment for making wood screw 木螺丝成套设备
complete equipment for manufacturing plywood 胶合板成套设备
complete equipment for salt refining 精制盐生产成套设备
complete equipment for shellac processing 虫胶加工成套设备
complete equipment of chemical plant 化工成套工厂
complete examination of water quality 水质全分析
complete exchange 完全交换
complete expansion 完全膨胀
complete expansion cycle 完全膨胀循环
complete-expansion diesel cycle 布雷顿循环
complete exploration line 总景勘探线
complete exponential model 完全指数模型
complete failure 完全故障
complete feed 全料
complete fertilizer 完全肥料
complete fixture line 整ябо卫生线路
complete flowchart 完整流程图
complete flowchart of soil classification 土分类完整流程图
complete flow of cash 现金流转全过程
complete flow process 整个流程
complete freezing 冰封;完全冰冻
complete Freund's adjuvant 完全佐剂
complete function series 完全函数序列;完备函数序列
complete fuse 完全熔合
complete fusion 完全熔接;完全熔化(物)
complete gas analyzer 气体全分析仪
complete gasification 完全汽化;完全气化
complete gasification gas 全气化煤气
complete gasification process 完全气化过程
complete glass fiber paper 全玻纤纸
complete graph 完整图;完全图;完备图
complete group 完全群
complete grouping 完全编组
complete handling line 全套搬运线
complete heat exchange 完全热交换
complete hiding 完全遮盖
complete hydraulic jump 完全水跃
complete ice coverage 封冻;冰封;完全冰封
complete image 全像;全图
complete inductance 全自感
complete induction 完全归纳法;数学归纳法
complete information 完整的资料
complete in place 全部竣工;就位定当
complete inspection 全面检查;全部检查
complete instantaneous phase measurement 瞬时完全相位测量
complete integral 完全原函数;完全积分
complete interaction analysis 整体相互作用分析
complete interchangeability 完全互换
complete inventory 全面盘存
complete isomeric change 完全异构变化
complete isomorphism 完全类质同象
complete joint penetration 整个焊缝熔透
complete jute spinning equipment 成套麻纺设备
complete knockdown export 全拆散式输出
complete lattice 完全格
complete life table 完全寿命表
complete lighting equipment 全套照明设备
complete line 整套线路;整个流程
complete linkage 完全连锁
complete liquefaction 完全液化
complete liquidation 全部清理
complete loop 闭合导线【测】
complete lubrication 完整油膜润滑
completely activated sludge oxidation process 完全活性污泥氧化法
completely additive family 完全加法族
completely automatic 全自动化的
completely automatic control 全自动控制
completely autotrophic nitrogen removal 完全自养脱氮

completely built-up area 全部建成地区
completely canonic(al) transformation 完全正则变换
completely consumed 完全消耗的
completely continuous linear transformation 全连续线性变换
completely differentiated intrusive body 完全分异岩体
completely dissolved 完全溶解的
completely elasto-plastic hysteresis 完全弹塑性滞回线
completely grouted rockbolt 全胶结式锚杆;砂浆全长黏结式锚杆
completely hydrated cement 完全水化水泥
completely inelastic collision 完全非弹性碰撞
completely integrated circuit 全集成电路
completely integrated thin-film circuit 全集成薄膜电路
completely knocked down 完全拆开
completely mixed activated sludge 完全混合活性污泥
completely mixed activated sludge process 完全混合活性污泥法
completely mixed activated sludge system 完全混合活性污泥系统
completely mixed aerated lagoon 完全混合曝气塘
completely mixed aeration 完全混合曝气
completely mixed aeration basin 完全混合曝气池
completely mixed aeration lagoon 完全混合曝气塘
completely mixed aeration system 完全混合曝气系统
completely mixed basin 完全混合池
completely mixed batch reactor 完全混合间歇式反应器
completely mixed biofilm reactor 完全混合生物膜反应器
completely mixed cell 完全混合室
completely mixed flow 完全混合水流;完全混合流
completely mixed flow reactor 完全混流式反应器
completely mixed stirred-tank reactor 完全混合搅拌池式反应器
completely mixed tank reactor 完全混合槽式反应器
completely monotonic 完全单调
completely nest 完全嵌套
completely no correlation 完全无相关
completely normal space 完全正规空间
completely overhauled 彻底检修的
completely penetrated well 完整井
completely penetrating artesian 完整承压井
completely penetrating gravity well 完整潜水井
completely perpendicular planes 完全正交平面
completely plant school 高水平苗木培育圃
completely random 完全随机
completely random design 完全随机设计
completely regular space 完全正则空间
completely reversed fatigue limit 全交变疲劳极限
completely reversed stress 对称性交变应力;完全反向应力
completely salified soil 超盐渍土
completely shrouded impeller 全闭式叶轮
completely specified decision 完全规范化决策
completely specified function 完全定义函数
completely submerged platform 全潜式平台
completely weathered 剧烈风化
completely weathered rock 完全风化岩石
completely welded and flame hardened crossing 全部焊接和淬火的辙叉
complete machine 成套计算机
complete manufacturing cost 全部制造成本
complete massive texture 整体块状结构
complete matrix ring 全矩阵环
complete metamorphosis 完全变态
complete metric space 完备度量空间
complete miscibility 完全溶合性;无限溶解;完全混溶性
complete-mix air system 完全混合空系统
complete mixing 完全混合
complete mixing activated sludge process 完全混合活性污泥法
complete mixing process 完全混合(方)法
complete mixing system 完全混合系统
complete-mix reactor 完全混合反应器
complete modulation 全调制;全调整;全调谐;全调幅
complete moment-resisting frame 完全抗弯框架
complete monopoly 完全垄断
complete motor type 配带电机型号
complete nappe 完整水舌
completeness 完备性;完整性;完全性;完全
completeness knock-down 整车解体出口
completeness of a model 模型完备性
completeness of aquifuge 隔水层完整性
completeness of combustion 燃烧完全度
completeness of deduction system 演绎系统完备性
completeness of modes 模的完全性
completeness theorem 完全性定理
complete nitritation 完全亚硝化
complete normed linear space 巴拿赫空间
complete obiteration 完全消失
complete of order 完全无序
complete oil analysis 全油分析
complete oil solvent extraction system 成套油脂浸出设备
complete operation 完全运算;完全操作;作业全程;完整运行;完整运算;完整操作
complete ordering 完全有序
complete orthogonal 完全正交
complete orthogonal system 完全正交系
complete orthonormal set 完全正交规范集
complete overhaul 全部检修;大修;全部仔细检修;全部拆卸检修
complete oxidation 完全氧化
complete oxidation process 完全氧化法
complete package 完整组件
complete penetraction and fusion in welding 全焊透
complete penetration 焊透;全熔透;全贯入;完全焊透
complete penetration butt weld 贯穿对焊
complete penetration well 完整井
complete period 全周期
complete picture 全貌
complete piping system 管系统
complete pivot(ing) 全主元(法)
complete plan 完整的计划;全套设计图
complete plane strain 完全平面应变
complete plant 成套设备;成套工厂
complete plants and equipment 全套设备
complete plants for cultivation 机耕成套设备
complete primitive function 完全原函数
complete production coefficient of impurities 污染物完全产生系数
complete production unit 成套(生产)设备
complete program(me) package 全套程序组件
complete project management 统筹规划管理法
complete proof map 套印图样
complete purification 完全净化
complete quadrangle 完全四边形
complete quadric combination method 完全二次型方根法
complete quadrilateral 完全四边形(的)
complete radiator 完全辐射体
complete randomized design 完全随机设计
complete rarefaction wave 全膨胀波
complete rationing system 包工制;包干制
complete reaction 完全反应
complete refrigerating equipment 成套冷冻设备
complete remission 完全缓解
complete resistance 完全抗性
complete responsibility for profits 利润包干
complete rest 全休
complete retraction 完全退缩;完全回缩
complete revision 全面修测
complete revolution 周转
complete roasting 完全焙烧
complete rotation 周转
complete routine 完整(例行)程序
complete running 完整运行
complete schematic diagram 总(线路)图
complete secondary sewage treatment 污水完全二级处理
complete section 全剖面
complete sectional view 全断面图
complete separation 完全性断离;完全分离
complete series 完整系列;完全系列
complete set 成套(机)组;全组;全系;全套;全集;完全系
complete set of 整套的
complete set of bill of lading 全套提单
complete set of control equipment 成套控制设备
complete set of directions 完全方向组
complete set of drawings 成套图纸
complete set of economic structural reform 配套的经济体制改革
complete set of equipment 整套设备;成套设备
complete set of files 成套锉刀
complete set of needle file 成套(小)锉刀
complete set of rice milling equipment 成套碾米设备
complete set of rolling mill 成套轧钢设备
complete set of variance 方差的完全组
complete set of wrench 成套扳手
complete sets of anti-corrosive equipment 成套防腐设备
complete sets of direct current equipment 直流成套设备
complete sets of equipment 成套设备
complete sets of farm machine 成套农机设备
complete sewage retention 污水完全停留
complete shadow 全影区
complete shear crack 完全切变裂缝
complete shell 满充壳层
complete shut-down 全关;停机
complete silk reeling equipment 成套缫丝设备
complete silk weaving equipment 成套丝织设备
complete sliding 整体滑动
complete sludge retention 污泥完全停留
complete solid solution 完全固溶体
complete solution 全解
complete space 完备空间
complete spare parts 全套配件;全套备件
complete spread 完整的(管道)施工队【给】
complete stability 完全稳定性
complete stabilization 完全稳定
complete state of stress in rock 岩石内全应力状态
complete stop 完全停止;完全停机;完全停车
complete storage system 完备存储系统
complete stowage plan 最终集装箱装载图
complete stress relief technique (岩石的)应力全部释放技术
complete structure model 整体结构模型
complete superstructure 全通船楼
complete survey 全面检验;全面调查;完整检验
complete survey of the well 钻孔全测
complete synchrone 全等时线
complete synchronism 完全同步
complete system 配套系统;完整系统;完整设备
complete tempering 完全退火
complete test 竣工试验;全面试验;完全试验
complete texture 整体结构
complete tillage 全套耕
complete time of oscillation 振荡周期
complete time of vibration 振动周期
complete titration curve 完全滴定曲线
complete tool equipment 全套工具装备
complete towel making equipment 成套毛巾设备
complete transduction 完全转导
complete treatment 全面处理;完全处理
complete treatment of sewage 污水深度处理;污水全部处理;污水(的)完全处理(包括初级和二级处理)
complete treatment plant 完全处理装置
complete treatment system 完全处理系统
complete trial 完全试验
complete turbulence 完全紊流;完全紊动
complete unification algorithm 完全一算法
complete unit 全套装置;全套设备;全套机组
complete veneer 全覆盖
complete verification 全部审查
complete water analysis 全量水分析;完全水分析;水(的)全分析;水样全指标分析
complete water quality analysis 水质全分析
complete water reuse 水完全回用
complete wood working machinery 成套木材加工机械
complete works 全集
completing task on time 按时完成工作
completion 竣工;完工;完成
completion acceptance 竣工验收;全部验收
completion acceptance of dredging works 疏浚

工程竣工验收
completion as scheduled 按期竣工
completion bit 完成位
completion bond 施工保证书;竣工保证书;承包保证书;工程质量保证书;完工(施工)保证书;完工承包保证书
completion by cuts 引割求全法
completion by stages 分期竣工
completion ceremony 竣工典礼
completion certificate 竣工证书;竣工证明
completion code 完成码
completion contract method 一次完工法
completion cost 完井费
completion cycle 时标
completion date 工程完工日期;竣工日期;完工日期;完成日期
completion depth 完井井深
completion drawing 竣工图
completion event 完工事件
completion final statements audit 竣工决算报表审计
completion fluid 成井液
completion fluid cost 完井液费
completion list 完工项目清单
completion method 完井方法
completion mixture 多元混合物
completion of construction 竣工
completion of discharge 卸载完毕
completion of jobs 批次产品完工
completion of well 钻井完成
completion of works 竣工;完工;工程竣工
completion on schedule 按期竣工
completion payment 竣工支付
completion period 工期
completion queue 完成排队
completion report 竣工报告;项目完成报告;完工报告
completion settlement 竣工结算
completion status 完成状态
completion test 竣工检验;竣工试验
completion time 完成时间;工期
completion tool 整体刀具
completion value 竣工价值
completion well date 完井日期
complex 联合装置;集合体;合成物;总厂;综合物;综合企业;综合建筑群;综合的;杂岩;多元的;复征;复形;复体;复数的;复合物;复合体;复合的;全套;配合物;络合物
complex abnormal curve 复变态曲线
complex absorption 全吸收
complex accelerator based on triethanolamine 三乙醇胺复合早强剂
complex achromatic system 复消色差系统
complex acid 络酸;配酸
complex action potential 复合动作电位
complex admittance 导纳复量;复数导纳
complex alloy 多元合金
complex alloy steel 合金钢;多元合金钢;多合金钢
complex alluvial fan coast 复式冲积扇海岸
complex alumin(i)um silicate 复杂铝硅酸盐
complex amines 复胺
complex amplitude 复值振幅;复值幅;复振幅
complex amplitude coefficient 复值振幅系数
complex-amplitude distribution 复式振幅分布
complex amplitude transmittance function 复振幅透过率函数
complex analysis 复分析;复变函数论
complex analysis procedure 综合分析法
complex analytic(al) curve 复解析曲线
complex analytic(al) function 复解析函数
complex analytic(al) structure 复解析结构
complex angle of incidence 复值入射角
complex anion 络阴离子;配阴离子
complexant 络合剂
complex anticlinal oil-gas field 复杂背斜油气田
complex antion 络合阴离子
complex arc 复式弧
complex argument 复自变量
complexation 络合作用;络合;复杂化
complexation adsorption 络合吸附
complexation chromatography 络合色谱法
complexation reaction 配位反应;络合反应
complexation-ultrafiltration coupling process 络合—超滤耦合法

complexation-ultrafiltration-nanofiltration coupling process 络合—超滤—纳滤耦合法
complex attribute 复型属性;复型表征;复数属性
complex attributes map 综合标志图
complex audit assignment 复杂的审计任务
complex autocorrelation 复自相关
complex automatic control system 综合自动控制系统
complex bacterium 复合菌
complex barchan dune 复合新月型沙丘
complex basin 复合盆地
complex belt conveyer 复合皮带运输机
complex biologic(al) reactor 复合生物反应器
complex brass 铜锌合金;特种铜;海军铜
complex building 复合建筑,综合楼
complex calcium lubricating grease 复合钙基润滑脂
complex calculus 复变微积分学
complex capital structure 复合资本结构
complex carbide 复合碳化物
complex catalysis 络合催化剂;络合催化;配位催化
complex cation 络(合)阳离子;配阳离子
complex cavity 复合窝洞
complex character 复杂性状;复合性状
complex characteristics of climate 气候的综合特点
complex chart 综合图
complex chemical reaction 复杂化学反应
complex chromatography 络合色谱
complex chromium ion 络合态铬离子
complex climatology 综合气候学
complex coacervation 复合凝聚
complex coagulant 复合混凝剂
complex coefficient 复系数
complex column packing 复合填充柱
complex compliance 络合柔量;复(数)柔量
complex component 复合构件
complex compound 络合(化合)物;配位(化合)物
complex compressibility of pool 油藏综合压缩系数
complex condition 复合条件
complex conduit 复合水道
complex conjugate 复共轭
complex conjugate function 复共轭函数
complex conjugate matrix 赫米特矩阵;复共轭矩阵
complex constant 复数型常数;复(数)常数
complex contaminant 复合污染物
complex contrast transfer function 复对比度传递函数
complex control 综合防治
complex control system 综合控制系统
complex coordinate 复坐标
complex correction 混合校正
complex correlation 复相关
complex covariance 复协方差
complex cover 复合覆盖
complex cross-bedding structure 复杂交错层理构造
complex cross correlation 复互相关
complex crossed-lamellar texture 复杂交错纹片结构
complex crystal 复晶体
complex crystalline glaze 复色结晶釉
complex-Cu reactive dye 络合铜活性染料
complex curve 线丛曲线;复(合)曲线
complex-curved glass 复合单面玻璃
complex cyanide 氰络合物
complex cycle 复杂循环
complex data 复型数据;复数值数据;复数数据;复数据
complex datahandling center 综合数据处理中心
complex declaration statement 复数说明语句
complex defect 复合缺陷
complex deformation 复变形
complex degree of construction site 建筑场地复杂程度
complex delta 复合三角洲
complex delta coast 复式三角洲海岸
complex deposit 多金属矿床;复合矿床
complex detector 复合探测器
complex development 综合开发
complex dielectric(al) constant 复数介电常;复介电常数
complex dielectric(al) permitivity 复电容率
complex die polisher 模具复合抛光机

complex differentiation 复微分
complex digestive solution 复合消解液
complex digital circuit 数字集成电路
complex displacement 复位移
complex display 复合显示
complex dissociation 络合离解
complex domain 复域
complex dune 复合沙丘
complex dye 络合染料;配合染料
complex dynamic(al) environment 复杂的动态环境
complex earthquake 复合地震
complexed cavity technique 复合腔技术
complexed metals wastewater 多金属废水
complexed regulation 混合式调节
complex effect 复合效应
complex elasticity 复弹性
complex elastic modulus 弹性复数模量;复数弹性模量
complex element 复元素
complex entity 复杂实体
complex envelope function 复包络函数
complex environment 复杂环境;复合环境
complex equilibrium diagram 复合平衡图
complexes 螯合物
complex ether 混合醚
complex exchange 全部交换
complex excitation 复合激振
complex experiment 复合试验
complex exponent(ial) 复指数
complex-exponential component 复指数分量
complex-exponential function 复指数函数
complex facility operator 全套设备操作人员
complex fault 复断层
complex ferrite 复合铁氧体
complex fiber 复合纤维
complex field vector 复数场矢量
complex film 复合膜
complex filter 复数滤波器
complex flat 复式住宅单元
complex flocculant 复合絮凝剂
complex fold 复褶皱;复式褶皱;复合褶皱
complex formability 配位度
complex formation method 配离子生成法
complex formation titration 络合滴定(法)
complex-forming ligand 络合配位
complex foundation 复杂地基
complex fraction 繁分数
complex frame 复式框架
complex frequency 复(合)频率
complex frequency response 复频反应特性
complex frequency response function 复频响应函数
complex function 复值函数;复合函数;复合官能;复合功能
complex Gaussian distribution 复高斯分布
complex geologic(al) structure 地质构造复杂
complex goal structure 复杂目的结构
complex gradient method 复梯度法
complex grease 复合润滑脂
complex group 复(杂)群
complex halo 复杂晕
complex harmonic current 复谐电流
complex harmonic quantity 谐和复量;复数谐量
complex harmonic voltage 复谐电压
complex heat transfer 复杂传热;复合传热
complex hydrograph 综合过程线;复合过程线;水文综合过程线
complex hydrothermal eruption 复合式水热喷发
complex hydroxy ion 配位羟离子
complex hyperbolic functions 复双曲线函数
complex impedance 复(数)阻抗
complex impedance spectroscopy 复合平面阻抗谱
complex index 复合指数
complex index of reflection 复反射率
complexing action 络合作用
complexing agent 络合剂;复合剂
complexing capacity 络合容量
complexing compound 金属配合物
complexing form 络合态
complexing ligand 络合配位体
complexing reagent 复式试剂
complexing separation 络合分离法
complex in involution 对合的线丛

complex initiation system 复合引发系统
complex inorganic pigment 复合无机颜料
complex instruction set computer 复杂指令系统计算机
complex interchange 复式立体交叉
complex inverse filter 复合反向滤波器
complexiometry 络合量法
complex ion 络(合)离子;复离子;配离子
complexion 络合剂
complexities of channel change 河槽变化复杂性
complexities of channel variation 河槽变化复杂性
complexity 综合性;错综性;复杂性;复杂度
complexity class 复杂性类
complexity of site 场地复杂性
complexity-stability theory 复杂性—稳定性理论
complex laboratory 综合实验室
complex labo(u)r 复杂劳动
complex land treatment system of wastewater 复合污水土地处理系统
complex layer 复合层
complex layer tunneling method 复杂地层掘进法
complex leachable core 复杂可溶芯
complex light amplitube 复光振幅
complex light correction method 复合光校正法
complex line 复线;复合管线
complex line drawer 复式拉线机
complex lipid 复合酯
complex liquid 复杂流体;复合液
complex locus 复合位点
complex logarithm 复对数
complex logic 复杂逻辑
complex low 多中心低压;复合低压
complex magnetic anomaly 复杂磁异常
complex magnetic flocculant 复合磁絮凝剂
complex magnetic permeability 复数磁导率;复合磁导率
complex manifold 复流形
complex map 综合图
complex mapping 综合制图;综合法测图
complex-marker 复标记符
complex material 复合材料
complex material mechanics 复合材料力学;复式材料力学
complex matrix 复矩阵;全矩阵
complex measure 复测度
complex membrane 复合膜
complex mercury ion 络合汞离子
complex method 复(合)型(方)法;复合形法;复合方式
complex mixture 复杂混合物
complex modulation 复合调制
complex modulus 复数模(量);复变模量
complex molecule 络分子;复杂分子;配分子
complex morphologic(al) 形态复杂的
complex multiplication 复数乘法
complex multiplier register 复数乘法寄存器
complex multiply register 复式乘法寄存器
complex network 综合网络;复合管网
complex normal stochastic process 复正态随机过程
complex notation 复数记法
complex nucleus 复合核
complex number 复型;复数
complex number field 复数域
complex number space 复数空间
complex number type 复数型
complex numeric data 复数数据
complex of blocks 街坊综合体
complex of curves 曲线丛
complex of external condition 外界条件总体
complex of geologic(al) origin 地质成因杂岩
complex of houses 住房综合体
complexometric titration 络合滴定(法);配位滴定(法)
complexometry 络合滴定(法);配位滴定(法)
complexonate 氨羧络酸盐
complexone 氨羧络(试)剂;络合滴定试剂
complex operator 复数(运)算符
complex optic(al) train 复合成像链
complexor 矢量
complex ore 综合矿石;复矿(石);复合矿(石)
complex organic molecule 络合有机分子
complex organism 超个体
complex orthogonal 复数正交

complex oscillation constant 复振荡常数
complex oxide pigment 复合氧化物颜料
complex pad 合金钢垫
complex parameter 复参数
complex parts 复杂零件
complex pattern 复模式
complex pellicle 复合膜
complex penetration 全贯入
complex permeability 复导磁率;复磁导率
complex permittivity 复数电容率;复介质常数
complex phase 复相
complex phase factor 复位相因子
complex phaser 复移相器
complex photographic material 复合感光材料
complex picture 复杂图形
complex pipe 复合管(道)
complex pipeline 复合管线
complex plane 复平面
complex plane analyzer 复平面分析器;矢量分析计算器
complex plastic flow 复数塑性流动;复合塑性流
complex plex structure 复杂丛结构
complex point 复点
complex polarization 复偏振
complex pollution 复合污染
complex polyelectrolyte 络合聚合电解质
complex potential 复位势;复势
complex power 相量功率;复数功率
complex power series 复幂级数
complex precipitation 络合沉淀
complex prescription 复方
complex process 多元过程;多相过程;复杂过程
complex product 复杂产品;复合乘
complex production 复杂生产
complex production cycle 复杂生产周期
complex projective plane 复射影平面
complex projective space 复射影空间
complex propagation vector 复传播矢量
complex pseudo-variable 复伪变量
complex purifying agent 复合净化剂
complex quantity 复数;复量
complex radical 复基
complex random process 复随机过程
complex random variable 复随机变量
complex ratio 复比
complex reaction 复杂反应
complex reactor 复式反应器
complex real system 复真实系统
complex record section of wavefront 前沿波带的复记录剖面
complex redundant system 复杂多余系统
complex reflector 复合反射器
complex reflex reaction 复合反射性反应
complex refractive index 复折射率
complex relative attenuation 复数相对衰减
complex relocatable expression 复(数)型浮动表达式;复合浮动表达式
complex representation 复数表示
complex resistivity 复电阻率
complex resolution of the identity 单位变换的复分解
complex response 复合响应
complex response method 复数反应法
complex rock 非均质岩石
complex roots 复根
complex salt 络(合)盐;复盐;配盐
complex sample 复杂试样
complex sampling unit 复合抽样单位
complex seas 复合浪
complex sea wave 复合浪
complex section 复杂零件
complex selection 综合选择
complex series 复(项)级数
complex set of hauling equipment 成套运输设备
complex shape 复杂形状
complex-shape orebody 形状复杂矿体
complex sharn 复杂矽卡岩
complex shear modulus 复数剪切模量;复合剪切模量
complex shears 复剪
complex silicate 复合硅酸盐
complex singularity 复奇点
complex sinusoidal current 复数正弦电流
complex sinusoidal quantity 复数正弦量

complex site 复杂场地
complex slope 复合边坡
complex soap grease 复合皂基润滑脂
complex softening agent 复式柔软剂
complex soil 复域土壤;复合土壤
complex solution 复合溶液;全解
complex source 复杂源
complex spatial filtering 复空间滤波
complex spectrum 复杂光谱;复谱;复(合)光谱
complex sphere 复球面
complex splitting 复杂类型分岔
complex static environment 复杂的静态环境
complex steel 多金钢;多元合金钢;合金钢
complex stimulated scattering 复杂受激散射
complex strain 复应变
complex strain wave 复合应变波
complex stress 综合应力;复杂应力;复合应力
complex stress function 复应力函数
complex structure 综合结构;复合结构
complex subtractive colo(u)rant mixture 复合减色法颜色混合
complex supercharger 复式增压器
complex surface 复曲面;复合地面
complex susceptibility 复合灵敏度
complex symbol 复(合)符号
complex symmetry 复合对称
complex synapse 复合突触
complex synaptic arrangement 复合突触排列
complex system 复杂系统;复系数;复合系统
complex target 复合目标
complex tariff 复式税则;复式税率;复合(关税)税则
complex tension 复合张力
complex tetrahedral phase 复发的四面体晶相
complex texture 复杂结构
complex tissue 复合组织
complex tombolo 复式陆连岛;复合连岛沙洲
complex tone 杂音;复(合)音
complex tone audiometer 复音叶力计;杂音测试器
complex topography 复杂地形
complex town planning 城市整体规划
complex transfer function 复传递函数
complex transformation 复变换
complex transmittance 复合透射比
complex truss 复式桁架
complex trust 复合信托
complex type 复数型
complex type stress pattern 复合应力图
complex unit 单位复数;复合单位
complex utility routine 综合性服务系统;复合实用程序的例行程序
complex value 复值
complex valued field amplitude 复值场振幅
complex valued function 复值函数
complex variable 复变数;复变量
complex variable function 复变函数
complex variable mapping method 复变绘图法
complex variable method in theory of elasticity 弹性广义复变函数方法
complex vector 复向量;复数矢量
complex vector space 复向量空间;复数域上的向量空间;复矢量空间
complex velocity 复速度
complex velocity of sound 复声速
complex viscosity 复数黏度
complex visibility function 复可见度函数
complex volcano 复火山;复合火山
complex wastewater 络合废水
complex water reducer 复合减水剂
complex wave 综合波;复合波;复波
complex wave function 复波函数
complex whip grafting 复舌接
complex Young's modulus 复数杨氏模量;复合杨氏模量
complex zone 综合区
compliance 可挠性;合格;依从;符合(性);柔软性;柔量;屈从;弯曲量;顺从;顺从(性)
compliance auditing 符合性审计
compliance audit report 符合性审计报告
compliance certificate 符合证明(书)
compliance constant 柔顺常数
compliance cost 合格成本
compliance delay 延期执行
compliance error 符合性上的错误
compliance in extension 拉伸柔量

compliance in shear 剪切柔量
compliance inspection 合格性检查
compliance modulus 柔性模量
compliance monitoring 合格性监测
compliance point 合格性监测点
compliance procedure 符合程序
compliance program(me) 合格性监测计划
compliance shape 复杂形状
compliance systems 符合环保法规的涂料体系
compliance tensor 柔性张量
compliance test 性能验证测试;符合性测试
compliance test apparatus 弹性测定器
compliance the contract 恪守合同
compliance torque 弹性变形力矩
compliance voltage 恒流制输出电压
compliance with schedule 符合进度表
compliance with statutes 恪守法令
compliance with the contract 合同的遵守
compliancy 柔量
compliant coating 符合环保要求的涂料
compliant platform 顺应式平台
compliant primer 符合环保法规的底漆
compliant technology 符合环保法规的
complicated boundary 复杂边界
complicated cross section 复杂的截面;复杂的断面
complicated degree of condition of mineral deposit-hydrogeology 矿床水文地质条件复杂程度
complicate degree of geologic(al) structure condition 地质构造条件复杂程度
complicate degree of geomorphologic(al) condition 地貌条件复杂程度
complicate degree of hydrogeologic(al) condition 水文地质条件复杂程度
complicate degree of water chemistry 水化学复杂程度
complicated function 复杂函数
complicated lithology type 岩性较复杂型
complicated molecule 复杂分子
complicated shape 复杂形状
complicated shape brick 异型砖
complicated shoal 复杂浅滩
complicated style in lithologic(al) character and geologic(al) structure 岩性及构造较复杂型
complication 复杂情况;混杂;复杂(化)
complier 编译程序
compluetic reaction 乏色曼反应
compluvium 房顶采光井;天花板上的方形通风孔;院内蓄水池(古罗马)
comply with 服从
comply with a formality 履行手续
comply with an agreement 履行合同
comply with the terms of contract 合同条款办理
compo 混合涂料;灰泥;工伤赔偿费;水泥砂浆
compo board 条木胶合板;树脂胶合板;纤维胶合板
compo bronze 粉冶青铜
compo bronze bearing 粉冶青铜轴承
compo-casting 混成砂铸造
compocrete 铺筑地面混合料(商品名)
compofloor 组合楼板
compoiste liner 复合衬砌
compole 极间极;换向极;整流极
compo mortar 混合砂浆;水泥砂浆;水泥石灰砂浆;石灰水泥砂浆
compo-mo(u)ld 混成砂铸型
componendo 合比定理
component 组元;组件;组分;组成部分;子星;构件;成员;成分;分图;分量;分力;部分;部件
component address 部件地址
componental movement 分运动
component analysis 组(成成)分分析;分量分析
component assembly 零件装配
component assembly machine 零件组装机;零件装配机;构件装配机
component bar-chart 分量条形图;分段条(线)图;分段条形图
component bridge 三用支桥
component building 组合建筑
component chamber 成分分析室
component characteristic 组件特性
component chart 元件表
component cleaning machine 部件洗净机
component colo(u)r 色成分
component concentration 组分浓度
component construction 预制建造;构件建造

component cost 部分成本;要素分析;要素成本
component current 分潮流
component day 分潮日
component density 元件密度;组件密度
component depreciation 分项折旧;房屋部件和设备的折旧
component discharge 分流量
component drawing 零件图
component effect 成分效应
component efficiency 局部效率;组件效率
component element 构成部分
component element of map 地图内容要素
component entry 部件入口;部件表目
component error 误差分量
component evaluation 组分评价
component failure 元件失效;部件失效
component failure impact analysis 部件故障影响分析
component family 部件族
component force 力的分量;分力
component goods 装配货物
component group 元件组
component hour 元件工作小时;分潮时
componential movement 组分运动
component identification 元件识别名
component library 部件图库;系统程序库
component life 元件寿命
component method 分部件估算造价法;构件估价法
component of acceleration 加速度分量;分加速度
component of additive 添加成分
component of alumin(i)um container 铝合金集装箱组件
component of a symmetric(al) system 对称系统的分量
component of cost 费用组成
component of current 电流轴线分量
component of deflection 挠度分量
component of displacement 位移分量
component of elasticity 弹性分量
component of flux linkage 磁链轴线分量
component of force 分力
component of gravity 重力分量
component of heavy compensation equipment 升沉补偿设备组成
component of interaction 交互成分
component of moment 力矩分量
component of petroleum 石油组成
component of price 价格组成
component of refuse 垃圾成分
component of rotation 转动分量;分转动
component of seabed thermal water 海底热水成分
component of strain 应变分量
component of the centrifugal force 离心力分量
component of the four displacements 四游仪
component of the six cardinal points 六合仪
component of variance 方差分量
component of vector 向量分量;分矢量;矢量分量
component of velocity 分速度;速度分量
component of vorticity 涡度分量
component of white noise 白噪声量
component panel 组合板;预制墙板
component parts 组件;零件;组合零件;构件;附件;分项;零配件;组成部分;元件;部件
component parts for assembly of a machine 台份
component percentage 分配百分比
component plate 分色板
component preparation machine 零件制备机
component prism 棱镜组件;棱镜部分;组合棱镜
component rancomization test 分量随机性检验
component reaction 组分反应
component review 部门审查
components assembly 部件
component short-circuit withstand ratings 组件承受短路能力
component sine wave 正弦波分量
components of covariance 协方差组分
components of drill rig 钻机部件
components of photoelectric(al) conversion 光电变换元件
components of stress 分应力
components of wastewater flows 废水流量的组成部分
component solubility parameter 组分可溶性参数

component solvent 混合溶剂
component species 组分物种
component specification 部件规格
component spinor 分旋量
component star 子星
component stress 分应力
component substances law 组分物质定律
component symbol 部件符号
component system 构件装配体系;构件组合系统
component target system 关键目标群
component test 零件检验;部件试验;部件测试
component test carrier 部件测试车
component testing 成分试验
component testing equipment 组件测试设备
component test memo 部件测试备忘录
component test set 元件测试装置
component test unit 部件测试装置
component tide 子潮;分潮
component transmission 传动部件
component type microcontroller 组合型微控制器
component under compression 受压构件
component under tension 受拉构件
component velocity 分速度
component voltage 分电压
component wave 分波
component weight of index 标志分权
component wire 芯线
component-wise product 按分量逐个作出的乘积
compo pipe 合金管;铅锡合金管
compose 构图;编写
compose of epsilon-type tectonic system 山字型构造成分
composer 编排器
compose rotational shear structural system 旋扭构造体系组成
composertron 作曲器
composing machines 铸排机
composing map 编图
composing room 组装车间;排字车间
composing rule 排版尺
compositae 菊科
composit compact 复合压块
composite 混合(料);合成物;构成;复合物;复合的
composite absorbent 合成吸收剂
composite absorber 复合式滤光片
composite abutment 组合式桥台
composite action 组成作用;组合作用
composite adhesive 合成黏合剂
composite air filter 混合式空气过滤器
composite algorithm 复合法
composite alluvial fan 复合冲积扇
composite analysis 组合分析;复合分析
composite arch 尖拱;花边尖拱;合成拱;复(合)拱
composite artwork 合成原图
composite assets goodwill 综合资产商誉
composite(backup) roll 组合式轧辊
composite balance 复合电秤
composite balance sheet 综合资产负债表;复合资产负债表
composite bar-diagram 复合条图
composite base 复合式基层
composite basin 复合盆地
composite batholith 复合岩基
composite beam (外包混凝土的)钢制楼板梁;劲性钢筋混凝土梁;结合梁;混成梁;组合梁;叠合梁;复合梁
composite beam bridge 联合梁桥;结合梁桥
composite bearing 多层轴瓦
composite bearing cage 复合轴承罩
composite bearing structure 复合式支承结构
composite bed 复合床;复合层
composite bed-sweeping 混合式扫床
composite belt system 组合传送带系统
composite biosorbent 复合生物吸附剂
composite block 线条网线混合板
composite block system 双信号闭塞制
composite board 合成板;组合板;复合板(材)
composite body 钢木组合车身;复式岩体;铁木混合结构车身
composite boiler 混合式锅炉;复合式锅炉
composite box beam 组合箱梁;组合箱形梁
composite brake block 复式闸瓦
composite brake drum 复合制动圆筒
composite brake shoe 合成闸瓦

composite braking 复合制动
composite breakwater 混合式防波堤;组合式防波堤;半直立式防波堤
composite brick 组合砖;复合砖
composite brick masonry 组合砖砌体
composite bridge 组合桥(梁)
composite(bridge) deck(ing) 组合桥面板
composite brush 复合电刷
composite budgetary estimate of single construction 单项工程综合概算
composite building 综合用途楼宇;混合建筑;多功能楼宇;组合建筑
composite building block module method 混合建筑区段模数法
composite building board 复合建筑板材
composite building member 复合建筑构件
composite building sheet 组合建筑板
composite building slab 复合建筑板
composite building unit 组合建筑单元
composite-built 混合建造的
composite-built vessel 组合建造船
composite bulkhead 组合舱壁
composite cable 混合多芯电缆;合成电缆;复合电缆
composite cage 组合保持架
composite can 复合罐
composite candle 复合陶瓷过滤芯
composite capital 组合柱头;组合柱颈
composite capital of Rome 罗马复合式柱头
composite car 组合车
composite card 组合卡片;复合卡
composite carrier 复合载体
composite casting 双金属铸件;双层金属铸造
composite catalog(ue) 总目录
composite catalyst 复合催化剂
composite cathode 合成阴极;复合阴极
composite cellulosic membrane 复合纤维膜
composite cement 复合水泥
composite ceramic cutting tool 复合陶瓷刀具
composite ceramic mo(u)ld 复合陶瓷型
composite chamber 多级燃烧室
composite character 合成特性
composite characteristic 合成特性曲线
composite chemicals 复合化合物
composite chord 组合弦杆
composite circuit 电报电话双用电路;混合电路;复合电路
composite coagulant aid 复合助凝剂
composite coal seam 复煤层
composite coast 复合海岸
composite coastline 复合海岸线
composite coating 复合物涂层;复合涂层
composite coefficient 合成系数;组合系数
composite cofferdam 混合式围堰;组合围堰
composite coil 组合带卷
composite colo(u)r 复合彩色
composite colo(u)r signal 复合彩色信号;全色信号
composite colo(u)r sync 复合彩色同步信号
composite column 劲性钢筋混凝土柱;混合柱;混合式柱子;混成柱;组合柱
composite columnar section 综合柱状(剖面)图
composite column-diagram 复式柱形图
composite column purification method 复合柱净化法
composite commodity 合成商品;复合商品
composite compact 复合坯块;复合粉粒料
composite component 复合组分;复合构件
composite compound unit 复合单元
composite compressive member 组合受压构件
composite concrete column 劲性混凝土柱
composite concrete flexural construction 组合式混凝土曲结构;组合式混凝土受弯构造
composite concrete flexural member 混凝土组合受弯构件;组合式混凝土受弯构件
composite concrete-steel girder 钢混凝土复合梁
composite concrete structure 混凝土组合结构
composite conditioning 综合调节
composite conductor 复合导线;复合导体
composite cone 混合锥
composite construction 劲性钢筋混凝土结构;混合(式)结构;混凝建筑;组合结构;复合结构
composite construction in steel and concrete 型钢混凝土混合结构
composite contact metal 复合电接触器材;复合电触头合金
composite contact of bed and vein 层脉复合接触带
composite container ship 组合式集装箱船
composite controlling voltage 复合控制电压
composite cooling 复合冷却
composite cost 工料合算单价;包工包料
composite cost of labo(u)r and material 工料合算价
composite cost of material and labo(u)r 工料合算价
composite craft 组合(式)船队
composite cross-section 复式横断面
composite crystal volume 组合晶体体积
composite curve 组合弧线
composite curve chart 复合曲线图
composite dam 混合式(土石)坝
composited circuit 复合线路
composite death rate 复合死亡率
composite deck cable stayed bridge 组合梁式斜拉桥
composite deformation 复合变形
composite demand 综合需求;复合需求
composite deposit 复合镀层;复合沉积层
composite depreciation 综合折旧
composite depreciation rate 综合折旧率
composite design 合成设计;综合设计
composited false colo(u)r photo 假彩色合成像片
composite diagram 复合图
composite die 拼合模
composite dielectric 复合质
composite die sections 组合式模具的构成元件
composite dyke 复合岩墙;复合岩床
composite display 复合显示器
composite distribution 复合分布
composite divider strip 混合分隔带;混合分车带
composite diving system 综合潜水系统
composite dolly 复式桩垫
composite door 组合门;复合门
composite drawing 全要素原图
composite duct 综合管道
composite earth(en) dam 组合式土坝
composite electrode 混合焊条;复合焊条
composite electrolyte 复合电解质
composite electronic navigational system 组合电子导航系统
composite engine 组合式发动机
composite entablature 混合柱顶盘;混合型柱顶盘(古罗马);混合式柱盘;混成型柱顶盘
composite environment index 综合环境指数
composite error 合成误差;综合误差
composite evaluation test 综合评定试验
composite experiment design 复合试验设计
composite explosive 混合炸药
composite fabric 复合组构;复合织物
composite fan structure 复合扇状构造
composite fault scarp 复合断层崖
composite feed 混合入选;不分级入选
composite feedback system 复合反馈系统
composite fiber 复合材料纤维
composite figure 组合图形
composite filling resin 复合充填树脂
composite filter 混成滤波器;组合滤波器;复式反滤层;复合滤波器
composite filter paper 复合滤纸
composite fire door 复合防火门
composite fire-proof and heat-insulation clothing 复合材料防火隔热服
composite fixing 综合定位
composite fixture 复合装置
composite flash 多道闪电
composite float 悬浮浮子;复式浮标
composite flocculant 复合絮凝剂
composite floor(board) 复合楼板;复合地板;组合楼板;组合地板
composite floor cover(ing) tile 组合地面砖
composite flooring 复合铺地材料;复合地板材
composite floor panel 复合楼板;复合地板
composite floor slab 复合楼板;组合楼板;复合地板
composite floor system 复合楼板系统;复合地板系统
composite foam 复合泡沫
composite fold 复合褶皱
composite folded slab 复合折板
composite force 合力;复合力
composite forecast 综合预报
composite forest 中林;复合林
composite foundation 复合地基
composite foundation pile 复合基桩
composite foundation with settlement-reducing piles 减沉复合疏桩基础
composite frame 组合桁架
composite framed structure 组合框架结构
composite fuel 组合燃料
composite function 合成函数;复合函数
composite function ceramics 复合功能陶瓷
composite general water quality vulnerability zone map 混合水质脆弱区总图
composite geometrical section 复式几何断面
composite geotextile 组合型土工织物
composite girder 混合大梁;混成大梁;组合大梁;钢与混凝土结合梁
composite girder bridge 叠合梁桥;结合梁桥;组合梁桥
composite girder with corbels 花篮梁
composite glacier 复式山谷冰
composite glaze 复合釉
composite gneiss 复片麻岩
composite goal 复合目的
composite gradient 合成坡度
composite gradient technique 复合梯度法
composite grain 复粒
composite graph 复合图
composite grating optic(al) differentiation 复合光栅光学微分
composite great circle sailing 混合大圆航行;混合大圆航法
composite green compact 复合压坯
composite ground 复合地基
composite ground wire with optic(al) fiber 带光纤的复合地线
composite group-flashing light 混合联闪光
composite grouping 复合分组
composite group occulting 混联明暗灯光
composite group-occulting light 混合联顿光
composite guideway 组合导轨
composite halo 复合晕
composite hard material 复合硬材料
composite hard metal base contacts 复合硬质合金电接触器材
composite harmonic motion 合成谐和运动
composite headway distribution 复合车间距分布
composite hologram 合成全息图;复合全息图
composite hydrogeologic(al) map 综合水文地质图
composite hydrograph 复合水文曲线;复合水文过程线
composite hypothesis 复合假设
composite import price index 综合进口价格指数
composite income 综合收益
composite income sheet 综合损益表
composite index 综合指数;复合指标;复合索引
composite index number 复合指数
composite index of coincident indicators 重合经济指标综合指数
composite index of lagging indicators 滞后经济指标综合指数
composite index of leading indicators 先行经济指标综合指数
composite insulation 复合绝缘材料
composite interface 复合材料界面
composite intermediate floor 复合中间(毛)地板
composite intrusion 复合侵入体
composite iron and steel 包钢铁
composite items 综合项目
composite joint 复合接头;混合连接;混合接头;组合联结
composite laminate 复合层压板
composite land development 综合土地开发
composite landscape 复成景观;复合景观
composite lattice 复合点阵
composite lava flow 复合熔岩流
composite layering 图案重叠控制
composite leg 复合引线
composite lens 复合透镜;复合镜头
composite life 综合年限;平均年限
composite life method 平均年限法
composite line 混成线路

composite lining 复合衬砌;组合衬砌
composite load 混合荷载
composite load bearing structure 复合受载结构
composite longitudinal ventilation 组合式纵向通风
composite macadam 复合碎石路
composite magma 复合岩浆
composite map 合成矿图;综合图
composite Markov process 复合马尔可夫过程
composite masonry wall 组合圬工墙
composite material 混合材;合成材料;组合材料;成层(合成)材料;复合材料
composite material container 混合材料集装箱
composite material film 复合材料薄膜
composite material structure 复合材料结构
composite material vessel 复合材料容器
composite member 组合件;组合构架;组合杆(件)
composite membrane 复合膜
composite merchandise index number 综合商品指数
composite metal 双金属;复合金属(材料)
composite metal plastic pipe 钢塑复合管道
composite-metal rod 双金属棒材
composite model 组合模型;复合模型
composite mortar 混合砂浆
composite mo(u)ld 多腔模;复合模具
composite mo(u)ld casting 复合型铸造
composite mo(u)lding 复合模塑;复合成型
composite move 复合动作
composite nailed beam 钉合实腹梁
composite nanofiltration membrane 复合纳滤膜
composite noise exposure index 总噪声暴露指数
composite nozzle 复合水口砖;复合喷嘴
composite number 合(成)数
composite of operand 操作数组合
composite ooid 复鲕
composite operator 复合运算符
composite optic(al) communication conductor 光通信复合导线
composite order 混合柱型;混合柱式;混成柱型;组合柱式
composite ore deposit of multiple genesis 多成因复成矿床
composite packaging 复合包装
composite panel 复合板
composite paper 复合纸
composite partition system 混合式隔墙做法;混合式隔断做法
composite pendulum 复摆
composite performance chart 复合性能图
composite phase refractory 多相耐火材料
composite phasor 合成相量
composite phosphor 复合磷光体
composite photograph 合成像片
composite picture 合成图像
composite picture signal 组合图像信号;复合彩色信号
composite pigment 复合颜料
composite pile 混成桩;组合桩;复合桩
composite piled foundation 复合桩基
composite pile in separate lengths 分段混合桩
composite pipe 复合管(道)
composite pipe line 综合管道
composite piston 合成材料活塞
composite plane 综合平面图;复合面
composite plane wall (多层材料的)复合墙
composite plan of pipelines 管道综合平面图
composite plasterboard 组合灰胶纸柏板
composite plastics 复合塑料
composite plate 多层电镀;组合板;复合电镀;全要素板
composite plywood 复合胶合板
composite pollution 复合污染
composite polymeric aluminum chloride 复合高分子氯化铝
composite polymeric flocculant 复合高分子絮凝剂
composite polysilicic acid flocculant 复合聚硅酸絮凝剂
composite Portland cement 复合硅酸盐水泥
composite position 全要素定位
composite post 混成柱
composite powder 复合粉(末)
composite power cable with optic(al) fibre 电力复合光缆

composite prestressed concrete beam 组合式预应力钢筋混凝土梁
composite prestressed concrete girder 组合式预应力钢筋混凝土大梁
composite price 综合价格
composite print 全色印样
composite prism 复合棱镜
composite probability 合成概率;复合概率
composite product 复合(材料)制品
composite profile 综合剖面;复式剖面
composite prognostic chart 综合形势预报图
composite projection 合成投影;组合投影
composite proof 全要素样图
composite propellant 合成推进剂;组合燃料
composite propeller 合成推进器
composite pulse 复合脉冲
composite quantity 合数
composite quartz 复合石英
composite rack 综合架
composite random terms 复合随机项
composite rate 综合率
composite rate of depreciation 混合折旧率
composite ratio 综合比率
composite reaction 复合反应
composite reed board 组合芦苇板
composite refractory 混合耐火材料;复合耐火材料
composite regulation 混合式调节
composite reinforced concrete cut-off wall 钢筋混凝土混合截水墙;钢筋混凝土混合防渗墙;钢筋混凝土混合齿墙
composite reinforced concrete structure 钢筋混凝土复合结构
composite resin filling material 复合树脂充填材料
composite resistance 组合阻力
composite resistor 复合电阻器
composite response spectrum 合成反应谱
composite result figure 综合成果图
composite rigid frame 组合钢架
composite ringer 双信号振铃
composite ripple marks 复合波痕
composite river 合成河;复源河;复性河;复式河道
composite rock 复合岩石;复合岩
composite rock fill and earth dam 混合土石坝
composite rockfill dam 复合堆石坝;组合式土石坝;组合堆石坝
composite rockfill earth dam 混合式土石坝
composite rocks 复合岩类
composite rod laser 复合棒激光器
composite roof(ing) 组合屋面;复合屋面;钢木合屋盖
composite roof(ing) slab 复合屋面板
composite roof truss 组合屋顶桁架;组合(屋)桁架
composite rope 复合缆
composite roughness 混合糙率;合成粗糙度
composite ruby laser 复合红宝石激光器
composites 综合性产品
composite sailing 混合航行法;混合大圆航行;混合大圆航法
composite salvage process 混合打捞法
composite sample 混合样品;混合(本);混合试样;组合试样;综合样本;复合试样;并合样品;并合试样
composite sampler 混合物取样器;混合采样器;组合取样器;复合(式)取(土)样器
composite sampling 混合取样;混合采样;组合采样;并合取样
composite sampling method of duplicate sample 付样组合法
composite sampling scheme 复合抽样方案
composite sampling technique 混合取样法
composite second order 复合二阶项
composite section 复合断面;集剖面;组合截面;复式横断面
composite section blanking die 组合式落料模
composite sediment 复合沉积
composite semi-submersible platform 组合型半潜式平台
composite sequence 复层序
composite set 组合设备;电报电话双用装置;收发两用机
composite shape 复合形式;组合形状
composite share 复合犁铧
composite sheet 复合板片
composite shield 合成屏蔽

composite ship 组合船队
composite signal(l)ing 合成信令
composite sill 复合岩体;复合岩床
composite simplex algorithm 复合单纯形法
composite sintered compact 层状烧结坯
composite slab 复合板;组合板
composite sleeper 合成轨枕
composite sliding surface 组合滑动面;复合滑动面
composite slip 混合滑动
composite slope 复式坡;复式边坡
composite soft material 复合软材料
composite soil 混合土(壤);混合基土
composite solid structure 复合坚实结构
composite solution method 混合溶液法
composite spectra 复合光谱
composite space truss 组合网架
composite spectrum 复合谱;复合光谱
composite spreader 组合式吊具
composite state 复合状态
composite statement 复合命题
composite statistical hypothesis 复合统计假设
composite steel 多层钢;复合钢;复层钢;双层钢
composite steel beam bridge 组合钢梁桥
composite steel box girder bridge with reinforced concrete slab 钢筋混凝土板面组合钢箱梁桥
composite steel bridge 复合钢桥
composite steel concrete bridge 钢混凝土组合桥梁
composite steel concrete column 劲性钢筋混凝土柱
composite steel concrete cut-off wall 钢筋混凝土混合截水墙;钢筋混凝土混合防渗墙;钢筋混凝土混合齿墙
composite steel-rubber spring 钢—胶复合弹簧
composite stranded wire 复合多股绞合线
composite-stratotype 复合层型
composite structural plane 复性结构面
composite structure 混合构造;组合结构;复合结构
composite style 组合形式
composite submarine cable 复合海底电缆
composite submerged hollow fibre membrane bioreactor 淹没式复合中空纤维膜生物反应器
composite submerged membrane bioreactor 淹没式复合膜生物反应器
composite supply 复合供给
composite supporting structure 混合支承结构
composite surface 合成曲面
composite surface of sliding 复式滑动面
composite suspension bridge 组合式悬索桥
composite symbol 复合符号
composite synchronizing signal 复合同步信号
composite synthetic(al) fibre 复合合成纤维
composite system 金银本位并用制;复合本位制
composite tape layer 复合带层
composite target 综合性指标
composite tariff system 复合税率制
composite team yard 综合性货场【铁】
composite technology 复合材料工艺
composite terminal 多种运输联合转运基地
composite terrane 复合地体
composite testing 复合试验
composite tie 组合轨枕
composite timber concrete bridge 木混凝土结合桥(梁);木混凝土合成桥(梁)
composite tolerance 复合公差
composite tooth form 混成齿形
composite topography 复循环地形
composite tower 复合塔;复合标
composite trace 合成地震记录线
composite track 混合航道;混合大圆航行;混合大圆航法
composite transport bill of lading 联运提单
composite trim 综合整理;综合修饰
composite triple beat 复合三阶差拍
composite trophic index 复合营养指数
composite truss 混成桁架;组合桁架;钢木屋架
composite truss bridge 组合式桁架桥
composite truss frame 组合桁架;混合构架
composite tubing 组合管
composite tubular mast 多节套管天线杆
composite turnover 综合周转率;综合周转额;复合周转数
composite type 复合型结晶纤维
composite type concrete pavement 复合式水泥

混凝土路面
composite type pavement structure 复合型路面结构
composite type rock fill dam 混合式堆石坝
composite underframe 复合结构的底架
composite unit 组合单元
composite unit graph 复合单位过程线
composite unit hydrograph 复合式单位过程线;混合单位过程线;复合单位过程线
composite variable 综合变量;复合变量
composite vector 合成向量;合成矢量
composite vein 复合(岩)脉
composite vessel 混合结构船;综合货船;铁木(结构)船
composite video signal 综合视频信号;复合视频符号
composite volcanic cone 复合火山锥
composite volcano 成层火山(岩)
composite wall 混合墙壁;组合墙;复合墙(体)
composite walling 组合墙
composite wall panel 复合墙板
composite wall reinforcing [parallel type] 组合墙体的加固(配平行钢筋)
composite waste 混合废物;混合废水
composite waste sampling 混合废物采样
composite wastewater overflow 混合废水溢流
composite wastewater sample 混合废水样品;混合废水样本;混合废水水样
composite water quality vulnerability zone map 混合水质脆弱区图
composite wave 合成波;复合波;波组;复式波
composite wave filter 复式滤波器
composite weld 密实焊缝;加强填密焊缝;紧密焊缝;组合焊缝
composite well 复合井
composite wheel 复合砂轮
composite window 组合窗
composite wire 双金属丝
composite wireline drill rod 复合式绳索取芯钻杆
composite wood 合成木材;复合木材;复合材
composite wood concrete bridge 木混凝土结合桥(梁);木混凝土合成桥(梁)
composite wood concrete deck 木混凝土组合桥面
composite working 收发混合运行
composite yard 综合性货场
composite zoning 综合区划
composition 介质名称和成分;合成;组合物;组成(物);构图;构成;复合(组成);配方;排版
composition action 复合作用
compositional maturity 组分成熟度;成分成熟度
compositional petrology 成分岩石学
compositional polarity 成分极性
compositional refinement 配方改进
composition analysis 成分分析
composition analysis instrument 成分分析仪器
composition and content of gas 瓦斯成分及含量
composition and structure of parent rocks 母岩的成分和构造
composition and structure of rock and soil 岩土成分与结构
composition backing 焊接垫板;焊剂垫材
composition beam 混合梁
composition bearing 组合轴承
composition bin 混合料仓
composition block floor 组合地板砌块
composition block layer 组合地板铺装工
composition board 组合板;纤维胶合板;复合镶板
composition brake block 复式闸瓦
composition brass 杂铜
composition bronze 复合青铜
composition caps 组合雷管
composition classify of sinter 泉华的分分类
composition classify of spring 温泉的化学成分分类
composition cloth 防水帆布
composition constant 组分恒定
composition containing guanidine compound 含胍化合物组分
composition core 组合芯件;合成芯件
composition cork 胶合软木
composition cost 建筑造价
composition crystal 孪晶;复晶
composition deed 债务和解协议
composition depreciation 组合折旧

composition details 建筑细节;建筑细部
composition diagram 组成图;合金平衡图;合金金相图
composition effect 复合效应
composition error 综合误差
composition face 接合面
composition factor 合成因子
composition fault 复合断层
composition fault scarp 复合断层崖
composition fibre 复合纤维
composition floor 组合地板
composition floor cover(ing) 组合地板;无缝地板
composition flooring finish 组合楼板装饰
composition floor layer 组合地板铺装工
composition focal mechanism 复合震源机制
composition glacier 复合冰川
composition gradient 合成梯度;组分梯度
composition hammer 合成树脂锤
composition history curve 组成变化曲线
composition in architecture 建筑构图;建筑成分
composition joint 铆焊并用接合;复合联结
composition map 编绘原图
composition material 组合材料;复合材料
composition metal 合金;高铜黄铜;复合金属
composition nail 钉板瓦的铜钉(英国)
composition nail of copper-zinc alloy 铜锌合金屋面板钉
composition of accelerations 加速度的合成
composition of acidizing fluid 酸液配方
composition of air 空气组成
composition of assets 资产构成
composition of atmosphere 大气组成
composition of batch 配料的组合
composition of cap core 帽岩成分
composition of coal petrology 煤岩成分
composition of concurrent forces 汇交力系的合成
composition of consumer demand 消费构成
composition of consumer goods 消费品需要结构
composition of couples 力偶合成
composition of deposits 沉积物组成;沉积物成分
composition of diapir core 底辟核成分
composition of digester gas 消化池气(体)成分
composition of displacement 位移合成
composition of dust 微尘成分
composition of filled vein 充填脉成分
composition of filling material of discontinuity 结构面充填物成分
composition of fixed asset 固定资产构成
composition of force in space 空间力系的合成
composition of forces 力系合成;力合成;力的合成
composition of forces in plane 平面力系的合成
composition of fumes 炉气成分
composition of fuzzy relations 模糊关系的合成
composition of gaseous phase 气体成分
composition of geothermal water 地下热水化学组分
composition of glass 玻璃成分
composition of hydrocarbon 烃类组成
composition of inventory 存货的构成
composition of investment 投资结构;投资构成
composition of leaching liquid from refuse 垃圾淋出液成分
composition of liquid phase 液相成分
composition of livestock herds 畜群结构
composition of mappings 映射的合成
composition of motions of a rigid body 刚体运动的合成
composition of ore 矿石组成
composition of parallel forces 平行力系的合成
composition of population 人口组成;人口构成
composition of probabilities 概率合成
composition of production technologies 生产技术构成
composition of proportions 比例构成
composition of refuse 垃圾组成
composition of regional population 地域人口构成
composition of relations 关系的合成
composition of resources available for commitment 可供承诺资金的组成
composition of river water 河水组分;河水组成
composition of rocks 岩石成分
composition of sample 样本构成
composition of sewage 污水成分;污水组分;污水组成

composition of shaft 井筒结构
composition of slope bottom 斜井平车道结构
composition of soil 土壤成分
composition of solid phase 固相成分
composition of solution 溶液组分;溶液组成
composition of source diapir 源层成分
composition of tectonic system 构造体系组成
composition of the ground 基let组成
composition of the railway system 铁路网结构
composition of trade 贸易组成
composition of vectors 向量(的)合成;矢量加(法);矢量合成
composition of velocities 速度(的)合成
composition of velocity law 速度合成定律
composition of wastewater 废水组成
composition of wastewater of blast furnace 高炉废水的组成
composition of well stream 充气石油成分
composition on the plane 平面上组合
composition on the plane-section 平剖面组合
composition on the section 剖面上组合
composition panel 组合板
composition parameter of drawbridge 吊桥结构参数
composition plane 复合(平)面;结合面;接合面
composition population 成分总体
composition profile 组成分布;成分分布
composition resistor 组合电阻器
composition roller 明胶墨辊
composition roof(ing) 组合屋面
composition roofing manufacture 组合屋面的制造;组合屋面的生产
composition roofing material 复合屋面材料
composition roofing shingle 组合屋面木片瓦
composition rule 合成规则
composition series 合成列;组成列
composition sheet 编绘原图
composition shingle 组合材料瓦;沥青瓦;组合屋面板
composition siding 复合墙面板材
composition surface 接合面
composition triangle 组成三角形
composition vertical seismic profile 合成垂直地震剖面
composition wave 组成波
composition window 组合窗
compositive rate of basic depreciation 综合基本折旧率
compositive rate of major repair depreciation 综合大修折旧率
compositive relation 复合关系
compositive use research 综合利用研究
compositivity 复合度
compositron 高速显字管
compost 混合涂料;灰泥;堆肥
compostable refuse 可堆肥垃圾
compost-applier 施堆肥机
compost grinder 堆肥粉碎机
compost heap 堆肥草垛;堆肥槽垛
composting 制堆肥
composting material 堆肥材料
composting method 速成堆肥法
composting of solid wastes 固体废物堆肥
composting pile 肥堆
composition of offences 违法即决
compostion on architecture 建筑构图
compost privy 堆肥厕所
compost rasping machine 堆肥捣碎机
compost shredder 堆肥捣碎机
compost temperature 堆肥温度
compost toilet 粪便垃圾槽
compound 混合物;化合物;合成物;复合物;复合(的)
compound abrasive 组合磨料
compound abutment 组合桥台
compound acinous gland 复泡状腺
compound addressing 复合寻址
compound adjustment 多级调整;复合调整
compound air compressor 复合式空气压缩机
compound alloy 合金钢
compound alluvial fan 复合冲积扇(形)
compound alternator 复励交流发电机;复激交流发电机
compound amount 复(利)本利(之和);本利和

compound amount factor 复利和因数;复合总量因素;本利和因数
compound amount of one 每元的复利终值
compound amount of one dollar per period 每期一元的本利和
compound amount of one per period 每期每元复利本利和
compound amount of one yuan 一元复利本利和
compound amount of one 每元复利本利和
compound analysis 复合分析
compound angle 复合角
compound annual rate of growth 复合增长年率;复合年增长率
compound antenna mast 复接天线杆
compound arbitrage 重复套汇;复合套利;复合套汇交易
compound arbitrate 重复裁定
compound arbitrated rate 复算汇率;复合汇兑价率;复合(裁定)汇率;复合比例汇率
compound arch 同心异跨联拱;合成拱;组合拱;复式拱;复合拱
compound ascendant fungi 复配优势菌
compound astigmatism 复性散光
compound attachment 复式刀架附件
compound automatic air switch 复合自动空气开关
compound average 综合平均值;综合平均数;复合平均数
compound ball bearing 组合滚珠轴承
compound beam 混合梁;混成梁;组合梁;组合架;迭合梁;拼装梁
compound bearing 组合滚珠轴承
compound bending 双向弯曲
compound bending die 复式弯压模
compound binomial distribution 复合二项分布
compound bioflocculant 复合型生物絮凝剂
compound body 复合体;混合体;合成体
compound bolster 复式承梁
compound boundary 复合边界
compound box 进刀箱;传动箱
compound breaker 复式破碎机
compound brush 金属碳混合电刷;铜碳电刷
compound bushing 充电绝缘物套管
compound cable 混合多芯电缆;复合电缆;分段组合电缆
compound calculation amplifier 复合运算放大器
compound cam 组合凸轮
compound casting 双金属铸件
compound catenary 复接吊架
compound catenary construction 复式悬链线结构
compound catenary equipment 双链形悬挂【铁】;双接触线链形悬挂
compound catenary suspension line 复式链形悬挂的接触导线
compound catenary system 复式悬链线吊装法
compound center drill without jacket 不带护套复合中心钻
compound centrifugal force 复离心力
compound chain 组合链系
compound chain of dune 复合沙丘链
compound chain system 复合链幕
compound chemical fertilizer 复合化肥
compound chimney 组合烟囱
compound chlorine dioxide 复二氧化氯
compound chlorine dioxide generator 复合二氧化氯反应器
compound chromatic harmony 复色谐调
compound circuit 复合电路
compound cloth 多层织物
compound clothoid 双螺线
compound coagulant 复合混凝剂
compound coast 复式海岸
compound coil 复合线圈
compound colo(u)r 调和色;混合色;复色
compound column 集柱;混成柱;组合柱
compound compact 复合压坯;复合压块;复合坯块
compound compression 多级压缩;双级压缩
compound compressor 多级压缩机;多级压气机;多缸压缩机;复式压缩机
compound condition 复合条件
compound constant 复合约束
compound construction 混合结构;混合建筑
compound contactor 复式萃取器
compound control action 复合控制作用
compound converter 复激变流机

compound correction 混合改正
compound correction value 混合改正值
compound course 航迹向;直航向
compound crane 复式起重机;复式门吊
compound crossing 复式交叉
compound crossing-shoal 复式过渡段浅滩
compound cross-section 复合截面;复合断面;复式断面
compound crusher 复式破碎机
compound cryosar 复合低温雪崩开关
compound crystal 变晶
compound curvature 空间曲率
compound curve 综合曲线;复曲线;复合曲线
compound cycle 复合循环
compound cycle operation 复合循环作业方式
compound cylinder 复合柱体
compound deal frame 复式排架
compound decay curve 复合衰变曲线
compound decision problem 复合决策问题
compound department 混合车间
compound determiner 复定因
compound die 复合(冲)模
compound discount 复(利)贴现
compound dislocation 复合位错;复合错位
compound distribution 复合分布
compound dredge(r) 混合挖泥机;混合挖泥船;复式挖泥机;复式挖泥船
compound drift correction of gravimeter 重力仪混合零点改正值
compound drive 复式传动
compound duties 混合税;复合(关)税
compound dynamo 复励直流发电机;复激发电机
compound ecological system 复合生态系统
compounded abrasive 复合磨料
compounded additive 复合附加物
compounded latex 加填料胶乳;复合乳胶;填充胶乳
compounded lubricating oil 合成润滑油
compounded oil 混合机油;复合油
compounded pigment 复合颜料
compounded plate spring 组合片簧
compounded resin 混合树脂
compounded rubber 复合橡胶;填料混炼胶
compound elastic scattering 复合核弹性散射
compound engine 复(合)式发动机
compound entry 复合分录;复合分类
compound epicyclic(al) reduction gear 两级行星减速器
compound equation of payment 分期摊还
compound erosion 复合型侵蚀
compound essence 合成香料
compound estuary 多汊河口;复杂河口
compound event 复合事件
compound excitation 复(式)励(磁);复激(励)
compound expansion 复膨胀;二级膨胀
compound expenses 复合费用
compound exponential distribution 复合指数分布
compound expression 复合表达式
compound extractor 复式萃取器
compound eyepiece 复式目镜;复合目镜
compound Fabry-Perot interferometer 复式法布里-珀罗干涉仪
compound factors 复本利系数
compound fault 复合断裂;复合断层
compound feed 配合饲料
compound feed-screw 复式刀架丝杠
compound feed-screw graduated collar 复刀架丝杠刻度圈
compound fertilizer 复合肥料
compound-filled 绝缘膏填充
compound filling resin 复合充填树脂
compound flexure 复合挠曲
compound flocculant 复合絮凝剂
compound floor 组合楼板
compound flooring 多层地板;复合地板
compound fold 复合褶皱
compound force 分力
compound formation 化合物形成
compound-formation chromatography 反应色谱法
compound fraction 繁分数
compound frame 组合肋骨
compoundfree beta alloy 不含化合物的贝塔合金
compound frequency distribution 复合频率分布
compound fruit 复合果

compound function 合成函数;复合函数
compound function city 多功能城市;综合性城市
compound function method 复合函数法
compound gap distribution 复合空档分布
compound gasket 组合垫片
compound ga(u)ge 真空压力表;真空压力(两用)计;复合规
compound Gauss objective 复合高斯物镜
compound gear 复(合)齿轮
compound gearing 复合传动装置
compound gear train 复式轮系
compound generator 复励发电机
compound girder 型钢大梁;合板梁;混成大梁;合成梁;组合(大)梁;叠合大梁
compound glacier 复合冰川
compound glass 多组分玻璃;多层玻璃;复合玻璃
compound glass optic(al) fiber 多组分玻璃光学纤维;复合玻璃光纤
compound graben 复合地堑
compound graphic(al) log 组合钻探剖面图
compound green compact 复合压坯
compound growth rate 复合增长率
compound harmonic function 复谐函数
compound harmonic motion 合成谐和运动
compound harness 复合吊综
compound head 复头状花序
compound heater 复合式炉
compound horn 复合喇叭
compound hydrograph 复合过程线
compound hypergeomertric distrbution 复合超几何分布
compound hypothesis 复合假设
compound hyrograph 复合水文曲线;复合水文过程线
compound IF statement 复合条件语句
compound impact crusher 复合型反击破碎机
compound impulse turbine 复式冲力涡轮机
compound indexing 复式分度(法)
compound indicator 复式指示计;复合指示剂
compound industry 合成工业;复合产业
compounding 合成;一次付款;按复利计算;配料;配方
compounding a debt 一次付清债务
compounding area of various tectonic systems 不同构造体系复合部位
compounding factor 复利系数
compounding factor for per annum 年金复利系数
compounding growth sequence 复合型生长层序
compounding ingredient 配合剂
compounding in parallel 并联混合
compounding in series 串联混合
compounding of structural system 构造体系复合
compounding of tectonic systems 构造体系的复合
compounding operation 配合工序
compounding period 复利计息周期
compounding technique 配合技术
compounding type of tectonic systems 构造体系复合类型
compounding variation 配方差异
compound injury 复合性损伤
compound inorganic polymeric flocculant 复合无机高分子絮凝剂
compound instruction 复合指令
compound interest 复利
compound interest amortization method 复利摊销法
compound interest annuity 复利年金
compound interest curve 复利曲线
compound interest estimate 复利估计值
compound interest factor 复利因子
compound interest law 复利法则
compound interest method 年金折旧法
compound interest method of amortization 复利摊销法
compound interest method of depreciation 折旧复利法;复利折旧法
compound interest table 复利表
compound interferometer 复式干涉仪
compound interferometer system 复合干涉仪系统
compound intersection 复合交叉(口);复式交叉(路口)
compound ion 复合离子
compound journal entry 复合日记账分录
compound kinematic chain 组合运动链

compound kinoform 复合开诺全息照片
compound ladders 复式梯线
compound landslide 混合式滑坡
compound lattice 复点阵
compound layerage 重复压条法
compound layering 波状压条
compound leaf 复叶
compound lens 复(式)透镜;复合透镜;复合镜头
compound lever 复(式)杠(杆)
compound light 复(合)光
compound lining 复合衬砌
compound link 复式链节
compound load 合成荷载;合成负荷
compound loading 总荷载
compound lock 复式闸门;复式船闸
compound locomotive 复涨式(蒸汽)机车;复式机车
compound logic element 多逻辑元件;复合逻辑元件
compound longitudinal sand-ridge 复合纵向沙垄
compound loudspeaker 复式扬声器
compound machine 复电机
compound magnet 复合磁铁
compound manometer 复式压力计;复式测压管
compound marble 杂色大理石;有彩斑的大理石
compound mastaba(h) 混合式古埃及长方形平顶斜坡坟墓
compound matrice 复合矩阵
compound meander 复合曲流
compound mechanism 复式机构
compound metal 合金
compound meter 复式流量计;复式量水器;复式流量计
compound meter of water 复合水表
compound microbial inoculant 复合微生物菌剂
compound microscope 复(式)显微镜;复合显微镜
compound middle lamella 复中层
compound mill 多仓磨(机);复式磨(机)
compound milling machine 双滚筒缩绒机
compound mineralizer 复合矿化剂
compound mode 组合方式
compound modulation 多重调制
compound molecule 复合分子
compound motion 复摆运动
compound motor 复激电动机;复绕电动机;复励电动机
compound mo(u)ld 多用途模具
compound movement 复杂运动
compound name 复合名
compound negative multinomial distribution 复合负多项式分布
compound normal distribution 复合正态分布
compound notch weir 复式槽口量水堰
compound nucleus 复合核
compound number 复名数
compound of tectonic system 构造体系的复合
compound oil 复合油
compound operation cycle 复合作业循环
compound option 复合选择
compound ore-containing structures 复合容矿构造
compound organic matter 复合有机物
compound output circuit 复合输出电路
compound ovary 复合子房
compound pair isotope fractionation 化合物对同位素分馏
compound particle 复合粒子
compound partition(wall) 复合隔墙;组合墙
compound pellet 复球粒
compound pendulum 复摆
compound perimeter 复合视野计
compound pier 集束桥墩;复合墩;簇柱;集墩;簇墩
compound pile 混成桩;组合桩
compound pillar 组合墩;组合(支)柱
compound pipe (有分叉的)复式管道;复合管(道)
compound pipeline 复合管线
compound pipes in parallel 并联复合管
compound pipes in series 串联复合管
compound planetary gear set 复式行星齿轮组
compound plate evapo(u)rator 复合板蒸发器
compound Poisson distribution 复合泊松分布
compound Poisson process 复合泊松过程
compound Portland cement 复合波特兰水泥

compound post 混成柱
compound presentation 复合先露
compound press tool 复式冲模;复合冲模
compound pressure and vacuum ga(u)ge 压力真空表;复合式真空压力表
compound pressure ga(u)ge 复式压力计;复合压力计
compound pressure-vacuum ga(u)ge 复式压力真空计
compound prism 混合棱柱
compound probability 合成概率;组合概率;复合概率
compound proportion 复比例
compound pulley 组合滑车;复(合)滑轮
compound pump 复式水泵;复式唧筒;复式泵
compound quadratic form 复合二次式
compound quantity 复合量
compound raceme 复总状花序
compound radical 复根
compound rate 复利率
compound rates of return 复(利)报酬率
compound ratio 复比
compound reaction turbine 复合反力式涡轮机
compound rectifying column 复式精馏塔;复合精馏塔
compound regeneration 复床再生
compound regenerative oven 复合再生炉
compound reinforcement 复式配筋;复式加筋;复筋
compound relay 复合继电器
compound resin 复合树脂
compound rest 复式刀架
compound rest attachment 复式刀架附件
compound rest bottom 复式滑动刀架的底部滑板
compound rest feed 复式刀架进给量
compound rest handle 复式刀架手柄
compound rest swivel 复式滑动刀架的转盘
compound rest top 复式滑动刀架的顶部滑板
compound revolving screen 复式回转筛
compound rhubarb powder 格雷戈里氏散
compound ripple 复式波痕
compound ripple mark 叠置波痕
compound river 汇流河;合流河;复源河
compound river channel 复式河槽
compound rod 组合杆
compound rubber 配合橡胶
compound sample 组合样品;混合样品;混合样(本)
compound sand cover 综合沙埋
compound scarp 复合(断)崖
compound screw 复式螺旋;复式螺钉
compound section 组合截面;复式截面
compound section weir 复式断面堰
compound self-excited diesel generator set 复式自激柴油发电机组
compound semi-annually 半年计算一次的复利;半年复利
compound semiconductor 化合物半导体
compound settlement 土地协议文书
compound shaft 复合井筒
compound shake 复合裂缝
compound shoal 复式浅滩
compound shoreline 合成海岸;复合滨线
compound shutter 复式快门
compound sieve plate 复筛板
compound silencer 复合式消声器
compound sintered compact 复合烧坯
compound size 复合大小
compound slide 复式滑动
compound slide rest 复式滑动刀架
compound slope 复合形坡
compound soap 复合皂
compound sound source 复声源
compound source 复声源
compound spring 复式弹簧
compound stage compression 复级压缩
compound staining 复合染色法
compound stanchion (钢的或木的)组合支柱;组合钢柱
compound starter 复合起重机
compound statement 复合语句;复合命题
compound statistical decision problem 复合统计决策问题
compound steam engine 复(合)式蒸汽机

compound steam hammer 双缸气锤
compound steam pump 复合式蒸汽泵
compound steam turbine 复式涡轮
compound steel 合金钢;复合钢;三层钢
compound steel beam structure 组合钢梁结构
compound stereomodel 复合立体模型
compound stream 合流河
compound stress 合成应力;综合应力;复合应力
compound structure 复合构造
compound substance 复合物质
compound sum 复利终值
compound surge tank 双室调压池;双室平衡罐
compound suspension 复链式悬挂装置
compound swinging type jaw crusher 复合摆动型颚式破碎机
compound swing jaw crusher 混合摆动颚式破碎机
compound swivel 复式转座
compound swivel arm 复式转臂
compound symbol 复合符号
compound system 复合体系
compound system of exploration line 复合勘探线
compound table 组合台;复合工作台
compound tackle 串联绞辘
compound tail 复合尾部
compound tariff 复合关税(税则)
compound tarifs 复合的塔立夫材积表
compound technique 复利技术
compound telescope 复合望远镜
compound temperature relay 复合热动继电器
compound tide 混合潮;合成潮;组合潮;复(合)潮
compound toggle lever stone crusher 双摇杆碎石机;复式肘节杆杆碎石机
compound tone 复合音
compound tool 复合刀具
compound tool rest 车床小刀架;复式刀架
compound train 复式轮系
compound trammel 多层转筒筛
compound transition curve 复合缓和曲线
compound trap 复合圈闭
compound trial balance 复合试算表
compound trommel 多段滚筒筛
compound truss 混成桁架;合成桁架;组合桁架
compound tube 复合管(道)
compound tube flowmeter 复合管气体流量计
compound tube mill 多仓管磨机
compound turbine 复式(反击)涡轮机;复式反击透平;复式反击汽轮机;复合透平机
compound turbine engine 复式涡轮发动机
compound turbo jet 双级压缩机涡轮喷气发动机
compound turboprop 双转子涡轮螺桨发动机
compound twin 李晶;复合李晶
compound twister 复捻机
compound-twist silk thread 复捻丝线
compound type apartment house 组合式住宅;复合式公寓
compound type floating crane 复合浮式起重机
compound type microcontroller 复合型微控制器
compound type wharf 复式码头
compound type wharf area 复式码头区
compound unit 组合单元
compound-vacuum ga(u)ge 复式压力真空计
compound valley glacier 复合谷冰川
compound value of an annuity 年金的复利值
compound valve 组合阀
compound vault 复式穹隆
compound vibrating feeder 复式振动加料器
compound vibration 复合振动
compound vibratory action pile driver 复振沉桩机
compound vortex 复合漩涡;复合式涡流
compound wall(ing) 组合墙;多层壁
compound water cyclone 多锥水力旋流器
compound water meter 复式水表
compound wave 复合波
compound web plate 组合式腹板;组合腹板
compound weir 复式断面堰
compound well 复管井;组合井(各高程直径不同)
compound winding 混合绕组;复(励)绕组;复激绕组;复(激)绕法
compound wire 复合导线
compound wound 复绕;复励;复激;复式励磁的
compound-wound current transformer 复绕式电流互感器
compound-wound generator 复激发电机;复励

发电机
compound-wound motor 复激电动机;复励电动机
compound-wound relay 复绕继电器
compound yield 复合收益
compound zero correction 混合零校正
comprador 康白度;买办
compreg 浸压木材;胶合木(材);层积塑料;本质层压材料
compregnate 浸压
compregnated wood 浸压木材;浸胶压缩木材;胶压木(材);胶合片料;胶合木材;木质层压材料;压缩木(材)
compreg web plate 胶合腹板
comprehend 理解
comprehension 理解;包括
comprehension tool 理解工具
comprehensive 综合的;广泛的;包括的;包含的;全面的
comprehensive agreement 全面协定
comprehensive analysis 综合分析;全面分析
comprehensive analysis method 综合分析法
comprehensive analysis of railway revenue 铁路收入综合分析
comprehensive appraisal 综合评价
comprehensive approach 综合方法;综合方案;综合处理;综合办法;通盘处理法
comprehensive area 综合区
comprehensive assessment 综合评价
comprehensive assessment method 综合评价(方)法
comprehensive atlas 综合地图集
comprehensive balance 综合平衡
comprehensive budget 综合预算;全面预算
comprehensive capacity of berth 泊位综合通过能力
comprehensive capacity of harbo(u)r 港口综合通过能力
comprehensive capacity of port 港口综合通过能力
comprehensive cartographic(al) standard 制图综合标准
comprehensive chart 详图
comprehensive colo(u)red map 综合彩色图
comprehensive communication planning 综合交通规则
comprehensive community plan 综合社区规划
comprehensive company 综合公司
comprehensive construction organization design 综合性施工组织设计
comprehensive curve of groundwater regime element 地下水动态要素综合曲线
comprehensive design 综合设计
comprehensive design of aids-to-navigation 航标综合设计
comprehensive development 综合开发
comprehensive development and utilization of coastal zone 海岸带综合开发与利用
comprehensive development area map 地区综合发展图
comprehensive development of economy 经济综合发展
comprehensive development of river-basin 河川流域综合开发
comprehensive development of water resources 水利资源综合开发
comprehensive development plan 综合建设规划;全面开发计划
comprehensive development program(me) 全面开发计划
comprehensive ecological risk 综合生态风险
comprehensive ecological risk index 综合生态风险指数
comprehensive economic equilibrium 综合经济平衡
comprehensive economic results 综合经济效益
comprehensive energy consumption 综合能耗
comprehensive engineering geology mapping 综合性工程地质测绘
Comprehensive Environmental Response, Compensation and Liability Act 综合环境影响,补偿和责任法令
comprehensive environmental science 综合环境学
comprehensive evaluation 综合评价(方)法;综合评价
comprehensive evaluation of environmental quality 环境质量综合指数
comprehensive evaluation of mineral exploration 矿床综合评价
comprehensive experiment 综合试验
comprehensive exploitation 综合开发
comprehensive exploitation plan 综合开发计划
comprehensive exploration of mine district 综合勘探
comprehensive external trade guarantee 境外贸易综合担保
comprehensive faunal province 综合性动物地理区
comprehensive general liability insurance 综合责任(总保)险
comprehensive geologic(al) logging 综合地质编录
comprehensive geologic(al) map of the survey area 测区综合地质图
comprehensive geophysical exploration method 综合物探法
comprehensive geophysical method 综合物探法
comprehensive geophysical survey 综合物探;综合地球物理勘探
comprehensive goods station 综合性货运站
comprehensive grasp of business 全面掌管业务
comprehensive harnessing 综合治理;综合利用;综合开发
comprehensive harnessing measure 综合整治措施
comprehensive health care 综合性保健
comprehensive hydrogeology maps 综合性水文地质图件
comprehensive improvement measure 综合治理措施;综合整治措施
comprehensive index of morphologic(al) complexity 形态复杂程度综合指标
comprehensive index of soil pollution 土壤污染综合指数
comprehensive index value 综合指标值
comprehensive indicator 综合指标
comprehensive industrial system 完整工业体系
comprehensive injecting water 综合压水
comprehensive injection mo(u)lding control system 综合注模控制系统
comprehensive institutional framework 综合体制
comprehensive insurance 综合保险
comprehensive integrated test 综合联调
comprehensive investigation 综合考察;综合勘察;综合调查;综合查勘
comprehensive investigation of river basin 流域综合勘察;流域综合勘查;流域综合查勘
comprehensive investigation ship 综合调查船
comprehensive lake ecosystem analyzer 综合湖泊生态系统分析器
comprehensive list 价目总表;明细表;一览表;综合目录
comprehensive lump-sum 合价包干
comprehensive management 综合经营;综合管理
comprehensive mandatory sanctions 全面的强制性制裁
comprehensive map 详图;综合(地)图
comprehensive marine exploration 海洋综合考察
comprehensive materials consumption quota 材料消耗综合定额
comprehensive measures 综合措施
comprehensive mechanism 综合机制
comprehensive mechanization 综合机械化;全盘机械化
comprehensive mechanized coal mining method 综合机械化采煤法
comprehensive method of protection 综合防护法
comprehensive model 整体模型;全局模型
comprehensive model for environmental impact assessment 环境影响综合评价模型;综合环境影响评价模型
comprehensive model test 整体模型试验
comprehensive national development plan 全国综合开发计划
comprehensive neotectonic map 新构造综合图
comprehensive package price 单价包干
comprehensive park 综合公园
comprehensive plan 整体计划;全面发展计划
comprehensive plan figure of mise-a-la-masse 充电法综合平面图
comprehensive plan figure of self-potential method 自然电场法综合平面图
comprehensive plan(ning) 总体规划;综合计划;综合规划;全面规划
comprehensive planning assistance 综合规划援助
comprehensive planning program(me) 全面规划方案
comprehensive policy 综合保险单
comprehensive pollution abatement 综合消除污染
comprehensive pollution index of water quality 水质综合污染指数
comprehensive prevention and control of water pollution 水污染综合防治
comprehensive prevention and cure 综合防治
comprehensive preventive health measure 综合防治措施
comprehensive profile of self-potential method 自然电场法综合平面图
comprehensive profiling figure of mise-a-la-masse 充电法综合剖面图
comprehensive profit rate 综合盈利率
comprehensive program(me) of action 综合行动方案
comprehensive programming 综合规划
comprehensive prospecting 综合找矿
comprehensive provisions 综合条款
comprehensive pumping system 综合抽气系统
comprehensive radio 全波无线电台
comprehensive reaction modulus of base 地基综合反应模量
comprehensive reclamation of river basin 流域综合治理
comprehensive redevelopment 综合性改建
comprehensive rehabilitation planning of urban environment 城市环境综合整治规划
comprehensive report 综合报告
comprehensive research 综合研究
comprehensive review 全面审查;全面复查
comprehensive risk 综合险
comprehensive river basin planning 综合流域规划
comprehensive sanction 全面制裁
comprehensive school 综合学校
comprehensive services 全面负责制;综合服务项目
comprehensive sewer separation 综合污水分流
comprehensive software package 综合软件包
comprehensive statistics 综合统计
comprehensive stormwater management 暴雨水综合管理
comprehensive study 综合研究
comprehensive summary 综合摘要
comprehensive sunshade 综合式遮阳
comprehensive survey 综合考察;综合勘察;综合调查
comprehensive survey of forest resources 林业资源普查
comprehensive survey of river basin 流域综合查勘
comprehensive survey of wide-area water pollution 大面积水污染综合调查
comprehensive survey vessel 综合调查船
comprehensive system 综合系统
comprehensive team yard 综合性货场【铁】
comprehensive test 综合试验;全面试验
comprehensive toxicity 综合毒性
comprehensive toxicity of wastewater 废水综合毒性
comprehensive transport study 总体运输研究
comprehensive treatment 综合治理;综合整治;综合处理
comprehensive treatment planning of urban environment 城市环境综合整治规划
comprehensive unit price 综合单价
comprehensive university 综合性大学
comprehensive urban development 城市综合开发
comprehensive use of timber 木材综合利用
comprehensive utilization 综合利用
comprehensive utilization benefit 综合利用效益
comprehensive utilization factor of solid wastes 固体废物综合利用率
comprehensive utilization of coal 煤炭综合利用
comprehensive utilization of sludge 污泥综合利用
comprehensive utilization of solid wastes 固体废物综合利用
comprehensive utilization of water resource 水资源综合利用
comprehensive utilization reward of solid wastes

固体废物综合利用奖
comprehensive wastewater management 污水综合管理
comprehensive wastewater management planning process 污水综合管理规划办法
comprehensive water pollution control 水污染综合控制;水污染综合防治
comprehensive water quality assessment 水质综合评价
comprehensive water quality management 水质综合管理
comprehensive water quality management planning 水质综合管理规划
comprehensive water transfer 综合调水
compreignacite 黄钾铀矿
compresol pile 灌注桩
compress 精简;压缩;敷布;受压;收缩器
compress data 压缩数据【计】
compressed 侧扁的
compressed air 压缩空气
compressed-air accumulator 压缩空气(器)储蓄器;压缩空气存储器
compressed-air aeration 压缩空气曝气
compressed-air agitation 压缩空气搅拌
compressed-air anchor winch 压缩空气起锚绞车
compressed-air assisted 压缩空气推动的;风动的
compressed-air bearing 压缩空气轴承;压气轴承
compressed-air blasting 压缩空气落煤
compressed air-boosted 压缩空气助推的
compressed-air bottle 压缩空气瓶
compressed-air brake 气闸;气闸;气凿;压缩空气制动器;压缩空气闸
compressed-air caisson 压缩沉箱;气压沉箱
compressed-air caisson foundation 气压沉箱基础
compressed-air chiseling hammer 气凿;风镐
compressed-air chuck 压缩空气卡盘;气压夹头;气动卡盘
compressed-air circuit breaker 压缩空气断路器
compressed-air cleaning 压缩空气吹洗;压缩空气除尘;气动净化;压缩空气清洗
compressed-air concrete placer 混凝土气压浇注机
compressed-air conduit conveyance 管道压气运输
compressed-air container 压缩空气容器
compressed-air control 压缩空气控制;压缩空气操纵器
compressed-air control system 压缩空气控制系统
compressed-air conveying system 压缩空气输送系统
compressed-air conveyer 气流输送机;压缩空气传送机
compressed-air cooler 压缩空气冷却器
compressed-air crane 气压起重机
compressed-air cylinder 压缩空气罐
compressed-air digger 风镐;风铲
compressed-air disease 减压病;沉箱病;气压沉箱病
compressed-air distributor 压缩空气分配器
compressed-air diver 压缩空气潜水
compressed-air diver procedure 压缩空气潜水方法
compressed-air diving profile 压缩空气潜水方案
compressed-air diving schedule 压缩空气潜水方案
compressed-air dressing machine 压气锻钎机
compressed-air drill 压缩气钻;风动凿岩机;气动凿岩机;风钻
compressed-air drilling 风动钻进;气动钻进
compressed-air drive 压缩空气驱动;压缩空气掘进;压气传动
compressed-air drive pump 压气驱动泵
compressed-air drying 压缩空气干燥
compressed-air duct 压气管道
compressed-air ejector 气压喷射器;压缩空气喷射器;压气喷射器
compressed-air elevator 压缩空气升降机;气动升降机
compressed-air engine 空气压缩机;压缩空气发动机;压缩空气断路器
compressed-air equipment 空气压缩机;压缩空气设备
compressed-air expander 压缩空气疏松器;压缩空气扩张器
compressed-air external vibrator 压缩空气外振动器
compressed-air feed motor 压缩空气进料电动机
compressed air for brake control 刹车器;操纵制动器压缩空气
compressed-air for drilling 钻孔用压缩空气
compressed-air foundation 气压沉箱基础;压气沉箱基础
compressed-air ga(u)ge 空气压力表
compressed-air gun 压缩空气枪
compressed-air hammer 气(动)锤;风铆机;风铆锤;压缩(空气)锤;风动锤
compressed-air hoist 压缩空气起重机;气动起重机;气动卷扬机;气动绞车
compressed-air horn 气笛
compressed-air hose 压缩空气软管
compressed-air illness 沉箱病;潜水员病;压缩空气病;压力病【救】;承压病
compressed-air impact hammer 风动冲击锤
compressed-air impact wrench 风动冲击扳钳
compressed-air in coal pits 煤矿矿井压缩空气
compressed-air inlet 压缩空气入口
compressed-air inlet pipe 压缩空气进入管
compressed-air inspirator 压缩空气吸入器
compressed-air installation 压缩空气装置;压缩空气设备
compressed-air internal vibrator 压缩空气插入式振捣器
compressed-air jack 风动千斤顶;气力千斤顶;气力起重器
compressed-air jet 压缩空气射流;压气喷射器
compressed-air junction manifold 压缩空气多向接头
compressed-air leg 气压支架
compressed-air lifting 压缩空气提升
compressed-air line 压缩空气管道
compressed-air lock 压缩空气闸;压缩空气塞
compressed-air machine 风动机(械)
compressed-air main 压缩空气干线
compressed-air mechanical tamper 气动夯
compressed-air method 压缩空气开凿法;压气法;气压法(隧道施工)
compressed-air method of tunnel 气压隧道开凿法
compressed-air method of tunnel(l)ing 隧道压缩空气开凿法;压缩空气开挖隧道法
compressed-air mixer 风动拌和机;气力搅拌机;压缩空气式拌和机;压缩空气搅拌机
compressed-air mixer-injector 压缩空气拌和机喷射器
compressed-air motor 压缩空气马达;风动马达;气动(电动)机
compressed-air nailing machine 压缩空气钉箱机
compressed-air operated valve 气动阀
compressed-air packing 压缩空气夯实
compressed-air paint apparatus 压缩空气油漆装置
compressed-air painting apparatus 压缩空气油漆机具;油漆喷涂器;压缩空气喷漆器械;压气喷漆装置
compressed-air painting gun 油漆喷枪;喷漆枪;喷漆器
compressed-air pick 压缩空气风镐;风镐
compressed-air pile chuck 压缩空气桩夹
compressed-air pile driver 压缩空气打桩机;风动打桩机
compressed-air pile hammer 压缩空气桩锤;压缩空气打桩机
compressed-air pipe 压缩空气管
compressed-air pipe line 压缩气管
compressed-air pipe(line) 压缩(空)气管(道)
compressed-air piping 压缩空气管系;压缩空气管道
compressed-air placer 气压浇灌机
compressed-air point 压缩空气凿尖
compressed-air power 压缩空气动力
compressed-air pressure 压缩空气压力
compressed-air pressure at face 工作面压气压力
compressed-air pulley block 压缩空气滑车
compressed-air pump 压缩空气泵
compressed-air pump drainage 压气泵排水
compressed-air pump station 空压机泵站
compressed-air quenching 压缩空气淬火
compressed-air ram 压缩空气夯锤;气动夯实;压缩空气冲压机
compressed-air rammer 压缩空气夯实机
compressed-air receiver 压缩空气罐;压缩空气储存器
compressed-air requirement at face 工作面需风量
compressed-air reservoir 风缸;压缩空气储[贮]存器;压缩空气储气罐;压缩空气储存罐
compressed-air riveting 气动铆接(法)
compressed-air riveting hammer 气动铆钉锤
compressed-air sampler 气动取样器
compressed-air scaling hammer 压缩空气敲锈锤
compressed-air shield 气压盾构
compressed-air shield driving 气动盾构掘进
compressed-air shot-blasting machine 喷丸机
compressed-air sickness 压缩空气病;减压病;压气病;压力病【救】;沉箱病
compressed-air sinking 气压下沉;气压打桩
compressed-air sliding door drive 压缩空气滑动门驱动(装置)
compressed-air soil rammer 气动夯土机
compressed-air spade 气动锄;风铲
compressed-air spraying 压缩空气喷涂
compressed-air starter 压缩空气启动机;压缩空气启动器
compressed-air station 压缩空气站
compressed-air suit 压缩空气潜水服
compressed-air supply 压缩空气供应;压缩空气供给;压气供应
compressed-air surge chamber 压气式调压室
compressed-air system 风动系统;压缩空气系统
compressed-air tamper 压气式夯具;压缩空气夯;压缩空气夯(具);风动夯(具);风动捣固机
compressed-air tank 压缩空气罐
compressed-air temperature 压缩空气温度
compressed-air thickness 压缩空气厚度
compressed-air tool 压缩空气工具
compressed-air transmission 压缩空气传动
compressed-air trowel 风动慢刀
compressed-air tube 压缩空气管(道)
compressed-air tunnel 气压隧道;压缩空气风道
compressed-air tunnel(l)ing 气压隧道施工
compressed-air type sprayer 压缩空气喷雾机
compressed-air unit 压缩空气机
compressed-air vessel 压缩空气容器
compressed-air vibrator 压缩空气振动器
compressed-air winch 压缩空气绞车;风动绞车
compressed-air wind tunnel 变密度风洞
compressed-air work 压缩空气工程
compressed amount 成型压缩量
compressed asbestos-cement panel 压制石棉水泥板
compressed asbestos-cement sheet 压制石棉水泥板
compressed asbestos fibre ointing with steel wire 进排气夹金属丝网橡胶石棉板
compressed asbestos high pressur metallic 夹钢丝高压石棉橡胶板
compressed asbestos sheet 压制石棉纸板;夹胶石棉板;加压石棉板;压紧的石棉胶板;石棉纸板
compressed asphalt 岩石(地)沥青路面;压制(地)沥青
compressed asphalt roaster 压制地沥青烘烤机
compressed asphalt tile 压制地沥青砖
compressed bar 受压钢筋;压杆;受压杆
compressed boom 受压的梁
compressed cement-slag brick 压制水泥矿渣砖
compressed chord (桁架中的)受压的弦杆
compressed code 压缩码
compressed column 受压的柱
compressed concrete 捣实混凝土;压制混凝土
compressed cork 压制软木;软木板
compressed corkboard 压制的软木板
compressed cork factory 压缩软木厂
compressed cork slab 压缩软木板
compressed data 压缩数据
compressed data storage system 压缩数据存储系统
compressed density function 压缩密度函数
compressed diagonal 受压斜杆
compressed edge 压缩边缘
compressed element 受压构件
compressed felt 压制毡
compressed fender 受压护舷
compressed fiberboard 压制纤维板
compressed flange (指梁中的)受压的翼缘
compressed flexible rail bond 压接式柔性导体接头
compressed format 压缩形式;压缩格式
compressed gas 压缩气体
compressed gas and liquid gas 压缩气体和液化气体

Compressed Gas Association 压缩气体制造业协会(美国)
compressed gas cut-out 压缩空气切断
compressed gas cylinder 压缩气瓶;压(缩)气筒
compressed-gas electrostatic generator 充气压型静电加速器
compressed gas forming machine 压缩气体成型机
compressed-gas generator 充气压型静电加速器
compressed gas installation 压缩煤气装置
compressed gas truck 煤气汽车
compressed hydrogen unit 压缩氢装瓶装置
compressed ignition 压缩点火
compressed-iron-core coil 铁粉芯线圈
compressed laminated wood 强化(层积)木材;压制多层板
compressed layered wood 分层压紧木材
compressed limit 压缩极限;压缩度;压缩界限
compressed magnetosphere 受压磁层
compressed material 压缩的材料
compressed member 受压构件
compressed memory 压缩储存
compressed mixture 密实混合料
compressed natural gas 压缩天然气
compressed natural rock asphalt 压制岩沥青
compressed pat kiln test 样饼试烧法
compressed pattern lumber 压缩制模木材
compressed peat 压缩的泥炭(土)
compressed pile 压制混凝土桩;压挤桩
compressed pipe 压缩管
compressed pits 压坑
compressed plate 受压板
compressed-pneumatic 气动压缩
compressed reinforcement 受压钢筋;抗压钢筋
compressed reinforcing 受压的加劲材料;受压的加筋材料
compressed ring 压环;汽铃;活塞平环
compressed rock asphalt 压制岩地沥青
compressed rock asphalt surfacing 压实天然沥青铺面
compressed rod 受压杆
compressed shear 压剪
compressed sponge 压缩海绵
compressed steel 压钢
compressed straw 压制禾杆
compressed straw fibre building slab 压制纤维板
compressed straw slab 马粪纸板;压制纸板;稻草板;草制纸板
compressed straw slab partition(wall) 压制稻草板隔墙
compressed tablet 压制片
compressed thickness 压缩厚度
compressed-time correlator 时间压缩相关器
compressed vegetable fiber 压制植物纤维
compressed water 受压水
compressed water pump 水泵
compressed width 受压宽度
compressed wood 压实木材;胶压木(材);胶合木材;压制木材;压缩木(材)
compressed zone 受压区;承压区;挤压构造带;承压带
compressed zone of wave 波密区
compressibility 敛缩性;可压缩性;压缩性;压缩率
compressibility burble 激波致扰气流区;激波后紊流;压缩性泡流
compressibility coefficient 压缩率;压缩(性)系数
compressibility degree 可压度
compressibility drag 压缩性阻力
compressibility error 压缩性误差
compressibility factor 压缩率;压缩因子;压缩因素;压缩因数;压缩系数
compressibility index 压缩指数
compressibility influence 压缩效应
compressibility modulus 压缩性模数;压缩模量
compressibility of formation water 地层水的压缩系数
compressibility of gas in place 地层天然气的压缩系数
compressibility of liquid 液体的压缩性
compressibility of petroleum in place 地层原油的压缩系数
compressibility of soil 土体可压缩性;土壤压缩性;土的压缩性
compressibility of wet suit 湿式潜水服压缩性
compressibility parameter 可压缩性参数

compressibility rank of soil 土的压缩性分级
compressibility ratio 压缩比(率);压短比
compressible 可压榨的;可压缩
compressible aerodynamics 可压缩空气动力学
compressible boundary layer 可压缩流体附面层;可压缩流体边界层
compressible cake 压缩性滤渣
compressible cascade flow 可压缩叶栅气流
compressible clay layer 压缩黏土层
compressible coefficient 可压缩系数
compressible fender 可压缩护舷
compressible flow 可压缩的流动;可压缩流;压缩流
compressible flow convergence 可压缩流会聚
compressible flow converging 可压缩流收缩
compressible flow divergence 压缩流扩大
compressible flow diverging 可压缩流扩张
compressible flow energy 压缩流量能量
compressible flow energy equation 压缩流量能量方程
compressible-flow principle 可压缩流原则
compressible fluid 可压缩流体
compressible fluid flow 可压缩流体流;可压缩的液流
compressible formation 可压缩土层;压缩性土层
compressible ground 可压缩土层;压缩性土层
compressible jet 可压缩(流体)射流
compressible liquid 可压缩液体
compressible material 可压缩的材料;压缩性材料
compressible medium 可压缩介质
compressible packing material 可压缩垫密材料;压缩性垫密材料
compressible soil 可压缩土(壤);压缩性土(壤)
compressible strata 压缩性土层
compressible stratum 可压缩层
compressible support 让压支架
compressible turbulence 可压缩性紊流
compressible volume 可压缩涡流
compressing 压紧
compressing apparatus 压缩机
compressing by pressure fusion 加压熔解
compressing diagonal 受压斜支柱
compressing dry granulation 干法压片
compressing force 压缩力
compressing pump 压送泵;增压泵
compressing strain 压缩应变
compressing strength 抗压强度
compression 紧缩;加压;密集;承压支承筒;扁度;受压气密性;压力
compressional area 受压面积
compressional bar 受压钢筋
compressional boundary 挤压边界
compressional chord 受压弦杆
compressional deep fracture 压性深断裂
compressional diagonal 受压斜杆
compressional dissected plane 纵向切割面
compressional face 受压面
compressional fault 挤压断层
compressional fissure 挤压裂缝
compressional flange 受压翼缘
compressional folding 挤压褶皱(作用)
compression algorithm 压缩算法
compressional hydromagnetic wave 磁流体压缩波
compressional layer 受压层
compressionally dilatational wave 胀缩波;疏密波;膨胀压缩波
compressional megasutures 挤压巨型接合带
compressional member 受压构件
compressional movement 挤压运动
compressional potential of wave 波压缩势
compressional stress 压缩应力
compressional vibration 压缩振动;纵振动
compressional viscosity 压缩黏度
compressional wave 压缩波
compressional zone 压缩区;受压区
compression amplifier 压缩放大器
compression-annealed pyrolytic graphite 压缩退火热解石墨
compression arch 受压拱
compression area 受压面积
compression arthralgia 加压关节痛
compression-bagging machine 压缩装袋机
compression bar 受压杆件;抗压钢筋;压杆;受弯钢筋;受压钢筋
compression bearing joint 承压缝;承接头

compression bed depth of foundation 地基压缩层深度
compression bib cock faucet 受压关闭咀水龙头
compression blasting 挤压爆破;压缩爆破
compression blow-by 从活塞上部空间漏气
compression boom (起重机的)受压桅杆;受压弦杆;受压起重臂
compression bulk density 压实松密度;压实容重
compression casting 压铸
compression chamber 压气室;加压室;加压舱;压缩室;气压匣;气压室
compression chord 承压弦杆;受压弦杆
compression cock 压缩旋塞
compression coefficient 压缩系数
compression coefficient of mud 泥浆压缩系数
compression coefficient of soil 土的压缩系数
compression column 承压柱
compression component 压缩构件
compression concrete 受压混凝土
compression conductor clamp 压接线夹
compression consolidation test 压缩(固结)试验
compression consolidation test of soil 土的压缩固结试验
compression contortion fault 压扭性断层
compression cord 折叠条纹
compression coupling 压缩联轴节;压紧联轴节;压紧管箍
compression cross-section 受压断面
compression cube 受压立方体(试块)
compression cup 压缩加油杯;润滑脂杯
compression curve 压缩曲线
compression cycle 压缩循环
compression cylinder 受压圆柱体(试块)
compression damage 受压损坏
compression deflection 压缩翘曲
compression deformation 压缩变形
compression diagonal 承压斜杆;受压斜杆;受压斜支柱
compression diagram 压缩图
compression dynamometer 压力测力计
compression edge 受压边缘
compression efficiency 压缩系数;压缩效率
compression engine 压缩机
compression equipment 压缩装置
compression exercise 加压锻炼
compression face 承压面;受压面
compression factor 压缩系数
compression failure 压缩破坏;压缩疲劳;压缩断裂;压力破坏;压毁;压环斜裂;压坏;受压破坏
compression faucet 压力龙头
compression fault 挤压断层;压性断层
compression feeder 压挤送料机
compression fibre 受压纤维
compression filter 压力过滤器
compression fire hydrant 压杆式消防栓;压紧型消防栓
compression fissure 挤压裂缝;压缩裂隙
compression fitting joint 压缩管件接头
compression flange 承压翼缘;补强凸缘;受压翼缘
compression fold 挤压褶皱构造
compression force 压力
compression forging 锻压
compression for laser pulse 光脉冲压缩
compression fracture 压裂
compression gasket 压缩密封垫
compression gasket joint 加压密封接合;压缩密封接头
compression gas refrigerator 压缩式燃气制冷机
compression ga(u)ge 压力表;汽缸压力表;压缩压力计;压缩压力表
compression glaze 压应力釉
compression glazing 挤压法装配玻璃
compression grade 加算坡度
compression grease cup 压缩加料滑脂杯;压力注油器
compression heat 压缩热
compression ignition 压缩点火
compression ignition engine 压燃式发动机;柴油机
compression index 压缩指数
compression index of soil 土的压缩指数
compression indicator 压缩指示器
compression inner ring 活塞环扩张器
compression joint 挤压节理;压力接合;压接(接头);压挤节理;压合连接;承压缝

compression knock 压缩爆燃
compression layer 压缩层
compression leak 漏气
compression leg 压柱
compression limit 压缩限度;压缩极限
compression link 受压连杆
compression load 抗压荷载;压缩荷载;压力荷载
compression manometer 压缩式压力计
compression material 压缩材料
compression member 受压构件;压杆;受压杆件
compression metamorphism 挤压变质作用
compression method 压缩法
compression modulus of aquifer frame 含水层骨架压缩模量
compression modulus of elasticity 变形模量
compression modulus of liquid volume 液体体积压缩模量
compression mo(u)ld 压型模;压铸模;压模
compression mo(u)lding 模压成型;压缩模塑;压塑;压模(成型);压力成型
compression mo(u)lding press 挤压成型机;压力成型机
compression mo(u)lding pressure 压缩模塑力
compressionn 压缩
compression nut 压紧螺母;压紧螺帽
compression of bandwidth 带宽压缩
compression of the earth 地球椭率;地球扁率
compression pad 压缩式防震垫
compression parallel to grain 平行于纹理的压缩;顺纹压力;顺木纹压缩;顺木纹压力;顺木纹受压
compression parallel to grain of timber member 木料顺纹压力
compression parallel to the grain 顺纹压缩
compression perpendicular to grain 横(木)纹压缩;横(木)纹压力;横(木)纹受压;垂直于文理的压缩;垂直木纹压力
compression perpendicular to grain of timber member 木料垂直于木纹压力
compression pile 受压桩
compression piston ring 活塞压环
compression plant 压缩机装置;压缩车间
compression plate 压力试验机座板;压板
compression plating 加压钢板固定
compression point 压缩点;受压点
compression pressure 压缩压力
compression process 压缩法
compression pump 压缩泵;压气泵;压力泵;压缩机
compression raker 受压斜桩
compression rating 增高压缩比
compression ratio 压缩率;压缩程度;压缩比(率);压力比
compression recovery rate 压缩复原率
compression refrigerating machine 压缩(式)制冷机
compression refrigeration 压缩(式)制冷
compression refrigeration cycle 压缩式制冷循环
compression refrigeration system 压缩制冷系统
compression refrigerator 压缩式制冷机
compression region 压缩层;受压区
compression regulator 压实度调节器
compression reinforcement 抗压钢筋;压缩加强件;受压钢筋
compression release 压缩漏气
compression release cock 减压旋塞
compression release lever 卸压杆
compression release shaft 卸压杆
compression release valve 泄压阀
compression relief cam 调压凸轮
compression relief cock 泄压开关;减压开关
compression resistance 抗压能力
compression resistance of shock absorber 减震器压缩阻力
compression resistance starter 加压变阻起动器
compression rib 受压肋
compression ring 受压环;活塞平环;压缩环;受压环
compression riveter 风动铆钉机
compression rivet squeezer 铆钉挤压器
compression rod 压杆
compression roll 压辊;(压路机的)驱动轮;压实辊
compression roller 压辊
compression rupture 受压破坏
compression seal 压力密封料;压封;压力密封接;压缩密封件;压力密件

compression seal gasket 压缩式密封垫
compression set 加压胶合;压缩变形;压缩变定;压力定型;压定;受压(残留)变形;压力定形
compression settling 压缩沉降
compression settling stage 压缩沉降阶段
compression shackle 压缩弹簧钩
compression shock 压缩冲击;压力冲击
compression shock wave 压缩激波
compression side 压侧(面)
compression silicified process 压力硅化法
compression sleeve 压管接头
compression specimen 受压试件
compression splice 受压拼接
compression sprayer 压力喷雾器
compression spring 压缩弹簧;压力弹簧;压簧
compression stage 压缩级
compression stalk rupture 压杆断裂
compression steel 抗压钢筋;受压钢筋
compression strain 压(缩)应变
compression stratum 压缩层
compression strength 抗压强度;耐压强度;压缩强度
compression strength class 抗压强度级别
compression strength parallel to grain 顺纹抗压强度
compression strength range 抗压强度范围
compression strength test 抗压强度试验
compression stress 压(缩)应力;受压应力
compression stress field 压力范围;压缩应力范围
compression stress relaxation 压缩应力松弛
compression stroke 压缩冲程
compression strut 抗压构件;压杆;受压支柱;受压支杆;受压支撑
compression subsidence 挤压沉陷
compression support skirt 承压支承筒
compression swelling 压膨胀
compression system 压缩系统;压缩设备;加压系统;压缩式系统
compression tank 压力水箱
compression technique 压缩技术
compression technique of light pulse 光脉冲压缩技术
compression temperature 压缩温度
compression-tension fatigue 压张疲劳
compression test 压力测定;抗压试验;抗压强度试验;加压试验;耐压试验;压缩试验;压力试验
compression tester 耐压试验机;压缩仪;压缩试验仪;压缩试验机
compression testing machine 压力试验机
compression test of loess 黄土压缩试验
compression test of soil 土的压缩试验
compression test of tubes 管子的抗压试验
compression test piece 受压试件
compression test specimen 抗压试件
compression thimble 压接环
compression time 压缩时间
compression track structure 压性追踪构造
compression/transfer mo(u)lding 压缩/转移模压机
compression travel 压缩行程
compression treatment 压缩处理
compression type 压缩类型
compression-type cock 压紧式水龙头
compression-type hydrant 压力式消火栓;压紧式消防栓;压紧式消防龙头;压力式消防栓
compression-type of connection 压接型连接
compression-type post fire 压杆式室外消火栓
compression-type post fire hydrant 压杆式室外消防栓
compression-type refrigerating cycle 压缩式制冷循环
compression-type refrigerating machine 压缩式冷冻机
compression-type refrigerating system 压缩式制冷系统
compression-type refrigeration system 压缩式冷冻系统
compression type refrigeration unit 压缩制冷机组
compression-type regularity tester 压缩式均匀度试验机
compression-type valve lifter 压力式起阀器
compression-type water chiller 压缩式冷水机组
compression valve 压力阀门;配气阀
compression wave 压缩波;激波;压力波

compression web 压送式输送带
compression web breaking 受压腹板破裂
compression web rupture 腹板受压破坏
compression with bending 弯曲压缩
compression wood 被压材;压缩木(材);受压木;生压木
compression yield point 压力屈服点
compression zone 压缩层;受压带
compression zone depth 沉降计算厚度
compression zone of concrete 混凝土受压区
compressive air 压缩空气;压气
compressive and torsion fracture 压扭性断裂
compressive apparatus 压缩仪
compressive bar 抗压钢筋
compressive belt 挤压带
compressive boom 受压杆
compressive chord 受压弦杆
compressive creep 压缩徐变;压缩蠕变
compressive cross-section 受压截面
compressive curvature (喷管的)压缩曲率
compressive curve 压缩曲线
compressive deformation 压缩应变;压缩变形;压变形;受压变形
compressive deformation stage of soil mass 土体压缩变形阶段
compressive elastic limit 压缩弹性极限
compressive failure 受压破坏
compressive fault 挤压断层;压性断层
compressive fissure 压缩裂隙
compressive flange 受压翼缘
compressive flow 压缩流
compressive fold 压缩褶皱
compressive force 抗压力;压(缩)力;受压应力
compressive force of concrete 混凝土压缩力
compressive formation 压缩性土层
compressive layer 受压层
compressive load(ing) 压力荷载;压缩荷载
compressive modulus 抗压模数;抗压模量;压缩模量
compressive non-linearity 非线性压缩;压缩非线性
compressive oscillation 压缩振荡
compressive parallel plate viscometer 压缩平行板黏度计
compressive plane 挤压面
compressive prestress 预压应力
compressive property 抗压性能
compressive raker 受压斜桩
compressive range 受压范围
compressive ratio 压缩比(率)
compressive ratio in cylinder 汽缸压缩比
compressive region 受压区;受压范围
compressive reinforcement 抗压钢筋;受压钢筋
compressive resilience 压缩回弹性
compressive resistance 压应力;压电阻;抗压强度;抗压阻力;耐压强度
compressive rigidity 抗压刚度;压缩刚度
compressive set 压力定型
compressive settling 压缩沉淀;压缩沉降
compressive shrinkage range 机械防缩机
compressive side 受压边
compressive size reduction 挤压粉碎
compressive specimen 受压试件
compressive stalk rupture 压杆破裂
compressive stiffness 抗压刚度
compressive strain 受压应变;压(缩)应变
compressive strength 可压强度;抗压强度;挤压强度;耐压强度;压缩强度;压力强度
compressive strength class 抗压强度分等
compressive strength of concrete 混凝土受压强度
compressive strength of rock 岩石的抗压强度
compressive strength parallel to grain 顺纹抗压强度
compressive strength perpendicular to grain 横纹承压强度
compressive strength range 抗压强度范围
compressive strength test 抗压强度试验
compressive strength test cube 抗压强度试验用立方试块
compressive strength tester 抗压强度仪
compressive stress 抗压应力;压(缩)应力;受压应力
compressive stress component 压应力分量
compressive stress field 抗压应力场
compressive stress ga(u)ge 压应力计

compressive stress of subgrade 路基面压应力
compressive structural plane 压性结构面
compressive substratum 地基压缩层
compressive surface stress toughening 表面压(缩)应力增韧
compressive tectonic zone 挤压构造带
compressive tectonite 挤压构造岩
compressive-tensile strength ratio 压—拉强度比
compressive test 抗压试验;受压试验
compressive test machine 抗压试验机
compressive ultimate strength 极限压缩强度;极限抗压强度
compressive vertical principal stress 垂直压缩主应力
compressive wave 压缩波;压力波
compressive web breaking 受压腹板开裂
compressive web failure 受压腹板破坏
compressive web rupture 受压腹板破坏
compressive yield point 抗压屈服点;压缩屈服点;受压屈服点
compressive zone 挤压带;受压带
compress mode 压缩方式
compresso-crushed zone 挤压破碎带
compressometer 压缩仪;压缩计;压气试验器;缩度计
compressor 空压机;压缩器;压缩机;压气设备
compressor air intake device 压气机进风装置
compressor air tank 空气压缩机储气罐;储气箱
compressor amplifier 频带压缩放大器
compressor and expander 压缩扩展器
compressor arm 压紧杆(打捆装置)
compressor assembly 压缩机组
compressor attachment 压缩机附件
compressor blade 压气机叶片
compressor blade row 压气机叶栅
compressor bleed 从压气机中排出气体;从压气机中抽气
compressor bleed governor 压气机放气调节器
compressor booster 升压压缩
compressor capacity 压气机容量
compressor cascade 压气机叶栅
compressor casing 压气机机壳
compressor cylinder 压缩机汽缸
compressor delivery temperature 压气机出口温度
compressor displacement 压缩机工作容量
compressor drive 压气机传动装置
compressor drum 压气机鼓轮
compressor efficiency 压缩机效率
compressor equipment depreciation apportion and overhaul charges 压气设备折旧摊销及大修费
compressor-expandor 压缩扩展器
compressor fan 压气通风机;鼓风机
compressor for freezer 冰柜压缩机
compressor for refrigeration 电冰箱压缩机
compressor governer 压缩机调节器;调压器
compressor gun 润滑油增压机;加油枪;润滑油冲压机
compressor head 压缩机端盖
compressor house 压缩房
compressor housing 压气机汽缸
compressor impeller 压缩机叶轮;压气机叶轮
compressor intake pressure 压气机进口压力
compressorless injection diesel(engine) 真空喷射式柴油发动机
compressor luboil 压缩机润滑油
compressor machine 压缩机
compressor map 压气机特性线图
compressor motor 空压机电机
compressor oil 压缩机油
compressor outlet pressure 压气机出口压力
compressor plant 空压机房;压缩机装置;压缩机车间;压缩机房;压气机房
compressor platform 压气机(多孔)放气平台
compressor power 压气机功率
compressor relief device 压缩机减压装置
compressor room 压气机房
compressor seal 压缩机密封
compressor section 压缩机单元
compressor set 压缩机组
compressor stall 压气机失速
compressor starting resistor 压气机起动电阻
compressor start interlock 压气机起动联锁
compressor station 压缩机站;压气站

compressor toggle switch 压气机扳钮开关
compressor turbine 压缩机透平;驱动压气机涡轮
compressor-turbine unit 燃气轮机组
compressor-type liquid chiller 压缩机式液冷机;压缩机式液体冷却器
compressor unit 压缩机组
compressor vacuum pump 压气机真空泵
compressor wire 预应力钢丝;内传力丝
compressor with multi-stage rotor 具有多级转子的压缩机
compresso-shear basin 压扭性盆地;压控性盆地
compresso-shear boundary 压剪边界
compresso-sheer deep fracture 压剪性深断裂
compresso-shear joint 压剪性节理
compresso-shear structural plane 压扭性结构面
compresso-shear structure plane 压剪性构造面
compresso-shear testing machine 压剪试验机
compress technique 压缩技术
compress theorem 压缩定理
compressure 压缩力
comprex 气波增压器
comprise 包括
compromise 中间物;折衷方案;妥协(方案);和解
compromise act 妥协条例
compromise alternative 折衷方案
compromise architecture 和解的建筑艺术
compromise balancing network 粗调(的)平衡网络
compromise boundary 谐和边界
compromise boundary cement texture 协和界面胶结物结构
compromise clause 妥协条款
compromised total loss 协议全损;协定全损
compromise faces 协和面
compromise form of architecture 建筑折衷形式
compromise joint 异型接头;异形轨;过渡接头
compromise joint bar 异形接头夹板
compromise network 折衷网络
compromise programming 制定综合平衡方案【给】
compromise proposal 折衷方案
compromise rail 异型钢轨;过渡轨
compromise settlement 协商解决
compromise sort 简易型平衡分类法;折衷分类
compromise strategy 调和对策
compromise tariff 调和关税
compromising emanation 泄露辐射
compromising joint 异型夹板
compromising joint bar 异型接头夹板
compromising pad 异型垫板
compromising rail 异型钢轨
comptograph 自动计算器
Compton absorption 康普顿吸收
Compton absorption coefficient 康普顿吸收系数
Compton cross section 康普顿截面
Compton-Debye effect 康普顿—德拜效应
Compton effect 康普顿效应
Compton electrometer 康普顿静电计
Compton meter 康普顿计
Compton polarimeter 康普顿偏振计
Compton process 康普顿过程
Compton radius 康普顿半径
Compton-Raman theory of scattering 康普顿—拉曼散射理论
Compton rule 康普顿规则
Compton scattering 康普顿散射
Compton shift 康普顿移动
Compton wavelength 康普顿波长
comptroller 审计员;审计委员会;审计师
compulsion type mixer 快速强制混料机
compulsive checking 强迫性检查
compulsive counting 强迫性计数
compulsive reaction 强迫反应
compulsory 强制性的
compulsory acquisition 强度征用
compulsory arbitration 强制仲裁;强制性制裁
compulsory arbitration settlement 强制仲裁解决
compulsory assessment 强迫纳税
compulsory audit 强制审计
compulsory certificate 强制性认证
compulsory circulation 强制流通
compulsory clause 强制条款
compulsory conciliation 强制和解
compulsory contribution 应交会费
compulsory course 必修课程

compulsory cultivation 强制耕作
compulsory discharge 强制卸货
compulsory execution 强制执行;强制性执行
compulsory fee 强制性规费
compulsory filtration rate 强制滤速;强制滤率
compulsory inscribing 强制内接
compulsory insurance 强制保险
compulsory land acquisition 征收土地;土地征用
compulsory licensing 强制性许可
compulsory loan 强制借款
compulsory maintenance system 强制保养制度
compulsory measure 强制性措施
compulsory mixer 螺旋桨叶式搅拌运输机;桨叶式搅拌机;桨叶拌和机;桨式混合机;桨式混合机;强制式搅拌机;强制式拌和机
compulsory payment 强制性支付
compulsory pilot 法定引航员;强制引航
compulsory purchase 强制购买
compulsory relocation 强制(性)迁移
compulsory reporting points 必报点
compulsory reserve 强制(性)准备金
compulsory sale 强制销售
compulsory sales by auction 强制性拍卖
compulsory sanction 强制性制裁
compulsory saving 强制储蓄
compulsory stop 被迫停止;强迫停车
compulsory type concrete mixer 强制型混凝土搅拌机
compulsory type mixer 螺旋桨叶式搅拌运输机;桨叶水混合机;桨叶式搅拌机
compulsory union membership 强制性工会会员资格
compulsory unloading 强制卸货
compulsory winding-up 强制解散清理
compund transfer 组合式换乘
compunication 计算机通信
compuscope 图形设计站
compus network 校园网
computability 可(计)算性
computable function 可算函数
computable number 可计算数
computable value 可计算的价值
computalk 计算机通话
computation 计算;计算机应用;推算
computational accuracy 计算精度
computational approach 计算方法
computational complexity 计算复杂性
computational condition 计算条件
computational efficiency 计算效率
computational error 计算误差
computational fluid dynamics 计算流体(动)力学
computational fluid dynamics language 计算流体动力学语言
computational geometry 计算几何学
computational image 计算图像
computational instability 计算不稳定性
computational item 计算项目
computational labo(u)r 计算劳动力
computational mathematics 计算数学
computational method 计算方法
computational method of foundation subsidence 地基沉降计算方法
computational method of linear algebra 线性代数计算算法
computational method of ore losses ratio 损失率计算方法
computational methods of probabilistically statistics 概率统计计算法
computational model 计算模型
computational photogrammetry 计算射影测量学
computational physics 计算物理
computational problem 计算(性)问题
computational procedure 计算过程;计算方法;计算程序;计算步骤
computational program(me) 计算程序
computational scheme 计算线图;计算线路
computational seismology 计算地震学
computational stability 计算稳定性
computational statement 计算语句
computational structural mechanics 计算结构力学
computational system 计算系统
computation and design division 计算与设计处
computation centre 计算中心
computation check 核算

computation diagram 计算图(表);计算曲线
computation hydraulics 计算水力学
computation in scalar mode 标量方式计算
computation line 推算路线;推算边
computation model 计算模型
computation model perturbation 计算模型摄动
computation module 计算组件;计算模型
computation of coordinate 坐标计算
computation of correction 改正计算
computation of critical distance 计算临界距离
computation of erosion 冲刷计算
computation of heat balance 热平衡计算
computation of oil and gas reserve 油气储量计算
computation of optic(al) system 光具组计算
computation of ore reserves 矿物储量计算
computation of quantities 数量计算
computation of rate of return 收益率计算
computation of reinforcement 补强计算
computation of scour 冲刷计算
computation of stability 整体稳定性计算
computation of ultimate load 极限荷载计算
computation of uniformly optimal plan 一致最优方案的计算
computation of wagon turnround time on the basis of correlating time spent 时间相关法
computation of wagon turnround time on the basis of correlating with number of serviceable wagons 车辆相关法
computation process 计算过程
computation process of deflection 钻孔偏斜计算方法
computation result of soil slope stability 土坡稳定性计算结果
computation room 计算室
computation sequence 计算顺序
computation sheet 计算表格
computation speed 计算速度
computation system of point-to-point cost 点到点成本计算法
computation table 计算表
computation time 计算时间
computation unit 运算单位;运算部件
computator 计算装置
compute bound 计算限度
compute compound interest 复利计算
computed altitude 计算高度
computed azimuth 计算方位(角)
computed bending stress 计算的弯曲应力
computed comound composition 计算的化合物组成
computed data 计算数据
computed data method 计算数据法
computed discharge 计算流量;推算流量;设计流量
computed entry table 计算项表
compute depreciation on an annual basis 逐年计算折旧
computed feed(ing) 计算进料;定量进料
computed GO TO statement 计算转向语句
computed high water 设计洪水
computed maximum load 计算的最大荷载
computed mode 操作方式
computed point of goods 货物的计算点
computed price 推算价格;推定价格
computed pulse shape 脉冲(的)计算波形
computed screen cut 计算的筛去量
computed span 桥涵计算跨径
computed speed 计算速度
computed strength 计算强度
computed tare 估计皮重
computed tensile stress 计算的张力应力
computed tidal height 计算潮位
computed tide height 计算潮位
computed tractive effort 计算牵引力
computed value 计算值
computed velocity 计算速度
computed velocity functions 计算速度函数
computed velocity of locomotive 机车计算速度
computed velocity reference 计算速度基准
computed water surface elevation 计算水面高程
computed width of goods 货物的计算宽度
computed zenith distance 计算顶距
compute mode 计算方式;运算状态;运算方法
compute price 计算价格
computer 计算者;计算员;计算机;电脑

computer adaptability 计算机可用性
computer adapter unit 计算机转接器;计算机适配器
computer administration system 计算机管理系统
computer administrative records 计算机辅助记录
computer-aid 计算机辅助的
computer-aid design 半电算化设计
computer-aided analysis 计算机辅助分析
computer-aided architectural design 电子计算机辅助建筑设计
computer-aided batcher plant 配有计算机的配料厂;配有计算机的拌和厂
computer-aided building design 计算机辅助建筑设计
computer-aided cartography 计算机辅助制图;计算机辅助地图制图
computer-aided colo(u)r separation 计算机辅助分色
computer-aided construction 计算机辅助施工
computer-aided design[CAD] 电子计算机辅助设计;计算机辅助设计;电算化辅助设计;利用计算机设计;用计算机(进行)设计
computer-aided design database 计算机辅助设计数据库
computer-aided design engineering 计算机辅助设计工程数据
computer-aided design experimental translator 计算机辅助设计实验性译码器
computer-aided design information software 计算机辅助设计信息软件;用计算机设计信息软件
computer-aided design language 计算机辅助设计语言
computer-aided design of diffused aeration system 扩散曝气系统计算机辅助设计
computer-aided design system 计算机辅助设计系统
computer-aided detection 用计算机进行探测
computer-aided diagnosis 计算机辅助诊断;用计算机诊断
computer-aided digitizer 计算机辅助数字化器
computer-aided dispatching system 计算机辅助行车调度系统
computer-aided drafting system 计算机辅助绘图系统
computer-aided engineering 计算机辅助工程
computer-aided engineering design system 计算机辅助工程设计系统
computer-aided experiment 计算机实验
computer-aided framework construction system 计算机辅助框架结构施工系统
computer-aided graphic expression 计算机辅助图形表示
computer-aided graphics package 计算机辅助图形软件包
computer-aided heating-up 计算机辅助烤窑
computer-aided information retrieval 计算机辅助情报检索
computer-aided inspection recording system 计算机辅助检验记录系统
computer-aided instruction 计算机辅助教学
computer-aided instruction network 计算机辅助教学网络
computer-aided instrument 计算机辅助仪表
computer-aided interpretation 计算机辅助判读技术
computer-aided learn 计算机辅助学习
computer-aided line drawing 计算机辅助绘图
computer-aided logic design 计算机辅助逻辑设计
computer-aided manufacture 计算机辅助制造
computer-aided manufacturing 电子计算机辅助生产;计算机辅助制造
computer-aided map compilation 计算机辅助编图
computer-aided mapper 计算机辅助绘图机
computer-aided mapping 计算机辅助绘图;计算机辅助制图
computer-aided method 计算机辅助方法
computer-aided navigation equipment 计算机助航设备
computer-aided navigation system 计算机(辅)助航(行)系统
computer-aided network design 计算机辅助的网络设计
computer-aided operation 计算机辅助操作
computer-aided performance analysis 计算机辅助性能分析

computer-aided planning 计算机辅助规划
computer-aided preplanning system for storage 计算机预配载系统
computer-aided production 计算机辅助生产
computer-aided production planning 计算机辅助生产规划
computer-aided programming 计算机辅助程序设计
computer-aided reliability estimation 计算机辅助可靠性估算(程序)
computer-aided ship design 计算机辅助船舶设计
computer-aided software engineering 计算机辅助软件工程
computer-aided stereoplotting 计算机辅助立体测图
computer-aided system hardware 计算机辅助系统硬件
computer-aided tanker operation 计算机辅助油轮操作
computer-aided test 计算机辅助测试
computer-aided test, inspection 计算机辅助测试与检测
computer-aided tool 计算机辅助工具
computer-aided traffic control 计算机辅助交通控制
computer-aided translation 计算机辅助翻译
computer-aid mapping 计算机辅助测图
computer-aids 计算机辅助设备
computer analog(ue) input 计算机模拟输入
computer analog(ue) input/output 计算机模拟输入/输出
computer analysis 计算机分析
computer animation 计算机动画
computer application 计算机应用
computer architecture 计算机功能结构;计算机体系结构
computer art 计算机技术
computer-assisted 计算机辅助的
computer-assisted cartography 计算机辅助制图;计算机辅助地图制图
computer-assisted classification 计算机辅助分类
computer-assisted colo(u)r separation 计算机辅助分色
computer-assisted command system 计算机辅助指令系统
computer-assisted design 计算机辅助设计
computer-assisted design system 计算机辅助设计系统
computer-assisted expert interrogation 计算机辅助专家咨询
computer-assisted information retrieval system 电子计算机辅助信息检索系统
computer-assisted instruction 计算机辅助指令
computer-assisted interpretation 计算机辅助判读技术
computer-assisted management 计算机辅助管理
computer-assisted mapping 计算机辅助制图;计算机辅助测图
computer-assisted plotting 计算机辅助绘图
computer-assisted programming 计算机辅助程序设计
computer-assisted reference 计算机辅助查阅
computer-assisted retrieval 计算机辅助检索
computer-assisted route selection system 计算机辅助选线系统
computer-assisted train monotoring 计算机辅助列车监控
computer augmented block system 计算机增大区段系统
computer automated test system 计算机自动测试系统
computer automated vessel 计算机自动化船
computer automatic measurement and control 计算机自动(化)测量和控制
computer-based 电算化的
computer-based management information system 管理信息计算机系统
computer-based numerical model(l)ing 基于计算机的数字模型;计算机数值模拟
computer-based supervisory control 计算机监控
computer-based training 计算机基本训练
computer block diagram 计算机方框图
computer board 计算台
computer browser 计算机浏览器

computer calculation 计算机计算
computer capacity 计算机容量；计算机能力；整机规模
computer card 计算机卡片
computer cartograph 计算机制图
computer cartography 计算机制图学；电子计算机制图
computer cell formulas 电子计算机程控配方
computer center 计算机中心；电子数据处理中心
computer check 计算机校对；计算机检查
computer chronograph 计时计算机
computer closed loop control 计算机闭环控制
computer cluster 计算机簇
computer code 计算机代码
computer coding 计算机编码
computer communication 计算机通信
computer communication network 计算机通信网(络)
computer-compatible method 计算机适用的方法
computer complex 计算装置；复合计算机
computer configuration 计算机配置
computer control 计算机控制
computer control communication 计算机控制通信
computer-controlled 计算机控制的
computer controlled marshalling yard 计算机控制编组站
computer controlled ship 计算机控制船
computer controlled telegraph switching 计算机控制自动转报
computer-controlled traffic 计算机控制交通
computer-controlled traffic signal 计算机控制交通信号
computer control of grate cooler 算式冷却机计算机控制
computer control of raw material mixing 原料配料计算机控制
computer control of rotary kiln 水泥回转窑计算机控制
computer control register 程序寄存器
computer control system 计算机控制系统
computer control unit 计算机控制器
computer cost model 计算机费用模式
computer data analysis 计算机数据分析
computer data base 计算机数据库
computer-dependent language 计算机专用语言
computer dependent program(me) 专用计算机程序
computer depth and signal strength digital readouts 计算机化深度和信号强度数字读数
computer design 计算机设计
computer design automation 计算机设计自动化
computer design language 计算机设计语言
computer-detectable 计算机能识别的
computer-distributed hierarchic processing 计算机分层分布处理
computer divide layer interpretation method 计算机分层解释法
computer draft 计算机制图
computer drafting aid system 计算机绘图辅助系统
computer-drawn map 计算机绘地图
computed tomography 计算机控制层析术
computer efficiency 计算机效率
computer element 计算机元件
computer environment 计算机环境
computer equation 计算机方程(式)
computer equipment 计算机设备
computerese 计算机语言；计算机术语
computer external device 计算机外设
computer facility 计算机设备
computer field system 计算机信息系统
computer file 计算机文件；计算机外存储器
computer game 计算机游戏
computer generalization 计算机制图综合
computer-generated filter 计算机产生的滤波器
computer-generated hologram 计算机输出的全息图；计算机产生的全息图
computer-generated holograph 计算机输出的全息图
computer-generated imagery 计算机输出的图像
computer-generated map 机绘地图
computer generation 计算机发展阶段
computer graphic(al) editing 计算机制图编辑
computer graphic(al) system 计算机制图系统
computer graphic(al) technique 计算机图示技术

computer graphics 计算机制图学；计算机制图；计算机图形学；计算机图示；计算机图解；电子计算机绘图；计算机图形图像
computer graphics interactive system 电子计算机图形对话系统
computer graphics interface 计算机图形接口
computer graphics metafile 计算机图形元文件
computer graphics module 计算机图形模块
computer graphics standard 计算机图形标准
computer graphics system 计算机制图系统
computer hardware 计算机硬件
computer hardware description language 计算机硬件描述语言
computer hardware structure 计算机硬件结构
computer hierarchy system 计算机分级系统
computer house 计算机房
computer image processing 计算机图形处理；计算机图像处理
computer-independent language 计算机通用语言；独立于计算机的语言
computer indicator 计算机指示器
computer in education 教育计算机
computer information unit 计算机信息元
computering center 电子计算机中心
computer input 计算机输入
computer input form microfilm 计算机微缩胶片输入；计算机缩放胶片输入
computer installation 计算机设备；计算机装置；计算(机)站；计算机设置
computer instruction 计算机指令
computer instruction code 计算机指令码
computer instruction system 计算机指令系统
computer integrated manufacturing 计算机集成制造
computer interface 计算机接口
computer interfaced 计算机接口的
computer interface unit 计算机接口装置；计算机接口部件
computer interlocking 计算机联锁
computer interpretation 计算机解释；计算机判读
computerite 计算机人员
computerization 装备电子计算机；电子计算机化；计算机化
computerization time 计算时间
computerized acquisition system 计算机化采集系统
computerized application of railway statistic(al) surveys 铁路各种统计调查中的计算机应用
computerized burden scale for cupola 冲天炉微机自动配铁装置
computerized calculation and interpretation 机器计算和解释
computerized cataloging system 计算机编目系统
computerized circulation system 计算机化流通系统
computerized colo(u)r control 电子计算机颜色控制
computerized construction 计算机化建筑业
computerized customer billing system 计算机化顾客收款制度
computerized data retrieval system 电子计算机数据检索系统
computerized design 电算化设计
computerized dispensing 电子计算机分配(配色)
computerized distribution model 计算机配水(管网)模型
computerized evaluation 电子计算机评价
computerized geographic(al) information system 地理信息计算机检索系统
computerized interpretation 计算机判读
computerized navigation set 计算机装备的导航设备；计算式导航设备
computerized numerical control 计算机数控系统
computer(ized) plotter 计算机绘图仪；计算机绘图机
computerized plotter 自动绘图仪
computerized safeguarding system 计算机防护系统
computerized simulation 用电子计算机模拟
computerized society 计算机化社会
computerized system 计算机处理系统
computerized telegraph switching equipment 计算机转报设备

computerized telephone 计算机电话机
computerized tomography 计算机化断层 X 射线摄影法；电脑化断层 X 射线照相法；电脑化 X 射线层析照相法；层析成像
computerized traffic signal control 交通信号计算机控制；计算机交通信号控制
computerized tunnel supervisory system 隧道计算机监控系统
computer key board 计算机键盘
computer language 计算机语言
computer language recorder 计算机语言记录器
computer leak detection 计算机检漏
computer-limited 受计算机限制
computer listing 用计算机作上市登记
computer logic 计算机逻辑
computer main-frame chassis 计算机中央处理单元
computer main frequency 计算机主频
computer maintainability 计算机可维护性
computer maintenance 计算机维护
computer-managed instruction 计算机管理教学
computer management system 计算机管理系统
computer manipulation 计算机处理
computer mapping 计算机制图
computer mapping package 计算机制图程序包
computer memory 存储器；计算机内存；计算机存储器
computer memory bank 计算机存储单元
computer memory drum 计算机存储磁鼓
computer micrographics 计算机微缩图形学
computer model 计算机模型；计算机模拟；电子计算机模式
computer modeling 计算机建模型
computer network 计算机网(络)
computer networking 计算机联网
computer network service 计算机网络服务
computer network system 计算机网络系统
computernik 计算机人员
computer on slice 单片组件式计算机
computer open-loop control 计算机开环控制
computer-operated electronics display 计算机操作电子显示
computer operating maintenance expenses 电子计算机运用维费
computer operation 计算机操作
computer optimization 计算机优化
computer optimization of river basin management 流域管理计算机优化
computer optimization package 计算机最佳组件；计算机最佳程序包
computer-oriented 面向计算机的；以计算机为基础的；电子计算机化
computer-oriented language 面向计算机语言
computer output 计算机输出
computer output in graphical form 计算机图形输出
computer output microfilm 计算机输出缩微胶片
computer pattern recognition technique 计算机自动识别技术
computer performance 计算机性能
computer-performed decision 计算机决策
computer peripheral 计算机外部设备
computer peripheral communication 计算机外围通信
computer-picked arrival of first break 计算机拾取初至时间
computer picture 计算机图像
computer pipelining 计算机流水线
computer plotting system 计算机绘图系统
computer potentiometer 计算机电位计
computer printout 计算机打印输出
computer process control 计算机过程控制；计算机处理控制
computer processing 计算机加工；计算机处理
computer product 计算机计算结果
computer program(me) 计算机(计算)程序
computer program(me) cost 计算机程序费用
computer program(me) library 计算机程序库
computer programming 计算机程序设计；计算机程序编制；计算机编程
computer programming language 计算机程序语言
computer programming of mix design 配合比设计的计算机程序
computer-rated 用电子计算机评定的
computer relay 计算机继电器

computer response 计算机应答
computer retrieval 计算机检索
computer revision 计算机改正
computer room 计算机室
computer run 计算机运行
computers aid design of heat exchananger 换热器的计算机辅助设计
computer science 计算机科学
computer security 计算机安全性
computer sensitive language 计算机可用语言
computer series 计算机系列
computer servo system 计算机伺服系统
computer shredder 计算机碎纸机
computer signal timing program(me) 信号配时计算机程序;计算机信号同步程序
computer signal timing system 计算机信号时间系统
computer-simulated earthquake hazard model 计算机模拟地震危害性模型
computer simulated plasma 等离子体模拟
computer simulation 计算机模拟;电子计算机模拟
computer simulation computation of carrying capacity at station 车站通过能力计算机模拟法
computer simulation diagram method 计算机模拟图解法
computer simulation method 计算机模拟法
computer simulation of heat exchanger networks 换热器网络的计算机模拟
computer software 计算机软件
computer space 计算机空间
computer station 计算(机)站
computer storage 记忆装置;计算机存储器;计算机储存器
computer store 计算机存储器;计算机储存器
computer structural analysis and design 计算机化的结构分析与设计
computer structure language 计算机结构语言
computer supervisory control system 计算机辅助计算机监控系统
computer-supported 计算机辅助的
computer-supported design system 计算机辅助设计系统
computer-supported plotter 计算机辅助绘图机
computer-supported programming 计算机辅助程序设计
computer support mapping 数控绘图
computer switching group 计算机转换器组
computer synergetic log 计算机综合显示测井图
computer system 计算机系统
computer system capability evaluation 计算机系统性能评价
computer system capability simulation 计算机系统性能模拟
computer system reliability 计算机系统可靠性
computer system simulation 计算机系统模拟
computer telephone exchange 计算机自动电话交换机
computer terminal 计算机终端
computer time 计算机时
computer traffic accident records 计算机交通事故记录
computer training field 计算机训练区法
computer transformer 计算机变量器
computer transistor 计算机晶体管
computer tree 计算机树
computer value 记算值
computer vision 计算机视觉
computer water distribution model 计算机配水模型
computer word 计算机字
computery 电脑系统
computing 计算的
computing area 计算面积
computing capacity 计算能力
computing centre 计算中心
computing centre for processing satellite tracking data 卫星计算中心
computing control system 运算控制系统
computing depreciation 计算折旧
computing differential 计算微分
computing element 计算元件
computing elevation 计算高程
computing error 计算误差
computing figure of hydraulic power 水力计算值

computing gear 计算机传动装置
computing impedance 运算阻抗
computing laboratory 计算实验室
computing loading estimates from non-point sources 非点污染源负荷估算
computing machine 计算机
computing machine room 计算机室
computing mechanics 计算力学
computing method 计算方法
computing method of residual oxygen 剩余氧计算法
computing method of stability failure of rock slope 岩质边坡稳定性计算方法
computing mode 算态
computing network 计算网络
computing of apparent susceptibility 视磁化率计算
computing parameter of groundwater age 地下水年龄计算参数
computing quantities 土方量计算
computing quantities of earthwork 土方量计算
computing rate 计算速率
computing scale 秋积尺
computing seismology 计算地震学
computing sight 瞄准用计算装置
computing span 计算跨径
computing speed 计算速度
computing statement 计算语句
computing store 计算存储器;计数存储器
computing system 计算系统
computing system coupler 计算系统的配线装置
computing technique 计算技术
computing technology 计算技术
computing terminal 计算终端
computing terminal system 计算终端系统
computing time 计算时间
computing weighing scales 带计算装置的秤
computron 计算机用多极电子管
comsat 通信卫星
Comsol 科姆索尔银铅焊料
Comstock process 康姆斯托克热压硬质合金法
comsumable melt 自耗电极熔化
conane 腺甾烷
conave setting 凹板调节
conbulker 集装箱散货(兼用)船
concamerate 覆以圆屋顶;做成圆拱形
concameration 拱或穹;房间;单元
concast 连续铸锭
concatenate 连锁;串级
concatenated code 链接码
concatenated data set 链接数据集;连续数据组;并置数据集
concatenated frequency changer 级联变频器;串级变频器
concatenated key 链接关键字;串联键;并置键;并置关键字
concatenated motor 链系电动机;级联电动机;串级电动机
concatenated tuple 串接元组
concatenate-range zone 并存延续带
concatenate rule 连接规则
concatenation 链状结合;链接;连锁;级联法;级联;串接;并置
concatenation character 并置字符;并置符号
concatenation control 串级调速
concatenation data set 并置数据集
concatenation error 并置错误
concatenation list structure 连接表结构
concatenation motor 链系电动机;串级电动机
concatenation of string 串的连接
concatenation operation 连接运算;并置运算
concatenation operator 并置(运)算符
concavation 凹度
concave 陷穴;圆锥破碎机的腔部;中凹(的);凹线;凹(入)的;凹面(物);凹处;凹板
concave adjusting bolt 凹板调节螺栓
concave adjusting lever 凹板间隙调节杆
concave adjustment shaft 凹板调整轴
concave angle 凹角
concave bank 冲刷河岸;凹岸
concave bank line 凹岸线
concave bar 凹板板条
concave barrel 凹形辊身
concave bead (釉面的)凹形小珠;凹面密封条;凹

式腻缝
concave beater 逐稿轮
concave bend frame 内弯肋骨
concave bit 凹心钻头;岩芯钻头;凹形钻头;凹形凿;凹面钻头
concave block 凹形支垫座(人字闸门)
concave bottom 凹底;凹板分组
concave bow 凹进(翘曲)
concave brick 凹形砖
concave buddle 固定凹圆形淘汰盘
concave cam 凹形凸轮
concave camber 凹线辊型;凹面;凹度
concave cave 凹洞
concave ceiling 凹进平顶;凹进顶棚;拱形层顶;坝垮间的拱板
concave chamfer 凹圆面
concave clearance 凹板间隙
concave clearance adjuster 凹板间隙调节机构
concave clearance lever 凹板间隙调节杆
concave coastline 凹岸线
concave collector 凹形接电器
concave-concave 两面凹的;凹凹;双凹(的)
concave-convex 一面凹一面凸的;凹凸(的)
concave-convex contact 凹凸接触
concave-convex game 凹凸对策
concave-convex lens 凹凸(透)镜
concave corner 凹面角
concave cross-bedding 凹面交错层理
concave cross-section 凹形(纵)断面
concave crown 凹顶
concave curvature 凹曲率
concave curve 凹(形)曲线
concave cutter 凹半圆刃铣刀;凹半圆成形铣刀
concave diffraction grating 凹(衍射)光栅
concave disk 凹面圆盘;凹光圆盘
concave disk colter 凹面圆犁刀
concave-down(ward) 下凹;凹向下的
concave downward utility curve 下凹效用曲线
concave edge 凹边
concave extension 凹板延长板
concave fault 凹断
concave feed plate 凹板喂入板
concave fillet weld 凹形角焊缝;凹面填角焊(缝);凹角焊
concave flank cam 凹腹凸轮
concave flow 凹岸水流
concave fold 凹褶
concave function 凹函数
concave game 凹对策
concave glass 凹面玻璃;凹镜
concave gradient 凹梯度
concave grate 凹板栅格;凹板筛
concave grating 凹面光栅
concave grating spectrometer 凹面光栅分光计;凹表光栅分光计
concave grid 凹格板;凹板筛
concave head bolt 凹头螺栓
concave head piston 凹顶活塞
concave hyperbolic contour 单叶双曲面轮廓
concave interface 凹界面
concave interlocking cutter 凹形联锁铣刀
concave joint 凹圆缝;凹圆缝;凹缝
concave key 凹面键
concave leaf 凹状叶
concave lens 凹(面)透镜;凹镜头
concave life 固定衬板寿命(破碎机)
concavely upward-facing tile roof 仰瓦屋面
concave meniscus 凹月形牙镜;凹液面
concave milling cutter 凹形铣刀
concave milling cutter for wood working 木工凹形铣刀
concave mirror 凹镶;凹(面)镜;凹面返光镜;凹镜
concave mounting bolt 凹板固定螺栓
concave of contact 接触带凹入部
concave of fault surface 断面凹
concave optic(al) tool 光学玻璃凹面磨具
concave package 凹面绕丝筒
concave perforated plate 凹形多孔板
concave plane 凹平面;凹底刨
concave plug bit 凹心钻头
concave polygon 凹多边形
concave polyhedron 凹多面体
concave programming 凹形规划法
concave programmingproblem 凹规划问题

concave protecting disk 凹面防护圆盘
concave radius 凹曲线半径
concave reflection grating 凹反射光栅
concave reflector 凹面反射器
concave region 凹域
concave relief 负向地貌
concave ridging body 短凹面起垄犁体
concave rim 凹形轮缘
concave rim wheel 槽缘轮
concave roll 凹面(轧)辊
concave rolling disk colter 凹面圆犁刀
concave roof 凹面屋顶
concave saw 碟式圆锯；凹面圆锯；凹口锯
concave-setting lever 凹板间隙调节杆
concave-shaped compact 凹面坯块
concave shear 凹形剪
concave shell 凹形薄壳
concave shore 凹岸
concave shoreline 凹岸线
concave side 凹岸一侧；凹(岸)面
concave slide 凹槽载片；凹玻片
concave slope 下凹形坡；凹(形)坡
concave solid cutter 凹半圆刃整体铣刀
concave spheric(al) mirror 凹球面(反射)镜
concave state 凹海岸国家
concave stream bank 凹河岸
concave sucking disc 吸盘状陷窝
concave surface 凹面
concave tile 凹瓦；牡瓦；底瓦
concave tooled joint 圆截面缝
concave-up(ward) 向上凹的；凹向上的；凹线
concave vein 凹脉
concave vertical curve 凹形竖曲线
concave washer 凹形垫块
concave weld 凹形角焊；凹(形)焊缝；凹面焊(缝)；轻型焊
concave wiring 暗线
concavex 凹凸面
concavity 凹性；凹曲(度)；凹度；凹处；熔洞
concavity slide 凹载玻片
concavo-concave 两面凹的；双凹形
concavo-convex 新月形；一面凹一面凸的；凹凸形(的)；凹凸的
concavo-plane 平凹形；凹底刨；一面凹一面平的
concealed 暗藏关门器
concealed air conditioner 暗装空调器
concealed arch 暗拱
concealed bedrock 隐蔽基岩；未露头基岩
concealed bend 视距不良的弯道
concealed bracket 暗托架
concealed cable 暗藏电缆
concealed carcasing 暗管
concealed ceiling grid system 暗龙骨吊顶系统
concealed cleat 暗固着楔；平接缝暗加固片
concealed closer 门的暗式开关
concealed closing mechanism 暗藏关闭机构
concealed coal field 隐伏煤田
concealed column 暗柱
concealed conduit 暗线管道
concealed convector 隐蔽式对流器
concealed damage 隐蔽损坏
concealed defect 隐蔽的缺陷
concealed display 隐蔽显示
concealed door 暗门
concealed door bolt 暗门锁；暗门闩
concealed dovetail 暗榫；隐蔽鸠尾榫
concealed downpipe 暗雨水管；暗水落管
concealed electric(al) wiring 暗式布(电)线
concealed erosion 潜蚀
concealed fastener 隐蔽式扣件
concealed fastening 隐蔽式锚固
concealed fencing 隐蔽式拦沙障
concealed fixing 隐式接合；隐蔽式装配
concealed flashing 暗的金属防雨板；暗泛水；暗挡条
concealed flashing piece 暗藏闪光装置
concealed gas cock 暗装煤气旋塞
concealed gas piping 隐蔽的煤气管道
concealed girder 隐蔽梁
concealed gutter 暗天沟；暗式雨水沟；隐蔽式天沟；暗檐沟
concealed heating 壁板式供暖；隐蔽式供暖；暗装供暖
concealed hinge 埋头；暗铰链；隐合页

concealed household 潜在户(口)
concealed illumination 暗藏式照明；无影照明
concealed-in-door closer 地垄；暗装门内关门器
concealed installation 隐蔽设备；暗藏设备；隐置设备；隐式装置(法)；暗设
concealed joint 交叠接合；隐藏接缝；盖板接合；暗接头；暗缝
concealed lamp 隐灯；隐藏灯
concealed-lamp sign 隐灯信号
concealed light 有灯槽的反光灯；暗灯
concealed lighting 隐蔽照明
concealed line 隐蔽线
concealed mutation 潜伏突变
concealed nail(ing) 暗钉；隐藏钉；藏钉；斜钉
concealed ore body 隐伏矿体
concealed outcrop 隐蔽露头
concealed pipe 暗管
concealed piping 暗管；暗装管道
concealed property 账外财产
concealed radiator 隐蔽式暖气片；隐藏式散热片；藏置(式)散热器；暗装散热器；暗炉片
concealed rafter 粗制椽；暗椽
concealed routing 木工暗铲口
concealed running board 内藏踏脚板
concealed safety step 内藏式安全踏脚板
concealed scale 隐藏式浴槽阶梯
concealed-screw rose 暗螺丝圆座盖
concealed shelving 隐蔽式壁架
concealed shoer with mixing valve 混合阀暗装淋浴器
concealed shower 暗装淋浴器；隐蔽式淋浴器
concealed shower with mixing valve 带有混合阀的暗装淋浴器
concealed solonchak 埋藏盐土
concealed space 隐蔽空间
concealed suspension system (吊件不露的)暗吊顶
concealed symmetry 隐蔽的对称性
concealed system of axes 隐蔽轴线系统
concealed tack 暗藏平头钉
concealed toilet paper holder 隐蔽式手纸架；暗手纸盒
concealed tube 暗管
concealed tubing 暗管
concealed-type purlin(e) hanger 暗式檩条托架
concealed valley 暗天沟
concealed valve 暗阀
concealed washer (用于墙板的)隐形垫圈
concealed water 埋藏水
concealed wire 暗(电)线
concealed wire chase 暗线槽
concealed wiring 暗(式布)线；隐蔽布线；布暗线；暗敷设
concealed works 隐蔽工程
conceal gutter 隐蔽式天沟
concealing the amount of income 匿报所得额
concealing wiring 暗线
concealment 匿报；隐藏；隐蔽；伪装
concealment analysis 伪装判读
concealment of fact 隐瞒事实真相
concealment paint 隐蔽色油漆；遮盖漆；覆面漆；暗漆
conceive 设想
conceived proposal 提议；设想方案
concenter wiring 管道内线路
concentrate 集中；集结；浓缩(物)；提浓物
concentrate all efforts 集中全力
concentrate bunker 精矿仓
concentrate colo(u)ring 母料着色
concentrated 浓的；密集
concentrated acid 浓酸
concentrated acid leaching 浓酸浸出
concentrated activated sludge 浓缩活性污泥；浓缩的活性污泥
concentrated air conditioning 集中空调
concentrated anchorage 集中锚固
concentrated blue 深蓝
concentrated brine 浓盐水
concentrated capacitance 集总电容
concentrated capacitor 集总电容器
concentrated capacity 集总容量
concentrated charge 集中装药
concentrated coefficient 浓缩系数

concentrated combustion 集中燃烧
concentrated cooling 集中供冷
concentrated demand 集中需水量
concentrated distribution 密集状分布
concentrated drainage 集中排水
concentrated dryer 铅锰催干剂；浓缩催干剂
concentrated earth movement 造山作用；造山运动
concentrated fall 集中落差
concentrated fall hydroelectric(al) development 坝后厂房式水力发电；集中落差式水力发电
concentrated fall type station 集中落差式电站
concentrated feeding stuff 精料
concentrated fertilizer 浓缩肥料
concentrated filling and emptying system 集中输水系统；集中灌水泄水系统
concentrated flow 集中水流；集中涌水
concentrated force 集中力；集中荷载；集中负载
concentrated glass colo(u)r 玻璃浓缩色料
concentrated heating 集中供暖；集中采暖(法)
concentrated hydrochloric acid 浓盐酸
concentrated information 集中信息
concentrated laser beams 激光聚光束
concentrated latex 浓缩胶乳
concentrated light 集束光
concentrated liquor 浓缩液
concentrated load(ing) 集中荷载；密集装车
concentrated-load stress 集中荷载应力
concentrated-load system 集中荷载系统
concentrated marketing 集中行销
concentrated message 集中信息；被集中消息
concentrated moving load 集中活动荷载
concentrated nitrate waste 浓缩硝酸盐废水
concentrated nitric acid 浓硝酸
concentrated parameter 集中参数
concentrated pigment dispersion 浓颜料分散体
concentrated pipe 集合管
concentrated point load 集中载重；集中荷载；集中负载
concentrated power feeding system 集中供电方式
concentrated power supply 集中供电
concentrated reaction 集中反应；集中反力
concentrated refrigerating 集中制冷
concentrated reinforcement 密集补强
concentrated reinforcement with curved surface 圆弧面密集补强
concentrated reinforcement with inclined surface 斜面密集补强
concentrated salting liquor 浓盐水溶液
concentrated sludge 浓缩污泥
concentrated slurry 浓料浆
concentrated solution 浓缩液；浓(缩)溶液
concentrated source 集中源
concentrated spring 集中弹簧
concentrated stay cable 集中拉索
concentrated stream 浓水流
concentrated stress of hatch corner 舱口角部应力集中
concentrated sulfuric acid 浓硫酸
concentrated sulphuric acid 浓硫酸
concentrated superphosphate 浓过磷酸钙；重过磷酸钙
concentrated support 集中支座；集中支承
concentrated suspension 浓缩悬浮液
concentrated vortex 集中涡流
concentrated vorticity 集中涡
concentrated water 浓(缩)水
concentrated water flow 集中的水流
concentrated wear 局部磨损
concentrated weight 集中荷重；集中载重
concentrated wells 集中布井
concentrated wheel load 集中轮载
concentrated winding 同心绕组；集中绕组；密集绕组；同心绕法
concentrate fodder 高于粗饲料
concentrate grade 精矿品位；精矿等级
concentrate handling 精矿处理
concentrate insulation cover 整体绝热层
concentrate ore grade 精矿品位
concentrate pump 浓缩泵
concentrates 精煤；精矿
concentrate sampling 精矿采样
concentrate tank 原液柜(泡沫灭火剂)；浓缩液舱柜

concentrate yield 精矿产率
concentrate zone theory 同心圆发展说
concentrating agent 浓缩剂
concentrating factor 集群因素;浓缩因素
concentrating fertilizer 浓缩肥料
concentrating function 浓缩功能
concentrating funds 集中资金
concentrating heating 集中供暖
concentrating in mound and pushing in one time 分堆集中
concentrating machine 选矿机
concentrating mill 选矿厂
concentrating of seismic activity 地震活动集中
concentrating pan 浓缩锅;浓缩罐
concentrating sludge 浓缩污泥
concentrating solar collector 聚光式太阳能集热器
concentrating table 精选摇床;选矿摇床;选矿台
concentrating tower 浓缩塔
concentration 精选;集中;集聚;浓缩作用;浓缩;浓度;密集度
concentration-activity relationship 浓度—活度关系
concentration basin 浓缩池;选矿槽
concentration by freezing 冷冻浓缩(法);冻结浓缩
concentration by refrigeration 冷冻浓缩(法)
concentration capacity of harbo(u)r 集运能力【港】
concentration capacity of port 集运能力【港】
concentration cell 浓差电池
concentration center 浓集中心
concentration Clark 浓集克拉克值
concentration class 聚集度
concentration coefficient 集中系数;浓缩系数;浓度系数
concentration coefficient method 浓缩系数法
concentration coefficient of electrode current 电极电流集中系数
concentration constant 浓度常数
concentration contour 浓度等值线
concentration control 浓度控制;改变浓缩度的调节
concentration control of pollutant 污染物浓度控制
concentration cross-section 浓度断面图
concentration cup 聚焦极
concentration current 浓差电流
concentration curve 涨洪段(过程线的);集流段;含沙量曲线;聚集曲线;浓度曲线
concentration cylinder 聚焦圆筒
concentration degree 集中度
concentration determination 浓度测定
concentration dialysis 浓差渗析
concentration difference 浓度差;浓差
concentration diffusion 浓度扩散;浓差扩散
concentration dilution test 浓缩稀释试验
concentration distribution 浓度分布
concentration distribution curve 浓度分布曲线
concentration distribution profile 浓度分布曲线
concentration-duration curve 浓度—历时曲线
concentration effect 浓度效应
concentration efficiency 浓缩效率
concentration ellipse 同心椭圆
concentration equilibrium 浓度平衡
concentration expressed in percentage by volume 体积百分浓度
concentration expressed in percentage by weight 重量百分浓度
concentration factor 精缩比;集中因素;集中因数;集中系数;浓缩系数;浓集系数;富集因素;富集系数
concentration gradient 浓度梯度
concentration guide 浓缩导槽
concentration index 集中指数;浓度指数
concentration index of suspended dust 悬浮尘埃浓度指数
concentration index of suspended smoke and soot 悬浮烟灰和烟的浓度指数
concentration indicator 浓度指示器
concentration inner zone 浓度内带
concentration intermediate zone 浓度中带
concentration key 集合按钮
concentration layout 集中布置
concentration lens 合聚透镜
concentration level 浓度水平;富集水平
concentration limb 集流段;涨洪段
concentration limit 极限浓度;浓度限度;下限浓度;最低检出浓度

concentration line 总线;公共线
concentration machine 精选机;选矿机
concentration medium 富集介质
concentration meter 浓度计
concentration method 浓缩法;浓集法;富集法
concentration multiple 浓缩倍数
concentration of aqua ammonia 氨水溶液浓度
concentration of beam 集束
concentration of capital 资本积累
concentration of carbon monoxide 一氧化碳浓度
concentration of channels 河槽汇流;河道汇流
concentration of chemical liquid for well cleaning 化学洗井液浓度
concentration of collector 捕收剂浓度
concentration of diamond 金刚石浓度
concentration of elements 元素浓度
concentration of entrained air 掺气浓度
concentration of floating dust 飘尘浓度
concentration of floating ice 浮冰密实度;浮冰结集
concentration of ground 地面外
concentration of harmful substance 有害物质浓度
concentration of industry 工业集中
concentration of inhibition 抑制浓度;抑制集中
concentration of ion 离子浓度
concentration of landholding 土地集中
concentration of mining 集中开采
concentration of nitrate nitrogen 硝态氮浓度
concentration of nitrogen oxides 氮氧化物的浓度
concentration of offer 集中报盘
concentration of particular pollutant 特性污染物浓度
concentration of poisons 毒料浓度
concentration of pollutant 污染物(的)浓度
concentration of population 人口密集;人口集中
concentration of production 生产集中
concentration of rail stresses 钢轨应力集中
concentration of rigidity 刚度集中
concentration of sewage 污水浓度
concentration of sludge 污泥浓度
concentration of soil solution 土壤溶液浓度
concentration of solution 溶液浓度
concentration of soot 煤烟浓度
concentration of stress 应力集中
concentration of suspended load 悬移质含沙量
concentration of suspended solids in mixed liquor 混合液悬浮固体浓度
concentration of tracer 示踪剂浓度
concentration of trade 贸易集中
concentration of volatile suspended solids in mixed liquor 混合液挥发悬浮固体含量
concentration of waste water disposal 污水排放浓度
concentration on engineering design 集中工程设计
concentration on ground 地面浓度;地表浓度
concentration outer zone 浓度外带
concentration pipe 集水管
concentration plant 选矿厂
concentration point 集流点
concentration point of stress 应力集中点
concentration polarization 浓差极化
concentration potential 浓集势;浓差电位
concentration profile 浓度剖视图;浓度分布图;浓度断面图;含沙量断面图
concentration quenching 浓缩淬灭;浓度淬火
concentration range 浓度范围
concentration ratio 集中率;浓缩比;选矿比;锥度比
concentration recovery 选矿回收率
concentration response curve 浓度反应曲线
concentration ring 集索圈
concentration sensitive detector 浓度敏感型检测器
concentration soulbility product 浓度容度积
concentration-spiral filtration dehydration 浓缩—螺滤脱水
concentration standard 浓度标准
concentration tank 沉淀池;浓缩池
concentration technique 浓缩技术
concentration test 浓缩试验;浓缩检验
concentration theory 汇流理论
concentration threshold 浓度极限
concentration time 集中时间;集流时间;浓缩时间;汇流时间;富集时间
concentration to control 控制浓度;防治浓度
concentration unit contour 浓度单位等值线

concentration upper limit 上限浓度
concentrative point-like distribution 集中点状分布
concentrative recharge 集中补给
concentrator 集中器;集中机;集线器;浓缩器;浓缩机;选煤机;选矿机
concentrator cell 聚光电池
concentrator marker 标识器
concentrator terminal buffer 集线器终端缓冲区
concentre wiring 管道内线路
concentric(al) 共心的;同心的;同心圆状
concentric(al) arc 同心弧
concentric(al) arch 同心拱
concentric(al) beam 同心梁
concentric(al) cable 同轴电缆
concentric(al) candle 同心电极弧光灯
concentric(al) chuck 同心夹盘;同心卡盘
concentric(al) circle 环形道路系统
concentric(al) circle diagram 同心圆图
concentric(al) circles 同心圆
concentric(al) circles theory 同心圆理论
concentric(al) city 同心圆城市
concentric(al) cleavage 同心劈理
concentric(al) cluster 共心发动机簇
concentric(al) coat 同心膜
concentric(al) coil 同心线圈
concentric(al) column load 集中柱荷载
concentric(al) compound trommel 同心复式转筒
concentric(al) conics 同心二次曲线
concentric(al) contraction 向心性缩小;向心收缩
concentric(al) converter 正口转炉
concentric(al) covering 同心药皮
concentric(al) crystalline zones 结晶筒
concentric(al) cylinder 聚焦圆筒
concentric(al) cylinder circuit 同轴电路
concentric(al) cylinder muffler 集筒式消声器;同心圆筒式消声器;套筒式消声器
concentric(al) cylinder rotation viscometry 同心圆筒旋转式测黏度法
concentric(al) cylinder visco(si)meter 同心圆筒式黏度计;同心筒式黏度计
concentric(al) die 同心模;轴对轴模
concentric(al) eclipse 同心食
concentric(al) exfoliation 同心剥离
concentric(al) faults 同心断层
concentric(al) feeder 同轴馈(电)线
concentric(al) float 环状浮子;环状浮标
concentric(al) fold 同心褶皱;同心褶曲
concentric(al) fractures 同心状断裂
concentric(al) girdle 同心环状的
concentric(al) groove 闭纹
concentric(al) halo 同心晕
concentric(al) jaw chuck 自动定心卡盘
concentric(al) joint 同心节理
concentric(al) lamellar 同心片状
concentric(al)-lay cable 同轴电缆
concentric(al) lay conductor 同心绞线
concentric(al) lens 同心透镜
concentric(al) line 公共线;同轴线;同轴电缆;同心线
concentric(al)-line oscillator 同轴线振荡器;同轴管振荡器
concentric(al) line resonator 同轴线谐振器
concentric(al) load 轴心荷载;中心荷载
concentric(al) locating 同心定位
concentrically built carbureter 同心式汽化器
concentric(al) mirror 共心反射镜
concentric(al) mirror system 共心反射系统
concentric(al) model 同心圆模型
concentric(al) nozzle 同心圆喷嘴
concentric(al) nylon bush 同心尼龙管
concentric(al) orifice plate 同心锐孔隔板
concentric(al) pattern (金刚石在钻头唇部的)同心圆排列式样;同心式
concentric(al) pencil 同心光束;同心线束
concentric(al) piston ring 同心活塞环
concentric(al) plan 同心圆平面
concentric(al) quadrics 同心二次曲面
concentric(al) reducer 同心锥形管;同心大小头;同心变径管
concentric(al) reducing 大小头
concentric(al) resonator 同轴共振器
concentric(al) ring air diffuser 同心环形空气散流器
concentric(al) ring infiltrometer 同心环渗透计;

同心环测渗仪;套环式测渗仪
concentric(al) ring pattern 密集环形图案
concentric(al) rings 环路;环形公路;同心圈;同心环
concentric(al) ring structures 同心环形山构造
concentric(al) route 环状道路;环形道路
concentric(al) scale 同心刻度
concentric(al) siphon 同心虹吸
concentric(al) sphere 同心球
concentric(al) squirrel cage mill 同心笼式粉磨碎机
concentric(al) stranded wire 同心绞线
concentric(al) structure 同心圆状构造;同心构造
concentric(al) system 共心系统
concentric(al) taper 同心锥形管;同心大小头;同心变径管
concentric(al) taper fitting 同心锥形件装配
concentric(al) taper pipe 同心锥形管
concentric(al) tendon 同心钢筋束;同心钢丝索
concentric(al) terminal shield 同心屏蔽筒
concentric(al) transmission line 同轴电缆
concentric(al) tree 同心树
concentric(al) tube 同心管
concentric(al) tube column 精密分镏柱;同心管(精密蒸馏)柱
concentric(al) tube heat exchanger 同心管热交换器
concentric(al) U bit 同心U形钻头
concentric(al) valve 同心阀
concentric(al) vascular bundle 同心维管束
concentric(al) weathering 球状风化;同心风化
concentric(al) winding 同心绕组
concentric(al) wire rope 同心式钢丝绳
concentric(al) wiring 集中布线;同心线布线;(带有接地带电外皮的)管道内线路
concentric(al) zone concept 同心圆论
concentric(al) zone model 同心环带模型
concentric(al) zone theory 同心圆发展理论;同心区理论
concentricity 同轴度;同心性;同心度;同心
concentricity checking fixture 同心度检查装置
concentricity ga(u)ge 同轴性检验规
concept 构思(建筑师或工程师的初步设想);概念
concept analysis 概念分析
concept books 丛书
concept development 概念发展
concept field 抽象字段
concept formation 概念形成
conception 立意;构思(建筑师或工程师的初步设想);概念(作用)
conceptional model 概念性模型
conception of per capita 人均观念
conception report 概念性报告;初步报告;方案研究
concept of expressive form 表达形式的概念
concept of form 形式的概念
concept of form and space 形状与空间的概念
concept of heat flow 热流概念
concept of model 概念模型
concept of non-reciprocity 非互惠概念
concept of settlement 沉降概念
concept of shape 形状的概念
concept of space 空间的概念
concept of style 型式的概念
concept of wholism 整体观念
concept phase 构思阶段;初步设计阶段;草图设计阶段;拟定阶段;设想阶段
concept plan 概要性规划图
conceptual 概念的
conceptual analysis 概念性分析;初步分析;方案研究;方案分析;概念分析
conceptual approach 概念性方法
conceptual art 概念艺术
conceptual attainment 概念获得
conceptual basis 基本原理
conceptual classification 概念分类
conceptual coordination 概念组配;概念协调;概念配位
conceptual cross-section 理想横剖面
conceptual dependency 概念依赖
conceptual design 方案设计;草图设计;构思设计;规划设计;概念(性)设计;初步(方案)设计;方案研究;方案(性)设计;方案设想
conceptual design and study 方案设计和论证
conceptual design stage 概念性设计阶段
conceptual development 方案研究;方案拟定;方案编制

conceptual drawing 概念图;方案图
conceptual engineering planning 规划及初步设计;工程初步规划
conceptual field 概念域
conceptual formation 概念形成
conceptual framework 概念性刚要;基本概念
conceptual internal mapping 方案内部安排
conceptual level 概念层;方案阶段
conceptual mode 概念模式
conceptual model 方案性模型;初设模型;概念(性)模型
conceptual modelling 概念建模
conceptual model of arched concrete dam 混凝土拱坝概念性模式
conceptual phase 初步设计阶段;规划阶段;构思阶段;概念阶段;草图阶段;方案设计阶段;方案阶段
conceptual plan 概要性规划图;概念图;构思性规划
conceptual planning 规划设计;概念性规划;轮廓性规划;方案性规划
conceptual prospect 概念的前景;方案的前景
conceptual reuse scheme 方案再用性设计
conceptual reuse system 构想再用系统
conceptual schema 概念模型;概念框图
conceptual scheme 概念性再用方案
conceptual seismic design 抗震概念设计
conceptual simulation model 概念性模拟模型
conceptual skill 理念上能力
conceptual stage 规划阶段
conceptual stage design 方案阶段设计
conceptual strategy 方案的战略
conceptual theory 概念性理论
conceptual view 方案视图
conceptualize 概念化
concern 商业企业
concerned 有关的
concert border 剧院舞台上口的灯光架;遮挡灯光架的横幕
concert chamber 演奏厅
concerted action 协同动作;一致行动;共同行动
concert hall 音乐厅;演奏厅
concertina amplifier 板阴输出放大器
concertina connection 手风琴式连接;缩叠式连接
concertina door 折叠式门;手风琴式折门
concertina door fitting 折叠式门装配
concertina folding door 手风琴式折叠门
concertina meander 蛇腹形河弯;蛇腹形河曲
concertina parition 手风琴式隔墙
concertina partition 折叠式隔墙
concertina wire 蛇腹形铁丝网
concert mechanism 协调机理
concert piano 大钢琴
concert pitch 标准音调
concession 矿山地质开采权;减让;优惠;租让地;租界;让步;特许权;特许
concession agreement 租让协议;租让合同;特许意向书;特许协议
concessionaire 小吃摊;租让合同承租人;租让合同;特许权所有者;特许权所有人;特许权获得者;特许权持有人
concessionaire granter 特许权所有人
concessional 特许的
concessional credit 通融信贷
concessional disposal 廉价处理
concessional loan 优惠贷款;特惠贷款
concessional rate 优惠价(格);让价
concession(al) terms 优惠条件
concession arrangement 特许安排
concessionary and contracting company 特许承包公司
concessionary grant 优惠条件特批
concessionary loan 优惠贷款
concessionary rate of tax 特惠税率
concessionary sale 优惠销售
concessionary terms 优惠条件
concession building 特许建筑
concession contract 租让合同
concessioner 特许权所有者;特许权所有人;特许权获得者
concession of tariff 关税减让
concessive aid 减让性援助
conch 海螺;贝珍珠;半圆形屋顶
concha 半圆形屋顶;半圆穹顶;穹内凹顶
Concha Ostreae 牡蛎

conchiform 贝壳状
conchite 霰石;方解石;泡霰石
conchitic 含有贝壳的
conchocelis stage 壳斑藻阶段
conchoid 螺旋线;蚌线
conchoidal 贝壳状的;贝壳状
conchoidal bevel 贝壳状斜边
conchoidal fracture 贝壳状破裂;贝壳状裂痕;贝壳状断口
conchoidal sandstone 贝壳状砂岩
conchoids 螺线管
conchology 贝类学
conchospiral 放射对数螺线
Conchostraca 贝甲目(节肢动物)
conciding indicator 同步指标
conciliation 和解;调解
conciliation agreement 调停协议
conciliation court 调解法庭
conciliation panel 协调小组
conclave 秘密会议;密室
conclude a bargain 订约;订合同;取得协议
conclude a contract 订立合约
conclude agreement 达成协议
conclude an agreement 签订协议
conclude angle 推断角
conclude a transaction 成交
conclude a treaty 订条约
conclude business after viewing samples 看样成交
conclude contract 订(立)合同;缔结合约
concluding line 绳梯中央绳
concluding note 结论
concluding report 总结报告
concluding stage 最后阶段;终期;完成阶段
conclusion 结论;总结;终结
conclusion of business 达成交易
conclusion of contract 合同缔结
conclusion of evaluation 评价结论
conclusion of report ratification 报告审批结论
conclusion of the contract 合同的订立
conclusions of testing 试验结论
conclusive 最后的;确定性的
conclusive evidence 决定性证据;最终证据;确证
conclusive evidence clause 确证条款
conclusive presumptions 无可异议的推定
conclusive test 决定性试验
concocter 策划者
concolo(u)r 同色
concolo(u)rous 单色的;同色的
concomitant 伴行的;伴随物;伴随的
concomitant disinfection 即时消毒
concomitant ion 共存离子
concomitant variable 相从变量;相伴变量;伴随变量
concomitant variation 相从变动
con-cooler shell 冷凝冷却器壳体
concord 谐和
concordance 和谐性;协调;用语索引;一致性;整合;词汇索引;便览;吻合;索引
concordance intrusive body 整合侵入体
concordance list 索引表
concordance program(me) 索引程序
concordancy 协调;整合;索引
concordant ages 一致年龄
concordant bedding 整合层面;平行层面;平行层理
concordant body 整合岩体
concordant cable 吻合预应力钢丝束;吻合索
concordant coast 纵式海岸
concordant coastline 纵式海岸线
concordant cross-section 吻合剖面;吻合断面;吻合(横)截面
concordant effect 吻合效应
concordant flow 协调流量
concordant fold 整合褶皱
concordant injection 整合岩体;整合灌入
concordant intrusion 整合侵入(体)
concordant line 吻合线
concordant microscope 吻合显微镜
concordant morphology 整合地貌
concordant pluton 整合(深成)岩体
concordant profile 吻合外形;吻合轮廓;吻合线(形);吻合剖面;吻合断面
concordant sample 协调样本
concordant shoreline 纵式滨线;顺向滨线
concordant tendon 吻合预应力筋;吻合力筋;吻合

钢丝束;无负弯矩值受拉钢筋
concordant twin 相似双生
concordant water-level 相应水位;对应水位
concordat 协定;契约
Concordia 一致线
Concordia diagram 一致线图
Concordia method 一致曲线法
concord universal ruling head 协和万能绘图头
concourse 集散厅;集合;中央广场;站厅;分配厅;人群;群集(场所)
concourse[美]中央大厅
concourse level 站厅层
concourse transfer 站厅换乘
concrescent teeth 结合牙
concresive 混凝土黏合剂
concrete 混凝土;固结的
concrete accelerator 混凝土速凝剂;混凝土快凝剂;混凝土促硬剂;混凝土促凝剂
concrete additive 混凝土外加剂;混凝土添加剂;混凝土附加剂
concrete additive powder 混凝土外加粉料
concrete adhesive 混凝土黏着剂
concrete adiabatic curing 混凝土绝热养护
concrete admixture 混凝土外加剂;混凝土附加剂
concrete advisory service 混凝土咨询服务
concrete aeratd with foam 泡沫混凝土
concrete agent 混凝土外加剂
concrete aggregate 混凝土集料;混凝土骨料
concrete aggregate composition 混凝土骨料成分
concrete aggregate feeding 混凝土集料供给;混凝土骨料供给
concrete aggregate grain 混凝土骨料颗粒
concrete aggregate heating 混凝土集料加热
concrete aggregate producer 混凝土生产者
concrete agitation 混凝土搅拌
concrete agitator 混凝土搅拌机
concrete agitator truck 混凝土搅拌运输车
concrete anchor 预应力混凝土锚具;预制混凝土构件中夹件
concrete anchorage wall 混凝土锚碇墙
concrete anchor block 混凝土锚碇块体
concrete anchoring 混凝土锚固
concrete anchoring wall 混凝土锚墙
Concrete and Construction Engineering 混凝土与建筑工程(英国期刊名)
concrete and crack control 混凝土和裂缝控制
concrete and mortar 混凝土和砂浆
concrete and reinforced concrete work 混凝土与钢筋混凝土工程
concrete anti-breaking machine 水泥抗折机
concrete anti-freezer 混凝土防冻剂
concrete appearance 混凝土外貌;混凝土外观
concret apron 混凝土散水;混凝土铺盖;码头前沿;混凝土护坦;混凝土护墙;混凝土底板
concrete arch bridge 混凝土拱桥
concrete arch dam 混凝土拱坝
concrete arch(ed girder) 混凝土拱梁
concrete architecture 混凝土建筑
concrete arch rib 混凝土拱肋
concrete area 混凝土面积;混凝土范围
concrete armo(u)r ring 混凝土护面
concrete armo(u)r unit 混凝土护面块
concrete article 混凝土制品
concrete astragal 混凝土半圆饰
concrete-as-you-go lining method 随挖随衬法
concrete as you go method 迅速加衬施工法
concrete attack 混凝土侵蚀
concrete backfill(ing) 混凝土回填
concrete backing 混凝土衬背
concrete baffle pier 混凝土消力墩
concrete bagging 混凝土装袋;袋装混凝土
concrete bagwork 袋装混凝土护岸工程
concrete ballast 混凝土用石渣;混凝土压载物;混凝土压载
concrete bar 钢筋;混凝土配筋
concrete-bar bending machine 钢筋弯曲机
concrete-bar drawer 钢筋拉伸机
concrete barge 钢筋混凝土趸船;水泥趸(船)
concrete barrel shell 混凝土筒形薄壳
concrete barrel vault 混凝土筒形拱顶
concrete barrow 混凝土手推车
concrete-bar straightening-cutting machine 钢筋调直切断机
concrete base(course) 混凝土基座;混凝土底层;混凝土基底;混凝土基础;混凝土基层;混凝土底座
concrete base laid at ground level 浇于地面上的混凝土
concrete base pad 水泥坐垫
concrete base slab 混凝土基础板
concrete basin 混凝土水池
concrete batch 混凝土一次制料量;混凝土配料
concrete batcher 混凝土配料器
concrete batching 混凝土配料
concrete batching and mixing tower 混凝土搅拌楼
concrete batching and mixing plant 混凝土配料拌和厂
concrete batching plant 混凝土分批搅拌设备;混凝土分批搅拌厂;间歇式混凝土搅拌站;混凝土配料车间;混凝土搅拌站;混凝土拌和设备;混凝土拌和楼;混凝土拌和厂
concrete batching scale 混凝土配料标尺
concrete bay 混凝土底板;现浇的混凝土面层
concrete bay subgrader 混凝土路基整平机
concrete beam 混凝土梁
concrete beam case 混凝土梁外壳
concrete beam encasement 混凝土梁饰面;混凝土梁包板
concrete beam encasure 混凝土梁套
concrete beam extension 混凝土墙支撑法
concrete beam haunching 混凝土梁腋
concrete beam making machine 混凝土梁制造机
concrete beam sheath coat 混凝土梁外壳涂层
concrete beam slipform device 混凝土梁滑模浇筑设备
concrete beam test mo(u)ld 混凝土棱柱体试模
concrete bearing system 混凝土支承系统
concrete bed 混凝土底脚;混凝土路床;混凝土基床;混凝土垫层;整体道床
concrete belfry 混凝土钟楼
concrete bell tower 混凝土钟楼
concrete bending stress 混凝土弯曲应力
concrete bent construction 混凝土构架结构;混凝土排架结构
concrete bin 混凝土储仓;混凝土储[贮]料斗
concrete binder 混凝土结合料
concrete biologic(al) shield(ing) 混凝土的生物屏蔽
concrete bit 混凝土钻头
concrete blanket 混凝土防水层
concrete blasting 混凝土面喷砂处理;混凝土爆破
concrete bleeding 混凝土泌水现象
concrete blinding 混凝土垫壳;混凝土模板;捣混凝土垫层;捣混凝土底板
concrete blinding coat 混凝土填充底层
concrete block 混凝土砌块;混凝土(方)块体;混凝土(方)块;混凝土垫块
concrete block and rock-mound breakwater 混凝土(方)块堆石防波堤
concrete block at pell-mell 抛筑的混凝土块体
concrete block bar support 支承钢筋的混凝土块
concrete block breakwater 混凝土(方)块防波堤;混凝土(方)块(体)防波堤
concrete block crushing testing machine 混凝土试块破坏试验机
concrete block cutter 混凝土(方)块切割机
concrete block for mooring post 系船柱混凝土块体;系泊块体
concrete block for shielding 防射线混凝土块
concrete block gravity quaywall 混凝土(方)块重力式岸壁
concrete block gravity wall 混凝土(方)块重力(式)墙
concrete block lining 混凝土(方)块支衬
concrete blockmaking 混凝土(砌)块制造
concrete block masonry arch bridge 混凝土砌块拱桥
concrete block masonry wall 混凝土砌块圬工墙
concrete block mo(u)lding machine 混凝土砌块成型机
concrete block-mound breakwater 混凝土(方)块斜坡式防波堤
concrete block pavement 混凝土预制块路面;混凝土砌块铺面;水泥混凝土预制块路面;混凝土(方)块路面
concrete block paving 混凝土(方)块铺砌路面
concrete block pell-mell 抛筑的块体
concrete block pitching 混凝土(方)块铺砌工程;混凝土(方)块护坡
concrete block press 混凝土块压制机
concrete block protection 混凝土(方)块护坡
concrete block quaywall 混凝土(方)块岸壁;方块码头
concrete block quay wall with relieving slab 带卸荷板方块码头
concrete block revetment 混凝土(方)块护岸
concrete block sound wall 混凝土(方)块挡声墙
concrete block splitter 混凝土(方)块切割机
concrete block wall 混凝土砌块墙
concrete block wall breakwater 方块防波堤
concrete block works 混凝土砌块工程;混凝土(方)块砌体
concrete blockwork structure 混凝土(方)块结构
concrete blockwork wall 混凝土(方)块体墙
concrete block yard 混凝土(方)块制造场;混凝土(方)块预制场;方块预制场;混凝土构件场
concrete blower 混凝土风力输送机
concrete boat 水泥船
concrete body 混凝土体
concrete bogie 混凝土运输车
concrete bond 混凝土抹面前的头道灰;混凝土黏结(层)
concrete bond beam block 混凝土结合梁块体
concrete bonding adhesive 混凝土黏结剂
concrete bonding agent 混凝土黏结剂
concrete bonding medium 混凝土黏结介质
concrete bonding plaster 混凝土黏着灰浆;混凝土黏着灰浆
concrete bond plaster 混凝土黏结灰泥;混凝土黏合灰泥
concrete borer 蛀石(海)虫
concrete bottom 混凝土基础;混凝土基层;混凝土底座;混凝土底面
concrete bottom boom 混凝土下弦(杆)
concrete bottom chord 混凝土下弦(杆)
concrete bound 水泥结的;混凝土结合的
concrete boundary wall 混凝土边界墙
concrete-bound macadam 水泥结合碎石路面
concrete-bound pavement 水泥结合碎石路面
concrete bowl 混凝土运载盘
concrete bowl lorry 运混凝土卡车
concrete box caisson 混凝土箱形沉箱
concrete box culvert 混凝土箱形涵洞;混凝土箱涵
concrete bracket 混凝土托座;混凝土牛腿
concrete breaker 混凝土凿碎机;混凝土破碎机;混凝土捣碎机;凿混凝土机
concrete breaker steel 混凝土破碎机钢锤
concrete breaking chisel 混凝土破碎凿
concrete breakwater 混凝土防波堤
concrete brick 混凝土砖;混凝土砌块;实心混凝土
concrete bridge 混凝土桥
concrete bridge pier 混凝土桥墩
concrete bucket 混凝土吊罐;混凝土吊斗;混凝土斗
concrete bucket lock 混凝土料斗闸门
concrete buggy 混凝土装卸车;混凝土运输车;混凝土用二轮车;混凝土手推车
concrete building 混凝土房屋
concrete building block 混凝土建筑砌块
concrete building brick 混凝土建筑砖
concrete building construction 混凝土房屋建筑;混凝土房屋构造
concrete building floor unit 预制混凝土地板单元
concrete building module 混凝土建筑砌块模数
concrete building tile 混凝土建筑瓦
concrete building unit 预制混凝土房屋构件;混凝土房屋构件;混凝土(建筑)构件;预制混凝土构件
concrete bunt on (竖井的)混凝土横撑杆
concrete burnt 混凝土灼伤
concrete cable cover 混凝土盖板电缆
concrete cable stayed bridge 混凝土斜拉桥
concrete caisson 混凝土沉箱;混凝土沉井
concrete caisson breakwater 混凝土沉箱防波堤
concrete caisson dolphin 混凝土沉箱靠船墩
concrete caisson quay wall 混凝土沉箱码头
concrete caisson sinking 混凝土沉箱下沉
concrete caisson yard 混凝土沉箱预制场
concrete canal lining 混凝土渠道衬砌
concrete cantilever 混凝土悬臂梁
concrete cap 混凝土帽梁;混凝土桩帽;混凝土柱帽;混凝土顶盖
concrete capping 混凝土压顶
concrete capping block 混凝土压檐块

concrete capping slab 混凝土压檐板
concrete capping tile 混凝土压檐砖
concrete carriageway 混凝土车行道
concrete cart 混凝土载运车;混凝土(运送)小车;混凝土手推车;运混凝土小车
concrete casing 混凝土蜗壳;混凝土外壳;混凝土饰面;混凝土面饰;混凝土浇面;混凝土衬物
concrete casting 浇捣混凝土
concrete casting system 混凝土浇筑系统
concrete cast in place 现浇混凝土
concrete cast in position 现(场)浇(筑)混凝土
concrete cast in situ 现浇混凝土
concrete catch-gutter 混凝土明沟
concrete ceiling 混凝土天棚;混凝土天花板;混凝土平顶;混凝土顶棚
concrete ceiling insert 混凝土顶板嵌入块;混凝土顶板嵌入件
concrete cellular block quay wall 空心方块码头
concrete cellular window 混凝土地下室窗户
concrete cement gun 水泥喷枪
concrete cement(ing agent) 混凝土胶凝剂
concrete channel unit 混凝土槽形构件
concrete check 混凝土配水闸;混凝土节制闸
concrete check dam 混凝土拦沙坝;混凝土谷坊
concrete chimney 混凝土烟囱
concrete chimney door 混凝土烟囱门
concrete chisel 混凝土凿
concrete church 混凝土教堂
concrete chute 混凝土输送槽;混凝土倾斜槽;混凝土浇槽;混凝土浇灌溜槽;混凝土急水槽;混凝土陡槽;运混凝土的溜槽
concrete chuting cable crane 槽送混凝土的缆索起重机
concrete cill 混凝土窗槛
concrete cinder 煤渣混凝土
concrete circular flat slab 混凝土圆形平板
concrete circular plate 混凝土圆板
concrete cladding 混凝土覆面
concrete cladding slab 混凝土覆面板
concrete cladding unit 混凝土覆面房屋
concrete class 混凝土等级
concrete coating 加混凝土涂层;混凝土表面涂层
concrete-coating interface 混凝土与涂料的界面
concrete code 混凝土标准;混凝土规范
concrete cofferdam 混凝土围堰
concrete collar 钢筋混凝土柱箍;混凝土柱环;混凝土系梁;混凝土圈梁
concrete colo(u)ring 混凝土着色
concrete column 混凝土柱
concrete community 基本群落
concrete compacted by jolting 振动夯实混凝土
concrete compacted element 振实混凝土构件
concrete compaction 混凝土捣实
concrete compactor 混凝土振实机;混凝土捣实机
concrete component 混凝土构件
concrete composition 混凝土组成;混凝土成分
concrete compression boom 混凝土受压吊杆
concrete compression chord 混凝土受压弦杆
concrete compression flange 混凝土受压翼缘
concrete compression machine 混凝土压力(试验)机
concrete compression ring 混凝土压缩环
concrete compression strength 混凝土抗压强度
concrete compressive strength 混凝土抗压强度
concrete compressive stress 混凝土压应力
concrete compressive test 混凝土抗压试验
concrete compressive zone 混凝土受压区域
concrete consistency 混凝土稠度
concrete-consistency meter 混凝土稠度计;混凝土黏度计
concrete consolidating 混凝土捣固;混凝土稠度
concrete construction 混凝土施工;混凝土结构;混凝土建筑;混凝土构造;混凝土工程
concrete constructional industry 混凝土建筑工业
concrete construction joint 混凝土施工缝
concrete construction type 混凝土建筑类型
concrete construction work 混凝土建筑工程
concrete container 混凝土容器
concrete containment structure 混凝土防护外壳
concrete contraction 混凝土收缩
concrete-conveying pipe 混凝土输送管
concrete conveyer 混凝土输送机
concrete cooling 混凝土冷却
concrete cooling equipment 混凝土冷却设备

concrete cooling method 混凝土冷却方法
concrete cooling pipe 混凝土冷却水管
concrete cooling plant 混凝土冷却设备;混凝土冷却工厂
concrete cooling system 混凝土降温设备
concrete cooling tower 混凝土冷却塔
concrete cope block 混凝土压顶块
concrete cope tile 混凝土压顶砖瓦
concrete coping 混凝土压顶;混凝土盖顶;混凝土顶盖
concrete coping drill 混凝土取芯钻
concrete coping slab 混凝土压顶板
concrete core 混凝土钻芯试体;混凝土筒体
concrete core-drilling testing apparatus 钻芯混凝土检测仪
concrete core slab 混凝土心板
concrete core unit 混凝土心板
concrete core wall 混凝土芯墙
concrete core wall type dam 混凝土芯墙式坝
concrete cornice 混凝土挑檐
concrete corrosion 混凝土腐蚀
concrete counterweight 混凝土平衡重
concrete cover(ing) 混凝土保护层;混凝土护面;混凝土盖
concrete covering for reinforcement 钢筋的混凝土保护层
concrete cover meter 混凝土保护层测量仪
concrete cradle 混凝土管座
concrete cradle bedding 混凝土基座;混凝土基底;混凝土承座基床
concrete creep 混凝土徐变;混凝土蠕变
concrete crew 混凝土作业队
concrete crib 混凝土格笼
concrete cribbing 混凝土箱格;混凝土框笼;混凝土框构筑物;混凝土垛式支架
concrete crib breakwater 混凝土箱格防波堤;混凝土木笼防波堤;混凝土框笼防波堤
concrete crib of construction 混凝土箱格结构
concrete crib retaining wall 混凝土格笼挡土墙
concrete cribwork 混凝土框构筑物
concrete cross 混凝土十字架
concrete cross-section 混凝土截面
concrete cross-tie 混凝土枕
concrete crushing strength 混凝土破碎强度
concrete cubage 混凝土容积
concrete cube 混凝土立方体(试块)
concrete cube compressive strength 混凝土立方块抗压强度
concrete cube crushing strength 混凝土立方体破碎强度
concrete cube strength 混凝土立方体强度
concrete cube test 混凝土立方块强度试验;混凝土立方体试验
concrete cubic yardage 混凝土立方体码数
concrete culvert 混凝土涵管;混凝土涵洞
concrete cupola 混凝土圆顶
concrete curb 混凝土路缘石
concrete curb mold 混凝土路缘石模板
concrete curb stone 混凝土侧石;混凝土路缘石;混凝土路边石
concrete curing 混凝土养护
concrete curing blanket 混凝土冬天养护遮毯;混凝土养护覆盖物;混凝土养护覆盖层;混凝土养护草垫;混凝土保温覆盖
concrete curing chamber 混凝土养护窑;混凝土养护室;混凝土养护舱
concrete curing composition 混凝土养护剂
concrete curing compound 混凝土养护剂;混凝土养护涂料
concrete curing mat 混凝土养护遮盖;混凝土养护席;混凝土养护盖垫
concrete curing membrane 混凝土养护薄膜
concrete curing paper 混凝土养护纸
concrete curing period 混凝土养护期
concrete curing solution 混凝土养护液
concrete curing technology 混凝土养护工艺
concrete curing tent 混凝土养护棚
concrete curing water 混凝土养护水
concrete curing yard 混凝土养护场
concrete curtain 混凝土幕
concrete curtain wall 混凝土幕墙
concrete curtain wall panel 混凝土幕墙板
concrete cushion 混凝土垫层
concrete cut-off 混凝土截水(心)墙

concrete cut-off abutment 混凝土防渗墙接头
concrete cut-off wall 混凝土隔墙;混凝土防渗墙;混凝土截水(心)墙
concrete cutter 混凝土切割机;混凝土切缝机;混凝土路面切缝机;水泥混凝土切割机;水泥混凝土切缝机
concrete cutting machine 混凝土切割机;混凝土切缝机
concrete cylinder 混凝土圆柱(试块)
concrete cylinder compressive strength 混凝土圆柱体抗压强度
concrete cylinder compressive strength test 混凝土圆柱体抗压强度试验
concrete cylinder crushing strength 混凝土圆柱体破碎强度
concrete cylinder test 混凝土圆柱体强度试验
concrete cylinder testing machine 混凝土圆柱体试验机
concrete cylinder wall 混凝土筒柱岸壁
concrete dam 混凝土坝
concrete dam storage closet 混凝土试件湿养护室
concrete deadman 混凝土锚墩;混凝土锚墙(拉岸壁板桩的锚桩);混凝土锚桩;混凝土地牛;混凝土地龙
concrete debris rack 混凝土拦污栅
concrete deck 混凝土桥面
concrete deep-water structure 深水混凝土结构
concrete deformation 混凝土变形
concrete degradation 混凝土侵蚀
concrete deliver(ing) car 混凝土运输车;混凝土输送车
concrete delivery 混凝土运送
concrete delivery mixer 混凝土搅拌运料车
concrete delivery pipe 混凝土输送管
concrete delivery truck 混凝土运送车;运输混凝土的卡车
concrete demolition work 混凝土拆毁工作
concrete densifier 混凝土增浓剂
concrete densifying admix(ture) 混凝土增密剂
concrete densifying agent 混凝土增密剂;混凝土密实剂
concrete density 混凝土密度
concrete deposit(e) 浇筑的混凝土混合物;混凝土浇筑体;混凝土浇灌体;混凝土灌筑体
concrete depositing 混凝土浇注
concrete design 混凝土(配合比)设计
concrete design criterion 混凝土设计标准
concrete destruction 混凝土破坏
concrete details 具体细节
concrete diagonal rod 混凝土斜杆;混凝土对角撑
concrete diaphragm 混凝土隔板
concrete diaphragm wall 地下连续墙
concreted in 注入混凝土
concrete discharge channel 混凝土卸料槽
concrete discharge gutter 混凝土卸料槽
concrete discharge pipe 混凝土卸料管
concrete disintegration 混凝土离析;混凝土崩解
concrete dispenser 混凝土摊铺器
concrete distance piece 混凝土隔片;混凝土定距块
concrete distress 混凝土损毁
concrete distributing plant 混凝土输送设备
concrete distributing tower 混凝土分配塔
concrete distributor 混凝土摊铺机;混凝土分配器;混凝土分布机
concrete dock lining wall 混凝土码头镶面墙
concrete dome 混凝土圆屋顶;混凝土穹窿
concrete door frame 混凝土门框
concrete drainage article 混凝土排水制品
concrete drainage channel 混凝土排水渠
concrete drainage goods 混凝土排水制品
concrete drainage gutter 混凝土排水沟
concrete drainage pipe 混凝土排水管
concrete drain pipe 混凝土排水管
concrete drain tile 混凝土排水瓦管
concrete dressing 混凝土琢面
concrete drill 混凝土钻孔器
concrete droppings 混凝土碎块
concrete dumper 混凝土翻斗车
concrete durability 混凝土耐久性
concrete edge beam 混凝土边梁
concrete edge stress 混凝土边缘应力
concrete edging 混凝土边框;混凝土镶边
concrete element 混凝土元件
concrete elevator 混凝土升降机;混凝土提升机

concrete-embedded 埋置在混凝土中的
concrete embedment 混凝土预埋件
concrete emulsion 混凝土乳浆
concrete encased 混凝土外壳的;混凝土封闭的
concrete encased steel beam 包混凝土钢梁
concrete encased steel column 包混凝土钢柱
concrete encased steel frame 包混凝土钢框架
concrete encasement 混凝土外壳;混凝土外包;混凝土套;混凝土包壳
concrete encasure 混凝土外壳;混凝土壳罩
concrete engineering 混凝土工程
concrete enveloped 混凝土包覆的
concrete equipment 浇灌混凝土设备;混凝土设备
concrete equivalence 混凝土当量
concrete expansion agent 混凝土膨胀剂
concrete expansion joint 混凝土伸缩缝
concrete expansive material 混凝土膨胀剂
concrete exposed to sea water 受海水作用的混凝土
concrete extrusion press 混凝土挤压机
concrete fabrication 混凝土装配;混凝土制备;混凝土配筋;混凝土制作;混凝土生产
concrete facade 混凝土房屋正面
concrete facade panel 混凝土房屋正面板
concrete face 混凝土面板;混凝土面
concrete face dam 混凝土面板坝
concrete face rockfill dam 混凝土面板堆石坝
concrete facing 混凝土饰面;混凝土面层;混凝土面板;混凝土护面
concrete failure 混凝土损坏
concrete farm 混凝土试验场
concrete farm building 混凝土农场建筑
concrete fascia wall 混凝土岸额墙;混凝土岸边线墙
concrete fatigue 混凝土疲劳
concrete feeding bucket valve 混凝土料斗闸门
concrete fence 混凝土栅栏;混凝土围篱
concrete fence picket 混凝土栅栏桩
concrete fence post 混凝土栅栏柱子;混凝土栅栏柱
concrete fence stake 混凝土栅栏小桩
concrete fencing picket 混凝土栅栏尖桩
concrete figure 具体数字;具体实例
concrete filled 混凝土填充的;填浇凝土的
concrete filled bag 混凝土袋
concrete filled caisson 混凝土充填沉箱
concrete filled steel pipe pile 混凝土钢管桩
concrete filled steel tube 钢管混凝土
concrete-filled tube column 混凝土填(塞)管柱;混凝土充填管柱
concrete filled with steel tubular 钢管混凝土
concrete filler block 混凝土填(充)块;混凝土衬块
concrete fillet 混凝土内补角
concrete filling 混凝土填料;混凝土灌注;混凝土填充
concrete filling pipe 混凝土填充管
concrete fill of drilled caisson 钻地沉箱的混凝土填方
concrete filter pipe 混凝土过滤管
concrete finish 混凝土修整;混凝土修面;混凝土饰面;混凝土抹面;混凝土表面涂层;混凝土表面光洁度
concrete finisher 混凝土整修机;混凝土修整机;混凝土修面机;混凝土抹面机;水泥混凝土路面修整机
concrete finishing 混凝土修整;混凝土表面磨光
concrete finishing machine 混凝土整面机;混凝土修整机;混凝土抹面机;混凝土表面粉光机;混凝土路面修整机械;混凝土表面加工机械
concrete finishing screed 混凝土刮尺
concrete fireproofing 混凝土防火性;混凝土防火层
concrete fitting 混凝土装配
concrete flag 混凝土(路面)板;薄混凝土路面板
concrete flag stone 混凝土铺路板
concrete flat tile 混凝土平瓦
concrete float 混凝土抹平器
concrete floating dock 水泥浮船坞
concrete floor 混凝土模板;混凝土楼面;混凝土楼盖;混凝土楼板;混凝土地坪;混凝土地面;混凝土底板
concrete floor beam 混凝土地板格栅
concrete floor cast(ing) 混凝土地板浇筑
concrete floor cover(ing) 混凝土地板面层
concrete floor dust 混凝土地面粉尘
concrete floor finish 混凝土地面修饰
concrete floor hardener 混凝土地面增硬剂;混凝土地面硬化剂;混凝土地面增强剂
concrete flooring 混凝土楼板;混凝土底座;混凝土底脚
concrete flooring tile 混凝土铺地砖
concrete floor member 混凝土地板构件
concrete floor paint 混凝土地板漆
concrete floor panel 混凝土地面镶板
concrete floor rib 混凝土地板肋
concrete floor screed(topping) 混凝土地板找平抹灰层
concrete floor sealer 混凝土填缝材料;混凝土密封材料
concrete floor slab 混凝土地板
concrete floor system 混凝土地板系统
concrete flowability 混凝土流动性
concrete flower trough 混凝土花盆
concrete flue 混凝土烟道
concrete flue block 混凝土烟道砌块
concrete fluidity 混凝土流动性
concrete fluidity test 混凝土流动性试验
concrete flume 混凝土运送道;混凝土渡槽
concrete foam 泡沫混凝土
concrete footing 混凝土底脚;混凝土基脚;混凝土基础;混凝土底座
concrete for carriageway markings 混凝土车道路标
concrete for civil defence structures 混凝土民防构筑物
concrete form 浇混凝土用的模板;混凝土壳子板;混凝土模板
concrete for marine structure 海工混凝土
concrete forming 混凝土成型
concrete forming circle 混凝土模板周转
concrete form oil 混凝土模板油
concrete formulation 混凝土配合比;混凝土配方;混凝土配料单
concrete formwork 混凝土模壳;混凝土模板作业;混凝土模板工程;混凝土模板
concrete for overbreakage 超挖衬砌
concrete for pressure grouting 压力灌浆混凝土
concrete foul water pipe 混凝土污水管
concrete foundation 混凝土基层;混凝土基础;混凝土底座
concrete foundation block 混凝土基(础)块(体)
concrete foundation pile 混凝土基础桩
concrete foundation wall 混凝土基础墙
concrete frame 混凝土排架;混凝土构架
concrete frame construction 混凝土框架结构
concrete framework 混凝土构架
concrete funnel 混凝土浇灌漏斗
concrete gable beam 混凝土山墙梁
concrete gasholder 混凝土储(贮)气柜;混凝土煤气库
concrete gate sill 混凝土闸门槛
concrete girder 混凝土大梁
concrete girder bridge 混凝土梁式桥
concrete girder for crane runway 混凝土行车梁
concrete glazing bar 混凝土玻璃格条
concrete goods 混凝土制品
concrete grade 混凝土(强度)等级
concrete grading 混凝土级配
concrete gravel 混凝土用砾石
concrete graving dock 混凝土干船坞
concrete gravity dam 混凝土重力坝
concrete gravity dockwall 混凝土重力式船坞墙
concrete gravity(drilling) platform 混凝土重力式钻探平台;混凝土重力式钻井平台
concrete gravity oil platform 混凝土重力式石油平台
concrete gravity overfall dam 重力式混凝土溢流坝
concrete gravity platform 混凝土重力式平台
concrete gravity quaywall 混凝土重力式岸壁
concrete gravity retaining wall 混凝土重力式挡土墙
concrete gray 混凝土灰色
concrete grillage 混凝土格框(基础);混凝土格床
concrete gripped wire 混凝土夹固的电线
concrete ground floor slab 混凝土底层地板
concrete grout 混凝土浆
concrete grouter 混凝土灌浆机
concrete grouting 混凝土灌浆
concrete grouting plant 混凝土喷注设备
concrete groyne 混凝土防波堤;混凝土丁坝
concrete guard rail 混凝土(防)护轨;混凝土防护栅;混凝土护栏
concrete guard wall 混凝土(防)护墙;混凝土挡墙
concrete guide wall 混凝土导墙
concrete gull(e)y 混凝土排水沟
concrete gun 混凝土喷枪
concrete gutter 混凝土檐槽;混凝土边沟
concrete handling 混凝土运输;混凝土操作;混凝土吊运
concrete hardcore 混凝土硬核;混凝土石块填料
concrete hardener 混凝土硬化剂;混凝土加硬剂
concrete hardening 混凝土硬结
concrete hardening agent 混凝土增强剂;混凝土硬化剂
concrete hauling 混凝土吊运
concrete hauling container 混凝土运送容器
concrete hauling unit 混凝土运输车辆
concrete haunching 混凝土镶边;混凝土起拱;混凝土加腋;混凝土梁腋
concrete haunting 加混凝土外壳
concrete-H beam composite pile 混凝土-工字钢组合桩
concrete head 浇筑混凝土的压力高度
concrete head wall 混凝土正墙
concrete heaping 混凝土浇筑
concrete hearth 混凝土火床
concrete high frequency vibrator 混凝土高频振动器
concrete highway 混凝土公路
concrete hinge 混凝土铰接承座;混凝土铰(接)
concrete hip tile 混凝土戗脊盖瓦
concrete hoist 混凝土升降机;混凝土卷扬机
concrete hollow block 混凝土空心砌块;空心混凝土块;混凝土空心(方)块;混凝土空气砖
concrete hollow block making 混凝土空心砌块制造
concrete hollow cellular block 空心混凝土格体
concrete hollow drill 混凝土钻孔器
concrete hollow filler(block) 混凝土空心填块
concrete hollow floor(slab) 混凝土空心楼板
concrete hollow mast 混凝土空心柱
concrete hollow panel 混凝土空心板
concrete hollow plank 混凝土空心板
concrete hollow pole 混凝土空心(电)杆
concrete hollow slab 混凝土空心板
concrete hollow slab flooring system 混凝土空心楼板系统
concrete hollow slab mo(u)lding machine 混凝土空心板成型机
concrete hollow ware 混凝土空心制品
concrete hopper 混凝土料斗;混凝土产品储[贮]斗
concrete hose 混凝土软管
concrete-image telemeter 双像测距仪
concrete impregnation agent 混凝土浸渍剂
concrete improvement 混凝土的改进
concrete improver 混凝土改进剂
concrete in 用混凝土浇入
concrete in bag 袋装混凝土
concrete inclinometer 混凝土倾角仪
concrete individual base 混凝土独立基础
concrete infilling panel 混凝土填充板
concrete infill(ing) unit 混凝土填充板
concrete in fluid state 流态混凝土
concrete in freezing weather 冰冻期的混凝土
concrete ingredient 混凝土成分
concrete injection unit 混凝土喷射机组;混凝土灌注机组
concrete injector 喷浆机
concrete-in-mass 大体积混凝土;大块混凝土
concrete-in-mass structure 大体积混凝土构筑物
concrete insert 混凝土插铁;混凝土预埋件;混凝土插件
concrete in situ 现(场)浇混凝土;现场灌筑混凝土
concrete inspection 混凝土检验
concrete inspection chamber 混凝土检查井
concrete integral waterproofing agent 混凝土整体防水剂
concrete integral water repellent admix(ture) 混凝土整体憎水外加剂
concrete integral water repelling agent 混凝土整体憎水剂
concrete intelligence 具体智能
concrete interface treating agent 混凝土界面处理剂
concrete interlocking pipe 混凝土嵌锁管子

concrete interlocking tile 混凝土镶嵌屋瓦;混凝土连锁瓦;混凝土联锁瓦
concrete intermediate floor 混凝土中间地板
concrete irrigation pipe 混凝土灌溉管道
concrete jacket 混凝土(柱)套;混凝土外皮;混凝土外层;混凝土壳
concrete jamb 混凝土门窗边框
concrete jetting machine 混凝土喷射机
concrete jetting machine set 混凝土喷射机组
concrete joint 混凝土接缝
concrete joint cleaner 路面清缝机(水泥混凝土)
concrete joint cutter 混凝土切缝机
concrete joint filling 混凝土填当;混凝土灌缝
concrete joint saw 路面锯缝机(水泥混凝土)
concrete joint sealer 路面填缝机(水泥混凝土)
concrete joint sealing compound 混凝土接缝密封合成物
concrete joist 混凝土密肋
concrete joist construction 混凝土小梁结构
concrete junction 混凝土连接处
concrete kerb 混凝土路缘石
concrete kerb mould 混凝土路缘石模板
concrete key stone 混凝土路边石
concrete key trench (土坝的)混凝土截水墙槽
concrete kiln curing 混凝土窑养护
concrete labor 具体劳动
concrete lab(oratory) 混凝土实验室
concrete laid in alternate bay 隔仓法摊铺混凝土(路面)
concrete lane 混凝土车道
concrete lath(ing) 混凝土板条
concrete lattice beam 混凝土格构梁
concrete lattice girder 混凝土格构大梁
concrete layer 混凝土层
concrete lead-lined cell 铅衬混凝土电解槽
concrete level(l)ing 混凝土找平
concrete level(l)ing course 混凝土摊铺机
concrete level(l)ing layer 混凝土找平层
concrete lift 混凝土起重机;混凝土升高层;混凝土浇筑层;混凝土(分)层;混凝土薄层
concrete lifting bucket 混凝土吊斗
concrete lighting column 混凝土照明柱
concrete lighting mast 混凝土照明(电)杆
concrete lighthouse 混凝土灯塔
concrete line 混凝土砌筑
concrete lined 混凝土衬里的;混凝土衬砌的
concrete-lined canal 混凝土衬砌(的)渠道
concrete-lined channel 混凝土衬砌(的)渠道
concrete-lined tunnel 混凝土衬筑隧洞;混凝土衬砌隧洞
concrete lining 混凝土衬砌;混凝土衬里
concrete lining wall 混凝土镶面墙
concrete lining works 混凝土衬筑工程;混凝土衬砌工程
concrete lintel 混凝土过梁
Concrete Industries Council 混凝土工业委员会(美国)
concrete load bearing system 混凝土承受荷载系统
concrete loader 混凝土装载机
concrete loading hopper 混凝土装料斗
concrete lock 混凝土(船)闸
concrete lock floor 混凝土船闸底板
concrete longitudinal cofferdam 混凝土纵向围堰
concrete lower chord 混凝土下弦
concrete machinery 混凝土机械
concrete made with crushed stone aggregate 碎石混凝土
concrete made with fine aggregate 细骨料混凝土
concrete maintenance 混凝土养护
concrete making 混凝土制备
concrete manhole 混凝土窨井;混凝土检查井
concrete manhole ring 混凝土检查井圈
concrete manufacture 混凝土制作
concrete manway 混凝土矿井井筒梯
concrete manway ring 混凝土矿井井筒梯圈
concrete marginal strip 混凝土路缘带;混凝土边带
concrete mark 混凝土强度等级
concrete masonry 混凝土(预制块)砌体;混凝土砌块;混凝土砌体
concrete masonry block 混凝土砌块
concrete masonry chimney 混凝土砌工烟囱
concrete masonry home 混凝土砌工住房
concrete masonry partition wall 混凝土砌工隔墙
concrete masonry unit 混凝土砌筑构件;混凝土

(墙)砌块
concrete masonry wall 混凝土砌工墙
concrete masonry window opening 混凝土砌工窗洞
concrete masonry work 混凝土砌工工程
concrete mast 混凝土桅杆;混凝土支柱
concrete master 混凝土含水率测量计
concrete mat(tress) 混凝土基础板;混凝土垫层;混凝土沉排
concrete mattress revetment 混凝土沉排护岸
concrete mattress roll 混凝土排辊
concrete meeting face 混凝土贴接面
concrete member jolt mo(u)lding machine 混凝土构件成型机
concrete membrane 混凝土罩面;混凝土薄层
concrete meter box 混凝土仪表箱
concrete micrometer 混凝土测微计
concrete mix 混凝土拌和物;混凝土拌和料
concrete mix design 混凝土级配设计;混凝土配合比设计
concrete mixed and placed in site 现拌现浇混凝土
concrete mix electric testing apparatus 混凝土混合料电测设备
concrete mixer 混凝土搅拌机;混凝土搅拌工厂;混凝土搅拌车;混凝土拌和机
concrete mixer drum 混凝土搅拌机圆筒
concrete mixer equipment depreciation apportion and overhaul charges 混凝土搅拌设备折旧摊销及大修费
concrete mixer heating attachment 混凝土搅拌预热装置;混凝土拌和机预热装置
concrete mixer-paver 混凝土搅拌铺(料)机
concrete mixer truck 混凝土拌和(汽)车;混凝土搅拌汽车
concrete mixer with cylindrical drum 有圆柱形鼓轮的混凝土搅拌机
concrete mixer with rotating drum 滚筒式混凝土搅拌机
concrete mixing 混凝土搅拌;混凝土拌制;混凝土拌和
concrete mixing "en route" 途中搅拌;途中拌和
concrete mixing center 混凝土搅拌厂
concrete mixing equipment 混凝土搅拌设备;混凝土浇灌设备;混凝土拌制设备
concrete mixing machine 混凝土搅拌机
concrete mixing on route 路拌混凝土;途中拌和
concrete mixing plant 混凝土搅拌设备;混凝土搅拌(工)厂;混凝土拌制设备;混凝土拌和设备;混凝土拌和车间;混凝土拌和厂
concrete mixing plate 混凝土搅拌板
concrete mixing station 混凝土搅拌站
concrete mixing temperature 混凝土搅拌温度
concrete mixing time 混凝土拌和时间
concrete mixing tower 混凝土搅拌塔
concrete mixing truck 混凝土搅拌运料车;混凝土搅拌车
concrete mixing vehicle 混凝土搅拌车
concrete mix proportion 混凝土配合(比)
concrete mix(ture) 混凝土混合物;混凝土混合料;混凝土掺和料
concrete mobile 混凝土汽车;混凝土流动配料搅拌车
concrete mobility 混凝土流动性
concrete model 混凝土模型;具体模型
concrete modification 混凝土修正;混凝土改正
concrete modifier 混凝土改进剂;混凝土调节剂
concrete modulus of elasticity 混凝土弹性模量
concrete moist curing 混凝土湿养护
concrete moist room 混凝土试样湿养护室;混凝土湿养护间;混凝土润湿室;混凝土润湿间;混凝土潮湿养护室
concrete mold or form release agent 混凝土模板脱模剂
concrete monolith 空心混凝土(大)方块;混凝土单块;混凝土大块体;混凝土沉井
concrete monolith breakwater 整体混凝土防波堤
concrete monolithic construction 混凝土整体性建筑
concrete monolith upright on rubble mound composite breakwater 巨型方块防波堤
concrete mortar 混凝土砂浆
concrete motor-track mixer 混凝土搅拌输送车
concrete motorway 混凝土汽车路

concrete mo(u)ld 混凝土模板
concrete mo(u)lding equipment factory 混凝土模板设备厂
concrete mo(u)ld oil 混凝土模板油
concrete mo(u)ld release oil 混凝土模板脱模油
concrete muff 混凝土套筒
concrete muntin 混凝土门窗中梃
concrete nail 混凝土钉;水泥钉
concrete nailing machine 混凝土射钉机
concrete narrow-ga(u)ge railcar 混凝土窄轨车
concrete normal block gravity wall 普通方块防波堤
concrete normal block quay wall 普通方块码头
concrete number 名数
concrete of dry consistency 干硬性混凝土
concrete offshore structure 海上混凝土结构物
concrete of high early strength 早强混凝土
concrete of lean mix 贫(灰)混凝土
concrete of low porosity 密实混凝土;低孔隙率混凝土;少孔隙混凝土
concrete of one consistency 等稠度混凝土
concrete of pump 泵送混凝土
concrete of stiff consistency 干硬性混凝土;低塌陷度混凝土
concrete of stiff consistency 干稠度性混凝土
concrete oil 模板油
concrete oil sump 混凝土油池;混凝土污水坑
concrete ordinary benchmark 混凝土普通水准标石
concrete orifice turnout 孔口式混凝土分水闸;孔口式混凝土分水阀;孔口式混凝土斗门
concrete ornamental string(course) 混凝土装饰线腿
concrete outfit 混凝土施工设备
concrete overflow dam 混凝土溢流坝
concrete overhang 混凝土板边悬空
concrete overlay 混凝土表层;混凝土罩面;水泥混凝土路面加厚层
concrete pad 混凝土座;混凝土基垫;混凝土垫层
concrete paint 混凝土涂料;混凝土(涂装)用(油)漆;混凝土面涂料;混凝土表面刷剧;刷混凝土墙用浆;刷混凝土墙涂料
concrete palisade fence 混凝土栏杆围篱
concrete panel fencing 混凝土条板围篱;混凝土板围篱
concrete panel with window opening 开有窗孔的混凝土墙板
concrete pantile 混凝土波形瓦
concrete parachute 混凝土巷道保险器;混凝土竖井防坠器
concrete parapet element 混凝土女儿墙附件
concrete pat 混凝土安定性试样
concrete patch 混凝土修补
concrete patching 混凝土填补;混凝土路面修补
concrete patch material 混凝土修补材料
concrete pavement 混凝土面层;混凝土路面;混凝土护面
concrete pavement block 混凝土路面试样块
concrete pavement curing 混凝土路面养护;混凝土地面养护
concrete pavement expansion joint cutter 混凝土路面切割机
concrete pavement finisher 混凝土路面整面机
concrete pavement joint 混凝土路面接缝
concrete pavement joint sealant 混凝土路面接缝密封膏
concrete pavement joint sealing compound 混凝土路面接缝密封合成物
concrete pavement pumping 混凝土路面抽吸现象
concrete pavement scaling 混凝土路面剥落
concrete pavement shattering machine 混凝土路面粉碎机
concrete pavement slab 混凝土铺面板;混凝土路面板
concrete pavement stress 混凝土路面应力
concrete pavement vibrator 混凝土铺路振动器
concrete paver 混凝土铺路机;混凝土摊铺机;混凝土路面铺筑机
concrete paving 混凝土面层;混凝土护面;铺筑混凝土路面
concrete paving block 铺路混凝土砌块
concrete paving equipment 混凝土路面机械
concrete paving flag 混凝土铺路石板
concrete paving form 混凝土铺路模板
concrete paving machine 混凝土路面机械

concrete paving sett 混凝土铺路小方石
concrete paving slab 混凝土铺板
concrete paving vibrator 混凝土路面振捣器
concrete pedestal 混凝土基座;混凝土基础
concrete pedestal of frame bent 座架排架混凝土底座
concrete penetration resistance tester 混凝土贯入阻力测定仪
concrete penetrometer 混凝土针入度仪;混凝土渗透仪
concrete perforated drain pipe 混凝土花管
concrete pergola 混凝土藤架;混凝土凉亭
concrete pier 混凝土码头;混凝土墩
concrete pile 混凝土桩
concrete piled pier 混凝土桩基承台
concrete pile driver 混凝土桩打桩机
concrete pile driving 打混凝土桩
concrete-pile follower 混凝土桩的送桩器;混凝土送桩机
concrete pile foundation 混凝土桩基(础)
concrete pile foundation structure 混凝土桩基结构
concrete pile jetting 混凝土桩水冲法;用水下冲法沉混凝土
concrete pile protection 混凝土板桩防护
concrete pile piling 混凝土板桩;混凝土(排)桩
concrete pillar 混凝土(支)柱;混凝土(支)墩;混凝土坡;混凝土标石
concrete pipe 混凝土(漏斗)管;混凝土导管(用于水下灌注混凝土)
concrete pipe bend 混凝土弯管
concrete-pipe composite pile 混凝土钢管组合桩
concrete pipe compression tester 混凝土管压力试验机
concrete pipe culvert 混凝土管洞;混凝土管涵
concrete pipe drain 混凝土排水管
concrete pipe laying unit 混凝土管埋置设备
concrete pipe making machine 混凝土管制造机
concrete pipe mould 混凝土管型模
concrete pipe pile 混凝土管桩
concrete pipe pipeline 混凝土管线
concrete pipe press 混凝土管压制机
concrete pipe press head 混凝土管压制机压头
concrete pipe rack 混凝土管架
concrete pipes cast in place 现场浇捣混凝土管
concrete pipe spinning machine 混凝土管离心制管机
concrete pipe works 混凝土管工厂
concrete piping 混凝土管
concrete pitching 混凝土护坡
concrete placeability 混凝土可砌筑性;混凝土可灌(注)性;混凝土和易性;混凝土和易度;混凝土的可浇筑性;混凝土的可浇置性
concrete placed by compressed air 压缩空气浇筑混凝土
concrete placed in situ 现浇混凝土
concrete placed in the work 混凝土浇筑
concrete placement 浇灌混凝土;浇捣混凝土;混凝土摊铺;混凝土浇筑;混凝土浇注;灌注混凝土
concrete placer 混凝土摊铺机;混凝土铺设机;混凝土喷射机;混凝土喷射机;混凝土浇注设备;混凝土浇注机;混凝土灌注机;水下混凝土浇注机
concrete placing 浇注混凝土;混凝土浇注
concrete placing boom 混凝土布料杆
concrete placing bucket 混凝土灌注桶
concrete placing by gravity 重力浇筑混凝土;混凝土浇筑
concrete placing chute 混凝土浇注槽;浇注溜槽
concrete placing hose 混凝土浇注管
concrete placing in site 现浇混凝土
concrete placing machine 混凝土浇注机
concrete placing plant 混凝土灌筑设备
concrete placing platform 混凝土浇注台
concrete placing sequence 混凝土浇灌程序
concrete placing skip 混凝土摊铺斗;混凝土浇筑料斗;混凝土浇注斗
concrete placing tower 混凝土浇注塔
concrete placing trestle 混凝土施工栈桥
concrete plain tile 纯混凝土屋瓦
concrete planer 混凝土路面整平机
concrete plank 混凝土(跳)板;预制空心楼板
concrete plank silo 混凝土板(筑的)筒仓
concrete plant 混凝土厂
concrete plasticizer 混凝土塑化剂
concrete plate 混凝土平板

concrete-platen interface 混凝土与压板的界面
concrete platform 混凝土平台
concrete plug 混凝土塞(封);混凝土封底
concrete plywood 胶合板制混凝土模板
concrete plywood for formwork 混凝土胶合板模板
concrete poker vibrator 插入式混凝土振捣器
concrete pole 混凝土支柱
concrete pond 混凝土储[贮]槽
concrete pontoon 混凝土平底船;混凝土浮船;混凝土囤船
concrete pool 混凝土(游泳)池
concrete porous pipe 多孔混凝土管
concrete portal frame 混凝土门式钢架
concrete post 混凝土柱(石)
concrete post shotcrete 混凝土柱支撑
concrete post-tensioning 后张法预应力混凝土
concrete pot for walls 空心混凝土砌块
concrete pouring 浇注混凝土;混凝土灌注
concrete pouring bucket 混凝土浇注吊桶
concrete pouring equipment 混凝土浇筑设备
concrete pouring machine 混凝土浇筑机械;混凝土浇注设备;混凝土浇注机
concrete pouring platform 混凝土浇注台
concrete power saw 混凝土锯缝机;混凝土动力锯(缝机)
concrete practice 混凝土实施技术;混凝土实施法
concrete precasting factory 混凝土预制工厂
concrete preservative 混凝土防腐剂
concrete pressure 混凝土压力
concrete pressure grouting machine 混凝土压力浇注机
concrete pressure pipe 混凝土压力管
concrete primary benchmark 混凝土基本水准标石
concrete primer 打底混凝土浆
concrete prism 混凝土棱柱体
concrete prismatic strength 混凝土棱柱体强度
concrete prism test 混凝土棱柱体试验
concrete probe 混凝土工作度探针
concrete product 混凝土(预)制品
concrete product factory 混凝土制品厂
concrete production 混凝土生产
concrete production plant 混凝土制品厂;混凝土拌制厂
concrete product maker 混凝土制品制造者;混凝土制品制造厂(商)
concrete product manufacturer 混凝土制品生产厂
concrete product mould 混凝土制品模具
concrete product plant 混凝土制品厂
concrete property 混凝土性质
concrete proportioning 混凝土配合(比)
concrete proportioning by trial method 用试验法配合混凝土成分
concrete proportioning by void-cement method 用隙灰比法配合混凝土成分
concrete proportioning by water-cement method 用水灰比配合混凝土成分
concrete protection fence 混凝土防护篱
concrete protection layer 混凝土保护层
concrete protection tent 混凝土防护棚
concrete protective agent 混凝土防护剂
concrete pump 混凝土泵;水下混凝土泵
concrete pumpability 混凝土可泵送性
concrete pumping 泵送混凝土;混凝土泵送
concrete pumping pipe 混凝土泵送管
concrete pump placing 混凝土泵浇筑;用混凝土泵浇注
concrete purlin(e) 混凝土檩条
concrete quality control 混凝土质量控制
concrete quality grade 混凝土质量分等;混凝土质量等级;混凝土性质分等
Concrete Quarterly 混凝土季刊(英国期刊名)
concrete quay 混凝土码头
concrete quaywall 混凝土岸壁
concreter 混凝土工(人)
concrete radiation shield 混凝土辐射屏蔽
concrete radiation shielding wall 混凝土防辐射墙
concrete raft 混凝土排基;混凝土筏基
concrete railroad tie 混凝土枕木
concrete railway sleeper 混凝土铁路轨枕;混凝土轨枕
concrete rainwater gutter 混凝土雨水檐槽
concrete rammer 混凝土振动器;混凝土振动机;混凝土振捣器;混凝土振捣机;混凝土夯实器;混凝土夯实机;混凝土夯捣器;混凝土夯捣器;混凝土捣实器

concrete reactor 混凝土芯反应堆;混凝土芯电抗器
concrete reactor vessel 混凝土反应堆压力容器
concrete rebound 混凝土回弹
concrete receiving hopper 混凝土受斗
concrete refuse water pipe 混凝土污水管;混凝土废水管
concrete re-handling 混凝土重新使用;混凝土再搅拌
concrete reinforcement 混凝土配筋
concrete reinforcement bar 粗钢筋
concrete reinforcement distance piece 混凝土钢筋间隔块
concrete reinforcement wire 混凝土配筋用钢筋;细钢筋;(混凝土中的)扎钢筋铅丝
concrete reinforcing bar 钢筋;混凝土(用)钢筋;混凝土配筋用钢筋
concrete reinforcing steel 混凝土加固钢筋
concrete reintegration 混凝土再捣实
concrete reintegration material 混凝土修补材料
concrete remixer 混凝土复拌机
concrete remover 混凝土脱模剂
concrete repair 混凝土修理
concrete repair material 混凝土修理材料
concrete replacement 混凝土重浇筑
concrete replacement method 混凝土置换法
concrete representation 点位精度
concrete reservoir 混凝土蓄水池
concrete resurfacing 混凝土面翻筑
concrete retaining wall 混凝土挡土墙
concrete retarder 混凝土缓凝剂
concrete retempering 混凝土重塑
concrete revetment 混凝土护岸
concrete rib 混凝土肋
concrete rib(bed)floor 混凝土密肋楼板
concrete rib(bed)slab 混凝土密肋(预制)楼板
concrete ribbon 混凝土运送带;混凝土跑道
concrete ridge tile 混凝土脊瓦
concrete ring 混凝土环
concrete ring beam 混凝土环梁
concrete-ring tensile test 环形混凝土试件抗拉实验
concrete road 混凝土(道)路
concrete roadbed 混凝土整体道床;混凝土路基
concrete road finisher 混凝土修平路面机;混凝土路面平整机
concrete road(inlet) gulley 混凝土道路进水口;混凝土道路集水沟
concrete road outlet 混凝土道路排水口
concrete road pavement 混凝土路面
concrete road paver 混凝土路面铺装机;混凝土(路面)铺路机;道路混凝土浇筑机
concrete rock-mouth breakwater 混凝土堆石防波堤
concrete roll 混凝土辊
concrete roller 混凝土碾压机
concrete rolling contact joint 混凝土滚动接头
concrete roof 混凝土屋顶;混凝土屋面;混凝土顶板
concrete roof deck 混凝土平屋顶
concrete roof gutter 混凝土屋顶檐沟;混凝土屋顶排水沟
concrete roofing slab 混凝土屋面板
concrete roofing tile 混凝土屋面瓦
concrete roofing tile machine 混凝土屋瓦机
concrete roofing tile tester 混凝土屋瓦试验机
concrete roof slab 混凝土屋面板
concrete roof tile 混凝土瓦;混凝土制屋瓦
concrete roof truss 混凝土屋架
concrete roughening unit 混凝土凿毛器
concreter's labo(u)rer 混凝土杂工
concrete runway 混凝土跑道
concrete rupture 混凝土破坏
concrete saddle (管道的)混凝土支座;混凝土托座;混凝土垫座;混凝土鞍座
concrete safety fence 混凝土安全栅栏;混凝土安全篱笆
concrete sample 混凝土试样;混凝土试件
concrete sand 混凝土用的砂;固结砂;胶结砂
concrete sand-lime building brick 灰砂混凝土建筑砖
concrete sandwich panel 混凝土夹层板
concrete sanitary building block 混凝土卫生建筑砌块模数
concrete sanitary unitized unit 混凝土卫生统一构件;混凝土卫生统一单元
concrete saw 混凝土锯;水泥混凝土锯缝机

concrete saw-tooth roof shell 混凝土锯齿形屋顶薄壳
concrete scaling 混凝土剥落
concrete scarp 混凝土残渣
concrete scrap 混凝土废料
concrete scraper 混凝土铲运机
concrete screed(coat) 混凝土找平层;混凝土整平
concrete-screed floating floor cover(ing) 混凝土刮板刮平楼板面层
concrete screeding 混凝土找平
concrete screed material 混凝土找平灰饼材料;混凝土定厚冲筋材料
concrete screed topping 混凝土刮板顶面刮平
concrete screw pile 混凝土螺旋桩
concrete screw-type mixer 混凝土螺旋搅拌机
concrete sculpture 混凝土雕塑
concrete scum slab 混凝土浮沫括板
concrete sealant 混凝土密封剂
concrete sealer 混凝土密封剂
concrete sealing 混凝土密封
concrete sealing hole 混凝土封孔
concrete seat 混凝土座位
concrete secondary liner 混凝土二次衬砌
concrete section 混凝土型材;混凝土剖面
concrete segment 混凝土管片
concrete segregation 混凝土离析
concrete selection 混凝土选择;混凝土取样
concrete separator 混凝土分隔物
concrete septic tank 混凝土化粪池
concrete setting 混凝土凝固;混凝土凝结
concrete sett paving 混凝土小块铺面
concrete sewage pipe 混凝土污水管
concrete sewer 混凝土污水管;混凝土下水道;混凝土排水管;混凝土管沟
concrete sewer pipe 混凝土污水管
concrete shaft 混凝土竖井
concrete shaped block 混凝土异形块体
concrete shapes 混凝土异形块体
concrete shear key 混凝土抗剪键
concrete shear wall 混凝土剪力墙;混凝土抗剪墙
concrete sheath coat 混凝土外壳;混凝土涂层
concrete sheet pile 混凝土板桩
concrete sheet pile breakwater 混凝土板桩防波堤
concrete sheet piling 混凝土板桩
concrete shell 混凝土壳体;混凝土薄壳
concrete shell cupola 混凝土薄壳圆顶
concrete shell design 混凝土薄壳设计
concrete shell dome 混凝土薄壳屋顶;混凝土薄壳穹顶
concrete shell pile 混凝土薄壳桩
concrete shelter 混凝土隐蔽所
concrete shield 混凝土罩面;混凝土屏蔽;混凝土衬砌层
concrete shielding wall 混凝土屏蔽墙;混凝土防护墙
concrete ship 混凝土船;水泥结构船;水泥船
concrete shoulder 混凝土路肩;混凝土拱座;混凝土挡肩(轨枕)
concrete shrinkage 混凝土收缩;混凝土干缩
concrete shutter 混凝土模板
concrete shuttering 混凝土模壳;混凝土模板
concrete sidewalk 混凝土人行道
concrete signal 具体信号
concrete sill 混凝土门槛;混凝土垫板;混凝土窗台
concrete silo 混凝土料仓
concrete sinker 混凝土沉锤
concrete siphon 混凝土虹吸管
concrete skeleton 混凝土骨架
concrete skeleton construction 混凝土骨架建筑
concrete skip 混凝土输送车
concrete skip hoist 混凝土料斗提升机
concrete skyscraper 混凝土摩天大楼
concrete slab 混凝土平板;混凝土面板;混凝土路面;混凝土板
concrete slab bridge 混凝土板桥
concrete slab compressive strength 混凝土板的抗压强度
concrete slab crushing strength 混凝土板的破碎强度
concrete slab facade 房屋正面混凝土板
concrete slab facing 混凝土板饰面;混凝土板面饰
concrete slab in situ 现场浇混凝土板
concrete slab lining 混凝土板衬里
concrete slab pavement 混凝土板护面
concrete slab press 混凝土板压机

concrete slab revetment 混凝土板护坡;水泥板护岸
concrete slabs and beams platform supported on piles 板梁式高桩码头
concrete slab sub-floor 混凝土板垫层地板
concrete slab track 混凝土板式轨道
concrete slab work 混凝土板护坡
concrete slatted floor 混凝土石板地面
concrete sleeper 混凝土(轨)枕
concrete sluice 混凝土节制闸
concrete slump cone 混凝土坍落度锥;混凝土坍落度筒
concrete slump test 混凝土坍落度试验
concrete snow 固结雪
concrete soffit 混凝土拱腹面
concrete solid block 混凝土实心块
concrete solid pier 混凝土实心桥墩
concrete solid tile 混凝土实心砖
concrete solvent 混凝土溶剂
concrete spacer(block) 混凝土垫块
concrete specification 混凝土说明;混凝土技术要求;混凝土规格;混凝土规范
concrete specimen 混凝土试样;混凝土试件
concrete spinning method 混凝土离心浇注法;混凝土离心制造法
concrete spiral casing 混凝土螺旋形外壳;混凝土蜗壳
concrete spiral stair(case) 混凝土螺旋楼梯
concrete splashing 混凝土喷射
concrete-spliced wood pile 混凝土套接木桩
concrete split duct 混凝土沟;混凝土分支管道;混凝土槽
concrete splitter 混凝土劈裂器;混凝土分离器
concrete spout(ing) 混凝土灌注;混凝土喷注;混凝土喷射;混凝土浇灌
concrete-spouting air-pressure 喷射工作风压
concrete-spouting bay 浇捣仓
concrete spouting equipment 混凝土喷射设备
concrete-spouting hydraulic pressure 喷射工作水压
concrete-spouting plant 混凝土浇灌设备;混凝土灌注设备
concrete-spouting plant depreciation apportion and overhaul charges 混凝土喷射设备折旧摊销及大修费
concrete-spouting technical parameter 喷射混凝土工艺参数
concrete sprayer 混凝土喷射机;混凝土喷布机
concrete-spraying 混凝土喷射
concrete-spraying machine 混凝土喷射机;混凝土喷布机
concrete-spraying stair(case) 喷射混凝土楼梯
concrete-spraying(staircase) step 喷射混凝土楼梯梯级;喷射混凝土楼梯踏步
concrete-spraying staircase tower 喷射混凝土楼梯水塔楼
concrete-spraying stand 喷射混凝土台架;喷射架
concrete-spraying steel 喷射混凝土中的钢筋
concrete-spraying strain 喷射混凝土的应变
concrete-spraying street(inlet) gulley 喷射混凝土街道进水口
concrete-spraying strength 喷射混凝土强度
concrete-spraying stress 喷射混凝土的应力
concrete-spraying string course 喷射混凝土束带层
concrete-spraying structural slab 喷射混凝土结构用板
concrete-spraying sub-floor 喷射混凝土底层地板
concrete-spraying subgrade paper 喷射混凝土用地基纸
concrete-spraying supporting medium 喷射混凝土支承介质
concrete-spraying surcharge 喷射混凝土过多
concrete-spraying surface 喷射混凝土表面
concrete-spraying sur faced iron pipe 喷射混凝土包裹的铁管
concrete-spraying surface finish 喷射混凝土表面修饰
concrete-spraying surface hardener 喷射混凝土表面硬化剂
concrete-spraying surface improver 喷射混凝土表面优化剂;喷射混凝土表面改进剂
concrete-spraying surface retardant 喷射混凝土表面缓凝剂
concrete-spraying surfacing 喷射混凝土路面

concrete-spraying surfacing joint sealing compound 喷射混凝土表面接缝密封合成物
concrete-spraying swimming pool 喷射混凝土游泳池
concrete-spraying table 喷射混凝土台架;混凝土喷射架
concrete-spraying tank 喷射混凝土储[贮]罐
concrete-spraying technician 喷射混凝土技师
concrete-spraying technologist 喷射混凝土工艺师
concrete-spraying technology 喷射混凝土工艺学
concrete-spraying television tower 喷射混凝土电视塔
concrete-spraying temperature 喷射混凝土温度
concrete-spraying tensile zone 喷射混凝土拉伸区
concrete-spraying tension 喷射混凝土的拉伸
concrete-spraying tension crack 喷射混凝土受拉开裂
concrete-spraying tension strength 喷射混凝土的抗拉强度
concrete-spraying test 喷射混凝土试验
concrete-spraying test beam 喷射混凝土试验梁
concrete-spraying(test)cube 喷射混凝土立方试块
concrete-spraying(test)cylinder 喷射混凝土圆柱体试件
concrete-spraying test hammer 喷射混凝土试验锤(回弹仪)
concrete-spraying testing machine 喷射混凝土试验机
concrete-spraying testing method 喷射混凝土试验方法
concrete-spraying texture 喷射混凝土组织
concrete-spraying thickness 喷射混凝土厚度
concrete-spraying threshold 喷射混凝土门槛
concrete-spraying tile 喷射混凝土瓦
concrete-spraying tile masonry wall 喷射混凝土砖坯工墙
concrete-spraying tile press 喷射混凝土瓦的压制机
concrete-spraying tile roof cover(ing) 喷射混凝土瓦屋顶铺设
concrete-spraying tile works 喷射混凝土瓦厂
concrete-spraying topping 喷射混凝土面层
concrete-spraying tower 喷射混凝土塔式建筑;喷射混凝土加料提升塔
concrete-spraying trimmer plank 喷射混凝土托梁板
concrete-spraying trough 喷射混凝土(输送)槽
concrete-spraying trough slab 喷射混凝土槽板
concrete-spraying type 喷射混凝土类型
concrete-spraying type of construction 喷射混凝土建筑类型
concrete-spraying umbrella 喷射混凝土顶盖
concrete-spraying unitized unit 喷射混凝土应用的单位
concrete-spraying vault 喷射混凝土穹隆
concrete-spraying wagon vault 喷射混凝土圆形穹隆
concrete-spraying wall 喷射混凝土墙
concrete-spraying wall block 喷射混凝土墙块
concrete-spraying wall pot 喷射混凝土墙洞
concrete-spraying ware 喷射混凝土制品
concrete-spraying ware factory 喷射混凝土制品厂
concrete-spraying waste pipe 喷射混凝土污水管;喷射混凝土废水管
concrete-spraying waterproofer 喷射混凝土防水剂
concrete-spraying waterproofing 喷射混凝土防水
concrete-spraying waterproofing powder 喷射混凝土防水粉
concrete-spraying water-reducing agent 喷射混凝土减水剂
concrete-spraying water tank 喷射混凝土水箱
concrete-spraying water waste preventer 喷射混凝土污水防止器
concrete-spraying wearing surface 喷射混凝土磨损表面
concrete-spraying web 喷射混凝土梁腹;喷射混凝土腹板
concrete-spraying wedge 喷射混凝土楔
concrete-spraying well ring 喷射混凝土井圈
concrete-spraying wet density 喷射混凝土湿密度
concrete-spraying window sill 喷射混凝土窗槛
concrete-spraying with cork aggregate 用软木作骨料的喷射混凝土
concrete-spraying with high early stability 具有超早期稳定性的喷射混凝土

concrete-spraying with high early strength 超早强喷射混凝土
concrete-spraying with large aggregate 大骨料喷射混凝土
concrete-spraying with oil addition 加油的喷射混凝土
concrete-spraying with synthetic resin dispersion 合成树脂悬浮液喷射混凝土
concrete-spraying with washed pumice gravel 水洗浮石骨料喷射混凝土
concrete-spraying work 喷射混凝土工程
concrete-spraying workability 喷射混凝土用工作度剂;喷射混凝土和易性;喷射混凝土工作度;喷射混凝土稠度
concrete-spraying workability agent 喷射混凝土用稠度剂;喷射混凝土用和易性剂
concrete-spraying yard inlet gulley 喷射混凝土工场进水沟
concrete spreader 混凝土布料机;混凝土撒布机;混凝土撒播机;混凝土平铺机;混凝土(路面)摊铺机;混凝土浇注机
concrete spreading 混凝土铺浇;混凝土摊铺;混凝土平仓
concrete spreading plant 混凝土散布机
concrete stair(case) 混凝土楼梯
concrete stand 混凝土架座;混凝土车架
concrete stationary time 拌和料静放时间
concrete stave 混凝桶板
concrete-steel 劲性钢筋;钢架混凝土;混凝土配筋;混凝土钢筋;钢筋钢
concrete-steel building 劲性钢筋混凝土建筑;劲性钢筋混凝土房屋;型钢混凝土房屋;钢筋混凝土建筑
concrete stiff consistency 混凝土干硬性
concrete stiffener 混凝土加劲筋
concrete stone 混凝土(方)块
concrete stop-end joint 混凝土施工缝
concrete strain 混凝土变形
concrete strain indicator 混凝土应变仪
concrete strength 混凝土强度
concrete stress 混凝土应力
concrete strip foundation 混凝土条形基础;混凝土带形基础
concrete structure 混凝土结构;混凝土建筑物;混凝土构造
concrete sub-base 混凝土底层;基层混凝土;混凝土下卧层
concrete sub-floor 混凝土毛楼;混凝土楼板;混凝土底楼
concrete sub-slab 混凝土底板;混凝土垫层
concrete support 混凝土支架
concrete support behind side wall 边墙后侧混凝土支护
concrete supporting medium 混凝土底座(基体)
concrete support plate 混凝土支撑板
concrete surcharge 混凝土超载
concrete surface 混凝土表面;混凝土地面
concrete surface channel 混凝土明沟
concrete surface dressing 混凝土琢面
concrete surface finishing 混凝土表面加工
concrete surface finishing machine 混凝土表面加工机械
concrete surface joint cutter 混凝土路面接缝切割机
concrete surfacer 混凝土表面处理器
concrete surface treatment 混凝土表面处理
concrete surfacing 混凝土面加工
concrete table 混凝土桌;混凝土平台;混凝土计算表格
concrete tank 混凝土储[贮]水池;混凝土水箱;混凝土油罐;混凝土水箱;混凝土槽
concrete taper pile 混凝土锥形柱
concrete technology 混凝土技术;混凝土工艺(学)
concrete temperature control 混凝土温控
concrete tensile crack 混凝土受力裂缝
concrete tensile stress 混凝土拉应力
concrete tensile zone 混凝土受力区
concrete terrazzo 混凝土水磨石;水磨石面混凝土
concrete test 混凝土试验
concrete test cube 混凝土试验立方块
concrete test(ing) cylinder 混凝土试验圆柱体
concrete test(ing) hammer 混凝土试验锤
concrete test(ing) machine 混凝土试验机
concrete test(ing) mo(u)ld 混凝土试(验)模
concrete testing mo(u)ld piece 混凝土试件
concrete test(ing) mo(u)ld tamper 混凝土试验捣固器
concrete test(ing) section 混凝土试验段
concrete tetrahedron 混凝土四面体
concrete tetrapod (防波堤上的)混凝土四角锥体块;(防波堤上的)混凝土四角块体
concrete thermal stress 混凝土温度应力
concrete thermometer 混凝土温度测量仪
concrete thrust block 混凝土抗推块;混凝土墩
concrete tie 混凝土枕;混凝土轨枕
concrete tile 混凝土烟道砖;混凝土瓦;水泥瓦
concrete tilt-up 混凝土现场浇注后竖起
concrete-timber composite pile 混凝土—木组合桩
concrete-timber pile 混凝土木桩;混凝土(和)木材(和)混合结构桩
concrete tip wagon 运混凝土倾卸车
concrete topping 混凝土面层;混凝土覆盖层
concrete tower 混凝土升运塔
concrete trackbed 混凝土道床
concrete track slab 混凝土轨枕板;轨枕板
concrete train 混凝土路面铺路机组;混凝土浇筑列车
concrete transfer car 混凝土转运车;混凝土运送车
concrete transfer pump 混凝土输送泵
concrete transport 混凝土运输
concrete transportation 混凝土运输
concrete transportation system 混凝土运输系统
concrete transporting equipment 混凝土运输设备
concrete tremie 混凝土导管(用于水下灌注混凝土)
concrete tribar 混凝土三柱体块;混凝土三棱体块
concrete truss 混凝土桁架
concrete tube 混凝土管
concrete tubing 混凝土管桩
concrete tubular pile 混凝土管桩
concrete unit 具体单位;混凝土预制件;混凝土心板
concrete unit masonry 混凝土(方)块砌体
concrete vault 混凝土掩蔽所;混凝土穹顶;混凝土地下室
concrete vessel 水泥船
concrete vibrating 混凝土振捣
concrete vibrating machine 混凝土振动器;混凝土振动机;混凝土振捣器;混凝土振捣机;混凝土夯实机
concrete vibrating screen 混凝土振动式找平器
concrete vibrating table 混凝土振动台
concrete vibration 混凝土振捣
concrete vibration method 混凝土振捣方法
concrete vibration stand 混凝土振动台
concrete vibrator 混凝土振动器;混凝土振动机;混凝土振捣器;混凝土振捣机;混凝土路面振动(夯实)器;混凝土捣实器;水泥混凝土振捣器
concrete vibratory machine 混凝土夯实机
concrete vibro-column pile 振动混凝土桩
concrete volumn 混凝土量
concrete waffle slab 混凝土格子板
concrete wall 混凝土墙
concrete wall anchorage 用混凝土墙锚碇
concrete wall block 混凝土墙块
concrete walling 混凝土墙砌体
concrete wall(ing) tile 混凝土壁砖
concrete wall panel 混凝土墙板
concrete wall type dam 混凝土芯墙式坝
concrete ware 混凝土制品;混凝土器皿
concrete warping 混凝土弯翘;混凝土卷曲;混凝土翘曲
concrete waste pipe 混凝土污水管
concrete water pipe 混凝土水管
concrete waterproofer 混凝土防水剂
concrete waterproofing 混凝土防水(性)
concrete waterproofing compound 混凝土防水剂
concrete waterproofing layer 混凝土防水层
concrete waterproof(ing) oil 混凝土防水油
concrete water tank 混凝土水箱
concrete wearing layer 混凝土耐磨板;混凝土耗层
concrete web 混凝土腹板
concrete weight 混凝土重量
concrete well foundation 混凝土井基础
concrete wharf 混凝土码头
concrete wheelbarrow 混凝土手推车
concrete window frame 混凝土窗框
concrete with artificial resin admix(ture) 合成树脂混凝土
concrete with cork aggregate 用软木为骨料的混凝土
concrete with entrained air 加气混凝土
concrete with entrance 加气混凝土
concrete with high early stability 早期高稳定混凝土
concrete with high early strength 早强混凝土
concrete with honeycombed spots 蜂窝混凝土
concrete with impregnation 浸渍混凝土
concrete with large aggregate 大骨料混凝土
concrete with oil addition 带油的混凝土
concrete workability 混凝土工作度;混凝土和易性;混凝土和易度
concrete works 混凝土工厂;混凝土工程
concreting 浇筑混凝土场地;浇灌混凝土;灌注混凝土
concreting aid 混凝土浇灌设备;混凝土浇灌工具
concreting air-lock 浇筑混凝土气闸
concreting bucket attachment 混凝土料桶附件
concreting crane 浇注混凝土用起重机
concreting curve 混凝土浇筑曲线
concreting deposit 混凝土浇筑物
concreting device 浇筑混凝土设备
concreting equipment 浇筑混凝土设备;混凝土筑设备;混凝土浇灌设备
concreting foil 混凝土浇注薄膜
concreting funnel 浇筑混凝土漏斗
concreting gang 混凝土工作队
concreting hopper 浇筑混凝土漏斗;浇灌混凝土料斗
concreting in cold-weather 混凝土冬季施工;混凝土冬季浇筑;冬季施工混凝土;冬季混凝土浇筑;冬季混凝土浇注;冬季混凝土浇灌
concreting in freezing weather 混凝土冬季施工;混凝土冬季浇筑;冻期浇注混凝土;冬季浇注混凝土;冬季混凝土浇筑;冰冻期混凝土施工
concreting in layers 分层灌浆混凝土
concreting in lifts 混凝土分层浇筑;分层浇筑混凝土;分层浇注混凝土;分层浇灌混凝土;分层混凝土;分层灌浆混凝土
concreting in rain 雨中浇注混凝土;雨中浇捣混凝土
concreting in water 水下浇筑混凝土;水下浇注混凝土;水下浇筑混凝土;水下灌筑混凝土;水下灌注混凝土
concreting job 混凝土施工现场
concreting line 混凝土浇注线;混凝土传送带
concreting machine 混凝土浇灌机械
concreting material 固结材料
concreting method 混凝土浇筑方法
concreting operation 混凝土灌筑
concreting outfit 混凝土施工设备
concreting output 浇灌混凝土完成量
concreting paper 混凝土浇筑用纸
concreting piston pump 浇灌混凝土活塞泵
concreting placing plant 混凝土浇筑设备
concreting plant 混凝土配制厂
concreting program(me) 混凝土浇筑计划;混凝土浇筑程序;混凝土浇注进度计划
concreting rate 混凝土浇筑速率
concreting scaffold(ing) 浇灌混凝土脚手架
concreting section 混凝土浇筑分段
concreting sequence 混凝土浇筑程序
concreting site 混凝土浇筑场地
concreting skip 浇混凝土用料车;浇混凝土用翻斗
concreting skip attachment 灌筑混凝土料车设施
concreting stage 混凝土浇筑阶段
concreting technique 混凝土浇筑技术
concreting temperature 混凝土浇筑温度
concreting tower 混凝土浇筑塔;混凝土浇灌塔
concreting train 混凝土浇筑机列
concreting under shelter 棚下浇筑混凝土
concretion 结核;凝块;(液体变为固体的)凝结;固结
concretionary 凝结物;已凝结的;凝固的
concretionary horizon 结核层;硬结地层;固结地层
concretionary layer 固结层
concretionary structure 结核状构造
concretionary texture 结核状结构
concretion of anhydrite 硬石膏凝结
concretion of lime 石灰质结硬
concretions of limestone 石灰岩夹渣;石灰包裹体
concretive 有凝固力的;凝结性的
concretor 混凝土工(人)
concurrence 并发;同时发生

concurrence between delivery and payment 交货与付款同时履行
concurrence chart 网络图
concurrency 并行性;并行
concurrent 共行的;共同的;共点的;共存的;重合的;并行的;并行
concurrent access 并行访问
concurrent activity 并行动作
concurrent asynchronous computer 并行异步计算机
concurrent authority 同等权力
concurrent axis 交叉轴;相交轴
concurrent boiler 直流锅炉;顺流锅炉
concurrent centrifuge 单向流离心机;顺流式离心机
concurrent completion 同时完工
concurrent computation 并行计算
concurrent computer 并行(操作)计算机;并联计算机
concurrent condition 相互依存条件
concurrent consideration 同时审议
concurrent control 并行控制;并发控制
concurrent control count 并行控制计数
concurrent control system 共行控制系统;并行控制系统
concurrent conversion 并行转换
concurrent curing 并发固化
concurrent delay 共同作用拖期
concurrent design 并行设计
concurrent discharge 对应流量;同时流量
concurrent disinfection 即时消毒
concurrent fired kiln 并流烧窑
concurrent flow 平行流(动)
concurrent flow mixer 同流混合器;顺流混合器
concurrent force 汇交力;共力点;共点力
concurrent heating 补(充加)热;同时加热
concurrent infection 同时传染
concurrent inhibition 半发抑制
concurrent input-output 同时输入—输出;并行输入—输出
concurrent insurance 共同保险
concurrent interruption 同时中断
concurrent jurisdiction 共同管辖权
concurrent leaching 同流浸出;顺流浸出
concurrent line 共点线
concurrently shared resource 并行共享资源
concurrently transmission 同时传送
concurrent microoperation 并行微操作
concurrent multiple terminal operation 并行多终端操作
concurrent operating control 并行操作控制
concurrent operating system 并行操作系统
concurrent operation 共行操作;并行运算;并行操作;同时操作
concurrent ownership 共同所有权
concurrent peripheral operation 并行外围设备操作;并行外部操作;同时外部操作
concurrent peripheral processing 并行外围处理;同时外围处理
concurrent planes 共点面
concurrent process 共行进程;并行进程
concurrent processing 共行处理;并行处理;同时处理
concurrent processor 并行加工程序;并行处理机;并行处理程序
concurrent program(me) 并行程序
concurrent program(me) design 并发程序设计
concurrent programming 并行程序设计
concurrent-range zone 共存延限带(生物地层单位)
concurrent realtime processing 并行实时处理
concurrent risk 共同风险
concurrent run unit 并行运行单位
concurrent scrubber 并行擦洗器
concurrent selection 同时选择
concurrent study 同时性观察研究
concurrent terms 相互一致条件
concurrent transaction 并行细目数据
concurrent transformation 共点变换
concurrent transmission 并行传输;同时传送
concurrent with discharge(payment of freight) 于卸货时支付(运费)
concurrent working 共行工作;并行工作
concussion 爆气冲击;震荡
concussion blasting 震动性;松动爆破

concussion burst 冲击爆破
concussion fracture 冲击破裂
concussion fuse 震动引信
concussion relief valve 冲击式安全阀
concussion spring 减振弹簧
concycle 共圆
concyclic(al) 共圆的
concyclic(al) points 共圆点
cond aluminium 导电铝合金
condary constraints 次要约束
condeep 深水混凝土结构
condeep platform 水下混凝平台
condemnation 征用;废弃
condemnation appraisal 征用估价
condemnation award 土地征用补偿
condemnation clause 征用条款
condemnation deformation spectrum 倒毁变形谱
condemnation threshold earthquake 倒毁阀限地震
condemnation value 征用价值
condemned 寿命已到的;可拆卸的;废弃的;报废的
condemned dwelling 被告为废弃的住宅(因拆迁等原因)
condemned road 废弃的道路
condemned stores 废品
condemnee 财产被征用者
condemner 征用财产者
condemning limit 报废(尺寸)界限
condemn limit 报废限度
condensability 浓缩能力;冷凝性;可凝缩性;缩缩性
condensable 可压缩;可凝结;可冷凝的
condensable vapo(u)r 可凝结的蒸气
condensance current 电容性电流
condensate 冷凝物;冷凝水;凝结物;凝结水
condensate booster pump 凝结水升压泵
condensate circuit 凝结水回路
condensate circulating water pipe 凝结水循环管
condensate collection line 凝结水管线
condensate collector 凝结液收集器
condensate control 凝结水控制
condensate depression 凝结水过冷(却);过冷度
condensate drain 冷凝排水
condensate drainage 冷凝水排除;凝结水排除
condensate draining 冷凝水排
condensate drain loop 疏水管路水封
condensate drain outlet 冷凝水排放口
condensate drain pan 凝结水盘
condensate extractor 轻质油回收装置
condensate filter 凝结水过滤器
condensate flash 冷凝液闪蒸
condensate gas 凝析气
condensate head tank 冷凝水高位槽
condensate knock-out pot 冷凝液分离罐;冷凝物分离罐
condensate line 凝结水管路
condensate liquid 冷凝液
condensate outlet 冷凝液(排)出口;凝结水引出管
condensate pipe 冷凝(水)管;凝结水管(道)
condensate pipework 冷凝管子工程
condensate polishing 冷凝液处理
condensate pump 冷凝液泵;冷凝水(排除)泵;冷凝(排水)泵;凝结水泵
condensate receiver 冷凝液受槽;冷凝液罐;冷凝水接受器
condensate recovery percentage 凝结水回收率
condensate reflux pump 凝结液回流泵
condensate return 凝结水回水;凝结水回流
condensate return collecting tank 凝结水回流集水池
condensate return pipe 冷凝水回水管;冷凝水回流管;凝结水回流管
condensate return pipework 冷凝回流管子工程
condensate return piping 冷凝水回流管道;冷凝回水管路;凝结水再循环管路
condensate return pump 冷凝(水)回水泵
condensate return system 冷凝水回收系统
condensate returrn pipe 冷凝水回水管
condensate scavenging installation 凝结水利用装置
condensate seal 冷凝水密封
condensate spill over pipe 凝结水溢流管
condensate storage tank 冷凝水储[贮]槽
condensate strainer 凝结水滤网
condensate stripper 冷凝解吸塔
condensate system 凝结水系统

condensate tank 冷凝槽;凝结水箱
condensate trap 阻汽器;疏水器;冷凝水井;冷凝液收集器;冷凝水隔汽具;冷凝水存水弯;冷凝槽
condensate treatment 凝结水处理
condensate unit 冷凝水排除机
condensate valve 凝结水阀
condensate water tunnel 冷凝水地下管道
condensate well 凝液井
condensating agent 缩合剂
condensating medium 凝结介质
condensation 冷凝(作用);结露;凝水;凝聚;凝结性;(气体变为液体的)凝结;致密;玻尔—爱因斯坦凝聚;缩合
condensation accelerator 缩合加速剂;缩合促进剂
condensation action 凝聚作用
condensation adiabat 冷凝绝热线;湿绝热
condensational wave 凝聚波;密波;缩波
condensation anomaly 凝聚异常
condensation by contact 接触冷凝
condensation by injection 喷射凝结
condensation catalyst 缩合催化剂
condensation chamber 冷凝箱
condensation channel 冷凝槽;凝结水槽
condensation coefficient 冷凝系数;凝聚系数
condensation compound 缩合物
condensation control 冷凝控制
condensation cooler 凝结水冷却器
condensation dampproofing 凝结防湿
condensation detector 结露检测
condensation efficiency 冷凝效率;凝结效率
condensation flue 冷凝管道
condensation free dwelling 无冷凝房屋
condensation groove 冷凝槽;凝结水槽
condensation gutter 冷凝水排水渠;冷凝水排水槽;凝结水槽;檐沟;集水沟
condensation heat 冷凝热
condensation in flues 烟道结露
condensation in hold dwelling 舱内汗湿
condensation layer 凝结层;云层
condensation level 凝结面;凝结高度
condensation method 冷凝法;凝结法;压缩法
condensation moisture 冷凝成分
condensation nucleus 凝聚块;凝结核
condensation nucleus counter 凝结核计数器
condensation nucleus size spectrometer 凝结核大小分光计
condensation of gas 煤气冷凝
condensation of moisture 冷凝;结露;水汽冷凝
condensation of vapo(u)r 气体冷凝
condensation plant 冷凝设备
condensation point 露点;冷凝点;凝(固)点
condensation polymer 缩聚体;凝聚物;缩(合)聚(合)物
condensation polymerization 缩聚(作用);缩聚合作用
condensation precipitation 凝结降雨量;凝结降水量;凝结沉降(作用)
condensation precipitation phenomenon 凝结沉淀现象
condensation precipitation tank 凝结沉淀(池)
condensation pressure 凝结气压
condensation-preventing chemicals 防冷凝化学品
condensation process 冷凝过程
condensation product 浓缩产品;冷凝产物;凝聚物
condensation pump 冷凝泵
condensation rainfall 凝结降雨量;凝结降水量
condensation rate 浓缩度
condensation reaction 缩合反应;缩尺反应
condensation region 凝聚区
condensation resin 缩聚树脂;缩合树脂
condensation return 凝结水回水
condensation return pipe 冷凝(水)回水管
condensation return piping 冷凝水回流管道
condensation return pump 冷凝(水)回水泵
condensation return riser 冷凝水回流立管
condensation sampling 冷凝取样
condensation sensor 冷凝探测器;冷凝传感器;结露传感器
condensation shock wave 冷激波;凝结激波
condensation side 冷凝边
condensation sink 冷凝水小集水槽(气窗上)
condensation sinking 冷凝水集水槽;冷凝集水槽;气窗上的冷凝水集水槽
condensation tank 冷凝水箱;凝结水箱

condensation temperature 冷凝温度;凝结温度;凝点;凝固温度;凝点;缩合温度
condensation temperature diagram 凝结温度图
condensation tendency 凝结趋向
condensation test 冷凝试验;并项检验
condensation theory 凝结理论
condensation trail 凝结尾迹
condensation trap 冷凝阱;冷凝阀;疏水器
condensation tray 冷凝盘
condensation trouble 结露问题
condensation trough 凝结水槽
condensation value 凝结值
condensation water 冷凝水;凝结水
condensation water discharge 冷凝水排出
condensation water filter 凝结水过滤器
condensation water quantity 凝结水量
condensation wave 密聚波;密波;缩波
condensation within structure 内部结露(围护结构)
condensation zone 聚合带;结露区
condensator 冷凝器
condense 冷凝;蓄电
condensed 密集的;稠合的
condensed 2-ring cyloalkane 稠合的双环烷烃
condensed 3-ring cycloalkane 稠合三环烷烃
condensed aromatics 稠合芳香烃
condensed asphalt 缩聚沥青
condensed azo-pigment 缩合偶氮颜料
condensed balance sheet 简明资产负债表
condensed cyloalkane 稠合的环烷烃
condensed deck 压缩卡片组
condensed fluid 冷凝液体;冷凝液
condensed income statement 简明损益表
condensed instruction deck 压缩指令卡片组;压缩指令汇卡
condensed material 冷凝物质
condensed matter 凝聚(态)物质
condensed mist separator 冷凝雾滴蒸汽机
condensed nuclear hydrocarbon 稠环烃
condensed nucleus 稠环;稠核
condensed oil 稠合油
condensed oil gas pool 凝析油气藏
condensed oil(gas)zone 凝析油带
condensed pack 压缩穿孔卡片叠
condensed phase 凝(聚)相;缩相
condensed planting 密植
condensed polycyclic(al)pigment 缩合型多环颜料
condensed profile 纵断面缩图
condensed projection 组合投影
condensed ring 稠环
condensed ring compound 稠环化合物
condensed ring system 稠环系
condensed section 简化路段
condensed silica fume 硅粉
condensed spark 电容火花
condensed specifications 简明技术规范
condensed star 凝聚星
condensed state 凝聚态
condensed steam 冷凝(水)蒸气;凝汽
condensed substance 凝聚物质
condensed system 冷凝系统;凝聚系统;凝聚(物)系;凝聚体系
condensed system phase rule 凝聚系相律
condensed-tannin gel 丹宁凝胶
condensed time 压缩时间
condensed wastewater 缩合废水
condensed water 冷凝水;凝结水
condensed water outlet 冷凝排水
condense lamp 聚光灯
condense liquor 冷凝液
condense pipe 冷凝水管
condenser 冷凝塔;冷凝器;聚光器;聚光镜;凝汽器;蓄电器;电容器;容电器
condenser antenna 电容器天线;容性天线
condenser aperture 聚光器孔径;聚光镜孔径
condenser armature 电容器板
condenser arragement 聚光器装置
condenser arrester 电容避雷器
condenser bank 电容器组
condenser block 电容器组;电容器盒
condenser bolometer 电容式测辐射热计
condenser box 电容器箱
condenser bracket 电容器架
condenser bushing 电容式套管
condenser capacity 电容器电容量
condenser cartridge 电容式拾音头
condenser casing 冷凝器外壳
condenser ceramics 电容器陶瓷
condenser charge 电容器充电
condenser checker 电容检验器
condenser circuit 电容器电路
condenser coating 电容器板
condenser coil 冷凝旋管;冷凝器盘管;凝汽盘管
condenser component 容抗;电容分量
condenser coupling 电容(器)耦合
condenser coupling amplifier 电容耦合放大器
condenser current 电容器电流
condenser depression 过冷度
condenser diode storage 电容器二极管存储器
condenser discharge 电容器放电
condenser discharge anemometer 电容放电风速计
condenser discharge blasting machine 电容器式放炮器
condenser discharge resistance welder 电容储[贮]能接触焊机;电容蓄能接触焊机
condenser discharge spot welding 电容储[贮]能点焊;电容储能点焊
condenser(discharge)spot-welding machine 电容器放电点焊机
condenser-discharge type exploder 电容式发爆器
condenser discharge welder 电容放电式电焊机
condenser divider 电容式分压器;电容分压器
condenser drain line 冷凝水排放管
condenser duty 凝汽器热负荷
condenser electroscope 电容器式验电器
condense return 冷凝回水
condense return pipe 冷凝回水管
condense return pump 冷凝回水泵
condenser fan 冷凝器风机
condenser filter 高通滤波器;电容滤波器
condenser flange 凝汽器颈部法兰
condenser focal plane 聚光镜(的)焦平面
condenser for solvent 溶剂冷凝器
condenser gasket 凝汽器垫密片
condenser ga(u)ge 凝汽器压力计
condenser input filter 电容器输入滤波器
condenser jacket 冷凝套
condenser kiln 冷凝干燥窑
condenser leakage 电容器漏泄
condenser leg(pipe) 冷凝器气压管
condenser lens 聚光镜
condenser lens system 聚光器透镜系统
condenser level control valve 凝汽器水位调节
condenser lightning arrester 电容(器)式避雷器
condenser loss 电容器损耗
condenser loudspeaker 电容式扬声器
condenser magnification 聚光镜放大率
condenser microphone 电容(式)话筒;电容(式)传声器
condenser mirror 聚光反射镜
condenser motor 电容式电动机
condenser oil 电容器油
condenser operation 调相运行
condenser optics 聚光镜光学零件
condenser paper 电容器纸
condenser partition wall 冷凝器隔板
condenser pickup 电容(式)拾音器
condenser pipe 冷凝管
condenser piping 凝汽器管道
condenser plate 电容器片
condenser potential device 电容式仪表用变压器
condenser pump 冷液泵;冷凝泵
condenser reactance 容抗
condenser-reboiler 冷凝器蒸发器
condenser receiver 电容式听筒
condenser resistance 电容器(的)电阻
condenser revolver 聚光镜旋转器;聚光镜旋转器
condenser roentgen meter 电容器伦琴计
condenser run motor 电容起动电动机
condenser-shaft rotation 电容器轴旋转
condenser sheet mica 电容器薄片云母
condenser shell 冷凝器壳体
condenser shunt type induction motor 电容分相式感应电动机
condenser shutoff valve 冷凝器截止阀
condenser spindle 电容器轴
condenser-start induction motor 电容起动感应电动机
condenser start motor 电容起动电动机
condenser stator 电容器定片
condenser storage 电容存储器
condenser stress corrosion cracking 冷凝器应力腐蚀开裂
condenser support 冷凝器架
condenser system 冷凝系统;聚光系统
condenser terminal post 电容器接线柱
condenser tester 电容(器)试验器
condenser tissue 电容器纸
condenser tissue machine 电容器纸机
condenser tissue paper 隔气纸
condenser-to-lens distance 聚光镜物镜的距离
condenser transmitter 电容送话器;电容式送话器
condenser-transmitter amplifier 电容式微音器的放大器
condenser tube 冷凝管;冷凝器管(子);凝结管
condenser tube friction 凝汽器管道摩阻
condenser tubing 冷凝管
condenser-type bushing 电容器式绝缘套;电容式进线套管
condenser-type spot welder 电容储能点焊机;电容储[贮]能点焊机
condenser-type terminal 电容式套管
condenser valve 凝汽阀
condenser voltage 电容器电压
condenser water 冷凝水
condenser water pipe 冷凝管
condenser water pump 冷凝器水泵
condensery 浓缩厂;炼乳厂
condenser yarn 废纺纱
condense utility 压缩实用程序
condense valve 冷凝水阀
condensifilter 冷凝滤器
condensing 聚光
condensing agent 冷凝剂
condensing apparatus 电容器;聚光器;冷凝器
condensing bleeder turbine 凝汽式抽汽涡轮机
condensing bulb 冷凝瓶
condensing coefficient 凝结放热系数
condensing coil 凝汽盘管;冷凝盘管
condensing electroscope 电容验电器
condensing engine 凝汽发动机
condensing engine set 凝汽机组
condensing equipment 凝汽设备
condensing flow 凝结流
condensing gas drive 凝结气驱
condensing heat rejection effect 冷凝段排热效能
condensing heat transfer 凝聚热传递
condensing lens 聚光镜
condensing lens system 聚光透镜系统
condensing liquor 凝结液
condensing locomotive 凝汽式机车
condensing mirror 聚光反射镜
condensing monochromator 聚光单色器
condensing objective 聚光物镜
condensing panel 聚光板
condensing plant 凝汽装置;冷凝设备;冷凝厂;凝汽设备
condensing power palnt 凝汽式发电机
condensing pressure 冷凝压力
condensing process 冷凝过程
condensing refrigerating effect 冷凝制冷效能
condensing ring 冷圈;激冷圈
condensing routine 压缩程序
condensing steam engine 冷凝式蒸汽机
condensing steam turbine 凝汽式涡轮机;凝汽式透平
condensing surface 冷凝面
condensing temperature 冷凝温度;液化温度
condensing tower 冷凝塔
condensing tube tongs 冷凝管夹
condensing turbine 凝汽式汽轮机
condensing turbine with extraction 抽气凝汽式汽轮机
condensing type electroscope 容电式验电器
condensing unit 冷凝机组
condensing vessel 冷凝器;复水槽
condensing water 凝结水
condensing water conduit 凝结水管(道)
condensing water pipe 凝结水管(道)
condensing water piping 凝结水管(道)
condensing worm 冷凝旋管
Condensite 康顿赛电瓷

condensive load(ing) 电容性负载;进相负荷
condensive resistance 电容抗
condie 老塘;(矿的)采空区
condiment 调味剂
condiment storage 调料库
con-dinozzle 收敛扩散形喷管
condition 荷载情况;状况;情况;情景;条件;使适应;调整
condition adjustment 条件平差
condition adjustment with parameters 附参数条件平差
condition adjustment with unknown 带未知数的条件平差
conditional acceptance 附有条件的认付;附条件承兑汇票;附带条件承兑汇票
conditional adsorption equilibrium constant 条件吸附平衡常数
conditional approval 附条件承认
conditional assemble function 条件汇编功能
conditional assembly instruction 条件汇编指令
conditional bill of lading 附有条件的提单
conditional bill of sale 附条件销售单据
conditional bond 附有条件的债券
conditional bond sales 附有条件的债券销售
conditional box 条件方框
conditional branch 条件转移;条件分支
conditional branch capability 条件转移权力
conditional branch instruction 条件转移指令
conditional breakpoint 条件断点
conditional clause 有条件的条款;暂行条款;条件子句
conditional climatology 条件气候学;室内人造气候学
conditional code 状态码
conditional coefficient 条件系数
conditional commitment 有条件的承担
conditional compilation 条件编译
conditional concept 条件概念
conditional conservation 条件保守法
conditional contract 有条件的契约;有条件的合同;暂行契约;附有条件的合同;暂行合约
conditional convergence 条件收敛
conditional covariance matrix 条件协方差矩阵
condition(al)curve 状态曲线
conditional delivery 附有条件的交货
conditional density function 条件密度函数
conditional disposition 有条件的处理
conditional distribution 条件分布
conditional dominance 条件显性
conditional dump 条件转储
conditional endorsement 附条件的背书;条件背书
conditional equality 条件等式
conditional equation 条件方程(式)
conditional equilibrium 有条件的平衡;条件平衡
conditional equilibrium constant 条件平衡常数
conditional error 条件误差
conditional expectation 条件期望
conditional expected value 条件期望值
conditional expression 条件表达式
conditional extraction constant 条件提取常数
conditional extreme value 条件极值
conditional extremum 条件极值
conditional failure rate 条件故障率
conditional fee estate 有限制条件的世袭家产
conditional financing 有条件的筹资;附(有)条件融资
conditional flag 条件标记
conditional forecast 条件预测
conditional formation constant 条件形成常数
conditional frequency 条件频率
conditional implication gate 隐含门
conditional import 有限进口
conditional income elasticity 条件收入弹性
conditional indicator 假定指标;条件性指示物
conditional indicator constant 条件性指示常数
conditional indorsement 附条件的背书
conditional inequality 条件不等式
conditional inhibiyion 条件性抑制
conditional instability 条件不稳定性;条件不稳定度
conditional instruction 条件指令
conditionality 条件性
conditional jump 条件转移
conditional jump instruction 条件转移指令
conditional lethal mutant 条件性致死诱变体
conditional lethal mutation 条件致死突变体
conditional likelihood 条件似然
conditional loan 有条件贷款;附条件的贷款
conditional log likelihood function 条件对数似然函数
conditionally acceptable dailty intake 条件性可接受的每日摄入量
conditionally compact game 条件紧对策
conditionally compact set 条件紧集
conditionally compact space 条件紧空间
conditionally convergent series 条件收敛级数
conditionally optimal pair 条件最优配对
conditionally optimal plan 条件最优方案
conditionally stable 条件稳定的
conditionally stable circuit 条件稳定电路
conditionally unbias(s)ed estimator 条件无偏估计量
conditionally unstable 条件性不稳定的
conditional macroexpansion 条件宏扩展
conditional macroexpression 条件宏表达式
conditional macroinstruction 条件宏指令
conditional match 条件颜色匹配
conditional mean 条件均值
conditional moment 条件矩
conditional Monte Carlo method 条件蒙特卡罗法
conditional name 条件名
conditional number 条件数
conditional observation 条件观测
conditional offer 附条件要约;附条件发价;附条件报价
conditional operation 条件运算
conditional order 附条件的订单;条件指令
conditional order unit 条件指令组
conditional plan 条件规划
conditional power source 条件电源
conditional power supply 条件电源
conditional power supply panel 条件电源屏
conditional precedent transaction 停止条件交易
conditional prefix 条件前缀
conditional prefix scope 条件前缀域
conditional probability 条件概率
conditional probability computer 条件概率计算机
conditional probability density 条件概率密度
conditional probability distribution 条件概率分布
conditional probability function 条件概率函数
conditional process 附有条件的加工
conditional profit 条件利润
conditional purchase 附有条件的购买
conditional reflex system 条件反射系统
conditional regression 条件回归
conditional relator 条件关系符
conditional reliability 条件可靠性
conditional request 带条件的请求
conditional resource 暂定资源
conditional return 条件返回
conditional right 附有条件的权利
conditional risk 条件风险
conditional sale 有条件的销售;附条件买卖
conditional sales 附条件销售;条件销售
conditional sales contract 有限制条件的销售合同;附有条件的销售合同
conditional sales of real property 有条件的不动产销售
conditional sample 条件抽样;条件采样
conditional sand 处理砂
conditional scheduling 条件调度
conditional secretion 条件反射性分泌
conditional security 附条件证券
conditional sentence 有条件的判决
conditional simulation 条件模拟
conditional Slutsky elasticity 斯鲁茨基弹性
conditional solubility product 条件性容度积
conditional stability 条件稳定性;条件稳定度
conditional stability constant 条件稳定常数
conditional standard error 条件标准误差
conditional statement 条件指令;条件语句
conditional statistic 条件统计量
conditional stop instruction 条件停机指令
conditional stop order 条件停机指令
conditional subscription 附有条件发行
conditional-sum adder 条件和加法器
conditional-sum logic 条件和逻辑
conditional survivor function 条件存活函数
conditional test 条件检验
conditional theorem 条件定理
conditional trade 有限贸易
conditional train 临时列车;专用列车
conditional transfer 条件转移
conditional transfer command 条件转移指令
conditional transfer of control 控制的条件转移
conditional unstability 条件不稳定性
conditional use 特许用途;特定条件的使用
conditional-variable 条件变量
conditional variance 条件方差
conditional weight 公量
conditional yield point 条件屈服点
conditionary periodic motion 条件周期运动
condition at operation 运行状况;工作状况
condition-based maintenance 视情维护;预防性维修;状态修理
condition bit 条件位
condition built-in function 条件内部函数
condition check 状态检查
condition clause 基本条款;条件子句;条件条款
condition code 状态码;条件码;特征码
condition code enable 可建立条件码
condition code register 条件码寄存器
condition concurrent 并存条款
condition curve 状态曲线
condition diagram 事故条件分析图;路况图;道路状况图
conditioned air discharge grill(e) 空调送风算子
conditioned air supply 空调供气
conditioned area 空调区;空调面积
conditioned ceiling 装有空调的天花板;装有空调的平顶;空气调节的天花板
conditioned connection 条件联系
conditioned line 高质通信线路
conditioned mo(u)lding sand 制备好的型砂
conditioned observation 条件观测
conditioned reflex 反射条件;人工条件反射;条件反射
conditioned response 条件反射反应
conditioned space 空调空间
conditioned weight 公重
condition equation of relative control 相对控制条件方程
conditioner 温度调节区(平板玻璃);(窑池的)冷却部;调理池;调节装置;调整器;调料槽;调理剂;调节人员;调节器;调节剂
conditioner channel 调节通路
conditioner roll 压扁辊
conditioner windrower 割晒压扁机
condition field 条件字段;条件域
condition for closing the horizon 圆周角条件;圆周闭合条件
condition for closure 闭合条件
condition for constrained annexation 强制附合条件
condition for construction 施工条件
condition for loan 贷款条件
condition for site planning 详细规划条件;建筑用地条件
condition for sum of angles 组合角条件
condition for validity 有效条件
condition guaranteed on arrival 保证到货时状态
condition in container 容器中状态
condition in damaged 受损状况
conditioning 木材天然干燥法;给湿;调湿调湿处理;调整;调湿;调湿;调理;调湿;条件作用;熟化
conditioning agent 调整剂;调节剂
conditioning cabinet 调温调湿箱
conditioning chamber 空气干湿调节室;加湿室;干湿度调节室;调节室
conditioning department 修整工段;清理工段
conditioning duct 调节风道
conditioning duct shaft 空调的通风井
conditioning equipment 调节设备
conditioning equipment room 空调设备室
conditioning house 生丝检验所
conditioning machine 给湿机;水分检查机
conditioning of air 空气调节
conditioning of cake 原丝筒回潮;原丝筒调节;原丝筒处理;空调装置
conditioning of goods 货物外装状况
conditioning of hay crop 干草预压
conditioning of road bed (铺砌路面前的)路基准备
conditioning of scrap 碎料分类;废料分类;废钢

料调节
conditioning of sludge 污泥调节
conditioning of wastes 废物调理
conditioning period 黏合剂黏固期;(下金刚石钻头时的)强力冲洗阶段;熟化期
conditioning plant 装修工厂
conditioning process 调节过程;调节处理
conditioning pulp 调和矿浆
conditioning section 平衡区
conditioning signal 调节信号
conditioning site 清理工段
conditioning station 装修站
conditioning sub-station 装修分站
conditioning system 空调系统;调节系统
conditioning system pipe 空调系统中的管子
conditioning tank 调节池;调节槽
conditioning tentering machine 给湿拉幅机
conditioning tower 空调塔;增湿塔
conditioning treatment 预备处理
conditioning unit 空调单元
conditioning water 处理水;调节水;空调水
conditioning yard 整坯场;整锭场;清理工段;坯修整工段
conditioning zone 空调区域;空调地带;调节区
condition in operating 操作条件
condition in the letter of credit 信用证的条件
condition line 条件线(空气调节);状态线(空气调节);温湿度状态线
condition list 条件表
condition loss 条件损失
condition map of groundwater development 地下水开采条件图
condition monitoring 状态监测
condition monitoring system 状况监测系统
condition name 条件名
condition-name condition 条件名条件
condition name description entry 条件名描述项
condition number 性态数
condition of affairs 清算资产负债表;财产状况说明书;事态
condition of atmosphere 大气状态
condition of average 比例分摊条件;比例分担条件
condition of balance 平衡条件;平衡条件
condition of bid 投标须知事项
condition of build mine surveying 建矿条件调查
condition of business 经营环境;企业环境
condition of car detained for repair 扣车条件【铁】
condition of carriage 承运条件
condition of compatibility 相容条件;适应条件;适合条件
condition of consistency 一致性条件
condition of consolidation 固化条件
condition of constraint 约束条件
condition of consumption 消费状况
condition of continuity 连续(性)条件
condition of contract 契约条件;合同条件
condition of convergence 收敛条件
condition of crystal growth 晶体生长条件
condition of cure 固化条件
condition of delivery 交货状态
condition of disaster 灾情
condition of drainage 排水条件
condition of equidistance 等距(离)条件
condition of equilibrium 平衡条件
condition of equivalence 等值条件;等效条件;等同条件;等价条件
condition of equivalency 等积条件
condition of execution of plan 计划执行情况
condition of experiment 试验条件
condition of exposure 露头条件
condition of finite character 有限特征条件
condition of flow 流动条件
condition of gas escape from spring 泉水气体逸出状况
condition of groundwater formation 地下水形成条件
condition of groundwater occurrence 地下水埋藏条件
condition of heat-treatment 热处理状态
condition of high temperature 高温状态
condition of industrial planning 工业规划条件
condition of integrability 可积条件
condition of intersection 交线条件

condition of inversion 逆转条件
condition of investment 投资条件
condition of leaky recharge 越流补给条件
condition of line 线路状态;线路状况
condition of loading 加载条件;荷载条件;荷载情况;负载情况
condition of mass flow continuity 质量流动连续条件
condition of mineral deposit inundation 矿床充水条件
condition of mo(u)lding 模塑条件;成型条件
condition of nourishment 营养补给条件
condition of obstruction to navigation 碍航情况
condition of orthogonality 正交条件
condition of passing through 逾越条件
condition of plasticity 塑性条件;塑性状态
condition of precedent 先决条件
condition of production 生产条件
condition of pure competition 纯粹竞争条件
condition of reduction 简化条件
condition of rest 静止状态
condition of rigidity 刚性条件
condition of sale 销售状况
condition of saline formation 盐渍化形成条件
condition of seawater intrusion 海水入侵状况
condition of service 服务条件;使用情况
condition of similarity 相似条件
condition of similitude 相似性条件
condition of spring utilization 泉水利用状况
condition of stability 稳定性条件
condition of static equilibrium 静平衡条件
condition of stiffness 硬刚性条件;劲度条件
condition of stone monument 标石完好情况
condition of storage 储[贮]存条件;储存条件
condition of support 支座状态;支承条件
condition of target 觇标完好情况
condition of target and monument 觇标及标石完好情况
condition of temperature 温度条件
condition of the contract 承包合同条件
condition of transport(ation) 运输条件
condition of uncertainty 不确定情况
condition of use 使用条件
condition of vertical infiltration 地表垂向入渗条件
condition of waste prevention of aquifer 含水层的防污条件
condition per cent 每期折旧后财产情况
condition point 曲线图状态点
condition precedent 先决条件
condition pseudo-variable 条件伪变量
condition sale 附(有)条件(的)销售;分期付款销售
conditions and exceptions 一切条款与免除条款
conditions at the arc 电弧状态
conditions change 条件变化
condition series 条件级数
conditions for commencement of works 开工条件
conditions for employment 聘用条件
conditions for glass formation 玻璃生产条件
conditions for particular applications 特定条件
conditions for site planning 总平面设计条件(如地形、地质、水文等)
condition shipping order 附条件的装货单
conditions of acceptance 验收条件
conditions of a loan 借款条件
conditions of auction sale 拍卖条件
conditions of bid 投标须知事项
conditions of borehole 钻孔条件
conditions of calcination 锻烧条件
conditions of carriage 货运条件
conditions of cartel 卡特尔条件
conditions of circumstance 环境条件
conditions of commercial sale 货物销售条件
conditions of compatibility 笑容条件
conditions of contract 合约须知事项;合同条款;合同条件;承包条件;契约条件
conditions of contract for electric(al) and mechanical works 电气与机械工程合同条件
conditions of contract for (works of) civil engineering construction 土木工程施工合同条件
conditions of determining solution of differential equation 微分方程定解条件
conditions of end fixity 端部固定条件
conditions of gas supply 供气条件
conditions of grant 批地条件

conditions of hire 租用条件
conditions of incoming water and sediment 来水来沙条件
conditions of labo(u)r 劳动条件
conditions of loading 荷载条件
conditions of mining assessment 矿山建设条件评议;开采条件评议
conditions of of statics 静力学条件
conditions of outer area of mining 矿区外部条件评议
conditions of oxidation and reduction 氧化还原条件
conditions of statics 静态条件
conditions of supply 供气条件
conditions of technologic(al)-economic of mine region 矿区经济地理条件
conditions of the bid 投标条件
conditions of the client/consultant model services agreement 业主咨询工程师标准服务协议书
conditions of the contract 承包合同条件
conditions of tow 拖曳条件
conditions of ultrasonic inspection 超声波探伤条件
conditions on shipboard 船上生活条件
condition space 条件空间
conditions precedent to commencement 开工的先决条件
conditions sheet 费率表
condition subsequent 但书
condition survey 状态检验;情况调查
condition switch 情况选择开关
condition test 条件测试
condition the hole 改善井眼
conditory 地下墓室;储藏室
condo 住户自有公寓
condominium 住户自有公寓;共有权;共管;私有共有(一种公寓产权的形式)
condominium conversion 改为住户自有公寓
condominium declaration 多户共有的公寓大楼主契约(美国)
condominium dwelling 住户自有公寓住宅
condominium housing 住户自有公寓住房
condominium management 公寓大楼的管理
condominium management agreement 公寓大楼的管理协议
condominium map 共有的公寓大楼详图
condominium mortgage insurance 住户自有公寓抵押金
condominium plan 共有的公寓大楼平面图
condominium project 住户自有公寓建设项目
condor 康多尔自动控制导航系统
Condorcet criterion 孔多塞标准
conduct 行为;指挥;办理
conductance 流导;电导;导电;传导性;导电力;传导度;传导;声导
conductance and susceptance measuring bridge 电导纳测量电桥
conductance coating 导电涂层
conductance factor 传导系数
conductance measurement 电导率测定;电导测定
conductance meter 电导仪
conductance monitoring 电导监测
conductance ratio 电导率;电导比;导电比
conductance relay 电导继电器;导纳继电器
conductance titration 电导滴定
conductance-variation method 电导变量法
conduct credit investigation 调查信用
conducted interference 自电源线来的干扰;传导干扰
conductibility 传导性;传导度
conductible finish 导电涂饰剂
conductimeter 导热计
conductimetric method 电导率测定法
conductimetric titrimetry 电导滴定分析法
conductimetry 电导分析法
conducting band 传导体
conducting bar 导条;导电棒
conducting bottom 传导炉底
conducting bridge 分路;分流;并联电阻;电阻电桥;电导电桥
conducting cell 电导电池
conducting channel 导电电路
conducting channel black 导电槽法炭黑
conducting coating 导电衬板;导电涂层;导电敷

层;传导层
conducting coefficient of water level of aquifer 浸水层水位传导系数
conducting-core heterofilament 导电芯型双组分长丝
conducting fiber 导电性纤维
conducting film 导电膜
conducting glass 导电玻璃
conducting hearth 导电炉底;传导敞炉
conducting island 导电微区
conducting layer 传导层;导电层
conducting liquid 导电液体
conducting magnetic iron 导电磁铁
conducting material 导热材料;导电材料
conducting medium 导电介质
conducting meter 电导计
conducting parenchyma 输导薄壁组织
conducting paste 导电胶
conducting period 导电周期
conducting polymer 导电聚合物
conducting power 传导性;传导能力;传导度;传导本领
conducting probe 电导探针;导电探针
conducting refractory material 导电难熔材料
conducting resin 导电树脂;导电胶
conducting ring 导电环
conducting roller 导辊
conducting salinometer 电导调混器
conducting salt 导电盐
conducting shell 传导层
conducting solution 导电溶液
conducting state 导通状态
conducting strip 导电片
conducting substrate 导电衬底
conducting tissue of wood 木材渗水组织
conducting water 电导水
conducting waveguide 馈电波导管
conducting wire 导线【电】
conduction 引流;导通;传导(性)
conduction angle 导电角
conduction band 导带;传导带
conduction block 传导阻滞
conduction carrier 传导载波
conduction cooling 导热冷却;传导冷却
conduction cooling bolometer 导热冷却测辐射热计;传导冷却测辐射热计
conduction current 船导电流;传导电流
conduction decrepitation thermometer 传导爆裂测温仪
conduction electron 船导电子;传导电子
conduction electronics 传导电子学
conduction error 导热误差
conduction field 传导场
conduction glaze 电导釉;导电釉
conduction heating 传导加热
conduction heating surface 传导热面积
conduction heat-transfer 传导传热
conduction hole 导电空穴;传导空穴
conduction level 电导能级
conduction loss bolometer 导热损耗测辐射热计;传导损失测辐射热计
conduction mechanism 传导机理
conduction of electricity 导电
conduction of heat 导热(作用);热传导
conduction parameter 传导参数
conduction pump 电导泵;电泵;传导泵
conduction state 电导状态
conduction surface 传热面;传导面积
conduction type furnace 感应式电炉
conductive 传导性的
conductive adhesive 导电胶
conductive body 导(电)体
conductive body mineral 导体矿物
conductive ceramics 导电陶瓷
conductive channel black 导电槽法炭黑
conductive coating 导电涂料;导电涂层;导电膜(层)
conductive contact 导电接头;导电触头
conductive coupler 电导耦合器
conductive coupling 直啮合;直接耦合;电导耦合;导偶
conductive discharge 导体放电
conductive-dot code 导电点代码
conductive earth 接地

conductive elastomer 导电弹性体
conductive fabric 导电织物
conductive floor cover(ing) 导电性地面覆盖物
conductive floor finish 传导性地面粉刷
conductive flooring 导电地面;防静电地面;防静电地板
conductive floor tile 传导性地面砖
conductive furnace black 导电炉(法炭)黑
conductive gasket 导电衬垫
conductive geothermal system 传导型地热系统
conductive glass 导电玻璃
conductive glaze 导电釉
conductive heat 传导热
conductive induction 电导感应
conductive ink 导电(油)墨
conductive lacquer 导电清漆
conductive layer 导电层;传导层
conductive liquid analog(ue) 水电比拟;导电液模拟设备
conductive lubricant 导电性润滑剂
conductively closed 闭合的;被屏蔽
conductive measurement 电导测定
conductive mortar 导电砂浆
conductive paint 导电涂料;导电漆
conductive paste 导电性胶;导电糊膏
conductive plastics 导电塑料
conductive polymeric composite 导电高分子复合材料
conductive primer 导电底漆
conductive rubber 导电橡胶;防静电橡胶
conductive rubber products 导电橡胶制品
conductive rubber slice 导电橡胶片
conductive rubber stick 导电橡胶棒
conductive sheet analogue 导电板模拟设备
conductive silver paste 导电银糊;导电银粉浆
conductive terrazzo 防静电水磨石
conductive tile 有传导性的砖瓦
conductive tissue 传导组织
conductive transfer 传导传输
conductive well 电导阱
conductivity 电导;导热率;导电率;导电性;传导系数;传导率;传导度
conductivity analyser 电导分析器
conductivity anomaly 电导率异常
conductivity apparatus 电导仪;导电仪;导电测定器
conductivity bridge 电导电桥;电传电桥;传导电桥
conductivity bronzes 导电青铜合金
conductivity cell 电导(率测定用)电池
conductivity coefficient 电导系数;导电系数;传导系数
conductivity connected charge coupled device 电导联结电荷耦合器件
conductivity construction 传导式结构
conductivity factor 导电系数
conductivity for heat 导热性
conductivity ga(u)ge 电导仪
conductivity glass 导电玻璃
conductivity indicator 导电率指示器
conductivity log curve 电导率测井曲线
conductivity measuring device 导电率测量装置
conductivity meter 电导率仪;电导率计;导电(率)表
conductivity mobility 导电迁移率
conductivity modulated rectifier 电导率调制整流器
conductivity modulation 电导率调制;电导率调变
conductivity modulation transistor 电导率调变晶体管
conductivity monitoring 电导监测
conductivity of an aperture 透孔性
conductivity of heat 导热作用
conductivity of medium 介质传导率
conductivity of rock 岩石电导率
conductivity of sphere 球体导电率
conductivity of wastewater 废水电导率
conductivity probe 电导探针;电导探示器
conductivity sensor 电导传感器
conductivity temperature depth recorder 温深电导记录仪
conductivity tensor 电导率张量
conductivity tester 电导试验器
conductivity theory 传导率理论
conductivity type flow meter 电导式流量计;电导流量计
conductivity water 校准电导水;导电水

conduct of a vessel 船舶航行情况
conduct of exploratory borehole 勘探孔的处理
conductograph 传导仪
conductometer 电导计;导热计;导电计;传导计;热导计
conductometer of heat 导热计
conductometric analysis 电导率分析法;电导分析法
conductometric end-point titration 电导终点滴定法
conductometric measurement 电导测量
conductometric method 电导法
conductometric particle counter 电导粒子计数器
conductometric sensor 电导传感器
conductometric sulfure dioxide analyzer 电导式二氧化硫分析器
conductometric titration 电导滴定法;电导滴定
conductometric titrimeter 电导滴定计
conductometry 电导滴定分析法;电导滴定法;电导测定法
conductor 列车长;响导者;导体;导水管;传导体;避雷针;前导子
conductor arrangement 布线;配线
conductor bond 钢轨导(电)接线;对头接
conductor box 泥浆槽
conductor cable 电缆
conductor casing 表层套管
conductor configuration 导线布置;布线;配线
conductor conflict 导线间距不符规定
conductor cross-section 导线截面
conductor element 导体元件
conductor glaze formation 导线雨淞;导线结冰
conductor head 水落斗
conductor hole 下导坑;下表层套管井眼
conductor hopper 水落斗
conductor installed in conduit 导线穿管敷设
conductor load 导线荷载
conductor loading 导线单位长度的总负载
conductor material 导电材料
conductor of sound 声音的传导体
conductor particulars 导体特性
conductor pattern 导线分布图
conductor pipe 孔口管【岩】;导管;表层套管
conductor propagation of bubble 磁泡导体传输
conductor rail 接触轨;导(向)轨;导电轨
conductor-rail anchor 接触轨锚;导电轨锚栓;导电轨固定器
conductor-rail bond 导电轨轨道接头
conductor-rail clamp 导电轨线夹
conductor-rail collector 第三轨受电器
conductor-rail cover 导电轨罩
conductor-rail insulator(with claw) 导电轨绝缘子
conductor-rail ramp 导电轨斜面
conductor-rail support 导电轨支撑
conductor-rail system 接触轨制;导电轨制
conductor resistance 导线电阻
conductor resistance test 导体电阻测量
conductor rope 导绳
conductor sag 导线垂度
conductor's cab(in) 列车员厢;列车员室
conductor screen 导体屏蔽
conductor selection 导线选择
conductor shielding 导体屏蔽
conductor skin effect 趋肤效应
conductor sleet formation 导线雨淞;导线结冰
conductor spacing 导线间隙
conductor string 套管;表层套管柱
conductor support box 导线支撑盒
conductor suspension clamp 导线悬垂线夹
conductor suspension clamp with hook 平行双耳鞍子
conductor system 乘务工作制
conductor turn 线匝
conductoscope 电导度计
conduct pipe 导管
conduct rail anchor 接触轨固定装置
conduct rail gap 接触轨间隙
conductron 光电导摄像管;导像管
conduct tests at selected plot 试点田
conduit 光学管;管道;电缆管道;导电管;槽道;暗线槽;渠道;输水管(道)
conduit bend 导管弯头
conduit body 管道连接件;管道本体
conduit box 配水井;配线箱;配线盒;接管井;接管盒;管道入孔;电缆管道分岔孔;导管接头盒;导

管分线匣
conduit breadth 管道宽度
conduit break-over 管线导通
conduit bridge 输水桥;排水渠桥;管线桥;管道桥
conduit bushing 导管套筒
conduit cable 管道电缆
conduit cast-in site 就地浇捣渠道
conduit clay pot 管道黏土槽
conduit cleaning 管线清理
conduit conductor rail 槽内接触轨
conduit conductor rail system 槽内接触轨制
conduit coupling 管路连接;电缆管道连接
conduit duct 地下电线管
conduit embedded in plaster 埋在灰浆中的管道
conduit entrance 风道进口;管道进口;水道进口
conduit ferrule 导槽
conduit fitter 管道装配工
conduit fittings 导管配件;输水管配件
conduit for prestressing steel arrangement 预应力钢筋孔道
conduit groove 管道沟槽
conduit hanger 管道吊架;管道吊钩
conduit head 储[贮]箱
conduit hollow clay(building)block 砌筑管道的空心黏土块
conduit jacking 顶管装置;顶管法;顶管
conduit joint 管线接头;管道接头;管道接口
conduit line 管道线
conduit liner 管道内衬;管道衬砌;管道衬里;管道衬层
conduit network 管道网
conduit on wall and ceiling 明敷管
conduit pipe 管道;地下线管;导(线)管;水管;输水管(道)
conduit pit 检查井;管道坑;探井
conduit plan 管道平面图
conduit pressure 管道压力
conduit resistance 水道摩阻(力)
conduit room 导管室
conduit run 电缆管道路线
conduit saddle 管托
conduit seal 线管壳
conduit section 管道断面;输水道断面
conduit slope 管道坡度
conduit spillway 溢水管;溢洪道
conduit storage 渠道畜水量
conduit support 立根台
conduit system 管道制;地下管道系统;地下导电轨制;导管系统;导管布线制;暗管装置;暗线系统;输水(管道)系统
conduit tile 管道砖
conduit tube 地下管线;导管;管道
conduit-type concrete sewer 导管型混凝土污水管
conduit-type sewer 导管型污水管
conduit-type sewer construction 污水管建造
conduit-type sewer work 导管型污水管工程
conduit under pressure 有压管道;压力管渠;压力管道
conduit valve 管道阀(门);水道阀门;输水管阀
conduit weather master system 诱导通风方式
conduit wiring 管内敷线;管内布线;管线穿管敷设
conduit work 管道工程;电线管道工程
condulet 导管节头;小导管;管中布线接线盒
condulet reducer 接线盒异径接头
conduloy 康杜洛镍铜合金
Condy's crystal 康狄晶体
cone 料钟;蘑菇状(用于固定集装箱);圆锥形;圆锥(体);牙轮;锥状地形;锥形(物);锥形筒;锥形滚柱轴承内座圈;锥体;风向袋;驳柱头
cone aerator 锥形曝气器;锥形曝气池;倒伞形曝气器
cone agitator 锥形搅拌机
cone anchor 流锚;海锚;浮锚
cone anchorage 锥形锚碇装置;锥形锚具;弗氏锚
cone anchorage device 锥形锚具
cone-and-jet cement mixer with hopper 带漏斗锥形射流式水泥搅拌机
cone-and-jet type cement mixer 漏斗射流式水泥搅拌器
cone-and-plate viscosimeter 平底锥形黏度计
cone-and-socket arrangement 锥窝接合装置
cone-and-socket joint 锥窝接头;锥窝接合
cone angle 锥角
cone angle banding 圆锥角环带法

cone antenna 锥形天线
cone-apex angle 锥顶角
cone area 锥面
cone assy 锥形装配;锥形支座;锥形零配件
cone axis 牙轮轴
cone-baffle classifier 锥形挡板分级机
cone balance method 漏斗均衡法
cone bearing 锥形轴承
cone-bearing casing hook 锥形滚子轴承套管打钩
cone bearing test 圆锥贯入试验;圆锥承重试验;圆锥承载试验;锥承试验
cone bearing test of subgrade 土基锥承试验
cone bearing value 圆锥承载力值
cone belt 三角皮带
cone bit 牙轮钻头;锥形(无岩芯)钻头
cone blender 锥形混合器
cone bobbin 宝塔纱
cone bolt 锥形螺栓
cone bottom 锥形底
cone bottom bucket 锥底式料桶
cone-bottom condenser 锥底冷凝器
cone-bottom gas-dispersion agitator 锥底气体弥散搅拌器
cone-bottom tank 锥形底储[贮]罐
cone brake 锥形制动器
cone buoy 锥形浮标
cone cable net 锥形电缆网
cone center 伞齿轮节锥顶
cone-channel condenser 锥形光路聚光器;光锥聚光器
cone-channel condenser optics 圆锥波道聚光器
cone chuck 锥形夹盘
cone classifier 圆锥分级机;锥形分级器;锥形分级机
cone closure autoclave 锥形密封式高压釜
cone clutch 锥形离合器;锥式离合器
cone collector 锥形骨料斗
cone compartment silo 锥形室搅拌库
cone core 宝塔筒管
cone coupler 锥形联轴器
cone coupling 锥形联轴
cone crusher 圆锥破碎机;圆磨;锥形压碎机;锥形碎石机;锥形破碎机
cone cup 锥形杯
cone cut 锥形掏槽
cone cut veneer 锥形卷刨(的)木刨片;锥形切削单板
cone-cylinder(body) 锥头(圆)柱体
cone cylinder model 锥柱体模型
cone data 锥形数据
coned dowel 锥形暗销
cone deformation study 锥状变形研究
cone delta 锥状三角洲;锥形三角洲;冲积锥
cone dewaterer 锥形脱水机
coned grinding wheel 圆锥磨轮
cone diaphragm 锥形膜片
cone dike 锥状岩墙;锥形岩席
cone discharging bucket 锥底式料桶
cone disc spring 碟锥形弹簧
cone dome 锥形圆屋顶;锥形穹顶
cone door type charging bucket 锥底式料桶
cone drive 锥轮传动;塔轮转动(装置)
cone-drum cyclorama 鼓形旋转的天幕
cone drums 铁炮
coned tent pile 帐篷型料堆;三角形断面料堆
cone duster 海格立斯除尘机;转笼式除尘机
cone edge 锥形边
cone effect 锥形效应
cone equivalent 测温锥相当温度
cone factor 锥头系数
cone fender 锥形护舷
cone fisher 锥形打捞器
cone for clover leaf plate 十字孔板用锥头
cone foundation 锥形基础
cone fracture 锥状断裂
cone friction(al)gear 摩擦减速器
cone friction brake 锥形摩擦闸
cone friction clutch 锥形摩擦离合器
cone friction coupling 锥形摩擦联轴节
cone friction gear 锥形摩擦传动装置
cone fusion test 测温锥试验
cone ga(u)ge 锥形量规;锥度量规
cone gear 圆锥齿轮;锥轮传动(装置);塔轮转动装置;塔轮齿轮;伞齿轮
cone governor 锥形调速器;调节器

cone granulator 锥形成粒机
cone grate 锥形格条
cone grinding 圆锥磨法
cone-headed 圆锥头的
cone-headed bolt 圆锥头螺栓;锥头螺栓
cone-head rivet 锥头铆钉
cone height 圆锥的高
cone height index 内焰(锥部)高度指数
cone hip(ped end)(四坡屋顶的)锥形屋脊
cone hip tile 斜脊锥形筒瓦;伐脊锥形筒瓦
cone impact test 锥击试验
cone-in-cone 叠锥;迭锥
cone-in-cone structure 叠锥构造
cone index 圆锥指数
cone influence of well 井的影响漏斗
cone in the jet 喷口内锥
cone-jet mixer 漏斗射流式搅拌器
cone joint 双锥管接合;圆锥接头
cone key 锥形键
conel 考涅尔铁镍铬合金
cone-lock 锥形锁止
cone-lock punch retainer 锥体锁止冲头护理
Conelrad 电磁波辐射管制
Conemauch 上石炭纪
Conemaugh series 科纳莫系【地】
cone method 圆锥筒试验;(测定混凝土坍落度的)圆锥法;锥体法
cone mill 斜轧穿孔机;锥形磨
cone milling cutter 锥形铣刀
cone neck rivet 圆锥铆钉
cone number 壮数
cone nut 锥形螺帽
cone nut tie 锥形螺帽拉杆
cone of aerial camera 航摄仪镜筒
cone of ambiguity 不定性锥区
cone of dejection 洪积锥;冲刷锥;冲积锥
cone of depression 降落漏斗;下降锥(体);下降漏斗;锥形沉陷;沉陷锥
cone of detritus 冲积锥
cone of drawdown 下降锥(体);锥形沉陷;降落漏斗;沉陷锥
cone of eruption 喷发锥
cone of exhaustion 降落漏斗;下降漏斗;排空漏斗
cone of experience 经验之塔
cone of explosion 爆炸锥
cone of friction 摩阻圆锥
cone of gears 齿轮锥
cone of groundwater depression 地下水(位)下降漏斗;地下水面下陷锥
cone of groundwater influence 地下水位影响漏斗
cone of growth 生长锥
cone of influence 降落漏斗;下降锥(体);下降漏斗;影响漏斗;锥形沉陷;沉陷锥
cone of influence of well 井的降落漏斗;井抽水时地下水位降落面
cone of intake 进水曲面
cone of intake of well 井的进水曲面
cone of light 散开光束;光锥(体);灯光锥形射束
cone of nulls 零锥
cone of origin 始锥
cone of photogrammetric(al)camera 航摄仪镜筒
cone of pressure relief 下降漏斗(地下水);降落漏斗
cone of pumping depression 抽水下降漏斗
cone of radiation 辐射锥(体)
cone of recharge 补给锥体;补给反漏斗
cone of revolution 回旋锥面
cone of silence 静区;圆锥形静区
cone of slope 边坡圆锥体;边坡锥体
cone of tangents 切锥
cone of visibility 能见锥区
cone of water table depression 地下水位漏斗线;地下水面下降漏斗;水位下降漏斗
cone out 锥形扩大(管道)
cone package 宝塔纱筒
cone pelletizer 圆锥造球机
cone penetration 圆锥贯入度;圆锥触探;锥体贯入度
cone penetration index 圆锥贯入指数;锥沉量
cone penetration number 锥沉量
cone penetration resistance 圆锥探点阻力
cone penetration stability test 锥贯入稳定性试验(测定沥青混合料用)
cone penetration test 圆锥触探试验;圆锥贯入试

验;圆锥贯入度试验;触探试验
cone penetration test with pore pressure measurement 孔压静力触探试验
cone penetrator 锥形硬度试头
cone penetrometer 锥探仪;圆锥流限仪;圆锥贯入仪;圆锥贯入器;圆锥触探仪;圆锥触探器;锥形透度计;锥形贯入仪;锥式液限(贯入)仪;触探仪
cone penetrometer for liquid limit test 锥式液限(贯入)仪;瓦氏平衡锥
cone pin 锥形销
cone pinion 小伞齿轮
cone pivot 锥形枢
cone plaque 测温锥底座
cone plate rheogoniometer 锥板式流尼旺戈仪
cone plate rheometer 锥板式流变性测定仪
cone plate visco(si)meter 锥板式黏度计
cone point screw 锥头螺钉
cone prism 锥体棱镜
cone pulley 圆锥形滑车;锥轮;宝塔(式滑)轮;塔轮
cone pulley drive 锥轮传动;塔轮传动
cone-pulley lathe 塔轮车床
cone pulley with steps 塔轮
cone pump 仓式泵
cone pyrometer 锥体高温计;测温锥
cone quartering 圆锥四分(取样)法;锥形堆四分法
cone radiometer 圆锥辐射计
cone reducer 铁炮式粗砂机
cone resistance 圆锥贯入阻力;锥体压入阻力;锥体贯入阻力;锥尖阻力
cone-resistance value 圆锥(贯入)阻力值
cone rock bit 凿岩钻头;锥形凿岩钻头;牙轮钻头
cone-roll piercing mill 蘑菇式穿孔机
cone roof 锥形顶
cone roof reservoir 锥形顶油罐
cone-roof tank 锥顶(储[贮])罐
cone sampler 锥形取样器
cone sampling 锥形堆取样法
cone scan autotracking 圆锥扫描自动跟踪系
cone screen 锥形筛
cone sensor 锥头传感器
cone settler 锥形沉淀箱
cone settling tank 锥形沉降槽
cone shaker 圆锥振动筛
cone-shaped 圆锥形的
cone-shaped beam 锥形光束
cone shaped charge 锥形装药
cone shaped circular storage hall 锥形圆储库
cone shaped cover 锥形罩;锥形盖
cone-shaped drill point 锥形钻刃
cone-shaped roll 锥形轧辊
cone-shaped sink hole 锥状落水洞
cone-shaped sounding apparatus 锥状测触探仪;锥形触探仪
cone shaped yarn 宝塔纱
cone-shape pitched slope at bridge abutment 桥头锥形护坡
cone sheet 锥形岩席;锥形片;锥形岩脉
cone shell 圆锥薄壳
cone shell foundation 锥形薄壳基础
cone shell stacking 锥壳堆料
cone shell stacking method 锥壳堆料法
cone shell stockpiling 锥壳形堆料法
cone shock 锥形激波
cone slip(ping) 锥体滑动
cone spacer 内隔圈
cone-spheroid 锥球体
cone spinning 锥形变薄旋压
cone spring 锥形弹簧
cone test 圆锥试验;圆锥法(测定混凝土坍落度的)
cone test for hardenability 锥体淬硬性实验
cone test method 圆锥试验法;圆锥测定法
cone thickener 圆锥形浓缩机
cone tile 屋脊弯(盖)瓦;戗脊锥形筒瓦;(用于四坡屋顶的)端头脊瓦
cone-to-cylinder junction 锥体与筒体连接
cone tool 锥形工具
cone-type 锥形的
cone-type bit deviation limits 牙轮钻头偏斜极限
cone-type collimator 锥形准直器
cone-type crusher 锥形碎石机;锥形破碎机;圆锥式破碎机
cone-type discharge grate 塔式卸料箅子
cone-type feeder 锥形给料器
cone-type feed reel 定心圆锥式开卷机
cone-type foundation 锥形基础
cone-type handle 喇叭形转向盘
cone-type liquid-film seal 锥形液膜密封
cone-type mixer 锥形混合机
cone-type mixer blender 锥形混合机
cone-type payoff reel 定心圆锥式开卷机
cone-type roller bearing rock 装有滚柱轴承的牙轮钻头
cone-type uncoiler 定心圆锥式开卷机
cone-type vault 锥形穹隆
cone valve 锥形阀(门);锥阀
cone visco(si)meter 锥形黏度计
cone vision 锥形视域
Conewangoan 科尼旺戈组
cone washer 锥形衬垫
cone-way valve 单向阀
cone wheel 锥形砂轮
cone winder 锥形筒子络纱机;宝塔筒子络纱机
cone winding machine 锥形筒子络纱机;宝塔筒子络纱机
cone-worm unit 球面涡轮减速机
cone worn out 钻头内锥面磨损
confectant 康费克坦
confectaurant 小吃店;点心店
confectionary 糖果点心店
confectioning 成型
Confederation of International Contractors Association 国际承包商联合会
Confederation of Swedish Trade Unions 瑞典工会联合会
confederation of trade union 行业协会
conference 会议;会谈;协商会议;商量协商
conference block 会议场所;会议场区;用于集会或会议的一组建筑物
conference building 会议大厦
conference call 会议电话;电话会议
conference cargo 协议货物
conference center 会议中心
conference circuit 会议电话电路
conference coach 会议车
conference communications 会议通信设备
conference control 灵敏度控制
conference for supplementing goods 补货会议
conference freight rate 协议运费率
conference hall 会议厅;会场
conference line vessel 参加海运同盟的班轮;同盟定期船
conference member 同盟会员
Conference of Experts on Climate and Mankind 气候和人类问题专家会议
conference of shipping 航运公会
Conference on Environment and Development 环境与发展会议
conference on production 生产会议
conference on the law of the sea 海洋法会议
conference rate 同盟费率;运价公会决定的费率;公会运费率
conference room 会议室
conference system 会议电话设备
conference table 会议桌
conference tariff 公会运价表;同盟费率表
conference telephone 会议电话
conference telephone control board 会议电话主机;会议电话控制台
conference telephone device 会议电话装置
conference telephone of administration's branch 局界会议电话
conference telephone subset 会议电话分机
conference telephone switch system 会议电话汇接机
conference telephone tandem board 会议电话汇接台
conference terms 公会条款
conference theater 会议剧场
conference TV 会议电视
conference unit 会议单元
conferva peat 丝状藻泥炭
confessional 忏悔室
confessionary 忏悔室
confetti 彩色雪花干扰
confidence 信心;信任;置信(度)
confidence band 置信范围;置信带
confidence belt 置信带
confidence bound 置信界限
confidence building measures 建立信任措施
confidence coefficient 可信系数;可靠系数;置信系数
confidence curve 置信曲线
confidence debt 信用借款
confidence density 置信密度
confidence distribution 置信分布
confidence distribution function 置信(度)分布函数
confidence distribution method 置信分布法
confidence factor 置信因数
confidence index 可靠性指标;置信指数
confidence inference 信用结论
confidence interval 信赖区间;置信区间;置信界限;置信范围
confidence interval of mean 均值的置信区间
confidence in the currency 货币信用
confidence level 可信赖程度;可信度(数值);可靠质量水平;可靠度;信赖级;信赖程度;置信水准;置信度;置信水平;置信级
confidence limit 可信反;可信界限;可靠极限;置信(界)限;置信极限
confidence limits for population mean 总体均值的置信界限
confidence probability 置信概率
confidence probability function 置信概率函数
confidence program(me) 可信度检测程序
confidence region 置信区域
confidential 机密的
confidential agreement 秘密协议
confidential data 机密资料
confidential documens 机密资料;机密文件;密件
confidential factoring 信贷保密保付代理
confidential information 保密情报
confidential interval 可靠区间
confidentiality 机密性;保密(性)
confidentiality agreement 机密性协定
confidential nature of contract 合同的保密
confidential papers 机密资料;机密文件
confidential phone 保密电话机
confidential project 保密工程
confidently 有把握地
configurated glass 印花玻璃;压花玻璃;图案玻璃
configuration 轮廓;结构;形状;形相;组成;组合;构型;格局;配置;配位;排列;位形;外形;设备配置
configurational asymmetry 形状不对称
configurational entropy 构型熵
configurational free energy 组态自由能
configurational symmetry 构型对称
configuration assignment unit 结构分配部件
configuration bit 结构位
configuration command 配置指令
configuration control 配置控制;配列控制;外形调整
configuration control register 配置控制寄存器
configuration coordinate model 构型配位模型
configuration counting series 构型计数级数
configuration dredging 整形挖泥;造型挖泥
configuration factor 形状因数
configuration frame 构成帧
configuration inspection 结构检查;外形检查
configuration interaction 组态相互作用
configuration line 轮廓线
configuration management 结构管理;构型管理
configuration map 结构图
configuration matrix 构型矩阵
configuration of air flow 气流形状
configuration of coal-rich belt 富煤带展布形式
configuration of earth 地形
configuration of electromagnetic double-dipole prospecting system 电磁偶极法测量装置
configuration of equilibrium 平衡组态
configuration of flow 流线谱
configuration of ground 地形
configuration of installation 装置布置
configuration of Loran transmitting stations 罗兰C发射台配置
configuration of receiving loop 接收线圈类型
configuration of sample (统计数学上的)样本构型;样本的构形
configuration of shoreline 岸线形状
configuration of surface 表面形态
configuration of system 系统构成
configuration of the ground 地貌
configuration of transmitting loop 发射线圈类型
configuration option 配置选择

configuration section 结构节;设备配置节
configuration space 构型空间
configuration stability 布局稳定性
configuration state 结构状态
configuration variable 形状变量
configurative change 构型变化
configurator 配置器;配置程序;设备组合程序
configured-off 非合用配置
configured-out 非通用的合用配置
configure switch 配置开关
confilm temporary curing 临时性养护薄膜(喷射于新浇混凝土表面上)
confine 磁场吸持;限制
confine bed 封闭层
confined and unconfined flow 有压无压流
confined aquifer 自流含水层;承压含水层;层间含水层;封闭含水层;非自由含水层
confined area 约束面积;承压水区
confined bed 封闭层;承压层
confined beds of aquifer 承压水层;封闭含水层
confined compression 侧限压缩
confined compression test 侧限压缩试验;侧限抗压试验
confined compressive strength 有限抗压强度;侧(向)限(制)抗压强度
confined compressive test 侧限压缩试验;有侧限压缩试验
confined concrete 约束混凝土;侧限混凝土
confined disposal facility 局限处理设备
confined eddy 固定漩涡
confined expansion 限制膨胀
confined explosion 约束爆炸
confined flow 封闭流;承压流;有压流;压力流;承压水流;受约束流
confined focus(s)ing 限界聚焦
confined geothermal system 封闭型地热系统
confined groundwater 自流地下水;承压地下水;受压地下水;封闭地下水;受约束的地下水
confined injection 填塞注浆
confined leaching 槽内浸滤
confined length 约束长(度)
confined lysimeter 封闭式渗漏测定计
confined O-ring joint 闭合环形接缝
confined plasma 可控型离子体
confined pot bearing 有侧限盆式支座
confined pressure 封闭压力;侧限压力
confined pressure effect 围压效应
confined pressure seepage flow 有压渗流
confined quarters 圈闭的角隅
confined reinforcement 约束钢筋
confined seepage flow 有压渗流
confined space 有限空间
confined state 被压缩状态
confined steady flow 受压稳流
confined stratum 隔水层;封闭层
confined stress 约束应力
confined swelling 有侧限膨胀
confined test 确认试验
confined-unconfined well 承压-无压井
confined water 自流水;承压水;层间水;受压水
confined water basin 承压水盆地
confined water head 承压水(水)头
confined waters 狭窄水域;狭浅水域
confined water well 承压水井
confinement 吸持;制约;磁场吸持;限制
confinement-contr olled seal 限压缩力密封
confinement of flow 约束导流
confinement pressure 侧限压力;(三轴试验的)侧压;(三轴试验的)围压
confines 范围
confining bed 承压层;阻水层;隔水底层;封闭(地)层;不透水层
confining fluid pressure 封闭液压
confining force 保持力
confining lateral pressure 横向侧限压力
confining lateral stress 横向侧限应力
confining layer 限制层;阻水层;隔水层;封闭层;不透水层
confining liquid 封闭液(体)
confining minor stress 侧限最小主应力
confining overlying bed 隔水顶层
confining pressure 周围压力;侧限压力;封闭压力;(三轴试验的)围压;围限压力
confining pressure effect 围压效应

confining region of reinforcement 钢筋约束区
confining stratum 不透水地层;限制地层;阻水层;隔水封闭层;隔水层;赋压地层;封闭地层;封闭层;不透水层
confining stress 侧限应力;包围应力
confining underlying bed 隔水底层
confirm 证实;保兑;确认
confirmability 可确定性
confirmation 证实;保兑;确认(书);确定
confirmation commission 保兑手续费
confirmation deed 确认书
confirmation hole 验证钻孔
confirmation in writing 书面确认
confirmation of order 订货确认书;订货承诺书;订单确认
confirmation of sales 销售确认书
confirmation product 确认样品
confirmation request 申请批准;查证请求;要求确认;请证
confirmation sheet 确认通知书
confirmation signalling 确认传信
confirmation test for guarantee 保证的确认考核
confirmation well 验证井
confirmatory check 验证检查
confirmatory measurement 核实性量度;核实性测量;验证性测量
confirmatory sales note 销售确认书
confirmatory sample 确认样本
confirmatory test 验证试验;证实(性)试验;确证试验
confirmed bank 保兑银行
confirmed credit 保兑信用证
confirmed documentary credit 保兑跟单信用证
confirmed irrevocable authority to purchase 保兑不可撤销委托购买证
confirmed irrevocable credit 保兑不可撤销信用证
confirmed letter of credit 保兑信用证;确认信用状
confirmed order 确认书;保付书
confirmed sample 确认样本
confirmed test 确定试验
confirming bank 保兑银行
confirming box 核实框
confirming charges 保兑手续费
confirming house 出口保兑行;保兑公司
confirming margin 保兑保证金
confirm order 确认订货
confirm your offer 确认你方的报盘
confiscate 没收
confiscated goods 没收货;没收货物;已没收货物
confiscation 征用;扣留;没收
confiscation of property 没收财产
confiscator 没收者
confiscatory taxation 没收性赋税
confix 连接起来
conflagrant 速燃的
conflagration 火灾;大火灾;爆燃
conflagration area 火灾区域;火灾面积;火灾地区
conflagration hazard 大火灾事故
conflagration region 火灾地区
conflagration zone 大火灾地带
conflation of words 词归并
Conflat seal 康弗拉特密封
conflex 包层钢
conflict 抵触;冲突
conflict alert 碰撞警告
conflict analysis 冲突分析
conflict analysis methodology 冲突分析方法论
conflict area 车流交汇冲突区;冲突区
conflict-cooperation nexus 抵触—协作关系
conflicting design 冲突设计
conflicting motion 不相容运动
conflicting operation 冲突性作业
conflicting route 敌对进路
conflicting signal 抵触信号;敌对信号【铁】;冲突信号
conflicting test 冲突试验
conflicting traffic 交汇车流;冲突车流
conflict management 冲突管理
conflict model 冲突模型
conflict objective 冲突目标
conflict of laws 冲突法;法律抵触;法律冲突
conflict point 冲突点
conflict process 冲突过程
conflict recognition 冲突识别

conflict regulation 冲突调节
conflict regulation theory 冲突调节理论
conflict resolution 冲突消除;冲突解决
conflict resolver 冲突分解
conflict rules 冲突法律
confluence 会合点;汇流;汇集;汇合点;汇合;河流汇合处;河道汇合点;合流点;合流;支流汇合处
confluence basin 汇流盆地
confluence ice 汇流冰
confluence of channels 水道汇合点
confluence of springs 泉集河
confluence reach regulation 交汇段整治
confluent 汇流的;汇合的;融合的
confluent angle 汇流角
confluent discharge 汇合流量
confluent filter 合流滤波器
confluent hypergeometric(al) function 合流超几何函数
confluent pitting 合流点蚀
confluent stream 合流河
confocal 共焦的
confocal backscatter laser velocimeter 共焦反向散射激光测速仪
confocal cavity 共焦腔
confocal cone 共焦(点圆)锥面
confocal conics 共焦二次曲线
confocal coordinates 共焦坐标
confocal curve 共焦曲线
confocal distance 共焦距
confocal ellipsoid 共焦椭圆面
confocal ends 共焦端面
confocal hyperbola 共焦(点)双曲线
confocal hyperboloid 共焦双曲面
confocal laser scanning microscope 共聚焦激光扫描显微镜
confocal lens line 透镜共焦线
confocal lens system 共焦透镜系统
confocal mirror 共焦反射镜
confocal mirror resonator 共焦反射镜谐振器
confocal mirror system 共焦反射镜系统
confocal optic(al) resonator 共焦光学谐振腔
confocal paraboloidal cooradinate 共焦抛物面坐标
confocal quardrics 共焦二次曲面
confocal resonance 共焦谐振
confocal resonator 共焦谐振器
confocal ring resonator 共焦环形谐振器
confocal ruby laser 共焦红宝石激光器
confocal spheric(al) laser resonator 共焦球形激光谐振腔
confocal spheric(al) reflector 共焦球面反射器
confocal strip laser resonator 共焦条状激光谐振腔
confocal unstable resonator 共焦式不稳定谐振腔
confocus 共焦
confocus cone 共焦锥面
conform 依从;遵照;遵守
conformability 一致性;整合性;适应
conformability of sleeve bearing 薄轴衬对轴和座的适应性
conformable bedding 整合层理
conformable contact 整合接触
conformable fault 整合断层
conformable injection 整合侵入
conformable matrix 共形矩阵;可相乘矩阵
conformable oil ring 弹簧胀圈油环
conformable optic(al) mask 符合光掩模
conformable stratum 整合岩层;整合地层
conformal 正形的;保角的;保角
conformal antenna 共形天线
conformal chart 正形投影图;正形投影海图;正形海图;等角投影海图
conformal coating 仿形涂层;保形涂料;保形(的)涂层;保角涂料
conformal condition 等角条件
conformal conic(al) projection 正形圆锥投影;等角圆锥投影
conformal cylindric(al) projection 圆柱正形投影;正形圆柱投影;等角圆柱投影
conformal double projection 正形双重投影
conformality 正形(性);保角性
conformal latitude 正形投影纬度
conformal longitude 等角投影经度
conformally equivalent 保角等价的
conformal map 保形变换图;保角映像

conformal mapping 共形映射;保形映射;保角映射
conformal mapping method 保角映射法;保角变换法
conformal parameter 保形参向
conformal polyconic(al) projection 正形多圆锥投影;等角多圆锥投影
conformal projection 正形投影;正形射影;保形射影;保角射影法
conformal reflection chart 保形反射图
conformal representation 保形表示法;保角表示法
conformal sphere 正形球体
conformal structure 保形结构
conformal transformation 正形变换;保形变换;保角变换
conformal wire grating 适形线栅
conformance efficiency 波及系数
conformance factor 波及系数
conformance to standard 标准一致
conformance with code 符合法规
conformation 适应;一致;形态;构造;符合
conformational analysis 构型分析
conformational change 形态变化;构造变化
conformational isomer 构像异构体
conformational theory 构型学说
conformation analysis 构像分析
conformation of ground 土地形态
conformation theory 构像理论
conformation trait 体型性状
conformed copy 符合要求的文件
conforming article 合格商品
conforming building 符合规范的房屋
conforming element 协调元;相容元
conforming imputation 一致的分配
conforming layer 均匀层
conforming plate 整形板
conforming structure 符合规范的结构
conformity 一致;整合;符合度;符合
conformity assessment 合格评定
conformity certificate 合格证;合格证明书;合格证明
conformity clause 一致性子句
conformity surveillance 合格监督
conformity with ISO 符合ISO标准
conform the specification to the requirement 符合所需规格
conform to actual situation 合乎实际
conform to conventional practice 依照习惯做法
conform to design specification 符合设计要求
conform to standard 合乎标准的
conform to the specification 符合规范规定
conform with realities 合乎实际
confounding 混淆(现象)
confounding bias 混杂偏性
confounding factor 混杂因素
confounding variable 混杂变量
confraternity 社团
confrication 磨碎
confriction 摩擦
confront 面对
confrontation 正视;对峙
confuence 河流汇合点
confused account 账目混乱
confused current 乱水
confused market 混乱的市场
confused sea 骇浪;道格拉斯九级浪;涨浪;暴涛
confused stage 混淆期
confused swell 乱涌;九级涌浪;暴涌
confused tone 色调紊乱
confusion 模糊
confusion reflector 扰乱反射体
confusion region 模糊区
conge 四分之一圆;拇指圆饰;圆弧凹形嵌边饰;凹(凸)形线脚
congeal 冷藏;凝集
congealed ground 冻土
congealed moisture 凝结水分
congealed solution 冻结溶液
congealer 冷却器;冷冻器;冷冻机;冷冻箱;冷藏器
congealing 凝固
congealing die 冻死
congealing point 冷冻点;凝聚点;凝固硬化点;凝固点;冻凝点;冻(结)点
congealing process 冷凝固化过程

congealing spray 喷雾冻凝法
congealment (液体变为固体的)凝结;冻结
congeal process 冷凝固化过程
congelating temperature 冻凝温度
congelation (液体变为固体的)凝结;冻凝;冻结(作用)
congelation agent 凝聚剂
congelation point thermometer 测定凝固点用温度计
congelation zone 冻结区
congelifraction 冰劈(作用);融冻泥流(作用);融冻崩解作用
congelisol 永冻土;永冻层
congeliturbation 融冻泥流(作用);融动作用
congen bill 航运公会木材提单
congeneric granite 同源花岗岩
congeneric industrial problem 同源的工业问题
congenial 相宜的;同性质的;适宜的
congested area 居民稠密地区;拥挤地区;稠密地区;人口稠密地区
congested flow 拥挤交通流
congested intersection control 拥塞交叉控制
congested map 详图
congested problem 密集问题
congested street 拥挤街道
congested traffic 交通拥挤;拥挤的交通
congested traffic area 交通拥挤地区
congested urban area 拥挤市区
congested use 频繁使用
congested with traffic 运输堵塞
congestion 挤港;集聚;密集;汇集;拥塞;拥挤交通;拥挤;阻塞(交通);堵塞(港口)
congestion call meter 忙呼叫计数表
congestion cost 混杂拥挤导致的费用
congestion degree (交通的)拥挤度
congestion detection 拥挤程度检测
congestion-free 无阻塞
congestion index 阻塞指数
congestion mode 拥塞程度型
congestion of bottom-hole zone 井底堵塞
congestion of reinforcement 钢筋密集
congestion of road traffic 道路拥挤(度)
congestion of traffic 车辆拥塞
congestion problem 密集问题
congestion surcharge 滞卸费;船舶拥挤附加费
congestion system 密集系统
congestive pterygium 赤膜
conginment invoice 送货单
conglagration 快速燃烧
conglobate 成团的;使成球形
conglobation 成团;团聚
conglobation reaction 成团反应
conglomerate 砾岩;跨行业公司;聚集物;聚成球形;集团企业;集成物;积聚成团;凝聚成团;密集体;圆砾岩;硬砾岩;多种行业联合的大企业;堆集物;成球形
conglomerate clay 砾岩土
conglomerated ice 密集冰
conglomerate dike 砾岩岩脉
conglomerate diversification 复合型多角经营
conglomerate facies 砾岩相
conglomerate fan 砾石扇
conglomerate financial statement 综合财务报表;联合财务报表;集合财务报表
conglomerate-gneiss 砾岩片麻岩
conglomerate merger 跨行业企业合并;多行业企业并合;企业集团合并
conglomerate mill 砾岩破碎机;砾石破碎机
conglomerate reservoir 砾岩储集层
conglomerate rock 砾岩
conglomerate silicosis 聚合性矽肺;聚合性石末沉着病
conglomerate stratum 砾石垫层
conglomerate takeover 集团企业接管
conglomerate test 砾石检验
conglomeratic 砾岩状;含砾的
conglomeratic mudstone 砾泥岩;类砾岩;副砾岩
conglomeratic sandstone 砾岩性砂岩
conglomeratic sandstone reservoir 砾质砂岩储集层
conglomeration 凝聚;密集型联合;堆集;团块
conglomeratlc texture 圆砾结构
conglutinant 黏合的
conglutinate 黏住

conglutination 胶着;黏附物;黏合;共凝聚作用
conglutination test 胶固试验
conglutinative complement adsorption test 胶固补体吸附试验
conglutinin 黏合素;团集素
Congo abyssal fan 刚果深海扇
Congo blue 刚果蓝
congo bo(a)rt 刚果金刚石
congo brown 刚果棕
Congo copal 刚果胶
Congo gum 刚果胶
congolian red fever 刚果红色热
congolite 菱锰铁方硼石
Congo pink 刚果粉红色
Congo red 茶红(色);刚果红
congo red test 刚果红试验
congo red test paper 刚果红试纸
congo rounds 球状刚果金刚石
Congo rubine 刚果玉红
congos 刚果红
congost 陡壁横谷
Congo stain 刚果红染剂
congou 中国红茶
Congo yellow 刚果黄
congregated district 人口稠密地区
congregate housing 集体住宅;集合公寓
congregation 聚集;集合;会合
congregational mosque 公理会清真寺
congregational mosque with antique columns 建有古式柱子的公理会清真寺
congregation plaza 集会广场
congregation-zone 集合带
congress 专业会议
congress building 议会大厦;议会大楼;国会大厦
congress canvas 刺绣十字布
congress house 议会大厦;议会大楼;国会大厦
congruence 目标一致;迭合;全等;同余
congruence axiom 全等公里
congruence class 同余类;同余组
congruence integer 同余整数
congruence of curves 曲线汇
congruence of first degree 一次同余
congruence of lines 线汇
congruence of matrices 矩阵相合;矩阵的相合
congruence relations 同余关系
congruency 迭合;全等;同余
congruent 相同的;符合的;同成分
congruent angle 全等角
congruent expression 同余式
congruent figures 叠合图形;全等图形
congruent generator 同余数生成程序
congruential generator 同余数生成程序
congruential method 乘同余数法
congruent innovation 一致创新
congruent matrix 相合矩阵
congruent melting 一致熔融;共熔;同质熔化
congruent-melting compound 一致熔融化合物
congruent melting point 一致熔融点;成分熔点;同成分熔点
congruent number 同余数
congruent point 叠合点
congruent segment 叠合线段
congruent transformation 等成分变化;全等转变;全等变换
congruent triangles 全等三角形
congruity 适合
congruity accuracy of gravimeter 重力仪一致性精度
congruity of parallel test 平行试验的符合性
congruous 一致的;协调的;适合的
congruous fold 相符褶皱
Coniacian 科尼亚斯阶【地】
conic(al) 圆锥形的;锥形的
conic(al) accumulation 圆锥状堆积
conic(al) anchor ring 锥形锚圈
conic(al) anchor wedge 锥形锚楔
conic(al) antenna 锥形天线
conic(al) arbor 圆锥柄轴
conic(al) ball mill 锥形球磨机
conic(al) barrel 锥形槽卷筒
conic(al) beaker 锥形烧杯
conic(al) beam 锥形束;锥形波束
conic(al) bearing 锥形轴承
conic(al) bed 锥形床

conic(al) bellows 圆锥形风箱
conic(al) blender 锥形混合筒;锥形混合机
conic(al) blender type dryer 锥形混合式干燥器
conic(al) body 锥形体
conic(al) boiler 锥形锅炉
conic(al) bolt 锥形螺栓
conic(al) bore 圆锥孔
conic(al) boring 锥孔镗削
conic(al) bottom 锥形底;斗底
conic(al) bottom bin 锥形底仓
conic(al) bottom concrete anchor bolt 锥底混凝土锚栓
conic(al) bottom hopper 锥底料斗
conic(al) bottom tank 锥底槽
conic(al) brake 锥形制动器
conic(al) breaker 锥形碎矿机
conic(al) broach roof 锥形屋顶;攒尖顶
conic(al) buoy 圆锥形浮标;锥形浮标
conic(al) bushing 锥形轴衬
conic(al) cable net 锥形电缆网
conic(al) cam 锥形凸轮
conic(al) camber 锥形弯曲
conic(al) center 锥形顶尖
conic(al) centrifuge tube 尖底离心管
conic(al) chain 锥轴链
conic(al) classifier 锥形分级器;锥形分级机
conic(al) collar 锥形轴环
conic(al) collector 锥形收集器
conic(al) collet 锥形有缝夹套
conic(al) column 锥形柱
conic(al) combustion chamber 圆锥形燃烧室
conic(al) combustor 锥形燃烧室
conic(al) contact 锥形接触
conic(al) contour 锥形
conic(al) convergent 锥形收敛的
conic(al) cowl 锥形风帽
conic(al) crossed slider chain 球面交叉滑块机构
conic(al) crusher 圆锥破碎机;锥形破碎机
conic(al) cup test 圆锥杯突试验
conic(al) cup value 锥形杯突深冲极限值
conic(al) cup wheel 锥形杯形砂轮
conic(al) curve 圆锥曲线;锥曲线
conic(al) cut 锥形掏槽
conic(al) cutter (加工砂轮的)圆锥形刀具;刀碗;锥形牙铣;锥形铣刀
conic(al) depression 锥状洼地;锥状陷部;锥形沉陷
conic(al) diaphragm 锥面薄片
conic(al) die 拉模
conic(al) diffuser 锥形扩散管;锥形扩压器
conic(al) discharge 供给锥形细流(喷油嘴)
conic(al) dispersion 弥散锥体;锥形分散器
conic(al) dome 圆锥(形)穹顶;锥形穹隆
conic(al) dowel 锥形销钉
conic(al) dowel pin 锥形定位销
conic(al) draft tube 锥形尾水管;锥形尾水管
conic(al) drawing die 锥形拉模
conic(al) drum 锥形滚筒
conic(al) drum centrifugal 锥鼓离心机
conic(al) drum hoist 圆锥绞筒提升机;圆锥卷筒绞车
conic(al) dry blender 锥形干混机
conic(al) embankment protection 锥体护坡
conic(al) envelope surface 锥形包络面
conic(al) equal-area projection 等积锥投影
conic(al) equi-distant projection 等距(离)圆锥投影
conic(al) equivalent projection 等积圆锥投影
conic(al) face wheel 圆锥形平面轮
conic(al) fibre 圆锥形纤维
conic(al) filling 锥体护坡
conic(al) fit 圆锥连接配合;锥度配合
conic(al) flask 锥形烧瓶
conic(al) flask with ground-in glass stopper 带磨砂玻塞烧瓶
conic(al) flask with stopper 带塞烧瓶
conic(al) float 锥形浮标
conic(al) floor 锥形底面
conic(al) flow 锥型流;锥形流
conic(al) fold 圆锥状褶皱;锥形褶皱
conic(al) folding 锥形褶皱
conic(al) foundation 锥形基础
conic(al) frustum 截头锥体
conic(al) ga(u)ge 锥度规
conic(al) gear 伞齿轮
conic(al) gear with curved teeth 曲线齿锥齿轮
conic(al) graduate 锥形量杯
conic(al) gravity platform 锥形重力平台
conic(al) grinding point 圆锥磨凿尖
conic(al) grip 锥形啮合;锥形柄
conic(al) groin 锥形穹棱;锥形交叉拱;圆锥交叉拱穹棱
conic(al) head 锥形(封)头
conic(al) head without straight flange 无折边锥形封头
conic(al) head bolt 圆锥头螺栓
conic(al) head rivet 截锥平头铆钉;圆锥头铆钉
conic(al) head without transition knuckle 无折边锥形封头
conic(al) head with straight flange 带折边的锥形封头
conic(al) heavy-medium separator 圆锥形重介质选矿机
conic(al) helimagnet 锥形螺旋磁体
conic(al) helix 锥形螺旋线
conic(al) hill 锥形丘
conic(al) hip 圆锥屋脊;锥形戗脊;锥形戗脊
conic(al) hipped end 锥形屋脊
conic(al) hip tile 锥形戗脊瓦
conic(al) hob 锥形滚刀
conic(al) hoisting drum 锥形卷扬机筒
conic(al) hopper door 圆锥形泥门【疏】
conic(al) horn lens 锥形喇叭透镜
conic(al) illumination 锥形照明
conic(al) indentation hardness 锥形低凹硬度
conic(al) indenter of durometer 硬度计锥形压头
conic(al) inner body 中心锥体
conic(al) irdome 圆锥形红外整流罩
conic(al) jet nozzle 锥形喷管
conic(al) journal 锥形轴颈
conic(al) key 锥形键
conic(al) lamp-shade 锥形灯罩
conic(al) light 锥形天窗;多面金字塔式天窗;多棱天窗
conic(al) liquid diversion baffle 锥形导叶板
conic(al) loaf 圆锥形丘
conic(al) mandrel 圆锥形轴棒;锥形心轴
conic(al) mandrel tester 锥形挠曲试验机
conic(al) mill 锥形磨(机)
conic(al) mixer 圆锥形搅拌机;锥形拌和机;锥形搅拌机;锥形混合器
conic(al) necked nut 锥颈螺母
conic(al) node 锥顶点;锥顶
conic(al) nose 锥形头部
conic(al) nozzle 圆锥形管嘴;锥形喷嘴
conic(al) nut 锥形螺母;锥形螺帽
conic(al) optic(al) fibre 锥度光纤
conic(al) orthomorphic projection 正形圆锥投影
conic(al) outlet 锥形卸料口
conic(al) packer 麻堆锥形压紧机构
conic(al) pendulum 锥摆
conic(al) pendulum governor 锥摆调节器
conic(al) penetrometer 锥形透度计;圆锥针入度仪;圆锥贯入度仪;锥形贯入仪
conic(al) pile 锥形桩;锥形火山岩体;锥形堆积体
conic(al) pinion differential gear 差动小伞齿轮
conic(al) pipe 圆锥管;锥形管
conic(al) pitching 锥体护坡
conic(al) pivot 锥形轴尖;锥形轴颈
conic(al) plug 锥形塞;(预应力混凝土的)圆锥形锚塞;锥形塞;锥形锚塞(预应力钢筋混凝土用);锥塞
conic(al) plug valve 锥形塞阀
conic(al) point 锥顶点
conic(al) polar 极二次曲线
conic(al) polarized light 锥形偏光
conic(al) prism 锥体棱镜
conic(al) probe 锥形探头
conic(al) projection 圆锥投影(法);圆柱投影;锥顶射影
conic(al) projection chart 圆锥投影地图
conic(al) projection map 圆锥投影地图
conic(al) projection with equal absolute values of distortion on boundary and mid parallel 边纬和中纬变形绝对值相等的圆锥投影
conic(al) projection with equal-area in definite region 定域等面积圆锥投影
conic(al) projection with two standard parallels 双标准纬线圆锥投影;圆锥正割投影;双标准纬线等积圆锥海图
conic(al) pulley 锥轮
conic(al) pulley drive 锥轮传动
conic(al) ray 锥光
conic(al) reducer 锥形齿轮减速器;锥齿轮减速器
conic(al) reducer without straight flange 无折边锥体变径段
conic(al) reducer with straight flange 带折边的锥形变径段
conic(al) refiner 锥形磨浆机
conic(al) reflector 锥形反射器
conic(al) ring 锥形环
conic(al) rivet 锥头铆钉
conic(al) rivet head 锥形铆钉头
conic(al) roll 锥形接缝条(铅皮屋面);(金属薄板屋顶咬口的)突肋形接缝;破碎机辊辗;(金属薄膜屋面咬口的)加固木条;锥形滚筒;锥形滚碎机
conic(al) roller 锥形滚柱;锥形滚轴
conic(al) roller thrust bearing 止推滚锥轴承
conic(al) roof 圆锥形屋顶;圆锥体屋顶;圆攒尖顶;锥形(屋)顶
conic(al) rotating screen 锥形转筒筛;锥形旋筛
conic(al) rotor type air separator 圆锥转子式选粉机
conic(al) sampler 圆锥式采样器
conic(al) scan 圆锥形扫描
conic(al) scanning 锥形扫描
conic(al) screen 多边筛;锥形滤(网);锥形格筛
conic(al) screen arrangement 锥形孔筛布置
conic(al) screen mill 锥形筛磨
conic(al) screw 锥形螺旋;锥形螺杆;锥形螺钉
conic(al) seal (contact) face 锥形密封面
conic(al) seat 锥形座
conic(al) seat valve 斜形座阀
conic(al) section 圆锥(形)截面;锥形截面;锥体部分;二次(圆锥)曲线
conic(al) selfprogramming controller 二次曲线自编程控制器
conic(al) settlement 锥形沉降
conic(al) settler 锥形澄清器
conic(al) settling tank 锥形沉降槽
conic(al) shank reamer 带柄锥形扩孔钻;带柄锥形绞刀
conic(al) shape 圆锥号型
conic(al) shell 圆锥(形)壳;锥形筒体;锥形壳(体)
conic(al) shell course 锥形锅炉筒
conic(al) shell foundation 锥形薄壳基础
conic(al) shell-reducer section 锥体变径段
conic(al) shell ring 锥形锅炉筒
conic(al) shock wave 锥形激波
conic(al) shoe 锥形桩头
conic(al) shutter 锥形快门
conic(al) side milling cutter 锥形侧铣刀
conic(al) sieve method 三角筛法
conic(al) skirt support 圆锥形裙座
conic(al) sleeve 锥形套筒;锥形套管
conic(al) sliding block 锥形滑块
conic(al) slope 锥坡
conic(al) socket 锥形套节;锥形套管;锥形凹槽
conic(al) spinner 锥形桨毂盖
conic(al) spiral fluted reamer 锥形螺旋槽绞刀
conic(al) spiral spring 锥形弹簧
conic(al) spiral valve spring 锥形螺旋阀簧
conic(al) spire 锥形塔尖;锥形尖顶装饰
conic(al) spray nozzle 锥形喷雾嘴
conic(al) spring 锥形盘簧;锥形弹簧
conic(al) steel body 锥形钢体
conic(al) stiffener 加强锥
conic(al) stockpile 锥形料堆
conic(al) straight fluted reamer 直槽锥形铰刀
conic(al) surface 锥面;锥式曲面
conic(al) suspension roof 锥形悬挂屋顶
conic(al) swing discharger 塔式转摆式卸料机
conic(al) tank 锥形水箱;锥形水塔;锥形顶储罐
conic(al) thread 锥形螺纹
conic(al) tilting mixer 圆锥形侧倾式搅拌机
conic(al) tip 圆锥顶点
conic(al) tire 锥形轮胎;锥形轮箍
conic(al) tooth 圆锥牙
conic(al) tower platform 锥形柱平台
conic(al) tread 锥形踏面
conic(al) trommel 锥形转筒筛;锥形滚筒筛

conic(al) trowel 锥形抹子
conic(al) tube 圆锥管；锥形管
conic(al) twin-screw extruder 锥形双螺杆挤塑机；锥形双螺杆挤出机
conic(al) type clarifier 圆锥形澄清池；锥形澄清池
conic(al) valve 锥面阀；锥形阀(门)
conic(al) vault 锥形穹顶
conic(al) vessel 锥形容器
conic(al) volcano 锥形火山
conic(al) wall nozzle 锥壁喷管
conic(al) water pipe 圆锥形水管
conic(al) weave fabric 锥形织物
conic(al) wedge 锥形楔
conic(al) wedge anchorage 锥形锚具
conic(al) wing valve 锥形圆盘导翼阀
conichalcite 砷钙铜石；砷钙铜矿
conicity 圆锥度；锥削度
conicity of model 模型锥度
conicograph 二次曲线规
conicoid 圆锥面；二次曲面；圆锥曲线
Coniconchia 锥壳钢
conicycle 浮水重量取样器
conidium 分生孢子
conifer 针叶植物；针叶(树)；松柏类植物
coniferals 针叶植物；松柏类
conifer balsam 针叶树脂
conifer garden 松柏园
coniferophyte 针叶植物；球果植物；松柏类植物
coniferous forest 针叶树林；针叶林
coniferous forest illimerized soil zone in subfrigid zone 亚寒带针叶林灰化土带
coniferous forest region 针叶林带
coniferous timber 针叶树材
coniferous tree 针叶树；松类树
coniflex gear 直齿锥齿轮
coniform 圆锥形的；锥形的
conifruticeta 针叶灌木群落；针叶灌木林
conilignosa 针叶木本群落；针叶木本林
conimeter 测角器；尘埃计；测尘器；测尘计
coning 形成圆锥形；锥形；锥堆四分法；锥度；成锥形的；水舌形成
coning and quartering 堆锥四分取样法
coning angle 圆锥角；锥旋角
coning in oil reservoir 油池漏斗
coning plume 锥形烟羽
coning quartering 锥形四分法
coning the well 水锥浸入油井
coniology 尘埃学；微尘学
coniometer 尘埃计算器
coniometer konimeter 灰尘计
coniophora cerebella 地窖粉孢革菌
coniosis 尘埃沉着病；粉尘病
coniscope 尘粒镜
conisilvae 针叶乔木群落；针叶乔木林
conisterium 力士润肤室(古希腊、古罗马体育场附属房间)
conjectural behavio(u)r 推测行为
conjectural variation 推测变化
conjecture 推测
conjoined manipulation 双手操作法
conjoining 交接
conjugacy 共轭性
conjugacy of the second kind 第二类共轭性
conjugate 复共轭
conjugate acid 共轭酸
conjugate angles 共轭角
conjugate axes 共轭轴(线)
conjugate base 共轭碱
conjugate beam 虚构梁；共轭梁
conjugate beam method 虚梁法；共轭梁法
conjugate branch 结合分支；共轭支路；共轭分支
conjugate centre 共轭中心
conjugate chord 共轭弦
conjugate cleavages 共轭劈理
conjugate complex 共轭复数
conjugate complex function 共轭复函数
conjugate complex numbers 共轭复数
conjugate conics 共轭二次曲线
conjugate constraint 共轭制约；共轭约束条件
conjugate control 同成控制
conjugate curve 共轭曲线；伯特兰曲线
conjugated acid 结合酸
conjugate(d) depth 共轭水深
conjugate deviation 同向偏斜

conjugated fatty acid 共轭脂肪酸
conjugated focus 共轭焦点
conjugate diameters 共轭直径
conjugate diametral planes of a quadric 二次曲面的共轭径面
conjugate direction method 共轭方向法
conjugate directions 共轭方向
conjugate direction search 共轭方向搜索
conjugate distance 共轭距
conjugate division 接合分裂
conjugate elements of a group 群的共轭元
conjugate equation 共轭方程
conjugate fault 共轭断层
conjugate fissures 共轭裂缝
conjugate focal planes 共轭焦面
conjugate foci 共轭点
conjugate folds 共轭褶皱
conjugate fold system 共轭褶皱系
conjugate fracture 共轭破裂；共轭断裂
conjugate fracture system 共轭破裂系
conjugate functions 共轭函数
conjugate gradient 共轭梯度
conjugate gradient method 合成梯度法；共轭斜量法；共轭梯度法
conjugate harmonic function 共轭调和函数
conjugate image 共轭像
conjugate image ray 同名光线
conjugate impedance 共轭阻抗
conjugate joint 共轭节理
conjugate kink-band 共轭膝折带
conjugate line 共轭线
conjugate linear operator 共轭线性算符(子)
conjugate lines in a conic(al) 二次曲面的共轭直线
conjugate lines in a quadric 二次曲面的共轭直线
conjugate matrix 相伴矩阵；共轭矩阵；伴随矩阵
conjugate numbers 共轭数
conjugate of function 函数的共轭值
conjugate operation 共轭运算
conjugate operator 共轭算子
conjugate phases 共轭相
conjugate photo points 共轭像点
conjugate plane 共轭面
conjugate point method 共轭点解释法
conjugate points 共轭(像)点
conjugate position 共轭位置
conjugate potential function 共轭位函数；共轭势函数
conjugate pressure 共轭压力
conjugate prior concept 共轭先验概念
conjugate ray 共轭射线；同名光线
conjugate reductant 共轭还原剂
conjugate redundant force 共轭多余力
conjugate section 共轭截面
conjugate shear angle 共轭剪切角
conjugate shearing stress 共轭剪切应力
conjugate shovel-run method 配合铲装法
conjugate solution 共轭解
conjugate stages 相配阶段；共轭阶段
conjugate stress 共轭应力
conjugate subgradient method 共轭次梯度法
conjugate system 共轭系
conjugate tangents 共轭切线
conjugate tensor 共轭张量
conjugate veins 共轭脉
conjugation 共轭性
conjugation line 连接线
conjugation reaction 结合反应
conjugation site 结合部位
conjugation tube 接合管
conjunct arc structure 联合弧形构造
conjunct-cycle geosyncline 跨旋回地槽
conjunction 联结；连接；连带；结合件；会合(点)；合时刻；合取；配合
conjunction(al) arc 联合弧
conjunction(al) wave 重复波；驻波
conjunction forecast 近期预测
conjunction of structures 构造体系联合
conjunction of tectonic systems 构造体系的联合
conjunctive form 逻辑积形
conjunctive kinetics 连接动力学
conjunctive path 连接通道；结合路线
conjunctive query 联合查询
conjunctive search 逻辑乘(法)探索；按与检索
conjunctive symbiosis 连接共生

conjunctive symbol 逻辑乘符号
conjunctive tissue 结合组织
conjunctive use 连接使用
conjunctive use of resources 资源结合应用
conjunct polymer 混合聚合物
conjuncture 局面；行情；事态
Conklin process 康克林选煤法
conloy 康洛铜铝合金
conn 指挥驾驶；指挥操舵；操舵
connarite 水硅镍矿
connate deposit 原生沉积
connate water 矿物油；埋藏水；化石水；原生水；封存水；天然水
connect 联系；连系；交接
connect box 连接框
connect collar 连接环；中间接环
connect data set to line 线路连接命令
connected axis 联动的
connected between borehole 井间速通
connected classification 联合分类
connected clique 连接集团
connected component 连通支；连通分支；连通分量
connected condition of karst 洞穴连通情况
connected directed graph 连通的有向图
connected domain 连通域
connected-element interferometer 连线干涉仪
connected footing 连接基脚；联合基础；连合底部；结合底脚
connected graph 连通图；连接图
connected head pile 接头桩
connected in parallel 并联的
connected in series 串联的
connected journal box 结合轴箱
connected load 装接容量；联结荷载；连接负载；接负荷；装料容量
connectedness 联络性；连通性；连通件
connected porosity 连通气孔率
connected position 通路位置
connected reference 连续访问；连接访问
connected region 连通区域
connected row set assignment 连通行集合分配
connected set 连通集
connected shaft 联动轴；联动的；连动轴
connected spline 连接样条
connected stack 烟囱连接段
connected state 连通状态
connected storage 连续存储
connected structure 接合结构
connected surface 连通曲面
connected system 连接系统；连通系统
connected traverse 连接导线【测】；通廊
connected yoke 系梁；连接横木；连接夹
connecter 连接器；联结件；连通器；连通器；联结管线；联结管道；连接物；连接头；连接器(开关)；连接片；连接符；结合器；结合环；接续器；接线端子；接头；接合物；接管；接插件；衔接；端子；插头插座；插塞与插座
connecter joint 连接器接头
Connecticut brownstone 康尼提卡特棕色石料(一种在美国初期广泛使用的深棕色建筑石料)
connecting 连通道
connecting action 联结作用
connecting angle 连接角(钢)；结合角钢；短角材
connecting arc 连弧线(两个圆形格箱的钢板桩弧线)
connecting arrangement 连接装置
connecting bar 接线柱；连系钢筋；连陆沙洲；连(接)杆；连岛沙洲
connecting beam 连系梁；连接梁；结合梁
connecting beam bridge 结合梁桥
connecting block 连接木块
connecting body 联结体
connecting bolt 连接螺栓
connecting box 连接器；接线盒；供料机通道口；连接箱
connecting branch 连接支管
connecting breakwater 连接式防波堤
connecting bridge 连接(接)桥；桥式连接器
connecting bus 连接母线
connecting busbar 连续汇流排
connecting cable 接合电缆；接线电缆
connecting canal 联渠；连接运河；引上通路
connecting chain 连接链；联结链
connecting channel 连接渠道；连接航道

connecting circuit 联系电路
connecting circuit between stations 站间联系电路
connecting circuit between yards 场间联系电路
connecting circuit of highway crossings within the station 站内道口联系电路
connecting circuit with a depot 机务联系电路
connecting circuit with automatic blocks 自动闭塞联系电路
connecting circuit with block instruments 半自动闭塞联系电路
connecting circuit with block signal(l)ing 区间联系电路
connecting circuit with semi-automatic block 半自动闭塞联系电路
connecting clamp 连接夹头；接头夹
connecting clamp for busbar 汇流排连接线夹
connecting cleat 连接夹板
connecting clip 结合扣
connecting conduct 导管
connecting conductor 连接导线【电】
connecting conduit 连接渠道；连接管(道)
connecting conveyer 连接输送机
connecting cord 连接塞绳
connecting corridor 连接(走)廊
connecting crossover 联络渡线
connecting curve 联络曲线；连接曲线；耦合曲线；缓和曲线
connecting cylinder 连接筒
connecting detail 连接细部
connecting diagram 接线图
connecting direction 连接方向
connecting door lock 套间门锁；连接房间门锁
connecting drawing 接线图
connecting duct 连接导管；连通沟槽；连接套管
connecting elbow 连接弯管
connecting fault 联通性断裂
connecting fitting 接续金具；横向连接件
connecting fixture 连接夹具
connecting flange 连接凸缘；连接法兰；联轴节
connecting footing 结合底盘；条形基础
connecting function 连通函数
connecting gallery 联络坑道；(隧道的)连络坑道；连接廊
connecting gangway 连接(的)引桥；连接的浮桥；连接引线
connecting gear 传送机构
connecting highway 连接公路
connecting hose 连接软管；软管；软管接头
connecting jack 接地塞孔
connecting joint 接头
connecting level line 符合水准路线
connecting lever 连接杆；连杆；拐臂
connecting line 联络线；联结线；连线；连接(路)线
connecting link 联结环节；连接链环；连接环节；杆；可拆链环
connecting loop 连接环
connecting means 连接方法
connecting method of counter with heat-supply network 热用户连接方式
connecting mole 连接式防波堤
connecting nipple 螺纹接套
connecting nut 连接螺母
connecting outlet 连接出口；接线盒
connecting panel 接线盘；接线板
connecting passage 联络通道；连接通道
connecting path 连接线路；连接道路
connecting piece 连接件；连接段；接套；接片
connecting pile 接合板桩
connecting pin 连接销钉；连接销
connecting pipe 联管；连系管；连通管；连接管道；连接管；结合管；导压管
connecting pipe line 连接管线
connecting piping 连接管
connecting place 连接点
connecting plate 接合板；联结板；连接板；紧固片；缓板
connecting platform 联络站台
connecting point for level(l)ing 水准联络点
connecting processing 连接处理
connecting railroad 联络铁路(线)
connecting railway 联络线；联络铁路(线)
connecting reinforcement 连接钢筋
connecting ring 连环
connecting rivet 连接铆钉

connecting road 连接路
connecting rod 联结杆；联杆；连(接)杆；结合杆；活塞杆
connecting rod aligner 连杆校正器
connecting rod and cap marks 连杆端盖装配标记
connecting rod and reach rod jaw 连杆与拉杆爪
connecting rod bearing 连杆轴承
connecting rod bearing cap 连杆轴承盖
connecting rod bearing reborer 连杆孔修刮机
connecting rod bearing reboring device 连杆轴孔镗削工具
connecting rod bearing saddle 连杆轴承轴瓦座
connecting rod bearing seat 连杆轴瓦座
connecting rod bearing shell 连杆轴瓦
connecting rod bearing shim 连杆轴承垫片
connecting rod bending tool 连杆校弯器
connecting rod big end bearing 连杆大头轴承
connecting rod body 连杆体
connecting rod bolt 连杆螺栓
connecting rod bottom-end bearing 连杆大头轴承
connecting rod bushing 连杆衬套
connecting rod bush lock 连杆瓦锁销
connecting rod cap 连杆轴承盖
connecting rod dipper 连杆油匙
connecting rod eye 连杆头孔
connecting rod for indication 表示连接杆
connecting rod guidance 十字头导承
connecting rod head 连杆端
connecting rod jig 连杆定位器
connecting rod lower bearing shell 连杆下轴瓦
connecting rod lower half bearing 连杆下瓦
connecting rod oil line 连杆油槽
connecting rod pin 活塞销；连杆销
connecting rod pressure 连杆方向的压力
connecting rod rest 连杆架
connecting rod shim 连杆垫片
connecting rod sleeve 连杆衬套
connecting rod small end bush 连杆小端轴瓦
connecting rod spurt hole 连杆润滑喷孔
connecting rod thrust 连杆方向的推力
connecting rod tip 连杆大端
connecting rod upper bearing shell 连杆上轴瓦
connecting rod with bearing shell 带轴瓦的连杆
connecting rod with oil passage 中空机油油道连杆
connecting rod wrench 连杆套筒扳手
connecting screw rod 连接螺杆
connecting shackle 连接卸扣；连接吊钩；锚链(连接)卸扣
connecting shaft 连轴
connecting sleeve 套节；连接套(筒)
connecting sleeve bolt 连接套筒螺栓
connecting sleeve pin 连接套销
connecting station 联轨站
connecting strand 联络索；联络丝
connecting strap 连接条；连接片
connecting structure 连接结构
connecting strut 连接支柱
connecting surface 连接表面
connecting tag 连接销
connecting taxiway 连接滑行道
connecting terminal 连接触点；接线端子；接头
connecting test 连通试验
connecting thread 联络丝
connecting track 引入线；联络线；联络铁路(线)；连络线；梯线【铁】
connecting traverse 连接导线【测】；附合导线【测】；双索吊梁
connecting tube 连系管；连接管(道)；连接管；结合管
connecting up 布线
connecting utricle 连接鞘
connecting valve 连接阀
connecting vessel 连通器
connecting wire 连接线；连接导线【电】
connecting yoke 连接横木
connect in parallel 并联连接
connect in series 串联连接
connection 联系；连接；连通；连接法；连法；连续；接合处；级联
connection angle 连接角(钢)；结合角钢；接合角钢
connection at ends of tracks 股道终端连接
connection band 接头箍
connection bar 连接杆
connection block 十八斗；交互斗

connection bolt 联结螺栓
connection box 连接箱；连接盒；接线箱；分线箱
connection bus 连接器汇流排
connection cable 接线电缆
connection canal 联渠；连接运河；连接通道；连接管
connection clamp 连接线夹
connection clip 结合扣
connection corridor 连接走廊
connection cover cutting pliers 剥线钳
connection cubicle 联络线开关柜
connection detail 连接细部；连接点详图
connection diagram 连线图；接线图
connection dimension 连接尺寸
connection eccentricity 接头偏心度
connection endpoint 连接终点
connection fittings 连接配件
connection flange 连接翼缘；连接法兰
connection force 连接力
connection foreman 管线工长
connection for gyrobattery 陀螺电池引线
connection gang 管线班
connection graph 连接图
connection hose 连接软管
connection identifier 连接标识符
connection in-parallel! 并接；并联
connection in series 串联
connection insulation 连接点绝缘
connection layout 布线装置
connectionless protocol 无连接协议
connection line 连接线
connection link 连接环；连接杆
connection load 连接荷载
connection matrix 连通矩阵；连接矩阵
connection multiplexing 连接多路转接
connection of circuits 电路的连接
connection of filter and impervious bottom bed 过滤器与隔水底板连接
connection of filter and impervious top bed 过滤器和隔水顶板连接
connection of safety valve to vessel 安全阀对容器的连接
connection of steam 消防汽车接头
connection operator 连接算符
connection orientation 管口方位图
connection oriented network layer protocol 面向连接的网络层协议
connection oriented protocol 面向连接协议
connection path 连接路
connection piece 连接件
connection pipe 连接管道
connection piping 连接管
connection plate 连接板
connection point 接点
connection point for orientation 定向连接点
connection point manager 结点管理
connection power 连接电源
connection quadrangle method 联系四边形法
connection rail 连接轨；转辙轨
connection rebar 连接钢筋
connection road 连接道路
connection rod 闸连杆
connection rose 电话机接线盒
connection screw 连接螺钉
connection shaft 连接轴；中间轴；天轴
connection sheet pile 连接板桩
connection signal 接通信号
connection size 连接尺寸(标准)
connection sleeve 连接套筒；连接套管
connection socket for aerial 天线插口
connections of polyphase circuits 多相电路联接
connection splitting 连接分割
connection strap 连接条；连接片；桥接线
connection survey 联系测量；连接测量
connection survey for shaft orientation 矿井定向联系测量
connection survey in mining panel 采区联系测量
connection terminal 连接终点；连接末端；接线端(面)板
connection test line 连接测试线
connection time 连接时间；接钻杆时间
connection to electric(al) supply 接电源设备；电源接通
connection triangle 连接三角形

connection triangle method 联系三角形法
connection tube 连接管道;接管
connection type of coordination polyhedron 配位多面体连接方式
connection under constraint 强制符合
connection valve 连接阀
connectitvity 连通
connective asymptotic stability 联结渐近稳定性
connective instability 联结不稳定性
connectively reachable digraph 联结可达有向图
connective operation 联结运算
connective reachability 联结可达性
connective stability 联结稳定性
connective suboptimality 联结次优性
connective ultimate boundedness 联结最终有界性
connective weld 联系焊缝
connectivity 连通性;连通度;联结性;连接性;接合性;贯通性
connectivity matrix 连接矩阵
connectivity number 连通数
connectivity pair 连通性偶;连通性对;连通度对
connector = connecter
connector assembly 接插件;插头座;插头连接
connector bank 终接器线弧
connector base 接线座
connector bend 连接(弯)头;连接弯管;接合弯头;弯(接)头;弯管
connector block 连接器插头块;连接件插头块
connector body 连接体
connector breakout 接线分线器
connector cable 电缆接头
connector contact 连接接触
connector for cantilever bracket 腕臂底座连接架
connector forehearth 过渡通路
connector for twin cntilever bracket 双腕臂底座连接架
connector frame 接插件框架
connector-induced optic(al) conductor loss 连接器引起的光导体损耗
connector insertion loss 连接器插入损耗
connector joint 连接器接头
connectorless 无连接器
connector link 接头连杆
connector lug 连接接头;接线衔套;接线头
connector motor magnet 回转电磁铁
connector name plate 接线指示牌
connector pin 接线插脚;插销;塞子
connector plug 连接插头;塞子;插头
connector relay 终接器继电器
connector ring 结合环
connector section 连接机构
connector shelf 终接器架
connector sleeve 连接套筒
connector socket 接线插座
connector splice 接头拼接板
connector sub 接头
connector wire 连接线
connect-out command 连出指令
connectron 气体放电管
connect signal 接线信号
connect time 连接时间
connect to earth 接地
connellite 羟硫氯铜石;铜氯矾
connerville meter 罗茨式煤气表;转子式煤气表
connexion 贸易上的往来关系;水陆联运
connexion ticket 联运票
conning tower 航行指挥塔;指挥台;军舰司令塔;驾驶指挥塔
connoisseur 鉴定家;行家
connon proof 防弹的
connor gate 缝隙浇口
connotative meaning 内涵意义
connwallite 翠绿砷铜石
conode 系线;共节点
conodont glacial stage 牙形石绝灭
conodontophoridia 牙形石类
conodrymium 常绿群落
conoid 圆锥形;圆锥体;劈锥曲面
conoidal dome 劈锥曲面穹顶
conoidal vault 锥形穹顶
conoidal vaulted 成锥形穹顶的
conoidal vaulting 锥面形穹顶;扇形锥状穹顶;尖拱顶
conoidal wave 浅水波;浅海波;椭圆余弦波;椭圆函数波
conoid process 锥形突
conoid roof 圆锥体屋顶;锥形屋顶
conoid shell 劈锥曲面壳
conoid vault 锥形穹顶
conophorium 针叶林群落
conoplain 山前侵蚀平原;山前平原
conormal 余法线;共法线
conoscope 晶体光轴同心圆观测器;锥光偏振仪;锥光镜
conoscope image 干涉图形
conoscopic observation 锥光观察
conpartment mill 多室磨碎机
Conpernik 康普尼克铁镍基导磁合金
conqinine 康奎宁
conqueror 康奎尔硅锰钢
conquest of nature 征服自然
Conrad discontinuity 康拉德面【地】;康拉德间断
Conrad-Limpach method 康拉德—林巴赫法
Conradson carbon value 康拉逊碳值
conradson method 残碳测定法
consanguineous 同源的
consanguineous association 同源组合
consanguinity 同源
conscience money 补交款;补缴款项
conscious error 已知误差
conscious parallel action 拟订相同价格的行动
consecutive 串行的
consecutive action 连续动作
consecutive ascending grades 连续上坡(道);连续升坡
consecutive collection 连续采集
consecutive computer 串行计算机;顺序操作计算机
consecutive data set 连续数据集
consecutive data set organization 连续数据集组织
consecutive days 连续(作业)日;连续装卸日(风雨和假日均计在内的装卸期)
consecutive descending grades 连续降坡;连续下坡(道)
consecutive digestion 连串消化
consecutive drift tube 连串漂移管;相接漂移管
consecutive earthquake 连续地震
consecutive entries 顺序连续记录
consecutive firing 连续爆破;串联爆破;顺序爆破
consecutive frame 相邻帧
consecutive grades 连续纵坡
consecutive hours 连续时(计算装卸期)
consecutive indexing 相邻指数
consecutive input 连续输入
consecutive lift 连续上升的浇筑层
consecutive line 连续行;相继行
consecutive machine method 流水机组法
consecutive mean 连续平均值;相继平均值;动态平均(值)
consecutive negative 连续航摄底片
consecutive number 邻数;连(续)号;编号
consecutive operation 顺序运行;顺序操作;连续运转;连续操作
consecutive organization 连续组织
consecutive photograph 连续航空像片
consecutive processing 连续处理;串行处理
consecutive-range zone 接续延限带
consecutive reaction 连续反应;连串反应
consecutive reading 相邻读数
consecutive readout 连续读出
consecutive return trip charter party 连续往返航次租船合同
consecutive run number 流水号
consecutive-sequence computer 串行操作计算机
consecutive shock 连续地震
consecutive single trip charter party 连续单航次租船合同
consecutive spilling 连续溢出法
consecutive storage 连续存储
consecutive train 连发列车
consecutive value 相邻值
consecutive voyage charter 连续航次租船
consecutive voyages 连续航次;连续航程
consecutive weather working days 连续晴天工作日
consecutive years 连年
Consenco classifier 自动卸载式干涉沉降分级机
consensual light reflex 互感性光反射;同感性光反射
consensus 协商一致;共同意见;多数人意见;同意
consensus method 一致法
consensus proceeding 一致程序
consensus standard 一致标准
consent 万能插口;塞孔
consent decree 同意法令
consent dividend 认定股利
consent election 同意选举
consent judgment 赞同判决
consent limit 允许限度
consent of surety 保证承诺;工程保证书
consent to build 同意建造;建筑许可证;施工执照
consent zoning 同意条件用地区划
consequence 结果;后果;推论
consequence analysis 后果分析;地震后果分析
consequence in decision tree 决策树中的结果
consequence of earthquake 地震后果
consequent 后项【数】;推论;顺向
consequent divide 顺向分水岭
consequent drainage 顺向排水(系);后成河;顺向水系;顺向河
consequent drainage system 顺向水系
consequential arc back 持续性逆弧
consequential commutation failure 连续换向失败
consequential damage 间接损害赔偿;间接损害
consequential file 相继文件
consequential loss 间接损失;从属损失
consequential loss insurance 灾后损失保险
consequent lake 顺向湖泊
consequent landform 顺地貌
consequent landslide 顺层滑坡
consequent logical 合乎逻辑的
consequent pole 交替极;间极;换向极;罩极;辅助极;屏蔽极;中间磁极
consequent-poles motor 交替磁极式电动机;变极式双速电动机
consequent-poles winding 交替磁极绕组;中间极绕组;罩极绕组
consequent river 顺向河
consequent stream 顺向河(流)
consequent succession 顺序演变
consequent valley 顺向河谷;顺向谷
consersion unit 反应设备
conservancy 自然资源保护区;保护
conservancy area 封山育林区;保护区;水土保持区;自然保护区
conservancy district 封山育林区;保护区
conservancy engineering 资源防护工程;水土保持工程;疏浚工程
conservancy matters (河道,港口的)疏浚维护工作
conservancy of sanitation 卫生管理
conservancy system 干厕;非冲水厕所;储存系统;非水运系统
conservancy tank 储存池
conservancy works 水利工程
conservation 资源保持;不灭;保留;保存;保藏;守恒
conservation amplitude wavelet processing of desired layer 目的层保振幅子波处理
conservation and collection of water 水的保护与采集
conservation area 自然保护区;封山育林区;保护区;水土保护(地)区;水土保持区
conservation area of wildlife 野生生物保护区
conservation design 保守设计
conservation district 保护区
conservation energy 节能
conservation farming 保持水土耕种
conservation field of force 保守力场
conservation force 守恒力
conservation law 资源保护法;守恒(定)律;生物保护法
conservation measure 保护措施
conservation method for the design in frozen soil 冻土地基保持法设计
conservation of ancient architecture 古建筑保护
conservation of angular momentum 角动量守恒
conservation of aquatic resources 水产资源保护
conservation of biologic(al) diversity 生物多样性的保护
conservation of biologic(al) life 生态保护
conservation of both moisture and soil 保持水土
conservation of classic(al) architecture 古建筑保护
conservation of climate 保护气候

conservation of energy 能量守恒;能量不灭
conservation of energy law 能量守恒定律
conservation of energy principle 能量守恒原理
conservation of freshwater 淡水保护
conservation of fuel 节约燃料
conservation of functional equation 泛函方程的守恒
conservation of groundwater 地下水保护
conservation of historic(al) buildings 古建筑保护
conservation of historic(al) cultural cities 历史文化名城保护
conservation of historic(al) landmarks and sites 文物古迹保护
conservation of historic(al) monument 古迹保护
conservation of human resources 人类资源保护
conservation of marine protection 海洋保护
conservation of mass 质量守恒
conservation of matter 物质守恒
conservation of moment of momentum 动量矩守恒
conservation of momentum 动能守恒
conservation of momentum momentum conservation 动量守恒
conservation of movement of the centre of gravity 重心运动的守恒
conservation of natural ecosystem 自然生态系保全
conservation of natural habitats 自然生境的保护
conservation of natural resources 自然资源保护
conservation of nature 自然保护
conservation of ocean protection 海洋保护
conservation of orbital symmetry 轨道对称守恒
conservation of probability 几率守恒
conservation of rate animal 珍稀动物保护
conservation of recreation resources 再生资源保护
conservation of resources 资源保护;储藏量守恒;保护资源;保护矿藏
conservation of scenic spots 风景名胜保护
conservation of soil and water 水土保持
conservation of storage 持恒蓄水
conservation of vorticity 涡旋度守恒;涡量守恒;涡度守恒
conservation of water 水体保护
conservation of water resources 水资源保护
conservation of waters 水资源保护
conservation of wildlife 野生生物保护
conservation plan 保护规划
conservation planning 保护计划
conservation plant 废料(利用)工厂;利用废料生产的工厂
conservation policy 保护政策
conservation pool 水库;蓄水池
conservation process 守恒过程
conservation program(me) 保护规划
conservation progress 水土保持的进展
conservation purpose 兴利蓄水目标;(需蓄水的)兴利目标
conservation rate 保持率
conservation regulations 保护条例
conservation regulator 恒定调节器
conservation relay 保护继电器
conservation reservoir 蓄水库;多年调节水库
conservation stocking 保守放养法
conservation storage 有效库容;蓄水库;兴利(蓄水)库容;储蓄水;持恒蓄水(量)
conservation storage reservoir 蓄水水库
conservation system 守恒系统
conservation tillage 水土保持耕作
conservation value 保护价值
conservation zone 资源保护区
conservatism 保守(性);稳健原则
conservatism convention 保守计算惯例
conservative 保守的;保藏的
conservative concentration 恒定浓度
conservative constituent 保守分量
conservative design 保守设计;保安设计
conservative designer 保守设计者
conservative element 保守元素
conservative estimate 保守估算;保守估计;偏低的估计
conservative figure 保守数字
conservative force 保守力
conservative force field 保守力场

conservative games manship 保守的对策策略
conservative grazing 合理放牧;适度放牧
conservatively 保守地
conservatively loaded bearing 保守负载轴承
conservative plate boundary 稳定板块边界
conservative pollutant 持恒污染物;保存性污染物
conservative prediction 保守的预测值
conservative principle 保守主义
conservative properties of air 空气的守恒性质
conservative property 守恒性质
conservative replacement 保守替换
conservative replication 保留复制
conservative scattering 守恒散射
conservative side slope 平缓边坡
conservative slope 平缓边坡
conservative solution 保守解
conservative substance 保存性物质
conservative substitution 保守替换
conservative system 保守系;守恒系统
conservative tracer 恒定程宗示
conservative treatment 保守处理
conservative use of funds 节约使用资金
conservative value 保守数值
conservative water quality 经常性水质;恒定水质
conservator 自然资源管理人员;储存器;保存器
conservatory 暖房;温室
conservatory of music 音乐学院
conserve biodiversity 保护生物多样性
conserved axial current 守恒轴矢流
conserved reserves 保有储量
conserved vector current 守恒矢量流
conserve regulation 保护条例
conserving agent 除腐剂;防腐剂;保存剂
conserving soil fertility 保持土壤肥力
conserving timber sleeper 木枕防腐处理
conshelf 大陆架
consider 考虑;认为
considerable latitude 相当大的幅度
considerable moisture 相同水分
considerable order 大量订购
consideration 对价;审议
consideration money 酬金;补偿金;报偿金
considerations in product design 考虑产品的设计
considered(dam) site 考虑的坝址;论证的坝址;设想的坝址
considered port site 考虑的港址
Considere pile 对称主筋桩
Considere system 康西多体系【建】
considering 鉴于
consign 托运
consign a letter to the post 付邮
consignation 寄存
consigned goods 承销品
consigned materials for processing 委托加工材料
consigned processing 委托加工
consigned receipt 承销收据
consigned to a foreign custom house broker 以进口地报关行为受货人
consigned to buyer 以购货人为受货人
consigned to collecting bank 以代收银行为受货人
consigned to issuing bank 以开证银行为受货人
consignee 代销人;承销商;承销人;受托人;受货人;收件人;收货人;收货单位;收件人
consignee's address 收货人地址
consignee's covers 提单信
consignee's letter of guarantee 收货人保证书
consignee's warehouse 收货人仓库
consigner 交付者;发货人;委托者;托运人;托销的货主
consigner and consignee 货主部门
consign for shipment 托运
consign goods by rail 货物交由铁路运送
consigning of freight 货物托运
consigning of luggages and parcels 行李包裹托运;行包托运
consignment 交货物;交付;寄售;一批货(物);一次交货量;代售;委托;拖运;托销
consignment account 寄销账目
consignment agent 寄售代理人;托运代理人
consignment application 托运申请书;托运申请单
consignment bill 火车运货契约;发货通知书;托运单
consignment business 寄销业务;委托买卖
consignment clause 指定代理人条款
consignment contract 寄售契约;寄售合同

consignment expenses 寄销费用
consignment guarantee money 委托保证金
consignment invoice 寄售发票;运货单
consignment inward 寄售收货
consignment note 寄售通知书;火车运货契约;发货通知书;发货通知;托运收据;托运单
consignment of goods 货物托运
consignment profit and loss 寄销损益
consignment purchase 委托收购
consignment record 寄售记录
consignment sale 寄售
consignment sheet 收货凭单
consignment stock 托销存货
consignment stock insurance 寄销货物保险
consignment store 委托行
consignment tax 承销税
consignment trade 寄售贸易
consignment warehouse 寄存仓库
consignment with lowest price limit 寄售最低限价
consignor 交付者;寄售人;发货人;委托人;拖运人;托运人;托运方;托销人;托销的货主
consistence 连贯性;浓度;相容性;一致(性);致密性;稠性;稠度
consistence coefficient 稠度系数
consistence degree 稠度大小
consistence limit 稠度极限
consistence measurement method 稠度测定方法
consistence meter 稠度仪;稠度计
consistence of composition 成分的一致性
consistence of equations 方程组的相容性
consistence range 稠度范围
consistency 坚松度;浓度;相容性;一致性;一贯性(原则);稠性;稠(密)度
consistency analyzer 稠度分析器
consistency check 相容性检验;一致性检验;一致性检查;稠度检查
consistency coefficient 稠度系数;一致性系数
consistency condition 相容条件
consistency constraint 一致性的约束
consistency controller 稠度控制器
consistency cup 稠度杯
consistency detector 稠度检测器
consistency equation 相容方程(式)
consistency error 一致性错误
consistency factor 稠度要素;稠度系数
consistency ga(u)ge 稠度仪;稠度计
consistency incorrect 稠度不合适
consistency index 稠度指数;稠度指标
consistency index property 稠度指数性质
consistency indicator 稠度指示器
consistency limit 稠度限度;稠度界限;稠度极限
consistency limit of soil 土的稠度界限
consistency meter 稠(密)度计;稠度仪
consistency of asphalt 沥青稠度
consistency of cement 水泥稠度
consistency of clay 黏土稠度
consistency of coefficients of correlation 相关系数的一致性
consistency of cohesive soil 黏性土稠度
consistency of colo(u)r 颜色的一致
consistency of composition 成分的一致
consistency of concrete 混凝土稠度
consistency of grease 润滑脂稠度
consistency of linear equations 线性方程组的相容性
consistency of mortar 砂浆稠度
consistency of soil 土的稠度
consistency of soil particle and cementing substances 土粒—孔隙体系的稠度
consistency operation 后台操作;一致性操作
consistency principle 一致原则
consistency probe 稠度探针;稠度探测仪
consistency profiles 轮廓法
consistency proof 相容性证明
consistency range 稠度范围
consistency regulator 稠度调节器
consistency routine 一致性程序
consistency state 稠度状态
consistency state of mud-stone flow 泥石流稠度
consistency test 一致性检验;稠度试验
consistency test of concrete 混凝土稠度试验
consistency test table 稠度试验台
consistency variable 稠度变量
consistent 相容;一贯

consistent approximation 相容逼近(法)
consistent biologic(al)regeneration 相容生物再生
consistent calibration 相容检定
consistent element 协调元;相容元
consistent equation 相容方程(式)
consistent estimate 相容估计;一致估计
consistent estimation 相容估计
consistent estimator 一致推算子;一致估计量
consistent fat 稠脂肪
consistent geometric(al) stiffness 相容几何刚度;一致几何刚度
consistent grease 黄油
consistent lubricant 黄油;滑脂;润滑脂
consistently ordered matrix 相容次序矩阵
consistent mass matrix 相容质量矩阵;一致质量矩阵
consistent matrix 相容矩阵
consistent nodal load 相容节点荷载;一致节点荷载
consistent order 相容次序;一致序
consistent policy 一贯方针;一贯的政策
consistent result 一致结果
consistent statistic 一致统计量
consistent stiffness matrix 相容刚度矩阵
consistent substitute 始终如一的置换
consistent test 相容检验
consistent vibration model 协和振动模型
consist fo vehicle 车辆编组
consistodyne 稠度调节器
consistometer 稠度仪;稠度计
consistometric value 稠度值
consistory 庄严的集会
consmopolitan speices 普生种
Consol 康索尔系统;扇形无线电指向标
consolan 康索兰
Consolan 康索兰系统;区域无线电信标
Consol beacon 康索尔信标;康索尔航标
Consol chart 康索尔图
console 落地式支架;落地柜;螺旋形柱;螺形支架;控制塔;牛腿【建】;支托;操作台;操作间;托架
console air conditioner 托架式空调器
console bracket 螺形支托;螺形托座
console buffer 控制台缓冲器;操作台缓冲器
console cabinet 控制室
console control circuit 控制台控制电路
console control desk 落地式控制台
console cornice 螺形支柱挑檐
console debugging 控制台调整程序;控制台调整
console directory 控制台显示牌
console display 控制显示(器);控制台显示(器);台式显示器
console documents copying machine 落地式复印机
console function 控制台功能
console keyboard 控制台键盘
console lift 舞台升降控制器;舞台升降机
console message 控制台信息
console model 落地式
console monitor typewriter 控制台监视(打字机)
console of centralized control 中心控制台
console operator 控制台操作员
console optical display plotter 控制台光显示绘图仪
console package 控制台部件
console panel 控制盘;操纵板
console printer 控制台打印机
console radio set 落地式收音机
console receiver 落地式收音机;落地式接收机
console room 控制室
console scope 控制台显示器
console screen 落地式屏风
console section 操纵台部分
console set 落地式收音机
console stack 控制台堆栈
console switch 控制台开关;操作开关
console table 蜗形腿狭台;墙装台案;靠墙小桌;螺形托脚桌案(古式家具)
console terminal 控制台终端
console trap 控制台陷阱
consolette 小型控制台
console type receiver 柜式接收机
console variable 控制台变量
consolidate 巩固;使合并;使固定
consolidated 加固的;固结的
consolidated account 合并账目;合并账户

consolidated annuities 统一公债
consolidated appropriation bill 统一拨款法案
consolidated bond 统一公债
consolidated cash budget 统一现金预算
consolidated company 附属公司
consolidated control 集中控制
consolidated debt 固定债务;合并债务;统一债务
consolidated depth 固结深度
consolidated design criterion and methology 统一设计标准和方法
consolidated-drained compression test 固结排水压缩试验
consolidated-drained direct shear test 慢剪试验;固结排水直剪试验
consolidated-drained shearing test 固结慢剪试验;固结排水剪切试验;固结排水剪力试验
consolidated drained test 固结排水试验
consolidated-drained triaxial compression test 固结排水三轴压缩试验
consolidated-drained triaxial test 固结排水三轴试验
consolidated entity 统一实体
consolidated failure 固结破坏
consolidated freight classification 统一运费分类表
consolidated fund 统一基金
consolidated fund service 固定资金;统一公债基金收支
consolidated goodwill 联合信誉
consolidated grouting 固结灌浆
consolidated hole 固结孔眼(向破碎岩石中注入胶结液)
consolidated ice 密集厚冰;冻连冰群
consolidated ice cover 固结冰盖层
consolidated immediate shear test 固结快剪试验
consolidated income sheet 综合损益表
consolidated income statement 汇总损益表
consolidated income tax return 所得税合并申报
consolidated industrial and commercial tax 工商业统一税;统一工商税
consolidated jail 拘留所
consolidated manning table 综合人员配备表
consolidated method of sand drains 砂桩加固法
consolidated mortgage 合并抵押
Consolidated Natural Gas 统一天然气公司(美国)
consolidated pack ice 固结积冰;固结浮冰(群);堆积厚冰
consolidated profit and loss 总损益
consolidated progressive income tax 统一累进所得税
consolidated progressive tax 统一(的)累进税
consolidated quick compression test 快速压缩试验;固结快压缩试验;固结块压缩试验
consolidated quick direct shear test 固结快剪直剪试验
consolidated quick shearing resistance 固结快剪强度
consolidated quick shear strength 固结快剪强度
consolidated quick shear test 固结快剪试验
consolidated quick shear value 固结快剪值
consolidated quick test 固结快速试验(土壤)
consolidated quick triaxial compression test 固结快速三轴压缩试验
consolidated rate 综合费率;整理发货手续费
consolidated report 综合报告
consolidated retained earnings statement 统一盈余留存表
consolidated return 汇总报告;合并收益;综合报告
consolidated ridge 联冰脊
consolidated rig (绞车、转盘、泵及发动机装在一平板上的)整体钻探设备
consolidated rock 固结岩石;坚固岩石
consolidated sand 固结砂
consolidated shipment 合并装运
consolidated silt 固结淤泥
consolidated sinking fund 统一偿债基金
consolidated slow compression test 慢速压缩试验;慢速固结压缩试验;固结慢速压缩试验
consolidated soil 固结土(壤)
consolidated statement 综合报表;合并报表
consolidated stock 统一公债
consolidated strength 固结强度
consolidated structure 凝团结构
consolidated subsoil 加固地基
consolidated surplus 结合盈余;合并盈余;总盈余

consolidated tax 统一货物税
consolidated thickness 固结厚度
consolidated trade catalog(ue) 商品总目录
consolidated triaxial compression test 固结三轴压缩试验
consolidated triaxial test 固结三轴试验
consolidated-undrained compression test 固结不排水压缩试验
consolidated-undrained shear 固结不排水剪切
consolidated-undrained shear test 固结不排水剪切试验;固结不排水剪力试验
consolidated-undrained test 固结不排水试验
consolidated-undrained triaxial compression test 固结不排水三轴压缩试验
consolidated-undrained triaxial test 固结不排水三轴试验
consolidated volume 固结体积
consolidated working fund 统一运用专款
consolidating 使固结
consolidating batch 压制成块的配合料
consolidating layer 固结层
consolidating pile 加固桩;强化桩(用以加固地的短桩)
consolidation 紧实;加固;集结;固结(作用);巩固期;捣固
consolidation apparatus 固结仪;渗压仪
consolidation box 联运箱;混装箱
consolidation building (集装箱码头的)装箱拆箱库
consolidation by displacement 挤淤法
consolidation by electroosmosis 电渗固结
consolidation by lease 租赁合并
consolidation by merger 吸收合并
consolidation by preloading method 预压排水固结法
consolidation by purchase 购买合并
consolidation by the vacuum method 真空加固法
consolidation by vacuum and surcharge preloading method 真空超载预压加固法
consolidation by vacuum preloading 真空预压加固
consolidation by vibrating 振动加固
consolidation by vibration 振动固结;振动捣实
consolidation cell 渗压仪;固结仪
consolidation chamber 渗压室;固结室
consolidation coefficient 固结系数
consolidation curve 固结曲线;渗压曲线
consolidation dam 护坝
consolidation deformation 固结变形
consolidation degree 固结度
consolidation depot 拼装站
consolidation device 固结设备
consolidation due to desiccation 干燥引起的固结
consolidation failure 固结(引起的)破坏
consolidation grouting 加固灌浆;用水泥浆加固;固结灌浆
consolidation grouting hole 固结灌浆孔
consolidation index 固结指数
consolidation line 固结曲线;渗压线
consolidation machine 固结仪
consolidation meter 固结仪
consolidation of collected data 整理所搜集的资料
consolidation of deposit 电积金属的熔凝
consolidation of earth dam 土坝压实;土坝固结
consolidation of foundation 地基夯实;地基固结
consolidation of geotechnical investigation data 整理勘察资料
consolidation of ground 软基加固
consolidation of permeability 渗流固结
consolidation of reclaimed earth 吹填土固结
consolidation of river bed 河床加固
consolidation of road surface 路面压实
consolidation of road surfacing 路面压实
consolidation of road surfacing under traffic 路面行车压实
consolidation of wooded areas 林区整理
consolidation parameter 固结参数
consolidation pile 固结桩
consolidation plant 夯碾设备
consolidation press 压密器;固结压缩机
consolidation pressure 固结压力;渗压力
consolidation rate parameter 固结速率参数
consolidation ratio 压密度;固结比;渗压比
consolidation research 固结研究
consolidation section 压实部分;固结部分

consolidation sedimentation 固结沉淀作用
consolidation service 集装箱作业;拼装作业
consolidation settlement 压密沉陷;固结沉陷;固结沉降
consolidation settling 集结沉积
consolidation shed (集装箱码头的)装箱拆箱库;拼装场
consolidation slow compression test 固结慢速压缩试验
consolidation slow shear test 固结慢剪试验
consolidation stage 固结阶段
consolidation state of soil 土的固结状态
consolidation station 拼装站
consolidation strength 固结强度
consolidation stress 固结应力
consolidation test 固结试验
consolidation test apparatus 渗压仪;固结仪
consolidation test device 固结试验装置
consolidation test device with electronic readout 带电测读数的固结仪
consolidation test progress 固结试验进程
consolidation test under constant loading rate 等速加荷固结试验
consolidation test under constant rate of strain 等应变率固结试验
consolidation test under controlled gradient 控制梯度固结试验
consolidation theory 压密理论;固结原理;固结理论
consolidation theory of soil 土的固结理论
consolidation time 固结时间
consolidation-time curve 固结时间曲线
consolidation type settlement 固结型沉降
consolidation vibrator 振实器
consolidation with cement grouting 水泥浆加固
consolidation within a specified time 限期整顿
consolidation works 加固工程;夯碾工作
consolidator 拼装承运商
consolidator's bill of lading 拼装承运商提单
consolidometer 固结(测量)仪;渗压仪
Consol(lattice) chart 康索尔海图
Consol sonne 康索兰
Consol station 康索尔台
Consol system 康索尔系统
consolute 互溶质;共溶性的
consolute component 互溶组分
consolute solution 会溶质溶液;互溶液
consolute temperature 临界溶解温度;混溶温度;会溶温度;互溶温度
consonance 谐和;协和;调和
consort 护航船
consortia bank 联营银行
consortium 联营企业;联合体;集团;(指国际的)财团;企业的合作
consortium bank 联营银行;联合银行(由其他银行入股组成的银行)
consortium lease 集团租赁
consortium of contractors 承包商联营体;承包者联合体;承包商联合体;承包人联合体;施工联合体
consortive group of flora 同生群
conspectus 线路示意图;一览表;大纲
conspicuous clause 醒目条款
conspicuous consumption 摆阔性消费;铺张浪费
conspicuous error 显著误差
conspicuous level 显著水平
conspicuous object 显著物标;显著目标
conspicuous object for navigation 航海显著标志
conspicuous tree 独立树
conspire 密谋策划
constac 自动电压稳定器
constance 恒定性
constancy 恒有度;恒定性;恒存度;不变性;稳定性
constancy class 恒有级
constancy of frequency 频率稳定性
constancy of heater motion 加热器恒定运动
constancy of level 能级不变
constancy of temperature 温度恒定
constancy of test condition 试验工况的稳定性
constancy of variance 方差恒定性
constancy of volume 容积恒定性;容积不变(性);体积稳定(性);体积不变(性)
constant 恒值;恒量;恒定;常数;常量
constant aberration 固定像差
constant acceleration 匀加速度;等加速度
constant action 恒常作用

constant address 常数地址
constant advance 固定超前
constant Air Volume system 定风量系统
constant altitude gamma survey 固定高度伽马测量
constant ambient pressure 周围恒压
constant amplitude 等幅
constant amplitude carrier 等幅载波
constant amplitude oscillation 恒幅摆动
constant amplitude sinusoid 等幅正弦波
constant amplitude system 等幅系统
constant-amplitude test 等幅应力疲劳试验
constantan 镍铜合金
constant angle 常量角
constant angle arch dam 恒角拱坝;定角拱坝;等中心角拱坝
constant angular acceleration 等角加速度
constant angular velocity 等角速度
constant aperture 固定孔径
constant area 常数区;常数存储区
constant area curve 恒定面积曲线
constant area flowmeter 固定面积流量计
constant axial velocity pattern 等轴速流型
constant band width filter 宽滤波器;恒定带宽滤波器;恒定带
constant bar 永外磁棒;永久磁棒
constant basis 恒定量
constant beamwidth 恒定束宽
constant bearing 方位不变
constant-bearing course 等方位航线;平行接近法
constant bearing navigation 常方位导航;平行接近法
constant bit rate 恒定比特率
constant block 常数块
constant boiling 恒沸
constant boiling binary mixture 恒沸二元混合物
constant boiling mixture 恒沸(点)混合物
constant boiling point 恒沸点
constant boiling solution 恒沸溶液
constant boiling ternary mixture 恒沸三元混合物
constant breadth cam 等宽凸轮
constant broadcast 常数传播
constant budget 固定预算;不变预算
constant capital 不变资本
constant center arch dam 等中心拱坝
constant channel 恒参信道
constant chord 固定弦
constant circulating oiling 等速环流润滑
constant circulation 定向循环;稳定循环
constant circulation flow pattern 等环量流型
constant clearance piston 恒定间隙活塞
constant clock 恒速计时器
constant coefficient 常系数
constant comparative force 恒定压实力
constant congelation soil 永久冻土
constant connection 固定联系
constant control system 恒定控制系统
constant cost 固定成本;不变成本
constant course line 恒向航线
constant creep speed 恒定徐变速度
constant critical flow 稳恒临界水流
constant cross-section 等截面;固定截面;固定断面
constant cross-section(al) arch 等截面拱;恒截面拱
constant current 恒流;恒定流;恒定电流;定常流;持续电流;稳定流
constant current anodizing process 恒定电流阳极氧化法
constant current arc welding power source 垂降特性弧焊电源
constant current characteristic 恒流特性
constant current charge 恒流充电;定电流充电
constant current condition 恒流条件
constant current contour 等流线
constant current control 定流控制
constant current DC potentiometer 恒流直流电位计
constant current discharge 恒流放电
constant current distribution 恒流配电
constant current dynamo 恒(流)发电机;直流(发)电机
constant current dynamotor 定流发电机
constant current filter 恒流滤波器
constant current generator 恒流发电机;定流发电机

constant current modulation 恒流调制;定流调制
constant current motor 恒流电动机;定流电动机
constant current potentiometry 恒电流电势法
constant current power supply 恒流电源;垂直下降特性电源
constant current process 恒电流过程
constant current regulator 恒流调节器;定流调整器
constant current source 恒流(电)源
constant current system 恒速常数据传输系统;恒流制
constant current titration 电位滴定
constant current transformer 恒流变压器
constant current welding machine 恒流电焊机
constant current welding source 恒流式焊接电源
constant curvature 常曲率
constant cutting speed 恒定切削速度
constant danger 固定火险性
constant data-rate system 恒速率数据传输系统
constant-deceleration orifice 等阻尼节流孔
constant declaration section 常数说明部分
constant declarator subscript 常数说明符下标
constant definition 常数定义
constant deflection test 恒定挠度试验;常挠度试验
constant-delay discriminator 恒定延时鉴频器;脉冲解码器
constant-delivery pump 定量(输送)泵
constant density contour 等密度线
constant density surface 等密度面
constant depth flume 定深水槽
constant depth sampler 定深取样器
constant depth sampling 定深取样
constant-depth tooth 等高齿
constant deviation 恒定自差;恒定偏差
constant deviation prism 恒偏向棱镜
constant deviation spectrograph 恒偏向摄谱仪
constant deviation spectroscope 恒偏分光镜;直视棱镜分光镜
constant diameter cam 等径凸轮
constant diameter elbow 等直径肘管
constant diameter growth 等径生长
constant dimension 常数维数
constant dimension column 等形柱;等截面柱
constant discharge 恒定排量;恒定流量;恒定出料量
constant discharge rate test 恒定排水速率试验
constant discharge test 定降深试验
constant displacement accumulator 等容蓄压器
constant displacement hydraulic motor 定排量液压马达
constant displacement motor 定量马达
constant displacement pump 定量泵;定量送料泵
constant displacement shaker 等幅抖动器
constant display system 常数显示方式
constant diversion spectroscope 等偏分光仪
constant dollar 不变美元;定值美元
constant dollar accounting 基元会计
constant dollar estimates 基元统计估值
constant dollar value 不变美元价值
constant drainage 接力排水
constant duty 常负荷;不变工况
constant duty cycle 恒定工作周期
constant duty engine 定负载发动机
constant edge 径向边;直边
constant effective stress 恒定有效应力
constant-effect model 定常效应模型
constant employment 固定职业
constant energy welding machine 恒功率电焊机
constant-enthalpy process 节流过程;等焓过程
constant entropy chart 等熵曲线图
constant entropy compression 等熵压缩
constant entropy expansion 等熵膨胀
constant equilibrium 恒定平衡;恒常平衡
constant error 恒(定)误差;系统误差;固定误差;常值误差;常数误差;常差
constant excitation 恒定激励;固定激励
constant expenses 经常费用;固定费(用);不变费用
constant exposure 正常曝光
constant extension speed telescopic cylinder 等速外伸的伸缩缸
constant factor 常因子
constant factor method 恒定系数法
constant failure period 常故障期
constant failure rate 偶然失效率;常值故障率
constant failure-rate period 正常工作时间
constant-fall rating 定落差水位—流量关系

constant false alarm rate 恒虚警率
constant fault 固定性故障
constant feeder 恒量喂料机;定量给料机
constant feed weigher 定量供料秤;定量给料秤
constant field 恒定场
constant field commutator motor 定激励整流式电动机
constant flexible test 等挠曲试验
constant flow 恒(定)流;定流量;定常流;常年性流水;不变流
constant-flow concrete pug mill mixer 连续式混凝土小型搅拌机
constant flow control 质调节
constant flow device 恒流量装置
constant flow equipment 稳流设备
constant flow grinding 稳流研磨;连续粉磨
constant flow lubrication 恒流润滑
constant flow machinery 连续性输送机械
constant flow mixer 稳定流动搅拌机
constant flow mixing 稳流搅拌
constant flow oxygen equipment 连续供氧设备
constant flow paddle pump 定量叶片泵
constant flow pump 恒定流量泵;定(排)量泵
constant flow screw-type mixer 稳流螺旋式拌和器
constant flux linkage theorem 磁链不变原理
constant force 恒力;力
constant force amplitude excitation 恒力幅激纵
constant force escapement 均力式擒纵机构
constant force liquid spring 恒力液体弹簧
constant force output 恒力输出
constant force spring 等拉力弹簧;恒力弹簧
constant fraction discriminator 恒定系数鉴别器
constant freezing point 恒冰点
constant frequency 恒频(率)
constant frequency control 恒定频率控制
constant frequency oscillator 恒定频率振荡器
constant frequency power supply 稳频电源
constant frequency variable dot method 固定频率变点法
constant function 常值函数;常函数
constant geometric(al) pitch 几何等距
constant geometry nozzle 固定几何形状喷管
constant glow potential 恒定弧电位
constant grade 持续上升
constant gradient 固定比降;固定梯度
constant gradient consolidation test 等梯度固结试验
constant-gradient synchrotron 等梯度同步加速器
constant gradient test 等梯度试验
constant-grid potential oscillator 恒定栅压振荡器;固定栅压振荡器
constant head 定水头;常水头;不变水头
constant head injection test 常水头注入试验
constant head meter 恒压头流量计
constant head orifice turnout 常水头孔口式配水闸;常水头孔口式(斗门)
constant head orifice-type feeder 恒压头孔口式加料器
constant head permeability test 恒水头渗透性试验;常水头渗透试验
constant head permeability test in laboratory 室内恒水头渗透性试验
constant head permeameter 恒水头渗透仪;定水头渗透仪;常水头渗透仪
constant head permeameter test 恒水头渗透性试验
constant head tank 恒水位箱;定位槽
constant head test 注水试验;常水头试验
constant head test apparatus 恒定水头渗透仪
constant heat oven 恒温炉
constant heat source 恒定热源
constant height balloon 定高气球
constant height platform 高度不变平台
constant humidity 恒(定)湿(度);稳定湿度
constant humidity cabinet 恒湿箱
constant humidity chamber 恒湿室
constant humidity system 恒湿系统
constant hydraulic radius channel 等水力半径渠道
constant identifier 常数标志符
constant impedance attenuator 恒阻抗衰减器
constant impedance circuit 定阻抗电路
constant impedance regulator 恒阻抗调节器
constant impedance unit 阻抗恒订单元
constant increment 常增量

constant increment integer 常增整变量
constant infiltration capacity 稳定下渗容量;稳定下渗能力;稳定入渗容量;稳定入渗能力
constant initial concentration 恒定起始浓度
constant injection rate method 恒定注射率法
constant input 常数输入
constant input concentration 恒定输入浓度
constant instruction 常数指令
constant intensity pyrometer 等光强高温计
constant interconnection matrix 常值关联矩阵
constant K filter 常 K 式滤波器
constant length 固定长度
constant length record 固定长度记录
constant length shift method 定长移位法
constant level 恒定水准;定油面;定水位;常液面
constant level balloon 定高气球
constant level carburettor 恒定油位化油器
constant level chart 等高面图
constant level control 恒定水平控制
constant level feed bin 恒定料位喂料仓
constant level feed tank 恒定料位喂料仓
constant level filter 定水位滤池
constant level lubrication 恒定油位润滑
constant level lubricator 恒定油位润滑器
constant level oiler 恒定油位器;恒定液位加油器
constant level regulator 恒位面调节器;恒定液面调节器
constant level splash system 定油面飞溅润滑系
constant-level tube 定长水准管
constant level valve 液面恒定阀
constant-linkage theorem 磁链守恒定律
constant load 恒载;定荷载;定(常)荷载
constant load balance 恒载天平
constant load limiter 定荷载限止器
constant load-rate tensile testing machine 恒速荷载强力试验机
constant load resistor 定荷电阻器
constant load rupture test 定荷载破坏试验
constant-load test 定载试验;常荷载下试验
constant loss 恒定损失;不变损耗
constant luminance principle 恒亮度原理
constant luminance transmission 恒定亮度传输
constantly acting load 恒定载荷
constantly changing 变化无常的
constant maintenance 日常维修;经常(性)养护
constant mark 固定标识;常数标记
constant mark and space code 固定标识和间隔代码
constant mark check method 固定标识检查法
constant mark space code 定标记间隔码
constant matrice 常数矩阵
constant mesh (齿轮的)经常啮合
constant mesh gear 固定啮合齿轮;常啮合齿数;常啮合齿轮
constant-mesh gear box 固定啮合齿轮变速箱;常啮合齿轮变速箱
constant mesh gearing 固定啮合传动
constant mesh transmission 常啮合式变速器
constant mesh type transmission 常啮合齿轮式变速箱
constant minor principal stress 恒定最小主应力
constant minor stress 恒定最小应力
constant moisture sampling[sample] 恒湿取样
constant molal overflow 恒摩尔回流
constant multiplier 恒定倍增器;标度因数
constant-multiplier coefficient unit 常数系数部件;常系数装置
constant natural mortality rate 恒定自然死亡率
constant net loss 恒定净损失;恒定净损耗
constant of aberration 光行差常数
constant of a meter 电度表的常数
constant of apsidal motion 拱线运动常数
constant of central tendency 集中常数
constant of channel maintenance 渠道养护常数
constant of composition 组分恒定
constant of dispersion 离散常数;离差常数
constant of earthquake epicentre 震中常数
constant of friction 摩擦系数;摩擦常数
constant of gravitation 引力常数;重力场常数
constant of gravity 重力常数
constant of integrating meter 积量计常数
constant of integration 积分常数
constant of level(l)ing staff 水准标尺常数
constant of linearity 线性恒定

constant of motion 运动常数
constant of pitch 齿节常数
constant of precession 岁差常数
constant of proportionality 比例常数
constant of reaction rate 反应速度常数
constant of river reoxygenation 河流复氧常数
constant of sedimentation 沉降常数
constant of the cone 圆锥常数
constant of the machine 电机常数
constant of universal gravitation 万有引力常数
constant open channel uniform flow 恒定明渠均匀流
constant optimum method 参数寻优法
constant outline teeth cutter 定型齿铣刀;成型齿铣刀
constant output 恒定输出
constant output amplifier 恒定输出放大镜
constant output oscillator 恒定输出振荡器
constant output pump 恒功率泵
constant overflow 恒溢流
constant parameter linear system 常系数线性系统
constant parametron 恒定参数器
constant parent regression 定亲回归
constant particle size 恒定粒度
constant penetration rate 恒钻速
constant period of drying 等干燥阶段
constant permeability alloy 恒磁导率合金
constant permeable alloy 恒导磁合金
constant phase 恒相
constant phase coherent illumination 恒相相干光照明
constant phase coherent imaging 恒相相干成像
constant phase shift 恒定相移
constant-phase-shifting network 常相移网络
constant pitch 定齿节
constant pitch airscrew 定距螺旋桨
constant pitch spacing mechanism 等距(离)间隔机构
constant point format 定点格式
constant pool 常数存储库;常数池
constant pore pressure 恒定孔隙水压力
constant position ratio 恒定状态速比
constant potential 具有平特性的;固定电位
constant-potential accelerator 等电势加速器;等电位加速器
constant potential generator 恒压发电机;定压发电机
constant potential mineral 恒电位矿物
constant potential modulation 恒压调制;定压调制
constant potential regulation 恒压调节
constant potential system 定压制
constant power braking 等功率制动
constant power generator 恒(定)功率发电机
constant power motor 恒定功率电动机
constant power process 恒功率法
constant power welding source 恒功率式焊接电源
constant pressure 恒(定)压(力);定芽;定压;等压力;稳定压力
constant pressure arrangement 定压装置
constant pressure calorimeter 恒压量热计;定压量热计
constant pressure cell 常压力盒
constant pressure centrifugal pump 恒压离心泵
constant pressure charging system 定压增压系统
constant pressure chart 等压面图
constant pressure combustion cycle 恒压燃烧循环
constant pressure combustion gas turbine 定压燃烧式燃气轮机
constant pressure combustor 定压燃烧室
constant-pressure cycle 定压循环
constant pressure cycle engine 定压式循环内燃机
constant pressure detector 定压式爆炸探测器
constant pressure device 恒压装置
constant pressure drop 定压降
constant pressure engine 定压发动机
constant pressure expansion valve 恒压膨胀活门;恒压膨胀阀
constant pressure filter 恒压过滤器
constant pressure filtration 恒压过滤
constant-pressure flow controller 稳定流量调节器;稳压流量控制器
constant pressure gas thermometer 定压气体温度计
constant-pressure gas turbine 恒压燃气轮机;定

压燃气轮机
constant pressure governor 恒压调节器
constant pressure injection 定压喷射
constant pressure line 等气压线
constant pressure liquid pump 恒压液泵
constant pressure map 等压图
constant pressure motor 等压内燃机
constant pressure point 恒压点;定压点
constant pressure pressure welding 恒压压力焊
constant pressure pump 恒压泵;恒定泵;定压泵
constant pressure pusher-type stock guide 定压推杆式导料装置
constant pressure regenerative cycle 定压回热循环;等压回热循环
constant pressure regulator 恒压调节器
constant pressure source 恒压源
constant-pressure specific heat 等压比热;定压比热
constant pressure surface 恒压面
constant pressure thrust chamber 等压推力室
constant pressure turbine 自由射流式水轮机;等压式水轮机
constant pressure unit(device) 定压装置
constant pressure valve 恒压阀
constant pressuro pattern flight 恒压型飞行
constant price 固定价格;不变价格
constant proportion 定比
constant proportional risk aversion 等比例的风险厌恶
constant Q interpolation 常数Q插值(法)
constant quantity 常量
constant radius arch dam 恒径拱坝;等半径拱坝;同心拱坝
constant rate 恒速(率);固定利率;常速率
constant rate creep 等速蠕变率
constant rate drying 恒速干燥
constant rate drying stage 等(速)干燥阶段
constant rate filter 等速过滤器
constant rate filtration 恒速过滤;等速过滤
constant rate injecting method 等速注入法
constant rate method of distillation 恒速蒸馏法
constant rate of flow 恒定流量
constant rate of loading 均匀加载;均匀加荷;等速加荷;常速加荷
constant rate of penetration method 等贯入度法
constant rate of penetration test 等速贯入试验;等贯入度试验
constant rate of production 常产量;稳定产量
constant rate of strain test 等应变率试验;常应变率试验
constant rate of traverse 恒定横移速率
constant rate of uplift test 等速上拔试验(桩工)
constant rate period 恒速率阶段
constant-rate period of drying 定速率干燥期;等速干燥阶段
constant ratio code 恒比码;定比(代)码
constant-ratio frequency changer 定比变频器
constant ratio mixing method 定比混合法
constant reaction blades 等反应度叶片
constant reaction design 等反应度的设计
constant region 恒定区
constant resistance dc potentiometer 恒定电阻直流电位计
constant resistance network 恒定电阻网络
constant retention period 恒定停留期
constant return 固定收益(率)
constant return activated sludge ratio 恒定回流活性污泥比
constant return to scale 比例固定收益
constant revolution 恒速转动
constant scanning 等速扫描
constant section 常数区域
constant section area 等截面积
constant section beam 截面恒定的梁;截面不变的梁
constant section conduit 等截面管
constant section pipe 等截面管
constant section tube 等截面管
constant segment method 等节法
constant series 置换序列
constant settlement 固定沉降
constant shear strain 等剪应变
constant shear viscosimeter 恒定剪力黏度计
constant shear viscosity 恒定剪切黏(滞)度

constant slope 一定斜度
constant solution 常数解
constant source 恒流电源
constant species 恒有种;恒存种
constant specific mass flow 等质量流率;等密流
constant spectral density 恒定谱密度
constant speed 恒速;(恒)定(转)速;等速度;等速
constant-speed airscrew 定速螺(旋)桨
constant speed chopper blade 恒速调制盘叶片
constant speed control 恒速控制;等速控制
constant speed cut 等速切削
constant speed drive 恒速传动;定速运转;常速驱动
constant speed engine 恒速发动机
constant speed governor 恒速调节器;定速调速器
constant speed model 等速型
constant speed motor 恒速电动机;定速电动机;等速马达;等速电动机
constant speed propeller 恒速螺旋桨;定速螺旋桨
constant speed ratio coupling 等速联轴器
constant speed reducer 等速减速器
constant speed regulator 等速调速器
constant speed running 定速运行
constant speed setting 恒速调节
constant speed steering 定速驾驶
constant speed unit 等速机构
constant speed universal joint 等角速万向节
constant spring 稳定泉
constant spring hanger 恒力弹簧支架
constant state region 等态区
constant storage 常数存储器
constant storage area 常数存储区
constant strain criterion 等应变准则
constant strain test 恒定应变试验
constant strain triangle 常应变三角形
constant strength 等强度
constant stress factor of safety 恒应力安全系数
constant stress test 恒定应力试验
constant sum game 常和博弈
constant super-heat line 等过热线
constant supply 经常供应
constant surveillance 连续监视
constant table 常数表
constant table entry 常数表登记项
constant taper screw 等螺距锥体螺杆
constant temperature 烘箱调节器;恒(定)温(度);定温;等温
constant temperature and humidity 恒温恒湿
constant temperature and humidity machine 恒温恒湿机
constant temperature and humidity system 恒温恒湿系统
constant temperature and humidity unit 恒温恒湿机组
constant temperature anemometer 恒温风速仪;恒温风速计
constant temperature automatic controller 恒温自动控制仪
constant temperature bath 恒温浴;恒温池
constant temperature cabinet 恒温箱;恒温室
constant temperature combustor 恒温燃烧室
constant temperature compression 等温压缩
constant temperature depth 常温深度
constant temperature electric(al) oven 恒温电炉;恒温电烘箱
constant temperature expansion 等温膨胀
constant temperature furnace 恒温炉
constant temperature line 等温线
constant temperature oven 恒温炉;恒温烘箱;恒温槽
constant temperature period 恒温期
constant temperature pressure welding 恒温压力焊
constant temperature process 等温过程
constant temperature regulator 恒温调节器
constant temperature room 恒温室
constant temperature system 恒温系统
constant temperature technique 恒温法
constant temperature water tank 恒温水箱
constant temperature workshop 恒温车间
constant tensile load joint test 恒定拉伸载荷接头试验
constant tension control 定张力控制
constant tension winch 恒张力绞车;恒定张力绞车;定张力绞车

constant term 常数项
constant term of conditional equation 条件方程式常数项
constant thickness arch 等厚度拱
constant-thickness arch(ed) dam 等厚拱坝
constant throughput 恒定输送量
constant throw turret lathe 等行程转塔刀架车床
constant thrust 恒定推力
constant tilt 固定倾斜
constant time-lag 恒定延时;定时限
constant tonnage feeder 定量给料器;定量给矿机
constant torque asynchronous motor 恒力矩异步电动机
constant torque control 定转矩控制
constant twist 等齐缠度
constant underflow 恒底流
constant vacuum carburetor 定真空汽化器
constant value 定值;等值;常数值
constant value control 恒定值控制;定值控制;定值调节
constant value line 等值线
constant vector 常向量;常数向量
constant velocity 恒速;恒定流速;定速;定流速;等速度;等速;常速度;稳定速度
constant velocity cardan joint 等速万向节
constant velocity control 等速控制
constant velocity domain 恒定速度范围
constant velocity grit channel 恒速沉沙渠;恒速沉沙池;恒速沉沙槽
constant-velocity joint 等角速万向节
constant velocity line 等速线
constant velocity motion 恒速运动
constant velocity recording 定速录音
constant-velocity scan 常速扫描
constant-velocity universal joint 定速万向节;等速万向节
constant vertical stress 恒定垂直应力
constant void ratio 恒定孔隙比
constant volatile-solid content 恒定挥发性固体含量
constant voltage 恒压(的);恒(定)电压
constant voltage anodizing process 恒定电压阳极氧化法
constant voltage charge 定压充电
constant voltage charger 等电压充电机
constant voltage device 恒压装置【电】;定电压装置
constant voltage direct current source 恒压直流电源
constant voltage dynamo 定压(发)电机
constant voltage generator 恒压发电机;定(电)压发电机
constant voltage modulation 定电压调制;稳压调节
constant voltage motor 恒(定电)压电动机;定压电动机
constant voltage motor generator 恒压电动机发电机
constant voltage power supply 恒压电源;稳压电源
constant voltage rectifier 恒压整流器
constant voltage regulation 恒电压调节
constant voltage regulator 恒压调节器【电】;稳压器
constant voltage source 恒压电源;恒定电压源
constant voltage system 等电压系统
constant voltage transformer 稳压器;恒压变压器
constant voltage transmission 定电压送电
constant voltage welding machine 恒压电焊机;恒电压焊机
constant voltage welding source 恒压式焊接电源
constant volume 等体积;等容(积)
constant volume air conditioning system 定风量空气调节系统
constant volume blower 定容式鼓风机
constant volume burning 定容燃烧
constant volume calorimeter 定容量热计
constant volume change 等容变化
constant volume combustion 定容燃烧
constant volume combustion gas turbine 等容燃烧式燃气轮机
constant volume curve 等容曲线
constant volume cycle 等(定)容循环;定容循环
constant volume cycle engine 等容循环发动机
constant volume feeder 恒体积加料器
constant volume gas thermometer 定容气体温度计

constant volume governor 等容量调节器
constant volume line 等体积线
constant volume process 等容过程
constant volume pump 恒容量泵
constant volume regenerative cycle 定容回热循环
constant volume sampler 等容积采样器
constant volume sampling 恒体积取样;定容采样;等体积取样;等容积取样;等容积采样
constant volume shear test 等体积剪切试验
constant volume specific heat 等容比热;定容比热
constant volume test 恒体积试验;定积试验;等容积试验;等积试验
constant wall pressure ring 等压环
constant water level 不变水位;定水位
constant water level method 常水头法
constant wave 等波幅
constant weight 恒重
constant weight feeder 恒重喂料机;恒重加料机;恒重给砂机;恒重给料机;定量给料机;定量喂料机;定量给砂机
constant white 沉淀硫酸钡;钡白
constant-width 等宽
constant wind 恒定风;稳定风
constant wire-feed mode 等速送丝方式
constant wire-feed system 等速送丝方式
constant word 常数字
constellation 星座(花饰);星区
constellation of Zodiac 黄道十二宫(符号);黄道宫
constituent 要素;组元;组分;组成物;组成部分;支量;构成部分;成分;分量
constituent company 原有公司;子公司;附属公司;分子公司
constituent corporation 子公司
constituent day 要素日;潮汐日;分潮日
constituent density of soil 土的组分比重
constituent element 组件;组分元素
constituent fiber of asbestos 石棉主体纤维
constituent grammar 构成文法
constituent hour 分潮时
constituent mineral 矿物成分
constituent model 组合模型
constituent number 分潮数;分潮号
constituent of soil 土壤成分
constituent of tidal current 潮流组分;分潮流
constituent of tides 潮汐组分;分潮
constituent ratio 构成比
constituent reaction 组分反应
constituent sampling 组分采样
constituents of septage 化粪污泥组分
constituents of wastewater 废水组分
constituents quality index 水质指示器
constituent suffix 分潮指数
constituent water 结构水
constituted authorities 合法当局
constituent of atmosphere 大气成分
constitutional 符合宪法的
constitutional ash 本体灰分
constitutional change 组分转变
constitutional details 结构零件
constitutional diagram 平衡图;组合图;组成图;状态图;相(位)图
constitutional equation 本构方程(式)
constitutional formula 结构式;组成式
constitutionality 符合宪法
constitutional provision 法规
constitutional supercooling 组分过冷
constitutional water 结构水;结构水;化合水
constitution diagram 金相图
constitution equation 基本方程(式)
constitution formula 结构式【化】
constitution of details 图幅接边
constitution of electric(al) analog(ue) machine 电模拟机结构
constitution of standard 标准的构成
constitution of the plant 装置组成
constitution water 化合水
constitutive behavio(u)r 本构性能
constitutive equation 基本方程(式);本构方程(式);结构方程
constitutive law 本构定理;本构定律
constitutive law of soil 土的本构关系;土的本构律
constitutive matrix 本构矩阵
constitutive model 本构模型
constitutive property 结构性质;组分性质

constitutive relation 本构关系;本质关系
constitutive relation of concrete 混凝土本构关系
constnat speed drive 等速传动
constrain 拘束;限制;约束
constrained acoustic(al) radiator 制约式声辐射器
constrained adjustment 约束平差
constrained annexation 强制附合
constrained arch 约束拱
constrained beam 约束梁;两端约束固定梁;固端梁
constrained centering device 强制对中器
constrained condition 约束条件
constrained coordinates 约束坐标
constrained damping layer 受约束的防潮层
constrained diameter 限制粒径
constrained distortion 强迫歪曲
constrained extremal problem 带约束极值问题
constrained factor 约束因素
constrained feed 强迫馈电
constrained force 约束力
constrained girder 约束大梁
constrained grade 限制坡度
constrained gravity model 有约束重力模型
constrained gyro 限制陀螺
constrained linear system 约束线性系统
constrained magnetization 强制磁化
constrained magnetization condition 强制磁化条件
constrained mechanism 受约束机构
constrained modulus 约束模量;侧限压缩模量;侧限模量
constrained modulus of soil 土的压缩模量
constrained motion 限制运动;约束运动;强制运动
constrained movement 约束运动
constrained operation 约束运行
constrained optimization 约束优化(数);约束(最)优化;条件最佳化
constrained optimization problem 有约束优化问题
constrained oscillation 制约振荡;强迫振荡;强迫摆动
constrained parameter 受约束参数
constrained regression model 受约束回归模型
constrained resources 限制性资源
constrained system 受约束系统
constrained vibration 限制振动;制约振动;强制振动;强迫振动
constrained yield point 条件屈服点
constrained yield stress 假定屈服应力;假定屈服限;条件屈服限
constraining condition 强制条件
constraining force 限定力;约束力
constraining moment 限制弯矩;约束弯矩;约束力矩
constraint 限制;强制;系统规定参数
constraint and variable 约束条件与变数
constraint condition 约束条件
constraint degree 拘束度
constraint equation 约束条件式;约束方程(式)
constraint factor 约束系数
constraint length 约束长(度);制约长度
constraint matrix 约束矩阵
constraint method 约束(方)法
constraint qualification 约束规范
constraint reasoning 约束推理
constraint term 待选项
constratum 木板地板(古罗马)
constricted channel 束窄水道;束窄的航道
constricted end joint 缩口接头
constricted end point 缩口接头
constricted flow 束窄水流
constricted free face 约束自由面
constricted furnace 缩腰炉膛
constricted paint sprayer 缩口喷漆头
constricted reach 束窄段
constricted section 束窄段;束狭段;收缩断面;收缩段
constricted tank 带挟喉的池容;带卡脖的池容
constricting 压缩
constricting nozzle 压缩喷嘴;收缩嘴
constricting works 束窄设施;束窄工程
constriction 压缩;紧窄感;缩窄;束集;收缩(部分);收敛
constriction correction 收缩修正
constriction meter 缩口测流计;缩口测流计
constriction obstacle 收缩障碍物

constriction of valley 河谷收束;河谷变狭
constriction-plate classifier 带有孔隔板的分级机;收缩板分级机
constriction project 束水工程
constriction resistance 集中电阻
constriction scour 束狭冲刷
constriction water meter 缩口水表
constrictive mark reaction 压痕反应
constrictor 压缩器;管导片;燃烧室收缩段;收缩物;收缩器;收敛式燃烧室
constrictor nozzle 收敛形喷管
constrict the money supply 紧缩资金投放
constringence 倒色散系数;收缩性
construct 建筑;建设;组成物;构成物;施工
constructability 施工可能性
constructable function 可构造函数
construct a graph 绘制曲线图
construct dilution 施工贫化
construct dilution ratio 施工贫化率
constructed 已建造的
constructed by the owner 业主建造的
constructed price 推定价格
constructed profile 剖面示意图;示意剖面图;示意断面图;施工剖面图
constructed rapid infiltration system 人工快速渗滤系统
constructed rate 推定运费率
constructed reed wetland 人工芦苇湿地
constructed time of station 建站时间
constructed upland-wetland wastewater treatment system 人工高地—湿地污水处理系统
constructed value 推定价值
constructed wetland 人工湿地
constructed wetland treating landfill leachate 人工湿地垃圾渗滤液
constructed wetland treatment system 人工湿地处理系统
constructer 建设者;施工人员
constructibility 可构造性
constructing contractor 建筑承包人;建筑承包商
constructing deblock 分块
constructing in train running interval 利用列车间隔施工
construction 结构;建筑;建设;营造;制作;织物结构;构造;施工
construction, installation costs 建筑安装工程费
construction access ramp 施工进场坡道
construction access road 施工交通路线;进场道路;施工通路;施工通道;施工路线;施工(进场)道路
construction accident 施工事故;建造事故
construction according to drawing 按图施工
construction account 建造账目;建筑账(户)
construction accounting 建筑会计
construction adit 施工横巷;施工横坑道
construction administration 营建管理;施工管理
construction advisory service 建筑咨询服务
construction aggregate 建筑集料;建筑骨料
construction ahead 前面施工
construction aid 施工辅助设施
constructional activities 建筑活动
constructional alloy steel 合金结构钢
constructional alumin(i)um 建筑铝
constructional and selected optimum development project 编制和选择最佳开发方案
constructional appliance 施工机具
constructional board 构造板
constructional class 构造等级
constructional climatology 建筑气候学
constructional column 构造柱
constructional concrete 建筑混凝土
constructional defect 建筑缺陷;施工缺陷
constructional deficiency 施工缺陷
constructional delta 堆积三角洲
constructional density 堆积密度
constructional detail 构造;结构详图;构造详图;施工详图
constructional drawing 结构图;制作图;构造图;施工图;施工详图
constructional element 构件
constructional engineering 建筑工程学;结构工程
constructional equipment 施工设备;施工机具;建筑设备
constructional equipment company 建筑设备公司

constructional error 安装误差
constructional excavator 建筑挖掘机
constructional expenditure forecast 建筑费用预测
constructional feature 构造特点
constructional form 建筑形式
constructional grade steel 建筑等级钢
constructional gradient 原始梯度
constructional height 建造高度
constructional industry 建筑工业
constructional installation 施工设备；施工机具
constructional iron 建筑钢；结构铁；建筑用钢铁；建筑用钢(材)
constructional iron works 铁结构工厂
constructional joint 构造缝
constructional laminate 建筑用层压板
constructional landform 堆积地形；构造地形
constructional limestone 结构灰岩
constructional loan agreement 建筑借款协议；建筑借款合同；建筑贷款协议；建筑贷款合同
construction all risks 施工一切险
constructional material 结构材料；建筑材料
constructional material dealer 建筑材料销售人；建筑材料商人
constructional material delivery 建筑材料运送
constructional material deposit 建筑材料储[贮]存
constructional material distributor 建筑材料分配器
constructional material engineer 建筑材料工程师
constructional material failure 建筑材料破坏
constructional material industry 建筑材料工业
constructional material machine 建筑材料机械
constructional material manufacturer 建筑材料制造者；建筑材料制造厂
constructional material market 建筑材料市场
constructional material processing 建筑材料加工
constructional material producer 建筑材料生产者；建筑材料生产厂
constructional material quality control 建筑材料质量控制
constructional material requirement 建筑材料规格；对建筑材料的要求
constructional material scale 建筑材料等级
constructional material show 建筑材料展览
constructional material standard 建筑材料标准
constructional material storage 建筑材料储[贮]存；建筑材料仓库
constructional material store 建筑材料商店；建筑材料堆栈
constructional material test 建筑材料试验
constructional material testing device 建筑材料试验设备
constructional material testing machine 建筑材料试验机
constructional measurement 施工测量
constructional mechanisms 施工机械
constructional member 构件；建筑构件
constructional metal 建筑用金属
constructional meteorology 建筑气象学；建造术物理学；建造术气象学
constructional of tectonic region 构造区结构
constructional organization design 施工组织设计
constructional physics 建筑物理(学)
constructional plain 构造平原；堆积平原
constructional plain coast 堆积平原海岸
constructional plan 建筑平面(图)
constructional plant 施工设备；施工机具；施工机械
constructional plastic profile 建筑塑料剖面图
constructional plastics 建筑塑料
constructional plastic tirm 建筑塑料线脚；建筑塑料修饰；建筑塑料贴面
constructional plastic unit 建筑塑料制品；建筑塑料单元
constructional principle 建筑原理
constructional profile 构造剖面图；建筑外形；建筑剖面图；建筑轮廓
constructional program(me) 施工计划
constructional reinforcement 构造配筋；构造钢筋；构造筋
constructional river 堆积河流
constructional rubber section 建筑橡胶部件
constructional rubber shape 建筑橡胶模型
constructional rubber trim 建筑橡胶镶饰；建筑橡胶贴面
constructional rubber unit 建筑橡胶单元
constructional safety regulations 施工安全条例

constructional sand 建筑用砂
constructional section 建筑型材；建筑剖面
constructional service 施工服务工作
constructional shape 建筑模型；建筑格式
constructional sketch 设计图
constructional speed 构造速度
constructional standard 建筑标准
constructional steel 结构钢；构造钢；建筑(结构)钢；构造筋
constructional steel profile 建筑型钢；建筑钢的剖面图
constructional steel section 建筑钢型材；建筑钢型材
constructional steel trim 建筑钢的镶饰
constructional steel unit 建筑钢构件；建筑钢单元
constructional stream 构造河流；堆积河流
constructional strength 船体强度
constructional stretch (钢丝绳的)结构伸长
constructional surface 堆积面
constructional system 建筑系统
constructional technique 施工技术
constructional terrace 构造阶地；堆积阶地
constructional testing production area 建设生产实验区
constructional timber 结构用木料；结构木材
constructional timber grade 建筑木材等级
constructional tin 建筑用锡
constructional tolerance 构造容许误差
constructional topographical relief 堆积地形起伏
constructional trim 建筑修整；建筑饰面
constructional unit 最近单元
constructional valley 堆积谷(地)
constructional-void porosity 原始堆积孔隙
constructional water 结构水
constructional waterproofing 构造防水
constructional work 建筑物；施工工作；施工工程
constructional zinc 建筑用锌
construction and acceptance code for reinforced concrete engineering 钢筋混凝土工程施工及验收规范
construction and addition 建筑与扩建；建筑和扩建
construction and demolition waste 建筑垃圾；施工和拆毁的废料
construction and erection drawing 施工安装图
construction and erection enterprise 建筑安装企业
construction and erection method 建筑与安装方法
construction and improvement of tailings dam 尾矿坝的建造与治理
construction and installation 施工安装
construction and installation enterprise 建筑安装企业
construction and installation investment 建筑安装工程投资
construction and installation works 建筑安装工程
construction and maintenance of fences 栅栏施工与保养
construction and repair 建造与修理
construction and road machine 建筑筑路机械
construction and use regulations 规定设计和运转的法则
construction apparatus 建筑机具
construction appliance 建筑机具
construction area 施工场地；施工区
construction assistance vehicle 建筑辅助(运输)车辆
construction authority 建筑权威；建筑工程局
construction authorization procedure 基建审批程序；施工核准程序
construction balance sheet 建筑资产负债表；基建资产负债表
construction bank 建设银行
construction banking 建设金融
construction base line 施工基线
construction bay 施工仓
construction bend 制作弯头
construction board 建筑板材
construction boat 工程船
construction boil 建筑箔片
construction bolt 施工用螺栓；安装螺栓；施工临时螺栓；架设螺栓
construction bond 建设公债；完工契约；完工保证书
construction boom 建筑起重机；建筑吊杆
construction bridge 施工桥；施工便桥
construction brigade 工程队；施工队

construction budget 建筑预算；基建预算；工程(投资)预算；施工预算
construction business 建筑业
construction butt joint 施工平接；构造平接
construction by overall planning 统筹法施工
construction by pipe jacking method (隧道、管道等的)顶进施工法
construction by swing 转体施工法；转体架桥法
construction calendar 施工日程表
construction camp 工棚；施工帐篷；施工临时房屋
construction capacity 施工能力
construction casing 施工套管
construction caution 施工注意事项
construction cement 建筑水泥
construction ceramics 建筑陶瓷
construction chart of roads and yards 道路堆场构造表
construction chemical 建筑用化学品
construction chemistry 建筑化学
construction claims 施工索赔
construction class 建筑物防火等级
construction classification 建筑分类；工程类别
construction clearance 建筑(接近)限界
construction clearance of special line for passenger train 客运专线铁路建筑接近限界
construction clearance of tunnel 隧道建筑限界
construction climatology 建筑气候学
construction code 建筑规范
construction company 营造厂；建筑公司；工程公司；工程单位；施工公司
construction company for machinery and equipment installations 设备安装公司
construction company for mechanical equipment installations 机械设备安装公司
construction component design 建筑构件设计
construction concession 建筑许可
construction concrete joint 混凝土施工缝
construction condition 建筑条件
construction contract 土木工程合同；建筑合同；建筑承包合同；建筑包工合同；工程建设合同；施工合同
construction contract award 建筑合同裁定书
construction contracting 施工承包
construction contract in process 在建工程合同
construction contractor 施工承包者；施工承包商；施工承包人
construction contrary to permit 违反许可规定的施工；违反许可证的施工
construction control 施工控制；施工管理
construction control measure 施工管理量测
construction control network 施工控制网
construction control network for building 建筑物平面控制网
construction coordinates 施工坐标
construction coordinate system 施工坐标系
construction copper 建筑用铜
construction corporation 建筑公司；工程公司
construction cost 建筑成本；(不含施工期利息的)静态投资；建筑造价；建筑工程费；建筑费用；建筑费；建筑成本；建造费(用)；建造成本；建设费用；造价；工程造价；工程费(用)
construction cost analysis 建筑成本分析
construction cost audit 建设成本的审计；施工成本审计
construction cost control 建筑费控制
construction cost engineering 建设工程造价
construction cost estimate 建筑费用预算书；建筑成本预算；工程费用预算；投资估算
construction cost estimating 建筑成本预算
construction cost index 建筑成本指数；施工费用指数
construction craft 施工工作船
construction crane 施工吊车；建筑起重机；施工起重机
construction crew 施工人员；施工队伍
construction crew shifting costs 施工队伍调遣费
construction cycle 建筑周期；建造周期；建设周期；施工周期
construction datum 施工基准面
construction debris 建筑垃圾；建筑废料
construction defect 建筑缺陷
construction demand of electricity 施工用电；施工需电量
construction density 建筑密度

construction department 施工科;建筑科;建筑部门;建筑工程处;工程处;施工处;施工部门
construction depth 施工深度;施工高度
construction design 施工组织设计;施工设计;修建设计
construction designer 施工设计者
construction details 结构细节;结构说明;建筑细节;建筑细部;施工详图;施工大样图
construction detour 施工便道;施工绕行便道;施工便线
construction diagram 施工图
construction diary 施工日志;施工日记
construction differential subsidies 建造亏损补贴
construction differential subsidy 造船差价补贴
construction difficulty 施工难度
construction dirt 建筑垃圾
construction discrepancy report 结构误差报告
construction dispatching 施工调度
construction dispatching plan 施工调度计划
construction diversion 施工导流
construction division 工区;工程处
construction dock 造船坞;(工程施工临时用的)施工码头
construction document 建筑文件;施工文件(包括施工图及说明)
construction documents design 施工图设计
construction documents design phase 施工图设计阶段
construction documents phase 施工说明书阶段;施工图设计阶段
construction dowel 施工定位销
construction drainage system 施工排水系统
construction drawing 施工图;建筑图;施工图
construction drawing budget 施工图预算
construction drawing design 施工图设计
construction drawing of bridge 桥梁施工图
construction drilling 施工钻探
construction economics 建筑经济学;施工经济学
construction effect 集群效应
construction effect of rock pressure 围岩压力的施工效应
construction element 结构部件;建筑部件;结构单元体
construction elevator 施工升降机
construction engineer 施工工程师
construction engineering 结构工程;建筑工程学;建筑工程;施工工程学
construction engineering corporation 建筑工程公司
construction engineering drilling 工程施工钻探
construction enterprise 工程公司;施工企业;建筑安装企业
construction equipment 施工设备
construction equipment and protection 施工设备及防护措施
construction equipment company 建筑设备公司
construction equipment item 施工设备项目
construction equipment manual 施工设备手册
construction equipment technical development 施工设备和技术的发展
construction error 施工误差
construction estimate 施工预算;施工图预算;施工图估算;施工估价
construction expenditures 结构费用;基建支出
construction expenditure forecast 建筑费用预测
construction expense added at night 夜间施工增加费
construction expense added in rainy season 雨季施工增加费
construction expense added in winter 冬季施工增加费
construction expense legality audit 建设费用合规性审计
construction experience 施工经验
construction expert 建筑专家
construction fabric 建筑用织物
construction facility 建筑设备;施工设施
construction fatalities 施工死亡事故
construction fault 施工差错
construction feature 结构特征
construction felt 施工用油毛毡
construction fibers 建筑用纤维
construction field 施工现场;建筑现场
construction financing 施工筹资;筹措建造资金;筹措建筑资金
construction finished 已完工程
construction firm 建筑公司;工程公司;工程单位
construction fit 施工容许误差
construction flexibility 建筑适应性;建筑机动性;施工灵活性
construction flood 施工期高水位
construction flow process 流水作业法;流水作业施工
construction forms 施工报表
construction formula 结构式
construction for non-productive purposes 非生产性建设
construction for productive purposes 生产性建设
construction fund 建筑资金;建设资金
construction gang 施工人员;铁路土方工程队
construction gap 施工缝
construction ga(u)ge 规划建筑线
construction ga(u)ge diagram 建筑接近限界图
construction ga(u)ge line 建筑接近限界线
construction general layout 施工总平面图
construction general plan 施工总平面图
construction glass 建筑玻璃
construction goal 建设目的;施工目的;施工目标
construction grade 结构级
construction grant program(me) 建筑资助计划
construction gravel 建筑砾石
construction grillage 整体式格床
construction ground 建筑场地
construction guide 施工准则
construction heading 施工导洞
construction headquarters 工程指挥部
construction heater 烤板;工地现场加热器;施工现场采暖器
construction height of bridge 桥梁建筑高度
construction hoist 施工起重机;施工吊车
construction home base 生活基地;生产基地
construction horizon 施工水平;施工地平
construction in compliance with design drawing 按图施工
construction inconvenience 施工不便
construction index 施工指标
construction industry 建筑业;建筑工业;建筑材料工业
Construction Industry Research and Information Association 建筑工业研究及情报协会
construction in freezing weather 冬季施工
construction in half-road width 路面半幅施工
construction in process 在建工程;在施工中;未完施工
construction in progress 在建工程;未完成的工程;施工中的工程
construction inquiry 建筑业调查
construction in rock 岩石施工
construction inspector 建筑师现场代表;施工检查员
construction-installation works 建筑安装工程
construction institute 建筑研究院
construction insulant 建筑绝缘材料
construction insulation grade material 建筑绝缘等级材料
construction insurance 建筑保险
construction insurance premium 工程保险费
construction interference 施工干扰;施工妨碍
construction "interval" 施工天窗【铁】
construction investment 建筑投资;建设投资;基建投资
construction iron 建筑用铁
constructionism 构成主义;构造论
construction item 建设项目;施工项目
construction item reference number 施工项目编号
construction job 施工任务
construction job sheet 施工任务单
construction joint 建筑缝;工作缝;施工缝;施工琅设备;伸缩缝
construction joint surface 工作缝缝面
construction joint tape 施工缝粘贴带
construction journal 施工杂志;施工日志;施工日记
construction labo(u)r 施工劳动力;建筑劳务
construction labo(u)rer 铁路土方工人
construction layout 施工放样;施工布置
construction level 施工基面标高;施工基面
construction licence 施工执照;施工许可证
construction lift (混凝土等的)浇筑层;水平施工缝
construction lighting 施工照明
construction lime 建筑石灰
construction line 建筑界线;建筑红线;作图线;辅助线;起始方位线;施工线
construction line survey 建筑红线测量
construction lining 整体式衬砌
construction load(ing) 施工荷载
construction loan 建筑贷款;建设贷款;施工贷款
construction loan agreement 建筑借款协议;建筑借款合同;建筑贷款协议;建筑贷款合同
construction loan draw 建设贷款的部分提取
construction location 施工地点
construction lumber 建筑木材
construction machine 建筑机械
construction machinery 建筑机器;建筑机械;施工机械
construction management 建筑施工管理;施工管理
construction management contract 施工管理合约;施工管理合同
construction management cost 施工管理费
construction management plan(ning) 施工组织设计
construction management research 施工管理研究
construction manager 施工经理
construction manpower 施工力量
construction map 建筑图;施工现场图;施工图;施工场地图
construction marker 施工标志
construction market 建筑施工市场
construction material 结构材料
construction material commodities 建筑原料矿产
construction material dealer 建筑材料销售人;建筑材料商(人)
construction material delivery 建筑材料运送
construction material deposit 建筑材料储[贮]存
construction material distributor 建筑材料商;建筑材料分配者
construction material engineer 建筑材料工程师
construction material failure 建筑材料破坏
construction material industry 建筑材料工业
construction material machine 建筑材料机械
construction material manufacturer 建筑材料制造者;建筑材料制造厂;建筑材料厂商
construction material market 建筑材料市场
construction material moisture 建筑材料湿度;建筑材料含水率
construction material processing 建筑材料加工
construction material producer 建筑材料生产者;建筑材料生产厂
construction material production 建筑材料生产
construction material quality 建筑材料质量
construction material quality control 建筑材料质量控制;建筑材料质量管理
construction material requirement 建筑材料规格;对建筑材料的要求
construction materials 建材;建筑材料
construction material sample 建筑材料样品
construction material scale 建筑材料等级
construction materials delivery 建筑材料送货;建筑材料交货
construction material show 建筑材料展览
construction materials of main parts 主要构件的结构材料
construction materials supplied by employer 支付品
construction material standard 建筑材料标准
construction material storage 建筑材料储[贮]存;建筑材料仓库
construction material store 建筑材料商店;建筑材料堆栈
construction material test 建筑材料试验
construction material testing device 建筑材料试验设备
construction material testing machine 建筑材料试验机
construction means 施工手段
construction mechanization 建筑机械化
construction member 建筑构件
construction metal plate 建筑金属板
construction metal sheet 建筑金属薄板
construction metamorphism 接力变质
construction method 工法;施工方法
construction methods and equipment 建筑方法

与设备
construction module 建筑模数
construction monitoring 施工监测
construction month 施工月
construction mortar 建筑砂浆
construction nail 建筑用钉
construction noise 建筑噪声;施工噪声
construction norm 施工定额
construction notes 施工记录
construction number 工程序号;工程号;工程代号;工程编号
construction of balance sheet 资产负债表结构
construction of barracks 工房修建
construction of building 房屋施工;房屋建筑;房屋建设
construction of buttress dam 支墩坝建造
construction of cable 电缆结构
construction of condition 条件结构
construction of contour lines 等高线描绘
construction of contract 合同结构;合同的主要内容;合同的编制方法
construction of cost 成本结构
construction of cross-section 横剖面的编制
construction of dams 水坝建筑
construction of dredge pump 泥泵构造
construction office 建筑施工办公室
construction of filter 过滤器结构
construction of fixtures 设备安装
construction of geologic(al) variable 地质变量的构成
construction of housing 住宅建设
construction of hydroelectric(al) station 水电站建设
construction of ice 冰结构
construction of infrastructural facility 基础设施建设
construction of line 线路架设
construction of magnitude 大型建筑(物)
construction of mine 矿山建设
construction of model 模型构造
construction of Mohr's circle 莫尔圆绘制
construction of new variables 新变量的构造
construction of platform 地坪的构造
construction of policy 保险单的解释
construction of pressure vessel 压力容器结构
construction of roads 道路施工
construction of roof 屋面构造
construction of sewers 管道施工
construction of shaft 竖井的建设
construction of shield 盾构施工法
construction of signal 造标
construction of small towns 小城镇建设
construction of stair(case) 楼梯施工
construction of switch 开关结构
construction of tanks 电解槽构造
Construction of the People's Republic of China 中华人民共和国宪法
construction of traffic 车流组成
construction of tube 管段制作
construction of tunnels 隧道掘进;隧道施工(法)
construction of window 窗户施工
construction of wire rope 钢丝绳的结构
construction of worker's dwellings 工人住房的修建
construction operation 施工(作业);施工工作;施工操作
construction operations and safety 施工操作与安全
construction order 施工订货(单);基建任务书;工程任务书;施工指令
construction organization 施工组织设计;施工单位
construction organization overhead 建设单位管理费
construction organization shifting costs 施工机构迁移费
construction overhead 施工杂费;施工间接费
construction paint 结构涂料;结构防护漆
construction paper 工作用纸
construction party 施工人员
construction payment 建筑造价
construction period 建筑(周期);建设周期;工期;施工周期;施工(工)期
construction period of unit project 单位工程施工工期

construction permit 施工执照;施工许可证
construction person 施工设计人员
construction personnel 施工人员
construction phase 建筑式样;建造阶段;建设阶段;施工阶段
construction phase-administration of the construction contract 施工阶段—施工合同执行的管理
construction pit 建筑基坑;施工坑;施工基坑
construction plan 结构图;构造图;建设计划;施工组织设计平面图;施工(平面)图;施工方案;施工布置图;施工计划
construction plane 基本结构平面;构造平面;工作平面;辅助平面
construction plan for individual project 单元工程施工进度计划
construction plan in section 区间施工作业安排
construction planning 建筑规划;施工规划;施工计划
construction planning legislation 建设计划立法
construction plant 工地施工厂房;建筑构件(工)厂;施工设备厂;施工机械;施工辅助工厂
construction plant costs 施工机械使用费
construction plastic film 建筑塑料薄膜
construction plastics 施工用塑料;建筑塑料
construction plastic sheeting 建筑塑料薄板
construction plate metal 建筑金属板
construction policy 建设政策;造船保单
construction possibility 施工可能性
construction practice 施工实践
construction preparation 施工准备;建设准备
construction preparation period 施工准备期
construction preparation stage 施工准备阶段
construction price 建筑造价
construction price index 建筑造价指数
construction price index figure 建筑价格指数表
construction price legislation 建筑造价规定;建筑价格规定
construction procedure 生产程序;施工过程;施工程序;施工步骤
construction product 建筑产品
construction profile 结构纵剖面图;船体纵剖面结构图
construction program(me) 施工方案;建设程序;建设计划;建设规划;施工计划;施工程序
construction programming 建设规划;施工规划
construction progress 工程进度;施工设备和技术的发展;施工进度
construction progress payment 按工程进度付款
construction progress report 施工进度报告
construction project 建设计划;基建计划;建筑设计;建筑工程;建设项目;建设工程;基建项目;工程建设项目;施工工程
construction project bidding 建设工程招标
construction project contract 建设工程承包合同
construction project management 施工项目管理;工程项目管理
construction project schedule network diagram 网络进度计划
construction project time management 施工进度管理
construction purpose 建设目的
construction put-in-place 建造量
construction quality 工程质量;施工质量
construction quantity 建筑工程量
construction quicklime 建筑生石灰
construction quota 施工定额
construction rate 建筑费率
construction recording 施工记录
construction regulation 施工计划条例;建筑章程
construction reinforcing bar 建筑用钢筋
construction reliability 结构可靠性
construction report 施工报告
construction research 施工研究
construction right-of-way 管带【给】
construction risk 施工(承包)风险(费)
construction road 工地道路;施工(用)道路;施工便道
construction ropeway 施工用架空索道
construction rules and regulations 建筑规则和条例
construction runoff 工地雨水径流
construction safety 施工安全
construction safety management 施工安全管理

construction safety regulations 施工安全规则
construction sand 建筑用砂
construction scale 建设规模
construction scar 施工遗迹;施工后的痕迹
construction schedule 建筑方案;施工日程表;施工(进度)计划;施工进度表
construction scheme 施工方案
construction sealant 建筑用密封剂;建筑用密封膏
construction season 施工季节
construction section 施工段;安装段
construction sequence 施工顺序;施工程序;施工步骤;修筑程序
construction shaft 施工井字架;施工竖井
construction shed 营造工房;施工棚
construction sheet 施工单
construction sheet metal 建筑金属薄板
construction ship 工程船
construction show 施工演示;建设展览
construction sign 施工标志
construction site 建筑现场;建筑工地;工地;施工现场;施工工地;施工地点;施工场地
construction site barrier 工地围栏
construction-site crane 建筑起重机
construction site dimensions 建筑场地尺寸
construction site facility 临时设施
construction site general layout 施工总布置图
construction site hoist 施工场地升降机;建筑工地起重机
construction site insurance premium 现场保险费
construction site lift 建筑工地升降机
construction site noise 建筑工地噪声;建筑施工噪声
construction site premium 工地奖金;工地补助费;施工场地额外费用
construction site security 施工现场安全措施
construction site service 工地设施
construction specification 施工技术要求;施工规范;施工说明(书)
Construction Specification Institute 施工规范协会
construction speed 建筑速度
construction stage 建筑施工阶段;建造阶段;建设阶段;施工阶段
construction stake 施工(标)桩;开挖桩
construction standard 施工规范;建筑规范
Construction Standard Committee 建筑标准委员会
construction standardization 施工标准化;建筑标准化;工程施工标准化
construction standard of waterway 航道建设标准
construction starting date 工程开工日期
construction starting report 开工报告
construction steel trestle 施工用钢栈桥
construction steps 施工程序;施工步骤
construction streamline method 施工流水作业法
construction superintendent 建设监理人员
construction supervising authority 施工监督官员;建筑监督权威
construction supervision 施工管理;建设监督;工程建设监理;工程管理;施工监理;施工监控;施工监督
construction supervision corporation 工程监理公司
construction supervision fee 工程监理费
construction supervision organization 监理单位
construction supervision personnel 工程监理人员
construction supervisor 建筑施工管理员;施工监理员
construction support base 施工生产基地;生产基地
construction survey 施工测量
construction survey for conventional shaft sinking method 普通凿井施工测量
construction survey for freezing method 冻结法施工测量
construction survey for open caisson 沉井施工测量
construction survey for shaft drilling 钻井施工测量
construction survey for well-sinking 沉井施工测量
construction surveying 施工测量
construction suspense account 基建暂记账
construction system 建筑系统
construction systematize 建筑体系化
construction tax 建筑税
construction team 工程队;施工队
construction technique 施工技术

construction technology 建筑术语汇集;建筑工艺学;施工技术;施工工艺
construction technology management 施工技术管理
construction term 施工术语;施工期限
construction timber 建筑用木材;建筑木料;建筑木材
construction time 建造时间;施工时间
construction time limit 建筑工期;工期
construction time to first generation 第一批机组发电(所需)工期
construction tolerance 施工(容许)误差;构造容许误差
construction tool 建筑工具;施工工具
construction tower 施工起重塔架
construction trade 施工行业;建筑业;施工工种
construction traffic 施工运输;施工期交通
construction train 建筑工程列车;建筑材料运输列车;工程列车;施工列车
construction trestle 脚手栈道;施工(用)栈桥;施工机桥
construction type flowmeter 压差流量计
construction uniformity 结构均匀性
construction unit 构件;结构单元;施工单位
construction vehicle 施工车辆
construction velocity 施工速度
construction ventilation 施工通风
construction ventilation of tunnel 隧道通风施工
construction vessel 工程船
construction waste 建筑废料
construction waterbar 施工止水条
construction water-level 施工水位
construction water-stage 施工水位
construction waterstop 施工截水墙;施工挡水板
construction way 施工手段;临时铁路
construction weight 结构重(量)
construction wharf (工程施工临时用的)施工码头
construction winch 建筑绞车
construction with logs 圆木建筑
construction without trestle 无支架施工
construction with ruler and compasses 直尺圆规作图法
construction wool 建筑用羊毛产品
construction work audit 基建工程审计
construction work element 分部工程
construction worker 建筑工人
construction work in freezing weather 冬季施工
construction working drawing 建筑施工图
construction work-in-process 在建工程
construction work in progress 进行中的建筑工程
construction work noise 建筑施工噪声
construction work policy 建筑工程保险单
construction work progress 施工进度
construction work quantity 建筑安装工程量
construction works 建筑工程;建设工程;工程建筑物;未完工程资金;施工工程;基建工程;建筑物
construction work starting 动工
construction work subelement 分项工程
construction work value 建筑安装工作量
construction wrench 大型安装扳手;大螺母扳手
construction yard 施工场地
construction yard premium 施工场地额外费用
construction zone 施工区域;施工区段
constructive 建设性的
constructive art 建筑艺术
constructive basin 建设型盆地
constructive boundary 建设性板块边缘
constructive clause 推定条款
constructive closing 推定结账
constructive comment 建设性意见;建设性建议
constructive delivery 推定交货
constructive delta 建设性三角洲;建设型三角洲;堆积型三角洲
constructive design 造型设计
constructive design system 造型设计系统
constructive detail 建筑细部
constructive discharge 推定解雇
constructive dishonour 推定拒付
constructive element 建筑构件
constructive estimation methods 结构估计法
constructive eviction 建筑遗弃;建设性退租(房产);建设性收回(房产)
constructive fault 推定过失
constructive geometry 构造几何

constructive height 建造高地
constructive interface 结构化界面
constructive interference 相长干涉;全息结构干涉
constructive legislation 建设立法
constructive margin 建设型边缘;增生性边缘
constructive mechanism 构造机构学
constructive metabolism 组成代谢
constructive metamorphism 接力变质;递增变质
constructive mortgage 法定抵押
constructive placement algorithm 构造性布局算法
constructive possession 法定所有权;推定占有
constructive production capacity stage 产能建设期
constructive project 施工项目
constructive receipts 推定收入
constructive suggestion 建设性意见;建设性建议
constructive total loss 推定全损
constructive total loss clause 推定全损条款
constructive trust 法定委托;法定信托
constructive value 推定价值
constructive wave 堆积浪;冲积波
constructivism 构成主义(其特点系以抽象及几何图案构图)
constructivist movement 构成派运动
constructivity 可构造性
construct losses 施工损失
construct losses ratio 施工损失率
construct on projection 展绘格网
constructor 建造者;建设者;制造者;构造者;构造程序;施工员;建筑人员;设计师
constructor's diary 施工员日志
constructor's mechanical plant 施工机械设备
constructor's railroad 工地铁路;工地铁道
constructor's railway 施工铁路;临时铁路;折移式轻便铁路;工地铁路;工地铁道;拆移式轻便铁路
consulage 领事签证费;签证手续费
consular fee 领事签证费;签证手续费
consular invoice 领事发票
consulate 领事馆
consulate general 总领事馆
consult 咨询(公司);查阅
consultancy 咨询业务
consultancy expert 咨询专家
consultancy service 咨询服务;咨询部门
consultant 咨询者;咨询人(员);顾问;查阅者;商议者
consultant agreement 咨询协议
consultant architect 顾问建筑师
consultant engineer 咨询工程师;顾问工程师
consultant engineering firm 咨询公司
consultant fee 咨询费;顾问费
consultant firm 咨询公司;顾问公司
consultant mission 顾问团
consultant paper 咨询文件
consultant program(me) 顾问程序
consultant service 咨询服务;咨询部门
consultants' office 顾问室
consultants' room 顾问室
consultation 会诊;协商;咨询;商议
consultation charges 咨询费;顾问费
consultation committee 咨询委员会
consultation office 咨询办公室
consultation on an equal footing 平等协商
consultation on the basis of equality 平等协商
consultative committee 协商委员会;咨询委员会;顾问委员会
Consultative Committee of International Telegraph and Telephone 国际电报电话咨询委员会
consultative conference 协商会议
consultative department 咨询部
consultative management 协议管理;协商管理
consultative panel 咨询(服务)小组
consulting 顾问资格的;参照
consulting agent 咨询代理
consulting architect 咨询建筑师;顾问建筑师
consulting center 咨询中心;顾问中心
consulting civil engineer 咨询土木工程师
consulting company 咨询公司;顾问公司
consulting engineer 顾问工程师
consulting engineering firm 顾问工程事务所;咨询工程公司
consulting fee 咨询费
consulting firm 咨询公司;顾问公司
consulting pilot 顾问引航员

consulting process 咨询程序
consulting room 咨询室;咨询处;诊(疗)室;顾问室
consulting service 咨询服务
consulting surveyor 验船顾问;顾问验船师
consultive committee 咨询委员会
consumable anode 自耗阳极
consumable arc-melting process 自耗电极电弧熔炼法
consumable article 消耗品
consumable electroarc furnace 自耗电弧炉
consumable electrode 自耗电极;熔化(电)极
consumable electrode cold-mold arc melting 自耗电极水冷坩埚电弧熔炼
consumable electrode-forming 自耗电极成型
consumable electrode furnace 自耗电极炉
consumable electrode melting 自耗电极熔炼
consumable electrode used only for heating 加热用焊条
consumable electrode vacuum furnace 真空自耗炉
consumable electrode welding 熔化极电弧焊
consumable guide electroslag welding 自耗定向电渣焊
consumable ingot 自耗弧熔锭
consumable insert 自耗嵌块
consumable item 消耗品;易耗件
consumable ledger 消费品总账
consumable material 消耗材料
consumable nozzle 熔嘴
consumables 消耗品;消耗材料;消费(商)品;船用备品;耗材
consumable store 建筑材料库;消耗品物料库
consumable supplies 库存易耗品
consumable tool and equipment 消耗工具和设备
consumable weight 可卸重量
consumably arc-melted ingot 自耗弧熔锭
consumably melted 自耗电极熔炼的
consume 浪费;耗费;消耗;消费;熔融消失
consumed boundary 消减边界
consumed cost 耗用成本
consumed energy 消耗能量
consumed man-hour of loading 装岩工时消耗
consumed oxygen 耗氧量
consumed plate boundary 消减板块边缘
consumed power 消耗功率
consumed power of loading 装岩动力消耗
consumed work 消耗功
consumer 消耗装置;消费者;消费户;用户;使用者
consumer action 消费者行为
consumer association 消费者联合会
consumer attitude 消费者态度
consumer behaviour model 顾客行为模型;顾客动向模型
consumer behavio(u)r research 消费者行为研究
consumer budget analysis 消费预算分析
consumer budgets 消费者预算
consumer buyer 用户
consumer buying expectations 消费者希望购买额
consumer calculator 家用计算器
consumer characteristics 消费者的特征
consumer city 消费城市
consumer commodity 消费品
consumer complaint 用户意见
consumer confidence 消费者信任
consumer contest 消费者有奖竞赛
consumer convenience goods 消费者方便物品
consumer cooperative 消费者合作社
consumer credit 消费信贷
consumer credit agency 消费信贷专业机构
consumer credit card slips 消费者信用卡保条
consumer credit control 消费信贷管理
consumer credit insurance 消费者信用保险
consumer credit protection act 消费信贷保护法
consumer decision 消费决策
consumer demand 消费(者)需求
consumer demand equation 消费需求方程
consumer demand theory 消费(者)需求理论
consumer density 用户密度
consumer develping country 消费的发展中国家
consumer durable goods 耐用消费品
consumer durables 耐用消费品
consumer equilibrium 消费者均衡
consumer excess burden 消费者额外负担
consumer expenditures 消费(者)支出

consumer expenditure-income pattern 消费支出收入模式
consumer finance company 消费信贷公司
consumer fraud division 消费者诈骗处理科
consumer gas piping 煤气用户管道
consumer goods 消费资料;消费者货物;生活资料;生产资料
consumer goods and services 消费品和劳务
consumer goods industries 消费品工业
consumer goods market 消费品市场
consumer heat inlet 热力入口
consumer indifference curve 消费者无差别曲线
consumering plate 消失板块;消减板块
consumering zone 消失带;消减带
consumer instalment credit 消费者分期付款信贷
consumer investment 消费者投资
consumerism 消费主义;用户第一主义;保护消费者利益主义
consumer items 消费品
consumerization 消费化
consumerization of information 信息的消费化
consumer jury 消费品评审组
consumer lending 消费者借贷
consumer loan 消费贷款
consumer loss 消费者损失
consumer macroeconomics 宏观消费经济学
consumer macrostructure 宏观消费结构
consumer main switch 用户总开关
consumer market 买方市场;消费市场
consumer mortgage credit 消费(者)抵押信用
consumer motivation 消费(者)动机
consumer movements 消费者运动
consumer non-durable goods 非耐用消费品
consumer non-durables 非耐用消费品;不耐久消费品
consumer organism 摄食生物
consumer orientated market 消费者目标市场
consumer orientation 消费者至上论
consumer package 消费包装;市售包装
consumer packaging 用户包装
consumer paint 市售涂料
consumer price 消费者价格;日用品价格;生活消费品价格
consumer price index 销售物价指数;消费(者)物价指数;消费(者)价格指数;消费品物价指数;消费品价格指数;日用品价格指数
consumer price index number 消费价格指数
consumer price inflation 消费品物价膨胀
consumer producer research 消费者产品调查
Consumer Product Safety Commission 消费品安全委员会
consumer protection 消费者保护
consumer purchase 消费品购买
consumer purchasing power 消费者购买力
consumer reporting agencies 消费者征信机构
consumer requirements 消费者要求条件
consumer research 消费者研究;消费品市场研究
consumer risk 买方风险
consumer risk point 消费者风险点
consumer sales 消费品销售
consumer sale disclosure statement 分期付款销售计划书
consumers' association 消费者协会
consumer's behaviou(u)r 消费者行为
consumer's choice 消费者选择
consumer's cooperative society 消费者协会;消费合作社
consumer's decision 消费决策
consumer's disposable income 消费者可支配收入
consumer sector 消费部门
consumer semi-durable 半耐用消费品
consumer service 消费部门;消费服务;消费性劳务
consumer's exceed demand function 消费者超需函数
consumer's expenditures 消费者开支
consumer('s) goods 消费者物品
consumer's installation 用户设备
consumer society 消费社会
consumer sovereignty 消费者主权
consumer spending 消费支出;消费者开支;消费性开支
consumer's plant 用户设备
consumer's power station 用户供电站

consumer's preference 消费者偏好
consumer's product 用于消费的产品
consumer's purchasing power 消费者(的)购买力
consumer's rent 消费者租金;消费者差额受益
consumer's requirements 消费者的需求
consumer's risk 消费(者)风险
consumer's service 用户服务设施
consumer's sovereignty 消费者主权;消费者至上
consumer's supply control 用户供电控制(装置)
consumer's surplus 消费者剩余
consumer's survey 消费者调查
consumer's terminal 用户供电站;用户终端
consumer strata 消费者阶层
consumer study 消费需求研究
consumer surplus 消费者差额受益;消费余量
consumer's wants 消费者的需求
consumer theory 消费行为理论
consumer thermal substation 用户热力站
Consumer Union 消费者联盟(美国)
consumer use test 消费者使用试验
consumer wants 消费者需要
consumer waste 生活垃圾
consuming city 消费城市
consuming electronic product 民用电子产品
consuming industries 消费产业部门
consuming plate 消失块
consuming public 消费公众
consuming zone 消失带
consummation 圆满成功;成就;完美
consummation of errors 平差法
consummatory act 耗尽
consumpition inflation 消费膨胀
consumption 耗量;消失作用;消减作用;消耗(量);消费(量)
consumption allowance 消费宽减额
consumption area of ozone depleting substances 臭氧消耗物质的消费领域
consumption boom 消费高涨
consumption capacity 消费能力
consumption capital 消费资金;消费(者)资本
consumption coefficient method 消费系数法
consumption control of material fuel and electric(al) power 材料燃料电力用量控制
consumption cost 消耗费用
consumption credit 消费信贷
consumption curve 耗水量曲线;消费曲线
consumption cycle 消费周期
consumption decrease 消耗缩减
consumption development trends 消费价值趋势
consumption diseconomies 消费方面的耗费增加
consumption duty 消费税
consumption economies 消费方面的节约
consumption entry 进口消费品申报单
consumption equation 消费方程
consumption expenditures 消费支出
consumption expenditure account 消费支出账户
consumption externality 消费外部性
consumption factor 消耗因数
consumption figure 消耗值
consumption function 消费函数
consumption function model 消费函数模型
consumption fund 消费基金
consumption fund source 消费基金来源
consumption goods 消费资料;消费品;生活资料;生产资料
consumption guarantees 消费保证
consumption-income ratio 消费收入比
consumption-income relation 消费收入关系
consumption-income schedule 消费收入表
consumption-use sequence 消费收入序列
consumption increase 消费增长
consumption increase rate 消费增长率
consumption indeference curve 消费无差异曲线
consumption index 消耗指数
consumption index of raw material 原材料消耗指数
consumption inflation 消费通货膨胀
consumption in port 在港耗量
consumption lag 消费滞差
consumption level 消费水平
consumption loan 消费贷款
consumption map 消耗地图
consumption meter 流量表;耗量表
consumption model 消费模式

consumption multiplier 消费乘数
consumptionn guild 消费协会
consumption norm 消耗定额
consumption norm of materials in building and installation 建筑安装材料消耗定额
consumption norm of subsidiary materials 辅助材料消耗定额
consumption of cement slurry 水泥浆用量
consumption of cooling water 冷却水耗量
consumption of corrosion inhibitor 缓蚀剂用量
consumption of energy 能量消耗
consumption of energy power 能量消耗
consumption of gasoline 汽油消耗
consumption of hydrochloric acid 盐酸用量
consumption of lubricating oil 润滑油耗量
consumption of materials 物资消耗;材料消耗(量);材料损耗量
consumption of petroleum 石油的消耗
consumption of power 动力消耗量
consumption of preflush 前置液用量
consumption of rinsing liquid 冲洗液消耗量
consumption of shots 钻粒消耗量
consumption of stabilizing 稳定剂用量
consumption of wash liquid 冲洗液消耗量
consumption of water 水的消耗;耗水量
consumption parameter 消费参数
consumption part of material life 消费性的物质生活
consumption path 消费途径
consumption patter 消费方式
consumption pattern 消费形态;消费模式
consumption peak 消耗高峰(期);用量高峰;最大消耗量;高峰消耗量
consumption per capita 每人用水量;每人消耗量
consumption per capita per day 每人每日用水量;每人每日消耗量;每人每日用(水)量
consumption per day 每日用水量
consumption per head 人均消耗(量);每人消费量;人均消费量
consumption per hour 每小时用水量
consumption point 消费场所
consumption possibility frontier 消费可能边缘
consumption possibility line 消费可能线
consumption product 消费产品
consumption quality 消费质量
consumption quantity index 消费(数)量指数
consumption quota of standard project 标准项目消费定额
consumption rate 耗电率;消耗率;消费率
consumption rate of electrodes 电极消耗比
consumption-related import 消费品输入
consumption residue 耗费残余物
consumption schedule 消费(图)表
consumption services 消费(性)劳务
consumption set 消费集
consumption standard 消费标准
consumption structure 消费结构
consumption tax 消费税
consumption technology 消费技术
consumption test 消耗量试验
consumption utility function 消费效用函数
consumptive 消耗性的
consumptive accumulation 消费性积累
consumptive power 消费力
consumptive refrigerated container 消耗冷藏集装箱
consumptive requirement 消耗需量
consumptive use 耗用水量;消耗性用途;消耗使用;消耗量;消费性用途
consumptive-use coefficient 消费性用途系数
consumptive use of crop 作物耗水量;田间耗水量
consumptive use of material 材料耗用量
consumptive use of water 耗水量;消费用水;消耗性用水
consumptive water 空耗水
consumptive water supply 消耗性用水
consumptive water use 耗水量
consutrode melting 自耗电极熔化
contacless smart card 无接触聪明卡
contact 联系;联络;接头(指找人联系);接触;相切;抵碰;初亏;触点;触头
contact absorption 接触吸附
contact acid 接触酸
contact action 接触作用;触杀作用

contact adhesive 压敏胶粘剂;接触黏结剂;接触黏合剂;接触胶粘剂;干粘剂;触压胶粘剂
contact aeration 接触曝气;接触掺气(污水)
contact aeration process 接触曝气法
contact aerator 接触曝气池
contact aerial 接触架空线
contact agent 接触剂
contact alignment 触点校准;触点调准
contact allergy 接触变应性
contact alloy 接点合金;接触器合金;电触头合金;触点合金
contact-altered rock 接触蚀变岩石;触变岩
contact anemometer 接触式风速表
contact angle 交会角;啮合角;润湿角;接触角
contact angle of air-mercury system 空气汞系统接触角
contact angle of air-oil system 空气油系统接触角
contact angle of air-water system 空气水系统接触角
contact angle of oil-water system 油水系统接触角
contact angle of saddle support 鞍座包角
contact angle of water-gas system 水—天然气系统接触角
contact angle of wetting 润湿接触角
contactant photo sensitizer 致光敏物质
contact application 触施
contact arc 接触弧;啮合弧
contact arc discharge 接触放电加工
contact arc starting 接触式引弧
contact area 接触区;接触面积
contact(area)pattern 接触斑点
contact area to ground 接地面积
contact arm 接触臂
contact assembly 接触簧片组
contact aureole 接触变质晕;接触变质带;热接触变质晕
contact bar 接触条;接触架
contact base 接触座
contact baseline measuring apparatus 接触式基线测量器械
contact basin 接触池
contact bed 接触(滤)床;分界岩层
contact bed method 接触床法
contact belt 接触带
contact binary 密接双星
contact biologic(al) filter 接触生物滤池
contact black 接触法炭黑
contact blade 触头片
contact block 接触块;光胶盘;触头组件;接点
contact blocking 光胶上盘
contact bond 接触连接;接触焊接
contact-bond adhesive 压敏胶粘剂;压敏黏结剂;压敏黏合剂;触压胶粘剂;接触黏合剂
contact bonding agent 接触胶粘剂
contact bonding medium 接触黏结介质
contact bounce 触点跳动;触点颤动
contact bounce elimination circuit 接触颤动消除电路
contact bounce time 触点回跳时间
contact breaker 刀式开关;接触(式)断路器;接触开关;接触断流器;闸刀开关;刀形开关
contact breaker cage 接触断路罩
contact brush 接触刷;接触(式)电刷;电刷
contact burn injury 接触烧伤
contact burnishing tool 接点清洁器;触点抛光器
contact bush 接触衬套
contact button 接触钮
contact capacity 接触容量
contact catalysis 接触催化剂
contact catalytic oxidation 接触催化氧化
contact ceiling 附顶天花板;无栅天花板;贴顶天花
contact cement 接触型黏结剂;接触型胶结剂;触压型胶粘剂
contact cementation 接触胶结
contact cementing agent 接触胶凝剂
contact cement method 接触黏合法
contact chamber 接触室;接触池;气浮接触室
contactchart 航空导航图
contact chatter 触点震颤;触点抖动
contact checker 接触式检验器
contact chip 接触片
contact chronometer 接触天文表;接触时计
contact clamp 接触夹板
contact clause 船壳撞损条款

contact clearance 触点隙
contact clip 接线夹;接触夹
contact clock 接触钟
contact closed 接点闭合
contact-closure 闭合接点
contact closure circuit 触点闭合线路
contact coefficient 接触系数
contact combination 触点组合
contact compliance 触点顺从度;触点柔度
contact condenser 接触冷凝器
contact configuration 接触布局
contact continuity 接触连续性
contact control 接触控制;触点控制
contact controller 接触式控制器
contact converter 接触式整流器;接触式换流器;接触式变流器
contact cooling 接触冷却
contact copy 接触晒印
contact copying 接触式复印方法;接触晒印
contact corrosion 接触腐蚀;隙间腐蚀
contact crystallization way 接触结晶作用方式
contact current 触点电流
contact current-carrying rating 触点额定通电容量
contact deasphalting method 接触脱沥青法
contact debt 借债
contact decolo(u)rization 接触脱色(法)
contact deposit 接触矿床
contact depth of tooth 接触齿高
contact dermatitis 接触性皮炎
contact desulfuriation 接触脱硫(法)
contact detector 接触(式)检波器
contact device 接触装置
contact digestion 接触消化
contact diode 点接触型二极管
contact discontinuity 不连续接触
contact discontinuity surface 接触间断面
contact disinfecting tank 接触消毒池
contact disinfection pond 接触消毒池
contact distillation 接触蒸馏
contact drop 接触电压降;触点电压降
contact drum 接触鼓(轮)
contact-effect relationship 接触—效应关系
contact electricity 接触电
contact electrode 接触式焊条;接触(式)电极
contact element 接触件
contact engaging and separating force 触点插拔力
contact erosion 接触面冲蚀
contact evapo(u)ration 接触蒸发
contact exposure 接触曝光
contact face 接触面
contact face of buffers 缓冲区接触面
contact factor 接触效率;接触系数
contact failure 接触破坏;黏结松脱
contact fault 接触故障;接触不良
contact file 触点清整锉
contact film 接触网(目)板;接触晒印胶片
contact filter 接触滤池
contact filtration 接触过滤
contact finger 接触指;接触片;触指;触头
contact fire-proofing 敷盖防火法
contact flocculation 接触絮凝(作用)
contact flocculator 接触絮凝池;接触絮凝器
contact flying 目力飞行
contact follow 接点跟随;触点追随;触点跟踪
contact force 接触力;触点压力
contact for flash unit 闪光灯触点
contact form area 接触模板面积
contact friction 接触摩擦
contact furnace 接触炉
contact fuse 着发引信;触发引信
contact gap 触点间隙
contact gasket width 垫片接触宽度
contact gear ratio 接触比
contact glass 接触镜;焦面玻璃(片);焦距玻璃;压片玻璃
contact glow discharge electrolysis 接触辉光放电电解
contact glue 触压胶
contact goniometer 接触测角仪;接触测角器
contact grill 接触烤板;煎板
contact grouting 接触面缝(隙)灌浆;接触面灌浆;接触灌浆(沉箱外围防水)
contact hardening 接触硬化

contact head 接触点;触头
contact heat exchanger 接触式热交换器
contact heating surface 接触加热面
contact heating system 接触加热系统
contact height 接触高度
contact herbicide 接触性除草剂
contact history 接触史
contact hyperfine interaction 接触超精细相互作用
contact infection 接触侵染
contact-infiltration metasomatism 接触渗透交代作用
contact(ing)angle 接触角
contacting area 接触面积
contacting biologic(al) filter 接触生物滤池
contacting line 啮合线(齿轮传动)
contacting profile 啮合齿廓
contacting surface 接触表面
contacting tower 接触塔
contacting with powdered carbon 用粉状炭接触
contact inhibition 接触抑制
contact initiated discharge machining 接触起火放电加工
contact input 接触输入
contact insecticide 接触杀虫剂;触杀剂
contact in series 串联
contact inspection 接触检测
contact insulation value 闪光触点绝缘值
contact interface 接触界面
contact interferometer 接触式干涉仪
contact interrupting capacity 触点切断能力;触点的切断能力
contact interval 接触间隔
contact investigation 接受调查;接触调查
contact jaw 接触钳钉;接触夹片;接触端;传电夹钳
contact joint 对接接头
contact junction 接点
contact laminating 接触层压
contact layer 接触层
contact leaching anti-fouling paint 接触渗毒型防污漆
contact leakage 接触面渗漏
contact length between the tool and chip 刀具切屑接触长度
contact length of tyre 轮胎接地长度
contact lens 接触(式)透镜;接触镜
contactless 无接点(的);无接触的
contactless control element 无接触控制元件;无触点控制元件
contactless ga(u)ge 非接触型测量仪
contactless operating volume meter 非接触式体积计
contactless resistivity measurement 非接触电阻率测量
contactless smart card 非接触式聪明卡
contactless tachometer 无接触转速计
contact level 符合水准仪;符合水准(器)
contact lever 接触杆
contact line 接触线
contact line bracket 电车线托架
contact line connection 馈线连接点;接触线连接
contact line inspection vehicle 接触线检查车
contact line protection 接触导线保护
contact load 接触荷载;滑移质;滚转负荷;触移质;触点负载;触点负荷;推移质
contact log 接触法测井;微测深法
contact logic 触点逻辑
contact loss 离线;接点损失;接触损耗
contact loss rate 离线率【铁】
contact lost 目标失踪
contact lysis 接触溶解
contact magnetic recording 接触式磁记录
contact maintenance 接触维修;直接维修
contact maker 开关;断续器;断路器
contact-making clock 触点电钟
contact making voltmeter 继电电压表;触点式伏特计
contact manometer 接触压力表
contact mark 接触标记
contact mast 接触网柱
contact material 筛滤材料;接触物质;接触材料;电接触器材;电接触材料
contact mechanism 接触装置

contact metal 接点金属;电触头合金;触头金属;触点金属
contact metamorphic 接触变质的
contact metamorphic action 接触变质(作用)
contact metamorphic belt 接触变质带
contact metamorphic facies series 接触变质相系
contact metamorphic metallization 接触变质成矿作用
contact metamorphic rock 接触变质岩
contact metamorphic zone 接触变质带
contact metamorphism 接触变质(作用)
contact metasomatic deposit 接触交代矿床
contact metasomatism 接触热变质作用;接触交代(变质)作用
contact method 接触法
contact method of sewage treatment 污水处理接触法
contact micrometer 接触测微器;接触测微计
contact microphone 接触(式)传声器
contact mine 触发水雷
contact mineral 接触变质矿物
contact-modulated amplifier 振簧式放大器
contact moisture 接触湿度;接触含水量
contact mo(u)lded reinforced thermosetting plastic laminate 触压模塑增强热固性层压材料
contact mo(u)lded glass-fiber-reinforced thermosetting resin flange 触压模塑玻纤增强热固性树脂法兰盘
contact mo(u)lding 接触成型;触压成型
contact network 接点网络;触点网络
contact network equipment 接触网设备
contact noise 接触噪声
contact normal 接触点法线
contact of earth 接地合闸;接地
contact of employment 雇佣合同
contact of engagement 雇佣合同
contact of the batholith 岩基接触
contactolite 接触变质岩
contact opened 接点断开
contact-operating lever 接点动作杆
contactor 接触器;断续器;触点;触头
contactor blower 接触器吹扫机
contactor board 接触器板
contactor control 控制接触器
contactor controller 触点控制器
contactor control system 触点控制系统
contactor operating coil 接触器运行线圈
contactor panel 接触器盘
contactor pump 混合泵
contactor segment 接触片
contactor sequence 接触器接触次序
contactor starter 接触器起动器
contactor switching starter 接触器开关起动器
contact oven 接触炉
contact over-travel 触点追随;触点超程
contact oxidation 接触氧化(法)
contact-oxidation for iron 除铁接触氧化法
contact oxidation method 接触氧化法
contact oxidation-oxidant oxidation process 接触氧化一氧化剂氧化工艺
contact oxidation pond 接触氧化塘
contact oxidation process 接触氧化过程;接触氧化工艺
contact oxidation tank 接触氧化池
contact panel 触点盘
contact parting time 触点断开时间
contact paste 接触浆剂;接触浆糊
contact pattern (钻头牙齿在井底的)接触印痕
contact percolation 接触渗滤
contact performance 接触性能
contact-period 接触周期;接触时间
contact person 联系人
contact pesticide 接触(杀虫)农药
contact pick-up 接触传感器
contact picture 接触面图
contact piece 接触片
contact pin 接片;触针;触销;插针
contact piping 接触管涌
contact piston 接触活塞
contact pitch 接点间隔
contact plane 接触面;触点平面
contact plate 接触(压平)板;焦面玻璃;焦距玻璃
contact plot 接触绘图
contact plunger 接触活塞

contact point 接点;接触点;触点
contact point bumper block 接触点防振垫块
contact point dresser 触点打磨机;白金打磨机
contact points and brushes tension ga(u)ge 触点及电刷电压计
contact points cleaning plate 断电触点清洁片
contact poison 接触(性)毒剂;接触毒物
contact poisoning 触杀作用
contact polyester 接触聚酯
contact positive 接触晒印正片
contact potential 接触电位;接触电势;触点电位
contact potential barrier 接触势垒
contact poyester 接触树脂
contact pressure 挤压力;接触压力;接触(面)压力;触(点)压力
contact pressure mo(u)lding 接触加压模塑
contact pressure resin 触压树脂
contact print 接触晒印;晒图
contact printer 接触印像;印相机;晒图机
contact printing 接触晒印
contact printing frame 接触晒像机
contact print map 接触晒印地图
contact probability 接触概率
contact probe 探头
contact problem 接触问题
contact process 接触(晒印)法;接触过程
contact protection 触点保护措施;防止接触
contact pyrometer 接触式高温计
contact quality 接触性能
contact rail 导电轨
contact rating 接点容量;触点额定值
contact ratio 啮合系数;重合度;重叠系数
contact reactions way 接触反应方式
contact reaction tank 接触反应槽
contact reactor 接触反应器
contact recording 接触式记录
contact rectification 接触式整流;接触精馏
contact rectifier 干式整流器
contact reflex 接触反射
contact reforming 接触重整
contact relation 接触关系
contact relation between intrusive body and country rocks 侵入体与围岩接触关系
contact relation of strata 地层接触关系
contact relay 触点式继电器
contact rerum 接触再蒸馏
contact resin 接触(成型)树脂
contact resistance 接触电阻;触点电阻
contact resistance meter 接触电阻测量仪表
contact ring 接触圈;输电环
contact rod 接触棒;导电棒
contact roller 接触滚轮;接触滚;压紧轮;滚轮
contact roughing filter 接触粗滤池
contact scanning 接触扫描
contact scour 接触冲刷
contact screen 接点荧光屏;接点屏蔽;接触网目片;接触网(目)板
contact screw 接触螺钉
contact seal 接触油封;接触式止水;接触密封;机械密封
contact section 接触段
contact sedimentation 接触沉淀(作用)
contact sedimentation method 接触沉淀法
contact sedimentation tank 接触沉降槽;接触沉淀池
contact segment 接触扇形体
contact sense 触点判定
contact sense module 接触读出模件
contact sensing 触点读入
contact sensitivity 接触敏感性
contact-separating force 接点分离拉力
contact separation 接点间隔;接点分离
contact sequence 接点接通顺序
contact series 接触序列;触点序列
contact servicing 直接维护
contact set 触点组
contact sheet 接触片
contact shoe 接触滑块;集电靴;触靴
contact signal 接触信号
contact's isolation 接触者隔离
contact site 接触部位
contact size 接触尺寸;插针尺寸
contact skate 接触滑座
contact-slide baseline measuring apparatus 接触滑动式基线测量器械

contact slider 滑动触点
contact slipper 接触滑块
contact socket 接触座
contact softener 接触软水器
contact solution 接触溶液
contact souring 接触面冲蚀
contact spacing 触点间距;插针间距
contact speed 滑行速度(飞机);碰撞速度
contact spike 导电棒
contact splice 接触拼接;互搭接合;钢筋搭接;搭接接头;搭接;接触并接
contact spring 接触泉;接触弹簧
contact spring assembly 触点弹簧装置
contact stabilization 接触稳定(作用)
contact stabilization activated sludge process 接触稳定活性污泥法
contact stabilization process 接触稳定过程;接触稳定法;吸附再生法
contact stabilization system 接触稳定系统
contact stabilization treatment 接触稳定处理
contact stain 接触污染;污色;污斑
contact stand 接触架
contact start-flying off 接触启动浮动停止
contact start stop 接触起停
contact stiffness 接触刚度
contact stimulus 接触刺激
contact strain meter 接触式应变片
contact strength 接触强度
contact stress 接触应力;接触面应力
contact stress transducer 接触应力传感器
contact strip 接触片;接触(滑)条;接触滑板
contact strip clamp 接触线夹
contact structure 接触构造
contact stud 接触钉
contact superheater 接触过热器
contact support 端钮支架
contact surface 接触(表)面
contact surface of a trickling filter 滴滤池的接触表面
contact surface of valve seat 阀座接触面
contact switch 接触开关;触簧开关
contact system 接点系统;接触系统;触点系统
contact system ga(u)ge 接触网界限
contact tach(e)ometer 触点式(自动归算)速测仪
contact tachymeter 触点式(自动归算)速测仪
contact tag 接片;接触金属箍
contact tank 接触罐;接触池
contact tempering 接触刚化
contact terminal 接触接点;触头
contact test 接触试验
contact thermal resistance 接触热阻
contact thermocouple 接触式热电偶
contact thermography 接触测温
contact thermometamorphic rocks 接触热变质岩
contact thermometer 接触温度计;接触式温度计
contact time 接触时间;触点动作时间
contact-time test 接触时间试验;接触率试验
contact tip 导电嘴
contact tongue 接触片
contact tower 接触塔
contact toxicity 接触毒性
contact transformation 接触变换;正则变换
contact transmission 接触传播
contact travel 触头行程;触点行程
contact treating 接触精制
contact tube 短焦距X射线管;导电(铜)管
contact twin 接触双晶
contact type 接触类型
contact type adhesive 接触型粘结剂;接触型胶结剂
contact type of grain 颗粒接触类型
contact-type operation (of explosive forming) 直接式爆炸成型
contact type solid air snap 接触式整体气动长规
contact-type strainometer 接触式应变仪
contact unit 接触装置
contact vibration 接触振动
contact vignetted screen 渐晕接触网目屏
contact voltage 接触电压;触点电压;人可接触的安全电压
contact voltage regulator 分级电位调整器
contact wear 接触磨损
contact welding 接触焊
contact wheel 滚极

contact wire 接触(导)线;电车线;滑接线
contact wire clamp 线夹
contact wire clip 接触线夹;定位线夹
contact wire connection clamp 接触线固定线夹
contact wire dropper clip 接触线吊弦线夹
contact wire for main line 正线接触线
contact wire gradient 接触线斜率
contact wire height 接触线高度
contact wire mid-point anchor 锚索
contact wire pre-sag 接触线预弛度
contact wires 交叉接触线
contact wire sag 接触线弛度
contact wire splice 接触线线头线夹
contact wire splice for wire crossing with cup-point screws 螺栓型交叉接触线接头线夹
contact wire splice with cup-point screws 螺栓紧固型接触线接头线夹
contact wire straightener 接触线整正器
contact wire suspension 接触导线悬挂
contact wire twister 接触线扭弯器
contact wire wear 接触线磨耗
contact zone 接触区;接触带
contagious disease 接触性传染病
contagious distribution 蔓延分布
contagious infection 接触(性)传染
contagious ward 传染病房
contain 整除;折合;包括;包含(物);容纳
contain a fire 围控
contain a lot of sand and clay 含有大量的沙子和黏土
contained concrete 自应力混凝土
contained O-ring joint 封闭圆环接点
contained plastic flow 限制塑性流动
contained sand island 围堰筑岛
contained spring 包容弹簧
contained text 被包含文本;包含的正文
contained underground burst 密封地下爆炸
container 集装箱;货柜;储[贮]器;储[贮]存器;罐体;储罐;包装物;包装容器;容器
container accounting 包装物核算
container asphaltic bitumen 含石油沥青基的沥青
container assembly 容器组件
container bag 集装袋
container bale capacity 集装箱包装容积
container bank 集装箱库;集装箱多层仓库
container barge 集装箱驳船
container/barge carrier 集装箱载驳两用船
container base 集装箱内陆中转站;集装箱基地
container berth 集装箱码头;集装箱泊位
container board 盒纸板
container box 集装箱
container bridge 集装箱装卸桥;集装箱吊桥;桥式集装箱起重机
container-bulker 集装箱散货船
container bulk ship 集装箱散货兼用船
container capacity 集装箱容量
container car 集装箱(专用)车
container cargo 集装箱货(物)
container cargo depot 集装箱货物仓库
container carrier 集装箱(运输)船;集装箱运输车
container-carrying capacity 装运集装箱能力
container-carrying ship 集装箱船
container-carrying trailer 集装箱拖车
container-carrying wagon 集装箱车辆
container case 容器外壳
container cell 集装箱井型架;集装箱格舱
container chassis 集装箱底盘车
container chassis for freight liner of JNR 国铁定期直达列车集装箱底盘车
container chassis side 集装箱底盘车侧
container cleaning 集装箱清洁处理
container coating 容器镀层
container compaction 容器压实
container conditioning 集装箱温度及湿度调节;集装箱内部调湿
container conservation 集装箱保存
container consortium 集装箱联合企业
container control system 集装箱控制系统;集装箱管理系统
container corner seating 集装箱角配件
container cost 容器成本
container cradle 容器摇架
container crane 集装箱起重机;岸壁吊车
container crane with rope trolley 缆车集装箱起重机
container crane with self-propelled trolley 自行小车集装箱起重机;自行缆车集装箱起重机
container crane with semi-rope trolley 半绳索式缆车集装箱起重机
container depot 货柜仓库;集装箱运站;集装箱拆装场
container detention charges 集装箱滞留费
container discharging place 集装箱卸货点
container disposition system 集装箱配置系统
container distribution center 集装箱分配中心
container equipment data exchange 集装箱设备数据交换
container equipment interchange 集装箱设备交换
container exit 集装箱出入口
container express 集装箱专用列车
container feeder port 集装箱支线港;集装箱喂给港
container feeder ship 集装箱集散(货)船
container feet 集装箱底脚
container ferry 集装箱渡船
container flat car 集装箱铁路平车
container fleet 集装箱船队
container flow 箱流
container forklift 集装箱叉车
container fork truck 集装箱叉车
container for refuse 垃圾箱;废料箱
container for wastes 废品箱
container freight liner 集装箱定期直达列车
container freight station 集装箱装箱拆箱库;集装箱中转站;集装箱货运站;集装箱拆装库;拆装箱库
container freight station cargo 集装箱货运站货
container freight station operation 集装箱货运站作业
container freight station to container freight station 集装箱站到站运输;站到站
container freight station to container yard 集装箱站到场运输;站到场(集装箱)
container freight station to door 集装箱站到门运输;站到门
container gantry crane 集装箱装卸桥;集装箱装卸机;集装箱龙门起重机
container glass 瓶罐玻璃
container handling 集装箱装卸
container handling crane 集装箱起重机
container handling facility 集装箱装卸设备
container handling gantry crane 集装箱装卸桥
container handling information system 集装箱管理信息系统
container handling machinery 集装箱装卸机械
container handling straddle carrier 集装箱跨(运)车
container handling system 集装箱装卸系统;集装箱卸方式
container handling vehicle 集装箱装卸车
container harbo(u)r 集装箱港
container height 集装箱高度
container hire 集装箱租金
container hoist 集装箱提升机
container hold 集装箱容积;集装箱舱
container hook 集装箱吊钩
container horizon 含油层
containering packing shed 拆装箱库
container interchange 集装箱互换
containerizable 可用集装箱运输的;可集装箱化的
containerizable cargo 可用集装箱运输的货物;适箱货
containerizable commodities 适箱货
containerization 集装箱化
containerize 集装箱化
containerized batch mixer 集装箱式分批混合机
containerized cargo 集装箱(化)货(物)
containerized freight 集装箱货(物)
containerized light aboard ship 集装箱化载驳货船
containerized refuse 集装垃圾
containerized traffic 集装箱运输;集装化运输
containerized transport 集装化运输;货捆运输;单元运输;束装运输
containerized transportation 集装箱化运输
container leasing company 集装箱出租公司
container length 集装箱长度
containerless casting 无模铸造
container liner 集装箱班轮;模腔圆圈;容器衬垫
container liner service to Hong Kong area by Shanghai COSCO (中远集团)上海—香港地区集装箱班轮航线
container lining 容器内壁保护层;容器衬里
container load 集装箱整箱货;集装箱荷重;整箱货
container load cargo 集装箱整箱货
container loader 集装箱装箱机
container loading 集装箱装箱作业
container loading list 集装箱装箱清单;装箱单
container load plan 集装箱装箱单
container load rate 整箱集装箱货运价
container lorry 集装箱运输车
container maintenance 集装箱的维修保养
container management system 集装箱管理系统
container marine line 集装箱海运公司
container marine line system 集装箱海上运输公司方式;集装箱搬运吊车方式
container marking 集装箱标记
container marshalling and storage area 集装箱整理存放场
container marshalling area 集装箱有效场地
container marshalling yard 集装箱编组场
container material 容器材料
container mixer 容器混合机
container number 集装箱箱号
container nursery 容器育苗苗圃
container of known size 已知大小的容器
container on flat car 集装箱上平板车;平板车装集装箱;铁路集装箱平车
container operation 集装箱业务
container operator 集装箱运输经营人
container/ore carrier 集装箱矿石两用船
container packing list 集装箱装箱清单
container packing shed 集装箱装箱拆箱库;集装箱拆装库
container paint 集装箱涂料;货柜涂料
container/pallet carrier 集装箱/托盘两用船
container park 集装箱堆场
container parking area 集装箱堆放场
container pier 集装箱码头
container planting 容器栽植
container platform trailer 集装箱平板车
container pool 集装箱堆放场
container portal crane 集装箱门座(式)起重机
container/rail car carrier 集装箱货车两用船
container rate 集装箱运费率
container refuse 集装垃圾
container rehandling list 集装箱翻箱清单
container rehandling report 集装箱翻箱报告
container release assembly 集装箱松紧装置
container released 集装箱已交付
container rental company 集装箱出租公司
container returned 集装箱已交还
container ring 模腔衬环;容器环
container rock 多孔储积岩;储集岩;熔岩
container roll ship 集装箱滚装船
container scale 计量罐
container seal 集装箱铅封
container securing device 集装箱栓固装置
container securing fitting 集装箱固定件
container semi-trailer 集装箱半挂车
container service 集装箱作业;集装箱运输;集装箱业务
container service charges 集装箱服务费
container service port 集装港
container ship 集装箱货轮;集装箱运货船;集装(箱)船;集装箱货运船
container shipping 集装箱船运;货柜车运
container side carrier 侧面集装箱叉式装卸车
container side fork lift truck 侧面集装箱叉式装卸车
container side loader 侧面集装箱叉式装卸车
container's lock 集装箱锁装置
container slot 集装箱箱位
container socket 集装箱承座
container soil volume 容器测土壤的容积
container spreader 集装箱吊具;集装箱吊架
container stacker 集装箱堆垛机
container stall 集装箱箱位;集装箱货架
container standardization 集装箱标准化
container station 集装箱(办理)站
container storage area 集装箱堆场
container stowage profile 集装箱装载剖面图
container stowage space 集装箱堆场
container straddle carrier 集装箱跨(运)车

container structure 集装箱结构
container stuffing 集装箱内装货
container sweat 箱汗(集装箱)
container system 集装箱系统
container team yard 集装箱货场
container terminal 集装箱装卸区;集装箱枢纽站;集装箱(枢纽)港;集装箱码头;集装箱(水陆交接)码头
container terminal capacity model 集装箱港口容量模型
container terminal facility 集装箱码头设施
container test 集装箱试验
container through loading net work 集装箱联运装卸网
container through traffic service 集装箱联运业务
container tractor 集装箱牵引车
container traffic 集装箱运输;集装箱运量
container trailer 集装箱挂车;集装箱底盘车
container/trailer carrier 集装箱拖车两用船
container transport 集装箱运输
container transport by loading more than one consignments of goods in one container 一箱多批运输;拼箱运输
container transport system 集装箱运输系统
container transport volume 集装箱发送量
container transshipment 集装箱转运
container truck 集装箱载运车;集装箱车
container turnaround days 集装箱周转天数
container unit train 集装箱专用列车;固定编组集装箱列车
container unloading list 集装箱卸货清单;卸箱清单
container vehicle 散装水泥罐车
container vessel 集装箱货轮;集装箱船
container wagon 集装箱车
container wall effect 器壁效应
container washing plant 集装箱冲洗厂
container weight 集装箱重量
container width 集装箱跨度
container yard 集装箱(后方)堆场;集装箱(车)场
container yard capacity 集装箱堆场能力
container yard cargo 集装箱场货
container yard operation 集装箱场作业
container yard to container freight station 场到站
container yard to door 场到门
container yard to yard 场到场
containing a fire 围控火势
containing asbestos fiber 含石棉纤维
containing hydrogen silicone oil 含氢硅油
containing mark 容量刻度
containing overflows 收集溢油
containing oxygen iron 含氧铁
containing vessel 安全壳
containing wall 防护外壁
containing water 含水
containment 拦集(水面浮油等);可容度;密封装置;密闭度;密闭度;置放装置;反应堆体;保护外壳;包容;包括;牵制
containment area 泥圹
containment area drainage 泥圹排水
containment area enclosure 泥圹围埝
containment grouting 抑制(周界)灌浆;内芯高压法灌浆;密封外壳法灌浆;周界灌浆
containment of solid waste 固态废物限制地区
containment procedure 抑制程度
containment shell 密封外壳
containment vessel 密闭壳;反应堆安全壳;保护壳;安全壳
contaminant 沾污物;沾染物;污染物;污垢物
contaminant accumulation 污染物累积
contaminant assessment 污染物评估
contaminant biomonitoring 污染物生物检测
contaminant classification 污染物分类
contaminant content 污染物含量
contaminant detection 污染物检测
contaminant effect 污染效应
contaminant flow 污染流
contaminant fluid 污染流体
contaminant-holding capacity of oil filter 滤油器纳垢容量
contaminant hydrology 污染水文学
contaminant ion 杂质离子
contaminant level 污染物水平
contaminant loading 污染物负荷(量)

contaminant migration theory 污染传播理论
contaminant plume 污染物烟缕
contaminant plume prediction 污染物烟缕预报
contaminant release 除垢
contaminant removal 污染物排除
contaminant separation 除垢
contaminant source characterization 污染源表征
contaminant source location 污染源位置
contaminant source release history 污染源排放史
contaminant transient flow field 污染物瞬时流场
contaminant transport 污染转移;污染物输移
contaminate 洗污;污染
contaminated 沾污了的;受污染的
contaminated aquifer 污染含水层
contaminated area 污染区;污染面积;污染地区
contaminated ballast bed 脏污道床
contaminated beet 污染的甜菜种子
contaminated by sewage 受下水道污染的
contaminated-catalyst 沾污的催化剂
contaminated effluent 污染流出液
contaminated environment 被污染的环境
contaminated fluid 污染液体
contaminated food 污染食物;污染的食物
contaminated food products 污染食品
contaminated groundwater 污染地下水
contaminated groundwater resources 污染地下水资源
contaminated land 受污染土地
contaminated-oil basin 污油池
contaminated oil pipe 污油管
contaminated oil settling tank 污水最终沉降槽
contaminated overburden 污染了的覆盖层
contaminated region 污染地区
contaminated river 污染河流
contaminated river water 污染(的)河水
contaminated rock 混杂岩
contaminated sediment 污染底泥
contaminated sediment management strategy 污染底泥管理对策
contaminated soil 污染(的)土壤;受污染土(壤)
contaminated stream 污染河流
contaminated water 脏水;染菌水;污染水
contaminated water remediation 污染水修复
contaminated waters 污染水域
contaminated zone 污染区;污染(地)带
contaminating air burst 污染性空中爆炸
contaminating material 污染物质
contaminating metal 杂质金属
contaminating radio activity 污染的放射性
contaminating sources 污染源
contamination 混杂作用;样品沾污;沾污;沾染;词的组合;污染
contamination accident 污染事故;沾污事故;污染灾难;污染偶发事件
contamination activity 污染放射性
contamination air 污染空气
contamination and remediation of groundwater 地下水污染治理
contamination and remediation of surface water 地表水污染治理
contamination by pesticides 农药污染
contamination coefficient 污染系数
contamination condensation 污染凝聚
contamination control 消除放射性措施;污染控制;污染管制;污染防治
contamination counter 测污器;污染计数器;污染计数计;污染计数器
contamination-decontamination experiment 污染去污试验
contamination degree 污染程度
contamination dispersal 污染物扩散
contamination dose 沾染剂量;污染剂量
contamination event 污染事件
contamination factor 污染因子;污染系数
contamination free 无污染(的)
contamination hazard 污染危害
contamination index 污染指数
contamination level and type 污染水平与分类
contamination limit 污染限度
contamination load 污染负荷量
contamination material 夹杂物
contamination meter 沾污测计计;污染剂量计
contamination monitor 沾染监测器;污染监测仪;污染监测器

contamination monitoring 污染监测
contamination of freshwater body 淡水水体污染
contamination of natural freshwater bodies 天然淡水水体污染
contamination of working fluid 工作液污染
contamination oil processing facility 污染油品处理设施
contamination oil reinjection device 污染油品再注入设备
contamination plume 羽状污染体
contamination precipitation 杂质沉淀
contamination rate 污染率
contamination risk analysis 污染风险分析
contamination sampling 污染物采样
contamination seal 防污染密封
contamination sources 污染源
contamination surface 污染表面
contamination-suspect area 可疑污染区
contamination tester 测污器;污染程度测定器;污染测试器
contaminative 沾染的;弄脏的;混染的;毒害的;掺染的;污染的
contaminative cargo 污染货物
contaminative-hydrogeologic(al) survey 污染水文地质调查
contaminator 污染物
contaminat sources 污染源
contango 交割限期日(英国交易所);交货延期费;延期金
contango dealing 延期交割交易
contango price of money 延期日息
contango rate 贴水率;升水率
contemplate 注视;预料;设想;企图;打算
contemporaneity 同时(期)性
contemporaneous breccia 同生角砾岩
contemporaneous conglomerate 同生砾岩
contemporaneous correlation 同期相关
contemporaneous covariance 同期协方差
contemporaneous deformation 同时变形;同生变形
contemporaneous dependence 同期相依
contemporaneous erosion 同时侵蚀;同生侵蚀;同期侵蚀
contemporaneous errors 同期误差
contemporaneous fault 同生断层
contemporaneous fold 同生褶皱
contemporaneous heterotopic facies 同时异相沉积
contemporaneous peat swamp 同生泥炭沼泽
contemporaneous structure 同沉积构造
contemporanrously uncorrelated variables 同期不相关变量
contemporary architecture 现代建筑
contemporary bearing wall 同期砌筑的承重墙;现代承重墙
contemporary ceramics 现代陶瓷
contemporary city 当代理想城市
contemporary comparision 同期比较法
contemporary gardens in China 中国近代花园
contemporary glaciation 现代冰川作用
contemporary glacier 现代冰川
contemporary issue 近期刊行出版物
contemporary records 当时记录;同期记录
contemporary survey 临时性调查
contemporary technique 现代工艺
contempt of court 藐视法庭
contend for markets 争夺市场
contending colo(u)r 斗彩
content 可容度;内装物;内容;含量;体积
content-addressable memory 联想存储[贮]器;按内容访问存储器
content addressed memory 相连存储器
content addressing 内容定址
content analysis 内容分析
content-based access method 按内容存取法
content by volume 单位体积含量;体积含量
content directory 内容一览表
content function 容度函数
content ga(u)ge 体积计;计量器;液位仪;液位计;液面计
content-independent address 内容独立地址
content indicator 内容指示器;内容显示器
contention 争用线路;争用;争论;冲突;通路争用
contention interval 争用时间间隔

contention method 争用法
contention mode 线路争夺状态
content map of slide radon 滑动氡气等值图
content meter 含量测定仪
content of air pollutant 空气污染物含量
content of alumina 三氧化二铝含量
content of aromatic hydrocarbon 芳烃含量
content of ash 灰分含量
content of ashes 灰分含量
content of bad-soluble salt 难溶盐含量
content of benzene-insoluble substance 苯不溶物含量
content of bitumen 沥青量
content of calcium oxide 氧化钙含量
content of carbon 炭的百分含量
content of cement in a cubic(al) meter 体量水泥系数
content of checking 检查内容
content of clay mineral 黏土矿物含量
content of clay particle 颗粒含量
content of consumption 消费内容
content of courses 教学内容
content of extractable organic matter 可抽提有机质含量
content of ferric oxide 三氧化二铁含量
content of ferrous oxide 氧化亚铁含量
content of fixed carbon 固定碳含量
content of free alumina 游离三氧化二铝含量
content of free ferric oxide 游离三氧化二铁含量
content of free silica 游离二氧化硅含量
content of glacial-marine sediment fraction 浮冰碎屑含量
content of harmful mineral of road metal test 道砟有害矿物含量试验
content of heavy placer mineral 重砂矿物含量
content of helium 氦含量
content of hydrobase-exchange 水化阳离子含量
content of information 信息量
content of investigation 勘察内容
content of K 4 radioactive 钾4 含量
content of La 138 镧 138 含量
content of Lu 176 镥 176 含量
content of lyso-soluble salt 易溶盐含量
content of magnesium oxide 氧化镁含量
content of manganese oxide 氧化锰含量
content of medium soluble salt 中溶盐含量
content of organic material in seawater 海水有机质含量
content of pelitic of road metal test 道砟含泥量试验
content of phosphorus pentoxide 五氧化二磷含量
content of platinum 铂含量
content of pollution 污染程度
content of potassium oxide 氧化钾含量
content of Rb 87 铷 87 含量
content of Re187 铼 187 含量
content of readily soluble salt 易溶盐含量
content of Rn222 氡 222 含量
content of saturated hydrocarbon 饱和烃含量
content of sesquioxide 倍半氧化物含量
content of silica 二氧化硅含量
content of Sm 147 钐 147 含量
content of sodium oxide 氧化钠含量
content of soft grain of road metal 道砟软弱颗粒含量
content of soft grain of road metal test 道砟软弱颗粒含量试验
content of Th 232 钍 232 含量
content of titanium oxide 氧化钛含量
content of total hydrocarbon 总烃含量
content of voids 孔隙的含量
content of volatile 挥发分百分含量
content of water 水百分含量
content of water-soluble salts 水溶盐含量
content percentage 含量百分率
content Ra 226 镭 226 含量
content ratio analyzer 相对含量分析器
content retrieval 内容检索
contents 目录
contents attribute 内容属性
contents broken 内部破损
contents conversion 内容转换
contents dictionary 目录字典
contents directory 内容目录;目录字典
contents exposed 内容暴露
contents hazard classification 房屋使用潜在危险性分级
contents leaking 货物漏出
contents lost 内容丢失
contents mildewed 内容发霉
contents of observation contents of reactive phase 活性相含量
contents retrieval 目录检索
contents rotten 内容腐烂
contents supervisor 内容监控器;目录管理程序
contents unknown 内容不明;货物内容不详
contents unknown clause 内容不详条款
conterminal 相连的
conterminous 邻接的;相连的
contest 竞赛
contestant 争执方
contest competition 争夺竞争
contested boundary 争议界;未定界限;未定界线
contesting parties 有争议双方
contesting party 诉讼当事人
context 前后关系;文脉;上下文
context check 上下文检查
context-dependent 上下文相关的
context-free 前后文无关的;上下文无关的
context indexing 关联变址
context searchig 上下文检索
context-sensitive 与上下文有关的;与前后文有关的
context sensitive menu 上下文有关的菜单
context switch 关联转换;上下文转换
context switching 关联转换;上下文转换
contextualism 文脉主义
contexture 织造;构造;组织
conticaster 连铸机
contigency interrupt 偶然中断
contigent 附条件而定的协定
contigent beneficiary 或有受益人;第二受益人
contigent fee 不可预见费
contigent fund 不可预见损失基金
contigent gain 不可预见收益
contigent interest 或有利益
contigent liability 不可预见的债务
contigent profit 不可预见利润
contigent remainder 或有剩余
contigent valuation method 应急评估法
contignation 梁系构架
contiguity 接触
contiguous 邻接的
contiguous angle 邻角
contiguous area 邻(接)区;毗邻地区
contiguous belt 接邻地带
contiguous chain 相接环形山链
contiguous crater 相接环形山
contiguous curve tangent 邻接曲线公切线
contiguous disk 邻接圆盘
contiguous file 连续文件
contiguous function 连接函数
contiguous graphics 相邻图形;邻接图形
contiguous interrupt 邻接中断
contiguous item 邻接项;相连项;相邻项
contiguous location 相连单元
contiguous model 相邻模型
contiguous obligation 随附债务;连带责任
contiguous operation 毗邻作业
contiguous river 毗邻河流
contiguous sheet 邻接图幅;相邻图幅
contiguous stratal membrane 邻接层膜
contiguous stratum membrane 邻接层膜
contiguous transmission loss 邻接传输损耗
contiguous work 毗邻作业
contiguous zone 邻接地带;接邻地带;毗邻(地)区;毗连(地)区
contilever span 挑出跨
continence 节制
continent 大陆
continental 大陆的
continental accretion 大陆增生;大陆外加作用
continental accretion period 大陆增生期
continental airmass 大陆(性)气团
continental anticyclone 大陆性高压;大陆(性)反气旋
continental apron 大陆裙
continental basin 大陆盆地
continental basin type 内陆盆地型
continental block 大陆块;大陆断块
continental borderland 大陆边地域;大陆边缘(地);大陆边界区;大陆边陲
continental brake hypothesis 大陆车阀说
continental breakup 大陆解体
continental bridge 陆桥;大陆桥
continental bridge hypothesis 陆桥假说
continental climate 大陆(性)气候
continental coast 大陆海岸
continental code 大陆电码
continental collision 大陆碰撞
continental condition 大陆条件
continental convergence 大陆会聚
continental convergence belt 大陆会聚带
continental crust 陆壳;大陆地壳
continental deposit 陆相沉积(物);大陆架矿床;大陆沉积(作用)
continental deposition 大陆条件
continental dispersion 大陆分离
continental displacement 大陆移迁;大陆位移;大陆迁移;大陆漂移
continental divide 大陆分水岭
continental drift 陆地运移;大陆位移;大陆迁移;大陆漂移
continental drift period 大陆漂移期
continental drift theory 大陆漂移(学)说;大陆漂移理论
continental earthquake 陆震;大陆地震
continental edge 大陆边缘
continental effect 大陆效应
continental electric(al) horsepower 公制马力(1 公制马力=0.986 马力)
continental embankment basin 大陆堤盆地
continental environment 大陆环境
continental epicentre 大陆震中
continental escarpment 陆崖
continental facies 陆相;大陆相
continental fauna 大陆动物区系
continental flysch formation 陆相复理石建造
continental fringe 陆缘;大陆边缘
continental geosyncline 大陆地槽
continental glacial cover 大陆冰川覆盖
continental glaciation 大陆冰川作用
continental glacier 大陆冰川
continental growth 大陆增长作用
continental heat flow 大陆热流
continental hemisphere 陆半球;大陆半球
continental high 大陆性高压
continental ice 大陆冰(川)
continental ice sheet 大陆冰原;大陆冰盖
Continental Illinois Corporation 伊利诺伊大陆公司(美国)
Continental Illinois National Bank 伊利诺伊大陆银行(美国)
continental interior 内陆
continental intermountainous basin type 内陆山间盆地型
continental island 陆连岛;大陆岛
continentality 大陆度
continentality glacier 大陆性冰川
continentality index 大陆度指数
continental lithosphere 大陆岩石圈
continental map 大陆图
continental margin 陆缘;大陆外缘部;大陆边缘
continental marginal arc 大陆边缘弧
continental marginal mountains 大陆边缘山脉
continental marginal terrain 陆缘区
continental margin deposit 大陆边缘沉积
continental-margin facies 大陆边缘相
continental marine terminal 海陆联运港;陆海联运港
continental mass 大陆块(体);大陆地块
continental mean heat flow 大陆平均热流
continental mean mantle flow 大陆平均地幔热流
continental migration 大陆迁移;大陆漂移
continental nucleus 陆核;地盾
continental ocean 大陆式海洋
continental-ocean(ic)interaction 海陆相互作用
continental plate 大陆块板;大陆板块
continental plateau 大陆台地;大陆高原
continental plate element 大陆板块单元
continental platform 大陆台地;大陆架
continental polar air mass 大陆极地气团
continental province coral reef 大陆区珊瑚礁

continental raised bog 高位沼泽
continental reef 大陆礁
continental resources 大陆架资源
continental rift 大陆裂谷
continental rifting 大陆断层活动
continental rift system 大陆裂谷系
continental rift zone 陆地断裂带
continental rise 陆隆;陆基;大陆(上)升;大陆隆起
continental rise apron 陆隆裙
continental rise cone 陆隆锥
continental rise deposit 陆隆沉积
continental river 内陆河(流)
continental rotation 大陆旋转
continental sea 陆表海;内(陆)海;大陆海
continental seating 不留中间过道的剧场座位
continental sedimentation 陆相沉积(物);大陆沉积作用;大陆沉积
continental sediments 大陆沉积物
continental segment 陆块;大陆块
continental separation 大陆分离
continental shelf 陆裙;大陆台地;大陆棚;大陆架
continental shelf deposit 大陆架沉积(物)
continental shelf oil 大陆架石油
continental shelf plain 大陆架平原
continental shelf sediment 大陆架沉积(物)
continental shelf slope 大陆架斜坡
continental shelf survey 大陆架调查
continental shelf topographic(al) survey 大陆架地形测量
continental shield 地盾;大陆盾;大陆地盾
continental shore plain 大陆架平原
continental shore terrace 大陆岸边台地
continental shoulder 大陆坡
continental slab 大陆板块
continental slope 大陆(斜)坡
continental slope deposit 大陆坡沉积
continental slope gravity anomaly 陆坡重力异常
continental sphere 陆圈
continental stream 内陆河(流)
continental subduction zone 大陆消减带
continental swamp soil 大陆沼泽土
continental talus 大陆阶地;大陆架外斜面
continental tectonics 大陆构造
continental terrace 大陆台地;大陆阶地;大陆岸边台地
continental tide 陆潮
Continental transform fault 大陆转换断层
continental transgression 陆侵;大陆沉积超覆
continental tropic(al) air 热带大陆气团
continental tropical airmass 大陆热带空气;大陆性热带气团
Continental United States 美国大陆;美国本土
continental uplift 大陆(上)升
continental volcanic deposit facies 陆上火山沉积相
continental volcanic formation 陆相火山岩建造
continental volcanic rock iron and copper-bearing formation 陆相火山岩含铁铜建造
continental water balance 大陆水量平衡
continental weathering 大陆风化作用
continental wind 大陆风
continent-bearing plate 大陆板块
continent-continent collision 大陆间碰撞;大陆—大陆碰撞
continent-continent interference 大陆间相互作用
continent cyclone 大陆性气旋
continent deposit 大陆沉积物
continent formation 大陆形成作用
continentization 大陆化作用
continent making movement 造陆运动
continent nucleus 大陆核
continent of Asia 亚洲大陆
continent-sea transition zone 海陆过渡带地貌
continent-splitting impact 大陆劈裂冲击
continent-wide 全洲的
continent-wide connected system 全洲联络(公路)网
contingence 偶然事故;相切;意外事故
contingence reserve 意外备用费
contingence sum 意外费用
contingence table 列联表
contingence table program(me) 列联表程序
contingencies estimating 不可预见费预算
contingency 不可预见(用);临时费用;列联;偶然性;偶然事故;相切;意外事件;意外事故

contingency account 应急账户;意外准备金账户
contingency allocation 意外开支拨款
contingency allowance 意外事故折让;预留时间;意外事项补偿;额外任务折让时间;不可预见费(用)
contingency approach 随机制宜
contingency appropriation 应急经费
contingency authority 意外开支权
contingency clause 偶发事故条款;应变条款
contingency coefficient 列联系数
contingency cost 预备费;不可预见费(用)
contingecy expropriation 紧急征用
contingency factor 临时因素;应急因素;意外因素;不可预测因素;事故因素
contingency fund 应急费用;预备费;不可预见费(用)
contingency interrupt 偶然(性)中断
contingency loading 意外荷载;事故荷载
contingency management 酌情管理
contingency management approach 权变管理方法
contingency measures 应急措施
contingency of reinforcement 强化过程系列
contingency plan 应变计划;临时计划;应急计划
contingency planning 应急部署
contingency procedure 偶然过程;应急过程
contingency relationship 权变关系
contingency reserve 应急费用;意外费用;意外准备金;储备经费;储备;应急储备金
contingency reserve fund 储备金
contingency strategies 权变战略
contingency sum 应急费用;意外费用;不可预见费(用)
contingency table 列联表;可数性状相关表;相依表(质量管理);质量性状相关表;关联表;并联表
contingency theory 权变理论
contingent annuity 或有年金
contingent assets 或有资产
contingent charges 或有费用
contingent credit facility 备用信贷
contingent depreciation 或有折旧
contingent duty 应变关税
contingent expenses 或有费用
contingent fund 应急基金;储备金
contingent import duty 附加关税
contingent interest 不确定利息
contingent issue of securities 特定发行
contingent items 不可预见项目
contingent liability 或有负责
contingent liability from endorsement 背书或有负债
contingent profit 或有利润
contingent reserve 应急准备金;意外事项准备金;意外事故准备
contingent survey 临时检验;临时检查
contingent survey of cargo gear 起货设备临时检验
continual command 连续指令
continual function with multiple maximum 多极值连续函数
continual function with single maximum 单极值连续函数
continual loading (consolidation) test 连续加荷固结试验
continual redemption sinking fund 不断的偿债基金
continual sharp focusing 连续清晰调焦
continual similarity of sediment discharge 输沙量连续相似
continuance 持续(时间)
continuance of observation 观测工作延续时间
continuance of tenancy 租借权继续
continuant 连续音;畸夹行列式
continuation 连续;继续;延长物;持续
continuation area 连续区;连接区
continuation card 延续卡片
continuation clause 连续条款;连保条件;继续条款;延误有效条款;展期条款
continuation code 连续码;继续码
continuation column 连续列
continuation day 交割限期日(英国交易所)
continuation line 连续行;继续行;延续行
continuation method 连续(方)法
continuation of gravity 重力延拓

continuation of the journey 续航
continuation order 连续次序
continuation property 连续特性
continuation rate 展期利率
continuation school 补习学校
continuation sheet 相邻图幅
continuation subset 连续子集
continue 接续;继续;延续
continue-any mode 连续任意方式
continue column 连续列;连续栏
continued 持续的
continued accumulation of dry matter 干物质继续积累
continued bond 延期偿付公债
continued control 继续控制
continued excitation 连续激发
continued fraction 连分数;连分式
continued-fraction approximation 连分数近似;连分式逼近
continued-fraction expansion 连分式展开式
continued-fraction method 连续分数法
continued growth 继续生长
continued insurance 继续保险
continued isolation 继续隔离
continued labelling 连续标记
continued multiplication 联乘法;连乘法
continued operation 连续操作
continued overflow 连续溢流;连续流(动)
continued overflow casting 连续溢流注浆法
continued product 连续乘积
continued ratio 连比
continued row 连续行;接续行
continued type speed control system 连续调速系统
continues flow system 连续流系统
continues flow tank 连续流水池
continues flow test 连续流试验
continue-specific mode 连续指定方式
continues period of water bursting 突水延续时间
continues running 连续运行
continues sterilization 连续消毒
continues throughput 连续输送量
continue to have 保留
continuing account 连续账户;结转账户;继续结转账户
continuing agreement 连续协议
continuing appropriation 连续经营;继续拨款;延续拨款
continuing assistance and grant-back right 技术继续援助和反馈权
continuing contract 连续合同;连续性契约
continuing cost 连续成本;继续结转成本
continuing expenditures 连续支出
continuing flow sedimentation tank 连续流沉淀池
continuing guaranty 连续保证书
continuing load test 持久荷载试验
continuing loan agreement 连续贷款协议
continuing market 继续交易市场
continuing planning process 连续规划过程
continuing project 连续的项目;持续的项目
continuing sources epidemic 续源流行
continuing the reproduction of labor power 延续劳动力再生产
continuity 连续(性);连贯性;延续性;导通;分镜头剧本
continuity and trend model 连续与趋势模式法
continuity bar 连续(钢)筋
continuity-bond 连续性轨道接头
continuity chart 连续性图
continuity check 连续性检查;连贯性校验
continuity coefficient 连续性系数
continuity condition 连续性条件
continuity condition for flow 水流的连续条件
continuity control 连续性调整
continuity data 连续性数据
continuity equation 连续性方程;连续方程(式)
continuity equation of diffusion substance 扩散物质连续方程
continuity equation of flow 水流连续方程
continuity equation of seepage 渗流连续性方程
continuity-fitting 连续性部件
continuity for a production set 生产集的连续性
continuity from above 上连续
continuity in the mean (square) 均方连续性

continuity law 连续性定律
continuity method 连续法
continuity model 连续模型
continuity of a bed 岩层的连续性;地层的连续性
continuity of acoustic(al) pressure 声压连续性
continuity of coating 涂层连续性
continuity of command 指令连续性
continuity of details 图幅接边
continuity of flow 水流的连续性
continuity of life 继续经营
continuity of life assumption 继续经营假设
continuity of pollutant mass 污染物质量连续性
continuity of sound pressure 声压连续性
continuity of space 空间的连续性
continuity of work 工作的连续性
continuity population 连续总体
continuity rank of rock mass 岩体连续性等级
continuity reinforcement(bar) 连续钢筋;连续配筋
continuity rod 连续杆
continuity stirrups 连续钢箍;连续箍筋
continuity stress 连续应力
continuity test 连续性试验;断路检查试验;持续性试验
continuity theorem 连续性定理
continuity type counter current rinsing 连续式逆流清洗
continum 连续区
continum index 连续指数
continuous 连续的;继续进行;不间断的;不断的
continuous absorption 连续吸收;连续吸附
continuous absorption coefficient 连续吸收系数
continuous access high speed camera 连续选速高速摄影机
continuous acid treating 连续酸处理
continuous acid-washing 连续酸洗
continuous acoustic(al) ceiling 连续吸音顶棚;连续隔声吊顶
continuous acting chlorinator 连续加氯机
continuous action atomizer 连续式迷雾器
continuous action refuse collection truck 连续式垃圾收集车
continuous action servomechanism 连续作用伺服机构
continuous activity 持续活性
continuous adsorption 连续吸收;连续吸附
continuous aeration 连续曝气
continuous ag(e)ing 连续老化;持续老化
continuous agitation 连续搅动
continuous air blowing 连续吹气氧化;塔式氧化
continuous air monitoring program(me) 空气连续监测规划
continuous air monitoring program(me) station 空气连续监测规划站
continuous air monitoring station 空气连续监测站
continuous air survey camera 连续航测摄影
continuous aluminising 连续渗铝
continuous amplitude filter 连续振幅滤波器
continuous analog 连续模拟
continuous analysis 连续分析
continuous analyzer 连续性分析仪;连续分析器
continuous analyzer for carbon monoxide 一氧化碳自动监测仪;一氧化碳连续监测仪
continuous analyzer for sulfur dioxide 二氧化硫自动分析仪;二氧化硫连续分布分析仪
continuous analyzer of nitrogen oxides 氮氧化物自动分析仪;氮氧化物连续分析仪
continuous analyzing system 连续分析系统
continuous anchor wall 连续锚碇墙
continuous and automatic monitoring system 连续自动监测系统
continuous and automatic monitoring system for atmospheric pollution 大气污染连续自动监测系统
continuous and automatic monitoring system for water pollution 水污染连续自动监测系统
continuous annealing 连续退火;连续(式)退火
continuous annealing furnace 连续式退火炉
continuous annealing lehr 连续式退火窑
continuous annealing oven 连续式退火炉
continuous annuity 连续年金;继续年金
continuous apron 连续护墙;连续挡板
continuous arch 连拱;连续拱
continuous arch bridge 连续拱桥
continuous arched dam 连拱坝

continuous arched girder 连续起拱大梁
continuous area 连续区
continuous ascending grades 连续升坡
continuous asphalt blowing 连续式沥青氧化
continuous asphalt mixing plant 连续式沥青混合料搅拌设备
continuous asphalt plant 连续式沥青制备设备;连续式沥青工厂
continuous assorting line 连续分选作业线
continuous atmosphere monitoring program(me) 连续大气监测计划
continuous audit 连续审计;经常审计;继续审计
continuous auger 连续式推运螺旋
continuous automatic timing water sampler 连续自动定时水采样器;连续自动定时采水样器
continuous automatic train control 连续列车自动控制
continuous automatic train-running control 列车运行持续自动控制
continuous automaton 连续自动机
continuous axle 连续轴
continuous azimuth method 连续方位角法
continuous background 连续背景
continuous band drier 链带式干燥机
continuous bar 连续(钢)筋
continuous bar reinforcement 连续钢筋配筋
continuous batcher 连续进料量斗
continuous batching by volume 按容量连续配料
continuous batching machine 自动连续卷布机
continuous batching plant 连续式配料设备
continuous beam 连续射束;连续梁;通梁
continuous beam bridge 连续梁桥
continuous beam linac 连续束直线加速器
continuous beam of light 持续光线
continuous beam on many supports 多跨连续梁
continuous beam supported at several points 多支点连续梁
continuous belt 转动皮带;传动皮带
continuous belt drum filter 连续带式转鼓真空过滤机
continuous belt weighing 连续传动带称量
continuous bending strength 连续弯曲强度
continuous bent 连续排架
continuous billet mill 连续式坯钢轧机;连续式钢坯轧机
continuous bioassay 连续流动生物测定
continuous biologic(al) phosphorous removal 连续生物除磷
continuous bit by bit 逐段连续的
continuous blanket spraying 连续全面喷雾
continuous blast 连续鼓风
continuous blast air 连续送风
continuous bleaching 连续漂白
continuous blending silo 连续式搅拌库
continuous block 连接块
continuous blowdown 连续排污
continuous blow-off 连续排污
continuous blow-off system 连续排污系统
continuous body 连续体
continuous boiling unit 连续煮沸器
continuous boil off machine 连续煮练机
continuous boss 穿透轴套;通孔套管
continuous box girder 连续箱式大梁;连续空心大梁;连续箱形梁
continuous brake 连续(制动)闸;连续制动器
continuous breaker 连续破碎机
continuous breasts 连续胸墙;连续窗下墙
continuous bridging layer 跨缝连续层
continuous bright-annealing furnace 连续光亮退火炉
continuous broaching machine 连续式拉床
continuous bucket ditcher 链斗式挖沟机;多斗挖沟机
continuous bucket elevator 链斗式提升机;连续斗式提升机;多斗(式)提升机;多斗连续提升机;斗式连续提升机;斗式连续升运器;无极链斗式提升机
continuous-bucket excavator 多斗(式)挖掘机
continuous budget(ing) 连续预算;延续预算
continuous bulkhead 连续舱壁
continuous bulk unloader 连续卸船机
continuous butt-weld mill 连续式炉焊管机组
continuous calender-feed system 连续压延—给料系统

continuous calender process 连续压延法
continuous canalisation 连续渠化;系列渠化;全河道渠化;梯级化
continuous capillary 连通毛细孔
continuous carbonization 连续碳化
continuous carrier 连续载体
continuous caster 连铸机
continuous casting (玻璃板的)连续压延法;连续铸锭;连续式铸造;连续浇铸;连续浇筑;连续浇注
continuous casting bay 连续铸锭跨
continuous casting machine 连续铸锭机;连续浇铸机
continuous casting process 连续浇筑法;连续浇注法
continuous casting ratio of steelmaking 炼钢连铸比
continuous ceiling 连续平顶;连续顶棚;连续铺板
continuous centrifugal dehydrator 连续式离心脱水机
continuous centrifugal drier 连续式离心脱水机;连续式离心干燥机
continuous centrifugal separator 连续式离心分离机;沉降螺旋卸料离心分离机
continuous centrifuge 连续离心机
continuous chain 连续链
continuous chamber kiln 连续式窑;长筒窑
continuous change 连续变速
continuous channel 连续通道;连续信道;连续渠道
continuous channelization 全河道渠化
continuous charge 连续充气
continuous charging grate 连续装煤炉箅
continuous chart 带形记录纸
continuous chemical analysis 连续化学分析
continuous chlorination 连续用氯消毒;连续氯化
continuous chromatography 连续色谱(法)
continuous circulation 连续循环
continuous clarifier 连续澄清装置
continuous clasp 连续卡环
continuous cleaning 连续清洗
continuous cleavage 连续劈理
continuous clock 连续时钟
continuous coal cutter 连续截煤机
continuous coking process 连续焦化过程
continuous columns 连续柱子
continuous combustion 连续燃烧
continuous combustion type 连续燃烧式
continuous comminution 连续破碎
continuous communication system 连续通讯系统
continuous compacting process 连续压粉法
continuous compaction 连续压制
continuous comparator 线性比较器
continuous compounder 连续加工配料机
continuous compounding 连续复利计算
continuous compound interest 连续复利
continuous concrete mixer 连续式混凝土搅拌机;连续式混凝土拌和机
continuous concrete spreader 混凝土连续铺筑机
continuous concrete transport 混凝土连续运输
continuous concrete wall (不设伸缩缝的)连续混凝土墙;地下连续墙
continuous concreting 连续浇筑混凝土;连续浇注混凝土;连续浇灌混凝土;连续灌筑混凝土;混凝土连续浇筑(法);混凝土连续浇注(法)
continuous concreting machine 混凝土连续灌筑机
continuous condition 连续条件
continuous construction 连续施工(法)
continuous consumption stream 连续消费流
continuous contact coking 连续接触焦化
continuous contactor 连续接触器;持续作用接触器
continuous contactor system 连续接触器系统
continuous-continuous casting 多炉连续铸钢
continuous contract 连续契约
continuous control 连续控制
continuous controller 连续作用调节器;连续式控制器
continuous control servomechanism 连续作用伺服系统
continuous converter 连续吹炼转炉
continuous converting method 连续吹炼法
continuous conveyer 连续传送器;连续式运输机;连续输送机
continuous conveyer belt method 流水传送带法
continuous cooling 连续冷却
continuous cooling transformation curve 连续

冷却转变曲线
continuous cooling transformation diagram 连续冷却转变图;连续冷却相变图
continuous co-orientation 连续定向
continuous core 连续芯墙
continuous core oven 连续烘芯炉
continuous core sampling 连续钻芯取样
continuous core wall 连续芯墙
continuous coring 连续取心(钻进)
continuous coring recovery 连续采取岩芯
continuous correction method 连续校正法
continuous correspondence 连续对应
continuous cost function 连续成本函数
continuous countercurrent ion(ic) exchange 连续逆流离子交换
continuous countercurrent leaching 连续逆流浸提
continuous countercurrent operation 连续对流操作
continuous countercurrent rinsing 连续式逆流清洗
continuous count station 连续计数站
continuous cracking 连续裂缝;连续裂纹
continuous credit 连续信贷法
continuous creep 连续蠕动;连续蠕变
continuous crescent dunes 新月形沙丘链
continuous cultivation 连续培养
continuous culture 连续培养(物)
continuous culture method 连续培养法
continuous current 连续(电)流;恒力电流;等幅电流
continuous current generator 直流发电机
continuous curvature nozzle 连续曲率喷管
continuous curve 连续曲线
continuous curve distance-time protection 平滑时限特性距离保护装置
continuous cut 连续切削;连续切割
continuous cylindric(al) shell 连续筒壳;连续的圆柱体薄壳
continuous data 连续(性)资料;连续数据
continuous days 连续天数;连续日
continuous decatizing machine 连续蒸呢机
continuous decomposable process 连续可分过程
continuous deflective separation 连续折流分离
continuous deformation 连续变形
continuous degradation 连续降解
continuous degree of discontinuity 结构面连续程度
continuous delivery 连续输水
continuous delta modulation 连续增量调制
continuous demand 连续需水量;全年需水量
continuous demineralization 连续脱盐
continuous demineralizer 连续除盐器
continuous demineralizing plant 连续纯水提取装置
continuous demolition 连续爆破
continuous density variation method 连续浓度变更法
continuous depositing 连续堆料
continuous descending grades 连续降坡
continuous desulfurization 连续脱硫
continuous deterministic process 连续确定性过程
continuous diaphragm type level detector 膜片式连续液位检测器
continuous diffusion 连续扩散
continuous digester series 连续压煮器组
continuous digitizing 连续数字化
continuous dilution 连续稀释
continuous discharge 连续放电;持续放电
continuous discharge stoker 连续除渣层燃炉
continuous discharge stoker grate 连续移动燃料炉排
continuous discontinuity 连续结构面
continuous discount(ing) 连续贴现
continuous disgester 连续蒸煮器
continuous disk winding 连续盘型绕组
continuous distillation 连续蒸馏
continuous distiller 连续蒸馏器
continuous distribution 连续分布
continuous distribution curve 连续分布曲线
continuous distribution function 连续分布函数
continuous double-helical gear 整齿人字齿轮
continuous draft 连续通风
continuous draft of surface water 地面水连续取水
continuous drainage 连续排水
continuous drainer 连续作用疏水器

continuous draw bar 连续拉杆
continuous drawing 连续拉制;不断头拉拔
continuous drawing machine 连续式拉丝机
continuous drawing process 连续拉引法
continuous drier 连续式干燥器;连续式干燥炉;连续式干燥机;连续干燥器;连续烘干机
continuous drier installation 连续干燥装置
continuous drilling machine 连续工作式钻机;连续工作式钻床
continuous drive 连续接合传动
continuous drive transmission 连续变速传动
continuous driving test 连续传动试验
continuous drying 连续式干燥
continuous drying stove 连续式干燥炉;连续烘干炉
continuous dust removal 连续除灰
continuous duty 连续值班;连续工况;连续负载;连续负荷;恒载连续运行方式;持续工作状态
continuous duty rating 连续负载额定值
continuous-duty relay 持续运行继电器
continuous dy(e)ing 连续染色
continuous dynamic(al) system 连续动态系统
continuous dynamic filter 连续动力过滤器
continuous dynamic thickener 连续动力浓缩池
continuous earth 永久接地
continuous echo sounding 连续回声测深(法)
continuous effect 连续效应
continuous elastic support 连续弹性支座;弹性连续支承体
continuous electrode 连续电极
continuous electrolytic chlorination 连续电解氯化
continuous electrophorus 持续起电盘
continuous elevator 效率升降机
continuous elutriation 连续淘洗
continuous emission 连续发射
continuous emission monitoring 连续排污监测
continuous emission spectrum 连续发射光谱
continuous engine 连续燃烧式发动机
continuous equation 连续性方程;连续方程(式)
continuous excavator 联斗挖掘机
continuous exciting source 连续激发源
continuous exclusive lane 连续式优先车辆专用车道
continuous exposure 连续曝光
continuous extension 连续延拓
continuous extraction 连续取料
continuous extraction unit 连续式萃取设施;连续式萃取设备
continuous extract ventilation unit 连续抽气通风设备;连续抽气通风单元
continuous extrusion 连续挤压
continuous extrusion machine 连续挤出机
continuous extrusion process 连续挤压法
continuous faced system 连续护面系统
continuous falling grade 连续下坡(道)
continuous falling gradient 连续下坡(道);长大下坡道
continuous feed 连续喂料;连续进料;连续加料
continuous feeder 连续送料器;连续进给装置;连续给纸装置
continuous feed incinerator 连续进料焚化炉
continuous feed welding 连续送丝电弧焊
continuous fender frame 连续护舷构架
continuous fiber 连续丝
continuous fiber pellet 连续纤维粒料
continuous fibre glass roving 连续纤维粗纱
continuous field ice 成片冰原冰
continuous field method 连续磁粉探伤法
continuous filament 长丝;连续纤维
continuous filament mat 连续纤维薄毡;连续玻璃纤原丝毡
continuous filament yarn 连续纤维线;复丝;长纤纱;长丝纱
continuous filled aperture array 连续满面天线阵
continuous fillet weld 连续贴角焊;连续(填)角焊缝
continuous film 连续膜
continuous film memory 连续膜存储器
continuous film processor 连续胶片冲洗机
continuous film projector 连续式电影放映机
continuous film scanner 均匀拉片扫描
continuous filter 连续(式过)滤池;连续过滤器;渗滤池
continuous filter press 连续式压滤机
continuous filtration 连续过滤
continuous firing 连续燃烧

continuous fishing zone 毗连渔区
continuous fixed base 连续固定底座
continuous flexible base 连续柔性底座
continuous flight auger 连续旋翼式螺钻;连续旋翼式螺旋钻
continuous flight power auger 连续旋翼式机动螺钻
continuous flooding 持续淹没
continuous floor 连续地板
continuous flooring 长地板
continuous flow 连续水流;连续流(动);连续车流
continuous flow analysis 连续流动分析
continuous flow analysis instrument 连续流动分析仪表
continuous flow bath 连续流浴盆
continuous flow bioassay 连续流动活体鉴定
continuous flow centrifugation 连续流动离心(法)
continuous flow chromatography 连续流动色谱(法)
continuous flow condition 连续流态
continuous flow constant level sequencing batch reactor 连续流动常序序批间歇式反应器
continuous flow conveyer 连续流动输送机
continuous flow cupola 连续出铁冲天炉
continuous flow diffusion model 连续流动扩散模型
continuous flow dryer 连续流动式干燥机
continuous flow enthalpimetry 连续流动热函测定法
continuous flow internal combustion engines 连流式内燃机
continuous flow irrigation 连续流灌溉
continuous flow method 连续流动法;连续流程法
continuous flow milk meter 连续式量乳计
continuous flow model 连续流(动)模型
continuous flow plant 连续流程设备
continuous flow process technology 流水作业工艺
continuous flow pug-mill mixer 连续式小型搅拌机;连续搅拌机
continuous flow pump 连续流泵;续流泵
continuous flow sampler 连续取样器
continuous flow sequencing batch reactor 连续流动序批间歇式反应器
continuous flow shortcut nitrification 连续流动短程硝化
continuous flow stirred tank aeration pond 连续流动搅拌池式曝气池
continuous flow stirred tank reactor 连续流动搅拌池式反应器;完全混合反应器
continuous flow stirred tank reactors in series model 连续流动搅拌池式反应器串联模拟
continuous flow system 连续流动系统
continuous flow system activated sludge process 连续流动升流式活性污泥法
continuous flow tank 连续流水池
continuous flow test 连续流动试验
continuous flow transport 连续流动运输
continuous flow ultra centrifuge 高速连续流动离心机
continuous flow ultrasonic reactor 连续流动超声波反应器
continuous fluid coker 流化床连续焦化设备
continuous flush 连续冲洗
continuous footing 连续基脚;连续底脚;条形基础
continuous force 持续力
continuous form 连续形式
continuous form card 连续(格式)卡片
continuous form line 地貌形态线
continuous forms 连续格式纸;连续打印纸;连续成型纸
continuous forms attachment 连续格式纸连接
continuous forms paper 连续格式纸
continuous forms stacker 连续格式纸堆积箱
continuous foundation 连续基础
continuous Fourier transform 连续傅里叶变换
continuous fractionation column 连续操作分馏塔
continuous frame 连续肋骨;连续框架;连续构架
continuous frame of reinforced concrete 钢筋混凝土连续框架
continuous frame steel structure 连续框架钢结构
continuous full-load run 连续满载运转
continuous function 连续函数;连续功能

continuous functional 连续泛函
continuous furnace 连续(式)炉;连续加热炉
continuous furnace run 连续开炉
continuous gallery machine 连续平巷采矿机
continuous game 连续对策
continuous gas carburizer 连续气体渗碳炉
continuous gas flow 连续气流
continuous gas lift 连续气举
continuous geometry 连续几何
continuous girder 连续(大)梁;单跨桁
continuous girder bridge 连续梁桥
continuous girder on elastic supports 弹性支承连续梁
continuous glass fiber 连续玻璃纤维
continuous glass filament 连续玻璃丝
continuous glass surface 连续玻璃表面
continuous gluing 连续胶合法
continuous going 连续运转;连续工作
continuous gradation 连续粒度组成;连续级配
continuous grade 连续纵坡;连续坡(度)
continuous gradient 连续梯度;连续坡道
continuous grading 连续粒度组成;连续级配;颗粒连续级配(土壤、砂砾等)
continuous grain growth 晶粒持续生长
continuous graphics 连续图形
continuous gravity concrete mixer 自落连续式混凝土搅拌机
continuous grazing 连续放牧;常年放牧;全季放牧
continuous grinder 连续研磨机
continuous grinder and polisher 连续研磨抛光装置
continuous grinding 连续研磨;连续粉磨;通磨
continuous grinding mill 连续式磨机
continuous half-pipe section machine 连续半管壳机
continuous handrail 统长扶手;连续扶手
continuous haulage 连续装运
continuous header 连续丁砖(砌合);连续过梁
continuous heading machine 掘进联合机
continuous heating 连续加热;连续供暖;连续采暖
continuous heat resistance 持续耐热性;耐热持久性
continuous heat-treating furnace 连续式热处理炉
continuous heavy duty service 连续重负载运行;连续重负荷运行
continuous heavy grade 连续陡坡;持续陡坡
continuous heavy gradient 连续陡坡
continuous heavy rain 连续性大雨;持续大雨
continuous helical auger 连续螺旋钻
continuous high chair (灌浆混凝土前临时支承钢筋的)钢丝托座
continuous hinge 钢琴铰链;长铰链;连续铰;连续合页
continuous hoist 连续提升机
continuous hollow flight screw conveyer 连续式刮板输送机
continuous homogenization 连续均化
continuous horizontal drier 连续式水平干燥机
continuous horizontal thin layer chromatography 连续水平薄层层析法
continuous horsepower rating 连续额定功率
continuous hot press 连续热压
continuous hot-strip mill 连续式带材热轧机
continuous hours 连续工作时数
continuous house 联排式多户住宅
continuous hunting 连续寻找;连续寻线
continuous ice sheet 大片冰层
continuous illumination 连续照明;连续光照
continuous image 连续象域【数】
continuous image processing 连续图形处理
continuous imagery 连续像取样;连续成像
continuous images 连续像
continuous impact test 连续冲击试验
continuous impost 连续拱座;连续拱脚垫块;连续拱墩
continuous indexing 连续分度法
continuous indicating 连续显示
continuous indicator 连续指示器
continuous industrial viscometer 连续式工业黏度计
continuous inflow 连续入流;连续流入;连续进水量
continuous information source 连续信源;连续情报源
continuous injection 连续注入

continuous in mean 均方连续
continuous innovation 连续创新
continuous in probability 概率连续
continuous insulation 连续绝缘;单质绝缘
continuous intake 连续摄入
continuous interstice 连续间隙
continuous interval function 连续区间函数
continuous interval survey in hole 环测法
continuous inventory 连续盘存;永续盘存
continuous irrigation 连续冲洗法
continuous item 连续项(目);连接项目
continuous joint 连续接缝;连续接头;连续缝
continuous kiln 连续作业窑;连续窑;连续式窑;渐进干燥窑
continuous laminating 连续层制;连续层压(法)
continuous land bucket dredge(r) 多斗式挖泥机
continuous layerage 连续压条法
continuous lehr 连续退火窑
continuous length 连续管线
continuous length of rails 焊接长钢轨
continuous lift 连续升降机
continuous light absorption 光(的)连续吸收
continuous light bands 带光源
continuous lighting 连续照明;均匀照明;统长窗采光
continuous lights 同开同关的联窗;连续照明;连续窗
continuous line 连续管线;实线
continuous line bucket system 连续戽斗链系统
continuous line lighting 带状照明
continuous line of shafting 贯通传动杆
continuous line recorder 带状记录器
continuous line source 连续线源
continuous line source model 连续线源模型
continuous liquid ejector 连续注液器
continuous liquid-liquid extracting system 连续液—液萃取系统
continuous liquid-liquid extractor 连续液—液萃取器
continuous load 连续荷载;连续负载;连续负荷;均匀加载;均匀加荷;分布荷载;分布负载
continuous load condition 连续荷载状态
continuous load current 持续荷载电流
continuous loader 连续连续装载机
continuous loading 连续加感;持续负载;持续负荷
continuous loading consolidation test 连续加荷固结试验
continuous loading test 连续加荷试验
continuous logic 连续逻辑
continuous longitudinal wave 连续纵波
continuous loop 连续环
continuously adjustable inductor 平滑调整电感线圈
continuously adjustable transformer 可变比变压器
continuously attended station 不断有人值勤的站
continuously differentiable function 连续可微分函数
continuously discharge 连续排出
continuously flow 连续排出量
continuously graded aggregate 连续级配集料;连续级配骨料
continuously impregnated compound 连续浸渍复合料
continuously integrated flow meter 连续累计流量计
continuously loaded cable 连续加感电缆
continuously loaded oedometer 连续加荷固结仪
continuously measuring control system 连续测量控制系统
continuously mechanical gravity dewatering 连续机械重力脱水
continuously modeling program(me) 连续式模拟模型程序
continuously moving hearth 连续移动炉床
continuously on-line instrument 连续在线仪表
continuously predicted data 连续预测数据
continuously queued station 连续排队站
continuously rated 按持续荷载计算的
continuously rated motor 连续额定运行电动机
continuously recharge 连续回充水
continuously recording sensor 连续记录的传感器
continuously reinforced 连续加固的;连续配筋的
continuously reinforced concrete 连续钢筋混凝土

continuously reinforced concrete carriageway 连续钢筋混凝土车行道
continuously reinforced concrete pavement 连续配筋混凝土路面
continuously reinforced concrete paving 连续配筋混凝土(路面)板
continuously reinforced slab 连续配筋混凝土(路面)板;连续钢筋混凝土板
continuously reinforcement(bar) 连续钢筋
continuously route 连续线路
continuously running 连续运转
continuously running duty 连续运行工作制度;连续运行方式;连续运行方法;连续工作方式
continuously running load 连续运行负荷
continuously running oscillation 连续振荡
continuously sensing control system 连续感受控制系统
continuously setting 连续沉降
continuously stirred sewage treatment reactor 连续搅拌式污水处理反应器
continuously stirred tank reactor 连续搅拌槽反应器
continuously variable control 均匀调整;连续可变调整
continuously variable delay line 连续可调延迟线;匀调延迟线
continuously variable thrust control 连续变推力控制
continuously welded rail 焊接长钢轨
continuously welded switches 无缝道岔
continuously welded turnout 无缝道岔
continuously writing camera 连续扫描摄影机
continuous machine 连续机
continuous(magnetic flux test) method 连续磁粉探伤法
continuous magnetic north reference 跟踪磁北基准
continuous magnetization method 连续磁化法
continuous magnification range 连续放大范围(光学仪器)
continuous manger 连续饲槽
continuous mapping 连续映射;连续测图
continuous mat 连续纤维毡
continuous maximum rating 最大连续功率;持续最大功率
continuous mean 连续平均值
continuous measurement 连续测量
continuous measuring ga(u)ge 连续式测厚计
continuous measuring hygrometer 连续测量湿度计
continuous medium 连续媒质;连续介质;弥散介质
continuous medium theory 连续介质理论
continuous melting 连续熔化
continuous melting method 连续熔融法
continuous melting process for optic(al) glass 光学玻璃连续熔制法
continuous membrane filtration 连续膜滤
continuous membranes 连续多层铺沥青和油毡;多毡多油防水层
continuous message 连续消息
continuous message source 连续信息源
continuous metal cast process 连铸法;连续铸法
continuous method 连续法
continuous microfiltration 连续微滤
continuous microfiltration membrane separation technology 连续微滤膜分离技术
continuous microfiltration membrane technology 连续微滤膜技术
continuous microfiltration technology 连续微滤技术
continuous microtubule 连续微管
continuous mill 连续研磨机
continuous milling 连续铣削
continuous miner 联合采煤机;联合采矿机;连续采煤机;采煤联合机
continuous ming 连续开采
continuous ming technology 连续开采工艺
continuous mining 连续采煤法
continuous mining machine 联合掘进机
continuous mixer 连续(式)搅拌机;连续式拌和机;连续混料器;连续混炼机;连续混合机;连续拌和机
continuous mixing 连续混合
continuous mixing plant 连续式搅拌设备;连续式

拌和设备;连续沥青混合料拌和设备
continuous mix plant 连续拌和设备;连续拌和厂
continuous mode 连续方式
continuous model 连续模型
continuous moderate rain 连续性中雨
continuous monitoring 连续监听;连续监视;连续监控;连续监测
continuous monitoring technique 连续监控技术
continuous monotonic increasing function 连续单调增量函数
continuous mortar mixer 连续式灰浆搅拌机
continuous motion 连续运动
continuous mo(u)ld-pouring machine 连续灌浆机
continuous mounting of rail 钢轨连续支承
continuous movement 连续运动;连续平滑运动
continuous multistage bleaching 连续多级漂白法
continuous multistage crystallizer 连续多级结晶器
continuous negative pressure suction 持续负压吸引
continuous(ness)law 连续性定律
continuous net 全面网
continuous noise 连续(性)噪声
continuous numerical method 连续数字法
continuous observation 连续观测
continuous oil flow 连续油流
continuous oiling 连续加油法
continuous on-stream monitoring of water quality 连续运转的水质监测
continuous operating 连续作业
continuous operation 连续作业(法);连续运转;连续运行;连续运算;连续行程;连续工作;连续操作
continuous operational process 连续操作过程
continuous operation system 流水作业系统;连续控制系统;连续操作系统
continuous operator 连续算子
continuous optimization program(me) 连续优化程序
continuous optimum 连续最优解
continuous ornament 连续式装饰;连续花饰
continuous oscillation 恒幅度摆动;连续振荡;等幅振荡
continuous oscillation magnetron 等幅振荡磁控管
continuous outcome variable 连续结果变量
continuous output 连续出量;持续输出;持续出力
continuous oven 连续炉
continuous overhead sprinkling 连续空中喷灌;进行空中喷灌
continuous oxidation 连续氧化
continuous oxidation assimilation 连续氧化同化(作用)
continuous pad-dyeing range 连续轧染机
continuous pallet-type sintering machine 连续台车式烧结机;带式烧结机
continuous panel between floors 楼板的连续嵌板
continuous paper 卷筒纸
continuous paper processor 连续像纸晒象机
continuous parabolic curve 连续抛物曲线
continuous parameter Markov chain 连续参数马尔可夫链
continuous parameter process 连续参数过程
continuous parameter time series 连续参数时间序列
continuous passing furnace 直通式炉
continuous path 连续通路
continuous path system 连续路线系统
continuous penetrating sounding machine 连续贯入触探机
continuous performance test 连续作业测验
continuous perfusion test 连续灌注试验
continuous permafrost 成片永冻层
continuous permafrost zone 连续永冻土带;连续多年冻土区;连续多年冻土带
continuous pesticide simulation model 连续农药模拟模型
continuous phase 连续相
continuous phase culture 连续周相培养
continuous phase frequency shift keying 连续相位频移键控
continuous photocathode 连续光电阴极
continuous photograph 连续(航) 像片
continuous photography 连续摄影
continuous physical inventory 连续实地盘存

continuous pickling 连续式酸洗
continuous piled wall 连续桩柱成墙
continuous pilot 延续式点火器;连续导向器;持续式指向灯
continuous pipe 连续管
continuous pipe without joint 无接头管道;无缝管道
continuous placing 连续浇筑
continuous plankton recorder 连续浮游生物自记仪;浮游生物连续采样器
continuous plant 连续式设备;连续操作装置
continuous plate 连续板
continuous plate sheet drawing process 连续拉引法(平板玻璃)
continuous point 连续点;相邻点
continuous point source 连续点(污染)源
continuous polisher 连续抛光机
continuous pollutant source 连续污染物源
continuous pollution source 连续污染源
continuous polycrystalline oxide fibre 多晶氧化物长丝
continuous polymerization 连续聚合(法)
continuous population 连续种群;连续群体;连续总体
continuous potentiometer 滑动触点电位器
continuous pour(ing) 连续浇筑;连续浇注;连续浇灌
continuous pour process 连续倾注法
continuous powder sample 连续粉末取样
continuous power 连续(运行)功率;连续发电;持续生产力
continuous power spectrum 连续功率谱
continuous prebaked anode cell 连续预焙阳极
continuous precipitation 连续析出;连续沉淀
continuous pressure filter 连续式压滤机;连续加压过滤器
continuous pressure operation 连续压按操作
continuous prestressed beam 连续预应力梁
continuous prestressed concrete frame 连续预应力混凝土框架
continuous prestressed concrete girder 连续预应力混凝土大梁
continuous printer 连续式打印机
continuous process 流水作业;连续制造程序;连续过程;连续法;连续成型工艺
continuous process for panels 连续制板工艺
continuous process industry 连续程序工业
continuous processing 连续处理
continuous processing line 连续作业线
continuous processing machine 连续冲洗机
continuous process method 连续步序法
continuous processor 连续加工配料机
continuous process plant 流水作业设备
continuous process system 连续生产系统
continuous production 流水生产;连续(性)生产
continuous production operation 连续生产作业;流水作业
continuous production operation sheet 流水作业图表
continuous production process 连续性作业
continuous production system 连续式生产体制
continuous profile 连续剖面
continuous profiling 连续剖面法
continuous program(me) 连续程序
continuous programming code 连续编码
continuous progression code 连续级数码
continuous proportioning 连续配料
continuous proportioning plant 连续式配料设备;连续配料车间
continuous pugmill 连续搅拌机
continuous pug mill mixer 连续搅拌机
continuous pulverizer 连续破碎机;连续粉磨机
continuous pumping test 连续泵送试验
continuous purlin(e) 连续檩条;连续檩
continuous pusher-type furnace 隧道窑;连续推进式加热炉;隧洞式炉;连续送料式炉
continuous quench and temper line 淬火回火连续作业线
continuous quick light 连续快闪光
continuous radiation 连续(光谱)辐射
continuous radiation scattering 连续辐射散射
continuous radioactivity measurement 放射性连续测量
continuous radio beacon 连续无线电信标

continuous radio time signal 连续无线电时号
continuous raft 连续筏基础
continuous rail frog 连续钢轨辙岔
continuous rain 连续雨;偏雨
continuous rainfall 连续降雨(量)
continuous random noise 连续无规噪声
continuous random process 连续随机过程
continuous random variable 连续(的)随机变量
continuous range 连续范围
continuous range audiometer 连续声ücklistening听力计
continuous rating 连续(运用)定额;连续运行功率;连续(运行)额定值;连续工作额定值;固定负载状态;连续(运转)额定值;持续运转额定容量;持续运转标称值;持续功率;长时运转额定值;长期运转额定值
continuous rating current 持续额定电流
continuous rating permitting overload 允许过载的持续额定(功率)值
continuous rating power 连续功率
continuous reaction 连续反应
continuous reaction series 连续反应系(列)
continuous reading 连续读数
continuous reading refractometer 连续读数折射计
continuous readout 连续读出
continuous ready electro-titration apparatus 连续读数式电滴定仪
continuous receive 连续接收
continuous recharge 连续回灌
continuous record 连续记录
continuous recorder 连续记录器
continuous recording curve of single component magnetic survey in borehole 井中单分量磁测连续记录曲线
continuous recording electronic manometer 连续记录电子测压计
continuous recording oscilloscope camera 连续记录示波器照相机
continuous rectification 连续精馏
continuous recycling 连续再循环
continuous refining process 连续精炼法
continuous refuse-burial machine 连续垃圾掩埋机;连续埋埋垃圾机
continuous regenerator 连通式蓄热室
continuous registration 连续登记
continuous regulation 不分档调节
continuous regulator 连续作用调节器
continuous reheating furnace 连续式加热炉
continuous reinforced 连续加固的
continuous reinforced concrete 连续配筋混凝土
continuous reinforced concrete beam 钢筋混凝土连续梁
continuous reinforced concrete girder 钢筋混凝土连续梁
continuous reinforced pavement 连续配筋路面
continuous reinforced slab 连续钢筋混凝土板
continuous reinforcement 连续强化
continuous repeater mill 连续式活套轧机
continuous revetment 连续式护岸;平顺护岸
continuous review 连续检查
continuous revision 连续修测
continuous rheostat 平滑调节变阻器
continuous rhythm 连续韵律
continuous ribbon 连续带饰
continuous rigid frame 连续刚性(构)架;连续刚架
continuous rigid frame bridge 连续刚架桥
continuous ring discharge 连续环形放电
continuous river 常年河流;常流河
continuous rock mass 连续岩体
continuous rod mill 连续式线材轧机
continuous rod reinforcement 连续钢筋配筋
continuous roller-hearth 连续式辊动炉底
continuous roller-hearth furnace 连续辊底式炉
continuous roll flooring 地板卷材
continuous roll-forming machine 连续式辊弯形机;连续式辊弯成形机
continuous rolling 连续轧制
continuous rolling process 连续压延法
continuous roll type filter 连续转动式过滤器
continuous roof-light 连续式屋顶采光
continuous rope 无极绳
continuous rope drive 连续绳索传动;无极绳传动
continuous rope drive system 连续绳轮传动制
continuous rope system 连续绳制
continuous rotary clarifier 连续回转净化器;连续

回转澄清池
continuous rotary filter 连续旋转过滤器;连续旋转过滤机;连续式回转过滤机
continuous rotation method 连续转动法(求稳定性的转动导数用)
continuous rotory filter 连续回转式滤机
continuous-row 长排的
continuous row fluorescent fixture 荧光灯带
continuous row lighting fixtures 连续式成排照明装置;连续式成排采光装置
continuous run 继续工作
continuous running 连续操作;连续生产;连续运转
continuous running duty 连续工作制度
continuous-running voltage 持续运行电压
continuous sample 连续样品
continuous sample method 连续取样法
continuous sampler 连续取样器;连续采样器
continuous sampler monitor 连续采样监测器
continuous sampling 连续取样;连续采样
continuous sampling method 连续采样法
continuous sampling of cores 连续取芯样
continuous sampling plan 连续抽样方案
continuous sand blast apparatus 连续喷砂装置
continuous sand-wash filter 连续洗砂过滤池
continuous sash 连续窗框
continuous sash window 连续框架窗;连续框格窗
continuous scan 连续扫描
continuous scanner 连续调分色机
continuous screw-type mixer 螺旋式连续拌和机
continuous seafloor sediment sampler 海底沉淀物连续取样器
continuous seamless tube rolling mill 连续式无缝管轧机
continuous sea protection 连续式护岸
continuous section 连续剖面
continuous seetp gradient 连续陡坡
continuous seismic profiler 连续地震剖面仪;海底浅地层剖面仪
continuous seismic profiling 连续地震剖面图;连续地震剖面法
continuous seismic source 连续震源
continuous seismic zone 连续地震带
continuous selfbaking electrode 连续自动烧成电极
continuous sensitivity control 灵敏度连续调整
continuous series 连续岩系;连续系列;连续数列;串联
continuous service 连续作业;连续运行;连续工作;连续服务
continuous service hydraulic power plant 持续运转水力发电厂
continuous servo 连续随动系统
continuous settlement 连续沉降量
continuous shaft kiln 连续(作业)竖窑
continuous sheet(ing) 长条片材
continuous sheet making 连续制板
continuous sheet micrometer 连续式钢板测厚计
continuous shell 连续薄壳
continuous shrinking machine 连续预缩机
continuous signal 连续信号
continuous silicon solar cell 连续硅太阳电池
continuous simulation 连续模拟;连续仿真
continuous single fillet welding 单面连续焊
continuous single pass welding 单面连续焊
continuous sintering 连续烧结
continuous siphon 连续虹吸
continuous slab 连续(楼)板
continuous slab deck 连续板面板
continuous slab-heating furnace 板坯连续加热炉
continuous slagging cupola 连续出渣冲天炉
continuous slagging spout 连续开炉
continuous sliding sash 连续(推)拉窗
continuous slope 连续斜坡
continuous slowing-down approximation 连续减速近似法
continuous sludge 连续沉淀物;液体中矿物颗粒连续沉淀物
continuous sludge removal 连续除泥
continuous sludge-removal tank 连续排泥沉淀器;连续排泥沉淀池
continuous sludge thickener 连续污泥浓缩池
continuous slurry wall 地下连续墙
continuous smelter 连续式熔窑
continuous smelting 连续熔炼
continuous snow cover(age) 连续积雪层;积雪层

continuous solid 连续光楔;连续固体
continuous solid concentration 连续固体浓度
continuous solid solution 连续固熔体
continuous solid-solution series 连续固溶体系列
continuous solution 连续解
continuous source 连续信源
continuous space 连续空间
continuous span 连续跨(度);连跨
continuous-span construction 匀连搭
continuous spandrel 连续式窗下墙;连续式路肩;连续拱肩
continuous spandrel walls 连续拱肩墙
continuous span structure 连续桥跨结构
continuous spectrum 连续谱(图);连续频谱;连续光谱
continuous speed 持续速度
continuous stacking method 连续堆料法
continuous start-stop operation 连续启停操作
continuous state material 连续相材料
continuous statement 连续语句
continuous static thrust 连续静正推入方式
continuous stationary reader 连续稳定阅读机;连续固定阅读器
continuous stationery 连续打印纸
continuous stationery reader 连续台式阅读器
continuous stations 相邻站
continuous stave pipe 连续引水管;连续桶板管
continuous steel sheet piling 连续钢板桩
continuous steel truss 连续钢桁架
continuous steel truss bridge 连续钢桁架桥
continuous steep grade 连续陡坡;持续陡坡
continuous sterilization 连续灭菌;持续灭菌法
continuous still 连续蒸馏器
continuous still battery 连续蒸馏
continuous stirred tank 连续流搅拌池;连续搅拌槽
continuous stitch welding 连接点焊
continuous stochastic process 连续随机过程
continuous stocking 连续放牧
continuous stone footing 条石连续底脚
continuous storm 连续暴雨
continuous stormwater pollution simulation system 连续暴雨水污染模拟系统
continuous strand 连续绳束;连续河岸
continuous strand annealing (线材的)多根连续退火
continuous strand mat 连续原丝薄毡;连续原丝毡
continuous strand sinter machine 带式烧结机
continuous strand-type furnace 单条薄带退火用的连续热处理炉
continuous streak 刀线
continuous stream 连续流;常年河(流);常流河
continuous streaming 连续流(动)
continuous stress 连续应力;持续应力
continuous string(er) 连续纵材;连续(的)楼梯梁;楼梯间连续外斜梁;(弯曲楼梯的)连续斜梁;连续纵梁
continuous strip 连续航片;连续板条;长条形的
continuous strip camera 连续(条带)摄影机
continuous strip galvanizing 带材的连续镀锌
continuous strip lacquering 带材的连续涂漆
continuous strip photograph 连续条形航摄片
continuous strip photography 连续航带摄影
continuous strip printer 连续航带印像机
continuous strip window 连续板条窗
continuous stroke measuring device 连续行程测量器
continuous suction unit 连续吸引器
continuous superstructure 连续式上层建筑
continuous supervision of trains 列车持续性监督
continuous support 连续支座;连续支承
continuous support type 连续支撑形式
continuous surface curve 连续水面曲线
continuous survey 连续检验
continuous survey system 连续检查制
continuous suspended ceiling 连续悬挂式天棚;连续悬挂式平顶
continuous suspended sediment record 悬移质泥沙连续测验记录
continuous suspension girder 连续悬梁
continuous suspension system 连续悬挂系统
continuous sweep 连续扫描
continuous syringe 连续注射器
continuous system 连续制;连续体系;连续(生产)系统

continuous system of physical inventory 连续实地盘存制度;永续实地盘存制度
continuous system simulation 连续系统仿真
continuous tank 连续式池窑
continuous taper tube 锥形管
continuous tapping 连续出铁
continuous tapping spout 连续出铁槽
continuous terrace 连续阶地
continuous test 连续试验
continuous T-frame bridge 连续T形结构桥;连续T形刚构桥
continuous thickener 连续浓缩装置
continuous thread 连续螺纹
continuous thunder and lightning 连续雷电
continuous tidal energy production 潮汐能连续生产
continuous time 连续(工作)时间
continuous time dynamic(al) system 连续时间动态系统
continuous time estimation 连续时间估计
continuous time Markov chain 连续时间马尔可夫链
continuous time Markovian motion 连续时间马尔可夫运动
continuous time multiple branching process 连续时间分支过程
continuous time-rated motor 连续运行电动机
continuous time sequence 连续时间序列
continuous time sequential decision process 连续时间的序贯决策过程
continuous time series 连续时间序列;连续时间系列
continuous time signal system 连续系统
continuous timing 连续测时法
continuous titration 连续滴定
continuous titrator 连续滴定仪
continuous tone 连续影调;连续(色)调
continuous tone controlled squelch system 连续单音控制静噪制
continuous tone copy 连续色调原稿;连续色调复制
continuous tone gray scale 连续色调灰度光楔
continuous tone image 连续(色)调图像
continuous tone object 连续色调物体
continuous tone original 连续色调原图;连续色调原版;全色原版
continuous tone photograph 连续色调像片
continuous tone photography 连续色调摄影
continuous tone printing 连续色调复印
continuous tone reproduction 连续色调复印
continuous tone system 连续单音制
continuous toolbar cultivator 整体横梁式中耕机
continuous tooth double-helical gear 连续齿人字齿轮
continuous trace camera 连续扫描摄影机
continuous track circuit 连续式轨道电路
continuous track shifter 连续移轨机;连续搬道机
continuous traction 持续牵引
continuous tractive effort 持续牵引力
continuous traffic 连续交通
continuous traffic wear 连续的车辆磨损
continuous train control system 列车持续控制系统
continuous train mill 连续式轧机
continuous transformation 连续函数
continuous transmitter 连续传真发送机
continuous transportation 连续运输
continuous treatment 连续处理
continuous trend 连续性趋向
continuous triangulation 连续三角测量
continuous triggered recording 连续触发记录
continuous truss 连续桁架
continuous truss bridge 连续桁架桥
continuous trussed girder 连续桁架式大梁
continuous tube process 连续管式吹塑法
continuous tuning 连续调谐
continuous tunnel furnace 隧道炉
continuous tunnel(l)ing machine 隧道连续施工机械
continuous twin-ribbon mixing conveyer 连续双螺条混合输送机
continuous twin-rotor paddle mixer 连续双转子桨式搅拌机
continuous two-stage twin-roller mixer 连续双辊混合机
continuous two-way slab 连续双向板

continuous type cab signal(l)ing 连续式机车信号
continuous type furnace 连续式加热炉
continuous ultraquick light 超快连闪光;连续超快闪光
continuous underpinning 连续托换(基础)
continuous univariate distribution 连续单变量分布
continuous unloader 连续卸货机
continuous vacuum drum filter 连续式回转真空过滤机
continuous vacuum filter 连续真空过滤器;连续真空过滤机
continuous value 连续值
continuous variable 连续变数;连续变量
continuous variable optic(al) attenuator 连续可变光衰减器
continuous variable programming 连续型规划;连续变量规划
continuous variable slope delta modulation 连续可变斜率增量调制
continuous variate 连续随机变量
continuous variation 连续变异;连续变化
continuous variation method 连续浓度变更法
continuous vector function 连续矢量函数
continuous velocity 持续速度
continuous velocity log(ging) 连续速度测井
continuous vent 连续排气管;连续通气管;连续风;通天通气管
continuous ventilating 连续通风
continuous ventilation 连续通风
continuous ventilator 连续通风器
continuous versus intermittent models 连续生产与间断生产模型的比较
continuous vertical drier 立式连续干燥机
continuous vertical drying stove 立式连续烘(干)炉
continuous vertical retort 连续直立式碳化炉
continuous very quick light 连续甚快闪光
continuous viaduct 连续高架桥
continuous viscosimetry 连续黏度测量法
continuous-vision window 通长窗
continuous voltage rise test 持续升压试验
continuous wall 连续墙
continuous wall drill 连续镐钻机
continuous wall drilling 连续墙钻进
continuous wall footing 连续墙基础
continuous waste 连续污水;公共污水;多源污水;连排污水
continuous waste and vent 连续式污水通风管;连续式污水及排臭气管
continuous water-level recorder 连续水位记录仪
continuous wave 连续波;等幅波
continuous wave generator 等幅波振荡器
continuous wave interference 连续波干扰
continuous wave jammer 连续波干扰器
continuous wave laser 连续波激光器
continuous wave laser action 连续波激光作用
continuous wave modulation 连续波调制
continuous wave nuclear magnetic resonance spectrometer 连续核磁共振谱仪
continuous wave optic(al) ranging system 连续波光学测距系统
continuous wave oscillator 连续波振荡器
continuous wave plasma arc pumping 等离子体电弧连续波泵浦
continuous wave radar 连续波雷达;等幅波雷达
continuous wave signal generator 等幅波信号发生器
continuous waves unmodulated 未调制等幅波
continuous wave telegraphy 等幅波电报
continuous wave tracking system 连续波跟踪系统
continuous wave transmitter 等幅波发射机
continuous wave tube 连续波振荡管
continuous wear 连续磨损
continuous weather maps 连续天气图
continuous weigher 连续称重器
continuous weighing 连续称重;连续称量
continuous weighing feeder 连续式定量供料器;连续计量喂料器;连续计量给料器;连续计量给料机
continuous weighing machine 连续称量机
continuous weld 连续焊(接);连续焊缝
continuous welded rail 连续焊接(钢)轨;无缝线路
continuous welded rail track 无缝线路
continuous welded rail track on bridge 桥上无缝线路
continuous welding 连续焊(接)
continuous welding joint 连续焊缝
continuous weld process 连续式炉焊管法
continuous whipstock 连续偏斜器【道】
continuous white noise 连续白噪声
continuous winding 连续绕组
continuous winding machine for pipe 连续缠绕制管机
continuous window 连续框架窗
continuous window spandrels 连续的窗下墙
continuous wire-belt eelvator 循环网带提升机
continuous wire mill 连续式线材轧机
continuous work 连续工作
continuous working 连续运转
continuous working period 持续工作时间
continuous working time 持续运行时间
continuous yarn 长丝
continuous zinc coating 连续镀锌
continuous zone refiner 连续区域精炼炉
continuous zone-refining apparatus 连续区域精炼设备
continuum 连续性;连续系统;连续介质;连续光谱;连续统【数】;连续体;连续区域;连续流;闭联集;群体连续体
continuum approach 连续法
continuum burst 连续谱爆发
continuum emission 连续谱发射
continuum energy spectrum 连续能谱
continuum fringe visibility 连续谱源条纹可见度
continuum mechanics 连续介质力学;连续体力学
continuum medium 连续介质
continuum medium approach 连续介质法
continuum medium mechanics 连续介质力学
continuum of states 连续态;连续的状态
continuum receiver 连续谱接收机
continuum source 连续谱源
continuum spectrum lamp 连续光谱灯
continuum theory 连续介质理论
contonier 法国梧桐
contorograph 表面图示仪
contort 扭弯;扭歪
contorted bed 扭曲层;褶皱层
contorted bedding 扭曲层理
contorted fold 扭曲褶皱
contorted mode 扭曲状水系模式
contortion 扭歪
contortion crack 扭曲裂缝
contortion fault 扭曲断层
contortion fissure 扭曲裂缝
contortion fold 扭曲褶皱
contour 等高线;弧面;周线;轮廓;外形
contour accentuation 加重轮廓;回路加重;勾边;提高(图像中物体)轮廓的明显性
contour account 抵消账户
contour acuity 轮廓锐度
contour analysis 轮廓分析;等值线分析;外形分析
contour area 等高线面积
contour bank 踏步式护堤;踏步式护岸;等高埂;等高堤
contour basin 等高洼地;等高(水)池
contour bench border-irrigation 等高畦灌;等高(田埂)灌溉
contour bender 成型弯曲装置
contour blasting 轮廓爆破;光面爆破
contour broach 外形拉削
contour cam 靠模;导形凸轮
contour chair 体型椅;人体体型椅
contour chart 等值线图
contour check 沿等高线种植;沿等高线栽培;等高田;等高畦;梯田
contour code 等高线电码
contour compilation map 等高线地图
contour control 轮廓控制
contour control system 外形控制系统
contour convergence 轮廓重合
contour correction 边界影响修正
contour cropping 等高种植;等高栽植
contour cultivation 等高种植;等高耕种;水平种植
contour current 等深流
contour current deposit 等深流沉积
contour current facies 等深流沉积相
contour curtain 可变换轮廓的幕帘
contour curve 等值曲线
contour cutter grinder 成型铣刀磨床
contour cutting 外形切削
contour diagram 等值线图;等密(度)图;等高线图
contour digitizing 等高线数字化
contour dimension 外形尺寸;外廓尺寸
contour ditch 等高沟
contour drafting 描绘等高线;等高线绘制
contour drainage 沿等高线修筑的雨水排除系统
contour drawing 轮廓图;描绘等高线;白描
contoured 波状外形的
contoured bathymetric(al) chart 等深线海(底地形)图
contoured chaplet 成型芯撑
contoured chill 成型冷铁
contoured container 特型集装箱
contoured densener 成型冷铁
contoured die 成型模
contoured field 梯田
contoured handle tinman's snips 环柄白铁剪刀
contour edition 等高线版
contoured land 坡地
contoured roll 成型轧辊
contoured roller 波形导辊
contoured sheet 异型板
contoured squeeze board 成型压头
contour effect 轮廓效应
contour-etching 外形腐蚀(加工)
contour extraction 轮廓抽取
contour extrusion 异型挤出
contour fabric 仿形织物
contour facing 靠模端面车削;仿形端面车削
contour farming 等高种植;等高栽植;等高耕作;等高耕种
contour finder 轮廓线检测器;探头取景器
contour flooding 等高沟淹灌
contour follower 靠模随动件;等高仪;仿形随动件
contour-following 跟踪等高线的
contour forging 模压;模锻
contour forming 形模成型
contour furrow 等高沟;田边沟;水平犁沟
contour furrow irrigation 等高沟灌法
contour ga(u)ge 轮廓仪;叶型规;仿形规;板规
contour gradient 等高坡线
contourgraph 等高仪;轮廓仪
contour grinder 仿形磨床
contour grinding machine 仿形磨床
contour hardening 轮廓淬火;表面轮廓热处理
contour hole 周边眼孔;周边炮孔
contour horizon 标准层位
contour-hypsocline generator 描绘等高线程序
contour information 等高线信息
contouring 轮廓;外形;地下测量;轮廓加工;磨球面;绘等高线;画轮廓;恒等线;等值线绘制;等高线绘制;测绘等高线;圈定法;圈定边界;外形修复
contouring accuracy 等高线描绘精度;等高线绘制精度;等高线测量精度
contouring control 轮廓控制;等值线控制
contouring machine 靠模机床;仿形机床
contouring mapping 等高线绘制
contouring method 等高线法
contouring note 等高线注记
contouring number 等高线注记
contouring program(me) 等值线程序
contour instrument 等高线笔
contour integral 围道积分
contour integration 围道积分
contour interpolation 等高线内插
contour interval 恒值间距;相邻等高线的高度差;等值线间距;等高线间隔;等高线距;等高线间距;等高线间隔;等距
contour irrigation 等高灌溉
contourite 等深积岩;等深流沉积物
contour lathe 仿形车床
contour-length method 等高线延长法
contour level 等高线水准测量;等高层
contour line 等高线;等值线;等强线;水平线;等高线
contour line map 等高线图;等高线地图
contourliner 等高线仪
contour machining 轮廓加工;成型加工;仿形加工
contour manuscript 等高线原图
contour map 围道线映射;轮廓图;等高线图;等值线图;等高线地图;等场强线图
contour map of alpha ray intensity 阿尔法强度等值图

contour map of apical bed of confined aquifer 承压水顶板等高线图
contour map of apparent absorption of radio wave penetration method 无线电波透视法视吸收系数等值线图
contour map of bottom bed of confined aquifer 含水层底板等高线图
contour map of bottom surface of coal seam 煤层底板等高线图
contour map of bottom surface of orebody 矿体底面等高线图
contour map of bursting water coefficient 突水系数等值线图
contour map of cross section of gamma 伽马断面等值图
contour map of Eman 埃玛等值图
contour map of fault surface 断层面等高线图
contour map of gamma-ray intensity 伽马强度等值图
contour map of groundwater 地下水等水位线图
contour map of heat releasing light 热释光等值图
contour map of ore body footwall 矿体底板等高线图
contour map of ore body roof 矿体顶板等高线图
contour map of overburden ratio 剥采比等值线图
contour map of polonium content 钋含量等值图
contour map of potassium content 钾含量等值图
contour map of shear(ing)stress 剪应力等值线图
contour map of stripping ratio of orebody 矿体剥离比等值线图
contour map of stripping ratio of ore deposit 矿床剥离比等值线图
contour map of Th/K ratio 钍钾等值图
contour map of thorium content 钍含量等值图
contour map of top surface of coal bed 煤层顶面等高线图
contour map of top surface of coal seam 煤层顶面等高线图
contour map of top surface orebody 矿体顶面等高线图
contour map of track density 径迹密度等值图
contour map of U/K ratio 铀钾等值图
contour map of uranium content 铀含量等值图
contour map of U/Th ratio 铀—钍等值图
contour map of vertical cross-section of radon 氡纵断面等值图
contour map program(me) 等值线图程序
contour method 等值线法
contour microclimate 地形(性)小气候
contour milling 等高走刀曲面仿形法
contour milling machine 靠模铣床
contour mode 等高线图形
contour model 轮廓模型
contour module 轮廓模数
contour number 等高线海拔注记; 等高线标数
contour-number pantograph 等高线注记缩放器
contour of anomalies 等异常线
contour of constant geometric(al) accuracy 等精度曲线
contour of equal perceived noisiness 等噪线
contour of equal travel time 等流时线
contour of groove 轧槽轮廓
contour of noise 噪声等值线
contour of recharge 补给边界
contour of river channel 河槽形态
contour of valley 河谷形态
contour of water table 水位等高线; 水面等高线; 等水位线
contour outline 略图
contour overlay 等高线叠置片
contour package 等高线程序组
contour painting 白描; 白功
contour patch 按炉子侵蚀外形修炉
contour pattern 恒值线图案; 等高线图案
contour paver 轮廓摊铺机; 仿形铺料机
contour peening 喷丸成型
contour pen (画等高线用的)曲线笔
contour plan 地形图; 等高线(平面)图; 等高层
contour plane 等高面; 等值面
contour planting 等高种植
contour plate 缘周板; 靠模等值面; 护模板; 靠模样板; 压型板; 成型吹板; 仿形样板
contour plots 等高线区划图
contour ploughing 等高耕作

contour plowing 等高耕种
contour prism 等角棱镜
contour projector 轮廓投影仪; 断面投影仪
contour quenching 轮廓淬火; 外形淬火
contour race 等高水道
contour recording 等强录音
contour representation 等高线表象
contour ridger 等高起垄机
contour ring 凸轮环
contour roll forming 轮廓滚形
contour rolling mill 轮廓轧机
contour sawing machine 仿形锯床
contour section of quintic-dip longitudinal sounding 五极纵轴测深等断面图
contour segment 等高线部分
contour size of the unit(s) 设备外形尺寸
contour sketching 描绘等高线
contour slices 等高线分层
contours of noise 噪声等值线
contours of water table 地下水位等高线; 水位等高线; 水平等高线; 水面等高线
contour specification 等高线规格
contour spinning 靠模旋压
contour squeeze 成型压头
contour squeeze mo(u)lding machine 成型压头造型机
contour stitcher 仿形缝合机
contour stripping 等高开采; 等高剥离
contour surface 等值面
contour survey 区界调查
contour terrace 等高梯田; 等高阶地; 水平阶地; 水平梯田
contour tillage 等高(线)耕作
contour tolerance 外形公差
contour tracing 轮廓跟踪
contour trench 水平沟
contour turning 轮廓车削; 仿形车削
contour value 等高线注记
contour weld 特形焊接
contour welding 绕焊
contour wheel 仿形轮
contra-account 对销账户
contra-angel 反角
contra-angle handpiece 弯机头; 弯车头
contra-angle porte-polisher 反角握柄磨光器
contra-aperture 对口
contra-arithmetic series 反算术级数
contra-asset account 资产对销账户
contra balance 对消余额; 反余额
contraband 禁运品; 走私货; 非法买卖; 私货
contraband goods 禁运品; 非法买卖品; 违禁品
contraband of import 进口违禁品; 非法进口
contraband of trade 违禁贸易
contra bossing 轴包架整流罩
contraclockwise 逆时针方向的; 逆时针旋转的
contra coater 逆向辊式涂布机
contra credit 贷方对消; 背对背信用证; 反信用
contract 合约; 合同; 承包合同; 承包; 契约; 契据; 聘书
contract accounting 合同说明
contract acquired property 合同所规定的性能
contract administration 合同执行的监理; (建筑师的)承包业务管理; 合同管理
contract agreement 合同协议书; 合同书; 承包工程协议书; 通汇契约
contract a loan 举债
contract amount 合同总额; 承包工程量; 承包额
contract at discretion 任意合同
contract authorization 签订合同权; 合同授权; 订约授权
contract award 合同签订; 发包; 合同裁决(书)
contract award date 合同签订日期
contract awarded 中标合同
contract-awarding procedure 合同签订手续
contract bargaining 契约谈判
contract bond 承包保证书; 合同担保书; 合同保证金
contract-bond requirement 承包合约的要求
contract bonus system 承包奖金制
contract builder 营造厂商
contract by competition 竞争合同
contract by deed 契约形式的合同
contract cancellation 合同取消; 取消合同
contract carrier 合同承运人; 包租的运输工具
contract certificate 立约凭证
contract change analysis 合同更改分析

contract change notice 合同更改通知
contract change notification 合同更改通知
contract change proposal 合同更改建议
contract change request 合同更改(的)请求; 请求更改合同
contract change suggestion 合同更改建议
contract channel 收缩通道
contract clause 合同条款
contract condition 合同条款; 合同条件
contract construction 包工建筑; 承包施工; 承包工程; 发包施工; 发包工程; 包工工程
contract construction by general builders 一般建筑者的承包建筑
contract contingency 承包合同预备费
contract control 合同检查; 合同监督; 建筑施工管理
contract cost 承包工程成本
contract cost work sheet 合同收入及有关成本分列表
contract currency 合同货币
contract curve 契约曲线
contract date 签约日期; 履约期限; 合同期限; 合同(订立)日期; 契约履行期限
contract definition phase 合同确定阶段; 签订合同阶段
contract delay penalty 延误合同罚款
contract deposit paid 已付合同保证金
contract deposit received 已收合同保证金; 已收合同存款
contract deposits paid 已付合同存款
contract deposits received 已收合同保证金
contract different lines of work 分业承包
contract disposal 收缩处理
contract documents 合同书; 合同文件; 承包文件; 承包契约
contract drawing 承包施工图; 发包图纸; 发包图样; 施工图
contract drilling 承包钻探; 承包钻井
contract duty 合同义务
contracted 包办的
contracted a disease 染病
contracted air duct 收敛形进气道
contracted basis 出包方式
contracted calculation 简算法; 简便算法; 简便计算
contracted channel 束窄水道; 束窄的航道; 束狭河道; 收敛形管道
contracted code sonde 发送电码无线电探空仪
contracted cross-section 剖切缩图; 截面缩图; 收缩断面; 缩狭断面
contracted division 简除法
contracted drawing 订合同图纸; 缩图; 缩绘
contracted fairway 束窄的航道
contracted flow 束窄水流; 收缩流
contracted jet 收缩射流; (水或气的)收缩喷注
contracted load 合用负载
contracted notation 简略符号
contracted opening 束水孔
contracted-opening discharge measurement 缩孔流量测定法
contracted opening method 缩孔法
contracted orifice 收缩孔口
contracted quota of wage in every hundred Yuan output value 百元产值工资含量包干
contracted responsibility system of editors 编辑承包责任制
contracted Riemann-Christoffel tensor 缩并的曲率张量
contracted section 缩窄段; 收缩截面; 收缩断面; 收缩段
contracted speed 合同航速
contracted stable region 收缩稳定区
contracted waterway 收缩水道; 束窄水道
contracted weir 缩口堰; 收缩堰
contracted width 收缩宽度
contracted words 缩写词汇
contract effective date 合同生效日期
contract engineer 合同工程师
contract evidenced by a writing 书面证明的合同
contract expired 合同期满
contract expiry 合同期满
contract fee 合同费
contract file 承包文件; 合同案卷
contract financing 契约融资
contract footage rate payment 按进尺包干计价
contract for assembly 来件装配合同

contract for borrowings 借用合同
contract for carriage 运输契约;运输合同
contract for carrying out design and survey works 设计勘察合同
contract for delivery 运货合同
contract for delivery of goods by installment 分期交货的买卖合同;分期付款的买卖合同
contract for goods 订货
contract for higher than the norm 超目标承包
contract for installation work 安装工程合同
contract for job 包工
contract for labo(u)r 劳务合同
contract for labo(u)r and materials 包工包料
contract form 合同格式
contract for outlay 经费包干
contract for processing work 加工承揽合同
contract for projects 承包工程
contract for purchase 购货合同;承买
contract for scientific and technical cooperation 科技协作合同
contract for service 劳务合同;服务合同
contract for supply of labo(u)r 提供劳务合同
contract for the delivery of goods by installments 分批交货的买卖合同
contract for the entire management of 包揽
contract for the expenses of erecting single building 栋号包干
contract for the payment of goods by instalment 分期付款的买卖合同
contract for work 承揽(包工);包工
contract gained 中标合同
contract goods 契约货品
contract horizon 视界缩小
contractibility 压缩性;收缩性
contractibility rate 收缩率
contractible 能收缩的
contractible graph 可缩的图
contractibleness 压缩性;收缩性
contractile 能收缩的
contractile root 收缩根
contractile skin 结合膜
contractile vacuole 收缩泡;伸缩泡
contractility 收缩性
contract implementation 合同执行;合同履行
contract in 保证承担义务
contracting 承包
contracting agency 承包经理处;承包代理人
contracting band brake 外带式制动器
contracting band clutch 缩带离合器
contracting brake 带闸;抱闸;抱闸式制动器;收缩式闸
contracting business 承包业;营造厂
contracting chill 收缩冷铁模
contracting chuck 筒夹;弹簧筒夹
contracting combine 缔约联合企业;建筑联合企业
contracting company 出包公司;承包公司
contracting crack 收缩裂缝
contracting current 收缩水流
contracting dam 束水坝
contracting dike 缩窄堤
contracting duct flow 束狭渠道水流;收敛管道内的流动
contracting earth 地球收缩说
contracting engineering project 承包工程项目
contracting enterprise 承包企业
contracting-expanding nozzle 拉瓦尔喷管;超音速喷管
contracting firm 承包公司;建筑合同
contracting hypothesis 地球收缩说
contracting industry 承包工业
contracting job 包工工作
contracting market 承包市场
contracting model 收缩宇宙模型
contracting nozzle 收缩喷嘴
contracting officer 甲方(工程)代表;甲方合同代表;发包人员;工程合同监督官(代表政府或发包方);缔约官员
contracting of the earth 地球的收缩
contracting packaged deal projects 承包整套工程项目
contracting parties 订约双方当事人;签约各方
contracting party 立约者;合同有关方(面);合同当事人;缔约方;契约当事人
contracting reach 缩窄段;收缩(河)段

contracting riverbed 束窄河床
contracting states 缔约国
contracting theory 地球收缩说
contract in restraint of trade 限制营业合同
contract instruments 合同文件
contract interest 约定利息
contract in writing 书面合同
contraction 简略字;压缩;缩窄;缩小;缩短;缩并;收缩作用;收缩
contractionary effect 紧缩效应
contraction at fracture 断裂处颈缩
contraction beading method 玻璃球收缩法
contraction cavity 缩孔;收缩孔洞
contraction coefficient 收缩系数
contraction crack 收缩破裂;收缩裂缝
contraction cracking 收缩裂缝
contraction distortion 收缩变形
contraction fault 收缩断层
contraction fissure 收缩裂纹;收缩裂缝
contraction fit 冷缩配合
contraction ga(u)ge 缩规;收缩尺
contraction hypothesis 收缩说;收缩假说
contraction in area 面积收缩
contraction in length 长度缩短
contraction joint (混凝土路面收缩开裂的)控制缝;缩缝;收缩节理;收缩接缝;收缩缝;伸缩接合
contraction joint grouting 收缩缝灌浆
contraction limit 缩限;收缩极限
contraction loss 截面收缩损失;缩窄损失;缩口损失;收缩损失
contraction mapping 收缩映像
contraction measures 紧缩措施
contraction of area 面缩;面积收缩
contraction of credit 紧缩信用
contraction of (cross sectional) area 断面收缩(率)
contraction of indicates 指标的短缩
contraction of jet 射流(束)收缩
contraction of mass concrete 大体积混凝土体积收缩
contraction of nozzle 喷嘴束流
contraction of tensor 张量缩并
contraction of visual field 视野缩小
contraction party 契约当事人
contraction percentage 收缩百分率
contraction percentage of area 断面收缩率
contraction phase 收缩相;收缩阶段
contraction pulsation 收缩脉动
contraction ratio (气流的)压缩系数;锥度比;收缩系数;收缩率;收缩比
contraction ratio of insulating gel 绝缘胶收缩率
contraction ring of uterus 班都氏环
contraction rule 铸尺;缩写规则
contraction scale 缩小比例尺
contraction section 收敛部分
contraction shape 收敛形状
contraction strain 收缩应变
contraction stress 收缩应力
contraction theory 冷缩论;收缩说
contraction works 缩窄建筑物;缩窄工程;缩狭工程;束窄工程;束水工程
contract item 合同项目;合同条款;付款项目说明
contract item material list 合同项目材料单
contractive clay 收缩黏土
contractive coefficient 收缩系数
contractive colo(u)r 似缩色
contractive crack 收缩裂缝
contractive criterion 收敛准则
contractive-dilative 收缩膨胀的
contractive domain 收敛域
contractive factor 收敛因子
contractive index 收缩指数;收敛指数
contractive joint 收缩缝
contractive limit 收缩极限
contractive pressure 缩压
contractive principle 收敛原理
contractive soil 收缩性土
contractive strain 收缩应变
contractive stress 收缩应力
contractive test 收缩试验
contractive velocity 收敛速度
contract journal 合同日记账
contract labo(u)r 契约劳工;包工;合同工;契约劳工;限定雇主劳工

contract labo(u)r agreement 包工合同
contract language 合同语言
contract law 合同法
contract leasing of governmental enterprises 出租国有企业合同
contract ledgers 合同分类账
contract letting 合同订立;承包订约;合同奖金;签订合同
contract lien 约定留置权
contract life 合同有效期(限)
contract limit 合同范围
contract lump sum 承包总额
contract mail service 约定邮件运输
contract maintenance 合同维护;承包养护;包修
contract management 承包管理
contract management range 合同管理范围
contract management region 合同管理范围
contract management system 合同管理制
contract manager 合同管理人;承包经理;合同经理
contract market 合同市场;期货市场
contract measurement and payment 合同的检查与支附
contract modification 合同修订
contract negotiation 合同谈判
contract No. 合同号
contract note 买卖合同;合同说明;买卖合同;合同书;契约书
contract note of sales 贸易契约;发售约书
contract number 合同号
contract objective 合同标的
contract of adhesion 同意加入合同
contract of affreightment 海运契约;运输契约;租船契约;租船合同;包运合同
contract of apprenticeship 学徒合同
contract of arbitration 仲裁合同
contract of bill 票据合同
contract of carriage 运输契约;运输合同;运货合同
contract of carriage by sea 海运合同;海上货物运输契约
contract of cooperation 协作合同
contract of cost and expenses 监督开支合同
contract of design 设计合同
contract of employment 劳务合同;聘约
contract of establishment 企业组建合同
contract of guarantee 担保合同;保证合同
contract of guaranty 担保合同;保证合同
contract of hire of labo(u)r and service 雇工合同
contract of indemnity 索赔合同;赔偿合同
contract of insurance 保险契约;保险合同
contract of lease 租赁契约;租赁合同
contract of lease of property 财产租赁合同
contract of life insurance 人身保险合同
contract of loan 借贷合同
contract of loss and compensation 损失补偿合同
contract of marine insurance 海上保险契约
contract of on-carriage 续运合同(集装箱)
contract of payment 支付合同
contract of price 契约价格
contract of privately performed work 私营工程合同
contract of project 工程合同
contract of readjustable 可调价合同
contract of renovation 更新合同
contract of sale of a building 建筑物买卖合同
contract of sales 卖船合同;买卖合同;销售合同;定约销售
contract of service 劳务合同;服务契约
contract of tenancy 租赁契约;租赁合同;租借契约
contract operating responsibility system 承包经营责任制
contract operation 合同操作;承包运转
contractor 立(契)约人;订约人;承包者;承包商;承包人;承包工厂;承包方;包工单位;契约商;收敛部分
contractor'a superintendence 承包商的监督
contractor consolidation of previous investigations 承包商整理已有勘察资料
contractor contingency 承包者预备费;承包商预备费;承包人预备费
contractor furnished and equipped 承包商供应和备份的;承包者供应装备的;承包商供应装备的;承包人供应装备的
contractor furnished equipment 承包者供应的设备;承包商供应的设备;承包人供应的设备

contractor furnished property 承包人所提供的性能;承包者提供的器材;承包商提供的器材;承包人提供的器材
contractor office staff 承包商办公人员
contractor overhead 公司管理费
contractor project 承包工程
contractor rail 施工铁路
contractor's account 承包商账户
contractor's affidavit 承包人承诺声明(书)
contractors affidavit 承包者宣誓书;承包商宣誓书;承包人宣誓书
contractor's agent 承包者代理人;建筑承包商;承包者代表;承包商代表;承包人代表
contractor's all risk insurance 建筑工程综合险
contractor's all risks insurance 建筑工程一切险
contractor's association 承包商协会
contractor's business 包工工程
contractors corporation 承包公司
contractor's default 承包者违约;承包商违约;承包人违约
contractor's drawings 承包者的图纸;承包商的图纸;承包人的图纸
contractor's equipment 承包者设备;承包商设备;承包人设备;包工设备
contractor's estimate 承包估算
contractor's facility 承包者的设施;承包商的设施;承包人的设施
contractor's fee 承包者的酬金;承包商的酬金;承包人的酬金
contractors' furnished equipment 承包人配备的设备;承包者配备的设备;承包商配备的设备
contractors' furnished property 承包人所提供的性能;承包商所提供的性能;承包人所提供的性能
contractor's general responsibilities 承包者一般责任;承包商一般责任;承包人一般责任
contractor's guarantee 承包者的担保;承包商的担保;承包人的担保
contractor's hole 承包者钻井;承包商钻井;承包人钻井
contractor's indirect cost 承包者的间接费用;承包商的间接费用;承包人的间接费用
contractor site office 承包人工地办公室
contractor's liability 承包者的责任;承包商的责任;承包人的责任
contractor's liability insurance 承包人义务保险;承包者责任保险;承包商责任保险;承包人责任保险
contractors name plate 承包者招牌;承包商招牌;承包人招牌
contractor's option 承包者的选择权;承包商的选择权;承包人的选择权
contractor's overhead 承包人的管理杂项开支
contractor's plant 承包者设备;承包商设备;承包人设备;包工设备;施工设备;施工机具
contractor's price 投标价格
contractor's profit 承包人的利益;承包者的利润;承包商的利润;承包人的利润
contractor's proposal 投标书;报价方案
contractor's representative 承包者代表;承包商代表;承包人代表
contractor's risks 承包者的风险;承包商的风险;承包人的风险
contractor's spread 施工设备场地;施工机械场地
contractor's staff 承包者的工作者;承包者的工作人员;承包商的工作商员;承包人的工作者;承包人的工作人员
contractor's total maximum liability 承包人的最高赔偿额
contractor's track 施工临时路线
contractors types 承包者类型;承包商类型;承包人类型
contractor supervisor 承包人监督者;工地监工(员)
contract(or) surveyor 工地监工(员)
contractor track 施工轨道
contractor winch 施工卷扬机
contractor work 承包工程
contractor workshop 工地加工车间
contract overrun 合同超价
contract overrun penalty 合同超期罚款
contract package 一揽子合同;整套合同
contract packaged deal projects 承包整套工程项目
contract particulars 合同项目;合同细节;合同条款

contract payment bond 合同付款保证书
contract penalty 合同罚款
contract performance 合同的执行;执行合同
contract performance bond 合同执行保证书
contract-performed maintenance 承包养护;包修;按合同规定的维修
contract period 合同期限
contract period of validity 合同有效期
contract plan 造船契约图
contract plant 合同工厂
contract price 合同价(格);承包价;发包价格
contract price adjustment 合同价格调整
contract processing service 合同加工服务
contract program(me) 承包方案
contract project 承包工程项目
contract proposal 承包建议书
contract provision 合同条款;合同规定
contract purchasing 期货购买
contract rate 合同运费率;合同运价
contract rate of interest 合同利率;约定利率
contract rate system 契约运费制度
contract records 合同案卷
contract renewal 变更合同
contract rent 约定地租;合同租金
contract report 合同报告
contract requirement 合同要求
contract research 研究合同
contract responsibility 合同责任
contract revenue 承包工程收益;契约收入
contract review 合同评审
contract rig 合同规定设备
contract risk 合同风险
Contract Rules for Construction Services 建筑合同法(规)
contract's agent 承包人代表
contract sale 合同销售
contract section 收缩断面
contract serial number 合同序号
contracts manager 合同管理人
contract specification 施工说明书;发包说明;合同规定;合同说明书;承包文件;合同文件
contract speed 约定速度
contract stevedore 合同装卸工
contract stipulated price 合同规定价格
contract stipulation 合同规定
contract style 承包方式
contract sum 合同总价;承包额
contract supervision 合同检查;合同监督;合同管理
contract surveyor 工地监理工程人员;承包工程监督员
contract system 合同制;发包制;包工制;包干办法
contract system of internal research subjects 课题承包制
contract system of scientific research 科研合同制
contract system on fiscal revenue and expenditures 财政收支包干制
contract tax 合同税;契税
contract technical report 合同技术报告
contract termination 合同终止
contract terms 合同条款;合同条件;契约条款
contract the line of credit 紧缩信贷
contract thrift institution 强制储蓄金融机构
contract tillage 承包耕田
contract time 合同工期;承包工期;合同期间;承包期限
contract time period 合同期限
contractual acknowledgement 契约上的认可
contract(ual) agreement 合同性协议(书);契约协议
contractual arrangement 合同安排
contractual capacity 缔约能力
contractual claims 合同规定的索赔
contractual commitment 合同承诺
contractual damage 合同引起的损害
contractual delivery 合同交货
contractual dividend 合同规定的股利;契约规定股利
contractual duty 契约规定的义务
contractual engagement 合同义务
contractual fines 契约上的罚金
contractual guarantee 合同规定的担保
contractual input 合同项下的投入
contractual interest 明息
contractual interface 合同接口

contractual investment 契约性投资
contractual joint(ad)venture 合作经营;契约式合资企业;契约式合资经营
contractual joint undertaking 契约式合营
contractual language 契约文字
contractual law 契约法
contractual liability 合同规定义务;合同责任
contractual management responsibility system 经营承包责任制;承包经营责任制
contractual obligation 合同规定的责任;契约规定的义务;合同义务
contractual party 合约当事人
contractual payment 合同规定的付款
contractual personnel 合同承办人
contractual practice 合同惯例
contractual procedure handling 缔约程序
contractuail purchase 合同采购
contractual quota 分摊(金)额
contractual regulation 合同条例
contractual reserve 契约规定准备金;契约规定储备金
contractual revenue 契约上的收益
contractual right 合同权利
contractual risk 合同风险
contractual route 契约规定航线
contractual service 订约服务
contractual specification 合同规定
contractual technical requirement 合同技术要求
contractual terms and conditions 合同条款及条件
contractual to investment 对投资的摊款
contractual to the fund 基金的分摊额
contractual transfer 合同转让
contractual usage 合同惯例
contractual value 契约上的价值
contractual venture 合作项目
contract underrun 合同低价
contract unit 合同的一部分;承包单位
contractura 斜面柱(柱径收缩)
contracture 收分
contract value 承包工程价格
contract version 合同文本
contract void 无效合同
contract voidable 可以撤销的合同
contract wage 合同工资
contract warranty 合同担保
contract with competitive bidding and secured by mortgage 招标抵押承包
contract without compensation 无偿合同
contract witness 合同见证人
contract worker 合同工
contract works 发包工程;包工(工程);承包工程;承包项目
contracurrent system 逆屈光系统
contra-cyclic(al) measures 反周期措施
contra debit 借方对销
contradict 相冲突
contradiction 矛盾;永假式;抵触
contradiction of benefit 利益冲突
contradictory 对立
contradictory propositions 矛盾问题
contradistinction 显然相反;区别;对比(鉴定)
contradistinguish 对比区别
contra-doping 反掺杂
contra-entry 对销分录
contrafissure 对裂
contraflail row cleaner 反击甩刀式行间清理器
contraflexure 反(向)弯曲;反向曲线变换点;反挠曲
contraflexure point 反挠点;反弯点
contrafloe lane 逆流车道
contraflow 逆流;回流;暂时电流;对流;反(向)流(动);额外电流;逆向车流;反向电流
contraflow condenser 逆流式冷凝器
contraflow cooler 对流式冷却器
contraflow gravel washer 逆流洗砾(石)机;反冲洗石机;逆流式洗砾机;反流洗砾机
contraflow heat exchanger 逆流式热水交换器
contraflow paddle type mixer 对流桨式搅拌混合机
contraflow regeneration 逆流再生
contraflow washer 逆流洗衣机
contragradience 逆步;反步
contragradient transformation 反步变换
contraguided stern 整流艉
contra guide rudder 导式整流舵
contra-harmonic mean 反调和平均数;反倒数平

均数
contrail 凝结尾流;逆增轨迹;转换轨迹
contrail-formation graph 凝结尾迹形成图
contra-incision 对口切开
contrainjection 逆向喷射;反向喷射
contrainjector 反向喷射器
contrajet 反射流
contralateral 对侧的
contralateral reflex 对侧反射
contra-liability account 负债抵销账户;负债抵销科目
contrapedal curve 逆垂趾曲线
contrapolarization 反极化作用;反极化
contraposed coast 叠置海岸
contraposed coastline 叠置海岸线
contraposed shoreline 叠置海岸线
contraposition 对置
contrapositive 逆反式;对置
contrapositive rule 逆反式规则;倒置规则
contraprop 导叶;反向旋转螺旋桨;同轴反转式螺旋桨
contra-propeller 整流舵;反向旋转螺旋桨;整流(螺旋)推进器
contraption 奇妙装置
contra-rotating 倒转
contra-rotating agitator 对转式搅动器
contra-rotating airscrew 反向旋转螺旋桨
contra-rotating drum 双转式双滚筒
contra-rotating propellers 对转螺旋桨;同轴正反转螺旋桨
contrarotation 反向转动;反(向)旋(转)
contra rudder 整流舵;导叶舵
contrary class 反组
contrary dipping direction 倾向相反
contrary event 对立事件
contrary flexure 反(弯)曲
contrary proposition 逆命题【数】;相反命题;反对命题
contrary sign 相反符号;异号;反号
contrary-turning propeller 对转推进器
contrary wind 逆风
contra solem 反日运行气流
contrast 对照,对立;对比物;对比(度);衬比;差异;反衬(度);反差
contrast amplification 对比放大
contrast between black and white 黑白对比
contrast between light and shade 明暗对比
contrast between solid and void 虚实对比
contrast coefficient 反差系数
contrast colo(u)r 对比色;对照色;反衬色
contrast colo(u)r accent 对比色突出
contrast control 对比度调整;对比度调节;反衬调整;反差控制
contrast decay 对比衰减;反衬度降低;反差衰退;反差衰减
contrast developer 强反差显影剂
contrast difference 衬度差
contraste 互逆风
contrast effect 反比效应;反差效果
contrast-enhanced image 反差增强图像
contrast enhancement 对比度增强;衬度增强;反差增强
contrast enhancement factor 反差增强因数
contrast enhancement layer 反差增强层
contrast factor 对比系数;反衬因子;反差系数
contrast figure of U/Rn anomaly 铀氡异常对比图
contrast filter 反衬滤色片;反差滤色片;反差滤光片;强反差滤光镜
contrast formation 反差形成
contrast function 反差函数
contrast gloss 对比光泽
contrast grade 衬度;反差度;反差等级;反差级别
contrast hue 对比色调;对比色彩
contrast image 反差像;反差图像;强反差图像
contrast improvement 反差加强
contrast in black-white transitions 明暗反差
contrast index 反差指数
contrast index curve 反差指数曲线
contrast index meter 反差指数计
contrasting analysis 对比分析
contrasting colo(u)r 对比色
contrasting pavement 异种路面;异色路面
contrasting signal 反衬度信号
contrast law 对比定律

contrast light compensating 反差光补偿
contrast lighting 衬托照明
contrast map 衬值图
contrast measurement 反差测量
contrast-medium 对比剂
contrast method of between exploration and mining data 探采资料对比法
contrast micrometer 对衬测微计;反差测微计
contrast modulation 反差调制
contrast of brightness 亮度对比
contrast of filter 滤光片的衬度
contrast of fringes 条纹反差
contrast of hue 色相对比
contrast of tone 色调对比
contrast paper 硬性像纸
contrast perceptibility 衬比感受性函数
contrast perception 衬比感觉;反差感觉
contrast photometer 对比光度计;反差光度计;比较式光度计
contrast potentiometer 对比度调整电位计
contrast range 对比度(范围);反差范围
contrast rating 对比度系数
contrast ratio 对比率;对比度(系数);反衬率;反差比
contrast reducing 减小反差;压缩反差
contrast reduction 对比率缩小;反差降低;反差减弱
contrast reduction factor 衬度降低因数
contrast relief 对照性地形
contrast rendering 反差再现
contrast rendering factor 对比显现因数
contrast rendition 反衬度重现;反差再现
contrast rendition factor 对比度再现因数
contrast reversal 反差反转
contrast scale 反差标度
contrast section 对照断面
contrast sensitivity 对比灵敏度;对比感受性;衬比灵敏度;反衬灵敏度;反差灵敏度
contrast sensitivity f eye 眼的反差敏感度
contrast stain 对比染剂
contrast staining 对比染色法
contrast stretch 衬度伸展;反差增强;反差增大;反差伸展
contrast stretched image 反差扩展图像
contrast stretching 对比度扩展;反差扩展
contrast strip 反差条
contrast study 对比研究
contrast test 对比试验
contrast threshold 最小敏感限;阈值反衬;对比阈;衬比阈
contrast transfer function 衬比传递函数;反差转换函数;反差传递函数
contrast transmission 衬比传输;反差传递
contrast transmission factor 衬比传递因数
contrast transmission function 衬比传输函数
contrast transmittance 反差透明比;反差表达能力
contrast type photometer 对比光度计
contrast value 反差值
contrasty 硬调的;反差大的
contrasuggestibility 反暗示性
contra-suggestion 反暗示
contrate 横齿的;横齿
contrate gear 横齿轮;端面齿轮
contrate wheel 冕状齿轮;横齿轮;端面齿轮
contra-turning propeller 整流推进器;同心套轴的两个对转螺旋桨
contra-type stern 整流艉
contravalency 共价
contravalid 反有效
contra valuation account 计价对销账户
contravane 逆向导叶;倒装小齿轮
contravariance 抗变性;反变性;反变
contravariant 抗变量;反变式;反变量
contravariant component of vector 矢量逆变分量
contravariant coordinate 逆变坐标;反变坐标
contravariant functor 反变函子
contravariant index 逆变指标;反变指标
contravariant system 反变系统
contravariant tensor 逆变张量;反变张量;曲率张量
contravariant vector 逆变向量;反变矢量
contravention to treaty 违反契约;违反合约;违反合同
Contra wire 铜镍合金丝
contra-wound 反绕
contra-wound helix circuit 对绕螺线

contra-wound helix coupler 反螺旋线耦合器
contrbution clause 损失分摊条款
contribute 捐助;贡献;摊派
contributed capital 已投股本;捐赠资本;投入资本;投入的资本;实缴股本
contributed surplus 资本公积金
contribute their respective shares of capital 缴足认购资本
contribute the land usage right as parts of its investment 以土地使用权入股
contribute the same amounts of funds and materials 平摊资金和物资
contributing area 产流面积
contributing area of run-off 产流区(域);产流面积
contributing region 影响区;供水区;供给区;补给区;水源区
contributing region of run-off 产流区(域)
contribution 捐献;稿件;负担保险损失;分配;分担额;分担(保险);边际收益
contribution by segment 分部贡献
contribution clause 分担条款
contribution factor 辅助系数
contribution in general average 共同海损分摊
contribution in kind 实物捐献
contribution margin 贡献毛利;边际贡献
contribution margin analysis 贡献毛利分析;边际利润分析;边际贡献分析
contribution margin per composite unit 每个组合单位的贡献毛利
contribution margin ratio 边际收益率;边际利润率;边际贡献率
contribution of property 财产捐献
contribution pool 边际贡献总额
contribution ratio 利量比率(利润销货额比率);边际贡献率
contribution theory 分担理论
contribution theory of taxation 课税的分担理论
contribution to profit 利润收益
contribution value 分摊值
contributive cell 贡献单元
contributor 贡献者;分摊人
contributor fault 货主或受害人自身过失
contributory 负连带偿还责任的人;分担责任;费用分担
contributory cause 连带原因;附带原因
contributory data 辅助数据
contributory dead load 有关恒载;分任恒载
contributory evidence 辅助数据
contributory factor 影响因素;辅助系数
contributory interest 分摊利益
contributory mortgage 分担抵押
contributory negligence 共同过失
contributory pension 已捐赠养老金;分摊养老金
contributory population 适度的人口;分担费用人口
contributory reinsurance 分摊再保险
contributory street 广场出入街道
contributory value 分摊价值
contributory zone 供水区
contributre the social working day 提供社会劳动日
contrivance 发明;设计比较估算
contrived scarcity 人为的稀缺性
contriver 发明家;设计者
contrlled avalanche rectifier 可控雪崩整流器
control 控制;监督;管制;对照物;操纵;防治;统制;调节;施业控制
control ability 控制能力;操纵性
control abstraction 控制抽象化
control accelerometer 控制加速表
control access 控制通路;进口控制
control account 控制账户;总账
control accuracy 控制准确度;控制精(确)度
control action 控制作用;控制行动;控制动作
control actuator 舵机
control agent 控制媒介;控制剂;调节体
control air valve 控制空气阀
control algorithm 控制算法
control amplifier 控制放大器;调整放大器
control analysis 控制分析
control and communication engineering 控制与通信工程
control and data acquisition system 控制和数据采集系统
control and display unit 控制与显示装置

control and management of pollution of water body 水体污染源控制和管理
control and management of water body pollution 水体污染源控制和管理
control and measureing unit 测控装置
control and monitor console 监控台
control and monitoring software 控制监测软件
control and prevention of water pollution 水污染防治
control and primary scavenging pump 控制泵和主动回油泵
control and test 控制和测试
control and timing unit 控制与计时装置
control animal 对照动物
control antenna 控制天线;监听天线
control apparatus 调整器;调节装置
control application 控制应用
control area 控制区;控制面积;控制范围;控制地区;禁区;对照区
control area of observation net 观测网控制面积
control area split 控制区分割
control arm 驾驶杆;操纵杆;操纵臂
control assembly 控制组件
control automation system 控制自动化系统
control azimuth 控制方位
control ball 控制球
control band 控制区域;控制波段
control bar 控制杆
control base 控制面;控制基线;检核基线
control behavior 控制状态
control bench board 控制斜台
control biofilter 控制生物滤池
control bit 控制位
control block 控制块;控制程序块;控制部件;消力墩
control block event 控制封锁事件
control board 控制(仪表)板;控制屏;控制盘;控制牌;操纵台;操纵盘;配电盘;配电板(控制盘)
control board for cooling unit 油冷装置操作盘
control board of shield 盾构操纵台
control boc for cooling water pump 冷却水泵控制箱
control booster 操纵系统助力器
control box 控制箱;控制台;控制柜;交通控制箱;操作箱;操纵台
control box for air conditioner 空调机控制箱
control box for chilled water pump 冷冻水泵控制箱
control box for concourse air return/exhaust fan 站厅回排风机控制箱
control box for cooling tower 冷却塔控制箱
control box for excalator 扶梯控制箱
control box for fan 风机控制箱
control box for fresh air fan 全新风机控制箱
control box for impulse fan 推力风机控制箱
control box for platform air return/axhaust fan 站台层回排风机控制箱
control box for small system fan 小系统风机控制箱
control box for tunnel fan 隧道风机控制箱
control box waste water pump 废水泵控制箱
control braking 有控制的制动
control break 控制中止;控制中断;控制改变
control break level 控制改变级
control budget 投资限额
control-budget variance 控制预算差异
control building 管制室;控制楼;调度室
control buoy 可控浮标
control bus 控制总线
control bus comparator 控制总线比较器
control button 控制钮;控制按钮;操纵按钮;调整按钮
control by drainage 用排水控制
control by exception 例外控制
control cab 操纵室;操纵控制室
control cabin 控制室;驾驶舱;操纵室
control cabinet 操纵柜;操纵板;控制柜
control cable 操作索;控制索;控制电缆;操纵索
control cable duct 控制电缆管道
control cable gallery 控制电缆廊道
control cable stop 操纵缆绳制动销
control cable to gun 接焊枪控制电缆
control cable tunnel 控制电缆地道;控制电缆隧洞;控制电缆隧道
control cam 控制凸轮

control camera 控制摄影机
control capability 控制能力
control capacitance 控制电容
control capstan 操纵绞盘
control card 控制卡片
control card specification 控制卡片说明;控制卡片规格
control card stack 控制卡片箱
control carrier transmission 控制载波传输
control casing 控制箱
control casing head 套管控制头;套管头
control catchment 控制流域;对照流域
control cathode 控制阴极
control cell 控制单元
control centre 控制中心;调度中心;控制台;指挥中心;指挥所;调度所;操纵室
control center of foreign exchange 外汇管理中心
control center of specific disease 特种病防治中心
control chain 控制链
control change 控制改变;控制变换
control channel 控制信道;控制通道
control character 控制字符【计】;控制符(号);操作符
control characteristic 控制特性
control chart 检查图表;监督进度表;控制图(表);检验图表;质量评估图;管理图(表)
control chart for analysing data 解析质量评估图
control chart for data analysis 用控制图分析资料
control chart method 控制图法;检验图法;管理框图法;按流程法
control chemical analysis 控制化学分析
control circuit 控制电路;控制回路
control circuit for magnetic starter 磁力起动器的控制电路
control circuitry 控制电路系统;控制电路图
control clerk 数据控制员
control clock 母钟;标准钟
control cluster 仪表板;控制钮板
control cock 控制旋塞
control code 控制码
control-code display 控制显示码
control column 控制卷;控制杆
control combination 控制组合
control command 控制指令;控制命令;操纵指令
control command data 控制命令数据
control command program(me) 控制命令程序
control commerce 市场控制商
control committee 控制委员会
control compartment 控制室
control component 控制元件
control computer 控制用计算机;控制计算机
control computer subsystem 控制计算机子系统
control concept 控制概念
control concurrence 控制并行性
control console 控制装置总成架;控制台;控制盘;控制柜;控制板;仪表操作台;调度台;操纵台;操纵控制点
control contactor 控制接触器
control cooled rail 缓冷法制钢轨
control cord 控制索
control correction vector 控制校正向量
control cost 管理费
control counter 控制计算器;控制计数器
control crank 驾驶杆;操纵手柄;操纵臂
control criterion 控制(质量)准则;控制标准
control cross-section 控制断面
control cubicle 控制台室;控制柜;操纵室
control current or voltage 控制电流或控制电压
control curve 控制弯段;控制曲线;对比曲线
control cycle 控制周期;管理周期
control cylinder 控制筒;控制风缸
control cylinder gasket 控制气筒垫密片
control cylinder-rod 控制杆
control damper 气流控制阀;调节风门;调节板
control data 控制数字;控制数据
control data card 三角点成果表
control data item 控制数据项
control data list 控制资料表
control data name 控制数据名
control data table 控制数据表
control day 控制日
control decision 控制决策
control density 控制网密度
control department 控制部;管理部门

control desk 控制台;控制塔;操纵台
control desk element 控制台单元
control detailed planning 控制性详细规划
control deviation 控制偏差
control device 操纵装置;控制装置
control device for handrail breakage 扶手带断代保护装置
control diagram 控制原理图;控制图(表)
control diagram language 控制图形语言
control display 控制显示(器)
control documents 控制文件
control domain 控制域
control drive 控制传动
control drive mechanism 控制传导机构
control drum 操纵鼓轮
control earthquake 控制地震
control earthquake value 控制地震值
control eccentric 控制偏心轮
control echo 控制回声
control effectiveness 操纵有效性
control efficiency 控制效率;控制能力
control electrode 控制电极
control electronics 控制电子学
control element 控制元素;控制元件;控制单元
control elevation 控制标高;控制(点)高程
control end 调度端
control engine 控制发动机
control engineer 调度工程师
control engineering 控制(点)工程;管理工程
control engineering of water pollution 水污染防治工程
control equation 控制方程
control equipment 控制仪器;调度设备;控制设备
controlet 控制机构
control experiment 对照实验;控制试验;检验实验
control extension 控制扩展法;控制点加密
control facility 操纵系统设备
control factor 控制因数;控制系数
control factor unit 控制因素单位
control feel 操纵感觉
Control Fiber 树皮碎屑堵漏剂(商品名)
control field 控制域
control field sorting 控制域分类
control figure 控制数字
control figures drill 数控钻机
control file 控制文件
control fill 控制填方
control fin 操纵舵;操纵片
control fitting 控制配件
control flange 控制法兰;检查法兰
control flight 控制航线飞行
control float 控制浮子
control flow 控制屏;控制流程
control flow computer 控制流计算机
control flow optimized program(me) 控制流程优化程序
control flume 计量槽;控制槽;测控水槽;测流槽
control fluoride emissions 氟化物排放控制
control footing 控制总数;控制小计;控制合计
control footing group 控制尾栏
control force system 操纵力系
control form 施业控制表
control format 控制格式
control format item 控制格式项
control for outfall of workshop 车间排放口治理
control function 控制职能;控制函数;控制功能;控制操作
control fuse 控制保险丝
control gallery 控制廊道
control gap 可控间隙;调整间隙
control gas 控制气
control gate 控制栅板;(场效应管的)控制栅;控制闸(门);控制门;节制闸门
control ga(u)ge 控制量规;检验量规;样规
control gear 控制装置;控制机械;自动调整仪;操纵装置
control generator 控制振荡器;控制发电机
control graduating spring 手控制递开弹簧
control grain size 控制粒径
control gravimeter point 重力控制点
control gravimetric base 重力测量检核基线
control grid 控制栅极;(场效应管的)控制栅
control-grid bias 控制栅偏压
control-grid plate transconductance 控制栅—阳

极跨导
control group 控制组;控制栏;对照组
control handle 控制旋扭;操纵手柄;操纵杆;操纵柄;操纵把手
control hand picking 检查性手选
control handwheel 操纵盘
control head 井口装置;调节头;控制头
control heading 控制头栏
control heading group 控制头栏
control height 高程控制
control hierarchy 控制级别;控制分层;控制层次;多级控制;递阶控制
control hole 控制孔;标志孔
control horn 控制杆
control house 控制室;管理所
control hysteresis 控制滞后【电】
control identifier 控制标识符
control impedance 控制阻抗
control in advance 提前控制
control index setting 控制指数调整;给定值调整
control information 控制信息;校正信息;调节信息
controling appliance 控制设备
controling device 控制设备
controling equipment 控制设备
controling facility 控制设施;控制设备
controling parasite 天敌
control input 控制输入
control instruction 控制指令
control instruction counter 控制指令计数器
control instruction transfer 控制指令转移
control instrument 控制仪表;控制仪器;操纵器械
control instrumentation 调节仪表
control interface 控制接口
control interlock 操纵链锁
control interval 控制时段;控制间隔;调节间隔
control interval password 控制区间口令
control interval split 控制间隔分割
control investment 控制投资
control item 控制项
control jet 控制射流
control joint 调节缝;控制接合点;控制(接)缝;伸缩缝
control joint block 控制缝块;控制缝块
control joint grouting 控制(接)缝灌浆;收缩缝灌浆
control key 控制键
control key switch 控制钥匙开关
control knob 控制旋扭;控制手柄;操纵按钮
controllability 控制性能;控制能力;可控(制)性;可监督性;可调节性;可控性;操纵性
controllability condition 可控性条件
controllability criterion 可控性标准
controllability index 可控性指数
controllability matrix 可控性矩阵
controllability principle 可控制性原则
controllability test 操纵性试验
controllable 可控制的;可调节的
controllable area nozzle 可调截面喷管
controllable burden 可控(制)费用;可控负荷
controllable contribution 可控贡献额
controllable cost 可控(制)成本
controllable coupler 可控耦合器
controllable diffraction grating 可控衍射光栅
controllable expenses 可控(制)费用
controllable factor 控制因子;可控制因素
controllable factory overhead variance 可控制造费用差异
controllable fixed cost 可控固定资本
controllable function 可控函数
controllable impedance 可调电感
controllable interest 可控权益
controllable investment 可控(制)投资
controllable liquid-crystal display device 可控液晶显示装置
controllable locomotive 支配机车【铁】
controllable nose intake 头部可调进气口
controllable pair 可控偶;可控对
controllable performance 可控业绩
controllable pitch airscrew 可调变距螺旋桨
controllable pitch fan 变距风扇
controllable pitch propeller 螺距可变的螺旋桨;可调螺距螺旋桨;变桨叶;可变距推进器;可变螺距螺浆;调距螺旋桨
controllable reactor 可控反应堆
controllable size wet suit 可调节湿式潜水服

controllable spark gap 可控火花隙
controllable spillway 可控溢洪道
controllable spoiler 可控扰流片
controllable state representation 可控状态表示方式
controllable storage 可控储量
controllable system 可控系统
controllable thrust 可调推力;可变推力
controllable variable 可控变量
controllable variance 可控(制)的差异;可控差异
control laboratory 检验室;化验室;控制实验室
control lag 滞后操纵;调整延迟;调节滞后;调节延迟
control lamp 控制灯
control latitude 控制纬度
control law 控制规律
control lead 操纵导管
controlled 被调整的;受控
controlled-access highway 控制进入的道路;进口控制道路
controlled accessibility 受控存取性
controlled access path 受控存取通路
controlled access road 控制进入路
controlled access route 全封闭线路
controlled accounts 明细账户;被控驭账户
controlled-air incinerator 控制空气焚化炉
controlled air space 控制空域
controlled(-angle) drilling 定向钻进;控制角钻进
controlled arc welding 可控电弧焊接
controlled area 监管区
controlled association 控制联想
controlled atmosphere arc welding 充气式电弧焊
controlled atmosphere (制造粒金刚石钻头时马弗炉中的)惰性气体;受控制的空气;受控气氛;受控炉气
controlled atmosphere arc welding 充气室电弧焊
controlled atmosphere furnace 可控气氛炉;保护气体炉
controlled atmosphere (gas) storage 气调储[贮]藏
controlled atmosphere heat treatment 可控气氛热处理
controlled atmosphere kiln 气氛窑
controlled atmosphere spectrochemical analysis 控制气氛光谱分析
controlled atmosphere storage 控气冷藏(法)
controlled avalanche device 可控雪崩器件
controlled avalanche rectifier 受控雪崩整流器
controlled bevel 精确倒边
controlled blasting 控制爆破
controlled blasting technique 受控爆破技术
controlled block 受控闭锁
controlled burning 控制燃烧;反控燃烧;受控燃烧
controlled by directional hole 定向孔
controlled by directional well 定向井
controlled cancel 控制消除;受控取消
controlled carrier 可控载波
controlled carrier modulation 可控载波调制;受控载波调制
controlled carrier system 控制载波系统;可控载波系统;受控载波系统
controlled chain growth polymerization 链增长受控聚合
controlled chilling 控制激冷
controlled circulation 强制循环
controlled clock 被控制电钟
controlled computational experiment 受控的计算试验
controlled concentration 控制浓度
controlled concrete 均匀控制混凝土;受控混凝土
controlled condition 控制条件;一定条件;受控条件
controlled condition system 条件控制系统
controlled construction 受控(制)的施工
controlled cooling 控制冷却
controlled cooling treatment 控制冷处理
controlled corporation 分公司;受控制公司
controlled cracking 控制(的)裂缝;开裂控制
controlled crest 设有调节水位装置的坝顶;闸门堰顶(设有调节水位装置的坝顶)
controlled crest spillway 控制设有调节水位装置的坝顶溢洪道
controlled crest wasteway 控制设有调节水位装置的坝顶溢洪道
controlled crossing 交通管理交叉口
controlled curve 控制曲线

controlled degradation 可控降解
controlled depletion polymer 受控溶耗性聚合物
controlled device 被控制装置;受控装置
controlled diameter 限制粒径
controlled differential 锁止式差速器;控制的差动装置
controlled-directed fishing 瞄准捕捞
controlled directional solidification 定向凝固
controlled discharge 控制泄放;计划泄放
controlled-discharge door concrete spreader 控制出料的混凝土铺筑机
controlled-discharge wagon 可控卸载车辆
controlled disposal of radioactive effluent 放射性流出物控制处置
controlled district 控制区段
controlled doping 控制掺杂;受控掺杂
controlled dumping 有控制的倾废
controlled economy 统制经济
controlled ecosystem pollution 可控生态系统污染
controlled ecosystem population 可控生态系统群落
controlled emergency braking 控制的紧急制动
controlled energy-flow machine 可控式高速锤
controlled entry to system procedure 受控的系统过程入口
controlled environmental release 可控环境释放
controlled environment house 自动控制环境畜舍
controlled exchange system 控制汇率制度
controlled expansion coefficient alloy 定膨胀合金
controlled experiments 可控试验
controlled explosion 控制爆破
controlled figures of plan 计划控制数
controlled fill 控制填方
controlled filling experiment 控制注水试验
controlled filling of a reservoir 水库控制蓄水
controlled filter 控制过滤器
controlled filtration 控制过滤
controlled float 控制的汇率浮动
controlled flooding 控制漫灌
controlled flow silo 受控料流均化库
controlled footage 控制进尺;最大进尺
controlled foreign trade 统制对外贸易
controlled funding 养老金受控筹资法
controlled funding pensions 养老金受控筹资法
controlled fusion 受控核聚变
controlled fusion reactor 受控聚变堆
controlled-gravity stowing 自流充填;水砂充填;水力充填
controlled head gap 控制头间隙
controlled house 自动控制畜舍
controlled humidity 控制湿度
controlled humidity drying 控制湿度下的干燥
controlled injury zone 控制危害区
controlled intersection 控制式交叉器
controlled intersection junction 信号控制交叉
controlled item 可控项目
controlled landfilling 有控制的垃圾填埋
controlled leak 受控漏泄
controlled magnetic core reactor 磁芯控制扼流圈;磁芯控制电抗器
controlled map 精确地图
controlled marker 可调指示器
controlled medium 被控媒质
controlled member 控制对象;受控件
controlled mercury-arc rectifier 可控汞弧整流器
controlled microcrystallization 微晶化
controlled mine 可控雷
controlled monopoly 控制性垄断
controlled mosaic 控制镶嵌;有控制点镶嵌图
controlled mosaicity method 控制镶嵌法
controlled motion 控制运动
controlled movement 控制运动
controlled object 控制对象;被控对象
controlled oscillator 控制振荡器
controlled oxidation 受控氧化作用
controlled parameter 受控参数
controlled parking area 控制停车地区
controlled parking zone 控制停车控制区;控制停车地区;规定停车地段
controlled pedestrian crossing 信号控制人行道;有控制的人行横道
controlled pedestrian crosswalk 受控制人行道横道
controlled pitch airscrew 双距螺旋桨
controlled plant 被控车间;受控对象

controlled plate conductance 跨导
controlled point 被控点
controlled pollution 可控污染
controlled polymerization rate 受控聚合速度
controlled pore ceramics 控制气孔陶瓷
controlled postmortem program(me) 受控检错程序
controlled potential electrolysis 可控电势电解
controlled potential method 控制电势法
controlled potential pulse 控制电位脉冲
controlled pressure grouting 控制压力灌浆
controlled price 控制价格;统制价格
controlled process 被控过程
controlled production 可调节产量
controlled quantity 控制量;可控数量;受控量
controlled radial steering 可控转动式舵机
controlled rate genlock 受控速度台从锁相
controlled rate of penetration 渗透控制率
controlled rectifier 可控(电压)整流器
controlled release 受控释放
controlled release disinfectant 可控释放消毒剂
controlled release of energy 能量的受控释放
controlled release pesticide 可控释放农药
controlled release technology 控释技术;可控释放技术
controlled resol tion 可控分辨率
controlled resolution 可控分辨率
controlled rewriting 受控重写
controlled seeding 受控籽晶法
controlled sender 受控发送器
controlled sheave 可控皮带轮
controlled silica 可控硅
controlled single-way channel 单行控制航道
controlled solidification 控制固化
controlled source 被控源
controlled speed 控制航速(港区)
controlled speed axle generator 恒速车轴发电机
controlled spillway 可调节溢流堰;浸入式溢流堰;活动堰;潜式堰;有控制设施的溢洪道
controlled spin 被控螺旋
controlled stage 控制级;被控级
controlled start 可控起动
controlled state 管理状态
controlled storage 控制储水(量);受控存储器
controlled storage allocation 受控存储(区)分配
controlled strain test 应变控制试验
controlled-stress load test 应力控制荷载试验
controlled stress rheometer 应力受控流变仪
controlled-stress test 应力控制试验
controlled substance 管制物质
controlled surface 限制表面
controlled suspension 刚性可调整的悬架;变刚度悬架
controlled swirl scavenging system 可控涡流清洗系统
controlled system 可控系统;被控系统;受控系统
controlled temperature 控制温度
controlled temperature cabinet 调温箱
controlled test 控制试验
controlled texture polymer 受控结构聚合物
controlled thermal cracking 温控裂纹试验
controlled thermal decomposition 受热热分解
controlled thermal severity test 控制性热力严重度试验;控制性热力破坏性试验
controlled thermohygrostat 可控恒温恒湿器
controlled thermonuclear reactor 受控聚变堆
controlled tip 废物堆填区;控制垃圾堆
controlled tipping 废物填埋场;受控倾倒法
controlled tipping site 废物堆填区
controlled-transfer welding 可控过渡电弧焊
controlled tungsten-arc welding 自动控制弧长的钨极电弧焊
controlled variability 受到控制的易变性
controlled variable 控制变量;控制量;可变因素;被控变量;被调参数;调节变量;受控变量
controlled ventilation window 可调通风窗
controlled vocabulary 可控制词表;规范化词表;受控词汇表
controlled volume pump 可控排量泵;计量泵
controlled vortex flow 控制涡流
controlled water 被控制的水;受控水体
controlled waters pollution 可控水体污染
controlled water supply 有控制的供水
controlled weir 潜坝;可调节溢流堰;活动堰;有节制溢流堰
controlled well drilling 控制钻井
controller 控制器;检验员;检查员;信号控制器;管理员;调度员;操纵器;擎链器;调节器
controller assembly 信号机组
controller buffer 控制缓冲器
controller bus 控制器总线
controller cage 控制器罩
controller case 控制器箱
controller configuration facility 控制器配置程序
controller conventional design 控制器常规设计
controller cover 控制器箱盖
controller creation parameter table 控制器生成参数表
controller cylinder 控制器圆筒
controller data 控制器数据
controller drum 控制器滚筒
controller function 控制器功能;控制器操作
controller gain 控制器增益;调节器增益
controller green time 信号机绿灯时间
controller handle 控制器把手
controller interface unit 控制器接口
controller interruption 控制器中断
controller of currency 通货检查员
controller of filter 滤池控制器
controller pilot valve 控制器导阀
controller's department 主计部;财务部;审外局
controllership 主计官权责
controller storage save 控制器存储器保留区;控制存储器保存
control level 控制水准;控制水平;控制级;控制电平;管理水平
control level indicator 控制级指示符
control level of surface contamination 表面污染控制水平
control lever 操纵杆
control lever boot 控制杆保护罩
control lever bracket 控制杆架
control lever clamp 控制杆夹
control lever for power-drive 动力驱动离合器操纵杆
control lever fulcrum 控制杆支点
control lever guide 导杆
control lever housing 控制杆罩
control lever sleeve 控制杆衬套
control library 控制(数据)库
control limit 控制限度;控制界限;控制极限;管理范围
control limit switch 控制极限开关;极限控制开关
control line 控制行;控制线路;防火线
controlling account 统驭账户
controlling agent 控制剂
controlling and distributing station 控制分配站
controlling and feedback with evaluation 评核管制与反馈
controlling apparatus 控制装置;控制器
controlling application program(me) 控制应用程序
controlling authority 监控组;监控当局
controlling beam 操纵波束
controlling body 监控组;监控当局
controlling borehole 控制孔
controlling buoy 控制浮筒
controlling circuit 控制电路
controlling company 总公司;管理公司;控制公司
controlling-company accounting 控制公司会计
controlling condition 控制条件
controlling corporation 控制公司
controlling current 控制电流
controlling decision 控制的决策
controlling degree 控制程度
controlling depth 控制深度;控制水深;可航最小水深;可航最大吃水
controlling device 控制装置;调节装置
controlling dimension 控制尺寸
controlling electric(al) clock 主控电钟
controlling element 控制设备元件;控制要素;控制构件
controlling elevation 控制标高
controlling entity 控制实体
controlling equipment 控制仪表;施控装置;施控系统
controlling expenses 控制费用
controlling factor 控制因素
controlling factor of organic matter evolution 控制有机质演化的因素
controlling factor of variation 变化控制因素
controlling field 控制场
controlling force 控制力
controlling gear 控制机构
controlling hold 重点货舱
controlling hole 控制钻孔[岩]
controlling humidity level 湿度水液控制
controlling influence 控制效应
controlling instrument 控制仪表
controlling instrumentation 控制仪表
controlling interest 控制(性)权益;控制股权(权益);可控制权益;统制股权
controlling ledger 控制分类账;统制分类账;统驭分类账;统驭分类账
controlling lever 控制杆
controlling line 控制线
controlling machine 控制机
controlling magnetic field 可调磁场;施控磁场
controlling manhole 控制人孔;控制井
controlling means 控制方法
controlling mechanism 操作机构
controlling member 控制机构
controlling moment 控制力矩
controlling obstacle 受制障碍
controlling of dredge-cut depth 挖槽深度控制
controlling of dredge-cut width 挖槽宽度控制
controlling of operating cost 运营成本控制
controlling organization 控制机关;控制机构
controlling parameter 控制参数
controlling point 控制点
controlling pontoon 控制浮筒
controlling record 控制记录;统驭记录
controlling ridge 挡水堤
controlling rod 控制杆
controlling rope 控制索
controlling seawater intrusion 控制海水侵入
controlling section 控制区间;控制区段
controlling signal valve 控制信号阀
controlling spring 控制弹簧
controlling stock 控制股票
controlling switch 控制开关
controlling system 施控系统
controlling the price 平抑价格
controlling torque 控制转矩
controlling unit 控制环节
controlling value 控制值
controlling valve 控制阀
controlling wastewater 废水控制
controlling wood protection 控制木材保护(法)
controlling zone 控制区
control load 控制荷载
control location 控制位置;控制地段
control logic 控制逻辑
control loop 控制环(路);控制循环;控制回路;驾驶系统;操纵系统
control magnet 控制磁铁
control manhole 检修孔
control man-made land subsidence 控制人为地面沉降
control mark(ing) 控制标记;控制符(号);控制标(志)
control master 检查工长
control matrix network 控制矩阵网络
control measurement 检查测量
control measure 控制措施;管理措施
control measures of soil erosion 土壤侵蚀治理措施
control mechanism 控制机理;控制机构;操纵机构
control memory 控制存储器
control message 控制信息;控制信号
control-message display 控制信息显示器;控制信号显示
control meter 控制表
control-metering point 系统控制记录点
control meter selector 控制仪表选择钮
control method 控制方法;检查法;对比方法
control methods for pipe lines 管路控制方法
control mode 控制模式;控制方式;控制方法
control mode equipment 控制动作装置
control model 控制模型
control module 控制组件;控制模件;控制部件
control module area 控制模块区

control moment 操纵力矩
control monitor unit 监控装置
control mosaic 控制点镶嵌图
control motor 控制马达;控制电动机
control net(work) 控制网【测】
control network density 控制网密度
control network for deformation observation 变形观测控制网
control network for monitoring deformation 变形监测网
control network for monitoring slide 滑坡监测网
control network of height 高程控制网
control nozzle 操纵喷管
control number 控制数;控制号
control nut 控制螺母
control objective 控制目标
control observation 控制观测;检差观测
control of a boundary 边界层控制
control of access 进入控制;进口控制;出入口控制;通道管理
control of aerial photography 航空摄影控制
control of agricultural pest 农业病虫害防治
control of air pollution 空气污染防治
control of air pollution sources 空气污染源控制
control of automobile exhaust gases 汽车尾气污染控制
control of blasting geology 爆破地质指导
control of cash payment 现金使用控制;现金管理
control of consumption 消费控制
control of core loss 岩芯遗失控制
control of crossfall 路拱高差控制
control of desert 沙漠治理
control of dioxide emission 二氧化硫排放控制
control of dust 集尘
control of economy 经济管理
control of electromagnetic radiation 电磁波辐射管制
control of expenses 费用控制(管理)
control office 控制局
control of fine particle 微粒控制
control of flood 控制洪水;洪水控制;防洪
control of fluoric gases by absorption with alkali 碱吸收法控制含氟废气
control of fluorine-containing gases 含氟废气治理
control of fluorine gases by absorption with water 水吸收法控制含氟废气
control of formation pressure 地层压力控制
control of gaseous contaminant 气态污染物控制
control of gaseous pollutant 气态污染物控制
control of grouts and grouting operations 灌浆控制与灌浆作业
control of headwater 水源控制
control of heat generation 发热量控制
control of high-pressure well 高压井控制
control of hydrogen sulfide 硫化氢污染控制
control of hydrogen sulfide by dry-oxidation process 干式氧化法控制硫化氢
control of hydrogen sulfide ferric oxide 氧化铁法控制硫化氢
control of impulse 推力控制
control of inflation 通货膨胀控制
control of insects 害虫防治;害虫控制
control of intersection of planes 交面控制
control of inventory 存货控制
control of inversion 变换控制
control of investment 投资控制
control of lending and deposit balance 借差控制
control of lesion 防止损害
control of levels 高程控制
control of marine pollution 海洋污染控制;海洋污染监控
control of merchandise 商品管理
control of metal temperature 金属温度控制
control of nematode 线虫的防治
control of nitrogen oxides 氮氧化物污染控制
control of NO_x by absorption oxidation 氧化吸收法脱氮
control of NO_x by absorption-oxidation process 吸收氧化法脱氮
control of NO_x by absorption with acid 酸吸收法脱氮
control of NO_x by absorption with alkali 碱吸收法脱氮

control of NO_x by adsorption process with activated carbon 活性炭吸附法脱氮
control of NO_x by adsorption process with silica gel 硅胶吸附法脱氮
control of NO_x by adsorption-reduction process 吸收还原法脱氮
control of NO_x by adsorption with water 水吸收法脱氮
control of NO_x by selective catalytic reduction 选择性催化还原法脱氮
control of ore deposits by compounding of tectonic system 构造体系复合控矿
control of ore deposits by multiorder of tectonic system 构造体系多级控矿
control of organic waste gas 有机废气处理
control of organic waste gas by thermal combustion 热力燃烧法净化有机废气
control of overcasting geology 剥离地质指导
control of pesticide 农药控制
control of phase sequence 相序控制
control of point sources 点污染源控制
control of pollution act 污染防止法
control of pollution sources 污染源控制
control of pressure 压力控制
control of price 价格管制
control of product 产品控制
control of production geology 生产地质指导
control of pumping up of underground water 地下水抽出控制
control of purchases 采购控制
control of purity 纯度(的)控制
control of raw material homogenizing 原料均化控制
control of reflux ratio 回流比控制
control of scrap 废品(料)的控制
control of setups 机器安装的调整
control of signal 信号控制
control of soil erosion by planting 控制土壤侵蚀的种植
control of soil pollution 土壤污染控制;土壤污染监测
control of sort procedure 分类过程的控制
control of source 源控制;水源控制
control of stamping 印章管理
control of static load pressure 控制静态负荷压力
control of stream 河流控制
control of stress 应力控制
control of substance hazardous to health 有害健康物质的控制
control of system 系统控制
control of thrust orientation 推力定向操纵
control of traffic 交通控制
control of variance 方差控制
control of ventilation 通风控制
control of warehouse temperature and moisture 仓库温度湿度管理
control of water-level 水位控制
control of weather 人工控制天气
control of weed 除草;杂草防治
control-oil pressure 控制油压力
control on consumer credit 消费信贷管理
control on settlement 结算管制
control opening 控制开口
control operating rod 操纵拉杆
control operation 控制操作
control operator's terminal 控制操作员终端
control optimization program(me) 控制最优化程序;控制最佳化程序
control order 控制指令
control organ 控制机构
control-orientation 侧重管理
control-oriented function 面向控制的功能
control-oriented microcomputer 面向控制的微型计算机
control orifice 控制节流口
control overrun 超限控制;控制越限
control over-supply 操纵供给
control packet 控制分组
control pair 控制线对
control panel 控制台;控制盘;控制(面)板;接线盘;接线板;调度台;电控设备;插接板;操纵台;操纵板;配线板;配电盘;驾驶控制台
control panel of gas supply 供气控制台
control panel with push-button 按钮控制板

control part 控制部件
control pass 控制孔型
control path 控制通路
control pedal 控制踏板
control pedestal 控制座;操纵台
control pen 控制笔
control period 控制时期;控制时间;控制时段;控制期(间);调节时间
control phase 控制相位
control(photo)strip 构架航线
control pick-up 操纵传感器
control pipe 控制管
control piston 控制活塞;调节活塞
control piston follower 调节活塞
control piston nut 控制活塞螺母
control pitch 改变车叶距
control plan 控制计划;管理计划
control platform 控制平台
control plotting 控制点展绘
control point 控制点;路线控制点;检测点;对照点
control point adjustment 控制点调整;给定器
control-point coordinates 控制点坐标
control point design 控制点设计
control point distribution for aerial triangulation strip 航线网布置
control point distribution for block aerotriangulation 区域网布点
control point distribution for control strip 控制航线布点【测】
control point files 控制点文件
control point for rectification 纠正点
control point interval along strips 航向控制点跨度
control point interval cross strips 旁向控制点跨度
control point list 控制点成果表
control point of alteration zone 蚀变带控制点
control point of boundary between different mapping units 填图单位界线控制点
control point of contact zone of intrusive body 岩体接触带控制点
control point of engineering in mountainous region 山地工程控制点
control point of fault zone 断层带控制点
control point of fold-hinge 褶皱枢纽控制点
control point of hydrographic(al)survey 海控点
control point of marker bed 标志层控制点
control point of mineralization zone 矿化带控制点
control point of ore body 矿体控制点
control point of photograph 像片控制点
control point of picture 相片控制点
control point of road 路线控制点【道】
control point of sedimentary facies change 岩相变化控制点
control point of shaft 近井点
control point of track 路线控制点【铁】
control point of vein 岩脉控制点
control(point)photograph 控制(像)片
control point selector 控制点转换开关
control point set 控制点装定
control point survey 控制点测量
control policy 防治方针
control port 控制油口
control portion 控制部件;控制段
control position indicator 控制示位器
control potentiometer 控制电位计
control power 控制权
control power supply 控制电源
control precision 控制精(密)度
control pressure 控制压力
control principle 防治原则
control procedure 控制程序
control process 监督过程
control processor 控制处理机
control processor state 控制处理机状态
control product 控制产品
control production 控制生产
control profile 控制断面图
control program(me) 控制程序;管理计划;防治规划
control program(me)command 控制程序命令
control program(me)parameter list 控制程序参数表
control program(me)privilege class 控制程序优先级
control program(me)read 控制程序读数

control protocol 控制协议
control pulse 控制脉冲
control punch 控制穿孔;标志孔
control(push)knob 控制按钮
control rack 操纵台;操纵板
control radius 控制半径
control range 控制范围
control-rate characteristic 控制速率特性
control ratio 控制比值;闸流管控压比
control reach 控制段
control reading 检核读数
control read-only memory 控制只读存储器
control reception 验收
control reconnaissance sheet 勘测检核样图
control region 控制区域;控制区
control register 控制寄存器
control relay 可控继电器;控制继电器;监测继电器
control report 控制报告
control resistor 控制电阻器;可调电阻器
control response 控制响应;控制灵敏度;控制反应;对控制指令的反应;操纵反应
control reversal 操纵反效
control risk 控制风险
control rocket 制导火箭
control rod 控制棒;控制杆;操纵杆
control rod actuator 控制杆传动机构
control rod drive 控制棒传动
control rod follower 控制棒导向装置
control rod guide 控制棒导向装置
control rod material 控制杆材料
control room 控制(台)室;仪表室;指挥舱;调度室;操作室;操纵室
control room block 控制室地段
control room building 控制室大楼
control routine 控制例行程序;控制代码
control run 操纵行程
control sample 校核试样;对照样品;对照试样;对比样;对比试片
control sampling 控制采样
control sampling inspection 抽样检查控制
control schedule 工程部分进度计划;工程进度控制
control scheme 控制方案
control section 控制区域;控制截面;控制节;控制断面;控制段;控制部分;检查段;管理部门
control segment 控制段
control sequence 控制序列
control sequence processing 控制序列处理
control sequence table 控制顺序表
control servo-tab 伺服调整片
control setting 操作钮安装
control set-up 控制装置
control shaft 控制轴
control sheet 控制图(表);校样图;坐标控制图
control signal 控制信号;操纵信号
control simulator 操纵模拟器
control sleeve 控制套筒
control SO$_2$ emisson 二氧化硫排放控制
control software specification 控制软件说明
control solution 校核液
control specification 控制说明书
control specimen 检查用的标准试样;校核试样;检验用的基准试样;对比试件
control spindle 分配轴
control spool 分配滑阀;分配滑阀
control stack 控制栈
control staff 管理人员;操纵人员
control stake 控制桩【测】
control stand 控制柱;控制台;检验台;操纵台
control standards 控制标准
control stand cover 操纵台罩
control state 控制(状)态
control statement 控制语句【计】
control statement analyzer 控制语句分析程序
control station 控制(测)站;管理所;调度站
control-station identification 控制点辨认
control step 控制步骤
control stick 控制杆;操纵杆
control storage 控制存储器
control storage save 控制存储器保存
control store 控制存储
control store literal 常数字段
control store monitor 控制存储监视器;存储监视器;存储地址监控器
control strategy 控制策略

control stress 控制应力
control stress for prestressing 张拉控制应力
control strip 控制条;控制航线【测】;骨架航线;构架航带
control structure 控制结构;控制建筑物;控制设施
control subarea 控制子区
control subassembly 调节子配件
control suction line 控制吸入管路
control sum 检查和
control supervision 调度监督
control supervisor 控制管理(程序)
control surface 控制面;操纵面
control surface actuator 舵机
control surface lock 操纵面夹板
control surface moment 控制面力矩
control surface pick-up 操纵面传感器
control surface tie-in 计算机操纵飞行
control surface torque 控制面力矩
control survey 控制测量;监视测量
control survey clasification 控制测量等级
control surveying of photograph 像片控制测量
control survey point 测量控制点
control survey station 测量控制点
control switch 控制电闸;控制开关;总开关;主令开关
control switching point 控制切换点;控制开关点;控制交换中心
control symbol 控制符(号)
control synchro 控制同步机;同步发送器
control system 控制制度;控制系统;控制方式;制导系统;操纵系统;防治体系;调整系统
control system damping 控制系统中的阻尼
control system design 控制系统(的)设计
control system dynamics 控制系统动力学
control system equipment 控制系统设备
control system feedback 控制系统反馈
control system flow charting 控制系统图示法
control system gear ratio 操纵系统传动比
control system of clutch 离合器操纵机构
control system of exhaust 废气净化系统
control system of exhaust gases 废气控制系统
control system of subsea drilling equipment 水下钻具控制系统
control systems engineering 控制系统工程
control system servo 控制系统动力传运装置
control system's logic 控制系统逻辑
control system test 控制系统测试
control tab 控制标记
control table 控制表
control tape 控制带
control tape mechanism 控制带机构
control task 控制任务
control technique 控制技术;检查技术;防治技术
control technology 控制技术;控制工艺
control technology of dust 尘粒控制技术
control technology of particulate 颗粒物控制技术
control telephone 调度电话
control template 控制模片
control terminal 控制终端(设备)
control test 控制试验;控制检验;检查试验;对照试验;对照实验;复查试验
control theory 控制(理)论
control through a legal device 通过法律手段的支配
control through budget 用预算管制
control through market forces 市场力量控制
control tide station 主要潮位站;主要潮流观测站;潮汐主要观测站
control timer 控制定时器;时间控制继电器
control tooth 限动齿
control torque 控制扭矩
control torrents 洪流控制
control total 控制总数;控制总额
control tower 控制塔;指挥塔;操纵塔;塔台
control-tower visibility 控制塔能见度;地面能见度
control trailer 控制拖车
control trajectory 控制轨线
control transfer 控制转移;转换操纵
control transfer instruction 转移指令
control transfer operation 控制转移操作
control transformer 控制变压器
control traverse of route 路线控制导线
control turn 控制线匝
control-type budget 控制式预算

control unit 控制装置;控制器;控制机组;控制单元;控制部件
control unit end condition 控制器结束条件
control unit with fault storage 故障记录控制单元
control valve 控制管;控制阀;节流管;节流阀;操纵阀;调节管;调节阀
control valve block 控制阀体
control valve bracket 控制阀托架
control valve breather 控制阀通气孔
control valve bush 控制阀衬套
control valve cap 控制阀盖
control valve dial 控制阀指度盘
control valve dial plate 控制阀刻度板
control valve for fluids 流体控制阀
control valve gland 操纵阀压盖
control valve manifold 控制阀支管;控制阀歧管
control valve of copying device 仿形装置控制阀
control valve of moderator level 慢化剂液位调节阀
control valve operating lever 控制阀操作杆
control valve piston 控制阀门活塞
control valve piston rod 控制阀活塞杆
control valve plug 控制阀塞
control valve spring 控制阀簧
control vane 操纵舵
control variable 控制变量
control variable name 控制变量名
control variance 差异控制
control variate method 控制变量法
control variety 对照品种
control verification 控制检定
control vessel 指挥船
control voltage 控制电压;操作电压
control volume 控制体积;控制容积;控制卷
control volume model 控制面模型
control wave 控制波
control waveform 控制电压波形
control weir 控制堰
control wheel 操纵轮;调整轮
control winding 控制线匝
control(wind)speed 控制风速
control(wind)velocity 控制风速
control wire 操纵索;控制线
control wiring 控制线路
Control Wool 石棉堵漏丝(商品名)
control work 整治工作;控制工作;调节工作;控制网布设
control works 防洪工程;控制工程
control zone 控制区域;控制区
control zone adapter module 控制模块
controsurge winding 浪涌屏蔽绕组;冲屏蔽绕组;防振屏蔽绕组
controversial question 有争议问题
controversy 争论
Contton-Mouton effect 康顿—穆顿效应
contumacy 蔑视法庭
contused wound 挫伤
contusion collar 挫伤轮
conularida 锥石类
conurbanization 城镇群
conurbation 大城市地区;城镇集聚区;具有卫星城的城市;集合城市;有卫星城镇;大市区;大城市;城镇区
conurbation transport 大城市圈运输(包括卫星城镇)
conus elasticus 弹性圆锥;弹力圆锥
convalescence 恢复期;休养
convalescent center 康复中心
convalescent centre for children 儿童康复中心
convalescent district 疗养区
convalescent home 儿童康复之家;疗养院;康复(医)院;休养院;休养所
convalescent home for children 儿童疗养院
convalescent hospital 疗养院
convalescent stage 恢复期
convas polishing wheel 帆布抛光轮
convected air 对流空气
convected heat 对流热
convected heat loss 对流热损失
convected heat transfer 对流热传递
convection air 对流空气
convectional aggregate 约定集料;约定骨料
convectional bank 对流管束
convectional bar reinforcement 常规配筋

convectional barrier 对流(传热)屏层
convectional boiler 对流式锅炉
convectional boring 一般取芯(钻进)
convectional breaker 对流式断路器
convectional cell 对流系统单元;对流窝;对流体;对流圈;对流环;对流核
convectional church 普通教堂;传统教堂
convectional circuit breaker 对流式断路器
convectional circulation 对流循环;对流性环流
convectional coagulation system 普通混凝系统
convectional coefficient 对流系数
convectional construction method 传统的施工方法
convectional cooling 对流冷却
convectional current 对流(气流);对流电流
convectional depreciation 传统折旧
convectional design 惯例设计;常规设计
convectional diagram 示意图
convectional digester 传统消化池
convectional dryer 对流干燥器
convectional drying 对流烘干;对流干燥
convectional engine 普通发动机
convectional equilibrium 对流平衡
convectional external balcony-access type of block 传统的外廊式建筑
convectional furnace 对流炉
convectional heat 对流热
convectional heater 对流式供暖器;对流式采暖器;对流加热器
convectional heating 对流式供暖;对流加热;对流采暖
convectional heating element 对流加热元件
convectional heat transfer 对流传热
convectional high-speed mixer 对流型高速混合器
convectional hobbing 逆向滚削
convectional hopper dredge(r) 普通自航式挖泥船
convectionalized 传统化的
convectional layer 对流层
convectional life estate 传统终身财产;常规终身财产
convectional lock 通用锁;普通锁
convectional loss 对流损耗
convectional memory 常规内存储存器
convectional mixing 对流混合
convectional modulus 对流模量
convectional mortar-bed method 传统的砂浆层法
convectional oven 对流烘炉
convectional overturn 对流下沉
convectional paint 普通油漆;普通涂料
convectional process 对流过程
convectional pumping unit 通用泵
convectional rain(fall) 对流性阵雨;对流(降)雨
convectional region 对流区(域)
convectional reinforced 按常规配筋的
convectional return 对流逆转
convectional section 对流区(域);对流段
convectional sharpening 普通刃磨
convectional shaving 纵向剃齿
convectional sign 习用符号
convectional snowmelt 对流融雪
convectional storm 对流性暴雨
convectional stoving 对流烘干
convectional superheater 对流式过热器
convectional theory 对流(学)说
convectional thickness 对流厚度
convectional transfer 对流传输
convectional trickling filter 传统生物滤池
convectional type needle valve 通用式针阀
convectional zone 对流层
convection burner 封闭燃烧器;对流燃烧器
convection cell 贝纳单体
convection coefficient 膜层散热系数
convection current 水流;气流;电流;对流;运流
convection diffusion 对流扩散
convection diffusion thermal conductivity cell 对流扩散式热导池
convection drift 运流漂移
convection hall 宴会厅
convection heating 对流加热
convection heating surface 对流加热表面
convection heating system 对流供热系统
convection of energy 能量对流
convection of heat 热的对流
convection of pollutant 污染运流
convection of sensible heat 可感热交换;可感热对流

convection oven 对流式烤炉;对流加热器
convection oven stoving 对流烘箱加温
convection section 对流部分
convective acceleration 对流加速度
convective activity 对流活动
convective cell 对流单体
convective circulation 对流(性)循环;对流性环流
convective cloud 对流云
convective-cloud-height diagram 对流云高度图解
convective condensation level 对流凝结高度
convective conductance 对流传导
convective cooling 对流冷却
convective core 对流核
convective current 对流气流;对流(环流);电流环流;气流环流
convective diffusion 对流扩散
convective diffusion equation 对流扩散方程
convective diffusion thermal conductivity cell 对流扩散式热导池
convective discharge 对流放电
convective dispersion 对流弥散
convective drying 对流烘干;对流干燥
convective effect 对流效应
convective envelope 对流包层
convective equilibrium 对流平衡
convective flow 对流(流)
convective furnace 对流加热炉
convective geothermal system 对流型地热系统
convective hear loss 对流热损耗
convective heat exchange 对流热交换
convective heat flux 对流热通量
convective heating 对流加热;对流式供暖;对流采暖
convective heating surface 对流传热面
convective heating system 对流供热系统
convective heat transfer 对流热传递
convective heat-transfer coefficient 对流传热系数
convective hydrothermal system 对流热水系统
convective instability 对流不稳定(性)
convective instability criterion 对流不稳定性判据
convective layer 对流层
convective lift 对流升高
convective lifting 对流抬升
convective loop 对流环路
convective mixing 对流混合
convective overshooting 对流过冲
convective overturn 湖水对流;对流翻腾
convective precipitation 对流(性)降水
convective pressure 对流压力
convective process 对流过程
convective rain 对流雨
convective rainfall 对流性阵雨;对流降雨
convective refraction 对流折射
convective regenerative cooling 对流再生冷却
convective region 对流区(域)
convective scale 对流尺度
convective shower 对流性阵雨
convective stability 对流稳定性
convective thunderstorm 对流性雷暴雨
convective transfer 对流转移
convective transport 热能对流传递量
convective velocity scale 对流速度尺度
convective weather 对流性天气
convective zone 对流带
convector 换流器;换流机;环流机;对流散热器;对流器;对流加热器;热空气循环对流加热器
convector fire 对流器加热
convector heater 对流式加热器
convector panel 对流器加热板
convector plate 对流垫板
convector radiator 对流(式)(散)热器;对流放热器
convectron 对流测偏仪
convelater 输送提升联合机
convelator 提升输送机
convene 召集
convener 会议召集人
convenience 生活设施
convenience foods 现成食品
convenience goods 大路货;方便商品;便利品;日用品
convenience motive 方便动机
convenience of customers 方便用户
convenience of path system 路线网方便性
convenience outlet 便利商店;简易式接线盒

convenience receptacle 插座
convenience sample 方便样本
convenience sampling 方便抽样;任意抽样
convenience store 自选商店;便利商店;方便店
conveniency 生活设施
convenient 便当的
convenient construction machine 通用结构电机
convenient goods 便利品
convenient sampling 方便性选样
convenient speed 适宜速度
convenienve receptacle 电源插座
convent 女修道院;女修道会
convent architecture 女修道院建筑
conventicler 集会场所
convention 协定(公约);习用性;习俗;习惯;规约;惯例;公约;常规;风俗习惯;条约
conventional 习用的;惯用的;传统的;常规的
conventional abundance 惯用丰度
conventional accounting 传统会计
conventional activated sludge 常规活性污泥
conventional activated sludge process 习用活性污泥法;常规活性污泥法;普通活性污泥法;通用活性污泥法
conventional activated sludge reactor 常规活性污泥反应器
conventional activated sludge system 常规活性污泥系统
conventional activity coefficient 通用活度系数
conventional aeration 传统曝气(法);常规曝气;普通曝气
conventional aggregate 惯用集料;惯用骨料;常规骨料;普通集料;普通骨料
conventional air aerobic digestion 传统空气好氧消化
conventional alphabet 通用字母
conventional anaerobic reactor 常规厌氧反应器
conventional barn 常规畜舍
conventional biologic(al) filter 常规生物滤池
conventional biologic(al) process 常规生物处理
conventional breakbulk ship 普通杂货船
conventional break hull ship 常规杂货船
conventional budding 常规芽接
conventional camera 传统摄像机
conventional cargo ship 常规杂货船;普通杂货船
conventional cargo terminal 普通杂货码头
conventional chemical process 传统化学法
conventional coagulation system 常规混凝系统
conventional collateral for loans 沿用贷款抵押品
conventional concrete 常规混凝土;普通混凝土
conventional construction 传统建筑;常规结构
conventional container ship 混载型集装箱船
conventional control 常规控制
conventional coordinates 标准坐标;通用坐标
conventional coordinate system 惯用坐标系;标准坐标系
conventional creep limit 规定徐变极限
conventional customs duties 协定关税
conventional data 通用数据
conventional depreciation 惯例折旧
conventional design 常规设计;通用设计
conventional design method 常规设计方法;通用设计方法
conventional diagram 示意图;关系示线图;习用图表;惯用图(表)
conventional diffused aeration tank 习用扩散曝气池;常规扩散曝气池;通用扩散曝气池
conventional digester 常规消耗池
conventional display 线性显示;常规显示
conventional dive 常规潜水
conventional diving 常规潜水
conventional diving operation 常规潜水作业
conventional door 常用门;普通门
conventional downflow filter 传统下流式滤池
conventional drawing 通用图
conventional drivage method 习用掘进法
conventional drying bed 常规干化床
conventional dump track 普通自卸汽车
conventional efficiency 约定效率
conventional elastic limit 协定弹限;常规弹限
conventional electrolysis 常规电解
conventional enclosed container 封闭型集装箱
conventional energy resource 常规能源
conventional energy source 常规能源
conventional equipment 常规设备;标准设备;普

通设备
conventional excavation 传统施工法;传统开挖法
conventional explosive 普通炸药
conventional expressing method 常规表示法
conventional figure 惯用图形;惯用数字代号
conventional force 集中力
conventional form 决算表;惯例式;传统格式
conventional Freundlich isotherm 通用弗兰德利吸附等温线
conventional fuel 常规燃料
conventional galvanizing 普通热镀锌
conventional gasoline engine 传统汽油发动机
conventional gear 常规渔具
conventional gravure 照相凹版
conventional grinding 常规研磨
conventional grouping 常规分组
conventional head 常规磁头
conventional heat treatment 普通热处理
conventional hobbing 普通滚削
conventional hotel 会议旅馆
conventional house 常规房屋;非预制房屋;传统房屋
conventional hull 常规船体
conventional industries 传统产业
conventional injection 钻杆注浆
conventional instability 对流性不稳定度
conventional interest rate 协定利率
conventional international law 国际协约法
conventional international origin 国际习用(原)点;国际惯用原点;国际协议原点
conventional international polar 国际惯用极
conventionalization 程式化(雕刻题材)
conventional jacket 习用夹套;整体夹套
conventional jointed track 传统式有接头的轨道
conventional law 协约法;契约法
conventional layout machine 普通划线机
conventional legend 图例
conventional letter 假设符号;惯用字符;代表符号
conventional linear strain 条件线应变
conventional liner terminal 普通定期班轮码头
conventional linework 普通线划图
conventional load(ing) 习用荷载;常用荷载;常遇荷载
conventional loan 普通贷款
conventional lot processing 常规批量处理
conventionally-powered ship 普通动力观测船
conventional management 因袭管理法
conventional mapping 常规测图
conventional measurement method 常规测定法
conventional measurement unit 习用计量单位
conventional measuring method 常测法
conventional memory 常规内存
conventional method 习惯做法;习用(方)法;惯用方法;惯例;传统方法;习用方法;常规方法;普通方法
conventional method of analysis 常规分析方法
conventional milling 逆铣;惯用铣法
conventional mining 常规采煤法
conventional money rate 协定利率
conventional mud 普通泥浆
conventional normal fault 标准正断层
conventional number 标志数
conventional of the bill 汇票的收款
conventional packed column 常规填充柱
conventional packing 习惯包装;通常包装
conventional paint 习惯用漆;传统漆
conventional place name 惯用地名
conventional plant 常规电站;常规电厂
conventional plumb point 假定天底点
conventional point sources 常规点污染源
conventional pollutant 常规污染物
conventional polypropylene biofilm carrier 常规聚丙烯生物膜载体
conventional practice 习惯做法
conventional price 协定价格
conventional printing 直接印花
conventional process 惯用方法;传统工艺;常规方法
conventional processing 常规处理
conventional program(me) 常规程序
conventional projection 习用投影;惯用投影
conventional radio service 常规无线电业务
conventional railway 常规铁路
conventional railway track 传统式铁路轨道

conventional rate 常规速率;通常利率
conventional reaction 通常反应
conventional reclamation treatment 常规回收处理
conventional rectification 常规纠正
conventional reflection seismic 常规反射地震
conventional refractory castable 普通耐火浇注料
Conventional Relating to Civil Liability in the Field of Maritime Carriage of Nuclear Material 国际海上核材料运输民事责任公约
conventional representation 传统表示法;习用表示法;惯用表示法
conventional resources 传统资源
conventional roving 并股无捻积纱
conventional sampling 常规采样
conventional saturation diving 常规饱和潜水
conventional sedimentation tank 常规沉淀池
conventional shaft sinking method 习用凿井方法
conventional ship 一般货船;常规(杂货)船;普通杂货船
conventional sign 习用符号;惯用符号;车辆标记;常规标志;标准图例;图例;通用符号;通用标志
conventional sign card 惯用符号表
conventional signs and abbreviations 通用海图图式
conventional single-phase continuously stirred-tank reactor 常规单段连续搅拌池式反应器
conventional species 常规物种
conventional spelling 惯用译名
conventional spinning 人工旋压法;普通旋压
conventional stabilization pond 常规稳定塘
conventional standard state 常规标准态
conventional steel section 普通型钢
conventional stormwater pollutant 平常暴雨水污染物
conventional stormwater pollutant control 常规暴雨水污染控制
conventional strain 名义应变;公称应变
conventional streamer 常规电缆
conventional study 常规调查
conventional substation 开敞式开关站;开敞式变电站;常规开关;常规变电站
conventional surface water treatment process 常规地表水处理法
conventional suspension 普通悬架
conventional switchgear 开敞式开关;常规开关
conventional switchgroup 敞开式组合电器;常规组合开关
conventional symbol 习用符号;惯用符号;图例;通用符号
conventional symbol card 惯用符号表
conventional symbolization 惯用符号
conventional synchrotron 常规型同步加速器
conventional system analysis and design 传统的系统分析和设计
conventional take-off landing 常规起飞降落
conventional tariff 协定税则;协定税率;协定关税(税则)
conventional tax 协定税额
conventional test 常规试验;常规测试;标准试验;普通试验
conventional tillage 常规耕作法
conventional tool 标准工具
conventional track support system 传统轨下基础
conventional transcription 惯用译名
conventional treatment 常规处理
conventional treatment process 常规处理过程
conventional trickling filter 习用滴滤池;常规滴滤池;标准滴滤池;通用滴滤池
conventional tunnel(l)ing 常规法隧道施工
conventional type 习用型;常规型
conventional type lubricant 习用润滑油
conventional type of sewage treatment 常规污水处理
conventional-type sprinkler 惯用式喷洒器
conventional unit system 常用单位制
conventional valve 习用阀
conventional vessel 常规容器;普通货船
conventional wall construction 双面墙结构(木框架结构夹衬板,两边覆以胶合板的墙体结构);常规结构
conventional waste water treatment 常规污水处理;常规废水处理(法)
conventional waste water treatment operation 常规污水处理运作

conventional waste water treatment plant 常规废水处理厂
conventional water pollution analysis index 常规水污染分析指标
conventional water treatment 常规水处理
conventional water treatment plant 常规水处理厂
conventional water treatment process 常规水处理工艺
conventional water treatment technique 常规水处理技术
convention area 公约区
Convention Concerning Protection of the World Cultureal and Natural Heritage 保护世界文化和自然遗产公约
conventioneer 到会人
Convention for Protection of the Mediterranean Sea Against Pollution 保护地中海免受污染公约
convention hall 会议厅;会议一宴会厅
convention method 习用(方)法
Convention of Facilitation of International Maritime Traffic 国际便利海上运输公约
convention of international trade 国际贸易惯例
Convention on Assistance in Case of a Nuclear Accident or Radiologic Emergency 核事故或辐射紧急情况援助公约
convention on civil liability for oil pollution damage 民事油污染事故责任公约
Convention on Early Notification of a Nuclear Accident 及早通报核事故公约
Convention on Fishing and Conservation of Aquatic Resources of High Seas 公海生物资源捕捞及养护公约
Convention on International Regulations for Preventing Collisions at Sea 国际海上避碰规则公约
Convention on International Trade in Endangered Species of Wild Fauna and Flora 濒危野生动植物物种国际贸易公约
Convention on Long-range Transboundary Air Pollution 远程越界空气污染公约
Convention on Maritime Mortgages and Liens 海上抵押权和留置权公约
convention on oil pollution 油污染公约
Convention on Prevention of Marine Pollution by Dumping Wastes and Other Matters into Ocean 防止倾倒废物及其他物质污染的海洋公约
Convention on the Continental Shelf 大陆架公约
Convention on the High Sea 公海公约
Convention on the Immunity of State-owned Vessel 国有船舶豁免权公约
Convention on the International Regime of Maritime Ports 国际海港制度公约
Convention on the Territorial Sea, the Contiguous Zone 领海与毗邻区公约
convention single tariff 协订单一税则
Convention to Combat Deserification 防治荒漠国际公约
conventual architecture 女修道院建筑
convectual building 女修道院房屋
convectual church 女修道院教堂
convectual community 女修道院社区
convectual kitchen 女修道院厨房
convectual limb 女修道院侧厅
convectual parlour 女修道院客厅
convergance 趋同作用
converge in probability 概率收敛
convergence 聚敛;聚合差;交会;会聚性;会聚(度);会合;汇合处;辐合;非周期性阻尼运动;趋同;收敛(点)
convergence acceleration 收敛性加速
convergence algorithm 收敛算法
convergence amplitude 会聚幅度
convergence-and-divergence 先收敛再发散
convergence angle 交向角;会聚角;收敛角
convergence attribute 收敛属性
convergence bolt 聚焦螺栓;会聚螺栓
convergence boundary 聚合型边界
convergence calculation 收敛计算
convergence circuit 会聚电路
convergence coil 聚集线圈;会聚线圈
convergence constant 收敛因子;收敛常数
convergence control 聚焦控制;聚焦调整;收敛控制
convergence correction 聚焦校正

convergence criterion 收敛准则;收敛性判据;收敛性判定准则;收敛(性)判别准则;收敛判据
convergence domain 收敛域
convergence electrode 会聚电极
convergence entrance 渐缩形进水口
convergence error 收敛差
convergence excess 辐辏过多
convergence exponent 收敛指数
convergence factor 收敛因子
convergence field 辐合场
convergence forward pruning 收敛正向修剪
convergence frequency 收敛频率
convergence ga(u)ge 下沉测量仪;沉陷测量仪
convergence indicator 收敛仪;顶底板闭合测示仪;收敛计
convergence in distribution 依分布收敛;分布收敛
convergence in mean 均值收敛;依平均值收敛;平均收敛
convergence in mean square 均方收敛
convergence in measure 依测度收敛
convergence in probability 依概率收敛;概率性收敛
convergence limit 会聚极限
convergence line 辐合线
convergence magnet 静会聚磁铁;会聚磁铁
convergence map 等容线图;等电位线图;收敛图
convergence measure 收敛测量
convergence method 逐次近似法;收敛法
convergence model 辐合模式
convergence nozzle 收敛形喷嘴
convergence of algorithm 算法的收敛
convergence of discrete approximation 离散逼近的收敛性
convergence of drainage 汇流
convergence of evidence 信息综合
convergence of generating function 母函数的收敛性;幂级数的收敛性
convergence of grid bearing 坐标纵线偏角
convergence of inner demands 内在要素汇集
convergence of meridian 子午线收敛角
convergence of power series 幂级数的收敛(性)
convergence of rays 光线会聚
convergence of series 级数的收敛性
convergence of truncation-error 截断误差收敛
convergence penalty function 收敛罚函数
convergence phase control 会聚电压相位调整
convergence plane 会聚平面
convergence point 收敛点
convergence pressure 会聚压力
convergence principle 收敛原则
convergence projection theory 会聚投射学说
convergence property 收敛性
convergence radius 收敛半径
convergence rate 收敛速率
convergence ratio 会聚比
convergence recorder 测录器
convergence reflex 聚合反射;集合反射;辐辏反射
convergence region 收敛区(域)
convergence sequence 收敛序列
convergence series 收敛级数
convergence surface 会聚面
convergence system 会聚系统;收敛系统
convergence test 会聚试验;辐向试验;收敛性判定;收敛试验
convergence theory 收敛理论
convergence thesis 趋同理论
convergence time curve 下沉量—时间曲线;收敛时间曲线
convergence voltage 会聚电压
convergence zone 会聚区;辐合区(域)
convergency 会聚性;会聚;辐合
convergency calculation 收敛计算
convergency constant 收敛常数
convergency error 收敛差
convergency measure 收敛测量
convergency point 辐合点
convergency theory 趋同理论
convergency value 收敛值
convergent 渐近分数;收敛子;收敛项;收敛的
convergent adaptation 趋同适应
convergent ages 收敛年龄
convergent anastomosing system 交织水系
convergent angle 交向角;会聚流;会聚角;收敛角
convergent area of main and branch waterways 干支流交汇水域
convergent beam 会聚光束;收敛束
convergent boundary 会聚边缘;收缩边界
convergent branching system 收敛分支系统
convergent camera 交向摄影机
convergent channel 会聚槽;收缩渠道;辐合状水道网
convergent character 趋同特性
convergent characteristic 趋同特性
convergent component 会聚元件
convergent conduit 收缩管道
convergent contour 收敛形
convergent control 收敛控制
convergent crustal consumption zone 地壳对接消减带
convergent-dilatation 收缩膨胀
convergent-divergence 收敛扩散
convergent-divergent 缩放;收缩膨胀的;收敛扩散的
convergent-divergent channel 收缩扩张管;拉瓦尔管;缩放形流道
convergent-divergent inlet 拉瓦尔喷管扩散段;渐缩放喷管
convergent-divergent nozzle 拉瓦尔喷管;超音速喷管;收敛扩散形喷管
convergent drainage 辐聚水系
convergent evolution 辐合进化;趋同进化
convergent flow 汇流;合流;收束水流;收敛(水)流
convergent ga(u)ge 收敛计
convergent homeomorphy 趋同同型
convergent homomorphism 趋同同型
convergent illumination 聚光照明
convergent improvement 辐合改进
convergent iterative procedure 收敛迭代程序
convergent lens 聚光(透)镜;会聚透镜;辐合透镜
convergent light 会聚光;汇聚光
convergent magnetic field 收敛磁场
convergent matrix 收敛矩阵
convergent mirror 会聚镜
convergent model datum 交向模型基准
convergent mode of motion 衰减运动
convergent mouthpiece 收缩管(嘴)
convergent movement 聚敛运动
convergent nozzle 渐缩喷嘴;收缩喷嘴;收敛喷嘴
convergent oblique 交向倾斜
convergent oscillation 减幅振动;减幅振荡;衰减振荡
convergent pencil 聚会光束;会聚光束
convergent pencil of rays 聚光束;集光角
convergent photograph 交向摄影像片
convergent photography 交向摄影
convergent pipe 异径管
convergent plate 聚敛(型)板块;聚合板块【地】
convergent plate boundary 聚敛板块边界
convergent point 会聚点
convergent polarized light 聚敛偏光
convergent position 交向位置
convergent precipitation 辐合降水
convergent reactor 收敛反应堆
convergent response 衰减响应;收敛响应
convergent section 喷管的收敛部分
convergent sequence 收敛序列
convergent series 收敛级数
convergent streaks 辐合条纹
convergent thinking 辐合思维
convergent tube 异径管
convergent wave 会聚波
converging action 会聚作用;收敛作用
converging beam 会聚波束
converging block post 回合线路所
converging breakwater 合抱防波堤;束口防波堤
converging channel 收束渠道
converging criterion 收敛性判据;收敛性判别准则
converging discharge channel 收缩型卸料通道
converging-diverging cones injector 锥式喷嘴
converging-diverging nozzle 拉瓦喷管;超音速喷管
converging duct 收缩管道
converging flow 渐缩流;会聚流;收缩水流;收束水流;收敛(水)流
converging flow into main channel 束水归槽
converging ion beam 会聚离子束
converging jetty 合抱防波堤;收束导堤
converging lens 会聚透镜
converging light 会聚光
converging magnetic mirror 会聚磁镜
converging meniscus 收敛凹凸透镜
converging mole 合抱防波堤;束口防波堤
converging nozzle 收缩喷嘴;收束管嘴
converging pier 合抱防波堤;束口防波堤
converging pipe 渐缩管;异径管
converging piping 渐缩管
converging point 收敛点
converging radius 收敛半径
converging ray 收敛射线
converging roads 汇合道路
converging transition 收缩渐变段
converging tube 渐缩(形)管;异径管;缩口(短)管
converging tubing 渐缩管
converging type cable-stayed bridge 伞形斜拉桥
convergometer 收敛计
conver lens 凸面镜
conversant with professional knowledge 精通业务的
conversation 会谈;会话
conversational compiler 会话编译程序
conversational entry 会话登记项
conversational system 会话系统
conversational time-sharing 对话式分时操作
conversational way 对话方式
conversation piece 室内风俗画
conversation room 谈话室
conversation test 交换测试
conversazione 学术座谈会
converse 逆命题【数】
converse deficiency 逆亏损
converse effect 逆效应
converse-Labo(u)rer formula 拉巴勒逆变换公式
converse-Laplacian formula 拉普拉斯逆变换公式
converse magnetostrictive effect 反磁致伸缩效应
converse piezoelectric(al) effect 逆压电效应
converse power protection 逆功率保护
converse power protection unit 逆功率保护装置
converse theorem 逆定理
conversion 交换;换位;换算;航向变换;作业变更;转换;转化;变交(段);折算;兑换;复律;反演;翻拍;变换
conversional diagram 示意图
conversion angle 转换角;半收敛差;半聚合差
conversion apparatus 转化设备
conversion at current price 按现价兑换
conversion at market price 按市价兑换
conversion at par 平价兑换
conversion between different wave numbers 不同波数变换
conversion board 坐标变换测绘板;变换测绘板
conversion burner 燃料替换燃烧炉
conversion by drift 漂移变换
conversion by retarding field 减速场变换
conversion characteristics 变换特性
conversion chart 换算图(表);换算表;转换表
conversion check 反校验;反向检查
conversion coating 电镀层;转化(型涂料);改性层(金属表面)
conversion code 变换码
conversion coefficient 换算系数;转换系数;变换因数;变换系数
conversion coefficient of heat 热量换算系数
conversion coefficient value 换算系数值
conversion coefficient value of spectrometer 能谱仪换算系数值
conversion conductance 变频互导;变频电导
conversion constant 换算常数;转换常数
conversion cost 加工成本;换算成本;变换成本
conversion curve 换算曲线;变换曲线
conversion detector 变频检查器
conversion device 转换装置;转换设备
conversion diagram 换算图表;转换特性曲线;变换特性曲线(图)
conversion effectiveness 转换系数
conversion efficiency 换能效率;转换效率;变频效率;变换效率
conversion electron 变换电子
conversion-electron emission 变换电子发射
conversion equation 换算公式;变换方程
conversion equipment 转换装置;转化设备;交换设备
conversion factor 兑换率;换算因子;换算因数;换

算系数;换算率;转换因子;转换因数;转换系数;
折算系数;折合因子;折合系数;再生系数;变流
因数;变频因子;变频因数;变换系数
conversion factor gamma value 换算系数伽马值
conversion factor in biomass 生物质转化系数
conversion factor of frazil slush 冰花折算系数
conversion factor of sheet glass 平板玻璃折算系数
conversion filter 转换滤色镜;变换滤光片
conversion filter for colo(u)r temperature 色温转换滤光器;色温变换滤光片
Conversion for the Protection of the Ozone Layer 保护臭氧层公约
conversion fraction 转换系数;转换部分;变换部分
conversion frequency 变换频率
conversion gain 转换增益;变频增益;变换增益
conversion gain coefficient 变频增益系数;变换增益系数
conversion gain tester 变换增益测试器
conversion gas-fired burner 煤气点火的转化燃烧器
conversion grade 加算坡度;换算坡度
conversion graph 换算图表
conversion in expression 表达式中的转换
conversion instruction 转换指令
conversion key 转换键
conversion kit 转换成套件
conversion length 换算长度
conversion line 变换线路
conversion loan 转换贷款
conversion loss 转换损失;转换损耗;变频损耗;变换损耗
conversion loss coefficient 变频损耗系数;变换损耗系数
conversion method 转化法
conversion of a loan 借款调换;贷款调换
conversion of angle 角度换算
conversion of a vessel 船舶改装
conversion of branch trial balance 分店试算表折算
conversion of cargo 非法交货;无单交货
conversion of course or bearing 向位换算
conversion of currency 货币兑换
conversion of energy 能量转换;能量转化
conversion of mode 波型转换
conversion of olefines 烯烃的转化
conversion of saline water 咸水淡化
conversion of seawater 海水的淡化;海水转化
conversion of stock 库存转换
conversion of stores 库存转换
conversion of subsidiary trial balance 子公司试算表换算
conversion of thermal energy 热能转换
conversion of timber 圆木锯解;锯材;圆木锯方;制材
conversion of unit 单位的换算;设备的重新安装
conversion of units of measurement 度量单位换算
conversion of wood to liquid fuel 木柴—液体燃料转换
conversion oil-fired burner 煤油点火的转化燃烧器
Conversion on Biologic(al) Diversity 生物多样性公约
conversion parity 兑换平价
conversion period 逆周期;反周期;变更期
conversion per pass 单程转化
conversion pig 炼钢生铁
conversion privilege 转换权利;权利转移
conversion process 转化过程;转化处理
conversion program(me) 变换程序;转换程序
conversion rate 换算率;转换速率;转换速度;转化速度;转化率;折算率;兑换率
conversion rate of food 饲料系数
conversion ratio 换算率;转换比值;转换比(率)
conversion reaction 转化反应
conversion resolution 转换分辨力
conversion rod 变换杆
conversion scale 换算尺度;换算比例尺;尺度换算(图)表
conversion spectrum 转换谱;转换(电子)光谱;变换(电子)光谱
conversion stage 变换级
conversion subroutine 转换子程序
conversion system 转换系统;翻拍方法;变更作业体制
conversion systems 图像转换法
conversion table 汇兑表;换位表;换算表;换算表
conversion temperature 转化温度
conversion test 转化试验
conversion time 转换时间
conversion to alternative fuels 转换到代用材料
conversion ton-kilometer 换算吨公里
conversion traffic volume 转换交通量
conversion transconductance 变频跨导;变频互导
conversion transformer 转换变压器
conversion trust funds 公约信托基金
conversion type roller 可更换滚筒;可更换辊轴
conversion unit 变换器;变换单元;变换机组
conversion valve 转向阀;转换阀
conversion voltage gain 变频电压增益;变换电压增益
conversion zone 转化段;反应段;反应层
convert 变换;折算
convert development environment 协同型开发环境
converted amount 折合数量
converted car length 车辆换算长度
converted cruiser 改装巡洋舰
converted data 数据变换
converted dwelling 改造住宅
converted flo(a)tation 多油浮选
converted grade 换算坡度
converted gradient of curve 曲线换算坡度
converted house trailer 改装的拖车住房
converted investment employed 折算投资占用
converted kerb 倒置式(路缘)石
converted length 折算长度
converted mechanism 协调机理
converted merchant ship 改装舰船
converted section 折算断面
converted ship 改装船
converted static load of wagon 换算静载(重)
converted steel 硬质钢;转炉钢;调质钢;渗碳钢
converted timber 锯(制木)材;锯制板;加工木材;粗加工材;成材
converted to an annual basis 以年为基础换算
converted ton-kilometers 换算周转量
converted ton kilometre cost 换算吨公里成本
converted ton kilometre liquidated revenue 换算吨公里清算收入
converted traffic 交通方式转换交通量;换乘交通量;变增交通量
converted traffic volume 转换交通量
converted turnover 换算周转量
converted turnover of railway transportation 铁路运输换算周转量
converted wave 转换波
converted wave processing 转换波处理
converted width 折算宽度
converter = convertor
converter air box 转炉风箱
converter air delivery system 转炉供风系统
converter arm 变流臂
converter blocking 换流器闭锁
converter blowing 转炉吹炼
converter body 转炉炉身
converter bottom 转炉炉底
converter bridge 换流器桥
converter charge 转炉炉料;吹炉炉料
converter charging platform 转炉加料台;吹炉加料台
converter circuit 变频(器)电路;变换器电路
converter coupling 转换联轴器
converter deblocking 换流器解锁;换流器导通
converter drain port 变扭器排油孔
converter equipment 变流设备
converter-fed motor 变流器供电电动机
converter gas 转炉(炉)气
converter heat ga(u)ge 换能器热力计
converter labyrinth port 变扭器迷宫口
converter lining 转炉炉衬
converter lockup 液力变矩器闭锁
converter logic 变换逻辑
converter mill 转炉车间
converter mouth 转炉炉口
converter noise 变频噪声;变换噪声
converter nose 转炉炉鼻
converter pig 转炉炼钢用生铁
converter plant 转炉车间;变频装置;变流装置
converter platform 转炉炉台
converter process 转炉炼钢(法)
converter punch tape 变换冲孔带
converter reactor 转换(反应)堆
converter residue 转换残渣
converter room 变流机室
converter shed 转炉厂
converter shell 转炉炉壳
converter slag 转炉炉渣
converter stage 变频级
converter station 换流站
converter steel 转炉钢
converter substation 换流分站
converter technique 转换技术;变频技术;变换技术;变换法
converter-transmitter 变流器发射机
converter trunnion ring 转炉托圈
converter tube 变频管
converter tuyere box 转炉风口箱
converter-type ladle 倾转式浇包
converter unit 变换器组;变换单元
converter unit firing control 换流器触发控制
convertibility 可逆性;可更换性;可兑换性;可变换性;互换性;自由兑换;转换性;转化性;转变性
convertibility of dollar into gold 美元兑换黄金
convertibility of expansion and shrinkage 胀缩可逆性
convertible 可转换的;可逆的;可交换的;可兑换的;可变换的;敞篷车;篷顶小客车
convertible account 可兑换账户
convertible attachment 可更换附件
convertible bank note 可兑换纸币;可兑换银行券
convertible body 活顶车(车顶可打开)
convertible bond 可转换债券;可换公司债;可兑现债券;可兑换债券;可调换债券
convertible bonds yields 可调换债券的收益;可兑换债券收益
convertible car 两用(汽)车;两用车辆;活顶汽车;折合式敞篷汽车;敞篷(小)汽车
convertible coating 交联型涂料;转化型涂料
convertible container ship 可变货柜船;集装箱两用船;杂货集装箱两用船
convertible coupe 活顶双门轿车
convertible crane 可更换装备(的)起重机;多用途起重机;通用起重机
convertible credit 自由外汇贷款
convertible currency 自由外汇;可(自由)兑换货币;可兑换的外币
convertible debenture 可转换债券;可换(公司)债券
convertible diesel-electric caterpillar excavator 柴油机—电动两用履带式挖土机
convertible double glazing 可转换双光泽
convertible drive 可转换驱动(器)
convertible equipment 可转换设备
convertible excavator 通用挖土机;通用挖掘机
convertible floating rate note 可转换浮动利率本票
convertible foreign exchange 现汇;国际货币
convertible front equipment 可更换的前端设备
convertible gas engine 可转换煤气内燃机
convertible gold note 可兑换金券
convertible husbandry 轮作经营
convertible insurance 转换保险
convertible lens 转换透镜
convertible loan 可兑换放款
convertible loan stock 可(兑)换信用债券
convertible matter 可变换的物质
convertible mortgage 可调换的抵押
convertible notes 可兑换货币
convertible objective 可换物镜
convertible paint 交联型涂料;转化型涂料
convertible paper 可兑换证券
convertible paper currency 可兑换纸币
convertible paper money 可兑现纸币
convertible preferred share 可转换优先股
convertible preferred stock 可换优先股;可兑换优先股
convertible price 兑换价格
convertible process 吹炼法
convertible Protar lens 可换普罗塔镜头
convertible revolving credit 可转换循环信用;可更换的循环信用证
convertible room 兼用起居室

convertible securities 可换证券
convertible securities adjustments 可转换证券调整
convertible sedan 活顶轿车
convertible ship 可改装的船(舶)
convertible shovel 两用铲;可变换铁铲;正反两用铲;正反铲挖土机;通用挖土机;通用挖掘机
convertible shovel-crane 挖掘起重两用机
convertible silver note 可兑换银券
convertible stock 可(转)股份;可兑换股票
convertible subordinated unsecured loan stock 可换股次级无抵押债务
convertible telephoto objective 可转换摄远物镜
convertible term assurance 可调换定期人寿保险
convertible transit level 经纬水准仪
convertible type 转化型
convertible unsymmetrical anastigmat 可变换的非对称消像散镜组
convertible vehicle 两用轮胎车;两用履带车;两用车辆
convertible vessel 可改装的船(舶)
convertin 转变加速素
converting 吹(炉)炼
converting area from depression into uplifting 下降转为上升区
converting area from uplifting into depression 上升转为下降区
converting cost 加工成本
converting furnace 吹炼炉;吹风氧化炉
converting plant 换流变电站;整流站
converting process 转炉炼钢(法);吹炼过程;吹炼法
converting sewage effluent 转化污水处理后出水
converting technique 变流技术
converting with air 空气吹炼
convert into dollars 折成美元
convertional method 矿山法(指隧道开挖)
convertiplane 推力换向式飞机
convertiport 垂直起降机场
convertor 换能器;换流器;转炉;转换器;转换程序;转化器;转化炉;再调制器;吹炉;吹风炉;变流器;变换器
convertor arm 换流臂
convertor bridge 换流桥
convertor connection 变流连接
convertor equipment 变流设备
convertor transformer 换流变压器;变流变压器
convert waste into useful material 利用废料
convert wastes into useful materials 废物利用
conservationist 自然资源保护主义者
convervative pollutant 恒定污染物
convervative value 恒定值
convex 凸状;凸圆的;凸形的;凸线;凸起;凸面;凸(出)的
convex and concave crepe 凹凸绉
convex and concave milling cutter 凸凹铣刀
convex angle 凸角
convex arch 内外三心拱;凸形拱
convex area 凸起表面
convex bank 突岸;凸岸
convex bar 凸岸边滩
convex bead 玻璃瓶子凸线条
convex bend 凸弯管
convex bend frame 外弯肋骨
convex billet 凸形钢坯
convex bit 中心凸出钻头
convex bottom 凸底(玻璃缺陷)
convex bow 凸起翘曲
convex box pass 凸底箱形孔型
convex camber 鼓出部
convex closure 凸包
convex combination 凸组合
convex-concave 凹凸形;凸凹(的)
convex concave shaped crusher plate 凹凸曲线型颚板
convex cone 凸锥
convex-convex 两面凸的;双凸
convex coping of a wall 墙的凸面压顶
convex corner 凸角
convex(cross-)section 凸形断面
convex crystal 凸晶
convex curvature 反挠度;凸曲率
convex curve 凸形曲线;凸曲线
convex cutter 凸形铣刀;凸半圆成型铣刀
convex edge 凸边

convex-end electrode 凸头形电极
convex fillet weld(ing) 凸面角焊(缝);凸(形)角焊缝
convex flank cam 凸腹凸轮
convex flood plain 凸形河漫滩
convex flow 凸岸水流
convex fracture 凸断口
convex fringe of crater 撞击坑凸边
convex glass 凸面玻璃;凸镜
convex gradient 凸梯度
convex head 凸形封头
convex head valve 凸形气门头
convex hull 凸形外壳;凸包
convex interface 凸界面
convex interlocking cutter 凸状联锁铣刀
convex iron 半圆铁
convexity 中凸;凸性;凸形;凸弯形;凸起;凸面体;凸度
convexity for a production set 生产集的凸性
convexity of character 外凸字体
convexity ratio 圆度比
convex joint 圆截凹缝;凸圆(接)缝;凸缝
convex lens 正透镜;凸透镜;凸面镜
convex lug 凸耳
convex milling cutter 凸半圆铣刀
convex mirror 凸(面)镜
convex mo(u)lding 凸纹线脚
convexo-concave 一面凸一面凹的;凹凸形;凸凹(的)
convexo-concave lens 凸凹透镜
convexo-concave seal face 凹凸密封面
convexo-convex 两面凸;双凸形;双凸(的)
convex of contact 接触凸出部
convexo-plane 凸平形的;一面凸一面平的;平凸形;凸平的
convex optic(al)tool 光学玻璃凸面磨具
convex perforated plate 凸形多孔板
convex plane 平凸形
convex polygon 凸多边形
convex polyhedral cone 凸多面锥体
convex polyhedral set 多面凸集
convex polyhedron 凸多面体
convex profile 背弧型面
convex programming 非线性规划;凸(形)规划(法)
convex programming problem 凸形规划问题
convex radius 凸曲线半径
convex relief 正向地貌
convex ridging body 长凸面起垄犁体
convex round mo(u)lding 凸圆线脚
convex rule 凸面卷尺
convex rule coating 钢卷尺涂料
convex set 凸集
convex shell 凸面壳(体)
convex shore 凸岸
convex side 凸岸一侧
convex slope 凸(形)坡
convex solid cutter 凸形整体铣刀
convex state 凸海岸国家
convex stream bank 凸河岸
convex surface 背弧面
convex table 凸面圆形淘汰盘
convex tile 盖瓦;牡瓦;凸瓦(俯瓦);凸筒瓦
convex tool 光学玻璃凸面磨具
convex tooled joint 凸圆接缝;凸圆缝
convex vertical curve 凸形竖曲线
convex weld 凸形焊缝;凸面焊;凸焊(焊)缝
convey 转让;传达
conveyance 流通;运送;运输工具;传递;传达;财产让与;让与证据;输送率;输送;输水率
conveyance by land 陆路运费
conveyance canal 输水渠
conveyance capacity 输水能力
conveyance channel 输水渠(道)
conveyance compartment 容器隔间
conveyance conduit 引水管道
conveyance culvert 输水涵洞
conveyance device 输液管
conveyance factor 运输因素;输水系数
conveyance fluid 运输流体;输送流体
conveyance loss 运输损失;运输损耗;输送损失;输水损失
conveyance of irrigation water 灌溉水的输送
conveyance of property 转让产权

conveyance of purified water 净水运输
conveyance of reuse water 再用水的输送
conveyance of turbid essence to the heart 浊气归心
conveyance power of water 输水能力
conveyancer 输送设备
conveyance ratio 输水系数;输水比
conveyance structure 输水建筑物
conveyance system 运输系统;输水系统;输水管线
conveyance tunnel 运输隧洞;运输隧道;输水隧洞;输水隧道
conveyancing 财产转让业务;不动产让与手续;让与证书制作业;物业交付文据制作
convey by air 用气运送
convey condition 车队行驶状况
convey cost 运送费
conveyed quantity 输送量
convey energy 输送能量
conveyer 运输器;运煤机;传送机;传送机;让与人;输送器;输送器
conveyer apron 输送带
conveyer auger 螺旋输送器
conveyer axle 输送器轴
conveyer ball joint 输送机窝球节
conveyer band 输送带;输送带
conveyer belt(ing) 运输(机皮)带;皮带运输机;皮带运输机;传送(器皮)带;输送带
conveyer belt frame 输送机带支架
conveyer belt furnace 带式传送炉
conveyer belt lehr 网带式隧道退火窑;履带式退火窑
conveyer-belt sorter 传输带分类机;传送带分类机
conveyer belt tripper 输送机卸料器
conveyer belt with sides 挡边输送机
conveyer boom 输送器架
conveyer bridge 输送机支架
conveyer bucket 运料斗;输送料斗
conveyer by compressed air 压缩空气传送
conveyer cable 输电线
conveyer canvas 帆布输送带
conveyer centrifuge 带式离心机
conveyer chain 搬运(机)链;输送(机)链
conveyer charging hopper 运输机装卸漏斗;运输机装料漏斗
conveyer coating 传送带涂装
conveyer crusher 输送轧碎机
conveyer delivery mechanism 传送机输送机构
conveyer depilator 传送带式脱毛机
conveyer drier 输送器干燥炉;输送带式干燥机
conveyer drive 输送器传动装置
conveyer driven shaft 动力输出轴
conveyer drive pinion 输送机传动小齿轮
conveyer drive ratchet wheel 输送机主动棘轮
conveyer drive shaft 输送器传动轴;输送机主动轴
conveyer drive shaft bush(ing) 输送机传动轴衬套
conveyer drive shaft end cover 输送机主动轴端盖
conveyer drying oven 带式干燥器;带式干燥机
conveyer end bearing 输送机端承
conveyer end bearing cover 输送机端承盖
conveyer end bearing cover bushing 输送机端承盖衬套
conveyer energy 输送能量
conveyer equipment 输送设备
conveyer feed distributor 带式撒肥机
conveyer filler 运输机装料工
conveyer flexible drive shaft 输送机挠性主动轴
conveyer for hot materials 热材料输送机
conveyer for silvering 带式镀银机
conveyer furnace 输送带式炉
conveyer gallery 皮带运输廊;输送机廊道;运输皮带廊道
conveyor grinding and polishing 连续磨光
conveyer house 输送机房
conveyer hopper 输送器漏斗
conveyer hopper bearing plate 输送机斗承板
conveyer hopper support 输送机煤斗支座
conveyer idler 输送带托辊
conveyer idler roller 输送带支持滚轮
conveyerization 运输机化;输送机化
conveyer jib 运输机壁
conveyer line 传送线;输送机线路
conveyer loader 运输装载机;输送带(式)装载机
conveyer loading hopper 运输载料斗;运输机装卸漏斗;运输机装料漏斗
conveyer pawl casing cover 输送机棘爪盖

conveyer pawl reverse 输送机棘爪回动法
conveyer pawl reverse pin 输送机棘爪回动销
conveyer picker 手选运输机;手选皮带;手选工人
conveyer pipe for material 风力送料管
conveyer pulley 输送机皮带轮
conveyer ratchet wheel 输送机棘轮
conveyer reverse cover 输送机回动盖
conveyer reverse quadrant 输送机回行扇形齿轮
conveyer reverse unit 输送机变向装置
conveyer roller 输送机滚柱
conveyer rolling machine 运输带鼓轮机
conveyer scale 传送带秤
conveyer screen 运输筛分机
conveyer screw 螺旋运输机;螺旋输送器;螺旋输送机
conveyer screw drive shaft 输送机螺旋主动轴
conveyer screw shaft 输送机螺旋轴
conveyer screw shaft coupling 输送机螺旋轴联轴节
conveyer screw thrust plate 输送螺旋止推板
conveyer slide support 输送机滑行托
conveyer stretcher 输送带张紧装置
conveyer system 流水作业系统;传送带流水作业法;输送机系统
conveyer table 转运台
conveyer train 输送机组;输带机组
conveyer trough 输送机承料槽
conveyer tunnel 输送带地道
conveyer trunk 传送管
conveyer type batcher scale 传送带式定量磅秤
conveyer type coating 带式挤涂;抽涂
conveyer type continuous furnace 输送带式连续炉
conveyer type furnace 输送带式加热炉
conveyer type molybdenum resistor furnace 传送带式钼电阻炉
conveyer type quench tank 输送带式淬火槽
conveyer type sample 皮带式取样器
conveyer type sand blast machine 输送机式喷砂机
conveyer type scale 输送带秤
conveyer type scraper 输送式刮泥机;输送器型刮泥机;输送刮泥器
conveyer type wagon unloader 皮带式卸车机
conveyer way 廊道;输水道
conveyer type weigher 带式计量秤
conveyer weigh meter 皮带输送机秤;传送带秤
conveyer worm 螺旋式输煤机
conveying 运输
conveying appliance 输送装置
conveying arrangement 运输装置
conveying belt 传送带;输送带
conveying bucket 输送斗
conveying canal 输水渠
conveying capacity 传送量;输送能力;输送量;输水能力
conveying chain 运输链
conveying channel 输水渠
conveying chute 输送溜槽;输送斜槽
conveying conduit 输送管(道)
conveying cost 运输费(用);运送费
conveying device 输送装置
conveying device for kiln 窑用输送设备
conveying distance 输送距离
conveying drier 运输干燥机
conveying efficiency 运输效率
conveying energy 输送能量
conveying en masse chain 埋刮链输送机;埋刮板输送机
conveying facility 输送设备
conveying fan 输送风机
conveying hose 输送管;输料软管;输送软管
conveying installation 传送装置
conveying line 传送线;输送线
conveying machinery 输送机械
conveying mechanism 输送机构
conveying medium 输送介质
conveying pipe 输送管(道)
conveying pipe line 输送管线
conveying plant 输送设备
conveying roller 传送滚筒
conveying screw 螺旋输送机;输送螺旋
conveying shute 输送槽
conveying spiral 螺旋输送机
conveying system 传送系统;输送系统

conveying track 输送段
conveying trough 溜槽;运输槽;传送槽;输送槽
conveying tube 输送管(道)
convey(ing) tunnel 输水隧洞;输水隧道
conveying velocity 传送速度;搬运速度
conveying vessel 仓式输送泵
conveying worm 螺旋输送器;输送蜗杆
conveyor = conveyer
convey tube 输送管(道)
convey tubing 输送管(道)
convey tunnel 运输机平巷;运输机坑道;输水隧洞;输送隧洞;输送隧道
conviction 确信
convince 确信
convinced test 确认试验
convocation 集会
convoke 召集
convolute 旋绕;盘旋形;盘旋面
convolute bedding 卷曲层理;旋转层理;旋绕层理;旋卷层理;包卷层理
convoluted bellow 卷绕波纹管
convoluted diaphragm 波纹膜片
convoluted seminiferous tubule 曲精细管
convolute lamination 旋卷纹理;包卷纹理
convolute structure 旋转构造;旋卷构造
convolution 卷折;卷旋;卷积;结合式;回旋;卷积;匝
convolutional ball 包卷球
convolutional code 卷积码
convolutional encoding 卷积编码
convolution code 卷积码;褶合编码
convolution filtered image 卷积滤波图像
convolution filtering 卷积滤波;褶积滤波
convolution filtering method 褶积滤波法
convolution integral 卷积积分;结合积分;回旋积分;褶积积分;折合积分
convolution integral model 卷积积分模型
convolution integration 回旋积分
convolution in wave number domain 波数域褶积
convolution matrix 褶积矩阵
convolution method 卷积法;褶合法
convolution of probability distribution 概率分布褶积;概率分布卷积
convolution operation 褶积运算
convolution output 卷积输出
convolution property 卷积特性
convolution result 褶积结果
convolution signal 卷积信号
convolution summation 卷积和
convolution theorem 卷积定理;褶积定理
convolution transform 褶积变换;褶合式变换;对合变换
convolve 盘旋
convolved Cauchy distribution 缠绕柯西分布
convolved distribution 缠绕分布
convolved normal distribution 缠绕正态分布
convolved Poisson's distribution 缠绕泊松分布
convolver output 卷积输出
convolvulus althaeoides 牵牛花
convoy 护队;护航线;护航
convoy assembly port 护航会集港
convoy condition 队列行驶
convoy(ing) ship 护航舰;护航船
convoy(ing) vessel 护航船
convoy pennant 护航三角识别旗
convoy vessel 护送船
convulsion 灾变
convulsionism 灾变论
Conwall bracket 康沃架(一种供暂时支承木脚手架、模壳的专利支架)
conweigh belt 装卸机
co-obligor 联合债务人
cook 厨师;烘烤固化
cook accounts 虚报账目;伪造账目
cook an account 虚报账目
Cook-Basmadjian method 科克—巴斯玛金法
cookbook 详尽说明书
cooked glue 热用胶
cooked mix(ture) 加热混合料
cooked mode 加工模式
Cooke elutriator 库克型淘析器
Cooke eyepiece 库克目镜
cookeite 锂绿泥石;细鳞云母
Cooke objective 库克物镜

cooker 炉灶;烤箱灶;火炉;蒸煮器;锅
cooker burner 灶面燃烧器
cooker cock 炉灶旋钮
cooker control unit 电灶控制设备
cooker hood 抽油烟机
Cooke triplet lens 库克三合透镜
Cooke triplet objective 库克三合物镜
Cooke unit 库克单位
cook house 厨房
cookie cutter 圆底形钢制沉箱
cooking 烘烤固化;烧煮;烹饪;焦化;(木材的)蒸煮处理;蒸缸熟化;熬炼;热炼
cooking apparatus 烹饪器具;烹饪器皿
cooking appliance 厨房工具;炊事用具
cooking banana 大蕉
cooking boiler 蒸煮锅
cooking degree 蒸煮度
cooking department 蒸煮车间
cooking kettle 蒸煮罐;热炼锅
cooking liquor 蒸煮液
cooking machinery 炊事机械
cooking place 烹饪场所;厨房
cooking-pot type 锅式
cooking purpose 烹调应用
cooking range 炉灶;厨灶
cooking room 厨房
cooking schedule 热炼规程
cooking smell 烹调气味
cooking snow 湿雪
cooking stove 炉灶;灶
cooking temperature 烧煮温度;热炼温度
cooking utensils 炊具;厨具;厨房用具
cooking vat 蒸煮桶
cooking vessel 烹调用器皿;陶罐
cooking ware 烹饪器皿
Cook inlet basin 库克湾盆地
cooklee 凹印
Cook lens 库克镜头
cook out sample 烘干样品
cook room 厨房
cook shop 小饭店;饭馆
cook's store 厨房仓库
Cook Strait 库克海峡
cook the accounts 造假账
cook up a report 捏造报告
cool 冷存
cool air 冷空气
cool-air blower 冷空气鼓风机;冷空气吹送器
cool-air curtain door 冷空气帘
cool-air damper 冷空气调节风门
cool-air duct 冷空气输送管
cool-air feed 供给冷气
cool-air hose 冷空气软管;冷风软管
coolant 载热质;载热剂;冷却液浴;冷却剂;冷却材料;冷媒;制冷液;散热剂
coolant apparatus 冷却液装置
coolant box 冷却水箱
coolant channel 致冷剂通道
coolant-charging system 冷却剂进给系统
coolant circuit 冷却剂回路
coolant circulation unit 冷却液循环装置
coolant cleaner 冷却液净化器
coolant coil 冷却蛇形管;冷却盘管
coolant compound 冷却剂
coolant condition 冷却剂条件
coolant connection 冷却液接口
coolant connector 冷却剂接通装置
coolant flow 冷却液流量
coolant-flow rate 冷却剂流速;冷却剂流率
coolant fluid 冷却液(流体)
coolant for wire drawing 拔丝冷却剂
coolant gap 冷却介质间隙
coolant inlet 冷却剂入口
coolant jacket 冷却套
coolant jet 冷却剂喷嘴
coolant liquid 冷却液
coolant material 冷却剂材料
coolant mechanism 冷却装置
coolant moderator 载热减速剂
coolant oil 冷却油
coolant outlet 冷却剂出口;冷却液出口管
coolant passage 冷却系统;冷却孔道
coolant pressure 冷却液压力
coolant pressurizing method 冷却液加压法

coolant pump 冷却液泵
coolant pump motor 冷却液泵电动机
coolant purification system 冷却剂净化装置
coolant refrigerating units 冷却液冷冻组合装置
coolant reservoir 冷却液储存器
coolant scrap 冷却废钢
coolant separator 冷却液铁屑分离器；冷却液清净器；冷却液分离器
coolant storage tank 冷却液储存箱
coolant system 冷却系统
coolant temperature 冷却液温度
coolant temperature ga(u)ge 冷却液温度表
coolant thermostat 冷却恒温槽；致冷恒温槽
coolant tube 冷却液管
coolant-water effluent 冷却废水
cool at low-temperature 低温冷却
cool bath 凉浴
cool belt 冷却骨料运输机胶带
cool cap 空气冷却保护罩；冷却护罩
cool chamber 冷藏间；冷藏室；冷藏箱
cool colo(u)r 冷色
cool compartment 通风舱
cool crack 冷却开裂
cool-down 冷却(下来)
cool-down rate 冷却速度；冷却速率
cool(-down)time 冷却时间
cool-dryness 凉燥
cooled cargo 冷藏货物
cooled cell radiometer 冷却元件辐射计
cooled(cld)冷却的
cooled coal gas 冷煤气
cooled freight 冷却货物
cooled goods 冷却货物
cooled infrared detector 冷却红外探测器
cooled in furnace 炉内冷却(的)
cooled maser 冷却微波激射器；冷却脉塞
cooled metal nozzle 冷却金属喷管
cooled motor 冷却式发动机
cooled parametric amplifier 致冷变量放大器
cooled partition 冷却隔板
cooled photocell 冷却型光电管
cooled rotor 冷却式转子
cooled shroud 冷却(水)套
cooled-tube pyrometer 冷却管式高温计
cooled turbine 冷却式透平
cooled water 冷却水
cooler 冷却装置；冷却器；冷却间；冷却机；冷凝器；冷凝机；冷冻机；冷藏箱；致冷装置
cooler air circuit 冷却机空气环路
cooler bin 冷却箱
cooler blender 冷却松砂机
cooler cargo 低温储[贮]藏货物(保持2～10摄氏度)
cooler casing 冷却器外壳
cooler condenser 冷凝器；冷却冷凝器
cooler crystallizer 冷却结晶器
cooler drum 冷却转鼓；冷却筒
cooler fan 冷却机鼓风机；冷风栅
cooler grate 冷却机箅子
cooler housing 冷却箱
cooler inlet 冷却器入口
cooler nail 冷却器钉
cooler pan 冷却盘
cooler-plated steel 加盖沸腾钢
cooler regulator 冷却调节器
cooler room 冷藏间
cooler screw 冷却机螺旋输送机
cooler seal 冷却水包封闭
cooler shrinkage 冷藏减缩率
cooler stress 冷却应力
cooler tension rolling mill 带钢张力冷轧机
cooler tube 冷却器管
cooler with gas pipe 火焰管冷却器
cooler with natural draught 自然通风冷却器
cool exchanger 冷交换器
cooley fund 贷款
Cooley-Tukey algorithm 库利—图基算法
cool flame 冷火焰
cool gas 冷气
cool house 冷藏室；低温室；低温温室(维持不结冰)
coolibah 棕色硬木(一种产于澳大利亚的木材)
coolie 苦力
co-oligomerization 共齐聚；共低聚
cooling 冷却技术；冷空气；降温

cooling absorber 冷却吸收器
cooling action 冷却
cooling after 随后的冷却
cooling after postweld heat treatment 焊接热处理之后的冷却
cooling agent 冷却介质；冷却剂
cooling air 冷却空气；冷气
cooling air circulation 冷却空气循环
cooling air circulation system 冷气循环装置
cooling air current 冷却空气流
cooling air curtain 冷风幕
cooling air jacket 气冷式套筒
cooling and heating heat pump 制冷供暖热泵
cooling apparatus 冷却装置；冷却设备；冷却器；冷藏装置；冷藏器具；冷藏器
cooling area 冷却面积
cooling arrangement 冷却系统布置图
cooling attachment 冷却附件
cooling back 中间冷却；循环水冷却法
cooling back installation 中间冷却装置；循环冷却装置
cooling baffle 冷却鱼鳞板；冷却隔火墙
cooling bank 冷床
cooling basin 冷却水槽
cooling bath 冷水浴
cooling battery 冷却管组；制冷管组
cooling bay 冷却池
cooling bed 冷床
cooling belt 冷却带
cooling belt system 冷却带系统
cooling blade 冷却(式)叶片
cooling blast 冷却气流
cooling block 冷却区
cooling blower 冷却风机
cooling box 冷却水箱
cooling brittleness 冷脆性
cooling by contact 接触冷却
cooling by natural convection 自然对流冷却
cooling by radiation 自然冷却
cooling by wetting 加湿冷却
cooling calender 冷却滚压机
cooling canal system for hot-water 热水冷却渠道系统
cooling canal system for hot-water reuse 热水再用中的冷却渠道系统
cooling cap 冷却保护罩
cooling capacity 冷却能力
cooling ca(u)lking 冷缩裂缝
cooling centre 冷水中心
cooling chamber 冷藏室；冷却室
cooling channel 冷却通风；冷却通道
cooling circuit 冷却回路
cooling circulating system for air conditioning 空调制冷水循环系统
cooling circulation 冷却循环
cooling coil 漏板蛇形冷却管；冷却旋管；冷却蛇形管；冷却盘管；冷却螺管；冷盘管；冷藏盘管
cooling coil(ed)pipe 冷却盘管；蛇形冷却器
cooling coil section 冷却盘管段
cooling collar 冷却套管
cooling column 冷水塔；冷却塔
cooling condenser 冷却冷凝器
cooling cone 冷却锥体
cooling converter 降温转炉
cooling conveyer 冷却运输机
cooling correction 冷却修正；冷却改正
cooling coupling 冷却装置接头
cooling crack 冷缩节理；冷却裂纹
cooling cracking 冷却开裂
cooling crystallizer 冷却结晶器
cooling crystallizing equipment 冷却结晶设备
cooling curve 冷却曲线；降温曲线
cooling cycle 冷却期
cooling cyclone 冷却旋流器
cooling cylinder 冷凝箱
cooling dam 冷却水包
cooling deformation 冷却变形
cooling degree day 冷却度日
cooling device 冷却设备；冷却装置
cooling dew point thermometer 冷却式露点温度计
cooling differentiation 冷却分异
cooling disc 冷却溜盘
cooling-down effect 冷却效应
cooling-down fan 散热风扇；冷却风扇

cooling-down grid 冷却帘格；冷却格栅
cooling-down jacket 冷却套
cooling-down load 冷却负荷
cooling-down looper 降温线环装置
cooling-down medium 冷却介质
cooling-down method 冷却方法
cooling-down(off)冷却
cooling-down operation 冷机操作
cooling-down period 冷却阶段
cooling draft 冷却通风；冷拔风
cooling drag 冷却(系统)阻力
cooling drum 回转式冷却器；冷却圆筒
cooling drum centrifuge 回转式冷却器离心机
cooling duct 冷却通道；冷却管(道)；冷却导管；空气通道
cooling effect 冷却作用；冷却效应
cooling effect of transpiration 蒸腾作用冷却效应
cooling efficiency 冷却效率
cooling efficiency of sweating 排汗冷却效率
cooling element 冷却元件
cooling end 冷却端；冷却部
cooling equipment 冷却设备
cooling equipment system 冷却设备系统
cooling facility 冷却设备；冷却装置
cooling fan 冷(却)风扇
cooling fan installation 冷却通风装置；冷却通风系统
cooling fin 冷却(叶)片；冷却肋片；散热(翅)片
cooling fissure 冷缩裂缝
cooling fixture 冷却型架
cooling flange 散热凸缘
cooling fluid 冷却液；散热液
cooling grid 冷气隔栅；箅条冷床
cooling hardness cast-iron 冷硬铸铁
cooling history 冷却史；冷却历程
cooling house 冷藏室
cooling in dust or sand 灰砂冷
cooling in heap 堆冷
cooling in packed formation 堆垛冷却
cooling in pile 堆冷
cooling in spaced formation 分离冷却
cooling installation 冷却设备
cooling jacket 冷却水套；冷却夹套
cooling jacket capacity 冷却套容积
cooling jet 冷却喷嘴
cooling joint 冷缩节理；收缩缝；冷却缝
cooling launder 冷却流槽
cooling liquation 冷却凝析
cooling liquid 冷却液；切削液；散热液
cooling load 冷却荷载；冷(却)负荷
cooling load from outdoor air 新风冷负荷
cooling load temperature 冷负荷温度
cooling loop 冷却回路
cooling manifold 冷却集管
cooling material 冷却材料
cooling means 冷却装置；冷却方法
cooling medium 冷却介质；冷却剂；冷媒；致冷剂
cooling medium consumption 冷却介质流量；冷却剂消耗量
cooling method 冷却法
cooling milk 牛奶冷冻
cooling mixture 冷却混合物
cooling modulus 冷却模数
cooling noise 冷却噪声
cooling nozzle 冷却喷嘴
cooling-off 冷却
cooling-off area 冷却面积
cooling-off coefficient 冷却系数
cooling-off curve 冷却曲线
cooling-off loss 冷却损失
cooling-off period 冷却周期；等待期(间)
cooling-off process 冷却过程
cooling-off rate 冷却速率
cooling-off surface 冷却表面
cooling-off zone 冷却区
cooling of gas 气体冷却
cooling of tank wall 池壁冷却
cooling of tyres 轮擂冷却
cooling oil pump 冷却油泵
cooling or heating jacket 冷却或加热夹套
cooling panel 冷却(防护)板
cooling passage 冷却液通道
cooling performance 冷却效果
cooling pipe 冷却管(道)

cooling pipe system 冷却管道系统
cooling plant 冷却装置;冷却设备;冷冻厂;冷藏库;降温设备;制冷厂
cooling plate 冷却板;散热片
cooling plate technique 凉板工艺
cooling plug 冷却水管塞
cooling pond 凉水池;冷却塘;冷却水池;冷却池
cooling power 冷却能力;冷却功率
cooling power anemometer 冷却功率式风速计
cooling process 冷却过程
cooling program(me) 冷却曲线
cooling pump 冷却泵
cooling rail 冷却轨
cooling range 冷却温降;冷却水温差;冷却幅度;冷却范围
cooling rate 冷却速度;冷却(速)率;降温速度;散热速率
cooling ratio 冷却比
cooling rib 散热片;冷却肋片
cooling ring 冷却圈;冷却环
cooling roll(er) 冷却辊
cooling room 冷却室;冷却间
cooling schedule 冷却制度
cooling section 冷却区;冷却段;冷却带
cooling shaft 冷却井
cooling shelf 冷却架
cooling shrinkage 冷却收缩
cooling silo 低温筒仓
cooling space 冷却区;收缩缝
cooling speed 冷却速率;冷却速度
cooling spiral 冷却螺管
cooling spray 冷却液água
cooling spray system 喷水冷却系统
cooling stack 冷却烟道;冷却塔;冷却竖管;自然通风冷却塔
cooling stave 立式冷却板
cooling strain 冷却应变;冷却变形
cooling stress 冷却应力
cooling substance 冷却剂
cooling surface 冷却面;冷表面;散热面
cooling system 冷缩说;冷却系统;降温系统;散热系统
cooling system capacity 冷却系统容量
cooling system for circulating water 循环冷却处理系统
cooling system of internal combustion engine 内燃机冷却系统
cooling system optimization factor 冷却系统优化因素
cooling table 热床
cooling tank 冷却箱;冷却罐;降温池
cooling temperature 冷却温度
cooling test 冷却试验
cooling time 冷却时间
cooling time clock 冷却记时钟表
cooling tower 凉水塔;冷水塔;冷却塔;冷却器
cooling tower condenser 冷却塔式凝汽器
cooling tower distribution system 冷却塔配水系统
cooling tower fan 凉水塔风机
cooling tower sancer 冷却塔底盘
cooling trap 冷阱
cooling treatment system for water circulation 水循环冷却处理系统
cooling tube 冷却管(道);散热管
cooling tunnel 冷却隧道;冷却廊道
cooling type 冷却方式
cooling unit 冷凝盘管;冷凝单元;冷却装置;冷却机组;冷气机组
cooling vapo(u)r 冷却蒸气
cooling velocity 冷却速度
cooling ventilation 通风降温
cooling wall 冷却壁
cooling waste water of high temperature 高温冷却弃水
cooling water 冷(却)水
cooling water circuit 冷却水回流
cooling water circulation 冷却水循环;冷(却)水环流
cooling water circulation system 冷却水循环系统
cooling water connection 冷却水接头
cooling water cycle 冷却水循环
cooling water discharge box 冷却水溢流箱
cooling water ex-pansion tank 冷却水膨胀箱
cooling water hose 冷却水软管

cooling water inlet 冷水进口;冷却水入口
cooling water inlet pipe 冷却水进水管
cooling water installation 冷却水装置
cooling water intake structure 冷却水进口结构
cooling water jacket 冷却水套;水冷式套筒
cooling water losses 冷却水损耗
cooling water model test 冷却水模型试验
cooling water outlet 冷(却)水出口
cooling water outlet and inlet 冷却水进出口
cooling water pipe 冷却水管
cooling water plant 冷却水厂
cooling water pump 冷却水泵
cooling water pumping plant 冷却水泵房
cooling water pumping station 冷却水泵站
cooling water return pumping plant 冷却水回水泵房
cooling water return station 冷却水回水泵站
cooling water supply 冷却水供应
cooling water supply system 冷却水供应系统
cooling water system 冷却水系统
cooling water tank 冷却水箱;冷却水塔;冷却水池
cooling water temperation ga(u)ge 冷却水温度计
cooling water tower 冷却水塔
cooling water treatment 冷却水处理
cooling water treatment of closed recirculation system 封闭式循环系统冷却水处理
cooling water treatment of once-through system 直流系统冷却水处理
cooling water-treatment of open circulating system 敞开式循环系统冷却水处理
cooling water tube 冷却水管
cooling work 冷却工作
cooling worm 冷却蛇形管;冷却盘管
cooling zone 冷却区;冷却带;冷却层;致冷层
cool isostatic compression 冷等压压制;冷等静压制
cool-light lamp 冷光灯
coolling connector 冷却装置接头
cool-off 冷却
coolometer 冷却率测定仪
cool recycle 循环冷却
cool-resistence 耐冷性(能)
cool season crops 冷季作物
cool spring 冷泉
cool storage 冷藏库
cool strain 冷却应变
cooltainer 冷藏集装箱
cool temperate zone 寒温带
cool time 淬火时间;冷却时间
cool tone 冷色调
cool-tube system 冷却管道系统
cool ventilation duct 冷却风道
cool water dweller 冷水栖居者
cool white 冷白色
coom 煤烟;煤灰;拱形顶板;深海谷
Coomassie brilliant blue 考马斯亮蓝
coombe 冲沟
coom ceiling 沿斜屋顶铺的顶棚;涂泥炭的拱形顶棚;隔音平顶;涂泥泥拱形顶棚
coon ceiling 涂灰泥天花板
Coons curved surface 孔斯曲面
coop 监牢(俚语);家禽养殖场;鸡笼
cooper 修筒工人;箍桶工人;缝包工;桶业制造者
cooperage 木桶制造业者;木桶(业);制桶业;残货修整费;桶货加箍费
co-operating 协同操作
cooperating installation 合作设施
cooperating investment 投资合伙
co-operating tide 协振潮
cooperation 联合运行;合作;协作
cooperation between banks and enterprises 银企合作
cooperation coefficient 协作指数
cooperation complex 共作用配合物
cooperation index 协同索引;协调系数
cooperation in the field of energy 能源合作
cooperation of concrete and steel 混凝土钢筋结合力;混凝土与钢筋的联合作用;混凝土钢筋结结力
cooperation of contractors 施工合作体
cooperation plan 协作计划
cooperation test 对照试验
cooperation with foreign countries in offshore oil 海洋石油对外合作
cooperative 合作社
cooperative advertising 联合广告
cooperative agency 合作机构
cooperative agreement 合作协定
cooperative apartment house 合作公寓;合建公寓楼
cooperative bank 合作银行
cooperative building society 合作建筑协会
cooperative business operation 合作经营
cooperative chain 合作连锁店
cooperative committee 协作委员会
cooperative communication 合作通信
cooperative compiled unit 合编单位
cooperative control 合力管理
cooperative credit association 联合信贷公司
cooperative design 合作设计
cooperative drainage 合作排水
cooperative economy 合作经济
cooperative effect 合作效应
cooperative enterprise 合作企业
cooperative equilibrium 卖主控制均势
cooperative exploration of offshore oil 海上石油合作勘探
cooperative housing 合作住宅;合作住房;集资合建住房
cooperative housing loan 合作住房贷款
cooperative housing mortgage insurance 合作住房抵押保险
cooperative housing society 合建住宅协会
cooperative international trade 协作贸易
cooperative management 合作经营
cooperative medical cure 合作医疗
cooperative medical station 合作医疗站
cooperative observer 合作观测员
cooperative ownership 合作住房所有权
cooperative phenomenon 集体现象;合作现象
cooperative production 合作生产
cooperative relationship 协作关系
cooperative research 合作研究
cooperative research program(me) 协作研究方案
cooperative sale 合作销售
cooperative shop 合作商店
cooperative slip 协调滑移
cooperative store 合作商店;合作社
cooperative target 合作目标
cooperative-type management 协同式管理制度
cooperator 合股人
coopered joint 桶式接缝
Cooper-Hewitt lamp 玻璃管汞弧灯
Cooperite 古波里特镍锆合金;硫铂矿
cooper's axe 桶工斧
cooper's bench plane 桶工工作台面;桶工用长刨
Cooper's loading 古柏氏荷载
cooper's rabbet 桶工用槽刨
co-op housing 合作住宅;合作住房
coordianted publishing 协作出版
coordimeter 自动坐标记录仪;直角坐标仪
coordinability 可协调性
coordinate 坐标;配位;配价
coordinate access array 协同存取数组;坐标存取列
coordinate adjustment 坐标平差
coordinate analyzer 坐标分析器
coordinate angle 坐标角
coordinate attribute 坐标属性
coordinate axes 坐标轴
coordinate axis 坐标轴
coordinate azimuth 坐标方位角
coordinate bond 配位键;配位价;配价键
coordinate closure 坐标闭合差
coordinate commission 协作委员会;协调委员会
coordinate comparator 坐标量度仪;坐标量测仪
coordinate complex salt 配位络盐;配价络盐
coordinate condition 坐标条件
coordinate conversion 坐标转换;坐标变换
coordinate conversion calculator 坐标换算器
coordinate conversion computer 坐标转换计算机
coordinate conversion error 坐标转换误差
coordinate conversion parameter 坐标变换参数
coordinate converter 坐标变换器
coordinate counter 坐标计数器
coordinate covalent bond 配位共价键
coordinate crystal 配位晶体;配位晶格
coordinate curve 坐标曲线

coordinate curved surface 坐标曲面
coordinate data 坐标数据
coordinate datareceiver 坐标数据接收机
coordinate datatransmitter 坐标数据发送机
coordinated bus service 联营公共汽车服务设施
coordinated complex 配位化合物
coordinated control 联动控制
coordinated control system 联动控制系统;联动调整系统;协调控制系统
coordinated decision 协调决策
coordinated development of regional economy 区域经济协调发展
coordinated distance 协调距离
coordinated drive 协调传动
coordinated evolutionalism 协同演化说
coordinated flow 协调流量
coordinate digitizer 坐标数字化转换器;坐标数字化仪
coordinate dimension 坐标尺寸
coordinate dimensioning 坐标尺寸确定(法)
coordinate display 坐标显示
coordinated metal complex 配价金属配合物
coordinated movement 协调运动
coordinated movement signal system 联动运行信号系统
coordinated operation 联合运用(一种水利工程操作方法)
coordinated planning 联动规划;协调规划
coordinated point 已知坐标点
coordinated polyhedra rule 配位多面体规则
coordinated price 协作价格
coordinated regional economic and social development 地区经济社会的协调发展
coordinated sets of measures 配套措施
coordinated signal 联动式信号
coordinated signal system 联动信号系统
coordinated system 联动系统;协作系统;坐标制
coordinated traffic signal system 联动交通信号控制
coordinated transformation matrix 坐标变换矩阵
coordinated transportation 联运;协作运输;综合运输;配合运输
coordinated transposition 交叉换位;配合换位
coordinated trial 协作试验
coordinated type 联动式
coordinated universal time 协调世界时
coordinated variation 坐标性变化
coordinated wage policy 协调的工资政策
coordinated water 配位水
coordinate equation 坐标方程
coordinate exercise 协调运动
coordinate figure 坐标值
coordinate frame 坐标格网
coordinate function 坐标函数
coordinate graph 坐标制图机
coordinate grid 坐标网格;坐标格网
coordinate grid method 坐标网格法
coordinate grid scale 坐标格网尺;方眼尺;特罗贝雪夫尺
coordinate index 协调指标
coordinate indexing 信息加下标;相关索引;坐标检索(法);对等检索
coordinate inspection machine 坐标检验机
coordinate line 坐标格网线
coordinate line of curvature 坐标曲率线
coordinate link(age) 配位键(合)
coordinate location device 坐标定位装置;纵横定位设备
coordinate machine 测绘仪;测绘器
coordinate measurement 坐标量测
coordinate measuring apparatus 坐标测量仪
coordinate measuring instrument 坐标量度仪;坐标量测仪;坐标测量仪
coordinate measuring machine 坐标测量机
coordinate method 坐标法
coordinate motif(pattern) 配位基型
coordinate multiplexing circuit 坐标复接电路
coordinate multiplex storage circuit 坐标复接存储电路
coordinate neighbo(u)rhood 坐标邻域
coordinate net 坐标网
coordinate note 坐标网说明
coordinate number 配位数
coordinate of central point of grid 网格中心点坐标
coordinate of details 细部坐标
coordinate of matching point 配合点坐标
coordinate of node 节点坐标
coordinate of non-nodal well 非节点井坐标
coordinate of water bursting point 突水点坐标
coordinate operator 坐标算子
coordinate pair 坐标对
coordinate paper 计算纸;坐标纸;方格绘图纸
coordinate perturbation 坐标摄动
coordinate plane 坐标平面
coordinate plotter 坐标绘图机
coordinate polymer 配位聚合物
coordinate potentiometer 坐标电位器
coordinate printer 坐标记录仪
coordinate protractor 坐标量角器
coordinate readback 坐标逆读
coordinate reading 坐标读数
coordinate record 坐标记录
coordinate recorder 坐标记录器
coordinate recording 坐标记录
coordinate recording unit 坐标记录装置
coordinate registrator 坐标记录仪
coordinate repeating system 坐标重复系统
coordinate retrieval 协同检索
coordinate scale 坐标标度
coordinate selector switch 坐标选择器开关
coordinate sequence 坐标序列
coordinate set 坐标集
coordinate setting 坐标定位;坐标测设
coordinate setting boring machine 坐标镗床
coordinate setting boring machine with tap-control 穿孔带控制的坐标镗床
coordinate setting drilling machine 坐标钻床
coordinate setting table 坐标工作台
coordinates for model 模型坐标
coordinates in space 空间坐标
coordinates of celestial sphere 天球坐标
coordinates of end point 端点坐标
coordinates of grid point 格点坐标
coordinates of image point 像点坐标
coordinates of motion 运动坐标
coordinates of photo point 像点坐标
coordinates of special point 特殊点坐标
coordinates of the pole 地极坐标
coordinate storage 矩阵(式)存储器;坐标存储器
coordinate string 坐标数字串
coordinate surface 坐标面
coordinate system 坐标系(统)
coordinate system conversion 坐标系统变换
coordinate system converter 坐标转换装置
coordinate system paper 方格厘米纸;坐标纸
coordinate table 坐标工作台
coordinate time 坐标时
coordinate titration 配位滴定(法)
coordinate transformation 坐标换算;坐标变形法;坐标变换
coordinate transformation of navigation 导航定位坐标转换
coordinate type adsorption 配价键型吸附
coordinate type potentiometer 坐标电位计
coordinate valence 配位价
coordinate valence force 配位价力
coordinate value 坐标值
coordinate value of equivalent point 等效点坐标值
coordinate zone 坐标带
coordinating atom 配位原子
coordinating body 协调机构
coordinating calculating center 坐标计算中心
coordinating catalysis 配位催化
coordinating catalyst 配位催化剂
Coordinating Committee on the Ozone Layer 臭氧层问题协调委员会
coordinating dimension 坐标尺寸
coordinating function 协调功能
coordinating group 协调小组
coordinating holes 同位孔
coordinating ion 配位离子
coordinating movement 交车运行(道路施工时);交叉运行
coordinating plane 坐标面;协调基准面
coordinating traffic signal 联动交通信号
coordination 进程协调;协作;协调;对等;配位;配价;同位;同等
co-ordination among specialized departments 专业化协作
coordination center 协调中心;调度中心
coordination chemistry 配位化学
coordination committee 协调委员会
coordination complex 配位化合物
coordination compound 配位化合物
coordination drawings 施工配合图
coordination form 协调式
coordination from of prospecting by mining 探采工作协调
coordination function 协调机能
coordination(ic) exception 例外联系
coordination isomer 配位异构体
coordination isomerism 配位异构
coordination lattice 配位晶格
coordination link 配位键
coordination linkage 配位键
coordination number 配位数
coordination of farming forestry and animal husbandry 农林牧结合
coordination of function 机能协调
coordination of production and marketing 产销平衡
coordination of reliability efforts 可靠性工作项目协调
coordination of transportation planning 交通规划协调
coordination of work 工程协调
coordination on site 现场协调
coordination polygon 配位多边形
coordination polyhedron 配位多面体
coordination polymer 配位聚合物;配位高聚物
coordination polymerization 配位聚合
coordination structure 配位结构
coordination theory 配位理论
coordination theory of development and environment 发展与环境协调论
coordination valence 配位价;配价
coordinative activity 配位活性
coordinatograph 坐标展点仪;坐标仪;坐标读数器;等位图;坐标制图器
coordinatograph for plummet observation 垂线观测坐标仪
coordinatograph photoplotter 带坐标仪的测图仪
coordinatometer 坐标尺
coordinator 联动控制器;连接成分;协调员;协调人;协调程序;坐标方位仪;调度员;配位仪;统筹人
coordinator architect 协作建筑师
coordinator of materials planning 材料计划联络人
coordinator of shipbuilding 造船调度员
coordinator routine 配位仪程序
coorongite 弹性藻沥青
co-oscillating partial tide 协振分潮
co-oscillating tide 协振潮
cooscillation 共振
coose 劣矿脉;贫矿脉
coouupational intoxication 职业性中毒
co-owner 共有者;共有人;共同所有人
co-ownership 联合债权人;共同所有权;共有
co-oxidation 共氧化
cooxidized alumina 共氧化(生成)的氧化铝
cop 圆锥形线圈;管纱;绕线轴
copal 快干树脂;柯巴(树)脂;苯乙烯树脂
copal adhesive 柯巴脂黏结剂
copal balsam 柯巴脂;苯乙烯软树脂
copal ca(u)lking compound 柯巴脂嵌缝合成物
copal cementing agent 柯巴脂胶凝剂
copal ester 柯巴酯
copal gum 透明树胶
copalic acid 黄脂酸
copalite 黄脂石;琥珀
copal mastic 柯巴玛琦脂
copal oil type varnish 柯巴脂油型清漆
copal resin 柯巴树脂
copal resin adhesive 柯巴树脂胶粘剂
copal resin bonding agent 柯巴树脂粘结剂
copal resin bonding medium 柯巴树脂黏结介质
copal resin compound 柯巴树脂化合物
copal resin sealing compound 柯巴树脂密封合成物

copal stopper 柯巴脂制止器
copal varnish 柯巴(脂)清漆;中油度清漆
Copan 考潘轴承合金
coparallel 上平行
coparallelism 上平行
coparties 共同当事人
copartner 合伙人;合股者
copartnership 合资公司;分红换股方式
Co/Pb-rich Mn nodule 富钴铅型锰结核
cop changer 自动加纤机构
cope 码头前沿;压顶【建】;遮盖物;盖顶;顶盖;墙帽;上型架;上砂箱;上部模型
cope and drag mount 上下型箱架
cope and drag pattern 两箱造型模;两片模
cope and drag print 垂直芯头
cope and drag set 上下型机组
cope block 压顶块
cope bollard (船闸的)闸顶系船栓
cope box 上砂箱
cope chisel 凿槽刀
cope cutter 切榫刀;切削雄榫器;钻头
coped 楔合的;接缝的;加压顶的;合缝的
coped beam 切口梁(以备联结其他构件)
coped joint 密合接缝;对缝接头;顶盖接头;合缝接头;搭接缝;暗缝
coped-lug type anode 挂耳式阳极
cope down 吊砂
cope elevation along wharf 码头前沿高程
cope flask 上砂箱
cope hole 上型箱孔
cope in steel beam 钢梁的顶层;削梁
Copel 考佩尔镍铜电阻合金
cope level 码头面高程;顶部高程
cope line 盖顶边线
cope match-plate pattern 上模板
Copenhagen water 标准海水
cope of project 工程范围
cope plate 托板;上模板
cope print 上芯头
coper green glaze 铜绿釉
Copernican 哥白尼体系
Copernican era 哥白尼纪
Copernican system 哥白尼体系
Copernicus 哥白尼卫星
Coperta glaze 釉面透明釉
copestone 最后完工的工作;压顶石;墙帽;盖石
cope tile 墙压顶瓦;顶盖瓦
cope wall 顶墙
cophasal modulation 同相调制
cophase 同相
cophase component 同相分量
cophased array 同相天线阵
cophased horizontal antenna 同相水平天线
cophased horizontal antenna array 同相水平天线阵
cophased vertical antenna 同相垂直天线
cophase excitation 同相激励
cophase supply 同相供电
cophase wave 同相波
copher 密码
copi 风化石膏
copiapite 叶绿矾
copier machine 复印机
copier monitor device 复印件监视装置
co-pilot 副引航员;自动驾驶仪;副驾驶员
coping 修磨;压顶【建】;盖梁;盖版(层);墩帽;挡轮坎(码头前沿);挡车槛(码头前沿);分割石板;墙压顶;墙帽
coping beam 压顶梁;帽梁
coping block 压顶(砌)块
coping bracket 墙压顶托架
coping brick 檐砖;履面砖;压顶砖
coping-cut 吊砂
coping down 挖砂
coping of abutment 台帽
coping of a vault 拱屋顶顶盖
coping saw 窄条手锯;弓形锯;手弓锯
coping slab 压顶板
coping stone 帽石;压顶石;盖石;盖顶石
coping stone of cesspool 污水池压顶石
coping tile 盖瓦
coping wall 码头顶胸墙;闸墙顶拦墙;坝顶栏墙
coping with inside wash 压顶内泛水
coping with outside wash 压顶双向泛水

coping with two-way wash 压顶双向泛水
copious cooling 深(度)冷(却)
copious current 强流
copious irrigation 漫灌;充分灌溉
copious oil supply 大量的润滑油供应
coplanar 共(平)面的;同一平面的;同(平)面
coplanar condition 共面条件
coplanar coordinate system 共面坐标系
coplanar displacement 共面位移
coplanar electrodes 共面电极
coplanar force 在同一平面的力;共面力;同平面内的力
co-planar forces 共面力系
coplanar force system 共面力系
coplanar grating 共面线栅;共面光栅
coplanar grid 共面删极
coplanar-grid valve 共面栅极管
coplanarity 共(平)面性;同面性
coplanar(ity) condition equation 共面条件方程
coplanarity equation 共面方程
coplanarity of orbits 轨道共面性
coplanar network 共平面网络
coplanar stabilizers 共面稳定器
coplanar strain 共(平)面应变;同(平)面应变
coplanar stress 共面应力;同平面应力
coplanar vector 共面向量;共面矢量
coplanar waveguide 共平面形波导
coplaner 共面
coplaner method 共面法
co-plasticizer 助塑剂;辅塑剂
Coplin jar 玻片染色缸
cop loader 自动加纤机构
copolar 共极的
copoly acrylate paste 丙烯酸酯共聚浆料
copoly alkylene oxide 共聚烯烃
copolycondensation 共缩聚(作用)
copolyester 共聚多酯
copolyether 共聚多醚
copolyether ester fibre 共聚醚酯纤维
copolymer 异分子聚合物;共聚物
copolymer fibre 共聚物纤维
copolymeric plastics 共聚塑料
copolymerization 异分子共聚作用;共聚作用;共聚(作用);共聚合反应
copolymer of acrylate and butadiene 丙烯酸酯与丁二烯共聚物
copolymer resin 共聚物乳胶漆;共聚树脂
co-polymers of vinyl acetate and maleic anhydride 醋酸乙烯酯—顺丁烯二酸酐共聚物
coppce-with standard system 中林作业法
copped joint 盖顶接头
copper 红铜;紫铜;铜钱;铜(红)色;铜币;深橙色
copper accelerated acetic acid salt spray test 铜催化乙酸盐雾试验
copper acetate 乙酸铜;醋酸铜
copper acetate arsenite 翡翠绿颜料
copper acetate senite 乙酸亚砷酸铜
copper acetoar senite 乙酰亚砷酸铜
copper aceto-arsenite 巴黎绿
Copper Age 铜器时代
copper alloy 铜合金
copper-alloy arc welding electrode 铜合金焊条
copper-alloy casting 铜合金铸件
copper-alloy pipe and fittings 铜合金管和配件
copper-alloy scrap 铜合金废料
copper-alloy steel 含铜合金钢;合金钢
copper-alloy trap 铜合金凝气筒;铜合金存水弯
copper-alloy wire 铜合金线
copper-alumin(i)um alloy 铜铝合金
copper-aluminium welding rod 铜铝焊条
copper amalgam 铜汞合金
copper amalgamating plate 铜混汞板
copper ammoniacal 铜铵合剂
copper anchor 铜锚
copper anode 铜阳极
copper antimonide 锑化铜
copper arc welding electrode 铜焊条
copper arsenate 砷酸铜
copper arsenite 亚砷酸铜
copperas 绿矾;七水(合)硫酸亚铁
copper-asbestos 铜包石棉
copper asbestos gasket 铜包石棉垫片
copper-asbestos-lined gasket 石棉衬里铜垫圈
copper-asbestos packing 铜包石棉衬垫

copper-asbestos ring 铜包石棉环
copperas black 皂矾黑颜料
copper ashes 铜渣
copperas red 绿矾铁红;干法铁
copper aventurine glass 铜金星玻璃
copper backing 铜垫片
copper bar 铜条;铜排;铜杆;铜棒
copper base alloy 铜基合金
copper base firction material 铜基合金摩擦材料
copper base matrix 铜基合金基体
copper base metal 铜基合金
copper basic carbonate 碱式碳酸铜;盐基碳酸铜
copper basic chlorite 碱式亚氯酸铜;盐基亚氯酸铜
copper-bearing 含铜的
copper-bearing alloy 含铜合金
copper-bearing concrete 含铜混凝土
copper-bearing formation 含铜建造
copper-bearing iron ore 铜铁矿石
copper-bearing sandstone 含铜砂岩
copper-bearing steel 铜钢;含铜钢
copper-bearing sulfur ore 铜硫矿石
copper-bearing wastewater 含铜废水
copper belt 紫铜带
copper bend 铜弯头;铜弯管
copper-beryllium 镀铜合金
copper beryllium alloy 铜铍合金
copper billet 铜条;铜块
copper binding-wire 铜扎线
copper bit 铜焊头;电烙铁;焊接器;紫铜烙铁
copper bit joint 铜焊接头
copper blasting cap 铜烙铁电爆器;铜起爆雷管
copper blight 铜色疫病
copper block 铜块
copper bloom 铜华
copper blue 铜蓝;天青蓝
copper blue copper sulfide 硫化铜
copper blue glaze 铜青釉
copper bolt 紫铜烙铁
copper bond 黄铜焊接;铜焊接合
copper-boned Rugalze with fish-roe crazing 铜骨鱼子纹汝釉
copper-boned Ruglaze without crazing 铜骨无纹汝釉
copper borate-coated silica 硼酸铜包覆二氧化硅
copper bottomed 包铜皮船底
copper bound 铜边(宋朝定窑瓷盘瓷碗铜质包边)
copper bowl 炉体铜碗
copper box roof gutter 筒箱形檐沟
copper brazing 铜(钎)焊
copper brittleness 铜脆
copper bromide 溴化铜
copper-bronze 铜—青铜薄片
copper-bronze paint 铜—青铜漆
copper-bronze powder 青铜粉
copper brush 铜丝布电刷
copper bullion 粗铜锭
copper burnishing wheel 铜抛光轮
copper busbar 铜(母)线
copper bush 铜轴套
copper cable 铜电缆
copper cadmium conductor 铜镉线
copper cadmium stranded conductor 镉铜绞线
copper carbonate 碱性碳酸铜;碳酸铜
copper cash 铜钱
copper casting 铜浇铸
copper cathode 电解铜
copper chloride 氯化铜
copper chromate 铬酸铜(木材防腐剂)
copper-chrome complex black 铜铬复合黑色颜料
copper chromite 含铜铬铁矿;亚铬酸铜
copper cistern 铜质储[贮]水器
copper citrate 柠檬酸铜
copper-clad 包铜(的);铜包
copper-clad alumin(i)um conductor 铜包铝(导)线
copper-clad aluminium wire 包铜铝线
copper-clad asbestos gasket 包铜石棉垫圈
copper-clad base plate 包铜底板
copper-clad connecting wire 镀铜连接线
copper-clad laminate 敷铜箔板;覆铜薄层压板
copper-clad nail 包铜钉
copper-clad panel 敷铜箔叠层板
copper-clad pitched roof 包铜皮的斜屋顶
copper-clad plate 敷铜板

copper-clad steel 包铜钢
copper-clad steel cable 铜包钢缆
copper-clad steel conductor 包铜钢丝;铜包钢缆
copper-clad steel wire 包铜钢丝;铜包钢缆;镀铜钢丝
copper-clad wire 包铜钢丝
copper clamp 铜扣钉;铜夹钳;铜箍
copper cleat 铜楔子;铜系缆墩;铜加劲条
copper-coated 镀铜的;铜包
copper-coated iron powder 包铜铁粉
copper-coated mild steel wire 镀铜软钢丝
copper-coated paper 涂铜纸
copper-coated stainless steel wire 镀铜不锈钢丝
copper-coated stitching flat wire 镀铜扁铁丝
copper coating 镀铜(的方法);涂铜
copper collar 铜环
copper colo(u)r cloth 古铜色布
copper colo(u)red 铜色的
copper commutator segment 铜换向片
copper compound 铜化合物
copper compound process 铜化合物法
copper concentrates 铜精矿
copper conductor 铜导体;铜导线
copper-constantan compensating conductor 铜—康铜补偿导线
copper-constantan thermocouple 铜—康铜热电偶
copper containing scrap 含铜金属废料
copper containing steel 含铜钢
copper-containing wastewater 含铜废水
copper content 含铜量;铜含量
copper converter 炼铜转炉;铜吹炉
copper converter gas 炼铜转炉炉气
copper converting 铜吹炼
copper cooling plate 铜冷却板
copper-copper-nickel compensating conductor 铜铜镍补偿导线
copper core 铜芯(线)
copper core concentrates 铜精矿
copper cored carbon 铜芯碳
copper core process 铜芯轧制法
copper corrosion inhibited grease 防铜腐蚀润滑脂
copper coulometer 铜板电量计
copper coupling 铜连接器;铜管接头
copper covered steel conductor 铜包钢缆
copper cross 铜四通(管);铜十字(接头)
copper crucible 铜坩埚
copper crusher 铜柱测压器
copper current 正电流
copper cuttings 铜屑;碎铜
copper cyanide 氰化铜
copper cylinder 测压铜柱
copper damper 铜阻尼器
copper dampproof(ing) course 铜防潮层
copper deactivator 铜减活剂
copper detonator 铜雷管
copper dioxide 二氧化铜
copper discharge pipe 铜排水管
copper dish gum 铜皿胶质
copper dish test 铜盆试验
copper distributed data interface 铜缆分布式数据接口
copper dowel 铜销钉;铜榫
copper drain(age) pipe 铜排水管
copper drainage tube 紫铜排水管
copper dryer 铜干料;铜催干剂
copper drill 空心钻
copper drop 铜阻电压降
copper dust disease 铜屑沉着病
coppered 镀铜的;包铜的
coppered carbon 镀铜碳
coppered joint 桶式接缝
coppered steel wire 镀铜钢丝
coppered wire 包铜丝
copper elbow 铜弯头;铜弯管
copper electrolytic refinery 铜电解精炼厂
copper electroplating wastewater 铜电镀废水
copper enamel 铜(胎)搪瓷;铜坯珐琅(铜胎搪瓷)
copper-engraving 铜雕刻
copper fabric 铜织物
copper facing 铜面饰;镀铜
copper fin 铜翅
copper fin(ned)pipe 铜制鳍管;铜散热器
copper fire box 铜板火箱

copper fitting 铜配件;铜接头
copper flashing(piece)铜(皮)泛水
copper flash smelting 铜闪速熔炼
copper fluoride 氟化(亚)铜
copper fluosilicate 铜的氟硅酸盐;硅氟化铜
copper foil 铜箔
copper foil-clad laminated sheet 敷铜薄层压板
copper foil laminate 敷铜板板;铜箔迭层板
copper form (镶焊钻头用的)铜片
copper fungicide 铜(杀菌)剂
copper furnace bell 钟形铜炉罩(烧结炉)
copper gasket 铜填料
copper glance 辉铜矿
copper-glass seal 铜与玻璃封焊
copper glazing 铜条镶嵌玻璃
copper gold 含铜金矿
copper granules 铜砾
copper-graphite alloy 铜石墨合金
copper-graphite composition 铜石墨制品
copper green 孔雀石绿颜料
copper gutter 铜(檐)槽(沟)
copper hammer 铜锤
copper head 铜头;黄斑(金属板镀层缺陷)
copper hydroxide 氢氧化铜
copper hydroxy carbonate 盐基碳酸铜
copper hydroxy chlorite 碱式亚氯酸铜;盐基亚氯酸铜
copper industry waste 炼铜工业废物
coppering 挂铜;镀铜
coppering bit 铜焊头
coppering hammer 大头锤;平头锤
coppering liquid 镀铜液
copper ingot 铜锭
copper in lubricating oil 润滑油的铜含量
copper insoluble metal 不溶于铜的金属
copper insoluble metal powder 不溶于铜的金属粉末
copper intensification 铜盐加厚法
copper ion 铜离子
copper-iron bimetal 铜—铁双金属
copperish 含铜的
copperized boliden salt 铜化加铬砷酸锌(防腐剂)
copperized iron powder 镀铜铁屑
copperized lead 含铜铅
copperizing 镀铜
copper jacket 铜套
copper-jacketed coil 铜屏蔽线圈
copper junk 铜废料
copper-lead alloy 铜铅合金
copper-lead bearing 铜铅轴承
copper-lead-bronze alloys 铜铅合金
copper-lead fabric sheet 铜铅织物板
copper-lead fiber glass 铜铅玻璃纤维复合板
copper-lead matte 铜铅锍
copper-lead sheet 铜铅薄板
copper-lead-tin 铜铅锡合金
copper light glazing 用防火分隔铜条装配玻璃;铜条装配窗玻璃;铜条镶嵌玻璃
copper-lined cistern 有衬里的储[贮]水器
copperlining 铜衬料
copper links 铜环(测梁连的标环)
copper liquor oil separator 铜氨液油分离器
copperlite 铜条装配法(一种防火玻璃装配方法)
copperlite glaze 铜釉
copper(lite)glazing 铜条装配玻璃;铜釉
copper loss 铜损(耗);铜耗
copper lustre 铜光泽(彩)
copper magnet alloy 磁性铜合金
copper-manganese 锰铜
copper-manganese alloy 锰铜合金
copper-manganese-nickle alloy 铜锰镍合金
copper master alloy 高铜合金;铜基中间合金
copper matrix 铜基体
copper matrix of sufficient porosity 多孔铜基体
copper matte 冰铜
copper mesh 铜丝网
copper metallurgy 炼铜
copper mica 云母铜矿
copper mill 轧铜厂
copper mining tailing 铜矿尾矿
copper-molybdenum 铜(环)钼合金
copper monoxide 一氧化铜
copper mo(u)ld 铜模
copper mo(u)ld base 铜模底

copper mo(u)ld wall 铜坩埚壁
copper nail 紫铜钉;铜钉
copper-nailed 钉铜钉的
copper-nailing 钉铜钉
copper naphthenate 环烷酸铜(一种木材防腐剂)
copper netting 铜丝网
copper-nicked converter 铜镍锍转炉
copper nickel 红砷镍矿
copper-nickel alloy 铜镍合金
copper-nickel-antimony-lead bronze 铜镍锑铅合金
copper-nickel deposit in ultra-basic rock 超基性岩铜镍面
copper-nickel-iron matte 铜镍铁锍
copper-nickel matte 铜镍锍
copper-nickel welding rod 铜镍焊条
copper-nickel-zinc alloys 铜镍锌合金
copper nickle 锌镍铜合金;德银;白铜
copper nitrate 硝酸铜
copper number 铜值
copper oleate 油酸铜
copper one-pipe 铜单管
copper one-pipe ring system 铜单管环系统
copper ore 铜矿(石)
copper ore of pyrite type 黄铁矿型铜矿石
copper orthoarsenite 原亚砷酸铜
copper oxide 一氧化铜;氧化(正)铜
copper oxide barrier film 氧化铜防护膜
copper-oxide photovoltaic cell 氧化铜光伏元件
copper oxyacetate 碱式乙酸铜;盐基乙酸铜
copper oxychloride 氯氧化铜
copper packing 铜垫
copper paint 铜质漆;防污漆;防虫漆;含铜涂料(木船防污漆)
copper panel 铜板
copper paper (铜与防水皱纹牛皮纸的)铜纸复合板
copper pentachlorophenate 五氯苯酚铜
copper peroxide 过氧化铜
copper phthalicyanine 酞菁蓝
copper phthalocyanine 铜酞菁
copper pigtail 柔韧铜辫
copper pipe 铜管
copper pipe and fittings 铜管和配件
copper pipe bend 铜管弯头
copper pipe coupling 铜管连接器;铜管管接头
copper pipe cross 铜四通管
copper pipe elbow 铜肘形管;铜弯管
copper pipe for radiator heating 散热器供暖用的铜管
copper pipe union 铜管联合会;铜管接头
copper piping 铜管
copperplate 薄铜板;紫铜板;铜版(印刷);铜版画;铜(凹)板
copper-plated 镀铜
copper-plated iron screw ring 镀紫铜铁吊环螺钉
copper-plated roller 镀铜油墨辊
copper-plated sash chain for door and window 门窗用镀铜吊链
copper-plated steel 镀铜钢板;包铜钢板
copperplate engraving 铜版画
copperplate etching 铜版画
copperplate ink 凹版油墨
copperplate paper 铜版纸
copperplate press 凹版印刷机
copper plating 镀铜
copper plating fog inhibitor 镀铜抑雾剂
copper plating liquid 镀铜液
copperply wire 镀铜钢丝
copper poisoning 铜中毒
copper pollutant 铜污染物
copper pollution 铜污染
copper porphyry deposit 斑岩铜矿床
copper powder 红铜粉;铜粉
copper precipitation unit 沉铜装置
copper printing roller 紫铜印染辊筒
copper product 铜制品;铜材
copper profile 铜型材
copper pyrite 黄铜矿
copper rainwater articles 铜制屋顶配件
copper rainwater goods 铜雨水制品
copper rainwater gutter 铜雨水槽
copper red 铜红
copper red glaze 铜红釉
copper removal 除铜
copper resinate 树脂酸铜

copper resist 铜光泽彩露花(陶瓷装饰方法)
copper resistor 铜电阻器
copper revetment 铜贴面
copper-rich matrix 富铜基体
copper rod 棒材铜;铜条;铜丝;铜杆;铜棒
copper rod gauze 铜丝网
copper roof cladding 铜屋顶覆盖(材料)
copper roof cover(ing) 铜屋顶覆盖物
copper roof gutter 铜屋顶檐沟
copper roofing 铜皮屋面;铜板屋面
copper roofing practice 铜屋顶实施
copper roofing sheet 铜皮屋面板;铜皮层面板
copper roof sheathing 铜屋顶衬板
copper rope 铜绳缆
copper ruby glass 铜红玻璃
copper rust 铜锈
copper salt 铜盐
copper sandoz 氧化低铜
copper scale 铜渣;铜屑
copper scrap 红铜杂件;废铜
copper seal 铜止水片;止水铜片
copper sealing strip 铜密封条
copper section 铜型材;型材铜
copper segment 扇形铜片
copper shale deposit 含铜页岩矿床
copper shape 铜型材
copper sheath 铜包皮
copper-sheathed cable 包铜电缆;无机绝缘电缆
copper sheathing 铜套
copper sheet 铜板;紫铜片;紫铜皮;薄(紫)铜板;铜片;铜皮
copper sheet plate 铜板
copper sheet roofing 铜皮屋面
copper shield 铜屏蔽
copper shingle 铜瓦片;铜板瓦
copper shot 铜球;铜弹
copper-silicon 硅铜
copper-silicon alloy 铜硅合金
copper-silicon welding rod 铜硅焊条
copper silmin 铜硅铝合金
copper silumin 含铜硬铝;铜硅铝明
copper slab 扁铜坯
copper slag 铜(矿)渣
copper slag block 铜渣块
copper slag brick 铜矿渣砖
copper slag sett 铜渣小方块
copper slag tile 铜渣瓦
copper slate 铜皮泛水;铜皮管脚泛水;铜板
copper sleeve 铜套管
copper sludge 铜泥
copper-slug 继电器的缓动铜套
copper smelter 铜熔炼炉
copper smelter slag 炼铜炉渣
copper smelting 炼铜
coppersmith 铜活工;铜器制造人;铜匠
coppersmith's hammer 铜匠锤
coppersmith shop 铜工车间
copper smithy 锻铜车间
copper soap 铜皂
copper solid section 铜实心型材
copper solubility 铜的溶解度;铜的可溶性
copper-soluble metal 铜溶金属
copper-soluble metal powder 铜溶金属粉末
copper space factor 铜线占空系数
copper sponge 海绵铜
copper spot 铜斑;铜色斑病
copper square bar 方铜棒
copper stain 铜(蚀)斑;铜绿污斑
copper staining 铜蚀污染;铜红扩散着色
copper stannopailadinite 铜锡钯
copper staybolt 铜撑栓
copper steel 含铜钢;铜钢
copper strip 铜带;铜条
copper strip bar copper 铜条
copper strip corrosion 铜条剥落腐蚀
copper-stripping autoclave 脱铜高压釜
copper stripping electrolysis 脱铜电解
copper-stripping reaction 除铜反应
copper strip roofing 铜带屋面
copper strip test 铜片腐蚀试验;铜带试验
copper sub-oxide 氧化亚铜
copper sulfate 硫酸铜
copper sulfate solution 硫酸铜溶液
copper sulfate test 硫酸铜试验

copper sulfate treatment 木材浸硫酸铜处理
copper sulphate 硫酸铜
copper sulphate test 硫酸铜水溶液腐蚀试验
copper sulphate treatment 浸硫酸铜处理(木材处理);木材浸硫酸铜处理
copper-sulphur alloy 铜硫合金
copper-surfaced 镀铜的;贴铜
copper-surfaced circuit 印刷电路
copper(sur)facing 铜贴面
copper suspension pulley 铜悬吊滑轮
copper tack 铜锹钉;铜平头钉;紫铜钉
copper-tape winding 铜带绕组
copper terminal clamp 铜线接头夹
copper thimble 铜护环
copper tie 铜拉杆
copper tile 铜瓦
copper-tin alloy 铜锡合金
copper-tin alloy plating fog inhibitor 镀铜锡合金抑雾剂
copper-tin amalgam 铜锡合金汞齐
copper-tin trap 铜锡存水弯
copper-tin trim 铜锡镶边;铜锡贴面
copper-tin tube ceiling rafiant heating panel 铜锡管平顶辐射供暖板
copper-tin unit 铜锡制品;铜锡单元
copper-tin valley gutter 铜锡屋顶斜沟
copper-tin washer 铜锡垫圈
copper-tin waste pipe 铜锡废水管
copper-tin water stop 铜锡阻水片;铜锡止水器
copper-tin wedge 铜锡楔
copper-tin welding rod 铜锡焊条
copper-tin wrought alloy 铜锡锻炼合金
copper-tipped 端部镀铜的
copper titanate 钛酸铜
copper to aluminum adapter bar 铜铝过渡板
copper-to-glass 铜玻璃封接
copper tube 紫铜管;铜管
copper tubing 铜管
copper-tungsten 铜钨合金
copper uranite 铜铀云母
copper vessel 铜制容器
copper vitriol 硫酸铜;蓝矾;胆矾
copper voltameter 铜(电解式)电量计
copper voltmeter 铜解伏安计
copper ware 铜器
copper water seal 止水铜片
copper waterstop 铜止水(条);止水铜片;铜止水带
copper wedge cake 热轧铜板用锭坯
copperweld 包铜钢丝
copper welding 黄铜焊接;铜焊
copper welding rod 铜焊条
copper weld wire 包铜钢丝
copper wheel 铜轮
copper wire 铜锡线;铜电线;镀铜钢丝;铜线;铜丝
copper-wire bar ingot 铜线锭
copper wire brush 红铜丝刷;钢丝刷
copper wire cloth 铜丝布
copper-wire gasket 铜线垫
copper wire gauze 铜丝网
copper wire loose anchor 锚碇铜丝
copper wire net 铜丝网
copper-wire rod 紫铜盘条
copper wire tie 锚碇铜丝
copper wool filter 紫铜毛滤清器
copper-worm 蛀虫
copper yellow 铜黄色(的)
copper-zinc accumulator 铜锌蓄电池
copper-zinc alloy 锌铜合金;铜锌合金
copper-zinc brazing mixture 铜锌焊药
copper-zinc solder 铜锌钎料
copper-zinc welding rod 铜锌焊条
coppice land 矮林地
coppice-method 矮林作业
coppice selection 矮林择伐作业
coppice stand 矮林林分
coppice system 矮林作业法
coppice with field crops system 矮林混林作物
coppice with standard 中林作业法
coppice(wood) 矮林;小灌木林;矮树丛;树丛;小树丛
coppicing 柳枝护岸;灌木护岸;矮林平荏
copping 植树护堤;遮檐
copping rail 成型轨
copping stone 帽石

copping tile 墙压顶瓦;顶盖瓦
coprecipitate of iron and manganese hydroxide 铁锰共沉淀物
coprecipitating antibody 协同沉淀抗体
coprecipitation 共沉淀(作用);共同沉淀
coprecipitation complex stabilizer 共沉淀复合稳定剂
coprecipitation method 共沉淀法
coprecipitator 共沉淀剂
coprocessor 协(同)处理器;辅处理机
coproduct 联产品;副产品
co-production 合作生产;副产品生产
coprogenic rock 粪生岩
coprogenic structure 粪化石构造
coprolite 磷钙土;粪粒体;粪(化)石
coprolithus 粪石
copromicrite 粪粒微晶灰岩
copromotor 共促进剂
coprophyte 嗜粪生物;养生植物
coprosparite 粪粒亮晶灰岩
coprostane 粪甾烷
coprostanone 粪甾烷酮
coprostene 粪甾烯
coprostenol 粪甾烯醇
coprostenone 粪甾烯酮
coprosterol 粪甾醇
coprozoon 粪生动物
cops 城齿
copse 小灌木林;小灌木林;杂树林;矮树丛;树丛
copse wood 小灌木林
copter 直升(飞)机
Coptic architecture 科普特式建筑
Coptic church 科普特教堂
Coptic glass 科普特式玻璃
Coptic monastery 科普特修道院
copulation orifice 交合口
copulatory organ 交合器
copulatory path 交合道
cop winder 卷纬机
copy 拷贝;摹写;描绘;一份;抄录;抄件;副本;复制件;复印件;复写;仿制(品);仿造(品)
copy and adjoin neighbo(u)ring portion of maps 地图抄接边
copy and correct 复制并改错
copy back 复原
copy board 原图板;晒图框;拷贝台
copy by scan(ning) 扫描复制
copy carving machine 仿形刻字机
copy check 副本检验;复制检查
copy control 拷贝控制;复制控制
copy counter 拷贝计数器;复制计数器
copy cycle 拷贝周期
copy documents 副本;单据副本
copy edit 审稿
copy editor 审稿人
copy engraving machine 仿形刻模铣床
copy error 复制误差;复制错误
copy facing 靠模切面
copyfitting 版面调整
copy function 拷贝操作
copy grinding machine 仿形磨床
copyhold 善本保有权;稿件夹;登录不动产保有权
copyholder 晒版架;原稿架;稿图架;稿件夹持器
copying 临摹;仿形切削;晒印;拷贝
copying and punching shear machine 仿形冲剪机
copying apparatus 复印机;靠模装置
copying attachment 靠模附件;仿形附件
copying autogenous cutter 仿形气割机
copying book 复写簿
copying boring bar 仿形镗杆
copying cam 模拟凸轮;协联凸轮
copying camera 复制照相机;复照仪
copying control 仿形装置
copying cutting 靠模切削
copying device 靠模装置;仿形装置
copying equipment 复制设备
copying error 仿形误差
copying feed 仿形进给
copying foreign models 仿造外国机器
copying frame 晒像框
copying hologram 复制全息图
copying ink 复印油墨
copying lamp 晒图灯
copying lathe 靠模车床;仿形车床

copying layer 感光层
copying lens 翻拍镜头
copying machine 复印机；仿形机床
copying material 复制材料
copying milling 靠模铣
copying objective 复制物镜
copying of binary tree 二叉树的复写
copying paper 复写纸
copying planer 靠模刨床；仿形刨床
copying press 复印机
copying process 复印法
copying program(me) 复制程序
copying projector 翻拍机
copying royalty 版税
copying screen 网线板
copying speed 拷贝速度
copying system 靠模系统；模拟系统；协联系统；仿形系统
copying template 靠模样板
copying tool 仿形刀具
copying turning 仿形车削
copying unit 仿形装置
copying valve 仿形阀
copy machine 伪制机床
copy mask 复制掩膜；二次主掩模
copy mass storage volume 复制海量存储卷
copy milling machine 靠模铣床
copy modification 复制修改
copy modification segment 复制修改段
copy-money 版税
copy of bid 标书副本
copy of certificate by customs 出口许可证海关回联单
copy of film 复制片
copy-on-reference 复制后引用
copy opacity 拷贝密度
copy operation 拷贝操作；复制操作
copy option 副本任选
copy pattern 复制图
copy polishing machine tool 仿形抛光机床
copy preparation 印刷品配页设备
copy quantity selector 拷贝数量选定器
copyright 著作权；版权；取得版权
copyright contracts 版权合同
copyright infringement suit 侵犯版权诉讼
copyright note 版权说明
copyright protection 版权保护
copyright reserved 版权所有
copyright royalty 版税
co-pyrolysis 混合高温分解
copy room 复印室
copy router 仿形镂床
copy rule 仿形尺
copy selector 拷贝选择器
copy separation 复制间隙；分离拷贝
copy truing 靠模整形
copy turning lathe 仿形车床
copy volume 拷贝卷；副本卷；复制卷
coque 煮沸
coquillage 贝壳花饰
coquille 薄曲面玻璃
coquimbite 针绿矾
coquina (建筑用的)介壳石；介壳灰岩；贝壳石灰石(产于美国南部,用于建筑)；贝壳灰岩
coquina shell 灰岩介壳；贝壳岩
coracle 小渔艇
coral 海石花；珊瑚色；珊瑚
coral aggregate 珊瑚集料；珊瑚骨料
coral atoll 珊瑚环礁
coral blush 珊瑚粉红色
coral bottom 珊瑚底
coral coast 珊瑚礁海岸；珊瑚海岸
coral concrete aggregate 珊瑚混凝土集料；珊瑚混凝土骨料
coral faunal province 珊瑚动物地理区
coral foreshore 珊瑚滩
coralgal 珊瑚藻灰岩；珊瑚(和藻类)沉积；珊瑚沉渣
coral graphite 珊瑚状细石墨
coral harbo(u)r 珊瑚港
coral head 珊瑚角；珊瑚岬
coralin 柯罗林(一种棕色线条晒图纸)
coral insect 珊瑚虫
coral island 珊瑚岛
coral knoll 珊瑚丘；珊瑚岬

coral-lac 红蜡(一种供楼面处治用的耐磨抗酸蜡)
coral limestone 珊瑚灰岩
coralline algae 珊瑚藻
coralline crag 珊瑚灰岩
coralline ware 珊瑚瓷；珊瑚色陶器
corallite 珊瑚石；珊瑚色大理石；珊瑚单体
coralloid 洞穴珊瑚状沉积；珊瑚状
corall-reef island 珊瑚礁岛
corall-sinter 珊瑚状泉华
corallum 珊瑚体
coral mud 珊瑚泥
coral ore 晨砂
coral pink 珊瑚粉红色
coral pinnacle 珊瑚尖峰
coral polyp 珊瑚虫
coral rag 硬珊瑚灰岩；珊瑚石灰岩
coral red 珊瑚色；珊瑚红(色)
coral reef 珊瑚(暗)礁
coral reef coast 珊瑚礁海岸
coral reef coastline 珊瑚礁海岸线
coral reef lagoon 珊瑚礁泻湖
coral reef oil pool 珊瑚礁油藏
coral reef platform 珊瑚礁台地
coral reef sea navigation 珊瑚礁海区航法
coral reef shoreline 珊瑚礁海岸线；珊瑚礁滨线
coral ring 珊瑚环状
coral rose 珊瑚玫瑰色
coral sand 珊瑚砂
coral sand beach 珊瑚砂滩
coral Sea 珊瑚海
coral sediment 珊瑚沉渣；珊瑚沉积
coral shoal 珊瑚洲；珊瑚暗礁
coral talus 珊瑚崖堆
coral type draghead 珊瑚型耙头
coral zone 珊瑚区
coramandel 装饰用木材
corange line 等潮差线
coravel 交叉相关视向速度仪
corbeau 乌鸦黑色
corbeil(le) 花篮雕饰；花篮饰；花篮或果篮雕饰
corbel 花篮形装饰；梁托；牛腿【建】；叠涩【建】；撑架；托臀【建】；托臂；突额
corbel arch 突拱(突出于立面的一种假拱)；挑出托拱；撑架拱；突拱(挑出拱石的拱)
corbel arched girder 托臂起拱大梁；突拱大梁
corbel arm 悬臂挑梁；拱
corbel back slab 悬挑板；悬臂板；引板；撑架悬臂板；岸板
corbel beam 托梁
corbel bracket arms cluster 一朵斗拱(古建筑)
corbel brick 托座砖；托架；突额砖；垫石；悬臂砖
corbel closet 墙式便桶；翅托式便桶
corbel course 翅托层；肱层；突腰线
corbel cupola 托臂圆顶；突额圆顶
corbel curved girder 托臂弯曲大梁
corbel dome 托臂圆屋顶
corbel gable 踏步式山墙
corbeling 撑架工程
corbel joint 挑头接合
corbel(led) arch 叠涩拱；突拱
corbelled brick construction 砖叠结构
corbelled brickwork 砖叠涩
corbelled coursing 叠涩砌法
corbelled out 挑出的；支托；锯齿状的
corbelling 用撑架托住；叠涩砌法；叠涩【建】；撑架结构；挑出圬工
corbel(ling) iron 悬挑铁支撑；悬挑铁件；翅托铁件
corbelling of bricks 砖托座(砖层挑出圬工)；砖叠涩
corbel masonry work 突额圬工工程
corbel mo(u)ld 挑出线脚；挑出的线脚
corbel out 挑头；支托；悬挑；挑出
corbel outwards 用梁托支出；用翅托挑出
corbel piece 支撑物；肱木；梁托；挑出块；承枕
corbel pin 悬挑铁件；翅托钉
corbel plate 托板；梁托垫板
corbel ring 柱环饰
corbel's method 柯柏尔计算法
corbel step gable 挑出砖牙压顶山墙
corbel steps 马头墙；山墙上踏步式压顶；挑出踏步
corbel table 挑檐；挑出面(层)；托梁挑檐；牛腿托砖
corbel-table frieze 突出墙面的雕带
corbel vault 托臂穹隆；托臂拱顶；挑檐穹隆；挑檐拱顶

corbel WC 挑式便桶
corbicula 粉筐
Corbicula fluminea 河蚬
corbie gable 阶形山墙；马头山墙；梯形山墙；踏步形山墙；阶梯状山墙
corbie step 梯形山墙顶；山形墙侧边梯形突出物
corbie step gable 挑出砖牙压顶山墙；梯形山墙；阶式山墙
Corbin 科宾铜合金
corbulin 柯布林(一种用于混凝土地面的防水剂)
corcass 河岸沼泽地
corchorus olitorius 长果种黄麻
corcovadite 花岗闪长粉岩
cord 木材堆单位；电缆；粗条纹；波筋；软电线；牵引线；条斑；绳(索)；塞绳
cord adjuster 塞绳调节器
cordage (以128立方英尺为单位测量的)木材总数(1立方英尺=0.0283168立方米)；缆索；考得数；纤维绳；细绳；索具；绳索
cordage fiber 制绳纤维
cordage oil 钢丝绳油；绳索润滑油
cordage room 索具库
cordage sling 绳索吊环
Cordaite-vitrite 科达树微镜煤
cordaitotelinite 科达木结构镜质体
cord belt 直纹带；钢丝径线带
cord braiding machine 编绳索机
cord breaker (轮胎的)缓冲层
cord carcass of tire 轮胎帘布层
cord chooser 塞绳选择器
cord circuit 绳电路；塞绳电路
cord circuit repeater 塞绳中继器；塞绳增音机
cord clip 索夹
cord cloth 灯芯绒
cord content 含绳量
cord control 绳索操纵
cord covering 绳索覆盖
cord design 线纹
cord dip latex 轮胎浸渍胶乳
Cordean 考杜导火线
cordeau 雷管线；爆炸导火线；爆炸导火索
cordeau-detonant 导火线
corded 起凸线；起棱纹
corded door 索系门；折叠百叶门
corded lava 绳状熔岩
corded way 棱条坡道；坡梯道
Cordelan fiber 柯台纶纤维
cordelle 拖缆；拖船索；纤绳
cordelling 拉纤
cordelova relief 凸纹墙纸
Corde plate 果尔达板块
corderoite 氯硫汞矿
cord fabric 帘布
cord fastener 塞绳接线柱
cord-foot 考得英尺；积英尺(美国)
cord-free plate glass 无条纹板玻璃
cord gimp 嵌芯狭辫带
cordgrass 带状草地
cord grip 塞绳结头
cord grommet 索环
cord hook 绳钩
cordierite 堇青石(角岩)
cordierite biotite gneiss 堇青黑云片麻岩
cordierite ceramics 堇青石陶瓷
cordierite garnet potash-feldspar and plagioclase gneiss 堇青石榴二长片麻岩
cordierite garnet potash-feldspar gneiss 堇青石榴钾长片麻岩
cordierite gneiss 堇青石片麻岩
cordierite granulite 堇青石麻粒岩
cordierite-mullite refractory 堇青石—莫来石耐火材料
cordierite muscovite gneiss 堇青石白云母片麻岩
cordierite porcelain 堇青石瓷
cordierite sillimanite hornfels 堇青石矽线石角岩
cordierite slate 堇青石板岩
cordierite stoneware 堇青石炻器
cordierite two mica gneiss 堇青石二云片麻岩
cordierite whiteware 堇青石白坯陶瓷制品
cordillera 连绵山脉；雁列山脉；山系；(并行的)山脉
Cordilleran-Andes Meridian tectonic belt 科迪勒拉安第斯经向构造带
Cordilleran brachiopod region 科迪勒拉腕足动物地理大区

Cordilleran floral region 科迪勒拉植物地理大区
Cordilleran geosyncline 科迪勒拉地槽
Cordilleran marine trough 科迪勒拉海槽
Cordilleran metallogenic belt 科迪勒拉成矿带
cording 圆形装饰(家具等);绳索;楞条织物;花边;饰边
cording diagram 拉线图;连接图;接线图;塞绳连接图
cordite 柯达无烟药;硝棉;线状火药;无烟线状火药;无烟火药
cord junction ring 集索圈
Cordkapp fender 考特开帕式护舷;绑扎橡胶护舷(考特开帕式护舷)
cordless telecommunication 无绳通信
cordless telephone 无绳电话
cordless telephone system 无绳电话系统
cord-mark decoration 绳纹装饰
Cord measure 考得量度
cordometer 木材体积测量计
cordon 交通计数区划线;封锁交通;飞檐层;带伤;交通封锁线;带形线条;封锁线;壁篱;区划分线(交通调查用)
cordon area 交通调查范围;交通调查(分)区;封锁区
cordon count 划区交通计量;划区交通调查;小区交通调查
cordon line 交通调查区界线;调查区境界线
cordon mark 警戒标志
cordonnier 警戒孔;相同位穿孔;同位穿孔
cordonnier system 纽带系统;同位孔系统
cordon traffic survey 境界出入调查
cordon training 单干形整枝
cord operated 绳索操作的
Cordova pink (一种产于美国得克萨斯州的)柯多瓦粉红色花岗岩
cord pendant 电灯吊线;软线吊线
cord pendant lamp 吊灯
cord-pour 线型浇注;索状浇注
cord-pull switch 拉线开关
cord read (沼泽地区铺路的)圆木道
cord retainer 塞绳保持器
cords 索线
cord seal(ing) 绳索密封
cord set 连接插头组件;连线装置
cord stay 绳拉条
cord strength 绳索强度
cord switch 凸纹电闸;拉线开关;悬挂开关
cord tension pulley 张纶滑轮
cord test jack 塞绳测试塞孔
cordtex 爆炸导火线;爆炸导火索
cord tire 帘布外胎
cord type teflon asbestos packing 绳型聚四氟乙烯石棉填料
cord tyre 帘布外胎
corduroy 木排路;灯芯绒
corduroy cloth 灯芯绒饰
corduroy effect 路面搓板效果
corduroy fabric 灯芯绒织物
corduroy mat 木排(圆木)铺面
corduroy road 木排路;木排道(路);横木杆道;圆木路(沼泽地区)
corduroy strake 垛式支架
cordwainer 鞋工
cord weight 绳锤
cord with weight 带平衡锤的吊绳
cord wood 成堆出售木材;柴捆;柴堆;层积材;堆积木材;小柴捆
cordwood arrangement 积木式排列
cordwood construction 积木式建筑
cordwood model 积木式微型组件
cordwood module 积木式微型组件
cordwood system 积木式
cordwood technique 积木式组装技术
cord woven fabric 线织物;绳索织物
cordycepic acid 虫草酸
cordy glass 有条纹玻璃
cordylite 氟碳酸钡铈矿;佩碳钡铈矿
core 芯;木髓;核心(部分);型芯(子);岩芯(子);筒形中心带;中心部分;中核;制芯;电缆芯线束;筒形膜;铁芯;绳芯
coreactant 共反应物;共反应剂
coreacting resin 共反应树脂
coreactivity 共反应性
core address 主存地址;磁芯存储地址
core adhesive 型芯黏合剂

core allocation 磁芯存储区分配
core allocator 主存分配程序
core analysis 岩芯分析
core and mantle texture 核幔结构
core and mineral core 岩芯与矿心
core arbor 芯骨架
core area 核心面积;核心(地)区;迎风面积;中心地带
core array 磁芯矩阵
core assembly 岩芯管组件;组芯(造型)
core assembly mo(u)ld 组芯铸型
core assembly process 组芯造型法
core baking 型芯烘干
core-baking oven 型芯烘干炉;泥芯干燥炉
core-balance protective system 磁势平衡保护系统;三芯电缆平衡保护装置
core-balance transformer 磁势平衡互感器
core band 内层带
core bank 磁芯体
core bar 型芯铁;芯铁;芯骨;芯杆
core barrel 型芯轴;管状型芯铁;芯管;岩芯钻筒;岩芯取样器;岩芯管;底质柱状采样管;取芯筒;取芯管
core-barrel bit 取岩芯钻头
core barrel head 岩芯管接头
core barrel head for single tube 单岩芯管接头
core barrel of double tube-rigid 双层双动岩芯管
core barrel of double tube-swivel type 单动双层岩芯管(内管不动);单动双层取芯器(内管不动)
core barrel of double tube-swivel type with core container 有岩芯容纳管的双层岩芯管
core barrel rod 岩芯管导杆;导向钻杆
core barrel stabilizer 岩芯管稳定器
core basket 岩芯打捞筒
core bedding frame 型芯头
core bench 制芯台
core binder 型芯黏合剂
core bit 空心钻头;岩芯钻(头);钻头;取芯钻头;取芯钻具
core bit connection 岩芯钻连接
core-blanket reactor 双区反应堆
core blister 基体结疤
core-block 坝心块体
core blockage 岩芯堵塞
core blow 泥芯气孔;砂芯气孔
core blower 型芯吹砂机;吹芯机
core blowing 吹芯
core blowing machine 型芯吹干机;吹芯机
core blow plate 吹芯吹砂板
core board 夹芯(胶合)板;型芯板;芯合板;芯板材;心板胶合板;车板板;厚芯合板
coreboard for building purposes 建筑用夹芯板
core bore 扩孔钻
core borer 岩芯机
core boring 岩芯钻探;岩芯钻进;取芯钻探
core borings 岩屑;岩粉;钻粉
core box 泥芯盒;型芯盒;芯盒;岩芯箱;岩芯盒;砂芯盒
core box bit 盒钻头
core box plane 型芯盒刨
core box sealer 芯盒密封垫
core box vent 芯盒排气槽;排气塞
core breakdown 型芯破碎
core breakdown property 型态溃散性
core breaker 岩芯折断器;岩芯提断器;除芯机;砂芯破碎机
core breaker spring 岩芯提断器卡簧
core bridging 关闭中心电极(发火栓)
core buffer 内存缓冲器;磁芯缓冲器
core-building machine 筒制机
core-built tire 筒制外胎
core-built tyre 筒制外胎
core cable 芯缆
core catcher 砂样防落簧;开口环提芯器;岩芯提芯器;岩芯取样器;岩芯卡簧;岩芯夹具;采芯器;取芯夹具
core catcher case 岩芯卡簧座;岩芯夹具座
core catcher gland 取岩芯爪压盖
core catcher length 岩芯抓长度
core catcher size 岩芯抓尺寸
core cavity 铸件内腔
core chamber 岩芯库
core chaplet 泥芯撑
core circumference 岩芯周长;芯部周长

core city 核心城(市);中心城市
core clad 磁芯敷层
core clamper 顶尖
core coating 型芯涂料
core collapsibility 型芯溃散性
core concrete 空心混凝土
core container 岩芯容纳管
content control 磁芯存储器内容控制
core cover 型芯涂料
core crab 芯骨架;主芯骨架
core crest (土坝的)芯墙顶(部)
core cross-section 岩芯断面
core crown (土坝的)芯墙顶(部)
core crush 芯子压环
core cutter 岩芯提取器;砂样制取器;取岩芯钻;岩芯取样器;岩芯切割筒;岩芯切割;取土样器;切样器;土样切割器
core cutting machine 钻芯型机
core cutting machine for concrete 混凝土心型钻机
cored 筒形的;空心的;装芯
cored bar 带芯棒
cored beam 已取芯样的混凝土梁;空腹梁;空心梁;取过钻芯试样的混凝土梁
cored block 空心块体;坝心块体
cored block masonry work 空心砌块圬工工程
cored (bore) hole 取芯钻孔
cored brick 含孔砖;空心砖
cored brick masonry work 空心砖圬工工程
cored calcium silicate brick 空心硅酸钙砖
cored calcium silicate slab 空心硅酸钙板
cored carbon 芯碳棒;贯芯碳条
cored ceiling 空心天花板;空心顶棚;空心楼板
cored cellular material 蜂窝状多孔材料
cored dendritic morphology 包心树枝晶形态
cored density 型芯透气性
cored department 制芯工段
cored drill bit 取岩芯钻头;空心钻头
cored driver 取芯棒
cored electrode 有芯电焊条;管状焊丝;药芯焊条
core depression 芯子凹陷
cored expanded cinder concrete block 空心膨胀煤渣混凝土砌块
core floor 空心楼板;空心地板
cored foamed slag concrete block 空心泡沫控制混凝土砌块;多孔泡沫矿渣混凝土砌块
cored footage 岩芯钻进进尺;取芯进尺
cored hole 型芯孔;混凝土预留中心孔口
core diameter 焊芯直径;焊条直径;岩芯直径
cored interval 岩芯示出的含矿层
core diode storage matrix 磁芯二极管存储矩阵
core divider 岩芯分隔标牌
cored lime-sand brick 空心灰砂砖
cored lime-sand slab 空心灰砂板
cored mo(u)ld 孔道模具
core documentation 岩芯编录
cored-out floor unit 抽芯楼板构件
cored panel 穿孔板
cored panel floor 空心镶板楼板
cored pattern 空心模型
cored pedestal pile 钻孔爆扩桩
cored plate 打眼板;穿孔板;心板
core drainage lock 堤下水闸
core drawer 芯盒翻转机
core drawing machine 芯盒翻转机
core dressing 型芯修整;型芯涂料
core drier 烘芯托架式箱;型芯烘炉
core drill 扩孔钻;空心钻;套料钻;岩芯钻;取芯钻具;取芯钻
core drill fittings 岩芯钻机附件
core drill for wood working 木工空心钻
core drilling 岩芯钻探;岩芯钻进;钻取岩芯;取芯钻法;取芯钻井;取芯钻进
core drilling exploration 取芯钻探
core drilling inspection 岩芯取样检验
core drilling machine 岩芯钻机
core drilling rig 岩芯钻机
core drilling with diamond 用金刚石钻取心
core drill machine 岩芯钻机
core-drill method 取芯钻探法;套料钻探法
core drill rig 岩芯钻机
core drill sample 岩芯样(品)
core drill sampling 钻取岩(芯)样
core drill test 钻孔取样试验;钻取岩芯样试验
core dryer 型芯干燥机

core drying furnace 型芯烘炉
cored-sand-lime brick 空心灰砂砖
cored screw 空心螺旋
cored section 筒形截面
cored slab 空心板;挖孔板
cored slab construction 空心板结构
cored soffit 多孔拱腹面
cored solder 带芯钎焊条
cored solder wire 空心焊丝;钎焊丝
cored steel 中空钢
cored structure 晶内核状偏析组织
cored tile 筒状瓦;坯心瓦;筒状瓦
coreduction 同时还原
core dump 内存转储;主存储器信息转储;存取器清除;磁芯信息转储;磁头信息转储
core dump memory clear 存储器清除
cored-up mo(u)ld 岩芯造型;组芯造型
cored wall 空芯墙
cored wall panel 空芯墙板
core dye 型芯染料
core earth dam 芯墙土坝
core electrode 包心焊条
core entry (岩芯管的)岩芯入口
core equipment 岩芯钻具;取芯设备
core erosion 岩芯冲蚀
core exposure 岩芯外露长度(介于钻头与岩芯管内末端之间的)
core extraction 岩芯采收率;岩芯采取率
core extractor 岩芯退取器;岩芯提取器;岩芯提断器
core extruder 挤芯机
core extruding machine 挤芯机
core fall off 岩芯脱落
core file 内存储文件
core filler 砂芯填料
core fin 芯飞边缺肉(铸件缺陷)
core fisher 岩芯捞取器
core flo(a)tation 漂芯;抬芯
core forging 空心件模锻;芯锻
core formation 成核
core former 型芯机;型芯成型工具
core foundantion 空心基础
core frame 芯框
core function 核心功能
core gap 芯板裂缝
core gate 型芯内浇口
coregionalized variables 协同区域化变量
core glass 芯玻璃;玻璃心
core grabber 岩芯爪
core graph 岩芯图;岩心图
core graphic system 核心图形系统
core grid 泥芯骨;芯铁
core grinding 岩芯磨损
core gripper 开口环提芯器;岩芯爪
core gripper case 岩芯卡簧座;岩芯夹具座
core ground mass 核心土石体
core grouting (由钻杆投卡料的)卡取岩芯;卡料
core gum 型芯粘合剂
core-halo model 核晕模型
core hammer bit 取芯凿岩钻;取芯风钻
core hardening 型芯硬化
core hardness 岩芯硬度
core hardness tester 砂芯硬度计
core head 岩芯钻头;电枢端板
core head of double tube-rigid type 双动双层岩芯管;双动双层取芯器
core heating 磁芯加热
core hitch 线芯钩
core hole 岩芯(钻)孔;铸孔
corehole extensometer 岩孔伸缩仪
core hook 吊芯钩
core house 岩芯库
core housing 核心住房
core image 内存映象;磁芯映像;磁芯存储器映像
core image dump 内存映象转储
core image library 内存映像程序库;磁芯(存储器)映像库
core-in-core structure 筒中筒结构
core indicator 型芯定位座
core-insert hollow-cone nozzle 插芯式空雾锥嘴
core insertion 芯型样品的嵌入
core insulation 芯线绝缘
core interpreting device 岩芯定向装置
core intersection 岩芯上下显示矿体厚度;岩芯断面
core interval 岩芯断面

core investigation 岩芯钻探
core iron 泥芯骨;型芯骨
core jam(ming) 岩芯堵塞
core jarring machine 泥芯落砂机;震实制芯机
core jig 型芯夹具
core joint flash 型芯披缝
core knockout 泥芯打出机;泥蕊打出机
core knock-out device 击芯机;出芯机
core knock-out machine 击芯机;出芯机
core-laden inner tube 岩芯内管
core laden split tube 岩芯半合管
core-leaving method 旁侧导芯法
core length 岩芯长度
core length of divided seam 分层岩芯长度
core length of roundtrip 回次岩芯长度
coreless armature 空心衔铁;无铁芯电枢
coreless cold elbowing 无芯冷弯管
coreless-type induction furnace 高频(电)炉;无(铁)心感应电炉
core library 岩芯库
core lift 抬芯
core lifter 岩芯提取器;砂样制取器;岩芯钻;开口环提芯器;岩芯提取器;取土样器;切样器;土样切割器
core lifter adapter 岩芯提断器接头
core-lifter case 岩芯提断器外壳
core-lifter wedge 滑动卡楔;取芯楔
core lifting 抬芯;取芯
core lightener 简化造型用型芯
core lighting 泛光照明
core load 磁芯负载
core locator 型芯定位座
core logging 岩芯记录;钻孔岩芯测井
core logic 磁芯逻辑
core loss 芯铁损失;岩芯损失;磁芯损耗;铁芯损失;铁损损耗;铁损;铁耗
core-loss component 铁耗分量
core loss on upper and lower bounds footage 顶末底初回次岩芯缺失量
core-loss test 铁耗试验
core machine 型芯机
coremaker's bench 制芯工作台
coremaking 制芯
core-making machine 制型芯机
core-making operation 制芯作业
core-mantle boundary 地核地幔界面;地核地幔边界
core mark 型芯记号
core marker 型芯定位座;岩芯标记器
coremat 芯层毡
core material 心材;型芯材料;芯层物料;堤心材料
core matrices 磁芯矩阵
core matrix memory 磁芯矩阵存储器
core memory 磁芯存储器;磁芯储存器
core memory address 磁芯存储器地址
core memory expansion 磁芯存储器扩展
core memory plane 磁芯存储板
core memory reentrant routine 内存复归程序
core memory register 磁芯存储寄存器
core memory resident 内存常驻区;磁芯存储器常驻区
core memory sense amplifier 磁芯存储读出放大器
core memory stack 磁芯存储体
core method 取岩芯法;筒制法
core method of tunnel construction 导坑法隧道施工(德国法);核心支持法;中央导坑式隧道施工法
core mixer 芯砂搅拌机
core mixture 型芯混合料
core moment 核矩;柱心力矩
core mo(u)ld 组芯铸型
core mo(u)lding 组芯造型
core mo(u)lding machine 制型芯机
core network 核心网络
core number 岩芯编号
core of a classic(al) cooperative game 古典合作对策的核心
core of anticline 背斜中心
core of a rope 绳芯
core of bank 堤心
core of cable 电缆芯
core of column 心柱;柱的核心部分(混凝土)
core of dam 土坝芯墙
core of earth dam 土坝芯墙
core of gypsum 消解石膏块

core of section 截面(核)心
core of soil 土芯样
core of syncline 向斜中心
core oil 泥芯油;芯油;制芯油
core orientation 岩芯定向
core orientation works 岩芯定向工作
core orienting device 岩芯定向装置
core oven 烘芯炉
core packing fraction 芯填充系数
core paint 型芯涂料浆
core panel 填心板
core paper 层在纸的芯纸
core parting line 型芯开边
core paste 型芯黏合膏;砂芯黏合膏
core pattern 芯模
core pendulum 摆动心
core phase 核震相
core picker 岩芯捞取器
core pick hammer 鹤嘴锤采岩芯器
core picking 采芯
core pin 芯杆;中心销;穿孔杆
core pipe 心管
core pit 岩芯坑
core plan 核心式平面图;磁芯面
core plane 磁芯屏蔽;磁芯板
core plate 芯板(压铸用);型芯板
core plunger (液压或螺旋的)岩芯推出器;退心接头
core-ply textured yarn 包芯合股变形纱
core point 核点
core pool 水力填土坝坝心
core position 内存位置
core positioner 下芯夹
corepressor 共阻遏物
core print 泥芯头;模样芯头(突);模样上的芯头;型芯座;砂芯头
core print seat 芯座
core print with register 定向芯头
core property of case hardened steel 表面热处理后钢中心的性质
core pull 型芯移出
core puller 型芯拉出器
core pull sequence 型芯移出程序
core punch (液压或螺旋的)岩芯推出器
core pusher plunger (液压或螺旋的)岩芯推出器
corequake 核震
corer 岩芯提取器;去芯器;去核机;取芯器
core rack 型芯架;岩芯架
core rack furnace 架式烘芯炉
core radiation sampling 岩芯辐射取样
core rail (连接扶手与栏杆的)扁铁
core raise 抬芯
core raised 漂芯
core ratio 孔洞率
core record 岩芯记录
core recovering 采芯
core recovery 岩芯收获率;岩芯获得率;钻孔岩芯采取
core recovery parameter 岩芯采取参数
core recovery(percent) 岩芯收取率;岩芯采取率
core recovery tool 岩芯采取工具
core refractiveness 型芯耐火性
core register 型芯定位座;芯头标记
core resident 磁芯存储器中的常驻区;常驻主存的
core resident area 磁芯存储器常驻区
core resident routine 磁芯存储器常驻程序;常驻主存程序
core resistance 湍流核心阻力
core reverse circulation drill 岩芯循环钻机
core rig 岩芯钻机
core rock 堤心石料;堤心石
core rod 模芯;芯棒
core rod cutter 岩芯棒切断机
core rod straightening machine 型芯铁条整直机
core roll-over machine 翻台式制芯机
core rope 型芯绳
core rope memory 奥尔逊存储器
core rope storage 磁芯线存储器
core run 岩芯采收率;岩芯采取率;取芯回次
core sag 沉芯;砂芯下沉
core sample 芯样;岩样;岩芯样(品);岩芯(试样);钻芯试样
core sample of sand 砂芯样品
core sampler 岩芯取样器;柱状采泥器;底质采集器
core sampling 岩芯取样;岩芯采样

core sand 型芯砂;芯砂;造芯混合料
core sand mixer 芯砂搅拌机
core sand mo(u)lding 型芯造型
core-sand mo(u)nd 组芯铸型
core saw 岩芯锯
|core section 反应堆活性区
core sense line 磁芯读出线
core separation 芯子分离
core separator 岩芯劈开器;岩芯劈开机
core setter 下芯机
core setting 装配泥芯;下芯
core setting ga(u)ge 下芯样板
core setting jig 型芯夹
core shack 岩芯棚;岩芯库房
core shadow 本影
core shanty 岩芯棚;岩芯库房
core shaped sample 柱状岩芯(芯)样
core sharing program(me) 磁芯共享控制程序
core sheet 芯层;中心层
core shelf 砂芯烘架
core shell 岩芯外皮;取芯内管(三层岩芯管);岩芯外壳
core shell coupling 岩芯卡簧座接头
core-shell emulsion particle 核壳颗粒乳液
core-shell polymer 核壳聚合物
core-shell structure 核壳结构
core shield 铁皮芯
core shift 错芯
core shooter 射芯机
core shooting 射芯
core shooting machine 射芯机
core shop 制芯工段
co-resident 共驻主存的
coresidual 同余
coresil 软木隔热垫(一种专卖品)
core size 岩芯尺寸;磁芯尺寸
core skeleton 型芯骨架
core sketch 岩芯素描图
core slicer 切割式井壁取芯器
core slope 芯墙倾斜(度)
cores-loss component 铁损分量
cores-loss test 铁损试验
core slurring 型芯粘合法
core snatcher 岩芯爪;岩芯提断器
core sort 磁芯分类
core spacing effect 中心间隔效应
core specimen 岩芯标本;钻芯试件
core specimen strength 钻芯试件强度
core spinning frame 包芯纺纱机
core splice adhesive 芯拼合胶粘剂
core splicing 芯子拼接
core splitter 岩芯切断器;岩芯劈开器;岩芯劈开机
core-splitting method 岩芯劈分法
core spring 岩芯卡簧
core spring adapter 岩芯提断器卡簧;卡簧接头
core spring carrier 岩芯卡簧座;岩芯夹具座
core spring case 岩芯卡簧座;岩芯夹具座
core-spun elastic yarn 包芯弹力纱
core-spun stretch yarn 包芯弹力丝
core-spun yarn 包芯纱;包芯纱
core stack 叠片磁芯;磁芯体
core stock 心坯;心轴;木门芯;门芯;厚芯合板;芯板料;芯板材
core stock coreboard 夹心碎料板
core stone 堤心石
core storage 岩芯库;磁芯存储器;磁芯储存器
core store 磁芯储存器
core strainer 滤渣芯片
core strength test 钻芯强度试验
core strickle 车芯板
core strickle template 车芯板
core structure 筒式结构
core sucking and filling system 冰心抽灌系统
core switch circuit 磁芯开关电路
core-switching circuit 磁芯换向电路
core system 核心式体系
core table 岩芯(置放)台
core tap 四方丝锥
core tape 带状磁芯;磁芯带
core temperature trip 铁芯温度跳闸
core template 型芯样板
core test 芯样试验;岩芯试验;钻芯试验;中心抽样检验
core testing 型芯表面硬度测定;钻芯测定法

core-testing circuit 磁芯测试电路
core theory 核心理论
core tongs 岩芯夹钳
core tool 取芯钻
core tool of long barrel 长筒取芯工具
core-to-sludge ratio 岩芯粉砂比
core transportation center 运输中心枢纽
core trap-ring 岩芯收集环
core tray 岩芯盘;岩芯箱
core trench 芯墙沟;芯墙截水沟;芯墙截水槽;芯墙基槽;堤坝芯墙下沟;坝心坑槽
core trench of dam 土坝芯墙基槽
core trimming 岩芯整理
core truck 送芯小车
core tube 岩芯(采样)管
core tube supported suspended structure 核心筒悬挂结构
core tube tap 岩芯管丝锥
core tuning 磁芯调谐
core turning lathe 型芯车床;制芯机床;制芯车床
core type 岩芯类型
core-type dam 芯墙(式土)坝
core-type induction furnace 有芯感应电炉
core type radiator 空心式放热器
core type reactor 内铁芯电抗器
core-up mo(u)ld 组芯铸型
core value 岩芯鉴定
core varnish 磁芯清漆;枢芯清漆
core vent 蜡芯出气孔;型芯排气塞;出气孔;排气塞
core venting 型芯排气塞
core vibrating unit 振动出芯机
core vibrator 振动出芯机
core wall 芯墙;岩芯墙
core wall dam 芯墙坝
core wall of dam 坝的芯墙;不透水坝芯墙;挡水坝芯墙
core wall type rockfill dam 芯墙式堆石坝
core wash 型芯涂料;岩芯冲蚀;涂芯型浆;涂芯型浆
core wash boring 取岩芯冲浆钻探
core welding-wire 焊芯
core well 取芯井
core wheel 轮心
core winding 芯绕组
core wire 焊条(铁)芯;芯线;芯钢丝;心线;电缆芯线;带芯铜线;铁丝芯骨
core wire for welding rod 焊条芯线
core wire straightening machine 芯骨校直机
core wrap 缆芯绕包层
core yarns 绳芯麻条
core zone 中心地带
corf 鱼笼;小型矿车;土筐;土篮;鱼笼;小型矿车;土框
corguide 康宁低耗光缆
Corhart 科尔哈特高级耐火材料;耐火材料
Corhart block 电熔莫来石—刚玉砖
Corhart standard 电熔莫来石—刚玉耐火砖
Corhart Zac-block 电熔锆刚玉砖
coriaceous 革质的
coriaria 马桑属
Co-rich Mn nodule 富钴型锰结核
Co-rich(Ni/Cu-poor) Mn nodule 富钴(贫镍铜)型锰结核
corindon 刚玉
coring 钻取土样;挖心;取芯样;晶界小晶体;黑心;岩芯钻进;岩芯取样;作型芯;钻取土样;枝晶偏析;(焊点边缘的)根须;包心偏析;取岩芯;取芯作业
coring apparatus 岩芯取器器;取岩芯设备
coring bit 取芯钻头
coring bit connection 岩芯钻头连接
coring crown 取芯钻头
coring device 岩芯取器器;取岩芯设备
coring footage 取芯进尺
coring interval 取芯井段
coring machine 岩芯钻机;取芯钻机
coring operation 钻取混凝土芯样
coring out 清理烟道(施工中);第一道粉刷(檐口或线脚粉刷中);清除烟道内灰垢
coring roller bit 取芯牙轮钻头
coring row 一排钻孔
coring segregation 晶内偏析
coring time 取芯钻进速度;取芯次数
coring tool 取芯钻;刮芯具;取芯具;取芯工具
coring tube 岩芯采样管
coring type turbodrill 取芯(式)涡轮钻具

coring-up station 下芯区
coring weight 取芯钻头压力
Corinthian 考林辛式建筑(古希腊)
Corinthian capital 考林辛式柱头
Corinthian column 考林辛柱
Corinthian entablature 考林辛柱顶盘;考林辛式顶盘
Corinthianesque base 考林辛式底脚
Corinthianesque capital 考林辛式柱帽
Corinthianesque column base 考林辛式柱础
Corinthianesque peripteral octastyle temple with fifteen columns on the flanks 考林辛两廊为15柱的八圆柱式神庙
Corinthianesque peripteral temple 考林辛式圆柱式神庙
Corinthianesque portico 考林辛式柱廊
Corinthian order 考林辛式柱型
Corinthian style 考林辛式
Coriolis acceleration 科里奥利加速度;互补加速度;复合向心加速度
Coriolis correction 科里奥利(力)校正
Coriolis deflection 科里奥利效应
Coriolis effect 科里奥利效应
Coriolis force 科氏力;科里奥利力;地球自转偏向力;地球偏转力;复合向心力
Coriolis influence 科里奥利影响
Coriolis operator 科里奥利算符
Coriolis parameter 科里奥利参数;地球自转偏向力参数
co-riparian states 共同沿岸国家
corite (一种专利用于防火楼面的)空心梁
corium 革片
cork 木栓;插塞;软木(塞);塞
cork article 软木制品
cork backing 软木背衬
cork bark 栓皮栎
cork base 软木基底;软木垫底;软木底层
cork based 软木垫底的
cork black 软木黑颜料
cork block 软木块
cork block board 软木块板
cork block for heating lagging 隔热软木块
cork block for heat insulation 隔热软木块
cork block sheet 软木薄板
cork block slab 软木块板
cork block wall 软木砌块墙
cork board 软木板;填缝软木板;木栓板
corkboard for ceiling 平顶软木板;顶棚软木板
corkboard for cold lagging 防寒软木板
corkboard for heating lagging 屋顶软木板
corkboard for heat insulation 屋顶软木板
cork borer 木塞穿孔器
cork boring machine 穿孔机
cork brick 软木块;多孔块;软木砖
cork buoy 软木浮标
cork cambium 木栓形成层
cork carpet 软木地板面;软木垫;软木地毯;软磨毡层
cork cell 软木元件
cork column 螺旋形柱
cork cortex 木栓皮层
cork cover(ing) 软木覆盖
cork covering cord 包软木的绳索
cork crust 木栓壳
cork drill 木塞钻孔器
cork dust 软木屑
corkdust concrete 软木屑混凝土
corker 木塞压紧机;压塞机
cork fender 护舷软木;软木碰球;软木碰垫;软木护舷材
cork filler 软木填料
cork fill(ing) 软木填充;软木填料
cork film 木栓膜
cork float 软木浮子
cork floor cover(ing) 软木地毡;软木地砖
cork floor finish 软木地板饰面
cork flooring 软木贴面;软木地面;软木地板;铺木地面
cork flooring tile 软木铺地砖
cork flour 软木粉
cork for heat insulation 隔热软木
cork for heat lagging 隔热软木
cork gasket 软木密(封)垫;软木垫圈;软木衬垫
cork ga(u)ge 塞(径)规
cork granule 软木粒

corking 木梁搁接法；堵塞缝道；开跨接；喷涂软木屑涂料；喷软木屑处理
cork-insulated hold 软木隔热舱
cork insulation 软木绝缘；软木绝热；软木隔热
corkite 磷铅铁矾
cork jacket 救生衣；软木救生衣
cork joint-filler 软木填缝料
cork-lagged 用软木板护壁的
cork lagging 用软木板护壁
cork life-belt 救生衣；软木救生带；软木救生衣
cork life buoy 软木救生圈
cork-light trim 空载纵倾
cork line 浮子网
corkline winch 浮子纲绞车
cork linoleum 软木油地砖
cork mat 软木垫
cork method 木塞法；软木塞法
cork mortar 软木砂浆
cork oak 软木橡木；栓皮栎
Corkoustic （一种专卖的）吸声材料
cork packing 软木垫
cork pad 软木垫
cork paint 掺软木屑涂料；软木（屑）涂料
cork paving sett 软木铺路小方块
cork pipe section 软木管段
corkpit 船尾部井型座舱
cork plug 软木塞
cork pneumoconiosis 软木尘肺
cork polishing wheel 软木抛光轮
cork powder 软木粉
cork pressor 木塞压紧器
cork product 软木制品
cork rope 浮子网
cork rubber 密封用软木橡胶；软木橡胶（密封用）；软木橡皮
corkscrew 螺旋状的；起软木塞的起子；螺旋拔塞器；螺丝起子；瓶塞钻；瓶塞起子；塞拔
corkscrew antenna 螺旋形天线
corkscrew auger method 螺旋钻（挖土）法；螺旋钻挖土法；螺旋钻进法
corkscrew column 螺旋形柱
corkscrew core 螺旋状岩芯
corkscrewing 在药包内掏雷管窝
corkscrew instability 螺旋不稳定性
corkscrew rule 螺旋法则；旋转法则
corkscrew spin 螺旋
corkscrew stair(case) 盘梯；螺旋楼梯；盘旋楼梯
corkscrew weave 螺旋纹织物
cork setting 软木小方块
cork setting asphalt 胶粘软木沥青；沥青黏结软木；粘贴软木沥青
cork shaving 软木薄片
cork sheet 软木纸；软木片；软木（薄）板
cork slab 软木板
cork slab for ceilings 顶棚用软木板
cork slab for cold insulation 隔冷软木板
cork slab for cold lagging 防寒软木板
cork slab for roof(ing) 屋顶软木用板
cork slab for walls 墙用软木板
cork spacer 软木垫片
cork stair tread 软木楼梯踏步
cork stopper 木塞；软木塞
cork strip 软木垫片
cork subfloor 软木毛地板
cork tackboard 软木布告板
cork tan 软木色
cork tar （混有木屑的）软木柏油
Cortex （一种专卖的）抗凝结涂料
cork tile 软木瓦；软木（铺）砖
cork tile foor(ing) 软木砖地板；软木块地板
cork tissue 木栓组织
cork tree 软木树；栓皮栎
cork underlay 软木垫层
cork wall 软木墙
cork wall tile 软木墙砖
cork wart 木栓瘤
cork washer 软木垫圈
cork wheel 软木轮
corkwood 轻木；软木（绒纤维）
cork worked by agglomeration 软木附聚加工
corky sinter 海绵状泉华
Corliss valve 柯立斯摆动阀
Corliss valve gear 柯立斯摆动阀装置
corm 球茎

cormophyte 茎叶植物
cormorant 鸬鹚
corn 小硬颗粒；玉蜀黍；谷物；谷粒(色)
corn belt 大农业地带
corn bin 粮箱
corn catcher 谷粒收集器
corn clad field 谷物种植地
corn cleaner 玉米清选机
corncob bit 锥形钻头（换径用）
corn crusher 玉米碾碎机
corn cutterbar 谷物切割器
cornean 隐晶岩
corned 呈粒状的
corneite 黑云angular
corner 隅（角）；转角；拐角；弓角；囤积居奇；墙角；囤积居奇；使有棱角
corner accessory 联测定标
Corner-Allen unit 科纳—阿伦单位
corner and connection pile 转角板桩；端角板桩
corner and junction pile 转角板桩
corner angle 外包角；棱角；顶角；边角钢
corner antenna 角反射器天线；角形天线
corner apartment 转角公寓
corner apartment unit 转角公寓单元
corner balcony 转角阳台
corner bar 角隅拉杆；角隅钢筋
corner bath 墙角浴盆
corner bath tub 转角浴缸
corner batt 隅角交点
corner bead 角焊缝；护角条；墙角线条；墙角护条
corner bead clip 护角夹；墙角线脚夹
corner beader 墙角圆线条；墙角护条
corner beam （钻塔底座的）角钢梁
corner bench 角凳
corner bend 角形弯曲
corner beveling 圆弯（波导管）
corner bit brace 角隅钻孔曲柄；角斜撑
corner block 墙角用空心砌块；角砖；内角加强三角条；转向滑车；墙角用空心砖（一端和两侧均为实心外面）
corner board 墙角（护）板
corner bond 墙角砌合
corner bonder 墙角砌合砖
corner bondstone 墙角砌合石
corner brace 角撑；对角撑；齿轮传动手摇钻；转角联系条；隅撑；斜撑
corner bracing 角斜撑；角撑铁；隅撑；斜撑
corner bracket 角托架；角架；转角牛腿
corner break(on concrete pavement) 隅裂；（混凝土路面的）板角损裂；角(隅)裂(缝)
corner brick 隅角砖；外角砖；角落砖
corner building 转角建筑
corner buttress 墙角支墩；墙角扶壁
corner cabinet 角柜；墙角柜
corner capital 角柱头；转角柱头；转角柱帽
corner casting 角(配)件；护角铸件；弯管铸造
corner chair 转角椅；角椅
corner chapel 转角小教堂
corner chimney 隅角烟囱
corner chisel 角凿
corner clamp 木工角夹；斜角夹板
corner clearance 转角视距
corner column 墙角柱；角柱；（钻塔底座的）角钢柱
corner condition 隅角条件
corner coordinates 图廓点坐标；图角坐标
corner corbel-bracket 转角斗拱
corner corbel-bracket set 角科斗拱
corner cramp （斜角接头的）扣钉
corner cube 角隅棱镜
corner cube mirror 直角反射镜；三直角锥反射镜
corner cube prism 直角棱柱体；直角棱镜；三直角锥棱镜；三面直角棱镜
corner cube reflector 三直角锥反射器
corner cupboard 三角形橱；角(隅式)柜；角隅式橱
corner curb radius 街角侧石半径；转角缘石半径
corner cut 内角加工；转角切除；切角
corner cutting 切角(的)
corner cutting tile 八角形瓷砖
corner damage 掉角
corner defect 缩边露角（漆病）
corner detail 角清晰度；图像角清晰度
corner dimension 夹角大小；弯头尺寸
corner door frame brick 墙角门框砖
corner drill 角钻；角轮手摇钻

cornered beam 抹角梁
cornered dragon-head ridge ornament 合角吻兽
cornered piston ring 棱边精车过的活塞环
cornered rat game 困鼠对策
corner effect 角效应；锐角效应
corner element 墙部件
cornerer 囤积居奇者
corner facing slab 墙角面板
corner facing tile 墙角面砖
corner fender 隅角防撞装置
corner fiducial mark 角框标
corner filler 贴角焊缝
corner fillet 贴角焊缝；填角块；角镶；圆角嵌条；隅缘线；贴角焊缝；角砖
corner finisher 边角修整器
corner-fired boiler 四角燃烧锅炉
corner fireplate 转角壁炉
corner firing 四角燃烧
corner firing burner 角式喷燃器
corner fittings 角(配)件；角端配件；护角铸件；箱角配件
corner flange weld 单卷边角焊缝；卷边角焊缝
corner flat 转角套间
corner focus 角聚焦
corner force 角隅力
corner foundation 角基础
corner frequency 转角频率；拐角频率；拐点频率
corner furniture 墙角家具
corner garden edging 转角公园边饰
corner gate 角形控制极
corner ghost 钢锭角鬼线；锭角鬼线
corner goods 囤积货物
corner grinding machine 磨角机
corner guard 护角铁；护角条；墙角护条
corner guide 箱角导向块
corner halving 角镶接；角嵌接；转角嵌接
corner head 角砖
corner header 墙角顶砖
corner hinge 角上铰链
corner house 转角房屋
corner illumination 角隅照明
cornering 回转；以角相接；转弯
cornering ability 转向性能；转弯能力
cornering at points 在道岔区进退两难
cornering force 回转力
cornering property 回转运动的性质
cornering ratio 横向力系数
cornering stress method 角点法
cornering tool 角隅车刀
corner insert 角形插头
corner insulator 转角绝缘子
corner iron 角铁
cornerite 室内隅角处金属条(防止抹灰粉裂纹)；金属墙角护条；护角金属镶条；包角
corner joint 角接接头；角接（头）；转角缝；弯头连接；弯管连接；弯管接头
corner knot 棱缘节；边棱节
corner lath 转角钢丝网板条
corner lavatory 角隅式盥洗室；角隅式洗手盆；三角洗手盆；三角洗面器
corner lavatory basin 墙角盥洗盆
corner lifter 角边拉底炮眼
corner lighting 角隅采光
corner link plate 墙角连接板
corner living unit 角隅起居室
corner load formula 角隅荷载公式
corner load(ing) (角)隅荷载
corner locked joint 梳齿接；角接
corner locking 直角组接
cornerlock joint 梳齿结合；企口连接；鸠尾榫接合；角锁接榫
corner loss 转弯损失
corner lot 转角地段
corner loudspeaker 角隅扬声器
corner mark 角标志；十字规矩线
corner metal 隅铁
corner mirror 隅角反射镜
corner mold 边角填块（梁柱）
corner mooring post 端角系缆桩
corner mo(u)lding 转角线脚
corner mullion 转角中梃
corner niche 角壁龛
corner nutrunner 角隅螺帽扳手
corner on the market 市场疲软；市场垄断

corner oriel 角隅凸肚窗
corner ornament 角隅装饰
corner pad 角部衬垫
corner panel 角板
corner pavilion 角亭
corner period 拐角周期
corner piece 角件;斗角加固块
corner pier 角柱石
corner pilaster 转角壁柱;转角半露方柱
corner pile 边桩;角桩;角偶板桩
corner pillar 角柱;转角柱墩;转角柱墩
corner pin 角钉;角插头
corner planting 角隅种植
corner plate 角撑铁;角板;角复板
corner plinth brick 墙角勒角砖
corner pocket 死角
corner point 角点;拐角点
corner-points method 角点法
corner pole 角柱;角砖;转角塔;(架电线的)转角架;转角杆
corner post 角柱,角支撑;转角塔;转角杆;转角窗柱;转角窗梃
corner post of balustrade 栏杆角柱
corner prism 隅角棱镜
corner punching machine 冲书角机
corner radiator 墙角散热器
corner radius 交叉口转角半径;转角半径;刀尖圆弧半径
corner rafter 墙角橡子
corner reflector 角(形)反射器
corner-reflector antenna 角形反射天线;角反射器天线
corner reflectors array 隅角反射器阵列
corner reinforcement (加固门框上角的)铁三角;转角处的加固钢筋;转角处的加固角钢
corner resolution 角隅分辨率;角清晰度;隅角分辨率;转角分辨率
corner return block 墙角用的空心砌块
corner rib 角肋
corner riveting hammer 角隅铆钉锤
corner roller 角辊
corner rounding (milling) cutter 圆角铣刀
corners and fillets 圆角和倒角
corners broken off main body 辊
corner section 转弯段
corner separator 角分离器
corner sheaf collector 集捆器
corner sheet pile 转角板桩;端角板桩
corner shop (住宅区附近的)小商店;街角商店
corner shower stall 角隅淋浴分隔间
corner sight angle 路口转角视角;转角视角
corner sink 墙角洗涤盆
corner slick 角刮子
corner slicker 抈角器
corner socket 角插座
corners of mouth 嘴角
corner spade 角形铲
corner stake 角桩
corner stay 角牵条
corner steel 角隅钢筋
corner step 墙角角踏步
corner sticks 十字规矩条
corner stiffener 角部加劲件;角部加固件
corner stone 角柱石;基石;隅石;奠基石;墙角基石
corner stone laying ceremony 奠基仪式
corner stone of a mining field 矿区墙角石
corner strap 角部连接条
corner stress 角隅应力
corner structure 角结构
corner stub 转角柱
corner support 角支撑
corner-supported rectangular slab 角隅支承的矩形板
corner tangent point 角隅切点
corner tank 舷侧angular水柜
corner test of association 相联性的象限检验法
corner the market 囤积居奇
corner thickness (混凝土路面的)板角厚度
corner tile 墙角瓦
corner tool 修角工具
corner tower 角隅塔楼;角楼
corner trap 舞台前活动拉门
corner trowel 角镘刀;阳角抹子;敞角泥刀;抹角泥刀
corner truss 角部桁架

corner tub 角隅式浴盆;角隅式洗手盆
corner tube boiler 角架式锅炉;角管式锅炉
corner turret 角隅炮塔;角塔;角楼
corner valve 角阀
corner vane 转角处导向叶片;导向叶片;导流叶(片)
corner-vane cascade 转弯导流叶栅
corner vertical 角柱
corner wash basin 墙角洗脸盆
corner waveguide 直角弯波导
corner wear 刀尖磨损
corner weld 角焊;角接焊缝
corner welding joint 角上焊接
corner window 角窗
corner zone lot 转角区土地
cornerwise 对角线的
cornetite 蓝磷铜矿
Cornet's forceps 盖片钳
corn exchange 粮谷交易所
corn-factor 捐客;谷物商
corn flakes 玉米片状铝粉
corn-flower 矢车菊
cornflower blue 浅蓝色
corn grinder 玉米粉碎机
cornice 雪檐;封檐;挑檐;上楣
cornice boarding 花檐板;钉挑檐板
cornice bracket 檐口托龛;挑檐托梁;上楣托座;沟槽支撑;挑檐托座
cornice casing 挑檐外包饰面
cornice facing 挑檐外包饰面
cornice gutter 外檐天沟
cornice lighting 沿墙线照明;檐板照明;挑檐照明
cornice of pedestal 座基上挑檐;台口线
cornice plane 鱼鳞纹面;鱼鳞面
cornice return 风檐板;挑檐的转向
cornice soffit 檐口平顶;挑檐底板
cornice stone 挑担石
cornice trim 檐口处理
cornice vent 檐底通风口
corniche 悬崖小路
corniferous rock 角页岩
corning 制成粒
corning glass 康宁玻璃;麻粒玻璃
corning-steuben 模制玻璃(一种供建筑和装饰用的专利产品)
cornish bit 扩孔刀;镗刀
Cornish boiler 内烟道锅炉;科尼斯(式)锅炉;单炉筒锅炉
Cornish-Fisher expansion 科尼斯—费雪展开式
Cornish granite 粗纹花白色花岗岩;科尼斯花岗岩
Cornish pump 科尼斯水泵
Cornish rolls 科尼斯辊碎机
Cornish single-flue boiler 科尼斯单烟道锅炉
Cornish slab 科尼斯石板
Cornith 考尼斯锰钢
corn-mash 粗钙质砂岩
cornmill 制粉机
cornoid 牛角线
corn oil fatty acid 玉米油脂肪酸
corn oil maize oil 玉米油
cornop gate 压边浇口
corn porter 谷物装卸工人
corn product waste 玉米产品废物
corn snow 春天粒雪;春融积雪
corn starch processing mill 玉米淀粉加工厂
corn starch production wastewater 玉米淀粉生产废水
cornstone 杂色石灰岩;玉米灰岩
corn tongs 宝石钳
corn trade clauses 谷物运输条款;谷物贸易条款
cornubianite 长英云母角岩
cornubite 羟种铜石
cornucopia (古希腊建筑的)丰饶角
Cornu formula 科尼尔公式
cornu inferius 下角
Cornu Jellet prism 柯尼尔—杰利特棱镜
Cornu mounting 柯尼尔棱镜装置
Cornu prism 柯尼尔棱镜
Cornu quartz prism 柯尼尔石英棱镜
Cornu's spiral 柯尼尔螺线;回旋曲线
cornwallite 墨绿砷铜矿
corn wet mill 玉米湿加工面粉厂
corn year 丰年
coro 唱诗房

coro alto 高位唱诗班席
coro conveyer 辊子输送机
corolla 花冠
corolla-like faults 花冠状断层
corolla lobes 冠瓣
corollary 推论
corollary equipment 配载设备;配套设备
corollary facility 配套设施
corolla tube 冠筒
corolliform 花冠状的
corollithic column 盘枝叶饰柱;叶饰柱身柱;叶饰柱身的柱子;叶饰柱(子)
corol lobe 冠片
coromandel 锡兰黑纹棕色硬木;东印度柿木
Coromant cut 科罗门脱型掏槽
coromat 包在管子外面防止腐蚀的玻璃丝层
Coromerc 苯汞乙二胺杀菌剂
corona 花冠;枝形吊灯;(教堂的)圆吊灯;花檐底板;飞檐的上部;焊点晕;晕边;月华;由臭氧引起的表面裂纹;光环;冠状物;冠;电晕;副花冠;飞檐上部;反应边;日华;挑檐滴水板;[复]coronae
corona breakdown 电晕击穿
corona counter 电晕放电计数管
corona current 电晕电流
corona current density 电晕电流密度
corona detector 电晕探测器
corona discharge 电晕放电
corona discharge curing 电晕放电固化(法)
corona discharge laser 电晕放电激光器
coronadite 锰铅矿
corona effect 电晕放电效应
corona electrode 电晕电极
corona electrode hammering 电晕极振打
corona emission 电晕放电;电晕发射
coronagraphic(al) technique 日冕技术
corona guard 电晕保护设备
coronal 冠状的;日冕的
coronal bright point 日冕亮点
coronal cloud 冕云
coronal condensation 日冕凝聚物;日冕凝聚区
coronal continuum 日冕连续光谱
coronal electrophotometer 日冕光电光度计
coronal enhancement 日冕增强区
coronal equilibrium 日冕平衡
coronal forbidden line 日冕禁线
coronal green line 日冕绿线
coronal heating 日冕加热
coronal hole 冕洞
coronal light 日冕光
coronal line 日冕谱线
coronal of the Galaxy 银冕
coronal optic(al) polarization 日冕光学偏振
corona loss 电晕损失;电晕损耗
coronal prominence 冕珥
coronal ray 日冕射线
coronal red line 日冕红线
coronal section 冠状面
coronal transient 日冕瞬变
corona lucis 圆环形烛灯圈(教堂用)
corona model 日冕模型
corona of the terrestrial atmosphere 地球大气上层地冕
corona power density 电晕功率密度
corona prevention 电晕防止
corona-proof cable 防电晕电缆
corona protection 电晕防护
corona-protective varnish 防电晕清漆
corona pulse analyzer 电晕脉冲分析器
corona radiata 辐射冠;放射冠
corona resistance 耐电晕性;耐电晕放电(击穿)能力;晕抗
corona resistant 电晕放电电阻;耐电晕放电
coronarium 挑檐装饰抹面(古罗马)
coronary 冠状的
coronary circulation 冠脉循环
coronary T wave 冠状 T 波
coronascope 日冕观测镜
corona shield 电晕屏蔽设施
corona stabilizer 电晕稳定器
corona-starting voltage 起始电晕电压
corona tester 电晕测试仪
corona texture 冠状结构;反应边结构
coronation cathedral (皇帝加冕的)大教堂
coronation chamber 加冕室

coronation church 加冕教堂
corona treatment 电晕处理
corona tube 电晕管
corona tube regulator 电晕管稳压器;电晕管调节器
corona unit 电晕装置;电晕器
corona unit cleaner 电晕器清洗器
corona voltage 电晕电压
corona voltmeter 电晕伏特计
corona zone 电晕区
Coronel 科罗内尔合金
coronet 门头线饰
coronet coupling 快速旋转式接插件
coronising 定形、柔软和着色工艺(整理玻璃纤维织物);连续热清洗;定纹工艺
coronite 反应边
coronium 白光环质
coronizing 连续热清洗;扩散镀锌;高温处理;定纹工艺
coronoid process 冠突
corotating intermeshing extruder 正转间隔挤出机
corotating self-wiping twin-screw extruder 正转自动卸脱双螺杆挤出机
corotating twin-screw mixing extruder 正转双螺杆混合挤出机
corotation 顺转
corotational departure 顺自转发射
corotation circle 共转圈
corotation radius 共转半径
corotation zone 共转带
corotron 电晕器
coroutine 联合程序;协同例程;协同程序;共行程序
corpocollinite 团块镜质体
corpohuminite 团块腐殖体
corpora quadrigemina 四叠体
corporate 共同的
corporate action 公司股东决定
corporate agency 企业
corporate assets 公司资产;公司资本
corporate body 法人团体;法人组织
corporate bond 公司债券
corporate capacity 法人资格;法人能力
corporate capital 公司资本
corporate charter 公司执照
corporate control 财团控制
corporate coverage 法人保险额
corporate deficit 公司亏损
corporate equity securttites 法人产权证券
corporate franchise 法人(公司)营业权
corporate income tax 公司所得税;企业所得税
corporate inhabitant tax 法人居住税
corporate investment 公司投资
corporate juridical person 公司法人;法人
corporate loan 公司贷款
corporate name 公司名称;商号
corporate non-financial enterprise 法人非金融企业
corporate organization 法人组织
corporate ownership rights 法人所有权
corporate powers 法人权限
corporate profit 公司利润
corporate representative 法人代表
corporate resolution 公司董事会决定;法人决议
corporate responsibility 共同责任
corporate seal 公司印章;法人印章
corporate shell 形式公司
corporate social responsibility 企业社会责任
corporate's ownership funds 企业自有资金
corporate spying 商业间谍活动
corporate structure 企业所属机构
corporate surplus 公司盈余
corporate town 自治城市
corporate trust 信托部
corporate veil 公司保障
corporation 联合集团;有限公司;股份有限公司;公司;公开公司
corporation accounting 公司会计
corporation attorney 公司法律顾问
corporation cock 接户旋塞;公司螺旋;断水旋塞;分水闸(门);分水栓;入户总闸(气、水)
corporation income tax 公司所得税
corporation law 公司法
corporation lawyer 公司法律顾问
Corporation of Lloyd's 劳氏船级社;劳埃德船级社
corporation profit tax 法人利润税
corporation readjustment 公司改组

corporation sole 单一法人
corporation stop 干管配气管;干管分水栓;接户断流阀;公司断流阀
corporation tax 法人税
corporeal hereditament 可继承有形不动产
corporeal movable 有形动产
corporeal property 有形资产;有形动产;有形财产
corporeal rights 房地产所有权
corposant 桅上电火;桅杆上端电火
corposclerotinite 团块菌质体
corpse 尸体
corpse gate (教堂的)墓门;可临时停留棺木的大门
corpsing (材料表面的)浅榫眼;(抹灰面层时的)凹疵
Corps of Engineers method 陆军工程部队设计法(美国)
Corps of Engineers of the United States Army 美国陆军工程师团
Corps of Engineers (or Marshall) asphalt mix-design method 陆军工程部队(或马歇尔)沥青混合料配合设计法(美国)
corpus 主体;集成;财产或投资本值;玻璃体;本金
corpus albicans 白体
Corpus Christ chapel 基督圣体节小教堂
Corpus Christ church 基督圣体节教堂
corpuscle 微粒
corpuscular beam 微粒束
corpuscular cloud 微粒云
corpuscular eclipse 微粒食
corpuscular emission 微粒发射
corpuscular nature of light 光的微粒性
corpuscular radiation 微粒辐射
corpuscular stream 微粒流
corpuscular theory 微粒子论;微粒学说
corpuscular theory of light 光的微粒理论
corpus geniculum 膝状体
corpus liberum 游离体
corpus of the laws and regulations 法规汇编
corpus restiformis 绳状体
corpus striatum 纹状体
corpus trapezoideum 斜方体
corrading river 下切河(流)
corrading stream 下切河(流);侵蚀河流
corral 畜栏;栅栏;把……聚集在一起;围成栅栏;牲畜栏;深水桩支承围栏
corraling a fire 消控火势;围控火势
corralling an oil slick 合围油膜
corral time 围控时间
corral-type dairy 围栏式乳牛场
corrasion 动力侵蚀作用
corrasion embayment 侵蚀河湾;侵蚀海湾;冲蚀河湾;冲蚀海湾
corrasion plain 海蚀平原;冲蚀平原
corrasion valley 刻蚀谷
correct 校正;正确的
correctable error 可校误差
correct action 调整作用
correct a deviation 矫正偏差
correct alignment 正切定线
correct and copy 校正复制
correct an error 纠偏
correct balance 正确平衡
correct combination of gears 齿轮正确组合
corrected 改正的
corrected accelerogram 校正后的加速度图
corrected azimuth 修正方位角
corrected bearing 校正方位
corrected clearance 校正清除率
corrected coefficient of determination 校正的测定系数
corrected data 校正数据;改正的数据
corrected density porosity 校正后的密度孔隙度
corrected depth 校正深度;(井的)修正(后)深度
corrected diffraction efficiency 校正衍射效率
corrected edition 订正版
corrected effective temperature 修正有效温度
corrected establishment 改正潮候时差;平均(月潮)高潮间隙
corrected factor 校正因素
corrected form 正字体
corrected function 校正函数
corrected gear 修正齿轮
corrected gross tonnage 修正总吨
corrected lens 校正透镜

corrected mean 修正均数
corrected multiple correlation coefficient 校正多元相关系数
corrected neutron porosity 校正后的中子孔隙度
corrected of dead time value 死时间校正
corrected oil 合格油
corrected omega 差奥米伽
corrected optics 校正过的光学系统
corrected output 修正功率
corrected picture 校正图像
corrected price indices 修正后的价格指数
corrected probit 校正概率单位
corrected rate 校正率;修正率
corrected reading 校正读数;改正读数
corrected resistivity 校正后的电阻率
corrected result 已修正的结果
corrected retention volume 校正保留体积
corrected screw 校正螺旋;校正螺丝
corrected sediment thickness 校正沉积厚度
corrected sextant altitude 真高度
corrected spectrum 校正频谱
corrected speed 校正车速;换算速度
corrected sum or squares 校正平方和
corrected technique 校正技术
corrected titer 校正滴定度
corrected tooth 修正齿
corrected total 冲销后总账
corrected value 校正值;改正值
corrected value of air humidity 气湿订正值
corrected value of cosmic(al) ray 宇宙射线修正值
corrected value of desquamation spectrum 剥谱校正值
corrected value of lake water 湖水修正值
corrected value of overburden 盖层校正值
corrected value of radon escape 氡逸出校正值
corrected value of seawater 海水校正值
corrected value of topography factor 地形系数修正值
corrected water-oil ratio curve method 修正的水油比曲线法
correct exposure 适当曝光
correct-exposure portion 正确曝光部分
correct-exposure region 曝光适当部分
correct grinding 精磨
correct grinding of tool 刀具正确刃磨
correcting 校正;改正
correcting behaviour 改正行为
correcting cam 校正凸轮
correcting circuit 校正电路
correcting code 校正码
correcting coil 校正线圈
correcting condition 校正条件
correcting current 修正电流
correcting deviation 校正偏差
correcting device 校正装置
correcting element 校正元件
correcting feature 校正性能;校正特性
correcting ingredient 校正物料;校正成分
correcting lens 校正透镜;改正透镜
correcting magnet 自差校正磁铁
correcting material 校正原料;校正物料
correcting mechanism 校准机构;校正机构
correcting mount 校正座;校正架
correcting plate 校正板;改正片;改正板
correcting projector 校正投影仪
correcting range 校正范围
correcting signal 校正信号;矫正信号
correcting unit 校正器;执行器
correcting variable 校正变量
correcting wedge 校正楔
correction 勘误;校正(值);校正数;修正值;修正(量);规正量;改正(量);订正;调正
correction advice 更正通知书
correction agitator 调整搅拌器
correctional screw 改正螺旋
correction and compensation device 校正补偿装置
correction angle 校正角;修正角
correction basin 校正池
correction bin 校正料仓
correction bunker 校正料仓
correction cam 校正凸轮;改正凸轮
correction chart 勘误表;校正表;修正表
correction code 审校符号
correction code check 校正码检查

correction coefficient 校正系数;修正系数;订正系数
correction coefficient of length of penetrometric pole 触探杆长度校正系数
correction computation 校正计算;修正计算
correction coordinates 改正坐标
correction copy 审校参考图
correction curve 校正曲线;修正曲线;补偿曲线
correction curve of elastic film 弹性膜校正曲线
correction data 校正数据;修正数据;修正表
correction deflection of vertical 铅垂误差改正量
correction device for tilt 倾斜校正装置
correction device of earth's curvature 地球曲率改正设备
correction down 减校正
correction equation 校正方程
correction factor 校正因子;校正因数;校正系数;修正因数;修正系数;改正因素;改正系数
correction factor for orientation 方向改正系数
correction factor for pressure 压力校正因子
correction factor of (mean) temperature difference 温度差修正因子
correction factor of ground wave phase 地波相位修正因子
correction filter 校正滤光片;校正滤光镜
correction flow rate equation 校正流量方程
correction fluid 改正液
correction for acceleration 加速度校正
correction for air density 空气密度校正
correction for alignment 定线修正;定线校正;定线改正
correction for altitude 海拔校正;高程校正;高程改正
correction for aneroid reading 气压计读数校正
correction for angularity 偏斜改正(测量断面)
correction for atmospheric refraction 大气折射改正;大气折射订正
correction for centering 归心校正;归心改正
correction for chord distance 弦距校正
correction for chronometer 表差
correction for clock 钟差
correction for compensation 补偿校正;补偿改正
correction for continuity 连续性校正
correction for curvature of plumb-line 垂线曲率校正
correction for curvature of star image path 星径曲率改正
correction for curvature of the earth 地球曲率改正
correction for curvature of the earth and refraction 地球曲率与折光差改正
correction for deflection of a suspension cable 悬索偏角改正
correction for deflection of vertical 垂线偏差改正
correction for depth 深度校正
correction for direction 方向校正;方向改正;方位校正
correction for eccentricity 偏心校正;偏心改正
correction for electric(al) wave propagation 电波传播改正
correction for elevation 标高校正
correction for excavation effect 挖掘效应校正
correction for free air 空间校正
correction for frequency deviation 频偏改正
correction for grade 倾斜改正;坡度校正;坡度改正
correction for gravity 重力校正
correction for grouping 归组校正
correction for inclination 倾斜校正
correction for inclination of sextant 六分仪倾斜角校正;六分仪倾斜角改正
correction for inclination of tape 坡度改正
correction for index error 指示差校正;指标差改正
correction for indirect effect 间接效应校正
correction for instrument azimuth 仪器方位校正
correction for isostasy 均衡补偿校正
correction for isostatic compensation 均衡补偿校正
correction for local variation 局部变化改正
correction for mean 均值校正
correction for meridian curvature 子午圈曲率校正;子午线曲率改正
correction formula 校正公式
correction for non-linear deformation 非线性变形改正
correction for observed value 观测值修正
correction for parallax 视差校正

correction for parallel curvature 平行圈曲率校正
correction for phase 相位校正
correction for pull 拉力校正
correction for rate of chronometer 表速校正
correction for reduction 归算校正
correction for reduction to centre 归心校正;归心改正
correction for reduction to the zenith 异位差
correction for refraction 折射校正;折光差改正
correction for relief 地势校正;投影差校正
correction for run of micrometer 测微器行差校正
correction for sag 悬链线改正;下陷改正;垂曲修正;垂曲校正;垂曲改正;垂度校正
correction for sea depth 海深改正
correction for semi-diameter 半径差
correction for settling 下沉校正
correction for skew normals 标高差改正
correction for slope 倾斜校正;倾斜改正
correction for sound velocity 声速改正
correction for temperature 温度改正
correction for the earth rotation 地球自转修正
correction for time-lag of radio electric(al) wave 电波传播时延改正
correction for transducer 换能器吃水改正
correction for transducer baseline 换能器基线改正
correction for unsymmetry of catenary 悬链线不对称改正
correction for variation of gravity 重力变化改正
correction from normal section to geodesic 截面差改正
correction function 校正函数
correction in altitude 高度改正量
correction items of magnetic measurement 磁测改正项目
correction line 改正线
correction liquor 调整液
correction maintenance 事后维修
correction map 现势图
correction matrix 校正矩阵
correction measure 校正措施
correction mechanism 修正机构
correction memo 更正通知书
correction meniscus 校正弯月透镜;校正凹凸透镜
correction method 校正法
correction middle latitude 中分纬度改正量
correction model 改正模型
correction number 修正数
correction of bearing 方位改正(量)
correction of channel 河槽治理;河槽整治
correction of chart 海图改正
correction of colo(u)r value 色值校正
correction of common calcium 普通钙校正
correction of common lead 普通铅校正
correction of defect 缺陷矫治
correction of depth 水深改正
correction of derivation 纠正偏差
correction of deviation 自差改正量
correction of error 误差校正
correction of field blow count 原位贯击数校正
correction of focal length 焦距改正
correction of gravity measurement of tide 重力潮汐改正
correction of gyrocompass 陀螺罗经改正量
correction of image coordinate 像坐标校正
correction of lag 滞后校正
correction of latitude 纬度校正(量);纬度改正(量)
correction of leeway 风压差改正量
correction of log interpretation 测井成果校正
correction of magnetic variation 磁变改正
correction of model deformation 模型变形校正
correction of mountain stream 山溪整治
correction of photo deformation 像片改正
correction of relief displacement 投影差改正
correction of residual attenuation value 校正剩余衰减值
correction of river 河流整治
correction of scale difference 档差改正(回声测深仪)
correction of sounder 测深仪校正;测深仪改正
correction of soundings 水深改正;测深改正
correction of strabismus 斜视校正
correction of stream 河流整治
correction of temperature 温度改正

correction of tidal zone 潮位分带改正
correction of track 轨道校正
correction of water-level 水位改正
correction of water level by analog(ue) method 模拟法水位改正
correction of water zoning 水位分带改正
correction of weighing 称量校正
correction of zero drift 零漂改正
correction of zero line 零位线改正
correction overlay 检索透明片
correction plate 校正板;改正板
correction plate carrier 改正平面框架
correction probability 修正概率
correction pulse 校正脉冲
correction rate of defect 缺陷矫治率
correction red copy 红色校样
correction retention time 校正保留时间
correction screw 校正螺钉
correction sheet 现势图
correction surface 校正面
correction symbol 审校符号
correction table 勘误表;校正表;改正(数)表
correction term 校正项;修正项
correction test 纠正试验
correction time 校正时间;修正时间
correction to average sea level 平均海平面订正
correction to mean sea level 平均海平面订正
correction to nominal length of tape 尺长改正
correction to program(me) 程序的校正;程序修改
correction to tension 拉力校正
correction to time signal 时号校正
correction up 加校正
correction value 改正值
correction value of altitude 高度修正值
correction value of atmosphere 大气修正值
correction value of Compton 康普顿修正值
correction value of spectral drift 谱漂移修正值
correction values of spheric(al) attenuation 球面衰减校正值
correction wedge 校准光楔;校正光楔;修正光楔
correction window 校正窗
corrective action 校正作用;纠正措施;校正动作;调整作用;改进措施;调整效应
corrective action measure 补救措施
corrective control model 校正控制模型
corrective delay 校正延迟
corrective dredging 整治性疏浚
corrective lag 校正延迟;调准延迟
corrective lens 校正镜(片)
corrective loop 修正回路
corrective maintenance 故障检修;更正维修;出错维修;出错复修;安全改进维护;设备维修;设备保养
corrective maintenance time 矫治性维修时间;出错维修时间
corrective material 校正原料
corrective measure 纠正措施;校正措施;整治措施;改正措施
corrective moment 修正力矩
corrective network 校正网络
corrective operation 补救操作
corrective pitting 修正缺陷
corrective price decline 调整性降价
corrective spectacle lens 校正镜片
corrective system 校正系统
corrective treatment 修正处理
correct level 校正水准;标准水准
correctly burned clinker 正常烧成的熟料
correctly identified model 校正识别模型
correct misprints 校正错字
correct mixture 标准(气体)混合物
correctness factor 校正系数
correctness of forecast 预报准确性;预报准确率;预报准确度
corrector 校正装置;校正者;校正线(路);校正算子;校正式;校正器;校正电路;校对员;修正器;改正镜;调整器
corrector cell 改正镜室
corrector formula 校正公式
corrector loop 校正环
corrector magnet 校正磁棒
corrector method 校正方法
corrector of magnetic compass 磁罗盘修正器
corrector plate 校正板;改正片;改正板

correctors and safety systems 瞬跳换日装置及调对和保险机构
correctostat paper 裱糊板
correct subset 正确子集；正码子集合
correct tension 正常拉力
correct that soil condition 改良那种土壤
correct time 准确（的）时间
correct timed ignition 正时点火
correct to 精确到……位数
correlate 联系数；相关物
correlated character 相关性状
correlated colo(u)r temperature 相关色温
correlated condition 相关条件
correlated data processor 相关数据处理机
correlated equation 相关方程
correlated inhibition 伴生抑制
correlated noise 相关噪声
correlated observation 相关观测
correlated process 相关过程
correlated sampling 相关抽样
correlated series 相关数列
correlated site 相关位置
correlated traits 相关性状
correlated variability 相关变异性
correlated variables 相关变量
correlated variation 相关变异
correlated water level 关系水位
correlate equation 联系数方程
correlation 交互作用；相互关系；相关张量；相关（性）；关联；对射变换；对射
correlation accuracy 相关精度
correlation analysis 相关分析
correlation analysis method 相关分析法
correlation analysis method of pal(a)eoclimate data 古气候信息相关分析法
correlation and regression 相关与回归
correlation and regression analysis 相关与回归分析
correlation and regression technique 相关与回归技术
correlation apparent feature 相关视要素
correlation behaviour 相关状态
correlation between exploration and exploitation data 深采对比法
correlation between mean and variance 均值与方差间的相关
correlation between series 序列相关
correlation between variables 变数间的相关
correlation calculation 相关计算
correlation chart 相关图
correlation coefficient 相关系数
correlation coefficient method 相关系数法
correlation coefficient of turbulence 湍流相关系数
correlation coefficient test 相关系数检验
correlation columnar section of quaternary system 第四系柱状对比图
correlation columnar section of strata 地层柱状对比图
correlation computer 相关计算机
correlation curve 相关图；相关曲线
correlation curve of balance element 均衡要素相关曲线
correlation curve of discharge 流量相关曲线
correlation curve of regime element 动态要素相关曲线
correlation curve of water-level 水位相关曲线
correlation data 相关资料；对比数据
correlation dependence 相互依存
correlation detection 相关检测
correlation detector 相关检测器；相关检波器
correlation device 相关器
correlation diagram 相关图
correlation direction finder 相关定向仪；相关测向台
correlation distances 相关距离
correlation echo processor 回波相关处理器
correlation echo sounder processor 回声相关处理器
correlation equation 相关方程
correlation factor 相关因子
correlation formula 相关公式
correlation from ranks 等级相关
correlation function 相关函数
correlation function analysis 相关函数分析
correlation function analyzer 相关函数分析仪

correlation gate width 相关门宽度
correlation index 相关指数；对比指标
correlation index value 对比指数值
correlation in space 空间对射变换
correlation interferometer 相关干涉仪
correlation length 相关长度；对比长度
correlation loss 相关损失
correlation map between exploration and exploitation data 探采对比图
correlation mask image 相关掩模图像
correlation mask technique 相关掩膜技术
correlation matrix 相关矩阵
correlation-measuring instrument 相关测量仪
correlation method 相关(方)法；(相关)对比法；比较法
correlation numbers 相关数
correlation observation 相关观测
correlation of attributes 品质相关
correlation of a well log 测井记录对比
correlation of coal-bearing strata 含煤地层对比
correlation of image 图像相关
correlation of imagery 图像相关
correlation of indices 指数相关；指标相关
correlation of linking wells 连井对比
correlation of ore-bearing strata 矿层对比图
correlation of orientation 方位相关
correlation of random noise 随机噪声相关
correlation of time series 时间数列相关
correlation orientation tracking and range system 相关取向跟踪测距系统
correlation parameter 相关参数
correlation precision 相关精度
correlation quality 相关质量
correlation radar 相关雷达
correlation-radial-velocities 交叉相关视向速度仪
correlation radius 相关半径
correlation ratio 相关比；(率)
correlation ratio method 相关比例法
correlation reception 相关接收
correlation refraction method 对比折射波法
correlation scatter diagram 相关散点图
correlation section 相关剖面
correlation sequence method 相关频数比值法
correlation spectrometer 关联分光计
correlation spectrum 相关光谱
correlation studies 相关研究
correlation table 相关表；对比表【地】；地层对比表
correlation technique 相关技术；相关方法
correlation theory 相关理论
correlation time 相关时间
correlation tracing and triangulation system 相关跟踪三角测量系统
correlation tracking and range 相关跟踪测距
correlation tracking and range system 相关跟踪测距系统
correlation tracking system 相关跟踪系统
correlation transformation 相关变换
correlation type 相关类型
correlation-type receiver 相关器
correlation water level 相关水位
correlative 关联词
correlative charges 相关费用
correlative curve 相关曲线
correlative differentiation 依存分化
correlative factor 相关因素
correlative figure 对射图形
correlative ga(u)ge 相关水位
correlative indexing 相关索引；相关检索
correlative investment 相关投资
correlative ratio 相关率
correlative value 对比值
correlative water rights (土地的)附有水权；相关水权
correlativity 相关性；相关程度
correlatogram 相关(曲线)图
correlatograph 相关图；相关函数计算记录器
correlator 环形解调器电路；相关仪；相关器；相关函数分析仪；关联子；乘积检波器；弱音响信号检测仪
correlator-servo loop 相关伺服回路
correlator-type leak detection device 相关型检漏仪
correlogram 相关图；相关曲线图
correlogram analysis 相关图分析；相关曲线分析
correlometer 相关计

correnie （一种产于阿伯丁的）亮橙红色花岗岩
corrensite 绿泥间蛭石
correspondence 相应；相当；对应；符合；通信
correspondence analysis 对应分析
correspondence between binary trees and forests 二叉树和森林之间的对应
correspondence course 函授课程
correspondence defining 定义对应
correspondence (match) 一致
correspondence principle 对应原理
correspondence school 函授学校
correspondence theorem 相似定理
correspondent 有商务关系的人；外来客户；通讯记者；商行；代理行；联行；客户
correspondent account 往来银行账户
correspondent agreement 通汇契约
correspondent bank 关系银行；代理银行；往来银行；通汇银行
correspondent letter of credit 通汇银行信用证
corresponding 相应的；对应的；对应
corresponding activity coefficient 对应活化系数
corresponding angle 对应角；同位角
corresponding bank 业务联系银行
corresponding change 对应变化
corresponding channel point selection 循经选穴法
corresponding coefficient 同位系数
corresponding concentration 对应浓度
corresponding crest stage 相应最高水位；相应洪峰水位；相应峰顶水位
corresponding data 相应的资料
corresponding depth 对应深度；对应水深
corresponding discharge 相应流量；相当流量
corresponding element 对应元件
corresponding epipolar ray 同名核线
corresponding equilibrium 对应平衡
corresponding equilibrium constant 对应平衡常数
corresponding grid method 直线化网络法
corresponding ground dimension 相应实地尺寸
corresponding image 对应影像
corresponding image points 对应影像点；同名像点
corresponding image ray 同名光线
corresponding images 对应影像点
corresponding intersected point method 合点法
corresponding line graphic(al) intersected method 共线图解交会法
corresponding lines 同位线
corresponding line transformation 共线变换
corresponding part 对应部分
corresponding percentage 相应的百分比
corresponding period 同期
corresponding period average method 同期平均法
corresponding point 相应点；对应点
corresponding projection ray 同名投影光线
corresponding ratio 对应比
corresponding reaction 对应反应
corresponding reaction rate 对应反应速率
corresponding sides 对应边
corresponding solid 对应固体
corresponding speed 相应速度
corresponding stage 相应阶段；相应水位
corresponding state 对应(状)态
corresponding subgrade soil design parameter 相应的地基土设计参数
corresponding temperature 对应温度；对比温度
corresponding variability 相应变异性
corresponding water level 相应水位
corridor 楼道；廊室；廊；狭长地带；甬道；走廊地带；走廊；过道；高速走廊；高速通道
corridor access type 通廊式平面布置
corridor approach 交通走廊进口
corridor control 高速走廊交通控制；高速通道交通控制；通道控制
corridor Diwa system 走廊地洼系【地】
corridor duct 走廊风道
corridor floor 走廊地板
corridor illumination 走廊照明
corridor lighting 走廊采光
corridor locker 走廊衣帽柜
corridor of high voltage electricity 高压线走廊
corridor on water 水廊
corridor pollution level 交通走廊污染水平
corridor stor(e)y 走廊楼层
corridor study 高速走廊调查；高速通道调查

corridor traffic 借道过境;高速走廊交通;高速通道交通;通道交通
corridor traffic assignment 通过交通分配
corridor train 有走廊相通的列车;各车厢有走廊相通的列车
corridor type basilica 单廊式长方形会堂
corridor wall 走廊墙壁
corrie 山腹凹地;圆形谷;冰坑;冰斗;山凹
corrie glacier 山腹冰川;冰斗冰川
corrie ice 碗状大冰块
corrie-lake 冰斗湖
corrigenda 正误表
corrigendum 勘误(表);[复] corrigenda
corroboration of an invention 发明人的确证
corrode 腐蚀;受损伤
corroded condition 腐蚀后状态
corroded crystal 熔蚀晶
corroded depression 溶蚀洼地
corroded fissure 溶蚀裂隙;溶蚀裂缝
corroded fossil 熔蚀化石
corroded funnel 溶蚀漏斗
corroded gypsum breccia 膏溶角砾岩
corroded mutant 腐蚀型突变体
corrodent 腐蚀剂;有腐蚀力的;锈蚀的
corrodibility 可腐蚀性;侵蚀性
corrodible 可腐蚀
corroding agent 腐蚀剂
corroding brittleness 酸洗脆性
corroding electrode 腐蚀电极
corroding inhibitor 阻锈剂
corroding lead 腐蚀铅
corroding proof 酸洗试验
Corrodkote test 涂膏耐蚀试验
corrod(o)kotec test 涂膏腐蚀试验
corrod(o)kote paste 金属腐蚀试验用的腐蚀膏
corrodokote test 涂膏密室耐蚀试验
Corronel 考拉聂尔钼铁合金;耐蚀镍钼铁合金
Corronil 考拉尼尔铜镍合金
Corronium 考拉尼姆合金
corrosscometer 测蚀计
corrosion 流蚀;刻蚀(作用);耐腐蚀硅钢;锈蚀;腐蚀;溶蚀;侵蚀;铁锈
corrosion allowance 锈蚀容许量;锈蚀厚度;允许磨蚀度;允许腐蚀度;腐蚀裕度;腐蚀余量;允许蚀余度;腐蚀限度;允许腐蚀厚度
corrosional pan karst 溶盘岩溶
corrosional plain 溶蚀平原
corrosional type 溶蚀作用类型
corrosion and crack control 腐蚀和裂缝控制
corrosion and deposit inhibitor 缓冲抑垢剂
corrosion ans scale inhibitor 缓蚀阻垢剂
corrosion at a seam 接缝腐蚀
corrosion attack by thermal water 热水腐蚀性侵蚀
corrosion behavio(u)r 腐蚀作用;腐蚀性能;腐蚀行为
corrosion border 熔蚀边;溶蚀边
corrosion by alkali vapo(u)r 碱蒸汽侵蚀
corrosion by gases 气体腐蚀;气蚀
corrosion by non-freezing solution 不冻液腐蚀
corrosion by sulphate 硫酸盐侵蚀
corrosion by thermal water 热水腐蚀
corrosion cavity 熔蚀孔
corrosion cell 腐蚀电池
corrosion cells within concrete 内部腐蚀电池
corrosion cell type 蜂窝型腐蚀
corrosion classify 腐蚀分类
corrosion coefficient 腐蚀系数
corrosion contaminant 腐蚀污染物;腐蚀沾染物
corrosion contamination 腐蚀(产生的)污染
corrosion control 腐蚀控制;腐蚀防治;腐蚀防止法;防锈蚀;防腐蚀;防腐方法
corrosion coupon 腐蚀试片
corrosion crack(ing) 腐蚀断裂;晶间开裂;腐蚀破裂
corrosion creep test 腐蚀蠕变试验
corrosion current 腐蚀电流
corrosion damage 腐蚀损坏
corrosion degree of casing well 管井腐蚀程度
corrosion department 防腐部门
corrosion due to welding 焊接腐蚀
corrosion effect 腐蚀效应
corrosion effect of air pollutants 大气腐蚀(作用)
corrosion embrittlement 腐蚀脆性;蚀脆化
corrosion environment 腐蚀介质
corrosion evolution 溶蚀评价

corrosion factor 腐蚀因素
corrosion fatigue 腐蚀疲劳
corrosion fatigue limit 腐蚀疲劳极限
corrosion-fatigue parameter 腐蚀疲劳参数
corrosion-fatigue test 锈蚀疲劳试验
corrosion fatigue testing machine 腐蚀疲劳试验机
corrosion figure 蚀图
corrosion film 腐蚀膜层
corrosion fracture 腐蚀断裂
corrosion free pump 耐蚀泵
corrosion ga(u)ge 腐蚀计
corrosion ga(u)ge point 腐蚀(检)测点
corrosion hazard 腐蚀危害
corrosion immunity region 腐蚀不敏感区
corrosion index 腐蚀指数;腐蚀深度指数
corrosion inhibior 防腐面层
corrosion inhibiter 阻蚀剂;防蚀剂
corrosion inhibiting 腐蚀抑制
corrosion-inhibiting admixture 抗腐蚀外加剂;抗蚀剂;阻锈剂;防侵蚀外加剂
corrosion-inhibiting fluid filler 防锈蚀液态填料
corrosion inhibiting mortar 抗腐砂浆
corrosion inhibiting pigment 防蚀颜料;缓蚀颜料
corrosion inhibiting primer 防腐蚀打底料
corrosion inhibiting slurry 抗蚀砂浆
corrosion inhibition 缓阻;缓蚀
corrosion-inhibitive 防蚀的
corrosion inhibitor 缓蚀剂;腐蚀抑制剂;防锈剂;抗腐蚀剂;减蚀剂;缓蚀剂;阻锈剂;阻蚀剂;腐蚀阻抑剂;腐蚀抑制剂;防蚀剂
corrosion inhibitor oil 防蚀油
corrosion inhibitor treated paper 防锈纸
corrosion intensity 溶蚀强度
corrosion loss 腐蚀损失;腐蚀损耗
corrosion mark 溶痕
corrosion mechanism 腐蚀机制;腐蚀机理
corrosion medium 侵蚀介质;[复] media
corrosion mitigation 腐蚀减轻
corrosion of casing well 管井腐蚀
corrosion of concrete pipe 混凝土管腐蚀
corrosion of metal 金属锈蚀
corrosion of pipe 管道腐蚀
corrosion of reinforcement 钢筋锈蚀
corrosion of sewer 排水管腐蚀
corrosion of steel 钢筋锈蚀;钢材锈蚀
corrosion peak 腐蚀尖峰
corrosion penetration 锈蚀深度
corrosion phenomenon 腐蚀现象
corrosion pit 腐蚀斑点;蚀坑
corrosion pitting 侵蚀点
corrosion pollutant 腐蚀污染物
corrosion pollution 腐蚀污染
corrosion potential 腐蚀电位;腐蚀电势
corrosion preventative 防腐蚀法
corrosion prevention 腐蚀防护;防(锈)蚀;防腐(蚀)
corrosion prevention for pipes 管道防腐蚀
corrosion prevention of sewers 管道防腐蚀
corrosion prevention quarry tile 防腐蚀缸砖
corrosion preventive 蚀防止;防蚀剂;防腐蚀法;防腐剂
corrosion preventive compound 防腐化合物
corrosion preventive procedure 防腐措施
corrosion probe 腐蚀探针
corrosion process 腐蚀作用
corrosion product 腐蚀产物
corrosion program(me) 防腐方案
corrosion promoter 增锈剂;腐蚀促进剂
corrosion-promoting 腐蚀促进
corrosion-proof 抗腐蚀(的);耐腐蚀;耐蚀的;不锈的
corrosion-proof cable 防蚀电缆
corrosion-proof coating 耐蚀面层
corrosion proof design 防腐蚀设计
corrosion proofing 防蚀处理
corrosion proof lined 衬耐腐蚀材料的
corrosion-proof metal 耐蚀金属
corrosion-proof paint 防蚀涂料
corrosion-proof paint for brass 黄铜防锈涂料
corrosion-proof steel 耐蚀钢
corrosion protecting lubricant 抗锈蚀润滑剂
corrosion protection 防锈;防(腐)蚀
corrosion protection agent 腐蚀防护剂
corrosion protection anode 保护阳极

corrosion protection blanket 腐蚀防护盖层;防腐蚀覆盖层
corrosion protection clinker tile 防腐蚀缸砖
corrosion protection coat(ing) 腐蚀防护涂层
corrosion protection composition 腐蚀防护组成物
corrosion protection foil 腐蚀保护薄膜
corrosion protection grout 腐蚀防护灌浆
corrosion protection mortar 腐蚀防护砂浆
corrosion protection paint 腐蚀防护油漆
corrosion protection pigment 腐蚀防护颜料
corrosion protection prime coat 腐蚀防护打底涂层
corrosion protection primer 腐蚀防护涂料
corrosion protection slurry 腐蚀防护稀浆
corrosion protective(agent) 腐蚀防护剂
corrosion protective covering 防蚀覆盖;防蚀被覆层
corrosion rate 锈蚀速率;腐蚀速率;腐蚀速度;腐蚀率;溶蚀率
corrosion reaction 腐蚀反应
corrosion region 腐蚀区
corrosion remover 除锈剂;防蚀剂;防腐剂
corrosion research 腐蚀研究
corrosion resistance 抗蚀性;抗腐蚀性;抗腐蚀能力;抗腐能力;耐蚀;耐(腐)蚀性
corrosion resistance measurement 耐腐蚀性测定
corrosion resistance to molten glass 抗玻璃液侵蚀性
corrosion-resistant 抗蚀;抗腐蚀的;耐腐的;耐腐蚀的
corrosion-resistant alloy 耐蚀合金
corrosion-resistant anodic coating 耐腐蚀阳极涂层
corrosion-resistant casing 不锈套管;耐腐蚀套管
corrosion-resistant cast iron 耐腐蚀铸铁
corrosion-resistant coating 防蚀层
corrosion-resistant concrete 耐腐蚀混凝土
corrosion-resistant low alloy steel 耐腐蚀低合金钢
corrosion-resistant mastic 耐腐玛琋脂
corrosion-resistant material 抗腐蚀材料;抗蚀性材料;耐(腐)蚀材料
corrosion-resistant metal 耐腐金属
corrosion-resistant metal and alloy 抗腐蚀金属和合金
corrosion-resistant oil 防锈蚀油
corrosion-resistant paint 防蚀涂料
corrosion-resistant plate 耐蚀钢板;不锈钢板
corrosion-resistant steel 耐(腐)蚀钢
corrosion-resist cast iron 耐蚀铸铁
corrosion resister 抗蚀剂;缓蚀剂
corrosion-resisting 抗腐蚀的;耐腐(的);耐蚀的
corrosion-resisting agent 防腐剂
corrosion-resisting agent for concrete 混凝土防锈剂
corrosion-resisting alloy 耐蚀合金
corrosion-resisting alumin(i)um alloy 耐(腐)蚀铝合金
corrosion-resisting bearing 耐磨蚀轴承;耐腐蚀轴承
corrosion-resisting material 耐蚀材料
corrosion-resisting metal 耐蚀金属
corrosion-resisting property 耐蚀性能
corrosion-resisting pump 耐蚀泵
corrosion-resisting steel 不锈钢;抗蚀钢;耐蚀钢;耐蚀钢
corrosion-resisting steel plate 耐蚀钢板
corrosion-resisting test 耐蚀试验
corrosion resistivity 耐腐蚀性;防蚀性
corrosion retarding agent 缓蚀剂
corrosion rig test 腐蚀试验台试验
corrosion rim 溶蚀边
corrosion(rust)preventive packing 防锈包装
corrosion scale 腐蚀鳞片;腐蚀等级;片状腐蚀产物
corrosion seawater 腐蚀海水
corrosion sensitivity 腐蚀敏感性
corrosion sound meter 超声波腐蚀测定仪
corrosion specimen 腐蚀标本
corrosion spool 腐蚀试片
corrosion stability 耐蚀性;抗蚀性;腐蚀稳定性
corrosion strength 耐蚀性;耐蚀强度;耐蚀力
corrosion stress 腐蚀应力
corrosion surface 熔蚀面;溶蚀面
corrosion survey 腐蚀检查
corrosion-tank finishing process 腐蚀法
corrosion target 牺牲电极;腐蚀电极

corrosion tendency 腐蚀趋势
corrosion test 耐腐蚀试验;溶蚀试验
corrosion troubles 腐蚀损坏;腐蚀干扰
corrosion type 腐蚀类型
corrosion under differential pressure 差压溶蚀
corrosion under differential temperature 差温溶蚀
corrosion under mechanical stress 机械应力溶蚀
corrosion under ordinary temperature and pressure 常温常压溶蚀
corrosion velocity 溶蚀速度
corrosion voltmeter 腐蚀伏特计
corrosion zone 溶蚀带
corrosion zone in fossil weathered crust 古风化壳溶蚀带
corrosiron 耐蚀硅钢
corrosive 腐蚀性的
corrosive acid 腐蚀酸
corrosive acid poisoning 强酸中毒
corrosive action 腐蚀作用
corrosive action of water 水的腐蚀作用
corrosive agent 腐蚀剂
corrosive alkaline poisoning 强碱中毒
corrosive atmosphere tester 腐蚀大气测试器
corrosive attach chamber 腐蚀促进室
corrosive attack 侵蚀;腐蚀
corrosive burn 腐蚀性灼伤
corrosive carbon dioxide 侵蚀性二氧化碳
corrosive cargo 腐蚀性货物
corrosive characteristic 腐蚀性
corrosive check 腐蚀裂缝
corrosive chemicals 化学腐蚀剂;腐蚀剂
corrosive compound 腐蚀性化合物
corrosive corporation 腐蚀性物品
corrosive effect 腐蚀效应
corrosive environment 腐蚀环境
corrosive flux 腐蚀性焊剂
corrosive goods 腐蚀性货物
corrosive impurity 腐蚀性杂质
corrosive inhibitor 防锈剂
corrosive liquor pump 腐蚀性液体泵
corrosive location 腐蚀部位
corrosive nature 腐蚀性能
corrosiveness 腐蚀作用;腐蚀性;腐蚀程度
corrosiveness of water 水的腐蚀性
corrosive phenocryst 熔蚀斑晶
corrosive pitting 点蚀
corrosive poison 腐蚀性毒物
corrosive potential 腐蚀电位;腐蚀电势
corrosives 腐蚀品
corrosive sample 腐蚀性试样
corrosive sampler 腐蚀性取样器
corrosive sewage 腐蚀性污水
corrosive shakes 腐蚀裂缝
corrosive solution 腐蚀性溶液
corrosive strength 耐蚀性
corrosive sublimate 氯化汞;升汞
corrosive substance 腐蚀性物质
corrosive velocity 腐蚀速度
corrosive waste 腐蚀性废水
corrosive water 侵蚀性水;有侵蚀性的水质;腐蚀性水
corrosive wear 腐蚀磨损;腐蚀耗损
corrosive wear test 腐蚀耗损试验
corrosive well 腐蚀井
corrosivity 腐蚀作用;腐蚀性;腐蚀度
corrosivity monitoring 腐蚀性检测
corrosivity of water 水的腐蚀性
corrosometer 腐蚀性测定计;腐蚀计;腐蚀测定计
corrugaled case 瓦楞纸箱
corrugate 成搓板状;成纹形的;起皱;起沟;使成波状
corrugate alumin(i)um plate 铝波纹板
corrugate core (胶合板的)波状夹芯
corrugated 皱折的;皱纹的;折皱的;波纹形的;波浪(形)的;起波浪的
corrugated absorbent panel 波形吸水板
corrugated absorber plate 波纹形吸热板
corrugated acoustic(al) panel 波形铝板;波形隔声板
corrugated alumin(i)um 波形屋顶铝板;波纹状铝
corrugated alumin(i)um roof cover(ing) 波形铝屋面
corrugated alumin(i)um roofing 瓦楞铝屋面;波形铝板屋面
corrugated alumin(i)um sheet 波形铝板;波纹铝板;瓦楞铝板
corrugated amplitude 波形波幅
corrugated anchor 波形锚;竹节式墙系杆和鸠尾形锚固件
corrugated and spiral rib pipe 螺旋肋式波纹管
corrugated arch 波形拱
corrugated arch roof 波形拱顶
corrugated-asbestos board 瓦楞石棉板;波形石棉板
corrugated asbestos cement 波形石棉水泥制品
corrugated-asbestos-cement board 波形石棉水泥板
corrugated-asbestos cement lumber 波形石棉水泥木料
corrugated-asbestos cement plate 波形石棉水泥板
corrugated-asbestos cement roofing 波形石棉水泥屋面板;瓦楞石棉水泥屋面
corrugated-asbestos-cement roof sheathing 波形石棉水泥屋面盖板
corrugated asbestos cement sheet 波形石棉水泥瓦;波形石棉水泥板;瓦楞石棉板;瓦楞石棉水泥屋面;瓦楞石棉水泥瓦;石棉水泥波形瓦
corrugated-asbestos cement sheet roofing 石棉水泥波形瓦屋面
corrugated-asbestos-cement siding 波形石棉水泥披叠板;波形石棉水泥壁板
corrugated-asbestos wall 波形石棉墙
corrugated asphalt board 波形沥青纸
corrugated asphalt felt 带楞沥青油毡
corrugated baffle 波纹挡板
corrugated bar 波形杆;波纹钢筋;竹节钢(筋)
corrugated barrel vault 波形筒拱;波形筒状穹顶
corrugated beam barrier 波形梁护栏
corrugated bearing 槽形轴承;梳状轴承
corrugated bedding 波皱层理;揉皱层理
corrugated bituminous board 波形沥青板
corrugated board 瓦垅纸;波纹(纸)板;波面纸板;瓦楞纸板;瓦楞纸(板)
corrugated board box 波楞纸板箱
corrugated board ink 瓦楞纸印刷油墨
corrugated buff 瓦楞形抛光轮
corrugated building paper 波形建筑用纸
corrugated bulb 皱纹灯泡
corrugated bulkhead 波形隔板;波形舱壁
corrugated canal gate 波形渠道闸门
corrugated carboard 波形薄纸板
corrugated cardboard 波状板;波纹纸板
corrugated catenary 波浪式悬链
corrugated cement asbestos board 波形石棉水泥板
corrugated cement asbestos panel 波形石棉水泥(镶)板
corrugated cement asbestos sheet 波形石棉水泥板
corrugated cement asbestos siding 波形石棉水泥披叠板
corrugated cement board roof cladding 波形石棉水泥屋面板
corrugated cement board roofing 波形石棉水泥屋面
corrugated ceramics 波纹状陶瓷
corrugated clay roof(ing) tile 波形屋顶黏土瓦
corrugated compensator 波纹管补偿器
corrugated concrete 波纹混凝土
corrugated concrete pavement 波纹混凝土路面
corrugated concrete slab 混凝土波纹板;波形混凝土板;波纹混凝土板
corrugated cone 分层锥
corrugated container 槽形壁集装箱
corrugated cooler 波纹板式冷却器
corrugated copper plate 波纹铜片
corrugated copper sheet 波形铜皮
corrugated core 波纹夹芯
corrugated cover 槽纹外层胶;波纹罩
corrugated covering 波纹状蒙皮
corrugated crushing roll 波辊压碎机;波形破碎辊
corrugated culvert 波纹铁皮涵洞;波纹铁管涵(洞);波纹管涵(洞);铁波纹管涵洞
corrugated cylinder 波纹圆柱
corrugated diaphragm 皱纹膜片;起纹膜片
corrugated drum 皱纹桶
corrugated element 波形单元

corrugated expansion joint 波形伸缩节;波纹涨缩管;波纹管补偿器
corrugated expansion pipe 波形膨胀(接)管
corrugated fabric 波形组织
corrugated fastener 转角连接用波纹铁片;波纹扣片;波纹扣件;波纹紧固钢带;波纹铁片(转角连接用)
corrugated fiberboard 波纹纤维板
corrugated fin 波状散热片;波纹状散热(翼)片
corrugated fire box 波纹火箱
corrugated flexible metal tube 波形柔性金属管;波纹金属软管
corrugated flexible pipe 波纹软管
corrugated flue 皱纹火管;波纹燃烧管
corrugated foil 波形箔
corrugated folded plate 波形折板
corrugated friction socket 波纹打捞筒
corrugated furnace 波形燃烧管;波纹火管
corrugated furnace boiler 波形炉胆锅炉
corrugated furnace tube 波纹炉膛管
corrugated furrow irrigation 密沟灌溉法
corrugated galvanized iron 镀锌波形铁
corrugated galvanized iron sheet 镀锌瓦楞铁皮;波形镀锌铁板(屋面)
corrugated galvanized iron sheet roofing 镀锌瓦楞铁皮屋面
corrugated galvanized sheet 镀锌瓦楞钢板
corrugated galvanized sheet iron 镀锌瓦楞铁皮;镀锌波纹铁皮
corrugated galvanized sheet roofing 镀锌波纹铁皮屋面;镀锌波纹铁皮屋顶
corrugated galvanized steel sheet 波形镀锌钢板屋面;波形镀锌钢板
corrugated gasket 波形垫片
corrugated gauze 波纹金属网
corrugated general-use paper 波形普通用纸
corrugated glass 波形玻璃;波纹玻璃
corrugated glass partition(wall) 波形玻璃隔墙
corrugated glass reinforced plastics roofing 波形玻璃钢屋面
corrugated glass roofing 波形玻璃屋面
corrugated grinding cone 波形磨锥
corrugated hatch cover 波形舱口盖
corrugated header 波纹集management;波纹管座
corrugated horn 波纹喇叭
corrugated insulator 波形隔热材料
corrugated iron 瓦垅薄钢板;波形铁;波纹铁;瓦垅铁;瓦楞铁
corrugated iron metal 波纹铁
corrugated iron pipe 波纹铁管
corrugated iron roof 波状铁皮屋面
corrugated iron roofing 瓦楞铁皮屋面
corrugated iron shed 瓦楞铁棚
corrugated iron sheet 波纹铁皮
corrugated iron-sheet tile 波型铁皮瓦
corrugated jaw crusher 颚式破碎机
corrugated jaw plate 波纹颚板
corrugated joint 波纹式接头
corrugated joint fastener 波形连接片
corrugated key 波形钥匙;凹凸钥匙
corrugated lath(ing) 波形(抹灰)板条
corrugated lens 波纹透镜
corrugated lens condenser 棱形聚光器;分层透镜聚光器
corrugated light-admitting board 波形透光板
corrugated light-admitting board roofing 波形透光顶盖
corrugated lining plate 波形衬板
corrugated lug 波状凸纹
corrugated material 波形材料
corrugated metal 波形铁;波形金属(板)
corrugated metal arch(ed) culvert pipe 波形金属拱形涵管
corrugated metal building sheet 波形金属建筑薄板
corrugated metal culvert 波形金属涵管
corrugated metal gasket 波纹金属垫片
corrugated metal joint ring 波纹金属接合填密环
corrugated metal lath(ing) 波形金属板条
corrugated metal pipe 金属波纹管;波形金属管;波纹金属管
corrugated metal pipe arch 波纹钢管拱
corrugated metal pipe culvert 波纹铁管涵(洞)
corrugated metal plate 波形金属板
corrugated metal roof cover(ing) 波形金属屋顶

盖板
corrugated metal sheath 波形金属挡板
corrugated metal sheet 波形金属(薄)板
corrugated metal sheet curved in two planes 两面弯曲的波形金属薄板
corrugated metal silo 波形金属筒仓
corrugated metal structure 波形金属结构
corrugated metal tunnel 波纹钢管隧道;波纹钢管廊道;波纹钢管地道
corrugated metal veneer ties 波形金属板系铁;波形金属板系件
corrugated mirror 波纹镜
corrugated mo(u)ld 波形铸模
corrugated nail 波型钉
corrugated nozzle 波状喷管
corrugated ornament 波形装饰
corrugated packing ring 波纹金属垫圈
corrugated panel 波形镶板;波形板
corrugated paper 波纹纸板;瓦楞纸
corrugated paper box 波状纸盒;瓦楞纸盒
corrugated perspex 波形透明塑料
corrugated pipe 瓦垄管;波纹管;瓦楞槽
corrugated pipe bend 波纹皱纹管弯头
corrugated plastic board 波形塑料板
corrugated plastic ceiling 波形塑料顶棚
corrugated plastic film 波形塑料薄膜
corrugated plastics 波形塑料
corrugated plastic sheet 波形塑料板
corrugated plastic sheeting 波形塑料片;波形塑料薄板;塑料波纹板
corrugated plastic sheet roofing 波形塑料瓦屋面
corrugated plate 皱褶板;波形板;波纹片;波支颚板;波纹板
corrugated plate intercepter 波纹板隔油池
corrugated plate interceptor 波纹板油水分离器;波纹板拦截器
corrugated plates scrubber 波纹板洗涤器
corrugated plexiglass 波形有机玻璃
corrugated plywood 瓦楞胶合板
corrugated polyester resin board 波形聚酯树脂板
corrugated polyester resin sheeting 波形聚酯树脂薄板
corrugated polyethylene fitting 波纹聚乙烯管件
corrugated polyethylene tubing 波纹聚乙烯管道
corrugated polyvinyl chloride film 波形聚氯乙烯薄膜
corrugated polyvinyl chloride sheeting 波形聚氯乙烯板
corrugated precipitator 百叶窗式除尘器
corrugated product 波形制品
corrugated profile 波形剖面图
corrugated rabbeted stop 波形门窗挡水条
corrugated rail bond 波纹导体接头
corrugated reducing sheet 波形渐缩薄板
corrugated rib 波形肋
corrugated ribbon 波纹带
corrugated ridge capping 波形屋脊帽盖
corrugated ring 波纹环
corrugated ripple mark 皱纹状波痕
corrugated road 搓板路
corrugated roll 槽纹辊筒
corrugated roll crusher 波纹状滚碎机
corrugated rolled glass 滚压瓦棱玻璃;波形压延玻璃
corrugated rolled glass partition 波形辊轧玻璃隔墙
corrugated rolled wired glass roof cover(ing) 波形铅丝网玻璃屋顶盖板
corrugated roller 瓦垄辊(瓦楞面镇压器)
corrugated roll washing machine 槽纹辊洗涤机
corrugated roof cladding 波形屋顶覆盖
corrugated roof glazing 波形玻璃屋面
corrugated roof(ing) 波纹屋顶;瓦楞屋顶;瓦楞屋面;波形箱顶;波形屋面(板);波形屋板屋面
corrugated roofing nail 波纹瓦屋面钉
corrugated roof insulation 波形屋顶隔热
corrugated roof insulation board 波形屋顶隔热板
corrugated roof insulation material 波形屋顶隔热材料
corrugated rooflight sheet 波形屋面采光板
corrugated roof sheathing 波形屋面衬板
corrugated roof sheet(ing) 波形屋面盖板
corrugated roof steel 波纹屋面钢;瓦楞屋面钢板
corrugated round iron 波形圆铁

corrugated rubber strip 波形橡胶条
corrugated screen wick 波纹丝网管芯
corrugated sealing piece 波形止水片
corrugated secondary scrubber 波形板二次洗涤器
corrugated section 波形剖面;波形截面
corrugated segment 波形钢管片
corrugated shape 波形形状;波形格式
corrugated sheath(ing) 波形衬板
corrugated sheet 波形片;波形板;波纹片;波纹薄板;波纹板;瓦楞铁皮;瓦楞槽;瓦楞板
corrugated sheet glass 瓦楞玻璃;波形玻璃
corrugated sheeting 波形薄板
corrugated sheet iron 波形铁皮;波纹铁皮;瓦楞铁皮;瓦楞白铁
corrugated sheet iron roofing 瓦楞铁皮屋面
corrugated sheet metal 波形金属薄板;瓦垄板
corrugated sheet-metal casing 焊制螺纹套管
corrugated sheet metal lath(ing) 波形金属(片)板条
corrugated sheet nail 波纹板钉
corrugated sheet rolling mill 波形板轧制厂
corrugated sheet roofing 波形板屋面
corrugated sheet steel 波形薄钢板;波形钢板
corrugated sheet zinc 波形锌皮
corrugated shell 波形薄壳;波形外板;波形壳;波纹形壳体
corrugated shell roof 波形薄壳屋顶
corrugated side 波形舷板
corrugated siding 波形墙板;波形挂板;波纹披叠板;波纹板壁
corrugated silo 波纹料仓
corrugated skin 波纹表皮
corrugated sound-absorbent panel 波形吸声(镶)板
corrugated sound-control panel 波形声控板;波形控制声音的板
corrugated square steel bar 方竹节钢条
corrugated steel 竹节钢筋;波形钢;波纹钢(板)
corrugated steel bar 波(纹)形钢(筋);竹节钢筋
corrugated steel culvert 波纹钢管涵洞
corrugated steel floor of bridge 波纹钢板桥面
corrugated steel glass 瓦棱玻璃;波形玻璃
corrugated steel plate 波纹钢板;瓦楞钢板
corrugated steel sheet 瓦垄钢板
corrugated steel sheeting 波纹钢板
corrugated steel shell diaphragm 波形钢壳心
corrugated strengthening 波形增固法
corrugated surface 路面波浪;冷硬表面;波支曲面;波纹(表)面;皱纹表面
corrugated tank 波形外壳
corrugated tie 波形拉杆
corrugated tile 波形瓦;波纹瓦
corrugated tile roof sheathing 波形黏土瓦屋顶衬板
corrugated tool 阶梯刨刀
corrugated tooth ring 波纹齿环
corrugated translucent sheet 波形半透明板
corrugated trim 波形镶边
corrugated tube 波形管;波纹管
corrugated tube valve 波纹管阀
corrugated tubing 波纹软管;波纹管
corrugated tunnel vault 波形隧道拱顶
corrugated type cast iron segment 波纹形铸铁管片
corrugated type cutter 波纹形磨轮
corrugated-type expansion joint 波形膨胀节
corrugated vault roof 波形圆拱顶;波形穹屋盖;波形穹屋顶
corrugated vessel 波形外板船
corrugated wall 波纹状墙壁
corrugated waveguide 皱波导;波形导管;波纹状波导管;波纹状波导;波纹波导
corrugated wear 波纹磨耗
corrugated web girder 波形腹板大梁
corrugated wire 刻痕钢丝;波纹钢丝
corrugated wire(d) glass 夹丝瓦楞玻璃;夹丝波形玻璃;波形夹丝玻璃
corrugated wood plug 波纹木塞
corrugated zincification sheet roofing 镀锌瓦楞铁皮屋面;镀锌波纹铁皮屋面;镀锌波纹铁皮屋顶
corrugated zinc plate 波形锌板
corrugated zinc roof cover 波形锌皮屋顶盖板
corrugater 波纹纸板机
corrugate sheath cable 波皱纹护套电缆;波纹护套电缆

corrugating 压制波纹;波纹板加工;波纹板冲压成型法
corrugating die 波纹成型模
corrugating line 波纹板辊压作业线
corrugating machine 波纹(纸板)机
corrugating paper machinery 波纹板机械
corrugating roll 波纹轧辊;瓦垄轧辊(轧制瓦楞板用)
corrugating roll mill 波纹板轧机
corrugation 压瓦垄;皱折;褶皱;折皱;灌水垄沟;(路面上的)搓板(现象);波纹状;波纹;波曲度;揉皱
corrugation infiltration 垄沟渗漏
corrugation irrigation 垄沟灌水;沟灌
corrugation machine 制波纹机
corrugation-method irrigation 垄沟灌溉法
corrugation of rail head 轨头波纹磨损
corrugation pitch 波峰间距
corrugation rolling mill 瓦垄轧机
corrugation texture 揉皱结构
corrugation web 槽形腹板
corrugator 瓦楞成型机;直压成型机;波纹板轧机;瓦垄板轧机
corrupt 腐化的;腐败的
corrupt data 不可靠数据
corrupted data 无效数据
corrupt element 腐化分子
corruption 腐化;腐败
corruptionist 行贿受贿分子;腐化分子
corrupt practice 行贿
corrupt transaction 舞弊行为
corsair 海盗船
corsehill (英国一种产于邓弗里斯的)红色砂岩
corselet 领盘【救】
corsham (英国,一种产于威尔特郡的)淡奶色石灰石
Corsican fir 欧洲枞木
Corsican larch 欧洲落叶松
Corsican pine 澳洲松
corsite 球状闪长岩
Corson alloy 科耳逊合金;铜镍硅合金
corso skirt 鞍形裙
cortate distance 垂足距
Cor-Ten 科尔坦耐大气腐蚀高强度钢
cortex 皮层
corteza del damajuhato fiber 达玛朱哈托韧皮纤维(巴西)
cortical layer 皮层
corticin(e)(用树胶胶成的)软木地毡;软木地毯
cortile 内院(意大利)
Corti's tunnel 螺旋器隧道
cortlanditite 普通角闪石橄榄岩
corundite 刚玉
corundolite 刚玉岩
corundum (金)刚砂;刚(玉)石
corundum alumin(i)um phosphate binder 磷酸铝刚玉胶结料
corundum bit 金刚砂钻头
corundum block 刚玉砖;刚玉质大砖
corundum brick 刚玉砖;刚玉耐火砖
corundum castables 刚玉质耐火质浇注料
corundum cloth 刚玉砂布
corundum crucible 刚玉坩埚
corundum deposit 刚玉矿床
corundum fillet 刚玉砂布带
corundumite 刚玉
corundum microsyenite 刚玉微晶正长岩
corundum-mullite ceramics 刚玉莫来石陶瓷
corundum-mullite porcelain 刚玉莫来石瓷
corundum plagioclasite 刚玉斜长岩
corundum porcelain 刚玉质瓷
corundum powder 金刚砂粉
corundum products 刚玉制品
corundum refractory 刚玉耐火材料
corundum refractory brick 金刚砂耐火砖
corundum refractory product 金刚砂耐火制品
corundum syenite 刚玉正长岩
corundum type structure 刚玉型结构
corundum wheel 刚玉砂轮
corve 矿场的轨道斗车
corvette 驱潜快艇;轻型护卫舰
corvic 聚氯乙烯树脂
corvusite 水复钒矿
corydalis green 紫堇色
cosa mica 致密钠云母
cosanic acids 二十级酸

cosanols 二十级醇
cosecant 余割
cosecant-squared antenna 余割平台型天线;平方余割天线
cosecant-squared beam 平方余割包线型射束
cosecant-squared diagram 平方余割形方向图
cosecant-squared dish 平方余割反射器
cosecant-squared pattern 平方余割形方向图
cosecant-squared weighting 余割平方加权
cosedimentation 同沉积作用;同沉积
coseismal 同震的
coseismal area 同震区
coseismal line 同震线
coseismic 同震的
coseismic area 同震区
coseismic circle 同震图;同震圈
coseismic curve 同震曲线
coseismic deformation 震时形变
coseismic effect 震时效应
coseismic line 等烈度线;同震线
coseismic movement 震时运动
coseismic strain 震时应变
coseismic zone 同震带
co-selector 补充选择器;补偿选择器
coseparation 共分离;同时分离
cosequence register 并列顺序寄存器
cosequent slope (岩层倾向与边坡坡向相同的)顺向坡
coset 复层组;傍集;伴集
coset code 伴集码
coset weight 伴集权
co-sharing state 共有资源国
co-signatory 联署人;联署者;共同署名者
co-signer 合签人;联署人
cosine 余弦
cosine capacitor 余弦电容器
cosine condition 余弦条件
cosine curve 余弦曲线
cosine curve distribution 余弦曲线分布
cosine distribution 余弦曲线分布
cosine emission law 余弦发射定律
cosine equalizer 余弦均衡器
cosine error 余弦误差
cosine formula 余弦公式
cosine formulas of spheric(al) triangle 球面三角余弦公式
cosine-fourth law 余弦照度定律;余弦四次方定律
cosine haversine formula 余弦半正矢公式
cosine impulse 余弦脉冲
cosine integral 余弦积分
cosine law 余弦定律
cosine modulation 余弦调制
cosine of direction(al) arm 方向导杆余弦角
cosine oscillations 余弦振荡
cosine series 余弦级数
cosine source 余弦律分布源
cosine-squared pulse 余弦平方脉冲
cosine transform 余弦变换
cosine wave 余弦波
cosine winding 余弦绕组
cosinusoidal intensity distribution 余弦强度分布
cosinusoidal motion 余弦运动
coslettise 磷化处理
coslettized 磷酸盐被膜防锈处理的;磷化处理的
coslettise 磷化
coslettizing 磷酸铁被膜防锈法;磷酸膜防(腐)蚀处理;磷化处理
Coslett process 钢铁防蚀的磷化处理法
cosmetic 化妆品;装饰性的
cosmetic appearance 装饰性外观
cosmetic(ci)ze 粉饰
cosmetic glassware 化妆用玻璃器皿
cosmetic-grade talc 化妆品级滑石
cosmetic improvement 外表改善;门面修饰
cosmetic industry 化妆品工业
cosmetic of computer cartography 计算机制图整饰
cosmetic plant for establishment 化妆品工厂
cosmetic wastewater 日用化工废水
cosmetic welding 盖面焊
cosmic(al) abundance 宇宙丰度
cosmic(al) abundance of element 元素的宇宙丰度
cosmic(al) age 宇宙年龄;宇宙暴露年龄

cosmic(al) background radiation 宇宙背景辐射
cosmic(al) cartography 宇宙制图学
cosmic(al) channel window width 宇宙道窗宽
cosmic(al) cloud 宇宙云
cosmic(al) dust 宇宙尘
cosmic(al) dust detector 宇宙尘埃检测器
cosmic(al) electrodynamics 宇宙电动力学
cosmic(al) environment 宇宙环境
cosmic(al) environmental space environment 航天环境
cosmic(al) expansion 宇宙膨胀
cosmic(al) gas 宇宙气
cosmic(al) geology 宇宙地质学
cosmic(al) iron 陨铁
cosmic(al) noise 射电噪声
cosmic(al) radiation 宇宙线;宇宙射线;宇宙辐射
cosmic(al) radiation value 宇宙辐射值
cosmic(al) radio wave 宇宙射电波
cosmic(al) ratio radiation 宇宙射电辐射
cosmic(al) ray 宇宙线;宇宙射线
cosmic(al) ray flux 宇宙射线通量
cosmic(al) ray influence 宇宙射线影响
cosmic(al) space 宇宙空间
cosmic(al) speed 宇宙速度
cosmic(al) water 宇宙水;大气水
cosmic(al) X-ray spectroscopy 宇宙 X 射线分光镜
cosmic(al) year 宇宙年
cosmic-ray background 宇宙线本底
cosmic-ray counter 宇宙射线粒子计数管
cosmic-ray meter 宇宙线测量计
cosmic-ray particle 宇宙线粒子
cosmic-ray shower 宇宙射线簇射
cosmic-ray source 宇宙线源
cosmic-ray telescope 宇宙线望远镜
cosmochemistry 宇宙化学;天体化学
cosmodrome 宇航基地;宇宙飞船发射场
cosmogenic elements 宇宙成因元素
cosmogenic hypothesis of mineralization 宇宙源成矿说
cosmogenic isotope 宇宙成因同位素
cosmogenic radioactive isotope 宇宙成因放射性同位素
cosmogenous 宇宙成因的
cosmogenous deposit 宇宙沉积
cosmogenous sediment 宇宙源沉积
cosmogeology 天体地质学
cosmogony 天体演化
cosmograd 太空城市
cosmoline 防腐油;涂防锈油
cosmological constant 宇宙学恒量;宇宙学常数
cosmological model 宇宙模型
cosmological principle 宇宙学原理
cosmology 宇宙学
cosmonaut 宇宙航行员;宇宙飞行员;宇航员
cosmonautics 航天;宇宙航行学
cosmopolis 国际性大城市;国际都市
cosmopolitan 普遍种
cosmopolitan speices 世界种
cosmopolitical 世界性的
Cosmos 大波斯菊
cosmos 菊花形
cosmosphere 天球仪
cosmotron 高能同步稳相加速器;同步稳相加速器
cosmozoan 宇宙物质;宇宙生物
cosolubilization 共增溶解作用
cosolvency 共溶性;潜溶性
cosolvent 共溶剂
cospace 共空间
co-spectrum 共谱;同相谱;同谱线;协相谱;余相谱
cosponsor 联合发起人;共同发起人
cosse green 豆荚绿色
cost 净到岸价格;价值;货价加运价和佣金;货价加运费和佣金;花费;卸货费在内的到岸价格条款;代价;船舱底交货的到岸价格;成本(加运费保险费价格);舱底交货的到岸价格;保险加空运费价格;班轮到岸条款;讼费
cost, insurance, freight, commission, exchange and interest 到岸价格加佣金,汇费及利息
cost, insurance, freight, commission and exchange 到岸价格加代理费和汇费
cost, insurance, freight, commission and interest 到岸价格加佣金和利息
cost, insurance, freight, interest and commission 到岸价格加利息和佣金

cost, insurance, freight, interest and exchange 到岸价格加利息和汇费
cost, insurance, freight and cleared 到岸价格加结关费用价
cost, insurance, freight and commission 到岸价格加佣金价;到岸价格加代理费
cost, insurance, freight and duty paid 到岸价格加关税价
cost, insurance, freight and exchange 到岸价格加汇费价
cost, insurance, freight and interest 到岸价格加利息价
cost, insurance, freight and landed 到岸价格加卸货费
cost, insurance, freight and landed terms 到岸价格加卸货价;到岸价格加卸货费条款
cost, insurance, freight and premium 到岸价格加保险费
cost, insurance, freight and war-risk 到岸价格加战争险;到岸价格加兵险
cost, insurance, freight cleared 到岸价格加报关所有手续费
cost, insurance, freight liner terms 到岸价格加班轮条件
cost, insurance, freight under ship's tackle 吊钩下交货的到岸价格
cost, insurance and freight London terms 到岸价格伦敦条款
co-stabilizer 共稳定剂
co-stabilizing lubricant 共稳润滑剂
costa bulb 舵前缘整流体
cost-access time product 价格存取时间积
cost account 成本账目;成本账(户);成本计算;费用计算
cost accountability center 责任成本中心
cost accountant 成本会计员
cost accounting 成本会计;成本核算
cost accounting analysis 成本核算分析
cost accounting by elements 要素别成本会计
cost accounting centre 成本核算中心
cost accounting division 成本会计科
cost accounting flow chart 成本会计流程图
cost accounting over-view 成本会计总观
cost accounting standard 成本会计准则;成本核算标准
cost accounting system 成本会计制度
cost accounts 成本类账户
cost accrued in construction contract 出包工程应付费用
cost accumulation accounts 成本类账户
cost adjustment 成本调整
cost advantage of economics 经济效益
costa equatorialis 赤道缘
cost-a-fixed-fee 成本加固定费
cost after split-off 分离后成本
cost after split-up 分离后加工成本
costalia 肋板
cost allocation 造价分配;成本分摊;费用分摊
costal margin 前缘
cost analysis 造价分析;成本分析;费用分析
cost analysis model 费用分析模式
cost analysis technique 成本分析技术
cost-and-benefit-sharing 共同分摊费用收益;成本与收益分摊
cost and delivery period 费用与支付期
cost and effectiveness trade-off criterion 费用和效能综合准则
cost and expense statement 成本费用表
cost and fee analysis 成本费用分析
cost and fee authenticity audit 成本费用真实性
cost-and-fee contracts 成本加费合同
cost and fee extent audit 成本费用范围审计
cost and fee management audit 成本费用管理审计
cost and freight 离岸价格加运费;离岸加运费价;货价加运价;货价加运费;货价和运价;运费在内价;成本加运费
cost and freight free out 船方不承担卸货费用;成本加运费除卸货费
cost and freight invoice 货价及运费发票
cost and freight price 成本加运费价格
cost and insurance 货价加保险费;成本加保险费
cost-and-percentage contracts 成本加百分比合同
cost-aperture analysis 造价口径分析

cost applicable to construction revenue 已完工程成本;已完成工程成本
cost applied account 实支成本账户
cost approach 成本处理;成本计算法
cost approach to value 房地产成本估价法
costate coordination 共态协调
cost audit 成本审计
cost base 成本基础
cost-based budgeting 按成本编制预算
cost based on actual batch purchasing price 分批实际进价成本
cost before split-up 分离前加工成本
cost-benefit 成本效益化
cost-benefit analysis 经济效益分析;成本利得分析;成本与收益分析;成本效益分析;成本收益分析;费用效益分析;费用收益分析;本利分析
cost-benefit analysis method 费用效益分析法
cost-benefit evaluation 费用效益评价;费用收益评价;费用经济效果评价
cost-benefit index 费用效益指标
cost-benefit indicator 经济效益指标
cost-benefit matrix analysis 损益矩阵分析法
cost-benefit ratio 投资收益比;费用效益比;费用收益比;投资效益比
cost-benefit relationship 成本—效益关系
cost-book 成本账
cost breakdown 成本细目;成本分类;成本分解;费用细目;费用剖析;合同金额(分项)明细表
cost budget 成本预算
cost calculation 成本计算
cost calculation for point to point of railway transport enterprise 铁路运输企业点到点成本核算
cost calculation statement 成本计算表
cost-capacity curve 成本—生产能力曲线
cost category 费用种类
cost center 成本项目;成本中心
cost classification 成本分类
cost clerk 成本记账员;成本管理员
cost code 工价编码引;价值编码;成本编码
cost coding 价值编码
cost coefficient 成本系数;费用系数
cost comparison 成本比较;费用比较
cost comparison approach 成本比较法
cost competitive 成本具竞争力的
cost concept 成本概念
cost-conscious 注意降低成本的
cost control 造价控制;工业成本控制;成本控制;成本管理;费用控制
cost control effectively 造价有效控制
cost control engineer 费用控制工程师
cost control figure 成本控制数字
cost control function 控制生产费用的职能
cost crude 成本油
cost cubic(al)metre 每立方米造价
cost curve 成本曲线;费用曲线
cost-cutting 降低成本
cost data 价格数据;成本资料;成本数据;费用数据
cost decrease 生产成本降低
cost decreasing rate 成本降低率
cost department 价格核定部门;成本会计处
cost depletion 成本耗损
cost differentials approach 差额成本法
cost distribution 成本分配;费用分配
costean 掘井(水力冲刷)勘探;井探;槽探;水力冲刷勘探
costeaning 井探;槽探
costeaning pit 探坑
cost economics 成本经济学
costeen 井探;槽探;水力冲刷勘探
cost-effective 良好经济效益;成本有效的;能降低成本的;划算的;有经济效益的
cost-effective means 成本有效法
cost-effective method 成本有效法;费用低廉的方法
cost-effectiveness 成本效果(的对比);成本有效性;成本效益;成本效率;费用有效度;费用效果
cost-effectiveness allowance 成本—效率容许量
cost-effectiveness analysis 成本—效益分析;技术经济分析;工程经济分析;成本效果分析;费用效能分析;费用效果分析
cost-effectiveness analysis method 工程经济分析法
cost-effectiveness model 成本效率模式
cost-effectiveness program(me) 经营效果研究

计划
cost-effectiveness ratio 费用效能比;费用效果比
cost-effective process 费用有效方法
cost-effective strategy 成本效益策略
cost efficiency 成本效率;投资效果
cost-efficiency analysis method 费用效率分析法
cost efficiency program(me) 经济效率研究计划
cost efficient 经济实用的
cost element 成本要素;费用因素
cost engineering 造价工程;成本管理
cost escalation 涨价;费用上涨
cost escalation clause 伸缩条款
cost estimate 造价估算;估价;估计费用;成本估算;成本估计;费用估算;费用概算
cost estimate and determine reasonably 造价合理计定
cost estimating 概算;成本预测;成本估算;成本估计
cost estimating procedure 成本估计程序
cost estimating relation 费用估计关系
cost estimating technique 概算技术
cost estimation 造价结算;造价估算;估价单;估价;概算;成本估算
cost estimation technique 成本估计技术
cost evaluation 成本估计
cost expansion path 费用扩大线
cost factor 造价影响因素;费用因子;费用因素
cost favor and exceptions 特许成本及例外事项
cost finding 成本估算;成本计算;费用计算
cost flow concept 成本流动观念
cost fluctuation 造价浮动;成本变动
cost for construction work 施工成本计算
cost for control 控制成本
cost for decision-making 决策费用
cost forecast 成本预测
cost for pumping 抽水费
cost for single productive work 单项生产工作费用
cost fraction 部分成本
cost-free 免费(的)
cost-freight-insurance 总运输费和保险费
cost freight price 交货收款价
cost function 价值函数;代价函数;成本函数;费用函数
cost function analysis 成本功能分析
cost function of sewage pipe 污水管道费用函数
cost function of unit process of water treatment 水处理单元过程费用函数
cost function of wastewater treatment plant 污水处理厂费用函数
cost function of water supply pipes 给水管道费用函数
costibite 硫锑钴矿
cost impact 成本影响;成本冲击
cost in business 营业成本;企业成本
cost incentive 价格刺激
cost incidental to wages 工资费用
cost in control 可控成本
cost incurred account 已发生成本账户;实际费用账;实际成本账
cost incurred in current period 本期发生成本
cost incurred method 发生成本法
cost index 价格指数;造价指标;成本指数;成本指标;费用指数
cost index number 成本指数
cost in economic accounting 经济核算成本
cost inflation 成本价格上涨;成本价格膨胀
cost information reporting 成本资料报表;费用数据报告
costing 劳动力统计;成本计算;成本会计;费用计算
costing by actual freight volume 按实际运量计费
costing by ship's displacement 按船泊排水量计费
costing details and allocations 成本计算的细节与分配
costing exercise 成本计算
costing of labo(u)r 劳务成本计算
costing of materials 材料成本计算
costing unit 成本单位
cost-insurance and freight 起岸价格;成本—保险费加运费;保险费加运费
cost-insurance and freight by plane 成本—保险加空运费价格
cost-insurance and freight duty paid 成本—保险费—运费加关税价
cost-insurance-freight 成本加保险费—运费价格

cost-insurance-freight and landed terms 成本加保险费—运费和卸货费价格条件
cost insurance freight inland waterway 内河到岸价格
cost in use 使用期费用
cost inventory 库存成本
cost investigation 成本分析
cost items 成本项目;费用项目
cost justifying 费用合算
cost keeping 成本核算
cost laid down 成本支出
cost ledger 成本分类账
cost less depreciation 成本减折旧
cost level 成本水平
cost limit 价格限制;造价限额
costly 昂贵的
costly goods 贵重货(物)
cost maintenance 养护费;保养费
cost management 成本管理
cost management dynamic 造价动态管理
cost management information system 成本信息系统
cost management throughout the construction process 建设全过程造价管理
cost map 生产成本图
cost mark-up 成本加价
cost matching income principle 费用与收入对应原则
cost memo 特项成本;成本通知单;转账凭单;费用通知单
cost method of inventory pricing 库存按成本计价法
cost-mindedness 费用意识
cost minimization 最低成本;成本极小化;生产成本最低化
cost-minimization analysis 最少费用分析
cost mode 成本模式
cost model 成本模型;费用模型
cost of a building 房屋造价
cost of acquiring technology 技术转让引进费
cost of acquisition 进货费用
cost of activity 作业成本
cost of alternation or repairs 改建或修理费用
cost of backfilling 回填土费
cost of borrowing 借款费用;借款成本
cost of borrowings issuance 借款的发放费用
cost of breaks 破损成本
cost of by-product sales 副产品销售成本
cost of capital 资本成本
cost of capital investment 基建投资费(用)
cost of carrying inventory 库存储备成本;存货保管成本
cost of charp tool 刀具成本
cost of civil engineering works 建筑工程造价;土木工程费用
cost of clearing 开垦费用
cost of common services 一般服务成本
cost of comparable products 可比产品成本
cost of completed works 完工工程成本
cost of compliance 合质成本
cost of construction 建筑成本;建造费(用);建设费(用);施工费(用)
cost of construction and assembly 建筑安装造价
cost of construction and equipment 建筑和设备费
cost of construction and installation 建筑安装造价
cost of construction equipment 施工设备费;施工机械费
cost of construction plant 施工设备费
cost of construction(project)工程造价
cost of constructor's mechanical plant 施工机械费
cost of contract 合同费
cost of cover 弥补费用
cost of debt 举债成本;债务成本
cost of delay 延误价值;延误费;误期费用
cost of demodeling 重新装修费用
cost of depreciation 折旧成本
cost of detailed estimation 预算造价
cost of development 开发费用;开发成本
cost of distribution 销售成本;分配费用
cost of drilling 钻探费用;钻井成本
cost of earthquake protection 防震费用
cost of effluent disposal 污水尾水处置费
cost of electric(al)energy 电能成本
cost of equipment 设备费(用)

cost of equipment installation 设备安装费
cost of equipment replacement 设备更新费(用)
cost of equity capital 普通股成本
cost of erection 建设费;架设费;安装费(用)
cost of erection work 安装工程费
cost of establishment 建设费用
cost of excavation 挖土费
cost of expenses 费用成本
cost of financing 筹资费用;财务费用
cost of floor space 地价成本;安装面积成本
cost of foreign exchange 换汇成本
cost of formation 造林费
cost of freight 运费
cost of freight service 货运成本
cost-offsetting advantage 抵补成本的优势
cost of fuels 燃料费
cost of fund 基金成本;资金成本
cost of further processing 再加工成本
cost of goods 商品成本
cost of goods available for sale 可供销售的商品成本;备销商品成本
cost of goods manufactured 产品成本
cost of goods on hand 库存商品成本
cost of goods purchased 进货成本
cost of goods sold 销售成本;销货成本
cost of goods sold on installment plan 分期付款销售成本
cost of grade separation 立交造价
cost of grading 土方修筑费;土地平整费
cost of handling and storing inventory 搬运与保管费用
cost of haulage 土方运距费
cost of heat at collective point 集核点热能成本
cost of heat at point of delivery 交付端热能成本
cost of human resources 人力资源成本
cost of initial estimation 估算造价
cost of installation 安装(用);安装成本;设置费;设备费(用)
cost of intangible assets 无形资产的成本
cost of inventory 盘存成本
cost of investigation 调查费用
cost of investment 投资费(用);投资成本
cost of labo(u)r 劳务费;人工成本
cost of labo(u)r turnover 人工周转成本
cost of labo(u)r used 人工费(用)
cost of land 地基费用;土地费用
cost of laying pipe 排管费
cost of layout 布置成本
cost of lease 租赁成本
cost of light 光源评价;照明费用
cost of living 生活水平;生活费(用)
cost of living allowance 生活指数津贴
cost of living clause 生活费用条款
cost of living escalation 生活费用的提高
cost of living factor 生活因素费用
cost of-living index 生活(费用)指数;物价指数
cost-of-living indexes 生活费用指数
cost of macromanagement 宏观管理费用
cost of maintenance 养护费(用);维修费(用);维护费(用);维持费(用)
cost of major repair 大修费(用)
cost of make and sell 制销成本
cost of management 管理费
cost of manufacture 制造成本;制成成本;造价
cost of market 成本或市价
cost of marketing 经销成本;销售费用;销货成本
cost of material 原材料成本;材料成本
cost of materials and inputs 材料及投入物成本
cost of merchandise sold 销货成本
cost of mining 可成本;采掘成本
cost of money 利息;资金成本
cost of new fixed assets 新增固定资产费用
cost of of work 工程成本费用
cost of operation 经营费用;运转费(用);运行费(用);作业成本;使用费;管理费
cost of organization 组织费
cost of overhaul 超程距离;超远成本;超距成本;超程运费;检修费;大修费(用);土方超运成本
cost of over-stock and stockpile items 超储积压物资
cost of passenger-kilometer 旅客成本
cost of passenger service 客运成本
cost of per-square-meter 每平方米造价
cost of planting 造林费

cost of pollution abatement 污染治理费用
cost of pollution protection 污染防护费用
cost of port development planned to take place 规划所要进行港口发展的费用
cost of possession 保有资产成本
cost of power 电费;燃料费
cost of power production 能量生产成本;电力生产成本
cost of preferred stock 优先股成本
cost of preinvestment studies 投资前研究费用
cost of price 成本价格
cost of primary estimation 概算造价
cost of prime implicant 初始隐含价值
cost of production 制造费(用);工本费;生产费(用);生产成本
cost of production budget 生产成本预算
cost of products 成品成本
cost of project implementation 项目建设费用
cost of pulling sheeting 拔板桩费
cost of pumping station 泵站抽水费
cost of quality 品质成本
cost of raw materials used 材料消耗成本
cost of recirculation 回流抽升费
cost of reconstruction work 改建工成本
cost of remedying defects 补贴缺陷的费用
cost of remodeling 重新装修成本
cost of removal 拆迁费(用);拆除费(用)
cost of removing equipment 移动设备成本
cost of repairs 修理费(用);修理成本
cost of repatriation 遣返费
cost of replacement 重置成本
cost of reproduction 再生产费用;再生产成本;重造成本;复制成本
cost of reproduction less depreciation 重造成本减折旧
cost of reproduction new value 新价值再生产成本
cost of sales 销售费用;销售成本;销货费;销货成本
cost of samples 样品费用;样本费
cost of selling 出售成本
cost-of-service 服务费(用);劳务费;服务成本
cost-of-service allocation 劳务成本分配
cost-of-service principle 服务费用原则
cost-of-service taxation principle 服务成本课税原则
cost of set-up 设置成本
cost of sewage chlorination 污水加氯消毒费
cost of sewage discharge 污水排放费
cost of sewage treatment 污水处置费
cost of sewer foundation 沟管基础费
cost of shared tubulars 管材摊消费
cost of sheet driving 打板桩费
cost of sludge disposal 污泥处置费
cost of sludge treatment 污泥处理费
cost of spoil 弃土费
cost of spoiled goods 废品成本
cost of steam heat 蒸汽热能成本
cost of stops 停车价值
cost of storage 储[贮]藏成本
cost of sulfur dioxide control 二氧化硫控制费用
cost of supervision 行政开支;管理费
cost of survey and design 勘察设计费
cost of suspension 暂时停工的费用
cost of tendering 投标经费;投标费用
cost of test 检验费用
cost of testing 试验费
cost of the energy at borehole 井口热能成本
cost of the investment 投资场所
cost of the works 工程造价
cost of tracer 示踪剂价格
cost of transfer 转让费(用)
cost of transportation 运输费(用)
cost of trench draining 沟槽排水费
cost of truck and transportation 运输费(用)
cost of unit working 单位工作成本
cost of upkeep 养护费(用);保养成本;保管费;日常养护费;日常维修费;维修(用);维修保养费(用)
cost of used fixed assets 已使用固定资产的费用
cost of vehicle operation 行车费(用)
cost of waste disposal 废弃物处置费用;废品处理费用;弃土费
cost of wasting 弃土费
cost of water 水价
cost of wear and tear 损耗费(用)

cost of winning 开采成本
cost of work performed 已完成工作量成本;已完成工程量成本
cost of works 工程造价
coston light 火焰信号;三色信号灯
coston signal 火焰信号
cost-optimal 最经济的;最合算的
cost optimization model 造价优化模型
cost-oriented pricing 根据费用定价;根据成本定价
costor silkworm 蓖麻蚕
cost outlay item 费用支出项目
cost out of control 不可控成本
cost outside business 营业外支出(费用)
cost over-run 成本超支;超限成本;超额费用
cost per benefit 单位效益造价
cost per byte 每字节价格
cost per cubic(al)foot 每立方英尺造价
cost per foot 每英尺(钻进)成本
cost performance 成本实效;价格性能;性能价格比;单位成本率;费用性能
cost performance analysis 成本与技术鉴定分析
cost performance ratio 价格性能比
cost per hole 每孔(钻进)成本
cost period 成本期
cost per kilowatt 每千瓦费用;单位千瓦成本
cost per kilowatt hour 单位电能造价
cost per locomotive-kilometer 机车一公里成本
cost per meter 海米成本
cost per mu 每亩成本
cost per square meter 单方造价
cost per ten thousand weight-ton-kilometer 每万总重吨公里成本
cost per thousand readers 每一千读者的收费标准
cost per ton-kilometre of goods 货物吨公里成本
cost per train-kilometre 列车公里成本
cost per unit 单位造价;单位费用;单位成本;单件成本
cost per unit recharge water volume 单位回灌量费用
cost per unit water volume 单位水量成本
cost per well 每口井(钻进)成本
cost plan(ning) 成本计划;造价规划;成本规划
cost-plus 附加价费;成本附加利润;成本附加报酬;按成本加收(管理费和利润等);成本加利润的;成本加费用
cost-plus-a-fixed contract 成本加附加合同
cost-plus-a-fixed-fee 成本加附加费;正价加附加费
cost-plus a profit 成本价加利
cost-plus award fees contracts 成本加酬金合同
cost-plus contract 累积成本合同;成本加酬契约;加酬合同;费用加利合同;变动价格合同;实报实销合同;成本加利润合同;成本加费用契约;成本加酬金合同;成本保利合同
cost-plus fee agreement 实报实销加管理费合同;加酬合同
cost-plus-fee contact 成本加酬金合同;成本加费用合同;加酬合同;正价加附加费合同;实费承包工程
cost-plus-fixed fee 成本加固定酬金;正价加固定附加费
cost-plus-fixed-fee contract 成本浮动酬金合同;正(价)加固定附加费合同;成本加固定费用合同;成本固定附加费合同;费用加固定手续费合同
cost-plus fluctuating fee contract 成本加浮动酬金合同;成本加浮动酬金合同;浮动利润合同;费用加浮动酬金合同
cost-plus incentive fee contract 成本加奖金合同
cost-plus-percentage contract 成本加提成契约;成本加比例报酬合同
cost-plus percentage fee 成本加一定比例费用
cost-plus percentage-of-cost contracts 成本加成本百分率合同
cost-plus price 成本加酬金价格;费用加利润;按中等开销加利润原则确定的价格;成本加费用价格
cost-plus-profit contract 成本加利润合同
cost-plus profit price 费用加利润价格
cost prediction 成本预测
cost preference 价格优惠
cost price 原价;成本价格;生产价格;生产成本
cost price differential 费用与价格差额
cost/price factor 费用价格因素
cost-price index 成本价格指数
cost procurement 订货成本
cost-profit ratio 成本利润率

cost-profit-volume analysis 成本—利润—产量分析
cost programming 造价规划
cost push 成本推动;生产成本提高
cost quota 成本定额
cost rate 成本率
cost rate per unit 单位成本率
cost rate per unit area 单位面积造价指标
cost record 成本账
cost record summary 开支总表;总开支
cost recovering 回收成本;收回成本
cost recovery 成本收回
cost recovery basis 以成本回收为基础
cost reduction 降低成本;成本降低
cost-reduction expenditures 降低成本的投资支出
cost reduction percentage of comparable products 可比产品成本降低率
cost reduction plan 降低成本计划
cost reduction program(me) 降低成本计划
cost reduction target on comparable products 可比产品成本降低任务
cost reference guide 造价参考指南
cost-reimbursable contract 成本结算方式合同
cost reimbursement-and-fee contracts 成本加酬金合同
cost reimbursement contract 成本加酬金合同;成本—补偿契约;成本补偿合同;费用补偿合同;补偿费合同
cost related factor 成本构成
cost rent 成本租金
cost report 成本报告;成本报表
cost reproduction 减少再生产
cost responsibility center 成本责任中心
cost-revenue analysis 成本收入分析
costrototype 同层型
cost saving 节约费用;省费用
cost-saving investment 节省成本的投资
cost schedule 成本明细表;成本测算表
cost segregation 成本分解
cost sensitive analysis 费用敏感分析
cost sensitivity 成本敏感性;费用灵敏度
cost sharing 分担费用;成本分担;费用分摊;费用分配;费用分担
cost-sharing formula 费用分摊办法
cost sheet 估价表;计算账单;成本账单;成本计算表;成本单
cost slope 成本斜率;费用增加率;费用斜率;费用坡度
costs of building work 建筑工程的造价
costs of construction work 建筑工程的造价
costs of litigation 诉讼费
costs of pollutant management 污染管理费
cost squeeze 压缩成本
cost stability 生产成本稳定
cost standard 成本标准
cost statement 成本计算表
cost stream 费用流程
cost-strength ratio 价强比;单价强度比
cost structure 成本结构
cost study 造价分析;成本分析
cost system 成本制度;成本计算方法
cost target contracts 成本目标合同
cost theory 费用说
cost-time curve 费用时间曲线
cost to replace remedial measure 更换矫正措施的成本
cost transfer 费用划拨
costume 服装
cost unit 成本核算单位;费用单位
cost unit ore reserve 单位储量成本
cost unit price 单位成本;成本单价
costunolide 广木香内酯
cost up 成本增高
cost valuation basis 成本估值基础
cost value 成本价值;费用价值
cost-value contract 成本价格合同
cost variance 成本差异
cost variance analysis 成本差异分析
cost variation clause 工程费变更条文
cost vector 成本向量;费用向量
cost versus effectiveness analysis 费用效能分析
cost-volume-profit analysis 成本—产销量—利润分析;本量利分析
cost-volume-profit assumption 本量利假定
cost-volume-profit projection 本量利预测

cosurfactant 辅助表面活性剂
cosy 温暖的;舒适的;保暖罩;保温套
cot 羊栏;茅舍;鸽棚;行军床;小屋;吊床;帆布床
cotamer 八聚物
cotangent 余切
cotangent scale 余切尺
cot bar (半圆形窗框的)弧形铁条;扇面亮子的弦向窗楞
cot bed 吊床
coteau 高原;高地;冰碛脊(法语)
cotectic line 同结线
cotemaker 制芯工人
cotenancy 共同租用
cotenant 合租承租人;共同租户
co-tensor 协张量
coterminal angles 共边角
cotidal 等潮的
cotidal and coranged 同潮时
cotidal and coranged chart 等潮差图
cotidal chart 等时潮;等潮线图;等潮(海)图;同潮(时)图
cotidal current chart 等潮流(时)图
cotidal current line 等潮流(时)线
cotidal hour 等潮时;同时潮;同潮时
cotidal line 等潮(时)线;同潮(时)线
cotidal line of partial tide 分潮同潮时线
cotidal map 等潮图
cotidal time 等潮时
cotinus 黄栌属
cotinus coggugria 栌木;黄栌
cotloft 谷仓阁楼
cotonier 法国梧桐
cotree 补树
cotruder 双螺杆挤出机
co-trustee 联合委托管理人
Cotswold n reversed polarity zone 科茨沃尔德反向极性带
Cotswold reversed polarity chron 科茨沃尔德反向极性时
Cotswold reversed polarity chronzone 科茨沃尔德反向极性时司带
cottage (郊外的)新式住宅;小型别墅;小屋;村舍;小屋;小别墅;村舍
cottage community 小型别墅区域(单元)
cottage furniture 乡村家具
cottage hospital 小医院;乡村医院;诊疗所
cottage industry 家庭手工业
cottage latch 小提升门;小门闩
cottage orne 粗面石砌小装宅(18～19世纪之间);粗面石砌小庄宅
cottage roof 小跨度屋盖;无桁架的小跨度屋盖
cottage suburb 小型别墅
cottaite 灰正长岩
cottar 接合榫;定缝钉
cotter 开尾销;开口销;接合榫;制销;定缝钉;插销
cotter bolt 锚栓;地脚螺栓;带销(螺)栓;牵条螺栓
cotter cutter 键槽铣刀
cottered joint 制销联轴节;(屋架中央主木与横木间的)双楔榫铰接
cotterel 锁销
cotter file 开槽锉
cotter hole 销钉孔(眼);扁销孔
cotterite 珠光石英
cotter joint 键接(合);销接合;扁销连接;开尾接合
cotter key 扁销键;扁键销;保险销
cotter mill 键槽铣刀;键槽铣刀
cotter mill cutter 键槽铣刀
cotter pin 扁销;开尾销;开口销
cotter-pin extractor 开口销分离机;开口销拔器
cotter pin flat pin 扁销
cotterpinning 装开尾销
cotter seat 气门弹簧座
cotter taper file 加工键槽用尖头锉
cotter volt 插销螺栓
cotterway 销槽
cotter with screw end 螺旋端锁销
cottle 外围挡板
cotton 棉纱;棉布
cotton and artificial silk mixture 棉纱人造丝混合织物
cotton and hemp rope 棉麻绳
cotton and rayon gimp 棉纱人造丝镶边带
cotton and wool mixture 棉毛混纺织物

cotton bagging 棉包用麻布
Cotton balance 科顿平衡仪;科顿磁秤
cotton ball 棉球
cotton ball clouds 棉球云
cotton ball tassel 棉球缨(球形边饰)
cotton base 棉花基地
cotton batting 棉絮
cotton belt 产棉区
cotton blue 棉染蓝
cotton blue staining solution 棉蓝染液
cotton blushing 硝化棉致发白
cotton canvas 棉帆布
cotton cloth 棉布
cotton cord 棉纱绳;八股棉纱绳
cotton-covered cable 纱包电缆
cotton covered wire 布电线;纱包线;纱包电线
cotton cover insulation 纱包绝缘
cotton cultivated areas 产棉区域
cotton drill 棉织斜纹布
cotton duck 棉织帆布;棉帆布
cotton duck tarpaulin 棉帆布盖布
cotton dust 棉尘
cotton dye 棉染料
Cotton effect 科顿效应
cotton-enamel covered wire 纱包漆包线;包纱漆包线
cottonette 棉毛混纺织物
cotton extractor 剥铃清棉机
cotton fabric 棉织物;棉织品
cotton fabric saturated with bitumen 沥青浸渍棉纤维屋面材料
cotton feeder 给棉机
cotton feeding machine 给棉机
cotton fiber content paper 含棉纤维纸
cotton fiber strength tester 棉纤维强力试验机
cotton filler 棉织衬垫
cotton-filling machine 充棉机
cotton floater 棉包橡皮包布和油包布
cotton flock 棉绒
cotton flock filler 棉绒填料
cotton-free dry 不黏棉纤维干;不黏尘土;不黏尘干;表干
cotton gauze 冷布
cotton gin 轧棉机
cotton ginning factory 轧棉厂
cotton gloves 棉纱手套
cotton goods 棉织品
cotton grey fabric 纯棉白坯布
cotton gum 紫树
cotton hair 棉线
cotton heaving line 棉纱撇缆
cotton hook 搬棉手钩
cottoning 絮状滚胶法;网纹状滚胶法
cotton hemp rope 棉麻绳
cotton-insulated wire 纱包线;纱包绝缘线
cotton insulation 棉纱绝缘
cotton insulation cable 纱包电缆
cotton lasting 厚实棉斜纹织物
cotton linters 棉绒
cotton loop bag fringe 棉圈球状边饰
cotton manufactured goods 棉织品
cotton mats 养护用棉毡;棉垫
cotton mill 棉纺厂;纱厂
Cotton-Mouton birefringence 科顿—莫顿双折射
Cotton-Mouton constant 科顿—莫顿常数
Cotton-Mouton effect 科顿—莫顿效应
cotton of short staple 短绒棉
cotton oil 棉(籽)油
cotton packing 棉填料
cotton paraffined rope 棉腊绳
cotton picker 采苗机
cotton picking receptacle 采棉室
cotton plant puller 拔棉柴机
cotton plug 棉塞
cotton polishing disk 棉絮抛光圆盘
cotton polish wheel 抛光棉轮
cotton powder 棉花火药
cotton pulp 棉浆液
cotton pulp black liquor 棉浆黑液
cotton rag 破旧衣服;棉质碎布;棉擦布
cotton region 棉区
cotton roll 棉卷
cotton roller 辊花机
cotton rope 棉纱绳

cotton rose 木芙蓉
cotton rose hibiscus leaf 芙蓉叶
cotton-seed cleaner 棉籽清选机
cotton-seed cleaning machine 棉籽清选机
cotton-seed hulls 棉籽壳堵漏材料
cotton-seed oil 棉子油;棉(籽)油
cotton seed oil fatty acid 棉籽油脂肪酸
cotton-seed oil semi-refined 半精炼棉籽油
cotton-seed oil stearin 棉籽油硬脂精;棉硬脂
cotton sewing thread in hank 棉绞线
cotton sewing thread on cone 宝塔线
cotton sewing thread on cop 管纱线
cotton sewing threads on cylinders 蜡筒线
cotton shirting 细棉布
cotton sleeving 纱包层
cotton soil 棉花土
Cottons patent frame 柯登氏平机
cotton spotted bollworm 翠纹金刚钻
cotton stalk puller 拔棉柴机
cotton swab 棉纱擦帚
cotton-tape 纱带
cotton textile 棉织品
cotton thread 棉线
cotton tie 钢皮带
cotton tire cord 轮胎帘布
cotton tree 木棉树
cotton twine 棉帆线
cotton varnished sleeve 黄蜡管
cotton velvet 棉绒
cotton velvet peeler 剥绒机
cotton warp linen 棉麻交织物
cotton waste 回花;废纱头;废棉纱头(堵漏用)
cotton waste filter 废棉滤器
cotton waste for cleaning 抹布
cotton wax 棉蜡
cotton wicking 捻缝棉条
cotton wood 棉白杨;杨木;三角叶杨;美洲白木
cotton wool 原棉;脱脂棉;棉絮;棉线;卫生棉
cotton-wool spots 棉絮状渗出点
cotton-wool tampon 棉塞
cotton yarn 棉纱
cotton yarn in skein 大扎绞棉纱
cottrell 电收尘器
cottrell chamber 电集尘室
cottrell dust 电收尘烟尘
Cottrell(dust) precipitator 科特雷尔除尘器
Cottrell effect 科特雷尔效应
Cottrell electric(al) precipitator 科特雷尔电分离器
Cottrell electrostatic precipitator 科特雷尔静电分离器
Cottrell moist precipitator 科特雷尔电除雾器
Cottrell pipe precipitator 科特雷尔管子静电集尘器
Cottrell precipitator 科特雷尔静电集尘器;科特雷尔静电沉降器;科特雷尔除尘器;静电除尘器
Cottrell rectifier 科特雷尔整流器
cottrell treater 电收尘器
cotty wool 次劣毛
cotunnite 氯铅矿
cotyle 臼
cotyloid receptacle 盘状花托
cotype 共型;副模式;全模标本
couch 抄纸压合;长沙发;伏辊;卧榻;榻;沙发椅
couch board 多层纸板
couchette 火车卧铺
couchette coach 卧铺客车
couch roll 伏辊
couch together 层叠
Coudé f-number 科德焦距比数;折轴焦距比数
Coudé focus 科德焦点;折轴(镜)焦点
Coudé mounting 折轴装置
Coudé reflector 折轴反射望远镜
Coudé refractor 折轴折射望远镜
Coudé spectrograph 科德摄谱仪;折轴摄谱仪
Coudé spectroscopy 折轴光谱学
Coudé system 科德系统
Coudé telescope 科德望远镜;折轴望远镜
Couette correction 库埃特校正系数
Couette flocculator 库埃特絮凝器
Couette flow 库埃特流动
Couette viscometer 库埃特黏度计
Couette viscosimeter 库尤特黏度计
coughing 专用排气口
coulability 铸造性
coulee 陆壁谷;斜壁谷;黏熔岩流;熔岩流;深冲沟;舌状泥石流体
coulisse 露天交易所;滑槽板;滑槽;穿堂门厅;侧面布景;槽柱
coulogravimetric analysis 库仑重量分析
couloir 通道(堆货用)
Coulomb 库仑
coulomb energy 库仑能
Coulomb equation for shear strength 库仑抗剪强度公式
Coulomb excitation 库仑激发
Coulomb field 库仑场
Coulomb force 库仑力
Coulombian field 库仑静电场
Coulombic attraction 库仑引力
Coulombic field 库仑静电场
Coulombic interaction 库仑相互作用
Coulombis standard formula 库仑标准公式
coulombmeter 电量计;库仑电量计;电量表;库仑计
Coulomb-Mohr shear failure theory 库仑—莫尔剪切破坏理论
coulomb potential 库仑电位;库仑势
coulomb repulsion 库仑斥力
Coulomb's attraction 库仑吸引
Coulomb's barrier 库仑位垒
Coulomb's basic equation 库仑基本方程
coulomb scattering 库仑散射
Coulomb's condition 库仑条件
Coulomb's damping 库仑阻尼
Coulomb's earth pressure theory 库仑土压力理论
Coulomb's effect 库仑效应
coulomb's efficiency 库仑效率
coulomb sensitivity 冲击灵敏度
Coulomb's envelope 库仑包络线
Coulomb's equation 库仑方程
Coulomb's failure envelope 库仑破坏包络线
Coulomb's formula 库仑公式
Coulomb's friction 库仑摩擦
Coulomb's interaction 库仑相互作用
Coulomb's law 库仑定律
Coulomb's law for magnetism 磁学的库仑定律
Coulomb's line 库仑线
Coulomb's rupture envelope 库仑破坏包络线
Coulomb's rupture line 库仑破坏线
Coulomb's soil failure prism 库仑土体破坏楔体
Coulomb's strength-envelope equation 库仑强度包络方程
Coulomb's theory 库仑理论
Coulomb's wedge theory 库仑楔体(土压理论)
coulomb-Terzagli criterion 库仑一太沙基破坏准则
coulometer 电量计;库仑计
coulometric analysis 库仑分析;电量分析
coulometric detector 库仑检测器
coulometric gravimetric analysis 库仑重量分析
coulometric hygrometer 库仑湿度计
coulometric method 电量测定法
coulometric titration 库仑滴定;电位滴定;电量滴定
coulometry 库仑分析法;电量(测定)法
Coulometry nitrogen oxides monitor 库仑法氮氧化物监测仪
Coulometry sulfur dioxide monitor 库仑法二氧化碳监测仪
coulopotentiography 库仑电势谱法
coulsonite 钒磁铁矿
co(u)lter 犁头(铁);犁刀(头);尖角沙嘴
Co(u)lter counter 库尔特(颗粒)计数器;悬浮液体颗粒组成电导测定仪
Coulmb's modulus 库仑模数;库仑模量
Coulumb's soil-failure prism 库仑土壤破坏棱柱体
coumaric acid 香豆酸
coumarone 香豆酮;氧茚;古玛隆;苯并呋喃
coumarone-indene resin 库玛隆树脂;香豆酮树脂;苯并呋喃—茚树脂
coumarone resin 香豆酮树脂;香茚树脂;古玛隆树脂;苯并呋喃(茚)树脂
council 会议;政务会;顾问班子;委员会
council chamber 会议室
Council for Mutual Economic Assistance 经济互助理事会;经互会
council for ports and harbo(u)r 港湾审议会
council house 会堂;会场
Council of Economic Adviser 经济顾问会议
Council of Housing Producers 住房开发委员会
Council on Environment(al) Quality 环境质量委员会(美国)
councilor 顾问;参赞
council school 公立学校;公办中等小学
counsel 法律顾问;辩护人
counsel for the defense 辩护律师
counsel(l)or 法律顾问;顾问;参赞
counsel(l)or of real estate 不动产法律顾问
count 统计
countability 可数性
countable 可数的
countable additivity 可列可加性
countable aggregate 可数集
countable function 可计数函数
countable number 可数数目
countable number of events 可计事件数
countable probability 可计算概率
countable set 可数集;可列集
countably additive 可数可加性
countably compact set 可数紧集
countably infinite subset 可数无穷子集
countably subadditive 可数次可加
count area 计数区
count attribute 计数属性
count block 计数区
count button 计数按钮
count code 计数码
count compensation 计数补偿
count cycle 计数循环
count detector 计数检测器
countdown 计时系统;逆序计数;应答脉冲比;询问无效率;倒读数;发射前的时间计算;漏失计数;计数损失;递减计数
countdown circuit 发射控制电路
countdown profile 操作程序表
count down sequence timer 递减顺序计时器
counted odo(u)r concentration 推定恶臭浓度
counted string 带计数的串;被计数的串
count enable 可计算的;可计数
counter 计算员;计数员;计数频率计;码子;相反的;柜台;定标器;船尾突出部;船舶凸面;反对;反
counteract 抵制
counteractant 冲消剂;恶臭冲消剂
counteracting force 反作用力
counteracting protective 反保护关税
counteracting valve type hydro-percussive tool 阀式反作用冲击器
counteraction 均衡作用;中和;对抗作用;抵抗;反作用;反诉
counteradvice 反对意见
counteragent 中和力;反作用剂;反向动作;反抗力
counter appliance 柜式用具(如煤气、家具等);柜式燃具;柜式煤气用具
counter approach 防御工事
counter arch 扶垛拱;反拱
counter-arched 有扶垛拱的(如护岸墙用)
counter arched revetment 反拱护壁
counter arm 计数器指针
counter-balance 平衡力;平衡锤;平衡块;平衡重(量);均衡(重);抵销;反压台;抗力;均衡块;相抵;砝码;补偿;配重;托盘天平;使平衡;使……达到均衡
counter-balance cylinder 平衡汽缸
counter-balanced brake 配重均衡制动
counter-balanced cage 平衡重笼
counter-balanced chute 平衡槽
counter-balanced door 均衡门
counter-balanced falling ball viscometer 平衡落球法粘度计
counter-balanced fork lift truck 平衡重式叉车
counter-balanced gate 均衡门;平衡(闸)门
counter-balanced sash 平衡式推拉门;平衡式上下推拉窗
counter-balanced scraper 带配重的刮板取料机
counter-balanced screen frame 平衡屏幕框架
counter-balanced shiplift 均衡式升船机
counter-balanced shutter door 平衡重闸门
counter-balanced spout 平衡导沟
counter-balanced sub-chute 平衡次斜槽
counter-balanced valve 背压反向阀
counter-balanced window 均衡窗;平衡窗
counter-balance moment 平衡力矩
counter-balance radius 平衡块伸距;平衡块伸出长度
counter-balance spooling (钢丝绳往滚筒上的)塔形平衡缠绕

counter-balance spring 补偿弹簧
counter-balance spring casing 平衡簧套
counter-balance system 平衡(重)系统
counter-balance valve 反平衡阀;背压阀;平衡阀
counter-balance weight 平衡重;砝码
counter-balancing 对重平衡
counter-balancing mass 抵衡质量
counter bar 定位尺;门窗外十字条;平衡锤杆
counter battens 屋面油毡压条;拼板防挠压条(固定在背面与板缝垂直);木地板木键;交叉压条;顺水条
counter bed 逆流床
counter bid 还价
counter bite 反咬合
counterblast 反气流;逆风
counter block 计数部件
counter blow 倒吹法
counterblow hammer 对击锤;锻锤;无砧座锤
counter bonification 反补偿
counterbore 埋头孔;锥(口)孔;沉孔;平头钻(头);平底扩孔钻;平底扩孔桩;镗孔;镗阶梯孔;扩孔(钻)
counterbore cutter 平底扩孔钻头
counterbored coupling 内流线型接头(冲洗液孔呈锥形)
counterbore guidance 扩孔钻导柱
counterbore guide 扩孔钻导柱
counterbore of the tool joint (钻杆接头的)划孔部分
counterboring 镗阶梯孔;扩钻
counterboring drill press 平底扩孔钻床
counter brace 转帆索;交叉撑;副斜杆;副对角撑;交叉对角撑;转帆索;副撑臂
counterbracing 交叉撑;副对角撑
counterbuffer 阻尼器
counter cable 反向缆索
counter-camber 预留弯度;预留拱度;反挠曲量
counter capacity 计数容量
counter ceiling (起隔声绝热作用的)悬吊平顶;设备吊顶;(起隔热、隔声等作用的)吊平顶;隔热吊顶;隔音吊顶
counter cell 反作用电池
counter center 反顶尖
counter chamber 计数室;计数管室
counterchange 交换;交互作用;互换
counter channel 计数器管道
countercharge 反诉;反控
countercheck 阻挡
counter-cheque 银行取款单
counter chronograph 计数式计时器
counter-circulation 反环流
counter-circulation wash boring 反循环洗孔钻进方法;反循环洗井钻进方法
counter-city 城市对抗
counter-claim 对等偿付债权;反索赔;反求偿;反诉
counter clamp 反向夹;双向夹紧装置;反向夹板
counter-clock 反时针的
counter-clockwise 逆时针的;逆时针方向的;反针方向的;反时针的;反时针
counter-clockwise airscrew 反时针螺旋桨
counter-clockwise angle 反时针角
counter-clockwise direction 逆时针方向;反针方向
counter-clockwise drift 反时针方向偏移
counter-clockwise motion 逆时针方向运动;反针方向运动
counter-clockwise polarized electromagnetic wave 逆时偏振电磁波
counter-clockwise polarized wave 反射针极化波
counter-clockwise polygon 逆时针多边形
counter-clockwise propeller 反时针旋转螺旋桨
counter-clockwise rotation 反时针回转;反时针(方向)旋转;逆时针(方向)旋转
counter-clockwise running 反时针转动
counter-clockwise twist 反手捻
counter coil 对流式盘管
counter collapse 抗倒塌
counter collapse skeleton 抗倒塌骨架
counter condition 不合要求
countercontact connection 来复式连接
counter control 计数器控制;用计算机检验
counter-controller 计数(器)控制器
counter-counter measures 电子反对抗;反电压对抗
counter counter-offer 反还盘

counter coupling 计数器耦合
counter crankshaft rotation 反曲轴转动方向的转动
counter-current 反向(电)流;反(电)流;逆(电)流;对流(法)
counter-current absorption 逆流吸附
counter-current action 逆流原理
counter-current agitation 逆流搅拌
counter-current balance 逆流平衡
counter-current boiler 逆流锅炉
counter-current braking 逆电流制动
counter-current capacitor 扼流电容器
counter-current cascade 逆流级联
counter-current centrifuging 逆流离心法
counter-current chromatography 反流色谱法;逆流色谱法
counter-current circuit 逆流循环
counter-current circulation 反循环;逆流循环
counter-current classification 逆流分级
counter-current classifier 对流式选分机;反流分级机;逆流式选粉机;逆流分级机
counter-current column 逆流喷淋塔
counter-current cooler 逆流冷却器;逆流冷却机
counter-current cooling tower 逆流式冷却塔
counter-current decantation 逆流倾析
counter-current decantation method 逆流倾析洗涤法
counter-current dispersion 逆向弥散
counter-current distribution 反流分布法;逆流分布
counter-current distribution apparatus 反流分布仪
counter-current distribution method 逆流分配法
counter-current drier 对流式干燥机;逆流式烘干机
counter-current drying 逆流干燥
counter-current drying system 逆流(式)干燥系统
counter-current efficiency 逆流效率
counter-current electrophoresis 对流电泳
counter-current elutriation 逆流澄析;逆流淘洗
counter-current extraction 逆流萃取
counter-current filtration washing 反流过滤洗涤
counter-current fired kiln 逆流窑
counter-current flow 逆向流动;逆流换热
counter-current fraction 逆流分级
counter-current furnace 逆流炉
counter-current gasification 逆流气化
counter-current heat exchange 逆流换热
counter-current heat exchanger 逆流热交换器
counter-current immunoelectrophoresis 对流免疫电泳
counter-current induced draft cooling tower 抽风逆流式冷却塔
counter-current ion exchange 逆流离子交换
counter-current ionphoresis 逆流离子电泳法
counter-current jet condenser 逆流淋凝器
counter-current leaching 逆流浸提
counter-current machine 逆流式浮选机
counter-current method 逆流方法
counter-current mixer 逆流式搅拌机;逆流混合机;逆流式拌和机;对流式拌和机
counter-current mixing plant 反流搅拌设备
counter-current operation 逆流操作
counter-current operation of multistage system 多级系统逆流操作
counter-current packed-bed filter 逆流填充层吸尘器;逆流填料层过滤器
counter-current pipe exchanger 反流管交换器
counter-current process 逆流过程
counter-current recycling 逆流循环
counter-current regeneration 逆流再生;对流再生
counter-current revoling-pan mixer 反流旋转拌和机
counter-current rotary dryer 逆流式转筒烘干机
counter-current scrubber 逆流式洗涤器
counter-current sizer 逆流分粒机
counter-current spiral heat exchanger 逆流螺旋热交换器
counter-current spray dryer 逆流喷雾干燥器
counter-current suspension preheater 逆流悬浮预热器
counter-current system 逆流系统
counter-current tower 逆流塔
counter-current treatment 对选矿法
counter-current tube and shell heat exchanger 逆流管壳式热交换器
counter-current ventilation 对流通风;反吹风滤袋清灰

counter-current vertical cooler 对流立式冷却机;逆流立式冷却器;逆流立式冷却机
counter-current washing 逆流洗涤
counter-current-wise 逆流地
counter curve 反(向)曲线
counter-curved axisymmetric(al) reducer 反向曲线旋转体变径段
counter-cyclic(al) action 反周期波动行动;反周期行为
counter-cyclic(al) tax 反向成本法;反危机税;反周期税
counter-cyclic(al) tendency 反周期倾向
countercylinder 副汽缸
counter dam 抗冲墙;抗蚀墙;护坝;副坝;前坝
counter dead time 计数器死时间;计数管死时间
counter decade 十进位定标器
counter-deed 反契据;反对证明
counter diagonal 副斜(拉)杆;交叉斜杆;交叉撑
counter dial 计数器刻度盘
counter-die 下模
counter-diffusion 逆扩散;传质异向;反扩散
counter dike 月堤;副堤
counter dopant 电子受体掺杂剂
counter-doping 反向掺杂
counter-down 脉冲分频器;分频器
counterdrain 河堤排水(沟);堤脚排水沟;副沟;背水坡脚排水;(坝渠岸堤的)背水面坡脚排水;漏水渠;截水渠;截水沟;副阴沟;对立面;辅助沟;背水坡脚排水(堤坝、渠岸背水面排除渗漏水用)
counterdraw 描图
counter drill 埋头钻
counter drive shaft 副传动轴
counteredge 固定刀刃;底刀刃
counter edging 边接边
counter efficiency 计数管效率
counter efflux 逆向冲采
counter electrode 极板;反电极
counter electromotive force 逆电动势
counter-electromotive-force inverter 反电动势换流器
counter-electromotive force relay 反电动势继电器
counter-electromotive force starter 反电动势起动器
counter electromotive force type regulator 反电动势式调整器
counter-electromotive regeneration 反电动势再生
counteretch 抗蚀性
counter evidence 反证
counterexample 反例
counter execution 互相履行债务
counterface 配合端面
counter fan system 辅助风机;二次风机
counterfeit 假冒;冒牌货;仿造;伪造品;伪造(的)
counterfeit articles 假货;冒牌商品
counterfeit products 仿造产品
counterfeit trademark 假冒商标;影射商标
counter field 反磁场;反向场
counterflange 假腿;反翼缘;反向凸缘;配对法兰;对接法兰
counterflange method 假翼缘法孔型设计;反翼缘法;反假翼缘法孔型设计
counter-flap hinge 盖板铰链(带中间连接片);背合铰链
counter flashing 帽盖泛水件;帽盖泛水;盖帽泛水;盖泛水;泛水盖帽;屋面金属防水条
counter flatter 矫正底模
counter flo(a)tation 反浮选
counterfloor 毛地板;粗地板(用于双层楼板);(楼板中的)垫板
counter-flow 逆流;对流(式);反向流(动);逆向车流;倒流
counter-flow combustor 逆流式燃烧室
counter-flow condensate 逆流冷凝水
counter-flow condenser 逆流凝汽器
counter-flow condition 回流条件
counter-flow cooling tower 对流式冷却塔;逆流式凉水塔;逆流式冷却塔
counter-flow cooling water 逆流冷却水
counter-flow cooling water system 逆流冷却水系统
counter-flow distribution 逆流分布
counter-flow draghead 对流耙头
counter-flow dryer 逆流式烘干机;逆流(式)干燥

器;对流式干燥机
counter-flow drilling 反向冲洗钻进
counter-flow drying 逆流干燥
counter-flow drying system 逆流式干燥系统;逆流式干燥系统
counter-flow engine 逆流蒸汽机
counter-flow furnace 逆流(式)炉
counter-flow heat exchanger 逆流式热交换器;对流热交换器
counter-flow jet condenser 逆流注水冷凝器
counter-flow method 反向冲洗方法
counter-flow mixer 逆流混料机;逆流式辊合机
counter-flow mixing 对流式拌和;逆流混合;逆流式拌和;逆流式搅拌
counter-flow of condensate 冷凝水逆流
counter-flow operation 逆流操作
counter-flow process 逆流过程
counter-flow rapid action mixer 逆流快速拌和机
counter-flow regeneration 逆流再生
counter-flow river 逆流河
counter-flow scavenging 逆流换气
counter-flow screen 反流筛
counter-flow scrubber 反流擦洗机
counter-flow shaft kiln 逆流式立窑
counter-flow solid bowl centrifuge 对流式无孔转鼓离心机
counter-flow stage 逆流级
counter-flow stream 逆流河
counter-flow stripping column 反向萃取塔
counter-flow thrust 反向推力
counter-flow tower 逆流塔
counter-flow tower drier 塔式对流干燥机
counter-flow unit 倒流装置
counter-flow washing 逆流洗涤
counter-flow water tube boiler 逆流水管锅炉
counter flush 反洗井
counterfoil (支票的)存根
counterfoil stub of a checkbook 票根
counter footrest 柜台踏脚板
counter for astronomic(al) observation 天文观测计数器
counter force 阻力;反力;对抗能力
counter for cylinder liner 汽缸衬垫扩孔
counter for petrol 加汽油泵的计量器
counterfort 支墩;围堰护桩;拉墙;拱柱;后扶垛;扶垛;扶壁式
counterfort abutment 扶壁式桥台;扶垛式桥台
counterfort dam 支墩坝;扶垛坝;扶壁式坝
counterfort drain 后扶垛排水沟
counterforted abutment 内支式桥墩;扶壁式桥台
counterforted lock wall 扶壁式闸墙
counterforted quay-wall 扶壁码头
counterforted retaining wall 扶垛式挡土墙;扶壁式挡土墙
counterforted type retaining 内支式挡土墙
counterforted wall 扶壁墙;扶壁结构
counterfort retaining wall 扶壁式挡土墙;后扶壁挡土墙;垛式挡土墙
counterfort type dam 支墩型坝
counterfort wall 后扶垛;扶壁式挡土墙;扶壁;扶垛墙;扶壁墙
counter gas 计数管填充气;计数管气体填料
counter gate 计数器门;平衡门
counter ga(u)ge 画榫线具;榫规
counter gear 分配轴齿轮;反转齿轮
counterglow 对照日;对日照
counter-grade siding 反坡岔线;避难线【铁】
counter guarantee 向保证人提供的担保;反担保函
counterguard 城墙堡垒;设障
counter hammer 对击锻锤
counter hand 计数器指针
counter head 平衡水头
counter heating non-smoke boiler 反烧式无烟茶炉
counter-inclined fault 反倾断层
counter-indemnity 背对背担保
counter indicate 计数器显示
counter-induction 反归纳
counter inhibit 计数(器)禁止(位)
counter initialization 计数器清除
counter-intuitive behavio(u)r 反直观行为
counter investment 相对投资
counter ion 反离子
counter-ion adsorption 反离子吸附
counter ion layer 反离子层

counter jib 后悬臂;平衡臂
counter jumper 店员
counter killer 计数器制动器
counter knob 计数器操纵柄
counter lamp 计数器信号灯
counterlath 副条
counter lathing (常规灰板条间的)嵌条;檩条;灰板木楔;(钉抹灰板条用的)横垫条;横垫条;板条子操作;双层灰条
counter leaf 反向叶片
counter letter 认赔书;返还保证书
counter letter of credit 相对信用证;对开信用证
counter life 计数管寿命
counter light 平顶照明;对面窗;逆光线;逆光
counter-light lens hood 物镜遮光罩
counter line 等高线
counter lock 平衡锁口;平衡导锁
counter locked die 平衡锁扣模;锁口模;双向配合模
counter loop 计数器回路
counter machine 计数器机器;反向计算机
counter-magnetic field 反(作用)磁场
counterman 店员
countermand 取消前令;取消订货;退订货
countermand of a cheque 撤回支票
countermand of an order 取消订单
countermand payment 止付
counter mark 戳记;副号;副标志
countermarketing 反销售
countermeasure 对策;防止措施;防范措施
countermeasure against earthquake disaster 防震对策
countermeasure design 对抗对策设计
countermeasure machine 运转计时机
countermeasurer 解答器
countermeasures for oil pollution 油污染防范措施
countermeasures sonar 对抗声呐
countermeasures set 电子对抗设备
counter mechanism 记数机构
counter memory 计数器存储
counter method of piloting 雷达岸形导航
counter mixing 逆流式拌和
counter model type 1 一型计数模型
countermodulation 反调制;负调制;解调
counter moment 均衡力矩;恢复力矩;反力矩;平衡力矩
counter monitor 计数器监控器;计数管监测器
counter motion 逆向运动;反向运动
counter-mounted cooking unit 柜式炊具设备;反向安装炊事装置
countermove 对抗手段;对策
countermure 附加墙;缀墙;拉墙;城墙副壁
counter murement 复墙内壁;副壁
counter note 伪钞
counter nozzle 反面喷嘴
counter nut 埋头螺帽;埋头螺母
counter offer 还盘;还价;复盘;返盘;反价;反要约
counter offer firm 还实盘
counter of nuclei 计核器
counteropening 对口切开
counter O reset 计数器清零
counter osmosis 反渗透作用
counter-osmosis device 反渗透器
counterpane 床罩
counter parking 对向停车
counterpart 相互匹配的一部分;补足物;配对(物);可替换部分;互换部件;交易对方;回头货币;相对订单(外汇订单);对应物;对应部分;对称物;副本;复本
counterpart fund 配套资金
counterpart personnel 对口人员
counterpart profile 相配内廓;配对齿廓
counterpart rack 铲形齿条
counterpart staff 对应人员
counterparty 对方;订约一方
counterpilaster 支墩;拱柱;扶垛;对向壁柱
counter plate 对面板
counter plateau 计数管坪
counterplot 预防措施;防止
counter plunger 反向柱塞
counter point 对照法;对点
counterpoise 砝码;平衡物;平衡砝码;平衡重;接地电线;地网;均重;衡重体;地网;平衡器;反力;平衡重量
counterpoise bridge 衡重吊桥;平衡式吊桥

counterpoise conductor 地下线
counterpoised conveyor belt 配重运输机带
counterpoised drawbridge 上开桥
counterpoised poke door 配重拨火门
counterpoise lift 橘棒;吊杆;平衡式提水工具
counterpoise spring 补偿弹簧
counterpoising 压坡
counterpoison 抗毒剂
counter potential 阻尼电位;反电位
counterpressure 均衡压力;负压强;负压力;平衡压力;反压(力);支力
counterpressure brake 均压闸;反压制动器;均压制动器
counterpressure casting 差压铸造
counterpressure cylinder 计数压力柱面;反压力圆筒
counterpressure die casting 反压压铸
counterpressure steam 背压式蒸汽
counterproductive 反生产
counterprofile 反侧面
counterpropagation 反向传播
counterpropagation beam 反向传输光束
counterpropagation laser pulse 反向传输激光脉冲
counterpropeller 反螺旋桨
counterproposal 反提案;反建议
counterproposition 反命题
counter pull 反拉力
counter pulley 计数轮;中间皮带轮
counter punch 冲孔机垫块;反向凸模
counter purchase 互购;产品回购;反向购买;反购买
counter-purchase credit exchange 相互购买信用交换所
counter race 对座圈
counter radiation 逆辐射;反辐射
counter rail 护轨;柜台栏杆;导轨
counter range 计数管区域
counter rapid action mixer 逆流式快速搅拌机
counter reaction 逆反应
counter readout 计数器读数
counterrecoil 复进
counterrecoil cylinder 复进筒
counterrecoil piston 复进活塞
counter recovery time 计数器恢复时间
counter-regional fault 反向区域断层
counter register 计数寄存器
counter regulation 反调节
counter-regulatory effect 逆调节效应
counter reservoir 平衡水库;对置(蓄)水库;反调节水库;反调节库容;平衡水池
counter response 计数器响应
counter revolution 反转
counter river 逆流河
counter-rotating 相反旋转;相对旋转
counter rotating airscrew 反向旋转螺旋桨
counter-rotating mass 反旋转质量
counter-rotating torque 恢复力矩
counter-rotating turbine 对转式双转子汽轮机
counter-rotating wave 反向旋转波
counterrotation 反转;反向旋转
counter-rotational departure 反自转发射
counter-rotational launch 反自转发射
counter-rotation condition 反向旋转条件
counter-rotatory illusion 反旋转错觉
counterrudder 整流舵
counter sample 照来样制成样品;对照样品;对等样品
counter scale 平衡重;平衡锤
counter scale with weight 案秤
counterscarp 城堡壕沟外岸
counterscarp wall 堡壕外护墙;城堡壕沟外护墙
counter screen 柜台矮帘;柜台矮围屏;柜台隔屏
countersea 逆行海流;逆浪
counter security 反担保
counter selection 反选择
counter septum 对隔壁
counter set 计数器置位
countershading 反荫蔽
countershaft 中间轴;逆转轴;对轴;对转轴;副轴;侧轴;平行轴;凡轮轴;天轴
countershaft bearing 中间轴承;副轴承;平衡轴承
countershaft bearing cap 副轴轴承盖
countershaft bearing cap gasket 副轴轴承盖垫密片
countershaft bearing cover 副轴轴承盖

countershaft bracket 副轴托架
countershaft clutch type 摩擦离合器副轴
countershaft cone 副轴宝塔轮
countershaft cone pulley 副轴宝塔轮
countershaft drive gear 副轴传动齿轮
countershaft fifth speed gear 中间轴五档齿轮
countershaft front or rear bearing 副轴前或后轴承
countershaft gear 副轴齿轮
countershaft gear bearing 副轴齿轮轴承
countershaft gear cluster 副轴齿轮组
countershaft gear cluster bearing 副轴齿轮组轴承
countershafting 副传动轴
countershaft intermediate gear 副轴中间齿轮
countershaft mechanism 副轴机构
countershaft mounting assy 中间轴支承总成
countershaft over drive gear 副轴超速齿轮
countershaft pulley 副轴滑轮
countershaft roller bearing 副轴滚柱轴承
countershaft sprocket 副轴链轮
countershaft unit 中间轴
countershear 反向剪力;反面剪切
counter shelf 柜台搁板
countersign 联名签署;会签;承认;副签;附署;附签;确认
counter signature 副签
counter signed bill of lading 副署提单
countersill 附加槛木
countersink 锚杆钻孔机(钻凿顶板锚杆孔用);尖底锪头孔;埋头孔;锥头孔;凹穴;尖底锪钻
countersink bit 埋头台(头)
countersink drill 埋头钻(锪钻);锥口钻;尖底锪钻
countersinker 埋头钻(头)
counter-sinkhead rivet 藏头铆钉
countersinking 钻埋头孔;锥形扩孔
countersinking bit 埋头钻;平头钻(头)
countersinking drill 埋头钻
countersinking drill press 尖底扩孔钻床
countersinking fixture 钻埋头孔夹具
countersinking of rivets 埋头铆钉
countersinking of the rivet holes 钻铆钉孔锥口
countersinking reamer 锪钻工具;埋头孔铰刀;埋头钻
counter sink screw pin shackle 平头螺栓卸扣
counter skylight 平顶照明
counter-sliding pile 抗滑桩
counter slope 挑出斜坡;反向坡度
counter solvent 反萃溶剂
counter spectrometer 计数(器)能谱计;计数器光谱计
counter spring 减振弹簧
counterstain 复染色
counter stay 船尾肋骨
countersteam 回汽
counter stern 悬伸艉;悬伸船尾
counter stop 触止
counter stream 逆流河
counterstreaming plasma 逆流等离子体
counter stress 相反应力;对应力
counterstrut 桁架抗压杆
countersuggestion 反暗示
counter suit 反诉
counter sun 反日
countersunk 埋头的;埋头孔;埋头;锥形孔;锥孔;沉孔
countersunk and chipped rivet 埋头铆钉
countersunk bolt 埋头螺钉;埋头螺栓;沉头螺栓
countersunk chequered head nail 埋头带槽钉
countersunk collar 埋头轴环
countersunk finishing washer 埋头垫圈
countersunk flat head bit 埋头平头钻头
countersunk flat head rivet 埋头头平顶铆钉
countersunk head 埋头钉头;埋头;埋头钉;沉头;暗钉头
countersunk head bolt 埋头螺栓
countersunk headed bolt 埋头螺栓;沉头螺栓
countersunk headed wood screw 埋头木螺钉
countersunk head machine screw 埋头机螺钉
countersunk head rivet 埋头铆钉;沉头铆钉
countersunk head screw 埋头螺钉
countersunk hole 埋头孔
countersunk not chipped rivet 半埋头铆钉;半沉头铆钉
countersunk nut 埋头螺母;埋头螺帽;半埋头螺母
countersunk oval head rivet 埋头圆顶铆钉

countersunk oval head screw 埋头圆顶螺钉
countersunk reamer 埋头(孔)铰刀
countersunk rivet 埋头铆钉;沉头铆钉
countersunk screw 埋头螺钉;埋头螺丝;沉头螺丝;沉头螺钉
countersunk seat 沉头座
countersunk shackle 埋头螺丝卸扣
countersunk spigot 埋头轴颈;锥形轴颈
countersunk square neck bolt 埋头方颈螺栓
countersunk washer 埋头垫圈
counter superelevation 反超高
counter switch 计数(器)开关;计时开关
counter tally 筹码表
counter tariff 对抗关税
counter telescope 计数(器)望远镜
counter-test 反向试验
counter thrust 反推力
counterthrust of soil 被动土压力
counter ticker 计数票
counter tide 逆潮(流);反潮
counter-tie 副系杆
counter timber 船尾肋骨
counter timer 时间间隔计(数)测(量)器;时间隔计测器
counter top 柜台(顶)面;柜台顶面;厨房台上
counter top basin 台式洗面器
countertop faucet-attached unit 厨房台上龙头式装置
counter top lavatory 嵌入式洗面器
countertop model 厨台型
countertop pour-though unit 厨房台上倒水式装置
countertorque 均衡扭矩;反力矩;反(抗)转矩
countertorque braking 反接线制动
countertrade 对等贸易;反向贸易;补偿贸易;往返贸易;反季(节)风
counter train 计量序列;指针回转齿轮机构
countertransference 逆转移
counter tube 计数管;计数放电管;电子注开关管
counter-twilight 对日照
countertwist 反捻
counter-type adder 累加型加法器
counter-type electronic timer 计数式电子计时器
counter-type frequency meter 数字频率计;计数式频率计
counter units 计数器(部件)
counter updating 计数器更新
counterurbanization 逆城市化
countervailing duty 反补贴税;反补税;反补偿税;平衡税
countervailing levy 反补贴税
countervane 导向(叶)片
counter-vault 反拱;倒拱;贴附穹顶
counter vein 交错(矿)脉
counter voltage 反电压
counter vortex 逆涡
counter wall 贴邻边墙
counter wedging 对拔楔法;双向楔牢;双向木楔法
counterweigh 用配重平衡;抵消;配重平衡
counterweight 平衡式吊臂;平衡重架子;平衡重安全装置;平衡筒;平衡锤;平衡重(量);配重体;平衡锤;平衡重(量);配重
counterweight arm 平衡锤臂
counterweight arm of bascule bridge 仰开桥衡重臂
counterweight balance 重锤式平衡;荷载平衡
counterweight block 平衡重滑块
counterweight block wall 衡重式方块码头
counterweight brake 配重制动器
counterweight cable 均衡缆索
counterweight catch 平衡重锁
counterweight cord 平衡重绳索
counterweighted balanced type spindle carrier 重锤平衡式接轴托架
counterweighted cage 平衡重笼
counterweighted window 垂直推拉窗;对重窗
counterweight embankment 反压台
counterweight fill 反压台;压重填土;平衡重填土;平衡压重填土
counterweight for tightening line 张力补偿架空线
counterweight for tightening overhead contact line 架空线张紧平衡块
counterweight guide 平衡重导轨
counterweight hanger 重锤平衡吊架
counterweight hoist 平衡重式水闸启闭机
counterweight lift 衡重式升船机;平衡重提升器

平衡重式升降机
counterweight line 衡重索;平衡重绳索
counterweight method 锤回单杆吊货法
counterweight pit (仰开桥的)衡重竖井
counterweight pulley 配重滑车
counterweight safety 电梯安全装置
counterweight scraper 衡重式铲运机
counterweight set 平衡重组
counterweight shaft (仰开桥的)衡重竖井
counterweight ship lift 平衡重式升船机
counterweight shutter 有平衡重的百叶窗
counterweight side 平衡重边
counterweight system 平衡重系统;平衡锤装卸方式;平衡锤系统
counterweight tower (仰开桥的)衡重桥墩台;(仰开桥的)衡重桥墩
counterweight weight 平衡锤
counterweight winder 平衡锤提升机
counter wheel 计数轮
counterwind 逆风;反信风
counter window 计数(管)窗
counter wood(en) ground 木地板木键
countess 屋顶石板瓦(20英寸×10英寸)(1英寸=0.0254米)
countess slate 标准石板瓦;屋顶石板瓦
count extend 计数扩展
count impulse 计数脉冲
count in 归入;包括;算入
count in different channels 分道计数
counting 计算;计数
counting apparatus 计数器
counting apparatus for bacteria 细菌计数器
counting assay 计数分析(法)
counting by naked eye 目视计数法
counting cell 计数池
counting chamber 计数室;计数池
counting channel 计数通路
counting clocktime 时钟计数时间
counting code 计数电码
counting control 计数控制
counting device 计算器;计数装置;计数器
counting dial 计数盘
counting diamond 金刚石计数器;计数金刚石
counting disk 计数盘
counting distribution 计数分布
counting down 脉冲频率分除法
counting(down)circuit 计算电路;计数电路;分频器
counting efficiency 计数效率
counting equipment 计数设备
counting error 计算误差
counting forward 顺向计算;顺向计数
counting frame 算盘
counting fringe 计数条纹
counting geometry 计数几何条件
counting glass 织物分析镜
counting grid 计数栅
counting head 计数管头
counting-house 会计室
counting information system 汇计信息系统
counting in reverse 逆向计数;反向计数
counting instrument 计数器;计量器
counting interval 计数间隔
counting loop 计数循环
counting loss 漏计数;计算失误;计数损失
counting machine 计算器;计算机;计数机
counting mechanism 计算装置;计数原理;计数机构
counting mechanism of a meter 电度表的计数器
counting method 计数法
counting of bacteria 细菌计数
counting period 计算间隔
counting-plate 计数盘
counting principle 计数原理;读数原理
counting process 计算过程;计数过程;计数法
counting property 计数特性
counting rate 计数率
counting-rate-difference feedback 速度差反馈
counting-rate measuring circuit 计数率测量电路
counting-rate meter 计数率计;计算速度测定器;脉冲频率计
counting rate of every channel 各道平均计数率
counting-rate versus voltage characteristic 计数率—电压特性
counting ratevoltage characteristic 坪特性曲线
counting region 计数区域;工作范围

counting register 计数器
counting relay 计数继电器
counting relay group 计数继电器群
counting ring 环式计算装置
counting scale 计数盘
counting slide 计数玻片
counting sort 计数分类
counting stage 计数级
counting statistics 计数统计
counting subroutine 计算子程序
counting the frequencies 频数次数计算
counting trigger 计数式触发器
counting tube 计数管；电子注开关管
counting type analog-to-digital converter 计算式模拟—数字变换器
counting unit 计数装置；计数单元
counting vial 计数杯
count-input pin 计数输入插头
count interrupt zero 零计数中断
count module N 按模 N 计数
count of yarn 纱线支数
count pulse 计数脉冲
count quadrate 计数样方
count rate 计数速率
count rate meter 计算速度测定器
count recording circuit 读数记录电路
countries having centrally planned economics 中央计划经济国家
Countries under the Agreement 协定国家
country bank 乡村银行
country builder 乡村建设者
country check 乡村支票
country church 乡村教堂
country club 乡村俱乐部；郊区俱乐部
country code 国名代码；国名代号；国家代码；国家代号
country code indicator 国家代码标志码
country-cut 实尺锯切的
country damage 内陆损害（海上货物保险附加险）；乡间损失；装船前货损
country foot path 乡村步行小道
country house 农村住宅；乡间别墅
country house'castle'乡村城堡住宅
country kiln 山区窑
country level 乡村水平
country mansion 宅邸；乡村大夏
country-of-birth basis 出生国
country of consignment 交付国
country of consumption 消费国
country of departure 起运国；启运国
country of destination 目的国；目的地国
country of dispatch 发货国
country of emigration 迁出国
country of exortation 输出国
country of immigration 迁入国
country of origin 来源国；原产国；原产地；起运国；启运国；生产国标志
country of payment 付款国
country of purchase 收购国
country of registration 船籍国；船舶登记国
country of residence 侨居国
country of sale 销售国
country of shipment 启运国
country park 郊野公园；乡村公园
country planning 农村规划
country quotas 国家配额；国别配额
country regional planning 县城规划
country road 乡间小路；乡间土路；乡间道路；乡村路；乡村道路
country rock 基岩；母岩；原岩；岩帮；主岩；废石；围岩
country rock alteration 围岩蚀变
country rock of orebody 矿体围岩
country's balance of trade 国家贸易差额
country seat 别墅；乡间住宅
country side 郊区；农村；乡村
countryside commission 乡村委员会
country tar 松焦油
country villa 乡间别墅
country-wide survey 全国范围的调查
count sampling 计数法采样
counts corrected for coincidence 符合校正计数
counts per-minute 每分钟记录；每分钟计数
counts per second 每秒钟计数；每秒计数

count table 计数表
count value 计数值
county boundary 县界
county diagram 州县区划示意图
county highway 县道；地方道路
county line 县界
county manager 县行政官
county map 县图
county master plan 县镇总体规划
county planning 县镇规划
county regional water quality control authority 县给水质量管理局
county-region planning 县城规划
county road 乡道；县镇地方道路；县公路；县级公路；县路（县道）；县道；地方道路
county route 县镇地方道路
county seat 县政府所在地；县城
county town 县城
county water and sewer authority 县给排水管理局
county where consigned 货物输出国；货物启运地
coupe 伐区
coupe-cabriolet 活顶车身（可由轿式改为敞式者）
couper's cabin at hump crest 驼峰连接员室
couplant 耦合介质；耦合剂；超声波法（一种非破损的检验方法）
couple 力偶；联结；耦合；偶联；偶合；一对；电偶；成对；匹配
couple axle 联动轴；连动轴
couple bar 联星
couple-close（用拉杆在脚部系住的）闭合对椽；成对椽子；紧耦合
couple-close roof 简易人字屋顶；闭合式对偶屋顶；三角架屋顶
couple computer 配合计算机
couple corrosion 电化腐蚀；双金属侵蚀
coupled 联结式；耦合的；成对的
coupled action 耦合作用
coupled amplitude 耦合振幅
coupled amplitude approximation 耦合振幅近似
coupled antenna 耦合天线
coupled apparatus 耦合器件
coupled arm 连接臂
coupled axle 成对轴
coupled axle-pin 联轴销；连轴销
coupled batter pile 叉桩
coupled beam 成对组合梁
coupled behavio(u)r 耦合作用
coupled bus 拖挂式大型客车
coupled camera 联挂摄影机
coupled cavity technique 耦合腔技术
coupled circuit 耦合电路
coupled colonnettes 成对的小柱
coupled column pier 双柱式桥墩
coupled columns 双(肢)柱；并柱；双柱；联用柱
coupled constant 耦合常数
coupled control 协调的调节
coupled electron 偶合电子
coupled engine 联轴式机车；联动发动机；成对发动机
coupled equation 耦合器；耦合方程
coupled equations 方程组
coupled exposure meter 联动曝光表
coupled factor 耦合系数
coupled field vector 耦合场矢量
coupled frame shear wall structure 框架剪力墙耦联结构
coupled harmonic oscillator 耦合谐振子系
coupled knee braces 夹板斜撑
coupled load 耦合负载
coupled matrix 耦合矩阵
coupled mode 耦合振荡模；耦合方式
coupled model 耦合模型
coupled moment 偶力矩
coupled motion 耦合运动
coupled ocean-atmosphere response experiment 海洋大气耦合响应实验
coupled optic(al) resonators 耦合光谐振腔
coupled oscillation hypothesis 成对振啸假说
coupled oscillator 耦合振子；耦合振荡器
coupled pair 耦合对
coupled pendulum 耦合摆
coupled pilasters 成对壁柱；双壁柱
coupled pin 连接器销；挂销

coupled planting 对植
coupled points 联动道岔
coupled pole 复合杆
coupled positions 复式交换机
coupled power 耦合功率
coupled process 耦合工艺
coupled rangefinder 联动测距器
coupled reaction 连接反应；耦合反应
coupled resonator 耦合谐振腔
coupled roof 双坡屋顶；人字形密椽屋顶；椽架屋顶
coupled running 重联运行
coupled shear wall 双肢剪力墙；耦合抗剪墙；并联剪力墙
coupled standing mo(u)ld 成对立模(法)
coupled state 耦合态
coupled structure 耦合结构
coupled substitution 耦合置换
coupled switch 联动开关
coupled system 联系系统；连系系统；耦联体系；耦合装置
coupled transition 耦合跃迁
coupled truck 拖拉机载重车；拖挂式载重(汽)车
coupled twin switch 双联开关
coupled-type generator 直联型发电机
coupled valve 联结阀
coupled vibration 耦合振动；互耦振动；相关振动
coupled vibration system 耦合振动系统
coupled vibrator 耦合振动器
coupled wave 耦合波
coupled wave equation 耦合方程
coupled wave theory 耦合理论
coupled wheel 联动轮；双轮
coupled window 成对窗；双扇窗；对窗
coupled zone 耦合区
couple mode 振型耦联；振型耦合
couple moment 力偶矩
couple of forces 力偶
couple of stability 稳性力偶
coupler 联轴节；联结装置；联结器；联结件；联结剂；连接器；可变电感耦合器；耦联器；耦合元件；耦合器；偶合剂；成色剂；车钩；分接器；匹配器
coupler and buffer test bench 车钩及缓冲器试验台
coupler bay of a duplicate-busbar system 双母线系统联络开关
coupler body 靠背壳体；车钩体
coupler box 连接器箱
coupler carrier 车钩托座；车钩托板
coupler casing 联结器套
coupler centering device 车钩复原装置
coupler centerline offset 车钩中心线偏移
coupler compressing grade 压钩坡【铁】
coupler draft casting 联结器牵挽挡铁
coupler draft spring support 车钩牵引簧托
coupler draw bar 车钩拉杆
couple resonator 耦合腔
coupler flange 联结器法兰
coupler girder 车钩大梁
coupler guard arm 自动车钩
coupler head 车钩头；车钩舌
coupler horn 自动车钩抵角
coupler knuckle 车钩关节
coupler knuckle lock 钩锁铁
coupler knuckle pin 车钩舌销
coupler knuckle thrower （机车的）钩舌锁铁
coupler latch 连接闩
coupler lock 自动闭锁
coupler loss 耦合损耗
coupler offset 连接器偏距
couple roof 对椽屋顶；椽子屋顶
couple roof construction 联檩屋架；檩架
coupler pivot pin 车钩枢销
coupler plug 连接插头
coupler pocket 车钩匣
coupler roof construction 联椽屋架
coupler seat 车钩座
coupler shank 车钩柄
coupler socket 车厢连接座
coupler switch 车钩开关
coupler test device 车钩检测仪
coupler tester 车钩检测仪
coupler yoke 车钩轭
coupler yoke key 钩尾销
couples 力偶
couple star 联星

couplet 耦联体;(由两根钻杆组成的)双根
couplet on pillar 楹联
coupling 联轴器;联结;联轴节;连接的;联管节;连接头;连接附件;接箍;耦合;偶联;相引;管接头;车钩;软管接头;配合;外接头
coupling adapter 连接器
coupling adjustment 耦合调整
coupling agent 耦联剂;耦合剂;黏结剂;成色剂
coupling amine 耦合用胺
coupling aperture 耦合孔
coupling arrangement 连接机构;连接装置
coupling attenuation 耦合衰减
coupling axle 连接轴
coupling band 联轴套;连接箍
coupling bar 连接杆;拉杆;牵引杆
coupling beam 连系梁
coupling between fields 场的耦合
coupling bit 耦合位
coupling block 联结块
coupling bolt 系紧螺栓;连接螺栓;联轴节螺栓;联结(器)螺栓;联轴螺栓;接合螺栓
coupling box 联轴(节)箱;电缆连接套管;联轴器箱;连接器箱;接线箱;分线箱
coupling bracket 联结器托架
coupling buckle 车钩环舌
coupling buffer workshop 车钩缓冲器检修间
coupling cable 耦合电缆
coupling capacitance 耦合电容
coupling capacitor 耦合电容器
coupling capillary electrophoresis to mass spectroscopy 毛细管电泳—质谱法联用
coupling casing 接箍套管
coupling chain 连接链;车钩连环
coupling change 联结器换向机构
coupling changement 联轴器换向机构;联结器换向机构
coupling clutch 牙嵌离合器
coupling cock 联结旋塞;连通开关
coupling coefficient 耦合系数;啮合系数
coupling coil 耦合线圈
coupling collar 扣环;连接环;结合环
coupling condenser 耦合电容器;偶合电容器
coupling cone 联接锥
coupling constant 耦合常数
coupling control 耦合控制
coupling crank 曲柄连接;双曲柄
coupling device 连接装置;连接器;连接装置;耦合装置;旋紧器
coupling dimension 连接尺寸
coupling disk 联结圆板;联结板
coupling drive 直接传动(的)
coupling effect 耦合效应
coupling efficiency 耦合效率
coupling element 耦合元件
coupling end 连接端
coupling face 连接面
coupling factor 耦合因子;耦合因数;偶联因子;偶合因子
coupling filter 耦合滤波器
coupling flange 联轴器法兰;联结翼板;联节轴凸缘;连接翼板;连接法兰(盘)
coupling for fixed pipe installations 固定管道接头
coupling for land vehicle 车辆连接器
coupling for the transmission of rotary motion 传递回转运动联结节
coupling frame 连接架
coupling frequency 耦合频率
coupling gap 耦合隙缝
coupling gas chromatography to Fourier transform infrared spectroscopy 气相色谱法—傅里叶红外光谱联用
coupling gas chromatography to mass spectroscopy 气相色谱法—质谱法联用
coupling gasket 接头垫片
coupling gear 联结装置;连接装置;车钩
coupling guard 联轴器罩
coupling half 半联合;半联结;半个法兰接头
coupling handle 连接器手柄
coupling head 连接头
coupling hole 耦合孔;耦合空穴
coupling hook 连接钩;连接叉头
coupling hose 联轴节软管;连接软管
coupling hose pipe 联结软管
coupling hysteresis effect 牵引效应

coupling impedance 耦合阻抗
coupling inductance 耦合电感
coupling inductor 耦合线圈;耦合电感器
coupling internal pressure strength 接箍抗内压强度
coupling interval 联结间期;偶联间期
coupling iris 耦合膜片;耦合光栏
coupling jack 拉合千斤顶;车钩锁铁
coupling joint 联轴器连接;接头;联轴节
coupling key 集合按钮
coupling knuckle pin 车钩关节销
coupling layer 耦合层
coupling length 接箍长度
coupling lever 离合器操纵杆
coupling lifter 接合提升机构
coupling link 联结杆;连接链;联杆;车钩连接
coupling loop 耦合回路;耦合环线;耦合环
coupling loss 耦合损失
coupling machine 管接头扭装机
coupling manufacturer 联轴器制造厂
coupling mechanism 联结机构;耦合机构;偶联机理
coupling media 耦合介质;耦合剂
coupling meter 耦合度测试器
coupling mirror 耦合镜
coupling modulation 耦合调制
coupling network 耦合网络;耦合回路
coupling nipple 车钩突榫;车钩突筍;软管用接头
coupling nominal diameter 接箍名义直径
coupling nut 联轴节螺母;连接螺母
coupling of circuits 电路耦合
coupling of definite amount of filling 定充量耦合器
coupling of locomotive at head of train 在列车头部挂钩
coupling of modes 振型耦联;振型耦合
coupling of motor vehicles 电动客车挂钩
coupling of pipes 管子连接件
coupling of shaft 联轴节
coupling of steel pipe 钢管接头
coupling of variable amount of filling 变充量耦合器
coupling of vehicles 车辆挂钩
coupling-out 耦合输出
coupling outside diameter 接箍外径
coupling oxidation flocculant 偶联氧化絮凝剂
coupling parameter 耦合参数
coupling paste 耦合剂;耦合膏
coupling pawl 联结器爪;联结卡子
coupling piece 联结器
coupling pin 连接销;连接枢;联结销;车钩销
coupling pipe 连接管
coupling plane 耦合面
coupling plank 楔合板
coupling plug 车钩插塞
coupling point 联结点
coupling polymerization 偶联聚合
coupling position 耦合位置
coupling power tight 机械紧固接头
coupling probe 耦合探针
coupling propeller shaft 传动轴接轴
coupling reactance 耦合电抗
coupling reaction 耦合反应;偶合反应
coupling resistance 耦合电阻
coupling resonance 耦合共振
coupling rigid 刚性联轴节
coupling ring 连接圈
coupling rod 联(结)杆
couplings 垂综
coupling scheme 耦合方案
coupling screw 螺旋联轴节;连接螺钉
coupling shaft 联结轴;连接轴
coupling shaft housing 接轴壳
coupling site 偶联部位
coupling sleeve 耦合套筒;联轴(节)套管;连接套管;套筒联轴节
coupling slot 耦合孔;耦合缝
coupling socket 球头联节
coupling solvent 耦合溶剂;共溶剂
coupling spectrum 耦合光谱
coupling speed 连接速度【铁】;车辆连挂速度
coupling spindle 联结(心)轴
coupling split 耦合分裂
coupling standard 接箍规格
coupling steel minimum limit strength 接箍钢材

最小极限强虞
coupling stress 耦合应力
coupling stub 耦合短截线
coupling tenon 接合榫
coupling term 耦合项
coupling thread joint strength 接箍丝扣连接强度
coupling thread root diameter 接箍扣根直径
coupling transformer 联络变压器;耦合变压器
coupling tube 耦合管
coupling type 接箍分类
coupling union sleeve 连接套管;套筒连接器
coupling unit 连接装置;耦合部分;耦合元件
coupling valve 联结阀
coupling variable 耦合变量
coupling weight 接箍重量
coupling with plain ends 光壁无螺纹接头
coupling with rubber pad 橡胶块联轴器
coupling with wrench flat 锁接箍
coupole 圆顶丘
coupon 利息单;金融证券;(测量腐蚀效应的)检验板;息票;采样管;附息票券;附单;试样块;试片;试棒
coupon advertising 附页广告
coupon bank debenture 附息票金融债券
coupon-based supply 凭票供应
coupon bond 息票(公司)债券;附息票债券
coupon origin-destination 起讫联票
coupon pack 附券商品
coupon payable 应付(债券)息票
coupon payments 息票支付
coupon rate 息票利率
coupon register 息票登记簿
coupon room 保险金库
coupons 取样片
coupon test 挂片试验
coupon ticket 联程票
coupon yield 息票收益率
Courbaril copal 柯巴里尔树脂
courier route 驿道
courier service 捷运快递服务
courier station 驿站
course 路线趋向;路程;矿层走向;科目;进程;航向;学科;过程;大横帆;方针;手续
course adjuster 航向调整器;航向调节器
course adjusting 航向调节
course again 航向复原
course and bearing indicator 航向—方位指示器
course and distance calculator 航向—距离计算器
course and distance indicator 航向—航程指示器
course and drift calculator 航向—偏移计算器
course and drift indicator 航向—偏移指示器;航向—偏航指示器;航路偏向指示器
course and speed 船位、航向和航速
course and speed calculator 航向—航速计算器
course-and-speed computer 航向—速度计算机;风速计算器
course and speed error 航向—航速误差
course angle 航向角
course back 飞回航向;返航航线
course beacon 指向标;定向无线电导航信号台
course bearing 航向方位;方位
course bond 丁砖层(砌合)
course brickwork 层砌砖工
course calculator 航路计算器
course changing quality 航向改变质量;改变航向性能
course changing rate 航向变换率
course computer 航向计算机
course control 航向调节
course converter 航向变换器;航路变换器
course correction 航向修正量;航向校正;方向校正量
course corrector 航向修正器
course cosine cam 航向余弦凸轮
coursed 成层的;层砌的;层铺的
coursed ashlar 成层琢石砌体;成层琢面;层砌琢石墙面;成层琢石;层砌条石;层砌方石
coursed block 层砌块体;层砌方块;方块层砌;片岩砌块
coursed block masonry 层砌块体圬工
coursed blockwork 层砌块圬工;混凝土预制块分层铺砌;层砌块体;层砌方块;层砌岩砌块;方块砌层
coursed brickwork 成层砌砖;成层砖砌体

coursed depth 层厚
course depth 墙体层高
course development system 课程开发系统
course deviate 航向偏移
course deviation indicator 航向偏离指示器;偏航指示器
coursed header work 不同尺寸的方石砌体
course director 航向指示器
coursed masonry 成层圬工;成层砌体;层砖砌体;层石砌体;分层(砌筑的)圬工
coursed pavement 铺装路面;层铺路面;成层铺面;成层路面
coursed paving 分层铺路;分层路面
coursed random 分层乱砌;成层乱砌
coursed random rubble 分层乱砌块石;成层乱砌蛮石圬工;分层毛石不规则砌体;分层乱石砌体
coursed random work 成层乱砌工程
coursed rockfill 成层填石;成层砌石;成层垒石;成层干砌块石;成层石堆砌;分层填石;分层抛石
coursed rubble 层砌毛石;成层毛石砌体;成层毛石;成层粗石圬工;成层粗石砌体
coursed rubble masonry 层砌毛石圬工;粗石垫底铺砌层;成层毛石工;分层毛石圬工;分层块石圬工
coursed rubble wall 分层毛石墙
coursed sand deposit 成层砂矿床
coursed sett paving 分层石块路面;成层小方石路面
coursed snecked rubble 成层杂乱毛石;成层乱石墙
coursed square rubble 层砌方块毛石;成层毛石;成层方块毛石(圬工);分层方块堆石
coursed stone pitching 粗石砌墙
coursed veneer 表面成层砌体
course error 航向误差
course feed 正常进给;正常给进
course granular andesite 粗粒安山岩
course heading servo 航向伺服系统
course indicating radio beacon 航线指示标
course indication 航向指示
course indicator 航向显示器;航向(偏离)指示器
course information 航向信息
course joint 行缝;成行缝;成层缝;层缝;底层接缝;(建筑物的)层间接缝
course joint mortar 层间接合砂浆
course-keeping ability 航向保持能力
course-keeping quality 航向稳定性
course keeping test 航向稳定性试验
course light 跑道指示灯;地面导航灯;航线灯标;导航灯
course line 航线【航空】;飞机航线
course linearity 航向的线性
course line computer 航线计算机
course ine deviation 航线偏差
course line deviation indicator 航向偏离指示器;航线偏差指示器
course line of great circle 大圆圈线;大圆航线
course line selecter 航线选择器
course made good 航迹向
course map tracer 航向图描绘仪
course mark 航向标记
course marker 航向标志器
course masonry 层砌琢石;层砌圬工
course modulation 航向调制
course of advance 计划航向;前进航向
course of an action 行为方案
course of blocks 砖的码砌序列;方块砌层
course of bonders 砌墙砖层;顶丁砖层
course of bonding headers 顶丁砖层
course of bondstones 系石层;束石层
course of bore 钻孔的进程
course of bricks 砖层
course of brick(work) 砖层
course of computation 计算过程
course of constant bearing 等角航线
course of currents 水流流向;水流流程
course of discharge 放电过程
course of exchange 兑换率;汇兑行情
course of exchange ratio 系数
course of fermentation 发酵过程
course of flight 航空线
course of floats 测流浮标;测流浮子(走向)
course offset 航向偏移
course of headers 满丁砖行;丁砖层
course of instruction 教学课程
course of manufacture 制造过程

course of outcrop 露头走向
course of reaction 反应过程
course of receiving 验收过程
course of river 河流线路;河道走向
course of stretcher 侧砖层;顺砖层
course of traverse 导线路线
course of working 加工过程;工作过程
course over the ground 航迹向;对地航向;地面上的航向
course plotter 航迹自绘仪
course position 导航区
course programmer 航线程序装置
course protractor 航向分角器
course pull 罗盘极化误差
course push 罗盘极化误差
course recommended 推荐航向
course recorder 航行记录器;航向记录器
course recording machine 航向记录机;航程记录仪;航程记录器
course recording paper 航向记录纸
course remote plotting 航向遥绘
course repeater 航向分罗经
course resolver 航向解算器
course scalloping 航向扇贝形现象
course selector 航向选择器
course sensitivity 航向灵敏度
course setter 定程器
course-setting compass 置定航向罗经;导航罗盘;航海罗盘(仪)
course-setting sight 航线指示器
course softening 航向灵敏度降低
course stability index 航向稳定性指数
course steadiness number 航向稳定特征数
course steering computer 航线操纵计算机
course to steer 操舵航向
course triangle 航行三角形
course-unstable ship 航向不稳定的船
courseware 课程软件
course width 航向角幅
course work 层砌砌筑
course writer 轨迹记录器
course yawing amplitude 航向偏荡角幅
coursing 顺砖层;侧砖层;分层铺设
coursing bubble 上升的气泡
coursing joint 共心缝(与拱同心但两层拱分开);分层砌砌(毛石砌墙);成行(砖)缝;成层缝;层缝(砌体)
court 裁判所
court architect 庭院建筑;宫廷建筑
court art 宫廷艺术
court attendant 法警
court building 庭院式建筑
court chapel 宫廷小教堂
court day 开庭日
court decision 法院裁决
court dock 封闭谷
court entrance 庭院入口
courtesy 特许转载
courtesy counter 服务台
courtesy desk 接待处
courtesy light 踏板照明灯
courtesy to brokers 经纪人酬金
court fee 诉讼费
court garden 宫廷花园
court grey 宫廷灰色
court height 天井高度
court house 内院式住宅;法院;县政府办公楼
Court Library at Vienna 维也纳宫廷图书馆
Court of Accounts of the Republic of France 法国审计法院
Court of Alberca 艾伯塔宫廷
court of appeal 上诉法院
court of bankruptcy 破产法庭
court of chancery 衡平法法院
court of claims 索赔法院
court of common pleas 民事上诉法庭
court of conciliation 调解法院
court of equity 衡平法法院
court of first instance 初审法庭
court of honour 荣誉法庭;纪念堂;纪念馆
court of justice 法庭;法院
court of last resort 最高上诉法院
Court of the Lions 英王(查理一世)宫廷
court of tribunal 民事审判庭

court painter 宫廷画家
court pavement 庭院道路
court plaster 薄橡皮膏布
court proceeding 法庭诉讼;法院审理程序
courtrail tiles (一种产自法国北部和比利时的)链锁瓦片
court reference procedure 法庭审查程序
courtroom 裁判室;审判室
courtship 求偶
court stable 宫廷马厩
court summons 法院传票
court theater 宫廷戏院
court ware 宫廷陶瓷器皿
courtyard 院子;庭院;天井
courtyard dwelling house 四合院住宅;四合院
courtyard economy 庭院经济
courtyard garden 院中花园
courtyard house 四合院(住宅);庭院式住宅;庭院花园
courtyard mosque 院中清真寺
courtyard of dwelling house 住宅庭园
courtyard wall 院墙
courtyard with building on the four sides 四合院
courtzilite 变沥青
cousinite 钼镁铀矿
cousins group 同等地位人员团体
coussinet 拱基石;基础垫层;爱奥尼亚冒头盖块
coustic(al) radiator 声辐射器
cousticity of white 白色灼剂
coustomary discount 折扣习惯
coutinuity of coating 涂层的连续性
Couwell method 科威尔法
covalence 共价
covalent 共价的
covalent binding theory 共价结合理论
covalent carbide 共价键碳化物
covalent complex 共价配合物
covalent compound 共价化合物
covalent crystal 共价晶体;共价(结)晶
covalent radius 共价半径
Covar 柯伐合金
covariability 共变性
covariance 协方差;协变性
covariance analysis 积差分析;协方差分析
covariance between inheritance and environment 遗传环境协方差
covariance for stratified sampling 分层抽样协方差
covariance function 积差函数;协方差函数;协变性函数
covariance law 协方差律
covariance matrix 协方差矩阵;协变矩阵;共差矩阵
covariance model 协方差模型
covariance parameter 协方差参数
covariance proportion 协方差比例
covariance response 协方差反应
covariance stationary process 协方差平稳过程
covarianee 共离散
covariant 协变(式的);协变量;共变式;随变
covariant calculation 协变计算
covariant component 协变分量
covariant curve 共变曲线
covariant equation 协变式方程
covariant functor 共变函子
covariant index 协变指标;共变指数
covariant of a curve 曲线的协变式
covariant tensor 协变张量
covariation 协变性
Cova's point 科瓦氏点
cove 小河口;小(海)湾;凹圆形;凹圆线;山凹
coveat 包退包换
cove base dividing strip 凹底分隔条
cove bit 内凹钻
cove box bit 岩芯箱车刀;型芯
cove bracketing 凹圆形(木)支架;隅折;凹圆托臂
cove ceiling 凹圆形顶棚;凹圆形平顶;拱形天花板;凹圆线顶棚;凹边圆顶棚
cove-charge 附加费
covective cooling tower 对流式冷却塔
cove culture 库湾养殖;湖湾养殖;河湾养殖;海湾养殖
coved and flat ceiling 凹圆线脚顶棚;凹圆线脚平顶;天弧拱
coved arch 凹角交叉拱;圆弧拱;大弧拱
coved base 凹圆形踢角板

coved bracket 凹圆饰托座;凹圆托臂
coved bracketing 凹圆托臂
coved ceiling 凹圆平顶;圆弧天花板;凹圆形顶棚;凹圆线顶棚;穹隆天花板
coved checker ceiling 凹圆线格子顶棚;四周凹圆的格子平顶
coved cornice 凹面檐口;凹面上楣
coved header brick 凹圆丁头砖
coved lattice ceiling 四周凹圆的细格平顶
coved lighting 穹顶泛光采光;泛光照明
coved mo(u)lding ceiling 凹圆形线脚顶棚
coved skirting 内凹踢脚板;凹圆踢脚线
coved vault 凹圆拱顶;凹圆穹顶
cove header 切头砖
cove header brick 凹圆形丁头砖
cove lighting 槽灯;凹圆暗槽灯;隐蔽照明
covelline 靛铜矿
covellite 靛铜矿
cove-mo(u)ld frame 有凹圆线脚的钢门樘
cove mo(u)lding 凹圆压条;凹圆线脚
covenant 订盟约;立契约;缔约;契约
covenantee 合同受约人;契约受益方
covenant not to compete 不参与竞争保证书(卖方作出在有限时间,范围内不与买方竞争的保证书)
covenant of right to convey 转让权协议
covenant of seisin 占有权契约
covenant of seizin 信托书
covenant of trust 地契
Covention Concerning Protection Against Hazard of Poison Arising from Benzene 防止苯毒害公约
coventry 径向梳刀
cove or coving (炉灶上的)顶口
cove plane 波动面
cover 机盖;面层;弥补;蒙皮;蒙;掩盖;遮盖物;遮盖;遮板;罩布;钢筋保护层;顶盖;抵偿;抵补;负担支付;阀盖;保温;保险项目;包皮;包括;封包;外壳;补进;抛补;涂层
coverage 绿地率;面层;有效距离;影响函数;作用距离;作用范围;覆盖面积;覆盖率;覆盖度;覆盖;服务区(域);保险总额;保额;拍摄范围;涂层;摄影面积
coverage angle 象场角
coverage area 到达范围;覆盖区(域);覆盖面积
coverage area of communication satellite 通信卫星覆盖区
coverage available 可控制的范围;覆盖效率;覆盖面
coverage contour 等场强曲线
coverage count 大范围运量观测
coverage count station 大范围运量观测站
coverage cycle 重复观测周期;覆盖周期
coverage diagram 可达范围图;制图区域示意图;覆盖区域图;反射特性曲线
coverage efficiency 覆盖效率
coverage electrode 涂剂焊条
coverage function 覆盖函数
coverage index 覆盖地区索引图
coverage interval 覆盖间隙
coverage monitoring range 覆盖监视范围
coverage of sample 样本范围
coverage of survey area 测区范围
coverage of taxation 课税范围
coverage power 遮盖力
coverage range 覆盖范围
coverage rate 覆盖速率;覆盖速度
coverage ratio 可建用地率;保付比率
coverage shed 仓库
coverage storage 仓库
coverage thickness 覆盖(层)厚度
cover aggregate 罩面骨料;盖面集料;盖面骨料;撒面集料;撒面骨料
cover aggregate for seal 封盖用面层骨料
coverall 连衣裤工作服
cover and cut method 盖挖法
cover annealing 罩式炉退火
cover area 覆盖面积
cover area of forest 森林覆被面积
cover bandage 盖的带束
cover bending machine 面起翘校正机;封面起翘校正机
cover bid 掩饰性投标;掩护性投标
cover blade 遮光叶片
cover block 隔板;盖砖;垫片;垫木;撑挡;护面块体;(混凝土的)盖板块体;保护层垫块

coverboard 盖板;犁壁覆土板
cover bolt 封盖螺栓
cover bush 泵盖衬套
cover capacity 覆盖能力
cover cap nut 螺帽母
cover carriage 揭盖机
cover carriage crane 揭盖起重机
cover chain welding 链节式药皮包丝焊接
cover charges 服务费(用);附加费
cover chip(ping)s 罩面石屑
cover clamp 盖夹;上盖压板
cover coat 釉面层
cover coat drywall compound 板墙覆盖涂料
cover coat enamel 面釉
cover core 盖芯
cover core print 盖芯芯头
cover cost 期货保证金
cover crane 地行揭盖吊车
cover crop 覆盖作物;遮盖作物
cover damp bags 外皮潮湿包(指货物)
cover deficit 弥补亏空
cover degree 覆盖度;植物覆盖度;盖度
cover-device 遮蔽物
cover die 凹压模;套模
cover dust 防尘罩
covered 遮盖着的;被遮盖的
covered ablation 间接消融;掩蔽消融;冰下消蚀;冰下消融;暗消融(冰下)
covered arc welding 手工电弧焊
covered area 建筑面积
covered barge 有盖舱口驳
covered bear 抛补空头
covered belt feeder 封闭式皮带喂料机
covered berth 有雨棚的泊位;遮蔽式泊位
covered burner 有罩燃烧器
covered by glaciers 冰川覆盖的
covered canal 加盖渠道;封闭式渠道
covered canopy 全遮式雨篷
covered car 棚车
covered ceiling 遮盖天花板;遮没的平顶;遮盖顶棚
covered coalfield 掩盖式煤田
covered conduit 封闭式水道;暗渠;暗管;暗沟
covered conduit flow 暗管流
covered conveyor system 包盖传送装置
covered distance 已走距离
covered ditch 加盖明沟
covered dollars 已抵补的美元
covered door 包盖的门;包皮门
covered drain(age) 地下排水;排水暗沟;暗排水沟
covered drainage ditch 暗排水沟
covered dry dock 有雨棚的干船坞
covered elastomeric yarn 包覆弹性纱
covered electrode 带药皮焊条;包剂焊条;包覆焊条;涂药焊条;涂料焊条;涂药焊条;色剂焊条
covered field 被覆盖地
covered filter 封闭式滤池
covered floor mo(u)ld 盖箱地面铸型
covered floor mo(u)lding 盖箱地面造型
covered gutter 暗沟
covered hopper car 漏斗棚车
covered in berth 室内船台
covered interest 提存利息
covered joint 暗接合;覆盖接合;复接合
covered karst 覆盖岩溶
covered knife switch 带盖闸刀开关
covered ladle 盖箱浇(注)包
covered land 被覆地
covered mall 有顶的走廊商场
covered mall building 商业廊亭;有顶的走廊商店建筑
covered mall center 商场;有顶购物中心
covered market(place) 有遮棚的市场;棚盖市场;有顶盖市场
covered metallic weldrod 涂焊药的金属焊条
covered multistage treatment reactor 暗式多级处理反应堆
covered open channel 无压暗渠;无压隧洞;无压隧道
covered parking 带顶停车场
covered passage 穿廊式通道
covered penstock 埋藏式压力水管
covered pipe 绝缘管;绝热管;保温管
covered plain 有覆盖平原;掩护平原;覆盖平原
covered plate 盖板

covered platform 带顶月台
covered position with a maturity gap 以满期缺口补进部位
covered pot 闭口坩埚
covered quay 有雨棚的顺岸码头
covered radiator 隐藏式散热器
covered railway station 有遮盖的火车站
covered reinforced concrete flume 封闭式钢筋混凝土渡槽
covered reserve 提存准备金;提存基金的准备
covered reservoir 有蓄水池;有蓄清水池;地下水池
covered rock 暗礁
covered rubber thread 包覆橡芯线
covered rubber yarn 包覆橡芯纱
covered shaft 有顶(的)通道
covered shating rink 有遮盖的溜冰场
covered shipbuilding berth 遮蔽式船台
covered shrine 有遮盖的神龛
covered sidewalk 骑楼式人行道
covered slip 室内船台
covered sludge drying bed 加罩污泥干燥床;加盖污泥干燥器
covered spectator's stand 有遮棚的观众席;屋顶瞭望台
covered storage 货栈
covered street-way 廊式街道;穿廊式街道
covered structure 覆盖结构
covered surface 制图区(域)
covered surge tank 封闭式调压井
covered swimming bath 有遮盖的游泳池
covered swimming pool 有遮盖的游泳池
covered terminal 遮蔽式码头
covered terrace 带顶平台
covered timber 粗加工木料;粗加工木材
covered timber bridge 走廊式木桥;有盖式的木桥;有屋面的木桥
covered truck 棚车
covered tube 绝缘管;绝热管
covered wagon 棚车
covered walk 廊式走道;林阴小径;蔓棚藤架阴道
covered walkway 穿廊式人行道
covered way 覆道;廊道;隐蔽的道路;暗道
covered wharf 有雨棚的顺岸码头
covered wire 绝缘线;被覆线;包线[电];皮线
covered with granulated cork 用粒状软木覆盖
covered with gravel 铺道砟
covered yarn 包线[电]
coverer 培土器;覆土器
cover excavation 盖挖法
cover fault 盖层断层
cover fillet 盖缝嵌条;压缝嵌条(顶棚、天花板及墙板接缝处);压缝条;盖条布;盖条
cover flange 封头法兰
cover flap 匣盖(盒式百叶窗)
cover flaps at the seat's front 座位前端活动盖板
cover flashing 披水;泛水;盖(帽)泛水;披水板
cover foil 覆盖金属箔
cover fold(ing) 盖层褶皱(作用)
cover for cable connection 电缆接线罩
cover for draft 票据抵补
cover for fusions 熔体覆盖物
cover for lines of credit 信贷限额保证
cover for seepage pit 渗流井盖
cover fracture 盖层断裂
cover furnace 罩式炉
cover gasket 盖垫片
cover glass 显微镜盖玻片;压片玻璃;盖玻片;盖玻璃;防护玻璃罩;玻璃护罩;玻璃盖片;保护玻璃;盖片
cover-glass forceps 盖片钳
cover-glass ga(u)ge 盖玻片测厚计
cover-glass thickness 盖玻片厚度
cover grate 遮盖炉算;遮盖格栅
cover guide 泵盖导槽
cover half 定模(压铸用)
cover handle 盖柄
cover head 顶封头
cover height 覆盖高度
coverhole 勘探孔;超前探水钻孔
covering 卖空补进;罩面;覆盖物;包覆材料;铺筑面层;(焊条的)涂药;补进;抛补;涂层
covering board 覆盖板;护舷贴面板;舷缘板
covering body 培土器
covering by seed paper 种子纸育草法

covering cap 遮盖罩
covering capacity 遮盖能力;覆盖能力
covering cloth 覆盖布
covering comain 覆盖域
covering contract 补进契约
covering curve 覆盖曲线
covering depth 保护层厚度;覆盖(层)厚度;覆土深度
covering dressed soil 覆盖容土
covering entry 详尽分录
covering fabric 覆盖织物
covering fillet 木条;盖缝木条
covering flux 覆盖熔剂;涂层;覆盖层(指熔剂)
covering foil 覆盖箔叶
covering fold 覆盖次数
covering fold map number 覆盖次数图幅号
covering for fire protection 防火保护层
covering grate 护栅
covering harrow 复土耙
covering in scale tiles 瓦片屋面
covering layer 覆盖层;保护层
covering length 覆盖长度
covering letter 详函;伴书;通知书
covering machine 包线机;包纱机
covering material 罩面材料;遮盖材料;覆面材料;涂料
covering note 承担保证书;认保单
covering of an electrode 焊条的药皮
covering of a slope 斜坡的覆盖物
covering of bars 埋置深度
covering of joint 压缝板;压缝条;接头覆盖
covering of load 路面
covering of piping 管道覆盖层;管顶覆盖
covering of reinforced concrete 钢筋混凝土保护层
covering of road 道路铺面
covering of roadway 道路铺面
covering of superstructure 上层建筑甲板覆盖层
covering-over 遮没;盖住;搭接
covering pipe 管子覆盖
covering plank 覆盖板
covering plate 盖板;罩面板
covering power 遮盖能力;覆盖能力;电镀能力;遮盖力;覆盖力;拍摄范围;视力作用范围
covering reagent 遮蔽剂
covering remittance 补进汇款
covering rock 覆盖岩
coverings for partition 间壁
covering slab 盖板
covering space 覆盖空间
covering spring 洞穴灯;暗泉
covering strake 遮盖列板
covering strata age 盖层时代
covering stratum 覆盖层;盖层
covering surface 覆盖(曲)面
covering tape 被覆带;包覆带;绝缘包布
covering tube 管子覆盖
covering up work 封盖工作
covering warrant 收款书
covering woven fabric 覆盖织物
covering yarn 包芯纱
cover insurance 投保保险
cover iron 护铁(木工刨刀)
cover layer 保护层
coverlet 被单;床罩
cover lid 罩布;床罩
coverllite 铜蓝
cover load pressure 覆盖岩层压力
cover mass 覆盖物
cover material 盖料;覆盖物料;覆盖料(保护地面用的木屑、砾石、纸等)
covermeter 钢筋覆盖层厚度测定仪(钢筋混凝土);面层测厚仪
cover missing 外皮失落(指货物)
cover mo(u)ld(ing) 压缝线脚;盖缝条;盖缝线脚;前模;压缝条贴边;盖模
covernote 暂保单;保险证(明)书;承保单
cover nut 螺母
cove ro coving 凹圆饰
cover of all risks 投保"一切险"
coverof a set of dependencies 从属集的覆盖
cover of cloud 云量;云层
cover of dust collector 集尘器上盖
cover of dust tight 防尘盖
cover of electrode 焊条的药皮

cover of joint 接头覆盖
cover of main bars 主筋保护层
cover of payment 抵补支付款
cover of piping 管顶覆盖
cover of pond 池盖
cover of unconformable strata 不整合盖层
cover of waterproof 防雨盖
cover over 遮没;覆盖
cover over reinforcing steel 钢架保护层
cover paper 书面纸;封皮纸;包皮纸;封面纸
cover pass 盖面焊道
cover piece 罩壳
cover plant 地被植物;覆盖植物;植被;草类植物
cover plate 压缝条;接头板;路面覆盖;面板;翼缘板;盖片;盖板;防护玻璃
cover plate for manhole 人孔盖板
cover plate joint 鱼尾板接合
cover plate splice 盖板拼接
cover plate to switch 开关盖板
cover pot 闭口坩埚
cover price 购入价格
cover printing 表面涂染;罩印
cover protectometer 钢筋保护层测定仪
cover ratio 绿地率
cover ring 套圈;盖圈
cover rock 盖层
cover run 盖面焊道
cover rusty 桶盖生锈
covers 抵偿金;暗礁群;篷(养护混凝土用)
cover sand 覆盖砂
covers and uncovers 干出礁
cover screenings 罩面石屑
cover screw 盖用螺钉
coversed sine 余矢
cover shadow 遮蔽
coversheet 层压板面板
cover sheet of plastics 塑料贴面板透明保护层
coversine 余矢
coversion objective 可换物镜
cover slab 面板;镶面板;压缝条;盖片;盖板
cover slab method 盖板法
cover slip 盖(玻)片
coverspun yarn 包芯纱
cover stone 盖石;盖面石(料);砌面石
cover strip 压缝条;盖(缝)条;防蚀镶片
covert 隐蔽密丛;涂布层
cover tap bolt 盖螺栓
coverted gradient of tunnel 隧道换算坡度
cover tender 掩饰性投标;掩护性投标
cover the cost 支付费用
cover the deficit 弥补亏损;弥补亏空
cover the hole 盖上井口
cover the loss 弥补损失
cover the position 补进
cover the shortage 弥补不足
cover thickness 覆盖厚度
covertible open side planer 活动支架单臂刨床
cover tile 盖瓦;保护层垫块;盖砖
cover time 覆盖时间
cover tire 外胎
covert mandatory plan 变相的指令性计划
covert to reinforcement 钢筋保护层
covert trade barrier 变相贸易壁垒
coverture 包覆
cover type 植被型;植被类型;覆盖类型
cover unit 封面印刷机
cover-up 罩面石料;隐蔽工事;撒布用石料;墙缝盖板条
cover wagon 有篷料车
cover weld 压缝焊;外焊
cover wing 遮光叶片
cover wire 外层钢丝
cover yoke 顶盖轭架
cove shirting 凹圆踢脚板
cove tile 地板与墙连接处圆角砖
covey 食品室;餐具室
covibration 共振现象
coving 墙顶藏灯凹槽;小海湾;圆阴角;凹圆线
covite 暗霞正长岩
cow 牛;通气帽;整流罩;炸弹型车身
Coward diagram 科沃德图(采矿)
CO-warning system 一氧化碳报警系统
cowcatcher 机车排障器;救助网;排障装置;(机车前的)排障装置

cow-day 放牧日
cowdi pine 淡黄棕色松木(新西兰);澳洲贝壳杉
cow glue 牛骨胶
cow hair 牛毛
co-wheel 齿轮
cow hide 牛皮
cow hitch 双合套;双合结
cowhorn bins 半圆形集料架
cowhorn picture 角雕画
cowhouse 牛舍;牛栅
cowing 罩
cowl 风帽;伞形帽;伞形风帽
cowl cover 通风斗罩;风斗罩
Cowles 考雷斯铜铝合金
Cowles furnace 考雷斯电炉
cowlesite 刃沸石
cowl flap 鱼鳞板;整流罩通风片
cowl head ventilator 喇叭式风斗;通风帽
cowl hood 缸头罩;汽缸头罩
cowling 机罩;发动机整流罩;通风罩;通风帽;整流罩;炸弹型车身
cowling flap 整流罩通风片
cowling flap control 整流罩通风片控制
cowling gill control 发动机鱼鳞片控制
cowling mount 整流罩架
cowling shutter 整流罩裙片
cowl lamp 边灯
cowl lamp bulb 罩灯泡
cowl lamp case 罩灯壳
cowllight 车头边灯
cowl panel shell 车头罩板壳
cowl ventilator 通风帽;带帽通风器;车罩通风器;车头(罩)通风器;风帽通风器
cowl ventilator adjuster link 车罩通风器调整杆
cowl ventilator brace 车头罩通风器拉条
cownose 端部成半圆形的砖
co-worker 合作者;同事
co-woven fabric 混纺织物;混织布
cow pen 室外牛栏
cowper stove 二通蓄热式热风炉;拷贝式热风炉
cowper with internal combustion 外燃室热风炉
cowri 淡黄褐色松木(新西兰)
Cowrie 南方贝壳杉;拷里树胶
cowrie pattern 贝纹
cowrie shells 玛瑙璃贝壳
cows 保护棱角用的排桩
cowshed 牛棚
cowshed waste 牛舍废水;牛场废水
cowshee 考斯风
cow's lip 牛皮块
cowsshee 卡活斯风
cow sucker (吊在钢丝绳端部用以加速下降的)重砣
cowtail 低级粗毛
cow-to-can milking room 直接装桶挤奶室
cow vetch 草藤
coxa vera 基前节
Cox chart 柯克斯图
coxcomb 梳形物;梳齿板;锯齿板;鸡冠状纹;梳状纹
Coxiella burnetii 贝纳特氏立克次体
coxswain's box 掌舵柜
cox(wain) 舵工
coyote blast(ing) 大量爆炸;大量爆破;峒室爆破;大爆破
coyote hole 药室;装药平巷;鼠洞式孔;山狗洞(爆破的一种)
coyoteite 柯水硫钠铁矿
coyote shot 碉室爆破弹药
coyote tunnel(l)ing method 装药平巷法;装药平硐法
coziness 舒适
CPM[Critical Path Method] 关键路线法
C-power supply 丙电源
CPT[Cone Penetration Test] 静力触探试验
CQC[Complete Quadratic Cube] 完全二次型方根法
crab 宽波段雷达干扰台;活动绞车;蟹抓式起重机;大芯骨架;起重小车;偏航;排头;偏流角
crab angle 航偏角;风流压差;漂流角;偏斜角;偏航角;偏流角
crabber 捕蟹船
crabber's eye knot 蟹眼结
crabbing 横漂;煮呢;整理工作
crabbing machine 煮呢机
crab boat 捕蟹船

crab bolt 地脚螺栓;锚栓;板座栓;板栓座;板螺栓
crab bucket 滑车戽斗;抓岩机抓斗;抓斗
crab capstan 起重绞盘
crab claw 蟹爪
crab claw open-type sampler 蟹爪敞口式取样器
crab crane 蟹爪式起重机;钳式吊(车)
crab derrick 移动式起重机;起重绞车架
crab dredge 扒网
crab drift 向下风横漂
crab frame 缆车架;桁车架
crab gathering arm loader 蟹爪式装载机
crab girder 主架大梁;起重机架
crab grass 马唐
crab grate 拦蟹栅
crab green glaze 蟹青釉
crab index 偏流角指针
crab-like 蟹足状
crab locomotive 装有绞车的电机车
crab process 壳型铸造法
crab rooting 露根
crab rope 绞车拉绳
crab runway 起重机轨道;吊车道
crab's foot crackle 蟹爪裂纹釉
crab-shape tank furnace 蟹形窑
crab to leeward 漂向下风
crab travel(l)ing winch 钳式移动绞车
crab traversing 起重机移动
crab trolley type bridge crane 载重小车式起重机
crab trolley type wall crane 绞车式壁装起重机
crab winch 起重绞车;卷扬机;绞杠;移动绞车;管扳子
crab with hook 带钩吊车
crabwood 沙果木;山楂木
crachin 蒙雨天气
crack 裂纹;裂隙;裂痕;裂缝;裂化;开裂;隙缝;砸开;缝隙;崩裂;敲碎;微开
crackability 可裂化性
crack and fissure 裂隙
crack appearance 裂纹外观;开裂的外观
crack a prospect 推销成功
crack arrest 止裂
crack-arrest capability 止裂能力
crack arrester 止裂器;止裂铆;阻裂材;防裂肘板
crack-arrest technique 防裂技术
crack arrest temperature 止裂温度;破裂停止温度
crack arrest test 止裂试验
crack a valve 略开阀(门)
crack both on body and glaze 坯釉均裂
crack branching 裂纹分支;裂纹分叉
crack bridging 裂纹密集;裂纹跨接;裂纹桥接
crack bridging ability 裂缝愈合能力
crack-cleaning machine 清缝机
crack closure 裂缝闭合
crack control 裂缝控制;防裂
crack control measure 防裂措施
crack control reinforcement 控制裂缝钢筋;防止裂缝钢筋;防裂配筋
crack count 裂缝统计
crack cut(vertical fissure cut) 龟裂掏槽
crack deflection toughening 裂纹偏转增韧;裂纹偏转韧化
crack detection 裂隙检查;裂纹探伤;裂纹检验;裂纹检查;探伤法;探伤
crack detector 裂纹探测仪;裂纹探测仪;裂缝探测器;探伤器
crack directional filtering 狭缝方向滤波
crack due to settlement 沉陷裂缝;沉陷引起的裂缝;沉降开裂
crack due to shearing force 剪切裂缝;剪力裂缝
cracked 裂化的;开裂的;断裂的
cracked ammonium 分解氨
cracked asphalt 热裂(地)沥青;裂化石油沥青;裂化(地)沥青
cracked brick 裂缝砖
cracked-carbon resistor 炭粉电阻器
cracked condition 开裂状态
cracked edge 开裂边缘
cracked fuel oil 裂化重油
cracked gas 裂解气;裂化气
crack ga(u)ge 测缝计
cracked gasoline 裂化汽油;分解汽油
crack edge 开裂的边缘;裂边;炸边
cracked grains 碎米
cracked-ice pattern 冰裂纹

cracked oil 裂解油;裂化油
cracked permanent deformation 龟裂永久变形
cracked pitch 裂化沥青
cracked position 很小的开度位置
cracked rail web 轨腰裂纹
cracked residue 裂解残渣;裂化残渣
cracked residuum 裂化残油
cracked section 裂缝断面;裂缝区段;开裂区段;开裂断面
cracked section check 开裂断面校核
cracked slab 裂缝水泥混凝土板
cracked snow surface 开裂雪面
cracked spirit 裂化汽油
cracked still 裂化炉
cracked tar 裂化焦油
cracked tension zone 开裂受拉区;张裂区
cracked valve 略开阀(门)
cracked zone 裂缝区;开裂区
crack end plate 裂纹终止板
cracker 裂化设备;裂化器(装置);裂化炉;碾碎辊;捣碎锤;闯入者;崩溃;破裂机;破产;碎裂器;砂岩中巨大的钙结核
cracker cell 裂解炉
cracker jack grinding wheel dresser 精密磨轮打磨机
crack extension 裂纹延伸;裂纹扩张;裂纹扩展;裂缝扩展
crack extension force 裂纹扩展力;裂纹开展力;裂缝扩展力
crack extension rate 断裂伸长率
crack factor 开裂系数
crack failure 裂纹破坏
crack fatigue propagation rate 裂纹疲劳扩展速率
crack filler 填缝(材)料
crack filling 裂缝充填(物);填缝
crack formation 裂纹形成;裂缝形成
crack formula 裂缝公式
crackfree 无裂纹的;无裂缝的
crack front 裂纹前缘
crackgas 裂化气
crack grain 裂纹表面
crack grouting 灌缝;裂缝灌注
crack growth 裂纹增长;裂纹延伸;裂纹长大;裂口生长;裂缝增长;裂纹扩大
crack growth behavior 裂纹增长特性
crack growth rate 裂纹扩展速度
crack growth resistance 抗裂纹生长性
crack hammer 地质锤
crack healing 裂纹愈合
crack impregnation test 裂纹浸渍试验
crack in body 坯裂
crack inducer strip 条带形诱缝设置
cracking 裂纹;裂开;裂解;裂化(作用);开裂;自发破裂;龟裂;断裂;穿透开裂;破裂;网状裂纹
cracking capacity 裂化设备生产量
cracking case 裂化反应器
cracking coil 裂化蛇(形)管
cracking control 裂缝控制;开裂控制
cracking-corrosion cycle 裂蚀—腐蚀循环
cracking detection 裂纹试验
cracking distillation 裂解蒸馏
cracking due to corrosion 腐蚀开裂
cracking due to shrinkage 收缩开裂
cracking energy 开裂能
cracking furnace 裂解炉;裂化炉
cracking gas 裂解气
cracking gas of petroleum 石油裂解气
cracking gasoline 裂化汽油
cracking in compression specimen 受压试件开裂
cracking index 开裂指数
cracking in paint 油漆开裂
cracking in plaster 灰浆开裂
cracking lacquer 裂纹漆
cracking limit state 开裂限制状态
cracking load 裂缝荷载;开裂荷载;破坏荷载
cracking loading 开裂荷载
cracking method 加热分解法
cracking moment 开裂弯矩;开裂力矩;形成裂缝弯矩
cracking moment capacity 开裂力矩能量
cracking of concrete 混凝土开裂
cracking-off 剥脱
cracking-off and edge melting machine 爆口烘口机

cracking-off machine 爆口机
cracking of molecules 分子破裂
cracking of petroleum 石油裂解;石油裂化
cracking of plaster 抹灰面开裂
cracking of walls 墙(壁)开裂;墙裂墙开;墙开裂
cracking on impact 冲击开裂
cracking pattern 裂纹图形;裂片图
cracking plant 裂化设备;裂化厂
cracking pressure 开启压力;启开压力
cracking process 裂化工艺;裂化法;干馏法;热裂法
cracking ratio 开裂率
cracking reaction 裂化反应
cracking resistance 抗裂纹性;耐开裂性
cracking resistant coefficient 抗裂系数
cracking ring 抗裂性试验环
cracking ring of concrete 混凝土抗裂性试验环
cracking risk 开裂危险
crackings 脆脂
cracking still 裂化蒸馏残油;裂化炉;裂化罐
cracking strain 开裂应变
cracking strength 开裂强度
cracking stress 开裂应力
cracking strip 防裂筋
cracking susceptibility 裂纹敏感性
cracking tendency 开裂趋势
cracking test 裂纹试验;抗裂试验;卷解试验;开裂试验;龟裂试验
cracking test for automatic welding 自动焊抗裂试验
cracking through surfacing 穿过面层开裂
cracking time 开裂时间
cracking tool 敲脱工具
cracking torque 开裂扭矩
cracking unit 裂化装置
crack initiation 裂纹开始;裂纹开裂;裂纹发生;起裂
crack initiation energy 裂纹初生的能量;裂纹发生的能量
crack injection 裂纹灌注
crack in tension 拉裂;受拉开裂
crack interval 裂缝间距;裂缝间隔
crack lacquer 裂纹漆
crackle 开片;产生裂纹;爆裂声;网状细裂纹;碎纹;碎裂花纹
crackle coating 碎纹涂料
crackled 裂纹的;炸裂花纹
crackled finish 裂纹饰面
crackled glass 有裂纹的玻璃;粗玻璃;碎纹饰玻璃
crackled glassware 碎纹式玻璃器皿
crackled lacquer 裂纹漆
crackle glaze 开片釉;纹片釉;碎纹釉
crack length 裂纹长度;裂缝长度;缝隙长度
crackle paint 裂纹涂料;裂纹漆
crackle test 裂纹检验;变压器油湿度检查
crackle varnish 裂纹清漆
crackling 破裂音;网状细裂纹;碎裂花纹
crackling arc 响弧
crack lip 断裂先端
crack loading 破裂荷载;裂缝荷载
crack measurement 裂缝观测
crack-measuring ga(u)ge 测裂仪(表)
crack meter 裂纹探测仪;裂缝探测仪;裂缝探测器;测裂缝仪;测缝计;测缝仪;超声波探伤器;超声波裂缝探测仪;探伤仪
crack monitoring 裂缝监测
crack observation 裂缝观测
crack-off 爆口
crack-off iron 敲脱刀
crack of weld 焊部裂纹
crack oil 裂化残油
crack on 满帆前进
crack opening 裂缝;裂缝宽(度);裂缝开度
crack opening displacement 裂纹张口位移;裂纹张开位移
crack opening displacement test 裂纹张开位移试验;裂纹张开位移测试
Crack Opening Distance[COD] 裂纹张开距离;化学需氧量
crack parameter 裂缝参数
crack pattern 裂纹形状;裂隙形式;裂缝形式;裂缝类型;裂缝(分布)图;裂缝分布情况
crack-per-pass 单程裂化量
crack pouring 灌缝
crack pouring paste 灌缝浆
crack-preventing 抗裂;防裂

crack-proof 防裂
crack propagation 裂隙开展；裂纹蔓延；裂纹扩展；裂纹开展；裂化增长；裂纹扩展
crack propagation energy 裂纹扩展能量
crack propagation rate 裂纹扩展率
crack reflection 裂缝反射
crack-reinforced concrete 防缝钢筋混凝土
crack resistance 抗裂性；抗裂度
crack resistance force 裂缝扩展阻力
crack resistance test of structural member 构件抗裂度检验
crack-resistant concrete 开裂混凝土
crack-ressitant 抗裂
crack restraint 裂缝抑制
crack resulting from load 荷载产生的裂缝
cracks accelerative extending and failure stage 裂纹加速扩展阶段
crack safety 抗裂度
crack sealer 封缝料
crack sealing 填缝；封缝；裂隙填封
crack sensitivity 裂缝敏感性；裂纹敏感性
crack sharpness 裂纹清晰度
cracks in brickwork 砖砌体裂缝
cracks increasing and extending stage 裂纹稳定扩展阶段
cracks owing to wall settlement 墙身沉陷裂缝
crack spacing 裂缝间距
crack speed 裂纹扩展速度
crack started technique 裂纹开裂技术
crack starter 落锤抗裂（试验）；开裂起始点
crack starter test 落锤抗断试验
crack stopper 防裂装置；防裂措施
crack stress 开裂应力；裂缝应力
crack survey 裂缝调查
crack test 抗裂试验
crack tester 裂缝试验机；抗裂试验机
crack tip 裂纹尖端；裂纹末端；裂缝尖端；缝尖
crack tip opening displacement 裂纹尖端张开位移
crack tip opening displacement method 裂纹顶端张开位移法
crack to beam depth ratio 裂缝与梁深比
crack treatment 裂缝处理
crack type 裂缝类型
crack vent 板边裂纹
crack wall 开裂墙
crack wander 裂缝移动
crack water 裂隙水；裂缝水
crackwax 防裂蜡
crack widening 裂缝加宽
crack width 裂缝宽（度）；缝隙宽度
crack-width curve 裂缝宽度曲线
cracky 裂缝多的；多裂缝的
crack zone 裂缝区
cradle for wagon 车辆摇枕
cradle 锯齿形圆刀；绞车架；活动支船架；活动船台；摇枕；摇床；支墩；支承垫块；吊架；刀具溜板；槽形支座；编排框架；保护网；板条框；鞍座；纹板架；托架；托板；淘汰机；吊篮
cradle and roller 运船小车；托架及滚轴
cradle angle （摇架的）摇角
cradle base 定子移动框架
cradle bedding 管道基层；管道垫块基座
cradle drifter 架式凿岩机；柱架式凿岩机
cradle dump 翻笼
cradle dynamometer 平衡式测功器
cradle feeder 盘式喂料器；盘式喂料机；盘式给料机；托架喂料机
cradle feet 支座
cradle frame 定子移动框架；炮架型车架
cradle guard 保护网
cradle head 转动关节
cradle housing （全齿切齿机的）摇台箱
cradle iron 摇振铁架
cradle knoll 根拔坑凸地
cradle-land 发源地
cradle machine 吊架机具
cradle of a pump 泵的底座
cradle plate （船下水滑道的）支架滑木；支架板；摇动板；炉座底板
cradle roll 托卷辊
cradle roof 支架式屋顶；支撑式屋顶；筒形屋顶
cradle scaffold 悬挂（式）脚手架；吊脚手架；挑出式脚手架；吊篮
cradle to grave management 由始至终的管理

cradle type stand 叉簧式支撑；叉簧式托架；叉簧式台架
cradle vault 筒形穹隆；筒形穹顶；筒形拱顶
cradling 支架支撑；支架；吊篮作业；吊顶木活；板条垫木；弧顶架；挡雪板
cradling member 支架杆件
cradling piece （壁炉烟囱与防火地板系梁间的）短格栅；支架杆（件）
cradling roof 架支屋顶；有支撑架的屋顶；用架支撑屋顶
craft 行业；专业；工种
craft brother 同行
craft clause 驳船条款
craft code 船舶代号
craft loss 船舶损失；船舶沉没
craft paper 牛皮纸；不透水纸；包装用牛皮纸
Craftplug 空心橡胶堵塞物（商品名，一种专利建筑材料）
craft railway 斜面升船道
craft risks 驳船险；驳运险
craft room 工艺室；手工艺室
craftsman 工匠；技工；手艺（工）人
craftsmanship 技巧；技能；工艺；工人技能；工匠技能；手艺
craft system 船舶运作组织
crafttone 透明材料
craft tug 港作拖轮
craft union 技艺工会；同业工会；行业工会
crag 峭壁；岩石碎片；陡崖；砂质泥灰岩
crag and tail 鼻山尾
craggan 陶制缸
cragged 崎岖的
craggy 多岩地区
Craigleith stone 克莱格雷斯石
Craigockhart basalt 克雷格玄武岩
Craig's formula 克赖格洪水强度公式
cram 填塞；夹钳
Cramer-Rao efficiency 克拉黑犬－拉奥有效性
Cramer-Rao inequality 克拉黑犬－拉奥不等式
Cramer's rule 克拉黑犬法则
Cramer's test 克拉默试验
Cramer-Tchebychev inequality 克拉黑犬－切贝谢夫不等式
cram frame 夹持器；虎钳；弓形夹；绑丝夹钳
crammed stripe 密经条纹
crammer 填塞物
cramming 堵塞管子；管子堵塞
cramoisle 绯红色布
cramoisy 绯红色布；深红色布
cramp 扣片；扣钉；夹线板；夹紧器；扎钉；扎锔（用以连接、固定相邻石块）；铁搭
cramp bar 夹紧杆；支杆
cramped bolt 夹紧的螺钉
cramped construction 紧凑结构
cramped flue 狭窄烟道
crampet 墙钩
cramp folding machine 折边机
cramp frame 弓形夹
cramping frame 虎钳
cramping table 制品固定板
cramp iron 弓形夹铁；扒锔子；扎钉；两爪钉；夹子；门形夹铁；蚂蟥钉；扎锔（用以连接、固定相邻石块）；铁钩；铁扒钉
cramp joint 钳接
cramp lapping （无齿隙的）压紧研齿
crampo(o)n 石棉瓦尾部的铆钉（防止风掀起）；金属爪；夹钳式起重机；起重吊钩；起重钩；吊钩夹
cramp ring 扣环
cramp run in lead 扒锔灌铅
cramshell shovel 抓斗挖土机
cranage 吊车费；起重机使用费；起重的使用（费）；起重机的能力
cranated 纯锯齿形的
cranberry 大果越橘
cranberry glass 带青紫色的透明红色玻璃；茶色玻璃
crance 箍钩；桅箍
crance iron 船首斜桅桅端铁箍圈
crandall 小锤；琢石锤；凿石锤；琢石锥
crandalled dressing 精雕细刻
crandalled stone 琢石
crandallite 纤磷钙铝石
crane 虹吸器；行车；吊车；起重机；升降架
crane accessories 起重机附件
craneage 吊车工时；起重机工时

crane aisle 吊车走道
crane and transporter load 起重运输机械荷载
crane attachment 升降设备
crane balance on slope 坡地上起重机的平衡
crane barge 浮吊；起重（机）船；起重机驳船；平底船（式）起重机
crane base 起重机支座；起重机底脚
crane bay 吊车间距；吊车跨
crane beam 行车梁；吊车梁；起重机梁
crane beam surveying 吊车梁测量
crane boom 吊杆；吊车起重杆；吊车臂；起重（机吊）架；起重（机）臂
crane bracket 吊车牛腿
crane brake 起重机制动闸；起重机制动器
crane bridge 起重桥；起重机桥（架）
crane bridge rail 桥式起重机轨道；桥式起重机钢轨
crane bucket 起重吊桶
crane buffer 吊车缓冲器
crane cab(in) 起重机驾驶室；吊车操作室；起重机驾驶室；起重机操纵室
crane cable 起重机索；吊索；承载索（起重机）
crane capability 起重能力
crane capacity 起重机（提升）能力；起重机承载能力
crane car 汽车起重机；起重（机）车
crane carriage 横行小车；桥工起重小车；起重小车
crane carrier 起重车；起重机载运车
crane center 起重机中心
crane chain 起重机链
crane column 吊车柱
crane control compartment 起重机驾驶室；起重机操纵室
crane controller 起重控制器
crane crab 绞车；起重机小车
crane crossing 交叉道岔【铁】
crane deck 起重行车桥面
crane/derrick barge 起重机/吊杆驳船
crane driver 吊车驾驶员；起重机驾驶员
crane driver's cabin 起重机手操纵室
crane electric lifting magnet 起重电气起重磁盘；电气磁力起重机
crane equipment 起重设备
crane-equipped 装有起重机的
crane erection 吊车安装；起重机安装
crane-excavator 挖掘－起重机；起重开挖两用机
crane factory 起重机厂
crane fall 吊车索；起重机吊索
crane falling plate 起重机夯板
crane fitter 起重机装配工
crane for placing stoplogs 叠梁闸门起门机；叠梁闸门启闭机
crane frame 起重机机架
crane gantry 起重机门架
crane girder 行车大梁；轨道梁（起重机）；吊车梁；吊车大梁；起重机行车大梁；起重机梁
crane grab 起重机抓斗
crane grease 起重机润滑剂
crane hinge 保险铰链；保险合叶
crane hoist 回旋式起重机；移动式起重机；吊车悬臂
crane hook 吊钩；起重（机吊）钩
crane hook block 起重机吊钩滑轮
crane hook coverage 起重机吊钩工作范围
crane hook return block 起重机吊钩回行滑轮
crane jib 起重机臂
crane ladder 吊车梯
crane ladle 吊车浇包；吊包
crane legs 起重机腿
crane lifting capacity 起重机起重能力
crane lifting service 起吊工作
crane lighter 有起重机的驳船
crane link 吊车吊架
crane load 吊车起重量；吊车载重；起重机起重量；起重机荷载
crane loading 用起重机吊装；起重机装运；起重机起吊；起重机吊装
crane locomotive 起重机车
crane machinery 起重机械
crane magnet 起重磁铁；电磁吸盘
crane man 起重机手；摘挂钩工人；吊车工；起重机指挥者；起重机驾驶员；天车工
crane man's house 起重机驾驶室；起重机操纵室
crane mast 起重机桅杆
crane motor 吊车电动机；起重（机）电动机
crane movement 起重机运转
crane navvy 挖土机；铲运机；上铲式单斗挖土机

crane oiler 起重机加油工
crane on cable railway 缆车起重机
crane-operated open clinker storage 敞开式吊车操作的熟料储库
crane-operated vibrator 起重振捣器;起重机操作的振捣器
crane operator 起重机手;吊车驾驶员;起重机驾驶员
crane operator's eye level 吊车工眼睛高度
crane output 起重能力;起重机生产率
crane pedestal 起重机底座
crane piece 起重机部件
crane pile driver 起重打桩机
crane pillar 起重机支柱;起重机转柱
crane plant 起重机上墩下水设备
crane platform 吊车平台;起重机台;起重机平台
crane portal 行车门座;门式行车;起重机门座
crane post 转臂式起重桅杆;起重机主柱;起重机支柱
crane pulley 起重机滑车
crane pump 起重机液压泵
crane radius 起重机(悬臂)伸距;起重机伸臂活动半径;起重机起吊半径;起重(机工作)半径
crane rail 吊车轨(道);起重机轨(道);起重机钢轨
crane rails used as neutral wire(grounding) 利用吊车轨道作接零线
crane rating 吊车载重量;吊车额定;额定起重量;起重机载重量;起重机额定重量;起重机定额
crane-rigged 起重机操纵的
crane rigger 吊车驾驶员;起重机指挥者
crane rope 起重机吊索;吊车钢丝绳;起重(机用)钢绳
crane runner 吊车驾驶员
crane running 起重机走动
crane runway 起重机滑道;起重机走道;起重机轨道;天车滑道;行车滑道
crane-runway girder 行车梁;起重机(行)车大梁
crane saver 料垛送进装置
crane scale 吊车衡
crane shaft 起重机主柱;起重机轴;起重机支柱
crane ship 浮吊;起重船;水上起重机
crane shovel 单斗挖掘机;挖土起重两用机
crane skip 吊车箕斗
crane slewing gear 起重机回旋装置
crane slope indicator 起重机坡度指示器
crane spreader 起重铺料机
crane stability 起重机稳定性
crane stake 起重机柱
crane stalk 起重机支柱;起动机柱
crane superstructure 起重机上部结构
crane switch 起重机开关
crane system 起重机装载方式
crane tamping plate 起重机夯板
crane tipping skip 起重机翻斗车
crane tower 起重吊塔
crane tpwer 塔式吊车
crane track 码头起重机轨道;起重机轨道
crane track height 起重机轨道标高
crane travel 起重机移距
crane travel(l)ing gear 起重机行走机构
crane travel(l)ing limiter 起重机行程限位器
crane travel mechanism 起重机运行机构
crane trestle 起重架;起重机栈桥
crane trolley 起重绞车;吊机滑车;起重小车;起重机行车;起重机小车;起重机滑车;起重吊车
crane trolley wire 行车接触导线;起重机接触导线
crane truck 卡车起重机;车,车装起重机;车载起重机;汽车起重机;汽车吊
crane-type loader 转臂式装载机;起重机式装载机
crane upper structure 起重机上部结构
crane wagon 起重车
crane way 码头起重机轨道;起重机轨道
crane way extension 带轨悬臂梁
crane weigher 吊钩称;吊车秤;起重秤
crane wheel 吊车车轮
crane width 起重机宽度
crane winch 吊机绞车;起重绞车;起重机绞车
crane with dipper attachment 铲门式起重机;铲斗式起重机
crane with double lever jib 四连杆(式)伸臂起重机
crane with fixed boom 固定吊杆起重机
crane with non-slewing jib 固定吊杆起重机
crane work 起重工作
Cranham (一种专利的)具有隔离性能的)空心陶砖
crank 卷片曲柄;匣钵垫板;拐肘;格子托架;曲拐

曲柄
crank-and-flywheel pump 曲轴飞轮泵
crank and pin 曲桩一销
crank and rocker mechanism 曲柄摇杆机构
crank angle 曲柄(转)角
crank arm 曲柄(臂);曲臂;起动手柄
crank arm operator (窗的)曲臂开关器
crank arrangement 曲柄布置
crank auger 手摇钻;曲柄螺旋钻
crank axis 曲柄轴线
crank axle 曲柄轴
crank bearing 曲柄轴承;曲柄装置
crank bearing casing 曲柄轴承罩
crank bearing liner 曲柄轴瓦
crank bearing oil seal 曲柄轴承护油圈
crank block 曲柄滑块
crank boring machine 曲柄镗床
crank boss 曲柄毂
crank brace 曲柄钻;钻孔器;手摇钻
crank brass 曲柄颈轴承铜衬
crank capstan 曲棒绞盘
crankcase 曲轴箱;曲柄箱
crankcase bearing 主轴承;曲轴轴承
crankcase bearing seat 曲轴轴承座
crankcase breather 曲轴箱通气管;曲柄箱通气管
crankcase center main section 主曲轴箱中部
crankcase compartment 曲轴箱分隔室
crankcase compression 曲轴箱压缩
crankcase conditioning oil 曲轴箱冲洗用油
crankcase cover plate 曲轴箱盖板
crankcase dilution 曲柄箱稀薄化
crankcase front main section 主曲轴箱前部
crankcase guard 曲轴箱护罩
crankcase inspection cover 曲轴箱检查盖
crankcase lube pressure 曲轴箱润滑油压
crankcase lube system 曲轴箱润滑系统
crankcase lubrication 曲轴箱润滑
crankcase oil foaming test 曲轴箱用油起沫试验
crankcase pressure 曲轴箱压力
crankcase scavenged engine 曲轴箱扫气二冲程发动机
crankcase scavenging 曲柄箱换气法
crankcase section 曲轴箱舱
crankcase stud 曲轴箱双头螺栓
crankcase sump 曲轴箱(油)槽
crank chain 曲柄链系
crank chamber 曲柄箱
crank cheek 曲柄颊板
crank circle 曲柄圆
crank closer 合绳机
crank connecting link 曲柄连杆
crank counter balance 曲柄平衡
crank disk including pin 带销曲柄盘
crank drive 曲柄传动
crank-driven 曲柄传动的
crank duster 手摇喷粉机
crankease rear main section 主曲轴箱后部
cranked 装有曲柄的;曲柄形的
cranked axle of plow 犁轮弯壁轴
cranked bearing 曲柄轴承
cranked dog 弯曲搭扣;曲柄形扒钉
cranked fish-plate 异形鱼尾钣;曲柄接合板
cranked flat iron 弯曲扁铁
cranked frame trailer 车架后部垂下的拖车
cranked gear 曲柄传动装置
cranked platform trailer 凹式平板挂车
cranked rod 曲柄杆
cranked screwdriver 弯成曲柄状的螺丝起子
cranked sheet 弯曲(覆盖)片;石棉水泥管瓦
cranked slab 曲拐板
cranked slab stair(case) 板式楼梯(无边梁及托梁的整体板楼梯)
cranked slot 曲槽
cranked string 曲柄状楼梯梁
cranked tire valve 轮胎气嘴
cranked trailer 车架后部垂下的拖车
cranked wrench 曲柄扳手
crank effort 曲柄回转力(矩)
cranker 操纵曲柄;手摇曲柄
crank-fastening angle 曲柄角度的调整定位
crank for electrode 电极移动曲柄
crank fulling mill 曲柄缩绒机
crank gear 曲柄传动装置
crank guidance 曲柄导槽

crank guide 曲柄月牙板;曲柄导向装置
crank-guide way 曲柄导向槽
crank hammer with fixed cylinder 定缸曲柄气锤
crank handle 摇把;曲柄执手;曲柄摇把;曲柄把手;手摇(曲)柄
crank handle embroidery 手摇曲柄绣花机
crank handle with offset lever 偏杆摇把
crank handle with straight lever 直杆摇把
crank head 曲柄头架
cranking 摇动;转动曲柄
cranking bar 摇把
cranking lever 曲柄启动杆;曲柄起动杆;起动曲柄
cranking loss 摇转损失
cranking motor 启动马达;启动电机
cranking speed 起动速度
cranking system 起动系统
cranking test 大电流试验
crank journal 曲柄轴颈
crank journal neck 曲柄轴颈
crankless engine 无曲柄发动机
crankless press 无曲柄压力机
crank lever 曲柄
crank lever escapement 曲柄式擒纵机构
crank link 曲柄连杆
crank mechanism 曲柄机构
crank metal 曲柄颈轴承合金
crank-motion 曲柄运动
crank-o-matic grinding machine 全自动曲柄磨床
crank-operated hammer 曲柄锤
crank-operated press 曲柄操纵的压床
crankpin 曲柄轴颈;曲柄销
crankpin angle 曲柄销配置角度
crankpin bearing 连杆轴承;曲柄销支承;曲柄连杆支承;曲柄销轴承
crankpin bush 曲柄销衬套
crankpin grease 曲柄销润脂
crankpin grinder 曲柄销磨床
crankpin lathe 曲柄销车床
crankpin metal 曲柄衬套合金
crankpin path 曲柄圆
crankpin returning tool 连杆轴颈修整工具;曲柄销修整工具
crankpin roller 曲柄销滚柱
crankpin seat 曲柄销座
crankpin spool 曲柄销猫头
crankpin step 曲柄销轴瓦
crankplaner 曲柄龙门刨床;曲轴刨床
crank planing machine 曲柄刨床
crank portion of shaft 轴的曲拐部分
crank press 曲柄式压机;曲柄压机;曲柄压床
crank pump 曲柄泵
crank radius 曲臂
crank rocker 曲柄摇杆
crankshaft 曲轴;曲柄轴
crankshaft alignment 曲轴轴对准器
crankshaft arm 曲轴臂
crankshaft axle 曲轴
crankshaft balancer 曲轴平衡器
crankshaft balance weight 曲轴平衡重
crankshaft bearing 曲轴(轴)承
crankshaft bearing metal 曲(柄)轴轴承合金
crankshaft breather 曲轴箱通气装置
crankshaft case 曲轴箱
crankshaft casing 曲轴箱
crankshaft chain 曲轴链系
crankshaft cheek 曲轴板
crankshaft collar 曲轴轴环
crankshaft connecting rod motor 曲轴连杆马达
crankshaft connecting rod system 曲轴连杆机构
crankshaft drive 曲柄驱动(的);曲轴传动装置;曲轴传动
crankshaft drive screen 曲轴驱动筛
crankshaft effort 曲轴回转力矩
crankshaft end thrust 曲轴端部推力
crankshaft flange 曲轴法兰盘
crankshaft flashbutt welding 曲轴电阻弧花压焊
crankshaft front bearing seal 曲轴前轴承油封
crankshaft front seal 曲轴轴前端密封(装置)
crankshaft ga(u)ge 曲轴轴颈测量器
crankshaft gear 曲轴传动装置;曲轴齿轮
crankshaft governor 曲轴调正器
crankshaft grinder 曲轴磨床
crankshaft harmonic balancer 曲轴谐和平衡器
crankshaft housing 曲轴箱

crankshaft impulse neutralizer 曲轴冲力平衡器
crankshaft journal 曲轴轴颈
crankshaft key 曲轴键
crankshaft lathe 曲轴车床
crankshaft milling machine 曲轴铣床
crankshaft oil hole 曲轴油孔
crankshaft oil seal retainer 曲柄轴护油圈
crankshaft oil slinger 曲轴甩油环
crankshaft oilway 曲轴油路
crankshaft pilot bearing 曲轴(引)导轴承
crankshaft pin 曲轴销;曲轴颈
crankshaft press 曲轴压力机;曲柄式压床
crankshaft pulley 曲轴皮带轮
crankshaft returning tool cutter 曲轴铰光用刀
crankshaft sprocket 曲轴链轮
crankshaft starting claw 曲轴起动爪
crankshaft starting jaw 曲轴起动爪
crankshaft straightening press 曲轴压直机
crankshaft throw 曲轴弯程
crankshaft thrust bearing 曲轴推力轴承
crankshaft timing gear 曲轴定时齿轮
crankshaft toothed wheel 曲轴齿轮
crankshaft twisting machine 曲轴扭转加工机床
crankshaft up 曲轴回转
crankshaft vibration damper 曲轴减震器
crankshaft with integral timing geat 带正齿轮的曲柄轴
crank shank lathe 曲柄车床
crank shaper 曲柄牛头刨床
crank shears 曲柄式剪切机
crank side shift 曲柄端换挡
crank slotter 曲柄插床
crank starter 曲柄起动器
crank strander 筐篮式捻股机
cranks working in opposite directions 双向运转的曲轴
crank tap 曲柄钻丝锥
crank throw 曲柄行程;曲柄弯程;曲柄半径
crank-type power unit 曲柄执行机构;曲柄动力机构
crank type press 曲柄式冲床
crank up 曲柄回转
crank vessel 易倾斜的船
crank web 曲柄连接板;曲柄颈;曲(柄)臂
crank wheel 曲柄轮
crank with drag link 带牵引杆的曲柄
cranky 歪倒的;扭折的缆索;弯曲的
crannog (爱尔兰在湖边建的)沼泽城堡
cranny (墙上的)裂缝
cranse 箍钩;桅箍
cranse iron 船首斜桅桅端铁箍圈
crantsite 水钒钠钙石
crapaudine 竖轴支承
crapaudine door 中轴转门;中心支轴旋开门;支轴转门
crape cotton 绉棉线;绉棉布
crape myrtle 紫薇
crape paper 绉纸
crape rubber 绉橡胶
crape wool 绉呢;绉毛线
craseology 液体混合论
crash 总崩溃(指市场);倒下;粗布;暴跌;碰撞;碰碎
crash accelerometer 应急加速度计
crash-a-head maneuver 全速倒车急转
crash ambulance 失事救援车(飞机)
crash-astern maneuver 全速正车急转全速正车
crash bar 推杆(太平门五金)
crash barrier 路中护栏;防撞护栏;碰撞护栏
crash beacon 带降落伞的紧急自动发报机
crash boat 救生艇;抢救快艇
crash change of speeds 变换速度时的撞击
crash collision 防撞衬垫
crash cost 应急成本;临界费用;应急费用;突击施工成本
crash course 速成班
crash cushion 防碰衬垫;防撞缓冲器
crash dive 紧急下潜
crashed critical path 紧缩关键路线
crashed time 特急时间;紧急期限;赶工时间
crashee 破产资本家
crash finish 粗布纹装饰
crash helmet 安全帽;防撞头盔;防护帽;保护头盔
crash-locator beacon 应急定位信标;失事定位信标(飞机)
crash pad 临时住宿处;防震垫

crash point 临界点
crash program(me) 速成计划;应急计划;应急措施;紧急计划;紧急工程项目;紧急方案;紧急措施
crash project 速成计划;紧急计划;紧急工程项目
crash proof 防撞击
crash reversal 紧急倒车
crash roll 防震垫
crash safety 撞击安全装置
crash schedule 应急日程表;突击进度安排
crash sensor 碰撞探测器
crash stop 突然停住;急速停车
crash test 破损试验;碰撞试验
crash time 临界时间;紧急限期;紧急期限;应急时间;特急时间
crash-towel 粗毛巾
crash truck 事故救护车;抢救车;失事救援车(飞机)
crash wag(g)on 失事救援车(飞机);抢救车
crashworthiness 抗撞性能;耐碰撞性能
crasnozem 红土;红壤
crassiclarite 厚壁孢子微亮煤
crassicutinite 厚角质体
crassidurite 厚壁孢子微陪煤
crassisporinite 厚壁孢子体
crassula 眉条
cratch 饲料架
crate 木条板箱;格框板条箱;大篓;板纸箱;板条箱;板条框;条板箱
crated goods 篓装货
crated weight 装箱重量;装箱后毛重
crater 火(山)口;花脸;弧坑(焊接时);陷孔;陷坑;月牙凹;圆形坑;焰口;焊接终点;放电痕;爆坑穴;盆状凹地;喷火口;露底缩孔
crater and lava flow map 火山口熔岩流分布图
crater blasting 漏斗爆破
crater bloom 凹痕(油漆);永久霜花;凹陷霜花
crater cavity 火口气孔
crater cone 火山口锥
crater control 缩弧控制
crater crack 焊槽裂纹;焊坑裂纹;火口裂纹;弧坑裂纹;缩孔裂纹
crater cuts 漏斗形掏槽
crater filler 焊口填充料;填弧坑
crater filling 填弧坑
crater geometry relations 陷弧几何关系(指冲击钻进时,陷口容积、钻头几何形状同岩石阻力三参数关系)
crater glaze 弧坑釉
crateric fumarole 火山口里的喷汽孔
crateriform 漏斗状的;喷火口状
crateriform depression 杯口状凹陷
cratering 起麻点;不黏着(涂料);漆膜起皮;露底缩孔
cratering effect 焰口效应;成坑效应
crater island 火山岛
craterkin 小火山口
crater king 金坑
crater lake 火山(口)湖;火口湖
crater lake scenic spot 天池风景区
crater lamp 环孔灯;点源录影灯
craterlet 小坑;小火山口
crater lip 爆破漏斗上边缘
crater magma 火山口岩浆
crater of elevation 高海拔火山口
crater rays 环形山辐射纹
crater rim 火山口边缘;火口沿
crater semi-diameter 漏斗半径
crater terrace 火山口阶地
crater wall 火口山墙;火(山)口壁
craticular stage 不活动期
crating and sorting area 分类包装室
crating and sorting room 分类包装室
crating and sorting space 分类包装室
crating charges 篓费(装箱篓费);装箱费
craton 克拉通;古陆核;稳定地块
craton basin 克拉通盆地
cratonic terrane 克拉通地体
cratonization 克拉通化作用
cratonizational stage 克拉通化阶段
cravat 烟囱内罩帽
cravenette 雨衣;防水布
crawer crane 爬行吊机;爬行吊车
crawing peg exchange rate 爬行固定汇率
crawl 围栏(浅水中);履带式;养鱼槛;点动;覆盖不均(油漆);蠕动;爬行
crawler 履带装置;履带式机械;履带车;履带牵引装置;履带牵引车;履带轮;履带车;爬行物
crawler asphalt paver 履带式沥青混凝土摊铺机
crawler attachment 履带牵引装置;履带附件
crawler baby bulldozer 履带式小型推土机
crawler bearing length 履带接地长度
crawler belt 履带
crawler bogie 履带台车
crawler bucket elevator loader 履带式升降装载机
crawler bucket ladder excavator 履带式链斗挖土机;履带式链斗挖掘机;履带式多斗挖土机;履带式多斗挖掘机
crawler bucket loader 履带式多斗装载机
crawler bulldozer 履带式推土机;履带式摊土机;履带式平土机
crawler clutch 履带式车离合器
crawler coalcutter 履带式截煤机
crawler concrete mixer 履带式混凝土搅拌机
crawler contact area 履带接触面积
crawler conveyer 履带式传送机
crawler crane 履带式起重机;履带起重机;履带吊(车)
crawler dozer 履带式推土机
crawler-drawn 履带拖拉机牵引
crawler drill 履带式钻机;履带式开山机
crawler drill jumbo 履带式钻车
crawler drive chain 履带驱动链
crawler drive clutch 履带式车驱动离合器
crawler dump wagon 履带式倾卸车
crawler excavator 履带式挖掘机;履带挖掘机
crawler face shovel 履带式正向铲土机
crawler flip-over bucket loader 履带斗式高位装载机
crawler foot shaft 履带式车转向轴
crawler foot shaft sprocket 履带式车转向轴链轮
crawler foot shaft sprocket yoke 履带式转向轴链轮架
crawler for towing wheel scrapers 拖曳轮式括板的履带式车
crawler frame 履带支架
crawler fronet-end loader 履带式前端装载机
crawler gate 履带式闸门
crawler height 履带高度
crawler jumbo 履带式台车;巨型履带式车
crawler lane 上坡车道;低速车道;爬坡车道
crawler loader 履带式铲车;履带式装载机;履带式装运机;履带式装料机
crawler machine 履带式牵引机
crawler mechanical shovel 履带式挖土机
crawler-mounted 装有履带的;履带式的
crawler-mounted asphalt finisher 履带式沥青路面整修机
crawler-mounted clamshell 履带式抓岩机
crawler-mounted crane 装有履带的起重机
crawler-mounted device 履带行走装置
crawler-mounted diesel excavator-crane 履带式柴油发动挖土起重机
crawler-mounted dragline 履带式索铲
crawler-mounted drill 拖拉机装钻机
crawler-mounted excavator 履带式挖土机;履带式挖掘机
crawler-mounted excavator-crane 履带式挖土起重机
crawler-mounted loader 履带式装车机
crawler-mounted piling rig 履带式打桩机
crawler-mounted pipe factory 履带式流动制管机
crawler-mounted power shovel 履带式铲土机
crawler-mounted tower crane 履带塔式起重机
crawler overhead loader 后倾式履带装载机;履带式高架装载机
crawler overshot loader 上倾式履带装载机
crawler pivot shaft 履带支承轴
crawler-powered scraper 履带动力括土机;履带动力铲运机
crawler-pulled 履带拖拉机牵引
crawler pusher 履带式推运机
crawler revolving crane 履带式旋转吊车
crawler revolving shovel 履带式旋转铲土机
crawler rocker shovel 履带式翻铲装载机
crawler roller 履带式压辊机
crawler scraper 履带式刮土机;履带式铲运机
crawler-scraper rig 履带式刮土机;履带式铲土机
crawler shoe 履带带板
crawler shovel 履带式挖土机;履带式挖掘机;履带式电铲;履带式铲土机

crawler shovel loader 履带式铲土装载机;履带式铲土装卸机;履带式铲土挖掘机
crawler side excavator 侧式挖沟机
crawler sprocket 履带链轮齿
crawler steam excavator 履带式汽动挖土机
crawler steering 履带转向(装置)
crawler swamp shoe 湿地履带板
crawler take-up 履带松紧调节装置
crawler tank car 履带式油槽车
crawler tension 履带张力;履带绷紧
crawler tower crane 履带式塔(形)吊(车)
crawler track 履带式行进装置;履带传动
crawler tractor 履带式推土机;履带式牵引机;履带式拖拉机;复带式拖拉机
crawler tractor attachment 履带式拖拉机附件
crawler tractor crane 履带式拖拉机吊车
crawler tractor-drawn scraper 履带式拖拉机拖曳铲运机
crawler tractor fitted with hydraulic angle dozer 装有液压斜铲的履带式拖拉机;履带式液压斜铲推土机
crawler-tractor-mounted bulldozer 履带式拖拉机推土机
crawler tractor-mounted front end loader 履带式拖拉机前置装载机
crawler-tractor-mounted shovel 履带式拖拉铲土机
crawler tractor with (bull) dozer 带推土机的履带式拖拉机
crawler trailer 履带(式)拖车
crawler tread 履带式行进装置;履带传动;履带传带
crawler truck 履带式货车;履带车
crawler type 履带式的
crawler type excavator 履带式挖土机
crawler type grader 履带式平土机
crawler-type motor grader 履带式自动平地机
crawler type tractor 履带式拖拉机
crawler-type truck 履带式运货车
crawler type vehicle 铺轨车;履带式牵引车
crawler unit 履带行走装置;履带行进装置
crawler universal excavator 履带式万能挖土机
crawler vehicle 履带式车辆
crawler vibrator 履带式振动机
crawler wagon 履带式运料车
crawler wagon drill 履带式钻探车
crawler wheel 履带轮
crawler-wheel excavator 履带式轮斗挖掘机
crawler-wheel trencher 履带式挖沟机
crawless pump 粪便污水移送泵
Crawley midget miner 克劳雷薄煤层采煤机
crawling 滚釉;表面涂布不均;表面起皱或隆起;涂膜疙瘩;缩边(漆病);烧缩
crawling and squeezing out 错动和挤出
crawling board 爬行板(有刻痕的木板);屋顶作业防滑板
crawling effect 蠕动效应
crawling gear 爬行变速齿轮
crawling of glazes 淌釉
crawling paint 起皱(纹)油漆
crawling pattern 蠕动图形
crawling peg 小幅度调整汇率;浮动钉柱;蠕动钉住;爬行式的汇率调整
crawling trace 爬行迹
crawling traction 履带牵引
crawl space (净空小的)房屋底层;低矮空间;管线空间;矮设备层;蠕动空间;爬行空间
crawl speed 慢车速度;蠕行速度
crawl tread tractor 履带式传动拖拉机
crawl trench 半通行管沟
crawlway 矮设备层;爬行管道
crayon 彩色铅笔;彩色粉笔;粉笔;炭棒;颜色笔
crayon drawing 粉笔画;蜡笔画
crayonsauce 彩色调料
craze 发丝裂缝(混凝土结构中);纤维状裂缝;细裂纹;银纹;出细裂缝
craze crack 裂纹微缝;发丝裂纹
craze resistance 抗细裂性
crazing 裂纹;隙裂;细纹开裂;细缝开裂;银纹;微裂纹
crazing crack 细微裂缝;细微裂纹
crazing glaze 皲裂(陶瓷);细裂纹釉
crazing of plaster 抹灰面开裂;抹灰面发裂
crazing resistance 抗釉裂;抗裂性
crazy pavement 乱铺石板小径;乱铺路面;乱石纹路面;碎纹石路
crazy paving 乱石纹路面;乱石铺路;散乱(的)片石铺砌;衬层裂缝
crazy paving path 水纹路;错铺路
crazy pitching 变坡护岸
crazy ship 应废而勉强航行的船;摇晃不定的船
creak detector 擦音检测器
cream 胶状(的);奶黄色;米色;化妆品;膏状物;膏;乳脂;乳酪色;水浆
cream base 膏基
cream body 乳酪色坯体
cream buff 淡黄至灰黄色;浅牛皮色
cream-colored 米色的
cream colo(u)r 米色;乳白(色)
cream-colo(u)red ware 淡黄色器皿
cream-digging rate 高峰抓货效率
cream emulsion 乳液油
cream engobe 淡黄色釉底料;淡黄色化妆土
creamery 奶油制造厂;乳酪厂
creamery butter 工厂制黄油
cream glaze 米色釉
cream ice 初冰;冰花;松软冰
creaming 乳油化;乳白化;涂敷脂
creaming agent 成乳油剂
cream laid 米色直纹纸;白条纸
creamlike suspension 乳状悬浮液
cream of latex 胶乳
cream of lime 石灰乳浊液;石灰浆;石灰膏
cream of tartar 重酒石酸铵
cream paper 乳白纸
cream solder 焊糊;糊状钎料;乳酪焊剂
creamware 乳酪色器皿;米色陶器
cream white 牙白色;乳白色;乳白
cream wove 米色布纹纸
creamy 奶油色的;米色的
creamy colo(u)r 奶油色
creamy white glaze 奶白釉
creasability 可折性
crease 皱褶;皱痕;褶痕;皱折;折痕;折叠;变皱
creased pipe 褶形管
creased pipe bend 褶缝弯管;折皱弯管
crease-flexible over test 折叠试验
crease in stocks 库存增加
crease(line) 折缝;折线
crease-proof 防皱的;不皱的
crease-proof finishing 防皱加工
creaser 压折缝的器具
crease recovery 折皱回复度
crease-resist 防皱
crease resistance 抗皱性;抗折叠性(能)
crease resistant fabric 防皱织物
crease resistant finish 耐皱涂层;耐折处理剂
crease-shrinkage 皱缩
creases of glacier 冰川裂隙
crease structure 皱纹(状)构造
crease tile 脊瓦
creaseyite 硅铁铜铅石
creashy peat 高沥青泥炭
creasing 折缝;挑檐;平瓦出檐
creasing course 烟囱压顶;逐层挑出的砖墙;逐层排出的砖墙
creasing iron 捻缝凿
creasing machine 压波纹机;褶边机;折边机
creasing resistance 抗折叠性(能)
creasing stake 弯管架
creasy 有折痕的
creasy surface 皱纹的表面;(铸件的)皱纹表皮
create 创造;产生
create a monopoly 形成垄断
create a more highly skilled workforce 提高劳动者素质
create a precedent 开先例
created symbol 引入符号
creating new varieties of plants 创造作物新品种
creating nonstandard label 生成非标准标号
creating segment 建立图段
creation 创造;编成过程
creation data 编成的日期
creation date 建立日期;编成日期;生成日期
creation facility program(me) 建立程序;生成设备程序
creation of impact sound 撞击声的发生
creation of vacuum 抽真空
creation operator 产生算符
creative 创作的
creative design material 造型材料
creative engineering 创造工程(学)
creative engineering program(me) 创造性工程开发计划
creative evolution 创造进化论
creative idea 创见
creative initiative 首创精神
creativeness 创造性
creative playground 儿童公园;儿童游乐场
creative selection 创造性选择
creative view-point 创见
creativity 创造性
creativity mobilization technique 创造性运用技术;创造性动员方法
creature 生物
creche 托儿所;孤儿院
credence 可信度;信任;供桌;食器柜(文艺复兴时期);祭台
credentials 证书
credenza 储藏柜
credibility 可信;可靠;信誉;信用;确实性;侵蚀性
credibility gap 信用差距
credibility interval 可信区间
credible 可信的
credit 信用;信任;信贷
creditability 可信性;可抵免性;信用程度
creditable tax 可抵免税款
credit account 赊购账户;欠账
credit advice 进账报单
credit agency 信贷机构;借贷所;信用咨询公司;信用调查机构
credit agent 信任代理人
credit agreement 信贷协议;信贷协定
credit allocation 贷方分配
credit amount 信贷金额;信用保证金额;赊账金额
credit amount exceeded 超支信用证金额
credit amounts 信贷额
credit analysis 信用分析
credit analysis firm 信用分析公司
credit and collection expenses 放账和收账费用
credit and loan 信贷
credit application 贷款申请登记表;信贷申请表
credit application form 信用贷款申请单
credit association 信用社;信贷联合
credit availability 信贷可获量
credit balance 结欠(金额);贷差
credit bank 信用银行;信贷银行
credit bank balance 票据交换后银行贷差
credit based on personal property 动产信用
credit based on real property 不动产信用
credit beneficiary 信用证收益人
credit bill 信用汇票
credit boycott 拒付贷款
credit bureau 信用资料社;信用咨询公司
credit business 赊售
credit capital 信贷资本
credit card 记账卡片;信用卡
credit card call 信用卡电话
credit card finance charges 信用卡财务费
credit card system 记账卡片系统;信用卡系统
credit ceiling 信贷最高限额
credit center 信贷中心
credit circulation 信用流通
credit classification 信用分类
credit clearing division 信用清算组
credit column 贷方栏
credit commitment 信贷承诺
credit competition 信贷竞争
credit construction 建设信贷
credit contract 信用购买契约
credit contraction 信用紧缩
credit control 信贷管理;信用管制;信贷控制
credit control act 信贷管制法
credit cooperative 信用合作社
credit cooperatives 信用社
credit country 债权国
credit creation 信用创造;信用产生
credit crisis 信用危机
credit crunch 信用恐慌
credit currency 信用通货;信用货币
credit cycle 信用循环
credit decision 信贷决策
credit department 信贷部(门);信用调查部
credit documents 信用证书;信用证券;信用单证
credit economy 信用经济

credit expansion 信用膨胀;信用扩张;信贷扩张
credit expired 信用证逾期
credit export 出口信贷公司(荷兰)
credit extending policy 信贷方针;融资方针
credit facility 信用透支;信用设施;信用措施
credit file 信用调查档案
credit first 信用第一
credit for returned goods 退货冲减
credit fraud 信用欺诈
credit freeze 信用冻结
credit fund 信贷基金;信贷资金
credit gap 信用差距
credit giver 借款人;授信人
credit guarantee 信贷担保;信贷保证;贷款信用担保
credit guarantee system 信用保证制度
credit guaranty 信贷担保
credit guild 信贷同业公会
credit inflation 信用膨胀;信贷膨胀
credit in foreign exchange 外汇信贷
credit information 信用资料;信用调查报告;信用调查;资信情况
credit information service 征信机构
crediting period 计入期
crediting period-fixed 固定计入期
crediting period revewable 可更新计入期
credit inquiry 信用调查;商业信用调查
credit inquiry division 信用咨询组
credit institution 信贷机构
credit instruments 信用票据;信用工具;信贷文件
credit insurance 信用保险
credit insurance premium 信用保险费
credit interchange 信用情报交换
credit interest 存息
credit investigation 偿还能力调查;资信调查
credit issue 信贷发行
credit item 贷方项目
credit leverage 信贷杠杆
credit life insurance 信贷人寿保险
credit limit 信贷限额;信贷额度
credit line 信用限额;信贷额度;信贷额度;贷款限额;透支限额
credit loan 信用借款;信用贷款
credit loss 赊账损失
credit man 客户信用调查员;信用调查员
credit management 信用管理;信贷管理
credit manager 信贷经理
credit market 信贷市场
credit memo 信贷凭证;贷款凭单
credit memo for purchase returns 进货退回通知单
credit memo for sales returns 销货退回通知单
credit memorandum for sales return 销货退回通知单
credit money 信用货币
credit multiplier 信贷乘数
credit note 信用票据;退款单;退货单
credit number 发送信件数
Credit of Agency for International Development 美援信用证
credit of bankruptcy 破产债权
credit of favorable conditions 条件优惠的信贷
credit of the currency 货币信用
credit operation 信贷业务
creditor 贷方;借方;债主;债权人
creditor bank 债权银行
creditor beneficiary 债权受益人
creditor country 债权国
creditor in a bankrupt estate 破产债权人
creditor investor 债权投资人
creditor nation 债权国
creditor of bankruptcy 破产(户之)债权人
creditor's bill 债权索赔清单;债权人开出票据
creditor's composition 偿还议定书
creditor's creditor 债权人的债权人
creditor's equity 行使扣押权的债权人;债主权益
creditor side 付项
creditor's ledger 债权人分户账
creditors of a bankrupt estate 破产债权人
creditor's petition 债权人破产申请书
creditor's rights 债权
credit package 一揽子贷款
credit paper 信用证券;信用票据
credit period 信用期限
credit picture character 贷方图像字符
credit policy 信贷政策;信用政策;信用保险单

credit position 资信情况
credit price 赊销价格;赊卖价格
credit purchasing 赊购
credit race 信贷竞争
credit rating 信用能力;信用定额;信贷分类;客户资信分类;客户信贷分类;信用评级;信贷配给;资信评价
credit ratio 信用比率
credit receipts and payments 信贷收支(平衡)
credit receiver 信贷接受人
credit reference 信用备询
credit relations 信贷关系
credit remittance 信用汇款
credit repayment 偿还贷款
credit report 信用报告
credit reporting agencies 信用报告机构
credit requirement 信贷需要
credit reserve 信贷储备
credit restraint 信贷节制
credit restriction 信贷限制;信贷节制
credit risk 信用风险;信贷风险
credit risk insurance 信用风险保险
credit sale 信用买卖;赊销
credit sale agreement 分期收款销货协议;赊销合同
credit sale invoice 赊销发票
credit screening 客户资信筛选
credit screw 信用紧缩
credit security 信用保证
credit side 贷方
credit society 信贷机构;信贷所
credit squeeze 信用紧缩;信贷紧缩;压缩信用
credit squeeze measure 紧缩信用措施
credit squeezing 信用紧缩
credit standard 信用标准
credit stand-by 备用信贷
credit standing 信誉;信用状况;信用名誉;信用地位;信贷价值;偿债能力
credit status 信用状况
credit supervision 信贷监督
credit supported by other credit 信贷支持的信贷
credit swap 信用换汇
credit symbol 贷方符号
credit system 信贷系统;信用制度;赊购制度
credit system service 记账制业务
credit terms 信贷条件;贷款条款;贷款条件;赊销付款条件
credit theory 信用论
credit ticket 储值票
credit tranche 信用部分贷款(国际货币基金组织);信贷份额
credit transaction 信用交易;赊账交易;信贷交易
credit transfer 信用转账;信用转让
credit union 储贷互济会;贷款互助会;信用组合;信用(合作)社;信贷协会
credit union investment pools 信用合作社投资总汇
credit user 信用证收益人
credit verification 信用核对
credit volume 信贷额
credit with soft clauses 信用证软性条款
credit with telegraphic(al) transfer reimbursement clause 带有电报索汇条款的信用证
credit worthiness 信誉;信贷价值
creditworthy 资信可靠度
crednerite 锰铜矿
creedite 铝氟石膏;氟铝石膏
Creed receiving perforator 克利德凿孔机
creek 小湾;小河口;小河;小海湾;溪;支流;短汉;潮汐沟;通海小河
creek wall 小河河堤
creel 经轴架;筒子架
creel stand 工字轮梁;排线架
creen rack 拦污栅栅条
creep 滑坍;滑塌;打滑;蠕动;蠕变;潜伸;频率漂移;爬行;塑性变形;冷流变
creepage 渗水;漏电;徐变;蠕动;溯流
creepage current 爬电电流
creepage discharge 潜流放电
creepage distance 爬行距离;爬电距离
creepage length 爬电距离
creepage of the coupler 接头滑扣
creepage path 爬电距离
creep and stress rupture test 蠕变持久试验
creep angle (前轮的)漂移角

creep at sliding 滑动徐变;滑动蠕变
creep behavio(u)r 徐变特性;蠕动特性;潜移特性;爬行特性;滑坍特性;蠕变行为
creep behavio(u)r test 蠕变性状试验
creep breaking strength 徐变断裂强度;蠕变断裂强度
creep buckling 蠕变压曲
creep capacity 徐变能力;蠕动能力;潜移能力
creep cell 蠕变盒;蠕变传感器
creep characteristic 徐变特性;蠕动特性;潜移特性;爬行特性;滑坍特性
creep coefficient 徐变系数;蠕变系数
creep collapse 蠕变毁坏
creep compliance 徐变屈服;蠕变柔量
creep condition 蠕变条件
creep constant 蠕变常数
creep correction factor 徐变修正系数
creep crack 徐变裂缝;蠕变裂缝
creep curve 徐变曲线;蠕变曲线
creep damage factor 蠕变寿命消耗率;蠕变破坏因素
creep deflection 徐变挠曲;塑性挠曲;蠕变挠曲;非弹性挠曲;蠕变挠度
creep deflexion 徐变挠曲;塑性挠曲;蠕变挠曲;非弹性挠曲
creep deformation 徐变变形;蠕动变形;蠕变变形
creep deformation tester 蠕变(变形)试验仪
creep design curve 蠕变设计曲线
creep displacement meter 蠕变位移仪
creep distance 爬距
creep effect 徐变效应;蠕动效应;蠕变效应
creep elongation 蠕变伸长
creeper 卷叶式浮雕;慢速连续输送机;定速运送器;葡匐砖;葡匐植物;爬行者;爬行物;爬车器;铁齿板(防滑用);深海钩
creeper attachment 缓行器
creeper crane 移动式装吊车;爬行式吊车;履带(式)起重机;移动式吊车
creeper derrick 爬行吊车;履带式起重机
creeper land 爬行道
creepers 蔓藤浮雕
creeper speed-propelled carrier 慢速自行底盘
creeper tractor 履带式拖拉机
creeper travel(l)er 徐升式起重机;爬升式起重机;履升式起重机;爬行车
creeper tread 履带板
creeper truck 履带车
creeper type tractor 爬行式牵引车
creeper wagon 履带式转臂起重机
creep extension 蠕变延伸
creep factor 蠕变因数;蠕变系数
creep failure 徐变破坏;蠕变破坏
creep failure criterion 蠕变失效准则
creep fatigue 蠕变疲劳
creep fault 蠕动断层
creep feed 间歇进给
creep feed ginder 缓进磨床
creep fiber 蠕变纤维
creep fluidity 蠕变流度;蠕变流动性
creep forming 蠕变成型
creep fracture 蠕动破裂;蠕变破坏;蠕变断裂
creep gradient 徐变梯度
creep growth 徐变增长;蠕变增长
creepie-peepie 便携式电视摄像机
creepies (铺屋面瓦片时用的)爬行板
creep increase 徐变增长
creep index 蠕变指数
creeping 涂料淌流;冷流变;蠕动;蠕变;慢速移动;滚釉;蠕升;潜移作用;漂移;爬行打滑;溯流;烧缩
creeping bent 蔓生草
creeping boat 探海艇
creeping covers 葡匐覆盖植物
creeping crack 蠕变断裂;蠕变裂纹
creeping discharge 沿面放电;蠕变放电
creeping distance 蠕变距离;沿面放电路径
creeping film 蠕动膜
creeping fire 慢行弱火;伏地火
creeping flange 缩缩法兰
creeping flow 蠕流(变);蠕动流
creeping force 爬行力
creeping inflation 慢而持续的通货膨胀;不知不觉的通货膨胀
creeping motion 滑动;慢速运动;蠕动;蠕波;爬动
creeping of grease 润滑脂蠕升

creeping of track 爬行【铁】
creeping pressure 徐变压力;蠕动压力
creeping rafter 伏椽;短椽;爬伏椽
creeping random method 缓移随机法
creeping resistance 防爬阻力
creeping slickenside 蠕动擦痕面
creeping slip 蠕变滑动
creeping slip fault 蠕滑断裂
creeping slope wash 蠕动坡积物
creeping speed factor 徐变速率系数
creeping-strain 蠕变应变
creeping strength 徐变强度
creeping trench 爬行管道(人可以进去里面进行检查)
creeping underground stem 地下横走茎
creeping wave 蠕波
creeping weasel 爬行式探测仪
creep instability 徐变失稳;蠕变失稳
creep lamella 蠕动层;蠕动薄层;蠕变层;蠕变薄层
creepless 无徐变的
creepless concrete 无徐变混凝土
creep life 蠕变寿命
creep limit 徐变极限;蠕变极限
creep limiting value 蠕变极限值
creep line 蠕流线;蠕动线
creep load 蠕变荷载
creep mechanism 徐变机理;蠕变机制;蠕变机理
creep meter 徐变仪;蠕变计;蠕变仪
creep model 蠕变模型
creep modulus 徐变模量;蠕变模量
creep observation 徐变观测;蠕变观测
creep of belt 皮带滑动
creep of concrete 混凝土徐变;混凝土蠕变
creep of frequency 频漂
creep of mantle material 地幔物质蠕动
creep of mortar 灰浆蠕变
creep of plastics 塑料蠕变
creep of rail 钢轨爬位
creep of rock 岩石蠕变
creep of soil 土层蠕动
creep of steel 钢筋徐变;钢的徐变
creep of steel reinforcement 钢筋徐舒
creep of the rails 轨道的爬行
creep path length 爬径长度(指渗流);渗流爬径长度
creep phenomenon 蠕变现象
creep point 蠕变点;屈服点
creep Poisson's ratio 蠕变泊松比
creep pressure 蠕变压力
creep process 徐变过程;蠕变过程
creep property 蠕变特性
creep rate 徐变速率;蠕动率;蠕变(速)率;爬坡速率
creep rate-time plot 蠕变率时间图
creep ratio 徐变比;蠕动比;蠕变比;溯流比
creep recovery 徐变恢复;徐变回复;徐变后效;蠕变恢复;蠕变复原
creep relaxation 蠕变松弛
creep resistance 抗蠕流性;抗蠕变性;抗蠕变力;蠕滑阻力;蠕变阻力;蠕变抗力
creep-resistant 抗蠕变的;抗蠕变的
creep-resistant alloy 抗蠕变合金;耐蠕变合金
creep-resistant steel 耐蠕变钢
creep-resisting 抗蠕变的;蠕变的;抗蠕动的
creep rupture 徐变破坏;徐变断裂;蠕变破坏;蠕变断裂
creep rupture life 持久寿命;蠕变破断寿命
creep rupture strength 徐变破坏强度;徐变断裂强度;持久强度;蠕变破坏强度;蠕变断裂强度
creep rupture test 持久试验;蠕变破坏试验;徐变破坏试验;持久强度试验;蠕变断裂试验
creep settlement 蠕变沉降
creep slide 碎岩滑动
creep slip 蠕变滑动
creep slippage 蠕变滑动
creep speed 慢行速率;徐变速率;徐变速度;最低速度;蠕变速度;爬行速度;爬坡速率
creep spring 保险弹簧
creep stage 蠕变阶段
creep state 蠕变状态
creep strain 徐动应变;蠕动应变;徐变应变;蠕变应变;蠕变(变形)
creep strain recovery 徐变应变回复;蠕变应变回复
creep strain-time plot 蠕变应变时间图

creep strength 抗蠕变强度;蠕变强度
creep stress 徐变应力;蠕变应力
creep temperature 蠕变温度
creep test 耐蠕变试验;徐变试验;蠕变试验
creep tester 蠕变试验机;蠕变实验机
creep testing machine 蠕变试验机
creep test machine 徐变试验机;蠕变试验机
creep test under deviated stress 偏应力蠕变试验
creep theory 徐变理论;蠕变理论
creep time curve 徐变时间曲线;蠕变时间曲线
creep traveler 爬行吊车
creep trench 匍匐沟槽
creep type 蠕动方式
creep under variable stress 变应力蠕变
creep value 蠕变值;蠕变差
creep variation 徐变差
creep velocity 蠕变速度
creep viscosity 蠕变黏度
creepwash 土壤滑泻;泥流;滑坡
creep wrinkle 蠕动皱纹
creep zone 屈服区
Creetex (一种专卖的用于防水面层的)塑性沥青结合料
creetown (产于苏格兰的)淡灰色花岗岩
cremains 骨灰
cremart 草胺磷(除草剂)
cremation 火葬
cremation certificate 销毁证明书
cremation urn 骨灰瓮;骨灰坛
cremation urn pit 骨灰瓮坑穴;骨灰墓
cremator 烧垃圾的人;垃圾焚化炉;焚尸炉;焚化炉
crematorium 焚尸炉;垃圾焚化炉;火葬场
crematory 垃圾焚化炉;火葬场;焚化的
crematory system 焚化系统;焚化处理
crematory system of sewage disposal 污物焚化处理系统;污泥焚化处理;污废物焚化处置法
cremnion 石生植物群落
Cremona's method 克列莫纳桁架计算法
Cremona's polygon of forces 克列莫纳多边形
Cremona's stress diagram 克列莫纳桁架内力图解法
Cremo(r)ne bolt 落地窗旋转插销;门窗长插销;长插销;通天插销
crenate 纯锯齿形的
crenation 钝锯齿状;纯锯齿形;缩皱
crenelet 小堡眼;(要塞上的)瞭望孔
crenel(l)ate 设雉堞;锯齿状的
crenel(l)ated 雉堞;锯齿状的;带枪眼工事;城垛式;带枪眼工事
crenel(l)ated bridge parapet 雉堞状桥墙;雉堞状桥栏
crenel(l)ated mo(u)lding 锯(齿)形线脚;牙齿花饰;齿花饰
crenel(l)ated pattern 钝锯齿形
crenel(l)ation 锯齿状物;雉堞
crenel(le) 枪眼;炮眼城垛;齿状城墙的凹口;堡眼
crenic acid 克连酸
crenitic 泉华化
crenogenic meromixis 半混合层
Crenothric 铁细菌
crenulate coast 锯齿形海岸
crenulate coastline 锯齿形海岸线
crenulated bedding 锯齿状层理
crenulate shoreline 锯齿形海岸线
crenulation 小圆齿;细褶皱;微钝锯齿状
crenulation cleavage 锯齿形劈理;滑劈理;褶劈理
crenulation lineation 褶纹线理
creosol 木焦油酚;(作防腐剂用的)木焦油醇
creosol pitch 甲氧甲酚树脂
creosote 克鲁素油;杂酚(油)
creosote bleaching oil 杂酚油漂白油
creosote bush 石炭酸灌木
creosote-coal-tar mixture 杂酚油—煤焦油混合物
creosote-coal-tar solution 杂酚油煤焦油溶液
creosote-contaminated water 杂酚油污染水
creosoted 浸防腐油的;防腐的;浸杂酚油防腐
creosoted pile 用杂酚油防腐处理过的木桩;浸(杂酚)油防腐木桩;油浸(木)桩;杂酚浸过的桩
creosoted timber 浸渍防腐木材;用杂酚油防腐处理过的木材;油浸(防腐)木材;杂酚浸过的木(材)
creosoted-wood pile 油浸防腐木桩;杂酚油浸防腐木桩;杂酚浸过的木桩
creosote emulsion 杂酚油乳剂;乳化防腐油

creosote oil 重质煤馏油;木馏油;杂酚油
creosote-petroleum solution 杂酚油石油溶液
creosote primer 煤焦油基涂层底料;木材防腐打底油
creosote pure 纯酚油
creosote treatment 杂酚油浸处理
creosote-type preservative 杂酚油型防腐剂
creosoting 杂酚油防腐处理;杂酚油浸渍
creosoting cylinder 油浸木材圆筒
creosoting of wood 木材杂酚油
creosoting post 浸涂杂酚油的防腐木柱
creosoting primer 煤焦油基涂层底料
creosoting process 注油法;杂酚油防腐(处理法)
creosoting timber sleeper 木枕防腐处理
crepe 皱纹纸
crepe-backed satin 绉缎
creped 绉缩的
creped cotton 绉线;绉布
crepe de chine 双绉
creped paper 绉纸
creped wool 绉呢
crepe paper 皱纹纸
crepe rubber 皱纹薄橡皮板
crepe-singles 单丝绉线
crepe varnish 皱纹清漆
crepidoma 台阶式基础
creping tissue 薄皱纹纸
creping treatment 绉缩处理
crepitation 捻发音;气蚀碎裂现象(混凝土)
crepon finish 皱纹漆
crepuscular arch 曙暮辉弧
crepuscular ray 曙暮辉光;日落辉
cresceleration 幂次加速度;按幂级数变化的加速度;速度规律性变化
crescendo 渐强
crescent 镰形;新月状;新月体;新月饰;月牙形;月牙饰;月牙卡铁;月牙槽;半月体
crescent arch 新月形拱;镰刀形拱;月牙(形)拱
crescent beach 新月形海滩
crescent beam 月梁;月牙形梁
crescent cracking 月(牙)形开裂;推挤裂缝
crescent cross-bedding structure 新月形交错原理构造
crescent current marks 新月形水流痕
crescent firmer chisel 半圆形木工凿
crescent formation 新月体形成
crescent gear motor 内啮合齿轮马达
crescent gear pump 内啮合齿轮泵
crescentic 新月形的
crescent(ic) dune 新月形沙丘
crescentic formed penciling 新月形条斑
crescent(ic) lake 牛轭湖;新月湖;弓形湖
crescentic moraine 新月形冰碛(层)
crescentic reef 新月形礁
crescentic sand-ridge 新月形沙垄
crescentic scour mark 新月形冲刷痕;新月形擦痕
crescentic-shaped bar 新月形沙洲
crescent lake 弓形湖
crescent moon 蛾眉月
crescent moon ornament 蛾眉月形装饰
crescent oblique rib 镶新月形料棱;镶新月形斜肋
crescent point bar 新月形凸岩边滩
crescent pump 月(牙)形齿轮泵
crescent rib bifurcation 月牙形肋岔管
crescent ripple 新月形波痕
crescent roof 月牙形屋盖
crescent roof truss 月牙形屋盖桁架
crescent scraper bucket 半圆形铲
crescent screen 月牙筛
crescent(-shaped) 新月形
crescent-shaped dike 半月形堤
crescent-shaped land (内啮合齿轮泵中的)月牙形间隙
crescent-shaped lever 月牙杆
crescent-shaped point bar 新月形凸岸边滩
crescent-shaped separator 月牙形隔板
crescent spacer 月牙形隔板
crescent sun 蛾眉日
crescent superposed folds 新月形叠加褶皱
crescent truss 镰刀形桁架;月牙(形)桁架;弓形桁架
crescent-type arch 月牙形拱
crescent-type arch bridge 月牙形(桁架)拱桥
crescent wrench 弯头扳手;可调扳手
crescumulate texture 强化堆积结构

cresol 甲(苯)酚;混合甲酚
cresol-formaldehyde resin 甲酚甲醛树脂
cresol pitch 甲酚树脂
cresol plastic 甲酚塑料
cresol red 邻甲酚磺酞;甲酚红
cresol resin 甲酚树脂
cresol resin adhesive 甲酚脂黏合剂
cresol saponatus 煤酚皂
cresol soap 来苏儿;甲酚皂
Crespi lining 克里斯皮炉衬
cresset 明火标灯;标灯;号灯
cress green 水芹绿色
cressing 锻头
crest 冠;带有图案的玻璃器皿;峰;嵌值;驼峰峰顶;凸出处;坡顶
crestal fracture 脊部断裂
crestal injection 外部注气
crestal plane 脊面
crest ammeter 峰值电流计
crest amplitude 最高振幅
crest angle 波峰角
crest arc chord 弧形坝顶的弦长
crest armo(u)r 堤顶护面
crestatron 高压行波管
crest block 顶块;顶部浇筑块
crest board 屋脊饰板
crest clearance (螺杆螺纹的)顶部间隙;端部间隙;齿端部间隙;齿顶间隙
crest cloud 盔云;驻云【气】;山帽云【气】
crest contraction 堰顶收缩;顶约束
crest control 溢洪堰顶控制;溢洪道顶控制;堰顶控制
crest control equipment 溢流堰顶控制设备
crest-control weir 溢流堰;溢流坝
crest curve 堰顶曲线;凸(形)曲线
crest depth 堰顶水深
crest development 最高发展;洪峰形成
crest discharge 洪峰流量;堰顶流量;过顶流量;峰值流量
crested wheatgrass 麦穗草
crest element fail 堤顶构件失效
crest elevation 堰顶高程;顶高;顶部高程;堤顶高程;坝顶高程
crest elevation of dam 坝顶高程
crest factor 振幅因数;波形系数;波峰因数;波顶因数
crest flow 洪峰流量;溢流;过顶流量
crest forecast 洪峰预报;最高水位预报;峰水位预报
crest forward anode voltage 最大阳极正向电压
crest freeboard 堰顶超高;坝顶超高
crest freeboard of dam 坝顶超高
crest gate 溢洪道闸(门);堰顶(闸)门;坝顶闸门
crest ga(u)ge 最高水位水尺;峰顶水尺
crest-ga(u)ge indicator 洪峰水位指示器
crest graben 脊部地堑
crest height 波峰高度
crest height of wave 波顶高度
cresting 屋顶装饰;脊瓦竖饰;饰饰(扶手);脊饰
crest interval 波峰间隔
crest inverse anode voltage 最大阳极反向电压
crest length 堰顶长度;(坝、堰的)顶长;顶部宽度;顶部长度(堤坝等);波峰长度;坝顶长度
crest level 堰顶高程;顶部高程;堤顶高程;坝顶高程
crest level of dike 坝顶高程
crest light 潮峰灯
crest line 脊线;顶线;峰(顶)线;分水线;山脊线
crest line of anticline 背斜的分水线
crest line of wave 波峰线
crest load 高量负荷
crestmorite 单斜硅钙石
crest of berm 滩肩脊顶
crest of core 坝心顶部
crest of corrugation (路面的)搓板脊脊
crest of dam 堤顶;坝顶
crest of diversion weir 导流堰顶
crest of hump 驼峰峰顶
crest of long period wave 长周期波峰
crest of overfall 溢流堰顶
crest of overflow 溢流堰顶
crest of peak 峰顶【铁】
crest of screw thread 螺纹牙顶
crest of slope 坡顶;顶部
crest of spillway 溢洪道顶(部)

crest of thread 螺纹牙顶;齿顶
crest of tide 高潮;潮峰
crest of wave 波顶;浪头;波脊;波峰
crest of weir 堰顶;堰口
crest overflow 坝顶溢流
crest over the weir 堰上液层高度
crest plane 脊面
crest platform 峰顶平台【铁】
crest point 顶点
crest pressure 波峰压力
crest profile 堰顶剖面(曲线);峰顶纵剖面;波缝纵剖面
crest province 海岭脊区
crest radius 坝顶曲线半径
crest reduction 洪峰削减(量)
crest ridge 冠脊(瓦)
crest segment 脊段;顶部;峰段;峰部
crest slab 顶板
crest speed 标定速度;名义速率;名义税率;名义车速;常见正常车速;标定速度(指在道路的一个区段中,驾驶员在没有交通干扰的情况下所能达到的行驶速率)
crest spillway 坝顶溢洪道
crest stage 洪峰水位;最高水位;顶峰阶段;峰水位;峰顶阶段
crest-stage indicator 洪峰水位指示器;最高水位计
crest-stage reduction 洪峰消减
crest structure 顶部结构(常指斜坡式防波堤等)
crest surface 脊面
crest table 墙帽
crest-thicken fold 顶厚褶皱
crest tile 屋脊饰瓦;脊瓦;装饰性脊瓦;屋脊瓦
crest-to-clearance distance 驼峰峰顶至计算停车点间距离
crest truncation 尖端钝化;削峰
crest value 颠值;峰值
crest vertical curve 凸形竖曲线
crest voltage 最大电压;峰值电压
crest voltmeter 峰值伏特计;峰值电压表
crest wall 堤顶墙;防浪(胸)墙
crest welder 波峰焊机
crest width 洪峰宽度;顶宽;顶部宽度;堤顶宽度;坝顶宽度
crest working reverse voltage 反向工作峰值电压
cresyl blue 焦油蓝
cresyl diphenyl phosphate 磷酸甲苯联苯酯;磷酸甲苯二苯酯
cresylic acid 甲酚和二甲酚的混合物
cresylite 甲苯炸药
cresyl mercury cyanide 氰化甲苯基汞
cresyl para-toluenesulphonate 对甲苯磺酸甲苯酯
creta 白垩
cretaceous 白垩的
cretaceous oils 白垩纪石油
Cretaceous period 白垩纪
Cretaceous system 白垩系
Cretaceous-Tertiary-Quaternary mixed polarity subchronzone 白垩—第三纪—第四纪混合极性超时间带
Cretaceous-Tertiary-Quaternary mixed polarity superchron 白垩—第三纪—第四纪混合极性超时
Cretacic 白垩系;白垩统
Cretan architecture 克里特式建筑;克里特建筑学
Cretan column 克里特式柱(型)
crete-o-lux 玻璃砖(一种专利玻璃砖产品)
Cretes prism 克里特棱镜
Crete trough 克里特海槽
creteway 踏出的小径;轨道
cretification 白垩化
cretonne 印花装饰布;大花型瑰丽印花装饰布
crevasse 裂隙;裂口;(municipality 的)裂口;大裂缝;冰隙;冰河裂隙;双峰谐振;双峰共振
crevasse channel 决口水道
crevasse crack 龟裂;裂缝
crevasse curve 裂隙曲线
crevasse deposit 冰隙沉积
crevassed glacier 裂隙冰川
crevasse filling 裂隙填充;裂隙充填;裂隙冰水沉积
crevasse hoar 冰隙白霜
crevasse repair (堤的)堵口;堵决口
crevasses of glacier 冰隙
crevasse-splay 决口扇(沉积)
crevasse splay facies 决口扇相
crevasse splay-natural levee 决口自然堤

crevasse splays 决口扇形滩
crevasses repair 决定修复
crevice 裂隙;隙缝;缝隙
crevice brush 裂缝刷
crevice corrosion 裂隙腐蚀;裂缝腐蚀;隙间腐蚀;缝隙腐蚀
creviced formation 微裂缝地层
creviced rock (岩石的)裂隙层
crevice formatiom (岩石的)裂隙层;裂隙岩层
crevice karst 裂隙卡斯特
crevice oil 裂隙油
crevice-searcher 裂隙研究
crevice-water 裂隙水
crew 小队;工队;船员;乘员组;乘务员;配员;水手
crew accommodation 船员住宿舱
crew and decompression 工班与减压
crew and effects 船员和行李
crew boat 联络艇;交通艇
crew-carrying superstructure 船员居住的上层建筑
crew changing at turnaround depot system 驻班制【铁】
crewel 松捻双股细绒线刺绣装饰
crew employment contract 船员雇用合同
crew escape system 船员逃生系统
crew gangway 船员步桥
crew hut 人员宿舍
crew in charge of construction 施工单位
crew key switch 列车乘务人员钥匙开关
crew key switch function 乘务员钥匙开关功能
crew landing permit 船员登陆证
crew list 船员名单;船员名册
crew list of shipsman 船员名册
crew member rest room 乘务员休息室
crew motor trolley 道班轨道车
crew of logging 测井队别
crew of steel fixers 钢结构工人组
crew pooling system of passenger train 轮乘制【铁】;旅客列车轮乘制
crew replacement 船员更动
crew rest room 工作人员休息室
crew's accommodation 船员舱;水手舱
crew safety system 船员安全系统
crew's book 船员名单;船员名册
crew's customs declaration 报小关
crew's effects 船员个人物品
crew's galley 船员厨房
crew's identification paper 船员身份证明书
crew's insurance certificate 海员保险证书;船员保险证书
crew's launch 船员交通艇
crew's messroom 船员餐室
crew's quarter 船员居住区;船员舱;水手舱
crew's room 船室
crew's space 船员舱;水手舱
crew station design 地面站设计
crew's wage 船员工资
crew's water closet 船员厕所
crew working system of passenger train 旅客列车乘务制度
Crédit Suisse 瑞士信贷银行
criador 克利亚德风
crib 轨枕盒【铁】;井框;井壁基环;木框;木垛;格排;儿童床;砂型套框;叠木框
crib and pile dike 木笼木桩堤
crib-ballast cleaner 枕间道砟清筛机
crib bed 基础始盘
cribber 支撑物
cribbing 木垛(支座);垛架;撑垫材;排节;竖井木框板壁;竖井环板壁
cribbing and matting 木底架
cribble 粗筛;粗粉;分离筛
cribbled 具有筛形孔的;带有小点的;圆点饰面
crib block 叠木块
crib breakwater 木笼防波堤
crib bucket 栅条式挖掘铲斗
crib check dam 木笼拦沙坝;木笼土坊
crib chock 枕木垛
crib cloth 粗筛布
crib cofferdam 木笼围堰;叠木围堰
crib consolidator 枕间捣固机
crib dam 框笼填石坝;木笼(填石)坝;木框坝;垛坝;石坝
crib dam filled with stone 木笼填石坝;填石木笼坝
crib filter 筐式过滤器

crib foundation 木笼基础
crib groin 木笼(填石)丁坝
crib groyne 木笼填石丁坝
cribling 叠木
crib member 木笼组件;叠木框组件
crib pier 木笼(桥)墩;叠木支座
crib protection 木垛支护
crib retaining wall 框格式挡土墙;格式式挡土墙(混凝土);格架式挡土墙;垛式挡土墙;填石木笼挡土墙
crib retaining wall built with precast reinforced concrete member 预制钢筋混凝土石笼挡土墙;预制钢筋混凝土格笼式挡土墙
cribriform 筛状的
cribriform plate 筛状板
cribrum 筛;[复]cribra
crib spur 木笼丁坝
crib type cofferdam 框格式围堰;木笼围堰
crib weir 木笼堰;木框堰
cribwork 木笼(作业);垛式支架;叠木框;石笼框梁;木笼围堰桥墩;木笼井架混凝土桥墩;木笼填石;木笼构筑物(填石);木(格)笼;格式式构造物
cribwork quay wall 木笼码头岸壁
cribwork retaining wall 木笼挡土墙
cribwork wall 木笼岸壁
crichtonite 尖钛铁矿;锶铁钛矿
crick 小型螺丝千斤顶(顶脚手架用);高丘陵
cricket 木制矮垫脚凳;防热屋顶;(烟囱后的)泻水假屋顶;斜沟小屋顶;板球
cricket bag 板球包
cricket cap flashing 鞍盖式披水板;鞍盖式泛水板
cricket cap flashing piece 鞍盖式披水板件;鞍盖式泛水板件
cricket counter flashing 鞍式披水板
cricket counter flashing piece 鞍式披水板件
cricket court 板球场
cricoid 轮形(的)
cricondenbar 临界冷凝压力
cricondentherm 临界冷凝温度
Criddle's erosion equation 克里特尔侵蚀方程
criggling 炭质页岩
crime against damage to environment 危害环境罪
crime against polluting environment 污染环境罪
crime of breaking dike 决水罪
crime of dereliction of public duty 渎职罪
crime of destroying mineral resources 破坏矿产资源罪
crime of destroying scenic spot 破坏名胜古迹罪
crime of illegal fishing fishery 非法捕捞水产品罪
crime of illegally cutting down and denuding forest or other woodlands 盗伐、滥伐林木罪
crime relating to environment pollution 环境污染罪;公害罪
criminal 违法的
criminal liability 刑事责任
criminal responsibility 刑事责任能力
crimp 卷曲;折边;弯皱;凸缘;使卷曲
crimp cloth 泡绉纱
crimp connection 压接连接
crimped 波纹状
crimped angle 曲贴角钢
crimped-center disk 平顶球面靶片
crimped connector 钳压管
crimped copper 皱铜;硬铜
crimped fiber 卷曲纤维;波形纤维
crimped fibre 波形纤维
crimped finish 皱纹状修饰
crimped lock 接线柱[电]
crimped section 弯曲断面
crimped steel fiber 压皱钢纤维;脆性钢纤维
crimped stiffener 变形肋条
crimped web stiffener 梁腹变形肋条;梁腹变形加固件
crimped wire 波纹钢筋;压波钢丝;皱纹钢丝;脆性钢丝;波形钢丝;波形钢筋
crimped wire screen 皱纹金属丝网格
crimper 褶皱器;折波机;压折器;卷曲机;钢筋折皱机;刻压机;卷缩机;卷边机;压紧钳;折缝机;钢筋弯折机;弯皱器;弯折机
crimping 卷边;咬口;咬边;皱痕压制;皱缩;波形皱缩;弯边
crimping machine 折边机;轧瓦楞机;拷花机;咭轧机
crimping plate process 卷板加工

crimping pliers 雷管夹钳;卷绕钳;卷边钳
crimping press 卷圆压力机;波纹压力机
crimping roll 槽纹辊;深槽纹辊
crimping roller 肋纹滚压机
crimple 皱;折缝
crimp mesh 波形钢丝网
crimp pliers 卷钳
crimp-proof 防皱的;不皱的
crimp-proof finish 防皱整理
crimp seal 锯齿形焊缝
crimp weave 皱纹织造
crimp wire net(ting) 波形钢丝网
crimson 紫红;绯红(类颜料);变成深红色;深红色
crimson antimony 锑朱颜料;锑红颜料
crimson foliage 红叶树
crimson lake 克里木森色淀;梅红;西洋红;绯红色淀
crimson purple glaze 玫瑰紫釉
crimson toner 绯红色原
crimsony 绯红色
crinanite 橄沸粒玄岩
cringle 索眼;索圈
crinkle 皱;缩卷
crinkle-crankle 蛇形墙(18世纪)
crinkle crepe 泡泡绉;泡沫绉
crinkled 皱折的
crinkled bedding 卷曲层理;皱纹层理;皱痕层理;褶曲层理
crinkle finish 波纹面饰;皱纹(面)漆
crinkle finishing 末道砗
crinkling 皱折;起皱(纹)
crinkling lamina 波状纹层
crinoidal 海百合灰石
crinoidal limestone 海百合灰岩
Crinoidea 海百合纲
crinoline 护舱围板
crinophagy 粒溶作用
criosphinx 狮身羊头
cripple 沼土;搭脚架
crippled leapfrog test 记忆部件连续检查试验;踏步检验
crippled version 残缺版本
cripple rafter 搭接椽
cripple scaffold 扶臂脚手架;扶壁脚手架
cripple stud 短构件
cripple timber 变形木材
cripple window 跛窗
crippling 局部压屈损坏(指钢梁);局部失稳破坏;角撑;断裂
crippling load 弯曲荷载;破坏荷载;临界失稳荷载;临界(断裂)荷载;压临界荷载;断裂荷载
crippling loading 临界荷载;断裂荷载;断裂负载
crippling strain 折曲应变
crippling stress 临界应力;折曲应力;断裂应力;曲折应力
criptome(te)r 遮盖力计
crisis 骤退;极期;[复]crises
crisis freshwater resources 淡水资源危机
crisis motivation 救灾式的激励
crisis of confidence 信任危机
crisis theory of overinvestment 投资过多危机论
crisp 脆的;起皱
crispate 皱波状的
crispation 卷缩
crispature 皱纹
crispen 使图像轮廓鲜明
crispening circuit 图像轮廓加重电路
crispening current 钩边电流
crispness 脆碎性
criss 脊瓦模具
criss-cross 交叉线;交叉的;相互交叉的;成十字形;方格形;方格;十字形(的);交错;交叉成十字形;纵横交错;十字号
criss-cross arrangement of moving stair(case) 自动楼梯交叉布置
criss-cross escalators 剪刀式自动扶梯
criss-cross eyepiece 十字线目镜
criss-cross method 计方格法;方格计数法
criss-cross mixer 交叉拌和机;十字板拌和机
criss-cross mixing 纵横混合;十字混合
criss-cross motion 交叉运动;交错运动
criss-cross pattern 十字形图案;十字形花样;网纹形
criss-cross sampling 方格取样(法)
criss-cross structure 方格构造;十字结构
criss-cross traffic 纵横交叉的交通

crista marginalis 帽缘
crista ntertrochanterica 转子间嵴
cristianite 钙长石
Cristite 克利斯蒂特合金
cristobalite 白硅石;方石英;方晶石
Cristoffel symbol 克里斯托弗尔符号
cristol glass 模制玻璃组件(一种专利产品)
criteria and standard 规范和标准
criterion 准则;标准;判据;判断标准;判定法;判别条件;判别式;评价准则;[复]criteria
criterion and standard division 规范和标准分类
criterion continuous concentration 标准连续浓度
criterion for aircraft noise 飞机噪声标准
criterion for application of one way street 采用单向交通的条件
criterion for appraising economic results 经济效果(的)评价标准
criterion for bed form 床面形态判别准则
criterion for comparing alternative 比较方案的标准
criterion for electric(al) deisgn and installation 电气设计安装规范
criterion for evaluation 评定标准
criterion for evolution 考核指标
criterion for faulting 断层标志
criterion for measuring objective achievement 目标成就的衡量标准
criterion for noise 噪声判据;噪声评价准则;噪声(评价)标准
criterion for noise control 噪声控制(评价)标准
criterion for recognition of tectonic element 构造单元划分依据
criterion for reservation assessment 自然保护区评价标准
criterion for super-fluidity 超流动性判据
criterion for wheel climbing 脱轨公式
criterion function 判别函数
criterion group 标准组
criterion maximum concentration 标准最大浓度
criterion numeral 衡准数
criterion of bank stability 岸坡稳定性标准
criterion of buckling 折曲准则
criterion of choice 选择的准则
criterion of classification 分类标准;分类原则
criterion of conduct 行为准则;行为守则
criterion of control quality 品质指标
criterion of convergence 收敛判据
criterion of curve 曲线准则
criterion of degeneracy 退化判别式
criterion of failure 破坏准则
criterion of ntependence 独立性准则
criterion of intensity 烈度标准
criterion of inverse probability 反概率准则
criterion of likelihood 似然性准则
criterion of optimality 最佳化准则
criterion of perfect fidelity 高逼真度准则
criterion of pessimism 悲观准则
criterion of realism 现实主义标准
criterion of record(ing) evolution 记录评价标准
criterion of rupture 破坏准则
criterion of selection 选取标准
criterion of sensitivity 感光标准
criterion of similarity 相似判数;相似判据
criterion of streamline motion 流线型运动准则
criterion of strengthening 加固准则
criterion of water log control 排涝标准
criterion path scheduling 判别路径表
criterion pollutant 标准污染物
criterion score 标准分数
criterion search 标准检索
criterion stability 准则稳定性
criterion variables 标准变数
criterion well 基准孔;基准井
critesistor 热敏电阻
crith 克瑞(1克瑞等于0.0986克);气体重量单位
critical 临界的;决定性的;重要的;关键的
critical absorption 临界吸收
critical absorption wavelength 临界吸收波长
critical acceleration 临界角加速度
critical activity 关键工序;紧要作业;紧急工作
critical adjustment 临界调整
critical age 临界年龄
critical air blast 最低鼓风量
critical altitude 临界高度

critical amplitude 临界振幅
critical analysis 临界分析
critical angle 临界角
critical angle of boom 吊杆临界角
critical angle of incidence 临界入射角
critical angle of refraction 临界折射角
critical angle of total reflection 全反射临界角
critical angle of track detector 径迹探测器临界角
critical angle refractometer 临界角折射计
critical angle refractometry 临界角测折射法；临界角测折射率术
critical annealing 临界退火
critical anode voltage 临界阳极电压
critical aperture 临界孔径
critical approach speed 临界进口道车速
critical area 临界区域；临界区；临界面积
critical area of extraction 临界回采面积（使地表某点达到最大可能沉降量的面积）；临界开采区
critical argument 临界幅角
critical aspect ratio 临界长径比
critical assembly 临界装置
critical attenuation 临界衰减
critical band 临界频带
critical bar 险滩
critical bearing capacity of pile 桩的临界承载力
critical bed density 临界床密度
critical bed depth 临界床高(度)
critical blocking temperature 临界阻挡温度
critical blocking volume 临界阻挡体积
critical boiling 临界沸腾
critical bottom slope 临界底斜率
critical bottom velocity 临界(河)底速度
critical branch 紧急供电设备
critical break rate 临界破损率
critical buckling load 临界压曲荷载；压曲临界荷载
critical capillary height 临界毛细高度
critical cavitation 临界汽蚀现象；临界空化现象
critical cell concentration 临界细胞浓度
critical charge 临界负荷
critical circle 临界(滑)圆；临界图
critical circle of slope 土坡临界圆
critical closing pressure 临界闭合压
critical coagulant concentration 临界混凝剂浓度
critical coefficient 临界系数
critical coefficient of cavitation 临界汽蚀系数；临界空化系数
critical coefficient of water bursting 临界突水系数
critical cohesion value 临界凝聚值
critical coincidence frequency 临界符合频率
critical cold point 临界冷点
critical component 关键部件
critical compression pressure 临界压缩压力
critical compression ratio 临界压缩比
critical compressive force 临界压强；临界压力
critical compressive stress 临界压应力
critical concentration 临界浓度
critical concentration range 临界浓度范围
critical condensated temperature 临界凝结温度
critical condensation pressure 临界冷凝压力
critical condensation temperature 临界冷凝温度
critical condition 临界状态；临界条件；临界情况；极限条件
critical confining pressure 临界侧向压力
critical consolidate temperature 临界会溶温度
critical constant 临界恒量；临界常数
critical content residue of soil moisture 临界土壤残余水分
critical cooling rate 临界冷却速度
critical coupling 临界耦合；中肯耦合
critical course and depth monitor 临界航向与深度监视器
critical covered load 临界盖重
critical crack extension force 临界裂缝扩展力
critical crack length 临界裂纹长度
critical crack-opening displacement 临界裂纹张开位移
critical crack size 临界裂纹尺寸
critical crazing strain 临界微裂应变
critical cross-section 临界截面；关键断面
critical cross-section area 临界截面面积
critical crystal nucleus 临界晶核
critical current 临界电流
critical current density 临界电流密度
critical curve 临界曲线

critical damage 临界损坏；临界破坏；严重损坏；严重损害；严重破坏；致命损伤
critical damper 临界阻尼器
critical damp(ing) 临界阻尼；临界衰减；临界减幅
critical damping attenuation 临界阻尼衰减
critical damping factor 临界阻尼因子
critical damping ratio 临界阻尼比
critical damping resistance 临界阻尼电阻
critical damping seismograph 临界阻尼地震仪
critical damping value 临界阻尼值
critical dark-period 临界暗期
critical day length 临界昼长；临界日长度
critical defect 临界缺点；严重缺陷；致命缺陷
critical defective 临界缺陷
critical deficit 临界亏空额
critical definition 临界清晰度
critical deformation 临界变形
critical degree of pigmentation 临界着色程度；临界颜料添加量
critical degree of saturation 临界饱和度；极限充水程度
critical density 临界(交通)密度；极限密度
critical density of fill 回填的临界密度
critical dependence 临界依赖关系
critical depth 临界水深；临界深度
critical depth discharge measurement 临界深度流量测定
critical depth discharge measurement method 临界深度流量测定法
critical depth flume 临界量水槽
critical depth line 临界水深线
critical depth meter 临界量水槽
critical depth of evapo(u)ration 极限蒸发深度
critical depth of flow 临界水流深度
critical depth of foundation 开挖临界深度
critical depth of groundwater 地下水临界深度
critical depth of phreatic water 潜水临界深度
critical depth of underground water 地下水临界深度
critical depth of water 临界水深
critical depth of wave base 波蚀临界深度
critical design review 关键性设计的检查
critical deviator stress 临界偏应力
critical diagram 判定图
critical diameter 临界直径
critical diameter of explosive 药包临界直径
critical dielectric(al) flux density 临界电通量密度
critical dimension 临界尺寸；极限尺寸
critical dip 临界倾角
critical discharge 临界流量
critical displacement 临界位移
critical dissolved oxygen 临界溶解氧
critical distance 临界距离
critical division 临界分裂
critical docking draft 坞内临界吃水
critical drag (force) 临界阻力；临界拖力
critical draught 临界吃水
critical drought period 临界干旱期
critical dry period 临界枯水期
critical duration 临界持续时间
critical earth pressure 临界土压力
critical ecosystem 临界生态系统
critical edge pressure 临塑荷载；临时边界压力；临界荷载；临界边缘压力
critical effect 临界效应
critical electric(al) field 临界电场
critical elevation 临界高程；制高高程
critical endurance 临界持续时间
critical energy 临界能量
critical energy release rate 临界能量释放率
critical engineering problem 关键性技术问题；工程技术关键问题
critical environmental area 临界环境区
critical environmental management 临界环境管理
critical environmental pollution 临界环境污染
critical environmental problem 临界环境问题
critical equation 临界方程
critical equatorial velocity 临界赤道速度
critical error 临界误差；关键性误差
critical error angle 临界误差角
critical evaluation 临界状态评价；临界评价
critical examination 关键检验
critical exit gradient 临界逸出坡降
critical experiment 临界试验；临界实验

critical exponent 临界指数
critical exposure pathway 关键照射途径
critical extension ratio 临界伸长比
critical external pressure 临界外压
critical factor 临界因素；主要因素；关键因素
critical factor analysis 关键因素分析
critical failure 致命故障
critical field 临界(磁)场；标准范围
critical flicker frequency 临界闪烁频率；临界闪光频率
critical flo(a)tation gradient 临界浮动梯度
critical flocculation concentration 临界絮凝浓度
critical flocculation temperature 临界絮凝温度
critical flocculation volume 临界絮凝体积
critical flow 临界流量；临界水流；临界流(动)
critical flow criterion 临界流态判别准则
critical flow depth 临界水流深度
critical flow diagram 临界流计算图
critical flowmeter 临界流量计
critical flow nozzle 临界截面喷管
critical flow prover 临界流量计
critical flow rate 临界流量
critical flow state 临界流态
critical flow velocity 临界流速
critical fluidization 临界流态化
critical flutter speed 临界颤振速率
critical flux 临界通量
critical focus 临界焦点
critical force 临界力
critical form (结构的)危形
critical form structure 危形结构
critical foundation depth 地基临界深度
critical fraction defective 临界不良率
critical frequency 临界频率；截止频率；熔结临界频率
critical frequency band 临界频带
critical friction velocity 临界摩阻流速；临界摩擦流速；泥沙起动流速
critical frost strength 极限冻结强度
critical Froude number 临界弗劳德数
critical function 判定函数
critical fusion frequency 临界停闪频率；临界融合频率；临界熔解频率
critical gap 临界空挡
critical geologic(al) discontinuity 临界地质不连续面；危险的地质不连续面
critical gradient 临界梯度；临界坡降；临界坡度
critical graph 临界图
critical ground motion 临界地面运动
critical group 关键人群组
critical growth rate 临界生长速率
critical growth velocity 临界生长速度
critical hardening 临界淬火
critical hardening temperature 临界淬硬温度
critical head 临界水头
critical head loss 临界水头损失
critical heat 临界热；相变热；转化热
critical heat flux 临界热流
critical height 临界(开挖)高度
critical height of face 开挖面极限高度
critical height of gas column 临界气柱高度
critical height of oil column 临界油柱高度
critical height of roadbed on soft soils 软土路堤极限高度
critical height of slope 土坡临界高度
critical high temperature 临界高温
critical humidity 临界湿度
critical hydraulic gradient 临界水力梯度；临界水力坡降
critical illumination 临界照明；临界照度
critical impact velocity 临界冲击速度
critical inclination 临界倾角
critical industry 尖端技术部门
critical information 关键信息
critical inlet pressure 临界入口压力
critical input 关键性投入
critical intensity 临界强度
critical interface tension 临界表面张力
critical intersection 关键交叉口
critical intersection control 临界路口控制；关键交叉口控制
critical interval 临界范围
critical isotherm 临界等温线
critical isothermal 临界等温线

criticality 临界(状况);临界(性)
criticality safety 临界安全
critical job 关键工程
critical lane 紧急备用车道;关键车道
critical lane detection 临界车道检测;关键车道检测
critical lane technique 关键车道法
critical lapse rate 临界直减率;关键衰减率
critical layer 临界层
critical leaching rate 临界渗出率
critical length 临界长度
critical length of grade 临界坡(道)长(度)
critical level 临界面;临界高度;临界水平
critical level of escape 逸逸临界高度
critical limit 临界(极)限
critical line 临界线;紧急线
critical link 临界环节;限制性环节
critical liquid water content 临界液态含水率
critical load approach 临界负荷法
critical-load design 临界负荷设计
critical load factor 临界负荷系数
critical load(ing) 临界荷载;临界负载;临界负荷;临界装载量;破坏负载;断裂负载
critical load(ing) level 临界负荷水平
critical load of pile 桩的临界荷载
critical locus 临界线
critical loop 关键环路
critical loss 临界损失
critically corrected lens 精密校正透镜
critically damped circuit 临界阻尼回路
critically damped seismograph 临时阻尼地震仪;临界阻尼地震仪
critically ill 垂危
critically safe 临界安全的
critical Mach number 临界马赫数
critical magnetic field 临界磁场
critical magnetic flux density 临界磁通密度
critical mass 临界质量
critical mass-ratio 临界质量比
critical material 临时用材料;紧缺材料;限制分配的材料;应配材料;应急物资;重要原料;关键物资;关键材料;供应紧张的材料;欠缺材料
critical maximum 临界最大值;临界极大值
critical maximum concentration 临界最大浓度
critical melting point 临界熔点
critical metacentric height 临界稳心高度
critical method 临界法
critical micelle concentration 临界微胶粒浓度;临界胶团浓度;临界胶束浓度
critical microoperation 临界微操作;关键微操作
critical mode 临界简正波
critical moisture 临界水分;临界含水率
critical moisture content 临界水分含量;临界含水率
critical moisture point 水分临界点
critical moment 临界时限;临界(力)矩;决定性时刻;紧要关头
critical movement 关键运行;关键流向
critical node 临界节点
critical normal pressure 临界法向压力
critical normal stress 临界法向应力
critical nucleus 临界晶核
critical number 临界值
critical numerical value 临界数值
critical oil content 临界含油量
critical opalescence 临界乳光
critical operation 临界状下下运行;临界操作
critical outlet diameter 卸料口临界直径
critical oxygen tension 临界氧张力
critical parameter 临界参数
critical parameter of flow 临界流动参数
critical part 要害部位;主要机件
critical particle size 临界粒径
critical path 临界通路;最短路线;关键线路;关键路线;关键路径
critical-path accounting 统筹法会计
critical-path analysis 关键路线分析;统筹分析
critical path diagram 最优路线法;统筹方法;施工组织设计优选法
critical path method 关键工序线路法;临界路线法;紧急线法;最优路线法;最短路径法;关键路线法;关键路径法;关键路线方法;统筹(方)法;施工组织设计优选法
critical path method schedule 关键线进度计划;关键路线法进度表

critical path planning 关键路线规划
critical path scheduling 关键路线图;关键路线制表;关键路径调度
critical path technique 施工组织设计优选法
critical path test generation 临界通路测试产生法
critical path time 临界通路时间
critical period 临界期;关键时期
critical period of expansion 膨胀危险期
critical period of growth 临界生长期
critical period of post-exposure 接触后的临界期
critical pH 临界酸碱度;临界pH值
critical phase 关键阶段
critical phenomenon 临界现象
critical phosphorus concentration 临界磷浓度
critical pigment volume concentration 临界颜料体积浓度
critical plane 关键面;临界面
critical plastic load 临塑荷载
critical point 临界点;要点;转折点;驻点;关键时刻
critical point of declination 赤纬临界点
critical points on the ring fabrication 支承环的制造要点
critical point tester 相变点测定器
critical point theory 临界点理论
critical pollutant 关键性污染物
critical pollutant concentration 临界污染物浓度
critical pollution soruce 临界污染源
critical pollution source area 临界污染源区
critical pore 临界孔隙
critical porosity 临界孔隙率;临时孔隙(度);临界孔隙度
critical position 临界状态;临界位置;要害部位
critical potential 临界(电)位;临界电势
critical power 临界功率
critical pressure 临塑压力;临界压强;临界压力
critical pressure in buckling 临界失稳压力
critical pressure ratio 临界压力比
critical probability 临界概率
critical process 关键工序
critical property 临界性质
critical pump rate 临界泵抽水率
critical quantity 临界量
critical quenching rate 临界淬火速度
critical queuing distance 临界排队距离
critical race 临界竞争
critical radiant flux 临界辐射流
critical radius 临界半径
critical radius for accretion 临界吸积半径
critical rain 临界(降)雨量
critical rainfall 临界雨量
critical rainfall intensity 临界降雨强度
critical range 临界程;临界区;临界距离;临界范围
critical range of Reynolds number 雷诺数的临界(值)范围
critical rate 临界(速)率
critical rate of salinity decrease 临界盐度降低率
critical ratio 临界比(率)
critical Rayleigh number 临界瑞利数
critical reach 关键河段;容泄水流最底溶氧度
critical reaction 危急反应
critical reactor 临界(状态反应)堆
critical real-time processing 临界实时处理
critical receiving water condition 临界纳污水体条件
critical reference surface 关键基面
critical reflection 临界反射
critical refraction 临界折射
critical regeneration 临界再生
critical region 临界区域;临界区;关键地区;判域;判别区域
critical relative humidity 临界相对压力
critical relative moisture 临界相对湿度
critical resistance 临界电阻
critical resonator 临界谐振腔
critical restabilization concentration 临界再稳定浓度
critical return velocity 临界返速
critical revolutions 临界转数
critical Reynolds number 临界雷诺德数
critical rotational speed 临界转速
critical roughness 临界糙度
critical runoff 临界径流量
critical salt concentration 临界盐浓度
critical sampling interval 临界抽样区间

critical sampling rate 临界抽样率
critical saturation 临界饱和度
critical scouring velocity 临界冲刷速度;临界冲刷流度
critical section 临界截面;临界区段;临界区;断面;临界段;限制区间;关键河段;危险截面;危险断面
critical section for moment 力矩控制截面
critical seepage velocity 临界渗透速度
critical sequence 关键程序
critical settlement criterion 临界沉降标准
critical settling point 临界沉淀点
critical sharpness 临界锐度
critical shear(ing) stress 临界切应力;临界剪(切)应力
critical shear strain energy theory 最大剪(切)应变能量理论
critical shoal area 危险浅滩区
critical shortage 严重短少
critical shortage of 奇缺
critical situation 紧急情况
critical situation notification 临界状态通知;紧急状态通知
critical size 临界体积;临界大小;临界尺寸
critical slide 临界滑动
critical slide surface 临界滑动面
critical slip circle 临界滑弧;临界滑圆;临界滑动圆(弧)
critical slit width 临界缝宽
critical slope 临界坡度;临界比降;休止角
critical slope angle 临界坡角
critical social value 临界社会价值
critical solidification rate 临界凝固速度
critical solubility 临界溶解度
critical solution temperature 临界溶液温度
critical span length 临界跨距
critical specific speed 临界比转速
critical speed 临界转速;临界速率;临界速度;临界车速
critical speed of mill 磨机临界转速
critical speed of revolutions 临界转速
critical speed of second order 第二临界转速
critical speed of shaft 轴的临界转速
critical spinning temperature 临界拉丝温度
critical stability 临界稳定性
critical stabilization concentration 临界稳定浓度
critical stage 临界水位;决定性阶段;病危期;危险期
critical standard penetration blow counts 临界标准贯入击数
critical state 临界(状)态
critical state energy theory 临界状态能量理论
critical state line 临界状态线;临界物态线
critical state locus 临界状态轨迹
critical state model 临界状态模型
critical state of flow 临界流态
critical state soil mechanics 临界状态土力学
critical state strength 临界状态强度
critical statistics 临界统计
critical steel ratio 最小配筋率;最小钢筋比
critical steel stress 钢筋极限强度
critical stirring speed 临界搅拌速度
critical storm runoff 临界暴雨径流
critical storm water flow 临界暴雨径流量
critical strain 临界应变;临界变形
critical strain energy release rate 临界应变能释放率
critical strength 临界强度
critical stress 极限应力;临界应力
critical stress circle 极限应力图
critical stress for fracture 断裂的临界应力
critical stress fracture toughness 临界应力断裂韧度
critical stress intensity 临界应力强度
critical stress intensity factor 临界应力强度系数;断裂韧性;临界应力强度因子
critical stress ratio 极限应力比
critical structural section 结构危险断面;结构临界断面
critical suction head 临界吸出高度
critical surface 临界(滑动)面
critical surface tension 临界表面张力
critical survival temperature 临界生存温度
critical system 临界系统
critical table 临界表;判定表

critical technique 关键性技术
critical temperature 临界温度
critical temperature difference 临界温度差
critical temperature of a plasma 等离子体的临界温度
critical temperature range 临界温度范围
critical temperature resister 临界温度电阻器
critical temperature thermistor 临界温度热敏电阻器
critical test 决定性考验
critical test concentration 临界试验浓度
critical thickness 临界厚度;中肯厚度
critical thickness of oil film 临界润滑膜厚度
critical throat area 喷嘴临界面面积
critical throat section 喷嘴临界截面
critical thrust 极限推力
critical time of gate closure 关闭闸门的临界时间
critical tractive force 临界推移力;临界牵引力;泥沙临界流速
critical tractive power 临界推移力;临界牵引力
critical tractive stress 临界推移应力
critical tractive velocity 临界起动流速;起动流速
critical traffic density 临界交通密度
critical transfer pathway 关键转移途径
critical value 临界(数)值;界值
critical value for buckling 临界曲率值
critical value of Reynolds number 临界雷诺德数
critical value signally 极限信号
critical value signally device 极限信号装置
critical velocity 临界流速;临界速度
critical velocity of sediment movement 泥沙运动临界速度
critical velocity of super-flow 超流动临界速度
critical vibration 临界振动
critical viewing distance 临界明辨距离
critical viscosity 临界黏度
critical void ratio 临界孔隙率;临界孔隙比;极限孔隙比
critical void ratio of soil 土的临界孔隙比
critical voltage 临界电压
critical volume 临界体积;临界容积;临界量
critical volume fraction 临界体积分数
critical wastewater demand 临界污水需求量
critical water point 水分临界点
critical water quality 临界水质
critical water quality characteristic 临界水质特征
critical wave length 临界波长
critical wave number 临界波数
critical weight on drill string 钻具临界钻压
critical wind speed 临界风速
critical wind speed of flutter 颤振临界风速
critical wind-speed of overturning a train 列车临界翻车风速
critical wind-speed of overturning single car 单车临界翻车风速
critical wind velocity 临界风速
critical withstand voltage 临界耐电压
critical work 临界功
critical zone 临界区(域);临界带;危险地区;危险地带
critrcal inversion density 临界反转密度
crivapo(u)rbar 临界蒸气压力
crivetz 克里维次风
crizzle 裂子;表面微裂纹;表面粗糙;平板玻璃表面裂纹,皱纹
crizzle skin 裂纹层
crizzling 表面缺陷
croaker 黄花鱼;石首鱼
Crobalt 克拉巴尔特铬钴钨钢
Crocar 克拉卡铬钒钴硅钢
Crocco's equation 克罗柯方程
Crocco-type nozzle 克罗柯氏喷管;单侧壁喷管
croceic acid 溯羟萘磺酸
croceous 藏红花色
crochet 编织器
crochet galloon machine 钩编织带机
crochet hook 钩针
crocheting machine 钩编机
crochet-look fabric 仿钩编织物
crochet machine 钩编机
crochet needle 钩针
crochet warp-knitting machine 钩编经编机
crociate 藏红花
crocidolite 钠闪石石棉;硅酸钠石棉;青石棉

crocidolite asbestos 青石棉
crocidolite-asbestos deposit 兰石棉矿床
crock 下水瓦管;破旧的汽车;瓦罐;瓦管;坛子
Crockatt magnetic separator 克罗克特型磁选机
crocked finish 歪口(玻璃制品缺陷)
crockery 瓦罐;陶器;瓦器
crocket 卷叶饰凸雕(哥特式建筑);卷叶(形花)饰
crocket capital 卷叶饰柱头;卷叶形柱头
Crockett separator 克劳凯特带式磁选机
crocking 摩擦掉色;泅色;不明显的起霜
crock tile 釉面沟瓦
crocodile 鳄鱼皮纹;平板运输机
crocodile clip 鳄鱼(嘴)夹;颚式夹子;弹簧线夹
crocodile hand lever shearing machine 鳄鱼式手柄剪床
crocodile shearing machine 杠杆式剪切机;鳄式剪床;鳄口(杠杆)剪切机
crocodile shears 杠杆(式)剪切机;鳄鱼剪;鳄式剪床;鳄口剪切机
crocodile skin 鳄鱼皮(缺陷)
crocodile spanner 鳄头扳手
crocodile squeezer 杠杆式弯曲机;鳄式压轧机;鳄式压挤机;鳄式挤压机;颚式挤压机
crocodile truck 鳄鱼式敞车
crocodile-type cutting machine 鳄式剪切机
crocodile wrench 管扳手;鳄形扳手
crocodiling 表面龟裂(粉饰、油漆);路面龟裂;龟裂;鳄纹装饰法;鳄纹;深度细裂纹
crocoisite 红色天然铬酸铅矿
crocoite 铬铅矿
crocus 橘黄色;秸红色;氧化铁研磨粉;研磨粉;藏红花;抛光粉
crocus abrasive 氧化铁抛光粉;氧化铁磨料
crocus cloth 磨粉布;细砂布;氧化铁(细)砂布;粗袋布
crocus martis 擦粉;抛光粉
crocus of antimony 锑藏红
crocus of mars 擦粉;抛光粉
Crodi 克拉迪铬钨锰钢
croft 小农场;地下通道;小田地
crofting 亚麻漂白
Croixian 库拉辛统【地】
crolite 陶瓷绝缘材料;陶瓷绝热材料
Croloy 铬基合金;铬钼耐热合金钢
Croma 克拉马铬锰钢
Cromadur 克拉马杜尔铬锰钒钢
Cromal 克拉马尔铝合金
cromalin 铝(合金)电镀法
cromalin colo(u)r proofing system 色层叠合打样法
cromaltite 黑榴霓霞岩
Croman 克拉曼铬锰钼铬硅钢
Cromansil 克拉曼西尔铬锰硅钢
crome metal 金属铬
Cromerian interglacial stage 克罗默尔间冰期
cromflex oil piston ring 钢制带胀圈的活塞环
cromlech (史前时代的)大石台;(史前时代的)环列石柱
cromn shaft 冠茎
Cromovan 克拉莫尔铬钼钒钢
Cromwell current 克伦威尔潜流
cronak method 常温溶液浸渍法
cronak process 常温溶液浸渍法;锌版防蚀处理
Cronapress 克罗纳转换法
Cronifer 克拉尼弗镍铬合金
Croning method 克郎宁壳模铸造法
Croning process 克郎宁壳模铸造法
cronite 镍铬耐热合金(因科镍)
Cronix 克拉尼克斯镍铬合金
cronstedtite 绿锥石;克铁蛇纹石
crook 挠曲;曲柄牧草棍;弯曲(材)
crooked alignment 曲折定线
crooked bevel 扭曲斜边
crooked chisel 铁钎
crooked curve 复曲线
crooked dealing 不正当交易
crooked drill pipe 印模;弯钻杆
crooked hole 斜孔;歪斜井
crooked hole country 斜孔弯斜地区
crooked iron 弯头捻缝凿
crooked leg 弯腿
crooked line 弯线
crooked nail 钩头道钉
crookedness 曲率;弯曲曲率

crooked road 弯路
crooked timber 弯曲木(材);弯木(料)
crooked well 偏斜井
Crookes dark space 阴极暗区;第二阴极暗区
Crookes glass 克鲁克斯玻璃
crookesite 硒铊银铜矿
Crookes radiometer 克鲁克斯辐射计
Crookes tube 克鲁克斯阴极射线管
crooking of the tubes 炉管弯曲
crook rafter 曲椽
crook tongs 弓形开度钳
crook warp 径向翘曲
crook warping 板材(上凸)纵向挠曲
croop property 蠕变性(能)
crop 露头;剪切;农作物;弧拱;叶板饰;收成
crop and cobble flying shears 端头废料飞剪
crop and cobble shears 端头废料剪切机
crop and root 总去
crop basal area 林作底面积
crop beam 船闸门框顶梁
crop coal 露头煤
crop cut 突然发生
crop cutter 刨草机
crop damage 作物伤害
crop density 林作密度
crop disposal 清除切头
crop disposal bin 废料收集坑
crop dusting 农药喷射器
crope 尖顶饰(旧称)
crop elevator 谷物升运器
crop end 去尾;切头;麻梢
crop end pusher 切料头推出机
crop failure 作物损坏
crop fair and refit 部分取下修复后再装上
crop forecast 收成预测
crop growing period 作物生长期
crop handling 清理切边
crop height 林作高
crop insurance 收成保险
crop insurance scheme 收成保险计划
crop irrigation requirement 作物灌溉需水量
cropland 连续轮作地;农田水利工程;农田;作物地;耕作地;耕地;田地
crop loss 农作物损失
crop loss compensation 农田青苗补偿;青苗补偿
crop map 农作物图
crop micrometeorology 作物微气象学
crop out 露出;出现;出露地表;露头
cropped 裁切不正
cropped ear 短耳
crop ped location 耕作区
cropped soil 耕作土(壤)
cropper 截断器;剪料头机;剪边机;修剪机;收获机
cropper machine 剪料头机
cropping 截弃;剪料头;修剪;勾画;切料(头);图片裁切;收割;门窗框角突肘;剪边;拱楔肩
cropping cut 碎边剪切
cropping die 切边模
cropping index 种植指数;复种指数
cropping machine 切料头机
cropping-out 露头地表
cropping-out of the groundwater 地下水露头(处)
cropping pattern 耕作形式
cropping principle 收获原理
crop(ping)shears 剪床;切线剪;切料头机;剪料头机
cropping system 耕作制度
crop plant 作物
crop planting machine 农作物种植机
crop-processing equipment 谷物加工设备
crop prospect 收成估计
crop protection 庄稼保护;植物保护
crop rate 出栏率
crop residue 残茬;作物残体;收获残余物
crop residue management 残茬处理
crop rotation 轮作作物;轮作制;轮种
crop sailure 歉收
crops for rotation 茬口
crop share rent 收成比例地租
crop smothering 草害
crop succession 轮作顺序
crop tree 主林木;主伐木
crop-tree thinning 主林木疏伐

crop up 突然发生
crop water requirement 作物需水量
crop yield 农作物产量
croquet 槌球
crosette (承邻石用,有肩拱块的)肩位;拱楔肩;门窗上缘的筋形突出部;螺旋形支柱
crosier 拐杖
cross -nailed material 交叉钉住的材料;交叉口;逆浪;横的;横插进去;直角器;四通;十字形;十字路;十字交叉
cross absorption 交叉吸收
cross absorption test 交叉吸收试验
cross acceptance 相互承付;交互承兑
cross accident 交叉口交通事故
cross adaptation 交叉适应
cross adding 交叉加算
cross addition 交叉相加
cross adit 石门
cross adjustment 横向调整
cross agglutination test 交叉凝集试验
cross ahead of the other 横过他船船首
cross air blasting 侧吹风
cross air duct 交叉通风管道
cross aisle 交叉甬道;交叉通道;横过道
cross ampere-turn 交磁安匝;正交安匝
cross-anchored bogie 横向加固转向架
cross and bevan cellulose 氯化法分离的纤维素
cross and top slides 纵横刀架滑板
cross-angle of air and gas streams 空气燃气流相交角
cross arch 交叉拱
cross arched vault(ing) 交叉穹窿
cross area of mud-stone flow body 泥石流体横断面积
cross arm 横担;横臂;线担;托架
crossarm brace 紧固物
cross arm pin 横担销针
cross-arm settlement ga(u)ge 十字臂沉降仪;十字臂沉降计;十字臂沉降管
cross arrangement 交叉排列
cross assembler 交叉汇编程序
cross assembly 交叉汇编
cross-assignment 交叉分配
cross auger 麻花钻
cross axis 横坐标轴
cross axle 十字轴
cross-axle undercarriage 十字轴式起落架
cross baffle 折流板
cross band 木纹交叉黏结;花边条;交叉频带;内层交叉单板;直交单板;中板
crossband beacon 交叉频率问答器;交叉频带信标
crossband-beacon system 交叉频带信标制
cross-banded lumber veneered board 木纹交叉的胶合板
crossband gap 镶板离缝
cross banding 交叉排列;交叉结合;频率交联;直交单板层;横纹饰带;横纹板层(胶合板中)
cross banding veneer 内层薄板
crossband principle 多频(率)收发原理;不同频率收发原理
crossband veneer 中板;胶合板逆纹内层;薄板
cross bar 四通管;十字形横挡;十字格版栏门门;横撑木;人孔撑架;四通管;十字杆件;横杆;拦车横木;箱带;挺杆
crossbar addressed dot matrix 正交线寻址点矩阵
crossbar automatic exchange system 纵横制自动交换方式
crossbar automatic telephone system 纵横自动电话交换机
cross bar centers 横杆中心距
crossbar connector 纵横制接线机
crossbar conveyer 横杆链式输送机;横杆输送机
cross bar file 交叉锉
cross bar handle 十字把手
crossbaring 安顶梁
crossbar knockout 横杆顶出
crossbar micrometer 十字丝测微计
crossbar of the mo(u)lding box 箱筋
crossbar over frames 隔肋棒
crossbar selector 纵横制选择器
crossbar slings 十字栅门;交叉杆吊具
crossbar switch 交叉开关;纵横开关;纵横制交换机;纵横式交换机
crossbar system 纵横系统;交叉制;纵横制

crossbar telephone center 纵横制自动电话局
crossbar telephone switching system 纵横制电话交换系统
crossbar transformer 纵横变换器
crossbar transition 交叉变换
cross bath 十字形槽
cross battens 交叉格;交叉压条;交叉条板;交叉板条
cross bay 横向间距;横跨
cross beam 主横梁;中间横梁;交叉梁;横肋梁;横撑;辅助横梁;交叉十字架;横梁;横支梁;横臂;顶梁;十字支座;十字梁
cross beam lifting roller 横梁升移辊
cross beam system 交叉梁系
cross bearer 横梁;桥铺板搁条;炉栅搁条;支持炉格的横杆
cross bearing 交会;交叉探向;交叉方位;交叉定位;侧面连测
cross beat 交叉脉动;交叉差拍
cross beater mill 交叉冲击磨
cross-bed 交错层
cross-bedded 有交错层理的
cross-bedded rock 有交错层理的岩石
cross-bedding 交错铺砌;交错衬垫;十字纹基床;交错层【地】;交错层(理)
cross bedding structure 交错层理构造
cross below a river 河底隧道
cross belt 交叉带
cross belt conveyer 横向皮带输送机;横向胶带输送机
cross belt separator 交叉皮带选换机
crossbend 交叉弯曲
cross bending 横向支撑;横向联系杆;交错弯曲;横向弯曲
cross bending strength 抗折强度
cross-bending stress 横(向)弯(曲)应力
cross-bending test 横向弯曲试验
cross-bend test 抗折试验
cross berth 横向泊位
cross bevel 交叉棱面;十字花纹的斜边
crossbill 交啄鸟
crossbill joint 交啄接头;叉口接合
cross binding 横向联系杆;横向连接;横向支撑;横向弯曲;交叉联
cross bit 十字扁铲;十字钻头
cross bit cutting edge 十字钻头刃口
cross bite 反咬合
cross bite occlusion 交叉牙合
cross bitt 系缆桩横杆;交叉缆桩;十字系缆桩
cross-bladed chisel bit 十字形钻(头)
cross-blast explosion pot 正交喷吹灭弧盒
cross-blast oil circuit-breaker 正交喷吹油断路器
cross bleed 交叉排气
cross block 十字形房屋
cross block layout 过街坊式布置;穿过街坊式布置;穿过式布置
cross board 配电开关板;交叉板;转换配电箱
cross board hut 转换配电盒
crossbolster 横向架梁
cross bolt 横销
cross-bolt lock 横销锁;横插销
cross bond 交联键;交错砌合;交叉扎线;交叉砌合
cross bonding box 交叉互联盒
cross-booking 交叉登记工时
cross bore 横孔
cross boring 横向镗孔
cross box 交叉式选择框;交叉(分线)箱;横连接箱;转换箱
cross brace 肋板;交叉(斜)撑;剪力撑;剪刀撑;横拉条;(横(刀)撑;十字刀撑;十字撑(条)
cross-braced bogie 带横斜撑转向架
cross-braced end frame 剪刀撑加固的端框
cross braced frame 横支架
cross bracing 横联交叉支条;交叉支托;交叉撑;剪刀撑系统;横撑
crossbracing 副对角撑
cross bracing of the bottom chord 下弦剪刀撑
cross bracing of the tope chord 上弦剪刀撑
cross bracing of timber longitudinal beam 木纵梁间的交叉撑
cross bracing wire 横拉线
cross braided rope 穿插编缆
cross brake 十字形弯
cross branch 四通管;十字管
cross break 横裂;横向破裂;横向开裂;横纹开裂

(木材);横折
cross breaking 横(向)断(裂)
cross-breaking strength 抗折强度;挠曲强度;木横纹破坏强度
cross-breaking strength of wood 木横纹破坏强度
cross-breaking test 横向挠曲试验
cross-bridge 天桥;交联桥;横桥
cross-bridge cycle theory 横桥循环学说
cross-bridge system 横桥系统
cross bridging 鱼骨形撑;交叉支杆;交叉撑;剪刀撑;格栅十字撑
cross brush 正交电刷
cross-bubble 十字管气泡水准仪
cross-buck sign 道口警告标志;交叉口标志
cross bud grafting 重复芽接
cross building 十字形平面房屋
cross bulkhead 横舱壁
cross bulk sign 交叉警告标(平交道口)
cross bunker 横向燃料舱;横煤舱
cross-bunt on 横向楔入;横向撑木;横向撑杆
cross calibration 交互标定;交叉标定
cross call 交叉呼叫;交叉调用;相互调用
cross cap 十字分配料道
cross chain 双锚单交叉;十字链
cross-channel 横向水道;交叉流道;横渡海峡的
cross channel vessel 英吉利海峡渡船
cross chart 交叉图;穿越图
cross check 交互(检查)核对;交叉校验;交叉检验;反复核计;画线支票;核对工会授权卡;相互校验;相互核核;相互核对;相互检验;相互检查;用不同方法校核;交叉核对
cross-checking 重复检验
cross chipper hammer 爪形冲击锤
cross chisel 掏槽凿;十字凿
cross-chopping bit 十字形钻头(破碎孔底岩柱用)
cross church 十字形教堂
cross circulation 交叉循环
cross classification chart 交叉分类表
cross cliff 横断崖;岩坝
cross clinker drag 横向熟料耙
cross coating 交叉涂覆法;十字涂覆法
cross cock 四通阀
cross cocking 雄榫接合;木榫接合
cross cogging 装榫
cross-coherence 交叉相干
cross coil 交叉线圈
cross coil instrument 交叉线圈式仪表
cross collar 环形
cross colo(u)r 串色;色失真;颜色失真
cross-colo(u)r noise 亮度串色
cross column 十字形柱
cross-comparison 交叉比较
cross compile 交叉编译
cross compiler 交叉编译(程序)
cross-compiling and assembling 交叉编译和汇编
cross component 横向分量;横向分力;侧向分量;侧向分力
cross component force 侧向分力
cross compound 并列多缸式
cross compound air compressor 并列复式空气压缩机
cross-compound arrangement 双轴结构;双轴布置
cross compound blowing engine 复式鼓风机
cross compound compressor 并列复式压气机
cross compound cycle 并联复式循环
cross compound engine 并联复式蒸汽机
cross compound gas turbine 并联复式汽轮机
cross compound locomotive 双缸复合机车
cross compound steam turbine 并联式汽轮机
cross compound turbine 横列式高低压涡轮机;横列高低压透平;横列式高低压汽轮机;双轴式汽轮机;并联复式涡轮机
cross compound unit 双轴机组
cross-computer software 交叉计算机软件
cross conduit 横向暗管
cross configuration 交叉配置
cross conformation 交叉构造
cross conjugation 交叉共轭
cross connected generator 正交磁场发电机
cross connected sheet 交叉连接片
cross-connecting area 交接区
cross-connecting board 纵横制交换台
cross connecting circuit 交接电路
cross connecting distribution 交接配线

cross connecting iron clamp 交叉铁夹
cross connecting piece 万向节十字头
cross connection 线条交叉;四通连接;四通管接头;双重接头连接;交叉连接;交叉接头;交叉接合;连通;跨接;横向连接;横贯通道;水管连通(饮用水与工业水串通);十字接头
cross-connection in piping system 上下水管道系连通
cross connection pattern 交叉连接图
cross connection sheet 交叉连接片
cross-connexion field 跨接线排
cross contactor 错流离心萃取机
cross-contamination 交叉污染(物)
cross control rod 十字形控制杆
cross conveyer 交叉输送器;横向输送器;交叉输送带;横向输送机;横向运输机;横向输送机;横式运输机
cross conveyer jack 横向运输机千斤顶
cross conveyer with turner 横向输送翻板机
cross core 胶合板层间的纹理交叉
cross corking 木榫接合
cross-correlation 交互相关;互相关;交互作用;交叉相关;互相关联
cross-correlation detection 互相关探测;互相关检测;相关检波
cross correlation flowmeter 互相关流量计
cross-correlation function 交叉相关函数;互相关函数
cross-correlation interferometer 互相关干涉仪
cross-correlation radar 交叉相关雷达
cross-correlation receiver 互相关接收机
cross correlator 互相关器
cross correlogram 交叉相关图;互相关图
cross-country 越野地方;越野的;断绝地
cross-country cargo carrier 越野载重(汽)车
cross-country chassis 越野车底盘
cross-country crane 越野吊车
cross-country mill 穿梭式轧机
cross-country minibus 越野小型汽车;越野面包车
cross-country motorcycle 越野摩托车
cross-country operation 越野行动
cross-country performance 越野性能;越野行动
cross-country power 越野性能;长程行车能力
cross-country route 横贯全国线路;横贯河谷线路
cross-country run(ning) 越野行驶
cross-country sample 全国性样本
cross-country service 越野运行
cross-country track 越野路线
cross-country truck 越野载重(汽)车
cross-country tyre 越野轮胎
cross-country vehicle 越野汽车
cross coupling 互耦;交叉耦合
cross-coupling coefficient 交叉耦合系数
cross-coupling correction value 交叉耦合校正值
cross-coupling effect 交叉耦合效应
cross-coupling optic(al) beam 交叉耦合光束
cross course 交叉航向
cross covariance 互协方差;互积差
cross-covariance function 交叉协方差函数;互协方差函数;交互协方差函数
cross crack 交叉裂纹;交叉裂缝;横裂纹
cross crank 前后向曲轴
cross cultivation 交叉耕作
cross culvert 横向涵洞
cross-cupola 四角圆顶小阁;角阁;交叉圆顶阁
cross currency risk 交叉通货风险
cross curreney interest accrual 交叉通货利息的增加额
cross current 交叉水流;交叉流(量);横向流(动);横向环流;横向海流;横流;正交流;错流;涡流;交错流;横向水流;临流
cross-current cooler 逆流冷却器;逆流冷却机
cross-current extraction 串联萃取
cross current gravity separator 涡流重力分离器
cross-current intensity 横流强度
cross-current mark 横流标
cross-current velocity 横向流速
cross curve 十字线
cross curved plate 拱波
cross curves of stability 稳(定)性交叉曲线
cross-cut 横导坑;联络坑道;捷径;龙锯;交切;交叉消隐;交叉纹路;横巷;横向切割;横向坑道;横割;横断;正交;穿脉坑道;横切
cross-cut adhesion test (试验涂膜黏结性的)棋格划痕试验
cross-cut arch saw blade 木工拉锯锯条
cross-cut bit 十字形钻(头)
cross-cut chisel 削凿;尖錾;扁錾;横切錾;十字形凿刀;横切凿
cross-cut cylinder bur 横切柱形牙钻
cross-cut end(of a beam) 端头表面;横切端头
cross-cut file 交叉纹锉;横割纹锉;双纹锉
cross-cut frame saw 横锯用锯床
cross-cut(ing) chisel 横切凿
cross cut-off wall 横(向)隔墙
cross-cut sail 横缝帆
cross-cut saw 横锯;横截锯;龙锯;截锯;横割锯
cross-cut saw blade 横割锯片
cross-cut saw machine 横割锯机
cross-cut shears 横剪机
cross-cutter 横向切割机;横切割器;裁切机
cross-cutter track 采板机导向装置
cross-cut test 划格法附着力试验;划格试验
cross-cutting 截件子;横截;造材;横切;横(向)切割
cross-cutting chisel 窄凿
cross-cutting machine 横向剪毛机
cross-cutting relationship 互切关系
cross dam 横坝;厂坝
cross damping 正交阻尼
cross deal 交叉交易
cross debt 相互抵销的债务;彼此可以抵销的债务
cross-defanlt clause 交叉弃约条款
cross default 交互不履行
cross default debt 三角债
cross detecting 交叉检测
cross-development 交叉开发
cross development system 交叉开发系统
cross diaphragm 横撑架
cross diaphragm cross frame 横隔板
cross diffusion 交叉扩散
cross dike 拦坝坝;拦海闸;横堤
cross dimensions 断面尺寸
cross direction 横向
cross direction of paper 纸张横向
cross discipline subject 交叉学科
cross disking 交叉耙
cross-disperser 横向色散器
cross distortion 交叉失真;横向扭曲
cross-distribution 交叉分布
cross ditch 横沟
cross division line 十字分划线
Cross-Doland method 克劳斯-杜兰法
cross-domain 跨越地域
cross-domain communication 跨越地域通信;跨域通信
cross-domain key 跨越地域键
cross-domain link 跨(越地)域链路
cross-domain logic(al) unit 跨域逻辑单元
cross-domain resource 跨(越地)域资源;交叉域资源
cross-domain resource management 跨域资源管理
cross-domain resource manager 跨(越地)域资源管理程序;交叉域资源管理程序
cross-domain subarea link 跨(越地)域分区链路;交叉域子区连接
cross-dome 四角圆顶小阁;交叉圆顶阁
cross-domed church 十字形拱顶教堂
cross draft gas producer 平吸式煤气发生炉
cross drain 横向排水沟;横向泥浆渠;横渠;间隔水沟;横贯水渠;横向排水管
cross drainage 交叉排水(沟);横向排水;过街沟
cross drainage cut-off 横跨流域取直
cross drainage method 横向排水法
cross draught kiln 横焰窑
cross drawing 交互开立;相互开发抵用票据
cross drill bit 十字钻头
cross drilling 横向钻孔
cross drilling attachment 横向钻孔附件
cross-drilling bit 十字形钻(头)
cross drive 交叉传动;横向传送
cross driver 十字螺丝起子;十字螺丝刀
cross drum boiler 横汽包锅炉;横鼓式锅炉
cross duct 交叉管道
cross dunnage 双层货垫
cross-dyed 多色染织品
cross-dyed fabric 交染织物
crossed 交叉的

crossed acoustic(al) response 交叉声反应
crossed arch 十字碹
crossed-arm pantograph 双臂受电弓
crossed axes angle 轴交角
crossed axes shaving method 交叉轴剃齿法
crossed axial plane dispersion 交叉光轴面色散
crossed axis 相错轴
crossed belt 交叉皮带
crossed check 画线支票;画线转账支票;平行线支票
crossed cheque 画线支票;不记名支票
crossed circles 双交圆
crossed classification 交叉分类
crossed-coil antenna 交叉环形天线
crossed control 交叉控制系统
crossed-core type transformer 交叉铁芯式变压器
crossed crank mechanism 交叉曲柄机构
crossed crown 十字碹
crossed cylinders 交叉圆柱镜
crossed double bond 横交双键
cross eddy 横向漩涡
crossed electric(al) and magnetic field 正交电磁场
crossed electric(al) magnet fields mass spectrometer 正交电磁场质谱仪
crossed field amplifier 正交场放大器
crossed field device 正交场器件
crossed field ga(u)ge 正交场电离真空计
crossed-field spectrometer 交叉场分光仪
crossed field tube 正交场管
cross-edged bit 十字形钻(头)
cross-edged drill 十字形钻具
crossed girdle 交叉环带
crossed grating 交叉光栅
crossed grating interferometer 交叉光栅干涉仪
crossed grating spectrograph 交叉光栅摄谱仪
crossed grating spectrum 交叉光栅谱
crossed grid 十字格
crossed-grid matrix 交叉网格矩阵
crossed grooves 交叉槽
crossed helical gear 交叉轴螺旋齿轮
crossed-lamellar texture 交错纹片结构
crossed masking 交叉掩蔽
crossed Nicols 正交偏光镜;正交偏光;正交偏振;正交尼科耳棱镜
crossed parallelogram 反向转动曲柄;平行四边形
crossed pits 十字纹孔对
crossed polar system 参差极化系统
crossed prism 正交棱镜
crossed products 交叉乘积
crossed sea 交叉浪;混浪
crossed spectra 交叉光谱
crossed static electric(al) and magnetic field 正交静电磁场
crossed strain 交错应变
crossed street 相交街道
crossed twinning 十字双晶
crossed weight index number 交叉加权指数
crossed wire 交叉接触线
cross effect 交叉效应
cross elasticity 交叉弹性
cross-elasticity of demand 交叉需求弹性;需求的交叉弹性
cross elevator 交叉升降机;横向升运机
cross-entry 抵销记入;交叉进入;横向巷道;石门
cross-equation error correlation 交叉方程误差相关
crosser 垫木;垫木;绳索机
crosser pipe 架空管道
crossette 门耳;窗耳;螺旋状支柱;突肩;柱顶过梁凸线脚;咬合接头;拱圈耳状线脚
cross exchange 交叉汇兑;通过第三国的汇兑;转汇
cross experimental design 正交试验设计
cross extrustion 横向挤压
cross fade 交叉混合叠像渐变;平滑转换
crossfade linearity 线性匀滑转换
cross-fading 交叉混合
cross fall 路拱高差;横斜度;横向坡降;横向坡度;横高差;横坡;断面高差;路面横向坡度;路拱横坡
crossfall controller 横坡定坡器
cross fault 横断层
cross feed 交叉馈电;交叉进给;交叉供电;交叉给进;横向喂入;横向输送;横向进刀;横向加料;进刀;串音;串馈;砂轮横移进给量
cross feed carbon 交叉进给印字带
cross feed control lever 横向进刀控制手柄

cross feed device 横向进给装置
cross feed motor 交叉馈电式电动机
cross feed screw 横进刀丝扛;横进刀螺杆
cross feed screw crank 横向进刀螺杆曲柄
cross feed screw dial 横向进刀螺杆指示盘;横进刀螺杆指示盘
cross feed screw handle 横向进刀丝杠手柄
cross feed system 交叉馈电系统;交叉进给系统
cross feed valve 交叉送水阀;交互送水阀
cross fiber 横行纤维
cross-fibered wood 扭纹木材;斜纹木材
cross field 交叉场;正交场
cross-field amplifier 交磁放大器;交磁放大机
cross-field bias 交叉场偏磁
cross field generator 交叉电场信号发生器;正交磁场发电机
cross field multiplier 正交场乘法器
cross field type electron beam multiplier 交磁场乘法器
cross file 椭圆锉
cross fire 交叉图形;交叉火力;信道间干扰;串扰电流;交叉火焰;串扰;串流;串报
cross-fired burner 横火焰燃烧炉;交射火焰灯头;交叉火喷灯
cross-fired furnace 横火焰窑
cross-fired tank furnace 横火焰池窑
crossfirer 联焰管
cross-firing 交叉提款
cross fissure 横裂缝
cross fitting 十字头
cross-five method 满五计数法
cross flame 横火焰
crossflame tube 联焰管
cross flight 横刮板
cross-flight photography 构架航线摄影
cross flooding fitting 横倾调整装置;横贯注水装置
cross-flow 交叉流(量);交叉气流;交叉流动;叉流式;横向气流;错流;正交流动
cross-flow absorber 错流吸收器;错流吸附器
cross-flow blower 贯流式风机
cross-flow cake filtration membrane 横向流滤饼过滤膜
cross flow clean room 水平层流式洁净室
cross-flow component 横向流动分量
cross-flow convection 交叉气流对流;横向对流
cross-flow cooling tower 交叉流式冷却塔;横流式冷却塔
cross-flow fan 贯流式通风机;横流式风机
cross-flow filtration 横向流过滤;横向流超滤
cross-flow heat exchanger 叉流式热交换器;交叉流式热交换器
cross-flow induced draft cooling tower 抽风横流冷却塔
cross-flow membrane bioreactor 横向流膜生物反应器
cross-flow microfiltration flux 横向流微滤通量
cross-flow nucleation scrubber 错流核晶洗涤器
cross-flow packed scrubber 错流填充洗涤器
cross-flow radiator 横流式散热器;横流散热器
cross-flow rate 漫流流速
cross-flow regenerator 横流式回热器;横流回热器;交叉流回热器
cross-flow shaft kiln 简式窑;横流窑
cross-flow surfactant-based ultrafiltration 横向流表面活性剂超滤
cross-flow tower 横流塔
cross-flow tray 单溢流塔板
cross-flow turbine 双击式水轮机
cross-flow type cooling tower 贯流式冷却塔
cross flow valve tray 横流式浮阀塔盘
cross-flow velocity 横流速度
cross-flow vibration 横向振动
cross-flow zone refiner 纵横流动区域精炼炉;正交流动区域精炼炉
cross flux 交叉磁通
cross flying 切割线飞行
cross flying and control-network flying 切割线控制网飞行
cross-folded 交叉褶皱的;交叉折叠的
cross-fold(ing) 交错褶皱;横褶皱;横跨褶皱;叠加褶皱
crossfoot 交叉结算;交叉计算;横尺
cross footing 交叉验算法;交叉合计;交叉总计
cross-footing test 交叉验算法测试

cross force 横向力
cross-forge 横锻
cross form 横向模板
cross fracture 横裂缝(木材缺陷);横向破裂;横向断裂;横断裂
cross frame 十字框架;交叉架;横撑架;交叉连架
cross frame portal 交叉式门座
cross framing 抗风横撑
cross frogs 交叉辙叉
cross front 镜头横移装置
cross furring 垫出的抹灰条;小龙骨;交叉龙骨;横龙骨;横钉板条(在格栅下);双层灰条
cross furring ceiling 横向垫高天花板
cross furrow 横沟
cross-gamma scale 交叉灰度
cross-gang 横向平巷
cross gap 交叉间隙
cross garnet 丁字形蝶铰;T形铰链
cross-garnet butt 十字铰链
cross-garnet hinge T形铰链
cross gasket 十字系索
cross gate 交叉拦栅门;横浇道;横交口;斜平巷
cross ga(u)ntry 跨轨信号架
cross geologic(al) section 地质横剖面图
cross girder 交叉大梁;横(大)梁;横托梁
cross-grafted copolymer 交叉接枝共聚物
cross grain 斜木纹;逆木纹;(木材的)交叉斜(木)纹;逆纹(理);木料上的斜纹;横纹;(木材的)斜行纹理;斜纹(理)
cross-grained 纹理不规则的;扭丝(的);逆纹;垂直木纹的;斜纹(的);有斜纹和横纹纹理的
cross-grained float 逆纹镘;横纹木抹子
cross grained rock 交错层砂岩
cross-grained slate 交叉粒状瓦
cross-grained timber 逆纹木材
cross grained wood 扭纹木材
cross grain modulus 横纹(弹性)模量
cross grain wood 斜纹木材
cross-grating 交叉光栅
cross grid 十字格栅
cross groove 横槽;十字形凹槽
cross grooved cam 交叉槽凸轮
cross-grooved flag(stone) 企口铺路石板
cross-grooved floor(ing)(finish) 企口地(面)砖
cross growth 横向生长
cross guarantees 交叉担保
cross-guide coupler 十字形定向耦合器
cross gulley 交叉排水沟
cross gully 横间排水沟
cross-hair cursor 十字(准线光)标
cross-hair eyepiece 十字线目镜;十字丝目镜
cross-hair reticle 十字丝
cross-hair ring 十字丝环
cross hairs 叉丝;十字准线;十字(照准)丝;叉丝;十字瞄准线;瞄准线
cross-halved joint 十字半搭接
cross handle 横把手;十字手柄
cross hatch 剖面线;交叉(阴)影线;交叉划线;桁磨网纹;网状纹;双向影线;画叉影线
cross hatch adhesion 划格法附着力
cross-hatched 用交叉线画成阴影的;画交叉阴影线的
cross hatching 交叉晕线;横梁线;断面线;十字影线;网状线;绘制阴影线
cross-hatching area 交叉晕线面积
cross-hatch pattern 网状光栅;方格测试图;棋盘格测试图
cross hatch signal generator 栅形场振荡器
cross hatch tape adhesion 划格胶带法附着力
cross-haul 横向运送
cross-hauling 横向搬运;对流运输;跨越搬运;横向运土;横向装材;横向滚卷
cross-haul traffic 对流运输
crosshead 十字结联轴节;十字(板)头
crosshead arm 十字头臂
crosshead beam 十字头连杆
crosshead bollard 十字系缆桩
crosshead bottom shoe 十字头下瓦
crosshead cotter 十字头销
crosshead die 直角模具
crosshead end 十字头端
crosshead engine 带滑块的发动机;十字头式发动机
crosshead extension rod 十字头伸长杆
crosshead extrusion of thermoplastics 热塑性塑料丁字模头挤出

crosshead gib 十字头扁栓
crosshead guide 十字头导板
crosshead guide bar 十字头导杆
crosshead heater 十字头加热器
cross head(ing) 横ми工作区间通道;小标题;丁字头;十字头;横向坑道;横坑道;工作区通道;联络巷道
crosshead inspection head 十字头观察口盖
crosshead link 十字头连杆
crosshead lipper shoe 十字头滑块
cross-head of a piston rod 活塞杆十字头
crosshead oven 十字头烘箱
crosshead pin 十字头销
crosshead pin bushing 十字头销套
crosshead rivet 十字头铆钉
cross head shear test 十字板剪切试验;十字板剪力试验
crosshead shoe 十字(头)滑块;十字头滑板
crosshead side shoe 十字头边瓦
crosshead top shoe 十字头上瓦
cross-head type walking mechanism 十字头式行动机构
crosshead wrist pin 十字头关节销
cross hedge 交叉对冲
cross hedging 交叉套头交易;交叉避险
cross hitch 单花钩结
cross-hole method 跨孔法
cross-hole-seismic 跨孔地震法的;孔间地震法的
cross hole seismic profile method 井间地震剖面法
cross hole seismic profile seismograph 井间地震剖面法地震仪
cross-hole shear wave velocity measurement 跨孔剪切波速法
cross-hole shooting 跨孔法
cross-hole sonic measurement 跨孔声波测量
cross hole technique 交叉孔法
cross-hole testing 跨孔实验
cross-identification 交叉判读
cross-ignition tube 联焰管
cross-impact analysis 交叉影响分析;相互影响分析;交互影响分析
cross impact matrix 交叉影响矩阵
cross impact matrix method 交互影响矩阵法
cross impost 四角拱墩(柱)
cross in circle 十字圈
cross induction 交磁感应
cross infection 交叉感染;交叉传染;相互传染
crossing 路口;跨接;交叉沙洲;交叉口;交叉部;交叉;过河航道;过渡河段;割切;道口【铁】;分叉;浅槽段(给);十字路口;十字交叉(刷涂法)
crossing above 上交叉
crossing accident 道口事故
crossing ahead of the other 横越他船船首
crossing and passing of trains 列车会让
crossing angle 交会角(度);交叉角;辙叉角;铁路的辙叉角
crossing angle of wires 钢丝(的)交咬角
crossing approach 道口引坡
crossing ar grade 平面交叉
crossing a traffic lane 穿越分隔航道
crossing at triangle 三角形交叉(口);三岔交岔口
crossing bar 跨河沙埂;门闩;横沙洲;横浅滩;过渡段浅滩;横杆
crossing barrier 交叉拦路木;交叉口栅门
crossing below 下交叉;铁路在下交叉
crossing bridging 格栅斜撑
crossing buoy 过闸浮标
crossing capacity 交叉口通行能力;交叉口通过能力
crossing channel 过河航道
crossing clamp 十字线夹
crossing compression 交叉压迫
crossing conflicts 交叉冲突点
crossing counter 交叉计数器
crossing curved blades snips 双弯刃剪
crossing depth 过渡段水深
crossing design 交叉口设计
crossing discharge rate 交叉口驶出率
crossing disturbance 交叉干扰
crossing flagman 道口看守员
crossing frogs 交叉辙叉
crossing gate 交叉拦栅门;道口遮断器;道口栏木;道口栏杆;岔口遮断器;铁路与道路交叉口的栅门
crossing guy 十字拉线

crossing insulator 跨越绝缘子
crossing keeper 道口看守员
crossing ledger 横杆
crossing lift 架式举升机
crossing loop 会让站
crossing mark 过河标
crossing nose 辙叉心轨尖端
crossing of dislocation 位错的割切
crossing off 删去
crossing of lines 路线交叉;线路交叉
crossing of routes 进路交叉;车站进路交叉
crossing of the rolls 轧辊轴线偏斜
crossing of tracks 线路交叉
crossing of walls 墙连接;墙交叉
crossing of wires 线的交叉
crossing over 交叉
crossing permit 穿许可证
crossing pier 角柱
crossing pipe 交叉管道;虹吸管
crossing plates 道口板
crossing point 交点
crossing pole 跨越杆
crossing power supply 交叉供电
crossing protection 道口防护;平交道口防护
crossing reach 过渡(河)段
crossing rod 分经棒
crossing route 穿越路线
crossing rule 横交规则
crossing runway 交叉跑道
crossing sequence 交叉序列
crossing shoal 横浅滩;过渡段浅滩
crossing-shoal with staggered pools 深槽交错过渡浅滩
crossing sign 行人过街标志;交叉(路)口标志;道口标志
crossing signal 交叉道口信号;道口信号
crossing site 穿许地点
crossing situation 交叉相遇局面
crossing sleeve 交叉(套)筒
crossing slide 横向滑移
crossing square (教堂的十字形耳堂)交点广场;道路交叉口广场
crossing station 交点里程;会让站
crossing stone 道口石
crossing streams 叉流
crossing structure 跨越结构;交叉建筑物;穿许结构
crossing the equator 跨越赤道
crossing the line 跨越赤道
crossing timber 长轨枕栅门
crossing tower 四角塔楼;角塔
crossing untying 交叉疏解
crossing vault 四角拱顶
crossing vessels 交叉相遇船
crossing watcher 道口看守员
crossing watchman 道口看守员
crossing with separate turning lane 专用转弯车道交叉口
crossing zero 交零点
cross-intensity function 交叉强度函数
cross-interaction 交叉相互作用
cross interchange 十字立体交叉
cross interference 交调干扰;交叉干扰
cross interference signal 交叉干扰信号
cross intersection 十字形交叉
cross iron 十字铁
crossite 青铝闪石
cross jet breaker 横向喷油灭弧开关
cross joint(ing) 交接缝;交错节理;交叉连接;交叉连接;垂直接缝;立梁缝;横缝;横(断)节理;错箱;四通;竖(焊)缝;十字接头
cross joint ruling pen 滑行墨线笔
cross key 交叉键;交叉关键字
crosskill roller 横齿环星轮碾压器
cross knurled 网状滚花
cross lamellar structure 交错层理构造
cross laminate 直交层积;交叉层压板
cross-laminated 交叉层积;交叉层压的
cross lamination 交错纹理
crosslap 十字搭接
crosslap joint 对开十字接;十字搭接;十字平接头;交叉搭接头;对接十字缝
cross lay(ing) 交叉钢绳捻向;交叉捻(绕);导向捻;交叉设置
cross lead screw 横向丝杆
cross lead screw nut 横向丝杆螺母
cross ledger 脚手架横杆
crosslet 小十字形
cross levee 横堤
cross level 横向巷道;横向水准仪;横向水准器;横向校正;直交水准器;十字(形)水准器;轨顶高差;轨道水平
cross level deviation 轨道横向水平偏差
cross-level(l)ing 横断面水准测量
cross level variation 轨道水平变化
cross liability 交叉责任(计算船舶碰撞事故赔偿责任的一种方法);相互责任;相互责任
cross liability insurance 交叉负责保险
cross license 交换许可证;相互特许
cross license contract 交换使用专利的合同;互换许可证合同
cross licensing 交换许可证;相互特许
cross-licensing agreement 交换许可协议
cross light 交叉光线;十字光
cross lighting 交叉采光
cross-like 十字形的
cross limb 十字形体的四肢
cross limb wrench 十字轴式套管扳手
cross-line glass screen 玻璃网屏
cross-line graticule 十字线分划板
cross-line plane 十字线平面
cross lines 交叉缆;十字丝;交叉线路;交叉线;交叉线;正交线
cross lines of sounding 测深检查线
cross link(ag) 交键;交联;交叉联结;横向耦合;单耳连接器;双耳连接器;交叉耦合;交联;横向连接
cross link assy 交叉连接组合
cross linked 交联(的);交叉绑结
cross-linked bentonite 交联膨润土
cross-linked coupling 十字形联轴器
cross-linked gel 交联凝胶
cross-linked high degree polyethylene 交联高密度聚乙烯
cross linked high polymer 交联高聚物
cross-linked index 交联指数
cross-linked polyethylene 交联聚乙烯
cross-linked polymer 交联聚合物
cross-linked polythylene/alumin(i)um/cross-linked polyethylene pressure pipe 交联聚乙烯/铝/交联聚乙烯压力管
cross-linked structure 交联结构
cross linker 交联剂
cross linking 交联;交键
cross-linking agent 交联剂;发低促进剂
cross-linking coating 交联型涂料
cross-linking of polymers 聚合物交键
cross-linking of resin 树脂交联度
cross-linking plasticizer 交联性增塑剂
cross-linking reaction 交联反应
cross lock 栓
cross locking 横锁
cross louver shielding 固定百叶窗
crossly wound multilayered cylinder by flat ribbon 扁平钢带错绕式多层圆筒
cross magnetic field 交叉磁场
cross magnetization 正交磁化
cross magnetizing 交叉磁化;正交磁化
cross-magnetizing ampere-turn 交轴磁势
cross magnetizing effect 正交磁化效应
cross mains 交叉干管
cross manifold 交叉管道
cross mark 交叉痕迹;桁磨网纹;十字痕迹;十字标记
cross marker 侧位限航灯标
cross marking light 侧位限航灯标
cross masking 越边掩蔽;超掩蔽
cross matched drill 十字形钻(头)
cross-measure drift 穿层石门
cross medium 交错介质
cross member 交叉杆;叉形杆件;横向构件;横(构)件;横板;线担;底梁(集装箱);底桁梁;十字梁
cross member brace 横挡撑板
cross member clinch unt 横梁扭紧螺母
cross member gusset 横梁角撑板
cross member reinforcement 横梁加强(板)
cross memory service lock 交叉存储服务锁
Cross method 克劳斯法;弯矩分配法
Cross method of continuous beam 设计连续梁的克劳斯法
cross-migration 交互渗移
cross milled 交叉滚花
cross modal 交叉形式
cross modulation 交扰调制;交叉调制
cross-moment method 交叉矩法
cross motion 横向运动
cross-mounted bit 十字钻头
cross mounted condenser 横向布置凝汽器
cross mouthed chisel 十字头凿;十字冲(击)钻
cross mouthed drill 十字头钻
cross mud conveyer 横向运泥机
cross nailing 交叉钉
cross needle indicator 交叉指示器
cross network 交叉网络
cross neutralization 推挽式中和;交叉中和
cross neutralization test 交叉中和试验
cross Nicol prism 正交尼科耳棱镜
cross Nicols 正交尼科耳棱镜
cross nogging 格栅撑;交叉支撑;交叉系木
cross-notching 对开槽
cross-off 交叉支条;交叉支撑;危险区堵墙
cross off account 销账
cross off an account 销账
cross offer 交错报价;交叉发盘;交叉发价;交叉报价
cross order 交叉订货;相互订货
cross-out 取消;删去
cross-out file 取消注册的档案
cross-over 跨插;交换型;线岔【电】;渡线【铁】;跨过;穿越;跨接;横越;上方交叉飞过
cross-over analysis 交叉分析
cross-over assembly 转换装置
cross-over bend 跨越弯头;换向弯头
cross-over block 绝缘垫块
cross-over circulation 井的回洗;逆循环
cross-over coil (变压器的)圆柱形线圈;饼式线圈
cross-over connection 转换管
cross-over crown line (在交叉口或超高处的)转移拱顶线
cross-over culvert 分流口处廊道
cross-over design 交叉设计
cross-over discount rate 交叉贴现率
cross-over distortion 窜渡失真;交界失真;交越失真
cross-over experiment 交叉型实验;交叉试验
cross-over fabric 横条花纹织物
cross-over flange 跨接法兰
cross-over frequency 过渡频率;区分频率;跨越率;交界频率;交叠频率;交叉频率;分频频率;分隔频率
cross-over gasoline valve 重叠(的)汽油阀
cross-over insulator 横跨绝缘子
cross-over joint 换向连接器;换向接头
cross-over junction 并纯
cross-over lane 转换车道
cross-over line 有渡线联结的线路;转换管线;重叠的输送管
cross-over main 桥管
cross-over model 交叉模型
cross-over network 跨接网络;分频网络
cross-over pipe 容气管;横通管;过桥管;连通管;立体交叉管;架空管
cross-over point 交点;交叉点;相交点;立体交叉点;跨越点
cross-over point error curve of line and cross line 直线切割线交点误差曲线
cross-over position 零位;中间位置
cross-over potential 临界电位
cross-over road 转线路;十字路;交叉路;渡车道;道岔;上跨立交道路;转线路;上跨式立体交叉道路
cross-over section 渡线段
cross-over shoe 十字靴
cross-over spiral 过渡纹;交绕螺旋
cross-over sub 换向连接器;换向接头
cross-over switch 转辙器
cross-over tee 十字头;转换三通
cross-over track 渡线;有渡线联结的线路;渡线【铁】;转线轨道
cross-over tube 架空管
cross-over type 交换型
cross-over value 交换值;互换值
cross-over valve 三通阀;转换阀;过桥阀;交换阀;隔断阀
cross-over velocity 横越速度

cross-over winding 分层绕组
cross-pane hammer 横头锤
cross panel stiffness 横纹劲度
cross parachute 十字形伞
cross parity check 纵横奇偶校验
cross parking 垂直停车场
cross-partial derivative 交叉偏导数
cross-partition 交叉分割(指管理数学)
cross partition wall 横隔墙
cross passage 横通道
cross-passage tunnel 横向联络隧道
cross passageway 横向通道
cross-path loss 横向电流损耗
cross pattern 十字纹
cross-peen 十字锤头;横锤头;斧锤;十字尖锤;(锤的)横头
cross-peen hammer 斧锤;横头锤;十字锤头;十字尖(头)锤
cross-peen sledge 大锤;双手锤
cross-peen sledge hammer 十字尖锤
cross-periodogram 交叉周期图
cross phase modulation 交叉相位调制
cross piece 过梁;连接板;绞盘横杆;横木;横挡;方向架;十字架;十字管头
cross pier 十字形柱;十字形墩
cross-piled loading 横向堆垛法
cross pin 十字头销;插销;插销
cross pin type joint 十字轴形节头;万向节(头)
cross pipe 交叉管;四通管接头;四通管;十字形管;十字管(接头)
cross pipeline 穿许管线
cross plane 横切面;横剖面
cross-plane permeability 横面渗透性
cross-platform 交叉站台
cross-platform interchange 换站台换乘
cross-plot 交会图;综合图表
cross-plot method 交会图法
cross plug bit 十字形无岩芯钻头
cross-ply 交叉排布;斜纹轮胎
crossply tire 十字轮胎
cross point 十字头;交换单元;交叉点;相交点;跨越点
crosspointer 交叉指针;双针
crosspointer indicator 双针式指示器;交叉指针式指示器
crosspointer instrument 双针式测量仪表
cross pointing 绳尾编成尖形
cross point noise 相交点噪声
crosspoint relay 交叉点继电器
crosspoint relay matrix 交叉点继电器矩阵
cross polar component 交叉极化分量
cross-polarization 交叉偏振;交叉极化;横向偏振;正交偏振
cross-polarization discrimination 交叉极化鉴别度
cross-polarization photography 交叉偏振摄影
cross polarizer plate 正交极化板
cross pole 交叉磁极
cross poling 横撑板
cross-pollution 交叉污染
cross-polutant 交叉污染物
cross-power density spectrum 互功密度谱
cross power spectrum 交叉功率谱;互功率谱
cross price elasticity 交叉价格弹性
cross-prism telescopic resonator 交叉棱镜望远镜腔
cross product 外积;向量积;叉积;叉乘;矢积
cross-product matrix 交叉乘积矩阵
cross-product term 交叉乘积项
cross profile 横断面;截面;横剖面图;横面;横截面
cross profile of terrace 阶地横剖面
cross-program(me) 交错程序
cross-protection 交互保护;交叉口防护;交叉保护作用;横向保护;防止碰线
cross proving 相互检查
cross pulse 交叉脉冲
cross-pumped laser 交叉抽运激光器;交叉泵浦激光器
cross-pumping 交叉泵送
cross purchase 交互购买
cross-purchase agreement 交互购买协议
cross pusher 横向出钢机
cross quarters 十字形花纹(花格窗中)
cross question 盘问
cross rail 交叉横挡;横梁;交叉横栏;横(导)轨;十字围栏

cross rail clamp 横向导轨夹
cross rail elevating device 横导轨升降装置
cross rail elevating mechanism 横导轨升降机构
cross rail head 横导轨进刀架
cross rail side gear box 横导轨侧面齿轮箱
cross range 横向;侧向
cross rate 交叉汇率;套汇率
cross rate of exchange 套算汇率
cross ratio 交比;重比;非调和比
cross reacting material 交叉反应物质
cross reaction 交叉反应
cross receptance 交叉动柔度
cross recess(ed) screw 十字槽头螺钉;十字凹口螺丝
cross recess head screw 十字槽头螺钉
cross reducing on both outlet 出口全缩四通
cross reducing on one outlet 出口单缩四通
cross reducing on one run and both outlets 二线出口异径四通
cross reeded (两面肋条花纹交叉的)压制玻璃
cross reference 交叉引证;交叉引用;交叉关系;交叉参照;交叉对照;互见条目;相(互)对照;相互参照;参照法;前后参照
cross reference dictionary 交叉引用字典
cross reference index 参照索引
cross reference list 相互对照表
cross reference logy 成组加工技术
cross reference number 参照号码
cross reference program(me) 交叉引用程序
cross reference table 交叉引用表
cross referencing 交叉访问
cross-regression 截面回归
cross reinforcement 双向配筋;横向钢筋;对角钢筋
cross-relation effect 交叉弛豫效应
cross relaxation 交叉弛豫
cross-relaxation line shape 交叉弛豫线形
cross-relaxation line width 交叉弛豫线宽
cross-resistance 交叉抵抗力;交叉抗药性
cross-resonance 交叉共振
cross response 交叉响应
cross rib 横肋;交叉肋
cross-ribbed 有横肋的;设横肋的;带有交叉肋的
cross-ridge roof 十字脊屋顶
cross ridge tile 交叉脊瓦
cross ring 十字环
cross ripple 交错波痕
cross-river buoy 过河标
cross-river traffic 过江交通;过河交通
cross-riveting 交错铆接;十字形铆接;交互铆接;交叉铆接
cross road 横路;十字形交叉【道】;十字路;十字路口;横街;交叉(道)路
cross-rod 横向钢筋
cross-rod of concrete 混凝土横向钢筋
cross-rod reinforcement 横向钢筋
cross roll 横轧辊;斜置轧辊
cross-roller bit 十字形牙轮钻头;十字形齿轮钻头
cross rolling 斜轧;横轧
cross-rolling mill 斜轧机;横轧机
cross-roll slewing ring 交叉滚轧转盘;交叉滚轧转环
cross-roll straightening machine 斜辊横矫直机
cross route 交叉道;交叉路(线)
cross-ruling screen 十字丝网目板
cross runner 横向格栅;次格栅
cross running knot 蟹眼结
cross scavenging 横流换气法
cross scoring 横切
cross scratching 抹灰面交叉划痕
cross screen 交叉滤色镜
cross screw 左右交叉螺纹
cross scribe 十字划痕
cross sea 交叉波;横浪;横渡海洋;恶浪
cross seam 十字缝;楼板交叉接缝;楼板横直接缝
cross-section 截面;叉(面)部分;横切面;横剖面图;横剖面;横截面;横断面;剖面法
cross-sectional 分类排列
cross-sectional aggregation 截面归并
cross-sectional analysis 截面分析
cross-sectional area 横截面(面)积;横断面(面)积;断面面积
cross-sectional area of soil sample 土样横截面积
cross-sectional area of tensile reinforcement 受

拉钢筋断面面积
cross-sectional area of waterway 水道断面面积
cross-sectional bed area 床层横截面积
cross-sectional coefficient of channel 航道断面系数
cross-sectional coefficient of navigation channel 航道断面系数
cross-sectional consumption function 截面消费函数
cross-sectional curve 截面曲线
cross-sectional data 横断面数据;剖面数据
cross-sectional deformation 截面变形
cross-sectional design 横断面设计
cross-sectional dimension 横面尺寸;横截面尺寸;横断面尺寸;截面积尺寸
cross-sectional drawing 断面图;横截面图;横断面图
cross-sectional elasticities 截面弹性
cross-sectional error component 截面误差分量
cross-sectional estimator 截面估计量
cross-sectional factor of channel 航道断面系数
cross-sectional line 剖面线
cross-sectional loading 横截面单位面积载荷
cross-sectional maximum concentration 断面最大浓度
cross-sectional method 断面法
cross-sectional microrelation 截面微观关系
cross-sectional monitor 横断面监测装置
cross-sectional non-uniformity 断面不均匀性
cross-sectional observation 截面观测值
cross-sectional of channel 河槽横断面;渠道截面;渠道横截面
cross-sectional of river 河流横剖面
cross-sectional paper 计算纸;横断面纸
cross-sectional passenger volume 断面客流量
cross-sectional profile 横断面
cross-sectional quality 横断面特性
cross-sectional random sample 截面随机样本
cross-sectional ratio 断面比
cross-sectional regression 截面回归
cross-sectional sample 截面样品;截面样本
cross-sectional scale 截面标度
cross-sectional shape 横断面形状;截面形状;断面形状
cross-sectional sheet 横断面纸
cross-sectional size of water collecting gallery 集水廊道断面尺寸
cross-sectional size of well table 井筒断面尺寸
cross-sectional study 横断面调查
cross-sectional survey 截面调查;横断面测量;断面测量
cross-sectional template 横截面样板
cross-sectional variable 截面变量
cross-sectional view 横截面图;断面景象;横断面图
cross-sectional view strength 横断面视图
cross-sectional volume 断面流量
cross-section analysis 典型分析
cross-section area 切面面积
cross-section area of a cave 洞穴横截面积
cross-section area of nozzle 喷嘴截面面积
cross-section area of soil column 土柱截面面积
cross-section at crown 拱冠断面
cross-section bar 十字铁
cross-section cone bit 十字形牙轮钻头
cross-section container 断面集装箱
cross-section contraction action 断面收缩作用
cross-section cutter 十字形牙轮钻头;十字形切割器
cross-section diagram 断面图
cross-sectioned 作出断面的
cross-section for tunnel finish 隧道装修剖面
cross-sectioning 取截面;画剖面图;断面测量;横切;横断面测量
cross-section interval 剖面间隔
cross-section iron 十字铁;十字配筋;剖面图
cross-section line stake 断面桩【测】
cross-section measurement 断面测量
cross-section method 断面法
cross-section method fro location of line 横断面选线
cross-section method of railway location 横断面选线
cross-section model test 断面模型试验
cross-section of air-gap 气隙截面
cross-section of array 阵列断面

cross-section of bridge 桥梁横断面
cross-section of channel 渠道横截面
cross-section of conductor 导线截面
cross-section of discharge 过水横断面
cross-section of dry dock 干坞横剖面
cross-section of engineering geologic(al) 工程地质横断面图
cross-section of exploration tunnel 探矿坑道断面
cross-section of free passage in checkwork 格子体有效断面;格子体流通断面
cross-section of heading 导坑断面
cross-section of load 负荷截面
cross-section of road 道路横断面(图)
cross-section of side drain 沟身断面
cross-section of subgrade 路基断面
cross-section of the rail 钢轨剖面
cross-section of wall 墙身断面
cross-section of waterway 水道横截面
cross-section paper 横断面纸;坐标纸;方格纸;方格绘图纸
cross-section point 剖面点
cross-section property 横断面特性
cross-section recorder 剖面记录器
cross-section reduction 断面收缩
cross-section selector 剖面选择器
cross-section shape of karst conduit 通道横截面形态
cross-section sheet 方格纸
cross-section size of observation tunnel 观测平硐断面尺寸
cross-section sketch 截面草图;断面草图
cross-section survey 典型调查
cross-section surveying 横断面测量
cross-section templet 横断面样板(路面);路拱板
cross-section through shaft 竖井横截面
cross-section type 断面类型
cross-section under compression 受压横断面
cross-section underground structure 大断面地下建筑物
cross-section velocity distribution 断面流速分布
cross-section view 剖视图
cross seizing (绳索的)十字合扎
cross sensitivity 交互灵敏度;交叉灵敏度;横向灵敏度;正交灵敏度;正交测斜灵敏度
cross sensitization 交叉致敏反应
cross-set 使错开
cross shaft 横轴
cross-shaft bracket 横轴托架
cross shaft bushing 横轴衬套
cross shaft guidance 横轴导承
cross shaft guide 横轴导承
cross shaft head 棱镜测角器
cross shaft press 曲轴横放压力机
cross shake 横裂;辐裂
cross shape 十字形
cross shape bollard 十字系缆柱
cross-shape column 十字形柱
cross-shape core 十字形型芯
cross-shaped abutment 十字桥台
cross-shaped budding 十字
cross-shaped plan 十字形平面(图)
cross-shaped transfer 十字形换乘
cross-shaped transverse dredging method 十字形横移挖泥法
cross-shape ground plan 十字形底层平面
cross-shape joint 交叉连接
cross-shape pier 十字形支柱;十字形支墩
cross-shape support 十字形支柱
cross shoal 横向浅滩
cross shooting 交叉爆破(地震勘探)
cross shunt 交叉分路
cross shunt push-pull circuit 交叉分路推挽回路
cross shuttle belt conveyer 横向梭式皮带输送机
cross sill 横槛;轨枕
cross simulator 交叉模拟程序
cross slab 横(隔)板
cross-sleeper 轨枕;横轨枕
cross sleeve 交叉筒
cross slide 交叉滑路;横向拖板;横向滑板;十字滑轨
cross slide carriage 横刀架
cross slide circular table 横向滑板回转工作台
cross slide coupling 十字滑块联轴器
cross slide feed cam 横向滑板进给凸轮
cross slide feed mechanism 横向滑板进刀机构

cross slide movement 横向滑板运动
cross slider 交叉滑块
cross slide rotary table 横向滑板回转工作台
cross slide screw 横滑板螺杆
cross slide system 交叉滑动系统
cross slide table 横滑板工作台;纵横移动工作台
cross slide tool box 横滑板刀具箱
cross slide unit 横滑组
cross slide way 横向导轨
cross sliding type 横移式
cross slip 交叉滑移
cross-slip region 横向滑动区
cross slope 横坡
cross-slope furrow irrigation 横向沟灌
cross-slope reference 侧倾基准面
cross-slope system of drainage 横坡排水系统
cross slot 十字形榫槽;十字(形)槽
cross snapping 横向掰断
cross socket 十字束节
cross spacer 横向隔离物
cross spale 临时撑木
cross-span 横跨面
cross-span drop bracket 定位索定位支座
cross-span eye clamp 软横跨环型线夹
cross-span tensioning spring 软横跨张力补偿器
cross-span wire 横跨索;定位索【铁】
cross-span wire clamp 定位环线夹
cross spectral density 交叉谱线密度;正交谱线密度;互谱密度
cross-spectral purity 交叉光谱纯度
cross spectrum 交叉谱;交叉频谱;互谱图
cross speed 横向分速率
cross spider 交叉线
cross spindle 横进刀导螺杆
cross spread 正交排列;十字排列
cross spring 交叉斜缆;交叉倒缆弹簧
cross springer 穹隆横肋;交叉穹肋
cross-spring suspension 交叉弹簧悬挂
cross stacking 交叉堆垛法;横向堆垛法
cross staff 照准仪;方向架;十字杆;十字测天仪;直角(照准)仪;直角十字杆【测】
cross staff head 直角器;十字架;棱镜测角器
cross station platform 跨车站站台;跨车站月台
cross stay 交叉支撑;交叉拉索;横向支撑;横向撑条;横撑(条);对角撑
cross steering 横转向装置
cross steering device 横转向装置
cross-stitched belt 交叉接头带
cross-stitched canvas 刺绣十字布
cross storage 交叉存储
cross-stratification 交错层理
cross-stratum 交错层
cross-streak method 交叉划线法
cross stream component 横向分量;横向分力
cross(stream)flow 横向流(动)
cross stream velocity component 横向分速度
cross-street 横向街道
cross striation 横纹
cross strut 交叉撑;剪刀撑;横挡;横撑
cross-subarea 跨越分区;交叉子区
cross-subarea link 跨越分区链路;交叉子区连接
cross-subsidization 交叉补贴
cross subsidizing 交叉补贴
cross substitutability 互置换
cross substitution 交叉置换
cross-substitution effect 交叉替代效应
cross superposed folds 横跨叠加褶皱
cross support 十字形柱
cross support arm 横支杆
cross suspension 横向悬挂
cross suspension preheater 格子式悬浮预热器
cross swell 乱涌;交叉涌;横涌;横向海浪
cross-switch 十字开关
cross system of heat exchanger 十字形热交换器
cross tail butt 十字铰链
crosstalk 交调失真;交乘效应;交叉干扰;交互干扰;串音(耦合);地震道相互影响;磁场转印;串话;串线;串扰;串道;色度亮度干扰
crosstalk attenuation 串音衰减;串话衰减
crosstalk attenuation tester 串音衰耗测试仪
crosstalk between spiral grooves 槽间串音
crosstalk compensation 串音补偿
crosstalk coupling 串扰耦合;串话耦合
crosstalk coupling tracer 串音耦合图示仪

crosstalk effect 串音效应
crosstalk image 串音像
crosstalk interface 串音干扰
crosstalk interference 串话干扰
crosstalk level 串音电平;串话电平
crosstalk loss 串音衰减;串话衰减;串话耦合
crosstalk measuring set 串音测量器;串话测试器
crosstalk meter 串音测试器
crosstalk signal 串话信号
crosstalk suppression 串扰抑制
crosstalk suppression filter 串音抑制滤波器
crosstalk unit 串扰单位
crosstalk volume 串音功率
cross tee 倒T形轻型钢构件;十字管
cross-ten method 满十计数法
cross tension test 十字形拉力试验
cross-term 截项
cross test level 纵横水准仪
cross texture 十字形结构
cross threading 卡住螺纹(拧螺眼过紧)
cross tide 横向潮流
crosstie 横向栏杆;横木;枕木;轨距联杆;撑杆;交叉系杆;横向拉杆;轨枕;垫木;横层
crosstie wall 横挡壁
cross tilt 交叉倾斜
cross timber 横木;横挡;起锚机挡板
cross tongue 横向加固板(木构件接头处);横舌榫
cross tool carriage 横刀架
cross tool head 横刀架
cross tower 错流塔
crosstown 横贯城市的
crosstown tunnel 越城隧道
cross track 交叉跟踪;十字跟踪
cross-track distance 航迹偏移距离
cross-track error 偏离航迹误差;跨轨误差
cross-tracking 十字跟踪
cross track passage 平过道【铁】
cross-track resolution 横向分辨率
cross tractive 横裂缝
cross trade 买空卖空
cross traffic 交叉车流;横向交通
cross trail error 横向尾流误差
cross training 交叉训练
cross transfer conveyer 横向运输机;横向运送机
cross transom 横撑;连杆
cross transverse 横向运动
cross transverse feed gear 横向进刀装置
cross travel 横移
cross travel speed 小车运行速度
cross traverse feed 横向进给
cross tree 桅顶横桁
cross trench 横沟
cross tripping 交叉断路(法)
cross tube 四通管
cross tube boiler 横管锅炉
cross turbine 并联复式透平
cross turret 横刀塔
cross twill 芦席斜纹
cross-type 十字形
cross type bit 十字形钎头
cross type crossover 交叉式线岔
cross type dowel 十字形暗销
cross type intersection 十字交汇
cross-type junction terminal 十字形枢纽
cross under 交叉;穿跨
cross under-current 横向底流
cross-under pipe 跨接管
cross valley 横谷
cross valve 四通阀;转换阀;转向阀;十字阀
cross variable 参变数
crossvariance 交互方差
cross variance function 互方差函数
cross variogram 交叉变差函数
cross vault 交叉穹隆;十字穹顶;十字拱
cross vaulting 交叉筒拱;交叉拱顶
cross vein 交错(矿)脉;横脉
cross-ventilated 穿堂风的
cross ventilation 十字通风;穿越通风;对流通风;垂直通风;穿堂(通)风;前后通风
cross view 横剖图;串像
cross-viscosity coefficient 第二黏性系数
crosswalk 过街人行道;人行横道;人行道;不同预算分类之间的相互关系
crosswalk line 人行横道线

crosswalk sign 人行横道标志
cross wall 交叉墙;隔墙;横(隔)墙;横壁;隔板;冰坎;石坎
cross wall bearing construction 横墙承重结构
cross wall construction 横墙(承重)结构
cross wall load bearing 横墙承重
cross wall temperature curve 横墙温度曲线
cross-wall type of construction 采用横向承重墙的构造
cross-wall unit 横墙单元
crossware 交叉件
cross wave 三角浪
cross-way 交叉路;横路;十字路口
crossway conflicts 交叉冲突点
cross weaved screen 布纹式滤网
cross-weave procedure 横向运条法;横向摆动焊接法
cross web 交合丝
cross wedge rolling 楔横轧
cross-weighted Fisher indexes 费歇尔交叉加权指数
cross-weighted index 交叉加权指数
cross weir 横堰
cross weld 横向焊接;十字交叉焊缝
cross-welded polished wire(d)plate 横向焊接的钢丝网磨光玻璃板
cross-welded wire(d)glass 横向焊接嵌丝玻璃;横向焊接钢丝网夹心玻璃
cross weld(ing) 横向焊接;横向焊缝
cross welt(ed seam) 横盖墙条;横向折边缝;金属板咬口;横咬口缝
cross wind 侧面风;逆风;横风;侧向风;侧风
cross-wind compensation 侧面风力补偿
cross-wind component 空气动力侧向力量
cross-wind diffusion 横风扩散
cross winder 交叉卷绕络纱机
cross-wind integratal concentration 侧向积分浓度
cross-wind landing 侧风降落
cross window 交叉窗
cross-wind sign 横风标志
cross-wire micrometer 十字丝测微器
cross wires 瞄准器;交叉线;叉丝;十字丝;十字交叉线;十字照准丝
cross-wire suspension 复并悬挂
cross-wire weld 十字焊
crosswise 交叉(的);横向的;交叉向;横向地;横地;成十字形;十字状
crosswise distribution 交叉分布;横向分布
crosswise movement 横向动作
crosswise movement of saddle 鞍架横向运动
crosswise reinforcement 横向加强;横向钢筋
crosswise rigidity 横向刚度
crosswise slide 横向滑动
crosswise stretching crack 横向伸延缝
crosswise welding 交叉焊接
crosswise wire strander 叉式绞线机
cross with side branch 侧口四通
crosswork 垂直走向采掘
cross-wound package 交叉卷绕卷装
crotal 红棕色
crotarbital 丁烯比妥
crotch 岔口;叉状物;叉架;叉杆;分叉处;弯脚钉;弯钩
crotchet 小钩;钩状物;叉柱;叉架
crotch stress 叉口应力
crotch swirl V形涡发
crotch veneer (取自树杈的)镶板;Y叉木制薄片;(取自树叉的)镶板
crotch weld 楔接焊接;楔接锻接
crotchwood 树叉木
crotoniazid 丁烯烟肼
crotonic acid 丁烯酸;巴豆酸
crotonic aldehyde 巴豆醛
crotonicnitrile 丁烯腈
crotonism 巴豆中毒
croton oil 巴豆油
croton oil collodion 巴豆油火棉胶
croton oil poisoning 巴豆中毒
crotonol 巴豆油醇
croton resin 巴豆脂
Crotorite 克罗托里特铜镍铝合金
crotoxyphos 丁烯磷
crotyl bromide 巴豆基溴
croute calcaire 钙积层

Crout reduction 克劳特化简法
Crova wavelength 克罗瓦波长
crow 铁棍;撬棍;克劳钙钢;起货钩;撬杠;铁橇
crowbar 铁棍;短路器;电撬;带爪尖的撬棍;撬棍;撬杆;起货钩;铁撬(棍)
crowbar circuit 消弧电路
crowbar connection 撬杠拨正管材连接
crowbar protection 急剧短路保护装置;撬杆保护
crowbar switch 撬杆开关
crow crane 料耙起重机
crowd 急速前进;拥挤;一堆
crowd chain 推进链;推进栏
crowd cylinder 反铲臂转折油缸
crowd density 拥挤密度
crowded downtown area 繁华商业区;闹市区
crowded pinion 小齿轮组
crowder 沟渠扫污机;沟渠清污机
crowding 挤压效应;拥挤;皱纹形成
crowding action (钻压过大金刚石切入过深的)挤压作用
crowding barrow 二轮手推车
crowding effect 聚集效应;拥挤效应;群集效应
crowding gear 液压传送装置;液压传动装置;液压传递装置
crowding level 聚集度;密集度
crowd lever 推拉操纵杆;掘进操纵杆
crowd mechanism 掘进机构
crowd pinpion 小齿轮组;(挖土机压力机件的)齿轮
crowd ram 推压活塞
crowd rope 掘进缆绳
crowd shovel 正向机械铲;正铲挖掘机
crowd shovel attachment 正铲挖掘机附件
crowd shovel fitting 正铲挖掘机配件
crowd sprocket 挖掘机推压链轮;推压链轮
crowd walking speed 人流速度
crowed 人群
crowed sedimentation 拥挤沉淀
crowed shovel 正铲
Crower process 克劳方法
crow-flight path 直线程
crow-fly distance 直线距离;直线飞行距离
crowfoot 岩石纹理;图中尺寸的箭头;三角钉;接线箱的接栓;联结扒钉;鸡爪吊;防滑三脚架;图中标尺寸箭头;三脚架
crowfoot bar 撬棍;拔钉器;起钉撬棍
crowfoot checking 爪式细裂纹
crowfooted 踏步形的
crowfoot elevator 打捞工具
crowfoot packer 爪形镇压器
crow gable 阶形山墙;马头山墙;踏步形山墙
crowing 隆起
crown 路拱;隆起部(分);冕状齿轮;滑车头;熔炉的顶部;轧辊凸面;管道内顶;冠顶;冠部;冠;顶拱;顶;顶;大臧;大顶梁;舱口空间;副花冠;凸度;树冠
crown adjustment (轧制过程中的)辊型调整
crown-area index 树冠面积指数
crown bar 炉顶支柱;拱顶梁;拱顶(撑)杆;洞顶木;顶梁;顶杆;导坑主纵梁
crown bar bracket 顶杆托架
crown bar link 顶杆链节
crown bar link bracket 顶杆系条托架
crown bar link pin 顶杆联杆销
crown bar on arch top 拱顶带钩托梁
crown bar pin 顶杆销
crown base 冠基
crown beam 天车木架
crown bed 华盖床
crown bevel gear 差动器侧面伞齿轮
crown bit 环状钻头;取芯钻头
crown blast 窑顶吹火
crown block 拱顶石;定滑轮;顶部滑轮;拱冠石;(吊杆的)顶滑车;起重机顶部滑车组;天车
crown block floor 天车台
crown bottom 球形筛底;拱形底板
crown brass 轮轴铜衬
crown brick 楔形砖;拱顶砖;大碹砖
crown bud 冠芽
crown budding 冠芽接
crown canopy 树冠层
crown cantilever adjustment 拱冠悬臂梁校正
crown cantilever analysis 拱冠悬臂梁分析(法)
crown cantilever beam 拱冠悬臂梁
crown cap 四开图纸

crown cell 锌锰碱性干电池
crown class 树冠级干;树冠级
crown closure 郁闭;树冠(遮蔽度);树冠郁闭
crown colicitor 政府律师
crown compound 冠状化合物
crown control 轧辊表面凸度控制
crown cork 冠形瓶塞
crown cork finish 冠形瓶口
crown cornice 大屋檐
crown corrosion 拱顶腐蚀
crown corrosion of sewer 排水管管顶腐蚀
crown course 顶层;铺上屋脊处的那道屋面材料;路拱曲线
crown cover 圆锯护罩;林冠;树冠覆盖
crown cover degree 树冠覆盖度
crown cross-section 树冠面;冠状剖面;顶部横剖面
crown curve 路拱曲线
crown cutter 冠状绞刀
crown daisy 筒篙
crown deck 拱形桥面;拱形底板;拱形舱面
crown density 树冠郁闭度;树冠密度
crown depth 树冠厚
crown development 波峰状发展
crown die 钻头铸型;钻头压模
crown displacement 拱顶位移
crown ditch 截水沟;天沟
crown drill 活头钻具;可拆卸钻头的钻;顶钻
crown drip 植物顶部滴下的雨水;冠滴
crown drop 窟滴
crowned 有路拱的
crowned barrel 弧面辊身
crowned bit 塔形钻头
crowned drill 顶钻
crowned drill point 顶钻钻尖
crown edge 凸缘
crowned pulley 钻机钢绳滑轮;凸面皮带轮;凸面滑轮;天车轮
crowned road 有路拱的道路;起拱路面;起拱道路
crowned roller 桶形滚柱
crowned roller path 滚柱轴承的凸形滚道
crowned section 拱形断面;凸形断面
crowned surface 有路拱的面层
crown elevation 顶高;顶部高程
crown ether 冠醚
crown feature 凸形特征;拱起特征
crown filler 优质填料;上等填料
crown filling 封拱
crown finish 冠形盖瓶口
crown flint glass 冕火石玻璃
crown for chilled shot 钻粒钻头;钢砂钻头
crown fork 冕状叉
crown gate 进水闸(门);引水闸门;冠闸门;上闸首闸门;上闸门
crown gear 冕状齿轮;冠(状)齿轮;差动器侧面伞齿轮;侧面伞齿轮;平面齿轮;伞形齿轮
crown gear coupling 冕状齿联轴节
crown glass 冕(牌)玻璃;优质光学玻璃;无铅玻璃;上等(厚)玻璃
crown glass lens 冕牌玻璃透镜
crown glass prism 冕牌玻璃棱镜
crown grafting 接接
crown head 冠头
crown height 路拱高;隆起高度;冠高;拱顶高度
crown hinge 拱冠铰;拱顶铰;顶铰
crown hog 顶弯;顶拱
crown holder 钻头座
crown imperial 冠顶
crowning 拱起;防锚咬底法;板材中心部分增厚;凸起;桶形齿
crowning curve 冕状曲线
crowning fire 树冠火
crowning pulley 凸面皮带轮
crowning set 中凹度磨削装置;凸面加工装置
crowning stone (山墙的)顶石
crown interception 树冠截留
crown joint 拱顶接头;拱顶接缝
crown knot 绳端结
crown lagging 支拱木(隧道拱顶部);封顶模板
crown layer 树冠层
crown leather 上等皮革
crown length 树冠厚度;树冠长度;顶部长度;冠长
crown length ratio 冠长树比高
crown lens 冕透镜
crown level 顶部高程

crown life 钻头寿命;钻头使用期限
crown-like top 冤状顶部
crown line 路拱线;隆起线;中心线;穹顶线
crown line of vault 穹顶线
crown lining 冠状衬砌;冠状衬垫
crown mean diameter 树冠平均直径
crown member 顶部构件
crown miter gate 上闸首人字闸门
crown mo(u)ld 钻头造型;钻头压模
crown mo(u)lding 装饰压条;装饰板条;冠顶饰
crown nut 槽顶螺母;槽顶螺帽
crown of a double bottom 双层底舱顶
crown of arch 拱冠;拱顶
crown of a tank 水柜顶
crown of beam 梁拱
crown of core 型芯顶部
crown of corrugation 波纹的顶
crown of levee 堤顶
crown of overfall 溢流堰顶
crown of overflow 溢流堰顶
crown of pavement 路面拱度;路拱
crown of piston 活塞顶
crown of road 路拱
crown of sewer 污染管管顶
crown of sidewalk 人行道顶
crown of tooth 齿顶
crown optic(al) glass 冕牌光学玻璃
crown percent(ratio) 树冠率
crown piece (墙上的)垫木;支撑短木;墙上垫木
crown plate 支柱冠板;支柱顶板;栋梁;梁枕;梁垫(顶)板;梁垫;柱顶垫板;顶板
crown platform 钻塔顶工作台
crown post 屋架中柱;人字架中柱;桁架中柱
crown process 冕牌玻璃制作法;冕牌操作过程;圆盘制板法
crown pulley 凸面皮带轮;(钻机的)钢绳滑轮;天车滑轮
crown radius 圆球半径;碟形半径
crown rail-bond 拱形轨隙联结器
crown remover 除冠器
crown-riding 沿路拱车道行驶
crown sag 顶部下垂;顶部下沉
crown saw 圆筒锯;圆孔锯;筒形锯
crown scarp 凹形崖
crown scissors 冠剪
crown section 拱冠断面;拱顶截面;拱顶段;拱顶截面
crown-section of tunnel 隧道拱截面
crown setting 钻头镶嵌
crown settle measurement 顶部下沉测量
crown shaving 一般剃齿
crown sheave (钻机的)冠轮;天车轮;天车滑轮
crown sling stay 顶板吊撑
crown sling stay eye bolt 活节撑眼螺栓
crown sling stay eye rivet 顶部吊撑眼铆钉
crowns nest 天车台
crown stay 梁枕;顶撑
crown steeple 冠状尖塔
crown-step gable 梯级山墙顶
crown-step 阶式山墙
crown stone 拱顶石;山墙顶石
crown structure 顶部结构;顶部构造
crown support 拱状保险丝
crown sur face 顶面
crown tap 顶盖
crown tile 脊瓦;冠瓦;顶瓦
crown top of burner 灯帽
crown tray 梁枕;梁垫;栋梁
crown-tree 双锯面对开材
crown vault 冠状拱顶
crown vent(ing) 冕式透气
crown wall 放浪墙;胸墙;顶墙;堤顶胸墙;防波堤胸墙
crown weir 存水弯溢水面
crown wheel 冕状齿轮;冠状齿轮;差动器侧面伞齿轮;侧面伞形齿轮;天车滑轮
crown wheel escapement 冕状轮擒纵机构
crown width 顶部宽度;顶宽;堤顶宽度
crown wire 冠丝(钢丝绳外围与滑轮沟接触的钢丝);表面钢丝;(钢丝绳的)外层钢丝
crow oot crack 鸟爪形裂纹;爪形皱裂;爪状裂纹;爪形裂缝;皱裂
crow-quill pen 鸭嘴笔
crow'sfoot(ing) 爪形皱纹(油漆);爪状晶纹
crow's foot mark 爪纹

crow's nest 街道交通岗亭;交通岗亭;架空交通(指示)岗亭;桅上瞭望塔;桅斗;守望台;(瞭望用的)小高台
crow step 挑出踏步;阶形山墙;翅状踏步;梯形山墙顶
crow(step)gable 阶式山墙;梯形山墙
crow stepping 阶形(山墙)
crowstone 阶形山墙(压)顶石
crow twill 三上一下斜纹
croy 丁坝;护岸挑木栅;护岸设施;突堤
Croydon ball valve 克罗顿式球阀
croystron 固态器件
croze 凿槽工具;栓槽
Crozet basin 克罗泽海盆
crozier 拐杖
crozzle 黏结变形砖;过烧砖;过火砖;烧成焦渣
crozzling 鳄鱼皮(状裂纹)
crozzling coal 炼焦煤
cruces pilorum 毛交叉
crucial bandage 十字绷带
crucial experiment 决定性实验
crucial moment 紧要关头
crucial test 决定性试验;决定性实验;判决性试验
crucian carp 鲫鱼;欧鲫
cruciate dichotomy 十字形二歧分枝式
crucible 炉缸;坩埚;熔缸
crucible belly 熔罐腹
crucible cast steel 坩埚钢;坩埚铸钢
crucible cover 坩埚盖
crucible die 坩埚模型
crucible disc 坩埚片
crucible for casting 技工用坩埚
crucible furnace 坩埚炉
crucible holder 坩埚座
crucible kiln 坩埚窑
crucibleless technique 无坩埚技术
crucible melting 坩埚熔化;坩埚炉熔炼
crucible method 坩埚法
crucible oven 坩埚炉
crucible steel 坩埚钢
crucible tilting furnace 倾动式坩埚炉
crucible tongs 坩埚钳
crucible triangle 坩埚用三角
crucible trolley 坩埚推车
crucible-type salt bath furnace 坩埚盐浴炉
crucible zone 熔层
cruciferous column 十字架饰柱;十字柱
cruciferous plan 十字形平面(图)
crucifix (耶稣钉在十字架上的)图像;十字架
cruciform 交叉形的;交叉形;十字形的
cruciform bit 十字形钻(头)
cruciform block 十字形大厦
cruciform bollard 十字形系船柱;十字形缆桩;十字头系船柱
cruciform breakwater 十字形防波堤
cruciform bridge 十字形桥梁
cruciform building 十字形建筑
cruciform centralized church 十字形教堂
cruciform centrally-planned church 十字形教堂
cruciform checker 十字形格子砖
cruciform church 十字形教堂;十字架状的教堂
cruciform clay cutter 十字形黏土冲切器
cruciform column 十字架柱
cruciform construction 十字形结构
cruciform core 十字形型芯;十字形泥芯;十字形铁芯
cruciform cracking test 十字(丝)接头抗裂试验
cruciform cross-section 十字形截面
cruciform crypt 十字形地窟
cruciform-domed church 十字形拱顶教堂
cruciform effect 十字形效果
cruciform girder 十字横梁
cruciform groin 十字形丁坝
cruciform ground plan 十字形底层平面图
cruciform groyne 十字形丁坝
cruciform gusset plate 十字节点板
cruciform iron 十字铁;十字配筋
cruciform joint 十字接头
cruciform mass 十字形体
cruciform mass of a building 十字形屋体
cruciform member 十字形横梁
cruciform of rudders 十字形舵
cruciform pattern 十字形式样;十字形图案
cruciform pier 十字形柱墩
cruciform plan 十字形平面(图)

cruciform screw driver 十字形螺丝起子
cruciform skyscraper 十字形摩天楼
cruciform slot head 十字槽螺丝帽
cruciform slot screw 十字槽螺钉
cruciform stern 十字艉
cruciform support 十字形支柱
cruciform symmetry 十字(形)对称
cruciform test specimen 十字焊缝试样
crucishaped 十字形的
cruck 叉柱;曲木屋架;帐幕式棚架木杆
cruck construction 树枝建筑(一种以树为屋脊,枝为屋面的原始建筑)
cruck house 帐篷式房屋
cruddy 透光不均匀的
crude 界面污物;毛糙的;粗紫胶;粗制香叶醇;粗褐煤蜡;粗糙;腐蚀沉淀物;未经加工的
crude aggregate 天然骨料
crude alabaster 粗雪花石膏
crude alcohol 粗酒精
crude aluminium 粗铝
crude ammonia liquor 粗氨液
crude analysis 原油分析
crude and careless style of work 粗枝大叶作风
crude arsenic 粗砷
crude asbestos 原料石棉;粗石棉;天然石棉
crude asphalt 原生(石油)沥青;粗制(地)沥青;粗沥青;生沥青
crude assay 原油鉴定法
crude benzene 粗苯
crude birth rate 毛总出生率;粗出生率;普通出生率
crude blister copper 粗泡铜
crude brine 原盐水;粗盐水
crude carbolic acid 粗酚
crude carrier 原油油轮
crude coal tar oil 粗煤溜油
crude copper 粗铜;泡铜
crude dam 天然坝
crude data 原始资料;原始数据
crude death rate 毛总死亡率;粗死亡率;普通死亡率
crude density 粗密度
crude drugs 药材
crude earthenware 粗陶器
crude emulsion 粗制乳液;天然乳液;天然乳胶
crude estimate 框算;粗略估算;粗略估计
crude ethylene 粗乙烯
crude-fibre crop 麻类作物
crude filter 粗过滤器
crude filtration 粗过滤
crude foul water 原废水
crude fuel 原燃料
crude gas 原煤气;粗煤气
crude gasoline 粗汽油
crude geraniol 粗制香叶醇
crude glass 粗玻璃
crude grease 粗油脂
crude hydrocarbon 天然碳氢化合物
crude iron 粗铁
crude line 原油输送管线;石油输送管线
crude magnesium 粗镁
crude map 临时版地图
crude masonry(work) 粗圬工
crude material 原(材)料
crude metal 粗金属
crude method 粗略法
crude mica 生料云母
crude montan wax 粗褐煤蜡
crude mortality 粗死亡率
crude mortality rate 普通死亡率
crude naphtha 粗石脑油;粗挥发油
crude naphthalene 工业萘
crude oil 原油
crude oil automobile 柴油汽车
crude oil carrier 原油船
crude oil cracking 原油裂解;原油裂化
crude oil engine 原油发动机;柴油发动机
crude oil film 原油(薄)膜
crude oil gas 天然石油裂化气
crude oil in reservoir 油层原油
crude oil processing plant 原油的炼油装置
crude oil products 原油产品
crude oil pump 原油泵
crude oil pumping station 原油泵房
crude oil storage tank 原油储罐
crude oil transportation 原油输送

crude oil unit 常减压装置；原油蒸馏装置
crude oil washing 原油洗舱
crude opinion 轻率的意见
crude ore 原矿石；粗矿；未处理矿石
crude-ore bin 原矿仓
crude petroleum 原油
crude phenol 粗酚
crude pig iron 沟铁（高炉）
crude pipe line 原油输油管
crude pottery 粗陶
crude processing 原油加工
crude producer gas 未净化的发生炉煤气
crude product 半制成品；半成品；粗品
crude production 原油产量
crude pyroligneous acid 粗木醋液
crude receiving terminal 石油接收码头
crude refuse 原废物
crude refuse water 原废水
crude refuse water crude foul water 原污水
crude rock 未加工石料；粗石料
crude rubber 粗橡胶；生橡胶
crude salt 原盐
crude sand 原砂
crude scale 鳞状蜡；粗石蜡
crude separation 粗分离
crude sewage 未处理污水；原污水
crude sewage pump 原污水泵
crude shale oil 原页岩油
crude silk 半脱胶丝
crude sludge 原污泥；未处理的泥沙
crude stabilizer 原油稳定塔
crude steel 未清理钢；粗钢
crude still 原油蒸馏锅；石油蒸馏锅
crude storage 原油（油）库；原油储藏
crude storage capacity 原油库容；原油储存能力
crude structure 原粗结构
crude sugar 原糖
crude sugar solution 原糖液；粗糖液
crude survival rate 概略存活率
crude talc ore 滑石原矿
crude tall oil 原松浆油；粗妥尔油
crude tar 原油接收站；生柏油；粗焦油（沥青）；粗柏油
crude terminal 原始收受站；粗特立尼达沥青
crude unloading terminal 石油卸货码头
crude wastewater 粗废水；未处理废水；原（状）废水
crude woodwork machinery 粗木工机械
crude zinc 粗锌
cruise 林地勘测；巡游；低速巡行
cruise base 旅游基地
cruise duration 续航时间
cruise liner 巡洋班轮
cruise-oriented 有关旅游客运的
cruise propulsion unit 巡航状态发动机
cruiser 警车；巡洋舰；大型快船
cruise radius 船舶续航力
cruiser dock 大型钢筋混凝土浮坞
cruiser stern 巡洋舰型船尾
cruise ship 旅游客仓；旅游船
cruise terminal 旅游码头
cruise train 旅游列车
cruise waterway 旅游水道
cruising 巡航
cruising craft 游艇
cruising depth 巡航深度
cruising duration 续航时间
cruising engine 主发动机
cruising gear 自动超高速传动装置
cruising radius 巡航半径；续航力；续航半径
cruising range 巡航范围；航程；巡航航程；巡航半径；续航距离
cruising rate 巡航速率
cruising rating 巡航速率
cruising speed 巡行速度；巡航速度
cruising thrust 巡航状态发动机推力；主发动机推力
cruising turbine 船用涡轮机
cruising velocity 巡航速度
cruising waterway 旅游水道
cruising way 慢车道
cruising yacht 巡航快艇
cruled felt 卷边毡
crumb 屑粒；小片；碎片；团粒
crumb boss 工地炊事员或其他负责人（俚语）
crumber （挖沟机的）整边机，清沟器，清沟机，清沟铲

crumb gang 清沟班
crumbing crew 清沟班
crumbing shoe 碾铧板
crumble 崩碎；崩溃；切碎；破碎；碎土
crumble away 崩解；剥落；脱落；碎裂
crumble coal 碎裂煤；松散煤
crumble peat 碎粒泥煤；松散泥煤；松散泥炭
crumble structure 团粒构造
crumble texture 团粒状结构
crumbling 岩块剥落；崩溃；崩解
crumbling away 粉碎；剥落
crumbling iron 破碎的铁
crumbling rock 酥碎岩石；松软岩石；风化岩石；崩解岩石
crumbling soil 团粒状土壤；碎屑状土壤
crumb party 清沟班
crumb rubber 粒状橡胶
crumbs 粒状橡胶
crumb screw 清沟班
crumb structure 团块状构造
crumb texture 团粒结构；团块状结构
crumby 团块状
crumby soil 屑粒土；团块状土壤
crummeridian altitude 近子午圈高度
crummeridian zenith distance 近子午圈天顶距
crummy 守车（货客运列车）；餐车（货车）
crump 岩石受压移动
crumple 弄破；变皱
crumpled 皱的；揉的
crumpled alumin(i)um foil 皱铝箔
crumpled texture 揉皱结构
crumple pattern 皱纹图案；皱纹
crumple zone of intrados 内弧揉皱带
crumpling 揉皱作用；起皱；盘曲皱纹
crumpling zone 揉皱带
crump out 铲出沟底松土
crunch seal 陶瓷封接
crunodal cissoid 结点蔓叶线
crunodal cubic(al) curve 结点三次曲线
crunode 结点；叉点；分支
crura membranacea 膜脚
crurls 卷曲木丝
Crusader 克鲁萨特（一种专卖的防水墙用的硬化液体）
crusader architecture 十字军式建筑
crusader castle 十字军城堡
crusader type church 十字军式教堂
cruse product 粗制品
crush 挤压；挤进；碾；压坏；顶板下沉；粉碎；塌箱；砂到碎裂；压碎
crushability 破碎性
crushability factor 碎性系数
crush barrier 防挤栏杆
crush borer texture 碎边结构
crush breccia 压碎角砾岩
crush conglomerate 压碎砾岩
crush cost 赶工成本
crush dresser 砂轮整形工具
crush dressing 磨轮表面修饰
crushed 轧碎的；压碎的；碾碎的
crushed aggregate 轧碎集料；轧碎骨料；破碎集料；碎石集料
crushed aggregate spreader 碎石撒布机
crushed ballast concrete 道砟混凝土
crushed basalt stone 玄武岩碎石
crushed belt 破碎带
crushed bort 碎粒包尔特金刚石
crushed boulder 碎漂石
crushed brick 碎砖
crushed brick concrete 碎砖混凝土
crushed brick concrete filler slab 碎砖混凝土填充块
crushed brick concrete wall slab 碎砖混凝土墙板
crushed broken stone 粗碎石
crushed chip(ping)s 碎屑
crushed cinder 碎制炉渣
crushed cinder sand 碎制矿渣
crushed coarse aggregate 压碎粗集料；压碎粗骨料
crushed coke 碎焦炭
crushed concrete 压碎的混凝土
crushed(concreting)sand 轧制的混凝土用砂；石砂；混凝土细集料
crushed conglomerate 天然聚集砾石
crushed dolomite 碎白云石

crushed dust 渣尘；石砂；石尘
crushed expanded cinder 碎膨胀炉渣；碎膨胀矿渣
crushed feldspar 碎粒长石
crushed fine aggregate 轧制的细集料；轧制的细骨料
crushed flast-furnace slag 碎高炉渣
crushed flast-furnace slag sand 炉渣砂；矿渣砂；高炉渣碎粒砂
crushed foamed blast-furnace slag 碎高炉泡沫矿渣
crushed glass 碾碎的玻璃；玻璃渣；粗碎玻璃；玻璃细粒
crushed granite 碎花岗岩；麻石子；花岗岩石子
crushed granite sand 花岗岩人工砂
crushed granulated blast-furnace slag 轧制的粒状高炉渣
crushed granulated cinder 碾轧成的粒状炉渣
crushed granulated slag 轧制的粒状炉渣；轧制的粒状矿渣
crushed gravel 轧碎砾石；碎卵石；碎砾（石）
crushed gravel aggregate 碎砾石集料；碎砾石骨料
crushed gravel sand 压碎砂砾；碎砾石砂
crushed gypsum 轧碎石膏；初碎石膏
crushed hard rock 碎硬石；碎石块
crushed head 轨头压溃
crushed ice 碎冰
crushed lava 碎的熔岩
crushed limestone 碎石灰石
crushed limestone sand 碾轧成的石灰石砂
crushed lump slag 轧成的块状矿渣；碎炉渣
crushed marble 碎大理石
crushed material 碎粒料；碾碎的材料
crushed news 废新闻纸
crushed ore 碎矿石
crushed ore feeder 碎矿石喂料机；碎矿石给料机；碎矿石给矿机
crushed porphyry 碎斑岩
crushed porphyry concrete 碎斑岩混凝土
crushed product 碾轧产品
crushed quartzite 碎粒石英石
crushed rock 压碎岩石；轧石；破碎岩体；碎岩石；碎石
crushed rock aggregate 碎石集料；碎石骨料
crushed-rock surfacing 碎石面层
crushed rock wrapped in wire mesh 铁丝网石笼
crushed-run macadam 统货碎石
crushed-run rock 未筛碎石；统货碎石
crushed sand 人工砂；轧细砂；轧碎砂；轧制砂；轧石残渣；碎石砂；石粉
crushed scrap anode butt 碎阳极残头
crushed shot 碎粒
crushed slag 碎火山灰岩；炉渣块；矿渣碎石；轧碎矿渣；碎渣；碎熔渣；碎矿渣
crushed slag sand 炉渣砂；矿渣砂
crushed slate 板岩碎块
crushed steel 碎钢粉
crushed stone 轧石；破碎石块；碎石
crushed-stone aggregate 碎石集料；碎石骨料
crushed-stone base 碎石基础；碎石基层；碎石垫层
crushed-stone base course 碎石底层
crushed-stone bed 碎石基层；碎石基床
crushed-stone concrete 碎石混凝土
crushed-stone course 碎石层
crushed-stone drainage layer 碎石排水层
crushed-stone grading 碎石级配
crushed-stone level(l)ing course 碎石填平层
crushed-stone pavement 碎石铺砌层；碎石路面
crushed-stone plant 碎石厂
crushed-stone road 碎石路
crushed-stone sampling 碎石取样
crushed-stone sand 石轧砂；石砂
crushed-stone soil 用碎石加固的土
crushed-stone stockpile 碎石堆场
crushed-stone stratum 碎石层
crushed zone 压碎带；破碎带
crusher 轧石机；碎石机；破碎机；漏斗式破碎机；挤压机；压碎机；轧碎机；粗磨机；砂轮刀
crusher ball 研磨体；轧碎机钢球
crusher block 耐压垫块
crusher board 土块细碎板
crusher car 碎矿车
crusher chamber 破碎机房
crusher chute 轧碎机
crusher circuit 破碎流程

crusher department 破碎车间
crusher drive 破碎机传动方式
crusher drive gear 轧碎机传动齿轮
crusher drive pinion 轧碎机传动小齿轮
crusher drive shaft 轧碎机主动轴；破碎机主动轴
crusher dust 轧碎残渣；石粉
crusher feed 破碎机喂料
crusher feeder 轧碎石就的供料器；碎石供料机
crusher feed hopper 轧碎机的供料门斗
crusher ga(u)ge 压缩压力计
crusher intermediate floor 破碎机中间平台
crusher jaw 轧碎机颚板；破碎机颚板
crusher mouth 轧碎机进料口
crusher oil 压碎机用油
crusher opening 轧碎机进料口；轧碎机出料口
crusher plate 轧碎板
crusher roll 粉末轧辊；破碎辊；辊式轧碎机
crusher-run 统货碎石；机碎石；未筛（轧）碎石（料）
crusher-run aggregate 未筛（分的）机碎骨料；机碎集料；机碎骨料；破碎料；未筛碎石集料；未筛机碎集料
crusher-run base 机碎骨料垫层
crusher-run material 机碎材料
crusher-run product 破碎机产品；（未过筛的）碾碎材料
crusher-run rock 破碎机石块；未筛（轧）碎石（料）
crusher-run stone 未筛分机轧碎石
crusher-run stone aggregate 未筛分轧骨料
crusher-sampler 破碎取样联合机
crusher sand 轧石砂；扎石砂
crusher screenings 轧制砂；石砂；人造砂
crusher setting 排料口间隙；排料口（调定）开度
crusher size 破碎机最大接受粒度
crusher-thrower 破碎抛料机
crusher unit 轧碎机组
crusher wall 轧碎机轮壁
crush-form dresser 磨轮压刮整型器；磨轮压刮整型机
crushing 压碎；破碎粒度；压损；容器瘪痕；破碎；砂轮修整
crushing action 破碎作用
crushing and grinding machinery 破碎粉磨机械
crushing and screening plant 轧石筛分厂；破碎筛分设备；破碎筛分厂；碎石筛石厂
crushing and screening tower 破碎筛分塔
crushing apparatus 压碎器
crushing appliance 破碎设备
crushing ball 捣碎锤
crushing bo(a)rt 破碎金刚石
crushing bortz 金刚石砂粒；破碎金刚石
crushing boulders 球磨石球；球磨机用石球
crushing bowl 破碎壳
crushing capacity 轧碎能力
crushing cavity 破碎室；破碎腔
crushing chamber 破碎室
crushing circuit 破碎流程
crushing clamp 挤压钳；压碎夹
crushing compartment 破碎仓
crushing concave 圆锥破碎机的破碎面
crushing cone 锥形轧辊；轧碎机锥体；破碎锥头；破碎圆锥
crushing device 破碎装置
crushing engine 轧石机；轧碎机；破碎机
crushing equipment 轧碎设备；碎石设备
crushing fold 压碎复褶曲
crushing force 挤压力；压轧力；压裂力
crushing head 破碎圆锥；破碎机锥头
crushing in closed circuit 闭路破碎
crushing in single pass 一次破碎；单程破碎
crushing installation 破碎装置
crushing jaw 破碎颚板；颚板
crushing limit 抗破碎强度
crushing load 破坏荷载；碾压负荷；压碎荷载；压碎负荷；压毁荷载；断裂荷载；断裂负载
crushing machine 压碎机；轧碎机；轧石机；破碎机
crushing mantle 破碎锥体的锰钢壳
crushing mechanics 破碎力学
crushing mechanism 破碎装置
crushing member 破碎头；破碎颚；破碎板
crushing method 压碎法
crushing mill 轧石厂；碎石厂；压碎机；粗磨
crushing movement 破碎动作
crushing of method 破碎方法

crushing of rail head 轨头压溃
crushing of refuse 垃圾破碎
crushing of solid wastes 固体废物破碎
crushing operation 碾轧操作
crushing output 轧碎产出量
crushing pad 压碎板
crushing plant 联合碎石机组；轧碎机；轧石厂；破碎设备；碎石设备；碎石场；碎石厂
crushing plate 臼板
crushing pocket 碎石仓；碎矿仓
crushing pressure 破碎压力
crushing process 轧碎过程；碾轧过程；破碎工艺
crushing proof load 破碎管载重
crushing rate 破碎率
crushing rate of gasket 垫圈的挤压率
crushing resistance 抗碾性
crushing ring 粉碎轮腔
crushing roll 压碎滚筒；轧碎滚筒；辊筒压碎机；破碎辊；滚筒压碎机
crushing roller 辊式破碎机；对辊破碎机
crushing roll segment 轧辊箍
crushing rolls for chip(ping)s production 石屑压碎机
crushing roll shell 压辊罩；破碎机滚筒护套
crushing sample 压碎样品
crushing-screening 破碎筛分
crushing section 破碎阶段
crushing space 破碎腔
crushing stage 压碎阶段；压碎台
crushing station 压碎站
crushing steps 破碎级数
crushing strain 压（毁）应变；破碎应变
crushing strength 抗（压）碎强度；挤压强度；碾压强度；耐压强度；压缩强度；压碎强度；压毁强度；破碎强度；破坏强度
crushing strength across grain 横纹抗压强度
crushing strength across the grain 粒间横压强度
crushing strength-age relationship 抗碎强度—龄期关系（混凝土）
crushing strength at elastic limit 弹性极限的抗压强度
crushing strength at maximum load 最大压碎强度
crushing strength class 抗碎强度级别
crushing strength of grain 颗粒压碎强度
crushing strength parallel to the grain 粒间纵压强度；顺纹抗压强度
crushing strength range 抗碎强度域；抗碎强度范围
crushing strength test 抗碎强度试验
crushing strength under high temperature 高温耐压强度
crushing stress 挤压应力；压碎应力；压（毁）应力；压坏（压毁）应力；破碎应力
crushing strip 增隙带
crushing test 压碎试验；压裂试验；压毁试验；破碎试验
crushing tool 压碎工具
crushing type 破碎型式
crushing value 压碎值；破碎值
crushing value of coarse aggregate 粗集料的压碎值；粗骨料的压碎值
crushing wheel 刀碗
crushing zone 破碎区
crush injury 挤压伤；压轧伤
crush load 超员；破碎荷载
crush plane 压碎面
crush plate 木框边框保护条；破碎板
crush roller 破碎辊
crush room 休息室；剧场休息室
crush-run aggregate 机碎集料；机碎骨料
crush seal 挤压密封；压挤密封
crush stress 压应力
crush strip 防跑火沟
crush texture 压碎结构
crush time 赶工时间
crush trauma 挤压伤
crusing radius 观察半径
crusker 破碎机
crust 黑皮；细白砂；硬表面；硬表层；铸件黑皮；地壳；外皮
crustacean 甲壳动物
crustaceous 壳质的；壳质的
crustal abundance 克拉克
crustal architecture and structures 地壳的结构构造
crustal block 地（壳）块

crustal contamination 地壳混染作用
crustal cupola 壳块
crustal deformation 地壳变形；地壳变动
crustal dynamics 地壳动力学
crustal earthquake 地壳地震
crustal fracture 壳断裂；地壳断裂
crustal garneliferous ultabasic rock 地壳型石榴石超基性岩
crustal heat flow 地壳热流
crustal inclination 地壳倾斜
crustal isostatic process by continental glaciers 大陆冰川地壳均衡过程
crustal longitudinal wave 地壳纵波
crustal low-velocity zone 地壳低速带
crustal movement 地壳运动
crustal plate 地壳板块
crustal province 地壳省
crustal recoil 地壳隆起
crustal seismic velocity 地壳地震波速度
crustal shock 地壳震动
crustal spreading 地壳扩张
crustal standing wave 地壳驻波
crustal strain 地壳应变
crustal stress 地（壳）应力
crustal structure 地壳构造
crustal subsidence 地壳下沉
crustal tectonic 地壳构造
crustal texture 地壳结构
crustal transversal wave 地壳横波
crustal type 地壳类型
crustal uplift 地壳隆起
crustal warping 地壳翘曲
crustal wave 地壳波浪
crustal-wave movement 地壳波浪运动
crustal-wave system 地壳波浪系统
crusta petrosa 白垩质
crust block 渣壳块；凝渣
crust breaker 浮渣破碎机（器）；表层碎土耙；打壳机
crust-breaking chips 打壳锤头
crust breaking hammer carrier 打壳锤架
crust breccia 断层角砾岩
crust brecciatic texture 压碎角砾结构
crust deformation measurement 地壳形变测量
crust-density magma 壳源岩浆
crust-derived granite 壳源花岗岩
crust forming agent 表皮形成剂
crust green 蟹壳青（色釉名）
crustified cement 枯壳状胶结；丛生胶结
crustified cement texture 丛生胶结构造
crusting 结硬皮
crust-mantle boundary 地壳地幔边界
crust-mantle interface 地壳地幔界面
crust-mantle mix 地壳地幔混合物
crust-mantle mixed granite 壳—幔混合花岗岩
crust-mantle system 地壳地幔系统
crust of cement 水泥表层
crust of iron 铁渣；氧化皮
crust of weathering 风化壳
crust ore 皮壳矿石；壳矿
crustose 壳状的；结壳岩；皮壳状
crust pulsation 地壳脉动
crust soil 结壳土壤
crusty structure 皮壳状构造
crut 短煤巷
crutch 拐杖；桨叉；桨架；支柱；船尾肘材；叉柱；叉杆；桅座叉柱
crutcher 螺旋式拌和机；螺旋（桨）搅拌器；螺旋（桨）搅拌机；搅和机；调料锅
crutch hole 桨架孔
crutching 搅和；防蝇去毛
crutching pan 搅和锅
crutch key 旋塞手柄；（螺旋式水龙头顶部呈T形的）塞子
crux commissa T字形十字架
crux decussate X形十字架
crux immissa 拉丁十字架
crux of the matter 问题的关键
crwon forest 王室林
crymic 冻原
cryoagglutinin 冷沉淀凝集素
cryo baffle 低温障板
cryobiology 低温生物学
cryocable 低温电缆

cryochem 冷冻干燥法;冰干法
cryochemical 低温化学的
cryochemical method 低温化学法
cryochemistry 低温化学
cryochem process 低温化学法
cryoconcentration 低温浓缩
cryoconite 冰尘
cryoconite cover 冰雪覆盖
cryoconite hole 冰雪孔;冰尘穴
cryoconite horizon 冰雪层
cryoconite strip 冰雪带
cryo-current comparator 低温电流比较仪
cryodrying 低温干燥
cryoelectronics 低温电子元件学;低温电子学
cryo-equipment 低温装置
cryoextractor 冷冻摘出器
cryofixation 冰冻固定
cryofocusing 低温聚焦
cryoforming 低温成型加工法;冷冻成型加工
cryofragmentising 低温碎裂法
cryogen 冷冻剂;制冷剂;低温流体;低温粉碎
cryogenerator 低温发生器;低温发电机;深冷制冷器
cryogenic 制冷的;低温(学)的;深冷的
cryogenic accelerator 低温加速器
cryogenic accelerometer 低温加速表
cryogenically cooled parametric amplifier 低温制冷参放;低温冷却参量放大器
cryogenic application 低温应用
cryogenic bearing 低温轴承
cryogenic bolometer 低温测辐射热计
cryogenic box 低温箱
cryogenic camera 制冷照相机
cryogenic coil 低温线圈
cryogenic compressor 低温压缩机
cryogenic computer 低温计算机
cryogenic conductor 超导体
cryogenic coolant 冷冻剂;低温制冷剂;低温冷剂
cryogenic cooling 低温制冷;低温冷却
cryogenic coupler 低温连接器
cryogenic device 低温装置;低温器件
cryogenic effect 低温效应
cryogenic electromagnet 低温电磁铁
cryomagnetic 磁制冷的
cryogenic engine 低温发动机
cryogenic engineering 低温工程
cryogenic equipment 低温装置;低温设备
cryogenic film 低温膜
cryogenic fluid 低温溶液;低温流体
cryogenic fluid pump 冷剂泵;低温流体泵;深冷流体泵
cryogenic forming 冷冻成型加工
cryogenic fragmentation 低温碎裂
cryogenic freezing 低温冰冻
cryogenic gas converter 低温液体汽化器
cryogenic gravity meter 超低温重力计
cryogenic grinding 冷冻研磨;低温研磨;深冷粉碎
cryogenic gyro(scope) 低温陀螺(仪)
cryogenic heater 低温加热器
cryogenic heat exchanger 低温换热器
cryogenic inertial sensor 低温惯性传感器
cryogenic insulation 低温绝缘(材料)
cryogenic lake 冰雪湖
cryogenic laser fusion fuel 冷冻激光聚变燃料
cryogenic liquid 低温冷却液
cryogenic liquid pump 低温液体泵
cryogenic liquid tank lorry 低温液体槽车
cryogenic magnet 低温(超导)磁铁
cryogenic magnetometer 超异磁力仪
cryogenic material 低温材料
cryogenic meadow-forest soil 冰冻草甸森林土
cryogenic measurement 低温测量
cryogenic memory 低温存储器
cryogenic period 低温冻结期;冰雪时期
cryogenic plasma 低温离子体
cryogenic preparation 低温制剂
cryogenic process 低温法;深度冷冻
cryogenic processing 低温处理
cryogenic propellant 低温推进剂
cryogenic property 低温性质;低温性能
cryogenic pump 冷凝泵;低温(抽气)泵;深冷泵
cryogenic pumping 冷凝排气;低温排气;低温抽运
cryogenic quenching 低温淬火
cryogenic recycling 低温再循环
cryogenic refrigerator 低温制冷器
cryogenic rocket engine 低温火箭发动机

cryogenics 低温(试验法);低温实验法;低温(材料)学
cryogenic sampler 冻结取(土)样器
cryogenic satellite telescope 低温卫星望远镜
cryogenic seal 低温密封
cryogenic stage 低温燃料级
cryogenic steel 低温用钢
cryogenic storage container 冷藏箱
cryogenic storage system 冷冻储[贮]藏系统;冷藏系统;低温储[贮]藏系统
cryogenic store 低温存储器
cryogenic stripping 冷冻脱漆
cryogenic superconductor 低温超导体;深冷超导体
cryogenic surface 低温体表面
cryogenic system 制冷系统;低温装置;低温系统;制冷装置系统;制冷装置
cryogenic tank 低温储[贮]槽
cryogenic target 冷冻靶
cryogenic technique 低温技术
cryogenic technology 低技术
cryogenic temperature 冷冻温度;极低温;低温温度;低温度;深沟温度
cryogenic temperature detector 低温检测器
cryogenic visco(si)meter 低温黏度计
cryogenin 冷却剂
cryogeny 低温学
cryogetter pump 低温吸气泵
cryohydrate 低共熔冰盐结晶;冰盐;食盐水防冻剂
cryohydric point 冰盐点
cryolaccolith 水岩盖
cryolac number 冰凝值
cryolite 冰晶石
cryolite deposit 冰晶石矿床
cryolite glass 冰晶石玻璃;冰晶石玻璃;冰晶玻璃;乳白玻璃
cryolith 人造冰晶石
cryolithionite 锂冰晶石
cryology 河海冰冻学;海冰学;冻岩学;冰岩学;冰雪学;冰雪水文学;冰雪冻土学;冰雪学;冰川学
cryoluminescence 冷致发光
cryolysis 冷冻溶解法;冷冻破坏法
cryomagnetic 磁制冷的
cryometer 低温温度计;低温计;低温表;深冷温度计
cryomicroscope 低温显微镜
cryomite 小型低温制冷器
cryomorphology 寒冷气候地貌学;冻土地貌学
cryonetics 低温学;低温技术
cryopanel 低温操纵盘;低温板
cryopanel array 低温板抽气装置
cryopathy 寒冷病
cryopedology 冻土学;冻结学
cryopedometer 低温表
cryophilic algae 雪生藻类
cryophilic microorganism 喜低温微生物
cryophilic mineral 绿鳞云母质矿物
cryophore 低温技术
cryophorus 冰凝器;凝冰器
cryophyllite 绿鳞云母
cryophysics 低温物理(学)
cryophyte 冰雪植物
cryoplanation 冻融夷平面
cryoplankton 冰雪浮游生物
cryoplate 低温抽气板;深冷抽气面
cryoplate array 低温板组
cryoprecipitability 冷沉淀性
cryoprecipitation 冷冻沉淀法;冷沉淀反应
cryopreservation 低温储[贮]藏
cryoprobe 冷冻探子;冷冻探针
cryoprotectant 防冻剂
cryoprotective 防冷冻的
cryoprotective agent 低温防护剂
cryoprotector 低温防护剂
cryopulverization 深冷粉碎
cryopulverizer 深冷粉碎机
cryopump 冷凝泵;低温(抽气)泵;深冷抽吸
cryopumping 冷凝排气;低温排气
cryopumping array 低温抽气装置
cryoresistive transmission line 低温超导输电线路
cryosar 雪崩复合低温开关;低温雪崩开关
cryoscope 冻点测定仪;低温测定器;冰点测定器;冰点测定计
cryoscopic constant 冰点降低常数
cryoscopic method 凝固点降低法;冰点(降低)法
cryoscopic solvent 冰点降低溶剂

cryoscopy 冰点降低(测定)法;冰点测定法
cryosel 冰盐
cryosistor 低温晶体管
cryosixtor 冷阻管
cryosorption 低温吸着;低温吸附;深冷吸附
cryosorption pump 冷凝吸附泵;低温吸着泵
cryosphere 永冻层;低温层;冰雪圈;冰雪圈;冰冻层
cryostat 制冷器;低温恒温器;超导线圈低温箱;深冷恒温器
cryosublimation trap 低温升华阱
cryosurgery room 低温外科室
cryotectonics 冰碛构造的;冰川构造的;冰成构造的
cryothermal treatment 冷处理;冰冷处理
cryotrap 冷凝阱;低温冷阱
cryotron 冷子管;冷持元件;低温管;低温电子管
cryotron gate 冷子管低温闸门;低温闸门
cryotronics 低温电子学
cryoturbation 冻融搅动;冻裂搅动(作用);冰冻翻浆;融冻泥流(作用)
cryoturbation structure 冻裂搅动构造
cryoultramicrotome 冷冻超薄切片机;冰冻超薄切片机
cryoultramicrotomy 冰冻超薄切片术
cryovacuum pump 低温泵
crypt 土窖;教堂地下室;地穴;地窖;穹顶地窖
cryptal 桉油萜醛
cryptalline 洞穴的
cryptic(al) colo(u)ring 保护色
cryptic colo(u)r(ation) 隐蔽色
cryptic mimicry 迷彩;掩饰物;伪装
cryptic suture 隐蔽缝合
cryptic suture line 隐伏缝合线
crypto ancillary unit 辅助密码装置
cryptobiolith 隐生物岩
cryptocenter 密码中心
cryptoclastic 隐屑的
cryptoclastic rock 隐屑岩
cryptoclastic texture 隐屑结构
cryptoclimatology 室内小气候学
crypto control unit 密码控制器
cryptocorpocollinite 隐团块镜质体
crypto-crystal 隐晶
cryptocrystalline 隐晶质;潜晶(质)
cryptocrystalline cement texture 隐晶质胶结物结构
cryptocrystalline graphite 隐晶石墨
cryptocrystalline graphite ore 隐晶质石墨
cryptocrystalline rock 隐晶岩
cryptocrystalline texture 隐晶质结构
crypto-depression 潜洼
cryptoexosporinite 隐孢子外壁体
cryptoexplosion crater 隐爆火山口
cryptoexplosive structure 隐爆炸构造
cryptofluorescence 隐晶荧光
Cryptogamia 隐花植物(纲)
cryptogelocilinite 隐胶质镜质体
cryptogene 成因不明(岩石)
cryptogenic 隐生的;隐成的;成因不明的
cryptograined 隐粒的;隐晶粒状
cryptograined texture 隐粒结构
cryptogram 密码
cryptogram residue class 密文剩余类
cryptogranitic 隐晶花岗状
cryptograph 密码机
cryptographic 加密的
cryptographically coded card 隐图编码卡片
cryptographic communication 密码通信
cryptographic equipment 密码装置
cryptographic protocol 密匙
cryptographic system 密码记录系统
cryptographic texture 隐晶文像结构
cryptography 加密术
cryptoguard 密码保护
cryptohalite 方氟硅铵石
cryptointosporinite 隐孢子内壁体
cryptolepine 白叶藤碱
cryptology 保密学
cryptomaceral 稳显微组分
cryptomagmatic deposit 隐岩浆矿床
cryptomagmatic mineral deposit 隐岩浆矿床
cryptomelane 锰钾矿;隐钾锰矿
cryptomere 隐微粒;隐晶岩
cryptomeria 柳杉
cryptomeria bark ash 柳杉皮灰
cryptomeria bark roofing 铺杉木皮的屋面

cryptomerism 隐微粒现象
cryptomerous 细晶质的；隐晶质的；隐晶的
cryptomerous rock 隐晶岩
crypto me(te)r 遮盖力计；遮盖力测定仪(涂料)；遮盖力(测定)计(涂料)；涂料遮盖力计
crypto module 密码模块
cryptomorphic 隐形的
crypto mull 隐腐熟腐殖质
cryptoolitic 隐鲕状
cryptoperthite 隐纹长石
cryptoportmus 古罗马地道
cryptor(h)eic 潜伏水系的；地下河水系的；潜流的
cryptosciascope 克鲁克管
cryptosecurity 保密措施
cryptostructure 隐构造
cryptotelinite 隐结构镜质体
cryptothermal deposit 隐温矿床
cryptovitrodetrinite 隐镜屑体
cryptovolcanic 潜火山的
cryptovolcanic earthquake 潜火山地震
cryptovolcanic structure 隐火山构造；潜火山构造
cryptovolcanic type 隐火山型
cryptovolcanism 潜火山作用
Cryptozoic 前寒武系的【地】
cry room 隔音室；不透声室
cryscope 冻点测定仪
crysrock table tops 胶石桌面
crysrock tiles 胶石砖
crystadyne 晶体振荡检波器
crystal 晶体；结晶氧化钡；结晶物；结晶体；结晶的；水晶玻璃；水晶
crystal active material 晶体激活材料
crystal activity 晶体活动性
crystal aerugo 乙酸铜
crystal aggregate 结晶集合体
crystal aggregation 结晶聚集
crystal amplifier 晶体管放大器；晶体放大器
crystal analysis 晶体分析；结晶分析
crystal angle 晶体角
crystal anisotropy 晶体各向异性
crystal anisotropy constant 晶体各向异性常数
crystal asymmetry 晶体不对称
crystalator 水晶自动楼梯；玻璃栏杆自动楼梯
crystal axial angle 晶轴角
crystal axial line 晶轴线
crystal axial plane 晶轴面
crystal axis 晶轴；晶体轴；结晶轴
crystal-bar 晶棒
crystal barrier 晶体势垒
crystal binding 晶体键联
crystal blank 晶体坯；坯晶
crystal block section 晶体检波部分
crystal body 晶体
crystal boundary 晶体外形；晶粒间界
crystal boundary corrosion 晶界腐蚀
crystal boundary migration 晶界迁移
crystal bring-up 晶体培养
crystal calibrator 晶体校准器
crystal can relay 晶体密封继电器
crystal carbonate 晶碱
crystal cartridge 晶体支架；晶体换能头
crystal cast 晶体印模
crystal chandelier 枝形吊灯；水晶吊灯
crystal checked 晶体稳定；晶体检定
crystal checker 晶体检验器
crystal-chemical formula of minerals 矿物晶体化学式
crystal chemistry 晶体化学；结体化学；结晶化学
crystal chronometer 石英天文钟
crystal clarity 晶体透明度
crystal class 晶类
crystal cleavage 晶体开裂
crystal clock 晶体钟；石英钟
crystal colloid theory 结晶胶粒理论
crystal combination 合晶
crystal conduction 晶体导电
crystal conduction counter 晶体传导计数管
crystal constant 晶体常数
crystal control 晶体控制
crystal-controlled 晶体控制
crystal-controlled dc servomotor 晶控直流伺服电动机
crystal-controlled oscillator 晶体控制振荡器；晶控振荡器

crystal-controlled receiver 晶体稳频接收机；晶体控制接收机
crystal-controlled transmitter 晶体稳频发射机；晶控发射机
crystal converter 晶体变频器
crystal coordinate axis 结晶学坐标轴
crystal corundum 结晶刚玉
crystal counter 晶体探测器；晶体计数器
crystal crystallizer 晶体结晶器
crystal current 晶体电流
crystal cut 晶体切片；晶体截片
crystal cutter 晶体割截器；压电刻纹头
crystal cutting machine 晶体切割机
crystal defect 晶体缺陷
crystal desalination 结晶脱盐
crystal detector 晶体探测器；晶体检波器
crystal differentiation 结晶分析作用
crystal diffraction 晶体衍射
crystal diffraction spectrometer 布拉格分光计
crystal diode 晶体二极管；半导体二极管
crystal diode power meter 晶体二极管功率计
crystal direction 结晶定向
crystal dislocation 晶体位错；晶体错位
crystal domain 晶畴
crystal drive 晶体激励
crystal druse 晶簇
crystal dynamics 点阵动力学
crystal edge 晶缘；晶棱；晶边
crystal edging machine 晶体切割机
crystal electric(al) clock 石英电钟
crystal electrostriction 晶体电致伸缩
crystal element 石英晶体元件
crystal enamel 晶纹瓷漆
crystal exciter 晶体激振器
crystal face 晶面
crystal face symbol 晶面符号
crystal field 晶场
crystal field effects 晶体场效应
crystal field parameter 晶场参数
crystal field spectrum 晶场光谱
crystal field splitting 晶体场分裂
crystal field stabilization energy 晶体场稳定能；晶体场稳定化能
crystal field theory 晶体场论；晶体场理论；晶场理论
crystal filed 晶场
crystal filter 晶体滤波器
crystal filter heterodyne analyzer 装有晶体滤波器的差频分析器
crystal finish 结晶纹饰(面)
crystal fixed bed 结晶固定床
crystal flambe glaze 结晶花釉
crystal form 晶状；晶形；结晶晶形；结晶单形
crystal for non-linear optics 非线性光学晶体
crystal fragment 晶屑
crystal frequency control unit 晶频控制装置
crystal frequency indicator 晶体频率指示器；晶体变频指示器
crystal frequency multiplier 晶体倍频器
crystal fundamental frequency 晶体基(本)频(率)
crystal gain 晶体增益；晶体放大系数
crystal galvanometer 晶体检流计
crystal gate 晶体符合线路
crystal glass 晶质玻璃；晶体玻璃；结晶体玻璃；结晶玻璃；富铅玻璃；水晶玻璃
crystal gliding 晶体滑动
crystal goniometer 晶体测角仪
crystal grain growth 晶粒长大
crystal grating 晶体光栅；晶格
crystal group 晶族；晶群
crystal grower 晶体生长器；单晶生长器
crystal growing 晶体生长
crystal growing section 晶体生长段
crystal growth 晶体成长；结晶增长；单晶生长
crystal growth from liquid phase 液相生长
crystal growth furnace 单晶炉
crystal growth in space 空间晶体生长
crystal growth method 晶体生长法
crystal habit 晶体习性；结晶惯态；结晶形态；结晶习性；结晶习惯；结晶惯态
crystal hoarfrost 结晶(白)霜
crystal hydrophone 晶体水声器
crystal ice 结晶冰；悬冰
crystal impedance meter 晶体阻抗计

crystal imperfection 晶体不完整；无定形晶
crystal imprint 晶体印痕
crystal index 密勒指数
crystal indicatrix meter method 晶体光率计法
crystal ingot 晶锭
crystal kit 晶体接收机成套零件
crystal lamp 水晶灯
crystal laser 晶体激光器
crystal lattice 晶体晶格构；晶体点阵；晶格；结晶格子
crystal lattice asymmetry 晶体点阵不对称性；晶格不对称
crystal lattice energy density 晶体点阵能量密度；晶格能(量)密度
crystal lattice filter 晶格滤波器
crystalline 晶状；晶质的；结晶状；结晶质；结晶体；结晶的；水晶的
crystalline agent 结晶剂
crystalline aggregate 晶质集合体；结晶(质)集料
crystalline aggregate of phenol-calcium 酚钙矿巢
crystalline alumina 结晶矾土
crystalline aluminium chloride 结晶氯化铝
crystalline amyloses 结晶直链淀粉
crystalline anisotropy 晶体各向异性
crystalline band 结晶谱带
crystalline basement 结晶基底
crystalline basement of Pre-Cambrian 前寒武纪结晶基底
crystalline bloom 结晶霜；起霜
crystalline bond 结晶连结
crystalline break 晶体断口
crystalline calcium hydroxide 结晶氢氧化钙
crystalline carbonate rock 结晶碳酸盐岩
crystalline cement texture 结晶质胶结物结构
crystalline ceramics 晶态陶瓷；结晶陶瓷；透明陶瓷
crystalline chloral hydrated chloral 水合氯醛
crystalline cleavage plane 晶体的解理面
crystalline clock 晶体钟
crystalline complex 晶体复合物
crystalline concretion 结晶架
crystalline constant 结晶常数
crystalline crack 结晶裂纹
crystalline cylinder 结晶筒
crystalline dichroic effect 晶体的二色效应
crystalline element determination method 结晶要素测定法
crystalline erosion index 结晶性侵蚀指标
crystalline field 晶场
crystalline flake graphite 鳞片石墨；结晶鳞片石墨
crystalline flaky graphite deposit 晶质鳞片状石墨矿床
crystalline flambe glaze 结晶铜红釉
crystalline form 晶形；结晶形状
crystalline fracture 晶状断口；晶体断裂；结晶状断口；结晶断面；结晶断口
crystalline glaze 结晶釉
crystalline glaze tile 结晶釉瓦
crystalline grain 晶粒
crystalline grain fracture 晶粒断口
crystalline-granular texture 结晶粒状结构
crystalline granular texture of carbonate rocks 碳酸盐岩晶粒结构
crystalline graphite ore 晶质石墨矿石
crystalline growth velocity 结晶速度
crystalline gypsum 结晶石膏
crystalline heaven 水晶天
crystalline host lattice 晶体基质点阵
crystalline host material 晶体基质材料
crystalline humor 晶状物
crystalline hydration product 结晶水化物
crystalline hyposulfite 结晶硫代硫酸钠
crystalline ice 结晶冰
crystalline imperfection 晶体(晶格)缺陷；晶体点阵缺陷
crystalline inclusion 结晶质包裹体
crystalline inhomogeneity 结晶不均匀性
crystalline ionic radius 结晶离子半径
crystalline lens 晶状体；水晶体
crystalline limestone 结晶(石)灰岩
crystalline material 晶体材料；结晶材料
crystalline matrix 结晶母体；结晶基体
crystalline metamorphism 结晶变质
crystalline ooid 结晶鲕
crystalline order 结晶序
crystalline overgrowth 结晶附生

crystalline particle 结晶颗粒
crystalline pattern 结晶式样
crystalline perfection 结晶完整性
crystalline phase 晶形；结晶相
crystalline plastics 结晶塑料
crystalline plutonic rock 结晶深成岩
crystalline polymer 结晶聚合物
crystalline-porphyritic 结晶斑状
crystalline powder 结晶性粉末
crystalline quartz 晶状石英
crystalline range 析晶温度范围
crystalline rate 结晶速率
crystalline region 晶区
crystalline rock 晶质岩；结晶岩(石)
crystalline schist 结晶片岩
crystalline shale 结晶页岩
crystalline shape 晶形
crystalline silica 结晶态氧化硅；结晶二氧化硅
crystalline silicon 晶体硅
crystalline skeleton 结晶骨架
crystalline slag 结晶矿渣
crystalline slate 结晶板岩
crystalline slurry 结晶浆液
crystalline solder glass 结晶焊接玻璃
crystalline solid 晶体物质；结晶性固体；晶体；结晶固体
crystalline solid phase 晶状固相
crystalline state 晶态；结晶(状)态
crystalline structure 晶体结构；结晶结构
crystalline surface texture 结晶状表面形态
crystalline texture 晶粒结构；玻晶结构
crystalline water 结晶水
crystalline water-soluble 晶体水溶性
crystalling 雏晶
crystallinic acid 结晶酸
crystallinity 晶性；结晶性；结晶度
crystallinity of crystal 晶体的结晶程度
crystallinoclastic 晶质碎屑
crystallinoclastic rock 晶屑岩
crystallinohyaline 玻基斑状
crystal-liquid pair 晶体—液体对
crystallisation 结晶
crystallite 雏晶；微晶
crystallite aggregate 晶粒聚集体
crystallite hypothesis 微晶子假说
crystal-lithic psammitic texture 晶岩屑砂状结构
crystal-lithic tuff 晶岩屑凝灰岩
crystallitic texture 皱晶结构
crystallizability 可结晶性；结晶性
crystallizable 可结晶的
crystallizable fragment 可结晶分段
crystallization 镜面化；晶化；结霜材料；结晶作用；结晶化；结晶；析晶
crystallization by cooling 冷却结晶作用
crystallization capacity 结晶容量
crystallization center 晶核；结晶中心
crystallization differentiation 结晶分异作用；结晶分异；结晶分化；分离结晶(作用)
crystallization-differentiation deposit 结晶分异矿床
crystallization fabric 结晶组构
crystallization face 析晶面
crystallization finish 油漆结晶纹饰；结晶状饰面
crystallization form solution 液体结晶作用
crystallization form vapo(u)r 气体结晶作用
crystallization from high temperature solution 从高温溶液结晶
crystallization from liquid 从液相结晶
crystallization from melt 从熔体结晶；熔体结晶(作用)
crystallization from vapo(u)r phase 从气相结晶
crystallization index 结晶指数
crystallization in painting 喷漆中的结晶物
crystallization kinetics 结晶动力学
crystallization mechanism 结晶机理
crystallization path 析晶路程
crystallization period 结晶阶段
crystallization point 结晶点
crystallization power 结晶能力
crystallization pressure 结晶压力
crystallization process 结晶法；结晶除银法
crystallization stage 结晶阶段
crystallization system 晶系
crystallization temperature 结晶温度
crystallization test 晶化试验

crystallization theory 结晶理论
crystallization time 结晶时间
crystallization volume 结晶容积
crystallization water 结晶水
crystallized dolomite 晶体白云石
crystallized finish 结晶状饰面；油漆结晶纹饰
crystallized glass fiber 微晶玻璃纤维
crystallized hydrate 结晶水化物
crystallized product 结晶产物
crystallized slag 结晶渣
crystallized verdigris 结晶铜绿
crystallizer 结晶器
crystallizer pan 结晶盘
crystallizer tank 结晶槽
crystallizing 晶化
crystallizing agent 结晶剂
crystallizing dish 结晶皿
crystallizing dish with spout 结晶皿
crystallizing evapo(u)rator 结晶蒸发器
crystallizing finish 晶纹饰；晶纹(罩面)漆
crystallizing lacquer 自干晶纹(喷)漆
crystallizing layer 结晶层
crystallizing magmatic melt 正在结晶的熔浆
crystallo axis 晶轴
crystalloblast 变(余斑)晶
crystalloblastesis way 变质结晶作用方式
crystalloblastic 变晶(质)的
crystalloblastic fabric 变晶组构
crystalloblastic ooid 变晶鲕
crystalloblastic order 变晶次序
crystalloblastic process 变晶作用
crystalloblastic sequence 变晶顺序
crystalloblastic series 变晶系列
crystalloblastic texture 变晶结构
crystallochemical analysis 结晶化学分析
crystallochemical classification of minerals 矿物晶体化学分类
crystalloclastic psammitic texture 晶屑砂状结构
crystalloclastic tuff 晶屑凝灰岩
crystallogeny 结体发生学；结晶发生学
crystallogram 晶体照片；晶体衍射图；晶体绕射图；结晶衍射图；结晶绕射图
crystallograph 检晶器
crystallographic(al) component 结晶组分
crystallo-graphic(al) axis 晶轴；结晶轴
crystallographic(al) apparatus 结晶测验器
crystallographic index 结晶指数
crystallographic misorientation 结晶错位
crystallographic(al) notation 晶面表示法
crystallographic orientation 定向结晶
crystallographic(al) plane 晶面；晶体学平面
crystallographic(al) shear 结晶剪切
crystallographic(al) system determination 晶系测定
crystallography 晶体照相术；晶体学；结晶学；检晶仪
crystalloid 晶体(的)；结晶的；类晶体；晶质；结晶样体；凝晶质；拟晶体；似晶质
crystalloid body 晶样体
crystalloid osmotic pressure 晶体渗透压
crystalloid solution 晶体溶液
crystallology 晶体学；结晶构造学
crystallo-luminescence 晶体发光；结晶发光
crystallomagnetism 晶体磁学
crystallometer 检晶器
crystallo-optics 结晶光学
crystallophysics 晶体物理学
crystalloplastic corrugated structure 塑变柔皱构造
crystal marker oscillator 晶体标志振荡器；晶体标志发生器
crystal marker switch 晶体指示开关
crystal maser 晶体脉塞
crystal matching 晶体匹配
crystal method 晶体法
crystal microphone 晶体传声器；压电式传声器
crystal mixer 晶体混频器
crystal mixer radio-meter 晶体混频辐射计
crystal mixing chamber 晶体混频腔
crystal model 晶格模型
crystal modulator 晶体调制器
crystal monochromatized radiation 晶体致单色化辐射
crystal monochromator 晶体单色器
crystal morphology 晶体形态(学)
crystal mount 晶函；检波头

crystal mush 晶粥
crystal needle 晶针；针状结晶
crystal nucleus 晶核
crystal of high activity 易激晶体；高活性晶体
crystalon 刚晶
crystal optics 晶体光学
crystal orientater 晶体光轴定向仪
crystal orientation 晶体定向；结晶取向；结晶定向
crystal orientation device 晶体光轴定向器
crystal oscillator 晶体振荡器
crystal-oscillator watch 石英电子手表
crystal osmotic pressure 晶体渗透压
crystal oven 晶体炉；晶体恒温器
crystal palace 水晶宫；用玻璃和钢架建造的建筑
crystal paper varnish 墙纸用晶纹清漆
crystal paradox 结晶佯谬
crystal pattern 晶体图案
crystal perfection 晶体完整性
crystal phase 晶形；晶相
crystal photoelectric cell 晶体光电池
crystal photoelement 晶体光电元件
crystal physics 晶体物理学
crystal pick-up 晶体拾音器；压电拾音器
crystal pick-up system 晶体拾音系统
crystal plane 晶面
crystal plasticity 晶质塑性
crystal plate 晶片；压电晶片；片状晶体
crystal powdered method 晶体粉末法
crystal probe 晶体探头；晶体传感器
crystal product 结晶成品
crystal projection 晶体投影
crystal puller 拉晶机
crystal pulling 拉(单)晶(法)；晶体提拉法；单晶拉制
crystal pulling furnace 拉晶炉
crystal-pulling machine 晶体(拉)机
crystal quartz 结晶石英
crystal radio 晶体接收机
crystal ratio 晶体整流系数
crystal receptacle 藏晶体
crystal rectifier 晶体整流器；晶体检波器；半导体二极管
crystal reflection 晶体反射
crystal reflective frequency probe 晶体射频探头
crystal reflector 晶体反射器
crystal refractometer 晶体折射计
crystal refractoscope 晶体折射镜
crystal region 结晶区
crystal resonator 晶体谐振器
crystal restoration 晶体恢复
crystal ribbon 片状晶体
crystal rock 晶质岩
crystal-router algorithm 晶状路线算法
crystals 结晶乙酸柏木酯
crystal sac 晶囊
crystal sacrophagus 水晶棺
crystal sand 晶砂
crystal scintillation counter 晶体闪烁计数器
crystal section 晶体截面；晶体断面
crystal seeds 晶种
crystal sender 晶体发送器
crystal separation 结晶分离
crystal set 晶体接收机；晶体检波接收机
crystal sheet glass 晶质玻璃；晶体玻璃；结晶玻璃薄片；结晶玻璃板；三厚玻璃(5～6毫米厚的窗玻璃)；橱窗玻璃
crystal shutter 晶体闸；晶体保护器
crystal silica sand 石英砂
crystal size 结晶粒度
crystal skeleton 晶体间架
crystal slip 晶体滑移
crystals of Venus 乙酸铜
crystal space grating 晶体空间光栅
crystal spectrometer 晶体光谱计；晶体分光仪；布拉格分光计
crystal spectroscopy 晶体光谱学
crystal stabilized transmitter 晶体稳定发射机
crystal stock 连晶
crystal storage 晶体存储器
crystal strain 结晶应变
crystal structure 晶体构造
crystal structure analysis 晶体结构分析
crystal structure determination 晶体结构测定法
crystal structure analysis package 晶体结构分析

程序包
crystal substrate 结晶基面;结晶基底
crystal surface 晶体表面
crystal surface area 晶体表面积
crystal switch 晶体开关
crystal symbol 晶面符号
crystal symmetry 晶体对称性
crystal system 晶系
crystal test set 晶体测试设备
crystal texture 晶体结构;晶体组织
crystal thermostat 石英恒温器
crystal thickness 晶体厚度
crystal-tipped 晶头
crystal transducer 晶体换能器
crystal triode 晶体三极管
crystal tuff lava 晶屑凝灰熔岩
crystal twin 孪晶
crystal unit 晶体装置;石英稳定器
crystalux (一种用于混凝土路面有透镜作用的)玻璃砖
crystal vessel 石墨容器
crystal video detector 晶体视频探测器
crystal video receiver 晶体视频接收机
crystal violet 结晶紫;晶紫
crystal violet method (测定氯的)结晶紫法
crystal violet oxalate 草酸结晶紫
crystal violet sulfate 硫酸结晶紫
crystal-vitric tuff 晶玻屑凝灰岩
crystal-vitric tuffaceous texture 晶玻屑凝灰结构
crystal voltmeter 晶体检波伏特计
crystal wafer 晶(体)片
crystal water 结晶水
crystal wavemeter 石英波长计
crystal whisker 晶须
crystal zone 晶带
crystal zone electron diffraction pattern 晶带电子衍射图
crystobalite 白石英;白硅石
crystolon 人造碳化硅磨料
crystopal 高度闪光的薄层材料;不透明;玻璃状
crytochimate 房屋小气候
cryto-crystalline texture 微晶结构
crytogram 暗号
crytograph 密码机
CS Askania gravimeter CS 形阿斯卡尼亚重力仪
C-scroll C 字形涡卷装饰(家具)
csiklovaite 硒硫碲铋矿
csing with inserted joints 插接套管
C-stage 丙阶段
C-stage resin 不熔(化)阶段树脂;不溶酚醛树脂;丙阶树脂
C-stud profile C 字形龙骨
ctatering effect 陷口效应
Ctenopharyngodon idellus 草鱼
ctesiphon arch 悬链曲线拱
C-type crane C 字形起重机
C-type graphite 重叠状石墨
cual-duct terminal unit 冷热分管终端混合箱(空调)
cualstibite 水锑铝铜石
Cuba architecture 古巴建筑
cubage 建筑体积;求体积法;求容积法
cubage of excavation 挖方
cubanite 方黄铜矿
Cuban mahogany 古巴红木;古巴桃花心木
cubature 容量法;(测量混凝土空气量的)容量法;求体积法;求容积法
cubature formula 求体积公式
cubby 小房(间);小杂物间;小壁橱
cubbyhole 舒适的地方;鸽笼式文件夹;分类格
cube 立方(体);正六面体;方;三次方
Cu-bearing lead ore 铜铅矿石
Cu-bearing Mo ore 铜钼矿石
Cu-bearing zinc ore 铜锌矿石
cube beam splitter 立方分束器
cube capital 立方体柱头;方块式柱头
cube concrete 立方体混凝土
cube concrete specimen 立方体混凝土块
cube concrete strength 立方体混凝土抗压强度
cube concrete test 混凝土立方体试验;混凝土立方试块试验;混凝土立方块(体)强度试验
cube concrete test specimen 立方体混凝土块;混凝土立方试块;混凝土立方试件
cube corner array 立体角反射器阵列
cube corner prism 直角棱柱体
cube corner reflector 角反射器
cube corner retroreflector 立方隅角反向反射器
cube crushing strength 立方压毁强度;立方体压碎强度;抗压立方强度
cube crushing strength test 立方体抗压强度试验
cube crushing test 立方试块抗压试验
cube-edge method 立方体棱法
cube foot 立方英尺
cube globe 立体球形灯罩
cube impact strength 立方抗冲击强度
cube-in-air method 空气中方块试验法
cube inch 立方英寸
cube method 立方体法(估算房屋造价);立方估价法
cube mixer 立方形搅拌机;方筒拌和器;方筒拌和机
cube mo(u)ld 立方体试模(混凝土)
cube ore 毒铁矿
cube pavement 方块料路面
cube paving sett 铺路方石块;路面方料;铺路面方石
cuber 制块机;切下机;打包机;制粒机
cubernetics 控制论
cube root 立方根;三次根
cube root colo(u)r difference equation 立方根色差方程
cube sett 方块石(铺路用);小方石;方块毛石
cube sett paving 方块石铺砌
cube spar 硬石膏
cube strength 立方体试块(抗压)强度;立方体试件(抗压)强度;立方体强度
cube strength at 28 days 混凝土立方试块 28 天强度
cube strength test 立方块强度试验
cube-surface coil 立方面线圈
cube test 立方体(强度)试验;抗压试验(混凝土立方体试验)
cube test specimen 立方试块
cubex 双向性硅钢片
cubic(al) 立方的;正六面体;体积;三次的;三次
cubic(al) acoustic(al) ceiling 立方体隔声天花板
cubic(al) acoustic(al) tile 立方体隔声板
cubic(al) aggregate 立方形集料;立方形骨料;立方体碎石骨料
cubic(al) algebra 立方代数
cubic(al) antenna 立方天线
cubic(al) anvil 立方压砧
cubic(al) array 三维阵列
cubic(al) axial system 立方轴系
cubic(al) axis 立方轴
cubic(al) block capital 斗状柱头;方块柱头
cubic(al) body-centered 体心立方
cubic(al) body-centered lattice 立方体格子
cubic(al) body-centered structure 立方体心结构
cubic(al) capacity 容积量;立体容积;立方容量;立方容积;容量;容积
cubic(al) capacity of lifeboat 救生艇容量;救生艇容积
cubic(al) cavity 立方形空腔谐振器
cubic(al) cell 立方晶胞
cubic(al) centimeter 立方厘米;立方公分
cubic(al) centimeter gradation 立方厘米分度
cubic(al) centimeter per hour 立方厘米/小时
cubic(al) centimetre 立方厘米
cubic(al) cleavage 立方体解理
cubic(al) closedpacked structure 立方密堆积结构
cubic(al) closed packing 立方密堆积;立方紧密堆积
cubic(al) closest packing 立方最紧密填充;立方最紧密堆积;立方密装(晶格)
cubic(al) compressive strength 立方体抗压强度
cubic(al) container 方容器;方罐
cubic(al) content 建筑物体积;立方容量;立方容积;建筑体积;容积
cubic(al) contents of a building 建筑容积
cubic(al) convergence 三次收敛
cubic(al) conversion 体积换算;容积换算
cubic(al) crushing strength 立方体试块压碎强度;立方体试块破坏强度;立方体试件压碎强度;立方体试件破坏强度
cubic(al) crystal 立方晶体
cubic(al) crystal structure 立方晶体结构
cubic(al) crystal system 立方晶系
cubic(al) curve 立方次曲线;三次曲线
cubic(al) curvilinear regression 三次曲线回归
cubic(al) decimeter 立方分米
cubic(al) deformation 体积变形
cubic(al) determinant 立方行列式
cubic(al) diamond 立方体金刚石
cubic(al) dilatation 体膨胀;体积膨胀
cubic(al) dimension 体积码;容积码;立方尺寸
cubic(al) displacement 排水吨位
cubic(al) distortion 三次畸变
cubic(al) effect 立体感
cubic(al) equation 立方次方程;三次方程(式)
cubic(al) equation problem 三次方程式问题
cubic(al) expansion 体(积)膨胀
cubic(al) face-centered 立方面心的;体心立方
cubic(al) factor (碎片的)立方体系数
cubic(al) feet per hour 立方英尺/小时
cubic(al) feet per minute 立方英尺/分;每分钟立方英尺
cubic(al) feet per second 立方英尺/秒;每秒立方英尺;秒立方尺
cubic(al) feet per second per day 每秒立方英尺日
cubic(al) fit 三次拟合
cubic(al) foot 立方(英)尺
cubic(al) foot method 立方英尺法(用于估算房屋造价)
cubic(al) form 立方形
cubic(al) function 立方次函数;三次函数
cubic(al) graph 三次图
cubic(al) hinge ultra high pressure and high temperature device 铰链式六压砧超高压高温装置
cubic(al) hyperbola 三次双曲线
cubic(al) inch 立方(英)寸
cubic(al) interpolation 三次内插;三次插值
cubic(al) kilometer 立方公里
cubic(al) lattice 立方晶格;立方点阵
cubic(al) martensite 立方马氏体
cubic(al) mass of a building 房屋的体量
cubic(al) material 立方(形)颗粒材料;按体积计量的材料
cubic(al) measure 体积;容量;立方量度
cubic(al) measurement 容积吨位;体积
cubic(al) meter 立方米;立方公尺
cubic(al) meter of base pit 基坑土量
cubic(al) meter of earth 土方
cubic(al) meter of earth and stone 土石方
cubic(al) meter of site digging 井场挖方
cubic(al) meter of site filling 井场填方
cubic(al) meter of stone 石方
cubic(al) meters per day 立方米/日
cubic(al) meters per second 立方米每秒;立方米/秒;秒立方米
cubic(al) meters per hour 立方米/小时
cubic(al) meters per minute 立方米/分
cubic(al) millimeter 立方毫米
cubic(al) modification 立方体饰形
cubic(al) niter 硝酸钠
cubic(al) notation 立方表示法
cubic(al) number 立方数
cubic(al) packing 立方填充;立方堆积
cubic(al) parabola 立方抛物线;三次抛物线;抛物挠线
cubic(al) parabola curve 三次抛物线曲线
cubic(al) parsec 立方秒差距
cubic(al) particle 立方体状颗粒
cubic(al) piece(of aggregate) (骨料的)立方形颗粒
cubic(al) piston displacement 活塞位移容积
cubic(al) plane 立方平面;立方面
cubic(al) polar 极三次曲线
cubic(al) polynomial 立方次多项式;三次多项式
cubic(al) powder 方体炸药
cubic(al) product 以体积计量的产品
cubic(al) receiver 立方律接收机
cubic(al) resistance 体电阻
cubic(al) root 立方根
cubic(al) shear box 立方体匣式剪切仪
cubic(al) space lattice 立方空间点阵
cubic(al) spline 三次样条
cubic(al) spline function 三次样条函数;三次仿样函数
cubic(al) strain 容积应变;体积张力;体积应变;体积胁变
cubic(al) structure 立方结构
cubic(al) surface 三次曲面
cubic(al) symmetry 立体对称
cubic(al) system 正规立方晶体;等轴晶系;立方晶系

cubic(al)ton 立方吨
cubic(al)transformation 三次变换
cubic(al)triaxial test apparatus 真三轴试验仪
cubic(al)wool 块状棉
cubic(al)yard 立方码;立方英码
cubic(al)yardage 体积;容量
cubic-equation transition 三次缓和曲线
cubic feet per minute 每分钟立方英尺
cubicity factor (碎石的)立方体系数
cubic-lattice cell 立方点阵晶胞
cubicle 密封配电盘;小(卧)室;小房(间);隔间;分隔间;配电装置的栅
cubicle aggregate 立方形粒料;立方形集料;立方形骨料
cubicle control panel 柜式控制板
cubicle switch (组合)室内开关;组合开关;室内用配电箱
cubicle switchboard 开关柜;组合开关板
cubicle switchgear 组合开关装置
cubicle system 箱形结构系统
cubic-shaped 立方形的
cubic-shaped aggregate 立方形集料;立方形骨料
cubiculo (剧院的)小包厢;隔间;小卧室
cubiculum 卧室(古罗马)
cubiform 立方形的
cubiform capital 方块柱帽
cubina (古罗马住宅中的)厨房
cubing 建筑体积计算;以体积计量;以体积计;求体积;体积估价法;体积测量
cubing machine 制粒机
cubing plant 码垛装置
cubing system 码垛系统
cubism 立体派图画作风(美术上);立体派
cubist 立体派艺术家
cubist architecture 立体派建筑艺术
cubist style 立体派形式
cubit 腕尺
cubitage 体积估价法
cubitainer 方容器;方罐
cubitape distance meter 微波测距仪
cuboctahedron 立方八面体
cubo-cubic(al)transformation 六次变换
cuboid 矩体;长方体;平行六边形
cuboid house 立方形房屋
cuboids-element method 长方体元法
cubond 铜焊剂
cubooctahedron 立方八面体
cubraloy 铝青铜粉末冶金
Cuccia coupler 电子耦合器
cuckoo 杜鹃
cuckoo shot 长壁工作面辅助挑顶炮眼
cucumber green 瓜皮绿
cucumber green glaze 瓜皮绿釉
cucumber of the sea 海参
cucumber tree 锐叶木兰
cucurbitula 吸罐
cudbear 地衣赤染料萃
cuddy 小室;小房(间);碗柜;配餐室(船上)
cue 长队;尾接组队
cue card 提示卡
cue circuit 指令电路;辅助电路
cue kit 指令跳光发生器
cue light 彩色信号灯
cue mark 指示标记
cue patch applicator 指令跳光带黏合器
cue-response query 询问反应标志;信息标号应答询问
cuesta 岊丘;单斜脊;单面山
cufenium 库弗尼厄姆铜镍铁合金
Cuferco 库非可铜铁铅合金
cuff 封(套)
cuffed brim hat 箍边帽
cuffing infiltration 袖套状浸润
Cui-gong ware 崖公窑
cuiller 鳞瓣
cuing scale 指示尺
cuisine 厨房;餐馆
cul-de-four 圆拱顶;炉灶后墙
cul de lampe 灯垂饰
cul-de-sac(street) 死胡同;尽头路;尽端路;陷凹;单一出口水道;袋状海湾;冰间死水道;铁路死岔;死巷
culdoscope 陷凹镜
culdoscopy 陷凹镜检查

culicidal 灭蚊的
culicide 灭蚊药;杀蚊剂
culiculus 蜗卷叶梗
culina 古罗马厨房
culinary 厨房(古罗马);厨房用的
culinary ware 厨房器皿
culinary water tank 食用水舱
cull 等外减量;等外材;残胶;废品砖;剔除之物;剔出;等外料
cullamix 库拉密斯(一种专卖的彩色水泥和骨料)
culled forest 择伐林
culled wood 废木材
cullender 滤器;漏锅;筛
Cullender isochronal method 克伦台尔等时法
cullet 废(碎)玻璃;碎废玻璃;碎玻璃片;碎玻璃(疤痕)
cullet adding 熟料含率
cullet bin 碎波仓
cullet bunker 碎玻璃仓
cullet catcher 碎玻璃桶;碎玻璃承接器;碎玻璃箱
cullet chute 碎玻璃溜槽
cullet conveyer 碎玻璃传送机
cullet conveyor chain 碎玻璃传送链
cullet crusher 废玻璃料破碎机;碎玻璃料破碎机
cullet cut 磨割伤;碎玻璃划伤
cullet-free batch 不含碎玻璃的配合料
cullet impression 碎玻璃压痕
cullet pan 碎玻璃承接盘
cullet picker 碎玻璃料拣选机
cullet ratio 碎玻璃比
cullet silo 碎玻璃仓
cullet tray 碎玻璃承接盘
cullet washer 碎玻璃清洗机
cullet yard 碎玻璃堆场
cull factor 立木减量系数;等外减量系数
Cullian 非洲之星
culling 分类
culling chute 鉴定栏
culling level 淘汰标准
cullis 舞台侧面布景;屋顶排水天沟;(使闸门滑动的)滑缝;(使闸门滑动的)沟柱;滑槽板;穿堂门厅;屋檐排水沟
cull-land 新开垦土地
cull lumber 等外材
cull of brick 劣等砖;过选砖;光端砖
culls 渣滓;选余的东西;废品
cull sleeper 不合格轨枕
cull-tie 剔出枕木;不合格轨枕;废枕木
cull tree 等外树;不良立木
cullum floor (由耐火黏土砖或瓷砖和钢筋组成的)防火地面
culm 劣质煤;空心秆;煤屑;小块无烟煤;竹麦秆;无烟煤砾;碳质页岩;煤茭
culm and gob banks 劣质煤炭废料堆
Culmann construction of soil pressure 土压力库尔曼图解法
Culman-Ritter method of dissection 库尔曼一里特(求桁架内力的)截面法
Culman's construction 库尔曼图解
Culman's graphical construction 库尔曼(图解)法
Culman's line 库尔曼线
Culman's method 库尔曼法
Culman's moment area 库尔曼弯矩面积法
Culman's procedure for computing earth pressure 库尔曼土压计算法
culm clay 煤泥;矸子土
culminant star 中(天)星
culminate 达到顶点
culminating point 极点;顶点;坡折点;最高点
culmination 积顶点;轴隆区;中天;褶轴顶点;褶升区;褶隆区;高点;顶点;过顶点;背斜最高部
culmination altitude 中天高度
culmination method 中天法
culmophyre 聚斑岩
cultch 碎屑;垃圾
cultellation 测点降低
culti-cutter 果园草地耕耘铲
cultigen 栽培种
cult image 神像;供礼拜的肖像;供礼拜的塑像
cultimulcher 耙地松土器;碎土松土器
cultipacker 碎土镇压器
cultivable land 可耕地;适耕地
cultivar 栽培品系
cultivar name 农作物名称

cultivate 开垦;中耕;栽培
cultivated area 农地;耕地;栽培面积;耕作区;耕地面积
cultivated area immersed 浸没耕地面积
cultivated coast 耕地海岸
cultivated field 耕地;田涛
cultivated form 栽培类型
cultivated grassland 栽培草地
cultivated land 农田;耕作地;耕地
cultivated plant 栽培植物
cultivated slope 耕作坡地
cultivated soil 熟土;熟地;农地;耕地;耕作土(壤);耕种土(壤)
cultivated species 栽培种
cultivated takyr 耕种龟裂土
cultivated variety 栽培品种
cultivated vegetation type 栽培植被型
cultivating implement 中耕除草机
cultivating machine 耕耘机
cultivating rotor 耕耘滚筒刀
cultivation 开垦;栽培化;栽培;培养法
cultivation density 种植密度
cultivation method 栽培法;耕作法;培育法;培养法
cultivation of sea bottom 海底养殖
cultivation of soil 土壤熟化
cultivation shifting 改变耕作方法
cultivation techniques 栽
cultivator 松土除草机;中耕机;耕耘机;耕地机
cultivator steel 耕耘机锄铲
cultivator with torsion(al)spring tines 扭力弹齿式
cult room 礼拜室
cult statue 圣像;神像;供礼拜的肖像;供礼拜的塑像
cultural activity center 文化活动中心
cultural and commercial zone 文化商业区
cultural and recreation park 文化休憩公园
cultural and scientific qualities 科学文化素质
cultural area 栽培区(域);耕作区域;耕作面积;文化区
cultural building 文化建筑;文化馆
cultural center 文化中心
cultural character 培养特性
cultural control 耕作防治
cultural deposit 文化堆积物
cultural detail 人工地物
cultural district 文教区
cultural documentary 科教片
cultural-educational zone 文教区
cultural exchange 文化交流
cultural exchange center 文化交流中心
cultural facility 文化设施
cultural hall 文化会堂
cultural heritage 文物古迹;文化遗产
cultural hydraulic engineering 农业水利工程
cultural indicator 文化指标
cultural landscape 文化景观
cultural landscape protected area 文化景观保护区
culturally advanced city 文明城市
culturallyinduced erosion 耕作砍伐导致的冲蚀
culturally sensitive design 文化特色区设计
cultural measure 培育措施
cultural milieu 文化背景
cultural movement 文化运动
cultural operation 培育措施
cultural organization 文化机构
cultural palace 文化宫
cultural park 文化公园
cultural plan 造林计划
cultural plaza 文化广场
cultural practices 栽培技术
cultural property 培养性质;文物
cultural relic 文物
cultural-social environment 文化社会环境
cultural structure of urban population 城市人口文化构成
cultural trait 培养特性
culture 地物;人工建筑(海图标志);培养物;文化
culture board 人工地物版
culture bottle 培养瓶
culture butter 培养黄油
culture community 栽培群落;人工群落
cultured fished 养殖鱼
culture dish 培养皿
cultured pearl 养殖珍珠;人工培养珍珠
cultured pearl necklace 人工养殖项链

culture drawing 地物与注记原图
culture environment 文化环境
culture flash 培养基烧瓶
culture flask 培养瓶
culture fluid 培养液
culture hall 文化会堂
culture in the open 露地培养
culture in the special economic zones 特区文化
culture jar 培养液瓶
culture medium 培养基
culture method 培养法
culture of bacteria 培养细菌
culture of cotton 棉花栽培
culture of low entropy 低熵文化
culture of microorganism 微生物培养法
culture of plants 植物栽培
culture of poverty hypothesis 贫困文化假设
culture pan 营养钵
culture plate 地物版;培养皿
culture preservation 菌种保藏
culture solution 培养液
culture symbol 地物符号
culture technics 栽培技术
culture technique 培养技术
culture test tube 培养试管
culture time 培养时间
culture valuables 文化珍品
culture vessel 培养容器
culver 斑鸠
Culverhouse 鸽笼
culvert 涵洞;地下电缆管道;输水廊道
culvertail 楔形榫;鸠尾榫
culvert aqueduct 涵洞式渡槽;管式涵洞
culvert body 涵洞身;涵洞洞身
culvert box 涵箱
culvert brick 涵洞砖
culvert capacity 涵洞过水能力
culvert diversion 涵洞导流
culvert drop 涵道跌落
culvert end wall 涵洞端墙;涵洞端壁
culvert filling system 长廊道充水系统
culvert-flow discharge measurement 涵洞流量测定(法);涵管测流法
culvert for railway 铁路涵洞
culvert gallery 输水廊道
culvert gate 涵管闸门
culvert grade 涵底坡度
culverting 建造涵洞
culvert inlet 涵洞进水口
culvert inlet and outlet 涵洞进出口;涵洞洞口
culvert inlet with flared wing wall 八字翼墙洞口
culvert intake 涵洞进水口
culvert opening 涵洞孔径
culvert outflow 涵洞出流
culvert outlet 涵洞出水口
culvert pipe 涵筒;涵管
culvert-pipe discharge measurement 涵管流量测定(法);涵管测流法
culvert propulsion 涵洞顶进
culvert siphon 倒虹吸涵洞
culvert site 涵址
culvert tube 涵管
culvert under floor 闸底涵管;闸底涵洞
culvert valve 廊道阀门;船闸充水阀门
culvert wall 涵洞边墙
culvert without top-fill 明涵
culvert with steep grade 陡坡涵洞
culvert with top-fill 暗涵;暗管
cum all 附有一切权利;附各项权利
cum all rights 附各项权利
cumalong 备焊机具
cumar 库玛隆树脂
cumar gum 库玛隆树胶;香豆树脂
cumaric acid 香豆酸
cumarone 香豆酮;氧茚
cumar resin 聚库玛隆树脂;香豆树脂
cumberlandite 钛铁长橄岩
Cumberland slate 坎伯兰板岩(英国)
cumbersome 笨重的
cumbersome cargo 笨重货物
cum bonus 附有红利
cumbraite 倍斑安山岩
cum call 催缴款通知(单);附催缴款项通知单;附带催缴股款

cum commission 连同佣金
cum coupon 附有息票;附息票券;附息票的股票;附利息票
cum distribution 附分配权
cum dividend 附(有)红利;附有股利
cum drawing 附抽签权;附带提存
cumeat 楔形晶体
cumec 立方米秒
cumene 枯烯;异丙(基)苯
cumene hydroperoxide 氢过氧化枯烯
cumengite 锥氯铜铅矿
cum interest 附(有)利息
cummingtonite 镁铁闪石
cummingtonite leptynite 镁铁闪石变粒岩
cummingtonite schist 镁铁闪石片岩
cummunity pollution 城市污染
cum new 附新股
cum right 附有购买新股票的权利;附带认股权;附带权利
cumsec 立方米/秒
cum sole 反气旋性
cum testament annex 附证明
cumulant 累积量
cumulant function 累积函数
cumulant generating function 累积量母函数
cumular sharolith 团粒
cumulate 累积的;积球雏晶;堆积岩
cumulate complexe 堆积杂岩
cumulated downward 向下累积
cumulated population 累积种群
cumulated service-time 累积服务时间
cumulated temperature 累积温度
cumulated upward 向上累积
cumulate texture 堆积结构
cumulation 累积;积存
cumulation accumulation 累计
cumulation incidence rate 累计发病率
cumulative 累积的;加重的
cumulative acceleration 累积加速度
cumulative accumulation 累积堆积量
cumulative action 累积作用;聚能作用;蓄积作用
cumulative amount of creep 蠕动量
cumulative amount of stick-slip 累积黏滑量
cumulative and control chart 累积和控制图
cumulative annual rate of growth 年累积增长率
cumulative batcher 累积配料计量器;累积配料分批箱
cumulative batching 累积配料
cumulative benefit and loss curve 累积收益和亏损曲线
cumulative benefit curve 累积收益曲线
cumulative bin conveyor belt 累积料仓输送带
cumulative binomial distribution 累积二项分布
cumulative bonus 累积红利
cumulative capital stock 累积股利的股票;累积股本
cumulative catch 累积鱼获量
cumulative causation model 累积因果模型
cumulative chain yield 链式反应累计产额
cumulative chart 累积图表
cumulative circular pitch 圆周累计齿节
cumulative coefficient 蓄积系数
cumulative compound 积复绕
cumulative compound excitation 积复励
cumulative compound generator 积复激发电机
cumulative compound machine 复绕电机
cumulative compound motor 积复激电动机
cumulative compound winding 积复激绕组
cumulative concentration 累积浓度
cumulative constant 总常数
cumulative correction 累积校正(值);总校正(值)
cumulative cost 累计成本;累积费用
cumulative cost curve 累积费用曲线
cumulative curve 累计曲线;累积曲线;积累曲线;增长曲线
cumulative curve chart 累积曲线图
cumulative curve of screening analysis 累积筛分曲线
cumulative curve of storage 蓄水累积曲线
cumulative damage 累积损伤;累积破坏;累积破坏
cumulative data 累积数据
cumulative decay spectrum 累积衰减谱
cumulative deflection curve 累计变形曲线
cumulative deformation 累积变形(值)

cumulative demand meter 累积最大需量电度表;累积需量计
cumulative demonstrated reserves 累计探明储量
cumulative departure 累积偏差
cumulative deviation 累积偏差
cumulative diacharge 总流量
cumulative diagram 累积曲线图;综合图解
cumulative disbursement 累计拨款额
cumulative discharge 累计流量
cumulative displacement 累积位移
cumulative distribution 累积数分布;累积分布
cumulative distribution curve 累积分布曲线;累积值曲线
cumulative distribution function 累积分布函数;积累分布函数
cumulative dividend 累计股息
cumulative dose 累积剂量;积累剂量
cumulative dosimeter 累积剂量计
cumulative dust index 接触粉尘累积值
cumulative effect 累积效应;聚能效应;蓄积作用;总效果
cumulative effect of shaking 振动累积效应;振动积累效应
cumulative effort function 累积作用函数
cumulative energy 累积能量
cumulative error 叠加误差;累计误差;累积误差;累积差(值)
cumulative evapotranspiration and precipitation 蒸发蒸腾的积累及降水量
cumulative excitation 累积激发
cumulative exposure index 累积接触指数;累积暴露指数
cumulative factor 累加因子;累积因素;累积因素
cumulative failure 累积破坏
cumulative failure probability 累积失效概率;累积故障概率
cumulative fatigue damage law 疲劳累积损伤定律
cumulative feet drilled 累计钻探进尺
cumulative float curve 累积浮物曲线;浮物曲线
cumulative flow curve 累积流量曲线
cumulative fluid production 累计流体产量
cumulative forecasting 累积预测法
cumulative frequency 累计频率;累积频数;累积频率;积累频率;累累频次
cumulative frequency chart 累积频数(曲线)图
cumulative frequency curve 累积频数曲线
cumulative frequency curve of tidal level 潮位累积频率曲线
cumulative frequency curve of tide level 潮位累积频率曲线
cumulative frequency diagram 累积频数(曲线)图
cumulative frequency distribution 累积频数分配;累积频数分布
cumulative frequency function 累积频率函数
cumulative frequency of generation 累积出现频率
cumulative frequency polygon 累积频数(曲线)图
cumulative frequency-strain plot 累积频率应变图
cumulative frequency table 累积频数表
cumulative fuel consumption 总耗油量
cumulative gas 累计产气量
cumulative gas production 累积产气量
cumulative grading curve 累积级配曲线;累积斜曲线;累积粒度曲线;累积级配曲线
cumulative grain size curve 粒级累积曲线;粒级累积曲线
cumulative grid detector 累积式栅极检波器
cumulative grid rectifier 累积式栅极整流器
cumulative growth 累积增长
cumulative heat production 累计热能产量
cumulative histogram 累积直方图
cumulative hours 累积装置时间;累积时间;积累时数
cumulative hunting 累积性振动
cumulative index(ing) 累积索引;累积变址;多重变址
cumulative interest 累加利息
cumulative intoxication 蓄积性中毒
cumulative ionization 雪崩电离
cumulative letter of credit 可累积信用证;可积累使用的信用证;保兑信用证
cumulative maximum requirement 累计产量最大需求
cumulative mean 累加平均;累积平均(数)
cumulative metering 累积计量

cumulative method 累积法
cumulative mortality rate 累计死亡率
cumulative normal curve 累积正态曲线
cumulative normal distribution 累积正态分布
cumulative normal probability function 累积正态概率函数
cumulative number frequency 累积数字频率
cumulative observation 累积观测
cumulative oil production 累积产油量
cumulative orbital payload 轨道总有效负载
cumulative output 累积出力；累计出力
cumulative output value 累计产值
cumulative oversize distribution curve 筛上物累计分布曲线
cumulative particle size distribution 累积粒度分布
cumulative passing 累计通过量
cumulative percentage 累计百分率；累积率；累积百分数
cumulative percentage passing 累计过筛百分数
cumulative percentage passing sieve 过筛累计百分率
cumulative percentage retained 累积筛余百分数
cumulative percent frequency 累积百分频率
cumulative physical recovery 累计物理开采量
cumulative pitch error 螺距累积误差；周节累积误差
cumulative plastic damage 累积塑性破坏
cumulative plastic deformation ratio 累积塑性变形率
cumulative plastic failure 累积塑性破坏
cumulative poinson 蓄积性毒物
cumulative poison 累积性毒物
cumulative poisoning 累积(性)中毒；蓄积性中毒
cumulative Poisson distribution 累积泊松分布
cumulative pore-size distribution 累积孔径分布
cumulative pore volume 累积孔体积
cumulative precipitation 累积雨量
cumulative precipitation ga(u)ge 累积降水量计
cumulative preference shares(stock) 累积优先股
cumulative probability 累积概率
cumulative probability distribution 累计概率分配；累积概率分布
cumulative probability function 累计概率函数
cumulative probability plot 累积概率图
cumulative process 累积过程
cumulative production 累计产量；累积产量；总产量
cumulative production to year end 年产量
cumulative profit 已累缴利润
cumulative profit and tax 已累缴利税总额
cumulative provision 累积条款
cumulative radiaiton effect 累积辐射效应
cumulative radiation dose 累积辐射剂量；积累辐射剂量
cumulative rain ga(u)ge 累积雨量器
cumulative recovery 累计采油量
cumulative regression 累积回归曲线
cumulative reliability 总可靠性；总可靠度
cumulative remainder 累积剩余
cumulative requirement 累计产量需求
cumulative residue curve 累积筛余曲线
cumulative risk 累计危险性
cumulative root mean square function 累积均方值函数
cumulative runoff 累积径流(量)
cumulative runoff diagram 累积径流曲线(图)
cumulative sample 累积样品
cumulative screen analysis 累积筛析
cumulative seismic strain 累积地震应变
cumulative service function 累积服务函数
cumulative settlement 累积沉降(量)
cumulative siltation 累积性淤积
cumulative sinking fund 累积偿债基金
cumulative site occupation function 累积点占有函数
cumulative size curve 累积粒度曲线
cumulative size distribution curve 累积粒度分布曲线
cumulative spectra 累积谱
cumulative spectral density 累积(光)谱密度
cumulative speed 累积速度
cumulative speed distribution curve 累计速度分布曲线；累积速度分布曲线；累积车速分布曲线
cumulative stability constant 累积稳定常数
cumulative statistic(al) sound level 累积统计声级
cumulative stock 累积股份；累积股份

cumulative strain 累积应变
cumulative stress 累积应力；积累应力
cumulative surface site distribution 累积表面点分布
cumulative surface site distribution function 累积表面点分布函数
cumulative switching off 累积断开
cumulative tax 已累缴税额
cumulative temperature 积温
cumulative test of fish 鱼类蓄积试验
cumulative throughflow 累计流量
cumulative time 累计时间；累积时间；总时间
cumulative time metering 累时计量
cumulative time series 累积时间系列
cumulative total 累计数；累积总数
cumulative toxicity 累积毒性；蓄积毒性
cumulative toxicity test 蓄积毒性试验
cumulative tripping 累积式跳闸
cumulative-type wire-drawing 蓄丝式拉丝机
cumulative undersize distribution curve 筛下物累积分布曲线
cumulative vibration 累积振动
cumulative volume 累积体积；总体积
cumulative volume curve 累积容积曲线
cumulative water encroachment 累积水浸量
cumulative water-oil ratio curve method 累积水油比曲线法
cumulative water production 累积产水量
cumulative weighing 累计称量；累积称(重)量
cumulative weight 累计重量；累积重量；累积量
cumulative weight batcher 累加分批称量器
cumulative weight percentage 累计重量百分比
cumulative yield 累计产率
cumulative yield curve 累积屈服曲线
cumulator 累积器
cumulocirrus 卷积云
cumulonimbus 积雨云；雷暴云
cumulo-nimbus arcus 弧状积雨云
cumulonimbus calvus cloud 秃积雨云
cumulonimbus cloud 积雨云
cumulonimbus incus 砧状积雨云
cumulose 炭质堆积的
cumulose deposit 碳质堆积土；有机沉积(物)；碳(素)质堆积层
cumulose soil 植积土；高腐殖质淤泥土；堆积土；腐殖(质有机)土；腐泥土；腐积土(壤)；碳质土
cumulosol 泥炭质土
cumulo-stratus 积层云
cumulo valuation 积云评估
cumulus 积云；堆积；丘群；晴天积云
cumulus cloud 积云
cumulus congestus cloud 浓积云
cumulus convection 积云对流
cumulus crystal 积云晶体
cumulus humilis 淡积云
cumyl hydroperoxide 枯基过氧化氢；异丙苯过氧化氢
cumyl phenol 枯基苯酚
cun 市寸(1 市寸=1/30 米)
cundy 小巷(道)
cuneate 楔形的
cuneatic arch 楔体拱；楔形砌拱
cuneiform 楔形条石；楔形碑；楔状的；楔形的；梯形的
cuneiform arch 楔形砌拱
cuneoid 圆(形)锥体
cunette 子沟；河岸加固工事；河岸加固板桩；干壕底沟；底沟
cunette-shaped section 底沟形断面
cuneus 一组梯形座排(古剧场)
cunic 库尼克铜镍合金
cunico 铜镍钴合金
cunico alloy 铜镍钴永磁合金
cuniculus 矮的地下通道
cunife 铜镍铁合金
cunife magnet alloy 坎尼夫磁合金；铜镍铁永磁合金
cuniman 铜锰镍合金
cunisil 铜镍硅高强度合金
Cu-Ni-rich Mn nodule 富铜镍型锰结核
Cu-Ni sulfide ore 硫化铜镍矿石
Cunningham correction factor 坎宁安校正系数
Cunningham correction(value) 坎宁安修正值；坎宁安校正值
cuno(oil) filter 叠片转动式滤油器
cup 绞盘轴座；锥形滚柱轴承外座圈；挂筒；冲盂；杯

状物；杯状凹地；杯形翘曲；杯；求并运算；受器
cupal 包铜的铝薄板
cup-and-ball joint 插口接合；球窝接头；球窝接口；球窝接合；球窝(关)节；球铰；关节状节理
cup-and-ball visco(si)meter 杯球黏度计
cup-and-cone 杯锥
cup-and-cone arrangement 钟头式装料装置
cup-and-cone bearing 钢碗滚锥式轴承；对开径向止推轴承
cup-and-cone fracture 杯锥形断口；杯形凹凸断口
cup-and-pipe discharge 杯管式排料装置
cup-and-pipe draw 杯管式排料装置
cup anemometer 旋杯式风速计；风杯风速计；风杯风速表；杯形风力计
cup bar 杯端杆
cup barometer 杯式气压计
cup base 圆钢柱的杯形基座
cup bearing 木插口；杯型承口
cupboard 小橱；柜子；橱柜；碗柜
cupboard button 橱柜旋钮
cupboard catch 橱柜门闩；碗橱开关
cup boarding 橱柜
cupboard latch 碗橱扣拴
cupboard lock 橱柜锁
cupboard turn 小橱柜门闩
cupboard wall 嵌装小柜的墙；嵌装碗橱的墙
cup burner 杯状燃烧器
cup chuck 钟形卡盘；杯形卡盘
cup-cone fracture 杯锥形断口
cup-conveyer type potato planter 杯送式马铃薯种植机
cup coral 单个珊瑚；凹圆状珊瑚
cup core 杯形铁芯
cup-cross anemometer 转杯风速表；旋杯式风速计；杯形风力计
cup current meter 转杯流速仪
cup dolly 圆形抵座(夹卡筒形工件用)；杯状顶铆模
cup-drawing test 深拉试验
cupel 灰吹盘
cup-electrode technique 杯形电极技术
cupellation 灰吹法
cupel(lation) furnace 灰吹炉；提银炉
cup emery wheel 杯形砂轮
cup escutcheon 碟形覆板
cupeth 波纹电缆护套
cup-feed drill 杯式排种的条插机
cup-feed mechanism 杯式排种装置
cup-feed potato ptanter 杯送式马铃薯种植机
cupferron 铜铁试剂
cup flow figure 杯溢法流动指数
cup flow test 杯模式流动试验；杯漏法流动试验
cup-forming line 杯类成型线
cup fracture 杯状断口；杯形断口
cup frilling 杯状竖割
cup frills 杯状竖沟
cupful 满杯
cup grease 黄油；油杯润滑脂；钙皂(基)润滑脂；稠黏润滑油；罐结润滑脂膏；杯滑脂
cup grinding wheel 杯形砂轮
cup head 半圆头；半球头；杯形机头
cup head bolt 半圆头螺栓
cup-headed bolt 杯状圆头螺栓
cup-headed nail 半球头钉；杯头钉
cup headed rivet 圆头铆钉
cup head pin 半圆头销
cup head rivet 半圆头铆钉；半球头铆钉；半球铆钉
cup head screw 杯状半圆头螺钉；圆头钉；半圆头螺钉
cup head wood screw 圆头螺钉；圆木铆钉；圆头木螺丝；半圆头木螺钉
cupholder 茶杯座；杯形绝缘子螺脚；绝缘子螺杆(线路用)
cup hook 丝杆吊钩
cupid's darts 发金红石
cup insulator 杯形绝缘子；杯式绝缘子
cup irrigation 盘状灌溉
cup-jewel 托钻
cup joint 套接
cup leather 皮碗
cup leather for air pump 气泵皮碗
cup leather for balance piston 平衡活塞皮碗
cup leather packing 皮碗填密；皮碗密封
cup meter 旋杯式流速仪；旋杯式海流计
cup method 杯封法

cup motor 杯形电机
cup nut 杯形螺母
cup of gasholder lift 储气罐塔节下挂圈
cupola 岩钟;钟状火山;穹顶;圆(屋)顶;旋转炮塔;倒拱形地板;熔铁炉;穹隆顶;拱顶(壳);拱顶面;拱顶棚;鼓形墙上的拱顶;平面呈多边形的穹隆;平面呈多角形的穹隆;圆拱形;圆拱檐口;圆拱缘饰;圆墙上的拱顶;轴对称圆顶
cupola above a square 方屋顶上的圆顶
cupola block 拱形块;扇形块;楔形块;化铁炉(异型耐火)砖
cupola blower 冲天炉鼓风机
cupola bottom 冲天炉炉底;拱形底
cupola brick 炉顶楔形桩;化铁炉砖;扇形砖;炉顶用楔形砖
cupola charge 冲天炉炉料
cupola charger 冲天炉加料机
cupola charging 冲天炉加料
cupola charging crane 冲天炉的装料吊车
cupola charging machine 冲天炉加料机
cupola cleaning door 冲天炉点火孔;冲天炉工作门
cupola coke 铸造焦炭
cupola column 冲天炉炉腿
cupola control equipment 冲天炉控制装置
cupola crown 圆拱顶
cupola crucible 冲天炉炉缸
cupola dam 穹隆形坝;双曲拱坝;圆顶形坝
cupola drag 清炉装置
cupola drop 冲天炉珠滴
cupola dust arrester 冲天炉灭火集尘器
cupola edge 圆拱边沿;圆拱边缘
cupola emission treatment 化炉铁排放物处理
cupola floor 倒拱形地板
cupola forehearth 冲天炉前炉
cupola foundation 冲天炉炉基
cupola fume collector 冲天炉灭火炉尘器
cupola furnace 冲天炉;化铁炉
cupola furnace bottom 冲天化铁炉炉底
cupola furnace rod 冲天化铁炉杆
cupola gun 冲天炉喷浆枪
cupola gun mix 冲天炉喷补料
cupola hearth 冲天炉炉底;冲天炉炉底
cupola hypothesis of mineralization 岩钟成矿说
cupola iron 化炉铁
cupola iron block 化铁炉砌块
cupola leg 冲天炉炉腿
cupola lighter 冲天炉点火器
cupola lighting torch 冲天炉点火器
cupola lining 冲天炉炉衬
cupola loader 圆筒装料机
cupola malleable iron 化铁炉可锻铸铁
cupola melter 冲天炉熔化工;大炉工
cupola melting 冲天炉熔化
cupola patching gun 冲天炉喷浆枪
cupola pier 圆拱墩
cupola receiver 冲天炉前炉
cupola shaft 冲天炉(炉)身
cupola shell 冲天炉炉壳;拱形壳
cupola slag 冲天(炉)炉渣;熔铁炉渣
cupola slag cement 化铁炉炉渣水泥
cupola slag hole block 冲天炉出渣口成型砖
cupola slag sulphated cement 石膏化铁炉炉渣水泥
cupola spark arrester 冲天炉灭火集尘器
cupola spout 冲天炉出铁槽
cupola stack 冲天炉烟囱
cupola stress 拱顶应力
cupola tender 冲天炉工长
cupola well 冲天炉炉缸
cupola working bottom 冲天炉炉底
cupolette 小型冲天炉
cup out 作出带有端面的深槽
cup packing 杯形填密法;皮碗填密;皮碗式密封;皮碗密封;杯形填密法
cupped 横向翘曲的;杯状的
cupped washer 凹垫圈
cuppiness 杯锥状蘑菇头断裂
cupping 胀凸(金属皮);翘成杯形;浇釉;木材干燥翘曲;横向翘曲;干缩翘起(木料);冲盂;采脂;杯状凹陷;杯形挤压;杯吸法(过程);瓦形弯
cupping ductility value 压延系数
cupping glass 拔火罐;吸罐
cupping machine 拉伸试验机;深拉压力机
cupping of gasholder 储气罐塔节起挂
cupping press 引伸压力机;深拉压力机
cupping setting up 受脂法
cup(ping)test 杯突试验;压凹试验;埃里克森试验
cupping tester 杯突试验仪
cupping testing machine 杯突试验机;深拉试验机
cupping value 杯突值;深拉系数
cup point 杯形端
cup point screw 杯形圆头螺钉
cup point setscrew 尖端止动螺钉
cup product 上积
cuppy 杯形的
cuppy wire 杯锥状断口钢丝
cupralith 库普拉利司铜锂合金
cupralium 库普拉利铜铝合金
cuprammonium rayon 铜铵(人造)丝
cuprammonium washing tower for removing CO 铜洗塔
cupreous 铜(色)的;铜赤色
cupreous sandstone deposit 含铜砂岩矿床
cupressene 柏烯
cupressin 柏树油
cupressoid pits 柏型纹孔
cupressus oil 柏树油
cuprian melanterite 含铜水绿矾
cupric 含铜的;正铜;高价铜;二价铜的
cupric acetate 乙酸(正)铜;正乙酸铜;醋酸铜
cupric arsemte 希尔绿
cupric arsenite 亚砷酸铜
cupric bichromate 重铬酸铜
cupric bioxalate 草酸氢铜
cupric cemented steel 渗铜钢
cupric chloramine 氯胺铜
cupric chloride 氯化铜
cupric compound 铜化合物
cupric cyanide 氰化铜
cupric fluoride 氟化铜
cupric meta-arsenite 偏亚砷酸铜
cupric nitrate 硝酸铜
cupric oxalate 草酸铜
cupric oxide 氧化铜
cupric salt 铜盐
cupric silicate 硅酸铜
cupric sulfate 硫酸铜;正硫酸铜
cupric sulfate crystal 结晶硫酸铜
cupric sulfide 硫化铜
cupric sulfophenate 苯磺酸铜
cupric sulphate 硫酸铜
cupric sulphide 硫化铜;铜蓝
cupric thiocyanate 硫氰酸铜
cupriferous dye 含铜染料
cup ring 胀圈;皮碗
cup ring nut 皮碗上的环形螺帽
Cuprinol 环烷酸铜(一种木材防腐剂)
cuprisone 铜试剂
cuprite 赤铜矿
cupro alloy 铜合金
cuproarquerite 铜银汞膏
cuprobismutite 辉铋铜矿
cuprobond 硫酸铜处理;镀铜
cuprocalcite 孔雀石;赤铜方解石
cuprocopiapite 铜叶绿矾
cuprodescloizite 钒铜铅矿
cuprodine 镀铜
cuproferrite 铜绿矾
cuprohalloysite 铜埃洛石
cuprolead 铅铜合金;铜铅合金料
cupromanganese 铜锰合金
cupron 铜试剂
cupro-nickel 镍铜合金;铜镍合金
cupro-nickel constantan 铜镍合金
cupropavonite 硫铋铜银铅矿
cuproplatinum 铜铂矿
cuprorivaite 硅铜钙石白
cupro silicon 铜硅合金
cuprosklodowskite 硅铜铀矿
cuprospinel 铜铁尖晶石
cuprostibite 锑铊铜矿
cuprotyngstite 铜钨华
cuprous 亚铜的
cuprous acetylide 乙炔化亚铜
cuprous chloride 氯化亚铜
cuprous cyanide 氰化亚铜
cuprous fluoride 氟化亚铜
cuprous hydroxide 氢氧化亚铜
cuprous iodide 碘化亚铜
cuprous oxide 一氧化二铜;氧化亚铜;氧化低(亚)铜
cuprous oxide photocell 氧化亚铜光电池
cuprous-oxide rectifier 氧化亚铜整流器
cuprous sulfate 硫酸亚铜
cuprum 铜
cup screen 圆柱筛;杯筛
cup seal 杯形封接
cup seaming machine 包缝机
cup shake 风裂;轮裂;木材环裂;环裂
cup shake wood 环裂木材
cup-shaped 成杯形;杯状的
cup-shaped bucket 杯状戽斗;杯形戽斗
cup-shaped hammer 铆钉锤
cup-shaped insulator 杯状绝缘体
cup shaped rotor 杯式离心头
cup-shaped sleeker 杯形堰刀
cup-shaped sprocket wheel 杯形链轮
cup shell 杯形件
cup spring 板簧;盘形弹簧
cup-square bolt 杯形方螺栓;半圆头方螺栓
cup stand 杯托
cup trimming machine 修杯机
cup type anemometer 杯式风速表
cup-type bristle brush wheel 杯型洁齿刷;杯形鬃刷轮
cup-type current meter 旋杯式流速仪;旋杯式海流计
cup type electrostatic paint sprayer 旋杯式静电喷涂机
cup-type elevator 杯式升运器
cup-type nozzle 杯形喷嘴
cup-type pycnometer 杯式比重瓶
cupule 杯形器;壳斗
cup valve 钟形阀;杯状阀;杯形阀;皮碗阀
cup vane assembly 转子(旋杯流速仪);风杯(转杯风速表)
cup visco(si)meter 杯式黏度计;黏度杯
cup warp 横向翘曲;环状翘曲;杯式翘曲
cup washer 环裂垫圈;杯状垫圈;杯形(螺栓)垫圈;杯形衬垫
cup weld 带盖板焊缝
cup wheel 杯形砂轮;杯形磨轮
cup wheel printer 杯型打印机
cup-wheel type potato planter 杯轮式马铃薯种植机
cup with cover 带盖杯
cup with high handle 高柄杯(瓷器名)
cup with picture of lofty gentlemen 高士杯(瓷器名)
curability 硫化性能;可切削性;固化性能
curable depreciation 可弥补的贬值
curative concentration 固化剂浓度
curative spring 治疗泉
curb 路缘(石);路牙;井栏;井框垛盘;井口锁口圈;道牙石;道牙;挡水面浇筑块;边缘墙胎;边饰;侧石
curb and gutter 缘石街沟;平侧石
curb angle 边角钢
curb arrangement 缘石排铺成行;路缘石准线
curb bar 路缘杆
curb below ground 路面(镶)边石(不高于路面的缘石)
curb bend 拐角条砖(釉面砖的配件砖)
curb bit 大勒缰
curb box 人行道地面水阀门箱;阀箱;阀井;路缘盒
curb break 人行道的缺口
curb brick 路缘砖
curb capacity 沿人行道停车容量;沿侧石停车容量
curb chain 马嚼式链条;制动链;锁链
curb clay tile 镶边陶块(屋顶)
curb cock 路边水栓;井内的关断阀;路缘旋塞;引水龙头
curb cock box 路缘旋塞盒;用户止水阀铁箱
curb collection 路边收集(垃圾)
curb concrete 拱脚托梁
curb corner 路缘转角
curb course 池底边砖
curb curve 路缘曲线
curb cutback 路缘后移
curb cuts 路沿下陷处
curb dealing 场外交易
curbed 有路缘石的
curb edger 路缘石施工工具;路缘修边器
curb elevation 路缘标高

curber 铺路缘石机;铺侧石机
curb exchange 场外交易;场外汇兑
curb face 路缘面
curb fender 路栏;壁炉周边装饰;路缘护栏
curb finish 路缘修整
curb footway 有缘石的人行道
curb for fire place 壁炉槛
curb form 路缘模板
curb gable tile 缘墙的山墙盖瓦
curb girder 侧梁
curb grab(bing) equipment 路边抓斗式设备
curb grade 路缘坡度
curb guard 护角(线条)
curbing 木井框支架;护轮槛;抑限制;做井槛;铺设缘石;设置路缘石;设路缘石
curbing inflation 控制通货膨胀
curbing machine 路缘石(自动)铺设机;设路缘设备
curbing paver 路缘石铺设机
curb inlet 路缘进水口;道牙进水口
curb inlet basin 路缘进水井
curb joint (复折形屋顶变坡处的)水平接缝
curb lane 路缘车道;路边车道;外侧车道;路缘车道
curb level 路缘水平;路缘石标高;路缘高度
curb level at building 建筑物前路边石标高
curb lighting 路边照明
curb line 路缘线;路边线
curb loading zone 路旁上下车带
curb machine 路缘机
curb market 路边市场
curb marking 人行道标牌;路侧标牌;路缘标示
curb mold 路缘线脚;混凝土路边石浇模
curb opening 道路边的开口
curb-opening inlet 路缘(侧)进水口;街道侧石进水口
curb outlet 路缘排水口;牙排水口;道牙排水口
curb parking 路缘(侧)停车;路边停车
curb parking capacity 路缘(侧)停车容量
curb parking space 路边停车场
curb parking spacing 路缘(侧)停车间距
curb parking taboo 路缘(侧)禁止停车
curb pins 内外夹
curb plate 天窗(复折屋顶两斜面相交处);侧板;缘墙板;复折屋顶变坡处的檩条;(圆形结构的)顶部墙板;边缘墙板
curb press 渗漏筒式榨油机;混凝土路缘石预制机
curb rafter 上排椽子(双折屋顶);侧椽;复折椽木
curb ramp 缘石坡道;路边斜坡;路缘坡道
curb return 交叉口(转向处的)弧形路缘石;转角路缘石
curb ring 转车台
curb ring crane 转盘起重机
curb roll (复折屋顶两斜面的)交线处卷瓦;复折接缝;(复折形屋顶变坡处的)水平接缝
curb roof 孟莎屋顶;折线屋面;复折(式)屋顶;复斜屋顶;变坡屋顶
curb separator 有路缘石的分车岛
curb shut-off 井内关断阀;用户管阀(设在路边)
curbside 街道边;路边
curbside loading 路边装卸
curb side strip 路侧带
curb space 路缘地带
curb space inventory 路缘地带使用清册
curb space map 路缘地带现状图
curb speculation 控制投机活动
curbstone 阶条石;路缘石;侧石;栏石;井圈;道牙石
curbstone finish 缘石修面
curbstone of a well 井口侧石
curb stop 路缘关闭阀;(设在路边的)用户管阀;井内关断阀
curb string 封闭式楼梯斜梁;全包楼梯斜梁
curb stringer 复折楼梯斜梁
curb tile 复折屋顶折线瓦;边瓦
curb-to-curb crossing distance 侧石到侧石距离
curb to curb width 车行道宽度
curb tool 路缘修饰器;路缘瓦刀
curb valve 井内关断阀;用户管阀(设在路边)
curb-valve box 人行道(水)阀门箱
curb wall 缘墙;围护墙
curcas oil 麻风树油
circuit contact 保险丝接点
circuit layer network 电路层网络
curcuma 姜黄
curcumenol 莪术醇
curd 凝乳;液体凝结物

curdle 凝结
curdled milk 凝乳
curdle effect (瓷釉的)凝结效果
curdling 乳液;凝胶化;凝固;胶化;凝化
curdy precipitate 凝乳状沉淀物
cure 矿泉疗养池;烘焙;固化;处置;熟化
cure accelerator 硫化促进剂;固化促进剂
cure box 高温烘焙机
cure concrete road 养护混凝土路
cure cycle 固化周期;熟化周期
cure dag 蒸煮室
cured by atmospheric-pressure steam 常压蒸汽养护(混凝土)
cured by atmospheric steam 低压蒸汽养护(混凝土)
cured by autoclaving 高压蒸汽养护(混凝土)
cured by low-pressure steam 低压蒸汽养护(混凝土)
cured panel 固化(刨花)板
cured resin 凝固树脂;硬树脂;固化树脂
cured strength 养护后强度
cured tobacco 烟叶加工
cure electric(al) monitoring 固化电测监控
curer 焙固机
cure rate 熟化速率
cure retardation 固化减慢;熟化减慢
cure retarder 固化阻滞剂;熟化阻滞剂
cure schedule 固化制度;固化规程;熟化规程
cure shrinkage 固化收缩(率)
cure stress 固化应力
cure system 固化装置;固化体系;熟化装置;熟化方法
curet 刮匙
cure the concrete 养护混凝土
cure time 熟化时间
curetonite 磷钛铝钡石
curette 刮匙
cure window 固化窗
curf 切缝;锯痕;开槽沟
Curfew 柯秀(一种专利滑门齿轮)
curia (罗马政府的)议会厅;议会厅(古罗马)
curiage 居里数;居里强度
Curie 居里(放射性强度单位)
Curie balance 居里秤
Curie constant 居里恒量;居里常数
Curie cut 居里截割
Curie cut crystal 居里截式晶体
Curie equivalent 居里当量
curiegram 居里图
curie-hour 居里小时
Curie isotherm 居里等温线
Curie isotherm surface depth map 居里面深度图
Curie law 居里定律
curienite 钒铅铀矿
Curie plot 居里图
Curie point 居里温度(点);居里点
Curie point depth 居里点深度
Curie point isotherm 居里点等温线
Curie point pyrolyzer 居里点热解器
Curie point temperature 居里点温度
Curie temperature 居里温度
Curie temperature of magnetic core 磁芯居里温度
Curie temperature scale 居里温度;居里温标
curing 硫化处理;养护;固着;保藏处理;熟化
curing accelerator 硫化促进剂;固化促进剂
curing agent 硫化剂;硬化剂;养护剂;固化剂;塑化剂;熟化剂
curing agent blush 固化剂泛白
curing agent for epoxy resin coating 环氧树脂漆固化剂
curing area 养护场(混凝土件);熟化区
curing bed 养护床
curing blanket 养护毯(混凝土);养护覆盖层;混凝土养护覆盖材料
curing by covering mats 草垫养护
curing by infrared radiation 红外(线)养护;红外(线)辐射养护
curing by ponding 池负养护;池式养护;围水养生;围水养护(混凝土);水池养护
curing by sprinkling 喷水养护;洒水养护
curing by steam 蒸汽养生;蒸汽养护
curing capacity 养护性能
curing catalyst 固化催化剂
curing chamber 硫化室;养护室;保养室;熟化室

curing composition 混凝土养护剂
curing compound 养护剂;混凝土养护剂;化学养护液;化学养护剂;养护剂
curing compound sprayer 养护液喷洒器
curing concrete 养护混凝土
curing condition 养护条件
curing cover 养护覆盖
curing cycle 烘培循环;养护(周)期(混凝土)
curing degree 固化度;熟化程度
curing delay 养护延时;静置期;静停期
curing department 硫化车间
curing enclosure for steaming 蒸汽养护罩
curing fixture 养护设备
curing humidity 养护湿度
curing in moisture 湿固化
curing in water 水中熟化;水中固化
curing kiln (混凝土蒸汽的)养护床;养护窑;蒸汽养护窑
curing machine 养护机;焙烘机
curing mat 养护席;养护覆盖(物)
curing material 硬化材料;养护材料
curing mechanism 固化机理;熟化机理
curing membrane 养育薄膜;湿治薄膜;养护薄膜
curing membrane equipment 混凝土养生薄膜洒布设备;混凝土养护薄膜洒布设备
curing of concrete 混凝土养护
curing of concrete units 混凝土砌块蒸汽养护
curing of test specimen 试件养护
curing of the mortar 砂浆养生
curing oven 硫化炉;固化炉
curing overlay 养护覆盖物
curing paper 混凝土养护纸;养护(用)纸;养护防水纸
curing period 熟成期;养护周期(混凝土);养护期;固化周期
curing power 熟成能力(沥青乳液)
curing pressure 养护压办
curing process 养护过程(混凝土);养护法;熟成过程(沥青乳胶);熟成法;固化处理
curingprocess rate 硫化速度
curing property 养护性质
curing quality 养护性质
curing rack 养护架(混凝土块);混凝土养护格栅
curing rate 养护速率(混凝土);熟化速率(沥青乳胶);熟成速率(沥青乳胶);固化速率
curing reaction 固化反应
curing rig 养护设备
curing room 硫化室;养护室;保养室;熟化室
curing schedule 养护制度
curing seal 养护密封装置
curing shed 养护棚
curing solution 养护液(混凝土)
curing speed 固化速度
curing strain 硬化变形
curing system 固化系统
curing tank 养护箱;养护池
curing temperature 硫化温度;养护温度;固化温度;熟化温度
curing tent 路面养护棚
curing test 养护试验
curing time 养护期;熟化期;熟成期(沥青乳胶);固化时间;固化时间;熟化时间;熟化期
curing title 分离的所有权
curing water 湿治水
curing yard 养护场(混凝土制品)
curio 珍品;古物;古玩;古董
curio dealer 古董商
curiosa 珍品;珍本
curio shelves 多宝格;博古架
curiosity 珍品
curiosity shop 古玩店
curite 板铅铀矿
curl 卷曲(度);卷边;旋量;旋度;波形纹;波纹材
curlator 揉搓式磨浆机
curl-de-sack 尽端路(尽端有回车道);袋形死巷
curled bedding 卷曲层理
curled clouds 云卷饰;卷云线脚
curled edge 波浪边
curled glass fiber 卷曲玻璃纤维
curled hair 毛毡;发卷
curled spindle 滚压齿纹摘锭
curl equation 旋度方程
curler 盘卷机
Curlew 柯勒物(一种防火滑门)

curley cannel 鸟眼状长烟煤
curl field 旋场
curlicue 卷曲装饰
curling 翘曲;扭曲;卷边;卷缩;卷花;毛绒卷曲;板材卷边;缩釉
curling action 卷曲作用
curling-cracking 卷曲裂开
curling of flooring 楼地面翘曲
curling of slab 板翘曲
curling of slabs 板的卷曲
curling punch 卷边阳模;卷边冲头
curling round the roll 缠辊
curling side guide 卷边侧导板
curling stress 弯翘应力;扭曲应力;翘曲应力
curling stuff 旋卷物
curling thickness 卷边厚度
curling wheel 卷边转盘;卷边转盘
curl of vector 向量旋度;矢量(的)旋度
curl operator 旋度算符
curl type 卷曲花纹
curl up 旋绕
curl wreath 旋转波纹材
curly bedding 卷曲层理
curly brackets 波形括号
curlycue 卷曲装饰
curly figure 羽状纹;皱状纹;波状纹理
curly grain 木节;卷(曲)纹(理);漩涡纹(理);皱纹(理);木瘤
curly pearlite 索拜珠光体
curly schist 卷曲片岩
curly veneer 蜗形镶面板;涡形饰纹镶面板;涡纹饰镶面板
cur-off device 断流器
curragh 沼泽地
curred link band 链圈表带
curren and voltage converter 电流电压变换器
curren cy 流通时间;流传;货币;当前值;通货
currency account 活期存款账户
currency account balance 经常收入差额
currency adjustment factor 货币(值)调整因素;货币(值)调整率;币值调整因素
currency adjustment surcharge 币值调整附加费
Currency and Bank Notes Act 通货和银行钞票法例
currency appreciation 货币升值;币值升值
currency arbitrage 套汇
currency assets 流动资金
currency balance 货币结余
currency bloc 通货集团
currency borrowed 借款的货币单位
currency chart file 当前图形文件
currency circulation 货币流通
currency clause 货币条款;汇款条款;汇兑条款;通货条款
currency coin 通货
currency collector 集电装置
currency control 货币管制
currency conversion 货币兑换
currency decay 电流衰减
currency deflation 通货紧缩
currency depreciation 货币贬值
currency depreciations and appreciations 币值升降
currency devaluation 通货贬值;货币贬值
currency drain 货币外流
currency drawing 通用图
currency electrode 电流电极
currency exchange 货币兑换;汇兑
currency exchange at the border 边境外币兑换;边境兑换
currency failure 电流故障
currency float 测流杆
currency fluctuation 币值变动
currency from irregular sources 迷走电流;漏泄电流
currency-impulse switch 脉冲开关
currency indicator 流通量指示器;当前值指示器;当前位置指示器
currency inflation 货币膨胀;通货膨胀
currency investment 短期投资
currency issued 投放现金
currency liability 短期负债;通货贷款
currency liquidity 通货流动性
currency loan 现金贷款;通货贷款
currency margin 现行限度;经营毛利

currency note 流通券;现款;现钞(尤指美钞);通货票据
currency of account 记账货币
currency of agreement 协议书规定的货币
currency of contract 合同规定货币;契约货币
currency of over valuation 币值高估
currency of payment 支付账;支付货币
currency of settlement 结算货币
currency overvaluation 币值高估
currency parity 货币平价
currency payment 经常性开支
currency pointer 当前值指针
currency premium 本期保险金
currency primitive attribute values 当前图元属性值
currency proportion 货币比例
currency rate 通货汇率
currency realignment 货币调整;币值调整
currency reform 货币改革;币制改革
currency regulation 外汇管理条例
currency reserve 货币储备
currency restriction 货币限制
currency revaluation 货币升值
currency stable unit 币值稳定单位
currency surcharge 货币贬值附加费;币值附加费
currency system 币制
currency transfer 货币调拨
currency unfit for circulation 不适宜流通的货币
currency value is firm 币值巩固
current 流动的;电流;当前的;水流
current account 流水账;经营项目;经常项目;交互计算;活期账户;往来账户
current account balance sheet 流动资金负债表;经常项目平衡表
current account credit 往来账贷方
current account margin 经常项目收支
current accretion 当年生长量;冲击量
current-activity stack 当前活动栈
current actuated leakage protector 电流起动型漏电保护器
current actuated rudder 襟翼舵
current address 现地址;当前地址
current algebra 流代数
current amplification 电流放大
current amplification degree 电流放大率
current amplification factor 电流放大倍数
current amplifier 电流放大器;电流放大电路
current amplifier tube 电流放大管
current and temperature recording system 海流温度记录仪
current angle 流压角差
current annual 连年的
current annual expenditures 本年度支出
current annual increment 连年(生长量);当年生长量
current annual revenue 本年度收入
current anomaly 电流异常
current anti-node 电流波腹
current appropriation 本期拨款;本年度预算经费
current arrangement 现时安排
current assets 流动资产
current assets audit 流动资产审计
current assets cycle 流动资产循环
current assets ratio 流动资产比率
current assets turnover 流动资产周转率
current associated with tsunami 海啸潮流
current attenuation 电流衰减
current audit 日常审计
current average daily traffic 现有平均日交通量
current averaging 电流平均
current backlog 当前积压
current balance 经常项目差额;经常收支;电流平衡;电流秤
current balance of payments 经常项目收支平衡
current balance relay 电流平衡式继电器
current balance type current differential relay 电流平衡式差动电流继电器;差动平衡式电流继电器
current beam position 当前电子束位置
current-bedded 交叉层理的;波状层理的
current bedding 流水层理;交错层理;波状层理;水流层理
current bill of lading 现时提单
current block number 当前块号
current boost 气流加压
current bottle 海流瓶;测流瓶;漂(流)瓶

current breaker 断流器;电流开关
current budget 本期预算
current business year 本营业年度
current capacity 载流量;电流负荷
current capital 流动资金;流动资本
current capital allocated by government 国家拨入流动资金
current capital management 流动资金管理
current capital quota 流动资金定额
current capital turnover 流动资金周转
current carrier 载流子
current-carrying capacity 载流(容)量;载流能力
current-carrying carrier 载流承流素
current-carrying conductor 载流线
current-carrying factor 载流系数
current-carrying part 载流部件;导电部分
current-carrying plate 导电板
current-carrying shaft 载流轴
current channel register 当前通道寄存器
current chart 海流图;潮流图;水流图
current check 例行校验;例行检查;日常检查
current chopper 斩波电流
current circuit 电流回路;电流电路;串联电路
current circuit break protection 电流回路断线保护(装置)
current city 现代城市
current code 现行规范
current coefficient 海流流动系数;潮流系数
current coil 电流线圈
current coincidence system 电流重合法;电流一致制
current collecting 集电的
current-collecting device 集流装置
current collection 受点
current collector 集流器;集电器;集电弓;受电头;受电器
current-collector bow 受电弓
current-collector for mine locomotives 矿用机车接电器
current colo(u)r 当前颜色
current column 等离子流束
current comparator 电流比较器
current comparison system 电流比较式
current compensational ground distance relay 电流补偿式接地远距继电器
current component 海流分量;电流分量
current condition 水流条件
current-conducting 导电的
current-conducting plate 导电板
current-conducting rod 导电棒
current connect group 当前连接组
current connection tab 导电接头
current connection tab assembly 导电接头总成
current constant 电流常数;潮流常数
current consumption 耗电量;电流消耗;电流耗用
current contact nozzle 导电嘴
current continuity 电流连续性
current control 日常检查
current-controlled 电流控制的;气流控制的
current-controlled current source 电流控制电源
current-controlled oscillator 电流控制振荡器
current-controlled switch 电流控制开关
current-controlled voltage source 电流控制电压源
current controller 电流控制器
current control reports 本期控制报告
current conversion locomotive 多电流制电力机车
current converter 换流器;电流转换器
current conveyance 电流通过
current coordinates 流动坐标
current cost 市值;现行价格;现行费用;现行成本;现时成本;本期成本;按市价计算的费用;市价;时价
current cost accounting 现行成本会计;时价会计
current cost of fund 资金的现时成本
current coupling 电流耦合
current course table 航向流压差表
current crescent 新月形水流痕
current cursor address 当前光标地址
current curve 流速曲线;水流曲线
current cycle 潮流周期;潮流循环
current damper 电流阻尼器
current data 新近资料;现时资料;现时数据
current data processing 现行数据处理;现时资料处理

current debt 短期债务
current decay 电流减弱
current deflecting 水流导向
current deflector 挑流构筑;水流导向设备
current demand 电流需量;当期需要
current density 流密度;电流密度;气流密度
current deposit 活期存款
current derating 电流额定值的下降
current determination by float 浮标测流
current device 测流装置
current diagram 海流图解;海流图;洋流图;潮流图
current difference 潮流差;潮流比数
current differential relay 电流差动继电器;电流差动保护
current differential relaying system 电流差动继电制
current dip 电流谷点
current direction 流向
current direction at bridge site 桥位水流方向
current direction indicator 流向指示器;流向仪
current direction measurement 流向测量
current direction meter 流向测定仪
current direction relay 电流方向继电器
current disease 流行性(疾)病
current-displacement motor 深槽感应电动机;深槽电动机
current distribution 电流分配;电流分布
current distribution meter 电流分布测量器
current divider 分流器
current division ratio 电流分配系数
current dollar 现值美元
current dollar estimate 现价美元估算值
current dollar value 美元时值
current domestic value 现行国内价格
current drag 水流拖曳
current drain 耗用电流;电流耗用;电耗
current drainage 排流
current drainage cabinet 排流柜
current drainage cable 排流电缆
current drainage connecting cabinet 排流接线柜
current drainage net 排流网
current drainage rail 排流轨
current drainage terminal box 排流端子箱
current drift 流程
current driver 电流驱动器
current drogue 测流浮子;测流浮标
current drying 电流干燥法
current earnings 现时盈余
current edition 现行版
current effect 水流影响
current efficiency 电流效率
current efficiency determination method 电流效率测定法
current electrode 供电电极
current element 电流元件
current ellipse 潮流椭圆
current energy conversion 海流能转换
current energy conversion commercialization 海流能转换商业化
current energy conversion development 海流能转换开发
current energy conversion installation 海流能转换装置
current energy conversion technology 海流能转换技术
current entry 日常分录
current equalization 均流
current equalizer 均流线;均流器
current estimate 现行估计值;最新估算
current event 当前事件
current expenditures 经费;经常开支;经常费(用);本期支出;本期费用;日常费用
current expenses 经常开支;杂费;日常费用
current factor 潮流系数
current failure alarm 电流中断报警设备;停电报警信号
current fast operating protection with delay 延时电流速断保护
current feed 电流馈接;电流馈电
current feed antenna 电流馈接天线
current feedback 电流反馈
current feedback amplifier 电流反馈放大器
current feedback circuit 电流反馈电路
current fender 防冲设备

current field 流场
current file 目前文件;当前文件
current fiscal year 本会计年度;本财政年度
current float 测流浮子;测流浮标
current flow 电流;水流
current flow angle 通角
current flow condition 载流状态
current fluctuation 电流波动
current focusing log 电流集中测井
current foldback circuit 过流保护电路
current follower 电流跟踪器
current force 潮流力;水流力
current force coefficient 潮流力系数;水流力系数
current force in beam current 横流水流力
current foreign exchange rate 现行汇率
current-free field 无电流场
current fund 流动基金;流动资金
current gain 电流增益
current-gain cut-off 电流增益截止
current general 通用
current generation 当前阶段
current generator 电流发生器
current grading 电流分级
current groundwater quality risk-assessment method 现行地下水水质风险评价法
current growth 当年枝;当年生长
current hogging 电流错乱
current hogging logic circuit 电流参差逻辑电路
current hour 最大潮流间隙
current ideal-standard cost 现行理想的标准成本
current-illumination characteristic 光电特性
current-illumination curve 电流照度曲线
current-illumination relation 电流照度关系式
current impulse 电流脉冲
current income 本期收益
current indicator 电流指示器;水流指示器;示流器
current indicator lamp 示流灯
current in estuary 河口水流
current information 海流资料;现势资料;现时资料;现时信息
current in middle wire 中线电流
current input 经常性投入;现时投入;电流输入
current in shallow water 浅海海流
current instruction code 现行指令码
current instruction word 现行指令字
current intangibles 流动无形资产;本期无形资产
current integration 电流积分
current integrator 电积分器
current intelligence 现势资料;动态情报
current intensity 电流强度
current interest 通行利息
current international transaction 经常国际交易(即经常项目交易)
current interrupter 电流断续器
current interruption 电流断路
current invisibleitem 经常无形项目;流动无形项目
current item 经常项目
current job 现职
current knife 电流刀
current lamination 流水纹理;波状层理;流层;波状纹理;波状纹层;水流纹理
current law 现行法律
current layer 当前层
current lead 电线
current leakage 电流漏泄
current length 当前长度
currentless 去激励的
current level 现时水平
current liability 流动负债
current-limited circuit 限流电路
current limiter 限流器;电流限制器
current-limiting circuit 限流电路
current-limiting fuse 限流熔丝
current-limiting inductance 限流电感
current-limiting inductor 限流线圈
current-limiting overcurrent protective device 限流过载保护装置
current-limiting protection 限流保护
current-limiting reactor 串联电抗器
current-limiting resistance 限流电阻
current-limiting resistor 限流电阻器
current limiting starter 限流起动器
current-limit relay 限流继电器
current limit switch 限流开关

current line 流速线;刻度绳;计程仪;测流索;测程仪绳;同潮流时线
current lineation 裂线理;水流线理
current line-type 当前线形
current loading 电流负载;潮流荷载
current location 当前位置
current location reference 现用单元参考
current log 计程仪;测流仪
current loop 电流波腹;当前循环
current loop interface 电流波腹界面
current-loss instrument 井漏仪
current lubrication 流动润滑(作用)
current luminous flux characteristic 光电特性
currently attainable standard 现时可达标准
currently effective code 现行规范
currently effective standard 现行标准
current maintenance 日常维护;经常维修;日常保养;维持电流
current map 现行地图
current margin 现行限度;工作电流范围;电流容限
current mark 波纹;波痕;水流痕(迹)
current market price 现行市(场)价(格);现时市价
current market value 现行市(场)价(值)
current maturity 本年到期的长期债务
current maturity of long-term debt 一年内到期的长期债务
current measurement 电流测量;水流测定
current measuring device 水流测定设备
current measuring equipment 水流测定装置
current measuring plant 水流测定设备
current meter 水流计;电流表;安培表;流速仪;流速表;流速计;海流计;电流计;冲流量计;测流仪
current-meter calibration 流速仪率定;流速仪检定
current-meter discharge measurement 流速仪测流(法)
current-meter measuring cross-section 流速仪测流断面
current-meter method 流速仪测流法
current-meter observation 流速仪测量
current-meter rating 流速仪率定;流速仪检定
current-meter rating flume 流速仪率定(水)槽;流速仪检定(水)槽
current method 流动方法(资产及负债评估方法);现行方法
current mirror 电流反射镜
current mirror circuit 电流镜电路
current mode digital-to-analog converter 电流型数模转换器
current mode logic 电流型逻辑
current mode logic circuit 电流型逻辑电路
current mode switch 电流型开关
current moment 电流矩
current monetary assets 本期货币流动资金;本期货币流动资产
current money 通用货币;通行货币;通货
current money unit 通用货币单位
current month delivery 当月交货
current multiplication 电流放大;电流倍增
current multiplication type 电流倍增型
current multiplication type transistor 电流倍增型晶体管
current negative feedback 电流负反馈
current net income 本期纯收益
current node 电流波节
current noise 电流噪声
current non-monetary assets 非货币性流动资产
current number 当前数
current number of maintenance wagons 检修货车现有数
current observation 近期观察;验流;测流;水流观测
current observation station 水流观测站
current of air 气流
current of commutation 换向电流
current-off upset allowance 无电顶锻余量;无电顶锻留量
current of higher density 泥质流;密度流;异重流
current of run-unit 过程的现行记录
current of set type indicator 当前系类型指示器;当前络类型指示符
current of the earth 大地电流
current of traffic 交通车流;船流;车流;车(船)流向
current of water 水流
current-on contact 开关触点电流
current-on upset allowance 有电顶锻余量

current operating performance basis 本期经营成果基础
current operating performance concept 本期经营成果观念
current operation 经营业务;经常性业务;目前业务
current operation expenditures 日常营业费
current operation symbol 现用符号
current operator 电流算符;当前算符
current order 即时指令;现时指令
current outlay cost 现付成本
current outlays 经常性支出
current output 输出电流
current output amplifier 电流输出放大器
current output state 当前输出态
current paper 近期论文;交流资料
current parent 当前母体
current pass-book 来往存折
current path 电流通路;气流通路
current pattern 流型
current payment 经常性开支;即期付款;经常性支付;经常性付款
current period 现期;比较期
current phase 电流相位
current phasor 电流矢量
current pickup 取流
current pickup shoe 集电靴
current plane lineation 分裂面线理
current pointer 当前指针
current pole 测流杆
current population 现有人口
current position 当前位置
current position analysis 流动(财务)状况分析
current position ratio 流动财务状况比率
current post 现职
current power generation 海流发电
current practice 现行惯例;现行工作制度;现行办法
current pressure 流压;水流压力
current price 现(行)价(格);市价;时价
current-price estimate 现价估算值
current priority indicator 现行优先次序指示器
current probe 电流探针
current probe amplifier 电流探头放大器
current processor priority 处理机现行优先级
current product cost 本期产品成本
current production 当前产式;当前产品;日常生产
current profile 流速纵断面
current profit and loss 本期损益
current protection 电流保护装置
current protector 电流保护器
current pulse 电流脉冲
current pulser 电流脉冲发生器
current pumping 电流抽运
current purchasing power 现实购买力;现时购买力
current range 电流范围
current rate 当日汇率;成交价;现行汇率;现行费率;现价;气流强度;现行兑换率
current rate method 现行汇率法
current rate of interest 现行利率
current rate translation method 现行汇率折算法
current rating 现行定额;电流额定值;额定电流
current ratio 流动资产率;流动比率;现行比率;电流比;电流变换系数
current-ratio regulator 电流比自动调节器
current ratio relay 电流比继电器
current record 电流记录
current rectifier 整流器;电流整流器
current regime 气流状态;电流状态;流况;水流状态;水流状况
current regulation 现行条令;现行规章;现行规则;现行条例
current regulator 电流调整器;整流器;电流调节器;稳流器
current relay 继电器;电流继电器
current release level 当前释放层
current repair 例行修理;经常修理;经常性修理;小修(理);现场修理;日常修理;日常维修
current replacement cost 现时重置成本
current requirements 现行要求
current reserve 经常储备;药剂周转储备金;周期转储备量
current resolver 电流分解器
current resonance 电流谐振;并联谐振
current resonance frequency 电流谐振频率
current resources 现时资源

current response 电流响应
current responsive circuit 电流响应电路
current retard 减流坝
current retarded dam 挡流坝
current return 当年收益;本期报酬率
current return cable 回流电缆
current revenue 当期收入;现收益
current reverser 电流换向开关;电流倒向开关
current reversing key 电流换向开关
current ripple(mark) 波痕;流水波痕;流痕;波纹;沙纹
current rips 流激涟波;洋流小浪日常;潮流波
current rise time 电流上升时间
current rose 流向频率分布图;流红图;海流图;海流玫瑰图;洋流玫瑰图;潮流玫瑰图;水流频率(玫瑰)图;水流(玫瑰)图
current rosette 潮流玫瑰图
current running repairs 运转中修理
current rush 电流骤增;冲激电流
current-rushing point 顶冲点
current sailing 潮流航行计算法(修正风流压差)
current sand capacity carried 海流携砂量
current saturation 电流饱和
current school statistics 经常教育统计资料
current scour 水流冲刷
current season 本季度
current sector number 当前扇面号
current selected switch 电流选择开关
current selling price 现行售价;现价
current sensing device 电流反应装置
current-sensitive colo(u)r screen 电流灵敏彩色屏
current-sensitive single gun colo(u)r display tube 电流灵敏单枪彩色管
current sensitivity 电流灵敏度
current sensor 海流探测设备;海流传感器
current sensor unit 海流测量装置
current shade 流行色
current sharing resistors 电流分配电阻
current-sheath model 电流鞘模型
current sheet 电流片;电流层;曲流层
current shifting 水流漂移;水流摆动
current shipment 近期装运
current ship's maintenance project 现行船舶维修计划
current signal indicator 电流信号指示器
current sink logic 电流吸收逻辑
current situation 现状
current slicing 电流分割
current slot 当前时间段
current source 电流源
current source circuit 电流源电路
current source for measurement 测量电源
current source inverter 电流源变器
current source logic 电流源逻辑
current square meter 平方刻度电流表
current stabilization 电流稳定
current stabilizer 电流镇定器;稳流装置;稳流器
current stack frame 当前栈帧;当前栈框架
current stake 水流标桩
current standard 现行标准
current standard cost 现行标准成本
current state 目前状态;当前状态
current statistics 现行统计数字
current status 现状;当前状态
current steering 电流导引
current steering logic 电流导引逻辑电路
current strength 电流强度
current stretcher 电流波形扩展器
current subsidies 经常补助费
current subsidy 流动补助金
current supply 供电;电(流)源
current supply device 电源设备
current surplus 本期盈余
current survey 流速测量;水流观测
current switch 电流开关
current switching mode logic 电流开关逻辑
current switching mode logic circuit 电流开关逻辑电路
current system 电流制
current table 潮汐表;潮流表;潮汐预报表;潮汐潮落表
current take-off 电流切断
current tap 分插座;插口灯座;复式灯座;分接头(电、水);分插口

current task 当前任务
current task table 当前任务表
current tax 本期税收
current tax payment act 现期纳税法
current tax system 现行税制
current technique 现代化技术
current technology 现代技术
current tectonic stress field 现代构造应力场
current terminal 电流端子
current term net loss 本期净损失
current term net profit 本期净益
current terms 现值
current term settlement 本期决算;本期净损益
current throttling type rotor 深槽式转子
current time 当时;实时
current-time curve 电流时间曲线
current-to-voltage converter 电流电压变换器
current track maintenance 线路经常维修
current tractive force 水流推移力
current traffic 现行交通量;现行运量;现时运量;现时交通量
current transaction 流动交易;经常项目交易;经常交易
current transaction account 经常项目交易账户
current transfer 经常性转移支付
current transfer order 现行转移指令
current transfer ratio 电流传输比
current transformer 换流器;互感器;中间变流器;电流互感器;电流变压器;变流器
current transformer comparator 电流互感器比较仪;变流器比较仪
current transformer disconnecting protection 电流互感器断线保护(装置)
current transformer phase angle 电流互感器相角
current trend 目前趋势;现行趋向;当前动向
current triangle 电流三角形
current trigger capability 电流触发性能指数
current-triggered type 电流触发型
current-type flowmeter 流速式流量计;速度式流量计
current-type telemeter 电流式遥测装置;电流式遥测仪;电流式遥测计
current undertaking 日常业务
current-using apparatus 电气设备
current value 现行值;现时值;当前值;通用值
current-value accounting 时值会计法;时价会计
current velocity 流速;水流速度
current velocity profile 海流流速分布
current voltage characteristic 伏安特性(曲线);电流电压特性
current-voltage characteristic curve 电流电压特性曲线
current-voltage curve 电流触电压曲线
current-voltage dual 电流电压对偶
current voltage follower 瞬时电压跟随器
current-voltage regulator 电流电压调节器
current voyage 本航次;本次航行
current wagon statistical number 货车现有数统计
current water quality standard 现行水质标准
current wave 电流波
current waveform 电流波形
current-wave laser 电流波激光器
current wavelength characteristic 电流与波长的关系特性
current weather 现在天气
current weigher balance 电流秤
current winding 电流绕组
currentwise 顺流(向的)
current with fixed base 固定式接电器
current year 当年(度);本年度
current year appropriation from treasury 本年预算拨款
current year precipitation 当年降水
current year runoff 当年径流量
current year's price 当年价格
current year's tax levy 本年度税收
current yield 经常收益率;现时产量;月产量;电流效率;本期收益;日产量
current zero 电流零点
curriculum plan 课程计划
curriculum program(me) 课程计划
curriculum vitae 履历(表)
currier 制革工人

curriery 制革厂
curring 卷边加工
currycomb 耙式整坡器
curry wool 低级粗毛
cursive 草写体
cursor 游视指针;游标;转动臂;光标【计】;标示器;透明指针片
cursor address 光标地址
cursor address register 光标地址寄存器
cursor comparator 光标比较器
cursor control 光标控制
cursor control key 光标控制键
cursor controller 光标控制键
cursor control register 光标控制寄存器
cursor data 光标数据
cursor format 光标格式
cursor home 光标位置;光标回零
cursor key 光标键
cursor left 光标左移
cursor line 光标线
cursor location 光标位置
cursor mark 游标符号
cursor memory 光标存储器
cursor movement 光标移动
cursor name 游标名
cursor position 光标位置
cursor positioning 光标定位
cursor register 光标寄存器
cursor remove 光标移动
cursor right 光标右移
cursor symbol 指示标
cursor tracking 光标跟踪
cursor up down 光标上下
cursory treatment 粗略论述
curstable 纹石束带层;有线脚的石砌层
curtail 削减;缩短;(楼板底级踏步的)曲线板头
curtail credits to enterprises 削减贷款
curtailed crossover 缩短渡线
curtailed inspection 抽样检查
curtailed ladders 缩短梯线
curtailed rail 缩短轨
curtailed works 简字体
curtail expenditures 节约支出;缩减开支
curtailing 缩短
curtailment 收缩;截断点;削减
curtailment diagram 截断钢筋的抵抗弯矩包络图
curtailment of bank facility 紧缩银根
curtailment of drilling 限制钻进工作量
curtailment of expenditures 经费节减
curtailment of investment 紧缩投资
curtailment of service 减少班次
curtail outlay 紧缩开支
curtail step 楼梯起步挑出踏步;起步梯级(指一端或两端为半圆形或漩涡形者);卷形踏步
curtain 帘幕;帘;幕;底釉起绉;石幔
curtain and festoon chain heat transfer system 花环链热交换系统
curtain aperture 调光孔径
curtain arch 幕拱;间壁拱
curtain baffle 挡帘
curtain block 托砖
curtain board 拦火抽风幕;遮板;防火幕屏
curtain box 窗帘匣;窗帘盒
curtain breakwater 透空式防波堤
curtain cable rail 横向帘式电缆栏
curtain cable rail stand 横铁帘式电缆栏杆
curtain cable system 帘式电缆系统
curtain chain 幕状链;垂挂链
curtain-coated finish 幕涂涂料;幕涂涂层
curtain coater 帘式涂料器;帘式涂覆器;帘式淋涂器
curtain coating 淋涂(法);帘式淋涂;幕式淋涂;屏幕涂层;屏蔽涂层;喷涂(法)
curtain coating method 喷涂法
curtain dam 卷帘坝;活动闸门坝;(有水平转轴的)闸门式水坝;防渗坝
curtain depth 帷幕深度
curtain door 幕门;防火幕门
curtain drain 闸门式泄水系统;截流沟;截流排水沟;截流排水道;排渗帷幕;帷幕式排渗;帷幕排水道
curtain effect 幕效应
curtain flow coater 帘式淋涂机;幕式淋涂机
curtain glaze 皱纹釉
curtain grout hole 帷幕灌浆孔
curtain grouting 幕式灌浆(防渗漏);灌浆帷幕;防渗(幕)灌浆;帷幕灌浆
curtain grouting process 帷幕灌浆法
curtain hardware 窗帘小五金
curtain hole 勘探孔;帷幕灌浆孔
curtain hung system 垂挂链系统
curtaining 消漆;漆膜垂落;流挂(漆病);溅模;垂落;漆面流坠
curtain jet 幕状射流
curtain kiln 帐幕干燥窑
curtain lath 百叶帘幕板条
curtain-like adhesion 幕状黏连
curtain line 舞台幕下垂线;大幕线;帷幕线
curtain masonry wall 幕墙;护墙
curtain of piles 桩排;板桩排(用于河岸或松软土壤);桩围堰
curtain of sheet pile 板桩截水墙
curtain of sheet piling 板桩截水墙
curtain of timber 木板桩岸壁
curtain of willow 柳枝浮坝;挂柳
curtain painting 幕帘涂装
curtain plate 外缘板
curtain-pole retarder 水帘滞流装置
curtain rail 帘轨;帘轨;窗帘(导)轨
curtain ring 帘圈;门窗帘圈;窗帘圈
curtain rod 帘辊;挂帘杆;窗帘辊;窗帘棒
curtain runway 窗帘滑轨
curtains 云状花纹表面
curtain set 台幕索具设备
curtain shutter 帘式快门
curtain slat 百叶帘幕板条
curtain step 楼梯第一踏步
curtain track 窗帘(滑)轨;拖拉帷幕轨;帘滑轮轨
curtain-type furnace 气幕式炉;气帘式加热炉
curtain wall 截水墙;幕墙;悬墙;吊墙;挡火管墙;屏风受热面;屏风管墙;帷(幕)墙
curtain wall between windows 窗间隔墙
curtain-wall block 具有围墙的建筑
curtain-wall breast panel 窗腰隔墙
curtain-wall building 具有围墙的建筑
curtain-wall construction 幕墙结构
curtain-wall enamel 建筑用搪瓷壁板
curtain-wall frame 围墙框架
curtain-wall framework 隔墙骨架
curtain-wall in all direction 各方向的隔墙
curtain walling 悬墙;安装隔墙
curtain wall panel 围墙嵌板;幕墙板
curtain-wall sealing material 围墙密封材料
curtain-wall semi-breast panel 半窗腰隔墙
curtain wall type breakwater 幕墙式防波堤
curtain weir 卷帘堰
curtain width 帷幕宽度
curtain zone 拉幕区
curtate 卡片划分;卡片横区;卡片部分
curtate cycloid 长幅圆滚线
curtate distance 黄道面投影距离
curtate epicycloid 长摆外摆线
curtilage 院子;住宅庭园;宅地;庭园
Curtis 柯蒂斯铝合金
Curtis-Doisy unit 柯蒂斯—多伊西单位
Curtis nozzle 扩放喷管;柯蒂斯喷管
Curtiss 柯蒂斯铝合金
Curtis stage 柯蒂斯级
Curtis steam turbine 冲击式汽轮机
Curtis turbine 柯蒂斯汽轮机;复速级汽轮机
Curtis wheel 柯蒂斯叶轮;复速式叶轮
Curtis-wheel turbine 柯蒂斯式涡轮机
Curtius reaction 库尔修斯反应
curtmeter 曲度计
curupay 黑纹深红色硬木(一种产于中美洲的木材)
curvatural radius of circular features 环形体曲率半径
curvature 弧度;直线性系数;曲率;曲度;弯曲率;弯曲部分;弯曲
curvature and refraction correction 两差改正;地球曲率和折光差改正
curvature angle 曲率角
curvature-area method 曲率面积法
curvature at first yield 初始屈服曲率
curvature change 曲率变化
curvature coefficient 曲率容许量
curvature correction 星径曲率改正;曲线校正;曲率校正;曲率改正;弯道土方计算修正值
curvature design 曲率设计
curvature difference 曲率差
curvature distortion 屈曲变形
curvature ductility 曲率延性
curvature factor 急弯系数;曲率系数;曲率容许量;曲流系数
curvature friction 曲面摩擦;弯曲拐折摩阻力损失值(预应力混凝土后张法钢筋)
curvature generator 曲面加工装置
curvature hyperopia 曲度远视
curvature indicatrix 曲率指示线
curvature invariant 曲率不变量
curvature line 弯曲线条;曲线;曲线线路
curvature loss 曲率损失
curvature measure 曲率量测
curvature measurement 曲率测量
curvature method 曲度法
curvature of a coast 海岸弯度
curvature of a conic(al) 二次曲线的曲率;二次曲率
curvature of a curve 曲线的曲率
curvature of a normal section 法截线曲率
curvature of beam 梁的曲率
curvature of curve 曲线曲率
curvature of face 面曲率
curvature of fault surface 断层面弯曲度
curvature of Gothic pass 弧菱形孔型半径
curvature of ground 地面曲率
curvature of hairline 曲周
curvature of image field 像场曲率
curvature of meridian 子午线曲率
curvature of parallel 平行圈曲率;纬线弯曲;纬圈曲率
curvature of path 光路曲率
curvature of space 空间曲率
curvature of the earth 地球曲率
curvature of the plumb-line 垂线曲率
curvature of tunnel 隧道曲率
curvature of vertical 垂线曲率
curvature of vessel surface 容器表面的曲率
curvature of wave function 波函数曲率
curvature radiation 曲率辐射
curvature radius 曲率半径;弯曲半径
curvature radius of bend 弯道曲率半径
curvature radius of channel 航道弯曲半径;航道曲率半径
curvature radius of track 线路曲率半径
curvature resistance 曲线阻力
curvature scalar 曲率标量
curvature tolerance 吻合度
curvature vector 曲率向量;曲率矢量
curve 玫瑰线;弧形;曲线(图表)
curve adjusting 曲线整正;整正曲线
curve analyser 曲线分析曲
curve analyzer 曲线分析曲
curve begin post 曲线起点标
curve bend 弧形弯管;弯段;弯道
curve block 弯石(砌)块
curve board 曲线指示牌;曲线板
curve body 单曲线
curve break 曲线拐点;曲线(的)转折点
curve center angle 曲线中心角
curve characteristic 曲线特性
curve chart 曲线图
curve compensation 曲线上的纵坡折减;曲线(坡度)折减;曲线补偿;(曲线路段的)坡度折减率
curve compensation ratio 曲线折减率
curve composition 曲线组合
curve controlling point 曲线控制点
curve convergence 曲线收敛
curve coordinates 曲线坐标
curve correlation method of dip log 曲线对比(方)法
curve curvature 曲线曲率
curve cut-off 裁弯取直
curve cutting face 曲线切剖面
curved 弧形的;曲线的;弯曲状
curved approaching rail 曲导轨
curved arch bridge 弯拱桥
curved armed pulley 曲辐带轮
curve data 曲线数据
curved backward 向后弯曲
curved baffle 曲面挡板
curved bale chute 曲形草捆滑槽
curve(d) bar 曲杆
curved barrier of rock 弯曲岩层遮挡

curved batter 曲面墙(一侧直墙,一侧曲面);曲斜面墙;曲斜撑;曲式横撑;弯曲倾度
curved beam 曲梁;弯枭
curved beam bridge 曲线梁桥;曲梁桥
curved beam with eccentric boundaries 偏心曲梁
curved bearing plate 弧形支座
curved bend 弧形管
curved bender 弧形顶弯器
curved blade 弯曲叶片
curved blades snips 弯刃剪
curved blade turbine type agitator 曲面透平式搅拌器
curved block 弯曲建筑;曲线型建筑
curved board 弧形板
curved board insulation 绝热弧形板
curved bone drill 弯骨钻
curved bottom chord 曲线形下弦杆
curve(d) boundary 曲线边界
curved box girder bridge 箱形梁曲线桥;曲线箱形梁桥
curved breakwater 弧形防波堤
curved brick 曲面砖
curved bridge 曲线桥;弯桥
curved building 弯曲建筑;曲线型房屋;平面弯曲的建筑
curved buttress dam 弧形(面板)支墩坝
curved cantilever 弓形腕臂
curved carpentry 曲线木作
curved catenary 垂曲线
curved cathode 曲面阴极
curved channel 弯曲水道;弯曲河道;弯曲河槽;弯曲航道
curved channel curvature 弯道曲率
curved chord 曲弦
curved chord truss 折弦桁架;曲线桁架;曲率桁架
curved chord truss bridge 折弦桁架桥
curved clasp 弯形卡环
curved closure plate 曲形封闭板
curved closure rail 曲合龙轨
curved coastline 弯曲海岸线
curved coefficient 弯曲系数
curved compact 曲面坯块
curved concave head 曲线型凹衬板
curved concrete roof 曲面混凝土屋顶
curved conduit 弯道
curved conveyer 曲线回路输送机
curved corrugated alumin(i)um 弯曲的波纹铝片
curved corrugated sheet 曲面薄板;瓦垄薄板
curved crevasses 弯曲裂纹;弯曲裂缝
curved crossing 曲线交叉
curved cross-section arch bridge 双曲线拱桥
curved crystal spectrograph 弯晶摄谱仪
curved crystal spectrometer 弯晶分光计
curved curb 曲面路缘石
curved cutting blade 弯曲切片
curved cutting face 曲切削面
curved dam 弧形坝
curved diagram 弯曲图
curved diamond 曲线菱形交叉
curved drawing instrument 曲线板
curved earthenware pipe 弧形瓦管;曲瓦管;弯曲瓦管
curved easement 曲线缓和
curve degree 曲线度数
curved element 弯曲构件
curved elevator 弯梃
curve design 曲线设计
curved face 弯曲面;曲面
curved fan 曲扇状流
curved fault 曲断层;弯曲断层
curved fault surface 曲断层面
curved fibre 柔性纤维;弯曲纤维
curved flow 曲线流(动);弯曲水流
curved forceps 弯钳
curved frame 曲架;曲肋骨
curved frog 曲线辙叉
curved gas-water interface 弯曲的汽水界面
curved girder 弯曲大梁;曲梁
curved glass 曲面玻璃;弯形玻璃
curved gravity dam 弧形重力坝
curved grillage girder bridge 曲线格子梁桥
curved ground plan 弯曲底层平面图
curved guide 曲面导轨
curved height 曲面矢高

curved holder for earth wire 地线弯卡
curved hole 弯曲孔
curved holographic(al) optic(al) element 曲面全息光学元件
curve diagram 曲线图
curved diamond rail 曲线尖轨
curved intersection 曲线交叉
curve distance 曲距
curved jack 弧形千斤顶;曲面千斤顶;弯曲千斤顶
curved jack rafter 小弯缘
curved jack technique 曲面千斤顶法
curved jet 曲线喷流
curved laminated member 层积曲木构件
curved laminated wood 曲形胶合板;曲形层积木材;曲面形多层胶合木
curved lead (道岔的)曲导程
curved lead rail 曲合龙轨
curved lever 成型杆
curved line 曲线
curved liner 弧形衬板
curved link 弧形链节
curved lip strike 锁舌弯碰板
curved member 弯曲构件
curved mesh 曲线网(格);网纹弯曲
curved mirror 曲面镜
curved-mirror cavity 球面反射镜腔
curved-mirror interferometer 曲面反射镜干涉仪
curved-mirror resonator 曲面镜共振腔
curved mole 弧形海堤;弧形防波堤
curved mo(u)lding 弯曲(形)线脚
curved names 弧形字列
curved needle holders 弯头持针钳
curved oil-water interface 弯油的油水界面
curved panel 曲形胶合板
curved path 曲线航迹;曲射线路径
curved-path error 光路弯曲误差
curved pediment 弧形山墙
curved physical surface 自然曲面
curved pipe 弯管
curved plan 弯曲平面图(房屋)
curved plane 曲面
curved plank roof 曲板屋顶;弧形板条屋顶
curved plate 曲面板
curved plate casting machine 弧形铅版铸造机
curved plate retaining wall 拱式挡土墙
curved plywood 曲面胶合板
curved points 曲线道岔
curved polisher 曲面抛光器
curved polyhedron 曲多面体
curved portion 弯曲孔段;弯曲井段
curved probe 弯探子
curved profile 内凹轮廓;曲线轮廓
curved profiled bar 弯型杆
curved projection 弯翘;反卷
curved pyramidal roof 弧形锥顶
curved quadrilateral 曲边四边形
curved radiator 弧形散热器
curved rafter 弯椽
curved rail 弯轨
curved ramp 弯子·(楼梯扶手的弯曲部分)
curve-drawing ammeter 自动记录安培计;曲线记录式电流表
curve drawing instrument 绘曲线仪
curve(d) reach 弯曲(河)段
curved reflection holographic(al) optic(al) element 曲面反射式全息光学元件
curved rib 曲肋
curved-rib truss 曲肋桁架;曲弦桁架;曲率桁架
curved rigid-type frog 曲线型固定辙叉
curved ripple mark 扭曲波痕
curved rod 弯杆
curved roof(ing) 曲面屋顶;弧形屋顶;曲线屋顶
curved ruler 曲线板;曲规
curved runner 弯滑刀式开沟器
curved scale 弧形刻度
curved scissors 弯剪
curved scrape 弯形刮痕
curved scraper 凹磨圆形刮刀
curved screen 弧形筛;曲面银幕
curved secondary cooling zone 弧形二次冷却区
curved section 曲线段
curved servomotor 环形接力器
curved sheet 弧形薄板
curved shell 曲线型薄壳;薄拱壳

curved shock 弯曲激波
curved slab 曲板
curved sliding door 曲面移门
curved space 弯曲空间
curved space-time 弯曲时空
curved spectrum 弯曲频谱
curved spit 弯曲沙嘴
curved split switch 曲线尖轨转辙器
curved spoke 曲辐
curved spoke shaver 曲刃刨刀
curved spring washer 弧形弹簧垫圈
curved stair(case) 弧形楼梯
curved steady arm 曲定位器
curved steady tube 弓形定位管
curved steel box girder bridge 曲线箱形钢梁桥
curved substrate 曲面基板
curved surface 曲面
curved surface projection 曲面投影
curved surface reduced to horizontal plane 曲化平面
curved surface trimming 曲面修整
curved switch 曲线转辙器(折叠式导轨);曲线道岔
curved switch point 曲线尖轨
curved switch rail 曲线尖轨
curved tempered glass 弯型钢化玻璃
curved thermometer 曲管温度计
curved thin-shell roof 弯形薄壳屋顶
curved timber 弯曲木材
curved tongue 曲线尖轨
curved-tooth bevel gear generator 弧齿伞齿轮铣齿机
curved tooth clutch 弧形齿离合器
curved top chord 曲线形上弦杆
curved track 曲线轨道;曲线线路;曲线股道
curved track open route 地面曲线段
curved track tunne 隧道内曲线段
curved trash rack 弧形拦污栅
curved triangle 曲边三角形
curved-trough flow device 转弯流槽;弯曲流槽
curved tunnel 曲线隧道
curved turbine type agitator 弯叶涡轮式搅拌器
curved viaduct 曲线高架桥
curved vortex line 曲涡线
curved vortex sheet 曲涡流层
curved wall 弧形墙;曲壁
curved wall effect 曲壁效应
curved weir 弧形堰
curved well 弧形井;弯曲井
curved work 曲面件
curve element 曲线要素
curve element in image identification correlation method 图形识别对比法曲线元素
curve end post 曲线终点标;曲面部分
curve enlargement 曲线放宽;曲线放大
curve family 曲线簇;一组曲线
curve final set(ting) 终凝曲线
curve fit(ting) 按曲线选择经验公式;确定适合曲线;曲线配合;曲线拟合(法);配线;适线法;选配曲线
curve fitting compaction 曲线拟合精简数据法;曲线近似压缩法
curve fitting method 曲线拟合法
curve fitting program(me) 曲线拟合程序
curve-fitting routine 曲线拟合程序
curve fitting technique 曲线拟合技术
curve flank 弯翼
curve flattening 曲线平整
curve follower 曲线阅读器;曲线输出机;曲线跟踪器;曲线跟随器
curve following 曲线跟踪
curve following stylus 描绘曲线笔尖;曲线描绘针
curve forming rest 弧形刀架;车弧形刀架
curve ga(u)ge 曲线规
curve generating 曲面成型
curve generator 金刚钻加工透镜机;粗磨球面透镜铣磨机;波形发生器;曲线产生器
curve going bogies (吊车的)曲线行走转向架
curve graph 曲线图
curve grease 曲轨滑油
curve guard rail 曲线护轨
curve hand 曲线转向
curve improvement 曲线改造
curve in river 河弯
curve in space 三向曲线;空间曲线

curve interpolation 曲线插值(法)
curve in the track 铁路曲线
curve length 曲线线长;曲线长(度)
curve-lined harrow 曲齿耙
curve-line graph 曲线直线图
curve lining 整正曲线
curve lining calculator 曲线整正计算器
curve lubricator 曲线涂油器
curve maintenance 曲线维修
curve-matching method 曲线拟合法;曲线对比(方)法
curve-matching method by computer 计算机曲线拟合法
curvemeter 曲率计
curve model 曲线模型
curve motion 曲线运动
curve name of well logging 测井曲线名称
curve-negotiating 可协调成曲线的
curve-negotiating belt conveyer 曲线协调的皮带运输机
curve-negotiating bogies 曲线协调的转向架
curve-negotiating S/R machine 曲线运动型有轨巷道堆垛机
curve negotiation 曲线通过;通过曲线
curve negotiation speed 曲线通过速度
curve of acousilog and density log 声波和密度测井曲线
curve of alignment 定线曲线
curve of area of transverse section 横剖面面积曲线
curve of axial dipole-dipole sounding 轴向偶极测深曲线
curve of azimuthal dipole-dipole sounding 方位偶极测深曲线
curve of borehole 钻孔弯曲
curve of buoyancy 浮力曲线
curve of candle-power 光强分布曲线
curve of center of gravity of water plane 漂心曲线
curve of combined three electrode sounding 联合三极测深曲线
curve of constant bearing 恒方位线;等方位线
curve of constant torsion 定挠率曲线
curve of corresponding discharge 相应流量曲线;相当流量曲线
curve of deficiency zero 无亏曲线
curve of depression 降落曲线;水位降落曲线
curve of discharge 流量曲线
curve of discharge-drawdown 流量降深关系曲线
curve of displacement 排水量曲线
curve of electric(al) sounding 电测深法曲线
curve of equal bearing 恒方位线;等方向角曲线
curve of equal intensity 等烈度线
curve of equal parallax 等视差曲线
curve of equal pressure 等压线
curve of equal pulsation intensity 等脉动强度曲线
curve of equal settlement 等沉陷曲线
curve of equal velocity 流速等值线;等速曲线
curve of equatorial dipole-dipole sounding 赤道偶极测深曲线
curve of error 误差曲线
curve of fatigue 疲劳曲线
curve of frequency distribution 频率分布曲线
curve of frequency sounding 频率测深曲线图
curve of Gauss 高斯曲线
curve of gradient array of single electrode 单极梯度曲线
curve of guarantee rate 保证率曲线
curve of guarantee rate of mining quantity 开采量保证率曲线
curve of guarantee rate of precipitation 降水量保证率曲线
curve of hardening 硬化曲线
curve of heat logging 热测井曲线
curve of hopper capacity 泥舱容量曲线
curve of hydrogeophysical logging 水文物探测井曲线
curve of induced polarization at near source 近场源激电曲线
curve of induced polarization charging array 激发极化充电曲线
curve of induced polarization combined profiling 激电联剖曲线
curve of induced polarization method 激电法曲线

curve of induced polarization mid-gradient array 激电中梯曲线
curve of induced polarization sounding 激电测深曲线
curve of initial set(ting) 初凝曲线
curve of input 输入曲线
curve of light distribution 光强分布曲线;光分布曲线
curve of liquid resistively logging in well 井液电阻率测井曲线
curve of loads 荷载曲线;负荷曲线;载重曲线
curve of longitudinal metacenter 纵稳定心曲线
curve of luminous intensity distribution 光强分布曲线
curve of magnetic polar wandering 磁极移动曲线
curve of magnetization 磁化曲线
curve of magnetotelluric method 大地电磁法曲线图
curve of metacenter 稳心曲线
curve of meteorologic(al) element change 气象要素变化过程曲线
curve of minimum radius 最小半径曲线
curve of opposite sense 反向曲线
curve of order 2 二阶曲线
curve of output 输出曲线
curve of parallel dipole-dipole sounding 平行偶极测深曲线
curve of percolating velocity-time 渗透速度时间关系曲线
curve of perpendicular dipole-dipole sounding 垂直偶极测深曲线
curve of potential induced in transient field method 瞬变场法感应电势曲线
curve of product curve 等产量曲线
curve of profiling method 电剖面法曲线
curve of pumpage-time 抽水量历时曲线
curve of radial dipole-dipole sounding 径向偶极测深曲线
curve of random error 偶然误差曲线
curve of reference value fro airborne sound 空气声隔声参考曲线
curve of reference values for impact sound 撞击声隔声参考曲线
curve of reflection 反射曲线
curve of relation 相关曲线
curve of relative eustatic sea level 海平面相对升降曲线
curve of righting arm 复原力臂曲线;稳性力臂曲线
curve of river flow 河水流量过程曲线
curve of river quality change 河水水质变化曲线
curve of rolling neat step shaft 轧齐曲线
curve of same sense 同向曲线
curve of schlumberger sounding method 对称四极测深曲线
curve of set(ting) 凝结曲线
curve of shearing force 剪力曲线
curve of size distribution 粒度分布曲线
curve of sliding 滑坡曲线;滑动线;滑动弧;崩滑线
curve of specific well yield-drawdown 单位涌水量降深关系曲线
curve of spectrum 光谱曲线
curve of spontaneous potential logging 自然电位测井曲线
curve of spring flow 泉水流量过程曲线
curve of stability 稳定性曲线;稳定曲线
curve of static stability 静稳(定)性曲线
curve of stowage 堆存曲线
curve of surface subsidence change 地面沉降量变化曲线
curve of surface water level change 地表水位变化曲线
curve of S zone S 区曲线
curve of temperature 升温曲线
curve of temperature measuring in drill hole 钻孔测温曲线
curve of temperature variation 温度变化曲线;温度变动曲线
curve of three-electrode sounding 三极测深曲线
curve of time equation 时差曲线
curve of tons per inch immersion 每英寸吃水吨数曲线;英寸吨位线
curve of total wave 全波曲线
curve of traction characteristic 牵引特性曲线

curve of transverse metacenter 横稳心曲线
curve of turnout 导曲线
curve of velocity logging 流速测井曲线
curve of water consumption 耗水量曲线;用水(量)曲线
curve of water content change in aeration zone 包气带含水量变化曲线
curve of water erosion 水侵蚀曲线
curve of water plane area 水线面面积曲线
curve of water supply 供水曲线;给水曲线
curve of wave zone 波区曲线
curve of weights 船重曲线
curve of zero velocity 零速度线
curve on river 弯曲河段
curve passage 通过曲线
curve passing speed 曲线通过速度
curve-pattern compaction 曲线式样精简数据法;曲线图形压缩
curve plate 曲板
curve plotter 绘图仪;绘图器;曲线描绘器;曲线绘图仪;曲线绘图机
curve plotting 绘曲线;点绘曲线;曲线标绘
curve point 曲线(起)点
curve pointer 曲线指示器
curve post 曲线标
curve profile 曲线形式
curve radius 曲线半径;曲率半径
curve ranging 曲线测设
curve reader 曲线读出器
curve resistance 非线性电阻;曲线阻力;弯道阻力
curve resistance coefficient 弯曲阻力系数
curver of area of midship section 船体中段剖面面积曲线
curve ruler 曲线板;曲线(定)规
curves 圆括号
curve segment 曲线段
curve setting 定曲线;曲线测设
curve sign 曲线标(志);弯道标志
curves of equal horizontal magnetic force 等水平磁力曲线
curves of equal magnetic dip 等磁倾角曲线图
curves of equal settlement 均匀沉降曲线;均等沉降曲线
curves of equal vertical magnetic force 等垂直磁力曲线图
curves of form 船体型线图
curves of liner 衬砌曲线
curves of magnetic dip 等磁倾角曲线图
curves of total magnetic force 等磁力曲线图
curves of total magnetic intensity 等磁力曲线图
curve-speed plaque 弯道减速标志牌
curve spiral 螺形曲线
curves ruler 曲线尺
curve steady arm 软定位器
curve steepness 曲线陡度
curve stone 曲线标
curve strengthening 曲线加强
curve super-elevation 曲线超高
curve-surfaced side slipway 横移变坡滑道
curve table 曲线表
curve template 曲线板
curve test 曲线检验;曲线测定;曲度检验
curve tester 曲线测定器;曲板检验器
curve theory 曲线理论
curve time distance 时间距离(关系)曲线
curve-tined harrow 弯齿耙
curve to curve 相连曲线点
curve-tooth bevel gear 弧齿伞齿轮
curve top 弯顶
curve to spiral point 圆缓点
curve to tangent point 圆直点
curve tracer 波形描绘器;波形记录器;曲线描绘器
curve tracing 曲线作图;曲线画法;画曲线;曲线描迹法;曲线描绘;曲线跟踪
curve type of power function 幂函数曲线型
curve union 曲线并集
curve versine 曲线正矢
curve widening 曲线加宽;平曲线加宽;弯段加宽;弯道加宽
curviline 曲线
curvilineal 曲线的
curvilineal coordinates of curve 曲线坐标
curvilineal drawing die 弧形拉模
curvilineal regression 曲线回归

curvilineal tracery 曲线窗格
curvilineal trend 曲线趋势
curvilinear 曲线状;曲线的
curvilinear air classifier 曲线式空气分级机
curvilinear angle 曲线角
curvilinear asymptote 渐近曲线
curvilinear axis 曲线轴
curvilinear boundary 曲线边界
curvilinear cone 曲线型纸盆
curvilinear coordinates 曲线坐标
curvilinear correlation 曲线相关
curvilinear detail 曲线细部
curvilinear distortion 曲线形畸变;曲线畸变
curvilinear equation 曲线方程
curvilinear figure 曲线形;曲线图形
curvilinear figure program(me) 曲线图程序
curvilinear flow 曲线流(动);弯曲水流
curvilinear function 曲线函数
curvilinear gable 弧形的顶山墙;山墙顶部坡度的几何曲线
curvilinear integral 线积分;曲线积分
curvilinear interpolation 曲线内插(法);曲线插值(法)
curvilinear model 曲线型模型
curvilinear motion 曲线运动
curvilinear net(work) 曲线网(格)
curvilinear ornament 曲线饰
curvilinear orthogonal coordinates 曲线正交坐标;正交曲线坐标
curvilinear path 曲线轨迹
curvilinear pediment 弧形山墙
curvilinear regression 曲线回归
curvilinear regression and correlation 回归和相关曲线
curvilinear regression of secondary degree 二次曲线的回归
curvilinear regulation circuit 曲调电路
curvilinear road 迂回路
curvilinear scale 不等分度标尺
curvilinear simple regression 曲线型简单回归
curvilinear street system 弯曲迂回的街道系统
curvilinear style 曲线式
curvilinear tile 曲线瓦
curvilinear tracery 曲线窗花格
curvilinear transformation 曲线变换
curvilinear translation 曲线平移
curvilinear travel 曲线行程
curvilinear trend 曲线趋势
curvimeter 曲线计
curving 弯曲
curving beach 弯曲海岸
curvity 曲率
curvometer 曲线计;曲线(长度)仪;曲率仪;曲线计
curzerene 莪术烯
curzerenone 莪术酮
cusec 秒立方英尺;立方英尺/秒流量
cuselite 云辉斑岩
cushion 换土垫层;胶垫;减振垫;缓冲垫;桩垫;直浇口下储铁池;垫子;垫;床垫;衬层;软垫;弹性垫;台垫;缓冲
cushion-back carpet 衬垫毡
cushion basin 消力池
cushion blasting 缓冲爆破(法);气垫爆破
cushion block 桩顶垫层;垫块;替打(垫块)
cushion board 由额垫板;檐垫板
cushion cap 诺曼式柱头;垫式柱头;桩平台;承台
cushion capital 罗曼式柱头;垫块状柱头;方块式柱头
cushion car 气垫列车
cushion circuit 气垫电路
cushion coat 垫层;软垫涂层
cushion course 垫层;软垫层
cushion cylinder 气垫缸
cushion dashpot 缓冲缸
cushion disk 弹簧圆板;弹性圆盘
cushion disc spider 弹性万向节叉
cushion dozer 减震推土机
cushioned bulldozer 缓冲式推土铲
cushioned frieze 枕垫雕饰带;凸状雕饰带
cushioned hammer 垫锤
cushioned swing check valve 带胶垫摆式止回阀
cushioned underframe car 底架带减震器的车辆
cushion face 承击面
cushion feeder 振动式供料器;振动式给料器;弹簧推车器
cushion frame 弹性坐位架
cushion gas 垫层气
cushion guardrail 缓冲护栏
cushion head 桩垫头;缓冲头;桩头帽
cushion hitch 缓冲连接装置
cushion hitch spring 拖车减震弹簧
cushion idler 缓冲滚轮
cushioning 软衬垫;减震;缓冲(作用);弹性压密;弹性垫层;缓冲软垫;缓冲器;弹性填层
cushioning ability 减震能力;缓冲能力
cushioning action 缓冲作用
cushioning capacity 减震能力;缓冲能力
cushioning device 缓和装置;缓冲装置
cushioning effect 减震作用;缓冲作用;缓冲效应;垫层作用;垫层效应;衬垫作用
cushioning fender 缓冲装置;缓冲护木;缓冲防护器;缓冲挡板
cushioning material 铸型补强剂;衬垫材料;容让性材料;弹性垫材
cushion kiln 气垫窑
cushion layer 减震层;缓冲层;垫层
cushion material 衬垫材料;缓冲材料;垫料
cushion movement 减震运转
cushion of air 气垫
cushion packaging 缓冲包装
cushion packing 防震衬垫
cushion piece 垫块;垫木
cushion pile 缓冲桩;垫桩
cushion plant 垫状植物
cushion planting 缓冲栽植
cushion plate 缓冲板
cushion plunger 缓冲柱塞
cushion pool 储备池;储备槽;静水池;缓冲池;水垫消力池;消力池
cushion push block 减震块;缓冲垫块
cushion rafter 垫椽;辅助椽(子)
cushion rubber 胶圈;垫层橡胶
cushion seat 软(垫)座;减震座
cushion-shaped distortion 枕形畸变
cushion shooting 缓冲爆破(法);气垫爆破
cushion socket 防震(管)座;防震插座;弹簧插座
cushion space 缓冲空间
cushion spring 座垫弹簧
cushion steam 垫气
cushion stress factor 垫层应力系数
cushion support 让压支架
cushion tank 缓冲水箱
cushion tire 软心轮胎
cushion valve 弹性阀;减震阀;缓冲阀
cusiloy 库西洛铜硅合金
cusp 两曲线的交会点;尖头;(曲线相交的)尖角饰;尖端;尖点;尖瓣;歧点;滩嘴
cuspate 三角状
cuspate bar 角形沙坝;尖沙洲;三角形沙坝;三角沙嘴;三角沙洲;三角沙坝
cuspate barrier 三角沙槛
cuspate beach 尖头海滩
cuspate delta 尖形三角洲;尖头三角洲;三角洲
cuspate fold 尖角褶皱
cuspate foreland 尖头前陆;三角岬;三角岬
cuspate foreland coast 三角岬海岸
cuspate reef 尖头礁
cuspate ripple mark 三角形波痕
cuspate sand key 尖头小沙岛
cuspate spit 尖(头)沙嘴
cusp beach 海滩嘴
cusped arch 尖拱
cusped edge 尖棱
cusped index 尖点指数
cusped locus 尖点轨迹
cusped point 尖点
cusped profile 尖头翼
cuspidal 尖型的
cuspidal cissoid 尖点蔓叶线
cuspidal cubic 尖点三次线
cuspidal edge 尖棱
cuspidal locus 尖点轨迹
cuspidal point 尖点;歧点
cuspidated 带尖头的
cuspidation 尖形饰;(用弧线相交的)尖角装饰
cusping 形成水锥;形成水舌
cuspis septalis 隔瓣
cusp locus 尖点轨迹
cusp of curve 曲线尖端
cusp of the first species 第一种歧点
cusp of the second species 第二种歧点
cusp point 峰点
cusps 月角
cussome 檐下(石)望板
custard glass 奶黄色不透明玻璃
custatic regression 海退变化
custodial area 保管面积
custodial warehouse 保管仓库
custodian 房产管理人;保管人;看守人
custodian fee 管理费;保管费
custodian fund 保管基金
custodian service 保管业务
custodis 楔形砖建烟囱法
custody 保管
custody account bookkeeping 保管账簿记
custody account charges 保管费用
custody bill of lading 存栈提单
custody charges 保管费
custody of arrival 到达保管
custody of drawings 图纸的保管
custody transfer 转输;密闭输送;运输监护(液化气)
custody transfer point 转输点
custom 习俗;习惯;常规的;风俗习惯
customable 可征收关税的
custom and practice 习惯与惯例
custom approach 惯例做法;惯例方法
customary 惯例的;常例的
customary average 照常规办理单独海损免赔额
customary business practice 商业惯例
customary commercial practice 商业惯例
customary consultation 例行磋商
customary deductions 习惯磨耗
customary discount 例行折扣;常例折扣
customary dispatch 习惯装卸速度
customary freight unit 习惯运费单位
customary international law 习惯国际法;国际惯例
customary law 习惯法(规);惯例法;不成文法
customary money 习惯货币
customary nomenclature 习惯命名法
customary packing 习惯包装
customary price 习惯性的价格;习惯价格
customary pricing 习惯订价法
customary procedure 习惯程序
customary quick despatch 迅速发送;习惯装卸速度;习惯快速装卸;港口习惯快速装卸
customary rent 通常租税
customary risk 惯常险;惯用险别
customary route 习惯航线
customary rule 习惯规则;惯例
customary tare 习惯皮重;惯用包装
customary tenure 习惯保有
customary tenure estates 习惯保有财产
customary voyage 习惯航线
custom billform 顾客账单
custom builder 承建商
custom-built 现场制造;定建(的);定制的;非定型的;按顾客制作的
custom-built form (work) 定制模板
custom-built home (按照用户要求建造的)承建的住宅
custom-built power unit 非标准动力头
custom calibration 委托校准
custom clearance permit 结关通行证
custom coating 定制涂料
custom colo(u)r 委托颜色
custom control 海关监督;关税控制
Custom Cooperation Council 关税协作理事会
custom cow pool 委托养牛场
custom-designed 非定型设计的
custom-designed package 按订货设计的包装
custom directory 海关标示牌
custom district 关区
custom-engineered system 顾客监督系统
customer 领图室;客户;卖主;耗电器;消耗器;用户;主客户;主顾;顾客;服务对象;使用部门
customer affairs 顾客事务
customer automated teller 客户自动出纳机
customer circulation 顾客(车辆)流通量
customer classification 客户分类
customer communication 向顾客提供信息
customer complaint 客户意见
customer connected load 装接容量

customer costing 顾客成本计算
customer density 用户密度
customer engineer 常规工程
customer first 一切为用户服务
customer friendly design 人性化设计
customer information control system 客户信息控制系统;用户信息控制系统
customer inspection 用户检查
customer list 客户表;用户表
customer loss ratio 顾客损失率
customer management 客户管理
customer mentality 消费者心理因素;消费心理因素
customer needs and wants 顾客需求
customer-oriented 以顾客为中心
customer own and maintenance communication equipment 顾客拥有并维护的通信设备
customer parking area 顾客停车处;顾客停车场
customer parking building 顾客停车建筑;顾客车库
customer penetration 用气普及率
customer premises equipment 用户驻地设备
customer premises network 用户驻地网
customer profitability analysis 客户盈利分析
customer rate 顾客运费率
customer risk 客户风险
customer's acceptance liability 客户的信用证负债;客户承兑负债
customer's account 客户账
customer's account discounted 已贴现应收客户款;客户款已贴现;客(户)账贴现;顾客账贴现
customer's approval 客户认可
customer's credit balance 客户账贷余;客户账贷差;客户保证金
customers' deposit 客户保证金;客户存款;客户定金
customer set-up 用户安装
customer set-up product 用户安装产品
customer's final acceptance check 用户的最终验收
customer's guarantee liability 客户保证负债
customer's installation 用户设备
customer's ledger 客户分户账;客户分类账
customer's letter of credit liability 客户信用证负债;客户的承兑负债
customer's order 客户订货单
customer's plant 用户设备
customer's service 顾客记录;顾客服务;服务至上
customer's statement 客户结账单;客户结单;客户对账单
customer's subscriber 用户
customer's training 客户培训;用户培训
custom fabricated 按订货条款制造的
custom grade 正品;常规级;定制等级
custom harbo(u)r 关税港
custom house 海关(办公楼)
custom house broker 报关行;报关办者;报关代办人
custom-immigration-quarantine 检关签证检疫
custom integrated circuit 非标准集成电路
custom invoice 报关单
customization procedure 定制过程
customize 定做;定制
customized computer 定做型计算机
customized image 定制映像
customizing 按用户要求制做
custom kiln drying 特约窑干
custom-made 定制(品);定做的;非定型的
custom-made forms 定制模板
custom-made map 普通地图
custom-made shade 定制的遮阳棚
custom manifest 海关申报单
custom millwork 特约订货;定做木活;特约订货;定做的加工件
customoffice 海关办公处
custom of merchant 习惯商法
custom of port 港口习惯;港口惯例;商港习惯
custom of trade 贸易常规
custom processing service 委托加工服务
custom report 顾客报告
customs 海关(办公楼)
customs administration 海关局
customs and habits 风俗习惯
customs appraisal 关税评定
customs appraised value 海关估价
customs appraiser 海关检查员
customs area 海关境界

customs assigned number 海关指定号码
customs auction 海关拍卖
customs authorities 海关当局
customs barrier 关税壁垒;关卡
customs bill of entry 海关进口船舶名单
customs bills of entry 海关关册
customs boarding officer 登轮关员
customs boat 海关艇;海关船
customs bond 海关保税;海关罚款
customs bonded warehouse 海关关栈仓库
customs boundary 海关疆界
customs broker 海关代理人;海关代理行;报关经纪人
customs building 海关大楼
customs bureau 税务局
customs certificate 海关凭证
customs classification procedure 结关手续
customs clearance 结关;海关结关;海关放行;海关出口许可证;验关;出口结关;清关
customs clearance area 海关境界
customs clearance charges 结关费
customs clearance date 结关日
customs clearance formality 结关手续
customs clearance procedures 通关手续
customs clearance statistics 结关或报关统计
customs clearing charges 结关费
customs clearing figure 报关数字
customs clearing procedure for export 出口报关手续
customs code 海关法
customs commissioner 海关关长
customs consignee 海关委托人
customs control area 海关境界
Customs Convention on Containers 集装箱海关公约;集装箱关税公约
customs cooperation council 海关合作理事会
Customs Co-operation Council Nomenclature 海关合作理事会税则商品分类目录
customs cost accounting 海关成本会计
Customs Court 海关法庭(美国)
customs cover 关封
customs cruiser 海关巡逻艇
customs custody 海关保管
customs debenture 关税报关;海关退税凭单;报关单
customs decision 海关处分通知书
customs declaration 海关申报(单);报关单;申报关税
customs declaration entry 进口报关单
customs declaration form 海关申报单
customs declaration made at the time of entry 入境申报单
customs depot 海关货栈
customs detention 海关扣留
customs drawback 进口货物再出口时的关税退税;海关退税
customs due 关税
customs duty 进出口海关税;海关关税;关税
customs duty rate 关税税率
custom section 定制型材
customs entry 进口报关手续;进口报关;海关登记
customs entry form 进口报关格式
customs examination 海关检查
customs examination area 海关检查场所
customs examination list 海关验关单
customs examiner 海关检查人员
customs fee 海关规费;海关费
customs form 海关格式
customs formalities 报关手续
customs formalities and requirements 海关手续和规定
customs free depot 海关免税仓库
customs goods 课税品
customs guarantee system 海关担保制度
customs guard 海关稽查员;税务缉私人员;水上警察
customs guard vessel 海关缉私船
customs hall 海关大厅
custom shop 顾客商店;定制商店
customs house 海关办公楼;海关
customs house broker 海关经纪人
customs house formalities 海关手续
customs house officer 海关人员
customs house report 海关报告

customs import tariff 海关进口税则
customs inspection 海关检查;验关
customs inspection post 海关检查站
customs invoice 海关关票;海关发票
customs launch 海关小艇
customs law 海关法
custom smelter 加工熔炼厂
customs nomenclature 海关税则分类
customs number 海关编号
customs office 海关
customs officer 海关工作人员;海关稽查员;关务员;税务缉私人员;水上警察
customs of trade 贸易惯例
customs operation and procedure 海关作业和手续
customs pass 海关通行证
customs patrol inspector 海关巡逻检查员;海关巡查员
customs patrol vessel 海关巡逻船
customs permit 海关许可证
customsplant 中心选矿厂
customs policy 关税政策
customs port 课税港;海关港;由海关经营的港口
customs preventive officers 海关缉私官员
customs procedures and norms 海关手续和准则
customs quarantine control 海关检疫
customs quota 海关配额
customs receipt 海关收据
customs re-entry permit 海关再进口凭证
customs regulations 海关规章;海关法则
customs report 海关进口船舶名单
customs revenue 海关税收;抄关
customs route 海关水路
customs seal 海关加封;海关封条;关封
customs sealing device 关封装置
customs search 海关检查;抄关
customs searching party 海关检查组
customs service 海关(业务)
customs shed 海关(验)货棚;海关卡
Customs Simplification Act 海关简化法(美国)
customs specification 海关清单
customs station 分支关
customs storage 海关仓库;保税关栈;保税仓库
customs supervision charges 海关监管费
customs supervision zone 海关监督区
customs surveyor 海关检查员
customs tare 海关规定皮重
customs tariff 海关税则;关税(税)率;税则
customs transit documents 海税过境单据;海关过境单
customs union 海关联盟;关税同盟;关税报告表
customs user fee 海关使用费
customs valuation 海关估价
Customs Valuation Code 海关估价规约
customs value 海关价值
customs warehouse 海关仓库;保税关栈;保税仓库
customs warrant 海关栈单;海关免税提货单;海关仓库货物出仓单;保税提货许可证
customs waters 海关水域
custom system 用户系统
custom-tailor 定做;定制
custons shed 海关仓库
custos 管理者;看守人;保管人;[复]custodes
cut 垅沟(疏浚);粒度级分;克特;开凿;横切矿脉;油分;印版版;琢磨花样;钩车(车组);刀伤;采掘带;切取倍分;切口;挖方;挖槽;掏槽;采伐;切削;切割;切裁
cutability 可切性
cut across 两船交叉航驶
cut across the bar 切滩
cut-admittance matrix 割容许矩阵
cut a groove 开轧槽
cut a melon 分派红利
cut-and-built platform 蚀积台地
cut-and-carry 连挖带运的
cut-and-carry die 拖件前进的连续模
cut-and-carry tool 挖运工具
cut-and-carve 切开
cut-and-cover 大开槽施工法;明挖;边挖边填;明挖回填;原土回填;挖方与填方;随挖随填;随挖随盖
cut-and-cover construction method 明挖法施工
cut-and-cover method 基坑法;明挖回填法;明挖法;随挖随填(施工)法
cut-and-cover section 挖方填方段;开挖覆土段;

明挖回填地段;大开挖段
cut-and-cover tank 半填半挖水池
cut-and-cover technique 大开挖施工法;敞开式隧道施工法
cut-and-cover tunnel 明洞;明挖隧洞;明挖隧道;大开挖隧道
cut-and-fill 移挖作填;挖方和填方;半挖半填;随挖随填;山腰上半挖半填;移挖半填;挖填作填;冲淤(作用);冲刷和填积;边冲边淤;挖填方;随挖随盖
cut-and-fill at center stake 中桩填挖高度
cut-and-fill balance 土方平衡
cut-and-fill balancing on cross-section 横断面上的挖填平衡
cut-and-fill estimate 土方计算
cut-and-fill excavation 挖方和填方;移挖作填;随挖随填
cut-and-fill location 挖填交点;挖填方分界线
cut-and-fill method 随挖随填(施工)法
cut-and-fill program(me) 填挖计划;冲刷及淤积计划
cut-and-fill section 半挖半填(式)截面;半挖半填(式)断面;半填半挖(横)断面
cut-and-fill slope 半挖半填的斜坡;半挖半填的边坡;半填半挖斜坡
cut-and-fill structure 冲淤构造
cut-and-fill system 分层充填开采法
cut-and-fill terrace 半挖半填的梯田
cut-and-full tank 半填半挖水池
cut-and-load machine 挖掘装载机
cut-and-mitered bead 隅角线条;偶角圆线条
cut-and-mitered hip 斜切戗脊;切割屋瓦与斜脊斜面接合
cut-and-mitered string 楼梯竖板与搁板斜接的明楼梯斜梁;隅角侧木;偶角侧木
cut-and-mitered valley 切割斜沟斜面接合
cut-and-replacement 换土法
cut-and-run 弃锚开航
cut-and-stored for later use 切碎储[贮]藏
cut-and-strip 刻剥薄膜
cut-and-trial 逐步接近
cut-and-trial method 逐次渐近法;试验法;逐步渐近法;试算法
cut-and-trial process 逐次渐近法
cut-and-try method 反复试验法;试算法;试凑法;渐次逼近法;逐次逼近法;尝试法;试探法
cut-and-try process 连续接近法;累次近似法;逐次接近法;试探法
cut-and-try work 试凑工作;实验工作
cut-and-valley 切割屋瓦与斜沟斜面接合
cut angle 交角
cutanit 刀具硬质合金;碳化物硬质合金
cutaway 切掉;切去隅角的;切去部分的;剖开立体图;剖面
cutaway boat 橇形船首
cutaway disk 缺口圆盘
cutaway disk harrow 带切盘的圆盘耙
cutaway disk packer 带缺口圆盘的镇压器
cutaway drawing 剖面图;断面图;剖开立体图
cutaway mine bit 十字钻头
cutaway section 剖面
cutaway view 剖视图;剖开立体图;内部接线图;局部剖视图
cutback 料液面侵蚀;液面线侵蚀;低熔点沥青混合物;后移;稀释产物;稀释;兑稀;反逆作用;轻制
cutback asphalt 稀释(地)沥青;轻制沥青;轻制地沥青
cutback asphalt binder 轻制石油沥青黏合料;轻制沥青混合料;轻制沥青黏合料
cutback asphalt emulsion 稀释沥青乳浊液;稀释沥青乳胶
cutback asphaltic bitumen 稀释沥青;轻制沥青
cutback asphaltic bitumen distillation apparatus 稀释沥青蒸馏装置
cutback asphaltic bitumen emulsion 稀释沥青乳胶;稀释沥青乳化剂
cutback bitumen 稀释沥青;轻制沥青
cutback blasting 挖方爆破
cutback coal tar 轻制煤(焦油)沥青
cutback cost 降低成本
cutback group 轻制沥青混合料
cutback irrigation 二次变量灌水
cutback pitch 轻制地沥青混合料;轻制焦油沥脂
cutback principle 轻制沥青法
cutback product 轻制产品;馏出产品;稀释产品

cutback rate 中转运费率
cutback road tar 稀释焦油沥青;稀释柏油
cutback road tar for cold repairs 冷修用稀释柏油
cutback side plate 切短的侧板
cutback stream 后期流量
cutback tar 稀释焦油;轻制溚;轻制柏油;回配沥青;轻制焦油沥青;稀释产物
cut balance 开口摆轮
cutbank 冲刷(形成的)陡岸;挖蚀岸;河流陡岸;切岸;切换台;视频切换台
cut bay 跨度;桥的跨度
cut bedding plane landslide 切层滑坡
cut boss 工长(操纵重型挖土机的,俚语)
cut bracket (边缘有线脚装饰的)牛腿
cut brick 切砖;砍砖
cut cable 电缆段;电缆端
cut capacity 切割容度
cutch 儿茶(色)
cutcherry 种植园事务所
cut core 截割铁芯;半环形铁芯;切面铁芯
cut corner for sight line 截角(路口)
cut deal 中厚板
cut decoration 踢花;剔红(朱红雕漆)
cut depth 挖土深度
cut diameter 分割粒径
cut-diameter method 分割直径法
cut diamond 磨琢金刚石
cut ditch 沉淀槽
cut dock 挖入式港池
cut-down 消减;向下挖;减价
cut-down a derrick 拆卸钻塔
cut-down and merge 裁并
cut-down a ship 改低船体
cut-down expenditures 紧缩开支
cut-down expenses 压缩开支
cut-down of staff 减员;裁员
cut-down the consumption of raw materials 降低原料消耗
cut-down to grade 挖到设计标高
cut drain 明沟;排水沟
cut drill 铣制钻头
cut edge 切割边;路堑边
cut-edge incidence matrix 割边关联矩阵
cut-edge plate 荷叶边盘子
cut engineered timber 锯好的木材;尺寸已加工的木材
cuter life 刀具寿命
CU-test 固结不排水三轴压缩试验
cut file 截锉;木锉(刀)
cut-fill 半挖半填;挖填方;随挖随填
cut-fill adjustment 土石方调配
cut-fill balance 土石方调配
cut-fill contact 挖方与填土连接处
cut-fill design curve 填挖方设计曲线
cut-fill height 填挖高度
cut-fill section 半填半挖式横断面;半挖半填(式)截面;半填半挖(式)断面;半填半挖断面;半填半堑;半堤半堑
cut-fill subgrade 半填半挖路基
cut-fill transition 挖填方调度(平衡);土方挖填调配;移挖作填;土方调配
cut-fill transition diagram 土方调配图
cut-fill transition program(me) 土方调配计划
cut film 切片;散张胶片
cutfit 备用工具
cut flight spiral 齿形螺旋
cut flower 插瓶花;切花
cut flower garden 切花园
cut flush 封面与书芯切齐
cut foilage 叶饰
cut form 规格切纸;单页;切开的格式纸;散页方式
cut-forms mode 切开格式纸方式
cut foundry pattern 塑料木模
cut gear 铣齿轮;切制的齿轮
cut glass 刻花的厚玻璃(器皿);刻花玻璃;雕花玻璃;车料玻璃;水晶玻璃
cut-glass decoration 刻玻璃装饰法
cut glaze 釉层太薄
cut glaze decoration 刻釉装饰法
cut grade design 割线设计
cut grafting 切接
cut graining 括制木纹法
cut growth resistance 切口增长阻抗
cut hay 割下的干草

cut hole 开炮孔;打(炮)眼;钻孔【岩】;掏槽(炮)眼
cut house 分类专业小组
cuticle 角质层
cuticle liptobiolith 角质残植煤
cuticoclarite 微角质亮煤
cuticoclarodurite 微角质亮暗煤;微角质暗亮煤
cuticodurite 微角质暗煤
cuticular layer 表皮层
cuticular transpiration 角质层蒸腾;表皮蒸腾
cutie 携带式辐射能测定仪
cut-image prism 分像棱镜
cut-image range finder 切像测距仪;切合测距仪;分像测距仪
cut-in 接通;标绘位置线;切入;角质;表皮质;开始工作
cut in a network 网络剖线
cut-in bottom 梁底板
cut-in-channel deposit 切入水道沉积
cuting 路堑
cut-in insulator 插入的绝缘子
cutinite 角质体
cutinite coal 角质煤
cutinite-collinite 角质无结构体
cut-in method 插接法
cut-in notes 文间注释
cut-in operation 接入工作
cut-in point 开始工作点;接通点
cut-in rates 费率降低
cut-in relay 接入(式)继电器
cut-in sand 冲砂
cut-in signalling 插入信号法
cut-in voltage 接通电压;互联电压;临界电压;开启电压
cutis marmorata 大理石色皮
cutite 微角质煤
cut joint 雄榫;凸榫;锯榫;接缝
Cutlasta alloy 卡特拉斯塔合金
cut length 切割长度;定尺长度
cutlery 刀具;刀剑
cutlery chest 餐具柜
cutlery stainless steel 刀具用不锈钢
cutlery steel 刀具钢;刃具钢
cutlery-type stainless steel 刀具不锈钢
cutlet 切片
cutlift 割草装载机
cut-line 垄沟线;切割线;图注;强韧细绳;铅铸窗的施工大样;图例;插图说明(指图下文字说明)
cut line deviation 切割线偏差
cut lining 挖方边衬
cut lock 凿制门梃的锁;插钉;插销
cut log 块状原木
cut lots 抽签
cut mark 分劈墨印;切割标记
cut material 裁制材料
cut matrix 割矩阵
cut member 裁切构件
cut meter 切削速度指示计
cutmixer 斩拌机
cut nail 楔形钉;方钉;切(制)钉
cut-nippers 剪丝钳
cut nippers with spring 弹簧剪钳
cut number 切断序数
cut-off 断火;停给;截止;截频;截流;捷水道;止水带;遮断装置;关机;隔断器;短捷航道;采板;裁弯取直;取直;切开;停电;缩径量;林间防火线;截断;切断
cut-off adjustment 截断调整;限界电平调整
cut-off amplification factor 截止放大因数;极限放大因数
cut-off amplitude 脉冲幅值
cut-off angle 截止角;截光角;遮光角
cut-off angular frequency 截止角频率
cut-off apron 护底截水墙;截水墙护坦;截水墙护底
cut-off attenuater 截止衰减器;极限衰减器
cut-off bar 阻栅杆
cut-off basin 封闭台地;闭合流域
cut-off bias 截止偏压
cut-off blanket 截水层;隔离层
cut-off buttress 横向防渗墙
cut-off characteristic 截止特性
cut-off check 裁切裂纹
cut-off circuit 截止电路
cut-off clutch 分离离合器
cut-off cock 断闭旋塞

cut-off collar 管口箍;管颈;防渗环;截水环
cut-off command 关闭发动机信号
cut-off computer 开断计算机
cut-off concentration 停车密度
cut-off condition 关机条件
cut-off control 裁切控制装置
cut-off corner 切角
cut-off current 截止电流
cut-off curtain 截水帷幕
cut-off curve 轧齐曲线;膨胀曲线
cut-off cycle 开断周期
cut-off cylinder 截圆柱
cut-off dam 截流坝
cut-off date 截止日期;截止期(限)
cut-off device 截流设备;截流器;截水器;断流器
cut-off diameter 截止直径
cut-off die 切边模;剪模
cut-off dike 截水堤;截流堤
cut-off ditch 截水沟
cut-off drain 截渗排水;截水沟
cut-off effect 截止效应
cut-off elevation 断桩高程
cut-off energy 截止能量;门限能量
cut-off equipment 截流装置
cut-off error 截断误差
cut-off filter 截止滤光片;截止滤波器
cut-off filter glass 截止型滤光玻璃
cut-off floor 采板台;采板层楼面;采板层
cut-off for lighting 遮光
cut-off frequency 截止频率
cut-off frequency of duct 管道截止频率
cut-off governing 停气调节
cut-off governor 断汽调节器
cut-off grade 边际品位(矿业);品位下限;边界品位
cut-off grinder 砂轮截断机
cut-off groove 截流槽
cut-off high 切断高压【气】
cut-off interval 停车间隔;停车时间
cut-off jack 串联切断塞孔
cut-off key 切断电键
cut-off knife 裁刀
cut-off lake 割断湖;牛轭湖;裁弯湖
cut-off lens 切割镜片
cut-off level 截止电平;限制电平
cut-off limiting 截止限制;截止限幅
cut-off line 闭合支线;截止线
cut-off low 切断低压【气】
cut-off machine 切断机
cut-off man 玻璃(切)裁工;采板工(人);切割工
cut-off meander 截弯河套;割ει曲流;裁弯曲流
cut-off meander spur 离堆山;曲流环绕岛
cut-off method 截止法;切角法
cut-off mode 截止模式
cut-off mo(u)ld 溢出式(铸)塑模
cut-off of pile 截桩
cut-off of river 裁弯取直(河道)
cut-off ore ratio 边界含矿率
cut-off peak 截止峰值
cut-off pile 截水板桩;截断桩
cut-off piling 截断桩;截水板桩;阻水板桩;隔水板桩
cut-off plane 截流面;截流面
cut-off plate 炉门;节汽门;闸板;斗仓的闸板;风挡
cut-off point 截止点;截断点;熄火点;分界点;投资截止点
cut-off polarizer 截止偏振器
cut-off procedure 截止程序
cut-off pulse height 脉冲幅值;脉冲幅度限
cut-off push 断路器按钮
cut-off radiation 辐射阀
cut-off rate 允许最低回收率;舍弃回收率
cut-off rate of return 临界收益率;能接受的最低收益率;报酬的舍弃率
cut-off ratio 断路比;裁弯取直率;等压容积比;等压膨胀比;裁弯取直率;裁弯(取直)比
cut-off relay 断离继电器;截止继电器;隔离继电器;断路继电器;切断继电器
cut-off relay circuit 断路继电器电路
cut-off rigidity 截止刚度
cut-off rubbers 停机橡皮
cut-off rule 分割线
cut-off saw 截土锯;截断锯;横截锯;切断锯
cut-off section 裁弯取直段
cut-off shears 截断机
cut-off sheet pile wall 截断板桩壁;板桩截水墙

cut-off shell 断电器壳
cut-off shot 哑炮;迟爆炮;拒爆炮(孔);瞎炮
cut-off signal 断开信号;停车信号
cut-off size 分离粒径;分离粒度
cut-off slide 横U形架;膨胀滑阀
cut-off slide valve 断流滑阀
cut-off slope 截止斜率
cut-off solenoid 切断螺管线圈
cut-off speed 截止速度
cut-off spring 断开弹簧
cut-off statement 截止报表
cut-off stop 门框做成的槛
cut-off supply 停止供水;停止供气;停止供电
cut-off switch 断路开关;断路接触器;断流开关;切断开关
cut-off system 防渗系统
cut-off table 切砖机;切泥坯机
cut-off tap 关闭龙头;关闭旋塞
cut-off test 停车试验
cut-off the edge 切边
cut-off the electricity supply 切断电源
cut-off time 截止时间
cut-off to grade 切至基准线;切到基准线
cut-off tool holder 车刀支架
cut-off trench 拦墙沟;隔墙沟;防渗隔离槽;截水墙;截水沟;截水槽;截水槽;齿槽(截水)
cut-off trench of puddle clay 黏土防渗槽
cut-off value 截止值
cut-off valve 断流阀;停汽阀;截止阀;截流阀;关闭阀;断流闸;停止阀
cut-off velocity 截止速度;停车速度
cut-off voltage 截止电压;终止电压;临界电压
cut-off wall 拦墙;防渗墙;截水墙;堰板;隔水墙;隔墙;齿墙
cut-off wall efficiency 防渗墙有效率;防渗墙防渗能力
cut-off water 渗透雨水;截断供水;渗透水
cut-off wavelength 截止波长
cut-off wheel 切断(磨)轮;切割砂轮
cut-off works 裁弯取直(工程);裁弯工程
cut oil 乳化油
cut-open view 剖视图
cut orientation 切割方向
cut-out 线路断开装置;断电器;切口线对;切断
cut-out board 断流板
cut-out box 安全箱;熔断器匣;断流器箱;保险盒;闸盒
cut-out carving 剔雕
cut-out case 保险盒;熔线盒
cut-out cock 截流旋塞门;切断旋塞;截断塞门
cut-out current coil 断电器电流线圈
cut-out device 安全保险装置;安全开关
cut-out fuse 保险装置;断流熔丝
cut-out governing 喷嘴调节
cut-out halftone 突出主体网点版
cut-out key 切断电键
cut-out overflow 溢水口
cut-out pedal 分离踏板
cut-out plug 断流栓;断流塞
cut-out relay 断流器;断路继电器
cut-out relay armature 断流继电器附件
cut-out relay armature spring 断流继电器电转子弹簧
cut-out relay contact bracket 断流器接触点托架
cut-out relay shunt coil 断流器并联线圈
cut-out shears 带材局部切除机
cut-out spring 分开弹簧
cut-out switch 截止开关;断路器;断路开关;切断开关
cut-out tap 关闭旋塞
cut-out valve 截止阀;断流阀;停止阀
cut-out voltage 截止电压
cut-out voltage coil 断电器电流线圈
cut-out yield 切块产量
cut-over 接入;开通;切换;采伐
cut-over land 原始森林大部分已伐除的土地
cut-over pasture 森林伐除后牧地
cut paint 兑稀用漆
cut-parabolic(al) antenna 切割抛物面天线
cut paraboloid 截抛物面
cut paraboloid reflector 截抛物面反射器
cut pattern 掏槽孔布置形式
cut payment 扣款
cut penetration 切割深度

cut pile 剪绒;割绒
cut-pile carpet 剪绒地毯;割绒地毯
cut pile tufting machine 簇绒割绒机
cut plane 截平面;剖面
cut-point 分离点;馏出温度;截点;割点;分馏温度;分馏界限;分馏点
cut-point ash 分界灰分
cut-point graph 割点图
cut/power relationship 分隔/功率关系
cut price 削价
cut price offer 廉价出售
cut product 切削制品
cut pulley 工作冲程皮带轮
cutrake 割草搂草机
cut rate 减价;减费
cut regulation 采伐调节
cut release speed 车辆出口速度
cut researching method 逐段检验法
cut ridge 截幅;截峰;限幅
cut rock 硬岩钻进;打硬岩
cut roof 削去屋脊的坡屋顶;平台式屋顶;无屋脊坡屋顶;盝顶
cut rubble 方正石块
cut sample 开挖的试样
cut sand 均质砂
cut searching method 逐段寻找法
cut section 路堑断面;分割区段;挖土断面;挖方断面
cut-section of a track circuit 轨道电路分割
cut-set 割集
cut-set code 割集码
cut-set equation 割集方程
cut-set matrix 割集矩阵
cut-set matrix equation 割集矩阵方程
cut-set saturation algorithm 割集饱和算法
cut sheet 开挖图;切片;挖方表
cut shoot 伐条
cut-short 捷径
cut shot 掏槽炮眼
cut side 挖土面;挖土边
cut size 特定尺寸;基本尺寸;极限粒径;裁成尺寸;分级粒度;切裁尺寸;切裁规格
cut size plate 平板玻璃切裁尺寸
cut slide valve 膨胀滑阀
cut slope 开挖坡度;路堑边坡
cut slope angle 路堑边坡角;采掘边坡角
cut spike 大方钉;普通狗头道钉;钩头道钉
cut splay 斜切(砖)
cut spring 分离弹簧
cut square 直角开挖
cut-squaring method 分格取样法
cut staple 短纤维
cut staple sliver 短切毛条
cut steam 断汽
cut steel nail 方钉
cut stone 石粉;切割石(块);毛石;琢石;方石
cut stone coping 石压顶
cut-stone masonry 琢石圬工
cut-stone paving 方石路面
cut-stone veneer 凿石镶面
cut stone work 琢石工程
cut string 开式楼梯梁;木楼梯露明小梁;明楼梯梁
cut stringer 连三角的楼梯梁
cut-string stair(case) 隐蔽来楼梯;露明梁楼梯(采用连三角斜梁的)楼梯
cut strip 采伐带
cutstuff 切割成较小的用料;碎木片
cuttage grafting 插接
cutted thread 车削螺纹
cutter 刻纹头;刻花工人;截切机;截断器;断路机;截板机;铰刀【疏】;剪断机;小渔船;独桅快船;刀具;裁剪机;方尾双桨救生艇;软砖(指可切削及雕刻的软砖);刀具;倾斜节理;切削器;切削工具;切割工具;切割膜腔;切断膜腔;切断机;切刀;气割枪;切割器
cutter adapter 刀具接头
cutter-and-reamer grinder 铣刀铰刀磨床
cutter and rubber 切刀;磨砖;范砖;刀割器
cutter and tool grinder 工具磨床
cutter and tool grinding machine 工具磨床
cutter angular slide 刀具斜滑板
cutter arbor 铣刀杆;刀具心轴
cutter arbor speed 刀具轴转速
cutter back-off 让刀;抬刀
cutter back ring 铰刀大环

cutter bar 截煤机截盘;刀杆;切割器
cutter bar and wedge machine 截煤机楔落煤机
cutterbar assembly 切割器总成
cutterbar clutch 切割器传动离合器
cutterbar control lever 切割器起落操纵杆
cutterbar drive 切割器的传动装置
cutterbar for high cut 高割型切割器
cutterbar for low cut 低割型切割器
cutterbar for middle cut 中割型切割器
cutterbar lead 切割器前斜伸量
cutterbar mower 切割器式割草机
cutter beater 击碎切割刀轮
cutter bed 截煤机底座
cutter bit 切削钻头
cutter bit sharpener 截煤机截齿磨锐机
cutter blade 铰刀片;铣刀片;刀片;切土刮刀;切裁刀;丝锥刃瓣
cutter blade for pipe 管子切割刀片
cutter blade setting angle 刀片装置角
cutter blank 刀坯
cutter block 组合铣刀;刀座;刀头
cutter body 割刀外壳;刀体;切碎装置外壳
cutter box 箱形刀架
cutter break-in 铰刀切入
cutter chain 截煤机截链
cutter chain excavator 截链式挖掘机;割链式挖掘机
cutter chain pick box 截链齿座
cutter change factor 齿轮刀具移距系数;齿轮刀具变位系数
cutter chuck 铣刀卡盘;弹簧卡盘
cutter clearance ga(u)ge 铣刀量隙规;刀具后角规
cutter collar 铰刀轴环
cutter collet 刀具弹簧套筒夹头
cutter cost 切割费用
cutter-creaser 冲压折痕机
cutter diameter compensation 刀具直径补偿
cutter disc 切削刀盘
cutter dredge(r) 绞吸(式)挖泥船;旋转式挖泥船;旋桨式挖泥船;切削挖土机;绞刀式挖泥船
cutter drill 切刀钻
cutter drive 铰刀驱动装置
cutter drive governor 铰刀驱动控制器
cutter drive power 铰刀驱动功率
cutter driver 铰刀机
cutter excavator 切削挖土机
cutter finishing machine 刀具研磨机
cutter for fluting twist drill 麻花钻槽铣刀
cutter for screw plate 螺丝板加工刀具;螺丝扳牙加工工具
cutter frame 刀架
cutter ga(u)ge 刀具样板
cutter gear 齿轮刀具
cutter grinder 切削刀具磨床;工具车床
cutter grinding 工具磨削
cutter grinding machine 刀具磨床;刃磨机床
cutter guide 刀具导轨
cutter gummer 截煤机除粉器
cutter head 切土头;截煤机头部;铰刀头;铰刀头;铣头;刀头;刀盘;刀架;切头;切碎装置;切碎器;切割机头;切割刀盘;刀具刀头;镗头
cutter-head barker 削片式剥皮机;滚刀式剥皮机
cutter-head cross-feed slide 刀架横向进刀滑板
cutter-head dredge(r) 链轮式挖土机;链轮式挖泥船;链轮式疏浚机;绞刀式挖泥船;吸扬式挖泥船;铣轮式挖土机;铣轮式挖泥机
cutter-head grinder 铣刀盘刃磨机
cutter-head pipe line dredge(r) 绞吸(式)挖泥船
cutter-head suction dredge(r) 吸扬式挖泥船;绞刀吸扬式挖泥船
cutter-head type suction dredge(r) 绞吸(式)挖泥船
cutter holder 刀夹;刀杆
cutter hub 铰刀轮毂
cutter interference 刀具干涉
cutter jib 截煤机截盘
cutter ladder 铰刀桥架
cutter ladder gantry 铰刀桥架吊架
cutter ladder winch 铰刀架绞车
cutterless sand-sucker 吸砂船
cutterless suction dredge(r) 无绞刀吸扬挖泥船
cutter life 刀具使用寿命;切削刃寿命
cutter lifter 切割挖掘机
cutter lifting 抬刀
cutter link 切齿链节

cutter loader 切割装载机;截装联合机;切碎装载机
cutter loader bar 联合采煤机截盘
cutter loader bed 联合采煤机底座
cutter loader colter 采煤康拜因破煤犁
cutter loader jib 采油康拜因截盘
cutter loader shearer 采矿康拜因立截盘;刨装机
cutter loader shearing jib 采矿康拜因立截盘
cutter location tape 刀位指令带
cutter making machine 割刀制造机
cutter mark 切(削刀)痕
cutter material 刀具材料
cutter-mixer agitator 切碎混合搅拌装置
cutter mixer pump 带切碎和混合器的泵
cutter module 刀具模数
cutter motor 截煤机电动机
cutter-mounting ring 钻头牙轮部分
cutter offset calculation declaration 刀具补偿计算说明
cutter of gear wheel 齿轮刀具
cutter oil ring 刮油环
cutter path 刀具轨迹
cutter pick 截煤机截齿
cutter pilot 刀具导杆
cutter pin 牙轮销;割刀销
cutter pitch diameter 齿轮刀具分度圆直径
cutter point 刀锋
cutter position indicator 铰刀位置指示仪
cutter pump 带切碎器的泵
cut terrace 浪蚀阶地;岩石阶地
cutter radius offset 刀具半径偏置
cutter relieving 让刀;抬刀
cutter rod 刀杆
cutter roller 刀辊;齿轮辊碎机;齿辊破碎机
cutter-rower 割晒机
cutter saddle 刀架;鞍式刀架
cutter's bay 切裁间
cutter's diamond 金刚石刀
cutter service platform 铰刀维修平台
cutter shaft 切削轴;铰刀轴;刀轴;切碎装置轴
cutter shaft coupling 铰刀轴联轴器
cutter shaft power 铰刀轴功率
cutter sharpener 刀具磨刀器;刀具磨床;刀具;刃磨器
cutter's lath 切割桌用尺
cutter slide 刀具滑板
cutter slit 切削槽
cutter spacing 刀位;刀具限位
cutter specification 刀具规格说明
cutter speed 铰刀转速;刀头转速;刀速
cutter spindle 铣刀轴;镗杆
cutter spindle sleeve 刀轴套
cutter's station 切砖台
cutter's straight edge 切割桌用尺
cutter's table 切桌;切割桌
cutter's table ruler 切割桌上刻度尺
cutter station 切坯台
cutter stock 沥青稀释油
cutter suction 绞吸
cutter-suction dredge(r) 绞吸(式)挖泥船;切吸式挖泥机
cutter suction dredge(r) with rotating bucket wheels 带旋转斗轮的绞吸挖泥船
cutter switch 截煤机起动器
cutter teeth 铰刀齿;牙轮轮齿
cutter thrower 切碎抛送机
cutter tilt 刀倾角
cutter tool holder 刀具夹杆
cutter tooth 刀齿
cutter tooth harrow 刀齿耙
cutter tooth number 铣刀齿数
cutter tooth top 刀齿齿顶
cutter torque 刀具扭矩
cutter track 铰刀轨迹
cutter trench 防火沟
cutter truing ga(u)ge 刀盘跳动检查仪
cutter unloader 切碎卸载机
cutter wearing 铰刀磨损
cutter wheel 滚刀轮;大刀盘(盾构切削用);切割轮
cutter windrower 割晒机
cuttery-marking 釉面刀痕
cut the melon 分配额外红利
cut thread 切割螺纹;方(橡)胶皮;方切橡胶线
cut thread tap 不磨牙丝锥
cut-throat competition 残酷竞争;恶性竞争;剧烈竞争
cut-through 联络巷道;贯通;开凿;切透;挖通;凿通
cut tide 涨潮区段
cutting 切削;切割;切裁;砍伐;砍;开挖;金属极气保护电弧切割;截土;雕刻;下锯;插条;采伐量;挖掘;挖方;挖除表土
cutting ability 切削能力
cutting across of routes 进路交叉
cutting action 切削作用;切割作用
cutting agent 磨料;切削材料
cutting alloy 刀具合金
cutting and bending of reinforcement 钢筋的断配和弯曲
cutting and bending reinforcement 钢筋加工
cutting and chopping block 砧板;菜板;案板
cutting and creasing press 冲压折痕机
cutting and fill 移挖充填;冲刷及淤积;充填法开采;边冲边填
cutting and forming shop 切割加工车间
cutting and grinding device for sliding carbon block 碳滑板切割打磨装置
cutting angle 刀面角;切削角;切割角
cutting angle adjustment 切角调节
cutting apparatus 切割设备
cutting arc 切削弧
cutting area 伐区
cutting at ground level 低斩
cutting attachment 割炬附件
cutting auger 切碎螺旋;切割螺旋
cutting back 稀释;修剪;稀释回配;短截
cutting back pruning 短截修
cutting below grade 地面以下的开采;地面以下的采掘
cutting bench 锯material台;切碎台
cutting bit 钻头;切刀
cutting blade 切刀边;刀片;遮光叶片;切削(叶)片;切(削)刀(片)
cutting-blank 采伐迹地
cutting block 凿眼垫墩(帆布冲眼用具)
cutting blowpipe 割炬;切割吹管;气割炬;喷割器
cutting board 刀板
cutting box 磨制楔形拱砖的盒
cutting brick 斩砖
cutting budget 采伐一览表
cutting buffer 停机橡皮
cutting burner 切割喷枪
cutting cam 切削凸轮
cutting-carrying capacity (冲洗液的)排粉能力
cutting ceramics 陶瓷刀具
cutting chain 割链
cutting chamber 割刀舱
cutting chin 刀刃;切削刃
cutting chisel 刻纹针
cutting chute 分群狭道;排屑槽
cutting circle 切削圆
cutting clamp 裁剪夹子
cutting class 采伐等级
cutting clearance 隙角
cutting compound 切削油;切削液;切削润滑液
cutting condition 切削条件
cutting coolant 切削冷却液
cutting corner 切角
cutting corral 分群栏
cutting coulter 犁刀
cutting curb 开挖路缘槽;沉箱刃脚;沉井筒脚
cutting current 切割电流
cutting cycle 采伐(间隔)期
cutting cylinder 裁纸圆筒;切割滚筒
cutting damage 切削损伤
cutting density in geomorphy 地貌切割密度
cutting depth 开挖深度;切削深度;切削厚度
cutting device 切裁装置
cutting diamond 玻璃割刀;金刚石刀;切裁玻璃用金刚石刀
cutting die 落料模;冲裁模;板牙
cutting dimension 标线尺寸
cutting direction 切削方向
cutting disc 圆盘刀;切削盘;切割沙片
cutting disk 切削盘;切割沙片
cutting distance 切齿安装距
cutting down 缩分;磨光;扩大井筒;减少;打光;磨掉;弱切削加工;缩短;表面平整加工
cutting down capital construction scale 压缩基建规模

cutting down expenditures and increasing income 节支增收
cutting-down of bed 落底
cutting down of tunnel bed 隧道落底
cutting drawing 切割图
cutting drilling 切削钻
cutting edge 切割边缘;铲刀刃;剪刃;刀刃;刀口;刃口;刃脚;刀脚;刀缘;切削刃;切削刀;(盾构的)切口环;切边
cutting edge angle 刃口角(度)
cutting edge at bottom of well casing 沉井筒脚刀口
cutting edge normal plane 主切刃法截面
cutting edge of a caisson 沉井刃脚
cutting edge of box frame 箱涵刃角
cutting edge of bucket 斗口
cutting edge of shield 盾构刀口;盾构切削口
cutting edge rounding 切削刃钝化
cutting effect of slag 炉渣侵蚀作用
cutting efficiency 切削效率
cutting effort 切削力
cutting electrode 切割(用)电极
cutting element 切削元件
cutting end 切削刃
cutting energy 切削能量
cutting engineering 切割工程
cutting equipment 切割设备
cutting face 开挖面;切削面
cutting feed 进刀
cutting ferrule 卡套
cutting finish for parts without drawing 无图零件切削面光洁度
cutting flame 切割焰
cutting flare 切割不齐
cutting float 刮刀
cutting flow 截流
cutting fluid 润切液;切削(润滑)液
cutting fluid additive 切削液添加剂
cutting for balance of ground 刷方减载
cutting force 切削力;切割力
cutting frame 切架;切割桌
cutting ga(u)ge 切割规;线勒子;切削规
cutting graftige 插接
cutting grain 切削颗粒
cutting grassland 打草地
cutting gun 割枪
cutting hammer 圬工锤;切断锤;破石锤
cutting hardness 切削硬度;切削强度
cutting head 联合采掘机刀盘;刻纹头;刻图头;切削头;切割头部
cutting height 开挖深度
cutting height control 切割高度调节杆
cutting-in 标线;齐边涂刷;隔线;遮挡漆涂;冲入;(切深孔型中的)切入
cutting inarching 插靠接
cutting-in line (不同颜色油漆的)交界线
cutting-in speed 起动转数;起动速度
cutting-in tool 切进刀
cutting iron 刮缝凿;刮缝刀;刨铁
cutting jet (水力开挖法的)开挖水射;切割射流
cutting knife 刻刀;锄刀;刻刀;刻刀器
cutting lay 切削层
cutting length 切削长度
cutting length of bucket chain 斗链着底距
cutting length of the first arrival 初至互切除长度
cutting life 刀具寿命
cutting lip 钻刃;钻口;钻唇;切削刃;切削刀
cutting liquid 切削液
cutting liquid wastewater containing oil 含油切削液废水
cutting list 截断表(钢筋);材料清单;木材规格表;下料表(钢筋或木材等);切削加工单;配筋表;下料清单;木料清单
cutting-list structural timber 按锯料单加工的木材
cutting loading machine 采矿康拜因
cutting loss 切削损失;切裁损耗
cutting lubricant 切削油;切削(润滑)液;切削冷却润滑液
cutting machine 块料切割机;下料压力机;钢筋切断机;裁切机;裁剪机;切纸机;切割器;切割机;挖沟机
cutting machine for corrugated sheet 波瓦切边机
cutting machine tool 切削机床
cutting management 采伐管理

cutting mark (浆纱的)分批墨印;裁剪画线
cutting method 切割方法
cutting milling 切削机
cutting moment of torque 切削扭矩
cutting mosaic 切割镶嵌
cutting needle 切割针
cutting nippers 剪丝钳;老虎钳;克丝钳;剪钳;电工手钳
cutting nozzle 割嘴
cutting of bars 钢筋切断
cutting of boring 岩芯钻孔的岩粉
cutting off 切开;切断;断路;断流
cutting offal 屠宰废弃物
cutting-off and centering machine 切断定心机
cutting-off and facing machine 切断车面机
cutting-off blade 切刀
cutting-off bushing 下料环
cutting-off by sheet pile wall 利用板桩墙断开
cutting-off lathe 截断车床;切断车床
cutting-off machine 切断机床
cutting-off male flower cluster 断蕾
cutting-off of pile 截桩
cutting-off of short-circuit 切断短路
cutting-off pile 截断桩;隔水板桩
cutting-off platform 裁板平台
cutting-off process 切断过程
cutting-off table 截砖台;切泥台
cutting-off tool 割刀
cutting-off tool rest 车刀架
cutting-off wheel 切断轮;切断砂轮
cutting of glass 划玻璃;雕刻玻璃
cutting of grooves 开挖沟;开挖槽
cutting of mud by gas 泥浆气侵
cutting of reinforcement 切割配筋;切割钢筋
cutting of stones 石料开采
cutting of trenches 开挖沟槽
cutting of wages 工资削减
cutting oil 切削液;切削(润滑)油
cutting oil additive 切削油添加剂
cutting oil separator 切削油分离器
cutting oil waste 切削油废水
cutting operation 剪切工序;切削操作
cutting optimizer 优选切割装置
cutting out 凿出;切开;切断;切除;割出;断路;断流;短路
cutting-out bar 扁头钎
cutting-out disease wood 剔除病木
cutting-out inward growing branches 修剪向里生长的枝条
cutting-out of bend 裁弯取直
cutting-out of rivets 切断铆钉
cutting-out press 冲切机;切断(压力)机
cutting output 切削量
cutting-out wire 切断用金属丝
cutting oxygen 切割氧
cutting pantograph 缩放刻度仪
cutting part 切割部分
cutting paste 切削用润滑冷却剂;切削润滑冷却剂
cutting pipe 切割管
cutting plan 采伐计划
cutting plane (制图学上的)破断面;割平面;切削平面;切割面;切断面
cutting plane algorithm 割平面算法
cutting-plane line 破裂线;破断面线
cutting plane method 割平面法
cutting plate 板材切割
cutting pliers 钳子;扁嘴钳;剪钳;割线钳;钢丝钳
cutting plow 切土犁
cutting point 切割点;刻点;刀口;切削点;刀刃
cutting-point test 刀刃试验
cutting position 掏槽位置
cutting profile 切削面
cutting prongs 切削牙齿
cutting propagation 插条繁殖;插条
cutting property 可切削性;切削能力
cutting punch 剪切冲头
cutting quality 切削能力
cutting range 修剪高度
cutting rate 切削率;刻纹速率;机械钻速;切削速度
cutting ratio 切屑厚度比;切削厚度比;切削比
cutting reducer 减薄液;等量减薄液
cutting resistance 抗切割性;抗切断强度;切削阻力;切削力;切削抗力
cutting ring 环刀;切断环

cutting ring method 切刀法
cutting ring sampler 环刀取样器
cutting room 裁切室
cutting rotor (松土拌和机的)粉碎砖筒;碎土转筒
cuttings 金属屑;切屑
cutting scar 剪切疤
cuttings chute 切屑接收器;收屑器
cuttings conveyer 煤粉输送器;切屑输送器
cutting scrap 切屑
cutting screed 切割用刮板;刮平用样板(混凝土);喷射混凝土整平刮板
cutting screen 切割筛
cutting section 采伐段;采伐点
cutting series 采伐列区
cutting session deed 割让证书(财产、权力等)
cutting shoe 桩管刃脚;刃脚;切土嘴;切土管头;(厚壁管前端的)切入靴
cutting shop 磨刻车间
cutting shore 刃脚
cutting shoulder 切口上缘;切口边
cutting side grafting 插腹接
cutting sieve 切割筛
cutting size 出刀尺寸;切削尺寸
cuttings laden mud fluid 携带大量岩屑的泥浆液;携带大量岩屑的泥浆流
cutting sleeve 卡套
cuttings lifting speed 岩屑上升速度
cuttings lifting velocity 岩屑升速
cutting slope 路堑边坡;挖方坡度;切削斜度
cuttings number 岩屑编号
cuttings of boring 钻屑
cutting solution 切削液
cutting specification 切削规范
cutting speed 机械钻速;切(削)速(度);切割速度
cuttings pick-up pipe 岩粉排出管
cuttings pick-up tube 岩粉排出管
cuttings pit 刀屑坑;切屑接收器;收屑器
cuttings returning rate 岩屑上返速度
cuttings sampling 岩粉采样;取岩屑
cuttings shape 岩屑形状
cuttings shoot 出屑槽;承屑盘;切屑坑
cuttings sliding down speed 岩屑下滑速度
cutting steam 切割气流
cuttings thickness 岩屑厚度
cutting stone 切裁刀
cutting stones 切削金刚石
cuttings transport ratio 岩屑携带比(率)
cutting strength 抗切割强度;切削强度
cutting strip 停机橡皮
cutting stroke 剪切行程;切削行程
cutting strokes per minute 每分钟切削行程
cutting stylus 刻针;刻纹刀
cutting surface 切削表面
cutting surface method 割曲面法
cuttings weight 岩屑比重
cutting system 锯切分级法
cutting table 切割桌;切割台
cutting tectonic 切割的构造层
cutting teeth 切齿
cutting test 切断检验
cutting thickness 切削厚度
cutting thread 车螺纹;分条经线
cutting through 刻纹过深;刻纹过度
cutting time of the first arrival 初至互切除时间
cutting tip 刻针;割尖;切削刀片;切削部分;切割喷嘴;气割嘴
cutting to length 按长度切割
cutting tolerance 切割公差
cutting tool 刻刀;刀具工具;刀具;刃具;切削工具;切削刀具
cutting-tool angle 刀具角度
cutting-tool coolant oil 切削工具冷却油
cutting-tool damage detection device 刀具损伤探测器
cutting-tool engineering 刀具技术
cutting-tool lubricant 刀具冷却润滑剂
cutting-tool steel 刀具钢;切削工具钢
cutting-tool wear 刀具磨损
cutting tooth 切削齿
cutting torch 割切焰;割炬;切割矩;切割(焰)吹管;切割炬
cutting to size 按尺寸切割
cutting trajectory 切削轨线
cutting tray 切屑盘

cutting tube 切气管
cutting type coring bit 切削型取芯钻头
cutting-type tunneling machine 切削式掘进机
cutting under water 水下开挖;水下挖土;水下切割(氢氧焰)
cutting unit 锯切单位
cutting-up 划切
cutting-up line 剪切作业线
cutting value 切削能力
cutting wallboard 切割墙板
cutting wheel 切削盘;切削轮;切割(砂)轮;切割刀轮
cutting whip 边斜削插条
cutting width 切削宽度
cutting window 开口(在套管上)
cutting wing 遮光叶片
cutting wire 割绳(砖工)
cutting wood 插条
cutting work 挖方工程;切削加工;开挖工程
cuttler 折布机
cuttling 折布;复原工序
cut to grade 挖到设计标高
cut-to-length 按长度截取;定长剪切
cut-to-length line 定尺剪切机组;带材定尺剪切作业线
cut-to-length sizes machine 定长切断机
cut to line 开挖到指定水平线;开挖到指定标高;挖到规定标高
cut-to-measure 按尺寸下料
cut-to-order 裁截
cut-to-tie ratio 断连比
cut transformation 割变换
cut tree 分割树
cut true and square 细心琢磨成圆;加工精确
cut tube 截断管头
cut up 船头破浪处
cut-up arrangement 分散布置;分配配置
cut-up test 切开试验
cut-up unit 剪切机组
cut value 切割容度
cut value of P-value data P值数据切除值
cut velvet 割绒
cut veneer 切制的镶面层板
cut voltage 割电压
cut washer 方形垫圈
cutwater 桥下防护架;尖船头;船首破浪材;分水角;分水尖;突堤端尖形披水;(突堤端的)尖形披水
cutwater bow 飞剪型船首;飞剪式船首
cut-widening 路基加宽;路堑加宽
cut width 底唇截面宽度;采宽
cutwork 雕绣;切削工作
cutwork embroidery 雕绣
cut-wound core 连续带绕铁芯
cut yards 开挖立方码
cuvette 小池;比色杯;透明小容器
cuzticite 黄碲铁石
C-washer 开口垫圈
C-wire 丙线
CWR[Continuous Welded Rail] 无缝线路;连续焊接轨
cyamelide 氰白
cyan 青色
cyanalcohol 氰醇
cyanaloc 氰基树脂(防水剂)
cyanamide 氨腈(类);氨基(化)氰;氰化氨;氰胺
cyanamide process 氰氨法
cyanate 氰酸酯;氰酸盐
cyanazine 氰乙酸肼
cyan brightness 青色明度
cyan group 氰基
cyanhydrin 氰醇
cyanic acid 氰酸
cyanicide 消除氰化物质
cyanidation 氰化法
cyanidation vat 氰化槽
cyanid-contamated plating solution 含氰电镀废水
cyanide 氰化物
cyanide attack bacteria 氰化物分解细菌
cyanide bath 氰化溶化槽
cyanide complex 氰化物络合物
cyanide-concentrated plating solution 浓氰化物电镀液
cyanide concentration 氰化物浓度

cyanide-containing waste 含氰废水
cyanide-containing wastewater 含氰化物废水
cyanide copper 氰化物电解铜;氰化物电镀铜
cyanide decomposition by ozone 氰化物的臭氧分解
cyanide deposit 氰化物淀积
cyanided steel 氰化钢
cyanide effluent treatment 氰化物污水处理;氰化物废水处理
cyanide elimination method 氰化物消除法;氰化物脱除法
cyanide gauze 氰化纱布
cyanide hardener 氰化物表面硬化剂
cyanide ion 氰化物离子;氰根离子
cyanideless electro-plating 无氰电镀
cyanide monitor 检氰装置;检氰仪器;氰化物监测仪
cyanide oxidation 氰化物氧化
cyanide plating wastewater 含氰化物电镀废水;氰化物电镀废水
cyanide poisoning 氰(化物)中毒
cyanide pollutant 氰化物污染物
cyanide pollution 氰化物污染
cyanide potassium 氰化钾
cyanide process 氰化(物)法;气体氰化法
cyanide pulp 氰化物矿泥
cyanide slime 氰化物微粒
cyanide waste water 含氰废水
cyanilide 苯胺腈
cyanin(e) 花青;喹啉蓝染料
cyanin(e) acid blue 花青酸性蓝
cyanin(e) blue 花青蓝;酞菁蓝颜料
cyanin(e) dyes 花青染料
cyanin(e) navy blue 花青海军蓝
cyanite 蓝晶石
cyanite biotite gneiss 蓝晶石黑云片麻岩
cyanite biotite quartz schist 蓝晶石黑云母石英片岩
cyanite-biotite schist 蓝晶石黑云母片岩
cyanite dimicaceous schist 蓝晶石二云母片岩
cyanite eclogite 蓝晶石榴辉岩
cyanite garnet potash-feldspar and plagioclase gneiss 蓝晶石榴二长片麻岩
cyanite garnet potash-feldspar gneiss 蓝晶石榴钾长片麻岩
cyanite muscovite gneiss 蓝晶石白云母片麻岩
cyanite muscovite quartz schist 蓝晶石白云母石英片岩
cyanite muscovite schist 蓝晶石白云母片岩
cyanite-quartz schist 蓝晶石石英片岩
cyanite schist 蓝晶石片岩
cyanite two mica gneiss 蓝晶石二云片麻岩
cyanite two mica quartz schist 蓝晶石二云母石英片岩
cyan layer 青色层
cyanme green 酞菁绿颜料
cyanoacetamide 氰基乙酰胺
cyanoacetic acid 氰基乙酸
cyanoacetylene 丙炔腈
cyanoacrylate 氰基丙烯酸酯;丙烯酸酯
cyanobenzene 苯甲腈
cyanocarbon 氰基烃
cyanochroite 钾蓝矾
cyanocobalamine 氰钴胺
cyanocomplex 氰配合物
cyanocrystallin 蓝晶质
cyanoethanoic acid 氰基乙酸
cyanoethyl cellulose 氰乙基纤维素
cyanoform 氰仿
cyanogas 放氰毒粉;氰钙粉
cyanogen 氰
cyanogen band 氰带
cyanogen compound 氰化合物
cyanogen-containing waste(substance) 含氰废物
cyanogen-containing waste water 含氰废水
cyanogenetic glycoside plant poisoning 含氰苷植物中毒
cyanogen fluoride 氰化氟
cyanogen iodide 氰化碘
cyanogens chloride 氯化氰
cyanoglycoside-containing plant poisoning 含氰苷植物中毒
cyano-group 氰基
cyanometer (测海洋或天空蓝度的)蓝度计;(测海洋或天空蓝度的)蓝度表;天蓝计
cyanometry 天色测量;天蓝计量

cyanophenyl 苯腈
cyanophoric 氰基的
Cyanophyceae 蓝藻纲
cyanophycin granule 藻青素颗粒
cyanophyllite 紫铜铝锑矿
Cyanophyta 蓝藻门
Cyanophyte 蓝藻植物
cyanoplatinate 铂氰化物
cyanopropyl 氰丙基
cyanopropyl silicone 氰丙基硅酮
cyanosensor 氰基传感器
cyanosite 五水硫酸铜
cyanotrichite 绒铜矾
cyanotype 蓝晒法;晒蓝图
cyanotype copy 蓝色印样
cyanuramide 蜜胺;氰脲酰胺
cyanurate 氰脲酸酯
cyanuric acid 氰脲酸;三聚氰酸
cyanuric amide 氰脲酰胺
cyanuric chloride 氰脲酰氯
cyanuric triamide 氰脲(三)酰胺
cyanurin 氰脲蓝
cyathozooid 不完全个员
cyberculture 电子计算机影响下的文化
cybernated society 计算机化社会
cybernation 计算机控制
cybernetician 控制论专家;控制论学者
cyberneti(ci)st 控制论专家;控制论学者
cybernetic model 控制论模型
cybernetic model of decision-making 决策的控制论模式
cybernetics 控制论
cybernetics culture 控制论优化
cybernetics system 控制论系统
Cyber service unit 斯伦贝谢公司数控测井仪
cyboma 集散微晶
cyborg 半机械人
cybotactic state 群聚态
Cycadales 苏铁纲(古植物)
cyc-arc welding 圆环电弧焊;自动电弧焊
cyclane 环烷
cyclanone 环烷酮
cyclator 凝聚沉淀装置
cycle 轮转;循环;自行车;周波;分批成本会计;使循环
cycle-amplitude relation 周幅关系
cycle annealing 循环退火
cycle billing 循环报表
cycle billing system 循环开单制
cyclecar 小汽车;小型机动车;三轮小汽车
cycle check 循环校验;周期检验
cycle climax 周期顶极
cycle compound 环化合物
cycle control 循环控制;定期控制
cycle coordinate method 坐标轮换法
cycle count 循环盘点;循环盘存;循环查点存货
cycle counter 转速表;周期计数器;周期计量器;周波表;循环计数器;频率计
cycle criterion 循环准则;循环判据;重复循环(总)次数
cycle crushing 闭路破碎
cycle defrost system 循环除霜系统
cycle delay selector 循环延迟选择器
cycle diagram 周期图
cycled interrupt 循环式中断;周期性中断
cycled precipitation during peak season 峰期旋回降雨量
cycled recondition 周期性检修
cycle duration 循环周期
cycle efficiency 循环效率
cycle epiphytotic 周期性植物流行病
cycle event 周期事件;周期过程
cycle failure 超饱和周期;失效周期
cycle fluid 循环工质
cycle for map revision 地图更新周期
cycle-free 变周期
cycle-free algorithm for network optimization 不定周期网络优化算法
cycle generator 交变频率发生器
cyclegraph technique 周期图示法
cycle index 循环指数;循环指标;循环次序;循环次数
cycle index counter 循环次数计数器
cycle in graph 图形中的回路
cycle in precipitation 降水周期

cycle inventory 周期存货;循环盘存
cycle length 掘进长度;循环时间;周期时间
cycle length change 周期时长变换
cycle length time 周期时长
cycle life 循环寿命
cycle limit 循环极限
cycle load(ing) 周期(性)加荷
cycle locking 周期同步;周期连锁
cyclelog (自动化的)程序控制器;程序调整器
cycle maintenance 周期维修
cycle matching 周波重合
cycle matrix 循环矩阵
cycle matroid 圈拟阵
cycle mobil-electrobath 循环流动式电解浴
cyclenes 环烯
cycle number 循环数
cycle of bending 挠曲周期
cycle of blasting 爆破循环
cycle of business operations 业务循环
cycle of carbon 碳循环
cycle of concentration 浓缩倍数
cycle of denudation 侵蚀循环;侵蚀旋回
cycle of deposition 沉积循环
cycle of detection 检查周期
cycle of development 发育循环
cycle of drawing-chamber 打炉周期
cycle of dry-and-wet years 干湿年分循环;多水少水年循环
cycle of engine 发动机循环
cycle of erosion 地貌循环;侵蚀循环;侵蚀旋回;侵蚀轮回
cycle of fluctuation 涨落周期;涨落循环;波动周期;起伏周期
cycle of flushing 涨落循环
cycle of fluvial erosion 河流侵蚀循环;河流冲积周期
cycle of freezing and thawing 冻融循环;冰融循环
cycle of fund movement 资金流动周期
cycle of groundwater circulation 地下水循环周期
cycle of infection 侵染循环
cycle of loading 加载周期;加荷周期
cycle of loads 荷载循环
cycle of magnetization 磁化循环
cycle of marine 海蚀旋回
cycle of marine erosion 海洋侵蚀循环;海蚀周期;海蚀循环
cycle of motion 运动循环
cycle of nitrogen 氮循环
cycle of operation 运行周期;工作循环;操作周期;操作循环;经营循环
cycle of oxygen 氧循环
cycle of regeneration 再生循环
cycle of repairs 修理周期
cycle of reproduction 再生产周期
cycle of river erosion 河蚀周期
cycle of sea current erosion 海流侵蚀循环
cycle of sedimentation 沉积循环;沉积旋回
cycle of stabilization 稳定循环
cycle of storage 蓄水周期;充水循环
cycle of stress 应力周期
cycle of stress reversal 应力交变周期;应力反复周期
cycle of sulfur 硫循环
cycle of the sun 太阳周
cycle of vibration 振动循环;振动周期
cycle operation 循环作业;周期操作
cycle output 周期生产率
cycle park 自行车停放处;摩托车停放处;两轮车停放处
cycle parking area 自行车停车场
cycle path 自行车(专用)道
cycle path edging 自行车道加边;三轮车车道加边;摩托车道加边;脚踏车道加边
cycle period 循环周期;操作周期
cycle per minute 周分
cycle per second 每秒周数;周/秒
cycle process 循环过程
cycle productivity 周期生产率
cycle program(me) 循环程序
cycle program(me) control 循环程序控制
cycle program(me) counter 循环程序计数器
cycler 循环控制装置;周期计
cycle-race cour 自行车比赛场
cycle race track 自行车跑道;自行车(比)赛路道

cycle racing track 自行车赛跑道
cycle range 循环范围
cycle rank 回路秩
cycle rate counter 周期计数器;周期计量器
cycle reset 循环计数器复原;循环复位
cycle road 自行车专用路
cycle room 脚踏车房
cycle route 自行车道
cycles 打击次数(桩)
cycle selection 周期选择
cycle-shared memory 周期共用存储器
cycle shift 循环移位
cycle skip(ping) 周波跳跃;跳周;跳步记录
cycle slip 滑周
cycles of concentration 浓缩倍数
cycles of decomposition 分解循环
cycles of stress reversal 反复应力周期;反复应力循环
cycle solution 循环解法
cycle specificity 循环特征
cycles per minute 每分钟周数;周/分
cycle(s) per second 每秒周数;每秒周期数
cycle stage 循环阶段
cycle stand 自行车架;自行车场;三轮车场;摩托车场;脚踏车场;脚踏车架
cycle stealing 周期性窃用
cycle steal mode 周期挪用方式
cycle stock 循环油料
cycles to failure 周期加荷至断裂;疲劳破坏的循环次数
cycle strain 周期应变
cycle stress ratio 循环应力比
cycle strip on carriageway 车行道上的自行车道
cycle system 环流系统
cycle tendency 循环趋势
cycle test 循环试验;往复试验;测试序列
cycle theory 循环学说
cycle time 换向时间;循环作业时间;循环时间;周期时间;周期
cycle time of excavator 挖掘机作业循环时间
cycle timer 循环计时器;循环定时器;周期替续器
cycle tire vulcanizing press 力车胎硫化机
cycle track 自行车小路;自行车道
cycle track crossing 自行车跑道交叉
cycle traffic 自行车交通
cycle turnover 周转的周期
cycle-type network 循环网路;循环式网络
cycle underpass 下穿式自行车道
cycle vector 圈向量
cycle water yield 旋回涌水量
cycleway 自行车道
cycleway edging 脚踏车道加边
cycleway network 自行车道网
cycleweld 合成树脂黏结剂;合成树脂结合剂
cyclic(al) 环的;循环的;周期的
cyclic(al) access 循环存取
cyclic(al) action 环化作用
cyclic(al) activated sludge process 循环式活性污泥法
cyclic(al) activated sludge reactor 循环式活性污泥反应器
cyclic(al) activated sludge system 循环式活性污泥系统
cyclic(al) activated sludge technology 循环式活性污泥法
cyclic(al) activation 循环活化
cyclic(al) A/D converter 循环数模/数转换器
cyclic(al) addition 环状加成;环化加成;成环加成
cyclic(al) addressing 循环寻址
cyclic(al) adjustment 循环变动调整
cyclic(al) annealing 周期退火
cyclic(al) average method 循环平均法
cyclic(al) base 环状碱
cyclic(al) bedding 旋回层理
cyclic(al) bedding structure 旋回层理构造
cyclic(al) binary code 循环二进制码
cyclic(al) bond 环内键
cyclic(al) buffer 循环缓冲器
cyclic(al) burner 循环燃烧器
cyclic(al) carousel system 循环轮盘系统
cyclic(al) carry 循环进位
cyclic(al) change 循环变化
cyclic(al) check 循环校验;周期检查
cyclic(al) checker 循环检验

cyclic(al) code 循环码
cyclic(al) compound 环状化合物;环行化合物
cyclic(al) compression 反复压缩
cyclic(al) connection 循环连通性
cyclic(al) constraint 循环约制;循环约束条件;循环约束
cyclic(al) control system 闭环控制系统
cyclic(al) coordinate 循环坐标
cyclic(al) criterion 循环判据
cyclic(al) current method 回路电流法
cyclic(al) curve 周期性曲线
cyclic(al) deformation 周期性变形
cyclic(al) degradation 周期衰化
cyclic(al) depletion 周期性消融(冰川);周期性耗竭;周期性退水;周期亏耗
cyclic(al) deposit 旋回沉积;周期性沉积
cyclic(al) deterministic model 循环决定模型
cyclic(al) deviation 循环离差
cyclic(al) dimer 环状二聚物
cyclic(al) diolefine 环二烯
cyclic(al) diolefin(e) resin 环二烯属树脂
cyclic(al) disease development 周期性发病
cyclic(al) disequilibrium 循环性支付差额
cyclic(al) docking 周期性进坞
cyclic(al) downturn 周期性下降
cyclic(al) drying 循环干燥
cyclic(al) duty 周期性工作
cyclic(al) economic recovery 周期性经济复苏
cyclic(al) economics crises 周期性经济危机
cyclic(al) effect 循环效率
cyclic(al) equation 循环方程
cyclic(al) error 循环误差;周期误差
cyclic(al) extension 反复伸长
cyclic(al) factor 循环因素
cyclic(al) fatigue 循环疲劳
cyclic(al) fatigue stress 周期疲劳应力
cyclic(al) feeding 周期性馈送;周期输送
cyclic(al) feed water heater 循环给水加热器
cyclic(al) field 循环场;周期场
cyclic(al) flow profile 周期流量图
cyclic(al) fluctuation 循环变动周期性波动;周期性波动;循环变动
cyclic(al) force 循环力;周期力
cyclic(al) fracture 旋回裂隙
cyclic(al) frequency 圆频率
cyclic(al) function 循环函数
cyclic(al) group 循环群
cyclic(al) heat test 周期加热试验
cyclic(al) hydrocarbon 环烃
cyclic(al) identity 循环恒等式
cyclic(al) index 循环指数;周期性指标
cyclic(al) indication 周期性循环指示数字
cyclic(al) indicator 环指标
cyclic(al) infection 循环性侵染;周期性感染;周期性传染
cyclic(al) inflation 周期性通货膨胀
cyclic(al) integrator 循环积分器
cyclic(al) interrupt 循环中断
cyclic(al) irregularity 周期不规则性
cyclic(al) iterative method 循环迭代法
cyclic(al) ketone resin 环酮树脂
cyclic(al) layering 旋回层理
cyclic(al) load-elution test 循环吸附—解吸试验
cyclic(al) load(ing) 循环加荷;周期(性)荷载;周期性负载;周期性加荷;反复荷载;交变荷载;交变负载
cyclic(al) loading treatment 反复负荷处理法
cyclic(al) load test 周期荷载试验
cyclically connected directed-graph 循环连通有向图
cyclically connected graph 循环连通图
cyclically perverse 经济周期反常现象
cyclic(al) magnetization 循环磁化
cyclic(al) matrix 循环矩阵
cyclic(al) measuring device 巡回检测装置
cyclic(al) memory 循环存储器
cyclic(al) method 循环法
cyclic(al) mining 循环开采
cyclic(al) mobility 循环流动性
cyclic(al) module 循环模
cyclic(al) motion 周期运动
cyclic(al) movement 循环变动;周期运动
cyclic(al) nick 循环裂点
cyclic(al) olefine 环烯

cyclic(al) olefinic bond 环内双键
cyclic(al) oligomer 环状齐聚物；环状低聚物
cyclic(al) operation 周期运行
cyclic(al) order 循环次序
cyclic(al) oscillation 循环性振动；周期性振动；周期性振荡；周期性波动
cyclic(al) oven 循环炉
cyclic(al) patterns 循环类型
cyclic(al) peak 循环波峰
cyclic(al) permeability 正常磁导率；周期磁导率
cyclic(al) permutation 循环置换；循环排列
cyclic(al) permuted code 单位间距代(码)；循环置换
cyclic(al) photophosphorylation 循环光合磷酸化作用
cyclic(al) pipe 循环管
cyclic(al) pitch control 周期变距操纵
cyclic(al) point group 循环点群
cyclic(al) polyolefine 环状结构聚烯
cyclic(al) pressure strength 周期压力强度
cyclic(al) process 循环过程
cyclic(al) pumping 循环抽取
cyclic(al) quaternary compound 环状四级取代化合物
cyclic(al) queues 循环排列
cyclic(al) queuing network 循环排队网络
cyclic(al) reaction 循环反应
cyclic(al) readout 循环读出；周期性读出；周期读出
cyclic(al) recovery 周期回升；周期复原
cyclic(al) redundancy character 循环冗余码符号
cyclic(al) redundancy check 循环冗余校验
cyclic(al) redundancy check code 循环冗余校验码
cyclic(al) regeneration 连续再生；循环再生
cyclic(al) representation 循环表示
cyclic(al) revision 定期修测
cyclic(al) river-bed change 周期性河床变化
cyclic(al) river-bed variation 周期性河床变化
cyclic(al) salt/fog expose 循环盐雾紫外线照射
cyclic(al) scanning system 循环检查制；循环检查系统
cyclic(al) sedimentation 旋回沉积作用；循环沉积作用
cyclic(al) selection 周期性选择
cyclic(al) sequence 商业循环序列；旋回层序
cyclic(al) shear stress ratio 循环剪应力比
cyclic(al) shear test 循环剪切试验
cyclic(al) shift 环形移位；循环移位
cyclic(al) simple shear test 循环单剪试验；周期单剪试验
cyclic(al) space 循环空间
cyclic(al) sterility 循环不稳性
cyclic(al) stock 敏感性股票
cyclic(al) storage 循环蓄水；循环存储器；循环储存；周期蓄水；周期库容
cyclic(al) storage access 循环存储器存取
cyclic(al) strain 循环应变
cyclic(al) strain softening 循环应变软化；周期应变软化
cyclic(al) stress 往复应力；循环应力；旋转冲击应力；周期应力
cyclic(al) stria 旋回条纹
cyclic(al) substituent 环式取代基
cyclic(al) supply 循环给水
cyclic(al) supply pipe 循环给水管
cyclic(al) surface 圆纹曲面
cyclic(al) surge 周期性涌波
cyclic(al) swing 周期性波动
cyclic(al) swings in demand 需求的周期性波动
cyclic(al) system 周期系(统)
cyclic(al) tangential loading 周期性切向加荷
cyclic(al) temperature change 周期性温度变化
cyclic(al) tension 反复张力
cyclic(al) term 周期项
cyclic(al) terrace 旋回阶地
cyclic(al) test 循环试验；重复测验
cyclic(al) thermomechanical system 循环式热力系统
cyclic(al) three-dimensional stretching 反复三向拉伸
cyclic(al) time 循环时间
cyclic(al) torsional shear test 循环扭剪试验
cyclic(al) train 周转轮系

cyclic(al) transfer 循环传送
cyclic(al) transformation 循环变换
cyclic(al) trend 循环趋势；周期性趋向
cyclic(al) triaxial test 循环三轴试验；周期加载三轴试验；周期加荷三轴试验；动力三轴试验
cyclic(al) tripolymer 环状三聚物
cyclic(al) twin 轮式双晶
cyclic(al) unemployment 周期性失业
cyclic(al) unit 中间负荷机组
cyclic(al) upturn and downturn 周期性上升和下降
cyclic(al) variable 循环变量
cyclic(al) variation 周期(性)变化；周期性变动
cyclic(al) variation of stresses 应力周期变化
cyclic(al) voltammetric sludge 循环式伏安污泥
cyclic(al) voltammetry 循环伏安法
cyclic(al) wetting 循环加湿
cyclic(al) wetting and drying 循环干湿法
cyclicamine group 环胺族
cyclicity 周期性
cyclics 环状化合物
cyclide 四次圆纹曲面
cycling 循环的；循环操作；周期工作的；反应堆功率循环
cycling-car 小型机动车
cycling circuit 循环电路
cycling life test 循环寿命试验；闪烁寿命试验
cycling load 循环负荷；周期性变负荷
cycling period 循环周期
cycling plant 循环装置
cycling pool 循环库
cycling process 循环过程
cycling service 周期性运行
cycling solenoid valve 周期工作的电磁阀；周期电磁阀
cycling stadium 自行车赛场
cycling test 循环负荷试验
cycling time 循环时间；周期时间
cycling tube 自行车隧道
cycling way 慢车道；非机动车道
cycling zone of siphon 倒虹吸管循环带
cyclis stress 交变应力
cyclist traffic 自行车交通
cyclitol 环多醇；环醇(类)
cyclization 环化作用；成环(作用)
cyclization process 环化过程
cyclizative condensation 环化缩合
cyclized fibre 环化纤维
cyclized rubber 环化橡胶
cyclized rubber resin 环化橡胶树脂
cyclizing agent 环化剂
cyclo 出租机动三轮车；三轮出租汽车
cyclo-addition 环化加成
cycloaddition reaction 环化加成反应
cycloalcohol 环式醇
cycloaliphatic epoxy resin 脂环族环氧树脂
cyclo alkane 环烷
cycloalkanes 环烷类
cycloal kanoate 环烷醇盐
cycloalkanoates 环烷金属化合物
cycloalkylation 环烷化；成环烷化作用
cycloamination 环胺化；成环胺化作用
cyclobutadiene 环丁二烯
cyclobutane 环丁烯；环丁烷
cyclobutane-carboxylic acid 环丁烷羧酸
cyclobutanol 环丁醇
cyclobutanone 环丁(烷)酮
cyclobutyl 环丁基
cyclobuty lene 环丁烯
cyclobutyrol 环丁酸醇
cyclo-cell 旋流浮选机
cyclocompound 环状化合物
cycloconverter 循环换流器
cyclocytic of stomata 轮裂气孔
cyclodehydration 环化脱水作用
cyclodos 发送电子转换开关
cyclogenesis 气旋作用；气旋生成；气旋发生
cyclogram 周期图
cyclograph 轮转全景照相机；金属硬度测量仪；画圆器；圆弧规；测定金属硬度的电子仪器；涡流式感应图示仪
cyclohexadiene 环己二烯
cyclohexane-carboxylic acid 环己烷羧酸
cyclohexane dimethanol succinate 丁二酸环己烷二甲醇酯

cyclohexanediol 环己二醇
cyclohexane sulfamic acid 环己烷氨基磺酸
cyclohexan-hexol 肌醇；环己六醇
cyclohexanone-formaldehyde resin 环己酮甲醛树脂
cyclohexanone industry wastewater 环己酮工业废水
cyclohexanone oxime 环己酮肟
cyclohexanone peroxide 过氧化环己酮
cyclohexanone resin 环己酮树脂
cyclohexene 环己烯
cycloheximide 放线菌酮
cyclohexitol 环己六醇类化合物
cyclohexlmide 环己亚胺
cyclohexyl 环己基
cyclohexyl acetate 乙酸环己酯
cyclohexylbenzene 苯基环己烷
cyclohexyl bromide 环己基溴
cyclohexyl chloride 环己基氯
cyclohexyl hydroperoxide 环己基过氧化氢
cyclohexyl iodide 环己基碘
cyclohexyl lactate 乳酸环己酯
cyclohexyl methacrylate 甲基丙烯酸环己酯
cyclohexyl oxalate 草酸环己酯
cyclohexyl phthalate 苯二甲酸环己酯
cyclohexyl searate 硬脂酸环己酯
cycloid 旋轮线；圆滚线；摆线；圈状的
cycloidal arch 圆形拱；圆滚线拱；摆线拱
cycloidal blade propeller 摆线叶片推进器
cycloidal blower 摆旋鼓风机
cycloidal cam 摆线凸轮
cycloidal chop 摆弧限制板
cycloidal curve 摆线曲线
cycloidal cylinder 圆形柱；摆线柱面；摆线圆柱
cycloidal deflection 成摆线的偏转
cycloidal gas meter 摆线式气表
cycloidal gear 摆线齿轮
cycloidal gear grinding machine 摆线齿轮磨齿机
cycloidal gear hob 摆线齿轮滚刀
cycloidal gearing 摆线啮合
cycloidal gear reducing motor 摆线齿轮减速电动机
cycloidal gear teeth 摆线轮齿
cycloidal helical motion 摆线螺旋运动
cycloidal mass spectrometer 摆线质谱仪
cycloidal motion 摆线运动
cycloidal needle wheel type motor 摆线针轮电动机
cycloidal path 摆线轨迹
cycloidal pendulum 圆滚摆；摆线摆
cycloidal pinwheel reduction box 摆线针轮减速箱
cycloidal planetary gear speed reducer 行星摆线针轮减速机；摆式行星齿轮减速机
cycloidal plate 圆形板
cycloidal propeller 直翼推进器；摆线型螺旋桨；平旋推进器
cycloidal propeller ship 平旋推进器船
cycloidal pump 摆旋泵；摆线泵
cycloidal teeth 摆线齿
cycloidal toothing 摆线啮合
cycloidal type rotary index units 摆线式回转分度台
cycloidal wave 圆形波；摆线波
cycloid gear grinding machine 摆线磨齿机
cycloid gear hydraulic motor 摆线齿轮油液压马达
cycloid knob action 锁执手摆线作用
cycloid motion 摆线运动
cycloid reduction gear 摆线减速齿轮
cycloid scale 圈鳞
cycloid system gear 摆线齿轮
cycloid tooth 摆线齿
cycloid wheel tooth 摆线轮齿
cycloinverter 双向离子变频器
cyclol 环醇
cyclolysis 气旋消失；气旋消除
cyclom 跳字计数器
cyclomatic number 圈数
cyclometer 示数仪表；测圆弧器；路码表；路程计；里程计；里程表；记转器；自行车计程仪；转数计；周期计；测圆弧器；跳字计数表
cyclometer counter 跳字计数器；数字显示式计量仪器
cyclometry 测圆法
cyclomorphosis 周期变形

cyclone 环酮;旋流器;旋风(器);旋风分离器;低气压区;除泥器;气旋
cyclone air separator 旋风式选粉机
cyclone blocking 旋风筒堵塞
cyclone breather 旋流通气器
cyclone burner 旋流式燃烧器;旋风炉
cyclone cellar 地下避风室;防风地下室;防风地窖;避风窖
cyclone chamber 旋风室
cyclone clarifier 回旋澄清器;旋液澄清器;旋流净化器;旋流分级器;旋流澄清器
cyclone classifier 旋风分选器;旋风分级器;水力旋流分离器
cyclone cluster 旋流器组
cyclone collector 降棉筒;旋风收集器;气旋集沉器;旋风集尘器;离心除尘器
cyclone combustor 旋风炉
cyclone course 气旋移动路径方向
cyclone dust catcher 旋风除尘器
cyclone dust collector 旋风吸尘器;旋风集尘器;旋风聚尘器;旋风(式)除尘器
cyclone dust extractor 旋风(式)除尘器
cyclone dust separator 旋风分尘器;气旋除沉器
cyclone filter 旋风(过)滤器;气旋过滤器
cyclone front 气旋前锋
cyclone furnace 旋风炉;气旋炉
cyclone gas washer 旋流式气体洗涤器
cyclone granular bed filter 旋风式颗粒层集尘器
cyclone grit washer 旋流洗砂器
cyclone hydraulic separator 旋流分离器
cyclone mill 旋风磨碎机
cyclone mist eliminator 旋风除雾器
cyclone of dynamic(al) origin 动力气旋
cyclone path 气旋路径
cyclone precipitation 旋风除尘;气旋性降雨
cyclone preheater 旋风预热器
cyclone preheater kiln 旋风预热器窖
cyclone pump 旋流泵
cyclone rain 气旋性降雨
cyclone receiver 旋风式收集器;旋风式集料器;旋风式除尘器
cyclone(re)circulating air separator 旋风式循环空气选粉机
cyclone-recovery skimmer 旋流回收式撇油器
cyclone scrubber 旋风洗涤器;旋风集尘器;旋涤气器;旋风除尘器
cyclone separating device 旋风分离器
cyclone separation 回旋分离器;漩涡分离;旋流分离
cyclone separator 旋(流)分(级)器;旋流分级机;旋风(式)分离器;旋风除尘器
cyclone settler 旋流收尘器;旋流沉淀器;旋风分离器
cyclone shaft preheater kiln 旋风立筒预热器窖
cyclone smelting method 漩涡熔炼法
cyclone spreader 旋风式撒布器
cyclone steam separator 旋风汽水分离器
cyclone suspension preheater 旋风式悬浮预热器
cyclone thickener 浓缩旋流器;旋风增稠器
cyclone tower 预热器塔
cyclone track 气旋路径
cyclone tube 旋风输送管;旋风分离管
cyclone-type after-condenser 涡流式二次冷凝器
cyclone-type air separator 旋风式选粉机
cyclone-type collector 回旋式集尘机;旋风式集尘机
cyclone-type electrostatic spraying apparatus 旋风式静电喷涂装置
cyclone washer 旋风洗衣机;旋风洗涤器
cyclone water-film scrubber 旋风水膜除尘器
cyclonic 气旋族
cyclonic center 气旋中心
cyclonic circulation 气旋式环流
cyclonic collector 气旋式集尘器
cyclonic coolant separator-inertial 惯性旋流冷却油分离器
cyclonic curvature 旋风曲度;气旋曲度
cyclonic droplet collector 旋风微滴收集器
cyclonic front 低气压前锋;气旋前锋
cyclonic model 气旋式模型
cyclonic motion 气旋式运动
cyclonic precipitation 低气压性降雨;气旋降水转
cyclonic rain 气旋(性)雨
cyclonic spray scrubber 旋流喷射洗涤器;旋风喷淋洗涤器;中心喷雾旋风除尘器
cyclonic spray tower 旋风喷雾塔;旋风喷淋塔
cyclonic storm 气旋性风暴;气旋性暴雨;气旋性暴风

cyclonic storm surge 气旋性风暴潮
cyclonic swirl 气漩涡流
cyclonic thunderstorm 气旋雷暴
cyclonic vorticity 气漩涡度
cyclonic wave 气旋波
cyclonic whirl 旋风形涡流
cyclonic wind 气旋风
cyclonite 黑索金炸药
cyclo-nitrifying filter 循环硝化滤池;环流硝化滤池
cyclonocope 气旋中心指示盘
cyclonograph 气旋中心指示盘
cyclonome 旋转式扫描器
cyclononanic acid 环壬烷酸
cyclonoscope 飓风中心测定仪
cyclo-octadiene 环辛二烯
cyclooctane 环辛烷
cycloolefine hydrocarbon 环烯烃
cycloparaffin 环烷
cycloparafine hydrocarbon 环烷烃
cyclopean 圆形城墙(古希腊);巨石堆积的;乱石堆;蛮石堆;蛮石
cyclopean aggregate 蛮石集料;蛮石骨料;大块石集料;大块石骨料
cyclopean block 巨型毛石方块;巨石块;大块石
cyclopean concrete 块石混凝土;毛石混凝土;蛮石混凝土;埋石混凝土;大块石混凝土
cyclopean masonry 毛石砌体;巨石工程;毛石圬工;蛮石圬工;蛮石砌体
cyclopean masonry dam 浆砌蛮石坝;浆砌块石坝;蛮石圬工坝
cyclopean masonry work 蛮石工程;乱石工程;巨石工程
cyclopean riprap 乱石堆;乱石层;大块石抛石体
cyclopean rubble masonry 毛石圬工;毛石砌体;块石砌体
cyclopean rustication 蛮石贴面(工程)
cyclopean-tadienized oil 环戊二烯改性油
cyclopean wall 乱石墙;蛮石墙
cyclopentadiene 环戊二烯
cyclopentadienyl 环戊二烯基
cyclopentadienyl anion 茂基阴离子
cyclopentadienyl potassium 环戊二烯合钾
cyclopentane 环戊烷
cyclopentane-carboxylic acid 环戊烷羧酸
cyclopentanol 环戊醇
cyclopentanone 环戊酮
cyclopentene 环戊烯
cyclopentenone 环戊烯酮
cyclopentenyl 环戊烯基
cyclopentyl 环戊基
cyclopentyl alcohol 环戊醇
cyclophon 电子注开关管
cyclophone 旋调管
cyclophoria 旋转隐斜视
cyclophosphamide 环磷酰胺
cyclopian 蛮石
cyclopite 钙长石
cyclopneumatic separator 旋风分离器
cyclopoly condensation 环构缩聚(作用)
cyclopolymer 环状聚合物
cyclopolymerization 环化聚合(作用);环构聚合(作用)
cyclopolyolefin 环聚烯烃;环多烯烃
cyclopropane 环丙烷
cyclopropane-carboxylic acid 环丙烷羧酸
cyclopropanyl 环丙烯基
cyclopropene 环丙烯
cyclopropenyl radical 环丙烯基
cyclopropyl 环丙基
cyclorama 弧形天幕;硬天幕;半圆形透视背景;舞台背后弧形背景幕布
cyclorectifier 循环整流器
cyclo reducer 行星针轮减速机
cyclorubber 环化橡胶
cycloscope 转速仪;转速计;转数计;里程计
cyclosilicate 聚合硅酸盐;环硅酸盐
cyclosizer 超微粒湿式旋流分级器
cyclosteel process 循环式直接炼钢法
cyclostrophic 旋衡;气转
cyclostrophic flow 旋衡流
cyclostrophic wind 旋衡风;旋风
cyclostyle 系列柱式;环形柱廊
cyclosymmetric function 循环对称函数
cyclotetraphosphate 环四磷酸盐

cyclothem 节奏层;旋回层【地】;韵律层
cyclothermic sedimentation 旋回沉积作用
cyclotomic 分圆
cyclotomic equation 割圆方程
cyclotomic polynomial 分圆多项式
cyclotomy 割圆法;分圆法;割圆术
cyclotrimethylene trinitramine 环三次甲基三硝基胺;旋风炸药
cyclotron 回转加速度;回旋加速器
cyclotron absorption 回旋加速吸收
cyclotron damping 回旋加速阻尼
cyclotron emission 回旋加速发射
cyclotron frequency 回转频率;回旋(加速)频率
cyclotron harmonic wave 回旋加速谐波
cyclotronic mass spectrometer 回旋质谱仪
cyclotronic resonance mass spectrometer 回旋共振质谱仪;回旋共振质谱计
cyclotron instability 回旋加速不稳定性;回旋不稳定性
cyclotron magnet 回旋加速磁铁
cyclotron maser 回旋微波激射器;回旋脉塞
cyclotron radiation 回旋(加速)辐射
cyclotron resonance 回旋共振
cyclotron resonance heating 回旋共振加热
cyclotron scattering 回旋散射
cyclotron turnover 回旋反转
cyclotron wave tube 回旋波管
cycloversion desulfurization 固定床矾土催化脱硫法
cycloversion reforming process 固定床矾土催化重整过程
cycloweld injection method 环氧树脂铸模法
cyclowollastonite 环硅灰石
cyconvin 聚氯乙烯—ABS掺混料
cyhexatin 环己锡;氢氧化三环己锡
cylic(al) 环状的
cylinder 镜筒;加热储水罐;唧筒;内径(数据库用);锡林滚筒;圆柱(状)体;圆柱;圆筒;印辊;压力水柜;桩基钢筒;柱体;柱面;辊筒;钢瓶;缸;磁道柱面;储气瓶;查找区域;发动机汽缸;气瓶;汽缸(排量);筒体
cylinder action 汽缸作用
cylinder actuator 油缸促动器
cylinder address 磁道柱面地址
cylinder alignment 缸的找正
cylinder and valve for power chuck 动力卡盘用缸及阀
cylinder area 缸面积
cylinder armature 鼓形电枢
cylinder arrangement 油缸布置;汽缸布置
cylinder assembly 液压缸总成;液压杆组件
cylinder axis 圆柱轴
cylinder back cover 汽缸底
cylinder baffle 汽缸外壳;汽缸导流片
cylinder bar (脱粒机的)滚筒夹条
cylinder barrel 汽缸筒
cylinder base 机缸座;汽缸座
cylinder base flange 汽缸座凸缘
cylinder battery 气瓶组
cylinder beater 逐稿轮
cylinder bit 圆柱钻
cylinder block 机体;汽缸组;汽缸排;汽缸座;汽缸体
cylinder block cracking 汽缸裂缝
cylinder block oil hole 汽缸组机油孔
cylinder body 筒形物;缸体;汽缸体
cylinder body bushing boring-machine 缸体轴瓦镗床
cylinder bolt 缸盖螺栓
cylinder bolting flange 汽缸螺栓凸缘
cylinder bore 缸(内)径;汽缸内径
cylinder bore ga(u)ge 汽缸内径规
cylinder borer 镗缸机
cylinder boring and honing machine 汽缸镗磨床
cylinder boring bar 汽缸镗杆
cylinder boring machine 汽缸镗床;镗缸机
cylinder bottom 缸底;汽缸底
cylinder bottom head 缸底盖;汽缸底盖
cylinder bottom steam pipe 缸底汽管;汽缸底汽管
cylinder bracket 汽缸托架
cylinder brushing machine 圆筒刷布机
cylinder bush 汽缸套;汽缸衬
cylinder bushing 汽缸衬筒
cylinder cage 板条式滚筒
cylinder caisson 沉井;开口沉箱;圆柱形沉箱;圆

筒形沉箱;圆筒沉箱;管柱
cylinder cam 筒形凸轮
cylinder cap 汽缸盖;汽缸盖
cylinder capacity 汽缸(工作)容量;汽缸工作容积
cylinder capacity modulation 汽缸容量调节
cylinder cap gadket 汽缸盖垫密片
cylinder carrier hay loader 捡拾滚筒式装干草机
cylinder casing 汽缸套
cylinder chamber 汽缸内腔
cylinder chopper 滚筒式切碎机
cylinder clearance 汽缸间隙
cylinder clothing 汽缸外罩
cylinder cock 汽缸通汽旋塞
cylinder column 管柱
cylinder complete 汽缸全套
cylinder complete unit 汽缸全套
cylinder compression strength 圆柱体抗压强度
cylinder compression tester 汽缸压缩试验器
cylinder compressive strength 圆柱体(试件)抗压强度(混凝土)
cylinder compressometer 圆柱压缩仪
cylinder concave clearance 滚筒凹板间隙
cylinder console typewriter 柱形控制台打字机
cylinder core 圆柱体岩心
cylinder corn sheller 滚筒式玉米脱粒机
cylinder cover 汽缸盖
cylinder crushing strength 圆柱强度;圆柱体压碎强度;圆柱体试件抗压强度
cylinder cushion 缸的缓冲装置
cylinder cut (隧道爆破的)圆筒式掘孔;柱面开挖;桶形掏槽
cylinder cutoff 脱粒滚筒阻稿板
cylinder cut-out 在汽缸里不爆发
cylinder cutter 滚筒刀式切碎机
cylinder cutter type chaff 滚筒刀式茎杆切碎机
cylinder cutting head 滚筒式切刀
cylinder cutting mechanism 滚筒式切碎装置
cylinder data 圆柱面数据
cylinder deck 汽缸体上部
cylinder diameter 缸径;汽缸内径;汽缸直径
cylinder discharge valve 汽缸排气阀
cylinder disk plow 垂直圆盘犁
cylinder displacement 汽缸排(气)量;汽缸(工作)容积
cylinder displacement volume 汽缸排量容积
cylinder double layer sand siftering machine 滚筒式双层筛砂机
cylinder drainage 汽缸放水口
cylinder drainage cock 缸排放塞门
cylinder drainage receiver 集汽水活门
cylinder drain cock 汽缸放水旋塞
cylinder drawback lock 弹子门锁
cylinder drawing process 圆筒拉制法(平板玻璃)
cylinder dryer 转筒烘干机;转筒干燥器;干燥机器;筒式干燥机
cylinder drive 滚筒的传动装置
cylinder drying 筒式干燥
cylinder end 缸头;汽缸头;汽缸底
cylinder end cap 缸端盖
cylinder end cover 缸端盖
cylinder end seal 缸头密封;汽缸头密封
cylinder escapement 工字轮式擒纵机构;筒形擒纵机构
cylinder expansion 汽缸膨胀
cylinder fault 缺柱面
cylinder feed beater 滚筒喂入轮
cylinder feed pipe (针形阀的)控制管;圆筒给水管;汽缸泄水管
cylinder fitting 汽缸附件
cylinder flat bed machine 凸版平台印刷机
cylinder foot 汽缸脚猫爪
cylinder(forming) machine 圆网成型机
cylinder frame 汽缸座
cylinder frame fit 汽缸座架配合
cylinder front cover 缸前盖
cylinder gas 储气瓶中的压缩气体;钢瓶气体
cylinder gasket 汽缸垫
cylinder-gate (木轮机的)圆筒闸门;圆柱形门
cylinder-gate intake 圆筒闸门进水口
cylinder ga(u)ge 圆柱体量规;缸径规;汽缸(内径)量规
cylinder glass 圆筒法平板玻璃;筒状玻璃量器;筒形玻璃
cylinder governor 脱粒滚筒调速器

cylinder grader 筒式种子精选机
cylinder grain 圆柱形火药柱
cylinder grate 筛式凹板
cylinder grinder 汽缸磨床
cylinder grinding machine 汽缸磨床
cylinder grinding wheel 筒形砂轮
cylinder hay loader 滚筒式装干草机
cylinder head 汽缸体盖;汽缸盖
cylinder head anchor pin 缸盖固定螺钉;汽缸盖固定螺钉
cylinder head bolt 缸盖螺栓;汽缸盖螺栓
cylinder head bolt and nut 缸盖的螺栓及螺母
cylinder head cap 汽缸前盖
cylinder head casing 汽缸盖罩
cylinder head clamp 汽缸头压板
cylinder head cover 汽缸(头)罩;缸盖罩;缸前盖
cylinder head end 汽缸盖端部
cylinder head fixed stud 汽缸盖固定螺栓
cylinder head fixing stud 缸盖固定螺栓
cylinder head gasket 汽缸垫片;汽缸床;汽缸衬垫
cylinder head jack nut 汽缸盖起卸螺母
cylinder head lock 圆筒销子锁
cylinder head oil inlet 汽缸盖进油口
cylinder head outlet connection 汽缸头出水接头
cylinder head plate 缸盖底板
cylinder head plug 汽缸盖塞
cylinder head pocket 汽缸盖凹顶
cylinder head stud 汽缸盖柱螺栓
cylinder head water jacket plug 汽缸盖水套塞
cylinder head water nozzle 缸盖喷水嘴;汽缸盖喷水嘴
cylinder head with antechamber 带预燃室的缸盖
cylinder hone 汽缸珩磨头
cylinder hone grinder 汽缸珩磨机
cylinder in-block 汽缸组
cylinder inner diameter 缸内径
cylinder inside diameter 缸内径
cylinder iron 缸用铸铁
cylinder jacket 缸套;汽缸(夹)套
cylinder jaw 轴环
cylinder jetty 圆墩(式)突式码头
cylinder key 圆柱形钥匙
cylinder knife 滚筒式切碎机刀片
cylinder knob 筒形把手
cylinder lagging 汽缸保温套
cylinder lapping machine 汽缸研磨机
cylinder latch 弹簧锁;弹子锁
cylinder lathe 汽缸车床
cylinder lawn mower attachment 滚筒式草坪割草装置
cylinder lid 汽缸盖
cylinder liner 汽缸衬筒;汽缸(衬)套;汽缸衬(垫)
cylinder liner diameter 汽缸衬套内径
cylinder liner packing ring 缸衬胀圈;汽缸压衬圈
cylinder liner puller 汽缸套拉出器
cylinder liner service tool 汽缸套修理工具
cylinder lining 汽缸套
cylinder lock 圆筒销子锁
cylinder lubrication cock 汽缸润滑旋塞
cylinder manifold 集流管;汇流排
cylinder method 圆柱注水泥法
cylinder microtome 圆柱型切片机
cylinder mill 圆筒碾磨机;管磨机
cylinder mixer 筒形混合机
cylinder mortice lock 圆柱形插锁
cylinder night latch 圆筒式弹簧碰锁;圆筒碰锁
cylinder number 环数;柱面数;柱面编号;圈数
cylinder of concrete 混凝土圆柱体(试件)
cylinder offset 汽缸偏置
cylinder oil 发动机用润滑油;汽缸润滑油
cylinder oiling 汽缸加油法
cylinder oil pipe 汽缸油管
cylinder oil pressure detector 缸体油压探测器
cylinder open caisson 筒形沉井
cylinder packing ring 缸套密封环;汽缸套密封环
cylinder paddle spreader 叶轮式撒布机
cylinder passage 汽缸通道
cylinder permeameter 圆形渗透计
cylinder pick-up 滚筒式捡拾器
cylinder pier 圆筒形码头;圆筒墩;圆柱形(桥)墩基
cylinder pile 圆柱(形)桩;圆筒桩;管柱
cylinder pile foundation 管桩基础
cylinder pile wharf 管桩码头
cylinder plug 汽缸塞

cylinder plug ga(u)ge 缸径塞规
cylinder port 汽缸通道;汽缸口
cylinder port block 中立油缸口关闭
cylinder post 管柱
cylinder press 滚筒印刷机
cylinder pressure 汽缸压力
cylinder pressure ga(u)ge 汽缸压力表
cylinder printing-machine 滚筒印刷机
cylinder process 吹筒法;平板玻璃拉管生产法
cylinder-processing machine 滚筒处理机
cylinder pulse 同位标脉冲
cylinder rack 气瓶集装架
cylinder rake 滚筒式搂草机
cylinder rake bar loader 滚筒搂耙式装干草机
cylinder ratio 汽缸容积比
cylinder reboring machine 汽缸镗床;镗缸机
cylinder reel rake 滚筒式搂草机
cylinder regulator 氧气筒调节器
cylinder relief cock 汽缸减压旋塞
cylinder rim night latch 圆筒式弹簧碰锁
cylinder ring 弹子锁芯垫圈;汽缸垫
cylinder roaster 筒式焙烧炉
cylinder roasting 回转窑内煅烧
cylinder rod 活塞杆
cylinder rod end 活塞杆端头
cylinder saddle 鞍形汽缸座
cylinder sampler 圆柱体试样
cylinder sand cooling machine 滚筒式型砂冷却机
cylinder sander 筒形打磨机;转鼓打磨机
cylinder sanding machine 筒形铺砂机
cylinder saw 筒形锯;桶形锯;圆筒锯;圆孔锯
cylinders castin-pair 铸成成对的汽缸
cylinder scoring 拉缸
cylinder screen 圆网
cylinder screw 弹子锁芯螺丝
cylinder seal 汽缸密封装置;汽缸密封垫圈
cylinder seat 汽缸座
cylinder segment 圆桶桶壁块
cylinder segment roof 柱段形屋顶
cylinder setting plate 压力水缸固定板
cylinder shaft 脱粒滚筒轴
cylinder shaping machine 汽缸刨床
cylinder sheller 滚筒式玉米脱粒机
cylinder shield 圆筒形挡(水)板
cylinder side cover 缸体侧罩
cylinder sieve 圆筒筛
cylinders in line 直排汽缸
cylinder sizing machine 烘筒式浆砂机
cylinder slasher 烘筒式浆砂机
cylinder sleeve 缸套;汽缸套(筒)
cylinder sorter 圆筒式分级机
cylinder specimen 圆柱形(混凝土)试件;圆柱体试件(混凝土)
cylinder splitting test 圆柱体(试件)开裂试验
cylinder spool 间距套管
cylinder stock 汽缸油(原料)
cylinder strength 圆柱体(抗压)强度
cylinder strength of concrete 混凝土圆柱体强度
cylinder stroke 汽缸冲程
cylinder stuffing box 汽缸盘锥根盒
cylinder suction valve 汽缸进气阀
cylinder support 圆柱体支座;圆柱体支承
cylinder support yoke 圆柱形支承轭状物;圆柱形支承轭架
cylinder surge chamber 圆筒式调压室
cylinder system 筒式热水采暖系统
cylinder teaseling machine 滚筒式起绒机
cylinder temperature indicator 汽缸温度指示器
cylinder test 圆柱形试件抗压试验(混凝土);圆柱体试验;圆柱体(试块)试验;圆柱(抗压强度)试验
cylinder tester 圆柱体(试件)试验机(混凝土)
cylinder testing machine 圆柱体(试件)试验机(混凝土)
cylinder theory 圆筒理论(拱坝设计)
cylinder tipping device 翻瓶机
cylinder top head 缸上盖
cylinder top steam pipe 缸上汽管
cylinder truck 钢瓶运输车
cylinder twist 圆柱扭曲
cylinder type absorber 伸缩筒型避震器
cylinder-type chaff-cutter 滚筒式铡草机
cylinder-type crusher 圆柱式碎石机
cylinder-type cutter 滚筒刀式切碎机

cylinder-type cutterhead 滚筒式切碎器
cylinder type jetty 圆筒式丁坝;圆筒式码头;圆筒式丁堤;沉井式码头
cylinder typemagnetic thin film memory 圆筒形薄膜存储器
cylinder type of lock valve 筒式闸阀
cylinder-type pipe 筒形管
cylinder-type shredder 滚筒式茎杆切碎器
cylinder-type side delivery rake 滚筒式侧向搂草机
cylinder under casing 汽缸底箱
cylinder vacuum cleaner 筒式吸尘器
cylinder valve 圆筒(气瓶)阀;汽缸阀;气瓶阀
cylinder volume 汽缸容积
cylinder wall 汽缸壁
cylinder wall lamp 圆筒形壁灯
cylinder wall temperature 汽缸壁温度
cylinder wall thickness 汽缸壁厚度
cylinder water inlet connection 汽缸进水接头
cylinder wear 印版滚筒磨损;汽缸磨损
cylinder weir 圆筒堰;筒形堰
cylinder well 液压缸座套(装于升降机井坑底部,用于安装液压缸)
cylinder wharf 管柱码头;墩座码头;筒柱式码头
cylinder wheel 管形砂轮
cylinder wheel with dovetail groove 带槽筒形砂轮
cylinder with differential effect 差动缸
cylinder wrench 圆筒扳手;管子扳手;握管器
cylindric(al) 圆柱(体)的;圆筒形的;柱状的;柱面的;圆柱形的
cylindric(al) arch 圆筒形拱
cylindric(al) arch dam 圆柱拱坝;圆筒形拱坝;圆筒拱坝
cylindric(al) armature 鼓形电枢
cylindric(al) array 圆筒形天线阵
cylindric(al) attachment 柱面加工夹具
cylindric(al) auger 筒形钻
cylindric(al) ball mill 圆筒球磨机
cylindrical bar 圆柱形钢筋
cylindric(al) barrage 圆筒式闸坝;圆筒形坝;滚筒式闸坝;滚筒式坝闸;带有圆筒形闸门的活动坝
cylindrical batch mill 滚筒形间歇式磨碎机;滚筒形间歇式磨坊
cylindric(al) bearing 弧形支座;圆筒轴承;滚柱轴承
cylindrical bending 筒状弯曲
cylindric(al) bit 圆柱形钻头
cylindric(al) block (预应力混凝土的)锚碇块;(预应力混凝土的)锚碇体系块
cylindric(al) bloom 大圆钢坯
cylindric(al) boiler 罐形锅炉;筒形锅炉
cylindric(al) boring 镗圆筒孔;镗汽缸内孔
cylindric(al) bow 圆柱形船头
cylindric(al) building 圆柱形建筑(物);圆形建筑(物)
cylindric(al) build package 圆柱形卷装
cylindric(al) buoy 圆筒形浮筒;圆筒形浮标;罐形浮标
cylindric(al) caisson 圆柱形沉箱;圆筒形沉箱
cylindrical campanile 圆柱形钟楼
cylindric(al) cast 圆柱状管型
cylindric(al) ceiling 筒形顶篷;筒形顶棚
cylindric(al) centerless-type grinder 外圆无心磨床
cylindric(al) center-type grinder 外圆中心磨床
cylindric(al) chart 圆柱形(投影)海图
cylindrical chimney 圆柱形烟囱
cylindric(al) coefficient 棱形系数;圆柱形系数
cylindric(al) cofferdam 圆柱形围堰;圆筒形围堰
cylindric(al) column 筒柱
cylindric(al) combustion chamber 圆柱形燃烧室
cylindric(al) compact 柱状坯块
cylindric(al) compressive strength 圆柱体抗压强度
cylindrical concrete cofferdam 环形混凝土围堰
cylindrical concrete column 圆混凝土柱;混凝土圆柱
cylindrical concrete manway 圆柱形混凝土人行巷道
cylindrical concrete pipe 混凝土圆管
cylindric(al) concrete shell 筒形混凝土(薄)壳
cylindrical concrete shell roof 筒形混凝土薄壳屋顶
cylindric(al) concrete test specimen 圆柱形混凝土试样;圆柱形混凝土试件
cylindric(al) condenser 筒形冷凝器
cylindric(al) coordinates 柱坐标;柱面坐标;圆柱坐标

cylindric(al) coordinates robot 圆柱坐标机器人
cylindric(al) coordinate system 圆柱坐标系(统);柱面坐标系
cylindrical core 圆柱形型芯
cylindric(al) coupling 筒形联轴节
cylindric(al) cross staff 圆柱直角器
cylindrical cup wheel 圆筒杯形砂轮
cylindric(al) cut 圆柱形陶槽
cylindric(al) cutting unit 滚筒式切碎装置
cylindric(al) domain 圆柱域
cylindrical donjon 圆柱形城堡主塔
cylindric(al) dryer 圆筒形干燥器;筒形干燥机;筒式烘干机
cylindric(al) drill 圆柱形钻头;柱钻;筒形钻;套料钻
cylindric(al) drum 圆柱形鼓筒
cylindric(al) drum hoist 圆筒卷筒绞车
cylindric(al) elevator 滚筒式升运器
cylindric(al) equal-area projection 圆柱等积投影;等积圆柱投影
cylindric(al) equal-spaced projection 等距(离)圆柱投影
cylindric(al) equidistant projection 等距(离)圆柱投影
cylindric(al) equilibrium 柱体平衡
cylindric(al) equivalent projection 等积圆柱投影
cylindric(al) face 圆柱面荧光屏
cylindric(al) fault 圆柱状断层;圆柱断层;圆柱面断层
cylindric(al) fit 筒形配合
cylindric(al) floating diametrically loaded fender 圆柱形浮动式(径向受力)护舷
cylindric(al) floating fender 圆筒浮动式护舷
cylindric(al) fold 圆柱形褶皱
cylindric(al) function 柱(面)函数;圆柱函数
cylindric(al) furnace 圆筒炉;筒式炉
cylindrical gas holder 圆筒形储气罐
cylindric(al) gate 圆柱形门
cylindric(al) ga(u)ge 缸径规
cylindrical glass tower 圆柱形玻璃塔
cylindric(al) granary 圆筒式粮舱
cylindric(al) grate 辊式炉算
cylindric(al) grate ball mill 格子式圆筒形球磨机
cylindric(al) gravel filter 筒式卵石过滤器
cylindric(al) grinder 外圆磨床
cylindric(al) grinding 外圆磨削
cylindric(al) grinding machine 外圆磨床
cylindrical grinding point 圆柱形磨碎凿尖
cylindric(al) ground type bunker 落地式圆筒(料仓)
cylindric(al) harmonic function 圆柱谐函数
cylindric(al) harmonics 圆柱调和函数;柱谐函数;贝塞尔函数
cylindric(al) heat exchanger 圆筒式热交换器
cylindric(al) helical spring 柱形螺旋弹簧
cylindric(al) helix 圆柱螺旋线;柱面螺旋线
cylindrical hinge 圆柱形铰链
cylindrical hip(ped end) 柱状斜脊屋顶端部
cylindric(al) holder 圆筒形储气罐
cylindric(al) hopper door 圆柱泥门
cylindric(al) hydrophone 圆筒式水下检波器;筒式海洋检波器
cylindric(al) illuminance 柱面照度
cylindric(al) insulation 管形保温
cylindric(al) insulator 筒形绝缘套
cylindrical interlocking pipe 镶嵌式圆管;链锁式圆管
cylindric(al) intersecting vault 筒形交叉穹顶
cylindricality 柱面性
cylindric(al) jack 圆柱形千斤顶;圆筒形千斤顶;圆筒式千斤顶
cylindric(al) jaw 柱面量爪
cylindric(al) journal 圆柱轴颈
cylindrical keep 圆柱形城堡主塔(中世纪);要塞
cylindric(al) ladle 鼓形浇包
cylindric(al) lapping 外圆研磨
cylindric(al) leg type 支柱
cylindric(al) lens 圆柱(形透)镜;柱面透镜
cylindric(al) lens raster 柱面透镜光栅板
cylindric(al) lenticulation 圆柱形光栅
cylindric(al) level 管状水准器
cylindric(al) lock 圆柱形锁;圆筒门形锁
cylindric(al) luminance 柱面照度
cylindrically balanced valve 柱形平衡阀
cylindrically based body 圆柱尾物体
cylindrically curved glass 单曲面玻璃

cylindric(al) magnetic thin film memory 管状磁膜存储器
cylindrical manhole 圆柱形人孔
cylindrical manway 圆柱形人行道巷道
cylindric(al) map projection 圆柱投影
cylindric(al) mill 圆筒碾磨机;圆筒磨;管磨
cylindric(al) milling cutter with coarse teeth 粗齿圆柱形铣刀
cylindric(al) mixer 圆筒形搅拌机
cylindrical mixing pan 圆筒形拌和盘
cylindrical mould 柱状模
cylindric(al) nodulizer 成球筒
cylindrical nut 圆柱形螺母
cylindrical object 圆柱体
cylindric(al) open caisson foundation 井筒基础
cylindric(al) passive screen 圆筒形被动筛
cylindric(al) perspective 圆柱透视(法)
cylindric(al) photoreactor 柱面光反应器
cylindrical piend 圆柱形屋脊
cylindric(al) pier 圆柱形桥墩;圆柱(形)墩;筒形桥墩
cylindric(al) pile 圆柱(形)桩;圆柱;筒(形)桩
cylindric(al) pin 筒形销
cylindrical pipe 圆筒管
cylindrical plate-steel roller 圆柱形钢板滚柱
cylindric(al) plug 缸径规
cylindric(al) plug ga(u)ge 圆柱测径规
cylindric(al) polar coordinates 柱面极坐标
cylindric(al) polisher 柱面磨光器
cylindric(al) pond 圆筒形水池
cylindric(al) post 圆柱
cylindric(al) press 滚筒式压呢机
cylindrical prestressed concrete shell 预应力钢筋混凝土
cylindric(al) projection 圆柱投影
cylindric(al) reamer 圆柱铰刀
cylindric(al) reducer 正齿轮减速器
cylindrical reinforced concrete shell 圆筒形钢筋混凝土薄壳
cylindrical reinforced shell 钢筋混凝土圆筒形薄壳
cylindrical reinforcement bar 圆钢筋
cylindrical reinforcement rod 圆钢筋
cylindric(al) roaster 回转窑
cylindric(al) rock-powder collector 筒形岩粉积聚管
cylindrical rod 圆棒
cylindrical rod web 圆杆网件;圆杆构成的网状构件
cylindric(al) roller 滚柱(测量齿厚或节圆直径用)
cylindric(al) roller bearing 短圆柱滚子轴承;圆柱滚动轴承;滚柱轴承
cylindric(al) roller bearing with long rollers 长滚柱轴承
cylindric(al) roller thrust bearing 圆柱形滚柱推力轴承
cylindric(al) rotary kiln 回转窑
cylindric(al) rotor 鼓形转子
cylindrical-rotor machine 非突磁极机
cylindric(al) rubber fender 圆柱形橡胶碰垫;圆柱形橡胶护舷;圆柱形橡胶防撞装置;圆筒形橡胶护舷
cylindric(al) rubber fender ring 圆柱形橡胶防撞圈
cylindric(al) sample 圆柱体试样
cylindrical screen 圆筒转动筛;圆筒筛;圆筒状屏
cylindric(al) screen feeder 笼形供料器;圆筒喂料机
cylindric(al) screening surface 筒形筛面
cylindric(al) section 筒形段
cylindric(al) sedimentation tank 圆(筒)形沉淀池
cylindric(al) settler 圆柱形沉降槽
cylindric(al) shaft 管柱
cylindric(al) shaft foundation 管柱基础
cylindric(al) shell 圆柱(形薄)壳;圆筒形壳;柱状壳体;筒形薄壳;筒壳
cylindric(al) shell roof 筒壳屋顶
cylindrical shell sandwich panel 筒状薄壳用三夹板
cylindric(al) shell section 筒节
cylindric(al) shield 筒状盾构
cylindric(al) sieve 圆(筒)形格筛;圆筒筛
cylindrical silo 筒仓
cylindric(al) skirt support 圆筒形裙座
cylindric(al) slide (土坡的)圆筒形滑动
cylindric(al) sluice gate 圆筒形(泄水)闸门
cylindric(al) solenoid 筒形螺线管
cylindric(al) space 柱形空间
cylindric(al) specimen 圆柱形(混凝土)试件;圆

柱试件
cylindric(al) specimen strength 圆柱体试件强度
cylindric(al) spiral 螺旋线；圆柱螺线
cylindric(al) spring 柱形弹簧
cylindric(al) spur gear 圆柱正齿轮
cylindrical steel column 圆钢柱
cylindric(al) stereographic(al) projection 极射赤面圆柱投影
cylindric(al) structure 滚筒状构造；圆筒结构
cylindric(al) surface 圆柱面；柱(形曲)面
cylindric(al) surface(d) screen 柱面屏
cylindric(al) symmetry 圆柱形对称
cylindric(al) tank 圆筒形水箱；圆筒形水池；圆柱形立罐；圆筒形箱
cylindric(al) taper drum 圆锥颈桶
cylindrical thread 圆柱形螺纹
cylindrical three-layers panel 筒状三夹板
cylindric(al) tile 筒瓦
cylindric(al) transverse projection 横圆柱投影
cylindric(al) trommel 滚筒筛
cylindrical tube 圆管
cylindric(al) turning 外圆车削
cylindric(al) type conveyer 圆筒形封闭式胶带运输机
cylindric(al) type electrolytic condenser 管形电解电容器
cylindric(al) type filter 圆筒形滤器
cylindric(al) valve 柱形阀；筒形阀
cylindric(al) vault 圆柱穹隆；半圆拱(顶)；筒形穹隆；筒拱
cylindric(al) vent 圆柱状孔道；圆柱形孔道
cylindric(al) vessel 圆柱形容器；圆筒形容器
cylindric(al) vibration plate 圆形振动板
cylindric(al) vortex 圆筒漩涡；柱状漩涡；柱形涡流
cylindric(al) vortex sheet 圆柱形涡流层
cylindric(al) washer 洗涤筒；圆筒形洗涤器
cylindrical washing screen 圆筒形冲洗筛
cylindric(al) wave 柱面波；柱波；柱波面
cylindric(al) wharf 筒柱式码头
cylindric(al) wheel 筒形轮
cylindric(al) wire 圆线(材)
cylindric(al) worm gears 筒形蜗轮
cylindricity 圆柱度；柱面性
cylindricizing 对称比

cylindriod 椭圆筒
cylindrite 圆柱锡矿
cylindroconic(al) 圆锥形的
cylindroconic(al) ball mill 圆锥形球磨机
cylindroconic(al) drum 柱锥混合式卷筒
cylindro-cylindric 高、跨不等的筒形体交叉的
cylindroid 拟圆柱面；圆柱状体；柱形面；正椭圆柱；椭圆柱
cylindrometer 柱径计
cylindro-spheric groin 高、跨不等球体与筒形交叉拱
cylix 高脚双柄大酒杯
cylohexane 环己烷
cylpeb 圆柱棒(粉碎材料用)；圆柱(钢)棒；小型碾一磨钢段；钢段
cylpebs grinding media 钢段研磨介质
cyma 浪纹线脚；反曲线(饰)
cyma downstream slab 下游反向曲线石板(保护桥墩用)
cyma inversa 里反曲线
cyma recta 上凹下凸的波状花边；混枭线脚；枭混线【建】；表反曲线【建】
cyma reversa 枭混线脚；上凸下凹的波状花边；里反曲线；反混线【建】
cymaric acid 磁麻酸
cymatium 拱顶花边；反曲线状；古典建筑柱顶装饰线脚；拱顶花边
cymatium recta 陶立克式波状花边的
cymatium reversa 爱奥尼亚式波状花边
cyma weir 反弧形堰
cymbia 圆嵌条；圆抹角；圆楞条；(绕柱身的)线脚条；绕柱身的线脚条
cymbiform 船形的
cymbiform leaf 船形叶
cymene 甲基异丙基苯；花烃；对异丙基甲苯
cymogene 粗丁烷
cymograph 线脚测绘器；轮廓线仪；反曲线仪；记录器；自记波频计；转筒记录器；自记波长计
cymometer 波频计；波长计；频率仪
cymomotive force 波动势
cymoscope 检波器；振荡指示器
cymrite 铝硅钡石；钡铝沸石
cynaosite 胆矾
cynarine 朝鲜蓟酸
cynoper 合成硫化汞

Cynthia Suberba Ritter 赤海鞘
C-yoke squeezer 钳式压铆机
cypenamine 苯环戊胺
cypher 押码；暗号
cyphering 译成密码
cypraea macula 紫贻贝
cypress 柏树；柏木
cypress oil 柏树油
cypress-wood column 柏木柱子
Cyprian green 地绿色
Cyprids 金星幼虫
cyprus umber 绿褐色棕；生赭石色
cyrilovite 水磷铁钠石
cyrostyle 圆形凸出的门廊
cyrtoceracone 弓角石壳
cyrtolite 曲晶石
cyrtometer 圆量尺；曲面测量计；曲度计
cyrtometry 曲面测量法
cyrtostyle 圆形凸出的门廊；柱子排列成向外突出的弧形的回廊
Cyrus's tomb 塞鲁士墓(波斯国王)
cystathionine 丙氨酸丁氨酸硫醚
cysteamine 半胱胺
cysteine 半胱氨酸；半光氨酸
cysteine hydrochloride 半胱氨酸盐酸盐
cysteine sulfinic acid 半胱亚磺酸
cysteinyl 半胱氨酰
cysteinyl glycine 半胱氨酰甘氨酸
cysteinylglycine sodium iodide 半胱氨酰甘氨酸碘化钠
cystohepatic triangle 卡洛氏三角
cystolith 钟乳体
cystolon 人造碳化硅(研磨用)
cytac 罗兰 C 导航系统
cytaster 星体
cytoplasmic filament 细质丝
cytoskeletal filament 支架微丝
Czapek-Dox solution 察—多二氏溶液
Czapek's agar 察氏琼脂
Czechoslovakia architecture 捷克斯洛伐克建筑
Czechoslovakia flexible pavement design method 捷克柔性路面设计法
C-Z loading 中—活载

D

dab 涂擦;轻拍;轻敷
dabbed finish 细凿琢面;细剁斧面
dabbed mortar 砂浆饼块
dabber 涂清漆刷(圆端头);硬毛刷;半球状软毛刷;涂漆刷
dabbing 琢凿石面;黏结砂岩;圬工勾缝;圬工冲孔;(有凉台的)高平房;敷擦石面;灰泥抛毛;用锤修琢;石面凿毛
dabble 溅湿;润湿;喷洒
dabble in stocks 经营部分股票
dabbling 灌注的
dable speculation 小试投机
da capo 重复信号;重发信号
Dacca cotton 达卡棉(孟加拉国)
Dacca silk 达卡无捻绣花丝线
dachiardite 环晶石
Dachprism 达赫棱镜
Dacian (stage) 达斯阶【地】
dacite 英安岩;石英质中长石
dacite liparite 英安流纹岩
dacite porphyry 英安斑岩
dacite tuff 英安凝灰岩
dacitic tuff 英安质凝灰岩
dacitic tuff lava 英安质凝灰熔岩
dacitic welded agglomerate 英安质熔结集块岩
dacitic welded breccia 英安质熔结角砾岩
dacitic welded tuff 英安质熔结凝灰岩
dacker 停滞空气
da copo 重发信号
Dacrolon 达克罗纶(美国)
Dacron 涤纶
dactylite 指形晶
dactylitic 指形晶状(的)
dactylolysis 指脱落
dactylopore 指孔
daddock 烂木头
dado 护墙板;护壁(板);柱墩;裙板;墙裙;台度;嵌槽接头
dado base 墙裙下部;基座底部;墙角
dado cap(ping) 墙裙压顶线;护壁板压顶线;护壁板压顶线;墙裙帽;柱头
dado frame 台度框架;护壁板架
dado framing 墙裙板框;护壁板
dado head 开槽刀具;开槽头;开槽工具;组合刀头
dado head machine 开槽刀具;开槽机
dadoing 刨槽
dado joint 嵌槽接头;护壁板接缝;企口连接;墙裙缝;垂直凹槽接缝
dado mo(u)lding 台基上部线脚;墙腰板线条(上部装饰);护壁板线脚;护壁木条(饰);墙裙线脚;护墙板压条
dado plane 开平底槽刨;开槽刨;沟槽刨;平槽刨;平底槽刨
dado rail 护壁靠椅栏;护壁(板顶)木条;装在墙裙上的木(轨)条;墙裙木条
dado tile 护壁贴砖
dadsonite 达硫锑铅矿
daedal(ian) 迷宫式的
Daedalus 代达罗斯(希腊神话中的建筑师和雕刻师)
Daelen mill 戴伦轧机
daemon 守护神
daffodil(e) 黄水仙;水仙花(色的)
daffodil(e) yellow 水仙黄色;镉黄色;淡镉黄颜料
dag 石墨粉
daga 高原洼地
dagger 剑形符;斜向构件;止滑木;短剑;十字形构件
dagger board 活动披水板(小船用);中插板【船】
dagger fiber 达格尔叶纤维
dagger knee 斜向肘材;十字时材
dagger mark 钢叉记
dagger shore 撑柱
dagoba 舍利子塔(佛教的);印度佛骨堂
dagobas 印度早期坟墓
Daguan glaze 大观釉
Dagu datum 大沽零点
Dagu glaciation 大沽冰期
Dagu-Lushan interglacial stage 大沽—庐山间冰期【地】
Dahlgren polar monitoring service 达尔格连地极监测服务
dahlia 大丽花;天竺牡丹
dahllite 碳酸磷灰石
Dahl tube 达尔测流管
Dahua ware 德化瓷
dahurian larch 落叶松(木);黄花松
daia (沙漠中的)小洼地
Daido's planetary mill 大通式单辊行星轧机
daiflon 戴氟隆树脂;聚三氟氯乙烯树脂
daily 每日的;日常的
daily aberration 周日光行差
daily account 日计表
daily accounting 日记账
daily accounting report 会计日报表
daily advance 日掘进(量);日进展;日进尺(钻探)
daily air temperature cycle 气温日循环
daily allowance 每日津贴;日津贴
daily amount 日量
daily amplitude 日振幅;日较差;日变幅
daily analysis 日常分析
daily and shift traffic plans 调度日班计划
daily and shift traffic working plan 日班计划
daily audit 日常审计
daily average 日;日平均(数)
daily average maximum allowable concentration 日平均最高容许浓度
daily average maximum concentration 日平均最高浓度
daily average maximum permissible concentration 日平均最高允许浓度
daily average number of car unloading 日均装车数
daily average permissible concentration 日平均容许浓度
daily average precipitation 日平均降水量
daily average rainfall 日平均降雨量
daily average traffic 日平均交通量
daily average value 日平均值
daily balance 每日余额;日结算余额
daily balance method 每日余额计算法
daily boring report 钻探日报表;钻探记录
daily burning cycle 日燃烧周期
daily capacity 每日生产量;日生产量
daily car requisition plan 日要车计划
daily cash balance book 现金余额日记账
daily cash ratio 每日现金需要比率
daily cash report 现金日报
daily change 日变化
daily change of air pollution 空气污染日变化
daily change of SO$_2$ concentration 二氧化硫浓度日变化
daily charter 按日租船
daily check 每日检查
daily climate chart 每日天气图
daily commuter 每天月票乘客;每天通勤乘客
daily commuting sphere 日产上班范围
daily compensation 日补偿
daily compensation period 日补偿期
daily construction reports 施工日报表
daily consumption 每日消费量;每日耗用量;日(消)耗量
daily course 日变程
daily cover 逐日覆盖
daily crude capacity 日原料量
daily cycle 昼夜周期
daily demand 日需量
daily difference of solid tides 固体潮的日差值
daily directional maximum sectional passenger volume 全日单向最高断面流量
daily discharge 日流量
daily diversity power 日参差功率
daily dose 一日量;日剂量
daily drainage 日排水量
daily dredged yardage 日挖方量
daily drilling progress 日进尺(钻探)
daily drilling report 钻探日报表;钻探记录;日报(钻探)
daily dry weather flow 日污水量
daily duration curve 日持续曲线
daily earning 日工资
daily environmental load 日环境负荷
daily evapo(u)ration discharge 日蒸发量
daily expenses 每日开支;每日费用
daily extremes 日极值
daily flood peak 日洪峰(流量)
daily flow 每日流量;日(径)流量;日产量
daily flow capacity 每日输送能力
daily flow volume 每日输送量
daily fluctuation 日涨落;日变化;日变程
daily fluctuation of water-level 水位的日变幅
daily fluctuation in soil temperature 土温的日变化
daily footage 日进尺
daily force report 劳动力日报表
daily forecast 日预报
daily gain 日增重
daily gain rate 日增重率
daily gains 每日获利
daily groundwater regime 地下水日动态
daily high tide 日高潮位
daily hours of operation 每日工作小时数
daily ice supply capacity per berth 泊位日供冰能力
daily increase 日增加量
daily increasing rate of chlorine content 氯离子含量日增率
daily inequality of tides 日潮不等
daily information 每日情报
daily inspection 例行测试;经常检查;每日检查;日常检验;日常检修;日(常)检查
daily instalment 按日摊付;每日分期付款
daily intelligence digest 每日情报摘要
daily interest 每日利息;日息
daily keying element 日变键控单元
daily kilometerage of locomotive 机车日公里
daily labo(u)r report 人工日报单
daily life 日常生活
daily limit of price changes 每日价位变动限幅;每日价格变动限幅
daily living sphere 日常生活范围
daily load 昼夜负载;日负载;日负荷
daily load curve 日负载曲线;日负荷曲线
daily load diagram 昼夜负荷图
daily load factor 日荷载率;日负载系数;日负荷率;日负荷率
daily load fluctuation 日负荷变动
daily load predication 日负荷预测
daily locking water volume 日过闸耗水量
daily logging 流水账记录
daily logging printer 流水账打印机
daily loss 日损失
daily magnetic change 周日磁变
daily maintenance 每日保养;日常维修;日常保养
daily maintenance task 每日保养工作
daily maintenance work 日常维护
daily making 每日产量
daily mass loading 每日总量负荷
daily materials and spoilage report 材料废料日报单;材料废料日报表
daily maximum 日最大量;日最大(值)
daily maximum air temperature 日最高气温
daily mean 逐日平均;按天平均
daily mean air temperature 日平均气温
daily mean discharge 日平均流量
daily mean flow 日平均流量
daily means 按日平均
daily mean sea level 日平均海平面
daily mean temperature 日平均温度
daily mean value 日均值
daily memorandum 每日备忘录
daily minimum 日最小量;日最小(值)
daily minimum air temperature 日最低气温
daily necessaries[necessities] 日用(必需)品;生活必需品
daily needs 日常需要

daily operating cost 每日营运成本
daily operating repair 日常检修
daily output 每日产量;日排出量;日出料量;日出力;日产量
daily output per serviceable car 运用车日产量
daily output ratio 出料率
daily paid staff 按日领工资的工作人员
daily parity 当日平价
daily passenger transport plan of station 车站旅客输送日计划
daily passenger volume 全日客运量
daily pay 计日工资;日(给)工资
daily peak 日洪峰;日高峰
daily peaking operation 日内调峰运行(水电站);日峰荷调节
daily peak load 日峰荷
daily per capita consumption 每人每日消耗量;每人每日耗(水)量
daily plan 日计划
daily plan of passenger transportation 旅客运输日常计划
daily pondage 日调节(库容);水池的日调节容量
daily porcelain ware 日用陶瓷
daily precipitation 日降雨量;日降水量
daily premium 每日贴水;每日升水;每日保(险)费;日保费
daily price limit 每日价格限幅;日价限幅
daily procelain ware 日用瓷器
daily process management 日常工艺管理
daily product 日产量
daily production 日产量
daily progress 日进度
daily progression rate 日增长率;日进度率
daily pursuits 日常事务
daily rainfall 每日降雨量;每日降水量;日(降)雨量;日降水量
daily range 逐日差程;日量程;日较差;日范围;日差程;日变幅
daily rate 日速;日差率
daily rate of flow 日流量
daily rate of pay 日工资率
daily receipts 日(常)收入
daily record 日志
daily record of construction 施工日志;施工日记
daily regime 生活制度
daily regulating reservoir 日调节水库
daily regulation 日调节
daily regulation reservoir 日调节水库
daily relay 每日换班
daily rental rate 日租率
daily report 日报单;日报(表)
daily report of cash income and expenses 现金收支日报
daily retardation of tides 潮汐推迟时间;日推迟时间;潮迟
daily return 日报
daily rotation 每日转动
daily routine 例行工作;日常业务;日常工作
daily run 每日运转
daily running plan 全天开行计划
daily runoff 日径流量
daily sales 日营业额;日销售额
daily sample 逐日取样
daily sea level 一日内测潮所得的海平面
daily service 日用泵
daily service report 每日业务报告
daily service tank 给水柜;常用油柜;日用水柜;日用水舱
daily sheet 每日工作记录;日报表
daily shipping report 航运日报
daily special drawing rights rate 特别提款权的逐日汇率
daily statement 每日账表;日报表
daily statement of account 每日账表;每日对账单
daily status 每日状况
daily stock of freight car 货车保有量
daily stock of serviceable cars 运用车日保有量
daily storage 日储藏的;日调节水库
daily storage plant 日调节电站
daily subsistence allowance 每日生活津贴
daily supplies 日常供给
daily supply 日常供应
daily supply tank 给水柜;常用油柜;日用水柜;日常供应罐

daily tally report 理货日报表
daily technical service fee 日技术服务费
daily temperature range 气温日较差
daily thermal wave 周日热波动
daily throughput 日输送量
daily tidal cycle 周日潮汐循环
daily time card 每日工作时间登记卡;日工作时间登记卡
daily time report 工作时间日报
daily totalized flow volume 日总计水流量
daily total passenger volume 日客运总量
daily total sales value 每日销售总值
daily traffic density 日交通密度
daily traffic pattern 交通量日变化图;日交通量变化图
daily traffic volume 每日交通量;日交通量
daily traffic working plan 运输工作日计划
daily transaction reporting 每日处理报告;日常处理报告
daily trial balance 日计表;试算表
daily tunnel footage 隧道开挖日进尺
daily turnover 日周转量
daily type of tide 日潮型
daily unbalance coefficient of cargo handled at the port 日吞吐量不平衡系数
daily unbalance factor 日不均系数
daily variation 每日变动;周日变化;日(内)变动;日变化;日变程
daily variation coefficient 每日变化系数;每日变动系数;日变化系数;日变动系数
daily variation factor 日变化系数;交通量日变化系数
daily variation of surface temperature 地表温度的日变化
daily vehicle-kilometer 车日行程
daily volume 日交通量
daily volume variation 交通量日变化图
daily wage 日工资;计日工资
daily wage rate 日工资率
daily wagon requisition plan 日要车计划
daily wants 日常需要
daily water consumption 每日用水量;日用水量
daily water consumption per capita 每人每日用水量
daily water demand 日需水量
daily water flow 日流量
daily wave 周日波动
daily weather chart 每天天气图;每日天气图
daily weather map 每天天气图
daily work 每日作业
daily working capacity 日生产能力;日产量
daily work load 每日工作量;日工作量
daily work report 每日工作报告单;工作日报单
daily yardage 日掘土方量;逐日掘土方量
daily yield 日产量
dairi 日本天皇住所
dairy 牛乳场;牛奶铺;牛奶房;牛奶场;制酪场
Dairy bronze 戴利(黄)铜
dairy building 牛奶场建筑;奶酪生产车间
dairy cattle 奶牛
dairy farm 奶场;制酪场;乳牛场
dairy-farm slurry 牛奶场排污;制酪场排污;奶牛场淤泥
dairy husbandry 乳酪业
dairy industry 奶品工业;制酪工业;乳品业
dairy industry wastewater 制酪工业废水
dairying 奶品制造业
dairy manure 制酪场肥料
dairy plant waste 奶品厂废物;制酪厂废物
dairy processing factory 乳酪加工厂
dairy product 乳产品
dairy stock 奶牛
dairy wastewater 牛奶场废水;制酪废水
dais 王座;高座;高背椅;法座;讲台;演坛
daisy chain 菊(花)链
daisy chain bus 菊瓣式链接总线
daisy chained priority mechanism 菊花全连式优先权机构
daisy chian 雏菊花环
daisy Mae 澳大利亚测角计
daisy wheel 菊(花)轮
Daitophor 大东福尔荧光增白剂
Daiwabo 大和纺粘胶短纤维(日)
Daiwabo Polyno 大和纺玻利诺高湿模粘胶纤维

(日)
dak bungalow (印度驿站的)旅舍
Dalamar yellow 达拉马黄颜料
dalan 有屋盖的阳台;(波斯或印度建筑中的)走廊
Dalby's theorem 达尔比定理
dale 小谷;溪谷;峪;装饰板材;排水管(道);水槽;山谷
Dalejan (stage) 达列杰阶【地】
d'Alembert's characteristic 达朗贝尔性质
d'Alembert's principle 达朗贝尔原理
Dale water-locating instrument 代尔水位探测仪
Dalian chih 大连纸
dallage 大理石地面;大理石层面
dallan 阳台(有屋盖的波斯、印度建筑)
dalle 装饰板
dalles 峡谷急流
Dall tube 多尔管
dallying receiver 空转接收机
Dalmatian wet-tability 多尔玛型润湿性
dalmeny basalt 徽斑橄玄岩
dal soil 泥炭土
dalta function 德尔塔函数
Dalton's alloy 道尔顿铋铅锡易熔合金
Dalton's law 道尔顿定律
Dalton's law of partial pressures 道尔顿分压定律
Dalton's overhead eccentric type jaw crusher 道尔顿高架颚式破碎机
Dalton's temperature scale 道尔顿温标
dalton 道尔顿(质量单位)
dalyite 硅钾锆石
dam 拦泥坝;密闭墙;筑堤堵水;堤;坝;轻便水箱;水坝;伸缩缝盖板
dam abutment 坝座;坝头岸坡;坝肩
dam accessories 坝体附属结构物
damage 毁坏;灾害;惩罚性赔款;残损;破坏;赔偿金;损失;损伤;损坏;伤害
damageability 易损坏性;易破坏性
damage accident 损坏事故
damage ag(e)ing 伤损寿命
damage and rupture 损伤与断裂
damage and shortage report 货损货差报告;短损(货物)报告单
damage assessment 损坏估价;损坏查定
damage assessment routine 故障恢复(例行)程序;故障估价例行程序
damage awards 损坏赔偿金
damage begond repair 损坏不能修的
damage by air pollution 空气污染损害
damage by drought 干旱损害
damage by fire 火灾损害
damage by fores-insects 森林虫害
damage by frost 霜(冻损)害;霜冻破坏;霜冻毁坏
damage by fume 烟气损害;烟害
damage by insect 虫害
damage by lightning 闪电害
damage by lightning strikes 雷击害
damage by nuclear weapon 核武器损伤
damage by severe frost 严重霜害
damage by snow 积雪损害
damage by storm 暴风雨损害
damage by water 水损(货物受水渍损坏)
damage by wind 风害
damage capability 破坏能力
damage cargo clerk 理赔员;理残损
damage cargo list 货损单
damage caused by accident 事故损坏
damage caused by atmospheric pollution 大气污染危害
damage caused by fire 火灾损失
damage caused by wind-storm 风害
damage certificate 残损证明(书)
damage claim 赔偿损害要求;损失索赔;损害赔偿要求;损害索赔;索赔要求
damage clause 残损条款
damage coefficient 破损系数;损系数
damage condition 残损情况
damage contour map 震害等值线图
damage control 船损管制;防损控制;损伤控制;损害管制
damage control bill 船损管制部署表
damage control flooding 船损管制灌注
damage control locker 应急器材柜;损管器材柜;

水损调整横倾装置
damage control plan 海损控制示意图
damage control tender 防损车
damage control truck 防损车
damage cost factor 破坏价值因素;破坏代价系数
damage criterion 破坏准则;危害准则
damage crops 损害庄稼
damage curve 损坏曲线;损耗曲线
damaged 受损
damaged and spoiled commodities 残损变质商品
damaged area 烧毁面积
damaged article 残品
damaged beyond repair 毁坏;彻底毁坏;无法修复;损坏难修;破损不能修理(的程度)
damaged cargo 受损货物
damaged cargo report 货物残损报告单;受损货物报告书
damaged condition 损失情况;损坏程度;受损情况
damaged drawing 已损坏图形;被损坏图形
damage deformation spectrum 破坏变形谱
damaged goods 残损货品;损坏货物;受损货物
damage distribution 震害分布
damaged length 烧毁长度
damaged market value 受损后市价;残损品市价;受损货物的市价
damaged non-conforming building 损坏的违章建筑
damage done 造成的损害
damage done clause 碰撞损害条款;损害赔偿条款
damaged party 受害方损坏
damaged premise 损毁的房屋
damaged property 受损财产
damaged rail 损伤钢轨
damaged ship 受损船
damaged stability 破舱稳性
damaged stability calculation report 破舱稳性计算书
damage due to driving 打桩引起的损害
damage due to humidity 潮湿引起的损坏
damage due to thaw(ing) 解冻引起的损坏
damaged value 受损后价值
damaged vehicle 已损毁客车
damaged wagon 已损毁货车
damage earthquake 破坏性地震
damage effect 震害效应;破坏效应;损伤效应
damage evaluation of environment(al) pollution 环境污染事故评价
damage fastness 抗损性;耐损性;耐损度
damage feature 破坏特点
damage-frequency curve 损害频率曲线
damage function 损害函数
damage identification 破坏识别
damage in collision 碰撞损失
damage increment 震害增量
damage index 震害指数;损坏指数
damage insurance 损失保险
damage intensity 震害率
damage in transit 运途中受损
damage items 损坏项目
damage length 损伤长度
damage line 破损线;损坏线
damage line test 损伤线试验
damage liquidate 赔偿金
damage mechanics 破坏力学
damage note 损坏通知单
damage of a structure 建筑物的损坏
damage of building 建筑物受损
damage of materials 材料受损;材料损害
damage of pressure 压损
damage on environment 致害环境
damage-only accident 仅因意外事故造成的损坏
damage parameter 破坏参数
damage pattern 破坏形式
damage phenomenon 震害现象
damage plan 损坏情况简图
damage portion 残损部分
damage potential 预计最大灾害;震害势;破坏潜势;破坏能力
damage potential index 潜在事故指数
damage pressure 破坏(性)压力
damage probability function 破坏概率分布函数
damage probability matrix 破坏概率矩阵

damage radius 破坏半径
damage rate 破坏率;损失率;损伤率
damage ratio 破坏率;损伤率;损坏率;受灾率
damage records 船损记录
damage repair 海损修理;损伤修理;事故修理
damage report 损坏报告书
damage resistance 免损力;耐损性;耐损度
damage resisting 抗磨损的
damage risk contours 危险线
damage risk criterion 损害风险标准;损害风险准则;失听危险标准
damages 赔款;赔偿费
damages for default 违约赔偿费
damages for delay 延迟损失费;误期赔偿费
damages for detection 延期损失赔偿
damages for detention 迟误损失
damage ship 海损船舶
damages liquidated 违约罚金;赔偿金
damage stability 海损稳定性
damage state 破坏状态
damage statistics 震害统计
damages to non-conforming use 不符合规定使用房屋的损坏
damage suit 损坏赔偿诉讼
damage survey 船损检验;危害性测量;损失调查;损坏检查;损伤调查;事故鉴定;事故检查;事故调查;破坏现场调查
damage threshold 损伤阈值;损坏阈;伤害阈值
damage threshold earthquake 破坏阈限地震
damage threshold earthquake intensity 破坏阈限地震烈度
damage to buildings 建筑物损坏
damage to crops 危害作物
damage to crops by drought 旱情
damage tolerance 损伤容限
damage tolerant design 破损设计
damage to person and property 人身与财产损害
damage to persons 人身损害
damage to prestige 损坏威信;损害威信
damage to property 损坏财产
damage to the accuracy 精度受损
damage to the environment 损害环境
damage to works 对工程的损害
damage vessel 海损船舶
damage vulnerable 易破坏的
damage wear 磨蚀造成的损伤
damaging deposition 严重淤积;严重淀积;严重电积;严重沉积
damaging effect 破坏作用;事故效应
damaging flood 灾害性洪水
damaging impact 危害性冲击;破坏性冲击
damaging range 危害范围;损害范围
damaging scour 严重冲刷
damaging sedimentation 严重淤积;严重淀积;严重电积;严重沉积
damaging stress 破损应力;破坏应力;损伤应力
dam appurtenance 坝体附属结构物
dam area power supply 坝区供电
damascene 金属镶嵌(的)
damascened 用波形花纹装饰的
damascening 金属镶嵌的,波形花纹的
Damascus blade 大马士革钢刀
Damascus bronze 大马士革铅锡青铜
Damascus steel 大马士革钢
Damascus ware 大马士革器皿(一种高硅质施釉陶瓷)
damask 锦缎色;红玫瑰色
Damask-hung 金属镶饰的挂件;绸缎悬饰
Damaxine 高级磷青铜;大马新磷铜
dam axis 坝轴线
dam back 坝的背水面
dam baffle 堰板
dam base 坝基
dam batter 坝坡
dam behavio(u)r 大坝特性
dam behavio(u)r measurement 坝的性能测定;坝的工况测定
dam block 坝块
dam body 坝体,坝身
dambo soil 坦泊土
dam break(ing) 坝决;决堤;水坝破坏;水坝溃决;溃坝;垮坝;坝(的)溃决
dam-breaking flood 溃坝洪水
dam-break wave 溃坝波

dam-building contract 建坝承包合同
dam-bursting 水坝溃决;溃坝(的);垮坝(的)
dam-bursting flood 溃坝洪水
dam-collapse flood 垮坝洪水
dam concrete 造坝混凝土
dam concrete work 大坝混凝土工程
dam construction 坝基施工;筑堤技术
dam construction technique 筑坝技术;坝工技术
dam construction works 筑堤工程
dam cooler 冷却水包
dam core 坝心;坝核;坝的芯墙
dam crest 坝顶
dam crest gallery 坝顶廊道
dam crest gate 坝顶(闸)门
dam crest length 坝顶长度
dam crest level 坝顶高程
dam crest road 坝顶道路
dam crest spillway 坝顶溢洪道
dam crest wall 坝顶防浪墙
dam crest width 坝顶宽度
dam-crossing pendulum pulley block 过坝摆滑轮组
dam-crossing tonnage 过坝运量
dam-crossing traffic 过坝交通
dam-crossing transportation mode 过坝运输方式
dam crossing transshipment 过坝转运
dam crossover by inertia 惯性过坝
dam damage 大坝损坏
dam deflection 坝体挠度;坝体变位
dam deformation 坝体变位
dam deformation observation 大坝变形观测
dam deformation survey 大坝变形测量
dam design 坝的设计
dam deterioration 大坝损坏
dam dislocation 坝体变位
dam face 堤坝上游面;堤坝承压面;堤坝承水面;坝面;坝的迎水面;坝的承水面
dam failure 坝溃;溃坝;大坝失事;坝失事;坝溃决;坝决;水坝破坏;水坝溃决;垮坝
dam for water supply 供水用坝
dam foundation 坝基;大坝基础
dam foundation height 坝基高程
dam foundation investigation 坝基勘察
dam foundation treatment 坝基处理
dam foundation underpinning 坝基置换
dam foundation width 坝基宽度
dam front water level 坝前水位
dam generalist 坝工专家
dam geology 坝地质(状况);大坝地质
dam heel 坝踵;坝的上游坡脚
dam height 坝高
dam heightening 堤坝加高;坝体加高;坝的加高
dam in 筑坝堵(拦)水;围坝堵水
dam inspection 大坝检查
damkjernite 辉云碱煌岩
Damkohler number 丹姆克尔数
dam lake 坝湖;人工湖
dam length 坝长
damless intaking 无坝引水
dam lift 坝的壅水高度
dam location 坝址;坝位;坝的位置
dam lower reaches slope 坝下游坡度
Damman cold asphaltic concrete 石屑冷拌沥青;冷拌拉起混凝土
dammar 达玛树脂
dammar gum 达玛(树)胶
dammarolic acid 达玛醇酸;达玛(树)脂酸
dammar penak 达玛琥珀色树脂
dammar resin 达玛树脂
dammar resin varnish 达玛树脂清漆
dammar varnish 达玛胶漆(玻璃装饰用);达玛(树脂)清漆
dammar wax 达玛蜡
dammarylic acid 达玛基酸
dam mass 坝体
dam measurement 水坝观测
dammed intaking 有坝引水
dammed lake 堰塞湖
dammed region 壅水区
dammed-up 拦蓄
dammed water 坝闸堵水;壅水
dammer 达玛树脂
Damm faience 丹姆陶器(德国施铅釉)
damming 蓄水;修筑隔墙;壅水;筑坝(拦)

damming area 壅水区
damming by reclamation 吹填筑坝
damming by sand bags 用砂袋筑坝(壅水)
damming length 壅水长度
damming limit 壅水界限;壅水极限;壅水范围
damming lock 拦江闸;蓄水闸;壅水(船)闸
damming of stream 河流堰塞
damming range 壅水范围
damming ring 挡料圈
damming shield 挡水防护体
damming step 壅水梯级
damming structure 筑坝壅水结构;壅水结构
damming-up 筑坝壅水
dam monolith 坝块
dam-needle 止水栅条
damning wire 拉筋
damnum absque injuria (申请赔偿中的)不能认可的损害
dam off 开坝放水
dam of gravity type 重力坝
dam of spoil area 抛泥场围埝
damourite 细鳞白云母;变白云母;水细鳞白云母
damourite-schist 水云母片岩
damouritization 水云母化作用
dam out 筑坝排水;筑坝拦水
damp 矿内毒气;矿井瓦斯;减振;压火;阻塞;潮气;阻尼;减幅
damp actuator 风门控制器;挡板控制器
damp air 高湿度空气;潮湿空气
damp and stick method 湿粘法
damp atmosphere 高湿度空气;湿大气
dampboard 隔音板
damp box 消振箱
damp check 防潮挡板
damp condition 衰减条件
damp course 防湿层;防潮层
damp diffusion 潮气扩散;水气扩散
damp down 浇水;封火
damp-dry 烘得半干
damped 阻尼的;衰减的
damped alternating current 减幅交流电;衰减交流电
damped angular frequency 衰减角频率
damped balance 减震天平;阻尼天平
damped ballistic galvanometer 阻尼冲击检流计
damped capacity 阻尼量
damped circular frequency 有阻尼圆频率
damped electron oscillator 衰减电子振荡器
damped exponential 减衰指数
damped force 阻尼力
damped forced vibration 有阻尼的受迫振动;有阻尼的强迫振动
damped-free vibration 有阻尼自由振动
damped harmonic motion 阻尼和谐运动
damped harmonic oscillator 有阻尼的谐振荡器
damped harmonic system 减幅线性振荡系统
damped impedance 阻尼阻抗;衰减阻抗
damped inductance 阻尼电感
damped lake 堰塞湖
damped least square method 阻尼最小二乘法
damped magnet 阻尼磁铁
damped motion 阻尼运动
damped natural circular frequency 有阻尼自振圆频率
damped natural frequency 有阻尼自振频率;有阻尼的固有频率;阻尼固有(振动)频率
damped oscillation 减幅振动;减幅振荡;阻尼振荡;阻尼摆动;衰减振荡
damped oscillator 阻尼振荡器
damped oscillatory component 衰减振荡成分
damped oscillatory mode of motion 阻尼振荡运动
damped pendulum 有阻尼摆;阻尼摆
damped periodic(al) element 阻尼摆动部分
damped periodic(al) instrument 阻尼稳定式仪表
damped potential index 阻尼势指数
damped pressure ga(u)ge 带缓冲器的压力计
damped resistor 阻尼电阻
damped seiche 阻尼驻波;阻尼水面波动;阻尼假潮
damped sinusoid 阻尼正弦曲线
damped sinusoidal quantity 阻尼正弦量;减衰正弦量
damped system 阻尼系统
damped vibration 有阻尼振动;阻尼振动;衰减振动
damped wave 减幅波;阻尼波;衰减波

damped wave detector 减幅波检波器
damp effect 阻尼效应
dampen 阻抑;阻尼;变湿
dampener 润湿辊;水辊;湿润器;湿润滚筒
dampening 返潮;润湿;润版
dampening chamber 湿润室
dampening effect 减振作用
dampening machine 增湿机
dampening material 阻尼材料
dampening rate of wave 波浪衰减速率;波浪减弱率
dampening roll 润湿辊
dampening roller 润版辊;水辊
dampening solution 润湿液
dampening system 润湿装置
dampening velocity of wave 波浪衰减速度
dampen out 减弱掉
damper 节气阀;减震器;减速器;烟道挡板;阻尼器;阻颤器;挡板;潮湿器;风门;风阀;防振锤;避震器;气流调节器;水辊
damper adjustment 挡板调节
damper applicator roller 抑制敷料滚筒;湿润剂施加辊
damper block 闸板砖
damper box 风挡箱
damper brake 制动闸
damper circuit 阻尼电路
damper coefficient 阻尼系数
damper control 挡板控制;风门控制装置
damper counterweight 减震器平衡锤
damper cylinder 减震筒
damper cylinder arm 减震筒臂
damper cylinder body 减震筒体
damper cylinder piston 减震筒活塞
damper frame 阻尼框;闸架
damper gear 阻尼装置;制动装置;缓冲装置
damper guard 减震器防护装置;制动器防护装置
damper leg 缓冲支柱
damper link 减震器连杆
damper lug 挡板拉耳
damper of chimney 烟囱风门
damper of smoke 挡烟活门
damper pedal 阻尼踏板
damper pipe union nut 减震器管联结螺母
damper plate 节流板;挡板;风挡
damper plate valve 翻板阀
damper register 减震器标示
damper regulator 调节阀;风门调节器;气闸调节器
damper rigging lever 抽阀开关杆
damper rigging rod pin 挡板杆销
damper shaft 减震器轴
damper shaft arm 减震器轴节;减震器轴臂
damper spring 缓冲器弹簧
damper tile 闸板砖
damper tube 阻尼管
damper valve 减震阀;调节阀
damper vibrator roller 减震滚筒;湿润剂摆动传递辊
damper weight 挡板平衡锤;平衡锤
damper with lockable actuator 手动风阀
damper with remote control device 遥控风门机构
damper workshop 减震器车间
Dampexe 丹帕克斯(一种墙面防潮漆)
damp goods 湿水货
damp grain 潮湿谷物
damp haze 湿霾
damp heat test 湿热试验
damping 减幅;阻尼;减振;减震;回潮;消声;润湿;润版;衰减;受潮;湿润(阻尼);湿润(加工)
damping action 减振作用;阻尼作用;制动作用
damping adjustment 阻尼调整
damping air 阻尼空气
damping air blower 阻尼风机
damping air diverter 阻尼风分流器
damping apparatus 减震装置;阻尼装置;阻尼器
damping area 减振面
damping arrangement 节制设备;减震装置
damping balance 阻尼天平
damping bar 扒渣钩
damping bearing of highway bridge 公路桥梁减震支座
damping broadening 阻尼致宽
damping by friction 摩擦减震
damping capacity 减震能量;减能能力;吸能能力;吸振能力;吸振能力;阻尼能力;衰减能量
damping chamber 减振室

damping characteristic 阻尼特性
damping circuit 阻尼电路;衰减电路
damping coating 阻尼涂料;防潮涂层
damping coefficient 减振系数;阻尼因数;阻尼系数;阻力系数
damping coefficient value 阻尼系数值
damping coil 阻尼线圈
damping cone 阻尼锥体
damping constant 减幅常数;阻尼常数;阻尼常量;衰减常数;衰减常量
damping construction 阻尼结构
damping control 加振控制器;阻尼控制;阻尼调整;湿润控制
damping control surface 阻尼控制面
damping coupling 阻尼耦合
damping course 阻尼层;衰减层;减幅层;减振层
damping current 阻尼电流
damping curve 阻尼曲线
damping decrement 减幅量;阻尼减幅
damping device 减阻装置;减震装置;减振装置;阻尼装置;阻尼器;制动装置;湿润装置
damping disc 减震盘;减振盘
damping disk 减振盘
damping distribution 阻尼分布
damping down 休风;压火
damping due to rotation 旋转阻尼
damping effect 阻尼效应;减震作用;减振效应;消能作用;制动效应;制动效能;制动效力;制动效果;衰减作用
damping element 减震元件
damping equipment 阻尼设备
damping error 阻尼误差;第二类冲击误差
damping factor 减幅因数;减幅系数;阻尼因数;阻尼系数;衰减因子;衰减因素;衰减因数
damping factor of wave 波的衰减系数
damping fin 阻尼片
damping fluid 减震液;减振液;缓冲液
damping foam material 阻尼泡沫材料
damping force 减振力;阻尼力
damping force vector 阻尼力矢量
damping for rolling oscillations 滚转振动阻尼
damping frame 阻尼架
damping impulse 阻尼脉冲
damping index 减振指数
damping influence coefficient 阻尼影响系数
damping in roll 滚动阻尼
damping key 阻尼电键
damping layer 阻尼层
damping least square method 阻尼最小二乘法
damping log(arithmic) decrement 阻尼对数衰减率
damping machine 阻尼机;给湿机;防潮电机;喷雾机;调温器;调湿机
damping magnet 阻尼磁铁;阻尼磁体;制动磁铁
damping material 缓冲材料;减振材料;阻尼材料;防潮材料
damping matrix 阻尼矩阵
damping measurement 阻尼测量
damping mechanism 阻尼机构
damping medium 阻尼介质
damping membrane 楼层或平屋顶的防潮层
damping model 阻尼模型
damping modulus 阻尼模量
damping moment 阻尼力矩;衰减矩
damping moment coefficient 阻尼力矩系数
damping of concrete 混凝土的吸湿性
damping-off 立枯病
damping of pulsation 脉动阻尼
damping of the oscillation of the spring 簧震阻尼
damping of wave 波浪阻尼;波浪衰减;波浪上涌段
damping oil 阻尼油
damping oil vessel 阻尼油罐
damping orthogonality 阻尼正交性
damping oscillation 阻尼振荡
damping pad 加湿垫
damping pad holder 加湿垫托架
damping parameter 阻尼系数;阻尼参变数
damping period 激后复原期;阻尼振荡周期;阻尼时间;阻尼期
damping piston 减震活塞
damping plant 阻尼设备
damping power 阻尼能力
damping pressure roller 加湿压力滚筒;湿润压力辊
damping property 阻尼性质
damping pulse 阻尼脉冲

damping quality 阻尼性
damping radiation 阻尼辐射
damping rate 阻尼(速)率
damping ratio 阻尼率;减振比;阻尼比
damping resistance 减震阻力;阻尼阻力;阻尼电阻;衰减电阻
damping resistor 阻尼电阻器
damping ring 阻尼环
damping roll 润湿辊
damping screen 整流栅;稳水栅
damping seiche 阻尼假潮
damping slab 阻尼板
damping solution 润湿液;润版药水
damping spring 减震弹簧;阻尼弹簧;防振弹簧
damping strength 阻尼强度
damping structure 阻尼结构
damping system of flexible structure 挠性结构减震系统
damping test 减震试验;减振试验
damping time 阻尼时间
damping time constant 缓冲时间常数
damping to half amplitude 半幅阻尼
damping torque 制动转矩
damping transient 衰减瞬变量
damping tube 阻尼管
damping turn 阻尼线匝
damping unit 减震装置;湿润装置
damping value 阻尼值
damping vibration 阻尼振动
damping vibration attenuation 阻尼减振
damping wall 防潮墙
damping washer 减震垫圈;减振衬垫
damping water 润湿液;润版药水
damping wave 阻尼波
damping weight 阻尼重量
damping width 阻尼宽度
damping width selector 加湿宽度选择器;湿润宽度选定器
damping winding 阻尼绕组
damping wire 减震拉筋
damp location 潮湿区域;潮湿部位
damp marsh 多水沼泽;湿沼泽
damp mass 软材
dampness 湿度;含水率;潮湿含水率;潮湿
dampness in dwelling 房屋潮湿度
dampness penetration 建筑受潮;湿气浸透
damposcope 坝址试探
dam power plant 蓄水发电厂;坝式发电站;大坝发电站
dam power station 坝式厂房
damp production 生产停滞
dampproof 防潮;防水;防(潮)湿的
dampproof and acoustic(al) panel ceiling 防潮吸声板吊顶
dampproof and acoustic(al) tile ceiling 防潮吸声板吊顶
dampproof coating 防潮涂料;防潮(面)层
dampproof concrete 不透水混凝土
dampproof corrugated board 防潮波纹纸板
dampproof course 避潮层;防水层;防潮层
dampproof foundation 防潮基础;防湿基础
dampproof fungicidal paint 防潮防霉漆
dampproof ground floor 防湿地面
dampproofing 防潮(的);防潮处理;防潮材料
dampproofing admixture 防潮剂
dampproofing agent (混凝土)防吸湿外加剂
dampproofing and waterproofing of above-ground rising structure 地面上建筑物的防潮与防水
dampproofing basement 防潮地下室
dampproofing coating 防湿涂层;防湿面层
dampproofing concrete 防潮混凝土
dampproofing course 防潮层;防湿层
dampproofing foundation 防潮基础;防潮地基
dampproofing ground floor 防潮底层地坪
dampproofing insulation 防湿绝缘
dampproofing machine 防潮电机
dampproofing material 防潮材料
dampproofing membrane 防潮(薄)膜
dampproofing of wall 墙身防潮
dampproofing packing 防潮包装
dampproofing product 防潮制品;防潮产品
dampproofing sheeting 防潮片材
dampproofing wall 防潮墙;防潮墙;防洪墙

dampproof insulation 防潮绝缘;隔潮
dampproof masonry 防潮圬工
dampproof material 防潮材料
dampproof membrane 防水膜;防水层
dampproofness 防湿性
dampproof paint 防潮漆
dampproof sheeting 为地下室外墙设防潮层;(地下室墙周的)防潮毡材
dampproof slab 防潮板
dampproof tile 防潮砖
damp-resistant compound 防潮化合物
damp room 潮湿养护室;潮湿的房间;雾室
damp room dampproofing 养护室防潮;潮湿室防潮
damp room light fitting 养护室照明装置;潮湿室照明装置
damp room light fixture 养护室照明设备;潮湿室照明设备
damp room luminaire fixture 养护室照明设备;潮湿室照明设备
damp room partition (wall) 养护室隔墙;潮湿室隔墙
damp room services 养护室设施;潮湿室设施
damp sand 阴湿地;湿砂
damp sawdust 湿锯木屑
damp sheet 风帘
damp-snow avalanche 湿雪崩
damp storage 湿度养护
damp-storage closet (养护混凝土用的)湿治室;(养护混凝土用的)高湿度养护室;(混凝土)湿养护室
damp tolerant plant 耐湿植物
damp-wood termite 湿木白蚁
dampy 潮湿的
dam raising 坝体加高
dam-reservoir interaction 坝水库相互作用
dam roadway 坝上道路
dam root (山傍的)坝基;齿槛
dam safety evaluation 大坝安全评价
dam section 坝段
dam sheeting 坝面护板
dam shell 坝壳
dam sill 闸槛
dam site 坝址
dam site appraisal 坝址评价
dam site exploration 坝址勘探
dam site investigation 坝址勘察;坝址勘查;坝址勘测
dam site plan 坝址平面图
dam site reconnaissance 坝址踏勘
dam site resettlement 坝区移民安置
dam site selection 坝址选择
dam site survey 坝址勘察;坝址勘测;坝址测量
dam slope 坝坡
dam spillway 堤坝溢洪道
dam spillway channel 溢流堤泄水槽
dam spring 水库源流;水库水源
dam steps 水坝梯级
dam strengthening 坝体加固;坝的加固
dam survey 大坝测量;坝测量
dam thickness 坝的厚度
dam toe 坝趾;坝脚;坝的下游坡脚
dam top 堤顶;坝顶
dam top height 坝顶高程
dam top width 坝顶宽度
dam type 坝型
dam type hydroelectric(al) plant 堤坝式水电站
dam type hydroelectric(al) power plant 蓄水式水电站
dam type hydroelectric(al) station 堤坝式水电站
dam type pouring basin 闸板式浇口杯
dam type pouring ladle 闸板式挡渣浇包
dam type power plant 水坝式发电站;靠坝式水电站;堰堤式发电厂;堰堤式(水力)发电厂;坝式发电厂
dam type power station 堤坝式水电站;靠坝式水电站;蓄水式水电站
dam up 挡起;拦阻;壅高;抑制;封闭
dam upper reaches slope 坝上游坡度
dam up ravine 闸山沟
dam vicinity plan 坝区平面图
dam volume 坝体积
dam wall 坝墙
dam warden 坝上管理人员
dam weight 坝体重量

dam width 坝宽
dam with attendance 有看护的坝
dam with core wall 心墙坝
dam with inclined core 斜(心)墙坝
dam without attendance 无看护的坝
dam without overflow 非溢流坝
dam with segmental-headed counterforts 圆弧形拱柱坝
dan 空中吊运车;小型滑车;担(中国和东南亚国家的一种重量单位,1 担 =50 千克)
danagered-off area 封闭的危险区
danaite 钴毒砂
danalite 铍榴石
dan anchor 浮标锚
danbaite 丹巴矿
dan buoy 标志浮标;临时(信号)浮标;临时浮筒;信号浮标;标识浮标;小浮标
danburite 硅酸硼钙石;赛黄晶
danby 炭质页岩
danced stair(case) 转弯踏步
dance floor 舞池地面;舞池
dance floor lighting 舞池照明
dance hall 舞厅
dancer roll 导毡辊;浮动滚筒;跳动辊;张力调节辊;调节辊
dancers 楼梯(俚语)
dancery 舞厅;跳舞厅
dancette 曲折线脚;锯齿饰【建】;曲折图案;曲折饰
dancette mo(u)lding 锯齿形线脚;曲折线脚
dancing dervish 尘旋
dancing devil 尘旋
dancing floor 舞池
dancing hall 舞厅
dancing pulley 均衡轮;调整轮
dancing saloon 舞厅
dancing steps 均衡梯级;均衡踏步;均衡阶步;楼梯旋步;转弯踏步;扇形踏步
dancing winders 盘梯;均衡(斜)踏步;转弯踏步
Danckwerts model 丹克韦尔兹模型
Dandelion 丹得来昂铅基轴承合金
dandelion 蒲公英
Dandelion metal 丹得来昂轴承合金;铅基锑锡承合金;铅基白合金
dandered coal 天然焦炭
Dando brick 丹多盛钢桶用砖
dandy 小型沥青喷布机;小型沥青喷洒机;渔船用拖网绞盘;二轮小车;双轮小车
dandy dancer 铁路道班工人;季节流动工
dandy fever 登革热
dandy finisher 末道粗纱机
dandy note 出库证书
dandy reducer 二道粗纱机
dandy roll (印透明纹的)柱辊;压胶辊
dandy rover 末道粗纱机
dandy roving 末道粗纱
dandy winch 拖网绞车
danemorite 锰铁闪石
danforth anchor 燕尾锚(具);长爪锚
danforth anchor of high holding power 丹福尔大抓力锚
danger 危险(物)
danger alarming 危险报警
danger alarm system 危险报警系统
danger angle 危险角;水平方向危险角
danger area 险区;危险区(域);危恶地
danger area in the Pacific 太平洋危险区
danger arrow 危险标志
danger beacon 危险标灯
danger bearing 危险(界限)方位
danger bearing alarm indicator 危险方位报警指示器
danger bearing transmitter 警戒发射机
danger board 火灾报警板;火险告示牌;火险报导板;危险信号牌;危险符号
danger buoy 危险位置指示浮标;示警浮标
danger button 紧急按钮
danger circle 危险圆
danger class 火险(等)级;危险等级
danger factor 危险因素
danger goods team yard 危险品货场
danger index 火险指数;火险指标;危险指数
danger label 危险标记
danger level 危险水位;危险水平;危险程度
danger light 告警信号灯光;危险信号灯

danger limit 危险界限;危险极限
danger line 警戒线;警戒水位;危险(界)线
danger mark 危险标志
danger message 危险通报
danger meter 火险计算尺
danger money 危险(工作)津贴;危险工作的额外报酬
danger number 危险量指数
danger of collision 碰撞危险
danger of ignition 着火危险
danger of soil blowing 风蚀的危险
danger of sparking 点火危险;打火花危险
dangerous accident 险性事故;事故苗子
dangerous area 危险区(域)
dangerous articles 危险货物;危险品
dangerous articles package 危险品包装
dangerous bend 危险弯段
dangerous building 危险房屋;危房
dangerous cargo 危险品;危险货物
Dangerous Cargo Act 危险货物法案
dangerous cargo container 危险品集装箱
dangerous cargo detection 危险货物探测
dangerous cargo list 危险货物清单
dangerous cargo on deck 甲板危险品
dangerous cargo terminal 危险品码头
dangerous cargo warehouse 危险品(仓)库
dangerous channel 危险水道
dangerous chemicals 危险化学品
dangerous circle analysis 极限圆分析
dangerous coast 危险海岸
danger(ous) coefficient 危险系数
dangerous condition 危险状态
dangerous (cross-)section 危险截面
dangerous cylinder 危险圆柱面
dangerous deck 危险甲板
dangerous drug cupboard 危险药品柜
Dangerous Drugs Act 危险药品条例
dangerous for anchoring and fishing 对锚泊和捕鱼有危险
dangerous goods 危险(物)品;危险货物
dangerous goods anchorage 危险品锚地
dangerous goods code 危险品规则
dangerous goods parking track 危险货物停留线
dangerous goods regulations 危险品规则
dangerous goods warehouse 危险品(仓)库
dangerous hill 险坡
dangerous industrial district 危险品工业区
dangerous intersection 危险交叉口
dangerous line 警戒线
dangerously explosive 极具爆炸的
dangerous mark 危险品标志;危险货物标志
dangerousness 危险
dangerous occupation 危险职业
dangerous oil 易燃石油产品
dangerous passage 险滩;危险水道
dangerous passage of reef pattern 礁石险滩
dangerous period 危险期
dangerous point 危险点
dangerous pressure 危险压力
dangerous quadrant 危险象限
dangerous reef 险礁
dangerous rock 险礁
dangerous section 险段;危险路段;危险断面;危险地段
dangerous semicircle 危险半圆
dangerous signal 危险信号
dangerous space 危险区(域)
dangerous stage 危险水位
dangerous stretch of road 危险路段
dangerous structure 危险建筑;危险结构
dangerous substance 危险物质
dangerous table 危险品隔离表
dangerous to health 对健康有害的
dangerous to man and animals 对人、牲畜有危险
dangerous voltage 危险电压
dangerous when wet 忌水危险品;遇水燃烧物
dangerous work 危险工作
dangerous wreck 碍航沉船(在水下不到十拓的沉船)
dangerous zone 危险地区;危险地带;险区;危险区(域)
danger point 危险点;危险地点
danger position 危险位置
danger protecting function 保险功能

danger range 危险距离
danger scale 危险率
danger sector 危险区段
danger sign 危险标(志)
danger signal 危险信号
danger signal buoy 危险信号浮标
danger signal lighting 危险信号灯
danger situation 危险状况
dangers of navigation 航行危险
dangers of seas 海上(意外)危险
danger sounding 危险探测;危险界限水深
danger space 防火间隔
danger table 火险计算表;危险表
danger threshold 安全限值;危险阈;危险极限
danger to life 生命危险
danger to navigation 航行危险物
danger warning 危险警告
danger warning light 危险警告灯
danger warning sign 危险警告牌;危险警告标志
danger warning signal 危险警告信号
danger zone 危险地带
dangler 悬摆物
dangling bond 悬空键;悬挂键
Dangyangyu kiln 当阳峪窑
Dangyangyu ware 当阳峪窑
Danian (stage) 丹尼亚阶【地】
Daniell cell 丹尼尔电池
Daniell flow point (method) 丹尼尔流点法
Daniell hygrometer 丹尼尔湿度计
Danish cultivalor 丹麦式中耕机
Danish running line trolling 丹麦式循环曳绳钓
Danish tong 丹麦鱼夹
Danish trawler 底拖网
Danjion astrolabe 丹容等高仪
Danjion impersonal prismatic astrolabe 丹容超人差棱镜等高仪
Danjion prismatic astrolabe 丹容棱镜等高仪
dank 潮湿的;黑色碳质页岩;煤页岩
dan layer 布标船;小浮标敷设艇;设标船
dan line 浮标绳
Danneberg kiln 丹奈伯格窑(环形窑)
Dannemora 丹内马拉高速钢
dannemorite 锰铁闪石
Danner process 丹纳法;水平拉管法
Dano composting plant 丹诺混合肥料厂
dan ship 浮标工作船;标志船
danshui shell carving picture 淡水贝雕画
dansite 氯镁芒硝
dant 煤母;低级(软)煤
dan tow 浮标绳
Dantzig oak 但泽橡木
Dantzing-Wolfe algorithm 旦泽格-瓦尔夫算法
Danube rule 多瑙河吨位法
Danxia landform 丹霞地貌
Danxia type 丹霞式
dao 拓木;菲律宾木
daomanusite 道马矿
dap 槽口;凹口;弹跳
daphne red 瑞香红色
daphnetin 白瑞香素
daphnia 水蚤
daphniphylline 虎皮楠碱
daphnism 白瑞香中毒
daphnite 铁绿泥石
daposit of title deeds 权利证书托存
dapped joint 互嵌接合;嵌槽接头
dapped joint connection 齿结合
dapped shoulder joint 互嵌肩接合
dapper 圆锯
dapping 挖槽(嵌接);刻槽;开槽
dappled 杂色的;有斑点的;有花纹的
dappled glaze 斑纹釉
dapt 榫眼
dar 住处;(印度和波斯建筑中的)门口
daraf 拉法(法拉的倒数)
daraller 计日工(人)
daranide 二氯苯磺胺
darapiockite 锆锰大隅石
darapskite 硫酸钠硝石;钠硝矾
darby 木杠;抹子;镘尺;混凝土刮板;刮片;刮杠;刮尺;长镘刀
darby float 刮尺;木杠;刮平尺
darbying 表面刮光(水泥砂浆或混凝土);木杠刮平
darby slicker 刮杠

D'Arcet metal 达塞特铋铅锡低熔点合金
Darcet's alloy 达塞特易熔合金
Darcy 达西(多孔介质渗透力单位)
Darcy number 1 第一达西数
Darcy number 2 第二达西数
Darcy's approximation 达西近似值
Darcy scale non-reactive transport equation 达西标度非活性物质运移方程
Darcy scale reactive transport equation 达西标度活性物质运移方程
Darcy's equation 达西方程
Darcy's flow regime 达西流态
Darcy's formula 达西公式
Darcy's friction factor 达西摩擦系数
Darcy's law 达西定律
Darcy's percolation velocity 达西渗透速度
Darcy's unit 达西单位
Darcy's velocity 达西速度
Darcy's viscosity 达西黏度
Darcy-Weisbach coefficient 第一达西数
Darcy-Weisbach equation 达西—威斯巴奇方程
Darcy-Weisbach formula 达西—威斯巴奇公式
Darcy-Weisbach friction factor 达西—威斯巴奇摩擦系数
Darcy-Weisbach roughness coefficient 达西—威斯巴奇粗糙系数
Darelle 达列尔防火黏胶短纤维
darg 任务;工作;产量定额
dark 暗黑;日暮
dark adaptation 对黑暗的适应性;暗(光)适应
dark adaptometer 暗适应计
dark air-curing 室内吹干
dark-and-light 浓淡;深浅
Dakar Port 达喀尔港(西非)
dark atom 暗原子;无放射性原子
dark background 黑暗背景
dark background luminance 暗背景亮度
dark band 暗带
dark-band effect 暗带效应
dark blind 严密遮光物;防光的
dark blue 藏青色;暗蓝色;深蓝色;暗青色
dark bottle 黑瓶(测浮游植物生产量用)
dark brown 黑棕色;棕褐色;茶色;暗棕色;铁棕(色)
dark-brown glass 茶色玻璃
dark brown matter 暗褐色物质
dark-brown soil 暗棕(色)土
dark burn 暗烧;疲劳;烧暗(荧光屏发光效率降低)
dark burn fatigue 发光衰退
dark cardinal 暗深红色
dark cathode 暗色阴极
dark change 暗换场
dark charge 暗电荷
dark cherry heat 暗樱红热
dark cloud 暗云
dark colo(u)r 不艳色;暗色;深色
dark-colo(u)red 深色的;暗色的;黑色的;深黑色的
dark-colo(u)red constituent 暗色组分
dark-colo(u)red extrusive rock 黑色喷出岩
dark-colo(u)red intrusive rock 黑色侵入体
dark-colo(u)red mineral 暗色矿物
dark-colo(u)red pavement 黑色路面
dark-colo(u)red soil 暗色土
dark-colo(u)red stocks 暗色原料
dark companion 暗伴星;暗伴天体
dark conductance 暗电导
dark conduction 暗电导
dark conductivity 暗电导率;无照导电性
dark contrast 暗反差
dark current 电极暗电流;暗流;暗电流;无照电流
dark current noise 暗电流噪声;暗(电)流噪声
dark-current pulse 暗电流脉冲
dark current shot noise 暗电流散粒噪声
dark currentvoltage curve 暗电流电压曲线
dark decay 暗衰退
dark degree of pleochroic halo 多色晕深度
dark discharge 暗放电;无光放电
dark distiller's grain 暗酒糟
dark dome 暗拱
dark door 不漏光门;遮光门
dark dust nebula 暗尘星云
darken 变暗;扑灭火焰
darkened filter media 黑化过滤介质
darkened sand 黑化砂

darkening 发暗;变黑
darkening plant 遮暗设备
darkening rate 变暗速率
darkening tendency 变暗倾向
darkening towards the limb 临边昏暗
darkening wavelength 变暗波长
darker shade 暗(色)调
dark exchange 外汇黑市
dark eye piece 目镜色片
dark face 灰色荧光屏;暗面
dark factice 暗黑胶;深胶
dark fibers 暗纤维
dark field 暗视域;暗视野;暗(视)场;暗区
dark-field amplitude mode 暗场振幅模式
dark-field colo(u)r immersion method 暗视野彩色油浸法
dark-field condenser 暗场聚光器;暗视野聚光器;暗视野聚光镜
dark-field elements 暗视野配件
dark-field illumination 暗视野映光法;暗场照明
dark-field image 暗视野像;暗场(图)像
dark-field interferogram mode 暗场干涉模式
dark-field line 暗场线
dark-field method 暗视野法;暗场法
dark-field microscope 超暗视微镜;暗场显微镜
dark-field microscopy 暗视场显微观察;暗视野镜检;暗场显微技术
dark-field of view 暗视场
dark-field reflection microscope 暗场反射显微镜
dark-field stop 暗视野光阑
dark flame 暗焰;无光焰
darkflex 吸收敷层
dark flocculus 暗谱斑
dark forest soil 暗色森林土
dark frame effect 黑框效应
dark Fraunhofer line 夫琅和费吸收线
dark galactic nebula 暗银河星云
dark grain 暗酒糟
dark-green 暗绿(的);墨绿;暗绿;深绿色
dark-grey 黑灰色的;暗灰色;深灰色
dark ground 暗视场
dark-ground illumination 暗场照明
dark-ground illuminator 暗场照明器
dark ground method of observation 暗场观察法
dark ground microscope 暗(视)场显微镜
dark heart 暗色心材
dark heat 光谱的红外部分;暗(红)热
dark heater 旁暗阴极加热器
dark-heat radiation 红外(线)辐射;暗热辐射
dark hole 黑洞;造成中心黑影
dark horse 夜班警卫员;守夜人员
dark-hued 暗黑色的
dark humus soil 暗色腐殖土
dark infra-red oven 暗红外线炉
darkish 微暗的;浅黑的
darkish brown 暗褐色
darkish red 暗红色(的);暗红
dark jalousie 百叶窗;遮光(百叶)窗
dark laminae 暗色纹层
dark lane 暗线;暗带
dark lantern 遮光提灯
dark layer 暗色层
dark light 不可见光;暗光
dark-light bottle method 黑白瓶法
dark lightning 暗闪电
dark limb 暗边缘
dark line 暗线
dark-line detection 暗线探测
dark-line spectrum 暗线光谱
dark luboil 黑色润滑油
dark maroon 深栗色
dark matter 暗物质
dark mildew 煤烟病;烟霉病
dark mineral 铁镁矿物;暗色矿物
dark mirror 暗镜
dark mottle 暗日芒
dark nebula 暗星云
darkness 盲度;黑暗;暗度
darkness acuity 暗敏度
darkness tremor 暗光眼球震颤
dark nilas 色黑而薄的冰
dark noise 暗噪声;暗流噪声
dark-noise signal 暗噪声信号
dark of the moon 月暗期;无月光期

dark oil 暗色油;深色油
dark part of flame 黑火头
dark peat 暗色泥炭
dark period 阴影周期
dark phase contrast 暗相差
dark picture area 图像暗区
dark pine 樟木
dark plaster 黑石膏
dark-plate method 黑色晕渣法
dark point section 暗点剖面
dark pulse 暗脉冲
dark radiation 暗辐射
dark ray 暗射线
dark reaction 黑暗反应;暗反应
dark reaction time 暗反应时间
dark reactivation 暗复活作用
dark red 赭色的;暗红(的);暗红色(的)
dark red ferrallitic soil 暗红色铁铝土
dark red ferralsol 暗红色铁铝土
dark red filter 暗红色滤光纸
dark red heat 暗红(炽)热
dark-red silver ore 深红银矿;深红硫锑矿
dark red spectacle glass 深红眼镜玻璃
dark region 暗区
dark repair 黑暗修复;暗修复
dark resistance 暗电阻;无照电阻
dark resistivity 暗电阻率
dark room 暗房;暗室
darkroom accessory 暗室辅助设备
darkroom camera 暗室式照相机;暗室制版照相机
darkroom clock 暗室钟;暗室定时器
darkroom equipment 暗室设备
darkroom filter 暗室滤光器;安全灯滤光片
darkroom fog 暗室灰雾
darkroom illumination 暗室照明
darkroom lamp 暗室灯
darkroom light 暗室灯
darkroom lighting 暗室照明
darkroom loading 暗室加载
dark room ophthalmic lamp 暗室(检)眼灯
darkroom processor 暗室显影机
darkroom technique 暗室技术
dark room test 暗室试验
darkroom thermometer 暗室温度计
darkroom timer 暗室定时器
darkroom trailer 活动暗室
dark ruby ore 深红银矿
dark satellite 暗卫星
dark sector 阴影扇(形)区
dark segment 暗弓
dark semi-fusain 暗色半丝炭
dark shadow 暗影
dark signal 暗信号
dark signature 暗标记
dark sky background 夜天背景
dark slatted blind 百叶窗
dark slide 遮光(滑)板;暗匣压平板
dark-smoke 黑烟;浓黑烟;深灰色
dark soil 深色土壤
dark sorrounds 黑暗环境
dark space 暗区
dark speck 暗斑
dark sphagnum peat 暗色水藓泥炭
dark spot 黑斑;暗点;暗斑
dark stage 暗摄影棚
dark star 暗星
dark tint face 暗淡面
dark-tint screen 中灰滤光屏
dark tonal anomaly 暗色调异常
dark tone 暗色调;暗调
dark trace 暗迹
dark-trace screen 暗线荧光屏;暗迹(荧光)屏
dark trace tube 暗迹阴极射线管;暗迹管;暗迹电子射线管
dark voltage 暗电压
dark window 遮光窗
dark yellow 暗黄色
dark yellow filter 深黄色滤光片
Darley Dale stone 达利戴尔石
darlington 复合晶体管
Darlington amplifier 达林顿放大器
Darlington connection 达林顿接法
darlistor 复合可控硅
darmar kuching 达玛库青树脂

darmold 石墨浆涂料
D'Arsonval galvanometer 达松伐耳电流计
D'Arsonval movement 达松伐耳传动机构
dart 箭头(装)饰;矢状饰【建】;射针;舌阀
dart and egg 蛋与簇形装饰
dart configuration 后尾式构型
Dartmoor granite 达特穆尔花岗岩
dart test 投掷试验
dart union 活络管子节
dart valve 突进阀;吊桶自动放水口;舌阀
dart(-valve) bailer 舌阀捞砂筒;(带突板球阀的)捞砂筒
Darwin ellipsoid 达尔文椭球
Darwinian fitness 达尔文适合度
Darwinian selection 达尔文选择
Darwin turning indicator 达尔文旋转式指示器
darya 河流
dash 阴影线;飞溅
dash adjustment 杠杆调整
dash against 撞上
dash-and-dot line 点画(虚)线
dash board 遮水板;控制板;挡泥板;仪表板;车前挡泥板;遮雨板;挡水板;防浪板(放在船头);防溅板;表板;仪表盘;操纵盘;防波板
dashboard light 仪表(板)灯
dashboard running jeams 木工工作裤(俚语)
dash-bond coat 浇涂(打底)层;水泥砂浆粉面层;砂浆涂层;水泥砂浆打底;浇涂黏结层
dash coat 浇泼涂层油;浇涂层;砂浆涂层
dash control 缓冲控制;按钮控制;按钮操纵
dash current 冲击电流;超值电流
dash-dot-line 点画线
dash-dotted line 画点线;点画线
dashed and dotted line 点画虚线
dashed area 晕线面积;阴影(线)部分
dashed contour 虚线等高线
dashed contour line 虚线等高线
dashed line 阴影线;虚线;短画(虚)线
dashed-line contour plot 虚线等高线图
dashed-line plotting 绘虚线
dasher 遮泥(水)板;挡水板;挡泥板;反波板
dasher block 单饼小滑车
dash finish 浇泼饰面;浇泼粉面;浇泼饰面;浇涂饰面
dashing lines (车身的)运动速率线
dashlamp 仪表板(照明)灯
dashlight 仪表板灯
dash number 零件编号
dash out 冲出;涂掉
dash pannel 仪表板
dash plate 缓冲板;挡水板;挡浪板
dashpot 阻尼器;消振器;缓冲器;缓冲筒
dashpot check valve 缓冲单向阀
dashpot governor 带缓冲器的调速器
dashpot piston 减震活塞
dashpot rotameter 带有阻尼器的转子流量计
dash receiver 接收板
dash unit 仪表板(装置)
dash valve 冲击阀
Dasikong Cun type pottery 大司空村类型陶器
dask adjustment 缓冲调节
dason (防潮层用的)沥青粗麻布
dasymeter 炉热消耗计;气体密度测定仪;气体成分测定仪
dasymetric map 范围密度法地图
dasymetric technique 区域统计图法
data 资料;数据
data accepted flag 数据接收标志
data access 数据存取
data access method 数据访问法;数据存取方式;数据存取法
data access path 数据访问路径;数据存取通路
data-access register 数据访问寄存器;数据存取寄存器
data access system 数据访问系统;数据存取系统
data accumulation 数据累积;数据积累
data accumulator 数据累加器;数据存储器
data acquisition 资料收集;数据收集;数据获取;数据采集
data acquisition and check system 资料收集与核对系统
data acquisition and control system 数据收集与控制系统;数据采集和控制系统
data acquisition and distribution 数据收集与分配
data acquisition and interpretation system 数据

收集和整理系统
data acquisition and monitoring equipment 数据采集和监测设备
data acquisition and monitoring equipment for computer 计算机的数据收集与监控设备
data acquisition and processing 数据采集与处理
data acquisition and processing system 数据收集与处理系统
data acquisition and recording system 数据收集与记录系统
data acquisition and reduction system 数据收集与简化系统
data acquisition camera 数据采集照相机
data acquisition center 数据收集中心
data acquisition computer 数据收集计算机
data acquisition controller 数据收集控制器
data acquisition equipment 数据采集设备
data acquisition form 数据收集形式
data acquisition instrument 数据采集记录器
data acquisition logging system 数据收集记录系统
data acquisition platform 数据收集站;数据获取平台
data acquisition probe 数据采集探测
data acquisition stage 数据获取级
data-acquisition statement 数据接收站
data acquisition system 数据搜索系统;数据获取系统;数据获得系统;数据采集系统
data acquisition unit 数据收集装置
data adapter unit 数据转接器
data address 数据地址
data-addressed memory 数据编址存储器;数据地址存储
data administrator 数据管理员
data age 取数时间;取数据时间
data agent 数据代理终端
data-aided loop 数据辅助环
data alignment 数据对齐;数据调整
data alignment attribute 数据调整属性
data alignment network 数据调整网络
data analysis 资料分析;数据分析
data analysis center 资料分析中心;数据分析中心
data analysis console 数据分析控制台
data analysis facility 数据分析设备
data analysis of coal log 煤田测井资料分析
data analysis/preliminary image interpretation 资料分析图像初译
data analysis system 数据分析系统
data analysis technique 数据分析技术
data area 数据区域;数据范围
data area number 数据区编号
data array 数据组;数据阵列
data association message 数据相关信息
data assurance unit 确信数据接收部件
data at seal 轴封处数据
data attribute 数据属性
data attribute list 数据属性表
data automation system 数据自动处理系统
data available 可用数据
data average 数据平均挡
data bandling system 数据处理设备
data bank 资料库;数据(总)库;数据资料库;数据库数据存储器;数据库集
data bank handler 数据库程序
data banking 数据储存
data bank parameter 数据库参数
data base 基本数据;资料库;数据库;数据基;数据单元
database abstraction 抽象数据库;数据库抽象化
database access language 数据库存取语言
database administrator 数据库管理员;数据库管理程序
database application system 数据库应用系统
database availability 数据库有效率
database consistence 数据库一致性
database consistency 数据库一致性
database control 数据库控制程序
database data model 数据库数据模型
database data name 数据库数据名
database definition language 数据库定义语言
database description 数据库描述
database description entry 数据库描述项
database design 数据库设计
database directory 数据库目录
data-based microinstruction 以数据为基础的微指令
data-based microinstruction cycle 以数据为基础的微指令周期
database engine 数据库引擎
database exception condition 数据库异常条件
database generator 数据库生成程序
database graph software 数据库绘图软件
database handler 数据库程序
database identifier 数据库标识符
database initialization 数据库预置;数据库创建;数据库初始化
database interpretation 数据库翻译
database key 数据库(关键)码
database key item 数据库码项
database machine 数据库机
database management 数据库管理
database management system 文献数据管理系统;数据库管理系统
database management system computer 数据库管理系统计算机
database manager 数据库管理员
database mapping 数据库制图
database model 数据库模型
database modifier 数据库修饰符
database parameter 数据库参数
database portability 数据库可移植性
database practitioner 数据库设计员
database procedure 数据库过程;数据库程序
database result wizard 数据库结果向导
database server 数据库服务器
database storage 数据库存储器
database subsystem 数据库子系统
database system 数据库系统
database utility 数据库实用程序
data bit 数据位;数据区
data block 数据块
data board 记录板
data book 资料工具书;参考书;参考资料手册;标准产品手册;数据手册;清单;数据表
data bridge 数据驾驶台
data bridge simulator 航海模拟训练仪
data buffer 数据缓冲器
data bulk 数据数量
data buoy 数据收集浮标站;数据浮标
data buoy system 数据浮标系统
data bus 数据总线
data bus coupler 数据总线耦合器
data cable 数据(同步)传输电缆
data capacity 信息容量;数据容量
data capsule 数据容器
data capture 数据获取
data capture system 资料收集系统;数据收集系统
data card 数据卡片
data carrier 数据载子;数据载波;数据(记录)载体;数据记录媒体;数据记录介质
data carrier detector 数据载波检波器
data carrier storage 数据载体存储器
data cell 数据单元
data center 数据中心;资料中心
data chain(ing) 数据(链)取;数据链接
data chamber 数据记录装置
data change proposal 数据更改建议
data change table 数据变化表
data channel 数据信道;数据通道
data channel converter 数据信道转换器
data channel cycle 数据通道周期
data channel cycle stealing 数据通道周期挪用
data channel interrupt 数据通道中断
data channel line printer 数据通道行式打印机
data channel multiplexer 数据通道复用器
data check 数据校验
data check list 数据检查表
data circuit 数据线路;数据电路
data circuit equipment 数据电路设备
data circuit switching system 数据电路交换系统
data circuit terminating equipment 数据电路终接设备
data circuit transparency 数据电路透明性
data class 数据类
data clause 数据子句
data clustering 数据分组
data code 数据代码;数据编码系统
data code indexing 数据码检索
data code translation 数据代码转换;数据代码翻译
data collected in karst investigation 岩溶调查数据采集
data collecting equipment 数据采集设备
data collecting of saline soil 盐渍土的数据采集
data collecting platform 数据采集平台
data collecting system 数据收集系统
data collection 资料收集;数据收集;数据集合;数据汇集;数据采集
data collection form 数据收集形式
data collection method 数据收集方法
data collection of observation net and points 观测网点数据采集
data collection of well cleaning 洗井工作数据采集
data collection platform 数据收集站;数据采集平台;数据采集平台
data collection satellite 数据收集卫星
data collection system 资料收集系统
data collector 数据收集器
data collector terminal 数据收集终端
data colo(u)r system 数据彩色系统
data communication 数据通信;数据电信;数据传输
data communication adapter unit 数据通信转接器
data communication channel 数据通信信道
data communication equipment 数据通信设备;数据传输设备
data communication input buffer 数据通信输入缓冲器
data communication link 数据通信线路
data communication monitor 数据通信监视器;数据通信监督端
data communication multiplexer 数据通信多路复用器
data communication network 数据通信网(络);数据传输网
data communication processor 数据通信处理机
data communication protocol 数据通信协议
data communication service 数据通信业务
data communication station 数据通信站
data communication system 数据通信系统
data communication terminal 数据通信终端
data compaction 数据压缩;数据精简;数据紧缩
data comparator 数据比较仪
data compatibility 数据兼容性
data compensation routine 数据补偿程序
data compilation 数据搜集
data complexity 数据复杂性
data component 数据成分
data compression 数据压缩
data concentration formatting 数据集中格式
data concentrator 数据集中器;数据汇集器
data condensation 数据压缩
data condensation algorithm 数据压缩算法
data condition code 数据条件码
data configuration 数据构形
data consistency 数据一致性
data control block 数据控制块
data control interval 数据控制间隔
data control system 数据控制系统
data control unit 数据控制装置
data conversion 数据转换;数据简化;数据换算;数据归纳;数据处理;数据变换
data conversion feature 数据转换特性
data conversion hardware 数据换算硬件
data conversion line 数据转换线
data conversion receiver 数据变换接收机
data conversion system 数据转换系统;数据变换系统
data converted to English system 换算成英制资料
data converter 数据转换器;数据转换程序;数据变换器
data copy 数据副本
data correction 数据校正
data correlation 数据相关
data counts 数据计数
data coupler 数据耦合器
data curve 试验测定曲线
data declaration statement 数据说明语句
data definition 数据定义
data definition mode 数据定义方式
data degradation 数据消减
data delimiter 数据分隔符;数据定义符;数据定界符
data demand 数据要求;数据请求
data demand protocol 数据要求协议
data description 数据描述

data description entry 数据描述(记入)项
data description language 数据描述语言
data description program(me) 数据描述程序
data descriptor 数据描述符
data design 数据设计
data design layout 数据设计方案
data device 数据装置
data dictionary 数据字典
data digital handling 数据数字处理
data-directed input-output 直接数据输入输出
data-directed output 数据定向输出
data directed transmission 数据定向传输
data direction register 数据流向寄存器
data display 数据显示
data display center 数据显示中心
data display controller 数据显示控制器
data display equipment 数据显示器
data display system 数据显示系统
data display unit 数据显示器
data distribution centre 数据分配中心
data distribution list 数据分配表
data distribution plan 数据分配计划
data distribution system 数据分配系统
data division 数据部分
data drawing list 资料图纸清单
data dump 数据转换;数据转储
data editing 数据编辑
data editing system 数据编辑系统
data edition 数据编辑
data element 数据类目;数据(单)元
data encoder 数据编码器
data encoding system 数据译码系统;数据编码系统
data encryption 数据加密
data end command 数据结束指令
data entity 数据实体
data entry 数据输入;数据进入项;数据登记项
data entry keyboard 数据输入键盘
data error rate 数据差错率
data evaluation 数据评估;数据估计
data evaluation technique 数据评价法
data exception 数据异常
data exchange 数据交换
data exchange program(me) 数据换算程序
data exchanger 数据交换器;数据交换机
data exchange service 数据交换业务
data exchange system 数据交换系统
data exchange unit 数据交换装置
data extend block 数据扩充块
data extent block 数据扩充块
data extractor 数据分离器
data feedback 数据反馈
data fetch 取数据;数据取出
data field 数据区
data file 固定数据文件;数据文件;数据外存储器
data file converter 数据文件转换
data file generation 数据文件生成
data film 数据胶片
data filtering 数据筛选;数据滤波;数据过滤
data filtering network 数据滤波网络
data fitting 数据拟合
data flow 数据流
data flowchart 数据流程图
data flow computer 数据流计算机
data flow diagram 数据流程图
data flow instruction 数据流指令
data flow language 数据流语言
data flow operator 数据流操作符
data flow procedure 数据流过程
data form 资料记录表
data format 数据格式
data formatting 数据格式编排
data for reference 参考数据
data forwarding 数据前送
data frame 数据帧
data gathering 资料收集;数据收集;数据汇集;收集资料
data gathering method 数据收集方法
data gathering panel 数据收集盘
data gathering set 数据收集设备
data gathering system 数据收集系统
data given in tabular form 表列资料;表列数据
datagram 数据图;数据报
datagram service 数据报业务
datagraphics 数据图形

data handbook 数据手册
data handler 数据自动处理器;数据(信息)处理器
data handling 信息处理;资料整理;资料整编;数据处理
data-handling capacity 数据处理能力;信息处理容量;数据处理容量
data-handling center 数据处理中心
data-handling component 数据转换元件;数据处理元件
data-handling equipment 数据处理装置;数据处理设备
data-handling program(me) 数据处理程序
data-handling routine 数据处理程序
data-handling system 数据处理系统
data head 数据头
data header 数据标题
data hierarchy 数据分级结构;数据层次
data highway 数据信息通路;数据高速通路
data hold 数据保持装置;数据保持
data identification 资料鉴定;数据识别;数据鉴定
data-in 数据输入器;数据输入;输入数据
data index 数据索引
data information line 数据信息线路
data information signal generator 数据信息信号发生器;数据信号发生器
data initialization statement 数据预置语句
data-initiated control 数据启动控制
data input 数据输入
data input bus 数据输入总线
data input pin 数据输入端
data input procedure 数据输入法
data input statement 数据输入语句
data input supervisor 数据输入监控器
data inputting methodology 数据输入分类法
data-input unit 数据输入器
data inscriber 数据记录器
data inserter 数据输入器;数据插入程序;数据插入
data insertion converter 数据插入变换器
data insertion system 数据插入系统
data inspection station 数据检查台
data integrity 数据完整性
data interchange 数据互换
data interface 数据界面
data interpretation of well testing 测试资料解释方法
data inversion 数据反演
data item 数据项;数据细目;数变项
data item validation 数据项验证
data key 数据键
datal 日薪
data layout 数据格式;数据打印格式
data level 数据级
data level access 数据级存取
data library 数据库
data library handler 数据库程序
data library parameter 数据库参数
data line 数据(传输)线
data line interface 数据线路接口
data line terminal 数据线路终端
data line terminating equipment 数据线路终端设备
data link 数据自动传输装置;数据(自动)传输器;数据中继器;数据率;数据链路;数据传输线
data link communicator 数据传输发信机
data link configuration 数据链路配制
data link connection identifier 数据链路连接标识
data link connector 数据链路连接器
data link control 数据链路控制;数据传输控制
data link control procedure 高级数据链路控制规程
data link escape 数据通信换码;数据传送换码
data link escape character 数据通信换码字符
data link system 数据传输系统
data link terminal 数据传输线路终端
data list 数据(列)表
data list structure 数据表结构
dataller 计日工
datalling 爆落顶板;挑顶
data location 数据单元
data locking 数据联锁
data logger 数据巡回检测装置;数值记录表;数据巡回检测器;数据输出器;数据记录器;数据记录表
data logger and verifier 数据记录器和核对器
data logger checker 数据输出校验器
data logger operation(al) desk 数据记录工作台

data-logging 数据自动记录;数据(资料)记录;数据巡回检测
data-logging device 数据记录装置
data-logging equipment 数据记录装置;动力装置参数自记仪;数据巡回检测装置
data-logging instrument 数据采集记录器
data-logging machine 数据记录机
data-logging-recording equipment 数据存入和记录设备
data-logging system 参数自动记录系统;数据自动记录系统;数据巡回检测系统
data-logging technique 数据记录技术
data management 数据管理
data management and analysis system 数据管理与分析系统
data management common code 数据管理公共代码
data management function 数据管理功能
data management program(me) 数据管理程序
data management system 数据管理系统
data manipulation 数据处理;数据操作
data manipulation language 数据操作语言
data manipulation register 数据操作寄存器
data map 资料图;数据图
data materialization 数据实体化
datamation 自动处理数据;自动数据处理;数据自动化;数据化;数据(处理)自动化
data matrix 数据矩阵
data measuring system 数据测量系统
data medium 数据(记录)载体;数据(记录)媒体
data message switching system 数据通路交换系统;数据报文交换系统
data migration 数据迁移;数据进位
data mode 数据状态;数据传送方式
data model 数据模型
data modeling 数据模拟
data modem 数据调制解调器
data monitoring system 数据监控系统
data multiplexer 数据多路复用器
data multiplexing system 数据多路复用系统
data name 数据名(称)
data network 数据网(络)
data network identification code 数据网络码
data networking 数据联网
data noise 偶然误差;速度测量误差
data number 数据编号
data of arrival 到达日期
data of component 分量记录数据
data of cross-section(al) comparison 截面对比资料
data of dark-line spectrum 暗线光谱数据
data of datum logging 数据收集日期
data of engineering construction 工程建筑资料
data of estimate 预算资料
data of fold elements 褶皱要素数据
data of groundwater development 地下水开采数据
data of groundwater regime observation 地下水动态观测数据
data of groundwater resources 地下水资源评价数据采集
data of inner orientation 相片内方位元素
data of mineral deposit-hydrogeology 矿床水文地质数据
data of mineral exploration engineering 探矿工程资料
data of observation 观测数据
data of other troubles 其他事故数据
data of outer orientation 相片外方位元素;外方位元素
data of reconnaissance and survey 勘测资料
data of regional geochemistry and geophysics 区域物化探资料
data of self-purification 自净作用数据
data of sticking 卡钻事故数据
data of strain measurement 应变测量数据
data of tectonic stress field 构造应力场数据
data of test engineering arrangement 试验工程布置数据
data of testing project 测试工程数据
data of water resources planning 水资源规划数据
data of well completion 完井数据
data of well structure 水井结构数据
data on environmentally significant chemicals

network 主要环境化学品数据网
data on flows 流量数据
data operation control complex 数据作业控制综合设施
data optimizing computer 数据最佳计算机;数据优化计算机
data ordering 数据整理
data organization 数据组织
data-oriented testing 面向数据的检验
data origination 资料来源;数据起点;数据来源;数据初(始)加工
dataout 输出数据;抹去数据
data output 数据输出
data output gate 数据输出门
data output level 数据输出电平
data overrun 数据过量运行
data ownership 数据谱系关系
data packet 数据分组;数据包
data panel 数据传送分配板;数据处理盘
data path 数据通路
data pattern 数据模式
data phase 数据阶段
dataphone 数据送话器;数据电话机
dataphone digital system 数据电话数字系统
dataphone service 数据电话业务
data pick-up 数据传感器
data pilot 数据驾驶仪
data plate 技术规格标牌;铭牌
data playback system 数据回放系统
data plotter 数据绘图仪;数据标绘仪;数据标绘器
data point 数据点
data pointer 数据指针
data pool 数据源;数据库
data positioning 数据定位
data potentiometer 数据输出分压器
data power 数据电源系统
data preparation 资料准备;数据准备
data preprocessing 数据预处理
data presentation 数据显示
data presentation device 数据表达装置
data presentation technique 数据显示技术
data printer 数据打印机
data privacy 数据保密
data processing 资料整理;资料整编;资料处理;数据加工;数据处理
data processing activity 数据处理活动
data processing algorithm 数据处理算法
data processing and interpretation 数据处理和解释推断
data processing center 数据处理中心
data processing computer 数据处理计算机
data processing device 数据处理装置
data processing element 数据处理单元
data-processing equipment 数据处理装置;数据处理设备
data processing installation 数据处理设备
data-processing inventory 数据处理清单
data processing machine 数据处理机
data processing manager 数据资料处理负责人
data processing method 数据处理(方)法
data processing of gravity anomaly 重力异常数据处理
data processing of magnetic anomaly 磁异常数据处理
data processing of surveying and mapping 测绘数据处理
data processing recorder 数据处理记录器
data processing sequence 数据处理序列
data processing stage 数据处理阶段
data processing system 数据处理系统
data processing technique 数据处理技术
data-processing terminal equipment 数据处理终端设备
data processing time 数据处理时间
data-processing unit 数据处理装置;数据处理设备;数据处理部件
data processor 数据处理器;数据处理机
data profile 数据剖面
data protection 数据保护
data pulse 数据脉冲
data purification 数据提纯;数据精练;数据精化
data qualification 数据限制;数据条件;数据分类
data quality control 资料质量控制
data quality monitoring system 数据质量监控系统

data radar 数据雷达
data rate 信息率;数据(数)率
data reader 数据读出器
data reading system 数据读数系统
data readout setup 数据读出装置
data ready 数据收发准备状态
data ready flag 数据就绪标记
data realignment 数据重定位
data realm 数据区域
data rearrangement 数据重排列
data recall 数据复示
data receiver 数据接收机
data recorded 记录号
data recorder 数据记录器
data record format 数据记录格式
data record(ing) 读数记录;数据(自动)记录
data recording amplifier 数据记录放大器
data recording capability 数据记录能力
data recording head 数据记录头
data recording processing 数据记录处理
data recording system 数据自动记录系统
data record size 数据记录长度
data recovery 信息整理
data reducer 数据简化器;数据变换器
data reduction 资料整理;资料折算;资料归算;资料归纳;数据整删;数据整理;数据压缩;数据简化;数据换算;数据归纳;数据处理
data reduction and distribution 数据处理与分布
data reduction and processing 数据简化和处理装置
data reduction center 数据处理中心
data reduction input program(me) 数据简化输入程序
data reduction program(me) 数据压缩程序;数据简化程序
data reduction subroutine 数据压缩子程序
data reduction subsystem 数据压缩子系统;数据压缩程序
data reduction system 数据简化系统
data register 数据寄存器
data-rejection 数据剔除;数据淘汰
data relay 数据中转;数据中继
data relay center 数据转送中心
data relay satellite 数据中继卫星
data relay satellite system 数据中继卫星系统
data reliability 资料可靠性;数据可靠性
data repeater 数据重发器;数据信号放大器;数据传送放大器
data report 资料报告;数据报告
data repository 资料馆
data representation 数据表示法
data reproduction 数据再生;数据复制
data restructuring 数据再构成
data retrieval 资料检索;数据检索
data retrieval system 数据检索系统
data revision 数据修正
data safety 数据安全
data safety guard 数据保护
data sailing 数据航行
data sample 样本资料
data sampling 数据抽样
data sampling switch 数据采样开关
data sampling unit 数据取样仪
data save mode 数据保存方式
data search 收集资料
data segment 数据段
data separation 信息区分;数据区分
data separator 数据区分符号
data sequence 数据序列
data service units 信道服务单元;数据服务单元
data set 数据组;数据装置;数据器;数据集;数据传输转换器;数据传输设备;数据传输机
data set access 数据集存取
data set adapter 数据传输机适配器
data set allocation 数据集分配
data set clocking 数据组定时;数据装置同步;数传机同步
data set coupler 数据集连接器
data set entry 数据集入口
data set identification 数据集识别
data set information 数据集信息
data set in-out 数据集输出入
data set instruction 数据集指令
data set label 数据集标号

data set migration 数据集迁移
data set name 数据集名
data set organization 数据集结构
data set ready 数据装置就绪;数据集就绪
data set reference number 数据集引用号
data set sequence number 数据集顺序号
data set serial number 数据集序列号
data set utility 数据集实用程序
data sharing 数据共享
data sheet 记录表;一览表;单页资料;数据一览表;数据(记录)表;数据单
data shift right 数据右移
data signal 数据信号
data signal(l)ing 数据信号传输
data signal(l)ing rate 数据信号速率;数据(信号)传输率;数据通信速度
data signal(l)ing rate transparency 数据通信速率透明性
data simulation 数据模拟
data sink 数据接收终端机;数据接收器;数据汇接器
data size 数据量
data slit 数据间隙
data slot 数据时间片
data smoothing 数据修匀;数据平滑;数据加工
data smoothing network 数据平滑网络;数据滤波网络
data snooping 数据探测法
data sorting system 数据分拣设备
data source 资料来源;数据源;数据发送器
data source and sink 数据收发装置
data sources 数据来源
data spacing error 数据间距误差
data specification 数据指定
data speed 数据速度
data stabilization 数据稳定
data stack 数据栈
data staging 数据分级
data station 数据站
data status word 数据状态字
data storage 信息存储(器);数据存储(器);数据储存(器)
data storage and retrieval 数据储存和检索
data storage and retrieval system 数据存储与检索系统
data storage area 数据存储区
data storage device 数据存储装置;数据存储器
data storage equipment 数据存储设备
data storage register 数据存储自动记录
data storage system 数据存储系统
data storage unit 数据存储装置
data store block 数据存储块
data stream 数据流
data strobe pulse 数据选通脉冲
data structure 数据结构
data structure choice 数据结构选择
data structure diagram 数据结构图
data structured system development method 数据结构化系统开发方法
data structure mapping 数据结构变换
data structure tree 数据结构树
data sublanguage 数据子语言
data submodel 数据子模型
data subset 数据子集
data-switch(ing) 数据转接;数据转换;数据交换
data-switching center 数据转换中心;数据交换中心
data-switching exchange 数据交换机
data synchro 数据自整机机;数据同步器
data synchronization 数据同步
data synchronization unit 数据同步装置;数据同步器
data system interface 数据系统接口
data systems integration 数据系统集合
data system specifications 数据系统规范
data table block 数据表块
data table of geodip 地层产状数据列表
data table of inclination 井斜数据列表
data table of interpretation result for dip log 地层倾角测井解释成果数据列表
data tablet 数据图形输入板
data taken signal 取数据信号
data take-off 数据消除
data tape 数据(记录)带
Data Telecommunication Center 全国银行电汇电脑中心(日本)

data telemetry 数据遥测(术)
data telemetry system 数据遥测系统
data telephone circuit 数据电话电路
data telex 数据电传
data terminal 数据终端
data terminal equipment 数据终端设备
data terminal message compiler and transmission system 数据终端信息自动编码和传输系统
data terminal ready 数据终端就绪
data terminal unit 数据终端装置
data time 数据存取时间
data to leaf analysis 叶片分析数据
data track 数据(磁)道
data tract 边界四址已定的土地
data traffic director 数据传输控制器
data transcription 数据转录;数据副本
data transcription equipment 数据转录设备
data transducer 数据传感器
data transfer 数据转移;数据转换;数据传送;数据传输;数据传递
data transfer interface controller 数据转换接口控制器
data transfer rate 数据传送速度;数据(传送)率;数据传递速度
data transfer system 数据转换系统
data transfer unit 数据转换装置
data transformation 数据变换
data translating system 数据变换系统;数据变化系统;数据翻译系统
data transmission 资料发送;资料传送;数据发射;数据传送;数据传输
data transmission and switching 数据传输与转换
data transmission block 数据传输块
data transmission channel 数据传输信道
data transmission device 数据传输装置
data transmission echoing unit 数据传输反回装置
data transmission efficiency 数据传输效率
data transmission interface 数据传接口
data transmission line 数据传输线
data transmission link 数据传输线路
data transmission package 数据传输装置
data transmission ratio 数据传输率
data transmission setup 数据传输装置
data transmission speed 数据传输速度
data transmission system 数据传输系统;数据传递系统
data transmission terminal 数据传输终端
data transmission terminal equipment 数据传输终端设备
data transmission testing set 数据传送系统测试器
data transmission unit 数据传输装置
data transmission utilization measure 数据传输效率
data transmission video display unit 数据传输视频显示器
data transmitting equipment 数据发送设备
data treatment 资料处理;数据处理
data trend 数据动向系统
datatron 数据处理机
data type 数据类型
data type conversion 数据类型转换
data type of a variable 变量(的)数据类型
data under voice 低于音频的数据
data unit 数据装置;数据机;数据发送部件;数据单元
data use identifier 数据使用标识符
data validation 数据有效性;数据校验;数据核实
data validity 数据真实性;数据有效性
data vector 数据向量
data verification 数据验证;数据检验
data verifying program(me) 数据校验程序
data volume 数据卷宗;数据(数)量
data way model 数据通路模型
data well 资料井
data window 数据窗
data wire 数据线
data withdrawal 数据抽出;数据撤出
data word 数据字
da(t)cha 俄罗斯乡间别墅
date 日期
date and time 日期和时间
date bill 远期票据
datebook 记事册
date closing 结算日(期);结算
date commenced 开始日期

date-compiled 编译完成日期
dated 保险日期
dated consumption claims 标时消费请求权
dated date 有效执行日期
dated deposit 定期存款
dated earned surplus 注明日期公积金
date declaration 宣布日期
date draft 定期汇票;发票日后付款汇票
dated stock 有具体偿还期债券
date due 满期;到期日
date for inspection 检查日期
date for inspection and testing 检查和检验日期
date for latest delivery of bids by bidder 投标人投标日期
date for testing 检验日期
date forward 预填日期
date indicator 日历指示器
date issued 发出日期
Datel circuit 得泰尔线路
date-letter 瓷器制造日期印记
date line 国际日期变更线;日界线
date mark 制造日期标记;瓷器制造日期印记;日期戳(记);日戳
date of acceptance 承兑日期
date of acquisition 需求日期
date of agreement 协定日期;合约签订日期;合同生效日期;协议日期
date of a letter 发函日期
date of appraisal 鉴定日期
date of approval investigation report 调查报告批准时间
date of approval of the contract 批准合同之日
date of availability 有效期限
date of balance sheet 资产负债表日期
date of built 建造日期
date of certification 证明日期
date of clearance 结关日期;票据交换日期
date of closing 结算日
date of commencement of the work 开工日期;建筑开工日期
date of complete of discharging 卸空日期
date of completion 竣工日期;完工日期
date of contract 合同签订日期
date of contraction 签约日期
date of data processing 资料处理日期
date of declaration 宣布日(期)
date of delivery 交货(日)期;交割日期;交船日期
date of departure 开航日期;启航日期
date of despatch 发运日期
date of discharge (义务等的)履行期限;卸货日(期);(债务)清偿期限
date of discharging 卸货日(期)
date of draft 汇票日期;出票日期
date of enforcement 实施日期
date of entry declaration 申报进口日期
date of entry into force 生效日期
date of evaluation 评价日期
date of exigibility of the obligation 债务到期日
date of expiration 有效日期
date of expiry 截止日期;有效日期;终止日期
date of extension 延长日期
date of finish investigation planning 调查工作设计完成时间
date of finish investigation report finished 调查报告完成时间
date of full availability 交付使用日期
date of installation required 要求安装日期
date of investigation work 已有调查工作日期
date of issuance 发单日期
date of issue 出票日期;发证日
date of landing 卸货日(期);卸岸日期
date of last repairs 上次检修日期
date of launching 下水日期
date of locating point 定点日期
date of location 定孔位日期
date of logging 测井日期
date of loss 失踪日期
date of manufacture 制造日期;出厂日期;生产日期
date of maturity 到期日;成熟日期;票据到期日期
date of maturity of the obligation 债务到期日
date of mining begins 投产时间
date of new edition 新版日期
date of non-conformity 违章日期
date of original register 原始登记日

date of payment 支付日期;付款日期
date of performance 履行日期
date of postmark 邮戳日期
date of printing 印刷日期
date of production 出厂日期
date of publication 出版日期
date of recharge 回灌日期
date of record 截止过户日期
date of retirement (固定资产)报废日期;报废期
date of sailing 开航日期
date of seeding 播期
date of settlement 决算日期
date of shipment 装运日期;付运日期;启运日期
date of shipping 装船期限
date of standard implementation 标准实施日期
date of substantial completion 工程完成交付使用日期;实际完工日期
date of survey 检验日期;测量日期
date of tax levied 纳税日期
date of termination 合同终止期;终止日
date of test(ing) 试验日期
date of testing ground established 试验场建立日期
date of validity 有效日期
date of value 起息日;起算利息日期
date of well cleaning 洗井日期
date on dock 进坞日期
date on which the claim become due 应履行债务日期
date on which the claim becomes due 债务到期日
date palm 枣椰树
date payable (股票)派息日
dater 日期戳子
date rate 日工资率
date received 收到日期
date sent to library 入库日期
date set for the opening of tenders 开标日期
date stamp 日期戳
date terms 日期条件
date tested 试验日期
date time group 日期时间组;时序分组
date to which payment becomes due 付款到期日
datholite[datolite, datolith]硅硼钙石;色球石
Datien bay 达连湾
dating 记日期;年代测定;信贷延期;断定年代;定年龄;测定年代
dating ahead 日期填早
dating backward 倒填日期;日期填迟
dating forward 开票日期提前;日期填早
dating method of groundwater 地下水年龄测定法
dating pulse 整步脉冲
dating routine 记日程序
dating subroutine 记日期子程序
dative bond 配价键
datometer 单日历表
datum 基准;基点;已知数;资料;测站零点
datum axis 基准轴(线);基线
datum axle 底座轴
datum bed 标准层
datum benchmark 水准基点
datum body 基准体
datum-centered ellipsoid 局部最佳拟合椭圆;参数共轴椭球
datum course 起始航向
datum deviation 基准偏差
datum diameter 基准直径
datum dimension 基面线尺寸;基面点尺寸
datum drag 基本阻力
datum drift 基准漂移;基准偏差
datum elevation 基准高程
datum error 基准误差
datum for reduction of sounding 测深折算基点
datum grade 基准标高
datum hole 基(准)孔
datum horizon 标准层位(作构造等高线用);标志层;基准地层
datum level 深度基准面;水准线;零电平;基准液面;基(准)面;高程基准(面);基准水平面
datum-limit protection 上下界防护
datum line 基(准)线;基照线
datum line system 基准线系统
datum mark 基准点(标高);基准标志;标高;水准点
datum marker buoy 基准面位置浮标;基准面标志浮标

datum mark level 基准点标高
datum mark of level(l)ing 水准原点
datum node 基准节点;参考结点
datum of soundings 测深基(准)面
datum of tidal level 潮位零点;潮位基准面
datum peg 基准桩
datum plane 基准水平(面);假设零位面;假设零点;基准(平)面;基(线)面
datum plane of ga(u)ge 测站基面;水尺零点
datum plane of identical atmospheric pressure 参考等压面
datum plane position 基准平面位置
datum point 基准点;固定点【测】;参考点
datum position parameter 大地基准定位参数
datum profile 基准齿廓
datum quantity 基准量;参考量
datum ray 基准光线
datum reference 参考基准(面)
datum setting 基准线确定;基准位置装定
datum speed 标称速度
datum sphere 基准球面
datum static 基准面静校正
datum station peg 基准测站标桩
datum surface 基(准)面;数据基准面
datum surface coordinate 基准面坐标
datum survey 基础资料调查
datum target 基准目标
datum temperature 基准温度
datum time 起始时间
datum transformation 大地基准变换
datum value 基准值
datum water level 水准面;基准水平面;基准水位;基点水位;水准零点;水位零点
datum zero 基准零点;图表零点
datura innoxia 毛曼陀罗
daub 涂抹;斧凿面;底涂;底色;打底色;粗抹面(层);胶泥;抹纸筋灰;抹泥;抹胶;打泥底;粗灰泥;粗灰;草土泥;皮革底色
daub coat 漆革第一层涂料
Daubenton's plane 道本顿平面
dauber 涂抹工具;泥水工;抿墙工人;抹工;涂抹者
daubing 凿纹;涂料;炉衬局部修理;灰泥;粗抹涂料;材料;泥灰抛毛;粗抹灰泥;石面凿毛;琢凿石面;灰面拉毛
daubing mud 搪料
daub of leather 皮革打底色
daubreeite 铋土
daubreelite 陨硫铬铁矿
Daubresse prism 道布莱斯棱镜
dauby 胶粘的
daugh 耐火黏土;软耐火黏土
daugher house (回教的)闺房
daughter 子系;子体;子核
daughter average 女儿平均(值)法
daughter board 子插件
daughter company 子公司
daughter crystal bearing inclusion 含子矿物包裹体
daughter mineral 子矿物
daughter product 衰变产物;子产物
daughter radioactivity 子体放射性
daughter radioisotope 派生放射性同位素
daughter ship 子驳
daughter substance 子系物质
daughter system 子系统
dauk 密实黏土砂岩;黑色碳质页岩;亚黏土;砂质黏土
daul distribution 双轨销售
daunialite 蒙脱石;硅质蒙脱岩;沉积膨润土
Dauphine law 多菲定律;道芬律
Dauphine twin 道芬双晶
Dauphine twin law 道芬双晶律
dauphinite 镁钛矿
davanite 硅钾钛石
Davcra cell 达夫克拉喷气式浮选机
davenport 两用沙发;活动小书桌;长沙发;书桌;沙发床
davenport bed 坐卧两用沙发
davetyn 达维廷呢
davey hook 滑钩
David graph 戴维图
davidite 镧铀钛矿;铀钛磁铁矿;铈铀钛铁矿
Davidon-Flecher-Powell variable metric method 大卫顿-佛拉歇-鲍威变尺度法
daviesite 柱氯铅矿
Davignon 达维南金铜铝合金

Davis bit 齿状钻头
Davis bronze 戴维斯镍青铜
Davis-Bruning colo(u)rimeter 戴维斯-布鲁宁比色计
Davis equation 戴维斯方程
Davis-Gibson filter 戴维斯-吉伯逊滤色器
Davis magnetic tester 戴维斯磁性测定仪
davisonite 板磷钙铝石
Davis raft 戴维斯排
Davisson coordinate 戴维森坐标
Davis Strait 戴维斯海峡
Davis table 戴维斯方位表
davit 挂艇架;吊(艇)柱;吊艇杆
davit arm 吊艇架臂
davit bearing 吊艇柱座承
davit bend 吊艇柱弯头
davit bollard 吊艇柱缆桩
davit bust 吊艇柱座承
davit cleat 吊艇柱羊角;吊艇挽耳
davit collar 吊艇柱挽耳
davit craft 小吊艇
davit crane 吊艇起重机
davit cranse 吊艇柱座承
davit guy 吊柱牵条;吊艇柱牵索;吊杆牵索
davit keeper 吊艇柱座承
davit launched inflatable lifecraft 吊放式气胀式救生筏
davit-launched life raft 吊放式救生筏
davit pedestal 吊艇柱基座
davit ring 吊艇柱座承
davit shoe 吊艇柱基座
davit socket 吊艇柱基座
davit span 吊艇柱头部连动系;吊艇杆跨索
davit spreader 吊艇柱顶环
davit stand 吊艇柱座
davit tackle 吊艇架绞辘
davit winch 吊艇绞车
davit winch interlocking device 吊艇绞车连锁装置
davreuxite 水硅铝锰石
Davy lamp 德氏安全灯
davyne(e) 钾钙霞石
Davy safety lamp 达维安全汽油灯
Dawes' limit 道斯极限
dawk (印度供旅游者住宿的)配凉台高平房;(含有矿脉的)黏质砂岩;黑色碳质页岩
dawnay's lintel 耐火黏土过梁(负荷楼板用的)
dawn effect 曙光效应
dawn pink 曙粉红色
dawn redwood 水杉
dawn shift 清晨班
dawn side 黎明侧
dawnside magnetosphere 晨侧磁层
dawnside magnetotail 晨侧磁层
dawsonite 碳钠铝石
Dawson's bronze 道氏青铜;道森轴承青铜
Daxi culture pottery 大溪文化陶器
d-axis 直轴(的)
d-axis subtransient reactance 直轴起始瞬态电抗
daxk box 暗箱
day 窗格
daya 积水洼地;小落水洞
day airglow 白昼大气辉光
day-and-date device 双历装置
day and night 昼夜
day-and-night changed bright zone 昼夜变化的强光带
day-and-night image 日夜成像
day and night telescope 船用夜视望远镜
day and night-time reconnaissance 日夜勘测
Dayao 大窑
day arc 昼弧
day arrangement 地面设备
day average precision 日平均精度;日内精度
day beacon 昼标;无灯航标
day beacon mark 昼标
day beacon range 昼用叠标
day beacon signal 昼标
day bed 躺椅
day bill 定期票据
day blindness 昼盲
day book 日志;日记簿
day call 日拆
day capacity 日生产量
day-care center 日托中心;白昼护理中心

day care facility 日托机构
day care home 日托所
day center 正午;白天场所
day clock 日钟
day coach 硬席车厢
day counter 周历计数器
day current 日发电量
day-date 双历
day-degree 日度
day disc 周历盘
day drift 通达地面的平巷
day finger 周历拨爪
day flow 昼间流量;日流量
day flow of sewage 污水日流量
day free of frost 无霜日
day fuel tank 日用量油箱
day gate 格子门
dayglow 白天气辉;日辉
day grate 安装在保险库内部的格栅
day hand 周历针
day hospital 专收门诊病人的医院;日间医院
Dayi kilns 大邑窑
day-in and day-out dependability 连续工作可靠性
day indicator 周历盘
day jessamine 昼花夜香树
day labo(u)r 临时工;计日劳工;计日工(人);工作日;短工;日工;散工
day labo(u)r construction 按日计酬施工
day labo(u)rer 按日计酬的散工;计日工作者
day labo(u)r system 按日计工制;日工制
day latch 白日上锁
day length 光照强度;日照时间;日长度
daylight 日光;白昼;压板开档;昼光;白天光照;天光
daylight base 日光灯管座
daylight blue 日光蓝颜料
daylight calculation 天然采光计算
daylight change 日照时间变化
daylight changing 白光换片
daylight clearance 空气隙
daylight colo(u)r film 白光彩色胶片;日光型彩色片
daylight control 日光控制
daylight controller 日光控制器
daylight dome 日照穹顶;日光圆顶
daylight driving 白昼行车
daylighted 日光反射的;日光照明的
daylighted from both sides 从两侧采光
daylight effect 白昼效应;日光作用;日光效应
daylight enlarger printer 日光放大印像机
daylight exposure 白光曝光
daylight factor 日照系数;白日光度;昼光因数;昼光系数;采光系数;日照因素;日照因数;日光因数;日光系数;天然采光系数
daylight filling 白昼装片
daylight film 白光胶片;日光型彩色胶片
daylight filter 昼光滤色镜;日光滤色镜;日光滤色器;日光滤光片
daylight fireball 白昼火流星
daylight fluorescent ink 荧光油墨
daylight fluorescent paint 日光型荧光涂料
daylight fluorescent pigment 荧光颜料;日光型荧光颜料
daylight from both sides 两侧开窗的
daylight glass 昼光玻璃;日光(过滤)玻璃
daylight illumination 日光照明;天然采光
daylight incandescent lamp 日光白炽灯
daylighting 自然采光;(易产生顺层滑坡的)陡倾高角斜坡面;采光;天然采光;日光
daylighting area 采光面积
daylighting band 采光带
daylighting by the roof 采光设计原理
daylighting coefficient 采光系数
daylighting curve 日光曲线
daylighting design 采光设计
daylighting design principle 屋顶采光
day lighting open 采光口
daylighting skylight 采光天窗
daylight lamp 日光灯
daylight loading cartridge 明室装暗盒
daylight loading cassette 明室装暗盒
daylight observation 白天观测
daylight of turnover 周转天数
daylight of turnover of current funds 流动资金周转天数
daylight opening 压机压板间距;(压机的)压板间

距;采光口;(压机的)台面距
daylight operation 亮室操作
daylight overdraft 日间透支
daylight period 白天光照时间
daylight photometry 日光光度学
daylight position 白昼位置
daylight prediction 日照预测
daylight ratio 日光系数;天然照明率;天然采光率
daylight room 明室
daylight saving meridian 日光节约子午线
daylight saving noon 日光节约正午
daylight saving time 经济时;夏令时(间);日光节约时间;夏时制;夏季时间
daylight signal 白昼信号;(交通)色灯信号
daylight signalling light 日间信号灯
day light signalling mirror 日光反照通信镜
daylight size (玻璃装配后的)透光尺寸
daylight slits 日光光隙
daylight source 自然光源
daylight speed 日光灯光度
daylight starter 日光灯启动器
daylight stream 白昼流星群
daylight target map 白天作战目标图
daylight time 白昼时间
daylight transparency 白天透明度
daylight type 日光型
daylight type colo(u)r film 日光型彩色胶片
daylight use in tunnel 隧道日光利用
daylight vision 明视觉;昼视觉;昼光觉;白昼视觉
daylight width 采光宽度;日照宽度;采光面积(宽度计量法)
day lily 黄花菜
day load 日(间)负荷
day loan 日拆;一天贷款
day man 计日工(人);日班工人
day manning 无夜勤的执勤制
day mark 昼(间)标;不发光航标;白昼标志;无灯航标
day mark buoy 不发光浮标
day maturity 到期日
day mineral 地面矿
day neutral 光照钝感的
day neutral plant 光期钝感植物
day-night average sound level 日平均声级
day-night cycle 昼夜周期;全日循环
day-night effect 日夜效应
day night equivalent sound level 昼夜等效声级
day-night rhythm 昼夜节律;昼夜节奏
day night switching equipment 昼夜转换装置;昼夜转换设备
day-night temperature 昼夜温度
day nursery 日间托儿所;托儿所
day of clear sky 晴空日
day of due date 满期日;到期日
day of embarkation 上船日期
day off 休假日;工休日;休息日
day of hail 雹日
day of maturity 满期日
day of maximum rainfall 最大降雨日;最大降水日
day of rain 下雨天
day of reckoning 结账日(期)
day of shipment price 装运日价格
day of snow 雪日
day of snow-lying 积雪日
day of supply 日供应量
day-of-week unbalance factor 周日不均衡系数
day-of-week variation 周内日变化
day of year 积日
day oil tank 日用量油箱
day old to death system 终生制
day on demurrage 超过装卸期限的日期;超过停泊日期数
day order 即日指示;当日定货
day output 日处理量;日产量
day pair 日工;日班工人;日班
day parker 长期停车者
day parking 日间停车场
day position 白昼位置
day quick setting 周历快速调整
day range 白昼作用距离
day rate 日工资率
day rate plan 日给制
day recording 计日
day release 休工培训日

day roller 周历盘
day room 文娱室;休息室
days after acceptance 承兑后若干日(交款)
days after date 出票后……天
days after sight 见票后……天
days after treatment 处理后的日数
day sample 逐日取样
day school 日校
day shape 号型
day shift 白(昼)班;日班
day shift goods dispatcher 日勤货运调度员
day shift system 日勤制
day side 向阳侧;白昼侧
day signal 昼间信号;不发光航标
day signal light 日间信号灯
days in transit 在途天数
days inventory ratio 每日存货比率
days man 零工
days of average inventory 平均存货期
days of average inventory on hand 平均库存天数
days of continuous rainfall 连续降雨天数;连续降雨日数;连续降水天数;连续降水日数
days of demurrage 滞留天数;停泊延期
days of disability 不能上班的天数
days of grace 宽延期;宽限日(期);优惠期;延期宽限日;恩惠日
days of heating period 采暖期天数
days of operation 运转天数
days of precipitation all over the year 全年降水天数
days of rainfall 降水天数
days of runoff producing 产流天数
days of steady rain 连续降雨天数;连续降雨日数;连续降水天数;连续降水日数
days of yearly sandstorm 年沙暴日数
days on demurrage 滞期日数
day('s) pay 计日工资
days purpose 装卸许可天数
day's receivables 平均应收账款日期
day's run 周日航程
day stone 自然状态露头;岩石露头;外露岩石;露头岩石
days to sell inventory 存货周转天数
day-supply reservoir 按日供应水库;按日供应仓库
day-supply tank 按日供应油箱;按日供应水箱
day('s) wages 计日工资
days when available 可使用的日数
day's work 每日工程;周日航行定位工作
days worked 工作日数
day's work joint 日班交接
day('s) work rate 计日工资
day system 三班(工作)制
daytaler 临时工;记日工;散工
day taller 计日工(人);临时工
day tank 间歇作业池窑;日用量油箱;日池窑;日常供应罐
day ticket 当日有效票
daytime 白天;白昼;日间
daytime approach 白日进场
daytime fog 日间雾
daytime home 白天养育所;日天收容所;白天疗养所
daytime light 昼灯;长明灯标;白昼灯标
daytime load 日间负荷
daytime meteor 白昼流星
daytime ozone profile 臭氧的日分布曲线
daytime population 昼间人口;白天人口;流动人口;日间人口
daytime radiance 白昼辐射率
daytime range 白昼作用距离
daytime resolution 白昼分辨率
daytime sensitivity 白昼灵敏度
daytime shift 日班
daytime signal 日间信号
daytime sky background 白昼天空背景
daytime sphere 昼间活动范围区
daytime stream 白昼流星群
daytime traffic 白天交通;日间交通
daytime train 白昼流星余迹
daytime transparency 白昼天空透明度
daytime treatment center 日间治疗中心
daytime visibility 白天能见度
day-to-day account 流水账
day-to-day activity 日常活动
day-to-day business 日常业务

day to day communications 日常通信
day-to-day goods 日用品
day-to-day investigation 日常调查
day-to-day issue 每日的供应
day-to-day loan 按日放款;日拆;通知放款
day-to-day loss control 日常(燃料)损耗控制
day-to-day maintenance 日常维护;日常保养
day-to-day management 日常管理
day-to-day money 暂借款;按日借款
day-to-day operations 日常经营
day-to-day posting 日常讨账
day-to-day production 日常生产
day-to-day repair 日常维修
day-to-day swaps 一日对冲
day-to-day test 每天试验;日夜连续试验;日常检验
day-to-day traffic working plan 运输工作日常计划
day-to-day variation 逐日变化
day-to-day variation factor 交通量日变化系数
day-to-day work 日常工作
day trade 即日买卖
day trading 日间交易
day turn 日班
day value 白天值
day visibility 白天能见度;白昼能见度
day wage system 日给制
day-wage work 计日工作;日工;散工
day water 地面水
day water of mine 矿山地面水
day with fog 雾天
day without frost 无霜日
day with rain 下雨天
day with snow 雪日
day with wind of gale force 大风日
daywork 零工;散工;计日工作;日工
day worker 临时工
day work joint 日工作缝;日施工缝
day work rate 计日工资单价;每日单价
day work schedule 计日工作表
day work system 计时制;计日工作制;日工制;按日计酬制
dazomet 棉隆
dazzle 目眩;眩目;眩感
dazzle-free 不眩光;不炫耀
dazzle-free glass 不眩玻璃;防眩光玻璃
dazzle lamp 车头灯;强光前灯
dazzle light 强光;眩目灯光
dazzle lighting 眩彩灯光;强烈照明
dazzle paint(ing) 伪装色彩;伪装漆
dazzling white 眩目白色;炽白光
D-bit 镶片钻头
D-cable 半圆导线双芯电缆
DC arc welder 直流电焊机
DC arc welding 直流电弧焊
D. C. casting 半连铸
DC-exponent DC 指数
DC generator 直流发电机
D crack(ing) D 密集裂纹
D-cracking of concrete 混凝土表面细裂纹;混凝土的 D 形裂缝
DC-squid magnetometer DC 超导磁力仪
DC-to-DC converter 逆变器
DC-to-DC inverter 逆变器
DC welding 直流电焊
de-accentuation 去加重
deaccentuator 校平器;减加重线路;平滑器
deacidification 除酸作用;脱酸(作用)
de-acidification plant 脱酸装置
deacification 除酸
deactivate 减活化;去活化
deactivated 使失效;使不活动
deactivated silica gel 减活化硅胶
deactivation 减活化(作用);灭活(作用);化学钝化法;消除放射性污染;惰性化;非活动化;不激活;不活动;使失效;去启动;去激作用;去激励;去活化(作用)
deactivation capacity 减活能力
deactivation column 减活化柱
deactivation constant 钝化常数
deactivation of a response plan 响应计划的撤销
deactivation of support 载体的钝化
deactivation of water 水的减活化作用;水的除氧;水脱氧
deactivator 减活化剂;钝化剂
deactuate 退动

deactuate pressure 反工作压力
dead 静的
dead abutment 隐蔽式桥台;止推轴承;固定支座
dead account 呆账
dead ahead 正前方
dead air 静止空气;静气;不流通空气;闭塞空气;停滞空气;死空气
dead-air compartment (空芯墙内的)闭塞空间;闭塞空气间
dead-air insulation 静止空气绝缘;闭塞空气绝缘
dead-air space 闭塞空间;静气保温空间;封闭空间;闭塞空气间
dead-air spot 空气滞留点
dead-air void 死空隙;闭塞空隙
dead alloy 稳定的合金;低碳合金
dead and down 死倒木
dead and dry face 干枯面
dead angle 死角
dead animal processing plant 死动物处理场
dead annealed 完全退火的
dead annealing 全退火
dead arc 残留岛弧
dead arcade 死连拱廊道;(一端不通的)连拱廊道;封闭拱廊
dead area 遮蔽面积;断面不受力部分;死滞区;死水区;死角区;死region
dead arm 枯枝;死岔道
dead ash 无放射性尘埃
dead assets 报废资产;无用资产;死资产
dead astern 正后方
dead axle 静轴;定轴;从动轴;不转轴;不动车轴;被动轴
dead band 空段;静区;静带;盲带;非灵敏区;不灵敏区;不工作区;无反应区;死区;死谱带;输出不变区
dead-band regulator 静区调节器;非线性调节器
dead bank 停煤燃烧
dead banking 长期封炉
dead bargain 长廉价交易;卖价极便宜的交易
dead-beam pass 闭口梁形轧槽
dead beat 晕线仪;不摆;赖账不还;信誉不好者;速示;直进式
deadbeat compass 不摆罗经;速示罗盘;无振荡罗经
deadbeat controller 命中型控制器
deadbeat discharge 非周期放电
deadbeat escapement 不摆擒纵机;直进式擒纵机构
deadbeat galvanometer 不摆检流计;不摆电流计;速示电流计
deadbeat instrument 不摆式仪表
deadbeat measurement 不摆测量;速示测试
deadbeat meter 不摆仪表;速示仪表
deadbeat pendulum 无周期摆
deadbeat response 不摆;速示响应
deadbeat stability 非周期稳定性
dead belt 静区;无风带
dead block 缓冲板;(化丝机的)固定卷筒
dead bolt 固定栓;无弹簧锁闩;锁定插销;锁紧螺栓(没有弹簧);矩形截面锁舌
dead bolt lock 无弹簧锁闩;弹子门锁
dead brake lever 固定闸杆
dead brake lever guide 固定闸杆导承
dead branch of river 死河汊
dead bright 磨光的;抛光
dead broken 完全破产
dead burn 僵烧;死烧;透彻煅烧
dead-burned 烧僵的;僵烧釉的;死烧的
dead-burned bauxite 重烧矾土熟料;死烧矾土
dead-burned calcium sulphate 僵烧硫酸钙
dead-burned dolomite 僵烧白云石;烧结白云石;烧僵白云石
dead-burned dolomite grains 冶金白云石砂
dead-burned gypsum 硬石膏;烧(结)石膏;烧僵的石膏(抹灰用的无水石膏);死烧石膏
dead-burned magnesia 僵烧镁砂;死烧氧化镁
dead-burned magnesite 僵烧镁矿(砂);死烧镁砂
dead-burned plaster 僵烧石膏
dead-burned refractory dolomite 烧结的耐火白云石;烧僵的耐火白云石;死烧耐火白云石
dead burn(ing) 僵烧;僵燃;死角燃烧;烧焦;硬烧;死烧的;烧结
dead-burnt 僵烧的;烧出釉的
dead-burnt gypsum 硬石膏;死石膏;无水石膏
dead-burnt lime 煅烧透彻的石灰
dead-burnt magnesite 重烧镁

dead calm 无风;完全无风
dead canal 废运河
dead capital 不生息资本
dead card 停用卡片
dead cast 钢丝盘卷均匀
dead center 静点;哑点;车床尾顶尖;止点;死点
dead center ignition 死点发火
dead center lathe 死顶尖车床
dead center mark 死点记号
dead center of lathe 车床尾顶尖
dead channel 干枯河道;残流河段;废渠;废河槽;死水支渠
dead charges 间接费用;非生产性费用
dead cheque 无效支票
dead circuit 无电电路;空路;死电路
dead coil 无效线圈
dead-cold chilling 激冷完全淬火
dead colo(u)r 呆色;不鲜明的色彩;暗色;油画底色
dead colo(u)ring 底图样
dead commission 固定佣金;固定经纪费
dead contact 空接点;空触点;开路接点;开路触点;闲接点
dead corner 死拐角
dead crystal 死晶体;失效晶体
dead dam 废坝;无效坝;死坝
dead deafening 隔声
dead dipping 金属酸浸除锈处理
dead door 门外遮板;假门;挡风暴遮板;(船上挡风暴的)门外假板
dead-drawn 强拉的
dead-drawn wire 冷拔钢丝;多次拉拔钢丝;强拉钢丝
dead dull 无光
dead earth 直通地;固定接点;固定接地;完全接地
deaden 消除亮度;减弱
dead end 空端;盲端;闭塞的一头;闭端;死水头;死端(头);短于六英尺(1英尺=0.3048米)的木板;紧密封闭
dead-end anchor 终端锚定;尽端站
dead-end anchorage 固定(端)锚具;固定(端)锚固件
dead-end areas in corridors 走廊终端面积
dead-end clamp 耐张线夹;终锚线夹;终端线夹
dead-end clamp with cup-point screws 螺栓顶紧式接触线终端锚固线夹
dead-end corridor 袋形走道;尽端(式)走廊;一端不通行的走廊
dead-end(ed) (路的)尽头
dead-ended piping 尽端管线
dead-ended railroad station 铁路终点站
dead-ended shed 一头不通的棚;一头不通的车库
dead-end effect 空端效应;空匝效应
dead-end feeder 终端馈线
dead-end fire driveway 尽头式消防车道
dead-end flue 死头烟道;假烟道
dead-end fracture 不连通裂缝
dead-end guy 终端拉线
dead-end hangar 一头不通的(飞)机库;尽端式飞机库
dead-end insulator 耐胀绝缘子;耐张绝缘子;耐拉绝缘子
dead-end loss 空匝(损耗)闲匝损失
dead-end main 干管尽端;尽头管;终端干管;闷头主管;死头(干)管
dead-end passenger terminal 尽头客运枢纽
dead-end path 断头路
dead-end piping 尽端管线
dead-end platform 尽头站台
dead-end pole 终端杆
dead-end pore 不连通孔隙
dead-end pressure 支点压力;死点压力
dead-end railroad station 尽端式火车站
dead-end road 死胡同;尽端路
dead-end sewer 下水道尽端;污水管尽端
dead-end shed 尽端式飞机库
dead-end siding 尽头(支)线【铁】;尽头侧线;死岔线;尽头披盖板;岔线尽头
dead-end station 尽头式车站;尽端点火车站
dead-end street 实巷;独头街;尽头街道;断头路;袋形路;死胡同;死头街道(即此路不通)
dead-end switch 空端开关;终端开关
dead-end termination 死锚
dead-end tie 导线(的)终端扎结
dead-end tower 终端塔架

dead-end track 尽头轨道
dead-end tunnel 终端隧洞;终端隧道
dead-end turn-back track 尽头折返线
dead-end weld 死端焊
deadener 隔音材料;消声器;隔音涂料
deadening 减弱;消音;消声处理;下降;吸声;隔音(作用);隔音材料;失光材料
deadening agent 消光剂
deadening dress (吸声的)粗面修琢
deadening dressing (吸声的)粗面修琢;吸声琢面
deadening effect 隔音效果
deadening fabric 隔音布
deadening felt 隔音纸;隔音毡
deadening felt paper 隔音毡纸
deaden the way 阻碍航道;停止前进(航行)
dead expenses 非生产性费用
dead eye 孔板伸缩节;眼床;轴承眼圈;三眼滑轮;孔板伸缩节;穿眼木滑车(接索用);钢丝绳绳环;三眼木饼;三眼辘轳;三孔滑车
dead eye hitch 双合套结
dead-fall 翻板陷阱;翻斗机
dead fallow 绝对休闲的
dead fallowing 秋耕休闲
dead fault 死断层
deadfender 消声器;隔声器
dead file 扁三角锉;废文件;停用文件;失效存储器
dead finish 普通饰面;无光泽抛光
dead flat 平光;无光
dead flat body 船中平行体;平行船中体
dead flat hammer 矫平面锤;平正锤
dead floor 废地板;粗地板(用于双层楼板);楼板中的垫板
dead flue 废(弃)烟道;作废烟道
dead form 古地形
dead freight 空舱运费;空舱费(用)
dead front 不带电面板;不带电一面;不带电一边;空正面;正面不带电的部件;死面
dead front switchboard 面板无接线的配电盘;不带电一面开关板
dead front type switchboard 固定面板式配电盘
dead fuel 死可燃物
dead fuel moisture 死可燃物含水量
dead furnace 死炉
dead furrow 堵头沟;死(水)沟
dead gear 惰齿轮
dead geyser 死间歇泉
dead glacier 停滞冰川;死冰川;化石冰川;不动冰川
dead glass 滞流玻璃;不(流)动玻璃液
dead granite 特种花岗岩
dead graphite 不含铀(块)石墨
dead grate 固定炉箅
dead grate area 炉箅不工作部分
dead groove 闭合式轧槽
dead ground 盲区;直通地;不通视地区;无矿岩层;无矿地层;死角区;死角区
dead guy 固定缆;固定牵索;固定缆风;固定拉杆
dead halt 完全停机;完全停车
dead handle 常闭式自动停车把手;常闭式安全把手
dead hard steel 高硬钢;极硬钢;特硬钢;高强(度)钢
deadhead 木块锚标;木质系船标;木浮标;系缆(木)桩;铸件冒头;顶尖桩;死头;锚浮标;床尾后顶尖座;系船柱;尾架
deadhead car 空放车辆
deadhead coiler 轴心进线的卷线机
deadheading 回空(行车);空载运行;空载返航
deadhead kilometers for dispatch 调度空驶里程
deadhead kilometers for passenger 接客空驶里程
deadhead pressure 零流量压力
deadhead resistance 废阻力
deadhead train 回空列车;排空列车
dead hearth generator 死膛发生炉
dead hedge 柴垒
dead hole 陷阱;盲孔;盲井;炸后残眼;(爆炸后的)残眼;不通孔;死孔;死洞;炮窝子
dead horizon 死土层
dead horse 预付工资;旧债
dead house 太平间;停尸房
dead idler 张紧轮;空转惰轮
dead influence 限期影响
deading 保热套
dead in line 轴线重合;配置于一线
dead in water condition 在水面静止
dead joint 固定连接;不可分连接;死连接

dead knife 固定刀片;底刀
dead knot 死结;木料的腐节;朽节;(木材的)腐节;死节
dead lake 死水湖
dead landslide 死滑坡
dead latch 可锁住的插销;紧锁销;安全门栓
dead layer 死层
dead leaf 固定门扉;定叶片
dead leg 盲管段;单流管;死水支管;热水管静止端节
dead length 截止长度
dead level (含量极少或不含硬焦油脂的)屋面沥青;绝对标高;零电平;空层;绝对水平;绝对高程;静态电平
dead level asphalt 沥青层
dead level roof(ing) 平(屋)顶
dead lever 固定杆
dead lever trunk 空层中继线;备用段干线
dead light (船上的)舷窗;门的厚玻璃;舷窗内盖;舷窗外盖;舷窗风暴series;木窗遮板;船上固定采光玻璃;亮瓦;固定(舷)窗;固定天窗
dead lime 过性石灰;失效(的)石灰
deadline 最后期限;截止时间;截止期(限);短等通管;不可逾越的界限;空线;静线;截止行;截止线;截止日期;闲置线路;不流通管道;安全界线;停电线路;死线;安全界限;限期
deadline anchor 死绳锚定
deadline date 截止日期
deadline delivery date 交货最后期限;交货截止期限
deadline effect 截止时间作用
deadline factor (车辆的)失修因素
deadline for acceptance of tenders 投标最后期限
deadline for delayed delivery 延期交货的截止期限
deadline for receipt of bids 接受投标的截止时间
deadline for receipt of tenders 接受投标的截止时间
deadline for shipment 交货期限;装运期限
deadline for submission of bids 投标提交的最后期限;投标截止日期
deadline for submission of tenders 投标截止时间;投标截止日期
deadline game 不可越界对策;不可逾越界线对策
deadline of bid 投标截止日期
deadline of payment 最后付款期限
deadline scheduling 时限调度
deadline sheave 死绳固定轮
deadline stresses 死绳拉力
dead load 静载;静荷载;静负载;静负荷;净负载;净负荷;净负荷;结构自重;恒(荷)载;总载重量;自重;自飘的束帆索;死荷载;底负载;底负荷;本底负载
dead-load coefficient of container 集装箱自重系数
dead-load deflection 自重挠度;静载挠曲;静载挠度;固定静负荷挠曲
dead-loaded safety valve 重锤式安全阀
dead-load lever 起重杆
dead-load moment 静力力矩;静载弯矩;自重力矩
dead-load of derrick 钻塔自重;钻塔的静荷载
dead load power 静载动力
dead-load stress 恒载应力;静载应力
dead-load thrust line 恒载压力线
dead loan 延滞贷金;死债
deadlock 闭锁器;单闩锁;闭锁;停滞;停顿;死锁;停止作业
deadlock avoidance 避免死锁
deadlocking lath 双料门;闭锁插销
deadlock property 死锁特性
dead locomotive 无火机车【铁】
dead loss 空匦损耗;净亏损额;(与负载无关的)固定损失;纯损失;纯损耗
dead low water 最低低潮
deadly compound 致死化合物
deadly condition 非常情况
deadly poisonous compound 剧毒化合物
deadly temperature 致死温度
dead main 空载线(路)
dead man 拉杆锚桩
deadman 叉杆;锚墩;锚定桩;锚定物;桩橛;地牛;地龙;锚回绳;松飘的束帆索;栓桩
deadman abort switch 常闭式紧急停车开关
deadman anchorage 锚定块;埋入式锚定;地龙
deadman brake 常闭式制动器;常闭式保险制动
deadman control 常闭式保险制动;停车闸;制动
deadman controller 叉杆控制器
deadman device 驾驶员失知制动装置;事故自动制动装置
deadman hole 绑绳锚桩孔
deadman honeycombed wall 地龙蜂窝墙
deadman method 锚回单杆吊货法;松飘的束帆索
deadman's brake 安全制动器
deadman's eye 锚定桩观察孔
deadman's handle 驾驶员安全装置;自回位手柄;带安全钮的手柄;常闭式自动停车把手;安全(警惕)手柄
deadman wall 地龙墙
dead margin 边缘枯萎
dead masonry wall 无门窗的墙;砖石砌筑的实墙
dead-matter 无机物质
dead-melted steel 镇静钢;全脱氧钢
dead memory 静止存储器
dead metal 废玻璃液
dead mild steel 极软(碳)钢;镇静钢;低碳钢;超低碳钢
dead milled rubber 重压橡皮
dead milling 重压
dead money 闲置资金;死钱
dead motor 关闭的电动机
dead neap 极小潮;最低(小)潮
deadness 衰萎
dead netting 裸网
dead nip stopper 活砧式掣链器
dead number 空号
dead oil 防腐油;石油渣油
dead parking 空车停放场;空车停放;空车停放处;空车停车处
dead pass 空轧孔型
dead pickling 呆液酸洗
deadpile 死堆
dead-piled 蒸馏石油的残油;无隔垫堆放的;静压胶合板;静压胶合的
dead pile yarn 埋头绒头纱
dead plant part 植物残体
dead plaster 过烧石膏;无水硫酸钙
dead plate 隔热板;炉前挡炭板;冷却板;障热板;固定(冷却)板
dead pocket 料仓内死角;静水区
dead point 静点;哑点;止点;死点
dead point of conversion gain 变换增益的死点
dead point of stroke 冲程死点
dead point of traction 牵引死点
dead position 失效位置
dead-pressed explosive 压死的炸药
dead program(me) 停用程序
dead pulley 惰(滑)轮
dead quartz 贫矿石英
dead rail 非测重钢轨
dead rear axle 固定式后轴
dead reckoning 积算航法;积算船位;盲航法;航位推算法;航路绘算;盲航法;航迹推算;行迹推算法;船位置简单推测法;船位积算推算(法);船位积算;推算定位;推算船位;推测航行法
dead-reckoning analog(ue) indicator 航迹推算模拟指示器
dead-reckoning analysis 船位推算分析
dead-reckoning analyzer 航迹推算分析器;航迹分析器【船】;航迹推算装置
dead-reckoning analyzer indicator 航迹绘算仪;船位推算分析指数器
dead-reckoning computer 航位推算计算机;航迹推算计算机
dead-reckoning distance 积算航程
dead-reckoning equipment 航迹推算器;航迹记录器【船】;航迹推算装置
dead-reckoning indicator 航迹推算指示器;航迹分析指示器
dead-reckoning plot 航迹推算作图
dead-reckoning position 积算船位;船位推算位置;无风无流的推算船位;推算船位;航位推算装置
dead-reckoning recorder 航迹记录器【船】
dead-reckoning time 已推算时间;推算时间
dead-reckoning tracer 航迹推算描绘仪;航迹记录器【船】;航迹绘算器;航迹推算描绘仪
dead-reckoning track (line) 航位推算航迹
dead region 料仓内死区;死区
dead rent (矿山、采石场)固定年租金;矿山固定租金;固定租金
dead resistance 寄生电阻;消耗电阻;吸收电阻;整流电阻
dead ring 绝缘环;紧固环

dead rise 底部升高量(航空、航海);船底斜高;船底斜度
deadrise boat 直框架结构船
dead rise of ship bottom 船底横向斜度
dead rising 木船船底升高
dead river channel (河流的)旧道;河流故道;废河道
dead roasting 死烧
dead rock 废石(头);脉石
dead-rolled rubber 重压橡皮
dead room 吸声室;静室;消音室;消声室;沉寂室
dead rope 固定索;手拉缆
dead runner 盲浇口
deads 井下废石;脉石;围岩
dead sand 原砂
Dead Sea 死海
Dead Sea asphalt 犹太沥青;死海沥青
dead season 寒季;淡季
dead section 空段;备用段;无电段;死区段
dead section of track circuit 轨道电路死区段
dead security 固定担保器
dead security mortgage 固定抵押品
dead segment 无效整流片
dead setting steel 去氧钢;脱氧钢
dead sheave 桅头滑槽
dead ship 严重失灵船;死船
dead shore 横撑木;支撑柱;固墙撑木;固定竖撑;撑木
dead-shore needle 固墙竖撑;围墙用撑;支柱横撑
dead shoring 静撑柱(仅承被撑构件的自重);门形支撑;固墙埋撑
dead short 完全短路
dead-short circuit 完全短路;全短路
dead size 定规尺寸
dead slack water 极缓流
dead slot 空槽
dead slow (speed) 微速;最低航速;最低速度
dead small 细末;粉末;(矿石的)小块料
dead smooth 极平滑;极光滑
dead smooth cut file 光锉
dead smooth equalling file 光扁锉
dead smooth file 极细锉;光锉
dead smooth knife file 光刀锉
dead smooth mill saw file 光锯锉
dead soaking pit 均热坑;加热保温坑
dead soft 极软退火;全退火;极软的
dead-soft annealing 极软退火
dead-soft steel 极软(碳)钢;低碳钢
dead-soft temper 极软退火处理
dead soil 死土
dead soil covering 枯枝落叶层;死地被物
dead sounding 隔声层;消减声响;隔音层
dead space 亏舱;静区;盲区;不工作区;弃位【船】;无用空间;无信号区;无效区;死隙;死水域;死区;死空间;死角
dead-space characteristic 静区特性
dead-space error 静区误差
dead spindle 静轴
dead spot 静点;哑点;死点
dead spot of traction 牵引死点
dead spring 压下弹簧;失效弹簧
dead square 准正方形
dead stackyard 死堆场
dead stall 小块煤
deadstart 静启动
dead state 停滞状态;死态
dead steam 废汽;废气;乏汽
dead steel 软钢;全镇静钢;全脱氧钢
dead stock 农具;农机;呆滞存货;呆滞资金;呆滞商品
dead stop 完全停止;突然停车
dead stop end point method 永停终点法
dead stop process 终点停止滴定法
dead stop titration 永停滴定法
dead storage 垫底库容;滞销货;储库死角;备用库存;无效储[贮]水量;死库容;死堆场
dead storage level 死水位
dead storage of sedimentation 泥沙淤积形成的死库容;泥沙淤积垫底库容;长期沉积库容
dead stream branch 淤塞河川的支流;淤塞的河流;死河汊
dead stream channel 河流故道
dead stress 静应力;呆应力
dead stroke 冲程死区;不反跳;死冲程
dead stroke hammer 不反跳弹簧锤;死冲锤;无反

跳击锤
dead swell 微涌
dead terminal 尽头站;终站;失效终端
dead tide 小潮;最低潮;死潮
dead time 空转时间;空载时间;静寂时间;截止时间;滞后时间;呆滞时间;纯时滞;不动时间;停滞时间;停歇时间;停工时间;停顿时间;死时间
dead time correction 死时间改正
dead time correction of X-ray detector X 射线探测器死时间校正
dead time delay 纯时滞
dead time in injection mo(u)lding 注塑静止时间
dead time loss 空载损耗;空载时间损耗
dead time of governor 调节器静止时间
dead-top 死梢
dead track 无电区段;无弹性轨道
dead true 完全平衡的
dead tunnel 死巷道
dead turn 空匝;无效线匝
dead valley 干谷;死谷
dead volume 死体积;死容积
dead wall 闷墙;隔音墙;挡墙;暗墙;无门窗的墙;无窗墙;实墙
dead water 静水;不流通水;停滞水;死水;死河汊;积水
dead water-level 死水位
dead water mechanical weighing method 静水力学称量法
dead water region 静水区域
dead water stage 死水位
dead water zone 死水区
deadweight 固定荷重;本身自重;静(载)重;总载重量;自重;重载;船舶载重量
deadweight anchor 重块式锚
deadweight brake 平衡重块制动器
deadweight capacity 总载重量;负载容量;载重能力;载重吨位
deadweight cargo 重量货(物)
deadweight cargo capacity 货物载重吨位
deadweight cargo tonnage 载货重量吨
deadweight carrying capacity 静载量;载重量
deadweight charter 满载租船合同
deadweight coefficient of container 集装箱自重系数
deadweight debt 不能减轻的负债
deadweight displacement coefficient 载重排水量系数
deadweight loss 额外损失;额外损耗
deadweight machine 基准测力机
deadweight moment 自重力矩
deadweight of structure 建筑物自重
deadweight of vessel 船舶载重量
deadweight piston page 重锤压力计
deadweight piston pressure ga(u)ge 底重式活塞压力计
deadweight ratio 载重排水量系数
deadweight roller 自重碾压机
deadweight safety valve 静荷载安全阀;荷重式安全阀;自重压力安全阀;杠杆式安全阀;平衡重块安全阀;自压安全阀;重锤式安全阀
deadweight scale 载重表尺;载重量标度表;载重标尺
deadweight stress 自重应力
deadweight tensioning 平衡重块绷紧
deadweight tester 真重校表仪
deadweight testing machine 重锤式万能试验仪
deadweight ton(nage) 重量吨;净吨;吨位;空载排水吨;载重吨位;总载重量;总吨位;载重量
deadweight tonnage of vessel 船舶载重量
deadweight-type regulator 静重式调节器
deadweight valve 重量保安阀
dead well 吸水井;枯井;废井;暗井;死井;渗水井
dead wheel 固定齿轮
dead white 无光泽涂料;无光泽漆
dead white paint 无光白漆
dead-wind 逆风
dead window 盲窗;隔音窗;假窗;死窗
dead wire 无电导线;已不通电流的电线;固定索;不载电导线
dead wood 沉木;沉材;力材;枯木;朽木;呆木;无用的人员(俚语);龙骨帮木;船首鳍;钝材;船尾鳍
dead wood fence 粗巨木栅栏
dead works 重活;非直接性生产工作;干舷;船体水上部分;上层建筑;水线上部建筑

dead zone 闭塞区;空区;静区;盲区;滞区;非灵敏区;不灵敏区;不工作区;死(水)区;死(区)域;死带;输出不变区
dead zone non-linear regulator 死区非线性调节器
dead-zone regulator 死区调节器
dead zone unit 静区装置;静区部件;不工作单元
DEAE-cellulose 二乙氨基乙基纤维素
deaerate 空气排除;除去空气;脱气;脱泡
deaerated concrete 去气混凝土;脱气混凝土;真空混凝土;除气混凝土
deaerated water 脱气水
deaerating chamber 除气室;脱气室
deaerating feed heater 除氧给水加热器
deaerating heater 除氧加热器;脱气加热器
deaerating layer 稀气层;稳压层;通气层;除气层
deaerating mixer (混凝土)排气搅拌器;(混凝土)排气搅拌机
deaerating plant 除氧装置;脱气装置
deaerating tank 脱气槽
deaeration 除(空)气;抽气;排气;脱气(法);脱泡;去气
deaeration agent 脱气剂
deaeration system 除气系统
deaerator 去气桶;去气器;消泡器;除氧器;除气器;除空气机;脱氧器;脱气塔;脱气器;脱泡机
deaerator extraction pump 除氧器凝水抽出泵
deaering 去空气法;排气
deaerization plant 除氧器;除气器
deaf and dumb school 聋哑学校
deafener 减音器;消音器;隔音器
deafening 止响装置;隔声装置;隔音材料;止响物;隔音装置;隔音止响;隔音层
deafening device 隔音装置
deafness percent 听力损失率
deaf ore 含矿脉壁泥
deagglomeration 解附聚(作用);松团作用
deaggregating effect 解聚集作用;解聚集效应
de-aired brick 除气砖;去气砖
de-aired clay 除气黏土;去气黏土
de-aired concrete 去气混凝土
de-aired water 脱气水
de-airing 除气;去气
de-airing auger 脱气挤泥机
de-airing extruder 除气挤压机;真空挤泥机;去气挤压机
de-airing machine 除气器;脱气器
de-airing mixer 真空搅拌机
de-airing of clay 泥料真空处理
de-airing pug mill 真空捏土机;真空练泥机
de-airing tower 除气塔
Deaister 多伊斯特旋风
deal 交易;买卖;厚窄板;松木板
deal at arm's length 彼此独立的交易
deal board 板材;松木板;杉板
deal butt 厚窄板头
dealcoholizing 脱醇
dealed coring tool 密封取芯工具
deal ends 短松板;短厚窄板;短板材
dealer 经销商;自营商;商人
dealer aids 协助推销人;推销员
dealer contest 竞销
dealer delivery materials 交货人
dealer helps 推销员
dealer inventory 流通库存
dealer price 批发价(格)
dealer's license 经销商执照
dealer tie-in 销售商广告;商人介入
deal floor(ing) 松木地板;板条地板
deal frame 木锯架;往复式排架;松木屋架;木框架;板材锯架
deal frame-saw 固定排锯
dealing future 期货
dealing in future 期货交易
dealing in securities 证券交易
dealing in stock and shares 股票经营
dealing risks 交易风险
dealing slip 交易记录单
dealing with digitized cartographic(al) data 数字地图信息处理
dealing with workers' appeals 接受工人申诉
dealkalization 脱碱作用;脱碱
dealkalize 脱碱
de-alkylation 去烃(作用);脱烷基化
deallocate 解除分配;重新分配;去分配

deallocation 存储单元分配;重新分配(地址)
deallocation of devices 设备的重新分配;设备的去配
deal on credit 信用交易
deal porter 搬运木材工人
deals 松料
deals and battens 板条和薄板
deals and boards 板材
deals board 厚板
deals ship 木板运输船
deal sub-floor 松木毛地板
dealumination 脱铝(作用)
dealwood 针叶木材;松木
deambulatory 回廊;教堂中的回廊通道
deambulatory vault 回廊拱顶
deaminating 脱氨基(作用)
deamination 脱氨基(作用)
deaminizing 脱氨基(作用)
deamochore 携布植物
deamplification factor 折减系数
deamplification signal 衰减信号
dean 教务主任;教务长
deanamorphoser 反变形镜头
Dean and Stark apparatus 迪安—斯脱克水分测定仪;迪安—斯脱克设备
dean-end layout 死头式布管
deanery 院长办公室
Dean number 迪安数;狄恩数
deanol 丹醇
deaquation 脱水作用
dear 昂贵的
dearer-money policy 高息政策
dearing spool 安装轴承的间距套管
dear money 高利率资金;紧缩银根;高息资金;高息借款;高利率;高利(贷)款项
dearness allowance 物价津贴
dearomatization 脱芳构化
dearsenification 脱砷
deas 两用凳(可作桌子用)
de-ash 除灰
de-ashed fuel 脱灰燃料
de-ashing 清灰;去灰分;脱灰作用
de-ashing device 脱灰设备
de-ashing fuel 脱灰燃料
deasphalted oil 脱沥青油
deasphalt(ing) 脱沥青
deasplaltene 脱沥青质
death annuity 抚恤年金
death assemblage 死亡组合
deathbed 尸体床
death board 尸体板
death by burning 烧死
death by suffocation 窒息死亡
death calm 死静
death control 死亡控制
death count 死亡数
death due to trussing up 捆绑死
death from automobile 车祸死亡
death from lightning 雷击死
death insurance 死亡保险
deathnium center 复合中心
death phase 死亡期;衰老期
death point 止点;冲程止点;死亡点
death rate 死亡率
death rate of urban population 城市人口死亡率
death ratio 死亡比(值)
death registration 死亡登记
deaths per fire 每起火灾造成的死亡
deathtrap 不安全建筑(物);危险区(域);危险场所
death watch beetle 根死甲虫;侵蚀木材的小甲虫
deaw hook 钩舌
debacle 凌汛泛滥;溃裂浮冰;解冻;融冰流;山崩
debalance of preparation and winning work 采掘失调
deballast 减载
deballasting 卸压载;卸压舱水
Debaltsevo polarity superchronzone 迪贝尔塞沃极性超时间带
Debaltsevo polarity superzone 迪贝尔塞沃极性超带;迪贝尔塞沃极性超时
debanker 破埂器
Debao-type phosphate deposit 德保式磷矿床
debarkation 上岸
debarking (木材的)去皮

debase 质量变坏
debasement 降低成色;贬值
debatable land 有争议的土地
debatable time 有争议时间;受损时间
debending roll 板直辊
debenture 海关退税凭单;信用债券;无担保债券
debenture account 债券账户
debenture bond 公司债券;信用(公司)债券
debenture capital 信用(借贷)资本;借入资本;债券资本
debenture certificate 信用债券;债票
debenture convertible at market price 按市价折合的债券
debenture holder 信用债券持有人
debenture in foreign currency 以外币发行的公司信用债券
debenture interest 借款股份股息;债券利息
debenture issue 债券发行
debenture stock 信用债券;信用贷款股份;信用股票;公司债券;借款股票;借款股份
debenture transfer 债券转让
debenture trust 债券信托
debenture trust deed 信用债券
debenzolization 脱苯
debenzolization of absorber 吸附脱苯
debenzolization of oil 洗油脱苯
debenzylation 脱苄基(作用)
debility 无活力
debit 借记;借方(账户);收方
debit advice 借记通知书;借贷通知单;出账报单;欠款报单
debit and credit 借贷
debit and credit accounting system 借贷记账法
debit and credit conventions 借贷法则
debit and credit guarantee 借贷担保
debit and credit memos 借贷凭单
debit balance 借余;借方余额;借方差额;借差
debit bank 借方银行
debit bank balance 票据交换后银行借差
debit card 签账卡;提款卡
debit column 借方栏
debit-credit bookkeeping method 借贷记账法
debit-credit mechanism 借贷机制
debit-credit plan 借贷记账法
debit customer 借方欠户;债务者
debit entry 借项(科目);借方分录
debiteuse 耐火槽;槽子砖;土制浮标
debiteuse bubble (玻璃熔制过程中槽子砖产生的)气泡;槽子气泡
debiteuse kiln 槽子砖窑
debiteuse method 有槽引上法(平板玻璃)
debiteuse seed 槽子气泡
debit instrument 欠据;欠单
debit interest 欠息
debit item 借项(科目)
debit memorandum 借记通知书;借项通知单;借据
debit note 借项通知单;借款清单;借记通知书;借方通知;借方票据;欠款通知
debitor interest 借方应付利息
debit row 借方列
debit schedule 发料清单
debit side 收方
debit voucher 借单
deblock 解块;分离数据块
de-blocking 解块;解封闭;消除结块;从字串中分离出;程序分块;数据分块
de-blocking factor 去封闭因子
de-blooming 去荧光
de-blurring 去模糊
de-blurring filter 去模糊滤光器
de-blurring method 去模糊法
debonded tendon 消除黏结的预应力筋
debonder 剥离器
debond(ing) 脱胶;解离;脱黏;去结合
debonding unit 分离装置
deboost 阻尼
debooster 限幅器;限制器;限动器;增压限制器
deboost phase 制动阶段
debossed 雕刻压印花纹
Debot effect 德博特效应
debouch 河口;使流出
debouchment 河口
debouchure 地下河出口
debounce logic 防反跳逻辑

Debreu-Gale-Nikaido theorem 德布勒—盖尔—尼卡多定理
debris 烂物;建筑碎料;含有泥沙贝壳石等的冰块;洗涤残余物;岩(石碎)屑;堆积物;残骸;弃渣;尾矿;瓦砾(堆);碎屑石;碎片(垃圾)
debris avalanche 泥石流;岩屑崩塌;岩屑崩落;塌方;碎石崩落
debris barrier 漂流物沉淀池;拦污设施;拦污设备;拦水坝;拦沙坝;碎渣隔拦
debris basin 拦沙池;储[贮]砂库;储[贮]砂池;砂砾池;沉沙池
debris bed load 河底泥砂;河底冲积物;推移质底沙
debris bed sand 推移质底沙
debris burning fire 垃圾焚烧用火;烧荒用火
debris chamber 沉沙池
debris clearing cost 清除残损物费用
debris cone 岩屑锥;岩屑堆;冲积锥;冲积扇;冲出锥;冰川岩屑锥;砂砾锥
debris dam 谷坊;拦水坝;拦沙谷坊;拦沙坝;冲积埂;冲积堤;冲积坝;碎石坝
debris deflector 漂流物导向设施
debris deposit 冲积物
debris dump 垃圾堆;岩屑堆;废渣堆;尾矿堆
debris fall 岩屑崩坍;岩屑崩塌;岩屑崩落
debris fence 拦石坝
debris-filled 填有碎石的;碎石填筑(的)
debris flow 泥石流;岩屑流;碎石流
debris flow deposit 泥石流沉积
debris flow sediment 泥石流沉积物
debris from demolition 建筑碎(石)料;拆房瓦砾
debris glacier 崩落冰川
debris guard 拦污装置;碎石防挡装置;碎片护罩装置;碎片防护装置;碎片防护板;碎片挡板装置
debris ice 堆积冰
debris-intercepting dam 拦沙坝
debris-laden flow 泥石流;挟沙水流;岩屑流
debris-laden ice 岩屑冰;冰碛冰
debris-laden river 夹大量砂砾河流
debris-laden stream 泥石流;岩屑流;大量挟沙河流;含大量砾石的河流;挟大量砂砾河流;含砂(砾)河流;含大量砂砾的河流
debris-laden water 含砂水
debris law 尾矿法
debris line 冲积线;冲痕;碎屑堆积线
debris loading means 装渣工具
debris log interval 岩屑录井间距
debris of oxide 锈皮
debris plain 岩屑平原
debris protection 拦沙设备
debris residuum 岩屑风化壳
debris retaining structure 拦沙建筑物
debris sample 块石样
debris slide 碎石滑动;岩屑滑动;砂砾滑坡;岩石滑动
debris soil 碎石土
debris storage basin 储[贮]砂库;储砂库;蓄泥沙水库;沉沙区;放淤区
debris storage capacity 拦沙能力;沉沙容积
debris tank 沉沙池
debristling machine 刨毛机
debris trap 漂浮物截留设备;沉沙井
debris unit 拦污设施;拦污设备
debris utilization 废品利用
De Broglie's equation 德布罗意关系式;德布罗意方程
De Broglie's frequency 德布罗意频率
De Broglie's relation 德布罗意关系式
De Broglie's theory 德布罗意理论
De Broglie's wave 德布罗意波
De Broglie's wavelength 德布罗意波长
Debrun candle 直角弧光灯
De Brun-Van Eckstein rearrangement 德布隆-范埃克施泰因重排【化】
debt 债务;负债;欠款
debt at call 即期债务
debt balance 负债余额
debt cancellation 取消债务
debt capacity 借债能力
debt capital 借入资本
debt capitalisation 债务资本化
debt ceiling 债务(最高)限额
debt chain 三角债
debt collector 收债人
debt-consolidators 协助清理债务人

debt coverage ratio 收入债务比率;债务抵补率
debt discount 债务折扣;债务贴现
debtee 债主;债权人
debt-equity ratio 负债与资产比(率);债务权益比率;债务股本比;债务对产权比率;负债产权比率
debt-equity swap 债务与股本交换
debt factor 债务系数(指债务总额对资产总额的比率);负债系数
debt financing 借贷筹资;负债融资;举债筹资(以发行债券或期票筹集资金);借款融资;借款资;以举债方式筹措资金;债务周
debt for environmental protection swap 环境保护交换债务
debt-for-nature swap 债务与自然掉换
debt funded 有担保公债
debt instrument 债务证券
debt interest 欠息
debt limit 债务限额
debt limit clause 债务限制条款
debt management 债务管理
debt memo 借项凭单
debt note 借款通知书
debt of hono(u)r 信誉借款;信用贷款;信用借款
debtor 借方;债务人;欠债人;欠户
debtor bank 票据交换轧差银行
debtor corporation 债务公司
debtor country 债务国
debtor-creditor relationship 借贷关系
debtor nation 债务国
debtor's ledger 应收账款分类账
debtor's petition 债务人破产申请书
debtors sharing an obligation 分担债务的各债务人
debtor summons 债务人传票
debt overdue 过期债款
debt payable 应付债款
debt paying ability 偿债能力
debt payment 应付债款
debt-pooler 协助清理债务人;帮助清理债务人
debt proportion 负债比重
debt ratio 债务比率;负债(比)率
debt receivable 应收账款
debt refunding 债务偿还
debt relief 免除债务;免偿债务;债务免除
debt remain 剩余债款
debt remittance 减免债务
debt rescheduling 重订还债期限
debt restructuring 债务重订
debt-ridden 负债累累
debts are rolling up 债台高筑
debt service 借款服务处;债务支付;债务本付息;偿债;债务清偿服务
debt-service constant 债务清偿常数
debt-service figures 债务偿还数字
debt-service fund 偿债基金;债务支付基金
debt-service fund requirements 应提偿基金
debt-service obligation 债务偿付义务
debt-service payments 债务偿还的开支
debt-service ratio 外债偿还率;债务偿还比率
debt-service requirement 负债资金需要量
debt serving 借款服务处
debt side 借方
debts of equal degree 同等级债权
debts of the deceased paid 已偿还遗债
debts-provable in bankruptcy 可证明的破产债务;破产时认可的债务;破产时可认债务
debt subject to statutory limit 法定限额的债务
debt tables 债务表
debt-to-assets ratio 负债与资产比(率)
debt to equipment ratio 债务与装备比
debt to export ratio 负债对出口比率
debt-to-income ratio 月债务与收入的比分
debt to net worth ratio 债务与财产净值比;负债与净值比;负债对净值比率
debt-to-total assets ratio 债务对资产总额比率;负债对资产总额比率
debt trap 债务陷阱
debt warrant bond 附买回条款债券
debug 排除错误;消除缺点;查错;排除(计算机)故障
debug aids 调试工具;调试辅助程序
debug command 调试指令;调试命令
debug function 调试功能
debugged program(me) 调试程序
debugged routine 调试程序
debugger 故障消除器;查错程序

debugging 故障(的)排除;排除故障;排除错误;调试
debugging activity 调试活动
debugging-aid facility 辅助调试设施
debugging-aid program(me) 调试辅助程序
debugging-aid routine 排故障程序
debugging aids 调试工具
debugging expenses 折卸搬迁费用
debugging facility 调试功能
debugging module 查错模块
debugging on-line 联线调试;联机调试;联机程序调整;在线调试
debugging packet 调试程序包
debugging period 早期失效期
debugging phase 排除故障阶段
debugging process 测试过程
debugging program(me) 调整程序;调试程序
debugging routine 调试程序
debugging system 调试系统
debugging tool 调试工具
debugging utility 调试实用程序
debug machine 调试机
debug macroinstruction 排障宏指令;调试宏指令
debug monitor 调试监督程序
debug on-line 联机排除错误;联机检错
debug phase 调试阶段
debug switch 调试开关
debug system 调试系统
de-bunched beam 去聚束;散聚束
de-bunched(beam)pulse 散聚脉冲;去聚束脉冲
de-buncher 去聚器;散束器
debunching 弥散;散失
debunching action 散束作用
debunching correction 散束校正
debunching effect 散聚效应
debunching force 分散力;散束力
deburr(ing) 修边;打毛刺;清除飞边;去(除)毛刺;去毛边;清理毛刺
deburring blade 去毛刺刀片
deburring file 倒角锉
deburring machine 除草籽机;修边机;去毛刺机
deburring operation 去毛刺工序
deburring reamer 去毛刺手铰刀
deburring tool 清理毛口工具;去毛边工具
deburring wheel 清理毛口齿轮
debutanization 丁烷馏除
debutanizer 丁烷馏除器
debutylizing 脱丁基(作用)
Debye 德拜(电偶极矩单位)
Debye-Falkenhagen effect 德拜—法尔肯哈根效应
Debye force 诱导力
Debye-Hukel equation 德拜—休克尔方程
Debye-Hukel model 德拜—休克尔模型
Debye-Hukel theory 德拜—休克尔理论
Debye-Jauncey scattering 德拜—姜西散射
Debye's camera 德拜照相机
Debye-Scherrer method 德拜—谢列尔 X 射线粉末照相法
Debye-Scherrer spectrum 德拜—谢列尔射线谱
Debye's crystallogram 德拜晶体衍射图
Debye's dispersion formula 德拜色散公式
Debye-Sears ultrasonic cell 德拜—席尔斯超声成像
Debye's effect 德拜效应
Debye's equation 德拜方程
Debye's equation for polarization 德拜积化方程
Debye's frequency 德拜频率
Debye's length 德拜屏蔽距离;德拜长度
Debye's model 德拜模型
Debye's pattern 德拜图
Debye's potentials 德拜势
Debye's radius 德拜半径
Debye's relaxation time 德拜弛豫时间
Debye's ring 德拜晶体衍射图
Debye's Scherrer camera 德拜—谢乐摄影机
Debye's Scherrer method 德拜—谢乐法
Debye's Scherrer photograph 德拜—谢乐摄影
Debye's shielding distance 德拜屏蔽距离
Debye's specific heat 德拜比热
Debye's sphere 德拜球
Debye's temperature 德拜温度
Debye's theory 德拜学说
Debye's unit 德拜单位
Debye-Waller factor 德拜—瓦勒因数
Debye wave 特超声
decaborane 癸硼烷

decabromodiphenyl ether 十溴联苯醚;十溴二苯醚
decade 旬;十年;十卷;十进(制)的;十进位;十个一组;十倍
decade box 十进(位)箱
decade carry 十进制进位
decade computer 十进制计算机
decade counter 十进(制)计数器
decade counter circuit 十进制计数电路
decade counting tube 十进计数管
decade counting unit 十进制计数装置;十进计数器单元
decade dial 十进位表盘
decade divider 十进(制)除法器
decadence 颓废
decadence style 颓废派型
decadent art 颓废派艺术
decadent school 颓废派
decadent style 颓废派艺术形式
decadent wave 减幅波;阻尼波;衰减波
decade rate multiplier 十进制比率乘数
decade ring 十进制计数环;十进计数环
decade scalar 十进算量;十进标量
decade scaler 十进制计数器;十进位换算电路;十进位换电路;十进位定标器;十刻度(器)
decade scaling 十进制刻度;十进换算
decade switch 十进位开关
decade unit 十进制器件;十进制单元;十进仪器(十进电阻箱或十进电感箱);十进数器件
decadic signal 十进位信号
decadiene 癸二烯
decadienoic acid 癸二烯酸
decafentin 癸磷锡
decagon 十面体;十角形;十边形
decagonal 十角形的
decagram(me) 十克
decahedron 十面体
decahydronaphthalene 萘烷;十氢化萘
decal 贴花印花法;贴花法(陶瓷装饰法);移画印花法
decalage 差倾角
decalcification 除石灰质作用;脱钙(作用)
decalcify 脱钙
decal comania 贴花膜;移画印花图案;贴花法(陶瓷装饰法)
decal comania method 移画印花法;贴花法
decal comania paper 印刷模写纸
decal comania process 贴花印刷法
decal comania transfer 贴花纸转印
decal comania transferring 贴花转印法
decalescence 吸热
decaline 十氢化萘
decaliter 十升
decalol 萘烷醇
decal paper for ceramic ware 陶瓷贴花纸
decalso 合成硅铝酸盐无机离子交换剂
decal tissue 贴花纸
decalvant 除毛的
decameter 十米
decametre wave 十米波
decan 黄道十度分度
decanal 教堂内坛南侧的
decanal side 教堂内坛南面
decane 癸烷
decane dicarboxylic acid 癸烷二羧酸
decanedioic acid 癸二酸
decanedioyl 癸二酰
decanewton 十牛顿
decanic acid 癸酸
decannulation 除套管法
decanol 癸醇
decanone 癸酮
decanoyl 癸酰(基)
decanoyl acetaldehyde 癸酰乙醛
decanoyl acetaldehyde sodium hydrosulfite 癸酰乙醛亚硫酸氢钠
decanoyl chloride 癸酰氯
decanoyl peroxide 过氧化二癸酰
decant 满流浇注;倒入;倒包浇注;滗;移注
decantation 移注;沉淀(分取)法;沉淀池;撤除;滗析;倾析(法);倾滤;撤除;倾注;撤离法
decantation method 沉淀法;撤离法
decantation test 静沉试验;洗涤(倾析)试验;倾析试验;颗粒洗涤试验
decantation tube 移液管

decantation washing 倾注洗涤
decanted solution 倾析溶液
decanter 缓倾器;沉淀分取器;细颈盛水瓶;滗析器;滗水器;倾注(洗涤)器;倾析器
decanter centrifuge 沉降式离心机
decanting 缓慢降压法
decanting glass 倾析瓶
decanting point 注入点
decanting tank 澄清罐
decanting valve 倾析阀;滗析阀
decantion 缓倾法
decantor 油水分离器
decapacitation 除能
decapacitation factor 出能因子
decapitating hook 断胎头钩
decapitator 断头器
decapsulation 无套模
decarbonate 除去二氧化碳
decarbonater 除碳酸气塔;脱碳酸气塔
decarbonating time formula 脱碳时间公式
decarbonating zone 除碳区;除碳带;除碳层
decarbonation 除去碳酸;除去二氧化碳;脱碳酸盐化作用;脱碳化作用;脱二氧化碳作用;去碳作用
decarbonation reaction 脱二氧化碳反应
decarbonation zone 分解带(碳酸盐)
decarbonization 除碳法;脱碳(酸)作用;脱碳
decarbonization preventing coating 防脱碳涂层
decarbonize 脱碳;除碳;除去碳素;脱焦炭
decarbonized biofilter 除碳生物滤池
decarbonized bottoms 除碳残渣
decarbonized steel 低碳钢
decarbonizer 除碳剂;脱碳剂
decarbonizing 脱碳(层)
decarboxylation 脱羧基作用
decarburation 脱碳作用
decarburisation slag 脱碳炉渣;脱碳矿渣
decarburization 脱碳;除碳(法);脱碳化作用
decarburization slag 脱碳矿渣
decarburize 除碳
decarburized cast iron shot 脱碳铸铁粒
decarburized depth 脱碳深度
decarburized layer 脱碳层
decarburized structure 脱碳组织
decarburizer 脱碳剂
decarburizing 脱碳作用
decarburizing annealing 脱碳退火
decarburizing atmosphere 脱碳气氛
decarburizing reaction 脱碳反应
decartelization 非卡特尔化
decastyle 十柱式房屋;十柱柱式式;十柱式门廊
decastyle building 十柱式建筑
decastyle portico 十柱式门廊
decastyle temple 十柱式庙宇
decastylos 十柱式建筑
decasualization 非临时工化
decasualization of labo(u)r 稳定就业
decasualize 消除临时工
decationize 除(去)阳离子
decationizing 除阳离子(作用)
decatizer 汽蒸机
decatone 水性涂料(涂料的一种)
decatron 十进制计算管;十进制计数管;十进位计数(放电)管;十进管
decatron scaler 十进管定标器
decatyl alcohol 癸醇
decauville 窄轨铁路;轻便铁路
decauville motor tractor 窄轨铁路电动牵引机;轻便铁路电动牵引机
decauville plant 窄轨铁路及设备
decauville portable railway 窄轨轻便铁路
decauville railroad 窄轨铁路
decauville railway 窄轨铁路;窄轨轻便铁道;轻便铁路;轻便铁道
decauville track 窄轨铁路
decauville truck 窄轨料车;窄轨斗车;轻轨料车;轻轨斗车;(窄轨的)轻轨轨道车
decauville tub 窄轨斗车;轻轨料车;轻轨斗车
decauville wagon 小型斗车;窄轨斗车;窄轨料车;轻轨斗车
decay 余辉;糟朽;腐朽;腐烂;衰减
decay accident 腐坏事故
decay angle 衰变角
decay area 减弱区;朽坏区;风浪衰减区;无风区;

衰减区
decay branching ratio 衰变分支比
decay-causing fungi 致腐菌
decay chain 衰变链
decay characteristic 衰减特性
decay characteristic phosphor 磷光体衰变特性；荧光屏余辉特性
decay characteristics 衰变特性
decay coefficient 裂变系数；衰减系数；衰耗系数
decay constant 裂变常数；衰减常数；衰减常量；衰变常数
decay curve 裂变曲线；衰减曲线；衰变曲线
decay curve analysis 衰变曲线分析
decay curve method 衰减曲线法
decay daughter 衰变子体；衰变产物
decay decomposition 化学风化(作用)
decay distance 减弱距离；消衰延程；衰减距离
decay distance of wave 波浪消衰延程
decayed 朽坏的；腐败
decayed knot 木材朽节；朽木；朽节；(木材的)腐节
decayed rock 风化岩(层)；风化岩(石)
decayed sack 腐烂包
decayed tie 腐朽枕木；腐蚀枕木
decayed timber 糟朽木材；腐朽木料
decay energy 衰变能量
decay factor 衰减因子；衰减因素；衰减因数；衰减系数；衰变因素
decay family 放射性系
decay heat 衰变热
decay heat cooler 衰变冷却器
decaying conduction current 衰减传导电流
decaying current 减幅电流
decaying excrement 腐败排泄物
decaying exponential 衰减指数
decaying organic matter 腐烂有机物
decaying pulse 衰变脉冲
decaying shock wave 衰减激波
decaying transient 衰减瞬态
decaying wave 衰减波；衰变波；减幅波
decay instability 衰变不稳定性
decay interaction 衰变相互作用
decay lag 衰变迟滞
decay law 衰减(定)律；衰变定律
decay method 衰减法
decay mode 衰变(类)型；衰变方式
decay of a disturbance 扰动的逐渐减弱
decay of air pollutant 空气污染物的逐渐减少
decay of lumber 木材腐朽
decay of luminescene 发光衰变
decay of organic matter 有机物质分解
decay of pollutant 污染物分解
decay-of-power 功率降低
decay of radioactivity 放射性衰变
decay of sound 声波的衰减；声波的减弱
decay of turbulence 湍流衰减
decay of waves 波浪衰减；波浪逐渐平息
decay of wood 木材腐烂
decay ooze 腐泥
decay oscillation 衰变振荡
decay parameter 衰变参数
decay pattern 混响衰减图
decay period 衰减(周)期；衰变期
decay phase 衰减相位；衰变相
decay photon 衰变光子
decay prevention 防朽
decay process 腐败过程；衰变过程
decay product 分解产物；衰变产物
decay property 衰变性能
decay radiation dose 衰变辐射剂量
decay rate 腐败率；衰减速率；衰减率；衰变(速)率
decay rate of demagnetizing field 退磁场衰减速率
decay ratio 衰变率
decay ratio method of tritium 氚衰减比率法
decay reservoir 衰变库
decay resistance 抗朽性；耐腐性
decay scheme 衰变图
decay sequence 衰变系列
decay series 放射性系；衰变族；衰变系
decay tank 冷却槽；老化池；放射性废物冷却槽
decay time 衰落时间；衰减时间；衰变时间；衰变期
decay time of scintillation 起伏衰落时间
decay type 衰变类型
decay wave 减波幅
Decca 台卡

Decca chain 台卡链；台卡电台链；台卡导航网
Decca chart 台卡位置线图；台卡双曲线导航图；台卡海图
Decca coverage 台卡导航覆盖区
Decca data sheet 台卡活页资料
Decca fixing 台卡定位
Decca grid 台卡格网
Decca Hi-Fix equipment 小型台卡导航仪
Decca homing fixing 台卡导向定位
Decca homing technique 台卡导向技术
Decca indicator 台卡指示器
Decca lane 台卡导航航线
Decca lane identification 台卡巷识别
Decca lattice chart 台卡图网海图
Decca long range area coverage 台卡远程覆盖区
Decca navigation 台卡导航
Decca navigation system 台卡导航系统
Decca navigator 台卡导航仪
Deccan basalt 德干玄武岩
Deccan trap 德干玄武岩
Decca positioning 台卡定位
Decca positioning system 台卡定位系统
Decca radio system 台卡无线电导航系统
Decca receiver 台卡接收机
Decca sheet 台卡活页资料
Decca suffix letter 台卡系统选台字母
Decca system 台卡导航系统
Decca tracing ranging navigation system 台卡跟踪测距导航系统
Decca track plotter 台卡航迹记录器
Decca zone 台卡区
decedents' estates 遗产
decelerability 减速能力
decelerate 减低
decelerated motion 减速运动
decelerated plasmoid 慢化等离子粒团
decelerating conveyer 减速运输机
decelerating economic growth 经济发展速度放慢
decelerating effect 减速作用
decelerating electrostatic field 减速静电场
decelerating field 减速场
decelerating flow 减速流(动)
decelerating force curve 减速力曲线
decelerating phase 减速阶段
decelerating screen 减速栅
decelerating wave 减速波
deceleration 减(加)速度；负加速度
deceleration area 减速区段
deceleration characteristic 减速特性
deceleration check valve 带单向阀的减速阀
deceleration curve 减速曲线
deceleration lane 减速车道
deceleration method 制动法
deceleration of electrons 电子制动
deceleration parameter 减速因子
deceleration performance 减速性能
deceleration radiation 减速辐射
deceleration rate 减速率
deceleration-speed curve 减速曲线
deceleration system 减速系统
deceleration test 减速度试验
deceleration time 减速时间；制动时间(型)
deceleration valve 减速阀
decelerative force 减速力
decelerator 减速装置；减速因子；减速器；减速剂；减速电极；缓动装置；延时器
decelerometer 减速仪；减速(度)计
deceleron 减速副翼
December bonus 年终奖金
December solstice 冬至(点)
de-cementation 去胶结作用；脱胶结作用
decency 风化
decene 癸烯
decene dicarboxylic acid 癸烯二羧酸
decenedioic acid 癸烯二酸
decennial 十年一次的
decennium 十年
decent 下坡
decenter 拆卸模拱；离心
decentered 偏心的
decentered lens 偏轴透镜
decentering 落架；拆卸手架；拆除模架；拆除减速机；拆卸拱架；拆拱模；拆除拱架
decentering distortion 偏心畸变

decentralise 分散；撤离中心点
decentralised control system 分散控制系统
decentralization 离心布局；向外疏散；分散化；分散(法)；分权化；非集中化；偏离中心；疏散
decentralization layout 分散布置
decentralization of authority 分权
decentralization of authority and responsibility 分权分责
decentralization of power 分权制
decentralization of power and interest 权力和利益的分散
decentralization of power and transfer of profits 放权让利
decentralization of responsibility 分散负责；分权负责；分层负责
decentralization principle 分散原则
decentralize 分散；撤离中心点；分散经营
decentralized accounting 分散制会计；分权制会计
decentralized arbiter 分散判优器
decentralized branch accounting 分店独立会计制
decentralized budgeting 分权化预算法
decentralized businesses 分散经营企业
decentralized channel 离散通道
decentralized commodity markets 分散的商品市场
decentralized company 分权制公司
decentralized computer network 非集中式计算机网络
decentralized computer system 分散式计算机系统
decentralized control 局部控制；分散控制；非集中控制
decentralized controller 分散控制器
decentralized controller design 分散控制器设计
decentralized control signalling 分散开支信号
decentralized control structure 分散控制结构
decentralized control system 局部控制系统；分散控制系统
decentralized copy station 分散复印站
decentralized data processing 分散(型)数据处理；非集中式数据处理
decentralized decision making 分权决策
decentralized desalination 分散脱盐
decentralized estimator 分散估计器
decentralized exchange economy 分散的市场经济
decentralized funds 分散使用资金
decentralized gain matrix 分散增益矩阵
decentralized heat-supply 分散供热
decentralized information system 分布信息系统
decentralized innovative unit 分权创新单位
decentralized input 分散输入
decentralise interlocking 非集中联锁
decentralised issue system 分散发行制
decentralized management 分层管理；分散经营；分散管理；分权管理
decentralized management of wastewater 分散式污水管理
decentralized management organization 分权的管理组织
decentralized management pattern 分权管理模式
decentralized material storage 分散原料库
decentralized multiaccess 分散多路存取
decentralized network 分布网络；非集中式网络
decentralized optimization 分散最优化
decentralized organization 分权制组织
decentralized planning 分散计划
decentralized planning control 分散计划管理
decentralized planning department 分散计划部门
decentralized processing 分散处理；分布处理
decentralized production control 分散生产管理
decentralized profit control system 分散的利润控制系统；分权的利润管理系统
decentralized regulator 分散调整器
decentralized reserve system 分散准备制
decentralized sewerage system 分散式下水道系统
decentralized solar power station 分散型太阳能电站
decentralized statistical system 分散型统计制度
decentralized statistics 分散统计
decentralized stochastic control 分散随机控制
decentralized structure 分散化结构
decentralized summary 分散汇总；非集中汇总
decentralized system 局部系统；分散制；分散系数；分权(体)制
decentralized wastewater treatment 污水分散处理
decentralize power on minor issues 小权分散

decentralizing management on a household basis 分户经营
decentral statistics 非集中统计
decentration 不共心;偏心
decentre 离(中)心
decentring 拆拱模
deception 掩饰
deception equipment 干扰施放装置
deceptive conformity 假整合
deceptive fold 假褶皱
deceration 除蜡法
decertification 否定代表资格
decertify 吊销执照;收回证件
de-channel(l)ing 去沟道效应
dechenite 红钒铅矿
dechlorinate 除氯;去氯
dechlorinating agent 脱氯剂
dechlorination 除氯;脱氯(作用)
dechlorining agent 除氯剂
decholesterinization 除胆甾醇作用
decholesterolization 除胆甾醇作用
dechromisation 除铬
deciacre 分公亩(1 分公亩 =1/10 公亩)
deci-ampere balance 分安培秤
decibar 分巴
decibel 分贝(音强单位、电平单位)
decibel above one volt 伏特分贝
decibel adjusted 调整分贝
decibel based on one milliwatt 基于一毫瓦的分贝数
decibel calculator 分贝计算图表
decibel gain 分贝增益
decibel level 测噪声标准
decibel-log frequency characteristic 分贝对数频率特性
decibel-loss 分贝衰减
decibel measurement 分贝测定
decibel meter 分贝计;分贝表
decibel notation 分贝符
decibel potentiometer 分贝电位计
decibel recorder 分贝记录器
decibels above l milliwatt 毫瓦分贝
decibels above picowatt[DBP]微微瓦分贝
decibels above reference coupling 超过基准耦合度的分贝数
decibels above reference noise 超过基准噪声的分贝数;超出参考噪声分贝数
decibel scale 分贝刻度;分贝标度
decibel scale for sound pressure level 声压级分贝标度
decibels relative to one volt 伏特分贝
deciboyle 分波义耳(压强单位)
decidability 可判定性;可决定性
decidable proposition 可决策命题
decidable subclass 可判定子类
decide against somebody 判决某人败诉
decide as circumstances require 依照情况而定
decided site of station 车站选址
decide in favor of somebody 判决某人胜诉
decide on awards through discussion 评奖
decide to order 决定订购
deciding factor 决定性因素
deciduous 阔叶;脱落的
deciduous broadleaf forest 落叶阔叶林
deciduous broadleaf scrub 落叶阔叶灌木丛
deciduous forest 落叶(树)林
deciduous hardwood tree 落叶硬材树
deciduous oak 落叶栎
deciduous plant 落叶植物
deciduous shrub 落叶灌木
deciduous tree 落叶树
deciduous wood 落叶树木材
deicer 防结冰装置;去冰器
decigramme 分克
decile 十分位数
decile interval 十分位距
deciliter 立方分米;分升;分合
decilog 常用对数的十分之一;分对数
decimal 十进制的;十进小数;十进位
decimal accumulator 十位累加器;十进制累加器
decimal adder 十进制加法器
decimal addition 十进制加法
decimal address 十进制地址
decimal adjust accumulator 十进制调整累加器

decimal adjust accumulator instruction 十进制调整累加器指令
decimal adjust instruction 十进制调整指令
decimal alignment 十进制对准;十进制对位;十进制定位
decimal arithmetic 十进制算术(运算)
decimal arithmetic(al) capability 十进制运算能力
decimal arithmetic(al) instruction 十进制算术指令
decimal arithmetic(al) operation 十进制(算术)运算
decimal arithmetic(al) trap mask 十进制算术运算自陷屏蔽
decimal attenuator 十进衰减器
decimal attribute 十进制属性
decimal balance 十进天平
decimal base 十进制记数制
decimal candle 十进位烛光(即国际烛光)
decimal carry 十进制进位
decimal classification 十进(制)分类法
decimal code 十进制(代)码;十进码;十进代码
decimal-coded 十进编码(的)
decimal-coded digit 十进制编码数;十进编码数字
decimal coded system 十进制编码系统
decimal computation 十进制计算
decimal computer 十进制计算机;十进位计算机
decimal constant 十进制常数
decimal conversion 十进制转换
decimal counter 十进制计数器
decimal currency 十进制通货
decimal decoder 十进制译码器
decimal determination 十进制
decimal digit 十进制数(字);十进码
decimal divide exception 十进制除法异常
decimal division 十进制除法;十进位主题分类
decimal equivalent 小数位等量;用小数表示的分数值;等值小数
decimal equivalent table 十进位等值表
decimal feature 十进制运算机构;十进制特点
decimal fixed-point constant 十进位定点常数;十进制定点常数
decimal floating-point constant 十进制浮点常数;十进位浮点常数
decimal floating point data 十进制浮点数据
decimal fraction 小数;十进(制)小数;十进制分数
decimal fraction format 十进制小数格式
decimal fraction notation 十进位小数标记符号
decimal ga(u)ge 十进制规
decimal index system 十进制系统
decimal integer 十进制整数
decimalism 十进法
decimalize 换算成十进制
decimal location 十进制数位;十进制单元
decimal marker 十进制(小数点)标记
decimal measure 十进制
decimal multiple 十进倍数
decimal multiplier 十进制乘法器
decimal normalization 十进制规格化
decimal notation 十进制记数法;十进位记数法
decimal number 小数;十进制数;十进小数;十进位数
decimal number base 十进制小数基
decimal number format 十进制数格式
decimal numbering system 十进制数制
decimal number system 十进制记数制;十(小)数制
decimal numeral 十进制数的
decimal numeration 十进位记数法
decimal numeration system 十进制数制;十进记数制
decimal operator 十进制运算符
decimal overflow 十进制数溢出
decimal overflow exception 十进制数溢出异常
decimal overflow trap 十进制数溢出自陷
decimal part 小数部分;十进制小数部分
decimal picture data 数字字符数据;十进制图像数据
decimal place 小数位
decimal point 小数点;十进(制)小数点;十进位小数点
decimal point alignment 小数点对位;小数点调准;十进制对位
decimal point positioner 小数定位器
decimal product generator 十进制乘积发生器
decimal radix 十进制基数

decimal representation 十进制表示
decimal rheostat 十进变阻器
decimal ringtype adder 十进环式加法器
decimals 小数
decimal scale 十进制;十进计数法
decimal selection 十进(制)选择
decimal sequence 十进序列
decimal storage system 十进制存储系统
decimal string 十进制数串
decimal subtracter 十进制减法器
decimal subtraction 十进制减法
decimal symbol 十进制(小数点)符号
decimal system 十进制系统;十进(位)制;十进数字法
decimal tabulator key 十进制制表键
decimal time 十进计时表
decimal-to-binary 十翻二
decimal-to-binary conversion 十进制到二进制转换
decimal value 十进制数值
decimal variable divider 十进制可分频器
decimation in time 时域疏样
decimeter 分米
decimeter continuum 分米波连续谱
decimeter height finder 分米波测高计
decimeter radio 分米波无线电通信
decimeter range 分米波段
decimeter test equipment 分米波测试器
decimeter wave 分米波
decimeter-wave antenna 分米波天线
decimeter-wave continuum 分米波连续谱
decimeter-wave mixer 分米波段混频器
decimeter-wave radar 分米波雷达
decimeter-wave transmitter 分米波发射机
decimetre 分米
decimetric wave oscillator 分米波振荡器
decimilligram 十分之一毫克
decimillimeter 丝米(1 丝米 =0.0001 米)
decimosexto 十六开纸
decine 癸炔
decineper 分奈
decipher 译码;破译密码
decipherable 翻译得出的;辨认得出的
decipherable map 易读地图
decipherator 译码器;译码机
decipherer 译码器;译码机
deciphering 解密码;解密【计】;译码
decision accounting 判定计算
decision action 决策行动
decisional management school 决策管理学派
decisional roles 决策方面的角色
decision and action analysis 决策和活动分析
decision analysis 决策分析
decision analysis process 决策分析过程
decision-assisting analysis 决策支援分析
decision band method 决策带法;决策层次作业评价法
decision behavio(u)r 决策行为
decision block 决策块;判定块
decision boundary 判别边界
decision box 决策箱;决策块;判定框(图)
decision center 决策中心
decision circuit 判定电路
decision computer 决策计算机
decision cost 决策费用
decision criterion 决策判据;决策标准;决策(衡量)准则;决策基准;抉择准则
decision data 决策资料
decision design 决策设计;优选设计
decision element 解算元件;判定元素
decision evaluation 决策估计
decision-event chain 决策事件链锁
decision factor 决策因子
decision feedback system 决策反馈系统;判定反馈系统
decision fork 决策图权
decision-form approach 决策表法
decision formulation 决策配制
decision function 决策函数;判定函数
decision game 决策模拟推断;决策博奕
decision gate 决策门
decision grid chart 判定网格表
decision information system 决策资讯系统
decision instant 判定瞬间
decision instruction 决策指令;判断指令;判定指

令;判别指令;条件转移
decision integrator 决策记分器;判定积分器
decision in the face of risk 冒险性的决策;风险决策
decision lag 决策时滞;决策时间滞差.
decision loop 决策环
decision-maker 决策者;决策人员;判定装置
decision-making 决策;作决定;作出判定;判定
decision-making accounting 决策会计
decision-making authority 决策权;决策当局
decision-making body 决策机构
decision-making complex 决策复合体
decision-making continuum 决策持续性
decision-making management 决策管理
decision-making package 综合决策
decision-making period 决策周期
decision-making power 自主权
decision-making power in operation and management 经营管理自主权
decision-making power of enterprise 企业自主权
decision-making process 决策过程;决策程序
decision-making steps 决策步骤
decision-making system 表决系统;决策系统;判定系统
decision-making technique 决策技术;决策方法
decision-making types 决策的类型
decision-making uncertainty 不确定型决策
decision-making under certainty 肯定情况下的决策;在确定情况下制订决策;确定条件下的决策
decision-making under risks 冒险性决策;风险下的决策
decision-making under uncertainty 易变情况下的决策;在不确定情况下制订决策
decision-making with experimentation 借助试验的决策
decision mapping 决策图
decision marking 标识决定
decision marking system 标识判定系统
decision matrix 决策矩阵
decision mechanism 决策机制;决策机理;判定机构
decision model 决策模式;决策模型
decision model base 判定模型库
decision model in traditional decision theory 传统决策理论中的决策模型
decision network 决策网络
decision node 决策(结)点
decision of maximization 最大化判定
decision of the court 法院裁决
decision on award 决标;定标
decision package 决策组;决策单元
decision parameter 决策参数
decision phase 决策阶段(价值工程工作计划的阶段之一)
decision plan 判定方案
decision point 决策点
decision problem 决定问题;决策问题;真伪判定问题
decision procedure 决策程序;判定方法;判定程序
decision process 决定过程;决策过程;作出决定的过程;判定过程
decision region 决策区域
decision right of equipment replacements and technical innovations 设备更新改造自主权
decision right of main repair cost 大修支出自主权
decision rule 决策规则;判决规则;判定规则
decision schedule 判定方案
decision scheme 判定方案
decision science 决策科学
decision scoring table 决策计算表
decision set 决策集
decision situation 决策状况;决策形势;决策情况
decision space 决策空间;判定域
decision split 决策分解
decision statement 决策陈述
decision steps 决策步骤
decision strategy 决策战略
decision structure 决策结构
decision support for water and environmental system 水环境决策支持系统
decision support system 决策支持系统;决策辅助系统
decision table 决策表;判定表
decision table compiler 判定表编译程序
decision table processor 判定表处理程序
decision table processor mode 决策表处理模式

decision test 决策检验
decision theoretic analysis 决策理论性分析
decision-theoretic approach 决策理论方法;判别推理方法
decision-theoretic approach to pattern recognition 模式识别的判定理论法
decision theory 决策(理)论;优选理论;判定(理)论
decision thesis of probability 概率的决策论点
decision to proceed with the project 工程项目实施的决定
decision tree 决策宗谱;决策图表;决策体系;决策树;决策程序流程图;树形决策图
decision-tree analysis 决策体系分析;决策树分析;分层决策分析;分层次决策分析;决策分析
decision-tree approach 决策树法
decision-tree method 决策体系法
decision under certainty 确定型决策
decision under conflict 竞争决策
decision under risks 风险型决策
decision under uncertainty 非确定型决策;不确定型决策
decision unit 决策单位
decision value 决策值;判决值
decision variable 决策变量
decisive 决定性的
decisive effect 决定性作用
decisive experiment 判决性试验
decisive influence 决定性影响
decisive model of organization 决策组织模型
decisive set 决定性集
decit (信息量的)十进单位
deck 楞堞;楞堞;甲板绕组;甲板磨石;甲板;机芯;面板;装车台;车身底板;舱面;平台
deck abstract logbook 舱面摘要日记
deck ankles 船员踝肿
deck appliance 甲板用具;甲板器械;舱面设备;舱面器械
deck auxiliaries 甲板属具
deck barge 甲板驳
deck base 筛框
deck beam 甲板(横)梁;顶棚梁;上承(式)架
deck bench 甲板长椅;舱面长椅
deck board 货架顶板;装饰盖板
deck bolt 甲板螺栓;平圆柱头螺栓
deck bottom rail 桥面板下轨梁;顶棚纵向梁
deck bridge 跨线桥;甲板桥楼;上承式桥;上承桥
deck broom 甲板刷;甲板扫帚
deck bucket 甲板水桶;太平桶
deck bung hole 甲板木塞孔
deck bunker scuttle 甲板上煤舱口
deck cadet 实习驾驶室
deck camber 甲板梁拱
deck cant 屋面平台撑木;平屋顶泛水;撑木
deck cantilever bridge 上承式悬臂桥
deck cargo 甲板货;舱面货(物)
deck cargo certificate 舱面货物证明书
deck cargo insurance 舱面货物保险
deck cargo rates 甲板货运费率
deck cargo shipper's account 舱面货物由货主担风险
deck carling 甲板端短纵梁
deck chair 船甲板躺椅;甲板用椅;折叠(式)躺椅;躺椅
deck charge 分段装火药;分段装药(法)
deck chest 甲板用具箱
deck chock (安置救生艇用)垫木;导缆口
deck clause 舱面条款
deck clearance 桥面净空
deck cleat 甲板系缆耳
deck clip 扣伴;紧固件;屋盖板夹片
deck collar 甲板唇口
deck composition 甲板涂料;甲板敷料
deck compression chambers 甲板加压舱
deck construction 码头岸面构造;码头面板;地板构造;桥面构造;平顶结构
deck coordinator 飞机降落区协调员
deck covering 舱面覆盖层;甲板覆盖层
deck covering plan 甲板被覆图
deck crane 甲板起重机;甲板吊车;船用起重机;船上起重机
deck crane operation 甲板起重机装卸作业
deck crew 舱面人员
deck curb 屋面(边)缘栏;盖板边饰;平屋顶凸缘
deck dam 平板坝

deck decompression chamber 甲板减压舱
deck delivery 油轮主油管从油泵室到甲板出口的一段
deck department 驾驶部门;甲板部
deck door 后层门
deck dormer 平屋顶无窗;平屋顶老虎窗;屋顶天窗
deck dowel 甲板木栓
deck drain valve 甲板泄水阀
deck duty 舱面值班
decked barge 舱口驳;平台甲板平底船
decked boat 甲板艇;甲板船
decked dobby 多针孔纹板多臂机
decked explosion 分层爆炸
decked lifeboat 浮箱式救生艇;水密双层底救生艇
decked rockfill dam 面板堆石坝
decked ship 甲板船
decked structure 叠瓦结构
decked vessel 甲板船
decken 推覆体
deck enamel 地板瓷砖
deck end roller 甲板端滚轮;导缆器
deck engineer 管理甲板机械的人员
decken structure 叠瓦构造
decker 甲板船;稠料器;分层装置;脱水机
deck erection 甲板上层建筑(物);舱面建筑物;舱面构筑物
deck expansion joint 桥面伸缩缝
deck factor D 筛分长度系数 D;层位系数(以顶层筛为1)
deck factor T 筛板孔形系数 T
deck fairleader 甲板道缆孔
deck fitting 甲板固定设备;甲板绑扎设备;甲板舾装;舱面属具
deck flange 甲板唇口
deck floor 平楼盖;平楼板;平台甲板
deck foam system 车载泡沫系统
deck form 平台式钢模板;甲板模板;平台模板
deck framing 桥面构架;甲板构架
deck gang 舱面人员
deck gear 甲板机械;舱面器械
deck girder 甲板纵材;面板支梁;码头面大梁;上承式大梁
deck-girder bridge 上承梁式桥
deck-girder dam 板梁坝
deck-girder steel dam 板梁式钢坝
deck glass 甲板天窗;铺面玻璃砖
deck glass roof 玻璃砖平台屋顶;玻璃平顶
deck gun 车载消防炮
deck hand 舱面水手;舱面人员;水手;甲板水手
deck hawse 甲板锚链孔
deckhead 舱口空间;舱顶甲板
deck height 甲板型高;甲板(间)高度
deck highway bridge 上承式公路桥
deck hook 甲板肘材(木船);甲板钩
deck house 甲板室[船];甲板舱室;舱面(舱)室
deck hull's eye 甲板天窗
deck hurricane house 甲板上搭的木屋
deck ice 船舶积冰
decking 甲板被覆;码头面板;装卸矿车;支撑板;盖板;叠放;分段装药法;桥面板;桥梁车行道;铺面;铺板;棚现
decking bridge continuous for temperature effect 温度效应连续桥面桥;温度桥面连续梁桥
decking chain 捆木链
decking engine 装罐发动机
decking gear 装罐机
decking ram 矿床捣锤
decking slab 上承板
decking system 屋顶下部结构系统
deck initiation 启动平台
deck inventory 甲板属具清单
deck iron 甲板铁;捻缝削刀
deck joint 面板接缝
deck ladder 甲板梯
deck lamp 甲板灯;煤水车灯
deck lashing container lock 甲板集装箱系固件
deck lashing for shiptainer 船用集装箱固定装置
deck lashing turnbuckle 甲板捆绑法兰螺丝;甲板捆绑凸缘螺丝
deck lashing wire system 甲板系索式栓固系统
deckle 稳纸框
deckle board 框板
deckle edge 纸边
deckle-edged 毛边的

deckless buttress dam 非平板式支墩坝
deckle strap 框带；定边带
deck light 甲板天窗
deck line 甲板线
deck load 甲板货；甲板负荷；面板荷载
deck loading 分段装药法
deck load per square metre 每平方米舱面负荷
deck lock 后层门锁
deck locker 甲板杂物柜
deck log 航行日志；航海日志
deck logbook 航行日志
deck longitudinal 甲板纵材
deck machinery 甲板机械；舱面机械；平台(式)机械
deck man 舱面装卸工人
deck module 桥面组件
deck name 卡片组名
deck of a goods shed 货棚装车站台
deck officer 驾驶员【船】；船舶驾驶员
deck-on-hip 四坡屋顶上的平屋顶；四坡平屋顶
deck opening 甲板开口
deck outfit 舱面设备
deck paint 瓷砖铺地面；甲板涂料；甲板漆；耐磨地面涂料；高度耐机械磨损漆
deck passage 搭乘统舱；统舱
deck passenger 甲板旅客；统舱旅客
deck pillar 甲板支柱
deck pipe 锚链管；车载消防炮
deck piping 舱面管系
deck piping control panel 舱面管路控制台(油船上)
deck plan 甲板平面图
deck plank(ing) 木甲板(条)
deck plate 钢甲板；盖板；波纹钢板；瓦垄钢板；台面板
deck plate beam 上承板梁
deck plate girder 上承板梁
deck plate girder bridge 上承式板梁桥
deck plating 钢甲板
deck plug 甲板木栓；甲板木塞
deck pontoon 甲板箱体
deck post 顶棚柱
deck prime coat(ing) 甲板第一道涂层；房屋楼面第一道涂层
deck prism light 甲板采光棱镜
deck profile grade 桥面纵坡
deck protection 甲板防护
deck pump 甲板泵；舱面泵
deck quoits 绳圈
deck rail 顶棚纵向梁
deck rate 舱面运费率；舱面货运费率
deck reinforcement 面板钢筋
deck ring 甲板地铃
deck rod lashing system 甲板拉杆式栓固系统
deck roller 打捆台禾谷输送辊
deck rolling crane 舱面移动起重机
deck roof 平屋顶；平台式屋顶；台式屋顶；晒台
deck rudder bracket 甲板止舵楔
deck rudder stops 甲板舵掣
deck sash 顶棚窗；顶棚架
deck scow 甲板(货)驳；平底船
deck screen 多层筛；振动筛组
deck scrubber 甲板刷；洗刷甲板的人
deck scuttle 甲板煤舱口；甲板煤舱孔
deck seamanship 船艺
deck seat 甲板长椅；舱面长椅
deck securing fitting 甲板固定件；甲板绑扎固定用具
deck sheathing 甲板覆盖层
deck sheer 甲板脊弧
deck slab 翼板；桥面板；上承板
deck socket 甲板承窝
deck soffit 上承拱复
deck span 上承式桥跨结构；上承式孔；桥跨径；桥面跨度
deck spillway 盖面溢洪道；溢流板；盖板溢洪道
deck sprinkler 甲板洒水装置
deck sprinkler pipe 甲板洒水管道
deck sprinkle system 甲板洒水系统
deck stanchion 甲板支柱
deck step 后层梯蹬
deck store 甲板部物料；帆缆材料室
deck store keeper 甲板部物料管理员
deck storeroom 甲板部物料间
deck-stowed container 放在甲板上的集装箱
deck straddling crane 跨式甲板起重机

deck strake 甲板列板
deck strength construction 甲板加强结构
deck stringer (桥梁车行道的)上承式纵梁；甲板边板；码头面纵梁；行车道纵梁；桥面纵梁
deck strong back 甲板强力背材；甲板纵梁
deck structure 桥体结构；舱面建筑；上承结构
deck structure of arch 拱上结构
deck stuffing box 稳舵盒
deck sundry 甲板用具
deck superstructure 甲板上层建筑(物)
deck surface 上承式桥面
deck surfacing 桥上铺面
deck switch 面板开关；同轴开关
deck system 甲板系统
deck tackle 舱面绞辘
deck tank 甲板水柜；舱面水柜
deck torpedo tube 水上鱼雷发射管
deck trailer 平板拖挂车
deck trainee 见习水手
deck transom 甲板艄梁(方尾船)
deck transverse 甲板横材(油轮)
deck truss 盘架(摇床)；上承(式)桁架
deck truss bridge 上承式桁架桥
deck tube 甲板电线套管
deck turntable 上承式转车台(铁路车站调车场)
deck twin crane 并列式甲板起重机
deck type 上承式
deck type barge 甲板驳
deck type bridge 上承式桥
deck type form 平台式模板
deck valve 甲板阀
deck wash hose 甲板冲洗软管；冲洗甲板软管
deck washing pipe 甲板冲洗管
deck washing pump 甲板冲洗泵
deck watch 精密航海表；甲板值班；船表；舱面值班
deck watchman 停泊时值班水手
deck waterproofing 平台防水
deck water seal 甲板水封
deck welding 重力焊(接)
deck width 甲板宽度
declarable 可申请交税的
declarant 申请人
declaration 公告；报关单；启运通知书(保险)；声明；申报
declaration and ratification of estimated reserves 储量报批
declaration and registration of pollutant discharge 污染物排放申报登记
declaration and registration system of pollutant discharge 污染物排放申报登记制度
declaration date 通知日
declaration day 记息日
declaration for export 出口报关单
declaration for exportation 出口报单
declaration for import 进口申报；进口报关单
declaration for importation 进口报关单
declaration form 报关表；申报单；申报表
declaration for warehousing 进栈申报单
declaration house 报关行
declaration insurance 通知保险
declaration inwards 进口报关单；入境报关单
declaration list 申报单；申报表
declaration of avoidance 宣告无效
declaration of blockade 封锁宣言
declaration of clearance 结关申报单
declaration of condominium 多户共有公寓大楼的主契据
declaration of dividend 宣布分配股息
declaration of entrance 入港申报(单)
declaration of estimated tax 估计税额申报
declaration of forfeiture 没收公告
declaration of hatches 各舱口报关单；舱口报关单
declaration of inability to pay 宣告无力偿债
declaration of inventorship 发明人的声明
declaration of restrictions 限制声明
Declaration of the Hague 海牙宣言
Declaration of United Nations Conference Human Environment 联合国人类环境会议宣言
declaration of value 价值说明；价值公告；申报价值
Declaration on the Human Environment 人类环境宣言
Declaration on the Protection of the Atmosphere 保护大气宣言

declaration outwards 出口报关单
declaration policy 船名未定保险单；长期有效保险单
declarative array 说明数组
declarative macro instruction 说明宏指令
declarative operation 说明性操作
declarative operation code 说明操作码
declaratives 说明部分
declarative statement 说明语句
declarator 说明符
declarator definition 说明符定义
declarator name 说明符名称
declarator subscript 说明符注脚
declaratory judgement 法院对原告拥有合法权力的裁决
declare 陈述；申报
declare a dividend 通告分红
declare at the custom 报关
declare bankruptcy 宣告破产
declared calorific value of gas 法定燃气热值
declared capital 法定资本；申报资本(额)；设定资本；设定股本
declared depth 公布的水深
declared dividend 已宣布分配的股息
declare deadweight 申报载重吨位
declare defeasance 宣告无效
declared efficiency 标称效率
declared goods 已报关货物；报关货物
declare discharging/loading port 宣布装/卸港
declared profit 已宣布利润
declared value 申报价格；声明价格；申报价值；设定价值
declared value for carriage 运费价格申报；申报货载价值
declared weight 申报重量
declare general average 宣布共同海损
declare goods for duty 货物报税
declare port option 宣布选择港口
declarer 说明词
declaring at customs 报关
declassification 解密【计】
declassified cost 重分类成本
declassified information 解密资料
declension 倾斜
declinate 差角
declinating point 罗盘修正台；磁偏点
declinating station 磁偏点
declination 下倾；赤纬；倾斜；偏斜
declinational inequality 赤纬不等
declinational tide 赤纬潮；赤道潮(汐)
declination angle 偏(向)角
declination arc 磁偏角弧
declination axis 赤纬轴
declination change in one hour 每小时赤纬变化量
declination chart 磁倾图；磁偏图
declination circle 赤纬圈；赤纬度盘
declination compass 磁偏针；磁偏计
declination curve 偏向曲线
declination D 陷偏角 D
declination diagram 三北偏角图
declination difference 赤纬差
declination error 罗盘变动误差；偏角误差
declination flying 偏向飞行
declination needle 磁偏指针
declination of compass 磁偏角
declination of grid north 格网(真北)偏角
declination of magnetic needle 磁偏角
declination of zenith 天顶赤纬
declination parallel 赤纬圈
declination variometer 磁偏变感仪；磁偏计
declinator 磁偏仪；磁偏计；赤纬仪；侧边固定器；坡度仪；偏角计；偏差器
declinature 拒保
decline 下倾
decline an offer 不接受报价
decline an order 谢绝订货
decline characteristics 递减特性
decline curve 下降曲线；递减曲线；衰退曲线
declined risk 拒受的保险业务；拒收的风险
decline in brightness 亮度消减
decline in economic usefulness 经济效用降低
decline in job 就业率下降
decline in judgement 判断力下降
decline in ozone 臭氧逐渐减少
decline in pressure 压力下降

decline in price 跌价;物价下跌
decline in production 生产衰退
decline loan 拒绝贷款
decline offer 谢绝发盘
decline of groundwater level 地下水位下降
decline of piezometric surface 水压面下降
decline of production 产量下降
decline of real growth rate 实际增长水平下降
decline of sale 销路不佳
decline of stress 应力下降
decline of underground water level 地下水位下降
decline of water-level 水位下降
decline of water table 地下水位下降;潜水面下降;水位下落;水位下降;水位降低
decline of well 油井衰退;油井枯竭
decline period of water table 水位下降期
decline phase 下落期;衰亡期;衰退期
decline rate 衰退速度
decline-rate filtration 减速过滤
decline section 下坡段
decline stage of production 产量递减期
decling phase 衰减相;衰减阶段
declinimeter 磁倾仪
declining 年金递减土地估价法;看跌;低落
declining average variable cost 平均变动成本下降
declining balance depreciation 余额递减法折旧
declining balance method 下跌平衡法;差额递减法;余额递减法
declining balance method of depreciation 折旧余额递减法
declining earth pushing process 下坡推土法
declining earth shoveling process 下坡铲土法
declining growth phase 减速增殖期
declining industries 夕阳产业
declining marginal efficiency of capital 资本边际收益递减
declining neighbo(u)rhood 衰退邻里
declining rate filtration 递降速过滤
declining stage 衰退期
declining volcanic activity 火山活动逐渐消失
declinometer 磁倾仪;磁倾计;磁偏计;方位计;倾斜表;坡度仪;偏角仪;偏角计
declivate 降坡的
declivent 降坡线;降坡的
declivitous 降坡的
declivity 降坡;下斜;倾斜面;倾斜;梯度
declivity observation 倾斜观测
declivity of ways 下水滑道倾斜度数
declivity rate of main body of building 建筑物主体倾斜率
declivity survey 倾斜测量
declutch 分离离合器;分离;分开离合器;脱开;松闸
declutch bearing 离合器轴承
declutch carrier 分离座架
declutch driving clutch 分离传动离合器
declutcher control lever 分离杆
declutching 脱开离合器
declutching lever 分离杆
declutching safety device 使离合器分离的安全装置
declutch shaft 分离轴
declutch shaft bearing gasket 分离轴轴承盖垫密片
declutch shaft carrier 分离轴座架
declutch shaft carrier gasket 分离轴座架垫密片
declutch shaft gear 分离轴齿轮
declutch shift lever 换挡操纵杆
declutch shift shaft 分离移动轴;拨叉轴
declutch shift shaft lock ball 拨叉轴紧锁钢球
declutch sliding clutch 分离滑动离合器
Deco 德可素工具钢
decoagulant 反絮(凝)剂
de-coat(ing) 去除覆盖层;除去涂层;脱漆
Decobra (alloy) 德可布拉铜镍锌合金
decoction crack 干燥裂隙;干燥裂纹
decoction made from powder 煮散
decodability 可解码性
decodable code 可解码
decode 解码;译码;译出指令
decoded mode 译码方式
decoded operation 译码操作
decoded signal 译码信号
decoded trigger pulse 译码触发脉冲
decode logic 译码逻辑
decode operation 译码操作
decoder 解码器;译码器;译码机

decoder connector 译码机连接器
decoder matrix 译码矩阵
decoding 译码;译出指令
decoding algorithm 译码算法
decoding circuit 译码电路
decoding constraint length 解码约束长度
decoding gate 译码门
decoding information 译码信号
decoding logic 译码逻辑
decoding matrix 译码矩阵
decoding network 译码网络
decoding relay 译码继电器
decoding room 译码室
decoding scheme 译码电路
decoding system 译码系统
decoding table 解码表;译码表
decohere 散屑
de-coherence 去相干
decoherer 散屑器
decohesion 黏着力的去除;减聚力
decoil 开卷
decoiler 开卷机;拆卷机
decoiler roll 开卷机带材板直辊
decoiling equipment 开卷装置
decoking 除焦;脱焦
decollate 分割
decollation 断螺顶
decollator 多联纸分理机;拆散器
decollement 挤离(作用);滑脱(构造);构造脱顶;浮褶(作用);剥离术;脱顶构造;脱底
decollement fold 挤离褶皱;滑脱褶皱;浮褶皱;表层褶皱;脱顶褶皱;脱底褶皱
decollement zone 滑脱构造带
decollimation 平行性破坏;去平行性
decolo(u)r 脱色
decolo(u)rant 脱色剂;褪色剂
decolo(u)ration 脱色
decolo(u)red water 变色水;脱色水
decolo(u)rimeter 脱色计
decolo(u)ring agent 脱色剂
decolo(u)riser 漂白剂;脱色剂
decolo(u)rite 多孔阴离子交换树脂
decolo(u)rization 漂白作用;脱色(作用);褪色(作用)
decolo(u)rization by chemical oxidation 化学氧化法脱色
decolo(u)rization efficiency 脱色效率
decolo(u)rization flocculant 脱色絮凝剂
decolo(u)rization index 脱色率
decolo(u)rization kinetics 脱色动力学
decolo(u)rization of wastewater 废水脱色
decolo(u)rize 脱色
decolo(u)rizer 漂白剂;脱色剂
decolo(u)rizing 脱色
decolo(u)rizing agent 退色剂
decolo(u)rizing carbon 脱色炭
decolo(u)r power 脱色力
decolo(u)r ratio 脱色率
decometer 台卡计
decommissioned waste 退役废物
decommissioning 退出运行
decommutation 反互换
decommutator 反转换器;反互换器
decompacting 松散
decompaction 振松
decompaction number 脱压实数
decompensation 代偿失调
de-complementize 去补体
decomplexation 解配合
decomplier 非编译
decomposability 可分解性;分解性能
decomposability attribute 可分解的属性
decomposable 可分解的
decomposable and crystalline erosion 分解结晶复合性侵蚀
decomposable and crystalline erosion index 分解结晶复合性侵蚀指标
decomposable erosion 分解性侵蚀
decomposable erosion index 分解性侵蚀指标
decomposable game 可分解对策
decomposable matrix 可分解矩
decomposable model 可分解模型
decomposable operator 可分解算子
decomposable organic matter 可分解有机质

decomposable process 分解过程
decomposable production system 可分解的产生式系统
decomposable system 可拆体系
decompose 分解;离解
decomposed 风化的
decomposed coal 风化煤
decomposed erosion 分解性侵蚀
decomposed explosive 分解炸药
decomposed granite 细砂土;风化花岗岩
decomposed out crop 风化分解露头
decomposed product 分解产物
decomposed rock 风化岩石;风化岩(层)
decomposer 分解物;分解者;分解器;分解剂;分解槽
decomposer organism 分解生物
decomposing agent 分解剂
decomposing furnace 分解炉
decomposing pot 分解釜
decomposition 离解作用;解体;还原作用;腐烂(作用);腐解;分解(作用);分解产物
decomposition aggregation analysis 分解集结分析
decomposition aggregation model 分解聚合模式
decomposition algorithm 分解算法
decomposition axiom 分解公理
decomposition by bombs 罐分解
decomposition by chlorination 氯化法分解
decomposition by combustion 燃烧分解(法)
decomposition by fusion 熔融分解(法)
decomposition by sealed tubes 密封管分解
decomposition by sintering 烧结分解
decomposition chart 分解图
decomposition coefficient 分解系数
decomposition combustion 分解燃烧
decomposition constant 分解常数
decomposition-coordination 分解协调
decomposition course 分解过程
decomposition curve 分解曲线
decomposition decay 分解风化
decomposition effect 分解作用
decomposition efficiency 分解效率
decomposition electric(al) tension 分解电压
decomposition extent 分解程度
decomposition flame 分解火焰
decomposition formula 分解公式
decomposition gas 分解气体
decomposition group 分解群
decomposition heat 分解热
decomposition in linear programming 线性规划中的分解
decomposition in sealed vessel 封闭分解法
decomposition intensity 分解强度
decomposition mechanism 分解机理
decomposition method 分解法
decomposition of forces 力(的)分解
decomposition of fraction 分式分解
decomposition of ink 油墨变质
decomposition of light 分解三棱色
decomposition of plastic waste 废塑料分解
decomposition of query 查询分解
decomposition of relation schema 关系模式的分解
decomposition of the sample 样品分解
decomposition of track 线路松动
decomposition of variation 变差分解
decomposition of wastewater 废水分解;污水分解
decomposition of water 水的分解
decomposition of white light 白光分解
decomposition peak 分解峰
decomposition point 分解点
decomposition point of distillation 蒸馏分解点
decomposition potential 分解电势
decomposition pressure 分解压(力)
decomposition principle 分解原理
decomposition product 分解产物
decomposition rate 分解率
decomposition reaction 分解反应
decomposition rule 分解规则
decomposition run 分解试验
decomposition stage 分解阶段
decomposition temperature 分解温度
decomposition texture 分解结构
decomposition theorem 分解定理
decomposition value 分解值
decomposition vibration 分解振动
decomposition voltage 分解电压;分解电势

decomposition water 分解水
decomposition with acid 酸分解
decomposition with aqua regia 王水分解
decomposition with H_2SO_4 硫酸分解
decomposition with H_3PO_4 磷酸分解
decomposition with HBr 氢溴酸分解
decomposition with HCl 盐酸分解
decomposition with $HClO_4$ 高氯酸分解
decomposition with HI 氢碘酸分解
decomposition with HNO_3 硝酸分解
decomposition with hydrofluoric acid 氢氟酸分解
decomposition with mixed acid 混合酸分解
decomposition with mixture of ammonium salts 混合铵盐分解
decomposition with mixture of H_2SO_4 and H_3PO_4 硫酸磷酸分解
decomposition with mixture of H_2SO_4 and HF 硫酸氢氟酸分解
decomposition with mixture of HCl and H_2O_2 盐酸过氧化氢分解
decomposition with mixture of HCl and $KClO_3$ 盐酸氯酸钾分解
decomposition with mixture of $HClO_4$ and HCl 高氯酸盐酸分解
decomposition with mixture of $HClO_4$ and HF 高氯酸氢氟酸分解
decomposition with mixture of HNO_3 and HF 硝酸氢氟酸分解
decomposition with overlapping 重叠分解
decomposition wit mixture of H_2O_2 and H_2SO_4 过氧化氢硫酸分解
decompound 多回分裂
decompounded motor 差复励电动机
decompounding winding 差复激绕组
decompressed curve 减压曲线
decompressed zone 减压区
decompression 解压；降压；减压(术)；潜水减压
decompression and recompression loop 卸载再压环
decompression cabin 减压室
decompression cabinet 减压箱
decompression chamber 降压室；减压间；减压箱；减压室；减压舱
decompression curve 减压曲线；卸荷曲线
decompression device 减压装置
decompression disorder 减压病
decompression dive 减压潜水
decompression drainage 减压引流
decompression illness 减压病
decompression induced supersaturation 减压引起的过饱和
decompression lever 减压杆
decompression load 减压荷载；消压荷载
decompression modulus 卸压弹性模量；卸荷模量
decompression moment 减压力矩；失压力矩
decompression release lever 减压杆
decompression retractor 减压(术)牵开器
decompression schedule 减压表
decompression sickness 减压病；潜水员病
decompression sphere 球形减压室；球形减压舱
decompression stage 减压站；减压阶段
decompression station 减压站
decompression stop 减压站；减压停留
decompression syndrome 减压综合症
decompression table 降压表
decompression time 减压时间
decompression time table 降压时间表
decompression wave 减压波
decompressor 减压装置；减压器；膨胀机
deconcentration 分散(布局)
deconcentrator 去浓器；消浓器；分线器；分散器；反浓缩器
decongestion 缓解拥挤
deconjugation 早期解离
deconstruction 解构【建】
deconstructivism 解构主义
decontaminability of surface 表面去污性
decontaminant 纯化剂；防污剂；去污剂
decontaminated agent 净化剂
decontaminating agent 净化剂；去污剂
decontaminating apparatus 净化器；消毒设备；去污装置；去污仪
decontaminating column 净化柱；提纯柱；去污染柱
decontaminating device 去污设备

decontaminating equipment 净化设备；去污设备
decontaminating installation 除污装置
decontaminating plant 去污设备
decontaminating unit 去污设备
decontamination 净化；消除污染；纯化；清除污染；清除放射性污染；排除污染；去杂质；去污作用；去污染；去污
decontamination area 净化区；消除污染区域；去污面积
decontamination center 净化中心；防放射污染中心；防放射污染站
decontamination chamber 净化室
decontamination cycle 净化循环；消除污染循环
decontamination device 净化装置；消除污染设施；清除放射性设备
decontamination facility 清污设备
decontamination factor 净化因数；纯化系数；除污系数；去污因子；去污因数；去污系数
decontamination filter 净化滤池
decontamination fluid 消除污染流体
decontamination index 净化指数；净化指标；去污指数
decontamination of feces 人类粪便无害化处理
decontamination of human excreta 人类粪便无害化处理
decontamination of materials 物料净化
decontamination of water 水的净化
decontamination plant 净化装置
decontamination procedure 净化程序
decontamination rate of urban refuse 城市垃圾无害处理率
decontamination shower 除污染淋浴；洁净吹淋(装置)
decontamination system 净化系统
decontamination waste 去污废物
decontrol 解除管制
deconvolution 解卷积；重叠合法；反旋卷；反(卷)褶积
deconvolution after stack 迭后反褶积
deconvolution filter 反卷积滤波器
deconvolution section of downwave 下行波反褶积剖面
deconvolution section of upwave 上行波反褶积剖面
deco-polymer flooring (以环氧树脂为底的)叠合楼面；以环氧树脂为底的叠合接面
decopper(ing) 脱铜；除铜
decoppering agent 除铜剂
decopperized lead bullion 除铜粗铅
decor 舞台艺术；舞台布景；彩绘；室内装饰(布置)
Decora 德可拉铬锰钼钒钢
decorate 装潢
decorated arch 装饰(性)拱；装饰(式)拱
decorated architecture 盛饰建筑；尖拱式建筑
decorated archivolt 美化的拱门饰；拱门饰
decorated archway 点景牌楼
decorated area 装饰过的面积；装饰面积；装饰范围
decorated border 边缘饰
decorated bracket 雀替【建】
decorated ceiling 装饰过的天花板；装饰过的顶棚；装饰天花(板)
decorated design 装饰设计
decorated door 装饰过的门
decorated gateway 牌楼
decorated gypsum 装饰石膏
decorated pottery 彩陶
decorated pulp cement brick 饰面纸浆水泥板
decorated style 哥特式建筑形式；尖拱式建筑形式；装饰形式；盛饰建筑形式(英国哥特式建筑之第二阶段)
decorated surface 装饰表面
decorated tie-beam 额枋【建】
decorated tile 装饰瓷砖；彩绘瓦；装饰瓦
decorated with battlements 用雉堞墙装饰；装饰雉堞墙
decorated with status 用塑像装饰；用雕像装饰
decorating 美化；彩绘
decorating art 装饰艺术
decorating cladding material 覆面装饰材料
decorating contractor 装饰承包人
decorating dislocation 缀饰位错
decorating fire 彩烧；彩烤；烤花
decorating kiln 烤花窑
decorating lehr 烤花窑

decorating machine 印花机；涂装机
decorating machinery 装潢机械
decorating media 装潢材料
decorating medium 装饰材料
decorating seaming 锁边(缝帆法)
decorating shop 彩绘车间
decorating with colo(u)rful light 张灯结彩
decoration 装饰(物)；装潢；饰品
decoration art 装饰艺术
decoration baking 彩烧
decoration carved on paste 雕花装饰
decoration cement 装饰水泥
decoration cost 装饰费用；彩绘费用
decoration fire 彩烧
decoration firing 彩烧
decoration firing lehr 烤花窑；彩烧窑
decoration glass 饰面玻璃；装饰玻璃
decoration kiln 烤花窑
decoration method 装饰法；染色法
decoration method of dislocation 位错缀饰法
decoration of porcelain 瓷器装饰
decoration sheet 装饰板
decoration stone 装饰石板
decoration with liquid gold 描金
decoration works 装修工程
decorative 装饰的
decorative acoustic(al) gypsum 装饰隔音石膏镶板；吸声石膏装饰板
decorative aggregate 装饰性集料；装饰性骨料；装饰用集料
decorative alumin(i)um 装饰用铝件；装潢用铝件
decorative anodized alumin(i)um 装饰的阳极化铝
decorative appearance 经过装饰的形象；经过装饰的外观；外观装饰
decorative application 装饰施工；装修作业
decorative arch 装饰(性)拱
decorative architecture 装饰建筑
decorative archivolt 装饰性拱门
decorative area 装修面积；装饰面积；装饰范围
decorative art 装饰艺术
decorative article 装饰制品；彩饰制品
decorative artificial stone 装饰用人造石；人造装饰石
decorative band 装饰带条
decorative barrel vault 装饰桶拱；装饰筒拱
decorative block 装饰(性)砌块；饰面砌块
decorative board 装饰板
decorative bond 图案砌合；装饰砌合(法)；装潢砌合；盛饰砌合
decorative bracket 装饰(性)托座
decorative brick 装饰(用)砖
decorative brick masonry(work) 装饰砖砌体
decorative brickwork 装饰砖工
decorative building unit 装饰性建筑制品；装饰性房屋用制品
decorative capacity 装饰功能
decorative cast block 装饰用铸块
decorative cast concrete block 装饰性混凝土块
decorative cast concrete product 装饰性混凝土制品
decorative cast concrete tile 装饰性浇制混凝土砖
decorative cast product 装饰性浇制品
decorative cast stone 装饰性人造石
decorative ceiling(board) 装饰天花(板)；装饰性顶棚(板)
decorative ceiling sheet 装饰性顶棚薄板；装饰天花(板)
decorative ceiling tile 装饰顶棚瓦；装饰顶棚花砖；装饰顶棚瓷砖
decorative cladding 装饰包层
decorative coat 装饰性美术涂装
decorative coating 装饰性面层；装饰性粉刷；装饰漆；美饰涂层；装饰涂层；装饰性罩面；装饰(性)涂料；装潢贴面
decorative cold water paint 装饰用冷水涂料
decorative colo(u)r painting 彩画
decorative column 装饰性柱子；装饰柱
decorative composition floor(ing) 装饰性拼花地板；装饰拼合地板
decorative concrete 装饰用混凝土；装饰(性)混凝土
decorative concrete product 装饰性混凝土制品
decorative concrete tile 装饰性混凝土砖；装饰性混凝土薄板
decorative cylinder 门簪

decorative door 装饰门
decorative effect 装饰效果
decorative element 装饰构件;装潢元件;装饰元素
decorative emboss(ment) 装饰(性)浮雕
decorative fabric 装饰织物
decorative faced block 装饰砌块;饰面砌块
decorative feature 装饰特征
decorative felt(ed fabric) 装饰用毡;装饰用毛织物
decorative finish 终饰;面层装饰
decorative fittings 装饰(性)小五金;装饰配件
decorative fixture 装饰性装置;装饰(性)设备
decorative floor cover(ing) 装饰性地板楼板面层;装饰性地板面层;地面装饰;地板装饰
decorative floor finish 装饰性地板面层
decorative floor(ing) 装饰性楼板
decorative foil 装饰用金属薄件;装饰用箔;叶形装饰
decorative form 装饰形式
decorative gable 装饰(性)山墙
decorative glass 有装饰图案的玻璃;装饰玻璃
decorative glass block 装饰性玻璃块;装饰玻璃砖
decorative glass brick 装饰性玻璃砖
decorative glass tile 装饰玻璃砖
decorative glued laminated slab 装饰胶合板
decorative glued laminated wood 装饰用胶合层积材;装饰用胶合层积板
decorative glued plywood 装饰性胶合板;饰面胶合板
decorative grill(e) 装饰性格栅
decorative gypsum board 装饰(性)石膏板
decorative gypsum board for lay-in installation 嵌装式石膏装饰板
decorative hardboard 装饰硬质纤维板
decorative hardware 金属装饰件;装饰小五金
decorative heavy ceramics 装饰性大型陶瓷
decorative hung ceiling 装饰性吊件
decorative illumination 装饰照明
decorative in-situ floor(ing) 装饰性现浇地板;装饰性现场浇制地板;无缝装饰板
decorative iron 装饰(性)铁件
decorative ironwork 装饰(性)铁制品
decorative ironwork technique 装饰性铁制品工艺
decorative jamb linings 装饰性门窗侧板
decorative joint 装饰性接合;装饰节点
decorative jointless floor(ing) 装饰性无缝地板;无接缝装饰地板
decorative laminate 装饰(壁)板;壁纸;装饰性贴面板;装饰层压板;塑料贴面板;饰面板
decorative laminated board 装饰性层压板;装饰性胶合板
decorative lamination 饰面
decorative lamp 装饰灯
decorative lattice 装饰性花格
decorative layer 装饰层
decorative light fitting 装饰性照明设备;装饰照明设施
decorative light fixture 装饰性照明设备
decorative lighting 装饰照明
decorative lighting equipment 街道装饰照明设备
decorative link 装饰铰链;装饰链环
decoraive lock 装饰(性)锁
decorative luminaire (fixture) 装饰性照明装置
decorative marble 装潢的大理石
decorative masonry bond 装饰性砖石砌合
decorative material 装潢材料;装饰材料;饰面材料
decorative melamine laminate 装饰三聚氰胺层压板
decorative metal 金属装饰件;装饰金属
decorative model(l)ed coat 装饰成有立体感的面层
decorative model(l)ed stucowork 装饰成有立体感的拉毛粉刷
decorative mortar 抹面砂浆;装饰砂浆
decorative motif 装饰主题
decorative motif taken from nature 取自自然界的装饰主题
decorative nail 装饰用钉
decorative nails on door leaf 门钉
decorative niche 装饰(性)壁龛
decorative openwork window 漏窗
decorative overhang 挂落
decorative overlay 层压板装饰贴面纸;装饰面层;饰面
decorative paint 装饰保护漆;装饰(用油)漆;装饰性油漆;装饰性色漆

decorative painting 装饰彩绘;装饰性油画;装饰性印刷;装饰(性)涂漆;彩画
decorative panel 装饰(镶)板;装饰墙板;装饰板材
decorative patent stone 装饰用人造石
decorative pattern 装饰(性)图案;花纹
decorative pattern brickwork 装饰性拼花砖工
decorative pavilion 装饰亭
decorative paving 装饰性铺砌;装饰性铺面
decorative perforation 装饰性漏窗;装饰性孔眼
decorative period 盛饰时期;盛饰时代
decorative plant 观赏植物;装饰植物
decorative plaster 装饰性抹灰
decorative plaster board 装饰性石膏板
decorative plastic board 塑料装饰板;装饰塑料板
decorative plastic-faced board 装饰用塑料贴面板
decorative plastic-faced wall board 装饰性塑料贴面墙板
decorative plastic film 装饰用塑料薄膜
decorative plastic laminate 塑料装饰层压板
decorative plastics 装饰塑料
decorative plastic sheet(ing) 装饰用塑料薄板
decorative plywood 装饰用胶合板
decorative pool 装饰性水池
decorative porcelain 彩瓷
decorative portal 装饰性入口
decorative power 装饰效力
decorative precast concrete block 装饰性预制混凝土块
decorative precast concrete product 装饰性预制混凝土制品
decorative precast concrete tile 装饰性预制混凝土砖;装饰性预制混凝土薄板
decorative printing 印铁
decorative profile 装饰型材
decorative property 装饰特性
decorative quality 装饰质量
decorative railing 装饰性围栏
decorative reconstructed stone 装饰用人造石
decorative rib 装饰性肋
decorative ridge tile 装饰性脊瓦
decorative rowlock paving 装饰性竖铺面
decorative screen 装饰屏
decorative seamless floor(ing) 装饰性无缝楼面
decorative sett paving 装饰性石块(镶花)铺砌;装饰性石块(图案)铺砌
decorative sheet 装饰性薄板;装饰片材
decorative sintered glass 装饰烧结玻璃
decorative sliced veneer 装饰性贴面板
decorative steel 装饰性钢件
decorative structural ceramics 装饰性结构陶瓷
decorative structure 装饰(性)结构
decorative style 装饰风格
decorative surface 装饰面
decorative suspended ceiling 装饰性吊顶
decorative tablet 装饰匾;装饰碑
decorative tie 装饰性柱箍;装饰性拉杆
decorative tile 装饰瓷砖
decorative timber veneer 装饰性贴面板
decorative touch 装饰格调
decorative town gateway 装饰性乡镇入口
decorative trim 装饰性修整
decorative tunnel vault 装饰隧道拱
decorative turret 装饰用小塔
decorative unit 装饰单元
decorative vault 装饰拱顶
decorative veneer 装饰贴面板;装饰单板;饰面薄板
decorative wagon vault 装饰性筒形拱顶
decorative wall 装饰(性)墙
decorative wall bracket 装饰性墙上托架;装饰性墙上牛腿
decorative wall lamp 装饰(照明)壁灯
decorative water(-carried) paint 装饰用水基涂料
decorative web 彩饰面纸
decorative window 装饰窗
decorative window curtain 装饰窗帘
decorative window frame 装饰窗框
decorative window screening 装饰窗纱
decorative wire(d) glass 装饰嵌丝玻璃
decorative wood veneer 饰面薄板
decorative work 装饰作业;装饰工作;装潢工件
decorator 制景人员;装修工;装饰师;装饰家;装饰工人
decorator's hammer 装饰用锤
decorator's lath 装饰者用的抹灰板条

decore 除芯
decoring 除芯;出芯
decoring device 除芯机
decorrelation 解相关;去相关
decorrelator 解相关器;解联器;去相关器
decorticate 去皮(质);脱壳
decortication 剥外皮(法)
decorticator 剥麻机;剥壳机;脱壳机;去(韧)皮机
decor tile 花砖
decount 缺除
decouple 去耦(合);退耦
decoupled approximation 解耦近似法
decoupled subsystem 解耦的子系统
decoupling 解耦;解开;摘钩;分离;脱扣;退耦;去耦(合)
decoupling and shunting 解体调车
decoupling capacitor 去耦电容器
decoupling circuit 去耦电路
decoupling control 解耦控制
decoupling factor 解耦因数;解耦系数
decoupling stock 保险性库存
Decousto 德寇斯托隔声、隔热材料
decoy 假目标
decoy attack 假目标进攻
decoy discrimination 假目标辨别
decoy discrimination radar 目标鉴别雷达
decoyl 己基
decoy return 假目标回波信号;假目标反射信号
decoy ship 伪装商船的军舰
decrating machine 出箱机
decrease 亏短数;减少;降低
decrease by degree 递减
decreased logarithmic phase 对数减少期
decreased overhead 削减管理费
decrease fold 降低倍数
decrease in book value 账面价值减少
decrease in brightness 亮度消减
decrease in cost 费用的减少
decrease in dip 倾角减小
decrease in population 人口减少
decrease in speed 减速比
decrease in value 降低价值
decrease liabilities 减少负债
decrease of definition 清晰度降低
decrease of discharge 流量递减
decrease of discharge from spring 泉水流量减小值
decrease of frequency 频率下降
decrease of income tax 所得税额减少
decrease of light 光度降低
decrease of load 降低负荷
decrease of pumping discharge 抽水量减小
decrease of rainfall 降水量减少
decrease of river flow 河水流量减小值
decrease of sensitivity 敏感度降低
decrease of specific mining quantity 单位开采下降值
decrease of strength 强度递减
decrease of vegetation cover 植被面积减少
decrease of water quantity in river 河流水量减少
decrease progressively 递减
decrease quantum of cost 成本降低额
decrease rate of cost 成本降低率
decrease the disposal of pollutants 消减污物处置
decrease the hole diameter 缩小钻孔直径
decrease train dwelling time on station 压缩列车停站时间
decrease train transit time in section 压缩区间运行时间
decreasing amplitude 降幅;减幅
decreasing annuity 递减年金
decreasing axial pressure fracture 周围减压断裂试验
decreasing axial pressure fracture test 轴向减压断裂试验
decreasing cost 递减成本;成本递减
decreasing credit 紧缩贷款
decreasing failure rate 递减失效率;初值故障率
decreasing forward wave 衰减前向波
decreasing function 降函数;下降函数;递减函数;单调非增函数
decreasing hazard rate 下降几率
decreasing lead screw 减螺距螺杆
decreasing loss of wash liquid 减少冲洗液漏失

decreasing of standard deviation 标准差降低
decreasing opportunity cost 递减机会成本
decreasing order of magnitude 由大而小排列；递减顺序；递减排列
decreasing pressure 递减压力
decreasing profit 收益递减
decreasing progression 递减级数
decreasing rate of energy source consumption 能耗降低率
decreasing rate of material spending 材料消耗降低率
decreasing rate of salinity 矿化度的递减率
decreasing returns 递减报酬
decreasing returns to scale 规模收益递减
decreasing seepage of polluted water from ground surface 减少地面污水入渗
decreasing sequence 下降序列；递减序列
decreasing series 递减级数
decreasing taxa 衰落中的生物分类群
decreasing utility 递减效用
decreasing vibration 降振
decreasing waves 减幅波
decree 法令；法规；判决
decree of completion of the bankruptcy proceeding 破产终结判决
decree of court 法院裁决
decree of distribution 产权分配裁决
decrement 减(缩)；减(量)；压缩量；缩减；衰减量
decremental arc 渐缩环形山弧
decremental chain 渐缩环形山链
decrement angle 衰减角
decrementation 分级卸载
decrement curve 减幅曲线
decrement factor 减缩系数
decrement jump 减量转移
decrementless conduction 不减量传导
decrement load 减量装入
decrement measurement 衰减量测量
decrement of foundation strength 地基强度衰减
decrement of ground strength 地基强度衰减
decrement of oscillation 振荡衰减量
decrement of velocity 减速
decrement pin 专用引线
decrement rate 减少率
decrement rate of ion concentration 离子浓度的递减率
decremeter 减缩量计；减幅仪；减计；衰减计；衰减测量器；衰变计
decrepigraph 爆裂图
decrepitation 爆裂(作用)；烧裂；烧爆(作用)
decrepitation apparatus 爆裂仪
decrepitation curve 爆裂曲线
decrepitation frequency elevation 爆裂频次高度
decrepitation halo diagram 爆裂晕图
decrepitation intensity 爆裂温度；爆裂强度
decrepitation method 爆裂法
decrepitation pulse times 爆裂脉冲数
decrescence 渐减
decrescendo murmur 渐递减杂音
decrescendo type 渐退型
decrescent 渐减的
decrescent function 渐减函数
decrustation 脱皮；去污沉积物
decrustation pliers 剥线钳
decrusting 刮洗
decrusting of a filter 清除滤料垢壳；滤池刮洗
decrypt 解码；翻译密码
decrystallization 解晶(作用)
dectaphone 漏水探知器
decticous 颚化的
Dectra 远程长波导航设备
deculator 纸浆排气装置
decumbent 垂下的
decurl 弄平整
decurler 卷曲消除器
decurrent 向下的；下延的
decursively pinnate 下延羽状的
decurvature 下弯
decussate 十字形的
decussate leaf 十字形对出叶
decussating fibers 交叉纤维
decussationes tegmenti 被盖交叉
decyanation 脱氰(作用)
decyclic acid 癸酸
decyclization 解环(作用)；脱环；去环作用

decyl 癸基
decyl acetate 乙酸癸酯
decyl alcohol 癸醇
decyl aldehyde 癸醛
decyl amide 癸酰胺
decyl amine 癸胺
decyl bromide 癸基溴
decyl butyl phthalate 邻苯二甲酸癸丁酯
decylene 癸烯
decylenic acid 癸烯酸
decylic acid 癸酸
decyl iodide 癸基碘
decyl phenyl ether 癸基苯醚
decyltrichlorosilane 癸基三氯硅烷
decyl-trimethylsilicane 癸基三甲基硅
decyne 癸炔
decyne carboxylic acid 癸炔羧酸
decynic acid 癸炔酸
Dedekind cut 狄德金分割
dedendum 齿根(高)
dedendum angle (齿轮的)齿根角
dedendum circle (齿轮的)齿根圆
dedendum cone 齿根锥(圆锥齿轮啮合)
dedendum line 齿根线
dedendum line of contact 齿根接触线
dedial temperature 平均温度
dedicated 专用的
dedicated autonomous module 专用自主模件
dedicated autonomous unit 专用自主部件
dedicated berth 专用泊位
dedicated channel 专用信道
dedicated circuit 专线
dedicated computer 专用计算机
dedicated connection 专用连线；专用连接
dedicated data 专用数据
dedicated data set 专用数据集
dedicated executive 专用执行程序
dedicated function 专用功能
dedicated line 专用线(路)；专用铁路；专用铁道
dedicated memory 专用存储区；特用存储区
dedicated network 专用网(络)
dedicated pin 专用引线
dedicated processor 专用处理机
dedicated space 存储空间
dedicated storage 专用存储区
dedicated system 专用系统
dedicated tape reader 专用程序带输入机
dedicated terminal 专用码头
dedication 题辞
dedication of ceremony 献词仪式；供献仪式
dedifferentiation 间变；反分化；去分化
dedition 让渡
dedolomitisation 脱白云石化作用；去白云石化作用；脱白云作用
de-drossing 掏渣
deduce 减低；演绎
deduced space 减去的位置；减去的空间
deduce rule 演绎规则
deduct 减去；折减
deduct a percentage from a sum of money 提成
deduct a percentage from profit 利润提成
deducted space 减除吨位(船舶丈量)
deduct expenses 扣除费用；削减费用
deduct from the salary 扣薪(年终分红用)；罚薪
deductible 可扣除的；可扣除的；可减税额；免赔额
deductible allowance 可扣除免税项目
deductible average 绝对免赔额
deductible clause 可减条款；减除条款
deductible contributions 可扣除借款
deductible cost 减除成本
deductible expenditures 可扣除支出
deductible expenditures-business 可扣除营业支出
deductible expenses 可扣除费用；可减免费用
deductible franchise 扣减免赔额；扣除的免赔额；绝对免赔额
deductible items 可减(免)项目
deductible loan 可扣除的借款
deductible loss 可扣除损失；可减损失；可减少的损失
deductible medical expenses 可扣除医疗费用
deductible medicines and drugs 可扣除药费
deductible money 扣款
deductible moving expenses 可扣除迁移费
deductible space 可减吨位

deductible tax 可扣除(的)税款
deducting payments 扣除借支
deduction 扣减；扣除(项目)；扣除(额)；结论；减账；减除数；演绎；折扣(额)；推论；推断
deduction exemption certificate 免扣预提税证
deduction from gross income 可从毛收入中扣除的项目；总收入中扣除数
deduction from income 可从收入中扣除的项目
deduction from income tax 所得税减免(额)
deduction from net income 可从净收入中扣除的项目；净收益中扣除数
deduction method 扣除法
deduction of the tax-exempt items 扣除免税项目
deduction result figure in induced polarization method 激电法推断成果图
deductions from net income 净收入往年减除数
deduction theorem 演绎定理；归约定理
deduction tree 演绎树
deductios and exemptions 扣除和豁免
deductive 推论的
deductive alternate 降低标价的标单；换料降标价
deductive approach 演绎法；推论(方)法
deductive coverage clause 可减保额条款
deductive database 演绎型数据库
deductive generation 演绎综合法
deductive inference 演绎推论；演绎推理
deductive logic 演绎逻辑
deductive method 演绎法；推论(方)法
deductive model 演绎模型
deductive operation on structured object 结构化事物的演绎运算
deductive procedure 演绎过程
deductive proof 演绎证明
deductive reasoning 演绎推论
deductive simulation 演绎模拟
deductive simulator method 演绎模拟法
deductive value 扣减价格
deduct losses from the total receipts 从总收入中减去损失
deduct money 扣钱；扣款
deduct prepayment 扣除预支(款)
dedust 脱尘
dedust by rapping 振打清灰
deduster 除尘器；脱泥机；脱尘机
dedust(ing) 除灰；除尘
dedusting box 除尘室
dedusting by filtration 过滤除尘(法)
dedusting by washing 洗涤除尘(法)
dedusting coefficient 除尘系数
dedusting cyclone 旋风集尘器
dedusting efficiency 除尘效率
dedusting equipment 除尘设备
dedusting filter 收尘器
dedusting of exit gas 废气除尘
dedusting of waste gas 废气集尘
dedusting plant 除尘厂
dedusting ventilation 去尘通风
dedust mask 防尘口罩
dee 拖链槽
deed 立契转让财产；合同；契约；契据
deed books 契据登记簿
deed box 契约箱；文件(保险)箱
deed fire 桅上电火
deed indented 锯齿边契约
deed in trust 财产托管信托书
de-editor program(me) 反编辑程序
deed money escrow 有付款条件的合同
deed of appropriation 使用权证明书
deed of arrangement 债务人与债权人的和解方案
deed of assignment 转让契约
deed of bargain and sale 卖契
deed of confirmation 确认契据；更正契据
deed of contact 全约证书
deed of gift 自动转让财产证书；赠与证书
deed of grant 财产转让契据
deed of indemnity 赔偿证书；赔偿契约
deed of mortgage 典契；抵押契约
deed of partnership 合伙契约
deed of purchase 卖据；卖契
deed of release 绝契
deed of sale 卖契；卖据
deed of security 担保契约；保证书
deed of settlement 协议证书；财产托管书

deed of surrender 将终身产权与继承权合并的证书;交产权契据
deed of transfer 转让证书;转让契约
deed of trust 信托契约;信托契据
deed release 转产契
deed restriction 契约限定;使用限制权
deed tax 契税
deed warranty 契约担保(房地产)
de-electrifying 去电
de-electronation 去电子(作用)
Deeley friction machine 迪ү摩擦机
deemanation 驱射气
deemed paid credit 信得过信贷
de-emphasis 减加重;信号还原;去加重
de-emphasis circuit 去加重电路
de-emphasis network 反预斜网络;去加重网络
de-emphasis parts of an apparatus 去加重器件
deemulsibility 乳化分解性
de-emulsification 破乳化(作用);脱乳化
de-enameling 除瓷
de-energize 解除激励;去能;切断电流;断电;去激励
deenergized 不带电(荷)的
deenergized period 释放周期;(继电器的)释放期间
deenergizing 断电路;去激励
deenergizing by short-circuit 短路释放
de-energizing circuit 去激电路;消除激励电路
deentrainment 除去雾沫;脱去夹带
deentrainment column 除雾沫柱
deentrainment filter 除雾沫过滤器
deentrainment tower 除雾塔
deep acoustic(al) fields 深海声场
deep admixture stabilization 深层拌和法
deep aeration tank 深曝气池
deep air cell 深型充气浮选机;深气升浮选机
deep anticyclone 深厚气旋;深厚(暖)高压
deep-arc piling 拱形板桩
deep area 深水区
deep bar 深鼠笼条
deep-bar cage winding 深槽鼠笼式绕组
deep-bar effect 深槽效应
deep-bar motor 深槽鼠笼式电动机
deep basement 深基
deep basin 深水盆地
deep basin gas trap 深盆气圈闭
deep bay 深水湾
deep bead 下扯窗盖条;深楞;挡风板;窗挡风条
deep beam 深梁;厚板;桁板横梁;主梁;高截面梁;深梁
deep beam slab 厚梁板
deep-bed dryer 厚层谷物干燥机
deep-bed filter 厚层滤器;深床滤池
deep-bed filtration 滤床过滤(池);袋式过滤;深床砂滤;深床过滤(法)
deep-bed granular filter 深床颗粒滤机
deep-bed ground water 深层地下水
deep bed stirring method gunite pile 深层搅拌法喷浆桩
deep bend 挡板板;深弯
deep blasting 深层爆炸;深层爆破
deep blue-black 深蓝黑色
deep blue-drown 深蓝褐色
deep blue glaze 深蓝釉
deep blue-green 深蓝绿色
deep blue-grey 深蓝灰色
deep blue-violet 深蓝紫色
deep bore 深孔;深井
deep bore hole 深(钻)孔;深炮眼
deep boring 深孔凿岩;深钻;深孔钻进;深层钻探
deep breaking up 深度破坏;深挖削
deep brown 深褐色
deep brown-black 深褐黑色
deep brown-blue 深褐蓝色
deep brown-green 深褐绿色
deep brown-grey 深褐灰色
deep brown-red 深褐红色
deep brown-violet 深褐紫色
deep brown-yellow 深褐黄色
deep building pit 深基坑;深挖坑
deep bulkhead 深舱壁
deep burialism 深埋作用
deep buried bimetal benchmark 深埋双金属标【测】
deep buried steel-pipe benchmark 深埋钢管标

deep-buried tunnel 深埋隧道
deep-cage-bar rotor 深鼠笼转子
deep camber 深弯度
deep carburizing 深度渗碳
deep-casting 深层采水
deep-cell machine 深室浮选机
deep cementing 深度渗碳
deep cement mixing 深层水泥拌和
deep center 深中心
deep channel 深水航道;深水槽;(航道的)深槽
deep channel section 深槽断面
deep circulating hydrothermal solution 深部循环热液
deep circulation 深水环流;深层环流
deep cleaning 深度清洗;深度清扫;深度净化
deep coal deposits 深层煤矿
deep coefficient 深度系数;深层系数
deep colo(u)r 纯色;深色
deep colo(u)red gold 赤金
deep compaction 深度压实;深层压实;深层加密(法)【岩】深入振实(法)
deep compaction method 深层压实法
deep concavities 深坑
deep-cone thickener 深锥形浓缩池
deep consolidation 深度加固;深层加固
deep controller 深度控制器
deep conversion 深度变换
deep cooling 深冷;深度冷却
deep cooling basin 深水型冷却池
deep cooling pond 深水型冷却池
deep cooling pool 深水型冷却池
deep cream 深奶油色
deep creep 深层蠕动
deep cultivator 深松土中耕机
deep culture 深层培养
deep current 深层流
deep curve 深度曲线
deep cut(ting) 深刻(槽);深切削;立剖;立锯;深刻蚀;深锯;深挖;深堑;深磨;深开挖;深部开采
deep dean ooze 深海软泥
deep densification 深层加密;深层加实
deep depression 深低气压
deep depth 深埋
deep depth sonar 深海声呐;深水声呐
deep dewatering at the mine 矿山深部疏干
deep digger 深挖土机;深松土器
deep digging 深松土;深松(法);深刨(法);深耕(法)
deep-digging dredge(r) 深挖式挖泥船
deep discount bond 大幅度贴现债券;巨额折价债券;高折价债券
deep dish paving 深碟式路面摊铺;深碟式路面铺设
deep dish steering wheel 深碟式驾驶盘
deep disintegrating 深度瓦解;深度分解;深度放松
deep diver 深水潜水员
deep diversion 深层引水;深层取水
deep diving 大深度潜水;深(水)潜水
deep-diving device 深潜装备
deep-diving experiment 深潜试验
deep-diving simulator 深潜模拟器
deep-diving submersible 深潜器
deep diving suit 深水潜水服
deep diving system 深潜系统
deep-diving trial ship 深潜试验船
deep-diving unit 深潜装备
deep-diving vehicle 深潜器
deep-diving vessel 深潜(潜水)船
deep dome 深部穹隆
deep-domed piston 凸顶形活塞
deep dormancy 深休眠
deep-draft 深吃水;吃水深的;深水的
deep-draft barge 吃水深的驳船;深水驳船
deep-draft channel 深水水道;深水河槽;深水航道
deep-draft harbo(u)r 深水港
deep-draft lock 深(吃)水船闸
deep-draft navigation 深水航运
deep-draft navigation channel 深水航道
deep-draft port 深水港
deep-draft project 深水通航工程
deep-draft route 深吃水航路
deep-draft vessel 深吃水船(舶);吃水深的船(舶);深水货轮;深水船舶
deep-draft water course 深水水道

deep-draft waterway 深水航道;深水水道
deep draught 满载吃水(船舶)
deep-draught route 深水水道;深水航线
deep-draught vessel 深吃水船(舶)
deep-draught waterway 深水水道
deep-draw 深拉
deep drawing (板材的)深冲成型;全回火;深延伸;深拉(延);深冲压
deep-drawing die 深压模
deep-drawing material 深冲材料
deep-drawing quality (板材的)深冲性
deep-drawing steel 深拉钢
deep draw mo(u)ld(ing) 深冲压模;深冲压成型
deep-drawn 深拉;深冲;深长的
deep-drawn sheet 深冲压的薄板
deep dredging 深浚
deep drill hole 深(钻)孔
deep drilling 深钻;深层凿岩;深层钻探
deep drilling cement 深油井水泥
deep-drilling engineering 深孔钻探工程;深孔凿岩工程
deep-drilling equipment 深孔钻探设备;深孔凿岩设备;深钻设备
deep-drilling rig 深孔钻机
deep drop frame semi-trailer 低底盘半拖车
deep dry gas zone 深部干气带
deep earthquake 深震
deep easterlies 赤道东风带
deep ecology 深层生态学
deep effect 深度效应;深度效果
deep embedment 深埋
deep embedment of sheet pile 板桩埋深
deepen 加深;深化
deepened beam 加深梁;加厚梁
deepened cross-section 加深的横截面
deepened web 加厚腹板
deep energy level 深能级
deep engraving 深雕刻
deepening 加深;向下侵蚀;挖深
deepening of capital 密集资本;资本深化;资本密集
deepening of riverbed 河床加深
deepening of river bottom 河底加深
deepening of the stream floor 疏浚河底;加深河底
deepening project 疏浚工程
deepening stage 加深阶段
deepen the channel of a river 加深河道
deeper-marine clastic deposit 深海碎屑沉积
deep erosion 强烈侵蚀;强度侵蚀;深向侵蚀;深向冲刷;深风化
deeper reflex 深层反射
deepest loadline 最大限度载重线
deepest subdivision loadline 最深分舱载重线
deep etch 平凹版腐蚀;深腐蚀;深度浸蚀
deep-etch digitizing 槽纹跟踪数字化
deep etched digitizing method 腐蚀槽纹数字化法
deep-etching 深蚀刻(法);腐蚀制版法;强腐蚀;深侵蚀
deep-etching paste 深蚀糊;深蚀膏
deep-etch lacquer 平凹版用基漆
deep-etch method 腐蚀法
deep-etch plate 平凹版
deep-etch printing 平凹版印刷
deep-etch process 凹版腐蚀法
deep etch test 深蚀试验
deep excavation 深(开)挖
deep extrusion method 深层挤密法
deep-eye 深海电视照相机
deep factor 深度因素;深度因数;深层系数
deep fade 强衰落
deep fat thermometer 熬油温度计
deep feed flight section 深螺纹喂料段
deep fill 深填(土);深填(方)
deep fillet welding 深角焊
deep filter 深滤器;深度过滤器
deep floe system 深部水流系统
deep floor 加强肋板;加强筋板;加肋楼板
deep flow 深层(水)流
deep focus 深(震)源
deep-focus earthquake 深震;深源地震
deep-focus earthquake plane 深源地震面
deep-focus of earthquake 深震源
deep-focus shake 深源震
deep foundation 深(埋)基础
deep foundation method 深基础施工法

deep foundation pier 深基础桥墩;深基础码头
deep fractural zone representing the Benioff zone of the West Pacific Island arcs 西太平洋岛弧毕鸟夫带深断裂系
deep fracture 深(大)断裂
deep fractures in marginal depression 边缘坳陷的深断裂
deep fractures system 深断裂系
deep fracture zone 深破裂带;深断裂带
deep frame 加强肋骨;深肋骨
deep frame swing jaw crusher 深肋骨架可动颚板破碎机
deep framing 宽肋骨架(横骨架式);加强肋骨架;深肋骨架
deep free diving 大深度自由潜水
deep freeze 低温快速冷藏
deep freezer 低温快速冷藏箱;冰柜;深度冷冻器;速冻冷库
deep freezing 冷处理;快冻处理;冰冷处理;速冷(处理);深度冷冻;深冻处理
deep-freezing plant 速冻装置
deep-furrow 深槽水沟;排灌水沟
deep gas 深层气
deep gas abundance zone 深层气富集带
deep gas below oil reservoir 油层深层气
deep gas survey 深层气体测量
deep geologic(al) mapping 深层地质填图
deep girder 深(大)梁
deep gloss 深邃光泽
deep-going vessel 远距离航行船
deep gold 深金色
deep gradient 深度梯度
deep gravel 深部砾石
deep green 深绿色
deep green-black 深绿黑色
deep green-brown 深绿褐色
deep green-grey 深绿灰色
deep greenhouse 地下温室
deep green pigment 深绿颜料
deep green-yellow 深绿黄色
deep grey 深灰色
deep grey-black 深灰黑色
deep grey-blue 深灰蓝色
deep grey-brown 深灰褐色
deep grey-green 深灰绿色
deep grey-red 深灰红色
deep grey-violet 深灰紫色
deep grey-white 深灰白色
deep grey-yellow 深灰黄色
deep-groove (ball) bearing 深槽滚珠轴承
deep ground water 深部地下水
deep grouting 深孔灌浆
deep gulch 深冲沟
deep handrail 深扶手
deep-hardening 深度淬火
deep heavy beam 重型深梁
deep hexagonal nut 深六角螺母
deep hibernation 深度冬眠
deep hold (cargo) ship 深舱货船
deep hole blasting 深孔爆破
deep hole block 深孔石块
deep hole boring 深孔钻探
deep hole drill 深孔钻头;深孔钻机;深炮眼凿岩机
deep hole drilling 深孔钻探;深孔凿岩
deep hole drilling machine 深孔钻机;深孔钻床
deep hole driving 天井深孔掘进法
deep hole excavating method 深孔开挖法
deep hole gamma ray survey 深孔伽马测量
deep hole grouting 深孔灌浆
deep hole infusion 深孔注水
deep hole method 深炮眼崩矿法
deep hole prospecting 深孔钻探;深孔勘探
deep hole seismic detection 深孔地震探测
deep hole seismometer 深孔地震仪;深孔地震计;深井地震仪;深井地震计
deep hole spectrum survey 深孔能谱测量
deep hole wagon drill 深孔汽车式钻机
deep hole work 深孔作业
deep holography 深全息术
deep ice 深冰
deep illumination optic(al) fiber 深部照明光纤
deep in debt 债台高筑
deep Indian red pigment 深印度红颜料
deep indicator 深度指示仪;深度指示器;深度计

deep inelastic transfer 拟裂变
deeping 深挖;深刻;顺纹锯开木材
deeping erosion ratio 强度侵蚀度
deep injection of wastes 废物的深灌注
deep injection well 深注水井
deep inland sea 深内陆海
deep intake 深式进水口
deep intaking 深层取水
deep intervention 深度干涉
deep in the money 极具实值
deep investigation induction log 深感应测井
deep investigation induction log curve 深感应测井曲线
deep investigation induction log plot 深探测感应测井图
deep investigation resistivity log plot 深探测电阻率测井图
deep karst 深喀斯特;深(层)岩溶;深部岩溶
deep lake 深水湖泊;深湖
deep lake deposit 深湖沉积
deep lake facies 深湖相
deep lateral log 深侧向测井
deep lattice(d) girder 深的格构大梁
deep layer 深层
deep layer aeration 深层曝气;深槽曝气
deep lead 深海砂矿
deep level 深能级;深层次;深层(的);深部层位
deep level excavation 深层开挖
deep level grillage 低桩承台
deep lift 深层
deep lift asphalt construction 深层沥青施工
deep lift asphalt pavement 加厚的沥青路面;深层沥青铺面
deep lift construction 深层构造
deep-lime mixing method 深层石灰搅拌法
deep-litter 厚垫草
deep-litter house 厚垫草鸡舍
deep-litter nest 厚垫草产卵箱
deep-litter system 厚垫草法
deep loading system of hopper 泥舱深装舱系统;深层装舱系统
deep load-line 最大限度载重线
deeply buried 深层的;深埋的
deeply embedded foundation 深埋基础
deeply eroded river bed 深蚀河床
deeplying 深埋的;深藏的
deeplying foundation 深埋基础
deeply recessed joint pointing 深凹勾缝
deep manhole 深井式人孔
deep meniscus 深弯月形
deep meter 深度计;深度表
deep mining 深井开采
deep-mixed pile 深层搅拌桩
deep mixing method 深层搅拌(法);深层化学拌和法
deep mixing pile 深层搅拌桩
deep model(l)ed concrete 深纹饰混凝土
deep molten bath 深熔池
deep-moored instrument station 深海系留仪器站
deepness 深度
deep non-saturation bell-diving research 非饱和潜钟深潜
deep nut 深螺母
deep ocean channel 深海声道
deep ocean environment 深海环境
deep ocean exploitation technique 深海开发技术
deep ocean exploration 大洋深测
deep ocean location 大洋定位;深海定位
deep ocean mining environment study 深海采矿环境研究
deep ocean mooring 大洋系泊;深海系泊
deep ocean pore water 深海间隙水
deep ocean power source 深海能源
deep ocean survey vehicle 深海探测器
deep ocean technology 深海技术
deep ocean temperature 大洋温度
deep ocean work boat 深海作业船
deep oil 深成油
deep oligotrophic lake 深水贫营养湖
deep open pit 凹陷露天矿
deep-operating vehicle 深潜潜水器
deep outlet 泄水底孔;深式泄水孔
deep pan apron conveyer 深斗裙式输送机;深斗板式输送机

deep pay 深部产油层
deep penetrating fault 深穿断裂
deep penetration 圆锥针入度;深贯入;深穿透
deep penetration bomb 深孔炸弹
deep penetration electrode 深熔焊条
deep penetration test 圆锥贯入度试验;深层触探试验
deep penetration weld 深熔焊
deep penetration welding 深熔焊接
deep penetration welding electrode 深熔焊条
deep percolation 深向渗透;深向渗水;深(向)渗滤;深向渗流;深层渗流;深层渗水;深(层)渗滤;深层渗漏;深层渗流;深部渗透;深部渗水;深部渗漏;深部渗流
deep percolation loss 深渗(透)损失
deep phreatic cave 深潜流洞
deep phreatic water 深井水
deep pier 深基础柱;深基础墩
deep pile 深基础桩
deep-pit cage house 深坑笼养鸡舍
deep-pit latrine 深坑厕所
deep-pit sewage pump 深井污水泵
deep plate 深盘
deep plate girder 深板大梁
deep-plough 深层犁;种植犁;深耕犁
deep ploughing and intense cultivation 深耕细作
deep-pocket classifier 深槽分级机
deep point 深度点
deep pool 深潭;深水池
deep pool section 深槽河段
deep probe 深度探头
deep-processed glass 深加工玻璃
deep production 深层开采(量)
deep profile 深度剖面
deep prospecting 深井勘探;深层钻探;深层勘探;深层勘察;深部勘探
deep pumped well 深泵井
deep pumping 深层抽水
deep punching 深冲
deep punching steel 深冲钢
deep quest 深水探测
deep range 深度范围
deep ratio 深度比
deep ravine 深冲沟
deep-recessed part 深腔零件
deep reciprocation pump 往复式深井泵
deep recorder 深度记录器
deep recording device 深度记录仪
deep red 深红(色的)
deep red-drown 深红褐色
deep red filter 深红色滤光镜
deep red-grey 深红灰色
deep red-violet 深红紫色
deep red wood work 朱红木雕
deep red-yellow 深红黄色
deep reflex 深层反射
deep refrigeration 深度冷冻
deep regional aquifer system 区域深层含水系统
deep research vehicle 深海考察(潜水)器
deep reservoir 深部热储
deep return current 深度回流;深度反向电流
deep-ribbed slab 深肋(T形梁)板
deep-rock temperature 深部岩体温度
deep-rooted crop 深根(性)作物
deep-rooted plant 深根植物
deep rule 深度法则
deep salty water pollution 深层咸水污染
deep sampler 深层取样器
deep sampling 深挖取样;深层取样
deep sands 深部含油砂层
deep-sawing 剖锯;深锯
deep scattering layer 音波(深海)散射层;深水散射层;深散射层;深海散射层;深海扩散层
deep scraping 深撇泡沫
deep screen 管井滤水管
deep sea 远海深水;深海;远海
deep-sea anchor dredge 深海锚泊
deep-sea anchoring winch 深海抛锚绞车
deep-sea barge 深海驳
deep-sea basin 深海盆地;深渊
deep-sea bathometer 深海测深仪
deep-sea bathythermograph 深海水深温度自记仪;深海自记温度仪
deep-sea bathythermometer 深海自记温度计

deep-seabed 深海(海)床
deep-sea benthos 深海底栖生物
deep-sea benthos 深海底栖生物
deep-sea berth 深水泊位
deep-sea bottom 深海底
deep-sea cable 深海电缆;深水电缆
deep-sea camera 深海照相机;深海摄影机
deep-sea canyon 深水峡谷;深海槽
deep-sea channel 深海水道(海底峡谷);深海航道;深海谷;深海槽
deep-sea circulation 深海环流
deep-sea cone 深海锥
deep-sea core 深海岩芯
deep-sea core drilling 深海取芯钻进
deep-sea course 深海航路
deep-sea current 深水洋流;深海流
deep-sea deposit 深海矿藏;深海沉积(物)
deep-sea detection ship 深海探测船
deep-sea disposal of wastes 废物的深海处置
deep-sea diver 深海潜水员
deep-sea dive work 重潜水作业
deep-sea diving 深海潜水
deep-sea diving dress 深海潜水服
deep-sea diving outfit 深海潜水设备;深海潜水工具
deep-sea diving suit 重型潜水服
deep-sea drag dredge(r) 深海拖斗采矿船
deep-sea dredging 深海疏浚
deep-sea drilling project 深海钻探计划
deep-sea drill of deep-sea 深海钻机
deep-sea earthquake 深海地震
deep-sea ecology 深海生态学
deep-sea expedition 深海调查
deep-sea facies 深海相
deep-sea fan 浊积扇;深海扇
deep-sea fauna 深海动物群;深海动物区系
deep-sea fishery 远海渔业;深海渔业;远洋渔业
deep-sea fishes 深海鱼类
deep-sea floor 深海海底
deep-sea furrow 深海沟
deep-sea graben 深海地堑
deep-sea hydraulic dredge 深海水力挖矿机
deep seal 深水封;深密封;(合流制排水管的)密封;深存水弯(一种反虹吸存水弯)
deep-sea lane 深海航路
deep-sea lead 深海铅测锤;深水测深锤;深海测深锤
deep-sea lead line 深海测深绳
deep-sea line 深海测深绳
deep-seal trap 防虹吸存水弯;深度密封存水弯;深存水弯(一种反虹吸存水弯)
deep seam 凹缝
deep-sea mining ship 深海采矿船
deep-sea moor buoy array 深水系泊浮筒阵列;深水系泊浮筒系统
deep-sea moored buoy 深海系留浮标
deep-sea mooring 深海系泊
deep-sea mud 深海泥
deep-sea multiple sampler 深水复式采样器
deep-sea ooze 深海软泥
deep-sea operation 深海作业
deep-sea photography 深海摄影
deep-sea plain 深海平原
deep-sea port 深水港
deep-sea propagation 深海传播
deep-sea research 深海调查船
deep-sea research vehicle 深海调查潜水器
deep-sea resources 深海资源
deep-sea reversing thermometer 深海倒转温度表;深水反转温度计
deep-sea robot 深海遥测设备
deep-sea salvage tug 海上救助拖轮
deep-sea sampler 深水采样器
deep-sea sand 深海沉砂
deep-sea sediment 深海沉积(物)
deep-sea sedimentary soil 深海沉积土
deep-sea simulation 深海模拟
deep-sea sounding 深海测深
deep-sea sounding apparatus 深海测深装置
deep-sea submarine cable 深海海底电缆
deep-sea submergence test 深海下潜试验
deep-sea surveying ship 深海探测船
deep-sea system 深海系统
deep-sea technology 深海技术
deep-seated 深位的;深嵌的;深部的
deep-seated blowhole 内部气孔;深位汽泡

deep seated burning 深层燃烧
deep-seated clay flowage 深部黏土流
deep-seated deposit 深藏矿床;深部矿床
deep-seated different in movement 深层分异运动
deep-seated dike 深成岩脉
deep-seated dispersion 深部分散
deep-seated environment 深部环境
deep-seated fault 深部断层
deep-seated fire 深位火
deep-seated fumarole 深源喷气孔
deep-seated geothermal fluid 深部地热流体
deep-seated geothermal system 深部地热系统
deep-seated grouting 深层灌浆
deep-seated openings 深埋隧洞
deep-seated rock 深层岩;深成岩
deep-seated steam 深源蒸汽
deep-seated tunnel 深埋隧道
deep-seated type 深嵌类型;深埋类型;深层类型;深部类型
deep-seated volcano 深海火山;深成火山
deep-seated weathering 深层风化
deep-sea terrace 深海台地;深海阶地
deep-sea thermometer 海底温度计;深海温度计
deep-sea tide 深海潮汐
deep-sea trade 远洋贸易
deep-sea trawler 远洋拖网渔轮
deep-sea trawl winch 深海拖网绞车
deep-sea trench 深海沟
deep-sea trough 深海槽
deep-sea tsunami 深海海啸
deep-sea tug 远洋拖轮
deep-sea velocimeter 深水流速计;深水海流计;深海流速仪
deep-sea velocity meter 深海流速仪
deep-sea wave 深海波
deep-sea wave meter 深海波浪计;深海波浪计
deep-sea wave recorder 深水波浪计
deep-sea winch 深水绞车
deep-sea work boat 深海作业船
deep-sea working 深海作业
deep-sea zone 深海区
deep sedimentary basin gas 深沉积盆地气
deep seeding 深播
deep seepage 深向渗透;深向渗水;深向渗滤;深向渗流;深层渗透;深层渗水;深层渗漏;深层渗流;深部渗透;深部渗水;深部渗漏;深部渗流
deep seism 深海地震
deep seismic sounding 深(地)震测深
deep seismic survey 深海地震测量
deep seismic zone 深震带
deep-set 深陷的
deep settlement 深层沉陷
deep settlement ga(u)ge 深层沉降仪;深层沉降计
deep shade type 深罩型灯
deep shadow 深(阴)影;深调
deep shaft 深井
deep shaft aeration 深井曝气
deep shaft aeration process 深坑曝气法
deep shaft system 深井系统
deep shakes 深环裂
deep shallow 投影
deep sheet 深衡薄板
deep slab 厚板
deep sleek 抓痕
deep slide 深层滑坡
deep slip 深水港池;深层滑动
deep-slot 深槽
deep-slot induction motor 深槽感应电动机
deep-slot motor 深槽感应电动机
deep-slot squirrel-cage motor 深槽鼠笼式电动机
deep-slot squirrel-cage induction motor 深槽鼠笼式感应电动机
deep slotting 深切槽
deep slow-flowage karst 深部级流岩溶
deep slow moving zone 深部缓流带
deep sluice 底部水闸;底部闸门;底部泄水隧洞
deep sluice section 底部水闸剖面(图);底部水闸截面
deep socket wrench 长套管型套筒扳手
deep soil 厚层土壤;深层土(壤)
deep soil agitating process 深层搅拌法
deep sonde 深水探测器
deep sound channel 深海声道
deep sound duct 深海声道

deep sound fields 深海声场
deep-sounding 深水测深
deep-sounding apparatus 测深器;测深仪;深层触探设备;深层触探器;深水测深仪;深水测深器
deep-sounding lead 深海测深锤;深水测深锤
deep-sounding machine 测深机;深海测深机
deep-sounding test 深层触探试验
deep soundpath phenomenon 深海声道现象
deep sound scattering layer 深海声波扩散层
deep source gas 深源气
deep space 外层空间;深外层空间;深空;深海空间
deep space instrumentation system 深空探测系统
deep-space laser communication 深空激光通信
deep-space laser tracking system 深空激光跟踪系统
deep space net 太空跟踪网
deep space network 太空跟踪网;深空跟踪网
deep-space probe 太空探测器
deep spout 供料机料盒
deep spring 晚春;深泉
deeps-staggered crossing-shoal 深槽交错过渡浅滩
deeps-staggered shoal 交错浅滩;深槽交错的浅滩
deep stabilization 深层稳定
deep stamping sheet 深冲薄板
deep state 深能态
deep stowage 纵深堆货法;深舱堆装法
deep strata 深水层
deep stratum 深层
deep stratum of clay 深层黏土
deep-strength asphalt pavement 深层高强度沥青路面;加厚的高强度沥青路面
deep structure 深层构造
deep submergence rescue vehicle 深潜救生艇
deep submergence rescue vessel 深潜救生船
deep submergence search vehicle 深潜搜索器
deep submergence search vessel 深海潜水船;深海调查船
deep submersible photographic(al) system 深潜照相系统
deep subsidence zone 深凹带
deep subsoil 深层底土
deep subsoil water 深层土壤水
deep suction apparatus 深吸水器
deep suction share 大垂直间隙的犁铧
deep suction well 深吸水井
deep tank 深水柜;深(水)舱;深舱油柜;深舱水柜
deep tank aeration/flo(a)tation system 深槽曝气/气浮系统
deep tank capacity 深舱容积
deep tectono-geochemistry 深部构造地球化学
deep telerecording unit 深水电视录像装置
deep temperature 深部温度
deep test 深测试验;深度测定;深层试验
deep thermal water 深层热水
deep-throat coping 深度遮檐
deep tillage 深耕;深翻
deep-tillage bottom 翻耕犁体
deep tint 深度着色
deep tooth form 长齿高制
deep tooth gear 长齿高制齿轮
deep-towed magnetometer 深海拖曳磁力仪
deep-towed system 深海拖缆系统
deep trades 赤道东风带;深厚信风带
deep trap 深陷阱
deep treatment process 深度处理法
deep trench 深河槽;深冲沟
deep trench latrine 深沟厕所
deep trial pit 深试坑
deep-trough idler 深槽托辊
deep-tube railway 地下铁道
deep-tube well 深管井
deep tunnel 深埋隧道
deep-turbidite trap 深海浊积岩圈闭
deep turbine pump 深井涡轮泵
deep underground disposal 深地层处置
deep underground structure 深埋地下结构
deep underwater nuclear counter 深水放射测定器
deep unmanned submersible 无人深潜器
deep valley 峡谷;深谷
deep violet 深紫色
deep violet-black 深紫黑色
deep violet-blue 深紫蓝色
deep violet-brown 深紫褐色
deep violet-grey 深紫灰色

deep violet-red 深紫红色
deep waisted vessel 高首尾船；深腰船
deep wall 厚墙
deep water 深水的；深水处；深水；深层水
deep-water aeration activated sludge process 深水曝气活性污泥法
deep-water amorphous facies 深水无定型相
deep-water area 深水域；深水区
deep-water ballast tank 深压载水舱
deep-water basin 深水港池
deep-water bell 深水潜水钟
deep-water berth 深水锚地；深水泊位
deep-water caisson 深水沉箱
deep-water case 深水情况
deep-water channel 深水航道；深水航槽
deep-water circulation 深水环流
deep-water cofferdam 深水围堰
deep-water current 深层流
deep-water delta 深水三角洲
deep-water depth 深水水深
deep-water dock 深水港坞
deep-water drilling 深水钻井
deep-water drilling operation 深水钻井作业
deep-water entrance 深水入口
deep-water fairway 深水航道
deep-water fauna 深水动物区系
deep-water flow 深层流
deep-water foundation 深水基础
deep-water harbo(u)r 深水港
deep-water herbaceous facies 深水草本相
deep-water hydrophone 深水探听器；深水探测器；深水检波器
deep-water isotopic current analyzer 同位素深海海流分析仪
deep-water layer 深水层
deep-water lead 深水测深锤
deep-water line 满载吃水线；最大载重水线
deep-water mass 深层水团
deep-water moorings 深水锚(泊)设备
deep-water muck 深水腐泥土
deep-water oil terminal 深水油库
deep-water pier 深水桥墩；深水码头
deep-water platform 深水平台；深水承台
deep-water port 深水港
deep-water power 深水波功率
deep-water quay 深水码头；深水岸壁
deep-water route 深水航线
deep-water sampler 深水取样器；深水采样器
deep-water sampling 深水取样
deep-water slip 深水港池
deep-water sonar 深海声呐
deep-water species 深水种类
deep-water table 深潜水位；水位位置；深层地下水位
deep-water-table condition 深水位条件
deep-water terminal 深水码头
deep-water test program(me) 深水试验计划
deep-water thermometer 深水温度计
deep-water transducer 深水换能器
deep-water trawl 深海拖网
deep-water turbidity deposit 深水浊流沉积
deep-water velocimeter 深水流速计；深水海流计
deep-water voyage 远洋航程
deep-water wave 深水波
deep-water wave height 深水波高
deep-water well 深(层)水井
deep-water wharf 深水码头
deep-water zone 深水域；深水区；深水带
deep wave 深水波
deep webbed 腹板很高的大梁；高腹板的
deep web shearer 深截式采煤机
deep welding 深焊
deep well 管井；深水井；深井
deep-well aeration 深井曝气
deep-well aeration activated sludge process 深井曝气活性污泥法
deep-well aeration biochemical process 深井曝气生化法
deep-well aeration method 深井曝气法
deep-well aeration tank 深井曝气池
deep-well array 深井台阵
deep-well backwater 深井回灌水
deep-well camera 深井照相机
deep-well dewatering 深井排水
deep-well disposal 深井处置；深井处头

deep-well drainage method 深井排水法
deep-well drilling 深井钻井
deep-well elevator 深井提升机；深井升降梯；深井吊梯
deep-well filter 管井滤水管；深井滤水管
deep-well impressed-current cathodic protection system 深井外施电流阴极保护系统
deep-well injection 深井灌注；深井注水
deep-well injection of hazardous waste 危险废物的深井灌注
deep-well jet pump 深井喷水泵；深井喷射泵
deep-well method 深井法
deep-well oil well cement 深井油井水泥
deep-well piston pump 深井活塞泵
deep-well plunger pump 深井柱塞泵；深井活塞泵
deep-well point 深井点
deep-well pump 潜水泵；深井抽水机；深井(水)泵
deep-well pumping 深井排水；深井抽水；深井泵疏干法
deep-well pumping unit 深井抽水机组；深井泵
deep-well reciprocating pump 往复式深井泵
deep-well rim 深轮辋
deep-well sampling pump 深井取样器
deep-well screen 管井滤水管
deep wells for foundation dewatering 地基降水用的深井
deep-well socket 长套筒
deep-well system 深井系统
deep-well treatment 深井处理
deep-well turbine 深井水轮机
deep-well turbine pump 深井涡轮泵；深井水轮泵；深井透平泵
deep-well type pump 深井式水泵
deep-well waste disposal 废物深井处置；深井废物处置
deep-well wastewater treatment 深井废水处理
deep-well water 深井水
deep-well water level measuring equipment 深井水位量测仪
deep-well water pump 深井水泵
deep-well working barrel 深井水泵
deep winding 深井提升
deep-winnowed-crestal trap 深海簸脊圈闭
deep-winnowed-flank trap 深海簸翼圈闭
deep-worked soil 深耕土
deep X-ray therapy room 深度X光治疗室
deep yellow 深黄色
deep yellow-brown 深黄褐色
deep yellow-green 深黄绿色
deep yellow-grey 深黄灰色
deep yellow-red 深黄红色
deep zone 深度分带
deer foot 鹿脚形斧柄
deerhorn antenna 鹿角天线
Deer Isle (一种粗纹理呈粉紫中灰色调的)缅甸大理石
Deerite 迪尔石
deer park 鹿苑
deeside 灰色花岗岩(一种产于挪威的花岗岩)
de-ethylation 脱乙基(作用)
deewan (穆斯林国家放有坐垫的)地面升起部分
de-excitation 灭励；灭磁；反励；去激(发)
de-excitation cross-section 去激发截面
de-excitation effect 去激发效应
de-excitation photon 去激发光子
de-excitation system 灭磁系统
de-excitation time-constant 灭磁时间常数
de-exciter 灭磁器
defaced coins 磨损残坏硬币
defacement 损坏路面
de-facto population 实际人口；事实人口
defalcate 盗用公款
defalcation 亏空额
defatting 脱脂
default 不履行(契约)；补缺；违约；违背契约；拖欠债款；拖欠；省缺；省略
default attribute 缺席属性；省缺属性
default choice device data 缺省选择设备数据
default data base 缺省数据库
default declaration 缺席说明；省缺说明
default department number 补缺部门号；缺席部门号
default drive 默认驱动(器)；默认的驱动器；缺省驱动(器)；缺省规则驱动

defaulted bonds 拖欠(清偿的)债券
defaulted debtor 违约债务人
defaulted interest 滞纳利息
defaulter 违约人；拖欠人；亏空公款者；缺席人；违约者
default file attribute 缺席文件属性
default fine 违约罚金
default for abnormality 反常性的缺省规则
default for attribute 缺省规则属性
default for data type 数据类型缺省规则
default for scope attribute 观察仪器属性缺省规则
default for storage class attribute 存储器种类属性缺省规则
default for transmission 传输缺省规则
default free securities 安全证券
default gateway 缺省网关
default group 补缺组；缺席组
default handler 缺省处理程序
defaulting behavio(u)r 违约行为
defaulting debtor 违约债务人
defaulting party 违约一方
default in investment 违约的投资
default interest 过期罚金
default interpretation 缺省说明；省略时解释
default judgement 缺席判决
default legend 缺省图注
default locator device data 缺省定位设备数据
default loss compensation reserve 违约损失赔偿准备金
default notice 不履行通知书
default of contractor 承包者违约；承包人违约
default of employer 业主违约
default of obligation 不履行债务
default of owner 业主违约
default of payment 拒绝付款；未付款；拖欠付款
default of pilot 引航员疏忽责任
default of tax payments 拖欠税款
default option 隐选择；缺席选项；缺省选项
default party 违约方
default point 预设点；收支平衡点
default ratio 预设比值；收支比值
default rule 省缺规则；省略规则
default set to 省缺时设置成……
default sign convention 符号约定缺省规则；省缺符号约定
default specification file 缺席说明文件；缺省说明文件
default strategy 空缺策略
default string device data 缺省字符串设备数据
default summons 因延迟给付而发出的传票；债务简易诉讼
default system control area 缺席系统控制区
default task 缺席任务；缺省任务
default valuator device data 缺省定值设备数据
default value 默认值；默契值；补缺值；缺省值；缺省值；省略值
defeasance 废止；废除契约的条款
defeasance clause 契约中的废止条款
defeasible 可废除
defeasible fee 可以废除的杂费
defeasible fee title 可取消的房地产继承权
defeasible title 可取消的地产所有权
defeated party 败诉方
defeated river 改道河流
defeated stream 改道河流；弃置河流；退化河流
defeathering machine 除羽毛机
defeathering unit 除毛机
defeat switch 消除开关
defeature 损坏外形
defecate 滤净
defecation 澄清；提净
defecator 过滤装置；澄清器；澄清槽
defect 亏损；亏量；病态；缺陷；缺损；缺点
defect absorption 缺陷吸收
defect center 缺陷中心
defect chemistry 缺陷化学
defect cluster 缺陷丛
defect compensation 工程缺陷补偿
defect concentration 缺陷密度
defect conduction 缺陷导电
defect deduction 缺陷扣计额
defect detecting test 缺陷检验；缺陷检查；探伤检查
defect detection 探伤
defect detector 故障检测器；探伤仪

defect echo 探伤回波
defect group 亏损群
defect in materials 在材料中的缺陷
defect in timber 木料缺陷
defect in vision 视距误差
defect in welds 在焊缝中的缺陷
defection 过失;变节
defection drilling 造斜钻进
defective 有缺陷的;不健全者;不合格线对;缺损的
defective aeration 土壤通气不良
defective brake 不良制动
defective brazing 有缺陷钎焊
defective check 不完整支票
defective coating 有缺陷涂占;不良涂层
defective concrete 有缺陷混凝土;不合格混凝土
defective coupling 不良耦合;不合格接头
defective equation 亏损方程
defective equipment 不良设备
defective gluing 有缺陷胶合
defective goods 次品;残品;不合格品
defective index 废品率
defective insulation 绝缘不良
defective item 亏损项目;不合格品;缺陷项目
defective joint 不良接头
defective material 次料
defective material report 不合格材料报告
defective materials 有缺陷原料;有缺陷材料
defective number 亏量;缺陷数
defective number chart 不合格品率控制图
defective pack 包装有缺点
defective package build 不良成型卷装
defective packing 不良包装;包装不良
defective pavement 有缺陷路面
defective per hundred unit 缺陷百分数
defective plumbing 有缺陷的卫生设备系统;有缺陷的室内管工
defective probability distribution 缺损概率分布
defective product 残品
defective product record 废品记录
defective quality 品质有缺点
defective rate 次品率
defective rate chart 不合格品率控制图
defective ringing 错误呼叫
defectives 次品
defective sample 残缺样本;不完全样本
defective semi-conductor 不良半导体
defective sleeper 失效轨枕;失效枕木
defective tie 失效枕木
defective tightness 不(够)紧密
defective title 不完整的产权
defective track 技术状态不良线路;故障磁道;不良线路
defective unit 不合格品;缺陷单元
defective value 亏(损)值;缺陷值
defective water 不纯(的)水
defective wood 缺陷(木)材
defective wool 缺陷羊毛
defective workmanship 工艺缺陷
defective works 不合格活计;有缺陷工程;不合格工程
defect lattice 缺陷晶格;缺陷点阵
defect marker 质量检查员;缺陷标记员
defect motion 缺陷运动
defect of complement system 补体系统缺陷
defect of gas accumulation of anode oxide film 阳极氧化膜蓄气性发花(起泡)
defect of images 图文缺陷;图像缺陷
defect of lacquer 漆疵病
defect of tissue (木材的)组织缺陷
defect of vision 视觉缺陷
defectogram 探伤图
defectometer 缺陷检测仪
defect on firing 烧成缺陷
defectoscope 探伤器(超声波);探伤仪(超声波)
defectoscopy 缺陷量测仪;探伤法
defect production 缺陷产生
defect repair 缺陷修补
defects assessment 缺陷评定
defect semiconductor 缺陷半导体
defects in forgings 锻件中的缺陷
defects in sash fixing 窗扇施工疵病
defects in sash work 窗扇施工疵病
defect sintering 缺陷烧结
defects in timber 木材缺陷;木材缺陷

defects in welds 焊接件中的缺陷
defects in window hardware fixing 窗五金安装疵病
defect skip 缺陷跳跃
defects liability 缺陷责任
defects liability certificate 缺陷责任证书
defects liability period 保用期;保修期;缺陷责任期
defects not discoverable 不能发现的缺点
defect solid solution 缺位固溶体
defects per unit 故障率;单位产品疵点数;缺陷率
defects prevention 缺点预防
defect structure 缺陷结构;缺陷构造
defect susceptible area 缺陷敏感区
defect system 缺陷分极法
defect ware 残次品
defence 防御;防卫;保卫
Defence Atomic Support Agency 国防原子支援局(美国)
defence attack 防御性灭火
defence communications system 国防通信系统
defence factory 兵工厂
Defence Mapping Agency 国防制图局(美国)
defence navigation satellite system 防导航卫星系统
defence need 防护要求
defence parapet 防御胸墙
defence problems 防敌航法
Defence Science Board 国防科学委员会
defence stand 防御点
defence transport 国防运输
defence wall 防御墙
defence work 防御工事;防御工程;防空工程
defended terrance 埋藏阶地
defender 防御者;辩护人;保护器
defending in-place 就地防御
defend in place strategy 就地保护战术
defend sea sovereignty 保卫领海权
defense = defence
defense factory 兵工厂
defense navigation satellite system 国防导航卫星系统
defense parapet 防御胸墙
defense wall 防御墙
defensible space (用建筑处理来加强邻里联系、防止偷盗和对付犯罪的)可防范空间;防御性空间
defensive gateway 加固大门;碉堡式城门
defensive hedge 防护栅栏;防护围墙
defensive investment 与竞争企业作抗衡的投资;防务投资;保护性投资
defensively armed merchant 武装商船
defensively equipped merchant ship 武装商船;武器商船
defensive operation mode 防御战斗程式
defensive portfolio 防御性资产组合
defensive pricing 防御性定价
defensive programming 防错性程序设计
defensive purchase 防务采购
defensive tower 防御塔楼;炮楼
defensive wall 雉堞墙;防御墙
defensive work 防御工事
defer 延迟;暂缓;递延
deference of longitude 经度差
deferent 输送物;输送的;导管;传送的;圆心轨迹;传送物
deferential rent 级差地租
deferment 迟延
deferment charges 逾期费
deferment delivery 推迟交货
deferment of a project 工程展期
defernite 戴碳钙石
defer payment 延期付款
deferred account 延期账户;递延账户
deferred action 延时动作
deferred address 延迟地址;递延地址;迟延地址
deferred addressing 延迟定址;递延寻址
deferred and prepaid expenses 待摊费用
deferred and rotational grazing 延迟式轮牧
deferred annuity 延期年金;递延年金
deferred assets 延期资产;递延资产
deferred assets audit 递延资产审计
deferred availability 递延的有效性
deferred bond 延期(付息)债券;延付债券
deferred charges 预付款;延期费(用);延迟费用;滚存费用;递延费用;待摊费用

deferred charges before amortization 摊提前递延费用
deferred check point restart 延迟检验点再启动
deferred compensation contract 延迟补偿合同
deferred concept of income tax allocation 分摊所得税的延期概念
deferred condition code 延迟条件码
deferred constraint 延迟约束
deferred construction 缓建;延期建设
deferred cost 递延成本
deferred credit 延期贷款;延期信用;递延贷款
deferred cycle 延迟周期
deferred debate 延期回扣
deferred debit items 债务项目滚存
deferred debt 递延借款
deferred deformation 延缓变形
deferred deformation of concrete 混凝土的延缓变形
deferred depreciation 递延折旧
deferred dividend 延付股息;递延股息
deferred duty 备用税率;备用费率
deferred echo 延迟回声
deferred entry 异步输入;延期输入;延迟入口;延迟输入
deferred exit 延期输出;延期出口;延迟输出;延迟出口
deferred expenses 延迟费用
deferred grazing 待期放牧
deferred gross profit 递延毛利
deferred income 延期收入;递延收入
deferred income tax 递延所得税
deferred input/output 延迟输入/输出
deferred insurance 递延保险费
deferred interest 延期利息;递延利息
deferred interest mortgage 延缓利息抵押贷款
deferred liabilities 预收款;延期负债;延迟负责;递延债务;递延负债
deferred life annuity 延期生命年金
deferred losses of prestress 延缓预应力损失;延缓水泥强度增长
deferred maintenance 逾期养护;延时性维修;延期维修;延缓维修;递延维修;延迟维持费
deferred maintenance time 延期维修时间
deferred mode 延迟态
deferred mount 延装装置
deferred overhead charges 延期间接费用
deferred payment 逾期税;延期付款;延付货款;迟付贷款;延期支付
deferred payment of letter of credit 延期付款信用证
deferred payment options 延期付款权利
deferred payment sale 分期付款销售;延期支付销售;赊销
deferred payment sales method 延期付款销货法
deferred payment trade 延期付款贸易
deferred payment transaction 延期付款交易
deferred performance liabilities 延期交货负债
deferred perpetuity 延期永续年金
deferred premium 递延保险费
deferred processing 延迟处理;延期处理
deferred profit on instalment sales 分期收款销货递延利润
deferred project 缓建项目;停缓建项目
deferred purchasing power 推迟购买力
deferred reaction 延迟反应
deferred rebate 延期折扣
deferred rebate system 延期回扣制(度)
deferred remuneration 递延酬金
deferred repairs 递延修理费
deferred repayment of capital and interest 延期还本付息
deferred restart 延期再启动;延迟再启动
deferred retirement benefit 延迟退休金
deferred revenue 延期收入
deferred rotation grazing 延迟轮牧
deferred shares 递延股
deferred shares stock 发起人股
deferred sight 延期票据
deferred sight letter of credit 延(期)付期信用证
deferred step restart 延迟步重新启动
deferred stock 延期付息股票
deferred strain of concrete 混凝土延缓变形
deferred swap accrual date 延迟互换交易权责日
deferred tax 递延税款

deferred tributary junction 延长支流汇点
deferred unapplied expenses 延期未分配费用
deferred update 延期更新;延迟修改;延迟更新
deferrization 除铁(作用);脱铁
deferrize 除铁
defer time 延期时间
defervescence 热退期
defervescent stage 热退期
defiber 分离线;分离纤维;脱纤维
defiberer 剥纤维机;剥韧皮机
defibering 分离成纤维;木屑合成纤维
defibering machine 纤维分离机
defibrated wood 纤维分离的木材
defibrater 木料碾碎机
defibration 纤维分离
defibrator 木料碾碎机;碎木机
defibrator-chemipulp 纤维分解化学制浆机
defibrator(-ter) 纤维分离机
defibre 分离纤维
defibrillation 除颤;去颤
defibrillator 电震发生器;除颤器;去颤器
defibrination 磨制木浆;磨浆
deficiency 亏数(变位);亏格;亏差;差量;不足数;不足量;不足额;缺陷;缺失
deficiency account 亏损账户;清算损失账
deficiency advance 预算不敷垫款
deficiency appropriation 弥补亏绌拨款;追加拨款;补拨款;拨补款
deficiency bills 补足票据
deficiency clause 缺额条款
deficiency curve 亏数曲线;亏率损曲线
deficiency estimates 亏损预算;亏损概算书;亏短概算书
deficiency in air supply 供气不足
deficiency index 亏指数;亏指标
deficiency in draft 风量不足;通风不足
deficiency in funds 缺乏资金
deficiency in reality with pseudo-excess symptoms 真虚假实
deficiency in water 水量不足
deficiency in weight 重量不足
deficiency judgement 差价裁决;裁定短额
deficiency letters 亏空通知书
deficiency modulus of groundwater 地下水亏损模数
deficiency of air 不足气流
deficiency of data 不完整数据
deficiency of elements 元素缺乏
deficiency of fluid with exuberant fire 津亏火炽
deficiency of men or stores 人员或物料不足
deficiency of wagon flow 车辆不足
deficiency of water 水资源不足;缺水
deficiency point 不符合标准的地点
deficiency report 故障报告
deficiency statement 清算损失账
deficient coupling 欠耦合
deficient data 不完整数据
deficient draft 缺水量;水深不足(航运)
deficient flow 流量差额;不足流量;欠缺流量
deficient number 亏数;亏量
deficient opacity 露黑
deficient pondage 不完全调节库容
deficient super elevation 欠超高
deficit 亏损(额);亏空额;亏短额;短缺;赤字;不敷
deficit account 亏损账户;亏欠账户
deficit at the beginning period 期初亏欠
deficit balance 赤字差额
deficit budget 赤字预算
deficit carried forward 亏损结转下期
deficit carried forward till extinction 亏损结转到结清为止
deficit clause 亏损条款
deficit coverage 弥补亏损
deficit-covering bond 用来弥补赤字的公债;赤字债券
deficit-covering financing 弥补赤字筹资
deficit-covering for the current 弥补近期赤字
deficit finance 赤字财政
deficit financing 赤字财政
deficit index 亏损指数
deficit in long-term capital 长期资本短缺
deficit on the books 账面赤字
deficit power 不足容量;不足功率
deficit reconciliation statement 亏损额计算书;亏损调节表
deficit spending 赤字开支
deficit statement 亏损表
deficit unit 亏损单位
deficit value 亏值
defilade 障碍物;隐蔽物
defile 峡谷;隘路;隧道;山中峡道
definability 可定义性
define 定义;确定
define area 定义范围
define byte 定义字节
define constant statement 定义常数语句
defined area 指定区(域)
defined array 定义数组
defined assets 限定资产
define data command 定义数据命令
defined attribute 定义属性
defined benefit plan 养老金的固定受益计划
defined-benefit plans 明确收益方案
defined coefficient 规定系数
defined contribution plan 养老金的固定缴款计划
defined external symbol 定义外部符号;外部定义符号
defined file 定义文件
defined instruction 定义指令
defined in years 年限
defined item 已定义项
defined pension plan 养老金的固定缴款计划
defined resources 确定资源
defined terms 规定的条件;定义词
defined variable 已定义变量;定义变量
defined word 定义字
define system 定义系统
define table 定义表
define tape file command 定义带文件命令
define-the-file 定义文件指令
define way-point 规定航路点
defining coal-forming prospect grade 圈定成煤远景分级
defining constant 定义常数
defining equation 定义方程
defining minerogenetic prospect grade 圈定成矿远景分级
defining occurrence 定义性出现
defining oil-gas accumulation prospect grade 圈定含油气远景分级
defining polynomial 定义多项式
defining postulate 定义公设
defining range 定义范围
defining record 定义记录
defining relation 定义关系
defining residual wavelet 剩余子波确定
defining scalar 定义标量
definite 肯定的;明确的;确定的
definite advice 确切通知
definite appropriation 定额费用;定额拨款
definite binomial coefficient 确定的二项式系数
definite cycle time 规定循环时间
definite decision 正式决定
definite designation 正式命名
definite division 定义除(法)
definited spare parts list 确定备件表
definite equation 定义方程
definite error 确定误差
definite event 确定的事件
definite extrapolation 有限外推
definite inquiry 确定询价
definite integral 定积分
definite invoice 正式发票;真实发票
definite kernel 确定核
definite machine 确定的时序机
definite magnetic time controller 磁力定时控制器
definite masses of congealed labor-time 一定量的凝固的劳动时间
definite melting point 固定熔点
definiteness 肯定性;确定性
definite order 正式保赔
definite plan report 详细设计报告
definite point 明置点;明显点
definite policy 确定保险单
definite project plan 详细工程设计
definite proportion 定比
definite purpose circuit breaker 专用断路器
definite quadratic form distribution 定二次型分布
definite quantity 定量
definite range law 定距法则
definite response 确切应答;确定应答;确定回答
definite Riemann integral 黎曼定积分
definite seasonal pattern 限定性季节模型
definite shape 一定形状
definite spare parts list 确定备件一览表
definite spot 离散斑点;分立斑点
definite stop 死止块
definite system 确定系统
definite time delay 定时滞后
definite time frequency relay 定时限周率继电器
definite time-lag 定时限
definite time limit distance relay 定时限远距继电器
definite time relay 定时(限)继电器
definite undertaking 明确承诺;确定承诺
definite value 定值
definite variability 一定变异性
definiting gate 导向挡板
definition 界定;定义;定界;分辨率;分辨力;分辨度
definitional part 定义部分
definition and classification 定义和分类
definition chart 分辨力测试图;清晰度测试卡
definition development program(me) 确定开拓方案
definition domain 定义域
definition dressing method 确定选矿方法
definition dressing technique 确定选矿技术
definition environment 定义环境
definition estimate 最终预算
definition flow-sheet of metallurgy 确定冶炼流程
definition flow-sheet of mineral dressing 确定选矿流程
definition in depth 景物的清晰度;景深
definition metallurgical method 确定冶炼方法
definition metallurgical technique 确定冶炼技术
definition mining method 确定采矿方法
definition mining technique 确定采矿技术
definition mode 定义方式
definition of air pollution 空气污染确定;空气污染的定义
definition of binary tree 二叉树的定义
definition of Boolean function 布尔函数的定义
definition of capital goods 资本财物的限定;资本财物的定义
definition of carrier clause 承运人定义条款
definition of colo(u)rs 分色
definition of constraints 约束条件限定
definition of contour 轮廓清晰度
definition of depth 景深清晰度
definition of image 影像清晰度
definition of mark image 各种标志清晰度
definition of telescope 望远镜清晰度
definition of term 条款解说
definition phase 技术设计阶段;定义(阶)段;初步设计阶段;方案论证阶段;确定任务阶段
definition power 分辨;清晰度
definition range 清晰范围
definitions of export quotations 出口价格条例
definition statement 定义语句
definition status 定义状态
definition table 定义表
definition test card 分辨力测试卡
definition wedge 清晰度测试楔形束
definitive
definitive answer 最后正式答复
definitive bond 正式证券
definitive estimate 确定性估算
definitive host 终宿主;终寄主
definitive instruction 定义指令
definitive judg(e)ment 最后判决
definitive loan 长期房地产租借的贷款
definitive money 限定货币
definitive survey stage 定测阶段
definitive text 最后文本;特定文本
definitive time 确定时间
definitive toxicity test 正式毒性试验
definitive variation 一定变异;确定变异
definitive weight 确定权重
deflagrability 爆燃性;暴燃性
deflagrable 易燃的
deflagrate 迅速燃烧;爆燃;暴燃
deflagrating mixture 爆燃混合物;暴燃混合物

deflagration 爆燃;暴燃(作用)
deflagration behavio(u)r 爆燃特性
deflagration fringe 爆燃区
deflagration front 爆燃的正面
deflagration phenomenon 爆燃现象
deflagration suppression 爆燃抑制
deflagration suppression concentration 抑爆浓度
deflagration to detonation transition 爆燃到爆轰的跃变
deflagration venting 爆燃通风
deflagration wave 爆波;暴燃波
deflagrator 爆燃器;突燃器
deflaker 高频疏解机
de-flashing 去飞边;去除毛刺
deflatable rubber tube 可紧缩的橡皮管
deflate 紧缩通货;抽出空气
deflate currency 紧缩通货
deflate currency and price 紧缩通货和压低物价
deflated condition 泄气状态
deflated tire 泄气轮胎
deflate the economy 紧缩经济
deflating index 紧缩指数
deflating price level 价格下降水平
deflating valve 放气嘴
deflation 紧缩通货;降阶;吹蚀;抽出空气;风蚀;放气;通货紧缩
deflation and inflation 紧缩与膨胀
deflation approach 紧缩法
deflationary effect 通货紧缩效应
deflationary gap 紧缩差额
deflationary monetary and fiscal policies 紧缩货币财政政策
deflationary spiral 紧缩的螺旋式上升
deflation basin 风蚀洼地
deflation column 风蚀柱
deflation flask 放气瓶
deflation opening 排气孔
deflation plain 风蚀平原
deflation plane 风蚀面
deflation policy 紧缩政策;通货紧缩政策
deflation quantity 风蚀量
deflation receptor 消气感受器
deflation ripple mark 风蚀波痕
deflation valley 风蚀谷
deflation valve 放气阀(门)
deflator 紧缩指数;紧缩因素;减缩指数;消除通货膨胀指数;平减指数
deflectable laser beam 可偏转激光束
deflect angle 偏移角
deflected air 偏流空气
deflected area 挠曲面积(原位与挠曲线之间的面积)
deflected ascent 斜升
deflected beam 偏转光束
deflected borehole 偏斜孔
deflected burning 偏火
deflected cable 偏转钢缆
deflected cable technique 弯曲钢筋束的工艺
deflected coil 偏转线圈
deflected drilling 偏斜(孔)钻进
deflected electron beam 偏转电子束
deflected pile 偏正桩;偏位桩
deflected sheet wall 变形的板墙
deflected strand technique 弯曲钢筋束的工艺
deflected stream 折弯水流;挑流
deflected stream limits of dike 坝的挑冲范围
deflected tendon 弯曲钢筋束;弓形受力筋;受弯力筋;预应力弯曲钢筋;挠曲钢键
deflected-thrust engine 抵力变向发动机
deflected well 定向井;偏斜井
deflect(ing) 偏转;造斜【岩】
deflecting angle of traveller 钢丝圈偏倾角
deflecting bar 转向杆;带钩杆;皮带挂杆
deflecting bit 造斜钻头;偏斜钻头
deflecting bucket 鼻坎反弧段;挑檐平顶;挑流厗斗;挑流弧坎;挑流鼻坎
deflecting cam 偏曲凸轮;脱扣凸轮
deflecting cavity 致偏谐振腔
deflecting coating 折向喷嘴
deflecting coil 致偏线圈;偏转线圈
deflecting core 偏斜孔岩芯
deflecting couple 转矩;偏力力偶
deflecting damper 偏流气闸
deflecting device 致偏装置;偏转装置

deflecting drill tool 钻孔偏斜钻具
deflecting electrode 致偏电极;偏转极
deflecting field 致偏场;偏转场
deflecting force 造斜力;偏转力;偏向力
deflecting gate 导向挡板
deflecting index of hole 钻孔偏斜指标
deflecting magnet 致偏磁体;致偏磁铁
deflecting method 偏转法
deflecting needle nozzle 偏向针形喷嘴;偏向针形管嘴
deflecting of strands (预应力混凝土中的)弯起钢缆
deflecting piece 转向杆
deflecting plate 遮护板;隔板;转向板;反击板;偏转板
deflecting plug 造斜塞
deflecting prism 转折棱镜;致偏棱镜
deflecting pulley 转向滑轮
deflecting rod 转向杆
deflecting roller 转向轮
deflecting roundel 偏光镜
deflecting shoe 偏斜靴【岩】
deflecting system 致偏系统
deflecting tools 造斜工具
deflecting torque 转矩;偏力力矩
deflecting vanes 折流叶片
deflecting voltage 致偏电压
deflecting wedge 造斜楔;偏心楔;偏斜器
deflecting well 斜井;偏斜井
deflecting yoke 偏转系统
deflection 挠曲;挠度;折转;折流;方向角;变位;偏转(光);偏移;偏斜;偏向;偏射;偏离;偏差角;弯沉【道】
deflectional instrument 指针偏转式仪表
deflection amplifier 偏转放大器
deflection anemometer 偏转风速表
deflection angle 角变位;转向角;致偏角;变位角;偏转角;偏斜角;偏向角;偏(离)角
deflection angle method 偏角法
deflection angle of scan mirror 扫描反射镜偏转角
deflection angle of traverse leg 导线边偏角
deflection angle traverse 偏角导线
deflection anomaly 偏差异常
deflection at break 裂断变位
deflection at rupture 破裂变位
deflection basin 倾斜盆地;弯沉盆
deflection beam 弯沉仪(用于路面弯沉测量)
deflection bit 校准钻头;造斜钻头
deflection board 方向修正板
deflection bowl 弯沉盆
deflection center 偏转中心
deflection change 方向角变换
deflection chassis 偏转部分
deflection coefficient 弯沉系数
deflection coil 偏转线圈
deflection command 偏转指令
deflection computer 前置量计算机
deflection contour (路面)等挠沉线;挠度等值线;(路面)等弯沉线
deflection correction 方向修正量
deflection criterion 挠度规定;变位准则;偏移准则
deflection cup 弯沉杯
deflection current 偏转电流
deflection current amplifier 致偏电流放大器
deflection curve 挠度曲线;变值曲线;变位曲线;偏离曲线;弯沉曲线
deflection defocusing 偏转散焦
deflection dial (贯入度计的)荷载刻度盘
deflection difference 集火量;分火量
deflection distance 偏距
deflection distortion 致偏失真
deflection ductility (factor) 挠度延性系数
deflection eclipse 偏拍
deflection efficiency 偏转效率
deflection equation 弯沉计算公式
deflection error 方向误差
deflection factor 偏转因数
deflection field 偏转场
deflection focusing 偏转聚焦
deflection fold 转向褶皱
deflection force 偏转力
deflection force for run-in period 磨合运转时期偏移力
deflection force of earth 地球自转偏向力
deflection force of earth rotation 地球自转偏向力

deflection formula 挠度公式
deflection-free shell 等应力壳体
deflection front 偏转峰
deflection ga(u)ge 挠曲测量仪;挠度计;偏角仪
deflection inclinometer 测斜仪
deflection indicator 偏移计;挠曲仪;挠度计;偏斜指示器;偏航指示器;偏度计;偏度表
deflection line method 变位曲线法
deflection measurement 挠度测定;弯沉量测
deflection meter 挠曲测量计
deflection method 造斜钻孔法;变位法;偏转法;移法;位移法(分析超静定结构的基本方法之一)
deflection method of measurement 偏差测量法
deflection-modulated indicator 调幅指示器
deflection modulation 偏向调制
deflection moment 偏转力矩
deflection multiplier 偏差倍增器
deflection nozzle 偏转管嘴
deflection observation 挠度观测
deflection of a girder 钢梁挠曲
deflection of beam 梁的挠曲;梁的挠度
deflection of bridge span 桥梁挠度
deflection of flange 法兰挠度
deflection of hole 钻孔偏斜
deflection of light 光线偏折
deflection of light beam 光束偏转
deflection of pipe line 管线的弯扁
deflection of plumb-line 垂线偏差
deflection of rail 钢轨挠曲
deflection of shuttle 货叉下挠度
deflection of spring 弹簧挠度
deflection of the bit 钻头偏离钻孔位置
deflection of the optic(al) path 光路偏转
deflection of vertical 垂线偏差;铅垂偏差
deflection of wave 波浪反射;波浪反折
deflection plane 偏转平面
deflection plate 偏转板
deflection plug 造斜塞
deflection point 变位点
deflection polarity of an oscilloscope 示波器偏转极性
deflection probable error 方向公算偏差
deflection procedure 弯沉法
deflection rate 造斜率
deflection scale 修正标尺;方向标度;偏转标度;偏移标度
deflection sensitivity 偏转灵敏度
deflection separator 折流分离器
deflection signal 偏转信号
deflection surface 挠曲面
deflection survey 挠度测量
deflection system 偏转装置
deflection table of plumb-line 垂线偏差表
deflection temperature 挠曲温度
deflection temperature under load 荷重畸变温度
deflection test 挠度试验;变曲试验;弯曲试验;弯沉试验【道】
deflection theory 挠曲理论;挠度理论;变位(理)论
deflection time 偏转时间
deflection tools 造斜工具
deflection torque 偏转力矩
deflection tube 导向管
deflection under loading 荷载挠度
deflection valve 偏转管
deflection wedge 校偏楔;造斜楔
deflective area 挠曲面积
deflective force 偏转力;偏向力
deflective plumbing 缺乏卫生设施系统
deflectivity 可弯性;偏向
deflectogram 挠度图;道路弯沉图;弯沉图
deflectograph 弯沉仪(用于路面弯沉测量)
deflectometer 挠度计;偏斜计;挠度计;偏差计;弯度计;弯沉仪(用于路面弯沉测量);挺度计
deflector 弧形导流器;转向装置;转向器;致偏器;致偏板;折转器;折流板;遮护板;造斜器;导向装置;导流器;导流片;偏转装置;偏转仪;偏向器;偏转装置;偏导器
deflector angle 偏导角
deflector apron 导向(盖)板;导向(挡)板
deflector bucket 鼻坎反弧段;挑流厗斗;挑流弧坎;挑流鼻坎
deflector control 导向器操纵杆
deflector device 防偏装置
deflector gate 导向挡板

deflector gear 偏流器
deflector jet 导流喷射式;偏流喷射式
deflector pinch roll 夹送导辊
deflector plate 折向板;偏转板;导流板;转向板;导向器;偏流板
deflector rod 拨杆
deflector roll 导辊
deflector tube 偏转管
deflector-type nozzle 导流式喷嘴
deflector-type separator 折板型除沫器
deflector-type sprinker 溅水盘式喷头
deflector wedge ring 造斜楔定位环;变向器固定环
deflectoscope 缺陷检查仪
deflectoscope for wire rope 钢丝绳断丝测定仪
deflect treatment 隐患处理
deflexion = deflection
deflexion angle 偏转角;偏斜角
deflexion curve 挠度曲线
deflexion curve radius 挠曲曲线半径
deflexion fold 转向褶皱
defloccualant 抗絮凝剂;解絮凝剂;胶体稳定剂;悬浮剂;反絮(凝)剂;反团聚剂;反凝聚剂
deflocculant dispersant 分散剂
deflocculate 反絮凝剂;散凝剂
deflocculated castable 散凝浇筑料
deflocculated colloid 不凝聚胶体
deflocculated graphite 胶态石墨;悬浮石墨
deflocculated sludge 抗絮凝污泥;反絮凝污泥
deflocculating 去凝作用;散凝作用
deflocculating agent 抗絮凝剂;胶体稳定剂;黏土悬浮剂;分散剂;反絮(凝)剂;反凝聚剂;散凝剂
deflocculation 抗絮凝(作用);解絮凝;絮凝分散现象;反团聚作用;反凝絮(作用)
deflocculator 抗絮凝器;抗絮凝离心机;悬浮剂;反絮(凝)离心机;反絮凝剂;反流态化
defluidization 流态化作用停滞;反流态化
defluorinated phosphate 脱氟磷肥
defluorinated stone 脱氟瓷石
defluorination 除氟作用;脱氟(作用)
de-flux 去焊(药)剂
defluxion 流下
de-foam 去泡沫
defoamant 去泡(沫)剂;消泡剂
defoamer (agent) 消沫剂;除泡剂;消泡剂;去沫剂;脱泡剂
defoaming 去泡沫;消泡作用;除沫
defoaming agent 抗泡剂;去(泡)沫剂;消泡剂;消沫剂
defoaming device 破沫设施
defoaming plate 除沫板
defoaming tank 除泡箱
defocus 去聚焦;散焦
defocused beam 散焦射束;散焦电子束
defocused code 散焦编码
defocused image 离焦像;散焦像
defocused impulse response 散焦脉冲响应
defocused lens 散焦透镜
defocused reflector 离焦反射镜
defocused speckle photography 散焦斑点照相术
defocused spot 散焦斑点
defocused system 散焦系统
defocusing amount 离焦量
defocusing coefficient 散焦系数
defocusing effect 离焦效应;散焦效应
defocusing technique 散焦技术
defocusing tolerance 离焦容限
defocussing 离焦
defogger 除水(汽)装置;驱雾器
defogging 去灰雾
defogging duct 防水汽导管
defoliant 落叶剂;枯叶剂;脱叶剂
defoliate plant 落叶植物
defoliation 去叶
defoliator 除叶器
deforcement 非法占有;非法侵占他人财产
deforest 毁林
deforestation 滥伐(森林);毁林;采伐森林;伐去森林;伐除森林;清除森林;森林砍伐
deforestation of land reclamation 毁林开荒
deform 变形
deformability 可变形性;加工性;变形性;变形(能)力
deformability due to bending 挠曲导致的变形性
deformability meter 应变仪;变形计;变形测定器

deformability modulus 变形模量
deformability of rock 岩石的可变性
deformable 可变形的;易变形的
deformable band 变形带
deformable bar 柔性杆
deformable body 可变形体;变形体;柔体
deformable body mechanics 变形体力学
deformable cross arm 变形横担
deformable damper 变形阻尼器
deformable event 变形事件
deformable mirror 可变形反射镜
deformable model 可变形的模型
deformable pore medium 变形多孔介质
deformable porous medium 变形多孔隙介质
deformable raft 变形浮箱
deformable rod 柔性杆
deformable zone 变形带
deformation 应变;残废;变形(纹);变位
deformation action 变形作用
deformational analysis 变形分析
deformational eustatism 地动性海面升降
deformational resilience 变形回弹
deformation(al) strain energy 变形应变能
deformation(al) stress 变形应力
deformation(al) structure 变形构造
deformational trap 变形圈闭
deformation amplitude 变形幅度
deformation analysis 变形分析
deformation and cataclasis ways 变形和碎裂作用方式
deformation angle 变形角
deformation area 变形区
deformation at failure 破坏时变形
deformation axis 变形力作用轴
deformation band 变形(条)带
deformation behavio(u)r 变形特性;变形状态;变形特征
deformation boundary condition 变形边界条件;边界变形条件
deformation character of rock 岩石的变形特征
deformation coefficient 畸变系数;变形系数
deformation coefficient of rock mass 岩体变形系数
deformation compatibility condition 变形协调条件
deformation computation 变形计算
deformation condition 变形条件
deformation control 变形控制
deformation coordinates 变形坐标
deformation correction 变形校正
deformation crack 变形裂缝
deformation cross method 变形交会测量法
deformation curtain 变形幕
deformation curve 变形曲线
deformation cycle 变形旋回
deformation defect 变形缺陷
deformation development 变形显影
deformation direction 变形方向
deformation distribution function 变形分布函数
deformation distribution method 变形分配法
deformation domain 变形域
deformation drag 变形阻力;变形阻抗;变形(拖)曳力
deformation due to bending 弯曲变形
deformation due to relief 投影差
deformation due to seepage 渗透变形
deformation due to shear 剪切变形;剪力应变
deformation due to welding 焊接变形
deformation during burning 焚烧过程中的变形
deformation dynamic inspection 应变动测
deformation effect 变形影响;变形效果
deformation ellipsoid 应变椭球(体);变形椭球体
deformation energy 变形能(量)
deformation equation 变形方程(式)
deformation eutectic 变形低共熔物
deformation fabric 变形组构【地】
deformation failure 变形失效
deformation form 变形方式
deformation fracture 变形破裂
deformation-free zone 固定区(域)
deformation front 变形前锋
deformation ga(u)ge 变量量具;变形计
deformation gradient 变形梯度
deformation gravel 变形砾石
deformation image 变形像

deformation imaging process 变形像记录法
deformation incised meander 变形深切曲流
deformation increment 变形增量;变形递增
deformation influential factor of rock 岩石变形的影响因素
deformation intensity 变形强度
deformation joint 变形缝
deformation lake 变形湖
deformation lamella 变形纹层;变形壳层
deformation limit 变形极限
deformation limit state 变形极限状态
deformation locus 变形轨迹
deformation map 变形图
deformation matrix 变形矩阵
deformation measurement 变形测定
deformation mechanics 变形体力学
deformation mechanism 变形机理
deformation mechanism map 变形机理图
deformation meter 变形测定计
deformation method 变形法
deformation modulus 变形模数;变形模量
deformation modulus of rock 岩石的变形模量
deformation monitoring network 变形监测网
deformation mountain 变形山
deformation observation 变形观测
deformation of bearing 方向变形
deformation of concrete 混凝土变形
deformation of image 像变形;图像畸形;图像变形
deformation of mirror cylinder 镜筒变形
deformation of riverbed 河床演变;河床变形;河床变迁
deformation of river channel 河槽变形
deformation of rock 岩石变形;岩层变形
deformation of surrounding rock 围岩内部变形
deformation of terrace 阶地变形
deformation of transformer winding 变压器绕组变形
deformation of wave 波浪变形;波的变形
deformation ooid 变形鲕
deformation orientation 变形取向
deformation parameter test 变形参数试验
deformation path 变形路径;变形历程;变形迹线
deformation pattern 变形图(像);变形模式
deformation plan 变形图
deformation plance 变形面
deformation plane 应变面
deformation point 变形软化温度;变形点
deformation polarizability 变形极化率
deformation polarization 变形极化率
deformation polyethylene liner 变形聚乙烯衬管
deformation potential 变形势;变形电位
deformation pressure 变形压力;单位变形力
deformation pressure of surroundings 变形围岩压力
deformation process 变形过程
deformation property 变形性质;变形能力
deformation range 变形范围
deformation rate 变形(速)率
deformation ratio 变形比
deformation recovery 变形恢复;变形回复
deformation resilience 变形回弹
deformation resistance 抗变形性能;变形阻力;变形抗力
deformation resistance curve 变形阻力曲线
deformation resolution 变形分辨率
deformation retraction 变形收缩
deformation seismograph 变形地震仪
deformation sequence 变形序列;变形顺序
deformation set 变定
deformation solid dynamics 变形固体动力学
deformation spectrum 变形谱
deformation stage of rock 岩石变形阶段
deformation standard 变形稳定标准
deformation state 变形状态
deformation structure 变形构造;变形结构
deformation style 变形形式
deformation survey 变形测量
deformation temperature 变形温度
deformation temperature of ash T_1 灰变形温度 T_1
deformation tester 变形测试器
deformation test(ing) 变形试验
deformation test of rock 岩石变形试验
deformation test of rock mass 岩体变形试验方法
deformation texture 变形结构

deformation theory 变形理论
deformation thermograph 变形温度计
deformation thermometer 变形温度计;变形温度表
deformation tie bar 变形拉杆
deformation-time curve 位移—时间曲线
deformation transducer 变形传感器
deformation twin 变形孪晶
deformation type 变形种类
deformation under load 荷重变形;荷载变形
deformation under load test 加载变形试验;荷载变形试验
deformation value 变形量
deformation velocity 变形速度
deformation verification of foundation 地基变形验算
deformation vibration 变形振动;变角振动
deformation wave 变形波
deformation work 应变功;变形功
deformation zone 变形区
deformed 变形的;异形的
deformed area 变形区
deformed axle 废轴
deformed bar 竹节钢筋;异形棒钢;变形钢筋;螺纹钢筋;异形钢筋;规律变形钢筋
deformed bars 各式钢筋
deformed bedding 变形层理
deformed caisson 异形沉箱
deformed channel steel 异形槽钢
deformed concrete caisson 异形混凝土沉箱
deformed cross-bedding 变形交错层(理)
deformed development 畸形发展;变形显影
deformed flat steel 异形扁钢
deformed fossils 变形化石
deformed hexagon 异形六角形;异形六边形
deformed hot rolled high yield bar 热轧变形高强钢筋
deformed hot rolled mild steel bar 热轧变形低碳钢筋
deformed ice 畸形堆冰
deformed kiln shell 筒体变形
deformed metal plate 轧边金属板;异形金属板;变形金属板材
deformed nucleus 变形核
deformed oolites 变形鲕粒
deformed pebble 变形砾石
deformed pipe 异形管;变形管
deformed plate 混凝土嵌缝板;凹凸板;变形板
deformed prestressed concrete steel wire 预应力混凝土异形钢丝;预应力混凝土结构用刻痕钢丝
deformed prestressing steel wire 异形预应力钢丝
deformed reduction spots 变形退色斑
deformed reinforcement 螺纹钢筋;变形钢筋;竹节钢筋
deformed reinforcing bar 异形钢筋;竹节钢筋;变形钢筋
deformed reinforcing steel 变形钢筋
deformed rolling 周期断面轧制
deformed round steel bar 圆竹节钢筋
deformed sheet pile 异形钢板桩
deformed steel 变形钢材
deformed steel bar 异形钢筋;竹节钢筋;变形钢筋;异形钢筋;周期断面钢材;螺纹钢
deformed steel reinforcement 变形钢筋
deformed tie bar 变形系筋;变形系杆;变形(钢)筋)拉杆
deformed Tor-steel 螺纹托尔钢;变形托尔钢
deformed tube 异形管
deformed tunnel 病害隧道
deformed typeface 变形字体
deformed ware 变形的器皿
deformed wire 异形(预应力)钢丝;变形钢丝
deformer agent 除沫剂
deformeter 应变仪;应变计;变形计;变形(度)仪;变形测定仪;变形测定计
deforming action 变形作用
deforming agent 除沫剂;变形剂
deforming alloy 变形合金
deforming effect 变形影响;变形效果
deforming force 变形力
deforming groove 周期断面轧槽
deforming lens 变形透镜
deforming rate 变形速率
deforming wedge 变形楔
deformity 变形

deformograph 变形图
deformographic storage display tube 变形成像存储显示管
defraud the revenue 漏税
defray(ment) 支付;付给;付出
defray the expenses 支给费用
defreeze 解冻
defrost 解冻;除霜
defroster 融冰器;去霜器;除雾器;除霜器;出霜器;防冻设备;防冻器
defroster blower 除霜鼓风机
defroster for vehicle 车辆除霜器
defroster (of vehicle window) 车窗玻璃除霜器
defrost header 除霜集管
defrosting 接触线除霜;解冻;除霜
defrosting compound 防霜冻剂
defrosting cycle 除霜循环
defrosting device 隔霜装置
defrosting measures 隔霜措施
defrosting pan 接水盘
defrosting pipe 除霜管
defrosting tray 接水盘
defrosting unit 隔霜装置
defrost water 除霜;防冻;去冰用水
defrother 消泡器;消泡剂;除泡器
de-frothing agent 去泡剂
defruit 异步回波滤除
defruiter equipment 反干扰设备
defruiting 异步回波滤除
deft pottery 荷兰白釉蓝彩陶器
defueling 取出燃料(指放射性燃料);泄出存油;二次加油;二次充气
defueling equipment 放油设备
defueling pump 抽油泵
defueling water cleanup system 二次注水净化系统
defunct company 已停业公司;已不存在的公司;停业公司
defurring 除垢
defying departure order 抗拒出港指令
degame wood 棕黄色硬木(一种产于西印度群岛的木材)
degas 脱气;去气
degaser 脱气剂
degasification 除气作用;脱气(作用);去气作用;去气法
degasified steel 镇静钢;除气钢;脱气钢
degasifier 除气器;脱气剂
degasifying agent 除气剂;脱气剂
degasity 脱气
degassed crude oil 脱气原油
degasser 去气器;吸气剂;除气器;除气剂;脱气装置;脱气塔;脱气器
degassing 除(毒)气;脱气;去气
degassing agent 消泡剂;脱气剂
degassing anneal 除气退火
degassing apparatus 除气设备
degassing flux 除气(熔)剂
degassing gun 去气枪
degassing installation 脱气装置
degassing mo(u)ld 排气塑模
degassing of eluent 洗脱液除气
degassing sprinkler 脱气式洒水车;脱气式喷洒器
degassing system 除气系统;抽空装置;脱气装置
degassing tank 脱气罐
degassing tower 脱气塔
degassing unit 除气装置
de-gauss 去磁
degausser 去磁器;去磁机;去磁扼流圈;去磁电路;退磁扼流圈
degaussing 去磁;消磁;退磁
degaussing apparatus 退磁设备
degaussing belt 消磁带
degaussing cable 去磁电缆;消磁电缆
degaussing coil 去磁线圈;消磁线圈;消磁绕组
degaussing component 消磁元件
degaussing facility 消磁设施
degaussing field 去磁场
degaussing range 消磁站;消磁观测场
degaussing ship 消磁船
degaussing vessel 消磁船;防磁性水雷的消磁船
degelatinizing 脱胶
degelling agent 脱胶剂
degenarative process 衰退过程

degeneracy 简并性;简并度;衰退
degeneracy collapse 简并坍缩
degeneracy concentration 简并浓度
degeneracy factor 简并因子
degeneracy in linear programming 线性规划的退化
degeneracy of energy level 能级简并
degeneracy operator 退化算子
degeneracy pressure 简并压力
degeneracy semi-conductor 简并半导体
degeneracy vibration 简并振动
degenerate 简并
degenerate amplifier 简并放大器
degenerate code 简并密码;简并编码
degenerate configuration 简并组态;退化位形
degenerate conic 可约二次曲线;退化二次曲线
degenerate continuum 简并连续流;简并闭联集
degenerate critical point 退化临界点
degenerated 退化的;变质的;变性的
degenerated bankruptcy 衰退倒闭
degenerated curve 退化曲线
degenerated form 退化类型
degenerate distribution 凝聚分布;单点分布;退化分布
degenerate distribution function 退化分布函数
degenerated organic wastes 变质的有机废料
degenerate doublet 简并模对
degenerated soil 退化土(壤)
degenerated type 消退型;退化型
degenerate energy level 简并能级
degenerate feasible solution 退化可行解
degenerate flow 退落水流
degenerate function 退化函数
degenerate gas 简并气体
degenerate junction 简并结
degenerate linear system 退化线性系统
degenerate matrix 退化矩阵
degenerate matter 简并物质
degenerate mode 简并模
degenerate normal distribution 退化正态分布
degenerate parametric amplifier 简并参量放大器
degenerate plasma 简并等离子体
degenerate process 退化过程
degenerate quadratic form 退化二次型
degenerate quadric 退化二次曲面
degenerate regression 退化回归
degenerate rotary tidal waves system 退化旋转潮波系统
degenerate segment 简并线段
degenerate semiconductor 简并半导体;衰减半导体
degenerate series 退化序列
degenerate specimen 简并样品
degenerate state 简并态;退化(状)态
degenerate tissue 退化组织
degenerate waveguide modes 简并波导模
degeneration 简并(化);负回授;腐化;变质;变性;退化(作用)
degeneration factor 负反馈系数;退化系数
degeneration mode 退化振荡模
degenerative 变质的
degenerative amplifier 负反馈放大器
degenerative atrophy 变性萎缩
degenerative character 退化性状;退化特征
degenerative feedback 负回授;负反馈
degenerative index 变性指数
degenerative recrystallization 衰退再结晶(作用)
degenerative stabilizer 负反馈稳定器
degenerative type frozen soil 退化型冻土
degerming 除菌
degerming agent 除菌剂
degeroite 硅铁石
deghosting 反虚反射
deglabration 变秃
Deglacial 高峰后冰期(北美);冰消期
deglacial period 冰消期
deglaciation 冰消(作用);冰河消退而露出的区域;冰川消退;冰川减退(作用)
de-glassing 去玻璃处理
deglitcher 限变器
deglossing liquid 消光液
deglued bone 脱胶骨
deglycerizing 除去甘油

de gourdi 低温素烧法(法国)
degradability 可降解性
degradable carbon 可降解碳;降解碳
degradable organic carbon 可降解有机碳
degradable organic material 可降解有机物质
degradable pollutant 可降解污染物;可降解分解物;可分解污染物
degradable substance 可降解物质
degradable substance concentration 可降解物质浓度
degradated failure 渐衰失效;渐衰故障
degradated spheroidization 球化衰退
degradation 陵削(作用);裂解;降解;降级;降格;减退;减低;河底刷深;河床消蚀;递降;递减作用;冲刷;剥蚀;退化;衰化;衰变;刷深
degradation ability 降解能力
degradation additive 降解添加剂
degradation along downstream reach of reservoir 水库下游沿程冲刷
degradation bacteria 降解菌
degradation below dam 坝下河底刷深;坝下冲刷
degradation by light 光致降解;光降解(作用)
degradation capability 降解能力
degradation catalyst 降解催化剂
degradation characteristic 降解特性;退化特征;退化特性
degradation coefficient 降解系数
degradation degree 冲刷程度
degradation factor 降级因数;递降因子;递降因数;递降系数
degradation failure 劣化故障;老化故障;渐衰失效;渐衰故障;缓慢失效;逐步失效;恶劣故障;退化(型)失效;退化事故;退化故障
degradation index 冲刷指标;衰化指数;衰变指数
degradation in size 粒度减少;磨细(作用);粉碎
degradation kinetics 降解动力学
degradation level 老化平面;剥蚀层;破裂平面
degradation model 劣化模型
degradation of contaminant 污染物降解(作用)
degradation of contrast 反差衰减
degradation of drinking water quality 饮用水水质恶化
degradation of energy 能的退降;能的递降
degradation of food chain 食物链的降解
degradation of groundwater quality 地下水水质恶化
degradation of level 水位下降
degradation of low-water 低水位消落
degradation of organic matter 有机质的降解作用;有机物降解(作用)
degradation of pesticide 农药降解
degradation of pollutant(substance) 污染物(质)降解(作用)
degradation of river 河流刷深
degradation of soil 土壤(退化)作用
degradation of stream 河流刷深
degradation of stream channel 河槽刷深;河槽冲刷
degradation of water quality 水质恶化
degradation parameter 衰化参数;衰变参数
degradation particle 粉碎颗粒
degradation pathway 降解途径
degradation peak 递减峰
degradation process 降解过程
degradation product 降解产物;递降分解产物
degradation rate 降解速率
degradation rate constant 降解速率常数
degradation reaction 降解反应
degradation recrystallization 衰退再结晶(作用)
degradation settlement 降解沉降
degradation terrace 剥蚀阶地
degradation testing 老化试验
degradation time 降解时间
degradation with ag(e)ing 老化降解
degradation zone of organic mater 有机质降解作用带
degrade 降级;降等;渐浅色调;递降
degraded 递降
degraded alkali soil 变质碱土
degraded beam 减速束;慢化束
degraded black-earth soil 变质的黑(钙)土壤
degraded chernozem 退化黑钙土
degraded copy 降级原稿
degraded illite 退化伊利石
degraded image 降质图像

degraded operation 降级运营
degraded products 次品
degraded pulse 递降脉冲
degraded radiation 递降辐射
degraded recovery 降级恢复
degraded soil 退化土(壤)
degraded water 变质水
degrader 降能器
degrade sewage 降解污水
degrading absorber 降低能量用吸收体
degrading bacteria 降解菌
degrading channel 冲刷河槽;刷深河槽
degrading characteristic 降解特性
degrading hydrocarbons 退化碳氢化合物
degrading impulse response 递降脉冲响应
degrading recrystallization 退变重结晶作用
degrading river 下切河(流);刷深河流
degrading soil 退化土(壤)
degrading stream 消蚀河溪;下切河(流);刷深河流;刷深河川
degrading valley 夷低谷
degradofusinite 氧化丝质体
degradomicrinite 氧化微粒体
de-grains 去除粒面
de-granitization way 去花岗岩化方式
degras 油鞣余物;氧化鱼油;羊毛脂;天然羊毛脂
degras acid 废脂酸
degras stearin 废脂硬脂酸
degrease 除油脂;除油污;去油脂;去油污
degreaser 除油剂;脱脂剂;去油器;去油剂
degreaser for alumin(i)um powder 铝粉脱脂剂
degreasing 脱脂;除油;去油脂;去油(污)
degreasing agent 除油剂;脱脂剂
degreasing effect 除油脂效果
degreasing installation 除油装置
degreasing of vapo(u)r 蒸气的脱脂
degreasing plant 脱脂设备;脱脂车间
degreasing solution 去油溶液;脱脂溶液
degreasing solvent 脱脂溶剂;脱垢溶剂
degree 度(数);程度
degree Celsius 摄氏度
degree cence 粉化(程)度
degree Centigrade 摄氏度
degree Clarke 克拉克度
degree-day 度·日;逐日温度
degree-day factor 度日因数
degree-day method 度日法;融雪系数法;温度天数法
degree-day value 度日值;温度日值
degree Engler 恩格勒度
degree estimate 根据同类型已建工程费用估算费用
degree exposure (混凝土的)曝露程度
degree for rectification 纠正自由度
degree grade of hydrogeologic(al) condition 水文地质条件程度等级
degree hour 温度时数
degree invariant transformation 次数不变换
degree Kelvin 开氏温标;开氏度;开尔文(温标);绝对温度
degree of abrasion 磨耗程度;剥蚀度
degree of absorption 吸附度
degree of accuracy 精确度;精(密)度;准确度
degree of accuracy of datum 数据可靠程度
degree of acidity 酸度
degree of activity 活(性)度
degree of adaptability 配合度
degree of administered-trade 贸易控制程度
degree of admission 进气度
degree of adoption 采用程度
degree of aeration 含气度;掺气度;曝气度;透气度
degree of ag(e)ing 老化程度
degree of aggregation 聚集度;聚合程度;集结度;团聚度
degree of aggressiveness 侵蚀度
degree of alkalinity 碱度
degree of an angle 角的度数
degree of angle 角度
degree of approximation 近似(程)度;逼近度
degree of arc 弧度
degree of a relation 关系的度
degree of association 缔合度
degree of asymmetry 不对称度
degree of automation 自动化程度
degree of balance 平衡度

degree of base saturation 盐基饱和度
degree of Baume 波美度
degree of bed development 河床发展程度;河床发育程度
degree of belief 可信度
degree of berth occupancy 泊位利用率
degree of blackness 黑度
degree of blame 过失程度
degree of boost 增压程度
degree of branching 支化度
degree of breaking 轧碎程度;破坏程度
degree of brightness 光亮程度
degree of bringing all farmland under irrigation 农田水利化程度
degree of brittleness 脆性程度
degree of burning 燃烧度
degree of calcinations 分解率
degree of canalization 渠化程度
degree of carbonation 碳化程度
degree of cavitation 空蚀度;空化程度;汽蚀(程)度;涡空度
degree of cementation 胶结度
degree of cementation of discontinuity 结构面胶结程度
degree of centigrade 百分度(数);摄氏度数
degree of chemisorption 化学吸附度
degree of choking-up 堵塞程度
degree of clarification 净化(程)度;澄清程度
degree of clay brick burning 砖的烧透程度
degree of cleaning 净化程度;清洗程度
degree of clogging 堵塞程度
degree of closeness 接近程度
degree of closenness 疏密度
degree of cloudiness 云度
degree of coalification 煤化程度
degree of coal oxidation 煤氧化程度
degree of coherence 相干度
degree of cohesiveness 黏聚度
degree of cold work 冷加工(材料变形)程度
degree of column utilization 柱利用度
degree of compaction 压缩比;紧密度;密实度;夯实度;压实度;致密度
degree of compatibility 混容度;相容度
degree of completion method 完工进度计算法
degree of compressibility 可压(缩程)度
degree of compression 压缩度;压实度
degree of concentration of control 控制集中度;集中控制度
degree of concrete maturity 混凝土成熟度
degree of confidence 信度;置信度
degree of confinement 填封度(装炸药)
degree of confirmation 可确定程度
degree of consistency 均匀度;稠度
degree of consolidation 固结率;固结(程)度
degree of controllability 操纵度
degree of convergence 收敛(程)度
degree of convergency 收敛程度
degree of convexity 凸度
degree of cooperation 合作程度
degree of core completion 岩芯完整程度
degree of correlation 相关程度
degree of coupling 耦合度
degree of cover 覆盖度
degree of coverage 覆盖率级别
degree of creaming 膏化度
degree of crosslinking 交联度
degree of crushing 破碎程度
degree of crystallinity 结晶度
degree of crystallization 结晶度
degree of cure 固化度;熟化程序
degree of current rectification 整流度
degree of curvature 曲率;曲度;曲面(程)度
degree of curve 曲线方程次数;曲线度数;曲度;弯曲度
degree of damage 破坏程度;损失程度;损坏范围;损害程度
degree of damping 阻尼(程)度
degree of dampness 潮湿程度
degree of decentralization 分散程度
degree of deflection 偏转度
degree of degeneracy 退化次数
degree of degeneration 退化系数
degree of demonstration 证实程度
degree of density 浓度;密实度;密度率;稠密度;

不透明度;疏密度
degree of depth 深度
degree of determination 可决程度
degree of deviation 偏离度;偏差度
degree of diaphaneity 透明度级别
degree of differentiation 分化度
degree of difficulty 难度
degree of dilution 稀释度
degree of dip 倾斜度
degree of disequilibrium 不平衡度
degree of disintegration (岩土的)分解度
degree of disorder 无序度
degree of dispersion 弥散度;分散度;色散度
degree of dissociation 分离度;离解度;解析度;解离度
degree of distortion 畸变度;变形程度;奇变度;失真度
degree of disturbance 扰动度
degree of dominance 显性度
degree of drainage 排水率
degree of drying 干燥程度
degree of ductility 延性度
degree of dullness 暗淡程度
degree of durability 耐久度
degree of dustiness 含尘度
degree of dyeing 着色程度;上色率
degree of eccentricity 偏心度
degree of efflorescence 起霜(程)度
degree of elevation 仰角
degree of environment(al) hazard 环境危害程度
degree of equation 方程式幂次;方程次数
degree of esterification 酯化度
degree of etherification 醚化度
degree of exactitude 精确度
degree of excitation 激发度
degree of exhaustion 抽气度;抽空度
degree of expansion 膨胀度
degree of exploration 勘探程度
degree of exploration is not enough 地质勘探程度不够
degree of explosion safety 爆破安全等级
degree of extraction 回收率;回采率
degree of Fahrenheit 华氏温度;华氏度数
degree of fastness 坚牢程度
degree of fill 填充度
degree of filling 充填程度
degree of filtration 过滤度
degree of financial leverage 财务杠杆作用盈利度
degree of fineness 细度
degree of finish 精加工度;光洁度;肥育(程)度
degree of fireproof 耐火度
degree of firmness 坚密度
degree of fixation 固定程度
degree of flatting 消光度
degree of flexibility 挠度;柔韧度
degree of flocculation 絮凝度
degree of fluctuation 起伏程度
degree of fractionation 分馏精确度
degree of fragmentation 破碎度
degree of freedom 自由度;维
degree of freedom of motion 动作自由度
degree of freeze 冰冻度
degree of fullness 充满度
degree of functionality 官能度
degree of geologic(al) complexity 地质复杂程度
degree of geologic(al) study 地质研究程度
degree of geologic(al) work 地质工作程度
degree of gloaasiness 光亮程度
degree of gradation 级配度
degree of graphitization 石墨化程度
degree of grinding 研磨程度;轧浆细度
degree of hardness 硬化程度;加固程度;硬度
degree of heat 热度
degree of homogeneity 均匀度;齐次度
degree of humidity 湿度
degree of hydration 水化程度
degree of hydrogen-ion capacity 黏土矿物亲水的程度
degree of hydrolysis 水解率;水解度
degree of importance 重要程度
degree of impregnation 浸渍度;浸染程度;注入程度
degree of incidence 关联度
degree of inclination 倾斜度;偏转度;偏斜度

degree of indeterminacy 超静定次数;不定度
degree of indeterminates 超静定次数
degree of induration 硬化(程)度;固结度
degree of industrial accumulation 工业积累程度
degree of inflation 充气度
degree of instability 不稳定度
degree of integration 综合程度
degree of integrity 完整性;完整程度;完善程度
degree of investigation of the construction site 勘察区研究的程度
degree of ion(ic) exchange 离子交换程度
degree of ionization 电离度
degree of irregularity 不匀度;不平整度;不均匀度;不规则程度
degree of isolation 分离程度
degree of isomerization 异构化度
degree of iteration 迭代次数
degree of karstification 岩溶率
degree of kiln filling 窑填充率
degree of latitude 纬度
degree of leakage 漏率
degree of level 水平度
degree of leverage 杠杆程度
degree of liberation 解离程度
degree of light 光的强弱
degree of lignification 木质化程度
degree of lime saturation 石灰饱和度
degree of liquefaction 液化度
degree of longitude 经度
degree of long range order 长程有序度
degree of looseness 松散度
degree of management sophistication 管理熟练程度
degree of mapping 映射度
degree of mattness 粗糙程度
degree of maturation (水泥的)成熟度
degree of mechanization 机械化程度
degree of mechanization of cargo handling 装卸作业机械化程度
degree of membership 隶属度
degree of metameris 同色异谱程度
degree of metamerism 位变异构度
degree of metamorphism 变质程度
degree of mineralization 矿化度
degree of mineralization of groundwater 地下水矿化度
degree of mixing 混合均匀度
degree of mobilization 流动度
degree of modulation 调制度
degree of moisture 湿度
degree of multiprogramming 多道程序数;多道程序设计的级;多道程序设计的道数
degree of nitration 硝代程度;硝化程度
degree of non-flammability 抗火焰度
degree of opalescence 乳光度
degree of order 有序度
degree of orientation 定向程度;取向度
degree of overlapping 重叠度
degree of oxidation 氧化(程)度
degree of packing 聚集程度;紧实度;装药程度;填实度
degree of pass through 通行程度等级
degree of penetration 渗透度
degree of pick 拾墨度
degree of pigmentation 颜料加入量;着色度
degree of pitch 倾斜度
degree of planeness 平整度;平面度
degree of plane separation 平面分离度
degree of plasticity 塑性度
degree of plasticization 增塑度
degree of plugging 堵塞程度
degree of polarity 极性度
degree of polarization 偏振度
degree of polishing 精碾程度
degree of pollution 污染(程)度
degree of pollution wave transformation 污染波衰减度
degree of polygonality 多边度
degree of polymerization 聚合度
degree of polynomial 多项式(的)次数
degree of pore filling 孔隙充填程度
degree of porosity 孔隙度;多孔度
degree of precision 精细程度;精确度;精密度
degree of predictability 可预报程度

degree of prestress 预应力度
degree of probability 概率度
degree of prospecting 勘探程度
degree of protection 保护等级
degree of pulverization 粉碎(程)度
degree of purification 净化(程)度
degree of purity 纯度
degree of putrescibility 可腐化性
degree of pyritization 黄铁矿矿化度
degree of quality 质量好坏程度
degree of ramification 分歧度
degree of ramming 紧实度
degree of reaction 反应度
degree of Reaumur 列氏(温)度
degree of reduction 还原程度
degree of redundancy 剩余度;静不定度;赘余度
degree of reflection 反射度
degree of regulation 控制程度;调整程度;(径流的)调节度
degree of reiability 可靠性度量
degree of relation 关系度
degree of reliability 可靠程度
degree of remo(u)lding 重塑度;扰动度
degree of reproducibility 再现度
degree of reservation 保留程度
degree of residual occupation 编组站台剩余程度
degree of resistance 耐久度;耐力程度
degree of restraint 拘束程度;限制程度;约束度;制约程度
degree of reverberation 混响度
degree of reversibility 可逆程度
degree of risk 风险程度;危险度
degree of rock decay 岩石风化程度
degree of roughness 粗糙(程)度
degree of rounding of grains 颗粒圆度
degree of roundness 圆度
degree of rusting 锈蚀程度;生锈程度
degree of safety 安全系数;安全(程)度;强度储备
degree of safety against rupture 防裂安全度
degree of salinity 盐渍化程度
degree of saturation 浸透程度;饱和度
degree of saturation at intersection 交叉口饱和度
degree of saturation grade of soil 土的饱和度分级
degree of saturation of ice 饱冰度
degree of saturation of soil 土的饱和度
degree of scatter 分散程度
degree of security 安全度
degree of seismic intensity 地震烈度级
degree of seismicity 震度;地震活动(程)度
degree of self-sufficiency 自给率
degree of sensitivity 灵敏度(比)
degree of separation 分离度
degree of settling of traffic paint 路面漆标稳定度
degree of short-range order 短程有序度
degree of shrinkage 收缩率;收缩度
degree of significance 重要程度
degree of sintering 烧结程度
degree of sinuosity 蜿蜒度
degree of size breakage 破碎比
degree of size reduction 粒度减少程度;表面积破碎比
degree of skewness 偏斜度
degree of slenderness 细长度
degree of slope 坡度;倾斜度
degree of smoothness 光洁度
degree of soil fertility 土壤肥力(等级)
degree of sorting 分选(程)度
degree of spatial correlation 空间相关程度
degree of sphericity 球形度
degree of sphericity of particle 颗粒球形度
degree of stability 稳定(程)度
degree of staining 着色度;污染度
degree of start-stop distortion 启动停止失真度
degree of statical indeterminacy 超静定次数
degree of stretch 拉伸度
degree of subdivision 细分散程度
degree of submergence 淹没度
degree of submersion 淹没度
degree of substitution 代换的次数;取代度
degree of supercooling 过冷(程)度
degree of superheat 过热(程)度
degree of supersaturation 过饱和度
degree of swelling 溶胀度;膨胀量;膨胀度;泡胀度
degree of temper 回火度;钢化度

degree of the complexity of product 产品复杂程度
degree of the stability price 价格稳定程度
degree of tightness 紧密度;密封度
degree of tilt 倾斜度
degree of torsion 扭转度
degree of traffic jam 交通阻塞度
degree of transparency 透明度
degree of treatment 处理程度
degree of trend surface 趋势面次数
degree of turbidity 混浊度;浑浊度;透过率
degree of turbulence 紊流程度;紊动度;湍流度
degree of turn 回转度
degree of Twaddell 特沃德尔度(比重单位)
degree of twisting 扭曲度
degree of unbalance 不平衡度
degree of unbalancedness 不平衡度
degree of uncertainty 不确定性程度
degree of undercooling 过冷度
degree of uniformity 均匀(程)度;一致程度
degree of unsaturation 不饱和度
degree of unsolvability 不可解度
degree of urbanization 城市化指数;城市化水平
degree of utility 效用等级
degree of vacuum 真空度
degree of variation 变异(程)度;变化程度
degree of vegetation 植被度;森林覆盖程度
degree of viscosity 黏度
degree of visual 视度
degree of water adsorption 吸水率
degree of water-resistance 防水程度
degree of wear 磨损程度;磨蚀度
degree of weathering 风化(强)度;风化程度
degree of wetness 潮湿程度
degree of wetting 润湿度
degree of yellowness 泛黄度
degree parallel 斜平行垫铁
degree scale 刻度;调节标度
degrees of frost 霜度
degree variance of gravity-anomaly 重力异常阶方差
degritting 除砂
de Gua's rule 德古阿法则
de-gumming 去胶
dehalogenation reaction 脱卤反应
de Hass-Schubnikov oscillation 德哈斯—舒尼科夫振动
de Hass-Van Alphen effect 德哈斯—范阿耳芬效应
de Hass-Van Alphen oscillation 德哈斯—范阿耳芬振动
dehexanizer 脱己烷塔
dehiscence 裂开
dehiscent by lid 盖裂的
Dehne filter 德恩型过滤机;达恩过滤机
dehorn 换刺
Dehottay freezing method 狄霍太冻结凿井法
Dehottay process 狄霍太打竖井法
dehrnite 碱磷灰石
dehumidification 脱水;减湿(作用);除湿;除潮;湿度降低;去湿
dehumidification of gas 天然气除湿
dehumidification plant 除湿设备
dehumidification system for cargo hold 货舱除湿装置
dehumidified 减湿的
dehumidifier 吸湿器;降湿器;减湿器;干燥器;除湿机;减湿器;去湿装置;去湿剂
dehumidifier coil 减湿盘管;干燥盘管
dehumidify 减湿;去湿
dehumidifying 减湿的;除湿
dehumidifying agent 减湿剂
dehumidifying capacity 减湿能力
dehumidifying cooling 减湿冷却
dehumidifying effect 减湿作用
dehumidifying heater 除潮加热器
de-humidifying plant 除湿设备
dehumidifying system 减湿系统
dehumidity unit 除湿机
dehumidizer 除湿器;去湿剂
dehwxanizing column 脱己烷塔
dehydrant 脱水剂
dehydrate 除水;烘干机;脱水(物)
dehydrate consolidation settlement 脱水固结沉降
dehydrated 脱水的

dehydrated alcohol 绝对酒精;无水乙醇;脱水酒精
dehydrated castor oil 脱水蓖麻油
dehydrated castor oil alkyd 脱水蓖麻油醇酸树脂
dehydrated castor oil fatty 脱水蓖麻油脂肪酸
dehydrated crude (oil) 脱水原油
dehydrated lime 氧化钙;生石灰
dehydrated oil 脱水油
dehydrated preservation 脱水保藏法
dehydrated tar 去水煤沥青;脱水焦油(沥青)
dehydrated water-gas tar 脱水水煤气焦油
dehydrater 干燥器;脱水机
dehydrating 脱水
dehydrating agent 干燥剂;脱水剂
dehydrating press 脱水压(力)机
dehydrating temperature 脱水温度
dehydrating tower 脱水塔
dehydration 除水;脱水作用;失水;去水作用
dehydration box 脱水箱
dehydration fever 脱水热
dehydration figure 脱水像
dehydration of air 空气脱水
dehydration of sludge 污泥脱水
dehydration peak 脱水峰
dehydration property 脱水性能
dehydration rate 脱水率
dehydration reaction 脱水反应
dehydration stage 脱水阶段
dehydration tank 脱水槽
dehydration temperaturre 脱水温度
dehydration test 脱水试验
dehydration zone 干燥带
dehydrator 干燥剂;烘干机;干燥器;除水器;除湿机;脱水器;脱水机
dehydrite 高氯酸镁
dehydroabietic acid 脱氢松香酸;脱氢枞酸
dehydroabietyl amine 脱氢松香胺;脱氢枞胺
dehydroacetic acid 脱氢乙酸
dehydrobenzene 苯炔;脱氢苯
dehydrocarbylation 脱烃基(作用)
dehydrochlorination 脱氯化氢(作用)
dehydrocyclization 环化脱氢;脱氢环化作用
dehydrofreezing 脱水冷冻(法)
dehydrogenated agent 脱氢剂
dehydrogenated oil 脱氢油
dehydrogenated rosin 脱氢松香
dehydrogenation 除氢;脱氢(作用)
dehydrogenation drying equipment 脱水干燥设备
dehydroisomerization 脱氢异构化作用
de-hydrolysis 去水作用
dehydrolyzing agent 脱水剂;脱水剂
dehydropregnenolone acetate 醋酸双烯醇酮
dehydroxylated face 脱羟基表面
dehydroxylated surface 无羟基表面
dehydroxylation 脱羟基
deicer 除水剂;除冰设备;除冰剂;防(水)结冰器;防冰加热器;防冻剂;防冰装置;防冰装置;碎冰装置
deicer chemical 抗冻化学品
deicer scaling 抗冻剂引起的剥落
deicing 除冰;防冻;碎冰装置;去冰
deicing agent 除冰剂;防冻剂;防冰剂;化冰剂
deicing and snow melting equipment 防冰与融雪设备
deicing apparatus 除冰装置
deicing chemicals 化冰药品
deicing equipment 防冰装置
deicing fluid 防冻液
deicing gear 防冰装置
deicing hand pump 防冰系统手摇泵
deicing liquid 除冰液
deicing salt 防冻盐;防冰;去冰盐
deicing salts 防冻盐类
deicing salt solution 防冻盐水;防冻盐溶液
deicing sealant 防冰填料料
deicing sluice 排冰闸(门);排冰道
deicing solvent 除冰溶剂
deicing system 去冰系统
deicing work 防冻工作
deihayelite 片硅碱钙石
deincrustant 除垢剂;除垢剂
deindustrialization 限制工业化;非工业化
de-inhibition 去除抑止
deinking 去墨剂;脱墨;脱黑
deinking process 脱黑过程
deinking waste(water) 去黑废水

deinsectization 灭昆虫
de-ion circuit breaker 消电离断路器
de-ion fuse 除离子熔断器;去电离熔丝
de-ionised water 去离子水
de-ionizater 去离子器
de-ionization 去离子(作用);消电离;消除电离作用;除盐;除离子作用;反电离作用;脱离子(作用)
de-ionization deexciter 消电离灭磁器
de-ionization potential 消电离电压
de-ionize 脱去离子;去电离
de-ionized water 去离子水;无电离子水;脱离子水
de-ionizer 去离子装置;去离子器;脱离子剂
de-ionizing 除去离子
de-ionizing agent 消电离剂
de-ionizing column 去离子柱
de-ionizing potential 去电离电位
de-iron 去铁
de-ironing 除铁
de-ironing separator 除铁器
Deister phase 德斯特阶段
deiterosomatic rock 再生岩
dejection cone 洪积锥;冲积锥;冲积锥
dejective loess 冲积黄土
de-jure population 常住人口
deka 十进的
deka-ampere balance (1~100 安培的)安培秤;十进安培秤
dekad 一旬;十天
dekagram(me) 十克
dekalbite 纯透辉石
dekaliter 十升
dekametre 十米
dekastere 十立方公尺
dekatron 十进计数管
dekatron calculator 十进管计算器
del 倒三角形
Delabole slate 德拉博尔石板
Delaborne prism 德拉博尔内棱镜
Delact's alloy 德拉格特合金
Delafield's hematoxylin stain 德拉菲尔德氏苏木精染剂
delafossite 赤钢铁矿;铜铁矿
delaine 毛棉布料
Delambre's analogy 德郎布尔相似式
delaminate 层离;脱层
delaminated clay 剥片瓷
delaminated joint 分层插接接缝
delaminated kaolin 剥片瓷;片状高岭土
delaminating 层脱
delamination 裂为薄层;离层;成层分离;层离;分层(剥离);起鳞;脱胶;脱层
delamination of plywood 胶合板开胶
delamination strength 脱胶强度
Delanium 人造石墨
delanium graphite 高纯度压缩石墨;人造石墨
de la Rue and Miller's law 德拉鲁和弥勒定律
de la Tour method 德拉图尔法
delatynite 德雷特琥珀
Delaunay element 德朗奈要素;德朗奈根数
Delaunay orbit element 德洛内轨道元素
Delaunay theory 德朗奈理论
De Laval centrifuge 蝶式离心机;德拉瓦尔离心机
De Laval centri-therm evapo(u)rator 德拉瓦尔离心加热式蒸发器
De Laval nozzle 德拉瓦尔喷管;缩放喷管
De Laval spool dryer 德拉瓦尔式筒管丝烘干机
De Laval turbine 德拉瓦尔汽轮机;单级冲动式涡轮机;单级冲动式透平
de-Lavaud process 离心铸管法
Delaware Aqueduct 德拉华水渠
Delawave effect 德雷伏效应
delay 延误;延缓;延迟;延爆段;滞后;迟延;拖延
delay action 延缓作用;延迟引起损害;延迟动作
delay action cap 定时雷管
delay-action circuit breaker 延时开关
delay-action detonator 缓爆雷管;缓发雷管;迟发雷管;定时(引爆)雷管;延发雷管
delay-action exploder 定时雷管
delay-action firing 延迟点火;迟发爆破
delay-action fuse 延时熔断器
delay-action relay 延迟动作继电器
delay-action stage 延效期;(炸药的)迟发期;迟爆期
delay-action switch 延迟动作开关
delay aeration 延时曝气

delay aeration system 延时曝气系统
delay allowance 延误外加时间；过程空裕；容许延迟；时延容许量
delay ambiguity function 延迟模糊函数
delay angle 延迟角；滞后角
delay base 迟缓接续制
delay bias 延迟偏差
delay blasting 延迟爆破；迟期爆破；缓爆；迟发爆炸；迟缓爆炸；延迟爆炸
delay-blasting cap 延爆雷管；延时雷管；缓发雷管；缓爆雷管
delay cable 延迟线；延迟电缆
delay cap 缓爆雷管；延发雷管；迟发雷管；延爆雷管
delay cell 延时元件；延迟脉冲
delay circuit 延时电路；延迟电路
delay coefficient 延迟系数
delay compensation 延迟补偿
delay component 滞后环节
delay connecting system 迟缓接续制
delay constant 时延常数
delay control 延迟控制；延迟调节
delay counter 延时计数器；延迟(线)计数器；时延计数器
delay crystal 延迟晶体
delay detonator 迟发雷管；延时雷管；延发雷管；延爆雷管；迟爆雷管
delay-differential equation 延误微分方程
delay disconnect signal 延迟切断信号
delay distortion 相延失真；延迟失真；延迟偏差；迟延畸变；时延畸变
delay distribution 延误分布
delay Doppler mapping 相延多普勒测图
delayed 迟发的
delayed action 延缓作用；延迟作用；滞后作用；延时器；延期作用；迟效作用
delayed action device 延迟装置
delayed-action electromechanical treadle 机电延迟动作脚踏板
delayed-action lifting device 缓动提升装置；平稳起落装置
delayed action time of discharge 喷射滞后时间
delayed alarm 延时报警；延迟信号装置；延迟警报
delayed alarm relay 延迟警报继电器
delayed alpha particle 缓发阿尔法粒子
delayed automatic gain control 延迟自动增益控制
delayed bit 延迟位
delayed blanking signal 延迟熄灭信号
delayed blast(ing) 延时爆炸；延迟爆破；延迟爆炸
delayed blasting cap 缓爆雷管
delayed bursting water 滞后突水
delayed carry 延迟进位
delayed-channel amplifier 延迟信号放大器；延迟通道放大器
delayed claim 迟索的赔款
delayed coagulant 迟发凝固
delayed coincidence 延迟符合；推迟符合
delayed coincidence method 延迟符合法
delayed coke 延迟焦炭；延迟焦化
delayed coking 延迟焦化过程
delayed combustion 滞后燃烧
delayed compression 次压缩
delayed consolidation 迟滞固结
delayed cracking 后期开裂；延迟裂纹
delayed crazing 后期龟裂
delayed critical 缓发中子临界
delayed deformation 延迟变形
delayed delivery 延期交货；延迟交割；延迟交货；延迟交付；延迟发送
delayed delivery cost 延期交货费用
delayed density dependence 密度依存
delayed development 延迟开发
delayed development well 扩边井
delayed device 延时装置
delayed dominance 缓发显性；迟发显性
delayed drop 平稳下降；缓慢下降
delayed effect 延迟效应
delayed elastic deformation 延迟弹性变形；滞弹性变形
delayed elastic effect 滞弹性效应
delayed elasticity 滞弹性；迟弹性；弹性后效
delayed elastic recovery 滞后弹性恢复
delayed expansion of cement 水泥的延迟膨胀
delayed explosibility 延期爆炸性
delayed explosion 延迟爆炸；迟发爆炸

delayed exponential response 延迟指数响应
delayed fallout 延缓沉降
delayed feedback audiometer 延迟反馈测听仪
delayed fill 滞后充填
delayed finish 延迟整修处理；延迟完成
delayed finishing 延时抹面
delayed firing 滞后点火
delayed flow 滞后径流
delayed fluorescence 缓发荧光；延迟荧光；迟滞荧光
delayed fracture 延迟断裂
delayed heat 延迟热；迟发热；剩余热
delayed ignition 延迟点火
delayed implantation 延迟植入
delayed impulse 延迟脉冲
delayed income credit 过期账收入
delayed income debt 过期账支出
delayed infiltration 延迟下渗；延迟入渗
delayed initiation 延迟起爆；延迟激发
delayed luminescence 延迟发光
delayed maturity 延迟成熟
delayed migration 迟延移行
delayed mixing 延迟拌和
delayed network 延迟网络
delayed neutron 缓发中子
delayed neutron accounting 缓发中子计数
delayed neutron fission product monitor 缓发中子裂变产物监测器
delayed neutron fraction 缓发中子比
delayed neutron leached-hull monitor 缓冲中子包壳监测器
delayed occlusion 关闭迟延
delayed opening 开放迟延；延迟开启
delayed output equipment 延迟输出设备
delayed payment 延误支付；延期付款；延迟付款
delayed phenomenon of groundwater pollution transport 地下水污染迁移的延后现象
delayed phosphorescence 延迟磷光
delayed plane position indicator 延时式平面位置指示器
delayed plan position indicator 延时平面位置显示器
delayed posting 延期过账
delayed project 拖延施工项目
delayed proton 缓发质子
delayed pulse 延迟脉冲
delayed quenching 预冷淬火；延时淬火
delayed radiation activity 缓发放射性
delayed radiation effect 缓发辐射效应；迟发辐射效应
delayed reaction 延迟反应；滞后反应；迟延反应
delayed reactivity 缓发反应性
delayed reclosure 延迟重合闸
delayed reconveyance 延期返还财产托管信托书
delayed recovery 延迟恢复
delayed recovery after effect 滞后回弹后效
delayed recurrent event 延迟的循环事件
delayed reflex 延缓反射
delayed relay 缓动继电器
delayed renewal process 延迟的更新过程
delayed repeater satellite 延迟式中断卫星
delayed-request mode 延迟申请方式；延迟请求方式
delayed response 延时响应；延时出动；延迟响应；延迟回答；迟发反应
delayed-response transducer system 延迟响应换能器系统
delayed run-off 延滞径流；滞后径流
delayed scan(ing) 延迟扫描
delayed seed 二次气泡
delayed sequence 延误序列；延迟序列
delayed settlement 延时沉降；次固结沉降量
delayed shipment 延期装运；延期装船；迟延装船
delayed signal 延迟信号；延时信号
delayed silicosis 晚发矽肺
delayed slaking 后期消解
delayed start 延迟起动
delayed starting 拖延起动
delayed subsidence 延迟沉降
delayed subsurface runoff 延迟壤中流；延迟表层流
delayed surcharge 延迟卸货附加费；延迟附加费(运费)
delayed sweep 延迟扫描；延时扫描
delayed test 延误检验
delayed time 延迟时间

delayed time of regime peak 动态峰值的滞后时间
delayed time of the works 工期延误
delayed time processing 延时处理；延迟处理
delayed time system 延时计系统
delayed toxicity 缓发毒性；延迟中毒；潜伏毒性
delayed transaction 延迟型更新
delayed trigger 延迟触发器；延迟触发脉冲
delayed tripping 延时脱扣；延迟跳闸
delayed type 缓发型
delayed union 连接延缓；延迟连接
delayed visco-elastic response 延缓滞弹性效应
delayed yield 延迟屈服；滞后屈服时间；迟后给水
delay element 延迟元件
delay equalization 延迟均衡；时延均衡
delay equalizer 延期均衡器；延迟均衡器；时延均衡器
delayer 缓燃剂；延时器；延迟器
delay error 延迟误差
delay factor 延迟系数
delay fault 延迟故障
delay filling 采后充填
delay filtering 延迟滤波
delay-finish 延迟整修处理(水泥混凝土路面)
delay flip-flop 延迟触发器
delay function 延迟函数
delay fuse 延发(电)信管；迟爆引线
delay gate generator 延迟门脉冲发生器
delay generator 延迟振荡器；延迟发生器
delay ground distance 滞后地面距离
delay igniter 延点火装置；延迟点火装置
delay in completion 竣工拖延
delay in delivery 交货延误；交货逾期；延迟交付；延迟发放；延迟传送
delay inequality 时延差
delaying elasticity 滞弹性
delaying multivibrator 延迟多谐振荡器
delaying state 延迟状态
delay in payment 延期付款；延迟付款
delay in placing 浇筑间歇；延迟浇筑
delay in purchase at consumer sector 顾客采购延误
delay in recruitment 延期征聘；延迟征聘
delay in settlement 逾期结算
delay in shipment 延迟装船
delay instruction 延迟指令
delay interest 延迟利息
delay line 延时线；延迟线
delay line canceller 延迟线消除器
delay line capacity 延迟线容量
delay line clock 延迟线时钟
delay line ejector 延迟线引出装置
delay line helix 螺旋延迟线；慢波螺旋；延迟螺旋线
delay-line inflector 延迟线偏转器
delay line memory 延迟线存储器
delay line oscillator 延迟线振荡器
delay line register 延迟线寄存器
delay line storage 延迟线存储器
delay line store 延迟线存储器
delay lock 延迟锁定
delay-lock loop 迟延锁定环
delay medium 延迟媒质；延迟介质
delay mixing 延迟拌和
delay mode 延迟模式
delay model 延误模型
delay modulation 延迟调制
delay multivibrator 延时多谐振荡器；延迟多谐振荡器
delay of firing 发火延迟
delay of material 原材料延误
delay of swing-response 应舵时间
delay on bearing 方位延迟
delay operator 延迟运算子
delay or temporary storage 延迟或暂时存储
delay pattern 延迟模式；起爆顺序
delay payment 延付
delay period 延滞期；延发时间；延迟期
delay period of explosion 爆破延发时间
delay policy 延迟策略
delay powder 延时炸药；延期炸药；定时炸药
delay primer 迟发起爆药包
delay probability 等待概率
delay programming 延迟程序设计
delay-pulse generator 延迟脉冲发生器
delay-pulse oscillator 延迟脉冲振荡器

delay rate 延误率
delay ratio 慢波比;延误比率
delay recorder 延时记录器
delay relay 延时继电器
delay release 延迟缓解
delay repair 延迟修理
delay resolution 延迟分辨率
delay routine 延迟程序
delays by subcontractors 分包商的误期
delay screen 持久荧光屏
delay set counter 延时位置计数器
delay shot firing counter 迟发爆破计数器
delay signal 延缓信号;滞后信号
delays of drawings 图纸误期
delay spread 延迟扩展
delays to traffic 阻滞交通
delay system 滞留系统
delay table 中间辊道
delay tag number 迟发标号
delay through obstructions 受阻延误
delay time 延误时间;延时;滞后时间;迟滞时间;迟延时间;时延
delay time at stop 滞站时间
delay time distribution 延迟时间分布
delay time indication 晚点表示
delay time of booster 继爆管延期时间
delay time of electric(al) detonator 雷管延期时间
delay time of explosion 爆破延发时间
delay timer 延时器
delay-time register 延迟时间寄存器
delay tracker 延迟跟踪器
delay tracking 延迟跟踪
delay unit 延时元件;延迟装置;延迟元件;延迟部件;传送延迟器
delay vector 延迟向量
delay voltage 延迟电压
delay working 延缓接续制
delcarant 报关人
Delcom vernier 带游标电感比较仪
Delco Remy distributor 笛尔可莱姆分配器
del credere 买主资力保证;担保还款;信用担保
del credere agent 信用担保代理人;担保付款代理人
del credere agreement 保付货价合同
del credere commission 保证收取货款佣金;保收货款佣金
de-lead(ing)除铅;退铅;脱铅;去铅
deleading reagent 驱铅剂
deleading therapy 除铅疗法
deleafing 铝粉脱漂浮
delegacy 代表团
delegate 代表
delegated power 被授予的权利
delegation 委托;代表团
delegation and revocation 授权和撤回
delegation of authority 专家代表团;权力下放;授权
delegation of authority and responsibility 授予权责
delegation of authority by type of wotk 依工作种类授权
Delepine reaction 德莱平反应
delessite 铁叶绿泥石
delete 作废;勾销;取消;删去
delete bit 删除位
delete capability 删除容量
delete character 抹掉(字)符;作废(字)符;删除(字)符
delete code 删错码
deleted clause 删除条款
deleted data 删除数据
deleted data mark 删除数据符号
deleted entry 删除(登记)项
deleted file 注销文件;删除文件
deleted marker 删除标志
deleted neighbourhood 删除邻域
deleted representation 删除表示
deleted species 残遗种;残留种
delete flag 删除标记
delete key 删除键
delete list 删除表
delete mark 删除符号
delete program(me)删除程序
deleterious 有毒的
deleterious agent 有毒物
deleterious effluent 有毒废水;有害废水

deleterious gas 有害气体
deleterious material 有害材料
deleterious matter 有害物(质)
deleterious particle content in aggregate 骨料有害颗粒含量
deleterious reaction 有害反应
deleterious salts 有害盐类
deleterious substance 有害物(质)
deleterious waste 有害废物;有害物(质);有毒废物
deleterious water 有害的水
deleting segment 删除图段
deletion 勾销;缺失;删去;删掉;删除部分
deletion from deque 从双向队删去;从双排队删去
deletion from doubly linked list 从双链接表删去
deletion from front 从前端删去
deletion from linear list 从线性表删去
deletion from stack 从堆栈删去
deletion from tree 从树删去
deletion from two dimensional list 从二维表删去
deletion model 删除模型;修测图板
deletion of exclusion 免除责任条款的删除
deletion of items 项目删除
deletion of stack 堆栈的删去
deletion record 删改记录;删除用新记录;删除记录
deletion trace 标描过渡透写图
delf 排流器;荷兰白釉蓝彩陶器;出水沟;薄矿层;海堤排水道
Delft continuous sampler 德尔夫特连续取样器
Delft metabolic model 德尔夫特代谢模型
Delft pottery 上釉陶器
Delft ware 德福特陶器(乳白釉石灰陶器)
Delhi boil 德里疖
deliberate 推敲
deliberate a case 评议案例
deliberate act 故意行为
deliberate breaching (堤口的)扒口
deliberate division 计划分洪;有准备的分洪;人工分洪
deliberate reconnaissance 计划勘测;周密选线;周密踏勘
deliberation 审议
delicate 精致的;精密的;精美的
delicate adjustment 精确调制;精密调整;精调(整)
delicate cargo 精致货
delicate colo(u)r 淡(颜)色
delicate comprehension of art 对艺术有敏感的鉴赏力
delicate flute cast 细槽铸型
delicate operation 谨慎操作
delicate question 棘手问题;必须谨慎处理的问题;微妙问题
delicatessen 现成食品店;熟食店
delicatessen store 熟食店
Deli coupling 德利联轴节
delictual claim 侵权行为引起的债权
delictual damage 侵权行为性质的损害
delictual liability 不法行为的责任
delignification 木素脱除;水致侵蚀作用;从木材中脱去木质素;去木质素作用
delime 脱灰
deliminate 分离成层
deliming 脱灰
deliming agent 脱灰剂
delimit 划界;指定界限;定界
delimit a boundary line 划定边界
delimitation 立界;定界;分离成层;分界;区划;区分
delimitation of bridge superstructure 桥梁建筑限界
delimiter 限制符;限定器;专用定义符;定界符;分界符;分隔符
delimiter line 定界行
delimiter macro 定界宏指令
delimit fishing area 划定捕鱼区
delimiting character 定界字符
delimiting the territorial sea 划定领海界限
delineascope 幻灯
delineate 描外形;画轮廓;勾划外形
delineation 描绘外形;划界;线条画;勾划轮廓;白描;圈定
delineation marking 路面划线标示;划线标示
delineation of island 表示路线和安全岛等的轮廓线
delineation of ore deposits 矿体圈定
delineation unit 描绘单元

delineator 路线反光标记;道路标线;描画器;绘图员
delinquency date 拖欠最后期限
delinquency rate on loans 贷款拖欠率
delinquency ratio 借款拖欠率
delinquent 过失者
delinquent account 逾期账户;拖欠账款
delinquent account receivable 逾期未收账款
delinquent instalment 拖欠分期付款
delinquent-loan ratio 拖欠贷款率
delinquent mortgage 拖欠抵押贷款
delinquent party 违约一方;违约当事人
delinquent receivables 拖欠应收款项
delinquent special assessment 滞纳特赋
delinquent tax 滞纳税款;拖欠税款
delinter 剥绒机
delinting machine 剥绒机
deliquate 冲淡
deliquation 冲淡
deliquency 拖欠债款
deliquesce 货物反潮;潮解溶化
deliquescence 可溶解性;潮解(性);溶化性
deliquescence salt 潮解性盐
deliquescent 吸水液化;易潮解的;潮解的
deliquescent agent 潮解剂
deliquescent effect 潮解作用
deliquescent gas 溶解气体
deliquescent material 溶解料;溶解剂
deliquescent matter 易潮物质;易潮解物质;溶解物(质)
deliquescent salt 潮解性盐;溶解盐
deliquescent solid 溶解固体
deliquium 潮解物
delist(ing)解除上市资格;停止上市
delitessen store 现成副食品店
deliver 引渡;给料;递交;传递;投送;投递;送交;输送
deliverability 可采度;供应能力;输送能力
deliverability of gas 天然气回采量
deliverable end item 可交付使用的结尾项目
deliverable grade 可交付品级
deliverable service 移位服务
deliverable state 适宜交收情形
deliver delay 交付滞留
delivered 已交付;业已交付
delivered after duty paid 完税后交货
delivered alongside 交到船边
delivered and received car yard 交接场
delivered and received track 交接线【铁】
delivered at frontier 边境交货价格条件;边境交货(价)
delivered at plant 到厂
delivered at the job 工场交货;工地交货
delivered carrier 交货承运人
delivered condition 交货状态
delivered cost 输水成本;交货成本;输水费用
delivered duty paid 完税后交货
delivered ex-quay 目的港码头交货
delivered ex-ship 目的港船上交货
delivered from quay 码头交货
delivered heat 供热
delivered horse power 收到马力;输出马力
delivered in person 专人递送
delivered on board the ship 交到船上
delivered payload capability 装载(净重)能力
delivered power 输出功率
delivered price 交货价格;到货价(格)
delivered quantity 交货数量
delivered site 交货地点
delivered sound 完好交货
delivered weight 交货(时)重量;卸货重量;到货重量
deliverer 交付者
deliver from godown 出栈;出货
deliver goods 发货
deliver goods to the customers 送货上门
delivering date estimating 交货日期估计
delivering device 卸货设备;发货设备
delivering gear 进给装置;进刀机构
delivering imported water 输送引进水体
delivering of goods 发货
deliver over 交付
deliver the goods 如期交货
delivery 交货(额);交割;交付;交船;供料;供给;供电量;传递;分送;发送;取模;排气量;释放;生产
delivery advice 交货单

delivery against bill of lading 凭提单交货
delivery against letter of credit 凭信用证交货
delivery against payment 付款交货
delivery air chamber 压缩空气室
delivery and acceptance certificate 交货验收证明书
delivery and customs agent 提货报关代理人
|delivery and measuring box 放水水量水箱
delivery and redelivery 交船与还船
delivery and receiving certificate 货载授受证
delivery and shipment 交货与装运
delivery angle 输送带倾角
delivery area 交货范围
delivery at frontier 指定边境地点交货(价)
delivery auger 分配螺旋
delivery-based priority principle 交货期优先原则
delivery belt conveyer 传递带式输送机
delivery bolt and nut 输送端螺栓和螺母
delivery book 交货簿;送货簿
delivery boom 递送桅杆;递送吊杆
delivery box 水表箱;放水箱;配水箱
delivery bridle 出口张紧装置
delivery by compressed air 气压输送
delivery canal 引水渠;输水渠
delivery capacity 排(水)量;交货额;排(送)量;输送能力;生产能力;生产额
delivery car 交货车;送料车;送货车
delivery certificate 交货证(明)书;出厂证(书)
delivery chamber 压缩室
delivery channel 输出信道;输出流道
delivery charges 交货费;交付费用;送货到户费
delivery check valve 送油节止阀
delivery chute 传递滑槽;输送滑槽
delivery clack 增压活门;增压阀;排出阀
delivery clause 交接条款(租船合同);交货条款
delivery clerk 票据分配员
delivery cock 放水旋塞;泄放旋塞;清洗旋塞;排污旋塞
delivery conduit 送风道;输送渠道;输送管(道);输水管(道)
delivery cone 压力喷嘴;排水锥管
delivery confirmation 确认发送
delivery-conscious 认真交货的
delivery contract 供货合同
delivery control 交货期限控制
delivery curve 供水曲线
delivery cycle 交货周期;交付周期
delivery cylinder 传送滚筒
delivery date 交货(日)期;交船日期
delivery date estimating 估计交货(日)期
delivery date rule 内格累氏规律
delivery day 交货日
delivery desk 借书处
delivery discharge 输送水量
delivery distance 供应距离;递送距离
delivery duty paid 指定目的地完税后交货(价)
delivery efficiency 输出效率
delivery end 卸料端;出料端(指水、气、汽、油等)出口
delivery end drive 卸载端驱动装置(输送机)
delivery endplate 出料端板
delivery end stop 传送制动器
delivery expenses 销货运费;运输费(用)
delivery exponent 流量曲线指数
delivery ex-warehouse 仓库交货
delivery facility 交付装置
delivery failure 交货误期;供货误期
delivery fee 送货到户费
delivery flash 分液瓶
delivery floor 供货楼层
delivery flushing flow 输出流量
delivery form 送货单
delivery formality 交货手续;提货手续
delivery free 送货到户费已付
delivery gate 放水门;供料门;斗门;配水闸(门)
delivery ga(u)ge 终轧尺寸;高压压力表;出口压力表;排出压力表
delivery guide 出口导板
delivery guide box 出口导板盒
delivery harbo(u)r 交货港(口)
delivery head 压送水头;供水水头;递送水头;提升水头;水头;输送水头;输水水头;输出水头
delivery head lift 输送压头
delivery hose 供水水带;出水软管;喷射软管;输送软管;输水软管;输出软管
delivery hose adapter 供水水带接口
delivery hose coupling 供水水带接口
delivery hose line 供水水带线
delivery information 询问交货
delivery in instalment 分期交货
delivery inspection 出厂检验
delivery installation 交付装置
delivery instrument 交收文件
delivery lag time 交货滞后时间
delivery lift 升ани高度;扬水高度;压升高度;输送水头;输水水头;输水高度
delivery line 导出管;输送管路
delivery long pipe 高压长管
delivery lorry 运输载货汽车
delivery loss 输水损失
delivery made by river barge 河边转运
delivery main 输送总管;输送干管;输水总管;输水主管;输水干管
deliveryman 送货人
delivery manifold 供水汇管;进水干管;供水管线;排气歧管
delivery mark 分液刻度
delivery mechanism 传送装置;传送机构;输送装置
delivery mistake 误交付
delivery month 交收月;交割月
delivery note 交货证(明)书;交货单;送货单
delivery notice 交收通知;交货通知
delivery nozzle 输出喷嘴
delivery of a deed 交付契据
delivery of canal 渠道输水量
delivery of cargo 交货
delivery of container 集装箱交付
delivery of empty cars 排空
delivery of energy 能量供给
delivery of fan 风扇功率;风扇输送
delivery of freight 货物交付
delivery of goods 交货;货物发送;供货;发货
delivery of goods by installments 分批交货;分批送货
delivery of international inputs 提供国际投入
delivery of luggages and parcels 行李和包裹的支付;行李包裹交付
delivery of material 运送材料
delivery of palletized cement 集装水泥的交货
delivery of profit 解缴盈利
delivery of pump 泵送量;泵排水量;泵功率;泵(的)供水量;水泵排(出)量
delivery of the bill of exchange 提交汇票
delivery of towline 拖缆的递送
delivery on arrival 货到交付;货到即提
delivery on board 船上交货
delivery on call 见票交货;根据买方通知发货;通知交货
delivery on consignment 托销
delivery on deck 甲板交货
delivery on demand 按需供水;见票交货
delivery on field 就地交货;现场交货;原地交货;原地交付
delivery on payment 付款交货
delivery on rail 货车交付
delivery on request 见票交货
delivery on term 定期交货;定期交付
delivery on the spot 现场交货
delivery operator 交付人员
delivery order 交货单;交割单;小提单;出栈凭证;出栈凭单;出货单;提(货)单;提货单位
delivery order of airfreight 空运提单
delivery orifice 出水孔;放流孔;送气孔;输出孔口
delivery outlet 排出口;输出口
delivery part 卸货的一侧;出料侧
delivery performance report 交货执行情况报告
delivery period 交货期(限);供水期
delivery phase 输送相
delivery pipe 送水管;输送管(道);输水管(道);输出管
delivery pipe branch 输水支管;输气支管
delivery pipe line 排水管;给水管路;送水管线;输送管线;输水管线
delivery pipe of pump 水泵输水管
delivery plant 交付装置
delivery plate 给料盘
delivery point 交货(地)点;交船地点
delivery points 交收点
delivery port 到货港;交货港(口);输送口;输出港
delivery pressure 递送压力;排气压力;排出压力;输送压力
delivery pressure head 供水压头
delivery pressure of air compressor 空压机输出压力
delivery price 交收价
delivery pump 给料泵;排油泵;送气泵;输送泵
delivery quality 交货质量
delivery ramp 发货斜台;发货坡道
delivery rate 给料速度
delivery rate of erosion 冲刷输沙率
delivery ratio 交货率
delivery receipt 交货收据;交货回单;送货回单
delivery-receiving 交接场
delivery-receiving station 交接站【铁】
delivery-receiving track 交接线【铁】
delivery record 交货记录
delivery record book 交货记录簿
delivery reel 松卷机
delivery regulator 进给调节器;给料调节器
delivery risk 交割风险
delivery roll 导辊
delivery roller 传送滚轮
delivery room 交货室;卸货间;产房
delivery schedule 交货时间表
delivery sequence 交货顺序
delivery settlement 实物交割
delivery shaft 输送井
delivery sheet 送料板
delivery side guide 传送侧导板
delivery side of pump 水泵输出端
delivery space 扩散室(离心泵);排气容积
delivery specification 供应条款
delivery speed 出口速度;排气速度
delivery spring 输出弹簧
delivery stall 货摊
delivery state 交货状态
delivery station 交货站;交货(地)点;发货站
delivery status 投送状态
delivery stroke 输送冲程;输出行程
delivery system 运载装置;供料系统;输送系统;收纸系统
delivery taken by river barge 河边转运
delivery tank 出水池
delivery temperature 排气温度
delivery tension roll 出口张紧辊
delivery term 交货期限
delivery terminal 交货终点
delivery terms 交货条件
delivery the goods 交货
delivery time 流出时间;交货时间;交割期;交船时间;引渡时间;传送时间
delivery to domicile 交货到户
delivery tolerance 喂料公差(偏离集料和沥青配比值)
delivery tongue 送料舌板
delivery track 供料支线;输送轨道
delivery tray 传送托盘
delivery trial 交船试航
delivery trip 交船航次
delivery truck 送货车;厢式送货汽车;载货汽车
delivery tube 输送管(道)
delivery tunnel 递送隧洞;传送隧道;输送隧洞;输送隧道;输水隧洞;输水隧道
delivery twist 出口翻钢导板
delivery unit 交付单位
delivery value 通话能力;输送能力
delivery valve 导出阀;出油阀;放水阀;排气阀;输送阀;输出阀
delivery van 厢式送货车;送货(汽)车;送货(篷)车
delivery vehicle 运输工具
delivery volume (泵的)流量;供风能力;抽水量;通风能力;输送容积
delivery wagon 递送车;输送车
delivery weight 交付重量
delivery wheel 传送轮
delivery window 传递窗
delivery work 输出功
delivery workers' salaries 运输工人工薪
dell 河谷谷地;小溪谷;峡谷;幽谷;出水沟
dellaite 羟硅钙石
Della Robbia ware 德拉罗伯陶器(硬质白色和光壳釉的艺术陶器)

dellenite 流纹英安岩
Dellinger fadeout 德林格尔渐弱现象
Dellinger phenomenon 德林格尔现象
dells 险滩;峡谷;急水滩
Delmonte tar 德尔蒙泰焦油;德尔蒙泰柏油
Delmontian (stage) 德尔蒙特阶
deloader 卸载器
deloading 减载
delocalization 离域;非定域作用;不定位
delocalization energy 非定域能
delocalized bond 非定域键
delocalized-bond model 非定域键模型
delocalized molecular orbital 离域分子轨道
De Long barge 狄龙式自开式驳
De Long wharf 狄龙式码头
del operator 倒三角形算子
delorenzite 钛铁铀钇矿
de Lorme 代洛姆式屋顶构造
delousing 灭虱
delph 堤后排水沟
Delphi forecast 德尔菲预测法
Delphi forecasting method 德尔菲预测法
Delphi method 德尔菲(方)法
delphinine 翠雀宁;翠雀碱
delphinium blue 翠雀蓝
delphinorum columnae (古罗马竞技场内有雕像的)双柱
Delphi technique 德尔菲技术
Delphi technique method 德尔菲技术法
delph stone 德尔菲岩石(一种产于英国约克郡的岩石)
Delrac 特拉克导航系统
Delrac system 特拉克导航制
Delrama 变形镜头系统
delrioite 水钒锶钙矿
delta 变化微量;半选输出差;三角洲;三角形
delta accumulation soil 三角洲堆积土
delta air chuch 三爪气动卡盘
Delta alloy 铜锌合金
delta antenna 三角形天线
delta area 三角洲地区
delta arm 三角洲支汊;三角洲汊河
delta bar 河口三角州;三角洲坝
delta basin 三角洲型盆地
delta bedding 三角洲层理
delta branch 三角洲支汊;三角洲河汊
delta building 三角洲形成(过程)
delta channel 三角洲水道;三角洲河汊
delta circuit 电路
delta clock 再启动时钟;变形钟
delta coast 三角洲海岸
delta cone 三角洲冲积锥
delta connected 三角形连接的
delta connected motor 三角形连接电动机
delta connection 星三角形接法;三角形连接(法);三角形接线;三角形接法
delta connector 三角形接端
delta current 三角形电流;三角接法电流
delta cycle 三角洲旋回
delta dam 三角洲坝
delta-delta connection 双三角形接(线)法;三角一三角形接法
delta deposit(e) 三角洲沉积土;三角洲沉积(物)
delta development 三角洲发育
delta electromotive force 三角形连接电动势
delta estuary regulation 三角洲河口整治
delta facies 三角洲相
delta facies sequence 三角洲相序
delta fan 扇状三角洲;三角洲冲积扇
deltafication 三角洲形成(过程)
delta field 三角形磁场
delta fill 三角洲填土
delta flat 三角洲洼地平地;三角洲洼地公寓
delta formation 三角洲形成(过程)
delta front 三角洲前缘
delta front deposit 三角洲前缘沉积
delta front facies 三角洲前缘相
delta front shape deposit 三角洲前缘斜坡沉积
delta front sheet sand 三角洲前缘席状砂
delta gasket closure 三角垫密封
delta geosyncline 三角洲地槽;外枝准地槽
delta growth 三角洲淤长
deltaic 三角洲的
deltaic bedding 三角洲层理

deltaic cone 三角洲锥;三角洲(冲积)锥
deltaic cross bedding 三角洲交错层
deltaic deposit 三角洲沉积(物)
deltaic embankment 三角洲路堤
deltaic environment 三角洲环境
deltaic estuary 三角洲型河口
deltaic fan 扇状三角洲;三角洲冲积扇
deltaic herbaceous facies 三角洲草本相
deltaic model 三角洲模式
deltaic plain 三角洲平原
deltaic plain deposit 三角洲平原沉积
deltaic region 三角洲地区
deltaic river mouth 三角洲河口
deltaic sediment 三角洲沉积
deltaic sedimentation model 三角洲沉积模式
deltaic sediments 三角洲沉积物
deltaic-sheet trap 三角洲席状砂圈闭
deltaic strand facies 海滨三角洲相
deltaic tract 三角洲段;三角洲地区
delta impulse function 三角形脉冲函数
deltaite 混钙银星石
delta lake 三角洲湖
delta lobe 三角洲朵体
deltalogy 三角洲学
delta-marginal plain 三角洲边缘平原
delta marsh deposit 三角洲沼泽沉积
delta matching 三角形搭接
delta matching antenna 三角形匹配天线
deltamax 铁镍薄板
delta measurement 三角洲测量
delta metal 黄铜合金;耐蚀高强黄铜
delta model 三角洲模式
delta modulation 增量调制;调制;三角调制
delta modulation coding 增量调制编码
delta modulation system 增量调制系统
delta moraine 冰接三角洲;三角洲冰碛
delta network 三角形网(络);三角形电路
delta plain 三角洲平原
delta plain facies 三角洲平原相
delta plain swamp 三角洲平原沼泽地
delta point 三角洲前缘
delta progradation 三角洲增长
delta project 三角洲整治计划
delta regime 三角洲体系
delta region 三角洲地区
delta-ring 三角形密封圈;三角垫
delta ring sealing gasket 三角形密封垫
delta river branch 三角洲河汊
delta river mouth 三角洲河口
delta sand trap 三角洲砂体圈闭
delta sediment 三角洲沉积
delta sediments 三角洲沉积物
delta sequence 三角洲层序
delta signal 半选输出信号差
delta-star connection 三角星形接线法
delta-star transformation 三角星形接线变换
delta succession 三角洲序列
delta survey 三角洲测量
delta terrace 三角洲阶地
delta trade 三角贸易
delta-type antenna 三角形天线
delta voltage 三角连接法线电压;三角接线电压
delta winding 三角形绕组
delta-winged plane 三角形飞机
delta-Y transformation 星形三角变换
delthyrium 柄孔;三角孔
deltic method 自相关比较波形法
deltidium 柄孔盖
deltohedron 四角三四面体
deltoid (al) 三棱的;三角形的;三角洲的
deltoidal fan 三角洲冲积扇
deltoid branch 三角洲支汊;三角洲河汊
deltoid plate 三棱板
deltology 三角洲学
delubrum 古代罗马的寺庙;圣殿或庙宇(古罗马)
deluge 洪水;淹没;大洪水;泛滥(暴雨);暴雨
deluge collection pond 蓄水池
deluge foam-water sprinkler system 雨淋喷水泡沫联用灭火系统
deluge gun 上炮
deluge proof motor 防水(式)电动机
deluge sprinkler 大水滴喷头
deluge sprinkler system 集水喷洒体系;雨淋喷水灭火系统

deluge system 人工喷泉系统;喷水灭火系统;浸水系统;集水系统;雨淋(灭火)系统
deluge valve 喷水控制阀门;集水阀门
deluge valve unit 雨淋阀组
delusion of foreign influence 被控制感
delusterant 消光剂
delustered nylon 消光尼龙
deluster(ing) 消光;清除光泽;去光泽
delustering pigment 消光颜料
delustre 去光泽;褪光
delustred coating 消光涂层;低光泽涂层
delustring 除去光泽
delustring coating material 低光泽涂料;消光涂料
deluvial placer 坡积砂矿
deluvial plain 洪积平原
deluvium 坡积物
de luxe 豪华的;最高级的
de luxe coach 高级客车
de luxe fabric 高光泽电光整理布
de luxe model 豪华型;高级型
de luxe shower 优质淋浴器;上等淋浴器;豪华型淋浴器
de luxe train 特别快车
de luxe type hotal 豪华型旅馆
de luxe volume 精装本
de luxe yacht 豪华游艇
delvauxite 胶磷铁矿
delve 凹地
demagging 除镁;脱镁
de-magnetism 去磁;退磁
de-magnetization 灭磁;消磁;退磁;去磁
de-magnetization curve 去磁曲线
de-magnetization factor 去磁系数
de-magnetization force 去磁力
de-magnetization loss 去磁损失
de-magnetization methods 退磁方法
de-magnetize 减磁;去磁
de-magnetized state 去磁状态
de-magnetizer 去磁装置;去磁器;退磁装置
de-magnetizing action 消磁作用
de-magnetizing ampere turns 去磁安匝
de-magnetizing apparatus 退磁设备
de-magnetizing band 去磁线圈
de-magnetizing effect 去磁效应;消磁作用
de-magnetizing energy 去磁能
de-magnetizing factor 去磁因数;消磁系数;消磁率;退磁因数;退磁系数
de-magnetizing force 消磁力
de-magnetizing resistance 去磁电阻;退磁电阻
demagnification 缩倍
demagnifier 缩倍器
demagnifying telescope 缩倍望远镜
demal 分玛
demand 需要(订立协议);需量系数;要价
demand activated bus system 按需配车公交系统
demand-actuated governor 文丘里调压器
demand aeration tank 按需曝气池
demand aeration tank-intermittent aeration tank process 按需曝气池间歇曝气池组合工艺
demand a lower price 压价
demand analysis 需求分析
demand and supply 需求与供应;需求与供给
demand-and-supply-curve framework 供应成绩结构;供需曲线结构
demand and supply model 需求与供给模型
demand and supply of energy 能源的需求与供应
demand an indemnity 要求赔款
demand-assessment period 最大负荷估计周期
demand-assignment 按需分配
demand assignment multiple access 动态分配多址
demand-assignment time-division multiple-access 按需分配时分多址
demand bill 限期票据;现期汇票;即期汇票
demand-capacity control 需求通行能力控制
demand carried forward 结转需求
demand carry-over 累积需求
demand certificate 活期存单
demand characteristics 需求的特性
demand charges rate 装机容量收费率;安装机容量收费率
demand check 即期支票
demand compensation 要求赔偿;索赔
demand competition 需求竞争

demand control 过载控制
demand-curbing measure 压低需求措施
demand curve 输水量曲线;负荷曲线;需要量曲线;需求曲线
demand curve of environment(al) quality 环境质量的需求曲线
demand debt 即期债务
demand-deficient unemployment 需求不足型失业
demand density 需求密度
demand depletion 需求缩减
demand deposit 即期存款;活期存款
demand detector 需求检测器
demand differential pricing 需求差异订价法
demand draft 即期汇票;汇票
demand elasticity 需求伸缩性;需求弹性
demand element 需求项目
demand estimation 需要量估计;需求估计
demand exceeds supply 求过于供
demand expansibility 需求扩张(性)
demand factor 需求率;需用因素;需求率;需要因素;需求因素;需求系数;需量系数
demand fetching 按需求读数
demand for bank funds 要求银行贷款
demand for capital goods 资本货物需求
demand for currency 现金需求
demand forecast(ing) 需求预测
demand for funds 资金需求
demand for labo(u)r 劳动需求
demand for payment 要求付款
demand for service 服务需求量
demand function 需求函数
demand goods 需求货物
demand grain and money from production teams 摊派粮款
demand graph 要求图形;请求图形
demand growth 需求增长
demand hydrograph 需用水量过程线;需用电量过程线;需要流量过程图;需求流量过程曲线
demand increase 需求增长
demand inflation 需求膨胀
demanding quantity of compressed air 压缩空气需要量
demand interval 需用时限
demand liabilitise 即期负债
demand limiter 需量限制器;电流限制器
demand loan 活期贷款;活期放款
demand loan and overdraft 活期放款及透支
demand loan department 活期贷款部
demand loan secured 活期担保贷款
demand management 需求管理
demand mass curve 需要量累积曲线;需水量累积曲线;需求累积曲线;需电量累积曲线
demand meter 负荷煤气表;需量表;最大负荷测试器
demand model 需求模式
demand mortgage loan 活期抵押放款;活期抵押贷款
demand note 见票即付票据;即期票据;支付申请书
demand notice 需求预告
demand obligation 现期债务;即期债务
demand of consumption 消费需求
demand of energy 能量需要
demand on processor 对中央处理机的需求
demand operation 按需操作
demand-oriented pricing 需求定价法
demand paging 请求调页
demand payment of a debt 要账
demand peak 峰值荷载
demand performance 催告履行义务
demand point 消费点;消费场合;需求点
demand pointer 用电量指针
demand pressure 需求压力;需求牵引力
demand price 需求价格
demand pricing 需求定价法
demand processing 立即处理;按需处理;请求处理
demand promissory note 即期本票
demand pull 需求压力;需求拉动;需求膨胀
demand-pull inflation 需求引起的通货膨胀;需求带动性的通货膨胀;需求促成的通货膨胀
demand pusher indicator 按钮指示器
demand quantity 需要量
demand quantity of material 物资需求量
demand rate 现期汇率;需求率;需量标准
demand reading 按需读出

demand recorder 需量记录器
demand register 需量记录器
demand-responsive transportation system 叶车运输系统
demand retrenchment 需求紧缩
demand saturation 需求饱和;充分满足要求
demand schedule 需求(一览)表
demand school 需求学派
demand season 需求季节
demand service 即时业务;即时处理
demand shift 需求变动
demand shifts from income 因收入增减引起的需求变动
demand side 需求学派
demand-side fluctuation 需求波动
demand-side subsidy 求方补助
demand sign 指挥信号
demand signal 指令信号
demand special privileges 索取特权
demand stability 需求稳定性
demand staging 按需传送;请求分段
demand strategy of mineral commodities 矿产需求对策
demand structure 需求结构
demand-supply ratio 需水供水比
demand surplus 需求过剩
demand system 自动供氧系统
demand system of irrigation 灌溉供水制度
demand-type mask 供气量自动调节式防毒面具
demand usage time 要求使用期
demand variable 需求变量
demand view of ecologic(al) economics 生态经济需求观
demand viscous water tanker-mixer 自动调节黏稠水的水罐车混合装置
demand working 立即接通
demand writing 按需写出
demand zero page 请求零页面
demanganization 除锰(法);脱锰
demantoid 翠榴石
demarcate boundaries 标定边界
demarcated section 测流断面;施测断面
demarcated site 监测站所在地段;测流站址;测流现场
demarcate on the ground 实地定界
demarcation 划界;分界;区划
demarcation line 界线;分界线
demarcation point 分界点
demarcation potential 分界电位
demarcation strip 分配板;分界条
demarcation survey 标界测量
demarcation widened 限界加宽
De Marre formula 德马尔公式
demasking 解蔽
demasking agent 可剥性涂料的脱漆剂;解蔽剂;暴露剂
dematerialized multiformity 非物质的多样性
dematron 代玛管
dematured 变性的
Dember effect 丹伯效应
Deme (现代希腊的、古希腊雅典的)市区
Demerara abyssal plain 德梅拉拉深海平原
demerit 过失
demerit sampling plan 权数抽样方案
demersal 沿底的;底栖的;底层的
demersal carnivore 底栖食肉动物
demersal population 底栖种群
demersal resources 底渔业资源
demersal species 底栖种
demeshing 脱离啮合
demesmaekerite 硒铜铅铀矿
demesne 领域;领地;范围;地产
demethylating 脱甲基作用
demethylation 反甲基化
demeton mixed 内吸磷混合物
demi-bastion (有一个正面和一个侧面的)防御工事
demi-bath 脚盆;半浴缸;坐浴(浴)盆
demicellization 解胶束
demi-column 半(露)柱;嵌墙柱
demi-double strength window glass 中厚窗玻璃(3毫米厚)
demi-double thickness 中厚(3毫米厚的窗玻璃)
demi-double thickness sheet glass 中厚窗玻璃(3毫米厚)

demi-hunter 半双盖表
demi-hydraulic fill dam 半冲填坝
demiinflation 半膨胀
demijohn 细颈大坛;用柳条笼装的细口瓶;坛;酸坛
demilitarized zone 非军事区
demilune 半月(体)
demi-metope (陶立克式式檐壁转角三拢板外侧的)半幅板;(陶立克式角柱上雕带外侧的)半块雕饰板
demimonstrosity 半畸形
demineralization 除盐(作用);软水作用;(水的)软化;脱矿质(作用);去矿质;去矿化作用
demineralization of water 水的脱矿化;水(的)软化
demineralize 去矿化;脱矿质
demineralized water 去离子水;软(化)水;脱盐水;脱矿质水
demineralized water circulation pump 软水循环泵
demineralized water tank 软水槽
demineralized water treatment device 软水装置
demineralizer 淡化器;除盐装置;软化器;脱矿质剂;水软化器;去矿化装置
demineralizing equipment 除盐装置
demineralizing plant 脱矿质装置
Deming circle 德明循环
demi-relief 浅浮雕;半浮雕
demi-relievo 很低浮雕;半浮雕
demi-rilievo 半浮雕
demise 遗赠;让与
demise charter 空船租船合同
demise charter party 空船租船契约;转让船租合同;转让船租合同
demise clause 过户条款;空船条款
demi-section 半剖视;半刻面;半节
demised premises 转让土地;转让房地产
demisemi 四分之一的
demise owner 二船东
demising clause 房地产租赁契约中的条款
demising partition 临时隔墙;临时隔墙;商场临时隔断
demising wall 间隔墙;让位墙
demister 除雾装置;除雾器;除沫器;去雾器
demister entrainment 液滴分离器
demister (of vehicle window) 车窗玻璃除霜器
demitint 中间(色)调
demi-tub 坐浴(浴)盆
demixing 分裂;分离;分层;反混合;配合料分层;脱混
demixing point 分层点
Demjanov rearrangement 德姆雅诺夫重排
demobilize 遣散
democratic management of enterprise 企业民主管理
demode 解码
demodifier 恢复符
demoding circuit 解码电路
demodulating system 解调系统
demodulation 解调;检波;反调制;反调幅;调整波形;去调幅
demodulation function 解调函数
demodulator 检波器;反调制器;反调幅器;去调幅器;解调器
demodulator amplifier 反调制放大器
demodulator filter 解调滤波器
demodulator oscillator 解调振荡器;反调制振荡器
demodulator probe 检波头;检波点
demogram 人口统计图
demogrant 负所得税
demographer 人口统计学家
demographic(al) analysis 人口分析
demographic(al) census 人口调查
demographic(al) change 人口变迁
demographic(al) characteristic 人口统计特征;人口统计特性曲线
demographic(al) composition 人口构成
demographic(al) data 人口统计资料
demographic(al) explosion 人口激增;人口爆炸
demographic(al) historic geography 人口历史地理
demographic(al) parameter 人口统计参数
demographic(al) policy 人口政策
demographic(al) projection 人口预测
demographic(al) situation 人口状况
demographic(al) statistics 人口统计
demographic(al) structure 人口(分布)结构
demographic(al) study 人口统计学研究
demographic(al) survey 人口调查

demographic(al) theory 人口理论
demographic(al) training 人口统计培训
demographic(al) transition 人口变化
demographic(al) trend 人口统计趋势
demographic(al) yearbook 人口统计年鉴
demographics 人口统计
demography 人口学;人口统计(学)
De Moivre's theorem 德摩弗定理
demolding 脱模
demolish 拆毁;拆除
demolishable 宜拆毁的;可拆毁的
demolish a building 拆除建筑物
demolished concrete 破坏的混凝土;拆毁的混凝土;废弃混凝土
demolisher 承包拆屋者;拆除机
demolishing 销毁;拆毁;拆除
demolishing and rebuilding 原拆原建
demolishing and resetting office 拆迁办(公室)
demolishing work 拆除工程
demolishment 销毁;拆毁;拆除
demolition 拆毁;拆除
demolition accident 爆破事故
demolition and breaking hammer 拆毁破碎用锤
demolition and construction ratio 拆建比
demolition and drill hammer 拆毁钻孔用锤
demolition and relocation 拆迁
demolition area 拆迁区
demolition ball 破坏球;爆破球
demolition bit 破坏钻头
demolition blast 炸毁;爆破
demolition blasting 拆除爆破
demolition chamber 爆破室
demolition chip(ping)s 拆毁的碎片
demolition concrete 拆毁的混凝土
demolition contract 解除合同;拆毁合同
demolition contractor 拆毁房屋承包者;拆毁房屋承包人
demolition cost 拆毁成本;拆除费用;拆毁费
demolition crew 拆毁房屋工程队
demolition device 拆毁设备
demolition expenses 拆毁费
demolition firm 拆毁工程公司
demolition gang 拆毁工程队
demolition grant program(me) 拆房补助计划
demolition hammer 冲锤
demolition loss 拆毁损失
demolition of back-court housing 拆毁后院房屋
demolition of buildings 拆毁房屋
demolition of tank 罐拆毁
demolition operation 拆毁作业
demolition order 拆除法令;拆迁规程
demolition party 拆毁工程承包队
demolition permission 销毁许可;拆毁许可(证)
demolition permit 拆屋执照
demolition pick (hammer) 拆毁用的鹤嘴锄;拆毁风镐
demolition press 破坏压榨机;拆毁印刷机;拆毁压榨机
demolition project 拆毁工程;拆毁方案
demolition rate 拆除率
demolition report 拆除申请书
demolition rubbish 拆毁废墟;拆房废料
demolition rubbish brick 拆毁废墟砖
demolition rubble 房屋拆毁后的废物堆
demolition scheme 拆毁计划;拆毁方案
demolition site 拆屋现场;拆毁地点
demolition spoil 瓦砾堆
demolition tool (混凝土路面的)捣碎器;击碎器;捣碎机;打桩机;拆毁工具
demolition waste 瓦砾堆;拆屋垃圾;工地废渣料
demolition wastewater 工地废水
demolition with explosives 爆炸拆除法
demolition work 拆毁(作业);拆毁(工作);拆除工程
demonetization 失去通货资格
demonstrated commodities 探明资源
demonstrated reserves 探明储量
demonstrate farm 示范农场
demonstration 论证;证明
demonstration area 试验区;示范区
demonstration building 示范建筑;示范馆;样板建筑
demonstration cities program(me) 示范城市建设计划

demonstration effect 示范效益;示范效果
demonstration forest 示范林
demonstration jack 演示插孔
demonstration kitchen 示范厨房
demonstration model 示范典型
demonstration of the type 型号说明
demonstration plant 样板厂;示范装置;示范工厂;示范电站;示范车;试验厂
demonstration plot 示范区;树木示范小区
demonstration pond 示范塘
demonstration project 示范(建筑)项目;示范工程(项目)
demonstration project grant 示范工程项目的拨款;示范建设项目的拨款
demonstration projects in the field of the environment 环境领域示范项目
demonstration pyrometer 直示高温计
demonstration reactor 示范堆
demonstration stage of scheme 方案论证阶段
demonstration system 显示系统
demonstration test 示范试验
demonstration unit 示范装置
demonstrative project 试验工程;示范计划
demonstrator 演示者;示者;解说员;示教器;示教板
demo plant 样板厂;试验厂
De Morgan's laws 德摩根定律
De Morgan's rule 德摩根法则
De Morgan's test 德摩根检验准则
De Morgan's theorem 德摩根定理
demorphism 风化变质作用
demorphismus 岩石分解
demo(u)ld 拆模
demo(u)lding 拆模(板);脱模
demo(u)lding agent 拆模剂;脱模剂
demount 拆卸;拆开
demountability 可拆卸性
demountable 可拆换的;可卸下的;可拆除;活的
demountable axle 可拆卸轴
demountable bearing 可拆卸轴承
demountable bit 可拆卸钻头
demountable bucket dredge(r) 拆装式链斗挖泥船
demountable building 可拆卸建筑物
demountable cargo container 拆装式货物集装箱;拆卸式货物集装箱
demountable connection 可拆卸的连接;可拆卸接点
demountable container 可拆集装箱
demountable cutter dredge(r) 拆装式绞吸挖泥船
demountable ditcher 可拆卸挖沟机
demountable division wall 可拆隔墙;活动隔墙;活动隔断
demountable floor 活动地板
demountable joint 可拆卸接合;可拆卸接头
demountable mast 可拆卸的桅杆
demountable partition 可拆式隔断;可拆隔墙;灵活隔断;可拆隔断;活动隔墙;活动隔断;装卸式隔断
demountable rim 可卸轮缘
demountable road-rail tanker 可拆卸的公路铁路两用油车;可拆卸的公路铁路两用水箱
demountable shoe 活动桩靴;活动柱脚;活动蹄形物;活动垫座;活动垫板
demountable stage 可卸式舞台;装配式舞台
demountable suction dredge(r) 拆装式吸扬挖泥船
demountable tank 可拆卸油箱;可拆卸水箱
demountable tube 可拆卸管
demountable walkway 活动走道
Dempster positive ray analysis 登普斯脱阳离子射线分析
demulsibility 乳化分解性;反乳化性;反乳化率;反乳比率
demulsibility agent 抗乳化剂
demulsibility test (乳化沥青的)反乳化度试验
demulsification 反乳化(作用);乳液破坏;破乳
demulsification number 反乳化值
demulsification test 反乳化试验
demulsifier 解乳剂;反乳化性;反乳化剂;乳液澄清器;破乳剂
demulsify 反乳化
demulsifying 反乳化作用
demulsifying agent 反乳化剂;乳液破坏剂;破乳剂
demulsifying compound 反乳化性;反乳化剂
demulsionfication 脱乳作用

demultiplex 信号分离;多路分配
demultiplexer 信号分离器;多路分解器;倍减器
demultiplex filter 多路分解滤波器
demultiplexing 解编;多路化;多路分解;分成多路(处理)
demultiplexing circuit 分离信号电路
demultiplication 倍减
demultiplier 递减器;倍减器
demur 表示异议
demurrage 延滞费;延时停车费;延期时间(指船舶在港期);延期罚款;装运滞期费;装卸误期费;滞留期;滞留费;船舶滞期费;误期费用;拖延时间;拖延费
demurrage and despatch 滞期费和速遣费
demurrage charges 延期滞延费;延期停泊费;延期费(用);滞期费;滞留费;滞港费
demurrage days 滞期天数;滞期日数
demurrage lien 滞期留置权
demurrage money 延期费(用);滞期费
demurrage of ship 船舶滞延时间
demurrage same despatch 滞期费和速遣费同等
demy 四开图纸;四开裁
demy check folio 四开图纸
den 兽穴;休息室;洞窟;动物园的兽笼
Denaby powder 铵硝化钾炸药
denary 十进制的
denary logarithm 以十为底的对数;常用对数
denary notation 十进制记数法
denary scale 十进制;十进法
denationalization 非国有化
denationalize 变成私营
denatonium benzoate 地那铵苯甲酸盐
denaturant 变性剂
denaturation 变性(作用);变性过程
denaturation map 变性图
denaturation of nuclear fuel 核燃料中毒
denaturation temperature 解链温度;变性温度
denaturation theroy 变性理论
denatured 变性的
denatured alcohol 工业酒精;变性酒精;变性醇
denatured gradient gel electrophoresis 改性梯度凝胶电泳
denatured salt 变性盐
denatured test 变性试验
denaturization 变性(作用)
dendrachy 树状分级
dendriform 树枝形
dendrimer 树枝状高分子
dendrite 枝晶;枝蔓状晶体;枝蔓晶;松树石;松林石;树枝状晶体;树枝形晶体;树枝石
dendrite arm 晶枝
dendrite formation 枝状晶形成
dendrite morphology 枝状形态
dendrite radius 枝状晶体半径
dendritic 枝状;枝晶的;树枝状的
dendritic copper powder 枝状结晶铜粉
dendritic crystal 枝(状)晶体;枝(蔓)晶体;树枝状冰晶
dendritic crystallization 枝蔓晶体;树枝状结晶
dendritic decoration 树枝状装饰
dendritic drainage pattern 枝状水系;羽状水系;树枝状模式
dendritic drainage system 羽状水系;树枝状水系;树枝形排水系统
dendritic dressing 树枝状装饰
dendritic glacier 树枝状冰川
dendritic growth 枝晶体生长
dendritic lake 多汊湖(泊)
dendritic morphology 枝(蔓)状结构
dendritic pattern 树枝形
dendritic powder 枝状(金属)粉末;树枝状粉末
dendritic ridge mo(u)ld 枝状脊模
dendritic river system 羽状水系;枝状水系;树枝状水系
dendritic segregation 枝状偏析;枝晶偏析;树枝状偏析
dendritic silicon 枝状硅
dendritic snow crystal 枝状雪晶
dendritic structure 树枝状组织;树(枝)状结构;树枝状构造
dendritic tufa 树枝状石灰华
dendritic valleys 树枝状谷
dendrocalamus affinis 麻竹
dendrocalamus giganteus 大麻竹

dendrochronology 年轮学;树木年代学
dendroclimatology 林业气候学;年轮气候学
dendrogram 树状图表;树形图
dendrograph 树径记录仪
dendrohydrology 年轮水文学
dendroid 树枝状
Dendroidea 树形笔石目
dendroid tufa 树枝状石灰华
dendroid venation 树桩脉序
dendrolite 树木化石
dendrology 树木学
dendrometer 测树器
dendrophysis 枝状拟侧丝;树状隔丝
dene 海边沙质地带;有林溪谷;绿林溪谷
Dengfeng group 登峰群【地】
dengue 登革热
dengue fever virus 登革热病毒
denial of facts 否认事实
de-nib 除粗粒
denidation 落床
denier 但尼尔;旦;(化学纤维规定长9000米、重1克为1旦)
denier balance 但尼尔天平
Denil fish pass 斜槽式鱼道
denim 粗斜棉布;劳动布
Denine 德奈恩钨钢
Denison core barrel 丹尼森(三层)岩芯管(取原状土样用)
Denison ga(u)ge 丹尼森发条规
Denison motor 轴向回转柱塞式液压电动机;丹尼森液压电动机
Denison pump 丹尼森轴向活塞泵;丹尼森泵
Denison sampler 丹尼森取土器
denitration 脱硝
denitridation 脱氮化层
denitriding 脱氮
denitrification 反硝化(作用);去氮作用;去氮法;脱硝作用;脱氮(作用)
denitrification characteristic 反硝化特性
denitrification coefficient 反硝化系数
denitrification denitrogen 反硝化脱氮
denitrification dynamics 反硝化动力学
denitrification inhibitor 反硝化抑制剂
denitrification in soils 土壤的脱氮作用
denitrification polyphosphorous 反硝化除磷
denitrification potential 反硝化势
denitrification process 脱氮过程
denitrification rate 反硝化速率
denitrification rate constant 反硝化速率常数
denitrification reaction 反硝化反应
denitrification sludge 反硝化污泥
denitrification system 反硝化系统
denitrifier 脱氮菌
denitrify 脱氮
denitrifying agent 反硝化剂;脱硝剂;脱氮剂
denitrifying bacterium 反硝化(细)菌
denitrifying biofilter 反硝化生物滤池
denitrifying filter 脱氮滤池;反硝化滤池
denitrifying fluidized bed reactor 反硝化流化床反应器
denitrifying microorganism 反硝化微生物
denitrogenate 除氮
denitrogenation 脱氮(法);除氮法;去氮法;脱氮作用
denitrogenation method 脱氮法
denitromethane 二硝甲烷
denivellation 湖面波动;湖面变化
Denki Kagaku Kogyo Kabushiki 电气化学工业公司
Denmark architecture 丹麦建筑
Denmark Strait 丹麦海峡
denninfite 碲锌锰石
dennison 可移动的高秤
denoise 降噪;消除干扰
de-nol tablet 德诺片
denominate number 名数
denominate quantity 名称量
denomination 命名;名目;名称
denominational number system 名数制
denominational value 面值
denomination value 票面价值
denominator 一般标准;共同特性;分母
denominator-weighted average 分母加权平均数
De Nora cell 德诺拉电解槽
denormalization 反向规格化

denotation 表示方法;起爆;外延
denotational semantics 标志语义
denotation token 标志记号
denotative meaning 外延意义
denote 指示
denoter 标志牌;标志符
denoting stamp 证明已纳税印花
denounce 通告废除
de novo synthesis 从头合成
denoxing 消除氮的氧化物
dens 齿节
Denscast 登岛卡特镍铬合金
dense 浓的;密实图;密(实)的;密集(的);颜色深暗的;致密的;稠(密)的
dense aggregate 密实集料;密实骨料;密集团聚体
dense aggregate concrete 普通混凝土;密实混凝土
dense air refrigeration cycle 逆布雷顿循环
dense air system 高压空气制冷系统
dense area 致密区
dense array 密集台阵
dense bacterial membrane 浓菌膜
dense barium crown glass 重钡冕玻璃
dense barium flint glass 重钡火石玻璃
dense bed 密实层
dense-bed column 紧密床柱
dense binary code 密集二进码
dense bitumen macadam 密级配沥青碎石路
dense black 浓黑;稠黑;非颗粒炉黑
dense block 致密砌块
dense body 密体
dense brick 高密度砖
dense bronze 致密青铜
dense-burned 密烧的
dense cake 致密结块
dense cargo 密度大的货物
dense charge 紧密炉料
dense chrome oxide 致密氧化铬
dense chrome oxide refractory 致密氧化铬耐火材料
dense cloud 密云
dense-coated macadam 密级配表面处治碎石路
dense communication network 交通网稠密
dense concrete 密实混凝土;重混凝土
dense concrete block 密实混凝土砌块
dense concrete wall 密实混凝土墙
dense condition 密实状态
dense conducting layer 强导电层
dense core 致密中心
dense crop 密林
dense crowd 稠密人流
dense crown 含钡重量玻璃
dense crown glass 重冕玻璃
dense degree of discontinuity 结构面密集程度
dense dolomite 致密白云岩
dense drilling 密网钻进
dense elastiomeric compression seal gasket 密实弹性体压缩密封垫
dense fissure zone 强裂隙带
dense flint glass 重燧石玻璃;重火石玻璃
dense fog 零级能见度;浓雾;大雾
dense form 致密型
dense formation 致密地层
dense gas chromatography 高密度气相色谱法
dense gradation 密级配
dense-graded 密级配的
dense-graded aggregate 密级配集料;密级配骨料
dense-graded asphalt concrete 密级配沥青混凝土
dense-graded bituminous concrete 密级配沥青混凝土
dense-graded bituminous surfacing 密级配沥青面层;密实级配沥青路面
dense-graded mineral aggregate 密级配矿物集料;密级配矿物骨料
dense-graded mix 密级配混合料
dense-graded pavement 密级配路面
dense grain 密纹
dense gravel-sand mixture 密砾砂混合物
dense index 密集索引;稠密索引
dense in itself 自稠密的
dense-in-itself-set 自密集
dense ionization 稠密电离
dense joint zone 节理密集带
dense lanthanum flint glass 重镧火石玻璃
dense layer 密集层;致密层;密实层

dense liquid 重液
dense liquid foundation 稠密液体地基
dense liquid subgrade 文克勒地基;稠密液体路基;稠密液体地基
densely graded 密级配的
densely graded aggregate 密级配骨料;密级配集料
densely graded mixture 密级配拌和料
densely inhabited district 居住拥挤地区;人口集(地)区;人口集中地区;人口高密度区;人口稠密地区
densely packed crystal 密集晶体
densely populated 人口稠密的
densely populated area 人口稠密地区;稠密居住区
densely populated urban area 人口稠密地区
densely settled area 稠密居住区;人口密集(地)区;居民稠密区
densely vegetated area 植物茂密地区
densely wooded 繁密的林木
dense manure 浓肥料
dense manure treatment 浓肥料处理
dense map continuous 密集连续图形
dense material 密致材料
dense matrix 密集矩阵;稠密矩阵
dense matter 致密物质
dense-media process 重介质选矿(法)
dense-media separator 重介质分离器
dense medium 重悬浮液;重介质;稠介质
dense medium bath 重介质分选箱;重介质分选槽
dense medium jig 重介质跳汰机
dense medium separation 重介选
dense medium treatment 稠介质处理
dense medium vibrating sluice 重介质振动溜槽
dense medium washing 重介质洗选
dense metal brine 含大量金属的卤水
dense microtremor network 密集微震台网
dense mineral 重矿物
dense mineral aggregate 密实矿物集料;密实矿物骨料
dense negative 密度大底片
densener 激冷材料
dense non-aqueous phase liquid 稠非水相液体
dense packing 紧密填筑;紧密夯实
dense parts 密致零件
dense patch 密斑
dense phase 浓相;密相
dense-phase fluidized bed 密相流化床
dense-phase pneumatic conveying 密相气力输送
dense-phase transporting system 密相(气流)输送系统
dense phosphate crown glass 重磷冕玻璃
dense powder 致密粉末
dense projection 致密突起
dense refractory 致密耐火材料
dense rock 致密岩
dense rubber 致密橡胶
dense sand 密(实)砂
dense sand-gravel mixture 密砂砾混合物
dense sediment 密度大的泥沙;重沉积物
dense set 稠密集
dense sky cover 密集云层
dense sludge 浓密冰泥;海面乱冰
dense smoke 浓烟
dense soda 重碱
dense sowing 密播
dense sowing in line 密条播
dense-sparse structure 疏密构造
dense stand 密林;密集林分
dense state 密实状态
dense structure 致密构造
dense subgraph 稠密子图
dense subset 稠密子集
dense tar macadam 密级配焦油沥青碎石路
dense tar surfacing 密级配焦油沥青面层;密级配焦油沥青路面
dense texture 紧密结构;密实结构
dense traffic 繁密交通
dense traffic line 繁忙干线
dense tubular system 密管系统
dense wood 紧密木材;密纹木(材)
dense zircon block 致密锆英石砖
dense zircon brick 致密锆英石砖
dense zircon refractory 致密锆英石耐火材料
dense zone 波密带
densification 加密;挤压法;密(实)化;压实稠化;

压实;致密;增浓;高密度化;稠化过程;稠化
densification by explosion 爆破压实
densification control point 加密控制点
densification kinetics 致密化动力学
densification network 加密三角网;加密控制网
densification point 加密点【测】
densification process 增密工艺
densification rate 致密速率
densification test 加密试验;击实试验
densified control network 加密控制网
densified hardboard 高密度硬质纤维板
densified impregnated wood 浸压木板;增强
densified laminated wood 密实化胶合木;高密层压木板;强化层积木材;硬化层压木板
densified plywood 压实胶合板;浸胶合板
densified powder 密化粉末
densified system containing homogeneously arranged ultrafine particles 均布超细颗粒的致密体系
densified with small particles 微粒填充
densified with small particles construction material 微粉密实的
densified wood 浸胶木材;密实木材;高密度木材
densified wood door 高密度制造的门
densifier 脱水机;浓缩器;浓缩剂;密实剂;密化器;增浓器;增密器;稠化器;稠化剂
densify 密实;增浓
densifying agent (水泥)增密剂;(水泥)添加剂;稠化混合剂
densilog 密度测井
densilog curve 密度测井曲线
densilog instrument 密度测井仪
densimeter 光密度计;密度计;液体比重计;稠密度计;比重计
densimetric flow 异重流
densimetric Froude number 密度中心弗罗德数
densimetric representation 密度表示法
densimetric velocity 密度中心速度
densimetry 密度法
densinite 密屑体
densi-tensimeter 密度压力计
densith of petroleum 石油的留度
densi(to)meter 显像密度计;密度计;液体密度计;深浅计
densi(to)meter aperture 光密度计光孔
densi(to)meter scale 光密度计刻度
densi(to)metric analysis 密度测量分析
densi(to)metric control 密度控制
densi(to)metric measurement 密度测量
densi(to)metric method 密度法
densi(to)metry 密度计量学;密度测定法;显像密度测定;显微测密术;测密度法
density 密度;致密度;增密;稠密(度)
density airspeed 密度空速
density altitude 密度高度
density analysis of lineament 线性体密度分析
density anomaly 密度异常;密度反常
density arm 密度臂
density bottle 密度瓶;密度测定瓶;比重瓶
density by volume 体密度
density change 密度换相
density change equation 密度变化方程
density change method 密度变化测定法
density channel 密度风洞
density chart 灰度图
density coefficient 密度系数
density component 浓度组分
density concentration 密度浓度
density contrast 密度反差
density control 密实度控制(控制现场浇筑混凝土的密实度);密度控制
density-controlled discharge 密度控制排料
density controller 密度控制器;车流密度控制机
density controlling material 密度调节材料(调节水泥浆或泥浆的密度)
density control unit 浓度控制装置;密度控制装置
density control valve 密度控制阀
density correction 密度修正量;密度校正;密度改善
density current 密度流;混浊流;异重流;重流
density current bed 异重流底层;密度流沉积
density current deposit 密度流沉积
density current in estuary 河口密度流
density current in estuary formed salt water wedge 盐水楔异重流

density current in reservoir 水库异重流
density current model 异重流模型
density curve 密度曲线
density cycle 密度周期
density data 密度数据
density decrease 密度降低
density defect 密度不足
density dependence of growth 密度对生长的相关性
density dependent 密度制约的
density dependent factor 密度制约因素;密度有关因子;密度有关因素;依赖于密度的因素
density depth gradient of concrete 混凝土密度—深度梯度
density difference function 密度差函数
density difference of positive 正片反差
density distribution 密度分布
density distribution function 密度分布函数
density distribution gradient 密度分布梯度
density distribution method 密度分布法
density district 建筑密度限制区;密集区域;密集建筑;密度区域
density drop 密集雾滴
density effect 密集效应;密度效应;异重效应
density error 密度误差
density-exposure curve 曝光密度曲线;密度曝光量曲线
density factor 密度系数
density flow 密度流;异重(水)流;重流
density flow model 异重流模型
density fluctuation 密度涨落;密度起伏;密度变化
density fluid 密度液
density fog 密雾
density fractionation 密度分离法
density function 密度函数
density ga(u)ge 密度计;密度测定器
density gradient 密度梯度
density gradient centrifugation 密度梯度离心分离作用;密度梯度离心(法)
density gradient column 密度气压柱
density gradient separation 密度梯度分离(法)
density-hygrometer method 湿度密度计法
density imbalance 密度不平衡
density impulse 密度冲量
density increase rate 密度增长率
density-independent 非密度制约的
density-independent factor 密度无关因素;密度无关因子;非密度制约因数
density index 相对密度;密度指数
density indicating meter 密度指示计
density indicator 密度指示器;密度计
density interface 密度界面
density interferometer 密度干涉仪
density latitude 灰度范围
density level 密度级
density log calibration value 密度测井刻度值
density log calibrator 密度刻度器
density logger 井下密度测定仪;密度测井仪
density log(ging) 密度测井
density map 密度图
density matrix 密度矩阵
density measurement 密度测量
density measuring instrument 密度计量仪
density meter 密度计;密度表;含泥率计;比重计
density-modulated beam 密度调制光束
density modulation 密度调制
density moisture relation of soil 土的密实度与含量关系
density-neutron method 密度中子测井法
density of aids allocation 设标密度
density of aids-to-navigation installed 设标密度
density of bentonite solution 膨润土胶液密度
density of bore holes 钻孔密度
density of building 建筑(物)密度
density of canopy 树冠覆盖密度
density of cement 水泥密度
density of chain system 链条悬挂密度
density of charge 装药密度
density of charging current 充电电流密度
density of clean sandstone 纯砂岩的密度
density of coal formation 煤层的密度
density of coal seam 煤层密度
density of coal zone 煤层密度
density of colo(u)r 彩色密度

density of compaction 夯实密度
density of compensation 补偿密度
density of container 集装箱重度
density of control 控制网加密
density of cover 植被覆盖密度
density of deposited material 淤积物密度
density of dispersed shale 分散泥质的密度
density of dissection 切割密度
density of distribution 分配密度
density of donor 施主密度
density of dry soil 干土密度
density of dust particles 真密度
density of energy state 能态密度
density of exploratory grid 勘探网度
density of explosive 炸药密度
density of faults 断裂密度;断层密度
density of field 场强;场密度
density of field energy 场能量密度
density of fill 填筑密度
density of fluid in pore 孔隙中流体密度
density of flux change 磁通量变化密度
density of formation water 地层水密度
density of frazil slush 冰花密度
density of freight traffic 货运密度;货运交通密度【道】
density of freight traffic by water 水运货运密度
density of gas 天然气的密度
density of gases 气体密度
density of ga(u)ging station 水文站网密度
density of geographic(al) names 地名注记密度
density of geologic(al) observation point 地质点密度
density of goods flow 货流密度
density of habitable rooms 居室密度
density of hydrogeologic(al) observation points 水位地质观测点密度
density of hydrologic(al) network 水文站网密度
density of imperfection 不完整性密度
density of infection 危害度
density of inhabitation 居住人口密度
density of integration 集成度
density of karst development 岩溶发育密度
density of limestone 石灰岩密度
density of line network 线网密度
density of line of magnetic force 磁力线密度
density of lines of forces 力线密度
density of living floor area 居住面积密度
density of load 荷载密集度
density of long-term observation points 长期观测点密度
density of marked fish 标志鱼密度
density of matrix 岩石骨架密度
density of measuring points 测点密度
density of modes 模密度
density of moist air 湿空气密度
density of network 网密度
density of neutrons 中子密度
density of object structure 目标结构密度
density of observation points 观测点密度
density of occupancy 居住密度
density of occupation 居住密度
density of ore-forming fluid 成矿流体密度
density of packing 填充密度
density of parametric(al) curve 参数曲线密度
density of passenger flow 人流密度
density of passenger traffic 旅客运输密度;客运密度
density of passenger transportation 客运密度
density of pigment 颜料密度
density of plantation 造林密度
density of points 点密度
density of pollutant 污染物密度
density of population 人口密度
density of prospecting net 勘探网密度
density of railway net 铁路网密度
density of railway traffic 铁路运输密度
density of refuse 垃圾密度;废物密度
density of registered inhabitants 居住人口密度
density of reservoir 储集层密度
density of residential floor area 居住建筑面积密度
density of road network 道路网密度
density of route 路线密度
density of runoff 径流密度
density of sampling points 取样点密度

density of sand 砂土的密实(程)度
density of saturation soil 饱和土密度
density of sea-ice 海冰密度
density of seawater 海水密度
density of sediment before compaction 沉积物压实前密度
density of settlements 居民点密度
density of ship flow 船流密度
density of shops 商业网密度
density of sludge 污泥密度
density of snow 雪密度
density of soil 土的密度
density of soil constituents 土的各种成分密度
density of soil grain 土粒密度
density of soil in place 原位土密度
density of soil particle 土粒密度
density of solid particle 固体颗粒密度
density of soundings 测深点密度
density of source rock 烃源岩的密度
density of spectral line 谱线黑度
density of sprinkling 洒水密度
density of states 态密度
density of station network 站网密度
density of subgrade 路基密度
density of submerged soil 浸没土(的)密度
density of suspended liquid 悬浮液密度
density of target structure 目标结构密度
density of test points 试验点密度
density of total passing tonnage 总重密度
density of traffic 交通密度;航行密度;运量密度
density of travel 出行密度
density of treating agent 处理剂浓度
density of vegetation cover 植被密度
density of wagon flow 车流密度
density of wastewater 废水密度
density of water 水的密度
density of water transport 水运密度
density of well spacing 井网密度
density of wind-sand current 风沙流密度
density of wood substance 木材密实度
density operator expansion 密度算符展开
density or erectness 密度或直立程序
density parameter 密度参数
density porosity 密度孔隙度
density porosity of wet clay 湿黏土的密度孔隙度
density portion 密度分布
density probe 浓度探测器;密度探针;密度探测器;同位素探测仪
density profile (沿板厚方向的)密度分布
density range 密度范围
density ratio 浓度比;密度比
density recorder 密度记录器
density reference scale 密度参考标度
density-resistivity method 密度—电阻率(测井)法
density rule 密度标准(木材分类标准);密度规则;密度定则
density scale 密度标度;密度(比例)
density separation 密度分离法
density shale 泥岩的密度
density shift 密度变化
density sliced image 密度分割图像
density slicer 密度分割仪
density slicing 密度分割;密变分割
density slicing device 密度分割仪
density slicing method 密度分割法
density slicing process 密度分割法
density sounded area 测深点加密区
density specific impulse 容积比冲
density-speed relationship 密度车速关系
density-spread 密度分布
density step-procedure 密度分级法
density step tablet 密度阶变图;密度分级片
density stratification 密度分层(作用);密度层
density stratified flow 异重流;成层异重流;成层密度流;分层异重流
density structure of city 城市密度结构
density test 密(实)度试验
density tool 密度测井仪
density transfer 密度调剂
density transformation 密度换算定则
density transition layer 密度跃层
density transition zone 密度跃层;密度变化区
density transmitter 密度传递器
density tunnel 高压风道

density type analyzer 密度分析仪
density underflow 底层异重流;潜异重流
density value of equivalent trace 当量径迹密度值
density value of track 径迹密度值
density variation 密度差异
density (water) stratification 异重水层
density wave 密度波
density wave distribution multiplexer 密集波分复用系统
density wave theory 密度波理论
density wedge 密度梯尺;密度光楔
density wedge filter 可变密度滤光片
density-wooded area 密林区
densofacies 变质相【地】
densogram 密度曲线图
densograph 密度坐标图;密度曲线自动描绘仪;密度曲线;黑度曲线
densometer 感光测定计
densus 浓密
dent 印痕;凹损;凹痕;凹部
dental alloy 补齿合金
dental alloy powder 补齿合金粉末
dental arch 齿弓
dental articulator 咬合架
dentalation 齿状结构
dental baffle 齿形挡板
dental base acrylic 干齿粉
dental brush wheel 洁齿刷轮
dental center 牙科中心
dental clinic 牙病防治所;牙科诊室
dental concrete 填塘混凝土
dental coupling 齿形联轴节
dental flask press 型盒压机
dental formula 齿式;凿式
dental fovea 齿突凹
dental laboratory lathe 技工打磨机
dental material 牙科材料
dental mottling 斑釉病
dental plaque 斑牙
dental porcelain 牙科瓷
dental room 牙室
dental star 齿星
dental surgery 牙科
dental swager 锤造器
dental treatment (坝基的)补缝填坑处理
dental treatment cubicle 牙科手术间
dental treatment of dam foundation 坝基补缝填坑处理
dental treatment room 牙科诊室
dental work 堵漏作业
dentary 齿骨
dentate 齿状的;配位基
dentate adsorption 配位基吸附
dentated bucket lip 齿形挑斗式消力槛;(溢洪道的)挑流齿槛
dentated sill 齿形消力槛;齿(形)槛(水工消能设备)
dentate fascia 齿状回
dentate incised 具齿状缺刻的
dentate layer 齿状层
dentate line 齿状线
dentate sclerite 齿片
dented spring leaf 凹形弹簧片
dented stilling basin 有齿槛消能池;有齿槛消力池
dentel 齿饰【建】
denter 穿筘刀
denticidal capsule 齿裂蒴果
denticle 小齿;齿饰【建】
denticle frieze 横齿饰带;丛齿饰;齿形装饰;齿形雕带
denticular 齿形装饰的
denticular cornice 齿形装饰的檐口;齿形装饰的挑檐
denticulate 齿状装饰的
denticulated 齿形(装饰)的
denticulated mo(u)lding 齿饰线脚
denticulation 小牙饰
denticulation corona 有齿饰的檐板;齿边檐板;齿饰挑檐滴水板
dentiform 齿状的
dentil 檐下齿形装饰;齿形的;小齿;齿形装饰【建】
dentil band 齿饰模板;齿饰线脚;齿饰带【建】
dentil cornice 齿饰挑檐
dentil course 齿形突腰线;齿形挑砖砌层;齿饰线脚

dentil edge 齿状边饰
dentil frieze 横齿饰带;丛齿饰;齿形雕带;齿饰中楣
denting 穿筘
dentist office 牙医室
dentistry 牙科
dentition 齿系
dentoid 齿状的
dent remover 去凹器
dent resistance 抗压痕性
dentrite arm spacing 晶枝间隙
denture acrylic 干托粉
denture clutch 齿轮离合器
denture plastics 补齿塑料
dentures 补齿
denudation 滥伐;剥脱;剥蚀(作用);剥裸;剥露
denudational fault 剥离断层
denudational intensity of orebody 矿体剥蚀程度
denudation coast 剥蚀(河)岸;剥蚀(海)岸
denudation factor 剥蚀因素
denudation intensity 剥蚀强度
denudation level 侵蚀表面;毁坏表面;风化表面;冲刷表面;剥蚀表面
denudation mountain 剥蚀山
denudation of lake 湖蚀作用
denudation plain 剥蚀平原
denudation plane 剥蚀面
denudation process 剥蚀过程
denudation rate 剥蚀(速)率
denudation ratio 剥蚀比
denudation slope 剥蚀坡
denudation soil 剥蚀土壤
denudation terrace 侵蚀地坪;剥蚀岩层;剥蚀阶地
denuded area 已砍伐的林地;剥蚀地
denuded mountain 荒秃山
denuded oil 解吸油
denuded quadrat 除光样方
denuded soil 剥蚀土壤
denuded zone 剥蚀区
denude of fire-fighting force 剥夺消防力量
denuder 溶蚀器
denumerable aggregate 可数集
denumerable function 可数序列
denumerable outcome variable 可数结果变量
denumerable process 可数过程
denumerable sample space 可数样本空间
denumerable set 可列集
denunciation 依约废止;废止合同;退约;声明条约无效
denunciation clause 废弃条款;废除条款;退约条款
denunciation of treaty 条约的废止;条约的废弃
Denver agitator 丹弗型搅拌机
Denver cell 丹弗型(底吹式)浮选机
Denver conditioner 丹弗型矿浆条件箱
Denver hydroclassifier 丹弗型水力分级机
Denver jig 丹弗型跳汰机
Denver mineral jig 丹弗型选矿跳汰机
Denver smaze 丹弗烟霾
Denver unit cell 丹弗型单室浮选机
deny 毛织物
denya 西非硬木(其中的一种)
deodar 雪松
deodar cedar 雪松木材
deodo(u)rant 去臭味剂;除臭剂;除臭的;防臭剂;脱臭剂
deodo(u)rant cream 除臭膏
deodo(u)rant liquid 除臭液
deodo(u)rant soap 除臭(肥)皂
deodo(u)rant solution 除臭液
deodo(u)ration 脱臭
deodo(u)riser 除臭器
deodo(u)rization 除臭过程;除臭(作用)
deodo(u)rization by adsorption 臭味吸附
deodo(u)rization by flame oxidation 火焰氧化法脱臭
deodo(u)rization of air 空气除臭
deodo(u)rization of water 水的除臭
deodo(u)rize 除去臭气;防臭
deodo(u)rizer 解臭剂;除臭器;除臭剂;除臭机;防臭剂;脱臭剂;脱臭机;脱臭剂
deodo(u)rized kerosine 脱臭煤油
deodo(u)rizing 去臭;脱臭
deodo(u)rizing equipment 恶臭处理装置
deodo(u)rizing material 去臭剂;除臭剂;除臭材料
deodo(u)rizing power 除臭能力

deoil 除油;去油
deoiler 除油器
deoiling 脱油
deoiling skimmer 撇油池
deomestication of microorganism 微生物的驯化
deontology 职责学
deorbit 离开轨道;脱离轨道
deores 向下的
deores cone 倒锥体
deoscillator 减振器
deoxidant 脱氧剂
deoxidation 还原;除酸
deoxidation method 脱氧(方)法
deoxidation sphere 漂白斑
deoxidize 除酸
deoxidized copper 脱氧铜
deoxidized steel 去氧钢;脱氧钢
deoxidizer 除氧剂;脱氧剂
deoxidizing agent 脱氧剂
deoxidizing slag 脱氧渣
deoxygenated area 缺氧地区;缺氧地带
deoxygenated waters 脱氧性水体
deoxygenated zone 缺氧地区;缺氧地带
deoxygenation 去氧;脱氧合(作用);脱氧(作用)
deoxygenation coefficient 脱氧系数
deoxygenation constant 脱氧常数
deoxygenization 除氧
deoxygen of the boiler feed water 锅炉给水除氧
deoxy-rhodo-etioporphyrin 脱氧玫红初卟啉
depaint 脱漆
depair 拆开对偶
Depal 德帕尔铝合金
depart 离开;飞出;起程
depart from 离开
departing passenger 离站旅客
department 车间;部门
department account 部门分类账
departmental 部门的
departmental account 分类账;分部账户
departmental accounting 分部会计
departmental budget 分部(门)预算;部门预算
departmental burden 分部制造费(用)
departmental burden rate method 分部负荷率法
departmental capital 分部资本
departmental charges 分部费用
departmental committee 专门委员会
departmental cos(ting) 分部成本;部门成本核算
departmental estimates 分部概算;部门概算
departmental expenses 分部费用
departmental expense allocation sheet 分部费用分配表;部门费用分配表
departmental expense analysis sheet 部门费用分析表
departmental expense rate 部门费用率
departmental financial organization 部门财务组织
departmental financial revenue and expenditures 部门财务收支
departmental force 部分设计组
departmental hour method 分部时间分配法
departmental hour rate 分部时间率
departmental hour rate method 分部小时率法
departmental income statement 部门收益表(损益计算书)
departmentalization 分车间核算;分组组织
departmentalization of overhead 管理费用分车间核算
departmental job order cost accounting 部门分批成本会计
departmental manufacturing overhead rate 分部制造费用率
departmental operation 分部作业;分部经营
departmental profit 分部利益
departmental rate of gross margin 部门毛利率
departmental standard 部部标准
departmental store 百货商店
departmental stores account 分部材料账户
departmental summary 部门累计
departmental system 分科制度;分部制度
departmental system of management 分部门管理体制
departmental trading accounts 分部营业账
departmental unit cost 分部单位成本;部门单位成本
departmentation 划分部门

departmentation specialty stores 专卖公司
department block 百货大厦
department building 百货商店
department chief architect 主任建筑师
department chief engineer 主任工程师
department cost 车间成本
department cost accounting 车间成本核算
department estimates 各部门概算
department finance 部门财务
department foreman 车间领班
department for treatment of raw material 备料车间
department head 部门领导
department in charge 部门主管
departmentized supermarket 分部门的超级市场
department manager 主管经理;部门经理
department margin 部门盈余
department of acupuncture and moxibustion 针灸科
Department of Agriculture 农业部
Department of Air Force 空军部
department of architecture 建筑系;建筑科;建筑部门
Department of Building and Housing 住房建设部
Department of Commerce 商业部
department of cost 成本部门
department of defence specifications 国防部缩微规范
Department of Defense 国防部
Department of Dentistry 牙科
Department of Energy 能源部
Department of Forestry 林业部
department of gynecology and obstetrics 妇产科
Department of Health and Human Service 健康和人类服务部
Department of Housing and Urban Development 住房及城市开发部(美国)
Department of Interior 内政部;内务部
Department of International Economic and Social Affairs 国际经济和社会事务部
Department of Justice 司法部
Department of Labo(u)r 劳工部;劳动部
department of medical administration 医务部
department of navy 海军部
department of ophthalmology 眼科
department of paediatrics 儿科;小儿科
department of public works 公共工程部门
department of real estate 房地产管理部门
Department of Scientific and Industrial Research 科学及工业研究署
department of stomatology 口腔科
Department of Technical Science 科学技术部
Department of Textiles 纺织部
Department of the Army 陆军部(美国)
Department of the Environment 环境署;环境部
Department of Trade 贸易部
Department of Trade and Industry 工贸部(美国)
Department of Transport and Industry 运输及工业部
Department of Transportation 美国运输部;运输部
Department of Transportation design pressure 美国运输部规定的设计压力
Department of Transportation system of labeling 运输部标志系统
Department of Treasury 财政部
department of water administration 水行政主管部门
Department of Water Quality 水质管理局
department of water use 用水部门
department of weights and measures 计量部门
department operation 分科作业
department organization 部门组织
department over-head 车间间接费用
department public works 市政工程局;市政工程处
department rule of environment 环境部门规章
departments engaged in intellectual production 非物质生产部门
department standard 部门标准
department statistics 部门统计
department store 百货商店;商店
department store passenger elevator 百货商店的顾客电梯
department store passenger lift 百货公司载客电梯
department substation 车间变电站

department supervisor 车间主任
department variance on overhead 部门间接费差异
department write-up 部门组织细则
department yard 发车场
departure 离开;离港;开船;横距;东西距;出港;起程点;启程;偏离;始航;始点
departure angle 出发角
departure cell 电力配谱指示
departure course 始航向
departure curve 离开曲线;偏离曲线;偏差曲线
departure dangerous section 出发危险地段
departure draft 离港吃水
departure frequency 发车频率
departure from isochronism 偏离等航线
departure from mean value 偏离平均值
departure from schedule 偏离进度表
departure from symmetry 偏离对称
departure gate 登机口
departure hall 机场候机楼
departure indication 离去表示
departure indicator 发车表示器;偏离指示
departure indicator circuit 发车表示器电路;偏离指示器
departure interval 发车间隔
departure line 出发线【铁】
departure line from depot 出段线【铁】
departure lounge 候机室;候车室;登机室
departure of tidal hour 潮时偏差
departure of train 列车开行
departure platform 出发站台;发车站台
departure point 出发点;起点;起航点;始航点
departure position 始航船位
departure process 离散过程
departure quarantine inspection 出口检疫
departure report 出港申请书
departure-returning form engine shed line 机车出入库线
departure route 发车进路
departure runway 起飞跑道
departure set of sidings 出发线束;出发线群
departure signal 出发信号;发车信号
departure time 出发时间;撤离时间;起飞时刻
departure time indicator 发车时间指示器
departure track 出发线【铁】;发车线【铁】
departure track indicator 发车线路表示器
departure yard 出发场【铁】;发车场;出发车场
depauperate colony 衰落群体
depegram 露点图
dependability 可依赖性;可信赖性;可靠性;信托
dependability of experience 经验可靠性
dependable 可靠的
dependable capacity 可靠容量;可靠库容
dependable crystal filter 恒特性水晶滤波器
dependable discharge 可靠(泄)流量;保证流量
dependable energy 可靠电能;保证出力
dependable flow 可靠流量;保证流量
dependable flow rule curve 保证供水调度线
dependable hydroelectric(al) capacity 可靠水电容量;保证水力发电量
dependable inflow 可靠来水量;保证入流量
dependable output 可靠出力;保证出力
dependable power 可靠功率;保证功率
dependable yield 可靠产(水)量;保证水量;保证产(水)量
dependence 相依性;相关(性);依赖性;依靠;依从关系;从属
dependence among equations 方程相关
dependence effect 依存效应
dependence function 相关函数
dependence graph 依赖关系图
dependence list 相关表
dependence method 依数法
dependence on foreign trade 贸易依存度
dependence on import 依赖进口
dependence relation 相依关系
dependence relationship 依存关系
dependence structure 依附结构
dependence test 依赖性试验
dependency 建筑物的一翼;从属;附属建筑(房屋);附属国
dependency allowance 抚养津贴
dependency coefficient 负担系数
dependency number 相关数
dependency preservation 依赖保持;从属性保护

dependency relation 相关关系
dependency relationship 相依关系;依赖关系
dependent 从属的
dependent action 相依行动
dependent agents 非独立代理人
dependent block 附属建筑
dependent building 附属建筑物
dependent contract 附属契约;附属合同
dependent control 从属控制
dependent coordinates 相关坐标
dependent country 附属国;附庸国
dependent covenants 相依契约
dependent deed 附属契约
dependent economy 从属经济
dependent equation 相关方程
dependent equatorial coordinates 第一赤道坐标
dependent equatorial coordinate system 第一赤道坐标系;时角赤道坐标系统
dependent error 相关误差;非独立误差
dependent event 相依事件;相关事件
dependent factor 依存因子
dependent failure 相关故障;生产误差故障
dependent function 相依函数
dependent functions 相关函数组
dependent information of evaluation 评价依据资料
dependent linear equation 相关线性方程
dependent mobile home 挂带的活动住房
dependent network 非独立网【测】
dependent observation 非独立观测
dependent on soil temperature 取决于土壤温度
dependent population 被抚养人口
dependent program(me) 相关程序
dependent random event 相依随机事件
dependent resurvey 不独立重测
dependent sample 相依样本
dependent signal(l)er 从属信号机
dependent time lag 相依时滞
dependent time-lag relay 变时限继电器;变时滞继电器
dependent trailer coach 汽车拖的活动房子
dependent triangulation chain 附合三角锁
dependent type 连接式
dependent unit 附属单元
dependent variable 因变数;应变数;应变量;因变量;附条件的变数;他变量
dependent variation in general dense 总体相依变化
dependent variation in local range 局部相依变化
depending air and water 依靠空气和水
depending on the city and serving the city 依托城市,服务城市
depending on the condition of the soil 取决于土壤的情况
depend on each other for existence 互相依存
depend on import 依靠进口
Depeq 德佩克风
depergelation 永冻层融化(过程)
depersonalized repair method 混装修理法
depeter 粉石齿面;仿斧鏨石;粉石凿面;碎石面饰(灰砂未干时,压碎石入内的一种墙壁面饰法)
Dephanox process 德法诺克斯工艺
dephased current 移相电流
dephasing 移相
dephasing measure 移相量度
dephenolization 脱酚(作用)
dephenolizer 脱酚剂
dephenolizing plant 除酚设备;除酚车间
Dephi method 德尔菲法
dephlegmating column 分馏柱
dephlegmation 分凝(作用);分馏
dephlegmator 蒸馏器;分缩器;分凝器;分馏器
dephlogistication 脱燃素(作用)
dephophorization 脱磷
dephosphorization 除磷
dephosphorize 脱磷
dephosphorized pig iron 脱磷生铁
dephosphorizing 脱磷
dephosphorylation 去磷作用;脱磷酸
depiction in outline 勾划轮廓
depigment 脱颜料
depigmentation 脱颜料(作用)
depiler 进料台;分送机

de-piling crane 叠板卸垛吊车
de-piling equipment 储料卸出台
deplate 除镀层;退镀
depletable resources 有限的资源;可耗竭资源;可耗尽资源
depleted 废弃的
depleted brine 废盐水
depleted cost 耗余成本;已(折)耗成本
depleted electrolyte 废电解液
depleted filtrate 废滤液
depleted fuel 贫化燃料
depleted material 贫化材料
depleted uranium 贫化铀
depleted well 枯竭井;耗竭井
deplete semiconductor 耗尽型半导体
depleting water source 消耗型水源地
depletion 亏水;枯竭;减液;耗损;折耗;贫化;损耗
depletion accounting 折耗会计
depletion allowance 可免税耗减;备抵折耗;备抵耗减
depletionary phase 枯竭阶段
depletion assets 耗损资产
depletion coefficient 疏干系数
depletion cost 耗损成本;耗竭成本;折耗成本
depletion curve 枯竭曲线;亏水曲线;亏耗曲线;退水曲线
depletion deduction 由于矿藏量衰竭而减税
depletion device 耗尽型器件
depletion-drive pool 消耗气驱油藏
depletion expenses 耗减费用;折耗费用;损耗费(用)
depletion flocculation 减液絮凝作用
depletion for wasting assets 消耗资产折耗
depletion hydrograph 亏水过程线;亏耗水文曲线(图)
depletion layer 减压层;耗尽层
depletion load 耗尽型负载
depletion mode 耗尽型
depletion of aquifer 含水层枯竭
depletion of energy 能源危机;能源枯竭;能量枯竭
depletion of foreign exchange reserve 外汇储备耗尽
depletion of groundwater 地下水亏耗;地下水的消耗
depletion of magnitudes 数量减少
depletion of ozone layer 臭氧层消耗;臭氧层耗尽
depletion of snow cover 积雪消融
depletion of soil 土壤流失
depletion of stock 资源枯竭
depletion rate 亏耗率
depletion rate of a year 年退水量
depletion region 耗尽区
depletion reserve 耗减准备
depletion response 减液反应
depletion stabilization 减液稳定作用
depletion stage 采竭阶段
depletion test 排空试验
depletion to additive 添加剂的消耗
depletion type channel 耗尽型沟道
depletion yield 枯竭性抽水量;疏干开采量
deploid 扁方二十四面体
deploy 布设;配备;疏开
deploy equipment to utmost capacity 拼设备
deploy fire-fight equipment 部署消防设备
deploy funds 配置资金
deploying plan of salvage operation 打捞施工展布图
deployment 调度;采纳;部署;布设;调配
deployment analysis 部署分析
deployment for dredging commencement 开工展布
deployment of labor force 调配劳动力
deployment of manpower 劳动力调配
deployment system 调度系统
depocenter 沉积中心
depolariser 去极剂;消偏振器;消偏(振)镜;消偏光振镜
depolarization 去极化(作用);消偏振(作用);除极(化)
depolarization band 退极化谱带
depolarization effect 去偏振效应
depolarization factor 去极化系数;消偏振因素;退极化因子
depolarization field 退极化场

depolarization phase 除极期
depolarize 去偏振;去磁;退极化
depolarizer 去极化器;去极(化)剂;消偏振器;消偏振镜;消偏镜;消偏光振镜;退极化剂
depolarizing factor 去极化因子
depolish 变粗糙
depolished glass 磨砂玻璃;毛玻璃;漫射玻璃
depollution 去污作用;消除污染
depollution of environment 环境去污
depollution of water 水的去污染;水的净化
depolymerization 聚解;解聚(作用)
depolymerized rubber 解聚橡胶
depopulate 人口减少
depopulation 人口减少;疏散人口
deport 卖空投机;移运
deportation 驱逐出境
deposed from office 免职
depositary 储[贮]藏所;保管人
deposit(e) 寄存;蕴藏量;存款;沉着物;沉积(物);沉淀(物);熔敷金属;托管;水柜
deposit(e) account 定期存款;存款账户
deposit(e) accumulation 沉积物的累积;沉积物的聚积
deposit(e) accumulation test system 积垢试验系统
deposit(e) allowance 存款利息
deposit(e) area 弃土区;排泥区
deposit(e) as collateral 以存款作担保
deposit(e) at call 即期存款;通知存款
deposit(e) at notice 通知存款
deposit(e) attack 沉积侵蚀;沉积腐蚀
deposit(e) book 存款簿
deposit(e) brine 沉积型卤水
deposit(e) builder 附着物
deposit(e) buildup 沉积物的聚积
deposit(e) capital ratio 存款与资本比率
deposit(e) capping 矿帽
deposit(e) carbon 沉积碳
deposit(e) certificate 存款存折
deposit(e) chamber 沉淀槽
deposit(e) combined account 综合存款方案
deposit(e) concrete 浇筑混凝土
deposit(e) control 沉积物控制
deposit(e) control agent 沉积物控制剂
deposit(e) corrosion 沉积腐蚀
deposited 沉积的
deposited alumin(i)um conductor 铝蒸发导体;铝淀积导体
deposited bank 收受存款银行
deposited carbon resistor 沉积炭电阻器;炭膜电阻器
deposited coating 沉积涂层
deposited copper 沉积铜
deposited drawings 工程图副本;备用图纸
deposit(e) development system 矿床开拓系统
deposited film 沉积膜
deposited island 沉积小岛
deposited material 沉积物;沉积泥沙;沉淀物
deposited metal 沉积金属;堆焊金属;熔接金属;熔敷金属;溶敷金属;喷镀层
deposited metal test specimen 熔敷金属试件
deposited on top of clay layers 沉积于黏土层面上的
deposited on top of sand layers 沉积于砂层面上的
deposit(e) dose 沉积剂量
deposited particulate matter 沉积的颗粒物质
deposit(e) drainage 矿床疏干
deposited reserve 存储备金
deposited sediment 淤积泥沙
deposited silt 淤泥;淤积物;沉积泥沙
deposited soil 沉积土壤
deposited sum 总存款
deposit(e) due from bank's account 应收银行到期存款
deposit(e) due to other banks 同业存款
deposited weathering crust 堆积型风化壳
deposit(e) fill material 放置填充材料;填方填充材料;倾填方材料;抛填方材料
deposit(e) for bidding documents 投标(文件)押金
deposit(e) for consumption 消费保证金
deposit(e) formation 沉积层
deposit(e) for safe custody 保护存款

deposit(e) for security 保证金
deposit(e) for tax payments 纳税准备存款;纳税住所纳税准备存款
deposit(e)-free 无沉淀物
deposit(e) from employees 职工存款
deposit(e) fund 周转金;存款基金
deposit(e) ga(u)ge 落灰计;降尘测定器;降尘计;沉积计;沉淀测量仪
deposit(e) ground 弃土区;排泥区
deposit(e) in a fictitious name 化名存款
deposit(e) in current account 往来账户存款
deposit(e) inducement 存款吸收
deposit(e) in foreign currency 外币存款
deposit(e) in-situ 原地淤积;原地沉积
deposit(e) institution 存款机构
deposit(e) interest rate 存款利率
deposit(e) in trust 信托存款
deposit(e) investigation 淤积调查
deposit(e) ledger 存款总账
deposit(e) line 平均存(款)余额
deposit(e) loan 以存款作抵押的贷款
deposit(e) loan ratio 存贷比率
deposit(e) metal 焊缝金属;熔化金属
deposit(e) modeling method 矿床模型法
deposit(e) modifier 沉积物改性剂
deposit(e) of asphaltic bitumen 浇灌沥青
deposit(e) of compound origin 复合成因矿床
deposit(e) of flue dust 烟尘沉积
deposit(e) of saline lake 盐湖矿床
deposit(e) of scale 水垢沉积(物);水垢
deposit(e) of suspended sediment 悬沙沉积
deposit(e) of the award 提交仲裁裁决
deposit(e) of the deep sea 深海沉积
deposit(e) of title-deeds 土地证抵押;权利证书托存
deposit(e) of withhold tax 代扣预提税
deposit(e) on contracts 合同保证金
deposit(e) on long-term lease 长期租贷预收款
deposit(e) paid on construction works 建设工程预付款
deposit(e) pass-book 存折
deposit(e) premium 预付保(险)费;预付保证金
deposit(e) rate 沉积速率;沉积速度;存款利率
deposit(e) receipt 定金收据;存(款)单;存款收据;保证金收据;保管收据;送款薄
deposit(e) received 预收款;已收存款
deposit(e) received on sale 预收货款
deposit(e)/refund system 押金退款制度
deposit(e) released 退还准备金;退还的押金
deposit(e) releases 退还保证金
deposit(e) reserves nuclear counting 矿床储量核算
deposit(e) retained 库存储备金;扣留的押金;扣存准备金;扣存押金
deposit(e) site of sludge (泥石流)停淤场
deposit(e) slip 送款单;存款单
deposit(e) slip book 存款条簿
deposit(e) steel 熔敷钢
deposit(e) structure 沉积构造
deposit(e) teller 收款员
deposit(e) ticket 存款单
deposit(e) turnover 存款周转
deposit(e) with landlord 押租
deposit(e) with public utility corporate 公用事业保证金
deposit(e) works 民间工程;非政府资助的工程
depositing concrete 浇注混凝土;浇筑混凝土
depositing dock 单坞墙式浮(船)坞
depositing fill 抛堆土石料
depositing floating dock 升托式浮船坞;沉降式浮船坞
depositing ground 垃圾区;倒泥区
depositing procedure 浇灌过程
depositing reservoir 澄清池;沉沙库;沉淀池
depositing-reworking current 沉积—改造水流
depositing site 沉积塘;沉积池;吹填区
depositing substrate 堆积场基底
depositing tank 沉积池;沉积槽;沉淀池;沉淀槽
depositing the fill(ing)的沉积块
depositing underwater concrete in bags 水下袋内浇进混凝土
deposit insurance act 存款保险法案
deposition 结垢;沉灰;沉着;沉积作用;沉积率;沉积量;沉积(法);沉淀;笔录供词;喷镀
deposition above dam 坝上游淤积

deposition ages 沉积年龄
depositional 沉积作用的;沉积的
depositional architecture pattern 沉积构成样式
depositional basin 沉积盆地;沉积池
depositional boundary 沉积边界
depositional clay 沉积黏土
depositional cycle 沉积循环;沉积旋回
depositional detritus remanent magnetization 沉积碎屑剩余磁化强度
depositional dip 沉积倾角
depositional environment 沉积环境
depositional environment of carbonate reservoir 碳酸盐储层沉积环境
depositional environment of coal formation 含煤岩系沉积环境
depositional environment of organic matter 有机质沉积环境
depositional environment of sand body 砂体沉积环境
depositional fabric 沉积组构;沉淀组构
depositional facies 沉积相
depositional force 沉积力
depositional form 沉积地形
depositional framework 沉积格架
depositional gradient 原始坡度;自然坡度;沉积坡度;天然(沉积)坡度
depositional interface 沉积间面
depositional material 沉积物
depositional rate 沉积速率;淤积速率
depositional rate method of sedimentary rock volume 体积沉积速率法
depositional remanent magnetism 沉积剩余磁性
depositional remanent magnetization 沉积剩余磁化;沉积剩磁
depositional ridge 沉积脊
depositional sequence 沉积层序
depositional setting 沉积背景
depositional soil 沉积土
depositional speed 淤积速度
depositional strike 沉积走向;侧向连续沉积
depositional system tract 沉积体系域
depositional topography 沉积地形
depositional trough 沉积沟;沉积槽
depositional velocity 淤积速度
deposition and conservation of organic matter 有机质的沉积和保存
deposition and erosion 淤积和冲刷
deposition and erosion control 沉积与腐蚀控制
deposition area 堆积区
deposition barrier 沉积障
deposition bluff 沉积(堤)岸
deposition burner 沉淀燃烧器
deposition by flood 洪水淤积;洪水沉积
deposition by wind 风积
deposition centre 沉积中心
deposition condition 沉积条件
deposition constant 沉降常数
deposition efficiency 焊敷效率;沉积效率;熔敷效率;熔敷系数
deposition fabric 沉积结构
deposition factor 沉积因素
deposition interface 沉积界面
deposition layer 淤积层
deposition mode 沉积方式
deposition model 沉积模式
deposition of coal particle 煤粒沉淀
deposition of flood current 洪流沉积作用
deposition of groundwater 地下水沉积作用
deposition of lake 湖泊沉积作用
deposition of paraffin 石蜡沉积
deposition of particles 质点沉积;粒子沉积;颗粒沉积
deposition of sediments 泥沙填塞
deposition of silt 淤泥沉淀;泥沙淤塞;泥沙沉积
deposition of suspended load 悬移质淤积
deposition of suspended material 悬移质淤积
deposition of tools 工具储备
deposition of wind 风的沉积作用
deposition potential 沉积电势
deposition process 沉积(工艺)过程;沉淀法;沉淀法
deposition rate 沉淀率;熔敷速度;焊着率(单位时间消耗堆焊金属的重量)
deposition reaction 沉积反应

deposition sedimentation 沉积沉淀
deposition sequence 焊着次序;焊着程序;熔敷顺序
deposition shoreline 沉降滨线
deposition speed 沉积速度
deposition tester 沉积试验器
deposition thickness 淤积厚度;沉积厚度
deposition time 沉积时间
deposition velocity 沉降速度;沉积速度
deposit metal 堆焊金属
depositor 拉条成型机;存款人;沉积器
depositorial clay 沉积土
depository 存放处;储存所;仓库;保管人(员)
depository dish 储藏盒
depository documents 存放单据
depository receipt 存单
depository receipt procedures 财政收入缴存手续
depository receipt system 财政收入缴存制度
depository right 存货权
depository trust company 储[贮]存信托公司;保管信托公司
deposits and savings 存储
deposits model 矿床模型
deposits of one-year maturity 一年为期的定期存款
deposits of six-month maturity 半年定期存款;半年为期的定期存款
deposits within the company 企业内部存款
deposit-taking institutions 接受存款机构
depositum 淡秋石
depot 机车库[铁];储[贮]存;储[贮]藏所;站段;储存;储藏库;车站;车辆段;仓库;仓储设施
depot and warehouse 仓库栈房
depot and warehouse sites 仓库栈房土地
depot broadcasting system 车辆段广播系统
depot control center 车辆段控制中心
depot driver 机务段司机
depot equipment 维修基地设备
depot fat 储脂
depot forming adjuvant 储存佐剂
depot fuel 库存燃料
depot negative return current system 车辆段负回流系统
depot plan 货场计划
depot repair 修理基地;修理厂
depots for commercial oil 商业石油库
depot ship 母舰;供应维修工作船;补给修理船
depot spare part 库存备件
depot superintendent 罐区主任
depot test run 车厂运行测试
depot-to-depot through goods train 区段货物列车
depot track 段管线;入库线
depot train controller 车辆段列车管理员
depot tug 港作拖轮
depot workshop equipment 车辆段设备
depreciable assets 应折旧资产
depreciable basis 折旧基数
depreciable cost 应折旧成本;应折旧价值
depreciable fixed assets 应力折旧的固定资产
depreciable life 折旧年限;使用寿命;应折旧年限
depreciable property 应折旧财产
depreciated cost 已折成本
depreciated currency 贬值通货;贬值货币
depreciated money 贬值货币
depreciated value 资产账面价值;折旧余值;已折旧价值;折余(价)值
depreciation 减值;折旧
depreciation account 折旧账户
depreciation accounting 折旧会计
depreciation accounting on a group basis 分组折旧法
depreciation accrual rate 折旧增值率
depreciation-age-life method 寿限折旧法
depreciation allowance 折旧费;可免税的折旧;折旧金提成
depreciation allowed 可提折旧额
depreciation and amortization 折旧及摊提
depreciation and apportion charges of unit footage of tunneling 单位进尺折旧摊销费
depreciation-annuity method 年金折旧法;折旧年金法;按年金折旧法
depreciation appraisal 折旧估价
depreciation-appraisal method 估价折旧法;估定折旧法;折旧估价法
depreciation-appropriation method 盈余拨抵折旧法

depreciation arbitrary method 任意折旧法;折旧任意决定法
depreciation assets 折旧资产
depreciation base 折旧基数;折旧基价
depreciation budget 折旧预算
depreciation by kind 分类折旧
depreciation-changing percentage of cost less scrap 余值变动折旧
depreciation-changing percentage of cost less scrap method 按成本减残值变率折旧法
depreciation charges 折扣;折旧费
depreciation charges of equipment 设备折旧费
depreciation charges per period 每期折旧费用
depreciation-charging percentage of cost less scrap method 折旧原值减残值百分法
depreciation-composite life method 综合平均年(限)折旧法;折旧综合平均寿命法;平均年限折旧法
depreciation costs 折旧费(用)
depreciation credit 折旧税收抵免
depreciation deduction 折旧减免额
depreciation-double declining balance method 双重余额递减折旧法;加倍定率递减折旧法
depreciation during construction 建筑期间折旧
depreciation-equal-annual payment method 年金折旧法
depreciation expenses 折旧费(用)
depreciation expense of house 房屋折旧费
depreciation factor 减光补偿系数;折旧因子;折旧系数;折旧率
depreciation fees from land funds 土地资金折旧费
depreciation-fifty percent method 折旧五成法;五成折旧法
depreciation-fixed instalment method 分期定额折旧法;折旧分期定额法
depreciation-fixed percentage of cost 成本定率折旧
depreciation-fixed percentage of diminishing value 定率减值折旧
depreciation-fixed percentage of diminishing value method 折旧照递减值定率百分法
depreciation fund 折旧基金;折旧费
depreciation-gross earning 总收益提成折旧
depreciation-gross earning method 总收益提成折旧法;折旧全年总收入法
depreciation-insurance method 折旧预储保险金法;按保险金额折旧法
depreciation-inventory method 盘存折旧法
depreciation-job method 分批折旧法
depreciation-maintenance method 折旧根据修缮费用计算法;维修折旧法
depreciation method 折旧(方)法
depreciation method of fixed percentage on cost 按成本的固定百分比折旧法
depreciation method of uniformity varying amounts 变数均匀折旧法;统一变数折旧法;统一变额折旧法
depreciation of equipment 设备折旧
depreciation of fixed assets 固定资产折旧
depreciation of machinery 机械折旧
depreciation of major repair 大修(理)折旧
depreciation of mechanical plant 机械折旧
depreciation of plant equipment 厂场设备折旧
depreciation of the year 本年度折旧
depreciation of value 贬值
depreciation on franchises 专利折旧
depreciation on property 财产折旧
depreciation on replacement value 重置价值折旧;按重置价值折旧
depreciation-percentage of original cost method 原始成本定率折旧法;折旧原始成本定率法;折旧按原值百分计算法
depreciation period 折旧期
depreciation price 折旧价
depreciation provision 备抵折旧
depreciation quota 折旧定额
depreciation rate 折旧率
depreciation rate of fixed assets 固定资产折旧率
depreciation recapture 折旧回收
depreciation reducing instalment method 分期递减折旧法;折旧分期递减法
depreciation-replacement method 重置折旧法
depreciation reserve 折旧准备(金);折旧准备(回收)

depreciation reserve building 建筑折旧储备金
depreciation reserve furniture and fixture 家具设施折旧储[贮]备金
depreciation reserve of delivery equipment 运输机械折旧储[贮]备金
depreciation reserve ratio 折旧准备率;折旧储[贮]备比率
depreciation revenue method 营业收入折旧法;收入折旧法
depreciation salvage value 折旧残值
depreciation-service capacity method 按生产能力折旧法
depreciation-service output method 按生产数量折旧法;折旧产量法
depreciation-sinking fund method 折旧偿债基金法
depreciation straight line 平均折旧
depreciation-straight-line method 直线折旧法;折旧直线法
depreciation study 折旧研究
depreciation-sum of expected life method 预计年限总数折旧法
depreciation table 折旧累计法
depreciation-uniformity varying amounts method 统一变额折旧法
depreciation-unit cost method 折旧单位成本法
depreciation-unit method 个别折旧法
depreciation-unit of product method 单位产量折旧法
depreciation value 折旧价(值)
depreciation-working hour 工作时数折旧
depreciation-working hour method 按工作时数折旧法
de-preservation 解除保护
depress 拉下;压下;推下
depressant prescription 降剂
depressed 凹下的
depressed arch 小矢高拱;四心拱;低(圆)拱;垂拱;扁平拱;(拱心在起拱线下的)平坦拱;坦圆拱
depressed area 工商业萧条的地区;萧条地区;低洼地;不景气地区;洼地
depressed basin 沉降盆地
depressed block 陷落地块
depressed center car (运大件的)凹形车
depressed center flat car 元宝车;凹底平车
depressed center flat wagon 低凹平车
depressed center wagon 凹底车
depressed cladding fiber 下陷包层光纤
depressed coast 沉陷海岸;沉降海岸
depressed collector 降压收集极
depressed conductor-rail 压下的接触轨
depressed-deck car 凹底车
depressed design 压花
depressed floor 下沉式地板
depressed flute cast 凹槽流痕
depressed fracture 凹陷断裂
depressed freeway 路堑式高速公路;堑式高速公路
depressed frog point 降低的辙叉尖端
depressed goods section 低货位
depressed highway 下穿路;堑式公路
depressed joint 钢轨低接头;低接头
depressed market 萧条市场
depressed nappe 下凹式水舌
depressed olfactory bulbar response 被抑制的嗅球反应
depressed open cutting 低于地面的明开挖
depressed panel method 倾斜拼合模板;凸边格形模板
depressed pole 俯极
depressed region 下降区
depressed road(way) 低于地面的道路;堑式道路;路堑;下穿路;低堑道路
depressed sewer 下穿污水管;倒虹式下水道
depressed sewer with two pipes 具有两管的下穿污水管
depressed track 低陷铁路;低线路
depressed trough 凹槽;路槽
depressed water-table 降落的地下水位;降低的地下水位
depressed weir 低堰;低坝
depressed well 凹槽
depressed well car (运大件的)凹形车
depressed zone 低洼地
depresses key 按钮
depresses sewer 倒虹吸污水管
depressing agent 浮选抑制剂

depressing effect 压制效应
depressing of the draw bar 压下引砖
depressing roll 升降辊
depressing spring 溢泉
depressing table 支撑辊道
depression 萧条;阻抑;低压;低度;抽空区;抽空;沉降;沉降;拗陷;凹陷;凹坑;凹地;洼地
depression angle 垂度;沉降角;俯(视)角;伏角;倾角;偏角
depression area 沉陷带;沉降区
depression basin 拗陷盆地
depression belt 低(气)压带
depression box 压力室
depression coast 下降海岸;下沉海岸;沉降海岸
depression cone 沉陷锥;水位降落漏斗;降水漏斗
depression cone method 降落漏斗法
depression contour 洼地等高线
depression crater 陷落喷火口
depression curve 降落曲线;降低曲线;下降曲线;凹曲线;浸润曲线
depression curve equation of confined water 承压水降落曲线方程
depression curve of confined water 承压水降落曲线
depression curve of phreatic water 潜水浸润曲线
depression detention 洼地滞洪
depression earthquake 陷落地震
depression effect 抑制效应
depression head 降落水头;下降水头
depression in shipping 航运业不景气;航运萧条
depression land 凹地
depression line 示坡线;示波线
depression of continent 大陆下沉
depression of contour 降坡线
depression of dew point 露点温(度)差
depression of freezing point 冰点降低
depression of ground 地面沉降
depression of horizon 俯视角
depression of joint 接头下陷
depression of land 陆凹;内陆洼地
depression of mediastinum 纵隔凹陷
depression of order 降阶法;阶次法
depression of rail joint 钢轨接头下垂
depression of sea horizon 海地平(线)俯角
depression of shore horizon 岸线地平俯角
depression of streets 降低街道
depression of supports 支座下沉;支座沉陷;支座沉降
depression of surface level 液面降低;液位控制降低;平面凹陷
depression of the dew point 温度露点差;温度与露点温度带
depression of the sea horizon 眼高差
depression of the wet bulb 干湿球温度差
depression of trade 贸易衰退
depression of water table 地下水位的降落
depression of wet-bulb 干湿球温差
depression of zero-reading 零点下降;冰点下降
depression pole 凹陷极
depression position finder 俯角取景器;竖基尺视距仪
depression range finder 竖基尺视距仪
depression shoreline 沉降岸线
depression signal 低气压信号
depression spring 溢泉;低洼泉;低地泉
depression storage 坑洼储水;坑洼蓄水;洼地蓄水(量);洼地储水(量);填洼水量
depression tank 减压箱;真空箱
depressive earthquake 陷落地震
depressor 缓冲机;压器;压峰剂;揿压器
depressor reflex 减压反射
depressor substance 减压物质
depress the pole 向赤道方向航行
depreter 碎石面饰;粉石齿面;洗碎石饰面;洗麻石刷面【建】
depretor 斩假石面
deprivation 脱除
deproceduring 非过程化
depropagation 链断裂(作用)
depropanizator 脱丙烷塔
depropanize 丙烷馏除
depropanizer 脱丙烷塔;丙烷馏除器
depropanizing column 脱丙烷塔
depropenizer 脱丙烷塔

deproteinized rubber 除蛋白橡胶
depth 进深;厚度;船深;深度
depth adjuster 深度调节器
depth-adjusting screw 耕深调节螺杆
depth aeration 深水曝气
depth alongside 码头边水深
depth alongside wharf 码头前沿水深
depth-area curve 面深曲线;雨深—面积关系曲线;雨量—面积关系曲线
depth-area-duration analysis 雨深—面积—历时分析;雨量—面积—历时分析
depth-area-duration curve 雨深—面积—历时关系曲线
depth-area formula 雨量—面积公式
depth-area relationship 面深关系;面积—深度关系
depth at chart datum 基准面以下深度
depth at quay 码头前沿水深
depth attachment 测深附件
depth-beam ratio 梁的高度与跨长之比
depth behind apron 港口陆域纵深
depth behind the port apron 港口陆域纵深
depth bomb 深水炸弹
depth bound 深度限度
depth calliper 深度卡规
depth capacity 最大深度
depth charge 深水炸弹;深水炸药包
depth chart 海深海图;测深图;水深图
depth clarifier 深层澄清池
depth classification 深度分类
depth clearance (航道的)附加水深;(航道等的)富余水深
depth clipping 深度剪取
depth compensated transducer 深度补偿换能器
depth compensated wet suit 深度补偿湿式潜水服
depth contour 莫霍面深度;等(水)深线;等海深线
depth contour chart 等深线海(底地形)图
depth contour map 等高线地图
depth control 深度控制;高度调节
depth control cam 耕深调节凸轮
depth control device 耕深调节器;深度控制装置
depth controller 深度控制器
depth control of dredge-cut 挖槽深度控制
depth control pipeline 深浅调节导管
depth control skid 仿形滑板
depth conversion 深度变换
depth correction 校正深度
depth creep 深层蠕动
depth crowning 齿高修形
depth-current meter 深水自记流速仪;深水自记海流计;深层(自记)海流计
depth-current recorder 深水自记流速仪;深水自记海流计;深层(自记)海流器
depth curve 进深曲线;等深线;深度曲线
depth datum 深度基准(面)
depth deficiency 水深不足
depth determination 深度测定
depth determined from cable mark 由电缆上的深度记号确定深度
depth determined from corrected curves 由曲线经校正后确定深度
depth determined from the roll of depth inversion 由深度转换轮确定深度
depth-determining sonar 测深声呐;回声探深仪;回声测深声呐
depth development 深层显影剂
depth deviation indicator 深度变化指示器
depth dial ga(u)ge 度盘式深度计
depth difference 景深差别
depth-diffusion process 深结扩散工艺
depth direction 景方向
depth discrimination 景深鉴别
depth distribution 深度分布
depth dose 深度剂量
depth dose distribution 深度剂量分布
depth drill 深孔钻
depth-duration curve 深—时曲线;雨深—历时(关系)曲线;雨量—历时关系曲线
depth effect 深度效应;深度效果
depth encoder 深度编码器
depth exponent 深度指数
depth face 铺块厚度
depth factor 埋深因素;深度因素;深度因子;深度系数
depth factor of slope 土坡深度系数

depth-feed mechanism 径向进给机构
depth figure 深度注记;深度数值
depth filtration 深床过滤
depth finder 回声探深仪;回音探深仪;测深仪;测深器;测深计
depth finder recorder 测深仪记录器
depth-first 深度优先
depth-first minimax procedure 深度优先最小最大过程
depth-first procedure 深度优先过程
depth-first search 深度优先搜索
depth float 深水浮子;水下浮标;双浮标;深水浮标
depth for siltation 备淤深度
depth ga(u)ge 高度计;检潮标;测深仪;测深量尺;测深规;水尺;深度计;深度规;深度表
depth ga(u)ge clearance 齿深规间隙
depth ga(u)ge setting 齿深规整齿
depth ga(u)ge tool 深度计仪器
depth gradient 深度梯度
depth-hardness curve 淬透性曲线
depth hoar 浓霜
depth ice 底冰;深水(底)冰
depth index 深度指标
depth indicator 刻痕深度指示器;回声测深仪;下潜深度指示器;深度指示仪;深度指示器;深度尺;深度表
depth indicator of shaft 竖井深度计
depth information pulse 深度信息脉冲
depthing tool 测啮合深度工具
depth-integrated method 积深法
depth-integrated sample 积深式水样
depth-integrated sampler 积深式采样器
depth-integrated sampling 积深式采样法
depth-integrated sediment sample 积深泥沙试样;含沙量积深试样
depth-integrating sampler 深度累积式取样器;积深式采样器(测含沙量)
depth-integrating sediment sample 随深度取的沙样
depth-integrating sediment sampler 随深度取沙器
depth-integration sampling 积深法采样;全深集总采样
depth interpretation 深度推断
depth limitation device 挖深限制装置
depth line 深度线
depth magnification 景深放大率
depth manometer 测深压力表;深度压力计
depth marker 深度标志
depth measurement 深度测量;测量水深;深度测量
depth measurement meter 测深仪
depth measurer 测深器
depth-measuring device 测(孔)深设备
depth-measuring vertical 测深垂线;测深垂丝
depth meter 深度(测量)计
depth micrometer 深度千分尺;深度测微计
depth migration 垂直洄游;深度偏移
depth migration operator 深度偏移算子
depth minimax search sequence 纵型最大最小探索次序
depth mo(u)lded 型深
depth no-motion 无流深度
depth number 深度注记
depth of abnormal section for well logging curve 测井曲线异常段深度
depth of a column 柱埋深度
depth of aeration tank 曝气池池深
depth of a flag 旗宽
depth of allowable aquifer dewatering 含水层允许疏干深度
depth of anomaly emerged 异常现象出现的孔深
depth of apical plate of aquifer 含水层顶板埋深
depth of arch 拱深;拱高;拱的厚度
depth of a room 房间的深度;房间的进深
depth of artificial recharge engineering 回灌工程深度
depth of a shaft 竖井深度;矿井深度
depth of ballast 道砟厚度
depth of bank over roof of culvert 涵洞上填方高度
depth of beam 梁深;梁(的)高度
depth of bearing stratum 持力层埋深
depth of bed 料层厚度
depth of bigger rat hole 大鼠洞深
depth of blasting segment 爆破段深度

depth of bore 钻孔深度
depth of borehole 钻孔深度
depth of borehole after pumping 抽水后的孔深
depth of borehole before pumping 抽水前的孔深
depth of borings 孔深【岩】;钻探深度
depth of bottom layer of aquifer 含水层底板埋深
depth of bottom of water collecting gallery 集水廊道底深
depth of breaking 碎浪深度;破波深度;破碎深度;碎波水深
depth of break of continental shelf 大陆架坡折水深
depth of burial 埋藏深度
depth of burning 燃烧程度
depth of burying 埋设深度;埋入深度;埋地深度
depth of camber 反拱高;上拱高
depth of carbonation 碳化浓度
depth of case (注水泥的)固结层厚度
depth of channel 航道水深
depth of chill 冷硬深度
depth of coiled tubing 问卷油管下入深度
depth of colo(u)r 色浓度
depth of colo(u)r saturation 色彩饱和度
depth of compaction 压实深度
depth of compensation 补偿深度
depth of compression 受压层深度
depth of compressive zone 受压区高度
depth of confined water top 承压水顶板埋深
depth of constant temperature 恒温层深度
depth of constricted section 收缩断面水深
depth of construction 建造深度
depth of court 天井的进深
depth of cover 上覆岩层厚度;覆盖深度;覆盖层厚度
depth of cracks 地裂缝深度
depth of cup 压痕深(度)
depth of current penetration 电流透入深度
depth of cut 路堑深度;开挖深度;锯割深度;进刀深度;切屑厚度;切土深度;切割深度;挖掘深度;挖方深度
depth of cut-off 截水墙深度
depth of cut-off wall 截水墙深度
depth of cutting 切削深度
depth of defreeze layer 解冻深度
depth of deposit 沉积厚度
depth of deposition 存泥深度
depth of depression 低压深度
depth of diameter changing 变径位置
depth of diapir top 底辟核顶部埋深
depth of discharge 放电深度
depth of dissection 切割深度
depth of disturbance 扰动深度
depth of ditch 沟深
depth of dock 坞深
depth of draining 排沟深度
depth of dredge-cut 挖槽深度
depth of drill hole sealing 封孔深度
depth of drilling terminal 终孔深度
depth of drill tool drop 钻具陷落深度
depth of earth fracture 裂缝深度
depth of effective fractured zone 有效破裂带深度
depth of elevation 取样深度标高
depth of embankment 埋置深度
depth of embedment 埋深;埋入深度;(桩)入土深度
depth of engagement 接合深度;啮合深度;衔接深度
depth of erosion 河岸冲刷深度;剥蚀深度;侵蚀深度
depth of evapo(u)ration 蒸发深度;蒸发量
depth of excavation 开挖深度;基坑深度;挖方深度
depth of experimental mining area 试采区水深
depth of experimental part 试验段深度
depth of exploration 勘探深度
depth of extra-dredging 超挖深度【疏】
depth of fall 沉下深度
depth of field 场深;景深;视野深度;视场深度
depth of field calculation 景深计算
depth of field indicator 景深指示器
depth of field scale 景深刻度;景深表
depth of fill(ing) 填土高度;路堤深度;填土深度;填高;填方高度
depth of finish 面漆的色泽深度;面漆厚度;肉头;涂饰深度;涂膜明晰度
depth of fish top 鱼顶井深
depth of fissure 裂缝的深度

depth of flo(a)tation 浮体吃水深度;漂浮深度
depth of flood 淹水深度
depth of floor 肋板高度
depth of flow 流动深度;水流深度
depth of focal range 焦距
depth of focus 聚焦深度;焦深;震源深度;震深
depth-of-focus table 焦深表
depth of folding 褶皱作用深度
depth of footing 大方脚埋深
depth of footing base 基础深度
depth of foundation 基础深度;基础铺置深度;基础埋置深度;填土高度
depth of foundation embankment 基础埋置深度
depth of foundation embedment 基础嵌入深度
depth of foundation on frozen soil 冻土的基础埋深
depth of foundation pit 基坑深度
depth of fractured zone of high-pressure water 高压水破裂带深度
depth of fractured zone of mine-pressure 矿山压力破裂带深度
depth of frame 肋骨深度
depth of free point 卡钻深度
depth of freeze layer 冻结深度
depth of freezing 冻结深度;冻结浓度;冰冻深度
depth of friction(al) influence 摩擦影响深度
depth of frostline 冰冻深度
depth of frost penetration 冻结深度;冻结浓度;冰冻深度
depth of frozen-thaw layer 融冻层深度
depth of fusion 融熔深度;(焊接)熔深;熔化深度
depth of gallery 坑道深度
depth of gas lift valve 气举阀下入深度
depth of girder 梁高
depth of gravel stuffing from top to bottom 填砾段起止深度
depth of groove 槽深;凹槽深度
depth of groundwater 地下水深度
depth of groundwater occurrence 地下水埋藏深度
depth of grouting hole 注浆孔深
depth of hardening 淬硬深度
depth of hardening zone 淬硬层深度
depth of hold 货舱深度;船舱深度
depth of hole 钻孔深度
depth of hole well coring 井壁取芯深度
depth of image 鲜映性
depth of image space 像方景深
depth of immersed ice 水浸冰厚
depth of immersion 浸液深度;浸水深度;浸湿深度
depth of impact crater 撞击坑深度
depth of impression 刻痕深度(布氏硬度试验);印痕深度;压痕深(度);切屑厚度
depth of indentation 印痕深度;压印深度;压痕深(度);低凹深度
depth of infiltration by water pouring 灌水入渗深度
depth of instrument burial 仪器埋置深度
depth of interfit 牙轮齿排互插深度
depth of intrusion 侵入深度
depth of invasion 侵入深度
depth of investigation 深测深度
depth of isostasy 均衡深度
depth of isostatic compensation 均衡补偿深度
depth of karst development region 岩溶发育区深度
depth of kerf 掏槽深度;水口深度(钻头)
depth of lake 湖泊深度
depth of landing nipple 加油管入岸深度
depth of liner hanger 尾管挂入深度
depth of liquefaction layer 液化层深度
depth of lot 地段进深(与街道垂直深入度)
depth of main detachment 主滑脱面深度
depth of mass tone 主色浓度;浓色浓度
depth of measurement 测点深度
depth of metal 玻璃液深度
depth of modulation 调制深度
depth of Moho-discontinuity 莫霍面深度【地】
depth of moisture variation zone 湿度变动带深度
depth of mouse hole 小鼠洞深
depth of navigable channel 通航水深
depth of neutralization 中性化深度
depth of nude well length from top to bottom 裸井段的起止深度
depth of object space 物景深

depth of observation shaft 观测竖井深度
depth of observation well 观测孔深度
depth of occurrence 揭示深度
depth of open pit 露采场深度
depth of origin 震源深度
depth of overburden 积土深度;积土厚度;覆盖(土)厚度;覆盖(层)厚度;覆土深度;剥离厚度
depth of overlying quaternary system 第四系覆盖层厚度
depth of packing 夯实深度
depth of page 版心
depth of parallelism 平行度
depth of parenthesis nesting 圆括号的嵌套深度
depth of parking stall 车位深度
depth of passageway 走廊进深
depth of pavement 路面厚度;铺面厚度
depth of penetration 桩的入土长度;针入深度;贯入深度;贯穿深度;穿透深度;插入深度;透入深度;渗透深度
depth of phreatic water (地下水)潜水埋深
depth of phreatic water level change zone 潜水水位变动带深度
depth of piston 活塞高度
depth of pit of water infiltration test 渗入试坑深度
depth of plastering 抹灰厚度;粉刷层厚度
depth of plowing 耕作深度;翻耕深度
depth of plunger 柱塞深度;泵的悬挂深度
depth of ponor 落水洞深度
depth of precipitation 降水深度
depth of primary migration 初次运移的深度
depth of primary migration peak 初次运移高峰深度
depth of processing 加工深度;操作深度
depth of pumping sections 取水段深度
depth of pump installation 水泵下置深度
depth of pump tube installation 泵管下置深度
depth of quenching 淬火深度
depth of rabbet 槽深
depth of radial pipe 辐射管深度
depth of rail 钢轨高度
depth of rainfall 降雨深度
depth of rear yard 后院的进深
depth of rebate 槽深
depth of recess 凹下深度
depth of recursion 递归深度
depth of roadbed 道床厚度
depth of rock 岩石厚度
depth of rock stratum 岩层厚度
depth of room 居室进深
depth of round 一次爆破循环进尺;炮眼组深度
depth of runoff 径流深度
depth of runoff zone 径流带深度
depth of salt-fresh water interface under sea level 咸淡水界面在海平面以下
depth of sample point 采样点深度
depth of scene 景深
depth of scour 冲刷深度
depth of scraping coal 刮煤深度
depth of screen tube installation 滤水管下置深度
depth of seam change 换层深度
depth of seam decision 判层深度
depth of seam division 分层深度
depth of seasonal creep 季节性体积变化深度
depth of seawater 海水深度
depth of seismic focus 震源深度
depth of seismic marker horizon 地震标准层深度
depth of sensor 传感器沉放深度
depth of setting (柱的)埋入深度
depth of shade 色调深度
depth of shiplift 升船机的水深
depth of shoal 浅滩水深
depth of shock 震源深度;地震深度
depth of shotcrete 喷层厚度
depth of side tracking 侧钻深度
depth of slot 槽片深度
depth of snow 积雪深(度)
depth of socket 套筒深度;套管深度
depth of soil 土层深度;土壤深度
depth of soil cover 覆土厚度
depth of soil-freezing 土壤冻结深度
depth of soil overlying bedrock 岩基上覆土层厚度
depth of soils 土壤深处
depth of source layer 源层埋深
depth of sphere center 球心埋深

depth of stratum 地层深度;地层厚度;岩层厚度
depth of stripping 剥离深度
depth of structural section 结构截面厚度;结构截面高度
depth of submergence 淹没深度
depth of supporting course 持力层埋深
depth of surcharge 堰顶溢水深度
depth of surface constant temperature zone 地表恒温带深度
depth of surface detention 积涝深度;地面滞留深度
depth of tank 池深
depth of temperature measurement 测温深度
depth of tension crack 受拉裂缝深度
depth of test pit 试坑深度
depth of test well 试验孔深度
depth of thaw 解冻深度
depth of thaw-collapse pit 融陷坑深度
depth of the bridge plug 桥塞下入深度
depth of the draw bar 引砖深度
depth of the fifth layer's top 第五层顶深
depth of the fourth layer's top 第四层顶深
depth of the ground stress measuring hole 地应力测量孔深度
depth of the market 市场深度
depth of the navigable channel 通航河道深度;航道水深
depth of the n th layer's top 第 N 层顶深
depth of the packer 封隔器下入深度
depth of the second layer's top 第二层顶深
depth of the side fussion zone 侧面熔深
depth of the sixth layer's top 第六层顶深
depth of the third layer's top 第三层顶深
depth of thread 螺纹深度;螺纹高度
depth of throat 弯喉深度
depth of tooth 齿高
depth of top of sheet 板体顶端埋深
depth of tracer injection 示踪剂注入深度
depth of transversal section 横断面高度
depth of trap seal 水封深度
depth of trench 缆沟深度
depth of truss 桁架高度
depth of tunnel 隧道埋置深度;隧道埋深
depth of underground observation laboratory 地下观测室深度
depth of vault 穹顶厚度
depth of velocity modulation 速度调制深度
depth of vena contraction 收缩断面水深
depth of wash liquid loss 冲洗液漏失深度
depth of waste disposal stratum 排污层深度
depth of water 水深;水封深度
depth of water-bearing section from top to bottom 含水段起止深度
depth of water entering tube installation 进水管下置深度
depth of water flowing over spillway 溢流堰水头
depth of water flowing over weir 溢流堰水头;堰顶溢流水深
depth of water in channel 渠道水深
depth of water layer 水层深度
depth of water on sill 槛上水深
depth of water seal 水封高度
depth of water sealing 止水深度
depth of water table 地下水埋深;到地下水位的深度
depth of wear 磨损深度;磨蚀深度;磨损厚度
depth of weathered zone 风化带深度
depth of weathering 风化深度
depth of weld 焊接深度
depth of well after well cleaning 洗井后的井深
depth of well before well cleaning 洗井前的井深
depth of well casing installation 井管下置深度
depth-o-matic 自动调位的液压机构
depthometer 测深仪;深度计;深度尺;深度表
depth on sill 坞口水深
depth over-dredging 超挖深度【疏】
depth over width ratio 深度宽度比
depth over-width ratio of memory 存储器深度宽度比
depth parameter 深度参数
depth penetration 深度贯穿
depth penetration of hardness test 硬度试验压入深度
depth perception 景深感(觉);深度感
depth point 深度点

depth pressure plot 深度—压力曲线
depth probe 深度探头
depth profile 深度(纵)剖面
depth queuing 景深层次
depth range 水深范围;深度范围
depth ratio 梁的高度与跨长之比;埋深比;充满度;深宽比;深度比
depth record 深度记录
depth recorder 深度记录仪;深度计;记录式声测测试仪;记录式深度计;回声测深仪;自记测深仪;水深记录器;深度指示仪;深度指示器;深度记录器
depth recording device 深度记录仪
depth resolution 深度分辨率
depth round-off 深度化整
depth rule 深度法则
depth scale 测深标尺;深度标尺;深度比例(尺)
depth scan(ning) sonar 深度扫描声呐;目标深度测量声呐
depth screw gear 耕深调节螺杆
depth section chart 深度剖面图
depth segment 深度段
depth shape 倾斜地形
depth signal mark 水深信号标
depth signal pole 水深信号杆
depth signal rod 水深信号杆
depth signal staff 水深信号杆
depth slide ga(u)ge 滑动式测深计
depth sort 深度排序
depth sound 测深
depth sounded 测得水深
depth sounder 回声测深计;测深仪;测深计;测深机
depth sounding 测深;测量水深
depth-sounding pipe 测深管
depth sounding sonar 测深声呐
depth sounding velocity 测深声速
depth-span ratio 高跨比;高度与跨度之比
depth stop 限深器;限深规
depth telemeter 深度遥测仪
depth-thickness ratio 高厚比
depth-time curve 深度—时间相关曲线
depth-to-discharge relation 水深—流量关系
depth-to-span ratio 高度跨度比;高跨比
depth to the bottom of the magnetic body 磁性体下底埋深
depth to the magnetic body 磁性体埋深
depth to the top of the magnetic body 磁性体上顶埋深
depth-to-width ratio 高阔比;高宽比;深宽比
depth tube spacing 纵向管距
depth-type filtration 深度过滤
depth under ground 地下深度
depth under keel 船底龙骨下水深
depth variation 水深变化
depth-velocity curve 水深—流速关系曲线;深度—速度曲线;流速垂直分布曲线;水位—流速曲线
depth vernier 深度游标尺
depth-volume curve 深度—容积关系曲线
depth warning device 深度报警装置
depth wheel 耕深调节轮
depth-width ratio 高阔比;高宽比;高度与宽度之比
depth zero of life 无生物深度
depth zone 深度(分)带
depth zone of earth 地球深带
depulization 灭蚤
depurant 净化器;净化剂;纯化剂
depurant tank 净化剂储[贮]罐
depurate 滤清;提纯
depurated 纯化的
depuration 滤清;纯化
depuration rate coefficient 排除速率系数
depurative 纯化的
depurator 净化剂;净化装置;净化器;纯化剂
deputation 委任代理;委派
depute 委任代理
deputy 代理人;代表;副职;副手
deputy chairman 副会长
deputy commissioner 副局长
deputy commissioner of customs 海关副税务司
deputy consul 副领事
deputy district head 副区长
deputy general manager 副总经理
deputy manager 副总裁;副经理
deputy mining surveyor 矿山副测量员

de-pyrogenation 去除热原法
Deqing ware 德清窑
dequeuc 退出队列
dequeue 解队;出列
deragger 破布拣除器
derail 脱轨;离开原定进程;开关;出轨装置;出轨
derailable by hand 人力脱轨器
derail block 块型脱轨器
derailer 脱轨装置;脱轨器
derail indicator 脱轨标志
derailing switch 开关器
derailleur 变速齿轮传动机构
derail(ment)脱轨;出轨
derailment coefficient 出轨系数
derailment limit 车轮脱轨极限
derailment protection 脱轨保护
derailment test 脱轨试验
derail of single-point 单辙轨脱轨器
derandomization 解随机处理
deranged drainage 冰碛平原水系
deranged drainage pattern 扰乱水系
deranged mode 紊乱水系模式
derangement 重排列;紊乱
derate 降额;降低额定值
derated fire load 降级火灾荷载
derated standard level 标准失效率
derating 降低额定值;减额;定额降低;额定值降低
derating coefficient 减额系数
derating level 减额等级
deration 取消定额分配
deratization 熏舱;灭鼠;除鼠
deratization certificate 灭鼠证书;熏舱证书
deratization exemption certificate 免予熏舱证书;免予灭鼠证书
deratting 除鼠
Deraup oscillator 德劳振荡器
derby 泥板;金属块;刮尺;木杠
Derby china 德拜瓷器
derby float 块形镘刀,(粉刷用)整平刮尺;整平标尺
derbylite 锑钛钛矿;锑钛铁矿
Derby Railway Technical Centre 达比铁路中心(英国)
derby red 镉铅红;铬红
derbyshire spar 晶石
derby slicker 刮平镘板
derby spar 萤石
derealization 现实解体;现实感丧失
dereeler 卷料开卷机
dereferencing 非关联化
deregenerative feedback 非再生反馈
derelict 海退遗地;海上遗弃物;漂流物
derelict building 被遗弃建筑物;危房
dereliction 被弃物
dereliction of duty 失职
derelict land 荒废地;冲积地;冲击地;废(弃)地;弃耕地
derepression 除抑制作用
dereption 水下剥蚀(作用)
dereserve 解除保留
deresinsate 脱树脂;脱胶脂
Deriaz turbine 斜流式水轮机;德立亚式水轮机
deriberite 板石
derichment 反富集
deriming valve 解冻阀
Deri motor 德里电动机
Deri repulsion motor 德里推斥电动机
derivant 衍生物[化];衍生物
derivate moment 导数矩
derivating agent 衍生剂
derivating post 改型支柱;改型支撑
derivating screw 改型螺钉
derivation 衍生;衍化;导引;求解运算;求导(数);派生;推导
derivation action 导数作用;微商作用
derivation angle 顶角
derivation graph 导引图
derivation income 派生所得
derivation of a fuzzifying function 模糊函数的导数
derivation of contour 等高线描绘
derivation of equation 方程式推导
derivation of formula 公式推导
derivation of reduction rules 归约规则的推导
derivation of wave climate 波候推算
derivation tree 导出树

derivative 诱导剂;引处物;衍生物[化];衍化物;导数;从变量;变形;派生的;微商衍生物;微商
derivative absorption spectroscopy 导数吸收光谱法
derivative action 导数作用
derivative air permissible pollution concentration 导出空气污染容许浓度
derivative coefficient 诱导系数
derivative control 一次微分控制;导数调节
derivative demand 派生需求
derivative departments 衍生部门
derivative differential thermal analysis 微商差热分析;微商差热分析
derivative dose equivalent limit 导出剂量当量限值
derivative equalizer 微分均衡器
derivative feedback 微分反馈
derivative form 派生形式
derivative index 派生指数
derivative interest 派生利息
derivative limit of surface contamination 表面污染导出极限
derivative magma 次生岩浆
derivative map 派生地图
derivative method 诱导法;引申法;衍生(物)法
derivative network 微分网络
derivative of gravity 重力导数
derivative of higher order 高阶导数
derivative of tar 柏油的衍生物
derivative on the left 左导数
derivative on the right 右导数
derivative polarography 导数极谱(法)
derivative proportional-integral control 坐标导数积分控制
derivative ratio method 导数比值法
derivative revenue 派生收益
derivative rock 衍生岩;导生岩;沉积岩
derivative spectrometry 导数光谱学
derivative spectrophotometric 导数分光光度法
derivative spectrum 导数光谱
derivative statistics 整理后统计数字
derivative structure 衍生结构;派生构造;派生结构
derivative thermodilatometry 微商热膨胀法
derivative thermogravimetric curve 微商热重曲线
derivative thermogravimetry 微商热重法
derivative thermometric titration 导数温度滴定
derivative time 微分时间
derivative time constant 微分时间常数
derivatography 微商热谱法;微商热谱法
derivator 导数装置
derive 导出;派生
derived air concentration 推定空气浓度
derived circuit 导出电路;分支电路
derived current 分支电流;分路电流
derived curve 导(出)曲线
derived data 计算资料;派生数据;推导数据
derived data item 导出数据项
derived data type 导电数据类型
derived envelope 导出包络
derived fossil 次生化石
derived from 采自
derived function 导(出)函数
derived graph 导出图
derived gust velocity 导出阵风速度
derived high polymer 衍生高聚物
derived investigation level 导出调查水平
derived key 推导关键词
derived liability 分赔责任
derived limit 推定极限
derived lipid 衍生脂类
derived map 资料地图
derived number 导数
derived point 引点
derived product 衍生产品;副产品
derived quantity 导出量
derived quantity of gravity 重力导出量
derived record 导出记录
derived relation 导出关系
derived rule 导出规则
derived savanna 衍生稀树干草原
derived scale 推导比例尺
derived series 导出群列;导出级数
derived set 导集
derived silt 导来泥沙
derived soil 残留土壤

derived sound system 导出声系统
derived statistics 派生的统计数字
derived structure 衍生结构
derived table 导出表
derived tide 派生潮
derived unit 导出单位
derived value 衍生价值;派生价值
derived wave 派生波
derivometer 测偏仪
Deri winding 德里绕组
derma 真皮
dermabrasion 擦皮法
dermadine 木菌素
dermal appendage 表层附属物
dermal fold 表层褶曲
dermal toxicity 皮肤中毒
dermarsal 底栖(鱼)
dermarsal fish 底栖鱼
Dermas 德马斯(一种铺地面的乳化沥青料)
dermateen 人造革;布质伪皮;漆布
dermatergosis 职业性皮肤病
dermatine 人造皮革
dermatitis 皮炎
dermatitis solaris 日晒性皮炎
dermatological department 皮肤科
dermatology 皮肤科
dermatosis 皮肤病
dermatosome 微纤维素
Dermitron 高频电流镀层测厚仪
dern 门框;门槛
derocker 除石机
derodo 摩擦材料
derogate 减损;毁损
derogatory 减阶的
deroofing 蚀顶
deroofing eruption 蚀顶喷发
derotation 反旋
derred payment 逾期付款
derrick 井架;架式起重机;钻塔;动臂起重机;吊杆起重机;吊杆;浮吊;人字起重机;起重(井)架;起重杆;桅杆起重机;塔式井架;塔架
derrick-and-ring 带配件钻杆
derrick assembly 井架安装
derrick band 吊杆箍
derrick barge 浮式起重机;浮动起重机;浮吊;起重(驳)船
derrick base 钻塔基础
derrick block 起重滑车
derrick board 井架板;台板
derrick boat 浮式起重机;起重船
derrick boom 吊臂;吊杆;起重机吊杆;起重臂
derrick boom rest 吊杆托架
derrick brace 钻塔支撑;井架支撑;起重支撑
derrick bullwheel 起重机转盘
derrick cap piece 吊杆头装置
derrick car 转臂吊车;汽车(式)起重机;起重车
derrick cargo handling 吊杆装卸
derrick cellar 井口圆井;圆井;方井
derrick cornice 起重塔顶挑檐
derrick crab 动臂起重机
derrick crane 斜撑起重机;人字起重机;转臂吊机;转臂(式)起重机;轨道吊车;动臂起重机;吊杆式起重机;扒杆(式)起重机;桅式起重机;桅杆转臂起重机;桅杆(式)起重机
derrick crown 钻塔顶(框);井架顶;钻塔天车台
derrick equipment 钻塔设备
derrick failure 钻塔损坏
derrick fall 吊货索
derrick flatform 钻塔平台
derrick floor 井场;钻台
derrick footing 井架基础;钻塔基础
derrick foundation 井架基础;钻塔基础
derrick foundation post 井架柱;井架基础柱;钻塔基础底梁
derrick girder 井架大梁
derrick girt 钻塔围梁
derrick grillage 钻井架格排
derrick guy 桅杆起重机缆风;钻塔绑缆;吊杆支索;吊杆稳杆;起重机缆索;起重机缆风
derrick guy winch 吊杆稳绞车
derrick hay stack 转臂式干草堆垛机
derrick head 吊杆头
derrick head block 吊货滑车;吊杆头滑车
derrick head cargo block 吊杆头起货滑车

derrick head fitting 吊杆头装置
derrick heel block 吊杆底导向滑车
derrick heel eye 吊杆叉头
derrick height 钻塔高度
derricking 改变起重臂倾角
derricking boom 改变起重臂倾角的吊杆
derricking crane 转臂起重机;俯仰起重机
derricking guy 起重机缆索;起重机缆风
derricking jib crane 转臂(式)起重机;人字式转臂起重机;伸臂式变幅起重机
derricking limiter 臂架变幅限位器
derricking mechanism 变幅机构
derricking motion 起重动作
derricksing rope 臂架钢丝绳
derrick irons 钻塔的金属部件(天车、游动滑车、钩环、螺栓、锻件等)
derrick jib 转臂
derrick jib crane 转臂(式)起重机
derrick kingpost 起重桅杆;起重(杆);起重吊杆柱
derrick leg 井架支腿;井架支柱;起重支架
derrick leg clamp 钻塔腿夹
derrick loop 吊杆箍
derrick man 钻塔工;井架工;架子工;塔项工(人)
derrick man elevator 活动工作台
derrick mast 起重把杆;船吊桅杆;扒杆;起重桅(杆);起重(扒)杆;起货桅杆
derrick monkey 塔顶工(人)
derrick motion 起重动作
derrick panel 塔板
derrick pendant 吊货索
derrick platform 钻井平台
derrick pole 起重桅(杆);起重杆
derrick post 将军柱;转臂起重机柱;吊柱;船吊桅杆;人字起重机柱
derrick post ventilator 吊杆柱通风筒
derrick pulley 钻塔滑车;天车轮
derrick rope 起重索;起重机钢索
derrick safety belt 井架安全皮带
derrick ship 吊杆起重船
derrick shoe 吊杆承座
derrick sill 钻塔底梁
derrick skinner 塔顶工(人)
derrick socket 吊杆承座
derrick stone 巨石(块);吊石;大石块;大块石;粗石块;粗块石
derrick stool 吊杆台
derrick substructure 井架底座;钻塔底座
derrick support 钻塔底座
derrick system 吊杆装置
derrick table 吊杆台
derrick test 吊杆负荷试验
derrick tower 桁架式;起重塔架;起重机架;起重吊(机)塔
derrick tower gantry 转臂塔式起重机架;门塔式起重台架
derrick truck 吊车
derrick type 井架类型
derrick type steel tapes 架式钢卷尺
derrick wagon crane 车座人字(形)起重机
derrick wheel (在司钻台操纵动力机加速器的)手轮
derrick winch 吊杆绞车;起重卷扬机
derrick worker 起重工
derriksite 多硒铜铀矿
derust agent 除锈剂
derusted 去锈的;除锈的
deruster 除锈机
derusting 除锈;去锈
derusting agent 除锈剂
derusting by sandblast 喷砂法除锈
derusting grade 除锈等级
derusting tool 除锈工具
Deruta ware 德鲁塔陶器(用珍珠云母及金属光泽装饰的陶器)
derv 柴油机车辆;柴油(英国)
derv fuel 重型车辆用的柴油
Deryagin number 德亚金数
desactivation 消除放射性沾染
desalinated water 除盐水
desalination 淡化;除盐;脱盐作用
desalination apparatus 淡化设备;脱盐设备
desalination by direct freezing 直接冷冻淡化
desalination by electrodialysis 电渗析法淡化
desalination by ion(ic) exchange 离子交换海水淡化

desalination by reverse osmosis 反渗透淡化
desalination by solvent extraction 溶剂提取脱盐
desalination factor 脱盐因素
desalination membrane 淡化膜;脱盐膜
desalination of seawater 海水脱盐;海水淡化
desalination plant 海水淡化(工)厂;淡化装置;淡化厂;脱盐装置;脱盐车间;脱盐厂
desalination process 脱盐法
desalination ship 制造淡水船
desalination system 脱盐系统
desalinization 淡化;脱盐
desalted 脱盐的
desalted water 淡化水;脱盐水
desalter 脱盐装置;脱盐器;脱盐剂
desalter brine 脱盐装置含盐污水
desalt(ing) 海水淡化;淡化;脱盐(作用)
desalting agent 脱盐剂
desalting environment 海水淡水化环境;脱盐环境
desalting kit 海水淡化器
desalting of seawater 海水脱盐
desalting of water 水的脱盐;水的除盐
desalting period 脱盐期
desalting plant 海水淡化设备;淡化装置
desalting process 除盐过程
desalting ratio 脱盐率
desalting ship 海水淡化船
desalting strength 脱盐强度
desalting technology 除盐工业技术;脱盐技术
desalting unit 脱盐装置
desample 解样
desampling 反抽样
desander 除砂器;去砂器
desanding 去砂;除砂;沉砂
desanding device 除砂装置;分砂装置
desanding screen 除砂器
Desargue's theorem 德扎尔格定理
desaspidine 地沙匹定
desaturated colo(u)r 不饱和色
desaturation 减饱和(作用);稀化;冲淡;饱和度减少;脱饱和;褪彰
desaturation curve 减饱和曲线
desaturator 稀释剂
desautelsite 羟碳锰镁石
De Sauty method 德绍蒂法
desaxe 异轴性;轴心偏移
descale 除锈;除鳞
descaler 除锈剂;除鳞机;除垢器
descaler liquid 除垢液
descale rust 脱鳞锈
descaling 船舶除垢;除锈;除去锈垢;除去锅垢;除鳞;除黑(氧化)皮;除垢除锈法;除垢;清除黑鳞;脱除锅垢;去氧化皮
descaling agent 除垢剂
descaling bath 酸洗槽
descaling box 氧化皮冲除箱
descaling capability 去垢能力
descaling machine 除鳞机;清理机;氧化皮打磨机
descaling mill 除鳞机
descaling spray 喷嘴组;除鳞喷嘴组
descaling stand 除鳞机座
descaling system 除鳞系统
descaling unit 除鳞装置
Descartes ray 笛卡尔光线
Descartes rule of signs 笛卡尔符号法则
descend 降下;降落;下降;下行
descendance of blocks 分程序的后代
descendant 递降;衰变产物
descendant action 分支动作
descend a rapid 下滩
descendent action 分枝动作
descending 降序;向下的;下行性的;下行(的);下落;下降的;递降
descending air 下降空气
descending air current 下降气流;下沉气流
descending axis 根系
descending barge 下行驳船;下水航驳
descending boat 下行船(舶)
descending branch 下垂枝;降弧;弹道降弧;下降段
descending chain 降序列
descending chromatography 下行色谱法;下行色层分离法
descending control 下行控制
descending current 下降流
descending curve 下降曲线

descending developing method 下行展开法
descending development method 下行展开法
descending difference 降差数
descending face 下降割面
descending flow 下降径流
descending flow of sap 下行液流
descending grade 降坡;下坡;下降坡度
descending infection 下行感染
descending inhibitory action 下行抑制作用
descending inhibitory path 下行抑制径路
descending inhibitory reticular projection 下行抑制性网状投射
descending key 降序键;递降键
descending lifeline 缓降器
descending line 入水绳
descending liquid 向下流动的液体;下行液体
descending method 下行法
descending mining 下行开采
descending node 降交点
descending node of satellite orbit 卫星轨道降交点
descending order 下降顺序;降序(排列);下降次序;递减顺序;递减次序
descending phase 下降相
descending plate 下倾板块;下板块;俯冲板块
descending powers 降幂
descending ramp 下坡道
descending secretion theory 下分泌说
descending sequence 降序序列
descending slope 下坡;下降斜率
descending sort 降序排列
descending spring 溢出泉;下降泉
descending stage 由上而下分段(灌浆)
descending temperature 下降温度
descending tracts 下行束
descending velocity 下行流速
descending vertical angle 俯角
descending vertical flue 下降式竖式烟道
descending water 下降水
descending wave 下降波
descend of regional groundwater level 区域性地下水位下降
descension theory 下降溶液成矿说
descent 降低沉陷
descent control device 缓降器
descent direction 下降方向
descent method 下降法
descent of regional ground water table 区域性地下位水下降
descent path 下降路径
descent through cloud 穿云下降
descent time 下潜时间
descent trajectory 下降段
descloizite 羟钒锌铅石
descrambler 解扰(码)器;反倒频器
descrete element 分立元件
describe 描述;描绘
described circle 外接圆
described function 描述函数
describe environment 描述环境
describer 列车指示牌
describing function 描述函数;描迹函数
describing function method 描述函数法
description 描述;摘要;文字描述;说明
description entry 描述项
description for drawing up standard 标准编制说明
description list 描述表
description note 说明注记
description of commodity 商品说明书
description of consequence 后果的说明
description of construction method 施工方法说明
description of construction progress 工程进度图表;工程进度说明;工程形象进度
description of core 岩芯描述
description of data evaluation 资料评价说明
description of design 设计说明
description of execution method 施工方法说明
description of final design 最终设计说明书
description of finishes 饰面说明
description of goods 货名
description of items 项目说明
description of job practice 施工方法说明
description of materials 土层描述;材料说明;材料品种
description of patent specifications 专利说明书
description of points 点之记;点的说明
description of profile 剖面描述
description of project 工程说明书;工程描述
description of property 资产项目;房地产项目;物品叙述
description of sample 样品描述;试样描述
description of station point 测点描述
description of stratum and rock type 地层岩性描述
description of structural construction 构造说明书
description of subrountine 子程序使用说明书
description of territory 业务地区说明
description of the goods 货物详单
description of the invention 发明说明书;发明的简要说明
description of the process 工艺说明
description of topography 地志
description of track damages 线路病害情况
description of wave loading 波浪荷载描述
description of works 工程描述
description point 取向标记
descriptive abstract 说明摘要;说明提要
descriptive astronomy 描述(性)天文学
descriptive bill 附说明的账单
descriptive billing 开出附说明的账单
descriptive catalog(ue) 附有说明的目录;附说明目录
descriptive climatology 描述性气候学
descriptive crystallography 描述(性)晶体学
descriptive data 说明注记
descriptive decision making 描述性决策
descriptive decision theory 描述性决策理论;叙述性决策理论
descriptive definition 描述性定义
descriptive ecology 描述性生态学
descriptive financial statement 附有说明的损益计算书
descriptive geometry 画法几何;投影几何(学)
descriptive hydrology 记述水文学
descriptive index 描述性指数
descriptive information 兵要地志
descriptive labeling 品种简介标签
descriptive meteorology 描述性气象学
descriptive method 记叙法
descriptive mineralogy 描述(性)矿物学
descriptive model 描述(性)模型
descriptive morphology 描述性形态学
descriptive name 说明注记
descriptive oceanography 描述性海洋学;海况学
descriptive palaeontology 描述性古生物学
descriptive petrology 描述(性)岩石学
descriptive procedure 描述性过程
descriptive process 描述性工艺过程
descriptive programming 描述性程序设计
descriptive provision 描述性规定
descriptive report 说明注记
descriptive research 描述性研究
descriptive sampling survey 描述性抽样调查
descriptive schedule 明细表;一览表;附说明图表
descriptive seismography 描述性地震学
descriptive series 描述性数列
descriptive simulation model 描述性模拟模型;概念性模拟模型
descriptive statement 说明注记
descriptive statistic 描述性统计
descriptive statistical method 描述性统计法
descriptive statistics 描述(性)统计学;叙述统计
descriptive stratigraphy 描述性地层学
descriptive survey 描述性调查
descriptive term 说明项
descriptive text 说明注记
descriptive transformation 画法变换
descriptor 解说符;描述信息;描述词
descriptor approach 描述符方法
descriptor for a segment 段的描述符
descruotuve statu statistical method 描述性计方法
desealant 封闭层防剥离药剂
deseam 气炬烧剥
deseamer 火焰清理机;焊瘤清除器;焊缝修整机;焊缝清理机;气柜烧剥机
deseaming 气炬烧剥;表面修整;气柜烧割;气炬切割
deseaming machine 焊缝清理机;修整锭面机;凿整锭面机
deseasonalization 消除季节性因素
deseasonalized data 消除季节变动后的数据
deseasonalized series 消除季节变动后数列
deseasonalizing 消除季节变动影响
desecrated church 被供作俗用的教堂
deseeder 亚麻梳籽装置
deseeding machine 亚麻脱粒机
desegregation 消除种族隔离
desending facilitory reticular projection 下行易化性网状投射
desensibilization 灵敏度降低;减敏化;去敏化
desensitisation 去敏作用;降低灵敏度;减敏现象
desensitivity 倒灵敏度
desensitization 降低灵敏度;减敏现象;去敏作用
desensitize 减感;脱敏
desensitizer 减敏剂
desensitizing agent 减感剂
desensitizing solution 减感液
deserialize 串并转换;并行化
deserializer 串并转换器;串并行变换器;并行器;并串行转换电路
deserpentinization 脱蛇纹石化
deserpidine 地舍平
desert 荒漠;赏罚;沙漠
desert animal 沙漠动物
desert area 沙漠地区
desert belt 荒芜地带;荒漠地带;沙漠带
desert climate 荒漠气候;沙漠性气候;沙漠气候
desert concrete 粗料混凝土;(用少量水泥作为胶结材料,大块地方粗石作为集料的)混凝土
desert creep 沙漠蔓延
desert crust 荒漠卵石覆盖层
desert deposit 沙漠沉积
desert devil 尘旋;沙卷风
desert ecosystem 荒漠生态系统
deserted island 荒岛
desert environment 沙漠环境
deserter 私自离船员
desert erosion feature 荒漠侵蚀地形
desert establishment 野外科学试验场
desert facies 沙漠相
desert gray soil 灰漠土
desertification 荒漠化;土地沙漠化;沙漠化
desertification control 治沙;沙漠治理;沙漠化控制
desertization 自然沙漠化;土地沙漠化;沙漠化
desert lake deposit 沙漠湖泊沉积
desert lake facies 沙漠湖泊相
desert locusts 沙漠蝗虫
desert paint 沙漠漆
desert patina 沙漠漆
desert pavement 荒漠卵石覆盖层;沙漠卵石覆盖层;沙漠砾石表层
desert peneplain 联合山麓侵蚀面
desert plain panfan 联合山麓侵蚀面
desert plant 沙漠植物
desert plateau 沙漠高原
desert railway 沙漠铁路
desert residence 沙漠(地区)住房;沙漠居住
desert scrub 沙漠灌丛
desert soil 漠土;荒漠土(壤);沙漠土(壤)
desert spoon 点心匙
desert spread 沙漠扩展
desert steppe 荒漠草原;沙漠草原
desert steppe soil 荒漠草原土
desert storm 尘暴;沙暴
desert survey 沙漠研究
desert test 沙漠试验
desert varnish 沙漠(岩)漆
desert vegetation 沙漠植被
desert wind 沙漠风
desert zone 荒漠地带;沙漠带
Desgu gasification method 德士古气化法
Deshaw process 德肖电解液除锈法
desheathing 取下外壳
de-shielding 去屏蔽
desication fissure 干缠裂缝
desiccant 去水分的;去湿的;干燥的;吸湿剂;干燥剂
desiccant cartridge 干燥剂包
desiccant salt 脱水盐
desiccant salt filling machine 干燥剂填装机
desiccate 干燥产物
desiccated alum 干燥铝矾
desiccated concrete 脱水混凝土
desiccated wood 干(燥)木材

desiccating action 脱水作用;干燥作用
desiccating agent 吸湿剂;干燥剂
desiccating machine 去湿机
desiccation 空气干燥法;干燥作用;干燥(法);干缩;干化(作用);除湿;变旱
desiccation breccia 干裂角砾岩
desiccation chamber 空气干燥室
desiccation conglomerate 干裂砾岩
desiccation crack 泥裂;龟裂;干缩裂隙;干裂(缝)
desiccation cracking 干缩裂缝
desiccation fissure 干缩裂隙;干缩裂缝
desiccation joint 干缩节理;干裂节理
desiccation of wood 木材干燥(法)
desiccation polygon 干裂多边土
desiccative 干燥剂;干燥的
desiccator 料浆蒸发器;干燥器;干燥剂;防潮砂;保干器;收湿器
desiccator diaphragm 干燥膜
desiccator plate 干燥器板
desiccator with porcelain plate 带瓷板干燥器
desicchlora 无水粒状高氯酸钡(干燥剂)
design 款式;图样;图案;设计
design, construction divided contracting 设计施工分离式承包
design abbreviation 设计符号
designability 设计可能性;可设计性;结构性
designable reserves 设计储量
design acceleration 设计加速度
design acceptability 设计的合格性
design accuracy 设计精度
design action 设计行动
design activity 设计活动
design adequacy 设计完备度
design administration 设计管理
design advance 设计发展
design aerodynamics 设计空气动力学
design agreement 设计协议
design aids 设计参考资料;辅助设计工具;设计工具
design aids set 整套设计工具
design air freezing index 设计空气冻结指数
design alteration 设计变更
design alternation 设计修改
design alternative 设计方案
design altitude 设计高程;设计高度
design analysis 设计分析
design-and-build contract 设计和施工合同
design and construct firm 设计和建筑公司
design and construction 设计与施工
design and construction firm 设计与施工公司
design and construction of port 港口设计和施工
design and drawing office 设计室
design and layout of streets 街道设计和布置
design annex 设计附件
design appendix 设计附件
design approach 设计途径;设计方案;设计程序
design approval drawing 批准的设计图
design architect 设计建筑师
design area of sprinkler system 喷淋系统的设计面积
design assembly 设计装配图
design assumption 设计假设;设计假定
designate 代称;标志;任命
designated 一定的;特指的
designated anchorage 指定锚地
designated area 指定(区域);指定地区;特定地区
designated bank 指定银行
designated berth 指定泊位
designated currency 指定货币
designated deposit 指定存款
designated design strength 标称设计强度
designated distance 一定距离
designated diversion area 指定分洪区
designated elevation 控制高程
designated frequency 指定频率
designated function of a city 城市性质
designated length 标称长度
designated mooring 指定停泊区;指定锚泊区
designated number of layer soil 一定土层
designated number of times 一定次数
designated operational entity 指定经营实体
designated order turn-around system 指定的订单转换系统
designated overhaul point 指定大修站
designated person 指定人员;指定代理人
designated personal 指定代理人;指定人员
designated port 指定港(口)
designated project 指定项目
designated real estate broker 指定的公司不动产代理人
designated real estate instructor 房地产讲师
designated size 合格粒度;选定粒度
designated symbol 指示符号
designated term of works 指定工期
designated value 指定值;特值;特定值
designated volume 一定体积
designated water body 指定水域;指定水体
designated water quality standard 指定水质标准
designated waters 指定水域;指定水体
designated weight 一定重量
designation 命名;名称;指派;钢号;符号名称;符号表示;标志数据;标识
designational expression 命名表达式
designation card 名称卡;标示卡;标识卡(片)
designation equipment 标志设备
designation hole 标志孔
designation index 标准指数
designation indicator 目标指示器;标识指示器
designation marks 书帖标记
designation number 指示数字;指定数;赋值数;标准指数;标志数(字)
designation numeral 指定数
designation of a structure 结构物标志;建筑物标志
designation of colo(u)rless optic(al) glass 无色光学玻璃牌号
designation of drawing 图纸名称
designation of goods 货品标志
designation of Omega station 欧米伽台名
designation of rock 岩石描述
designation punch 标志孔
designation punching 标记穿孔
designation strip 名牌;标(示)条;标识条;铭牌
designation system 标志系统;标志法
designator 命名符;指示符;标志符
designator for Loran A station pair 罗兰A台对名称
designatory term 专用术语
design automation 自动设计;设计自动化
design automation of digital system 数字系统设计自动化
design automation system 设计自动化系统
design automation system of bridges 桥梁自动化设计系统
design basic acceleration of ground motion 设计基本地震加速度
design basis 设计依据;设计基础
design basis accident 设计基本事故
design basis data 设计基础数据
design basis earthquake 设计依据地震;设计基本地震
design basis for investigation 勘察设计依据
design basis natural event 设计所依据的自然条件
design bearing pressure 设计支承压力
design bedding 图案花坛
design bending moment 设计弯矩
design bolt load 设计的螺栓荷载
design book 设计说明书
design bottom 设计河底
design brief 设计草图;设计概要;设计任务书
design budget 设计预算
design build experience 设计施工经验
design build process 设计兼施工
design by deformation 按变形设计
design by modulus system 模数制设计
design by stress analysis 按应力分析设计
design calculation 构造计算;设计计算
design California bearing ratio 设计加州承载比(又称设计的加利福尼亚承载比)
design capacity 设计能力;设计容量;设计能量;设计功率
design center 设计中心
design certified value 设计保证值
design change 设计的更改;设计变更
design change analysis 设备更改分析
design change notice 设计更改通知
design change proposal 更改设计建议
design change request 设计更改要求
design change summary 设计更改一览
design change verification 设计更改检验
design change work order 更改设计操作规程
design channel elevation 设计航道高程
design characteristic period of ground motion 设计特征周期
design charges 设计费
design chart 设计图(表)
design check 设计检查
design circuit 设计电路
design code 设计规章;设计规则;设计规范;设计规程
design coefficient 设计系数
design commission 设计委托
design comparison 设计比较
design competition 设计竞争;设计竞赛;设计比较
design compression ratio 设计压缩比
design computation 设计计算
design concentration 设计浓度
design concept 设计概念;设计思想;设计思路;设计构想;设计构思;设计方案
design concept change 设计概念的改变
design condition 计算条件;设计条件
design conference 设计会议
design consideration 设计上的考虑;设计考虑;设计中考虑的问题;设计考虑事项;设计根据;设计依据;设计要点;设计构想;设计构思
design constant 结构常数;设计常数
design constraint 设计控制条件;设计制约条件;设计约束条件
design construct 设计兼施工
design construct contract 设计(兼)施工合同;设计施工合约
design construction contract 设计施工合约;设计施工合同
design construction team 设计施工小组
design consultant 设计顾问
design consultation 设计咨询
design contest 设计比赛
design contract 设计合同
design control 设计控制;设计管理
design control drawing 设计控制图表
design control specification 设计检验规范
design coordinate 设计坐标
design cost 设计成本
design crest level 设计路拱高程;设计坝顶高程;设计定高程;设计(坝)顶标高
design criteria for large dams 大坝设计准则
design criterion 设计准则;设计规范;设计标准
design current speed 设计水流速度
design current velocity 设计流速
design curve 设计曲线
design cycle 设计周期;设计流程
design data 设计数据(资料)
design data gathering 设计资料收集
design datasheet 设计数据单;设计资料图表
design decision 设计定案
design deficiency 设计缺陷
design deflection value 设计弯沉值
design density of sprinkler system 喷淋系统的设计密度
design department 设计室;设计部门
design depth 设计深度
design details 设计细则;设计细节;设计细部
design development 设计草图加工;草图加工;技术设计
design development drawing 技术设计图
design development phase 建筑二期技术业务;技术设计阶段;初步设计阶段
design development shop 设计试制车间
design development year 设计水平年
design deviation 设计偏差
design diagram 设计(计)算图;设计图表
design dilution 设计贫化
design dilution ratio 设计贫化率
design discharge 设计流量
design discharge of bridge 桥梁设计流量
design discharge of pumping station 泵站设计排水量
design displacement 设计排水量
design document(ation) 设计文件;设计资料
design draft 设计图样;设计草图
design drawing 设计图(纸)
design dredge-cut 设计挖槽
design dredge level 设计疏浚标高;设计浚挖高程
design duty 设计生产能力

design earthquake 设计地震
design earthquake intensity 地震设计烈度;设防烈度
designed 设计的
designed air velocity 设计风速
designed by 设计人
designed capacity 设计能力
designed carbon dioxide level 设计二氧化碳度
designed channel 设计的河槽
designed composition 设计组成;设计成分
designed cut depth 设计挖深
designed damming level 设计挡水位
designed depth 设计水深;设计深度
designed depth of dredging 设计挖深
designed depth of navigation channel 航道设计水深
designed dimension 设计尺寸
designed dimension of channel 设计航道尺度;设计航道尺寸
designed discharge 设计排水量;设计排放量;设计流量
designed distance 设计距离
designed draft 设计吃水
designed draft of typical ship 标准船舶设计吃水
designed draft of typical vessel 标准船舶设计吃水
designed drag 设计纵倾
designed dredging speed 设计挖泥航速
designed duty of water 设计用水量
designed efficiency 设计效率
designed elevation 设计标高;设计高程
designed elevation mark 设计标高标志点
designed estimation 设计概算
designed experiment 预定试验
designed filling 设计充满度
designed flood 设计洪水
designed flood frequency 设计洪水频率
designed flow 设计流量
designed flow frequency 设计洪水频率
designed flow per second 设计秒流量
designed flow velocity 设计流速
designed for capacity of n tonnages 设计吞吐量为 n 吨
designed full load draft 计划满载吃水;设计满载吃水
designed full load line 设计满载吃水线
designed grade 设计坡度
designed gradient line 设计坡度线
designed head 设计水头
designed highest navigable stage 设计最高通航水位
designed highest navigable water-level 设计最高通航水位
designed horse-power 设计马力
designed input 设计输入量
designed life 设计寿命
designed load 设计荷载;设计负荷
designed loading 设计(时采用的)负荷
designed load point 设计负荷点
designed longitudinal grade 设计纵坡
designed lowest navigable stage 设计最低通航水位
designed lowest navigable water-level 设计最低通航水位
designed maximum navigable discharge 设计最大通航流量
designed minimum navigable discharge 设计最小通航流量
designed mining time 设计开采时间
designed mining yield 设计开采量
designed mix 设计配(合)比
designed navigable water-level 设计通航水位
designed navigation water level at downstream of dam 坝下设计通航水位
designed offence 指出的违约行为
designed output 计划产量;设计输出量
designed paper 设计图纸
designed parameter 设计参数
designed passenger flow 设计客流量
designed period 设计期限
designed person 指定代理人
designed position 设计位置
designed pressure 设计压力
designed productive capacity 设计生产能力
designed pumping discharge 设计的抽水量

designed reliability 设计耐久性;设计可靠性;设计可靠度
designed rule 设计规则
designed service life 设计使用年限
designed shopping center 设计的购物中心
designed slope 设计边坡
designed speed 设计速度;设计航速
designed stages of navigable channel 航道设计水位
designed system 设计的系统
designed temperature 设计温度
designed temperature difference 设计温差
designed tide level 设计潮位
designed traffic volume 设计交通量
designed trim 船设计纵倾;设计纵倾;设计吃水差
designed type 设计型号
designed value 特定值
designed velocity 设计速度;设计流速
designed water consumption 设计用水量
designed water-level 设计水位
designed waterline 设计水线(面)
designed water quality 设计水质
designed working life 设计寿命;设计使用年限
design effect 设计效果
design effort 设计工作;设计计划
design elevation 设计高程;设计标高
design engineering 设计工程
design engineering inspection 设计工程的检验
design engineering show 设计工程展览会(场)
design engineering system approach 工程设计系统分析
design engineering test 设计工程试验
design entrusting 设计委托
designer 设计者;设计师;设计人
designer colo(u)rs 舞台设计用的水色漆
designer-in-charge 设计(总)负责人
designer in chief 总设计师
designer of formal gardens and parks 园林设计师
design error 设计误差
design error failure 设计故障
designer's check list 设计师校核表
designer's load water line 设计载重水线;设计荷载水线
design estimate 设计估算;设计概算
design estimating index 设计概算指标
design estimating norm 设计概算定额
design evaluation test 设计鉴定试验
design examination 设计审查
design example 设计举例;样板设计
design external loads 设计外载
design factor 设计因子;设计因素;设计因数;设计系数
design factor of safety 设计安全系数
design failure 设计故障
design fault 设计故障;设计错误
design feature 结构特点;设计特征;设计特点
design fee 设计费
design file 设计文件
design filter rate 设计滤速
design fixed number of year 设计年限
design fixed type test 设计定型试验
design flexibility 设计灵活性
design flood 设计洪水
design flood discharge 设计最大流量;设计洪水流量
design flood flow 设计洪水流量
design flood frequency for bridge 桥梁设计洪水频率
design flood hydrograph 设计洪水过程线
design flood inflow 设计洪水进流量;设计洪水入流量
design flood level 设计洪水位
design flood occurrence 设计洪水出现率
design flood peak 设计洪峰
design flood peak discharge 设计洪峰流量
design flood stage 设计洪水位
design flow 设计交通量;设计流量
design force 设计外力
design force spectrum 设计荷载谱
design for durability 耐久性设计
design for earthquake 抗震设计
design for earthquake-proof 防震设计
design formula 设计公式
design for protection against frost action 抗冰冻保护设计;抗霜冻保护设计
design for stabilization pond 稳定塘设计
design foundation 设计基础
design framework 设计结构;设计骨架;设计框架
design freezing index 设计冰冻指数
design frequency 设计频率
design frequency of drainage 排水设计重现期
design fundamental 设计依据;设计基础
design goal 设计目标;设计目的
design gross weight 设计总重量
design ground motion 设计地面运动
design group 设计组
design guide 设计准则
design guideline 设计指南;设计方针
design handbook 设计手册
design head 设计水头;设计主任;设计负责人
design head of pumping station 泵站设计扬程
design heat consumption per hour 设计每小时耗热量
design heat load 设计热负荷
design heat loss 热损失设计
design height 设计波高
design high water-level 设计高水位;设计高潮位
design hour 设计小时
design hourly volume 设计小时交通量
design humidity 设计湿度
design hydrograph 设计水文过程线
design hypothesis 设计假设;设计假定
design idea 设计意图
design improvement program(me) 设计改进程序
design in detail 详细设计
designing 设计(方案)
designing aim 调查目的
designing capacity 原设计能力
designing drains with flow net 用流网设计排水沟
designing engineer 设计工程师
designing factor 设计因子;设计因素;设计因数
designing firm 设计事务所
designing for soil conditions 根据土质条件设计
designing for the disabled 残疾人设计;方便残疾人设计
designing institute 设计院
designing load 设计荷载;设计负荷
designing period of sewage (treatment) plants 污水厂设计年限
designing plan 调查计划
designing rainfall intensity 设计降雨强度
design ingredient ore 设计配矿
designing room 设计室
designing scheme 设计方案
designing task 调查任务
designing water level 原设计水位
design in intaglio 凹线纹
design in reinforced concrete 钢筋混凝土设计
design institute 设计单位
Design Institute for Emergency Relief System 紧急救援系统设计院
design instruction 设计说明书
design intensity 设计烈度
design interception 设计截流量
design item 设计项目
design-it-yourself system 设计组装系统;按组装原理设计系统
design lane 设计车道
design latitude 设计幅度;设计范围
design layout 设计布置
design liaison 设计联络;设计代表
design liaison group 设计联络组
design liaison meeting 设计联络会议
design life 设计(使用)寿命;设计使用期;设计保固期限;设计使用周期;设计(使用)年限
design life of structure 结构物设计使用期限
design lift 设计压实层厚
design limit 设计极限
design limit load 设计极限荷载
design line 设计线
design load 荷载设计值;设计荷载;设计载重;设计负荷
design load factor 设计荷载因数;设计荷载系数;设计负荷系数
design load safety factor 设计荷载安全系数
design losses 设计损失
design losses ratio 设计损失率
design low water-level 设计枯水位;设计低水位

design manual 设计手册
design margin 临界设计
design margin evaluation test 结构储备能力评价试验;结构安全系数试验
design mass of S/R machine 有轨巷道堆垛设计重量
design matrix 设计矩阵
design memorandum 设计记录
design method 设计(方)法
design methodology 设计方法学
design mix 配料设计;设计混合料
design mixture 配合比设计
design moment 设计力矩;设计弯矩
design monitoring 设计检验
design motif 设计主题
design number 设计编号
design objective 设计目的;设计目标
design of apron 护坦设计
design of armo(u)r 护面块设计
design of base 基础设计
design of beam 梁的设计
design of blast 爆破设计
design of blockwall 块体墙设计
design of bolted flange connection 螺栓法兰连接设计
design of breakwater head 堤头设计
design of bridge crossing 桥渡设计
design of buildings 房屋设计
design of buried structure 埋入式结构设计
design of caisson wall 沉箱壁的设计
design of cellular wall 格型墙设计
design of cementing casing 固井设计
design of coal mine 煤矿设计
design of code 编码设计
design of composite foundation 复合地基设计
design of composite ground 复合地基设计
design of concrete lining 混凝土衬砌设计
design of connections 联结设计;连接设计
design of core 堤心石设计
design of crest structure 胸墙设计
design of discharge system 出水系统设计
design of double anchored wall 双锚碇墙的设计
design of dredge pump 泥泵设计
design of dry dock floor 干坞底板设计
design of dry dock sump 干坞集水坑设计
design of dry dock walls 干坞坞墙设计
design of elevation (城市道路)竖向设计
design of elevation of subgrade 路基设计高程
design of equal bearing capacity 等强度设计
design of equal stability 等稳定设计
design of experiments 实验设计
design of farm work and weaving 耕织图
design office 设计室;设计事务所
design of fire separation 防火分隔设计
design of flexible dolphin 柔性墩的设计
design of gravity structure 重力式结构设计
design of grooves 孔形设计
design of holding characters in both hands 棒字图
design of lining 衬砌设计
design of load bearing element 承载构件的设计
design of lock structure 船闸结构设计
design of longitudinal section 纵断面设计
design of marks 标志设计
design of mine 矿井设计
design of mix 配合比设计
design of mix proportion 配合比设计
design of mixture 配料设计
design of monolith wall 沉井壁的设计
design of mooring point structure 系泊荷点结构的设计
design of municipal water distribution system 城市配水网设计
design of organizing construction 施工组织设计
design of passes 孔形设计
design of pile foundation by K-method 桩基计算 K 值法
design of pile foundation by M-method 桩基计算 M 值法
design of pipe system 管网设计
design of plane 平面设计
design of plant 成套设备设计;设备设计
design of process(ing) flow 处理流程的设计;工艺流程设计

design of pumphouse 泵房设计
design of railway location 铁路选线设计
design of report 报告的设计
design of retaining wall 挡土墙设计
design of rigid dolphin 刚性墩的设计
design of roads 道路设计
design of salvage work 打捞工程设计
design of sample 样本设计
design of section 定出断面
design of seepage control measures 控制渗流措施设计
design of service offered 提供服务的设计
design of statistic(al) inquiry 统计调查设计
design of steel joist 钢托梁设计
design of steel structure 钢结构设计
design of supporting structure 支承结构的设计
design of the plant 工厂设计
design of throat 车站咽喉设计
design of toe 堤脚设计
design of tubes 沉管设计
design of under layer 垫层设计
design of variety test plot 品种试验小区图
design of vertical alignment 纵断面设计
design of vertical vessels skirt 裙座的设计
design of virtual instrument system 虚拟仪器系统设计
design of water and drainage 给排水设计
design of water quality monitoring system 水质监测系统设计
design of working drawing of a capital construction project 基本建设项目施工图设计
design of working organization 施工组织设计
designograph 设计图解(法)
design operation 设计业务;设计工作
design option 设计方案
design order 计划(任务)书;设计任务书
design organization 设计单位
design-oriented 与设计有关的
design outline 设计草图
design paper 设计图纸;绘图纸;打样纸;设计文件
design paper's duplication 设计文件复制
design parameter 设计参数
design parameters of ground motion 设计地震动参数
design patent 工业专利设计;设计专利权
design-pattern limitation 计算限度
design payload 设计有效荷载
design performance 设计性能
design period 设计期间;设计周期;设计期限;设计阶段
design personnel 设计人员
design phase 设计阶段
design philosophy 设计原则;设计原理;设计思想
design pipe diameter 设计管径
design plan 设计详图;设计平面图;设计方案;设计草图
design point 计算点;设计状态;设计工况;设计点
design power 设计功率
design power efficiency 设计工况效率
design power loading 设计动力荷载
design precept 设计方案;设计原则;设计任务书
design precipitation 设计(降)雨量;设计降水量
design pressure 设计压力
design principle 设计方针;设计原则;设计原理
design problem 设计任务;分配的任务
design procedure 设计方法;设计过程;设计程序;设计步骤
design process 设计过程
design processing 设计图案处理
design procurement 提供设计
design product 设计产品
design production department 生产计划科
design profession 设计职业;设计业务
design program(me) 设计任务书;设计计划书;设计程序;设计方案
design programmer 程序设计员
design project 设计项目
design promoter 设计项目倡议人
design proof cycle 设计检查周期
design proposal 设计建议;设计方案
design quality 设计质量
design quantity 设计用量
design rainfall 设计雨量
design rate 设计流量

design ratio 设计比例
design recurrence interval 设计重现期
design reference period 设计基准期
design reference year 设计水平年
design reliability 设计可靠性
design report 设计报告
design requirement drawing 设计要求的图纸
design requirements 设计要求
design reserves 设计储量
design return period 设计重现期
design review 产品试验;设计审查;设计评审;设计检查;设计复审
design review conference 设计评审会议
design revision 设计修改
design revision notice 设计更改通知
design right 设计权
design roller 滚花辊
design roller coating 辊涂压花法
design rule 设计规则
design running speed of passenger train in section 路段旅客列车设计行车速度
designs act 工业样品法
design safety 设计安全
design safety factor 设计安全系数
design safety limit 设计安全限
design schedule 设计进度(表);设计计划表;设计表格
design scheme 设计方案
design sea level 设计海面高程;设计潮位
design section 设计组;设计室;设计路段;设计截面;设计断面;设计部门
design seismic coefficient 设计震动系数;设计地震系数
design seismic force 设计震动力;设计地震力
design seismicity 抗震设计烈度;设计烈度
design seismic load 设计地震荷载
design series 设计系列
design sheets 设计(说明)书
design ship 设计船型
design sight distance 设计视距
design simulation 设计模拟
design size 设计尺寸
design sketch 设计草图
design solution 设计方案
design specification 设计任务书;设计说明(书);设计规范;设计规程;设计要求
design specification for building 建筑设计规范
design specification for reinforced concrete structure 钢筋混凝土结构设计规范
design spectrum 设计谱
design speed 设计(行车)速度;设计车速;设计航速
design speed of car 设计车速
design stage 设计水位;设计阶段
design stage of constructive map 施工图设计阶段
design standard 设计规范;设计标准
design standard for large dams 大坝设计规范
design standard manual 设计标准手册
design standard of waterway project 航道工程设计标准
design standard specification 设计标准技术规范
design starting point 设计出发点
design stop order 设计停止指示
design storm 设计暴雨(径流量)
design strategy 设计策略
design strength 许用应力;设计(基准)强度
design stress 设计应力
design studio 设计室
design subgrade factor 设计地基反力系数
design subgrade modulus 设计地基反力模量
design subgrade strength 设计路基强度值
design summary 设计一览表
design supplement 设计附件;设计补充文件
design survey 工程设计测量
design system 设计系统
design table 设计图表;设计表格
design task 设计任务
design team 设计组
design technology 设计工艺
design temperature 设计温度
design temperature difference 设计温差
design tension of anchor 锚杆设计拉力
design test 鉴定试验
design theory 设计理论
design thermal efficiency 设计热效率

design thickness 限度厚度;设计厚度
design thrust 设计推力
design tide level 设计潮位
design to cost goal 目标成本设计
design traffic capacity 设计通行能力
design traffic number 设计交通量
design tsunami 设计海啸
design type calculation of beat exchanger 换热器的设计型计算
design ultimate load 极限设计荷载;设计极限荷载
design unit 设计单位
design value 结构参数;计算值;设计(数)值;设计参数值
design value for strength of materials 材料强度设计值
design value of action 作用设计值
design value of load 荷载设计值
design variable 设计参数;设计变量
design vehicle 设计汽车;设计车辆
design velocity 构造速度
design velocity of flow 设计流速
design verification 设计验证
design verification test 设计验证试验
design vessel 设计(标准)船型
design vessel type and size 设计船型
design volume 设计交通量
design water head 设计水头
design water-level 设计水位;设计水面(高程)
design water stage 设计水位
design wave 设计波(浪)
design wave climate 设计波候
design wave condition 设计的波浪条件;设计波况
design wave height 设计波高
design wave parameter 设计波参数
design weight 限定重量;设计重量
design wheel load 设计轮载;设计轮压
design wind 设计风力
design wind pressure 设计风压
design wind speed 设计风速
design wind velocity 设计风速
design with nature 适应自然的设计
design with proper forethought 深谋远虑的筹划
design work 设计工作
design working pressure 设计资用压力(工作资用的压力值)
design year 设计使用周期;设计年度
desilicate 除硅
desilication 硅氧淋失;除硅;脱硅(作用);去硅反应
desilicification 脱硅
desilker 除须器
desilter 滤水池;除砂器;沉砂池;砂井;砂阱
desilt(ing) 沉砂;放淤;清淤;排淤;脱泥
desilting area 沉砂面积;沉淀面积
desilting basin 沉砂池;沉淀池;放淤区
desilting canal 放淤渠(道);排淤渠(道);排砂渠(道)
desilting channel 放淤槽;过滤槽
desilting funnel 排砂漏斗
desilting installation 沉砂设施
desilting sluice 冲淤闸(门);排砂闸(门)
desilting strip 脱泥地带;过淤地带;沉砂畦条;放淤畦条
desilting works 除砂工程(如沉砂池等);沉砂工程;放淤工程;防砂工程;排砂设施
desiltor 脱泥机
desilver 除银;提银
desilvered lead 除银铅
desilverization 除银
desilverizing kettle 除银锅
desingnation 牌号
desintegrator 笼型粉碎
desintering 清理
desirability 客观需要
desirable criterion 必要标准
desirable environment 合适的环境
desirable property of a model 模型的期望性质
desired cut 理想切削
desired effect 预期影响;预期效果
desired form 期望形式
desired ground point 所需地面点
desired ground zero 预期爆心投影点
desired impact 要求达到的效果
desired input 输入量的希望值
desired length 要求长度

desired output 输出量的希望值
desired performance 预期完成
desired-profit point 预期利润点
desired result 预期效果;预期收益
desired speed 理想转速
desired strength 理想强度;理论强度;预期强度;预定强度;要求强度;所需强度
desired thickness 要求厚度
desired to undesired signal ratio 期望信号与不期望信号之比
desired track 计划航迹;航向;预计航迹
desired value 预期值;指标值;给定值;期望值;期待值;所需值
desire for goods 购货欲望
desire for mobility 对机动性的要求
desire line 希望线;期望线
desire line chart 期望路线图
Desi silk 德西丝(孟加拉)
desist and refrain 不动产查封令
de Sitter space 德西特空间
desize 脱浆;退浆
desizer 脱油炉
desizing 脱浆;除浸剂;退浆
desizing/dyeing wastewater 脱浆/染色废水
desizing mangle 脱浆轧布机
desizing wastewater 脱浆废水
desk 窑车上耐火材料面层;办公桌;书桌
desk blade 盘形刀片
desk calculating machine 台式计算机
desk calculator 台式计算器
desk centrifuge 台式离心机
desk check 桌上检验;部件校验;台式校验;台面检验
desk checking 手工检查
desk computer 台式计算机
desk control unit 桌式控制台;桌式操纵台
desk engineer 内业工程师
deskew 抗扭斜
de-skewing 去偏斜
de-skew processing 去斜处理
desk fan 桌式电扇;台式风扇;台扇
desk frame 台架
de-skilling 降低作业的技术程度
desk jobber 向厂商直接进货的零售商
desk keyboard 桌式控制台
desk lamp 台灯
desklighting 台灯
desk lock 书桌锁;办公室桌锁
desk machine 台式计算机;台式机器
desk structure 桌状构造
desk-survey computer 测地计算机
desk switchboard 台式配电盘;台式配电板
desk telephone set 桌机;台式电话机
desk-top computer 台上计算机;台式计算装置
desk-top machine 台式计算机
desktop management interface 桌面管理接口
desk type control panel 台式控制屏;台式控制板;台式插接板
desk-type machine 桌式控制台
desk-type relay 座式继电器
desk work 科室工作
deslag 除渣
deslagging 除渣;倒渣;排渣
deslicking 防滑
deslicking material 防滑材料
deslicking tile 防滑瓷砖
deslicking treatment 防滑处理
deslime 除矿泥
deslimer 脱泥机
desliming 除(去)矿泥;脱泥
desliming classifier 脱泥分级机
desliming cone 脱泥圆锥分级机
desliming screen 脱矿泥筛
desliming sieve bend 脱泥弧形筛
de-sludge 污泥排除;除去污泥;清除淤泥;除渣;清除油泥
de-sludger 排污泥器;除泥设备
de-sludging 除渣;清除泥
de-sludging agent 去泥剂;脱泥剂
de-sludging mechanism 除垢机构
de-sludging operation 除渣操作
de-sludging separator 除垢器
de-slurrying 湿法脱除细粒
de-slurrying by screens 筛去细粒
desmic surface 连锁曲面

desmid earth 硅藻土
Desmidiaceae 鼓藻科
Desmids 鼓藻(类)
desmine 辉沸石
desmocollinite 基质镜质体
desmodur 聚氨基甲酸酯类黏合剂
Des Moinesian 德莫阶【地】
desmose 连接纤丝
desmos-emicollinite 基质半镜质体
desmosite 条带绿板岩
desmosome 桥粒
desmotrope 稳变异构体
desmotropism 稳变异构体现象
desmurgia 绷扎法
desmurgy 绷扎法
desmutting 除去附着物
desolation 荒地;废墟
de-solde-ring gun 去焊枪
de-solvation 去溶剂作用;去溶剂化;退溶
desolventizer-toaster 脱溶剂器-蒸炒缸
desorber 解吸器
desorex 活性炭
desorption 解吸作用;解吸附作用;解吸;清除吸附气体;脱附;退吸(作用)
desorption agent 脱附剂
desorption by displacing 置换脱附
desorption by evacuation 降压脱附
desorption by heating 升温脱附
desorption by stripping 汽提脱附
desorption coefficient 解吸系数
desorption column 解吸塔
desorption curve 解吸(作用)曲线;消退曲线
desorption factor 解吸因子
desorption isotherm 解吸等温线;脱附等温线
desorption of gas 气体的解吸
desorption of moisture 脱湿(作用)
desorption of pesticide 农药解吸
desorption process 解吸过程
desorption rate 解吸速率;脱附速率
desorption solution 解吸溶液
desorption solvent 解吸溶剂
despatch = dispatch
despatch barrel 发送室
despatch boat 通信船
despatch days 发送日数
despatch discharging only 速遣费仅在卸货时计付
despatcher 调度员
despatch in bags 袋装发运
despatch in bulk 散装发运
despatching barrel 发送室
despatching center 调度中心;发货中心
despatching control 发货处
despatching crossover 调度渡线
despatching list 发货单
despatching of goods 发货
despatching of luggage 行李发出
despatching plan 发货安排
despatching program(me) 发货安排
despatching sheet 发货单
despatching shuttled through block trains 固定车底循环运用的列车
despatch inquiry 发出询价
despatch in sacks 袋装发运
despatch loading only 速遣费仅在装货时计付
despatch money 速遣费
despatch note 包裹附单
despatch of cargo 货物发运
despatch of goods 货物发运
despatch of ship 船舶速遣
despatch room 调度室;分发室;发送室
despatch station 出发站;发送站
despiker 峰尖校平设备
despiker circuit 尖峰展平电路;尖峰平滑电路;削峰电路
despiking 削峰
despiking circuit 反尖峰电路
despiking resistance 削峰电阻;阻尼电阻
despin 反自转
despin mechanism 消旋装置
despin(ning) 停止旋转
despotic network 主钟控制网
despujolsite 钙锰矾
despumation 除去浮渣;除沫法;排除杂质
despun 反自转

despun antenna 消旋天线;反自转天线
despun motor 反自转电动机;反旋转电动机
despun platform 反自转台;反旋转台
desquamation 鳞状剥落;脱屑
desquamation of surface soil 表土剥离
dessert fruit 餐后水果
dessert spoon 中匙
dessin 线条;图案
destabilization 稳定作用;失稳;去稳定化作用
destabilization of particles 颗粒脱稳
destabilization speculation 使市场不稳定的投机
destabilize the market 扰乱市场
destabilizing effect 去稳效应;失稳作用
destabilizing factor 不安定因素
destabilizing moment 破坏力矩
destabilizing speculation 不稳定的投机
de-stacker 拆堆垛机
destacking 卸垛
destage 离台
destaging error 离台错;降级误差;不分级错
destaticization 消除静电(作用)
destaticizer 除静电器;除静电剂;脱静电剂;去静电器
destaticizing 静电消除
destaticizing of paper 纸张静电消除
de-stemmer 去梗机
desterilization 解除货币管制
desterilization of reserves 准备金恢复使用
desterilize 恢复使用
desticker 分离器
destination 目的地;目标;需要点(运输问题中心需要点);预定用途;预定目标;终点;船到终点
destination address 结果地址;接收站地址;目的地址;收报地址
destination address field 目的地址字段
destination address field prime 接收站地址主字段;目的地址字段加小撇;目的地址主字段
destination addressing mode 目的定址方式
destination address register 目的地址寄存器
destination airport 终点航(空)站
destination and distance sign 终点和距离标志
destination bill 目的地提单
destination board (公共汽车的)路线指示牌;指示牌;指路牌
destination button 终端按钮
destination case 目的地角色
destination category 目的范畴
destination code 接收站代码;目的(地)指示代码;目的地代码;目的地编码;目标代码
destination code base 指定编码基数
destination contract 目的地契约
destination control statement 目的地管制声明
destination control table 目的地控制表
destination element field 目的地元字段
destination field 目的地字段
destination file 结果文件;目标文件
destination flag 目的地国旗
destination goal 目的
destination host 目的主机
destination indicator 到站指示牌;目的地表示器
destination mark 目的港标志;目的地标志
destination mask 目的地指示掩码
destination memory 目的存储器
destination memory location 目的存储器单元
destination network 目的地网络
destination operand 目的操作数
destination packet 目的包
destination point 终点;到达点
destination port 目的港
destination queue 目的(地)队列
destination railway 到达铁路
destination register 目的(地)寄存器
destination release 到达地放行证书
destination sign 目的地指示标志
destination station 目的站
destination survey 目的地(交通量)调查
destination terminal 到达终点
destination warning marker 终端警告标志
destination workspace register 目的工作区寄存器
destination zone 目的区;终点区
destine 指定
destined station 指定车站
destinker 去味器
de-stratification 去层理作用;消层
de-stress 放松应力;钢轨应力放款;放散(温度)应力

de-stressed zone 去应力带;应力释放带
de-stressing 应力放散;去应力
de-striping 去条带
destroy by rush of water 冲毁
destroy chain 消链
destroyed by fire 毁于火灾;焚毁
destroyed slope surface 坡面被破坏
destroyer 驱逐舰;破碎机
destroyer depot ship 驱逐舰母舰
destroyer escort 护卫舰
destroyer stern 驱逐舰船尾型
destroying 损坏;毁灭;毁坏
destroying substance 破坏性物质
destroy plants and trees 毁坏庄稼和树木
destroy the crop 毁掉作物
destroy weed growth 破坏性杂草生长
destruct 自毁;摧毁
destructibility 破坏力;破坏性
destruction 破坏
destructional delta 侵蚀三角洲
destructional landscape 侵蚀景观;破坏性景观
destructional plain 侵蚀平原
destructional terrace 侵蚀阶地
destruction belt 破坏带
destruction level 破损水准仪;剥蚀水平面
destruction of balance 破坏平衡
destruction of bond 黏结破坏
destruction of buildings 建筑物破坏
destruction of explosives 销毁炸药
destruction of ice 冰裂
destruction of insulation 绝缘损坏
destruction of organic material 有机质破坏;破坏有机质
destruction of ozone layer 臭氧层破坏
destruction of pool and oil-gas redistribution 油气藏破坏和油气再分布
destruction of soil 土壤破坏(作用)
destruction of stock 销毁库存
destruction plain 侵蚀平原;破坏平原
destruction plant 垃圾处理厂
destruction region 破坏区
destruction surface 破坏路面
destruction test 毁坏试验;破坏性试验
destruction zone 破坏带
destruction zoning of apical or bottom bed 顶底板破坏分带
destructive 破坏(性)的
destructive addition 破坏式加法
destructive attack of vehicle 车辆对路面的有害冲击
destructive breakdown 破坏性击穿
destructive capacity 破坏能力
destructive combustion 毁坏性燃烧
destructive competition 破坏性竞争
destructive cursor 破坏性光标
destructive delta 侵蚀型三角洲;破坏性三角洲
destructive deodo(u)rization 分解性除臭
destructive disease 毁灭性病害
destructive distillation 毁馏;分解(破坏)蒸馏
destructive distillation of wood 木材干馏
destructive distillation turpentine 干馏松节油
destructive earthquake 破坏性地震
destructive effect 破坏作用;破坏性;破坏效应
destructive examination 破坏性检查
destructive expansion 破坏性膨胀
destructive experiment 破坏试验
destructive exploitation of species 物种的破坏性开发
destructive fire 破坏性的火
destructive force 破坏力
destructive insect 毁灭性虫灾;害虫
destructive inspection 破坏性检查;破坏检验
destructive interference 相消干扰;破坏性干扰
destructively distilled pine oil 干馏松油
destructively distilled wood turpentine 干馏木松节油
destructive malfunction 破坏性故障
destructive mechanism 自爆装置
destructive metabolism 破坏性代谢(作用)
destructive metamorphism 破坏变质作用
destructive method 有损探伤法
destructive motion 破坏运动
destructive movement 破坏性运动;有害运动
destructiveness 破坏性;毁灭性;破坏程度

destructive pest 毁灭性虫灾;害虫
destructive phenomenon of apical or bottom bed 顶底板破坏现象
destructive plate margin 破坏性板块边缘
destructive poisons 毁坏性毒物
destructive power 破坏力
destructive read 抹去读数
destructive reading 抹掉信息的读出;破坏性读数
destructive readout 抹掉信息的读出;破坏(性)读出;破坏信息读出
destructive shock 破坏性震动
destructive soil-inhabiting insect 地下害虫
destructive test 破损试验;击穿试验;断裂试验;爆破试验;破坏(性)试验;破坏检查
destructive testing 破损试验
destructive testing method 破坏性检验法
destructive testing of concrete 混凝土的破损试验
destructive vibration 破坏性震动
destructive wave 破坏性波浪
destruct line 自毁线
destructor 垃圾焚化炉;焚烧炉;废料焚化炉;破坏装置;破坏器
destructor plant 垃圾焚化厂;垃圾处理厂
destructor room 垃圾焚化间
destruct system 自毁系统
destructure plate boundary 破坏性板块边缘
desublimation 消升华(作用);去升华作用;凝结作用
desuetude 废绝;失效
desulfation 脱硫酸盐作用
desulfidation 除硫;去硫
desulfogypsum 脱硫石膏
desulfurated 脱硫的
desulfurater 脱硫剂
desulfurating 脱硫的
desulfurating agent 脱硫剂
desulfuration 去硫(作用);脱硫
desulfuridation 脱硫酸盐作用
desulfuring installation 脱硫装置
desulfurization 除硫;脱硫作用;去硫
desulfurization by activated carbon 活性炭脱硫
desulfurization by active carbon 活性炭脱硫
desulfurization by electronic beam irradiation 电子束辐照烟气脱硫
desulfurization by fluidized bed combustion 流化床燃烧脱硫
desulfurization by high gradient magnetic separator 高磁场梯度磁力分选脱硫
desulfurization by limestone injection process 石灰石喷入法脱硫
desulfurization from exhaust gas 排气脱硫
desulfurization from fuel oil 燃料油脱硫
desulfurization gypsum 脱硫用石膏
desulfurization method 脱硫法
desulfurization of flue gas 烟道气脱硫
desulfurization of fuel oil 燃料油脱硫
desulfurization purification 脱硫净化
desulfurization refinery 脱硫精炼厂
desulfurization unit 脱硫单元
desulfurized fuel 脱硫燃料
desulfurizer 脱硫装置;脱硫塔;脱硫器;脱硫剂;脱硫槽
desulfurizing agent 脱硫剂
desulfurizing furnace 脱硫炉
desulfurizing tower 脱硫塔
desulfurizing unit 脱硫装置
desulphidation 脱硫作用;去硫作用
desulphurater 脱硫剂
desulphuration 脱硫
desulphurization 脱硫(作用)
desulphurization bacteria 脱硫菌
desulphurization of fuel 燃料脱硫(作用)
desulphurize 除硫
desulphurizing roasting 脱硫焙烧
desuperheat 减温;过热后冷却
desuperheated steam 减温蒸汽
desuperheater 减热器;蒸汽冷却器;过热降低器
desuperheater spray 减温器喷水
desuperheating 过热下降
desuperheating station 减温器
desuperposition 反叠加
deswelling 消溶胀(作用)
desyl 二苯乙酮基
de-symmetry 偏位
de-synchronization 去同步化作用

desynchronize 同步破坏
de-synchronizing 去同步；失步
detach 卸下；除去；车辆脱钩；分下；派遣
detachability 可分离性；可拆性；脱渣性；脱除性
detachable 可拆(卸)的
detachable axle 可卸轴
detachable bearing 活动轴承
detachable bit 可(拆)卸钻头；可(拆)卸式钻头；活(络)钻头；活动钻头
detachable blade 可折桨叶；可拆卸螺旋桨叶；可拆桨叶；可拆卸页片
detachable bubble horizon sextant 可卸气泡水准六分仪
detachable bucket 可卸叶片；拆换式戽斗
detachable bulkhead 活动舱壁
detachable chain 活络链
detachable coil 可卸线圈
detachable column 单立柱
detachable detector 可拆卸探测器
detachable device 可拆卸装置
detachable ditcher 可拆卸反向铲挖土机
detachable drill head 可卸式钻头；可卸式钎头；活钎头
detachable fitting 可拆卸配件；可拆接头
detachable fixing 可拆卸(式)装配
detachable head of cylinder 可拆汽缸盖
detachable-hook chain 钩头链
detachable joint 可卸连接；可拆接合
detachable key 活络键
detachable lever bracket 分离杆托架
detachable link 可拆链环；散合式连接链环
detachable-link chain 钩头链
detachable magazine 可拆卸暗盒
detachable mount 可卸式架
detachable needle 可拆卸针头
detachable panel 可拆卸面板
detachable parts 可拆式部件
detachable plate 隔舱壁上装的可移动式铁门
detachable plugboard 可卸插接板
detachable point seat 节伸式阀座；活络阀座
detachable point shovel 换尖式挖土铲
detachable propeller 可扩螺旋桨；组合推进器
detachables 可卸部件
detachable shoe 活动垫座
detachable shutter 可卸的关闭器；可卸的百叶窗
detachable speaker box 分体式音箱
detachable specular reflector 可拆卸吸顶灯；可拆卸反光镜；活动镜面反射器
detachable tank 可拆卸油箱；可拆卸水箱
detachable tow-bar 可拆卸的牵引杆
detachable trencher 可拆卸挖沟机
detachable tungsten carbide insert bit 碳化钨合金补强的可拆卸式钻头
detachable valve seat 可拆下的阀座
detachable viewfinder 可卸检影器
detachable wagon fittings 车辆脱钩设备
detachable warrant 可分割认股权证
detachable wheel 可卸下的轮
detach coupons to 发给利息票
detached access 独立式出入口
detached arch core 挤离背斜心
detached binary 不接双星
detached block 飞来峰；外来岩块
detached breakwater 分离式防波堤；离岸防波堤；孤立防波堤；独立式防波堤；岛式防波堤；岛堤；分立防波堤；不接岸防波堤
detached building 独立(式)房屋
detached chimney 独立(大)烟囱
detached cloud 碎云
detached coefficient form 分离系数形式
detached coefficient tableau 分离系数表
detached column 独石柱；独立柱；单立柱；单独柱
detached contact method 分离接点法
detached core 挤离褶曲岩心
detached core of syncline 挤离向斜心
detached duty 临时任务
detached dwelling 独立式住宅
detached garage 独立车库
detached harbo(u)r 岛式港(口)
detached house 独立式住宅
detached house quarter 独立式住宅地段
detached island 孤(立)岛；分立岛；分离岛
detached keel and keelson 分离龙骨
detached keyboard 分离式键盘

detached-lever escapement
detached mass 孤残岩块；飞来峰；外来岩块
detached mole 孤立防波堤；岛(式)防波堤
detached palace 离宫
detached peninsula 独立半岛
detached pier 岛式码头；岛埠头；独立(支)墩
detached privy 户外厕所
detached process 分派进程
detached reef 分立礁
detached rock 孤(立)礁；分立礁
detached shaft 离合轴
detached shock wave 脱体激波
detached single family dwelling (一个家庭的)独立住宅；独门独户住宅
detached soil material 松散土质
detached statuary 圆雕；独立雕像
detached stratus 分裂层云
detached structure 分离建筑；独立结构
detached superstructure 局部船上层建筑；局部船楼；岛式上层建筑
detached vehicle 已分解客车
detached wagon 已分解货车
detached wharf 岛式码头
detached work 前哨工事
detacher 拆卸器
detach ice 除冰(块)
detaching 分离；分开；卸下
detaching and attaching of cars 摘挂调车
detaching cam 分离凸轮
detaching efficiency 分离效率
detaching hook 脱钩；分离钩；触板脱钩
detaching motion 分离运动
detachment 脱离；分离；可分件；摘车
detachment allowance 释放容许量
detachment fault 挤离断层；滑脱断层；脱顶断层；推覆体底基逆掩断层
detachment gravity slide 脱顶重力滑动
detachment of car 车辆摘钩
detachment of devices 设备的卸下
detachment of interest 无利可图
detachment of pontoon line 拆开挖泥浮管线
detach wagons 分解货车
detackifier 防粘剂；脱粘剂
detail 零件；详细情况；细图；细部；大样；逐一处理
detail account 明细账；明细分类账
detail assembly 细部装配
detail assembly template 细部装配样板
detail bit 说明位
detail budget 细目预算
detail budgeting 明细预算
detail chart 详细流程图
detail construction 结构详图；细部构造
detail contrast 明细对比；细节对比度
detail coordinate point 细部坐标点
detail cost analysis 详细造价分析
detail design 零件设计；详图设计；细部设计；详细设计
detail designator 详细标记
detail drawing 详图；加工详图；明细图；细部图；大样图；分图；施工图；设计详图
detail drawing of parts 详细零件图
detail drawing of reinforcement 配筋图
detailed 详细的
detailed account 明细账；清点材料；清单
detailed analysis 详细分析
detailed appropriation 详细经费
detailed audit 详细审计
detailed audit report 详细审计报告
detailed balance 细致平衡
detailed budget 详细预算
detailed card 细目卡(片)
detailed catalogue 细目
detailed coastline 详细海岸线
detailed construction 细部结构
detailed construction schedule 施工进度计划
detailed cost estimate 详细概算
detailed coverage 详细论述
detailed data 详细资料；详细数据
detailed description 详细说明
detailed description of the invention 发明的详细说明
detailed design 具体设计；详细设计；施工图设计；细节设计；细部设计(阶段)
detailed design stage 详细设计阶段；细部设计阶段

detailed drawing 施工图；大样(图)；零件图；详图
detailed engineering 细节设计；工程细节；施工细节；施工设计
detailed engineering rules and regulations 工程细则
detailed estimate 详细概算；设计预算
detailed estimate (based on working drawing) 施工图预算
detailed estimate norm 预算定额
detailed estimate (of construction cost) 施工预算；详细预算；工程造价详细预算；工程建筑费用详细估价
detailed estimation 工程预算；详细概算书
detailed evaluation of technical economics 详细技术经济评价
detailed examination 精密调查；详细检查
detailed explanation 详细说明
detailed exploration 详细勘探；详细勘察
detailed exploratory report 详勘报告
detailed exploratory well 详探井
detailed field data 详细现场资料
detailed final sorting 细分
detailed geotechnical investigation 详细勘探；详细勘察
detailed gravity survey 重力详查
detailed hydrogeologic(al) investigation 水文地质详细勘察
detailed inspection 详细检查
detailed instruction 详细说明
detailed interview 详细面谈法
detailed investigation 结构详图；详细调查；详查；详细勘探
detailed layout planning 详细布置计划
detailed ledger 明细分类账；详细分类账
detailed list 明细表；详细清单；清单
detailedly controlled 达到详细控制
detailed map 明细图；详图
detailed mineral investigation 矿产详查
detailed operating procedure 作业程序细则；操作程序细则
detailed operating[operation] schedule 详细作业计划；细部作业计划
detailed packing list 详细装箱单
detailed particulars 详细说明
detailed perpetual inventory records 永续盘存明细记录
detailed plan 零件图；详细平面；详细计划；详细规划图；细部图；碎部图
detailed planning 具体规划；详细规划；细部规划
detailed planning stage 选择场址勘察阶段
detailed price list 详细价目表
detailed profile 详细纵断面图
detailed program(me) 具体方案；详细(工程)计划；细部规划
detailed project design 工程详细设计
detailed project report 工程详细报告
detailed prospecting 详细调查；详细勘探；详细勘察；详查
detailed prospecting for construct 详查最终
detailed prospecting report 详查报告
detailed provisions 详细条款
detailed radar image 精细雷达影像；高分辨率雷达影像
detailed record 详细记录
Detailed Regulations for Implementation of the Law of the People's Republic of China on Prevention and Control of Water Pollution 中华人民共和国水污染控制实施细则；中华人民共和国水污染防治实施细则
detailed report 详细报告
detailed requirement for fuel oil 燃料油的具体要求
detailed rules and regulations 细则
Detailed Rules and Regulations for the Implementation of the Forest Law of the People's Republic of China 中华人民共和国森林实施细则
detailed rules and regulations of supervision 监理工作细则
detailed schedule 进度计划明细表；明细进度表
detailed schedule of special fund 专用基金明细表
detailed scope of supply 详细供应范围
detailed section 详细剖面；细部截面
detailed shoreline 详细海岸线
detailed sketch 详细绘图；详图
detailed soil map 土壤详图

detailed soil survey 土壤详测
detailed specification 详细说明书;详细规格(书)
detailed standard of current capital 流动资金明细定额
detailed statement 明细表
detailed statement of accounts 账户明细表
detailed study 详细研究
detailed subsidiary records 辅助明细记录
detailed supporting statement 明细附表
detailed survey 细部测量;碎部测量;碎部测图;详细勘测;详查,详测;构造调查
detailed survey of pollution sources 污染源详查
detailed system description 详细系统说明
detailed technical requirement 详细技术要求
detailed technical specification 技术规格一览;详细技术说明书
detailed test objective 详细的试验目的
detailed test of mineral dressing ability 详细可选性试验
detailed topography 地形碎部;碎部地形
detailed use of commodity 详细用途
detailed ways and means 详细办法
detail engineering 详细工程;详图工程
detail enhancer 细节放大器;清晰度增强器
detailer 细部设计员;大样设计者;大样设计员;推销员;施工设计人员
detail estimate sheet 工程细目估价单
detail file 明细档案;详细文件资料;细目文件;细目文档;说明资料
detail flowchart 明细流程图;详细流程图
detail fracture of rail head 轨头微裂纹
detail geologic(al) map based on remote sensing 遥感详细地质图
detail geotechnical investigation 工程地质详细勘察
detail index mark 详细索引标志
detailing 详细设计;细节;大样设计
detailing design 细部设计
detailing of parts 零件设计
detail item 细目表
detail labo(u)r 局部工人
detail layout plan 详细布置图
detail ledger 明细账;明细分类账
detail ledger audit 明细分类账审计
detail log 详测曲线图
detailloss 清晰度降低
detail man 推销员
detail map store 地图供应库
detail network 加密网;详细网示(工程进度表);详细网络图;测图三角网
detail of construction 零件图;施工详图
detail of design 设计详图;设计(图样)
detail of facade 立面细部(房屋);立面细节
detail of joint 接缝详图
detail of manhole 检查井大样图
detail of treaty 协议细节
detail operating schedule 小日程计划;详细作业(进度)计划
detail paper 绘大样图用的图纸;描图纸;底图纸;大样图纸
detail pen 粗线笔
detail pen for double lines 划双线笔
detail planting design 种植大样图
detail plate 线划板【图】
detail plotting 细部测图
detail point 细部点;地物点;碎部点
detail point description 碎步点记载手薄
detail printing 细目打印
detail prospecting stage 详细勘探阶段
detail prospecting stage of evaluation 评价性详探阶段
detail record 明细记录
detail rendering 细节再现
detail reproduction 细部再现力
detail requirement 详细规格;详细要求
details 详细报告;细节
details as per attached list 详见附表
detail schedule 详细进度计划(表);小日程计划;详细日程计划
details of apron 垂裙大样
details of basin support 洗脸盆托架详图
details of case 案构
details of construction 结构详图;结构零件;构造详图;构造细节

details of design 设计细则;设计细节;设计细部
details of seismic design 抗震构造措施
details of treaty 合同细节
details of warning sign of overhead electric(al) danger (集装箱)登箱顶触电警告标记
details of work items 工程项目明细表
detail specification 详细规范;详细说明书;详细规格
detail status report 详细情况报告
details to follow 详细内容后告;详情随后邮寄
detail surveying 详细勘测;细部测量
detail survey of tunnel 隧道细部测量
detail survey stage 详细调查阶段
detail tape 细目带
detail test 分部试验
detail time 详细时间
detail triangulation 图根三角测量
detained goods 扣留货物
detained pending investigation 扣留待查
detainee 被扣押者
detainer 扣押者;扣留者;非法占有
detain flood 滞洪
detaining zone 滞流区
detainment 扣留;拖延
detainment of the cargo 扣留货物
detajmium bitartrate 地他铵重酒石酸盐
detar 脱焦油
detarrer 除焦油器;脱焦油设备
detarring 除焦油
detarring plant 脱焦油设施
det-cord 导爆线
det drill 熔化穿孔机
dete 潜艇雷达
detearing 沥水;沥干;除余漆
detect 查明;发觉;检出;探测
detectability 可检取性;可检查性;可检测性;可检波性;鉴别率;检验能力;检测限;检测能力;检波能力;整流能力;探测能力
detectable 可探测;可检波
detectable element 可检出元素;可检测元素;可检测像素
detectable error 可检测误差
detectable group 可检出组;可检测(像素)组
detectable quantities of smoke 烟的可探测量
detectable rate 检出率
detectable trap 可探测的圈闭
detectable type fire detector 可拆式火灾探测器
detectagraph 窃听器
detectaphone 监听设备;监听器;窃听器;探漏器
detected error 检测出的错误
detected fault 检测到的故障
detected signal 已检波的信号;探测信号
detect(ing) 检测
detecting action 检波作用
detecting block 检测块
detecting customer's needs 探求顾客需要
detecting device 探测装置
detecting element 灵敏元件;检测端;检测元件;检波元件;传感元件;探测元件
detecting engine troubles 检查发动机故障
detecting equipment 探伤设备
detecting foil 箔探测器
detecting grating 检波光栅
detecting head 检波头;探测头
detecting instrument 检测仪器;检测仪表;检波仪器;探测仪(器);探测系统传感器
detecting layer 探测层
detecting limit 检测极限
detecting of bottom characteristics 底质探测
detecting piece 检测片
detecting-range sonar 探测与测距声呐
detecting sand level 探砂面
detecting sensitivity 检测灵敏度
detecting system 检测系统
detecting threshold 检测阈
detecting turning wheel mechanism 探伤转轮机构
detection 检测;检波;拖延;检出;探测
detection and alarm system 探测与报警系统
detection and assessment of photochromism 光致变色的检验和评定
detection and measurement technique 检测与测量技术;检测技术
detection array 探测器阵(列)
detection block 检测块
detection by reed switch 干簧管检测

detection capability 检测能力
detection characteristic 检波特性
detection check valve 检测单向阀
detection circuit 传感电路;探测电路
detection coefficient 检波系数
detection criterion 检定准则;检测准则
detection curve 检测曲线
detection device 检测装置;检测设备
detection distortion 检波失真
detection-effectiveness parameter 觉察力参数
detection efficiency 检波效率;探测效率
detection equipment 检测装置;探测指数;探测设备
detection facility 检测设施
detection function 检测函数
detection index 检测指数;探测指数
detection limit 检出限(界);检测(极)限;探测极限
detection mode 检测方式
detection model 检测模型
detection monitoring program(me) 探查性监测计划
detection of acetate 醋酸根检验
detection of axle cracks 车轴裂纹探伤
detection of defects 探伤
detection of elements 元件的检测
detection of error 检查错误;错误的检测;发现错误
detection of fraud 检查舞弊
detection of infrared radiation 红外(线)辐射探测
detection of interaction effects 交互作用影响的检测
detection of odo(u)rs 气味检索
detection of oil pollution 油污染物测定
detection of parallelism 并行性检测
detection of radiation 辐射探测
detection of toxic-organic compound 毒性有机化合物检出
detection phase 探测阶段
detection piece 检测片
detection pressure 探测压力
detection probability 检测概率
detection process 检波过程
detection radius 检测半径
detection range 探测距离;探测范围
detection rate 检出率;探获率
detection record 探伤记录
detection resolution 检出分辨率
detection response value 检测响应值
detection search 检验搜索
detection sensitivity 检测灵敏度;探测灵敏度
detection signaling 检查信号装置
detection standard 探伤标准
detection system 检测系统;探测系统
detection theory 检测论
detection time 探测时间
detective bandwidth 探测带宽
detective camera 侦察摄影机
detective piece 检测片
detective rate of pollutant 污染物检出率
detective time constant 探测时间常数
detectivity 可探测率;检测能力;探测能力;探测率;探测灵敏度
detector 鉴定器;检验器;检测器;检波器;验电盘;整流器;感察器;传感元件;探头;探测装置;探测元件;探测仪(器);探测设备;探测器;随动机构
detector area 探测器灵敏面积
detector array 检测器阵列;探测器阵(列)
detector aspect 检测器形式
detector assemblage with cylindric(al) shape 圆柱状探测器组合
detector assemblage with square shape 方形探测器组合
detector balanced bias 检波器平衡偏压
detector bar 道岔锁闭杆
detector cell 探测器
detector cell volume 检测器池体积
detector check valve 探测器止回阀
detector check valve meter 探测器止回阀的水表
detector coefficient 检波器系数
detector condition monitoring 探测器状况监视
detector converter 检波变频器
detector coupling 检波器耦合
detector coverage 探测器覆盖率
detector diode 检波器二极管
detector efficiency 检波器效率;探测器效率
detector element 探测元件

detector field 探测器试场
detector for bearing parts 轴承零件探伤机
detector frequency meter 检波频率计
detector guard 检测器保护装置
detector head 探测头
detector holder 检测器支座
detector lamp 检漏灯
detector material 探测器材料
detector noise 检测器噪声;检波器噪声
detector noise discrimination 探测器噪声滤波器
detector noise-limited operation 检波器噪声限制作用
detector of adhesive quality 胶接质量检验仪
detector of deflects 探伤仪
detector of photon 光子探测器
detector of smoke 烟雾探测器
detector of tritium contamination 氚气污染探测器
detector of underground metal pipe and wire 地下金属管线探测仪
detector pad 车辆检数板
detector response 检测器响应
detector sensitivity 检测器灵敏度
detector stage 检波级
detector tester 探测器;检验仪
detector threshold sensitivity 探测器阈值灵敏度
detector tube 检气管;检测管;检波管
detector tube method 检测管法
detector type 探测器类型
detector with fixed value 定值探测器
detector with rolling grating 滚动光栅检测头
detector with up and down structure 上下结构探测器
detectoscope 探伤仪;探伤器;探伤法;水中探音器;水下探测器;潜艇探测器
detectoscope at sea 海中信号检测器
detemperature rate 降温率
detensioning (预应力钢筋混凝土的)解除张力
detent 插销;扳手;棘爪(掣);门扣;制动器;定位凹槽;掣轮器;擒纵机构
detent adaptor 锁接头
detent bridle 拉挡限动杆
detent catch 锁闩;锁门
detent eccentric 拉挡偏心销
detent escapement 天文钟擒纵机构
detention 扣押;扣留;禁闭;滞留;停滞
detention area 滞洪区;分洪区
detention basin 滞水区;拦洪水库;滞洪水库;滞洪区
detention building 拘留所
detention by ice 冰封阻留
detention center 集中营
detention charges 滞留费
detention clause 滞留条款;滞留条款
detention dam 拦砂坝;拦洪坝;拦河坝;滞流坝;滞洪坝
detention door 安全门
detention fee 滞留费
detention home 青少年教养院;少年教养院
detention hospital 隔离病院;传染病医院
detention of sewage 污水沉淀池
detention of ship 滞留船舶
detention of storm flow 暴雨滞留
detention of vessel or cargo 扣留船(只)或货(物)
detention period 延迟时期;滞留周期;滞留时期;持续时期;停留时期;停留时间
detention period in grit chamber 沉砂池停留时间
detention reservoir 拦洪水库;蓄洪水库;滞洪水(库);调节水库
detention screen 滞流屏;隔离屏;安全纱窗;加重型纱窗(拘留所用)
detention storage 拦洪库容;滞洪蓄水;滞洪库容;土壤滞留蓄水量
detention structure 蓄洪设施;蓄洪建筑物;蓄洪构筑物
detention surface 阻滞表面
detention tank 滞流沉淀池;滞留池;滞留槽;污水沉淀池;停留池
detention ticket 扣留物品凭单
detention time 滞留时间;滞后时间;持续时间;迟滞时间;停留时间
detention time of car in transit without resorting 无调中转车停留时间
detention time of car in transit with resorting 有调中转车停留时间

detention volume 拦洪容量
detention window 防盗窗;安全窗
detent mechanism 制动机构;定位器
detent pin 止动销
detent plate 制动器板
detent plate stop 带齿的制动板
detent plug 制动箱;止销
detent torque 起动转矩
deter 阻止
detergence 净化力;去污力;去垢
detergency 净化力;洗净性;除垢能力;去污性;去污力;去垢性
detergency mechanism 去污机理
detergency promoter 助洗剂
detergency test 去污染试验
detergent 流质去污剂;洗涤液;洗净剂;洗洁剂;洗涤剂;除污剂;除垢剂;去污剂;去污的;去垢剂;清洁剂;脱垢剂
detergent action 去垢作用
detergent additive 去垢添加剂
detergent alkylate 十二烷基苯
detergent application 清洗剂使用
detergent builder 助洗剂
detergent compound 洗涤剂化合物
detergent dispersant 洗涤分散剂;清洁分散剂
detergent effect 净化效果;去污效果;去垢作用
detergent for household use 民用洗涤剂
detergent line 清洗管路
detergent oil 净化油;洗涤油;去垢油
detergent pollution 洗涤剂污染
detergent power 去垢本领
detergent resistance 耐洗涤剂性;耐洗涤剂腐蚀
detergent spray truck 洗净剂喷布车
detergent waste 洗涤剂废水
detergent zeolite 沸石洗涤剂
deteriorate 劣化;退化
deteriorated area 荒芜区;城市功能恶化区
deteriorated building 劣化建筑
deteriorated dwelling house 破旧住宅
deteriorated grassland 退化草地
deteriorated insulator 老化绝缘子;失效绝缘子
deteriorated residential quarter 破旧住宅区
deteriorating housing 破落住房
deteriorating hysteresis structure 退化滞变结构
deteriorating neighborhood 破落街区
deteriorating restoring force 退化恢复力
deteriorating supplies 易变质军需品
deterioration 劣化;恶化;变质;变坏;破落;退化变质;退化;损蚀;损坏
deterioration agent 对品质有害溶剂
deterioration failure 磨损故障;变质失效
deterioration in accuracy 降低精度
deterioration indicator 连年续作业;恶化指标
deterioration level 退化程度
deterioration mechanism 退化机理
deterioration of a case 变证
deterioration of business conditions 商情恶化
deterioration of cement 水泥的变质
deterioration of environment 环境退化;环境恶化;环境变坏
deterioration of groundwater quality 地下水质恶化
deterioration of infiltration condition in aeration zone 曝气带入渗条件变差
deterioration of inventory 存货耗损
deterioration of pipe joint 管道接口腐朽
deterioration of roofing felts 屋面油毡老化
deterioration of water quality 水质恶化
deterioration of waterway 航道恶化
deterioration of wood pile 木桩腐烂
deterioration rate 退化速率
deteriorative 恶化的;变坏的
deteriotate 变坏
determinability 可确定性
determinable 可测定
determinable error 可测误差
determinable fee 有终止可能的财产
determinable income 固定收入
determinable liabilities 可确定的负债
determinable losses 可决定的损失
determinacy 确切性;确定性
determinacy case 确定性情况
determinacy counterpart 确定性副本
determinacy model 确定性模型

determinand 被测定物
determinant 决定性的;决定簇;行列式;定子;待测物
determinantal expansion 行列式展开式
determinant assay 决定簇测定
determinant attributes 限定性特征
determinant divisor 行列式因子
determinant expansion 行列式展开(式)
determinant factor 行列式因子(式)
determinant of a matrix 矩阵的行列式
determinant of coefficient 系数行列式
determinant of matrix 矩阵行列式
determinant of transformation 变换行列式
determinant rank 行列式秩
determinants of consumer demand 消费者需求的决定因素
determinants of demand 决定需求的因素
determinate boundary 已定界
determinate coefficient 规定系数
determinate depth of borehole 确定孔深
determinate error 预计误差
determinate fault 确定性故障
determinate growth 有限生长
determinate machine 判定机
determinate method of indeterminate coefficient 待定系数确定方法
determinate model 确定性模型
determinateness 定值
determinate service 固定劳务
determinate solution 确定解
determinate structure 静定结构
determinate system 确定系统;确定体系
determinate variation 定向变异
determination 测定(法)
determination absorbance method 测定吸光度法
determination by reduction 还原测定
determination coefficient 决定系数;测定系数
determination data 决定数据;测定数据
determination limit 测定限(度)
determination method 测定方法;确定的手段
determination method of groundwater velocity 地下水流速测定法
determination of acetylene 乙炔测定
determination of activity 活性测定
determination of agglutination titer 凝集价测定
determination of azimuth 方位角测定
determination of carbon monoxide 一氧化碳测定
determination of clock correction 钟差测定
determination of control point 控制点测定
determination of coordinate 坐标确定
determination of decision criteria 决策标准的确定
determination of direction and range dodar 导达
determination of epicenter 震中测定
determination of epicentre 确定震中
determination of extractives 浸出物测定
determination of fineness 细度测定
determination of flashing point 闪点测定
determination of focus 焦距测定
determination of hardness 硬度测定
determination of hydrophysics property of soil sample 土样水理性质测定
determination of ignition point 燃点测定
determination of latitude 纬度测定
determination of line direction 线路定向
determination of longitude 经度测定
determination of map size 地图宽廓尺寸的确定
determination of minerogenetic solution ice point 成矿溶液冰点的测定
determination of moisture 水分(含量)测定
determination of nitric oxide 一氧化氮测定
determination of optimum bias 最佳偏压测定
determination of orbit 轨道测定
determination of output 功率确定
determination of oxygen 氧的测定
determination of permeable coefficient 渗透系数测定
determination of position 位置测定
determination of pressure of collapsibility coefficient 测定湿陷系数压力
determination of price 价格决定
determination of protection forests 防护林确定法
determination of rating curve 关系曲线确定
determination of real flow velocity 实际流速测定法
determination of relation curve 关系曲线确定;确

定关系曲线
determination of soyabean oil 豆油鉴定;豆油测定
determination of standard base 标准碱测定
determination of standing point 确定站立点
determination of the mission 确定任务
determination of thermal stress 热应力量测法
determination of the roof shape 屋顶形式的决定
determination of tilt 倾角测定
determination of tinting strength 着色力测定;白色颜料消色力测定
determination of total base 总碱量测定
determination of trace 痕量测定
determination of unsteady flow parameter 非稳定流参数的确定
determination of vertical dimension 垂直距离测定
determination of yield 产量测定;采伐量测定法
determination on site 现场决定
determination subsoil dynamic(al) parameters 地基动力参数测试
determination theory of economic development 经济发展决定论
determination vibration 定数振动;确定性振动
determination wave function 行列式波函数
determinative bacteriology 鉴定细菌学
determinative factor 决定因素
determinator 决定因素;测定仪;测定器
determine 决定;确定;判定
determine by votes 投票决定
determined cumulative runoff 测定累积径流
determined measures 坚决措施
determined periodic(al) income 定期固定收入
determine evaluation strategy 决定评价方法
determine production schedule 决定生产日程进度表
determiner 决定因素
determining alternative course 替代方案的决定
determining alternative project 替代方案的决定
determining cost of labo(u)r 设计的人工成本
determining cost of layout 决定布置成本
determining factor 决定(性)因素
determining function 决定函数;母函数
determining isotope composition 实测同位素成分
determining particle-size distribution of soil 土的颗粒大小分布测定
determining production standard for all operation 决定各种生产作业标准
determining quantities of production runs 决定生产量
determining radon method by mountainous project 山地工程测氡法
determining radon method by water sample in hole 井中水样测氡法
determining radon method in drill 钻孔测氡法
determining radon method in shallow hole 浅孔测氡法
determining rate-controlling step 速率控制步骤
determining ratio cost to selling price 确定成本对售价的比率
determining source 决定源
determining tendency 推定趋势
determining the items of work 决定工作项目
determining the position 船位测定
determining work method 决定工作方法
determining work method for each work area 决定各工作区的工作方法
determinism 决定论;因果律
deterministic 确定性的
deterministic algorithm 确定性算法
deterministic analysis 定值分析;确定性分析
deterministic analysis method 确定性分析法;确定性分析法
deterministic appraisal approach 决定性评价法
deterministic approach 确定法;决定法;定数法
deterministic assessment of agricultural non-point source pollution 确定性农业非点源污染评估
deterministic automation 确定性自动机;决定性自动机
deterministic base 确定的依据;定数基础
deterministic channel 确定信道
deterministic component 确定性分量;确定性成分;确定性部分
deterministic data 定值数据
deterministic decision process 确定型决策过程

deterministic demand 确定性需求
deterministic design method 定值设计法
deterministic diagram 确定性状态图
deterministic distribution 确定性分布
deterministic dynamic(al) programming 确定性动态规划
deterministic dynamic(al) response analysis 定值动力反应分析
deterministic dynamic(al) system 确定性动态系统
deterministic erasing stack automaton 确定的抹除堆栈自动机
deterministic evaluation 定值估计方法
deterministic expression 确定的表达式
deterministic finite automaton 决定性有限自动机;确定的有穷自动机
deterministic fire model 确定火灾模型
deterministic forecast 因果预报;确定性预报
deterministic function 定值函数;确定性函数;确定的函数
deterministic geochemical model 确定性地球化学模型
deterministic intensity function 定值强度函数
deterministic interconnection 确定性关联
deterministic inventory model 确定性库存模型
deterministic language 确定性语言
deterministic linear bounded automaton 确定的线性界限自动机
deterministic mode 确定性模式
deterministic model 确定性模型
deterministic non-erasing stack automaton 确定的非抹除堆栈自动机
deterministic output 定值输出
deterministic parsing 确定性分析
deterministic policy 决定性策略
deterministic-probabilistic contaminant transport model 确定性概率统计污染物输移模型
deterministic problem 确定性问题
deterministic process 确定(性)过程;判断过程
deterministic push-down automato 确定的下推自动机
deterministic queuing system 确定性排水系统
deterministic retrieval 确定(性)检索;判定型检索
deterministic routing 确定性路径选择
deterministic schedule 确定性调度
deterministic simulation 定值模拟;确定性模拟;确定性仿真
deterministic simulation model 确定性模拟模型
deterministic source 确定源
deterministic stack automaton 确定的堆栈自动机
deterministic standard 确定性标准
deterministic system 确定性系统
deterministic technique 确定性技术
deterministic topdown grammar 确定性自顶向下文法
deterministic vibration 确定性振动
deterring type median 阻车(进入)式中央分隔带
detersive 洗涤剂;去污剂;去污的;清净剂
deter vandalism 防止破坏行为
Dethridge metre 德里季自记流量计
de-tin 除锡;脱锡;去锡
detinned scrap 除锡切边;除锡废边料
de-tinning 去锡作用;去除锡层
detinning apparatus 除锡装置
Det norske Veritas 挪威船级社
detonable 可爆炸的;易爆的
detonate 引爆;爆破;起爆
detonated dynamite 起爆炸药
detonate tube 雷管;起爆(雷)管
detonating agent 引爆剂;发爆剂;起爆剂
detonating cap 雷管;工业雷管;发火帽;起爆(雷)管
detonating capacity 引爆能力;起爆能力
detonating cartridge 引爆药卷;响墩
detonating charge 起爆装药;起爆药包;引爆药
detonating circuit 引爆电路
detonating combustion 爆震燃烧
detonating composition 起爆药;起爆剂;起爆成分
detonating compound 起爆药
detonating compression pressure 爆燃压缩压力
detonating cord 火药导线;引爆线;导火线;导爆索;导爆索;传爆线;爆炸引线
detonating cord initiation 导爆索起爆
detonating device 起爆装置
detonating equipment 引爆装置;爆破设备;起爆设备

detonating explosive 爆破炸药;爆轰炸药;起爆(炸)药
detonating fuel 爆震燃料
detonating fuse 爆炸信线;引爆线;导爆线;导爆索;传爆线;爆炸引火线;爆炸导火索;爆破线;爆管导火线;起爆引信
detonating fuse blasting 导爆索起爆
detonating gas 爆鸣气
detonating mixture 混合炸药
detonating net 火药导线网
detonating pedestal pile 爆扩桩
detonating powder 起爆药
detonating pressure 爆炸压力
detonating primer 引爆雷管;起爆药包;起爆器
detonating ram 爆炸锤
detonating rammer 爆炸式夯锤
detonating ram pile driver 爆炸撞锤打桩机
detonating rate 引爆速率
detonating relay 继爆器;继爆管;起爆继电器
detonating signal 响爆信号;爆响信号;爆裂信号;起爆信号
detonating slab 防弹墙
detonating string 引爆线;导爆索
detonating tube initiation 导爆管起爆
detonating velocity 爆轰速度
detonating wave 爆震波;爆轰波
detonation 起爆;引爆;爆炸;爆燃;爆鸣;爆轰;突爆
detonation action 爆轰作用
detonation agent 爆轰剂
detonation cap 雷管
detonation coating 爆炸涂覆;爆炸涂层
detonation combustion 爆轰燃烧
detonation consumption per cubic entity rock 每立方米原岩雷管消耗量
detonation consumption per meter 每米坑道雷管消耗量
detonation distance 诱爆距离
detonation energy 爆轰能量;爆轰能
detonation explosive 爆轰炸药
detonation forming 爆炸成型
detonation front 激波面;爆炸前沿;爆轰波前锋
detonation fuse 起爆引线
detonation gas 爆轰气体
detonation gun 爆震喷枪
detonation hazard 爆轰危害
detonation heat 爆轰热
detonation hopper 爆破漏斗
detonation impedance 爆炸阻抗
detonation indicator 点火指示器;爆燃指示器
detonation inhibitor 防爆剂
detonation limit for mixture gases 可燃混合气的爆轰范围
detonation meter 爆震仪
detonation mixture 爆轰混合物
detonation parameter 爆炸参数;爆破参数
detonation performance 爆炸性能
detonation plating 爆炸镀金法
detonation point 爆震点;爆炸点
detonation power 引爆能力;爆炸(能)力
detonation pressure 爆炸压力;爆轰压力;爆发压力
detonation property 爆轰(特)性;起爆性
detonation ram 爆炸锤
detonation range 爆轰范围
detonation rate 起爆速率
detonation reaction 起爆反应
detonation sealer 阻爆器
detonation sensitivity 爆轰感度
detonation sound 爆炸声
detonation speed 起爆速度
detonation spraying 爆震喷涂
detonation stability 爆轰稳定性
detonation state 起爆态
detonation suppression 防爆剂
detonation temperature 爆轰温度;起爆温度
detonation train 传爆系统;传爆道;起爆引火线
detonation velocity 爆炸速度;爆速;爆燃速度;爆破速度;爆轰速度
detonation wave 爆震波;爆炸波;爆破波;爆轰波
detonator 雷管;响墩;引信;引燃剂;引爆剂;发爆器;发爆剂;爆炸雷管;爆破雷管;爆鸣器;爆剂;起爆装置;起爆器;起爆管;起爆(雷)管;起爆剂
detonator cap 雷管帽盖
detonator cartridge 炸药筒;炸药包
detonator circuit 信管电路;引发电路

detonator crimper 雷管钳
detonator device 爆炸装置
detonator fuse bridge 雷管熔丝电桥;雷管电桥线
detonators connected in parallel 并联雷管
detonators connected in series 串联雷管
detonator signal 爆炸信号
detonator tube 雷管壳
detonator wire 雷管脚线;电雷管导线
detonics 爆炸学
detorsion 扭转矫正术,反扭转
detour 迂回(路);便道;绕行线;绕行道(环形公路)
detour arrow sign 迂回线指向标志;迂回路指向标志
detour behavio(u)r 迂回行为
detour bridge 临时便桥;便(道)桥
detour direction 绕道指向
detour drift 迂回坑道
detoured train 迂回列车
detouring section 绕行地段
detouring section of both line 双绕地段
detouring section of single line 单绕地段
detour line 迂回线
detour marker 迂回线指示标
detour phase effect 迂回相位效应
detour plan (施工期的)绕行路平面图;绕行路计划
detour road 便道,便路;临时便道;环行路;迂回路;绕行路
detour route 迂回进路;迂回航线
detour sign 绕行标志;迂回路标志
detour temporary access road 临时便道
detoxicant 解毒素;解毒剂
detoxification of pesticide 农药解毒作用
detrainment ramp 紧急门
detrending 消除长期趋势
detrial lime tuff 碎屑石灰凝灰岩
detribalization 非部族化
detrimental change 有害变化
detrimental effect 有害作用;不良影响;不利影响
detrimental expansion 有害膨胀,不良膨胀作用
detrimental impurity 有害杂质
detrimental reflection sound 有害反射气
detrimental resistance 有害阻力
detrimental settlement 有害沉降;危害沉降
detrimental soil 有害土壤;不稳定土(层)
detrimental sponginess 有害海绵状结构
detrimental subsidence 危害沉陷
detriotor 方形沉砂池
detrital 由碎屑形成的;碎屑状(的)
detrital cone 岩屑堆;碎屑堆
detrital deposit 碎屑沉积(物)
detrital dolomite 碎屑白云岩
detrital dump 岩屑堆
detrital fan 冲积扇;碎屑扇形地
detrital mineral 碎屑矿物
detrital mineral mater 碎屑矿物质
detrital ore deposit 碎屑矿床
detrital ratio 碎屑比
detrital remanent magnetization 碎屑剩余磁化强度
detrital reservoir 碎屑储油层
detrital rock 碎屑岩
detrital sand 岩屑砂;碎屑砂
detrital sediment 碎屑沉积(物)
detrital slope 岩屑(边)坡
detrital tuff 碎屑凝灰岩
detrition 磨损,磨耗
detritomicrinite 碎屑微粒体
detritus 碎屑;杂粒;残屑;碎(岩)屑;岩屑(腐质);碎石
detritus avalanche 岩崩;泥石崩坍
detritus chamber 沉砂箱;窨井;大颗粒沉砂池;沉渣池;沉污槽;沉砂池;沉砂池(下水道)
detritus chamber or pit 窨井;(下水道的)沉淀池
detritus clearing 清渣
detritus equipment 制造设备;破碎设备;碎石设备
detritus food chain 腐屑食物链
detritus intrusion 岩屑入侵
detritus pit 窨井;沉污槽;沉砂池;沉砂池(下水道)
detritus rubbish 岩屑堆;废屑垃圾;碎屑垃圾
detritus slide 碎岩滑动;面土滑动
detritus stream 泥(石)流
detritus tank 废渣池;窨井;一级处理沉淀池;沉渣池;沉污槽;(下水道的)沉砂池
detritus zone 碎屑带;岩屑地带
detrogelinite 碎屑凝胶体

Detroit electric(al) furnace 底特律电炉
Detroit rocking furnace 窄可倾炉;底特律摇摆式电炉
detrusion 剪切变形;冲移;(木材的)顺纹剪切
detrusion ratio 剪切比
detune 解谐;失谐
detuned circuit 失谐电路
detuner 解调器;排气减音器
detuning 去谐;失谐,失调
detuning capacitor 离调电容器;失调电容器
detuning phenomenon 失调现象
detuning stub 解谐短截线
detwinning 双晶消除
detwisted 解捻的
de-update 重更新
deurbanization 逆城市化
Deutche Bundes patent 德国专利
Deutch equation 多依奇方程
deuterate 重水合物
deuterated thermal water 含氘热水
deuterated water 重水
deuteric 岩浆后期的
deuteric ore deposit 岩浆后期矿床
deuteric salinization 后期盐化作用
deuterium 重氢
deuterium and oxygen-18 method 氘和氧-18 测定法
deuterium arc lamp 氘弧灯
deuterium fluoride 氟化氘
deuterium lamp 氘灯
deuterium-moderated reactor 重水中化堆
deuterium oxide 氧化氘;重水
deuterium thyratron 充氘闸流管
deuterium-tritide 氚化氘
deuterium-uranium reactor 重水铀反应堆
deutero-aetioporphyrin 次初卟啉
deuteroartose 次阿托
deuterogaikum 小地旋回【地】
deuterogamy 再配合
deuterogene 后生岩;后期生成;次生岩(石)
deuterogene rock 后成岩
deuterogenic 后期生成的
deuterogenic component 后生组分
deuterogenous 后期生成的
deuteromorphic 后生变形的
deuterosomatic 再成岩
deutomerite 后节
deutoplasm 副浆
deutoscolex 次生头节
Deutsche Industrie Normen 德国工业标准
Deutsche Industrie Normen colo(u)r difference equation 德国工业标准色差方程式
Deutsche Industrie Normen colo(u)r system 德国工业标准表色系统
Deutsche Industrie Normen test method 德国工业标准试验方法
Deutsche Mark 德国马克
Devacchis process 第瓦齐斯冶矿法
Deval abrasion test 狄法尔(双筒)磨耗试验;双筒式磨耗试验
Deval abrasion tester 狄法尔磨耗试验机
Deval abrasion testing machine 狄法尔磨耗试验机;双筒磨耗试验机
Deval attrition machine 狄法尔磨损试验机
Deval attrition test 狄法尔磨损试验
Deval machine 狄法尔磨耗试验机;双筒磨耗试验机
Deval rattler 狄法尔(双筒)磨耗(试验)机
Deval test 狄法尔磨损试验
Deval testing machine 狄法尔磨损试验机
devaluation 贬值
devaluation of currency 货币贬值
devaluation of dollar in terms of gold 美元对黄金的贬值
devaluation of US dollar 美元贬值
devaluation surcharge 贬值附加费
devanning 拆箱
devanning report 出箱报告表;(集装箱的)拆箱(报)单
devanture 锌华凝结器;蒸锌炉冷凝器
devapo(u)ration 蒸汽凝结;蒸气凝聚
devapo(u)ration membrane 渗透气化膜
devapo(u)rizer 蒸发冷却器;清洁器
Devarda's alloy 迪氏铝铜锌合金;德瓦达铜铝锌合金
Devard's alloy 戴瓦合金

devastate 毁坏
devastated land 灾区;受灾土地
devastated stream 荒溪;荒废河流
devastating earthquake 破坏性地震
devastating flood 毁灭性洪水
de Vaucouleurs-Sandage classification 伏古勒-桑德奇分类
develling (抹灰底面上的)刮粗
develop 开拓(采矿);洗片;展开;导出;发展;研制
developable curved surface 可展曲面
developable head 可开发水头
developable hydropower 可开发的水力发电
developable surface 可展(曲)面;展开面
develop a fault 发生故障
develop area 开发区
develop a sphere onto a plane 把球面投影到平面
develop distortedly 畸形发展
developed area 发展区;已开发土地;展开面积;发达地区
developed at top speed 高速发展
developed blade area 展开叶片面积
developed blank 改进型坯料
developed commercial reserves 开发工业储量
developed country 发达国家
developed curve 展开曲线
developed dimensions 展开尺寸
developed drawing 展开图
developed dye 显色染料
developed elevation 展开立面
developed film 显影胶片
developed floodplain 已垦殖的洪泛平原;已开发的洪泛平原;已发育的洪泛平原
developed geothermal field 已开发的地热田
developed grain 显影粒子
developed head 开发水头
developed image 已显像;放大照片
developed land 已开发土地
developed length 展开长度
developed nations 发达国家
developed ore 可靠储量;开拓储量;开采矿石
developed plantform 展开平面形
developed power 发出功率
developed pressure 产生的压力
developed project 开发规划;开发工程
developed quarter 已开发地区
developed region 发达地区
developed representation 展开图示法
developed reserve 开拓储量;可靠储量
developed resources of groundwater 地下水开采资源
developed river stretch 开发河段
developed section 展开截面;发育段;开发段
developed site 已开发地区
developed surface 展开面
developed surface flaw 发展的表面缺陷
developed view 展开图
developed water 开发水
developed water power 开发出水力(发电)
developed width 展开宽度
developer 建设者;开发者;显影液;显影剂;显像剂;显示剂;显色剂;展开剂;放样员;放样工(人)
developer streaks 显影剂条纹
develop field 开发油田
develop financial resources 开辟财源
develop form nothing 从无到有
develop heat 产生热
develop high technical proficiency 达到高度的技术熟练程度
develop in a complementary way 相辅相成的发展
develop in an unhealthy way 不能健康发育
developing 显影;衍变
developing agent 显影剂;显色剂;展开剂
developing and fixing 显影及定影
developing apron 冲片胶带
developing area 正在开发的地区;发展区
developing coefficient of live load reservation 预留活载发展系数
developing cost 开发成本
developing countries' rate 发展中国家优惠关税税率
developing country 发展中国家
developing direction of crack 裂缝延展方向
developing economy 发展中经济

developing effect 显影效果
developing equipment 冲洗机
developing favorable export items 发展具有优越性的出口产品
developing fire 蔓延中的火
developing ink 显影墨
developing instruction 显影指南
developing island 发展中岛国
developing machine 显影机;洗片机
developing new market 开拓新市场
developing of photography 摄影处理
developing outfit 显影设备
developing-out paper 显像纸
developing process 显像;发展过程
developing reserves 开采储量
developing solution 显影液
developing speed 显影速度
developing stage 开采阶段
developing state of the art 技术发展水平
developing talents 人才开发
developing tank 显影桶;显影缸;展开罐;展开槽
developing target market 开发市场
developing test 研制试验
developing tray 显影盘
developing velocity 显影速度
develop in proportion 按比例发展
develop in spiral 螺旋式发展
development 研制;矿井开拓;开发;显影;显像;显色;演变;展开;冲洗;发展;发育;推导
development aid 开发援助;发展援助
developmental 开发的
developmental age 发展年龄
developmental anomaly 发育异常
developmental condition 发育条件
developmental crisis 发展性危机
developmental curve 发育
developmental index 发育指数
developmental indication 发育标志
developmental item 开发项目
developmental lag 发展迟滞
developmental marketing 开发性销售
development along hillside 傍山展线
developmental patterns and lifestyles 发展模式和生活方式
developmental period 发育期
developmental project 改良土地规划
developmental research 发展研究
developmental stage 发育阶段
developmental stage of mountain chain 山脉发育阶段
developmental stage of river valley 河谷发育阶段
developmental technology of new energy resources 新能源开发技术
developmental threshold 发育临界
developmental type frozen soil 发展型冻土
developmental zero 发育零点
development and driving of coal mines 煤矿开拓掘进
development and reform of science of education 教育科学发展与改革
development and test 研制与试验
development and use of freshwater resources 淡水资源的开发和利用
development area 开发(地)区;建筑区;展开面积;发展区
development area of tourism geology 旅游地质开发区
development arrest 发育阻滞
development assistance 发展援助
Development Assistance Committee 开发援助委员会
development assistance group 发展援助集团
development bank 开发银行
development batch 分批显影
development board 开发委员会;开发厅;开发局;开发部;计划委员会
development bond 开发债券
development bond stress 锚固黏结应力
development borehole 详细勘探钻孔;详勘钻孔
development by leaps and bounds 跳跃式发展
development capacity 开发容量
development capital 开发资金
development center 显影中心
development condition 显影条件

development contract 开发合同
development contrast 显影反差;显色对比度
development cooperation 发展合作
development corporation 开发公司
development cost 开发费用;开发成本;发展费用
development credits 开发售货
development curve 演变曲线;发展曲线
development cycle 开发周期;发展周期
development decision rules 发展决策准则
development design 规划设计
development district 政府鼓励工业投资的地区;失业严重地区;发展区;开发区
development drawing 展开图;采掘图
development drift 开拓平巷;开拓坑道
development drilling 开拓钻孔;详细勘探钻井
development economics 开发经济学;发展经济学
development effect 显影效应
development effort 开发计划;研制计划;研制工作;发展工作
development engine 研制中的发动机
development engineer 发展研究工程师
development engineering 新结构设计
development engineering management system 开发工程管理系统
development environment(al) model 开发环境模型
development equipment 显影设备;洗井装置;洗井设备
development error 显像误差
development expenditures 发展费用支出
development expenses 开发费用;发展费用
development exploration 发展勘探
development facility 研究机构
development feature of growth fault 生长断层发育特征
development financing 开发资助;开发融资;发展资金的供应
development fog 显像雾翳
development framework 发展体系
development fund 开发计划;开发基金;发展基金
development geology of oil-gas field 油气田开发地质
development group 开发小组
development health examination 发育健康检查
development impact fee 开发影响费用
development index 发展指数
development institution 开发机构
development intensity 采油强度
development investigation 发展调查
development length 延伸长度
development loan 开发贷款
development loan fund 开发贷款基金
development machine 掘进机
development mechanical kit 显影器
development method 开拓方法;平展法
development model 发展研制模型
development natural resources 开发自然资源
development new mines 增建新矿山
development of a conic(al) projection surface 圆锥投影面展开
development of a function 函数展开
development of aquatic living resources 水产资源开发
development of a well 油田的开发;水井的增量
development of barrages and locks in river 河流的梯级开发
development of coast 海岸动向
development of colo(u)r 显色
development of cones 锥面展开
development of economy 经济发展
development of engine power 发动机功率的提高
development of faculty 能力开发
development of formula 公式展开
development of gas 放出气体;放出毒气
development of heat 发生热力;热的产生;生热
development of high technology 高科技开发
development of highway 公路运输开拓
development of hydroelectric(al) resources 水力资源开发
development of land 土地开发
development of line 展线
development of mineral deposits 矿床开拓
development of natural resources 自然资源开发
development of new products 研制新产品

development of new technology 新技术开发
development of ocean 海洋开发
development of open pit 露天矿开采
development of peat land 泥炭土的发展
development of railroad transport 铁路运输开拓
development of renewable resources 可再生资源开发
development of resources 资源开发
development of right-of-way 管带开拓
development of rigidity (水泥浆、水泥砂浆、混凝土的)凝固
development of river pattern 河型发育
development of seaside 沿岸开发
development of seaside area 沿岸开发区
development of shoreline 岸线扩展
development of slope cable lift 斜坡卷扬开拓
development of soffit 拱腹图的展开
development of soils 土地开发;土壤发育
development of solar energy 太阳能开发
development of strength 强度的形成;强度增长
development of style 风格的形成;风格的发展;风格的成长
development of submarine resources 海底资源开发
development of technology 技术开发
development of waste land 开垦荒地
development of water resources 开发水资源
development of work 工作开展
development opening 开拓巷道;开拓工程
development organization 建设单位
development-oriented resettlement 开发型移民
development-oriented shareholding enterprises 开发型股份企业
development paper 像纸
development path 发展途径
development pattern 发展格局
development period 开发期;研制期;发展期
development personnel 发展人才
development phase 发育阶段
development plan(ning) (土地的)开发规划;开发计划;发展计划;发展规划;开展计划;建设规划;研制计划
development plan scheme 发展规划纲要
development potential 发展潜力
development priority area 优先开发地区
development procedure 开发程序
development process 开发过程;显色法;发育过程
development program(me) 开拓方案;发展方案;开发方案;建设规划;研制计划;改进计划;发展计划;发展规划;发展纲要;开发计划
development project 开发计划;开发项目;发展规划;发展项目;发展(单元)计划
development promotion area 鼓励开发区
development quotient 发展商数
development range 显像范围
development ratio 采掘比
development records 开发记录
development report 研究报告
development representation 展开图
development research 开发研究
development reserves 开拓矿量
development restriction area 开发限制区域
development retardation 发育迟缓
development rights 开发权
development rights transfer 开发权转让
development road 开拓道路
development rock 开拓矿石
development running 研制性运行
development scheme 建设方案
development security facility 发展担保手段;发展担保基金
development shop 试制车间;试验车间
development speed and growth rate 发展速度和增长率
development stage 开发阶段;显影过程;研制阶段;发展期;发展阶段
development stage enterprise 开创阶段企业
development stage of relief 地形发展阶段
development standard 发育标准
development strategy 发展策略;发展战略
development support library 开发程序库
development system 开发系统;研制(生产)系统;研制(开采)方法
development system step 系统研制步序

development tank 显影罐;显影槽;展开罐
development tax 开发税;发展税
development temperature 显影温度
development tendency 发展趋向
development test 新产品试验;研究试验;试制品试验;试采井
development test station 实用化试验台
development time 开发时间;显影时间;研制时间
development tool 开发手段;开发工具
development toxicity 发育毒性
development traffic 发展交通(量)
development traffic volume 开发交通量
development trench 发展趋势
development trough 开拓堑沟
development tube 显影管
development type 试验样机;开采型;研制型;试验样品
development wastewater 显影废水
development well 开发井;详探井
development well classification 开发井类别
development without destruction 无害发展
development work 开拓工程;开发工作;研制工作;研究工作;采准工作;试制工作
development works in stone 岩巷掘进;脉外开拓
development zone 开发区;发展区
develop new tools 研制新农具
develop pressure 产生压力
develop resources 开发资源
develop the equipment to a high pitch of efficiency 把设备效率发挥到很高程度
develop to finality 显影完毕
develop urban and rural markets 开拓城乡市场
Devereaux agitator 代伐若搅拌机
devest 撤资
deviant 不正常者;偏移值;偏离标准的;偏差值
deviascope 自差校正试验仪
deviate 偏斜;偏向;偏移(航线)
deviated drilling 偏斜钻进
deviated flow 偏斜水流;偏向流动
deviated line 偏离规定路线
deviated rate 优惠利率;优惠保险费
deviated track 岔线【铁】
deviated traffic 偏离规定路线
deviate from 离开
deviating force 自差力;偏转力;偏向力
deviating hole 斜钻孔;偏斜孔
deviating prism 转向棱镜【测】;偏折棱镜
deviating wedge 偏向光楔
deviating well 偏斜井
deviation 离差;差;变差;绕航;偏转;偏折;偏移;偏位;偏离;偏差(数)
deviation absorption 近临界频率吸收;偏移吸收
deviation adjustment 偏差调整
deviational survey 偏角测量
deviation angle 偏移角;偏差角
deviation beacon 偏离标志;偏差指向标
deviation between map size and nominal size 图廓尺寸与理论尺寸较差
deviation calculator 偏差计算器
deviation calibration curve 偏差校准曲线
deviation card 自差图;磁罗盘偏差记录卡片
deviation changing rate 井斜变化率
deviation channel 航道改道
deviation clause 变更航程条款;绕航条款;偏航条款
deviation coefficient 自差系数;天然气偏差系数
deviation compensation device 自差补偿装置
deviation compensator 偏差补偿器
deviation computer 偏差计算机
deviation constant 偏差常数
deviation control 偏差控制
deviation correction 纠斜【岩】
deviation curve 自差曲线;偏航曲线;误差曲线
deviation detector 偏差测定器
deviation diagram 自差曲线
deviation dispersion 漂移
deviation display 偏差显示
deviation equalizer 偏差均衡器
deviation factor 压缩因数;偏差系数
deviation frequency 偏移频率
deviation from average 离均差
deviation from datum 正负零偏差
deviation from exact shape 变形
deviation from isochronism 等时性误差
deviation from mean 离均差;均值偏差;均值离差

deviation from par 平均差幅;平价差幅
deviation from plumb 垂直偏差
deviation from pulse flatness 脉冲平滑性偏移
deviation from the mean 离均差;距平
deviation from voyage route 偏离航程
deviation hole 偏斜井
deviation indicator 偏差指示器
deviation in level 平整度偏差
deviation in position 位置偏差
deviation intensity of azimuth 方位角弯曲强度【测】
deviation log book 自差记录簿
deviation matrix 差值矩阵
deviation measurement 偏差测量
deviation moment 惯性离心力矩
deviation of directivity 方向偏差
deviation of equiphase surface 等相位面偏差
deviation of geoid 大地水准面差距
deviation of level bubble 水准气泡偏差
deviation of light 光偏差
deviation of magnetic compass 磁罗经自差
deviation of market price 偏离市场价格
deviation of plumb-line 垂线偏差(分量)
deviation of reading 示值偏差
deviation of relative dispersion from normal 相对色散偏离值
deviation of the contract 违约事件
deviation of the vertical 竖直线偏差
deviation of transposition 换位偏差
deviation permit 偏离许可
deviation point 偏差点
deviation prism 偏折棱镜
deviation product sums 离差积和
deviation range 偏移范围;偏差范围
deviation ratio 偏移系数
deviation recorder 偏差记录仪
deviation report 绕航报告;偏航报告
deviation scale 偏移刻度
deviation section 差值剖面
deviation sensitivity 漂移灵敏度;偏位灵敏度;偏差灵敏度
deviation shaft 差轴
deviation standard 偏差标准
deviation stress 偏应力
deviation table 自差表;磁罗盘偏差表
deviation to left 偏左
deviation to right 偏右
deviation track 便道
deviation wedge 偏差调整楔(X射线带钢测厚仪部件)
deviator 偏置器;偏量;致偏装置;变向装置;偏向装置
deviatoric state of stress 偏应力状态
deviatoric state of stress distortion 体积不变而形状变化时的应力状态
deviatoric strain 偏应变
deviatoric stress 主(要)应力差(土工试验);轴差应力(三轴压缩试验最大主应力与最小主应力之差);偏应力
deviatoric stress tensor 偏应力张量
deviatoric tensor 偏张量
deviatoric tensor of strain 应变偏(张)量
deviatoric tensor of stress 应力偏(张)量
deviator strain 偏应变
deviator stress 主应力差;主要应力差(土工试验);轴差应力(三轴压缩试验最大主应力与最小主应力之差);偏应力;偏差压力
device 计谋;机构;仪表;装置;器械;器件;设施;设备
device acknowledge 设备应答
device address 设备地址
device allocation 设备分配
device allocation routine 设备分配程序
device and rule 分散控制
device assignment command 设备赋值命令
device assignment information 设备分配信息
device attachment 设备连接
device availability 设备效率;设备利用率;设备可用性
device backup 设备后援;设备备份
device base control block 设备基本控制块
device busy 设备忙
device busy control 设备忙碌控制
device category 设备种类
device character control 设备字符控制
device characteristics table 设备特性表格

device check 设备检查
device class 设备类别
device code 设备(代)码
device control 装置控制;设备控制
device control block 设备控制块
device control cell 设备控制单元
device control character 外围设备控制符;设备控制(字)符
device controller 设备控制器
device control register 设备控制寄存器
device control table 设备控制表
device control unit 设备控制装置;设备控制器
device coordinate 设备坐标
device correspondence table 设备对应表
device data block 设备数据块
device-dependent 与设备有关的;设备相关的
device-dependent program(me) 设备相关程序
device descriptor block 设备描述块
device descriptor module 设备描述模块
device diagnostic program(me) 设备诊断程序
device driver 设备驱动器;设备驱动机构;设备驱动程序
device efficiency 设备效率
device end 设备结束
device failure 设备失效
device field 设备字段
device flag 设备标志;设备标示器
device for automatic power 自动功率调整器
device for closing 闭锁装置
device for demonstrating capillarity 毛细装置
device for drying and filtering 干燥过滤器
device for line variation 变线仪
device for measuring rail wear 钢轨磨耗测量器
device for stopping buffer action 车钩缓冲停止器
device handler 设备驱动器;设备处理器;设备处理程序
device identification 设备标志;设备标识
device identifier 设备标识符
device independence 设备无依性;设备独立性
device-independent 装置独立性;与设备无关的
device-independent graphics software 独立于(绘图)设备的制图软件
device-independent program(me) 设备独立程序
device inoperable 设备不能动作
device input format 设备输入格式
device input format block 设备输入格式块
device input queue 设备输入队列
device interface module 设备接口模块
device interrupt 设备中断
device interrupt vector table 设备中断向量表
device level 设备级
device lifetime 装置寿命
device line 设备信息行;设备线路
device manipulation 设备操纵
device message handler 设备信息处理程序;设备消息处理程序
device name 设备名字;设备名(称)
device number 设备号
device object program(me) 设备目标程序;设备结果程序
device of dust separation 除尘设备
device oriented 面向设备的
device output format 设备输出格式
device output format block 设备输出格式块
device page 设备页面
device parameter list 设备参数(列)表
device polling 设备轮询
device pool management 设备库管理;设备池管理
device power supply 设备电源
device program(me) language 设备程序语言
device queue 设备序列;设备排队;设备队列
device ready 设备准备好状态;设备就绪
device reconfiguration 设备再组合;设备结构变换
device register 设备寄存器
device reserve 设备保留
device reserve word 设备保留字
device resolution 设备确定
device select code 设备选择码
device selection bit 设备选择位
device selection check 设备选择校验
device select line 设备选择线路
device selector 设备选择器
device selector logic 设备选择逻辑
device service routine 设备服务程序

device service task 设备服务任务
devices for absorbing regenerative electric(al) energy 再生电能吸收装置
device source program(me) 设备源程序
device space 存储空间;设备空间
device spanning 设备跨展;设备跨越
device status address 设备状态地址
device status byte 设备状态字节
device status field 设备状态字段;设备状态域
device status register 设备状态寄存器
device status word 设备状态字
device support routine 设备支持例行程序
device switching unit 设备转换装置;设备开关装置
device test instruction 设备测试指令
device to eliminate 'running back' 止回装置
device tolerance 设备容差
device to measure the flow of water 计量水流装置
device type 设备类型
device type logical unit 设备类型逻辑部件
device vector table 设备向量表
device waiting queue 设备等待队列
device work queue 设备运行队列;设备工作队列
devil 加热焊料的小炉子;木甲板缝;划毛抹子;尘旋风;扯碎机;扯碎
devil bolt 假螺栓
devil date 魔日
devil float 刮痕抹子;钉抹;钉镘板;划毛抹子;带钉抹子;常钉抹子
devilline 钙铜矾
devilling 刻槽;划毛;底灰划毛
devil liquor 废液
devil's brew 硝化甘油
devil's claw 螺旋锚链掣
devil's pitchfork 叉式打捞工具
devil's steel hand 叉式打捞工具
devil water 废水
deviometer 航向偏差指示器;斜视计;偏航指示器;偏差计
devious 不正当的
devise 不动产遗赠
devisee 遗赠受益人
deviser 发明者;发明家;设计者
devision device 除法装置
devisor 遗赠人
devitalizing effect 早衰效应;促老效应;衰弱效应
devitrification 去玻璃化作用;析晶;反玻璃化;脱玻作用;脱玻璃化作用;脱玻化;透明消失;失透;失去博览光泽
devitrification glaze 失透釉
devitrification inclusion 脱玻包裹体
devitrification nuclei 析晶晶核
devitrification of glass 玻璃析晶;玻璃透明消失;玻璃闷光
devitrification stone 析晶结石
devitrified bunchy texture 脱玻束状结构
devitrified comb structure 脱玻梳状构造
devitrified cryptocrystalline texture 脱玻隐晶结构
devitrified crystalline texture 脱玻雏晶结构
devitrified drusy texture 脱玻晶簇构造
devitrified felsitic texture 脱玻霏细结构
devitrified glass 闷光玻璃;微晶玻璃;失透玻璃
devitrified glass fiber 微晶玻璃纤维
devitrified micropoikilitic texture 脱玻微嵌晶结构
devitrified slag 失去光泽的渣
devitrified spherolitic texture 脱玻球粒结构
devitrified structure 脱玻构造
devitrified texture 脱玻结构
devitrified variooitic texture 脱玻球颗结构
devitrify 析晶;脱玻作用;失去博览光泽
devitrifying glass 不透明玻璃
devitrifying solder 失透性焊剂
devitrite 失透石
devitroceram 微晶玻璃;玻璃陶瓷
Devitro ceramics 德维特罗陶瓷;玻璃陶瓷
devoid of directionality 缺乏方向性
devolatilization 去挥发分作用;脱挥发作用;脱挥发分
devolatilization reaction 脱挥发反应
devolatilizer 抑制挥发剂;脱挥发器
de-voltage 失压【电】
devolution 崩坍;崩塌
devolution of agreement 移交协议
devolution of powers 放权

Devonian glacial stage 泥盆纪冰期
Devonian igneous rocks 泥盆纪火成岩
Devonian limestone 泥盆纪石灰岩
Devonian period 泥盆纪【地】
Devonian system 泥盆系
devulcanization 反硫化;脱硫化;去硫化
devulcanizer 反硫化器
dew 露水;结露数
Dewar bottle 真空瓶;杜瓦瓶
Deward 德瓦特锰钼钢
Dewar flask 杜瓦瓶;杜尔瓶皿;真空瓶
Dewar type container 杜瓦容器
Dewar vacuum vessel 杜瓦真空瓶
Dewar vessel 杜瓦瓶
dewater 增稠;除水;脱水
dewaterability 脱水性
dewaterability characteristics of sludge 污泥脱水性能
de-water curve 脱水曲线
dewatered sludge 干化污泥
dewatered sludge' dried sludge 脱水污泥
dewaterer 除水器;脱水器;脱水机
dewatering 除水;脱水;降低地下水位法;降低地下水位;泄水;放水;排水;疏水;疏干
dewatering agent 脱水剂
dewatering and classifying tank 分级脱水箱;分级脱水槽
dewatering and drainage engineering 降排水工程
dewatering area 疏干区
dewatering bin 脱水仓;脱水装置
dewatering blow-down tank 脱水排污舱
dewatering borehole 疏干井
dewatering boring 排水钻孔
dewatering box 脱水箱;脱水仓
dewatering bucket 排水吊桶;疏浚戽斗
dewatering bunker 脱水仓
dewatering by belt press filter 带式压滤机脱水
dewatering by centrifuge 离心式脱水机脱水
dewatering by plate frame press filter 板框式压滤机脱水
dewatering by roll 辊轧式脱水机脱水
dewatering by wells 井点降水
dewatering centrifuge 离心式脱水机;脱水离心机
dewatering channel 放水渠;脱水槽
dewatering coefficient 疏干系数;脱水系数
dewatering conduit 泄水底孔;放水管;排水管(道);排水底孔;脱水管
dewatering cone 脱水圆锥分级机
dewatering conveyer 脱水传送带
dewatering detour heading 迂回排水施工法
dewatering drift 迂回排水施工法
dewatering drum 脱水转鼓
dewatering elevator 脱水提升机
dewatering equipment 排水设备
dewatering excavation 排水工程
dewatering facility 脱水设施
dewatering factor 疏干因素
dewatering filter 脱水滤器;脱水滤池
dewatering filtration 脱水过滤
dewatering flight conveyer 洗矿槽脱水传送器
dewatering gallery 排水廊道
dewatering hole 排水孔;排水井点
dewatering in filter press 压滤机滤脱水
dewatering installation 降低地下水位设备
dewatering lane 疏干巷道
dewatering level 疏干水平
dewatering method 排水施工法
dewatering milliscreens 脱水细筛滤网
dewatering of activated sludge 活性污泥脱水
dewatering of cofferdam 围堰排水
dewatering of dry dock 干坞排水
dewatering of excavation 开挖降水;开挖中的排水
dewatering of sand 砂脱水
dewatering of sewage sludge 污水污泥脱水;污泥脱水
dewatering of sludge 污泥脱水
dewatering of vacuum filter 真空过滤机脱水
dewatering oil 脱水石油
dewatering opening 泄流孔
dewatering orifice 泄流孔;放水孔
dewatering outlet 放水口;放水管
dewatering period 脱水段
dewatering pit 排水坑
dewatering plant 脱水装置;脱水工厂;脱水车间

dewatering plastic sheet 排水塑料板
dewatering potential of sludge 污泥脱水势
dewatering preloading 降水预压
dewatering press 脱水压榨机
dewatering principle 降水原理
dewatering pump 排水泵;脱水泵
dewatering radius 疏干半径
dewatering screen 脱水筛
dewatering shaker 振动脱水机;脱水摇动筛
dewatering sieve bend 脱水弧形筛
dewatering silo 脱水筒仓;脱水谷仓
dewatering sludge 脱水污泥
dewatering stabilization 脱水加固
dewatering sump 集水坑;集水井;集水槽;排水井
dewatering system 降水系统
dewatering tank 排水井;脱水池;脱水槽
dewatering time 疏干时间
dewatering type 疏干型
dewatering unit 脱水装置;脱水器
dewatering vibrator 振动脱水机
dewatering volume 疏干体积
dewatering water volume 疏干水量体积
dewatering way 排水道
dewatering wheel 脱水轮
dewatering with gallery and borehole 巷道钻孔结合疏干
dewatering with mining 并行疏干
dewatering zone 烘干带;脱水带
dewaterizer 脱水器;脱水机
de-wax 除蜡;脱蜡
dewaxed dam(m)ar 脱蜡达玛树脂
dewaxed oil 脱蜡油
dewaxed shellac 脱蜡虫胶
dewaxing 排蜡;脱蜡(过程)
dewaxing casting 失蜡铸造
dewaxing plant 脱蜡装置;脱蜡工厂;脱蜡车间;除蜡设备
dewaxing-sintering furnace 排蜡烧结炉
dewaxing with urea 用尿素脱蜡;尿素除蜡
dew blown 露水膨胀
dew cap 露罩;露冠;防露盖
dew caution 结露告警
dew cell 湿敏元件
dew claw 悬爪;残留趾
dew detection lamp 结露检测指示灯
dewdrop 露珠
dewdrop glass 露珠玻璃;水滴(花样)玻璃;失透玻璃
dew drop slot 梨形槽
dew duster 湿式喷粉机
deweeding oil 除莠油
dewetting 反湿润;反浸润;外向湿润;去湿
Dewey decimal classification 杜威十进制分类(法)
Dewey decimal system 杜威十进制系统
deweylite 水蛇纹石
dew ga(u)ge 露量表
dew indicator 结露指示器
dewindtite 磷铅铀矿;粒磷铅铀矿
dewing 结露
dewing machine 给湿机;喷雾机
De Witte relation 德威特关系
dewlap 集露片
dew point 露点;结露度
dew point boundary 露点边界
dew point composition 露点组成
dew point control 露点控制
dew point curve 露点曲线
dew point deficit 露点温(度)差;温度露点差
dew point depression 露点温(度)差;露点降低
dew point determination 露点技术
dew point diagram 露点图(表)
dew point formula 露点公式
dew point hygrometer 露点湿度计;露点湿度表
dew point indicator 露点指示器
dew point measurement 露点测定
dew point pressure 露点压力
dew point recorder 露点记录器
dew point rise 露点升高
dew point spread 露点差
dew point temperature 露点温度
dew point testing 露点测试
dew pond 露塘;鱼鳞坑;存水池
Dewrance's alloy 迪尤兰斯合金;迪朗斯轴承合金

dew retting 露浸
dew valve 滴水阀
dex 布里格
dexbolt method 二次减径螺钉镦锻法
dexion 德新体系(一种预制单元构造体系)
d-exponent d 指数
dexterity test 敏捷性测定
dextoxification 解毒(作用)
dextrad 向右侧
dextral dislocation 右旋错动
dextral displacement 右旋位移
dextral drag fold 右拖褶曲
dextral fault 右旋断层;右侧断层
dextral fold 右旋褶曲
dextral horizontal displacement 右旋水平位移;右旋水平层移
dextrally tilting swing 顺时针翘倾摆动
dextral motion 右旋运动
dextral offset 右旋位移;右旋断错
dextral-separation fault 右旋错位断层
dextral slip on fault 断层的右旋滑动
dextral strike slip 右旋走向滑动
dextral strike-slip movement 右旋平移运动
dextral torsion 右行扭动
dextral translation 右行直移
dextral wrench fault 右旋平移断层
dextriferron 糊精铁
dextrin 糊精
dextrine gum 糊精胶
dextrin-fuchsine sulfite 糊精品红亚硫酸盐
dextrin glue 糊精胶
dextrinize 糊精化
dextro-compound 右旋(化合)物
dextrogyre 右旋物
dextro-levo 右旋左旋
dextroposition 右移位
dextrorotary 向右旋转的;右旋的
dextrorotary quartz 右旋石英
dextrorotation 向右旋转
dextrorotatory 向右旋转的
dextrorsal curve 右挠曲线
dextrotatory compound 右旋化合物
dextroversion 右转
dey-containing wastewater 含染料废水
Deyoung-Axford model 德扬—阿克斯福德模型
dezinc 脱锌
dezincification 除锌(作用);脱锌(作用)
dezincify 除锌;脱锌;去锌
dezincing 除锌
dezincked lead 除锌铅
de-zinkify 去锌;脱锌
dezocine 地佐辛
décor 舞台装置
D-glass 低介电玻璃
D-grade wood D 级木材
dhanari column 幢
D-handle shovel D 形柄手铲
D-herringbone track 箭翎型编组站
D-horizon 丁层;下覆岩土层;D 层(土)
DHV[design hour volume]设计小时交通量
Diabaig group 迪亚拜群【地】
diabantite 辉绿泥石
diabase 辉绿岩
diabase aplite 辉绿细晶岩
diabase chip(ping)s 辉绿岩屑
diabase facies 辉绿岩相
diabase for cast stone 铸石用辉绿岩
diabase for cement 水泥用辉绿岩
diabase group 辉绿岩类
diabase porphyry 辉绿斑岩
diabase traprock 辉绿暗色岩
diabasic 辉绿质的
diabasic lining 辉绿岩衬里
diabasic texture 辉绿结构
diabasic tuff 辉绿凝灰岩
diabatic 传热的;非绝热的
diabatic process 传热过程
diabatic wind profile 非绝热风速廓线
diablastic 筛状结构的
diablastic texture 筛状变晶结构
diabled ship 失去航行能力的船
diaboleite 羟氯铜铅矿
diabolo 对三角形(货物标志)
diaborit 孕镶式钻头(瑞典)

diabrochometamorphism way 浸透变质作用方式
diacaustic 折射散焦线
diacetamide 二乙酰基胺
diacetanilide 二乙酰苯胺
diacetate 双乙酸盐
diacetate fibre 二醋酸纤维
diacetic acid 二乙酰乙酸
diacetic ether 乙酰乙酸乙酯
diacetin 甘油二乙酸酯
diacetone 乙酰丙酮;双丙酮
diacetone acrylamide 双丙酮丙烯酰胺
diacetone alcohol 双丙酮醇
diacetoxyl 二乙酸基
diacetyl 联乙酰
diacetyl acetic ester 二乙酰乙酸酯
diacetyl acetone 二乙酰基丙酮
diacetyl amide 二乙酰胺
diacetyl-carbinol 二乙酰甲醇
diacetyldioxime 丁二酮肟
diacetyldioxime spectrophotometry 丁二酮肟分光光度法
diacetyl dioxyphenylisatin 二乙酰酚靛红
diacetyl di-sulfide 二乙酰二硫
diacetylmonoxime 丁酮肟
diacetyl peroxide 过氧化二乙酰
diacetyl succinic acid 二乙酰丁二酸
diacholestane 重排胆甾烷
diachronism 历时性
diachronous 年序堆积层
diacid 二(元)酸
diacid amide 二酰胺
diacidic base 二元碱
diacid wastewater 二酸废水
diaclase 正方断裂线;构造裂缝
diaclinal 横向褶曲河
diaclinal river 顺斜道逆斜河
diaclinal stream 顺斜道逆斜河
diaconicon (古代教堂中的)圣器室
diaconicum (古代教堂中的)圣器
diacope 重切伤
diacoustics 折声学
diacranterian 横隔齿列的
diacranteric 横隔齿列的
diacrete 硅藻土混凝土
diacritical 半临界值的
diacritical current 半临界值电流
diacritical point 间临界点
diacritical sign 区别记号
diactine 二放体
diactor 直接自动调整器
diacyl peroxide 过氧化二酰;二酰基过氧化物
diacytic type 直轴式
diad 二重轴;二重的;二素组;二联体;二价原子;二价元素;二价基;二合一;二分子;二分体;二单元组;双位二进制
diad axis 二重轴;二次对称轴
diad derivative thermogravimetric curve 二次微商热重曲线
diad derivative thermogravimetry 二次微商热重法
diadelphous 二体的
diadic 二重轴;二素组;二价的;双值
diadic operator 双值算子
diadochite 磷铁华;磷铁矾
diadromous 越区回游
diagenesis 成岩(作用)
diagenetic 成岩作用的
diagenetic breccia 成岩角砾岩
diagenetic concretion 成岩结核
diagenetic differentiation 成岩分异作用
diagenetic dolomite 成岩白云岩
diagenetic dolostone 成岩白云岩
diagenetic fissure 成岩裂隙
diagenetic fracture 成岩裂隙
diagenetic gas 成岩气
diagenetic gas zone 成岩气带
diagenetic gelification 成岩凝胶化作用
diagenetic metamorphism 成岩变质作用
diagenetic Mn nodule 成岩作用锰结核
diagenetic stage 成岩作用阶段
diagenetic trap 成岩圈闭
diagenetic water 成岩水
diagenism 成岩(作用);沉积变质作用
diagentic process 成岩过程
diageotropism 横向地性

diaglyph 凹刻;凹雕
diaglyphic 凹雕的
diaglyphic ornament 凹刻装饰;凹雕装饰
diaglyphic work 凹雕(作品)
diagnosis 判断;特征简介
diagnosis in species description 物种描术鉴定要点
diagnosis of activated sludge process 活性污泥工艺诊断
diagnosis of damage 伤害诊断
diagnosis refrigeration 冷却分析
diagnosis status word 诊断情况码
diagnostic ability 可诊断性
diagnostic capability 诊断本领【计】
diagnostic check 诊断校验;固着检验
diagnostic dose 诊断剂量
diagnostic element 诊断性元素
diagnostic factor 诊断因子
diagnostic fossil 标准化石;特征化石
diagnostic function 诊断功能
diagnostic indicators 诊断性指标
diagnostic logout 诊断记录
diagnostic message 诊断信息
diagnostic mineral 指相矿物;标志矿物;特征矿物
diagnostic program(me)诊断程序
diagnostic robot 诊断机器人
diagnostic routine 诊断程序
diagnostics 编译出错信息
diagnostic surface horizon 诊断表层土
diagnostic test 鉴别检验;诊断(性)试验;诊断检验;诊断检查;诊断测验
diagnostic tool 诊断工具;分析工具
diagnostic tree 诊断树
diagnostic X-rays 诊断 X 光
diagometer 电导计
diagonal 中线斜;对顶线;对角线的
diagonal abutment pier 斜岸墩
diagonal arch 对角拱
diagonal band 斜角小带;对角(钢筋带)
diagonal bar 斜向钢筋;斜列砂坝;斜拉杆;斜(钢)筋;斜(撑)杆;对角钢筋
diagonal bars tread 斜交条纹
diagonal baulk fish pass 斜槽式鱼道
diagonal beam 对角梁;斜梁
diagonal beam floor 斜梁地板;斜梁楼盖结构
diagonal-beam pass 斜置钢梁孔型
diagonal bedding 斜层理
diagonal bedding structure 斜层理构造
diagonal bollard tie 对角系船柱拉杆
diagonal bond 斜置丁砖皮;人字形砌合;人字形面砖排列;对角黏合;斜角砌合;对角式砌合法;对角砌合
diagonal bond raking bond 斜向砌合
diagonal brace 斜拉条;斜撑(杆);对角拉条;对角撑;斜支撑;剪刀撑;对角斜撑
diagonal bracing 斜角力撑;斜(支)撑;对角支撑;对角系杆;对角拉条
diagonal bridging 斜支杆;剪刀撑;斜撑;对角支撑
diagonal brush 斜向电刷
diagonal built 斜铺法
diagonal buttress 斜撑墙;斜扶壁;转角扶垛;转角扶壁
diagonal buttressing pier 斜支墩;斜扶柱
diagonal cable (预应力混凝土中的)斜拉钢缆;(预应力混凝土中的)对角钢缆
diagonal cable bridge 斜缆桥
diagonal carvel built 斜平镶舢板
diagonal cassette slab floor 斜镶板楼盖;斜镶板地板
diagonal check 对角线检查
diagonal coffered soffit 具有斜方格花纹的拱腹
diagonal coffer (slab) floor 斜镶板楼盖;具有斜方格花纹的地板
diagonal compression 主压力;斜向压力
diagonal compression failure 斜压破坏
diagonal compression stress 斜向压应力
diagonal conducting wall generator 斜导电壁发电机
diagonal conjugate 对角径
diagonal cored slab floor 具有斜方格花纹的地板
diagonal crack(ing)斜裂纹(展开);斜向开裂;斜(向)裂缝;对角裂缝
diagonal crossed 对角交
diagonal cut 对角线切削
diagonal cut joint 斜开口

diagonal cut sail 斜接缝帆
diagonal cutter 对角切割器
diagonal cutting nippers 斜嘴钳
diagonal cutting pliers 斜切手钳；斜口钳
diagonal decomposition 对角线分解
diagonal direction 对角线方向
diagonal displacement 对角线位移
diagonal-dominant matrix 对角占优势阵
diagonal drain 斜水沟；斜渠
diagonal draw bar 斜拉杆
diagonal dross scar 斜线道（浮法玻璃缺陷）
diagonal edge 直角边
diagonal element 对角元素
diagonal entry 交叉巷道；(矩阵的) 对角元素；对角线元素；对角平巷
diagonal expansion 交叉扩展；斜向发展
diagonal eye(piece) 棱镜目镜；折轴目镜；对角目镜
diagonal face 倾斜工作面
diagonal fault 剪力断层；斜(向)断层
diagonal finishing beam 斜滚式整平梁（修整混凝土路面用）
diagonal fissure 斜交裂隙
diagonal flange method 斜配孔型轧制法
diagonal flooring 斜铺地板
diagonal flow angle 对流角
diagonal flow compressor 斜流式压缩器
diagonal flow generator 斜流通风式发电机
diagonal flow pump 混流泵
diagonal flow turbine 斜流式水轮机
diagonal form 对角线形式
diagonal frame 斜框架；斜构架
diagonal game 对角线对策
diagonal glide plane 对角滑移平面
diagonal grain 制材斜纹；斜纹理；斜（木）纹；对角（线）纹理；沿长向斜纹理（木材）
diagonal-grid system 斜交格构体系
diagonal harrowing 斜向耙地
diagonal hobbing 对角线滚削
diagonal hole pattern 菱孔型
diagonal homing 对角导向
diagonal horn 对角线极化喇叭；对角喇叭
diagonal horn antenna 对角喇叭天线
diagonal in compression 斜向受压；受压斜杆
diagonal in tension 斜向受拉；受拉斜杆
diagonalizable matrix 可对角化矩阵
diagonalization 作成对角线；对角（线）化
diagonalization of matrix 矩阵对角化
diagonal jib 斜接缝三角帆
diagonal joint 斜节理；斜(削)接头
diagonal junction of flat bars 扁钢斜接
diagonal laying 斜置
diagonal ligament 管间斜距
diagonal line 对角线
diagonally braced frame 对角拉条底架
diagonally dominant 对角优势的
diagonally dominant matrix 对角优势矩阵
diagonally dominant system of equation 对角优势方程组
diagonally isotone mapping 对角保序映射
diagonally m-block tridiagonal matrix 对角 m 块三对角矩阵
diagonal masonry bond 斜向圬工砌合；对角接合
diagonal masonry pattern 斜向圬工图案
diagonal matrix 对角(线)矩阵
diagonal member 斜(向)构件；斜(撑)杆；对角斜
diagonal member head 斜撑头
diagonal member truss 三角孔桁架；腹杆桁架
diagonal micro instruction 对角线型微指令
diagonal mirror 对角镜
diagonal network 对角线网络
diagonal of a matrix 矩阵对角线
diagonal pair 对角对
diagonal panel slab floor 具有斜方格花纹的地板
diagonal parking 斜列停车；斜角停车
diagonal parking lane 斜列(式)停车道
diagonal pass 斜置孔型；对角孔型
diagonal pattern 斜纹图案；斜纹花型；斜砌图案
diagonal paving 斜(向)铺砌；对角铺砌
diagonal perspective 斜透视
diagonal pillar 斜支柱
diagonal pitch 交错铆距；斜凿纹；斜心距；斜铆距；斜间距；斜齿纹；对角间距
diagonal placed 斜置
diagonal plane 对角(平)面

diagonal plane of a polyhedron 多面体的对角面
diagonal planking 斜镶舢板；斜接式铺板
diagonal plywood 斜纹胶合板
diagonal process 对角线方法
diagonal ramp 斜向坡道；对角向匝道
diagonal rectangular grid floor 斜长方格地板；菱形格地板
diagonal reinforcement 弯起钢筋；弯曲钢筋；斜(向)钢筋；斜筋
diagonal rib 斜肋(条)
diagonal ribbed tile 斜肋瓦
diagonal rib of wall-arch 墙拱斜肋
diagonal ribs 交叉肋(条)；交叉(弯)肋；对角肋
diagonal ridge for hip roof 垂脊
diagonal ridge gable and hip roof 戗脊
diagonal rod 斜杆
diagonal rod head 斜拉杆头
diagonal roller beam 斜滚式整平梁（修整混凝土路面用）
diagonal roofing 斜铺板材屋面
diagonal ruling 对角晕线【测】
diagonal scale 斜线（比例）尺；斜分比例尺；对角线刻度；对角线标度
diagonal screed 斜刮板
diagonal screed finisher 斜刮板式修整器
diagonal section 对角截面
diagonal series 对角线级数
diagonal set 对角集合
diagonal shaving 对角线剃齿
diagonal shear member 受剪对角斜杆
diagonal sheathing 斜铺衬板；斜角覆盖层；斜护板
diagonal siding 斜墙；斜壁
diagonal slating 斜铺板；斜盖瓦；对角铺瓦做法；对角铺砌
diagonal slip fault 斜滑断层
diagonal slot 钻粒钻头斜水口
diagonal square grid 斜方孔格栅；菱形孔格栅；斜交方格
diagonal square matrix 对角(线)方阵
diagonal stacking 斜堆
diagonal staggered pigeon hole packing 井字码法
diagonal stanchion 斜支柱
diagonal standing wave 斜向驻波；三向驻波
diagonal stay 斜支索；斜撑条；对角拉撑
diagonal stock 脱方轧件
diagonal straight flange method 直腿型钢斜轧法
diagonal stratification 交错层；假层理
diagonal stray 对角支撑
diagonal street 斜向街道；斜街；斜交道路；对角线街道
diagonal strut 斜角撑杆；斜撑；对角(线) 支撑；受压斜杆；对角撑；斜支柱；托脚
diagonal subfloor 斜铺毛地板
diagonal surface 对角线曲面
diagonal system 斜置孔型系统；斜列系统
diagonal tensile stress 斜拉应力；主拉应力
diagonal tension 斜(向)拉力；对角张力；主拉力
diagonal tension bar 斜向抗拉钢筋
diagonal tension crack 斜拉裂缝
diagonal tension failure 斜拉破坏；剪切破坏
diagonal (tension) stress 斜拉应力；主拉应力
diagonal tie 斜拉索；斜拉杆；对角拉条
diagonal tie of a dome 圆顶斜拉条
diagonal tie plate 甲板底斜衬条；斜牵板
diagonal tile 斜铺瓦；斜铺面砖；铺面砖嵌条
diagonal topping 斜切顶
diagonal transfer 对角变换
diagonal triangle 对边三角形
diagonal trussing 斜置桁架；对角桁架
diagonal truss rod 对角桁架杆
diagonal truss tube 斜桁架式外筒
diagonal tube 斜管
diagonal turbine 斜流涡轮机
diagonal valley 斜谷
diagonal ventilation 对流式通风；对角式通风（法）
diagonal voltage bridge 对角电压电桥
diagonal waffle floor 斜镶格楼盖
diagonal waffle slab floor 具有斜方格的地板
diagonal warping 斜(弯)翘曲
diagonal web member 桁架斜杆；斜腹杆
diagonal winch 斜置式(蒸汽)绞车
diagonal-wire placed 斜置
diagonal wiring 对角拉线

diagonaly body break 辊身对角折断
diagram 图(表)
diagram factor 功图因子
diagram flow 操作程序图
diagram for lighting feeder 照明干线系统图
diagram language 图表语言
diagram mapping 调制要图
diagrammatic 图解的
diagrammatic arrangement 原则性布置；示意性布置；概略布置；图式布置；示意布置图
diagrammatic curve 图示曲线；图解曲线；图表曲线
diagrammatic decomposition 图表分解
diagrammatic decomposition method 图解分解法；图表分解法
diagrammatic determination 图解测定
diagrammatic diagram 简图
diagrammatic drawing 略图；简图；草图；示意图
diagrammatic layout 原理图；示意布局
diagrammatic map 图表
diagrammatic plan 图解平面；示意平面(图)
diagrammatic presentation 用图表示；图示
diagrammatic profile 图解剖面
diagrammatic representation 用图表示；图形表示；图示；图解表示(法)
diagrammatic representation of input-output 投入产出表
diagrammatic section 图示剖面；图解剖面
diagrammatic sectional drawing 断面草图
diagrammatic sign 图形标志
diagrammatic sketch 草图；简图；草简；示意图
diagrammatic view 简图；图解视图
diagrammatization 用图表示；做成图表
diagrammatize 做成图表；用图解法表示
diagram method 图解法；图表法
diagram of adjoining chart 邻图索引
diagram of boiler control 锅炉控制原理图
diagram of component forces 分力图
diagram of computational program(me) 计算程序框图
diagram of connection 简图；草图
diagram of curve 曲线图示
diagram of daily traffic variation 日交通量变化图
diagram of errors 误差图解
diagram of fault system 断层系统图
diagram of foliations 面理图解
diagram of forgeability behavio(u)r 塑性图
diagram of gears 传动装置图
diagram of horizontal water pressures 水平压力（分布）图
diagram of hourly traffic variation 时交通量变化图
diagram of link 链接的图式
diagram of list 列表的图式
diagram of monthly traffic variation 月交通量变化图
diagram of node 节点的图式
diagram of observation and experiment 观测
diagram of operation(al) procedure 操作程序图
diagram of power supply system 供电系统图
diagram of principal stress axes 主应力轴图解
diagram of quintic-dipole longitudinal sounding 五极纵轴测深曲线图
diagram of radioactive result 放射性成果图件
diagram of sounding curve 电测深曲线图
diagram of spectrum analysis 频谱分析图
diagram of stresses 应力图
diagram of structural information 结构信息的图式
diagram of tectonic stress field 构造应力场图解
diagram of terminal connection 接头端子连接图
diagram of the plane of horizon 地平平面图
diagram of the situation 对势图
diagram of velocities 速度图
diagram of wiring 接线线路图
diagram of work 示功图
diagram on thee plane of the equator 赤道平面图
diagram on the plane of the celestial equator 天赤道平面图；时系图；时间图
diagram on the plane of the celestial meridian 子午线平面图
diagram on the plane of the equinoctial 时系图
diagram reinforcement 弯起钢筋
diagram showing the chemically zonation model of elements 元素化学分带模式图
diagram showing the distribution of elements 元素分布图

diagram showing the geochemical cycle of elements 元素地球化学循环图
diagram showing the geochemical evolution of elements 元素地球化学演变图
diagram showing the main geochemical processes and their pathway of elements 元素主要地球化学作用及途径图
diagram showing the partition coefficient as a function of oxygen fugacity 分配系数与晶体生长函数关系图
diagram showing the ratio of elements 元素比值图
diagram showing the relation between pressure and density 压力与密度关系图
diagram showing the relation between temperature and depth 温度与深度关系图
diagram showing the relationship between temperature and oxygen fugacity 温度与氧逸度关系图
diagram showing the relative change in the concentration of elements 元素浓度相对变化图
diagram showing the solubility as a function of pH 溶解度与pH函数关系图
diagram showing the stability field of components 化合物稳定场图
diagram showing the variation in abundance of elements 元素丰度变化图
diagram showing the variation trend of element 元素变化趋向图
diagram showing the vertical distribution of element 元素垂直分布图
diagrams of rocks 岩石学图件
diagram stage 图解阶段
diagram tacheometer 图解式测距仪；图解视距仪
diagraph 扩大绘图器；机械仿形仪；描字器；作图器；制图仪；分度尺；放大绘图仪；仿型仪；仿形仪
diagraphy 作图法
diagrid 斜眼格栅；斜眼格筛；斜格
diagrid floor 双向格式网架楼板；斜网格楼板；斜肋楼板；斜交网格型楼板
diagrid system 斜交格构体系
diakinesis 终变期
diakinesis stage 终变期
dial 号(码)盘；信号度盘；针盘；度盘；分划；分度盘；拨号盘；表盘；标度盘；调节控制盘；数字盘；示数盘
dial analogue indicator 标度盘模拟指示器
dial arc 刻度弧
dial-a-ride service 电话叫车服务
dial-a-ride system 电话叫车系统
dial balance 刻度盘天平；圆盘指针式天平；表盘秤；标度盘秤
dial barometer 槽式气压计；气压指示计；气压表
dial batcher 刻度盘计量器
dial bench ga(u)ge 台上指示表
dial cable 表盘绳
dial caliper ga(u)ge 带表内卡规
dial calipers 带表卡尺
dial central office 自动电话总局
dial clock work 计数器的钟表结构
dial comparator micrometer 度盘式比较千分尺
dial compass 刻度罗盘；刻度规
dial condenser 度盘式可变电容器；度盘式电容器
dial control 度盘控制
dial cord 刻度盘线
dial counter 刻度盘(式)计数器；指针式计数器；指针读数；表盘式计数器
dial depth ga(u)ge 深度千分表
dial-digital thermometer 度盘式数字温度计
dial disc 刻度圆盘
dial division 度盘划分
dial drum 刻度凸轮；刻度筒
dial dynamometer 度盘式测力计
dialed circuit 拨号电路
dialed digit 拨号数字
dial exchange 自动交换区；拨号交换机
dial extensimeter 刻度盘变形仪；测微引伸仪
dial extensometer 针盘式伸长计；测微伸缩仪
dial face 刻度盘面
dial flange 刻度盘座
dial foot 表盘面脚；表盘固定销
dial for number setting 数字安置盘
dial ga(u)ge 量表；刻度计；刻度尺；英制千分表；指示表；直读式测厚仪；度盘式指示器；测微仪；千分表
dial ga(u)ge comparator 带有千分度的比较仪；度盘规比较器
dial ga(u)ge micrometer 带表千分尺
dial ga(u)ge stand 指数表架；千分表架
dial gear 度盘传动装置
dial holder 仪表架；千分表架
dial illumination 刻度照明；(仪器的)刻度盘照明；仪表照明；度盘照明
dial-in 拨入
dial-in cluster 拨入群控器
dial indicator 刻度指针；刻度盘指示器；度盘(式)指示器；拨号盘(速度)指示器；标度盘指示器；百分表；千分表；温度指示器
dial indicator strain ga(u)ge 针盘式应变计
dialine 棕色线晒图
dialing 拨号；调谐
dialing adapter 字盘式转接器
dialing in 拨入
dialing system 拨号系统
dial inside micrometer 表内径千分尺
dial instrument 指针式仪表；度盘式仪表；表盘式仪表；标度盘仪表
dialkene 二烯烃
dial key setting 信号度盘时间键定位
dialkyl 二烃基
dialkyl amine 二烃基胺
dialkylate 二烃基化合物
dialkylene 二烃基
dialkyl mercury 二烃基汞
dialkyl peroxide 过氧化二烷基
dialkylphosphinate 二烃基亚磷酸酯
dialkyl pyrophosphoric acid 二烷基焦磷酸
dialkyl sulfates 二烃(基)硫酸盐
dialkyl sulphosuccinic salt 磺基琥珀酸二烷基酯
dialkyl tin oxide 氧化二烷基锡
dialkyl tin sulphide trimer 硫化二烷基锡(环)三聚体
dialkyphosphate 二烃基磷酸盐
diallage 异剥石
diallagite 异剥辉石岩
dial lamp 刻度盘灯；度盘灯
dial light 度盘照明
dialling 罗盘导线测量
Diallist 迪阿里司特镍铝钴铁合金
dial lock 字码锁；转字锁；度盘锁档；标度锁档
Diallocs 戴洛陶瓷
Dialloy 戴洛伊硬质合金
Di-alloy 迪合金
diallyl amine 二烯丙基胺
diallyl chlorendate 氯菌酸二烯丙酯
diallyl cyanamide 二烯丙基氰胺
diallyl disulfide 二硫化二烯丙基
diallyl fumarate 富马酸二烯丙酯
diallyl isophthalate 间苯二甲酸二烯丙酯；异苯二甲酸烯丙酯
diallyl maleate 马来酸二烯丙酯
diallyl oxalate 草酸二烯丙酯
diallyl phthalate 邻苯二甲酸二烯丙酯树脂
diallyl phthalate resin 聚邻苯二甲酸二烯丙酯树脂
diallyls 聚二烯丙酯(类)
diallyl sulfide 烯丙基化硫
diallyphthalate 邻苯二甲酸二丙烯
dial measuring ga(u)ge 标度盘量规
dial mercury sensing thermometer 盘式水银传感温度计
dial micrometer 厚度千分表；直读式测微仪；度盘测微计
dial needle 刻度盘指针
dial-needle heat-efficiency meter 度盘指针式热效率测定仪
dial office 自动电话局
dial of meter 度盘；仪表刻度盘
dial of vertical adjustable screw 垂直可调丝杠
dialog(ue) box 对话框
dial operation 度盘控制
dial-out 拨出
dial padlock 号码挂锁；拨号挂锁
dial pattern bridge 插塞式电阻电桥
dialphanyl phthalate 邻苯二甲酸二脂族醇酯
dial plate 指针板；拨号盘；表盘；标度盘
dial pointer 度盘指针；标度盘指针
dial private branch exchange 自动小交换机
dial pulse 拨号脉冲
dial reading 读数
dial register 指针式计量装置；度盘式记录器；拨号计数器
dial scale 度盘刻度；带表盘秤；刻度盘
dial screw 度盘螺钉
dial selector 十进制步进式选择器
dial service 自动电话业务
dial sheet 度盘座
dial sheet ga(u)ge 厚薄指示表
dial snap ga(u)ge 指示卡规；带表卡规；千分表卡规
dial speed 信号度盘转速；拨号速度
dial supervision 图盘式监视
dial switch 拨码盘式开关；盘式开关
dial system 自动电话系统
dial system installation 自动电话装置
dial system tandem operation 自动转接的操作
dial telephone 自动电话
dial test indicator 杠杆千分表；度盘式指示器
dial thermometer 刻度盘式温度计；指针(式)温度计；度盘式温度计；表盘式温度计
dial thickness ga(u)ge 针盘式测厚规；表盘式厚度规
dial thread indicator 螺纹指示盘；乱扣盘
dial through busy 拨入忙线
dial transfer machine 度盘式回转连续自动工作机床
dial type 度盘式
dial type cylinder ga(u)ge 圆盘式缸径规
dial type feed mechanism 转盘式加料器
dial type indicator 度盘式指示器
dial type resistor 度盘式电阻器
dial type rheostat 圆盘式变阻器
dial type scale 刻度盘；圆弧尺；字盘秤；标度
dial unit 信号度盘
dial-up 拨码；拨号
dial-up line 交换线路
dial-up networking 拨号连接上网
dial-up phone line 拨码电话线
dial-up telemetric system 拨号遥测系统
dial vapo(u)r-pressure thermometer 盘式蒸汽压力温度计
dial window 刻度窗
dialycerin 双甘油
dialysate 淡水流；透析液
dialyse 析离体；透析
dialysing membrane 渗析膜
dialysis 透析；渗析(分析)；渗透分析
dialysis membrane 渗析膜
dialysis treatment 渗析处理
dialysis unit 渗析单元
dialyte lens 分离透镜
dialytic membrane 渗析膜
dialyzable 可透析的
dialyzate 渗析液；渗透液
dialyzator 渗析器
dialyzer 透析器；渗析器；渗析膜
dialyzing shell 渗析壳
diamagnet 抗磁体
diamagnetic 抗磁(性)的；逆磁性的；反磁性的
diamagnetic alloy 抗磁合金
diamagnetic anisotropy 抗磁的各向异性
diamagnetic body 抗磁性；反磁体
diamagnetic effect 抗磁效应
diamagnetic Faraday effect 抗磁性法拉第效应
diamagnetic gas 抗磁的气体
diamagnetic gaussmeter 抗磁磁强计
diamagnetic ion 抗磁的离子
diamagnetic loop 抗磁圈
diamagnetic material 抗磁(性)材料；反磁材料
diamagnetic moment 逆磁矩
diamagnetic plasma 抗磁等离子体
diamagnetic probe 抗磁探针
diamagnetic resonance 回旋共振
diamagnetic response 抗磁感应
diamagnetic shield 抗磁屏蔽
diamagnetic shielding 抗磁性屏蔽；反磁性屏蔽
diamagnetic shift 反磁位移
diamagnetic state 抗磁态
diamagnetic substance 抗磁(物)质；反磁物质
diamagnetic susceptibility 抗磁(磁)化率
diamagnetism 抗磁性；逆磁性；反磁性
diamagnetism mineral 逆磁性矿物
diamant 玻璃刀；金刚刀
diamante 闪光片
diamantin(e) 金刚铝；铁铝氧耐火材料；金刚烷；似

金刚石
Diamantine fracture zone 迪亚曼蒂纳破裂带
Diamantine trench 迪亚曼蒂纳海沟
diamantini 薄片碎玻璃;鳞片碎玻璃
diamertrical 直径的
diameter 直径;透镜放大倍数
diameter at bottom of thread 螺纹底径
diameter at location of gasket load reaction 垫片荷载作用位置处的直径
diameter at smaller end 细端直径
diameter breast high 胸径
diameter class 直径级
diameter class of wood 直径级木
diameter clearance 径向间隙;直径留隙
diameter control 直径控制
diameter deformation 径向变形;横向变形
diameter difference of cross section 断面直径差
diameter dimensioning 直径(尺寸)标注
diameter diminution 直径缩小
diameter enlargement 径节放度
diameter enlarging 放肩
diameter flat jack process 径向液压枕法
diameter group 比直径
diameter increment 加粗生长;直径增量
diameter index safety system 直径指数安全系统
diameter length 直径长度
diameter limit 径级择伐;直径(级)界限;下限直径
diameter of a doline 岩溶漏斗直径
diameter of advance 回转圈最大直径
diameter of anchor bolt hole 地脚螺栓孔直径
diameter of aperture 孔径
diameter of axle 轴径
diameter of bolt 螺栓直径
diameter of bolt hole 螺栓孔直径
diameter of borehole mouth 开孔直径;开孔直径
diameter of boss 轮毂直径
diameter of brake drum 刹车鼓直径
diameter of center hole 中心孔直径
diameter of centre hole in bolts 螺栓中心圆直径
diameter of charge 药包直径
diameter of conduit 通道直径
diameter of diapir core 底辟核直经
diameter of distant hole 终孔直径【岩】
diameter of drill pipe 钻杆直径
diameter of electrode 焊条直径
diameter of equivalent circle 当量圆直径
diameter of explosion cartridge 药包直径
diameter of finally formed hole 终孔直径【岩】
diameter of grain 粒径
diameter of gravel stuffing 填砾直径
diameter of grouting hole 注浆孔径
diameter of hammer 锤直径
diameter of hole 孔径【岩】
diameter of horizontal circle 水平度盘直径
diameter of impact crater 撞击坑直经
diameter of impeller 叶轮直径
diameter of inflow tube 进水管直径
diameter of inlet 吸入口直径
diameter of invaded zone 侵入带直径
diameter of karstic stony column 石柱直径
diameter of observation shaft 观测竖井直径
diameter of opening 孔口直径;洞口直径;孔径
diameter of outflow tube 出水管直径
diameter of outlet 排气口直径
diameter of padding steel reinforcement 垫筋直径
diameter of particle 颗粒直径
diameter of penetrometer 锥底直径
diameter of percolation zone 渗透带直径
diameter of pipe 管径
diameter of plunger 柱塞直径
diameter of ponor 落水洞直径
diameter of pump tube 泵管直径
diameter of radial pipe 辐射管直径
diameter of rock shield 石盾直径
diameter of rope 缆索直经
diameter of sampler 取样器直径
diameter of screen tube 滤水孔直径
diameter of screw 螺旋直径;螺纹(标称)直径
diameter of shot-hole 炮眼直径
diameter of slot 槽孔直径
diameter of soil column 土柱直径
diameter of soil particles 等容粒径
diameter of stack opening 烟囱上口直径
diameter of stalactite 钟乳石直径

diameter of stalagmite 石笋直径
diameter of taper 直径锥度
diameter of the core 钻进岩芯直径
diameter of threads 螺纹直径
diameter of tube 管径
diameter of twined wire 缠丝直径
diameter of vane 十字板头直径
diameter of water collecting pipe 集水管直径
diameter of water ga(u)ge glass 测水管直径
diameter of water pipe 水管直径
diameter of wheel 轮径;车轮直径
diameter of wind pipe 风管直径
diameter of wire 钢丝(网)粗度;电线直径
diameter of work 工件直径
diameter over bark 连皮直径
diameter pitch 径节;直径节距;整节距
diameter plane 直径面
diameter quotient 直径率
diameter ratio 内外径比;直径比
diameter run-out 径向跳动
diameter sizer 径选机;按直径分级机
diameters of dissolved crystal 被溶晶体直径
diameter strain 径向应变;横向应变
diameter tape 直径卷尺
diameter testing ga(u)ge 量径规
diameter tolerance 直径公差
diameter track 直径线
diameter transversa 横径
diameter under bark 去皮直径
diametral 直径的
diametral compression test 径向受压试验;劈裂(抗拉)试验
diametral crushing test 直径压碎试验
diametral curve 沿径曲线;直径曲线
diametral gap 直径径向公差
diametral pitch (齿轮的)径距;径向角变量;径节
diametral pitch system 径节系统
diametral plane 直径(平面);径向平面
diametral plane of a sphere 球径平面
diametral prism 第二正方柱
diametral pyramid 第二正方柱
diametral quotient 直径系数
diametral surface 直径曲面
diametral system 径节制
diametral voltage 正负对向电压
diametral winding 整(节)距绕组;全节矩绕组
diametric(al) 正方的;直径方向的
diametric(al) arrangement of brushes 电刷的径向布置
diametric(al) compression 径向压缩;对径压缩
diametric(al) connection 径向连接;沿径连接;对角接线
diametric(al) curve 沿径曲线;径曲线
diametrically opposite 直径上对置的
diametric(al) pitch 整节距;全(节)距
diametric(al) projection 等角轴测投影;径点互射影
diametric(al) ratio of flange 法兰径比
diametric(al) rectifier 全波整流器
diametric(al) tappings 对径分接
diametric(al) voltage 对径电压
diamictite 混杂陆源沉积岩;混积岩;杂岩
diamicton 杂层[地];双层砌体;(古罗马建筑中填碎料的)空芯墙;混杂陆源沉积物
diamide 联氨;二酰胺;二酰氨
diamido 二酰氨基
diamidogen 联氨;二胺
diamine 二胺化物;双胺染料
diamine fast violet BBN 双胺坚牢紫 BBN
diamino 二氨基
diamino acid 二氨基(甲)酸
diaminobenzene 苯二胺
diaminobenzidine tetrahydrochloride 二氨基联苯胺盐酸盐
diaminocaproic acid 二氨基己酸
diamino dicarboxylic acid 二氨基二羧酸
diaminodiphenyl 二氨基联苯
diaminodiphenyl methane 二氨基二苯甲烷
diamino-diphenyl sulphone 二氨二苯砜
diaminogen dye 双胺精染料
diaminomaleonitrile 二氨基顺丁烯二腈
diamino monocarboxylic acid 二氨基羧酸
diamino-monophosphatide 二氨磷脂
diaminonaphthalene 二氨基萘
diaminophenol 二氨基苯酚

diaminophenol dihydrochloride 二氨基酚盐酸盐
diamino-phenylacetic acid 二氨基苯乙酸
diaminopimelate 二氨基庚二酸
diaminopimelic acid 二氨基庚二酸盐;二氨基庚二酸
diaminopropyl tetramethylene diamine 精胺
diamino sulfonic acid 二氨基磺酸
diaminotoluene 二氨基甲苯
diaminotoluene dihydrochloride 二氨基甲苯盐酸
diaminovaleric acid 二氨基戊酸
diaminuria 二胺尿
diammonium hydrogen phosphate 磷酸氢二铵
diammonium phosphate 磷酸二铵
diamond 菱形组合;菱纹呢;金刚石;钻石
diamond abrasive 钻石磨蚀剂
diamond and carbide tool boring machine 金刚石和硬质合金刀具镗床
diamond antenna 菱形天线
diamond anvil 金刚石砧
diamond-anvil cell 金刚石砧压槽
diamond bar 菱形凸纹钢筋;菱形钢筋构件;菱形钢筋;多角形钢筋
diamond beads 金刚石串珠
diamond bearing 金刚石轴承;含金刚石的
diamond-bearing gravel 含金刚石砂砾
diamond bit 金刚石钻头;金刚石车刀;金刚(钻)钻头;尖凿
diamond bit boring machine 金刚石钻探机
diamond bit drilling 金刚石钻头钻进
diamond bit returned 回收的金刚石钻
diamond bit selection guide 金刚石钻头选择指南
diamond bit setting 金刚石钻头镶嵌
diamond blade 金刚石刀片
diamond blade bits 金刚石刀片钻头
diamond bonded wheel 金刚石结合砂轮
diamond boring 金刚石钻进;金刚石镗孔
diamond brace 十字形拉条
diamond brake shoe 金刚石闸瓦
diamond bronze 赛金刚青铜(铜88%、铝10%和硅2%)
diamond-charged ring tool 金刚石环形刀具
diamond check 菱形格子
diamond chip(ping)s 金刚石支承履;金刚石片;金刚石碎屑
Diamond chrome dye 迪阿蒙德铬媒染料
diamond circular saw 金刚石圆锯;金刚砂圆锯
diamond circular saw blade 金刚石圆切片刀
diamond cleavage 金刚石解理
diamond cleaving 金刚石劈开
diamond cloth 斜方块织物;菱纹织物;斜纹布
diamond coating 金刚石涂层
diamond cold cathode luminescence 金刚石冷阴极发光
diamond compact 金刚石粉烧结体
diamond composition 金刚石制品
diamond concentration 金刚石浓度
diamond cone 金刚钻锥体;金刚石锥
diamond content 金刚石含量
diamond core 金刚石钻进(取得的)岩芯
diamond core barrel 岩芯筒体
diamond core bit 金刚石取芯钻头
diamond core boring bit 金刚石空心钻头;金刚石岩芯钻头;金刚石取心钻头
diamond core drill 金刚石岩芯钻机;金刚石取芯钻机;金刚石钻岩机
diamond core drilling 金刚石岩芯钻进
diamond core head 金刚石岩芯钻头
diamond coring 金刚石取芯
diamond coring bit 金刚石钻头
diamond count (钻头上的)金刚石粒数;(钻头上的)金刚石颗数
diamond counter 金刚石计数器
diamond crossing 菱形(立体)交叉;菱形道叉;棱形交叉
diamond crossing frog 菱形交叉辙叉
diamond crossing on straight track 直线轨道间菱形交叉
diamond crossing with movable points 可动心轨菱形交叉
diamond crossing with one curved track 曲直线轨道间菱形交叉
diamond crossing with single slip ramp 单式匝道菱形交叉

diamond crown 金刚石钻顶;金刚石凸板
diamond crown bit 金刚石钻头
diamond crystal counter 金刚石晶体计数器
diamond cubic 菱形立方体
diamond cubic structure 金刚石立方结构
diamond custting machine 金刚石切割机
diamond cut lug fusion-cast refractory 斜注切割电熔耐火材料
diamond cutter 金刚石切割刀;金刚石刀具;金刚石车刀;钻石玻璃刀;玻璃(切割)刀
diamond cutter held upright 直立式金刚石刀架
diamond cutting 金刚石切割
diamond cutting disk 金刚石切割盘
diamond cutting edges successively appear 金刚石切削刃陆续露出
diamond cutting saws 金刚石锯片
diamond decking 防滑甲板材
diamond deposit 金刚石矿床
diamond die 金刚钻拉丝模;金刚石(拉)模
diamond die polishing machine 金刚石拉丝模抛光机
diamond disc 金刚石小圆锯片
diamond disk 金刚石小圆锯片
diamond drag bit 金刚石刮刀钻头
diamond drawing plate 金刚石拉模
diamond drawn wire 金刚石(模)拉丝
diamond dresser 金刚石整形器;金刚石修整工具;金刚石修整笔
diamond dressing 金刚石整修
diamond dressing roller 金刚石修整滚轮
diamond drill 金刚石钻岩机;金刚石钻探机;金刚石钻(孔)机;地质钻机
diamond drill bit 金刚石钻头
diamond drill bit matrix 金刚石钻头胎体
diamond drill boring 金刚石钻进
diamond drill core 金刚石钻头钻取的岩芯
diamond drill cover 金刚石钻进钻孔
diamond drill crew 金刚石钻机机组人员
diamond drilled well 金刚石凿井
diamond driller 金刚石钻进钻工
diamond drill hole 金刚石钻孔;菱形钻孔洞
diamond drill hole probe 金刚石钻孔探测管
diamond drilling 金刚石钻钻探;金刚石钻探;金刚石钻进;钻石凿井
diamond drilling machine 金刚石钻机
diamond drilling outfit 金刚石钻进设备
diamond drilling rig 钻石凿井机
diamond drilling system 金刚石钻进方法
diamond drill pipe 金刚石钻进钻杆;金刚石钻进管材
diamond drill rig 钻石凿井机;金刚石钻(探)机;金刚石凿井机
diamond drill rod 金刚石钻进钻杆
diamond dust 金刚石粉(末);金刚砂;金刚粉;冰晶粉;冰晶尘
diamond expanded metal 菱形网眼钢板
diamond exposure 金刚石出another
diamond fabric 斜方块织物;菱花织物;斜纹织物
diamond field 金刚石矿区;金刚石产地
diamond film 金刚石膜
diamond finishing bur 钻石精修钻
diamond for carbide and steel 磨碳化物和钢用金刚石品种
diamond form 金刚石成型修整器
diamond fragments 劣质碎粒金刚石
diamond frame saw blades 金刚石框架锯条
diamond fret 方块形花饰的线脚;菱形回纹饰;斜方形浮雕花饰;斜方回纹饰
diamond frieze 菱形雕带
diamond girder 菱腹大梁
diamond glass cutter 金刚石玻璃刀
diamond grade 金刚石品级;金刚石类别;金刚石级别;金刚石等级
diamond grains 金刚砂;金刚石细晶粒
diamond grains for metal bond 金属结合剂用金刚石
diamond grains for metal bond dressing tool 修整工具用金刚石(颗粒)
diamond grains for metal bond saw 锯用金刚石
Diamond gray 灰色花岗岩(一种产于美国明尼苏达州的花岗岩)
diamond green 金刚绿
diamond green lake 孔雀石绿色淀颜料;金刚绿色淀颜料

diamond grid sampling 菱形网采样
diamond grinding plate 金刚石磨盘
diamond grinding tools for optic(al) industry 金刚石光学磨具
diamond grinding wheel 金刚石砂轮
diamond grinding wheel dresser 金刚石砂轮整形器
diamond grindstone 金刚石砂轮;金刚石磨石
diamond grit 金刚石砂粒;金刚石磨料粒;金刚石粒度;金刚砂
diamond hammer 菱形锤;金刚石冲头
diamond hand tool 金刚石手工具
diamond hardness 金刚钻石角锥硬度
diamond head 金刚石钻头
diamond-head buttress(ed) dam 方头支墩坝;大头(撑墙)坝
diamond heatsink 菱形散热片
diamond held trailing 移动式金刚石刀架
diamond held with firm grid 带固定夹的金刚石刀架
diamond holder 金刚石夹具
diamond horseshoe 二楼楼厅
diamondierous 含金刚石的
diamond-impregnated bit 孕镶金刚石钻头
diamond-impregnated blade 金刚砂刀片
diamond-impregnated composition 镶嵌金刚石的制品
diamond-impregnated tool 金刚石掺结磨具
diamond-impregnated tungsten carbide blade 孕镶金刚石的碳化钨刮刀
diamond-impregnated wheel 金刚石磨轮
diamond indenter 金刚石压痕计
diamonding (木材不均匀干缩引起的)菱形变形
diamond in looped end 轮菱
diamond insert 菱形插头
diamond inserts stabilizer 金刚石稳定器
diamond-inset tool 金刚石钻具
diamond instrument 钻石器械
diamond interchange 菱形道路交汇处;菱形立体交叉;菱形(互通式)立交
diamondite 碳化钨(硬质)合金;赛金刚石合金
diamond joint 菱形砌缝
diamond jumper 十字头凿子
diamond knife 金刚石刃;金刚石刀
diamond knot 握索结
diamond-knurled 交叉滚花
diamond knurls 金刚石滚花刀
diamond lap 金刚石研磨器
diamond lapping machine 金刚石研磨机
diamond lath 菱形板条;菱形带条
diamond lathe 金刚石车床
diamond lattice 金刚石晶格;金刚石点阵
diamond-like carbon 类金刚石碳
diamond liner 菱形衬板
diamond loss 金钢石消耗;金刚石耗量
diamond loss per bit 每个金刚石(损)耗量
diamond loss per foot drilled 每英尺钻进的金刚石损耗量
diamond loss per meter 每米金钢石消耗
diamond lustre 金刚石光泽;金刚石光彩
diamond marking 金刚石压花
diamond matching 菱形拼板;菱形花饰镶拼;菱形嵌缝板条
diamond mesh 菱形孔眼;菱形筛眼;菱形筛孔;菱形钢板网;菱形(孔)金属网
diamond mesh flat expanded metal 金属条板拉展菱形网
diamond mesh grating 菱形筛眼格栅
diamond-mesh lath 菱形钢板网;菱形丝网板条
diamond-mesh metal lath 菱形孔金属拉网
diamond-mesh reflector 菱形网状反射器;菱形栅反射器
diamond mesh wire lath 菱形筛眼金属丝网
diamond micro-powder 金刚石微粉
diamond milling tool 金刚石铣盘
diamond mining 菱形钻采矿
diamond mortar 冲击钵
diamond motif 菱形主题装饰;菱形装饰主题
diamond mo(u)lding 菱纹线脚;钉头线脚
diamond mounted point 金刚石磨头
diamond needle 金刚石真空吸笔
diamond nib 金刚(钻)钻头
diamond of error 误差菱形
diamond oil drill bit 金刚石油钻头

diamond opposed anvil 金刚石压砧
diamond orientation 金刚石定向
diamond ornament 菱形饰
diamond pan 淘洗金刚石用淘盘
diamond panel 菱形镶板
diamond particle bit 细粒金刚石钻头
diamond pass 菱形孔型
diamond paste 金刚石(研)磨膏;金刚石磨浆
diamond pattern 菱形图案;(钻头底唇的)金刚石排列形式;金刚石模式;菱形花形
diamond-patterned 有菱形图案的
diamond-patterned glass 有菱形图案的玻璃
diamond pattern knurling 金刚石滚花
diamond pattern liner plate 菱形凸纹衬板
diamond pavement 菱形铺砌路面;菱形铺砌地面
diamond paving 菱形铺砌
diamond pendulum 菱形飞摆
diamond penetrator 金刚石压头
diamond penetrator hardness 维氏金刚石硬度
diamond pick-up tube 金刚石真空吸笔
diamond pipe 金刚石真空吸笔
diamond pit 金刚石刀头
diamond plate 菱形板;金刚石板;花钢板;钻石砣盘;防滑钢板
diamond plug bit 金刚石不取芯钻头
diamond point 金刚石吸笔;金刚石头;金刚石尖点;金刚石笔/钻石刻刀;钻石砂石针;斜方形轨道(交)叉点
diamond-point chisel 菱形錾;金刚石尖头凿
diamond-pointed bit 菱形钻头;金刚石尖头钻头
diamond-pointed rustication 菱形粗琢
diamond-point engraving 金刚石磨头;金刚石笔刻磨;用金刚石雕刻的玻璃表面图饰;金刚石雕刻(图案)
diamond-point tile 菱形尖瓦;金刚石钻头瓦
diamond-point tool 菱形尖端切割工具
diamond polishing 金刚砂抛光
diamond polishing powder 金刚石抛光粉
diamond powder 金刚石粉(末)
diamond pressure 每颗金刚石上的压力
diamond pyramid hardness 金刚钻角锥硬度;金刚石锥体硬度;金刚石角锥硬度;维氏金刚石硬度
diamond-pyramid hardness test 金刚石锥体硬度试验
diamond quality 金刚石质量
diamond reamer 金刚石钻头扩孔器;金刚石铰刀
diamond recovery 金刚石回收
diamond ring effect 金刚石环效应
diamond rivet 菱形铆钉
diamond riveting 菱形铆接
diamond rock drill crown 硬岩金刚石钻头
diamonds 菱形孔型
diamond salvage 金刚石回收
diamond saw 钻石(齿圆)锯;金刚石切割轮;金刚石锯;金刚石砂锯;圆盘式金刚石锯
diamond saw blade 金刚石圆锯片
diamond-sawn (圆盘式的)金刚石锯锯口
diamond saw (splitter) 金刚石圆锯
diamond scale 金刚石天平
diamond scrap 金刚石次料
diamond screen 金刚石(分级)筛
diamond screen set 金刚石(分级)筛组
diamond scribing 金刚石划片
diamond separator 金刚石分选机
diamond set 镶有金刚石的
diamond-set bit 金刚石钻头
diamond-set casing shoe 镶有金刚石的套管靴
diamond-set core bit 金刚石岩芯钻头
diamond-set inserts 金刚石镶嵌块
diamond set reaming shell 金刚石扩孔器
diamond-set ring 金刚石镶的(扩孔)环
diamond setter 金刚石镶嵌工;钻石安装工
diamond setting pattern 金刚石镶嵌模式
diamond-set tool 金刚石镶的工具
diamond shape 菱形形状;斜方形
diamond-shaped 菱形的
diamond-shaped core 菱形模芯
diamond-shaped damper 菱形叶片调节阀
diamond-shaped ground plan 底层菱形平面
diamond-shaped knurling 菱形滚花
diamond-shaped motif 菱形主题(装饰)
diamond-shaped pane of glass 菱形玻璃片
diamond-shaped pattern of well hole spacing 菱形井网;钻孔的菱形布置

diamond-shaped plan 菱形平面
diamond-shaped pylon 钻石形塔架
diamond-shaped slab 菱形(混凝土)板
diamond-shaped slate 菱形石板瓦
diamond-shaped tower 菱形塔
diamond-shaped window pane 菱形窗玻璃
diamond shoe 金刚石管靴钻头
diamond size 钻石粒度
diamond slate 菱形瓦
diamond slicing disk 金刚石切割片
diamond sliding 金刚石刻划
diamond slitting wheel 金刚石切割砂轮;金刚石磨头切割轮;镶金刚石锯盘
diamond spar 刚石;刚玉
diamonds per carat 每克拉金刚石粒数
diamond sphero-conic(al) penetrator 金刚石球锥贯入仪(用于路面硬度试验)
diamond spinning tool 尖头旋压工具
diamond-square passes 菱方系统孔型
diamond stick 金刚石油石;金刚石磨条
diamond stop 菱形光栏
diamond strength tester 金刚石强度计
diamond structure 金刚石结构
diamond styli 金刚石触笔
diamond stylus 金刚石触针
diamond substitute 金刚石代用品
diamond target 菱形靶标
diamond thin film 金刚石薄膜
diamond tool 金刚石工具;金刚石(车)刀;钻石工具;钻石针头
diamond tool holder 金刚石刀夹
diamond tool lathe 金刚石刀车床
diamond truss 菱形桁架
diamond turning 金刚石车削
diamond type 菱形式
diamond-type coil 菱形线圈
diamond-type diesel engine 菱形柴油机
diamond vault 网格拱顶;带有凹槽的肋穹隆
diamond washer 菱形垫圈;金刚石洗选机;瓦楞垫圈;波形瓦垫圈
diamond wear 金刚石磨损
diamond weave coil 菱形编织线圈
diamond wheel 金刚石砂轮;钻石砂轮
diamond wheel dresser 金刚石磨轮整形机
diamond winding 菱形绕组;模(型)绕组
diamond window 金刚石窗口
diamond wire drawing die 金刚石拉丝模
diamond wire saw 金刚石绳锯
diamond wooden bar 菱形木花格
diamond work 菱形砌筑;菱形砌块(圬工)
diamond worn flat 金刚石磨平
diamond woven fabric 菱形织物;斜纹布
diamond yield rate 金刚石转化率
diamonite water treatment agent 硅藻土水处理剂
diamorphine 海洛因
diamox 醋唑磺胺
diamy 表镶金刚石钻头(瑞典)
diamyl 二戊基
diamylamine 二戊胺
diamyl benzene 二戊苯
diamyl disulfide 二戊基二硫
diamylene 癸烯
diamyl ester 二戊酯
diamyl ether 二戊醚
diamyl phenol 二戊基酚
diamyl sulfide 二戊基硫醚
dian 双酚A
dianegative 透明底片
Diane pigment 戴恩颜料
dian epoxy resin 双酚A环氧树脂
dianilazurin(e) 双苯胺青染料
dianil bordeaux 双苯胺枣红染料
dianiline 双苯胺
dianion 二价阳离子
dianisidine orange 联茴香胺橙
dianthus chinensin 石竹
dianthus garden 石竹
Diaoyudao tectonic knot 钓鱼岛构造结【地】
diapause 休止;滞育
diaped tendon 偏心钢筋束
diaper 菱形组饰;菱形(花纹)织物;菱纹
diaper bond 席纹砌合;棋盘式砌合;菱形砌合;交错砌合
diapered type brick paver 菱形铺地砖

diaper ornament 平面图案装饰
diaper wall 围护墙
diaper work 棋盘式砌合;菱形图案砌合;菱形组饰;菱形花纹(装饰)
diaper work of bricklaying 菱形砖砌图案;菱形图案砌砖工作;砖砌的菱形图案
diapgragm of the weight indicator 指重表传压膜;指重表传感片
diaphaneity 透明性;透明度;透彻度
diaphanometer 透明度仪;透明(度)计
diaphanometry 透明度测定法
diaphanoscope 彻照器;透照镜
diaphanoscopy 透照术
diaphanotheca 透明层
diaphanous 透明的
diaphanous ink 透明油墨
diaphax 敏感X线片
diaphone 雾中信号笛;雾笛
diaphonic paper 薄彩色纸
diaphonics 不调和音节
diaphorite 辉锑铅银矿;异辉锑铅银矿
diaphototropic 横向光性的
diaphragm 横隔(板);截水墙;膜(片);横隔梁;振动膜;遮光板;光圈;隔板;隔膜;隔光器;隔仓板;电池隔板;槽壁;薄膜
diaphragm accelerometer 膜式加速(度)计;膜盒加速度计
diaphragm-actuated 隔膜改动的
diaphragm-actuated regulator 薄膜调节器
diaphragm adjusting lever 光阑调节杆
diaphragm and piston pump 膜式和活塞式水泵
diaphragm and piston-type pulsator 薄膜和活塞式脉动器
diaphragm aperture 光阑孔径
diaphragm aperture measurement 光阑孔径测量
diaphragm asbestos cloth 隔膜石棉布
diaphragmatic prosthesis 人工横膈
diaphragmatic waveguide 膜片加载波导;盘荷波导
diaphragm automatic valve 自动放件阀
diaphragm baffle 挡气膜;薄膜挡圈
diaphragm ball valve 隔膜球阀
diaphragm barometer 膜片式气压计;膜盒压力表
diaphragm blast ga(u)ge 膜片式风压表
diaphragm box level controller 膜盒式料位控制器;气压型鼓膜液面控制器
diaphragm burner 辐射板燃烧器
diaphragm capsule 膜盒
diaphragm capsule type level ga(u)ge 膜盒式液位计
diaphragm case 送话器盒
diaphragm cell 有横墙的格仓;鼓形格体;隔墙式格体;隔膜电池;扇形格体
diaphragm cellular cofferdam 格形隔板围堰(由两排连续弧形板桩构成)
diaphragm chamber 薄膜盒
diaphragm chamber type rate-of-rise detector 薄膜盒温升率探测器
diaphragm chuck 膜板式夹头;薄膜式夹具;薄板式夹头
diaphragm compressor 膜(片)式压缩机;隔膜式压缩机;隔膜式压气机;隔板式增压器
diaphragm contractor's pump 膜式施工用泵
diaphragm control knob 光阑控制组
diaphragm current 隔膜电流
diaphragm dam 有隔墙的坝;有防渗墙的坝;隔墙坝
diaphragm deluge velve 隔膜雨淋阀
diaphragm draught ga(u)ge 膜式通风计
diaphragm-ejector amplifier 膜片喷射放大器
diaphragm electrolytic cell 隔膜电解槽
diaphragm expansion joint 膜片胀缩接合
diaphragm feeder 隔膜给料器
diaphragm field 光阑视场
diaphragm filter press 隔膜压滤机
diaphragm float 横隔浮标;(流速仪率定车上的)隔板浮子
diaphragm for excess pressure head 超压膜
diaphragm forming 隔膜法成型
diaphragm for penetration of bus 母线穿墙隔板
diaphragm fuel pump 膜片(式)燃料泵
diaphragm gas meter 膜盒式气压表;隔膜煤气计
diaphragm ga(u)ge 膜(片)式压力计;膜盒真空计;膜(盒)式压力计;薄膜压力计;薄膜式压力盒
diaphragm governor 薄膜调节器
diaphragm grate 盘形浇口

diaphragm horn 膜片喇叭;电动高音雾号
diaphragm in bulbous bow 球鼻首隔板
diaphragm index 光阑指数
diaphragm jig 隔膜跳汰机
diaphragm knapsack sprayer 背负式膜片泵喷雾机
diaphragm level ga(u)ge 膜片式液位计
diaphragm logic element 薄膜逻辑元件
diaphragm manometer 膜片(式)压力表;隔膜压力计
diaphragm meter 干式煤气表;膜片(式)计量器;隔膜流量表;薄膜式煤气表
diaphragm metering pump 膜式计量泵
diaphragm motor 膜片阀控制电动机;光阑驱动电动机
diaphragm mo(u)lding 膜压造型
diaphragm mo(u)lding machine 膜压造型机
diaphragm opening 光阑孔径
diaphragm-operated 薄膜传动
diaphragm-operated bin material-level indicator 薄膜操作式料仓料位指示器
diaphragm-operating mechanism 薄膜致动机构
diaphragm operation 隔膜操作
diaphragm orifice 光阑孔;隔膜孔口;板孔
diaphragm packing ring 迷宫式隔板汽封环
diaphragm paper 隔膜纸
diaphragm plate 孔板;横隔板;隔片;隔(膜)板
diaphragm plate fish pass 隔板式鱼道
diaphragm pressure 薄膜压力
diaphragm pressure ga(u)ge 膜片式压力计;薄膜压力计
diaphragm pressure metering 薄膜压力调节
diaphragm process 隔膜法
diaphragm pulsator 薄膜式脉动器
diaphragm pump 隔膜式抽水机;膜板式泵;膜片;隔膜(式)泵;膜式水泵
diaphragm quaywall 地下连续墙码头岸壁
diaphragm range 光圈变动范围
diaphragm regulating valve 膜片调节阀
diaphragm regulator 薄膜调节器;膜片调节器
diaphragm relay 膜片式继电器
diaphragm retaining wall 地下连续墙;帷幕截水墙;帷幕挡土墙
diaphragm retrofit 隔膜法改型
diaphragm scale 光圈分划
diaphragm screen 平筛
diaphragm seal 膜片(式)密封;密封片;密封膜;薄膜密封
diaphragm sealed displacement transmitter 密封隔片容积式液位传感器
diaphragm seal flowmeter 膜片式密封流量计
diaphragm seal pressure ga(u)ge 隔膜密封压力计
diaphragm setting 光阑定位
diaphragm shell mo(u)lding machine 膜压壳型机
diaphragm shutter 光阑快门
diaphragm-spot 栏影圈
diaphragm sprayer 膜片泵喷雾机
diaphragm spring 膜片弹簧;隔膜簧
diaphragm spring clutch 膜片弹簧式离合器
diaphragm strain ga(u)ge 膜(片)式应变计
diaphragm stress 膜应力
diaphragm switch 薄片开关
diaphragm tank 隔膜水箱
diaphragm transducer 膜式转换器;膜式传感器
diaphragm-transmitted sound 膜片振动音
diaphragm type 膜(片)式;隔膜式
diaphragm type accelerometer 膜片式加速表;膜盒式加速表
diaphragm type accumulator 膜片式储能器
diaphragm type carburettor 薄膜式汽化器
diaphragm type cellular cofferdam 隔墙式格型围堰
diaphragm type compressor 膜片式压缩机;薄膜式压缩机
diaphragm type differential pressure ga(u)ge 膜片式差压计
diaphragm type filter press 隔膜式压滤机
diaphragm type fuel pump 膜片式(燃)油泵
diaphragm type gas meter 膜片式气量计;膜片式煤气表
diaphragm type jig 隔膜式跳汰机
diaphragm type level meter 膜片式液位计
diaphragm type metering pump 膜片式计量泵
diaphragm type micromanometer 膜片式微压计

diaphragm type paint sprayer 膜片式喷漆器
diaphragm type positive displacement metering pump 隔膜型正排量计量泵
diaphragm type power brake 膜片式动力制动器
diaphragm type pressure balance detector 膜片式压力平衡检测器
diaphragm type pressure ga(u)ge 膜片式压力计;膜盒式压力计;膜盒式压力表;薄膜式压力盒;薄膜式压力计
diaphragm type pressure sensor 膜片式压力传感器
diaphragm type sludge pump 膜片式泥浆泵;隔膜式污泥泵
diaphragm type washbox 隔膜式跳汰机
diaphragm type wind pressure ga(u)ge 膜片式风压表
diaphragm vacuum ga(u)ge 膜片式真空计
diaphragm valve 膜式阀;膜片阀;隔膜阀;薄膜活门
diaphragm wall 隔水墙;隔墙;地下连续墙;堤坝防渗墙
diaphragm wall counterfort 地下连续墙扶壁
diaphragm walling 地下连续墙
diaphragm wall with multibracing 多支撑地下连续墙
diaphragm with lifting scoops 带扬料板的隔仓板
diaphragm works 地下连续墙工程
diaphragram lens 光阑透镜
diaphthoresis 退化变质作用
diaphthorite 退化变质岩
diapir 挤入褶皱
diapirc structure 底辟构造
diapire 底辟【地】
diapir fold 刺穿褶皱
diapir granite 底辟花岗岩
diapiric dome 底辟穹丘
diapiric fold 挤入褶皱;底辟褶皱
diapiric intrusion 底辟侵入
diapiric salt 底辟盐丘
diapiric uprise 底辟上升
diapir pool 刺穿油藏
diapir trap 刺穿圈闭
diapositive 透明像片正片;反底片;透明正片;幻灯片
diapositive film 透明正片
diapositive lever 夹皮弹簧杆
diapositive printer 透明正片晒像机
diara 河心沙洲
diarch 二原形
diarrhea 腹泻
diarsenide 二砷化物
diarsenite 二(亚)砷酸盐
diaryl 二芳基
diarylamine 二芳基胺
diaryl arsenious acid 二芳基亚胂酸
diaryl arsine oxyhalide 二芳基胂化氧卤
diarylhydrazine rearrangement 二芳基肼重排作用
diaryl hydroxy arsine 二芳基羟胂
diarylide yellow 乙酰乙酰芳胺黄;双乙酰芳胺黄
diaryl phosphate 磷酸二芳酯
diary method 日记法
diary of the reporting household 报告户日记账
Dias 二叠系【地】
diaschistic 二分的
diaschistic dyke rock 二分脉岩;分浆脉岩
diaschistic rock 二分岩
diaschistite 分浆岩;二芬岩;二分岩
diascope 彻影器;反射幻灯机;透射映画器
diasonograph 超声波诊断仪
diaspore 硬水铝石;硬水铝矿;一水硬铝石;传播体;水铝石;水矾石
diaspore clay 硬铝土黏土;水矾土
diaspore-kaolinite ore 硬水铝石高岭石岩矿芒
diaspore refractory product 水铝石耐火材料制品
diasporite 硬水铝石;水矾土
diasporometer 旋转光楔对【测】;偏离计
diastatite 角闪石
diastem 小间断;沉积暂停期
diasterane 重排甾烷
diasterene 重排甾烯
diastereoisomeric 非对映异构的
diastereomer 非对映体
diastereomeric 非对映的
diastereomeric form 非对映形
diastereotopic 非对映异构的
diastimeter 测距仪;距离测定计

diastole 舒张期
diastrophic activity 地壳运动
diastrophic coast 错动海岸
diastrophic cycle 地壳运动旋回
diastrophic eustatism 构造性海面升降运动;地壳变动(海面升降)
diastrophic lake 陷落湖
diastrophism 地壳运动;地壳变动
diastrophometer 测畸形仪
diastyle 三倍柱径双列柱门廊;三倍柱底径空距柱廊;柱间间隔为三米的;三柱径式
diatectite 高度深熔岩
diatexis 高度深熔作用
diatexis way 高级深熔作用方式
diatexite 高度深熔岩
diathermal 透热的;热导的
diathermancy 传热性;透热性
diathermaneity 透热性
diathermanous 透热辐射的
diathermanous body 导热体;透热体
diathermic coagulation 透热电凝法
diathermic heating 高频加热
diathermic mirror 冷光镜
diathermic wall 绝热壁
diathermocoagulation 透热电凝法
diathermometer 导热计
diathermous envelope 透热包壳
diathermy 透热法
diathermy apparatus 透热器
diathermy machine 透热机
diathyrum (古希腊住宅的)门道
dia-titanit 钛钨硬质合金
diatom 硅藻
Diatomacae 硅藻纲
diatomaceous 硅藻(土)的
diatomaceous brick 硅藻土砖
diatomaceous calcite 硅藻方解石
diatomaceous chert 硅藻燧石
diatomaceous clay 硅藻黏土
diatomaceous concrete 硅藻土混凝土
diatomaceous earth 硅藻土
diatomaceous earth filter 硅藻土滤层;硅藻土过滤器;硅藻土(过)滤池
diatomaceous ooze 硅藻软泥
diatomaceous shale 硅藻页岩
diatomaceous soil 硅藻土
diatomaceous support 硅藻土载体
Diatoma hiemale 冬生等片藻
diatom analysis 硅藻分析
diatom biological index 硅藻生物指数
diatom bloom 硅藻水花
diatom brick 硅藻土砖
diatom-earth 硅藻土;云母
diatomic 二价的
diatomic acid 二价酸
diatomic alcohol 二元醇
diatomic concrete 硅藻土混凝土
diatomic filter 硅藻土滤层;硅藻土过滤器;硅藻土(过)滤池
diatomic ooze 硅藻软泥
diatomics-in-molecule method 分子中双原子法
diatom insulator 硅藻土绝缘器
diatomite 硅藻岩;硅藻土
diatomite asbestos plaster 硅藻土石棉灰浆
diatomite brick 硅藻土砖
diatomite deposit 硅藻土矿床
diatomite filter 硅藻土滤层;硅藻土过滤器;硅藻土(过)滤池
diatomite filter-aid 硅藻土助滤池
diatomite filtration 硅藻土过滤
diatomite in powder 硅藻土粉
diatomite insulating block 硅藻土保温块
diatomite insulating brick 硅藻土保温砖
diatomite insulating layer 硅藻土保温层
diatomite insulator 硅藻土保温材料
diatomite support 硅藻土支护;硅藻土载体
diatoms 硅藻类
diatom-sapropel 硅藻腐泥
Diatom W 白色硅藻土载体
diatoni 古希腊建筑的贯穿砌合石、突隅石;突隅石
diatonous 突出墙面的砖或石块
dia-tool 镶有金刚石的工具
diatreme 火山角砾岩筒;火山道;火山爆发口;火山爆发筒

diatrine 浸渍电缆纸化合物
diatropism 离向性;横向性
Diatto surface-contact system 迪阿托面接触系统
diauxic growth 二峰生长
diauxic growth curve 二峰生长曲线;二次生长曲线
diaxial crystal 二轴晶
di-axis door 偏心门
diazamine brilliant black 重氮胺亮黑
diazanyl 二氮烷基
diazene 二氮烯
diazenyl 二氮烯基
diazepam tablet 安定片
diazete 二氮杂环丁二烯
diazetidine 二氮杂环丁烷
diazetine 二氮杂环丁烯
diazidoethane 二叠氮基乙烷
diazine 二氮杂苯
diazo 重氮(基)
diazoacetate 重氮基乙酸盐;重氮基醋酸盐;重氮基醋酸酯
diazoaminobenzene 重氮胺基苯
diazoamino compounds 重氮氨基化合物
diazobenzene 重氮基苯
diazo benzene acid 苯重氮酸
diazobenzene hydroxide 氢氧化重氮苯
diazobenzene sulfonic acid 重氮苯磺酸
diazo brilliant green 重氮亮绿
diazo colo(u)r 重氮染料
diazo compositor 重氮合成仪
diazo compound 重氮化合物
diazo copier 重氮复印机
diazo copying 重氮盐晒图
diazo copying process 重氮盐晒图工艺
diazocyanide 重氮氰化物
diazodinitrophenol 重氮二硝基苯酚
diazo document copying machine 重氮资料复印机
diazo dye 显色染料;重氮染料
diazo film 重氮胶卷;重氮感光片
diazogen blue 重氮精蓝
diazoic acid 重氮酸
diazoimide 叠氮酸
diazo indigo blue 重氮靛蓝
diazo-ketones 重氮酮
diazol 重氮盐
diazol bordeaux 重氮盐枣红
diazol colo(u)r 显色盐染料
diazole 二唑
diazol fast orange 重氮盐坚牢橙
diazol light orange 重氮盐浅橙
diazol light yellow 重氮盐浅黄
diazols 重氮盐类
diazoma 环绕古罗马圆形露天剧场的休息平台
diazomethane 重氮甲烷
diazomethylbenzenesulfonamide 重氮甲基苯磺酰胺
diazo microfilm 重氮缩微片
diazonitrophenol 重氮硝基酚
diazonium 重氮(基)
diazonium compounds 重氮化合物
diazonium salt 重氮盐
diazooxonor leucine 重氮氧代己氨酸
diazo paper 重氮晒图纸;重氮复印纸
diazo paper raw base 重氮晒图原纸
diazoparaffins 重氮烷
diazo positives 重氮正片
diazo-print 重氮晒图;重氮复印
diazo printer 重氮复印机
diazo printing apparatus 重氮晒印装置
diazo process 熏晒图重氮方法;重氮照相法;重氮(基)
diazo pseudo-colo(u)r composition 假彩色合成重氮法
diazo-reaction 重氮化反应
diazo resin 重氮树脂
diazosalicylic acid 重氮水杨酸
diazosalt 重氮盐
diazo scribing coating film 重氮刻图片
diazosoplit 重氮分解物
diazo stabilizer 重氮稳定剂
diazo stripping film 重氮撕膜片
diazosulfanilic 对重氮苯磺酸
diazosulfide 苯并噻二唑
diazotability 重氮化本领
diazotable dye 重氮染料

diazotate 重氮酸盐
diazo-thermography 重氮热敏成相法
diazotization 重氮化
diazotization process 重氮化处理
diazotization titration method 重氮化滴定法
diazotizing colo(u)rs 重氮染料
diazotizing dyes 重氮染料
diazotype 重氮盐晒图；重氮盐成相法
diazotype process 重氮盐印相法
dibasic 二取代的；二碱的
dibasic acid 二元酸
dibasic acid ester 二元酸酯
dibasic alcohol 二元醇
dibasic aminoacid 氨基二酸
dibasic magnesium phosphate 磷酸氢镁
dibber 戳孔器
dibble 点播器
dibble-dabble 试算
dibbler 栽植挖穴机；挖穴手铲
dibbling 戳孔器造林
dibbling machine 作穴机
dibenzalacetone 二亚苄基丙酮
dibenzanthracene 二苯并蒽
dibenzanthron green 双苯绕蒽酮绿
dibenz-dibutyl anthraquinol 二苯二丁蒽夹二酚
dibenzenazoresorcin 二苯偶氮间苯二酚
dibenzhydroxamic acid 二苯苯氧肟酸
dibenzimide 二苯甲酰亚胺
dibenzofuran 二苯并呋喃
dibenzothiazine 吩噻嗪
dibenzothiazyl-dimethyl-thiourea 二苯并噻唑基二甲基硫脲
dibenzothiophene 二苯并噻吩
dibenzo thioxine 吩噻噁
dibenzoyl 苯偶酰
dibenzoyl disulfide 二苯甲酰二硫
dibenzoyl-ethylene 二苯甲酰乙烯
dibenzoyl hydrazine 二苯甲酰肼
dibenzoyl ketone 二苯甲酰基甲酮
dibenzoyl methane 二苯甲酰甲烷
dibenzoyl peroxide 过氧化二苯甲酰
dibenzoyl thiourea 二苯甲酰硫脲
dibenzphenanthrene 二苯并菲
dibenzyl amine 二苄胺
dibenzyl aniline 二苄苯胺
dibenzylchlorethamine 苄氯乙胺
dibenzyl disulfide 二苄二硫
dibenzyl ether 二苄醚；苄基醚
dibenzyl fumarate 富马酸二苄酯
dibenzylidene 二亚苄基
dibenzyline 苯氧苄胺
dibenzyl ketone 二苄基甲酮
dibenzyl maleate 马来酸二苄酯
dibenzyl oxalate 草酸二苯酯
dibenzyl prtoxydicarbonate 过氧化二碳酸二苄酯
dibenzyl sulfide 二苄硫醚；二苄基硫
dibhole 井底排水孔
dibit 二位组；二位二进制数；双位
dibit encoding 两位编码
diborane 二硼烷
diboride 二硼化物
diboson 双玻色子
dibrom 二溴磷
dibromide 二溴化物
dibrominated 二溴化的
dibromizated 二溴化的
dibromoacetic acid 二溴乙酸
dibromoacetone 二溴丙酮
dibromo-acetyl bromide 二溴乙酰溴
dibromoacetylene 二溴化乙炔
dibromobenzene 二溴苯
dibromobenzoic acid 二溴苯甲酸
dibromobutane 二溴丁烷
dibromo-butyric acid 二溴丁酸
dibromodhenyl silane 二苯基二溴(甲)硅烷
dibromodifluoro methane 二溴二氟甲烷
dibromodiphenyl ether 二溴二苯醚
dibromo ether 二溴醚
dibromoethyl alcohol 二溴乙醇
dibromoketone 二溴丁酮
dibromomaleic acid 二溴马来酸
dibromomalonic acid 二溴丙二酸
dibromomethane 二溴甲烷
dibromomethylethylketone 二溴乙基甲酮

dibromonitromethane 二溴硝基甲烷
dibromophakellin 二溴扇形海绵素
dibromophenyl-propionic acid 二溴苯基丙酸
dibromopropanal 二溴丙醛
dibromopropane 二溴丙烷
dibromopyruvic acid 二溴丙酮酸
dibromo-succinic acid 二溴丁二酸
dibucaine 二丁卡因
dibusadol 地布沙多
dibutene 二丁烯
dibutoline sulfate 地布托林硫酸盐
dibutoxy ethyl sebacate 癸二酸二乙氧基乙酯
dibutyl 二丁基
dibutyl acetamide 二丁基乙酰胺
dibutyl amine 二丁胺；二丁胺
dibutyl butyl phosphonate 丁膦酸二丁酯
dibutyl carbitol 二丁基卡必醇
dibutyl cyanamide 二丁氰胺
dibutyl disulfide 二丁二硫
dibutyl ether 二丁醚
dibutyl formamide 二丁基甲酰胺
dibutyl fumarate 富马酸二丁酯
dibutyl hydroxy toluene 二丁基羟(基)甲苯
dibutyl maleate 马来酸二丁酯
dibutyl malonate 丙二酸二丁酯
dibutyl oxalate 草酯二丁酯；草酸二丁酯
dibutyl phosphate 磷酸二丁酯
dibutyl phthalate 邻苯二甲酸二庚酯；邻苯二甲酸二丁酯
dibutyl phthalate plasticizer 邻苯二甲酸二丁酯增润剂
dibutyl sebacate 癸二酸二丁酯
dibutyl sulfide 二丁硫
dibutyl sulfone 二丁砜
dibutyl sulphide 丁硫醚
dibutyltin bis 双月桂硫醇二丁锡
dibutyltin bis monoalkylmaleate 双一烷基马来酸二丁锡
dibutyltin dilaurate 二丁基二锡
dibutyltin laurate-maleate 月桂酸马来酸二丁锡
dibutyltin maleate 马来酸二丁锡
dibutyltin mercaptan 硫醇二丁锡
dibutyl tinoxide 氧化二丁锡
dibutyltinsulphide 硫化二丁锡
Dical 迪卡尔铜硅合金
dicalcium aluminate hydrate 水化铝酸二钙
dicalcium ferrite 铁酸二钙
dicalcium phosphate 磷酸二钙
dicalcium phosphate dihydrate 磷酸二钙二水合物
dicalcium silicate 重钙酸钙；硅酸二钙
dicalcium silicate hydrate 水化硅酸二钙
dicarbocyanine 二碳花青
dicarbonate 重碳酸盐；碳酸氢盐
dicarbonating zone 重碳酸盐带
dicarboxylcellulose 二羧基纤维素
dicarboxylic acid 二羧酸
dicarboxylic ester 二羧酸酯
dicasterium (古建筑中的)法庭
dice 小方块；玻璃龟裂；碎粒；碎块
dice block 流液洞侧砖
diced line 黑白相同铁路符号
dice-pattern 简单几何花纹
dicer 切粒机；切块机
dichan 气化性防锈剂
dichastasis 自行分裂
dichlone 二氯萘醌
dichloralide 二氯交脂
dichloramine 二氯胺
dichlorated 二氯化的
dichlorethylene 二氯乙烯
dichloride 二氯化物
dichloriphenyl phosphate 磷酸二氯苯酯
dichloro acetaldehyde 二氯乙醛
dichloro acetamide 二氯乙酰胺
dichloroacetic acid 二氯乙酸
dichloroacetone 二氯丙酮
dichloroacetylene 二氯代乙炔
dichloroaniline 二氯苯胺
dichloroanthraquinone 二氯蒽醌
dichlorobenzaldehyde 二氯苯甲醛
dichlorobenzene 二氯苯
dichlorobenzoic acid 二氯苯甲酸
dichlorobenzoyl peroxide 过氧化二氯苯甲酰
dichlorobutane 二氯丁烷

dichlorobutylene 二氯丁烯
dichloro compound 二氯化合物
dichlorodibromomethane 二氯二溴甲烷
dichlorodifluoromethane 二氯二氟甲烷；二氟二氯甲烷
dichloro-diiodomethane 二氯二碘甲烷
dichlorodinitromethane 二氯二硝基甲烷
dichloro-dipropyl ether 二氯二丙醚
dichloroethane 二氯乙烷
dichloroethane poisoning 二氯乙烷中毒
dichloroethanol 二氯乙醇
dichloro ether 二氯代醚
dichlorofluoromethane 二氯氟甲烷
dichlorohydrin 二氯丙醇
dichloroiodomethane 二氯碘甲烷
dichloroisoproterenol 二氯特诺
dichloromethane 二氯甲烷
dichloromethyl ether 双氯甲醚
dichloromethyl phenylsilane 甲基苯基二氯(甲)硅
dichloronaphtalene 二氯萘
dichloronitrobenzene 二氯硝基苯
dichloronitroethane 二氯硝乙烷
dichlorophenamide 二氯苯磺胺
dichlorophenarsine 二氯苯胂
dichlorophenol 二氯苯酚
dichlorophenoxyacetic acid 二氯苯氧基乙酸
dichlorophenyl sulfonic acid 二氯苯肼磺酸
dichloroprene 二氯丁二烯
dichloropropane 二氯丙烷
dichloropropanol 二氯丙醇
dichloropropionanilide 二氯丙酰苯胺
dichloropropionic acid 二氯丙酸
dichloropropylene 二氯丙烯
dichlorosilane 二氯甲硅烷
dichloro-succinic acid 二氯代丁二酸
dichlorosulfonphthalein 二氯磺酞
dichlorotetra-fluoroethane 二氯四氟乙烷
dichlorotetraglycol 二氯四甘醇
dichlorotoluene 二氯甲苯
dichloroxylenol 二氯二甲酚
dichlorphenoxyacetic acid 二氯苯氧基乙酸
Dichograptidae 均分笔石科
dichotomic mode 叉状水系模式
dichotomic variable 二分变量
dichotomizing 对分(检索)；对半检索
dichotomizing search 对分检索；二分法检索
dichotomous choice method 二项选择法
dichotomous classification 两分法分类
dichotomous key 分叉式检索表
dichotomous population 二项总体；二分总体
dichotomous search 两分搜索；两分搜寻；分组探索
dichotomous variable 两分变量
dichotomous venation 叉状脉序
dichotomy 两分法；均分；二歧式；二分(法)
dichotomy method 两分法
dichotomy of variables 变量的均分
dichotomy system 二分系统
dichotomy testing system 二分区测试系统
dichotopodium 重叉生式
dichotriaene 重出三叉体
dichroic 二向色性(的)；二向色(的)；二色性的
dichroic absorption 二向色性吸收
dichroic beam splitter 分色光束分离器；二向分色镜；二色光束分离器；二(向)色分光镜
dichroic cross 十字形分色镜
dichroic crystal beamsplitter 二(向)色晶体分光器
dichroic filter 选色镜；二向色滤光片
dichroic fog 二色(性)灰雾
dichroic glass 变色玻璃；两色玻璃；双色玻璃
dichroic layer 分色层；二向色层
dichroic magnifier 二向色放大镜
dichroic mirror 分色镜；二向色(反射)镜；二色镜
dichroic mirror splitter 分色镜分离器
dichroic polarizer 二向色偏振器
dichroic prism 二向色棱镜
dichroic reflector 分色反射镜；二向色反射镜
dichroic reflector lamp 分光反射器灯
dichroic sheet 二色片
dichroism 分色性；分光特性；二(向)色性
dichroite 堇青石
dichromat 二色觉者
dichromate 重铬酸盐
dichromated gelatin 重铬酸盐明胶；重铬明胶
dichromated gelatin film 重铬酸盐明胶板；重铬

明胶膜;双色胶片
dichromate ion 重铬酸根离子
dichromate oxidation method 重铬酸盐氧化法
dichromate oxidizability 重铬酸盐氧化性
dichromate oxygen consumed 重铬酸盐耗氧量
dichromate oxygen demand 重铬酸盐需氧量
dichromate treatment 重铬酸盐处理(法);重铬酸盐表面处理;重铬酸处理
dichromate value 重铬酸盐值
dichromatic 二色(性)的
dichromatic dye 双色染料
dichromatic vision 二色视觉
dichromatism 二色性
dichromatopsia 二色视
dichromic acid 重铬酸
dichroscope 二(向)色镜
dicing 切割
dicing cutter 小块切割机
dicing saw 划片机
Dickens formula (for runoff) 狄更斯(径流)公式
dicker 小生意
Dicke' radiometer 狄克辐射计
Dicke' receiver 狄克接收机
Dicke's radiometer 狄克辐射计
Dicke' switch 狄克开关
dickey 小艇尾室
dickinsonite 绿磷锰矿;磷碱锰石
dickite 地开石;迪开石
dicky 小艇尾室;汽车后部备用的折叠小椅
diclinic 双斜的
dicoaster 双星
dicode 双码
dicode signal 双码信号
dicord system 双塞绳制
dicotyledon 双子叶植物
Dicrolene 迪克罗纶聚酯纤维(阿根廷)
dicrotia 二波脉现象
dicrotic 二波脉的
dicrotic notch 降中峡;重搏切迹
dicrotic pulse 重搏;二波脉
dicrotic seiche 双节假潮
dicrotic wave 降中波;重脉波;重搏波
dicrotism 重脉;二波脉(现象)
dicroton 二聚丁烯醛
dicroton aldehyde 二聚丁烯醛
dictabelt 录影带
dictaphone 录音(电话)机
dictating machine 记录录音机;指令机
dicthenone 二乙烯酮
dictionary catalog 词典式目录
dictionary of place name 地名辞典
dictionary pattern 基准图型
dictionary storage and retrieval 词典存储和检索
dictograph 窃听录音机
dictyonema bed 网笔石层
dictyonite 网状混合岩
dictyotheton 花格网孔设计;希腊砖石网状砌体
dicumyl peroxide 过氧化二枯基
dicyan 氰
dicyanamide 二氰胺
dicyandiamide 氰基胍;双氰胺
dicyanine 双花青
dicyanobutylene 二氰基丁烯
dicyanogen 氰
dicyclic(al) crinoids 双环海百合
dicyclic solution 双循环解法
dicyclohexyl amine 二环己基胺
dicyclohexyl amine nitrite 亚硝酸二环己胺;二环己胺亚硝酸盐
dicyclohexyl ketone 二环己基甲酮
dicyclohexyl maleate 马来酸二环己酯
dicyclohexyl oxalate 草酸二环己酯
dicyclohexyl peroxydicarbonate 过氧化二碳酸二环己酯
dicyclohexyl phthalate 邻苯二甲酸二环己酯
dicyclopentadiene 二聚环戊二烯;双茂;双环戊二烯
dicyclopentadiene dioxide 二氧化二聚环戊二烯
dicyclopentylmethane 双环戊基甲烷
diddle-net 抄网
dideoxy 双脱氧法
did price index 标价指数
didymium 钕镨
didymium glass 钕镨玻璃
didymium glass standard 钕镨玻璃标准

didymolite 钙蓝石
die 模嘴;模子;模具;硬模;印模;铸模;管心;代型;冲模;冲垫;扳牙;塑模;螺丝扳(牙)
die accessories 压模附件
die adapter 模接头
die adaptor 模具套环
die and mill engraving 钢芯雕刻
die and profile shaping machine 刨模机
die angle 模口角度
die approach 拉拔模入口
die approach angle 模孔入口锥角;模孔变形锥(模孔第二部分)
die arrangement 模具;压模装置
die assembly 模具组合;模具(总成);压模装置;压模装配
die attachment 模片固定;小片连接
die away 消失;渐弱;降低压力
die-away test 衰减试验
dieback 模孔出口锥;顶死;顶梢枯死;树木顶梢枯死;顶枯病
die backer 凹模支撑圈
die backing plate 模具垫板
die barrel 压模衬筒
die base 口型座板
die bearing 模口支承面;模口部分
die bed 模座;底模
die bell 拉模入口锥
die blank 模具坯料
die block 螺丝扳(牙);模座;模块;模具坯料;模度坯料;滑块;滑板;底模;板牙;凹模固定板;塑模承套
die block hardener 冲模硬化剂
die body 模体
die bolt cutter 板牙切丝机
die bonder 芯片焊接机
die bonding 模片键合
die bonding jig 管心焊接夹
die box 拉模盒;板牙切头
die box mo(u)ld 真空造型阴模
die bushing 模具镶套
die button 模具叶状模槽
die carrier 模架;水压机模座
die case 拉模套;铸模盒
die cast 金属模铸造;模铸;压模;压铸(件)
die cast bearing 模铸轴承
die-cast chassis 压铸底盘
die cast dies 压铸模
die caste 模具等级
die cast furniture 模铸空铅
die casting 压铸件;压铸法;模铸(件);模铸法;模铸材料;压铸(法);压模铸造;压模铸件;硬模铸造;压力铸造
die-casting alloy 压铸合金
die-casting element 模铸元件
die-casting furnace 保温炉
die-casting machine 模铸机;压铸机;压力铸制器;钢模压铸机
die-casting machine for motor rotor 电机转子压铸机
die-casting material 压铸产品;模铸金属
die-casting metal 压铸金属;压铸轴承合金
die-casting mo(u)lding 铸压成型
die-cast machine 钢模造型机
die cavity 模穴;模腔;模槽;阴模
die center 模子定心盘;模具定心装置
die chaser 螺丝扳(牙);扳牙梳刀
Dieckman condensation 狄克曼缩合作用
die clamp 模夹
die clamping cylinder 装夹模具用汽缸
die clearance 模(具)间隙;模(板)间隙;冲裁间隙
die clicker 冲压裁剪机;冲裁剪机
die closing mechanism 合型机构
die-closing swager 闭式模旋转锻机
die closing unit 合型机构
die closure 模具闭合量
die coating 模压涂层
diecoefficient 介电系数
die collar 母模上模;打捞环
die contact area 夹头接触面
die coupling 打捞母锥;打捞接头
diectional emissivity 定向发射率
die cushion 模具缓冲装置;模具的气垫;模垫;缓冲器
die cut 冲切;模切

die-cut carton 模切纸盒
die cutter 冲压裁剪机
die cutting 冲压裁剪法;冲切;模切
die-cutting machine 冲压裁剪机
die drawing 模拉
die drawing ring 拉模孔壁磨损圈
die edge 模具镶块
die entrance 拉模入口锥;拉模的入口喇叭
dieer 方粒切粒机
die exit angle 模孔出口锥
die face land 模具面刃口
die-face pelletizing 模面切割造粒
die-filling 压模装料;装模
die fill ratio 模装填比;压模充填比;体积压实比
die finish 精整模;精加工模
die forging 模型锻造;模锻(件);热模锻
die forging air hammer 模锻空气锤
die forging machine 模锻机
die for left hand thread 左螺纹扳牙
die-formed 模压的;冲压的
die-formed parts 模压成型件
die for metric fine thread 公制细(牙)螺纹板牙
die forming 模压成型
die for pipe thread 管螺纹板牙
die for right-hand thread 右螺纹扳牙
die for special purpose 专用模
die for taper thread 锥螺纹板牙
die friction 压模摩擦
die gap 模隙
die half 半型
die hammer 模锻锤;印号机
die handle 板牙架
die head 模头;冲模;扳牙(切)头
die head bush 板牙头夹套
die head chaser 板牙头螺纹梳刀
die-head rivet 冲垫铆钉
die heating zone 模头加热区
die height 模子闭合高度;模转闭合高度
die hob 标准(扳牙)丝锥
die hobbing 模压制模法
die hobbing press 模压制模压力机
die holder 模座;扳牙扳手;凹模固定板
die hole 模孔;模槽
die impression 刻印模;模槽
dieing machine 模压机;高速精密冲床;板料自动压力机
dieing out press 冲模压机;冲床;冲压压力机
die insert 模具银块;模具镶块;模衬;压模嵌入件
die joint 分模面
die land 塑模型槽
die layout 锻模设计图
dieldrin 狄氏剂
dielectric 绝缘材料;介质绝缘材料;介质的;电介质(的)
dielectric(al) absorption 介质吸收;介电吸收
dielectric(al) absorption constant 介质吸收常数;电介质吸收常数
dielectric(al) amplifier 电介质放大器
dielectric(al) anisotropy 介电各向异性
dielectric(al) antenna 电介质天线
dielectric(al) beamsplitter 介质分光镜
dielectric(al) body 绝缘体;介电体;电介质体
dielectric(al) breakdown 绝缘击穿;介质破环;介电击穿
dielectric(al) breakdown test 绝缘击穿试验
dielectric(al) breakdown voltage 电介质击穿电压
dielectric(al) capacitance 介电常数
dielectric(al) capacity 电容;介电容量
dielectric(al) capcitance 电容率
dielectric(al) ceramics 电介质陶瓷
dielectric(al) circuit 介质电路
dielectric(al) coated mirror 镀介质镜
dielectric(al) coating 介质膜
dielectric(al) coefficient 介电系数
dielectric(al) conductance 介质电导
dielectric(al) constant 介电系数;介电常数
dielectric(al) constant of rock 岩石介电常数
dielectric(al) core baker 高频烘芯炉
dielectric(al) core oven 高频烘芯炉
dielectric(al) crystal 介质晶体;介电晶体;电介质晶体
dielectric(al) diode 电介质二极管
dielectric(al) dispersion 介电色散;介电耗散;介

电分散
dielectric(al) dissipation factor 介质损耗因数
dielectric(al) dissipation factor test 介质损耗因数试验
dielectric(al) dissipation fraction 介电损耗率
dielectric(al) dissipator factor 介电损耗因子
dielectric(al) drier 介电干燥机;高频(加热)干燥机;高频干燥炉
dielectric(al) drying 高频烘烤
dielectric(al) ellipsoid 介质椭球
dielectric(al) embossing 介电压花;高频压花
dielectric(al) factor 介电功率因数
dielectric(al) fatigue 介质疲劳
dielectric(al) filling 介电填料
dielectric(al) film 绝缘膜;介质(薄)膜;介电薄膜
dielectric(al) fluid 介电液体
dielectric(al) (flux) density 电通量密度
dielectric(al) force 击穿力
dielectric(al) gas 介质气体
dielectric(al) glass 低介电玻璃
dielectric(al) gradient 介电(常数)梯度
dielectric(al) guide 介质波导管
dielectric(al) heat 介电热
dielectric(al) heat curing 介电热固化
dielectric(al) heating 介电加热;高频加热;电介质加热;电感加热
dielectric(al) induction 介电感应
dielectric(al) isolation 介质隔离
dielectric(al) layer 介质膜;介质层
dielectric(al) lens 介质透镜;电介质透镜
dielectric(al) -lens antenna 介质透镜天线
dielectric(al) level 绝缘水平;介电水平
dielectric(al) load 介电负荷
dielectric(al) -loaded linac 介质加载直线加速器
dielectric(al) log 介电测井
dielectric(al) logger 介电测井仪
dielectric(al) loss 介质损耗;电介质损耗
dielectric(al) loss angle 介质损耗角
dielectric(al) loss (angle) tangent 介质损耗角正切
dielectric(al) loss factor 介电损耗因子
dielectric(al) matching plate 电介质匹配板
dielectric(al) material 介质材料;电介质材料
dielectric(al) medium 介电媒质;电介质
dielectric(al) mirror 介质反射镜
dielectric(al) oil 介电油
dielectric(al) optic(al) waveguide 介质光波导
dielectric(al) permeability 介质传导率
dielectric(al) phase angle 介电相角
dielectric(al) phase difference 介电相差
dielectric(al) phenomenon 介电现象
dielectric(al) planar waveguide 介质平面波导
dielectric(al) plastics 介电塑料
dielectric(al) polarization 介质极化;介电极化
dielectric(al) power factor 介质功率因数
dielectric(al) preheating 介电预热
dielectric(al) process 电介质处理过程
dielectric(al) property 介电(特)性
dielectric(al) radiation 电介质辐射
dielectric(al) radiator 介质天线;介质辐射器
dielectric(al) radom 介电整流罩
dielectric(al) relaxation 介电松弛
dielectric(al) relaxation time 介电弛豫时间
dielectric(al) -rod antenna 电介质杆天线
dielectric(al) rod waveguide 介质棒波导
dielectric(al) saturation 介电饱和
dielectric(al) separation 介质分离;电介质分选
dielectric(al) separation method 介电分离法
dielectric(al) separator 介电分离器
dielectric(al) shielding 介质屏蔽
dielectric(al) -slab 介质板;解相过程
dielectric(al) -slab waveguide 介质板波导
dielectric(al) solution 电介质溶液
dielectric(al) strenght test 绝缘强度试验
dielectric(al) strength 绝缘(介质)强度;介电强度;击穿强度;击穿力;电气强度;电介质强度
dielectric(al) strength tester 耐压测试器
dielectric(al) stress 电介质应力
dielectric(al) substance 电介质;介电质;绝缘物质
dielectric(al) surface 介质表面
dielectric(al) susceptibility 介质磁化率;电介质极化率
dielectric(al) -tape camera 介质带摄像机
dielectric(al) target 介电屏幕
dielectric(al) tensor 介电张量

dielectric(al) test 介质性能试验;介电性能试验
dielectric(al) test set 高频绝缘试验器
dielectric(al) thin film filter 介质薄膜滤波器
dielectric(al) track detector 电介质径迹探测器
dielectric(al) tube 介电管
dielectric(al) tube resonator 介质管谐振器
dielectric(al) type moisture meter 电介质湿度计;电容式湿度计
dielectric(al) varnish 绝缘清漆
dielectric(al) waveguide 介质波导
dielectric(al) waveguide mode 介质波导型
dielectric(al) wedge 电介质劈
dielectric(al) wire 介质导线;电介质线
dielectrolysis 电解渗入法
dielectrometer 介质测试器;电介质测试器
dielectrometry 电测量学
dielectronic recombination 双电子复合
dielectrophoresis 介电电泳
dieless drawing 无模拉拔
dietric resistanse 介质电阻
die level 凹模上面高度
die life 模具寿命
die lifter 起模装置
die line 模痕;模具划痕
die liner 模衬;压模衬里
die lining 压模衬片
diel movement 昼夜活动
die locked chain 扣锁环链
die locking force 合型力
Diels-Ader reaction 狄尔斯—阿德耳反应;六元环合成反应;二烯合成
die lubricant 模具涂料;模具润滑剂;压模润滑剂;铸模润滑剂
die made by low-melting point alloy 低熔合金模
die maker 制锻模工
die making 制钢模法
die making machine 刻锻模机
die making shop 模具制造(工)厂;模具车间
die mark 拉模划痕
diem basis 按日租赁
die metal 模铸金属
die mill 模具铣刀
die milling machine 钢模刻纹机
die misalignment 模具调整不良
die modifications 模具改良
die mo(u)lding 模塑造型;模片键合
Dien 迪恩聚丁二烯纤维(荷兰)
diene 二烯
diene polymerization 二烯系聚合作用
diene resin 二烯树脂
diene rubber 二烯橡胶
diene series 二烯系
diene synthesis 二烯合成
diene synthetics 二烯基合成橡胶
diene value 二烯值
die nipple 公锥;打捞公锥;丝锥接套
dienoic acid 二烯酸
dienol 二烯醇
dienone 二烯酮
die nut 螺母状板牙;六角(形)扳牙;钢板螺母
die of auger machine 制螺旋钻机的模
die of stamp 捣矿砧
die opening 夹头开度;极间距;模腔;模槽;电弧间隙
die orifice 模孔;拉丝孔模;模口
die out 灭绝;消失
die pad 冲模垫
die pantograph machine 钢模缩小机
die parallel 拉模孔圆柱形部分
die parting face 分模面
die parting line 分模线
die piece 模具拼块;拼合模块
die piercing 拉模钻孔
die-pin 凹模顶杆;托杆
die plate 拉模板;印模;模板;压模台板
die plate insert 凹模镶块
die preheating furnace 模具预热炉
die pressing 模压(法);压属法;片材造型
die processing unit 模具加工机
die profile 拉模孔型
die profile radius 凹模圆角半径
die proof 模具灌铅检验法(检查模腔精度用)
die quenching 模压淬火;模具淬火;压模淬火
die radius 拉深模模口圆角半径;模口半径
die reduction angle 模孔压缩锥

die relief angle 模孔出口锥
dierite 无水硫酸钙;烧石膏
die rolled section 周期断面
die rolling 周期断面轧制
die rolling mill 周期断面型材轧机
dies ad quem 截止日
dies a quo 起算日
dies blank 模度坯料
die scalping 拉模剥光
dies cavity 型腔
Diescher elongator 迪舍尔轧管机;狄舍尔轧管机
Diescher mill 迪舍尔轧机;狄舍尔轧机
Diescher process 迪舍尔法;狄舍尔法
die score 拉模划线
die scratch 拉模划痕
die seat 模座
die section 拼合模块
die segment 压模元件;组合模块;拼合模块
diesel 柴油机;狄塞尔内燃机
diesel air compressor 柴油机空气压缩机
diesel alternator 柴油发电机
diesel anchor winch 柴油锚固卷扬机;柴油锚固绞车
diesel and diesel electric(al) locomotive terminal 内燃和内燃电力机务段
diesel boat 柴油(机)船
diesel bucket excavator 柴油斗式挖掘机;柴油斗式挖土机
diesel car 柴油汽车;柴油机动车
diesel card 柴油发动机示功图
diesel combustion 柴油机内的燃料燃烧
diesel-compressor aggregate 柴油压缩机组
diesel concrete mixer 柴油混凝土搅拌机
diesel container ship 柴油机集装箱船
diesel crane 柴油起重机
diesel crawler excavator 履带式内燃挖掘机
diesel crawler face shovel 履带式内燃正铲挖土机
diesel crawler shovel 履带式内燃挖土机
diesel crawler tractor 履带式柴油牵引车;柴油履带式拖拉机
diesel cycle 柴油循环机;柴油机循环
diesel cycle engine 柴油循环发动机
diesel cylinder oil 柴油汽缸油
diesel deposits 柴油机积炭
diesel direct drive 柴油机直接传动
diesel dope 柴油添加剂;柴油机燃料添加剂
diesel dredge(r) 柴油机挖泥船;柴油挖泥机
diesel-driven 柴油机驱动(的)
diesel-driven auger 柴油机驱动的输送螺旋
diesel-driven chain bucket 柴油机传动链斗
diesel-driven direct current welder 柴油驱动直流焊机
diesel-driven fire pump 柴油机驱动的消防泵
diesel-driven generator 柴油驱动发电机
diesel-driven hand-guided two-wheel dozer 柴油驱动人工控制双轮推土机
diesel-driven pump 柴油机驱动泵
diesel-driven rig 柴油机驱动的钻机
diesel-driven ship 柴油机驱动的船
diesel-driven single centrifugal pump 柴油驱动离心泵
diesel-driven single stage centrifugal pump 柴油机驱动单极离心泵
diesel-driven track-laying crane 柴油机驱动铺轨机
diesel-driven water pump 柴油机驱动离心水泵
diesel driving plant 柴油机驱动发电厂
diesel dumper 柴油机倾斜(垃圾)车
diesel dynamo 柴油机直流发电机
diesel dynamo aggregate 柴油机直流发电机组
Diesel effect 狄塞尔效应
diesel-electric(al) 柴油机拖动发电的
diesel-electric(al) bus 内燃电力车;内燃电动车;柴油电气公共汽车
diesel-electric(al) dredge(r) 柴油电动挖泥船
diesel-electric(al) drive 柴油机电气传动装置;柴油机电力驱动;柴油机电力传动;柴油电动驱动装置
diesel-electric(al) driven bucket chain 柴油机电力驱动斗链;柴油机电力传动斗链
diesel-electric(al) emergency set 备用柴油机发电机组
diesel-electric(al) engine 柴油发电机
diesel-electric(al) floating crane 内燃电力浮式起重机
diesel-electric(al) generating equipment 柴油

发电机设备
diesel-electric(al) generator set 柴油发电机组
diesel-electric(al) locomotive 内燃电力传动机车;内燃电动传动机车;柴油电力机车;柴油电动机车
diesel-electric(al) mobile crane 柴油发电汽车吊;柴油发电起重车
diesel-electric(al) mobile power shovel 柴油电力移动式电铲
diesel-electric(al) motor ship 柴油电动机船
diesel-electric(al) plant 柴油机发电厂
diesel-electric(al) power plant 柴油机发电装置
diesel-electric(al) power station 柴油(机)发电厂
diesel-electric(al) power unit 柴油电力机组
diesel-electric(al) propelled trailing suction hopper dredge(r) 柴油电动耙吸式挖泥船
diesel-electric(al) railcar 柴油电动机车;柴油电动轨道车
diesel-electric(al) railway car 柴油电动机车
diesel-electric(al) remote-control unit 柴油电动摇控装置
diesel-electric(al) ship 柴油机电动船
diesel-electric(al) shovel 柴油电动挖掘机;柴油电动铲土机
diesel-electric(al) system 柴油电动系统
diesel engine 柴油(发动)机;狄塞尔发动机
diesel engined 柴油机拖动的
diesel engine driven centrifugal pump 柴油机驱动离心泵
diesel engine driven direct current arc welding machine 柴油机驱动直流弧焊机
diesel engine driven hammer 柴油机驱动锤
diesel engine driven mixed-flow pump 混流式柴油机泵
diesel engine driven self-priming pump 柴油机自吸泵
diesel engine dynamometer 柴油机测力计
diesel engine exhaust gas 柴油发动机排气
diesel engine exhaust pyrometer 柴油机排气高温计
diesel engine for automobile 车用柴油机
diesel engine fork truck 柴油机叉车
diesel engine fuel oil 柴油机燃料油
diesel engine generating set 柴油机发电机组
diesel engine generator 柴油发电机
diesel engine oil 柴油机油
diesel engine oil additive 柴油机油添加剂
diesel engine over-speed relay 柴油机超速继电器
diesel engine pile hammer 柴油机桩锤
diesel engine power 柴油机功率
diesel engine reversing gear 柴油发动机回动装置
diesel engine road vehicle 柴油机道路车辆
diesel-engine road vehicle 柴油机车辆
diesel engine rock drill 内燃凿岩机
diesel engine three-wheel(ed) roller 柴油机三轮压路机
diesel engine vehicle 柴油机车辆
diesel engine with air cell 空气室式柴油机
diesel engine with antechamber 预燃室式柴油机
diesel engine with direct injection 直接喷射式柴油机
diesel engine with precombustion chamber 预热燃烧室式柴油机;预燃室(式)柴油机
diesel exhaust 柴油机废气
diesel exhaust standard 柴油机排气标准
diesel field loco(motive) 工地内燃机车
diesel floating crane 内燃浮式起重机
diesel fork lift truck 内燃叉车
diesel fuel 柴油机燃料;柴油
diesel fuel additives 柴油添加剂
diesel fuel carbon 柴油机燃料的残炭
diesel fuel grades 柴油的等级
diesel-fuelled vehicle 柴油车辆
diesel fuel oil 重柴油
diesel fuel specification 柴油燃料规格
diesel fuel system 柴油机燃料系统
diesel fuel water and sediment 柴油中的水和沉积物
diesel generating set 柴油发电机组
diesel generating station 柴油(机)发电站
diesel generator 柴油发电机
diesel generator cooling water pump 柴油发电机冷却水泵
diesel generator set 内燃发电机组;柴油发电机组

diesel governor setting 柴油机调速器的调整
diesel hammer 柴油机汽锤;柴油(打)桩锤;柴油锤
diesel hammer pile driving frame 柴油机锤打桩架
diesel high compression engine 高压柴油机
diesel hydraulic dredge(r) 柴油液压挖泥船
diesel-hydraulic-driven bucket chain 柴油机液压传动斗链
diesel hydraulic (-transmission) locomotive 内燃液力传动机车;液力传动内燃机车;液力传动柴油机车
diesel hydromechanical locomotive 内燃液力机械传动机车;液力传动内燃机车
diesel index 柴油指数
dieseline kilometrage 内燃化线路公里里程
dieseling 柴油机压缩爆燃
diesel injection pump 柴油机喷射泵;柴油射流泵
diesel injector 柴油机喷油器
dieselisation 内燃化
dieselization 内燃化;柴油机化
dieselized kilometrage 内燃化线路里程
diesel knock 柴油机爆震(声)
diesel liner 柴油机汽缸衬套
diesel liner wear 柴油机汽缸衬套磨损
diesel loader 柴油装载机
diesel locomotive 内燃机车;柴油机车;狄塞尔机车
diesel locomotive crane 内燃铁路起重机
diesel locomotive terminal 内燃机机务段
diesel lube(oil) 柴油机润滑油
diesel lubricating oil 柴油机润滑油
diesel mechanic 柴油机技工
diesel-mechanical drive 柴油机机械传动(装置)
diesel-mechanical locomotive 机械传动内燃机
diesel mine locomotive 矿用柴油机车
diesel motor 柴油发动机;狄塞尔发动机
diesel motor roller 柴油压路机;柴油碾压机
diesel multiple unit 内燃车组
diesel multiple-unit operation 多机组内燃机运转;多机组内燃机操作
diesel multiple unit set 柴油机组
diesel narrow-ga(u)ge loco(motive) 窄轨内燃机车
diesel number 柴油值
diesel oil 柴油
diesel oil cement 加有表面活性剂的柴油水泥(地热井用);柴油水泥
diesel-oil electric(al) generator wagon 柴油发电车
diesel oil purifier 柴油净化器
diesel oil separator 柴油分离器
diesel oil transfer pump 柴油输送泵
diesel-operated vibrator 柴油机操纵的振动器
diesel particulate traps 柴油机粒子吸附器
diesel pile driver 柴油打桩机
diesel pile-driving engine 柴油机打桩机
diesel pile-driving plant 柴油机打桩设备
diesel pile extractor 柴油拔桩机
diesel pile hammer 柴油机打桩锤;内燃打桩机;柴油(打)桩锤
diesel pile puller 柴油机拔桩机
diesel plant 柴油发电机
diesel power 柴油发电机
diesel-powered fire pump 柴油机消防泵
diesel-powered truck 柴油机车;柴油货车
diesel-powered vehicle 柴油机驱动车辆;柴油车辆
diesel power generating set 柴油发电机组
diesel power plant 柴油机动力装置;柴油发电站;柴油发电厂
diesel power station 柴油发电厂
diesel power unit 柴油动力机组
diesel propulsion engine 柴油推进机
diesel pump 柴油泵
diesel railcar 柴油机有轨自动车;柴油机压道车
diesel rig 柴油钻机
diesel road roller 柴油压路机
diesel roller 柴油碾压机
diesel scavenging air 柴油机废气
diesel self-sprinkling pump set 柴油自喷泵组
diesel set 柴油机发电机组
diesel ship 柴油机船
diesel shovel 柴油(机)铲;柴油挖掘机;柴油铲(土机)
diesel shuttle dumper 柴油倾卸车
diesel smokometer 柴油机烟浓度计
diesel soot emission 煤烟发生量
diesel starting fuel 柴油机起动燃料

diesel stock 柴油机车辆
diesel suction dredge(r) 柴油吸泥机
diesel tandem roller 柴油车串列压路机;柴油车串联压路机
diesel traction 柴油机牵引;内燃(机)牵引
diesel traction interference 内燃牵引干扰
diesel traction railway 内燃牵引铁路
diesel traction unit 内燃牵引机
diesel tractor 内燃拖拉机;内燃机牵引车;柴油拖拉机
diesel tractor-compressor 柴油拖拉机式压缩机
diesel tractor crane 柴油拖拉机式汽车吊
diesel train 内燃牵引列车;内燃机列车
diesel tramcar 内燃电车
diesel truck 柴油运货车;柴油机卡车;柴油车辆;柴油(卡)车
diesel tug 柴油机拖轮
diesel underground locomotive 井下用柴油机车
diesel universal compressor tractor 柴油万能压气机式拖拉机
diesel universal excavator 柴油万能挖掘机
diesel wheeled tractor 柴油轮式拖拉机
diesel winch 柴油绞车;柴油卷扬机
die separation 模片隔开;管心切割
die set 模组(熔模铸造用);冲模定位架;成套冲模
die set height adjustment 模具高度调节量
die sets for presses 冲床用成套冲模
die setting 模具(的)安装
die setting press 模具装配压力机;装配冲模床
dies gratia 恩惠日
die shank 模柄
die shoe 模瓦;模托;模板板;下模座;冲模底板
dies holder 模板座
die shop 模具厂;制模车间
die shrinkage 模片收缩率;模后收缩;脱模后收缩
die side wall 模壁面
die sinker 靠模铣床;模具铣床;铣模机;制模工;刻模机
die sinker's file 精密工具加工用锉;模具锉刀
die sinking 刻模;开模;制阴模;模具型腔的加工
die sinking cutter 刻锻模刀具;凹模铣刀
die sinking electric(al) spark machine 电火花刻模机
die sinking machine 刻锻模机;雕模铣床
die sinking milling machine 凹模铣床
diesis 双十字符
die sleeve 模套
die slide 模下压板;模具滑移装置
die slot 模缝
die slotting machine 冲模插床
dies non 法定休息日;法定假期
die space 模腔;模槽;型腔
die spacer block 模具定位块
die spacing 夹钳距离
Diespeker 狄尔斯佩克(一种耐燃楼板构造)
dies plate tap bolt 模板紧固螺栓
die spotter 调试冲模用压力机
diespotting press 修整冲模压力机
die square 枋子;方柱形木材;方材;中枋;正四方块;大方木;方子(约100毫米×100毫米的方木材)
die-squared 枋子;方木
die-squared log 方梭原木
dies redrawing 重复拉模
dies scalping 精整冲裁
die-stamped circuit 冲压电路
die stamping 模压;压花;压(凹)凸印刷;冲压;凹凸印刷;凹版印刷;热模烫印
die-stamping machine 钢模压印机
die stamping press 凹印印刷机
die-stamp marking 打印标记
die stand 拉模板座
die steel 模具钢;板模钢
diester 二(元酸)酯
diester oil 合成双酯润滑油
die stock 螺丝扳手;扳牙扳手;螺丝绞板;螺丝;扳牙绞手
die stock set 成套扳牙架;全套螺丝绞板
die storage rack 模具存放架
Diesulforming 脱硫重整
die surface 模面
die swell ratio 挤出胀大比;模口膨胀比
diet 饮食;膳食
die tap 板牙丝锥
dietary standard 膳食标准;饮食标准

Dieterici equation of state 狄特利希物态方程
Dieterici's equation 狄特利希方程
Dietert centrifuge 迪塔特式离心机
Dietert tester 迪塔特式硬度计
diethacetic acid 二乙基乙酸
diethadione 地沙双酮
die thickness 凹模厚度
diethonolamine 二乙醇胺
diethoxy acetic acid 二乙氧基乙酸
diethoxymethane 二乙氧基乙烷
die throat diameter 模孔直径
diethy-aceto oxalate 草酸二乙酯
diethyl 二乙基(的)
diethylacetaldehyde 二乙基乙醛
diethylacetic acid 二乙基乙酸
diethyl acetoacetic ester 二乙基乙酰乙酸酯
diethylacetonitrile 二乙基乙腈
diethylacetylene 二乙基乙炔
diethylallylacetamide 二乙烯丙醋胺
diethyl alumin(i)um chloride 氯化二乙基铝
diethylamine 二乙胺
diethylamine hydrochloride 二乙胺盐酸盐
diethylamine phosphate 二乙胺磷酸盐
diethylamine solution 二乙胺溶液
diethylaminoethyl cellulose 二乙(基)氨基乙基纤维素
diethylaniline 二乙苯胺
diethylaniline hydrochloride 二乙苯胺盐酸盐
diethylbenzene 二乙(基)苯
diethyl carbinol 二乙基甲醇
diethyl carbonate 碳酸二乙酯
diethyldithiophosphoric acid 二乙基二硫代磷酸
diethylene diamine 对二氮已环
diethylene glycol 二乙二醇;二甘醇
diethylene glycol diethyl ether 二乙二醇二乙醚
diethylene glycol dilaurate 二乙二醇双酯
diethylene glycol dimethyl ether 二乙二醇二甲醚;二甘醇二甲醚
diethylene glycol dinitrate 二甘醇二硝酸酯
diethylene glycol distearate 二甘醇硬脂酸酯
diethylene glycol monoacrylate 二乙二醇单丙烯酸酯
diethylene glycol monobutyl ether 丁基卡必醇;二乙醇单丁醚
diethylene glycol monoethyl ether 卡必醇;二乙二醇单乙醚
diethylene glycol monomethyl ether 甲基卡必醇
diethylene triamine 二乙撑三胺
diethylester 二乙酯
diethyl ether 二乙醚
diethylformamide 二乙基二酰胺
diethyl fumarate 反丁烯二酸二乙酯
diethyl fumarate diisocyanate 反丁烯二酸二乙酯二异氰酸酯
diethyl ketone 戊酮-3
diethyl maleate 马来酸二乙酯
diethyl malonate 二乙基丙二酸酯;丙二酸二乙酯
diethyl malonic acid 二乙基丙二酸
diethyl malonic ester 二乙基丙二酸酯
diethyl malonylurea 巴比士
diethylnitrosamine 二乙基亚硝胺
diethyloxadicarbocyanine iodide 二乙基噁二碳化青碘化物
diethyl oxalate 草酸二乙酯
diethyl oxaloacetate 草乙酸二乙酯
diethyl oxaloacetic ester 二乙基草乙酸酯
diethyl peroxide 过氧化二乙基
diethyl phosphite 亚磷酸二乙酯
diethyl phthalate 邻苯二甲酸二乙酯
diethyl-p-nitrophenyl phosphate 对硝基苯磷酸二乙酯;磷酸二乙基对硝基苯酯
diethyl-p-phenylenediamine sulfate 二乙基对苯二胺硫酸盐
diethyl-p-tlouidine 二乙基对甲苯胺
diethyl sebacate 癸二酸二乙酯
diethyl succinate 琥珀酸二乙酯;丁二酸二乙酯
diethyl succinic acid 二乙基丁二酸
diethyl sulfate 硫酸二乙酯
diethyl sulfide 二乙硫
diethyl thionamic acid 二乙胺基磺胺
diethyl tin sulphide trimer 硫化二乙基锡三聚体
diethyltoluamide 二乙甲苯酰胺
die tie-plate 模具垫板
dietitian 营养学家

diet kitchen 营养厨房;病号厨房
die tongs 牙钳
die tool grinding machine 模具工具磨床
dietrichite 锰铁锌矾;锌铝矾
die turning gear 塑模旋转装置
die typing 反印法
dietzeite 碘铬钙石
die wall 模壁;斜墙
die wall effect 模壁效应
die wall finish 模壁光洁度
die wall friction 模壁磨擦
die wall life 模壁寿命
die wall lubricant 模壁润滑剂
die wall lubrication 模壁润滑
die wall surface 模壁面
die wear 模具磨损
die weld(ing) 模(锻)焊;横锻接;冲模堆焊
die width 模具宽度
die yoke 压模滑架
di-fan antenna 双扇形天线
difebarbamate 苯巴氨酯
difference 异点;差值;差异;差数;差分;差额;差(别)
difference account 盈亏通知书
difference address operand 差分地址操作数
difference altitude 高差
difference amplifier 差频信号放大器;差分放大器
difference approximation 差分近似
difference between both ends 两端直径差
difference between commercial purchasing and selling prices 商业进销差价
difference between cost and book value 成本与账面价值差异
difference between fore and back sighting distances 前后视差距
difference between hydrostatic(al) end forces 流体静压轴向力差
difference between log's readings 计程仪读数差
difference between means 平均数差数
difference between town and country 城乡差别
difference channel 差分电路;差动声道
difference coefficient 差商;差分系数
difference correction 差分校正
difference correction method 差分校正法
difference counter 差值计数器
difference current 差动电流
difference curve 减差曲线;差异曲线;差分曲线;温差曲线
difference detector 差值检波器;差分检波器
difference-differential equation 差(分)微分方程;微分差分方程
difference diode 差分二极管
difference engine 差分机
difference equation 差分方程
difference expression 差分表示式;差分表达式
difference formula 差分公式
difference frequency 差频
difference frequency effect 差频效应
difference frequency fluctuation 差频起伏
difference frequency generator 差频发生器
difference frequency term 差频项
difference function 差函数
difference gamma sampling 差值法伽马取样
difference gamma spectrum sampling 差值法伽马谱取样
difference ga(u)ge 极限量规;测差规
difference grid 差分网格
difference grid point 差分格点
difference image 差值图像
difference in altitude 高程差
difference in attitude of beds 地层产状差异
difference in deformation 变形(作用)差异
difference in depth modulation 调制深度差
difference in elevation 高(程)差
difference in floor lever of sieve tray 筛板表面水平偏差
difference in gradients 坡度差
difference in height 高(度)差
difference in level 高程差;标高差
difference in magmatic activity 岩浆活动差异
difference in metamorphism 变质作用差异
difference in opinion 分歧意见
difference in pressure 压力差
difference in regional price levels 地区差价
difference in riser level 升气管顶面水平偏差

difference in settlement 沉降差
difference in shrinkage 收缩差
difference in temperature 温(度)差
difference interpolation 差分内插法
difference in water levels 水位差
difference isopleth of distances 等距(离)差曲线
difference length operand 差分长度操作数
difference limen difference test 强度辨差阈差异试验
difference limen test 辨差试验;强度辨差阈试验
difference matrix 差分矩阵
difference measurement 差分测量
difference method 差值法;差分法;差别法
difference mode gain 差模增益;差分信号增益
difference modes 差分模式
difference of absolute cost 绝对成本差异;绝对成本差额
difference of a function 函数差
difference of altitude 高度差
difference of annual air temperature variation 年气温变幅
difference of brightness 亮度差
difference of coal seam thickness 煤层厚度差值
difference of comparative cost 比较成本差异
difference of curvature 曲率差
difference of daily temperature 日气温变幅
difference of elevation 高(程)差
difference of elevation between indoor and outdoor 室内外标高差
difference of elevation between inside and outside 室内外标高差
difference of elevations 标高差
difference of focus 焦点差
difference of forward and backward observation 往返测较差
difference of ground height 地面高差
difference of height 高差
difference of latitude 纬距;纬(度)差
difference of level 高程差;水位差;水位变幅;水平差
difference of level surface 水准面差
difference of longitude 经距;经差
difference of luminance 亮度差
difference of magnification 等倍差;放大率差
difference of meridianal parts 纬度渐长率差
difference of meridional parts 纬差
difference of moisture yearly variance 土壤年内湿度变差
difference of observation 观测值差
difference of phase 相位差
difference of polynomial 多项式的差
difference of potential 位能差;势差;电位差;电势差
difference of principal stress 主应力差
difference of rail at joint 钢轨接头上下错牙
difference of readings 读数差
difference of readings at the opposite orientation 转向差
difference of readings in the same orientation 同向差
difference of sample means 样品平均数差
difference of set 集差
difference of subgraph 子图差
difference of tidal height 潮高差
difference of two measures 两次测量结果之差
difference of two sets of coordinates 两组坐标较差
difference of water head 水头差
difference of water head between the grouting heavy curtain inside and outside 帷幕内外水头差
difference of water head between water blocking before and after 堵水前后水头差
difference of water level 水位差
difference of water-level variation 水位变幅
difference of yearly earth temperature 年地温差
difference operator 差分算子
difference pattern parameters of migration 偏移差分格式参数
difference preference 不同偏好
difference pressure flow meter 差压式流量计
difference-product 差积
difference quotient 差商
difference radio-frequency mixer 射频差信号混频器
difference report 差分报告;差别报告

difference resonance 差共振
differences among means 许多平均数间的差数
difference sampling converted coefficient 差值取样换算系数
differences between means 两个平均数间的差数
difference scheme 差分格式
difference sequence 差分序列
difference set 差集(合)
difference-set code 差集码
difference signal 差信号
difference spectrophotometer 吸收分光光度计
difference spectrum 差光谱;示差光谱
difference spiral 差较螺线;差分螺线
difference string 差分串
difference structure 差分结构
difference table 差分表
difference tone 差音
difference type pickup electrode 异形信号电极
differencing 差分化
differencing scheme 差分格式
differentail arc lamp 差接电弧灯
different amounts of phosphorus 不同磷量
different application at different depths 分层施肥法
different crystallite 各种晶出体
different cultivars 不同品种
different depths of plowing 不同耕作深度
different design 设计不符
different elevation 不同高程;不同标高
different fumigator 各式熏烟清毒器
different head ratio 不同水头比
different hodograph 差异时距曲线
differentiability 可微分性
differentiability from the right 从右边可微
differentiability of a pointwise supremum 点态上确界的可微性
differentiability property 可微性性质
differentiable arc 可微弧
differentiable atlas 可微图
differentiable function 可微(分)函数
differentiable manifold 微分流形
differential 差致的;差速机构;示差;差接;差动(器);差动(的);差异;分化的;微分(的);示差的
differential aberration 较差光行差;光行差较差;差动光行差
differential ablation 差异消融
differential absorption 吸收上的差异;区别吸收
differential absorption dye laser radar 差动吸收染料激光雷达
differential absorption rate 微分吸收率
differential absorption ratio 示差吸收比
differential accounting 差量会计
differential accumulator 差动蓄水器;差动储器
differential acting hammer 差动式桩锤
differential acting pile hammer 复动打桩机
differential acting pump 差动水泵
differential acting steam hammer 差动蒸汽锤;差动(式)汽锤;差动(式)蒸汽锤
differential acting type 差动式
differential actinometer 示差曝光表
differential action 差动(作用)
differential adjusting wrench 分速器轴承调整扳手;分速分速器轴承调整扳手
differential adjustment 微差平差法
differential adsorption 鉴别吸附
differential aeration 差动充气
differential aeration cell 差动充气电池
differential affinity 差别亲和力
differential air pressure type 差动气压机
differential air thermometer 示差空气温度计
differential altimeter statoscope 高差仪
differential ammeter 差动安培表
differential amplifier 差示放大器;差动放大器;分差放大器;微分放大器
differential amplifier multiplier 差分放大乘法器
differential analyser 差热分析仪
differential analysis 差值分析;差异分析;差热分析;差动分析;微分分析;示差分析
differential analysis method 差异分析法
differential analyzer 差动分析器;差分仪;微分分析仪
differential anatexis 分异深熔作用
differential and integral calculus 微积分学;微积分
differential and integral equation program(me) 微分和积分方程程序
differential annealing 局部退火
differential annealing process 局部退火深拉延法
differential anode conductance 动态微分阳极电导
differential anode resistance 动态阳极电阻
differential apitude test battery 鉴别能力倾向的成套测验
differential approximation 差分逼近
differential arc regulator 差联电弧调节器
differential assembly 差动总成
differential atmospheric absorption 较差大气吸收;大气吸收较差
differential attachment 差分装置;差动机构;分速器十字头销
differential axle 半轴
differential ballistic wind 差动弹道风
differential bathygraph 差动自记测温仪
differential bearing 差速器轴承;差动轴承
differential bearing adjustinng nut 差动轴承调整螺母
differential bearing cone 分速器轴承锥
differential bearing cup 差速器轴承杯
differential bearing indicator 差动方位指示器
differential benefit 差别利润
differential bevel gear 差动伞齿轮
differential bit 差动式钻头
differential block 差动绞辘;差动滑轮;差动滑车;差别滑轮;神仙葫芦
differential block structure 差异性断块构造
differential bolt 差动螺栓
differential booster 差接增压机;差动升压机
differential bracket 差动轮架
differential brake 差速制动器;差动制动器;差动闸
differential bridge 差接电桥
differential bridge type spectrometer 差示电桥式分光计
differential busbar ring 差动式汇流环
differential calculus 微分(学)
differential calorimeter 示差量热器
differential calorimetry 示差量热术
differential camber 差异反拱
differential cancellation 差值消除法
differential capacitance 微分电容
differential capacitance acceleration transducer 差容式加速度传感器
differential capacitor 差动电容器
differential capstan 差异绞盘;差动绞盘
differential capsule 差压膜盒
differential carrier 差速器座架;差速器壳;差速搬运车
differential carrier cap 差速器座架盖;分速器座架盖
differential carrier gasket 差速器座架垫密片
differential case 差动齿轮箱;分速器箱
differential case oil collector drum 分速器箱集油鼓
differential catalogue 较差星表
differential centrifugation 示差离心法;差速离心(分离);示差离心分离;特异离心
differential chain block 差动滑轮组;差动(式)链滑车;差动链(式)滑车;差动绞辘;差动滑车组;神仙葫芦
differential chain hoist 差动链式起重机;差动链式吊车
differential change 微变
differential change gear 差动变换齿轮
differential chart 差值图
differential chemical reactor 微分化学反应器
differential chromatography 差示色谱法
differential circuit 差动电路;微分电路
differential coating 双面差厚涂法;双面差厚涂镀
differential coefficient 导函数;差动系数;差动系数;微商;微分系数
differential coherent detection 差分相干检测;差动相干检测
differential coil 差动线圈
differential colo(u)rimeter 色差仪
differential colo(u)rimetry 示差比色法
differential compaction 差异压实(作用);分异致密
differential comparator 差分比较器;差动比较器
differential comparator amplifier 差分比较放大器
differential compartment 示差分隔间
differential compensating gear 差速机补正齿轮
differential compensation 差额补偿;差动补偿
differential compound 差绕复激;差复励
differential compound generator 差(绕)复激发电机;差复励发电机
differential compound motor 差(绕)复激电动机;差复绕电动机;差复励电动机
differential compound winding 差复励绕组;差复激绕组
differential computing potentiometer 微分计算电位器
differential concatenation 差级联;反向串联
differential concentration 浓度差
differential condensation 差异冷凝
differential condenser 差异冷凝器;差动电容器
differential conductivity recorder 差热导记录仪
differential conductometric titration 差示电导滴定
differential cone bearing 差速器锥形轴承
differential cone bearing cap 差速器锥形轴承盖
differential cone bearing cup 差速器锥形轴承杯
differential connection 差速器接合
differential consolidation settlement 差异固结沉降
differential contact-type air plug ga(u)ge 差动接触式气动塞规
differential contraction 差异收缩
differential control 差动控制;微分控制
differential controller 差压调节器;差动控制器
differential cooling 差示冷却
differential correction 较差改正;差动修正
differential corrosion 差异溶蚀
differential cost 差异成本
differential counter 差动计数器
differential count(ing) 分类计数
differential cross 差速器十字头;差动十字轴
differential cross pin 差速器装置
differential cross-section 微分截面
differential crown wheel 差速器侧面伞齿轮
differential current 差动电流;差别电流
differential current protection 差动电流保护装置
differential current relay 差流继电器
differential curve 微分曲线
differential cylinder 差动油缸;差动汽缸
differential declines 不同衰减
differential deformation 不均匀变形;差异变形
differential delay 较差时延;差值延迟;差分延迟
differential-density saparation 比重分选法
differential depletion 差动衰减
differential detection 差分检波
differential detector 差示检测器;差接检波器;差动探测器;差动指示器
differential determination 较差测定
differential device 差动装置
differential diagnosis 鉴别诊断
differential diaphragm 差压膜板;差压隔膜
differential-difference equation 差分微分方程
differential dike-rock 分异墙岩;分异脉岩
differential dilatometer 示差热膨胀仪;示差热膨胀计;示差膨胀仪
differential dilatometry 差示热膨胀量测法
differential dimensional change 尺寸差变
differential discriminator 鉴差计;差分甄别器;差分鉴别器;差动式鉴频器;差动式鉴别器
differential disintegration 优先破碎
differential displacement 差动位移;分异位移
differential distance system 距离差制;双曲线定位制;双曲线定位系统
differential distillation 微分蒸馏
differential divisor 微分因子
differential Doppler 微分多普勒
differential double column 差示双柱
differential draft ga(u)ge 差示通风计
differential draught ga(u)ge 差压通风计
differential drive 差速(器)传动;差动装置;差动传动
differential drive gear 差速(器)传动齿轮
differential drive gear thrust block 差速传动齿轮止推块
differential drive gear thrust screw 差速器环齿轮推力螺钉
differential drive pinion 差速器传动小齿轮
differential drive pinion bearing 差速器主动小齿轮轴承
differential drive pinion bearing cage 差速器主动小齿轮轴承罩
differential drive pinion bearing lock 差速器主动齿轮轴承锁
differential drive pinion carrier 差速器主动小齿轮座架

differential drive shaft coupling 差速器传动轴联轴节;分速器传动轴联轴节
differential drive tachometer 差动传动式转速表
differential dry-pipe valve 差动式干汽管阀门
differential duplex 差动双工
differential duplex circuit 差动双工电路
differential duplex system 差动双工制
differential duty 差别关税
differential dynamometer 差动式功率计;差动测力计
differential earning 差额收益;差额收入
differential earnings from land 土地级差收益
differential ebuliometer 差示沸点计
differential effect 差动效应
differential electromagnet 差动电磁铁
differential electrometer 差动静电计;差动电位计
differential elements thumbline 微量恒向线
differential engine 差动式发动机
differential entrapment 差异聚集
differential epicyclic gearing 差动调转齿轮装置
differential equation 微分方程
differential equation analog(ue) computer 微分方程模拟计算机
differential equation computer 微分方程计算机
differential equation of composite type 复合型微分方程
differential equation of first order 一阶微分方程
differential equation of geodesic 大地线微分方程
differential equation of higher order 高阶微分方程
differential equation of motion 运动微分方程
differential equation solver 微分方程计算机
differential equation with constant coefficient 常系数微分方程
differential erosion 差异侵蚀;差别侵蚀;分异侵蚀;不等侵蚀
differential excitation 差励;差分激励;差动激励;差动激磁
differential-excited generator 差励发电机
differential-excited welding generator 差激电焊发电机
differential expansion 差胀;差异膨胀;不均匀膨胀
differential expansion indicator 差胀指示器;差异膨胀指示器
differential expression 微分式
differential factor 差动因数;绕组系数
differential fault 剪断层
differential feed 差动进给
differential feedback method 差动回授法;差动反馈法
differential-field motor 他激差绕直流电动机
differential-field series motor 串激差绕直流电动机
differential file 勘误文件
differential filter 差接滤波器;差动滤波器
differential firing 换层燃烧
differential flexure 较差弯沉
differential flo(a)tation 优先浮选;差别浮选;分异浮选
differential flow detector 微分流动检测器
differential flowmeter 差示流量计;差动流量计
differential focusing 差动式调焦
differential force 不均匀力
differential force system 不均匀力系
differential form 微分形式
differential four-channel gamma-ray spectrometer 差动四通道伽马射线能谱仪
differential frequency circuit 差频电路
differential frequency meter 差示频率计
differential frequency relay 差周继电器
differential friction 摩擦差异
differential friction(al) effect 摩擦差微效应
differential frost heave 不均匀冻胀
differential fusion 分异熔融
differential gain 微分增益;微分增量;差额所得
differential gain control 微分增益旋钮;微分增益控制
differential gain controller 微分增益控制器
differential galvanometer 差示电流计;差绕电流计;差动检流计
differential game 差分博弈
differential gap 不可调间隙
differential gap controller 差动间隙控制器
differential ga(u)ge 差动(式)压力计;压差计;差压计;差压表;微分(气压)计;示差流速计
differential gear 差速齿轮;差动装置;差动(齿)轮;差别齿轮;分速轮
differential gear adder 差动齿轮加法器
differential gear box 差动齿轮箱
differential gear for screw conveyer 螺旋输送机的差动装置
differential gearing 差动齿轮装置
differential gear ratio 差动齿轮传动比
differential gear shift sleeve 差动式变速器套筒
differential gear slipmeter 差动齿轮转差计
differential gear train 差动齿轮机构
differential generator 差动自整角机;差动振荡器;差动式传感器;差动式伴感器;微分发生器
differential geometric factor 微分几何因子
differential geometry 微分几何(学)
differential global positioning system 差转全球定位系统
differential governor 差动调节器
differential grinding 选择性研磨;优先研磨
differential hammer 差动式打桩锤
differential hardening 阶段硬化;差致硬化;差别淬火
differential head 差异压头;差动头;不等压头;水头差;示差水头
differential head of filter 滤池压差
differential heating 局部加热;差温加热;差别加热(法)
differential heating curve 差热曲线
differential heat of adsorption 微分吸附热
differential heat of dilution 稀释热;微分冲淡热
differential heat of solution 微分溶解热
differential heat of sorption 微分吸收热
differential heat treatment 局部热处理
differential heave 不均匀隆胀
differential height 高差
differential hogging 差异翘曲;差异挠度;差异反拱
differential hoist 差动式卷扬机;差动绞辘;差动复滑车;神仙胡芦
differential hole ga(u)ge 楔形验孔规
differential holographic interferometry 差分全息干涉计量术
differential host 鉴别寄主
differential housing 差速机壳;差动齿轮箱
differential housing cover 差速器箱体盖;分速箱壳
differential-hydration separation 差示水化分离
differential hydrostatic(al) pressure 静水压(力)差
differential idle gear 差速器空转齿轮
differential idler gear shaft 差速器空转齿轮轴
differential illumination 不均匀照明
differential image 微分图像
differential imaging equation 成像微分方程
differential income 差额收入
differential incomes of land 土地级差收益
differential increment 差额增量
differential index 不均匀指数
differential indexing 差动分度法
differential indexing centre 差动分度中心
differential indexing head 差动分度头
differential inequality 微分不等式
differential input 差分输入;差动输入
differential input shaft 差动输入轴
differential instrument 差动(式)仪表
differential intake 差动式取水口;差动式进水口
differential integral method 微分积分法
differential intensive active area in West China 西部幅度差异性活动区
differential interference 差分干涉
differential interference contrast 微分干涉反衬
differential interference contrast microscope 微分干涉反衬显微镜
differential interference microscope 差分干涉显微镜
differential interferometry 差分干涉测量法
differential invariant 微分不变式
differential ionization chamber 微分电离室
differential ion pump 差动电离泵
differential iron tester 差动铁质测试器
differential jack 差动千斤顶
differential keying 差动键控
differential laser gyro 差动激光陀螺
differential laser interferometer 差动激光干涉仪
differential lateral 差分侧位偏移
differential law 微分法则
differential leakage 电枢齿磁漏
differential leakage flux 差漏磁通
differential leakage reactance 差漏抗
differential leak detector 差动检漏器
differential level(l)ing 差动水准测量;微差水准测量;水准测量;水平测量;高程差测量
differential lever 差速杠杆;差动控制杆
differential liberation 差异释出
differential light emitting diode indicator 示差发光二极管指示器
differential lighting 不均匀照明
differential light scattering analyzer 示差光散射分析仪
differential linearity 差分(直)线性
differential liquid level ga(u)ge 差式液位计
differential lock 差速器闭锁机构;差速器锁;差速机锁
differential locking jaw 差速器夹爪
differential low oil pressure warning signal 差速器低油压警告信号
differential lubricant 差速器用润滑油
differentially coherent transmission system 差动相干传输系统
differentially compensated regulator 差动式补偿稳压器
differentially compound-wound machine 差复励电机
differentially compound wound motor 差复激电动机
differentially driven engine 差速传动发动机
differentially protected circuit breaker 差动保护断路器
differentially value-equivalent 可微分值等价
differentially wound 差绕
differentially woundbell 差绕电铃
differentially wound motor 差绕电动机
differential magnetic susceptibility 可逆磁化率;增值磁化率
differential manometer 压差计;差压计;差压机;差动压力计;差动压力表;微压计;水差计;示差压力计
differential manometry 差示测压术;示差测压术
differential marking system 差覆厚面标志方法
differential mass curve 差积曲线
differential master gear 差速器主齿轮;差速器大被动齿轮
differential matrix 微分矩阵
differential measurement 较差测量;微分测量;微差量测
differential measuring instrument 差动式测量仪表
differential mechanism 差速器;差速机构;差动机器
differential medium 鉴别培养基
differential megnetometer 差动磁强计
differential meter 差压计;差动仪(表)
differential method 差示法;差动法;微分法
differential methods for field point layout 微分法全野外布点
differential methods of measurement 差示测量法;微差测定法
differential methods of photogrammetry 微分法测图;微分测图法
differential micromanometer 差示微压计;差动式微压计
differential micrometer 差示测微计
differential micrometer dial 标度盘
differential microphone 差动式话筒;差动式传声器
differential milliammeter 差动式毫安计
differential modal attenuation 模模减差
differential modal dalay 模时延差
differential mode 异态
differential mode interference 差模干扰
differential mode signal 差分模信号
differential modulation 差异调制;差分调制
differential monitor 差动监控器
differential mortality 差别死亡率
differential motion 差速运动;差动(运动)
differential motor 差绕电动机;差动式电动机;微分马达
differential mounting plate 差速器固定板
differential movement 差异运动;微差移动
differential nut 差动螺母
differential nutation 较差章动
differential observation 较差观测
differential of a arc 微弧
differential of air temperature 空气温差
differential of area 面积元(素)

differential of higher order 高阶微分
differential oil screen 差速器油滤网
differential omega 差欧米伽
differential omega system 差欧米伽系统
differential operation 差示操作;微分运算
differential operation(al) amplifier 差分运算放大器
differential operational amplifier 微分运算放大器
differential operator 微分算子;微分算符
differential parallax 视差较
differential parallelogram 微分平行四边形
differential parameter 微分参数
differential partial condensation 部分冷凝的蒸馏分离
differential pendulum cam 差动摆式凸轮
differential permeability 动磁导率
differential permittivity 微分介质常数
differential phase 差分相位;微分相位
differential phase inversion modulation system 反相差分调制方式
differential phasemodulation 差动式调相
differential phase-shift keying 差分相移键控;差分移相键控;差动相移键控技术
differential phase test 相移失真试验
differential photo-electronic spectrometer 差动光电子谱仪
differential photo-electrooptic(al) refractometer 差动电光学折射计
differential photographic method 较差照相方法
differential photography 缝隙摄影
differential photometry 较差光度测量;较差测光;差示光度计量;差示光度测量;差示测光;示差测光
differential piece-rate system 差别计件工资率制度;分等计件工资制
differential pinion 差速器小(伞)齿轮
differential pinion case 差速器小齿轮箱
differential pinion cross 差速器小齿轮十字轴
differential pinion shaft 差速器主动齿轮轴
differential pinion spider 差速器小齿轮十字轴
differential piston 差径活塞;差动活塞
differential piston compressor 差动活塞式压缩机
differential piston ga(u)ge 差动活塞式压力计
differential piston pump 差动活塞泵
differential planet gear 差动行星齿轮
differential plunger pump 差动(式)柱塞泵;差动式活塞
differential point 差别点
differential polarogram 差示极谱
differential polarography 差示极谱法
differential polynomial 微分多项法
differential potential stripping analysis 差示电位溶出分析
differential potential titration 差示电位滴定(法)
differential potentiotitration 差示电位滴定(法)
differential precession 较差岁差
differential pressure 压力落差;压力降;压(力)差;差异压;差压;差动压力;不均匀压强;不均匀压力
differential pressure amplifier 差压放大器
differential pressure cell 差压传感器
differential pressure controller 压力调节器;压差控制器;压差调节器;压差控制器;剩余压力调节器;剩余压差调节器
differential pressure control mechanism 压力差调节机构
differential pressure control valve 差压调节阀
differential pressure device 差压(式)装置
differential pressure flowmeter 压力流量计;压差流速计;压差流量计;差式流量计
differential pressure fuel valve 压差油阀
differential pressure ga(u)ge 差动(式)压力计;差压流量计;差式压力计;差式压力表;差式压力计;差示压力表;差压流表;微差压力计;压差计
differential pressure indicator 差压指示计;差压(式)指示计
differential pressure meter 差式压力计;差式压力计;差示压力表;差(动)式压力表;差压计
differential pressure multiplier 压差乘子
differential pressure pickup 压差传感器;差压传感器
differential pressure recorder 差压记录器
differential pressure recorder controller 差压记录控制器
differential pressure regulator 差压调节器
differential pressure sticking 压差卡钻

differential pressure switch 压差开关;差压开关
differential pressure transducer 压差传感器;差压变换器
differential pressure type airspeed indicator 差动压力式风速计
differential pressure type flowmeter 差压式流量计
differential pressure type level detector 差压式液位检测器
differential pressure type level ga(u)ge 差压式液位计
differential pressure under production 生产压差
differential pressure valve 煤气自动阀;差压阀
differential process 差动过程
differential products 认识差异产品
differential profit ratio 差额利润比率
differential projected element 微分投影元素
differential protection 差动保护(装置)
differential protection cable 差动保护电缆
differential protection for cable line 电缆线路差动保护
differential protection relay 差动保护继电器
differential protective system 差动保护装置;差动保护系统
differential pulley 差动滑车(轮);差别滑轮
differential pulley block 差动链滑车;差动滑车组;差别滑轮组
differential pulley purchase 神仙葫芦;差动绞辘
differential pulse 微分脉冲
differential pulse anode solvation voltammetry 微分脉冲阳极溶出伏安法
differential pulse anodic stripping voltammetry 差示脉冲阳极提溶伏安法
differential pulse code modulation 差异脉码调制;差分脉码调制
differential pulse code modulation quantizer 差分脉冲编码调制量化器
differential pulse polarography 差示脉冲极谱(法);微分脉冲极谱法
differential pump 差压泵;差动(式)泵
differential pumping 差动泵浦
differential purchase 差动绞辘;差动复滑车
differential quadrature phase shift keying 差分四相移键控
differential quantizer 差值量化器
differential quenching 阶差淬火;差温淬火
differential quotient 微商;微分系数
differential rate 差别运价率;差别工资率;特价
differential reaction rate 微分反应速度
differential reactor 差式反应器
differential recorder 压差记录器
differential rectification 缝隙纠正;微分纠正
differential rectifier 微分纠正仪
differential reduction gear 差动式减速齿轮器;差动式齿轮减速器
differential refraction 较差折射
differential refraction detector 差示折光检测器;示差折光检测器
differential refractive index detector 差示折光(率)检测器
differential refractometer 差示折射计;差示折光计;差动式折射计;示差折光计
differential refractometer detector 差示折光(计)检测器
differential regulating transformer 差接可调变压器
differential regulator 差动调节器
differential reinforcement 区别性强化
differential relay 差动继电器;差别继电器
differential relaying panel 母差保护屏
differential release valve 差压缓解阀
differential resistance 微分电阻
differential resistance type instrumentation 动态电阻仪;差动电阻式仪器;微分电阻仪
differential response 不同反应;微分反应
differential returns 差别报酬
differential revenue 差别收入
differential reversing gear 差动式回动装置
differential ring gear 差速器环齿轮
differential rock bolt extensometer 差动式岩石锚杆引伸仪
differential roll 差压辊;差速对辊机
differential rotation 较差自转;微分动
differential route 运费低廉的交通线;特价线路
differential scale 差动天平
differential scanning calorimeter 示差扫描量

计;差示扫描量热仪;差示扫描量热器;差示扫描量热计;差动式扫描量热仪
differential scanning calorimetric curve 差示扫描量热曲线
differential scanning calorimetry 差示扫描量热法
differential scintigram 微分闪烁图
differential screen 差动筛
differential screen analysis 微分筛析
differential screw 差动装置螺钉;差动螺旋
differential screw gearing 差动螺旋传动装置
differential screw jack 差动(螺旋)千斤顶;差动螺旋起重器
differential sedimentation 差示沉降
differential segment 相异区段;差别区段
differential selection 特异选择
differential selsyn 差动自整角机;差动自动同步机
differential selsyn motor 差动自动同步电机
differential sensitivity 差示灵敏度;差动灵敏度
differential sensor 差动式传感器
differential separation 差动分离
differential series winding 差接串励绕组
differential serration 错齿
differential servo 差动伺服机构
differential settlement 沉降差;差异沉降;不均匀下沉;不均匀沉箱;不均匀沉陷;不均匀沉降
differential settlement of structure 结构物差异沉降
differential settling 速差沉淀
differential shaft 差动轴
differential sheave 差动轴滑轮
differential shift 差动相移
differential shock mounting plate 差速器减振固定板
differential shrinkage 差分收缩;差异收缩;不均匀收缩;收缩微
differential shunt motor 差并励电动机;差分并极电机
differential shut of water 分层止水
differential side adjusting tool 差速器壳调整工具
differential side bearing 差速器壳轴承
differential side case 差速器侧箱
differential side gear 差速器侧齿轮;差速器被动齿轮;差速半轴齿轮
differential sign 微分符号
differential site occupation function 差异点占有函数
differential size frequency curve 微分粒度频率曲线
differential sliding clutch 分速滑动离合器
differential sliding clutch shift shaft 差速器主动齿轮横轴
differential sliding valve 差动滑阀
differential slope coefficient 不同斜率系数
differential smoothing device 微分平滑装置
differential soil movement 不均匀土体运动;不均匀土壤运动
differential solubility technique 差示溶解度法
differential spectrometry 差示光度法
differential spectrophotometer 差示分光光度计
differential spectrophotometric method 差示分光光度法
differential spectrophotometry 差示分光光度法;差比分分光光度学
differential spectroscopy 差示光谱学;差示光谱法
differential spectrum 差示光谱;微分谱
differential speed 差速(的)
differential speed disk cutter 差速圆盘剪切机
differential speed regulation 差动调速
differential speed roll 差速辊
differential speed sensor 速差传感器
differential spider 差速器十字轴;差速机十字轴
differential spider gear 差速器架齿轮
differential spider pinion 分速架小齿轮
differential spur drive gear 差速器传动正齿轮
differential spur drive pinion 差速器传动小正齿轮
differential spur drive pinion cross shaft 差速器主动齿轮横轴
differential staining 鉴别染色法;对比着色
differential star catalogue 较差星表
differential static pressure 静差压
differential steam calorimeter 差示量热计;差示蒸汽量热器
differential steering 差速转向;差动式转向;差动操纵
differential steering brake 差速转向制动
differential steering device 差动转向机构

differential stratification 微差层
differential stress 差(异)应力
differential subsidy 差价补贴
differential supercharged engine 差速增压发动机
differential supercharger 差动增压器
differential surface heating 表面不均匀加热
differential surface site distribution 差异表面点分布
differential surface site distribution function 差异表面点分布函数
differential surge chamber 差动式调压室
differential surge tank 差压调压塔；差动式调压箱；差动式调压塔
differential susceptibility 不同感受性；微分磁化率
differential symbol 微分符号
differential synchro 差接自动同步机；差级同步机；差动同步；差动自动同步机；差动同步机
differential synchro amplifier 差动同步放大器
differential synchro motor 差动同步电动机
differential synchronizer 差动同步机
differential synchro receiver 差动同步接收机
differential synchro transmitter 差动同步发射机
differential system 混合线圈；差动装置；差动制；差动系统；微分(方程)组
differential table 差分表
differential tackle 差速绞辘；差动滑车组；差动滑(车)轮；差别滑轮；神仙葫芦
differential tariff 差别税则；差别费率
differential temperature 差异温度
differential temperature log 微差井温测井图
differential thermal analysis 温差分析；差(示)热分析；热差分析；微热分析；示差热分析
differential thermal analysis apparatus 差热分析设备
differential thermal analysis curve 差热分析曲线；温度自记曲线
differential thermal analysis meter 差示热分析仪
differential thermal analysis pattern 差热分析图
differential thermal analysis thermogram 差热分析曲线图
differential thermal analyzer 差(示)热分析仪
differential thermal apparatus 差热分析设备
differential thermal calcination 差热煅烧
differential thermal curve 差热温度曲线；温差温度曲线；微差温度曲线
differential thermal gravimetric analysis 差热重量分析；微分热解重量分析
differential thermal measuring range 差热量程
differential thermal solarimeter 差热日射计
differential thermocouple 差作用温差电偶；差示热电偶；差动热电偶
differential thermogram 差示热分析图
differential thermogravimetric analysis 差示热重分析；差示热解重量分析；差热失重分析；微分热解重量分析
differential thermogravimetry 差热重量法
differential thermometer 差示温度计；微差温度表；双金属温度计；示差温度计
differential thermometric titration 差示温度滴定法
differential thread 差动螺纹
differential threshold 差阈；辨别阈
differential thrust block pin 差速器止推座销
differential titration 差示滴定
differential tolerance to Pb 对铅有不同耐性
differential torque of stereoscopic pair 立体像对旋角差
differential transducer 差动传感器
differential transformer 差示变压器；差接变压器；差动变压器
differential-transformer transducer 差接变压器换能器
differential transmission 差速传动；差动传动
differential transmitter 差动式变送器；差动发送器
differential transmitting linkage 差动式传动机构
differential transmitting selsyn 差动自整角机发送机
differential transparency of atmosphere 不同层次的大气透明度
differential trap 差热式阻汽器
differential turn 差动转向
differential type dry pipe valve 差动型干式阀
differential type galvanometer 差动式检流计
differential type gate 差分型门
differential type relay 差动式继电器

differential ultrasonic velocimeter 差示超声测速仪
differential unit 差动单位；分速器
differential U tube 差示U形管
differential vacuum ga(u)ge 压差式真空计；差示真空规
differential value 微分值
differential valve 差压阀；差动阀
differential vane 差动阀
differential vapour pressure thermometer 差示蒸汽压力温度计
differential variable condenser 差动可变电容器
differential viscosity 差示黏度
differential voltage 差分电压；差动电压
differential voltmeter 差动式伏特计；差动式电压表
differential wages 差别工资
differential water capacity 分层持水量
differential water pressure 水压差
differential wave 差动信号
differential weathering 差químico风化(作用)；分异风化
differential weight distribution 微分重量分布(函数)
differential weights 不等权数
differential weir 差动堰
differential winch 差动卷扬机
differential wind 差动风
differential winder 差速卷绕机
differential winding 差动绕法
differential windlass 差动(式)卷扬机；差动辘炉；差动绞盘；差动滑车
differential wound field 差绕磁场
differential wound motor 差励电动机
differentiate 分异岩；区别；求微分；求导；判别
differentiated 分化型
differentiated dike 分异岩脉
differentiated filament yarn 变异长丝
differentiated layering 分异层理
differentiated rock 分异岩；分化脉岩
differentiated sill 分异岩床
differentiated treatment 差别待遇
differentiate intrusive 分异侵入体
differentiate mark of active fault 活断层判别标志
differentiate rate 特定的运费率(铁路)
differentiating amplifier 差分放大器；差动放大器
differentiating circuit 辨析电路；微分电路
differentiating effect 辨impl效应；区分效应
differentiating network 微分电路
differentiating pulse 微分脉冲
differentiating solvent 辨别溶剂；区分性溶剂
differentiating time constant 微分时间常数
differentiation 演变；分异(作用)；分化(作用)；取导数；区别；微分法
differentiation center 分化中心
differentiation index 分异指数
differentiation of a matrix 矩阵的微分
differentiation of conditioned reflex 条件反射分化
differentiation of magma 岩浆分异(作用)
differentiation of random process 随机过程的微分
differentiation of sediment 泥砂鉴别；泥砂分选
differentiation of symptoms and signs 辩证
differentiation period 分化期
differentiation phase 分化相
differentiator 鉴别者；差示装置；示差器；差动装置；差动轮；差动电路；微分元件；微分器
different kinds of liquid solvents 不同种类的溶剂
different level 不同高程；不同标高
different levels of potassium 不同水平施钾
different levels of productive forces 不同层次生产力
different line 各种线
different moisture levels 各种湿度
different N concentration 不同氮浓度
different oxygen concentrations 不同氧浓度
different poultry farming 各类养鸡场
different quality 品质不符
different-sized original 不同尺寸原稿
different size of autodrinker 各种自动饮水器
different soil-forming processes 不同的成土过程
different soils 不同土壤
different spacings 不同行距
different state 各种状态
different surface 各种面
different temperature point 各种温度点
different temperatures 不同温度

different time observation 异时观测
differflange beam 宽缘工字(钢)梁；宽翼梁；宽缘梁；不等缘工字梁；不等翼梁
differflange I-beam 不等翼工字梁
differ flange (rolled) section joist 宽缘工字梁
differing site conditions 现场变异条件；不同的现场条件
differs form symbol 不等符号
difficult area 碍航区
difficult communication 可听度差；难于通信；通信困难
difficult country 丘陵地区；难通行地区；丘陵地带
difficult country ground 丘陵地区
difficult foundation 难处理地基
difficult-fusible ash 难熔灰分
difficult ground 难通行地区；(难处理的)软弱土层；软弱地层
difficultly falling roof 难冒落顶板
difficultly selfcombustion 不易自燃
difficult particle of screening 难筛粒
difficult section 困难区间；难行路段
difficult soluble 难溶解
difficult soluble compound 难溶性化合物
difficult stretch of road 难行路段
difficult-to-cut alloy 难加工合金
difficult-to-machine material 难加工材料
difficult-to-screen material 难筛材料
difficult-to-settle particle 难沉颗粒
difficulty 难度
difficulty degree of detection 检出的难易程度
difficulty in decannulation 拔管困难
difficulty of control 难控性
difficulty to control 难控制性
diffluence 分流
diffluent 易溶的；液化的；分液的；分流性
diffluent river 出渗河(流)；分流支流
diffluent stream 出渗河(流)；分流支流
difform 形状不同的
diffract 绕射
diffracted beam 衍射束；衍射束；绕射射束
diffracted contrast 衍射反差
diffracted integrated migration 绕射积分偏移
diffracted intensity 衍射强度
diffracted light 衍射光；绕射光
diffracted maximum 衍射极限
diffracted pattern 衍射花样
diffracted P wave 衍射P波；绕射初波(地震纵波)；绕射P波
diffracted ray 衍射光线；衍射(射)线；绕射线
diffracted-scaly 龟裂鳞片状的
diffracted spectrum 衍射光谱
diffracted wave 衍射波；绕射波
diffracted wave vector 衍射波矢量
diffracting edge 衍射边
diffracting mask 衍射掩模
diffracting medium 衍射介质
diffracting object 衍射体
diffracting opening 衍射孔
diffracting power 衍射本领
diffraction 衍射(作用)；绕射(作用)
diffractional field 衍射场
diffractional lattice 衍射点阵
diffractional pulse height discriminator 脉冲振幅选择器
diffraction analysis 衍射分析
diffraction angle 衍射角；绕射角
diffraction aperture 衍射孔径
diffraction approach 衍射法
diffraction area 绕射面积
diffraction-based merit function 衍射评价函数
diffraction blur circle 衍射弥散圆
diffraction blurring 衍射模糊
diffraction broadening 衍射增宽
diffraction by a circular aperture 圆孔衍射
diffraction by aperture 孔径衍射
diffraction by slit 缝隙衍射
diffraction circle of confu-sion 衍射模糊圈
diffraction coefficient 衍射系数；绕射系数
diffraction coefficient of wave 波浪绕射系数
diffraction cone 衍射锥
diffraction contrast 衍射对比；衍射衬度
diffraction corona 衍射晕
diffraction coupling 衍射耦合
diffraction cross-section 衍射截面

diffraction cross-section area 绕射截面积
diffraction crystallography 衍射结晶学
diffraction current 衍射流
diffraction curve 衍射曲线
diffraction decoupling 衍射去耦合
diffraction depth migration 绕射深度偏移
diffraction diagram 衍射图;绕射图
diffraction direction 衍射方向
diffraction disc 衍射圆面
diffraction edge 衍射边缘
diffraction effect 衍射效应
diffraction efficiency 衍射效率
diffraction efficiency of hologram 全息图的衍射效率
diffraction fringe 衍射条纹;绕射条纹
diffraction grating 衍射光栅;绕射光栅
diffraction-grating inter-ferometer 衍射光栅干涉仪
diffraction-grating replica 衍射光栅复制
diffraction-grating spectral order 衍射光栅光谱级
diffraction group 衍射群
diffraction halo 衍射晕
diffraction image 衍射(影)像
diffraction image of double star point 双星点衍射像
diffraction image point 衍射像点
diffraction instrument 衍射计【物】
diffraction integral 衍射积分
diffraction intensity standard 衍射强度标准
diffraction lens 衍射透镜
diffraction limit 衍射极限
diffraction-limited devergence 衍射极限发散度
diffraction-limited diameter 衍射极限直径
diffraction-limited lens 衍射限透镜;受衍射限制镜头
diffraction-limited mirror 衍射限镜
diffraction-limited mode 衍射限模式
diffraction-limited optics device 衍射限光装置
diffraction line 衍射线
diffraction loss 衍射损失;衍射损耗;绕射损耗
diffraction mark 衍射符号
diffraction maximum 衍射极限;衍射极大;衍射峰
diffraction method 衍射法
diffraction microscopy 衍射显微术
diffraction modulation transfer function 衍射调制传递函数
diffraction of light 光(的)衍射
diffraction of ocean(ic) waves 海浪绕射
diffraction of sound 声音绕射
diffraction of sound wave 声波衍射
diffraction of water waves 波浪衍射
diffraction of wave 波的绕射
diffraction optics 衍射光学
diffraction optics head-up bisplay 衍射光学平视显示器
diffraction order 衍射级
diffraction oscillation 衍射振荡
diffraction pattern 衍射图(样);衍射图(像);衍射花样;绕射图像
diffraction peak 衍射极限;衍射峰
diffraction phenomenon 绕射现象
diffraction picture 衍射像片
diffraction plane 衍射平面
diffraction point 衍射点
diffraction-produced hot spots 衍射产生的热点
diffraction propagation 绕射传播
diffraction ray 衍射线
diffraction reflection 衍射反射
diffraction region 衍射区
diffraction ring 衍射光环
diffraction scan 绕射扫描
diffraction scattering 衍射散射;绕射散射
diffraction screen 衍射屏
diffraction spacing 衍射间距
diffraction spectrometer 衍射分光计
diffraction spectroscope 衍射分光镜
diffraction spectrum 衍射光谱;绕射频谱;绕射光谱
diffraction spot 衍射斑点
diffraction spot radius 衍射光斑半径
diffraction symmetry 衍射对称
diffraction theory 衍射原理;衍射理论
diffraction wave 衍射波;绕射波
diffraction wedge 衍射楔
diffraction X-ray tube 衍射用 X 射线管
diffractive spillover 衍射溢失

diffracto-ga(u)ge 衍射应变仪
diffractogram 衍射图
diffractometer 衍射仪;衍射器;衍射计【物】;衍射表;绕射计
diffractometry 衍射学
diffuence 扩流区
diffusance 扩散度;扩散性
diffusant 扩散杂质;扩散剂
diffusant particle 扩散源质点
diffusate 扩散物;弥散物;渗出液
diffuse 扩散漫射;弥漫(性的);弥漫(的);铺散的
diffuse background 扩散背景
diffuse branch 喷雾水枪
diffuse circulation 扩散循环
diffuse constant 扩散常数;扩散常量
diffuse-cutting filter 漫散减光滤光片
diffused 扩散的;漫射的
diffused absorption band 漫吸收谱带
diffused aeration 扩展曝气;扩散(式)曝气法
diffused aeration system 扩散式曝气系统
diffused air 扩散空气
diffused air aeration 扩散(空气)曝气;扩散掺气;空气扩散曝气;鼓风曝气;散气曝气
diffused air aeration system 扩散空气曝气系统
diffused air agitation 扩散空气搅拌
diffused air pipe 空气扩散管
diffused air plate 空气扩散板
diffused air tank 曝气池(污水处理);扩散空气池;扩散空气槽
diffused air water 扩散空气曝气水
diffused area 传播地区
diffused aurora 漫射极光
diffused band 漫射谱带
diffused band absorption 漫射光带吸收
diffused base 扩散基极
diffused blur circle 漫射模糊圈
diffused bond 漫反射带
diffused calcification 弥漫性钙化
diffused capacitor 扩散电容器
diffused capacity 扩散电容
diffused channel waveguide 扩散通道波导
diffused coating 扩散涂料;扩散涂层;扩散渗镀
diffused coherent illumination 漫射相干照明
diffused-collector method 集电极扩散法
diffused daylight 漫射天然光;漫射日光
diffused density 漫射(光)密度
diffused diffraction 漫散射衍射
diffused enlarger 漫射式放大机
diffuse density 散射密度
diffused feflectance spec-troscopy 漫反射光谱法
diffused feflectivity 漫反射系数
diffused filter 漫射式滤色镜;漫射滤光器
diffused fringe 漫散条纹
diffused function determination 弥漫功能测定
diffused growth 弥散生长;漫散生长
diffused halo 漫射晕
diffused heat radiation 漫射热辐射
diffused holography 漫射全息照相
diffused illumination 散射照明;弥散光线照明法;漫射(光)照明;扩散照明
diffused illumination hologram 漫射照明全息图
diffused illumination imaging system 漫射照明成像系统
diffused insolation data 漫射日照数据
diffused junction 扩散结
diffused-junction detector 扩散结探测器
diffused kinetochore 漫散着丝粒
diffused layer resistance 扩散层电阻
diffused light 弥散光;漫射光(线)
diffused lighting 弥漫(光)照明;漫射照明;漫射光照明;漫射光均匀照明
diffused light source 漫射光源;扩散光源
diffused line 模糊谱线
diffused matter 弥漫物质
diffused maximum 漫射峰
diffused measuring mark 漫射测标
diffused-meltback 扩散反复熔炼法
diffused microdensitometer 漫射式微密度计
diffused object 漫射物体
diffused object beam 漫射物体光束
diffused optic(al) waveguide 扩散光波导
diffuse double layer 扩散双电层;分散双电层
diffused parameter 扩散参数
diffused pinch effect 漫掐缩效应

diffused pollution 扩散污染
diffused pollution loading 扩散污染负荷
diffused pollution source 扩散污染源
diffused radiation 漫(辐)射
diffused radiation field 漫辐射场
diffused radiation form 漫辐射形态
diffused radio emission 漫射电辐射
diffused ray 漫射(光)线
diffused reflectance 漫反射系数;漫反射率;漫反射度;漫反射比
diffused reflecting power 漫反射率;漫反射本领
diffused reflecting target 漫反射目标
diffused reflection 漫反射
diffused reflection coefficient 漫反射系数
diffused reflection factor 漫反射系数
diffused reflection glass 漫反射玻璃
diffused reflection photo meter 漫反射光度计
diffused reflection spectrum 漫反射光谱
diffused reflector 漫(反)射罩;漫反射体;漫反射器;漫反射镜
diffused refraction 漫折射
diffused resistor 扩散电阻
diffused ring 漫射环
diffused scattering 漫散射
diffused scattering background 漫散射背景
diffused scattering field 漫射场
diffused scattering surface 漫散射面
diffused screen 漫屏
diffused sedimentation 扩散式沉淀
diffused series 漫线系;漫射系
diffused sideband 漫散边带
diffused sky radiaiton 漫天空辐射
diffused sound 漫射声;散音;漫音
diffused sound reduction 减低噪声
diffused source of pollution 扩散污染源
diffused spectrum 漫(射)光谱
diffused spot 漫(射)斑
diffused streak 弥散条纹
diffused surface 漫射面;扩散面
diffused surface water 地面扩散水
diffused surround 漫射晕
diffused transmission 扩散透射(漫透射);漫透射;漫射传输
diffused transmission density 漫透射密度;慢透射密度
diffused transmission factor 漫透射系数;漫射因数;漫射透光系数
diffused transmissivity 漫透射率;漫射透射比;漫射透光度
diffused transmittance 漫(射)透射率;漫射透光率
diffused type 弥漫型
diffused vibration 发散振动
diffused wall cavity 漫射壁腔
diffused wave 漫射波
diffused waveguide 扩散波导
diffused window 漫射窗
diffused X-ray reflection 漫射 X 线反射
diffuse efficiency 扩散效率
diffuse-enhanced spectrum 漫强谱
diffuse equation 扩散方程
diffuse field distance 扩散场距离
diffuse finish 漫射面加工
diffuse front 弱散锋
diffuse layer 扩散层
diffuse layer model 扩散层模型
diffuse light 扩散光;散光;漫射光
diffuse light luminaire 漫射光照明器
diffusely scattered neutrons 漫散射中子
diffusely transmitting 漫射发射
diffuse masking 扩散掩蔽法
diffuse nebulae 不规则星云
diffuseness 漫射
diffuse oslar radiation 太阳散射辐射
diffuse plate 扩散板
diffuse polluting source 分散污染源
diffuse-porous 扩散的;多孔的;散孔
diffuse porous wood (木材的)散孔材
diffuser 扩压器;扩散体;扩散器;扩散рано道;扩散管;扩散段;空气散流器;浸提器;浸提体;浸取器;进气道;漫射体;漫射器;洗料池;偏光片;散射体;散射器;散气器;散流器;散光罩
diffuse radiation 扩散辐射
diffuser air supply 散流器送风
diffuser area ratio 散气面积比

diffuser as mixers 扩散混合器
diffuser casing pump 扩压泵；扩散泵
diffuser chamber 扩散(器)室
diffuser cone 扩散锥(体)
diffuser dome 空气扩散罩
diffuser drag 扩散器阻力
diffuser efficiency 扩压器效率
diffuse reflectance 扩散反射率
diffuse reflectance spectroscopy 漫反射分光法
diffuse-reflecting power 漫反射能力
diffuse-reflection 漫反射；扩散反射
diffuse-reflection factor 漫反射系数
diffuse-reflective cavity 漫反射腔
diffuse resinite 细分散树脂体
diffuser feeding 散流器送风
diffuser for sewage discharge 污水排放扩散管
diffuser grid 扩散器栅；栅形扩散器；导流栅
diffuser inlet 扩压器入口
diffuser intake 扩压器进口
diffuser lip 扩压管唇部
diffuser loss 扩散损失
diffuser nozzle 扩散式喷嘴；多用水枪
diffuser pipe 扩散管
diffuser plate 扩散(掺气)板
diffuser priming of pump 扩散(器)式水泵；扩散器启动泵
diffuser rating 扩散器额定值；扩散器率定
diffuser ratio 扩散器比例
diffuser screen 散射屏；散躲屏
diffuser spray nozzle 喷雾水枪
diffuser system 扩散器系统
diffuser tube 扩散管
diffuser tube in tank 油箱中扩散管；水箱中扩散管
diffuser-type centrifugal pump 扩散式离心泵
diffuser-type of shade 扩散式遮罩
diffuser valve 扩散阀
diffuser vane 散气片；扩压器内导向叶片；扩散叶轮；扩散(器)叶片
diffuse seepage 扩散渗流
diffuse skylight 天空漫射光
diffuse sky radiation 天空漫射；天空扩散辐射
diffuse solar radiation 太阳漫射辐射
diffuse sound 漫射声音；扩散声(音)
diffuse sound control 扩散声控制
diffuse sound field 声音扩散场；扩散声场
diffuse sound reduction 扩散声衰减
diffuse sound transmission 扩散声传递
diffuse spectrum 绕射光谱
diffuse transformation 扩散相变
diffuse transmission 扩散渗透；扩散传输
diffuse transmission factor 扩散传输因数
diffuse urban runoff pollution management 城市径流污染弥散管理
diffusibility 散播力；弥漫性；扩散性；扩散率；扩散本领
diffusibility of gases 气体扩散率
diffusibity 弥散性
diffusible 可扩散的
diffusible calcium 可扩散性钙
diffusible electrolyte 扩散性电解质
diffusing action 扩压作用
diffusing agent 扩散剂
diffusing background 漫射本底
diffusing block 扩散块；扩散板
diffusing capacity 弥散量
diffusing ceiling 照明天花板；照明顶棚；漫射光(照明)天花板
diffusing ceiling system 天花板漫射照明系统
diffusing coupler 扩散型成色剂
diffusing disc 漫射柔光器
diffusing disk 漫射圈；漫射片；漫反射斑
diffusing extinction of arc 扩散灭弧
diffusing fitting 漫射灯具
diffusing glass 毛玻璃；漫射玻璃；散光玻璃
diffusing glass screen 漫射玻璃屏
diffusing globe 漫射器；漫散球
diffusing heat treatment 扩散热处理
diffusing illumination 散光照明
diffusing lens 漫射透镜
diffusing media 漫射媒质
diffusing mixer 扩散混合器
diffusing object 扩散物体
diffusing panel 漫射板；散光镶板；散光嵌板
diffusing pattern 扩散方式

diffusing phenomenon 扩散现象；弥散现象
diffusing pit 扩散坑
diffusing plastic sheet 散光塑料膜；散光塑料板；漫光塑料膜；漫光塑料板
diffusing relief valve 散流安全阀
diffusing screen 漫射遮光装置；漫射(遮光)屏；散射屏
diffusing scroll 扩散涡管
diffusing sheet 散射屏
diffusing source 漫射(能)源
diffusing surface 扩散面；漫射(表)面；散光表面
diffusing tissue 渗滤布
diffusing type steam valve 扩压式汽阀
diffusing unit 散光器；光线扩散器
diffusing wall 照明壁；漫射光照明墙(壁)；散光墙体
diffusing well 地下水补给井；扩散井
diffusiometer 扩散率测定器；弥散率测定器；气体扩散计；渗滤计
diffusion 扩压；扩散；弥散；漫射；渗滤
diffusion-activation doubly controlled reaction 活化扩散双重控制反应
diffusion aerator 扩散通气器；扩散曝气器
diffusional bond 扩散结合
diffusional creep 扩散蠕变
diffusional decay 扩散衰变
diffusion-alloying 扩散合金化
diffusion along dislocation-pipe 位错管线扩散
diffusional resistance 扩散阻力
diffusion altitude 漫射宽容度
diffusional transfer 扩散性转移
diffusion analogy 扩散模拟；扩散类比(法)；扩散比拟(法)
diffusion analysis 扩散分析
diffusion and dilution of pollutant in river 河水污染物扩散稀释
diffusion annealing 扩散退火；均匀退火
diffusion anomaly 扩散异常
diffusion anti-fouling paint 接触渗出型防污漆
diffusion approximation 扩散近似
diffusion area 扩散面积
diffusion attachment 漫射附件
diffusion aureoles 扩散环
diffusion barrier 扩散阻挡层；扩散阻碍；扩散膜；阻扩散剂；阻扩散层
diffusion block 扩散式消力墩；扩散块
diffusion bonding 扩散黏结(法)；扩散压合；扩散连接；扩散结合
diffusion boundary layer 扩散边界层
diffusion bounded tip 扩散黏结漏嘴
diffusion brazing 扩散硬钎焊
diffusion bucket 扩散式消力斗
diffusion capacitance 扩散电容
diffusion cell 扩散池
diffusion chamber 扩散箱；扩散盒
diffusion charging 扩散荷电
diffusion circulation 扩散循环
diffusion cloud chamber 扩散云室
diffusion coating 扩散镀层；扩散涂层；扩散包覆层
diffusion coefficient 扩散系数；扩散率；漫射系数
diffusion coefficient of grain boundary 晶粒边界扩散系数
diffusion coefficient of particle 颗粒扩散系数
diffusion coefficient of vacancy 空位扩散系数
diffusion coefficient value 扩散系数值
diffusion colo(u)ration 扩散着色
diffusion combustion 扩散(式)燃烧
diffusion component 扩散成分
diffusion conductance 扩散电导
diffusion constant 扩散常数；扩散常量；弥漫常数
diffusion control 扩散控制
diffusion-controlled growth 扩散控制生长
diffusion-controlled reaction 扩散控制反应；受控扩散反应
diffusion-controlled reaction rate 受控扩散反应速率
diffusion-controlled release disinfect 受控扩散释放消毒剂
diffusion cooling 扩散冷却；发散冷却
diffusion couple 扩散偶
diffusion creep 扩散蠕变；分散蠕变
diffusion current 扩散电流
diffusion current density 扩散流密度
diffusion cyclone 扩散式旋风吸尘器
diffusion depth 扩散深度
diffusion diagram 扩散图解

diffusion dialysis 扩散渗析
diffusion disc 散射圈
diffusion disk 漫射片；漫射盘
diffusion distance 扩散距离
diffusion effect 扩散作用；扩散效应
diffusion electrode 扩散电极
diffusion enlarger 漫射放大机
diffusion equation 扩散方程；漫射方程
diffusion equilibrium 扩散平衡
diffusion factor 扩压因子；扩散因素；漫射因数；漫射系数；车队离散系数
diffusion fastness 扩散阻力
diffusion film 扩散膜
diffusion filter 扩散滤波器
diffusion flame 扩散焰；层流焰
diffusion flame burner 扩散式燃烧器；光焰燃烧器
diffusion flame firing 扩散火焰煅烧
diffusion flow 扩散流
diffusion flux 扩散通量
diffusion function 弥散功能
diffusion furnace 扩散炉
diffusion glass 散射玻璃；散光玻璃；漫射玻璃
diffusion gradient 扩散梯度
diffusion halation 扩散光晕
diffusion halo 扩散晕；漫射光晕
diffusion hardening 扩散硬化处理；金属表面渗碳
diffusion heat 扩散热
diffusion humidification 扩散增湿作用
diffusion humidity 扩散湿度
diffusion hygrometer 扩散湿度计；扩散湿度表
diffusion hypothesis 扩散假说
diffusion index 扩散指数；景气动向指数
diffusion inductance 扩散电感
diffusion jump 扩散式水跃
diffusion kernel 扩散核
diffusion knife 切刀
diffusion lamp 漫射灯泡
diffusion layer 扩散层
diffusion length 扩散距离；扩散长度
diffusionless transformation 无扩散型相变
diffusion-limited aggregation model 扩散限制聚集模型
diffusion mask 扩散掩蔽
diffusion measurement 扩散测量
diffusion mechanism 扩散机理
diffusion membrane 扩散膜
diffusion method 扩散法
diffusion mixer 扩散混合器
diffusion mixing 扩散式混合
diffusion model 扩散模型
diffusion moisture 扩散湿气
diffusion mo(u)lding 扩散造型
diffusion network(ing) 扩散网络
diffusion normalizing 高温扩散正火
diffusion nozzle 扩散喷嘴
diffusion number 扩散数
diffusion of atmospheric pollution 大气污染扩散
diffusion of contaminant 污染物扩散
diffusion of heat 热扩散(作用)
diffusion of image point 像点模糊
diffusion of innovations 新产品渗透理论
diffusion of light 光的漫射
diffusion of lower molecular weight hydrocarbon 低分子量烃类扩散
diffusion of noxious substance 有害物扩散
diffusion of pollutant 污染物扩散
diffusion of pollution 污染扩散
diffusion of property 产权分散
diffusion of solids 固体扩散
diffusion of water vapo(u)r 水蒸气扩散
diffusion of wealth theory 分散财富说
diffusion optic(al) fiber 弥散光纤
diffusion parameter 扩散参数
diffusion path 扩散途径；扩散程
diffusion pattern 扩散方式
diffusion photometer 漫射光度计
diffusion pit 扩散坑
diffusion plate 扩散板
diffusion plate test 扩散板试验
diffusion potential 扩散电位；扩散电势
diffusion pressure 扩散压力
diffusion prior to distribution 扩散先验分布
diffusion process 扩散(转印)法；扩散过程
diffusion profile 扩散曲线；扩散分布图

diffusion pump 扩散泵
diffusion radiation 扩散放射;漫射辐射
diffusion rate 扩散(速)率
diffusion ratio 扩散比(率)
diffusion region 扩散区
diffusion research 扩散研究
diffusion resistance 扩散阻力;扩散电阻
diffusion ring 扩散环
diffusion silicon pressure sensor 扩散硅压力传感器
diffusion sintering 扩散烧结
diffusion source of water pollution 扩散水污染源
diffusion speed 扩散速率
diffusion stack 扩散柱
diffusion still 扩散蒸馏釜
diffusion streak 扩散条纹;扩散痕色;疵纹(镀层面上的)
diffusion technique 扩散技术
diffusion technology 扩散工艺
diffusion temperature 扩散温度
diffusion theory 扩散理论
diffusion-tight 紧密不扩散的
diffusion transfer 扩散转印;扩散转移;漫射转移
diffusion transfer process 扩散转印法;扩散转移过程;扩散传送过程
diffusion transfer reversal process 扩散转移反转法
diffusion transformation 扩散转变;扩散型相变
diffusion treatment 浸泡处理;扩散处理
diffusion trench 消能沟;消能槽
diffusion-type enlarger 漫射光放大机
diffusion-type junction 扩散结
diffusion value 漫射值
diffusion vane 扩散叶片;扩散(流)叶轮
diffusion velocity 扩散速度
diffusion-ventilated 扩散通风的
diffusion ventilation 扩散通风
diffusion wall 扩散壁
diffusion washing 扩散洗涤
diffusion welding 扩散焊(接)
diffusion well 扩散井;地下水补给井
diffusion zone 扩散带范围
diffusive 扩散的
diffusive capacity 扩散能力
diffusive climb 扩散攀移
diffusive flow 物料扩散流动
diffusive force 扩散力
diffusive metasomatism 矿散交代作用
diffusive metasomatism way 扩散交代作用方式
diffusive property 漫射特性
diffusivity 扩散性;扩散系数;扩散率;弥漫性;热量扩散率
diffusivity equation 扩散方程
diffusor 浸提器
diffusor entry 扩散器入口
diffussion problem 扩散问题
difinite bud 定芽
difinition determiningfocal length method 清晰度定焦法
difluoride 二氟化物
difluorinated 二氟化的
difluorizated 二氟化的
difluoro-benzene 二氟化苯;二氟代苯
difluorobromomethane 二氟一溴甲烷
difluorochlobromo-methane 二氟一氯一溴甲烷
difluoro compound 二氟化合物
difluorodibromomethane 二氟二溴甲烷
difluoro ester 二氟酯
difluoro ether 二氟醚
difluoroethylene 二氟乙烯
difluoromethane 二氟甲烷
difluorophosphoric acid 二氟基磷酸
difumarate 富马酸氢盐
difunctional 双作用(式)
difusaire and difusic 百叶通风器(一种通风器)
dig 开凿;小坑(光学玻璃缺陷);采掘
digallic acid 丹宁酸;丹宁
dig-and-turn time 挖掘回转时间
dig a well 打井;挖井
dig down 挖倒
dig-down pit 下陷坑;下沉坑
digenesis 世代交替
digenesis evolution stage of organic matter 有机质成岩作用演化阶段
digenetic euhedral texture 成岩自形晶结构

digenetic mineral 成岩矿物
dignenite 蓝辉铜矿
digest 汇集;消化法;摘要;提要;水解液
digested piggery wastewater 猪场废水消化液
digested security 已吸收证券
digested sludge 消化污泥;消化过的污泥
digested sludge drawoff 消化污泥排除
digester 浸煮器;加热消化池;化污染池;消化器;消化剂;消化池;煮解器;蒸煮器;蒸煮罐;热压蒸溜器;酸解器;熟化器;熟化槽
digester access manhole 消化池出入人孔
digester acid fermentation 消化池酸性发酵
digester coil 化污器盘管;消化盘管
digester draft tube mixer 消化池引流搅拌器
digester gas 沼气
digester gas circulation mixer 消化气(体)循环搅拌器
digester gas production 消化池气(体)产量
digester gas utilization 消化池气(体)利用
digester heat exchanger 消化池热交换器
digester heating 消化池加热
digester liquor drawoff 消化池液体排除管
digester mixing 消化池搅拌
digester recirculation suction 消化池循环唧泥(管)
digester room 蒸煮锅;蒸煮车间
digestibility 可消化性;消化性;消化率
digestible 可消化的
digesting compartment 消化室;污泥消化池
digestion 浸煮;浸提;加热溶解;消解;消化作用;消化;煮解;排泄【生】
digestion cake 酸解固相物
digestion chamber 消化池;消化室
digestion coefficient 消化系数
digestion of sludge 污泥消化
digestion period 消化时间;消化期
digestion process 消化过程;消化处理
digestion residue 溶出残渣
digestion slime 溶出残渣
digestion tank 腐化池;老化罐;化污池;化粪池;消化罐;消化池;隐化池;煮解罐;熟化槽
digestion tank heating 化粪池加热;腐化池加热
digestion temperature 消解温度
digestion time 消化时间;消解时间
digestion tower 污水处理塔
digestion trial 消化试验
digestion water tank 消化水池
digestive canal 消化道
digestive disease 消化系(统)疾病
digestive efficiency 消化效率
digestive gas 消化气体
digestive gas storage tank of sludge 污泥消化气罐
digestive solution 消化液
digestive system 消化系统
digestive treatment 消化处理
digest liquor overflow 消化池液体溢流
digest of statistics 统计摘要
digestor 浸煮器;化污池;消化器;消化剂;消化池;煮解器;酸解器;熟化器
digest sewage sludge 消化污水污泥
digger 掘凿器;采掘机;挖土机;挖掘机;勺斗
digger blade 挖掘铲
digger body 翻耕犁体
digger bottom 翻耕犁体
digger chain 挖掘链
digger dredge(r) 采掘式挖泥船
digger plough 挖沟犁;犁式挖沟机;开沟犁;机犁式挖沟机
digger plow 犁式挖沟机
digger tooth adapter 铲齿接头
digger wasp 地蜂
digging 挖掘;采掘
digging ability 挖掘能力
digging action 挖掘作用
digging angle 切削角;切入角;挖掘角度
digging area 挖掘区
digging arm loader 立爪扒渣机
digging attachment 挖掘附件
digging auger 挖穴螺旋钻
digging bit 矛式钻头;刮刀钻头
digging bucket 挖土铲斗;挖掘铲斗
digging cable 挖掘线
digging coefficient 挖掘系数

digging cycle 挖掘循环
digging depth 掘凿深度;掘削深度;挖掘深度
digging depth line 开挖深度线
digging diagram 挖掘图
digging down 地面以上挖土
digging dredge 挖掘机;挖泥船
digging drum 挖掘机绞车滚筒
digging face 挖掘面
digging flight 挖掘机斗导架;挖掘机斗梯状支架
digging force 挖掘(能)力
digging fork 挖掘叉
digging head 转动头
digging height 露天矿阶段高度;采掘高度;挖掘深度
digging in depth 挖掘深度
digging in height 挖掘高度
digging ladder 链斗式挖掘机;挖掘机斗导架
digging line 开挖界线;挖土动力索;开挖深度线
digging line boom point sheave 铲斗拉绳吊杆端滑(车)轮
digging machine 挖土机;挖掘机
digging machinery 挖掘机械
digging method 开挖方法;挖掘方法;挖掘办法
digging motion 挖掘运动
digging out of barge 驳船清舱作业;清舱作业
digging pile 挖孔桩
digging platform 挖土平台
digging plough 挖沟犁
digging point 铲刃;挖掘地点
digging position 挖掘位置
digging prong 掘斗刃齿;挖土刀齿
digging radius 挖掘(机)半径
digging rake 掘土耙
digging ram 挖掘锤
digging reel 挖掘转子
digging rope 铲斗拉绳
diggings 矿山
digging shovel 挖掘铲;铲土机
digging site 挖掘现场;挖掘场地
digging spade 掘锹
digging speed 挖掘摆幅;挖掘速度
digging teeth 钻头齿刃
digging time 挖掘时间
digging tools 挖掘工具
digging well 挖掘井;掘井
digging wheel 圆盘挖掘机;挖沟器转子;挖掘圆盘
digging width 挖掘宽度
digicon 数字像管
digifax 数模
digigrid 数字化器
digilock 数字同步
digilogue channel 数模信道
digilogue circuit 数模电路
digimatic calipers 电子数显卡尺
digimatic depth ga(u)ge 电子数显深度尺
digimatic height ga(u)ge 电子数显高度尺
digimatic micrometer 数显千分尺
digimer 数字式万用表
dig in 掘进;挖土掩盖
dig-in basin 挖入式港池
digiplot 数字作图
Digiralt 数字雷达测高计
digit 计数单位;一ং宽;二进位;位数;数码;十进数的位
digit absorber 号位吸收机;消位器;数字吸收器
digit-absorbing selector 数字吸收选择器;脉冲吸收(选择)器
digit absorption 消位;数字吸收
digital 数字的;数值式
digital acquisition system 数字采集系统
digital adaptive technique 数字自适应技术
digital adder 数字加法器
digital alarm transmitter 数字警报传送器
digital all-sky camera 数字全天候相机
digital-analog(ue) multiplication 数字模拟乘法器
digital-analog(ue) conversion circuit 数字模拟转换电路
digital-analog(ue) converter 数字模拟转换器;数字模拟变换器
digital-analog(ue) decoder 数字模拟译码器
digital-analog(ue) type 数字模拟型
digital analyzer 数据分析器;数字分析仪
digital angular position 数字角位置
digital announcement device 数字式通知装置
digital approximation 数值近似(法);数值逼近

digital area photometer 数字面积光度计
digital arithmetic(al) centre 数字运算中心
digital atmospheric model 数字大气模型
digital audio 数字声音文件
Digital Audio Visual Council 数字视听委员会
digital automat 数字自动机
digital automatic focus 数字自动聚焦
digital automatic frequency control 数字式自动频率控制
digital automatic map compilation 数据化自动编图
digital automatic tracking and ranging 数字自动跟踪与测距
digital automation 数字自动装置;数字式自动化
digital automation device 数字式自动装置
digital automatization 数字式自动化;数字计算机自动化
digital autopilot 数字化自动驾驶仪
digital back-up 数字回溯(法);数字后援
digital bench scale 数字台秤
digital bit 数字位
digital block 数字区段
digital block flip-flop 数字式阻塞双稳态触发器
digital block inverting amplifier 数字式阻塞倒相放大器
digital cable map 数字管线地图
digital calculator 数字计算器
digital camera 数字式摄像机
digital cartography 数字制图学;数字地图制图
digital channel link 数字信道连接
digital character 数据化字符
digital character representation 数字化字符表示
digital circuit 计算线路;数字线路;数字电路
digital circuit family 数字电路族
digital circuit multiplication system 数字话路扩容系统
digital circuitry 数字式电路
digital classification program(me) 数字化分类程序
digital clock 计数脉冲;数字(时)钟
digital cloud map 数字云图
digital coast line generator 数字海岸发生器
digital code 数字(代)码;数码
digital code converter 数字电码变换器
digital coder 数字编码器
digital code wheel 数字代码盘
digital coding 数字编码
digital coding of voice 声音(的)数字编码
digital coefficient 数字系数
digital coefficient unit 数字式系数单元
digital combiner 数字组合器
digital command 数字指令
digital command system 数字指令系统
digital communication 数字通信
digital communication system 数字通信系统;数字传输系统
digital comparator 数字比较器
digital complement 按位(的)补码
digital computation 数字计算
digital computer 数字(电子)计算机;数字电脑;数学式计算机
digital computer holography 数字计算机全息术
digital computer programming 数字计算机程序编制
digital computer system 数字计算机系统
digital computer system of central control 中央控制数字化系统
digital computing system 数字计算系统
digital connection 数字链接
digital console 数字式控制台
digital control 数字控制;数控
digital control computer 数字(控制)计算机;数控计算机
digital control design language 数字式控制设计语言
digital controlled drawing 数控绘图
digital controller 数字(化)控制器
digital control system 数字控制系统
digital control table 数控绘图桌
digital conversion 数字转换;数字变换
digital conversion receiver 数字变换接收机
digital converter 数字转换器;数字变换器
digital coordinate output 数字坐标输出
digital coordination 数字配位
digital cordless telephone 数字无绳电话

digital correlation 数字相关
digital correlation kit 数字相关器
digital correlator 数字相关器
digital counter 数字计数器
digital cross connect equipment 数字交叉连接设备
digital current meter 数字海流计
digital current recorder 数字海流计
digital data 数字资料;数字信息;数字数据;数据
digital data acquisition and processing system 数字资料收集和处理系统
digital data acquisition system 数字数据收集系统
digital data broadcast system 数字数据发播系统
digital data communication 数字数据通信
digital data compaction technique 数据压缩技术
digital data computer 数字数据计算机
digital data converter 数字数据变换器
digital data display system 数字数据显示系统
digital data distributor 数字数据分配器
digital data group 数字数据群
digital data handling 数据处理
digital data modulation system 数字信息调制系统
digital data network 数字数据网
digital data packet 数字数据包
digital data process 数字数据处理
digital data processing equipment 数字数据处理设备
digital data processing system 数字数据处理系统;数据处理系统
digital data processor 数字数据处理机
digital data receiver 数字数据接收机
digital data recorder 数字数据记录器;数字式记录器
digital data reproduction system 数据复制系统
digital data service 数字信息业务;数字数据系统
digital data system 数字数据系统
digital data tape 数字数据带
digital data terminal 数字数据终端
digital data transceiver 数字数据收发机
digital data transmission 数字数据传输
digital data transmitter 数字数据传输器
digital data wash 数字数据消除
digital decade counter 数字十进制计数器
digital decoder 数字译码器
digital deep seismograph 数字式深层地震仪
digital deflection 数字偏差
digital delay generator 数字式延迟发生器
digital demultiplexer 数字分接器
digital demultiplexing 数字分接
digital demultiplier 数字倍减器
digital detector 计数探测器;断续探测器
digital determinant 数字行列式
digital device 数字装置
digital dial 数字标度盘
digital differential analyzer 数字微分分析机;数值积分器
digital display 数字显示(器);数字(式)显示;数控显示
digital display circuit 数字显示电路
digital display generator 数字显示信号发生器
digital display unit 数字显示装置
digital distribution frame 数字分配架
digital down counter 数字溢出计数器
digital drawing 数控绘图
digital drum 数字(磁)鼓
digital dust measuring apparatus 数字粉尘测量仪
digital earphone 数字耳机
digital echo sounder 数字回声测深仪
digital editing 数字编辑
digital electric-hydraulic control system 数字式电液控制系统
digital electronic universal computing engine 数字式电子通用计算机
digital elevation model 数字高程模型
digital enable 数字启动
digital encoder 数字编码器
digital encoder handbook 数字编码器手册
digital enhancement 数字增强
digital equation 数字方程
digital equipment 数字式设备
digital error monitoring subsystem 数字式误差监控分系统
digital event recorder 数字事件记录器
digital exchange program(me) 数字交换程序
digital exchanger 数字交换机

digital expansion system 数字扩展系统
digital feedback 数字反馈
digital filling 数字填充
digital filter 数字滤波器
digital filtering 数字筛选作用;数字滤波
digital filter method 数字滤波法
digital flowmeter 数字流量计
digital force transducer 数字式力传感器
digital form 数字形式
digital format 数字形式
digital framing structure 数字图像帧结构
digital frequency analyser 数字式频率分析器
digital frequency analyzer 数字频率分析器
digital frequency display 数字式频率显示
digital frequency divider 数字分频器
digital frequency meter 数字(式)频率计
digital frequency sounding instrument 数字频率测深仪
digital function 数字式功能部件;数字功能
digital function generator 数字函数发生器
digital gamma ray survey 数字伽马测量
digital ga(u)ge 数字量规;数字量测仪
digital gaussmeter 数字高斯计
digital geophone 数字地震检波器
digital graphic(al) processing 数字图形处理
digital graphics processing 图形数据处理
digital handling 数字处理
digital hologram 数字全息图
digital home network 数字家用网
digital image 数字影像;数字图像
digital image correlation 数字图像相关
digital image processing 数字图像处理;数字图像加工
digital image processing products 数字图像处理产品
digital image processing system 数字图像处理系统
digital image processing technique 数字图像处理技术;数字图像处理方法
digital image recorder 数字图像记录仪
digital image scan(ning) and plotting system 数字图像扫描记录系统
digital image tube 数字像管
digital impulse 数字脉冲
digital inclinometer 数字(式)测斜计
digital incremental plotter 数字增量绘图仪
digital indexed light deflector 数字光偏转器
digital indicating system 数字显示系统
digital indicator 数字指示器;数字显示仪;数字显示器
digital induced polarization instrument with dual frequency 双频道数字激电仪
digital inertial navigation system 数字式惯性导航系统
digital information 数字信息;数据
digital information detection 数字信息检查
digital information display 数字信息显示
digital information display system 数字信息显示系统
digital information display unit 数字信息显示器
digital information processing 数字信息处理
digital information processing system 数字信息处理系统
digital input 数字输入
digital input base 数字数据库
digital input data 数字输入数据
digital instrument 数字式仪表
digital instrument for elelctric(al) method 数字电法仪
digital integrated circuit 数字集成电路
digital integrated circuit element 数字集成电路元件
digital integrating circuit 数字积分电路
digital integration 数字积分
digital integration computer 数字积分计算机
digital integrator 数字积分器
digital interpolation 数字内插
digitalization 数字化
digitalizer 数字化装置;数字(变换)器
digital keyboard 数字键盘
digital keypunch 数字键控孔机
digital knob 数码钮
digitall analog(ue) conversion 数字模拟转换
digital laser beam deflector 数字激光束偏转器

digital line 数字注记线
digital linear measuring system 光栅式线位移测量装置
digital link 数字链路
digital logger 数字测井仪
digital logging system 文字记录测井仪
digital logic 数字逻辑
digital logic routine 数字逻辑线路
digital logic system 数字系统
digital loop carrier 数字环路载波
digital loop movie 数字循环移动
digital low-pass filtering 数字低通滤波
digitally controlled machine 数字控制机
digitally programmable modem 可编程序数字调制解调器
digitally programmable servo 数字程序化的伺服机构
digitally recording offshore tide ga(u)ge 数字记录式岸用验潮仪
digital machine 数字计算机
digital magnetic intensity instrument 数字式磁性强度仪
digital map 数字地图
digital mapping 数字(化)测图
digital mapping software 数字制图软件
digital mapping system 数字测图系统
digital mapping technique 数字化制图技术
digital map revision 数字地图修测
digital measuring apparatus 数字测量仪
digital measuring instrument 数字式测量仪
digital memory 数字存储器;数字储存器
digital message entry device 数字信息输入装置
digital message link 数字化信息线路
digital message terminal 数字信息终端
digital method 数字(方)法
digital microcircuit 数字微波电路
digital micrometer 数字测微计
digital microphone 数字传声器
digital microwave communication 数字微波通信
digital microwave radio 数字微波无线电
digital microwave relay system 数字微波中继系统;数字式微波接力通信系统
digital model 数字模型
digital model test 数字模型试验
digital modem 数字调制解调器
digital modulation 数字调制
digital module 数字微型组件
digital mosaic 数字镶嵌
digital movies from three-dimensional seismic 三维地震资料的数字电影显示
digital multibeam scan(ning) sonar 数字多波束扫描声呐
digital multibeam steering array 数字多波束基阵
digital multimeter 数字(式)万用表;数字多用表
digital multiplex equipment 数字复用设备
digital multiplexer 数字(信号)复接器
digital multiplex hierarchy 数字复用系列;数字复用体系
digital multiplexing 数字复接
digital multiplier 数字乘法装置;数字乘法器
digital multipurpose wind speed ga(u)ge 多功能数字风速仪
digital name 数字代号
digital navigation set 数字式导航设备
digital network architecture 数字网络体系
digital number 数字
digital ocean-bottom seismograph 数字式海底地震仪
digital ocean data acquisition system 海洋数字资料收集系统
digital ohmmeter 数字式欧姆表
digital operation 数字运算;数学应用
digital operational circuit 数字运算电路
digital optical tracking set 数字光学跟踪装置
digital orientation 数字定向
digital output 数字输出
digital output timer 数字输出计时器
digital packet 数字包裹
digital paging system 数字寻呼系统
digital path 数字通道
digital phase detector 数字相位检测器
digital phase-locked loop 数字锁相环
digital phasometer 数字相位计
digital photogrammetric plotting 数字化摄影测图

digital photogrammetry 数字摄影测量学
digital photography 数字摄影
digital photomosaic 数字像片镶图
digital picture 数字图像
digital picture input tablet 数字化图形输入板
digital plot 数字绘图
digital plotter 数字绘图仪;数字描绘器;数字绘图机;数控绘图机
digital plotting system 数字绘图系统
digital point plotter 数据点标绘器
digital position information 数字位置信息
digital potentiometer 数字(式)电位计
digital potentiometer system 数字分压器制式;数字电位计制式
digital presentation 数字显示
digital-presentation counter 数字显示计算器;数字显示计数器
digital printer 数字印字机;数字打印机
digital process 间断过程;不连续过程
digital process control 数字顺序控制;数字过程控制;数字程序控制
digital process control system 数字过程控制系统
digital processing 数字加工;数字处理
digital processing and control unit 数字与控制装置
digital processing unit 数字处理装置;数字处理器
digital profile data 数字断面数据
digital program(me) controlled switch 数字程控交换机
digital protection device 数字式保护装置
digital pulse 位脉冲
digital pulse duration 数字脉冲持续时间
digital pulse sequence 数字脉冲序列
digital quantity 数字值
digital quantizer 数字转换器;数字量化器
digital radar landmass simulation 数字雷达地面模拟系统
digital radio meter 数字式无线电仪表
digital radio section 数字无线段
digital ramp 数字化台
digital rate 数字速率
digital ratiometer 数字比率表
digital ratio path 数字无线通道
digital ratio system 数字无线系统
digital readout 数字读出(装置)
digital readout device 数字读出仪
digital readout oscilloscope 数字读出示波器
digital readout photometer 数字读出光度计
digital readout pressure indicator 数字显示压力指示器
digital readout system 数字读出系统
digital readout timer 数字读出计时器
digital receiving station 数字接收站
digital recorder 数字自记器;数字记录仪;数字记录器
digital record(ing) 数字(式)记录
digital recording analyzer 数字式记录分析器
digital recording equipment 数字记录设备
digital recording process 数字记录过程
digital recording system 数字记录系统
digital rectification 数字纠正
digital reflection hologram 数字反射全息图
digital regenerator 数字再生器
digital remote unit 数字式遥控装置
digital representation 数字表示(法)
digital resistance strain indicator 数字电阻应变仪
digital resolution 数字分辨率
digital resolver 数字解算器
digital restitution 数字测图
digital restitution instrument 数字测图仪
digital result 数字结果
digital rheostat 数字变阻器
digital roll 数字卷筒
digital salinity temperature and depth instrument 数字温盐深测量仪
digital scale 数字刻度;数字比例尺
digital scan 数字扫描
digital scanner 数字扫描器
digital section 数字段
digital seismic system 数字式地震记录系统
digital seismogram 数字地震图
digital seismograph 数字(记录)地震仪
digital seismometer 数字地震计
digital selective call equipment 数字选择性呼叫设备
digital selective calling 数字选择性呼叫
digital sensor 数字传感器
digital sequence integrity 数字序列完整性
digital servo 数字伺服系统
digital servomechanism 数字伺服机构
digital servosystem 数字跟踪系统
digital signal 数字信号
digital signaling system 数字信号系统
digital signal(l)ing 数字信令
digital signal processor 数字信号处理器
digital signature 数字签名
digital simulated analog(ue) computer 数字仿真模拟计算机
digital simulation 数字式仿拟;数字模拟;数字仿真
digital simulation language 数字模拟语言
digital simulation technique 数字模拟技术
digital simulator 数字式模拟器;数字模拟装置
digital simulator computer system 数字模拟计算机系统
digital simultaneous voice and data 声音和数据同时传输协议
digital situation model 数字位置模型
digital slope model 数字坡度模型
digital smoothing function 数字光滑函数
digital snow map 数字雪盖地图
digital solution of water quality model 数值解水质模型
digital sonar 数字式声呐
digital sort 数字式分类
digital spectral analysis 数字谱分析
digital speech coding 数字语音编码
digital stepping recorder 数字步进式记录器
digital stereocompilation 数字立体测图
digital storage 数字存储器
digital storage system 数字存储系统
digital submarine cable 数字海底电缆
digital subscriber loop 数字用户环
digital subset 数字子集
digital subtracter 数字减法器
digital surface model 数字地形模型;数字表面模型
digital switch 数字开关
digital switching 直接数字交换
digital synchronometer 数字比时同步指示器
digital system 数字化系统
digital system of photographic(al) acquisition 摄影测量数字获取系统
digital system quantizing error 数字系统量化误差
digital tachometer 数字式转速计
digital tape 数字带
digital target characteristic 数码目标特性
digital team 数字化小组
digital technique 数字(化)技术
digital telemetering 数字遥测;数字式远距离测量
digital telemetry 数字遥测术
digital telemetry register 数字遥测寄存器
digital telemetry system 数字遥测系统
digital telemetry unit 数字遥测装置
digital telephone network 数字电话网
digital television 数字电视;数控电视
digital television converter 数字电视转换器
digital television monitor 数字电视检测器
digital television system 数字电视系统
digital template 数字模片
digital terminal 数字终端
digital terrain model 数字地形模型;数字地面模型
digital terrain module 数字地形模数
digital thematic classification 数字信息按专题分类
digital theodolite 数字经纬仪
digital thermister unit 数字热敏电阻装置
digital thermocouple unit 数字热电偶装置
digital thickness ga(u)ge 数字式测厚仪
digital time dissemination 数字时间发播
digital-to-analog(ue) channel 数模信道
digital-to-analog(ue) circuit 数模电路
digital-to-analog(ue) conversion 数模转换
digital-to-analog(ue) converter 数模转换器;数模换能器
digital-to-analog(ue) decoder 数模译码器
digital-to-analog(ue) ladder 数模转换阶梯
digital-to-analog(ue) simulation 数模转模拟器
digital topographic(al) system 地形数字化系统
digital-to-video display 数字视频显示(器)
digital tracing machine 数控绘图机

digital traffic control 数字计算交通控制;数符交通控制
digital transducer 数字转换器;数字传感器
digital transit 数字经纬仪
digital translator 数字变换器
digital transmission 数字传输
digital transmission densitometer 数字式透射密度计
digital transmission system 数字传输系统
digital transmitter 数字发送器
digital type torquemeter 数字式转矩计
digital up counter 数字溢出计数器
digital vertical deflection 数字垂线偏转
digital video 数字视频
digital video broadcast 数字图像广播
digital video disc 数字(激光)视盘
digital visual panel 目视读数面板
digital voltage encoder 数字电压译码器
digital voltmeter 数字(式)伏特计;数字(式)电压表
digital watch 数字显示式电子表
digital water-stage recorder 数字自记水位仪;数字水位计
digital wavefront measuring interferometer 数字波面测量干涉仪
digitalyer 模拟数字变换器
digit area 数码面积
digitate 分指状
digitated pinnate 掌状羽状的
digitate drainage pattern 鸟足状水系;掌状水系;分指状水系
digitate margin of delta 指状三角洲趾;三角洲前趾
digitation 指状分散作用
digit backup 数字后备电路
digit bit pulse 数字二进位脉冲
digit buffer 位缓冲器
digit by digit method 逐位法
digit capacity 位数
digit check 数字校验
digit-coded voice 数字编码声音
digit column 数位列
digit compression 数字压缩;数位压缩;数位减缩
digit count 位数计算;数字计数
digit counter 数位计数器
digit decimal 单个整数数字;十进制数字
digit drive pulse 数字驱动脉冲
digit driver 数字驱动器;数字激励器
digit duration 数字脉冲宽度
digit electrode 数位电极
digit emitter 数字发送器
digit group 数组
digit impulse 数位脉冲
digitiser = digitizer
digitising board 数字化器
digitising on-line system 联机数字化系统
digitising table 数字化器
digitization 数字化
digitization method 数字化方法
digitized 数字化跟踪头
digitized acceleration 数字化加速度
digitized cartography 数字化制图
digitized coordinates 数字坐标
digitized data 数字化数据
digitized echo-sounding 数字记录回声测深
digitized error 数字化误差
digitized file 数字化文件
digitized image 数字化影像;数字化图像
digitized image processing 数字化图像处理
digitized information 数字化信息
digitized instrumentation 数字化仪器
digitized layout 数字化的设计图;数字化版图
digitized mapping 数字化测图
digitized model 数字化模型
digitized photography 数字化摄影
digitized platform 数字化平台
digitized playback system 数字化回放系统
digitized point data 数字化点数据
digitized processing 数字化过程
digitized ramp 数字化台
digitized signal 数字化信息
digitized speech technology 数字化语言技术
digitized strong seismometer 数字化强震仪
digitized surface 数字化面
digitized table 数字化平台
digitized terrain data 数字化地形数据

digitizer 数字器;数字化转换器;数字化仪;数字化设备;十字准线光标数字转化器;三维数字转化器
digitizer format 数字化器型号
digitizer stage 数字化阶段
digitizing accuracy 数字化精度
digitizing board 数字化器
digitizing equipment 数字化器
digitizing machine 数字化器
digitizing scale 数字化比例尺
digitizing station 数字化台
digitizing trainer 数据化教学设备
digit key strip 计数电键片
digit layout 数位配置;数的配置
digit layout parameter 数位配置参数
digit line 位线;数位线
digit order number 数位
digit pair 位偶
digit path 数字道
digit period 数字(信号)周期
digit place 一位数的位置
digit plane 数位面
digit plane driver 数位面驱动器
digit position 一位数的位置;数字位(置);数位位置
digit pulse 位脉冲;数字脉冲
digit punch 数字穿孔;数字穿孔(机)
digit rate 数字率
digit readout micrometer 读数千分尺
digit register spots 数字寄存点
digitron 数字(指示)管;数字(读出)辉光管
digits delay 数位延迟
digit selector 位选择器
digit spacing 数位容量
digit specifier 数字区分符
digits recellens 扳机状指
digit switching 数字交换
digit-symbol generator 数字符产生器
digit synchronization 数字同步
digit tape 数字带
digit time 数字(信号)周期;数字时间
digit time slot 数位时隙
digit track 数字道
digit transfer bus 数字信号传送总线
digit transfer trunk 数字信号传送总线
digitus medius 中指
digit wave form 数位波形
digit wheel 符号轮;数字轮
digitwise operation 按位运算;按数字运算
digiverter 数字转换器
digivolt 数字式电压表
diglyceride 二脂酸甘油酯
diglycerin 一缩二甘油
diglycerol 双甘油
diglycidyl ether 二环氧甘油醚
diglycidyl terephthalate 对苯二甲酸二缩水丙酯;对苯二甲酸二环氧丙酯
diglycol 二甘醇
diglycol aldehyde 二甘醇醛
diglycolamine 二甘醇胺
diglycolic acid 二甘醇酸
diglycol phthalate 邻苯二甲酸二甘醇酯
diglycol stearate 二甘醇硬脂酸酯;二甘醇硬脂酸盐
diglyme 二甘醇二甲醚
diglyph 双陇槽;双槽排挡;双槽面;双槽板(突出的装饰board,带双槽)
dignity of courtroom 法律尊严
digonal axis 二角轴
digram 二字母组
digraph 有向图;双图
digraph representation of a structure 结构的有向图表示法
digroup 数字基群
digs 住所(俚语);寓所
digtato-pinnate 掌状羽状的
digue 防浪堤
dig up 挖松
dihalide 二卤化物
dihedral 两面角;角形的;二面的
dihedral angle 二面角
dihedral angle reflector 二面角反射器
dihedral corner reflector 二面角反射器
dihedral group 二面簇
dihedral reflector 二面反射器;二面反射镜
dihedron 二面体
diheptal base 十四脚管底

dihexagonal 复六方的;双六角棱镜
dihexagonal bipyramid 复六方双锥
dihexagonal prism 复六方柱
dihexagonal pyramid 复六方单锥
dihexahedron 复六方面体;双六面体
dihexyl phthalate 邻苯二甲酸二己酯
Dihl porcelain 迪尔硬瓷器
dihydrate 二水氯化钡;二水合物
dihydrate dolomitic lime 双水化(镁)石灰
dihydrate gypsum 二水石膏
dihydrazone 二腙
dihydric acid 二价酸
dihydric alcohol 二元醇
dihydric phenol 二元酚
dihydric phosphate 磷酸二氢盐
dihydric salt 二氢盐
dihydride 二氢化物
dihydroabietyl phthalate 邻苯二甲酸二氢化松香醇酯
dihydro bromide 氨基乙基异硫脲二溴氢酸盐
dihydrocoumarin 苯并二氢吡喃酮
dihydrodiketo-anthracene 蒽醌
dihydrofolic acid 二氢叶酸
dihydroke toacridine 吖啶酮
dihydrol 二聚水;双水分子
dihydroterpinyl acetate 二氢乙酸松油酯
dihydroxy acetic acid 二羟基乙酸
dihydroxy alcohol 二元醇
dihydroxyaluminum aminoacetate 氨乙酸甘羟铝
dihydroxy phalophenone 酚酞
diimide 二酰亚胺
diimine 二亚胺
diimino succinonitrile 二亚氨基丁二腈
di-interstitials 双填隙(子);双填充子
diiodide 二碘化物
diiodo-acetic acid 二碘乙酸
diiodoaniline 二碘苯胺
diiodo-benzene 二碘苯
diiodofluorescein 二碘荧光素
diiodoform 二碘仿
diiodomethane 二碘甲烷
di-iron phosphide 磷化二铁
di-iso-amyl 二异戊基
di-iso-amylamine 二异戊胺
di-iso-amyl phenylphosphonate 苯基磷酸异二戊酯
diisobutyl 二异丁基
diisobutyl alumin(i)um chloride 氯化二异丁铝
diisobutyl carbinol 二异丁基甲醇
diisobutylene 二异丁烯
diisobutyl ketone 二异丁酮
di-iso-butylmanice 二异丁胺
diisobutyl phthalate 邻苯二甲酸二异丁酯
diisocyanate 二异氰酸酯;二异氰酸盐
diisodecyl phthalate 邻苯二甲酸异癸酯;邻苯二甲酸二异葵酯
diisooctyl azelate 壬二酸二异辛酯
di-iso-octyldiphenylamine 二异辛基二苯胺
diisooctyl phthalate 邻苯二甲酸二异辛酯
diisooctyl sebacate 癸二酸二异辛酯
di-iso-octyl succinate 琥珀酸二异丁酯
diisopropanolamine 二异丙醇胺
diisopropyl 二异丙基
diisopropylamine 二异丙胺
diisopropylbenzene 二异丙苯
diisopropylcarbinol 二异丙基甲醇
diisopropyl ether 异丙醚;二异丙醚
diisopropylethylamine 二异丙基乙胺
diisopropyl fluorophosphate 氟代磷酸二异丙酯;二异丙基氟磷酸
diisopropylideneacetone 二异亚丙基丙酮
diisopropyl ketone 二异丙基甲酮
diisopropyl peroxydicarbonate 过氧化二碳酸二异丙酯
dikaka 固定沙丘
dike 岩墙;岩脉;堤;防波堤围堤;排水道;透水木笼堤;透水格笼堤;塘堤岸
dike area 堤防区
dike batter 堤前斜坡
dike body 堤体;坝身
dike bottom protection 坝的护底
dike breach 决堤;堤(坝)决口
dike breaching 堤防溃决
dike breaking 堤防溃决
dike building 堤防工程

dike burst 决堤;堤决口;堤防溃决
dike cleaner 沟渠清理机
dike consolidation by reclamation 吹填固堤
dike cutter 开槽刀
diked 围堤的
dike-dam 护堤;堤坝;护坡
diked area 筑堤区域;障碍区;围堤区域;围堤面积;圩区
dike design 坝的设计
diked land 围堤地;堤防地;圩地
diked marsh 垸田;堤围泽地;圩地
dike drainage lock 围堤泄水闸
diked reach 有坝河段
dike failure 决堤;堤决口;堤防溃决
dike field 坝田
dike fill 堤坝填土
dike foot 堤脚
dike footing 堤防底脚
dike fortifying project 护堤工程
dike head 坝头
dike keeper 护堤员;巡堤员
dike ledge 戗台;坡腰平台
dikelet 小岩墙;小岩脉
dike lock 堤闸;堤坝闸门;坝闸
dike maintenance 堤防维修;堤防维护
dike management 堤防管理
dike master 筑堤工人
dike opening 堤口
dike path 沿堤小路
diker 挖渠机;筑堤工人;筑堤机
dike ramp 下河斜坡道
dike reeve 护堤员;巡堤员
dike rehabilitation 复堤(工程)
dike rock 脉岩;围岩
dike root 坝根
dike-root caving 搜根
dikes and dams 堤坝
dike-satellite 围脉
dike set 岩脉组
dikes filling of solid bitumen 固体沥青岩墙
dike slope 堤(前斜)坡
dike strengthening by colmation 放淤固堤
dike summit 堤顶
dike swarm 岩墙群;岩脉群
dike system 堤防系统;堤示
diketen(e) 双烯酮
diketoalcohol 二酮醇
diketone 二酮
dike top 堤顶
dike-type cofferdam 土工围堰
dike-type seal 栏框式密封
dike way 沿堤小路
dike width 堤宽
Di kiln 帝窑
diking 围堤;开沟排水;筑(围)堤
diking system 堤防系统
dikites 墙岩;脉岩;半深成岩
diktyonite 网状混合岩
dilantin 大伦丁;二苯乙内酰脲
dilapidated building 倒塌的房屋;破烂建筑物;危险房屋;老朽房屋;将塌房屋
dilapidated housing 危房
dilapidation 残破不堪;财产荒废;崩塌;崩落;倾圮;倾毁
dilatability 膨胀性;膨胀率;伸缩度
dilatable 可膨胀的
dilatable soil 膨胀土
dilatancy 扩张性;扩容现象;搅胀性;剪胀性;逆触变性;膨胀性;膨胀变形
dilatancy clay 膨胀黏土
dilatancy criterion 剪胀准则(非黏性土剪力变形)
dilatancy diffusion model 膨胀扩散模型
dilatancy effect 膨胀效应
dilatancy genesis of earthquake 扩容成因
dilatancy hardening 扩容硬化
dilatancy model 扩容模型
dilatancy of soil 泥土膨胀性;土的膨胀性
dilatancy of subgrade 路基膨胀
dilatancy test 手搓试验
dilatant 胀流体;膨胀物
dilatant consistency 膨胀性稠度
dilatant dispersion 胀流型分散体
dilatant flow 胀性流动;胀流型流动
dilatant fluid 胀流型流体;膨胀性流体

dilatant plastisol 膨胀性增塑溶胶;膨胀性增塑糊
dilatant thickener 膨胀增稠器
dilatation 扩张术;剪胀(性);膨胀(性);膨胀作用;膨胀度;膨胀比(率);膨胀;松胀
dilatational 膨胀性的
dilatational disturbance 膨胀扰动
dilatational strain energy 膨胀应变能
dilatational strain energy gradient 膨胀应变能量梯度
dilatational viscosity 第二黏度;膨胀黏度
dilatational wave 疏密波;膨胀波
dilatation clay 膨胀性黏土
dilatation coefficient of viscosity 黏性膨胀系数
dilatation constant 膨胀常数;膨胀常量
dilatation device 膨胀装置
dilatation diffraction 扩散衍射
dilatation fissure 膨胀裂隙;膨胀裂缝
dilatation joint 钢轨间隙;膨胀缝;伸缩缝
dilatation soil 膨胀性土(壤)
dilatation source 膨胀源
dilatation strain meter 膨胀应变计
dilatation stress factor 膨胀应力系数
dilatation test 膨胀试验(如水泥等)
dilatation wave 扩散波
dilated 扩张的
dilating device 膨胀伸缩机构
dilation 扩张术;剪胀;膨胀度;伸缩
dilational vein 扩张脉
dilation fissure 膨胀裂隙
dilation point 剪胀点
dilation pressure 膨胀压
dilation strength 剪胀强度
dilation test 膨胀试验
dilation value 剪胀量
dilative 剪胀性的
dilative soil 剪胀性土;膨胀性土(壤);膨胀土
dilatometer 径向松胀仪;触头膨胀计;侧胀仪;膨胀仪;膨胀计
dilatometer measurement 膨胀计测定
dilatometer test 膨胀度试验
dilatometric change 膨胀变化
dilatometric curve 膨胀曲线
dilatometric method 测膨胀法
dilatometric softening point 变形软化温度;变形点
dilatometric test 膨胀仪检验
dilatometry 膨胀测量术;膨胀测量法;膨胀测定术;膨胀测定法
dilator 扩张器;膨胀箱
dilauroyl peroxide 过氧化十二酰
dilaurylamine 二月桂胺
dilazep 地拉齐普
dilemma zone 进退两难区【铁】
di-lens 介质透镜
dilettante 艺术爱好者
diligence 公共马车
dilinoleic acid 二亚油酸
dilipoxanthine 二脂黄质
dill 草茴香
dillinite 水铝高岭石
dilly 斜井拖运系统;小型货车;小车;平板车
dilly hole (中型的)中间沉淀池;方钻杆鼠洞
dillying 筛上冲洗
diloxanide 二氯尼特
dilsh 劣质薄煤层;黑黏土
diluent 稀释剂;稀料;增量剂;冲淡剂
diluent air 加稀空气;稀释空气;二次空气
diluent cooling 液冷冷却;喷液冷却
diluent gas 稀释气体
diluent material 冲淡剂
diluent naphtha 石脑油溶剂
diluent solvent 稀释溶剂
dilutability 可稀释性;可稀释度;稀释度
dilute 稀释的
dilute acid washing 稀酸洗涤
dilute alkali 稀碱
dilute alkali immersion 浸稀碱液
dilute alloy 低合金
dilute-alloy martensite 低合金马氏体
dilute aquatic system 稀释水系统
dilute aqueous solution 稀释水溶液
diluted 稀释的
diluted acid 稀酸
diluted $CaCl_2$ solution return pump 稀氯化钙回流泵

diluted concentration 稀释浓度
diluted debris flow 稀性泥石流
diluted factor approach 稀释因子法
diluted lubricant 稀释润滑剂
diluted mineral acid 稀释无机酸;稀释矿物酸
diluted paint 稀漆;调合漆
diluted sludge volume index 稀释污泥体积指数
diluted slurry 稀释料浆
diluted soluble oil 稀释乳化油
diluted solution 稀释溶液
diluted water 冲淡水
dilutee 非熟练工(人)
dilute hydrochloric acid 稀盐酸
dilute organic-water mixture 稀释有机物—水混合物
dilute phase 稀相
dilute phase bed 稀相床
dilute pour point 稀释倾点;稀释流动点
dilute power 稀释力
dilute salt 稀盐
dilute sewage 稀释污水
dilute sewerage 稀释污水
dilute solution 稀溶液
dilute sulfuric acid storage tank 稀硫酸储[贮]槽
dilute sulphuric acid 稀硫酸
dilute swine wastewater 稀释猪猡污水
dilute volume 稀释量
dilute waste 稀释废水
dilute wastewater 稀释污水
dilute water 稀释水
diluting agent 稀料;冲淡剂;稀释剂
diluting effect 稀释作用
diluting factor 稀释因子
diluting installation for hopper 泥舱稀释装置
diluting or packing of goods 掺杂使假
diluting polluted water by artificial recharge 人工回灌冲淡已污染水质
diluting power 稀释力
diluting tank 稀释池
diluting the latex 稀释胶剂
dilution 稀(释)度;股权收益减损;淡度;冲淡(度)
dilution agent 稀释剂
dilution air 混入烟气的空气;变稀的空气
dilution analysis 稀释分析
dilution and concentration test 稀释浓缩试验
dilution and diffusion 稀释扩散
dilution and disposal 冲稀排放
dilution and seeding method 稀释与接种法
dilution approach 稀释法
dilution capacity 稀释能力
dilution coefficient of dissolved carbon 溶解无机碳稀释系数
dilution colo(u)rimetry 稀释比色法
dilution discharge 稀释排放
dilution effect 稀释效应
dilution efficiency 稀释效率;冲淡效率
dilution end point 稀释终点
dilution factor 稀释因子;稀释系数;稀释倍数
dilution gas 稀释气体
dilution ga(u)ging 浓度测流速法;稀释测量;稀释测流法
dilution heat 稀释热;冲淡热
dilution limit 稀释限度;稀释极限
dilution mechanism 稀释机制
dilution metering 稀释测定;冲淡测定法
dilution method 浓度测流速法;稀释法
dilution of ore 矿石贫化率
dilution of precision 精度稀释
dilution of seawater 海水淡化
dilution of sewage 污水稀释
dilution of water bodies 水体稀释
dilution oxidation pond 稀释氧化塘
dilution phenomenon 稀释现象
dilution prediction 稀释预测
dilution process 稀释作用
dilution property test 稀释性试验
dilution rate 稀释(速)率;稀释比例
dilution ratio 径污比;稀释比(率);冲淡率
dilution ratio of ore 矿石贫化率
dilution ratio of water 水稀释比
dilution requirement 稀释要求;稀释需水量
dilution solution 稀释液
dilution stability 稀释稳定性;稀释安定性
dilution test 稀释(量)试验

dilution titer 稀释滴度
dilution tube 稀释管(测爆仪用)
dilution ventilation 稀释通风
dilution visco(si)meter 稀释黏度计
dilution water 稀释水
diluvial 洪积的
diluvial alluvial fan zone 洪积冲积扇地带
diluvial clay 洪积黏土
diluvial deposit 洪积物;洪积层
Diluvial epoch 洪积世【地】
diluvial fan 洪积扇
diluvial formation 洪积层
diluvial layer 洪积土层
diluvial loam 洪积壤土
diluvial soil 洪积土
diluvian 洪积的
diluvian placer 洪积砂矿床
Diluvium 洪积统【地】
diluvium 洪积物;洪积层
diluvium gravel 洪积砾石
dilvar ferronickel 铁镍合金
Dilver(alloy) 迪维尔合金
dim 暗淡的
dimargarin 二珠酯
dim design 图案不清晰(压花玻璃缺陷)
dimefadane 二甲法登
dimenhydrinate 乘晕宁
dimension 尺寸;量纲;量度;面积;线度;因次;大小;尺度;尺寸标注实体;维(数);维度
dimension after drop down 降维后维数
dimensional 量纲的;尺寸的
dimensional accuracy 量纲尺寸;尺寸准确度;尺寸精度;尺寸的准确性
dimensional analysis 量纲分析;因次分析;尺度分析;维量分析
dimensional analyzer 维量分析器
dimensional chain 尺寸链
dimensional change 尺寸变化
dimensional characteristic 尺寸特性
dimensional chart 尺寸图
dimensional check 规格尺寸检查;尺寸校核
dimensional check record 尺寸检验记录
dimensional check report 尺寸检验报告
dimensional constant 量纲常数
dimensional control 尺寸检验;尺寸控制
dimensional coordination 构件尺寸配合;尺寸协调;尺寸配合;空间配合关系;结构布置图
dimensional data 尺寸数据
dimensional deflection 尺寸偏差
dimensional deviation 尺寸偏差
dimensional discrepancy 尺寸上的差异;尺寸误差;尺寸偏差
dimensional drawing 轮廓图;有尺寸图纸;注明尺寸的图;尺寸图;标有尺寸的图;有尺度图
dimensional effect 尺寸效应
dimensional equation 量纲方程;因次方程
dimensional error 尺寸误差
dimensional figure 尺寸图;尺寸数字
dimensional formula 量纲公式;因次公式
dimensional framework 模数制
dimensional ga(u)ging 尺寸检查
dimensional group 量纲群
dimensional homogeneity 尺度均一性;尺寸均一性
dimensional inspection 尺寸检验;尺寸检查
dimensional inspection report 尺寸检查报告
dimensional interchangeability 尺度互换性;尺寸互换性
dimensionality 维数
dimensionality reduction 降维
dimensional limit 尺寸限制
dimensional limitation 尺寸限制
dimensional line 注尺寸的线;尺度线;尺寸线
dimensional lumber 成材
dimensionally accurate 量纲准确的;尺寸准确的
dimensionally coordinated 尺寸上调整协调的
dimensionally homogeneous 量纲齐次的
dimensionally stable 尺寸稳定的
dimensionally stable anode 形稳性稳定电极;尺寸不变阳极
dimensionally stable electrode 形稳性稳定电极
dimensionally stable film 尺寸稳定胶片
dimensionally stable material 尺寸稳定材料
dimensional map 标有尺寸的地图
dimensional method 量纲法;因次(理论)法

dimensional metrology 尺寸测量法
dimensional non-homogeneity 量纲非齐次性
dimensional of bearing plate 承压板尺寸
dimensional orientation 三度定向;确定空间坐标;空间定向;空间方位;形体排列
dimensional output 公称输出功率
dimensional paper 尺度纸
dimensional precision 尺寸精度
dimensional proportion 尺寸比例
dimensional quantity 有因次量;有量纲量
dimensional range 尺寸范围
dimensional ratio 量纲比
dimensional reference system 尺寸参考系统
dimensional regularization 量纲正规化;量纲规则化
dimensional relation 量纲关系;因次关系
dimensional relationship 量纲关系;尺寸关系
dimensional requirement 尺寸要求
dimensional resonace 形状共振
dimensional resonance 形状共振
dimensional restrictions 尺度限制
dimensional scale 尺度比例;尺寸比例
dimensional scaled model 比例模型
dimensional ship 标准船型
dimensional similarity 量纲相似性
dimensional sketch 轮廓草图;尺寸简图
dimensional sound 立体音响
dimensional stability 形稳性;尺度稳定性;尺寸稳定性;尺寸恒定性;外形稳定性
dimensional standard 尺寸标准
dimensional standard specification 尺寸标准规范
dimensional stone 规格石料
dimensional table 尺寸项目表
dimensional timber 规格木材
dimensional tolerance 尺寸容差;尺寸公差;容许尺寸误差
dimensional transformation 量纲变换
dimensional unit 量纲单位;因次单位;尺寸单位;维量单位
dimensional variation 尺寸偏差;尺寸变化
dimension attribute 维数属性
dimension book 工程尺寸记录册
dimension bound 维数界
dimension characteristic recognition in two-dimensional data 二维资料中维特征识别
dimension chart 轮廓尺寸图
dimension control 尺寸控制
dimension cut shingle 按尺寸切割的木屋面板
dimension diagram 轮廓尺寸图
dimensioned drawing 尺度图;轮廓图;有尺寸图纸;尺寸图
dimensioned non-homogeneity 非齐次性
dimensioned timber 规格木材
dimensioned water level 规定水位
dimension figure 表示尺寸的数字;尺寸图
dimensioning 量尺寸;拟定尺寸;选定尺寸;尺寸计算;参数选定;标(注)尺寸
dimensioning of joint(s) 榫头划线;接榫划线;定榫头的尺寸;节点尺寸的确定
dimensioning of locks 测锁法
dimensioning rule (工程图的)尺寸标示规则
dimensioning ship 标准船型;设计船型
dimensionless 无因次的;无量纲;无尺寸的
dimensionless activity quantity 无量纲活度量
dimensionless coefficient 无因次系数
dimensionless current velocity distribution 无因次流速分布
dimensionless current velocity profile 无因次流速分布
dimensionless curve 无因次曲线
dimensionless displacement 无量纲位移
dimensionless factor 无维因数;无量纲因子;无量纲因素;无量纲系数
dimensionless frequency 无量纲频率
dimensionless group 无量纲参数
dimensionless Henry constant 无量纲亨利常数
dimensionless hydrograph 无因次过程线;无量纲过程线
dimensionless measure 无因次量纲
dimensionless number 无因次数;无名数;无量纲数;无尺度数
dimensionless parameter 无因次参数;无维参数;无量(纲)参数
dimensionless performance 无因次特性线
dimensionless quantity 无因次量;无量纲量;无尺

度量
dimensionless rating curve 无因次水位—流量关系;无因次率定曲线
dimensionless response factor 无量纲反应因数;无量纲反应系数
dimensionless shape vector 无量纲形状失量
dimensionless time function 无量纲时间函数
dimensionless unit 无尺寸单位
dimensionless unit hydrograph 无因次单位过程线
dimensionless value 无量纲值
dimensionless variable 无量纲变量
dimensionless variance 无量纲方差
dimension limit 极限尺寸
dimension limit indicator 极限尺寸指示器
dimension limit system 极限尺寸制;公差制度
dimension line 尺寸(标注)线;标注实体;表示尺寸的线
dimension line arc 尺寸标注弧线
dimension lumber 规格木材;分级木材;标准尺寸木材
dimension of anomaly 异常规模
dimension of a simplex 单纯形的维数
dimension of a vector space 向量空间的维数
dimension of chart 图幅
dimension of connection 连接尺寸
dimension of drilling rig 钻机尺寸
dimension of human figure 人体尺度
dimension of image 像幅
dimension of larger earthquake 较大地震震级
dimension of mineral shapes 矿物形状的量纲
dimension of navigational channel at low stages 枯水航道尺度
dimension of picture 图像纵横比;图像尺寸;像幅
dimension of plastic zone 塑性区尺寸
dimension of principal earthquake 主震震级
dimension of sample 土样尺寸;试样规格
dimension of sliding mass 滑坡体规模
dimension of test pit 试坑尺寸
dimension of the earth 地球大小
dimension of wall 墙身尺寸
dimension of waves 海浪幅度
dimension paper 尺度纸
dimension relation 因次关系;量纲关系
dimensions 规模;尺寸规格
dimensions and figure of the earth 地球的尺度和形状
dimensions and mechanical property of the earth 地球的尺度和力学性质
dimension saw 方板锯;浅切锯
dimension scale 尺寸比例;比例直尺
dimension's chart 尺寸图;外廓尺寸图
dimension select 维数选择
dimension shingle 标准木瓦;标准木板;规格木瓦;规格墙面板
dimensions of aeration tank 曝气池尺寸
dimensions of channel 航道尺度
dimensions of concrete 混凝土量度;混凝土尺寸
dimensions of debris flow 泥石流规模
dimensions of door opening 门洞尺寸;箱门开口尺寸
dimensions of drill shank 钎尾规格
dimensions of drydock 干船坞的尺度
dimensions of karst 岩溶形态尺寸
dimensions of lifeboat 救生艇尺度
dimensions of opening 洞口尺寸
dimensions of orebody 矿体大小
dimensions of shiplift 升船机的尺寸
dimensions of submerged member 水下构件尺寸
dimensions of suction and discharge flanges 进出口法兰尺寸
dimensions of the platform 平台尺寸
dimensions of window opening 窗洞(口)尺寸
dimension staff 规格木料;标准尺寸木料
dimension standard 尺寸标准;连接尺寸标准
dimension statement 维数语句
dimension stock 规格木料;规格(板)材;规格(木)材;标准尺寸木材
dimension stone 标准尺寸石料;规格石料;琢石;规定尺寸的琢石;分级石块;按规格分级的石料
dimension stuff 规格木板
dimension style 尺寸标注类型;标注类型
dimension suffix 维数后缀
dimension text 尺寸(标注)文字
dimension theory 维数理论

dimension timber 建筑木料;建筑木材
dimension variable 尺寸标注变量;标注变量
dimension word 维数字
dimensoinless quantity 无维量
dimer 二聚物;二聚体
dimercaprol 二硫基丙醇
dimercaptoethyl maleate 马来酸二巯基乙酯
dimercurousammonium 氨汞基
dimeric 二部组成的
dimeric compounds 二聚化合物
dimeric dibasic acid 二聚酸
dimeric vinyl acetylene 二聚乙基乙炔
dimeric water 二聚水
dimerisation 二聚作用
dimerization 二聚作用
dimerization addition 加成二聚合作用
dimerization reaction 二聚反应
dimerous 二部组成的
dime store 廉价商品店;一角商店(卖便宜货的商店)
dimetalation 二金属取代作用
dimetallic 二原子金属的
dimetasomation way 双交代作用方式
dimetasomatism 双交代作用
dimethachlon 纹枯利
dimethadione 二甲双酮
dimethicone 二甲聚硅氧烷;二甲基硅油
dimethiodal sodium 二碘甲磺钠
dimethoxanate 二甲氧酯
dimethoxy 二甲氧基
dimethoxyaniline 二甲氧基苯胺
dimethoxy-ethane 二甲氧基乙烷
dimethoxymethane 二甲氧基甲烷
dimethoxyphenylacetic acid 二甲氧苯乙酸
dimethoxysuccinic acid 二甲氧基琥珀酸
dimethyl 二甲基
dimethyl acetamide 二甲基乙酰胺
dimethyl acetic acid 二甲乙酸
dimethyl acetoacetic ester 二甲基乙酰乙酸酯
dimethyl acetophenone 二甲基苯乙酮
dimethyl amination 二甲基胺化作用
dimethyl amine 二甲(基)胺
dimethylamine agueous solution 二甲胺水溶液
dimethylamine solution 二甲胺溶液
dimethylaminoazobenzene 二甲氨基偶氮苯
dimethyl ammonium chloride 二甲氯化铵
dimethyl ammonium formate 二甲基甲酸铵
dimethylan 二甲蓝
dimethylane 二甲蓝
dimethylaniline 二甲苯胺
dimethylaniline hydrochloride 二甲苯胺盐酸盐
dimethyl benzene 二甲苯
dimethylbenzidine 二甲基联苯胺
dimethylbenzoyl 二甲苯酰
dimethyl butadiene 二甲基丁二烯
dimethyl butadiene rubber 二甲丁二烯橡胶
dimethylbutane 二甲基丁烷
dimethyl butene 二甲基丁烯
dimethyl cabonate 碳酸二甲酯
dimethyl chlorendate 氯菌酸二甲酯
dimethylchlorosilane 二甲基氯硅烷
dimethyl cyclohexyl phthalate 邻苯二甲酸二(甲基环己)酯
dimethyldiallylammonium chloride 二甲基二烯丙基氯化铵
dimethyl dichlorosilane 二甲基二氯硅烷
dimethyl diketone 丁二酮;双乙酰
dimethyl dioxane 二甲基二氧乙环;二甲基二恶烷
dimethylene 二亚甲基
dimethyl ester 二甲酯
dimethyleterahydronaphthalene 二甲基四氢化萘
dimethyl ethanolamine 二甲乙醇胺
dimethyl ethenyl carbinol 二甲乙烯基甲醇
dimethyl ether 二甲醚
dimethyl ethyl carbinol 二甲乙基甲醇
dimethyl formamide 二甲基甲酰胺
dimethyl glyoxime 丁二酮肟;二甲乙二肟
dimethyl glyoxime spectrophotometry 丁二酮肟分光光度法
dimethylhexane 二甲基己烷
dimethylhydroxybenzene 二甲苯酚
dimethyl isobutyryl chloride 二甲基异丁酰氯
dimethyl isophthalate 间苯二甲酸二甲酯;异苯二甲酸二甲酯
dimethylisopro panolamine 二甲基异丙醇胺

dimethyl itaconate 衣康酸二甲酯
dimethyl ketone 二甲基酮
dimethyl malenate 丙二酸二甲酯
dimethyl malonic acid 二甲基丙二酸
dimethyl malonic ester 二甲基丙二酸酯
dimethyl naphthidine 二甲基联萘胺
dimethyl nitrosamine 二甲基亚硝胺
dimethylol ethylene urea resin 二羟甲基乙烯脲树脂
dimethyl oxalate 草酸二甲酯
dimethyl phenol 二甲(苯)酚
dimethylphenylene diamine 二甲苯二胺
dimethyl phthalate 邻苯二甲酸二甲酯
dimethylpiperazine tartrate 赖塞托
dimethyl polysiloxane 二甲(基)聚氧烷
dimethyl-p-phenylenediamine 二甲基对苯二胺
dimethyl-p-phenylenediamine sulfate 二甲基对苯二胺硫酸盐
dimethyl propane 二甲基丙烷
dimethyl sebacate 癸二酸二甲酯
dimethylselenide 二甲基硒
dimethyl silane 二甲(基)甲硅烷
dimethyl-silicane 二甲基硅烷
dimethyl silicone oil 二甲基硅酮油
dimethyl-silicone-polymer fluid 二甲基硅氧烷聚合液
dimethyl silicone rubber 二甲基硅橡胶
dimethyl siloxane 二甲基硅氧烷
dimethyl silscone 二甲基硅酮
dimethyl succinate 琥珀酸二甲酯;丁二酸二甲酯
dimethyl succinic acid 二甲丁二酸
dimethyl sulfate 硫酸二甲酯
dimethyl sulfate test 硫酸二甲酯检验(法)
dimethyl sulfide 二甲基硫醚
dimethyl sulfide and metha-nethiomethane 二甲硫
dimethyl sulfonium methylide 二甲基亚甲基硫
dimethyl sulfourea 二甲基硫脲
dimethyl sulfoxide 二甲基亚砜
dimethyl sulfoxonium methylide 二甲基亚甲基硫氧
dimethyl terephthalate 对酞酸二甲酯;对苯二甲酸二甲酯
dimethyl thiourea 二甲基硫脲
dimethyl tin oxide 氧化二甲基锡
dimethyl urea 二甲脲
dimethyl yellow 二甲基黄
dimethyl zinc 二甲锌
dimethyoxy succinic acid 二甲氧基丁二酸
dimeticone 二甲硅油
dimetric 正方形
dimetric drawing 正二轴测图
Dimet wire 迪梅特线
dim filament 暗灯丝
dimicaceous schist 二云母片岩
dimictic lake 双季对流混合湖
dimidiate 折半;对开的;对分的;对半的;二分;二(等)分
diminazene aceturate 重氮氨苯脒乙酰甘氨酸盐
diminish 降低;缩小
diminished angle 缩角
diminished arch 扁平拱;弧形拱;平圆拱
diminished block 缩减闭段
diminished column 窄柱
diminished image 缩小像
diminished pipe 缩窄管;缩径管
diminished radix 基数减1
diminished radix complement 基数减1补码;基数反码
diminished rule 柱径渐减的柱样板
diminished scale 蜗线尺;缩尺
diminished shaft 锥形柱;柱径渐减的柱身
diminished stile 窄门梃;不等宽门边梃;不等宽门窗边梃
diminished utility 效应递减
diminisher 减光器
diminish(ing) 减少
diminishing assets 递耗资产
diminishing balance depreciation 折旧分摊递减法
diminishing balance method 余额递减法
diminishing balance system 递减分摊制度
diminishing cost 降低成本;逐步降低的成本
diminishing courses 渐收瓦列;尺寸递减层;递减行距瓦层

diminishing course work 依次缩小行间距离的挂瓦法
diminishing efficiency 效率递减
diminishing line 缩线
diminishing marginal productivity 边际生产率递减
diminishing marginal utility 边际效用递减
diminishing method 递减法
diminishing piece 锥形件;锥形管;接合器;(联结不同直径管子的)变径配件
diminishing pipe 缩径管;锥形管;缩窄管
diminishing productivity 生产率递减
diminishing relative value 相对价值递减
diminishing return 报酬渐减;报酬减少;报酬递减;收益递减;递减输出
diminishing returns from increasing intensity 密度增加而收益下降
diminishing returns to scale 规模报酬递减
diminishing rule 柱径渐减的柱样板;仿形尺
diminishing scale 缩小的比例(尺);缩尺
diminishing socket 缩小套节
diminishing step 转向踏步;扇形踏步
diminishing stile 不等宽门窗边梃;收截面门梃(通常在上部装玻璃处收小)
diminishing stop bevel 圆棱
diminishing strakes 渐薄船壳板
diminishing stuff 渐薄船壳板
diminishing substitution possibility 替代可能性递减
diminishing tower 小塔(建筑装饰用)
diminishing utility 效用递减
diminish of accuracy 降低精度
diminish pipe 渐缩管
diminish piping 渐缩管
diminish rates on cost method 变率递减法
diminish tube 渐缩管
diminish tubing 渐缩管
diminuendo 渐弱
diminution 变小;变细;缩减
diminution of backwater effect 回水影响衰减
diminution of employment 缩减雇员
diminution of roots 缩根法
diminutive current meter 小型流速仪
diminutive tower 装饰性角楼
diminutive turbine 小型水轮机;微型水轮机
dimiodol 多米奥醇
dimity 格条麻纱;仿麻布
dim lamp 磨砂灯泡
dim letter 图案不清晰(压花玻璃缺陷)
dim light 小光灯;暗光灯;暗淡光(线)
dim light bulb 小光灯泡
dim light circuit 微亮电路
dimmed headlight 暗光头灯
dimmed illumination 城市车灯;小光灯;小灯光
dimmed light 暗光;微亮;微光
dimmer 减光器;小光灯;光强调节器;光度调整器;变光器;调光器
dimmer coil 减光线圈
dimmer control 亮度调节;减光控制
dimmer controller 减光控制器
dimmer device 遮光器
dimmerfoehn 迪默焚风
dimmer glass 减光玻璃
dimmer knob 遮光控制钮
dimmer room 调光室;剧院调光室
dimmer sweep trace 扫描轨迹
dimmer switch 减光器开关;调光开关
dimmer unit 减光器
dimming 发蒙(无光);发霉
dimming installation 减光设施
dimming resistor 调光电阻
dimming rheostat 亮度调节器
dimmish 暗淡
dimness 暗淡
Di-Mol 迪-钼尔钼高速钢
dimolecular 二分子(的);双分子
dimorph 同质异象变体
dimorphic 二型的;二形的;二态的;同质二形体
dimorphism 岩石分解;风化变质作用;二形;二态现象;双形(现象);双晶现象
dimorphite 硫砷矿
dimorphous 二形的;二态的
dim-out 遮光;灯光管制;灯光暗淡;半灯火管制
dim-out cap 灯光管制帽
dim-out installation 灯光管制装置

Dimpalloy 迪帕洛伊银焊料合金
dim pattern 图案模糊
dimple 定位窝；表面微凹；凹座；凹痕；浅凹
dimpled jacket 蜂窝型整体夹套
dimpled plate 波纹板
dimple-plank isolating-oil pool 波纹板隔油池
dimple rock 硅石
dimple roller 波纹辊
dimple spring 小洼泉；凹地涌泉；洼坑泉
dimpling 嵌线条；加波纹；造窝；沉头孔压形；凹陷
Dimroth condenser 迪姆罗回流冷凝管
dim switch 减光开关
dimyristyl ether 双肉豆蔻醚
dina 第纳干扰器
di-n-amylamine 二戊胺
Dinantian 狄南统【地】
Dinantian (stage) 狄南阶【地】
dinaphthenate 二环环烷酸盐
dinaphthyl methane 二萘甲烷
Dinaric Alps 迪那里克阿尔卑斯山脉
dinas 砂石；硅石
Dinas brick 戴那砖；硅(石)砖；灰砂砖
dinas firebrick 硅石耐火砖
di-n-butylacetic acid 二丁基乙酸；二丁基醋酸
di-n-butyl carbonate 碳酸二正丁酯
di-n-butyl phthalate 邻苯二甲酸二正丁酯
dinbutyl tin dichloride 二正丁基二氯化锡
diner 餐车
dinergate 兵蚁
dineric interspace 二液相界面
Dines anemometer 丹斯风速计
Dines pressure anemograph 达因风压计
Diness anemometer 达因风向风速计；达因风向风速表
dinette 进餐凹室；小餐厅桌椅
dinette set (厨房旁边的)小餐厅桌椅
dinette table 餐桌(厨房用)
dineuric 二轴突的
dineutron 双中子
dinex 二硝环己酚
ding 氧乙炔粉体枪；勾缝
dinge 表面凹陷
dingey 小型救生艇或供应艇；小艇；小舢板；船上小舢板
dingey cockboat 折叠式救生艇
dinghy 舰上小艇；小型救生艇或供应艇；小舢板；小船；折叠式救生艇；船载划艇；船上小舢板
dinging 一次抹面；钩缝；单层粗抹灰
dinging hammer 平板锤
dingle 小排水沟；雨棚；封闭谷地；幽谷；防风暴门；储藏间
dingot 直熔锭
dingot cleaning 直熔锭清理
dingot metal 直熔金属锭
dingot regulus 直熔锭块
dingot scalping 直熔锭修整
dings 板材的弯折
ding slope 弯曲斜度
Dings magnetic separator 丁斯磁力分离器
dingus 小机件；小装置
Ding ware 定窑器
ding white ware 定白
dingy 颜色灰暗；暗褐色；色彩暗淡
Dingyao-type 定窑系
Dingzhou kiln 鼎州窑
di-n-hexylamine 二己胺
DIN index DIN 感光度
dining alcove 进餐凹室
dining area with terrace 有阳台的进餐地方
dining car 餐车
dining carriage 餐车
dining corner 就餐角；用餐地方
dining hall 餐厅；饭厅
dining kitchen 兼用膳的厨房；兼容小餐厅的厨房；厨房兼餐室
dining nook 进餐凹室；餐座
dining place 用餐地方
dining recess 进餐凹室
dining room 餐室；饭厅；食堂
dining saloon 食堂；饭厅；餐厅
dining table 餐桌
dining terrace 室外餐厅；就餐露台；进餐平台
dinitrate 二硝酸盐
dinitration 二硝化作用

dinitroaniline 二硝基苯胺
dinitrobenzal-dehyde 二硝基苯甲醛
dinitrobenzene 二硝(基)苯
dinitrobenzoic acid 二硝基苯甲酸
dinitrobenzoic acid piperidine salt 二硝基苯甲酸六氢吡啶盐
dinitrobenzoyl chloride 二硝基苯酰氯
dinitrobenzoylene urea 二硝基邻苯甲酰亚脲
dinitrocellulose 二硝酸纤维素
dinitrochlorobenzene 二硝基氯(化)苯
dinitrochlorobenzol 二硝基氯苯
dinitrodiazophenol 二硝基重氮酚
dinitrodiphenyl carbazide 二硝基二苯碳酰二肼
dinitrodiphenyl ether 二硝基二苯醚
dinitrofluorobene 二硝基氟苯
dinitrofluorobenzene 二硝基氟苯
dinitrogen heterophenanthrene 二氮杂菲
dinitrogen pentoxide 硝酐
dinitroglycerine explosive 二硝基甘油炸药
dinitroglycol 二硝基乙二醇
dinitrohydroquinone acetate 二硝基氢醌乙酸酯
dinitrohydroxyazo 二硝基羟偶氮
dinitromethane 二硝基甲烷
dinitronaphthalene 二硝基萘
dinitro naphthalene disulfonic acid 二硝基萘二磺酸
dinitro naphthalene sulfonic acid 二硝基萘磺酸
dinitrophenol 二硝基酚钠；二硝基苯酚
dinitrophenolate 二硝基酚
dinitrophenylation 二硝基苯基化
dinitrophenylhydrazine 二硝基苯肼
dinitrosalicylic acid 二硝基水杨酸
dinitrotoluene 二硝基甲苯(炸药)
dink 无子女双职工
dink(e)y 窄轨小机车；小型(调车)机车
dink(e)y locomotive 窄轨机车；运土用的窄轨机车；运石用的窄轨机车；轻便机车
dinking 空心冲
dinking machine 平压切断机
dinking machine shear die 冲切机
dinky 小型起重机(俚语)
dinky skinner 轻便机车驾驶员(俚语)
dinner cloth 餐桌台布
dinner clothes 餐服
dinner lift 食梯
dinner pail 饭盒
dinner-party expenses 接待费
dinner service 餐具
dinner-set 成套餐具
dinner table 餐桌
dinner wagon 带轮的食品输送架；食品输送车
dinnerware 餐具
di-n-nonyl sebacate 癸二酸二壬酯
dino block 扭 S 形块体
di-n-octylamine 二辛胺
di-n-octyl phenylphosphonate 苯基膦酸二正辛酯
di-n-octyl phthalate 邻苯二甲酸二正辛酯
di-n-octyl sebacate 癸二酸二辛酯
dinoctyl tin dichloride 二正辛基二氯化锡
dinonyl phthalate 邻苯二甲酸二壬酯
Dinosaur 恐龙属
dinosaur block 扭 S 形块体
dinosaurs extinction 恐龙绝灭
dinoseis 气爆震源
di-n-propylamine 二丙胺
dint 印痕；压出痕迹；凹痕
dinting 卧底
Diocletian's palace at Split (位于南斯拉夫斯普利特的)戴克里森宫
dioctadecyl terephthalate 对苯二甲酸双酯
dioctahedral 二八面体的晶体结构
dioctahedral layer 二八面体层
dioctahedron 双八面体
dioctron 交叉电磁场微波放大器
dioctyl-acetic acid 二辛基乙酸
dioctyl azelate 壬二酸二辛酯
dioctyl phthalate 邻苯二甲酸二辛酯；邻苯二甲酸二丁辛脂
dioctyl phthalate wastewater 邻苯二甲酸二辛酯废水
dioctyl sebacate 癸二酸二辛酯
dioctyl sodium sulphosuccinate 丁二酸二(异)辛酯磺酸钠
dioctyl sulpho-succinate 磺基丁二酸二辛酯

dioctyl terephthalate 对苯二甲酸二辛酯
diode 二极管；半导体二极管
diode-capacitor storage 管容电存储器
diode coupled bipolar memory cell 二极管耦合双极存取单元
diode coupler 二级耦合器
diode current capacity 运载能力；车辆运载能力
diode demodulator 二级解调器
diode drop 二极管降压
diode forward voltage 二极管正向电压
diode gate 二极管门电路
diode image converter 二级变像器
diode laser 激光二极管
diode logic 二极逻辑
diode luminescence 二级发光
diode matrix 二级矩阵
diode microwave oscillator 二极管微波振荡器
diode rectification 二极管整流
diode rectifier 二极管整流器
diode sensor 二级传感器
diode transistor logic 二极晶体管逻辑
diode trigger 触发二极管
diode voltage gate circuit 二极管门电路
diode voltage regulation 二极管稳压器
diodide 二碘化物
diodone 碘沃酮
d-i offset system 图像直接转换胶印法
diofrequency heating dryer 射频感应加热干燥
diogenite 紫苏辉石无球粒陨石；奥长古铜无球粒陨石
diol 二醇
dioleate 二油酸脂
diolefin(e) 二烯化合物；二烯(烃)
diolefin hydrogenation 二烯加氢
diolefinic acid 二烯酸
diolefins 二烯属(烃)
diolein 二油精
Diolen 迪奥纶聚酯纤维
dioleyl phosphite 亚磷酸二油醇酯
dionic recorder 导电度记录仪
dionic tester 导电度仪
diopside 透辉石
diopside amphibole plagioclase granulite 透辉角闪斜长麻粒岩
diopside amphibolite 透辉角闪岩
diopside glass 透辉石玻璃
diopside hornfels 透辉石角岩
diopside hypersthene hornfels 透辉石紫苏辉石角岩
diopside-iadeite 玉质透辉石
diopside marble 透辉石大理岩
diopside peridotite 透辉石橄榄岩
diopside picrite 透辉石苦橄岩
diopside porcelain 透辉石质瓷
diopside-tremolite marble 透辉闪大理岩
diopsidite 透辉石岩
dioptase 透视石
diopter 折光度；照准仪；觇孔；屈光度；视度
diopter cylinder 视度筒
diopter difference 视度差
diopter dividing ring 视度分划圈
diopter hair 瞄准器游丝
diopter lens 视度透镜
diopter reduction 照准归心改正
diopter regulation 视度调节
diopter ring 视度圈
diopter rotating screw 视度转螺
diopter scale 折光标；照准仪分划
diopter thread 瞄准线
diopter value 视度值
dioptoeikonometer 镜片影像计
dioptometer 屈光度测定器
dioptra 望筒
dioptre 折光度；屈光度
dioptric 折射(的)；折光的；光线折射的
dioptric antenna 折光天线
dioptric apparatus 折光器；屈光器
dioptric element 折光元件
dioptric glass 光线屈折镜
dioptric imaging 折射成像
dioptric lens 折光透镜；屈光透镜
dioptric light 折射光
dioptric lighthouse 折射光灯塔
dioptric lighting 折射灯(光)；折光灯

dioptric power 焦度;折光本领
dioptrics 折射光学;屈光学
dioptric strength 焦度
dioptric system 折射光学系统;折光系统;屈光组;屈光系统
dioptric type 曲折式
dioptrometer 折光度计;视度计
diorama 透视画
diorientation migration 双向迁移
diorite 闪绿岩;闪长(细晶)岩
diorite andesite rock 闪长-安山岩
diorite group 闪长岩类
diorite iron and copper-bearing formation 闪长岩含铁铜建造
diorite porphyrite 闪绿斑岩;闪长玢岩
dioritic 闪长岩的
dioritic aplite 闪长细晶岩
dioritic appinite 闪长暗拼岩
dioritic gabbro 闪长辉长岩
dioritic pegmatite 闪长伟晶岩
dioritite 闪长细晶岩
dioxalate 草酸氢盐
dioxalate of potassa 草酸氢钾
dioxane 二氧杂环己烷;二氧六环
dioxazine dyes 二恶嗪染料
dioxide 二氧化物
dioxide of silicon 二氧化硅
dioxin 二恶英
dioxolone 二氧杂环戊烯酮
dioxydichloride 二氯二氧化物
dioxysulfide 一硫二氧化物
dip 浸泡;船首下沉;弛度;凹片;倾斜;泡浸
dip angle 俯角;倾角
dip angle of fault surface 断层面倾角
dip angle of fracture surface 断裂面倾角
dip angle of hole 钻孔倾角
dip angle of orebody 矿体倾角
dip angle of plate 板体倾角
dip angle of shaft 采井倾角
dip application 浸涂施工
dip-arrow map 油田构造图
dipartite 分成几部分的
dip at high angle 急倾斜(大于15度);陡倾角
dip at low angle 缓倾斜;缓倾角
dip at right angle 急倾斜(大于15度)
dip blasting 倾向爆破
dip-braze 铜浸焊
dip brazing 浸渍铜焊;浸浴钎焊;浸沾钎焊;硬浸焊;沉浸钎焊
dip calculations near borehole 井旁地层倾角计算
dip calorizing 浸率处理;浸镀铝
dip can 选择器
dip chart 磁倾角曲线图
di-p-chlorophenoxy-methane 二对氯苯氧基甲烷
dip circle 俯角圈【测】;磁倾仪;测斜仪
dip cleat 倾斜割理
dip coat 浸涂涂层
dip-coated electrode 浸涂焊条
dip coater 浸涂机
dip coating 浸涂涂料;浸浴涂敷;浸漆;浸入涂层;浸涂涂层;浸渍敷层;浸涂(法);浸涂涂层
dip coating deposition 浸没法沉积镀膜
dip colo(u)ring 浸染
dip compass 矿山罗盘;倾斜仪
dip-corrected map 原位地层图;变动前地质图;倾角校正图
dip correction 俯角校正;俯角改正;伏角差
dip correction height of eye correction 眼高差改正
dip counter 负荷计数管
dip crossing (道路的)向下倾斜交叉
dip detector 浸入式探测器
dip (direction) 倾向;倾斜方向
dip direction of ore body 矿体倾向
dip direction of polarized body 极化体倾向
dip direction of rock body 岩体的倾向
dipentene 二(聚)戊烯;松油精芑烯
dip equator 磁赤道
dip face 沿倾斜方向的工作面;浸渍面
dip fault 倾斜断层;倾向断层
dip-feed drum drier 封闭式供料滚筒干燥机
dip-feed lubrication 浸浴润滑
dip-foot spring 倾向面山足泉
dip forming 浸渍成型;蘸塑

dip galvanizing 浸渍镀锌
dip gas water interface 倾斜的气水界面
dip ga(u)ge 磁倾计;垂直计;垂度规
dip geologic(al) profile 倾向地质剖面图
dip-grained wood 波浪纹木材;浸润纹(理)木材
dip groin 浸没折流坝;浸没挑流坝;斜面丁坝
diphase 两相;二相
diphase cleaning 双相清洗
diphase equilibrium 两相平衡
diphase mcrality 双相死亡现象
diphaser 两相(交流)发电机
diphase rubber 两相橡胶
diphasic 双相(性)的
diphasic action potential 两相性动作电位
diphasic potential 两相电位;双相电位
diphasic pulse 双相脉冲
diphasic titration 两相滴定法
diphasic variation 二相变异
dip hatch 计量孔
diphead 下山;沿岩层倾斜面向下开掘巷道
diphenazyl 二苯甲酰乙烷
diphenol 联苯酚
diphenol cabazide 联苯酚卡巴肼
diphenol diacetate 双乙酸联苯酯
diphenolic acid 双酚酸
diphenol propane 二酚基丙烷
diphenychlorarsine 二苯氯化胂
diphenyl 联苯;二苯基
diphenyl-3-thiosemicarbazide 二苯胺基硫脲
diphenyl acetylene 二苯乙炔
diphenylamine 二苯胺(杀虫剂)
diphenylamine-arsine chloride 二苯胺氯胂
diphenylamine carboxylic acid 二苯胺羧酸
diphenylamine-chloro-arsine 二苯胺氯胂
diphenylamine derivatives 二苯胺衍生物
diphenylamine hydrochloride 二苯胺盐酸盐
diphenylamine sulfate 硫酸二苯胺;二苯胺化硫酸
diphenylamine-sulfonate 二苯胺磺酸盐
diphenylamine sulfonic acid 二苯胺磺酸
diphenylamine sulfonic acid sodium salt 二苯胺磺酸钠
diphenylamino-azo-m-benzene sulfonic acid 二苯胺偶氮间苯磺酸
diphenylaminocyanarsine 二苯胺氰胂
diphenylarsine 二苯胂
diphenyl arsine sulfide 二苯胂基硫
diphenyl arsinic acid 二苯胂酸
diphenylbenzidine 二苯胺联苯
diphenylbenzidine sulfonic acid 二苯联苯胺磺酸
diphenyl boiler 联苯锅炉
diphenyl carbamyl chloride 二苯(替)氨甲酰氯;二苯氨(基)甲酰氯
diphenylcarbazide 二苯碳酰二肼;二苯基卡巴肼
diphenyl carbene 二苯基卡宾
diphenyl carbinol 二苯基甲醇
diphenyl carbonate 碳酸二苯酯
diphenyl chloroarisine 二苯氯胂
diphenyl cresyl phosphate 磷酸二苯基(一)甲苯酯
diphenyl cyanarsine 二苯代胂腈
diphenyl cyanoarsine 二苯胂基氰;二苯氰化胂
diphenyl decyl phosphite 亚磷酸二苯基一癸酯
diphenyl dichloromethane 二苯二氯甲烷
diphenyl dichlorosilane 二苯基二氯硅烷
diphenyldiimide 偶氮苯
diphenyl dimethyl ethane 二苯基二甲基乙烷
diphenyl diphenoxysilicane 二苯二苯氧基硅
diphenyl disulfide 二苯二硫
diphenyl dye 二苯基染料
diphenylene 二亚苯基
diphenylene-oxide 二苯并呋喃
diphenyl ester 二苯酯
diphenyl ether 二苯醚;苯醚
diphenyletherformalde-hyde resin 二苯醚甲醛树脂
diphenyl ethylene 二苯(基代)乙烯
diphenylformamide 二苯基甲酰胺
diphenylglycollic acid 二苯乙醇酸
diphenylglyoxal 苯偶酰
diphenyl guanidine 二苯胍
diphenyl guanidine carbonate 碳酸二苯胍
diphenylhydantoin 二苯乙内酰脲
diphenyl-imidazolone 二苯咪唑酮
diphenyl ketene 二苯乙烯酮
diphenyl ketone 二苯甲酮;苯酰苯;苯酮
diphenyl ketoxime 二苯甲酮肟

diphenyl malonate 丙二酸二苯酯
diphenyl methane 二苯(基)甲烷
diphenylmethane dyes 二苯甲烷染料
diphenylmethane group 二苯甲烷基
diphenylmethanes 二苯甲烷类
diphenyl methanol 二苯基甲醇
diphenylmethylation 二苯甲基化作用
diphenyl methyl phosphonate 甲基膦酸二苯酯
diphenyl mono-o-xenyl phosphate 磷酸二苯基邻联苯酯
diphenyl-m-tolylmethane 二苯基间甲苯基甲烷
diphenylol propane 双酚A;二苯酚基丙烷
diphenylol propane epoxy resin 双酚A型环氧树脂
diphenyl oxide 二苯基氧
diphenyloxide cycle 二苯醚循环
diphenyl oxide glass cloth plate 二苯醚玻璃布板
diphenyl-para-phenylene diamine 二苯对苯二胺
diphenyl phosphate 磷酸二苯酯
diphenyl phthalate 邻苯二甲酸二苯酯
diphenyl-p-phenylenediamine 二苯对苯二胺
diphenylsemicarbazide 二苯氨基脲
diphenyl silanediol 二苯基硅二醇
diphenyl silanodiol 二苯基(甲)硅烷二醇
diphenyl silicone diisocyanate 二苯基甲硅烷二异氰酸酯
diphenyl succinic acid 二苯丁二酸
diphenyl sulfone 二苯砜
diphenyl sulphide 苯硫醚
diphenyl thiourea 二苯(基)硫脲
diphenyl tinoxide 氧化二苯锡;二苯基氧化锡
diphenyl tridecyl phosphate 亚磷酸二苯基十三烷酯
diphenyl urea 二苯脲
diphenyl xylenyl phosphate 磷酸二苯基二甲苯酯
diphonia 双音
diphosgene 聚光气;双光气
diphosphate 二磷酸盐
diphosphoglyceric acid 二磷酸甘油酸
diphthalate 苯二酸氢盐
Diphyl 狄菲尔换热剂
dipicolinic acid 吡啶二羧酸
dip inductor 磁倾感应器
dip iron 集脂铲
dip joint 倾向节理
diplane 双平面
dipleg 浸入管
diplex 同向双工;双通路;双倍
diplexer 天线分离滤波器;共用天线耦合器;两倍半传机;同向双工器;天线共用器;双信伴传机
diplex generator 双频发生器;双工振荡器
diplex operation 双工工作
diplex radio transmission 双工无线电传输
diplex reception 同向双工接收
diplex sender 同向双工发送器
diplex telegraphy 单向双路电报
dip lineation 倾向线理
diplinthius 两砖厚墙(古罗马)
dip load 倾斜荷载
diplocardia 双芯
diplodization 二倍化
diploe 板障
dip log 地层倾角测井(仪);地层产状测井;倾角钻井记录
diplogen 氘
diplogenesis 叠生成因
diplogentic 叠生的
diplogram 重复X射线照片
diploid 二重(的);二倍(体);倍数的;偏方复十二面体
diploidization 二倍化
diploid nucleus 二倍核
diploid state 二倍状态
diplokaryon 二倍核
diplolc type 板障型
diplomatic clause 外交人员租用房屋条款
diplomatic conference on maritime law 海事法外交会议
diplomatic edition 仿真本
diplomatic enclave 外交飞地;他国飞地
diplomatic immunity 外汇豁免
diplomatic language 外交辞令
diplomatic mission 使官
diplomatic parlance 外交辞令
diplomatic practice 外交惯例

diplomatic privileges 外交特权
diplomatic protocol 外交礼节
diplomatic row 使馆区
diplomatic visa 外交签证
diplont 二倍体
diplopia test 复视试验
diplopiometer 复视计
diplosome 双心体
diplotene 双线期
dip lubricating system 浸入润滑系统
dipmeter 岩层产状仪；钻孔地下水记录仪；地层倾角测量仪；地层倾角测井仪；倾斜仪；倾角（测量）仪；测倾仪；浸入式探测器；磁倾角测量仪
dipmeter calibration value 地层倾角测井仪刻度值
dipmeter calibrator 地层倾角测井刻度器
dipmeter log 倾斜计测井记录
dipmeter survey 地层倾斜测井；倾斜仪测量
dip method 浸渍法
dip mo(u)ld 下压模
dip mo(u)lding 浸渍模塑；浸渍成型；蘸塑
dip-moveout correction 斜动时差校正
dip needle 倾度计；测斜仪；磁倾仪；磁倾计
dip needle work 磁法勘探
dip net 捞鱼网；抄网
dipnictide 二磷族元素化物
dip nozzle 伸入管
dip of bed 地层（的）倾斜
dip of fault surface 断层面倾向
dip of formation 地层倾角
dip of fracture surface 断裂面倾向
dip of gasholder lift 储气罐爬梯倾角
dip of horizon 眼高差
dip of load curve 负荷曲线峡谷
dip of lode 矿脉俯角
dip of needle 磁（针）倾角
dip of orebody 矿体倾向
dip of rock layer 岩层斜度
dip of sea horizon 海地平（线）俯角
dip of shore horizon 岸线坡度；岸线俯角差
dip of stratum 地层倾斜
dip of structure 构造倾角
dip of the bait 拍子浸入深度
dip of the horizon 视地平俯角；地平俯角
dip of the intermediate principal stress 中间主应力倾角
dip of the magnetic body 磁性体倾角
dip of the maximum principal stress 最大主应力倾角
dip of the minimum principal stress 最小主应力倾角
dip of the track 径迹深度
dip oil-water interface 倾斜的油水界面
dipolar 偶极的；两极的
dipolar agent 极性溶剂
dipolar bond 偶极键
dipolar coordinates 双极坐标
dipolar ion 偶极离子
dipolar loss 偶极损耗
dipolar moment 偶极矩
dipolar polarizability 偶极子极化率
dipolar section 双极性剖面
dipolar site 偶极部位
dipole 偶极（子）；对称振子；双极
dipole absorption 偶极子吸收
dipole antenna 偶极天线；半波振子天线
dipole array 偶极天线阵
dipole attraction 偶极引力
dipole broadening 偶极加宽
dipole-dipole force 取向力
dipole-dipole profiling curve 偶极剖面曲线
dipole-dipole sounding 偶极测深
dipole-dipole sounding method 偶极测深法
dipole disk feed 偶极圆盘馈电
dipole dislocation 偶极位错
dipole effect field 偶极作用场
dipole-elastic loss 偶极弹性损耗；高弹偶极损耗
dipole feed 偶极馈源
dipole field 偶极（子）场
dipole lattice 偶极晶格
dipole layer 偶极层；双电荷层
dipole-layer mode 双极层模
dipole magnetic field 偶极磁场
dipole matrix element 偶极矩阵元
dipole mode 对称振子振荡模

dipole modulation 偶极调制
dipole moment 偶极矩；磁偶极矩
dipole oscillator 双极振荡器
dipole polarization 偶极子极化；定向极化
dipole profiling method 偶极剖面法
dipole radiation 偶极辐射
dipole radiation pattern 偶极子辐射图样
dipole radiator 偶极子辐射器
dipole-radical loss 侧基偶极损耗
dipole relaxation 偶极子弛豫
dipole sound field 偶极子声场
dipole source 偶极子源
dipole transition 偶极跃迁
dipolymer 二聚物；二聚体
dipolymer reactor 二聚反应器
di(polyoxyethylene) hydroxymethyl phosphonate 羟甲基膦酸二聚氧乙烯酯
diponium bromide 地泼溴铵
dipotassium hydrogen phosphate 磷酸氢二钾
dipout furnace 双用熔炉
dip package 浸涂包装
dipped article 浸渍制品
dipped bean 近光
dipped brass 蚀刻黄铜
dipped electrode 浸液电极
dipped finish 斜口（玻璃制品缺陷）
dipped goods 无缝制品
dipped rail joint 钢轨低接头
dipped yarn 浸油麻丝
Dippel's oil 地柏油；骨焦油（指一种污染饮水的物质）
dipper 浸渍法；近距车灯；汲器（铲斗）；扉斗；显影液槽；铲斗；施釉工；勺（取）；勺（斗）
dipper arm 铲斗柄；铲臂；勺斗柄
dipper arm wire 斗柄推压缆
dipper backing equipment 铲斗反铲装置
dipper boom 铲斗吊杆
dipper bucket door 铲斗底（卸）门
dipper bucket dredge(r) 单斗挖掘船；铲斗（式）挖泥船
dipper bucket ram 铲斗冲锤
dipper bucket tooth 铲斗齿
dipper capacity 斗容（量）；铲斗容量
dipper crowding gear 铲斗推压装置
dipper door 汲器可开的底；铲斗活底（门）
dipper dredge(r) 杓斗挖泥机；斗式挖泥机；单斗挖泥机；单斗挖泥机；铲扬式挖泥机；铲斗（式）挖泥船；挖泥铲斗；铲扬式挖泥船
dipper dredge(r)'s outreach 铲斗挖泥船舷外伸距
dipper factor 满斗系数
dipper frame 铲斗支架
dipper handle 杓柄；铲斗柄；铲臂；挖掘机（杓）斗杆；勺柄
dipper hoisting gear 铲斗提升装置
dipper lip 杓斗刃口；斗缘；铲斗刃口；铲斗斗唇；勺斗刃口
dipper machine 铲斗机
dipper opening 斗孔
dipper rock-breaking gear 铲斗碎石装置
dipper sheave block 铲斗滑车组
dipper shovel 动力挖土机
dipper stick 水位指示器；量杆；量尺；铲斗杆；（挖掘机的）杓斗柄；斗柄；铲斗柄；挖掘斗柄柱
dipper stick ram 旋转挖掘机的冲压件
dipper stick sleeve 铲斗连杆套筒
dipper teeth 铲斗刀；铲斗齿；斗齿；杓斗齿
dipper trip 铲斗门开头；铲斗开合器
dipper turning gear 铲斗机回转装置
dipper (type) dredge(r) 铲斗（式）挖泥船
dip pile 斜桩
dipping 浸渍；浸釉；浸涂（药皮）；倾头现象【船】；艏沉
dipping asdic 声呐；吊放式声呐
dipping basket 浸入式电石篮
dipping bath 浸浴
dipping coat 浸渍涂层
dipping coil 浸渍线圈
dipping coil primary meana 电磁式检测设备
dipping compass 矿用磁倾仪；磁倾仪；测斜仪；倾度仪
dipping compound 浸涂混合料
dipping consistence 浸染浓度；浸染稠度；浸渍稠度
dipping consistency 浸染稠度
dipping cylinder method 圆筒内陷法

dipping dike 俯头丁坝；斜面丁坝
dipping distance 灯光初隐距离
dipping engobe 浸釉底料；浸没上釉底料
dipping fault plane 倾滑断层面
dipping film method 浸膜法
dipping glazing 浸渍上釉；浸没上釉
dipping groin 俯头丁坝
dipping groyne 俯头丁坝
dipping hatch 量油孔
dipping height 艏沉深度
dipping lacquer 浸漆；浸没上漆
dipping layer 倾斜层
dipping line 量油尺
dipping lug 前倾纵帆
dipping maceration tank 浸渍槽
dipping machine 浸机
dipping method 浸泡法；浸渍（方）法
dipping method of timber treatment 木材浸渍处理法
dipping needle 磁倾针
dipping needle instrument 倾斜自差调整器；倾差校正仪
dipping of anomaly body 异常体倾向
dipping paint 浸漆
dipping polish 浸渍抛光
dipping process 浸渍过程
dipping pyrometer 插入式高温计
dipping refractometer 浸液析射计；浸没式折射计；蘸液折射计
dipping reservoir 倾斜油田
dipping rod 测深尺
dipping solution 浸渍溶液
dipping sonar 吊放式声呐；投吊式声呐
dipping stick 测深尺
dipping stone drift 下倾岩巷
dipping stratum 倾斜地层
dipping switch 近光开关；减光器开关
dipping tank 浸渍槽；浸涂槽；酸洗槽
dipping tape 量油尺
dipping test 浸渍试验
dipping toward down stream 倾向下游
dipping toward upstream 倾向上游
dipping treatment 浸渍处理
dipping varnish 浸渍清漆；浸涂清漆
dipping vat 浸涂槽
dipping viscosity 浸渍黏滞度
dipping weight 浸釉量
dip pipe 液封管；倾斜管；浸渍管；内筒；下垂管
dip plain 倾斜平原
dip plane 倾斜平面
dip plating 浸镀
dip pole 地磁极；磁极
dip pole strength 磁极强度
dip polishing 浸渍上光；浸渍抛光
dip-primed 浸渍法清漆打底的
dip process 保护层沉浸涂镀法
dip-proof burner 防扼喷燃器
dippy twist 转落螺旋
dip quench 浸入淬火
dip reading 倾角读数
dip reduction 眼高差
dip reversal 倾斜逆转
dip river 倾向河
dip rod 探条；料钎；量油杆；浸量尺；液面测定杆；测水深尺；玻璃液取样铁棒；水位指示器
dip roll coater 浸辊涂布机
dip rope 解链引绳
dipropyl 二丙基
dipropyl-acetic acid 二丙基乙酸
dipropyl amine 二丙胺
dipropyl barbituric acid 二丙基巴比土酸
dipropyl disulfide 二丙基二硫醚
dipropylene glycol 一缩二丙二醇；二丙撑二醇
dipropylene glycol monomethyl ether 二丙二醇单甲醚
dipropyl-ethyl-phenyl-silicane 二丙基乙基苄基甲硅烷
dipropyl lead 二丙基铅
dipropyl maleate 马来酸二丙酯
dipropyl malona-te 丙二酸二丙酯
dipropyl oxalate 草酸二丙酯
dipropyl peroxydicarbonate 过氧化二碳酸二丙酯
dipropyl phthalate 邻苯二甲酸二丙酯
dipropyl sebacate 癸二酸二丙酯

dipropyl sulfide 二丙硫醚
dipropyl sulfone 二丙砜
dipropytene glycol monosahcylate 一水杨酸一缩二丙二醇酯
diproqualone 地普喹酮
dip rose diagram 倾向玫瑰图
diprosopia 双面
dipro sopy 双面
diproton 双质子
diprotonic acid 二元酸
dip sampler 浸入取样器;插入取样器
dip seal 液封
dipsea lead 深水砣
dip section 倾向剖面
dip separation 倾向离距;倾向隔距;倾向距
dips(e)y 渔钓上软木浮子;软木浮子
dip shift 倾向移位;倾向移距
dip shooting 倾向爆破
dipsie 渔钓上软木浮子
dip slip 倾向滑距;倾向滑动
dip-slip drag 倾向滑动拖曳
dip slip fault 倾向滑断层;倾滑断裂
dip slip reverse fault 倾向滑逆断层
dip slip source mechanism 倾(向)滑震源机制
dip slope 反坡;倾(向)坡
dip-solder 浸焊
dip soldering 浸焊;浸入焊接;软浸焊
dip solution 浸溶;浸蚀雕刻
dipstick 量油尺;量杆;浸量尺;水位指示器;油尺;浸量杆
dipstick gauging 用尺测深
dipstick metering 用尺测深;尺量
dipstone trap 浸石板式存水弯
dip stool 计量管
dip stream 倾向河
dip-strike symbol 倾角走向符号
dip switch 斜队道;(汽车大灯的)变光开关
dip tank 浸渍槽;浸釉槽;浸油槽;浸浆桶;淋冰槽
dipteral 带两个翼部的房屋;四周双列柱廊式建筑;双列柱廊的
dipteral building 双柱廊建筑;双列柱廊建筑
dipteral temple 双列柱廊式神庙
dipteros 有两个翼部的寺庙;四周双列柱廊式建筑;双柱廊建筑
dip test 降落试验;倾角测定
dip the flag 行点旗礼
dip throw 倾向落差
dip time 浸渍时间
dip transfer metal-arc welding method 浸渍变换金属(电)弧焊接法
dip trap 浸没式水封
dip tube 出料管
diptych 双连画;双幅折叠画板;双幅联画雕刻
dipulse 双脉冲
dipulse system 双脉冲方式
dip up 挖开
dip valley 倾向谷
dip vat 浸渍筒
dipvergence 垂直发散度;双目垂直角差
dip wall dam 斜墙坝
dip weight 测锤;测量海洋深度锤;测海深度锤;深度锤
dip working 沿倾斜方向向下开采
dipylon 并置双门;双门的大门
dipyramid 双棱锥(体)
dipyre 钙钠柱石;针柱石
dipyre slate 钙钠柱石板岩
dipyrite 针柱石;磁黄铁矿
Dirac beta function 脉冲函数;狄拉克的尔塔函数
Dirac comb 狄拉克梳
Dirac comb function 狄拉克梳状函数
Dirac covariant 狄拉克协变量
Dirac delta function 脉冲函数;狄拉克的尔塔函数
Dirac electron theory 狄拉克理论
Dirac equation 狄拉克方程
Dirac fields 狄拉克场
Dirac function 狄拉克函数
Dirac gamma algebra 狄拉克伽马代数
Dirac-Jordan cosmology 狄拉克—约旦宇宙论
Dirac matrix 狄拉克矩阵
Dirac moment 狄拉克磁矩
Dirac monopole 狄拉克单极
Dirac particle 狄拉克粒子
Dirac quantization 狄拉克量子化

Dirac's constant 狄拉克常数
Dirac theory 狄拉克理论
Dirac wave function 狄拉克波函数
dircet mineral cycling theory 直接矿物循环理论
dirctional circuit-breaker 方向性断路器
direct 指挥;直接的
direct access 直接取数;直接进口路;直接借入;直接存取;随机取数;随机访问;随机存取
direct access capability 直接存取能力
direct access communication channel 直接存取通信信道
direct access device file control block 直接存取设备文件控制块
direct access display channel 直接存取显示通道
direct access file 直接存取文件
direct access inquiry 直接(存取)询问
direct access library 直接存取程序库;随机存取库
direct access method 直接存取方式;直接存取法
direct access mode 直接存取方式
direct access programming system 直接存取程序设计系统
direct access sorting 直接存取分类
direct access storage 直接存取存储器
direct access storage device 直接存取装置
direct access storage device erase 清除直接存取存储设备
direct access storage facility 直接存取设备
direct access volume 直接存取媒体
direct acting 直接作用
direct acting brake 直接作用式制动机
direct acting carcinogen 直接致癌原;直接致瘤物
direct acting control valve 直接作用控制阀
direct acting current meter 直流电表
direct acting design 直接传动设计
direct acting engine 直接作动发动机;直接驱动发动机
direct acting factor 直接影响因素
direct acting finder 直接作用的旋转式选择器
direct acting gas thermostat 直接推动式煤气恒温计;自动式煤气恒温器
direct acting hoist 直接作用(的)提升机
direct acting load 直接作用荷载
direct acting mutagen 直接致突变物
direct acting pump 直接作用泵;直接推动式水泵;直接联动泵;直接传动泵;直动泵
direct acting reciprocating pump 直接联动往复泵
direct acting recording instrument 直接传动记录仪表
direct acting regulating valve 直动调节阀
direct acting shock absorber 直接作用式减振器;套筒式减振器
direct acting steam engine 直接式蒸汽机
direct acting steam pump 直接式蒸汽泵
direct acting valve 直动阀
direct action 直接行动;直接作用;直接动作
direct action avalanche 即发性雪崩;风暴引起的雪崩
direct-action pressure controller 直接作用式压力调节器
direct activation 直接活化
direct activities 直接业务
direct actuating relay 直接作用式继电器
direct adaptation 直接适应(作用)
direct add instruction 直接相加指令
direct addition 直接加法
direct additional charges 直航附加费
direct address 一级地址;直接地址
direct address code 直接地址码
direct addressing 一级寻址;一级选址;一级编址;直接寻址;直接选址;直接定址
direct addressing method 直接寻址方式
direct addressing mode 直接寻址模
direct address processing 直接寻址处理
direct address relocation 直接浮动地址
direct adjacent channel interference 邻近通道直接干扰;直接相邻波道干扰
direct affect 直接影响
direct agglutination reaction 直接凝集反应
direct agglutination test 直接凝集试验
direct air conditioning system 直流式空气调节系统
direct air cycle 直接空气循环
direct air die casting 低压铸造
direct air injection die-casting machine 直接气压式热室压铸机

direct air route 直达航空线
direct allocation 直接分配
direct allocation method 直接分配法
direct allocation of goods by the factory 就厂直拨
direct alongside activities 直接换装作业
direct ammonia recovery 氨气直接回收法
direct amplification 直接放大
direct analogy 直接模拟;直接类比;正相似
direct analysis 直接分析
direct and indirect process cost 直接与间接过程成本
direct and inverse development in electric(al) prospecting 电法勘探的正反演
direct and inverse position computation 位置正反算
direct and reversed observation 往测与返测
direct angle 直接角;正镜角
direct aperture antenna 直接式孔径天线
direct application of international standard 国际标准的直接应用
direct approach 直接法
direct arbitrage 两地套汇;直接套利;直接套汇
direct arc 直接电弧(熔铸)
direct arc furnace 直接电弧炉
direct arc melting furnace 直接电弧熔炼炉
direct arc melting method 直接电弧熔炼法
direct area 直接灌区
direct arrival wave attenuation factor computation 直达波衰减系数计算
direct arrival wave spectrum analysis 直达波频谱分析
direct ascent 直接上升
direct assay 直接测定
direct-attachment clip 直接固定卡
direct attack 直接射水
direct attractive range 直接吸引范围
direct axis 纵轴(的);纵向轴线;纵向的;轴向的;直轴(的);顺轴(的)
direct axis circuit 直轴电路;顺轴电路
direct axis component 直轴分量
direct axis reactance 顺轴电抗
direct axis subtransient reactance 纵轴次瞬变电抗
direct axis synchronous reactance 纵轴同步电抗;直轴同步电抗
direct axis transient electrodynamic(al) potential 纵轴瞬变电动势
direct axis transient reactance 纵轴瞬变电抗
direct-babbitted connecting rod 大头直接浇铸巴比合金的连杆
direct band 正向带
direct band gap 竖直带隙
direct bank guarantee 直接银行保证
direct bank protection 直接护岸
direct barter 直接易货;直接以货易货
direct bearing 直立支承;直支承物;直接影响;直接引导方位;直接意义;直式支座;导向轴承;导向支承
direct bearing foundation 直接基础
direct benefit 直接效益;直接受益
direct bilge suction 舱底独立吸管
direct bill of lading 直运提单;直达提单
direct biphasic reaction 直接双相反应
direct blending method 直接混合法
direct boiling 直接汽化
direct bonded basic brick 直接结合碱性耐火砖
direct bonded rate 直接结合率
direct bonded refractory 直接结合耐火材料
direct bonding 直接结合;直接接合
direct broadcasting satellite 直播卫星
direct buffering mode 直接缓冲方式
direct built-in system 直接埋设式
direct burial cable 直埋电缆
direct burial method 直埋法
direct buried cable 直埋电缆
direct bury cable 直埋式电缆
direct business 直接买卖
direct call 直接通话;直接呼叫
direct calorimetry 直接测热(法)
direct capacitance (两导体间的)静电容;部分电容
direct capitalization 直接资本还原率
direct cargo 直达货(物)
direct cascade reuse 直接串联再用
direct casting 直接浇铸;顶铸;上铸
direct cause 直接起因

direct cause of death 直接死因
direct census 直接计数
direct centrifugal flo(a)tation method 直接离心浮集法
direct challenge system 直接询问系统
direct characteristic 方位特征
direct charges 直接费(用)
direct charge-off method 直接冲销法
direct chemical attack 直接化学腐蚀
direct chill casting 直接激冷铸造；半连续浇铸
direct chill process 直接冷铸法；半连续铸造法
direct chlorination 直接氯化作用
direct chopper 直接短切机
direct chopping 直接短切
direct circuit 直通线路；直通电路；直接电路
direct circulation 直接循环；正循环；正环流
direct circulation washing 正循环冲洗【岩】
direct circulation (washing) drilling 正循环冲洗钻进【岩】
direct cleaning 工件为阴极的电解清洗
direct closing method 直接结算法
direct coal fired kiln 直接喂煤回转窑
direct coal firing mill 直接燃烧系统的煤磨
direct code 绝对码；直接码
direct coding 切口编码法
direct colo(u)rimetry 直接色度测量学；直接比色法
direct colo(u)r print 着色印刷
direct colo(u)r separation 直接分色(法)
direct combustion 直接燃烧
direct combustion black 直接燃烧灯黑
direct communication 直接通信；直达通信
direct communication link 直达通信线路
direct comparison 直接比较
direct comparison method 直接比较法
direct comparison of characteristics 直接比较特点
direct compatibility 直接兼容性
direct compensation 直接赔偿
direct componental movement 直接组分运动
direct composition 直接成分
direct compression 直接压力；单纯受压
direct compression process 直接压片法
direct compressive stress 直接压应力；直向压应力
direct computation of coordinates 坐标正算
direct condenser 回流冷凝器；回流冷凝管
direct connect 直接接合
direct connected 直接连接的；直接接合的；直接传动(的)
direct connected exciter 直连励磁机
direct connected motor 直连电动机
direct-connected mower 悬挂式割草机
direct connected pump 直连泵；直联泵
direct connection 直通；直联；直接连接
direct connection design 立体交叉的直接联系设计
direct connection interchange 直连式立(体)交(叉)
direct consideration 直接报酬
direct consignment 直接托付；直接交货
direct construction cost 直接施工费；直接工程费(用)
direct consumption 直接消耗；直接消费
direct consumption coefficient 直接消耗系数
direct consumption coefficient of environment(al) resources 环境资源直接消耗系数
direct consumption coefficient of water resources 水资源直接消耗系数
direct consumption tax 直接消费税
direct contact 直接接触
direct contact condenser 接触式冷凝器；混合式凝汽器；组合式冷凝器
direct contact cycle system 直接接触循环系统
direct contact desuper-heater 喷水减温器
direct contact exposure 直接接触曝光
direct contact heater 接触式加热器
direct contact heat exchanger 混合式换热器；混合式热交换器；直接接触式换热器；直接接触换热器；气水混合式换热器
direct contact membrane distillation 直接接触膜蒸馏
direct contact method 直接接触法
direct contact seal 接触型密封
direct contact transmission 直接接触传播
direct contact type 直接接触式
direct continuation 直接续租
direct contract 直接合同

direct contract stipulation 合同的直接规定
direct control 直接控制
direct control connection 直接控制连接
direct control facility 直接控制机构
direct control microprogram(me) 直接控制微程序
direct control microprogramming 直接控制微程序设计
direct control valve 直接控制阀
direct conversion reactor 直接(热)转换反应器
direct cooler 直接冷却器
direct cooling 直接冷却
direct cooling system 直接冷却系统
direct cooling water 直接冷却水
direct copy 机械靠模；直接晒印
direct copying 接触晒印
direct correlation 直接相关；正相关
direct corrosion 直接腐蚀
direct cost 可变成本；直接工程费(用)；直接费用；直接费；直接成本(包括材料工时在内)
direct cost calculation method 直接成本计算法
direct costing 可变成本法；直接成本计算；直接成本法
direct costing income statement 直接成本收益表
direct cost method 直接成本法
direct cost of exploratory drift 坑探直接成本
direct cost of project 工程直接成本
direct cost pricing 按直接成本定价
direct cost reduction 直接费用的减少
direct counter 正比计数器
direct counting method 直接计数法
direct couple 直接耦合
direct coupled 直接耦合的；直接连接的；直接联结的；直接耦合的
direct coupled amplification 直接耦合放大
direct coupled amplifier 直耦式放大器；直接耦合放大器
direct coupled circuit 直接耦合电路
direct coupled computer 直接耦合计算机
direct coupled exciter 直连式励磁机
direct coupled generator 直连式发电机；直连励磁机
direct coupled power amplifier tube 直接耦合功率放大管
direct coupled roll 直接传动辊
direct coupled (steam) turbine 直接传动式汽轮机
direct coupling 直接轴联节；直接耦联；直接耦合；直接联轴节；直接连接的；直接接合
direct coupling exciter 直连励磁机
direct coupling motor converter 连轴电动换流机
direct coupling system 直接耦合方式
direct course 直航向
direct course made good 直航向
direct cranking starter 直接转动的起动器
direct cross-connection 直接交叉连接
direct crushed sampling 直接压坏试样法
direct current [DC] 直流(电)
direct current-alternating-current converter 直流交流变换器
direct current ammeter with shunt 带分流器的直流电流表
direct current amplifier 直流放大器；直流电流放大器
direct current arc 直流电弧
direct current arc welder 直流弧焊机
direct current arc welding 直流电弧焊(接)
direct current bridge 直流电桥
direct current cables 直流电缆
direct current circuit 直流电路
direct current code 直流电码
direct current commutating machine 整流式直流电机
direct current component 直流分量
direct current compound generator 复励直流电动机
direct-current controllable reactor 饱和扼流圈
direct current dielectric(al) separation method 直流介电分离法
direct current discharge 直流放电
direct current drive 直流电驱动
direct current dynamo 直流(发)电机
direct current electric(al) locomotive 直流电力机车
direct current electric(al) traction 直流电力牵引；直流电气牵引

direct current electric(al) traction system 直流电力牵引制
direct current electrode negative 负直流电极
direct current electrode positive 正直流电极
direct current equipment 直流电设备
direct current excited synchronous motor 直线励磁同步电动机
direct current form factor 整流电流波形因素
direct current generator 直流发电机
direct current heating 电加热
direct current high-speed circuit-breaker 直流快速断路器
direct current high-voltage system 高压直流系统
direct current induced polarization 直流激发极化法
direct current keying 直流键控
direct current load 直流负载
direct current machine for electric(al) mining shovel 电矿铲用直流电机
direct current main 直流主电源
direct current method 直流法(勘探)
direct current motor 直流电(动)机
direct-current motor control 电动机电子控制
direct current motor driven pump 直流电动机驱动泵
direct current of glass 玻璃液的生产流
direct current panel 直流屏
direct current pantograph 直流受电器
direct current plasma torch 直流等离子体喷枪
direct current power 直流功率
direct current power supply 直流电源
direct current power supply panel 直流电源屏
direct current power supply system 直流供电制
direct current pulse 直流脉冲
direct current relay 直流继电器
direct current restoration 钳位
direct current series motor 串激直流电动机
direct current servomotor 直流伺服发动机
direct current source 直流电源
direct current stabilizer 直流稳压器
direct current supply 直流供电；直流电源
direct current switchgear 直流开关柜
direct current tachogenerator 直流测速发电机
direct current testing of cables 电缆的直流测试
direct current track circuit 直流轨道电路
direct current traction 直流牵引
direct current traction system 直流电力牵引
direct current transformer 直流变压器
direct current voltage 直流电压
direct current welder 直流电焊机
direct current welding 直流电焊
direct current working voltage 直流工作电压
direct cut attachment 直接切碎装置
direct cut grass and pickup attachment 直接切碎收集装置
direct cut operation 硬切换
direct cycle 直接空气循环
direct cycle integral boiling reactor 直接循环一体化沸水堆
direct cycle reactor 直接循环反应堆
direct cylinder 直接热水罐
direct damage 直接损失；直接损害
direct data 直接资料
direct data channel 直接数据通道
direct data checking 直接数据检核
direct data entry 直接数据输入
direct data organizatioin 直接数据组织
direct data processing 直接数据处理
direct data set 直接数据集
direct daylight factor 直接采光系数
direct daylighting 直接采光
direct dealing 直接交易
direct debiting 直接借入
direct debt 直接负债
direct deflection method 直偏法；直接偏转法
direct delayed reaction 直接迟缓反应
direct delivery 直接交货；直接交付
direct demand 直接需求
direct demonstration 直接阐明
direct department 直接部门
direct departmental expenses 直接部门费用；分部直接费用
direct department cost 直接部门成本
direct depth sounding 直接测深

direct derivation 直接推导
direct derivative 直接派生的
direct descendant 直接分支
direct design 直接设计
direct desulfurization 直接脱硫
direct detection intensity modulation 直接检测强度调制
direct detection receiver 直接检波式接收机
direct determination 直接测定
direct development 直接显影
direct development in electric(al) prospecting 电法勘探的正演
direct development model in electric(al) prospecting 电法勘探正演模型
direct development of spectrophotometry 直接显色分光光度法
direct dial(l)ing 直接选择
direct dial(l)ing system 直接拨号系统
direct die 正向模
direct diffused light 漫射光源;漫射灯;直接漫射光
direct diffused lighting 直接漫射(光)照明
direct diffusion 直接扩散
direct digital control 直接数字(程序)控制
direct digital controller 直接数字控制仪
direct digital control system 直接数字控制系统
direct digital encoder 直接数字编码器
direct digital interface 直接数字接口
direct digital transfer 直接数字交换
direct digitizing seismograph 直接读数地震仪
direct dipping process 直接浸渍法
direct dip-reading chart 真实倾角读数图
direct discharge of spoil 直接排泥
direct discharging 直接卸货
direct displacement 直接排代
direct disposal of spoil 直接排泥
direct dissemination 直接播散
direct distance 直航距离
direct distance dial(l)ing 长途自动拨号;长途直拨;长途直(接)拨(号)电话;直接长途拨号;直接远距离拨号
direct distance dial network 直接长途拨号网
direct distance measurement 直接测距
direct distribution 直接分配
direct distribution method 直接分配法
direct diving 直接潜水
direct division 直接分裂
direct dosing 直接配料;配料后直接入磨
direct draft 直接汇票
direct drainage 直接排水
direct draught 正压下
direct dredging 直接疏浚
direct drilling cost 直接单位成本
direct drive 直接转动;直接驱动;直接激励;直接传动(的)
direct drive acceleration 直接传动加加速度
direct drive auger 直接传动式输送螺旋
direct drive clutch 直接挡离合器
direct drive cyclometer 直接传动的转数表
direct drive dial 直接传动度盘
direct drive diesel ship 由柴油机驱动的船
direct drive engine 直接驱动发动机
direct drive fixed mount 直接传动固定架
direct drive gear 直接传动
direct drive locomotive 直接驱动的机车
direct drive machine 直接传动的机器
direct drive motor 直接驱动电机
direct driven airscrew 直接驱动螺旋桨
direct driven shears 直接传动式剪切机
direct driven synchro 直接传动同步机
direct drive of machine 机械直接传动
direct drive propeller 直接传动螺旋桨
direct drive pump 直接传动泵
direct drive shift valve 直接传动(液力)变速阀
direct drive shift valve accumulator 直接传动(液力)蓄油器
direct drive speed 直接传动速度
direct drive system 机械式传动系统
direct drive transmission 机械式传动系统
direct drive unit 直接传动装置
direct drive vibration machine 直接传动振动机
direct driving clutch 直接传动离合器
direct dry 直接干燥
direct drying 直接加热干燥
direct dumping 直接卸料;直接浇灌;直接灌注法

direct dye(stuff) 直接染料
direct earnings of parent company 母公司直接收益
direct earth capacitance 接地电容
direct ecotoxicological effect 直接生物毒性效应
directed 定向的
directed aerial photography 定向航空摄影
directed angle 有向角
directed application 直接施用
directed away edge 有向离开边
directed beam 定向射线;定向光束
directed circuit 有向回路
directed distance 有向距离
directed drilling 定向钻井;定向钻进
directed drilling technique 定向钻进技术
directed dynamic(al) pressure 单向动压
directed edge 定向边;有向边
directed-edge train 有向边序列
directed education 定向培养
directed energy 单向能量
directed Euler line 有向欧拉线
directed fiber preforming equipment 定向纤维预成型设备
directed forward edge 有向进入边
directed graph 有向图;方向图
directed interesterification 定向酯交换
directed line 有向直线;定向线
directed line segment 有向线段
directed-path 有向路
directed pressure 定向压力
directed property 有向性
directed radiation 定向辐射
directed ray 指向射线;定向射线
directed research 定向研究
directed scan 控制扫描;受控扫描
directed seeding 直播(种)
directed set 有向集
directed shear test with large shear box 大型直剪试验
directed sowing 直播(种)
directed spraying 定向喷雾
directed spray type combustion chamber 定向喷射式燃烧室
directed tree 定向树形(网络);方向树;有向树
directed-tree graph 有向树图
directed-tree matrix 有向树矩阵
direct effect 直接效应;直接效果
direct electric(al) arc furnace 直接电弧炉
direct electric(al) heating 直接电热;直接电动供暖
direct electric(al) process 正淬火法;电阻直接加热处理
direct electric(al) starting system 直流电力起动系统;直流电动起动系统
direct electrography 直接电照相
direct electromotive force 直流电动势
direct electrophotography printer 直接电子照相印刷机
direct elimination method 直接消去法
direct embedded cable 直埋电缆
direct employed labo(u)r 直接雇工
direct energy conversion 能量直接转换;直接能量转换
direct energy input 能量直接输入
direct engineering cost 直接工程费(用)
direct enlargement 直接放大
direct entry mode 直接记入方式
direct environment 直接环境
direct esterification 直接酯化
direct estimate 直接估算;直接估计
direct evapo(u)rator 直冷式蒸发器;直接式蒸发器
direct evidence 直接证据
direct exchange 直接交换;直接汇兑
direct exchange contract 直接汇兑合约
direct exchange interaction 直接交换作用
direct exchange of arbitrage 直接套汇
direct exchange quotation 外汇直接报价
direct excitation 直接激励
direct expansion 直接膨胀;直接发泡
direct expansion coil 直接蒸发盘管;直接膨胀盘管
direct expansion cooling 直接蒸发冷却;直接膨胀冷却
direct expansion evapo(u)rator 直接蒸发的蒸发器
direct expansion refrigeration 直接制冷系统;直接膨胀制冷

direct expansion system 直接蒸发系统;直接膨胀式
direct expenditures 直接列入支出;直接财政支出
direct expenses 直接费(用)
direct expense audit 直接费审计
direct expense cost 直接费用成本
direct expense escalation 按房产的直接支出自动调整租费条款
direct export 直接出口
direct export and import 直接出口和进口
direct extrusion 直接挤压(法);正挤压;顺挤法
direct fastening indiscrete fastener 不分开式扣件
direct fed coil 直馈式线圈
direct feed 直接馈电;直接供电;直接传送工件
direct feedback 直接反馈
direct feedback system 直接反馈系统
direct feeder 直接馈路
direct feed incinerator 直接进料焚化炉
direct feeding system with negative feeder 带回流线的直接供电方式
direct feeding system with return cable 带回流线的直接供电方式
direct feed mill 直接喂煤系统的磨机
direct field 直接声场
direct file 直接文件
direct file access 直接文件存取
direct filtration 直接过滤
direct finance for 直接为……筹措资金
direct financing 直接供应资金;直接筹资
direct financing leases 直接融资租赁
direct finder 直接定位仪
direct fire 明火;活火头;活火(头)直接烧;直接烧;直接点火
direct-fired 活火烧了的;直接火加热的
direct-fired air heater 直热式空气加热器
direct-fired appliance 直接燃烧式用具
direct-fired boiler 直吹式制氧锅炉
direct-fired coil furnace 直接焰烧的热处理炉
direct-fired drier 明火直烧干燥机;直接火烘干燥机;直烧干燥器
direct-fired evapo(u)rator 直烧蒸发器
direct-fired furnace 火焰直接加热炉;明焰窑;直接加热的热处理炉
direct-fired heater 直接火力加热器;明火炉;直热式空气加热器
direct-fired installation 直吹式燃烧系统
direct-fired kettle 直接火烤锅
direct-fired kiln 明焰窑
direct-fired lithium-bromide absorption type refrigerating machine 直燃式溴化锂吸收式制冷机
direct-fired mull 直吹式制粉系统磨煤机
direct-fired oven 直接(式)烤箱;内热式烤箱
direct-fired rotary drier 直烧旋转干燥器
direct-fired scale-less of free heating furnace 敞焰少无氧化加热炉
direct fired system 直吹式制粉系统
direct fire heating 明火加热;活火加热;直接火加热的;直火加热
direct fire pressure 直接消防压力
direct fire suppression 直接扑火;直接灭火
direct fire suppression method 直接灭火法
direct firing 明火烧成;直接(燃)烧;直接火加热的;直吹式燃烧
direct firing coal bowl mill 无中间煤粉仓直接燃烧的碗煤磨
direct firing coal mill 无中间煤粉仓直接燃烧的煤磨
direct firing coal system 无中间煤粉仓直接燃烧系统
direct firing furnace 明火加热炉;直焰炉;直接火力加热器
direct flame 直接焰
direct flame atomic absorption method 直接火焰原子吸收法
direct flame boiler 直焰式锅炉
direct flame incineration 直火焚烧炉
direct flexure 简单弯曲;单纯弯曲;纯弯曲
direct float 流向(测验)浮子
direct flo(a)tation 直接浮选
direct flooding system 直接注水系统
direct flow 恒稳流;直进流;直接流通;直接流量;直接径流;顺流
direct flow demineralization 直流式脱盐
direct flow hollow spool valve 直流式空心阀;顺流式空心阀
direct flow water system 直流水系统

direct flushing 正循环冲洗【岩】;正洗井;正冲件
direct flush valve 直接冲洗阀
direct foreign exchange quotation 直接标价法
direct foreign investment 直接对外投资
direct foundation 直接基础
direct freight 直接航运
direct frequency modulation 直接调频
direct frequency modulation system 直接调频制
direct from drum cold emulsion sprayer 圆桶式冷乳液直接喷射装置
direct from drum type sprayer 圆桶式直接喷射装置
direct gain 直接受益
direct gaseous reduction 直接气体还原
direct gas heating 直接煤气供热
direct gasoline 直接分流汽油
direct gear drive 单独传动
direct-geared 直接啮合;齿轮直接传动
direct geothermal steam 天然地热蒸汽
direct glare 直射炫目光线;直接眩光
direct glazing 无窗框安装法;直接装配玻璃
direct glue-down 直接粘贴
direct goods 直接财产
direct grants 直接补贴
direct graphite furnace method 直接石墨炉法
direct gravity culvert 直接重力涵洞
direct gravure coater 直接槽辊涂布机
direct grinding mill 直吹磨机
direct group 直通线群
direct guarantee 直接担保
direct halftone 直接挂网片
direct harm done by purely man-made factors 人为的直接破坏
direct heat drier 直接加热烘干机;直接加热干燥器
direct heated 直接加热
direct heated cathode 直热式阴极
direct heated thermistor 直热式热敏电阻;直热式热变电阻器
direct heater 直接供暖器
direct heat exchange 直接换热
direct heating 直接加热;直接供暖
direct heating apparatus 直接加热装置;直接加热器;用火加热的加热器
direct heating device 直接供热设备;直接采暖设备
direct heating hot water supply system 直接加热热水供应系统
direct heating surface 直接供暖面
direct heating system 直接供热系统;直接采暖系统
direct heating unit 直接采暖器
direct heat loss 直接热损失
direct heat source 直接热源
direct high 直接档
direct holding 直接持股
direct hologram 直接全息图
direct holography 直接全息摄影
direct hose stream 直接软管水流
direct hot water system 直接热水系统
direct humidifier 直接增湿器
direct illumination 直接照明;垂直照明
direct image 直射影像;直接像
direct-image offset system 图像直接转换胶印法
direct imagery 直接成像
direct imaging 直接成像
directimaging mass analyzer 直接成像质谱仪
direct imaging optics 直接成像系统
direct impact 直碰
direct impact amplifier 对冲式放大器
direct impact type air speed indicator 直冲式风速计
direct impedance 直接阻抗
direct import 直接进口
direct impulse 直射脉冲;正向脉冲;探测脉冲
direct incineration 直接焚烧;直接焚化
direct indent system 直接订货制度
direct indexation 直接指数化
direct indexing 直接分度法;直接变址
direct index plate 直接分度盘
direct index plate pin 直接分度盘销
direct indicating compass 直读罗盘;直读罗经
direct indicator 直接指标
direct-indirect heating 半间接暖气装置
direct-indirect lighting 漫射光照明
direct inductive coupling 直接电感耦合
direct infection 直接传染

direct infection diesel unit 柴油直接喷射器
direct infection nozzle 直接喷射式喷油器
direct infeed 直接切入磨法;直接横向进磨
direct influence 直接影响
directing curve 有向曲线
directing graph 有向曲线
directing operation 控制程序
direct ingot 直熔锭
directing point 基准点
directing property 定向性
directing sign 指向标
directing spray 直接喷洒
directing staff 标桩
directing vane 导向叶
directing works 导流设施;导流建筑物
direct inhibition 直接抑制
direct injection 直接注入(法);直接射入;直接喷油;定向注射
direct injection carburetor 直接喷射式汽化器
direct injection cylinder head 直射式汽缸盖
direct injection diesel engine 直接喷射式柴油机
direct injection engine 直接喷射式发动机
direct injection method 直接喷射法
direct injection mo(u)lding 直接注射造型
direct injection pump 直接喷射式燃料泵
direct injury cost 直接损害费用
direct in-line 直线叫入
direct in-plant reuse 厂内直接再用
direct input 直接输入
direct input circuit 直接输入电路
direct input table 直接投入表
direct insert 直接插入的
direct insert routine 开型子程序;直接插入例行程序
direct insert subroutine 开型子程序;直接插入子程序
direct instruction 直接(地址)指令
direct insurance 直接保险
direct integration 直接积分
direct interpolation 直接内插法;直接插值法
direct interpretation 直接解译
direct interpretation key 直接解译标志
direct interpretation method 直判法;直接解释法
direct interview 直接问询调查
direct investment 直接投资
direct investment abroad 国外直接投资
direct investment dividends 直接投资红利
direct investment flow 直接投资流动
direct investment fund 直接投资基金
direct investment interest 直接投资利息
direct inward dial(l)ing 对内拨号;直接向内拨号;互拨内线
direction 方向;方位
directional 定向的;方向的
directional accuracy 方向精度
directional action 定向作用
directional adjustment 方向平差
directional aids 定向器材;定向航标;定向(辅助)设备
directional amplitude modulation 定向调幅
directional angle 方向角;方位角
directional antenna 定向天线;指向天线
directional arm 方向导杆
directional arrow 指向箭头
directional averaging method 方向平均法
directional baffle 指向障板
directional beacon 指向信标;定向信标;定向导标;无线电定向信标
directional beam 定向射束(无线电);定向波束
directional beam beacon 定向无线电航标
directional bearing 测向方位
directional bit 定向钻头
directional blasting 定向爆破
directional blasting closure 定向爆破截流
directional blasting damming 定向爆破筑堤;定向爆破筑坝
directional carrier 定向搬运车
directional casting 定向柱状晶铸造
directional census 交通方向调查
directional changing 反转的
directional channelization 交通定向渠化
directional characteristic 方位特性;方向性图形;方向特性;定向特征
directional clinograph 方位测斜仪
directional clutch 定向离合器

directional coefficient 方向系数
directional coil 定向线圈
directional collector 定向收集器
directional command activated sonobuoy system 定向指令声呐浮标系统
directional comparison microwave relay 方向比较微波继电器
directional component 方向分量;方向数
directional connection plug 反转转接插头
directional contactor 可逆接触器;换向接触器
directional control 方向控制
directional controller 方向控制器
directional control valve 方向控制阀;定向控制阀
directional cooling 定向冷却
directional correction 方向校正
directional correlation 方向相关
directional cosine 方向余弦
directional cosine vector 方向余弦矢量
directional counter 定向计数器
directional coupler 定向耦合器
directional coupling 方向性耦合
directional cross 定向交叉(口)
directional crossing 定向式立(体)交(叉)
directional crystal 柱状晶体;定向结晶
directional current meter 定向海流计;定向流速仪
directional current protection 方向电流保护装置
directional damper 方向阻尼器;偏搬阻尼器
directional data 方向数据
directional derivative 方向微商;方向导数
directional design-hour volume 定向设计小时交通量
directional detector 方向探测器;定向探测器;定向检测器
directional deviation of borings 钻探定向偏差;钻孔定向偏离;钻孔定向偏差
directional diagram 方向图
directional differential protection 方向差动保护
directional diffuse reflection 定向漫反射
directional diffusion 方向性扩散;定向扩散
directional direct current relay 定向直流继电器
directional distance relay 定向距离继电器
directional distortion 方向变形
directional distributed coefficient 方向分布系数
directional distribution 定向分布;方向分配;方向分布;按向分布
directional divergence 方向性扩散
directional division multiplexing 方向分割复用
directional drilling 定向钻井;定向钻进;定向钻孔
directional drilling head 定向钻进作用的回转器
directional drilling tool 定向钻进器具
directional duality principle 定向的对偶性原理
directional dust ga(u)ge 直接集尘器
directional earth-leakage relay 对地漏电方向继电器
directional echo sounding 定向回声测深(法)
directional effect 方向效应;方位效应
directional element 方向元件
directional error 方向误差;定向误差
directional exit sign 指向出口标记
directional explosion 定向爆炸;定向爆破
directional explosive echo-ranging 定向爆炸回声测距
directional feature of earthquake 地震(的)方向性
directional film 定向膜
directional filter 分带网络;方向滤波器;定向滤波器
directional filtering 定向滤波
directional filtering image with partial coherent light 部分相干光方向滤波图像
directional finder 定向仪;定向器
directional finding 方位测定
directional finding compass 定向罗盘
directional fixed grill(e) 定向固定格栅
directional float 定向浮子
directional float tray 浮舌塔盘
directional flow 流向
directional flow fitting 导向设施;导向管件
directional fluctuation 方向起伏
directional frequency modulation 定向调频
directional frequency response characteristic 方位频率感应特性
directional function 指向性函数
directional gain 指向性指数;指向性增益
directional ga(u)ge 方位指示仪
directional ground relay 接地方向继电器;定向接

地继电器
directional gyrocompass 方位陀螺仪
directional gyro(scope) 定向陀螺仪;方位陀螺仪;陀螺方向仪;航向陀螺仪;定向陀螺(仪)
directional gyroscope reference 方向陀螺基准
directional high frequency antenna 定向高频天线
directional hole 定向(钻)孔
directional homing 定向归航
directional hydraulic conductivity 定向导水性;定向导水率
directional hydrophone 方向性水听器;指向水听器;定向水听器
directional impedance relay 方向阻抗继电器;定向阻抗继电器
directional index 方向指数;指向性指数
directional indicating device 方向指示仪;方向指示器
directional indicator 方向指示仪;方向指示器
directional information 方向信息
directional inlet fitting (能调整水流方向和流量的)进水口配件
directional instrument 方向仪
directional interchange 定向(式)立(体)交(叉)
directional intersection 定向交叉(口)
directional island (道路上的)导向岛,方向岛;渠化(交通)岛
directionality 指向(性);方向性
directionality effect 定向(性)效应
directional lane 导流车道
directional light 定向灯标;方向灯
directional lighting 指物照明;定向照明;方向(性)照明
directional light source 定向光源
directional line 方向线
directional-listening sonobuoy 定向声呐浮标
directional log 方向录井
directional main 方向转接器
directional measurement 定向测量
directional method 方向(观测)法
directional method of adjustment 方向观测平差法
directional microphone 指向传声器;定向传声器
directional migration 定向迁移
directional mirror 定向反射镜
directional modulation 定向调制
directional moment of meridian 子午方向力矩
directional negative phase-sequence current protection 负序电流方向保护(装置)
directional numbers 方向数
directional of blast orifice 风口方向
directional of development 发展方向
directional of lay 敷设方向
directional of source force 发震力方向
directional of wind 风向
directional overcurrent protecter 方向过电流保护装置
directional overcurrent relay 定向过流继电器
directional parameter 方向参数
directional passenger flow 分向客流
directional pattern 辐射方向图;方向性图;指向性图
directional pavement marking 路面指向标示;路面导向标志
directional permeability 方向渗透率
directional phase changer 定向相位转换器
directional phase shifter 方向性移相器;定向移相器
directional pin-point blasting 定向抛掷爆破
directional pipe 定向钻杆
directional pole 定向杆
directional polish 方向性抛光痕;抛光不足
directional post 导向柱
directional power meter 方向性功率表;定向功率计
directional power protection 方向功率保护装置;功率方向保护(装置)
directional power relay 定向功率继电器
directional precision 方向精度
directional preponderance 优势偏向
directional pressure 定向压力
directional probe 方向测针
directional property 方向性;指向特性;各向异性;定向性
directional protection with carriercurrent blocking 载波闭锁方向保护(装置);高频闭锁方向保护(装置)
directional pull 直牵引力
directional radio 定向无线电;无线电定向

directional radio beacon 定向无线电指向标;定向无线电信标
directional radio navigational beacon 定向无线电导航信号台
directional radio range 定向无线电叠标
directional radio station 定向无线电信号台
directional ramp 定向式匝道
directional ranging 方向测距
directional ratios 方向数
directional reading 方位角读数
directional reading hardness test 直接单读硬度试验
directional receiver 定向(信号)接收机
directional reception 定向接收
directional rectifier 方向整流器
directional reflectance 定向反射率
directional reflection 定向反射
directional reflection characteristic 方位反射特性
directional relay 极化继电器;定向中继;方向继电器;定向继电器
directional response 方位感应
directional response pattern 指向性图案
directional roadway marking 路面指向标示;道路指向标志
directional rod 定向钻杆
directional scanning 定向扫描
directional screen 定向银幕
directional selection 定向选择
directional selectivity 方向选择性
directional sensor 方向传感器
directional separation 上下行分开
directional separation filter 分向滤波器
directional separator 车道指向分隔带;方向分隔带
directional shear cell 定向剪力盒
directional shooting 定向爆炸
directional shooting closure 定向爆破截流
directional shrinkage 定向的缩
directional sign 指向标志;指路标(志);方向标志;方向标识
directional silicon steel 各向异性硅钢
directional silicon steel strip 各向异性硅钢片
directional slide 方向滑槽
directional solidification 定向凝固;定向结晶;顺序凝固
directional sonar 指向性声呐
directional specimen 定向标本
directional spectral density 方向性谱密度函数
directional spectrum 方向谱
directional split 方向划分;方向分路;流向分配
directional split of traffic volume 交通量对向划分率;交通量定向划分率
directional spreading function 方向分布函数
directional stability 航向稳定性;方向稳定性
directional stability of beam 光束方向稳定性;光束方向稳定度
directional stabilizer 方向指示仪;方向指示器
directional structure 定向构造;方向构造;指向构造
directional survey(ing) 方向测量;井斜测量
directional surveying of a well 井的方位测量
directional survey method 方位测井法
directional survey with magnetic compass 地磁场定向测量法
directional swell 定向的胀
directional switch 方向开关
directional technique 定向技术
directional theodolite 方向经纬仪
directional time 定向时间
directional tire 方向性轮胎
directional tool 定向钻具
directional traffic 定向交通
directional traffic power source 方向电源
directional traffic power supply 方向电源
directional transformed area 方向转换区
directional transit 方向经纬仪
directional transmitter 定向发射机;无线电定向发射台
directional traverse 方向线测定
directional-type interchange 定向型互通立交
directional-type tricone bit 定向型三牙轮钻头
directional valve 换向阀;定向阀;方向阀
directional variable 方向变量;方位变量
directional variation 定向变异
directional vector 方向向量
directional vibration 定向振动

directional wattmeter 定向瓦特表
directional well 定向井
directional well cementing 定向井注水泥
directional well surveying 定向井测量
directional wireless installation 定向无线电装置
directional yaw stability 方向安定性
directional zero phase-sequence current protection 零序电流方向保护(装置)
directional zoning 方向分区
direction and magnitude of force 力(的)矢量
direction and pathway of migration 运移方向和通道
direction and revolution indicator 轮机转向转数指示器
direction arm 指路牌;指向牌
direction arrow 行车箭头标
direction beacon 定向导标
direction bearing 探向方位;探测方位
direction board 指示牌
direction calculation 直接计算
direction calculation method 直接计算法
direction center 引导中心;制导中心
direction-changing equipment 改向装置
direction clutch 定向离合器
direction control damper 航向控制的阻尼器
direction control of tunnel 隧道导向控制
direction-determining board 正方案
direction finder 定向仪;测向仪;测向器;测向计;无线电罗盘;无线电罗经;探向器
direction finder antenna 无线电测向天线;测向器天线
direction finder-bearing indicator 测向仪方位指示器
direction finder deviation 测向仪自差;测向仪偏差
direction finder repeater 测向仪复示器
direction finder set 测向机
direction finding 定向;测向;方位测定;探向
direction finding antenna 测向天线
direction finding antenna system 测向天线系统
direction finding apparatus 测向仪
direction finding chart 导航图
direction finding compass 测向罗盘
direction finding device 测向装置
direction finding facility approach 定向设备进近
direction finding net(work) 测向网
direction finding receiver 测向接收机
direction finding sensitivity 测向灵敏度
direction finding station 定向台;测向台
direction finding system 测向系统
direction float 流向(测验)浮子;流向浮标;定向浮子
direction for safe use 安全使用说明书
direction for use 使用说明
direction-free 无方向性的
direction ga(u)ge 测角计
direction-giving beacon 指向信标
direction-independent 方向独立的
direction-independent radar 方向无关雷达
direction indicating device 示向器
direction indicator 方向指示器;航向陀螺仪;方位指示器
direction indicator for vehicle 车辆方向指示器
direction instrument 万向仪
direct ionization 非本征光电发射
direction jet grouting 定喷
directionless pressure 无(定)向压力
direction light 方向灯
direction limits switch (电梯的)保险开关
direction-listening device 声波定向器
direction load cast 定向压模
direction marking 行车指向箭头
direction meter 定向器
direction method of measuring horizontal angles 水平角方向观测法
direction of acceleration 加速度方向
direction of a curve 曲线的方向
direction of angle of deviation 偏离角方向;偏差角方向
direction of angular deviation 角偏离方向;角偏差方向
direction of approach 进场方向
direction of arrow 箭头方向
direction of Burger's vector 布格矢量方向
direction of camera lens 镜头方位
direction of cardinal wind 常风向

direction of commerce 货物流向
direction of compaction 驱动方向
direction of cross-current 横流方向
direction of current 流向；电流方向
direction of deflection 偏斜方向；偏斜方位
direction of deformation 变形方向
direction of dip 倾斜方向
direction of dip of strata 岩层走向
direction of displacement of interesting point 交点位移方向
direction of drilling 钻进方向
direction of dune movement 沙丘移动方向
direction of earthquake 地震方向
direction of easy axis 易磁化方向
direction of evacuation 疏散方向
direction of exposure 受风影响
direction of extinction 消失方向；消光方向
direction of extrusion 挤压方向
direction of fall 下坡方向
direction of feed 进料方向；进给方向
direction of flight 摄影航向
direction of flow 流向；流体流向；流程方向
direction of force 力的方向
direction of friction 摩擦（力）方向
direction of gales 强风向
direction of gate 大门方向
direction of geostress 地应力方向
direction of goods flow 货流方向
direction of greater goods flow 货流较大方向
direction of groundwater flow 地下水流向
direction of groyne 折流坝方向；丁坝方向
direction of illumination 照明方向
direction of imposed force 外力作用方向
direction of landslide axle 滑坡轴方位
direction of lay 捻向
direction of light 光的方向
direction of load application 加载方向
direction of loading 荷载方向；负荷方向
direction of long diameter oblique to the flow 长轴与水流方向斜交
direction of long diameter parallel to coast 长轴与岸线方向平行
direction of long diameter parallel to the flow 长轴与水流方向平行
direction of long diameter perpendicular to the flow 长轴与水流方向垂直
direction of longitudinally tectonic migration 纵向构造迁移方向
direction of loop 环的方位
direction of main stress 主应力方向
direction of management 经营方向
direction of map 图件说明
direction of maximum variation of ore body 矿体最大变化方向
direction of measurement 沿量度方向
direction of migration 洄游方向；迁移方向
direction of most severe wave attack 最强浪向
direction of motion 运动方向
direction of observation line 观测线方向
direction of observation line for test 试验观测线方位
direction of optic(al) axis 摄影方向
direction of plumb-line 垂线方向
direction of polarization 极化方向
direction of pressing 压穿方向
direction of prevailing wind 常风向
direction of principal curvature 主曲率方向
direction of propagation 传播方向
direction of regime observation line 动态观测线方位
direction of relative movement 相对运动方向
direction of river 河流流向
direction of rotation 旋转方向；转动方向
direction of scan(ning) 扫描方向
direction of search 搜索方向
direction of shock 振动方向
direction of simple shear 单剪运动方向
direction of sound 声音方向
direction of spiral 螺旋方向
direction of spring distribution 泉群分布方向
direction of strain 应变方向
direction of strata 岩层走向；走向断层；地层走向
direction of stream 河流流向；水流方向
direction of strike 走向（矿层或地层）
direction of strong wind 强风向
direction of structural lineament 构造线方向
direction of take off 起飞方向
direction of the applied force 加力方向
direction of the applied load 所加荷载方向；加载方向
direction of the borehole 钻孔方向
direction of the intermediate principal stress 中间主应力方位角
direction of the maximum principal stress 最大主应力方位角
direction of the minimum principal stress 最小主应力方位角
direction of throughput 输送方向
direction of tide 潮流方向
direction of tilt 倾斜方向
direction of trade 贸易方向
direction of traffic 行车方向；疏散交通
direction of transversely tectonic migration 横向构造迁移方向
direction of travel 旅行方向；走行方向；前进方向
direction of variogram 变异方向
direction of view 观测方向
direction of wave advance 波浪推进方向
direction of wave propagation 波（浪）传播方向
direction of wave travel 波浪前进方向；波（的）传播方向
direction of weft 信号旗方向；求教信号方向
direction of welding 焊接方向；焊接方位
direction of working 回采方向
direction opposite this tectonic region 相对本构造区的方向
direction oriented decrossing 方向别疏解
direction peg 导向桩
direction permeability 定向渗透率
direction plate 指示牌
direction post 指向柱；指向牌
direction radiation 定向辐射
direction-range method 测向距离法
direction reduction mortgage 逐期还本减息抵押贷款
direction relay 定向继电器
direction rose 流向图
directions 用法
direction selector valve 方向性选择开关
direction-sense 定向性
direction sense movement 定向性运动
directions for use 用法说明
direction signal light 转向灯
direction signal switch 转向信号开关
directions of use 用法指导；用法说明
direction stability 偏航安定性
directions test 指导测验
direction sub 定向接头
direction switch 升降开关
direction tower 定向塔；定位塔
direction traffic 定向交通
direction vector 方向矢量
direct irrigation 直接灌溉
direct irrigation area 直接灌溉面积
direct isotope effect 直接同位素效应
direct iteration 直接迭代；正迭代（法）
directive 指向（的）；指令；指挥仪；定向的
directive action 定向作用
directive beacon 无线电定向信标
directive charge 定向炸药包
directive command 指示指令
directive construction organization design 指导性施工组织设计
directive development of environment 环境定向发展
directive diagram 指向图；方向图
directive effect 方向作用；方向效应
directive efficiency 定向效率
directive erosion 定向侵蚀
directive factor 定向系数
directive feed 定向耦合器
directive force 指引力；指向力；定向力
directive force of magnetism 磁性指向力
directive gain 指向性增益；定向增益；方向增益
directive guidance 方向性指导
directive gyro 定向陀螺仪
directive gyroscope 定向陀螺仪
directive joint 定向接头
directive meter 单转向计数器
directive moment 指向力矩
directive of wave 波向
directive pattern 指向性图
directive plan 指令性计划
directive power 定向功率
directive property 指向性
directive radio beacon 定向无线电指向标；无线电指向标
directive reception 定向接收
directive rule 规程
directive sending 定向发送
directives of construction 施工指令
directive statement 指示语句；指令语句；定位语句
directive structure 定向结构
directive tool 定向钻具
directive to work 工作指示
directive transmission 定向发送
directive variation 方向性变化
directive wave 直达波
directive wheel 导向轮
directivity 指向性；方向性（系数）
directivity angle 辐射仰角
directivity characteristic 方向特性曲线
directivity diagram 辐射方位曲线；方向图
directivity factor 指向系数；指定系数；方向系数；指向性因数；定向系数；方向性因子；方向性因数
directivity function 指向性函数；方向性函数
directivity gain 指向性因数
directivity index 指向性指数；定向指数；方向性指数
directivity pattern 指向性图案
direct jet solid stream 直射充实水柱
direct jump 直接转移；直接水跃
direct key 直接判读标志
direct labo(u)r 直接（人）工；直接雇佣劳动力；直接劳动；非熟练工；生产工人
direct labo(u)r and indirect labo(u)r 直接劳动与间接劳动
direct labo(u)r budget 直接人工预算
direct labo(u)r cost 直接人工费；直接工值
direct labo(u)r cost budget 直接人工成本预算
direct labo(u)r cost method 直接人工成本法
direct labo(u)r expenses 直接人工费用
direct labo(u)r expense budget 直接人工费用预算
direct labo(u)r hour 直接人工小时
direct labo(u)r hour rate 直接人工小时率
direct labo(u)r standard 标准时间
direct labo(u)r time 直接工时
direct laying 电缆直接敷设
direct leasing 直接租借
direct lending 直接借出；直接贷款
direct level(l)ing 直接测高程；直接水准测量；直接高程测量
direct liability 直接负债
direct light 直接光（线）；直射光
direct light illuminant 直接光照明器
direct lighting 直射照明；直接照射；直接照明；直接采光
direct-light luminaire 直射光照明
direct lightning stroke 直接雷击
direct lightning stroke over-voltage 直接雷击过电压
direct light reflex 直接光反射
direct limit 正向极限
direct line 直通行；直通线路；直（达）线；定向线
direct linear transformation 直接线性变换
direct link 直接连接
direct linkage roll feed 直接杠
direct liquefaction 直接液化
direct listening sonar 噪声测向声呐
direct lithography 直接平印
direct load 直接装载；直接寄存；直接荷载
direct loading 直接装运法；直接装入
direct loading and unloading 直接装卸
direct load loss 直接负荷损失
direct loan 直接贷款
direct location mode 直接定位方式
direct loss 直接（火灾）损失
direct luminaire 直接照明器；直照光源
directly buried installation 直埋敷设
directly consumable products 可直接消费的产品
directly controlled traction equipment 直接控制的牵引设备
directly cooled machine 直接冷却电机

directly coupled 直接耦合的
directly fed antenna 直馈式天线
directly fired kiln 直烧窑；直接加热炉
directly fired (pressure) vessel 直接火焰加热压力容器
directly heated 直热式的
directly heated oven 直接(式)烤箱；内热式烤箱
directly lead borehole for dewatering 直通式疏干钻孔
directly produce 直接产生
directly proportional 直接比的；有直接比例的；成正比(例)的
directly proportional to 与……成正比
directly reduce 直接归约
directly reduce to 直接归纳到……
directly related entertainment 直接有关的款待
directly related family 直系家庭
directly seating 直接承垫
directly surveillance network 进程遥信网络
directly surveillance subsection 进程遥信分区
directly viewed infrared image tube 直观红外成像管
directly visible area 直接能见区
direct magnetizing effect 直轴磁化作用；直轴磁化效应
direct mail 直接邮寄
direct maintenance 直接维修
direct manner 直接法
direct manufacturing expenses 直接生产费用
direct manufacturing process 直接生产过程
direct mapping 直接映射；直接变换
direct mark 直接标志
direct marketing 直接行销
direct marketing at the factory door 工厂门市部直接销售
direct marketing by integration 一体化直接销售
direct marketing by personal solicitation 向个人直接销售
direct material 基本材料；直接(原)材料
direct material budget 直接原材料费用预算
direct material consumed 耗用直接原料
direct material cost 直接原料成本；直接材料成本
direct material cost method 按直接原材料成本法
direct material purchases budget 直接原料购买预算
direct materials basis 直接物料法
direct measurement 直接量测；直接计量；直接度量；直接测量；直接测定；往测
direct-melt process 直熔法拉丝；池熔法拉丝；池窑拉丝工艺
direct memory access 存储器；直接存储器存取；存储器直接访问；存储器直接存取
direct memory access channel 存储器直接存取通道
direct memory access data transfer 存储器直接存取数据传送
direct memory access transfer 存储器直接存取传送
direct memory address 直接存储地址
direct memory execution 直接内存执行
direct metal 直接金属
direct metamorphosis 直接变态
direct method 直接(教学)法
direct method heat balance 正热平衡法
direct method of current account 往来账户直接法
direct method of isolation 直接分离法
direct method of triangularization 三角化直接法
direct microscopic method 直接显微镜法
direct mobility 原点导纳；直接导纳
direct mode 直接方式；正向模
direct model analysis 直接模型分析
direct mode operation 联机作业法
direct moment distribution 直接力矩分配法
direct monetary loss 直接货币损失
direct motion 直接运动；顺行
direct motor drive 电动机直接传动
direct-motor-driven 单电动机传动的
direct mount 直接定位卡；直接包装
direct-mounted 悬挂式的；直接装上的
direct mouth 直接(入海)河口
direct multiplex control 直接多工控制
direct nailing 直接钉入；正面钉合；垂直钉面
direct negotiation 直接协商
directness 直率；直捷；直接(性)

direct noise 直接噪声
direct noise amplifier 直接噪声放大器
direct numerical control 直接数字(程序)控制；直接数值控制；直接数控；群控
direct numerical integral 直接数值积分
direct obligation 直接责任；直接偿付
direct observation 直接观测；直接调查
direct observation method 直接观察法
direct observation through microscope 显微镜直接观测
direct of unit footage of tunneling 单位进尺直接成本
direct oil detection 直接探测石油
Directoire style 法国革命时代的风格(过渡的拟古派风格)
direct on enamel 直控搪瓷
direct on enameling 直接上釉
direct on-line processor 直接联线处理机；直接联机处理机
direct on-line starter 直接联线启动器
direct on-line switching 直接起动；直接合闸
direct on starter 直接起动器
direct on starting 直接起动
direct operated 直接操纵
direct operating cost 直接生产费用；直接经营费用；直接操作费
direct operating expenses 直接操作费用
direct operation 直接作业；直接传动；直接操作
direct operational features 直接营运设施
direct optic(al) measurement 直接光学测定法
direct option 直接选择
director 理事；引向器；引导程序【计】；主任；指挥者；指挥仪；指挥器；指挥机；指导者；管理者；董事；定向器；定向偶极子；导向偶极子引向器；导控器；操纵仪表；无源定向偶极子；所长
directorate 管理局；董事会
Directorate of General Water Engineering 普通水工工程理事会(英国)
Directorate of Overseas Survey 海外测量局(英国)
direct orbit 顺行轨道
director board 董事会
director circle 准圆；切距圆
director coil 指示器线圈；探测线圈
director cone of a ruled surface 直纹面的准锥面
direct order 直接订货
director dipole 引向振子
director element 引向单元；导向元件
direct organization 直接结构；直接编制
direct organization file 直接编制文件
director general 总监(工)
direct orientation 直接定向法
director liability 董事职责
director meter 指挥器记数器
director of hospital 医院院长
director of project management 工程管理经验
director of site management 工地主任
directors building 行政管理楼
director's console 指挥台；导演纵台
directorship 指挥职能
director-sight system 指挥仪瞄准系统
director's office 厂长室
directors of lower echelon of the board 初级董事会
director sphere 准球面
director's railcar 管理人员专用铁道车辆
director's report 董事会报告
director's room 董事室
director system 指挥制；指挥系统
director telescope 指示望远镜；光学瞄准器
director-type computer 指挥仪型计算机
director valve 主导汽门；主导阀；导向阀
direct or with transshipment 直运或装船
directory 号码簿；号码表；一览表；人名录；索引簿；手册
directory area 目录区
directory board 指示牌
directory capacity 目录容量
directory control entry 目录控制项
directory data set 目录数据集
directory entry 目录登记项
directory file 目录文件；目录档案
directory key 目录关键字
directory (list) 目录
directory maintenance 目录保护；目录管理程序

directory management 目录(表)管理
directory manager 目录管理程序
directory member entry 目录成员项
directory name 目录名
directory of importers and exporters 进出口商行名录
directory operator 查号台话务员
directory record 目录记录
directory retrieval 目录检索
directory routing 目录式路由选择；目录式路径选择；表式路由选择
directory search 目录搜索；目录查找
directory sign 指示标志；导向标志
directory structure 目录结构
directory system 索引系统
directory trust 指示信托
direct output 直接输出
direct output writer 直接输出程序
direct outward dial(l)ing 直接对外拨号；直接向外部拨号
direct over-current release device 直流过流释放装置
direct overhead 直接管理费
direct overhead accounts 直接管理费账户
direct oxidation 直接氧化
direct oxidation leaching 直接氧化浸出
direct oxidation process 直接氧化法
direct oxidation reaction 直接氧化反应
direct packing 直接包装
direct paper 直接汇票
direct particle 初生粒子
direct passage of current 直接通电
direct patenting 线材拉后直接退火；直接韧化处理；直接拉后退火
direct path 直接通路
direct path wave 直接波
direct payment 直接支付；直接付款
direct permeability apparatus 直接渗透仪
direct personal selling 人员直销
direct personnel expenses 人员直接费用；直接个人费用；直接人员薪金
direct photodetection 直接光电探测
direct photodetection receiver 直接光探测接收器
direct photographic(al) method 直接复照法
direct photography 直接拍摄法
direct pick-up 直接拾波；直接摄影；直接摄像；直接录音；直接传感；广播室广播；正常极性下启动电流
direct pick-up receiver 转播用接收机
direct piecework system 直接计件制
direct piezoelectric(al) effect 正压电效应
direct piezoelectricity 直接压电现象
direct placement 直接安排
direct placement of securities 直接安排(私自推销)债券
direct plummet observation 正锤线观测
direct-plunger elevator 柱塞直接驱动的升降机
direct pointing 直接瞄准
direct point repeater 直接中继器
direct pollutant discharge coefficient 直接污染物排放系数
direct port 直达港
direct position of telescope 正镜测量；正镜【测】
direct positive 直接正像；直接接触正片
direct posting 直接过账
direct pouring 直接浇铸
direct power acting push sweep rake 直接动力作用推扫耙
direct power feeding system 直接供电方式
direct power generator 直接发电机
direct precipitation 直接降水(量)
direct preparation 直接配制
direct pressure 直接压力
direct pressure closing 直压法
direct pressure hotchamber machine 直接气压式热室压铸机
direct pressure wave 直接压力波
direct price 直接价格
direct pricing 直接计价
direct printer 直接印刷机
direct printing 直接印刷；直接印染；直接印花
direct printing decorative board 直接印刷装饰板
direct probe analysis 直接测试法
direct problem 正算问题

direct process 直接(冶炼)法
direct product code 直积码
direct production coefficient of impurities 污染物直接产生系数
direct product of sets 直积集
direct products 直接产品
direct program(me) control 直接程序控制
direct projection 直接投影
direct projectivity 正射影变换
direct prompt reaction 直接迅速反应
direct proof 直接证明
direct property damage 直接财产损失
direct proportional effect 正比例效应
direct proportion(ality) 正比(例)
direct prospecting indication 直接找矿标志
direct prot injection system 直接进口喷射系统
direct pull method 直接拉伸法
direct pulse 直达脉冲;探测脉冲
direct pulverized sub-bituminous coal firing system 直接成粉亚烟煤燃烧装置
direct purchase 直接采购
direct quench aging 热浴时效处理
direct quenching 直接淬火
direct quench system 直接急冷系统
direct question 直接询问
direct quotation 直接估价;直接标价
direct radial plot 直接辐射三角测量
direct radial ray 直接辐射线
direct radial triangulation 直接辐射三角测量
direct radiant heater 直接辐射式供暖器
direct radiant heating appliance 直接辐射供热用具
direct radiation 直接辐射
direct radiator 直接传热炉片
direct raising 直接打捞【救】
direct raising method 直接打捞法
direct ramp 直接式匝道
direct rate 直接费率
direct rate of exchange 直接汇率
direct ratio 直接比;正比(例)
direct ray 直接辐射波;直射线
direct rays of the sun 太阳直射光线
direct reaction 直接反应
direct reaction to light 直接光反应
direct reader 直接示值;直接读数
direct reading 直接阅读;直接读数;直读式;直接读出
direct reading analytical balance 直读式分析天平
direct reading analyzer 直读式分析器
direct reading balance 直读天平
direct reading calculator 直接读数式计算机
direct reading capillary chart 直接读数毛细管作用图
direct reading compass 直读罗盘;直读罗经;磁罗盘
direct reading current meter 直读式海流计
direct reading direction finder 直接指示探向器
direct reading ga(u)ge 直读表
direct reading instrument 直读式(测试)仪器;直读式(测试)仪表
direct reading liquid level ga(u)ge 直读液面计
direct reading manometer 直读式压力计
direct reading meter 直读计;直读定标器
direct reading method 直读法;直接读数法
direct reading micrometer 直读式测微仪
direct reading oil level ga(u)ge 直读式油位计
direct reading phase analyzer 直读式相位分析器
direct reading pH meter 直读式 pH 计
direct-reading spectrochemical methods 光谱化学分析直接读数法
direct reading spectrograph 直读式分光计;直读光谱分析仪
direct reading spectrometer 直读式光谱仪
direct reading tach(e)ometer 直读式转速表;直读式视距仪;直读式测速仪
direct reading temperature 直接读数温度
direct reading test method 直读试验法
direct reading thermometer 直读式温度计
direct reading torque spanner 直读扭矩扳钳;直读扭矩扳钳
direct reading torque wrench 直读扭矩扳手;直读扭矩扳钳
direct readout 直接发送;直接读出
direct readout infrared radiometer 直接读出红外辐射计

direct receiver 直放式接收机
direct recharge 直接补给
direct recombination 直接复合
direct recording 直接记录(的)
direct recording current meter 直读自记海流计
direct recording film 直接记录胶片
direct recording seismograph 直接记录地震仪
direct recording system 直接记录系统;直接记录方式
direct rectification 直接法纠正
direct red 直接大红
direct reduced iron 直接还原铁
direct reduction 直接还原;直接归约
direct reduction mortgage 分期偿还抵押贷款
direct reduction process 直接还原法
direct reductive analysis 直接还原分析
direct reference address 直接访问地址;直接参考地址
direct reflectance 单向反射比
direct reflection 直接反射;定向反射
direct refrigerating system 直接制冷系统;直接冷冻系统
direct relation telemeter 正比遥测计
direct release 直接释放
direct release brake 直接缓解式制动机
direct remittance 顺汇
direct repeating system 直接复示系统
direct replacement 直接置换
direct reprint 直接翻印
direct reproduction 直接复印
direct requirement 直接需要
direct resistance furnace 直接电阻炉
direct resistance heating 直接电阻加热
direct return scheme 异程式布置;直接回水系统;直接回流系统
direct return system 直接回水系统;异程回水式系统;直接回水制
direct reuse 直接再用;直接回用
direct reversal film 直接反转片
directrix 准线
directrix curve 准曲线
directrix plane 准平面
direct roll coating 直接涂布法;同向辊涂法
direct rolling action 垂直压力
direct rolling mill 无锭轧机
direct-rope haulage 直接钢丝索
direct rotational direction 正旋转方向
direct route 直接路由;直接航线;直航;往测路线
direct route powder reduction 直接粉末还原
direct routing 直接路由选择
directroy shared 共用目录
direct rule 直接规则
direct run 往测
direct runoff 地表径流;直接径流量;暴雨径流
direct runoff hydrograph 直接径流过程线;地面径流过程(曲)线;地表径流过程线
direct runoff rate 地层径流量
direct sailing 直航
direct sailing vessel 直航船
direct sale 直接销售;直接出售
direct sale clause 直接售货条款
direct sale price 直接销售价
direct sales comparison approach 直接销售比较法
direct sampling 直接取样;直接采样
direct sandflow filtration 直接砂流过滤
direct saturation method 直接饱和法
direct savings 直接节省额
direct scanning 直接扫描
direct scanning camera 直接扫描摄影机
direct screening 直接加网
direct screening colo(u)r separation 直接分色加网
direct screening method 直接加网分色法
direct-seal valve 盘形封接管
direct search 直接搜索;直接检索
direct search method 直接搜索法
direct secondary distress alerting network 直接第二遇险报警系统
direct securities 直接证券;直接担保机
direct seeding 直播
direct segmentation 直接分裂
direct selection 直接选择
direct selection system 直接选择方式
direct selector 简单调谐旋钮;直接选择器

direct sequence spread spectrum 直接序列扩展频谱
direct sequential file 直接顺序文件
direct service water heater 给水管直接加热器
direct servicing 直接操作
direct shear apparatus 应力式直剪仪;直接剪切仪;直接剪力仪;直剪仪;纯剪切仪
direct shear method 直接剪切法;直剪法
direct shear strength 直剪强度
direct shear test 直接剪切试验;直接剪力试验;直剪试验
direct shipment 直接运输
direct shipment order 直接装运通知
direct ship-to-wagon operation 直接卸船装车
direct ship-to-wagon working 直接卸船装车;船车直接换装
direct shock wave 正冲波
direct short 短路
direct sight telescope 直接瞄准望远镜
direct simple shear test 直接单剪试验
direct smear 直接涂片
direct smelting in converters 直接吹炼法
direct smoke 定向烟
direct solar radiation 太阳直射辐射;太阳直接辐射
direct solicitation 直接招工
direct solid-liquid separation 直接固液分离
direct solution 直接解
direct sound 直达声
direct sound field 直接声场;直达声场
direct sounding 直接声测法
direct sounding transmission 直接传送探测数据
direct sound transmission 声音直接传播
direct space-heating furnace 直接采暖炉
direct spindle drive 主轴直接传动
direct spinning 正旋
direct splice 直接拼接
direct spot welding 双面点焊
direct staining 直接染色
direct standardization 直接法标准化
direct starting 直接起动;直接启动
direct stationary 顺留
direct steam 新气;活气;直接蒸汽
direct steam generator 直接蒸汽发生器
direct steel process 一步炼钢法;直接炼钢法
direct stereoscopic view 直接立体观察
direct stereoscopic vision 直接立体观察
direct stiffness method 直接劲度法;直接刚度法
direct store expenses 商店直接储[贮]存费用
direct strain 直接应变;正应变
direct strain apparatus 应变式直剪仪
direct strain detector 直接应变仪
direct stratification 原始层理;原生层理
direct stress 直接应力;正应力;法向应力
direct stress machine 直接应力试验机
direct subsidy 直接补贴
direct substitution 直接替代;直接取代
direct subtract instruction 直接相减指令
direct sub-transient reactance 直接次瞬态电抗
direct sunlight illumination 直射阳光照度
direct sun rays 直接紫外线
direct supply 直流电源
direct supply at fixed location 定点供应
direct supply reservoir 直接供水水库
direct supporting type of feed control 直接给进调节器
direct surface runoff 地表径流量;直接(地面)径流
direct survey 直接检验
direct suspension 直接悬挂
direct suspension construction 直接悬挂结构
direct sweep second 中心秒针
direct switching-in 直接合闸
direct switching starter 直接开关启动器
direct symbol recognition 直接符号识别
direct system 直接系统
direct system of refrigeration 直接制冷系统
direct tax 直接税
direct taxation 直接赋税
direct tax credit 税收直接抵免
direct taxes and charges on corporation 公司直接税及应付费用
direct taxes and charges on person 个人直接税
direct tax on consumers 直接消费税
direct technologic(al) income 直接技术收益
direct telephone exchange 直拨电话交换台

direct telescope	直视望远镜;正镜【测】
direct tempering	直接回火
direct template theory	直接模板学说
direct tensile strength	直接抗拉强度
direct tensile stress	直接拉应力;法向拉应力
direct tension	简单受拉;直接拉力
direct tension strength	直接拉力强度
direct through lane	直行车道
direct through-connection equipment	分路机
direct thrust	直接推力
direct thrust A-frame	轴向荷载 A 形井架
direct tide	直接潮汐;上中天潮
direct titration	直接滴定
direct-to-ground capacity	对地电容
direct to line	直接接到线路
direct toning	直接调色(法)
direct-to-plate	直接制版法
direct torque	直接转矩
direct-to-scale	按规定比例尺
direct toxicity assessment	直接毒性评估
direct trade	直接贸易
direct traffic	直达运输
direct transaction	直接交易
direct transaction in business	直接通商
direct transfer	直接换装
direct transform	直接变换
direct transformation	直接变换
direct translator	直接转换器;声谱显示仪
direct transmission	直接传递;悬滴法;直接通信;直接输送;直接传送;直接传输;直接传动;直达通信;正透射;单向透射
direct transmission density method	直接透射密度法
direct transmission factor	直接透射系数;直接传输系数;单向透射率
direct transmission method	直接透射法
direct transmission system	直接发送系统;直接传送系统
direct transport	直达运输
direct transshipment	直接作业;直接换装
direct trip	直接脱扣
direct trip circuit-breaker	直接跳闸断路器
direct trunk	直通中继线
direct ultraviolet fluorometric method	直接紫外荧光法
direct undertaking	自建
direct undertaking works	自建工程
direct utility	直接效用
direct utility and indirect utility	直接效用和间接效用
direct vapo(u)rization technique for solution	溶液直接蒸发技术
direct variable cost	直接变动成本
direct vented type	密闭式燃(气用)具
direct verification	由客户直接退回查账员的对账回单
direct vernier	正游标;顺游标;顺读游标尺
direct vessel	直达船
direct viable count	直接活菌计数
direct view display	直接显示
direct viewfinder	直接取景器
direct viewing	直视;直接观看;直观
direct viewing storage tube	直观式存储管
direct viewing tube	直观显像管
direct view kinescope	直观显像管
direct vision	直视
direct vision erecting prism	直视正像物镜
direct vision finder	直视检景器
direct vision method	直视显影法;直接目测法;直观法
direct vision nephoscope	直视测云镜
direct vision objective prism	直视物方棱镜;直视物端棱镜
direct vision port	直视缝;窥视缝
direct vision prism	直视棱镜
direct vision spectroscope	直视(棱镜)分光镜
direct vision storage tube	直接检视式储存管
direct vision viewfinder	直观取景器
direct visual tracing	直观追踪
direct voltage	直流电压;正向电压
direct wage	直接工资
direct warmer	直接供暖设备
direct warming appliance	直接采暖用具
direct wave	直接(辐射)波
direct welded connection	直接焊接头
direct welded joint	直接焊接头
direct welding	双面点焊
direct wet chopper	直接湿短切机
direct wheel load	直接车轮荷载
direct wire circuit	单线电路
direct wire link	直接导线连接
direct write-off method	直接冲销法
direct write-off of bad debt	坏账直接冲销
direct writing galvanometer	直写电流计
direct writing oscillograph	直接记录示波器
direct writing recorder	直接记录器
direct yield	直接收益;直接产量
dire poverty	赤贫
Dirichlet boundary condition	狄利克雷边界条件
Dirichlet condition	狄利克雷条件
Dirichlet problem	第一类边值问题;狄利克雷问题
Dirichlet series	狄利克雷级数
Dirichlet test for convergence	狄利克雷收敛判别法
Dirichlet transform	狄利克雷变换
dirigation	功能锻炼
dirigibility	可控制性
dirigible balloon	飞艇;飞船;可操纵气球
dirigible wheel	操向轮
diriment	使无效的
dirk	小刀
D-iron	圆角槽铁
dirt	矿物杂质;泥土(污物);油垢;尘埃;污垢
dirt and foreign matter	杂质
dirt bailer	钻泥提桶;钻泥砂泵
dirt band	岩石夹层;夹石层;冰川碎石带;污(浊冰)层;污积带
dirt bed	夹石层;夹矸层;泥土层;淤泥层;有机质层;污层
dirt bench	污染层
dirtboard	挡泥板
dirt bucket	装土斗
dirt catcher	除尘器
dirt cloud	尘土烟云
dirt collection	积尘
dirt collector	吸尘器
dirt cone	泥雪碎屑堆
dirt-depreciation factor	积尘折减系数
dirt disposal	余土处理;废石处理;垃圾处理
dirt excluder	防尘条;收泥器
dirt-excluding groove	积垢槽;出渣槽
dirt filter	滤尘器;污物滤器;污物滤池
dirt floor(ing)	泥土铺地坪;泥土地面
dirt-holding capacity	截污能力;截活能力
dirt loader	装岩机
dirt money	污浊货作业津贴
dirt mover	运土机(械);推土机;土方机械;挖掘机
dirtness	污染度;污秽
dirt optics	不洁光学装置
dirt parting	夹矸
dirt path	(无路面的)土路
dirt percentage	含杂率
dirt pickup	积尘;吸垢
dirt pickup resistance	防集尘性;防积垢性
dirt pits	尘点
dirt pocket	泥箱(除尘袋);泥袋;尘坑;除尘袋
dirt pocket trap	沉泥井;沉泥井存水弯
dirtproof	耐脏的;防(灰)尘的
dirtproof boot	防尘套
dirtproof sleeve	防污套
dirt-repellent	防垢的;防脏物;防尘物;防污剂
dirt-repelling	防尘
dirt resistance	耐污性;耐灰性
dirt retention	积污;积尘;积尘性
dirt road	泥(泞)路;泥(土)路;(无路面的)土路;天然土路
dirt screed	匀泥尺
dirt screen	尘土分离筛
dirt screening	土筛分;土过筛;除尘筛
dirt seal	防尘封
dirt seal housing	防尘箱
dirt seam	劣质煤层
dirt separator	除尘器
dirt settling	涂膜积垢
dirt shroud	防尘罩
dirt slip	黏土脉
dirt spot	污斑
dirt streak	带色条纹
dirt track	赛车跑道;煤屑铺跑道;泥铺跑道;煤渣跑道;(无路面的)土路
dirt trap	污染阱;不纯净的阱;集渣器;集污器;集尘器;挡渣器;排污阀;收泥器;焦渣器
dirt wagon	垃圾车
dirt wall	土墙;泥墙
dirt well	污井
dirty aggregate	未清洗集料;脏污集料
dirty air	污浊空气
dirty ballast	沉淀柜;污油柜
dirty bill of lading	附有某些条件的提单;不清洁提(货)单;不洁提单
dirty bond	非固定价格公债;不洁公债
dirty cargo	油污货物;不洁货物;污脏货物;污染(性)货物
dirty casting	夹杂物铸件
dirty channel	泥砂河槽
Dirty City	肮脏城(美国纽约市的俚称)
dirty coal	脏煤;高灰分煤
dirty colo(u)r	浊色
dirty complexion	面垢
dirty deal	肮脏交易
dirty environment	多尘环境;污秽环境
dirty factor	污垢系数
dirty finish	粗糙瓶口
dirty floating	肮脏的浮动
dirty galley	毛样
dirty ice	脏冰
dirty manuscript	毛稿
dirty money	重活薪金;码头工人装卸污臭货物的额外工资;装卸污染货物所增加的装卸费;脏污费
dirty mud sump	污染泥浆池
dirty oil tank	废油箱
dirty oil trade	重油运输
dirty read	错读
dirty sand	泥(质)砂层;淤积砂;脏砂层;污砂
dirty ship	运油船;重油轮;重油船
dirty snowball model	脏雪球模型
dirty steam	脏蒸汽
dirty steel	不洁钢
dirty tanker	原油油轮;重油油船;不洁油轮
dirty vessel	不清洁船
dirty ware	粘脏器皿
dirty water	污水
dirty-water ditch	污水沟
dirty water meter	污水流量计
dirty water pump	污水泵
dirty weather	坏天气
disability	劳动能力丧失;无资格;伤残
disability benefit	伤残补助;残废偿金
disability clause	伤残条款
disability glare	阻塞眩光;致盲强光
disability insurance	残废(保)险;伤残保险
disability of vehicle	车辆报废
disability payment	残废救济金
disability pension	残废抚恤金
disability rate	报废率;伤残率
disability status	无劳动力能力
disabled	禁止的;残废;报废的
disabled interruption	被禁中断;被禁止的中断
disabled module	禁止中断模块;不可中断模块
disabled motor switch	电动机故障断路器
disabled page fault	禁用缺页;不可中断页故障
disabled person	残疾人
disabled switch	禁用开关
disabled vehicle	不能行使的车辆
disabled worker	残废工人
disable instruction	非法指令【计】;不能执行的指令
disablement	残废
disablement before retirement	退休年龄前丧失劳动力
disablement benefit	残废救济金
disable people's elevator	残疾人电梯
disabler	功能停止装置;功能失效
disable statement	断开语句
disable switch	禁止开关
disable word	断开字
disabling injury	造成缺勤的工伤
disabling injury frequency	伤残频率
disabling injury frequency rate	造成缺勤的工伤发生率
disabling injury index	造成缺勤的工伤指数
disabling injury severity rate	工伤缺勤率;伤残严重程度率
disabling pulse	禁止脉冲;截止脉冲;封闭脉冲;闭

塞脉冲
disabling signal 截止信号
disabling statutes 禁止财产转移法令
disaccharide 双糖
disaccommodation 磁化率衰减;失去调节
disaccomodation 磁导率减落
disaccord 不一致
disacidify 除酸
disacryl 狄萨雷【化】
disadjust 失调
disadvantage 不利
disadvantaged group 贫困阶层
disadvantage factor 不利因子
disadvantage of competition 竞争的不利
disafforestation 毁林;伐尽森林
disaggregate 解开聚集
disaggregate approach 分解研究
disaggregated model 非集合模型;分类模型
disaggregate estimation of mineral resources 非总合式定量预测
disaggregate method 分解分析法
disaggregate model 分解模型;非综合模型
disaggregation 解集作用;分散作用
disaggregation matrix 不聚合矩阵
disaggregation model 非集合模型
disagreeable odo(u)r 难闻的气味
disagreeable taste 不良味道
disagreement 异议;不符合
disagreement set 不一致集合
disalignment 中心未对准;中线偏转;错位;不同轴;不同心(度);不平行度;不对准;偏离轴心;偏离中心线;失中;中心线偏离;定线不准
disalinement 中心未对准
disallowance 剔除
disallowed input/output interrupt 不允许输入出中断
Disa matic mo(u)lding machine 迪沙无箱射压自动造型机
disambiguating 二义性消除
disamenity 不愉快感;不舒适
disamenity cost 不舒适的代价
disappear 消失
disappearance 消失作用;缺少;失踪
disappearance of a legal person 法人消失
disappearance of outcrop 露头隐藏;露头缺失
disappearance potential spectroscopy 消隐电势谱
disappearance rate of foam 泡沫消失率
disappearing bed 暗式壁(柜)床;壁内折叠床
disappearing chain 隐式链;暗式链
disappearing filament 光学高温计
disappearing filament optic(al) pyrometer 隐丝光学高温计
disappearing filament pyrometer 隐丝式光测高温计;灯丝隐灭式高温计;隐丝高温计
disappearing finger auger 偏心扒杆式输送螺旋
disappearing highlight test 强光隐灭试验
disappearing ice 消融冰
disappearing ladder 折叠梯;软梯
disappearing river 消失河(流);伏流
disappearing stair(case) 阁楼梯子;隐式梯;暗楼梯;隐式楼梯
disappearing stream 消失河(流);地下河(流);伏流(河)
disappearing stream sinking 伏河
disappearing target 隐显目标
disappointment 失望
disappreciation 低估价值
disapprobation 不赞成
disapproval 不赞成
disarm 拆去发火件;拆除引信;排除发火装置
disarmed interrupt 拒绝中断;解除中断;截止中断
disarmed state 拒绝状态
disarm state 解除状态
disarrange 使失谐
disarrangement 紊乱;失谐
disassemble 拆卸;拆散;拆开;反汇编
disassemble mo(u)ld 开模
disassembler 反汇编程序
disassembling 反汇编
disassembling equipment 拆卸设备
disassembling flange 可拆卸法兰
disassembling operation 拆卸操作
disassembly 解体;拆开;分解;不正确装配

disassembly and assembly block 拆卸与安装专用木架
disassembly and assembly department 拆装部
disassembly procedure 拆卸程序
disaster 灾难;灾祸;灾害
disaster analysis 灾难分析
disaster area 灾区;受灾地区;失事区
disaster assistance centre 救灾中心
disaster box 保险盒
disaster caused by a windstorm 风灾
disaster caused by hail 雹灾
disaster clean-up operation 灾害清理工作;灾难清理作业
disaster control 紧急情况处理预案
disaster dump 灾难性转储[贮];灾难性转储;故障转储;大错转储[贮];大错转储;事故性转储[贮];事故性转储
disaster geology 灾害地质学
disaster handling 致命错误处理
disaster housing mortgage insurance 受灾住房的抵押保险
disaster loan 灾情贷款
disaster management 灾害管理
disaster mode 灾害模式
disaster planning 城市防灾规划;防灾规划
disaster preparedness 防灾准备
disaster preparedness condition 防灾准备状态
disaster prevention 防灾
disaster prevention design 防灾设计
disaster prevention facility of tunnel 隧道防灾设施
disaster prevention plan 防灾规划
disaster prevention planting 防灾栽植
disaster-prone area 容易受灾地区
disaster-prone region 易受灾区域;灾害多发区
disaster-proof building 防灾建筑
disaster reduction 减灾
disaster relief 灾难救援
disaster-resistant community 抗灾社区
disaster science 灾害学
disasters preparedness and relief 灾害防备和救济
disaster unit 抢险救援车
disaster valve 安全阀
disastrous 灾难性的;灾害的
disastrous accident 灾难性事故
disastrous effect 恶果
disastrous fire 火灾
disastrous flood 严重洪灾;灾害性洪水;大水灾
disastrous shock 烈震
disastrous shutdown 事故停车
disaturated 二饱和的甘油酯
disazo compound 二重氮化合物
disazo condensation pigment 偶氮缩合颜料;缩合型双偶氮颜料
disbalance 平衡差度;失衡
disbalance uranide method 不平衡铀系法
disbark 剥皮
disbarking machine 剥皮机
disbenefit 负效益
disboard 卸下
disbond 开胶;脱胶
disbonding 脱胶;(内聚力的)松释
disbranch 修剪树枝
disbudding 断角术
disburden 解除负担
disburse 拨款
disbursement 支付(额);支出费用;支出额;港口使费;付出款
disbursement account 支付账(目)
disbursement and settlement 费用结算
disbursement clause 费用条款
disbursement for customers 代客付出款项
disbursement from fund 动用基金
disbursement insurance 支付保险;费用保险(附属于船壳保险)
disbursement of dividends 支付股息
disbursements 港口费(用)
disbursement voucher 支付凭单;支付凭证;付款凭单
disbursement warranty 船舶费用附加保险
disburse money 付款
disbursing office 支付机构
disbursing officer 出纳主管
disc 轮盘;圆片;圆盘;盘(状);甩油盘
disc action 圆盘作用

discal 平圆盘(的);盘状的
discale 碎鳞
discaling pump 鳞屑冲洗泵
discaling roll 齿面轧辊;齿辊;碎鳞轧辊
Discaloy alloy 迪斯卡洛伊合金
discard 切头;挤压尾料;解雇;废品;不考虑;报废(件);抛弃物;排斥工人
discardable ballast 可弃压载
discard bin 废票箱
discarded article 废弃品
discarded dam site 放弃的坝址
discarded metal 废弃金属
discarded site 废弃坝址
discard electrode 焊条头
discarding change 忽略修改
discarding petal 扇形分离瓣
discard products recovery 废品回收
discard raffinate 废弃残液
discard weak plants 废弃瘦弱植株
disc atomiser 离心式喷雾圆盘
disc auger 盘式螺旋钻
disc bit 圆盘钻(头)
disc black 盘黑
disc block 盘法炭黑
disc brake 制动闸;制动盘;圆盘式制动;盘形制动器;盘式制动器
disc brake calliper 制动盘测径仪
disc breaker 盘式轧碎机
disc calculator 盘式计算机
disc cam 盘形凸轮
disc centrifugal process 盘式离心法
disc centrifuge 盘式离心机
disc chromalography 盘式色谱
disc-clinometer 碟式测绘仪
disc clutch 盘形离合器;盘式离合器
disc comparator 旋盘比色器
disc conveyer 盘式输送机
disc-coupled vibration 轮系振动
disc crank 盘形曲拐
disc crusher 盘式轧碎机;盘式压碎机;盘式破碎机
disc cup 杯状盘
disc cutter 圆片铣刀;圆片截煤机;圆片车刀;圆盘形切割器;滚刀;切断铣刀;盘形铣刀
disc-cutting unit 切削饼设备
disc diffuser 盘式扩散器
disc distributor 撒料盘
disc drier 盘式干燥机
disc electrophoresis 盘(状)电泳;盘式电泳法
disc electroporesis 盘状电泳
discending velocity 下降速度
discernible detail 可分辨碎部
discernible step 可识别级
discerning method 分辨法
discernment 见识,洞察力
discerp 扯脉
disc extruder 圆盘式挤压机
disc feeder 盘式喂料机;圆盘喂料机;盘式加料器;盘式加料机
disc filter 圆盘式料浆过滤器;盘形滤池;盘式滤池;盘式过滤器;盘式过滤机;盘滤机
disc floret 盘花
disc flowmeter 盘式流量计
disc flywheel 盘形飞轮
disc-footed pile 扩脚桩
disc friction 圆盘摩擦
disc friction loss 圆盘摩擦损失;圆盘摩擦损耗
disc frieze (具有圆盘状浮雕的)横饰带;圆盘花饰
disc front connection 盘前接线
disc gear cutter 盘形铣齿刀
disc granulator 成球盘;盘式成球机
disc grinder 圆盘磨(床);盘式研磨机
discharge earth volume 排土量
discharge 流(出)量;开除;解除;结清证明;基本流量;煤粉出口;卸载;卸土;卸货;卸荷;卸;泄出物;运量量;溢液;出炉;出料;发放;发染;清偿;起岸;排料;排量;排放;排出(物);排出流量
dischargeable capacity 有效容积;排送能力
discharge accident 喷射事故
discharge a contract 解除契约
discharge a debt 清偿债务
discharge afloat 水上卸货
discharge after water blocking 堵水后的排水量
discharge air 排风
discharge(air)chamber 排气室

discharge air shaft 出风井;通风竖管;通风井
discharge air slide 卸料空气斜槽
discharge alarm 放电警点
discharge amount 排放量
discharge analyzer 放电分析器
discharge anchorage 卸载锚地
discharge and suction hose 吸排气水软管
discharge angle 卸料角;出流角;出口角(度)
discharge apparatus 卸料装置
discharge area 解租地区;卸料口面积;泄水断面面积;过水(断面)面积;出口(截面)面积;出口(断面)面积;排泄区
discharge area of groundwater 地下水排泄区;地下水泛出区
discharge area of screen 筛料口面积
discharge arm 抛掷杆
discharge arrestor 放电避雷器
discharge at constant current 定流放电
discharge auger 卸料螺旋
discharge bar 排流条
discharge basin 泄水池
discharge bay 泄水池;出水池;放水池
discharge before water blocking 堵水前的排水量
discharge belt 卸载传送带
discharge belt conveyer 卸料胶带输送机
discharge berth 卸载泊位;卸货码头
discharge bibcock 出水龙头
discharge blade 排出叶片
discharge blower 风力吹送卸载器
discharge boom 卸料悬臂;卸料吊杆
discharge borehole 泄水钻孔
discharge box 卸料箱;出料箱;卸料箱
discharge box for the mud screen 泥筛出料箱
discharge branch 排水支管
discharge branch from manifold 废汽管的放出分导管
discharge bridge 放电测量电桥
discharge bucket 卸料斗
discharge button 卸料按钮;电容器放电按钮;放料按钮;放电按钮;排料按钮
discharge by pump 用泵卸车
discharge by pumping 倾卸;倾倒卸物
discharge by wash 冲刷放出
discharge calculation 流量计算;排水量计算
discharge canal 泄水渠(道);排水渠(道);排沙渠(道)
discharge capacity 流量(生产力);泄水能力;泄流能力;过水能力;工作能力;放电容量;排泄能力;排污能力;排流能力;排放能力;排放量;排出量;通行能力;通过能力
discharge capacity of output of ten-thousand-Yuan 万元产值排污量
discharge cargo form a wreck 从沉船上卸货;吊除余货【救】
discharge carrier 排放通道
discharge cart 卸料小车
discharge case 出口外壳
discharge cash 卸料容器
discharge casing 排水箱
discharge cell 放电室
discharge chamber 卸料仓;压缩室;放电室
discharge channel 排气管;排放通道;泄水渠(道);分离液出口;放出沟;放出槽;排水渠(道);排水河槽;排水道;排水槽;排沙渠(道)
discharge characteristic 流量特征;流量特性;排气特性
discharge chart 流量图表
discharge check ball 出口止回球
discharge chest 放出柜
discharge chute 卸料(斜)槽;卸料溜子;卸料(溜)槽;出料(溜)槽
discharge circuit 放电电路
discharge cock 潜式龙头;排放旋塞;放出旋塞;排水龙头;排放旋塞
discharge coefficient 流量系数;出流系数;放电系数;排放系数
discharge coefficient for orifice 节流孔流量系数
discharge coefficient of flow nozzle 喷嘴流量系数
discharge condition 流水条件;放电状态;排放条件;排放情况
discharge conduit 卸料管;泄水管(道);排泄管(道);排水渠(道);排水管(道)
discharge cone 卸料锥(体);泄水锥;出水锥形口;放泄喷嘴;排矿圆锥;排矿斗;排出喷嘴

discharge connection 出口管;放电接线
discharge connection delivery 输出管接头处流量
discharge contact 放电触点
discharge contactor 放电接触器
discharge control 流量控制
discharge control valve 出口控制阀
discharge conveyer 卸载输送器;卸载输送机;卸料输送机
discharge counter 放电记录器;放电计数管
discharge coupling 出口接口
discharge cover 卸渣门;排出端泵盖
discharge cradle 卸料翻斗
discharge cross-section 过水断面
discharge culvert 泄水廊道;泄水涵洞;排水涵洞
discharge current 放电电流;排出流
discharge current noise 放电电流噪声
discharge curve 流量曲线;放电曲线;水位流量关系曲线
discharged-area development 反转显影
discharged as loaded 原装原卸;原收原交
discharged bankrupt 已清偿债务的破产
discharged bill 付讫票据
discharge deficiency 排放不足;流量不足
discharge demand 流量需要
discharge density 喷放密度
discharge detector 放电检测器
discharge device 放电器;避雷器;卸料装置;避电器;喷放装置;排放器
discharged finish 卸货完毕
discharged for inefficiency 因无能被解职
discharged from prison 出狱
discharged head 递送水头
discharge diagram 流量(过程)图
discharge diameter 出口直径
discharged into 排入
discharge diode 放电二极管
discharge distance 排送距离
discharge distribution pipes 装舱管系
discharge ditch 排水沟
discharge ditch sweeper 排水沟疏浚机
discharge door 出料门;排料门
discharge drain cock 出水龙头
discharge duct 排气管;出料管;排水管垫层;排水管(道);排气管道
discharge duration 放水历时
discharge duration curve 流量历时曲线
discharge-duration relationship 流量历时关系
discharged water 排出水
discharged workers payroll 被解雇工人的工资
discharge electrode 有源电极;放电极;放电电极
discharge electrode rapping mechanism 放电极振荡机构
discharge-elevation relation 泄量高程关系
discharge-elevation relation curve 泄量高程关系曲线
discharge elevator 卸载升运器
discharge end 发射端;卸料端;出料端;排放装置末端
discharge-end block 窑口圈砖;出料端墙砌块
discharge-end refuse extraction chamber 排料端排渣室
discharge-end trunnion 卸料端中空轴;卸料端空心轴(颈)
discharge engineering 疏导工程
discharge equation 流量方程
discharge equipment 卸出装置
discharge excited 放电激发的
discharge factor 流量因数;排放系数
discharge fan 抽风扇;单向排气扇;排(气)风扇;排风机
discharge fee 排污费;排放税;排放费用
discharge filter 排气过滤器
discharge fitting 出水管接头
discharge flange 卸料管法兰;排水管法兰;排气管法兰
discharge flow 泄流;出流
discharge flow grate 卸料阀
discharge flow-off cock 出水龙头
discharge flow rate 排放水量
discharge flue 排烟管(道);排气道;排烟(道)通道
discharge formula 流量计算公式
discharge formula of two parallel canals 二条平行水渠流量公式
discharge formula of water collecting pipe un-

der riverbed 河床下集水管流量公式
discharge for river bed function 造床流量
discharge for unit width 单宽流量
discharge frame 卸料架
discharge frequency 放电频率
discharge frequency curve 流量频率曲线
discharge from safety and relief valves 由安全阀和泄压阀泄放
discharge front end 排料前端
discharge funnel 卸料漏斗
discharge gap 放电间隙
discharge gas 废气
discharge gate 出料口;卸料(阀)闸门;闸阀;出料门;排料口;排出闸门
discharge gate mortar mixer 活门出料式灰浆搅拌机
discharge gate rope 出料闸门悬索
discharge ga(u)ge 高压压力表;出水压力计;放电真空计;排出压力表
discharge grate 雨水口箅盖;雨水口算盖;进气算子;卸料算板
discharge grating 进气箅栅;卸料炉栅
discharge grid 雨水口栅盖;卸料格栅
discharge gutter 排水沟;排水槽
discharge head 出水压头;扬程;压力高度(压缩机);压力的高度;供水压头;出口压头;排气压头;排出扬程;输送压头;输送高度
discharge head curve (水泵的)流量扬程曲线;D-H曲线
discharge header 排出总管
discharge head of dredge pump 泥泵扬程
discharge-head relation 流量水头关系
discharge height 卸料高度;排放高度
discharge hole 卸料孔;泄流孔;泄放孔;出料口;出口(指水、气、汽、油等);成品出口;排出孔
discharge hopper 卸料斗;卸货斗;放料斗
discharge hose 泄水软管;出口胶管;排水胶管;送水软管
discharge hose assembly 排放软管组
discharge hydrograph 流量(水文)过程线
discharge hydrograph cutting method 流量过程线切割法
discharge hydrograph for characteristic year 特征年流量过程线
discharge impulse oscillator 脉冲发生器
discharge inception test 开始放电试验
discharge inception voltage 放电起始电压
discharge indicator 放电指示表
discharge in insulation 绝缘放电
discharge instrument 流量计
discharge intensity 卸料强度;单宽流量;排水强度;排气强度
discharge into the environment 排入环境中
discharge into the sea 排放入海
discharge into the water 排入水中
discharge jet 出口射流;喷嘴;喷射水束;喷射水舌;射水管
discharge jetty 卸货码头
discharge key 放电键
discharge lamp 放电管;放电灯(管)
discharge launder 卸载槽
discharge leader 废汽管;排出管
discharge leg of siphon 虹吸管流出端
discharge length 排送距离
discharge lever 释放杆
discharge liability 解除债务
discharge limit 排污限度
discharge line 排放管道;排水管路;卸出线;出口管线;排水线;排出管线;排出管路
discharge-line loss 泄水管道水头损失
discharge liquid 废液;排出液
discharge liquor 排出液
discharge load 放电负载
discharge location 排污点
discharge loss 卸油损耗;卸油损失;出口损失;出口损耗;排气损失
discharge manifold 集气管;卸出支管;排油歧管;(闸室的)排水道;排水总管;排气集合管
discharge manifold of lock chamber 闸室排水道
discharge mass 流量累积
discharge mass curve 流量累积曲线
discharge material 放电材料
discharge matter 排放物
discharge measurement 流量量测;流量测验;流量

测量;流量测定(值);测流;排水测量;排放测量
discharge mechanism 卸载装置;排泥机构
discharge modulus 流量模数;流量模量
discharge monitor 流量监测器
discharge monitoring 排放监测
discharge monitoring report 排污监测报告
discharge nozzle 油嘴;出料管;燃气短管;喷嘴;排气喷管;排放喷嘴;排放出喷嘴接管;排出喷嘴
discharge observation 流量观测
discharge obstacle 水流障碍;排泄障碍
discharge of a debt 清偿债务
discharge of an attachment 解除财产扣押
discharge of a party 解除汇票上签字人的义务
discharge of a screen 筛孔排出
discharge of a well 井出水量
discharge of brush 电刷放电
discharge of capacitor 电容器放电
discharge of claims in bankruptcy 破产程序中清偿债权
discharge of concrete 混凝土排出量
discharge of contract 履行合同;解除合同;取消合同
discharge of debt 免除债务
discharge of dewatering engineering 疏干工程排水量
discharge of effluent 出料
discharge of electricity 放电
discharge off 放电完毕
discharge of flood peak 洪峰流量
discharge off shore side 外档卸货【船】
discharge of groundwater 地下水排泄;地下水排放;地下水流(出)量
discharge of groundwater to the sea 地下水入海流量
discharge of liability on a bill of exchange 履行或解除汇票上的义务
discharge of mortgage 解除按揭
discharge of mud-stone flow 泥石流流量
discharge of opening 排水
discharge of opinion 放弃表示意见
discharge of overflow 溢水流量
discharge of permeability 渗透流量
discharge of pollution to underground water 排污地下水
discharge of pump 水泵溢流;泵的排水量;泵抽水量;水泵排水量;水泵出水量
discharge of pumping well 抽水孔流量
discharge of pumping well group 抽水孔组流量
discharge of rainfall 降雨流量
discharge of repayment 免除清偿
discharge of reporting and registering 污染物排放申报登记
discharge of river 河道流量
discharge of safety valve 安全阀泄放
discharge of sewage 污水排出量;污水流量;污染物排放
discharge of sewage effluent 污水处理后出水排放
discharge of sluice 闸门流量
discharge of solids 固体输移量;底沙输移量;推移质输移量;泄沙量;固体(径)流量;输砂量
discharge of tax obligation 撤消或免纳纳税义务
discharge of temperature 出厂温度
discharge of the fuel 卸燃料
discharge of the right of action by lapse of time 因时效丧失权益
discharge of undersize 筛下物料
discharge of untreated effluent 未处理过的污水排放;未处理废水排放
discharge of waste water 污水排放量
discharge of wastewater effluent 污水处理后出水排放
discharge of water 水流量;排水量
discharge of well 井的出水量
discharge one's duties 行使职权
discharge one's functions and powers 行使职权
discharge opening 卸料口;卸料孔;泄水;泄水孔;泄水涵洞;泄流孔;出料口;排水口
discharge orifice 排气孔;流水孔;泄水孔;泄流孔;排水孔;排泄孔;排出口;排出孔
discharge outlet 卸料口;卸料口;泄水口;泄水口;放水口;排料出路;排放口
discharge over the head end 高架端卸料
discharge passage 放水道
discharge path 放电路径
discharge pattern 流量类型;喷放形式;排泄方式

discharge period 放电持续时间;排放期
discharge permit 排放许可
discharge per unit width 单宽流量
discharge pipe 卸载管;卸料管;泄水管(道);压送管(道);排水管(道);排气管道;排泥管;排放管;排出管
discharge pipe elbow 卸料管子弯头
discharge pipe line 排水管线;排水管路;排放管线;排泥管线
discharge pipeline float 排泥管浮筒
discharge pipeline supporting frame 排泥管架
discharge pipe loss 排水管(压头)损失
discharge pipe part 出液管部件
discharge pipe system 排放管系
discharge pipework 排放管系
discharge piping 出水管系;排出管
discharge place 卸料场
discharge plasma 放电等离子体
discharge platform 卸货站台
discharge point 排水口;排放点;排污点;卸载点;地下水流出地面的泉眼;射出点
discharge port 卸料口;卸货港;泄水口;放水口;放出口;排水口
discharge potential 放电电位;放电电势
discharge prediction of industrial pollution sources 工业污染源排污预测
discharge pressure 供给压力;出口压力;排水压力;排气压力;输送压力
discharge pressure ga(u)ge 排气压力表
discharge prevention 排放预防
discharge printing 放电印刷
discharge process 放电过程
discharge-producing rain 造流降雨
discharge product 卸货产品
discharge property 排放性质
discharge pulse 放电脉冲
discharge pump 卸油泵;卸料泵;排泄泵;排水泵;排气泵;排放泵;排出泵;输送泵
discharge purge cock 出水龙头
discharge pusher 卸货推料器
discharge-quality-head model 流量—水质—水头模型
discharge quay 卸货码头
discharger 卸料器;卸货人;卸货机;卸放装置;溢出管;废物的原有者;放电器;排污设备;排放者;排放管;排出装置;污染者;推料机
discharge ram 卸料推杆
discharge rate 卸载速度;卸料速度;卸货效率;出口速度;放电(速)率;排泄量;排水率;排放速率;排出率;疏散率;释放率
discharge rate of effluent 废水排放率;排污率
discharge rating curve 流量率定曲线;水位—流量关系曲线
discharge rating table 流量率定表
discharge ratio 流量比;排污比;排放比
discharge ration 排污限额
discharge receipt 卸货收据
discharge record 流量资料;流量记录;排放记录
discharge recorder 流量记录仪;流量记录器
discharge reduction 卸货扣减;流量折减
discharge regime 流量变化情况
discharge regulating device 流量调节装置
discharge regulation 流量调节;径流调节
discharge regulator 流量调节器
discharge reporting and registering 排污申报登记
discharge required 需要流量
discharge resistance 放电电阻
discharge resistor 放电电阻
discharge resources 径流资源
discharger gap 火花(间)隙
discharge ring 基础环;转轮室中环;底环;出口环;前结圈
discharge roll 卸料滚筒
discharge roller 出料辊
discharge rotary valve 卸料滚动阀门;转子阀门
discharge rotation 卸货次序
discharge routing 流量演算
discharge scale ratio 流量比尺
discharge scow 卸货平底船
discharge screw 卸料绞刀
discharge screw conveyer 卸料螺旋输送机
discharge section 过水面积;过水断面;测流段;排水沟断面
discharge section(al) area 过水断面面积

discharge sectional line 测流断面
discharge section line 测流断面线
discharge sensor 流量传感器
discharge service van 压出阀
discharge sewer 排放污水管
discharge shaft 卸料竖管
discharge shutoff valve 排气截止阀
discharge shutter 卸料闸板
discharge side 出料侧;卸料侧
discharge side cartridge 排出侧筒形接头
discharge site 测流站;测流断面;测流段;测流地址;测流地点;排污点
discharge skin 放电表面层
discharge sluice 泄水闸
discharge source (of wastewater) 排污源
discharge space 放电空间
discharge spark 放电火花
discharge spout 漏嘴;喷口
discharge-stage curve 水位流量关系曲线
discharge-stage relation 水位流量关系
discharge standard 排放标准
discharge standard of water pollutant for dyeing and finishing of textile industry 纺织工业染整水污染排放标准
discharge standard of water pollutant for paper industry 造纸工业水污染物排放标准
discharge standards for regional pollutants 地区水污染排放物标准
discharge station 卸货站
discharge strength 排气强度
discharge stroke 排出冲程
discharge structure 泄水建筑(物);泄水构筑物
discharge sump 排水坑;排水池
discharge switch 放电换接器
discharge system 排气系统;放油系统;排水系统
discharge table 流量表(格)
discharge tank 出料罐
discharge temperature 排气温度;排出温度
discharge temperature of bituminous mixtures 沥青混合料出厂温度
discharge terminal 卸料站
discharge test 放电试验
discharge time 卸货时间;排水时间;卸料时间;放电时间;喷射时间
discharge time constant 放电时间常数
discharge time curve for characteristic year 特征年流量过程线
discharge time delay test 喷水时间延迟试验
discharge timer circuit 放电定时电路
discharge tongs 放电叉
discharge trough 卸料槽;排水槽;排矿槽
discharge tube 卸料管;泄水管(道);泄放管;闸流管;放电管;喷粉管
discharge-tube leak indicator 放电管泄漏指示器
discharge-tube noise generator 放电管噪声发生器
discharge-tube rectifier 充气管整流器
discharge tunnel 排泄隧洞;排放烟道;泄水隧洞
discharge uniformity 流量均匀性
discharge value 疏散流量
discharge valve 卸料阀;泄水阀(门);泄放阀;出水阀;放泄阀;排泄阀(门);排气阀(门);排出阀;输送阀
discharge valve deck 放泄阀盖
discharge valve for freeze-proof 防冻泄水阀
discharge valve of pump 水泵出水管阀
discharge valve seat 放泄阀座;放水阀座
discharge valve spindle 放泄阀轴
discharge valve stud 放出阀柱螺栓
discharge variability 流量变率
discharge variation 流量变化
discharge vehicle 卸载式载重汽车
discharge velocity 流速;泄水速度;泄流速度;出流速度;出口速度;排水速度;排气速度;排放速度;排出流速度
discharge-velocity curve 流速—流量关系曲线
discharge-velocity triangle 出口速度三角形
discharge vent pipe 安全排气管
discharge vessel 卸料容器
discharge vlave spring 放出阀簧
discharge voltage 放电电压
discharge voltage regulator 放电调压器
discharge water 排放水
discharge wharf 卸货码头
discharge wire 放电极丝;放电电缆
discharging 卸料;卸货作业;排泄(指水、气等);排放

discharging afloat 船上卸货
discharging agent 拔染剂;脱色剂;漂白剂
discharging arch 卸载拱;辅助拱;助拱;卸荷拱;门窗洞口边框;(码头堤岸等用的);减卸荷载拱过梁拱
discharging at wharf 靠码头卸货
discharging bay 卸货河弯
discharging belt conveyer 卸料用胶带输送机
discharging berth 卸货码头;卸货泊位
discharging bucket elevator 斗式提升卸货机
discharging capacity 卸货能力;泄水能力;泄流能力
discharging cargo work 卸货作业
discharging chain 卸载输送链
discharging charges 卸船费
discharging choke coil 放电扼流圈
discharging connection 可卸连接
discharging crew 卸货班
discharging criterion 排放基准
discharging current 泄放水流;放电电流;排出流
discharging current strength 放电强度
discharging curve 卸货曲线
discharging day 卸货日(期)
discharging dredge(r) 吸扬(式)挖泥船
discharging efficiency 卸货效率
discharging equipment 卸货装置;卸货设备
discharging expenses 卸货成本
discharging expense discharging charges 卸货费(用)
discharging facility 卸货设施
discharging for refloat 卸载脱浅
discharging friction 出口摩擦
discharging gang 卸货班
discharging gear 卸货装置
discharging gear release mechanism 放出机构
discharging hatch 卸货口【船】
discharging jet 喷出流
discharging method 卸装方法
discharging of brake 风闸放气
discharging pallet 释放叉
discharging party 卸货班
discharging period 卸货时间;卸货期
discharging permit 卸货许可证
discharging pit 卸料坑
discharging place 卸货地点;卸货处
discharging plant 卸货装置;卸货设备
discharging plate conveyer 卸料用板式输送机
discharging platform 卸料站台;卸货作业台;卸货站台;卸货平台
discharging point 卸货点
discharging pollutant 排放污染物
discharging port 卸货港
discharging port mark 卸货港标志
discharging quay 卸货码头;装卸码头
discharging rate 卸货效率;卸船率
discharging resistor 放电电阻器
discharging rod 避雷针;放电棒
discharging roller 释放滚轴
discharging sluice 泄水闸
discharging spring 金簧;有流体排出的泉
discharging team 卸货班
discharging time 卸料时间;卸货时间;出料时间
discharging time list 卸货时间表
discharging tongs 放电钳
discharging tube 泄水管(道);出水管
discharging valve 排水阀(门)
discharging vault 卸载拱顶
discharging weight 卸货重量
discharging wharf 卸货码头
discharging zone 排放带
disc harrow 圆盘(式)耙;圆盘式耙机;圆板耙路机
Dischinger prestressing method 迪斯钦格预应力法
disc hyophone 压电圆片式水下检波器
Disciflorae 盘花群类
discifloral 盘状花的
disciform 圆形的;盘状的
Discinacea 平圆贝超群【地】
disc insulated cable 垫圈绝缘电缆
disc insulator 盘式绝缘子
disciplinary room 惩戒室
discipline 学科;专业
discipline design engineers 专业设计工程师
disciplines of environment(al) science 环境科学学科
disclaim 不索赔

disclaimer 放弃
disclaimer of opinion 放弃意见书;放弃表达审计意见
disclaimer's statement 仅供参考声明
disclaim liability 否认责任
discl disk 阀叶
Disclean 狄思表面清洁剂
disc lightning arrester 盘式避雷器
disclosed principal 公开委托人
disclosure 揭示
disclosure of an invention 发明的公开
disclosure of relevant information 相关资料的表达
disclosure statement 发放贷款的书面说明
disc meter 盘式流量计
disc method 圆片法
disc mill 车轮轧机;圆盘式粉碎机;盘磨
disc milling cutter 盘形铣刀
disc mixer 盘式搅拌机
disc of confusion 弥散盘;弥散斑
disc of pump 泵轮
discoid 平圆形(物);圆盘的;平圆形(的);盘状的;盘形挖器;盘形的
discoidal 盘状
discoidal nodule 盘状结核
discoidal valve 盘形瓣膜
discoid dental scaler 匙形牙刮器
discoid meniscus 盘状半月板
Disco light 迪斯科灯
Disco loom 迪斯科无梭织带机
discolo(u)ration 变色;脱色作用;色斑
discolo(u)ration agent 脱色剂
discolo(u)ration material 变色材料
discolo(u)ration test 退色试验;漂白试验;变色试验
discolo(u)red and distorted area 变色和变形部位
discolo(u)red bearing 变色轴承
discolo(u)red clay 变色黏土;退色黏土
discolo(u)red clinker 变色熟料
discolo(u)red seawater 变色海水
discolo(u)red water 海水变色
discolo(u)ring agent 脱色剂
discolo(u)ring clay 漂白土;脱色黏土
discolo(u)ring water 变色水
discolo(u)rization 退色
discolo(u)rment 退色
discomfort 局促;不愉快感;不舒适感;不适
discomfort glare 耀眼的眩光;致眩强光;刺眼强光;不适眩光(一种不良照明现象);不舒适眩光
discomfort index 不(舒)适指数;不安指数;温湿指数
discomfort threshold 不舒适阈
discomfort zone 不舒适区
discommodity 负商品;无使用价值商品
discomposition 原子位移
discomposition effect 原子位移效应
discompressor 松压器
discone antenna 盘锥形天线
disconformity 假整合;不一致;不协调(性);平行不整合
disconformity index 非一致指数
disconjugacy 不共轭性
disconnect 折断;隔断;拆线;拆接;释放
disconnectable clutch 可解脱离合器;可拆离合器
disconnectable coupling 可卸接合
disconnect a clutch 摘开离合器
disconnected 分离的;非直接;脱开的
disconnected contact 空间接点
disconnected crawler track 不连贯的履带式轨道
disconnected digraph 不连通的有向图
disconnected elestromagnetometer separation method 间断式电磁仪分离法
disconnected fracture 不连通裂缝
disconnected graph 不连通图
disconnected metric space 不连通度量空间
disconnected pore 不连通孔隙
disconnected position 断路位置
disconnected set 不连通集
disconnected signal 切断信号
disconnected switch 隔离开关
disconnected vehicle 分解车辆
disconnecting 拆开
disconnecting apparatus and disengaging gear 拆卸器械
disconnecting chamber 检查井;隔离室

disconnecting gear 分离装置
disconnecting hook 脱钩;摘钩;分离钩
disconnecting knife-switch 脱开刀闸开关
disconnecting lever 分离杆
disconnecting link 摘环;闸刀;刀闸;分离环
disconnecting manhole 检查孔;观察孔;隔断探井;渠道的隔断探井
disconnecting mean 断路方法
disconnecting plug 断流塞
disconnecting protective panel 断电保护屏
disconnecting signal 拆线信号
disconnecting spring 分离弹簧
disconnecting stirrup 摘环;分离环
disconnecting switch 阻断开关;隔离开关;断路开关;切断开关
disconnecting switch reverser 隔离反向器
disconnecting trap 截流式封;截流存水弯头;存水弯
disconnecting valve 截止阀;闭锁阀
disconnection 解脱;断路;断开;拆线;拆开连接;拆开;拆接
disconnection fault 断电故障
disconnection of lines at no-load 空载倒闸
disconnection register 拆线寄存器;释放信号寄存器
disconnection switch 断路开关
disconnection test 断路试验
disconnection valve 断开阀
disconnect lamp 可拆灯泡
disconnector 拆开者;隔离开关;隔断器;断路器;切断开关
disconnector release 拆线器
disconnector with fuse 熔断丝隔离开关
disconnect signal 断开信号
disconnect time-out 超时断开
disconnect-type clutch 分离式离合器
discontiguous 不接触的
discontinous transformation group 不连续变换群
discontinuance 撤回诉讼;间断;中断
discontinuance of legal proceedings 中断诉讼
discontinuation 中止;中断;不连续
discontinue 中止;中断
discontinue an action 中断诉讼
discontinued operations 非连续性经营活动
discontinuities (defect) in forgings 在锻件中的缺陷
discontinuities in materials 在材料中的缺陷
discontinuities in welds 在焊缝中的缺陷
discontinuity 间断(性);骤变点;地震波速度突变面;不连续(性);突跃;速度跃变
discontinuity condition 不连续条件
discontinuity focusing 不连续震源
discontinuity function 不连续函数
discontinuity indication of defect 缺陷指示
discontinuity in the cooling ports 冷却管中的断裂
discontinuity layer 不连续层
discontinuity mechanics 不连续体力学;不连续介质力学
discontinuity medium mechanics 不连续介质动力学
discontinuity motion 不连续运动
discontinuity movement 不连续运动
discontinuity of current 水流不连续(性)
discontinuity of defect 缺陷不连续性
discontinuity of material 材料(的)不均匀性
discontinuity of rock 岩石不连续性
discontinuity of rock mass 岩体结构面
discontinuity of structure 结构不连续性
discontinuity of the first kind 第一类间断性
discontinuity point 间断点;不连续点
discontinuity pressure 不连续压力
discontinuity sliding surface 不连续滑动面
discontinuity stress 间断性应力;不连续(性)应力
discontinuity structural plane 不连续构造面
discontinuity surface 不连续面
discontinuity switch 切断开关
discontinuity waviness 结构面起伏度【地】
discontinuous 断续的;不连续的
discontinuous action 断续动作;不连续作用
discontinuous aggregate 间断级配集料;间断级配骨料
discontinuous amplifier 非连续放大器;不连续放大器
discontinuous approximation 对不连续函数的近似解
discontinuous band absorption 不连续光带吸收
discontinuous beam 不连续梁;不连续波速

discontinuous capacitance 不连续电路
discontinuous capillary pore 不连续毛细孔
discontinuous change 不连续变化
discontinuous composites 不连续复合材料
discontinuous construction 断离式构造【建】；不连续结构；隔声构造；非连续性施工；非连续构造
discontinuous control 间断调节
discontinuous controller action 不连续性调节器作用；不连续的调节器作用
discontinuous control system 离散控制系统；不连续控制系统；断续（作用的）控制系统
discontinuous conveyer 非连续式运输机
discontinuous coring 不连续岩芯
discontinuous cost function 非连续成本函数；不连续成本函数
discontinuous crack 不连续裂纹
discontinuous deformation 不连续变形
discontinuous detector 不连续检测器
discontinuous dipmeter 不连续记录地层倾角仪
discontinuous distribution 间断分布；不连续分配；不连续分布
discontinuous distribution function 不连续分布函数
discontinuous distribution of concrete 混凝土的不连续分配；混凝土的不连续分布
discontinuous double-roll process 间歇双辊压延法
discontinuous dynamic(al) system 不连续的动力学系统
discontinuous easement 要求间歇使用非已有的权利；间歇要求使用非已有的权利
discontinuous electro-phoresis 不连续电泳
discontinuous fatigue 不连续疲劳
discontinuous fibre 定长玻璃纤维
discontinuous filter 间歇式过滤器；周期式滤清器；不连续滤波器
discontinuous filtration 间歇式过滤
discontinuous flow 间断车流；不连续车流
discontinuous flow-concentration model 间断型流量－密度模型
discontinuous fluid 不连续运动
discontinuous fold 断续型褶皱
discontinuous frequency function 不连续频率函数
discontinuous function 间断函数；不连续函数
discontinuous game 不连续对策
discontinuous glass fiber 非连续玻璃纤维
discontinuous glide 不连续滑移
discontinuous going 间歇操作
discontinuous gradation 不连续级配
discontinuous gradient 不连续梯度
discontinuous grading 间断级配；不连续级配
discontinuous grain growth 不连续晶体生长
discontinuous granulometry 不连续颗粒分析；不连续颗粒测定法
discontinuous group 不连续群
discontinuous growth 不连续长大
discontinuous gully 不连续冲沟
discontinuous heating 间歇（式）供暖；定时供暖；不连续供暖
discontinuous hypotheses 非连续性假设
discontinuous image 不连续成像
discontinuous impost 不连续拱墩；间断拱墩
discontinuous interference 不连续干扰
discontinuous interstice （岩石的）不连续间隙
discontinuous jump 不连续转移
discontinuous layer 不连续层；不连通层
discontinuous layer of density 密度跃层
discontinuous load 间歇荷载；不连续负载；突变性负载；突变荷载
discontinuous loading 不连续加载
discontinuously distributed suction 不连续多孔吸除
discontinuously graded aggregate 间断级配骨料
discontinuously graded concrete 间断级配混凝土
discontinuous market 不连续证券市场
discontinuous Markov process 不连续马尔可夫过程
discontinuous mat 不连续纤维毡
discontinuous measure 不连续测量
discontinuous measurement 间断测量
discontinuous medium 不连续介质
discontinuous microelement grinder 间歇式微细离心粉碎机
discontinuous mining technology 间断开采工艺
discontinuous motion 间断运动；不连续运动

discontinuous noise 间断性噪声
discontinuous observation 不连续调查
discontinuous oil phase 不连续油相
discontinuous operation 间断作业；间断运行；间断工作
discontinuous oscillation 断续振荡
discontinuous periodic(al) motion 非连续周期运动；不连续周期运动
discontinuous permafrost 间断永冻土；间断永冻层；非连续性永久冻土；不连续多年冻土层
discontinuous phase 间断相；分散相；不连续相；被分散相
discontinuous phenomenon 不连续现象
discontinuous plane 不连续面
discontinuous porosity 不连续孔隙性
discontinuous precipitation 不连续析出；不连续沉淀
discontinuous process 不连续过程
discontinuous production 间歇生产
discontinuous quantity 不连续量
discontinuous rating 间断定额
discontinuous reaction series 不连续反应系列
discontinuous reception 不连续接收
discontinuous reflection 不连续反射
discontinuous rhythm 间断韵律
discontinuous rock mass 不连续岩体
discontinuous running 阶段作业；间歇运转；周期作业；断续运行
discontinuous save segment 不连续保存段
discontinuous segment 不连续段
discontinuous series 不连续系列；不连续数列；不连续级数
discontinuous shiftwork 不连续轮班工作制
discontinuous simultaneous technique 间歇联用技术
discontinuous solution 间断解；不连续解
discontinuous spectrum 间断光谱；不连续（频）谱；不连续（光）谱
discontinuous spectrum absorption 间断光谱吸收
discontinuous sterilization 间歇灭菌(法)
discontinuous structure 不连续组织
discontinuous surface 不连续面
discontinuous suspended acoustic(al) ceiling 间断式吸声吊顶
discontinuous synthesis 不连续合成
discontinuous system 离散系统；不连续系统
discontinuous transmission 不连续发送
discontinuous transport 间歇运输
discontinuous type 间断型
discontinuous-type regulator 间断式调节器
discontinuous variable 间断变量；非连续变量；不连续变数；不连续变量
discontinuous variation 断续变化；不连续变异；不连续变化；不连续变差
discontinuous warping 不连续式整经
discontinuous wave 断续波；非连续波
discontinuous wedge 不连续光楔；梯级光楔
discontinuous welding 断续焊
discontinuous yielding 间断屈服
discontinuum 不连续体
discontinuum mechanics 不连续介质力学
discontinuum medium mechanics 不连续介质动力学
discontrol of the market 放开市场
disco process 低温渗碳法
discord 不谐和(弦)；不和谐
discordance 不整合；不匀整；不和谐性
discordance intrusive body 不整合侵入体
discordance permutation 不和谐排列
discordant 不和谐的
discordant age 不谐和年龄
discordant basin 不整合盆地
discordant bedding 不整合层理；不平行层理
discordant coast 不匀整海边
discordant contact 不整合接触(面)
discordant drainage 不协调水系
discordant fold 不整合褶皱
discordant injection 不整合注入；不整合贯入
discordant intrusion 不整合侵入
discordant intrusive body 不整合侵入体
discordant junction 不平齐汇流
discordant permutation 不和谐排列
discordant pluton 不整合深成岩体
discordant sample 不和谐样本

discordant valley 不整合河谷
discordant value 不和谐值
discordogenic fault 分界断层
discorrosion lining cloth 防腐衬布
discotheque （放流行歌曲唱片供来客跳舞的）夜总会
discount 扣息；扣价；折扣；把票据贴现；让头；贴现利率
discountability 可贴现性
discount a bill 扣减期票
discountable 可折现；可贴现的；可打折扣的
discount account register 账款贴现簿
discount allowance 市场交易折扣
discount allowed 销货折价
discount and allowance 折扣和让价
discount bank debenture 贴现银行债券
discount basis 贴现基础
discount broker 不动产经纪人；贴现（票据）经纪人
discount cash book 贴现现金簿；贴现票据账簿
discount category 贴现分类
discount center 零售百货商店
discount charges 贴现费用
discount clerk 票据贴现人员；贴现人
discount coefficient 贴现系数
discount cost 折扣价格；折扣费用；贴现费用
discount coupon 减价券
discount curve 贴现曲线
discount debenture 贴现债券
discount difference 贴现率差数；贴现差额
discount earned 折扣收入；贴现收入
discounted account register 客账贴现簿；已贴现账款登记簿
discounted benefit/cost ratio 按现值法计算的效益成本率；用现值法计算的效益成本比率
discounted bill 已贴现票据；贴现票据
discounted cash 贴扣现金
discounted cash flow 贴现现金流量（表）；折现现金流量；现金流转贴现法；贴现值系列；贴扣现金流动
discounted cash flow analysis 现值分析法
discounted cash flow method 现金流量贴现法；资金流动折现评估法；按现值计算现金流量法
discounted credit 贴现信贷
discounted flow 贴现流量
discounted mortgage 减价出售抵押
discounted note 已贴现票据；贴现票据
discounted of account receivable 应收账款贴现
discounted payable period 贴现回收期
discounted payback method 按现值计算回收法
discounted payback period 贴现投资回收期
discounted present value 贴现后的现值
discounted price 折扣价格
discounted rate 贴现率
discounted rate of return 折现投资盈利率；贴现投资收益率
discounted value 折算值；折现值；贴现（价）值
discounter 廉价商店；贴现人
discount expenses 贴现费用
discount factor 折现率；贴现因子；折扣因素；贴现系数
discount for cash 贴现
discount for risks 风险贴现
discount from the price 货价折扣
discount granted 销货折价；给予折扣
discount house 贴现（商）行；贴现公司
discount income 贴现收入
discount(ing) 打折扣；贴现
discounting a bill 票据贴现
discounting bank 贴现银行
discounting methods 贴现法
discounting notes and acceptance receivable 应收票据与承兑汇票的贴现
discounting of account receivable 应收账款贴现
discounting of bill of exchange 汇票贴现
discounting rate 折现率；贴现率
discounting technique 贴现技术
discount interest 贴现利息
discount issue 发行折扣
discount lapsed 折扣失效
discount liability 折价债务
discount loss 折扣损失
discount margin 贴现率差
discount market 削价市况；贴现市场
discount method 贴现法

discount offered 允给折扣;提供折扣;提供贴现
discount on a promissory note 期票贴现
discount on bill bought 购入票据得贴现
discount on bill sold on condition of repurchase 附有回购条件票据的贴现
discount on bond 债券折扣;债券折价
discount on bond purchased 购入债券折价
discount on bonds payable 应付债券折价
discount on capital stock 股本折价
discount on check 贴现支票
discount on common stock account 普通股折价账
discount on debenture 公司信用债务折扣
discount on exchange 贴现(外汇);汇水
discount on note payable 应付票据贴现
discount on purchase 购买折扣;进货折扣
discount on returned sales 销货退回折扣;退货的销售折扣
discount on stock 股本折价
discount paper 贴现票据
discount period 折扣期限;贴现期(限)
discount point 货款贴现百分点
discount policy 贴现政策
discount preference 折现偏好
discount present value 贴现现值
discount price 打过折扣的价格
discount quotation 贴现行市;贴现行情
discount rate 减成率;折扣率;折减率;贴现(利)率;贴水率;折现率
discount rate of interest 贴现率
discount rate poilicy 贴现率政策
discount received 购货折扣
discount return rate 贴现回收率
discount sale 减价出售
discounts and allowances 折让
discount shop 廉价商店
discount store 廉价商店;折扣商店
discount table 贴现表
discount taken 获得折扣;已取得的折扣
discount teller 贴现人
discount to present value 折现;折算为现值
discount window 贴现窗口
discoupling 断开;切断
discourse of trade 贸易论
discovered curve 发现曲线
discovered new resources regions 发现新的资源基地
discovered new types of ore deposits 发现新的矿床类型
discovered reserve 已发现储量
discovery 显像;发现
discovery bonus 发现定金
discovery bore 探孔;探井
discovery boring 探孔;探井
discovery deep 发现海渊
discovery hole 发现孔
discovery of facts 披露事实
discovery of petroleum 发现石油
discovery point surveying 露头测量
discovery rate 发现率
discovery sampling 发现抽样
discovery ship 探险船
Discovery tablemount 迪斯卡弗里平顶海
discovery time 发现时间
discovery value method 发现价值法
discovery well 发现新油井;发现热田的钻井
disc parking 计时泊车;定时泊车;贴标签停车
disc pile 盘底桩;阔脚桩;扩底桩;空心阔脚桩;圆盘柱
disc punching 冲片
disc record 磁盘记录
discredited 无信用
discreet value 预估值
disc refiner 盘磨机
disc relay 盘形继电器
discrepance 差异;不一致
discrepancy 较差;互差;不一致;不相符;不符值;不符合;不法行为;偏差
discrepancy and claim clause 异议和索赔条款
discrepancy between observation sets 测回(互)差
discrepancy between twice collimation errors 两倍照准差;二倍照准误差偏差;二倍视准差互差;2C 较差
discrepancy in closing 闭合差【测】
discrepancy in elevation 高(程)差;高程不等值
discrepancy in position 位置偏差
discrepancy sheet 订正表
discrepancy switch 差速开关
discrepant item 分歧的项目
discrepitate 龟裂;布满裂纹
discrete 离散(的);分散;分立的;分离的;非连续(的)
discrete absorption 离散吸收;不连续吸收
discrete address 离散地址;非连续地址
discrete address communication system 离散地址通信系统
discrete aggregate 松散集料;松散骨料
discrete amount 个别量
discrete analog(ue) 离散模拟
discrete analysis 离散分析
discrete approximation 离散近似;离散逼近
discrete-area draught 普染要素清绘原图
discrete assembly 分立组件
discrete automat 离散自动机
discrete automaton 离散型自动机
discrete band 分立带
discrete bit optic(al) memory 打点式光存储器
discrete block 松散块体
discrete-carrier hologram 离散载体全息图
discrete cell 特定单元
discrete channel 离散(信息)通道
discrete choice 离散选择
discrete circuit 分立电路
discrete coefficient 离势系数
discrete command 离散指令;离散命令;断续指令;不连续指令
discrete component 离散构件;分立元件
discrete component amplifier 分立元件放大器
discrete component card 分立元件插件
discrete compounding 非连续复利计算
discrete conditional distribution 离散条件分布
discrete consumption stream 离散消费量流
discrete control 离散控制
discrete convolution 离散褶积
discrete coordinates 离散坐标
discrete correlation 离散相关
discrete cosine transform 离散余弦变换
discrete crack-bounded 裂隙分离边界
discrete crenulation cleavage 不连续褶劈理
discrete data 离散性资料;离散数据;分立数据;不连续资料
discrete decision process 非连续决策程序
discrete density levels 离散密度级
discrete depth plankton sampler 定深浮游生物采集器
discrete deterministic process 离散确定性过程;离散决定性过程
discrete device 分立器件;分离器件
discrete differential game 离散微分对策
discrete distribution 离散(型)分布;不连续分布
discrete distribution function 离散分布函数
discrete dynamic(al) system 离散动态系统
discrete ecotone 分离式生态过渡带
discrete element 分离单元;离散元
discrete energy 分立能量
discrete energy eignvalue 分立能量本征值
discrete energy levels 分立能级
discrete energy state 分立能态;分离能态
discrete error 个别误差;分立误差
discrete event simulation package 离散事件仿真程序包
discrete fiber 分离线
discrete field-stop aperture 分立视场光栏孔径;分立场阑孔径
discrete film zone 土壤水带
discrete fixed base 离散固定底座
discrete flaw 分散性疵点
discrete flexible base 离散柔性底座
discrete Fourier series 离散傅里叶级数
discrete Fourier transform 离散(型)傅里叶变换
discrete frequency 离散频率;不连续频率
discrete frequency generator 离散频率发生器
discrete function 离散函数
discrete hinge model 离散铰模型
discrete image 离散像点
discrete information source 离散信息源
discrete input 离散输入
discrete integral element method 离散单元法
discrete intelligent data acquisitor 分布式数据采集器
discrete interval theorem 分立间隔定理
discrete inversion Fourier transform 离散傅里叶逆变换
discrete ion 离散离子
discrete irregularity 单独不平顺
discrete kink-band 分立扭折带
discrete level 分离电平
discrete logarithmic energy spectrum curve 离散对数能谱曲线
discrete logarithmic normal distribution 离散对数正态分布
discrete logic 离散逻辑
discrete luminescent center 分立发光中心
discretely tunable infrared laser 不连续可调红外激光器
discrete Markov process 离散马尔可夫过程
discrete mass 离散质量
discrete material 松散材料
discrete mathematics 离散数学
discrete maximum principle 离散最大值原理;离散极大值原理
discrete media 松散介质;松散介体
discrete message 离散信息;离散消息
discrete message source 离散信号源
discrete method 非连续法
discrete model 离散模型
discrete mode spectrum 离散模谱
discrete multivariate distribution 离散多元分布
discreteness 离散性;目标相对于背景的明显度;不连续性
discrete normal distribution 离散正态分布
discrete optimization 离散最优化
discrete optimization problem 离散型最优化问题
discrete optimum 离散最优解
discrete-ordinate method 分立纵坐标法
discrete output 离散输出
discrete package 分立封装
discrete parameter 离散参数
discrete parameter Markov chain 离散参数马尔可夫链
discrete parameter process 离散参数过程
discrete parameter system 离散参数组;离散系(统);不连续参数组;不连续参系(统)
discrete parameter time series 离散参数时间序列
discrete Pareto distribution 离散帕累托分布
discrete part 分立元件;分立式器件
discrete particle 离散粒子;离散颗粒;分立质点
discrete peak 分立峰
discrete phase 离散相;不连续相;被分散相
discrete picture 离散图像
discrete planar epitaxial device 分立平面外延器件
discrete point 离散点;分立点
discrete point loads 不连续点荷载
discrete probability distribution 离散概率分布
discrete probability function 离散概率函数
discrete process of physical chemistry 物理化学分离法
discrete process of wet 湿研磨分离法
discrete programming 离散(型)规划
discrete pulse 离散脉冲;分离脉冲;不连续脉冲
discrete quantum 分立量子
discrete radiation spectrum 分立辐射谱
discrete radio source 分立射电源
discrete random error 离散型随机误差
discrete random nonlinear system 离散随机非线性系统
discrete random process 离散随机过程
discrete random variable 离散随机变量
discrete range 离散幅度;离散范围
discrete regulator 离散调整器
discrete relaxation 不连续松弛
discrete representation 离散表示(法)
discrete sample 离散样元
discrete sample analyzer 不连续采样分析仪;分散采样分析器
discrete sampling 离散取样;离散抽样;离散采样;分散取样;分立取样;非连续采样;不连续采样
discrete scale 离散标度
discrete search 离散搜索
discrete series 离散序列;离散数列;间断数列;不连续数列

discrete set 离散集
discrete settling 分散沉降
discrete shear zone 不连续剪切带
discrete signal 离散信号;分离信号
discrete simulation 离散模拟;离散仿真
discrete simulation model 分散模拟模型
discrete source 离散信源
discrete source model 离散声源模型
discrete space 离散空间
discrete space time 离散时空;分立时空
discrete spectrum 离散(频)谱;离散(光)谱;分立光谱;不连续谱
discrete spot 离散斑点
discrete state 离散状态;分离状态;不连续(状)态
discrete state material 分散相材料
discrete steps 不连续(的)阶段
discrete stochastic process 离散随机过程
discrete stochastic variable 离散随机变量
discrete structure 离散结构;个别结构
discrete switch and button 分离式开关和按钮
discrete symbol 散列符号
discrete system 离散体系
discrete time 离散时间
discrete time control 离散时间控制
discrete time dynamic(al) system 离散时间动态系统
discrete time Markovian motion 离散时间马尔可夫运动
discrete time queue 分散时间排队
discrete time series 离散时间序列;离散时间数列
discrete time system 离散时间系统
discrete topology 离散拓扑;离散结构
discrete transistor 分立(式)晶体管
discrete type 离散类型
discrete type data 离散型数据
discrete type random variable 离散型随机变量
discrete type transistor protection relay 分立式晶体管保护继电器
discrete uniform distribution 离散均匀分布
discrete unit 离群车辆
discrete value 离散值;非连续值;不连续值
discrete-valued function 离散值函数
discrete-valued objective 离散值目标函数
discrete-valued random variable 离散值随机变量
discrete variable 离散变数;离散变量;不连续变量
discrete variable method 离散变量法
discrete variable problem 离散型问题
discrete white noise 离散白噪声;白噪声
discrete winding 不连续绕组
discretion 离散性;自由裁量权;自行处理;处置权
discretional order 无条件的定货
discretionary account 任意账户
discretionary account transaction 任意账户交易
discretionary appropriation 任意盈余指拨
discretionary array method 选择阵列法
discretionary cost 决策差别成本;选择成本;自由处理成本
discretionary credit limit 自定信用项目
discretionary expense center 可支用费用中心
discretionary fixed cost 选择性固定(性)成本
discretionary funding 选择性拨款
discretionary funds 任意资金
discretionary income 任意支配所得;随意支配的收入
discretionary licensing 酌情发给许可证
discretionary limits 自主决定权限
discretionary order 任意处理的订单
discretionary policy 任意政策
discretionary profits 相互抉择利润
discretionary purchasing power 任意购买力;剩余购买力
discretionary spending 自行处理的开支
discretionary stabilization 相互抉择的稳定
discretionary trust 全权信托;全权托管
discretionary wiring (method) 随意布线(法);选择连接法;选择布线法
discretization 离散化;离散比
discretization error 离散化误差
discretization of continuous time system 连续时间系统离散化
discriminability 可分辨性;鉴频能力;鉴别力
discriminant 鉴别式;判别式
discriminant analysis 判别(式)分析
discriminant approach to pattern recognition 模式识别的鉴别法
discriminant equation 判别方程(式)
discriminant function 辨别函数;判断函数;判别函数
discriminant of quadratic form 二次方程判别式
discriminant parameter 判别参数
discriminant threshold 鉴别阈;判别阈
discriminate 区别
discriminate analysis model 判别分析模型
discriminate threshold 鉴别阈
discriminating breaker 逆流自动断路开关
discriminating circuit-breaker 流向鉴别断路器
discriminating conduct 辨别方法
discriminating cut-out 鉴相断路器;鉴频断路器;鉴别断路器
discriminating duty 差别关税;区别对待性关税
discriminating element 鉴别元件
discriminating function 判别函数
discriminating order 指令;判定指令;判别(指令)
discriminating power 辨别能力
discriminating pricing 差别定价
discriminating protective system 鉴别性保护系统;区域选择性保护系统
discriminating rate 差别费率
discriminating relay 鉴别继电器;谐振继电器
discriminating satellite exchange 装有区别机的支局
discriminating score 判别计量
discriminating selector 鉴别选择器;区域选择器
discriminating tariff 差别税率
discriminating threshold 甄别阈
discriminating value 判别值
discrimination 鉴别率;鉴别力;区分;区别;判别;条件转移;识别率
discrimination accuracy 鉴别精度
discrimination by income 按收入多少给予的差别对待
discrimination data processing system 鉴别数据处理系统;数据鉴别处理系统
discrimination error 鉴别误差
discrimination factor 差异系数;鉴别因子
discrimination filter 鉴别滤波器
discrimination function 判别函数
discrimination instruction 判定指令;判别指令
discrimination learning 辨别学习
discrimination output 鉴频输出;鉴别输出
discrimination process 鉴别过程
discrimination ratio 判别比
discrimination reversal 辨别颠倒
discrimination selecting machine 鉴别选择机
discrimination test 辨别测试
discrimination threshold 鉴别阈;鉴别灵敏度
discrimination time 辨别时间
discrimination training 辨别训练
discrimination treatment 歧视待遇
discriminative clause 歧视性条款
discriminative condition 歧视条件;不平等条件
discriminator 鉴频器;鉴别器;假信号抑制器;甄别器
discriminator circuit 鉴相位电路
discriminator peak separation 鉴频器谐振点间距
discriminator rate 倾轧费率
discriminator transformer 鉴频变压器
discriminator tube 鉴频管
discriminatory across-rates 差别套汇率
discriminatory analysis 判别分析
discriminatory analysis on Bayesian criterion 贝叶斯准则下的判别分析
discriminatory analysis on Fisher criterion 费歇尔准则下的判别分析
discriminatory critical value 判别分界值
discriminatory discharge 歧视性解雇
discriminatory exchange rate 差别汇率
discriminatory index 判别指标
discriminatory quota 歧视性配额
discriminatory score 判别得分
discriminatory selling 差别对待的销货
discriminatory tariff 差别对待的税则
discriminatory taxation 差别征税
discriptive item 描述项
disc rivet 圆盘铆钉;平头铆钉
disc roller 圆盘式压路机;圆盘式滚筒;滚轮
disc rotor 盘形叶轮
disc sander 圆盘磨光机;圆盘打磨机
disc-sanding 盘式打磨

disc saw 圆锯机
disc screen 盘(式)筛;盘式格栅
disc-seal 盘形封口(的)
disc separator 圆盘式分选器;圆盘式分离器
disc sewage screen 圆盘式污水滤网
disc-shaped cavity 盘形腔
disc shaving cutter 盘形剃齿刀
disc shutter 圆盘快门
disc signal 圆排号志
disc skimmer 盘式水上浮油回收装置
disc slicer 盘式切片机
disc spreader 圆盘式撒布机
disc spring 圆盘形弹簧;盘簧;螺旋弹簧
disc tablet counter 盘式数片机
disc tensioner 圆盘式张力器
disc tiller 重直多圆盘犁
disc turntable 舞台转盘(施转舞台的圆形地板部分)
disc type flaw 圆片形缺陷
disc type lightning arrester 盘形避雷器
disc type nodulizer 成球盘;盘式成球机
disc type power float 圆盘镘平板
disc type suction filter 盘式真空过滤器;盘式真空过滤机
discus 圆片;铁饼
discuss and determine wage-grade 评薪
discussion statement 讨论式决算表
discussive 消散的
disculient 消散的
disc valve 圆盘阀;碟阀;片状阀;盘阀
disc vibrator 圆盘振动器;盘式振捣器
discvision 光盘显像系统
disc water meter 盘式水量计;盘式水表
disc wheel 薄片砂轮;盘(形)轮
disc winding 盘式绕组
disc with outer teeth 带外齿的圆盘
disdrometer 雨滴测量器
disdropmeter 示滴仪
disease 疾病
disease caused by poisonous water 有毒水引起的疾病
disease caused by pollution 污染引起的疾病
disease caused by soil and water 水土病因
disease control 疾病控制;疾病管理
disease development 发病
diseased wood 病木
disease frequency 发病频次
disease index 疾病指数
disease notification 疾病报告
disease of agricultural plants 农作物疾病
disease of plants 植物病害
disease outbreak 疾病爆发
disease prevention 疾病预防;防病
disease-producing 致病的;发病的
disease producing germ 病原菌
disease resistance 抗病性;抗病力
disease sources 病源
disease time 发病时间
disease-treated seed 消毒种子
disease vector 疾病媒介(物)
disebacate 癸二酸氢盐
di-sec-octyl hexahydro phthalate 六氢化邻苯二甲酸二仲辛酯
diseconomy 不经济
diseconomy of scale 规模不经济
di-sectoctyl hexahydro phthalate 环己烷邻苯二甲酸二仲辛酯
disembarkation 下船;上岸
disembarked group 上陆类群
disembarked period 上陆期
disembarkment 下船;登陆
disembogue of river 河流出海
disembogue of stream 河流出海
disendow 剥夺基金
disengage 解约;解开;加除;拆卸;放开;脱离;使分离
disengage a clutch 摘开离合器
disengaged free 空线
disengaged (free) line 闲线
disengaged position 脱离状态
disengaged side 下风舷
disengaged vapo(u)r 释放汽
disengagement 脱离
disengagement area 蒸发面积
disengagement gear 解脱机构

disengagement of dredge pump 泥泵脱开
disengagement surface 蒸发面;汽水分界面
disengaging 空闲的
disengaging bar 分离杆
disengaging clutch 脱开式离合器
disengaging coupling 离合联轴节
disengaging device 排气装置;分离器;断路器;分离装置
disengaging device operating lever 分离机构操纵杆
disengaging device push rod 分离机构推杆
disengaging fork 离合器叉
disengaging fork lever 分离叉杆
disengaging gear 解脱机构,齿轮离合器;分离装置;分离机构;脱缆钩
disengaging hook 分离钩;脱缆钩
disengaging latch 分离链;分离掣爪
disengaging lever 分离杆
disengaging mechanism 分离机构
disengaging movement 分离运动;停车运动
disengaging of barge train 解拖驳队
disengaging pawl 分离爪
disengaging plate 脱离板
disengaging rod 分离杆
disengaging section 拆卸部分
disengaging shaft 分离轴
disengaging spring 分离弹簧
disentailing assurance 解除限定继承权保证
disentomb 发掘;从坟墓中挖出
disequilibrate 打破平衡
disequilibration 平衡不稳
disequilibrium 非均衡;不稳定;不平衡;平衡不稳;失去平衡
disequilibrium assemblage 不平衡组合
disequilibrium compaction 不平衡压实
disequilibrium economics 不均衡经济学
disequilibrium model 不均衡模型
disequilibrium price 不均衡价格
disequilibrium situation 不平稳状态
disequilibrium system 不均衡体系
diserimity 可识别性
diserophie tectono-geochemical field 亲硫构造地球化学场
disfigure 损伤外形
disfigurement 失形;破相;外形毁损;损形
disfigurement of surface 路面损坏;地表起伏
disforest(ation) 砍伐森林;毁林(辟田);伐除森林
disforesting 伐去森林
disfunction 功能失常
disfunctional 失去功用引的
disgerminator 除胚芽机
disgregation 分散
disguise 伪装
disguised cession 变相让与;变相割让
disguised support 暗中支持
disguised unemployment 变相失业
disgusting situation 恶劣情况
dish 截抛物面反射器;碟形天线;凹部;凹陷;抛物面反射器;盘形物
dish aeration 碟式曝气
dish aerator 碟式曝气器
dish and pillar structure 碟柱构造
dish antenna 截抛物面天线;抛物面天线
disharmonic 不谐和的
disharmonic faulting 不谐和断裂作用
disharmonic feature 不谐和地形
disharmonic fold 不谐和褶皱;不谐和褶曲;不协调褶皱
disharmonic folding 不协调褶皱作用;不谐和褶曲【地】
disharmonic habitat 不谐和生境
disharmonious habitat 不谐和生境
disharmony 不谐和;不调和
dish baffle 盘形挡板
dish brush 碟刷
dish cabinet 碗柜
dish cloth 抹布
dished 表面下凹;半球形的;半凹形的;凹状的
dished and ribbed 加助碟式
dished design 碟式
dished disk 碟形圆盘
dished end 碟形端板;碟形封头;盘形底
dished end plate boiler 碟形底板式锅炉;凸底锅炉
dished hotplate 盘形灶面板

dished-out 圆屋顶构架托座
dished perforated plate 碟形多孔板
dished riser 凹顶形冒口
dished roof 穹窿形屋顶
dished sluice gate 圆盘式泄水闸门;圆盘式冲沙闸门
dished spinner 碟形撒布轮
dished steel plate 碟形钢板
dished tank 碟形储[贮]槽
dished turn 碟形曲面的超高弯道
dished washer 盘形垫圈
dish emery wheel 碟形砂轮;凹盘轮
dish friction 圆盘摩擦
dish friction loss 圆盘摩擦损失;圆盘摩擦损耗
dish garden 盆景
dish gas holder 湿式气柜
dishing 形成凹面;大半径凹进成型;辐板压弯;凹圆槽;平底器皿下沉变形;凹陷
dishing of surface 土壤表面碟状沉陷
dishing press 碟形压机;车轮轮辐压弯机
dishing settlement 碟状沉陷
dishonest 不正当的
dishonest process 不真实过程
dishonor 拒绝兑付;拒绝承兑
dishonored bill of exchange 拒付汇票
dishonored certificate 拒付证书
dishonored check 拒付支票
dishonored note 拒付票据;拒付汇票
dishonour 拒绝付款;拒付;不名誉;不光彩
dishonour a bill 不支付期票;不承兑期票
dishonoured cheque 空头支票
dish pan 洗碟盆;洗碟池
dishpan antenna 盆式天线
dish plate 碟形板;弯边圆钢板
dish sander 盘式砂光机
dish-shaped depression 碟形坑
dish-shaped ewer 盘形壶
dish skimmer 碟形扒渣器
dish sprayer 盘式曝气器;盘式溅水器
dish structure 碟状构造
dish tubesheet 碟形管板
dish turner 木盘车工
dish type discharge grate 盘式卸料算子
dish type grinding wheel 碟形砂轮
dish-type liquid limit machine 碟式液限仪
dish type nodulizer 成球盘;盘式成球机
dish vacuum filter 碟式真空吸滤机
dishware 餐具;容器
dish-warmer 菜肴保温板
dish washer 洗碟机
dish washing agent 洗碟剂
dish washing detergent 餐具洗涤剂
dish washing machine 洗碗机;洗(碗)碟机
dish washing room 洗盘间
dish-washing sink(unit) 洗碗池;洗涤池;洗碟盆;洗碟池
dish water 洗碟(子)水
dish wheel disc wheel 碟形砂轮
dish with sky blue glaze 天蓝釉盘(瓷器名)
dish with white glaze 白釉多口盘(瓷器名)
disilane 乙硅烷
disilanoxy 乙硅烷氧基
disilanylene 亚乙硅烷基
disilanylthio 乙硅烷硫基
disilazanoxy 甲硅烷氨基甲硅烷氧基
disilicate 二硅酸盐
disilicide 二硅化物
disilkane 乙硅烷
disilmethylene 亚甲基二硅
disiloxane 二甲硅醚
disiloxanoxy 甲硅烷氧代甲硅烷氧基;甲硅醚氧基
disiloxanyl 甲硅烷氧代甲硅烷基;甲硅醚基
disiloxanyl amino 甲硅烷氧代甲硅烷氨基;甲硅醚氨基
disiloxanylthio 甲硅醚硫基
disilthianoxy 甲硅硫醚氧基
disilthianyl 甲硅硫醚基
disimpaction 嵌塞解除法
disincentive 抑制措施;无刺激
disincentive action 反激励作用
disincentives 不利于经济发展的因素
disincrustant 除垢剂;防水锈剂;去水垢
disincrustation 水垢溶化
disindection by chlorine dioxide 二氧化氯消毒

disinfect 灭菌;消毒
disinfectant 灭菌剂;消毒(杀菌)剂;消毒的
disinfectant byproduct 消毒剂副产物
disinfectant concentration 消毒剂浓度
disinfectant cup 消毒杯
Disinfectant/Disinfection Byproducts Rule 消毒剂与消毒副产品法令
disinfectant evaluation 消毒剂评价
disinfectant paint 杀菌涂料
disinfectant residual 消毒剂残留
disinfected sewage 经消毒污水
disinfected soap 消毒皂
disinfected waste 消毒废物
disinfected wastewater 消毒废水
disinfecting action 灭菌作用;消毒作用
disinfecting medicament and apparatus for silkworm 蚕用消毒药剂及用具
disinfecting pool 消毒池
disinfecting tank 消毒池
disinfection 灭菌作用;灭菌;消毒作用;杀菌;消毒
disinfection by ammonium chloride 氯化铵消毒;氯胺消毒
disinfection by bleaching powder 漂白粉消毒
disinfection by chlorination 氯化法消毒
disinfection by chlorine 加氯消毒
disinfection by hydrogen peroxide 过氧化氢消毒
disinfection by ozone 臭氧消毒
disinfection byproduct 消毒副产物;消毒副产品
disinfection by ultraviolet ray 紫外线消毒
disinfection chamber 消毒室;无菌室
disinfection combination process 消毒组合工艺
disinfection equipment 消毒设施;消毒设备
disinfectioner 消毒员
disinfection facility 消毒设施
disinfection kinetics 消毒动力学
disinfection method by liquid chlorine 液氯消毒法
disinfection method by ozone 臭氧消毒法
disinfection method by sodium-hypochlorite 次氯酸钠消毒法
disinfection method by ultraviolet ray 紫外线消毒法
disinfection of domestic water 生活用水消毒
disinfection of sewage 污水消毒
disinfection of water 给水消毒;水的消毒
disinfection of water supply 自来水消毒
disinfection of well water 井水消毒
disinfection plant 消毒厂;消毒设备
disinfection process 消毒工艺
disinfection room 消毒室
disinfection seed 消毒种子
disinfection soil 消毒土壤
disinfection with ultraviolet rays 紫外线消毒
disinfector 消毒器;消毒剂;消毒者
disinfestation 灭虫法
disinflation 通货紧缩
disinflation policy 通货紧缩政策
disinhibition 抑制解除
disinscrustant 除垢剂
disinsection 灭虫法
disinsectization 灭虫法
disintegrant 崩解剂
disintegrate 解体;分裂;切碎
disintegrated concrete 剥蚀混凝土
disintegrated earth 坍塌山体
disintegrated electrolytic powder 粉碎的电解粉末
disintegrated granite 崩解花岗岩;风化花岗岩
disintegrated rock 崩解性岩石
disintegrated soil 分散性土
disintegrating 分裂本领
disintegrating agent 分裂剂
disintegrating granite 崩解花岗岩
disintegrating method 分裂法
disintegrating mill 解磨机;破碎机
disintegrating process 风化过程
disintegrating salvage 解体打捞【救】
disintegrating scrubber 冲激式除尘器
disintegrating slag 水碎炉渣;崩解熔渣;崩解炉渣;崩解矿渣;碎渣
disintegration 解体;解磨(粉碎);分裂;分解;剥蚀;崩解作用;瓦解;脱变作用;碎磨;碎裂
disintegration by frost 冰冻裂解
disintegration chain 放射性系;衰变链
disintegration coefficient 衰变系数
disintegration constant 分解常数;衰变常数

disintegration family 放射性系
disintegration index 崩溃指数
disintegration of ballast 道砟破碎
disintegration of concrete 混凝土的破裂;混凝土(的)离析
disintegration of filament 灯丝烧毁;灯丝烧坏
disintegration of glaze 釉的碎解;釉的溶蚀
disintegration of measure 测度的积分分解
disintegration of plaster 抹面剥落
disintegration per minute 每分钟衰变数
disintegration product 衰变产物
disintegration series 放射性系
disintegration time 崩解时间
disintegration time limited 崩解时限
disintegration velocity 崩解速度
disintegration voltage 扩散电压;崩离电压;破坏电压
disintegration volume 崩解量
disintegration weathering 崩解;风化
disintegrative 崩解性
disintegrator 对辊破碎机;分离器;解磨式粉碎机;转笼磨碎机;轧石机;粉碎机;分裂因素;气体洗涤机;破碎机;碎裂器
disintegrator pump 破碎泵;喷散泵
disintegrator scrubber 粉碎洗涤器
disinterested appraisal 无偏见的评估;公平评估
disintergation 打碎
disintermediation 资金的游离;大量提款;逆中介;脱媒
disintoxication 解毒(作用)
disinvestment 负投资;停止投资
disjoining pressure 排斥压力;分离压;膨胀压力
disjoint 拆散;不相交的;不连续的;不连贯
disjoint circuit 不相接的回路
disjoint collection 不相交集族
disjointed 脱开的
disjointed complement 不相交余数
disjointed path 不相交路线
disjointed track 不相交迹线
disjoint elements 不相交元素
disjoint event 不相交事件
disjoint intervals 不相交区间
disjoint policy 不相交策略
disjoint range 分隔范围
disjoint sets 分离集;不相交集(合)
disjugate 不连合的
disjunct 断离的;不连接的
disjunction 析取;折断;分离;脱节
disjunction cleavage 分离劈理
disjunction fold 间断褶皱;脱节褶皱;碎裂褶皱
disjunction mark 基线
disjunctive absorption 分离性吸收
disjunctive fold 间断褶皱;脱节褶皱;碎裂褶皱
disjunctive form 逻辑和形
disjunctive Kriging estimator 析取克立格估计量
disjunctive Kriging method 析取克立格法
disjunctive Kriging variance 析取克立格方差
disjunctive normal form 析取范式
disjunctive path 分离路线
disjunctive search 析取检索;按或检索
disjunctor 开关
disk = disc
disk action 圆盘作用
disk-and-cup feeder 杯盘给料机
disk-and-drum turbine 盘鼓形汽轮机
disk-and-wheel integrator 盘轮式积分器
disk antenna 盘形天线
disk armature 圆盘衔铁;盘形电枢
disk atomizer 离心式喷雾圆盘
disk attrition mill 盘磨
disk bit 圆盘钻(头)
disk boot 圆盘开沟器
disk bottom hole packer 盘式井底封隔器
disk-bowl centrifuge 盘碗离心机
disk brake 圆盘制动器;碗盘闸;圆盘制动;盘形制动器;盘式制动器
disk-braked train 带制动盘的动力传动
disk breaker 盘式轧碎机
disk brushing machine 碟式刷洗机
disk-bursting test 轮盘破裂试验
disk canvas wheel 帆布抛光轮
disk capacitor 半圆形可变电容器
disk car loader 圆盘装车机
disk centrifuge 盘式离心机

disk chuck 花盘;平面卡盘
disk clutch 圆盘离合器;盘形离合器
disk coalcutter 盘式截煤机
disk coil 盘形线圈
disk colo(u)rimeter 圆盘比色计;盘式色度计
disk comparator 旋盘比色器
disk conveyer 圆盘输送机;盘式输送机
diskcopy 磁盘复制(件)
disk counter 表盘式计数器
disk coupling 柔性盘联轴节;盘形联轴节
disk crusher 圆盘压碎机;圆盘碎土机;盘式轧碎机;盘式压碎机;盘式破碎机
disk cultivator 圆盘碎石机;圆盘式碎土机
disk cutter 圆盘形切割器;圆盘式切割机;滚刀;盘形刀具;盘式截煤机
disk diaphragm 旋转式光阑
disk discharger 盘式放电器
disk dish 圆盘凹度
disk-disk rheometer 双盘式流变仪
disk-doughnut baffle 盘环形折流板
disk dowel 圆木暗销;圆木榫
disk drier 圆盘干燥器
disk dryer 圆盘式干燥器;盘式干燥机
disked 盘形的
disked bottom 碟形封头;碟形底盖
disked closure 碟形封头
disk emery cloth 圆盘磨光机
disk engine 回旋汽机;盘式发动机
diskette 柔性塑料碟盘
disk extractor 环形脱膜工具
disk facing 盘形衬片
disk fan 圆盘风扇;盘扇
disk feeder 圆盘(式)加料器;圆盘式给料器;圆盘给料机;转盘喂料机;盘式加料器;喂料盘
disk filter 盘式过滤机;圆盘(式)过滤器;盘式滤池
disk flywheel 圆盘式飞轮;盘式飞轮
disk format 磁盘储存格式
disk formation pile 扩底桩
disk friction 轮盘摩擦;圆盘摩擦
disk friction clutch 圆盘摩擦离合器
disk friction loss 轮盘摩擦损失
disk friction wheel 盘形摩擦轮
diskfruit wingnut 青钱柳;青钱枫柏
disk gap transmitter 盘隙发射机
disk ga(u)ge 圆盘规
disk gear 盘形齿轮
disk gear cutter 盘形齿轮刀具
disk generator 盘式发电机
disk grinder 磨圆盘砂轮;圆盘研磨机;圆盘磨(床);盘磨机
disk grinding machine 盘式平面磨床
disk grizzly 圆盘(滚轴)筛;辊动筛
disk-grizzly screen 圆盘铁栅筛
disk harrow 圆盘中耕机;圆盘式路耙;圆盘耙
disk hinged valve 蝶门自动阀
disk hob 圆盘滚刀
disk holder 代座压板
disk humidifier 圆盘式加湿器;盘形增湿器
disk impeller 轮盘搅拌器;盘式激动器
disking twice 两次耙地
disk insulator 盘式绝缘子
disk key 圆盘式判读样片
disk leather wheel 牛皮抛光轮
disklike 圆盘状的
disklike structure 盘状结构
disk material thrower 圆盘式抛料机
disk meter 盘式水表;圆盘计量器;转盘式表;盘式仪表;盘式水表;盘式流量计
disk mill 盘式磨粉机
disk mixer 圆盘搅拌机;盘式搅拌机
disk nail 圆盘钉
disk nozzle 可调圆片式喷嘴
disk nut 盘形螺母
disk of conic(al) profile 锥形叶轮
disk of constant stress 等强度叶轮
disk of constant thickness 等厚度叶轮
disk of variable profile 变截面轮盘
disk of variable thickness 变厚度轮盘
disk on rod type circuit 加馈同轴电路
disk paring plow 圆盘灭茬犁
disk pelletizer mixer 盘式造球混料机(烧结机二次混料机的一种)
disk piercer 盘式穿孔机
disk pile 阔脚桩;圆盘柱;盘头桩;盘脚桩;盘底桩

disk piston 盘形活塞
disk piston blower 盘形活塞鼓风机
disk planer 圆盘平地机
disk planimeter 圆盘求积仪;圆盘面积仪
disk plough 圆盘路犁
disk plug 平面螺栓塞;盘塞
disk polisher 圆盘抛光机
disk ratio 盘面比
disk recorder 表盘式记录器
disk refiner 圆盘精研机
disk retarder 盘式制动输机
disk ripper 圆盘松土器
disk rivet 圆盘铆钉;圆盘钉
disk roller 圆盘碾压机
disk rotor 盘式转子
disk ruling 圆盘式划线
disk sander 圆盘磨光机;圆盘打磨机;地板打磨器;砂轮磨光机
disk saw 圆(盘)锯;盘锯
disk scanner 析像盘
disk scraper 圆盘刮泥机
disk screen 圆盘筛;盘筛
disk seal 盘封;盘形封口(的)
disk-seal tube 盘封管
disk separator 盘式分离机
disk-shaped radiator 圆盘状辐射体
disk-shaped roll 盘形辊
disk shutter 盘式快门
disk sieve 盘形筛
disk size 磁盘容量
disk slide rule 计算盘;圆盘计算尺
disk slotting cutter 盘形插齿刀
disk solid-cone nozzle 可调圆片式实心雾锥喷嘴
disk spacer 圆隔板
disk spide roller 盘齿式镇压器
disk spring 圆片弹簧;碟形弹簧;盘簧
disk stroboscope 盘式频闪观察仪
disk telescope 日包望远镜
disk test 巴西法圆盘试验;劈裂法圆盘试验
disk track 盘径
disk type cutter 盘式打磨机
disk-type electrostatic sprayer 转盘式静电喷涂机
disk-type friction clutch 盘式摩擦离合器
disk-type gear milling cutter 盘形齿轮铣刀
disk-type magnetic separator 盘式磁选机
disk-type marker 圆盘式划行器
disk-type nodulizer 盘式球机
disk-type reamer 盘形铰刀
disk-type rectifier 干式整流器
disk-type reducing valve 盘式减压阀
disk-type relay 盘式继电器
disk-type retarder 盘式制动运输机
disk-type root cutter 圆盘式块根切碎机
disk-type rotor 盘形转子
disk-type water-meter 圆盘式水表;转盘式水表
disk valve 横轴式蝴蝶阀;圆盘阀;平板阀(门);片状阀;盘形阀
disk valve bailer 盘阀捞砂筒
disk visco(si)meter 盘式黏度计
disk-wall packer 盘形井壁封隔器
disk washer 盘式洗涤器
disk water meter 圆盘水表;转盘式水表
disk weeder 圆盘式除草机
disk wheel 盘形(砂)轮;盘形车轮;盘轮
disk winding 盘形绕组
disk with bent blades 带弯叶的圆盘
disk with camlike teeth 带凸形齿的圆盘
dislike point 嫌忌点
dislike value 嫌忌值;厌恶量
dislimn 变模糊;使轮廓模糊
disller's feed 烧酒糟饲料
disloading of sediment 底砂推移;清除泥砂
dislocate 使变位
dislocated segment 离位节数
dislocating inclined fault 倾斜断层错动
dislocation 离位;转位;转换位置;断错;断层;叠差;错位;变位;位错;脱位;脱臼
dislocation angle 移动角;错角;位错角
dislocation area 移动带;错动区;错动带
dislocation basin 断陷盆地;断层盆地
dislocation boundary 位错界面
dislocation breccia 断错角砾岩;断层角砾岩
dislocation climb 位错攀移
dislocation climb model 位错攀移模型

dislocation coalescence 位移聚结
dislocation crack 位错裂缝
dislocation creep 位错蠕变
dislocation damping 位错阻尼
dislocation density 位错密度
dislocation dipole 位错偶极子
dislocation dynamics 位错动力学
dislocation earthquake 断错地震;断层位移地震;断层地震
dislocation fault 错断断层
dislocation free 无位错
dislocation free crystal 无位错晶体
dislocation glide 位错滑移
dislocation habit 位错惯态
dislocation half-loop 位错半环
dislocation jog 位错割阶
dislocation line 乱位线;断层线;位错线
dislocation locking 位错锁住
dislocation loop 位错环
dislocation metamorphism 断错变质;断层变质作用;错动变质作用
dislocation motion 位错运动
dislocation mountain 断层山
dislocation movement 位错运动
dislocation mucleation 位错中心
dislocation network 位错网络
dislocation node 位错结
dislocation of optic(al) range-finder 光学测距机的失调
dislocation plateau 断层高原
dislocation ring 位错环
dislocation scattering mobility 位移散射迁移率
dislocation segment 位错段
dislocation solubility 位错溶度
dislocation source 位错源
dislocation structure 错位构造;位错结构
dislocation substructure 位错亚结构
dislocation tangle 位错缠结
dislocation theory 位错理论
dislocation-time function 位错时间函数
dislocation valley 断裂谷;断层谷
dislocation velocity grade of active fault 活断层错动速率分级
dislocation velocity of active fault 活断层错动速率
dislocation wall 位错墙;位错壁
dislocation wave 位移波
dislodge 离开住处;从住处迁出;去除
dislodged sludge 沉积泥渣
dislodger 沉积槽;沉淀器
dislodger settler 沉积槽
dislodging of sediment 底质推移;底砂推移;挖除泥砂
dismal 沼泽;湿地
dismantle 卸开;拆散;拆除
dismantled plate restoring spring 拆卸板回位弹簧
dismantled railroad 已拆毁铁路
dismantled railway 已拆毁铁路
dismantlement 拆卸;拆除
dismantlement of import tariffs 废除进口关税
dismantle representation 拆卸画法
dismantling 拆卸
dismantling and lifting in section 解体打捞【救】
dismantling bay 拆卸间
dismantling building 拆除房屋
dismantling device 拆装工具
dismantling equipment 拆卸设备
dismantling flange 滑动法兰(盘);滑动法兰(板)
dismantling gang 拆卸组
dismantling key 拆装用扳手
dismantling of false work 拆除脚手架
dismantling of shuttering 模板拆卸
dismantling operation 拆除作业;拆除操作
dismantling pit 拆卸坑
dismantling shaft 拆卸井
dismantling yard 拆船厂
dismantlling of the axle-boxes 轴箱拆卸
dismembered drainage 解体水系
dismembered geosyncline 解体地槽;残块地槽
dismembered river 解体河
dismembered river system 解体水系
dismembered stream 解体河;海侵河
dismemberment 解体
dismetrical line of junction terminal 枢纽直径线
dismicrite 扰动微晶灰岩;扰动泥晶灰岩;磐斑灰岩

dismiss 开除;解聘;撤职
dismissal 解雇
dismissal of appeal 驳回起诉
dismissal of legal action 驳回诉讼
dismissal pay 解雇费
dismissal wage 离职工资
dismiss and replace 撤换
dismiss compensation 辞退金
dismiss workers 辞退职工
dismount 卸下;拆卸
dismountability 可卸性;可拆性
dismountable 可拆卸的
dismountable antenna 可拆式天线
dismountable auger conveyer 可拆式输送螺旋
dismountable boat 拼装式船舶
dismountable connection 可拆卸连接;活络节点
dismountable cutter suction dredge(r) 组合式绞吸挖泥船
dismountable division wall 可卸隔墙;活动隔墙;灵活隔断
dismountable dredge(r) 组装式挖泥船;组合式挖泥船
dismountable element 可拆卸的部件
dismountable flange 可卸法兰;活动法兰(盘)
dismountable installation 可卸装置
dismountable joint 可拆卸接头
dismountable mast 可拆卸天线杆
dismountable package 可拆卸包装
dismountable partition 可拆卸隔板
dismountable partitioning 可移隔板;活动隔板
dismountable ship 组合船队;拼装式船舶
dismountable vessel 拼装式船舶
dismounted 可拆卸的
dismounted presentation 拆卸画法
dismounting 拆装
dismounting and cleaning flowline for bearings of shaft box 轴箱轴承分解清洗流水线
disnormality 非正态性
disobliteration 闭塞消失;闭塞复通
disodium calcium ethylene diamine 依地酸二钠钙
disodium ethylene-bisdithiocarbamate 代森钠
disodium hydrogen phosphate 磷酸二(氢)钠
disodium phosphate 磷酸氢二钠;磷酸二(氢)钠
disodium salt 二钠盐
disodium sulfide 硫化二钠
disome 二体
disorbit 离开轨道;出轨
disorbition 脱轨
disorder 异常;无序;无规律;紊乱
disordered alloy 无序合金
disordered chain propagation 不规则链增长
disordered environment 环境秩序混乱
disordered field 无序场
disordered fold surface 不规则折叠面
disordered lattice 无序晶格;无序点阵
disordered material 无序材料
disordered orientation 不规则取向;无序取向;无规则取向
disordered state 无序态
disordered structure 无序结构
disordering 无序化
disorderly 不规则的
disorderly closedown 意外停歇;故障停机;非正常停机
disorderly conduct 妨碍治安行为;妨害治安行为
disorder of carbon dioxide transport 二氧化碳运输障碍
disorder of CO_2 transport 二氧化碳运输障碍
disorders of social instinct 社会本能障碍
disorientation 乱取向;迷航;定向(力)障碍
disoxidation 减氧作用
disparity 差异;差别;非平价;不一致;不均等性
disparity error 不一致误差
disparity items 不同项目;不等式项
disparity pay 差别工资
dispatch 急件;调遣;发送;派遣;速办;疏运
dispatch algorithm 调度算法
dispatch boat 递送公文的船;通信船
dispatch case 文件袋
dispatch clerk 调度员
dispatch days 速遣日数
dispatch delay 调度延迟;出车准备
dispatch drive 急速驾驶
dispatch driving 快速行车

dispatch earnings 快卸效益
dispatcher 调度员;调度器;调度区段
dispatcher controlled area 调度区域
dispatcher control table 调度程序控制表
dispatcher primitive 调度员原语
dispatcher queue 调度排队;发送排队
dispatcher's control 调度控制
dispatcher's control table 调度控制表
dispatcher's instruction 调度命令
dispatcher's log 调度日志
dispatcher's meter 调度尺
dispatcher's office 调度所;调度室;发送室
dispatcher's rask 调度员任务
dispatcher's room 调度室
dispatcher's signal chart 调度信号图
dispatcher's supervision 调度监督
dispatcher's supervision equipment 调度表示设备
dispatcher telephone 调度电话(机)
dispatcher telephone device 调度电话装置
dispatching 调度;发送
dispatching and scheduling of drilling job 施工调度计划【岩】
dispatching arrangement 调度安排
dispatching bill 发货单
dispatching card 调度卡
dispatching center 调度中心;发送中心
dispatching circuit 调度电路
dispatching communication 调度通信
dispatching computer 调度计算机
dispatching control 发货处
dispatching crossover 调度渡线
dispatching cycle 调度周期
dispatching device 电梯自动停车器
dispatching diagram 调度图
dispatching engineer 调度工程师
dispatching equipment 调度设备
dispatching file 调度表
dispatching list 发货单;调度表
dispatching money 派遣费
dispatching of empty container 发放空箱(集装箱)
dispatching of goods 发货
dispatching of train 发车【铁】
dispatching plan 发货安排
dispatching point 调度点
dispatching priority 调度优先(权);调度优先(级)
dispatching program(me) 发货安排
dispatching room 调度室
dispatching sequence 调度顺序
dispatching station 调度站;调度所;调度室;发运站
dispatching supervision system 调度监督系统
dispatching system 分配系统
dispatching telephone control board 调度电话主机
dispatching telephone control panel 调度电话主机
dispatching telephone equipment 调度电话设备
dispatching telephone of administration's branch 局界调度电话【铁】
dispatching telephone subset 调度电话分机
dispatching telephone system 调度电话系统
dispatching tools 疏运工具
dispatch inquiry 发出询价
dispatch list 调度表
dispatch loop 循环发送
dispatch means 疏运工具
dispatch money 快装费;速遣费
dispatch network 调度网络
dispatch of bagged cement 水泥袋装发运
dispatch of cargo 货物发运
dispatch of goods 货物疏运;货物发运
dispatch of ship 船舶速遣
dispatch operator 调度员
dispatch room 调度室;分发室;发送室;急件室
dispatch station 出发站;发送站;输送站
dispatch station for operation of harbo(u)r 港区作业调度室
dispatch switchboard 调度交换台
dispatch system 调度系统
dispatch track 发车线【铁】
dispatch trains in reverse direction 反向发车
dispatch trains in straight direction 顺向发车
dispatch vessel 通信船
dispensability 可省略性
dispensable circuit 调配电路;调剂电路

dispensable mo(u)ld 一次型
dispensary 门诊部；医务所；药房；诊疗所；防治所
dispensation 分配(物)；配方
dispense 分送；分发；调剂
dispense point 加油站；分配站
dispenser 自动售货机；给料器；分送器；取料器；喷射发泡器；配料器；配料泵；配量器；配合器；调合器；撒布器；撒播器
dispenser cathode 储备式阴极；补给式阴极
dispenser for wrapping paper 包装纸分配器
dispense with royalties 免缴使用费
dispensing 配方
dispensing assembly 分配装配
dispensing bottle 投药瓶
dispensing bulk plant 加料区
dispensing depot 加料区；配料区
dispensing device 撒布装置
dispensing equipment 分配装置；配料设备
dispensing gas 弥散气体
dispensing mechanism 计量器
dispensing package 分量包装
dispensing point 加料点
dispensing pump 分配泵；配料泵
dispensing report 加料报告
dispensing section 分配段；分配班
dispensing service 加料业务
dispensing station 加料区；配料区
dispensing tank farm 加料罐区
dispensing test 分配试验；配料试验
dispensing unit 加油机组
dispergator 胶溶剂
dispersable vat dyes 分散型还原染料
dispersal 分散的；疏散
dispersal agent 扩散剂
dispersal area 疏散用地
dispersal colo(u)r 浆状色料；色料浆；散胶色料
dispersal hut 工作队临时棚屋
dispersal of pollutant 污染物扩散
dispersal of stocks 分散储料
dispersal on contour lines 在等高线上分散
dispersal oscillator 扩散振荡器
dispersal parking area (机场的)疏散区
dispersal pattern 分散模型；分散类型
dispersal potential 散布潜力
dispersal route 扩散途径；扩散路线；扩散路径
dispersal stores 分散储料
dispersal vessel 疏浚船
dispersancy 分散力
dispersant 料浆稀释剂；扩散剂；研磨助剂；砂浆稀释剂；分散剂
dispersant additive 分散剂
dispersant agent 分散剂
dispersant bottle 分散剂瓶
dispersant-to-oil ratio 分散剂与油比
dispersate 分散质
dispersator 分散搅拌机
disperse 弥散的；分散(的)；切碎；喷粉
disperse aerosol 分散型气溶胶
disperse coefficient value 放射系数值
dispersed 漫布
dispersed acetate dyes 分散性醋酸纤维染料
dispersed aeration 分散式曝气
dispersed aeration process 分散(式)曝气法
dispersed aerodrome 设施疏开布置的机场
dispersed air conditioning system 分散空调系统
dispersed air flo(a)tation 分散气浮法；曝气浮选(法)；分气浮选法
dispersed charge 分散药包
dispersed chromatin 分散分布染色质
dispersed component 分散组份
dispersed development 分散开发
dispersed dye stuff 分散染料
dispersed element 分散元素
dispersed fault 分散断层
dispersed filling and emptying system 分散输水系统
dispersed fresh water mud 细分散淡水泥浆
dispersed gas injection 分散注气
dispersed habitat 分散(向郊区)居民点
dispersed hard standing 分散的停机坪
dispersed heat sink 分散热穴
dispersed intelligence 分散智能；分布信息
dispersed layout 分散布局
dispersed medium 弥散剂；分散介质；分散媒

dispersed network phase 弥散相
dispersed organic matter 分散有机质
dispersed part 分散相；弥散质；分散质；分散部分
dispersed particle 分散粒子
dispersed phase 分散(内)相；不连续相
dispersed phase hardening 弥散硬化
dispersed plug flow 分散活塞流
dispersed residential pattern 分散居住式
dispersed resin 分散树脂
dispersed runway 疏开布置的跑道
dispersed settlement 自然村；稀疏集居；分散型聚落；散列式居民点
dispersed shrinkage 分散缩孔
dispersed soil 分散性土(壤)
dispersed structure 分散结构
dispersed substance 分散物质
dispersed sulfur dye 分散硫化染料
dispersed supply of heating 分散供热
dispersed system 分散系(统)
dispersed town 散置城镇；布局分散的城镇
dispersed unconformy 分散不整合
dispersed village 自然村
dispersed wave 分散波
disperse dyeing 分散染料染色
disperse dyes 分散染料
disperse element 稀散元素
disperse grass green 分散草绿
disperse medium 悬浮介质；分散介质
disperse metallic dyes 分散金属络合染料
disperse metals commodity 稀散金属矿产
disperse mill 分散研磨机；乳化机
disperse nodule 散状结核
disperse of elements 元素分散作用
disperse part 分散内相
disperse phase 弥分散相
disperse policy decision 分散决策
disperser 弥散器；蒸馏塔中的泡罩；粉碎器；分散器；喷粉器；调节浮选剂；色散器
disperse solid 悬浮体
disperse spectrometry 色散分光法
disperse state 扩散状态；弥散状态；悬浮状态；分散状态
disperse system 悬浮系；分散体系
disperse yellow 分散黄
dispersibility 分散性；分散能力
dispersible 可分散的
dispersimeter 弥散计；色散计
dispersing additive 分散添加剂
dispersing agent 乳化剂；(液体的)颗粒悬浮剂；扩散剂；悬浮剂；分散剂
dispersing aid 分散助剂
dispersing auxiliary 分散助剂
dispersing bulk plant 配料区
dispersing characteristic 分散特性
dispersing component 色散元件
dispersing depot 配料灌区
dispersing device 撒播装置
dispersing element 色散组元
dispersing equipment 分散设备
dispersing flight 扬料板
dispersing flux 磁力线分散
dispersing goosery 分散养鹅场
dispersing hen coop 分散鸡笼
dispersing honeycomb 分散蜂房
dispersing instrument 色散仪
dispersing lens 色散透镜
dispersing liquid 调剂液
dispersing machine 金属箔抛散装置
dispersing medium 分散介质；分散剂
dispersing mixer 分散混合器
dispersing plant 分散装置
dispersing power 分散能力
dispersing prism 色散棱镜
dispersing pump 配料泵
dispersing reflector 色散反射器
dispersing report 配料报告
dispersing resonator 色散共振腔
dispersing service 配料业务
dispersing spectromodulator 色散光谱调制器
dispersing tankage 配料(灌)容量
dispersing tank farm 配料灌区
dispersing-type cationic polyacrylamide 分散性阳离子聚丙烯酰胺
dispersing-type spectroradiometer 色散型分光辐射计
dispersing wooden beehive 分散蜂箱
dispersion 离散(度)；离差(度)；扩散；弥散；分散作用；分散状；分散(液)体；分散；频散；漂移；色散；散射；撒落作用
dispersion action 分散作用
dispersion adhesive 分散型黏合剂
dispersion agent 分散剂
dispersion analysis 弥散分析
dispersion and deposition of toxics 毒物的色散与沉淀
dispersion and scale inhibition 分散阻垢
dispersion angle 扩散角；色散角
dispersion angle for load distribution 荷载分布扩散角
dispersion apron 分散裙
dispersion attenuation 色散衰减
dispersion attenuation factor 色散衰减因数
dispersion-based plastic adhesive 分散性塑料黏合剂
dispersion capacity 分散能力
dispersion coating 分散型涂料
dispersion coefficient 离散系数；扩散系数；弥散系数；分散系数；色散系数
dispersion coefficient of convection 对流弥散系数
dispersion coefficient of hydrodynamics 水动力弥散系数
dispersion coefficient of reinforcement 配筋分散性系数
dispersion colloid 分散胶体
dispersion cone 分散锥；撒料锥体
dispersion constant 色散常数
dispersion contactor 分散接触器
dispersion cup 分散容器(土壤试验)
dispersion current 耗散电流
dispersion curve 折射率分布曲线；频散曲线；色散曲线
dispersion degree 离散程度；分散度；色散度
dispersion drier 撒料式烘干机
dispersion due to flow profile 流型扩散
dispersion dye 分散染料
dispersion effect 弥散效应；分散效应；频散效应；色散效应
dispersion element 色散元件
dispersion equation 扩散方程；频散方程；色散公式；色散方程
dispersion error 离散误差；散布误差
dispersion factor 弥散因数；车队离散系数；散射因子
dispersion fan 分散扇
dispersion field 分散场
dispersion flow 分散流
dispersion fluidized bed 散式流化床
dispersion flux 弥散通量
dispersion force 弥散力；分散力；色散力
dispersion formula 色散公式
dispersion frequency 色散频率
dispersion fuel 分散染料
dispersion function 分散函数
dispersion gradient 色散梯度
dispersion halo 分散晕
dispersion-hardened alloy 弥散硬化合金
dispersion-hardened material 弥散硬化材料；弥散化合金
dispersion hardening 弥散硬化；分散硬化
dispersion hardness 弥散硬度
dispersion index 离散指数；分散指数
dispersion in quality 质量不齐
dispersion in strength 强度的分散
dispersion in the air 在空气中扩散
dispersion ladder 散布梯尺
dispersion law 离散定律；色散律
dispersion-limited 色散限制
dispersion-limited operation 色散限制作用
dispersion line 色散谱线
dispersion machine 分散机
dispersion measure 色散量
dispersion measurement 分散测定；色散测定
dispersion medium 弥散介质；弥散介质；悬浮介质；分散介质；分散剂；频散介质；色散媒质；色散介质
dispersion method 弥散胶体法；分散法；标准离差法；色散法
dispersion mill 分散(研)磨机

dispersion model 弥散模型
dispersion model of groundwater age 地下水年龄的弥散模型
dispersion mushroom 分散蘑菇
dispersion network 弥散网络
dispersion of a random variable 随机变量的离差
dispersion of atmosphere 大气色散
dispersion of binomial distribution 二项分布的离势
dispersion of clay 黏土悬浮液;黏土分散剂
dispersion of colo(u)rs 色的分解
dispersion of difference scheme 差分格式频散
dispersion of distribution 分布宽度;分布的离中趋势;分布的离散度
dispersion of explosives 炸药的分布
dispersion of light 光的色散;分散的光
dispersion of numbers 数值离差;数值的离散;数值差量
dispersion of oil on sea 海水油弥散
dispersion of optic(al) axis 光轴色散
dispersion of seismic wave 地震波频散
dispersion of unemployment 失业率的差别
dispersion of velocity 速度弥散(度)
dispersion of wave 波的弥散;波的扩散;波的分散
dispersion of waveguide fiber 波导纤维色散
dispersion of waves 波弥散
dispersion paint 分散型涂料
dispersion parameter 扩散参数;分散度参数
dispersion part 分散部分
dispersion pattern 弥散形式;分散形式;分散模式;散布图
dispersion phase 分散外相
dispersion phenomenon 弥散现象
dispersion photometer 色散光度计
dispersion plate 空气分布板;分散板;撒料盘;撒料板
dispersion point 弥散点;色散点
dispersion polymerization 分散系聚合作用;分散聚合(作用)
dispersion prism 色散棱镜
dispersion process 分散过程
dispersion property 分散性能
dispersion radius 扩散半径
dispersion range 分散范围
dispersion rate 分散率
dispersion ratio 离散率;分散(比)率
dispersion relation 色散关系
dispersion resin 离散树脂;分散树脂;色散树脂
dispersion separator 撒料式选粉机
dispersion shift single mode fiber 色散位移单模光纤
dispersion sling 抛掷分散器
dispersion soil 分散土
dispersion spectrum 色散光谱
dispersion stabilizer 阻凝剂;分散稳定剂;防絮凝剂
dispersion state 悬浮状态;分散状态
dispersion-strengthened composite 弥散增强复合材料
dispersion-strengthened lead 弥散强化铅
dispersion-strengthened material 弥散强化材料
dispersion-strengthened metal 弥散强化的金属
dispersion-strengthening 弥散强化;弥散硬化
dispersion structure 分散结构
dispersion surface 色散曲面
dispersion system 离散系统;悬浮系;分散体系
dispersion system disperse system 分散物系
dispersion technology 分散体工艺学
dispersion tendency 离心趋离
dispersion tensor 弥散张量
dispersion test of water quality 水质弥散试验
dispersion texture 散体结构
dispersion theory 扩散理论;弥散理论;色散理论
dispersion theory of smoke 排烟扩散理论
dispersion toughencng 弥散韧化
dispersion train 分散列;分散带
dispersion train anomaly 分散流异常
dispersion type 弥散型
dispersion-type fuel element 分散型燃料组元
dispersion-unshifted single mode fiber 非色散位移单模光纤
dispersion variance 离散方差
dispersion wave 频散波
dispersion wavelength 色散波长

dispersion zone 分散带;频散带;散布区域
dispersity 弥散度;分散性;分散度;色散度
dispersive 分散的
dispersive ability 分散能力
dispersive action 分散作用
dispersive angle of tracer 示踪剂的弥散角
dispersive capacity 扩散能力;分散本领;色散本领
dispersive clay 浸水崩解黏土;分散性黏土;崩解性黏土
dispersive curve 频散曲线
dispersive delay line 色散延迟线
dispersive distribution 离散状分布
dispersive effect 扩散作用
dispersive element 分散性单元
dispersive filter 波散滤波器
dispersive function 频散函数
dispersive medium 弥散介质;悬浮介质;频散介质
dispersive modulator 色散调制器
dispersive movement of contaminant 污染物扩散运动
dispersive optical system 色散光学系统
dispersive pollutant transport 污染物扩散输移
dispersive power 分散本领;色散能力;色散率;色散本领
dispersive pressure 分散压力
dispersive process 分散法
dispersive Rayleih wave 频散瑞利波
dispersive replication 分散复制
dispersive soil 分散性土(壤)
dispersive spectrometer 色散分光计
dispersive spectrometry 色散光谱;色散测谱学
dispersive stress 分散应力
dispersive target 离散目标
dispersive ultrasonic delay line 色散超声延迟线
dispersive wave 弥散波;分散波
dispersivity 弥散性;弥散系数;分散率差;色散性
dispersivity quotient 微分色散
dispersoid 离散胶体;弥散质点;弥散体;分散胶体
dispersoid distribution 弥散质点分布
dispersoid particle 弥散质点
dispersoid particle size 弥散质点粒度
dispersoid size 弥散质点粒度
Dispersol dye 狄司潘素染料
dispersor 分散器;色散器
disperstiveness 分散性
disphenoid 复正方楔;双半面晶形
disphotic zone 弱光区;弱光带;弱光层
displace 换置
displaceable 可替换的
displaced anomaly 位移异常
displaced exponential distribution 移位指数分布
displaced farmer 农业移民
displaced foundation 换土地基;置换基础
displaced halo 位移晕
displaced ore body 移位矿体
displaced persons 难民
displaced plant 废弃工厂
displaced sediment 迁移沉积层
displaced soil 换土;被排挤的土
displaced stitch 波纹组织
displaced structure 拆迁建筑(物);搬迁房屋
displaced surface 位移面;推力面
displaced threshold 转移的跑道入口
displaced volume 排不容量
displaced water 排出水
displaced weight 排水重量
displacement 排水量;排代作用;排(出)量;易位;移置作用;移置;移位;移动;转移;置换作用;断错,错位;汽缸工作容积;位移;替位
displacement 2000 tons 排水两千吨
displacement 50000 tons 排水五万吨
displacement activity 替位活动;替换活动
displacement amplitude 位移振幅;位移幅值
displacement and scale effects 财政收入的移置和规模扩大效应
displacement angle 位移角;失配角
displacement angle at cylinder 缠绕筒体进角
displacement angle of dome 缠绕封头位角
displacement auger 置换螺(旋)钻
displacement blower 容积式增压器;容积式鼓风机
displacement boat 滑行艇;非滑行快艇;排水型艇;排水型船
displacement bound 极限位移;位移范围
displacement calculation 位移计算

displacement capacity 输送能力
displacement casting 排溢铸造
displacement chamber 汽缸工作容积
displacement chromatography 置换色层法;顶替色谱法;排代色谱法
displacement cliff 断崖
displacement coefficient 形状系数;排水量系数;位移系数
displacement compatibility 位移相容性;位移协调条件
displacement component 位移分量
displacement compressor 容积式压缩机;容积式压气机;排量式压气机
displacement cone 分料锥
displacement configuration 排油结构
displacement conservation 位移守恒
displacement contact 移位接触器;移位触发器
displacement continuity 位移连续性
displacement control 位移控制
displacement criterion 位移准则
displacement crosslinking 置换交联
displacement current 位移电流
displacement curve 沉降曲线;移动曲线;排水量曲线;位移曲线
displacement development 置换显影;顶替(展开)法;取代展开法;取代展开法
displacement diagram 变位曲线;位移图;变位图
displacement diaphragm 排量膜板
displacement direction of hanging wall 上盘位移方向
displacement discharge rate 排出量
displacement ductility 位移延性
displacement ductility factor 位移韧性系数
displacement effect 移位效应
displacement efficiency 置换效率
displacement element 位移元
displacement energy 位移能(量);置换能
displacement error 偏移误差;位移误差
displacement fault 平移断层;位移断层
displacement feeder 排代投配机
displacement field 变位场;位移场
displacement flow meter 容积式流量计
displacement fluid 顶替液
displacement flux 位移通量
displacement frequency spectrum 位移频率谱
displacement fully laden 满载排水量
displacement function 位移函数
displacement ga(u)ge 变位仪
displacement grouting 排水灌浆;替代灌浆
displacement gyroscope 位移陀螺仪
displacement head 汽缸工作室盖
displacement hull 排水体
displacement impedance 动刚度;位移阻抗
displacement indicator 位移指示器
displacement in earth mass 土体位移
displacement in position 位移
displacement inversion 位移转变
displacement knob 观测手轮;位移旋钮
displacement law 置位定律;放射性位移定律
displacement length coefficient 排水量船长系数
displacement length ratio 排水量船长比
displacement light 空载排水量
displacement limit 位移极限
displacement line 位移线
displacement load 位移荷载
displacement manometer 位移压差计
displacement measurement 位移量测;位移测量;位移测定
displacement-measurement procedure 位移测量法
displacement meter 变位(流量)计;容积(式)水表;活塞水表;浮子式液体比重计;排代式水表;位移式水表变位计;位移计
displacement method 挤淤法;变形法;置换法;变位法;排水法(浮游生物定量方法);位移法(分析超静定结构的基本方法之一)
displacement mobility 动柔度;位移导纳
displacement noise 排放噪声;位移噪声
displacement observation 位移观测
displacement of abutment 桥台位移;支座位移
displacement of aids 航标移位
displacement of compressor 压缩机活塞的行程容积
displacement of coordinate axis 移坐标轴;坐标

轴位移
displacement of discontinuity method 不连续位移法
displacement of dredge pump 泥泵排量
displacement of earthquake fault 地震断裂位移
displacement of families 家庭迁移;搬家
displacement of fault 断距
displacement of images 像点位移
displacement of inner contour of tunnel 隧道内轮廓位移
displacement of joint 结点位移;节点移动;节点位移
displacement of labo(u)r 替代劳工
displacement of observation points 观测点位移
displacement of pipe 管道位移
displacement of piston 活塞的行程容积
displacement of pulse porches 脉冲沿边沿差
displacement of pump 泵的排水量
displacement of rail joint 钢轨接头位移
displacement of rock 石块崩落
displacement of rock masses 岩体移动
displacement of roof rock 盖层排驱压力
displacement of ship 船舶排水量
displacement of soil 换土
displacement of the crank 更换曲柄
displacement of vessel 船舶排水量
displacement of water 排水量
displacement oil pump 旋转式油泵;排量油泵
displacement one hundred and fifty thousand tons 排水十五万吨
displacement one hundred thousand tons 排水十万吨
displacement operator 位移算子
displacement parameter 位移参数
displacement path method 位移轨迹法
displacement period 排代周期
displacement phase transition 位移相变
displacement pickup 位移传感器
displacement pile 挤压桩;压入桩;打入桩;排土桩;送入桩
displacement plan 输送计划
displacement plating 置换电镀;排代电镀
displacement point 位移点
displacement potential 位移势
displacement precipitation 置换沉淀
displacement pressure of roof rock 盖层排驱
displacement pulse 位移脉冲
displacement pump 容积泵;活塞泵;排量泵;排代泵;往复式水泵
displacement range 位移范围
displacement reaction 置换反映;排代反应
displacement resistance 位移阻力
displacement response coefficient 位移反应系数
displacement response factor 位移反应因数
displacement response spectrum 位移反应谱
displacement scale 载重标尺;排水量标尺
displacement seismograph 位移地震仪;位移测震仪
displacement seismometer 位移地震计
displacement sensitivity 位移灵敏度
displacement sensor 位移传感器
displacement series 置换序列;置换次序
displacement shape 位移形状
displacement sheet 排水量计算表
displacement shift 变位
displacement ship 排水型艇;排水型船
displacement slow rinse 再生液置换
displacement slurry 换浆
displacement spectral density 位移谱密度
displacement spectrum 位移谱
displacement stress 位移应力
displacement tensor 位移张量
displacement test 剥离检验(沥青膜);位移检验
displacement theory 大陆漂移(学)说
displacement thickness 取代厚度;排量厚度
displacement threshold 变位阈
displacement time 排水时间;排代时间
displacement-time curve 位移—时间关系曲线
displacement-time diagram 位移—时间关系图
displacement-time relationship 位移—时间关系
displacement titration 置换滴定;排代滴定法
displacement ton 排水吨
displacement tonnage 排水量;排水吨位;排水吨数
displacement transducer 变位传感器;位移传感器
displacement transmission coefficient 位移传递系数
displacement transmissivity 输送能力
displacement type ferro-electrics 位移型铁电体
displacement type liquid meter 根据排代原理运转的流量计
displacement type meter 容积式防护水表
displacement type seismometer 位移型地震仪;位移式地震检波器
displacement value 置换价;位移量
displacement vector 位移向量;位移矢量
displacement velocity 排代速度
displacement volume 置换体积;工作容量;被置换的体积;汽缸排量;排代容积
displacement wake 排水伴流
displacement water line 排水状态水线
displacement water meter 排代式水表
displacement wave 位移波
displacement well 排水砂井
displace meter 变位流量计
displace mud 压出泥浆;置换泥浆
displacer 排除物;定距垫块;代用品;蛮石;埋石;置换器;置换剂;平衡浮子;排代剂;排出物
displacer cone 出料锥体;出料锥斗
displacer level meter 沉筒液面计
displacer-type flowmeter 浮子式流量计
displacer-type meter 浮子式计
displace simulation 显示模拟
displacing fluid 置换液
displacive precipitation 推进沉淀作用
displacive transformation 位移型转变
display 显像;显示;展览;陈设;陈列;区头向量
displayable index 可显示索引
display access 显示访问;显示存取
display actuator 显示激励器
display adapter 显示适配器;显示适配部件;显示转接器
display adapting unit 显示适配部件
display advertisement 醒目广告
display alarm 显示报警器
display analysis console 显示分析控制台
display and control design 显示与控制系统的设计
display and control unit 显示与控制装置
display and edit station 编辑显示台
display animation 动画显示
display area 陈列区;展览区
display behavio(u)r 显示行为
display booth 显示小室;展出间隔
display buffer 显示缓冲区;显示缓冲器
display buffer storage 显示缓冲区存储器
display building 陈列馆
display cabinet 橱窗;陈列柜
display case 样品柜;(展览商品的)橱窗;陈列柜
display center 显示中心
display change 转换显示
display character 显示字符
display characters per frame 每帧显示字符数
display column 显示柱;显示列
display command 显示指令
display conservatory area 观赏温室区
display console 显示控制台;操作控制台
display control 显示控制
display controller 显示控制器
display copier 显示拷贝器;显示复印机
display cycle 显示周期
display data analysis 显示数据分析
display data module 显示数据模块
display device 显示装置;显示设备
display disc 指示度盘
display drawing 展览图;展示图
display editing system 画面编辑系统
display equipment 显示设备
displayer disc 指示度盘
display feature 显示特性
display field 显示字段;显示域;显示场;数据显示区
display file 显示文件
display format 显示形式
display frame 显示域;显示帧
display gallery 展览廊
display gamma 重显灰度
display glass 显示屏
display greenhouse area 观察温室区
display green time 显示绿灯时间
display group 显示组
display image 显示图像
display in digital form 数字显示
display information processor 显示信息处理机
displaying instrument 指示仪器
displaying items 展览品
display instrument 显示仪表
display item 显示量
display key 显示电键
display lamp 显示灯;指示灯
display lighting 陈列品照明
display line 显示行;显示线
display lock 显像固定
display logic unit 显示逻辑装置
display loss 可见度因数
display mark 显示记号
display memory 显示器储存容量
display menu 显示菜单;显示清单
display mode 显示模式;显示方式
display model 显示模型
display module 显示组件
display of combination of intersections 相交剖面的综合显示
display of seismic result 地震成果显示
display of slope distance 斜距显示
display of three-dimensional data 三维资料显示
display packing 显示方法
display panel 显示装置;显示屏;显示盘;显示板
display pedestal 显示器基座
display place 显示区
display plotter 显示绘图仪
display position 显示位置
display primary 显像三基色
display processing unit 显示处理器
display processor 显示处理机
display psychology 显示心理学
display rack 货架;陈列架
display range 显示范围
display refresher rate 显示更新速度
display region 显示区域
display register 显示寄存器
display roof 显影屏打样法
display room 商品陈列室
display routine 显示程序
display screen 显示屏
display sensibility 显示灵敏度
display sensitivity 指示灵敏度
display shelf 展览架
display size 显示尺寸
display software 显示软件
display space 显示区;显示空间
display stand 陈列架;陈列柜
display station 显示站
display storage tube 显示存储管
display surface 陈列面
display switch 显示开关
display system 显示系统
display technique 显示技术
display technology 显示技术
display terminal 终端显示装置
display tube 显像管;显示管;阴极射线管
display unit 显示装置;显示器;显示单元;显示部件
display window 显示窗口;展示窗玻璃;展览橱窗;橱窗(玻璃);陈列橱窗;商店橱窗
display windows with I louvered panel 橱窗带一扇百叶窗
display work 标题字排版
displuviatum (屋顶从房顶采光井向外倾斜的)中庭
disposable 可置换件
disposable articles and containers 可处理的物品和容器
disposable assessed income 可支配的税后收入
disposable bag 一次性(滤)袋
disposable buoyancy 可用有效浮力
disposable business income 企业可用收入
disposable capital 游资移动
disposable cartridge 限使用一次的小气瓶;一次性滤芯
disposable filter media 一次性滤材
disposable fire extinguisher 一次性灭火器
disposable funds 使用资金
disposable goods 可处理物品
disposable house 听任使用的房屋;可毁房屋
disposable income 税后可支配收入;可支配收入;可支配的个人收入;纳税后个人可用收入;税后

收入
disposable load 可卸荷载;自由载量
disposable packaging 一次性包装;不回收的包装
disposable pallet 非回收的纸盘;拆解式托盘
disposable personal income 可支配的个人收入;纳税后个人可用收入;个人可支配的收入
disposable plastic item 一次性塑料制品
disposable price 可支配价格
disposable profit 可支配利润
disposable real income 可支配实际收入
disposable syringe 限使用一次注射器
disposable type 任意型号
disposable value aggregate 可支配价值总额
disposable wave buoy 移动式波浪浮标
disposable weight 活动重量
disposal 处置;处理;配置
disposal and reclamation of solid wastes 固体废物处置和利用
disposal area 处理场地;弃土区;抛泥区;排泥区;挖泥处置区;疏浚泥砂倾倒区
disposal bag 垃圾袋
disposal by dilution 稀释处置
disposal capability 处理容量;废物处理能力
disposal capacity 处理容量;处理能力
disposal chute 倒垃圾的斜槽;抛泥槽
disposal compliance 遵照处置
disposal container 废物箱
disposal contractor 废物处理承保人
disposal cost 清理成本
disposal ditch 导水沟
disposal drain 配置排水管
disposal dump 垃圾倾倒处
disposal element 处理部件
disposal facility 处置设施
disposal field 处理场;废品处理场;水流吸收系统
disposal ground 抛泥区
disposal in the form of scrap 废料处理
disposal lift 有效升力
disposal load 活动荷载
disposal of activated sludge 活性污泥处置
disposal of donated shares 捐赠股处理
disposal of dredged material 疏浚土处理
disposal of excess water 积水排除;排除积涝
disposal of forfeits 没收物的处置
disposal of gas container 气体容器的处置
disposal of hazardous waste 危险废物处理
disposal of industrial solid waste 工业固体废物处置
disposal of industrial wastes 工业废物处置;处理工业废物
disposal of passengers caught injury by accident 旅客伤害事故处理
disposal of passengers suddenly caught sick or died by accident 旅客发生疾病或死亡处理
disposal of plant and equipment 厂房和设备的处理
disposal of radioactive wastes 放射性废物处置
disposal of refuse 垃圾处置;垃圾处理
disposal of refuse in land-fills 垃圾填埋处理
disposal of rubbish 垃圾处理
disposal of screenings 筛渣处置
disposal of sewage 污水处理;污水处置
disposal of sludge 污泥处置
disposal of sludge at riverbed 河床污泥处理
disposal of solid wastes 固体废物处理
disposal of spoil 出渣;弃土处理;排泥
disposal of spoil with belt conveyer 皮带机排泥
disposal of treasury shares cost basis 按成本计算库存股票的售价
disposal of wastes 粪便处置;废物处置;废物处理
disposal of wastewater 废水处置
disposal of water 水的处理
disposal on land 陆上废物处理
disposal point 卸载点;清理场地
disposal price 处理价格
disposal region 排泥区
disposal right of assets 资产处置权
disposal shaft 垃圾竖井
disposal site 废料堆放处;垃圾倾倒区;垃圾倾倒场;垃圾堆置场;垃圾放置场;埋藏场;废物处理场;弃土区;抛泥区;排泥区
disposal system 处置系统;处理系统
disposal tip 处理置场
disposal to land 埋入地下

disposal trench 处理污水用地沟;处置地沟;灌溉水沟;灌溉渠(道)
disposal unit 泔水处理器
disposal value of assets to be retired by project 旧产变卖价值
disposal well 污水渗井;处置井;处理井;渗井;排污水井
dispose 解决;配备;排列
dispose-all 垃圾斜道;垃圾箱;垃圾桶;垃圾竖井
dispose of remaining assets 分配剩余资产
dispose of without a license 非法销售
disposer 垃圾桶;处置器
disposing capacity of juristic person 法人的行为能力
disposing of surplus funds 多余资金的处理
disposition 交叉(线路);财产付与;部署;布置;布局;安排;倾向;配置;素质
disposition about world 关于世界的布局
disposition board 仪表板;配电盘
disposition center 调度中心
disposition code 处置码
disposition design 布置设计
disposition of equipment 设备利用;设备使用;设备处理(单)
disposition of goods 处理货物
disposition of installment obligations 分期付款债务的处理
disposition of lot 成批处理
disposition of net earnings 净收益的处理;净收入的处理
disposition of net income 净收益分配;净收入分配;净收入的处理
disposition of net profit 净利润的处理
disposition of non-conformity 不合格处置
disposition of property 财产处置
disposition of real estate statement 房地产处置声明
disposition of resources 变卖资源
disposition of standard cost variances 标准成本差异处理
disposition plan 布置平面;配置(平面)图;排列平面图
disposition plan of equipment 设备配置平面图
disposition price 城市更新用地售价
disposition processing 处理支配权;部署处理
disposition station 调度站
dispositive plate 透明板
dispossess 迁出;霸占
dispossession 征用;强占;没收
dispossess proceedings 剥夺起诉;搬迁诉讼
disproof 反证
disproportion 不相当;不相称;不均衡;不合比例;不成比例
disproportionality 比例失调
disproportionate 不相称的;比例失调的;歧化(作用)
disproportionate class numbers 不成比例组含量
disproportionately graded 级配不良的
disproportionate sampling 非比例抽样
disproportionate sampling with paired selections 配对选取非比例抽样
disproportionate stratified sampling 不等比例分层取样
disproportionate subclass numbers 不等比例子类数
disproportionate withdrawal 不均衡开采
disproportionation 不相称;歧化(作用)
disproportionation reaction 歧化反应
disputable 可争论的
dispute 纠纷;争议;争端
dispute among the people 民事纠纷;民间纠纷
dispute at law 法律争执
dispute concerning private rights 民事权利(的)争执
disputed area 争议区
disputed boundary 有争议边界
disputed invoices 有争议发票
disputed undefined boundary 未定界限;未定界线
dispute over water affairs 水事纠纷
disputing parties 争议当事人
disqualification 不合格;取消资格
disqualified goods 不合格物品;不合格的货物
disqualified product 不合格产品
disqualified upon inspection 检验不合格
disqualify 不合格

disquality 使不合格
disquisition 学术论文;专题论文
disregard 漠视;不理
disregistry 错合度
disrepair 破损;失修
disresonance 非谐振
disrupt 妨碍;使分裂
disrupt economic order 扰乱经济秩序
disrupted anomaly 脱节异常
disrupted bed 破坏层
disrupted drainage system 不规则水系
disrupted halo 脱节晕
disrupted horizon 变位层
disrupted terrane 分裂地体
disruption 击穿;分裂;破裂(作用);碎裂
disruption in the balance 破坏平衡
disruption of progress 进度打乱
disruption of wound 切口裂开
disruption potential 击穿电势
disruption test 断裂试验
disruption time 瓦解时间
disruptive 分裂的;爆裂的;破裂的;破坏性的
disruptive conduction 击穿导电;破裂电导
disruptive currency flow 破坏性货币流动
disruptive discharge 击穿放电;火花放电;破裂放电;破坏性放电
disruptive distance 击穿距离
disruptive effects 断裂效应;破坏性作用
disruptive expansion 分裂膨胀
disruptive explosive 烈性炸药;爆炸性炸药
disruptive field intensity 击穿电场强度
disruptive flow of capital 破坏性资本流动
disruptive gradient 击穿梯度
disruptiveness 破裂性
disruptive potential gradient 击穿电势陡度
disruptive pressure 击穿气压
disruptive spark 击穿火花
disruptive strength 击穿强度;电击穿强度;破坏强度
disruptive test 击穿试验;耐(电)压试验
disruptive voltage 击穿电压
disrupt the market 扰乱市场
dissatisfier 不满足因素
dissave 动用储蓄金
dissaving 超支;负储蓄;提取储蓄存款
dissect 剖析
dissected basin 切割盆地
dissected low plain 切割低平原
dissected plain 切割平原
dissected plane 切割面
dissected plateau 切割高原
dissected topography 切割地形
dissectible 挤切的
dissectible product 可分割的产品
dissecting 套色拼隔版
dissecting force 挤切力
dissecting hook 剥离钩
dissecting microscope 解剖显微镜
dissecting needle 分析针
dissecting room 解剖室
dissecting saw 割解锯
dissecting table 解剖台
dissecting valley 切割谷;切割盆地
dissection 分析作用;切割作用
dissection of heterogeneous distribution 非同质分布分解
dissector 析像管;分析器
dissector multiplier 析像管倍增器
dissector tube 图像分解器
disseisin 非法侵占不动产等
Dissel mean effective pressure 狄塞尔平均有效压力
dissembling inspection 拆卸检查
disseminate 分散卸货
disseminated 弥散性;侵染的【地】
disseminated Au ore 浸染状金矿石
disseminated deposit 分散油藏
disseminated ore 浸染状矿石
disseminated Pb-Zn ore 浸染状铅锌矿石
disseminated structure 浸染状构造
disseminated values 散状矿物
dissemination 传播(作用);分散度;发播;播散;侵染作用;散布(作用)
dissemination in the air 在空气中散播
dissemination of information 情报传播

dissemination of new technology 新技术推广
dissemination of radioactive effluent 放射性废水散布
dissemination of radioactive wastewater 放射性废水的浸染
dissemination of statistic(al) information 统计资料的散发
dissemination structure 侵染状构造
dissent 不赞成
dissenting opinion 反对意见
dissepiment 鳞片；鳞板；磷片；隔壁；水平隔
dissepimentarium 鳞板带
dissertation 论文；学术演讲；学术论文；报告
dissetible pin 挤切销
dissimilarity 异点；不相似(性)
dissimilar processor 异种处理机
dissimilar terms 不同类项
dissimilar welding rod 不同成分焊条；与被焊金属成分不同的焊条
dissimilar weld metal 不同金属的焊接；不同的焊接金属
dissimilation 异化作用
dissipated energy 消散能量
dissipated heat 散失热
dissipater 喷雾器
dissipating 消散
dissipating heat 散热
dissipation 耗散；消散；逸散；散逸(作用)
dissipation cost 浪费资金
dissipation effect 耗散作用
dissipation energy 离解能量；消能
dissipation factor 耗散因数；耗散系数；消散系数；损耗因子；损耗因素；损耗因数
dissipation function 黏性耗散函数；耗散函数；散逸函数
dissipation interaction 耗散相互作用
dissipationless line 无耗散线
dissipation line 耗散线
dissipation of energy 能量散逸；能量耗散；能力消散；消能
dissipation of fog 消雾；雾的消散
dissipation of heat 热(的)消散
dissipation of heat of hydration 水化热消散；水化热散逸
dissipation of kinetic energy 消能；动能消散
dissipation of noise 噪声消散
dissipation of pore water 孔隙水消散
dissipation of pore water pressure 孔隙(水)压力消散
dissipation of surface flow 面流消能
dissipation of wave energy by plantation 植物消波
dissipation percentage pore pressure 孔隙压力消散百分数
dissipation power 耗损功率
dissipation resistance 耗散阻力
dissipation resistivity 体积电阻率
dissipation test 耗散试验；消散试验；消能试验；散热试验
dissipation trail 耗散尾迹；消散尾迹
dissipative alternation 损耗衰减
dissipative attenuation 损耗衰减；散射性衰减
dissipative attenuator 耗散型消声器
dissipative cell 耗能元件
dissipative curve of pore pressure 孔压消散曲线
dissipative effect 消散作用
dissipative element 耗能元件
dissipative factor 耗散因数；耗散系数
dissipative force 耗散力
dissipative function 耗散函数
dissipative muffler 消声器
dissipative net 耗散网络
dissipative network 耗散网络
dissipative resistance 耗散阻力
dissipative sort(e)y 散逸层
dissipative structure 耗散结构
dissipative system 耗能系统
dissipative tunnel(l)ing 耗能隧道效应
dissipative work 消耗功
dissipator 配电盘；耗散器；消散器；消能建筑物；吸热部件
dissmilar skeins 不同丝绞
dissmilatory process 异化过程
dissociable 可分离

dissociate diet 分类进食
dissociated operation 完全分离操作
dissociating gas 离解气体
dissociation 离解作用；解离；分离(作用)；分解作用；分化变异；不相联；变异
dissociation cell 裂解炉
dissociation chemical interference 离解化学干扰
dissociation constant 离解常数；解离常数；电离常数；分离常数
dissociation curve 解离曲线
dissociation degree 离解(程)度
dissociation energy 离解能；分解能
dissociation energy of bond 键控解能
dissociation equilibrium 离解平衡
dissociation field effect 离解场强效应
dissociation index 解离指数
dissociation interference 分离性干扰
dissociation in water 水中离解
dissociation of molecules 分子离解
dissociation of water 水的离解
dissociation potential 离解电势
dissociation pressure 离解压力
dissociation process 离解过程
dissociation property 离解性质
dissociation reaction 离解反应
dissociation recombination 离解复合
dissociation temperature 离解温度
dissociation time 瓦解时间
dissociation yield 离解率；解离率
dissociative adsorption 离解吸附
dissociative capture 离解俘获
dissociative chemisorption 离解化学吸附
dissociative diffusion 离解扩散
dissociative disorder 分离性障碍
dissociative excitation 离解激发
dissociative mechanism 分解机理
dissociative reaction 分裂反应
dissociative recombination 离解复合
dissociator 离解子；分离器
dissogeny 两次性成熟
dissolubility 可溶(解)性；溶解性；溶(解)度
dissoluble 可溶解的；可分解的
dissoluble rock type 可溶岩类型
dissoluble well cementing 可溶性地层注水泥
dissolution 解除；溶解(作用)；溶化
dissolution amount of calcite 方解石溶解量
dissolution basin 溶蚀盆地
dissolution boiler 溶化锅
dissolution cavity 熔洞
dissolution cell 溶解电解槽
dissolution fissure 溶蚀裂隙；溶蚀裂缝
dissolution gallery 溶蚀通道
dissolution kinetics test method 溶解动力学试验方法
dissolution kinetics test velocity equation 溶解动力学试验速度方程
dissolution method 溶解法
dissolution of company 公司解散
dissolution of contract 解除合同
dissolution of partnership 合伙企业解散
dissolution of responsibility 免除责任
dissolution pore 溶蚀孔隙；溶孔
dissolution pore and cave 溶蚀孔穴
dissolution precipitation mechanism 溶解沉淀机理
dissolution rate 溶解速率；溶解速度
dissolution rate of calcite 方解石溶解速度
dissolution rate of dolomite 白云石溶解速度
dissolution reaction 溶解反应
dissolution reprecipitation material transfer mechanism 溶解沉析传质机理
dissolution speed 溶解速度
dissolution temperature 溶解温度
dissolution trace of core 岩芯溶蚀痕迹
dissolvability 可溶性；溶解性
dissolvable inorganic phosphorus 可溶性无机磷
dissolvable salt analysis of soil 土壤可溶盐分析
dissolvable starch 可溶性淀粉
dissolvable thallium 可溶性铊
dissolvant 溶媒；溶剂
dissolve 溶解；溶化
dissolve coefficient value 溶解系数值
dissolved 溶化的
dissolved acetylene 溶解乙炔

dissolved adsorbate 溶解吸附质
dissolved adsorbate concentration 溶解吸附质浓度
dissolved air flo(a)tation 溶气浮选(法)；溶解空气浮选(法)
dissolved air releasing valve 溶气释放阀
dissolved air tank 溶气罐
dissolved air vacuum flo(a)tation 溶气真空气浮法
dissolved air vessel 溶蚀气罐；气浮溶气罐
dissolved cadmium species 溶解辐射物种
dissolved carbon 溶解碳
dissolved cation 溶解阳离子
dissolved chloride activity 溶解氯化物活性
dissolved chlorine ion 溶解氯离子
dissolved concentration 溶解浓度
dissolved constituent 溶解成分
dissolved contaminant 溶解污染物
dissolved contaminant concentration 溶解污染物浓度
dissolved contaminant transport 溶解污染物输移
dissolved copper concentration 可溶性铜浓度
dissolved divalent cation concentration 溶解二价阳离子浓度
dissolved element isotopic 水中元素的同位素
dissolved form 溶解态
dissolved fuel cell 溶解燃料电池
dissolved gas 溶解气；溶解(的)气体
dissolved gas drive 内部气驱；溶解气(体)驱动
dissolved gas drive reservoir 溶气驱动油藏
dissolved gases 溶解气体
dissolved gas in high-pressure thermal water 高压热水溶气
dissolved gas in oil 油溶气
dissolved gas in water 水溶气
dissolved humic materials 溶解腐殖质
dissolved hydrogen 溶解氢
dissolved hydrophilic organic substance 溶解亲水性有机物
dissolved hydrophobic organic substance 溶解疏水性有机物
dissolved impurities 溶解杂质
dissolved inorganic nitrogen 溶解无机氮
dissolved inorganic phosphorus 溶解无机磷
dissolved inorganic salt 溶解无机盐
dissolved inorganic solids 溶解无机固体
dissolved inorganic substance 溶解无机物质
dissolved ion 溶解离子
dissolved iron 溶解(性)铁
dissolved ligand 溶解配位体
dissolved lignin 溶解木质素
dissolved lipids 溶解脂类
dissolved load 溶解荷载；溶解质；溶解量
dissolved matter 溶质；溶油浮选；溶解物(质)；溶化物
dissolved mercury species 溶解汞物种
dissolved metal concentration 溶解金属浓度
dissolved metal ion 溶解金属离子
dissolved mineral 溶解无机质；溶解矿物质
dissolved nitrogen 溶解氮
dissolved nutrient salts 溶解营养盐类
dissolved oil flo(a)tation 溶油浮选法
dissolved organic carbon 溶解有机碳
dissolved organic carbon concentration 溶解有机碳浓度
dissolved organic compound 溶解有机化合物
dissolved organic distribution 溶解有机物分布
dissolved organic halide 溶解有机卤素
dissolved organic material 溶解有机物
dissolved organic matter 溶解有机物
dissolved organic matter concentration 溶解有机物浓度
dissolved organic matter content 溶解有机物含量
dissolved organic molecule 溶解有机分子
dissolved organic nitrogen 溶解有机氮
dissolved organic phosphorus 溶解有机磷
dissolved organics removal 除溶解有机物
dissolved organic substance 溶解有机物
dissolved oxygen 溶解氧
dissolved oxygen analyzer for seawater 海水溶解氧测定仪
dissolved oxygen and chlorine ion 水中的溶解氧和氯离子
dissolved oxygen concentration 溶解氧浓度
dissolved oxygen consumption 溶解氧消耗量

dissolved oxygen consumption rate 溶解氧消耗速率
dissolved oxygen content 溶解氧含量
dissolved oxygen content in marine water 海水溶解氧含量
dissolved oxygen corrosion 溶解氧腐蚀
dissolved oxygen deficit 溶解氧饱和差
dissolved oxygen deficit contribution 溶解氧饱和差基值
dissolved oxygen demand 溶解氧需求量
dissolved oxygen depletion 溶解氧消耗
dissolved oxygen determination 溶解氧测定
dissolved oxygen device 溶解氧测定仪
dissolved oxygen distribution 溶解氧分布
dissolved oxygen effect 溶解氧效应
dissolved oxygen electrode 溶解氧电极
dissolved oxygen gas analyzer 溶解氧气体分析仪
dissolved oxygen level 溶解氧水平
dissolved oxygen meter 溶解氧计；溶解氧测定仪
dissolved oxygen monitor 溶解氧检测仪
dissolved oxygen profile 溶解氧分布曲线；溶解氧分布图
dissolved oxygen removal 除溶解氧
dissolved oxygen routing model 溶解氧演算模型
dissolved oxygen sag curve 溶解氧下垂曲线；溶解氧挠度曲线；氧垂曲线
dissolved oxygen saturation capacity 溶解氧饱和量
dissolved oxygen sensor 溶解氧传感器
dissolved oxygen standard 溶解氧标准
dissolved oxygen uptake rate 溶解氧摄取率
dissolved peak 溶峰
dissolved phase 溶解相
dissolved phase contaminant concentration 溶解相污染物浓度
dissolved phosphorus 溶解磷
dissolved pillar 溶柱
dissolved reactive phosphate 溶解活性磷酸盐
dissolved reactive phosphorus 溶解活性磷
dissolved salt 溶解盐
dissolved salt concentration 溶解盐浓度
dissolved salts 溶解盐类
dissolved sensor 溶解传感器
dissolved silicon 溶解硅
dissolved silicon dioxide 可溶性二氧化硅
dissolved sodium chloride 溶解氯化钠
dissolved solid 溶(解)质；溶解固体；溶油浮选
dissolved solid concentration 溶解固体浓度
dissolved solid content 溶解性固体量
dissolved solid yield 溶解质径流量
dissolved solute 溶解溶质
dissolved state 溶解态
dissolved substance 溶解物(质)
dissolved sulfide 溶解硫化物
dissolved sulfur 溶解硫
dissolved trace metal 溶解微量金属
dissolved zinc ion 溶解锌离子
dissolve gas component 溶解气体成分
dissolvent 有溶解能力的；溶媒；溶剂
dissolve on 裁撤
dissolve out 溶出
dissolver 溶解剂；溶解机；分解剂；溶解装置；溶解器
dissolving 溶解
dissolving-air tank 溶气罐
dissolving box 溶解箱
dissolving capacity 溶解能力
dissolving heat 溶解热
dissolving-out substance 溶出物
dissolving power 溶解力；溶解本领
dissolving pulp mill wastewater 溶解纸浆厂废水
dissolving state 溶解状态
dissolving sulphite pulp 溶解亚硫酸盐纸浆
dissolving tank 溶解池；溶解槽
dissonance 不和谐；不协调(性)
dissonance coil 非谐振消弧线圈
dissonant colo(u)rs 不协调色
dissymmetric(al) 不对称的
dissymmetric(al) dispersion of interference 干涉色不对称色散
dissymmetric(al) field 不对称场
dissymmetric(al) filter 不对称滤波器
dissymmetric(al) impedance 不对称阻抗
dissymmetric(al) magnetic field 不对称磁场
dissymmetric(al) mode 不对称模式

dissymmetric(al) molecule 不对称分子
dissymmetric(al) network 非对称网络；不对称网络
dissymmetric(al) peak 不对称峰
dissymmetric(al) polyphase system 不对称多相制
dissymmetric(al) structure 不对称结构
dissymmetric(al) transducer 不对称换能器；不对称变换器
dissymmetry 不相称；不对称(现象)
dissymmetry coefficient 非对称系数；不对称系数
dissymmetry factor 不对称因子；不对称性因数
distad 向远侧
distal bar 远砂坝
distal bar deposit 远砂坝沉积
distal end 离中心端；远心端
distal fan deposit 扇远端沉积
distal part 远部
distal portion 远侧部
distal storm deposit 远端风暴沉积
distal turbidite 远端蚀积岩
distance 路程；距离；航行距离；位距
distance accuracy 测距精度
distance across 横穿距(离)
distance adjustment 距离校正
distance amplitude compensation 距离振幅补偿
distance angle 求距角
distance apparatus 遥测仪(器)
distance azimuth measuring equipment 方位遥测设备
distance bar 横撑；限程杆；牵条；间距撑；定距撑条
distance between acting center 轴颈中心距
distance between active fault and site 场地距活动断裂距离
distance between adjacent flight lines 相邻航线间隔(航测)
distance between axial centers 轴心距
distance between backs of wheel flanges 轮对内侧距离
distance between backs of wheel rims 轮对内侧距离
distance between bearings 轴承间距离
distance between bogie centers 转向架中心距
distance between bogie centres 转向架中心距
distance between bogie pivots 车辆定距
distance between buffer centers 车辆全长
distance between buildings 建筑物间距
distance between buoys 浮标间距
distance between buttresses 支墩中心距
distance between center lines 中线间距(离)
distance between center lines of station 站间距离
distance between centers of area element and arbitrary offset middle point 任意偏离中点与面元中心之间的距离
distance between centers 中至中距离；中心距(离)；中心间距；顶尖距
distance between centers of crawlers 履带中心距(离)
distance between centers of freight turning rack 货物转向架支距
distance between centers of lines 线(路中心)间距；正线中心距离
distance between centers of tracks 铁路中线间距
distance between center to center 中至中距离；中心间距(离)
distance between center to center of bogies 转向架中心距
distance between (collecting) plates 板间距
distance between commutator segments 片间距离
distance between conductors 导体间距离
distance between coupler centers 车钩中心距
distance between cross line 切割线间距
distance between electrodes 电极间距(离)
distance between fiducials 框标距离
distance between girders 梁的间隔
distance between groin 丁坝间距
distance between groups of pumping well 抽水孔组间距离
distance between grouting hole 注浆孔间距
distance between hills 穴距；株距
distance between holes 孔距
distance between insides of rims 轮对内侧距离
distance between insides of tyres 轮箍内侧距
distance between joints 缝距
distance between lateral shot spread 横向炮排距离
distance between master and slave sensor 主从传感器距离
distance between mine and market 矿山距市场里程
distance between mine and station 矿山距火车站里程
distance between observation and reagent dropping point 观测点到投源点距离
distance between observation point and pumping well 观测点到抽水孔距离
distance between observation wells 观测孔间距
distance between (on) centers 中心间距(离)
distance between out to out 外侧间距
distance between padding steel reinforcement 垫筋距离
distance between parks 公园间距
distance between pins 摇枕中心距
distance between pivots 上心盘中心距
distance between pole tips 极端距离
distance between profiles 断面带间距
distance between pumping well and observation well 观测孔到抽水孔距离
distance between pumping wells 抽水孔间距离
distance between rail supports 钢轨支点间距
distance between reagent dropping well and receiving well 投源孔到接收孔距离
distance between rivets 铆钉间距
distance between rows 行距；排距；排间距离
distance between shafts 轴距
distance between source and sample 源样距
distance between stair(case) wells 楼梯井间距；楼梯间间距
distance between stations 站间距；站间距(离)；站间距距
distance between stiffening rings 加强圈间距
distance between stirrups 钢箍筋间距
distance between stops 停站距离
distance between tension equipment 拉紧装置间距
distance between the extremes of any of two more consecutive axles 最远轴距【机】
distance between the site and building material field 料场距工地距离
distance between the survey(ed) points 测点距离
distance between the toes of a weld 焊缝宽度
distance between track centers 股道间距
distance between tracks 线间距；股道间距
distance between transmitter and receiver 收发距
distance between trusses 桁架间距
distance between two distribution 两个分布的间距
distance between two goods turning racks 跨装支距
distance between two telescopic shuttles 货叉间距(中心距)
distance between vertical shot spread 垂向炮排距离
distance between welding wires 焊丝间距
distance between wells 井间距离
distance between zero point of mid-gradient curve 中梯曲线零值点间距
distance between zero points 零值点间距
distance block 定位垫块；隔板；定距环；定距(隔)块
distance bolt 定距螺栓
distance bound 距离限
distance braking 间隔制动【铁】
distance bush(ing) 定距衬套；间隔衬套
distance by engine revolutions per minute 主机每分转数航程
distance by engines revolution 主机转数航程
distance by log 计程仪航程
distance by radar 雷达距离
distance by stadimeter 测距仪距离
distance calibration 距离校准
distance carriage 长度滑架
distance check 距离校验
distance circle of position 距离船位线
distance coefficient 距离系数
distance collar 间隔轴环；间隔圈；隔环；定距环
distance conception 距离概念
distance condition 距离条件
distance control 远距离操纵；遥控；调节间距
distance-controlled boat 遥控艇

distance convergence 距离收敛
distance coolar 定距轴环
distance coolar receiving plate 定距轴环支承板
distance correction in Gaussian projection 高斯投影距离校正；高斯投影距离改正
distance cost 距离成本
distance covered 所走距离
distance decision function 距离判决函数
distance-difference measurement 双曲线测位制
distance display 距离显示
distance distortion 长度变形
distance distribution 距离分布
distance element 距离元（素）
distance error 距离误差
distance factor 里程系数
distance factor between air voids 加气混凝土的气孔间距系数
distance finder 测距计
distance finding 测距
distance finding station 测距（电）台；测距站
distance for inspection couples 拉钩检查距离【铁】
distance form centre to centre 中到中距离
distance form initial point 起点距
distance freight 海里程运费；超程运费
distance frequency relay 远距离周率继电器
distance from fault 断层距
distance from hole to shaft 孔至井距离
distance from locality of strain measurement to reference line 应变测量点至参考线距离
distance from midship to center of flo(a)tation 漂心距船中距离
distance from midship to centre of buoyancy 浮心距离中距离
distance from seawater intrusion 海水入侵距离
distance from source area 油源区距离
distance from the bank 岸边距
distance from the epicenter 震源距（离）
distance from the focus 震源距（离）
distance function 距离函数
distance gate 距离开关
distance ga(u)ge 距离测量仪；测远仪；测远器；测距仪；测距仪；测距规
distance ga(u)ge for sleepers 枕距规
distance ground control 稀疏控制
distance hardness （顶端淬火时的）距离硬度
distance-height indicator 距离高度指示器
distance holder 定位块
distance impedance protection 距离阻抗保护
distance in back of hole 孔后距
distance indicating automatic navigation equipment 自动测距导航设备；测距自动导航设备
distance indicating system 距离指示系统
distance indication 距离显示
distance indicator 示距天体
distance in front of hole 孔前距
distance interval 距离步长
distance light 远光
distance line 行动绳；堵漏毡吊索
distance link 定距链节
distance log 计程仪
distance-luminosity relation 距离光度关系
distance made good 航行距离；直航程；实际航程
distance mark 距离标志
distance marker 距离指点标
distance marking light 距离灯标；距离标线灯
distancemat 测距仪
distance measurement 距离测量；距离测定；测距；微波测距
distance measuring 测距；距离测量
distance measuring device 距离测量装置
distance measuring equipment 测距装置；测距仪；测距设备
distance measuring equipment indicator 测距装置指示器
distance measuring equipment interrogator 测距装置询问机
distance measuring equipment memory 测距装置存储器
distance measuring equipment transponder 测距装置应答机
distance measuring instrument 测距仪
distance measuring potentiometer 测距电位计
distance measuring relay 距离测量继电器

distance measuring system 测距定位系统
distance measuring theodolite 测距经纬仪
distance measuring wedge 视距测量光楔
distance meter 距离测量仪；测远仪；测距器；测距计
distance mode selector 距离方向选择器
distance modulus 距离模数
distance nut 隔垫用螺母
distance observation 远距（离）观测；测距
distance of closest point of approach 会遇最近距离
distance of critical dissolved oxygen 临界溶解氧距离
distance of dislocation glide 位错滑移距离
distance of dispersion 弥散距离
distance of dissemination 传播距离
distance of distinct vision 明视距离
distance of fiducial marks 框标距离
distance of gliding 滑移距离
distance of hydraulic jump 水跃距离；水跃长度
distance of lattice point 结点间距离
distance of leaky percolation 越流渗透距离
distance of migration 运移距离
distance of mining influence 开采影响距离
distance of nappe transporting 推覆体运移距离
distance of object when abeam 物标正横距离
distance of passenger flow 旅客流程
distance of payment after date 远期付款交单
distance of radar horizon 雷达地平（线）距离
distance of rays intersection 光线交会距离
distance of rectification zone 纠正带距
distance of relative movement 相对运动行程
distance of run 行程
distance of suffering wind 受风距离
distance of tray support to support 塔板支座间距
distance of tray to tray support 塔板至塔板支座的距离
distance of visibility 视地平距离
distance of visible horizon 视地平距离
distance of weir to inside surface of tank 溢流堰至容器内壁的距离
distance of wheel backs 轮背（内侧）距离
distance operation 遥控
distance out to out 外沿间距离
distance over the ground 实际对地航程
distance piece 定位（垫）块；垫铁；垫片；间隔物；隔片；隔板；定距片；定距环；定距（隔）块
distance plate 定距（隔）板；隔片；定位块
distance pole 标柱
distance post 路程标；里程标；分段标
distance propate rule 改正边长配赋法
distance protection 距离防护；距离保护；远距离保护；远距防护；阻抗保护系统
distance protection with carrier current blocking 载波闭锁距离保护（装置）；高频闭锁距离保护（装置）
distance pulse generator 里程脉冲传感器
distance pulse sensor 里程脉冲传感器
distancer 测距仪
distance range 距离范围；通达距离
distance rate 按距收费率（电话）
distance reading tach(e)ometer 遥测转速计；遥测转速表
distance receiver 航程显示器
distance receptor 距离感受器
distance recorder 距离记录仪；计距器；航程记录仪；航程记录器；航程计算器；遥测记录器
distance recording watch 记步记距两用表
distance regulation 远距调节
distance relay 距离继电器；阻抗继电器
distance ring 间distance环；隔环；定距环
distance rod 隔杆；推力杆
distance run 所走距离
distance run by train in departing from station 列车出站距离
distance run by train in entering into station 列车进站距离
distance scale 距离尺度；距离比例尺；焦距标度；线性比例尺；长度比
distance scene 远景
distance sensor 距离传感器
distance separation 建筑物防火间隔；分隔间距；防火间距
distance servo amplifier 距离伺服放大器
distance sets 测距成果组

distance setting 调节间距
distance setting ring 距离调整环
distance sign 距离标志；里程标志
distance signal 距离信号；距离标志；远距离信号
distance sink bolt 定位螺栓；牵条螺栓
distance sink tube 定距管
distance sleeve 隔离套筒
distance space 间距
distance stone 里程标
distance strap 定距块
distance sum measurement 距离和测定
distance sum measurement system 距离和测量制；椭圆定位制
distance table 里程表；航程表；测距成果表
distance test 距离试验
distance thermometer 远测温度计；遥测温度计；遥测温度表
distance tie bolt 定距系紧螺栓
distance-time curve 距离时间曲线
distance to airfield kilometerage 距机场里程
distance to closest point of approach 最接近点距离
distance-to-coupling measurement 测长
distance to docks kilometerage 距码头里程
distance to go 应驶距离
distance-to-go block 准移动闭塞；半移动闭塞
distance-to-go measurement 测长
distance tolerance 距离容许误差
distance to new course 从转舵到新航向的距离
distance to object 至目标距离
distance to the horizon from height of observer 测者视地平距离
distance tube 隔离套筒
distance-type 遥控式
distance-type of cooling thermometer 遥测空气温度计
distance velocity lag 距离速度延迟；传动延迟
distance view(ing) 远距离观察
distance visibility 能见度
distance washer 间距垫圈；定距垫圈
distant 远距离的
distant action instrument 遥测仪表
distant adjustment 遥调
distant admixture （沉积岩中的）特大碎屑
distant collision 远距离碰撞
distant control 远距离控制；遥控
distant control of signals and switches 号志与道叉的人工联动操作
distant control on floor 地面遥控【机】
distant drive 远距离驱动
distant earthquake 远距离地震；远震；远源地震
distant earthquake instrument 远震仪
distant exchange 远端电话局
distant field 远场
distant heating 远程供暖
distant hole 终孔
distant hybrid 远缘杂种
distant-indicating instrument 远距离指示仪表
distant indication 远距离信号；远程指示；距离显示；远距离显示；遥示
distant indicator 远距离指示器；遥测指示器
distant level indicator 远距离液位指示器
distant measurement 遥测
distant object 远距离目标
distant operation 远距离操作；遥控
distant point 远点
distant position indicator 远距离位置指示器
distant reading 遥测读数
distant-reading compass 遥示罗经
distant-reading inclinometer 远读测斜器
distant-reading manometer 远距传送压力表
distant-reading tach(e)ometer 遥测转速计
distant reception 远距离接收
distant recording instrument 远距离自记设备；遥测自记设备
distant recording system 遥测记录系统
distant regulator 遥控调节器
distant set 遥调
distant setting 遥调
distant shock 远震
distant sign 远距标志
distant signal 远距（离）信号；远程信号；预告信号【铁】
distant signal(l)er 预告信号机

distant signal mechanism 预告信号机构
distant surveillance 远距离观察
distant switching-in 远距离合闸
distant synchronization 远距离同步
distant thermometer 远距离温度表;遥测温度计;遥测温度表
distant transmitter 远程发射机
distant view 远视;远景
distant viewing 远距离观察
distant view photograph 全景照片
distant warning device 远距离警报装置;远距离报警装置
distant-water fishery 远洋渔业
distant water state 远海国家
distearyl thiodipropionate 硫代二丙酸二硬脂酯
Distec system 铝翼反射式显示系统
distegia (古希腊、罗马剧院布景房的)上层
distele 双中柱
distemper 胶画用的颜料或涂料;涂料刷;水溶性涂料;水浆涂料;水粉画颜料;刷墙水浆涂料;刷墙粉;色胶;色浆
distemperature 刷墙水浆涂料;刷墙粉
distemper brush 涂料刷;色胶刷;胶质涂料刷;画笔;长毛刷墙刷;调漆刷
distemper coat 胶质涂料面层
distemper colo(u)r 刷墙水浆涂料用色料
distempering 粉刷墙壁;刷浆;色浆涂刷
distemper paint 调和漆;刷墙水浆涂料
distensibility 扩张性;伸长性
distensible 可伸展的
distension 膨胀化;膨胀装置
distention reflex 膨胀反射
disthene 蓝晶石
disthene-mica schist 蓝晶石云母片岩
distichous 二列的
distil(l) 蒸馏(净化)
distilland 蒸馏母液;被蒸馏物
distillate 馏出液;馏出物;蒸馏液;蒸馏产物
distillate cooler 蒸馏冷却器
distillate desulfurization unit 蒸馏脱硫装置
distillate fraction 馏分
distillate fraction oil 馏出油
distillate fuel 馏出燃料
distillate fuel oil 馏出燃料油
distillate selective cracking process 柴油选择裂化工艺
distillating 蒸馏的
distillating boiler 蒸馏锅
distillating column 蒸馏塔
distillating still 蒸馏壶;蒸馏釜
distillating tank 蒸馏壶
distillating tower 蒸馏塔
distillating tray 蒸馏塔板
distillation 蒸馏(作用);蒸馏(法);抽出物
distillation apparatus 蒸馏器
distillation bench 蒸馏装置组
distillation chamber 蒸馏室
distillation coal 高沥青褐煤;干馏用煤
distillation curve 蒸馏曲线
distillation distilling tray 蒸馏塔板
distillation equipment 蒸馏设备
distillation flask 蒸馏(烧)瓶
distillation fraction 蒸馏馏分
distillation gas 蒸馏气(体)
distillation kettle 蒸馏釜
distillation loss 蒸馏损失
distillation membrane 蒸馏膜
distillation method 蒸馏法
distillation plate calculation 蒸馏塔板的计算
distillation pot 蒸馏釜
distillation process 蒸馏法
distillation range 蒸馏区间;蒸馏范围;沸腾范围
distillation refining 蒸馏精炼
distillation residue 蒸馏残渣
distillation retort 蒸馏瓶
distillation still 蒸馏釜
distillation test 蒸馏试验
distillation test of liquid asphalt 液体沥青分馏试验
distillation test of liquid tar 煤沥青分馏试验
distillation test residue 蒸馏后的残渣
distillation titration method 蒸馏滴定法
distillation-titrimetry 蒸馏滴定法
distillation total moisture yield 干馏总水分收率
distillation tube 蒸馏管

distillation under pressure 加压蒸馏
distillation zone 蒸馏区
distillatory 蒸馏(的)
distilled 蒸馏(过)的
distilled coconut fatty acid 馏出椰子脂肪酸
distilled crude carbolic acid 蒸馏粗碳酸
distilled crude cresylic acid 蒸馏粗甲苯基酸
distilled fatty acid 蒸馏脂肪酸
distilled grease olein 从废油脂馏出的甘油三酸酯
distilled linseed fatty acid 馏出的亚麻子脂肪酸
distilled tar 蒸馏焦油;蒸馏煤沥青
distilled temperature 馏出温度
distilled water 蒸馏水
distiller 蒸馏器
distiller's by-product 烧酒副产物
distiller's dried grains 干酒精
distiller's dried soluble 烧酒糟;干酒精
distiller's grain 酒糟
distiller's product 烧酒副产品
distiller's soluble 烧酒糟残液
distiller's spent grains 烧酒糟
distillery 酒厂;蒸馏室;蒸馏场;造酒厂
distillery refuse 烧酒业下脚料
distillery residues 酒厂废渣
distillery waste 酒厂废物
distillery waste(water) 酒厂污水;蒸馏厂废水
distilling 蒸馏作用;蒸馏的
distilling apparatus 蒸馏装置
distilling apparatus with ground on coil condene 磨口蛇形管蒸馏器
distilling condenser 蒸馏冷水器;蒸馏冷凝器
distilling flask 蒸馏烧瓶
distilling furnace 蒸馏炉
distilling industry 蒸馏业
distilling plant 蒸馏设备;造水装置
distilling ship 蒸馏水供给船
distilling tank 蒸馏壶
distilling tower 蒸馏塔
distilling tube 蒸馏管;分馏管
distilling tube with two bulbs 二球分馏管
distilling water 蒸馏水
distinct 明晰的;清晰的
distinct acid 特色酸
distinct beam condition 不同光束条件
distinct boundary 明显边界
distinct chemical species 特色化学物种
distinct eigen value 不重复本征值;不等本征值
distinct flocculence 明显絮状沉淀法
distinct hearing 明晰听觉
distinctive 鉴别性;有特色的;区别性的
distinctive appearance 特殊外形
distinctive colo(u)r 区别显著的颜色
distinctive emblem 区标志
distinctive feature 区别性特征
distinctive mark 辨认标记
distinctiveness 差别性;区别性
distinctive ringing 区别铃声
distinctive threshold value 区分阈值
distinctness 差别;清晰度
distinctness of image 鲜映性
distinctness-of-image gloss 像片清晰度光泽
distinct root 相异根;不等根
distinct variety 稳定品种;特征明显的变种
distinct vision 明晰视觉
distinct zero 相异零点;不重复零点;不等零点
distinguish 判别(显现)
distinguishable 可区别的
distinguished antique style 著名的古式风格
distinguished guests' waiting room 贵宾候车室
distinguished sequence 判别序列
distinguished stain 标志用有机色料
distinguished symbol 已发出的信号;判别符号
distinguished vertex 奇异点
distinguishing characteristic 明显特征
distinguishing colo(u)rs 区别显著的颜色
distinguishing criterion for metamorphic rock succession 变质岩地层层序判别准则
distinguishing feature 特征;特点
distinguishing flag 区别旗
distinguishing mark 识别标志
distinguishing pennant 识别尖旗
distinguishing reaction 区别反应
distinguishing sequence 辨别序列;区别时序;判别序列

distinguishing signal 区别信号;识别信号
distinguishing state 区别状态
distinguishing symbol 识别符号
distinguishing test 区别试验
distinguishing tree 区别树
distoceptor 距离感受器
distomat 测距仪
distometer 硅藻测定仪
distort 曲解;使变形
distorted 无畸变的
distorted alternating current 失真交变电流
distorted bedding 畸变层理;扭曲层理;歪曲层理
distorted bevel 扭曲边
distorted cash balance 失真的现金余额
distorted cross method 变形交会测量法
distorted crystal 歪晶
distorted deformation 扭曲变形
distorted degree 畸变度
distorted dial(l)ing 失真拨号
distorted electric(al) field 畸变电场
distorted element 畸变单元;变态单元
distorted geological section 变态地质剖面
distorted grid 扭曲格网
distorted image 畸变像
distorted image formation 畸变成像
distorted lattice 畸变点阵
distorted loop 变形线圈
distorted model 不等比模型;畸变模型;扭变模型;变态模型
distorted pattern 变形模
distorted peak 畸变峰
distorted region 畸变区域;畸变范围;失真区
distorted river model 变态河流模型
distorted scale 变态比例(尺)
distorted-scale model 比例变态模型
distorted shape 形状扭曲
distorted spectrum 畸变谱
distorted surface 畸变表面
distorted water 畸变水;变态水
distorted wave 畸变波;失真波
distorted waveform 失真波形
distorted wave method 畸变波法
distorted wave theory 扭曲波理论
distorted wire rope 畸变钢丝绳
distorter 畸变放大器;失调
distorting lens 变形镜头
distorting mirror 哈哈镜
distorting stress 扭(转)应力
distortion 畸变(像差);扭转;扭歪;扭曲;反常;变率;歪曲(作用);歪扭(作用);失真
distortional coefficient 畸变系数
distortional deformation 扭曲变形
distortion aligned phase 畸变排列相
distortion allowance 反扭曲量;变形余量
distortional point 扭变点
distortional stress 变形应力
distortional wave 剪切波;畸变波;横波;变形波;S波
distortion analyzer 失真分析器
distortion and bias 失真和偏移
distortion bridge 测失真用电桥;测畸变电桥
distortion centre 畸变中心
distortion characteristic 畸变特性
distortion compensation 畸变补偿;失真补偿
distortion correcting cam 畸变校正齿轮
distortion correction 畸变校正
distortion correction factor 畸变校正系数
distortion corrector 变形校正器
distortion curve 畸变曲线
distortion delay 畸变延迟;失真延迟
distortion due to feedback 反馈引起失真
distortion during quenching 淬火应变;淬火变形
distortion effect 畸变效应;失真效应
distortion elimination 畸变消除;失真消除
distortion energy 畸变能
distortion energy theory 歪能理论
distortion energy yield criterion 畸变能屈服条件
distortion error 畸变(误)差;失真误差
distortion factor 畸变率;畸变系数;畸变率;失真因数;失真系数;失真度
distortion factor meter 失真系数测试器;失真度表;畸变系数测量器
distortion factor of a wave 波形失真系数
distortion-free 无畸变
distortion-free correcting instrument 无畸变校正仪

distortion-free hologram 无畸变全息图
distortion-free image formation 无畸变成像
distortion-free virtual image 无畸变虚像
distortion function 畸变函数
distortion graph 畸变曲线图
distortion inaccuracy 畸变不准确度
distortion in angle 角度畸变
distortion indicator 失真指示器
distortion in the crystal lattice 晶格畸变
distortion isograms 等变形线
distortion lattice 畸变点阵
distortionless 无失真的;无畸变的
distortionless condition 无失真条件
distortionless line 无畸变线路
distortionless recording 无畸变记录
distortion limited operation 失真限制作用
distortion measuring equipment 畸变测量仪
distortion meter 失真(度)测试仪;畸变计
distortion model 畸变模型
distortion of bearing 方向变形
distortion of chart 海图变形
distortion of contact wire 接触线畸变
distortion of cross section 横截面畸变
distortion of field 磁场畸变
distortion of flow pattern 流型的畸变
distortion of handle and spout 歪嘴斜把
distortion of hatch 舱口变形
distortion of laminate 层合材料变形
distortion of lattice 晶格畸变
distortion of lens 物镜畸变差
distortion of map projection 地图投影变形
distortion of parallel lines 平行线变形
distortion of projection 投影变形
distortion of roughness 粗糙度的变形;糙率失真(模型试验中)
distortion of scale 尺度变形
distortion of slope 边坡变形
distortion of sound 声音失真
distortion of the mesh 网格变形
distortion of vessel 容器变形
distortion of wood(en) sash 木窗扇变形
distortion pad 失真衰减器
distortion pattern 畸变图
distortion power 畸变功率
distortion printing 畸变印刷
distortion pulse code 失真脉冲码
distortion range 畸变范围;失真范围
distortion rate 畸变率;畸变率
distortion ratio 变态率
distortion ratio of model 模型变率
distortion resistance 变形阻力;变形抗力
distortion set 失真测试器
distortion settlement 畸变沉降量;瞬时沉降
distortion shift 畸变位移
distortion structure 畸变结构
distortion temperature 变形温度
distortion tolerance 变形公差
distortion tolerance of track 线路变形公差
distortion transformer 失真变压器
distortion under heat 热扭变
distortion under load 负载作用下的畸变
distortion value 畸变值
distrail 耗散尾迹
distrain 扣押
distrained goods 扣押的货物
distrainee 财物被扣押者;财物被扣押的人;财产被扣押
distrainer 扣押他人财物的人;扣押人
distrainment 扣押财产
distraint 扣押财物;动产扣押
distress 扣押物;扣押财产;遇险;遇难;疲劳;事故
distress acknowledgment 遇险确认;遇难确认
distress altering 遇险报警;遇难报警
distress and rescue procedure 海难与救助程序
distress area 灾区;受灾地区
distress beacon 失事信标
distress call 海难紧急呼叫;遇险呼号;遇难呼号;求救呼号;遇险信号
distress cargo 急运货物
distress commodity 赔钱货
distress condition 损坏状态
distressed area 遭难地区;贫困地区;经济萧条区
distressed city 贫困城市
distressed goods 亏本出售的商品

distressed mortgage 无能偿付的抵押人;困难户抵押人
distressed property 要拍卖的抵押不动产
distressed structure 需要加固的结构;超载结构;变形结构
distressed zone 应力释放带;去应力带
distress flag 遇险旗;遇难旗
distress for rent 扣押货物充抵租金
distress freight 扣押货物充抵运费;削价运费;低微运费;低价运费;填载运费
distress frequency 呼救信号频率;遇险频率;遇难频率
distress goods 亏本出售的货物
distress in concrete 混凝土事故;混凝土(的)龟裂;混凝土缝隙
distressing 木器家具仿古装饰;放散温度应力;去应力
distress landing 危急降落(飞机)
distress light 遇险灯号;遇难灯号;遇险火号;遇难灯号;求救灯号
distress mechanism 破损机理
distress merchandise 廉价品;亏本(出售的)商品;赔钱货;跳楼货
distress phase 遇险阶段;遇难阶段
distress priority request message 遇险优先度申请电文;遇难优先度申请电文
distress procedure 遇险求救程序;遇难求救程序
distress radio call system 遇险无线电呼叫系统;遇难无线电呼叫系统
distress rates 低价运费率
distress relay 遇险转播;遇难转播
distress rocket 遇险火箭;遇难火箭
distress sale 公卖;廉价销售;扣押物拍卖;扣押货拍卖;扣押财物的出售
distress selling 廉价抛售;亏本出售
distress signal 呼救信号;遇险(呼救)信号;遇难(呼救)信号;求救信号
distress signal warning device 求援信号装置
distress warrant 财物扣押令;财产扣押令
distribond (含膨润土的)硅质黏土
distributable 可分配
distributable assets 可供分配的资产;可分配资产
distributable profit 可供分配的利润
distributary 叉河;分流;配水沟支流
distributary channel 分支河道;分流河道
distributary channel deposit 分流河道沉积
distributary channel-fill trap 分流河道砂圈闭
distributary channel levee 分流河道天然堤
distributary glacier 支流冰川;分支冰川;分叉冰川
distributary mouth 分流河口
distributary mouth bar 分流河口(砂)坝
distributary mouth bar deposit 分流河口坝沉积
distributary-mouth bar trap 分流河口砂坝圈闭
distributary of river 网状河流
distributary river 分流河
distributary stream 分流河;分汊河
distributary system 分流系统
distributary tongue 分支冰舌
distribute 被分配到财产的人;重新分配;拆版;分送;分派;分发;分布
distribute a bonus 分红;分发红利
distribute a dividend 分发股利
distribute assets 可供分配的资产
distributed 分布(式)的
distributed access system 分布式访问系统
distributed adaptive routing 分布式自适应路由选择
distributed agent approach 分散模块法
distributed air conditioner 分布式空调器
distributed amplifier 传输线放大器;分布(式)放大器
distributed application 分布式应用
distributed application program(me) 分布式应用程序
distributed arbitration program(me) 分布式仲裁
distributed area of tectonic layer 构造层分布地区
distributed array processor 分布式阵列处理机
distributed bar 分布钢筋
distributed block trial 分散区组试验
distributed body forces 分布的体积力
distributed Bragg reflector 分布布喇格反射器
distributed calculating environment 分布式计算环境

distributed capacitance 分布电容
distributed capacity 分布电容
distributed charge 分布电荷
distributed circuit 分布电路
distributed collector power plants 分布聚热太阳能电站
distributed communicaition processor 分布通信处理机
distributed communication 分配通信;分布式通信
distributed communication architecture 分布通信体系结构;分布式通信体系;分布式通信结构
distributed communication processor 分布式通信处理机
distributed component 分布元件
distributed component filter 分布元件滤波器
distributed computer 分布式计算机
distributed computer architecture 分布式计算机体系结构
distributed computer network 分布式计算机网络
distributed computer processing 分散制计算机处理
distributed computer system 分式计算机系统;分布(式)计算机系统
distributed computing 分布式计算(方式);分布(式)计算
distributed computing operating system 分布式计算操作系统
distributed condition of aquifer 含水层分布条件
distributed constant 分布式系数;分布常数
distributed constant filter 分布常数滤波器
distributed contaminant source 分布污染源
distributed control 分布式控制
distributed control module 分布式控制模块;分布式控制模件
distributed control system 集散控制系统;分散控制系统;分布式控制系统
distributed control unit 分布式控制单元
distributed core type transformer 分布铁芯变压器
distributed cost 已分配成本
distributed damping 分布阻尼
distributed data 分布式数据
distributed database 分布(式)数据库
distributed database management system 分布式数据库管理系统
distributed database network 分布数据库网络
distributed database system 分布式数据库系统
distributed database technique 分布式数据库技术
distributed data processing 分散(制)数据处理;分布(式)数据处理;分布(方式的)数据处理
distributed data processing network 分布式数据处理网络
distributed date 发行时间
distributed delay model 分布时延模型
distributed diagram 分布图
distributed direction of an underground river 暗河分布方向
distributed domain 分布域
distributed environment 分布式环境
distributed executive-like system 分布式类执行程序系统
distributed exhaust for flue gas 分散排烟
distributed expert system 分布式专家系统
distributed extent of aquifer 含水层分布范围
distributed factor income 已分配的要素收入
distributed fault 分布式故障
distributed feedback 分布反馈
distributed feedback laser 分布反馈激光器
distributed feedback semiconductor laser 分布反馈半导体激光器
distributed fiberoptic sensor 分布式光纤传感器
distributed fiber sensor 分布式光纤传感器
distributed file system 分布式文件系统
distributed force 分布力
distributed form of aquifer in space 含水层空间分布形式
distributed frame 分布式结构
distributed frame alginment signal 分散式帧定位信号
distributed free space 分布(式)自由空间;分布式空白区
distributed function 分布式功能
distributed function(al) computer system 功能分布(式)计算机系统
distributed function(al) microprocessor 分布式

功能微处理机
distributed gas 外供气
distributed graph 分布图
distributed graphic network 分布式图形网络
distributed host command facility 分布式主机命令设备
distributed indexed access method 分布式索引存取法
distributed inductance 分布电感
distributed insulator 配线绝缘子
distributed intelligence 分布信息;分布(式)智能
distributed intelligence microcomputer system 分布式智能微型计算机系统
distributed intelligence system 分布式智能系统;分布式灵活系统
distributed interconnection 分布式互联
distributed isomerous computer system 分布式异构型计算机系统
distributed lag 已分摊差滞;分配滞后;分配迟延;分布滞后
distributed lag model 分布滞后模型
distributed lag pattern 分布滞后模型
distributed lag regression 分布滞后回归
distributed lag studies 分布滞后研究
distributed law of collapse 塌陷洼地分布规律
distributed linear collector system 分布线型集热器系统
distributed load 均布荷载;分布荷载;分布负载;分布负荷
distributed logic 分布(式)逻辑
distributed magnet 分布磁铁;配置磁铁
distributed management facility 分布式管理设施
distributed mass 分布质量
distributed mass beam 质量均匀分布梁
distributed message switching system 分布式报文交换系统
distributed microcomputer 分布式微型计算机
distributed microcomputer network 分布式微型计算机网络
distributed model 分散模型;分布模型
distributed moment 分配弯矩;分配力矩;分布弯矩
distributed multilevel systems 分布多层系统
distributed multiprocessor 分布式多处理机
distributed multiprogrammed operating system 分布式多道程序操作系统
distributed network 分散式网络;分布(型)网络;分布(式)网络
distributed network management 分布式网络管理
distributed network system 分布式网络系统
distributed operating system 分布式操作系统
distributed organization 分布式组织
distributed packet switching 分布式包交换
distributed parallel logic theory 分布式并行逻辑理论
distributed paramagnetic amplifier 分布常数顺磁放大器
distributed parameter 分布参数
distributed parameter circuit 分布参数电路
distributed parameter delay line 分布参数延迟线
distributed parameter element 分布参数元件
distributed parameter generator 分布参数振荡器
distributed parameter integrated circuit 分布参数集成电路
distributed parameter line 分布参数线路
distributed parameter measuring equipment 分布参数测量设备
distributed parameter model 分布参数模型
distributed parameter network 分布参数网络
distributed parameter problem 分布参数问题
distributed parameter system 分布参数组;分布参数系统
distributed parameter system model 分布参数系统模型
distributed paramp 分布参量放大器
distributed Pitot tube 多点皮托管
distributed plan 分布图
distributed plotting system 分布式绘图系统
distributed polar rotor 隐极式转子;分布磁极绕组转子
distributed presentation management 分布式(消息)表示管理;分布式显示管理
distributed presentation services 分布式消息表示服务程序;分布式显示服务程序
distributed processing 分散处理;分布(式)处理

distributed processing control executive 分布式处理控制执行程序
distributed processing memory 分布处理存储器
distributed processing network 分布(式)处理网络
distributed processing structure 分布式处理结构
distributed processing system 分布(式)处理系统;分布式操作系统
distributed processor 分布式处理机
distributed processor system 分布式处理机系统
distributed profit 已分配利润
distributed query processing 分布式询问处理
distributed queue dual bus 分布式排队双总线
distributed quota of drilling depth 钻探进尺分布定额
distributed quota of observation line 观测路线分布定额
distributed reinforcing bar 分布钢筋
distributed resistance 分布阻力;分布电阻
distributed resistive suppressor 分布电阻抑制器
distributed rod 分布钢筋
distributed shunt conductance 分流电导;分布分流电导;分布并联电导
distributed software system 分布式软件系统
distributed source of pollution 分布污染源
distributed statement of service cost 服务成本分配表
distributed steel 分布钢筋;配筋
distributed stress 分布应力
distributed system 油系统;分配系统;分布(式)系统;分布(参数)系统;气系统;配电系统;水系统
distributed system environment 分布式系统环境
distributed system executive 分布式系统执行程序
distributed system license option 分布式系统特许可选项
distributed system network 分布式系统网络
distributed target 分布性目标
distributed test 分散试验
distributed thin-film waveguide 分布薄膜波导
distributed time-division multiple-access 分布式时分多址
distributed unit 发行单位
distributed volume 提供卷宗
distributed water quality modeling 分布水质模拟
distributed wave function 分布波函数
distributed weight 分配重量
distributed winding 分布绕组
distributed yielding 分布屈服
distributer 撒布器
distributing amplifier 分配放大器
distributing and regulating module 搅纵调速组件
distributing arm 分配杆
distributing auger 分配螺旋
distributing bar 分布钢筋
distributing belt 布料胶带
distributing bin 配料舱;配料仓
distributing board 配电板
distributing boom 布料杆
distributing box 交接箱;给料分配箱;分线匣;分线盒;配电盒
distributing bridge 工作便桥
distributing bucket (混凝土的) 布送斗;布料斗;配料斗;撒布料斗
distributing business 经销企业
distributing cabinet 配电盒;分线盒;分配电箱
distributing cable 配电电缆
distributing canal 分渠
distributing center 集散地;集散中心
distributing chamber 调压室;缓冲室
distributing channel 分支河道;配水渠;支渠
distributing chute 配料槽架;布料溜槽;撒布溜槽
distributing cock 分配旋塞
distributing coefficient 分配系数
distributing condition 分配情况
distributing conveyer 分配运输机
distributing damper 分配挡板
distributing device 布料器
distributing disc 撒料盘
distributing ditch 农渠;毛渠;配水渠
distributing edge 切断边
distributing factor 分配系数
distributing financing report forms 分布财务报表
distributing frame 配线架
distributing gutter 配水沟槽
distributing insulator 配线绝缘子

distributing lane 分流车道
distributing law 分配定律
distributing line 配水管线;分布线
distributing main 供水干线;配水干线;配水干渠;配水干管;配电干线
distributing manifold 分配歧管;分配管汇
distributing moment 分配力矩
distributing mouth bar 分支河口砂坝
distributing net(work) 配水(管)网;分配网络;配电网
distributing nozzle 导入喷嘴;分配喷嘴
distributing oblique gear 分配斜齿轮
distributing of security and bond 公司债票的分配
distributing operation 分配运算
distributing pipe 配水管道;分配管(道)
distributing plant 分销工厂
distributing plate 布料板;承重板;分配板;分料板;分布板;撒料盘
distributing point 馈电点
distributing pole 分线杆
distributing ramp 装卸台
distributing reinforcement 分布钢筋
distributing reservoir 配水池
distributing roll 分配辊
distributing roller 均布辊;匀料辊
distributing room 配电室;配电间
distributing screw 布送螺旋
distributing screw conveyer 螺旋摊铺器
distributing section 分配段
distributing shaft 分配轴
distributing shed 调运仓库
distributing slide valve 分配滑阀
distributing station 配电站
distributing steel 分布钢筋
distributing stor(e)y 配货楼层;发货楼层
distributing substation 配电变电站
distributing switch 电流分配器
distributing system 分配系统;配电系统;配水系统
distributing system of port 港口疏运系统
distributing tank 沥青喷布机;沥青分布机
distributing terminal assembly 终端配线部件
distributing trough 布料槽
distributing tube 分配管(道)
distributing valve 分油活门;分配阀;配水阀;配气阀
distributing valve chest 分配阀箱
distributing valve gasket 分配阀垫密片
distributing wagon 风动分渣车
distributing water pipe 配水管(道)
distributing well 配水井
distributing yard 分配车场
distribution 经销;销售;拆版;分配装置;分布;布局;配置;配料方法;配给;配电;撒布
distribution according to labo(u)r 按劳动分配
distribution according to needs 按需分配
distribution accounting machine 分配会计机
distribution accounts statistics 分配核算统计
distribution across the boundary layer 边界层内法向分布
distribution activities 储运作业
distributional area of metamorphic rock 变质岩分布区
distributional assumption 分布(的)假设
distributional effect 分配作用
distributional equity 分配的公平
distributional stability 分布稳定性
distributional weight 分配的权数
distribution amplifier 功率分配放大器;信号分配放大器;分配放大器;分布放大器
distribution analyser of acoustic(al) emission 声发射分布分析仪
distribution and cultivation 分布与栽培
distribution and fuse board 配电与保险丝盘
distribution anisotropy 分布各向异性
distribution area 分布区;分布面积;配线区;配水区域
distribution area of condensated gas pool 凝析气藏分布区
distribution area of oil and gas pool 油气藏分布区
distribution area of pure gas pool 纯气藏分布区
distribution area of saturated oil pool 饱和气油藏分布区
distribution area of undersaturated oil pool 未饱和气油藏分布区

distribution area of well group 井群分布面积
distribution bar 分布钢筋
distribution bar reinforcement 配力钢筋;(与主筋成直角的)细钢筋;分布钢筋
distribution based on manufacturing cost 根据制造成本分配
distribution basin 配水池
distribution blade 分散叶片
distribution block 配电板;接线盘;接线板;配线盘
distribution board 分线盘;分配盘;配线盘;配电屏;配电盘;配电板
distribution board for lighting 照明配电盘
distribution box 分线箱;分配箱;阀箱;配水箱;配水井;配水槽;配电箱;污水配水槽
distribution box for lighting mains 照明配电箱
distribution box for ships 船用配电箱
distribution bucket 布料料罐
distribution budget 分配预算
distribution business 分配业务
distribution cable 分支电缆;分配电缆;分布(式)电缆;配线电缆;配电电缆
distribution canal 配水渠
distribution capability 分配能力
distribution capacity 分配能力;配水量
distribution center 分配站;分货站;配电中心;货物储运中心;商品流通中心
distribution chain 销售过程
distribution chamber 分配室;配线人孔
distribution chamber for forced hot air 加压热气分配室
distribution channel 分销渠道;分配渠道;配线管(道);配线电缆管;配水渠;配水管;分配路线
distribution channel density 分销渠道密度
distribution characteristics 分布特征
distribution charges 销售费用
distribution chute 配料斜槽;分配料槽
distribution circuits 配电网
distribution clause 分配条款
distribution coefficient 分配系数;分布系数;比色系数
distribution coefficient model 分布系数模型
distribution company 煤气公司
distribution conduit 配水道
distribution constant 分配常数;分布恒量
distribution control 分布控制
distribution control system 分布控制系统
distribution cost 销售成本;调运费用;分配成本;发售成本;推销费用
distribution cost allocation 分配成本分摊
distribution cost analysis 分配(销售)成本分析
distribution cost budget 供销成本预算
distribution curve 级配曲线;分配(销售)曲线;分节曲线;分布曲线
distribution curve flux 配光曲线
distribution curve of daylight factor 采光系数分布曲线;天然照度系数曲线
distribution curve of flying path deviation 航迹偏离度分布曲线
distribution curve of pulse amplitude 脉冲幅度分布曲线
distribution curves of element 元素分布曲线图
distribution cutout 配电断流器
distribution decks 分配卡片叠
distribution density 分布密度
distribution determination theory 分配决定论
distribution diagram 分配图
distribution diagram of hydrostatic(al) pressure 静水压强分布图;静水压力分布图
distribution disk 配电盘
distribution ditch 配水渠
distribution duct 分配管(道)
distribution economics 分配经济学
distribution effect 分配效果
distribution efficiency 灌溉效率
distribution energy 分布能量
distribution entry 分配分录;分配表目;分布项目
distribution equilibrium 分配平衡;分布平衡
distribution equipment 分配设备
distribution error 分布误差
distribution expenses 摊销费用;分销费用
distribution facility 销售设施;调运设备
distribution factor 分配系数;分配率;分配因数;分布(式)系数
distribution factor method 分配系数法

distribution floor 配货楼层;发货楼层
distribution frame 配线架
distribution-free 非参数;分布自由;无分布
distribution-free inference 非参数推断
distribution-free method 自由分布法;非参数方法
distribution-free regression procedure 非参数回归方法
distribution-free statistic(al) method 非参数统计(方)法
distribution-free statistics 非参数统计;分布自由统计
distribution-free test of fit 拟合的非参数检验
distribution frequency 分布频率
distribution frequency of element 元素分布频率
distribution function 分配职能;分配(销售)函数;分配函数;分配功能;分布函数
distribution fuse panel 配线熔线盘
distribution gate 分水闸(门);配水闸(门)
distribution gear 分配机构;分配齿轮
distribution graph 分配曲线(图);分配过程线;分布曲线
distribution graph of isotope composition 同位素组分分布图
distribution grid 分布板;配水管网;配电网
distribution gutter 配水(沟)槽
distribution harbo(u)r 调运港
distribution header 分配总管;分配站
distribution holder 地区储气罐
distribution image 分布像
distribution impedance 分布阻抗
distribution in altitude 按高度分布
distribution in area 面上分布;按面积分布
distribution in a unified way 统一分配
distribution in a vertical 沿垂线分布
distribution independence 分布独立
distribution in depth 纵深配置
distribution index 分配指数
distribution-infrared method 分配红外线法
distribution in height 按高度分布
distribution in plane 平面内分布
distribution installation 配电装置
distribution in time 按时间分布
distribution isotherm 分配等温线
distribution kit 分布式配套元件
distribution klystron 分布作用速调管
distribution lane 分流车道
distribution law 分配(定)律;分布(定)律;分布规律
distribution law of karst 岩溶分布规律
distribution ledger 分配(分类)账;分布账
distribution library 分配(程序)库;分布库;提供程序库
distribution line 分布线;配水管线;配电线路;配电干线
distribution liquid 分配液(法)
distribution list 分配(列)表;分布清单;分布表
distribution load 分层荷载;分布荷载
distribution loss 分配损失
distribution main 配水总管;配水干线;配水干渠;配电干管
distribution manager 经销经理
distribution manifold 分配集箱;分配管汇;配水总管;配水歧管;配水母管
distribution map 装配详图(面砖、板材等);分布图;人机分布图
distribution map in exploratory engineering of ore deposit 矿床探矿工程分布图
distribution map in exploratory engineering of ore district 矿区探矿工程分布图
distribution map of deviation of the superficial sediments 表层沉积物离差分布图
distribution map of element content of manganese nodules 锰结核元素含量分布图
distribution map of engineering 工程布置平面图
distribution map of environment(al) diseases 环境疾病图
distribution map of foraminifera and ostracoda 沉积物中有孔虫、介形虫分布图
distribution map of heavy mineral 重矿物分布图
distribution map of heavy minerals in the superficial sediments 表层沉积物中重矿物含量分布图
distribution map of karst form 岩溶形态分布图
distribution map of logarithmic probability 对数概率分布图
distribution map of manganese nodules 锰结核分布图
distribution map of medium size of the superficial sediments 表层沉积物中值粒度分布图
distribution map of mineral deposits 矿产分布图
distribution map of mud in the superficial sediments 表层沉积物中泥含量分布图
distribution map of number of the foraminifera in the dry sample with the same weight 在同重干样中有孔虫个致分布图
distribution map of number of the ostracoda in the dry sample with the same weight 在同重干样中介形虫个致分布图
distribution map of organic matter in the sediments 沉积物中有机质含量分布图
distribution map of pollution sources 污染源分布图
distribution map of recent erosion area 现代剥蚀区分布图
distribution map of recent sedimentation area 现代沉积区分布图
distribution map of sand in the superficial sediments 表层沉积物中砂含量分布图
distribution map of skewness of the superficial sediments 表层沉积物偏态分布图
distribution map of spring 温泉分布图
distribution map of the clay mineral in the sediments 沉积物中黏土矿物含量分布图
distribution map of the grains of the heavy minerals in the superficial sediments 表层沉积物中重矿物颗粒含量分布图
distribution map of the spore-pollen and algae in the sediments 沉积物中孢粉藻类分布图
distribution marker 分配标识器
distribution mark-up 分配加价
distribution matrix 分布矩阵
distribution mechanism 推销机构
distribution memorandum 分配通知单
distribution method 分配(方)法
distribution mix 分配销售组合;分配渠道组合
distribution mode 分布格式;分布方式
distribution model 分布模型
distribution model of direction(al) data 方向数据的分布模型
distribution modulus 分配模量
distribution moment 分配矩
distribution net(work) 配水(管)网;配电网;销售网;分配(销售)网络;分配网络;分布网络
distribution network piping system 配水管网
distribution number 分配数;分布(系)数
distribution of adsorbate molecule 吸附质分子分布
distribution of adsorptive equilibrium constant 吸附平衡常数分布
distribution of agriculture 农业布局
distribution of bearing pressure 承压力分布
distribution of bending 弯曲分布
distribution of bending moment 弯矩分配
distribution of capital 资金分配
distribution of circulation 环流分布
distribution of clay 黏土矿物分布
distribution of clay mineral 黏土矿物分布
distribution of composition in fault rock 断层岩组分分布
distribution of concrete 混凝土运送;混凝土(从搅拌机到模板的)运输
distribution of contact pressure 接触压力分布
distribution of cost 费用分布
distribution of cracks 裂缝分布
distribution of daylight factor 采光系数分布
distribution of deformation 变形分配法
distribution of demand 用量分配;要求分布
distribution of disease 疾病分布
distribution of docking force 靠岸力分布
distribution of documents 分发文件
distribution of ductility demand 要求延性分布
distribution of earth quantities 土石方分配
distribution of element 元素分布
distribution of energy 能量分布
distribution of environmental noise level 环境噪声等级分布
distribution of errors 误差分布
distribution of estate 财产分配

distribution of expenses 费用分配
distribution of exploration net 勘探网间距
distribution of factor income 要素收入的分配
distribution of failures 故障分布
distribution of felling 采伐顺序
distribution of fire-extinguisher 灭火器的配置
distribution of gas 燃气分配;配气;煤气分配
distribution of glacial-marine sediment fraction 浮冰碎屑分布
distribution of goods 货物分配
distribution of grain size 粒度分级;粒度分布;颗粒级配
distribution of heat 热量分布
distribution of heat flow 热流分布
distribution of inclusion 包裹体分布
distribution of income 收入分配
distribution of income account 收入分配账户
distribution of industry 工业布局
distribution of light 光线分布
distribution of load 荷载分布;负荷分布
distribution of LPG 液化石油气供应
distribution of luminous intensity 光强度分布
distribution of material 物资调运
distribution of mean of sample 样本均值的分布
distribution of molecular speed 分子速率分布
distribution of molecular velocity 分子速度分布
distribution of momentum 动量分布
distribution of mooring force 系船力分布
distribution of n-alkane 正构烷烃分布
distribution of national income 国民收入分配
distribution of net profit 净利润的分配
distribution of normal vibration 简正振动分布
distribution of oil and gas field 油气田分布
distribution of oil and gas field by depth 油气田的深度分布
distribution of oil and gas field by geologic(al) age 油气田的时代分布
distribution of oil and gas field by lithology of reservoir 油气田的储集层岩性分布
distribution of oil and gas field by position in basin 油气田在盆地中位置的分布
distribution of ozone, precipitation 臭氧,降水分布
distribution of pathogen 病原分布
distribution of peaks 峰值分布
distribution of pipe 管线(的)布置;管路分布
distribution of population 人机分布
distribution of pores 微孔分布
distribution of pore size 孔径分径
distribution of power 电力分配
distribution of pressure 压力分布
distribution of prestress 预应力分布
distribution of production 生产布局
distribution of productive forces 生产布局
distribution of profit 利润分布
distribution of profit and loss 损益分配
distribution of property 财产分配
distribution of radiant heat 辐射热分布;辐散热的分布
distribution of rafter 橡的布置
distribution of rain 降雨量分布
distribution of rainfall throughout the year 全年降雨分布
distribution of rain intensity 降雨强度分布
distribution of reflux 回流分布
distribution of reinjection 再注入分配
distribution of residual property 剩余财产分配
distribution of resource 资源分配;资源分布
distribution of risks 危险分布;风险分配
distribution of saline soil 盐渍土的分布
distribution of salinity 盐度分布
distribution of salts 盐的分布
distribution of sewage 污水量分布
distribution of shadow 阴影的分布
distribution of ship 配船
distribution of sizes 粒度分配;大小分布
distribution of skewness 分配不匀;偏态分布
distribution of slip 滑动分布
distribution of soil 土壤分布
distribution of source rock in geologic(al) age and strata 烃源岩时代和层位分布
distribution of spectral intensity 谱强度分布
distribution of stations 车站分区;车站分布
distribution of stress 应力分布
distribution of subgrade reaction 路基反力分布;地基反力分布
distribution of suspended matter concentration 悬浮物浓度分布
distribution of taxation 赋税的分配
distribution of variance 方差分布
distribution of velocity 流速分布
distribution of wave energy 波能分布
distribution of wave height 波高分布
distribution of wavelength 波长分布
distribution of wave period 波周期分布
distribution of wealth 财富分配
distribution of wealth and income 财富和收入的分配
distribution of work 功分配
distribution on a reciprocal basis 交互分配法
distribution on entries 分配分录
distribution on gross profit 按毛利分配
distribution on manufacturing cost 按制造成本分配
distribution on reciprocal terms 互惠发行
distribution on selling price 按数量分配;按售价分配
distribution orifice 配水孔
distribution packaging 销售包装
distribution panel 配电盘;配电板
distribution parameter 分布参数
distribution path 分配销售途径
distribution pattern 分配方式;分配结构;分布形式;分布模式;分布类型;分布规律
distribution pattern of heavy placer prospecting 重砂测量布局方式
distribution percentage 分配百分率
distribution permutation 分布排列
distribution photometer 光度分布计
distribution pipe 配水管
distribution pipeline 分配管线
distribution planning 城市布局规划
distribution plan of material balance 物资平衡分配计划
distribution plant 销售工厂
distribution plate 布煤盘;撒料盘
distribution plenum 配风室
distribution point 配线点;配水点
distribution port 集散港;集散点;调运港
distribution pressure 输气压力;配水压力
distribution price 分配价
distribution probability 分布概率
distribution problem 分配问题
distribution program(me) 分配程序;提供程序
distribution range 分布幅;分布范围
distribution range of fold 褶皱展布范围
distribution rate 分配率
distribution rate constant 分布速率常数
distribution ratio 分配比(率);分布系数;提取系数
distribution region 分布区
distribution register of organic pollutant 有机污染物分布自记仪
distribution reinforcement 分配筋;分布(钢)筋;配筋
distribution relations 分配关系
distribution reservoir 分水库;配水箱;配水库;配水池
distribution resonance 分布谐振
distribution ring 撒料圈
distribution road 交通分散道;分散交通道路
distribution road system 分流道路系统
distribution rod 分布钢筋;配力钢筋
distribution series 分配数列;分布数列
distribution shaft 分配轴;配力轴
distribution shed 调运仓库
distribution sheet of job time 工作时间分配表
distribution sheet of stores 存料分配表;材料分配单
distribution sort 分配分类
distribution spider 分料槽
distribution stage 分配级
distribution state 分布状态
distribution statement of service 服务成本分配表
distribution statement of service cost 服务费用分配表
distribution station 储配站;分流站;配电站
distribution steel 分布钢筋;配力钢筋
distribution structure 配水闸(门)
distribution substation 小配电站;变电站;变电所;配电所;配电(分)站
distribution surface 分配面
distribution switchboard 配电盘
distribution switching metwork 分配接续网络
distribution system 配气系统;辐射状配电系统;分配制(度);分配系统;配水系统;配给系统;配电系统;调运系统
distribution system biostability 配水系统生物稳定性
distribution system component 配水系统组成部分
distribution system of gas 煤气分配系统
distribution system of water supply 自来水配水管网;配水管网
distribution system water quality 配水系统水质
distribution technique 调运技术
distribution temperature 分布温度
distribution theory 分布论
distribution thickness 涂抹厚度;撒布厚度
distribution tile 污水分配管;配水瓦管
distribution to income beneficiaries 分配给收益人的款项
distribution trade 批发贸易
distribution transformer 配电变压器
distribution transformer station 配电变电站
distribution valve 压力调节阀;分配阀;配水阀
distribution variance 分布方差
distribution voltage 配电电压
distribution volume 分布容积
distribution warehouse 流动仓库;商品流通仓库
distribution well 配水井
distribution width measurement 沥青洒布宽度测定
distribution winding 分布绕组
distribution wire 配电线
distribution with density 分布密度
distribution works 配水工程
distribution zone 配水区;驼峰调车区
distributive 分配的;分布的
distributive ability 喷油能力;撒布能力
distributive analysis 分配性分析
distributive and service trades 分配性和服务性行业
distributive bargaining 分配性谈判
distributive condition 分配条件
distributive cost 销售成本;发行成本
distributive effect 分配作用
distributive fault 阶状断层;阶梯形断层;分支断层;分支断裂
distributive function 分配职能
distributive industry 个体工业
distributive inequality 分配不等式
distributive judgement 分配评价
distributive justice 分配的公正
distributive lagging model 分布滞后模型
distributive lattice 分配格
distributive law 分配律;分布率
distributive operation 分配运算
distributive pairing 分配配对
distributive profit 可分配利润;经销利润
distributive province 分布区
distributive selection 分配选择
distributive trade 经销(行)业
distributive trade survey 销售调查
distributive value 分配权值;喷油量
distributivity 分配性
distributor 经销商;给料机;分线匣;分配凸轮;分配器;分配机;分流道路;分料器;分布器;发行人;布水器;配水器;配料装置;配电器;排出装置;洒布机;撒布器
distributor adjusting screw 分配调整器螺钉
distributor bar 喷布机喷管
distributor block 接线板;接线盒;分电器
distributor box 配电箱
distributor breaker arm 配电器断电臂
distributor breaker point 分电器断电接触点
distributor brush 配电器电刷
distributor buckle 配电器沟槽
distributor cam 分配凸轮;分电器凸轮
distributor case 分配箱
distributor channel 过渡通路
distributor circuit 分配电路
distributor cog wheel 分电器齿轮
distributor condenser 分电器电容器;配电器电容器
distributor contract 经销合同

distributor control arm 配电器控制臂
distributor discount 经销商折扣;批发折扣
distributor disk 分配盘;配电盘
distributor drive shaft coupling 分配器传动轴联轴节
distributor duct 分配器管道;配电干线管道
distributor elevation 导水机构安装高程
distributor gear 分配机构;分电器齿轮
distributor governor 分配器调节器
distributor housing 分电器外壳
distributor inquiries 经销商询价
distributor mechanism 分送器
distributor nozzle 旋转式水枪
distributor of commodity 商品经营者
distributor piston 分配器活塞
distributor plate 分线板;分电器盖;撒料盘
distributor point 配电器触点
distributor pump 配油泵
distributor road 分支道路;分流道路
distributor roller 传墨辊
distributor rotating disk 分电器回转盘
distributor rotor 分路器转子;分电器转子
distributor shaft 分电器轴;配电架
distributorship agreement 经销协议;经销契约;分配协议
distributor-spreader 联合洒布机
distributor's profit margin 经销商利润赚头;经销利润率
distributor's stock 经销商存货
distributor street 交通分流街道;分流街道
distributor switching 转换分销商
distributor system 交通分流道路系统;分流道路系统
distributor-trailer 拖挂式喷布机
distributor-transmitter 分配发送器
distributor truck 沥青铺布车;路面喷铺沥青卡车
distributor type fuel pump 分配式燃料泵
distributor vacuum brake spring 分配器真空闸弹簧
distributor vacuum control 分配器真空控制
distributor with herical blades 带螺旋形叶片的分布管
distributory river 汉河
distributory stream 汉河
district 林管区;地区;地方;大区;区域;区(段)
district 1 区一
district bank 区域银行
district board 区域界限
district boiler room 区域锅炉房
district center 地区中心;区中心
district centroid 小区形心
district chief 大队长
district communication 区段通信
district connector 区接线器
district cooling 分区冷却
district court 地方法院
district cross-cut 采区石门
district distributor 地区级分流交通道路;区域性输送线;区域性分流道路
district division 地方分区
District Engineer 分局局长(美国)
district engineer 工区工程师;总段工程师;工务总段长
District Engineer's Office 工程兵分局(美国)
district forecast 地区天气预报
district forester 林区长
district forest-office 林区署
district government 区政府
district governor 地区调压器
district haulage conveyer 采区运输机
district heating 局部加热;地区集中供热;分区供暖;区域(集中)供热;区域供暖;区域采暖
district heating and cooling center 地区供热供冷中心
district heating and cooling system 区域供热供冷系统
district heating cable 地区供热电缆
district heating duct 区供热管道;分区供暖管道
district heating line 区供热管道;分区供暖线
district heating network 分区供热系统;分区供暖系统
district heating plant 区供热厂;分区供暖设备;区域供暖锅炉房
district heating system 分区供暖系统;区域供热系统;区域供暖系统
district heat supply 区域供热
district highway 区公路
district holder 地区储气罐
district housing management bureau 区房管局
district local train 区段摘挂列车
district local trains pickup plan 摘挂列车甩挂车作业计划
district manager 地方经理
district office 林区署
district officer 营林所长
district of passenger flow 客流区段
district of quaternary strata 第四纪地层分区
district park 区公园
district passenger ticket 区段票
district plan 地区计划
district planning 分区规划
district pumping station 地区泵站
district pump station 区抽水站
district railway 地方铁路
district ranger 分区营林员
district-ranger system 林区保护员制度
district road 区道;区域性道路
district scale 地区规模
district school 区立学校;地区学校;区县学校
district sector 地区部门
district selector 选线机;地区选择器;区域选择器
district station 区段站【铁】
district surveyor 工区监工;工区测量员;工程检查员;地区工程检查员
district system 区域系统
district telephone network 市内电话网
district thermal heating 区域供热
district thickly inhabited 人口稠密地区
district through train 区段直通列车
district toll switching center 长地区中心局
district train 区段列车
district transfer train 区段小运转列车
district transformer station and distribution centre 地区配电所;地区变电所
district ventilation 采区通风
district wireless station 地区无线电台
distrofication 河湖污染
distrubution channel 分配流通渠道
disruptive distance 跳火距离
disturb 打扰;扰乱
disturbance 搅动;干扰;扰动;紊乱;失调
disturbance-accommodating controller 适应干扰控制器
disturbance allowance 干扰容差
disturbance at the springing 起拱边扰动
disturbance by livestock 牲畜破坏程度
disturbance case 扰动情况
disturbance covariance 扰动协方差
disturbance current 干扰电流
disturbance daily variation 磁扰日变化
disturbance degree 扰动度
disturbance factor 扰动因素
disturbance frequency 扰动频率
disturbance function 干扰函数;扰动函数
disturbance index 扰动指数
disturbance indicator 恶化指标;侵害征兆
disturbance in qi transformation 气化不利
disturbance of analysis 分析障碍
disturbance of circulation 循环紊乱
disturbance of plumb-line 垂线扰动
disturbance of set(ting) 凝结干扰
disturbance of traffic 运输扰乱
disturbance of value 价格扰动
disturbance patterns 扰动图形
disturbance potential function 扰动势函数
disturbance quantity 扰动量
disturbance ratio 扰动比
disturbance-solidification method 搅凝法
disturbance term 扰动项
disturbance variable 干扰变量
disturbance vector 扰动向量
disturbance velocity 扰动速度
disturbance wave 干扰波
disturbance zone 扰动区
disturbed ages 扰动年龄
disturbed area 干扰区;扰动区;受震区;受扰区
disturbed bedding 扰动层理
disturbed belt 错动带
disturbed body 受摄体
disturbed cell 被干扰单元
disturbed coordinates 受摄坐标
disturbed ecologic(al) equilibrium 生态平衡失调
disturbed flow 扰流
disturbed force 扰动力
disturbed forest 受扰林
disturbed fractured zone 扰动破碎带;受扰破裂区
disturbed harmonic process 被扰动的调和过程
disturbed ice 扰动冰
disturbed index 扰动指数
disturbed jet 受扰射流
disturbed motion 干扰运动;受扰运动;摄动运动
disturbed orbit 受摄轨道
disturbed-periodicity model 扰动周期性模型
disturbed profile 扰动剖面
disturbed sample 扰动样品;扰动土样;扰动试样;受干扰样品
disturbed sample of soil 扰动土样;非原状土样
disturbed sand 扰动砂
disturbed soil 扰动土(壤)
disturbed soil sample 非原状土样;扰动土样
disturbed strata 扰动岩层;扰动地层
disturbed test 干扰测试
disturbed track 扰动过轨道
disturbed-upper atmosphere 扰动高层大气
disturbed wave 扰动波
disturbing acceleration 干扰加速度
disturbing background 干扰背景
disturbing body 摄动体
disturbing calculation current of feeding section 供电臂干扰计算电流
disturbing current 干扰电流;串音电流
disturbing dredging 扰动疏浚
disturbing echo 干扰回波;受扰回(声)波
disturbing effect 摄动效应
disturbing force 干扰力;扰(动)力;摄动力
disturbing frequency 干扰频率
disturbing function 摄动函数
disturbing influence 扰动影响
disturbing moment 颠覆力矩;扰动力矩;倾覆力矩
disturbing phenomenon 扰动现象
disturbing potential 扰动位
disturbing potential function 扰动位函数
disturbing resonance 扰乱共振;扰动谐振;扰动共鸣
disturbing sample of soil 扰动土样
disturbing signal 干扰信号;扰动信号
disturbing the peace 扰乱治安
disturb output 干扰输出
distyle 双柱式(门廊);双柱式(的)
distyle in antis 双柱式门廊
distyle pier 双柱式桥墩
distyle pylon 双柱式桥塔
disubstituted compound 二元取代物
disuccinic acid peroxide 过氧化丁二酰
disulfide oil 二硫化物油
disulfo-cyanic acid 二聚硫代氰酸
disulfole 二噻环戊二烯
disulfoxide 二亚砜
disulphonic acid 二磺酸
disunion 不愈合;不统一
disuse 不用
disused canal 废弃运河
disused railway track 废弃线(路);废弃铁路线
disused well 废井
disutility 负效用;反效用
dis-utility of labo(u)r 无效劳动
disymmetry 两侧辐射对称;双对称
dit 小(孔)砂眼
ditactic 构型的双中心规整性;双中心规整性
ditan 二苯(基)甲烷
ditartronate 丙醇二酸氢盐
ditch 刻沟;开渠;沟壑;沟堑;沟;渠;排水明沟;水渠
ditch and bank for arresting sand 挡砂沟堤
ditch and barrier of obstructing sand 挡砂沟堤
ditch-and-trench excavator 开渠挖沟机;挖沟机
ditch blasting 开沟爆破;挖沟爆破
ditch by explosives 爆破开槽
ditch canal 平水运河
ditch check 明沟节制闸;沟中消能槛;沟挡板;防冲沟埂;灌渠斗门
ditch cheek 沟堰
ditch cleaner 清沟机;水沟清理机

ditch cleaning 清沟;清理沟渠
ditch cleaning bucket 清沟桶;清沟斗;清沟构
ditch cleaning machine 清沟机
ditch cleaning out 清理边沟
ditch conduit 沟埋式管道;明排管道;沟槽管道
ditch-corner excavating process 沟角开挖法
ditch crossing 涵洞
ditch cutter 挖沟机
ditch cut(ting) 挖沟;开沟
ditch depth 沟深
ditch digger 开沟机;挖沟机
ditch digging bucket 挖沟铲斗
ditch drainage 明渠排水;明沟排水;排水盲沟
ditch dredge(r) 清沟机;沟渠清理机
ditch ecosystem 明沟生态系统
ditch-end excavating process 沟端开挖法
ditcher 挖壕机;挖沟者;挖沟机
ditcher arm 高频振动臂
ditcher boom 挖掘机臂;挖沟(机)臂
ditcher bucket 挖沟机铲斗;斗式挖沟机
ditcher for laying cable 敷设电缆挖沟机
ditcher ladder 多斗式挖沟机
ditch erosion 沟蚀
ditcher pump 高频脉动泵
ditcher stick 挖掘机铲臂
ditch excavation 挖沟
ditch excavator 挖沟机
ditch for foundation 基槽;基坑
ditch grade 沟底坡度
ditching 挖沟;开沟;挖方支撑;甩开
ditching and trenching machine 挖沟机
ditching by explosives 爆破开槽
ditching by grader 用平地机挖沟
ditching car 为开挖土方而配备的铁路车辆
ditching excavator 挖沟机
ditching grab 挖沟机抓斗
ditching grader 开沟平路机
ditching machine 挖沟机
ditching of road 修路边沟
ditching plough 沟犁;挖沟犁
ditching plow 犁式挖沟机;开沟犁;挖沟犁
ditching scoop 挖掘铲
ditching shovel 反铲;挖掘铲
ditch irrigation 沟灌
ditch junction 沟道交接点
ditch lining 沟(渠)衬砌;沟(道)衬砌
ditch maintenance 沟渠养护
ditchman 挖沟机
ditch method 壕沟法
ditch of foundation 基坑
ditch on top of wall of portal 洞门墙顶水沟
ditch outlet 沟的出口
ditch oxidation 沟渠氧化;氧化沟
ditch planting 壕沟栽植
ditch planting way 壕沟栽植法
ditch plough 挖沟机;沟犁
ditch relief culvert 边沟泄水涵洞
ditch retard 沟内防冲挡板
ditch rider 灌区管理员
ditch sample 挖土取样;泥浆槽岩屑样品
ditch sampling 泥浆槽取样
ditch section 沟断面
ditch sett paving 小石块铺砌沟渠;沟内毛石铺砌
ditch shoe 清沟靴
ditch-side excavating process 沟侧开挖法
ditch slope 沟身边坡
ditch spoil 挖沟弃土
ditch sweeper 清沟机
ditch type brush aerator 沟型转刷曝气器
ditch water level 沟渠水位
diterpene 二萜;双萜
diterpenoids 双萜类
ditert-butylhydro quinone 二特丁基对苯二酚
di-tertbutyl peroxyphthalate 过氧化苯二甲酸二叔丁酯
ditertiary butyl peroxide 过氧化二叔丁基
ditetragon 双四边形
ditetragonal dipyramid 复四方双锥
ditetragonal prism 复正四方柱(棱镜);复四方柱
ditetragonal pyramid 复四方单锥
ditetrahedron 双四面体
Dithane manganese Manebdithane 代森锰
dithecal 二室的
dither 高频振动(器);高频脉动;高频颤动;抖动;颤抖
ditherable magnetron 颤动调谐磁控管
dither arm 高频颤动臂;颤动臂
dither effect 振动效应
dithering 颤动调谐;并列调色
dithering tuner 颤动调谐器
dither motor 高频振动(用)电动机
dither pump 高频振动泵
dither signal 抖动信号;颤动信号
dithiane 二噻烷
dithiazole 二噻唑
dithiene 二噻烯
dithiobutane 二硫代丁烷
dithiocarbamate 二硫代氨基甲酸酯(一种浮选剂)
dithiocarbamic acid 氨荒酸
dithiocarbonic acid 二硫代碳酸
dithiocyanogen 二硫代氰
dithiocyanomethane 二硫氰基甲烷
dithioglycol 二硫代乙醇
dithiolane 二噻茂烷
dithiole 二噻环戊二烯
dithionate 连二硫酸盐
dithionic acid 连二硫酸
dithionite 连二亚硫酸盐
dithionite critrate-bicarbonate 柠檬酸连二硫酸盐重碳酸盐
dithiophosphate 二硫代磷酸盐(一种浮选剂)
dithizone extraction method 双硫腙提取法
dithranol 蒽三酚
Ditisheim balance 迪蒂斯海姆摆轮
ditidante 二钛酸盐
ditolyl sulfone 二甲苯砜
ditridecyl phthalate 邻苯二甲酸双十三烷酯
ditriglyph 三陇板间距;三槽板间距(古希腊陶立克建筑上)
ditriglyphe 复排档
ditrigon 双三角形
ditrigonal dipyramid 复三方双锥
ditrigonal prism 复三角棱镜;复三方柱
ditrigonal pyramid 复三方单锥
ditrigonal scalenohedron 复三方偏三角面体
ditroite 方钠霞石正长岩
dittmarite 迪磷镁铵石
ditto machine 复印机
Dittrich's plug 迪特里希栓
Dittus-Boelter equation 第塔斯-波尔特方程
ditungsten carbide 碳化二钨
diural fluctuation 日变幅
diural range 日变幅
diuranate 重铀酸盐
Diurene diving suit 防热潜水服
diurnal 每日的;昼间的;白天活动的;日刊
diurnal aberration 周日光行差
diurnal acceleration 周日加速度
diurnal age 日潮不等潮龄
diurnal age of tide 日潮龄
diurnal amplitude 昼夜温差;昼夜变化幅度;日振幅;日较差;日变幅;日变程
diurnal and seasonal variation 昼夜和季节变化
diurnal apparent motion 周日视运动
diurnal arc 周日弧;地平上弧
diurnal change 昼夜变化;周日变化;日变化
diurnal circle 周日圈
diurnal clock rate 周日钟速
diurnal component 全日分潮
diurnal constituent 周日分潮;全日分潮
diurnal correction 日变改正
diurnal correction value 日变改正值
diurnal current 周日潮流;全日(周)潮流
diurnal curve method 昼夜曲线法
diurnal cycle 日循环;昼夜循环
diurnal drift 周日漂移
diurnal fluctuation 昼夜变动;日波动
diurnal force 日潮力
diurnal heating 昼夜加热
diurnal horizontal parallax 周日地平视差
diurnal inequality 日均差;日潮位差;日潮流速差;日潮差;日差
diurnal inequality of tides 日潮不等
diurnal lamination 日纹理
diurnal libration 周日天平动;周日秤动
diurnal load 日负荷
diurnal load diagram 昼夜负荷图;日负荷图
diurnal load factor 日荷载率;日负荷率
diurnal maximum 日最大量;日最大(值)
diurnal maximum tide 日最高潮
diurnal mean 日平均
diurnal mean value 日平均值
diurnal migration 昼夜迁移
diurnal minimum 日最小量;日最小(值)
diurnal motion 周日运动
diurnal nutation 周日障动
diurnal ozone profile 臭氧的白日分布曲线
diurnal parallax 周日视差;地心视差
diurnal parallel circle 周日平行圈
diurnal pattern 日间型
diurnal period 日周期
diurnal periodicity 昼现周期性;日周期性
diurnal phase change 相位日变化
diurnal range 每日潮高差;大日较差;大日潮差;日较差;日潮差
diurnal rhythm 昼夜节律;日周期节律
diurnal rotation 周日转动
diurnal temperature 昼夜温度
diurnal temperature change 每月气温变化;日气温变化
diurnal temperature range 温度日较差
diurnal thermal wave 周日热波动
diurnal tidal correction 周日潮汐校正;周日潮汐订正
diurnal tidal current 周日潮流;日潮流
diurnal tide 昼夜潮;周日潮;日周潮;日潮;全日潮
diurnal tide harbo(u)r 日潮港
diurnal tide-producing force 周日起潮力
diurnal type of tide 日潮型
diurnal variability 昼夜差异性
diurnal variation 昼夜变异;昼夜变化;周日变化;日变化;日变短
diurnal variation in earth's temperature 地温昼夜变化
diurnal variation in flow 流量日变化
diurnal variation method 日变法
diurnal vertical migration 昼夜垂直移动
diurnal wave 周日波动;日波
diurnal wind 日变风
divacancy 双空位;双空格点
divagation 改道;泛滥偏差
divagation channel 游移;改道河槽;摆动河道;摆动河床;迁移
divagative river 游移河
divagative river channel 游移河槽
divagative stream 游移河
divalent 二价的
divalent element 二价元素
divalent ion 二价离子;双价离子
divalent metal 二价金属
divalent radical 二价基
divan 长沙发
divan-bed 两用(长)沙发
divariant system 二变系
divaricatic acid 分枝地衣酸
divaricating channel 分叉河道
divarication 河流分叉;分支;分叉点
divarication of a river 河流分叉;河分流
dive 俯冲
dive air compressor 潜水空气压缩机
dive-bomber 俯冲轰炸机
dive brake 下潜制动装置;俯冲制动器
dive culvert 下潜涵洞;倒虹吸管;倒涵管
diveded axle 分离轴
diveded shield 分开屏蔽
dive flap 减速板;制动板
dive knife 潜水小刀
diver 矿浆密度计;潜水者;潜水员
diver boots 潜水靴
divercency 离散
diver communications 潜水员通信
diverge 偏离
divergence 离向运动;离散;扩展度;扩散(度);敞开;辐散;方向行驶;分散;分歧;分出;反常急剧上升;发散(性);发散(度);背离;趋异;歧异;歧离;偏离;散开;散度
divergence angle 扩张角;扩散角;开度角;张角;发散角
divergence area of impact-clastics 撞击碎屑散布面积
divergence boundary 离散型边界;发散界限
divergence curve chart 分歧曲线图

divergence error 离散误差
divergence expression 发散式
divergence factor 发射因数;发散因数
divergence field 辐散场
divergence half-angle 发散半角
divergence indicator 发散指标
divergence line 发散线
divergence loss 扩散损失;发射损耗;发散损失
divergence of a vector 矢量的散度
divergence of a vector-valued function 向量值函数的散度
divergence of pipe 管子的分支
divergence of price from value 价格背离价值
divergence of style 风格的歧异
divergence point 分歧点;分路点;发散点
divergence point of seismic fault 地震断裂分支点
divergence problem 发散问题
divergence rate 扩张度
divergence ratio 扩张比
divergence speed 发散速度
divergence theorem 散度定理
divergence theory 发散理论
divergence zone 扩散区;辐散区
divergencies of prices from values 价格脱离价值
divergency = divergence
divergency boundary 扩散边界
divergency point 分路点
divergent 岔开的;辐散的;偏斜的;偏离;散开的
divergent beam 发射束;发散线束
divergentbeam photography 发散的光束摄影术
divergent belt 发射带
divergent birth process 发散增延过程
divergent bore nozzle 扩散型喷嘴
divergent boundary 离散边缘;扩散边界
divergent branching system 发散分支系统
divergent bundle 发散电子束
divergent channel 扩张形流道;扩散形喷管;扩散渠道
divergent circulation 辐散环流
divergent component 发散部件
divergent cone 扩散锥(体)
divergent-convergent duct 扩散收缩管
divergent current 辐射流;分叉洋流
divergent cycle 发散周期
divergent cyclical pattern 分散循环形态
divergent die 分流模
divergent flow 扩散(水)流;辐散流
divergent function 分散功能;发散函数
divergent integral 发散积分
divergent jet guiding 发散喷注导路
divergent laser beam 发散激光束
divergent lens 发散透镜
divergent lens system 发散(的)透镜系统
divergent light 发散光
divergent lode 分散矿脉
divergent matrix 发散矩阵
divergent meniscus 发散弯月透镜;发散凹凸透镜
divergent meniscus lens 凹凸发散透镜
divergent mirror 发散镜
divergent nozzle 扩张型喷管;扩散式喷嘴;渐扩喷嘴
divergent nozzle pump 发散喷嘴扩散泵
divergent oscillation 增辐振动;增幅振荡
divergent parabola 发散抛物线
divergent pencil of rays 发射束
divergent photograph 离向摄影像片
divergent photography 离向摄影
divergent pipe 扩散管
divergent plate 离散板块;辐散板块;背离(型)板块
divergent plate boundary 离散板块边界;趋离板块界线
divergent radiation 发散辐射
divergent seismic reflection configuration 分叉地震反射结构
divergent sequence 发散序列
divergent series 发散级数
divergent solution 发散解
divergent spheric(al) wave 发散球面波
divergent strains 相异品系
divergent streams 分支流动
divergent structure 发散结构
divergent unconformity 成角不整合
divergent vein 分散矿脉
divergent wave 扩张波;船首波;发散波;(船首的)八字波

divergent wind 发散风
divergent zone 辐散区
diverger 发散部件
diverging 分流;发散
diverging belt sorter 皮带式分级机
diverging cone 扩张锥(体)
diverging conflicts 分流冲突点;分岔冲突点
diverging duct 喇叭形管;扩散水槽;锥形导管
diverging fault 分支断层;分叉断层;树枝状断层
diverging flow 扩散水流
diverging lane 分流车道
diverging lens 分散透镜;发散透镜
diverging light 发射光;发散光
diverging line 分歧线路
diverging meniscus 分散透镜;发散透明透镜
diverging meniscus lens 负弯月透镜;发散弯月形透镜
diverging nozzle 喇叭形管嘴;放射管嘴
diverging ocular 发散目镜;散光目镜
diverging of traffic flow 分散车流
diverging out belt grader 带式分级机
diverging pipe 扩张管;扩散管;渐缩管;渐粗管;锥形管
diverging piping 渐缩管
diverging ramp 分离匝道
diverging rays 发散线
diverging series 发散级数
diverging spheric(al) wave field 发散球面波场
diverging spoke width 发散幅宽度
diverging traffic 交通分流
diverging tube 扩散管;扩散管;渐缩管;锥形管
diverging tubing 渐缩管
diverging veined pore 支脉孔
diverging wave 扩张波;船首波;分散波;发散波;(船首的)八字波;散波
diverging yaw 增幅偏航
diver helper 潜水员助手;潜水信绳员
diveriform 形状不同的
diverless flowline connection 无潜水员输油管连接
diverless operation 无潜水员作业
diverless technique 无潜水员技术
diverless underwater repair 无潜水员水下修理工作
diverless well-head 无潜水员井口
diver method 沉降法;潜沉法
diver's air bottle 潜水储气罐
diver's barotitis 潜水员耳炎
diver's boots 潜水鞋
diver's breathing supply hose 潜水员呼吸供气管
diver's cap 潜水帽
diver's disease 潜水病
diver secondary class 二级潜水员
diverse crops 多茬庄稼
diverse crossing station 变更会车地点
diverse economy 多种经济
diverse farming 多种经营农场
diverse forms of ownership 多种所有制形式
diverse loading 调整装车
diver's emergency transfer system 潜水员应急转运系统
diverse overtaking station 变更越行地点
diver service boat 潜水工作艇;潜水工作船
diversfication of economy 经济多样化
diver's helmet 潜水头盔
diversification 经营多样化;多种经营;多样化;分散化
diversification consortium 多种经营企业
diversification investment 多样化投资
diversification of export 出口多样化
diversification of risks 分散风险;风险分散;风险多样化
diversification of sales 销售方式多样化
diversification of sports funds 体育资金多元化
diversified 各种各样的
diversified agricultural activities 农业多种经营
diversified agriculture 多种经营农场
diversified business 多种经营
diversified common stock fund 分散普通股基金
diversified company 多种经营公司
diversified economy 多种经营;多种经济
diversified economy assets 多经资产
diversified farm 多种经营农场
diversified functions of sports 体育的多元化功能
diversified industrial city 综合性工业城市
diversified investment 多样化投资

diversified load 总负荷;综合负荷;参差负荷;多变荷载;变化荷载
diversified management 多种经营
diversified manufacture 多样化生产
diversified operating strategy of railway enterprise 铁路企业多元化经营战略
diversified product 产品多样化
diversified program(me) of investment 投资(的)多样化规划
diversified slash disposal 多种伐区清理
diversified twill 变化斜纹
diversifier 分散经营者
diversiform 各式各样的
diversifying operation 多种经营
diversing-flow system 导流系统;导流设备
diversing track 岔线【铁】
diversing tube 锥形管
diversion 换向;引流;改道;分出;牵制;偏转
diversion and protection structure 导治建筑物
diversion angle 分流角
diversion aqueduct 引水渡槽;导流渡槽
diversion area 分水区;分流面积(用于地下水);分洪区
diversionary aerodrome 备降机场
diversion barrage 引水坝
diversion belt sizer 皮带式分级机
diversion blind drain 引水渗沟;引水盲沟
diversion bottom outlet 导流底孔
diversion box 分水箱;分流箱
diversion by alluviation 冲积分流
diversion by warping 放淤分流
diversion canal 引水渠;导流渠;分水运河;分水渠;分洪渠(道)
diversion canal development 引水式工程
diversion canal type of power plant 引水式电站
diversion certificate 转运证明书
diversion chamber 引水井;引流室;分水室;分水池
diversion channel 引水渠;导流渠;分水渠;分水道;分洪渠(道)
diversion channel closure 导流明渠截流;导流明渠封堵
diversion check flood 导流校核洪水
diversion chute 分料溜子
diversion clause 绕航条款
diversion closure 截流
diversion conduit 分水管
diversion construction 引水建筑物;引水构筑物;导流工程;分流建筑(物);分流工程;分洪建筑物
diversion curve 转移曲线
diversion cut 截水渠;泄洪口;溢水道;分水渠;分水槽;分洪渠(道)
diversion dam 拦河坝;引水坝;导水坝;导流坝;分水堰;分水坝;分流坝
diversion design flood 导流设计洪水
diversion dike 导流堤;分流堤;引水堤
diversion discharge 分流流量
diversion ditch 截水沟;引水沟;分水沟
diversion duty of water 分水灌溉率
diversion-flow system 导流系统
diversion flume 导(流)水槽
diversion gallery 转向坑道
diversion gate 导流闸门;分水闸(门);分泥板【疏】;分流闸门
diversion headwork 引水渠首工程;分水渠首工程;分流渠首工程;调水渠首工程
diversion hydraulic characteristics 导流水力学特性
diversion intake 导流进水口
diversion layout 导流布置
diversion manhole 分水检查井;分叉处人孔
diversion method 导流方法
diversion of a road 道路改移
diversion of complement 补体转向
diversion of floodwater 分洪
diversion of flow 水流转移;水流改向
diversion of highway 公路转向;分路
diversion of river 河流分流;河流改向;河流改道;河道支流;改河;导流;分流
diversion of river course 河流改向
diversion of road 道路分流
diversion of sediment 泥砂疏导;导砂;排砂
diversion of stream 改移河道
diversion of stream course 河流改道
diversion of traffic 交通改道;交通分流
diversion of water course 水路改道

diversion of waterway 水路改道
diversion outlet 导流底孔
diversion panel 导流屏
diversion passageway 分流通道
diversion pier 分水墩
diversion pipe 分料管
diversion plate 隔板
diversion power plant 引水式(发)电站
diversion privilege 绕航权
diversion procedure 导流程序
diversion project 引水工程;导流工程;调水工程
diversion ratio 分流(流量)比;分水率;分水比
diversion river course 河流改道
diversion road 曲折路;疏散道路
diversion scheme 导流方案
diversion sign 分路标志;分流标志
diversion sluice 分水闸(门)
diversion spillway 分水溢洪道
diversion structure 导流结构物;导流建筑物;引水建筑物;引水构筑物;分水建筑物;分流设施;分流建筑(物)
diversion system of hydropower station 水电站引水系统;水电站引水设备;水电站引水工程
diversion terrace 导水(防冲)阶地;导流台地;分水阶地
diversion trough 分水槽
diversion tunnel 分支隧道;引水隧洞;引水隧道;支隧道;导流隧洞;导流隧道;分水隧道;分流隧洞
diversion tunnel development 引水隧洞式电站
diversion type hydroelectric(al) power plant 引水式水电站
diversion type power station 引水式水电站;引水式电站
diversion type river power plant 河流引水式水力发电厂
diversion valve 引水阀;转向阀
diversion vane 导流屏
diversion waterway 引水道;分水道
diversion weir 引水堰;导流堰;分水堰;分流堰;渠首堰
diversion works 引水设施;引水建筑物;引水构物;引水工程;导流工程;分水建筑物;分水工程;分流设施;分流建筑(物);渠首工程
diversity 相异;多样化;参差;分集;不同;疏散性
diversity antenna 分集(式)天线
diversity branch 分集支路
diversity combiner 分集合并器
diversity condition 异性条件
diversity constant 差异常数;差异系数
diversity effect 参差效应;分集效应
diversity factor 差翼系数;差异因数;差异度;分散系数;分散率;不同时系数;不同时率;不均质系数;不均匀系数;不等率
diversity gain 分集增益
diversity gradient 多样性梯度
diversity index 经营多样化指数;相异性指数;多样性指数
diversity indices 多样比率
diversity of biocommunities 生物群落多样性
diversity of commodity 矿产资源变化度
diversity of geology 地质变化度
diversity of metallogenic types 多种成矿类型
diversity of types and size 型号的多样化
diversity pattern 多样形式
diversity radar 分集式雷达
diversity radio receiving set 分集无线电接收机
diversity ratio 差异比
diversity receiver 分集式接收机
diversity receiver-voter system 分集接收选择系统
diversity reception 分集接收(法)
diversity spacing 分集间隔
diversity stack 反比叠加
diversity system 分集制
diversity telemetry 分集遥测
diver's knife 潜水员刀
diver's ladder 潜水(员)梯
diver's lead weight 压铅
diver's linesman 潜水员助手(潜水员线路指示通信员)
diver's log book 潜水日志
diver's palsy 潜水员瘫痪
diver's paralysis 潜水病
diver's pump 手压潜水供气泵

diver's service boat 潜水工作艇;潜水工作船
diver's shoes 潜水鞋
diver's suit 潜水服
diver's underwear 潜水衬衣
diver's work 潜水作业
diver's wrist depth ga(u)ge 水深手表
diver's wristwatch 潜水手表
divert 变换方向;使转向
divert aerodrome 备降机场
divert current away from the bank 挑流离岸
diverted channel 分支河道
diverted flow 引出水流
diverted gully 改沟
diverted heat 不合格熔炼(炉次)
diverted river 转向河;改向河;改道河(流)
diverted stream 转向河;改向河;改道河
diverted switch 转换开关
diverted track 岔道【铁】
diverted traffic 改道运量;分流运量
diverted traffic volume 转移交通量
diverted tree 马刀树(即醉林)
diver tender 潜水员助手;潜水信绳员
diverter 析流器;转向器;电阻分流器;导流器;分路变阻器;分流管汇;分流(调节)器;分流电阻
diverter coil 分流线圈
diverter control 分流电阻调速
diverter pole charging set 分流电极充电机
diverter pole generator 分流(电)极发电机;分流(磁)极发电机
diverter relay 分流继电器
diverter valve 转换阀;分流器阀门;分流阀
diverter van 换向阀
diverting 改道;绕道
diverting agent 导引剂
diverting dam 引水坝;分水堰;分水坝
diverting device 分散装置
diverting dike 引水堤
diverting river 改道河(流)
diverting stream 改道河
diverting wall 导流墙;分水墙
diverting water 引水;调水;水量分流
diverting wave 船首波
diverting weir 分水堰
diverver gate 转向门
diver work 潜水施工;潜水工作
Divesian 牛津阶【地】
divesting of a right 丧失权利
divestiture 财权利等的剥夺
divestment 转让部分投资
dive-strafer 俯冲轰炸机
divetail slide 燕尾滑板
divide 划分;隔开;等分;分开;分隔
divide and conquer 分步解决
divide-by-10 被十除线路
divide by zero trap 被零除的俘获
divide check 除法校验
divide check exception 除法检验事故
divide cut canal 越岭运河
divide cut reservoir 越岭水库
divided axle 分开式前桥
divided axle-box 可分式轴箱
divided bar graph 分段条形图
divided bath 分格的槽子
divided battery 分组电池
divided beams 分光束
divided beam system 分束系统
divided bearing 分离式轴承
divided blast pipe 组成吹管
divided body 分隔的水带箱
divided channel 分隔航道
divided circle 刻度盘;圆度盘;度盘;分度盘;分度卷
divided circuit 分流电路
divided coarse aggregate 多粒级集料
divided conductor 分裂导线;分裂导体
divided-conductor protection 绕组支路断线保护
divided control runs 分开的操作行程
divided coverage 分担危险
divided crank 组合曲柄
divided crankcase 分离式曲柄箱
divided developer 双液显影
divided diamond 分菱形(货物标志)
divided difference 均差
divided difference operation 均差运算
divided dose 分剂量

divided fall 分段落差
divided fall type 分差式
divided-flow turbine 分流式涡轮机;分流式汽轮机
divided highway 分线公路;分道公路;分出公路;快慢车道分行公路(有分隔带的公路)
divided highway sign 公路分道标志
divided-iron core 分裂铁芯
divided-lane 分车道
divided-lane highway 分车道公路;有分车带的公路
divided light door 分格玻璃门
divided magnetic circuit 磁分路
divided manifold 分支(歧)管;分歧管;分路导管;歧管
divided material 松散材料
divided ownership 分割的所有权
divided pattern drum 分离式提花滚筒
divided pitch 分隔螺距
divided-plate oscillator 分瓣振荡器;分瓣阳极振荡管
divided policy 分红政策
divided propeller shaft 组合螺旋桨轴
divided pulse 分频脉冲
divided rate 不同比率
divided regenerator 分隔式蓄热室
divided reservoir 分间储筒
divided reset 分段重置;分段反馈
divided return duct 分支回流导管
divided rim 组合轮缘
divided road 分线路;分开行驶的道路
divided scale 分刻度;分划尺;(钢铁构件表面的)黑色氧化铁屑
divided seat ring type alarm valve 分座环式报警阀
divided shovel 组合式挖掘铲
divided skirt piston 导缘开缝活塞
divided slit scan 分划扫描
divided sphere ultra high pressure and high temperature device 分割球式超高压高温装置
divided tenon 夹口榫;双榫
divided tow-way highway 分隔行驶的双向公路
divided type disk wheel 组合轮辋式盘轮
divided under carriage 分体式起落架
divided uptake 分隔烟道
divided ventilation 分道通风
divided wheel 分度轮
divided winding 分流绕组;分开式绕组
divided-winding rotor 分开式绕组转子
divided yield 股息收益
divide equally 平分
divide into sections 分格划线
divide into several classes 分成若干类
divide line 分水线
divide migration 分水岭转移
dividend 红利;资本红利;被除数;摊还
dividend account 股息账户
dividend accumulation 累积红利
dividend address operand 被除数地址操作数
dividend bank account 银行股利专户
dividend book 股息簿
dividend check 股利支票
dividend coupon 股利券;股息息票
dividend cover 利润股息比率;净收益对普通股息比率;盈利对股息比率
dividend coverage 股利保证金
dividend credit 纳税时的股息扣除或抵免
dividend declared 议决分派的股利;已宣布股息
dividend earned 股利收益
dividend exclusion 计算所得税时股息不予列计
dividend exemption 股息免税额
dividend extra 额外股利
dividend for preference share holders 优先股利息
dividend fund 股息基金;股利基金
dividend in arrears 积欠股息
dividend income 股息收益
dividend in kind 实物股息
dividend length operand 被除数长度操作数
dividend mandate 股息委托书;股东委托公司将其股息存入银行的委托书
dividendo 分比定理
dividend off 无股息
dividend-on 带息
dividend on land shares 土地分红
dividend on shares 股息
dividend paid 已付股利
dividend payable 应付股息

dividend payable in capital stock 以股票支付的股息
dividend payable shares 付红利股票
dividend paying stock 付红利股票
dividend payment book 股利支付簿
dividend payment sheet 股利支付表
dividend payout percentage 股利支付率
dividend payout ratio 派息率
dividend-per-share presentation 以每股股息额表示
dividend rate 股息率;股金利率
dividend receivable 应收股息
dividend reserve 股息储备金
dividend rights 领取股息的权利
dividend scrip 股利期票
dividends receivable 应收股利
dividend stock 股利股份;股利股票
dividend string 被除数串
dividend tax 纯益税
dividend warrants 股息单;支付股利(保证金)
dividend with a source in China 来源于中国的股息
dividend yield 投资份额收益;每股股利与目前价格的比率;股息率
dividens 分脉
divide out 约去;除
divide ownership 分割所有权
divider 间隔物;线规;除数;除法器;分种器;分水器;分配器;分毛辊;分流管;分茎器;分禾器;分隔物;分隔器;切块机;双脚规
divider bow 分禾器弓形杆
divider cal(l)ipers 等分卡钳;画规
divider chain 分频器链
divider corrector 分配校正器
divider fender bar 分禾器挡禾杆
divider guard 分禾器挡杆
divider height adjuster 分禾器高度调节器
divider rod 分禾器杆
dividers 两脚规;分(线)规
dividers method 两脚规法
divider strip 分格(嵌)条;水磨石分格条;分隔条
divider stripper 分禾器翼板
divider tape 分条皮带
divider valve 分水阀
divider wall 分隔墙
divide shifting 分水岭转移
divide statement 除法语句
divide the work 企业劳动分工
divide time 除法时间
divide water shed 分水岭;分水界
dividing 刻划;分开;分界;分割;分度
dividing and cutting machine 分切机
dividing apparatus 分度装置
dividing arm 分度手柄
dividing attachment 分度附件
dividing box 终端套管;分配箱;分块机;分割箱
dividing breeching 二分水器
dividing broken line 分界折线
dividing bushing 隔离衬套
dividing carriage 分禾车
dividing circuit 除法电路
dividing control valve 分配控制阀
dividing crest 分水岭;分水脊
dividing dam 分流坝
dividing device 分度装置
dividing dial 分度盘
dividing dike 隔流堤;分流堤
dividing disk 分度盘
dividing engine 刻度机;分度机
dividing error 分度误差
dividing fence 分界线
dividing fillet 分类嵌条
dividing filter 分频滤波器;分路滤波器;分离式滤波器
dividing flow 分流
dividing frequency 分配频率;分割频率
dividing gear 分度齿轮
dividing head 分度头
dividing head center 分度头中心
dividing head chuck 分度头卡盘
dividing head driver 分度头传动轮
dividing head spindle 分度头主轴
dividing island 分车岛
dividing knife 分禾割刀
dividing line 界线;中和线;分箱线;分界线

dividing machine 刻度机
dividing maching 分度机
dividing mark of tectonic layer 构造层划分标志
dividing mechanism 分度机构
dividing mineralized zone 矿带划分
dividing network 分频网络;分配网络
dividing panel 分隔(板)
dividing partition 分隔式隔板
dividing pier 分水墩;隔墩
dividing plane 分划面
dividing plate 分度盘
dividing range 分水岭;分散范围
dividing reed 分经筘
dividing ridge 分水岭;分水脊
dividing rock fascia 岩相带划分
dividing rod 分度标杆
dividing roller 分毛辊
dividing shears 切分剪
dividing slate 夹层板岩;板岩夹层
dividing spindle 分度轴
dividing stationary tower crane 分立固定式塔吊
dividing strata 地层划分
dividing strip 分隔带;分隔条;分格(嵌)条;分车带
dividing stripe greening 分车带绿化
dividing strip for terrazzo 水磨石嵌条;水磨石分格条
dividing tectonic layer 构造层划分
dividing the tow 分拖
dividing up state-owned assets 分产
dividing vein 分脉
dividing wall 隔离壁;隔(水)墙;分水墙
dividing wall type 间壁式
dividing wall type heat exchange 间壁式换热
dividing wall type heat exchanger 间壁式换热器
dividing waterwall 双面水冷壁
dividing wheel 分纱轮;分度轮
dividing worm wheel 分度蜗轮
dividng ruler 刻度尺
dividual 分开的
divied duct 分路导管
divinding shears 纵切剪(切)机
diving 潜水;跳入;水深探测
diving accident 潜水事故
diving air hose 潜水空气软管
diving alarm 下潜信号
diving anoxia 潜水缺氧症
diving apparatus 潜水装具;潜水设备;潜水器具
diving armor 潜水衣;潜水盔
diving at altitude 高低潜水
diving ballast 下潜压载
diving bell 潜水钟;钟形潜水器
diving bell foundation 潜水钟式基础
diving bell operation 潜水钟作用
diving bell with platform 带平台的潜水钟
diving board 跳水板潜水箱;跳板(体育)
diving boat 潜水工作艇;潜水工作船
diving box 潜水箱
diving brake 俯冲减速器
diving breathing gas purity standard 潜水呼吸纯度标准
diving buoy 潜水浮标
diving cage 潜水吊笼
diving centrifugal fish pump 潜水式离心鱼泵
diving chamber 潜水舱
diving compartment 潜水舱
diving complex 成套潜水设备
diving cooler 浸入式冷却器
diving current 潜(水)流
diving decompression chamber 潜水减压舱
diving decompression sickness 潜水减压病
diving decompression speed 潜水减压速度
diving depot ship 潜水作业保障船;潜水母船
diving depth 潜水深度
diving depth limit 潜水深度极限;潜水极限深度
diving descent 潜水下潜
diving disease 潜水员病
diving downtime for some reason 因故停潜时间
diving dress 潜水衣;潜水服
diving engineering drill 潜水工程钻机
diving equipment 潜水装具;潜水装备;潜水设备
diving fall 潜水坠落
diving for supporting off shore operation 支援近海作业的潜水
diving gang 潜水作业班

diving gear 潜水装备;潜水用具
diving goods 潜水用具
diving group 潜水组
diving helmet 潜水帽;潜水盔
diving inspection before salvage 救捞作业前的潜水检查
diving inspection before salvage operation 探摸检查
diving inspector 潜水视察员
diving instrument 潜水器
diving intercommunicator 潜水对讲机
diving lock 潜水闭锁(装置)
diving medical security 潜水医务保证
diving medical support 潜水医务保证
diving medicine 潜水医学
diving motor 潜水电动机
diving operation 潜水作业
diving outfit 成套潜水设备;潜水设备
diving party 潜水作业班
diving physiology 潜水生理学
diving plan 潜水计划
diving plane 潜水舵
diving planning 制订潜水计划
diving plant 潜水设备
diving platform 跳台
diving pool 跳水池
diving program(me) 潜水计划
diving psychology 潜水心理学
diving pump 潜水泵
diving research ship 潜水研究船
diving ring and packing 带填料隔离器
diving rudder 潜水舵;水平舵
diving saucer 小型双人潜水器;碟形潜水器
diving ship 潜水工作艇;潜水工作船
diving simulator 模拟潜水装置
diving skip 潜水吊笼
diving slab 俯冲板块
diving stand 跳台
diving stand-by and support 潜水守护
diving suit 潜水衣;潜水服
diving supervisor 潜水监督
diving support vessel 潜水支援船
diving system 潜水系统
diving team 潜水作业班
diving team leader 潜水长
diving technician 潜水技术员
diving telephone 潜水电话
diving tool 潜水具
diving torpedo 深水炸弹
diving tower 跳台;跳塔
diving trim 潜水位势
diving under-ice 冰下潜水
diving wall 分隔墙
divining device 测探设备;探查设备;探矿设备
divining rod 探水树杈;测深竿;测深杆;探(矿)杖
divinity school 神学院
divinyl 联乙烯;二乙烯(基)
divinyl acetylene 二乙烯基乙炔
divinylbenzene 二乙烯基苯
divinyl rubber 丁二烯橡胶
divinylstyrenethermoelastoplasts 二乙烯苯乙烯热弹性塑料
divinyl triammonium 二乙烯三氨
divinyl vinylethylene 丁二烯
divisa (农场、农田、畜牧场等的)地界
divisibility 可约性;可分(割)性;可除尽性;晶体可劈性;解理性;整除性
divisible 可分割的;可除尽的
divisible benefits 可受益
divisible letter of credit 可分割(的)信用证
divisible staff 旋分刻度签
divisible train set 可分级列车组
division 火险区;划分;等分;单位;除(法);分裂;分划;分隔;分部;分(角度单位);部门;标度
divisional 分开的;分部的
divisional accounting 分部会计
divisional bonds 局部债券
divisional colo(u)r 分界色
divisional control 分部控制
divisional cost 分部成本
divisional court 地方法院
divisional forest-office 森林分区署
division algebra 可除代数
division algorithm 带余除法

divisional grade of engineering geology 工程地质区等级
division(al) island 分车岛;中央分车岛[道];分流岛
divisionalization 分部门制度;实现分部制
divisional line 分界线
divisional management 部门管理
divisional management level 分部管理水平
divisional management zone 分部管理地区
divisional method 分区法
divisional officer 营林区长
divisional organization 分区组织
divisional performance 分部业绩;分部经营成绩
divisional performance analysis 分部业绩分析
divisional performance report 分部业绩报告
divisional plane 分隔面;结构面;节理面;构造划分面;断层面;分割面;劈理面
divisional report 分部报告
divisional reporting 分部报表
divisional station of railway administration 局界站[铁]
divisional structural plane 分划性结构面
divisional symbol 分区代号
divisional system 分区制;分部门制度
divisional type 分区类型
divisional unit 分区单位
divisional work 分部工程
division bar 分隔杆;隔条;窗棂
division board 分隔
division boss 段长
division box (灌渠上的)分水装置;分水箱;配水闸(门)
division bulkhead 分(隔)舱壁
division buoy 分支浮标(在中洲下游)
division center 中心结构;分裂中心
division circle 刻度盘
division communication system 区段电话装置
division contract 分项承包(合同)
division contribution statement 分部贡献表
division controller 工区管理员
division converted to multiplication 除法转换为乘法
division delay 分裂延迟
division during sample preparation 试样制备时的缩分
division engineer 主任工程师;工务段长;工区工程师;区局局长(美国陆军工程师团)
division engineer's office 工区区局
division error 分度误差
division establishment right 机构设置权
division fence 分隔篱笆
division gate 分水闸(门)
division general manager 分部总经理
division head 隔仓板
division header 部首;部分头;部(分)标题
division in capital ratio 按资本比例分配
division income statement 分部收益表;部门利润表
division indicator 刻度指示器
division in proportion 按比例划分
division into lot 地段划分
division into parcels 地段划分
division island 分车岛;分隔(车)岛
division lamp 区划灯
division line 分划线;分隔线;分道线
division manager 分部经理
division map of balance calculation 均衡计算分区图
division map of metamorphism 变质分区图
division map showing prospective mineralization area 成矿远景区分区图
division marshal 支队长
division masonry wall 分隔圬工墙;防火圬工墙
division name 部名;部分名称
division of area 面积划分
division of authority and responsibility 权与责的划分
division of business 营业部
division of coal-bearing strata 含煤地层划分
division of common property 分配共同财产
division of construction 工务科;建筑科;建筑部门;工程科
division of design 设计分工
division of economic zones 经济区划
division of enterprise 企业分工
division of function 机能分工

division of geochemistry districts 地球化学区域划分
division of heading section 导坑断面分割
division of highway 公路分区
division of income 收益分配
division of labor and lines of specialities 分业分工
division of labor and lines of work 分业分工
division of labo(u)r 分工(制)
division of labo(u)r and specialization 分工与专业化
division of labour in society 社会分工
division of labo(u)r with individual responsibility 分工负责(制)
division of land 土地划分
division of land property 土地所有权的分配
division of net gains or loss 损益分配
division of net profit 净利润的分配
division of palaeobiogeographic(al) province 生物古地理区划
division of plate tectonic units 板块构造单元划分
division of power 权力划分
Division of Real Estate 房地局
division of relation 关系的划分
division of responsibility 责任分担;分工负责(制)
division of safe degree 安全程度分区
division of soil environmental capacity 土壤环境容量区域分异
division of space 空间划分
division of standard 标准单元
division of technical information 技术情报处
division of tectono-magmatic belt 构造岩浆带
division of tectono-metamorphic belt 构造变质带分区
division of tectono-sedimentary facies belt 构造岩相带分区
division of the graticule 分划板刻线
division of the revenue and expenditures boundaries 收支范围
division of units in regional mapping 区域填图单位划分
Division of Water Pollution Control 水污染控制局
division of wavefront 波阵面分割
division of work 分工
division operator 除法算符
division pier 墩座;分水墩
division plate 隔板;分隔板;分度板
division ratio 分频比;分割比;标度比
division ratio of bifurcated channel 汊道分流比
division ratio of branched channel 汊道分流比
division routine 除法程序
divisions dorsales 背侧股
division sign 分路标志
division stage 分裂期
division subroutine 除法子程序
division surface 界面;隔仓板;分隔面
division system 分部门制度
division under direct vision 直视分离术
division value 分格值;分度值
division wall 间隔墙;隔(离)墙;分隔墙;防火墙;双面水冷壁
divisive reorganization 分裂改组
divisor (道路上的)分车带;约数;因子;除数;分压自耦变压器;分压器;分隔器
divisor address operand 除数地址操作数
divisor length operand 除数长度操作数
divisor string 除数串
divisural line 开裂线
divorced cemen 离散渗碳体
divorced eutectic 离散共晶体
divorced from reality 脱离现实
divorced pearlite 粒状珠光体
divot 方块草皮;草皮(块);草皮层
divulsion 扯裂
Diwa metallogenic theory 地洼成矿理论
diwan 穆斯林法院议事厅;吸烟室
Di ware 帝窑
Diwa region 地洼区[地]
Diwa stage 地洼阶段
Diwa structural layer 地洼构造层
Diwa theory 地洼学说
Diwa-type coal-bearing formation 地洼型含煤建造
Diwa-type magmatic formation 地洼型岩浆建造
Diwa-type metallogenic formation 地洼型成矿建造

Diwa-type oil and gas-bearing formation 地洼型含油气建造
Diwa-type oil shale-bearing formation 地洼型含油页岩建造
dixanthogen 二磺原酸
dixanthyl urea 二磺质基脲
dixed-film process 固着膜法
dixenite 黑硅砷锰石
dixie cup 饮用纸杯;饮料纸杯
Dixoil bronze 迪克索尔青铜
Dixon ring 狄克松环
dizao group 重氮基
dizaomethyl 重氮甲基
dizziness 眩晕;头晕
dizziness caused by smoking 吸烟眩晕
dizziness due to heatstroke 冒暑眩晕
dizzy 发昏
djerfischerite 阴碱铜硫镍铁
djerfisherite 硫铁铜钾矿;硫铁钢钾矿;阴硫铜钾矿
D-joint D 节理[地]
Djuifian 朱伊夫阶[地]
dlignment tolerances of plate edges 板边校直公差
D-line crack D 形密集裂纹;D 裂缝
D-logE curve 感光胶片特征曲线
dnduction by current 电流感应
dneprovskite 纤锡矿
D-normality test 正态性 D 检验法
dnotenschiefer 留状岩
doab 汇流区;河间地段;河间冲积地;多砂黏土;砂质页岩
do-all 杂工
do-all tractor 万能牵引车
do alternately 交叉进行
do a stroke of business 做一笔生意
doat 腐朽(木材瑕疵);腐烂
dobbie 多臂机
dobbin 带磨干燥器
Dobbins-Camp BOD-DO [Biological Oxygen Demand-Dissolved Oxygen] water quality model of river 多宾斯-坎普 BOD-DO 河流水质量模型
Dobble nozzle 多布尔喷嘴
dobby 多臂机
dobby board 高级封面纸板
dobby machine 多臂机
dobby weave 小提花织物
dobie 裸露剂包;裸露爆破;黏土砖(坯);糊炮;砖坯;二次破碎
dobie blasting 裸露爆破
dobie man 爆破工
Dobrowolsky generator 多勃罗沃尔斯基三线发电机
dobschauite 辉铜镍矿
Dobson instrument 多布森分光计
Dobson meter 多布森分光计
Dobson primary standard 多布森一级标准
Dobson prop 多布森支柱
Dobson spectrophotometer 多布森分光光度计
Dobson spectrophotometer measurement 多布森分光光度测量
Dobson spectrophotometer station 多布森站;多布森分光光度测量站
Dobson support system 多布森支架
Dobson total ozone spectrophotometer 多布森臭氧全量分光光度计
Dobson unit 多布森单位
Dobson value 多布森值
do business in a small way 小本经营
do business on one's account 自营商业
dobutamine 多巴酚丁胺
dobying 泡痕;鼓泡;起泡
doc 附证件
Doccia porcelain 多西亚软瓷器(意大利)
dock 港池;码头;站台;船坞;车位[港];停泊处;截短
dock accommodation 港口码头设施;船坞设备
dockage 靠码头;进场费;码头(使用)费;码头捐;船坞停泊费;船坞设备;船渠费;入坞费;停泊费
dock and harbo(u)r 港湾;港口;船坞与港口
dock and harbo(u)r accommodation 港口码头设施;港口码头设备
dock and harbo(u)r authority 港务局;港务管理机构
dock and harbo(u)r engineering 港湾工程学;港口工程;港工
dock and harbo(u)r equipment 港口码头设备

dock and harbo(u)r facility 港口码头设施
dock and harbo(u)r installation 港口码头设备
dock and harbo(u)r management 港务管理
dock and town dues 码头和城市税；入港及入城税
dock apron 坞槛
dock arm 悬臂式坞壁小车；港池支汊
dock barge crane 港口趸船起重机
dock base station 码头基点
dock basin 系泊水域；闸门港；港池；闭式港池
dock block 倒装辘轳；码头龙骨墩；坞墩
dock board 港务局；短跳板
dock boss 码头搬运长
dock bottom 坞底
dock bumper 码头护木；装卸台车挡
dock caisson 箱形坞门
dock cell 码头格体；格体式靠船墩
dock chamber 船坞室；坞室
dock charges 码头(使用)费；船坞停泊费；入坞费
dock control house 码头管理室
dock crane 码头起重机；造船起重机；船坞起重机
dock crane track 码头起重机轨道
dock development 船坞开发
dock displacement 进坞排水量
dock drainage system 船坞排水系统
dock dues 码头税；码头捐；码头费；入坞费
dock dues and shipping 码头费和装货起运
docked vessel 系船泊；停靠在码头的船只
dock engineer 港口工程师
dock engineering 闸坞工程；船坞工程
docken silk 成绞生丝
dock entrance 进港航道；引航道；船坞入口；船坞进口；坞口
dock entrance channel 船坞进口航道
dock entrance lock 船坞进口闸门；船池进口船闸
dock entrance sill 船坞进口闸板
dock equipment 船坞设备
docker 船坞工人；码头工人；港口工人；船坞工作人员；(码头的)搬运工房；搬运工(人)（指码头作业工人）
docket 记入记事表；附加提要；附笺；标签；签条
dock expansion 码头扩充；船坞扩建
dock face 码头前沿
dock face line 码头前沿线
dock facility 船坞设备；坞修设备
dock fender 船坞碰垫；码头护木；码头防撞设施；码头防撞设备；防冲装置【港】；防冲设备
dock filling culvert 坞灌水涵管
dock-filling system 船坞灌水系统
dock floor 干坞底；船坞底板；坞底；停船甲板
dock floor profile 坞底板剖面
dockfront 码头区；码头前沿
dock frontage 码头前沿长度
dock gate 有闸港池；船坞(闸)门；船坞入口；坞门
dock gate opening 坞门开启
dockglass 大杯
dock group 坞群；码头群；港池群
dock hand 码头工人
dock hangar 修理棚【船】
dock harbo(u)r 船坞式港；封闭式港池；闭式港池；闭合式港(口)
dock harbo(u)r basin 闭式港池
dock head 船坞坞首；船坞水头；坞首
dock house 码头办公室；船员临时办公室
dock in 进坞
docking 进坞；截短；泊码头；入(船)坞；上墩【船】
docking accommodation 靠泊设施；靠泊设备；码头设施；码头设备；进坞设备；入坞设备；入(船)坞设施
docking adapter 对接接合器
docking and slipping 坞修和上排
docking and undocking maneuver 进出坞操纵
docking area 靠船区；系船区；碇泊区；泊船区
docking barge 趸船；浮码头
docking block 船坞干船坞垫船木；船坞中的船底垫块；船底垫块(船坞中的)；坞座墩木；坞座垫块；坞墩；稳船块；龙骨垫
docking bridge 船尾桥楼；尾桥【港】
docking bridge awning 尾桥楼天幕
docking bridge awning beam 尾桥楼天幕横梁
docking bridge awning boom 尾桥楼天幕直梁
docking bridge awning stanchion 尾桥楼天幕柱
docking capacity 泊船容量
docking charges 停泊费；进坞费
docking cone 对接锥体

docking deck 尾桥【港】
docking dues 停泊费
docking energy 靠船能(量)
docking facility 进坞设施；码头设施；码头设备；系泊设施；泊船设施；泊船设备；入坞设施
docking facsimile 入港航行的传真
docking fee 进坞费
docking force 靠船力
docking harbo(u)r 泊船设施；靠泊设备
docking impact 靠船撞击力；靠船冲击力
docking impact force 靠船冲击力
docking impact load(ing) 靠船冲击荷载；靠船冲击荷载；撞击力(船靠泊时)；船靠冲击冲
docking indent 进坞契约
docking instruction 进港指令
docking keel 坐坞龙骨；坞(座)龙骨
docking keel block 入船坞盘木；坞内龙骨垫；船内稳船垫木
docking mechanism 对接机构
docking operation 进坞作业
docking order 入坞修理通知书
docking plan 进坞图
docking rail 棚厂操作轨
docking repair 坞修；进坞修理
docking report 进坞报告
docking saw 截土锯；截头锯；吊截锯
docking shackle 卸船卸扣
docking ship 靠泊船；靠船船舶
docking sonar 靠岸测速仪
docking speed 靠船速度
docking stress 靠坞应力
docking structure 靠船建筑物
docking system 船舶靠泊系统
docking telegraph 进坞传令钟；系船传铃钟
docking television 入港航行的传真
docking time 入坞时间
docking velocity 靠船速度
docking vessel 靠泊船只；靠泊船舶；正在进行靠船作业的船只
docking winch 进坞绞车；系船绞车(靠码头用)；带缆绞车
dock installation 码头设备；船坞设备
dockisation 河道船坞设施；港口码头设施
dock ladder 码头爬梯
dock land 码头陆域；港区陆域
dock landing account 码头卸货账单
dock landing ship 海滩登陆舰；船坞登陆舰
dock leg elevator 伸臂式链斗卸船机
dock level(l)er 装货跳板；自动调节站台梯板；车位渡板
dock line 码头线
dockman 码头装卸工(人)；码头水手；码头人员；码头带缆工人；码头搬运工；船坞工人
dock manager 港区经理
dock marshalling yard 港区调车场；港区编组站
dockmaster 码头管理(人)员；船坞长；码头负责人；船坞主管；坞长
dockmaster building 码头管理办公楼
dock of closed construction 实体式码头
dock office 码头办公室
dock of open construction 透空式码头
dock of solid construction 实体式码头
dock operation 船坞营运
dock out 离坞
dock pass 进坞纳费收据；码头出入证；船坞出入证；出货许可证
dock pilot 港内引航员
dock price 出厂价(格)
dock pump 船坞水泵；坞泵
dock quay 港池式码头；码头
dock railing siding 港区铁路；港区铁道
dock railroad 港口铁路（美国）
dock railway 港区铁路；港区铁道；港口铁路（英国）
dock receipt 码头收据；码头收货凭证；收货单
dock rent 趸船码头租金
dock shed 码头前方货栈；码头前方仓库
dock sheet 码头装卸登记表
dock shelter 码头棚
dock shore 坞中支撑
dock shunting locomotive 港内调车机车
dockside 里档；码头一侧；码头区；码头前沿；码头(岸)边
dockside cargo 岸边货物
dockside cargo crane 岸边货物起重机；岸边货物吊机
dockside crane 码头起重机；岸边起重机；船坞起重机；岸壁起重机；坞边起重机
dockside crane rail 坞边起重机导轨
dockside doctrine of appropriation of water rights 港区
dockside fender 码头防冲装置
dockside gantry crane 岸壁门吊
dockside operation 码头作业
dockside road 码头道路
dockside service 码头作业；港口装卸作业
dockside switcher 港内调车机车
dockside travel(l)ing crane 码头前沿移动式起重机
dockside trial 系泊试验；系泊试车
dock sill 船坞门限；船坞(底)槛；坞(门)槛
dock sill level 坞门槛标高
dock siltation 码头淤积
dock slab 码头面板
dock slope protection 码头护坡
dock stevedorage 码头装载
dock stowage 码头装载
dock structure 船坞结构
dock superintendent 船坞长
dock system 港池系统
dock test 系泊试验；系岸试车
dock tonnage dues 码头吨位费；码头吨税
dock trial 码头试车；系泊试验；系泊试车
dock tug 码头拖轮；移泊拖轮；港内作业拖轮；港口作业拖轮
dock-type lock chamber 坞式船闸闸室
dock undertaking 码头公司
dock wall 码头岸壁；船坞墙；驳岸；坞墙
dock walloper 码头工人；装卸工(人)；码头短工
dock warehouse receipt 仓库收领证据
dock warehousing 仓库业
dock warrant 码头栈单；码头仓单
dock weight 码头重量记录单
dock weight note 码头重量记录单；入港货物通知单
dock wharf 港池
dock winch 码头绞车；船坞绞车
dock with intermediate gate 双坞室式船坞
dock work 码头工作；闸坞工程
dock worker 船坞工人
dock yard 造船所；造船厂；港区车场；厂；船舶修造场；船舶修造厂
dock yard over-haul 船修理厂；厂修【船】
docosandioic acid 二十二烷二酸
docosane 二十五烷；二十二烷
docosanol 二十二醇
docosene dicarboxylic acid 二十二烯二羧酸
docosenoic acid 二十二烯酸
docosoic acid 二十二烷酸
docrystalline 多晶质的
doctor 输送校正器调节机；油墨辊；医生；镀层修理器；辅助机构
doctor bar 定厚刮刀；定厚刮尺；刮条；刮棍；刮杆；刮棒；刮刀；涂胶刀；调节配件
doctor blade 刮浆刀；刮墨刀(片)；刮(胶)片；刮(浆)刀；刮棍；漆膜涂布器；涂胶刀
doctor-blade application 刮涂施工
doctor-blade casting 流延法注浆；带式浇筑
doctor blade method 刮涂法
doctor-blading process 流延法；刮片法
doctor coater 刮涂机
doctor coating 刮涂
doctoring of inks incorrect 刮墨不正确
doctor kiss coater 辊纸刮涂机；辊触刮涂机；轻触刮涂器
doctor kiss coating 刮涂；轻触刮涂
doctor knife 刮片
doctor knife coating 刮涂
doctor line 刮刀纹；条痕
doctor mark 条痕
doctor negative 低硫的
doctor of engineering 工程学博士；技术科学博士
doctor of science 理科博士
doctor positive 高硫的
doctor roll 匀浆辊；刮刀辊；涂漆辊；涂胶辊；涂布量控制辊；涂布辊；上浆辊
doctor roller 墨斗辊
doctor's case 需要医治的工伤
doctor scraper 涂胶刀
doctor scrapper 刮漆板；刮墨板
doctor's office 医务室；医师办公室

dolomitic karst 白云岩岩溶
dolomitic lime 含镁石灰;白云质石灰
dolomitic lime putty 含镁石灰浆
dolomitic limestone 白云质石灰岩;白云灰岩;白云石质灰岩;白云石质石灰岩;白云灰岩;白云化石灰岩
dolomitic marble 镁质大理石;白云(石质)大理石
dolomitic marl 白云石泥灰岩;含镁泥灰岩
dolomitic (quick) lime 高镁石灰;白云质石灰
dolomitic sand 白云(质)砂
dolomitic sandstone 白云质砂岩
dolomitic shale 白云质页岩
dolomitite 白云岩
dolomitization 白云石化(作用)
dolomitized bioclast limestone 白云石化生物屑灰岩
dolomitized bioherm limestone 白云石化生物礁灰岩
dolomitized biostratic limestone 白云石化生物层灰岩
dolomitized intraclast limestone 白云石化内碎屑灰岩
dolomitized limestone 白云石化灰岩
dolomitized lump limestone 白云石化团块灰岩
dolomitized oolitic limestone 白云石化鲕灰岩
dolomitized pellet limestone 白云石化球粒灰岩
Dolores beds 多洛雷斯层
doloresite 氧钒石
dolorudite 砾屑白云岩
Dolos 扭工字块体;杜勒斯块体
dolosiltite 粉砂屑白云岩
dolosparite 亮晶白云岩
dolostone 白云(灰)岩
dolphin 墩台;靠船墩;锚船桩群;护墩桩;护墩椿;海豚;系缆桩;系(靠)船墩;系船桩;系船浮筒;簇桩
dolphin beacon 簇桩航标
dolphin bed 墩式台座
dolphin before pile 防冲桩
dolphin berth 离岸停泊;海上停泊
dolphine striker 船首桅杆垂木
dolphin fast 系在缆桩的绳
dolphin of the mast 桅围
dolphin pier 墩式码头
dolphin pile moorings 系船簇桩
dolphin quay 墩式码头
dolphin structure 墩式结构
dolphin system 墩式系统
dolphin type berth 墩式泊位
dolphin type breasting structure 桩式靠船建筑物
dolphin type quay 墩式码头
dolphin wharf 墩式码头
Dolter surface-contact system 道尔式脱表面接触系统
domain 领域;晶域;晶畴;域;功能区;畴;版图
domain administrator 域管理员
domainal cleavage 域式劈理
domain boundary 磁畴界壁;畴界;畴壁
domain fabric 粒团组构;团粒组构
domain growth 畴的生长
domain hypothesis 功能区假说
domain maritime 领海权
domain mode 畴模式
domain motive 主要动机
domain name server 域名服务器
domain name system 域名系统
domain of a fuzzy relation 模糊关系的定义域
domain of attraction 吸引域
domain of convergence 收敛域
domain of definition 定义域
domain of dependence 依赖域
domain of feasible solution 可行解区域
domain of influence 影响域
domain of interpretation 解释域
domain of study 研究域
domain of walker 浮动范围
domain operator 区域运行程序;区域操作员
domain reversal 电畴反转
domain rotation 磁畴旋转;畴转动
domain stability 畴稳定性
domain structure 电畴结构;磁畴结构;畴结构
domain-switching process 磁畴开关过程
domain texture 畴线结构
domain theory 畴理论

domain tip 磁泡;畴尖
domain tip device 磁泡器件;畴尖器件
domain-tip memory 畴尖存储器
domain wall 畴壁;布洛赫壁
domain-wall coercivity 畴壁矫顽力
domain-wall mobility of magnetic bubble 磁泡畴壁迁移率
domain-wall movement 畴壁运动
domain-wall resonance 畴壁共振
domain-wall switching 畴壁翻转
domain-wall velocity 畴壁速度
domal 圆顶状的;穹隆状的
domal anticline zone 穹隆背斜群
domali 女儿墙;矮墙
domal structure 穹状构造
domal uplift 穹状隆起
domanial sampling 地区抽样
domatic 坡面的
dome 流线型罩;海底半球状突出地(水深在六百英尺以下);圆屋顶;圆穹顶;圆盖;圆顶丘;圆顶;整流罩;拱形圆顶;顶盖罩;导流罩;导流纳;封头;反映双面;半球形水团;穹丘;穹顶;坡面
dome above a square 方形建筑上的穹隆
dome aerator 圆罩式曝气器
dome and basin structure 穹隆和构造盆地
dome apex 穹隆顶点;圆顶顶点
dome bar 穹隆杆
dome barrel 汽包
dome base 汽室垫圈
dome-basin structure 穹盆构造
dome-block mountain system 穹状断块山系
dome bottom 半球形底
dome brick 拱砖;顶砖;土窑砖;楔形砖
dome cap 钟形罩
dome collar 钟形汽室垫圈
dome construction 穹形建筑
dome cornice 圆顶挑檐
dome crown 圆顶拱顶
dome cupola 拱圆顶
domed 拱凸;表面凸起;使成穹顶
dome dam 穹隆坝;圆穹顶坝;双曲拱坝
domed arch 馒头碹
domed arch dam 穹顶形拱坝
domed basilica church 圆顶长方形教堂;带有圆顶的巴西利卡式教堂
domed building 穹顶建筑;带有圆顶的建筑物
domed ceiling 拱形天花板
domed central-plan church 中央穹隆式教堂;带有穹顶的集中式教堂
domed chapel 穹隆作顶的小教堂
domed church 有穹隆的教堂
domed-cruciform church 圆顶十字形教堂
domed dam 穹形拱坝;双曲拱坝;穹隆坝
domed diminutive tower 圆顶小塔楼;带有穹顶的小塔楼
domed end 圆顶端
domed floor 半球形地板
domed form 穹隆状;半球状
domed hall 圆顶大厅;有圆顶的大厅
domed hall church 穹顶大厅教堂
domed-head stopper 塞头砖
dome diffuser 圆顶扩散器;钟铃形扩散器
domed mosque 圆顶伊斯兰教院;圆顶清真寺;带有圆顶的清真寺
domed mountain 钟形山;隆起山
domed nut 盖螺母
dome door stop 球形碰头
domed pavilion 圆顶亭
domed roof 穹隆屋顶;圆屋顶
domed roof-light 穹隆的顶窗;碟形穹隆的顶窗
dome-drum 圆屋(顶)座盘;穹顶的鼓座
domed shell 半球形薄壳;半圆形薄壳;球形壳(体)
domed slab 穹隆状板
domed square 方底穹隆
domed stadium 穹隆作顶的体育场
domed structure 穹隆作顶的建筑物;圆顶结构
domed style 穹隆风格建筑
domed tank roof 容器球形盖
domed temple 有穹隆的神庙
domed turret 圆顶小塔楼;有圆顶的小塔楼
dome dune 穹状沙丘
domed vault 隆起的拱顶地下室;穹形拱顶
domed volcano 穹形火山

dome edge 穹隆的边;圆顶边缘
dome eye 圆屋顶观察窗
dome flange 汽室凸缘
dome floor 半球形地板
dome form 穹隆形式;圆顶形式;穹顶模板
dome format 穹式结构
dome foundation 穹隆式基础
dome-geometry light source 半球形光源
dome head 球形封头;球面底
dome-head cylinder 半球形压缩室汽缸
dome-head piston 圆顶活塞
dome housing 圆顶居所
dome illumination 天棚照明
dome impost 穹隆柱墩;圆顶拱基
dome key 穹隆顶
dome kiln 馒头窑;圆屋窑;圆顶地下式石灰窑;圆顶窑
dome lamp 穹面灯;顶灯;天棚灯;穹顶灯
dome lamp fixture 天棚照明灯具
dome light 顶灯;座舱顶灯;穹面灯;穹顶天窗孔;天棚灯
dome like 穹顶状的
domelike structure 穹形状结构;穹隆状结构;穹顶状结构
dome-like upheaval 穹状隆起
dome manhole 圆顶人孔;储罐入孔
dome mountain 隆起山;穹形山;穹隆山
dome nut 圆盖螺母
dome of multiangular plan 多角穹隆;多角穹顶
dome of polygonal plan 平面为多边形的穹隆;多角穹隆;多边穹顶
dome of regulator 调节器帽
dome of rotational symmetry 轴对称穹隆
Dome of the Invalides 伤残者穹隆(1680～1691年建于巴黎的文艺复兴时期穹隆建筑)
Dome of the Rock 罗克穹隆(688～692年建于耶路撒冷的伊斯兰神殿,具有高穹隆)
dome on tambour 鼓形壁上的穹隆;鼓座上的圆顶
dome organ 钟形感器
dome pressure 气包压力
dome reflector 穹形反射器;穹面反射器
dome reservoir 穹隆状热储
Domerian (stage) 多麦尔阶【地】
dome ring 穹隆圈;穹隆环;圆顶环
dome roof 圆盖形屋顶;穹式屋顶
domes 月面拱形结构
dome segment 圆顶(分)片;穹隆分部
dome-shaped 圆丘状
dome-shaped bottom 穹隆形底;锅形底
dome-shaped contact 半球形接点
dome-shaped dam 穹形坝
dome-shaped floor 穹隆地板
dome-shaped form 穹隆状形式
dome-shaped iceberg 圆顶状冰山
dome-shaped reflector 半球形反射罩
dome-shaped roof 圆顶炉;穹隆状屋顶;圆(盖)屋顶
dome-shaped shell 穹隆状薄壳
dome-shaped slab 穹隆状板
dome shell 穹隆薄壳
dome silo 穹顶混凝土料仓
dome skylight 采光罩;圆顶天窗
dome slab 穹隆状板;圆顶板
dome slot 半球形缝
domestic 民用的;国内的
domestic accommodation 本国货
domestic activities 家庭活动;国内活动
domestic aerodrome 国内机场
domestic agent 国内代理商
domestic airport 国内航空港;国内机场
domestically associated enterprise 内联企业
domestic altar 家用圣台
domestic animal trade tax 牲畜交易税
domestic appliance 家用设备;家庭用具
domestic appliance burning town gas 烧管道煤气的家庭用具
domestic appliance circuit 家用电器线路
domestic arbitration award 国内仲裁裁决
domestic architecture 居住建筑;民房建筑;住宅建筑(学)
domestic assets 国内资产
domesticated stage 驯化阶段
domestication of aerobic-settling condition 好氧沉淀条件下驯化

domestication of anaerobic-settling condition 厌氧沉淀条件下驯化
domestic baseboard heating 室内踢脚板供暖;国产隔热板
domestic base plate heating 室内踢脚板供暖
domestic bathroom 家庭浴室
domestic bidder 国内投标人;本国投标者;本国报价人
domestic bill 国内票据
domestic block 住宅大厦;居住街坊
domestic boiler 家用锅炉;日用锅炉
domestic bond 国内债券
domestic bond issue 境内债券
domestic borrowing 内债
domestic branch 国内分行
domestic building 居住房屋;住宅(房屋);住房房屋
domestic building type 住宅大楼型
domestic burning appliance 家用炉子;家庭燃烧用具
domestic capital 本国资本
domestic capital as circulating funds 流动资金内资金部分
domestic capital used for purchase of rolling stock 机车车辆购置费中内资
domestic cargo 国内航线货物
domestic carriage 国内运输
domestic ceramics 家用陶瓷;日用陶瓷
domestic chapel 礼拜堂;小教堂
domestic chimney 住宅烟囱;普通烟囱
domestic clay brick block 砖砌住宅楼
domestic clay brick building 砖砌住宅楼
domestic climatology 生活气候学
domestic coke 家用焦炭;民用焦炭
domestic commerce 国内贸易
domestic communication 国内交通
domestic connection 家庭用水管连接
domestic consolidation 国内兼并
domestic construction 住宅施工;住宅建造;住宅建设
domestic construction material 国产建筑材料
domestic consumer 普通用户
domestic consumption 家庭需水量;民用耗水量;国内消费;生活用水量;生活消费量
domestic container 家庭垃圾容器;国内集装箱;国产集装箱
domestic content rate 本国自制率
domestic corporate issues 国内公司发行
domestic corporation 州内公司;国内公司
domestic cost 国内费用
domestic credit 国内信贷;国内信用
domestic credit expansion 国内信用增加额
domestic cullet 厂内回炉碎玻璃;工厂碎玻璃
domestic currency 本国货币
domestic currency bill 国币汇票
domestic debt 内债
domestic demand 国内需求;生活需要
domestic deposit 国内存款
domestic detergent 家用洗涤剂
domestic development of major technical equipment 重大技术装备国产化
domestic diesel oil 民用柴油
domestic discharge 生活排污
domestic dispute 内部争端
domestic draft 国内汇票
domestic earthenware 日用陶器
domestic economy 家庭经济;国内经济;本地经济
domestic electric(al) appliance 国内用电器(设备)
domestic electric(al) installation 家用电器装置
domestic electric(al) water heater 家用电热水器
domestic energy requirement 民用能源要求
domestic enterprise 国内企业
domestic exchange 国内汇兑
domestic exchange unsettled account 国内汇兑未清账户
domestic expenditures 国内支出
domestic expenses 家庭支出
domestic factory 家庭工场
domestic fecal sewage 家庭粪便污水;生活粪便污水
domestic filter 家用过滤器
domestic fine pottery 日用精陶
domestic fire 家庭炉灶
domestic flight 国内航线
domestic food waste 家庭食品垃圾;厨房垃圾

domestic fowl farming 家禽业
domestic fuel 家用燃料;民用燃料
domestic fuel gas 民用燃料气
domestic fuel oil 民用燃料油
domestic furnace 家用火炉
domestic furniture 家用家具
domestic garage 住宅汽车间;家用汽车间
domestic garbage 生活垃圾
domestic gas 民用(煤)气
domestic gas appliance 家用燃(气用)具;家用煤气设备;家庭煤气用具
domestic gas burning appliance 家庭煤气燃烧用具
domestic gas-fired incinerator 家用煤气焚化炉
domestic gas meter 家用煤气表
domestic gasoline 民用汽油
domestic glass 日用玻璃
domestic glassware 日用玻璃器皿
domestic goods 国货;本国货;本国产品
domestic grade 民用规格
domestic grade block 高档住宅;采样拼花地板的住宅
domestic heating 家用热水;家庭采暖;民用供热
domestic heating appliance 家用取暖设备
domestic heating installation 住宅采暖装置;家庭采暖装置
domestic heating oil 民用取暖油
domestic heating plant 家用采暖设备
domestic heating system 家用供热系统;家用采暖系统
domestic help 家务照顾
domestic hot water 家用热水;民用热水
domestic hot-water heater 家用热水器
domestic hot water system 住宅热水供热系统;家用热水(供热)系统;民用热水系统
domestic incinerator 生活垃圾焚化炉
domestic induction heater 家用感应加热器
domestic industrial district 手工业区
domestic industry 国内产业
domestic inflation 国内通货膨胀
domestic installation 生活用电设施;生活设施
domestic institution 境内机构
domestic instruments 家庭用具
domestic insurance 家计保险
domestic investment 国内投资
domestic kerosene 民用煤油
domestic kitchen 住宅厨房;家用厨房
domestic legal person 国内法人
domestic legislation 国内立法;国内法规
domestic letter of credit 国内信用证
domestic level 本地水平
domestic line 国内航线
domestic load 民用负荷;生活用电量;生活用电负荷
domestic loan 内债;国内贷款
domestic loom 脚踏木织机
domestic mains 用户管
domestic manufacture 国内制造工业;本国制造品
domestic-manufactured 本国产的
domestic map 本土地图
domestic marble 国产大理石
domestic market 国内市场
domestic means of payment 国内支付手段
domestic merchant 国内中间商
domestic meter 家用水表;用户水表
domestic money market 国内资金市场
domestic money order 国内汇票
domestic money order cashing 国内汇票兑现
domestic money order issuing 国内汇票发行
domestic needs 国内需求
domestic noise 住宅噪声;家庭噪声;生活噪声
domestic office 国内办事处
domestic oil 国产石油
domestic operation accident 国内航线飞行事故
domestic outside-type incinerator 民用室外型焚化炉
domestic passenger 国内旅客
domestic pesticide 家用杀虫剂
domestic photomap 本国领土地图
domestic pollution 公用生活污染;生活污染
domestic pollution of water body 水体生活污染
domestic pollution sources 生活污染源
domestic porcelain 日用瓷器
domestic potable water 生活饮用水
domestic preference 国内优先;国内优惠

domestic premises 居住用房屋;居住用房产
domestic price 国内价格
domestic price level 国内价格水平
domestic price of products 产品国内价格
domestic product 国内生产
domestic production 民用生产;国产(品);本国产品
domestic products 国货;土产(品)
domestic public frequency bands 国内公用频带
domestic pump 家用(水)泵
domestic quarter 国内旅客区;住宅区
domestic range 家用炉灶
domestic refuse 家庭垃圾;生活垃圾
domestic refuse compactor 生活垃圾压实机
domestic refuse disposal 生活垃圾的处置;家庭垃圾的处置;城市垃圾处置
domestic remittance 国内汇兑
domestic resources 本国资源
domestic reuse 生活再用
domestic robot 家用机器人
domestic room door 住宅内室门
domestic satellite [DOMSAT] 国内通信卫星
domestic saunabath 家庭桑拿浴室;家庭蒸汽浴室
domestic-scrap 厂内废料
domestic septic tank effluent 生活化粪池污水
domestic sewage 家庭污水;民用下水道(系统);生活污水
domestic sewage treatment 生活污水处理
domestic sewerage 家庭污水工程;民用排水系统;生活污水
domestic sewer network 生活污水管网
domestic sewer system 家庭污水(管)系统;生活污水管系统
domestic ship 本国船舶
domestic shipping 国内航运
domestic situation 国内形势
domestic skirting heating 家庭壁脚板供暖
domestic sludge 生活污泥
domestic smoke 家庭烟气
domestic softener 家用水软化剂
domestic solar water heater 家用太阳热水器;民用太阳能热水器
domestic solid fuel boiler 烧固体燃料的家用锅炉
domestic solid wastes 生活固体废物
domestic source of water pollution 水生活污染源
domestic stair(case) 家用楼梯
domestic standardization 国内标准化
domestic standard of water quality 国内水质标准
domestic stoneware 日用炻器
domestic storage financing 国内财政储存
domestic stove 家用火炉
domestic supply sewage 生活供水工程
domestic supply system 家庭供水系统
domestic tariff 国内税收
domestic tax 国内税收
domestic terminal building 国内候机室;国内航空站;国内乘客候机楼
domestic timber 国内木材
domestic trade 国内运输;国内贸易
domestic traffic 国内运输
domestic tranquility 生活安宁
domestic trust 国内赊售
domestic use 家用;家庭使用
domestic use of water 民用水;生活用水
domestic utility corridor 住宅实用走廊;家庭实用走廊;家用过道
domestic utility patio 住宅实用内院;家庭实用内院
domestic utility room 住宅杂用室;家庭杂作间;家庭杂用室
domestic utilization of gas 家庭用气;生活用气
domestic vein 同簇矿脉
domestic waiting room 国内候机室
domestic waste 家庭废弃物;生活废物;家庭污水;家庭垃圾
domestic waste(water) 家庭污水;家庭废水;生活污水;生活废水
domestic wastewater post-biotreatment 生活污水后生物处理
domestic wastewater reclamation 生活污水回收
domestic wastewater treatment 生活污水处理
domestic water 生活用水;家庭用水
domestic water demand 居民需水量;生活需水量
domestic water heater 用户热水器;用户加热器
domestic water heating 住宅热水采暖
domestic water meter 民用水表;用户水表

domestic water piping system 住宅水管系统
domestic water preheat storage 民用水预热储存
domestic water reuse system 生活回用水系统
domestic water shut-off 生活给水关闭
domestic water softener 家用饮水器
domestic water source 本地水源
domestic water supply 家庭供水水源;民用水(水)源;民用给水;生活水源;生活供水
domestic water supply quality 生活供水水质
domestic water supply system 家庭供水系统;生活供水系统
domestic water system 生活给水系统
domestic water use 生活用水
domestic window 国内售票窗口
domestic wood 国产木材
domestic workroom 住宅杂用室;家庭工作室
domestration of activated sludge 活性污泥驯化
dome stress 圆顶应力
dome structure 穹状构造;穹隆构造
dome surface 圆顶表面
dome system 拱形体系;混凝土料仓建造系统
dome theory 自然平衡拱理论
dome top 圆顶顶端
dome trap 穹隆圈闭
dome type 穹式
dome type distribution 穹隆形分布
dome type skylight 采光罩
dome whistle 钟形汽笛
dome with spire 有尖塔的穹顶
domeykite 砷铜矿
domiant weak acid 优势弱酸
domical 圆屋顶的;半球形穹顶的
domical architecture 圆顶式建筑
domical bottom 半圆底;半球底
domical floor 半球形地板
domical mound 半球形土墩;舍利(子)塔
domical shell 拱形薄壳
domical slab 穹隆形板
domical vault 内凹穹顶;圆穹顶;半球形穹顶
domicile 户籍;原籍;期票支付所
domicile bill 外埠付款票据
domiciled bill of exchange 外埠支付汇票
domiciled check 外地付款支票
domiciled credit 托付信用证
domiciled note 外地付款期票
domicile of origin 原籍
domicile register 户籍
domick 小石块
dominance 优势(度);超先;超过
dominance at locus 位点的显性
dominance of strategies 策略的优势
dominance relationship 支配关系
dominance strict 严格支配
dominancy 优势度
dominant 支配夫;占优势
dominant action 优先行动
dominant activator 主激活剂
dominant algae species 优势藻类物种
dominant alternatives 关键方案
dominant bank 主导河岸
dominant colo(u)r 主色(调)
dominant company 主要公司
dominant crop 优势作物;主要作物
dominant diagonal 对角占优势
dominant discharge 控制流量;主要流量;支配流量;造床流量
dominant dissolved species 溶解优势物种
dominant eigenvalue 主本征值
dominant estate 地产支配权
dominant fault line 主断层线
dominant firm 占有统治地位的公司
dominant flow 主要流量
dominant formative discharge 河道造床流量;主要造床流量;主要成槽流量;成槽流量
dominant frequency 优势频率;卓越频率;主频(率)
dominant function 控制函数
dominant future 支配性期货
dominant harmony 单色绘画;单色调配
dominant hue 主色相;主色调;支配(彩)色
dominant line 主测线
dominant mechanism 主要机理
dominant metal species 优势金属物种
dominant mode 基谐模式;主模;波基型

dominant mountain spur 高耸孤立丘
dominant negative diagonal 负对角占优
dominant obstacle allowance 最高障碍物容许限额
dominant organism 优势生物
dominant oxidation state 优势氧化态
dominant peak 主峰
dominant plant 主要作物
dominant pool method 留钢法
dominant population 优势种群
dominant shape 主形
dominant socials 优势种
dominant soluble speices 可溶性优势杂种
dominant species 优势种;优势类型;主要种类
dominant speices 优势种
dominant tenement 主地产;承役地
dominant term 主(要)项
dominant tree 优势树;支配木
dominant variable 控制变量
dominant vector 控制向量
dominant wave 主要波浪;主波;盛行波(浪)
dominant waveguide transmission mode 波导传输主模
dominant wavelength 主(要)波长
dominant wavelength of seismic wavelet 地震子波的主波长
dominant wind 盛行风
dominant wind direction 最多风向
dominate 优势;占优势;超出
dominated 中等木;中等的
dominated convergence theorem 控制收敛定理
dominating condition 控制条件
dominating influence 决定性影响;主要影响
dominating integral 优势积分
dominating series 优势级数
dominating wind 常风;盛行风
domination of input combinations 控制投入组合
domination of strategies 策略的支配;策略的超优
doming 成拱作用
doming-up of pavement 凸起路面
Dominican architecture 多米尼加式建筑
Dominican church 多米尼加教堂
Dominican mahogany 多米尼加红木
Dominican monastery 多米尼加修道院
Dominican novitiate monastery 多米尼加见习修道院
Dominican nunnery 多米尼加修女院;多米尼加女修道院
dominion 主权;领土
Domino effect 多米诺效应
domus (古罗马或中世纪时期的)住宅;房子
domy 圆(屋)顶的
Donal 多纳尔铝合金
Donaldson sludge density index 多纳尔逊污泥密度指数
donalt conveyer 袋式提升机
donarite 多纳炸药;安全炸药
donate 捐助;捐赠
donated account 捐赠账户
donated assets 捐增资产
donated capital 捐赠资本
donated fund 捐赠基金
donated land 捐赠地产
donated land reserve 捐赠地产储备
donated stock 捐赠股份
donated surplus 捐赠盈余
donated working capital 捐赠运用资本;捐赠营运资本
Donath chart 多纳斯图
donathite 四方铬铁矿
donation 捐赠;捐款;赞助
donation account 捐赠账户
donation tax 赠与税
donator 赠送人;给予体;给体;施主
Donau glacial stage 多瑙冰期【地】
Donau-Gunz interglacial stage 多瑙—贡兹间冰期【地】
donbassite 片硅铝石
Donders reduced eye 东德尔斯简约眼
done bit 操作位
donee 受赠人
donegan 工地活动厕所(俚语)
Donetsk polarity hyperchron 多涅茨克极性亚时
Donetsk polarity hyperchronzone 多涅茨克极性巨时间带

donga 峡谷;陡壁;陡岸干沟;冰峡谷;山峡
Donghai-Nanhai block-zone 东海—南海地块带(中国)
Dongola leather 东古拉革
Dongpo old land 洞鄱古陆
Dong ware 董窑
Dongyang wood carving 东阳木刻
doniker 巨形块石(俚语)
donjon 中世纪城堡的主塔;城堡主塔;城堡主楼
donk 亚黏土;脉壁黏土
donkey 移动式牵引机;辅助的;平衡重;上坡牵引车
donkey boiler 副锅炉;辅助锅炉
donkey boiler equipment 辅助锅炉设备
donkey check valve 辅助止回阀
donkey crane 辅助起重机;蒸汽吊车;小起重机
donkey doctor 发电设备的技师;修理工(俚语)
donkey drain pipe 副泄水管
donkey engine 卷扬机;绞盘;绞车;小汽机;移动式蒸汽机;辅助发动机;辅机;轻便发动机
donkey engine feed pump 辅助发动机添水泵
donkey fan 辅助鼓风机
donkey hoist 辅助升降机;辅助起重机
donkey-key 辅助发动机
donkeyman 辅助操作工;副司炉;停泊时机舱值班员
donkey pump 锅炉水泵;蒸汽往复(给水)泵;蒸汽泵;辅助(水)泵
donkey's breakfast 草席;草垫
donkey's ear 斜接插销;(钳口中的)插板
donkey stack 备用烟囱
donkey topsail 辅助上桅帆;四角斜桁帆
donkey work 辅助工作
Donnacona 多纳柯那(一种木纤维墙板)
Donnan effect 道南效应
Donnan equilibrium 道南平衡
Donnan potential 道南电位
Donnan's membrane equilibrium 道南膜平衡
donnayite 碳钇锶石
donnish 略暗褐色
Donoghue and Bostock centrifuge 多级碗式离心机
Donoghue equation 多诺休方程
donor 捐赠者;捐款人;供体;给体
Donora smog episode 多诺拉烟雾事件
donor atom 供电子原子
donor basin 供水流域;调出流域;施调流域
donor canal 供给管道
donor concentration 施主浓度
donor country 捐赠国;捐款国;援助国;资助国
donor dopant 电子供体掺杂剂;施主掺杂剂
donor impurity level 供体杂质含量;施主杂质含量
donor ion 供体离子;施主离子
donor levels 施主能级
donor limestone 输出灰岩
donor plaque 捐赠牌
donor principle 主式原则
donor region 供水区;调出区;水源区;施调区
donor selection 供体选择
donor site 供体部位
Donors smog incident 多诺拉烟雾事件
donor stream 原河流;供水河流;调出河流;施调河流
do not drop 小心掉落
do-nothing operation 空操作
do not load near inflammables 不能靠近易燃品
do not stake on top 勿放顶上
do not store in damp place 勿放湿处
do not turn over 不可倒置
Donovan's solution 碘化汞砷溶液
donpeacorite 斜方锰质辉石
Don polarity hyperzone 多涅茨克极性巨带
donsella (一种产于美洲的)红棕色硬木
don't-care 不管
don't-care point 不管点
don't cast 勿掷
don't cross 不准超车(交通管理)
don't log 不做日记
don't lose leaves in winter time 冬季不落叶
don't pass 不准超车
donut 环形室;大型垫圈;超环面粒子加速器;起落架轮距;环形混凝土垫块
donut roll 卷盘水带
donwhole recording probe 孔下记录探测器
donwhole recording system 孔下记录系统
doodad 小装饰品
doodlebug 小机动车;船上金砂精选厂;探条
doodlebug crew 地震钻井队

dook 斜井(巷);木屑;木塞
dooly 炸药(俚语)
doomed vessel 遇险船;遇难船;沉没船
do one's marketing 到市场做买卖
door abutment piece 门槛
door accessories 门的配件;门的附件;门上附件
door actuation mechanism 门驱动机构
door actuator 门机机构
door air reservoir 车门风缸
door and frame packaged unit 门与门框成套单元
door and window draught-proofing 门窗防风
door and window dust-proofing 门窗防尘
door and window greening 门窗绿化
door and window hardware 门窗及五金
door and window hook 门窗钩
door and window ironmongery 门窗及五金
door and window schedule 门窗(一览)表
door and window tax 进口税
door aperture 门洞;门道入口;门口
door arch 门(顶)拱;门楣
door area 全部门面积(包括门框)
door assembly 门的装配;整套门
door axis 门(中)轴线
door bar 门栓
door barge 开底泥驳
door bearing 门枕
door bell 门铃
door-bell and button system 门铃系统
door bevel 门梃斜边
door bolt 门栓;门(插)销;挂门钩
door bolt mechanism 门栓机构
door bottom rail 门扇下冒头
doorbrand 门栓;门板系木;铁板铰链
door breaker 重铁铤
door buck 门樘(外框)
door buffer 制止器;车门缓冲器
door bumper 门挡;制止器;制(动)器;车门软垫
door buster 广告商品
door butt 门铰链
door bypass function 车门旁路功能
door case 门框
door casing 贴脸;门框(饰);经装修的门框;门侧装修
door catch 门后夹;门插销;铁门销子;门扣
door centerline 门轴线
door centre line 门轴线
door chain 门链;安全门链
door check 门制止器;制(动)器;制门框侧板;自动关门器;关门器;自动缓闭器
door check and spring 止门弹簧
door cheek 门柱;门边梃;门(边)樘;门边框;门框侧板
door chime 门铃
door cill 门槛
door clamp 门扣
door class 门防火等级
door clearance 门的间隙(门与地板之间);门底净空;门空隙
door closed shaft kiln operation 立窑闭门操作
door close indication light 车门关闭指示灯
door closer 门制动器;缓闭器;自动关门器;关门器;闭门器
door closer bracket 关门器座
door closer-holder 闭门装置
door closer mounting 关门器装置
door closer-spring 门弹弓;鼠尾弹簧
door closing force 车门关紧力
door contact 开门亮;门控开关;门触点;门的接触开关
door control equipment 门鸠尾槽
door control homonuclear decoupling method 门控同核去耦法
door control panel 门控屏
door cornice 门楣
door covered with sheet iron on both sides 双面包铁皮门
door cross member (集装箱的)箱门横梁
door curtain 门帘
door-curtain rod 门帘杆
door damper 炉门挡板
door dike 电冰箱门突出的绝缘层;门封
door draught excluder 门框封条
door dressing 门贴脸;门筒子板

door drive operator 开门装置
door dynamo 发电机门
door edge member 门底缘材(集装箱)
door endurance 车门耐久性试验
door engine 门动力控制机构;自动关门机
door equipment 门的配件
door extractor 炉门机;摘门机
door factory 门厂
door finish 门边装修;门边装饰;门的修整
door finishing 门上装饰
door fitting 门的五金;门的附件(不包括门锁和铰链);门配件
door fixer 门止器
door fixture 门用零件
door fork 导边器
door frame 框槛;门樘(子);门框;窗框架
door frame anchor 门框锚固件
door frame fixing 门框的固定
door frame heater 门框式电热器
door frame junction 门框架接缝
door frame without rebate 无槽的门框
door frame with panels 有嵌板的门框
door frame with rebate 有槽门窗
door frame with transom 有门楣的门框
door frame (work) 门框架
door framing 门框
door furniture 门(金属)配件;门用小五金(件)
door gap 门缝
door gasket 门衬;门垫圈;门垫板
door gear 门具;门的开启装置;门装置
door glass 门玻璃
door glass channel 门玻璃槽
door glass knob 玻璃门拉手
door glazing 亮子
door grill(e) 门格栅;门(下)通风算子
door guide 门导轨
door hand 开门方向;门的左右摆向
door handle 门拉手;门把(手)
door handle fittings 门把小五金;门把手零件
door handle of porcelain 瓷制门把手
door hanger 门钩;挂门钩
door hardware 门金属配件;门(用)五金
door head 门头(梁);门上梃;门楣;门(框)上槛
door hinge 门铰链;门合页
door hinge pin 门铰销
door holder 门制止器;门钩;门定位器;门止;门扣;箱门搭扣件(集装箱);定门器;脚踏门掣
door holder with cushion 门的缓冲夹钩
door holding device 门的保持装置
door hold-open detection 保持门开启探测器
door hold-open device 保持门开启装置
door hood 门上雨篷;门斗
door hook 门钩
door indicator lamp "open/closing" 车门指示灯"开启/关闭"
door indicator lamp "out of order" 车门故障指示灯
door initiation 车门启动
door installation 门的安装
door interlock switch 门锁开关
door ironmongery 门五金
door jack 门夹具;装门架
door jamb 门(侧)柱;门梃;炉门垛;炉门侧柱;门樘子边梃;门(框)边框;门窗侧壁;门边樘;侧门壁;前墙砖垛
door joinery 门的装修;门的精细加工
door keeper 守门人;门卫
doorkeeper's house 门(守)卫室;门房
doorkeeper's lodge 门(守)卫室;门房;传达室
doorkeeper's room 门(守)卫室;门房;传达室
door key 门钥匙
door knob 门(锁)把手;门上球形捏手;门球;球形门拉手;球形(门)把手
door knob fittings 门拉手零件
door knob furniture 球形门拉手零件
door knob hardware 门拉手小五金;门拉手金属件
door knob transformer 门钮形转换器
door knob transition 门钮形转变;门钮变形换
door knocker 门环;门铍
door latch 门栓;门碰锁;门扣;门插销;排料启动器
door latch keeper 门锁栓扣座
door latch (with lock) 门栓锁
door lead with window 带窗门扇
door leaf 门栓;门扇
door leaf with sliding window 带活动的门扉

door ledge (拼板门的)横档;拼板门的横档
doorless 无门的
door light 门上镶玻璃面积;门采光面
door lining 门筒子板;内门框;门衬
door lintel 门楣;门过梁
door lock 门锁
door lock cylinder 门锁汽缸
door lock handle 门锁把手
door locking device 门锁装置
door lock rod 门锁杆
door lock rod bracket 门锁托架;锁杆托架(集装箱)
door lock striker 门锁撞针
door lock with stainless steel knob 不锈钢球形房门锁
door loop cutout switch 车门环路截断开关
door louver 门百页;门百叶
doorman 门侧服务员;门房
door mat 门前(擦鞋)棕垫;门口毡;门口擦脚条;擦鞋垫
door mechanism 门具;门的开启装置;门装置
door middle block 门扇中冒头
door middle rail 门扇中冒头
door mirror 门镜
door money 入场费
door mullion 门中立樘;(双门口的)中间立柱;门中梃
door muntin 门扇中梃
door nail 门簪;护帽钉;大头钉
door niche arch 门(顶)拱;门(龛)拱
door noise 门噪声;门静噪声
door of fire-check 防火门
door opener 开门服务员;侧门服务员;开门装置;开门器
door opening 门净空;开门;门框尺寸;车门开放;入口;门洞(口)
door opening degree 车门开度
door opening handle 开门拉手
door opening/releasing possibilities 车门开放/解锁的可能性
door opening size 门洞尺寸
door operator 司门;升降门操作器;(电梯门的)电动开关;门(的)自动开闭装置;升降机操作器
door panel 门心板
door panel arrangement 门的镶板设计
door panel-lining 门垫板;门板衬里
door panel retainer 门心板护圈
door pier 门墩(柱)
door pillar 门柱
door pivot 门(枢)轴;门(转)轴
door pivot brick 门枢轴砖
door plank 门板
door plate 门牌;户名牌
door plate pin washer 炉门板锁垫圈
door pocket 门套;滑轨门槽;推拉门箱;推拉门架;(活动门的)门囊
door post 门柱;门边樘;门边框;门边梃
door profile 门的组成件;门的外观;门贴脸
door pull 门拉手
door pull handle 门拉手
door rabbet 门槽口;门凹凸榫;门止槽;门边半槽
door rail 门扇冒头;门横木;门横档;冒头
door rebate 门槽口;门凹凸榫
door release 门开放信号;门动电开关
door release/open 车门解锁/开启
door releasing 车门解锁
door retainer 防火门自动控制装置
door reveal 外墙门侧;门侧
door roller (推拉门的)滚轮;拉门滚轮;门滚轴;门滚轮
doors 门点
door saddle 门口踏板;门槛
door safety mirror 门上安全镜
doors and windows repair workshop 门窗检修间
door sash 车门窗框
door schedule 建筑物门种一览表;门一览表;门规格表
door scraper 擦门人;擦门机
door screen 门纱;门屏;门帘;纱门
door seal gasket 箱门密封垫;箱门封条
door seal(ing) 门嵌缝密封;门封
door sealing fillet 门的封口条;门密封嵌条
door seals 闸门隔层装置
door section 门的剖面

door separation 隔墙门
doorset 门组合件
door shape 门形;门的形状
door sheet (集装箱)门板
door sheet of firebox 火箱门板
door shell 门壳
door shield 门挡
door silencer 门静噪器
door sill 阈;门槛
doors per side 侧门数
door spring 门(用)弹簧
door spring lock 门锁弹簧;门上弹簧锁
door stay 停门器;门制
door stead 门口
door step 门前踏步;门前石阶;门阶
door step service 门到门捷运作业
door stile 门梃;门扇边梃;门框条
door stone 门前铺石;门口铺石;门(槛)石;门槛处石踏步;门阶
door stop 止门(器);门框企口;门钩;门挡;门制止器;门制(动)器;门制;门碰头;窗风撑;车门止挡
doorstope strain ga(u)ge 孔底应变计
door stopper 门扎头
doorstopper method (量测岩石应力的)门阻塞器方法;门塞法【岩】
doorstopper rock stress measuring equipment 门塞式岩体应力测量设备
door straightener 门整直器
door strap 门挂板;门系板;门带
door strength 箱门强度
door strength test 箱门强度试验
door strip 撇水板;门槛嵌缝条;门周护条;门底挡雨线脚板
door stud 门边立木
door support 排料门底座
door sweep 门扇防风刷
door swing 开门方向;门的左右摆向
door switch 开门亮;门(控)开关;门(动)开关;门(触)开关
door system 门的布局;门安装
door threshold 门坎;门槛
door threshold plate 门坎铁板
door tidy 信孔盖
door to cfs (集装箱的)门到站
door to container freight station (集装箱的)门到站
door to container yard (集装箱的)门到场
door to cy (集装箱的)门到场
door-to-door 门到门
door-to-door car 送货用汽车
door-to-door container traffic 上门联运集装箱运输
door-to-door delivery 逐户分送
door-to-door selling 挨户推销
door-to-door service 上门服务;门对门运输;门到门运输;门到门服务;户到户货运业务
door-to-door system (集装箱的)门到门系统
door-to-door transportation 门对门运输;门到门运输
door-to-door transport of container 集装箱门到门运输
door to passenger compartment 通往客室门
door to passenger compartment with peep hole 带窥孔通往客室门
door top rail 门扇上冒头
door top rear view mirror 车门上端装置的后视镜
door-to-service 上门服务
door tree 门柱
door trigger 门开关
door trim 门头线;门(框)贴脸;门框饰
door trip 门开关;溢料门
door type cores 各型门芯
door-type sampler 侧门式取样器(取砂砾岩样品)
door unit 门配件
door unit contact system (升降机的)门联锁触点装置
door viewer 门窥视镜;窥镜;门窥视孔
doorway 门阀;入门;口门;门道
doorway zone 入口区
door window 落地窗
door window curtain 门窗帘
door window mullion 门窗竖框
door wing 门扇
door with sidelights 联窗门
door with window 门连窗

door yard 大门前庭;门前庭院
dopamine 多巴胺
dopant 掺质;掺杂物质;掺杂物;掺杂剂;掺杂
dopant material 掺杂材料
dope 绝缘和包缠;模型油;蒙布漆;掺杂(剂);掺入;涂布漆;上涂漆;上漆
dope adhesive 胶浆;涂布黏合剂
dope brushability 蒙布涂刷性
dope bucket 装封口油灰的斗(管子接头配件)
dope can 加油枪(发动机开动时);注油壶
doped 掺杂(过)的
doped atomized water 掺漆雾化水
doped chemical 掺杂剂
doped chemical element 掺杂元素
doped coating 涂漆包皮;加固涂料;含添加剂的涂料;涂料层;喷漆层
doped crystal 掺杂晶体
doped crystal interferometer 掺杂晶体干涉仪
doped crystal laser 掺杂晶体激光器
doped cutback (掺添加剂的)稀释沥青
doped diamond 掺杂金刚石
doped fabric 涂漆蒙布
doped fabric covering 涂漆蒙布面
doped fuel 含添加剂的燃料;含铅汽油
doped gadolinium gallium garnet 掺质钆镓石榴石
doped gasoline 防爆汽油
doped glass 掺杂玻璃
doped glass laser 掺杂玻璃激光器
doped glass rod 掺杂玻璃棒
doped mercury lamp 掺杂汞灯
doped modification 掺杂改性
doped oil 含添加剂的油;防爆润滑油
doped oxide diffusion 掺杂氧化物扩散
doped photoconductor detector 掺杂光电检测器
doped seed 掺杂籽晶
doped-silica cladded optic(al) fiber 掺杂石英包层光纤
doped silica glass 掺杂石英玻璃
doped-silica graded fiber 掺杂石英渐变型光纤
doped silicon 掺杂硅
doped single crystal 掺杂单晶
dope dye 调色染料
dope-dyed fibre 纺液染色纤维
dope dyeing 原液染色
dope for diesel fuels 柴油防爆剂
dope gang 包管道工(沥青保护层等)
dope machine 管子涂包沥青机;绝缘机
dope material 绝缘材料
doper 滑脂枪
dope room 喷漆间
dopes for gasoline 汽油防爆剂
dope textile shrink property 蒙布收缩性
dope transistor 掺杂质晶体管
dope-vapourizer 掺杂剂蒸发器
dope vector 内情向量;数组信息;信息向量
doping 施防水剂;浸渍;加蒙料(在燃料或油内);晶格掺杂;掺杂;涂漆;上涂料;加燃料(在燃料或油内)
doping accuracy 掺杂精度
doping agent 掺杂剂
doping compensation 掺杂质补偿
doping compositional superlattice 掺杂组分超晶格
doping content 掺杂浓度
doping density 掺杂密度
doping effect 掺杂效应
doping gradient 掺杂陡度
doping level 掺杂度
doping machine 包管子机
doping material 掺杂材料
doping method 掺杂方法
doping of binders 胶黏剂的增黏
doping of gasoline 汽油掺添加剂
doping profile 掺杂剖视图
doping superlattice 掺杂超晶格
Dopol 双逆流预热器
Dopol kiln 多波尔窑
Dopol preheater 多波尔预热器
Dopol preheater with precalciner 带预分解炉的多波尔型预热器
do poorly done work over again 返工
dopping 模子上油
Doppler accuracy 多普勒精度
Doppler broadening 多普勒致宽;多普勒(谱线)增宽;多普勒加宽法

Doppler bubble detector 多普勒气泡检测仪
Doppler contour 多普勒轮廓
Doppler core 多普勒核心
Doppler count 多普勒计数
Doppler counter 多普勒计数器
Doppler current meter 多普勒海流计
Doppler data 多普勒资料;多普勒数据
Doppler data translator 多普勒数据传输器
Doppler detector 多普勒检测器
Doppler displacement 多普勒位移
Doppler echo indicator 多普勒回波指示器
Doppler effect 多普勒效应
Doppler equation 多普勒方程
Doppler fixation 多普勒定位
Doppler-Fizeau effect 多普勒·斐索效应
Doppler flowmeter 多普勒流量计
Doppler frequency 多普勒频率
Doppler frequency shift 多普勒频移
Doppler frequency spectrum 多普勒频谱
Doppler generator test set 多普勒振荡器检测设备
Doppler indicator 多普勒指示器
Doppler inertial navigation(al) equipment 多普勒惯性导航装置
Doppler inertial system 多普勒惯性系统
Doppler invariance 多普勒不变性
dopplerite 灰色沥青;橡皮沥青;腐殖质块;(泥炭中的)腐殖凝胶;天然沥青;弹性沥青
Dopplerizer 多普勒效应器
Doppler laser velocimeter 多普勒激光流速计
Doppler log 多普勒计程仪
Doppler message 多普勒信息
Doppler meter 多普勒流量计
Doppler method 多普勒法
Doppler modulation 多普勒调制
Doppler motion 多普勒运动
Doppler navigation 多普勒导航
Doppler navigation(al) sonar 多普勒导航声呐
Doppler navigation(al) system 多普勒导航系统
Doppler navigator 多普勒导航仪
Doppler positioning 多普勒定位
Doppler positioning by the short arc method 多普勒短弧法定位
Doppler principle 多普勒原理
Doppler profile 多普勒轮廓
Doppler radar 多普勒雷达
Doppler-radar fix 多普勒雷达定位
Doppler range 多普勒测距系统
Doppler ranging 多普勒测距
Doppler ranging measuring system 多普勒测距系统
Doppler receiver 多普勒接收机
Doppler relation 多普勒关系
Doppler satellite geodetic network 多普勒卫星大地网
Doppler satellite network with short arcs 多普勒卫星短弧网
Doppler satellite observation 多普勒卫星观测
Doppler scattering 多普勒散射
Doppler shift 多普勒频移
Doppler shift measurement 多普勒频移测量
Doppler shift ocean-current meter 多普勒频移海流计;多普勒频移海流测速仪
Doppler ship's speedometer 多普勒船舶速度计;多普勒船舶速度表
Doppler single-point positioning 多普勒单点定位
Doppler sonar 多普勒声呐
Doppler sonar navigation(al) and docking system 多普勒声呐导航与系泊系统
Doppler sonar navigator 多普勒声呐导航仪
Doppler sonar system 多普勒声呐导系统
Doppler sonar transducer 多普勒声呐转换器
Doppler sonar very high frequency omnirange beacon 多普勒甚高频全向信标
Doppler speed log 多普勒船速计程仪
Doppler spread 多普勒频散
Doppler system 多普勒系统
Doppler system with double frequency 双频多普勒系统
Doppler system with single frequency 单频多普勒系统
Doppler tracking 多普勒跟踪
Doppler translation 多普勒移位
Doppler translocation 多普勒联线定位法

Doppler two-colo(u)r laser anemometer 多普勒双色激光风速计
Doppler ultrasonic scanner 多普勒超声波扫描器
Doppler velocimeter 多普勒流速计
Doppler velocity and position 多普勒速度和位置;多普勒测速和定位
Doppler velocity and position finder 多普勒测速和测位器
Doppler width 多普勒宽度
Doppler wind 多普勒风
Dopploy 多普勒铸铁
dorag dolomitization 混合白云石化作用
doran 多普勒测距系统
doranite 变菱沸石
Dora reinforced block floor 多拉加筋块材楼板
doras 壤土
Dorco filter 多尔科型真空过滤机
dore 节理缝
dore bullion bar 金银合金锭
doreite 钠粗面安山岩
dore metal 金银块;金银合金
dore silver 多尔银
dorfmanite 水磷氢钠石
Dorfner test 多夫纳试验;多尔夫纳应力测试验法(上釉制品)
dorian 陶立克式(希腊式建筑)
doria stripes 多利亚条子细布
doric 多立克体
Doric architrave 陶立克式楣梁;陶立克柱顶过梁
Doric base 陶立克式柱座;陶立克柱基础
Doric capital 陶立克柱头
Doric colonnade 陶立克柱廊
Doric column 陶立克柱
Doric-columned 采用陶立克式柱子的
Doric cornice 陶立克柱檐口
Doric cyma(tium) 陶立克波纹线脚;上凹下凸的双弧形线脚
Doric echinus 陶立克式凸圆线脚;陶立克柱头顶板下四分之一圆饰
Doric entablature 陶立克柱檐座;陶立克柱顶盘
Doric epistyle 陶立克式楣梁
Doric frieze 陶立克式檐壁;陶立克柱上飞檐过梁;陶立克柱上飞檐雕带;陶立克雕带
Doric order 陶立克型;陶立克柱式
Doric portico 陶立克门廊
Doric structure 陶立克式建筑;陶立克结构
Doric style 陶立克风格
Doric temple 陶立克(式)神庙
dormand 枕木;系梁;横梁;地梁
dormant 休止的
dormant account 不活动账户
dormant account balance 闲置账户结存
dormant application 休眠期施用
dormant assets 闲置资产
dormant bolt 暗门销;暗栓;暗(门)栓;埋头螺栓
dormant capital 游资;未利用资金
dormant company 不活跃公司
dormant economy 停滞的经济
dormant equipment 闲置设备
dormant fire 潜伏火灾
dormant geothermal system 休眠地热系统
dormant geyser 休眠间歇泉
dormant lock 暗销
dormant mine 海底水雷
dormant partner 匿名合伙人;匿名股东;幕后合伙人;隐名合伙人;出资而不参与经营的合伙人(隐名合伙人)
dormant partnership 匿名合伙营业
dormant period 休眠期
dormant processor 静止处理机
dormant pruning 休眠期修剪
dormant resource 未开发资源;潜在资源
dormant scale 自重天平
dormant scrap 报废器材
dormant screw 埋头螺钉
dormant season 休眠期
dormant-season irrigation 非生长季节灌溉
dormant-season streamflow 枯季径流
dormant spraying 休眠期喷射
dormant spring 休眠泉
dormant stage 休眠期
dormant state 待用状态
dormant tree 枕木;系梁;楣(梁;地梁;横梁
dormant volcano 休眠火山;熄火山

dormant window 老虎窗;屋顶天窗;天窗
dormer 老虎窗;屋顶(天)窗;屋顶采光窗
dormer bolt 老虎窗螺栓
dormer cheek 老虎窗两侧;老虎窗侧壁;屋顶窗侧壁;气楼侧壁
dormer covering roof 设有较多老虎窗的屋顶
dormer(ed) roof 设屋顶窗的屋顶;设老虎窗的屋顶;有老虎窗的屋顶
dormer-ventilator 屋顶通风窗
dormer-ventilator opening 屋顶通风窗口
dormer window 老虎窗;屋顶窗;天窗
dormette 躺椅
dormitory 集体宿舍;寝室;宿舍;郊外住宅区
dormitory area 宿舍区
dormitory block 宿舍楼;住宿区
dormitory building 宿舍楼
dormitory car 宿营车
dormitory house 宿舍
dormitory quarter area 宿舍区面积
dormitory ship 住宿船
dormitory suburb 卧城区;郊外住宅区;市郊住宅区
dormitory town 居住城镇
dormitory unit 宿舍单元
Dorn effect 多恩效应
dornick 小块石
dornveld 多刺高灌丛
doroid 半环形线圈
dorok (由双层油毡组成的)沥青平屋面系统
Dorotheenthal faience 多洛甇塔尔锡釉陶器
Dorr agitator 多尔(型)搅拌器;多尔(式)搅拌器;多尔(式)搅拌机
Dorr balanced-tray thickener 料盘式增稠器;多尔平衡盘式增稠器
Dorr balance tray thickener 多尔平衡盘式增稠器
Dorr bowl clarifier 多尔浮槽分级机
Dorr bowl classifier 多尔式浮槽分级机
Dorr bowl-rake clarifier 多尔浮槽耙式分级机
Dorr bowl-rake classifier 多尔式浮槽耙式分级机
Dorr clarifier 多尔清除淤泥器;多尔(型)澄清器;多尔分粒器;多尔沉淀器
Dorr classifier 选粒器;多尔型粒子分级器;多尔分级机;多尔式选粒器;多尔式分级机;多尔型分级机
Dorr clone 多尔式水力旋流器
Dorrclone classifier 多尔旋流分级器
Dorrco filter 多尔真空滤机;多尔滤浊机
Dorr continuous thickener 多尔连续式浓缩器
Dorrco sand washer 多尔选砂机
Dorr detritus 多尔沉砂池
Dorr duplex classifier 多尔双联分级机
Dorr fluo-solide reactor 多尔液态化反应器
Dorr-Oliver detritor 多尔—奥立夫沉砂池
Dorr-Oliver hydrocyclone 多尔—奥立佛水力旋流器
Dorr-Oliver siphon sizer 多尔—奥立佛虹吸分级机
Dorr rake classifier 多尔耙式分级机
Dorr thickener 多尔(型)增稠器;多尔浓缩池;多尔搅拌机;多尔式厚浆池;多尔沉降器;单室增稠器
Dorry machine 多莱陶瓷耐磨性试验设备;多莱陶瓷耐磨性试验机
dorsal 罩篷;叶底部;门上雨篷的支托;挂帘;门上雨罩;背(脊)的;背(侧)的;背(部)的
dorsal border 背侧缘
dorsal branch 背侧支
dorsal chord 脊索
dorsal column 背柱
dorsal commissure 背侧连合
dorsal cord 脊索
dorsal divisions 背侧股
dorsal fissure 背裂
dorsal funiculus 背柱;背侧固有束
dorsal group 背侧组
dorsal hump 背隆起
dorsal light response 背部光反应
dorsal line 背中线
dorsal lip 背唇
dorsal longitudinal fasciculus 背侧纵束
dorsal margin 背侧缘
dorsal median sulcus 背正中沟
dorsal mesocardium 背心系膜
dorsal nucleus 背核
dorsal position 背卧位
dorsal radix 后根
dorsal reflex 背反射
dorsal root 脊根
dorsal scale 背壳

dorsal segment 背段
dorsal stylet 背针
dorsal surface 背侧面
dorsal suture 背缝(线)
dorsal trace 背迹
dorsal view 背视图
dorse 门上雨罩;雨棚;遮檐
dorsiferous 背生的
dorsifixed 背着的
dorsiflexion 背(侧)屈
dorsol rays 背辐肋
dorsomedial nucleus 背内侧核
dorsoventral axis 背腹轴
dorso-ventrally compressed 背腹扁平的
dorsum 背部;山脊;宿舍;[复]dorsa
dorsum sellae 鞍背
dorter[dortour] 寺院宿舍;男修道院宿舍;寺院的卧室部分;僧寮
Dortmund method 多特蒙德单管提升法
Dortmund tank 多特蒙德罐;多特蒙德沉积槽;多特芒德(沉淀)池【给】
Dorvon 绝缘隔热(聚苯乙烯)塑料板
dory 小平底(鱼)船
dory anchor 沙地锚
dory-hand-liner 手钓母船
dory skiff 方尾平底渔船
dory trawler 小平底拖网船
dosage 剂量(测定);吸收剂量;用药量;用量投药配量;配药;配(人)量;配(料)量;适用量
dosage bunker 配料槽
dosage compensation 剂量补偿(作用)
dosage effect 剂量效应
dosage form 剂型
dosage monitoring survey design 投药剂量监测测量图
dosage of chemicals 化学药剂投加量
dosage plant 化学软水设备
dosage pump 剂量泵;计量泵
dosage pump set 馈料泵组
dosage rate 剂量率
dosalic 多硅铅质
dosciss bids 议标
Dosco miner 道斯科联合采煤机
dose 剂量;吸收剂量;一次剂量
dose counter 剂量计数器
dose-effect curve 剂量效应曲线;剂量反应曲线
dose-effect relationship 量效关系;剂量效应关系
dose equivalent 剂量当量
dose-equivalent commitment 剂量当量负担
dose-equivalent index 剂量当量指数;剂量当量指标
dose-equivalent limit 剂量当量极限;限制剂量当量;剂量当量限值
dose-equivalent rate 剂量当量率
dose-frequency relationship 剂量频率关系
dose glass 剂量玻璃
dose limit 剂量限度;剂量极限
dose limit equivalent concentration 剂量限度当量浓度
dose-meter 辐射剂量计;量筒
dose of flocculant 絮凝剂用量
dose of lime 石灰投量
dose of radioactivity 放射性剂量
doser 定量加料器;定量给料器;测剂量装置;测绘装置
dose rate 量率;剂量率;投料量
dose rate meter 剂量率计
dose rate monitor 放射率探测器
dose rate of radioactivity 放射性剂量率
dose rate out 出射剂量率
dose reduction factor 剂量减低系数
dose-response 剂量反应
dose-response assessment 剂量反应评定
dose-response correlation 剂量反应相关
dose-response curve 剂量响应曲线;剂量反应曲线
dose-response pattern 剂量反应模式
dose-response relationship 剂量反应关系
do shoddy work and use inferior material 偷工减料
dosifilm 胶片剂量计
dosimeter 剂量仪器;量筒;剂量计;液量计;测剂量装置;放射剂量计
dosimetric system 剂量制
dosimetry 剂量子;剂量学;剂量计量术;剂量测定法;放射剂量测定法

dosing 计量配料;定量加料;配料;配量;投料计量
dosing apparatus 投配装置;投配器
dosing appliance 配料设备
dosing box 投配箱
dosing chamber 配制室;配料池;污水量配池;投配室;投料计量室
dosing cock 投配龙头;投料控制
dosing control 投配龙头
dosing control system 投药剂量控制系统
dosing cycle 投配周期
dosing device 定量装置;配料设备;投配器
dosing equipment 喂料装置;投配设备
dosing instrument 配料设备
dosing interval 投配间隔
dosing mechanism 配量装置
dosing of reagent 试剂投配;试剂配制
dosing pump 加药泵;定量泵;比例泵;投配泵
dosing rate of sewage irrigation 污水农灌定额
dosing ratio 投配比
dosing reagent 试剂投配;试剂配制
dosing room 投配间
dosing rotor 喂料转盘
dosing rotor weigher 圆盘喂料秤;喂料转盘秤
dosing rotor weighfeeder 转子式定量给料秤
dosing scale 计量秤
dosing siphon 投配虹吸
dosing syphon 投配虹吸
dosing system 喷油剂量系统;配料系统
dosing tank 量箱;量斗;加药箱;污水量配池;投配箱;投配池(污水处理);投配槽
dosing tank in biologic(al) filter 生物滤池馈水池
dosing unit 定量加料装置;投配器;投配单元
dosiology 剂量学
dosis 剂量;一次剂量
doss 廉价寄宿房屋
dossal 墙幕;帐帘;艇背部;高椅背幔
dosser 背篓;帐帘;驮筐
dosseret 拱基垫块;副柱头
doss house 下等客栈;简陋小客栈
dossier 档案(材料)
dossoret 拜占廷式高柱帽
do statement 循环语句【计】
dosulepin 度硫平
dosulite 水硅锰石
dosy timber 腐朽木材
dot 小点;圆点;垫料;点(纹)
dot address 点地址
dot alloy mesa transistor 点接触合金台面型晶体管
dot and dab fixing 点状黏结固定
dot-and-dash line 点划(虚)线
dot-and-dash signal 点划信号
dot and dash technique 点划技术
dot angel 点状亮影
dot area 网点面积
dot area coverage 网点覆盖率
dot-bar generator 点条状图案信号发生器
dot beamsplitter 点式分光镜
dot-blur pattern 记忆矩阵
dot chart 点(阵)图;布点图
dot circuit 点形成电路
dot convention 打点习惯
dot count register 点计数寄存器
dot crawl 点蠕动
dot cutter 刻点仪
dot cycle 基本信号周期;点周期;点循环;打点周期
dot-dash graver 点划虚线刻绘仪
dot-dash line 点划线
dot-dash method 点划法
dot-dash mode 点划法
dot density 点密度;网点密度
dot density method 像素密度法
dot diagram 点式图示
dote 木材糟朽;腐朽物;腐化;腐败物
doted 腐朽的
dot engraving 网点刻蚀
dot enlargement 网点增大
dot etching 斑点腐蚀;网点蚀刻;网点腐蚀
dote timber 腐朽木材
dot file 点文件
dot fluorescent screen 点荧光屏
dot for dot 网点翻版点
dot format 点格式
dot frequency 点频率;短点频率
dot gain 网点增大

dot generation 点生成
dot generator 点信号发生器
dot graph 散点图
dot image 点像
dot interlacing 隔点扫描;跳点扫描
dot line 虚线;点线
dot map 点描法地图
dot mark 刻印标记;点标记
dot matrix 点阵(法);点(矩)阵
dot matrix character 点阵字符
dot matrix character generator 点阵字符发生器
dot matrix format 点阵格式
dot matrix method 点阵法
dot matrix plotter 点阵绘图仪
dot matrix printer 点阵打印机
dot matrix printer image 点阵打印图像
dot method 点值法;点方式
dot of light 光点
dot-on-dot register 精确套准
dot pattern 光点图形;点图(形)
dot pattern generator 点模式发生器;点信号发生器
dot pitch 点距
dot printer 点式打印机
dot product 点乘;标量积;数量积
dot punch 冲子
dot random mosaic 无规则的斑点马赛克
dot ratio 网点比例
dot recording system 点式记录仪
dot reduction 网(目)点缩小
dots 刮灰平面标点
dot scale 网点比例
dot screen 点状网目板
dot scriber 刻点仪
dot-sequential 点顺序制
dot-sequential image 逐点发送图像
dot-sequential method 逐点传送法;点序法
dot shank harrow support 点状柄耙支架
dot signal 点信号
dot signal generator 点状图案信号发生器;点信号发生器
dot size 网点大小;网点尺寸
dots per inch 英寸点数
dots recognition 网点识别
dot's sector 点弧区
dot system 微点法
dotted 打点
dotted arrow 虚(线)箭头
dotted curve 点曲线;虚曲线
dotted decimal notation 点分十进制记法
dotted leaf 斑叶
dotted line 虚线;点(虚)线
dotted line marker 点线时标
dotted pair 点对
dotted rule 点线
dotted swiss 点子花薄纱
dotter 刻点仪;划点器;点圆规;点标器
dotting impulse 点信号脉冲
dotting instrument 点线器;点绘仪
dotting machine 刻点仪
dotting needle 点孔针
dotting-on 类加;并加
dotting pen 点线笔
dotting punch 中心冲头;冲眼
dotting recorder 打点式记录器
dotting test 画点测验
dottling 平盘装窑法;平盘烧针装窑法
dot trio 三组圆点
dot type 打点式
dot weld(ing) 填补焊;点焊
dot wheel 划点转轮;骑缝线滚轮
doty 腐朽的
doty sleeper 腐朽枕木;腐蚀枕木
doty wood 腐朽木(材)
dou 斗(谷类测量单位,1 斗 = 10 升)
Douala Port 杜阿拉港(喀麦隆)
doubel standard 复本位制
double 两倍;加倍;翻一翻;二倍;倍;卧铺车双人包房;双重(的);双倍
double account form of balance sheet 复式资产负债表;双账式资产负债表
double account system 复式账户制;复式账;复式记账法;双账制
double-acting 往复的;双作用式;双动(式)
double-acting and spring return type 双动与弹簧回位型

double-acting atomizer 双动式弥雾机
double-acting brake 复动制动器
double-acting butt 双动对接铰
double-acting butt hinge 双向作用铰链
double-acting centrifugal pump 双作用离心泵
double-acting comber 双动式精梳机
double-acting compressed air cylinder 双作用压缩汽缸
double-acting compression 两面压制;双效压制
double-acting compressor 双作用压缩机;双动压气机;双动式压缩机
double-acting cylinder 双作用(汽)缸
double-acting damper 双动减振器
double-acting die 双向压模;双动模
double-acting dobby 复动式多臂机
double-acting door 摇门;弹簧门;双向门;双面摇门;双开(式)弹簧门;双动门
double-acting engine 双作用(式)发动机;双动发动机;双开门
double-acting floor spring 双面地弹簧
double-acting forcemeter 双作用式测力计;双动式测力计
double-acting frame 双向门框
double-acting gas engine 双作用式煤气机
double-acting gin 双动式轧花机
double-acting hammer 复动式锻锤;双动锤;双作用汽锤(蒸汽和空气)
double-acting hand pump 双动式手摇泵
double-acting hinge 双向合页;双动铰链;双向铰链
double-acting hydraulic jack 双动液压千斤顶;双向作用水力千斤顶
double-acting hydraulic ram 双向作用水力千斤顶
double-acting jack 双作用千斤顶
double-acting pawl 双动(棘)爪
double-acting pile hammer 双动式打桩机;复动桩锤
double-acting pump 双作用式泵;双动(式水)泵;吸压双作用泵
double-acting relay 双侧作用继动器
double-acting rotary jar 双作用回转振击器
double-acting shock absorber 双向减振器
double-acting single basin 双作用水池
double-acting sprayer 双动式喷雾机
double-acting spring butt 双动弹簧铰链;双向弹簧铰(链)
double-acting spring butt hinge 双面弹簧铰链
double-acting spring door 双面弹簧门
double-acting spring hinge 双向弹簧铰链
double-acting steam engine 双动式蒸汽机
double-acting steam hammer 双动式蒸汽锤;双动(式)汽锤;双作用蒸汽锤
double-acting steam pile hammer 往复式的复动蒸汽打桩机;双作用蒸汽机桩锤;双打蒸汽机桩锤
double-acting switch 双动电开关
double-acting valve type hydro-percussive tool 阀式双作用冲击器
double action 复动
double-action bellows 双动风箱
double-action black japanned spring hinge 黑漆双弹簧铰链
double-action clutch 双作用离合器
double-action die 双效冲模
double-action disc harrow 双功圆盘耙
double-action disk harrow 双排圆盘耙路机;双排式圆盘耙
double-action door 摇摆门;双向门
double-action double door 双扇双向门
double-action end cutter 双作用的端铣刀
double-action floor spring 双面地弹簧
double-action forming 双效成型
double-action hammer 动力锤;双作用锤
double-action hinge 双向合页
double-action hydraulic cylinder 双作用液压缸
double-action hydraulic press 双动冲压水压机
double-action jack 双作用千斤顶
double-action mechanical press 双动机械压力机
double-action pile hammer 双向动作的桩锤
double-action piston 两面作用活塞
double-action press 双动压力机;双动冲床
double-action pressing 双向压制;双面加压
double-action pump 双动式(水)泵
double-action spring bolt 双面弹簧插销
double-action spring butt 双动弹簧铰链

double-action spring butt hinge 双面弹簧铰链
double-action steam hammer 双动蒸汽锤
double active filter 双层活性滤池
double activity 双重活度;双重活动性
double adapter 双头同性接口
double address 双地址;二地址
double adjustment suspension 双调节悬挂系
double aging 双时效
double agitator 双搅拌器;双向搅拌机
double air pump 复式气泵
double alkali effect 双碱效应
double alkali scrubber process 双碱洗涤法
double alkoxide 金属复式醇盐
double-alloy steel 二元合金钢
double alternate 双(重)交错
double altitudes 相等高度
double ambulatory 双回廊
double amplitude 正负峰间振幅值;正负峰间的倍幅值;倍幅;全幅值;双(向)振幅;双幅
double amplitude-modulation multiplier 双重调幅乘法器
double amplitude peak 双幅度峰值
double anastigmat 双消像散透镜
double anchored sheet wall 双锚碇板墙
double anchored wall 双锚碇墙
double and single fastening 单双板固定法(一种木船船壳板装配法)
double angle 双角钢构件;双角钢杆件;双角铁
double angle bar 双角钢
double angle formula 倍角公式
double angle iron 双角铁
double angle mirror 复式直角镜
double angle section 双角型钢
double angle tee 双角T字钢
double angle tie 双角钢构成的系杆
double(-angle) V-belt 双面三角皮带
double angle web 双角钢腹杆
double annealing 二次退火;双重退火
double annual ring 双年轮
double anode 对阳极;双阳极
double antenna 偶极天线
double aortic knob 双主动脉球
double aperture lens 双孔透镜
double aperture slit 双狭缝
double application 分两次缴入
double arc bit 双弧钻头
double arch 双层拱
double arch bridge 双曲拱桥
double arched corbel table 双拱支承的平板(支承平板的牛腿呈拱形);双拱挑檐
double arched dam 双曲拱坝
double arches 二重拱;双窟
double archway 双重门过道;双重门拱廊;双(重)拱门
double arcing 双燃弧;双弧现象
double arc-melting 两次电弧熔炼
double arc technique 双电弧技术
double armature 双衔铁;双铠装;双电枢
double armature motor 双电枢电动机
double armature relay 双衔铁继电器
double arm draw bridge 双臂式开合桥
double arm kneader 双轴练泥机;双臂捏练机;双臂捏合机
double arm lever 双臂杠杆
double arm mixer/extruder 双轴混合/挤出机
double armo(u)red cable 双层铠装电缆
double armo(u)ring 加双重钢筋;加复筋;复式钢筋;双重铠装;双重护面
double arm pantograph 双臂受电弓
double arm press 双柱压力机
double arrow sign 双指向标志
double-articulated arch(ed) frame 双铰接拱架
double-articulated flat arch(ed) girder 双铰接平拱大梁
double-articulated flat parabolic arch(ed) girder 双铰接平抛物线拱大梁
double-articulated gable(d) frame 双铰接人字架
double-articulated rectangular frame 双铰接方框架
double-articulated reductor 二级链式减速器
double-articulated segmental arched girder 双铰接弓形拱大梁
double astrograph 双筒天体照相仪;双筒天体摄影仪
double atlas 双页地图集

double attenuator 双段衰减器
double awning 双层天幕
double axe 双刃斧
double axial flow turbine 双轴流式水轮机
double axial mixer 双轴搅拌混合器
double-axis symmetry 双轴对称
double azimuthal projection 双重方位投影
double back 稀疏织物背衬;双层面层做法;复式背衬
double back gear 双跨轮;双过轮;(车床的)双背齿轮装置
double backing 复合背衬
double back-substitution 双重回代
double bag filling machine 双嘴包装机
double balanced mixer 双平衡混频器
double balanced modulator 双平衡调制器
double ball catch 双球扣栓
double ball foot valve 双球脚阀
double ball pitman arm 双球连接杆臂
double bandmill 双带锯制材厂
double band sawmill 双带锯制材厂
double bank 双排梯架
double-banking 并靠(两船)
double bank radial engine 双排星型发动机
double bank switch 双触排开关
double bar and yoke method 双条双轭法
double bar ga(u)ge 双杆划线机
double bar link 双联扣;双档链环
double barrel 双筒枪;双管枪
double barreled drilling 双井筒钻进
double barrel hand pump 双筒手动泵
double barrel-mud gun 双筒泥炮
double base 双支承;双汇
double base phototransistor 双基极光电晶体管
double base powder 双基火药
double base propellant 双基推进剂
double base rocket propellant 双基火箭推进剂
double base transistor 双基极晶体管
double batten 双板条
double battery gin 双滚筒式轧花机
double bay 双跨(间)
double bead 双圆线脚;双珠饰(的)
double bead lap joint 双焊道搭接接头
double beam 双重梁;双(光)束
double beam cathode-ray tube 双束阴极射线管
double beam densitometer 双光束密度计
double beam double focusing mass spectrometer 双束双聚焦质谱仪;双束双聚焦质谱计
double beam electric-null infrared spectrometer 双光束电零点红外分光计
double beam far-infrared Fourier spectrometer 双光束远红外傅里叶光谱仪
double beam Fizeau fringes 双光束斐佐干涉条纹
double beam grating spectrophotometer 双光束光栅分光光度计
double beam interferometer 双光束干涉仪
double beam microspectrophotometer 双光束显微分光光度计
double beam optic(al) system 双束光学系统
double beam oscillograph 双束射线示波器
double beam oscilloscope 双踪示波器;双线示波器;双束示波器
double beam photometer 双(光)束光度计
double beam polarizer 双光束偏振器
double beam principle 双光束原理
double beam recording spectrophotometer 双光束记录分光光度计
double beam spectrometer 双束分光计
double beam spectrophotometer 双束分光光度计;双光束分光光度计
double beam system 双(光)束系统
double beam technique 双光束技术
double beam tube 双注管;双束射线管
double bearing rudder 双支承舵
doublebeat 双重差拍;双拍
double beat drop valve 双座阀
double beat sluice 双向(泄)水闸
double beat valve 双开阀
double becket bend 双编结
double bed 双人床;双床【给】
double bed(ded) room 双人床客房
double bed filter 双层滤池
double bed filter material 双层滤池滤料
double bed guest room 双人床客房
double bedroom 可容双人卧室

double Belgian mill 双列比利时式轧机
double bell 双碗形
double bell bend 双承弯头
double bell differential manometer 双钟罩压差计
double-bellied 开弓形栏杆小柱;双称端的
double bellow type spout 双层伸缩卸料嘴
double belt 双层皮带
double belt conveyer 压带输送机;双带式输送机
double-belted elevator 双带式升降机
double belt laminator 双皮带层压机
double belt press 双带式压机
double bend S形弯管;双弯(头);双弯(管);双弯;双编结
double-bend(ing) fitting 双弯(接)头
double bent 双跨排架
double bent blade 双曲叶片
double bent glass 双弯曲玻璃
double bent shell roof 双弯薄壳(屋)顶;双曲薄壳屋顶
double berth 双层铺
double beta decay 双重贝塔衰变
double bevel 双斜式;双斜边;双斜面
double-beveled edge 双斜边
double bevel groove K形坡口
double-bevel groove weld 双斜坡口焊;双斜边坡口焊(缝)
double bevelling 双斜面
double bevel scarve 双头斜榫
double bevel T joint 双斜T形接头
double bilateral turnout 对称双开道岔
double bin method 复棚法
double bit error 双位错误
double bitt 双柱系缆桩;双柱系缆柱;双柱系船桩;双缆柱
double-bitted axe 双刃斧
double-bitted key 双面齿钥匙
double (bituminous) surface treatment 沥青保护层多层处理;双层沥青表面处治
double black 双层面层做法
double blackwall hitch 双花钩结
double blade batch dispersion mixer 双叶片分批分散混合机
double bladed knife switch 双闸刀开关
double bladed plough 双铧犁
double bladed plow 双铧犁
double blade kneader 双桨捏合机
double blast 双重音汽笛
double blast pipe 双送风管
double-blind method 双盲法
double block 复滑车;双阻塞部件;双闸瓦;双轮滑轮;双轮滑车
double block brake 双闸瓦制动;双瓦闸;双块制动器
double blow heading 双击镦锻
double board platform (钻塔的)架二台;(钻塔的)副二层台
double bolection mo(u)lding 双镶板线脚;双凸式线脚
double bollard 双柱系缆柱;双柱系船柱;双柱系缆桩;双柱系船桩
double bolt 双螺栓
double bolt cutter 双轴切丝机
double bond 附条债券;双键
double bond isomerism 双键异构
double bond shift 双键位移
double bookkeeping method 复式记账法
double bottom 双夹底;双层(船)底
double bottom afterward 尾部双层底
double bottom cells 双层底之间的间隔
double bottom cellular 双层船底框架
double bottom compartment 双层底舱
double bottom forward 首部双层底
double bottom level 双底水准仪
double bottom open floor 双层底组合肋板
double bottom party 双层底清洁组
double bottom ship 双(层)底船
double bottom tank 双层底舱
double bottom tank top plating 双层底内底板
double bottom test 双层底水密试验
double bottom trailer combination 双联拖车
double bottom trailer system 双联拖车方式
double bounce calibration 双回波校准法
double bowed vessel 两头船
double bowline 水手结;双套结
double bowl sink 双盆洗涤池

double bowl vacuum centrifuge 双鼓真空离心过滤机
double bowstring truss 鱼形桁架；双弓弦桁架
double box 双箱式
double box collar 双母接头钻链
double boxed mullion 双匣窗梃；双腔式中梃；匣式中竖框
double box girder 双箱式大梁
double box girder bridge 双箱式梁桥
double box pile 双箱桩
double box section 双室箱形截面
double bracing 双支撑
double bracing system 双支撑系统
double brackets 双括号
double braided nylon hawser 双层尼龙缆
double braided polypropylene filament hawser 双层多丝丙伦缆
double braided polyester nylon rope 双层丙伦尼龙缆
double branch 双叉管
double branch anchor beam 双支下锚横梁
double branch bend 双支弯头
double branch elbow 双支弯头
double branch gate 双侧浇口
double branch pipe 双支管
double-breadth barge train lock 倍宽船队船闸
double-breadth lock 倍宽船闸
double break 双断
double break circuit breaker 复断器；双断断路器
double break contact 双开路接点；双断接点；桥接接点
double break double-make contact 双断双闭触点；桥接转换触点
double break jack 双断开关；双断塞孔
double break switch 双断开关
double breasted plough 双壁开沟犁
double brick 双砖
double brick wall 两砖墙；四九墙
double bridge 双(臂)电桥
double bridge probe 双桥探头
double bridge-type penetrometer 双桥式触探仪
double bridging 双交叉撑；格栅双行剪刀撑；两对交叉撑；剪刀撑；双剪力撑
double brilliant 双多面型
double broad 双排丝散光灯
double broken and double screened chip(ping)s 二次轧碎和二次过筛的碎屑；二次破碎和二次过筛的碎屑
double bucket 双地址；双存储位
double bucket elevator 双斗提升机
double bud 重芽
double budget system 复式预算制
double buff 双折布抛光轮
double buffer 双缓冲
double buffered data transfer 双缓冲数据传送
double buffing machine 复式抛光机
double builders hoist 双排建筑用吊车
double bulkhead 双层舱壁
double bull-nose brick 双圆角砖
double bull type column 双圆角式柱
double bunk 双层铺
double buoys mooring 双浮筒系泊；双浮标系泊
double buoys seaplane 双浮筒式水上飞机
double burned dolomite 二次煅烧的白云石
double burner 双喉管燃烧器
double bus 双重母线；双汇流排
double busbar 双母线
double-busbar system 双母线系统
double bus connection 双母线连线
double buttress 双支墩
double buttress dam 双支墩坝
double butt strap 双拼板(木结构)；双接贴板
double-butt-strap riveting 对接双面盖板铆接
double buttwelded 双重平焊接
double cabin 双间圆木屋
double cable crane 双缆索起重机
double cable plane 双索面
double cable release 双线快门开关
double cable ropeway 双缆架空索道
double cable suspension bridge 双索悬索桥
double cableway 双(向)索道；双向缆道
double cable winch 双索扬机
double-cage reinforcement 双笼式钢筋架
double-cage reinforcing 双笼钢筋

double cake winding 分股卷绕
double calcite plate 双方解石片
double-calculation 两次计算；复算
double cal(l)ipers 两用卡钳；内外(两用)卡钳
double camber 双弧线
double camera 双物镜摄影；双片摄影机
double cancellation 双系斜杆
double canoe 双体船
double cant 双斜面
double cant frame 双斜边肋骨
double cantilever anchor span 双悬臂锚孔
double cantilever beam bridge 双悬臂梁桥
double cantilever beams 双悬臂梁
double cantilever beam specimen 双悬臂梁试样
double cantilever beam clip-in displacement meter 双悬臂夹式引伸计
double cantilever crane 双悬臂起重机
double cantilever gantry crane 双悬臂装卸桥；双悬臂门式起重机；双悬臂龙门吊
double cantilever girder bridge 双悬臂梁桥
double cantilever pylon 双悬臂塔柱
double cantilever roof truss 双悬臂屋架
double cantilevers 双悬臂
double cantilever shell 双悬臂壳体；双悬臂薄壳
double cantilever swing bridge 双悬臂式旋桥；双悬臂旋开桥
double cap 大四开图纸；双大裁
double capacitor 双联电容器
double cap casing head 双套管接头
double capital 双柱(冒)头；双(层)柱头
double capstan 双头绞盘
double carbide 复合碳化物；三元碳化物
double carbon arc lamp 双碳极弧光灯
double carbon arc welding 间接碳弧焊
double carriageway motorway 双车道汽车路
double carrick bend 双花大绳接结
double carry 双重进位；双位进位
double casement fastener 双平开窗插销
double casement window 双层(开关)窗
double casing 双层机壳式
double casing bucket elevator 双外壳斗式提升机
double casing machine 双机壳电机；内外机座型电机
double casing motor 双层机壳式电机
double casting 双面注浆
double catenary construction 双悬链线结构
double catenary suspension 双链悬挂法
double caulker 水手结
double cavity joint 双空腔缝
double cavity mo(u)ld 双穴模；双联模
double ceiling 双层天花
double centering 双定心法
double-center theodolite 双中心经纬仪；复测经纬仪
double-center transit 双中心经纬仪
double centrifugal pump 双吸式离心泵
double chain 双链
double chain conveyer 双链输送机
double chain draw bench 双链式拉拔机
double chain grab 双索抓斗
double chain grab crane 双索抓斗起重机
double chain rehandling grab 双链装载抓斗
double chain sling 双吊货索链
double chain suspension bridge 双链式悬索桥
double chair 双人椅
double chamber bed 双室床[给]
double chambered drill 双斜面钻
double chamber furnace 双室炉
double chamber sand blast apparatus 复室喷砂装置
double chamber surge shaft 双室式调压室
double chamber tunnel kiln 双室式隧道窑
double channel 双沟道
double channel duplex 信道双工通信；双通道双工
double channel section 双槽断面
double channel simplex 双通道单工
double charges 双倍收费；双倍负载；双倍充电
double charge ion 双电荷离子
double charging bell 双料钟
double check 重复检验；双重校验；双重校对；复核
double check system 重复检验制；双重监视系统
double check valve 双向回弹阀门；双(作用)止回阀
double check-valve assembly 双逆止阀装置
double chime unit 双乐音门铃

double chisel drill bit 双錾头钻头；二字形钻头
double chlorination 二级氯化；二次加氯量(法)；两次加氯量(法)
double choir 双合唱诗班
double church 双层教堂
double circle readings 双刻度读数
double circle rotating crystal camera 双圆转动单晶相机
double circle theodolite 双刻度经纬仪
double circle transit 双刻度经纬仪
double circuit 加倍电路；双电路
double circuit lines on the same tower 同塔双回线
double circuit tower 双回路线塔；双回路线架
double circular current 双向环流
double circular mill 双圆锯机；双上下圆制材厂
double circular projection 双圆投影
double circular sawmill 双线圆锯制材厂；双上下圆锯制材厂
double circulation 双重循环
double clam gate 双扇弧形门
double classifier 双层筛
double claw flexible coupler 双爪式挠性联轴节
double clay pot 黏土空心砌块；空心黏土砌块
double-cleat ladder 双开扶手梯架；双行道宽梯；双缆坡道
double-clevis-base bracket for drop tube 立柱双压管底座
double clewed jib 蝴蝶三角帆
double closed tube 双闭管
double closing hoop 双闭箍
double cloth 双层织物
double cluster 双星团
double clutch 双离合器
double-clutching 两次踏离合器换挡法
double coagulation 双重凝聚；双重混凝
double coat 双重涂层
double coated electrode 双层药皮焊条
double coated paper 双层涂料纸
double coated tape 双面涂布带；两面涂胶带
double coated X-ray film 双涂层X射线照相片
double coater 双面辊涂机
double coating 双层粉饰；两道涂装；两道涂覆；两道抹面；双道涂料；双层涂层；双层粉饰
double code 对偶码
double cog 双面齿
double coil 双线圈
double-coiled lamp 双线绕式电灯
double-coiled relay 双线圈继电器
double coil filament 双线圈灯丝
double coil holder arrangement 双卷位开卷机
double coil method 双线圈方法
double coincidence of need 需要的双方吻合
double coincidence spectrometer 双符合谱仪
double coincidence switch 双重合开关；二重符合开关
double cold reduction 再冷轧
double cold reduction mill 再冷轧用轧机
double colonnade 柱廊；双列柱子
double column 双柱式
double column hydraulic planer 双柱液压刨床
double column ledger account 双栏分分类账账户
double column machine 双柱式机床
double column pier 双柱式桥墩
double column planer 双柱刨床
double column planer milling machine 双柱龙门铣床
double column planing machine 双柱龙门刨(床)
double column tariff 双栏税则
double column type 双柱式
double column type digital height ga(u)ge with dial 带刻度盘双柱数字高度计
double column vertical boring machine 双柱(立式镗床)；双(立)柱
double combination 拖头和拖车相连
double combing 复梳
double combustion chamber 双燃烧室
double commutator motor 双整流子电动机；双换向器电动机
double compartment septic tank 双室化粪池
double component 二元组分
double component blending bed 双组分预均化堆场

double composition grease 复合润滑脂
double compound turbojet 双轴涡轮喷气式发动机
double compression 双重压制
double concave 双凹形;双凹(的)
double concave glass 双凹透镜
double concave lens 双凹透镜
double concrete pump 双联混凝土泵;双体混凝土输送泵
double condenser 双聚光器
double conductor 双导线
double conduit 双孔水道;双孔涵洞
double cone 对顶圆锥;双锥区
double cone air classifier 双圆锥式空气分级机
double cone bit 双牙轮钻头
double cone blender 双锥形混合机;双锥形掺和机
double cone classifier 双锥选粉机;双锥形粒子分级
double cone clutch 双锥离合器
double cone crusher 双锥形破碎机
double cone gasket 双锥面垫圈
double cone impeller mixer 双锥形桨式混合机
double cone invert discharging type mixer 双锥反转出料式搅拌机
double cone mixer 双锥混合机
double cone mo(u)lding 诺曼底式建筑中拱的线脚;饰双锥体线脚;对锥线脚;顶锥线脚
double cone pile 双尖料堆,双锥体料堆
double cone seal ring 双锥密封环
double cone type cyclone 双锥型旋风筒
double conical rotary vessel 双锥转鼓
double conical rotary vessel with jacket 带夹套的双锥转鼓
double conical saw 双凸面圆锯
double connection 二重连接
double connector 双重接头;双向(螺纹)套管
double contact 双触点
double contact key 双触点电键
double continuous curve 正弦曲线
double contraction 双重收缩
double contraction pattern 双重收缩模
double control 双重控制(自动和人工)
double control switch 双控开关
double conversion system 双变换制
double converter 反并联接法;双换流器
double convex 双凸;两面凸
double convex arch bridge 双曲拱桥
double convex glass 双凸透镜
double convex lens 双凸透镜
double convex type 双凸型
double copy 对开图纸
double core barrel 双层岩芯管;双层取芯器
double core barrel sampler 回转取土器
double core plastic wire 双芯塑料电线
double core point 双芯头
double corkscrew 双螺旋打捞器;双捞牙(捞钢丝绳用)
double corner block 六面光混凝土砌块;双棱(角)混凝土砌块
double corridor 复廊
double corridor layout 双走道布局
double corrugated sheet 双波纹铁皮
double cotton-covered wire 双纱包线
double counting 重复计算
double counting method 双计数法
double couple 双力偶
double couple method 双力偶法
double couple model 双力偶模型
double couplers 双联夹
double coupling piece 复式连接件
double course 檐口双层瓦;双层(板);双(板)层
double course surface treatment 双层式表面处治;双层式表面处理
double coursing 双层铺板法
double cover butt joint 双盖板对接
double cover-coat enamel 双搪(瓷)
double covered type 双层掩盖型
double crane 双钩起重机
double crank 双曲柄
double crankless press 两点式无曲柄压力机
double crank lock 双曲柄闭塞器
double crank mechanism 双曲柄机构
double crank press 双曲轴压力机
double cropping index 复种指数
double cropping rice 双季稻
double cropping system 双季栽培制

double crop rice 双季稻
double cross 双横杆十字架;双交
double cross grain 木材交织纹
double cross head bollard 双十字头系缆柱
double crossing 双交叉
double crossing leading mark 首尾导标
double crossing over 双交替
double crossover 交叉渡线;双交叉
double cross timber 复式交叉(木)系杆
double cross welt 双咬口十字缝
double crown 对开图纸
double crown furnace 双碳窑炉;双拱窑炉
double crown knot 双冠形结节
double crown tank furnace 双拱顶池窑
double crucible 双层坩埚
double crucible method 双坩埚法
double crucible process 双坩埚法
double-crushed chip(ping)s 二次碾轧碎屑
double crusher 双顶圆锥轧碎机
double crust type 双壳型
double crypt 教堂双层墓窟;教堂双层地下室
double crystal diffractometer 双晶衍射计
double crystal spectrograph 双晶摄谱仪
double crystal spectrometer 双晶分光计
double cube 双立方体
double-cube room 双立方体室
double cup bearing 双排外夹环滚动轴承
double cup insulator 双碗式绝缘子
double currency 复本位
double current 双(电)流
double current bridge duplex 双流桥接式双工
double current furnace 双电流电炉
double current generator 双电流发电机;交(直)流发电机
double current locomotive 双电流制式电力机车
double current method 交直流法;双流式;双极性电流法
double current relay 双流式继电器
double current return to zero system 双流归零制
double current system 双电流制
double current transmission 双向电流传送
double curvature 双曲率
double curvature arch 双曲拱
double curvature arch aqueduct 双曲拱渡槽
double curvature arch dam 双曲拱坝
double curvature arch flume 双曲拱渡槽
double curvature shallow shell 双曲扁壳
double curvature shell 双曲(面)壳体;双曲薄壳
double curvature translation(al) shell 双曲率平移薄壳
double curved 双弯曲
double curved arch bridge 双曲拱桥
double curved arch dam 双曲拱坝
double curved bridge 双弯曲的桥梁
double curved cupola shell 双曲圆顶薄壳
double curved glass 双曲面玻璃
double curved rib 双曲肋
double curved shell 双曲扁壳
double curved spike grid 双曲的钉格板;钉格凸板
double curved surface 二重曲面;复曲面
double curved vault roof 双曲拱顶
double cut 双纹;双掏槽
double cut file 斜格锉;双纹锉;交叉滚花锉
double cut-off saw 双轴圆盘锯(床)
double cut planer 双向切削龙门刨
double cut plough 双层犁
double cut saw 双向切割锯;双割锯;双向(齿)锯
double cutting bands 双刃带锯
double cutting bandsaw 双刃带锯
double cutting drill 双刃钻
double cutting snips 轧剪;咬剪
double cycled signal 双周期信号
double cycling 双周期
double cyclone 双旋网捕焦器;双级旋风分离器
double cylinder 双筒
double cylinder engine 双缸(式)发动机
double cylinder hydraulic sounding machine 双缸液压触探机
double cylinder jacquard 双花筒提花机
double cylinder lock 双弹子锁
double cylinder stair(case) 双圆柱形楼梯
double cylinder type rotary kiln 双筒回转窑
double cylindrical boiler 双筒锅炉
doubled 复式的

double damask 八枚花缎
double danger angle 双倍危险角
double day tide 隔日潮;双日潮
double deal 二英寸厚板;二寸厚板(英制);双层木铺板
double deal frame 复式排锯
double deck 两层甲板船;双层(结构)
double deck airplane 双层飞机
double deck arch 双层拱
double deck bed 双层床
double deck blending and storage silo 双层搅拌储存库
double deck block 二次拉丝机
double deck bridge for highway and railway 公路铁路两用桥
double deck bus 双层公共汽车;双层叠瓦
double deck car-carrying wagon 双层汽车运输车
double deck classifier 双层分级机
double deck coach 双层车箱
double decked 双层的
double decked bridge 铁路公路两用桥;双层桥(梁);双层桁架桥;双层大桥
double decked bridge for railway and highway 铁路公路双层桥
double decked bus 双层公共汽车
double decked oven 双层炉
double decked ship 双甲板船
double decked wharf 双级式码头
double deck elevator 双层电梯
double decker 双层火车;双层公共汽车;双层装置;双层小公寓;双层甲板船;双层叠瓦;双层电车;双层床
double decker coach 双层客车
double decker train 双层火车
double deck floating roof 双层浮顶
double deck freeway 双层(式)高速干道;双层超速干道
double deck furnace 双层炉
double deck lift 双层电梯
double deck motorway 双层汽车道路
double deck pallet 双面托盘
double deck plough 双层犁
double deck pull rod 双层拉杆
double deck road 双层道路
double deck rotary 双转盘台
double deck screen 双层筛
double deck station 双层车站
double deck table 双层摇床
double deck type of crown block 双层天车
double deck vibrating screen 双层振动筛
double-declining balance 加倍余额递减
double-declining balance depreciation 双倍递减余额折旧
double-declining balance depreciation method 双倍递减余额折旧法
double-declining balance method 双倍余额递减法
double decomposition 复分解
double decomposition grease 复合解润滑脂
double deflection grille 双向挠曲格栅
double deflection indicator tube 双偏转指示管
double degeneracy 二重简并(度)
double demy 对开图纸
double density 倍密度;双(倍)密度
double-density encoding 倍密度编码
double-density format 倍密度记录格式;倍密度格式
double depth plough 双层犁
double detection 双重检波
double detection receiver 超音频外差式接收机;超外频外差式接收机;双检波接收机
double detector 双检测器
double detector check 双检测器单向阀
doubled flanged ring 双边钢领
doubled folded yarn 双股丝
double diamond interchange 双菱形(互通式)立体交叉
double diamond knot 双花菱形结节
double diaphragm cone 双隔膜圆锥分级机
double diaphragm pressure cell 双膜式土压力盒
double diaphragm pressure ga(u)ge 双膜式土压力盒
double diaphragm pump 双膜式泵
double dichotomy 二重二分法;双二分法
double differectial 双差速器
double difference 二重差分

double difference code 二异码
double differential 双差动
double differential fix resolution 双差固定解
double diffracted 二倍衍射的
double diffracting medium 双衍射介质
double diffraction 双衍射
double-diffuse 复漫射
double diffusion 双(重)扩散;双(向)扩散
double-diffusion method 双扩散法(木材防腐法)
double diffusion precipitation test 双向扩散沉淀试验
double diffusion treatment 双重扩散处理
double diffusive interface 双扩散面
double digit inflation 两位数的通货膨胀
double digitizing 第二次数字化
double-digitizer 重复数字化器
double dike closure 双戗堤截流
double dilution (method) 双(重)稀释(法)
double dipole 双偶极子
double dipping 二次浸釉
double directional focusing 双向聚焦
double directional working (on single track line) 双方向运行
double direction automatic block 双向运行自动闭塞【铁】;双向自动闭塞
double direction thrust ball bearing 双向推力球轴承
double disc electrophoresis 两向盘状电泳法;双盘电泳法
double discharge gear pump 双出口齿轮泵
double discharge spiral water turbine 双排量涡壳式水轮机
double discharge stabilization 双放电稳定
double disc harrow 双盘耙路机
double disking 复耙
double disk marker 双圆盘划行器
double disk opener 双圆盘开沟器
double disk turbine rotor 双轮盘涡轮转子
double disk winding 双盘型线圈
double dispersion 双色散
double displacement pump 双向往复泵
double distillation 重蒸馏
double distilled 重馏
double distilled water 重蒸馏水
double distributary estuary 双叉河口
double distribution 双重分布;双流分配
double distribution method 偶极法
double dock 双厢船坞
double doffer condensor 双道夫搓条机
double donut 大卷盘水带
double door 双重门;双扇门;双道门;双开门
double door bolt 双(门)插销;通天插销;天地销
double door drop-bottom bucket 对开底卸料斗
double door fittings 折门零配件;双扇门零配件
double door furniture 折门设备;双扇门设备;对开门设备
double door hardware 折门五金;折门零件;双扇门零件
double door leaves 双门扇
double dot dash line 双点划线
double dot display 双点显示;分叉点显示器
double-double cap 全开图纸
double-double crown 全开图纸
double double effect 双偶效应
double double folio 全开图纸
double double iron sheet 叠轧铁板
double double radio source 双双射电源
double doublet antenna 双偶极天线
double Dove scan prism 双道威扫描棱镜
double dovetail key (硬木制的)银锭榫;双燕尾榫(扣);双燕尾键;双鸠尾榫键
double draft 双联单
double draft bull block 二次拉丝机
double drain 双排水
double drainage 双向排水
double draining (搪瓷表面的)二次流浆
double draw bridge 双臂式仰开桥;双翼仰开桥;双翼开合桥
double drift 二星流;双偏流测风法
double drill double column 钻机用双支柱(坑道钻进用);双柱钻架
double drive 双传动式
double driver plate 双销拨盘
double driveway 双线公路

doubled-roller applicator 双辊式浸润器
double drop machine 双落料机
double dropping electrode 双滴电极
double-drum boiler 双汽包锅炉
double-drum cable control unit 双卷筒缆索控制装置;双筒电缆控制单元
double-drum (concrete) mixer 双(鼓)筒混凝土搅拌机
double-drum diesel hoist 双筒柴油卷扬机
double-drum diesel winch 双卷筒柴油绞车;双筒式柴油卷扬机
double-drum drier 双滚筒干燥机;双转鼓干燥器
double-drum dryer with top-feed 顶部加料式双滚筒干燥器
double-drum haulage engine 双滚筒运输机;绞车运输机
double-drum hoist 双筒起重机;双筒卷扬机;双筒绞车;双卷筒提升机;双滚筒提升机
double-drum hoisting gear 双鼓筒卷扬机装置
double-drum mixer 双(鼓)筒搅拌机
double-drum pav(i)er 双(鼓)筒铺路机
double-drum pile driving winch 双卷筒打桩绞车
double-drum roller 双筒振动压路机;双筒振动(路)碾
double-drum rope winch 双筒卷扬机;双筒绞车
double-drum steam winch 双卷筒蒸汽绞车
double-drum type mortar mixer 双筒式灰浆搅拌机
double-drum winch 双卷筒卷扬机;双卷筒起网机;双卷筒绞车;双滚筒绞车
double-drum windlass 双径滚筒绞盘
double drying drum 双干燥鼓
double dry plate clutch 双干板离合器
double duct system 双风道系统
double dunnage 双层货垫
double duo-mill 双二辊轧机
doubled-up door 折叠门
double duty 双重用途;两用(的)
double-duty double-acting hammer 两用双动锤
double duty plant 两用机械设备
double-duty ship 两用船
double dwelling 两间式建筑物
doubled yarn 并捻纱
double-ear fancy colo(u)r vase with lotus design 斗彩勾莲双耳樽
double ear knot 双耳结
double earth fault 双相接地故障;两相接地故障
double eaves 重檐(指两层之间加假屋檐)
double eaves course 檐口双层瓦;双檐层木板瓦;檐口垫瓦;双檐瓦层
double eaves roof 重檐屋顶
double ebb 复退潮;复落潮;复落潮流
double eccentric connector 双偏心连接器
double eccentricity 双偏心
double echo 双回波
double edged 双刃的
double edged skate 双边铁鞋
double edge grinder 双边研磨机
double edge grinding machine 双边磨边机
double edge milling cutter 二面刃铣刀
double edger 双圆锯裁边机;双裁边锯
double edging tool (修整混凝土板边缘的)双面规尺
double effect 双效(的)
double effect evapo(u)rator 双效蒸发器
double effect lithium-bromide absorption-type refrigerating machine 双效溴化锂吸收式制冷机
double egress frame 双向开门框
double elbow pipe S形管
double electrode layer 双电层
double elelctric(al) layer 双电层
double ellipsoidal tank 双椭圆形水箱
double elliptical cavity 双椭圆腔
double elliptical spring 双椭圆弹簧
double elliptical system 双重椭圆系
double embrace (用钢带止动鼓的)双抱
double end 双端引线
double end block plane 双向刨
double-end bolt 双头螺栓;柱螺栓
double-end(ed) 双端;双头(的)
double-ended abdominal retractors 双头腹壁牵开器
double-ended ball mill 双锥滚筒式球磨机;双头球磨机;双流式球磨机
double-ended boiler 双头锅炉;双炉门式锅炉

double-ended bolt 双头螺栓;双端螺栓
double-ended boring machine 双头镗床
double-ended box spanner 双头的套筒扳手
double-ended box wrench 双头的套筒扳手
double-ended charge 双端装药爆破筒
double-ended coating line 双端式镀膜线
double-ended control 双端控制
double-ended cord 两头塞绳
double-ended cylinder 双端口汽缸
double-ended floor 两端砌块的地板;构架楼面;构架楼板
double-ended grinder 双端磨床
double-ended locomotive 双头机车
double-ended machine 双滚筒联合采煤机
double-ended power transmitting tube 双端功率发射管
double-ended pressing 两端压制
double-ended probes 双头探针
double-ended retractors 双头牵开器
double-ended ring spanner 双头环形扳钳
double-ended ring wrench 双头环形扳手
double-ended ship 首尾同型船
double-ended swivel wrench 双头转体扳手
double-ended system 双端系统
double-ended tubular lamp 双头管形灯(泡)
double-ended turbine 双排汽口汽轮机
double-ended type 双端引线型
double-ended vessel 两头船
double-ended wrench 双头(死)扳手
double ender 两头船;法兰杆;双头扳手;首尾同型船
double end knife colter 翻输式直犁刀
double end mill 双端铣刀
double endorsement 双重背书
double end prestressing 两端张拉
double ends 双经
double end snap ga(u)ge 双头卡规
double end spanner 双头扳手
double end spark plug wrenches 两用火花扳手
double end tenoner 双端制榫机
double end trimmed 两端锯齐的;双端修饰的
double end wrench 双头扳手
double energy transient 双能量过渡过程
double entry 复式记账(法);两侧进气
double entry account 复式会计
double entry accounting system 复式簿记账法
double entry bookkeeping 复式簿记;复式记账
double entry card 双入口卡片
double entry impeller 双吸叶轮;双面进气叶轮
double entry method 复式记账法;又平巷采矿法
double entry system 复式会计制度
double epitaxial method 双外延法
double equal-angle cutter 对称双角铣刀
double equatorial 双筒赤道仪
double equilibrium diagram 二相平衡图
double error 双重错误
double error correction scheme 双磁道纠错法
double escrow 转手买卖地产
double esker 重叠蛇形丘
double exchange 双重交换
double exchange interaction 双重交换作用
double exhaust pipe 双尾气排放管
double expansion compound steam engine 两级膨胀复式蒸汽机
double expansion engine 双胀式蒸汽机;二级膨胀式发动机
double exponential distribution 二重指数分布;双指数分布
double exponential regression 二重指数回归
double exponential smoothing 二重指数平滑;二次指数平滑
double exposed 两次曝光的
double-exposed hologram 双曝光全息图;两次曝光全息照片;两次曝光全息图
double-exposed holographic interferometry 二次曝光全息干涉测量术
double exposure 两次曝光;二次曝光;双(重)曝光
double exposure device 两次曝光装置(重拍机构)
double exposure hologram 双曝光全息图;二次曝光全息图
double exposure holographic interferometry 两次曝光全息干涉测量;双曝光全息干涉法
double exposure holographic interferometry 两次曝光全息干涉量度学
double exposure holography 双曝光全息术;两

次曝光全息术
double exposure interferometry 两次曝光干涉量度学;双曝光干涉度量学
double extended shaft 双向外伸轴
double extension ladder 二节式可延伸梯
double exterior-corridor layout 双外廊布局
double extra dense flint 双超重火石玻璃
double extra heavy pipe 特加重管;双重厚壁管
double extra strong pipe 特强厚壁管;特加重管;特高强管;双重厚壁管;双加厚管
double eye end rod 双环杆
double fabric 重经组织
double face 对偶工作面
double face coating 双面涂布
double-faced 双面织物;两面用的双层布;双面的;双向线脚;两面修饰材料
double-faced adhesion band 双面胶带
double-faced architrave 双面线脚门头板
double-faced book rack 双面书架
double-faced hammer 双(击)面锤;无砧座锤
double-faced insulation 双贴面防热制品
double-faced partition wall tile 双面隔墙砖
double-faced satin 双面缎子织物;双面缎
double-faced secondary clock 双面子钟
double-faced sledge hammer 双面撬锤
double-faced slide gate valve 两面滑门阀
double-faced stack 双面书架
double-faced stock 两面光板材
double-faced tile 双面砖
double-faced ware 双釉面瓷器
double face jaw crusher 双面颚式破碎机
double face sign 双面标志
double face valve 双面阀
double face wall 双面墙
double face ware 双面搪瓷器
double factorial terms of trade 双因贸易条件
double fag(g)ot iron 双压挤熟铁
double failure 双故障
double fall system 联杆吊货法
double false ceiling 双层假天棚;双层假平顶
double fan 二联通风机;双风扇
double fan cut 双扇形掏槽
double fare 来回票价
double fastening 双排钉
double fault indicator relay 复式故障指示继电器
double feathering 双叶瓣饰
double fed asynchronous machine 双馈电式异步电机
double fed repulsion motor 双馈推斥电动机
double feed 双馈
double fee rate 双费率
double female 双雌接口
double female adaptor 双头内螺纹接合器
double figures 两位数
double file stream 双(面)列车流
double fillet lap joint 双面角焊缝搭接接头;双面搭焊接
double fillet welded joint 双面焊接接头;双面焊缝
double fillet welded T-joint 双面丁字形焊接
double filtration 二级过滤;双重过滤
double final drive 双末级传动
double finder telescope 双焦取景器
double finger pier 双突堤式码头
double firing 二次煅烧
double firing process 二次烧成工艺
double five-spot 双五点网
double flame furnace 返焰炉
double flange 双凸缘
double-flanged butt joint 双折边对接;双弯边对接接头
double-flanged butt weld 双弯边对接焊缝
double-flanged rail wheel 双缘钢轨轮
double-flanged reducer 双法兰渐缩管
double-flanged wheel 双面凸缘轮
double-flanged zed 双法兰外移管件;双法兰 Z 形管件
double flange roller 双凸缘压板机
double flange track roller 双边支重轮
double flange wheel 双缘轮
double flap clay pantile 双翼黏土波形瓦
double flap pantile 双翼波形瓦
double flap valve 双瓣阀
double flash distillation 两次急骤蒸馏法
double flash evapo(u)ration 二级闪蒸

double flash unit 两级扩容装置
double flat rod 双扁杆
double flattened strand 双层扁股钢丝绳
double flat yard 双向平面调车场
double Flemish bond 双面一顺一丁砌合;双(面)荷兰式砌合
double flight feed crew 双叶螺旋加料器;双头进料螺杆;双头螺旋加料器
double flight stair(case) 双梯段楼梯;双跑楼梯
double float 复式浮标;复浮子;双球浮子;(测流用的)双浮标;深水浮标
double float seaplane 双浮筒式水上飞机
double flood 复涨潮;双涨潮流
double floor 双层楼面;(格栅置在次梁上的)楼面构造;下有吊顶天花板;下有吊顶的楼板;双层楼板;双层(龙骨)地板
double floor covering 下有吊顶的楼板饰面
double floor finish 双层楼板饰面;双层地板饰面
double flooring 双层地板
double flow 双(排)汽
double flower 重瓣花
double flow exhaust 双流排汽
double flow induced-draft cooling tower 强制通风双流式冷却塔
double flow turbine 双流(式)涡轮(机);双流式透平;双流式汽轮机
double flue boiler 双烟道锅炉;双燃烧管锅炉;双炉胆锅炉
double fluid cell 两液电池;双液电池
double fluid coupling 闭式循环液力耦合器
double fluid epoxy primer 双液型环氧底漆
double-fluid theory 二流体理论
double fluked anchor 双爪锚
double-fluted bit 双槽钻头
double focus arrangement 双焦装置
double focusing 双聚焦
double focusing mass spectrograph 双聚焦质谱仪;双聚焦质谱计
double focusing mass spectrometer 双聚焦质谱仪
double focus interference microscope 双焦点干涉显微镜
double focus interferometer 双焦点干涉仪
double focus method 双焦法
double focus system 双聚焦系统
double fold disappearing stair(case) 双折隐梯
double folder 复式折页机
double folding door 双折门;两摺式摺门;双层摺门;两摺门
double folio 对开图纸
double folium 双叶线
double foolscap 对开图纸
double force 双力
double fore-and-aft planking 双层纵向木壳板
double forked strut 双叉式支撑
double format pavior 双开铺面砖
double formed crystals 双形晶
double foundation pit hoist 双基坑吊车
double frame 双幅字盘架;双层框架
double frame camera 双像幅照相机
double-framed floor 双重支承楼板;双梁系楼盖;双层框架楼板
double-framed generator 内外机座型发电机;双机座型发电机
double-framed roof 双承式屋盖;双坡式屋顶
double-framed wall 双重构架墙
double-framed window 双层窗
double frame hammer 双柱式锤
double framing 双倍用料的构架
double Frank's mirror 双法兰克反射镜
double freight operations 双重作业
double frequency 双频
double frequency beacon 双频信标
double frequency coding 双频率编码
double frequency generator 双频发电机
double frequency induced polarization instrument 双频道激电仪
double frequency inductor 双频感应器
double frequency oscillation 双频振荡;倍频振荡
double frequency oscillator 双频振荡器
double frequency sequence signal(l)ing system 双频顺序信号制
double frequency spectral density function 双重频率谱密度函数
double frequency table 二项频数表

double frieze balustrade 垂台钩栏
double frill girdle 双竖沟环割
double fringe 双条纹
double frit glaze 双熔块釉
double fritting 双熔块配釉
double frog 棱形交叉
double-frogged brick 双凹槽砖
double frontage 对穿地块
double frontage lot yard 双面临街院落
double-fronted lot 两面临街地产
double fuel engine 双燃料发动机
double function key 双功能键
double funnel 双层漏斗
double furnace boiler 双燃烧管锅炉;双炉膛锅炉
double furrow 闭垄
double furrow plough 双铧犁
double fuselage plane 双机身飞机
double gable roof 双山头屋顶;双山墙屋顶;双人字屋顶;双脊屋顶
double galvaized wire 加厚锌层镀锌钢丝
double galvanized wire 加厚镀锌钢丝
double gang disk harrow 双排圆盘耙路机
double gap 双火花间隙
double gap cavity 双隙缝谐振腔
double gas passage travel(l)ing grate 气体二次通过的炉算加热机
double gate 双闸;双选通;双门;双控制极
double gate elevator 双开门吊卡;双铰链提引器
double gate lock 双重门船闸;双门(式)船闸
double gate stop 双开门止门器;双开门门挡
double gateway 双重大门;双重门道
double ga(u)ge line 复式轨距线;双轨距线路
double Gauss derivatives 双高斯型物镜变形
double Gauss goniometer 双向测角仪
double Gauss lens 双高斯镜头
double Gauss lenses 双高斯透镜
double Gauss objective 双高斯型物镜
double Gauss type lens 双高斯式物镜
double gauze wire cloth 双股金属丝网布
double gear cone transmission 复式锥轮传动装置
double geared 双级齿轮传动
double-geared drive 二极齿轮传动;二级齿轮传动
double geared roller crusher 双齿辊破碎机
double girder bridge 双梁桥
double girder gantry crane 双梁门式起重机
double girder overhead crane 双梁桥式起重机
double girders 双(大)梁;并置大梁
double girder truss gantry crane 桁架结构的双梁龙门起重机
double glass 双层玻璃
double glass seal 双层玻璃气密封接
double glass window 双层玻璃窗
double glass wired glass machine 双加料夹丝玻璃成型机
double glazed window 双层玻璃窗
double glazed window frame 双层玻璃窗框
double glazed window unit 双层玻璃窗
double glaze unit 双层玻璃窗
double glazing 两次施釉;双面上光;双层玻璃
double glazing glass 中空玻璃;双层玻璃板
double glazing unit 双层玻璃单元;双层玻璃板;双层中空玻璃;双层玻璃窗
double-gourd vase 葫芦瓶
double governor 双重调节器
double graded structure 双粒序构造
double grading 双重特性;双重分级
double grating 双光栅
double grating monochromator 双光栅单色仪
double grating spectrograph 双光栅摄谱仪
double grid 双栅
double groove 双面坡口;双面槽
double groove joint 双面坡口接头
double groove weld 双面坡口焊缝
double ground fault 双线接地故障
double group 双值群
double grouser shoe 双齿履带板
double grouser track shoe 双锚定桩靴
double guest room 双人(客)房
double gulley 双集水沟
double gutter tile 双层沟瓦;双企口瓦
double gyre 双环流
double half-day tide 双半日潮
double half hitch 两半结
double halfwave filter 双半波滤片

double hammer breaker 双锤破碎机
double hand drilling 双人手工打眼；双人手工打孔
double handed saw 双人锯；双柄锯
double hand grip 双机械抓手
double handling 双重转运；双重处理；二次搬运；两次转运
double handling costs 二次搬运费
double-handling of material 物料的重复操作
double hanger 开式吊架(轴承用)
double hanging (roof) truss 立字桁架；双柱上撑式桁架
double harmonic oscillator 双谐振动器
double harvest rice 双季稻
double head automatic transmitter 双机头自动发报机
double head bitt 双柱头系缆柱
double head chaplet 双面型芯撑
double-headed axe 双头斧
double-headed camera 双头摄影机；立体摄影机
double-headed ceiling 双嵌线平顶
double-headed nail 双头钉；拼钉；双帽钉
double-headed rail 双头导轨；护墙板
double-headed spray gun 双口喷枪
double-headed train 双机牵引列车
double-headed tube bender 双头弯管机
double header 双头式列车；双击冷镦机；双木边梁；双车头牵引列车；复式大梁
double header boiler 双联箱横水管锅炉
double header curtain coating 双头幕式涂布
double head flat scraper 双头平刮刀
double head grinder 双头研磨机
double heading 双重牵引的；双开挖面掘进；双机(牵引)；双导洞掘进
double heading grade 双机坡度【铁】
double heading gradient 双机坡度【铁】
double heading locomotive 重联机车【铁】
double headings 双平行煤巷
double head rail 双头轨
double head spray gun 双头喷枪；双口喷枪
double head spur dike 双头丁坝
double head tyre 双缘轮胎
double head wrench 双头扳手
double hearth furnace 双层底炉
double heating 两次加热；二次加热
double heatsink diode 双重散热片二极管
double hedging 双重对冲
double helical bevel gear 人字伞齿轮
double helical double helix 双螺旋形的双螺旋
double helical double-reduction unit 两级人字齿轮减速器
double helical gear 人字齿轮
double helical gearing 人字形齿轮
double helical gear planer 人字齿轮刨床
double helical gear tooth 人字齿
double helical heater 双螺旋(线)灯丝
double helical mixer 双螺旋桨混合器
double helical spring 复式(螺旋)弹簧
double helical (spur) gear 双螺旋齿轮
double helical spurgear 人字形齿轮
double helix 双螺旋
double helix screw pump 双头螺旋螺旋泵
double helix structure 双股螺旋线
double helmet for piles driving 打双桩用的桩帽
double herringbone 双人字纹
double heterodyne system 双外差法
double heterojunction laser 双异质结激光器
double heterostructure 双异质结构
double high rolling mill 双辊式轧机
double high tide 双高潮
double high water 双高潮
double-hinged arch 两铰拱；双铰拱
double-hinged arch bridge 双铰拱桥
double-hinged arch(ed) frame 双铰拱架
double-hinged arch(ed) girder 双铰拱大梁
double-hinged arm 双枢杆
double-hinged door 双扇对开门
double-hinged flat arch(ed) girder 双铰平拱大梁；双铰平拱形梁
double-hinged flat parabolic arch(ed) girder 双铰平抛物线拱大梁
double-hinged frame 双铰框架
double-hinged gable(d) frame 双铰人字架
double-hinged gate 双合页门
double-hinged parabolic arch(ed) girder 双铰抛物线拱大梁；双铰抛物线拱形梁
double-hinged rectangular frame 双铰方框架；双铰矩形框架
double-hinged segmental arch(ed) girder 双铰弓形大梁
double-hinged swivel 复式转座
double-hinged swivel arm 复式转臂
double-hinged swivel wall drilling machine 复式转臂墙钻床
double hoist 双排吊车
double hole type nozzle 双孔式喷嘴
double hollow bit tongs 十字空心钳
double hollow clay (building) block 双孔空心黏土砖
double hook 双钩(吊钩)；双(抱)钩
double hook bolt lock 双弯钩栓锁
double Hooke's joint 双万向节
double hoop closing 双箍闭合
double hopper 双仓斗
double horizontal cordon 双平干矮树
double horizontal milling machine 复式卧铣床
double horn radiator 双号筒辐射器
double horse-shoe flame tank furnace 双马蹄焰池窑
double house 两间式建筑物；一对半分离房屋；拼连的两所房屋
double housing planer 双柱(式)龙门刨床；龙门刨床
double H steel column 双工字钢柱
double hull 双层船体；双层船壳
double hulled 双重机壳
double hull structure 双层船壳结构
double hump 双驼峰
double humped resonance 双峰谐振；双峰共振
double humped wave 双峰波
double hump effect 双峰效应
double humping 双溜放
double humping and double rolling 双推双溜【铁】
double humping facility 双溜放设备
double hump yard 双向驼峰编组场
double hung 双吊钩
double hung (counterweight) window 上下推拉窗；上下扯窗
double hung sash (window) 上下扯窗；双悬窗
double hung window 双侧平衡铊式吊窗；水下扯窗；双悬窗(框)；提拉窗
double hydrant 双出口消火栓
double hydrated lime 倍水化石灰
double hydraulic 双液压的
double hysteresis loop 双磁滞回线
double ignition 双点火
double image 重影；重像；双影；双(图)像
double image echo 双图像回波
double image effect picture 重影效应图像；双像效应图像
double image eyepiece 双像目镜
double image lens 双向透镜
double image-method 双像法
double image-method optic(al) plumbing 双像法光学垂准
double image micrometer 双像测微器
double image ocular 双像目镜
double image overlap tacheometer 双像重合测距仪
double image prism 双像棱镜
double image range finder 双像测距仪；双像测距机
double image signals 重影信号
double image tach(e)ometer 双像测距仪；双像视距仪
double image tachymeter 双像测距仪
double image zenith telescope 双像天顶仪
double impedance coupling 复阻抗耦合
double impeller breaker 双转子锤式破碎机
double impeller flo(a)tation cell 双轮式浮选机
double impeller impact breaker 双叶轮冲击破碎机；双推进器冲击破碎机
double impeller impact crusher 双转子反击破碎机
double impeller impact mill 双叶轮式冲击粉碎机；双转子反击式磨机
double imposition 复税
double impression method 锤击布氏硬度测定法
double impulse 成对脉冲
double incline 双斜面；驼峰调车场；驼峰(铁)
double inclined plane 双斜面
double indemnity 加倍赔偿；加倍补偿；双倍赔偿
double indemnity clause 人身意外死亡的双重赔偿条款
double index method 双指标法
double induction regulator 双电感式电压调整器
double injector 复式喷射器
double inlet system 双进样系统
double input mixer 双输入混频器
double insulated appliance 双重绝缘的设备
double insulated conductor 双绝缘导线
double insulating reed mat 双层绝缘芦苇垫
double insulation 双重绝缘
double insulator 双重绝缘子
double insurance 重复保险；双(重)保险
double insurance policy 双重保险单
double intake valve 双进气阀
double integral 重积分；二重积分
double integrating accelerometer 二重积分加速度计
double integrating gyro 双积分陀螺
double integrating gyroscope 双积分陀螺仪
double-integrating range unit 重积分测距装置
double integration (method) 两次积分法；二重积分法；重积分法；双积分法
double integrity tank 双壁储罐
double intended circle 双凹圆
double interferometer 双元干涉仪
double interlaced handle 双泥条绞合手柄；麻花形手柄
double-interlocking H-pile 双锁口H桩
double-interlocking Spanish tile 双连锁西班牙瓦
double-interlocking tile 双联锁瓦
double interpolation 二重插值；双内插
double intersecting truss 双向交叉桁架；复式交叉桁架
double intersection truss 复式交叉桁架
double intertie 复式交叉系杆
double ionization 二次电离
double ionization chamber 双电离室
double ionization ga(u)ge 双电离压力计
double iron 工字钢
double iron plane 双刃刨
double irradiation 双照射
double jack 双柱支架；双手(大)锤；八磅大锤；锻工锤；双柱钻架；双支柱；双塞尺
double jack bar 双柱式塔架；(凿岩机的)双支架
double jacket hose 双外层水带
double jack process 双千斤顶法
double jack rafter 复式小椽；双支点椽条；复式小梁
double jaw joint 双重十字轴式等速万向节
double jet carburetor 双嘴汽化器
double joint 宽缝；双接缝
double-jointed 双重关节的
double-jointed arm 双枢杆
double-jointed compasses 双接头圆规
double-jointed propeller shaft 双万向节传动轴
double-jointed swivel 复式转座
double-jointed swivel arm 复式转臂
double-jointed swivel wall drilling machine 复式转臂墙钻床
double jointing 双管焊接
double jointing machine 双管焊接机
double jointing plant 双管焊接厂
double jointing site 双管焊接场地
double jointing yard 双管焊接场地
double-joisted floor 双格栅楼板；双层龙骨地板
double junction 双联会合；双联结点；双交叉；十字岔管接头
double jute bag 双麻袋
double key 双头切口扳手
double K groove 双边K形坡口
double king post 双起重柱
double knife mower 双刀割草机
double knit fabric 双面针织物
double knitting wool 粗绒线
double knock 双报警
double knot 双结
double lacing 双缀条
double ladder 折叠梯；人字梯
double ladder dredge(r) 双吸管架挖泥船；双排多斗挖泥船
double laminated spring 椭圆弹簧；双弓弹簧

double lancet window (中间有竖框的)双尖窗；双尖顶窗
double lane 双线(航道)的；双车道
double lane bridge 双车道桥(梁)
double lane channel 双线航道
double lane highway 双车道公路
double lane road 双车道道路
double lane traffic 双车道交通
double lap 两面搭接；插接
double lap joint 双盖板搭接接头
double lap tile 双盖瓦；双搭瓦；普通瓦；双层叠瓦
double large 双大裁
double large post 双大邮裁
double lark's head 两层双合套
double lath 加厚(的灰)板条；双重板条；双面灰板条
double lathe 复式车床
double lathing 双层灰条；双缀条
double lay 双线铺设
double layer 电偶层；双电层；双层数；双层(次)偶层；重层；二次布料
double layer antireflection coating 双层增透膜；双层减反射膜
double layer aquifer 双层含水层
double layer brick 复合砖
double layer coat 双层保护层
double layer coating 双层膜
double layer construction 双层建筑；双层构造
double layer crust model 双层地壳模型
double layer diaphragm plate 双层隔仓板
double-layered 双(荧光)层的
double-layered space frame shell (采用立体构架的)双层壳体；空间框架薄壳
double-layered system 双层体系
double layer effect 双层效应
double layer filter 双层过滤器；双层(过)滤池
double layer grid 双层网架；双层网络
double layer interface theory of crystal growth 双层界面晶体生长理论
double layer lens 双层透镜
double layer lining 双层衬砌
double layer metalization 两层金属化
double layer of coating 双涂层
double layer of reinforcement 双重钢筋
double layer phosphor 双层磷光体
double layer potential 双层位
double layer settling tank 双层沉淀池
double layer theory 双层体理论(柔性路面)
double-layer-truss centering 叠桁式拱架
double layer wallboard 双层纸面石膏板；双层墙板
double layer winding 双层绕组；二层绕组
double lead 双引线
double-lead covering 双(层)铅包皮
double-leaded 两倍行距
double-lead process 包覆层铅皮法
double-lead screw 双螺旋式输送机；双管绞刀
double-leaf bascule bridge 双翼竖旋桥
double-leaf clay brick veneer 双页黏土砖饰面
double-leaf construction 双面建筑
double-leaf door 双翼门；双扇门
double-leaf gate 双叶闸门；双扇闸门
double-leaf hinged bottom bucket 对开底卸料斗
double-leaf hook-shaped gate 双扇形闸门
double-leaf hook-type gate 双扇形闸门
double-leaf level bridge 双层桥
double-leaf party wall 双层(共用)隔墙；双层隔断
double-leaf swing bridge 双翼平旋桥
double-leaf vertical lift gate 双叶垂直提升闸门
double lean-to roof 蝶形屋顶；内坡屋顶；(中间有天沟的)双坡屋顶；V形屋顶
double least squares method 二重最小平方法
double leaves door 双开门
double leg stirrup U形网箍
double length 双倍字长
double length accumulation 双倍长累加
double length arithmetic 双字长运算；双倍位运算
double length constant 双倍长常数
double length multiplication 双倍长乘法
double length number 双倍长数；双准数；倍长数
double length of drill pipe 双根钻杆
double length operation 双字长运算；双倍长运算
double length register 双倍长寄存器
double length word 双倍长字
double length working 双倍工作单元
double lens 双(合)透镜

double lens magnifier 双透镜放大镜
double letter 合成字【计】
double level bridge 双层桥
double level fibre forming 双层拉丝作业
double level geometry 双层作业线
double level(l)ing 双(转点)水准测量
double level mineral 双层矿物
double level operation 双层拉丝作业
double level polysilicon 双层多晶硅
double level road 双层式道路
double lever 双杠杆
double lever jib 象鼻架伸臂；四连杆式伸臂
double lever mechanism 双杆机构
double liabilities 双重负债
double liability 双重责任
double lift 双水级(船闸)；复动式；两级提升；二级提升；双水级船门；双线升船机
double lift cam 双升程凸轮；双阶凸轮；双级凸轮
double lift dock 双级船闸
double lifting gate 双层提升(闸)门
double lifting hooks 双提升吊钩
double lift lock 二级船闸
double light engines attached 单机重联
double limit 二重极限
double limiting 双向限幅
double line 重叠线；复线；双线(线路)；双线(段)；双轨
double linear stage pattern 复线生产结构
double line automatic block 双线自动闭锁
double line bridge 复线桥；双线桥
double line connection 双线联络线
double line cutter 双线刻刀
double line feature 双线地物
double line graduation 双线刻度
double line hanger 双线吊弦
double line level(l)ing 双转点水准测量；双线水准测量
double line method (延长直线的)双倒镜法
double line of pecks 双虚线
double line pen 双线笔；划双线笔
double line point 双线刻针
double line road 双线路【道】
double line scriber 双线刻刀
double line section 双线区段
double line semi-automatic block 双线半自动闭塞
double line sleeper bed sieve machine 双边枕底清筛机
double lines of locks 双线船闸
double line spectroscopic binary 双谱分光双星
double line symbolization 双线符号
double line tariff 复线税税率
double line traffic 复线交通；双线交通
double line tunnel 双线隧洞；双线隧道
double line working 双线运行
double lining 复合衬砌；二次衬砌；联合式衬砌
double-lining method 双层衬砌法
double-linked arch(ed) frame 双铰拱架
double-linked flat arch(ed) girder 双铰平拱大梁
double-linked flat parabolic arch(ed) girder 双铰平抛物线拱大梁
double-linked frame 双铰框架
double link equivalent-strain direct shear system 四联等应变直剪仪
double linking 双连
double link jib 组合臂架
double link jib in parallelogram 平行四边形组合臂架
double link suspension 双接触线链形悬挂
double link type luffing crane 四联杆式起重机
double link unloader 四联杆式卸船机；双悬臂卸船机
double lip countersink bit 带安全锥埋头钻头
double lip seal 双重密封
double load 双装药
double-loaded corridor type 内廊式
double lobe bucket 双叶斗
double lobe system 双波瓣天线系统
double-lock 双(厢)船闸；双(线)船闸
double-lockage 分两次过闸(船)
double-lock cross welt 双咬口十字缝
double-locked 双重锁定
double-locked flat seam 双咬口平接缝
double-locked standing seam 双咬口立缝
double-locking device 双锁装置

double locks 复线船闸
double-lock seam 双面盖板缝；双盖缝；卷边咬合接缝；双咬口；双锁卷边接合
double-lock welt 双锁盖条；双折边缝；双咬口
double-locomotive moving kilometers 重联机车走行公里【铁】
double logarithm 双对数
double logarithm coordinate 双对数坐标
double logarithm increase 双对数增加
double log frame 复式排锯
double-log scale 重对数比(例)尺
double-log washer 复式槽洗机
double longeron wing 双梁机翼
double loop 双列蛇形管圈
double loop antenna 双环天线
double loop for girders 大梁双吊箍
double loop pattern 双箕斗
low water 双低潮
double L-square 双用角尺
double L stair(case) 双 L 形楼梯；三跑楼梯
double magazine 双卷暗盒
double magnetic circuit 双磁路
double magnetic components 两种磁性成分
double magnifier 双重放大镜
double main 双干管
double main system 配水系统；排水系统；双干线系统；双干管制；双(干)管系统；双干管式
double make contact 双工作触点；双闭合接点
double male 双雄接口；双头配件；双头插销
double male and female 双凹凸式密封面
double male fitting 双凸形配件
double male reduction 简化插销
double male stacking fitting 双头式堆装件
double manifold with four plus four connectors 四加四的双向管道连接器
double manure salt 硫酸钾镁盐
double-margined door 重楗窗门；双幅门；双边饰门
double marking 双重标志
double mass curve 双累积曲线
double mast A 型桅杆
double mast S/R machine 双柱型有轨巷道堆垛起重机
double measure 双面处理；双面装修；双面有线脚的木工活；双面修饰
double measurement 双观测
double mechanical end face seal 双端面密封
double mechanical joint connecting piece 双机械接头连接件
double mechanical reed 双簧类
double mechanical seal 双端面机械密封
double mechanical trowel 双面机动镘刀
double media filter 双层介质滤池
double meeting rail 两面搭头的窗横档
double meeting stile 两面搭头的窗梃
double melted ingot 双熔锭
double melting point 双重熔点
double member 双肢吊杆
double membrane electrode 双膜电极
double membrane pneumatic system 双膜气压系统
double membrane process 双膜法
double meridian distance 倍子午距；倍横距
double message system 双信息系统
double metallic standard system 复金属本位制
double meter point 双米点法
double method 双重法
double methods of titration 两种滴定法
double microelectrode 二联微电极
double microscope 双(筒)显微镜
double milling machine 双轴铣床
double mirror 双反射镜
double mirror device 双镜装置
double mirror galvanometer 双镜式电流计
double mirror technique 双反射镜法
double mode 双模
double model 双模型
double model stereotemplate 双模型立体模片
double moding 跳模；双振荡型
double modulation 多重调制；双(重)调制
double module 双模
double modulus 双模数
double monastery 统一管理下的男女修道院
double monochromator 双单色仪
double monochromator arrangement 双单色仪

装置
double monochrometer 双单色仪
double motion agitator 双动搅拌器
double motion mixer 双动混合器
double motion paddle mixer 双轴式桨叶搅拌机；双轴式桨叶拌和机；双向桨叶拌和机；双动桨(式)混合器
double motor 双电动机
double mould 复合式线脚
double mould-broad ridging plough 双壁起垄犁
double mo(u)lding 复合式线脚
double movable jaw crusher 移动式双颚式轧碎机
double moving average 二重移动平均(数)
double muffle furnace 双层内罩炉
double nailing 双钉合
double name paper 双(记)名票据
double naved church 双中殿教堂
double negotiation 重复押汇
double nickel salt 硫酸镍铵
double nip 双接(口)
double nipple 双向短接管；双螺纹接套
double normalizing 两次正火
double nose 双塔轮式前端
double nose brick 双圆角砖
double notching 开槽叠接
double nozzle spray gun 双口喷枪
double nuclear resonance magnetometer 双重核共振磁力仪
double nut 双螺母
double objective 双透镜物镜
double objective microscope 双物镜显微镜
double oblique junction 双斜接合
double offset 双支距(测量)；二级起步时差；双乙字管；双效抵消；双效补偿；双偏置(管)；双登插管
double offset ring spanner 梅花(双头)扳手
double offset ring spanner black finished 黑色梅花扳手
double offset ring spanner chrome plated 镀铬梅花扳手
double offset screwdriver 双偏置螺丝起子
double oil seal 双油封
double open end spanner 双叉口扳手
double open end wrench 双又口扳手；双头(开口)扳手
double open end wrench black finished 黑色双头扳手
double open end wrench chrome plated 镀铬双头扳手
double opening 双孔的
double operand instruction 双操作数指令
double optic(al) square 双镜直角器
double option 双重选择权
double order for traversing tree 遍历树的双重次序
double O-ring rubber seal 橡胶双 O 形环密封
double or reverse calculation 两次计算或反向计算
double oscilation mode 双振荡模式
double oscillator 双式振荡器
double oscillograph 双电子束示波器
double osculating plane 二重密切面
double overlap 双重迭代
double-over test 折叠试验
double pack 双重包装
double packer 双堵塞器
double packer grouting 双塞灌浆
double paddle 两头桨
double-panelled door 双腹板门
double-pane safety glass 双层安全玻璃
double-pane sash 双格窗；两片玻璃的窗；两块玻璃的窗扇
double parabolic girder 双抛物线大梁
double parabolic slab 双曲板
double parallel distance 双倍间距；倍平行距
double parallelogram 双平行四边形
double parameter 双参数
double-park 与人行道平行的双排停车
double parking 双行停车；双列停车
double partition 空心隔墙；双重隔墙；双隔；双层隔墙；双层隔断
double pass(age) 双程
double pass coating 两道涂布
double pass cooler 空气二次通过的冷却机
double pass drier 双路干燥器；二次通过式干燥机；热风二次通过的烘干机
double pass (flow) tray 双流塔板

double pass grate 气体二次通过的炉算加热炉
double passing 双程
double pass machine 双加料夹丝玻璃成型机
double pass monochromator 双程单色仪
double pass-out turbine 双抽汽式汽轮机
double pass spectrometer 双程分光计
double patent 重复授予专利
double path system 双路系统
double peak anomaly 双峰异常
double peak logistic distribution map 双峰对数分布图
double peaks 双峰
double pedestal desk 两头沉书桌
double pendulum 双摆
double pendulum bearing 双摆(式)支座
double pendulum flap gate 摆锤式双翻板阀
double pent 双坡内倾屋顶；V 形屋顶
double pentagonal prism 双五角棱镜
double period section steel 复合周期断面钢材
double perspective cylindric(al) projection 双重透视圆柱投影
double petticoat insulator 双裙绝缘子
double petticoat porcelain insulator 双裙瓷绝缘子
double phase linear comparative method 两相直线对比法
double pick 双嘴锄；双嘴镐
double pick insertion 双尾引送
double pick-up tongs 双口锻工钳
double pier 加宽的突堤码头；复式突堤码头
double piercing process 两次穿孔法
double pile helmet 双桩帽
double piles 双柱系缆桩；双柱系船桩；双(排)桩
double piles driving 打双桩
double pillar 双列支柱
double pilot 双导频
double pin 双销钉
double pinion 双小齿轮
double-pinned arch(ed) frame 双铰拱架
double-pinned flat arch(ed) girder 双铰平拱大梁
double-pinned flat parabolic arch(ed) girder 双铰平抛物线拱大梁
double-pinned frame 双铰框架
double-pinned gable(d) frame 双铰人字架
double-pinned parabolic arch(ed) girder 双铰抛物线拱大梁
double-pinned rectangular frame 双铰方框架
double-pinned segmental arch(ed) girder 双铰弓形大梁
double pin sub 双公扣大小头
double pin sucker rods 两头公扣抽油杆
double pipe 套管式；套管；双重承管
double pipe clamp 复式管夹
double pipe condenser 套管(式)冷凝器
double pipe culvert 双管涵洞
double pipe dropping system 下降式双管系统；上分式双管(热水供暖)系统
double pipe exchanger 套管热交换器
double pipe heat exchanger 双管热交换器；套管式换热器
double pipe heating system 双管供暖系统；双管采暖系统
double pipe heat interchanger 套管热交换器
double (pipe) line 双管线
double piston 双活塞
double piston pump 双活塞泵
double pitch 双向坡；双面坡
double pitched 双坡的
double pitched beam 双坡梁
double pitched roof 双折线屋顶；折线形屋顶；双坡(面)屋顶
double pitched skylight 双坡天窗
double pitch roof 折线形屋顶；斜屋顶；人字屋顶；双坡屋面；双坡屋顶；两坡屋顶；复折形屋顶
double pitch skylight 两坡天窗
double pitch sodium paper 双沥青纯碱纸
double pivoted gate 双支枢式闸门；双枢轴闸门
double pivoted pattern 双轴尖式；双支枢式
double pivoted type 双轴尖式
double pivoting block 双轮转环滑车
double plains 双层平纹织物
double plane 二重面；榫舌刨
double plane harp cable stayed bridge 双索面竖琴形斜拉桥
double plane iron 双刨刀；带护铁(的)刨；双刀刨

double planetary-mixer 双行星式混合机
double plank wall 双层(木)板墙
double plate 叠板
double plate bulkhead 双板舱壁
double plate clutch 双(圆)片离合器
double plate girder 双层板梁；槽形梁
double plate grinding machine 双盘研磨机
double plate holder 双层暗盒
double plate rudder 复板舵
double plate semibalanced rudder 复板平衡舵
double plate wheel 复板轮
double plumbing method 双垂线法
double plumbing of shaft 竖井双垂线定向
double plumbing of single shaft 单竖井双重线对中
double ply 补强层【道】；双股；双层
double point 二重点
double pointed 双端
double-pointed nail 双头螺栓；双头尖钉；接合销(钉)
double-pointed pick 尖镐
double-pointed tack 双尖钉
double point gear 多点啮合齿轮
double point interpolation 双点内插法；双点插入法
double points 复式道岔
double point switch 双刀开关
double point thread chaser 叉形螺纹梳刀
double point threshold 两点差别阈
double point tool 双刃刀具
double point toothing 双点啮合
double pole 双刀；两极
double pole cut-out 双极断流器
double pole double throw switch 双极双投开关；双刀双掷开关
double pole front connection 双极正面连接
double pole fuse 双极熔断器
double pole mast 双腿井架
double pole on-off switch 双极开关
double pole pull cord type switch 双极拉线开关
double pole relay 双组触点式继电器；双极继电器
double pole scaffold 双排(柱)脚手架；双杆支脚手架
double pole single-throw 双刀单掷
double pole single throw switch 双极单投开关；双刀单掷开关
double pole switch 双极开关；双刀开关
double port 双口
double portio 双门廊
double post bollard 双柱缆桩
double post roof 双柱屋架
double post roof truss 双柱屋顶桁架
double post row brush dam 双排桩梢(料)坝
double post row crosswise brush type check dam 双排桩梢横编柴梢式拦砂坝
double post sign 双柱式标志牌
double pour 二次浇筑；二次涂刷
double pouring displacement technique 双金属浇注置换
double power series 二重幂级数
double precision 二倍精密度；双精度；双倍精(密)度
double precision arithmetic 双精确度运算；双倍精度运算
double precision complex type 双倍精度复数型
double precision constant 双倍精度常数
double precision exponent part 双精度阶部分
double precision floating macro-order 双精度浮点宏指令
double precision hardware 双倍精度硬件
double precision number 双精度数；双精度量；双倍长数
double precision numral 双精度数
double precision operation 双精度操作；双倍精度运算
double precision quantity 双倍精度量
double precision real data 双精度实数
double precision type 双精度实数型
double pressed brick 二次压实砖；两次压制砖
double pressed stearic acid 二压硬脂酸
double-press process 两次压制法
double pressure ga(u)ge 双压力计
double pressure system 双极施压法
double pressure-vessel conveyer 双压力容器输送机；双仓输送泵

double prestressing 双向预应力
double price 双重价格
double price basis 双重价格基础
double priority 双优先
double priority mode 双优先式
double prism 双棱镜
double prismatic square 直角棱镜;双棱镜直角器
double prism field glass 双棱镜望远镜
double prism monochromator 双棱镜单色仪
double prism optical square 双(棱镜)光学直角镜
double prism plummet 双棱镜垂标
double prism spectrograph 双棱镜摄谱仪
double prism spectrography 双棱镜摄谱法
double prism square 双棱镜直角器
double probe system 双探针系统
double product 二重和
double profile 复式断面
double profit 双重利润
double projected window 双凸窗
double projection 双重投影(法);双像投影
double projection direct-viewing stereoplotter 双投影直接观测立体测图仪
double projection instrument 双像投影仪
double projection plotter 双投影绘图仪
double projection stereo-plotter 双投影立体绘图仪
double projection stereoscopic plotting instrument 双像投影立体测图仪
double projection system 双重投影系统
double projector 双像投影器
double proportionate measurement 双比例量测法
double protractor 双斜量角器
double pulse excitation 双脉冲激励
double pulse modulation 双脉冲调制
double pulse recording 双脉冲记录(方式)
double pulse selection 双脉冲选择
double pulse station 双脉冲发射台
double pulsing 双脉冲列发射
double pulsing station 双脉冲电台
double pump 双水泵;双联泵;双缸泵
double pumping action 双抽运
double pump roll torque converter 双泵轮变压器
double punch 双穿孔
double punch and blank-column detection 双孔和无孔检测
double punching machine 复式冲压机;复式冲床
double punch layout 双排排样立模
double punch test 复穿孔试验
double purchase counterweight batten 复式吊杆
double purchase pulley 双铰链滑车
double purchase winch 双力绞车;双卷筒绞车
double purpose camera 两用摄影机;两用照相机
double purpose valve 双作用管
double push-pull 双推挽
double pyramid 对顶棱锥
double quarter turn 带有踏步的双直角转弯楼梯
double quarter-turn stair(case) with windows 带窗的双分式楼梯
double quartz prism spectroradiometer 双石英棱镜光谱辐计
double quenching 二重淬火;双重淬火;双液淬火
double queue 二重排队
double quire 双合唱班
double-quire church 双合唱班教堂
double-quirk(ed) bead 双槽(凸)圆线脚;双槽串珠线脚
double quotidian fever 双峰热
doubler 折叠机;乘二装置;二倍器;并条机;倍压器;倍频器;倍加器;双联拖车
double rabbet 双重凸凹榫接
double-rabbeted frame 双企口门框
double radio frequency sputtering 双射频溅射
double radio source 双射电源;射电双源
double raft foundation 双层筏式基础
double rail circuit 双轨条轨道电路
double rail logic 双线逻辑
double rail motor crab 双轨电动起重机
double rail motor hoist 双轨电动起重机
double rail retarder 双轨条杂式缓行器
double raised panel 双面凸凹镶板
double ram press 双柱塞压机
double ram type preventer 双闸板型防喷器;双闸板式防喷器
double range 双量程
double range voltmeter 双量程伏特计

double rate 双率计
double ratio 重比;复比
double ratio estimate 复比估计值
double ratio estimation 复比估计
double ratio estimator 复比估计量
double reaction 双重反应;双反应法
double read out 双倍读出;加倍读出
double rebated 双企口门框;内外双开门框
double rebated frame 双企口门框;双铲口门框
double rebated linings 双铲口门膛板
double reception 双(工)接收;双(倍)接收
double recipe 重方
double recoil method 双反冲洗法
double recoil system 双重后座装置
double record check 双记录校验
double recovery circuit 双回收回路
double-reduced tin plate 二次冷轧镀锡薄板;超薄马口铁
double reduction 双重减速
double reduction axle 复式减速轴
double reduction drive 两级减速驱动
double reduction final drive 双级最终传动
double reduction gear 双套减速齿轮(装置);双减速齿轮;复式减速齿轮
double reduction gear device 双级减速齿轮传动装置
double reduction geared turbine 两级减速船用汽轮机
double reduction gearing 两级减速装置;两级减速齿轮
double reduction motor 两级减速电动机
double reduction shaft 复式减速轴
double reduction spur pinion 复式减速小正齿轮
double reed mat 双层芦苇垫;双层草垫
double reeling frame 双面摇纱机
double refined iron 重炼铁
double refining 二重精炼
double reflecting stereoscope 双反光立体镜
double reflection 双反射
double refracting spar 冰洲石
double refraction 双折射
double refraction angle 双折射角
double refraction calcspar 双折射方解石
double refrigerant system 双制冷系统;二次制冷系统
double register 倍长寄存器
double regulating angle valve 双调节角阀
double regulating valve 双调节阀
double regulation 双重调节
double reheat steam turbine 二次(中间)再热式汽轮机;中间再热式汽轮机
double reinforced 双配筋(的)
double reinforced beam 配受压筋的梁;双重配筋梁;双重钢筋梁;双配筋梁;双筋梁;复筋梁
double reinforced concrete 双重配筋混凝土
double reinforced concrete beam 双重配筋混凝土梁
double reinforcement 复筋;双(重)配筋;双重钢筋;双面钢筋;双筋
double replacement 互换
double replica 二次复型
double resection in space 空间双点后方交会
double-residue (mixture) 二渣(混合料)
double resonance 双谐振;双共振
double resonance optic(al) parameter oscillation 双共振光量参振荡
double resonance spectrograph 双共振谱仪
double resonant optic(al) parameter oscillator 双共振光量参振荡器
double resonator 双(腔)谐振器
double response 双调谐
double response controller 双重作用调节器
double-return circuit power supply 双回路供电电源
double-return siphon 双网虹吸管;双弯虹吸管;乙字形存水弯;乙字形弯管
double-return stair(case) 合上双分式楼梯;双分式对折楼梯;双回行楼梯;双分平行楼梯;双合分式楼梯
double-return trap 乙字形存水弯
double-return valve 双止回阀
double reversal system 双互换制度
double reverse bend 折边双反折
double reversing mirror 双镜反像镜

double reversing mirror system 双反像平面镜系统
double revolving branch 双头水枪
doubler for glass fibre 玻璃纤维并捻机
double rhombic antenna 双菱形天线
double rib 双棱;双肋
double ribbon agitator 双螺带式搅拌器
double rice system 双轨价格制
double ridding bollard 双十字形系缆柱
double rig 双腿木钻塔
double right-angle prism 双直角棱镜
double ring ahead 加速前进【船】
double ring astern 加速后退【船】
double ring infiltrometer 双环渗透仪
double ring silo 双环库
double ripple 双波性
double rivet 双行铆钉
double riveted 双行铆接的
double riveted butt joint 双行铆钉对接
double riveted joint 双重铆钉铆合;双行铆接头;双行铆接缝;双排铆接头
double riveted lap joint 双行铆钉搭接
double riveting 双行铆接
double roasting 二重焙烧
double rocker bearing 双摇座;双桥梁摇座;双摇动支座
double rodded line 双转点路线
double rodded method 双转点法
double Roebel bar 双罗贝尔线棒;双换位线棒
double roll 双漩涡饰;双滚筒
double roll breaker 双滚筒碎石机;双滚筒破碎机;双辊破碎机
double roll crusher 双滚筒轧碎机;双滚筒碎石机;双滚筒破碎机;双辊破碎机;对辊破碎机
double roll curb tile 双横滚路缘石
double-rolled cast 压延锤痕玻璃
double-rolled glass 双辊压延玻璃
double roller 双(圆盘)滚轮;双压路机;对辊破碎机;双碾;双辊轴
double roller chock 双滑轮导缆器;双滚轮导缆钳
double roller (lever) escapement 双圆盘擒纵机构;双滚擒纵机构
double roller mill 双辊研磨机
double roller mixer 双辊混合器
double roller press 双辊辊压机
double roller safety action 双圆盘保险装置
double roller size applicator 双辊涂油
double roll feed 两边滚子进料
double rolling hump 双溜放驼峰
double rolling on double pushing track 双推双溜【铁】
double roll interlocking tile 双筒咬口瓦
double roll mansard tile 双滚折线式屋面瓦
double roll pelletizer 辊式压球机;对辊成球机
double roll press 双辊压力机
double roll roof 双坡层屋顶
double roll slow speed crusher 慢速双辊破碎机
double roll tile 双圆凸瓦
double roll toilet tissue fixture 双轴手纸架
double roll under-ridge tile 双卷边瓦
double roll verge tile 双卷边山墙突瓦;檐瓦
double Roman tile 加大罗马式瓦;复式罗马瓦;双筒咬口瓦
double roof(ing) 双重屋顶;双层屋顶;双屋面;桁条屋顶;多特征屋顶;复式屋顶
double room 双(人)床房;双床房;双房间客房
double root 重根;二重根
double rope aerial cableway 双缆式架空索道
double rope grab 双索抓斗;双绳抓斗
double rope grab crane 双索抓斗起重机
double rope rehandling grab 双索转载抓斗
double ropeway 双向索道;双缆索道
double rope winch 双索卷扬机
double rose 双玫瑰花型
double rotator mill 中卸粉磨;带烘干仓的中卸磨机
double rotor crusher 双转子(辊式)破碎机
double rotor hammer crusher 双转子锤式破碎机
double rotor impactor 双转子冲击式破碎机
double rotor impact pulverizer 双转子反击式破碎机
double round 双轮
double round nose bit 凸圆弧金刚石钻头(圆弧直径等于壁厚)
double row 双行座;双排
double row ball bearing 双列向心滚珠轴承

double row ball journal bearing 双列径向滚珠轴承
double row ball thrust bearing 双排滚珠止推轴承
double row bearing 双列(滚珠)轴承
double row cowshed 双列式牛舍
double row hog house 双列式猪舍
double row layout 双排排样;双列布置(机电)
double row pile cofferdam 双排板桩围堰
double row radial engine 双列辐射型发动机
double row rivet 双行铆钉
double row riveted joint 双行铆接头;双行铆接缝
double row riveting 双行铆接
double row seating 双行座
doublers 双联拖车方式
doubler twister 并捻机
double rub 往返擦洗
double ruling pen 双曲线笔
double run level(l)ing 双程水准测量
double run line 双程水准线;双程水准测量
double run method 双程测量法
double run procedure 双程测量法
doubler winder 并纱机
doubles 双联拖车;双粒级煤
double safety valve 双安全阀
double-sag crossing 双向凹形交叉
double salt 双盐(混合物);复盐
double sampling 两次取样检验;成双抽样;复式取样;复式抽样;二重抽样;二次质量检验;双(重抽)样
double sampling for stratification 分层(的)双重抽样
double sampling inspection 二重抽样检验
double sampling plan 二重抽样方案;双次验收抽样法
double sand catching basin 双三层截流井
double sash casement window 双开窗
double sash window 推拉窗;双扇窗户
double saw 复齿锯
double sawbuck 二十美元的钞票
double saw-tooth system 双锯齿形布置;交错式布置
double scale 二重刻度尺;二重标度;双刻度
double scale instrument 双标度仪表
double scale rod 双列刻度水准(标)尺
double scales 两排划分
double scattering 双散射
double scattering experiment 双散射实验
double scoop feeder 双勺喂料机;双斗给料机
double scoop type excavator 双斗式挖土机
double screed finisher 双样板整面机;双样板整修机;双找平修整器
double screed transverse finisher 双样板横向整面机;双样板横向修整机
double screen 双(涂层)屏
double screen coaxial cable 双画面同轴电缆
double screen saturator 双网型浸渍机
double screw bolt 螺柱
double screw conveyer 双管螺旋输送机;双管绞刀
double screwed bolt 双头螺栓
double screw extruder 双螺杆挤出机
double screw pump 双螺杆泵
double screw washer 双螺旋洗矿机
double sculling 单人双桨船
double seal 双(重)密封;双封层
double seal coat 双封(闭)层
double seal(ed) bearing 双封轴承
double seal manhole cover 人孔复式封盖
double seal porewater piezometer 双管封闭式孔隙水压力仪
double seam 双重咬口;双重卷口;双重接缝
double seam can 双接口罐;双接缝罐
double seamer 封罐机;双层卷边机
double seaming 双重卷口接缝;双锁边
double seated ball valve 双座球阀
double seat valve 双座阀
double secondary transformer 双副边变压器;双二次绕组变压器
double second difference 倍程二次差
double segmental baffle 双弓形折流板
double seismic zone 双震带
double selvedge 双边缘;卷边
double sequence 二重序列
double series 二重级数
double service 双用的;双效(的);两用的
double sewing line 重叠缝迹

double sextant 双(测量)六分仪
double shackle 双钩环
double shaft 双轴
double shaft crusher 双轴破碎机
double shafted spring spike 双杆式弹性道钉
double shaft hammer crusher 双轴锤式破碎机
double shaft mixer 双轴(式)搅拌机;双轴混合机;双辊搅拌机;双辊混合机
double shaft paddle mixer 双轴式桨叶搅拌机;双轴式桨叶拌和机;双向浆叶拌和机
double shaft preheater kiln 双立筒预热窑
double shaking screen 双层摇动筛
double shaping machine 复式牛头刨床
double shared double wheeled plow 双轮双铧犁
double shared plough 双铧犁
double shear 双面受剪;双(面)剪切;双剪(力)
double shear bolt 双剪螺栓
double shear deformation 双向剪切变形
double shear joint 对接铆接
double shear method 双面剪切法
double shear rivet 双剪铆钉
double shear riveted joint 双剪铆接;双剪铆(钉)接头;对接铆接
double shears 双面剪床
double shear test 双面剪切试验
double sheath 双极性层
double sheathed needle 双套针管
double sheave pulley 双槽阀
double shed jacquard 双梭口提花机
double shed porcelain insulator 双裙式瓷绝缘子
double sheepfoot roller 双联羊脚碾
double sheet bend 双编结
double sheet detector 多纸检测器
double sheet detector control 多纸检测器控制
double sheet ejector 多纸推出器
double sheet piling 双层板桩
double shell 双层薄壳
double shell recuperator 套筒式换热器
double shell rotary drier 双筒回转烘干机
double shell tile 双面层(空心)瓦
double shield enclosure 双层屏蔽罩
double shielding 双层保护
double shift 两班制;双班制
double shifting 双班刻度
double-shift work 两班制工作
double ship lifts 双线升船机
double shock intake 双激波进口
double shoe brake 蹄式制动器;双闸瓦制动
double shooting 套晒制版
double shot mo(u)lding 二极模塑造型;双注射模塑;双色注塑
double shouldered tie plate 双肩垫板
double shrinkage 双重收缩
double shrouded wheel 闭式叶轮
double shunt field relay 双闭磁路继电器
double shutter curtain 双窗帘
double-side band 双边带
double-sideband modulation 双边带调制
double-sideband receiver 双边带接收机
double-sideband suppressed-carrier modulation 抑制载波的双边带调制
double-sideband system 双边带制
double-sideband transmission 双边带传输
double-sideband transmission system 双边带传输制
double-sideband transmitter 双边带发射机
double-sided anchor dredge(r) 双侧锚式挖泥船
double-sided angle-reading system 双向读角系统
double-sided arm 双侧臂
double-sided bicycle stand 双面自行车架
double-sided book shelves 双面书架
double-sided coating 两面粉饰;双面涂布
double-sided copying 双面复印
double-sided function 双边函数
double-sided gear drive 双边传动
double-sided high speed press 双柱高速压力机
double-sided hood 双面拔气罩
double-sided impeller 双吸式叶轮
double-sided mosaic 双面嵌镶幕
double-sided partition (wall) 双面隔墙
double-sided pattern plate 双面模板
double-sided plate trimming shears 中厚板双边剪边剪切机
double-sided reinforced concrete jetty 二边钢筋混凝土突码头(靠船用)

double-sided socket 双向插座
double-sided sprayer 两面作用的喷雾机
double-sided staff 双面水准标尺
double-sided toothed plate 双侧齿板
double-side partition (wall) 二面空心隔墙
double-side shuttering 双面模板
double-side tipping wagon 两面自卸车;两轮运货马车;两侧倾斜车;双边倾卸车
double-side weir overflow 双侧堰溢流
double-side welding 双面焊接
double sighting 正倒镜分中法
double sigmoid-blade mixer 双曲拐叶片搅拌机
double sign 重号
double signal location 信号双置点;并置信号点
double silk-covered wire 双(层)丝包线
double sintering 二重烧结
double site principle 双位置原理
double six array 双六(单元)天线阵
double size 两次施胶;双层胶粘;双倍尺寸
double-sized brick 双倍(大的)砖
double-size pantile 大号波形瓦
double sizer 双齐边机
double skew notch 双斜槽口
double skin 重皮;双层
double-skin boat 双壳艇
double-skin construction 双(壳)层结构
double-skin curtain wall construction 双层幕墙结构
double-skin duct 双层管道
double-skin floor 双层楼板;双层屋顶(上层排水、下层吊顶);双壳层地板
double-skin partition (wall) 双面层隔墙
double-skin plate valve 双面板阀门
double-skin roof 双层屋顶
double skip 双翻斗
double skirting 榫接高踢脚板;双截墙裙;双层高踢脚板
double slab 派形板
double slab-type block 双平面建筑;双板式建筑
double slab-type building 双板式房屋
double slag process 双渣法
double sledge 双爬犁
double sleeper joint 双轨枕夹接
double slider coupling 十字滑块联轴器
double slider crank chain 双滑块曲柄链系
double slide valve 双滑阀
double slide-wire potentiometer 双滑线电势计
double sliding door 双向移门;双扇移门;双扇拉门
double sliding door installation 双扇滑门安装
double sliding gate 双向推拉门;双向滑门
double sliding table press 双滑台式压机
double slip casing bowl (可循环洗井的)双卡瓦打捞筒
double slip switch 复式交分道岔;双动转辙器
double slit 双缝(隙)
double slit interferometer 双缝干涉仪
double sloping bed 双面温床
double sloping roof 两坡屋顶
double slow sand filtration 双重慢砂滤
double slug tuner 双插芯式调谐器
double sluice gate 双层泄水闸(门);双层提升水闸;双层冲沙闸(门)
double small 双小裁
double smoothing 二次平滑法
double socket 双插座;双承口管
double socket pipe 双承口管
double solid laminated 双层实心叠层的
double solvent 复(合)溶剂
double solvent refining 双溶剂精炼
double sound-boarded floor 双层隔音地板;双层消音地板
double sound source 双声源
double source 双源
double span moorings 双跨底链的系船浮筒
double spans 双跨(度)
double span trussed roof structure 双跨屋架结构
double specimen oedometer test 双样固结试验
double spectroheliograph 双筒太阳单色光照相仪
double spectroprojector 双重光谱投射器
double speed pulley 两速滑车
double spheric(al) pendulum 双球面摆
double spike 双稀释剂
double spindle hammer breaker 双轴锤式破碎机

double spindle hammer crusher 双轴锤式破碎机
double spine saw 复齿锯
double spiral ramp 双螺旋斜坡道
double spiral stair(case) 双螺旋楼梯
double spiral turbine 双排量涡壳式水轮机
double spit 双沙嘴
double splayed coping 双坡压顶
double split flow 双分流
double spot technique 双光点技术
double spot tuning 双点调谐
double spread 双面涂布；双面涂胶
double spread gluing 双面涂胶
double spreading 两面涂胶
double spring 双叠板簧
double spring bottom oil can 复簧底油壶
double spring washer 双层弹簧垫圈
double spur grade 人字坡
double square 两用方尺
double square junction 垂直接合；T式接合
double squirrel cage motor 双鼠笼(式)电动机
double squirrel-cage rotor 双鼠笼转子
double squirrel-cage winding 双鼠笼绕组
double stable type 双稳型
double-stack container car 双层集装箱车
double-stack container wagon 双层集装箱货车
double-stack system 双通风管系统；双立管系统
double-stage 两级的
double stage compressor 二级(空气)压缩机；双级压缩机
double-stage cooler 两级冷却机
double-stage crushing 两级破碎
double-stage digester 双级消化池
double-stage driving 两级传动
double-stage grinding 两级粉磨
double-stage mixing plant 双阶式搅拌机
double-stage nitriding 二级氮化
double-stage open circuit crushing 两级开路破碎
double-stage proportioning 两级配料；两级计量
double-stage pump 两级泵；双级泵
double-stage super-charger 双级增压器
double-stage turbine 双级式透平
double-staggered checker 错列砖格子
double stain 二重染剂
double staining 二重染色法
double standard 复本位；双重标准；双倍宽度标准的(耐火砖)
double standard brick 倍宽砖
double standard country 复本位国
double standard interpolative method 双标准内插法
double standard seam 加力缝
double standard system 复本位制
double standing seam 双(盖条)立缝
double standing seam-type zinc roof 双盖立缝式镀锌板屋面
double standing welt 双盖立缝
double standing welt-type zinc roof 双盖立缝式镀锌板屋面
double stand rolling mill 双机座轧机
double star 双星
double star equal altitude method 东西星等高测时法
double star-guad cable 双星绞电缆
double star point 双星点
double-start screw 双头螺栓
double-start thread 双头螺纹
double-start thread screw 双线螺丝
double state 双重线项
double state compass 双态陀螺罗经
double steel comb joint 双梳式接缝
double steering 双转向
double stem threads 双头杆螺纹
double-step diffusion technique 二级扩散技术
double steps 双踏步木接点；双段接头；W形接头；两阶齿榫；双阶齿榫；双槽齿
double stirrup for girder 支梁双箍
double stochastic matrix 二重随机矩阵
double stopper arrangement 双棒浇注装置
double stopper hitch 双掣索结
double stor(e)y silo 双层库
double straight-line block 双直线式建筑；双板面式建筑
double straight-line building 双直线式建筑；双直线式房屋；双板面式建筑

double strand 双股的
double strand conveyer 双线传送装置
double strand drag conveyer 双线输送机；双路拖运机
double stranded polymer 双股聚合物
double strand mill 双线轧机
double strand roller chain 双股滚子链
double strand winder 分拉拉丝机
double strap 双均压环
double strap joint 双盖板接头
double strap lap joint 双盖板搭接接头
double strapped joint 双(面)盖板接头
double strap web joint 双面盖板接头
double stream amplifier 双电子束放大器
double stream cooler 双流式冷却器
double streamline 双流线型(钻杆接头)
double stream tube 双注管；双电子束射线管
double strength 高强度；强力；双料
double-strength glass 双料玻璃；加厚玻璃；高强度玻璃；厚玻璃(3～4毫米厚)；强力玻璃
double-strength pipe 高强度管
double-strength window glass 高强窗玻璃；双料窗玻璃
double stretcher 双顺砖；两皮顺砖
double strike-off method 两次刮平法(建筑混凝土路面)
double strings 双连绝缘子串
double string suspension preheater 双列式悬浮预热器
double stroke 二冲程；往复冲程；双冲程的
double stroke deep well pump 双冲程深井泵
double stroke engine 二冲程发动机
double strut 双柱
double strutted frame 双压杆框架；双压杆桁架；双支撑架
double strut trussed beam 双压杆桁架梁；双柱桁架梁；双支撑桁架梁
double stub tuner 双短线调谐器
double-style hanger 双排环
double suction 双吸口；双吸
double suction centrifugal pump 双吸离心泵
double suction compressor 双吸压缩机
double suction impeller 双吸口叶轮
double suction pump 双吸(式)泵；双吸(口)泵
double suction share 加宽垂直间隙的犁铧
double suction single stage volute type pump 双吸式单级螺旋泵
double-sulfate 硫酸盐复盐
double sum 二重和
double summation 二重求和法
double summation rule 二重求和规则
double summer time 双重夏令时
double super effect 拍频干扰效应；双差频效应
double superheterodyne receiver 双变频超外差式接收机
double superheterodyne reception 双超外差接收
double-super system 二次变频制
double supported rudder 双支承舵
double surfacer 双面刨光机
double surface treatment 双层表面处治(路面)；双层处理；双层表面处理
double swage hammer 双对击式镦锻机
double swage saw 双压料锯
double sweep tee 锐角三通(管)；双弯头三通；双弯头T形接头
double swing bridge 双翼平旋桥
double swing door 双向摆动门；双面摇门；双面弹簧门；双摆门
double swing frame 双摆门的门框
double swing gate 双摇摆门
double-swinging door 双扇弹簧门；双摆门
double swing lever mechanism 双摆四连杆机构
double swing-span bridge 双开平旋桥
double switch 两路开关；双又轨
double switch call 双转换呼叫
double switch turnout 对称道岔
double swivel barrel 双转环套筒
double swivel hook 双旋转钩
double-swiveling uncoiler and recoiler 并列回转式开卷卷取机
double swivel nozzle 双头喷嘴
double symmetrical 双对称的
double symmetrical cross-section 双对称断面
double symmetry 双对称

double system sound recording 声像同步录制系统
doublet 耦极天线；偶极子；二重态；二联微管；双(重)线；双重态；双合透镜
double tabular form 二项表格式；二次表式
double tailed test 二重切线
double tamping beam finisher 双夯实梁整面机；双夯实梁修整机
double taper 双楔形
double taper bit 双锥钻头；双斜刃钎头
double tapered roller 双锥面旋轮(旋压用)
double taper wedge 双斜楔
double target 双层测标
double target level(l)ing rod 双面水准标尺
double tariff 二重税；双重税；双率税则
double tariff system 复关税率制；双重税率制；双价制
double tariff (system) meter 双价电度表
double taxation 重复课税；复税制；双税(收)
double taxation relief 免除双重税收；重复税减免
double taxation relief agreement 重复课税减免协定
double tax system 两次税制
double tax treaty 双重(课)税协定
double T-beam 双梁上楼板挑出；双丁字梁；工字钢梁；双T形梁
double T bed 双丁支座；双T支座
doublet component 双透镜组件
doublet distribution 偶极分布
double tee 十字接管；π形的；四通
double tee beam 双T形梁；双丁梁
double tee floor(ing) system 工字梁地板；π形混凝土板地板
double tee floor (slab) 双T(形)楼板；π形地板
double tee form(work) 双T形模板
double tee frame 双T框架
double tee iron 工字铁；π形钢
double teem(ing) 双浇；重铸；重浇
double tee plate 双T形板
double tee prefabricated girder 双T形预制大梁
double tee roof slab 双T形屋面板
double tee slab 双T形板
double-tee slab π形(混凝土)板
double tempering 双回火
double template jigger 双刀旋坯机
double tenon 双(雄)榫；双(凸)榫
double tenon joint 双榫接合
double tension test 双重拉伸试验
double texture 双重结构
doublet-fine structure 双重线精细结构
double T floor(ing) system 工形梁地板；π形混凝土板地板
double T floor slab 双T地板；π形地板
doublet flow 偶极子流动
double T formwork 双T字形模壳
double T frame 双T(形)框架；双T形结构
double the angle on the bow 船首倍角法
double thermomechanical treatment 二级变形热处理；双重热机械处理
double thickiness 双面刨光机
double thickness 双厚
double thicknesser 双面压刨
double thickness sheet glass 双料窗玻璃
double thickness window glass 双料窗玻璃
double thread 双线螺纹；双头螺纹；双丝
double threaded screw 双纹螺栓；双头螺栓
double thread hob 双头滚刀
double thread method 双丝法
double thread screw 双纹螺纹；双头螺杆；双螺纹螺杆
double thread worm 双头蜗杆
double three-legged gravity escapment 双重三星轮重力式擒纵机
double threshold detection 双阈检测
double throat burner 双喉管燃烧器
double throw 双掷；双投
double-throw bolt 双程插销
double-throw circuit breaker 双投断路器
double-throw contact 切换触点；双投触点
double-throw crankshaft 双联曲柄轴；双臂曲轴
double-throw disconnecting switch 双投隔离开关
double-throw knife switch 双向闸刀开关
double-throw lock 双保险锁
double-throw switch 双向开关；倒顺刀开关；双掷开关；双投开关

double-thrust bearing 双向止推轴承;对向推力轴承
double tidal ebb 双落潮流
double tidall-flood 双涨潮流
double tide 复潮;双潮
double-tier partition (贯穿中间层楼的)隔墙;两层高隔墙;两层高隔断;双层框架间壁;双层框架隔墙
double timber lintel 双木过梁
double time 加班;双工资;额外加班(费)
doublet interval 双线间隔
double tires 双轮胎
double T-iron 工字铁;π形钢;工字钢;双T形铁
doublet magnifier 二重放大镜;双合放大镜
doublet objective 双胶物镜;双合物镜
double-toe rail fastener 双趾弹簧扣件
double toggle 双肘节;双肘板
double toggle chuck 双绞曲柄卡盘
double toggle crusher 双肘板颚式破碎机;双肘破碎机
double toggle jaw crusher 双肘板颚式破碎机
double tone (ink)双色版油墨;双色调油墨
double tone printing 双色版印刷
double tongued chock 双钳导缆钳
double tongue grafting 重舌接
double tooth 复齿锯
double toothed crusher 双齿辊破碎机
double tooth rake 二齿(灰)耙
double torsion machine 双列螺旋弹簧缠绕机;双列缠绕机
doublet oscillator 赫兹振荡器
double touch 双触键
double tower 双锥体
double towered 双塔的
double-towered gatehouse 两侧有双塔的城楼;双塔式闸室;双塔式大门;双塔式城楼
double-towered facade 双塔式立面(建筑);双塔正面
doublet peak split distance 双峰裂距
doublet T plate 双T形板
double T prefabricated girder 双T形预制大梁
doublet profile lens 双胶合轮廓投影物镜
double tracer technique 双示踪技术
double trace scope 双射线示波器
double traces oscilloscope 双踪示波器
double track 复线;双线(线路);双(线)轨道;双轨(的);双轨(道)
double track all-relay semi-automatic block system 复线继电半自动闭塞系统
double track automatic block system 复线自动闭塞系统;双线自动闭塞系统
double track bridge 复线铁桥;双轨铁路桥;双轨铁道桥;双轨(道)桥
double track cam 双面凸轮
double track circuit 双轨条轨道电路
double tracked railroad 双线铁路;双线铁道
double tracked railway 双线铁路;双线铁道
double tracked railway tunnel 双线铁路隧道;双线铁道隧道
double tracking 双线化;修建第二线
double tracking railway 复线铁路
double tracking tunnel 双轨隧道
double track insert 双线插入段
double track interpolation 双线插入段
double track kilometrage 双线公里里程
double track line 双线道路;双线(道路);双线铁路;双线铁道;双轨线路
double track main line 双线干线
double track packing machine 双列包装机
double track railroad 双线铁路;双轨铁道;双线铁路
double track railway 双线铁路;双轨铁道
double track section 双线区间;双线区段
double track skip hoist 双轨翻斗式绞车
double track sliding door 双轨推拉门
double track swing parting 双轨甩车调车场
double track system 双轨制
double track tunnel 双线隧洞;双线隧道
double T-rail 工字形钢轨
double trains attached 两列合并运行
double transept 十字形教堂的袖廊;十字形教堂的双耳堂
double transfer 两次转换正像;重复转记
double transit 二次中天
double transit oscillator 双渡越速调管

double transmission 双传动(装置);双波发送
double transmission method 双透射法
double trap movable weir 双阱式活动堰
double traverse method (延长直线的)双倒镜法
double treating 淬火回火处理;二次处理
doublet refractor 双折射器
double trenching 双层开沟
double triangle 六角星
double triangulated system 双三角结构;双(边)支撑;箭头形支撑
double trigger 双脉冲触发器
double trimmer 双托梁
double triple box 三肋双箱式
double trolley system 双线制
double T roof(ing)slab 双T(字)形屋面板
double truncated cone seal 双截头圆锥密封
double truncated exponential distribution 双截头指数分布
double T-section 工字形剖面;工字(形)断面
doublet separation 双线分离度
double T slab 双T形(梁)板
doublet soap 胰皂
doublet spectrum 双线光谱
double T-steel 工字钢
doublet structure 双重结构;双谱线结构
doublet throw 双掷
doublet trigger 双触发
double tube burner 双管燃烧管
double tube core barrel 取芯双管钻具;双管式岩芯管;双管取芯钻具;双层岩芯管;双层取芯器;双层取芯管
double tubed containment boom 双管围油栏
double tube double packer 双管双封隔器
double tube grouting method 双管灌浆法
double tube method 双管法
double tube permeameter 双管渗透计
double tube reactor 套管反应器
double tube rigid barrel 双动双层岩芯管;双动双层取芯器
double tube rigid type core barrel 双动双层岩芯管;双动双层取芯器
double tube sampler 双管取样器;双层取样器
double tubesheet 双管板
double tube single-packer 双管单封隔器
double tube swivel type core barrel 单动双层岩芯管(内管不动);单动双层取芯器(内管不动)
double tube waste heat boiler 双管废汽锅炉
double tubular shaft 双套管转轴
double tuckstone 双挂钩砖
double tumbler bearing 双摆动支座;双摇座
double tuned amplifier 双调谐放大器
double tuned circuit 双调谐电路
double tuned coupling 双调谐电路耦合
double tuned detector 双调谐检波器
double tuning 双调谐
double T unit 双T形构件
double turbine torque converter 双涡轮变矩器
double turn-back track 双折返线
double turn lock 双圈锁;双向水头船闸
double turnout 对称道岔;复式道岔;双叉轨;三开道岔;双分道岔
double turnout with diamond crossing 有菱形辙叉的双开道岔
double twin 复合孪晶
double-twin coil shank 带双弹簧圈的铲柄
double-twisting frame 倍捻捻线机
double-twisting machine 倍捻加捻机
double-twist twisting machine 倍捻捻线机
double type 复式的
double type filter 重合型滤波器
double type marshalling station 双向编组站【铁】
double type marshalling yard 双向编组场【铁】
double type soil tank 复式土壤蒸发器
double type trace heating 套管式伴随加热法
double U groove 双面U形坡口
double undulated preheater 双波纹板回转式预热器
double unit hydraulic pump 双单元组水力泵
double unit motor 双电动机机组
double unit system 双机制
double unit triaxial test apparatus 二联三轴试验仪
double unit type blow-out preventer 双单元爆裂防护器

double universal coupler 双万向(联轴)节
double up 加双绳(系缆时);加双绑(系缆时);双层连续抹灰(作业)
double up and secure 加双时挽紧
double up door 双开门;双道门
double vacancy 双空位
double valley 双向谷
double value 赔偿损失时的加倍付给
double value compensation 赔偿损失的加倍付给
double valuedness 双值性
double valve 双重阀;双联阀
double vane 双翼;双叶片;双舵
double vane attenuater 双片衰减器
double vane attenuator 双片衰减器
double variation method 双变法
double vault 双层地下室;复式穹顶;双重圆顶;双拱圆顶;双层穹顶
double V-butt joint 双面V形对(焊)接;双V形接头
double V-butt weld X形对接焊
double V-butt weld joint 双面V形对接焊
double vee 双V形的;X形的
double vee clutch 双锥离合器
double vee gutter 双V形槽
double velvet loom 双层丝绒织机
double vent pipe tile 双口管瓦管
double vernier 复游标;双游标
double V groove weld 双斜坡口焊;双斜边坡口焊缝
double V groove with root face 带纯边X形坡口
double V gutter 双V形槽
double vibrating roller 双(轮)振动压路机
double vibration 双振动
double vibrator 双振捣器
double vibratory roller 双轮振荡压路机
double vision 双视
double vision test 复视试验
double voltage connection 倍压连接
double voltage rating 双额定电压
double wall 双排墙;双层墙
double wall board layer 双墙板层
double wall breakwater 双重箱式防波堤
double wall buttress 双壁支撑
double wall cofferdam 双层板桩围堰;双壁围堰
double wall cofferdam with concrete fill 填充混凝土的双层板桩围堰
double wall construction 双面墙结构
double wall corrugated board 双层双面波纹纸板
double wall dredge pump 内套式泥泵
double-walled 双墙的;双层的
double-walled bell jar 双层烧结钟罩
double-walled heat exchanger 双壁冷却器
double-walled piston 闭式活塞
double-walled sheet-pile cofferdam 双板桩围堰
double-walled shell 双壁壳
double-walled steel waling 双层钢围图
double-walled tank 双壁罐
double-walled well 重壁井
double wall funnel 夹壁漏斗
double wall knot 双花绳端结
double wall partition 双层隔仓板
double wall sheet pile cofferdam 双排板桩围堰;双壁板桩围堰
double wall structure 双墙式结构
double wall timber piling cofferdam filled with puddle 双墙木板桩泥土填心围堰
double wall with concrete fill 混凝土填充墙
double warp 双经
double Warren truss 复式华伦桁架;复合华伦桁架
double wash(ing)stand 双漱洗座
double water flow model 双胶水流模式
double water glass 双(料)水玻璃
double water internal cooling 双水内冷
double wave 双(面)波;双(叠)波
double-wave detection 全波检波
double waved tube differential pressure ga(u)ge 双波纹管差压计
double way channel 双向航道;双线航道
double way connection 桥形连接;桥式接法
double way gradient 双坡
double way launching 双滑道下水
double way lock 双线船闸
double way lockage 双向过闸
double web 双肋式
double-webbed 双肋的
double-webbed beam 双腹梁

double-webbed girder 双腹板(桁)梁
double-webbed I-beam 双腹板工字梁
double-webbed plate arch 双腹板拱
double-webbed plate arch(ed) beam 双腹板拱梁
double-webbed T-beam 双腹板T形梁
double-webbed truss 双重腹杆系桁架
double-webbed wheel 复板轮
double-web bridge 双肋式桥梁
double-webbed section 双腹式断面
double wedge 菱形翼型;凸齿;双楔形;双面楔;菱形的(指剖面形状)
double-wedge aerofoil 菱形翼面
double-wedge boat 直框架结构船
double-wedge cut 双楔掏槽
double-wedge clamp 双楔垫老虎钳
double-wedge section 棱形截面
double weight paper 双重量纸
double-welded butt joint 双面焊对接接头;双面对焊接
double-welded joint 两面焊缝;双面焊接接头
double-welded lap joint 双面焊搭接接头
double weld(ed seam) 双盖缝
double welding 双管焊接
double welt 双咬口
double welted seam 双面盖缝
double welted standing seam 双面盖条立缝
double wharfage 加码头费(船舶未付码头费离港)
double wheel plow 双轮犁
double whip grafting 重舌接
double-whizzer classifier 二级离心式空气分级机
double winding alternator 双绕组交流发电机
double winding armature 双绕组电枢
double winding synchronous generator 双绕组同步发电机
double window 防风窗;双层(玻璃)窗;双层(平开)窗
double wing door 双簧门;双扇门;双侧门
double-wing(ed) building 有翼建筑(物);有两翼的建筑物
double wing pusher 双螺旋挤泥机
double wire 双丝
double wire-armo(u)red cable 双层铁线铠装电缆;双铁线式铠装电缆
double wire mask 双线掩模
double wire method 双线法
double wire operated semaphore 双线臂板信号机
double wire system 双线制
double-work 二重嫁接
double-working switches 双动道岔
double worm mixer 双螺旋搅拌机;双螺旋混合机
double worm pusher 双螺杆顶车机
double wound 双股的
double wound coil 双绕线圈
double wound relay 双线圈继电器
double wound rotor 双绕式转子
double-woven 双层织造的织物
double-woven-wire cloth 双层织造网筛布
double wrench 两头扳手
double Y 双支管(件);斜三通
double Y branch 双重的Y形管;叉管连接的支管
double Y connected rectifier 双星形连接整流器
double Z-arm mixer 双曲臂和面机
double Zepp antenna 双齐伯天线
double zigzag fold 双闭合褶皱
double zone gas producer 双层煤气发生器
double zone (pugmill) mixer 双叶片混料机
doublication 重合
doubling 夹胶;加倍;折回;重合;并线;并捻;并重
doubling an angle 倍角复测法
doubling and lapping machine 对折卷板机
doubling and plaiting machine 折布机
doubling and rolling machine 并卷布机;对折布机
doubling and tacking machine 折幅缝间机
doubling and twisting machine 并捻机
doubling back 双layer油漆
doubling calendar 重合研光机
doubling calender 重合轮压机
doubling circuit 倍增电极;倍压电路;倍频电路
doubling course 檐口垫瓦;出挑瓦层;双檐瓦层;双板(瓦)层
doubling dilution 二倍稀释
doubling dose 加倍剂量;倍加剂量

doubling effect 加倍作用;倍增现象
doubling fillet 三角垫木
doubling frame 并线机;并捻机
doubling method 倍加法
doubling on itself 作一百八十度转摺
doubling(-over) test 折叠试验
doubling piece 加强条;檐口垫瓦条;风檐板;双层木条;双边
doubling plate 加强钢板;双面盖板;双层板
doubling preventer clamp 两截桅接头防护铁
doubling prism 双像棱镜
doubling register 加倍寄存器;倍增寄存器;倍加寄存器
doubling roller 贴合辊
doubling the angle on the bow 倍增机头方位角
doubling thread 纺线
doubling time 加倍时间;重复时间;倍增时间;倍加时间
doubling time meter 倍增时间测量计;双重时间测量计;双时间仪;双倍计时表
doubling up 多户合住一套房
doubling winder 并丝机
doublures 封里材料装饰
doubly 加倍地
doubly balanced incomplete block design 二元平衡不完全区组设计
doubly charged ion 双倍带电离子
doubly cladded slab dielectric(al) waveguide 双包层条形介质波导
doubly clad optical fibre 双包层光学纤维
doubly curved 双弯曲的;双曲率弯曲的
doubly curved shell 双曲(面)壳体;双曲薄壳
doubly curved shell dome 双曲薄壳穹顶
doubly curved translational shell 平移双曲壳体
doubly degenerate 二度简并
doubly differentiable function 二阶可微函数
doubly diffuse density 双散射密度;双漫射密度
doubly dispersed spectrum 双色散光谱
doubly excited state 双重激发态
doubly fed series (or repulsion) motor 双馈串联电动机
doubly inked list 双重连接列表
doubly integrating accelerometer 二重积分加速度
doubly linked bond 双键结合
doubly linked circular list 双重联结环状列表
doubly linked linear list 双重联结线性列表
doubly perspective 二重透视
doubly plunging fold 双倾伏褶皱
doubly prestressed concrete 双向预应力混凝土
doubly re-entrant winding 双口式绕组
doubly refracting biplate 双折射双片
doubly refracting crystal 双折射晶体
doubly refracting medium 双折射媒质
doubly reinforced 双重加固的;配复筋的
doubly reinforced concrete (具有受压钢筋和受拉钢筋的)双筋混凝土;配复筋混凝土
doubly reinforced section 双筋截面
doubly resonance 双共振
doubly resonant oscillator 双共振振荡器
doubly ruled surface 双直线曲面;双母线曲面
doubly stochastic matrix 双随机矩阵
doubly uniform channel 双重均匀信道
doubtful assets 可疑资产
doubtful bed 可疑层
doubtful debt 可疑债务
doubtful debts provision 预提坏账准备
doubtful loan 迟付贷款;可疑放款
doubtful notes and accounts 呆账和票据
doubtful operation 原因不明的动作
doubtful sounding 可疑水深
doubtful title 可疑的土地财产所有权
Doucai contrasting colo(u)rs 斗彩
doucai vase with tendril vine 斗彩蔓草纹瓶(瓷器名)
douche 灌洗;灌水器;冲洗器;冲洗;送水器;葱形曲线饰
douche bath 喷射浴
douche can 冲洗桶
douche solution 灌洗剂
doucine 反曲线形装饰【建】
dough 料团;捏塑体;揉好的陶土
dough batch 打面机
doughboy table (下可储[贮]物的)桌子;发面箱
dough deformation 柔性变形

dough mill 调胶机;调面机
dough mixer 捏合机;和面机;粉料调水混合机
dough mo(u)lding 捏塑造型
dough mo(u)lding compound 团状模塑料
doughnut 混凝土柱环;环状物;环状结疤;环形(室);圆形图;超环面粒子加速器;热堆快中子转换器;扩大垫圈;钢筋混凝土柱箍(提升用);定位垫块;垫环;环形混凝土垫块
doughnut antenna 绕杆式天线
doughnut kiln 蒸笼窑
doughnut phenomenon 环状发展现象;"炸面饼"式的城市发展
doughnut ring 环形隔板
doughnut-shaped 环形的
doughnut-shaped pattern 环形喷束
doughnut tire 低压轮胎;超低压轮胎
dough salver 分粉托
doughter product 子代产物
doughy 柔软的
Douglas bag 道格拉斯气袋
Douglas fir 黄杉;花旗松【植】
douglasite 氯钾铁盐;绿钾铁盐
Douglas pine 美国松;黄杉
Douglas pouch 道格拉斯陷凹
Douglas powder 道格拉斯炸药
Douglas protractor 道格拉斯量脚规
Douglas scale 道格拉斯(波)浪(等)级
Douglas sea and swell scale 道氏波浪等级;道格拉斯浪涌级表级;道格拉斯海况和长浪尺度;道格拉斯(波)浪(等)级
Douglas wave scale 道氏波浪等级;道格拉斯(波)浪(等)级
douke 黏土质(材料);硬的;致密的;砂质黏土
Doulton's joint 道顿接头(一种湿土内雨水管接头)
Dounce homogenizer 杜恩斯匀浆器
doundry effluent 铸造车间的三废
doup 半综
dourfold block 四饼滑车
douse 急闭
douser (电影放映室用的)防火门;遮光板
dousing water 淬凝水
doustage grate cooler 两级型篦式冷却机
Douthitt diaphragm control 道施特光圈调节装置
Douzieme 道兹密
Douzieme ga(u)ge 道兹密卡规
dove 淡灰色;暗蓝色
dove catch 鸽形掣子
dove-colo(u)r 淡灰色
dove-cot 鸽舍;鸽棚
dove fret 鸠尾回纹饰
dove grey 鸽灰色
dove hinge 鸠尾铰
dove hole 鸠尾形孔
dove marble 淡灰大理石
Dove prism 多夫棱镜;德拉博尔内棱镜;道威棱镜;梯形棱镜;特夫棱镜
Dover pneumatic breakwater 多佛尔气压式防波堤
dovetail 鸠尾形;鸠尾;楔形榫(头);燕尾榫;磁极尾
dovetail anchor 燕尾锚(具)
dovetail anchor slot 燕尾锚定槽
dovetail baluster 楼梯的鸠尾栏杆
dovetail bit 燕尾钻
dovetail condenser 同轴调整电容器
dovetail cramp 燕尾夹;燕尾扒钉;双燕尾榫扣;鸠尾扣钳;(石或金属制的)银锭榫
dovetail cutter 油灰刀;燕尾(铣)刀
dovetail dowel 鸠尾合缝钉
dovetail feather 鸠尾(榫)滑键
dovetailed fillets 鸠尾楔形(木)条;燕尾楔形木条地板
dovetailed groove 梯形槽;燕尾榫槽
dovetailed growing and tonguing 鸠尾鸳鸯接头
dovetailed halving joint 半嵌鸠尾接
dovetailed joint 燕尾榫接
dovetailed lathing 鸠尾形固结板
dovetailed merion 鸠尾形雉堞;鸠尾形城齿
dovetailed side piece 楔形侧部
dovetailed way 燕尾导轨
dovetailer 鸠尾榫机;制榫机
dovetail fastening 鸠尾接合;鸽尾接合;插筒接合
dovetail feather 燕尾榫口;燕尾键;(硬木制的)银锭榫
dovetail feather joint 楔形榫键;鸠尾榫键;双楔形榫键;双鸠尾榫键

dovetail female 门鸠尾榫
dovetail form 燕尾形;楔形
dovetail groove 燕尾槽;燕尾接合
dovetail guide 燕尾导轨
dovetail half-lap 燕尾榫半搭接
dovetail half-lap joint 燕尾榫半叠接;半盖燕尾榫接
dovetail halved 半盖燕尾榫接
dovetail halved joint 燕尾对开接合;燕尾榫半搭接
dovetail halving 鸠尾榫制作;鸠尾榫嵌接;鸠尾对半结合;燕尾榫半搭接
dovetail hinge 鸠尾铰链;燕尾铰;燕尾合页
dovetailing 楔形结合;燕尾连接;燕尾接合;燕尾形表面刻痕
dovetailing machine 制榫机;削榫机
dovetailing plane 榫刨
dovetail joint 鸠尾榫接头;鸠尾榫接合;鸠尾结合;燕尾形接头;燕尾接合;半盖燕尾榫接
dovetail joist 燕尾龙骨
dovetail key 鸠尾键;燕尾键
dovetail key way 鸠尾键槽;燕尾键槽
dovetail lap joint 鸠尾互搭接合
dovetail lath 肋形(钢)板条;带肋网眼钢板
dovetail machine 制榫机;鸠尾榫机;燕尾开榫机
dovetail margin 燕尾榫板条
dovetail merlon 鸠尾雉蝶墙
dovetail miter 燕尾斜角接合;燕尾形斜面接合;暗马牙榫
dovetail mortise 燕尾形榫眼
dovetail mo(u)lding 鸠尾榫找线形,鸠尾饰线脚;(饰以鸠尾形的)浮雕线脚;鸠尾饰;三角鸠尾连锁式线脚
dovetail pile 燕尾桩
dovetail plane 燕尾刨;梯形槽刨
dovetail plate 榫眼板
dovetail rib 鸠尾肋;燕尾肋
dovetail saw 榫头锯;鸠尾锯;燕尾锯
dovetail sheeting 燕尾形片材;鸠尾形片材;鸠尾形模板;鸠尾榫锯;燕尾形板
dovetail slide 鸠尾导轨
dovetail slide bearing 燕尾(式)导轨
dovetail slot 鸠尾形槽道;鸠尾槽;燕尾槽
dovetail slot and anchor 燕尾槽锚固
dovetail tenon 鸠尾榫;燕尾榫
dovetail tie 鸠尾形条件
dovetail vaulting 鸠尾形拱顶
dovetail vernier caliper 测量燕尾槽专用游标卡尺;测定燕尾槽专用游标卡尺
dovtail halving 半盖燕尾榫接
Dow cell 道氏镁电解槽
Dow chemical extruded 道氏化学制剂挤压过的
dowdrift 下滑漂移
dowel 螺桩;两尖钉;连接筋(钢筋混凝土);夹缝钉;木销钉;合缝钉;轴销;钉;暗销接合钉;定缝销钉;传力杆;插铁;暗销;暗榫;榫钉;插筋
dowel action 销栓作用;销钉作用;传力杆作用;暗销作用
dowel bar 榫钉接合;传力杆(接缝);暗销接合;缝条;销钉;插筋;暗销筋;榫条;接头插筋
dowel bar assembly 传力杆组装件
dowel bar chair 传力杆支座
dowel bar reinforcement 插铁;插筋;连接筋;接缝筋
dowel bearing strength 榫钉支持强度
dowel bit 销钉钻(木工用);半圆形木钻;枸形钻(头)
dowel board 暗钉钉合板
dowel bolt 精确配合螺栓
dowel brick 受钉砖;木砖
dowel bush 定位套管
dowel cap 销钉帽
dowel crack 暗榫裂缝
dowel crown 桩冠
dowel cutter 接合钉切割器
dowel cutter centralizing pin 接合钉切割器定中心针
dowel cutter shaft 接合钉切割器杆
dowel deflection 销钉歪斜;传力杆挠曲(混凝土路面)
dowel driver 销钉打入工具;接合钉推进器
dowel hole 定位销孔;定缝销钉孔;插筋孔
doweling 用榫钉连接;设置传力杆;加销钉;放置传力杆;埋插筋
dowel installer 销钉缝;加榫钉连接;合钉接合;定缝钉接合;套接;用传力杆接合;暗销接合
dowel joint 榫钉接合;加榫合缝接缝;合钉接合;定缝

dowelled beam 键接组合梁;加键叠合梁;键接梁;键合梁
dowelled connection 用销钉连接的键接;暗销连接
dowelled edge 设暗销的边缘;设传力杆的板边
dowelled expansion joints 设置传力杆的伸胀缝
dowelled joint 销连接;加榫合缝接缝;插销接合;暗销结合;暗销接头;螺栓合合;接合;销钉接头;销钉接合;传力杆缝;暗销接合;企口
dowelled pin joint 暗销结合;暗销接头;暗销接合
dowelled timber beam 键合木梁
dowelled tongue and groove joint 有榫钉的企口缝;有传力销的企口接缝;有传力杆的企口接缝;有暗销的舌槽接缝
dowelled transverse expansion joints 设置传力杆的横向伸缩缝
dowelled transverse expansion or contraction joints 设置传力杆的横向伸缩缝
dowelled wooden beam 键合木梁
dowel(ling) jig 钻孔夹具;孔钻
dowel(l)ing joint 销钉接头;销钉接合
dowel load-transfer unit 传力杆荷载传递装置
dowel lubricant 伸缩缝润滑油;插筋润滑油;连接筋润滑剂;(无黏套钢筋的)伸缩缝润滑销
dowel lubrication 暗销润滑剂
dowel machine 传力杆设置机
dowel masonry 暗栓圬工
dowel pile 钉桩
dowel pin 接合栓销(合(缝)销;销钉;桩钉;柱销;定位销;定缝销钉;暗销;拼钉;拼合用销;两头尖钉;带倒刺的无头钉;木钉
dowel pin hole 销钉孔(眼)
dowel pin joint 销钉接头;销钉接合
dowel plate 销钉板
dowel rod 传力杆
dowel screw 双头螺钉;两头螺钉;螺纹栓
dowel shear 销钉断面剪力;传力杆剪力;合缝钢筋所受的剪力
dowel sleeve 连接筋套;销钉套筒
dowel socket 传力杆套筒
dowel spacer 间隔暗销
dowel spacing 传力杆间距
dowel steel 合缝钢条;传力钢条;传力杆
dowel-supporting assembly 传力杆支座
dowel template 连接筋样板
Dow-etch plates 道氏腐蚀法印版
Dowfume EB$_5$ 二氯丙烷混合剂(熏蒸杀虫剂)
Dowfume F 二氯乙烷混剂(熏蒸杀虫剂)
Dowicide G 五氯酚钠(木材防腐剂)
Dow-Jones index 道·琼斯指数
Dow-Jones transportation average 道·琼斯运输业股票价格平均指数
Dowling paper 道林纸
Dow metal 道氏镁铝合金;道氏金属
Dow metal alloy 道氏合金
Dow motionless mixer 道氏静止混合器
Dow moving bed process 道氏移动床法
down 开阔高地;故障停机;岸边沙丘;山岗
down cut milling 同向铣削
down-and-out option 见跌即止期权
down apron 后挡板
down-away time 释放时间
downbound 下行(的);下水
downbound boat 下行船(舶)
downbound cargo flow 下行货流
downbound commerce 下行运量
downbound entry time of lockage 下行进(船)闸时间
downbound exit time 下行出闸时间
downbound exit time of lockage 下水出闸时间
downbound fleet 下行船队
downbound journey 下水航程
downbound push-train 下行顶推船队
downbound ship 下行船(舶)
downbound tows 下行(驳)船队
downbound traffic 下行运量;下行交通;下水(货)运量
downbound traffic volume 下行运量;下水运量
down-bound train 下行列车
downbound transport 下行运量
downbound vessel 下行船(舶)
down buckling 向下弯曲
down by stern 尾倾
down by the head 船首纵倾;前倾;首倾【船】

down by the stern 后倾【船】;船尾纵倾
downcast 下落;下风井;垂直断眶;入风井;通气竖坑;通风(竖)坑;通风井
downcast air 进风井气流
downcast fault 下落断层
downcast shaft 进风井
downcast side 陷落侧;下落翼;断落侧
downcast ventilation 下向通风
downcast ventilator 下向通风机;供气通风筒
downcoast 下行海岸
downcoastal flow 近岸流
down coiler 地下卷取机
downcomer 泄水管;下气烟道;排气管;废水管;降水管;下水管;下降管;下导管;高炉下气道;排液管
down-coming wave 下射波
down Concordia intercept 一致曲线下交点
down-conversion 下变频
down converter 下送变换器;下变频器;向下变换器
down corner 精馏柱溢流管;溢流管
down corner leg 落水管支路
down counter 降计数器;逐减计数器
downcountry 向海边
down current 下电气流;下沉气流
downcut 叨下;向下侵蚀;下切
downcut shears 下切式剪切机
downcutting 顺铣;向下侵蚀;下切(作用);淘刷
downcutting erosion 向下侵蚀
downcutting method 下切法
downcutting river 刷深河流;下切河(流)
downcutting stream 刷深河流;下切河(流)
downcutting tool 立刨刀;插刀
down digging excavator 俯掘机
down dip 下倾
downdip block 下倾断块;沿倾向的下落断块;顺倾向的下降盘
down direction 下行方向
down-Doppler 下行多普勒
down draft 向下通风;倒焰;倒风;下曳气流;降低;下向通风;下沉气流;向下气流
down-draft carbureter 下吸式汽化器
down-draft carburetor 下吸式化油器
down-draft combustion process 强制通风燃烧法
down-draft drying zone 抽风干燥段
down-draft furnace 倒焰炉;倒风炉;倒烟炉
down-draft gas producer 下吸式煤气发生炉
down-draft hood 下吸式抽风罩
down-draft intermittent brick kiln 向下抽风间歇式砖窑
down-draft kiln 倒风窑;倒焰窑
down-draft sintering 吸风烧结
down-draft sintering machine 下流式烧结机;吸风烧结机
downdrag 桩负摩擦力;下拉荷载
downdrag load 下拉荷载(桩工)
down draught 向下抽风;下向气流;下吸式通风;(立管的)下流气流;倒灌(风);下泄;下向通风;下吸;倒烟;向下气流
down-draught calciner 倒焰式煅烧窑
down-draught carburetor 下吸式汽化器
down-draught cooler 抽风式冷却机
down-draught diverter 坠灌隔流阀;倒灌隔流网
down-draught fan 抽风机
down-draught grate 抽风式链算机;抽风带式焙烧机
down-draught intermittent brick kiln 向下抽风间歇式砖窑
down-draught kiln 倒焰窑
down-draught pot furnace 倒焰式坩埚窑
down draught producer 回流式发生炉;向下通风发生器;倒风发生器
down-draught type producer 回流式发生炉
downdraw 水位降落;水位下降
down-drawing tube process 垂直下拉管法
down-draw process 下拉法
downdrift 向下(游)漂移
down-drift direction 朝下游方向
down drilling 向下穿孔
downdrop 下落
down end 断头
down-ender 翻卷机
downer 断纸停机
downfall 下落;下降;突然大量降落(雨,雪等);坍方
downfall pipe 水落管
downfaulted basin 断陷盆地
down faulting 下落断层作用

down feed 下供式
down-feed distribution 下供式分配；下(行)给水
down-feed overhead system 上行下给式系统(暖通空调)
down-feed overhead two pipe system 上行下给式双管系统
down-feed riser 下给立管；下供式立管
down-feed riser system 下供式立管系统
down-feed screw 垂直丝杠
down-feed system 下供(式)系统；上行下给系统；向下给料系统；下给(式)系统；上分系统
downfield 低磁场
down-fired furnace 顶烧式炉
down-flop of coats 涂层的随角异色效应
down-flop property 随角异色性能
downflow 下流；下降气流
downflow anaerobic turbulent bed 下流厌氧紊动床
downflow apron 降液挡板
downflow bubble contact aerator 降流气泡接触曝气池
downflow clean room 垂直层流式洁净室
downflow contact bed 下流式接触床
downflow filter 下流式滤池
downflow filtering media type electronic air cleaner 静电一滤材并用型空气净化设备
downflow fixed bed 下流式固定床
downflow fixed bed reactor 下流固定床反应器
downflow fixed ffilm bioreactor 下流固定膜生物反应器
downflow flue 下降式烟气；下降式火道
downflow furnace 逆流式烟道采暖炉
downflow hanging sponge 下流悬挂海绵
downflow pipe 下水道；溢流管；下水管
downflow regeneration 顺流(式)再生
downflow stationary film 下流稳态膜
downflow-type central furnace 下部排气式集中采暖炉
downflow water source 下游水源
downflow weir 溢流堰
down fold 向斜槽
downfolding 由于断层作用而陷落；下折式
down gate 直浇口
downgoing plate 下沉板块；俯冲板块
down going wave section 下行波剖面
downgrade 下坡(的)；衰落的；下坡(度)；降低等级；低质量
down grade excavation 下坡开挖
downgrade flow 自流
down grading 品位降低
downhand electrode 平焊条
downhand position 平焊位置
downhand weld 平焊缝
downhand welding 俯焊；平焊(接)
downhauler 收帆索
downhaul utility capsule 下拖式潜水工作舱
downhill 下(山)坡；低质量材料
downhill casting 顶浇；上铸法
downhill conveying 下倾输送；下向输送
downhill conveyer 下倾输送机；下向输送机
downhill creep 移动滑坡
downhill grade 下坡；降坡
downhill loading 倾斜荷载；下向加荷
downhill path 下坡路
downhill pipeline 下倾的输送管线
downhill road 下坡路
downhill screening 倾斜筛选
downhill section 下坡段落；倾斜段落
downhill travel of rotary kiln 回转窑体向下窜
downhill welding 俯焊
downhole 底部井眼；孔内的；孔底的；井下(的)；井内的；井底的
downhole accelerograph 井下加速仪
downhole accelerometer 井下加速度计
downhole array 井下台阵
downhole blowout preventer 井下防喷器
downhole boring 井下钻进
downhole camera 井下照相机；井下录像机
down hole choke 井下油咀
downhole colo(u)r TV 井下彩电
downhole condition 井下条件；孔底条件；井底条件
downhole drill 潜孔凿岩机；下向凿岩机；人孔冲击钻机；潜孔钻
downhole driller 潜孔钻机

downhole drilling 向下钻孔；井下钻进；潜孔钻进
downhole (drilling) machine 潜孔钻机
downhole drilling motor 井下钻进马达
down hole electric(al) drill 孔底电钻钻具
downhole electrodrill 孔底电钻
downhole energy gathering equipment 井下能量集聚装置
downhole engine pump 井底水力活塞泵
downhole equipment 井下钻具；井下设备；钻孔配件
downhole explosion 孔内爆破
down hole flowing pressure 井底流动压力
downhole flowmeter 孔内流量计
down hole flow rate measuring 测井下流量
downhole formation 井下岩层
downhole gas burner 井底气体燃烧器
downhole geophysical logging 井下地球物理测井
downhole ground 井下接地
downhole hairpin heat exchanger 井下U形燃烧器
downhole hammer 潜孔锤；潜孔冲击钻机；井下振动锤
down hole hammer bit 潜孔锤钻头
downhole hammer drill 潜孔锤式钻机
downhole hammer drilling 潜孔锤钻进
downhole instrument specification 井下仪器规格
downhole logging 井下记录
downhole measurement 井下测量；井底测量
downhole method 下孔法
downhole motor 井底发动机
downhole motor drill 孔底动力钻机
downhole motor drilling 孔底动力(机)钻进
downhole motor tool 孔底动力钻具
downhole package 井下成套设备
downhole packer 井下封隔器
downhole partial reverse circulation 孔底局部反循环
downhole partial reverse circulation tool 孔底局部反循环钻具
downhole pendulum 井下摆
downhole percussion drill 冲击式潜孔钻机
downhole performance 井底状态；井下工作性能
downhole pressure 井下压力；(钻井时的)钻头压力
downhole pressure recorder hot air calibration bath 井下压力计空气浴
downhole pump 井下泵
downhole reamer 潜孔扩孔器
downhole recorder 孔底记录器
downhole recording system 孔底记录系统
downhole sample pump 井下取样泵
downhole sampler 井下取样器
down hole seismograph 地下测震仪
downhole seismometry 井下测震学
downhole separator 井下分离器
downhole shooting 下孔法
downhole solid sampler 井下固体取样器
downhole speed 下孔速度；下井速度
downhole super-drill 超级潜孔钻机
downhole survey 孔内测量
downhole telemetry 井下遥测技术
downhole telemetry system 井下遥测系统
downhole television 井下电视
downhole television camera 孔压电视录像机
downhole temperature 井下温度
downhole temperature instrument 井下测温仪
downhole trouble 孔内事故【岩】
downhole TV camera 井下电视录像机
downhole vibration drill 孔底振动钻具；潜孔振动钻机
downhole vibro-drilling 孔底振动钻进
down land 低地；山地牧场；丘陵地
down lead 引下线
down leakage layer 下越流层
downlight 顶棚小型隐蔽式灯；吸顶小灯；向下照光
down line 脚线；下行线路；下行(线)；沿铁路线；下游管线
down link 下行线路
down load(ing) 下载；卸荷；下装
down lock 下位锁
down main track 下行正线
down milling 顺铣
down payment 分期付款的定金
downpass 下向通道
down payment 交付定金；现付金额；预付款；预付定金；定金；分期付款的定金；首期付款；首笔现付金额；分期付款的首笔付款；首次付款
down payment paid 已付定金
down period 停工检修时间
down pipe 下水道；下流管；旁通管；水落管；下落管；雨水管；厕所喷水管的连接管
downpipe filter 水落管滤池
downpipe shoe 水落管足
down pointing borehole 斜向钻井
downpour 注下；大(暴)雨；倾盆大雨
downpour gate 底注浇口(内浇口向上倾斜)
down pressure 向下压力；钻头对孔底压力
down price clause 减价(合同)条款；减价(契约)条款
down pull 下拉力
down quenching 急冷
down ramp 下坡道；下坡道
downrange station 离开发射中心的测试站
down right 从上向下；直下
downriver 向河口处；向下游；向上游(的)；下游(的)；在下游；在河口处
down-run 下行制气
down runner 直浇口
down rush 下刷；下冲
downs 白垩山丘；丘陵草原；岗
down sand 丘砂
down sand-dune 砂丘；砂堆
Down's cell 道恩电解池
down service 重力输水设施
Down's formula 道恩公式(一种设计柔性路面厚度的古典公式)
down shift 降速变换；减速(汽车)；换慢挡；下移；接减速传动
downshifting 转换低速
downshift range 降速变换幅度；减速幅度
downshot-type furnace 倒烟炉
downside 下行方面；下侧(的)
down-sideband 下边带
downside risk 亏损的风险
down-size staffs and improve efficiency 减员增效
downslide 下滑
downslide surface 下滑面
down-slip fault 下落(滑)断层
downslope 降坡；下坡(的)
downslope displacement 下滑位移
downslope ripple 下移波痕
downslope slow 下坡气流
downslope time 电流下降时间
downslope wind 下坡风
down-splitting 下分岔
downspout 落水管；降液管；水落管；(阳台、平台等向外排水的)泄水嘴
downspout basket strainer 雨水管下水口箅子
downspout conductor 下流管；溢流管
downspout elbow 泄水弯头；雨水管(出水)弯头
downspout hanger 雨水管卡子
downspout leader 雨水竖管
downspouts for bridge surface water 桥面落水管
downspout shoe 雨水管出水弯头
downspout strainer 雨水管下水口箅子
downspout strap 雨水管固定铁条
Down's process 道恩法
downsprue 浇注系统；直浇口；直浇道
downstage 舞台前部
downstair(case) 往楼下；楼下(的)；在楼下
downstair(case) merger 向下并合
downstand beam 托梁；下翻梁
down state 停运状态
downstream 后续程序；向下运动；向下游；下游物流；下游(的)；下行的；朝向河的下游；顺流(而下)
downstream anchorage 下游停泊区
downstream-angled spur dike 下挑丁坝
downstream apron 下游护坦；防冲铺砌；防冲护坦；坝后防冲护底
downstream batter 下游坡；下游面坡度；(坝的)背水坡；背水面坡度
downstream bay (临近建筑物的)下游河段；下游池；下水池
downstream benefit 下游受益
downstream berthing area 下游停泊区
downstream boat 顺流船
downstream bottom 下游河床
downstream chain 惰链
downstream chamber 出口室
downstream cofferdam 下游围堰

downstream concrete apron 下游混凝土护坦;下游混凝土防冲铺砌
downstream course 下游段;顺流洄游路程
downstream cross section 下游断面
downstream deep 下深槽
downstream direction 顺流向
downstream dredging 顺流挖泥;顺流施工
downstream eel migration 鳗鲡顺流洄游
downstream end of forehearth 通路下游端
downstream equipment 下游设备
downstream erosion 下游冲刷
downstream extremity of the dock 码头下游端
downstream face 下游(坝)面;(坝的)背水面
downstream face of a weir 堰的背水面
downstream face of dam 坝的下游面
downstream facility 管线后段设施
downstream feeding 下游喂料
downstream fill (土石坝的)下游楔形体
downstream film 气流下流外延层
downstream fish migration 鱼群顺流洄游
downstream floating of bamboo or log raft 竹/木排流放
downstream floating of log raft 木排流放
downstream floor 防冲铺砌;下游护坦;防冲铺盖;防冲护坦
downstream flow 顺流向流动
downstream garage 下游停泊区
downstream gate head 下闸首
downstream gate sill 下游闸首门槛
downstream ga(u)ge 下游水深标尺
downstream guide wall 下游导墙
downstream heat exchanger 顺流热交换器
downstream highest navigable water level 下游最高通行水位
downstream injection system 顺流喷射系统
downstream installation 管线后段设施
downstream leaf 下游闸门门扇
downstream level 下游水位
downstream line 下游管线
downstream lock 下游船闸
downstream lock chamber 下游闸室
downstream lowest navigable water level 下游最低通行水位
downstream margin of unbroken ice cover 未破裂冰覆盖层的下游界限
downstream method 下游法
downstream migrant 顺流洄游鱼
downstream migration 降河洄游;顺流洄游
downstream migration of eel 鳗鲡顺流洄游
downstream migration of fish 鱼群顺流洄游
downstream moved distance 下移距离
downstream nose 顺流墩尖
downstream of pipeline 管线下游
downstream pier nosing 下游墩尖;闸墩头部;闸墩尾部;墩的下游端点
downstream pipe 下游管
downstream plant 管线后段设施
downstream pool 下深槽
downstream power plant 下游发电站;下游发电厂
downstream pressure 下游压力;阀后压力
down-stream pressure of orifice meter 测气孔板差压
downstream process 后续处理
downstream products 下游产品(制造业)
downstream products of float glass 浮法玻璃深加工制品
downstream profile of crown section 拱冠断面下游面轮廓
downstream protection 下游护坡;下游护岸;下游防护措施
downstream radius of crest 坝顶下游面半径
downstream reach 下游河段
downstream region 下游区
downstream reservoir 下游水库
downstream revetment 下游护岸
downstream sailing 下水行驶
downstream sand bank 下边滩
downstream section 下游区段
downstream shell 下游坝体;下游坝壳
downstream ship 顺流船
downstream side 下游边;下游面;下游侧
downstream side of the compressor 压缩机的压缩端
downstream slope 下游坡;下游面坡度;(坝的)背水坡;背水面坡度;坝的背水坡
downstream slope of dam 坝的下游面坡度
downstream spray pattern 顺喷
downstream stage 下游梯级;下游水位
downstream station 下游电站
downstream station arrangement 尾部式布置(地下式水电站)
downstream surface 下游水面
downstream surge chamber 下游调压室;下游调压井
downstream technique of biotechnology 生物工程下游技术
downstream temporary dam 下游临时坝
downstream toe 堤趾;坝趾;下游面坡脚;下游坝趾
downstream toe of dam 坝的下游坡脚
downstream traffic 下水运输;下水(货)运量
downstream transition 下游渐变段
downstream transitional section 下游衔接段
downstream transportation 下水运输
downstream tube 下游管
downstream vessel 顺流船
downstream water 下游水
downstream water level 下游水位
downstream waters 下游水域;下游河段
downstream wing wall 下游翼墙
down stroke 下(降)行程;头款;定金;下(行)冲程
down stroke hydraulic press 下冲式水压机
down stroke press 压头下降式压机
down structure deflection of the well 底层结构位移的竖井
down suction 垂直间隙
down surge 水面下降
down survey crystal volume 下测晶体体积
down sweep 下(向)扫描
downswing 下落
down symbol 降符号
downtake 降落管;下气道;下导气管
downtake superheater 锅后过热器
downtake tube 降液管;下导管
downtank 下流槽
down the country 向海边
down (the) dip 下向
down-the-hole 潜孔
down-the-hole corrosion 井内腐蚀
down-the-hole hammer 潜孔锤;潜孔冲击钻机
down-the-hole inspection 下井检查
down-the-hole rock drill 潜孔岩石钻机
down-the-hole treatment 井内处理
down-the-hole type of machine 井底射孔器
down the-hole type rock drill 潜孔钻岩机
down the river 朝向河的下游
down the wind 顺风的
down throat 下行咽喉
downthrow 坍陷;下落(翼);正断层;陷落地块;塌陷
downthrow block 陷落地块;下落地块【地】;下落盘【地】;下降断块【地】
downthrow fault 下落断层;正断层
downthrow side 下投侧;(断层的)下降盘;陷落侧;下侧面
downthrow wall 下降盘
downthrust 下冲
down-tick 股票价格下跌的符号
down-tilter 翻卷机
down timber 成片倒木
downtime 空转时间;修理时间;故障时间;非工作状态时间;不可工作时间;停机时间;检修时间;下落时间;闲故障时间;故障(停机)时间;窝工时间;停歇时间;停工时间;停工期;停产(时间)
downtime accumulator 故障时间累积器
downtime day 停工日;故障日
downtime of computer 计算机发生故障或操作错误的时间
downtime of machine 机械停工时间
downtime rate 停机时间率
downtime ratio 停工率;停机时间比率
down-to-basin fault 下落盆地断层;盆地侧下降断层
down-to-earth 切实地
down-to-floor lattice window 落地花格窗
down-to-ground openwork screen 落地罩
down to her marks 装到满载吃水
Downton Castle Sandstone 当唐堡砂岩
Downtonian series 当唐阶【地】
Downtonian stage 当唐阶【地】
Downton pump 邓通曲柄手摇泵;当通手摇泵;达温特曲柄式手摇泵
down total 停机总计
downtown (area) 市(中心)区;商业(中心)区;闹市区;市内;市中心商业区
downtown business 城市商业
downtown business district 城市商业区
downtown business spots 城市商业网点
downtown streets 街市
downtown terminal traffic 以市中心区为终点的交通;向市中心区的交通
down track 下行线;左线
down traffic 下行(交通)
down-train 下行列车
downtree 倒树
downtrend 下降趋势
down trip 下降行程(电梯);下降路程(电梯)
down upon edge 逼近所追赶的船
down ventilation 下向通风
down voltage minimum break 最小击穿电压
downward 向下(的)
downward adjustment 调低
downward air leg 下向式气腿
downward bias 向下偏倚;向下偏斜
downward borehole 斜孔(坑道钻进)
downward bowing 向下(弓形)弯曲
downward broom 向下帚形
downward bulge 向下凸出
downward call 向下调用
downward communication 下行信息交流;下行沟通
downward compatibility 向下兼容性
downward compatible 向下兼容的
downward construction method 逆作法
downward continuation 向下延拓
downward continuation method of data 数据向下延拓法
downward continuation of gravity anomalies 重力异常向下延拓
downward conveyer 向下输送机
downward conveying 向下输送
downward deflection 向下弯曲
downward discharge unit heater 下流式单位散热器
downward drifting 向下开挖
downward drilling 向下钻孔
downward erosion 向下冲刷
downward facing 面向下
downward filtration 下向过滤
downward flow 向下水流;下行流
downward flow flue 下降烟道
downward force 向下力
downward gradient 下降梯度
downward landscape 俯视景观
downward migration from source rock 自母岩向下运移
downward mining 下行开采
downward modulation 向下调制
downward movement 向下运动
downward ozone transport 臭氧向下移动;臭氧下沉
downward plunging 沉陷;下沉
downward-point groin 下挑丁坝
downward pressure 向下压强;向下压力
downward price 价格趋落
downward price trend 价格趋跌
downward radiation 向下辐射
downward scour 向下冲刷
downward shift distance 下移距离
downward sloping 向下倾斜
downward sloping hole 倾斜孔
downward stroke 下行冲程
downward system 下行系统
downward system of ventilation 向下通风系统;下向通风系统
downward tendency 下跌趋势
downward total radiation 向下全辐射
down-ward-to-the-ground of gravity value 空对地重力归算值
downward transport in the soil 在土壤中向下移动
downward ventilation 下降式通风
downward ventilation system 下向通风系统
downward weathering 向下风化
downward welding in the inclined position 下坡焊

downward welding in the vertical position 向下立焊
downward zero crossing 下跨零点
downwarp 下沉;反弯;拗弯;下翘;下凹
downwarped basin 拗陷盆地
downwarped extracontinental basin 陆外下陷盆地
downwarped lake 凹陷湖
downwarping 下翘;下曲;向下弯曲;下挠;拗陷;洼陷作用
downwarping belt 拗陷带
downwarping faulted belt 拗断带
downwarping faulted region 拗断区
downwarping faulted zone 拗断区
downwarping folded belt 拗褶带
downwarping folded region 拗褶区
downwarping movement 拗褶运动
downwarping region 拗陷区
downwarping structure 拗褶构造
Downwarping zone of East China sea shelf 中国东海陆架拗陷带
Downwarping zone of northern continent margin of south China Sea 中国南海北部陆缘拗陷带
Downwarping zone of southern continental margin of South China Sea 中国南海南部陆缘拗陷带
down wash 向下输送;下冲气流;污物下沉;冲刷物;下洗;下沉气流;冲掏
down washing 下层洗气;烟气下沉
downwash velocity 泄流冲刷速度;下沉速度;诱导速度;溢流下泄速度
downwelling 下降流;沉降流;水流下沉
downwelling current 下降流
down-wind 下(沉)风;下风(的);顺风向;顺风(的)
down-wind course 顺风航向
down-wind direction 顺风向
down-wind distance 下风距离;顺风距离
down-wind landing 顺风落地
down-wind leg 顺风边
down-wind source of pollution 顺风污染源
downy 丘陵起伏的
down zoning 降低密度区划;用地区划强化;降低人口密度的区域规划
Dow oscillator 道氏电子耦合式振荡器
dowrah jute 多拉黄麻
dowse 急冷;急闭;淬火
dowser 遮光板
dowsing 用测杆测地下水位
dowsing practitioner 用机械探寻(水源)的专业人员
dowsing rod 测水杆
dowsing water 淬凝水
Dowson gas 混合煤气;城市加煤气
dowtherm 高沸点有机溶液;导热姆(换热剂)
dowtherm boiler 导热姆热媒锅炉
dowty accelerator 加速顶
dowty accelerator-retarder 加减速顶
dowty controllable retarder 可控减速顶
Dowty prop 道梯液压支柱;道道型支柱
dowty retarder 减速顶
doxapram 多沙普仑
doxepin 多塞平;多滤平
doxofylline 多索茶碱
doxycycline wastewater 强力霉素废水
doyleite 督三水铝石
Doyle rule 多伊尔板积表(美国硬材厂商协会)
Doyle-scribner rule 多伊尔原木板积表
doze (木材的)腐朽
dozen file 一打卷宗;一打锉刀
dozer 推土机;推土铲
dozer apron 推土机挡板
dozer blade 推土机铲刀
dozer blade for felling work 伐木用推土机铲刀
dozer blade for rooting work 除根用推土机铲刀
dozer boss 推土机领班
dozer company 推土机班
dozer-equipped track-type tractor 装备履带式拖拉机的推土机;履带式推土机
dozer fitted to track-laying (type) tractor 装配铺轨式拖拉机的推土机;履带式推土机
dozer fitted to wheel tractor 装配轮式拖拉机的推土机;充气轮胎式推土机
dozer lift cylinder 推土机升举汽缸
dozer-loader 推土装载两用机
dozer ram 推土机撞锤

dozer shovel 推土挖掘机;推土铲
dozer shovel(l)ing 推土机铲挖
dozer-shovel unit 铲式推土机装置
dozer stabilizer 推土机稳定器
dozer with angling blade 角刀推土机;斜板推土机
dozing 用推土机清除;推土(作业)
dozing blade 推土机清除铲刀
dozing chamber 格筛室;筛矿室
dozing output 推土机清除产量
dozing tool 铲具;推土工具
dozy 快要腐烂的;烂的
dozzle 铸模补助注口
dpdt double pole double throw 双刀双掷
dq phasor 直角坐标相量
draa 臂形韵律层
drab 褐色斜纹布;褐色的;茶褐色;土褐色的
drabness 淡褐色
drab soil 黄褐土;褐色土
drachm(e) 打兰(古希腊的重量单位,1 打兰 = 0.5 盎司)
dracone 弹性拖油容器
draconitic month 交点月
draconitic revolution 交点周
draconitic year 交点年
draegerman 矿山救护人员
draft 待付汇款;草图;拉引;拉拔;减面率;汇票;一吊货;支票;抽力;吃水深度;吃水;草拟;草稿;付款通知书;付款通知单;法案;牵引;脱模斜度;图样;图案;画草图;穿综
draft advice 票据委托书
draftage 公耗
draft agreement 草约
draft air 流通空气;气流;通风气流
draft amidship 舯吃水
draft and loading monitor 吃水与装载监测器
draft and telegraphic(al) transfer 票汇及电汇
draft and telegraphic(al) transfer payable 应付汇票及电汇
draft angle 模锻斜度;烟囱锥度;拔模斜度;脱模角
draft anima 畜耕
draft animal 役畜
draft apparatus 通风设备
draft arm 牵引臂
draft articles 条文草案
draft attached 汇票随附
draft at thirty days sight 三十天期的汇票
draft bar 连接杆;牵引杆
draft bead 窗户止风条;防吸风压条
draft blast 送风
draft board 牛皮纸板
draft by negotiation 逆汇
draft capacity 排出风量;送风能力
draft chamber 气流室;通风室
draft channel kilometerage 拟建航道里程
draft check damper 气流调节门
draft chisel 琢边凿
draft contract 合同草案
draft control 抽力控制
draft control system 力调节系统
draft convention 协定草案
draft copy 摹绘图;草稿复制;草案拷贝
draft cupboard 通风柜
draft curtain 屋顶下热烟气隔板;风障;气流幕;屋顶挡烟隔板;通风(装置)隔板
draft customs nomenclature 关税分类草案
draft damper 气流闸;通风阀
draft deflector 挡风器
draft design 方案设计;通风设计;设计草图
draft device 牵引装置
draft difference (船首尾的)吃水差
draft differential 抽风压差
draft distributing box 分风箱
draft diverter 通风转向器
draft drawn against mortgage 抵押汇票
draft drawn against securities 担保汇票
draft drawn under travellers letter of credit 旅行信用状汇票
drafted budget 拟订预算;预算草案
drafted general budget 拟订总预算;总预算草案
drafted margin 块石琢边;细凿边框;琢边
drafted stone 琢边块石
draft engine 排表机,通风机
draft environment(al) impact statement 环境影响报告书草案

draft equipment 通风设备
drafter 练条机;描图器;制图机械;牵伸机构;起草人
drafter control program(me) 描图控制程序
draft exchange rate 票汇汇率
draft excluder 阻挡气流的门廊;阻挡气流的穿堂;气流阻挡设施;气流隔断
draft excluding threshold 气流阻挡阀
draft exclusion 排除穿堂风
draft extreme 龙骨吃水
draft fan 吸风机;无通风;无气流,无穿堂风;抽风机;风扇;排烟风扇;通风计;通风机;通(风)风扇
draft feeder highway kilometerage 拟建公路里程
draft feeder railway kilometerage 拟建支线里程
draft fillet 窗镶条
draft final statement 最终付款申请初稿
draft fire 逆风火
draft flue 排气烟道
draft force 牵引力
draft for collection 托收汇票
draft frame 牵引机架
draft-free building 不吸风建筑物
draft from reservoir 水库放水
draft from storage 水库泄放;水库放水
draft full load 满载(吃水)装载量
draft furnace 通风炉
draft gas pipeline meterage 拟设输气管长度
draft ga(u)ge 通风表;拉力计;吃水(指示)仪;吃水指示器;差式压力计;差式压力表;差示压力计;差示压力表;风压表;通风计
draft gear 牵引装置
draft head 吸出水头;吸出落差;尾水头;通风气流压力;吸水水头
draft hole 通风孔
draft hood 吸气罩;排气罩;烟囱风罩;通风罩;厨烟抽风罩
draft indicator 通风指示器;吃水指示仪;吃水指示器;通风计
draft in duplicate 双联汇票
drafting 意匠;作图;制图;牵曳;牵伸;起草;画草图;穿综
drafting ability 挽力
drafting accuracy 绘图精度
drafting board 绘图板
drafting chisel 琢边锤;琢石边的凿子
drafting data 绘图数据
drafting department 设计科
drafting device 绘图装置
drafting editing 清绘原图审校
drafting film 描图膜
drafting group of standard 标准编制组
drafting guide 分色样图
drafting ink 绘图墨(水)
drafting instrument 绘图仪器;制图仪器
drafting kit 盒装绘图仪
drafting line 取水量线
drafting machine 绘图机;自动制图机;制图机;分度规;平移角尺;平行运动分度规;平行尺
drafting office 制图室;绘图室
drafting pen 绘图笔
drafting pencil 绘图铅笔
drafting pit 试泵间
drafting room 绘图室;制图室;设计室
drafting scale 绘图比例(尺);曳引标度;制图尺
drafting set 盒式绘图仪;全套绘图仪
drafting standard 制图规程;制图标准
drafting stylus 刻尺
drafting system 绘图系统
drafting table 绘图桌
drafting technique 绘图方法
drafting template 制图模板
drafting yard 分隔牧场
draft intensity 通风强度
draft international standard 土样国际标准;国际制图标准;国际标准草案
draft issued 汇出汇票
draft-limited channel 吃水限制的水道
draft line 吃水(深度)线;牵引线
draft load 牵引力
draft lobby 阻风门廊;阻风穿堂
draft logout 汇票注销
draft loss 气流损失;通风损失(量);通风损耗(量)
draft machine 制图机;绘图机
draftman 制图人
draft mark 吃水(线)标志;水尺;船舶吃水尺;吃

水标(尺)
draft meter 吃水标(尺);水标尺
draft midship 船中(部)吃水
draft mo(u)lded 型吃水
draft mud 漂泥
draft note 草案
draft numerals 吃水线标志;吃水标(尺)
draft of construction 施工布置图
draft of the plan 计划草案
draft on demand 来取即付的汇票;见票即付票据;见票即付汇票;即付汇票
draft on sill 闸槛水深
draft or note payable to bearer 凭票即付持票人的汇票;凭票即付持票人
draft outline zoning plan 分区计划大纲草图
draft payable 待付汇款
draft payable at sight 见票即付汇票
draft payable with terms 有条件支付的汇票
draft pipe 吸出管;尾水管
draft pipe surge 尾水管涌浪
draft plan 计划草案;规划设想图
draft pond 抽水池
draft power 通风能力
draft preventer 阻风设施;阻风门廊
draftproof 无气密的;防风的;隔气的
draftproof barrier 阻风障体
draftproof door 阻风门;防风门;隔气门
draftproofing 阻(穿堂)风;防风
draftproof lobby 阻风门廊;隔气间
draft proposal 建议草案
draft provision 条款草案
draft rate 取水率;票汇汇率
draft recommendation of International Standardization Organization 国际标准推荐草案
draft regulation 法规草案
draft regulator 通风调节器;气流控制器;气压闸
draft report 非正式报告;报告草案;文件草稿
draft resolution 决议草案
draft responsible member 力调节器测量器;牵引力传感器
draft responsible system 力调节系统
draft retarder 通风减速板
draft roller 牵伸辊
draft room manual 制图手册
drafts and telegraphic transfers payable 应解汇款
draft scale 载重标尺
draftsman 起草人;拉拔工;描图员;绘图员;制图员;起草者
draftsman chief 绘图组长;制图组长
draftsmanship 绘图术;制图质量;制图(技)术
draftsman's scale 绘图员用比例尺
draftsman's set 盒式装图仪
draft standard 标准化标准草案;标准草案
draft standard for discussion 标准讨论稿
draft standard for examination 标准送审稿
draft stop 止火板;挡火物;升降止动器;拦火道;防吸风压条;防火门;顶棚防火板;挡火墙
draft stopping 安装通风挡板
draft survey 吃水检查
draft telpherage meterage 拟建架空索道长度
draft temperature 排风温度;通风温度;送风温度
draft test 通风试验
draft transmission line meterage 拟建输电线长度
draft transport pipeline meterage 拟建输送管道长度
draft tube 汲取管;吸入管;吸出管;牵引横梁;尾水管;通风管;牵引横柱;处于穿堂风状态
draft tube access gallery 尾水管进入廊道
draft tube aerator 导流式曝气器
draft tube bend 尾水管弯管段
draft tube bulkhead gate 检修闸门;平板闸门;尾水管闸门
draft tube channel barrier ditch 通风管渠式设障沟
draft tube deck 尾水管平台;尾水管盖板
draft tube diffuser 尾水管扩大段
draft tube efficiency 尾水管效率
draft tube elbow 尾水管肘管段
draft tube entrance 尾水管进口
draft tube exit 尾水管出口
draft tube floor 尾水管底板;尾水管层
draft tube gate 尾水管闸门;通风管闸门
draft tube gate chamber 尾水管闸门室
draft tube gate slot 尾水管闸门槽
draft tube inspection passage 尾水管检查通道

draft tube liner 尾水管里衬
draft tube loss 尾水管损失;吸出管损失
draft tube manhole 尾水管进入孔
draft tube mixer 抽吸式搅拌机
draft tube performance 尾水管性能
draft tube port 尾水管出口
draft tube pressure surge 尾水管压力脉动
draft tube roof 尾水管顶板
draft tube soffit 尾水管拱腹
draft tube support vane 汲取管支承板
draft tube surge 尾水管涌浪
draft tube wall 尾水管边墙
draft type 通风方式
draft without recourse 无追索权的汇票
draft with recourse 有追索权的汇票
drafty 通风的
drafty room 通风良好的房间
drag 拉引;拉拔;抗力;粘底拉裂;下模型;曳力;走锚;滞后;刮路器;船尾倾时首尾吃水差;牵引;缆绳拉海法;拖拉;拖集木;拖;扫海钢索;拖曳
drag acceleration 减速(度);阻力加速度;负加速度
dragaded cullet 水淬碎玻璃
drag anchor 流锚;海锚;浮锚
drag and drop 拖放
drag and suction device 耙吸装置
drag angle 制动角;耙角;拖曳角
drag antenna 下垂天线
drag arc 牵引弧
drag area 迎面力面积;阻力面积
drag arm 耙臂
drag auger 敞式自动装料螺旋
drag axis 阻力轴;阻力曲线
drag balance 流阻平衡
drag bar 连接杆;拉杆;牵引杆
drag bar conveyer 链板运输机
drag bit 磨削钻头;刮刀型钻头;割刀;锚;切削型钻头;翼状钻头;十字镐;刮刀式钻头
drag bit cutting 刮刀钻头钻进
drag bit drilling 刮刀钻头钻进
drag bit structure 刮刀钻头结构
drag boat 驳船;挖泥船;拖网(鱼)船;扫船;牵引船
drag-body flowmeter 阻力体流量计
drag bolt 拉紧螺栓
drag box 下型箱;下砂箱;牵引箱
drag brake 制动装置;减速板;棘轮闸;阻力闸
drag brake application 施行减速制动
drag brake system 阻力制动系统
drag braking 减速制动
drag bridle 带材张紧装置
drag broom 刮泥刀;刮路刷
drag bucket 拖斗;拉斗;反铲
drag-bucket sampler 拖筒式取样器;刮斗取样器
drag cable 拉绳;牵引绳;牵引(钢)索
drag calculation 阻力计算
drag carriage 拖运小车
drag center 阻力中心
drag chain 拉链;牵引链;拖链;刹车链
drag chain belt conveyer 拉链输送带
drag chain conveyer 刮板链式输送机;链(刮)式输送机;拉链输送机;链式运输机
drag chain cooler 拉链冷却机
drag-chisel bit 扁铲钻头
drag chute 刹车伞
drag classifier 拉曳分粒器;刮板式颗粒分级机;刮板分粒机;刮板分级机;耙式分级机
drag coefficient 曳力系数;阻力系数;水流力系数;阻力系数
drag coefficient value for circular cylinder 圆柱的阻力系数值
drag cone 制动锥;浮囊
drag conveyer 链动输送机;刮板(式)输送机;链刮式输送机;链板式输送机
drag-cup generator 拖杯式(测速)发电机
drag-cup induction machine 拖杯形感应电机
drag-cup induction motor 空心转子感应电动机
drag-cup motor 拖杯式电动机;拖杯形电动机
drag-cup tachogenerator 空心转子测速发电机;拖杯形转子测速发电机
drag-cup tachometer 拖杯式测速发电机
drag-cup type rotor motor 空心转子电动机
drag curve 阻力曲线
drag cut 连带回采
drag dampening device 阻力减摆器
drag data 阻力数据

drag dewaterer 刮板(式)脱水机
drag diagram 阻力图;拖力图
drag dipper 拖曳铲斗
drag direction 阻力方向
drag disk harrow 牵引式圆盘耙
drag distribution 阻力分布
drag down 下曳;下拉;挂低速挡;负摩擦(力)
drag due to shock wave 激波阻力
drag during brushing 刷涂时拉力
drag effect 曳引效应;牵制效应;拖曳效应
drag embedment anchor 埋入式锚
dragen beam 枪式小梁
drag energy 阻力能
drag experiment 阻力试验
drag factor 曳力系数
drag falls 拖缆
drag fault 拖曳断层
drag flask 下砂箱
drag-flight conveyer 刮板式输送机
drag-flight elevator 刮板升运器
drag flights 刮板式梯格
drag flow 有阻力流;沿压力方向的流动;正流
drag fold 牵引褶皱;拖曳褶皱
drag folds of fault 断盘牵引褶皱
drag force 空气阻力;迎面阻力;曳阻力;阻力;制动力;牵引力;拖(曳)力;拖(牵)力;(作用于水工建筑物的)水流力
drag (force) coefficient 阻力系数
drag-free satellite 无阻力卫星
drag friction 制动摩擦
drag gear 牵引装置
dragged 毛边的石块;被拉毛的
dragged finish 梳刷面饰
dragged form method 拉模法
dragged lubricant 带走的润滑油
dragged work 用钢齿饰石细工;钢齿琢石;石工修饰工作;钢齿加工过的石面
dragger 牵引机;拖网鱼船
dragging 摩擦;横丝错位;走锚;刮平;带走;打捞;打滑;不完全脱扣运动;牵引的;牵动;耙掘;拖动;扫海;扫床(指探测河床有无障碍物);拖曳
dragging alarm apparatus 走锚自动报警器
dragging anchor 拖锚
dragging beam 承托脊椽梁
dragging brake equipment 闸阻设备
dragging buoy 疏浚区浮标
dragging construction 拖拉施工
dragging-equipment detector 减速装置检测器
dragging method 拉法
dragging motion 携带运动
dragging movement 携带运动
dragging of anchors 走锚
dragging out 拖出舱角货;带出
dragging pin 牵引钩钉
dragging-shoe 曳板
dragging-slip 曳板
dragging tie 隅撑;角铁联系;角铁联结
dragging track 滑行轨道;滑木道;横木滑道;拖拉道
dragging-up 自动离职(俚语)
draggy sales 滞销
drag handling 操耙【疏】
drag harrow 宽齿耙;牵引耙;拖耙
drag head 耙吸式挖泥机掘头;(耙吸式挖泥船的)耙头
draghead adapter 耙头接头管
draghead cullet 水淬法制碎玻璃
draghead gantry 耙头吊架
draghead ladder 耙头架
draghead visor 耙头罩
draghead visor controller 耙头罩控制器
draghead wearing 耙头磨损
draghead with hydraulic jet (挖泥船的)带水力喷嘴的疏浚头
drag hinge 阻力铰;竖直铰链
drag hole 拖链孔
drag hook 拉钩
drag-in 带入
drag-in conveyer 拖曳输送机
drag increment 阻力增量;阻力增大
drag-irons 海底耙
drag keel 弧形龙骨
drag ladle 料勺;水淬
drag-ladled cullet 水淬(法)碎玻璃
drag level(l)ing course 整平层;刮平层

dragline 拉引线道;集材索;曳痕;拉索;系索;导索;切割波痕;绳斗电铲;索铲
dragline bucket 拉索戽斗;拉铲铲斗;吊铲抓斗;索铲铲斗
dragline bucket attachment 吊铲抓斗附件
dragline bucket boom 拉索铲斗吊杆
dragline bucket dredge(r) 牵索铲斗挖掘机
dragline bucket suspension 臂式拉索铲斗
dragline cable way 拉索挖掘索道
dragline cableway excavator 缆索拖铲挖土机;缆索式拖铲;吊斗挖掘机
dragline chain 牵引链
dragline conveyer 索铲挖土机;拉铲挖土机;链刮式输送机
dragline crane shovel 索斗铲
dragline dredge(r) 拉索挖泥船
dragline excavator 拉索(斗式)挖土机;拉铲挖掘机;拉铲铲斗;吊铲;(索斗)铲;索斗(式)挖土机;索斗(式)挖掘机
dragline go-devil 拉索清管
dragline machine 索铲
dragline pig 拉索清管
dragline scoop 牵索挖斗;索铲铲斗
dragline scraper 拉铲运载机;拉铲挖土机;拉铲挖掘机;吊铲;电耙;带式刮土机;拉铲(挖土机);索铲
dragline stripper 剥离电铲
dragline tower excavator 塔式挖铲;塔式拖铲挖土机
dragline track 牵出线【铁】
dragline-type shovel 拉索挖土机;索斗(式)挖土机;索斗(式)挖掘机
dragline yoke 拉索架;拉铲轭
drag link 拉杆;转向拉杆;牵引杆;偏心曲拐;拖链;双曲柄连杆机构
drag link ball 拉杆球
drag link ball nut 拉杆球形螺母
drag link ball seat 拉杆球座
drag link ball socket 拉杆球窝
drag link ball stud 拉杆球头柱螺栓
drag link bearing 拉杆支座
drag link body 拉杆体
drag link conveyer 刮板式输送器;刮板式传送器;链板式输送机;刮板式运输机;链刮式输送机
drag link dust cover 拉杆防尘罩
drag link dust cover pad 拉杆防尘盖垫
drag link dust seal 拉杆防尘盖
drag link end 拉杆端
drag link end plug 拉杆端塞
drag link front end grease fitting 转向纵拉杆前端油嘴
drag link grease retainer 拉杆护脂圈
drag link mechanism 拉杆机械装置;拉杆机构
drag link motion 快速退回运动
drag link safety plug 拉杆安全塞
drag link spring bumper 拉杆弹簧座
drag link spring seat 拉杆弹簧座
drag link steering rod 转向纵拉杆
drag loader shovel 牵引式装载机铲斗;手动刮土机
dragman 机铲工;刮路机铲工;拖网渔工;刮路机手;搬运工
drag mark 拖痕;水纹
dragmode 托运方式
drag model 阻力模型
drag net 拖网
drag of anchor 走锚
drag-off carriage 板料拖运小车
drag of film 薄膜的阻力
drag of kerf 切缝痕
drag of vessel 船尾纵倾度
dragoman 旅行向导
dragon 拨进;装甲曳引车;单舱游艇
dragon beam 支承脊橡梁;分角梁;(屋顶支持戗脊的)垫木
Dragon boat 龙舟
drag-on carriage 板料拖运小车
dragon design 龙纹
dragon-head gargoyle 螭首【建】
dragon-head main ridge ornament 正吻【建】
dragon-head ridge ornament 吻兽【建】
dragon piece 墙角挑檐;承托脊橡梁
dragon's blood 龙血红;红粉;麒麟血;清漆红颜料;龙血树脂
dragon skin glaze 龙皮釉
dragons pattern 和玺彩画

dragon spruce 云杉
dragon's tail 降交点
dragon's teeth 消能齿;消力齿;齿槛
dragon's tooth 消能墩;消力墩
dragon tanker 超大型油轮
dragon tie 角铁联系;支承脊橡梁;支撑脊缘梁;分角撑;隅撑
dragon wagon 牵引式货车
dragon-wreathed column 蟠龙柱
drag-out 带出液;被工件带出的酸洗液;酸洗废液
drag-over 横向自动拖送机
drag overall 总阻力
drag-over assembly 拖运机部件
drag-over mill 递色式轧机;二辊周期式薄板轧机
drag-over skid 转运装置
drag-over unit 移送机;拖运机
drag parameter 阻力参数
drag perturbation 阻力扰动
drag piece lost 掉剖刀片
drag pinch roll 拉紧咬送辊
drag planer 刮路机
drag plate 下型箱板
drag plate apron conveyer 链板裙式输送机
drag plough 牵引犁
drag plow 牵引犁
drag point 下型芯头
drag power 阻力动力;牵引力
drag rake 负角(切削);耙角
drag-reducing agent 减阻剂
drag reducting cowling 减阻罩
drag reduction 减阻
drag reduction agent 减阻剂
drag reduction method 减阻方法
drag resistance 空气阻力;车形阻力;牵引阻力;拖曳阻力
drag ring 摩擦环
drag rise 阻力增长
drag road-scraper 牵引式刮路机;牵引式铲运机
drag rod 拉杆;牵引杆
drag rope 牵(引)索;牵引绳;牵引钢丝绳;绳绳;拖拉钢丝绳
drag rotary detachable bit 折卸式刮刀钻头
drag sail 流锚;海锚;浮锚
drag saw 拉锯;狐尾锯;往复锯;横切锯
drag scraper 拖铲;牵引式铲运机;拉铲挖土机;拉铲挖掘机;刮运装置;刮削机;拖拉铲运机;拖拉铲土机;拖铲牵引式铲运机
drag scraper and loader 牵引式铲土和装载机
drag scraper bucket 牵引式铲斗;拖铲铲斗
drag scraper feeder 刮板给料器
drag scraper hoist 拖铲绞车;牵引式铲运起重机;扒矿绞车
drag scraper installation 牵引式铲运机装置
drag scraper machine 牵引式铲运机(械)
drag scraper with steeply inclined ramp 可作陡坡倾斜的牵引式铲运机
drag screen 刮板筛;耙式筛
drag seine 拖围网
drag-separating 依靠阻力分离
drag separation 阻力分离
drag sheet 流锚;海锚;浮锚
drag shoe 磨鞋;制动铁鞋;制动块
drag shovel 拉铲挖土机;拉铲挖掘机;反向(机械)铲;拖拉铲土机;拖拉铲运机;拖拉铲土机;拖铲挖土机;反铲(挖掘机);反(向)铲;无钢索内向挖土机
drag spot 浆疙瘩
drag spreader 牵引式碎石撒布机
drag spreader box 牵引式碎石撒布机箱
drag steering 拖物代舵
dragster 高速赛车
drag-stone mill 拖拉石滚式磨碎机;石滚磨;石辊式磨粉机
drag strut 阻力支柱;阻力撑杆
drag-suction dredge(r) 吸泥式挖泥船;耙吸(式)挖泥机;耙吸(式)挖泥船
drag-suction method 耙吸(挖泥法)
drag sweep 拖缆扫海法
drag technique 牵弧技术
dragtender 操耙手
drag theory 阻曳理论(设计混凝土路面钢筋的理论)
drag the sea 扫海
drag torque 曳力矩;拖曳转矩

drag-to-weight ratio (车辆的)拉力与重量之比;车辆拉力重量比
drag trencher 牵引挖沟机
drag truss 阻力桁架
drag turbine 摩擦涡轮机
drag twist 捞矛;扫孔器
drag-type bit 翼状钻头;刮刀(式)钻头;多刃(式)钻头
drag-type classifier 刮板式分选机;刮板式分粒机;刮板式分级机
drag-type feeder 刮板式给料机
drag-type scraper 拉索清管
drag-type tachometer 涡流转速计
drag unit 反张力装置;反拉装置
drag vector 阻力矢量
drag wedge 压紧楔;牵引楔
drag wire 迎面阻力张线;阻力张线;正面阻力张线
drain 沥干;枯竭;径流;耗用电流;引流物;注口;滴干;导液;放净口;排渣及放气;排污口;排污管;排水明沟;排空;排干;排放口;排除器;排除管;排除阀;外流;水沟
drainability 滤水能力;排水性;排水能力
drainability of soil 土壤排水能力
drainable 可排水的
drainable porosity 有效孔隙率;排水孔隙率
drainable sludge 可增稠污泥;可脱水污泥;易脱水污泥
drainable superheater 疏水型过热器
drainage 引流术;导液法;排油;排泄(指水、气等);排水装置;排水性;排水(法);疏水;疏干
drainage adit 排水廊道
drainage aggregate 排水集料;排水骨料
drainage and dewatering 排水与降水
drainage and filter 排水和倒滤层
drainage and irrigation 排灌
drainage and irrigation by mechanical electric(al) power 机电排灌
drainage and irrigation equipment 排灌设备
drainage and irrigation machinery 排灌机械
drainage and irrigation station 排灌站
drainage and waterproofing system of tunnel lining 隧道衬砌防排水图
drainage anomaly 水系异常
drainage appliance 排水设施
drainage area 流域(面积);集水面积;汇水面积;泄水面积;排水区域;排水流域;排水面积
drainage area boundary 流域边界
drainage area of well 井的排水面积
drainage area precipitation 流域降水
drainage article 排(雨)水用制品
drainage balance 排水平衡
drainage basin 流域(盆地);流域(排水盆地);集水池;排水盆地
drainage basin morphology 流域形态(学);流域地貌
drainage basin of groundwater 地下水流域
drainage basin topography 流域地形(学)
drainage basin yield 流域平均年径流量;流域产水量
drainage berm 坝趾滤水平台;排水棱柱体平台
drainage blanket 排水铺面;排水(垫)层
drainage board 雨水板;排水板;水系版
drainage borehole 放水孔
drainage bunker 泄水斗
drainage by air invasion 充气排水
drainage by compressed air 压缩空气排水
drainage by consolidation 固结排水
drainage by desiccation 疏干排水;疏干井水
drainage by drift 巷道排水
drainage by electroosmosis 电渗排水
drainage by filter wells 井点排水
drainage by gravity 重力排水
drainage by open channel 明渠排水
drainage by pumping station 泵站排水
drainage by sand piles 砂井排水
drainage by suction 吸收排水;负压排水
drainage by surcharge 超载排水;超载排水
drainage by vertical well 竖井排水
drainage by well points 井点排水
drainage by wells 井群排水
drainage canal 排水渠(道)
drainage canal excavated to aquifer 完整排水渠
drainage canal unexcavated to aquifer 非完整排水渠

drainage capacity 排水能力
drainage channel 河网;排水渠(道);排水沟渠;水系
drainage channel and pump for the maintenance of working condition 疏水系统(船坞)
drainage characteristic 流域特征(值);排水性能;排渗特征;水系特征
drainage cock 放水龙头;排水旋塞
drainage cock on strainer 过滤排水开关
drainage coefficient 径流系数;日排水量;排水系数;排放系数
drainage coil 分流线圈
drainage collector 排放物收集器;污水集水管;排水干管
drainage condition 排水条件
drainage condition of groundwater 地下水排泄条件
drainage conduit 排水渠(道);排水管(道)
drainage connection 排液接头;排液接管
drainage controller 疏水(调节)器
drainage correction 流出体积修正
drainage cost 排污费;排水费
drainage course 排水层
drainage crossings 交叉排水渠;交叉排水管
drainage culvert 排水阴沟;排水涵洞;疏水隧洞
drainage curtain 排水幕;竖向排水系统(混凝土坝体)
drainage cushion 排水垫层
drainage density 河网密度;排水密度;水系密度
drainage design 排水设计
drainage device 排水装置
drainage dip 排水浅洼
drainage discharge channel 分流渠
drainage distribution pattern 河网分布形式;水系分布形式
drainage district 排水系统管理机构;排水区(域)
drainage ditch 排水明沟;排水沟(渠);截流沟
drainage divide (line) 分水岭;流域分界线;分水界;分水脊;排水边线;地面分水岭;地面分水界
drainage drawing 水系原图
drainage drift 排水坑道
drainage drop 排水沟落差
drainage duct 排水管
drainage elevator 脱水提升机
drainage engineering 排水工程(学)
drainage equilibrium 排水平衡
drainage equipment 排水设备
drainage equipment capacity 排水设备能力
drainage excavation 排水开挖
drainage face 排水表面
drainage facility 排水设施
drainage facility of port area 港区排水设施
drainage facility 排水设备
drainage factor 排水因数;排水系数
drainage feature 水系地物
drainage fee 排污税
drainage field ditch 田间排水沟
drainage fill 排水沟填土;排水暗沟
drainage filter 排水滤体;排水滤层
drainage fitting 铸铁排水管零件;排水装备
drainage fixture unit 设备排水能力测定单位
drainage flow 冷泄水
drainage funnel 排水漏斗
drainage gallery 排水巷道;排水平峒;排水廊道;疏干巷道
drainage gang 沟工队
drainage ga(u)ge 渗漏计
drainage goods 雨水制品;排水制品
drainage grate 排水箅栅;排水格栅
drainage grating 拦污栅;雨水箅;排水箅;地面废水箅
drainage grid 排水格栅
drainage grommet 漏水垫圈
drainage ground water 排除地下水
drainage gutter 排水边沟
drainage hat 污水阱;污水井
drainage head 排水水头;排水区最高点;排水源头;排水区内最远或最高的地点
drainage hole 泄水孔;排水孔;排水口
drainage hole of foundation 地基排水孔
drainage in borehole 井下放水
drainage indicator 排水指示器;渗透仪
drainage information system 排水信息系统
drainage inlet 排水入口
drainage in open 明沟排水

drainage lake 出流湖;排水湖
drainage lateral 横向排水沟
drainage layer 滤层;排水层
drainage level 地下排水道;排水平巷
drainage line 雨水沟道;排水沟道;地下水沟道;流域界线;排水(干)线
drainage loading 排水加荷
drainage location 排污点
drainage lock 堤闸
drainage machine 排水机
drainage machinery 排水机械
drainage map 流域图;水系图
drainage method 水系法
drainage modulus 日径流深;排水模数;排水模量
drainage net(work) 河网;水系(网);雨水网;排水(管)网;排水(沟)网;地下水管网
drainage of a building 房屋的雨水排除
drainage of coal mines 煤矿排水
drainage of coal pits 煤矿矿井排水
drainage of exploratory tunnel 勘探坑道排水
drainage of foundation 基础排水;地基排水
drainage of leak 渗水排除
drainage of mine works 井巷排水
drainage of seepage 渗漏水排除
drainage of slip 回浆;排浆
drainage of the ground 地面排水
drainage of tunnel 隧道排水
drainage opening 泄水孔;排水孔
drainage outlet 排水出口
drainage party 沟工队
drainage passage 排水道
drainage path 排水路径
drainage path for excess pore water 超量孔隙排水通道
drainage pattern 河网形式;格状水系;排水系统类型;排水方式;水系形状;水系模式;水系(类)型;水系格式;水系布局
drainage perimeter 分水岭周长
drainage pipe 雨水管;排水管(道);放泄管;排污管
drainage pipe elbow 雨水管弯头
drainage pipe fitting 雨水管件
drainage pipe gallery 排水管坑道;排水管廊
drainage pipe line 排水管线;泄水管线;排水管(道)
drainage pipe system 雨水管系统;排水地下管系(统);排水管网
drainage pipework 雨水管网;排水(管)网;排水沟网;地下水管网
drainage piping 排水管网;排水沟网;排水系(统);排水管路;排水管垫层
drainage pit 排水井;排水坑
drainage plant 排水设备
drainage plate 水系版
drainage plow 开排水沟犁
drainage plug 放水塞;排泄塞
drainage point 排污点
drainage post 排流柱
drainage pot 污水阱;污水井
drainage pressure 排水压力
drainage product 雨水制品;排水制品
drainage product of concrete 混凝土排水构件
drainage project 排水计划;排水工程
drainage pump 排泄泵;排水泵;吸油泵;抽油泵;抽水泵;疏水泵
drainage pumping station 排水(泵)站
drainage quota 排水定额
drainage radius 供油半径;排油半径;排水半径
drainage radius of an oil well 油井的排水半径
drainage rate 析水速率;排污率
drainage ratio 径流系数
drainage receiver 冷凝罐
drainage recovery 排出水回收
drainage region 排水流域
drainage requirement 排水定额
drainage right of ways 排水用地;排水系统用地范围
drainage rights 排水权
drainage riser 排水立管
drainage sand drains 砂井排水
drainage sand mat 排水砂垫层
drainage scheme 排水系统方案;排水纲;水网
drainage screen 脱水筛
drainage sediment anomaly (stream sediment anomaly) 水系沉积物异常

drainage separation 水系版
drainage sewage canal 排污沟渠
drainage sewage reservoir 排污水库
drainage shaft 排水井
drainage shaft 泄水井;排水井;排水竖井
drainage slope 排水坡(度)
drainage slope fill 屋面排水坡找平层
drainage slope of quay area pavement 码头路面排水坡度
drainage sluice 排水闸(门)
drainage standard of industrial waste water 工业废水排放标准
drainage standard of radioactive waste water 放射性废水排放标准
drainage station 泵站;排水站
drainage stratum 排水层
drainage structure 排水建筑物;排水结构;排水构筑物
drainage sump 排水井;排水集水坑;泄水阱
drainage surface 降水面;排水表面
drainage surface water 排除地表水
drainage survey 水系调查;水系测量
drainage system 排水装置;下水道系统;河网;水网;排水(管)系统;排水管路;水系
drainage system design 排水系统设计
drainage tank 泄水柜;进水井
drainage team 沟工队
drainage terrace 排水阶地
drainage thread 挂线;排液线
drainage tile 排水瓦管
drainage time 析液时间;排水时间
drainage tool 排水工具
drainage to surface water 排入地表水体
drainage tray 排水槽
drainage trench 排水沟
drainage trench digger 排水沟挖掘机
drainage tube 引流管;排水管(道)
drainage tunnel 排水隧洞;排水隧道;排水廊道;排水暗沟
drainage-type lysimeter 排水式土壤蒸发仪
drainage valve 放水阀;排污阀;排水阀(门)
drainage vessel 排水容器
drainage ware 排水制品
drainage water 污水;排出水
drainage water quality 排放水质
drainage water quality standard 排放水质标准
drainage way 排水道
drainage well 污水阱;污水井;泄水井;残水井;排水(砂)井;渗水井
drainage wind 重力风
drainage with ditch 渠排
drainage within tunnel 隧道内部排水
drainage with well 井排水
drainage works 排水系统;排污工程;排水构筑物;排水工程(设施)
drain air space ratio 排水气隙比
drain area 排水面积
drain away 流尽
drain ball-valve 泄水球阀
drain basin 排水池
drain bin 排水斗
drain blockage 排水用的小方石
drain board 流水板;滴水板;排水板;泄水板
drain bolt 泄水孔塞
drain boring 排水钻孔
drain box 排水箱
drain break down 漏极击穿
drain bushing 放料漏板
drain cap 漏盘
drain cashing 空心注浆
drain chute 排水陡槽;溜槽;冲槽;排水斜管
drain cistern 泄水柜
drain cleaner 排水沟清理器;排水管清洗机
drain cock 旋塞;泄水孔塞;泄水管门;泄水阀(门);泄放塞;放水旋塞;排液旋塞;排气旋塞
drain cock body 泄水塞体
drain collector 排水干管
drain conductance 漏极电导率
drain conductivity 漏(极)电导率
drain conduit 排水管;出水渠;排水暗渠
drain connection 排泄管道;排水(连)接管;排水管接头
drain connector 排水管接头
drain contact 漏极接点

drain cooler 泄水冷却器
drain cover 渠盖;沟渠盖;沟盖(板)
drain cover slab 水沟盖板
drain cup 放泄漏斗
drain current 漏(极)电流
drain depth 排水管深度;排水沟深度
drain dias 漏偏置
drain diffusion 玻姆扩散
drain digger 排水沟挖掘机
drain discharge 排水(流)量
drain district 排泄区
drain ditch 泄水沟;排水沟
draindrawlayer 排水管铺设机;排水管铺管机
drain drop 排水落差
drained 已排水的;排水的
drained angle of internal friction 排水内摩擦角
drained bulk weight density of soil 土的水上重度
drained cavity 排水空腔
drained chemical board 排水化学板
drained field 排干田
drained glazing 有排水通道的玻璃窗
drained ground base 排水地基;干燥地基
drained hole 排水孔
drained joint 明沟排水;泄水节点;排水接头;排水接缝
drained load 排水加荷
drained load cycle 排水荷载循环
drained product 排出物
drained product collector 排出物收集器
drained repeated direct shear test 排水反复直剪试验
drained rod 通沟条
drained shear strength 排水剪切强度
drained shear test 慢剪试验;排水剪切试验;排水剪力试验;土壤慢剪试验;渗透剪力试验
drained test 排水(剪切)试验;排水(剪力)试验
drained triaxial test 排水三轴试验;渗透三轴试验
drain elbow 排水弯管
drain electrode 漏极
drain equipment 排水设备
drainer 滤干器;冷凝罐;泄水器;洗浆池;储[贮]浆池;放泄器;放泄机;排水工
drain ferret 排水管探测器
drain field 排放场
drain flute 排水槽
drain for rain water 雨水排水口
drain function 排污功能
drain gate 泄水门;排水闸(门)
drain grating 挡污栅;排泄孔;排水沟;排泄槽;排污栅;排水沟栅
drain gully 落水口
drain gutter 排水边坡;檐沟;天沟
drain hat 舣水井
drain header 排水集管
drain hole 排水孔;放玻璃水板;泄水孔;放水孔;放料口;放玻璃水口;排污孔
drainhole drilling 排水孔钻进
drain hose 排水软管
draining 放玻璃水;倾浆;排泄(指水、气等);排水作业;排去
draining adit 排水平洞
draining and flooding tank 浮沉水舱(潜艇)
draining aperture 排放孔
draining area 流域;排水面积
draining arrangement 排水布置
draining basin 流域(盆地)
draining blanket 排水面层
draining board 泄水板;槽式倾斜排水板;滴水板
draining borehole 放水钻孔
draining canal 排水渠(道)
draining channel 排水渠(道);排水沟
draining cock 放泄旋塞;泄水旋塞
draining conduit 排水渠(道)
draining ditch 排水沟
draining effect 穿流效应
draining equipment 泄水设备
draining funnel 放液漏斗
draining gang 排水工程队
draining grate 排水箅
draining gutter 排水沟
draining hole 放玻璃孔;排水孔
draining installation 泄水设备
draining line 排水沟道
draining modulus 排水模数

draining of foundation 地基排水
draining off the water 泄水
draining of the basin 池子放空
draining opening 泄水口
draining or intercepting flood water 排出或截留洪水
draining outfall 排水渠(道)
draining pipe 排水管
draining pipe elbow 排水管弯头
draining pipeline 排水管道;放水管线
draining pipe system 排水管道系统
draining plough 开排水沟犁;排水沟犁
draining plow 开排水沟犁
draining plug 泄水塞;排水塞
draining point 排水点;进水点
draining pump 排水泵
draining screen 脱水筛;排水滤网
draining shaft 排水竖井
draining sieve bend 脱水弧形筛
draining spade 排水用铲
draining sucker 排水吸附器
draining system 排水网
draining system diagram 排水系统示意图
draining tank 排液池;放玻璃液池
draining tap 泄水龙头
draining test of brush 涂刷黏度试验
draining time 排水时间
draining transformer 抗流变压器
draining valve 排水阀(门)
draining water-logged 排涝
draining well 排水井
draining well points 井点排水
drain interval 换油期限
drain junction 漏结
drain layer 排水管铺设机;排水管铺管机
drainless lake 无出流湖
drain line 放条纹;排水管路;排水管理线
drain logging 漏记
drain main 排水总管
drain manifold 泄水总管;排水总管
drain of boiler 锅炉排污
drain of capital 资金外流
drain off 流出;吸出;放空;排除;排出;流干
drain of foreign exchange 外汇外流
drain of reserve 储备外流(指外汇)
drain oil 放油孔
drain oil recovery equipment 废油回收装置
drain on cash 现金消耗;现金外流;现金枯竭
drain on manpower and material resources 人力物力的消耗
drain on supplies 脱销
drain opening 排水出口;泄水口
drain outlet 排污通道
drain output 漏极输出
drain pan 泄油箱;泄水盘;油底壳;滴水盘;放油盘;排水盘
drain period 换油周期;排水周期
drain pile 排水砂桩;砂井
drain pin 泄水孔塞
drain pipe 管沟;排水管(道);疏水管;泄水管
drain pipe line allowing infiltration 允许渗透的排水管线
drain pit 排水坑;水窝;水仓
drain plate 排水板
drain plug 泄水(孔)塞;放水塞;放水塞;排水塞;排泄塞;排污螺塞;排水管栓
drain plug for oil sump 油池放油塞
drain plug gasket 螺塞垫圈
drain port 污水井;排水口;排污孔
drain pot 舣水井
drain pressure 放出压力
drain pump 排污泵
drain region 漏区
drain resistance 漏电阻
drain rod 排水管疏通软杆
drain sample 排出的试样
drain sentinel 阴沟标志;阴沟标记
drain separator 卸料刮板;排液分离器;脱水器
drain sewer 排水阴沟
drain shoe 带承口排水管靴
drain silting 排水管淤积
drains in aeration tank 曝气池排空管
drain sleeve 冷凝管;排出套管
drain-source resistance 漏极源极电阻

drain spacing 排水(管)间隔;排水间距
drain spout 落水管;雨水管
drain stem 放泄杆
drain stopper 泄水塞;排水塞;排水管栓
drain-substrate junction 漏衬底结
drain sump 排水坑;脱水罐
drain system 疏水装置;排水装置
drain tank 疏水槽
drain tap 放水龙头
drain terminal 漏引出线;漏端子
drain test 排水系统检漏试验
drain testing 排放试验;排水管试验
drain the oil 放油
drain the transmission 从齿轮箱内放油
drain tile 排水陶管;排水瓦管
drain tile line 排水管线;管道
drain to reduce pressure 排水减压
drain trap 聚水器;存水弯(管);沉污井;沉淀池;放泄弯管;放水弯管;防臭弯管;排水器;排水防气瓣;疏水箱;疏水罐
drain trench 排水沟
drain-trench spade 排水沟铲
drain trough 疏水槽
drain trunk 溢流道;排水干管
drain tub 雨水承受器
drain tunnel 排水坑道;排水廊道
drain valve 放油阀;放泄阀;排污阀;排水阀(门)
drain valve ball 泄阀球
drain valve plug 放泄阀塞
drain voltage 漏电压
drain water 废水;排出废水;排出的水
drain way 排沟;泄水路
drain wire 排扰线;漏电引出线;加蔽线
drain yield 排水量
drain-zone chimney 土坝中央排水道
drake 浮石片
drake device 浮标式指示器;漂浮式指示器
Drake passage 德雷克海峡
Drake seaway 德雷克海路
dram 打兰(重量单位)
dramamine 乘晕宁
drama theater 话剧院
dramatic 显著的
dramatic decline 明显下降
dramatic growth 迅速增长
dramatic lighting 舞台照明
dram funnel 放液漏斗
dram shop 小酒店;小酒吧;酒吧间
drapabiltiy 铺覆性
drape 悬挂;隔音;盖层;覆盖;起皱纹;披盖
drapeable fabric 轻软的悬挂织物
draped 用帷装饰的
draped round rubber fender 横挂垂悬式防撞橡胶护舷
draped tendon 弓形受力筋;弯受力筋;偏心钢筋束
draped tension 偏心钢筋束
drape fold 披盖褶皱
drape forming 区域成型
drape mode 披盖模式
drape mo(u)ld 罩布模具
drape mo(u)lding 覆布模型
draper 带式输送器;布商;布面清选机
draper apron 帆布输送带
draper canvas 帆布输送带
Draper catalog 德雷伯星表
Draper Catalogue 德雷伯星表
draper cleaner 绒布选种机
draper conveyer 帆布输送带
Draper effect 德雷伯效应
draperies 帷幕
draper stop 输送带停转机构
draper-type 带式
draper-type pickup 带式检拾器
drapery 毛织物;装饰性窗帘
drapery hardware 帷幕五金
drapery panel 褶布饰镶板
drapery track 帘幕滑杆;帘幕滑竿
draping 悬挂;帷幕;吸声材料;吸声材料;隔音材料;隔声帘;披盖
draping deposit 披盖沉积
draping property 悬垂性(能)
drastic 烈性;激烈的
drastically reduce budget expenditures 紧缩预算支出

drastic change 急剧变化
drastic crack 深裂
drastic extraction 深度抽提
drastic reduction 大幅度削减
drastic variation 急剧变化
draughly workings 通风巷道
draught 一吊货;穿堂风;抽力;吃水(深度);草稿;草案;图案
draught aft 船尾吃水(深)
draught air 通风气流
draught allowance 拔模余量
draught animal 力畜;耕畜
draught area 通风面积
draught at midship 船中(部)吃水
draught bar 拉杆;牵引杆
draught bead 窗扫风条;窗挡风条;止水条
draught box 窗扫匣
draught capacity 通风能力
draught center 阻力中心
draught chamber 通风室
draught damper 气流闸;气流阀;通风阀
draught deflector 挡风器
draught distributing box 分风箱
draught diverter 排烟式折流板
draughted stone 琢边块石
draught-engine 排水机
draughter 制图者;描图器;制图机械
draught excluder 阻挡气流的门廊;阻挡气流的穿堂;气幕;气帘;封闭阀;门槛嵌缝条;防风条;气流隔断;排风器
draught excluder strip 密封条
draught excluding threshold 气流阻挡阀
draught exclusion 排除穿堂风
draught exclusion device 防风措施
draught fan 引风机;风机;通风风扇
draught fillet 防风条
draught flue 排气烟道
draught forward 船首吃水
draught-free 无气流;无风;无穿堂风
draught fully laden 满载吃水
draught-furnace 通风炉
draught ga(u)ge 抽力计;吃水指示仪;吃水指示器;差式压力计;差式压力表;差示压力计;差示压力表;风力计;通风计
draught head 吸水水头;吸出水头;吸出落差;挡风条;气压差;通风压头
draught height 吸出高度
draught hole 通风孔
draught hood 烟橱通风罩;通风罩;通风柜
draught horse 挽车马
draught indicator 船舶吃水测示仪;差式压力计;差式压力表;通风计
draughtiness 通风
draughting 绘制;制图
draughting accuracy 绘图精度
draughting board 制图板
draughting desk 绘图桌
draughting machine 制图机
draughting scale 制图室;曳引标度;绘图室;绘图比例尺
draughting up 鼓风增热
draughtless 无气流;无穿堂风
draught line 吃水深度线
draught link anchorage 拉杆固定点
draught lobby 阻风门廊;阻风穿堂;气流隔断
draught loss 压力损失;气流损失;通风损耗(量)
draught machine 制图机;绘图机
draught margin 通风裕量
draught mark 吃水深度标记;吃水标志;吃水标(尺)
draught on demand 来算即付的汇票
draught penetration 透风
draught per pass 每道压缩量;每道次压下量
draught pipe 通风管
draught power 畜力
draught pressure 压缩力;轧制(压)力
draught preventer 阻风设施;阻风门廊;气流隔断
draught-proof 无气流的;防风的
draught-proof barrier 风障
draught-proof door 阻风门;防风门
draught-proofing 阻穿堂风;防风
draught-proofing channel 防风槽
draught-proofing device 防风措施
draught protocol 议定书草案

draught regulator 气流调节器;抽力调节器;牵引力调节器;气压调节阀;通风调节器
draught retarder 通风阻流器
draught rod 窗扫棍
draught screen 风口挡
draught sensing link 自动调力拉杆
draughtsman 描图员;制图员;起草者
draughtsmanship 制图质量;制图(技)术;绘图术
draughtsmen's protractor 绘图量角器
draught spring 拉力弹簧
draught stabilizer 稳定通风板;稳定装置;气流稳定器
draught stop 挡火物;风挡;防风槽;拦火道;防火门;挡火墙;挡风物
draught strip 牵引条;挡风条
draught stripping 装挡风条
draught survey 水尺检验;水尺检量
draught tube 吸引管;引流管;导管;通风管
draught tube agitator 导管式搅拌装置
draught tube gate 通风管闸门
draught tube tunnel 通风坑道;管坑道
draught ventilation 自然通风
draughty 有过穿堂风的;通风(良好)的
Dravidian style 德维达风格
dravite 镁电气石
draw 拉延;拖拽;开立票据;汲(取);描绘;干涸河道;抽取;放落悬煤;放矿;蠕变影响;牵引;牵曳
drawability 拉深性;可拉(制)性;可拉(伸)性;回火性;压延性能
draw a bill 开发票;出票
draw a blank 打废井
drawable 拉出式
drawable currency 可提款货币
draw a charge 出炉
draw a conclusion 作出结论;推断
draw a design 打样
draw a money-order 开汇票
draw and buffer gear 缓冲车钩
draw a plan 草拟计划
draw arm 漏模架;起模架
draw a splice 解开插接
draw at scale 按比例尺制图;按比例尺描绘
drawback 退款
drawback collet 弹性夹头
drawback device 夹紧装置
drawback (for duties paid) 退税;退回已付税金;退还已缴税款
drawback lock 手拉锁;内开锁(内用手外用钥匙开的锁)
drawback mo(u)ld 对开金属型
drawback piston 回程活塞
draw back plate 推拉板
drawback rod 后拉杆
drawback spring 夹紧弹簧;回复弹簧;回动弹簧;回程游丝
drawback temperature 回火温度
draw bail 连接结构
draw bar 引砖;牵引砖;拉杆;沉砖;牵引杆
drawbar ball joint 牵引架球头关节
drawbar ball joint socket 牵引架球头关套节
drawbar box 拉杆箱
drawbar bubble 引砖(产生的)气泡;槽子气泡
drawbar capacity 挂钩能力;牵引杆受拉能力
drawbar carrier 拉杆托板
drawbar contour 引砖刻面
drawbar coupling 牵引装置
drawbar cradle 牵引杆托架
drawbar dynamometer 拉力表
drawbar efficiency 牵引动力装置效率
drawbar eye 拉杆眼
drawbar frame 拉杆箱
drawbar grader 拉杆平地机
drawbar guidance 拉杆导板
drawbar guide 拉杆导板;牵引杆导板
drawbar guide sleeve 拉杆导套
drawbar hook 拖钩
drawbar (horse) power 拉杆马力;牵引功率;牵引马力;挽钩牵引马力
drawbar jack 主牵引梁千斤顶
drawbar load 挂钩拉力
drawbar load sensing mechanism 牵引力传感器
drawbar machine 拉杆机
drawbar performance 牵引性能;牵引特性
drawbar plate 牵引杆板

drawbar plate brace 拉杆撑板
drawbar pocket 拉杆匣
drawbar profile 引砖刻面
drawbar pull 拉杆拉力;揽钩牵引力;拉杆牵引力;牵引杆拉力;挽钩牵引力;拖钩牵引力
drawbar pull-weight ratio 牵引杆拉力重量比
drawbar pump 拉杆泵
drawbar rib 引砖加强筋
drawbar ring 联结环
drawbar spiral coupling sleeve 牵引杆螺旋连接套管
drawbar spiral sleeve 牵引杆螺旋套管
drawbar spring 牵引杆弹簧
drawbar test 牵引试验
drawbar trailer 挂钩牵引拖车;拉杆拖车
drawbar yoke 拉杆轭
draw base line 画基线
draw bead 拉深压边筋;拉道;拉延筋
drawbeam 绞盘机
drawbench 拉拔(工作)台;拔管机;拉丝机;冷拔机;拉床
drawbench bed 拉拔机的机座
drawbench chain 拉拔机链条
draw-bit 松紧针钩
drawboard 盖缝板;带销的榫槽结合
draw bolt 小插销;调节松紧的销钉;紧固螺栓;接合螺栓;牵引螺栓;套环螺栓
drawbolt lock 内开锁(内用把手、外用钥匙开的锁)
drawbore 销孔钉;榫销孔(洞);钻销孔
drawbore pin 榫销;钻孔销
draw box 牵伸箱
draw bridge 开启桥;开合桥;开合板;活动桥;旋转吊桥;吊桥;铰链式仰开桥
drawbridge coupler 开合桥连接器
drawbridge inset of slope 斜井吊桥调车场
draw bucket 吊桶
draw by lot 抽签
draw by sight 草绘
draw chisel 板正錾
draw cock 小型排水阀;小型排气阀;小旋塞;排气栓
draw collet 拉杆式弹簧夹头
draw cord 信号绳
draw course 分离横列
draw curtain 拉帘;拉幕;扯帘
draw cut 拉切;拉剪(用张力卷筒);回程切削
drawcut type keyseater 键槽拉床
draw direction 提拉方向
draw door weir 垂直闸门堰;直升闸门堰;竖直闸门堰
drawdown 泄降;消落;下降;曳下;刮涂(膜);抽水降深;(水库的)水位泄降;水位下降;水位坡降;水位降深;水面下降;压力降;提取资金;水面降落
drawdown angle 陷落角;自卸角;放矿角
drawdown area 消落区
drawdown at inflection point 拐点处的降深
drawdown bar 涂膜涂布器;湿膜刮涂器
drawdown blade 涂膜涂布器;湿膜刮涂器
drawdown bottom pressure curve 井底压降曲线
drawdown card 刮涂卡片
drawdown cone 地下水下降漏斗;降落漏斗
drawdown cone of groundwater 地下水下降圆锥体
drawdown curve 地下水降落曲线;跌水曲线;浸润曲线;降落曲线;泄降曲线;抽水降深曲线;水位降落曲线;压降曲线
drawdown curve of groundwater 地下水下降漏斗曲线;地下水(位)降落曲线
drawdown curve of water-level 水位下降曲线
drawdown curve of water-stage 水位下降曲线
drawdown-discharge curve 泄降水位流量曲线
drawdown-distance curve 降深—距离曲线
drawdown-distance graphic(al) method 降深—距离图解法
drawdown-distance superimposed line method 降深—距离配线法
drawdown elevation 曳下高程
drawdown exploration 测液面勘探(法)
drawdown land 消落区(土地)
drawdown level 下降水平面;下降标高;消落水位(水库);动水位
drawdown of hole 钻孔水位降低
drawdown of level 水位消退;水位消落

drawdown of reservoir 水库泄降;水库消落
drawdown of water-level 水位消退;水位消落
drawdown of water level in observation well 观测孔的水位降深
drawdown of water level in pumping well 抽水孔的水位降深
drawdown of well 钻井水位降低
drawdown operation 下降操作
drawdown period 贷款支用期
drawdown period of reservoir 水库消落期
drawdown ratio 降落比;牵伸比;缩小比(例);水位下降比
drawdown rod 刮涂棒
drawdown stability chart 降落稳定图
drawdown surge 泄降涌浪
drawdown test 降压测试;降低液面试井法;涂料刮涂试验;骤降试验;抽水试验
drawdown the curtain 闭幕;放下窗帘
drawdown-time and distance curve 降深—时间距离曲线
drawdown-time and distance graphic(al) method 降深—时间距离图解法
drawdown-time and distance superimposed line method 降深—时间距离配线法
drawdown-time curve 降深—时间曲线;水位降深历时曲线
drawdown-time graphic(al) method 降深—时间图解法
drawdown-time superimposed line method 降深—时间配线法
drawdown value of matching point 配合点的降深值
drawdown water level 下降水平面
drawdown zone 消落区
draw draft 开汇票;出票
draw drapery 扯帘
draw drum 放线盘;牵引卷筒
draw edge radius 拉延凹模圆角半径
drawee 支付人;支付汇票人;(汇票的)付款人;票据付款人;受票人
drawee bank 付款行;汇票付款银行;票据付款行
drawee-company 汇票付款公司
drawer 拉装工;开(汇)票人;制图人;出票人;抽屉(柜);抽出;拔取工具
drawer/compiler 制图者
drawer dovetail 抽屉(燕尾)榫;互搭鸠尾榫
drawer-front dovetail 互搭鸠尾榫
drawer kicker 抽屉挡块
drawer knob 抽屉拉手
drawer lock 抽屉锁
drawer lock chisel 抽提锁凿子
drawer of a bill 出票人
drawer-paper 以自己为收款人的汇票发票人
drawer pull 抽提拉手
drawer roller 牵引滑轮;抽屉滚柱
drawer runner 抽提滑道;抽屉滑条;抽屉滑道
drawer shutter 瓣式快门
drawer slide 抽提滑轨;抽屉滑板;抽屉滑板
drawer slip 抽屉滑带
drawer stop 抽屉挡
drawer test 抽屉试验
drawer tuner 抽屉式频道选择器
drawer type drying stove 抽屉式烘干炉
drawer type shutter 闸门式快门
draw feature code 绘图特征码
draw filing 磨锉法
draw-flo(a)tation method 提拉悬浮法
draw for fixation 安装图
drawform 抽去式模板(用于浇灌特种饰面混凝土)
draw frame 并条机;起模框架;拉伸机
draw gate 推拉门;运河中水闸;闸门
draw gear 压延装置;车钩;牵引装置
drawgear length 牵引装置长度
drawgear spring plate 拉绳机弹簧板
drawhead 牵杆;拉床(机)机头
draw hoist 卷扬机
drawhole 拉拔模孔;拉模(孔);放矿口
draw-hook 拉钩;牵引钩
draw-hook bar 牵引钩装置
draw-hook pin 牵引钩销
draw-in 水位消退;水位消落;拉(缩)回;港湾式(停)车站;道路的港湾式停车站
draw-in air 引入空气
draw-in attachment 铲齿装置;卡套

draw-in bar 内拉杆
draw-in bolt 拉紧螺栓
draw-in chuck 弹簧夹头;内拉簧卡盘;内拉簧夹盘
draw-in collet 内拉簧套圈
draw inference 举一反三
drawing 拉拔;开发票;绘制;绘图;绘画;汇总图;抽(签);冲压成型;吃饱风;并条;拔丝;安装图;牵伸;脱箱;图样;图形;图画;提存;收回;拉制
drawing account 提款账目;业主提款账户;提款账;提存账户
drawing advice 票据委托书;提款通知
drawing appliance 绘图设备;绘图器具;绘图用具
drawing appliances and instruments 绘图工具和绘图仪器
drawing arm 绘图臂
drawing attachment 绘图附件
drawing-back 回火
drawing-back of pillars 后退式回采矿柱
drawing bar 引砖
drawing base 绘图片;绘图板
drawing basin 引上池
drawing belt conveyer 牵引式皮带运输机;卸料皮带输送机
drawing bench 拉床;制图桌
drawing block 拉丝卷筒;拉丝模板
drawing-out frame 拉丝模框架
drawing board 拉丝模板;绘图板;画板;制图板
drawing breakage 拉裂
drawing bristol 制图用光泽纸板
drawing by belt 皮带牵引
drawing by shortfall 短缺提款界限
drawing capstan 张紧卷筒
drawing card 裱糊板
drawing chamber 拉引室;引上室;成型室
drawing change 更改图纸
drawing change notice 图纸更改通知
drawing change request 更改图纸的要求
drawing change summary 图纸更改一览
drawing channel 引上通路
drawing chute 放矿漏斗
drawing cloth 描图布
drawing coefficient 拉延系数
drawing compasses 绘图用两脚规;绘图圆规;制图圆规
drawing compound 拉拔用润滑剂;拉拔用乳剂;金属拉丝用润滑剂
drawing curve 曲线板
drawing data 绘图数据
drawing date 编测时间
drawing datum 绘图基面
drawing deformation for a pass 拉拔道次变形量
drawing design 图纸设计
drawing desk 描绘台;绘图桌;制图桌
drawing detail pencil 打样铅笔
drawing device 制图设备;控制设备;绘图装置
drawing die 深冲模;拉(拔)模;拉延冲模
drawing die orifice 拔丝模孔口
drawing digitizer 图形数字化器
drawing door weir 垂直闸门(围)堰
drawing-down 锻延
draw-in gear 平面卡盘
drawing edge radius 拉延凹模圆角半径
drawing effect 回火作用
drawing enamel with copper body 铜胎画珐琅
drawing equipment 绘图装置;绘图机;制图仪
drawing equipment and material 绘图设备机材料
drawing eraser 绘图橡皮
drawing error 描画误差;绘图误差
drawing fan 抽风机
drawing film 拉延润滑膜
drawing fires 抽火
drawing floor 卸料平台;划线工作台
drawing for fixation 安装图
drawing for order 订货图纸
drawing from a bushing 漏板拉丝
drawing from bins 从料斗中卸料
drawing from rods 棒法拉丝
drawing from silos 从筒仓中卸料
drawing furnace 回火炉;退火炉
drawing gap 拉延间隙
drawing gear 平面卡盘
drawing grease 拉延用润滑脂;拉拔润滑脂
drawing grid 绘图网格;绘图格网
drawing-in 拉进;穿经

drawing-in and reeding 穿经和箝
drawing-in box 引线盒
drawing-in device 穿模装置
drawing induced colo(u)ration 拉引着色
drawing-in frame 穿综架;穿经机
drawing-in hook 穿综钩
drawing ink 绘图墨(水)
drawing insertion 图形插入
drawing in side sectional elevation 横剖面图
drawing instrument 绘图工具;绘图仪(器)
drawing interchange capability 图形交换能力
drawing interchange file 绘图交换文件
drawing interface 绘图接口
drawing in the credit branches 信用份额提款
drawing issued for construction 准备好的建筑设计;施工出图;成熟的建筑设计
drawing joint examination system 图纸会审制度
drawing key 底图
drawing kiln 引上窑室
drawing knife 刨子;绘图刮刀;制图(用)刮刀
drawing landslide 牵引滑坡
drawing line 线道;型线图
drawing line tester 直线描绘测试仪
drawing list 图纸目录;图纸清单
drawing load 拉拔力
drawing logo 图签
drawing machine 拉丝机;拉线机;拉拔机;引上机;制图机;抽丝机;拔丝机;起模机
drawing machine for quartz 晶体提拉机
drawing material 绘图材料
drawing mill 金属丝制造厂;拔丝厂;拉制车间;拉丝厂
drawing name 图名
drawing number 制图编号;设计编号;图纸编号;图号
drawing of bonds 抽签;提取债券
drawing of brick works 砌砖图
drawing of elevations and sections 制立面图和剖面图
drawing of excavation construction 开挖施工图
drawing of external shape 外形图
drawing-off 掉炉
drawing office 产品设计组;绘图室;制图室;设计室
drawing office apprentice 实习绘图员;绘图室实习生
drawing of partial enlargement 局部放大图
drawing of patterns 脱模
drawing of position to be detected 探伤部位图
drawing of sample 样板图;抽取试样
drawing of site 基地平面图;平面图;位置图
drawing of tools 工具图纸
drawing of tubes 拉管机
drawing oil 拉制用油
drawing on issue 发行的图纸
drawing operation 拉延工序
drawing original 清绘原图
drawing out 拉窄;拉细;拉伸;并条;取回;拉制;出车
drawing-out frame 拉模板
drawing out methane 抽放瓦斯
drawing paper 制图纸;绘图纸;图(画)纸
drawing passes 拉拔道次;拉模孔
drawing pen 制图笔;绘图笔;直线笔;鸭嘴笔
drawing pencil 绘图铅笔
drawing pen holder 绘图笔架
drawing pen nib 绘图笔尖
drawing pin 揿钉;绘图钉;图钉
drawing pit 引上窑室
drawing plate 描绘板;起模板
drawing plate apron conveyer 皮带输送机
drawing plate conveyer 卸料板式输送机
drawing point 划线针
drawing point carriage 绘图头骨架
drawing polyester film 绘图聚酯薄模
drawing pot 引上池
drawing practice 制图
drawing press 引伸压力机
drawing pressure 拉延力
drawing process 回火过程
drawing pump 汲水式水泵;吸入泵;抽出泵
drawing punch radius 拉延凸模圆角半径
drawing quality 拉制性;压延性能;深冲质量
drawing quality plate (适用控制成型的)低碳钢板
drawing rate 拉引率;引上率
drawing rate of real assets heavy repairing 固定

资产大修理提存率
drawing ratio 拉延比
drawing reproduction 复制图
drawing rights 提存权;提款权;提货权
drawing rights of international monetary fund 国际货币基金组织提款权
drawing rod type jack 拉杆式千斤顶
drawing room 客厅;绘图室;会客室;起居室;休息室
drawing rule 绘图(比例)尺
drawings and documents room 图档室
drawing scale 绘图比例(尺);制图比例尺
drawing screw conveyer 抽送式螺旋输送机
drawing section 引上工段
drawing set 绘图仪(器);绘画支架
drawing sheet 拉延用钢板;图纸
drawing shop 拔丝车间
drawing silk 画绢
drawing size 图幅
drawing small 磨损变细
drawing speed 拉引速度;拉延速度;拉拔速度;绘图速度;引上速度
drawing square grid (with Chinese unit of length "li")计里画方(中国长度计算单位)
drawing standard specifications 制图标准规范
drawing statement 绘图语句
drawing steel 冷拉钢
drawing stress 拉拔应力
drawing strickle 刮板
drawing strickle guide 导框
drawing studio 绘图室
drawing subsystem 绘图子系统
drawing summary 图纸一览
drawing surface 绘图面
drawing system 绘图系统
drawing table 绘图桌;绘图台;制图台
drawing tank 引上窑室
drawing technique 绘图术;绘图方法;制图技术
drawing temper 回火处理
drawing temperature 回火温度
drawing template 刮板
drawing test 拉制试验
drawing timber 回收支架;坑木回收
drawing tool 拉制工具;绘图用具
drawing trap 存水弯
drawing T-square 绘画丁字尺
drawing under the credit 按信用证开出汇票
drawing unit 绘图单元
drawing up 上炉
drawing up a protest 制成拒付证书
drawing velocity 拉拔速度;绘图速度
drawing water level 绘图水位
drawing with colo(u)red chalk 粉笔画
drawing with colo(u)red pencil 彩色铅笔画
drawing without resource clause 注明无追索权票据
drawing with pen and ink 钢笔画
drawing with two scales 双重比例尺作图
drawing work 拉延工艺
drawing worker 引上工
draw-in rod 拉入杆
draw-in runner 底抽浇口
draw-in system 管沟敷线体系;电缆拉入系统
draw-in ventilation 引入式通风
draw-in winding 穿入式绕组
draw knife (木工的)刮刀;拉刮刀;木工用刮刀;双柄拉刨;(两端有柄的)挎刀
draw knob 拉钮
draw lessons from 借鉴
drawlift 吸入泵(扬程不超过吸高)
draw machine 牵引机
draw mark 拉拔伤痕;深冲划伤
draw metal 拉制金属
draw money 提款
draw moneyorder 开汇票
draw money out of a bank 从银行取出款项
draw mo(u)ld 铸(塑)模;薄壁模(子)
drawn 拉伸的
drawn nail 起模钉
drawn and rolled tubes 拉制轧制管材
drawn at a very high-speed 高速拉制
drawn bar 冷拉钢筋
drawn bill 开出票据
drawn clause 出票条款;发票条款
drawn component 拉延件

drawn cylinder 无缝气瓶;无缝的
drawn cylinder glass 拉制的圆筒摊平玻璃
drawn fiber 冷拉纤维;并纺纤维
drawn glass 普通窗玻璃;拉制玻璃
drawn grader 牵引式平土机;牵引式平地机;拖式平地机
drawn grain 皱面
drawn-in scale 残余氧化皮层
drawn-in scratch 拔痕
drawn-in tandem 串列拖带
drawn into economic circulation 被投入到经济流通中
drawn lead 拉伸的铅条
drawn lead trap 用铅条密封;密封用铅条
drawn line 实线
drawn metal 拉制金属
draw n meters of water 吃水 n 米深
drawn mo(u)ld 铸塑模
drawn-off pipe 泄水管(道);卸料管
drawn on 硬背装订
drawn-out 时间拖延过长;延长;抽出
drawn-out wave 延伸波
drawn part 拉制件
drawn pipe 控制钢管;拉制管
drawn product 拔制产品;拉伸制品
drawn profile 冷拉的型材
drawn profile in stainless steel 冷拉的不锈钢型材
drawn right 提款权;提货权
drawn section 冷拉的型材
drawn section in stainless steel 冷拉的不锈钢型材
drawn shape 冷拉的型材;拔制型材;拉制型材
drawn shape in stainless steel 冷拉的不锈钢型材
drawn sheet glass 拉制平板玻璃
drawn steel 冷拉钢;拉制钢;拔制钢
drawn steel piping 拉制钢管
drawn steel tubing 拉制钢管
drawn symbol 形象符号
drawn thread 分离线
drawn trim 冷拉的镶边
drawn trim in stainless steel 冷拉的不锈钢镶边
drawn tube 拔制管材;拉制管
drawn tubing 拉制管
drawn turbine flo(a)tation 叶轮吸气上浮法
drawn unit 冷拉构件;冷拉的型材
drawn unit in stainless steel 冷拔的不锈钢构件
drawn-up 细高的
drawn wire 冷拉钢丝;冷拔钢丝;拉制钢丝
drawn without recourse 对出票人无追索权;无追索权
drawn work 二道抹灰层;二道墁抹罩在湿底灰上
draw of eyepiece 目镜管
draw-off 抽去;抽取;泄水(排除);引出;抽出
draw-off arrangement 放水装置
draw-off bin gate 储仓排料闸门
draw-off box 排料箱
draw-off cock 排气龙头;泄水龙头;泄放旋塞;放出开关;排放塞门
draw-off culvert 泄水涵洞
draw-off discharge 泄放流量;放水流量
draw-off gear 拉拔装置
draw-off hole 底钻孔
draw-off pan 泄液板;侧线出料塔盘
draw-off pipe 泄水管(道);排污管
draw-off plug 排出塞
draw-off point 泄水点
draw-off pump 抽出泵;抽取泵
draw-off shaft 蓄水井
draw-off tank 抽水罐
draw-off tap 放水龙头
draw off tower 泄水塔;取水塔
draw off tunnel 泄水隧洞
draw-off unit 抽水装置
draw-off valve 排气(阀);排汽阀;排水阀门;泄水阀门
draw on 临近;动用;签发
draw on a frame 脱框起模
draw on one's credit 凭信用借款
draw on pins 顶杆起模
draw on rollover 翻转起模
draw on scale 按比例制图;按比例尺描绘
draw on the future 预支
draw out 拉长;抽出;拖长
draw-out breaker 抽出式断路器
draw-out design 抽出式设计

draw-out metal-clad switchgear 拉出式铠装开关装置
draw-out relay 拉出式继电器;抽出式继电器
draw-out switchboard 抽出式开关板
draw-out track 牵出线【铁】
draw-out track at grade 坡度牵出线【铁】
draw over 蒸馏;拉下遮盖
draw payable after date 出票远期付款交单
draw peg 起模木棒;起拉螺钉
draw period 地下水的抽降期;(地下水的)抽降期
draw piece 拉制件;压延件
draw pin 活络销;插销;榫销
draw pipe 冷拔管;无缝管;拉制管
draw plate 起模板;拉模(板);牵引板
drawplate oven 拉板炉
draw plough 铲土犁;排土犁;拖挂式除雪犁;推土犁
draw point 装车点;卸矿点;放矿溜井;放矿点
draw power 抽运功率
draw punch 拉冲头
draw radius 冲模半径
draw resonance 拉伸共振
draw rest 平旋桥护座
draw ring 拉模板;牵引环
draw-rod method 玻棒拉丝法;棒法拉丝
draw roll 拉伸的金属板;拉伸辊;进料辊;紧缩辊;引入辊
draw roller 喂料机;喂料辊
draw roof 可吊挂的(活动)屋顶;(矿井坑道的)假顶
draw runner 滑条
draws 铸件孔穴
draw scraper 拉刮刀
draw screw 起模螺钉;起拉螺钉
draw shallow sea bottom seismograph 提取式浅海海底地震仪
draw shave 木工用刮刀;刨刀;刮土机;刮刀
draw span 开合桥跨;开合桥孔;仰开跨(度)
draw spring 缓冲弹簧;牵簧
draw-spring connecting rod 缓冲弹簧连接杆
draw stick 起模针;起模棒
draw stop 拉钮
draw stroke 漏模行程;起模行程
draw the accumulation fund 提取积累金
draw the temper 回火
draw-through heater 再循环供暖机组
draw tongs 紧线钳;扳导线钳
draw tongue 拖杆;拖车挂钩;车辆连接装置;拉杆;牵引拉杆
draw top 推拉台面
draw top table 推拉台面桌
draw to scale 照尺度制图;照尺度放样;照尺度翻样;按尺度制图;按比例绘制;按比例尺制图;按比例尺描绘
draw trials 取出试样
draw tube 镜筒;套筒
draw tube slide-trombone 拉管
draw-twist machine 拉伸加捻机
draw-up 拉起;拟订;画出;制订;草拟
draw up a check 开支票
draw up a clear statement 开具清单
draw up a contract 拟合同;草拟合同
draw up a deed of transfer 立转让楔;立卖契
draw up a development planning 编制发展规划
draw up a list 列列清单
draw up an agreement 订协议;订合同
draw up a plan 起草计划
draw upon 动用
draw vice 拉(线)钳
draw water 汲水
draw well 吊桶井;提水井;深井
draw winder 拉伸绕丝机
draw wire 拉成丝
draw works 旋转钻(进)绞车
draw works cat head 牵引式吊锚架;绞盘式吊锚架
draw works drum 绞车转筒
draw works engine 绞车
dray 运货马车;载重板车;载货马车;低架载重车;大车
drayage (短途的)陆运运费;马车运货费;运货马车费
drayage company (短途的)陆运公司
dray-chain conveyer 链板输送机
dray screen 乱物筛
dreadage 装载杂货选择权
dreading 装载杂货选择权

dreadmought (片状沥青和预制浮石块的)屋顶；厚呢；防火组合门
dream hole 天窗；塔身透光窗孔；(塔身上的)采光窗；(仓库等的)风窗；(仓库等的)气窗；(塔身上的)采光口
dreamland 幻境
Drechsel washer 德雷克泽尔煤气洗涤器；玻璃煤气洗涤器
dredge 拖曳式采样器；拖网；浚泥船；采泥船；采矿矿船；挖土机；挖泥船；挖掘机；水底采集器；疏浚工；疏浚船
dredge boat 采矿船；采金船；挖泥船
dredge capacity 吸泥能力
dredge cut 浚槽；疏浚槽；挖槽
dredge-cut alignment 挖槽定线
dredge-cut design 挖槽设计
dredge-cut setting out 挖槽放线
dredge-cut sideline control unit 挖槽边线控制装置
dredged area 已挖区；挖泥区域；挖出区；疏浚区
dredged berth 挖泥船台；挖泥泊位；挖成的泊位；疏浚的泊位
dredged bottom 挖成的河底；挖成的海底；挖泥海底；挖成的底部(水底)
dredged bulkhead 后挖式岸壁
dredged channel 挖深的航道；挖泥航道；挖成的航道；疏浚(的)航道；疏浚(的)航槽；已疏浚航道
dredged channel buoyage 航槽标
dredged inland port 内河港口疏浚
dredged level 挖泥高程；挖泥标高
dredged material 挖出物；疏浚材料；浚挖的泥砂；弃土；挖泥船挖出物；挖出的泥砂；疏浚弃土
dredged material assessment framework 疏浚土评价框架
dredged material disposal 疏浚土处置；疏浚土处理
dredged material dumping 卸泥
dredged matter 弃土；挖出的泥砂
dredged peat 挖出的泥炭土
dredged pit 挖泥坑
dredged profile 浚挖断面监测器
dredged spoil 挖出的废料；浚挖的泥沙；弃土；挖泥船挖出物；挖出的泥砂
dredged spoil basin 排泥塘
dredged spoil matter 疏浚弃土
dredged trench 挖泥槽
dredged trench built-in-place tunnel 挖槽现浇式隧道
dredge engineering survey 疏浚工程测量
dredge excavation 疏浚开挖
dredge for 捞回
dredge hopper 开底挖泥船；挖泥斗；挖泥船泥舱
dredge hull 挖泥船船身
dredge level 疏浚标高；挖土线标高
dredge master 浚工长
dredge muck 挖出的腐殖土；疏浚污泥
dredge pipe 吹泥管；排泥管；挖泥船出泥管
dredge-placed fill 挖泥填筑
dredge pump 泥(浆)泵；吸泥泵；排泥泵；排泥泵；污水泵；挖泥泵
dredge pump capacity 吸泥泵额定功率
dredge pump characteristic curve 泥泵特性曲线
dredge pump coupling 泥泵联轴器；泥泵联轴节
dredge pump discharge pressure 泥泵排泄压力
dredge pump drive 泥泵驱动
dredge pump efficiency 泥泵效率
dredge pump engine 泥泵主机
dredge pump flow rate 吸泥泵流量
dredge pump test 泥泵试验
dredge pump unit 泥泵装置
dredge pump wearing 泥泵磨耗
dredger 挖泥船
dredge(r) bow 挖泥船艏部
dredge(r) bucket 泥斗；挖泥机泥斗；挖泥船泥斗；挖泥(铲)斗
dredge(r) bucket (type) elevator 挖泥斗式提升机
dredge(r) contractor 挖泥承包者；挖泥承包人
dredge(r) cutter head 挖泥船绞刀
dredge(r) excavator 链斗式挖泥机；链斗式挖掘船；链斗式挖泥机
dredge(r) fill 吹填土；冲填土
dredge(r) fleet 挖泥船队
dredge(r) ladder 挖泥船桥梁架；挖泥臂架
dredge(r) load draft indicator 挖泥装载吃水指示仪
dredge(r) pipe line 挖泥船排泥管道；挖泥船出泥管道
dredge(r) pontoon 非自航式挖泥机
dredge(r) port 挖泥船港口
dredge(r) productivity 挖泥船生产率
dredge(r) profile monitor 挖泥船断面监测
dredge(r) scoop 挖泥戽斗
dredge(r) scraper 拖铲
dredge(r) ship 挖泥船
dredge(r) shovel 单斗挖泥机
dredge(r) spoil 挖泥船废料
dredge(r)'s production capacity 挖泥船生产能力
dredge(r)'s production rate 挖泥船生产率
dredge(r)-spud hoisting ram 挖泥船定位桩提升柱塞
dredge(r) spuds 挖泥船桩柱；挖泥船定位桩；挖泥船撑柱；吸泥船锚柱
dredge(r) stern 挖泥船艉部
dredge(r)'s unproductive downtime 挖泥船非生产性停歇
dredge(r) walking 挖泥船摆动移进
dredge(r) winch 挖泥机绞车
dredge sampling 吸泥采样；挖泥取样
dredge spoil 挖出的废料
dredge waterway 整治航道
dredge well 挖泥船链斗井；挖泥井
dredge work 挖泥工作；疏浚工作
dredging 干法涂塘；清淤；挖泥；挖掘法；疏浚
dredging against current 逆流挖泥
dredging assessment 疏浚评估
dredging assessments of marine borrow areas 海底取土区的污泥评估
dredging barge 泥驳
dredging bit 匙形钻(用于散粒土钻进)；匙形钻头
dredging bottom 挖泥河底；清淤河底
dredging box 戽斗；挖泥斗
dredging branch fleet 挖泥船支队
dredging bucket 挖泥机挖斗；挖(泥)斗；挖泥船挖斗；脱水戽斗
dredging buoy 挖泥浮标；疏浚(区)浮标
dredging by layers 分层挖泥
dredging capacity 挖泥能力；疏浚能力
dredging community 疏浚界
dredging computer 挖泥计算机
dredging construction 疏浚施工
dredging contractor 挖泥承包者；挖泥承包人；疏浚承包人
dredging control system 挖泥控制系统
dredging conveyer 捞挖输送机
dredging craft 挖泥船
dredging crane 挖泥起重机
dredging cutting head 吸泥切割头
dredging cycle 挖泥周期
dredging depth 浚挖深度；挖深；疏浚深度
dredging depth indicator 挖泥深度指示器
dredging depth tolerance 浚挖深度容许误差
dredging diagram 浚挖示意图；挖泥示意图
dredging engine 挖泥机；挖泥船
dredging engineering 疏浚工程
dredging equipment 挖泥设备；疏浚设备
dredging execution 疏浚施工
dredging exploration survey 疏浚勘测
dredging firm 挖泥公司；疏浚公司
dredging fleet 挖泥船队
dredging frequency 疏浚频率
dredging in layers 分层挖泥
dredging in sections 分段挖泥
dredging in strips 分条挖泥
dredging intensity chart 疏浚强度图
dredging knife 挖泥刀
dredging ladder 疏浚船链斗斗架；挖泥臂架
dredging machine 浚挖机；挖土机；挖泥机；挖泥船
dredging machinery 疏浚机械
dredging mark 疏浚标志
dredging material 挖出物；疏浚材料；疏浚土
dredging matter 疏浚土
dredging method 浚挖方法
dredging module 挖泥组件
dredging muck 疏浚土
dredging operation 浚挖作业；挖泥作业；挖泥操作；疏浚作业；疏浚施工；疏浚操作
dredging parameter 挖泥参数
dredging period 挖泥周期
dredging pipe(line) 排泥管(线)；吸泥管；疏浚管道
dredging pit 排泥坑
dredging plant 挖泥设备；疏浚设备
dredging plume behavio(u)r 挖泥卷流特性
dredging process 疏浚工艺
dredging profile 浚挖剖面
dredging project 疏浚工程
dredging project chart 疏浚工程图
dredging pump 挖泥机水泵；清淤水泵；泥浆泵；吸泥泵
dredging quantity 疏浚量
dredging remote control system 挖泥遥控装置
dredging rule 挖泥法则
dredging scoop 挖泥斗
dredging shovel 单斗挖泥机；铲斗挖泥；挖泥铲；疏浚铲斗
dredging signal light 挖泥信号灯
dredging site 挖泥地点；疏浚地点；浚挖区；挖泥区；挖泥工地；挖出区；疏浚工地；疏浚地段
dredging site condition 疏浚工况
dredging situation display 挖泥情况显示器
dredging soil 疏浚土
dredging spoil 挖出废料；疏浚废料
dredging sub-fleet 挖泥船支队
dredging survey 疏浚测量
dredging survey sheet 疏浚测图
dredging task 疏浚任务
dredging technique 疏浚工艺
dredging thickness 挖泥厚度
dredging time 挖泥时间
dredging tolerance 浚挖容许误差；回淤富裕深度；回淤富余深度
dredging took place within a silt screen 防淤帘范围以内污泥
dredging track 疏浚轨迹
dredging track plot 挖泥轨迹图
dredging trench 疏浚槽
dredging tube 吸泥管
dredging turning 拖锚掉头
dredging unit 挖泥设备
dredging well 挖泥链斗槽
dredging wheel 疏浚操纵轮
dredging width 挖泥宽度
dredging with current 顺流挖泥法
dredging with the current 顺流拖锚
dredging works 挖泥工作；疏浚工程
dreelite 杂石膏重晶石
dregginess 含渣量；沉淀物
dreggy 含渣的
dregs 渣滓；沉渣；渣子
dregs formation 渣滓形成
dreikanter 三棱石
drench 浸液；浸润；湿透
drenched corrosion time 溶浸时间
drencher 洒水器
drencher fire extinguisher 水幕式灭火器
drencher head 骤雨水头；暴雨水头；水幕喷头
drencher installation 水幕装置
drencher system 甲板洒水系统；喷水灭火系统(指一种手动洒水装置)；水幕系统；手动洒水装置；喷水系统；喷淋系统
drenching apparatus 灌水机
drenching installation 水幕装置；手动洒水装置
drench pit 脱灰槽；脱水槽
Drenthe stade 德伦特冰阶【地】
Dresbachian (stage) 德里斯巴赫阶【地】
Dresden green 德雷斯登绿(陶瓷彩料)
Dresden porcelain 德雷斯登瓷器(迈森瓷器的别名)
dress 修剪；洗矿；整饰；服饰；服饰
dressability of ore 矿石可选性
dress (a) bit 修理钻头
dress and personal adornment 服饰
dress a shop window 布置商店橱窗
dress a store window 布置商店橱窗
dress circle 花楼；戏院内的前排包厢；二楼楼厅；特等包厢
dressed 磨光了的；修琢过的；整修过的
dressed and headed 刨光并在端部作榫的
dressed and matched 整平配接；企口拼合；刨光镶接
dressed and matched boards 刨光企口板；刨光拼(合)板；刨光镶板；企口板
dressed and matched flooring 刨光镶地板；刨光企口地板
dressed ashlar 琢面方石
dressed board 刨光板

dressed brick 烧制前修饰加工砖坯;磨面砖;磨光砖
dressed fair face 加工过的平整面;光洁的石面;天然石装修
dressed full 挂满旗【船】
dressed full with up and down flags 满旗纵挂法
dressed granite meeting face 细花岗岩贴接面
dressed lumber 修整好的木材;刮平的木材;刨光木料;刨光木材;刨光板
dressed masonry 镶面砌体;细琢石圬工;细琢石砌体;琢面圬工;敷面圬工
dressed matchboard 刨光企口板;刨光拼合板
dressed natural stone 加工过的天然石料
dressed one side 单面修整
dressed particle board 饰面碎料板
dressed rock 羊背石【地】
dressed size 刨光(材)尺寸;修整后的尺寸;实际尺寸;刨光后的尺寸;净尺寸;成型尺寸
dressed stone 料石;修琢石;琢石
dressed stone pavement 细琢石路面
dressed stonework 加工过的石制件
dressed stuff 修饰过的材料;刨光木材;刨平的材料
dressed timber 刨光木料;刨光木材;刨光板
dressed tongue-and-groove flooring 刨光企口地板
dressed two sides 双面修整
dressed weight 整理后的重量
dressed with 铠装的
dressed with masthead flags 桅顶饰旗
dressed with natural stone 装修过的天然石料
dresser 浆纱机;化妆台;选矿工;修整器;修整工具;修饰工具;整修工具;整经机;雕琢工;打磨机;清理风鎼;梳妆台
Dresser coupling 无螺纹的管接头;德列斯连接器
Dresser flexible coupling 德列斯柔性联轴节
Dresser flexible joint 德列斯柔性接头
dresserite 水碳铝钡石
dress form 服装(人体)模型
dress grinding wheels 整修磨轮
dressing 加印版边条;修整;修饰;琢面;整修;打磨;敷料;包扎;湿整理;上浆
dressing and preparation engineering 装饰与制备工程
dressing bit 修整钻头
dressing bit tongs 钻头修理钳;多整钻头用夹钳
dressing block 钻头修理铁砧
dressing board 敷料板
dressing by flo(a)tation 浮选法
dressing by lathe 车削精加工
dressing by magnetic separation 磁选
dressing by screening 筛选
dressing by washing 洗选
dressing cab(in) 修饰过的小室;化妆室
dressing carriage 敷料车
dressing case 化妆箱
dressing composition 修饰用组成物;屋顶胶黏物质
dressing compound 沥青涂料;沥青覆盖物;表面处治混合料;涂敷沥青
dressing compound for Hausler type roof(ing) (木纤维水泥屋顶用的)木纤维水泥
dressing conveyer 分选输送带
dressing cost 选矿成本
dressing course 找平层
dressing cubicle 梳妆室;幕墙小室
dressing diamond 金刚石修整笔
dressing drum 敷料罐
dressing equipment 选矿设备;锻钎设备
dressing floor 整理车间
dressing forceps 敷料钳
dressing furrow 施肥沟
dressing hammer 修整锤;整面锤;敷面锤
dressing iron 刨刀护铁;琢石修边垫铁;铁挡块
dressing leather 饰革
dressing line 满旗绳
dressing locker 衣橱;衣帽柜
dressing-locker room 衣橱间
dressing machine 校正机;修整机;修钎机;锻钎机;去毛刺机;清粮机
dressing material 修整料;镶面材料
dressing of a casting 铸件的清理
dressing-off 清理
dressing of grinding wheel 磨轮的整修
dressing of steel ingots 钢锭的整修
dressing output 选矿产量

dressing paint 修饰用油漆;屋顶涂料
dressing plant 选矿装置
dressing plate 校正平台
dressing pliers 敷料镊
dressing procedure 选矿程度
dressing room 化妆室;修整工段;更衣室;更衣间;敷裹室
dressing room for man 男更衣室
dressing room for woman 女更衣室
dressings 覆盖层;表面处理;文艺复兴及其派生建筑正面上的线脚和装饰
dressing scissors 敷料剪
dressing ship 挂满旗的船
dressing shoe 清选机
dressing shop 清理车间
dressing sieve 筛粉筛
dressing stake 手砧
dressing stand 平整机座
dressing station 包扎所(多用于前线军营中)
dressing stick 校正工具;整形块;整形棒
dressing stone 镶面石
dressing table 镜台;化妆台;厨房桌;清理台;梳妆台
dressing tool 校正工具;修整器;修整工具;凿石工具
dressing trolley 敷料车
dressing work 选矿厂
dressing yard 整场场;整锭场;清整工段
dressing yield 选矿产率
Dressler kiln 德雷斯勒式隧道窑(隔焰式)
dress linen 衣料麻纱
dressmaker 时装店
dressmaker's dummy 木制人体模型
dressmaker's measure 裁缝用尺
dressmaker's shears 裁缝剪刀
dress one side 单面修整
dress plate 加固用钢板;盖板
dress pole 挂衣棍
dress rack 挂衣架
dress shield 防护外衣
dress ship 船舶旗布
dress ship overall 挂满旗【船】
dress smooth 修光
dress stand 服装人体模型
dress up 现场组装;预制拼装
dress water 选矿水
dreston 吸湿硬膏(即湿熟石膏)
Drew number 德鲁数
Drew's hypothermia apparatus 德鲁低温装置
drew-string bag 抽带式袋
drex 德莱克斯(纤维单位)
Drexed glass-bottle 德雷克塞尔煤气洗涤瓶
Drexel bottle 煤气洗涤瓶
dreyerite 德钒铋矿
drib 点滴
dribbing (道路的)零星修补;修整道路坑凹
dribble 滑滴;滴水沟;出水碎石沟;碎石沟;使滴下
dribble action 细流动作;滴落动作
dribble applicater 滴洒机
dribble blending 一滴一滴地混合
dribbled fuel 没有蒸发的燃料
dribble from jet 喷嘴残滴
dribble pipe 滴水管
drib(b)let 涓滴;岩饼
drib(b)let cone hornito 岩饼锥
dribbling 滴落;滴流;碎石冒落;修整道路坑凹;零星修补
dribbling diesel fuel 低黏度柴油燃料;低黏度柴油机燃料
dribbling nozzle 滴流喷嘴
dri-crete 空心混凝土块(其中的一种)
dried-air cooler 干空气冷却器
dried-air drier 干空气干燥机
dried alum 烧明矾;焦矾;干燥明矾
dried blood 干血胶;蛋白兰
dried blood adhesive 干血黏结剂;蛋白兰胶黏剂
dried body 干坯
dried brick 干砖坯
dried crystal varnish 烘干晶纹清漆
dried explosion-proof transformer 干式防爆变压器
dried exsiccant 干燥剂
dried extracted fermentation soluble 干燥的发酵浸出物
dried female insect 虫红
dried fish scrap or meal 干鱼粕或鱼粉
dried flower 干花

dried gel 干凝胶
dried glaze 干釉
dried ice 坚固冰
dried ice cream mix 冰淇淋粉
dried in shade 阴干
dried joint 过干接合
dried kernel 干棕仁
dried lake bed 干旱湖床
dried lemon peel 干柠檬皮
dried lily bulb 百合干
dried marine algae powder 海藻粉
dried-out 干涸的
dried oyster 牡蛎干
dried plant samples 植物干燥样品
dried product 干制品
dried sand 干砂
dried sludge 干(化)污泥
dried slurry 烘干料浆
dried strength 干燥强度;烘干强度
dried to touch 快干涂料
dried-up 干缩的
dried wood 干(木)材
driel mo(u)ld 干型
drier = dryer
drier absorption 催干剂失效
drier activator 助催干剂
drier coil 干燥盘管
drier conveyer 干燥机输送带
drier crusher 烘干破碎机
drier dissipation 催干剂降效
drier felt 干燥毡
drier-filter of the refrigerator 冰箱干燥过滤器
drierite 燥石膏;无水硫酸钙
drier mill 烘干破碎机;烘干磨
drier scum 阴沟中的浮渣;墙干燥后浮垢
drier sol 速干剂
drier soluble in drying oil 能溶于干性油中的干燥剂
drier white 陶器干燥泛白
dries 干出(海图图式)
drift 连通道;孔锤;铆钉孔冲;滑移;航迹偏移距离;航差;乡村道路;移动;沿脉巷道;烟流向;堆积物;导坑(开挖);打桩器;吹集;浮动;风压漂移;分步开挖;冰碛(堆积物);任意游走;飘失;飘踪;漂失;漂散木;漂流;漂积物;漂变;偏移;偏离;偏航;退套楔;铜冲头;水平巷道;漂移
drift action 漂移作用
driftage 流压差;漂流物;漂浮物;漂程;漂移
driftance 漂移度
drift anchor 流锚;海锚;浮锚
drift and speed indicator 偏航与航速指示器
drift angle 顶角【岩】;风压角;风流压差;漂(移)角;漂(流)角;偏流角;偏航角
drift angle build up 偏航增大
drift angle control 偏航角控制
drift angle indicator 顶角测量仪;偏航角指示器
drift angle of hole 钻孔偏差角
drift angle under wind/current pressure 风/流压偏角
drift ashore 漂向岸边
drift axis 回转的垂轴;漂移轴
drift band 软土夹层
drift barrier 拦浮栅;漂木屏障;漂浮物拦栅(河道中拦漂浮木料设施);漂浮物栅栏
drift barrier lake 冰碛堰塞湖
drift base 漂移基板
drift beam instability 漂移束不稳定性
drift bed 坡积(层)
drift boat 流网渔船;漂网(鱼)船
drift bolt 锚栓;系栓;系紧螺栓;穿钉;冲头螺栓;冲钉;退楔
drift borehole 偏差钻孔
drift boring 超前钻探
drift bottle 测流瓶;浮标瓶;瓶式浮标;漂流瓶
drift boulder 漂砾
drift canal 浮运运河
drift card 测流卡片
drift carrier 漂移载流子
drift channel 漂移沟道
drift clay 冰碛泥;漂砾泥;漂积黏土
drift coefficient 偏流系数
drift compensating 漂移补偿
drift compensation 漂移补偿
drift computer 偏差计算机
drift constant 漂移常数

drift control 积雪控制
drift-corrected amplifier 漂移校正放大器;漂移补偿放大器
drift correction 零点漂移改正;航差改正;漂移修正;漂移校正;漂移改正;偏流修正;偏流校正;偏航修正;偏航校正
drift correction of gravimeter 重力仪零点改正值
drift cross-bedding structure 迁移交错层理构造
drift current 缓慢水流;海洋中暖流;吹流;风吹流;漂移电流;漂流;表流
drift curve 漂移曲线;漂描曲线
drift cyclotron wave 漂移回旋波
drift dam 冰碛堰;冰碛坝
drift-dam lake 冰碛湖
drift-dammed lake 冰碛湖
drift deposit 冲积物;冰川沉积(物)
drift detritus 漂移石屑
drift deviation intensity 顶角弯曲强度
drift diameter 通孔直径
drift diameter ga(u)ge 通径规
drift-dissipative instability 漂移耗散不稳定性
drift drill 穿孔钻;穿孔取芯钻头
drift due to wind 吹流
drifted borehole 偏斜孔
drifted casing 整形套管
drifted material 洪积物;漂移物
drifted sand 流沙;飞沙
drifted soil 冰碛土;吹扬土
drift eliminator 除水器;漂浮物清除器;水滴挡板
drift epoch 流冰期;冰期;漂浮期
drifter 隧道工人;冲头;漂网渔船;架式钻机;架式风钻;架式(风动)凿岩机;巡回检查员;支柱式开山机;风钻工;漂网鱼船;漂流者;漂流物
drifter bar 凿岩机支架
drifter drill 架式凿岩机;风钻钻机;取芯钻(传孔钻);穿孔钻
drifter drill for underground 地下钻机
drifter fleet 漂网鱼船队
drifter hammer 架式凿岩机
drift error 漂移误差;偏流差
drifter-type machine 架式风钻
drift face 掘进工作面;掌子面;工作面前壁;平巷掘进工作面
drift fence 流动栅栏;两端开门的围栏
drift field 漂移电场
drift fisher 流网渔船;漂网(鱼)船
drift force 漂移力
drift for knocking out of tubes 管子穿孔器
drift frequency 漂移频率
drift ga(u)ge 风雪计;偏差计
drift glacier 吹流冰川
drift goods 漂流货物
drift heading 导坑;导洞
drift hole 偏差钻孔
drift hoop 箍檐
drift ice 流冰;浮冰;漂凌;漂冰;淌凌
drift index 侧移指数;房屋横向偏移指数
drift indicator 倾斜计;倾角计;量坡计;测斜器(钻孔);航差指示器;漂流指示器;偏流计;偏航指示器
drifting 走向水平掘进;穿孔调准;平巷掘进;漂流;漂积;漂浮;漂动
drifting automatic radio meteorologic(al) station 漂流自动无线电气象站
drifting automatic radio meteorologic(al) system 漂流自动无线电气象系统;自动无线电气象漂浮系统
drifting beach 移动海滩
drifting biomaterial 漂水植物
drifting buoy 漂流浮标
drifting buoy cooperation panel 漂流浮标站合作小组
drifting buoy (station) report 漂流浮标站报告
drifting continent 漂移大陆
drifting convergence 不稳定收敛;漂移收敛
drifting distance 流程
drifting dune 移动砂丘;移动砂垄
drifting dust 低吹尘;吹尘;飘尘
drifting grab 漂斗
drifting ice 浮冰;导冰
drifting locus 漂移轨迹
drifting machine 掘进设备;柱架式凿岩机;平巷掘进机;凿岩机
drifting-management 流动管理
drifting mine 漂流水雷

drifting moment 偏航力矩
drifting recording station 漂浮式自记(海洋)观测站
drifting sand 低吹砂;吹砂;漂砂
drifting-sand filter 漂砂快滤池
drifting snow 雪暴;低吹雪;吹雪
drifting test 穿孔试验
drift instability 漂移不稳定性
drift key 漂移栓;卸出斜铁
drift klystron 漂移速调管;偏移式速调管
drift lead 测走锚铅锤;漂流指示锤
drift leeway 风流压差;漂流角
drift length 漂移长度
drift level 漂移电平
drift limit 流速限定;流速界限
drift limitation 房屋横向偏移极限;漂移极限
drift line 主流线;漂移(曲)线
drift-line fishing 浮选延绳钩
drift log 流木;流放的木材;漂木;偏差钻孔;井斜测井
drift logging 井斜测井
drift loss 水的漂失
drift map 覆盖层地质图;冰山漂积海图;冰碛图
drift mardrel 量规(量套管等内径)
drift marker 偏差指示器
drift mathematical model 漂移数学模型
drift meter 偏移测量仪;偏流计;偏航计;漂移计;漂流测示器;偏差计
drift method 导坑法(隧道);导洞掘进法;漂移法;隧道导坑法
drift mine 地下采矿
drift mining 地下采矿;导洞开采
drift mobility 迁移率;漂移迁移率;漂移率
drift motion 漂移运动
drift net 流网;捕鱼流网;漂网
drift net fishing 流网作业
drift net method 流网法
drift observation 偏流观测
drift of a borehole 钻孔偏移
drift of continent 大陆漂移
drift of convergence 会聚漂移
drift of frequency 频率漂移
drift of gravimeter 重力仪零点位移
drift of plummet wire 垂球丝漂移
drift of stor(e)y 层间侧移
drift of the gyro 陀螺漂移
drift of tunnel 隧道偏差
drift path 漂流路径
drift peat 冰碛泥炭
Drift period 更新世冰期;冰雪碛纪
drift pin 紧配合销;尖冲钉;冲子;冲头;锥枢(销);穿孔销
drift-ping 锥塞
drift pipe 穿孔砧板;流管;漂移管
drift plate 漂移板块
drift plug 打入销;冲头;铅管矫直木棒
drift point 滑移点
drift pole 测流杆;漂流杆
drift prevention 积雪防治措施;防止积雪
drift punch 打孔器;锥形冲头;冲头
drift rate 漂移速度;漂移率;偏移率;偏移度
drift reamer 冲头铰刀
drift recorder 漂移记录器
drift region 漂移区
drift regulation 浅滩整治
drifts 作业分额;分额
drift sail 流锚;海锚
drift sand 沿岸漂沙;流沙;冲积沙;风积沙;漂沙;海岸沿岸漂沙
drift-sand filter 浮砂式滤池
drift-scan measurement 漂描测量
drift scanning 漂移扫描
drift scratch 冰碛擦痕;冰川擦痕
drift seine 流网
drift setting 航差调整
drift sheet 流锚;海锚;浮锚;冰碛层
drift sheet influence 冰碛层影响
drift sight 漂移角测定器
drift slope 坑道坡道
drift slot 斜形槽;出屑槽
drift snow 吹堆雪;风积雪;飘雪
drift soil 冰碛土;飘移土
drift soil landslide 推积土滑坡
drift space 漂移空间
drift spraying 飘雾喷射
drift stabilization 漂移补偿

drift station 流动台;漂浮观测站
drift storm 雪暴;暴风雪
drift subpulse 漂移次脉冲
drift surface 漂移面
drift surface landslide 堆积面滑坡
drift terrace 冲积阶地
drift test 扩孔试验;穿孔试验;扩张试验;校验内孔直径;冲头扩孔法钢材延性试验;冲孔试验
drift theory 异地成煤说;漂移(学)说;漂流理论
drift time 漂移时间;漂航时间
drift topography 冰碛地形
drift trajectory 漂流路线
drift trammel net 三层网流
drift tube 漂移管
drift tube mass spectrometry 漂移管质谱分析
drift turbulence 漂移湍动
drift vector 漂流矢量
drift velocity 迁移速度;漂移速度;漂流速度
drift voltage 漂移电压
drift water 挟砂水
drift wave 漂移波
drift way 流压差;导向井;平巷;平峒;漂程;漂流物;马道;流程;导坑;大车路;大车道;坑道
drift weed 漂浮(海)藻
drift wood 漂流木;流放的木材;浮木;水上浮木;漂(流)木;管流材
driftwood peat 漂流木泥炭;丢弃的泥炭
drikold 干冰
drilitic 干电解电容器
drill 练习;钻孔钻头;操练
drillability 可钻性;可钻度;钻进难易程度
drillability classification of rock 岩石可钻性分级
drillability index 可钻性指数;可钻度指数
drillability index of rock 岩石可钻性指数
drillability of rock 岩石可钻性
drillability test 钻探性测试
drill a blasting hole 打炮眼
drillable 可钻孔的
drillable depth 可钻深度
drillable metal casing 可钻掉的铝或镁合金套管
drillable packer 可钻掉的封隔器
drill adapter 夹钎器;钻杆卡头;钻杆夹头
drill and blast construction method 打眼放炮施工法
drill and blast method 钻爆法
drill and draw 充水排水操作;充水放水周期
drill and steel wire ga(u)ge 钻径及钢丝规
drill and turning tool holder 钻孔与车削两用刀夹
drill anvil block 凿岩机撞锤
drill around 钻越过
drill assembly 条播装置总成
drill attachment 钻机附件
drill attachment for excavators 挖掘机钻头附件
drill bar 钻杆
drill barge 钻探船;钻孔驳船
drill barrel head 钻的筒形头
drill barrel shoe 钻的筒形靴
drill base 钻机底座
drill bit 打孔钻头【岩】;扁钻;钎头;钻头
drill bit cutoff 钻头额定进尺数
drill bit cutting edge 钻头切口
drill bit diameter 钻头直径
drill bit edge 钻头刀口
drill bit ga(u)ge loss 钻头直径磨损(量)
drill bit grinder 钻头研磨机
drill bit grinding ga(u)ge 钻头研磨量规
drill bit life 钻头寿命
drill bit parameter 钻探钻头参数
drill bit type 钻探钻头类型
drill bit wear detector 钻头磨损检测器
drill blower 钻进除粉管;钻机吹粉器
drill bo(a)rt 钻探用金刚石
drill boat 钻探船;钻井船
drill boot 吊锤【岩】
drill bortz 钻用金刚石屑;钻探用金刚石
drill bow 手(摇)钻弓柄
drill break 钻进突变
drill building 钻头修理间
drill bush 钻套
drill cable 钻机钢丝绳;钻探用钢丝绳
drill capacity 钻机钻进深度;钻机能力;钻进能力
drill carriage 车钻;凿岩台架;钻轴支架;钻轴支撑;钻轴滑座;钻孔台车;钻架(车);手推凿岩机车
drill casing 凿岩机外壳

drillcat 装有压气机的履带式自行钻机
drill centering machine 拱架钻机
drill centers 钻中心孔
drill chamber 凿岩峒室
drill change 钻距；钎头更换；钎距
drill chuck 钻头夹具；钻箍；钻头卡(盘)；钻头夹盘；钻夹头
drill chuck barrel 锥钻轴套
drill chuck bushing 凿岩机钎尾套筒
drill chuck jaw 凿岩机夹钎爪；风钻夹钎爪
drill circuit 练习线路
drill collar 钻孔口；加长钻铤；钻铤；钻环
drill collar break off 断钻键
drill collar lubricants 钻铤丝扣油
drill collar stabilizer 钻环稳定器；钻铤稳定器
drill colter 条播开沟机
drill column 开山机支柱；钻孔柱状图；钻机柱架；钻机支柱
drill column jack 钻机顶柱
drill core 钻孔岩芯
drill-core sampling 岩芯钻探采样
drill coulter 条播开沟器
drill cutter 钻头铣刀
drill cuttings 钻末；钻孔岩屑
drill cylinder 凿岩机汽缸
drill density 钻机(布置)密度
drill derrick 钻塔；架铁；钻机起重机架
drill diamond 钻探用金刚石
drill director 钻进导向器；钻进导向管
drill doctor 钻头、工具检修工(俚语)；机具修理工
drill dog 钻探剖面
drill down to a point 打丢直径(钻探用语)
drill drift 磨薄机；卸钻套楔；凿岩平巷；冲杆
drill dust from boring 钻粉
drilled and driven pile 钻孔打入桩
drilled area 已钻地区
drilled as an alliterative to casing 接管法钻进
drilled burner 钻孔燃烧器
drilled caisson 钻孔墩；钻孔沉井；钻孔桩；管柱；钻孔沉箱
drilled casing 钻孔套管
drilled core crushing strength 钻孔岩芯抗碎强度
drilled dry 干钻的
drilled footage 钻探进尺
drill edge 钻头切削刀
drilled grout hole 钻成的灌浆孔；灌浆钻孔
drilled in 钻通的
drilled-in anchor ties 机钻锚杆
drilled-in caisson 钻地沉箱；钻入沉箱；钻孔沉井；钻孔沉箱
drilled-in caisson retaining wall 钻地沉箱挡土墙
drilled lining tube 钻孔套管；钻孔衬管
drilled meter 钻孔计量器
drilled nozzle 钻制喷管
drilled pier 钻孔桩；钻孔墩
drilled pile 钻孔桩
drilled raise 超前钻孔开凿开井法
drilled rod 钻杆
drilled shaft 钻孔桩；钻孔竖井；钻孔机轴
drilled wall 柱列式地下连续墙
drilled well 管井；凿井；钻(成)井
drill equipment 钻头附件
drill equipment depreciation apportion and overhaul charges 凿岩设备折旧摊销及大修费
driller 钻(探)工；钻(孔)机；钻床；重型凿岩机
driller-month 月·台【岩】
driller's discretion 泥钻酌定
driller's helper 辅助钻工
driller's log 钻探记录；钻井记录
driller's mud 钻探泥浆
driller's position 司钻位置
driller's station 钻井机站；钻井机位置
driller's tour report 钻机日钻孔报表；每日钻探记录单；司钻班报表；司钻班报表
drill exhaust 凿岩机排气装置
drill extractor 钻头提取器；钻头提取器；打捞钳；取钻器
drill feed 钻头进给装置；钻孔装置；钻机给进装置；凿岩机推进器
drill file 小细锉
drill finish 光钻
drill fittings 钻孔配件；钻机配件；钻机附件
drill fixture 钻机设备
drill floor 钻台

drill floor equipment 钻台设备
drill fluid 洗井液
drill flute 钻槽；槽钻
drill foreman 钻班班长
drill for quarrying 采石钻(孔)机
drill for underground 坑道钻进
drill frame 钻机(支)架；钻机机架
drill free 钻头空转
drill front head 凿岩机机头
drill furrow 条播沟
drill ga(u)ge 钻(径)规
drill gismo 吉斯莫型钻眼万能采掘机
drill goods 钻探用品
drill grinder 磨钎机；钻头磨床
drill grinding ga(u)ge 钻头磨削量规
drill grinding machine 钻头磨机；钻头磨床
drill groove 钻头构槽
drill ground 练兵场；操场
drill guard with dust bag 用吸尘箱防护的钻机
drill hall 教练房；操练房
drill hammer 钻锤；打眼锤；冲击钻头；手冲钻机；手冲击钻
drill-harrow 除草小耙
drill head 钻床床轴头箱
drill head drive gear 钻机回转器主动齿轮
drill headroom 钻机头部空间
drill heel 钻头根面
drill holder 钻头卡具；钻夹头；钻套；变径套
drill hole 井眼；钻孔【岩】；炮孔
drill hole burden 炮眼内炸药与自由间距
drill hole column 钻孔柱状图
drill hole columnar section 钻孔柱状(剖面)图
drill hole depth 钻孔深度；孔深【岩】
drill hole direction 钻孔方向
drill hole discharge 钻孔排放
drill hole exploration 钻孔勘探
drill hole flushing 钻孔
drill hole geologic(al) record 钻孔地质编录
drill hole grouting pile 钻孔灌注桩
drill hole inclination survey 钻孔测斜
drill hole layout map 钻孔设计图
drill hole loading test 钻孔荷载试验
drill hole log 钻孔岩芯记录
drill hole pattern 钻孔图型
drill hole quality inspecting rule 终孔质量验收制
drill hole returns 钻孔反流
drill hole sampling 钻孔取样；打眼取样法；钻孔采样
drill hole sealer 封孔器
drill hole spacing 钻孔间距
drill hole structure 钻孔结构
drill hole technical file 钻孔技术档案
drill hole testing technique in situ 钻孔原位测试技术
drill hole wall 钻孔壁
drill hole water level 钻孔水
drill hose 钻机用软管；凿岩机风管
drill hose connection 钻机软管接头帽；凿岩机连接管接头
drilling 钻眼作业；油气钻井；钻探；钻孔工作；钻进；凿孔；打眼
drilling accessories 钻探附件
drilling accident 钻探事故
drilling ahead 超前钻进
drilling and blasting 钻孔爆破；凿岩爆破
drilling and blasting data 钻孔与爆破数据
drilling and blasting method 钻爆法
drilling and blasting operation 钻眼爆破作业；钻孔爆破作业
drilling and blasting pile 爆破式灌注桩
drilling and blasting ship 钻探炸礁船
drilling and grinding machine 钻磨两用机床
drilling and tapping machine 钻孔和攻丝两用机床
drilling and wireline coring system 钻探绳索取芯系统
drilling angle 钻进角
drilling apparatus 钻进设备
drilling area 钻面
drilling attachment 凿岩机附件；钻头夹具；钻工夹具
drilling at the sea 海洋钻探；海上钻探
drilling auxiliary equipment 凿岩辅助设备
drilling barge 钻探(驳)船；钻井驳船
drilling beam 钻杆；摇摆梁
drilling bench 钻孔梯段；钻炮眼段；凿岩梯段
drilling bit 钻头；钻探钻头

drilling bit life 钎头寿命
drilling block 钻井装置
drilling boat 海底钻探船；海底钻井船
drilling borehole 钻孔【岩】
drilling breakers 岩屑；钻屑
drilling bucket 钻斗【岩】
drilling bucket boring machine 回转斗成孔机；回转斗钻孔机
drilling by drill carriage 用台车凿岩
drilling by flame 火力钻进
drilling by jetting method 水力冲刷钻进；射流法钻进
drilling cable 钻缆；钻探用绳(索)；钻井用钢丝绳
drilling campaign 钻探作业(周期)
drilling capacity 最大钻孔直径；最大钻进深度；钻探能力；钻孔容量；打钻能力；钻进能力
drilling carriage 钻车；凿岩台车
drilling character of rocks 岩石的可钻性
drilling chatter 钻进时振动
drilling clay 钻探用黏土
drilling column 钻孔(孔)柱
drilling condition 钻探条件；钻进情况
drilling construction 钻探施工
drilling contract 钻探合同
drilling contractor 钻探承包人
drilling control 钻进控制；钻探控制
drilling core 钻探芯样
drilling cost 钻探费用；钻探成本；凿岩成本
drilling cost per meter 每米钻井成本
drilling cramp 钻架
drilling crew 钻探班组；钻探工班；钻探队；钻井队
drilling crown 钻冠
drilling curve 钻速曲线
drilling cuttings 岩屑
drilling cuttings separator 钻屑分离器
drilling cycle 钻进循环
drilling department 钻探部门
drilling depth 进尺【岩】；钻孔深度；钻进深度
drilling derrick 钻塔；钻井起重机架
drilling derrick brace 钻井起重机架支撑；钻塔支撑
drilling derrick cellar 钻塔地下室；井架地下室
drilling derrick cornice 钻塔上檐口；井架上檐口
drilling derrick crane 钻塔起重机；钻塔吊车；井架起重机；井架吊车
drilling derrick crown 钻塔顶；井架顶
drilling derrick floor 钻塔楼板；井架楼板
drilling derrick foundation 钻塔基础；井架基础
drilling derrick foundation post 钻塔基柱；井架基柱
drilling derrick girder 钻塔围梁；井架围梁
drilling derrick grillage 钻塔基础格床；井架基础格床
drilling derrick leg 钻塔支柱；井架支柱
drilling derrick man 钻探井架工
drilling derrick platform 井架平台；钻塔平台
drilling derrick substructure 钻塔下部结构
drilling design 钻井设计
drilling developer 钻机研制者
drilling device 钻井装置
drilling diamond 钻探金刚石；黑金刚石
drilling direction 钻探方向
drilling doctor 钻机修理工
drilling dust 钻尘；岩粉；钻粉；钻屑
drilling dust extraction 钻粉排出
drilling duty 钻进能力
drilling effect 凿岩生产率
drilling efficiency 钻进效率
drilling engine 钻(探)机
drilling engineer 钻探工程师
drilling engineering 钻探工程
drilling equipment 钻探设备；钻井设备；钻孔设备；凿岩设备
drilling experimentation 钻进实验
drilling exploration report 钻探报告
drilling facility 钻孔设备
drilling factor 钻进系数
drilling failure 埋钻；钻井事故
drilling field 钻探工地
drilling field equipment 钻探工地设备
drilling finished well depth 完钻井深
drilling firm 钻探公司
drilling fluid 钻用冲洗液；钻液；钻探(用)泥浆；钻探流体；钻探护壁液；钻井泥浆

drilling fluid additive 钻井液添加剂
drilling fluid contamination 钻井液污染
drilling fluid loss control 钻井液漏失控制
drilling fluid pressure 钻井冲洗液压力
drilling footage 钻进进尺;进尺【岩】;钻进米数
drilling for conduits ducts 敷管钻进
drilling foreman 钻探技师;钻探工头
drilling for ground investigation 地面调查钻探
drilling for soil investigation 土壤调查钻探
drilling for tunnel exploration 坑探凿岩
drilling for ventilation ducts 敷设通风管钻进
drilling for water supply 给水钻探
drilling free 空转(钻进时)
drilling from floating vessels 在浮船上钻探
drilling gang 钻探队
drilling grade of rock 岩石可钻性等级
drilling grinding machine 钻磨两用机
drilling hammer 钻机冲锤;打眼锤
drilling head 钻进动力头;钻头头部
drilling hole diameter 钻眼直径
drilling hook 大钩;钻井钩;钻机吊钩;提钩
drilling hose 钻井软管
drilling hours log interval 钻时录井间距
drilling house 钻机皮带管
drilling implement 钻孔设备
drilling in 钻入
drilling inclination survey 钻孔测斜
drilling in coal 煤田钻探
drilling in cramped quarter 小断面钻进
drilling index 钻孔指数
drilling indicator 钻探取样重量指示器
drilling industry 钻探工业
drilling installation 钻探装置;钻探机
drilling instrument 钻探仪表;钻探工具
drilling-in unit 小钻机(矿层钻进专用)
drilling island 钻机平台;钻探平台
drilling jars 钻进振击器;钻用振击器
drilling jig 钻模;手摇钻
drilling jumbo 凿岩台车;轨道式钻机;钻探车;钻孔台车;钻架车
drilling length 钻孔深度;钻程长度
drilling level 钻进水平(地下钻进时);钻进层位
drilling life 钻探设备使用期限
drilling line 钻探用钢(丝)绳;钻绳
drilling load 钻压;钻头负荷
drilling location 孔位
drilling log 钻井记录
drilling machine 钻(孔)机;钻床;凿岩机
drilling machine column 钻床立柱
drilling machine for pile foundation 桩基钻机
drilling machine operator 钻工
drilling machinery 钻机机械
drilling machine spindle 钻轴
drilling machine upright 钻床柱
drilling machine with jointed arm 旋臂钻床;悬臂钻床
drilling machine with jointed arm pivotal 摇臂钻床
drilling machine with opposite drilling heads 对位钻孔机
drilling magnetometer 钻孔磁力仪
drilling management 钻探生产管理
drilling map 钻孔布置图
drilling mast 钻机桅杆
drilling master 钻探工长;钻探队
drilling material 钻进器材;钻进磨料
drilling meal 钻粉
drilling meterage 钻孔进尺
drilling method 钻探方法;钻进方法;条播法
drilling-mining combine exploration 探采结合勘探
drilling-mining combine of development phase 开拓阶段探采结合
drilling-mining combine of minor development phase 采准阶段探采结合
drilling-mining combine of stoping 回采阶段探采结合
drilling motion 钻具运动(钢绳冲击钻进时)
drilling motor 钻探电动机
drilling mud 钻渣;钻探(用)泥浆;钻泥;钻孔泥浆;钻井泥浆
drilling mud cake 钻泥皮
drilling mud jet 钻泥浆喷射
drilling mud materials 钻泥浆材料
drilling mud metering pump 钻探泥浆计量泵

drilling mud practices 钻探泥浆使用技术
drilling mud surfactant 钻探泥浆表面活化剂
drilling number 钻井编号
drilling obligation 钻探契约
drilling obstacle 钻探障碍(物)
drilling off 空转(钻进时)
drilling of glass 玻璃钻孔
drilling of tubesheet 管板钻孔
drilling on the bottom 孔底钻进;纯钻进
drilling on the rake 倾斜钻探
drilling on water region 水上钻探
drilling on waterways 水上钻探
drilling operating instruction 钻探操作规程
drilling operation 钻井作业;钻探作业;钻探操作;钻井操作
drilling operation planning 钻探施工计划
drilling out 钻碎
drilling out a cementing plug 钻水泥塞
drilling out cementing plug time 扫水泥塞时间
drilling outfit 钻探设备;钻井设备
drilling out of the cement plug 钻穿水泥填塞物
drilling parameter 钻探参数;钻井参数;钻进参数
drilling party 钻探队;钻井队
drilling pattern 钻孔图(式);钻探形式;钻孔布置形式;钻孔布置(图);凿岩方式;炮眼形式;炮眼布置(图);炮孔排列
drilling people 钻探者
drilling performance 钻探深入率
drilling performance diagram 钻探作业图
drilling performance per drill and shift 凿岩台班效率
drilling performance per driller and shift 凿岩工班效率
drilling period 建井周期
drilling pile 钻孔桩
drilling pipe 钻凿(井)管
drilling platform 钻井平台;海上钻探平台;海上钻井平台;海上井架;钻探平台;钻台;钻架台车;钻采平台
drilling point 钻(孔)点;钻口;打井位置
drilling pontoon 钻孔工作船
drilling position 孔位;钻孔位置;钻井位置
drilling practices 钻进规程
drilling preengineering 钻前工程
drilling press spindle 钻床主轴
drilling pressure (钻进时的)钻头压力;钻压
drilling pressure meter 钻压表
drilling process 钻井过程
drilling program(me) 钻探程序;钻探计划;钻探方案;钻井程序
drilling progress 钻探进度;钻井进度
drilling progress chart 钻探进度表
drilling quality 钻井工程质量
drilling range 钻探范围
drilling rate 钻进速度;钻探速率;钻探速度
drilling rate per rig-shift 台班效率【岩】
drilling rate per-year 台年效率
drilling recommendation 钻进规范
drilling record 钻探报表;钻探记录(报)表;钻探记录
drilling regulation 钻进规程
drilling report 钻探报告
drilling result 钻探结果
drilling returns 钻探冲洗液返流
drilling rig 钻塔;钻井设备;钻具;钻井装置;钻机;打井机;钻探机(具);钻探设备
drilling rig developer 钻机设计人员
drilling rod 钻杆;钎子
drilling rope 钻探用绳(索);钻探用钢丝绳;钻探绳索;钻井用钢丝绳
drillings 钻屑
drilling sample 钻探土样;钻探取样
drilling scow 钻探(驳)船
drilling sequence 钻孔顺序
drilling set 钻具组
drilling shaft 钻杆柱
drilling shift 钻探轮班;钻探班次
drilling ship 钻探船
drilling shock 钻进时振动
drilling shot-holes in the face 表面炸孔钻探
drilling site 孔位;钻孔位置;钻井工地;钻井位置
drilling site area 井场面积
drilling site arrangement 井场布置
drilling site length 井场长度
drilling site wide 井场宽度

drilling sludges 钻屑
drilling spacing 钻孔间距
drilling specification 钻探规范;钻探规程
drilling speed 钻进速度;钻探速度
drilling speed per round trip 回次钻速
drilling stand 钻床柱
drilling standard 钻探规范
drilling steel 钻探用钢(材);钢钎
drilling stem 钻杆(钢丝绳冲击)
drilling string 钻杆柱
drilling stroke 凿岩机冲程;冲击钻头行程
drilling superintendent 钻探指挥人
drilling superstructure 凿岩机台车上部结构
drilling survey 钻孔测量
drilling survey by boring 钻探
drilling synchronized to mucking 钻眼装岩平行作业
drilling system 钻探(工程)系统;钻探方法
drilling target 钻孔目的
drilling team 钻探队;钻井队
drilling technical management 钻探技术管理
drilling technical specification 钻探技术要求
drilling technique 钻进技术
drilling technology 钻探工艺;钻井技术;钻进工艺;凿岩工艺
drilling template 孔口盘;井口盘
drilling tender 钻探辅助船;钻井附属船
drilling test 钻孔试验;钻探试验
drilling thrust (钻进时的)钻头压力
drilling time 钻探时间;钻探施工期;钻进时间
drilling time break 钻探时间中断
drilling time logs 钻速曲线;钻进时间曲线;钻时记录
drilling time recorder 钻速记录仪;钻进时间记录仪
drilling to completion 钻到设计深度
drilling tool 击探钻具;钻探工具;钻孔刀具;钻具
drilling tool blocking 卡钻
drilling tool joint 钻探工具接头
drilling tool substitute 钻具接头
drilling tower 钻探塔架;井架;钻(井)塔
drilling trouble 钻探事故
drilling unify coding 统一编码
drilling unit 钻井装置;钻机机组
drilling up 钻carry
drilling up cement retainer 钻碎水泥塞
drilling valve 钻井用阀门
drilling vessel 钻探船;钻井船
drilling wastewater 钻井污水
drilling water 钻孔冲洗水;钻井用水
drilling weight 钻压;钻具重量
drilling well 钻井;钻探井筒;凿井
drilling well depth 钻井深度
drilling winch 钻机绞车
drilling with aerated water 用汽水钻孔
drilling with casing 跟管钻进
drilling with counterflow 反循环钻进
drilling with reversed circulation 反循环钻孔
drilling with sound vibration 声频振动钻进
drilling worker 钻工;钻探工人
drilling works 钻探工程;钻井工程;钻井操作
drilling yard 钻探现场
drill jib 斜村钻架;钻机托臂
drill jig 钻模
drill jig for air cylinder 风箱钻模
drill jig for gasket 垫圈钻模
drill jig for lathe 车床钻模
drill jumbo 凿岩台车;钻架车;轨道式钻;大型多头钻机
drill jumbo with two booms 二臂凿岩台车
drill key 拔钻楔铁
drill ladder 钻梯
drill lathe 钻床;卧式钻床
drill lip angle 钻刀角
drill lip clearance 钻刀余隙角
drill log 岩芯记录;钻杆柱状图;钻孔记录;钻孔编号;钻井柱状图
drill man 钻工
drill manual 训练手册
drillmaster 钻探技师;钻探工头;教练;训练员;高架式潜孔钻机
drill material 钻井磨料
drill member 钻机部件
drill mill 铣刀式钻头(消减孔内钻具用)
drill mobile 钻车;凿岩车

drill motor 钻机马达
drill-motor rotor vane 钻孔转子叶片
drill-motor totor 钻机马达转子
drill mounting 钻架;钻机装载
drill off 扫孔【岩】
drillometer 钻进仪表(负荷和钻速指示器)
drill out 钻开;钻掉;钻出;取出岩芯
drill over 钻开;过钻
drill pad 钻垫
drill pcint ga(u)ge 钻头角度规
drill peripheral speed 钻周速
drill pipe 钻管;钻杆
drill pipe annular return velocity 钻杆环空返速
drill pipe box and pin 钻杆公母扣;钻杆公母接头
drill pipe break off 断钻扦
drill pipe cement head 钻杆水泥头
drill pipe clamp 钻杆夹持器
drill pipe cone elevator 钻杆锥形提升器
drill pipe coupling 钻杆接箍;钻管连接器
drill pipe cutter 钻杆割刀;割管器(岩)
drill pipe float 钻杆浮鞋
drill pipe float valve 钻杆浮阀;钻管浮球阀
drill pipe frozen 卡钻
drill pipe ga(u)ge 钻杆螺纹规
drill pipe joints 钻管接头
drill pipe load 钻管荷重;钻管荷载
drill pipe lubricant 钻杆丝扣油
drill pipe pin 钻管销
drill pipe plug 钻杆胶塞
drill pipe protector 钻杆丝扣护箍
drill pipe slips 钻杆卡瓦
drill pipe stabilizer 钻杆稳定器
drill pipe stand 钻杆立根
drill pipe string 钻杆组;钻杆柱
drill pipe sub 钻杆代用品;钻杆接头
drill pipe thread 钻管螺纹
drill pipe thread dope 钻管螺纹润滑
drill pipe tongs 钻杆钳
drill pipe torque 钻杆扭矩
drill pipe twist-off 钻管拧紧
drill pipe wall thickness 钻杆壁厚
drill piston 凿岩机活塞
drill piston bars 凿岩机活塞杆
drill piston stroke 凿岩机活塞冲程
drill planter 条播器
drill planting 畦植
drill planting method 条植法
drill plate 钻模板
drill platform 钻井平台
drill plugger 凿岩机
drill point 钻头;钻刃;钻尖
drill point angle 钻尖角
drill pointer 钻尖
drill point grinder 钻头刃磨机
drill pointing machine 钻头磨尖机
drill post 钻机支柱
drill power feed 凿岩机自动推进
drill press 立式钻床;压钻机钻床;钻床
drill press grinding 钻床磨削
drill press parts 钻床零件
drill press sleeve 短钻套
drill press socket 长钻套
drill pressure 钻头负载
drill program(me) 钻进计划
drill progress 钻进过程
drill pump 钻探用泵
drill quality diamond 钻(探)用金刚石
drill rate 钻速
drill rate value 钻进速度值
drill reamer 钻铰复合刀具
drill record 岩芯记录;钻孔剖面;钻孔记录
drill regime 钻进规程
drill repair shop 修钎车间;钻机修理车间;凿岩机修配(车)间;凿岩机修理(车)间
drill rests 凿岩机架
drill rig 架式钻床;(成套的)钻进设备;钻架(车);凿岩机
drill rig drifter 架式凿岩机
drill rod 带孔棒;钻杆
drill rod bit 钻杆钻头(不带岩芯管的)
drill rod check valve 杆式止回阀
drill rod coupling 钻杆接头;钻杆接箍
drill rod drive quill 钻杆卡盘
drill rod joint 钻杆接头

drill rod life 钻杆寿命
drill rod rotation speed 钻杆回旋速度
drill rod size 钻杆尺寸
drill rod sticking 卡钻
drill rod string 钻杆柱
drill rotation ratchet 凿岩机旋转棘轮机构
drill runner 钻工;凿岩机操作员
drill sample 钻取岩(芯)样
drill sampler 钻杆式采样器
drills and mountings 打眼装药
drill school 训练学校
drill scout 轻型钻探机;轻型移动式钻机
drill series 钻机系列
drill set 一套钻头;钻探器具;钻具
drill setup 钻探装置;凿岩机装置
drill shack 钻塔棚;场房【岩】
drill shank 钻柄;钎尾;钎杆
drill sharpener 钻探刃磨;钻头整修器;钻头整修机;钻头磨尖器
drill sharpening 钻头整修;钻头刃磨
drill sharpening and shanking machine 钻头磨尖和磨细机
drill sharpening equipment 钻头整修器;钻头整修机
drill sharpening machine 磨钻头机;钻头刃磨机
drill sharpening practice 钻头修配作业
drill sharpening shop 修钎车间;钻机修理车间;钻头修配间
drill ship 海底钻探船;海底钻井船;钻井船
drill shoe 条播开沟器
drill single twist 单出屑槽钻
drill site 井位;钻探现场;钻场;凿岩地点;地盘(岩)
drill sleeve 钻头套筒;钻套;短钻套
drill sludge cake 钻泥块
drillsmith 钻头整修工;锻钎工
drill socket 钻头插口;钻头接套;长钻(头)套筒
drills operation 多机凿岩作业
drill speed 钻头转速;钻具转速
drill speeder 速钻器
drill spindle 钻机芯轴;钻轴
drill spindle support 钻轴支架
drill spindle with rack 带齿条长钻轴
drills steel sharpener 改钻机
drill stabilizer 钻具(下部)稳定器
drill stalling 钻机失速
drill stand 凿岩机架;钻机支架;手摇钻台架
drill steel 钻杆;钎子;钻(头)钢
drill steel extension 钻钢接长
drill steel extension diameter 钻钢接长直径
drill steel furnace 钎钢加热炉
drill steel holder 夹钎器
drill steel retainer 钢钎护圈;夹钎器
drill steel set 成套钻具;钻头组
drill steel sharpener 钻头磨锐机
drill steel shop 钻钎车间
drill steel support 钢钎支架
drill stem 钻柱;钻杆
drill stem bushing 钻杆螺丝缩接;钻盘补心;方钻杆卡瓦
drill stem bushing holder 钻杆螺丝缩接夹具
drill stem controlled down hole test valve 钻杆控制井底测试阀
drill stem logging 钻杆记录
drill stem tester 地层试验器
drill stem test(ing) 钻杆试井;钻杆试验;钻杆(地层)测试
drill stock 钻柄
drill stop 钻头定程停止器
drill string 钻具组;钻柱;钻杆组
drill string changing time 换钻具时间
drill string heave compensator 钻柱升沉补偿器
drill string torque 钻杆柱扭矩
drill string tripping 钻具组升降程序
drill studs 钻头镶嵌物
drill swaying 钻杆柱弯曲
drill tap 钻孔攻丝复合刀具
drill template 钻头径规
drill test 钻孔试验
drill the plug 钻碎塞堵
drill throttle valve 凿岩机节流阀
drill throttle valve handle 凿岩机节流阀手柄
drill thrust 钻削力;钻头(轴向)压力
drill time log 钻井记录

drill tip 钻尖
drill to design depth 钻至设计深度
drill tool 凿岩工具;钻具
drill tool for controlling hole deviation 控制钻孔偏斜钻具
drill tool sticking 卡钻
drill to predetermined point 钻到预定深度
drill tower 训练塔;钻架
drill tower brace 钻架支撑;钻塔支撑
drill tower cellar 钻塔地下室;钻架地下室
drill tower conrice 钻塔上飞檐;钻架上飞檐
drill tower crane 钻塔吊车;钻架吊车;钻井起重机
drill tower crown 钻塔塔顶;钻架塔顶
drill tower foundation 钻塔基础;钻架基础
drill tower foundation post 钻塔基柱;钻架基柱
drill tower girder 钻塔大梁;钻架大梁
drill tower grillage 钻塔基础格床;钻架基础格床
drill tower leg 钻塔支柱;钻架支柱
drill tower man 钻塔工作人员;钻架工作人员
drill tower substructure 钻塔下部结构;钻架下部结构
drill tripod 三脚钻塔;三脚钻架;三角钻架
drill truck 钻探车
drill tubing 钻探管材;钻管
drill unit 钻削动力头;钻探机组
drill up 钻完
drill upward 由下向上钻进(在坑道内)
drill valve chest 凿岩机阀门箱
drill wagon 钻车
drill water 钻进用水
drill water hose 凿岩机供水管
drill water line 钻探供水管路
drill water valve 凿岩机水阀门
drill way 孔道;钻出的孔
drill weave 斜纹布
drill well 管井
drill with casing tubes 套管钻进
drill with cylindrical shank 有圆柱形柄的钻
drill with ferrule 弓钻;方钻
drill with hexagonal shank 有六边形柄的钻
drill with Morse cone 莫氏圆锥钻
drill with Morse taper 莫氏锥度的钻
drill with slurry 用泥浆钻进
drill working-month 钻月
drill yard 铁路货车调车场
drimeter 含水量测定计;干度计;水分测定仪;水分测定计;湿度计
drinker 饮水器
drinkery 酒吧间;酒店
drinking and lodging tax 膳宿税
drinking bowl 饮水器
drinking cup 饮水器
drinking-cure 饮疗
drinking fountain 喷泉式引水龙头;饮用喷泉;饮水器;饮水龙头;自动饮水装置;自动饮水器;喷泉式饮水器
drinking fountain stand 饮水台
drinking glass 饮用玻璃杯
drinking jar 饮水缸
drinking paper 吸水纸
drinking-place 饮水供区
drinking spring 饮用泉
drinking tank 饮水柜
drinking test 喝水试验
drinking tolerance 饮用水中污染物容许量
drinking trouph 饮水器
drinking water 饮用水
drinking water administrator 饮用水管理者
drinking water biostability 饮用水生物稳定性
Drinking Water Conservation Zoning Regulation 饮用水分区保护法令
drinking water contaminant 饮用水污染物
drinking water cooler 饮用水冷却器
drinking water criterion 饮用水基准
drinking water criteria standard 饮用水水质基准标准
drinking water defluoridation 饮用水除氟
drinking water demand 饮用水需求量
drinking water disinfection 饮用水消毒
drinking water disinfection byproduct 饮用水消毒副产物
drinking water distribution system 饮用水配水系统
drinking water fluoridation 饮用水加氟;饮用水氟化

drinking water for livestock 牲畜饮用
drinking water issue 饮用水问题
drinking water microbial contamination 饮用水微生物污染
drinking water network 饮用水网
drinking water pool 饮水池
drinking water preparation 饮用水制备
drinking water pressure pipe 饮用水压力管
drinking water pump 饮用水泵
drinking water quality 饮用水水质
drinking water quality assessment 饮用水水质量评价;饮用水水质评估
drinking water quality limit 饮用水水质容限
drinking water quality monitoring 饮用水水质监测
drinking water quality standards 饮用水水质标准
drinking water reservoir 饮用水蓄水池;饮用水水库
drinking water sanitary standard 饮用水卫生标准
drinking water shortage 饮用水短缺
drinking water source 饮用水源;饮用水水源
drinking water source area 饮用水水源区
drinking water standard 饮用水(质量)标准
drinking water storage tank 饮用水蓄水池
drinking water supply 饮(用)水供应;饮用水供水;饮水供给
drinking water supply pipe 饮用水供水管
drinking water supply quality 饮用水供水水质
drinking water supply source 饮用水供水水源
drinking water supply system 饮用水供水系统
drinking water surveillance 饮用水监测
drinking water surveillance program(me) 饮用水监测程序
drinking water tank 饮用水水箱;饮用水箱;饮用水槽
drinking water test 饮用水检测
drinking water testing method 饮用水检查法;饮用水检测法
drinking water tolerance 饮用水含污染物容许量
drinking water treatment 饮用水处理
drinking water treatment plant 饮用水处理厂
drinking water treatment system 饮用水处理系统
drinking water well 饮用水井
drink wastewater 饮料废水
driography 无水胶印
drip 检油池;泄出器;滴注法;滴水(器);滴水(槽);滴流器;滴(流)口;滴出物;水滴;湿透;滴下
drip bead 珠状挂线
drip box 冷凝液箱;承油盘
drip cap 外窗台;遮檐;滴槽;滴水挑檐;外阳台
drip catcher 捕集器
drip chamber 沉淀室;排水室;排水池
drip channel 滴水槽
drip cock 滴水器旋塞;滴水开关;滴(降)栓;取样开关;排气旋塞
drip collector 采酸管
drip condenser 水淋冷凝器
drip cooler 点滴式冷却器;水淋式冷却器;水淋冷凝器
drip cooling plant 点滴式冷却设备
drip crater 锡滴坑(浮法玻璃缺陷)
drip culture 滴液培养
drip cup 收集器;油样收集器;承接杯
drip curtain 滴水帷幕;石灰华帷幕
drip drapery 滴水帷幕;石灰华帷幕
drip-dry 滴干
drip edge 滴水檐槽;滴水槽檐;滴水边缘
drip feed 逐滴供给;点滴注法;点滴注水;滴油润滑
drip feeder 滴给(加药)器
drip feed lubricator 滴给式注油器;滴油润滑器
drip feed oiling system 滴油润滑系统
drip feed oil lubricator 点滴注油器
drip feed valve 流滴供给阀;滴馈阀
drip flame time 滴液燃烧时间
drip flow 滴流
drip-free paint 无滴痕涂料
drip furnace 液体燃料炉
drip gasoline 液馏汽油
drip groove 滴水(凹)槽
drip hole 滴水孔;滴水洞;排孔;排水管(道)
drip impression 滴痕
drip irrigation 滴灌
drip irrigation and spray irrigation 滴灌和喷灌
drip irrigation emitter 滴灌器

drip irrigation technique and equipment 滴灌技术及其配套设备
drip irrigator 滴灌器
drip joint 金属片互搭接头;屋面金属片互搭接头
drip leg 冷凝液罐;滴油器;滴水管;滴水斗
drip line 滴水线
drip loop 滴水环;水落环管
drip lubrication 点滴润滑;滴油润滑法;滴油润滑
drip lubricator 滴油润滑器
drip melt 滴熔(法)
drip melting 悬熔;吹灰熔炼
drip melting apparatus 滴熔设备
drip mo(u)ld(ing) 滴水线(脚);石砌滴水槽;成型件;落水板;分水板
drip mo(u)lding pliers 成型件装拆钳
drip nozzle 滴嘴;滴水喷嘴
drip nozzle connection 滴嘴接头
drip oil 管道冷凝液
drip oiler 滴液式注油器
drippage 滴下的水
drip pan 滴(水)盘;接漆盘;油滴收集盘;滴(油)盘;滴(料)盘;存油盘;承屑盘;承水盘;酸样收集器;收集盘;盛液盘
dripper 沥干架;滴头
drip pin 滴料冲头
dripping 水滴声;流挂;浸渍滴淌;浸涂滴淌;液滴;水滴;滴下
dripping board 滴水板
dripping cup 酸样收集器;承油杯
dripping density 淋水密度
dripping-drop atomization 点滴喷雾
dripping eaves 滴水檐(口)
dripping fault 滴水断层
dripping moisture 冷结水;滴落水;冷凝水
dripping noise 滴水声
dripping pan 接滴盘;烘烤盘;承屑盘
drippings 跌落物;滴下物
dripping temperature 成滴温度
dripping tile 滴水瓦
dripping water 滴水
dripping wet 淋湿的
drip pipe 冷凝(液)泄出管;冷凝(水)泄出管;滴管;排出管
drip pocket 凝液收集匣;凝液收集袋;冷凝液收集袋
drip point 冷凝点
drippoint packing 点滴式填料
drip pot 冷凝液罐;滴水筒
drip-proof 防滴(漏)的;不透水的
drip-proof burner 防塞燃烧器;防滴喷燃器
drip-proof generator 防滴式发电机
drip-proof machine 防(滴)水电机;不滴油机器
drip-proof motor 防滴式电动机
drip-proof screen-protected machine 防滴罩防护式电机;防滴网罩保护式电机
drip-proof type 防滴型
drip-proof type induction motor 防滴式感应电动机
drip-proof type motor 防滴水式电动机
drip pump 小型(抽)水泵
drip ring 滴油圈;滴水环;润滑环
drip riser 冷凝液排放管
drips 料滴;粉块;收集液
dripshed 滴水棚
drip shield 防滴罩
drip sink 滴水盘;滴水盆
drip spot 滴迹
drip stock 冷凝液排放管
dripstone 檐溜石;滴水檐口;钟乳石;滴水石;石笋;钟乳状方解石;滴石类;石灰水檐
dripstone cave 滴水石洞穴
dripstone course 滴水石层
drip strip 易脱模板条;刷油脱模
drip tank 集漏油箱;滴液箱;收集罐
drip-tight 防滴的
drip tile 滴水
drip tip 滴水叶尖
drip torch 滴液点火器
drip trap 冷凝液罐;滴水收集器;滴阀
drip tray 滴水碟;接漆盘;接墨盘
drip trough 集水凹槽;滴水槽;疏水槽
drip tube 滴管
drip valve 泄水阀(门);(调节液压进的)针阀
drip-vessel 漏壶

drip watering 滴灌
drip water plate 淋水板
drivage 巷道掘进
drivage method 开凿方法;掘进方法;打桩方法
drivage ratio 掘进率
drive 驾驶(车辆);内驱力;原动机;沿脉平巷;传动;驱动;推进;散漂材
drive a roaring 生意好
drive assembly 传动装配
drive a well 打井
drive axis 主动轴
drive axle 驱动(车)轴;主动轴
drive back gear 倒车齿轮
drive band 护桩箍;桩箍
drive belt 传动皮带
drive belt adjuster 传动皮带调节器
drive belt housing 传动皮带罩
drive bit 打入式钻头
drive block 井架;带动销;打桩锤;套管锤
drive-block extension 吊锤加长段
drive bolt 传动螺栓
drive bushing 传动套管;转盘补心
drive cam 传动凸轮;主凸轮;导(动)凸轮
drive cap 引帽;引盖;导盖;锤帽;打头护帽
drive capstan 主导轴
drive case 后桥壳体
drive casing 打入套管
drive chain 主动链;传动链条
drive chain and belt identification 传动链和传动带的辨别
drive chain guard 传动链条罩
drive characteristic 激励特性;传动特性;驱动特性;调制特性
drive chuck 钻机立轴卡盘;传动卡盘
drive circuit 激励电路;策动电路;驱动电路
drive clamp 钻管夹;钻杆夹;(旋冲钻杆的)增重环;承击夹持器;打入管夹板
drive clutch 传动离合器
drive coil 激励线圈;传动线圈;驱动线圈
drive collar 桩箍;打入管接箍;套管帽
drive control device 传动控制装置
drive controller 传动控制装置
drive control relay 继电切换器;切换继电器
drive control unit 传动控制装置
drive current 激励电流;驱动电流
drive cylinder 传动缸
drive down 压低
drive down casing spear 震击式捞管器
drive down socket 震击式打捞筒
drive drum 传动卷筒;驱动压路机;驱动碾轮
drive element 驱动元件
drive end 轴的驱动端;主动端;驱动端
drive end bearing 主动端轴承
drive end felt washer guard 主动端毡垫圈护罩
drive end head 主动端盖板
drive end plate slinger 主动端抛油环
drive fit 牢配合;密配合;打入配合
drive flange 主动凸缘;传动法兰(盘)
drive force 主动力
drive friction plate 传动磨擦盘
drive gear 传动齿轮;驱动机构;驱动齿轮
drive (gear) box 传动齿轮箱;主动齿轮箱
drive gear carrier 主动齿轮箱
drive gear casing 主动齿轮箱
drive gear mechanism 传动齿轮机构
drive gear (wheel) 主动齿轮
drive hammer 吊锤【岩】;打入锤
drive-hammer extension 吊锤加长段
drive head 桩帽;打头;传动机构;承锤头(桩的、触探);套管帽;钻杆打头
drive head frame (输送机的)机头架
drive head yoke 压紧套箍(用以压入取样器)
drive hear 传动机构
drive home 敲进去;车送回家
drive housing 传动箱
drive idler gear 空转传动齿轮
drive-in 打入;敲进;路旁餐馆;汽车电影院;路旁旅社
drive-in bank 自动出纳银行;路旁汽车银行;汽车银行
drive-in business 路旁服务小店
drive-in cinema (可以把汽车开进位,并坐在汽车里看的)露天电影院
drive-in cotter 打入用楔
drive-in diffusion 主扩散

drive in/drive off ship 机动车辆专运船
drive-in establishment 路旁服务业;路旁服务企业
drive-in facility 路旁设施(如餐馆等);露天汽车电影院设施
drive-in full section 全断面掘进
drive-in refreshment stand 路边茶点亭;路边茶点站
drive-in restaurant 饮料亭;外送餐馆;快餐亭;茶点亭;汽车饭店;汽车餐馆
drive-in teller 自动银行出纳员;(开车进去方便服务的)出纳员
drive-in teller window 汽车银行出纳窗
drive-in theater 汽车剧场;汽车(电)影院;停车露天电影场
drive level 驱动电平
drive line 传动线;驱动管路;传动系统
driveline retarder 传动系统减速器
drive link 传动(联)杆;驱动链
drive lock ring 传动锁环
drive magnet 驱动磁铁
drive master 自动传动装置
drive mechanism 传动机构;驱动机械装置;驱动机理;驱动机构
drive motor 传动马达;驱动电(动)机
driven 从(传)动的;传动的;被驱动的
driven array 有源天线阵
driven ashore 被冲上岸
driven bevel pinion 从动伞齿轮
driven blocking oscillator 单稳多谐振荡器
driven bogie 从动台车
driven by investment demand 受投资需求拉动
driven caisson 管柱;打入式沉箱
driven cased pile 打入的现浇钢壳混凝土桩
driven casing 打桩套筒
driven cast-in-place pile 钢管混凝土现场灌注桩;就地灌注桩;灌注桩;沉管灌注桩;套管就地灌注桩
driven cast-in-situ pile 打入现浇混凝土桩;灌注桩(打入式);打入灌注桩
driven chain 传动链
driven disk 被动圆盘
driven end 从动端
driven fit 紧配合
driven fluid 被引射流体
driven friction disk 被动摩擦盘
driven friction ring 被动摩擦圈
driven gear 从动机构;从动(齿)轮;被动齿轮
driven gear shaft 从动齿轮轴;被动齿轮轴;随动齿轮轴
driven ground 埋管接地
driven-in pin 插入销
driven in-situ pile 打入式灌注桩
driven lever 从动杠杆
driven machine 从动机
driven mechanism 从动机构
driven member 从动(构)件;被动轮;从动构件
driven multivibrator 从动多谐振荡器;随动多谐振荡器
driven nozzle 高压远射程喷嘴
driven off by shaft 轴传动的
driven on rock 搁礁
driven on the butt 与次解挥垂交掘进
driven on the face 与主解挥垂交掘进
driven on the rake 倾斜打入
drive nozzle 压力喷嘴
driven pile 打入桩;锤入桩;锤击桩;入土桩
driven pile underpinning 打入桩托换
driven piling 打入
driven precast concrete pile 打入式预制混凝土桩
driven pulley 从动皮带轮;从动摩擦轮;从动滑轮;被动摩擦轮
driven radiator 有源天线;激励辐射天线
driven record 打桩记录
driven roller 从动辊
driven roller conveyer 辊式传动运输机;驱动式转子输送机
driven shaft 从动轴;被动轴
driven shaft bearing 被动轴承
driven shaft bearing cap 从动轴承盖
driven shaft bearing cap gasket 从动轴承盖垫密片
driven shaft bearing cone 从动轴承锥
driven shaft bearing cup 被动轴承杯
driven shaft gear 被动轴齿轮
driven shaft nut 从动轴螺母
driven shell pile 打入带壳桩;打入包壳桩

driven spindle 从动(转)轴
driven spring 复进簧
driven sprocket 从动链轮
driven steel pile 打入钢桩
driven sweep 驱动扫描
driven timber pile 打入木桩
driven torque 驱动扭矩
driven torus 从动环
driven transmitter 受激发射机
driven tube pile 打入式管桩
drive number 驱动器编号
driven underpinning 打入式托换工程
driven underpinning pile 打入式托换桩
driven vehicle 电动车辆
driven vertical pilot beam 打入垂直导向梁
driven well 管井;机井;(打)(人)井
driven well pump 钻井泵
driven wheel 从动轮;被动轮
driven wood(en) pile 打入木桩
drive oblique pile 打入式斜桩
drive off 驱散;馏出
drive of transverse slide 横向滑板传动
drive on/off 开上开下;轮渡式船舶
drive on/off ship 汽车轮渡
drive on/off system 开上开下方式
drive out 打掉
drive part 主动件;驱动部件;驱动部分
drive pattern 驱动(误差)图形;驱动斑纹
drive pile 打桩
drive pin 带动销;传动销
drive pinion 主动小齿轮;传动小齿轮;驱动小齿轮
drive pinion adjusting nut 传动小齿轮调节螺母;传动齿轮调整螺母
drive pinion and shaft 主动小齿轮及轴
drive pinion cone bearing 主动小齿轮锥形轴承
drive pinion cone bearing cup 传动小齿轮锥形轴承杯
drive pinion housing gasket 传动小齿轮壳垫密片
drive pinion oil seal 主动小齿轮油封
drive pipe 洛阳铲;护筒;主动钻杆;主动管;打入管;取样套管
drive pipe drilling 打管法钻进;套管钻进
drive pipe head 打入管头;套管打头
drive pipe ring 导管环;套管环
drive pipe shoe 导管鞋;套管靴
drive piping operation 下定向管操作
drive piston 传动活塞
drive point 策动点;驱动点
drive point penetrometer 打入式尖锥;打入式贯入器
drive power 推进力;牵引能力;驱动力
drive pulley 传动(滑)轮;主动(皮带)轮;驱动皮带轮;驱动滚筒
drive pulse 激励脉冲;触发脉冲;策动脉冲;驱动脉冲
drive-pulse generator 驱动脉冲发生器
drive quill 钻机立轴卡盘;钻杆卡盘
driver 螺丝起子;(汽车的)驾驶员;激励器;激励级;打入工具;传动箱;传动器;驱动机;策动器;驱动器;驱动极;驱动程序;司机(卡车)
drive ratio 传动比
driver axle 动轮轴
driver brake 主动轮制动器
driver brake oil cup 主动轮闸油环
driver chuck 拨盘
driver circuit 驱动电路
drive-reduction 驱力降低作用
driver element 驱动单元;驱动部件
driver-extractor vibrating hammer 打桩拔桩两用振动锤
driver-extractor vibratory hammer 打桩拔桩两用振动锤
driver gear 传动齿轮
driver guide arm 驱动导杆
driver information system 驾驶员(获取)信息系统
drive ring 导向环
driverless reversal operation 无人驾驶自动折返
driverless reversing of train 自动折返
driverless train 无人驾驶列车
driver level 驱动器电平
driver line 驱动线路
driver link 驱动链节
driver load line 激励级负载线
driver motor 主驱动电动机
drive roadway 打通道路

drive rod 钻机立轴;传动杆;驱动杆
drive-rod bushing 钻机立轴套筒
drive roll 主动辊;传动辊;驱动压路机;驱动碾轮;驱动滚轮
drive roller wheel 压路机主动轮
drive round 使传动
driver parking 自由停车
driver parking building 自由停车房建筑
driver plate 带动盘;(车床的)拨盘;驱动圆盘
driver pricking 打桩机测探
driver probing 打桩机测探
driver pulley 传动(机)皮带轮
driver ratio 传动比
driver routine 传送程序
driver's assistant 驾驶员助手
driver's automatic brake valve 驾驶员自动制动阀
driver's brake valve 驾驶员制动阀
driver's cab 驾驶室
driver's cab installation 驾驶员室安装
driver's cab interior finishing 驾驶员室内装修
driver's cab lighting 驾驶员室照明
driver's cab side door 驾驶室侧门
driver's cage 驾驶员室
driver's compartment 操纵室
driverscope 驾驶仪
driver's desk 驾驶员(操作)台
driver seat 驾驶座
driver's infrared viewer 驾驶员红外夜视仪
driver skew 驱动器时差
driver's licence 驾驶执照
driver's license 驾驶执照
driver's low light level viewer 驾驶电微光夜视仪
driver's oil 采矿工用油
driver speed 驱动器速率
driver spring 传动轮弹簧
driver's room 驾驶室
driver's safety device 驾驶员安全装置
driver's seat 驾驶员座位
driver stage 激励级
driver's tool 随车工具
driver's tool and accessories 随车工具及附件全套
driver sweep 触发(式)扫描
driver's weight adjustment 车架刚度调节器
driver swivel seat 驾驶员旋转座
driver training area 驾驶员训练场
driver training field 驾驶员练习场地
driver training simulator 驾驶员模拟操纵教学设备
driver transformer 激励变压器
driver trolley (有驾驶员座位的)电动缆车
driver unit 主振部分;推动器
drive sample 击入式土样;压入法取样
drive sampler 击入式取土器;压入法取样器;打入式取样器
drive sampling 钻进取样;打入式取样;推进取样;压入法取样
drive screw 敲入螺钉;螺纹钉;压入螺钉;传动螺杆;打入螺钉
drivescrew nail 螺纹钉;打入螺钉
drive section 传动装置
drive selection 驱动(器)选择
drive sense 驱动器检测
drive set 驱动装置
drive shaft 主动轴;传动(主)轴;驱动轴;驱动杆
drive shaft bearing 主动轴轴承
drive shaft of the camshaft 凸轮轴的主动轴
drive shaft safety strap 传动轴安全圈
drive shaft spline 传动主轴止转楔
drive shoe 桩靴;管靴;打入靴;卡箍;套筒靴;套管管靴
drive side 驱动侧;主动端;主动侧
drive side of tooth 工作齿面
drive signal 驱动信号
drive sleeve 传动轴套
drive soil sampler 打入式取样器
drive spindle 驱动杆
drive spring 主动弹簧
drive sprocket 履带驱动轮;主动链轮;驱动链轮
drive sprocket axle 主动(链)轮轴
drive stand 驱动机座
drive station 动力站
drive stopping relay 止动继电器
drive style 传动方式
drive system 传动系统
drive tcrque 传递力矩

drive the pile home 把桩打到底
drive the rivets 装铆钉
drive through 驾驶通过；驾驶穿过
drive-through access 通车便道
drive to maturity stage 趋于成熟阶段
drive to refusal 打到抗贯入
drive tube 击入式取土器；打入式取样器；打入式取土器
drive tumbler 驱动轮
drive type 驱动装置形式
drive-type grease fitting 压入式加油嘴
drive-type oil cup 压力加油器；压力加油杯
drive-type split-tube soil-sampling device 带半合管压入式取样器；带半合管压入式取土器；打入式开缝管取土器
drive unit 传动装置；动力单元；传动系；驱动装置；驱动设备；驱动机组
drive up 加速传动；抬高
driveway 行车道；公路；汽车道（路）；赶畜道；车道；快车道；机动车道
driveway greening 车行道绿化
drive wedge 打入式（造）斜楔
drive wheel 主动轮；驱动轮
drivie shaft housing 主动轴箱
driving 掘进；主动（的）
driving adaptability 驾驶适应性
driving agent 驱动装置
driving amplifier 激励放大器；驱动放大器
driving anchor 浮锚
driving and reversing mechanism 进退装置
driving angle 传动角度
driving area 转弯面积（汽车）
driving attachment 打桩附件
driving axis 主动轴
driving axle 主动轴；动轴；传动（车）轴；驱动轴；驱动桥
driving axle box 传动轴箱
driving axle housing 主动轴套
driving axle sleeve 主动轴套
driving band 护桩箍；桩箍；传动带；铁箍桩帽；木桩顶头钢箍
driving bar 传动杆
driving beam 远光
driving belt 传动（皮）带
driving bevel gear 主动伞齿轮
driving bevel pinion 主动小伞齿轮
driving blade 传动刀片
driving block 传动滑块
driving borehole for dewatering 打入式疏干钻孔
driving box 主动轴箱；驱动轮轴箱
driving bushing 打桩衬套
driving cab 驾驶室
driving cam 工作凸轮；传动凸轮；驱动凸轮
driving cap 桩帽；钢桩帽；打桩头；打桩帽；受锤桩帽
driving capacity 牵引能力
driving carriage 自行式车辆
driving center 传动毂
driving chain 主动链；传动链（条）
driving chain sprocket 主动链轮；传动链轮
driving characteristics of train 列车牵引性能
driving chuck 传动拨盘
driving circuit 激励电路；驱动电路
driving claw 主动爪
driving clock 转仪钟；驱动钟
driving clutch 传动离合器
driving cog 传动小齿轮；传动齿
driving condition 行车条件
driving cone 级轮；塔轮
driving cone pulley 主动锥形轮
driving console instrumentation 驾驶室控制装置
driving control 行车控制；传动控制
driving crank 传动曲柄
driving curve 打桩曲线
driving cycle 驱动循环；试用期
driving depth 打桩深度；打入深度
driving device 传动装置
driving direction 行车方向；驱动方向；驾驶方向；主动机械运动方向；传动方向
driving disc 传动盘
driving dog 传动止块；传动夹头；传动挡块；桃子夹头
driving drum 缆车绞索轮；传动鼓轮；驱动滚筒
driving efficiency 驾驶效率；驾驶技能；操纵效率
driving effort braking mode 动力制动方式

driving end 驱动端；主动端；传动端；扳手头
driving energy 打桩能量；驱动力
driving engine 动力机
driving error 传动误差
driving excitability 驾驶兴奋
driving face 开挖面；掘进面；压力面；（桨叶的）推力面
driving fatigue 驾驶疲劳
driving fid 插接木笔
driving fit 装配；打实（用锤等）；牢配合；紧配合；密合；打入配合
driving fit bolt 牢配合螺栓；密配合螺栓
driving flange 传动法兰；传动凸缘
driving fluid 工作液体；工作流体；传动流体
driving force 主动力；传动力；策动力；驱动力
driving force characteristic curve 汽车动力特性曲线
driving force factor 汽车动力因素
driving force for crystallization 结晶驱动力
driving force index 驱动力指数
driving force of heat-transfer 传热推动力
driving force of motor vehicle 汽车驱动力
driving formula 打桩公式
driving frame 打桩架
driving frequency 策动频率；驱动频率
driving friction ring 主动摩擦圈
driving gear 传动装置；传动齿轮；驱动装置；驱动齿轮
driving gear assembly 传动机构
driving gear box 主动齿轮箱；传动齿轮箱
driving gear key 主动齿轮键
driving gear shaft 主动齿轮轴
driving grout pile 打入式灌注桩
driving hammer 击锤；桩锤；打桩锤
driving head 桩帽；主动头；打桩头；桩锤；驱动头
driving helmet 桩盔；桩帽；受锤桩帽
driving hole 扳手孔
driving home 就位（部件）；打到应在的位置（桩）
driving hood 桩帽
driving index 传动指数
driving influence 传动力；驱动力
driving in stone 硬岩钻进
driving instructor 打桩指导员
driving intermediate gear 传动中间齿轮
driving into hard ground 打入硬质地面
driving iron 铁钻杆
driving jack 驱动千斤顶
driving key 打入键；传动键
driving lane 行车车道
driving lane width 行车道宽度
driving lever 主动杆；拨杆
driving licence 驾驶执照
driving light 行车灯光
driving machine 原动机；驱动机；打桩机；驱动装置
driving machinery 传动机械
driving magnet 传动磁铁；启动磁铁
driving mallet 捻缝锤；镶嵌锤
driving mechanism 主动机构；传动机械装置；传动机构；备用传动装置；驱动器；驱动机构
driving mechanism of shutter 驱动机件快门
driving medium 传动件；驱动介质
driving member 主动构件；传动构件；驱动构件
driving method 掘进方法
driving mode 传动方式
driving moment 转动力矩；驱动力矩
driving motion 传动
driving motor 驱动马达
driving nut 主动螺母；垫帽
driving obstacle 行车障碍
driving offence 违反行车规则
driving off tail rod 导杆传动
driving of pile 打桩
driving of plugs 钉木楔
driving of rivets 铆钉作业
driving of sheet piling 打板桩
driving of stakes 钉木橛
driving of working face 工作面推进
driving on broad front 宽面掘进
driving on the rake 倾斜掘进；倾斜打桩
driving on the right side 右侧行车
driving on the side 靠右（侧）行驶；靠右（侧）行车
driving operation 行车操作
driving oscillator 激励振荡器
driving output 打桩效率

driving part 主动（部）件；传动部件
driving pawl 棘轮爪；传动爪；推动爪
driving pile 锤击沉桩
driving pile abutment 打桩台
driving pin 带动销
driving pinion 主动小齿轮；传动小齿轮；齿轨机车主动小齿轮；驱动小齿轮
driving pipe ring 套管环
driving plant 打桩设备；打桩机；水上打桩设备
driving plate 带动盘；（车床的）拨盘；驱动圆盘；驱动板
driving-plate pin 拨盘插销
driving point 策动点
driving point admittance 驱动点导纳；策动点导纳
driving point characteristic 策动点特性
driving point function 策动点函数；驱动点函数
driving point impedance 策动点阻抗；驱动（原）点阻抗；驱动输入阻抗；激励点阻抗；供电点阻抗
driving point impedance function 驱动点阻抗函数
driving point mobility 驱动点导纳
driving power 励磁功率；激励功率；驱动能力；驱动功率；拖动力；推进力
driving power for dredge pump 泥泵驱动功率
driving preliminaries 打桩准备工作
driving pressure 驱动压力
driving pressure differential 驱动压差
driving process 打桩方法；打桩程序；打桩操作
driving profile 螺钉头部（旋纹）形式
driving psychology 驾驶心理
driving pulley 主动皮带轮；主动滑轮；主动滑车；传动皮带轮
driving pulse 触发脉冲；起动脉冲
driving pupil 学习驾驶员
driving rain 倾盆大雨；大风雨
driving-rain index 暴雨指数
driving-rain rose 暴雨玫瑰图
driving ram 传动柱塞
driving ratchet 主动棘轮
driving ratio 打桩效率；传动率
driving record 掘进记录；打桩记录
driving resistance 夯入阻力；行驶阻力；桩的打入阻力；打桩阻力；打入阻力；锤击阻力
driving resistance of a pile 打桩阻力
driving rig 打桩设备；打桩机架
driving rod 动力触探杆；传动杆
driving roll 驱动滚轮
driving rope 传动索
driving rope wheel 驱绳轮
driving route 行驶路线
driving sampling 击入取样
driving screw 主螺杆；导螺旋；导螺杆
driving section 掘进断面
driving section size 掘进断面尺寸
driving shaft 组合器轴；主动轴；传动轴；驱动轴
driving shaft dowel 传动轴定缝销钉
driving shaft motor 传动轴电机
driving shaft universal joint fork 传动轴万向节叉
driving shoe 桩靴；套管管靴；推进式管头
driving side 主动侧；传动边；驱动侧
driving side of tooth 轮齿主动侧
driving signal 驱动信号
driving simulator 操纵模拟装置
driving skill 驾驶技能
driving sleeve 传动轴套
driving snake 传动软轴
driving snow 暴风雪；大风雪
driving source 激励源
driving speed 行车速率；打入桩速度
driving spring 主动弹簧；传动弹簧
driving sprocket (wheel) 主动链轮
driving stage 驱动极
driving steam turbine 驱动汽轮机
driving stem 主动钻杆
driving strap 传动带
driving stress 桩打入地下时的冲击应力；打桩应力；沉桩应力
driving stroke 工作冲程
driving surface 工作面
driving swarm 分出群
driving sychro 主动自整角机
driving synchro 传动同步机；驱动同步机
driving system 转动系统；传动系统；驱动系统
driving test 试车试验（道路）；驾驶考试；打桩试验
driving test of pile 桩的贯入试验

driving time 驾驶时间
driving toothed gear 主动齿轮
driving torque 主动扭矩;传动转矩;传动力矩;驱动转矩
driving to stopping ratio 采掘比
driving-trailer 驱动拖车
driving truck 牵引车
driving tube (自流井的)竖管;桩管
driving unit 主振器;传动装置;驱动装置;驱动机组
driving unit enclosure 驱动装置罩
driving universal joint 传动万向节
driving up the pitch 逆倾斜掘进;仰斜掘进
driving velocity 打入桩速度
driving vessel 打桩船
driving vibration 打桩(引起的)震动
driving visibility 行车能见度
driving voltage 励磁电压;激励电压
driving washer 活动垫圈
driving water 传动水
driving wheel 主动轮;动轮;传动轮;驱动轮
driving wheel box 传动齿轮箱
driving wheel brake 主动轮闸
driving winding 激励绕组;策动绕组
driving with broad face 宽面掘进
driving with narrow face 窄面掘进
driving work 打桩(工程)
drivometer-equipped test car 装备驾驶仪的测试车
driwood mo(u)ldings 得来木压条
drizzle 毛毛雨;细雨
drizzle drop 毛毛雨滴
drizzling 细雨
drizzling rain 微雨;毛毛雨;细雨
drocarbil 槟榔胂胺
drogue 流锚;空中加油锥套;海锚;坠子;测流浮子;测流浮标;浮锚;浮标;风向(袋);捕鲸索浮标
drogue nozzle 锥套形喷嘴
drogue parachute 引导(降落)伞
droit 法定所得;权利
droke 灌木丛
drome 飞机场;飞机库
dromometer 速度计
dromos 通往地下古墓或古庙的通道;(进入埃及古墓前的)深长引道
dromotron 直隙微波加速器
dromotropic(al) action 变传导作用
dromotropic(al) effect 变导作用
drone 无人驾驶船舶
drone ship 遥控船
drong 通道(英国方言);弄堂;岩席
drooling 垂涎
droop 下垂(度);使下垂
droop birch 欧洲白桦
droop characteristic 下垂特性(曲线)
droop governor 下降调速器
drooping 下降
drooping belt stone 副子;垂带
drooping characteristic curve 下降特征曲线
drooping ridge 垂脊
drooping-ridge mounted animal ornament 垂兽【建】
droop line 下降(曲)线;垂线
droop setter 不等率调整装置
drop 落线;落距;落角间隙;楼面底部离天花板上边距离;枯萎病;下跌;跌水;洒水器;滴剂;拆版;分出点;脱落;水滴;湿透;冒出
drop accretion 雨水累积
drop accumulation soil 坠积土
drop accumulator 滴液聚集器
drop anchor 抛锚
drop annunciator 号牌式交换机;调« 报器;吊牌通报器;色盆降落信号指示器
drop apron 金属屋面滴水槽;檐口金属片滴水槽
drop arch 落地拱;垂拱(拱高度小于一半跨度者);平圆拱;矮碹
drop arm 转向(垂)臂;操纵杆
drop astern 落后【船】
drop a vessel 送过他船
drop-away current 脱扣电流
drop-away time 释放时间
drop-away voltage 释放电压
drop awning 下垂的遮阳篷
drop back downstream 顺流后退
drop ball 坠锤;重力冲击球;捣釜锤

drop ballast 可弃压载
drop-ball penetration test 落球贯入度试验
drop-ball test 落球试验
drop-ball visco(si)meter 落球式黏度仪
drop bar 接地棒;密封滑板;叠梁;落杆(水底凿岩的疏浚设备)
drop-bar grizzly 离合棒条筛;翻落式格筛
drop beam 垂梁
drop bed semi-trailer 低底盘半拖车
drop bed trailer 下垂式拖车
drop black 黑色料滴;椴骨炭;落黑;黑漆粒;珠状炭黑;上等黑料
drop black pigment 灰黑色颜料
drop board hammer 落锤
drop bolt 活节栓;吊螺栓
drop boring tool 落锤式钻探工具
drop bottle filler 滴剂装瓶机
drop bottom 活底;底卸式
drop bottom barge 活底式驳船
drop bottom bucket 落底式铲斗;活底铲斗;钟形底料罐;底开式料桶;活底料罐;活底(式)料斗
drop bottom car 底卸式矿车
drop bottom mine car 活底式矿车
drop bottom seal 自动斗门
drop bottom skip 底卸料斗
drop bow 点圆规
drop-bow compasses 弹簧点圆规
drop box 吊盒;矿浆缓流箱;开口垃圾箱
drop bracket 活动托架;定位单环;吊架
drop brake 自动下落制动器
drop bucket 吊斗
drop catcher 液滴捕集器;捕滴器
drop ceiling 下降平顶;低平顶;低吊顶;(能产生天篷低矮幻觉的)天篷装饰法;假吊顶
drop center rim 凹槽轮缘;深凹式轮辋
drop center wheel 凹槽轮
drop chaining 水平丈量
drop chisel 冲击式凿子;冲击式錾子
drop chisel rock breaker 落凿碎石机
drop chute 泥浆井;跌水(陡)槽;落落式溜槽;跌落急水槽;混凝土溜槽
drop cloth 盖布;苫布
drop compasses 小(圈)圆规
drop connection (连结检查井的)竖向排水管;跌水检查井;跌落连接;跌落井
drop cord 悬垂的照明电线
drop core 下落芯;爬芯
drop core sampling test 动力取芯试验
drop coring device 落锤式取样器
drop coring sampler 落锤式取样器
drop counter 计滴器;液滴计数器;滴剂计数器
drop crushing 落重破碎;坠锤破碎
drop curtain 吊幕;舞台水下活动的幕
drop curtain against smoke 挡烟垂壁
drop-dead halt 完全停机;完全停车;突然停机
drop delivery 下落输送
drop detector head 点滴检测器检测头
drop diameter 水滴直径
drop die 落锤锻模
drop door 落板装置;落板口;吊门;升降门
drop down 减弱;倒下;降落
drop-down curve 降水曲线;降落曲线;陡降曲线;下垂曲线;泄降曲线
drop-down function 降落函数
drop-down section 跌水段
drop drawer pull 下垂式抽屉拉手
drop dry 屋面预抛光;屋面玻璃
drop elbow 起柄弯头
drop ell L形块吊耳;吊耳L形块;带柄的水管弯头
drop-end car 前卸式矿车
drop energy 降落能
drop energy dissipation 跌水消能
drop energy dissipator 跌水消能工
drop-error 抛弃错误
drop escutcheon 下垂式钥匙孔盖
drop fault 下投断层
drop fault indicator relay 吊牌式故障指示继电器
drop feed carburizing 滴注式渗碳法
drop feed lubricator 滴油润滑器
drop feed oiler 滴油式加油器;滴油器
drop fill material 抛填料
drop fill rock 抛石;堆石;填石;抛抛石(料)
drop forge 落锤锻
drop-forged 锤锻;冲锻;落锻

drop forged shackle 模锻卸扣
drop forged steel bench vice 锻钢台虎钳
drop forging 冲锻;落锻;落锤锻造;模锻(件);锤锻
drop-forging die 落锻钢模
drop-frame trailer 低身拖车;低架拖车
drop-free potential 不降落电势
drop front 活动面板
drop gangway 活动引桥
drop gate 提升式闸门;直接(成型)浇口;垂直滑动闸门;升降式闸门
drop gate sluice 吊门水闸;沉放闸门式水闸
drop grate 除灰栅;倾倒炉算
drop guide 落料器
drop guide funnel 导料漏斗
drop hammer 自由落锤(打桩用);落锤;模锻锤;锻锤;单作用锤;单动锤;打桩落锤
drop hammer cable 落锤绳索
drop hammer die 落锤模
drop hammer driver 落锤打桩机
drop hammer for shattering old pavements 击碎旧路面用的落锤
drop hammer hardness of rock 岩石落锤硬度
drop hammer line 落锤绳索
drop hammer method 落锤法
drop hammer pile driver 落锤(式)打桩机;重力打桩机
drop hammer pile driving method 落锤打桩法
drop hammer pile driving plant 落锤式打桩设备
drop hammer stamping 落锤模锻法
drop hammer test 落锤试验
drop hammer tester 落锤试验机
drop handle 活络拉手;悬垂把手
drop hanger 降落钩;吊钩;起落吊钩;架空传动吊架
drop-hanger frame 吊架
drop hardness test 肖氏硬度试验
drop heart 悬锤心
drop height 跌落试验落高;跌落高度;冲击试验落锤高度
drop hole 落砂孔;加砂孔
drop hook 下垂钩;升降钩
drop-impact penetrometer 落锤冲击贯入仪;锤击贯入仪
drop impact test 跌落冲击试验
drop in 冒出;混入信息
drop in atmospheric(al) pressure 气压下降
drop in beam 垂降就位的梁;下挂梁
drop in delivery 降低泵的排量
drop indicator 液滴指示器;吊牌指示器
drop in girder 吊梁;悬梁
drop in head 水头落差
drop inlet 落底式进水口;堕式进水口;跌水进口;跌式进水口
drop-inlet culvert 跌落式进水口涵洞;有落底式进水口涵洞
drop-inlet dam 跌水进口涵箱式土坝
drop-inlet spillway 落底式进水口溢洪道;竖井(跌水)式溢洪道
drop in level 高(程)差;落差
drop in ozone 臭氧减少
drop in pressure 压力下降;压差
drop in pressure head 压强水头差降
drop in pressure water flow alarm 压降水流报警器
drop in price 落价;跌价
drop in substitute 随手可得的代用品
drop in temperature 降温;温度下降
drop into 开进
drop in truss 悬吊桁架
drop in winding 落入式绕组
drop keel 滑动龙骨;下垂龙骨
drop key 钥匙孔盖板
drop key plate 活络柄钥匙
drop leaf 活动翻板;翻板桌
drop-leaf table 折叠桌;悬板桌;有活动翻板的桌子;折页桌;活板桌
droplet 涓滴;小(液)滴;液滴;飞沫;表面滴斑;熔滴;微(液)滴
droplet burning time 液滴燃烧时间
droplet erosion 滴点(高度)冲蚀
droplet infection 飞沫传染
droplet nulei transmission 飞沫核传染
droplet separator 分滴器
droplet test 点滴试验

droplet transmission 飞沫传播
droplet under finish 瓶口下液滴
drop lifter 升降钩
droplight（上下滑动的）吊灯;（上下滑动的）吊窗
drop line 用具进气管;引入线
drop lip hook 安全吊货钩
drop-lock bond 保息债券;自变息债券
drop lock clause 降低锁住条款
drop lubrication 液滴润滑;滴油润滑(法)
drop lubricator 滴油(润滑)器
drop-machine brick 冲压机制砖
drop-machine silica brick 冲压成型的硅砖
dropmade column 砂柱
drop manhole 进人井;跌水井;跌落式窨井;跌落式检查井;跌落井;倒垃圾口
drop meter 滴量计
drop method 点滴法;滴入法;垂滴法
drop model of nucleus 核的液滴模型
drop monkey 模锻件活板车;模锻活扳手
drop mo(u)lding 下垂托板造型;镶板门下沉线脚;低线脚
drop nozzle 滴降式喷嘴
drop number 跌水系数
drop number method 量滴数法
drop of beam 材料试验机加载梁下压
drop of drill tools 钻具陷落
drop-off 浮移;陡坎;陡坡;摘下
drop-off to pickup ratio 返回系数;继电器返回系数
drop-off voltage 跌落电压
drop of potential 电位下降
drop of pressure 压力下降;压降
drop of stress 应力减小
drop of temperature 降温;温度下降;温度降落
drop of the flame 火焰缩短
drop of viscosity 黏度减小
drop oiler 液滴润滑;滴油器
drop ornament 悬吊装饰;吊(悬)装饰;垂饰
dropout 漏码;漏失(信息);回动;遮掉;涂去
drop out colo(u)r 打印油色
dropout compensator 信号失落补偿器
dropout count 斑点总数;丢失总数
dropout current 回动电流;开断电流;下降电流;去励磁电流
dropout error 遗失误差
dropout fuse 跳升式熔断器;跳开式熔断器
dropout halftone 高光区无网凸版
dropout information 丢失信息
dropout line 紧松放空线;放空线;排泄线
drop out of step 失去同步
dropout value 释放参数
dropout voltage 回动电压;开断电压;灭磁电压;去励磁电压
dropout wheel repair 落轮修【铁】
drop overboard 落到船外;抛出船外
droppable landing gear 可放下的起落架
droppable solid ballast 可抛固体压载
droppage 扣回款项
drop panel 无梁楼盖托板(在无梁楼盖结构中围绕柱顶加厚的托板);无梁楼盖加厚托板;托板
drop panel form 柱顶加厚托板用模板
drop pawl 绞盘安全掣;掣手
drop pebble 坠石
dropped bottom 瓶底下沉
dropped ceiling 吊天花板;吊平顶;吊顶棚
dropped core 残留岩芯
dropped curb 坡形路缘口
dropped dot 点虚线
dropped end 断头
dropped end detector 断头指示器
dropped girder 下垂桁架;顶承格栅
dropped girt 顶承配梁;(楼板结构下的)圈梁
dropped-line chart 断续线图
dropped-line exposure attachment 断续线曝光装置
dropped panel（无梁楼板与柱头接触的）加厚部分;下垂拖板;顶柱加厚托板
dropped punt 瓶底下沉
dropped seat 凹椅座
dropped throat 沉降式流液洞
drop pen 点圆规
drop penetration sounding 锤击贯入测探
drop penetration sounding appararus 落穿测深装置
drop penetration test(ing) 锤击贯入试验;动力贯入试验(试验土壤密实度等);落锤贯入试验;水滴渗透试验

dropper 料滴计数器;降压器;转臂收割机;真空阀;挂钩;吊弦;吊索;吊架;滴管;测重计（表面张力);分脉;骑马针
dropper assembly 吊弦装配图
dropper bottle 滴瓶
dropper clamp 滴管夹(钳)
dropper clip 整体吊弦
dropper clip for catenary wire 承力索吊弦线夹
dropper clip for messenger wire 承力索吊弦线夹
dropper clip for twin catenary wire 双承力索吊弦线夹
dropper clip for twin contact wire 双接触线吊弦线夹
dropper eye clamp 定位吊线环
dropper in tunnel 隧道整体吊弦
dropper key 吊弦键
dropper saddle 吊弦鞍子
dropper support loop 环形吊悬
dropper table 吊弦制作台
dropper thimble 吊弦套环
dropper wire 吊弦线
dropper wire clamp 吊弦线夹
drop pile hammer 落锤(式)打桩机;打桩(的)落锤
dropping 槽沉法;表面陷坑;艏落;首跌落【船】
dropping board 滴水板;粪板(鸡舍)
dropping board scraper 除粪机;耙粪器
dropping bottle 滴瓶;滴管式给药机
dropping characteristic 陡降外特性
dropping cut slice 水平切片
dropping electrode 滴液电极
dropping equipment 降压装置
dropping fraction 漏片
dropping funnel 滴液漏斗
dropping glass 滴瓶
dropping head 渐落水头;滴落水头
dropping height 落锤高度
dropping incidence of disease 发病率降低
dropping in hill 穴播;点播
dropping liquid 滴液
dropping mercury electrode 滴汞电极
dropping mercury electrode pulse polarography 滴汞电极脉冲极谱法
dropping moor 退抛双锚
dropping off barge 减驳
dropping of the charge 塌料
dropping of the tool 钻具下落
dropping pipette 滴液吸移管
dropping pit 粪坑
dropping point 滴点;滴油润滑点;初馏点
dropping rate of water-level 水位下降速率
dropping resistor 降压电阻(器)
droppings 滴下物
dropping section 升降段
dropping shadow 落影
dropping shaft dryer 下落式立筒烘干机
dropping tube 滴管
dropping valve 降压阀
dropping voltage generator 降压特性振荡器
dropping weight 强夯
dropping-weight compaction machine 落锤式压实机
dropping-weight method of compaction 落锤压实方法
drop pipe 下悬(喷)管;下落管;水井竖管;深井井管
drop pipe line 水井竖管;深井竖管
drop pit（整修机车底部的）地坑;修车坑;跌水坑;跌水井;渠道集水井;落轮坑
drop plate 卸渣炉篦;撒种设备
drop point 用沥青浇敷设屋面石棉板;下降点;滴点(沥青)
drop-point slating 斜铺板;斜盖瓦;对角铺砌石板瓦屋面;对角砌;对角瓦做法;蜂窝状铺石棉板（瓦）;吊脚铺石板
drop press 落锤;模锻压(力)机
drop preventing device for carriage 限速防坠装置【机】
drop print 下落芯头
drop procedures 脱落步骤
dropproof 防滴落
drop recorder 计滴器
drop relay 吊牌继电器;脱扣继电器
drop repeater 分转站;分接中断器;本地端接中继器

drop ring 球形活络把手;吊环
drop riser 下供竖管;下供立管
drop roller 递墨辊;递稿辊
drop rudder 低于龙骨的垂下舵;垂下舵
drops 冷珠;铁豆
drop sampler 沉落取样器
drop scene 吊下布景;吊悬布景工法;舞台吊幕
drop scenery 吊景
drop separator 集液器
drop sequence 空投程序
drop shaft 溜井;自重运输井筒;跌水（竖)井;沉井;倾卸轴
drop shaft foundation structure 开口沉井基础结构
drop shaft with cross walls 有隔墙的开口沉井
drop shipment 直达运货
drop shipper 向厂商直接供货的零售商;出口中间商;承运批发商
drop shutter 快门;下落快门
drop shutter test 坠落试验
drop-side 侧卸(的)
drop-sided body 侧卸车身
drop-sided car 侧卸式矿车
drop-sided trailer 栏板挂车
drop siding 互搭披叠板;互搭板;外墙用披叠壁板;外墙包层;外墙垂层护壁板
drop size 雨落大小;水滴大小
drop size distribution 水滴大小分布;滴谱
drop-slab construction 降(„滑)施工法
drop slide window 上下滑动吊窗
dropsonde 下投式探空仪
drop sorter 降落式选材装置
drop sounder 落锤测深
drops pen 点线笔
drop sphere visco(si)meter 落球黏滞度仪
drop spillway 跌水式溢洪道
drop stage 升降舞台
drop stamp 落锤
drop stamping 落锤锻造;热冲压
drop stone 钟乳状矿解石;滴石(钟乳状矿解石)
drop strake 合并列板;并(掉)列板;互搭箍条板;插窄条板
drop structure 吊架;拦砂坝;谷坊;跌水建筑物;跌水构筑物
dropsy 积液;积水
drop system 上行下给式供暖系统;水循环供暖系统;倾落式(煤炭装船设备的一种形式)
drop table 活动桌
drop tank 可投放邮箱;副油箱
drop tee 起柄三通;悬挂三通;T形吊耳
drop test 降落试验;落体试验;落锤试验;降压试验;降落式强度测定仪;降落伞投下试验;坠落试验;跌落试验;电压降试验;滴油试验;锤击试验;冲击试验;抛掷法试验;投放试验
drop testing machine 冲击试验机;落锤试验机
drop test of rail 钢轨落锤试验
drop the business 放弃交易
drop-throat 下沉式流液洞
drop tide 落潮
drop tracery 花格吊窗;下错窗
drop tube 专用吊柱
drop type instrument 投放式仪器
drop valve 用具旋塞;落座配汽阀;落阀
drop valve gear 坠阀装置
drop vent 吊风管口
drop wall 阶墙;跌水墙
drop water 跌水;放水
drop water drop corrosion test 静态水滴腐蚀试验
drop water head 跌水水头
drop weight 打桩锤;落锤
drop weight apparatus 滴重器
drop weight method 滴重法
drop weight tamper 重力夯
drop weight tear test 落锤扯裂试验
drop weight test 落(球)锤试验
drop well 跌水井
drop window 吊窗
drop wire 电源线;进户线;用户引入线
drop wire clamp 分线接线柱
dropwise condensation 珠状凝结;滴状凝结;滴状冷凝
drop worm 落蜗杆
drop zone 空投区
drosal part 背侧部
drosograph 露量表

drosometer 露量表;露量计
dross 结渣;灰渣;锡渣;氧化皮;杂质;浮渣;废渣;熔渣;溶渣;铁渣;撇渣
dross band 锡渣带;浮渣带
dross box 渣箱;掏渣箱
dross cement 浮渣水泥
dross coal 渣滓;铁渣;碎屑;矿渣;渣煤
dross extractor 清渣器;撇渣器
dross filter 滤渣芯片
dross fines 微细浮渣
dross inclusion 氧化夹杂物
drossing 撇渣
dross kettle 除渣锅
dross of metal 金属渣
dross spot 锡渣
dross tippler 浮渣翻车机
dross trap 浮渣捕集器;扒渣;撇渣
drossy 渣状
drossy coal 低级煤;碎煤
drought 干旱;短少
drought code 干旱指数
drought condition 旱情
drought control 抗旱
drought crack 旱裂
drought defying 抗旱
drought disaster 旱灾
drought disaster control forest 防旱林
drought early warning 干旱预警
drought-enduring 耐旱植物
drought-enduring crops 耐旱作物
drought-enduring plant 耐旱植物
drought fire 干旱引起的火灾
drought forecast(ing) 旱情预报;干旱预报
drought hardening 干旱锻炼
drought hardiness 耐旱性
drought index 干燥指标;干旱指数;干旱指标
droughtiness 干旱
drought injury 旱害
drought meadow 旱草甸
drought mitigation benefit 抗旱效益
drought period 枯水期;枯冰期;旱期;干旱期
drought preparedness 做好干旱准备
drought probability map 干旱可能性图
drought-prone region 易于干旱区域
drought relief 干旱救灾
drought resistance 抗旱性;耐旱性
drought-resistance measures 抗旱措施
drought-resistant crop 抗旱植物
drought-resistant plant 抗旱植物
drought resisting measure 抗旱措施
drought ring 旱轮
drought season 枯(水)季;旱季;干季
drought susceptible 不耐旱的
drought tolerance 耐旱性
drought-tolerant species 耐旱品种
drought warning 旱情预报
drought watch 旱情监视
droughty 干旱的
drought year 旱年;干旱年代
Drouily's method 控制电解粉末粒度法;德劳利控制电解粉末粒度法
drove 阔凿子;平凿
drove chisel 阔凿;平凿;石工平凿
droved dressing 短槽凿面(用阔凿修琢)
droved works 粗凿工作;粗凿工程
drove finish of stone 石料的短槽纹修琢
drover 漂网(鱼)船
drove road 不通行机动车的小路;大车路;大车道
drove way 不通行机动车的小路;大车路;大车道
drove work 粗凿工作;宽凿工;粗琢工作
drowned 淹死的
drowned atoll 潜环礁
drowned bar 沉溺沙洲
drowned coast 溺河海岸;溺岸;淹没岸;沉溺(海)岸;沉没(海)岸
drowned culvert 淹没式涵洞
drowned dam 潜坝
drowned estuary 溺河河口;沉溺河口
drowned flow 溺流;淹没流;潜流
drowned (hydraulic) jump 淹没水跃
drowned lime 消石灰;熟石灰
drowned longitudinal coast 沉溺纵向海岸
drowned(-out) hydraulic jump 淹没式水跃
drowned over-fall 沉溺式堰流

drowned pipe 浸没管;沉浸管
drowned pump 淹没的泵;沉没式泵;沉浸泵;潜水泵;深井泵
drowned reef 溺礁;沉没礁;暗礁
drowned river 溺河;隐入河;沉溺河
drowned river mouth 河口湾;沉溺河口;沉没河口
drowned shoreline 沉溺(海)岸线;沉溺滨线;沉没海岸线;沉没滨线
drowned spring 沉没泉;溺泉
drowned stream 淹没河;隐入河;沉溺河
drowned topography 海溺地形
drowned valley 溺谷;沉没(河)谷
drowned valley coast reef 溺谷海岸礁
drowned weir 溺堰;潜堰
drowning 沉没的;溺水;淹死;淹溺;淹没
drowning-out 淹没
drowning pipe 淹没进水管;潜管
drowning ratio 沉浸比率
drown sphere viscometer 平衡落球法黏度计
droxtal 过冷水冰滴
drub 敲打
drub coal 含煤页岩
Drude equation 德鲁特定律
Drude law 德鲁特定律
Drude' theory of conduction 德鲁特传导理论
drueckelement 测压体
drug 难销货物
drug and hospital equipment store 医药用品商店
drug controlled release material 药物控制释放材料
drug-fastness 耐药性
drugget 粗毛呢;棉毛地毯;粗毛地毯
druggist's drawer pull 带标签抽提拉手
drug grade 制剂
drug in the market 呆滞商品
drugmanite 水磷铁铅石
drug resistance 耐药性
drug storage 药品库
drug store 药品库;药房;药店;杂货店
drug sunchries 卫生用品
drug tolerance 耐药性
druids altar 牌坊
druid stone 平炉石
drum 卷盘;小炉口;圆筒;圆鼓屋顶;柱状物;滚筒;光电鼓;鼓状物;鼓形柱座;鼓式容器;鼓盘;磁鼓;测微鼓;绕线架;汽包;桶形
drum-and-scoop feeder 磨机中空轴内螺旋喂料装置
drum armature 鼓形电枢
drum axle 鼓轮轴
drum back-sight mechanism 鼓形表尺
drum barrow 放线车
drum boiler 汽包锅炉
drum brake 卷筒刹车;鼓形制动器;鼓轮刹车
drum broken and leaking 桶破漏
drum buffer 磁鼓缓冲器
drum buffer clear 磁鼓缓冲器清除
drum-built column 鼓形柱
drum-built pillar 鼓形柱;孤立的圆形屋顶柱
drum buoy 鼓形浮标;桶形浮标
drum cam 柱形凸轮;鼓形凸轮;凸轮轴
drum camera 鼓轮(式)摄影机
drum can buoy 桶形浮标
drum can package 桶包装
drum capacity 滚筒容量;鼓筒容量;卷筒容量;磁鼓容量
drum carrier 轨道搬运车
drum cell 鼓形电池
drum chart 自记仪器图纸;转筒记录纸
drum cleaner 鼓形清洗器
drum clutch 卷筒离合器
drum coating 转鼓涂布
drum cobber 圆筒选矿机
drum control 滚筒控制
drum controller 鼓形控制器;磁鼓控制器
drum conveyer 滚筒式升运器
drum cooler 冷却鼓;单筒冷却机;鼓式冷凝器
drum cradle 油桶架
drum curb 鼓形路缘(石);井圈;圆顶路缘石
drum cutter 滚筒式切料机
drum cylinder crystallzer 圆筒式结晶器
drum dam 弧形门堰;圆筒形闸门坝
drum development 滚筒显影
drum dial 鼓形度盘
drum diameter 卷筒直径;转筒直径;滚筒直径

drum digger 滚筒式挖掘机;斗轮(式)挖掘机
drum digger shield 圆筒形掘进盾构
drum digit path 磁鼓数字道
drum disc rectifier 鼓形干式整流器
drum drawing process 滚筒拉丝法
drum drawn mat 滚筒薄毡
drum drier 鼓式干燥器;筒式干燥器;干燥筒;转鼓式干燥器;滚桶干燥机
drum drive 卷筒传动;磁鼓驱动器;磁鼓驱动
drum driver 磁鼓驱动器
drum dump 磁鼓(信息)印出;磁鼓(信息)读出
drum elevator 滚筒式升运器
drum end 汽包封头;桶端
drum face 卷筒工作面
drum factor 卷筒系数
drum feed beater 滚筒喂入轮
drum filter 筒式过滤机;滤鼓式过滤器;滤鼓;转筒式过滤器;筒形过滤器;筒式滤筛
drum flange 滚筒凸缘;卷筒凸轮
drum float 鼓形浮子;鼓形浮标;鼓式测流浮标
drum for (asphaltic) bitumen 沥青混合料拌和用滚筒
drum for the plummet wire 垂球丝卷筒
drum furnace 圆筒炉;鼓形炉
drum gallery 鼓廊
drum gas meter (拌和机干燥筒的)气量计
drum gas velocity 沥青拌和机干燥筒
drum gate 弧形转动闸门;圆闸门;圆筒形闸门;圆阀(门);滚筒式闸门;鼓形闸门
drum gear 转鼓齿轮
drum glazer 滚筒式刨光机
drum guide 磁鼓导引装置
drum handler 搬桶机
drum head (两侧鼓状的)最低一级楼梯踏步;卷扬机平台;绞盘头;鼓轮盖;磁鼓磁头;汽包封头
drumhead man 绞盘手
drum high-voltage ceramic capacitor 鼓形压瓷电容器
drum hob 弧形齿顶花键滚刀;弧形齿顶花锭滚刀
drum hoist 卷筒(式)提升机;鼓筒式卷扬机;卷筒式绞机;卷筒式绞车;转鼓吊车;滚筒式卷扬机;滚筒式绞车
drum information assembler and dispatcher 磁鼓信息收集和分配器;磁鼓式信息收发器
drum insulator 鼓形绝缘子
drum interface 磁鼓接口
drum internals 汽包内部装置
drum ironelectric(al) porcelain capacitor 鼓形铁电瓷介质电容器
drum jack 线盘千斤顶
drum jacket 鼓轮套
drum ladle 封闭式金属桶
drum latency time 磁鼓等待时间
drum length 卷筒长度;整卷长度;滚筒长度
drum length of cable 电缆长度
drum lens 鼓形透镜
drum lighting lamp 油桶照明灯
drumlin 鼓丘;冰河堆集成的小山;冰丘
Drumm accumulator 德鲁姆蓄电池
drum magnetic separator 筒式磁选机
drum manhole 汽包观察孔
drum mark 磁鼓记录终端标记;磁鼓标记
drum melting furnace 鼓筒式熔炉
drum memory 存储磁鼓;磁鼓(记录);磁鼓(存储器)
drummer 煮茶工人(英国工地上);锻铁工人;旅行推销商
drummer discount 旅行推销商的折扣
drum mill 滚筒式粉碎机;圆筒式球磨机;转筒磨;滚筒式磨机
drumminess (支撑圆屋顶的)圆筒形墙壁的倾斜度
drumming 卷筒加工法;转筒加工法;发嗡声
drumming noise 振动噪声
drum mixer 圆筒混合器;圆筒混合大;转筒式拌和机;转鼓式拌和机;滚筒式拌和机;筒型混合机;滚动式拌和机;筒式搅拌器;鼓式搅拌机
drum mixer with reversing discharge 有可逆式卸货的圆筒混合机
drum mooring buoy 鼓形系船浮筒;筒形系船浮筒;桶形系船浮筒
drum motor 鼓形电动机
drum movement 鼓钟机心
drum nodulizer 成球筒;筒式成球机
drum number 磁鼓(号)数

drum of column 鼓状柱身；鼓状柱段；柱圆筒段
drum of dome 圆鼓屋顶；穹顶坐圈；穹顶的鼓座
drum of mixer 搅拌鼓
drum of roller 滚筒式压路机
drum of winch 绞车绕索筒；绞车卷(缆)筒；绞车滚筒
drum operation(al) system 磁鼓操作系统
drum packager 滚筒式包装机
drum panel(l)ing 空腔镶板结构；皮面门板
drum parity 磁鼓(的)奇偶检查
drum parity error 磁鼓奇偶错
drum pelletizer 滚筒式制粒机；圆筒制粒机
drum penning recorder 鼓形笔写记录器
drum plating 滚镀
drum plotter 滚筒式(自动)绘图机；滚筒绘图机
drum polisher 转鼓抛光机
drum polishing 转鼓抛光
drum preheater 滚筒式预热器
drum pressure 汽包压力
drum printer 鼓形打印机
drum printing 转筒印刷；转筒印花
drum processor 鼓轮型洗片机
drum pulse 磁鼓脉冲脉宽调制
drum pulverizer 转筒粉磨机
drum pump 回转(式)泵；油桶泵；转子泵；滚筒泵
drum rate 滚筒速率
drum reading 测微鼓读数
drum recorder 滚筒记录器；鼓形记录器；磁鼓记录器
drum recording 滚筒记录
drum reel 卷扬机滚筒；筒式卷绕机；筒式卷取机
drum revolution counter 滚筒旋转计数器
drum ring gear 搅拌筒大齿圈
drum rinser 洗桶机
drum roller 搅拌筒支承滚轮；滚筒式压路机；滚筒式碾压机；滚压机(修路用)
drum rolling 滚筒压平
drum rotations per minute 滚筒每分钟转数
drum rotor 鼓形转子
drum rupture 桶破裂
drum sander 转鼓打磨机；滚筒式砂光机；辊式磨光机；鼓形磨光机；砂鼓磨床
drum sanding 辊筒磨光
drum saw 圆筒锯；筒形锯
drum scanner 鼓形扫描设备；鼓形扫描机
drum scanning 滚筒扫描
drum screen 鼓筛；回转筛；圆筒筛；滚筒鱼栅；滚筒筛；滚动筛；鼓形筛(网)
drum screening 滚筒筛选；转筒筛选
drum separator 转鼓式分离器；筒式筛分机；筒形分离机；筒式选矿机；筒式筛分机；筒式分离机
drum separator with counter-rotating magnet 反转磁铁筒式磁选机
drum servo 鼓形伺服机构
drum setting-out machine 鼓形平展机
drum sextant 鼓轮六分仪
drum shaft 卷筒轴；绞筒轴；滚筒轴；成型鼓主轴；沉井
drum-shaped column 鼓形柱(座)
drum-shaped column base 鼓形柱座
drum-shaped metro passenger rolling stock 鼓形地铁电动客车
drum-shaped stone block 抱鼓石
drum shell 圆鼓形薄壳；滚筒罩；鼓形(薄)壳；鼓形壳；汽包筒身
drum shield 滚筒式盾构
drum shutter 鼓形快门；鼓形光闸
drum sieve 圆筒筛；转筒筛；鼓形筛
drum sifter 圆筒筛
drum skimmer 转鼓撇油器
drum slaker 石灰消化鼓
drum sluice 鼓形泄水闸门
drum snader 鼓形磨床
drum sorter 磁鼓分类程序
drum sorting 磁鼓排序；磁鼓分类(程序)
drum speed 转筒速度；滚筒转速；磁鼓速度脉宽调制器
drum spline 弧形齿顶花键滚刀
drum spline hob 弧形齿顶花键滚刀
drum spool 卷扬机滚筒
drum spreader 绕线架
drum's speed of rotation 球磨机转速
drum starter 鼓形起动器
drum steam generator 汽包式蒸汽锅炉
drum storage 圆筒储存器；磁鼓(存储器)

drum stub 汽包管接头
drum submergence 转筒淹没度；存水弯淹没
drum support 周边支撑
drum switch 鼓形开关
drum table 鼓形圆桌
drum tannage 转鼓鞣革
drum tape system 鼓带系统
drum tightening device 卷筒式张紧装置
drum to rope ratio 卷筒和钢丝绳(直径)比；绞筒钢丝绳直径比
drum-tower 鼓楼
drum transaction 磁鼓更新
drum transmitter 鼓形发送器
drum trap 鼓形罩；鼓形疏水器；鼓式凝气阀；鼓式存水弯
drum turbine 鼓形涡轮机
drum type 滚筒式
drum-type boiler 鼓筒式锅炉；汽包式锅炉；弯管锅炉
drum-type brake 鼓式制动器
drum-type building machine 鼓式胎成型机
drum-type cast coater 鼓式涂铸机
drum-type coiler 筒式卷绕机；筒式卷取机
drum-type concrete mixer 鼓形混凝土拌和机；转筒式混凝土搅拌机；转筒式混凝土拌和机
drum-type controller 鼓式控制器
drum-type cooler 鼓式冷却器；鼓式冷却机
drum-type drawbench 鼓式拉拔机
drum-type drier 鼓式烘干器；鼓式干燥器
drum-type drying 鼓式干燥法
drum-type electromagnetic separator 滚筒式电磁分离器；筒式磁选机
drum-type feeder 转筒给料机；鼓式喂料机；鼓式加料机
drum-type filter 鼓式过滤机
drum-type float 鼓式测流浮标
drum-type flying shears 鼓轮式飞剪机
drum-type gas meter 鼓式煤气表
drum-type gas scrubber 鼓式气体洗涤器
drum-type granulator 鼓式粒化器
drum-type hay loader 滚筒式装干草机
drum-type heater 鼓式加热器；鼓式加热机
drum-type heat exchanger 鼓式热交换器；圆筒式热交换器
drum-type indicater 鼓形指示器
drum-type kiln 筒式窑
drum-type ladle 鼓形浇包
drum-type line printer 鼓式宽行打印机
drum-type miller 鼓式铣床
drum-type milling machine 鼓形铣床
drum-type mixer 滚筒式拌和机；鼓式搅拌机；鼓式搅拌器；鼓式混合机；鼓式拌和机
drum-type pickler 滚筒式酸洗装置
drum-type pickling machine 滚筒式酸洗机
drum-type pick-up 滚筒式捡拾器
drum-type plotter 鼓式绘图机
drum-type pouring ladle 鼓形浇包
drum-type preheater 鼓式预热器
drum-type printer 鼓式打印机
drum-type reactor 鼓式反应器
drum-type recorder 鼓式记录器
drum-type root cutter 滚筒式块根切碎机
drum-type rotary dryer 滚筒式烘干机
drum-type rubber damper 鼓形橡胶减振器
drum-type sampler 鼓式采样器
drum-type screen 鼓式旋转筛；鼓式格栅
drum-type scrubber 鼓式洗涤器
drum-type separator 鼓式分离器
drum-type shears 滚筒式剪切机
drum-type slicer 鼓式切片机
drum-type spindle 滚筒式摘锭
drum-type spindle picker 筒筒摘锭式采棉机
drum-type starter 鼓式启动器
drum-type storage 鼓式存储器
drum-type stripper 滚筒脱模机
drum-type swath aerator 滚筒式条铺摊晒通风机
drum-type tow dryer 鼓式丝束烘缝机
drum-type trap 鼓式凝气阀；鼓式存水弯
drum-type tumbler 鼓式拌和机
drum-type turret lathe 回转式六角车床
drum-type uncoiler 鼓轮式开卷机
drum type vacuum filter 转鼓式真空过滤机；鼓式真空过滤器；鼓式真空过滤机
drum-type vulcanizing press 鼓式硫化机

drum-type washer 鼓式洗浆机；鼓式洗涤机
drum-type washing machine 鼓式洗涤机
drum-type water meter 鼓式水表
drum-type weir 鼓式闸门堰
drum unit 磁鼓机
drum up trade 奖励贸易
drum vacuum filter 圆筒真空过滤机；圆筒式(真空)过滤器
drum valve 鼓形阀
drum volume 搅拌筒几何容量
drum washer 圆筒洗矿机；滚筒洗矿机；滚筒式洗石机
drum washing 转鼓式洗涤
drum-water meter 鼓式水表
drum weir 弧形闸门；弧形门堰；圆筒堰；转筒堰；鼓形闸门堰
drum wheel 鼓轮
drum wicket 鼓形闸门
drum winch mounted on a vehicle 装置在车辆上的转筒式绞车
drum winder 转筒卷绕机；滚筒式提升机
drum winding 转筒卷绕(的)；卷扬；鼓形绕组；鼓形绕圈；鼓形绕法
drum with jacket 带有夹套的容器
drum with pins 钉齿式滚筒
drum with rasp bars 纹杆滚筒
drum with record sheet 附记录纸的圆筒
drum worm wheel 鼓形蜗杆蜗轮
drum-wound armature 鼓形(绕线)电枢
drum zero 测微鼓零分划
drunken saw 摇摆(圆)锯；宽口锯；开槽锯；切槽锯
drunken saw blade 斜轴锯
drunken sawing machine 切槽锯床
drunken thread 周期变距螺纹；不规则螺纹
drunkery 酒馆
drunkness error 螺纹导程周期误差
drunkometer 测醉器(交通安全)
druse 晶洞；晶簇
druse of calcites 方解石晶簇
Drusian foot 回教徒用尺
druss 土状煤
drusy 晶簇状
drusy cement texture 晶簇状胶结构结构
drusy coating 晶簇状外壳
drusy structure 晶簇(状)构造
druxey 腐木状态；朽心木材
druxiness 朽心木材；腐木状态
drwing list 图纸编目
dry 干的
dry abrasive cloth 干磨砂布
dry abrasive cutting 干磨削
dry abrasive paper 干磨砂纸
dry accumulator 干蓄电池
dry activated carbon method 干式活性炭法
dry adiabatic 干绝热
dry adiabatic atmosphere 干绝热大气
dry adiabatic change 干绝热变化
dry adiabatic lapse rate 干绝热直减率；干绝热减温率；干绝热递减率
dry adiabatic process 干绝热过程
dry adsorption 干吸附
dry after treatment 干养护
dry ag(e)ing 干老化
dry-agent fire tender 干粉灭火车
dry air 干(燥)空气
dry-air blast 干风
dry-air cooler 干空气冷却器
dry-air pump 干(燥空)气泵；干风泵
dry-air sterilizer 干风杀菌机
dry ammonia method 干氨法
dry analysis 干法试验；干(法)分析
dry and brittle 干脆
dry and wet bulb hygrometer 干湿球湿度计；干湿球温度表
dry annealing 干式回火
dry arch 防潮基拱；不加装饰的拱
dry area 干斑；采光井；房基通风井；干燥区域；房基空格墙空；地下室通风井
dry ash 干灰
dry ash digestion 干灰消解
dry ash-free 干燥无灰的
dry ash-free basis 干燥无灰基
dry ashing 干(法)灰化
dry assay 干样品；干法化验

dry automatic sprinkler system 干式自动喷水系统
dry avalanche 干雪崩
dry back boiler 背包式锅炉;干背式锅炉
dry back fill 干填
dry bag 干袋
dry bag isostatic compacting press 干袋等静压机
dry bag isostatic pressing 干袋法等静压成型
dry bag tooling 干袋模具
dry ballast 干石渣;干道碴;固体压载
dry ball mill 干式球磨机
dry barrel 干货琵琶桶
dry basin 干旱盆地
dry basis 折干计算;干燥基;干基(准);按干物质计算;按干计算
dry batch 一次搅拌的干拌物(混凝土等);混凝土一次搅拌的干拌物
dry batched aggregate 干混凝土;干的分批计量的骨料;干拌和集料
dry batched concrete 干料分批拌和混凝土;干拌(和)混凝土;干配料混凝土(按水比加水即可使用)
dry batched method 干拌法
dry-batch(ing) plant 干拌设备;干料称量装置
dry-batch material 干拌材料
dry-batch method 干料分拌法
dry batch mixer 干料混合机
dry batch system 混凝土干料分拌法
dry batch weight 一次搅拌的混合料干重量;干配料重量
dry battery 干电池(组)
dry beach 干海滩;潮漫滩;退潮时露出的海滩
dry bearing 干轴承
dry bed 干河床;污泥干燥床
drybilt 装配式木屋
dry blast 干燥送风;低湿度送风
dry blasting 干喷射处理;干喷砂处理
dry blast(ing) cleaning 干喷射处理;干喷砂清洗;干喷处理;干喷砂打净
dry blend colo(u)ring 干混着色
dry blender 干混机;干合料机
dry blend(ing) 干混合;干掺和;干混料;干搅和
dry blends 干合料
dry block(ing) 干式卡取(岩芯)
dry blowing 干蒸;干选
dry body 无釉陶器;干坯
dry body strength 干坯强度
dry-bond adhesive 压敏胶黏剂;接触黏结剂;接触黏合剂;干黏胶剂
dry bonding 干黏合
dry bonding strength 干燥的黏合强度;干砌合强度
dry bond macadam 干结碎石(路)
dry bond strength 干黏结强度
dry boring 干钻
dry bottom boiler 干底锅炉
dry bottom hopper 冷灰斗
dry bottoming 干铺(底)基层
dry bottom producer 干底发生器
dry bound crushed stone bases 干结碎石基层
dry bound macadam 干灌浆碎石路
dry box 干燥吸收剂
dry brick 干砌砖;无砂浆砌砖
dry brick building 干打垒;干砌砖房屋
dry brickwork 干砌砖工
dry bridge 跨线桥;旱桥;旱桥
dry brightness 干态明度;干亮度
dry bright polish 干的上光蜡
dry brush 干刷;干笔
dry brushing 干刷净;干刷光
dry bulb 温度计干球;湿度计的干球
dry bulb temperature 干球温度
dry bulb thermometer 干球(水银)温度计;普通温度计
dry bulb thermometry 干球温度
dry bulk cargo 干散货
dry bulk container 干散货集装箱
dry bulk density 干重度;毛体积干密度;干燥密度;干土密度;干么重;干毛(体积)密度
dry bulk density of aggregate 集料干重度
dry bulk density of sediment 泥沙干重度
dry bulker 干散货船
dry bulk specific gravity 干容积比重;干松比重
dry bull density 干体密度
dry burning coal 干燥燃煤;低结的不结块煤;贫煤;瘦煤

dry cabinet 烘干(小)室
dry calorimeter 干式量热器
dry camp 无饮水宿营地
dry capacity 遮盖力
dry capacity per batch 一拌干容量
dry card compass 干罗经
dry cargo 干货
dry cargo container 干货集装箱;杂货集装箱;通用集装箱
dry cargo hold 干货舱
dry cargo ship 干货船
dry cargo tramp 不定期干货船
dry casting 干铸法;干模铸件
dry cast method 干铸法
dry cast pipe 用干浇筑法制成的管;干铸管
dry cast process 干铸造法;干处理法(制混凝土管)
dry catchment 干收集
dry ceiling 石膏板天棚;石膏板顶棚;石灰板天花板
dry cell 干电瓶;干电池
dry-cell battery 干电池组
dry-cell cap light 干电池防爆安全帽灯
dry centrifugal collector 干式离心集尘器;干式离心除尘器
dry centrifugal dust remover 干式离心除尘器
dry certificate 干燥证明书
dry chamber 干舱
dry-charged battery 干充电电池
dry-charge smelting 干料熔炼
dry chemical extinguisher 小苏打灭火干粉;干式化学剂灭火
dry chemical extinguishing method 干式化学灭火法
dry chemical extinguishing system 干化学药品灭火系统;干粉灭火系统
dry chemical feeder 干式化学进料器
dry chemical (fire) extinguisher 干化学灭火器;化学干式灭火器;干式化学剂灭火器
dry chemical powder 干粉
dry chemical projectile 干粉灭火弹
dry chemicals 干化学药品;不含水化学品
dry chemical sonde 干式化学探空仪
dry chip(ping)s 干碎屑
dry-chlorination 干加氯法
dry chlorine 干氯
dry choked stone 干嵌石料;干结块石
dry churn 干法黄化鼓
dry classification 干法分级
dry classifier 干分级器
dry cleaner 干洗剂;干洗机
dry cleaning 干洗;干法选矿;干法精选
dry cleaning agent 干洗剂
dry cleaning building 干洗店
dry cleaning detergent 干洗剂
dry cleaning fluid 干洗流体
dry cleaning shop 干洗店;干洗房
dry climate 干燥气候;干旱气候
dry climate condition 干旱的气候条件
dry closet 茅坑;旱厕(所);干燥土室;便坑
dry closet receptacle 干厕所容器
dry closure 干封器
dry clutch 干式离合器
dry coal 干煤
dry coal preparation 干煤制备
dry coated 干法包膜的
dry coating procedure 干法包膜法
dry cobbing 干法磁选
dry cold bending 干冷弯(曲)
dry collection 干收集
dry colo(u)r 干色料;干色粉
dry colo(u)rant 干着色剂
dry colo(u)ring 干法着色
dry column-packing 干法装柱
dry combustion chamber boiler 干燃膛锅炉
dry combustion method 干烧法
dry compacted weight 密实干重;干压重度
dry compaction 偏(压)干压实
dry compass 干罗经
dry compass card 干罗盘
dry-compound cleaning method 干洗剂清洗法
dry compression chamber 干舱
dry compression strength 干抗压强度
dry compression test 干压强度试验
dry concentration 干选

dry concentrator 干式选矿机
dry concrete 稠混凝土;干(硬)性混凝土;干混凝土
dry concrete mix 干硬性混凝土混合料
dry condense pipe 干式冷凝管
dry condenser 干式聚尘器;干式电容器;干冷凝器
dry condense return 干式冷凝回水
dry connection 干接管
dry conservancy 干保护;干保存;旱厕(所);干燥土室
dry conservancy receptacle 干厕所容器;干保护插座
dry consistency[consistence] 干硬度;干硬(稠)度
dry constriction coefficient 干燥收缩系数
dry construction 干法施工;干作业施工;干(式)施工;干式建造;干地施工;干建筑;干打垒
dry construction method 干式建造法
dry construction partition 干作业隔断
dry contact 干式无电位接点;干触点
dry contact rectifier 干片接触整流器
dry conversion process 干法转化工艺
dry cooling 干式冷却
dry cooling tower 干冷却塔
dry copper 干铜;凹铜
dry core cable 空气纸绝缘电缆;空心纸绝缘电缆;干芯电缆
dry-core transformer 干芯变压器
dry corrosion 干(式)腐蚀;干(腐)蚀
dry course 干底层
dry crankcase 干式曲柄箱
dry criticality 干临界
dry crushed 干式破碎的
dry crushing 干法破碎
dry crushing mill 干(磨)碎机
dry crushing roll 干碎辊式破碎机
dry crust 硬结表土层;干硬壳层;干泥皮
dry cup 干吸杯
dry-cup method (在0~50%湿地范围内的)扩散测试法;干湿球法;干杯法
dry curing 空气养生;空气养护;干养护
dry curve 干淹
dry cutting 干挖
dry cyclone 干式旋风分离器
dry cylinder liner 干式汽缸衬套;干式汽缸衬套
dry damage 旱灾;旱害
dry-damp process 干捣法
dry decatizer 干蒸呢机
dry delta 干三角洲;冲积锥;冲积扇
dry denitrification 干法脱氮
dry density 干(燥)密度;干容重
dry density-moisture content graph 干密度—含水量图
dry density-moisture ratio 干密度与含水量比
dry density of expansive clay 膨胀黏性土的干密度
dry density of soil 土壤干密度;土的干密度
dry deposition 干沉降;干沉积
dry deposition velocity 干沉积速度
dry desiccant dehydration 干态干燥脱水作用
dry desiccant dehydrator 干燥剂脱水器
dry desulfurizing 干法脱硫
dry diazo copy 干法重氮复印
dry diazo processing 干法重氮复印
dry disc clutch 干盘式离合器;干式圆盘离合器
dry disk clutch 干盘(式)离合器;干式圆盘离合器
dry distillation 干馏
dry distillation of coal 煤干馏
dry distillation of wood 木材干馏
dry distiller's soluble 烧酒槽残液干燥物
dry divider 干料分流器
dry diving 干式潜水
dry diving dress 干式潜水服
dry diving suit 干式潜水服
dry dock 进坞;修船船坞;干(船)坞;船坞
dry dock accessories 干船坞附属设备
dry dock caisson 浮坞门;干船坞闸门
dry dock draft 干船坞吃水
dry dock drainage pump 干坞辅排水泵
dry dock engineering 船坞工程
dry dock floating arrangement 干船坞放水设备
dry dock flood device 干船坞放水设备
dry dock iceberg 船坞形冰山
dry dock(ing) 坞修;进干船坞
dry dock keel-block 干船坞龙骨墩
dry dock length 干船坞长度

dry dock lock 干坞坞闸
dry dock pump 干坞泵
dry dock type building 干船坞式建筑物
dry dock type structure 干船坞式建筑物
dry dock wall 干船坞墙
dry dock without walls 无坞墙的干坞
dry dredge(r) 陆地挖泥机；干式挖泥机
dry drill 风钻
dry drill cuttings 干法钻头刨屑
dry-drilled pile 干成孔灌注桩
dry drilling 干式钻机；干法打孔；干法打眼；干钻；干式钻进；干式凿岩；无冲洗液钻进
dry drilling method 干钻法
dry drilling system 干法钻孔操作体系；干法打眼操作体系
dry dross 铁渣
dry drum 蒸汽鼓
dryductor drill 干式吸尘凿岩机（英国）
dry dust arrester 干式除尘器
dry dust collection device 干式集尘装置；干式除尘装置
dry dust collector 干式集尘器；干式除尘器；干粉吸尘器
dry duster 干式除尘器
dry dust removal 干法除尘
dry dust remover 干式除尘器
dry dy(e)ing 干染色
dry earth dredger excavator 多铲式挖掘机
dry edge 干边
dry edging 枯边；边角缺陷
dryed sludge 干化污泥
dry electric(al) dust precipitator 干式电集尘器；干式电除尘器
dry electrolytic capacitor 干电解电容器
dry electrolytic rectifier 干电解整流器
dry electrostatic cleaner 干式静电除尘器
dry electrostatic precipitator 干式静电吸尘器；干式静电沉淀器；干式电滤器
dry element air cleaner 干电池空气滤清器
dry element cell 干电池
dry elutriation 空气选粒；干淘（洗）
dry elutriator 干淘机
dry emulsion 干乳胶
dry end 干端
dryer 烘干炉；烘干机；干燥设备；干燥炉；干燥工；干料；催干剂；喷气式单板干燥器
dryeration 干燥通风作业
dryeration process 干燥通风法
dryer car 烘干车
dryer crusher 烘干破碎机
dryer drum 烘干筒；干燥滚筒；干燥鼓
dryer for mineral wool slab 矿棉板烘干机
dryer mill 烘干磨；干燥机
dryer-pulveriser 干燥粉磨机
dryer take-off unit 干燥窑出料装置
dryer white 干坯冒霜
dry etching 干蚀刻
dry evapo(u)rator 干式蒸发器；干式蒸发机
dry excavation 干（开）挖
dry expansion chiller 干膨胀式冷却器
dry expansion evapo(u)rator 干式膨胀蒸发器；干膨胀式蒸发器
dry extract 干提取物
dry face 干割面
dry-face water content 面干含水率
dry fall-out 干散落物
dry fallow 干休耕
dry fan 风干器；风干机；风干扇
dry farm 旱田；旱地
dry farming 旱作（农业）；旱农；非灌溉农业
dry farming technique 旱作农业技术
dry farmland 旱田
dry feed 干式送料；干进料
dry-feed chlorination 干加氯法
dry-feed device 干投设备
dry-feed dossage 干加剂量
dry feeder 干料给料机；干进料器；干加药器；干加料器
dry feeding 干式给药
dry feed method 干加料法；干投（料）法
dry feel 干感
dry feet 建筑物的砖、石、混凝土地基
dry felt 干油毡
dry felted fabric for roofing 铺屋顶用马粪纸；铺屋顶用干油毛毡
dry felt jumbo roll 可卷油毛毡；干毡大卷
dry felt-laying 干铺油毡
dry felt manufacture 油毛毡制造；干毡制造
dry felt web 油毛毡卷筒；干毡织物
dry fill 干填土
dry filled 干填的
dry-filling fill method 干式充填法
dry filling filter 干燥填充过滤器；干式填充过滤器
dry film 干膜
dry film lubricant 干（燥）膜润滑剂
dry-film resist 感光胶膜；干片保护层
dry film test 干膜检验
dry film thickness 干（漆）膜厚度
dry filter 干滤器；干燥过滤器；干式滤光器；干式过滤器
dry filtration 干过滤（器）
dry-fine 干抛光
dry finish 干压光；干上光；干浆料
dry finished terne plate 不涂油镀铅锡合金钢板
dry finishing 干修；干态修坯
dry flashover voltage 干飞弧电压
dry flash test 干闪试验
dry flax 干亚麻
dryflex 抗腐石墨油漆；干折曲；干弯曲
dry flong 干纸型纸
dry flood 干洪流
dry flow 干流量
dry flue gas 干烟气
dry fluid centrifugal coupling 干液离心式连接器
dry fluid drive 干液力传动
dry flushing 干挤水法
dry-foam cleaning method 干泡沫清洗法
dry fog 干雾
dry food store 干食品库
dry foot 毛脚
dry forest zone 干旱林区
dry formed 干燥成型的
dry-formed fabric 干法成型织物
dry freeze 干冻(结)
dry friction 固体摩擦；干摩阻力；干摩擦(力)
dry friction damper 干摩擦阻尼器
dry friction damping 干摩擦阻尼
dry frozen ground 寒土；干冻土
dry fuel 固体燃料
dry-fuel engine 固体燃料发动机；固定燃料发动机
dry fumarole 干喷汽孔
dry furnace 烘干炉
dry gallon 干加仑
dry galvanizing 干（燥）法镀锌；干镀锌；干镀锌（法）
dry gas 干（煤）气；干气（体）
dry gas compressor 干式气压缩机
dry gas fuel 液化气体燃料
dry gas-holder 活塞式储气罐；干式储气罐；干式（储）气柜
dry gas meter 干式（煤）气表；干气体计量器
dry gas phase-out zone 干气消失带
dry gas purification 干气净化
dry gas purifier 干燥气体净化器；干气净化器
dry ga(u)ge 料勺
dry geothermal field 干地热田
dry geothermal-gradient heat source 干地热梯度热源
dry glazing 干装玻璃；干法施釉；无灰装玻璃
dry glazing dispenser 干法施釉器；干法撒釉器
dry glazing roof 干法装配屋顶玻璃；(不用油灰的)玻璃屋顶
dry gluing 干胶合
dry goods 干货
dry goods stores 纺织品类商店
dry grain 干谷粒
dry granular 干粒（一种釉料）
dry granulation 干法成粒
dry gravity stamp 干法重力式捣击机
dry grind 干磨
dry grinder 干碾机；干磨机
dry grinding 干磨（削）；干法研磨；干法粉碎；干法粉碎
dry grinding cylinder mill 干法圆筒磨机
dry grinding grate ball mill 干式格子型球磨机
dry grinding machine 干碾机；干磨机
dry grinding mill 干式磨机
dry grinding operation 干磨操作

dry grinding plant 干碾设备；干磨设备；干法粉磨设备
dry grit-blast cleaning 干砂喷射清理
dry ground 干法研磨；干法粉碎
dry-ground muscovite mica 干磨白云母粉
dry-ground whiting 干磨白垩粉
drygum waxed paper 干式蜡纸
dry harbo(u)r 浅水港；干涸港湾；低潮时干出或水较浅的港；潮漫滩
dry hard 烘干水分
dry hard clay 干硬黏土
dry hardness 干(态)硬度
dry hard of coats 涂层干硬
dry haze 干雾；干霾；飞尘
dry heat 干热
dry heat curing 干热固化
dry heat exposure 干热状态暴露
dry heating system 干采暖系统
dry heat sterilization 干热（空气）灭菌法
dry heat sterilizer 干热消毒器；干热灭菌器
dry heat vulcanization 干热硫化
dry hiding 干遮盖
dry hiding power 干遮盖力
dry high-pressure water ga(u)ge 高压干式水表
dry holder 活塞式储气罐；干式储气罐
dry hole 乏油井；干（钻）孔；干炮眼；干井；未见矿钻孔；上斜炮眼
dry hole plug 干井塞
dry hole ratio 干孔比
dry hot environment 干热环境
dry-hot-rock geothermal system 干热岩地热系统
dry hot vulcanization 热空气硫化
dry hot wind 干热风
dry hydrant 干消火栓；无水消防栓
dry hydrate 熟石灰粉
dry hydrated lime 氢氧化钙；建筑用石灰；建筑用钙质泥浆；干水化石灰
dry-hydrate type by-product lime 干水化物型副产品石灰
dry-hydrate type carbide lime 干水化物型碳化石灰
dry hydrogen 干氢
dry ice 固体二氧化碳（俗称干冰）；干冰
dry ice fire extinguisher 干冰灭火器
dry ice freezer 干冰制冷器
dry ideal sands and gravels 干的理想砂和砾石
dry impactor 干式撞击取样器；干式碰撞器
dry impingement 干式碰撞器
dry impinger 干式撞击取样器；干式碰撞器；干式冲击集尘器
dry inertial collector 干式惯性除尘器
drying 烘干；干燥；干出（落潮时露出）
drying action 干燥作用
drying agent 干燥剂
drying aid 助干剂
drying air duct 干燥空气输入管
drying-air temperature 干空气温度
drying air temperature control 干燥用空气温度调节器
drying air unit 干燥用空气加热器
drying alkyd resin 干性醇酸树脂
drying and heat setting stenter 干燥及热定型拉幅机
drying and mixing plant 烘干与拌和设备
drying and pulverizing unit 烘干粉磨设备
drying apparatus 烘干装置；烘干器；干燥设备；干燥器
drying area 干燥地区
drying arrangement 烘干装置；干燥装置
drying baker 烘干箱
drying balcony 晒干衣服的阳台
drying battery 化学干燥场
drying bed 干燥床；干燥场；干化床
drying bed system 干化床系统
drying behavio(u)r 干燥性能；干燥情况；干燥反应
drying board 干燥箱
drying bottle 干燥瓶
drying box 干燥箱
drying box by electric(al) heated blasting 电热鼓风干燥箱
drying breather 密封式呼吸器
drying bulb reading 干球温度读数
drying bulb temperature 干球温度
drying bulb temperature measuring method 干球温度

drying bulb thermometer 干球温度计;干球温度表
drying bulk density 干堆积密度
drying bulk weight 干重度
drying by evapo(u)ration 蒸发干燥;物理干燥法;挥发干燥
drying by fire 焙干
drying by hot-air jet 热空气喷射干燥
drying by induction heating 感应加热干燥
drying by oxidation 氧化干燥
drying by waste heat 利用余热干燥
drying cabinet (住宅内的)烘干柜;烘(衣)箱;烘干(小)室;干燥箱;干燥室
drying capacity 干燥能力
drying car 干燥车
drying centrifuge 干燥离心机
drying chamber 烘干室;烘干仓;干燥室;干燥罐
drying condition 干燥条件
drying constant 干燥常数
drying crack 干燥裂隙;干燥裂纹;干裂(缝)
drying creep 干燥徐变
drying cure 干燥措施
drying curve 干燥曲线
drying cycle 干燥周期
drying cylinder 干燥筒;干燥烘缸;干燥鼓
drying defect (木材的)干燥缺陷
drying drum 干燥滚筒;回转烘干筒;回转烘干机;干燥转筒
drying effect 干燥作用
drying equipment 干燥装置;干燥设备
drying fatty 燥油
drying fault 干燥故障
drying film machine 干膜机
drying filter 干滤色片;干滤器
drying finger 垫(板)条
drying frame 烘干架;干燥(框)架
drying grinding 粉磨烘干
drying grinding mill 干法粉磨机
drying ground 干燥场地
drying hearth 烘干炉
drying height 干出高度;潮间区
drying hopper 干燥式料斗
drying house 干燥室
drying incorrectly 干性调整不正确
drying in counter-current 逆流烘干
drying index 干燥指数
drying indicator test 干燥指示剂试验
drying inhibitor 缓干剂;阻干剂;干燥抑制剂
drying in parallel current 顺流烘干
drying installation 干燥设备
drying kiln 干燥窑;烘干窑
drying loft 干燥室;干燥棚
drying machinery 烘干机;干燥设备;干燥机械
drying mark 干燥斑痕;干燥(黑)斑
drying mechanism 干燥机理;干燥装置
drying medium 干燥介质
drying meter 测干仪;漆膜干燥计
drying method of oxidative deodorization 干法氧化脱臭
drying mixer 装有干燥设备的拌和机
drying mixer combination 干燥拌和两用机
drying mud yard 晒泥场
drying of ink 油墨干燥
drying of lake 湖水干涸
drying of screeds 饰面烘干
drying of sludge 污泥干化
drying of wheat in sacks 袋灌小麦的干燥
drying of wood 木材干燥(法)
drying oil 干性油
drying oil based ink 干性油底基油墨
drying oil binder 干性油黏剂;干性油黏结剂
drying oil ink 干性油油墨
drying oil vehicle 干性油载色剂
drying on basket 吊篮干燥
drying out 干燥;干化;失去水分;疏干
drying-out period 到达干燥周期
drying-out time 到达干燥期限
drying oven 烘箱;干燥(烘)炉;干燥箱;干燥炉
drying-packing 干包环割法
drying pallet 干草垫
drying pan 干燥盘
drying period 油漆干燥期
drying pipe 干燥管
drying pistol 干燥枪
drying plant 烘干装置;干燥装置;干燥设备;干燥车间
drying pot 烘罐
drying power 干燥性能;干燥(能)力
drying press 榨干机;干燥机
drying print 干印
drying process 干燥处理;干燥操作
drying property 干燥性
drying rack 干燥的支架;晾片架;干燥架
drying rate 干燥(速)率;干燥速度
drying reef 干出礁
drying regime 干燥情况
drying return pipe 干回流管
drying rock 干出礁
drying roll 烘燥滚筒;干燥辊道
drying room 烘房;洗衣烘干室;干燥室;干燥间
drying rot 干腐
drying schedule 干燥制度
drying screen 百叶窗式分离器
drying sensitivity 干燥敏感性
drying shed 干燥(凉)棚;干燥房
drying shoal 干出滩
drying shrinkage 干燥收缩(率);干缩率;干缩(量)
drying shrinkage behavio(u)r 干缩特性
drying shrinkage crack 干缩裂缝
drying shrinkage curve 干缩曲线
drying shrinkage limit 干缩限度
drying shrinkage stress 干缩应力
drying shrinkage value 干缩值
drying siccative oil 干性油
drying sieve 干筛
drying stage 晒台
drying storm 干风暴
drying stove 烘燥炉;烘干炉
drying strength 干燥强度
drying stress 干燥应力
drying system 干燥制度;干燥系统
drying temperature 干燥温度
drying tendency 干燥速度(混凝土)
drying test 干燥试验
drying through 干透
drying time 干燥时间;干化时间
drying time automatic recorder 干燥时间自动记录仪
drying time clock 干燥记时钟表
drying time meter 干燥时间测定器
drying tower 干燥塔
drying transverse strength 干燥抗弯强度
drying treatment 干燥处理
drying tube 干燥管
drying tumbler 烘燥转筒
drying tunnel 干燥隧道
drying unit 干燥机组
drying up 干燥
drying up deposit 干涸油藏
drying varnish 干清漆
drying weather discharge 旱季流量
drying weather flow 旱流污水;旱季流量
drying yard 干化场;晒场
drying zone 干燥区;干燥带
dry injury 旱害
dry ink 干油墨
dry installation 干式装置
dry instrument 干式仪表用的;室内仪表用
dry insulation 干式绝缘的
dry insulation transformer 干式绝缘变压器
dry in the air 晾干
dryin-up point 干燥点
dry inversion 干转变
dry iron 低硅生铁
dry jet mixing method 喷射搅拌法
dry joint 干接头(使用密封垫片的接头);干(缩)缝;干(式接)缝;扭接;虚焊(接头);易断接头;非黏合结缝;缺胶接头;干(砌)缝
dry-jointed 干砌接头
dry-ki 枯死树
dry kiln 干燥窑;干燥室;干燥炉;烘干窑
dry-laid 干砌的
dry laid masonry 干砌体;干砌(石)圬工;干砌块石砌体
dry-laid masonry dam 干砌石坝
dry-laid masonry wall 干砌墙
dry-laid rubble 干砌毛石
dry-laid rubble masonry 干砌毛石圬工
dry-laid stone 干砌石
dry lake 干(涸)湖
dry laminate 缺胶层合板
dry laminating machine 干式层压机
dry land 陆地;旱田;旱地;干燥地;干旱地区
dry-land afforestation 干旱地区造林
dry-land crop 旱地作物
dry-land cultivation 旱地栽培;旱地耕作
dry-land drilling 陆地钻探
dry-land farming 旱地耕耘法
dry-land operation 陆上作业;陆地钻探
dry-land rehabilitation 旱地复原
dry-land rice 陆稻
dry-land suction dredging plant 陆地吸泥挖土设备
dry layer 干(燥)层
dry laying 干铺砌
dry laying of felt 干铺油毡
dry-lean concrete 干贫混凝土
dry lean rolled concrete 干性贫碾压式混凝土
Drylene 德里纶聚乙烯单丝
dry lens 干式透镜
dry lighting storm 干雷暴
dry limestone process 干石灰石法
dry lining 干衬壁;干衬砌
dry loam sand 干壤土壤质砂土
dry loamy sand 干壤质砂
dry looper 干式圈环捆束机;干式活套捆束机
dry loss bulk density 干松散重度
dry loss unit weight 干松单位重
dry loss volume 干松体积
dry lot 饲养场;肥育地;牲畜饲养场
dry machine 干燥机
dry magnetic compass card 干罗盘
dry magnetic dressing 干磁选
dry magnetic particle 干磁粉
dry magnetic separation 干法磁选
dry magnetic separator 干法磁选机
dry manufacture (水泥的)干法生产
dry marshalling 干砌筑;干砌工作
dry-mash hopper 干的碎料饲槽
dry masonry 干砌砖墙;干砌圬工;干砌体;干砌石砌体;无砂浆砌筑;无浆砌筑;砖石干砌体
dry masonry dam 干砌圬工坝
dry masonry pitching 干砌石护坡
dry masonry protection of slope 干砌石护坡
dry masonry wall 不用灰浆的砖石墙;无砂浆砌筑墙体
dry masonry work 干式砌墙法
dry material 干料
dry material mixer 干料搅拌机
dry matter 干燥物质;干物重
dry matter content 干物质含量
dry measure 干容量单位;干衡器;干测;干量;不带偏见的测度
dry mechanical grading 干法机械筛分
dry metallurgy 火法冶金学
dry meter 干式煤气表;薄膜式煤气表
dry mid-latitude climate 中纬度干旱气候
dry mill 干式制粉厂;干式粉碎机;干磨机;干法粉碎
dry milling 干磨;干法破碎
dry-milling machine 干磨机
dry mineral matter free basis 干燥无矿物质基
dry mining 干法开采
dry mining oil or natural gas 干采石油或天然气
drymion 森林植被型
dry mix 硬拌(砂浆);干硬性拌和物;干(搅)拌;干燥料;干拌和物;干硬性混凝土拌和料;干拌;干混合
dry-mix concrete 干拌混凝土
dry mixed 干混合的;干(法搅)拌的
dry mixed concrete 干拌(和)混凝土
dry mixed material 干拌材料
dry mixed shotcrete 干拌喷射混凝土
dry mixer 干式混合器
dry mixing 干混合;干拌
dry mixing time 干混时间
dry mix process 干拌工艺;干拌步骤;干拌法
dry mix shotcrete 干拌喷射混凝土(一般在喷射口加水)
dry mixture 干拌混合料;干混合物;干硬性混合料
dry mixture concrete 干混合物混凝土
dry monsoon 干季风
dry mortar 干硬性水泥砂浆;硬练(砂浆);干砂浆
dry mortgage 无追索权的抵押;干抵押
dry-mo(u)lded 干塑的

dry-mo(u)lded tile 干塑瓦管
dry mounting 干式裱贴;干裱
dry mud 黏土粉
dry mulch 干燥层
dry multiplate clutch 干式多片离合器
dry natural gas 无水天然气;干燥天然气
dryness 干燥度;干物质含量;干旱程度
dryness factor 干燥率
dryness-fire 爆火
dryness fraction 蒸汽干度;干燥率;干燥度
dryness fraction of wet steam 湿蒸汽干燥度
dryness tester 干度检验器
dry neutralization 干中和
dry niche 干壁龛
dry nurse 保姆
dry nursery 旱秧田
dry objective 干(式)物镜
dry off 使干乳
dry-off oven 烘炉;干燥炉(窑);干燥炉(器);干燥炉(机立窑)
dry offset 干胶印
dry offset ink 干胶印油墨
dry offset plate 干胶印版
dry offset press 干胶印机
dry offset printing 干胶印印刷法
dry offset (process) 干胶印印刷法
dry of optimum 最优干度
dry oil 无水石油;干性油
dry operation 干操作
dry out 过早干燥;变干
dry-out period 干燥周期
dry-out sample 无水试样;干燥样品;无试样
dry-out time 干燥时间
dry oven 烘干箱
dry-packed concrete 夯实干硬性混凝土;干填混凝土;干夯混凝土;干硬性混凝土;干夯落度混凝土
dry pack(ing) 干装填充法;干型阻水;干法填装;干法填充;干衬垫;填塞干硬性混凝土;干填法;干填(砂浆);干夯实;干填(混凝土)
dry pack(ing) method (混凝土或砂浆的)干填法;干挤法
dry pail latrine 旱厕
dry paint film 干燥涂膜
dry painting 干涂法;粉末涂料涂装法
dry pan 干(式轮)碾机;干磨盘;干式盘磨机
dry pan mill 干磨石;干(式)盘磨机
dry parting 界砂;分型粉
dry partition 干筑隔墙
dry partition wall 干砌式隔墙
dry pass 盈式干粉碎机
dry paving 干砌面层;干铺砌;无(灰)浆铺砌
dry pelletizing method 干式造粒法
dry pendent sprinkler 干式下垂型喷头
dry period 枯水季;旱季;干季;干旱(周)期
dry permafrost 干永冻层
dry permeability 干透气性
dry photography 干法摄影
dry photopolymer film 干式光聚合物薄膜
dry pick 干拔
dry pick hammer 干式鹤嘴锤;干式风镐锤
dry pigment 干颜料
dry pile 干堆原电池
dry pipe 干(燥)管;干(汽)管
dry-pipe automatic drencher system 干管自动水幕系统
dry-pipe sprinkler system 干式喷水(灭火)系统;干管式消防喷水系统;干式喷淋系统
dry-pipe system 干管式消防喷水系统
dry-pipe underground 地下干管
dry-pipe valve 干式阀;干管阀
dry piston compressor 干活塞式压缩机
dry pit 干坑
dry pitching 干砌(护坡)
dry pitching stone 干砌块石
dry pit pump 干坑泵
dry-placed concrete 干填混凝土
dry-placed fill 干式充填;干填土
dry placer 干砂金矿
dry plankton 去水浮游生物
dry-plant design 干式工厂设计
dry plate 干片;干版;干板;镀锡薄板
dry plate clutch 干盘式(摩擦)离合器;干板(式)离合器
dry plate glass 底片玻璃

dry plate rectifier 干式整流器
dry plate reprography 干片复照
dry plate sheath 干板暗匣
dry point 干点;凹板雕刻针
drypoint etching 针刻腐蚀凹版
dry polishing 干磨光;干法抛光
dry powder 干(燥)粉;干(粉)料
dry powder appliance 干粉消防车
dry powder automatic forming hydraulic press 干粉自动成型液压机
dry powder blender 干粉混合机
dry powder colo(u)r 干色粉
dry powder extinguisher 干粉灭火器
dry powder extinguishing system 干粉灭火装置
dry powder fire-engine 干粉消防车
dry powder fire extinguisher 粉末灭火器;干粉灭火器;干粉灭火机
dry powder paint 干粉状涂料
dry preblend 干预混料
dry precipitation 干式除尘
dry precipitator 干式聚尘器
dry prefabricated construction 干式预制配装构造
dry prefabricated frame assembly method 干式装配施工法
dry premix 干预混料
dry pre-mixing 干式预先混合(水泥制造)
dry preparation 干法精选
dry preserved specimen 干制标本
dry press 干压机
dry press brick 干压砖
dry press brick machine 干压压砖机;干式制砖机
dry-pressed 干压成型的
dry-pressed brick 干压砖
dry-pressed hardboard 干法热压硬质纤维板
dry pressing 干压法;干(法)压制;干压成型
dry press method 干压法
dry press process 干压成型;干压法
dry pressure brick making machine 干压制砖机
dry pressure coating 干压包衣法
dry printing 干式印刷;干法印花
dry printing head 干式打印头
dry process 干(洗印)法;干工艺;干式法;干法制备;干法冶金;干法工艺;干法;干处理
dry process development 干式显影;干法显影
dry process enameling 干法搪瓷;干法涂搪
dry process kiln 干法窑
dry process kiln with cyclone preheater 旋风预热器干法窑
dry process kiln with shaft preheater 立窑预热器干法窑
dry process kiln with suspension preheater 悬浮预热器干法窑
dry process mat 干法毡
dry process of flume gas desulfur 干法烟气脱硫
dry process of shotcreting 干法喷射混凝土
dry process penetration macadam 干法渗入碎石铺路
dry process plain kiln 干法中空窑
dry process rotary kiln 干法旋窑水泥;干法回转窑;干燥作业回转窑
dry process type rotary kiln 干法式转窑;干燥作业回转窑
dry process with boiler 带余热锅炉的干法生产
dry-proof paper for silkworm 蚕用防干纸
dry pruning 修剪枯枝;修干枝
dry pulverizer 干磨机
dry puridication 干法净化
dry purification 干法提纯
dry purification process 干式净化过程
dry quart 干夸脱
dry quenching 空冷淬火
dry quenching of coke 干法熄焦
dry reaction 干(态)反应
dry reagent feeder 干式给药机
dry reclamation 干法再生
dry reconditioning 旧砂过筛处理
dry rectifier 干式整流器
dry rectifier locomotive 带有金属整流器的机车
dry reed contact 干簧(片)触点
dry reed relay 干(舌)簧继电器
dry regime 干燥状态
dry region 干燥区;干旱区
dry region relief 干燥区地貌
dry-relief offset 凸版胶印

dry-rendered 干炼的
dry rendering 干法炼油
dry reservoir 干热储
dry residue 干残渣
dry residue generator 干式乙炔发生器
dry resultant temperature (室内的)黑球温度
dry return (锅炉的)空回水管;回气管;干(式)回水;干式回火
dry return pipe 干式回水管;无水回管
dry rice 陆稻
dry riser 空立管;干式消防立管;干式竖管;无水立管(消防用);干立管
dry rising main pipe 干式消防立管
dry river bed 旱谷;干(涸)河床
dry river valley 干河谷
dry rock cutting 干法岩石切割
dry rock paving 干砌块石面层;干砌块石路面;干铺砌
dry rock paving 干石块铺砌
dry-rodded 干捣的
dry-rodded aggregate 捣实的骨料;干捣实的骨料
dry-rodded concrete 干碾压的混凝土
dry-rodded unit weight 干重度
dry-rodded volume 干捣(实)体积
dry-rodded weight 干捣(实)重量
dry rodding 干(碾)压;干捣(实);干捣(法)
dry-rolled 干碾(压)的
dry-rolled road surface 干碾压的路面
dry-rolled surface 干碾面层
dry rolling 干压;干碾
dry roofing felt 干燥屋顶油毛毡;铺屋面用干油毡
dry roofing glazing (不用油灰的)屋顶玻璃装配
dry room 干燥的房间
dry rot 干枯(木材腐朽);干枯木;干腐;干朽
dry rubbing 干打磨;脱色试验干搓
dry rubble 干砌毛石
dry rubble-block mass stone 干砌片石垛
dry rubble construction 干砌毛石构造;干砌毛石结构
dry rubble dam 干砌乱石坝
dry rubble fill 干填毛石;干砌毛石
dry rubble masonry 干砌片石;干砌乱石圬工;干砌块石圬工;干砌毛石圬工
dry run 空运转;预排演;干钻回次;干操作;试操作;空白试验;预检;干试车;干馏
dry running 干钻
dry-run tar 干馏焦油
dry runway 干跑道
dry saltery 干货店
dry salting 干盐化
dry salt lake 干盐湖
dry salt-pan phase 干盐盘相
dry sample 干(岩)样;干(试)样
dry sample barrel 干式取样筒;干式取样器;表土层取样器
dry sampler 干式取样器
dry sampling 干式取样
dry sand 干(型)砂;干砂(层)
dry sand cast 干砂铸法
dry sand casting 干砂浇铸;干型铸造
dry sand core 干砂型心
dry sand facing 干型涂料
dry sand mo(u)ld 干(砂)型
dry sand mo(u)lding 干砂造型
dry saturated steam 过热蒸汽;干蒸汽;干饱和蒸汽
dry saturated vapo(u)r 干饱和蒸汽
dry-scale disposal 干法排除氧化皮;干法除鳞
dry screening 干(法过)筛
dry scrubber 干法再生装置
dry scrubbing 干法擦洗
Drysdale ac polar potentiometer 德赖斯代尔交流极性电位计
Drysdale permeameter 德赖斯代尔磁导计
Drysdale potentiometer 德赖斯代尔电位计
drysdallite 硒钼矿
dry seal 无油密封
dry sealed cable pothead 干封电缆头
dry seal pipe thread 干封管螺纹
dry seal thread 气密螺纹
dry season 枯水季(节);枯(水)季;旱季;干旱期
dry season flow 枯水季水流;枯水季流量
dry season months 枯水月份
dry sediment 干泥沙
dry self-cooled transformer 干式自冷变压器

dry sensation 干燥感
dry sensitivity coefficient 干燥敏感系数
dry separation 干(法分)选
dry separator 干式选矿机;汽水分离器
dry shake 干撒抹面(水泥粉);干洒抹面;干撒
dry shake floor 干振法(混凝土楼地板耐磨层施工)
dry shaping 干法成型
dry shaver 干修面器
dry-shear test 干剪强度试验
dry sheet 干垫层;不黏结的油毡;(屋面上为防止沥青渗透的)不黏结轻质板
dry-sieve method 干筛法
dry sieving 干筛分;干筛;干法筛分;干法过筛
dry silver film 干式银盐胶卷;干银胶片
dry silver paper 干式银盐照相纸;干式银盐像纸
dry silver print 干式银盐印刷
dry silver printer 干式银盐复印机
dry silver process 干法银盐复印
dry sink 洗涤盘;晾碟架
dry sized 干加工尺寸
dry skid 干滑
dry-skidding test 干滑试验
dry slag 重矿渣;稠渣
dry slagging combustion chamber 固态排渣式燃烧室
dry slake 减弱;缓和;干燥
dry-slaked lime 干熟石灰
dry-slaked quicklime 干消化的生石灰
dry slaking 干崩解
dry sleeve 干缸套;干衬垫
dry sleeve cylinder 干套筒汽缸
dry slide 干滑道
dry slope 临陆坡;(坝的)背水坡
dry sludge 干污泥
dry snow 干雪
dry soap 干皂
dry soil 干(燥)土
dry soil density 干土密度
dry soil sample 干土样
dry soil sample of known weight 已知重量的干土样
dry solid cement 干固水泥
dry solid content 干固体含量
dry solids filtration 干物料过滤
dry solids mixing 干固体混合
dry solution 速干剂
dry solvent 无水溶剂
dry spell 连续干旱;干(旱)期;缺水期;无雨期
dry spinning 干纺亚麻
dry spot 干斑
dry spray 干喷;糖面喷涂法;砂状表面;喷逸;喷溅
dry sprayer 喷粉器
dry spread 干涂布量
dry sprinkler 干式喷头
dry sprinkler system 干管式消防喷水系统;干式喷水(灭火)系统
dry stability 干稳定性
dry stage 干燥阶段
dry stained 干渍
dry staining test 干泅色试验
dry stamp battery 干标记蓄电池
dry standpipe 干竖管;干式立管
dry standpipe system 干式立管系统
dry state 干燥状态
dry steam 干蒸汽
dry steam drum 干蒸汽包
dry steam field 干蒸汽田
dry steam geothermal electric-power generator 利用干蒸汽的地热电站
dry steam humidifier 干蒸汽加湿器
dry steam well 干蒸汽井
dry steel 低级钢
dry stiffness 干硬度
dry stock 经干燥处理木材
dry stone 干石
dry stone base 干(砌)石基层
dry stone drain 盲沟;干砌(石)排水沟
dry stone facing 干砌块石护面
dry stone fill 干填石
dry stone foundation 干砌石基础
dry stone lining 干砌石衬砌
dry stone masonry 干砌石圬工;干砌块石砌体
dry stone pitching 干砌(块)石护坡
dry stone pitching breakwater 干砌块石防波堤

dry stone wall 干砌(石)墙
dry stone walling 干砌块石
dry stone wall with moss-filled joints 干石墙缝中长满苔藓的墙
dry storage 干货栈;干储[贮]藏;干储存;风干储[贮]存
dry strainer 干式过滤器
dry streak plate 灰斑纹的镀锡薄钢板
dry stream bed 干河床
dry strength 耐干燥性;烘干强度;干(态)强度;干(态机械)强度;常态胶着力
dry strength test 干强度试验
dry stress 干燥应力
dry stripping 干脱漆;干剥膜
dry submersible 干式潜水器
dry subsea X-mass tree 干式海底采油树
dry substance 干燥物质
dry subtropic(al) climate 亚热带干旱气候
dry suction 干式吸尘
dry suit 干(式)潜水服
dry sulfide system 干硫化物系统
dry sump 干滑油槽;干沉淀池
dry sump lubrication 干燥润滑法;干式润滑
dry suspended solid 干悬浮固体
dry sweating room 蒸汽室
dry sweetening 干法脱硫
dry system 干式消防设备系统;干式系统
dry system lens 干镜头
dry system of construction 干法施工系统
dry system of curing 干式养护系统
dry tabling 干式摇床选矿
dry tack 半干黏性;干黏性
dry tack adhesion 成型黏着性
dry tack-free 表面干燥;指触干;不发黏;表干
dry-tamped concrete 干夯混凝土;干捣实混凝土
dry tamping 干击法(试件制备);干捣
dry tamping method 干捣法
dry tamping process 干捣法
dry tank 干物质存罐;干舱
dry tap 干抽
dry tape 接缝纸带;压敏胶带
dry-tapy fuel cell 干带燃料电池
dry technical of construction 干法施工技术
dry tempering 干式回火
dry tensile strength 干拉强度
dry tensile strength test 干拉强度试验
dry tensile test 干拉试验
dry test 干试验;干试法;干法检验;排泄试验;排空试验
dry test meter 干火表;干式煤气表
dry-thread raising 干法起绒
dry-through 全干
dry ticket 干票
dry tilling tractor 旱地拖拉机
dry to handle 漆膜完全干燥;指触干
dry-to-handle of film 漆膜可搬运干
dry ton 干吨(美国)
dry toner 墨粉
dry tongue 干(燥)舌
dry-to-non tacky 表干;指触干;不发黏干
dry to permit handling 干燥至可以装运
dry topping 干洒抹面;干撒抹面
dry-to-recoat by brushing 干后用刷子再涂抹
dry-to-recoat by spraying 干后用喷射法再涂抹
dry-to-recoat time 再涂干时间
dry-to-sand 打磨干
dry-to-touch time 指触干时间
dry tower 干塔
dry tract 干旱地带
dry transfer 干式转印
dry transfer submersible 干转移型潜水器
dry tropic(al) climate 干旱热带气候
dry tropic(al) forest 干燥热带森林
dry tropic(al) scrub forest 干燥热带灌木丛
dry tube mill 干(式)管磨机
dry tumbling 干法(滚)抛光;干法翻滚抛光法
dry tumbling barrel 干式清理滚筒
dry tundra 干燥冻土地带
dry-tuned gyro 干调谐陀螺仪
dry turning 干法车坯;干车
dry two-phase sand 干二相砂土
dry two-phase soil 干二相土
dry two-phase unit weight 干二相重度
dry two-phase unit weight of soil 土的干二相重度

dry-type 干式(的)
dry-type air cleaner 干式空气滤清器;干式空气净化设备
dry-type air cooler 干式空气冷却器
dry-type clutch 干式离合器
dry-type cooling system 干式冷却系统
dry-type cooling tower 干式冷却塔
dry-type dust collection system 干式收尘设备
dry-type electric(al) precipitator 干燥型电动聚尘器;干式电吸尘器
dry-type electrolytic capacitor 干式电解电容器
dry-type instrument transformer 干式仪表变压器
dry-type multidisc electromagnetic brake 干式多片电磁制动器
dry-type multidisc electromagnetic clutch 干式多片电磁离合器
dry-type rectifier 干片整流器
dry type tower abrasion mill 干式塔式磨粉机
dry-type transformer 空气冷却(式)变压器;干式变压器;气冷变压器
dry-type (water) meter 干式水表
dry type well method 干式水井法
dry-type wire drawing machine 干式拉丝机
dry unit weight 干重度
dry-up 非图文部分着墨
dry up a well 把井排空
dry upright sprinkler 干式直立型洒水喷头
dry-use adhesive 不防水胶;干用胶
dry vacuum 干式真空
dry vacuum airpump 干式真空气泵
dry vacuum pump 干式真空泵
dry valley 干谷
dry valve 无水的管阀门;干阀
dry valve arrestor 干阀避雷器
dry valve shaft 干阀井
dry vanning 干式淘选法
dry vent 干式通气管;干式通风管
dry Venturi scrubber 干式文丘里洗涤器
dry vibrating material 干式振动料
dry-volume measurement 干体积配料计量;干(料)容积计量;干容积测量法
dry voyage 干货航次
drywall 干(饰面内)墙;干(砌)墙;不用砂浆砌成的自承重毛石墙或方石墙;清水墙;无浆砌墙板;干(式)墙;板墙
drywall adhesive applicator 板墙嵌缝枪式涂胶器;干墙涂胶器
drywall adhesive gun 干墙注胶枪;板墙缝注胶枪
drywall construction 干作业墙体;干饰面内墙构造;干施工墙板构造;清水墙壁结构;干砌墙壁结构
drywall contractor 清水墙承包人;干砌墙的承包人
drywall finish 干作业内墙覆盖面板;内墙干饰面
drywall frame 干作业墙门框
drywall garden 墙壁花园
dry wall(ing) 干砌(墙);清水墙;拼装式墙体;干砌(护墙);无浆砌墙
drywall material 干墙用材料;内墙干饰面材料
drywall partition 用干材料筑成的隔墙;清水隔墙
drywall screw 干墙螺钉
drywall system 板墙系统;干墙系统
dry wash 干谷;干洗;干适淹(水底石在低潮时露在水面);干河床
dry-wash cleaner 干洗剂
dry washer 干洗机;干式选矿机;风力选矿机
dry washing of coal 煤干洗
dry waste 干垃圾
dry watchman 井场停钻值班员
dry weather 干燥天气
ddry-weather flow 旱季流量;旱天流量;干流量;旱季径流;枯水季水流;枯(水)季流量;家用污水量;晴天污水量(无雨水)
dry-weather runoff 旱季径流
dry wedging 干砌楔块;临时楔块;干楔固
dry-weighed batch 干重分拌
dry weight 干重;干燥重量
dry weight basis 干重组成
dry weighted batch 干重配料
dry weight of sludge mass 污泥质量干重
dry weight of soil 土的干重
dry weight of soil solids 土粒干重
dry weight of the aggregate 集料干重;骨料干重
dry weight of the casing 套管干重
dry weight rank method 干重分级法
dry weir 低潮干出渔堰

dry weld 易断接头
dry well 干井；排水井；渗井；泄水井
dry well cutting 干井挖掘
dry well pumping house 干室泵房
dry-wet bulb hygrometer 干湿球温度表
dry-wet cell type barge carrier 干湿舱型载驳货船
dry-wet combination diving suit 干湿两用潜水服
dry-wet cooling tower 干湿式冷却塔
dry-wet cyclone 干湿旋风除尘器
dry-wet type carrier 干舱载驳船
dry-wet-type cooling system 干湿式冷却系统
dry white lead 白干铅粉
dry wind 干燥风
dry winding 干法缠绕
dry wire drawing 干法拉丝
dry wood 烘干木材；干(燥)木材
dry-wood termite 干木白蚁
dry work 干作业(施工)；干式施工；干地施工
dry year 枯水年；旱年；干年；旱早年(代)；少水年
dry yield 干产量
dry zone 干旱地区
dry zone planning 干旱区计划
dry zone programming 干旱区计划
dry zone research 干旱区研究
dry zone study 干旱区研究
dry zone survey 干旱区调查
Drzwiecki theory 德兹威基理论
D-shaped spalling 半圆形剥落；半球形碎石片
D-shaped tower 半圆形塔楼；D形塔楼
dtention time of car 货车停留时间
d-tert-butyl-p-cresol 二特丁基对甲酚
D-trap D形隔气具；D形存水弯
duab(ing) 涂抹；涂料
duad 成对之物
dual 孪生；加倍；对偶(的)；二重(的)；二的；双重(的)；双数
dual accelerometer 双向加速度仪；双向加速度计；双向加速度表
dual access 双臂存取
dual action lift 双向升降机
dual address stack 双地址栈
dual agency 双重代理人
dual agent attack 双灭火剂灭火
dual-agent system 双剂联用灭火系统
dual air pump 复式空气泵；双联空气泵
dual alkali process of flue gas desulfurization 双碱流程
dual all integer 对偶全整型
dual amplification 双重放大
dual amplifier 双重放大器
dual antenna 双重特性天线
dual arc weld 双弧焊
dual arc welding 双弧焊接
dual-arm rotating jet device 双臂式旋转喷射装置
dual-arm rotating jet washer 双臂式旋转喷射冲洗装置
dual aspect convention 复式记账惯例
dual automatic direction-finder 双指针自动测向仪
dual axis compensator 复轴补偿器
dual axle ratios 双轴比
dualayer solution 双层溶液
dual basing-point system 双重基点制
dual-beam 双光束；双射线；双电子束
dual-beam betatron 双束电子感应加速器
dual-beam interferometer 双束干涉仪
dual-beam observation 双束观测
dual-beam oscilloscope 双(射)线示波器
dual-beam synchronoscope 双线同步示波器
dual-beam synchroscope 双电子束同步示波器
dual-beam system 双线示波器；双束制
dual-beam technique 双波束技术
dual-bed continuous contactor 双塔连续接触床
dual-bed dehumidifier 双床吸附减湿器
dual berth facility 有两个泊位的码头
dual block tie 双块混凝土轨枕
dual bore cluster drilling 双筒密集钻井
dual bracing 双重支撑
dual bracing system 双支撑体系；双(重)支撑系统
dual brake-control 双刹车控制
dual bucket unloader 双排链斗卸车机
dual budgets 复式预算
dual burner (可用二种燃料的)双燃烧器；双工质燃烧
dual bus 双总线

dual capacitor 双联电容器
dual-capacitor motor 双电容器式电动机
dual card 对偶卡片；对联卡片
dual carriageway 两块板车行道；复式公路；复式车行道；双幅式车行道；双车道
dual carriageway motorway 双车行道汽车路
dual carriageway (road) 双线车道；两块板道路；双幅式(道)路；双幅路
dual-car rotary dumper 串联双翻车机
dual cast cylinder 复铸汽缸
dual-catalyst 双功能催化剂
dual center 双主轴箱机床
dual chain bench 双链拔丝机
dual-chamber baler 双室式捡拾压捆机
dual chamber incinerator 双室焚化炉
dual chamber lock 双闸室船闸
dual channel 复式水槽；复式河槽；双渠；双(向)通道
dual channel amplifier 双通道放大器
dual channel analysis 双道分析
dual channel burner 双风道燃烧器；双风道喷嘴
dual channel collector 双接收器
dual channel controller 双通道控制器；双路控制器
dual channel device 双道装置
dual channel flame photometric detector 双道火焰光度控制器
dual channel oscilloscope 双束示波器；双电子示波器
dual channel receiver 双接收器
dual channel regulator 双渠道调节器；双路调节器
dual channel sampler 双分道采样器
dual channel side scan sonar 双道旁侧声呐
dual channels of distribution 双渠道推销
dual channel system 双通道系统
dual channel thermal infrared scanner 双通道热红外扫描仪
dual channel type 双沟道型
dual charge structure of tariff 运价的双费率结构
dual chartering 双重执照
dual circuit 对偶电路；双电路
dual circuitry check 双重电路校验
dual circulation 双循环
dual close-in pressure valve 二次关井工具
dual code 对偶码
dual column chromatography 双柱色谱(法)
dual combustion and denitration furnace process 双燃烧和除氮炉法
dual combustion cycle 双燃循环
dual combustion engine 双燃发动机
dual compartment septic tank 双间隔化粪池；双隔间化粪池
dual completion 双层完成
dual completion well 双油管完井
dual component 两用元件
dual compression 双向压缩
dual-compression trench roller 双轮沟槽压路机
dual compressor 双转子压气机
dual concentric drilling string 双层同心钻杆柱
dual concentric drill pipe 双层同心钻管；双壁钻杆
dual concrete pump 复式混凝土泵
dual-cone drum 双锥鼓
dual conomic structure 二元经济结构
dual control 对偶控制；(自动和人工的)双重控制；双杆控制
dual control system 复式控制系统
dual-conversion superheterodyne receiver 双变频超外差式接收机
dual coordinates 对偶坐标
dual cross-section 双断面
dual cure mechanism 复合固化机理；双固化机理
dual curvature contact lens 重弯隐形眼镜
dual cycle 混合循环
dual-cycle boiling-water reactor 双循环沸水堆
dual cycle operation system 往复循环方式
dual-cycle reactor plant 双循环反应堆发电厂
dual-cycle toilet 双档大便器
dual-cylinder locomotive 带有复涨式汽缸机车
dual delayed sweep 双重延迟扫描
dual densilog 双源密度测井
dual densilog curve 双源距密度测井曲线
dual density 双重记录密度；双密度
dual density device 双密度装置
dual design criterion 双重设计准则
dual detector 双向检波器；双检测器
dual-detector receiver 双检波接收机

dual diaphragm pump 双隔膜泵；复式隔膜泵
dual disc check valve 双盘防逆阀
dual disinfection 双消毒
dual display 双重显示；双显示器
dual disposal 混合处理；雨水与污水的混合处理；二元处理
dual disposal method 混合处理法
dual distribution system 双配水(配电)系统
dual diversity receiver 双(重)分集接收机
dual dot 对偶点
dual dot screen 对偶点网目板
dual drainage model 混合排放模型
dual drill rig 双钻机
dual drive 双重传动；双联传动装置；复式传动；双驱动(器)
dual drive conveyer 双传动输送机
dual drive of mill 磨机双传动
dual drive separator 双传动选粉机
dual drive way 双车道
dual-drum 双筒；双轮；双卷筒的
dual-drum grabbing crane 双卷筒抓斗起重机
dual-drum mixer 二重圆筒混合机；鼓拌和机；双筒搅拌机
dual-drum pav(i)er 双滚筒铺路机；双鼓式(混凝土)铺路机
dual-drum roller 双轮压路机；双轮路碾
dual-drum tumbler 双鼓式转鼓
dual drum vibratory roller 双筒振动压路机；双筒振动(路)碾
dual-drum winch 双卷筒绞车
dual-dual highway 四幅式公路；四车道公路；双复式(四车道)公路
dual duct 双管沟；双导管
dual duct air conditioning system 双风管空气调节系统；双风道空调系统
dual duct system 双风管空气调节系统；双风道系统；双风道方式
dual eccentric tower type grate 双偏心塔式箅子
dual economy 二元经济
dual effect control device 双效控制设备
dual effect of investment 投资的双重效益
dual electron image 二次电子图像
dual-element fuse 双成分熔断器
dual elevator 双台电梯
dual-emitter switch 双射极开关
dual emulsion 二元乳化液
dual engine 双联发动机
dual evaluation 双保价
dual evaluation clause 双保价条款
dual evapo(u)ration 二级蒸发
dual exchange market 双重外汇市场
dual exchange rate 双重汇率
dual exciting linear motor 双励磁线性电机
dual face 对偶面
dual feasible condition 对偶容许条件
dual feasible solution 对偶可行解
dual feed 双面进料；双路馈电；双端馈电
dual fiber cable 双光缆
dual fiber optic(al) jumper monomode 单模双芯跳线光纤
dual filter 对偶滤波器
dual-filter hydrophotometer 双色水下光度计
dual final drives 双轴主驱动
dual fired furnace 双燃料窑
dual flame detector 双火焰检测器
dual flank gear rolling tester 双面啮合齿轮检查仪
dual-flight stair(case) 双跑楼梯
dual flow 双流
dual-flow oil burner 双流式油喷燃器
dual flow path 双流路
dual flushing system 双挡冲洗系统
dual-frequency beacon 双频信标
dual-frequency converter 双频直流变换器
dual-frequency echo-sounder 双频回声测深仪
dual-frequency locomotive 混合频率型机车
dual-frequency motor 双频率电动机
dual-frequency sounder 双频测深仪
dual-frequency traction unit 混合频率牵引机组
dual fuel 混合燃料
dual-fuel burner 油气两用燃烧器
dual-fuel customer 双燃用户
dual-fuel engine 双(用)燃料发动机；使用两种燃料的发动机；双燃机
dual-fuel fired furnace 双燃料窑

dual-fuel system 双燃料系统
dual function 对偶函数;双重作用
dual furnace 两用燃烧器
dual-gap head 双缝磁头
dual gas-shield welding 双重气体保护焊
dual ga(u)ge line 双轨距线路
dual ga(u)ge railway 双轨距铁路
dual ga(u)ge system 双用测定器体系
dual gear 二联齿轮
dual girders 双(大)梁
dual glazing 双层玻璃窗
dual graph 对偶图
dual-grid tube 双栅管
dual group 对偶群
dual guided slush service valve 双导板泵阀
dual gyro pilot 复式自动操舵
dual-hardness rubber ring 双硬度橡胶圈
dual head girder 双顶梁
dual head grinder 双头研磨机
dual head nail 拼钉;双头钉
dual highway 复式公路;双幅式公路;双车道公路
dual hologram 双像全息照片;双全息图
dual-homed gateway 堡垒主机
dual hump 双驼峰
dual humping 双溜放
dual humping facility 双溜放设备
dual hydraulic(al) brake 双组液力制动器
dual hydraulic circuit brake system 双管路液压制动系统
dual ignition 双(重)点火;双塞点火;双火花塞点火;双磁电机点火
dual image processing system 双影像处理系统
dualin 双硝炸药
dual independent map encoding 双独立地图编码
dual index mode 二重变址方式
dual indicator 双针式指示器
dual induction curve 双感应测井曲线
dual induction-focused log 双感应聚焦测井
dual induction-focused log curve 双感应聚焦测井曲线
dual induction-focused log plot 双感应聚焦测井图
dual induction-laterolog 双感应八侧向测井
dual induction-laterolog 8 curve 双感应八侧向测井曲线
dual induction log 双感应测井
dual induction log plot 双感应测井图
dual induction-spheric(al) focused log 双感应球形聚焦测井
dual induction-spheric(al) focused log curve 双感应球形聚焦测井曲线
dual injection pump 复式注油泵
dual-in-line 双列直插
dual inline package 双列直插式组件;双排标准组合插件;双列直插式外壳;双列直插式封装
dual inline type 双列直插式
dual input controller 双输入控制器
dual inside and outside seal 双重内外密封
dual instruction flight 双人教练飞行
dual-in-tandem landing gear assembly 双串式着陆轮架装置
dual integral equation 对偶积分方程
dual interferometer 双干涉仪
dual ion beam detector 双离子束探测器
dualism 二性性;二重体;二元论
dualism concept of accounting 二元观念的会计
dualistic 二元论的
dualistic economy 二元经济
dualistic nature 二象性
dualistic society 二元社会
dualistic structure 二元结构
dualistic transformation 对偶变换
duality 对偶(性);二重性;二元性;二象性
duality gap 对偶间隙
duality law 对偶(性定)律
duality map 对偶映射
duality method 二重法
duality principle 对偶(性)原理;波粒二象性
duality theorem 对偶定理
duality theory 对偶理论
dualization 对偶(化);复线化;二元化
dualizing 极反演变换
dual lane 双线航道;双车道
dual laser 双波长激光器
dual lateral tile 双层支座砖

dual laterolog 双侧向测井
dual laterolog curve 双侧向测井曲线
dual lay 双线铺设
dual layer lining 双层衬里
dual layer solution 双层溶液
dual level sensing 双电平读出
dual level turret 双层转塔
dual lever 双杠杆
dual line 复线;双线(段)
dual linear potentiometer 双线电位计
dual linear programming 对偶线性规划
dual liner 双列分行机
dual lines of lock 双线船闸
dualling 加倍的;成对的
dual loader 双向装车机
dual loading system 往复装载系统;往复装载方式
dual load wheel 双负重轮
dual-lug yoke plate 双耳联板
dually completed well 双层完井
dual main system 双干管制;双干管系统;双干管式
dual mechanism 双重机制;双重机构
dual-media filler 双层滤料滤池;双层介质滤池
dual-medium 双滤料
dual-medium stratified filter bed 多滤料分层滤床
dual membrane process 双层滤膜工艺
dual mesh 对偶网格;对偶网格;二重网格;二重格网
dual metal casting 双金属离心铸管法
dual meter 双重表;两用表
dual microscope 双筒显微镜
dual mineral method 双矿物法
dual mode 双模;双重方法;两种方式;两种运输方式
dual-mode bus system 两用公共汽车系统
dual-mode control 复式控制;双式控制
dual-mode range acquisition system 双波形测距系统
dual mode system 两用交通系统
dual mode travelling wave tube 双模行波管
dual modulation 双调制
dual-motor drive 双机驱动
dual mo(u)ld 双模壳;复式线脚
dual nationality 双重国籍
dual nature 双重性
dual nature of investment 投资的双重性
dual network 互易网络;对偶网络;对偶电路;二元网络
dual non-linear decomposition 对偶非线性分解
dual non-linear programming 对偶非线性规划
dual normal log curve 双电位法测井曲线
dual nozzle spray gun 双嘴喷枪
dual offset 双重偏移;双重壁阶
dual of tangent space 对偶切空间
dual operation 对偶运算;对偶操作;二元运算;二元操作;双重控制(自动和人工);双重操纵;双态操作
dual operations of wagon 双重作业
dual optic(al) wavelength 双色光波长
dual optimal solution 对偶最优解
dual organic cation bentonite 双有机阳离子膨润土
dual orientation 双向取向
dual otter board system 双网板装备
Dualoy 杜阿洛姆钨合金
dual pack 双组分
dual packer 双打包机
dual-pattern light measurement 双区测光
dual peak 二重峰
dual performance rear axle 双速后桥
dual phase extraction 两相萃取
dual photometer 双光度计
dual plan 双重计划
dual plan for standard cost 标准成本的双行记账法
dual platen 双面板
dual polarization 双极化;双偏振
dual-polarization laser 双偏振激光器
dual-polarization oscillation 双偏振荡
dual-polarized ring laser 双偏振环形激光器
dual portal 双(入口大)门
dual port controller 双通道控制器
dual-ported memory 双端口存储器
dual-power aerated lagoon 双动力曝气塘
dual-power instrument 双倍率仪器
dual-power level lagoon system 双动力水平塘系统
dual-power multi-cellular lagoon 双动力式多格塘

dual-power multi-cellular lagoon system 双动力式多格塘系统
dual-pressure 双重压力
dual pressure chamber 高低压室
dual-pressure controller 高低压控制器
dual-pressure control switch 高低压控制开关
dual-pressure cycle 两段加压循环
dual-pressure preformer 弹簧模预压机
dual-pressure press 弹簧模压机;双向压力压机
dual price system 二部价格制
dual principle 对偶原理
dual problem 对偶问题
dual process 对偶过程
dual processing 二重处理
dual processing system 双信息处理系统
dual processor 双(信息)处理机
dual processor system 双信息处理机系统
dual producer 双发生器
dual-product line 双生产线
dual-product operation 两种产品输送
dual programming 对偶规划
dual projector 双投影器
dual property 二象性
dual pump 双缸泵;复式泵
dual pumper operation 泵浦车双干线供水
dual-pumping well 双层采油井
dual-purpose 兼用的;双重目的(的);双效(的);两用(的)
dual-purpose bucket 两用铲斗;两用戽斗
dual-purpose bulk carrier 矿石原油两用船
dual-purpose camera 两用摄影机
dual-purpose dolphin 靠船系船墩
dual-purpose flat files 两用扁锉
dual-purpose kerosene 两用煤油
dual-purpose machine 双用途机器
dual-purpose medium 两用介质
dual-purpose mixing plant 两用拌和设备;两用搅拌设备
dual-purpose nozzle 多用水枪
dual-purpose reactor 两用反应堆
dual-purpose roller 两用压路机;两用路碾
dual-purpose room 两用房间
dual-purpose screwdriver with plastic handle 两用塑料柄螺丝刀
dual-purpose sequential sampler 两(用)顺序取样器
dual-purpose sewer 合流阴沟;两用阴沟
dual-purpose ship 两用船
dual-purpose stove 两用炉
dual-purpose tinter 双色着色器
dual radioactive decay 多重放射性衰变
dual-rail renewal 成对换轨方法
dual range 双量程;双距离;双波段
dual range hoist 两挡卷扬机
dual range portable tester 轻便双程材料试验机
dual rank flip-flop 双列触发器
dual rank register 双列寄存器
dual raster television system 双光栅电视系统
dual-rate moon camera 月球双摄影机
dual-rate satellite camera 卫星双速摄影机
dual-rate spring 可变刚性弹簧
dual-rate system 双重运费制(运费同盟的两价运费制);双重费率制度
dual-ratio control 二级微动控制
dual-ratio reduction 双减速比机构
dual-ratio steering 双级操纵
dual rear tire 双后轮胎
dual rear wheel 双胎后轮
dual regulation 双重调节;双调
dual reheat 两次再热
dual relationship 对偶关系
dual relief valve 减压阀;保险阀;调压阀;双联安全阀
dual representation 双重代表权
dual reuse system 双再用系统
dual rheostat 双圈变阻器
dual ring adaptor 双环适配器
dual ring controller 双环(信号)控制机
dual riser system 双立管系统
dual road 复式车行道
dual rod puller 双杆提引器
dual roll breaker 双滚筒碎石机;双滚筒破碎机
dual roll crusher 双滚筒碎石机;双滚筒破碎机
dual roller 双作用压路机;双滚筒

dual rotors 双转子
dual rotor valve 双旋子阀门
dual runway 双线跑道
dual rupture disks 双防爆膜
dual sand catching basin 双层砂集水池
dualscatter laser velocimeter 双重散射激光测速计
dual-seal 双重密封
dual-seal tubing joint 双封型油管接头
dual sense amplifier 双重读数放大器
dual service 不同源供电;双源供电
dual set 对偶集
dual-ship operating 双船作业
dual simplex algorithm 对偶单纯形法
dual simplex method 对偶单纯形法
dual slide scanner 双联幻灯片析像器
dual slope analog/digital converter 双斜率模/数转换器
dual solvent process 双效溶剂萃取(油脂)法
dual space 对偶空间
dual spectrum 双光谱复印法
dual-stage granular activated carbon treatment 双级粒状活性炭处理
dual standard receiver 双标准接收机
dual-status tax year 双重纳税年度
dual stereoscope 双立体镜
dual-stream dry chemical nozzle 双射流干粉炮
dual string completion 双油管完井
dual structural system 双重结构系统
dual structure 对偶结构
dual structure of economy 双重经济结构
dual structure system 双重结构系统
dual supply system 双供水系统;双供电系统
dual surface treatment 双表面处理(路面)
dual symmetric amplitude 双关对称振幅
dual system 两用钻机;二重系统;双重分配电路;双管系统;双重体系;双机系统
dual system track 双线制线路【铁】
dual tabularform 二次表式
dual tandem 双串式
dual tandem landing gears 双串式(飞机)着陆装置
dual-tank type pump 双箱型泵
dual tariff 加重关税;双重关税
dual tariff system 复关税率制
dual temperature brine system 双盐水系统
dual temperature control 双重温度控制
dual temperature separation 双温分离
dual tensor 对偶张量
dual texture of flood plain 河漫滩二元结构
dual theorem 对偶理论
dual three-lane motorway 复式三车道汽车路
dual threshold 双阈(值)
dual thrust 双推力
dual-thrust motor 双推力发动机
dual tires 拼装轮胎;双(料)轮胎
dual tire truck 双轮胎货车
dual tire wheel 双胎车轮
dual-tone frequency 双音频率
dual-tone multi-frequency 双音多频
dual-tone multiple frequency 双音多频
dual toroidal yoke 双环形偏转线圈
dual tower facade 双塔立面;双塔的正面
dual tower suspension preheater 双塔式悬浮预热器
dual-trace amplifier 双迹放大器
dual-trace oscilloscope 双线示波器;双迹示波器
dual transformation 对偶变换
dual transmitter 两用发射机
dual traverse barograph 双杆气压计
dual tubes heat exchanger 双重管式换热器;双套管式换热器
dual tunnel 复线隧道;双向隧道
dual two-lane motorway 双幅式双车道高速公路
dual tyews 双料轮胎
dual type circuit 对偶电路
dual-type loading 双轮荷载
dual usage 对偶的运算
dual-use line 双用线
dual-use packaging 双重用途包装
dual valuation clause 双重价格条款
dual valve 复式阀;双(联)阀
dual-valve jetting tip 双气阀喷嘴
dual variable 对偶变量
dual vector 对偶矢
dual vector space 对偶向量空间

dual vent 成双并置通气管;两用通气管;两用通风管;双用通气管
dual vibrator 成对振动器
dual vibratory pile hammer 成对振动打桩锤
dual-voltage motor 双压电动机
dual-wall drill pipe 双壁钻杆
dual water scheme 双水源给水工程
dual water supply 分质供水
dual water supply system 饮用杂用双水系统;双给水系统
dual waveband infrared sensor 双波段红外传感器
dual-wavelength-double beam ultraviolet-visible spectrophotometer 双波长双光束紫外可见分光光度计
dual-wavelength spectrophotometer 双波长分光光度计
dual-wavelength spectrophotometry 双波长分光光度法
dual-wavelength thin layer chromatography scanner 双波长薄层色谱扫描器
dual well 双筒井
dual well completion 双井开采
dual wheel 双轮
dual wheel assembly 双轮装配
dual wheel loading 双轮荷载
dual wheel steering gear 双轮转向装置
dual wheel under carriage 双轮起落架(飞机)
dual wing Christmas tree 双翼采油树
dual worm extruder 双螺杆挤出机
dual yoke vent 双轭通气管
dual zone well 双层出油井
Duane-Hunt law 杜安·亨脱定律
Duane-Hunt limit 杜安·亨脱界限
Duan ink slab 端砚
Duanxi ink stone 端溪砚
Duatol dye 杜阿托尔染料
dub 扎;池塘;涂油
dubbed corner 切角
dubbed off 找平;刮去
dubbin 油液;塑化剂;防水油(脂)
dubbing 抹灰找平;打底;刮平;粉刷找平;防水油(脂);翻印;板边欠厚;塑化剂;皮革保护油
dubbing machine 配音机
dubbing out 整平;刮平;锤平;填塞;抹灰基层修整;填平
Dubbs cracking 杜布斯裂化法
dubiocrystalline 微隐晶质
dubious foundation 不可靠基础;不可靠地基
dubious value 可疑值;不可靠值
dubmsrine relief 海底地形
dub off 刮去;倒角
Duboir oscillograph 杜波依尔示波器
Du Bois balance 杜波阿天平
Dubrovin ga(u)ge 杜布罗文真空规
Ducal Palace 公爵住宅
Duchemin's formula 杜赤明公式
Duchesnian(stage) 杜契乃阶【地】
duchess 石板瓦(12英寸×24英寸)
duchess slate 岩板尺寸(24英寸×12英寸)
duck 粗布;轻帆布;水陆两用摩托车;水陆两用飞机;闪避
duck ant 白蚁
duck belting 帆布带
duckbill bit 鸭嘴形钻(头)
duckbill blade 鸭嘴式装载铲
duckbill loader 鸭嘴式装载机
duckbill nail 鸭嘴钉
duckbill pen 鸭嘴笔
duckbill pliers 鸭嘴钳
duckbill snips 鸭嘴钳;鸭嘴剪
duckbills type spillway 鸭嘴形溢洪道
duckbill weir 鸭嘴形堰
duck blue 靛蓝色
duck board 修屋面用步级板;罐顶人行栈桥;沟槽板道;挡雪板;挡水墙;(泥泞道上铺的)木板道
duck canvas 帆布
duck-egg porcelain 鸭蛋青瓷器
ducker 潜水人
Ducker-Prager(yield)criterion 德鲁克—普拉格准则
duck fabric 帆布织物;工业用织物
duck foot 直角弯头;支座;鸭掌式锄铲;箭形中耕锄铲
duckfoot bend 托座弯头;支撑弯管;端弯头
duckfoot cultivator 鸭掌式中耕机

duckfoot shovel 鸭脚锄铲
duckfoot sweep 鸭脚铲
duck green 鸭绿色
duck loom 帆布织机
duck propeller 全回转螺旋桨
duck roller 墨斗辊
duck's-egg green 鸭蛋青
ducks meat 浮萍
duck truck 水陆两用车
duckweed 浮萍(属)
duck work 管道系统
Duco cement 杜珂胶(一种粘贴应变片的黏合剂)
ducol-punched card 每列多孔卡片
Ducol steel 低锰结构钢
ducon 配合器
duct 孔道;空气通道;管(道);导流管道;导管;槽道;风管;风道;波道
duct, pipe, hole for cable laying 敷设电缆的槽、管、孔
duct adjusting screw 墨斗调节螺丝
duct air flow noise 管道气流噪声
Ductalloy 延性铸铁
duct attenuation 风道消音;管道消声
duct blade 墨斗刮刀
duct cable 管道电缆
duct channel 地沟
duct cleaner 电缆管道清洁器
duct condenser 耦合电容器
duct construction 管道施工
duct cover 管道盖;通道盖
duct design 管道设计;风道设计
ducted 管道中的
ducted air heater 管式热风炉
ducted body 中空体
ducted cable 管中缆索
ducted cooling 管道式冷机;导流冷却;隧管冷却
ducted fan 内外函喷气发动机;导管式通风机;导管风扇
ducted-fan jet engine 导管风扇式喷气发动机
ducted propeller 装有导流管的螺旋桨;槽道螺旋桨
ducted-propulsion unit 空气喷气推进装置
ducted rocket 火箭冲压喷气发动机
ducted warm air 管道热空气
duct entrance 地下管道入口
ducter 微阻计
duct fan 风道风机
duct fittings 管道配件
duct for electric(al) wiring 电气敷线管道;安装电线管道
duct for service 运用的管道;供应通道
duct friction loss 风道摩擦损失
duct furnace 烟道采暖炉;导管(暖气)炉
duct grouting 孔道灌浆;孔道压浆;导管灌浆
duct heater 热导管
duct height 波导高度
ductibility 延伸性
ductibility testing machine 延度试验机
ductible frame 延伸框架
ductile 小导管;韧性的
ductile alloy 展性合金
ductile base oil 黏性铺路油;优等延性油
ductile bead 韧性变形缘
ductile bed 塑性层
ductile-brittle transition 韧脆转变
ductile-brittle transition temperature 脆性转变温度
ductile building 延性建筑物
ductile cast iron 延性铸铁;球墨铸铁;可锻铸铁
ductile cast iron pipe 延性铸铁管
ductile cast iron roll 可锻铸铁辊;可锻铁铸辊
ductile cast iron segment 球墨铸铁管片
ductile concrete 延性混凝土
ductile crack 延性破裂
ductile crustal spreading 地壳延性扩张
ductile deformation 延性变形;柔性变形
ductile design 延性设计
ductile failure 延性破坏;延压破坏;韧性破坏
ductile failure mechanism 延性破坏机制
ductile fault 韧性断层
ductile fracture 变形断裂;纤维状断裂;延性破裂;延性断裂;延性断口;韧性破裂;塑性断裂
ductile iron 延性铁;韧性铁(可锻铸铁);球墨铸铁
ductile iron pipe 可延铸铁管;球墨铸铁管
ductile iron pipe joint restraint 延性推入接头
ductile iron restrainer 延性(铁)固定器

ductile iron roll 变性铸铁轧辊
ductile iron segment 球墨铸铁管片
ductile iron T-bolt 延性铁 T 形螺栓
ductile joint ring 可塑的接合环
ductile level 延性水平
ductile material 延(展)性材料;韧性材料
ductile metal 延性金属;韧性金属
ductile moment-resisting space frame 延性抗弯空间框架
ductileness 延性
ductile nickel-resist cast iron 高镍球墨铸铁
ductile reinforced concrete frame 延性钢筋混凝土框架
ductile response 延性反应
ductile rupture 韧性断裂
ductile shear belt 韧性剪切带
ductile shear zone 韧性剪切带
ductile stage 金属的延展阶段;延展阶段
ductile steel 韧(性)钢
ductile structure 延性结构
ductile test 韧性试验
ductile treatment 球化处理;塑性处理
ductile wrench-shear zone 平移韧性剪切带
ductile yield 延性屈服
ductilimeter 延性试验机;延性计;延度仪;延度计;展度计;伸缩计
ductilimeter test 韧性试验;反复弯曲试验
ductilimetry 测延术
ductility 可延展性;可塑性;延(展)性;延(伸)性;延(伸)度;柔软性;韧性
ductility coefficient 延性系数
ductility demand 延性要求
ductility-dip crack 失塑裂纹
ductility factor 延性因数;延性系数;延度系数
ductility limit 流限;屈服点
ductility machine 延性仪
ductility material 延性材料
ductility of bitumen 沥青延(伸)度
ductility potential 延性势
ductility ratio 延性比;延度比
ductility requirement 延性要求;延性需要量
ductility response spectrum 延性反应谱
ductility spectrum 延性谱
ductility test 延展试验;延性试验;延度试验
ductility tester 延性试验机;延度仪;延度试验机;延度计
ductility test machine 延性试验机
ductility test mo(u)ld 延性试验用模(沥青材料)
ductility traits 延性特征
ductility transition 韧性转变
ductility transition temperature 延性转变温度;塑性转变温度
ductility value 延度值;延性值
ductilometer 延性计;延度仪;延度计;展度计
ductilometer test 反复弯曲试验;反复挠曲试验
ducting 烟道;管道;风管;导管;导通
ducting for electric(al) wiring 电线布线管道
ducting system 管路系统
duct insert heater 风道加热器
duct insulation 导管保温材料
duct keel 箱形龙骨
duct knife 墨斗刮刀
ductless 无(导)管的
duct line 电缆管道
duct lining 管道衬砌;管道衬里;风道衬层;风道衬里
duct loss 管道损失
duct-mounted smoke detector 安装在管道里的感烟探测器
duct opening 管道开口
duct optic(al) cable 管道光缆
ductor 压辊;涂漆辊
ductor roller 水斗辊
duct piece 管片
duct pilot 控制回路
duct pressure loss 风道压力损失
duct propagation 波导传播
duct propulsion 喷气推进
duct resistance 风道阻力;风道摩阻
duct riser 管道提升井;竖向管道
duct roughness correction 风管粗糙校正值
duct run 电缆管道
duct sealing compound 电线管道封口胶
duct shaft 通风道竖井
duct sheet 制管薄板

duct silencer 导管消声器;风道消声装置
duct sound absorber 风道吸声装置
duct space 风道空洞
duct survey 管道测量
duct system 管道系统;导管系统;风道系统
duct temperature sensor 风管温度探测器
duct thermostat 温度调节器
duct thickness 波导厚度
duct trench 管道沟
duct type drier 管道式干燥机
duct type smoke detector 通风管式感烟探测器
ductule 小导管
ductuli aberran-tes 迷管
ductus 管
ductus semiciculari 膜半规管
ductus semi-circulares 半规管
duct velocity 通过管道的速度
duct velocity profile 管内流速分布图
duct ventilation 管道式通风
duct water distribution system 管道配水系统
duct way 管式通道;管道沟槽;管沟;导管
duct width 波道宽度;波导厚度;通路宽度
duct work 管道铺设;管道工程;管道敷设作业;风道工程;管道系统;管道设施;风管
duct wrap 管道包裹物
dud 假货
Duddell arc 杜德尔弧
Duddell galvanometer 杜德尔热检流计
Duddell oscillograph 杜德尔示波器;可动线圈式示波器
dudleyite 珍珠蛭石
due 应付款;到期(的);到付款日期
due authority 适当权限
due bank 收款银行
due bill 借票;到期汇票;借约(以服务抵偿)
due care 合格质量验收
due course 顺次;适当时候
due course of transit 适宜的运输途径
due date 满期日;应付款日期;支付日期;到期(之)日;到期日期;付款日期;票据到期日
due date for payment 付款截止日
due date of coupon 息票到期日
due despatch 尽责速递
due diligence 谨慎处理
due dispatch 尽责速递
due east 向东方;正东
due for correspondents account 往来账户欠款
due for debt 向……讨债
due form bank 托收款项
due from 应收(款项)
due from affiliated and associated companies 联属及营业欠款
due from agencies 代理商欠款
due from banks 银行欠款;应收银行款;同业存款
due from consignor 寄销人垫付款;应收寄销人款
due from officers and employees 职工欠款
due from other fund 其他基金欠款
due from subscribers 应收认缴股款
due fulfillment of the contract 合同的适当履行
due-in 待收
duel 决战
due north 正北
due of shipment 装运限期
due-on-sale 即期销售;即期出售
due-out 待发
due presentment 依限提出
due process 规定程序;法定诉讼程序
due reward 应有的奖励
dues 捐款;会费;税收
dues and taxes 捐税
due south 正南
due time 顺次;适当时候
due time of arrival 应当到达时间;正点到达时间
due to 应付
due to affiliated and associated companies 欠联属及联营企业款
due to agencies 欠代理店款
due to balance 你方账户(国际汇兑)
due to be paid 应支付的
due to consignor 应付寄销账款
due to correspondent's account 欠往来账户款
due to fiscal agent 欠财务代理人款
due to foreign branch 欠国外分支机构款
due to other funds 应付其他基金款

due value 预期值
due west 正西
dufault page 省略时页面
duff 枯枝落叶层;煤屑;煤粉;细(劣)煤;地面落叶层;地面腐殖质;残落物层;半腐殖质;半腐层;揉面
duff hygrometer 半腐层湿度计
duffle bag 水手袋;行旅袋;粗厚呢绒袋
duff mull 半腐解腐殖质
Dufour oscillograph 杜符示波器
dufrenite 绿磷铁矿
dufrenoysite 硫砷铅矿
duftite 砷铜铅石;砷铜铅矿
dug anchorage 地下锚固
dug and covered underground railway 先挖后填式地下铁道
dugganite 砷碲锌铅石
dug-iron 熟铁
dug method 掘井方法
dugout 独木舟;地下掩蔽部;地下室;防空壕;挖出
dugout earth 挖出的土
dugout way 路堑段
dugout well 挖土井
dug peat 挖泥炭
dugway 路堑段
dug well 掘井;大口井;挖土井;挖掘井
Duhamel integral 杜亚米尔积分
duhamelite 杜钒铜铅矿
Duhamel's integral 杜哈梅积分
Duhamel's theorem 杜哈梅定理
Duhem Margules equation 杜安·马古斯方程
Duhem's equation 杜安方程(式)
Duifang paper 对方纸
dukedom 公国
Dukes 杜克斯钨钢
Dukes-metal 杜克斯高电阻合金
dukey 提升平车
Dukler theory 杜克勒理论
dulateral track 双面积调制声迹
Duliray 杜里瑞镍铁合金
dull 减轻;钝的
dull agent 消光剂
dull art paper 暗美术纸
dull-banded coal 暗淡条带煤
dull bit 钝钻头;钝钎头
dull bit evaluation 钝钻头评定;钝钎头评定
dull black ink 无光油墨
dull brown coal 暗褐煤
dull cherry-red 淡樱红;暗樱桃红色
dull clarain 无光泽的亮煤
dull coal 暗(色)煤;暗(淡)煤
dull coated paper 暗涂布纸
dull colo(u)r 浊色;彩色不正
dull corrugated roll 暗瓦楞轧辊
dull deposit 毛面镀层
dull die 磨损(的)模具
dulled cutting edge 钝刻刀刃
dull edge 钝刃口;钝刀口
dull emitter 暗发射体
dull fibre 暗纤维
dull finish 消光修饰;无光泽修饰
dull-finish(ed) lacquer 消光整理喷漆
dull finish(ed) paper 无光纸
dull-finish(ed) sheet 毛面钢板;无光薄板
dull glass 暗淡玻璃;无光玻璃
dull glaze 无光釉
dull-glazed art paper 暗色美术纸
dull-glazed paper 暗色蜡光纸
dull-glazed tile 毛面的玻璃砖
dull gold 暗金黄色
dull grain 钝砂粒
dulling 磨钝;倒光;无光
dulling life 磨钝寿命
dull knife 钝刀
dull luster 呈阴暗色泽;暗淡光泽
dull market 市面萧条;市面冷淡
dullness 浊响;光泽不佳;低色品度的颜色;低彩度颜色;暗度;暗淡无光;无光(指釉面)
dull of sale 销路不佳
dull paint 无光漆
dull plate 无光镀锡薄钢板
dull polish 磨砂;研磨消光
dull polishing 无光研磨
dull purple 暗紫色

dull purplish black 暗绛紫色;茄紫色
dullrae (看不见光线的)电加热系统
dull-red 暗红色(的)
dull-red heat 暗红(炽)热
dull resonance 浊响
dull roll 毛面辊
dull rubbing 擦暗
dull sale 销售不畅;滞销
dull season 淡季
dull steel sheet 不光亮的钢板
dull surface 无光面
dull-surface paper 无光纸
dull varnish 暗色漆
dull weather 阴沉天气
Dulong and Petit's rule 杜隆与普蒂规则
Dulong number 埃克特数
Dulong-petit law 杜隆与普蒂定律
Dulong's formula 杜隆公式
Dultgen process 道尔金凹版法
dulux 精制的;高级的
duly admitted 依法准许
duly certified copy 经正式核证的副本
duly constituted authority 正式合法当局
dumb 低能终端;非机动的;不灵活
dumb antenna 解谐天线
dumb arch 假拱
Dumbar filter 登巴滤池
dumb barge 非自航驳(船);非机动驳船;排泥驳船;无风帆(无动力驳船);无帆船;拖船
dumb bell 哑铃
dumbbell depression 孪生低气压
dumbbell kiln 哑铃形窑
dumbbell pier 哑铃形桥墩
dumbbell resonator 凹腔谐振器
dumbbell-shaped curves of zero velocity 哑铃形零速度线
dumbbell-shaped package 哑铃形卷装
dumbbell slot 哑铃形槽
dumbbell specimen 哑铃形样板;哑铃形试验片
dumbbell tenement 纽约哑铃形平面(住房)
dumb-bidding 秘密以最低价格竞争;秘密投标
dumb buddle 固定圆形淘汰盘
dumb card 哑罗经
dumb chamber 无出口房间;闷室
dumb compass 哑罗经
dumb contact 空触点
dumb craft 非自航驳(船)
dumb iron 捻缝削刀;簧托架;修缝凿;钢板弹簧支架;边梁配件;填缝铁条
dumb lighter 非自航驳(船);非机动驳船
dumboard 隔音板
dumb scow 非自航驳(船)
dumb screen 筛孔堵塞的筛子
dumb sheave 滑槽;无饼滑车
dumb sound 哑音
dumb terminal 简易终端;哑终端;低能终端;不灵活终端
dumb vessel 非机动船
dumb waiter 自动旋转菜碟架;轻型运货升降机;旋转食品架;小件升降机;小型升降机;食梯
dumbwaiter door 运货升降机门
dumbwaiter hoistway 饭菜升降机竖井
dumbwaiter shaft 送菜吊车井;食梯井;送菜升降机井
dumb well 枯井;无水井;污水阱;污水井
Dumet 杜美合金;镀铜铁镍合金;代用铂金
Dumet wire 杜美(合金)丝;铜包镍铁线
Dumfries sandstone 邓弗里斯砂岩;二叠纪砂岩
dumi 矮灌丛
dummied 空轧过的(无压下轧制的)
dumming 空轧通过
dummy 空白样本;假设施;假程序;模造物;名义阴极;虚假物;修复体;平衡器;平衡机件;浮码头
dummy activity 假想作业;名目作业;虚作业;虚拟活动;虚过程
dummy airfield 伪装机场;假机场
dummy antenna 等效天线;假天线;仿真天线
dummy area 哑音区(混凝土)
dummy argument 假变元;虚(拟)变元;哑(自)变量;哑变元;伪自变数
dummy argument array 哑变元数组
dummy axis 虚轴
dummy bar 引锭杆

dummy barge 空驳;系船驳(船);趸船;垫档驳;浮箱;浮码头
dummy beach 消能滩
dummy bit 空位;空白比特;虚比特
dummy block 挤压垫;热压垫块
dummy building 假建筑
dummy bus 模拟母线
dummy car 缆车;卸料车;有机车设备的车厢
dummy casting 烫模铸件
dummy cell 空单元
dummy club 凸耳
dummy coil 虚设线圈;虚假线圈
dummy company 秘密代营企业
dummy compass 哑罗经
dummy conductor 无载导线
dummy control section 哑控制段
dummy corporation 影子公司
dummy coupling 临时连接器;空联节;暂用联轴节;仿真耦合
dummy deck 空卡片叠
dummy demand 假想(的)需求量
dummy diode 仿真二极管
dummy director 挂名董事
dummy door 假门
dummy doorway 假门道;假门口
dummy element 虚拟杆件
dummy exogenous variable 虚拟外生变量
dummy fill 试验性充水
dummy ga(u)ge 补偿应变计;无效应变片
dummy groove joint 假缝
dummy hole 假孔
dummy hose coupling 软管虚连接
dummy imaginary load 虚设荷载
dummy index 哑指标
dummy-index notation 哑指标记号
dummy indicator 等效指示器;仿真指示器
dummy industry 虚拟产业部门
dummying 虚阴极电镀;荒锻;预锻
dummy ingot 引锭
dummy input circuit 等效输入电路
dummy instruction 空操作(指令);伪指令【计】
dummy job 空头买卖
dummy joint 假结合;槽线;假接合;假(接)缝;半缝
dummy journal 假轴颈
dummy level 虚拟水平
dummy load 模拟电路;虚(荷)载;虚荷载;虚负载;等效荷载;仿真荷载;仿真负载;仿真负荷;等效负载;等效负载
dummy load method 虚负载法
dummy lock 插锁
dummy man 温度记录仪;热损失估测仪
dummy market 虚拟市场
dummy member 虚拟杆件
dummy-mesh current 等效回路电流
dummy model 虚设模型
dummy node 哑节点
dummy-node voltage 等效节点电压
dummy observation 虚拟观测值
dummy order 伪指令【计】
dummy panel 模拟板
dummy pass 空轧孔型
dummy pendulum 静摆;假摆
dummy pile 虚拟桩
dummy piston 假活塞;平衡活塞
dummy piston and retainer 平衡活塞与护圈
dummy pit 二次加料室
dummy plate 空白版;隔板
dummy plug 空塞子
dummy procedure 哑过程
dummy record 假记录
dummy register 空寄存器
dummy regression 虚拟回归
dummy resistance 假负载电阻
dummy ring 平衡环;填密环
dummy riser 暗冒口
dummy rivet 假铆钉;装合铆钉
dummy road 伪装道路;石梁平巷;盲巷
dummy-road packing 伪装道路夯实
dummy roll 空转辊;传动轧辊
dummy run 试操作
dummy section 虚节;哑节;哑段
dummy shaft 假轴;中间轴
dummy sheave 呆滑轮
dummy signal 模拟信号

dummy signal(l)er 模拟信号机
dummy sketch 引锭坯;无用板坯
dummy slider 嵌片
dummy spindle 假轴
dummy stand 空轧机座
dummy statement 空语句
dummy step 空转程序步
dummy strain ga(u)ge 补偿应变计;补偿应变片
dummy string 哑串
dummy subscript index 虚拟下标指标
dummy suffix 哑下标
dummy test 人体模型冲击试验
dummy tetrad 虚假四位二进制
dummy treatment 虚拟处理
dummy tube 等效电子管
dummy type contraction joint 假缝式收缩缝
dummy unit load method 单位虚荷载法
dummy variable 空变量;拟变数;虚拟变数;虚(拟)变量;虚假变数;哑(变)元;哑变量
dummy variable model 虚拟变量模型
dummy variable name 哑变量名
dummy variable regression 虚拟变量回归
dummy variable trap 虚拟变量陷阱
dummy window 假窗
dumontite 羟磷铅铀矿;水磷铀铅矿
Dumont's blue 杜蒙蓝颜料
Dumore 杜莫尔铬钼钢
dumortierite 蓝线石
dump 垃圾堆;内存信息转储;货堆;信息转储;转储(方法);清除计算机;倾卸;倾翻;切断电源;弃土堆;抛售(物资)
dump after update 更新后转储
dump analysis 卸出分析
dump and restart 断电后重新启动
dump area 废土场;倾倒区域;倾倒场;弃土场;垃圾场
dump articulated trailer 倾卸式半拖(挂)车
dump bailer 倾卸式泥浆泵;可倾式抽筒
dump bank (堤、路基的)倾卸边缘;堆卸滩地
dump barge 垃圾船;开店泥驳;垃圾驳(船);泥驳;自倾卸驳(船);倾卸驳船;抛石驳
dump-bed 俯卸式
dump-bed truck 底卸式货车
dump before update 更新前转储
dump bin 废料箱
dump body 翻卸车身;翻倾式车身;翻斗车身;倾翻车厢;倾卸式车箱
dump body bottom plate slope 倾卸车身底板斜度
dump body heating 倾卸车身加热
dump body hoist cylinder 倾卸车身升起汽缸
dump body position pad 倾卸车身位置灯光
dump body prop 倾卸车身支撑
dump body track 倾卸运货车
dump body trailer 自倾卸拖(挂)车;翻卸式拖车
dump body truck 货卸式货车;自行卸载的货运汽车
dump body warning light 倾卸车警告灯
dump bottom (排种装置等的)活动底
dump-bottom car 底卸车
dump box 倒泄筐
dump box machine 翻斗壳形机
dump box mo(u)lding 翻卸制壳法
dump bucket 卸(料)斗;装载斗头;翻斗;倾倒式加料桶
dump buggy 自倾双轮小车;倾卸小车
dump by tilting 倾卸
dump car 自动卸料车;翻斗汽车;翻斗小车;倾卸汽车
dump carrier 自动倾卸式拖车
dump cart 倾卸车;倾卸车;垃圾车
dump check 计算机工作检验;转储[贮]校验(内存信息);转储检验;大量倾卸
dump cinder car 渣罐车;倾卸渣车
dump clearance 卸载高度
dump condenser 事故凝汽器
dump crane 倾卸起重机
dump cylinder of floor 转动底板倾卸油缸
dump door 泥舱门
dumped 废弃的;抛填的
dumped deposit 速卸沉积
dumped earth fill 抛填土
dumped fill 废石堆;倾倒填料;倾倒填土;抛填土;抛填;堆填
dumped goods 倾销商品
dumped material 疏浚弃土

dumped packing 填料
dumped riprap 乱石堆;抛石(体);抛石护坡;抛石护岸
dumped rock embankment 抛石(路)堤;抛石坝
dumped rockfill 堆石;倾卸堆石;倾斜堆石;抛(填堆)石体
dumped rockfill dam 倾卸堆石坝
dumped stone riprap 倾卸抛石护岸;倾卸乱石堆
dumped truck 倾卸车
dumped weathered sand 抛填风化砂
dump energy 过剩电能;储存能量;抛弃能量;剩余能(量);剩余电量
dumper 卸货车;自动卸料车;自卸(汽)车;自动倾卸车;翻斗车;清洁工(人);倾卸者;倾卸式运料车;倾卸机;倾翻器;倾倒器
dumper car 自动倾卸汽车
dumper truck 自卸汽车
dump fastening 暗栓钉
dump grate 卸灰炉排;翻转炉排;倾卸炉箅
dump hoist 倾卸起重机
dump hopper 卸料(漏)斗;卸车坑;自卸漏斗;底开门漏斗;倾料车
dump in bay 海湾抛泥
dumping 倾销;倾倒;卸料;排土;商品倾销
dumping and filling on water 水上抛填
dumping angle 倾卸角(度)
dumping apron 溜槽闸门;卸载溜槽;卸料溜子;卸料溜槽
dumping area 垃圾场;卸泥区
dumping area survey 卸泥区测量
dumping at sea 海洋倾卸;海洋倾废
Dumping at Sea Act 海洋倾弃法(英国)
dumping at sea of hazardous waste 危险废物倾弃于海洋
dumping bar 翻倒式炉排
dumping barge 开底泥驳;泥驳;抛石船
dumping board 舱口下护板;倾卸板
dumping body 倾卸车身
dumping bucket 倾卸桶
dumping car 翻斗车;倾卸车;翻转式矿车
dumping cart 倾卸垃圾车
dumping cradle 翻锭座
dumping crane 卸料起重机
dumping device 倾卸装置;卸装置
dumping device for straddle carrier 跨立式转运工具的倾卸装置
dumping door 泥门
dumping dredged silt 疏浚弃土
dumping embankment from trestle 从栈桥上倾卸筑堤
dumping exchange duty 倾销税
dumping field 海外倾销市场
dumping fill 抛填
dumping gear 倾卸装置;翻转机构;倾翻机构
dumping grab (for rehandling) (转载用的)倾卸式挖土机
dumping grate 卸渣炉栅;振动式炉排
dumping ground 垃圾堆(积)场;垃圾倾倒场;垃圾场;倾卸场;倾销市场;抛泥区;抛泥场;排泥区;疏浚弃泥区;垃圾倾倒场
dumping height 卸料高度;卸载高度;倾卸高度;抛卸高度
dumping height at maximum radius 最大半径时的卸料高度
dumping hopper 翻卸斗
dumping intensity 抛投强度
dumping in thin fill 薄层填料
dumping in thin-layers 薄层填土;薄层填料
dumping in water 水中卸泥;水下卸泥
dumping lorry 翻斗矿车
dumping mark 卸泥标志
dumping mechanism 倾卸机械装置;倾卸机构
dumping of hazardous waste 危险废物的倾弃
dumping of overburden 覆土卸载
dumping of toxic waste 有毒废物的倾弃
dumping of wastes 倾卸废物
dumping place 垃圾场;垃圾堆(积)场;垃圾倾倒场;弃土区;抛泥场
dumping plank 舱口下护板
dumping plough 排土犁
dumping policy 倾销政策
dumping position 卸料场;垃圾倾卸场;倾卸位置
dumping price 倾销价格
dumping radius 卸载半径;排泥半径

dumping ram 倾卸油缸
dumping reach 卸载距离
dumping roller 翻车滚轮
dumping shaft skip 倾卸轴小车
dumping site 倾置场;垃圾倾倒场;卸料场;堆渣场;堆物场;堆料场;堆积场;废石倾卸场;氢弃场;弃土区;抛泥区;卸泥区;排泥区
dumping site at sea 海上抛泥区
dumping space 弃土区;排泥区
dumping time 卸料时间
dumping trough 卸料槽
dumping truck 自卸(汽)车;翻斗车;自卸卡车
dumping vessel 卸泥船
dumping wagon 翻斗车;倾卸(垃圾)车;垃圾车
dumping well 垃圾坑
dumping with tilting bucket 用翻斗车卸料
dumping zone 倾废区
dump in piles 分堆卸料;分堆堆料
dump in windrow 卸成行;卸成(长)堆
dump latch 可倾式锁栓
dump light 卸矿信号灯
dump line 排汽管路
dumpling 倒置盆形木块;中心土墩;矮土墩
dump list 卸出列表
dump lorry 倾卸车
dumpman 卸料工;翻车工
dump mechanism 倾卸(机械)装置
dump mode 转储方式
dump oil 桶装油
dump pile 卸料堆
dump pit 垃圾坑;废土坑;废料场;弃渣坑;弃碴坑;排土场;尾矿场;填坑
dump place 垃圾场
dump plate 卸渣炉排
dump platform 卸料站台;倾卸台
dump point (程序的)检查点;卸载[贮]点;转储点
dump position 倾料场;垃圾倾卸场
dump power 倾销功率;倾销电力;剩余功率;剩余电力
dump program(me) 转储程序;清除程序
dump pump 回油泵
dump rake 横向搂草机
dump ram 倾卸油缸
dump rod 抛掷杆;抛草杆
dump routine 信息转储程序;转储程序;清除程序
dump sample 已化验试样
dump scow 垃圾驳(船);卸料平底船;运泥船;自卸泥驳;抛石驳
dump semi-trailer 自动倾卸式半拖车
dump sheave 倾卸滑轮;倾卸滑车
dump site 垃圾堆集场
dump skip 装运斗;卸料斗;斗式升降机;翻车箕斗;翻斗提升机
dump steam 排汽
dump tank 溢水箱;倾料箱;排水箱;排料接受槽
dump test 微粗试验;顶锻试验
dump time 卸载时间
dump time of garage 车库出空时间
dumptor 倾卸式运料车
dump tower 浇筑塔;卸料塔;装料塔;灌注塔;堆料塔
dump tower of concrete 混凝土灌注塔
dump track 卸料车轨道;出渣线
dump trailer 自卸挂车;自卸拖车;自动倾卸式拖车;翻斗挂车;倾卸拖车
dump truck 自卸卡车;自卸汽车;翻斗汽车;自卸式货车;倾卸斗汽车;倾斜式卡车
dump-type 自卸式
dump valve 放卸阀;放泄阀;切断阀;事故排放阀
dump wagon 后翻式自卸车;自卸拖车;底卸卡车
dump well 污水井;垃圾坑
dumpy 定镜水准仪
dumpy level 定镜水准仪
dumpy tree 矮状树
dums 框架和外壳的位置名称
dun 微暗的;设防住所;暗褐色;讨债人
Dunagan analysis 杜纳根分析
Dunbar filter 邓巴滤池
Dunbar's absorption theory 顿巴吸收理论
Duncan-Chang model 邓肯·昌模型
dundasite 白铝铅矿;水碳铝铅石
dundy 接触变质煤
dune 山岗;沙丘(堆);沙垄
dune bedding 沙丘层理
dune buggy 沙滩(轻便汽)车

dune coast 沙丘海岸
dune complex 沙丘覆区
dune covered river bed 沙丘河床;有起伏沙丘的河床;沙垄河床
dune drift sand 沙丘流沙
dune fixation 沙丘固定
dune-fixing forest 固沙林
dune motion 沙丘运动
dune movement 沙丘运动
dune on the march 推进的沙丘
dune-phase traction 沙丘相推移
dune plain 沙丘平原
dune ridge 沙峰;沙(丘)脊
dune sand 沙丘砂
dune sand trap 砂丘砂体圈闭
dunes on bottom 河底沙丘
dune tracking 推移质沙波法测验
dune vegetation 沙丘植被
dune work 沙丘工作;沙丘工程
dun for debt 追债
dung 粪(肥)
dungannonite 刚玉闪长岩
dungarees 粗布工作服
dung-cart 粪车
dung crane 装粪吊车
dung depot 粪坑;粪窖;粪库
dungeon 地牢(仅在牢顶有出口);城堡主楼;土牢
dung fly 粪蝇
dunghill 粪堆
dung hole 污水坑
dung tub 粪桶
dungun iron ore charter 矿石租船契约
dung yard 积粪场;粪场
dunhamite 碲铅华
dunite 纯橄榄岩;纯橄榄石
Dunkard Series 邓卡德统【地】
dunking rolls 沉浸压辊
dunking sonar 吊式声呐;投吊式声呐
dunk rinsing 浸水清洗
Dunlop cornering force machine 邓禄普横向力机
dunnage 截余木材;货垫;垫货材;垫(货)板;垫衬物;垫舱(木料);衬垫;废料;填舱物料
dunnage bags 衬垫袋
dunnage batten 垫条
dunnage board 垫货板;垫舱板
dunnage cargo 垫底货(物);垫舱货物
dunnage charges 垫料费
dunnage gratings 垫货木格板
dunnage mat 垫(舱)席
dunnage paper 垫料纸
dunnage plank 垫货板
dunnage plate 垫舱板
dunnage tarpaulin 垫舱油布
dunnage wood 垫货板;垫舱板
Dunn anchor 双尖嘴无杆锚
dunn bass 夹矸;黏土质页岩
dunnet shale 油页岩
dunning notice 催款通知
dunnite D 型炸药
dunny 盥洗室;厕所
Dunoyer cell 杜诺叶光电池
Dunoyer's two-stage pump 迪努耶尔两级扩散泵
duns 煤矸石
dunsapie basalt 长橄辉斑玄武岩
dunstone 镁灰岩;杏仁细碧岩;硬质耐火黏土;底黏土
dunt 吹裂
dunter 花岗岩磨面工;石面抛光机
dunter machine 石面磨光机
dunting 重击;冷却开裂;风裂
dunting point 转化开裂温度
Dunville stone 邓维尔石(一种浅色细粒砂石)
duo 二部;双行影像系统
duobinary number 双二进制数
duobinary system 双二进制
duocards 双联梳棉机
duocentric motor 同心双转子电动机
duo-cone 高低音扬声器;双锥;双纸盒扬声器
duo-cone seal 双锥形密封
duo-cone tilting (type) mixer 双锥形可倾斜拌和机;双锥体倾斜混凝土混合器
Duocreme pigment 干涉吸收色颜料;双彩颜料
duode 感应驱动扬声器

duodecimal 十二进制的
duodecimal base 十二脚管座
duodecimal notation 十二进制(记数法)
duodecimal number 十二进制数字
duodecimal numeral 十二进制数字
duodecimal system 十二进位算法;十二进(位)制
duodecimo 十二开本
duo-directional current 双向电流
duo-directional relay 双向继电器
duodynatron 双负阻管;双打拿管
duo-faced hardboard 双面的硬木板
duofunctional catalyst 有两种功能的催化剂
duograph 双色网线板
duolaser 双激光器
duolateral 蜂房式的
duolateral coil 蜂房形线圈
duolateral winding 蜂巢绕组
Duolite 离子交换树脂
duo method 双页缩拍法
duo mill 二重式轧机;二辊(式)轧机
duomo 教区中的主要教堂;意大利教堂
duo-muffle furnace 二层马弗炉
Duo-pactor 杜欧式压实机(土坝、路基等用)
duopage 双面复制页
duo-pitched roof 双坡屋顶
duoplasmatron 双等离子体管;双等离子体发射器
duopoly 双卖主垄断;双头垄断(即市场某一商品由两家卖主垄断)
duoprimed word 次首字
duo-prism monochromator 双棱镜单色仪
duopsony 双买主垄断;双头垄断市场
duo-servobrake 双瓦自动闸;双力作用制动器
duo-sol extraction 双溶剂提取
duo-sol process 双溶剂润滑油精制过程
duostage grate cooler 阶梯形算式冷却器
duothem air 二次热空气
duotone 艺术品复制体;同色浓淡套印法;双色照片;双色套印法;双色调;双色版(印件)
duotone ink 双色版油墨
duo-tread tyre 双滚动面轮胎
duotricemary notation 三十二进制(记数法)
duo-trio flow table 二三重流程表
duo-trio test 二三检验
duotype 双色调网点印版
Duovac method 杜奥瓦克法
dupe 复制品
duped print 复制正片
dupe negative 复制底片
Duperrey's lines 杜佩里线
Duplay's syndrome 冻肩
duplet 对;电子对
duplet bond 偶键
duplex 成对;复式结构;复式的;二重(的);双炼;双联;双工;双倍器
duplex adapter 双路转接器;双口接头
duplex air ga(u)ge 双针气压计;双针风表
duplex air governor 复式空气调节器
duplex air pump 双缸气泵
duplex alloy 二相合金
duplex apartment 双户住宅;二联式公寓;跳层公寓单元(占有两层的公寓单元);公寓套房
duplex balance 双工平衡;双方向平衡
duplex balanced feeder 双对称平衡给料器
duplex ball bearing 成对双玻向心推力球轴承;双联球轴承
duplex basebar 复式基线杆尺
duplex base-line apparatus 双金属基线测量器械
duplex bearing 双轴承
duplex beater 双盘式打浆机
duplex boring machine 复式镗床
duplex bracket 复式托架
duplex breaker 复式破碎机;二重破碎机
duplex burner 双喉管式燃烧器;双油路喷嘴;双头燃烧器;双路燃油喷嘴;双路喷燃器
duplex cable 双芯电线;双芯电缆;双股电缆
duplex calculating machine 倍加计算机
duplex calculator 双制计算器
duplex carburetor 复式化油器;双联汽化器
duplex cavity 双腔谐振器
duplex chain 双联式链;双排(滚子)链;双滚柱链
duplex channel 双向信道;双向通道;双路通路;双工信道;双波道
duplex chilled 复合冷硬铬钼合金
duplex circuit 双工电路

duplex classifier 双联式分级机
duplex colo(u)r ink 双色版油墨
duplex communication 双路通信;双工通信
duplex communication system 双重通信方式
duplex compressor 双缸压缩机
duplex computer 双(联式)计算机;双(联)计算机
duplex computer system 双计算机系统
duplex continuous integral fin tube 双金属套片式翅片管
duplex control 复式控制;双工控制
duplex crane 双联吊车
duplex crank press 双曲轴压机
duplex crusher 复式破碎机;双重破碎机
duplex cyclone 双旋流器
duplex design tubular derrick 双管伸缩式钻塔
duplex dialling 双工拨号
duplex double-acting pump 复式双动水泵;双缸双作用泵;双缸复动泵
duplex double-action press 二重双动压机
duplex dwelling 二联式住宅;双户住宅
duplex earth shoveling process 双联铲运法
duplexed system 双套装置;双套系统
duplex engine 双活塞发动机
duplexer 天线转换开关;双工器;双工机
duplexercoupler 复式耦合器
duplexer of waveguide system 波导收发转换装置
duplex escapement 双重擒纵机构;双联擒纵机构
duplex facsimile equipment 双工传真机
duplex feedback amplifier 双重反馈放大器
duplex feeding 两路供电;双重馈电;双路馈电;双端馈电
duplex fibre 复合纤维
duplex film 双层膜
duplex film applicator 双膜厚涂布器
duplex filter 复式过滤器;双联滤机
duplex fitting 复式接头;双通接头;三通接头
duplex flax hackling machine 亚麻双面梳麻机
duplex fuel nozzle 双连式喷油嘴;双级燃料喷嘴
duplex furnace 双联炉
duplex ga(u)ge 双针计
duplex gear cutter 齿轮加工组合铣刀
duplex gear hobber 双轴滚齿机
duplex grain size 双重晶粒结构
duplex-headed nail 双头钉
duplex-headed pump 双水位泵
duplex head milling machine 双头铣床
duplex high speed data 双工高速数据
duplex horizontal miller 复式卧铣床
duplex horizontal sprayer 双管卧式喷雾器
duplex house 两套住房合成的住宅;两户合住的房子;联式房屋;双联式住宅;双户房屋
duplex impurity 复合夹杂
duplexing 二联法;双重;双联法;双工运行
duplex iron 两用钢(碱性炼钢炉及酸性炼钢炉都适用的生铁);双联铸铁
duplexite 硬沸石
duplexity 二重性
duplex jig 双室跳汰机;双式跳汰机
duplex lamp 双边灯
duplex lathe 复式车床
duplex line 双线(路)
duplex lock 双联锁;双铃鼓锁;双重钥匙锁
duplex mechanical pump 联式机械泵
duplex melting 双联熔炼
duplex metal 双金属
duplex method 双页复印法
duplex mill 双面磨粉机
duplex milling machine 双轴铣床
duplex mixer 双式混料机;双式混合机
duplex nails 双头钉
duplex nozzle 双重喷嘴式喷射头
duplex operation 双工制;双工运行;双工操作
duplex outlet 双插座接线盒;双插座
duplex paper 叠层纸;双层纸
duplex parts 成对部件
duplex pendulum seismograph 双摆式地震仪
duplex piston pump 双缸活塞泵
duplex planning 联式房屋(平面)布置;成双的(房屋)平面布置
duplex plate planer 复式板刨床
duplex plunger pump 双缸柱塞泵
duplex polarization antenna 双极化波共用天线
duplex power feed type A. C. commutator motor 并联馈电整流式交流电动机

duplex power pump 双缸曲柄泵
duplex practice 双联法
duplex pressure controller 复式压力控制器
duplex process 二联法;双炼法;双联法
duplex process for steel making 双联炼(钢)法
duplex pump 联式泵;双筒泵;双联(式)泵;双缸泵;双动机械泵;双水泵
duplex pump feeding 双泵合流
duplex pump governor 双缸泵调节器
duplex punch 二重冲压机
duplex purchase 差动绞辘;复式涡轮传动滑车;神仙葫芦
duplex radiophone 双向无线电话
duplex rake classifier 双联式分级机
duplex receptacle 双插座
duplex receptacle outlet 双孔插座出线口;双孔插座出口线
duplex reciprocating pump 双缸往复泵
duplex regulator 双联调节器
duplex roller chain 双滚子链
duplex running 双机运行
duplex set 双马达传动
duplex Shreiner calender 双层电光机
duplex signal(l)ing 双工信令
duplex slide rule 两面计算尺
duplex slip 双滑移
duplex sound track 双锯齿声迹
duplex-spiral classifier 双螺旋分级机
duplex spot weld 双点焊;双点焊接头
duplex spot welder 双点焊机
duplex spot welding 双点焊接
duplex spreadblade cutting method 复合双面切削法
duplex steam pistom pump 双缸蒸汽往复泵
duplex steel 二联钢;双炼钢
duplex strainer 双滤网;双层滤器
duplex structure 双冲构造
duplex system 双重系统;双工系统
duplex tandem compressor 双并列串压气机
duplex tooth taper 双重收缩齿
duplex transmission 两向传导;双重电信;双工运行;双工传输
duplex tube 双管
duplex turning point method 双转点法
duplex type 复式
duplex-type atomizer 复式喷嘴;复式喷雾器;复式喷雾口
duplex-type house 双户住宅;二联式住宅;并联式住宅
duplexure 分支回路
duplex vertical sprayer 双管立式喷雾器
duplex winding 串并联绕组;并绕;双(重)绕组
duplex wind tunnel 双试验段风洞
duplex wire 双芯导线
duplex-wire method 双线法
duplicability 再现性
duplicate 加倍的;一式两份;重复件;抄件;副账;复制品;复制的;复写;复件;双份
duplicate basebar 复式基线杆尺
duplicate bill of lading 副本提单;提单副本;提单抄本
duplicate block 重复区
duplicate-busbar 双母线
duplicate carriageway 复线道路
duplicate cavity plate 双腔型板
duplicate cheque 重发支票
duplicate clamping arrangement 双筒式夹紧装置
duplicate control 双重控制;复式控制
duplicate copy 副本;复本
duplicate copy of invoice 发票副本
duplicated adhesive 加倍活化黏合
duplicated capacity 重复容量
duplicated control 复式控制
duplicated drawing 复制图
duplicate deposit ticket 存款单副本
duplicate design 套用设计
duplicate design drawing 套用设计图
duplicated film 复制片
duplicated record 复制记录;备用记录;备份记录
duplicated system 重复系统
duplicated tunnel 复线隧道
duplicate error message 重复错误信息
duplicate feeder 第二馈路;双馈路
duplicate field 复制区

duplicate film 复制胶片；复印胶片
duplicate gear 双联齿轮
duplicate indicator 复制显示器
duplicate injection 重复注射
duplicate invoice 发票副联；发票副单
duplicate key 复制键
duplicate level line 双程水准线；双程水准测量
duplicate line 双红电路
duplicate lock 双线船闸；复线船闸
duplicate mass storage volume 复制海量存储卷
duplicate measurement 重复测量
duplicate name 重名
duplicate negative 复制底片
duplicate of draft 汇票副本
duplicate of way bill 铁路运单副本；运单副本
duplicate order 再订货
duplicate original 原本的副本；复制原版
duplicate parts 备件；配件
duplicate picture 复制像片
duplicate plate 复制像片；复制版
duplicate positive 复制正片
duplicate print 复制印样
duplicate production 成批生产
duplicate ratio 复比
duplicate reading 重复读数
duplicate recoiler 双位卷取机
duplicate record 重复记录
duplicate relay 备用继电器
duplicate routine 重穿程序；复制例行程序
duplicate rows 双行
duplicate sample 副样；副份样品；复样；平行样
duplicate sampling 重复抽样
duplicate scale 比例加倍；双刻度
duplicate slide 复制正片
duplicate sluice box 双输泥箱；双泄水箱
duplicate source 二重声源；双源
duplicate specimen 重复试样
duplicate stained 二重染色
duplicate supply 双路馈电；双电源
duplicate taxation 重复课税
duplicate test 重复试验；重复检验；复试；平行试验；替换测验
duplicate transparency 复制透明正片
duplicate volume 复制卷
duplicate warrant 副本凭单
duplicate wheel arrangement 成双联轮装置
duplicating 描摹复制；复制；仿形
duplicating attachment 复式附件
duplicating card punch 复制卡片穿孔
duplicating device 双套装置
duplicating experimental condition 使实验条件完全相同
duplicating lathe 仿形车床
duplicating machine 复制机
duplicating milling machine 复制铣床
duplicating model 翻制模具
duplicating paper 复写纸
duplicating prism 双括棱镜
duplicating room 复制室；复印室
duplicating unit 复制机
duplication 加倍；重复；复制；复印；倍加；双重；双摺
duplicational polyploid 重复多倍体
duplication check 重合校验；复校验；重复检验；双重检验（量）
duplication code 重复码
duplication factor 重复因子；重复因数；复制因数
duplication formula 加倍公式；倍量公式；倍角公式
duplication of drawing 图纸复制
duplication of imports 重复引进
duplication of labor 重复劳动
duplication of marking 双重标志
duplication of name 重名
duplication of natural condition 仿造天然环境
duplicator 靠模复制装置；油印机；复制装置；复印机；仿形装置；仿形机
duplicator ozone pollution 复印件臭污染
duplicator with semi-automatic master change 半自动换母板复制机；半自动更换原版复印机
duplicuils 复云
duplicatus cloud 叠层云
duplicident 重齿的
duplicitas 并胚
duplicity 二重性
duplicity theory 二元视学说

duplicity theory of vision 视觉双性理论；视觉二重性理论
Du pont abrasion resistance index 杜邦耐磨指数
Du pont flexural tester 杜邦弯曲试验仪
Du pont impact tester 杜邦冲击试验机
Dupont process 杜邦选矿法
Duppler's alloy 杜普勒镜用合金；杜普勒合金
Dupre equation 杜普雷方程
duprene rubber 氯丁橡胶
Dupry process 直接炼铁法
Dupuit assumptions 裘布意假设
Dupuit confined well formula 裘布意承压水井公式
Dupuit differential equation 裘布意微分方程
Dupuit equation 杜毕方程
Dupuit phreatic water well formula 裘布意潜水井公式
Dupuit planar flow formula of phreatic water 裘布意潜水平面流公式
Dupuit relation 杜普特关系式
Dupuits equation 裘布意方程
Durabil 杜拉比尔钢
durability 耐用性；耐用年限；耐久性；耐久力；耐久率
durability against pollution 耐污能力
durability factor 耐久性因素；耐久（性）系数
durability factor of concrete 混凝土耐久性系数
durability for demand elasticity 需求弹性的持久性
durability index 耐用（性）指数；耐久性指数；耐久率
durability line 耐久性曲线
durability of concrete 混凝土（的）耐久性
durability of wood 木材耐久性
durability period 耐用期
durability rating 耐久性等级
durability ratio 耐久比；相对耐久性
durability test 耐久（性）试验；疲劳试验
durable 坚固的；耐用的；耐久的
durable clause 耐用条款
durable consumer goods 耐用消费品
durable facility 耐用设备
durable goods 耐用品
durable goods price 耐用品价格
durable goods sector 生产耐用品的部门
durable hours 使用时间
durable life 使用期限
durable material 耐久性材料
durableness 耐用性
durable press 耐久压烫
durable pressed tumbler 钢化杯
durable-press resin 永压树脂
durables 耐用品
durable seal 耐久嵌缝
durable structure 耐久结构
durable wood 耐久材
durable years 耐用年限；使用年限
durabolin 苯丙酸去甲睾酮；苯丙酸诺龙
Durachome 杜拉钼铬硅合金铸铁
Duraflex 杜拉弗莱克斯青铜
durain 暗煤（块）；陪煤
durain peat 暗煤质泥炭
Durak (alloy) 杜拉克（压铸）锌基合金
dural 硬铝
dural alclad 包杜拉铝
Duralium 杜拉铝；镁铝合金
Duraloy 杜拉洛依耐蚀耐热铬钢
duralplat 锰镁合金被覆硬铝；包硬铝的铜板
dural plate 硬铝板
dural trace generalized oscilloscope 二踪通用示波器
Duralumin 杜拉铝；硬铝；飞机合金
duralumin alclad 包硬铝
duralumin(i)um 硬铝
duralumin plate 硬铝板
duramen 木心；心材；中心木质
Duramium 杜拉密高速钢
Durana 杜拉纳高强度黄铜
Duranal 杜拉纳尔铝镁合金
Durana metal 杜拉纳高强度黄铜；杜氏合金
durangite 橙砷钠石
Durangoan (stage) 杜兰戈阶【地】
Duranic 杜拉尼克铝（镍锰轻）合金
Duranickel 杜拉镍；杜拉非磁性耐蚀高强度镍铝合金
Duranodic finish 杜拉纳饰面
Duranol dye 杜南醇染料
duranusite 红硫砷矿
durapatite 多晶型磷灰石

Duraperm 杜拉帕姆硅铝铁合金
Durargid 黏磐旱成土
Duraspon 杜拉斯邦聚氨基甲酸乙酯纤维
duration 历时；延续性；涨落历时；工作时间；持续（时间）；持续（期间）；持续；期间；生存期
duration and frequency of observation 观测时间和频率
duration-area curve 历时—面积曲线
duration-area-depth relation 历时—面积—深度关系
duration change of dissolved oxygen in day and night 溶解氧昼夜变化
duration characteristic 持续时间特征
duration clause 期限条款
duration control 时间控制
duration curve 历时曲线
duration curve of stage 水位历时曲线
duration curve of tide level 潮位历时曲线
duration curve of water-level 水位历时曲线
duration-frequency method 保证率频率法
duration hydrograph 历时曲线
duration in scan(ning) 扫描持续时间；扫掠持续时间
duration load(ing) 持续荷载
duration of activity 作业持续时间
duration of agreement 协议有效期（限）
duration of back-pumping 回扬时数
duration of blast 吹炼期
duration of braking 制动时间
duration of breaker contact 断电器接触时间
duration of certain stages 一定发育阶段
duration of charging 装料时间；注水时间；充气时间；充电时间
duration of combustion 耐烧时间；燃烧时间
duration of contact 接触时间
duration of contract 合同有效期；契约有效期间
duration of crystallization 冰冻期
duration of curing 养护持续时间
duration of cycle 周期历时
duration of day 日照长度
duration of drought 干旱历时
duration of each burst of noise 每次噪声袭鸣时间
duration of each movement 每次活动时间
duration of earthquake 地震持续时间
duration of ebb (current) 落潮（流）历时；落潮持续时间；退潮历时
duration of ebb tide 落潮历时
duration of eclipse 掩食时间
duration of exciting 激发持续时间
duration of existence 存立期间
duration of exposure 暴露时间
duration of fall 落潮持续时间；落潮（流）历时；退潮历时；退潮持续时间；水位下降时间
duration of fault activity 活动持续时间
duration of feeding 供油延续时间
duration of fire resistance 耐火极限
duration of flood 涨潮（历）时；涨潮持续时间；高潮历时
duration of flood current 涨潮流历时；涨潮持续时间
duration of flood tide 涨潮历时
duration of flushing 涨落持续时间
duration of franchise 特许期限
duration of freeze 冰冻持续时间
duration of freezing 冻结持续时间
duration of frost-free period 无霜期
duration of grinding 粉磨（持续）时间
duration of ground 脉冲持续时间
duration of ground motion 地动持续时间
duration of guarantee period 保证期限
duration of guaranty 保证期（限）
duration of hearings 审理期限
duration of heat 加热持续时间
duration of heating 冶炼时间
duration of ice cover 封冻期；封冻历时
duration of ignition 点火时间
duration of illumination 光照延续
duration of impulse 脉冲持续时间
duration of inflow 入流历时
duration of insurance 保险期限
duration of inundation 淹没历时
duration of liability 责任期限
duration of load 复合持续时间；荷载作用时间；受载历时

duration of load application 负荷时间
duration of locomotive complete turn-round 机车全周转时间
duration of mixing 搅和(持续)时间
duration of navigation-obstruction 碍航历时
duration of occurrence of a berth 泊位占用时间
duration of operation 运行持续时间;作业持续时间
duration of oscillation 摆动时间
duration of photopic vision 明视持久度
duration of pollution 污染历时;污染持续时间
duration of precipitation 降水历时
duration of pumping 抽水时间
duration of rain 降雨历时
duration of rainfall 降雨历时;降雨持续时间
duration of rainfall excess 超渗(降)雨历时
duration of recharge 回灌时数
duration of recharging water detention underground 补充水地下停滞时间
duration of repair 检修停时
duration of rise 涨水历时;涨落持续时间;涨洪历时;涨潮(历)时;涨潮(持续)时间
duration of risk 危险持续期限
duration of risk clause 保险期间条款
duration of runs 展开时间;运转时间;运转历时;持续运转时间;持续运转额定容量
duration of service 服务期限;使用年限;设备使用年限
duration of shaking 振动持续时间
duration of shock pulse 冲击脉冲持续时间;冲击波宽度;冲击波持续时间
duration of short-circuit 短路持续时间
duration of spawning season 产卵期持续时间
duration of starting 起动持续时间
duration of staying on bottom 水底停留时间
duration of storm 风暴持续时间;暴雨持续时间
duration of storm (rainfall) 暴雨历时
duration of strong shaking 强震持续时间
duration of submergence 淹没历时
duration of submersion 淹没历时
duration of sunshine 日照历时;日照(持续)时间;日照长度
duration of super-pressure 超压持续时间
duration of test runs 试验持续时间
duration of the settlement 沉降历时
duration of tidal lift 涨潮历时
duration of tide 潮(汐)历时;潮期
duration of tunnel passage 过隧道历时
duration of twilight 晨昏蒙影时间
duration of unemployment 失业期限
duration of validity 有效时间;有效期间
duration of volcanic activity 火山活动延续时间
duration of voyage 航行历时;航程历时
duration of wavefront 波前宽度
duration of wind 风时
duration on a project 工程工期
duration performance 续航性能
duration response 持续时间响应
duration running 持久试验
duration scanning 扫描时间
duration selector 连续导线器;持续时间选择器
duration spring 持续作用弹簧
duration-strength factor 持续时间—强度因素;强度持续因素
duration stress 持续应力
duration test 运转试验
duration time 持续时间
duration time modulation 时间宽度调制
durative 持续
durax-cube pavement 嵌花式小方石路面
durax pavement 嵌花式小方石路面
durax paving 嵌花式小方石路面;嵌花式(小方石)铺砌
durax stone block 嵌花式小方石块
durbachite 暗云正长岩
durbar 接见大厅(印度)
Durbar 杜尔巴轴承合金
durchmusterung 星表
Durcilium 杜尔西利厄姆铝合金;杜尔西里铝合金;铜锰铝合金
Durco 杜尔柯合金;杜尔耐热耐蚀镍铬合金
Durehete 杜尔赫特抗蠕变钢;杜尔赫特铬钼钢
durene 杜烯
Duresco 杜拉斯科(一种注册的多种油漆的商品)
Durex 杜雷克斯铜锡合金;屋面油毡;烧结石墨青铜

Durex bearing alloy 杜雷克斯铅基铜镍烧结轴承合金
Durex bronze 杜雷克斯多孔石墨青铜;多孔石墨青铜
Durex iron 杜雷克斯多孔铁;多孔铁
Durham fitting 铸铁螺纹接头
Durham system 达勒姆系统
durian 毛荔枝;榴莲(树)
Durichlor 杜里科洛尔不锈钢
duricrust 硬壳;古土层;钙质壳
durifruticeta 硬叶灌木群落
duriherbosa 硬草本群落
durilignosa 硬叶树林
Durimet 杜里米特奥氏体不锈钢
during a drought 在干旱期间
during case 持续时间角色
during evolution 在进化过程中
during fabrication 在制造时
during-operation service 运转维护
during probation 在试用期间;试用期
during the late stages of developmet 在后期阶段
durinite 亚硫碳树脂;暗煤素质
durinode 硬结
durionising 电镀硬铬法
duripan 硬磐
duriprate 硬草草甸
Duriron 杜里隆高硅铸铁;杜里龙耐酸铸铁;杜里龙高硅钢
durishilvae 硬叶乔木群落
durite 微暗煤
duroclarite 微暗亮煤
Durodi 杜劳迪镍铬钼钢
Durogrip (一种专利的有突起点的)陶瓷防滑铺地砖
Duroid 杜劳特铬合金钢
durol 杜烯
Durolith 杜劳里斯锌基合金
durolok 聚氯乙烯酚醛树脂类黏合剂
durometer 硬度试验仪;硬度计;硬度测定器;钢轨硬度计
durometer hardness 计示硬度;肖氏硬度;压痕硬度
Duromit 杜勒密(一种水泥晶体地坪和路面)
Duronze 杜龙兹硅青铜;杜朗青铜
duroplastic 硬质塑料
duroscope 轻便硬度计
durotelovitrite 微暗煤质结构镜煤
durovitrite 微暗镜煤
Durox gas concrete 杜乐斯泡沫混凝土
Durville casting 杜威勒浇铸法
Durville pouring 翻炉浇注法
Durville process 德维尔翻炉铸造法
dusk 黄昏;薄阴;薄暮;薄明;变微黑;变微黑;暗紫灰色;微暗色;曙暮色
dusk blue 一种淡紫蓝色到淡紫色
dusk grey 暗灰色
duskiness 微黑;微暗
duskish 微黑;微暗
dusky 暗黑;微黑色
dusky belt 暗带
dusky brown 微棕色
dusky ring 暗环
dusky veil 暗纱
dussertite 绿砷钡铁石
dust 垃圾;灰尘;游尘;尘土;尘埃;粉末(尘埃);粉尘(比);微粒干粉;微尘
dust abatement 除尘
dustability index 粉尘指数
dust absorption 吸尘
dust absorption polishing machine 吸尘式抛光机
dust accommodating capacity 集尘能力
dust accumulation 灰尘聚集;粉尘聚集
dust accumulator 集尘器
dust adsorption 吸尘
dust after burner 后燃除尘器
dust agglomeration 粉尘结块
dust air mixture 含尘空气;粉尘和空气的混合物
dust allayer 除尘器;收尘器
dust allaying 抑尘的;防尘的;集尘;捕尘
dust allayment 集尘;捕尘
dust alleviation 除尘(工作);减尘(工作)
dust allowance 飞尘津贴

dust amount 粉尘含量
dust analysis 尘埃分析;粉尘分析
dust analyzer 尘埃分析仪
dust and fume monitor 煤尘和烟雾检测仪
dust and mist collector 尘雾收集器
dust and noise abatement 清除灰尘与噪声;减少灰尘与噪声
dust and poison filtering room 除尘滤毒室
dust arrester 集尘器;吸尘器;挡尘器;除尘器;捕尘器
dust-arrester installation 收尘装置;集尘装置
dust arrester plant 吸尘装置;集尘装置
dust arrest(ment) 吸尘;灰尘捕集;除尘
dust arrestor 吸尘器
dust atmosphere 粉尘环境
dust avalanche 尘崩
dust bag 吸尘袋;集尘袋;收尘袋
dustband (表的)防尘圈
dust base 粗粉剂
dust bin 垃圾箱;减尘箱;灰尘箱;灰仓;除尘室;尘(埃)箱
dust-bin chamber 垃圾桶室
dust binding composition for sweeping 除尘黏结剂
dust-binding oil 黏尘铺路油;铺路用油
dust-bin lift 垃圾桶提升机
dust-bin room 垃圾桶室;废物堆积室
dust blower 吹尘器
dust blue 土灰蓝色
dust board (有抽提的)橱隔板;挡尘板
dust boot 防尘罩;防尘套
dust-borne radioactivity 尘埃(传播的)放射性;放射性灰尘;放射性尘埃
dust-borne transmission 尘埃传播
dust bowl 尘暴(区);风沙中心;风沙浸蚀区;旱涸区(长期干旱及有尘暴的地区)
dust box 垃圾箱;集尘箱;积灰筒;捕尘箱
dust bunker 集尘箱;捕尘箱
dust burden 矿尘量;含尘量
dust burdening 含灰量
dust cage 集尘罩
dust canopy 防尘罩
dust cap 防尘罩;防尘盖
dust capacity 粉尘量;容尘量
dust cap adapter 防尘盖接头
dust capture 集尘;捕尘
dust car 废料车
dust carrier 粉末载体
dust carryout 尘粉飞扬
dust cart 垃圾车
dust carting 垃圾车
dust case 集尘室
dust catcher 吸尘器;除尘器;防尘器;捕尘器
dust catching 集尘
dust cell 除尘室
dust chamber 集尘室;除尘室;沉降室;尘室
dust chimney 排尘管
dust chute 垃圾道;微粉流道
dust cleaner 集尘室
dust cleaning apparatus 除尘装置
dust clinker 粉状熟料
dust cloud 尘云;尘雾
dust-cloud hypothesis 尘云假说
dust coal 煤末
dust coat 干法涂搪;风衣;防尘外壳;防尘衣
dust collecting 集尘
dust-collecting cyclone 旋风集尘器;旋风除尘装置;收尘旋风筒;旋流集尘器
dust-collecting equipment 集尘设备
dust-collecting facility 处尘装置
dust-collecting fan 吸尘器;真空吸尘器;除尘器
dust-collecting installation 集尘装置
dust-collecting part 集尘段
dust-collecting screw 集尘螺旋
dust-collecting sleeve 吸尘袋;收尘袋
dust-collecting system 集尘系统;粉尘收集系统;收尘系统;吸尘系统
dust collection 集尘;聚尘;吸尘;收尘
dust collection by sound wave 声波除尘
dust collection coefficient 除尘系数
dust collection efficiency 除尘效率;粉尘过滤效率
dust collection method 除尘方法
dust collection plant 除尘车间
dust collection system 集尘系统

dust collector 聚尘器;集尘室;集尘设备;集尘器;
    积尘器;吸尘器;除尘器;收尘器
dust collector for cupola 冲天炉专用除尘器
dust collector hood 集尘罩
dust collector scrubber 除尘洗涤器
dust collect plant 集尘装置
dust collect unit 集尘装置
dust colo(u)r 暗褐色;灰褐色
dust composition 粉尘组成
dust concentration 含尘浓度;粉尘浓度
dust concentration index 粉尘浓度指数
dust concentrator 吸尘器
dust condensing 粉尘沉降
dust condensing flue 沉炭烟道;沉灰烟道
dust connection 连接管道
dust constant head box 恒压集尘室
dust contained gas 含尘气体
dust contamination 粉尘污染
dust content 灰尘含量;含尘量;尘含量;粉尘含量
dust content intensity 含尘浓度
dust content of air 空气含尘量
dust content of exit gas 废气含尘量
dust control 尘埃控制;粉尘监控;防尘(的)
dust control barrier 防尘隔离物;防尘隔离板
dust control system 粉尘控制系统
dust control unit 除尘装置
dust-control with enclosure 封闭防尘
dust-control with water-jetting 喷水抑尘法
dust conveying 粉尘输送
dust conveyer 粉尘输送机
dust-cooled reactor 气流粉尘冷却反应堆
dust core 压粉铁芯
dust cork 粉末状软木
dust counter 尘度计;计尘器;尘量计;尘粒计数
    器;尘埃计数器;测尘器;测尘计;粉尘计数器
dust counting 尘粒计数
dust-counting microscope 计尘显微镜
dust cover 护封;护套;防尘罩;防尘外壳
dust cover box 防灰盒
dust crops with an insecticide 撒农药
dust cup 防尘罩;(空压机滤清器的)集尘环
dust curtain 防尘帘
dust deposit 积尘;尘土堆积;尘埃堆积;尘埃沉积;
    粉尘沉积
dust determination 粉尘测定
dust development 扬尘;粉末显影
dust devil 小尘暴;尘卷(风);尘暴
dust-devil effect 尘卷效应
dust dike 粉煤岩脉
dust diluent 粉剂填料
dust discharge 粉尘排放;排尘
dust discharger 卸灰器
dust discharging plate 排粉盘
dust disease 灰尘病;尘埃沉着病
dust distributor 喷粉嘴
dust dry 表干;不粘尘干燥
dust earth 尘土
dust ejector 粉尘喷射器
dust electric(al) resistivity 粉尘比电阻
dust emission 灰尘逸散;扬尘;尘埃排放
dust emission value 粉尘排放量
dust enclosure 集尘罩
dust engineering 防尘工程
dust entrainment 带走粉尘量
duster 空井;揩布;集尘器;除尘器;除尘机;喷粉
    器;洒粉器;撒粉器
duster cloth 抹布
duster cost 防尘外壳
dust-exhaust system 排尘系统
dust excluder 防尘器;防尘条;防尘器
dust excluding plate 防尘板
dust exclusion 防尘;除尘
dust exhaust apparatus 吸尘装置;排尘装置
dust exhauster 排尘器;吸尘器;除尘器;粉尘排放器
dust exhaust hood 吸尘罩
dust exhausting 排尘
dust exhausting device 排尘器
dust exhausting equipment 出尘设备;排尘设备
dust exhausting fan 除尘风扇;排尘风扇
dust exhausting hood 排尘器
dust exhaustor 排尘器
dust exhaust plant 吸尘设备;除尘设备
dust exploding 粉尘爆炸
dust explosion 煤粉爆炸;尘末爆炸;粉尘爆炸;微粒爆炸
dust explosion accident 煤尘爆炸事故
dust explosion hazard 煤尘爆炸危害
dust explosion testing gallery 粉尘爆炸试验坑道
dust exposure method 染尘法
dust extinction 尘埃消光量
dust extraction 抽尘
dust extraction fan 除尘风扇
dust extraction from exit air 从出口空气中除尘
dust extraction hood 除尘罩
dust extraction port 排灰口
dust extractor 灰尘分离器;除灰器;除尘器;抽尘
    器;捕尘器;收尘器
dust extrator 吸尘器
dust fall 落尘;降尘;灰尘沉降;粉尘沉降
dustfall jar 集尘杯;采尘器;降尘罐;降尘测定瓶
dustfall measurement 降尘量测
dustfall quantity 降尘量
dust fan 抽尘(风)扇
dust-fast 耐尘的
dust fastness 耐尘性
dust feedback 粉尘回收
dust figure 粉像
dust filter 滤池;滤尘器;灰尘过滤器;过滤式除尘
    器;除尘器;粉尘过滤器
dust filtration 滤尘
dust fired 粉末燃烧的;粉尘燃烧的;烧煤粉的
dust-firing 粉末燃烧发火
dust flo(a)tation 矿尘浮选
dust flow 尘流
dust flue 尘道
dust flush 粉尘倾泻
dust fog 尘雾
dust formation 灰尘的形成;尘源;撒粉
dust-free 防尘的;无尘的
dust-free air 无尘空气
dust-free area 防尘区
dust-free atmosphere 无尘空气
dust-free continuous ship unloader 无尘污连续
    卸船机
dust-free cooling 无尘冷却
dust-free drilling 无尘钻眼
dust-free dry 不粘尘干
dust-free dry time 不粘尘干时间
dust-free environment 无尘环境
dust-free gas 无尘煤气
dust free operation 无尘操作
dust-free plant 无尘车间
dust-free road 无尘路
dust-free time 无尘时间;表干时间
dust-free workshop 无尘车间
dust from cement factory 水泥工厂粉尘
dust from wind erosion 风蚀粉尘;风蚀尘埃
dust fuel 粉状燃料
dust fused on the clinker 熔融在熟料上的粉尘
dust ga(u)ge 滤尘网;测尘仪;粉尘计
dust gauze 滤灰尘网;防尘网
dust generation 扬尘
dust goggles 防尘眼镜;防尘护目镜
dust gold 金粉
dust grain 粉尘颗粒;尘粒
dust granule 粉粒剂
dust guard 除尘设备;防尘装置;防尘罩;防尘设
    备;防尘板
dust guard holder 防尘板座
dust guard plate 防尘板
dust guard ring 防尘挡圈
dust gun 粉末喷枪
dust handling unit 粉尘输送设备;粉尘处理设备
dust hazard 尘害;尘埃危险;粉尘危害
dust haze 尘霾
dust heap 灰堆;垃圾堆
dust helmet 除尘罩
dust holding 容尘量
dust holding capacity 容尘量
dust holding plant 滞尘植物
dust hole 垃圾坑;尘槽
dust hood 集尘柜;除尘罩;尘罩
dust hook 集尘罩
dust hopper 集灰桶;集尘斗;灰斗;灰仓
dust horizon 尘埃层顶;粉尘层顶
dust housing 集尘室
dust-ignition-proof 防粉尘着火
dust impurity 灰尘杂质;粉末杂质

dustiness 积灰甚厚;尘污
dustiness degree 含尘度
dustiness inex 含尘指数
dustiness of air pollution 空气污染度
dustiness of the air 空气污染度
dust infection 尘埃传染;粉尘传染
dusting 喷粉;清除灰尘;起砂;(路面磨损后造成
    的)起尘;喷干釉;涂粉
dusting beak 喷粉器
dusting bronze 揩金粉;敷青铜粉
dusting brush 撒粉刷;除尘刷
dusting clay 喷粉黏土
dusting clinker 粉化熟料
dusting finish 粉作面修饰
dusting ink 揩金(粉)油墨
dusting loss 粉尘损失
dusting machine 除尘机;喷粉机
dusting method 撒灰法;撒粉法
dusting (on) 撒粉
dusting roll 擦光辊
dusting with cement 撒水泥
dust jar 集尘瓶;集尘罐
dust keeper 集尘器;防尘装置
dust-laden 充满尘埃的
dust-laden air 含尘空气;尘暴;有尘空气
dust-laden gas 含尘气体;含尘炉气
dust lane 尘埃带
dust layer 粉尘层;防尘器;防尘层
dust laying 防尘的;防尘处治
dust laying agent 消尘剂;防尘剂
dust laying composition 除尘合成制剂
dust laying material 防尘(材)料
dust laying oil 灭尘油;防尘(涂)油
dust leg 集尘管
dustless 无尘的
dustless asbestos cloth 无尘石棉布
dustless asbestos gloves 无尘石棉手套
dustless carbon black 粒状炭黑;无尘炭黑
dustless drilling 无尘钻眼
dustless loading 无尘装载
dustless screenings 无灰筛屑;无尘石屑
dust-like 尘状的
dust load 积灰荷载;含尘量;尘埃浓度;积尘载重
dust loading 灰尘含量;尘量;尘埃浓度;粉尘负
    荷;粉尘含量
dust loess 粉黄土
dustman 清道工;清洁工(人)
dust mask 防尘面具;防尘面罩
dust measurement 粉尘测量
dust measuring instrument 测尘仪器
dust meter 尘量计
dust-methane-air-mixture 煤尘
dust minder 防尘提示装置
dust mixer 混尘器
dust monitor 灰尘监测器;烟尘浓度计;粉尘监测器
dust monitoring system 灰尘监测系统;粉尘监控
    系统
dust mop 长柄拖把
dust mulch 细土覆盖(层);地面防风幕
dust nodulizing process 粉尘球化方法;料粉成球法
dust nuisance 尘害
dust of mite 尘螨
dust of vegetable origins 植物性粉尘
dust palliative 减尘剂(道路用)
dust pan 畚箕;簸箕;除尘器
dustpan dormer 簸箕式屋顶天窗;簸箕形老虎窗
dustpan dredge(r) 畚箕式吸扬挖泥船;箕头式吸
    扬挖泥船
dust panel 抽屉隔板
dustpan head 吸盘头
dustpan-like 箕状
dustpan (suction) dredge(r) 吸盘挖泥船
dustpan (type of hydraulic) dredge(r) 畚箕式
    疏浚机
dust particle 尘粒;粉状粒子;粉尘微粒
dust particle size distribution 粉尘颗粒组成
dust particulate control 尘粒控制
dust pelletizing system 除尘系统
dust plume 尘缕
dust pocket 沉灰室;集尘室;储(贮)尘室
dust pollutant 粉尘污染物
dust pollution 尘污染;粉尘污染
dustpoof bulkhead 防尘舱壁
dust pot 灰斗

dust powder 撒布剂
dust-precipitating system 集尘系统
dust precipitation 聚尘;降尘;落尘;除尘;粉尘沉降
dust precipitator 收尘器;聚尘器;集尘器;除尘器;除尘室
dust pre cleaner 预收尘器
dust precollector 预收尘器
dust press 干法成型
dust-pressed process 粉尘加压方式
dust pressing 干粉形成
dust preventer 防尘剂
dust-preventing sealer 粉尘封口机
dust prevention 尘埃预防;防尘措施;防尘(的)
dust prevention planting 防尘栽植;防尘栽植
dust prevention sprinkler 防尘喷头
dust preventive 抑尘剂;除尘剂;尘埃预防(法);粉尘预防
dust preventive agent 避尘剂;防尘剂
dust process 干法形成
dust-producing 扬起飞尘;产生灰尘
dustproof 避尘的;防尘(的)
dustproof building 防尘建筑
dustproof case 防尘箱
dustproof ceiling 防尘天花板;防尘顶棚
dustproof electric(al) equipment 防尘式电气设备
dustproof enclosure 无尘区
dustproofer 防尘器;防尘剂
dustproof flap 防尘挡板
dustproof grab 防尘抓斗
dustproofing 去尘;防尘
dustproofing seal 防尘密封条
dustproof insulator 防尘绝缘子
dustproof lamp 防尘灯
dustproof lighting fitting 防尘照明装置
dustproof luminaire 防尘灯
dustproof machine 防尘型电机;防尘式电机;防尘机器
dustproof rubber strip 防尘橡皮条
dustproof strike 防尘锁舌磕片;防尘栓塞
dustproof switchgear 防尘式开关装置
dust protecting mask 防尘口罩
dust protection agent 抗尘剂;防尘剂
dust protection grummet 防尘罩
dust protection wall 防尘墙
dust rain 尘雨
dust ratio 含尘率
dust-reclaiming mill 灰尘处理设备
dust recovery 收尘;粉尘回收
dust recovery system 粉尘回收系统
dust removal 除尘
dust removal by gravity 重力除尘
dust removal efficiency 除尘效率;粉尘过滤效率
dust removal for pavement 路面除尘
dust removal installation 除尘设施
dust-removal system 除尘系统
dust remover 滤尘器;除尘器;除尘器
dust remove valve 排灰阀
dust removing filter 除尘滤器
dust-removing mechanical rectifying equipment 除尘用机械整流设备
dust removing plant 除尘装置;除尘设备
dust-removing preparation 除尘制剂
dust removing system 除尘系统
dust-repelling 驱尘;防尘
dust resistivity 粉尘电阻系数;粉尘比电阻
dust respirator 防尘面具;过滤式呼吸器;除尘口罩;防尘口罩
dust retaining capacity 捕尘能力
dust-retention effect of plant 绿地滞尘作用
dust return 粉尘回收
dust return bin 回灰斗;回灰仓
dust return metering device 回灰计量装置
dust return rate 回灰量
dust return scoop 回灰勺
dust ring 防尘环;前结圈
dustroad 土路;乡村道路
dust ruffle 防尘褶边
dust sample (空气洗空时得到的)岩粉样品
dust sampler 灰尘取样器;粉尘采样器
dust sampling 粉尘采样
dust sampling equipment 粉尘采样装置
dust sampling meter 粉尘采样器
dust sand 粉砂

dust scraper 刮板除尘机
dust scraping ring 刮尘圈
dust screen 除尘网;防尘网
dust scrubber 洗尘器;涤尘器;刷尘器
dust seal 防尘圈;防尘密封件;尘封;防尘密封条
dust sealing 防尘密封
dust separating 灰尘分离的
dust-separating appliance 灰尘分离装置
dust-separating equipment 除尘分离设备;灰尘分离器
dust separation 灰尘分离;除尘净化
dust-separation equipment 吸尘装置
dust separator 集尘器;灰尘分离器;岩粉分离器;除尘(分离)器;粉尘分离器
dust separators train 除尘器组
dust setting chamber 降尘室;粉尘沉降室
dust setting compartment 集尘室
dust setting pocket 降尘袋
dust settle 粉尘沉降器
dust settler 除尘器
dust settling chamber 集尘室;除尘室
dust settling compartment 降尘室;除尘室
dust settling pocket 除尘袋
dust shaft 垃圾槽(楼房倒垃圾用);垃圾井筒
dust shaker 打土机;除尘机
dust shield 放尘罩;防尘板
dustshield pocket 集尘袋
dust shower 尘雨
dust source 尘源
dust source control 尘源控制
dust spectacles 防尘眼镜
dust spot count test method 粉尘试验计数法
dust spot efficiency 粉尘捕捉率
dust stop 密封装置;防尘剂
dust storm 大灰尘;尘暴
dust stratification 积灰
dust suction 吸尘
dust suction installation 吸尘装置
dust suppression 降低粉尘浓度;粉尘抑制;防止起灰
dust suppression device 灭尘装置
dust suppression measures 防尘措施
dust suppression system 灭尘系统
dust tea 末茶
dust test 透尘试验;砂尘试验
dust test instrument 粉尘试验器
dust throw off 粉尘逸出
dust-tight 尘密;不透灰尘的;绝尘的;防尘(的)
dust-tight belt conveyer 封闭式皮带输送机
dust-tight construction 防尘构造;防尘建筑
dust-tight material 防尘材料
dust-tightness 密闭防尘;防尘性
dust-tight partition 防尘密封隔板
dust-tight type motor 防尘式电动机
dust trap 集灰器;集尘器;除尘器;捕灰器;捕尘器
dust trash meter 灰尘废料测定仪
dust tube 测尘管
dust turbidity 灰尘混浊度;尘浊度;粉尘浑浊度
dust velocity 集尘速度
dust washer 防尘垫圈;防尘挡圈
dust weight arrestance test method 粉尘试验重量法
dust well 垃圾坑;尘穴;尘坑;冰井
dust whirl 尘旋;尘卷风;粉尘漩涡
dust wind 尘风
dust wiper 防尘器;除尘器;粉尘去除器
dust wrapper 防尘罩;除尘套
dusty 起尘的;灰尘(多的)
dusty air 含尘空气
dusty aqua (blue) 土赤水蓝色
dusty aqua green 土灰水绿色
dusty blue 灰蓝色
dusty-butt 勤杂工;普通工
dusty cargo 有扬尘货;扬灰货物
dusty clinker 粉状熟料
dusty coat 砂状涂层
dusty crust 粉状结皮
dusty gas 含尘烟气;含尘气体
dusty green 灰绿色;土灰绿色
dusty jade green 土灰玉绿色
dusty line 粉线
dusty material 粉状材料;粉末(状)材料
dusty mauve 土灰紫红色

dusty money 灰尘费(装卸扬尘货物时的附加费)
dusty olive 土灰橄榄绿色
dusty orange 土灰橙色
dusty peach 土灰桃红色
dusty pink 土灰粉红色
dusty place 扬尘点
dusty rose 土灰玫瑰红色
dusty seed 灰泡
dusty spray 喷涂粉末
dusty turquois 土灰绿松石色
Dutch arch 砖砌平拱;荷兰式平拱
Dutch architecture 荷兰建筑
Dutch auction 减价拍买
Dutch barn 荷兰式谷仓;荷兰式(盖)棚
Dutch Baroque Cathedral of St Peter and St Paul 圣彼得—圣保罗的荷兰巴罗克式大教堂
Dutch bond 荷兰式砌合;荷兰式砌法
Dutch brick 荷兰式(炼)砖;高温烧结砖
Dutch cell 荷兰(式)三轴仪
Dutch cell test 荷兰式三轴试验
Dutch cleanser 去垢粉
Dutch clinker 荷兰(式)缸砖;铺路硬砖
Dutch colonial architecture 荷兰在北镁殖民地建筑风格
Dutch colonial house 荷兰殖民地房屋
Dutch cone 荷兰锥
Dutch cone penetration method 荷兰式静力触探试验
Dutch cone penetration test 荷兰式圆锥贯入试验;荷兰式静力触探试验
Dutch cone penetrometer 荷兰式圆锥贯入仪;荷兰式圆锥触探仪
Dutch cone test 荷兰圆锥试验
Dutch cyclone 荷兰式旋流器;旋波分离器
Dutch diaper bond 荷兰菱形花式砌砖
Dutch door 两截门;荷兰(式)门;上下可以分别开关的两截门
Dutch door bolt 两截式门门栓
Dutch draghead 荷兰耙头
Dutch garden 荷兰庭园
Dutch gilding 荷兰式涂金;荷兰金漆
Dutch gold 黄铜箔;荷兰(饰)金
Dutch Guilder 荷兰盾
Dutch lap 石板瓦两边搭接;荷兰式搭接
Dutch light 活动玻璃框(温室中);荷兰式活动玻璃框;荷兰式窗
Dutch liquid 荷兰液
dutchman 插入楔;填隙片;连接销;木塞块;补缺块;塞孔补缺的木料
Dutchman's log 漂物测速法
Dutch marble paper 荷兰大理石纹纸
Dutch mattress 荷兰式柴排;荷兰沉排;用芦苇和木材制成的柴排
Dutch metal 荷兰金箔;荷兰合金;低锌黄铜箔
dutch nose tool 宽头刀
Dutch oil 荷兰液
Dutch oven 荷兰式(壁)炉
dutch penetrometer 透度计
Dutch pink 荷兰桃红色;荷兰粉红
Dutch process 荷兰法
Dutch rasp 荷兰磨
Dutch Renaissance 荷兰文艺复兴
Dutch sieve bend 弧形筛
Dutch sounding 荷兰式触探法
Dutch staatsmijnen screen 荷兰国家标准筛
Dutch stove 荷兰火炉
Dutch tile 荷兰蓝色釉砖;彩砖;饰砖
Dutch white 荷兰白(颜料)
Dutch white metal 荷兰白色饰用合金;白色饰用合金
dutiable 应征税的;应付关税
dutiable article 应征税商品;应缴税物品
dutiable goods 有税品;应纳税货物;应纳税货品;征税货物
dutiable price 应征税价格;完税价格
dutiable stores 应缴税物料
dutiable value 应征税价值;完税价值;征税价值
duties 税捐;税金
duties on buyer's account 目的港码头交货
duttonite 羟钒石
duty 义务;职责;职务;职位;空占;灌溉水量;关税;工作(状态);代理人;班次;任务;税(收)
duty actually paid 实付税款
duty allowance 职务津贴;岗位津贴

duty and quota free 免税及无配额限制
duty appraiser 关税鉴定人员
duty assessment 课税
duty boat 值班艇
duty classification of a relay 继电器工作分类
duty cycle 有荷因数;作业周期;占空度;占空比;占空系数;(焊机的)暂载率;工作因数;工作循环;工作比;负载周期;负载比;频宽比
duty-cycle capacity 断续负载容量;断续负荷容量
duty-cycle operation 工作循环
duty-cycle rating 持续定额;反复使用定额
duty-cycle ratio 工作循环负载比
duty cyclometer 占空因数计;负载循环计
duty drawback 关税退款;退税
duty factor 利用因数;利用因数;占空因数;占空系数;工作因数;工作系数;负载周期;负载比
duty forward 预付关税
duty free 免税
duty free access to the market 免税进入市场
duty free articles 免税物品
duty free certificate 免税执照;免税证明(书);免税凭证
duty free entry 免税进口
duty-free goods 免税品;免税(进口)货物
duty free import 免税进口
duty free importation 免税进口
duty free imports 免税输入品
duty free material 免税材料
duty free quota list 免税限额表
duty free shop 免税(商)店
duty free slips 免税单
duty free storage 免税仓库
duty free territory 免税区
duty free zone 免税地带
duty horse power 有效马力;标称马力;额定马力;报关马力
duty in civil affairs 民事义务
duty insurance 关税保险
duty is imposed on 对……课税
duty limited import 进口限制
duty memo 税单;上税单
duty of averting or minimizing losses 防止和减轻损失的义务
duty of care 认真地尽责
duty of engine 发动机功能
duty officer 值班驾驶员;值班官员
duty of performance 履行义务
duty of pump 泵流量;泵的效率;泵的能率;泵出力
duty of water 灌溉(用水)率;灌溉需水量;供水流量;灌水定额
duty on imported goods 商品进口税
duty paid 进口完税价;目的港码头交货;已付(关)税;税款付讫;已完税;已纳税
duty-paid certificate 完税单;收税单
duty-paid goods 已完税货物
duty-paid price 完税货价;完税价格
duty-paid proof 完税凭证;完税证明
duty-paid terms 卖方交纳进口税的交货条件;通过海关后交货条件
duty parameter 工作状态参数
duty-paying value 纳税价值
duty plate 性能标志板
duty preiod 工作周期
duty proof 已完税;已纳税
duty pump 主用泵;常用泵
duty qualified independent inspector 正式有资格的独立检查人
duty quotas 海关限额;关税限额
duty rate 税率
duty ratio 占空因数;占空率;占空比;负载比;负荷比
duty receipt 纳税收据;纳税凭证;完税凭据
duty room 值班室
duty rosters 排班表
duty row(ing) boat 值勤艇
duty runway 值勤跑道
duty-shift commander 值班指挥员
duty unpaid 关税未付
duty water 灌溉额;灌溉用水;保证供水流量
duty work schedule 值班时间表
Duvan 杜钒钢
D-value 差值
dwang 转动杆;大螺母扳手;板条墙立筋间斜撑;支撑;撬棍;楼板格栅之间的支撑
dwarf 矮小的动物或植物;矮化

dwarf banana 粉蕉
dwarf branch 矮枝
dwarf capstan 轻便手动绞盘
dwarf cherry 矮木樱
dwarf door 两截门;矮门;小门(活动门的下半截门)
dwarfed crown 树冠矮化层
dwarfed enamel 薄釉质
dwarf elm 榆树;白榆
dwarf flower bed 矮生花坛
dwarf flowering cherry 郁李
dwarf form 矮生型
dwarf fruit 矮果树;棚枝果树
dwarf grass 矮牧草
dwarfing effect 矮化作用
dwarfing plant 矮化植物
dwarfing stock 短化砧木;矮化砧
dwarfing tree 矮性树
dwarfism 矮态
dwarf Japanese quince 日本贴梗海棠
dwarf partition 半截隔墙;矮间墙;短隔屏;半截隔断;矮隔墙;矮隔屏;矮隔断
dwarf partition wall 短隔墙
dwarf plant 矮生植物
dwarf plug 短塞
dwarf rafter 小椽
dwarf sequence 主星序
dwarf shoot 短枝;矮枝
dwarf sign 矮标号志
dwarf signal 小型信号机;矮柱信号机;矮型信号
dwarf signaller 矮型信号机
dwarf skylight 小天窗;矮天窗
dwarf thicket 矮树丛
dwarf tree limit 矮乔木界限
dwarf-tree orchard 矮生果园
dwarf trees 矮化林
dwarf tripod 矮三脚架
dwarf trivet 矮三脚架
dwarf wainscotting 矮护壁
dwarf wainscotting plate 矮护壁板
dwarf wall 女儿墙【建】;胸墙;矮(院)墙;矮隔屏;矮(挡土)墙
dwarf wave 矮波
dwell 居住;闭锁时间
dwell angle 凸轮回转角度
dweller 居住者;居民;住户
dwelling 农村住房;住处
dwelling area hygiene 居民区卫生
dwelling area noise 住宅(区)噪声
dwelling attached 联立式住宅
dwelling bathroom 住房浴室
dwelling building 住房房屋;住房房屋
dwelling burrow 居住潜穴(遗迹化石)
dwelling condition 居住条件
dwelling construction 居住建筑;住宅建设
dwelling density 居住密度;住宅密度
dwelling detached 分立式住宅
dwelling district 居住区;小区
dwelling duplex 双户住宅;二联式住宅
dwelling environment 居住环境
dwelling equipment 住房设备
dwelling expenses 居住费
dwelling facility 居住设施
dwelling group 居住群
dwelling house 独立式住宅;居住房屋;农村住房;住宅;住处
dwelling house comprehensive 住宅综合
dwelling house for staff and workers 职工住宅
dwelling house scale 住宅规模
dwelling inspection record 住房检查记录
dwelling kitchen 民居厨房;家用厨房
dwelling land rate 宅地地租率
dwelling level 居住水平
dwelling living room 住房起居室
dwelling occupancy 居住用途
dwelling on honeycomb 蜂窝式住宅;蜂窝式住所
dwelling place 住址;住所;住地;住处
dwelling room requirements 居室要求
dwelling size 住宅规模;户型
dwelling space 居住面积
dwelling standard 居住标准
dwelling structure 居住迹构造
dwelling tower 居住的高楼
dwelling townhouse 三幢以上并联住宅
dwelling trace 居住迹

dwelling unit 单户住宅;居住单元;住宅单元
dwelling unit interview 居住单元调查法
dwelling unit interview method 住户访问法;居住单元访问法
dwelling unit scale 居住单元规模
dwelling units for low-income families 低收入家庭的居住单元
dwell mark 停止痕
dwell pressure 保压压力
dwell tack 持续黏性
dwell time 静态时间;驻留时间;闭模时间;保压时间;保持最大塑性压力的时间;停滞时间;停站时间;停延时间;停留时间
dwell time at station 车站停站时分
dwell time table 列车停站时分表
Dwey 德怀风暴
Dwigh 德怀风暴
Dwight chart 德维特电压调整卡
Dwight-Lloyd machine 德怀特—劳埃德型焙烧炉
Dwight-Lloyd process 德威特—劳埃德法
dwindle 缩减
dwornikite 杜铁镍矾
Dwoy 德怀风暴
Dwyka tillite 德怀卡冰碛岩
dyad 二数;二价元素;二分体;并矢(量)
dyadic 二元的;二数;并矢式;双值;双积的
dyadic array 二元阵列;二元数组;二维阵列;二维数组
dyadic Boolean operation 二元布尔运算
dyadic Boolean operator 二元布尔算符
dyadic expansion 二进展开
dyadic formula 二元公式;双向公式
dyadic indicant 二元指示符
dyadic indication 二元指示
dyadic matrix 并矢矩阵
dyadic number 二进制数;二进数(字)
dyadic operation 二运算数运算;二元运算;二元操作;二操作数运算;双值运算;双运算数操作
dyadic operator 二元算子;二元算符;二元操作符
dyadic operction 二算子运算
dyadic rational 二进有理数
dyaman 计日工
Dyas 二叠系【地】
Dybln dye 迪布尔恩染料
dydrant 消防龙头
dyeability 可染性
dye absorption 染料渗透试验法(测定陶瓷气孔率)
dye adsorption test 吸红试验
dye backing 背色膜
dye beaker 染色烧杯
dye bleaching process 染漂法
dye cell 染料盒
dye coupling 染料耦合
dye crop 染料作物
dyed agent 着色剂
dyed cellulosic materials 着色纤维素材料
dye decolo(u)rization 染料脱色
dye destruction process 染料破坏法
dye developer 色素显影剂
dyed fluid 着色液体;染色液体
dyed gasoline 着色汽油
dyed glass yarn 染色玻璃纤维纱
dye diffusion 染色扩散
dye dilution curve determination 染料稀释曲线测定
dye dispersion study 染色离散研究
dyed nankeen 毛蓝土布
dyed scribe-coating 染色刻图膜
dye duration 颜料耐久性
dyed volume 着色水体;染色水体
dye exclusion test 染料排泄功能试验
dye-gelatin filter 染色明胶滤色片
dye-house 染色间;染坊
dye-house wastewater 印染厂废水
dy(e)ing 染色(品)
dy(e)ing and finishing 染整
dy(e)ing assistant 染色(辅)助剂
dy(e)ing flower 红
dy(e)ing machine 印染机
dy(e)ing method 染色法
dy(e)ing retouching process 染色分涂法
dy(e)ing wastewater 染色废水;印染废水
dye laser 染料激光器
dye layer 染色层

dyeline 熏制图;氨图;重氮复印法;染料线条
dyeline printer 重氮复印机
dye manufacturing wastewater 染料生产废水;染料工业废水
dye marker 染水标志;染色标记
dye mordant 媒染剂
dye penetrant 染色渗透液;染色渗透剂
dye penetrant inspection 着色检验;染色观察
dye penetration 染料渗透试验法(测定陶瓷气孔率)
dye penetration test 染色渗透试验
dye penetrator inspection 着色检验
dye pigment 染料性颜料
dye plume 染色卷流
dye print false colo(u)r composite 染印法假彩色合成
dyer 染色机
dye-retarding agent 染色减速剂
Dyer method 戴尔陶瓷插口成型法
dyes and dyestuffs 有机着色剂
dye sensitization 染料敏化
dye stain test 染料着色试验
dyestripping 撕膜
dyestuff 颜料;着色剂;染料
dye stuff industry 染料工业
dyestuff intermediate 染料中间体
dyestuff intermediate wastewater 染料中间体废水
dyestuff lake 染料池
dye stuff solution 染液
dyestuff suspension 染料悬浮体
dyestuff works 染料厂
dye test 染色试验
dye time concentration curve 着色浓度时间曲线
dye toning 染料调色法
dye tracer 染料示踪剂
dye tracing 色素示踪(用于测流、测波等)
dye under pressure 高压染色
dye-variant fibre 差异染色性纤维
dye vat 染缸
dyeware 颜料;染料
dye waste(water) 染料废水
dye wastewater treatment 染料废水处理
dye-weave polychromatic dyeing machine 彩色织花型染色机
dye woods 用于提染料的硬木
dye-works 染坊;染厂
dygoram 自差曲线图
dying 垂死的
dying away of vibration 振动衰减
dying of activated sludge 活性污泥干化
dying oscillation 减幅振荡;阻尼振荡;衰灭振荡;衰减振荡
dying out 死去;衰减(消失)
dyjord 湖底植物沉积
Dykanol 狄加拿尔
dyke = dike
dyke area 堤防区
dyke body 坝身
dyke boring 堤防钻探
dyke breach 堤坝裂口;决堤;堤坝破坏;堤坝决口
dyke building 堤防工程
dyke country 堤内泽地
dyked 堤坝的
dyke defect detecting 堤防隐患探测
dyke design 坝的设计
dyke failure 决堤;堤防溃决
dyke field 坝田
dyke fill 堤坝填土
dyke foot 堤脚
dyke fortifying project 护堤工程
dyke head 坝头
dykeland 圩地
dyke lock 堤坝闸门
dyke maintenance 堤防维修;堤坝维护
dyke management 堤防管理
dyke opening 堤口
dyker 筑堤机;筑堤工人;砌石(墙)工人
dyke reinforcement 堤防加固
dyke root 坝根
dyke slope 堤前斜坡;坡坡
dyke summit 堤顶
dyke swarm 岩脉群
dyke system 堤防系统
dyke top 堤顶

dyke width 堤宽
Dy-krome 戴克拉姆铬钼钒钢
Dymal 戴码尔铬锰钨钢
Dymaxion house (生活与机械检修合并的)房子
Dymetrol 戴梅特罗尔尼龙
Dynacote process 中子射线处理表面涂饰法
Dyna-drill 戴纳钻具
dynaflect 动力弯沉
dynaflector 动力弯沉(测定)仪
Dyna-Flex machine 动力弯沉(测定)仪
dynaflow 流体动力(传动)
Dynaforge 戴纳高速高能锻压机
dynaform 同轴开关
Dynaforming 金属爆炸成型法
dynagraph 记录式测力计;轨道试验器
Dynalens 消震镜头
dynalysor 消毒喷雾器
dynamafluidal 动力流体的
Dynamax 戴纳马克薄膜磁芯材料;镍钼铁合金
dymeter 扩力器;放大率计;倍率计(光学用)
dynametropolis 活力大都市;沿交通干线有计划发展起来的城市
dynamic(al) 动态的;动力学相似模;动力(学)的
dynamic(al) absorber 动力吸振体;动力减振器
dynamic(al) access 动态访问;动态存取
dynamic(al) accounting facility 动态统计手段
dynamic(al) accuracy 动态偏差;动态精(确)度;动力精度
dynamic(al) accuracy test system 动态精(确)度测试系统
dynamic(al) action 动态作用;动力作用
dynamic(al) address relocation 动态地址再分配;动态地址再定位
dynamic(al) address translation 动态地址转换;动态地址翻译
dynamic(al) address translation program(me) 动态地址翻译程序
dynamic(al) address translator 动态地址转换器;动态地址译码器
dynamic(al) adsorption 动态吸附
dynamic(al) adsorption breakthrough curve 动态吸附穿透曲线
dynamic(al) adsorption-desorption 动态吸附—脱附
dynamic(al) adsorption parameter 动力吸附参数
dynamic(al) air flow factor 气流动力系数
dynamic(al) air separator 动力空气选粉机
dynamic(al) allocation 动态(存储)分配
dynamic(al) allocation and deallocation 动态分配与去配;动态分配与解除分配
dynamic(al) allocation memory 动态分配存储器
dynamic(al) allocator 动态分配程序
dynaxion(al) analog(ue) device 动态模拟装置
dynamic(al) analogue test 动态模拟试验
dynamic(al) analogy 动态模拟;动力相似;动力模拟
dynamic(al) analysis 动态分析;动力分析
dynamic(al) analysis map of reservoir 储集层动态分析图
dynamic(al) analysis method 动态分析法
dynamic(al) analysis of heat exchanger 换热器的动态分析
dynamic(al) analysis of mineral resources reserved 矿产资源储备动态分析
dynamic(al) analyzer 动态分析程序;动态分析器
dynalens(al) and evolution of the lithosphere project 岩石圈动力学及演化计划
dynamic(al) and static universal testing machine 动静态万能试验机
dynamic(al) angle of internal friction 动内摩擦角
dynamic(al) angle of repose 动态休止角
dynamic(al) anodegrid characteristic curve 动态阳栅特性曲线
dynamic(al) anti-cyclone 动力性高压;动力反气旋
dynamic(al) area 动态区(域);动态(存储)区
dynamic(al) array 动态数组
dynamic(al) astronomy 动力天文学
dynamic(al) asynchronous logic circuit 动态异步逻辑电路
dynamic(al) axis 动力轴
dynamic(al) axis of flow 河流动力轴线;水流动力轴线
dynamic(al) axis of water flow 水流动力轴线
dymaxion(al) backout 动态取消
dynamic(al) balance 动(态)平衡;动(力)平衡

dynamic(al) balance level 动平衡标准
dynamic(al) balancer 动平衡(试验)机;动(力)平衡器
dynamic(al) balance running 动平衡试验
dynamic(al) balance test 动平衡试验
dynamic(al) balance tester 动平衡试验机
dynamic(al) balancing 动力平衡
dynamic(al) balancing machine 动平衡机
dynamic(al) balancing tester 动平衡试验机
dynamic(al) balancing test(ing) 动平衡试验
dynamic(al) ball-impact test 动力球冲击(硬度)试验
dynamic(al) ball-impact test method 动力落球冲击试验法;动态球击试验方法
dynamic(al) ball indentation test 落球硬度试验
dynamic(al) bearing capacity 动承载力
dynamic(al) behavio(u)r 动态(学)特性;动态机理;动(力)学性状;动力性能;动力特性;上动态;动态行为
dynamic(al) bottom 动态河床;动力河床
dynamic(al) boundary condition 动力边界条件
dynamic(al) brake 动力制动;发电制动
dynamic(al) braking 能耗制动;制动装置;动力制动;发电制动
dynamic(al) breakage 动力破裂
dynamic(al) breccia 构造角砾岩;动力角砾岩
dynamic(al) buckling 动力弯折;动态压曲;动力压屈(负荷影响下);动力压屈;动力失稳
dynamic(al) buffer allocation 动态缓冲器分配
dynamic(al) buffering 动态缓存;动态缓冲
dynamic(al) buoy 动浮力
dynamic(al) business analysis 业务动态分析
dynamic(al) calculating method 动态计算法
dynamic(al) calculation 动力计算
dynamic(al) calibration 动态校准
dynamic(al) calibration system 动态校准系统
dynamic(al) call 动态调用
dynamic(al) capacity 动态电容
dynamic(al) capillary pressure 动力毛管压力
dynamic(al) cataloging 动态分类
dynamic(al) catalogue 动态目录
dynamic(al) centering 动态中心调整
dynamic(al) ceramic membrane 动态陶瓷膜
dynamic(al) chamber 动态流过试验室
dynamic(al) change of temperature 温度的动力变化
dynamic(al) channel 动态通道
dynamic(al) channel assignment 动态信道分配
dynamic(al) characteristic (电弧的)自振特性;动态特征;动态特性;动力特性;负载特性曲线
dynamic(al) characteristic curve 动态特征曲线;动态特性曲线
dynamic(al) characteristic of arc 电弧动特性
dynamic(al) check 动听测试;动态校验;动态检验;动态检查
dynamic(al) circuit 动态电路
dynamic(al) climatology 动力气候学
dynamic(al) cluster 动态聚类
dynamic(al) cluster parameter 动态束参数;动态聚类参数
dynamic(al) coated method 动态涂覆法
dynamic(al) coating 动态涂覆
dynamic(al) coefficient 动力系数
dynamic(al) coefficient subgrade reaction 地基反力动力系数
dynamic(al) cohesion 动内聚力
dynamic(al) collection control 动态文献库控制
dynamic(al) compaction 动态成型;动力压实;动力夯实(法);动力固结法;强夯
dynamic(al) compaction method 强夯(实)法
dynamic(al) compensating device 动态补偿装置
dynamic(al) compliance 动态柔量
dynamic(al) component 动力分量
dynamic(al) compression test 动力压缩试验
dynamic(al) compressor 动压力压气机;动力压缩器
dynamic(al) compress tester 动力压缩测试器
dynamic(al) concentration of pollutant 动态污染物浓度
dynamic(al) condenser 动态冷凝器;同期调相机;同步补偿器
dynamic(al) condenser electrometer 动态电容器式静电计
dynamic(al) condition 动态条件;动态

dynamic(al) cone penetrometer 动力圆锥触探仪
dynamic(al) consolidation 动力夯实(法);动力固结(法);强力夯实;强夯(法)
dynamic(al) consolidation method 强夯加固法
dynamic(al) consolidation process 强夯(实)法
dynamic(al) constant 动态常数
dynamic(al) constraint 动态约束;动态限制
dynamic(al) contact force 动态接触力;受电弓动态接触力
dynamic(al) control 动态控制;动力控制
dynamic(al) control function 动态控制功能
dynamic(al) control module area 动态控制模块区
dynamic(al) control system 动态控制系统
dynamic(al) convection 动态对流;动力(性)对流
dynamic(al) convergence 动态会聚
dynamic(al) cooling 动力冷却
dynamic(al) coordination system 动态联动(信号)系统
dynamic(al) core allocation 动态存储分配
dynamic(al) correction 动(力)校正
dynamic(al) cosmology 动力学宇宙学
dynamic(al) costing 动态投资
dynamic(al)-coupled amplifer 动态耦合放大器
dynamic(al) coupling 电磁离合器
dynamic(al) course stability index 航向动稳定性指数
dynamic(al) creep 动态蠕变;动力蠕变
dynamic(al) crossfield photomultiplier 动态交叉场交电倍增器
dynamic(al) cross-interaction 动态交叉相互作用
dynamic(al) current 动稳电流;持续电流
dynamic(al) curve 动态特征曲线;动态特性曲线
dynamic(al) cycle 动力循环
dynamic(al) damper 消振器;动态减振器;动力阻尼器;动力消振器;动力减振器
dynamic(al) damping 动力减振
dynamic(al) data 动态资料;动态数据
dynamic(al) data exchange 动态数据交换
dynamic(al) data recording system 动态数据记录系统
dynamic(al) data set definition 动态数据集定义
dynamic(al) data structure 动态数据结构
dynamic(al) debugging 动态调试
dynamic(al) debugging routine 动态调试程序
dynamic(al) debug mode 动态调试方法
dynamic(al) decoupling 动态去耦
dynamic(al) deflection 振动变形;动(载)弯沉;动载挠度;动力弯曲;电动偏转;冲击挠度;deflection of spring 弹簧动挠度
dynamic(al) deformation 动力变形
dynamic(al) delivery head 总输水水头;动输水水头
dynamic(al) design 动力设计
dynamic(al) design analysis method 动态设计分析方法
dynamic(al) design approach 动态设计方法
dynamic(al) design criterion 动力设计准则
dynamic(al) desorption 动态脱附
dynamic(al) deterministic model 动态确定型
dynamic(al) deterministic-random mixed model 动态确定随机混合型
dynamic(al) detuner 动力减振器
dynamic(al) deviation 动态偏移;动态偏差
dynamic(al) device allocation 动态设备分配
dynamic(al) device reconfiguration 动态设备重新组合;动态设备重新配置
dynamic(al) differential calorimeter 动态热量计
dynamic(al) differential colo(u)rimeter 差动比色计;动差比色计
dynamic(al) differential colo(u)rimetry 差动比色法
dynamic(al) diffusion 动力扩散
dynamic(al) digital torque meter 动力数字式转矩计
dynamic(al) digitizer 动态数字化器
dynamic(al) digitizing 动态数字化
dynamic(al) dilatation 动力扩容作用
dynamic(al) discharge head 动泄水水头
dynamic(al) dislocation model 动力位错模型;动力错位模型
dynamic(al) dispatching 动态调度
dynamic(al) dispersivity 水动力弥散系数
dynamic(al) displacement 动(态)位移
dynamic(al) display 动态显示
dynamic(al) display image 动态显示图像

dynamic(al) dissipation 动力耗散
dynamic(al) distortion 动畸变
dynamic(al) document 动态资料;动态文献
dynamic(al) document space 动态文献空间
dynamic(al) draft of transducer 换能器动态改正
dynamic(al) draught head 动吸出水头
dynamic(al) drift 动态变化
dynamic(al) drive 电动驱动
dynamic(al) ductility 动力延展性;冲击韧性
dynamic(al) dump(ing) 动态打印;动态转储
dynamic(al) duty 动态工作
dynamic(al) eccentricity 动态偏心率
dynamic(al) ecology 动态生态学
dynamic(al) economic model 动态经济模型
dynamic(al) economics 动态经济学
dynamic(al) economy 动态经济
dynamic(al) ecosystem 动态生态系(统)
dynamic(al) effect 动态效应;动力影响;动力效应
dynamic(al) effective stress method 动力有效应力法
dynamic(al) effect of liquid core of the earth 地球液核动力效应
dynamic(al) efficency 动力效率
dynamic(al) elastic analysis 动力弹性分析
dynamic(al) elastic behavio(u)r 动(态)弹性特性;动弹性状态
dynamic(al) elastic limit 动(态)弹性极限
dynamic(al) elastic modulus 动(力)弹性模量;动基床系数
dynamic(al) electricity 动态电;动电学
dynamic(al) electrometer 振簧静电计
dynamic(al) electronic rail balance 动态电子轨道衡
dynamic(al) element 动态元件
dynamic(al) elevation 动力高
dynamic(al) elevation correction 动力高校正
dynamic(al) elevation number 动力高数
dynamic(al) elevation system 动力高系统
dynamic(al) ellipticity 力学椭率;动力学扁率
dynamic(al) ellipticity of the earth 地球动力扁率
dynamic(al) encompassing 动态包围
dynamic(al) end-effects 动态端部效应
dynamic(al) endurance test 动力耐久试验;变负载强度试验
dynamic(al) energy converter 动力换能器
dynamic(al) engineering 动力工程学
dynamic(al) engineering property in-situ 原位动力工程性质
dynamic(al) enhanced microelectrolysis 动力强化微电解法
dynamic(al) environment 动力环境
dynamic(al) environmental capacity 动环境容量
dynamic(al) equation 动力方程
dynamic(al) equation of equilibrium 动力平衡方程(式)
dynamic(al) equilibrium 动态平衡;动力平衡
dynamic(al) equinox 力学分点
dynamic(al) equipment policy 动态设备对策
dynamic(al) error 动态误差;动态误差
dynamic(al) error debug 动态错误排除
dynamic(al) evaluation 动态评价
dynamic(al) evolution 动力学演化
dynamic(al) exchange coefficient 动力传递系数
dynamic(al) experiment 动态实验
dynamic(al) extension viscosity 动态同轴伸长黏度
dynamic(al) factor 动载系数;动力因数;动力系数
dynamic(al) fatigue 动态疲劳;动力(性)疲劳
dynamic(al) fatigue test 动态疲劳试验;动力疲劳试验
dynamic(al) fault model 动态故障模型;动力断层模型;电动断层模型
dynamic(al) feature 动态特征
dynamic(al) field 变动场
dynamic(al) file organization 动态文件组织;动态文件结构
dynamic(al) finite-difference equation 动态有限差分方程
dynamic(al) flattening 力学扁率;动力学扁率
dynamic(al) flexural stiffness 动力弯曲劲度
dynamic(al) flip-flop 动态触发器
dynamic(al) flocculation 动力絮凝
dynamic(al) flow 流体动力
dynamic(al) flow diagram 动态流程图
dynamic(al) fluctuation of wheel-load 轮载的动力变化

dynamic(al) fluidity 动力流度
dynamic(al) focus 动焦点
dynamic(al) focusing 动态聚焦
dynamic(al) focusing control 动态自动聚焦调整
dynamic(al) force 动(态)力
dynamic(al) forecast 动态预报
dynamic(al) forecasting model with recusive compensation by grey numbers of identical dimension 等维灰数递补动态预报模型
dynamic(al) formula 动能公式
dynamic(al) foundation exploration 动态基础探查;动态基础探测
dynamic(al) fracture mechanics 动力断裂力学
dynamic(al) frequency allocation 动态频率分配
dynamic(al) frequency characteristic 动态频率特性
dynamic(al) frequency spectrum analyzer 动态频谱分析仪
dynamic(al) friction 动摩擦;动力(学)摩擦
dynamic(al) frictional angle 动摩擦角
dynamic(al) friction coefficient 动摩擦系数
dynamic(al) function of ecosystem 生态系统的动态功能
dynamic(al) function of money 货币的动态职能
dynamic(al) gate 动态门
dynamic(al) geodesy 动力大地测量学
dynamic(al) geology 动力地质学
dynamic(al) geomorphology 动力地貌学
dynamic(al) gradient 动力梯度
dynamic(al) grain refinement 动力晶粒细化法
dynamic(al) gravity measurement 动态重力测量
dynamic(al) gravity meter 动力重力仪
dynamic(al) gravity survey 动态重力测量
dynamic(al) handling 动态处理
dynamic(al) hardening 快速加热冷却淬火法
dynamic(al) hardness 刮刻硬度;冲撞硬度;冲击硬度;马尔特氏硬度
dynamic(al) hazard 动险态;动态冒险
dynamic(al) head 液压头;动(压)水头;动(力)水头;速度水头
dynamic(al) heat flux 动力热通量
dynamic(al) heating 动力增温;动力加热;动力冲温
dynamic(al) heeling angle 动横倾角
dynamic(al) height 力高;动力高度
dynamic(al)-height anomaly 动力高度偏差
dynamic(al) high-speed spectrograph 动态高速摄谱仪
dynamic(al) holdup 动态储量
dynamic(al) horsepower 净实马力;指示马力;传动马力
dynamic(al) host configuration protocol 动态主机配置协议
dynamic(al) hot pressing 动力热压
dynamic(al) hydrothermal process 水热合成法
dynamic(al) ice detector 动态结冰探测器
dynamic(al) ice pressure 流冰动压力
dynamic(al) image analysis 动态图像分析
dynamic(al) image compress coding 动态图像的压缩编码
dynamic(al) imbalance 动态不平衡
dynamic(al) imitation test 动态模拟试验
dynamic(al) impedance 动态阻抗
dynamic(al) increment 动力增量
dynamic(al) indentation 回跳法硬度试验;动力压凹
dynamic(al) indentation test 回弹法硬度试验;球印硬度试验
dynamic(al) index 动态索引;动力指数
dynamic(al) indirect addressing 动态间接寻址
dynamic(al) inductance 动态电感
dynamic(al) influence 动力影响
dynamic(al) information 动态信息;动态情报
dynamic(al) information dump(ing) 动态信息转储[贮];动态信息转储
dynamic(al) information service 动态情报服务
dynamic(al) information structure 动态信息结构
dynamic(al) information value 动态情报值
dynamic(al) inhalation system 动式吸入装置
dynamic(al) inorganic membrane 动态无机膜
dynamic(al) input-output analysis 投入产出动态分析
dynamic(al) input-output model 动态投入产出模型
dynamic(al) input-output table 动态的投入产出表

dynamic(al) instability 惯性不稳定性；动态不稳定性；动力不稳定性
dynamic(al) instruction 动态指令
dynamic(al) interaction 动力相互作用
dynamic(al) inventory model without set-up cost 不考虑设置成本的动态库存模型
dynamic(al) lag 动态滞后；动态时滞
dynamic(al) lagoon model 动力污水氧化塘模型
dynamic(al) lateral force of soil 动侧土压力
dynamic(al) lateral pressure coefficient 动态侧向压力系数
dynamic(al) lateral stiffness 动力侧向刚度
dynamic(al) leak test 动态检漏
dynamic(al) level 动力水面；动水位
dynamic(al) library 动态图书馆
dynamic(al) libration 动力(学)天平动
dynamic(al) life table 动态生命表
dynamic(al) lift 流动浮力；动浮托力
dynamic(al) limit of yield point 动力屈服点
dynamic(al) linear analysis 动力线性分析
dynamic(al) linker 动态连接器
dynamic(al) link(ing) 动态链接；动态连接
dynamic(al) load 动(荷)载；动态负载；动力荷载；动负载
dynamic(al) load coefficient 动(态)荷载系数
dynamic(al) load factor 动载系数；动(力)荷载系数
dynamic(al) load fluctuation 动载波动
dynamic(al) load for a loaded car 重车动载重
dynamic(al) loading 动力荷载；动压负荷运动；动态装入程序；动(态)荷载；冲击荷载
dynamic(al) loading and linking 动态装入与连接
dynamic(al) loading test 动荷载试验
dynamic(al) load line 动力荷载线
dynamic(al) load lining 动载重被覆
dynamic(al) load of car 货车动载重
dynamic(al) load per unit width 单宽坝体重量
dynamic(al) load simulator 动载模拟器
dynamic(al) load stress 动载应力
dynamic(al) load test 动载试验
dynamic(al) load test of pile 桩的动(力)荷载试验
dynamic(al) log 动态登录
dynamic(al) logic 动态逻辑
dynamic(al) logic circuit 动态逻辑电路
dynamic(al) loop 动态循环
dynamic(al) loop jump 动态循环转移
dynamic(al) loss 动态损失
dynamic(al) loudspeaker 电喇叭；电动(式)扬声器
dynamic(al) lubrication 动力润滑
dynamic(al) luminous sensitivity 动态光照灵敏度
dynamically balanced 做过动态平衡的
dynamically balanced system 动态平衡输水系统
dynamically descendent on-unit 动态子系中断单位
dynamically loaded beam 受动负载的梁
dynamically oriented evaluation 动态定向评价
dynamically positioned drilling ship 动力定位钻探船；动力定位钻井船
dynamically positioning 动力定位
dynamically positioning rig 动力定位钻探平台；动力定位钻井平台
dynamically positioning semi-submersible platform 动力定位半潜式平台
dynamically similar model 动力相似模型
dynamically stable 动力稳定的
dynamically tuned gyro 动态调谐陀螺
dynamically user microprogrammable machine 动态用户可编微程序计算机
dynamically viscous coefficient 动力黏度系数
dynamic(al) magnetic induction 动磁感应
dynamic(al) magnification 动力扩大系数；动力扩大率；动力扩大倍数；动力放大
dynamic(al) magnification coefficient 动放大系数
dynamic(al) magnification factor 动力放大因素；动力放大系数
dynamic(al) map 动态地图
dynamic(al) mapping 动态变换
dynamic(al) mapping method 动态制图法
dynamic(al) mapping procedure 动态制图法
dynamic(al) mapping system 动态制图系统
dynamic(al) mass spectrometer 动态质谱计
dynamic(al) matrix 动力矩阵
dynamic(al) mean sun 假想太阳
dynamic(al) measurement 动态测量
dynamic(al) measurements of pile 桩的动测法

dynamic(al) mechanical analysis 动态机械分析
dynamic(al) membrane 动态膜；动力膜
dynamic(al) membrane bioreactor 动态膜生物反应器
dynamic(al) membrane pressure method 动态膜压法
dynamic(al) memory 动(态)存储器
dynamic(al) memory allocation 动态存储(器)分配
dynamic(al) memory interface 动态存储器接口
dynamic(al) memory refresh 动态存储器刷新
dynamic(al) memory relocation 动态存储分配；动态存储(器)重新分配；动态存储器再定位
dynamic(al) metamorphism 动力变质(作用)
dynamic(al) meteorology 动力气象学
dynamic(al) meter 测力计；动力米
dynamic(al) method 动态法；动力法；变化场(法)
dynamic(al) method of relative permeability determination 动力法测定相对渗透率
dynamic(al) method of satellite geodesy 卫星大地测量动力法
dynamic(al) method of satellite orbit 卫星轨道动力法
dynamic(al) microelectrolysis 动态微电解法
dynamic(al) microelectrolysis-Fenton agent process 动态微电解—芬顿试剂法
dynamic(al) microphone 电动(式)传声器
dynamic(al) microprocessor 动态微处理机
dynamic(al) microprogramming 动态微程序设计
dynamic(al) microprogramming control 动态微程序控制
dynamic(al) mock-up 动力模型
dynamic(al) mode incremental digitizer 动态增量数字化器
dynamic(al) mode incremental digitizing 动态增量数字化
dynamic(al) model 力学模型；动态模型；动力(学)模型；动力模式
dynamic(al)-model(l)ing language 动力学建模语言
dynamic(al) model of competitive equilibrium 竞争性平衡的动态模型
dynamic(al) modulation 动态调制
dynamic(al) modulus 动态模量；动态模数；动模量；动力学弹性模数；动力模量
dynamic(al) modulus of elasticity 动(力)弹性模量；动弹性系数；动弹性模数；动(态)弹性模量
dynamic(al) modulus of shear 动剪切模量；动剪力模量
dynamic(al) monitoring 动测(动力监测)
dynamic(al) mount 减振架
dynamic(al) mounting ring 减振环形架
dynamic(al) muscle 动力肌
dynamic(al) noise suppressor 动噪声抑制器
dynamic(al) non-linear analysis 动力非线性分析
dynamic(al) normal stress 动正应力
dynamic(al) number 动力势差数
dynamic(al) oblateness 力学扁率
dynamic(al) oceanography 动力海洋学
dynamic(al) oedometer 动力固结仪
dynamic(al) oil-damper 油液减振装置
dynamic(al) olfactometry 动态测嗅法
dynamic(al) operation 动态操作
dynamic(al) optimization 动态(最)优化；动态最佳化
dynamic(al) optimum population 动态适度人口
dynamic(al) ordering procedure 动态排序过程
dynamic(al) overlay 动态覆盖
dynamic(al) overload 动态过载；动力过载
dynamic(al) overload control 动态超载控制
dynamic(al) oxidation 动氧化
dynamic(al) packing 动力密封垫
dynamic(al) panel 发电机盘
dynamic(al) parallax 力学视差；动力视差
dynamic(al) parameter 动态参数；动力参数
dynamic(al) parameters test of rock and soil 岩石动参数试验
dynamic(al) parametric test 动态参数测试
dynamic(al) partition 动态划分；动态分区；动态分割
dynamic(al) partition balancing 动态分区平衡
dynamic(al) partitioning 动态划分
dynamic(al) passes 振动压实遍数
dynamic(al) path 动态路径

dynamic(al) payoff period 动态返本期
dynamic(al) penetrating meter 动力触探仪
dynamic(al) penetration 动力触探
dynamic(al) penetration test 动力贯入试验；动力触探试验
dynamic(al) percolation model 动力学渗透模型
dynamic(al) performance 动态性能；动态特性
dynamic(al) permeability 动磁导率
dynamic(al) phase error 动态相位误差
dynamic(al) photo-elasticity 动力光(测)弹性学；动态光弹性(法)
dynamic(al) photo-elastic test 动力光弹性试验
dynamic(al) photopoint check 动光点检查
dynamic(al) pickup 动态拾振器；动态传感器；动力拾振器；电动式拾音器；电动拾声器
dynamic(al) pile-driving 动力打桩
dynamic(al) pile-driving formula 动力打桩公式；打桩动力公式
dynamic(al) pile-driving resistance 打桩动阻力
dynamic(al) pile formula 动力桩(荷载)公式；桩动力公式
dynamic(al) pile test 动力测桩
dynamic(al) pinch effect 动力箝缩效应
dynamic(al) planning 动态规划(法)
dynamic(al) plate impedance 动态阳极阻抗
dynamic(al) plate loading method 承载板动态施荷法
dynamic(al) plate resistance 动态板极电阻
dynamic(al) point resistance 动力探头阻力
dynamic(al) Poisson's ratio 动泊松比
dynamic(al) pool block 动态组合块；动态页面池块
dynamic(al) population 动态人口
dynamic(al) position 动力定位
dynamic(al) positioning 动态定位
dynamic(al) positioning accuracy 动力定位精度
dynamic(al) positioning and tracking system 船位与航迹显示装置
dynamic(al) positioning system 动态定位系统；动力定位系统
dynamic(al) positioning technique 动力定位技术
dynamic(al) potential 动电势
dynamic(al) precipitator 动力除尘器；动力沉淀器
dynamic(al) prediction 动力预报
dynamic(al) pressure 动压(推)力；动压强；冲击压力；速压头
dynamic(al) pressure apparatus 动态高压设备
dynamic(al) pressure computer 动压力计算机
dynamic(al) pressure log 水压计程仪
dynamic(al) pressure method 动压法
dynamic(al) pressure transducer 动压传感器
dynamic(al) pressure type flow meter 动压式流速计
dynamic(al) printout 动态打印(输出)
dynamic(al) priority 动态优先(级)
dynamic(al) priority allocation 动态优先权分配
dynamic(al) probabilistic inventory model 概率型动态库存模型
dynamic(al) probing test 动力触探试验
dynamic(al) problem 动态问题；动力问题
dynamic(al) problem check 动态问题检验；动态解题校验
dynamic(al) process 动态过程
dynamic(al) process creation 动态进程建立
dynamic(al) process deletion 动态进程取消
dynamic(al) process simulation 动态过程模拟
dynamic(al) program(me) 动程序
dynamic(al) program(me) loading 动态程序装入；程序动态装入
dynamic(al) program(me) relocation 动态程序浮动
dynamic(al) program(me) structure 动态程序结构
dynamic(al) programming 非线性规划；动态规划(法)；动态程序设计
dynamic(al) programming algorithm 动态规划算法
dynamic(al) programming and bounded measure 动态规划与有界测度
dynamic(al) programming formulation 动态规划形成
dynamic(al) property 自振特性；动态特性；动力(学)性质；动力特性
dynamic(al) property of rock and soil 岩土的动力性质
dynamic(al) quenching 动态淬火法

dynamic(al) random access memory 动态随机存取存储器
dynamic(al) random model 动态随机型
dynamic(al) range 密度范围;动态量程;动态范围;动力学研究范围
dynamic(al) ratios 动态比率
dynamic(al) reaction 动力反作用;动态反力
dynamic(al) real-time information processing system 动态实时信息处理系统
dynamic(al) rebound deflection 动态回弹弯沉;动载回弹弯沉
dynamic(al) reciprocity 动力互换
dynamic(al) reconfiguration 动态重新配置;动态再配置
dynamic(al) reconfiguration data set 动态重新配置数据集;动态再组合数据集
dynamic(al) recorder 动态记录器
dynamic(al) recovery 动态回复
dynamic(al) recrystallization 动态重结晶;动态再结晶
dynamic(al) reference model 动态参考模型
dynamic(al) reference system 动力学参考系
dynamic(al) reflectance spectroscopy 动态反射光谱学
dynamic(al) refocusing 动态再聚焦
dynamic(al) region 动态区
dynamic(al) region area 动态范围区
dynamic(al) register 动态寄存器
dynamic(al) regrouping 动态重组
dynamic(al) regulator 动态调节器
dynamic(al) rejuvenation 动力回春
dynamic(al) relation 动态关系
dynamic(al) release 动态释放
dynamic(al) reliability 动态可靠性;动力可靠性;动力可靠度
dynamic(al) relocation 动态再分配;动态再定位;动态浮动
dynamic(al) relocation memory 动态重新分配存储器;动态再定位存储器
dynamic(al) relocation program(me) 可动态移位程序;动态浮动程序
dynamic(al) remanent magnetization 动力剩余磁化强度
dynamic(al) remote sensing 动态遥感
dynamic(al) replacement 动力置换(法)
dynamic(al) replication 动态重复
dynamic(al) reproducer 电动拾声器
dynamic(al) reserve 动储量
dynamic(al) reservoir 动态热储
dynamic(al) reservoir model 热储动力模型
dynamic(al) resilience 动力回弹能;动力回弹性;动弹性回复
dynamic(al) resistance 动(力)阻力;抗动载强度;抗冲击强度;动(态)阻力;动态电阻
dynamic(al) resonance 动态谐振;动态共振;瞬态谐振
dynamic(al) resource 动态资源
dynamic(al) resource management 动态资源管理
dynamic(al) response 动态响应;动(态)反应;动(力)反应;动力效应
dynamic(al) response analysis 动力反应分析;动反应分析
dynamic(al) response approach 动力反应法
dynamic(al) response characteristic 动态响应特性
dynamic(al) response curve 动态响应曲线;动态特征曲线;动态特性曲线
dynamic(al) response factor 动态特性系数
dynamic(al) response spectrum 动响应谱
dynamic(al) response test 动态响应试验
dynamic(al) restructuring 动态再构成
dynamic(al) rest space 动休息区
dynamic(al) rheometer 动态流变仪
dynamic(al) righting lever 动稳性力臂;动复原力臂
dynamic(al) rigidity 动力刚性;动刚性
dynamic(al) rijuvenation 动力回春作用
dynamic(al) ripple 动态波纹
dynamic(al) road net 动态路网
dynamic(al) rock mechanics 岩石动力学
dynamic(al) role 能动性
dynamic(al) roll response 滚动扰动动力特性
dynamic(al) roughness 动力糙度
dynamic(al) route allocation 动态通路分配
dynamic(al) routine 动态程序
dynamic(al) rupture 动力破裂

dynamic(al) sampling 动态抽样
dynamic(al) satellite geodesy 卫星动力测地
dynamic(al) scattering 动态散射
dynamic(al) scattering device 动态散射器
dynamic(al) scheduling 动态调度
dynamic(al) scheduling simulator 动态调度模拟系统
dynamic(al) seal(packing) 动态密封
dynamic(al) search 动态搜索
dynamic(al) sediment 动力沉积物
dynamic(al) segment attribute 动态图块属性
dynamic(al) segment relocation 动态区段再定位
dynamic(al) sensitivity 动态灵敏度
dynamic(al) sensor 动态传感器
dynamic(al) sequential control 动(态)顺序控制
dynamic(al) series 动态数列
dynamic(al) set 动态组;动态集
dynamic(al) shearing force 动切力
dynamic(al) shearing stress 动剪应力
dynamic(al) shear modulus 动(力)剪力模量;动态剪切模量
dynamic(al) shearometer of mud 泥浆动切力计
dynamic(al) shear strength 动抗剪强度
dynamic(al) sheet of fixed assets 固定资产动态表
dynamic(al) shift register 动态移位寄存器
dynamic(al) similarity 动态相似;动力相似(性);动力比拟
dynamic(al) similarity principle 动力学相似原理
dynamic(al) similitude 动力相似性
dynamic(al) simple shear test 动单剪试验
dynamic(al) simulation model 动态模拟模型;动力模拟模型
dynamic(al) simulator 动态特征模拟;动态特性模拟;动力模拟器
dynamic(al) skew 动态偏斜
dynamic(al) slack 动态呆滞
dynamic(al) sludge age 动态污泥年龄
dynamic(al) sludge reverving process 动力污泥沉淀工艺
dynamic(al) soil 动力土壤
dynamic(al) soil-tunnel interaction 土(壤)隧道动力相互作用
dynamic(al) solution 动力学解法
dynamic(al) sounding 动力触探杆;动力探测;动力触探(试验);动力测深
dynamic(al) sounding apparatus 动力触探仪
dynamic(al) speaker 动圈式扬声器;电动扬声器
dynamic(al) spectrum 动态频谱
dynamic(al) spectrum analyzer 动态频谱分析器
dynamic(al) split generation 动态绿信比的产生
dynamic(al) split system 动态分离系统
dynamic(al) spring constant 动态弹簧常数
dynamic(al) spring rate 动力弹簧刚性
dynamic(al) stability 动稳性;动态稳定性;动态稳定度;动力学稳定性;动力稳定(性);动力稳定(度)
dynamic(al) stability characteristics 动稳定性特性
dynamic(al) stability control 动稳定性控制
dynamic(al) stability model 动稳定性模型
dynamic(al) state 动态
dynamic(al) state approach 动态状况研究
dynamic(al) statement 动态(报)表
dynamic(al) state of fixed assets 固定资产的动态
dynamic(al) state of reserves 储量变动
dynamic(al) stationing system 动力定位系统
dynamic(al) statistics of population 人口动态统计
dynamic(al) steady state 动力稳定态
dynamic(al) stereo photography 动态立体测量
dynamic(al) stereotype 动力定型
dynamic(al) stiffness 动态劲度;动态刚性;动(态)刚度;动(力)刚度
dynamic(al) stop 动态停机
dynamic(al) storage 动态存储(器)
dynamic(al) storage allocation 动态存储(区)分配;动态存储(器)分配
dynamic(al) storage area 动态存储区
dynamic(al) storage model 动态存储模型
dynamic(al) storage system 动态存储系统
dynamic(al) strain 动(态)应变;动(力)应变
dynamic(al) strain aging 动态应变时效
dynamic(al) strain amplifier 动态应变放大器
dynamic(al) strain indicator 动态应变仪
dynamic(al) strain meter 动态应变仪
dynamic(al) strainometer 动态应变仪

dynamic(al) stratigraphy 动力地层学
dynamic(al) stream 动力流
dynamic(al) strength 抗动载强度;抗冲击强度;动态强度;动力强度;冲击强度
dynamic(al) strength of material 材料动力强度
dynamic(al) strength of shearing 动抗剪强度
dynamic(al) strength of tensile 动抗拉强度
dynamic(al) stress 振动应力;动态应力;动(力)应力
dynamic(al) stress concentration 动态应力集中
dynamic(al) stress ratio 动应力比
dynamic(al) stress singularity 动力应力奇性
dynamic(al) structural analysis 动力结构分析
dynamic(al) structure 动态结构;动力结构
dynamic(al) subgrade reaction 动基床反力
dynamic(al) subroutine 动态子(例行)程序
dynamic(al) subsurface sounding 动力地下触探
dynamic(al) suction head 动吸升水头;动吸入水头;动吸(水)头
dynamic(al) suction lift 动吸程;动式吸入高度;动(力)吸升高度
dynamic(al) support system 动态支持系统;动态后援系统
dynamic(al) surface tension 动(态)表面张力
dynamic(al) suspension 可缓冲支承法
dynamic(al) system 脉冲制;活动系统;动态制;动态系统;动力系(统)
dynamic(al) system analysis 动态系统分析
dynamic(al) system maintenance 动态方式维修;动态方式维护
dynamic(al) system model 动态系统模型
dynamic(al) system of design 设计动态系统
dynamic(al) system of distribution 动分布系统
dynamic(al) system respose 动态系统频率特性
dynamic(al) system simulator 动力系统模拟器
dynamic(al) system synthesizer 动态系统综合装置;动态系统合成器
dynamic(al) tear test 动态撕裂试验;动态扯裂试验
dynamic(al) technique 动力技术
dynamic(al) temperature 动温
dynamic(al) temperature change 温度动态变化
dynamic(al) temperature correction 动力温度校正;温差校正;温差改正
dynamic(al) tensile strength 动态抗拉强度
dynamic(al) tensile test 动力张力试验
dynamic(al) test bed 动力试验床
dynamic(al) test for cargo gear 起货设备动负荷试验
dynamic(al) test(ing) 动态试验;动听测试;动态测试;动力试验;动测(动力监测);冲击试验
dynamic(al) testing machine 动态试验机;动力试验机
dynamic(al) testing technique 动力试验技术
dynamic(al) test of pile 桩的动力荷载试验
dynamic(al) test of soil 土的动力试验
dynamic(al) test stand 动力试验台
dynamic(al) theory 动态理论;动力(学)理论
dynamic(al) theory of tides 潮汐动力理论
dynamic(al) thermomechanometry 动态热机械法
dynamic(al) thickness 动力厚度
dynamic(al) threshold discharge 临界流量;起动流量
dynamic(al) time 力学时
dynamic(al) time scale 力学时标度
dynamic(al) topography 动力地形测量学
dynamic(al) torque 加速转矩;动转矩
dynamic(al) total enclosure chamber 动式全密闭室
dynamic(al) track ga(u)ge 动态轨距
dynamic(al) track irregularity 动态不平顺
dynamic(al) track response 轨道动力响应
dynamic(al) track stabilizer 动力稳定机
dynamic(al) traffic 动态交通
dynamic(al) training 动力性训练
dynamic(al) transaction backout 动态事项逆序操作
dynamic(al) transconductance 动跨导
dynamic(al) transducer 电动式换能器
dynamic(al) transfer system 动力传输系统
dynamic(al) transient pool area 动态暂驻组合区;动态瞬时页池区
dynamic(al) transmission 动力传输
dynamic(al) triaxial shear apparatus 动三轴剪力仪
dynamic(al) triaxial test 振动三轴试验;动力三

轴试验
dynamic(al) triaxial test of soil 土的动三轴试验
dynamic(al) trigger 动态触发器
dynamic(al) trim 动力调整
dynamic(al) trimming angle 动纵倾角
dynamic(al) trough 动力槽
dynamic(al) tube constant 电子管动态常数
dynamic(al) tuned gyroscope 动态调谐陀螺仪
]dynamic(al) tuning 动态调谐
dynamic(al) type 电动式
dynamic(al) ultra high pressure and high temperature method 动态超高压高温法
dynamic(al) ultra high pressure and high temperature process 动态超高压高温法
dynamic(al) unbalance 动态失衡;动(态)不平衡;动力失衡;动(力)不平衡
dynamic(al) unequilibrium 动力不平衡
dynamic(al) unit 动力单位
dynamic(al) unstability 动力失稳
dynamic(al) valve 动态阀;动力阀
dynamic(al) variable 动态变量;动力变量
dynamic(al) vehicle envelope 动态车辆包络线
dynamic(al) vertical 动垂线;视垂线
dynamic(al) vibration absorber 动力振动阻尼器;动力吸振器;动力减振器
dynamic(al) viscoelasticity 动态黏弹性
dynamic(al) viscoplasticity 动态黏塑性
dynamic(al) viscosity 绝对黏度;动黏滞度;动态黏(滞)度;动(力)黏滞性;动力黏滞率;动力黏(滞)度
dynamic(al) viscosity coefficient 动力黏滞(度)系数
dynamic(al) volumetric(al) method 动态容量法
dynamic(al) warming 动力增温
dynamic(al) water environmental capacity 动态水环境容量
dynamic(al) water environmental capacity of river network 河网动态水环境容量
dynamic(al) water level 动(力)水位
dynamic(al) water pressure 动水压(力);动水水压力;动孔隙水压力
dynamic(al) watershed model 动态流域模型
dynamic(al) water table 动态水位
dynamic(al) water table of back-pumping 回扬动水位
dynamic(al) wave 动力波
dynamic(al) weathering 动力风化
dynamic(al) wheel balancer 车轮动平衡机
dynamic(al) wheel demonstrator 车轮动平衡表示器
dynamic(al) wheel load 动轮载
dynamic(al) wheel-rail contact force 动态轮轨接触力
dynamic(al) window articulation 动态窗口连接
dynamic(al) window translation 动态窗口平移
dynamic(al) wind rose 风向风力动态图;风力玫瑰图;风力风向动态图
dynamic(al) work 动力作业
dynamic(al) writing range 动态记录范围
dynamic(al) yield strength 动态屈服强度
dynamic(al) Young's modulus 动力杨氏模量
dynamicist 力学家;力本论者;物力论塔
dynamicizer 串化器;动态器;动态逻辑转化元件
dynamics 运动学;动态学;动力学
dynamics of compressible fluids 气体动力学
dynamics of domain movement 电畴运动的动力学
dynamics of ecosystem 生态系统动力学
dynamics of exploited stocks 开发种群的数量动态
dynamics of intergroup conflict 小组间冲突动态性
dynamics of jets 射流动力学
dynamics of machinery 机械动力学
dynamics of management 管理动态
dynamics of marine structure 海洋结构动力学
dynamics of migration 移民动态
dynamics of ocean currents 海流动力学
dynamics of orbits 轨道动力学
dynamics of pollutant concentration 污染物浓度动态特征
dynamics of population 人口动态
dynamics of power 权力动态
dynamics of resedimentation 再沉积动力学
dynamics of rigid body 刚体动力学
dynamics of silt sedimentation 泥砂沉淀动力学
dynamics of stream 河流演变;河流动态

dynamics of structure 结构动力学
dynamics of variable mass 变质量动力学
dynamics of vehicle movements 车辆运动动力学
dynamics of water quality 水质动态
dynamics on the track 轨道上的动力学
dynamics property of minerals 矿物力学性质
dynamics similarity 动力相似性
dynamism 动力论
dynamite 烈性甘油炸药;黄色炸药;炸药;甘油炸药;代那买特炸药
dynamite cartridge 硝化甘油炸药包
dynamite charge 炸药装量;炸药装填
dynamite fume 爆炸烟雾
dynamite glycer 硝化甘油
dynamite magazine 火药库;炸药库
dynamite method 爆炸方法
dynamite punch pricker 插管棍
dynamites available 炸药可用性
dynamite source 炸药震源
dynamite stick 硝甘炸药卷
dynamiting 炸药爆破
Dynamitron 地那米加速器;高频高压加速器
dynamo 直流发电机;发电机
dynamobronze 电机青铜;耐蚀铝青铜;耐熔铝青铜;特殊耐磨铝青铜
dynamo brush 整流子电刷;电刷
dynamo condenser 调相器
dynamo cradle 发电机架
dynamo current meter 发电机式流速仪
dynamo door latch 发电机门栓
dynamo door latch pin 发电机门弹键销
dynamo drive jockey pulley 发电机导轮
dynamo effect 发电机效应
dynamo-electric(al) 电动的
dynamo-electric(al) amplifier 动电式放大器;电动放大器
dynamo-electric(al) amplifier generator 电动放大发电机
dynamo-electric(al) machine 电动发电机
dynamo-electric(al) motor 旋转换流机
dynamo engine 发电用发动机
dynamo exploder 点火机
dynamofluidal 动力流动的
dynamofluidal texture 动力流状结构
dynamo geomagnetic theory 地磁发电机理论
dynamogoniograph 磁偏角记录器
dynamo governor 发电机调节器
dynamo-graph 动力自记器;动力自计器
dynamo ignition 发电机点火
dynamo magneto 永磁发电机
dynamo-metamorphic rock 动力变质岩
dynamometamorphism 动力变质(作用)
dynamometer 电力功率仪;量力器;拉力计;功率计;电动式仪器;测力计;测功计;测力计;测力环;测力盒;测功器;测功计;放大率计;土壤阻力计
dynamometer ammeter 功率计式安培计
dynamometer brake 测功制动器
dynamometer car 动力试验车;动力测定车;测试车;测力试验车
dynamometer card 测力图
dynamometer check 测力试验
dynamometer engine test bed 发动机测功器试验台架
dynamometer for stretching the tape 张力功率计
dynamometer instrument 电力表
dynamometer life 台架试验寿命
dynamometer link 测力杆
dynamometer machine 测功机
dynamometer multiplier 功率计式乘法器
dynamometer prop 测功支柱;测力计支架
dynamometer ring 动力计环
dynamometer test 制动测功仪试验;测功试验
dynamometer-type instrument 电动式仪表;测力计型仪表
dynamometer-type multiplier 功率计式乘法器;测力计式乘法器
dynamometer wattmeter 功率计式瓦特计
dynamometric 测力的;计力的
dynamometric apparatus 动力测量装置
dynamometric device 功率测定装置;测力装置
dynamometric exchange 动力交换
dynamometric measuring apparatus 动力测量装置
dynamometry 计力法;测力法;测力术;测功法
dynamotor 电动发电机;旋转换流机

dynamo oil 电机油
dynamo output 发电机功率输出
dynamo output control 发电机输出调节器
dynamo room 发电机房
dynamoscope 动力测验器
dynamoscopy 动力测验法
dynamo sheet 电机硅钢片
dynamo sheet steel 电机用薄钢片
dynamostatic 动电产生静电的
dynamo stator core automatic welder 电机定子铁芯自动焊接机
dynamo steel 电机钢
dynamo steel sheet 电机(用)硅钢片;电机钢板
dynamo strap 发电机固定带
dynamo theory 发电机假说;发电机理论
dynamo-thermal 动热的
dynamothermal gradient 热动力梯度
dynamothermal metamorphism 动热变质(作用);热动力变质
dynamotor 电动直流发动机;电动(直流)发电机;电动机发电机组
dynamo wave 发电机波
dynapak method 高速高能锻造法
dynapak press 高能(高)速压机;高能束压机
dynapolis 动态发展城市;沿交通干线有计划发展起来的城市;干线城市;带形城市
dynaprop 动力支柱
Dyna pump 戴纳泵
dynastic order 朝代次序
dynasty 朝代;王朝
dynatherm 透热机
dynatron 打拿管
dynatron effect 打拿效应
dynatron frequency meter 负阻管频率计
dynatron oscillator 负阻管振荡器
dynatron pulse circuit 负阻式脉冲电路
dynaturtle 电动龟标
Dynavar (alloy) 戴纳瓦(恒弹性)合金
Dyna whirlpool process 代物重介质旋流选矿法
Dyna whirlpool separator 代纳旋流分离器
dyncmic tear test 动力扯裂试验
dyne 达因(力的单位)
dyne centimeter 尔格
dynectron 真空壳汞整流器
dynemeter 达因计
dynistor 负阻晶体管
dynode 中间极;电子倍增器电极;二次发射极;倍增(器)电极
dynode spots 倍增器电极斑点
dynode system 倍增系统
dynofiner 纸浆精磨机
dynomo-generator 电动发电机
dyostyle 双柱式
dyotron 超高频振荡三极管
dypingite 球碳镁石
dypnone 缩二苯乙酮
dysanalyte 铌钙钛矿
dysbarism 降压病;减压障碍;气压痛;气压病
dysbiosis 生态失调
dyscrasite 锑银矿
dyscrystalline 不良结晶质
dysdiemorrhysis 毛细管循环迟缓
dysentery with bloody stool 赤痢
dysfunction 机能障碍;机能失调;机能不良;功能障碍;功能异常;功能紊乱;功能失调;功能不良;损坏;失灵
dysfunction of sodium pump 钠泵功能障碍
dysgeogenous 不易风化的
dysintegration 整合障碍
dyskaryosis 核异常
dyskeratosis 不良角化
dyslysin 难溶素
dysmetria 辨距障碍;辨力不良
dysodile 挠性褐煤;硅藻腐泥褐煤
dysontogenesis 发育不良
dysosma 八角莲(属)
dysoxidizable 难氧化的
dysphotic 弱光性
dyspnea 气急
dysponderal 重量异常的
dysprosium ores 镝矿
dysraphism 闭合不全
dyssymbolia 构思障碍
dyssynergia 协同不能

dystaxia 行动失调
dystectic 高熔点的
dystectic mixture 高熔混合物
dyster 多伊斯特风
dystetic mixture 高熔混合物
dystome spar 硅硼钙石
dystonia 张力障碍
dystopia 非理想化的地方
dystrophic 无养分的
dystrophic environment 无滋养环境
dystrophic lake 腐殖营养湖；贫营养湖；无滋养湖；沼泽湖；泥塘
dystrophy 营养不良
dytetragonal prism 复正方柱棱镜
dytory 胶体泥浆
Dywidag method 狄威达克法
Dywidag system 迪威特施工法
dzhalindite 羟铟石

E

each of the contracting parties 缔约各方
each particular kind of product 每种特殊产品
each-pass own code 每遍扩充工作码
each person 每人
each run 每次行程
each time 每次
Eads Bridge 伊兹桥(1867~1874 年美国密西西比河上第一座用悬臂法建造的钢拱桥)
eager 激浪;河水上涨;怒潮;涌潮;潮水上涨
eagle 预制楼板组合件(一种注册的建筑材料产品);希腊古建筑正面上方的三角墙;鹰徽
eagle capital 鹰头柱头
eagle ewer 鹰壶
Eagle mounting 伊格尔装置
eagle nose pliers 鹰嘴钳
Eagle-Picher blister box 伊格尔·皮切尔型起泡试验箱
eagle shaped din 鹰鼎
eagle stone 鹰石
eaglewood 沉香木
eagor pororoca 怒潮
eagre 怒潮;涌潮;潮水上涨
eagroforestry 农林学;混农林业
eagro-sylvo-pastoral system 农林牧系统
eakerite 硅铝锡钙石
eakinsite 块辉锑铅矿
eakleite 硬硅钙石
ealing cap 封帽
Eamian interglacial epoch 埃姆间冰期
Eamian interglacial stage 埃姆间冰期
ean-shuin 艾绒
ear 耳状物;凸耳
ear corn 带穗玉米
ear cup 耳杯
ear defender 护耳器;耳护
ear drops 荷包牡丹属植物;灯笼海棠
eared 带耳的
eared nut 有耳螺母;翼形螺母
eared stop 带角光圈
ear handle 耳柄
earing 形成花边;皱耳;出耳子;凸耳;雪耳;花边
ear insulator 耳环绝缘子
earity 稀有度
ear lamp 耳形灯
earlandite 水碳氢钙石;水柠檬钙石
ear length 耳长
earlier completion 提前竣工
earlier failure period 初期故障期
earlier period 早期
earlier stage 前期
earliest arrival time 最早到达时间
earliest breakup 最早解冻日
earliest complete freezing 最早封冻日
earliest event occurrence time 最早项目完成时间;事件(的)最早发生时间;预计最早完成时间
earliest event time 事件最早时间
earliest expected date 最早预计日期
earliest final ice clearance 最早终冰日
earliest finish time 最早完成时间;最早结束时间
earliest finish time of activity 作业最早完成时间
earliest first appearance of ice 最早初冰日
earliest infection 最初侵染
earliest possible time of arrival 可能最早到达时间
earliest stages of cultivation 初期栽培阶段
earliest start 最早开工(日期);最早开始
earliest start data 最早开工日期
earliest starting time of activity 作业最早开始时间
earliest start time 最早开始时间;最早开工时间
ear lifter 谷穗导板
earliness of forecast 预报预见期;预报时效
Earlougher type curve 埃洛弗尔样板(解释)曲线
earlshannonite 褐磷锰高铁石
Earlumin 伊尔铝合金
Earlumin alumin(i)um alloy 伊尔铝合金
early acquisition 预早购置
early adolescence 青春期
early-age cracking 早期裂纹
early Alpine geosyncline 早阿尔卑斯期地槽

early Baroque 早期巴洛克式建筑艺术风格;早期变态式(装饰过分的)建筑
early blooming 早花
early bond 早期黏合;早期结合
early breakthrough 早期突破
early burning 早期火烧;危险期前用火
Early Caledonian geosyncline 早加里东期地槽【地】
Early Cambrian transgression 早寒武世海浸【地】
Early Carboniferous transgression 早石炭世海浸【地】
early cause 早期因素
early Christian architecture 早期基督教(会)建筑
early Christian church architecture 早期基督教堂式建筑
early Christian structures 早期基督教的建筑物
early commutation 超前换向
early compression 早期压缩;早期压力;初压(力);初始压(力)
early compressive strength 早期抗压强度;初始抗压强度
early concrete volume change 混凝土早期体积变化
early contact 先动触点
early contraction 早期收缩
early crack 早期裂缝
early cracking 早期开裂
Early Cretaceous climatic zonation 早白垩世气候分带
early curing period 早期养护周期;早期养护阶段
early cut-off 早断
early cut-off signal 提前切断信号
early days 前期
early death 早期死亡
early decline 初降
early delivery of goods 提前交货
early detection 早检测;早期探测(火灾发生时的探测)
early detonation 过早爆燃
early development 早期开发
early Devonian transgression 早泥盆世海浸
early diagenesis 早期成岩作用
early door 提早入场门(戏院)
early effect 早期效应;阿莱效应
early English architecture 早期英式建筑
early English cathedral style 早期英国天主教堂式建筑
early English church 早期英国式教堂
early English style 早期英式(建筑)
early English style base 早期英式建筑柱基
early English style window 早期英式建筑窗
Early equivalent circuit 阿莱型等效电路
early erosion 早期冲蚀
early evaluation 早期鉴定
early failure 早期失效;过早损坏;初期故障
early failure detection 早期故障检测;故障预检
early failure period 早期失效期
early fallout 早期放射性沉降物;早期沉降(物)
early fallow 早季休闲;春季休闲;春耕休闲(地)
early field life 油田开发初期
early finish date 最早结束时刻
early fire detection 早期火灾探测
early Florentine Renaissance (style) 早期佛罗伦萨文艺复兴风格
early food 初期饵料
early French Gothic (style) 早期法国哥特式建筑风格
early frost 早霜;初霜
early frost hidden 黑霜
early gate 前闸门
early generation material 早世代材料
early generation test 早代测验
Early generator 阿莱发生器
early Georgian style 早期乔治式【建】
early glacial epoch 早冰川寒冷期
early Gothic 早期哥特式【建】
early Gothic style church 早期哥特式教堂

early growth 前期生长
early hardening 首次淬火;早期硬化
early Hercynian geosyncline 早海西期地槽
early history 早期历史
early holes 早期施工的(钻)孔
early ignition 提前点火
early impoundment 初期蓄水
early in its development 初期
early injection 提前喷射
early Jurassic climatic zonation 早侏罗世气候分带
early leaf blight 早期枯叶病
early loading 最初的负载;早期加载
early magmatic ore deposit 早期岩浆矿床
early magmatic segregation 早期岩浆分凝作用
early maturing variety 早熟品种
early-mid-Triassic transgression 早中三叠世海浸【地】
early model 旧型号;初期型号
early morning boost 采暖清晨生火装置
Early Neocathaysian system 早期新华夏构造体系【地】
early offshore drilling 早期海上钻井
early offshore exploration 早期海上勘探
early origin theory of petroleum 早期生油说
early period prevalence 早期发生型
early Permian regression 早二叠世海退【地】
early Permian transgression 早二叠世海浸【地】
early phase-out 及早逐步淘汰
early Plantagenet style 早期金雀王朝式建筑
early plowing 初耕
early Pointed 早期英国尖顶建筑;早期英国尖拱式建筑
early pressure data 测压早期资料
early pressure information 测压早期资料
early pressure maintenance scheme 早期保持压力方案
early production installations 早期开发装置
early production period 开采初期
early production system 早期生产系统;早期开采系统
early profit 初期利润
early proterozoic glacial stage 早元古代冰期
early pruning 早期修剪
early radard warning 雷达预警
early radiation effect 早期辐射反应
early reaction 早期反应
early reflection 早期反射声
early Renaissance cupola 早期文艺复兴时的半球形屋顶
early Renaissance dome 早期文艺复兴时期圆屋顶
early Renaissance (style) 早期文艺复兴式【建】;早期文艺复兴时代建筑式
early retirement benefit 提前退休养老金
early Romanesque style 早期罗马式建筑
early sample 原始样品;原始样本
early sand control 早期防砂
early scoop 挖土铲斗
early (season) rice 早稻
early set(ting) 早凝
early setting cement 快凝水泥;早凝水泥;速凝水泥
early shipment 提早装船
early shrinkage 早期收缩
early simulation 早期模拟
early slag 初渣;初期渣
early spark 提前火花
early spectral type 早光谱型
early spring 早春
early spring irrigation 早春灌溉
early stability 早期稳定性
early stage 初期(阶段);早期
early stage cracking 早期形成的裂缝
early stage of failure 提早断裂期
early state 初期状态;初始状态
early state of stress 初期应力状态
early stiffening 过早变稠;早硬结
early Stone Age 旧石器时代
early strength 早期强度;初期强度;早强剂
early strength addition 早强添加剂

early strength admixture 早强（外加）剂；早强掺和剂
early strength cement 快硬水泥
early strength component (of concrete)（混凝土的）快硬剂；混凝土早强剂
early strength concrete 早强混凝土；速凝混凝土
early strengthened oil well cement 早强油井水泥
early strength gain 早期强度增长
early strength rockbolt 早强锚杆
early stress 初期应力
early summer 夏初
early summer or late spring 夏初或春末
early suppression fast-response 早期灭火快速反应
early suppression fast-response sprinkler 早期灭火快速反应洒水喷头
early termination 中止
early test 早期测验
early-time data 早期数据
early-time portion 早期段
early-time treatment 早期处理
early timing 超前成圈
early transient regime 早期不稳定阶段
early transplanting 早插
early uses 早期用途
early warning 预警；早期预警
early-warning device 早期报警装置
early-warning radar 远程警戒雷达
early-warning signal 预警信号
early-warning system 预（先）警（报）系统；早期警报系统
early-warning technique 预警技术
early-warning test （工程结构的）早期预警试验
early water breakthrough 早期见水
early watering 提早灌水
early Weichselian glacial epoch 早魏克塞尔冰期
early Wisconsin 早维斯康辛
early wood 幼材；春材；早材
early write 初期写入
early write cycle 初期写入周期
early Yanshanian geosyncline 早燕山期地槽【地】
early Yanshanian subcycle 早燕山亚旋回【地】
early youth 青年早期
earmark 明白指定（资金）用途；打上标；指定用途；打上记号；圈存资金
earmarked deposits for taxes 纳税准备金
earmarked fund quota 专用资金额
earmarked loan 专用贷款
earmarked surplus 指定用途盈余
earmarked tax 附加税
earmark fund to 拨款给……
ear marking 标以耳号；打耳标
earmarking of taxes 税款专用
ear muffs 减噪耳盖；护耳套；耳套
earned capacity value 劳动所得价值
earned capital 赚得资本
earned for ordinary stock 普通股收益
earned home payments account 受补偿支付房价的账户
earned income 劳动收入；已获收入；赚得收益；工资收入
earned income for ordinary stock 普通股能分到的收益
earned interest 已获利息；赚得利息
earned premium 满期保（险）费；已到期保险金；期满保险费；实收保险费
earned revenue 已获收入
earned surplus 营业盈余；已获盈余；赚得盈余
earnest 真实；定价金
earnest money 定金；保证金
earnest money received 已收保证金
earn foreign exchange 附加值高
earn index 获利指数
earning 所得；收入
earning after tax 扣除税金后净利
earning assets 有收益资产；收益资产
earning block 获利区块
earning capability 收入能力
earning capacity 利润率；营运吨位；收益能力；收益额；生产能力
earning capacity of traffic 运输量
earning chance 赚钱机会
earning coverage ratio 收益偿债能力比率
earning cycle 经营周期；营业循环

earning forecast 收益预测
earning life of vessel 船舶营运年限
earning observation 收益观察
earning on investment 投资收益
earning performance 盈利实况
earning power 盈利能力；赚钱能力；收益（能）力
earning power of real assets 投资利润率
earning quality 收益质量
earning rate 赚得率；收益率
earning ratio 收益率
earning report 收益报表
earning retained 收益留存
earnings 利润；盈利；业务收益；报酬；收益
earnings after tax 课税后利润；税后收益额
earnings before interest and tax 减除利息和税款前收益；付息及税前利润
earnings before tax 纳税前收益额；税前收益额
earnings capitalization method 收益资本化方法
earnings capitalized value 收益资本化价值
earnings-dividend ratio 收益股利比率
earnings on invested capital 投入资本盈利
earnings per share 每股收益；每股平均收益
earnings position 收益状况
earnings ratio 收益比率
earnings report 收益报告
earnings/rice ratio method 盈余价格比率方法
earnings standard 收益标准
earnings statement 收益表；业务收益表；损益计算书
earning surplus 营业盈余
earnings variability 收益变动性
earning well 收益井
earning yield 收益率
earnmark 专用拨款
earnmarking 专款专用
earnmarking of fund 资金专用
Earnshaw escapement 天文钟擒纵机构
ear nut 翼形螺帽
earphone 译意风；耳机；听筒
earphone adapter 耳机转接器
ear piece 耳件；耳机；耳承；受话盖
ear plug 防噪声耳塞
ear protection 护耳
ear-protection helmet 防声头盔
ear protector 护耳器
ear receiver 耳塞
earring cringle 帆耳眼环
earshot 听力范围
ears of wheat 小麦穗
ear spoon 耳匙
ear squeeze 中耳气压伤
earth 泥土；地球；土（壤）
earth age 地球年龄
earth-air current 地空电流；大地大气电流
earth-air heat pump 土壤空气热泵
earth albedo 地球反照率；地球反射率
earth alkali metal 碱土金属
earth anchor 地层锚杆；锚杆；根络；土锚（杆）；土层锚杆
earth anchorage 地锚固定
earth anchored cable system 地锚缆绳系统
earth anchored stay cable 地锚拉索
earth anchored suspension system 地锚式悬索体系
earth anchored system 地锚体系
earth and rockfill dam 土石混和坝；土石坝
earth antenna 接地天线；地下天线
earth arrester 接地火花隙避雷器；接地放电器；接地避雷器
earth arrestor 接地放电器
earth atmosphere 地球大气
earth-attenuation 地层衰减
earth attraction 地球引力
earth attraction radius 地球引力半径
earth auger 螺旋钻；螺旋形钻土器；泥土钻；麻花钻；钻土螺钻；地螺钻；取土螺钻；挖穴机；土（螺）钻
earth auger extension 地螺钻延伸杆
earth auger truck 钻土器卡车
earth backfill 回填土
earth backing 回填土；还土；覆土；培土；复土
earth back pressure 土壤反压力
earth bag 土包；砂袋；砂包
earth ballast 土道砟
earth ball cutting 带土团扦插

earth balsam 土香胶（脂）
earth bank 路堤；堤岸；土堤（岸）
earth banking 筑土埂；土埂
earth banking apparatus 筑埂器；筑埂机器
earth bar 地线棒；接地棒
earth base 地面基地；土基
earth based 地面的
earth based coordinates 地球坐标
earth based coordinate system 地球坐标系
earth based station 地面站
earth-baseplate resonance 地面底板共振
earth beam test 土梁试验；上梁试验
earth bearing strength 接地耐力；地耐力
earth bench 土工作台
earth berm 截水土堤
earth blanket （堆石坝的）黏土防渗面层
earth board 犁板（平路机铲刀）；翻土工具
earth body 土体
earth bolt 土锚
earth bore 地质钻孔
earth borer 螺旋钻机；钻土器；挖穴机；土钻
earth boring auger 麻花钻取土器；地钻；泥土钻
earth boring machine 平巷掘进机
earth boring outfit 钻土设备；钻土器具
earth boring tools 钻探工具
earth buckling 地壳弯曲
earth building 抗震建筑（物）；生土建筑
earth bulge 地面隆起
earth bus 接地母线
earth buttress 土支撑
earth canal 土渠
earth capacitance 对地电容；地电容
earth capacity 大地电容
earth carrier 运土筐
earth cementation 土的黏结作用；土的黏结性；土壤固化
earth centered ellipsoid 地心旋转椭圆体；地球质心椭球
earth centered inertial coordinate 地心惯性坐标
earth central angle 地球中心角
earth channel 土渠
earth chemical analysis 地球化学分析
earth circuit 接地回路
earth clamp 接地夹（子）；地线夹（头）
earth clamping machine 堆藏覆土机
earth clip 接地夹
earth closet 便坑；旱厕（所）；土厕；撒土厕所
earth closet receptacle 干土厕所容器
earth coal 泥煤；土褐煤
earth cofferdam 土围堰
earth coil 地磁感应线圈
earth colo(u)r 颜料土；矿物颜料
earth column 土柱
earth compacting machine 土壤压实机；土壤夯实机
earth compaction 土壤压实；土壤击实；土壤夯实
earth compact work 夯土工程
earth concept 土的概念
earth concrete 掺土混凝土
earth conducting cable 接地电线
earth conductivity 大地电导率
earth conductor 地线
earth conductor cable 接地电线
earth conduit 陶管；瓦管
earth connection 接地工程；接地线
earth connector 接地器
earth conservancy 干厕所；土壤保持
earth conservancy receptacle 干土厕所容器
earth consolidation 土壤固结（作用）
earth contact 接地
earth contact house 土中住宅
earth contact pressure 接地压力
earth-continuity conductor 接地连接导线
earth contraction 地球收缩
earth cord 接地软线
earth core （电缆的）接地心线；接地软线；地核
earth core rockfill dam 黏土心墙堆石坝
earth coupling 大地耦合
earth cover 土壤覆盖；土壤封面
earth-covered construction 覆土建筑
earth covering 返填土；覆土；覆盖土厚度；覆盖土层
earth-covering pressure 覆盖土压力
earth creep 滑坡；大地蠕动；大地蠕变；土体蠕动；土崩；坍方

earth crust 地壳
earth crustal warping 地壳翘曲
earth crust dip monitor 地倾斜观测
earth crust pulsation 地壳脉动
earth crust stress 地壳应力
earth crust structure 地壳构造
earth current 接地电流;泄地电流;地电(流);大地电流;土壤电流
earth current meter 泄地电流表;地电流测量器
earth current resistor 接地电流电阻器
earth current storm 大地电流暴
earth curvature 地球曲率
earth-curvature correction 地球曲率改正
earth curve 地壳弯
earth cutting 挖土
earth cutting ability 土壤切削能力
earth dam 土坝
earth dam ag(e)ing 土坝老化
earth dam compaction 土坝压实
earth-dammed reservoir 土蓄水池
earth dam on pervious foundation 透水地基上的土坝
earth dam paving 土坝护面
earth-damp concrete 水灰比恰好的混凝土;干硬性混凝土
earth dam with clay core 黏土心墙坝
earth dam with inclined clay core 黏土斜心墙(土)坝
earth day 地球日
earth deformation 地形变;土变形
earth densification 土壤密实;土壤压实;土壤夯实
earth deposit 沉积土层;弃土堆;土堆
earth-detective radar 探区雷达
earth detector 接地指示器;接地检测器;通地指示器
earth dike 土堤
earth displacement 土壤位移
earth ditch 土沟
earth drill 泥土钻;钻土器;钻土机;表土钻机;挖坑机;土钻
earth drill method 钻土法
earth-dry mortar strength test 硬练砂胶强度试验法
earth dynamic(al) ellipticity 地球力学椭率
earth dynamics 地球动力学
earthed 接地的
earthed brush 接地电刷
earthed cathode 接地阴极;阴极接地的
earthed circuit 接地电路;单线回路
earthed collecting plate 接地集尘板
earthed concentric wiring 接地同轴电缆
earthed concentric wiring system 接地同轴电缆制
earthed conductor 地导体
earthed contact 接地触点
earthed continuity conductor 接地导线
earthed network 接地网路
earthed neutral 接地中线;接地中点
earthed neutral conductor 接地中线导体
earthed overhead line 架空地线
earthed pole 接地杆
earthed system 接地装置;接地系统
earthed telegraphy 单线电报系
earth elasticity 地球弹性
earth electric(al) field 地电场
earth electricity 地电
earth electrode 地线;接地(电)极;地电极
earth electrode system 地电极系统
earth ellipsoid 地球椭圆体;地球椭球(体)
earth embankment 路堤;土堤;填土
earth embankment dam 土堤;土坝;路堤
earth embankment laying pipeline 土堤敷设管线
earth embankment stadium 有土堤的运动场
earthen 泥土的;土制的;陶制的
earthen basin 土池
earthen brick 黏土砖
earthen container 陶制容器
earthen flood bank 土筑防洪堤
earthen glaze 黏土釉;黄土釉
earthen kiln 土窑
earthen mound 土丘
earthen pipe 瓦管
earthen pit 土质泥浆池
earthen pond 泥池
earthen road (无路面的)土路
earthen settling tank 土沉淀池
earthen storage tank 陶制储[贮]罐

earthen structure 土工构筑物;土工构造物;土工建筑(物)
earthen tile 黏土瓦
earthen tile pavement 陶砖铺面
earthen tower 陶制蒸馏塔
earthenware 缸瓦器;瓦器;陶器;陶瓷;硬质精陶
earthenware body 陶坯体
earthenware cable cover 陶土电缆覆盖;陶制电缆护盖
earthenware clay 陶器黏土
earthenware container 陶制容器
earthenware deposit 陶粒黏土矿床
earthenware duct 陶路管;陶瓷管道
earthenware filter pipe 陶土滤水管;陶制滤管
earthenware glaze 陶器釉
earthenware glazed finish 陶土釉面修饰;陶面釉
earthenware jars 陶器罐
earthenware mosaic 陶土马赛克;陶土锦砖;马赛克;陶制马赛克
earthenware or glass carboy 陶质酸坛或玻璃瓶
earthenware paving block 铺地陶砖
earthenware paving tile 铺地陶砖
earthenware pipe 缸瓦管;瓦管;陶管
earthenware porcelain receptacle 陶瓷容器
earthenware ring 瓦圈
earthenware shale deposit 陶粒页岩矿床
earthenware slab 陶器板
earthenware small-sized mosaic 陶制小块马赛克
earthenware tile 黏砖;陶制面砖
earthenware tile pavement 陶面砖铺地;陶砖铺面
earthenware tower 陶制塔
earthenware vessel 陶制容器
earthenware wall tile 陶土墙面砖;陶制墙面砖
earth equatorial plane 地球赤道面
earth evolution model 地球演化模式
earth excavating machine 挖土机
earth excavation 挖土(方);土方开挖
earth excavation works 挖土工程
earth excavator 掘土机
earth expanding theory 地球膨胀理论
earth exploration satellite 地球勘测卫星
earth fall 滑坡;地崩;大塌方;土塌;土崩
earth family element 土族元素
earth fault 接地故障
earth fault current 接地故障电流
earth fault protection 接地故障保护(装置);接地保护;防止故障性接地
earth fault relay 接地故障继电器;接地继电器
earth feed ditch 引水土沟
earth-field magnetometer 地磁场磁强计
earth fill 填土(方);填方
earth-fill bag 土袋
earth-fill cofferdam 土围堰
earth-fill compaction inspection 回填土密实度检验
earth-fill dam 土坝;堆土坝;填土坝
earth-fill dam paving 土坝护面
earth-filled bag 填土袋;填土包
earth-filled dam 土坝
earth-filled stone arch bridge 填土实腹石拱桥
earth-filled straw bag cofferdam 装土草袋围堰
earth filling 填土处理
earth-fill puddle core dam 胶土核心坝
earth-fill structure 土筑结构
earth-fill timber dam 填土木坝
earth filter 硅藻土过滤器;硅藻土过滤池;土滤池;陶土滤器
earth filtering 地层滤波
earth filtering effect 地层滤波效应
earth fissure 地裂隙;地裂缝
earth-fixed Cartesian coordinate 固地笛卡尔坐标
earth-fixed coordinate system 地球固定坐标系统;地球定点坐标系统;地固坐标系(统)
earth-fixed satellite coordinate 地球固定卫星的坐标
earth-fixed satellite position 地球固定卫星的位置
earth flax 高级石棉;石质纤维
earth flood bank 防洪土堤
earth floor 黏土地面;泥地
earth flow 泥(石)流;土(石)流;土崩
earth formation 土的形成;土的构成
earth-forming element 构成地球的元素
earth foundation 土质地基;土基
earth-free 不接地(的)
earth-free system 不接地系统

earth free vibration 地球自由振动
earth furrow 垄沟
earth gas 天然气
earth grab(bing) bucket 挖土机抓斗;抓土斗
earth grade 土路基
earth gravimetry 地球重力学
earth gravitational model 地球重力位模型
earth gravity 地心引力
earth gravity determination 地球重力测定
earth gravity field 地球重力场
earth gravity model 地球重力位模型
earth grid 接地栅极
earth hauling 运土
earth heaving 地面隆起
earth history 地质历史(学);地史学
earth hoe 挖土锄
earth holography 大地全息摄影术
earth house 土屋;土房;土筑房
earth hummock 土丘;土岗;土堆
earth humus 腐殖土;土壤腐殖物质
earth impedance 对地阻抗
earth induction 地磁感应
earth induction compass 地磁感应罗盘;磁罗经
earth inductor 地球感应器;地磁感应仪;地磁感应器
earth inductor compass 地磁感应罗盘
earthing 接地;盖上;培土
earthing bar 接地棒
earthing blade 培土铲
earthing brush 接地电刷
earthing brusher equipment 接地装置【铁】;接地装置
earthing bus 接地干线
earthing busbar 接地母排
earthing cable 接地电缆
earthing clamp 接地夹子;接地环;接地端子
earthing clip 接地线夹;地线夹子
earthing conductor 接地导体
earthing connection 接地
earthing contact 接地触点
earthing cover(ing) 土被
earthing device 接地装置;码头静电接触装置
earthing equipment 接地装置
earthing fixture 接地卡具
earthing grip 接地端
earthing installation 接地装置
earthing isolator 接地隔离开关
earthing lead 接地电极线索;地线导管
earthing method of system neutral 系统中性点接地方式
earthing network 接地网
earthing of frame 接机座
earthing of lightning arrester 避雷器接地装置
earthing plate 接地平板;接地板
earthing pole 接地棒
earthing protection 接地保护
earthing relay 接地继电器
earthing resistor of neutral point 中性接地电阻器
earthing rod 接地极;接地棒;接地柱
earthing screw 接地螺钉
earthing switch 接地开关;接地转换开关;接地电闸;接地开关
earthing terminal 接地端(子);接地线端;接地夹子;接地电线接头
earthing transformer 接地变压器
earthing-up plow 培土犁;培土器
earth insulation 主绝缘
earth interior 地球内部
earth-ionosphere cavity resonance 地电离层空腔共振
earth ionosphere waveguide 地电离层波导
earth isolator 接地刀闸
earth kiln 土窑
earth lake 白土色淀
earth lamp 检漏灯(检查漏电用)
earth layering 地球分层
earth-layer propagation 地层传播
earth lead 接地线;接地导线
earth leakage 接地漏电;通地漏电
earth-leakage circuit-breaker 接地漏电断路器;接地保护断路器
earth-leakage current 漏地电流;接地漏电(电流);漏电电流
earth-leakage current circuit breaker 接地漏电断路器

earth-leakage detector 漏电探测器
earth-leakage indicator 漏地电流指示器
earth-leakage-meter 对地泄漏测量计
earth-leakage protection 接地漏电防护
earth leakage relay 对地泄漏信号继电器
earth-leakage trip 接地保护自动断路器
earth levee 土堤
earth level(l)er 平土器
earth light 地球反照(光);地光;大地光;新月灰光
earth line 接地(母)线;公共接地线;地线;地下电缆线路
earth-lined 表面抹土的
earth lining 抹面土层
earth-link box 地线盒
earth load 土重
earth loader 运土机;运土工;装土机
earth lodge 木架土屋
earth-loosening equipment 松土设备
earth loss 地层损耗
earth lug 接地连接板
earth lurch 地倾斜
earthly 地球的
earthly environment 地球环境;大地环境
earthly heat 地热
earth magnetic field 地磁场
earth magnetic field balance 地磁场平衡器
earth magnetic(field)effect 地磁场影响
earth magnetism navigation 地磁导航
earth magnetosphere 地球磁层
earth major radius 地球长半径
earth mantle 地幔
earth manual 土壤手册;土工手册
earth mass 土块;土堆;土体
earth mass calculation 土方计算
earth mat 接地网;地网
earth material 接地材料;泥土;泥砂土石;土料
earth materials laboratory 土壤材料试验室;土壤实验室
earth medium 大地介质
earth membrane 黏土防渗墙;黏土防渗层 media
earth metal 土金属
earth metal sheet 接地金属板(材)
earth meteorite crater 地球陨击坑
earth micropulsation 地壳微脉动
earth microtremor measurement 地脉动测
earth model 地球模型
earth model of lithologic(al) characters 岩性分析的地层模型
earth moist concrete 水灰比恰好的混凝土;干硬性混凝土
earth-moon system 地月系统
earth mortar 黏土砂浆;土灰浆
earth mound 土丘;土墩;土堆
earth movement 岩层移动;造山运动;地球运动;地壳运动
earth mover 运土机(械);堆土机;大型推土机;推土机;土方机械;大型挖土机
earth mover equipment 运泥工具
earthmover tire 运土机轮胎;挖土机轮胎;推土机轮胎
earthmoving 土方搬运;运土
earthmoving contracting firm 运土承包公司
earthmoving engineer 土方工程师
earthmoving equipment 运土机械;铲土运输机械;运土设备;堆土设备;推土设备
earthmoving for landscape purposes 以景观为目的的土方工程
earthmoving gear 运土传动装置
earthmoving job 土方工程
earthmoving machine 运土机(械);土方工程机械
earthmoving machinery 土方机械;铲运机械
earthmoving operation 运土工作;运土工序;运土作业
earthmoving plant 运土机械设备;运土机(械);运土设备
earthmoving scraper 铲运机;运土刮土机;运土铲土机;铲土运土机
earthmoving train 运土列车
earthmoving vehicle 铲土运输机械
earthmoving works 土方工程;土方搬运(工程)
earth mulch 土盖
earth network 接地网;地网
earth neutral system 中性点接地系统
earth noise 地噪声

earth observation satellite 地球观察卫星;地球观测卫星
earth observatory satellite 地球观测卫星
earth of casing 外壳接地
earth oil 原油;石油;地沥青
earth(o)meter 接地测量计;接地检查器;兆欧计;高阻表
earth(o)metering 接地测量
earth-orbit 绕地(球)
earth-orbital imagery 地球轨道成像
earth-orbital operation 绕地(球)轨道运行
earth-orbital photography 地球轨道摄影术
earth-orbital rendezvous 绕地(球)轨道会合
earth-orbiting satellite 地球轨道卫星;地球观测卫星
earth oscillation 地球振动
earth ozone layer 地球臭氧层
earth pad 土台;土基
earth path 土筑土路;(无路面的)土路
earth path indicator 地面航迹指示器
earth patriotism 对地球的忠心
earth-paved bridge 土桥
earth physics 大地物理学
earth pigment 矿物颜料;地质颜料;天然(矿物质)颜料
earth pile 土桩
earth piled hill 土山
earth pillar 土柱;土墩
earth pin 接地销针
earth pipe 瓦管
earth pit 地下温室;地下温度;地窖;取土坑;土坑
earth pitch 矿柏油;地沥青;软沥青;软地沥青;土沥青
earth placement 铺土
earth plasmasphere 地球等离子体层
earth plate 接地电极板;接地(导)板;支腿底座;地线铜线头;地线铜版
earth platform 土平台
earth plug 带接地触点的插头
earth point 接地点
earth polar system of coordinates 地极坐标系
earth poles 地极
earth potential 地电位;地电势
earth potential difference 对地的电位差;地电位差
earth potential field 地引力势场
earth pressure 岩层压力;地压;地层压力;土压(力)
earth pressure at-rest 静土压力;静止土压力
earth pressure balance 土压力平衡
earth pressure balanced shield machine 土压力平衡盾构机
earth pressure balance shield 土压(力)平衡盾构
earth pressure balance shield machine filled with betonite or other high efficiency blister 加泥式土压平衡盾构机
earth pressure balance shield tunneling 土压力平衡盾构施工
earth pressure cell 土压力计;土压力盒;土压(力)元件
earth pressure coefficient 土压力系数
earth pressure computation 土压力计算
earth pressure distribution 土压力分布
earth pressure during earthquake 地震土压(力)
earth pressure ga(u)ge 土压力计
earth pressure line 土压力线
earth pressure measurement 土压力测定
earth pressure measuring cell 土压力量测压力盒
earth pressure of cohesion soil 黏性土土压力
earth pressure of cohesiveless soil 无黏性土土压力
earth pressure probe 土压力计
earth pressure theory 土压力理论
earth pressure wedge 土压力楔(形体)
earth-probing radar 地质探测雷达
earth proof 抗震的
earth-proof construction 抗震建筑
earth protection block relay 接地保护闭锁继电器
earth protection installation 接地保护装置
earth puddle work 夯土工程
earth pulsation 大地脉动
earth pyramid 土柱
earthquake 地震
earthquake acceleration 地震加速度
earthquake accelerograph 地震加速度(记录)仪
earthquake action 地震作用

earthquake activity 地震活动(性)
earthquake allowance 地震裕度
earthquake analysis 地震分析
earthquake area 地震范围;地震区域
earthquake area map 地震面积图;地震范围图
earthquake attenuation 地震衰减
earthquake axis 震轴;地震轴
earthquake bedrock 地震基岩
earthquake behavior 抗震性能
earthquake belt 地震地带
earthquake belt intensity 场区烈度
earthquake bracing 地震安全支撑;抗震斜撑
earthquake calamity 地震灾害
earthquake capacity 抗震能力
earthquake catalogue 地震目录
earthquake catastrophe 震害;地震灾害
earthquake center 地震中心;地震源;震中;震源
earthquake coefficient 地震系数
earthquake coefficient method 地震系数法
earthquake consequence 地震后果
earthquake construction 抗震建筑;地震建筑
earthquake control 地震控制;地震防治
earthquake controlling fault 控震断裂
earthquake countermeasure 防震措施
earthquake country 地震区
earthquake cycle 地震周期
earthquake damage 地震震害;地震损失;震害;地震损害;地震破坏
earthquake damage effect 震害效应
earthquake damage increment 震害增量
earthquake damage index 震害指数
earthquake damage insurance 地震损失保险
earthquake damage ratio 地震破坏率
earthquake damage survey 震害调查
earthquake damage susceptibility 震害敏感性
earthquake damaging phenomenon 震害现象
earthquake data 地震资料
earthquake design 抗震设计;地震设计;防震结构设计
earthquake disaster 地震震害;地震灾害;地震危害(性)
earthquake disaster mitigation 减轻地震灾害
earthquake disaster potential 发生地震灾害可能性
earthquake disaster preparedness 震前防灾准备
earthquake dislocation 地震位错
earthquake dispersal area 地震疏散用地
earthquake displacement 地震位移
earthquake displacement field 地震位移场
earthquake distribution 地震分布
earthquake due to collapse 崩陷地震;塌陷地震
earthquake dynamic earth pressure 地震动土压力
earthquake earth pressure 地震时土压力
earthquake effect 地震影响;地震效应
earthquake effect field 地震效应场
earthquake energy 地震能量
earthquake engineering 抗震工程学;工程地震学;地震工程(学)
earthquake epicenter 震中;地震震中
earthquake excitation 地震激发
earthquake excitation factor 地震干扰因子
earthquake exploration 地震探查
earthquake factor 地震系数(重力加速度百分数)
earthquake failure 地震破坏
earthquake fault 地震断层
earthquake faulting 地震断层作用
earthquake fault plane 地震断层面
earthquake feature 抗震特征
earthquake-felt area 地震有感范围
earthquake filtering 地震滤波
earthquake fire 地震火灾
earthquake fire hazard 地震引起的火灾危险
earthquake foci 震源
earthquake focus 地震源
earthquake force 地震力
earthquake forecast 地震预测;地震预报
earthquake foreshock 前震
earthquake frequency 地震频度
earthquake frequency increase 地震频度增大
earthquake frequency spectrum 地震频谱
earthquake fundamental intensity 地震基本烈度
earthquake gas 地气
earthquake-generating mechanism 发震机制
earthquake ground motion 地震动
earthquake ground motion characteristics 地震

动力特性
earthquake hazard 震害;地震震害;地震危险性;地震危害(性)
earthquake hazard evaluation 地震危险分析
earthquake hazard map 地震危险(区划)图
earthquake hazard mitigation 地震危害减轻
earthquake hydrodynamic(al) force 地震动水作用力
earthquake hypocenter 震源
earthquake influence field 地震影响场
earthquake in intermit period 地震平静期
earthquake input 地震输入
earthquake insurance 地震保险
earthquake intensity 烈度;地震强度;地震烈度
earthquake intensity scale 地震烈度分级;地震强度计;地震烈度表
earthquake investigation 地震调查
earthquake light 地震光
earthquake list 地震目录
earthquake load(ing) 地震荷载;地震负荷;抗震荷载;地震土压(力)
earthquake location 地震位置
earthquake magnitude 震级;地震(震)级;地震等级
earthquake mechanics 地震力学
earthquake mechanism 地震机制
earthquake meter 地震记录仪
earthquake migration 地震迁移
earthquake magnitude 震级
earthquake model 地震模型
earthquake modification 地震变化
earthquake movement 震时活动
earthquake non-resistive building 非抗震建筑
earthquake non-resistive construction 非抗震结构
earthquake observatory 地震气象台;地震观测站
earthquake observing station 地震观测站
earthquake occurrence model 发震模型
earthquake of distant origin 远震;远源地震
earthquake of intermediate depth 中深地震
earthquake origin 震源
earthquake oscillation 地震激烈振荡;地震震荡
earthquake oscillation mode 地震震型
earthquake parameter 地震参数
earthquake participation factor 地震参与因数
earthquake period 地震周期;地震时期
earthquake periodicity 地震周期性
earthquake precautionary criterion 抗震设防标准
earthquake precautionary intensity 抗震设防烈度
earthquake precursor 地震先兆现象;地震前兆
earthquake prediction 震情预报;地震预测;地震预报
earthquake pregnant fault 孕震断裂
earthquake pregnant tectonic systems 孕震构造体系
earthquake premonitory phenomenon 地震预兆
earthquake prevention 地震预防
earthquake prevention subsidies 抗震加固补助费
earthquake process 地震过程
earthquake prone area 易发震区;地震活动区
earthquake prone region 地震多发地区;易发震区;多震区;地震危险区
earthquake-proof 抗震(的);抗地震;耐地震的;防(地)震的
earthquake-proof building 抗震建筑物
earthquake-proof construction 抗震结构;抗震建筑;抗震工程;抗震构造;避震建筑;防震建筑(物)
earthquake-proof design 耐震设计
earthquake-proof foundation 抗震基础;抗地震基础;防震基础
earthquake-proof(ing) technique 抗震技术
earthquake-proof joint 抗震接头;防震缝;耐震接头
earthquake proofness 抗震性
earthquake-proof site 抗地震试验场
earthquake-proof structure 抗震建筑(物)
earthquake propagation 地震传播
earthquake protection 抗震设防
earthquake province 震区
earthquake rattle 地震声
earthquake record 震波图;地震记录
earthquake recording method 地震记录法
earthquake recording system 地震记录系统
earthquake recurrence interval 地震重现间隔;地震重复间隔
earthquake recurrence map 地震重复性图

earthquake region 震区;地震区(域)
earthquake region map 地震区域图
earthquake relief headquarters 抗震救灾指挥部
earthquake resistance 抗地震性(能);抗震(的);抗地震
earthquake resistance design 抗地震设计
earthquake resistance feature 抗震特点
earthquake resistance of structure 结构物抗震性能
earthquake resistance structure 抗震结构
earthquake resistant 抗(地)震的;耐震的;防震的;耐地震的;防地震的
earthquake resistant analysis 抗震分析
earthquake resistant bearing 抗震支座
earthquake-resistant behavio(u)r 耐震性能;防震性能;抗震性能
earthquake resistant code 抗震规范
earthquake resistant component 抗震部件
earthquake resistant design 抗震设计
earthquake resistant design code 抗震设计规范;耐震设计规范;防震设计规范
earthquake resistant measure 抗震措施
earthquake resistant plate 抗震拉板
earthquake resistant regulation 抗震规则
earthquake resistant structure 抗震结构;耐震结构;防震结构
earthquake resistant wall 抗震墙
earthquake resistant wall column 抗震墙柱
earthquake-resisting 耐地震的
earthquake-resisting behavior 抗震性能
earthquake-resisting element 抗震构件
earthquake-resisting wall 抗震墙
earthquake resistivity 抗震性
earthquake response 地震响应;(建筑物的)地震反应特性
earthquake response analysis 地震反应分析
earthquake response function 地震反应函数
earthquake response spectrum 地震响应波谱曲线;地震反应谱
earthquake return period 地震重现期
earthquake risk 地震危险性
earthquake risk region 地震危险区
earthquake risk zoning 地震危险区划
earthquake rupture surface 地震破裂面
earthquake safety facility 防震安全设施
earthquake scale 地震级别;地震震级
earthquake scarp 地震崖
earthquake sea wave 地震海啸
earthquake seismology 测震学;天然地震学
earthquake separation 抗震缝
earthquake sequence 地震序列
earthquake series 震群;地震系列
earthquake shadow 震影
earthquake shelter 抗震棚
earthquake shock 地震冲击
earthquake simulating shaking table 模拟地震振动台
earthquake simulator 地震模拟器
earthquake slide 地震滑坡
earthquake smell 地味
earthquake sound 地震声;地震声;地声
earthquake source 震源;地震震源
earthquake source dynamics 震源动力学
earthquake source mechanism 震源机制;震源机理
earthquake spectrum 地震谱
earthquake station 地震测站
earthquake statistic(al) prediction 地震统计预报
earthquake statistics 地震统计学
earthquake stimulation 地震激发
earthquake strength 地震强度
earthquake stress 地震应力
earthquake subsidence 震陷(量)
earthquake survey 地震调查
earthquake swarm 震群;地震群;微震
earthquake tectonic line 地震构造线
earthquake tectonic movement 地震构造运动
earthquake telemetering system 地震遥测系统
earthquake telemetry 地震遥测术
earthquake trace 地震记录迹线
earthquake tremor 地震颤动;地颤动;微震
earthquake triggering 地震触发
earthquake triggering fault 发震断裂;发震断层;发展断裂
earthquake triggering part 发震部位
earthquake triggering tectonic systems 发震构

造体系
earthquake undersea bottom 海底地震
earthquake uplift 地震隆起
earthquake vibration 地震颤动
earthquake volume 震源体积;震量
earthquake warning 地震警报
earthquake warning apparatus 地震警报装置
earthquake warning system 地震警报系统
earthquake watch 地震监视
earthquake wave 地震海浪;地震波
earthquake weather 地震天气;地震前天气
earthquake zone 震区;地震区;地震(地)带
earth quantity 土方工程量;土方量
earth quantity determination 土方工程量测定;土方量测定
earth radiation 地球辐射(包括大气辐射);地面辐射;大地辐射
earth radiation budget 地球辐射收支;地球辐射平衡
earth radiation budget experiment 地球辐射收支试验
earth radio radiation 地球射电辐射
earth rammer 夯具;夯土机;土夯;夯土器
earth ramp 土匝道;土坡道
earth rate 地球自转速率
earth rate correction 地转速改正;地球转速修正
earth rate unit 地球角速度单位;地球速率单位
earth refilling 回填土
earth refining 陶土精制
earth reflection 地球反射;大地反射
earth reflection produced by ultrasonic inspection 超声波检验产生的大地反射
earth reservoir 土质水库;土水库;土水池
earth resistance 接地电阻;地(面)电阻;大地电阻
earth resistance tester 接地电阻测量仪
earth resistivity 大地电阻率;地电阻率
earth resistivity anomaly 地电阻率异常
earth resource data interactive processing system 地球资源数据人机联系处理系统
earth resource information 地球资源信息
earth resource observation and information system 地球资源观测与信息分析系统
earth resource observation satellite 地球资源观测卫星
earth resource observation satellite data center 地球资源观测卫星数据中心
earth resource observation system 地球资源观测系统
earth resources 地球资源
earth resource satellite image 地球资源卫星图像
earth resources satellite 地面资源卫星;地球资源卫星
earth resources technology satellite 资源技术卫星
earth resource technologic(al) satellite 地球资源技术卫星
earth resource technologic(al) satellite photo(graph) 地球资源技术卫星像片
earth resource technology satellite 地球资源技术卫星
earth resource terrestrial satellite 地球资源卫星
earth resource test satellite 地球资源试验卫星;地球资源观测卫星
earth response function 地响应函数
earth-retaining 挡土的
earth-retaining structure 挡土结构;挡土构筑物;挡土建筑物
earth-retaining support 挡土支护
earth-retaining wall 挡土墙
earth return 接地回线;接地回路;地(电)回路
earth return brush 接地回流刷
earth return circuit 接地回(流电)路;地回电路;单线送电大地返回电路
earth return system 地回路制;地线回路
earth revolution 地球公转
earth ridge 土垄
earthrise 地出
earth road 土路
earth roadbed 土路基
earth rock blasting 土石方爆破
earth rock cofferdam 土石围堰
earth rock dam 土石坝
earth rock excavation 土石方开挖
earth rockfill 土石填方
earth rockfill dam 土石坝

earth rock longitudinal cofferdam 土石方纵向围堰
earth rock transversal cofferdam 土石方横向围堰
earth rod 接地极;接地棒;地极
earth rotation aperture synthesis 地球自转综合
earth rotation parameter 地球自转参数
earth rotation rate 地球自转速率;地球转速
earth rotation synthesis array 地球自转综合天线阵
earth rotation theory 地球转动说
earth rubber 泥胶
earth runway 土跑道
earth satellite 地球卫星
earth satellite orbital image 地球卫星轨道像片
earth satellite rocket 地球卫星火箭
earth satellite vehicle rocket 地球卫星运载火箭
earth('s) axis 地轴
earth's center 地心
earth science 地学;地球科学
earth science laboratory 大地科学实验室
earth's compliance factor 地球顺从系数
earth's core 地心;地核
earth scraper 刮土机;铲运机;铲土机
earth screen 接地网;接地屏蔽;地网
earth screw 麻花钻;钻土螺钻;钻土机;取土样的麻花钻;挖穴钻;土钻
earth's crust 地壳
earth sculpture 地面刻蚀
earth sea dike 土海堤
earth's ellipticity 地球椭(圆)率;地球扁率
earth's energy budget 地球能量平衡
earth sensing relay 接地敏感继电器
earth sensor 地球传感器
earth's formation 岩层
earth's geomagnetism 地磁
earth's graticule 地理坐标网
earth shadow 地球影
earth shaking 极其重大的
earth shape 地球形状
earth shell 地壳
earth-sheltered home 土筑住宅;生土住宅
earth-sheltered housing 掩土住房
earth shield 接地屏蔽
earth shielded 接地屏蔽的;接地隔离
earth shift 地位移;地层移动
earth shine 地球反照(光);大地光
earth shock 地震
earth shoulder 未加固的路肩;土路肩
earth shovel 挖土铲
earth silicon 二氧化硅;硅石
earth's layers 地球圈层
earth slide 滑坡;大塌方;土体滑动;土滑(坡);土层滑动;(地层结构上的)塌方
earth slip 滑坡;大塌方;土体滑动;土滑(坡);土层滑动;(地层结构上的)塌方
earth slope 土坡
earth slope stability 土坡稳定(性)
earth slope stability analysis 土坡稳定性分析
earth's magnetic dip angle 地磁倾角
earth's magnetic field 地球磁场
earth's magnetism 地磁
earth's mean axis of rotation 地球平均旋转轴
earth's mean coordinate system 地球平坐标系
earth's nucleus 地核
earth solidification 土的凝固(作用);土壤固化
earth's orbit 地球(公转)轨道
earth sound 地声
earth-sound inspecting meter 地音探测仪
earth's ozone layer 地球臭氧层
earth spaceship 地球飞船
earth spheroid 地球扁球体
earth spin 地球自转
earth spur dike 土钉坝
earth's radiation balance 地球辐射平衡
earth's radius chart 地表放射线图
earth's reflected radiation drag 地面反射的辐射阻力
earth('s) rotation 地球自转
earth('s) rotation vector 地球自转矢量
earth's shadow 地影
earth's sphere 地球
earth's sphericity 地球球形度
earth('s) spheroid 地球椭球体
earth('s) spin vector 地球自转矢量
earth's stabilization 地球稳定
earth's surface 地球表土层;地球表面;地面;地表(面)

earth's survey satellite 地球观测卫星
earth stability 地面稳定
earth stabilizer 稳定土壤剂;土壤稳定剂
earthstar 地星
earth station 地球站;地面(通信)站;地面(电)台
earth-station polarizer 地球站极化器
earth-station receiving antenna 地球站接收天线
earth's temperature regime 地球热状态;地球的温度状态
earth's thermal balance 地球热量平衡
earth stone 贴地石座
earth storage 土油池
earth strain 地应变;地球应变
earth stranded wire 接地钢丝绳
earth stratosphere 地球平流层
earth structure 土工构造;地球结构;地球构造;土结构;土工结构物;土工建筑(物)
earth-substance 地球物质
earth supply ditch 土水渠
earth-supported structure 挡土结构;土支承结构(物)
earth-supporting structure 土支承结构(物)
earth surface 地球表面;地面;地表
earth surface collapse 地表塌陷
earth surface environment 地表环境
earth surface road 土铺路面
earth surface subsidence 地表沉陷
earth surface vibration drill 地表震动钻机
earth survey 土地测量
earth survey satellite 地球观测卫星
earth's water balance 地球水分平衡
earth's way 地球向角
earth swell 胀方
earth switch 接地开关;剩余电荷放电装置
earth synchronous orbit 地球同步轨道
earth synchronous satellite 地球同步卫星
earth system 地线系统
earth table (墙基、柱基、台基等的)贴地层;土底座;贴地石座
earth tank 饮畜塘;土水库;土水池
earth tap 接地抽头
earth tectoshphere 地球构造圈
earth temperature 地温
earth terminal 接地接线柱;接地端;地线接线柱;地线接头
earth termination network 接地端子网络
earth terrain camera 地表摄像机
earth test 接地试验;电路接地试验
earth tester 接地检查仪;接地电阻测试器
earth thermometer 地温计
earth thrust 土推力
earth tidal correction value 固体潮改正值
earth tide 陆潮;固体潮;地球潮汐;地潮
earth tide anomaly 固体潮异常
earth tide correction 固体潮校正
earth tide effect 固体潮影响
earth tide gravimeter 固体潮重力仪
earth tide gravity meter 地球潮汐重力仪
earth tide table 固体潮值表
earth tilt 地面倾斜
earth-to-moon laser ranging method 地月光测距法
earth transmission characteristic 地层透射特性
earth traverse 土的横向保护物
earth tremor 地震预兆;地层微脉动;地颤动;微震;微地震
earth tunnel 土质隧道;土层隧道
earth-type 陶制的
earth-type filter 陶制过滤器;陶类滤器
earth-type resin 土类树脂
earth up 作垄;培土
earth upper mantle 上地幔
earth volume 土方量
earth wall 土墙;土坯墙
earth wall bracing 土壁支撑
earth wall construction method 地下连续墙施工法
earth ward 向地球方向
earth ware pipe 陶土管
earthware tile pavement 陶砖铺面
earthwatch 地球观察
earth watch program(me) 全球环境监测计划
earth water 泥浆水;硬水
earth water proofing 水土防护;抗水土壤;防地下水渗漏
earth water scale-producing water 硬水
earth wave 地震波;地波
earth wax 地蜡
earth wind zone 地球上的风带
earth wire 接地线;地线
earth wire clamp 地线线夹
earth wire clamp for busbar 汇流排地线线夹
earth wire connecting clamp 接地连接线夹
earth wire of lightning arrester 避雷器接地线
earth wire of limited resistance 有限阻力的接地线
earth wire termination 架空地线下锚
earthwork 土方作业;土方(工程);土制艺术品;土石方工程;土工
earthwork adjustment 土方调配
earthwork balance 土方平衡
earthwork balance factor 土方平衡系数
earthwork balance sheet 土方平衡表
earthwork balancing chart 土方平衡表
earthwork blasting 土石方爆破
earthwork blasting procedure 土石方爆破法
earthwork cofferdam 土围堰
earthwork contract section 土方承包班组
earthwork control statistics 土方工程控制统计
earth work cross-section 土石方横断面图
earthwork dam 土坝;土方筑坝
earthwork design 土方工程设计
earthwork drawing 土方工程图
earthwork engineering 土方工程;岩土工程
earthworker's tools 土工工具
earthwork haul distance 土方运距
earthworking machinery 土方工程机械设备
earthwork machinery 土方工程机械
earthwork operation 土方工程工序;土方运算(工程);土(石)方工程作业
earthwork plan 土方计划;土方工程计划
earthwork quantity 土方工程量
earthworks balance 土方平衡
earthworks contract section 土方工程承包地段
earthwork side slope 土方边坡
earthwork site 土方工程现场
earthwork slope 土工边坡度
earthworks plant 土方工程设备
earthworks site 土方工程现场
earthwork vehicle 土方工程车辆;土方运输工具
earthworm biofilter 蚯蚓生物滤池
earthworm-motion wave 蠕行波
earthy 地电位的;土状;土壤的
earthy bauxite 土状铝土矿
earthy cobalt 土质钴;钴土
earthy dolomite 含土白云石
earthy fire (过低温度或过短时间烧成的)低温烧成;低温煅烧
earthy graphite 土状石墨
earthy gypsum 土(质)石膏;脆性石膏
earthy lime 土质石灰
earthy material 泥质材料;黏土;泥砂
earthy ore 土质矿石
earthy smelling substance 土臭物质
earthy spring 泥水泉;泥泉
earthy taste 土腥味
earthy tripolite 硅藻土
earthy variety of brown coal 褐煤的不同含土量
earthy vivianite 含土蓝铁矿
earthy water 硬水
earth zone 地球带
ear-trumpet 助听器
easamatic 简易自动式
easamatic power brake 简易自动制动闸;真空闸
ease 减轻;放宽;安逸
ease automatic 简易自动式
ease automatic power brake 简易自动制动闸
eased edge 倒棱边;缓弧边;小圆棱;光圆边缘
ease degradation organic matter 易降解洗涤剂
ease driving 平稳行车
ease her astern 低速后退
easel 绘画架;放大尺板;绘图台;画架;黑白架
easel mask 障板;放大框
easel painting 架上绘画
easel picture 画架上绘画;架上画
easel plane 承影面
easel size 承影尺寸
easel tilt 承影倾斜
easement 缓和;用土权;自然弯线;地役权;采光

权;附属建筑物;平顺;土地使用权;顺弯(使转折处平缓)
easement boundary 土地使用权范围
easement by implication 暗示通行权
easement curve 螺旋形缓和曲线;介曲线;缓和曲线;缓变曲线;过渡曲线;调制曲线
easement of necessity 必要通行权
easement of way 通路权
ease of control 容易控制程度
ease off 减低冲程;松些
ease of handling 手控便利
ease of ignition 易着火;易点燃性
ease of operation 易于操作
easer 油墨添加剂;辅助钻孔;辅助炮眼;掏槽爆破孔
easer hole 辅助钻孔
easer shot 辅助眼爆破
ease the grade 缓和坡度
ease the helm 回舵
ease turn 和缓转弯
easiest rolling car 最易行车车辆【铁】
easiest washing 极易选
easily accessible 易于进入的;便于接近的
easily cleanable 易洗的;易扫除干净的;易弄干净的
easily controlled grasses 容易防治的杂草
easily detectable trap 易探测圈闭
easily eroded bank material 易冲蚀的河岸质
easily falling roof 易冒落顶板
easily hydrated clay 易水化黏土
easily identification image point 明置像点
easily selfcombustion 易自燃
easily workable glass 容易加工的玻璃
easily-worn parts 易损零件
easiness of money market 市场利率低
easiness to milling 易磨性
easing 放松物;曲线缓和;松型;顺弯(使转折处平缓)
easing centers 缓移鹰架;缓移拱架
easing gear 卸物装置;卸货装置
easing lever 检查控制杆
easing ship 减轻船载;减轻船荷
easing the bit in 钻头轻压慢转钻井地层
easing the wedges 拆除支撑;易脱模块
easing valve 减荷阀;溢流阀;保险阀
easing wedge 松动楔块;辅助楔;易脱模块;对垫楔块;拆除架楔
east 东部
East Africa coast current 东非沿岸海流
East African bivalve province 东非双壳类地理区
East African copal 东非珂巴树脂
East African Graben 东非地堑
East African Meridian tectonic belt 东非经向构造带
East African rift seismotetonic zone 东非裂谷地震构造带
East African rift system 东非裂谷系
East African Rift Valley 东非裂谷
East American brachiopod realm 东美洲腕足动物地理区系
East Asian reflectional axis 东亚镜像反映中轴(带)
eastasy 海面进退
East Australia current 东澳大利亚海流
East Australian realm 东澳区系
eastbound 向东行驶
east by north 东偏北;朝东偏北
east by south 东偏南
east cardinal mark 东方位标
east Caroline basin 东加罗林海盆
east Caroline trough 东加罗林海槽
East China Sea 中国东海
East China Sea subsiding basin 中国东海沉降海盆
east coast 东海岸
east coast coal port 东海岸煤港
east coast magnetic anomaly 东海岸磁异常
east elevation 东立面(图)
east elongation 东距角
east end 东端;教堂东端
easter-betonite grease 酯一膨润土润滑脂
Easter I fracture zone 复活岛破裂带
easterity weather 东风气候
easterlies 东风带
easterly climate 东风气候
easterly component of an air current 气流的东西分量

easterly current 东风流
easterly hour angle 东时角
easterly jet 东风急流
easterly limit 东图廓
easterly trade wind 东贸易风
easterly wave 东风波
Easter Monday 复活节后的星期一(英国银行休业日)
eastern American realm 东美区系
eastern amplitude of the sun 日出没方位角
eastern apsis 教堂东端的半圆形凸出部分;教堂东端的半圆形殿
eastern bay 东方式海湾
eastern chapel 东方小教堂;东方小礼拜堂
eastern choir 东方教堂唱诗班;东方教堂式建筑
eastern choir tower 东方教堂唱诗班塔楼
eastern church 东方教堂式建筑
eastern church tower 东方教堂尖塔
eastern closet 东方式便器;蹲(式)便器
eastern crossing tower 东方交叉塔楼;东方教堂的中央尖塔
eastern crown 宗谱纹章图案
Eastern East China Sea-Western Xinzhu depression region 中国东海东部—新竹西部拗陷地带
eastern elongation 东距角;东大距
Eastern European platform 东欧地台
eastern exedra 东方的开敞式座谈间;东方的半圆形露天建筑;教堂东端的前廊
eastern greatest elongation 东大距
eastern hemisphere 东半球
eastern hemlock 加拿大铁杉
eastern Himalaya tectonic segment 东喜马拉雅构造段
eastern Indian walnut 山合欢木
easternism 东方主义
eastern larch 美洲落叶松
Eastern method 东方砌砖法(一种英国式砌砖法)
easternmost 最东的
eastern pediment 东方山墙上的檐饰;正面入口上方三角墙(古希腊建筑)
Eastern plane 法国梧桐
eastern quadrature 东方照
eastern quire 东方教会歌唱队;东方教堂式建筑
eastern quire tower 东方教会歌唱队塔楼
eastern red cedar 铅笔柏;北美圆柏;东方红杉
eastern spruce 美国红果云杉;东云杉
eastern standard time 东部标准时间
eastern suburbs 东郊
Eastern Tomb 东陵(清朝)
eastern W. C. pan 东方跨式便盆
Easter sepulchre 伊斯特尔坟墓
East European belemnite region 东欧箭石地理大区
East European coral region 东欧珊瑚地理大区
East European floral region 东欧植物地理大区
East European plate 东欧板块
East European platform transgression 东欧地台海浸
East European ware 东欧窑
East Gobi basin 东戈壁盆地
East Greenland current 东格陵兰海流
East Greenland marine trough 东格陵兰海槽
East Indian gum 东印度树胶
East Indian kino 东印度吉纳树胶
East Indian laurel 东印度月桂树
East Indian ocean 东印度洋
East Indian satinwood 东印度椴木
easting 向东航行;向东方航程;东行航程;东向分量;东西距;东航;朝东方向
east longitude 东经
east(ly) wind 东风
east magnetic component 东向磁力
east Manchuria old land 东满古陆
Eastman colo(u)r 依斯特曼彩色胶片
Eastman survey instrument 伊斯门测斜仪
Eastman whipstock turbine 伊斯门造斜涡轮
east Mediterranean bivalve subprovince 东地中海双壳类地理亚区
east Melanesian trench 东美拉尼西亚海沟
east northeast 东北东;北东东;东东北
Easto circulating sub 伊士托循环接头
east of Greenwich 格林威治东
eastonite 富镁黑;铁叶云母

Easton red 伊斯顿红
east orientation 东向
east Pacific basin 东太平洋海盆
east Pacific molluscan realm 东太平洋软体动物地理区系
east Pacific ocean 东太平洋
east Pacific ocean basin block 东太平洋海盆巨地块
east Pacific plate 东太平洋板块
east Pacific ridge 东太平洋海岭
east Pacific rise 东太平洋隆起;东太平洋海隆
east point 东点
east sea area 东部海域
east Siberian Sea 东西伯利亚海
eastside 东面
east southeast 南东东;东南东;东东南
eastward 向东
eastward component 东向分量
east-west 从东往西
east-west asymmetry 东西不对称(性)
east-west component 东西分量
east-west direction 东西向
east-west effect 东西效应
east-west level 主水准器(中星仪)
east-west line 东西线
east-west structural zone 东西构造带
east-west tectonic system 东西构造系
east wind drift 东风飘流
east window 教堂东端(主祭坛处)之窗
easy 便当的
easy access 交通方便;便于检修;容易接近
easy ahead 缓进
easy areas 可航行的
easy bench (井场的)休息长凳
easy bend 平缓弯子;平缓弯头;平缓弯道;慢弯管;半弯管
easy bilge 平顺底边
easy care finish 易保养涂饰剂
easy chair 单人沙发;安乐椅
easy-change transmission 易换速变速器
easy-clean(ing) hinge 长脚铰链;长脚合页;延伸窗铰;长翼铰链
easy cleavage 明显解劈理
easy credit 扩大信用
easy credit policy 放弃信贷政策
easy credit terms 放宽贷款条件
easy curve 平顺曲线;平缓弯度;平缓曲线
easy device 轻便装置;轻便仪表
easy digging 容易挖掘
easy disconnected suction tube 易拆式吸管
easy dome 简易穹顶;平缓圆顶;平缓圆屋顶
easy drilling 在易破碎岩层中钻进;轻快钻进
easy fatigability 易疲劳性
easy fired (过低温度或过短时间烧成的)低温烧成
easy flow 易流动
easy fluxing mix 易熔生料
easy-fusible ash 易熔灰分
easy grade 慢坡;缓坡;平缓坡度;平缓坡道
easy gradient 缓坡;平缓梯度;平缓坡度;平缓坡道
easy gradient for acceleration 加速坡度;加速缓坡
easy gradient section 缓坡地段
easy helm 回舵
easy instruction automatic computer 教学用自动计算机
easy jacking system 简便起重器
easy-machining steel 易加工的钢材
easy magnetizing axis 易磁化轴
easy maintenance 易于维修
easy money 银根松;低息资金;低息借款;低息贷款;低利贷款
easy off 容易拆卸的;易拆
easy-off flask 滑脱砂箱
easy of sale 销路好
easy-on 易装
easy-open tin can 易开白铁皮罐
easy payment 分期付款
easy processing channel black 易混槽法炭黑
easy processing resin 易加工树脂
easy purchase 分期付款购置
easy push fit 轻推配合;滑动配合;松推配合
easy ride knurling 简易曲线凸边
easy rolling car 易行车【铁】
easy rolling track 易行线
easy running 平滑运转

easy running fit 轻转配合
easy running track 易行线
easy servicing 小修(理);容易检修
easy slide fit 滑动配合;轻滑配合
easy-soluble compound 易溶性化合物
easy-soluble salt test 易溶盐试验
easy speed 平缓速度
easy spring 平缓弹簧;缓冲弹簧
easy squeeze blind rivet gun 拉铆枪
easy starboard 向右缓转
easy starter 简易起动装置
easy terms 不苛刻条件
easy-to-blend 易掺混的
easy-to-change 易于转换的
easy-to-drill formation 易钻地层
easy-to-handle 易于操纵的
easy to machine 便于加工
easy to raise 容易种植
easy-to-read 易读的
easy-to-screen material 易于筛选材料
easy-to-use 易于使用的
easy transport 容易运输
easy washing 易选
easy washing sample of coal 煤筒选样
eat away 锈坏;腐蚀
eat-back of chemical corrosion 化学腐蚀蔓延
eatery 餐馆
eath-rock-fill dam 堆石土坝
eathwork mechanical construction 土方机械化施工
eating 腐蚀;侵蚀作用;膳宿税
eating and drinking establishment 饮食店
eating away 蚀掉
eating house 饮食店;餐馆;饭馆
eating place 餐馆
eating room 食堂
eating-through 蚀穿
eat out 耗尽
eat up part of state revenue 冲减财政收入
eau-de-Javel 惹芙耳溶液
eau-de-Javelle 次氯酸盐消毒液
eau-de-Labarrque 拉巴腊溶液
eaves 屋檐
eaves batten 檐口挂瓦条
eaves bearer 檐口托架;檐口支座
eaves board 檐口垫瓦条;封檐板;连檐板;檐(口)板;风檐板
eaves catch 连檐板;檐垫条
eaves ceiling 廊檐顶棚;檐口平顶
eaves channel 檐槽
eaves cornice 檐口线脚
eaves course 檐口瓦层;双板层
eaves detail 檐口大样;屋檐详图
eaves drip 檐水滴落地;檐檐滴水
eaves drop 屋檐滴水;檐水滴流池
eaves fascia 檐口托板;(柱顶横梁的)横带;檐口板;封檐板
eaves fillet 檐瓦条木
eaves flashing 披水;挡水板;檐口披水;檐口泛水;屋檐泛水
eaves gallery 檐下廊道;檐廊
eaves girder 檐口大梁
eaves gutter 天沟;檐口滴水沟;檐沟
eaves gutter nozzle 檐沟导流管
eaves height 屋檐高度;檐高
eaves-hung fascia 挂檐板
eaves lath 屋檐板条;连檐板;檐口板条
eaves mo(u)lding 檐边;檐口模板;檐饰线条;檐饰线脚;檐口线脚
eaves lead 铅制屋檐水槽
eaves overhang 屋檐悬挑;屋檐出挑
eaves plate 檐口垫木;封檐板;檐口梁;檐口板;檐脚梁
eaves plate tile 檐板瓦
eaves plummet 屋檐铅垂线
eaves pole 檐瓦垫条;屋檐柱;披水条;檐口嵌条
eaves projection 檐口凸出物;檐口出挑
eaves purlin(e) 檐檩;桃檐枋
eaves rafter 檐椽(古建筑)
eaves soffit 檐腹板;挑檐平顶板;檐口托板
eaves strip 镶边板条
eaves strut 檐撑;檐口支撑
eaves-supporting post 擎檐柱
eaves tiebeam 檐枋

eaves tile 檐瓦;檐口瓦;勾头瓦;勾滴瓦;瓦挡
eaves trough 檐沟;檐槽
eaves trowel 檐口水沟;檐沟
eaves unit 屋檐单元
eaves vent 檐口通风口;屋檐下的通风孔
eaves wall 檐墙
Ebanoid (一种木块地板用的)专卖沥青
ebb 潮路;退落;退期;衰退
ebb anchor 落潮锚
ebb and flood 涨落潮
ebb and flood of tide 潮的涨落
ebb and flood tides 潮汐
ebb and flow 潮的涨落;进退(指潮水);起伏;盛衰;涨落(潮)
ebb and flow gate 涨落潮闸门;挡潮闸(门)
ebb and flow structure 涨落潮流构造;潮流涨落构造
ebb and flow zone 潮涨落地带
ebb axis 落潮流轴
ebb channel 落潮水道;落潮流沟槽;退潮航道
ebb current 落潮水流;落潮流;退潮流
ebb-current flow structure 落潮流构造
ebb dike 防潮堤
ebb dry 干出(落潮时露出)
ebb gate 落潮闸(门);退潮闸门
ebbing 沉陷
ebbing and flowing spring 潮汐泉;涨落泉
ebbing spring 退落泉
ebbing well 落水井;涨落井
ebb interval 落潮间隙;落潮流间隙;退潮时间间隙
ebb ratio 退潮回流比
ebb stream 落潮流
ebb strength 退潮强度;落潮流速;最大退潮流(速);最大落潮流
ebb surge 落潮涌浪;退潮涌浪
ebb tidal delta 退潮三角洲
ebb tidal deltaic deposit 退潮三角洲沉积
ebb tide 落潮;低潮;退潮
ebb tide channel 退潮航道
ebb tide current 落潮流;退潮流
ebb tide gate 落潮(时用的)闸(门);退潮闸门
ebb tide range 落潮潮差
ebb tide stream 落潮流
ebb tide volume 落潮量
Eberhard effect 埃伯哈德效应
Eberhardt effect 埃伯哈德效应
Eber's solution 埃贝尔溶液
Ebert-Fastie monochromator 艾勃特—法斯梯单色器
Ebert-Fastie mounting 艾勃特—法斯梯装置
Ebert-Fastie spectrograph 艾勃特—法斯梯摄谱仪
Ebert ion counter 埃伯特离子计数管
ebicon 电子轰击导电性
Eble type 埃贝尔方式
ebong wood 乌木;黑檀木
ebonite 胶木;硬质橡胶;硬质胶;硬皮;硬橡胶;硬化橡胶
ebonite base 胶木座;橡皮座
ebonite board 胶木板;硬橡胶板
ebonite bush 胶木套管;硬橡皮衬套;硬胶套管
ebonite bushing 硬橡胶轴衬
ebonite cell 胶木覆蔽电池
ebonite clad plate 硬胶蔽电池极板
ebonite driver 胶柄螺丝起子
ebonite knot 硬橡皮扣
ebonite monoblock 正铸胶壳
ebonite reed relay 胶木簧片继电器
ebonite socket 硬橡皮座
ebonite stud 硬胶皮螺帽
ebonite wax 乌木蜡
ebony 黑檀色的;黑檀木
ebony asbestos 黑石棉
ebony wood 黑檀木;乌木
eboulement 滑坡;坍塌
ebullated bed 流化床;沸腾床
ebullated bed dryer 沸腾床干燥器
ebullated bed reactor 流化床反应器
ebullator 流化器;沸腾器
ebullience 沸腾
ebullience chamber 鼓泡室;沸腾室
ebullient cooling 蒸发冷却
ebulliometer 沸点升高计;沸点酒精计;沸点(测定)计
ebulliometry 沸点测定

ebullioscope 酒精气压计;沸点升高计;沸点酒精计;沸点(测定)计
ebullioscopic equation 沸点升高公式
ebullioscopic solvent 沸点升高溶剂
ebullioscopy 沸点升高检查;沸点升高测定法;沸点上升法
ebullition 沸腾;气泡生成;起泡
ebullition cooling 气相冷却
eburicoic acid 齿孔酸
eburnean 象牙色的
eburneous 象牙样的
Eburonian cold epoch 埃布朗寒冷期
Eburonian glacial stage 埃布朗冰期
ecad 境型型;境范生物;适应型
Ecaussines stone 厄瓜新石
eccentered gun 偏心射孔器
eccentered screw pump 偏心螺杆泵
eccentering arm 偏心臂
eccentraliser = eccentralizer
eccentralizer 偏心器
eccentric 偏心的
eccentric action 偏心作用;偏心动作
eccentric adjuster 偏心调节装置
eccentric adjusting sleeve 偏心套筒;偏心调节套
eccentrically arranged valve 偏置阀
eccentrically bored spindle 偏心钻杆
eccentrically braced 偏心支撑的
eccentrically compressed 偏心受压的
eccentrically compressed column 偏心受压柱
eccentrically curved beam 偏心曲线梁
eccentrically loaded 承受偏心荷载的;偏心荷载的
eccentrically loaded column 偏心荷载柱
eccentrically loaded footing 偏心荷载基础
eccentrically loaded pile 偏心荷载桩
eccentrically loaded rotating shaft 偏心荷载转轴
eccentrically precompressed precast-concrete-pile 偏心预压混凝土预制桩
eccentrically rotating 偏心旋
eccentrically rotating mirror 偏心旋转镜
eccentrically stiffened 偏心加固的
eccentric anchor pin 偏心支销
eccentric-andpitman drive 偏心轮与连杆传动装置
eccentric angle 离心角;偏心角
eccentric angle measurement 偏心测角
eccentric annulus 偏心环空
eccentric anomaly 偏心异常;偏心像差;偏近点角
eccentric application of force 力的偏心应用;力的偏心施加
eccentric arm 偏心杆;偏心臂
eccentric axial load 偏心轴向荷载
eccentric axis 偏心轴
eccentric balance-weight 偏心平衡锤;偏心配重
eccentric bar 偏心杆
eccentric beam-column connection 偏心梁柱连接
eccentric bending moment 偏心弯矩
eccentric bit 偏心钻(头)
eccentric bit load 偏心钻头载荷
eccentric block 偏心块
eccentric bolt 偏心螺栓
eccentric brace 偏心支撑
eccentric breaker 偏心轧碎机
eccentric burner 偏心燃烧器;偏心喷燃管
eccentric bush 偏心轴衬
eccentric bushing 偏心轴套;偏心轴衬;偏心衬垫
eccentric cam 偏心凸轮
eccentric catch 偏心制动(装置)
eccentric center 偏心
eccentric chuck 偏心卡盘
eccentric circle 偏心圆
eccentric circle-in 偏心圈入
eccentric circle-out 偏心圈出
eccentric clamp 偏心自锁搭扣;偏心压板
eccentric clip 偏心夹环
eccentric coefficient 偏心系数
eccentric collar 偏心钻铤
eccentric compound 偏心分量
eccentric compression 偏心压缩;偏心压力;偏心受压
eccentric compression member 偏心受压构件
eccentric compression method 偏心受压法
eccentric conic(al) shell 偏心锥形筒体
eccentric connection 偏心联结;偏心接合
eccentric contraction 离心收缩

eccentric contraction 偏心校正
eccentric crank 偏心曲柄
eccentric crank mechanism 偏置曲柄机构;偏心曲柄机构
eccentric crankshaft 偏心曲柄轴
eccentric cutter holder 偏心刀架
eccentric cylinder rheometer 偏心圆筒式流变仪
eccentric cylinder rotary oil pump 偏心缸回转油泵
eccentric dipole 偏心偶极
eccentric direction 偏心方向
eccentric disc 偏心(圆)盘
eccentric discharge 偏心卸料
eccentric distance 偏心距离;偏心距
eccentric drill bit 偏心钻头
eccentric drive 偏心轮传动;偏心传动
eccentric driven pump 偏心泵
eccentric drive screen 偏心振动筛;偏心传动筛
eccentric error 偏心误差
eccentric error of alidade 照准部偏心差
eccentric error of circle 度盘偏心差
eccentric eruption 偏心喷溢
eccentric factor 离心因子
eccentric fas lift valve 偏心气举阀
eccentric fastening stud 偏心轮紧固柱螺栓
eccentric feed motion 偏心进给运动
eccentric fitting 偏心管接头;偏中心管件
eccentric force 偏心力
eccentric fork 偏心轮叉
eccentric foundation 偏心基础
eccentric gab 偏心轮叉子
eccentric gear 偏心机构;偏心(传动)装置;偏心(齿)轮
eccentric gear drive 偏心齿轮传动
eccentric gearing 偏心传动
eccentric grinder 偏心研磨机
eccentric groove 偏心纹
eccentric halo 偏心晕
eccentric impact 偏心碰撞
eccentric injection mandrel 偏心配水器
eccentricity 离心率;反常;偏心距(离);偏心(度)
eccentricity coefficient 偏心系数
eccentricity correction 偏心率改正
eccentricity correction device 偏心矫正装置
eccentricity factor 偏心度系数
eccentricity indicator 偏心计
eccentricity of alidade 照准仪偏心;照准部偏心
eccentricity of circle 度盘偏心
eccentricity of collimation axis 视准轴偏心
eccentricity of ellipse 椭圆偏心率
eccentricity of ellipsoid 椭球偏心率
eccentricity of instrument 仪器偏心;测站偏心
eccentricity of load 负荷偏心率
eccentricity of loading 荷载偏心距
eccentricity of orbit 轨道偏心率
eccentricity of rest 静止偏心距;静态偏心率
eccentricity of satellite orbit 卫星轨道偏心率
eccentricity of signal 照准点偏心;觇标偏心
eccentricity of spheroid of revolution 旋转椭球偏心率
eccentricity of the earth 地球偏心率
eccentricity of the earth ellipsoid 地球椭球偏心率
eccentricity of thrust 推力偏心矩
eccentricity pressure 偏心压力
eccentricity ratio 偏心率(率)比
eccentricity recorder 偏心记录仪
eccentricity tester 径向跳动检查仪;偏心距检查仪
eccentricity throw-out 偏心推出器
eccentricity tolerance 偏心度公差
eccentric jaw crusher 颚式偏心轧碎机;颚式偏心压碎机;颚式偏心碎石机;颚式偏心破碎机
eccentric joint 偏心接合
eccentric key 偏心轮销
eccentric knockout grid 偏心振动落砂机
eccentric lathe 偏心车床
eccentric latitude 偏心纬度
eccentric level 偏心水准
eccentric lever 偏心杠杆;偏心杆
eccentric liner 偏心轮衬面
eccentric link 偏心连杆
eccentric load 偏心载重;偏心钻压;偏心荷载;偏心负载
eccentric loaded column 偏心承载柱
eccentric loaded foundation 偏心承载基础
eccentric load factor 偏载系数

eccentric loading 偏心负载;偏心加载
eccentric mandrel 偏心杆
eccentric mass 偏心质量
eccentric mass-type vibration generator 偏心质量式激振器
eccentric mechanism 偏心机构
eccentric moment 偏心力矩
eccentric mooring buoy 偏心系船浮筒
eccentric motion 变角加速运动
eccentric-motion reversing gear 凸轮回行机构
eccentric nut 偏心螺母
eccentric object 偏心目标
eccentric observation 偏心观测
eccentric oiler 偏心注油器
eccentric ooid 偏心鲕
eccentric orbit 偏心轨道
eccentric orifice 偏心孔板
eccentric orifice with flange taps 法兰取压偏心孔板
eccentric pattern 螺旋状排列
eccentric pin 偏心销(钉)
eccentric pipe 偏心管
eccentric piston ring 偏心活塞环
eccentric position 偏心位置
eccentric press 偏心压机;偏心(式)压力机
eccentric pressure 偏心压力
eccentric production mandrel 偏心配产器
eccentric pulley 偏心皮带轮
eccentric pump 偏心轮式泵;偏心泵
eccentric radius 偏心半径
eccentric ratio 偏心比
eccentric rebel tool 偏心变向器
eccentric reducer 偏心渐变器;偏心锥形管;偏心异径管;偏心渐缩管;偏心大小头
eccentric reduction 偏心改正
eccentric ring 偏心环
eccentric ring emergency governor 偏心环式危急保安器
eccentric ring structure 偏心环形山构造
eccentric riveted joint 偏心铆接
eccentric rod 偏心棒;偏心杆
eccentric rod clamp 扁形夹持器
eccentric rod seat 偏心杆座
eccentric roller 偏心滚柱;偏心滚轮
eccentric roller rail bender 偏心滚子弯轨器
eccentric rotary pump 偏心泵
eccentric rotary shaker 偏心旋转式激振器
eccentric rotation 偏转
eccentric rotor engine 偏心转子发动机
eccentric rotor sliding vane compressor 偏心转子滑叶式压缩机
eccentric runner 偏心转子
eccentric screen 偏心筛
eccentric-screw breechblock 偏心闭锁机
eccentric sector 偏心扇形齿轮
eccentric segment 偏心扇形齿轮
eccentric set-up 偏心安置
eccentric shaft 偏心轴
eccentric shaft plunger 偏心轴柱销
eccentric shaft press 偏心曲柄压力机;偏心冲床;偏心轴压机
eccentric shaker table 偏心振动台
eccentric sheave 偏心皮带轮;偏心轮
eccentric sheave key way 偏心皮带轮键槽
eccentric sheave set 偏心滑轮制动螺钉
eccentric sighting mark 偏心觇标
eccentric signal 偏心信号;偏心觇标;偏心标志
eccentric sliding vane pump 偏心滑叶泵
eccentric socket 偏心套管
eccentric stabilizer 偏心稳定器
eccentric station 偏心站
eccentric stiffener 偏心加劲杆
eccentric strap 偏心固定夹;偏心外轮;偏心环
eccentric-strap cleaning plug 偏心环清除塞
eccentric-strap liner 偏心环衬
eccentric-strap oil cup 偏心环油杯
eccentric-strap oil pocket 偏心环油槽
eccentric-strap seat 偏心套环座
eccentric surveying point 偏心测站
eccentric swage 压料器
eccentric table joint 偏企口接合
eccentric tee 偏心异径三通
eccentric telescope 偏心望远镜
eccentric tendon 偏心预应力钢丝束;预应力偏心

钢丝束
eccentric tension 偏心受拉;偏心拉力
eccentric tensioner 偏心衬筒
eccentric tension member 偏心受拉构件
eccentric thermodynamic(al) type steam trap 偏心热动力疏水器
eccentric throw 偏心轮行程
eccentric thrust 偏心推力
eccentric tongs 偏心钳;偏心管钳
eccentric tumbling mixer 偏心滚动拌和机;偏心转筒式拌和器;偏心转筒式拌和机
eccentric turning 偏心车削
eccentric-type pick up 偏心式拾拾器
eccentric-type vibrating screen 偏心式振动筛
eccentric-type vibrator 偏心(式)振动器
eccentric under-reaming bit 管下扩眼偏心钻头;管鞋下扩眼偏心钻头
eccentric valve seat grinder 偏心阀座磨床
eccentric vibrating mill 偏心振动磨
eccentric vibrating screen 偏心振动筛
eccentric water distributor 偏心配水器
eccentric wear 偏心磨损;偏磨
eccentric weight 偏心重;偏心锤
eccentric weight drive shaking screen 偏重驱动摇动筛
eccentric weight shaft 偏心重轴;偏重轴
eccentric weight type vibrator 偏心重式振动器
eccentric wheel 偏心轮
eccentric yokes 偏心接箍
ecclesia 教室;教堂
ecclesiastical 教堂的
ecclesiastical architect 教堂建筑师
ecclesiastical architecture 教堂建筑学;教堂建筑术;宗教建筑
ecclesiastical basilica (教堂的)长方形会堂;教堂的长方形会堂
ecclesiastical building 教堂建筑;宗教房屋
ecclesiastical building style 教堂建筑风格;教堂式建筑
ecclesiastical Gothic (style) 教堂建筑的哥特式风格
ecclesiastical monument 神圣纪念碑;教堂纪念碑
ecclesiastical structure 教堂建筑结构
ecclesiological architecture 教堂建筑
ecclesiology 教堂建筑及装饰研究;教堂装饰学;教堂建筑学;教堂建筑术
Eccles-Jordan circuit 埃克勒斯—乔丹电路
Eccles-Jordan multivibrator 埃克勒斯—乔丹多频振荡器;双稳态多频振荡器
ecdemic 非地方性的
ecdemic disease 外来病
ecdemite 氯种铅矿
ecesis 定居;不稳定居
ecesis anomaly 定居
echange function 更换功能
echard 无效水量【植】;无效水分
echauguette 古堡角楼
echaust-gas analyzer 废气分析器
echea 助响瓮
echelette 红外(线)光栅
echelette grating 红外(线)光栅;小阶梯光栅
echelle 图纸比例尺;阶梯光栅
echelle grating 中阶梯光栅
echelle spectrograph 中阶梯摄谱仪;中阶梯光栅摄谱仪
echelogram 阶梯光谱照片
echelon 精度系列;阶梯式;雁行;排成梯队;梯列;梯队;梯次配置
echelon antenna 梯形天线
echelon clouds 梯状云
echelon condenser lens 阶梯聚光透镜
echelon device 阶梯光栅装置
echelon faults 雁行断层
echelon fold 雁行褶皱
echelon form 梯形
echelon fracture 雁行式裂缝
echelon grating 阶梯光栅
echelon lens 阶梯透镜;阶梯式透镜
echelon lens antenna 梯形透镜式天线
echelon-like veins 斜列式矿脉
echelon maintenance 分级维修
echelon matrix 阶梯矩阵;梯阵
echelonment 阶梯状;梯状配置;梯次配置
echelon operation (铺筑沥青混合料的)阶梯操作法
echelon pattern 雁行构造形式;雁列式

echelon prism 阶梯棱镜
echelon-sections 剖面阶梯型排序
echelon spectroscope 阶梯光栅分光仪
echelon strapping 阶梯式绕带；梯形绕带
echelon structure 阶梯构造；雁行构造
echelon wave 阶梯式波
echicaoutchin 艾奇树脂
echiceric acid 艾奇蜡酸
echinate 棘刺状
echinoderm 棘皮动物
Echinodermata 棘皮动物门
echinoid 海胆状的
echinoids 海胆类
echinus 钟形圆饰【建】；[复]echini
echinus and astragal 柱头凸圆线脚
echo 回音；回声；回波；反响
echo acquisition 回波捕捉
echo altimeter 雷达测高计；回声测高仪；回波高度计；回波测高计
echo amplifier 回声放大器；回波（信号）放大器；反射信号放大器
echo amplitude 回波振幅；反射波振幅
echo apparatus 回波测距仪
echo area 回波区；回波面积
echo arrival 回波初至
echo attenuation 回声衰减；回波衰减；反射衰减
echo bearing 回波方位
echo board 回波显示屏
echo box 回波谐振腔；回波箱；回波共振腔
echo-box actuator 回波箱激励器；回波谐振腔装置；回波空腔谐振器
echo cancellation 回波对消；重影消除；双回路对消
echocardiograph 超声波音波描记器
echo cavity resonator 回波空腔谐振器
echo chamber 回声室
echo channel 回声波道；回波波道
echo check(ing) 回送检测；回送检验；回读校验；回波校验；回波检验；返回检查
echo checking means 回送检验方式
echo check technique 回波检验技术
echo contour 回波等强线
echo depth sounder 回声测深仪；回波测深仪；回音测深仪；回波测深仪
echo depth sounding 回波测深（法）；回波测深
echo depth sounding sonar 回波测深仪；回声深声呐；超声波深度测定器；超波深度测定器
echo distortion 回波失真；回波畸变
echo duration 回波时间；回波宽度
echo ed signal 回波信号
echo effect 回声效应；回波效应
echo elimination 消除回声；消除回波；反射信号抑制；回声消除
echo eliminator 回波消除器
echoencephalography 超声波回波脑照相术
echo equalizer 回波均衡器
echo-fathom 回声尺
echo filter 回波滤波器
echo fixation 回波定位
echo flutter 回波颤动
echo frequency 回波频率
echogram 回声图；回波测深图；回声测深记录；回声测深(深度)记录图；回波图；回声图；音响测深图
echograph 回声测深（自动）记录仪；回波测深装置；回声测深仪；回声测深记录器；回波记录仪；鱼群探测器；音响测深自动记录仪
echography 回波描记术
echo image 回波图像；双像；回波像
echo impulse 回声脉冲；反射脉冲
echoing 回声现象；回波现象；反照现象
echoing characteristic 回波特性
echoing end reply 返回结束回答
echo intensity 回声强度；回波强度
echo interference 回波干扰
echo killer 回声抑制器；回波抑制器；反射(信号)抑制器
echo length 回波长度
echoless area 无回声区
echo level 回声级
echo liquid level instrument 液面回波探测仪
echo location 回响定位法；回声探测；回声定位(法)；回声测定；回波测距
echo location technique 回声定位技术
echo locator 回声定位器；回声(波)勘定器

echo log 回波测井
echo logical 回波逻辑的
echo loss 回声损耗
echo machine 回音设备
echo measurement 音响测量
echometer 回波测深仪；回声测深仪；回波测深仪；回声测深计；回声测深仪；回波计；音响测深仪；水声测深仪
echo method 回波法
echometry 测回声术
echo-peak 回波峰值
echo-phenomenon 模仿现象
echo pip 反射脉冲
echo-plex 返回传送
echo portico 回声柱廊；回声门廊
echo pulse 回脉冲；回波脉冲
echo-pulse receiver 回声脉冲接收器；回声脉冲接收机
echo radar 回波雷达
echo ranger 回波测距仪
echo ranging 回声定位；回波测距；回声测距(法)
echo-ranging apparatus 回声测距仪
echo-ranging detection 回声测距(法)
echo-ranging indicator 回波测距指示器
echo-ranging sonar 回声定位声呐；回声测距声呐
echo-ranging station 回声测距站
echo-ranging system 回声测距系统
echo receiver 回声接受器；回声接收器
echo recognition 回波识别
echo repeater unit 回波中继装置
echo resonator 回声谐振器
echo room 回声室
echo sampler 回声采样器；回波采样器
echo satellite 回波卫星
echo sign 回波符号
echo signal 目标信号；回声信号；反射信号
echo signal of sounder 测深仪回波信号
echo signals 重影信号
echosonogram 超声波回波图
echo sonoscope 回波式超声仪
echo sounder 回音测深仪；回声探测器；回声测深器；回波测深仪；超声波回声测深仪；音响测深机；测深仪
echo sounder calibration 回声测深仪校准
echo sounder machine 回声测深仪
echo sounder profile 回声测深剖面图
echo sounding 回声探测；回声测深（法）；回波探测；回波测深；音响测深法
echo-sounding apparatus 回波测深声器；回声测深仪
echo-sounding device 回波测深仪
echo-sounding frequency 回波测深频率
echo-sounding gear 回声探测仪
echo-sounding instrument 回波测深声器；回声测深仪
echo-sounding launch 回声测深艇
echo-sounding machine 回声测深仪
echo-sounding receiver 回声测深接收器
echo-sounding recorder 回声测深计
echo-sounding sonar 回声测深声呐
echo source 回声源
echo-splitting radar 回波分裂雷达
echo strength 回声强度；回波强度
echo strength indicator 回波强度指示器；反射信号强度指示器
echo studio 回声播音室
echo suppresser 回声遏制器；回波阻尼器；回波抑制器；反射抑制器；反射波抑制
echo suppression 回声抑制；回波抑制；回波信号（的）抑制
echo talker 回送干扰
echo test 回声测试
echo time 回声时间
echo timing 回波计时
echo trap 回波抑制装置；回波抑制设备；回波陷波器；回波滤波器；回波阱
Echo Wall 天坛回音壁
echo wave 回波；回声波
ecidio climate 微生态气候
Eckart-Young factor analysis 埃卡特·扬因素分析
Eck-atomizer 极超微粉碎机
eckermannite 镁钠闪石；氟镁钠闪石
Eckert die cast machine 埃克特立式压铸机
Eckert number 埃克特数
Eckert projection 艾克尔特投影
eclarite 艾辉铋铜铅矿

eclectic architecture 折衷式建筑；折衷主义建筑
eclectic-garden 折衷式庭园
eclecticism 折衷主义
eclectic structure 折衷主义建筑；折衷式建筑
eclipse 星掩蔽；月蚀；全蚀带
eclipse beginning 初亏
eclipsed conformation 遮蔽构型；重叠(式)构像
eclipsed form 重叠形；重叠式(指混凝土沉排形式)
eclipse duration 掩食时间
eclipse factor 阴影率
eclipse of the moon 月食
eclipse of the sun 日食；日蚀
eclipse period 掩蔽期
eclipse phase 隐蔽期
eclipse weather 日食天气
eclipse year 食年
eclipsing 重叠
eclipsing effect 重复效应；重叠效应
ecliptic 黄道
ecliptic armillary sphere 黄道经纬仪
ecliptic coordinates 黄道坐标
ecliptic coordinate system 黄道坐标系(统)
ecliptic diagram 黄道图；黄道带星图
ecliptic latitude 黄纬
ecliptic longitude 黄经
ecliptic map 黄道星图
ecliptic meteor 黄道流星
ecliptic node 黄道交点
ecliptic obliquity 黄道斜度
ecliptic of date 瞬时黄道
ecliptic plane 黄道平面；黄道面
ecliptic polar distance 黄极距
ecliptic pole 黄极
ecliptic stream 黄道流星雨
ecliptic system of coordinates 黄道坐标系(统)
eclogite 榴辉岩
eclogite facies 榴辉岩相
ecnephias 地中海的飑或雷暴；埃克内菲斯飑
ecoactivism 生态能动论
eco-activist 生态活动家
eco-agricultural system 生态农业系统
eco-agriculture 生态农业
eco-animal husbandry 生态畜牧业
eco-atmosphere 生态大气
eco-catastrophe 生态灾难
ecochronology 生态年代学
ecocide 生态灭绝
eco-city 生态城市
eco-city planning 生态城市规划
ecoclimate 生态气候
ecoclimate adaptation 生态气候适应
ecoclimate forecasting 生态气候预测；生产气候预测
ecoclimatic forecasting pests 害虫生态气候预测
ecoclimatology 生态气候学
ecocline 生态倾差；生态渐变群；生态差别；生态差型；生态变异
ecocrisis 生态危机
ecocycle 生态循环
ecocycle rule 生态循环规律
ecodeme 生态同类群
eco-demonstration region 生态示范区
ecoeconomics 生态经济学
ecoengineering 生态工程(学)
ecoengineering system 生态工程系统
eco-environment 生态环境
eco-environment(al) change 生态环境变化
eco-environment(al) effect 生态环境效应
eco-environment(al) establishment 生态环境建设
eco-environment(al) feature 生态环境特征
eco-environment(al) function 生态环境功能
eco-environment(al) index 生态环境指标
eco-environment(al) planning 生态环境规划
eco-environment(al) protection 生态环境保护
ecofactor 生态因子；生态因素
eco-forestry 生态林业
ecogenesis 生态种发生；生态发生
ecogenetics 生态遗传学
ecogeographical divergence 生态地理趋势
ecograph 生态图解
ecogroup 生态群
eco-herbary 生态草业
ecohydrological water management approach

生态水文学管理方法
ecohydrology 生态水文学
eco-industry park 生态工业园
ecolabelling 生态标志
Ecole Cistercienne (11世纪法国早期哥特式建筑的)西斯丁学派
E. coli 大肠杆菌
E. coli index 大肠杆菌指数
E. coli reverse mutation test 大肠杆菌回复突变试验
ecolofy cullet 回收碎玻璃
ecologic(al) absolute order 生态绝对命令
ecologic(al) accounting 生态会计学
ecologic(al) activity 生态活动
ecologic(al) adaptation 生态适应
ecologic(al) age 生态年龄
ecologic(al) agriculture 生态农业
ecologic(al) amplirute 生态幅度
ecologic(al) analysis 生态分析
ecologic(al) and habitat control of pest 害虫生态和环境控制
ecologic(al) approach 生态学方法
ecologic(al) architecture 符合生态学法则的建筑;生态建筑
ecologic(al) art 生态艺术
ecologic(al) aspects 生态概况
ecologic(al) assemblage 生态组合;生态集合体
ecologic(al) assessment 生态评价
ecologic(al) association 群丛;生态群落
ecologic(al) backlashes 生态冲击
ecologic(al) balance 生态平衡
ecologic(al) benefit 生态学方面的利益
ecologic(al) biogeography 生物生态地理学
ecologic(al) bonitation 生态多度计算
ecologic(al) botany 生态植物学;植物生态学
ecologic(al) building materials 生态建材
ecologic(al) capacity 生态容量
ecologic(al) carrying capacity 生态承载力
ecologic(al) change 生态变化
ecologic(al) character 生态性状;生态特性
ecologic(al) chemistry 生态化学
ecologic(al) classification 生态分类
ecologic(al) climatology 生态气候学
ecologic(al) community 生态群落
ecologic(al) compensation and amends 生态补偿与赔偿
ecologic(al) competition 生态竞争
ecologic(al) complex 生态综合体;生态复合体
ecologic(al) comprehensive assessment 生态综合评价
ecologic(al) comprehensive index 生态综合指数
ecologic(al) concentration 生态浓集
ecologic(al) conception of history 生态史观
ecologic(al) concrete 生态混凝土(用废料制成的混凝土)
ecologic(al) consciousness 生态意识
ecologic(al) consequence 生态影响;生态后果
ecologic(al) consideration 生态原因;生态考虑
ecologic(al) construction 生态建设
ecologic(al) control 生态控制
ecologic(al) correlation 生态学相关
ecologic(al) crime 破坏生态罪
ecologic(al) crisis 生态危机
ecologic(al) crop geography 生态作物地理学
ecologic(al) culture 生态文化
ecologic(al) cycle 生态循环
ecologic(al) damage 环境损害;生态损害;生态破坏
ecologic(al) degeneration 生态退化
ecologic(al) degradation 生态退化
ecologic(al) demand 生态需求
ecologic(al) density 生态密度
ecologic(al) design 生态设计
ecologic(al) disaster 环境灾害;生态灾难;生态灾害
ecologic(al) disaster area 生态灾区
ecologic(al) displacement 生态更替
ecologic(al) disruption 生态破坏
ecologic(al) distance 生态距离
ecologic(al) distribution 生态分布
ecologic(al) disturbance 生态失调
ecologic(al) divergence 生态分歧
ecologic(al) diversity 生态多样性;生态差别
ecologic(al) dominance 生态优势
ecologic(al) ecogenesis 生态史
ecologic(al) economic equilibrium 生态经济平衡

ecologic(al) economic planning 生态经济规划
ecologic(al) economics 生态经济学
ecologic(al) economic system 生态经济系统
ecologic(al) education 生态教育
ecologic(al) effect 生态影响;生态(学)效应
ecologic(al) effect assessment 生态效应评价
ecologic(al) effect for water pollution 水污染生态效应
ecologic(al) effect of harbo(u)r engineering 港口工程生态效应
ecologic(al) effect of harbo(u)r works 港口工程生态效应
ecologic(al) effects of atmospheric pollution 大气污染的生态效应
ecologic(al) efficiency 生态效率;生态等值
ecologic(al) engineering 生态工程(学)
ecologic(al) engineering system 生态工程系统
ecologic(al) entomology 生态昆虫学
ecologic(al) environment 生态环境
ecologic(al) environmental change 生态环境变化
ecologic(al) environmental data 生态环境资料
ecologic(al) environmental effect 生态环境效应
ecologic(al) environmental establishment 生态环境建设
ecologic(al) environmental feature 生态环境特征
ecologic(al) environmental function 生态环境功能
ecologic(al) environmental hydrogeology 生态环境水文地质学
ecologic(al) environmental index 生态环境指标
ecologic(al) environmental information 生态环境资料
ecologic(al) environmental planning 生态环境规划
ecologic(al) environmental rebuilding 生态环境重建
ecologic(al) environmental safety 生态环境安全
ecologic(al) equilibrium 生态平衡
ecologic(al) equivalence 生态等值;生态相当;生态均等
ecologic(al) equivalent species 生态等位种
ecologic(al) ethics 生态伦理学
ecologic(al) evaluation 生态学评价;生态评价
ecologic(al) evaluation criteria 生态评价标准
ecologic(al) evolution 生态进化
ecologic(al) fabric 生态织物
ecologic(al) facies 生态相
ecologic(al) factor 生态因子;生态因素
ecologic(al) failure 生态破坏
ecologic(al) fallacy 生态学谬论
ecologic(al) farm 生态农场
ecologic(al) floating-floor 生态浮床
ecologic(al) food chain 生态食物链
ecologic(al) forecasting 生态预测
ecologic(al) genetics 生态遗传学
ecologic(al) geobotany 生态地(球)植物学
ecologic(al) geography 生态地理学
ecologic(al) group 生态群(体);生态类群
ecologic(al) growth efficiency 生态生长效率
ecologic(al) hazard 生态危害
ecologic(al) health 生态健康
ecologic(al) home-ostasis 生态内稳定(现象)
ecologic(al) homeostasis 生态稳态
ecologic(al) hydrology 生态水文学
ecologic(al) imbalance 生态不平衡
ecologic(al) impact 生态影响;生态冲击
ecologic(al) impact assessment 生态影响评价
ecologic(al) imperialism 生态帝国主义
ecologic(al) indicator 生态指示物;生态表征;生态表型
ecologic(al) indicator plant 生态指标植物
ecologic(al) interaction 生态相互作用
ecologic(al) isolation 生态孤立;生态隔离
ecologic(al) kinetics 生态动力学
ecologic(al) law 生态学规律
ecologic(al) limit 生态限度
ecologically reliable water 生物安全用水
ecologically safe product 生态无害产品
ecologically sound characteristic 生态无害特性
ecologically sound product 生态无害产品
ecologically sound technology 生态无害技术
ecologic(al) manipulationd 生态操纵
ecologic(al) map 生物地理图;生态图;生态地理图
ecologic(al) master factor 生态主导因素
ecologic(al) model 生态系统模型;生态模型

ecologic(al) monitoring 生态监测
ecologic(al) monitoring strategy 生态监测对策
ecologic(al) monitoring unit 生态监测机构
ecologic(al) moral principle 生态道德原则
ecologic(al) moral standard 生态道德标准
ecologic(al) morphology 生态形态学
ecologic(al) mortality 生态死亡率
ecologic(al) niche 生态小生境;生态位;生态(适宜)环境;小生境;生境龛
ecologic(al) optimum 生态最适度
ecologic(al) outlook 生态观
ecologic(al) parameter 生态参数
ecologic(al) pessimism 生态悲观主义
ecologic(al) phase 生态相
ecologic(al) pholosophy 生态哲学
ecologic(al) physiology 生态生理学
ecologic(al) planning 生态规划
ecologic(al) plant anatomy 生态植物解剖学
ecologic(al) plant geography 生态植物地理学;植物生态地理学
ecologic(al) pollution limit 生态污染极限
ecologic(al) polymorphism 生态多态型现象
ecologic(al) pond system 生态塘系统
ecologic(al) pond-wetland system 生态池湿地系统
ecologic(al) potential 生态势
ecologic(al) prevention and treatment 生态防治
ecologic(al) process 生态过程
ecologic(al) protection 生态保护
ecologic(al) pyramid 生态金字塔
ecologic(al) race 生态族;生态宗(族)
ecologic(al) range 生态幅度
ecologic(al) reef 生态礁
ecologic(al) regime 生态状况
ecologic(al) regionalization 生态区划
ecologic(al) rehabilitation 生态复原
ecologic(al) relationship 生态关系
ecologic(al) replacement 生态替换
ecologic(al) resiliency 生态恢复力
ecologic(al) response 生态感应
ecologic(al) restoration 生态修复
ecologic(al) restoration technique 生态修复技术
ecologic(al) risk analysis 生态风险分析
ecologic(al) risk assessment 生态风险评价
ecologic(al) risk characterization 生态风险表征
ecologic(al) risk grade 生态风险级别
ecologic(al) risk index 生态风险指数
ecologic(al) river quality 生态河流水质
ecologic(al) science 生态学科学
ecologic(al) security 生态安全
ecologic(al) security status 生态安全状况
ecologic(al) seral method 生态系列法
ecologic(al) series 生态系列
ecologic(al) setting 生态背景
ecologic(al) simulation 生态模拟
ecologic(al) sociology 生态社会学
ecologic(al) speciation 生态性物种形式
ecologic(al) species group 生态物种组
ecologic(al) stability 生态稳定性
ecologic(al) statistical analysis 生态统计分析
ecologic(al) stimulation 生态刺激
ecologic(al) strain 生态菌株
ecologic(al) strategy 生态对策
ecologic(al) structure 生态结构
ecologic(al) succession 生态演替;生态序列;生态迁移;生态接续
ecologic(al) suitability 生态适应性
ecologic(al) survey method 生态调查法
ecologic(al) sustainability of river 河流生态可持续性
ecologic(al) synthesis 生态综合
ecologic(al) system 生态系统
ecologic(al) system engineering 生态系统工程
ecologic(al) technique 生态技术
ecologic(al) technology 生态技术学;生态工艺
ecologic(al) threshold 生态阈限;生态临界;生态极限
ecologic(al) tolerance 生态容限;生态耐性
ecologic(al) toxicity 生态毒性
ecologic(al) type 生态型
ecologic(al) value 生态价值
ecologic(al) variable 生物变数
ecologic(al) variety 生态变种
ecologic(al) vulnerability 生态脆弱性

ecologic(al) water 生态用水
ecologic(al) winter hardiness 生态耐寒性
ecologic(al) zoogeography 生态动物地理学
ecologist 生态学家;生态学工作者
ecologo-geographic(al) distribution 生态地理分布
ecologo-geography 生态地理学
ecology 均衡系统;环境适应学;生态(学)
ecology and analysis of trace contaminant 生态与微量元素污染物分析
ecology bank 生态银行
ecology group 生态群体
ecology movement 生态运动;生态活动
ecology of marine 海洋生态学
ecology of marine sediments 海洋沉积生态学
ecology of nature 自然生态学
ecology of relationship 关系生态学
ecology of river system 水系生态学
ecology reservoir 油藏生态学
ecology technology treatment system 生态技术处理系统
eco-mark 生态标记
eco-museum 生态博物馆
econocrete 经济混凝土
Economet 镍铬铁合金
econometric analysis 计量经济分析
econometric analysis model 计量经济分析模型
econometric approach 经济计量方法
econometric criteria 经济计量准则
econometric experimentation 经济计量试验工作
econometric forecasting 经济计量学预测
econometric forecasting method 经济计量学预报法
econometric forecasting model 经济计量预测模型
econometric function 经济计量函数
econometric gaming 经济计量对策模拟
econometrician 经济计量学家
econometric jargon 经济计量学术语
econometric method 经济计量方法;计量经济法
econometric methodology 经济计量学方法论
econometric model 经济计量模式;计量经济模型
econometric model analysis 经济计量模型分析
econometric modeling technique 经济计量模型建立技术
econometric model method 经济模型法;经济计量学法
econometric model simulation 经济计量模型模拟
econometric research method 经济计量研究方法
econometrics 经济计量学;经济测量学;计量经济学;估算经济学
econometric software package 经济计量软件包
econometric statistical inference 经济计量统计推论
econometric system 经济计量系统
econometric technique 经济计量技术
econometry 经济测量学
economic 经济的
economic accounting 经济核算
economic accounting of commercial enterprises 商业企业经济核算
economic accounting of group 班组经济核算
economic accounting system 经济核算制
economic accounting target 经济核算指标
economic active population 经济活动人口
economic activity 经济活动
economic activity analysis 经济活动分析
economic activity analysis decision support system 经济活动分析决策支持系统
economic adjustment 经济调整
economic advance 经济发展
economic advantage 经济优势
economic advice 经济方面的意见
economic advisability 经济合理性
economic advisor 经济顾问
economic affairs 经济事务
economic age-life method 经济寿命法
economic agent 经济代理人
economic aggregate 经济活动总量
economic agreement 经济协定
economic aid 经济援助
economic ailment 经济失调
economic air speed 经济风速
economic algae 经济藻类
economically advanced country 经济发达国家
economically coordinated region 经济协作区
economically dependent industrialenterprises 非独立核算工业企业
economically feasible 经济上可行的;经济上合算;经济可行的
economically practical 经济上切实可行的;经济上可行的
economically recoverable oil 可经济地开采的石油;经济上有开采价值的石油;工业性可采石油(储量)
economically sound thermal design 经济的声热设计
economically underdeveloped area 经济不发达地区
economically unsound project 经济上不合算项目
economic analysis 经济分析
economic analysis and accounting school 经济分析和会计学派
economic analysis evaluation 经济分析评价
economic analysis life 经济分析年限
economic and commercial data bank 经贸信息管理系统
economic and monetary union 经济和货币同盟
economic and population forecasting techniques 经济和人口预测技术
economic and practical 经济实惠
economic and sector analysis 经济和地区分析
economic and social benefits 经济和社会利益
Economic and Social Commission for Asia and the Pacific 亚太经济和社会委员会
economic and social determinant 经济和社会决定因素
economic and social development plan 经济和社会发展计划
economic and social development strategy 经济和社会发展战略
economic and social justification 经济和社会效益论证
economic and social justification for road development 道路建设的经济和社会合理性
economic and technical cooperation 经济技术协作;技术和经济合作
economic and technical development zone 技术经济开发区
economic and technologic(al) development zone 技术经济开发区
Economic and Trade Arbitration Commission of CCPTT 中国国际贸易促进会经贸仲裁委员会
economic and trade exhibitions 经济贸易展览会
economic appendage 经济附庸
economic application 经济应用
economic appraisal 经济评价;评估
economic arbitration 经济仲裁
economic area 经济区(域)
economic argument 经济论证
economic arteries 经济命脉
economic assessment 经济评价
economic assets 经济资产
economic assistance project 经济援助项目
economic associations 经济协会
economic atlas 经济地图集
economic atmosphere 经济情势
economic austerity 经济紧缩
economic average 经济平均值
economic background 经济背景
economic balance 经济平衡
economic barometer 经济情势观察指标;经济观测指标
economic base 基本经济;经济基础
economic base multiplier 经济基础乘数
economic basis 经济基础
economic batch determination 经济批量决定法
economic batch quantity 经济批量
economic bebefit for environmental protection 环境保护经济效益
economic behavio(u)r 经济行为
economic benefit 经济效益;经济利益
economic benefit indicator 经济效益指标;经济效果指标
economic benefit of environmental protection 环境经济效益
economic benefits of standardization 标准化经济效果
economic bloc 经济集团
economic blockade 经济封锁
economic boiler 经济式锅炉
economic boom 经济繁荣
economic boundaries 经济界限
economic breakdown 经济崩溃
economic business 经济业务
economic bust 经济崩溃
economic calculation at the factory level 厂级经济核算
economic calculation team 经济核算队
economic canon 经济法规
economic capability 经济势能;经济能力
economic capacity of land 土地经济容力
economic carburet(t)or 经济汽化器
economic case 经济案件
economic causality 经济因果律
economic center 经济中心
economic change 经济变革
economic channel 经济航道
economic chaos 经济混沌
economic character 经济性状;经济特性
economic characteristic 经济特征
economic circle 经济圈;经济界
economic circulative graph 经济循环图
economic city 经济城市
economic climate 经济晴雨表;经济情势
economic coal reserves 煤经济储量
economic code 经济法典
economic coefficient 经济系数
economic cohesion 经济内聚力
economic collaboration 经济合作
economic combination 经济联合
Economic Commission for Europe 欧洲经济委员会
economic community 经济界;经济共同体
Economic Community of West African States 西非国家经济共同体
economic comparison 经济比较
economic comparison between design proposals 设计方案经济比较
economic compensation 经济赔偿
economic competition 经济竞争
economic competition act 经济竞争法
economic complementariness 经济补助性
economic complex 经济综合体;经济联合
economic concentration 经济集中制
economic condition 经济情况
economic conflict 经济矛盾;经济冲突
economic conflict groups 经济冲突群
economic consequence 经济结果;经济后果
economic conservation 经济资源保护
economic consideration 经济利益;经济考虑
economic constitution 经济法
economic constraint 经济约束;经济制约
economic construction 经济建设
economic construction bonds 经济建设公债
economic continuity 经济连续性
economic continuous rating 经济连续出力
economic contract 经济合同
economic contract arbitration 经济合同仲裁
economic contract arbitration committee 经济合同仲裁委员会
economic contraction 经济紧缩;经济萎缩
Economic Contract Law of the People's Republic of China 中华人民共和国经济合同法
economic control 经济控制
economic control law 经济统制法
economic cooperation 经济合作;经济协作
economic cooperation administration 经济协作局
economic cooperation as joint ventures 合资经营
economic coordination 经济协调
economic coordination region 经济协作区
economic cost 经济费用(按影子价格计算的费用);经济成本
economic cost comparison 经济费用比较
Economic Council of Arab League 阿拉伯联盟经济理事会
economic counselor 经济参事
economic counselor's office 经济参赞处
economic creation potential 经济建设潜力;创收潜力
economic crisis 经济危机
economic criterion 经济准则;经济法则;经济标准
economic-criterion system 经济准则系统
economic crop 经济作物

economic cross-section 经济断面
economic current density 经济电流密度
economic curve 经济曲线
economic cut 经济开挖
economic cybernetics 经济控制论
economic cycle 经济周期;经济循环
economic cycle theory 经济周期理论
economic cyclical changes 经济周期波动
economic damage 经济损失
economic dam height 经济坝高
economic data 经济资料;经济数据
economic decision 经济性决策
economic decision-making 经济决策
economic decline 经济衰退
economic demonstration 经济论证
economic density 经济密度
economic dependence 经济依赖
economic dependency ratio 经济信赖率
economic depletion 经济耗竭
economic deposit area 经济的抛泥区
economic depreciation 经济折旧;经济贬值;实质折旧
economic depression 经济萧条
economic depth 经济深度
economic depth range 经济水深
economic design of wastewater treatment 经济型污水处理厂设计
economic despatching 经济调度
economic determination of mineral resources 矿产资源经济决策
economic development 经济开发;经济建设;经济发展
economic development accountancy 经济发展会计
economic development cycle 经济发展周期
Economic Development Institute 经济发展学院
economic development levy 经济开发税;经济发展税
economic development plan 经济发展计划
economic development strategy of coastal areas 沿海地区经济发展战略
economic development zone 经济开发区
economic diameter 经济直径
economic diameter of pipe (管道的)经济管径
economic dislocation 经济混乱
economic disparities 经济差异
economic dispatch 经济输送法;经济分配;经济调度
economic disposal area 经济的抛泥区
economic dispute cases 经济纠纷案件
economic distance 经济距离
economic distress 经济景气
economic division (用于财产增值的)经济划分
economic downturn 经济下降
economic drain 经济耗费
economic dredging depth 经济挖深
economic dredging speed 经济挖泥航速
economic dualism 经济二重性
economic durable years 经济的使用期;经济的耐用年限
economic dynamics 动能经济学;经济动态
economic effect 经济效果;经济效益效应;经济成效
economic effectiveness analysis 经济效益分析
economic effect of curvature 弯道的经济影响
economic effect of development 开发的经济效果
economic effect of geologic(al) working 地质工作经济效果
economic effect of project 工程经济效益
economic effect of road improvement 道路改建的经济效果
economic efficiency 经济性;经济效益;经济效率;经济效果
economic efficiency of commecial enterprises 商业企业经济效益
economic efficiency program(me) 经济效率研究计划
economic ends 经济目标
economic entity 经济组织;经济实体
economic environment 经济环境
economic equilibrium 经济平衡;供需平衡
economic establishments 经济事业
economic estrangement 经济的异化
economic evaluation 经济评价;经济评估;经济核算;经济估值
economic evaluation model 经济评价模型
economic evaluation of a project 项目经济估价
economic evaluation of ore deposits 矿床经济评价
economic evaluation of project 工程经济评价
economic evaluation of railway investment 铁路投资经济评价
economic evaluation technique 经济评价方法;经济核算技术
economic events 经济事态
economic expansion 经济扩张
economic expansion incentives 经济扩张动力
economic expansion rate 经济扩张率
economic experiment 经济试验
economic exposure 经济暴露
economic facing point lock 简易对向道岔锁闭器
economic factor 经济因素
economic feasibility 经济可行性;经济合理性
economic feasibility analysis 经济可行性分析
economic feasibility study 经济上可行性研究
economic feature 经济情况
economic fire-control theory 经济防火理论
economic flow rate 经济流量
economic fluctuation 经济波动
economic force 经济力量
economic forecast(ing) 经济预测
economic forecasting method 经济计量预测法
economic forecasting of new products 新产品经济预测
economic forest 经济林;用材林
economic form 经济形态
economic foundation 经济基础
economic freedom 经济自由
economic friction 经济摩擦
economic function 经济职能
economic gain 经济收益;经济成果;经济效果
economic gain or loss 经济损益
economic gantry crane 简易龙门起重机
economic gap 经济差距
economic geography 经济地理(学)
economic geology 经济地质学
economic goal 经济目标
economic goods 经济物品;经济商品;经济货物;经济财货
economic grade 经济坡度
economic gradient 经济坡度
economic grant 经济补助
economic groundwater yield 地下水经济产水量
economic growth 经济增长;经济成长
economic growth factor 经济增长系数
economic growth rate 经济增长率;经济成长率
economic growth rate in the near future 近期经济增长速度
economic growth theory 经济增长论
economic guideline 经济指标;经济方针
economic hauling distance 经济(的)运距;经济量程
economic hauling distance of soil 土方调配经济运距
economic head 经济扬程
economic height 经济高度
economic hierarchical model 经济递阶模型
economic hierarchy scheme 经济阶层结构模式
economic hinterland 经济腹地
economic hydraulic horse power 经济水马力
economic hypothesis 经济假设
economic imbalance 经济不平衡状态
economic impact 经济影响;经济冲击
economic impact analysis 经济影响分析
economic impact evaluation 经济影响评估
economic incentive 经济鼓励
economic income 实质收益
economic index number 经济指数
economic indication 经济表现
economic indicator 经济指数;经济指标
economic indicator of mineral commodities 矿产经济指标
economic indicator of mining 矿业经济指标
economic infiltration 经济渗透
economic information 经济信息;经济情报
economic infrastructure 经济基础设施
economic institution 经济机构
economic instrument of environment(al) management 环境管理的经济手段
economic integration (不同经济条件住房的)混合居住;经济整合;经济一体化
economic interest 经济利益
economic internal rate of return 经济内部收益率
economic internationalization 经济国际化
economic interpretation 经济解释;经济阐述
economic investigation for water transport 水运经济调查
economic investigationl economic research 经济调查
economic investment 经济投资
economic justification 经济合理性;经济论证
economic law 经济规律;经济法则;经济法(律)
economic law and regulation 经济法规
economic legislation 经济立法;经济法规
economic lever 经济杠杆
economic liberalism 经济自由主义
economic life 经济生活;经济年限;经济使用年限
economic life cycle forecast method 经济寿命预测法
economic lifeline 经济命脉
economic life of equipment 设备经济使用寿命
economic life time 经济寿命;经济使用年限;经济使用期限
economic limit 经济限制;经济极限
economic limitations 经济局限性
economic limit rate 经济极限产量
economic load 经济负载;经济负荷;经济负担
economic load dispatcher 经济负载调度装置
economic load factor 经济负荷因素
economic long cycle theory 经济长周期论
economic lose 经济损失
economic lots 经济划分
economic lot size 经济批量;最佳生产或订购的一次批量
economic lot-size analysis 经济批量规模分析
economic lot-size equation 经济批量方程
economic lot-size formula 经济批量公式
economic lot-size model 经济批量模型
economic lot sizes in purchasing 采购的经济批量规模
economic machinery drilling speed 经济机械钻速
economic magnitude 经济数量;经济量
economic man 经济人
economic management 经济经营;经济管理
economic management instrument 经济管理手段
economic manufacturing quantity 经济生产量
economic map 经济地图
economic mapping 经济制图
economic maturity 经济成熟度
economic means 经济手段;经济方法
economic measurement 经济计量
economic mineral 有用矿物;有经济价值的矿物
economic misfortunes 经济不幸
economic mobility 经济变动性
economic mobilization 经济总动员;经济动员
economic model 经济模型;经济模式
economic model analysis 经济模型分析
economic model construction 经济模型结构
economic model forecast 经济模型预测
economic momentum 经济势头
economic motive 经济动机
economic movement mechanism 经济运行机制
economic net present value 经济净现值
economic network 经济网
economic norms 经济指标
economic objective 经济目标
economic obsolescence 经济性淘汰;贬值;经济废弃
economic of free enterprise 自由企业经济
economic offset policy 经济补偿政策
economic of space 空间的节约;经济利用空间
economic operation 经济操作运行
economic operational mechanism 经济运行机制
economic operation model 经济运行模式
economic opportunities for 经济出路
economic optimization criterion 经济优化标准
economic order 经济秩序
economic order(ing) quantity 经济订货量;经济订购(批)量;最佳订货量;经济定购量;经济订单数量;最经济订货批量
economic order quantity method 经济采购量决定法
economic order size 经济购货批量;经济订货批量
economic ore 经济矿藏

economic organization 经济组织
economic organization rationalization 经济组织合理化
economic organization structure 经济组织结构
economic outlook 经济展望;经济前景
economic output 经济功率;经济产量
economic parameter 经济指标;经济参数
economic parasitism 经济上的依附
economic pattern 经济形态;经济模式
economic penetration 经济渗透
economic percentage of steel 钢筋的经济百分比;经济含钢率
economic performance 经济特性;经济实绩;经济成就
economic performance index 经济实效指标;经济绩效指标
economic performance indicators 经营成果指标
economic perspective 经济远景
economic phenomenon 经济现状;经济现象
economic philosophy 经济哲学
economic picture 经济状态
economic pipe size (管道的)经济管径
economic planning 经济计划;经济规划
economic planning administration 经济计划管理
Economic Planning Board 经济计划局(英国)
economic planning council 经济计划会议
economic planning for water transport 水运经济规划
economic plant 经济植物
economic pleasure 经济的享受
economic poison 实用毒剂
economic policy 经济政策
economic policy analysis 经济政策分析
economic policy environmental protection 环境保护经济政策
economic policy insurance 经济政策保险
economic policy-making 经济决策
economic pollution threshold 经济污染阈值
economic position 经济地位
economic possibility 经济可能性
economic postulate 经济假设
economic potential 经济潜力
economic power 经济功率
economic prediction 经济预测
economic premise 经济前提
economic pressures 经济压力
economic price 经济价格
economic principle 经济原则
economic principles of production 生产经济原则
economic problem 经济问题
economic process 经济过程
economic product 经济产品
economic production 经济生产
economic production life 经济开采期限
economic productivity 经济生产率
economic profile 经济概况
economic profit 经济利润
economic progress 经济进步
economic projection 经济预测
economic proportional relation 经济比例关系
economic prosperity 经济振兴
economic protectionism 经济保护主义
economic proximity 经济接近性
economic purchase quantity 经济购入量;经济购货量
economic purchasing quantity 经济订购(批)量
economic quantities 经济数量
economic ranking criteria 经济等级准则
economic rate of return 经济回收率
economic ratio 经济比率(指钢筋混凝土中钢筋和混凝土的适当比率);经济配筋率
economic rationality 经济合理性
economic ratio of reinforcement 经济配筋率;经济含钢率;最佳配筋率
economic ratio of reinforcement to concrete 混凝土经济配筋率
economic reality 经济现实
economic reasonableness 经济合理性
economic reasoning 经济推理
economic recession 经济衰退
economic reckoning 经济核算
economic reconstitution 经济重建
economic reconstruction and readjustment 经济重建和调整

economic recoverable ore reserves 经济可采储量
economic recovery 经济复原;经济复兴;经济复苏;经济采收率
economic reform 经济改革
economic region 经济区(域)
economic regional 经济区级
economic regulating mechanism 经济调节机制
economic regulatory mechanism 经济调节机制
economic rehabilitation 经济恢复;经济复兴善后
economic relations 经济关系
economic relationships 经济关系
economic rent 经济租金;经济地租;经济纯利;物产租金(不包括设备和服务)
economic reorganization 经济改组
economic research 经济研究;经济调研
economic research center 经济研究中心
economic research method 研究经济的方法
economic research of goods traffic 货运经济调查
economic research of passenger flow 客流经济调查
economic reserve 经济储量
economic resistance of heat transfer 经济热传导电阻
economic resources 经济资源;财源
economic responsibility 经济责任
economic responsibility system 经济责任制
economic restructuring 经济改组
economic result 经济效益;经济结果;经济效果
economic result indicators 经济效果指标
economic resurgence 经济复兴
economic retrenchment 经济紧缩
economic returns 经济收益
economic returns and social benefits 经济和社会利益
economic revitalization 经济复苏
economic rights and interests 经济权益
economic risk 经济风险
economic rivalry 经济竞争
economic ruin game 经济破产对策
economic running 经济运转
economics 经济因素;经济学
economic sanction 经济制裁
economic satisfaction 经济满足
economic scale 经济规模
economic scale factor 经济比例系数
economic scene 经济状态;经济实况
economic science 经济学;经济科学
economic section (构件的)经济截面;经济断面
economic section of bridge 桥梁经济断面
economic sector 经济成分
economic sector owned by the whole people 全民所有制经济
economic security 经济安全
economic selection cutting 经济择伐
economic selenology 经济月质学
economic self-sufficiency 经济自足
economic sense 经济概念
economic separation 经济割据
economic series 经济序列
economic service life 经济使用年限
economic setup 经济体制
economics for development 发展经济学
economic shape of cross-section 断面的经济形式
economic sheet 经济计算表
economic short trees 矮型经济林
economic signals 经济信号
economics in steel sheet pile 钢板桩的经济性
economic situation 经济形势
economic size 经济尺度;经济尺寸
economic societies 经济学会
economic sociology 经济社会学
economics of control 控制经济学;统制经济学
economics of energy 能量经济学;能源经济学;动能经济学
economics of flood control 防洪经济学
economics of grade separation 立交的经济性
economics of human capital 人力资本经济学
economics of innovation 技术革新经济学
economics of internal organization 内部组织经济学
economics of land resources 土地资源经济学
economics of learning 学习的经济效益
economics of management 管理经济(学)
economics of marine resources 海洋资源经济状态

economics of materials 材料经济学
economics of production 生产经济学
economics of scale 经济规模;经济比例;规模经济学;比例经济
economics of scale and scope in transport industry 运输业规模经济与范围经济
economics of wastewater ozonation 废水臭氧化的经济性
economics of water conservancy 水利经济学
economics of welfare 福利经济学
economic span 经济跨度;(构件的)经济跨度
economic span-length 经济跨度;(构件的)经济跨径
economic span of bridge 桥梁经济跨径
economic speed 经济速率;经济航速;经济车速
economic sphere 经济影响范围;经济活动区域
economic stability 经济稳定(性)
Economic Stabilization Act 经济稳定法
economic stabilization policy 经济稳定政策
economic stabilizer 经济稳定器
economic statistics 经济统计(学)
economic status 经济法规;经济地位
economic stimulus 经济鼓励;经济刺激
economic storage (水库的)经济库容
economic stragegy 经济战略
economic strategy goal system 经济战略目标系统
economic stratification 经济阶层的形成
economic strength 经济实力
economic stripping ratio 经济剥采比
economic structural equation 经济结构方程式
economic structural mode 经济体制模式
economic structure 经济体制;经济结构
economic structure shift 经济体制转机
economic study 经济研究;经济设计
economic superiority 经济优势
economic super power 经济超级大国
economic supervision 经济监督
economic support 经济支援
economic surplus 经济剩余;经济过剩
economic survey 经济调查
economic survey of maritime navigation 水运经济调查
economic system 经济制度;经济系统;经济体制;经济体系
economic system reform 经济体制改革
economic tactic 经济战术
economic take-off 经济起飞
economic target 经济目标
economic target system 经济指标体系
economic technical index 技术界限标定
economic technical system 经济技术系统
economic test 经济检验
economic theorem 经济定理
economic theory 经济理论
economic theory and method research 经济理论与方法研究
economic thermal resistance 经济热阻
economic thickness 经济厚度
economic threshold 经济限阈
economic ties 经济纽带
economic time series 经济时间数列
economic traits 经济特点
economic transaction 经济交易
economic transportability 经济输送距离
economic transport distance 运输经济距离
economic trees 经济林;用材林
economic trench 经济趋势;经济发展趋势
economic-type appraisal method 经济型评价法
economic unequality 经济失调;经济失衡
economic union 经济联盟
economic unit 经济单位
economic unity 经济统一
economic utility 经济效用
economic utilization 经济利用
economic value 经济价值;工业价值
economic value of project 工程经济价值
economic variable 经济变数;经济变量
economic velocity 经济速度;经济流速
economic viability 经济生存能力;经济可行性
economic war(fare) 经济战;经济竞争
economic water-oil ratio 经济水油比
economic water resources planning 经济水资源规划
economic water resources planning model 经济

水资源规划模型
economic waterway 经济航道
economic welfare 经济福利
economic wonder 经济奇迹
economic work 经济业务
economic worth 经济价值
economic worth of project 工程经济价值
economic yardstick 经济标准
economic yield 经济出水量；经济产量
economic zone 经济区(域)
economic zoning 经济区划
economiser ＝economizer
economising degrees of freedom 节约自由度
economism 经济主义
economist 经济学家；经济工作者；节约的人
economization 节约；节省；缩减法
economization rate of labo(u)r occupancy 劳动占用节约率
economize in raw materials 节约原料
economize on raw materials 降低原料消耗
economizer 炉水预热器；经济至上者；节约装置；节约器；节油器；节热器；省热器；省煤器
economize raw materials 节约原料
economizer bank 空气预热管；节热器排管；预热管
economizer body 省油器本体
economizer conic(al) valve 锥形省油器阀
economizer hopper passage plug 经济喷口油道塞
economizer relay 节电继电器
economizer surface 省煤器受热面
economizer valve 省油阀
economizing 节约的
economizing basin 节水池
economizing labo(u)r time 节约劳动力时间
economizing rate of exploration expenses 地勘费用节约率
economy 经济(性)；节约；缩减率
economy anatomy 经济剖析
economy brick 节约型砖；小型砖；经济砖
economy calculation 节约计算；经济计算
economy class 客机普通舱；经济舱
economy class room 经济舱
economy construction area of urban geology 城市地质经济建设区
economy flat plate stern 经济式船尾
economy grade 廉价级；经济级；低级；次级；便宜级
economy in consumption 节约消费
economy in oil consumption 节约石油消费
economy jet (节煤省气的)经济喷射器
economy load 经济负载
economy load curve 经济荷载曲线
economy of abundance 富裕经济
economy of control 控制点合理安排
economy of energy 能源经济；能源的经济节约
economy of forestry 林业经济
economy of gain 增重经济效益
economy of helper grades 辅助坡度的经济性
economy of housing construction 住宅建设经济
economy of large locomotives 大型机车的经济性
economy of manpower 节约劳动力
economy of material 节约用料
economy of plenty 富裕经济
economy of scale 规模经济
economy ought to be restructured in a package deal 一揽子经济体制改革
economy resistance 经济电阻
economy's dark continent 经济的黑暗大陆(指经济未开发的非洲)
economy sink unit 经济冲洗槽
economy's payments mechanism 经济的支付方法
economy test 经济效果评价；对经济效果的评价
economy treadle 经济型油门摇臂
economy-type carburettor 省油汽化器
economy wall 经济墙
econonic account 经济账户
econo-technical index 经济技术指标
econo-technical norm 经济技术指标
ecophene 生态变种反应
ecophenotype 生态表型
eco-pond 生态池塘
ecoregion 生态区
ecoscience 环境科学
ecospecies 生态种
ecosphere 生物域；生物大气层；生态域；生态圈；生态层
ecostate 无肋骨的
ecostratigraphic(al) classification 生态地层分类
ecostratigraphic(al) unit 生态地层单位
ecostratigraphy 生态地层学
ecosystem 生态系统；生态区系
ecosystem approach 生态系观点
ecosystem assessment 生态系统评价
ecosystem assessment index system 生态系统评价指标体系
ecosystem assessment model 生态系统评价模型
ecosystem balance 生态系统平衡
ecosystem characterization 生态特性
ecosystem component 生态系统要素
ecosystem condition 生态系统条件
ecosystem conservation group 生态系统养护组
ecosystem degradation 生态系统退化
ecosystem development 生态系统演化
ecosystem development strategy 生态系统开发战略；生态系统开发对策
ecosystem diversity 生态系统多样性
ecosystem dynamics 生态系统动力学；生态系动力学
ecosystem ecology 生态系统生态学；生态系生态学
ecosystem effect 生态系统效应
ecosystem energetics 生态系统力能学
ecosystem engineering 生态系统工程
ecosystem hierarchical model 生态递阶模型
ecosystem hydrologic simulator 生态系统水文模拟器
ecosystem interface 生态系(统)界面
ecosystem management 生态系统管理
ecosystem management theory 生态系统管理理论
ecosystem mapping 生态系统制图
ecosystem model 生态系模型
ecosystem modeling 生态系统模拟
ecosystem monitoring 生态系统监测
ecosystem of courtyard 庭院生态系统
ecosystem of natural water bodies 天然水域生态系统
ecosystem of water body 水域生态系(统)
ecosystem prediction 生态系预报；生态系统预测
ecosystem processes 生态系统的过程
ecosystem productivity 生态系统生产力
ecosystem protection strategy 生态系统保护对策
ecosystem rehabilitation 生态系复原
ecosystem response unit 生态系统反应器
ecosystem restoration 生态系统修复
ecosystem security 生态系统安全性
ecosystem service 生态系统服务
ecosystem simulating model 生态系统模拟模型
ecosystem stability 生态系统稳定性
ecosystem structure 生态系统结构
ecosystem structure and function 生态系结构与功能
ecosystem succession 生态系统演替
ecosystem transfer rate 生态系迁移率
ecosystem type 生态系统型；生态系统类型
ecosystem uncertainty analysis 生态系统不确定性分析
ecotechnique 生态技术
ecotechnology 生态无害技术；生态技术
ecotelemetry 生态遥测术
ecotonal community 交错区群落
ecotone 群落交错区；群落过渡区；生态过渡带
ecotope 生态区；生态环(境)
ecotopic 适应特殊生态的
eco-tourism 生态旅游
ecotoxic effect 生态毒理性效应
ecotoxicological effect 生态毒理性效应
ecotoxicological impact 生态毒理性影响
ecotoxicological test 生态毒理性测试
ecotoxicology 生态毒理学
ecotype 生态型；生态类型
ecotypical selection 生态变种选择
ecoulement 重力滑坡；崩坍
ecouvillon 硬刷子
ecouvillonnage 擦洗术
ecozone 生态区
ecronic 河口湾；港湾
ecru 灰黄色；浅褐色
ecru silk 半脱胶丝
Ecsaine 埃克萨伊纳含有聚酯的非织造织物
ectal layer 外生层
ectatic 扩张的
ectexis way 原地混合岩化方式；注入变熔作用方式
ectocrine 外界微量影响物
ectodynamic(al) soil 外动力土壤
ectodynamorphic 外动力型的
ecto-entad 由外向内
ectogene 外因
ectogeny 外生现象
ectogony 外生效应
ectomorph 瘦型体质
ectomorphy 瘦型体质
ectoplasm 外质
ectoscopy 外表检视法
ectosphere 外球
ectotheca 外壁
ectozoon 外寄生物
ectropometer 哑罗经；方位盘
ectype 副本；复制品
ecumenopolis 全球城市；普世城(指按某种理想规划的城市)；世界性都市
Edale shales 埃达尔页岩
edaphic 土壤的
edaphic climax 土壤演替顶极；土壤顶极
edaphic climax association 土壤演替顶极群丛
edaphic climax community 土壤演替顶极群落
edaphic community 土壤生成群落
edaphic condition 土壤条件
edaphic control 底土控制
edaphic ecotype 土壤生态型
edaphic factor 土壤因素
edaphic formation 土壤群系
edaphic indicator 土壤指示植物；土壤指示剂
edaphic race 土壤族
edaphic scale 土壤图表
edapho-climatic condition 风土条件
edapho-coenotic series 土壤植物群落系列；土壤植物群落分布区
edaphogenic succession 土壤发生演替
edaphology 应用土壤学；土壤学；土壤生态学；生态土壤学
edaphon(e) 土壤微生物(群)；土壤生物群落
edaphonekton 土壤水中栖居生物；土壤水生生物
edapho-phytocoenotic area 土壤植物群落区；土壤群落分布区
edatope 土壤环境
Eday Sandstone 艾台砂岩
Eddington approximation 爱丁顿近似
Eddington-Bar bier method 爱丁顿—巴比叶方法
Eddington limit 爱丁顿极限
Eddington luminosity 爱丁顿光度
Eddington standard model 爱丁顿标准模型
eddy 漩水；漩涡；涡度
eddy action 漩涡作用
eddy available potential energy 涡动有效势能
eddy axis 漩涡轴线；涡流轴线
eddy built bar 涡成砂坝
eddycard 涡流卡片
eddycard memory 涡流卡片存储器
eddycard store 涡流卡片存储器
eddy catching plate 涡流捕尘板
eddy conduction 涡流传导
eddy conductivity 涡流传导性；涡动传导性；涡动传导率
eddy correlation 涡动相关
eddy correlation method 涡度相关法；紊流交换法(估计蒸散量)
eddy current 漩涡水；漩涡电流；涡(电)流
eddy current anomaly 涡流损耗异常
eddy current brake 电磁离合器；涡流制动器；涡流制动(机)；涡流闸
eddy current braking 涡流制动法
eddy current braking magnet 涡流制动磁体
eddy current clutch 涡流离合器
eddy current coefficient 涡流系数
eddy current confinement 涡流约束
eddy current constant 涡流常数
eddy current coupler 涡流耦合器
eddy current damper 涡电流阻尼器
eddy current damping 涡流阻尼
eddy current disc 涡流盘
eddy current dynamometer 涡电流扭矩仪
eddy current effect 涡流作用；涡流效应
eddy current examination 涡流检测
eddy current factor 涡流系数
eddy current flaw detection 涡流探伤法

eddy current flowmeter 涡流流量计
eddy current ga(u)ge 涡流计
eddy current gravimeter 涡流重差计
eddy current heater 涡电流加热器
eddy current heating 感应加热;涡流加热;涡电流加热
eddy current inspection 涡流探伤;涡流检验;涡电流探伤
eddy current instrument 涡流测量仪器
eddy current loss 涡流损失;涡流损耗;漩流损失
eddy current motor 涡流电动机
eddy current rail brake 涡流缓行器;轨道涡流制动机
eddy current retarder 涡流阻尼器
eddy current revolution counter 涡流转速计;涡流式转速表
eddy current rotating brake 涡流旋转制动机
eddy current screen 涡流屏蔽
eddy current seismometer 涡流地震检波器
eddy current separator 涡流隔离板;涡流分离器
eddy currents in attraction type motor 吸引型电动机中的涡流
eddy current speed indicator 涡流速度指示表;涡流测速计
eddy currents ultrasonics 涡电流超声
eddy current tachometer 感应式转速表;涡流(式)转速计
eddy current test(ing) 涡流探伤;涡电流试验
eddy current testing equipment 涡流探伤机
eddy current thickness meter 涡流测厚仪;涡电流法测厚仪
eddy current transducer 涡流式传感器
eddy current-type dynamometer 涡电式测功器
eddy current type geophone 涡流式检波器
eddy current type retarder 涡流式减速器
eddy deformation 涡漩变形
eddy-deposited silt 漩涡沉积泥砂;涡流沉积淤泥
eddy diffusion 扰动扩散;涡(流)扩散;涡度扩散;涡动扩散;紊流扩散
eddy diffusion coefficient 涡流扩散系数;涡流扩散率
eddy diffusion in estuaries 河口涡流扩散
eddy diffusion in oceans 海洋涡流扩散
eddy diffusion term 涡流扩散项
eddy diffusivity 涡流扩散率;涡流扩散度;涡动扩散率;紊流扩散系数
eddy displacement currents 涡流位移电流
eddy effect 涡流效应
eddy effusion 涡流喷射
eddy energy 涡动能量;涡动能
eddy error 涡流误差
eddy field 涡漩场
eddy flow 旋流;涡流;紊流
eddy flux 涡流通量;涡流动;涡动通量;紊动通量
eddy flux evaporation formula 涡动通量蒸发公式
eddy forcing 涡动作用
eddy free flow 无漩涡水流
eddy-free front 流线型前身
eddy frequency 涡漩频率
eddy generation 涡流形成
eddy heat conduction 涡流热传导;涡动热传导;紊流热传导
eddy heat flux 涡流热传导
eddying 涡流形成
eddying flow 涡流流动
eddying kinematic viscosity 涡流运动黏度
eddying motion 涡流运动;涡漩运动
eddying resistance 涡流阻力
eddying shedding 涡流分离
eddying turbulence 涡漩紊流
eddy kinematic viscosity 涡流运动黏度
eddy kinetic energy 涡流动能;涡动动能
eddylike 涡流形的
eddy loss 涡流损失
eddy making 涡流形成
eddy-making resistance 产生涡流的阻力
eddy marking 涡流痕迹;涡痕
eddy mill 漩涡磨机;锅穴;凹洞(河床岩石上的);涡流式碾磨机;涡流粉碎机
eddy milling 漩涡研磨
eddy mixing 涡流混合
eddy motion 涡动
eddy plate 抗涡流板;整流板
eddy pump 涡流泵

eddy reaction 涡流反应
eddy region 漩涡区;漩涡地区
eddy resistance 漩涡阻力;涡旋阻力;涡流阻力;涡动阻力
eddy scale 漩涡尺度;涡流尺度
eddy separation 涡流分离
eddy separation flowmeter 涡流分离流量计
eddy shearing stress 涡动剪应力
eddy shedding flowmeter 涡流分离流量计
eddy size 涡漩尺度;涡流尺度
eddy spectrum 涡漩谱;涡流谱
Eddy's theorem 埃迪原理(计算拱上任意断面弯矩的定理);埃迪定理
eddy stress 雷诺应力;涡流应力;涡动应力
eddy thermal diffusivity 涡流热扩散系数
eddy trail 尾涡;回旋小径
eddy transfer 涡流传递
eddy transport 涡动输送
eddy turbulence 涡动
eddy unit 涡形器
eddy velocity 涡流速度;涡动速度
eddy viscosity 涡动黏滞性;涡动黏(滞)度;涡流黏滞性;涡流黏(滞)度
eddy viscosity coefficient 涡流黏滞(度)系数;涡动黏滞系数
eddy wake 涡流伴流
eddy washing machine 漩涡式清洁机
eddy water 漩涡水;涡流
eddy wind 小旋风;涡旋风
eddy with horizontal axis 横轴漩涡
eddy with vertical axis 立轴漩涡;竖轴漩涡
eddy zone 旋涡区;涡流区
Edelean process 爱迪林精炼法
edelforsite 不纯硅灰石
edema 浮肿;水肿;[复]edemas 或 edemata
edema due to unacclimatization 不服水土肿
Edenborn coiler 艾登堡氏卷取机
Edenborn reel 艾登堡氏线材卷取机
Edenian 伊登阶;艾登阶
edenite 浅闪石
Eder-Hecht wedge 埃德·赫克特光楔
edge 锯成棱边;交界处;徐徐移动;缘(边);主切削刃;刀口;材边;边缘;边(沿);板边;刃;矢线
edge abrasion 边缘磨损
edge absorption 边缘吸收
Edge Act 艾奇国际条例
edge action 边缘作用;边界作用
edge action point 边缘作用点
edge angale and edge radius 刃角和刃半径
edge analysis 轮廓分析;图像边缘分析
edge angle 棱角;边缘角;刃角;偏角
edge aquifer 边缘蓄水层
edge arch 边缘拱顶;边缘发券;边沿弧线;边拱
edge-arrissing residue 磨边残屑
edge as cut (板玻璃切断后的)毛边;切割毛边
edge away 尖灭;楔出
edge band 边条;边瓦
edge banding 做边条;有边条;边饰条带
edge bar 缘杆;边铁
edge bar reinforcement 板框加固钢筋;边框加固钢筋;边缘钢筋
edge batten 边饰板条
edge beam 边梁
edge beam of shallow shell 扁壳边梁
edge bedding 边缘垫层
edge bend 起拱面;起拱点;弯钩(状)
edge blurry 边缘模糊
edgeboard connector 卡片边沿连接器
edge bonding machine 封边机
edge boring 边缘钻探;边孔
edge bowl 碗形边器(无槽法生产平板玻璃)
edge break 边缘
edge brick 角(砌)砖;墙角砖
edge broken 边破
edge brushing 刮边
edge buckle 边缘卷曲;边扣环
edge butt joint 边缘对接;边抵正接合;抵头接合
edge calking 搭边捻缝
edge cam 端面凸轮;平面凸轮;凸轮盘
edge capacitance 边缘电容
edge capacity 边容量
edge card 边缘卡片;边缘插件板
edge-card connector 板边插接器
edge catcher 碎玻璃箱

edge-cell seed plate 槽口式排种盘
edge-chamfering machine 边缘倒棱机
edge circular saw 切边圆锯
edge clamp mounting 夹边安装
edge clearance 边间隙;边部间距
edge clearance angle 刃后角
edge clocked D-type register 脉冲边沿同步的D型触发器;边沿同步的D型触发器
edge coal group 边缘煤群
edge coated card 边缘涂层卡片;边沿涂层卡片;包边卡片
edge coded card 边缘编码卡
edge coil 扁绕线圈;边绕线圈
edge collision 擦边碰撞
edge column 边柱
edge compression 边缘压缩
edge compressive strength 棱边压缩强度
edge condition 边缘条件;边界条件
edge connectable 带插接边的
edge connector 印制板插头;印刷板插头;边缘连接器;边缘接头;边缘插接件
edge constraint 边界约束
edge contact 边缘接触
edge-contact-cooled rectifier 边缘接触冷却的整流器
edge contour 边缘外形
edge control device 边缘控制装置;板边控制装置
edge course 侧砌
edge coverage 边缘覆盖率
edge covering 边角涂覆性
edge crack 横裂隙;边缘裂纹;边缘裂缝;边裂(缝);板边裂纹
edge creep 边部变形
edge-crimped yarn 刀口卷曲变形纱
edge cross member 边缘横梁;边缘横杆
edge current 边缘电流
edge cut 切边
edge-cutting force 刃切割力
edged 加边
edge damage 破边
edge-dam of spreader 涂布机边挡板
edge data 图廓外资料
edge defect 缺棱
edge-defined film-fed growth 导膜生长法;边缘限定薄膜供料生长法
edge deleting device 磨边装置
edge delineation 路边标线
edge density enhancement 边缘密度增强
edge design 边缘设计
edge detection 边缘检测
edge detection filter 边缘检测滤波器
edge detector 边缘检测器
edge deterioration 边缘磨损
edge diffraction 边缘衍射
edge direction 边缘方向
edge disjoint circuit union 边不相接回路的并集
edge disjoint cut-set union 边不相接割集的并集
edge disjoint subgraph 边不相交子图
edge dislocation 行列位错;边缘位错;刃式位错;刃型位错
edge distance 边到边距离;边距
edge distance of rivets 铆钉边距
edge distance ratio 边距比
edge disturbance 边缘扰动;边缘干扰;边界干扰
edge doctor 边缘涂料刮刀
edge doctor of roll paint 辊涂机刮边刀
edge dowel 边缘连接筋;边缘插筋
edged plate 镶边平板;镶边干板
edge-dressing machine 板边加工机床
edge drop plate 侧落式槽口排种盘
edged timber 边饰木材
edge-edge contact 边边接触
edge effect 棱角效应;翼缘效应;边缘影响;边缘效应;边缘效果;边行效应;边界效应;边际效应
edge effect modulator 边缘效应调制器
edge effect of shell 壳体边界效应
edge-emitting diode 边发射二极管
edge-enhanced image 边缘增强图像
edge enhancement 轮廓增强;勾边;浮雕法(遥感图像);边缘增强
edge fabric stress 边缘纤维应力
edge-face contact 边面接触
edge failure 啃边;路面啃边;边缘损坏
edge fault 边部缺陷

edge file 刃锉
edge filter 流线式(过)滤器;边缘滤器
edge fine grinding 板边细磨
edge finish(ing) 边缘加工;立轧
edge firing 边缘点火法;圈烧
edge fit 接边
edge-fixed 边缘固定的
edge flake 边部变形
edge-flange joint 卷边对接头
edge flare 卷边对接;边缘闪光
edge-flexure vibration mode 棱边弯曲振动模式
edge flux 边缘磁通
edge focusing 边缘聚焦
edge fog 灰边;边缘黯影;边缘模糊;边缘灰雾;边沿灰雾
edge fold 折痕;折边
edge folder 叠边机
edge following 边界跟踪
edge force 边界力
edge form 混凝土地面隔条;板的边模;模盒边框挡板;混凝土地面隔板;边模板
edge forming 侧边成形
edge forms 边模(板)
edge fracture 边裂(缝)
edge-frame 内图廓【测】
edge fringing 边缘起纹效应
edge function 边缘函数
edge fusion 熔边
edge gain 边增益
edge gate 压色浇口;缝隙内浇口
edge girder 边缘大梁;边梁
edge glass 棱镜
edge gluer 单板横接机;边缘涂胶机
edge gluing 侧面胶合;边端胶结
edge gradient 边缘梯度;边沿梯度
edge gradient analysis 边缘梯度分析
edge gradient technique 边缘梯度分析技术
edge grain 四开木材;径向切纹;径切面;径面纹理;(木材的)直行纹理;直木纹;边纹;半径面
edge-grained 四开的
edge-grained lumber 四开木料;径锯木材;径向锯材;径开木料;径截板;四破木材
edge-grained shingle 径切纹理的木瓦;径切木板
edge graph 边缘图
edge grid 边格
edge grinding 磨边
edge grinding machine 磨边机;边缘研磨机(石料)
edge grip 边侧插接
edge grouting 周边灌浆
edge growth 边缘生长
edge guide 导边器
edge guide mode isolator 导纤模隔离器
edge guide of spreader 涂布机导边器
edge gumming 边胶接
edge hinge 明铰链;明合页
edge holder 拉边纠偏装置
edge-holding power 边缘握持能力
edge hook 边钩
edge horizontal member 棱边水平的构件
edge-illuminated scale 侧面照明度盘
edge illumination 边缘照明
edge illumination hologram 边缘照明全息图
edge in 渐渐逼近;侧入
edge inking machine 边沿着墨机
edge in with 渐渐移近
edge iron 角铁;护角铁;边缘铁;铁制边缘;花匠工具
edge isolation 边端绝缘;伸缩缝嵌条;边缘隔断
edge jitter 边缘跳动
edge joining adhesive 对接胶;边接粘胶剂
edge joint 边缘拼接;卷边接头;端接(接)头;端接头;边缘连接;边缘接缝;边缘焊接头;接边(头);拼缝;顺纹接缝
edge joint of flooring board 纵接地板
edge joint weld 边缘焊
edge knife 修边刀
edge knot 侧面节疤;边节
edge knurling machine 滚花机;滚边机
edge lease 油气田边缘租地
edge lift 翘边
edge-lift agitator 边管提升搅拌器
edge light 边缘照明;跑光
edge lighting 边缘折射;侧线折光
edge line 车行道边线;边线;边 线

edge line of carriageway 车行道边(缘)线
edge lining 镶边;钩边
edge lip 边缘凸起变形
edge-lit display 边光显示
edge loaded 边上负荷的
edge load(ing) 边缘荷载;刃口负荷
edge load stress 边载应力;边缘荷载应力
edge lot 沿边建筑基地;沿边建筑基地
edge machine 缝边机
edge mark 界标;材边标记
edge mask 图廓蒙片
edge matching 接边;边匹配
edge measurement 边缘测量
edge melting 熔边
edge melting machine 熔边机
edge member 边缘构件;边缘杆件
edge mill 轮碾压;轮碾式混砂机;轮碾磨;碾子;碾碎机;边碾;轮碾机
edge milling 轧边;侧压下
edge milling machine 铣边机
edge moment 边缘弯矩
edge mo(u)lding 镶边线条;边缘线饰;边缘线脚
edge nailing 阴钉;侧钉;边钉(法)
edge notch 边沿切口
edge notch card 边缘切口卡片
edge-notched 切边
edge-notched card 边缘凹口卡片
edge nucleation 晶棱成核
edge number 胶片编号;边缘标号;片边号码
edge of a dihedral angle 二面角的棱
edge of a pulse 脉冲边缘
edge of a wood 森林界
edge of continental shelf 大陆架(外)缘
edge of ditch 沟缘;沟沿
edge of forest 林缘
edge of foundation 基础边缘
edge of fracture 断裂面边缘;破裂面边缘
edge of gallery floor 底板边缘地带
edge of knife 刀口
edge of Mach's cone 马赫锥母线
edge of pavement 路缘石;路面边缘
edge of punch 冲头缘
edge of regression 脊线;回归边缘
edge of shake 裂纹板头
edge of shovel 铲刃
edge of ski 滑雪刀
edge of stream 射流界限
edge of the format 图廓
edge of tool 刃口
edge of tunnel roof 顶板边沿部
edge of vertical transition 垂直过渡边缘
edge of work 工件边缘;工件边缘
edge of wound 创缘
edge-on object 侧向天体
edge-on spiral 侧向漩涡星系
edge orientation 菱边定向(金刚石);晶棱定向;切削刃定向
edge peeling 掉边;脱边
edge phenomenon 边沿现象;边缘现象
edge pick stitching machine 边缝缝合机
edge plane 边刨
edge planer 修边刨床;刨边机
edge planing 边刨
edge planing machine 刨边机
edge plate 护角板;镶边板;边板
edge polishing 边抛光;刷边
edge positioner 边缘定位器
edge preparation 边缘预加工;接边加工;修边准备;边缘整理;板边加工;坡口加工
edge pressure 边缘压力
edge printer 边缘印字机
edge printing 边缘印字;边沿印出;边沿曝光
edge profile 边的轮廓;边侧面
edge protection 边缘加固;边缘防护;边缘保护
edge protection bead 护角压条
edge protection strip 护角条
edge protection with rubber 边上用橡皮保护
edge protector 护边物
edge pull 嵌入平拉门边框的把手;推拉边(门窗)
edge pulse 边缘脉冲
edge pumping 板边抽吸
edge-punched card 边缘穿孔卡(片)
edge-punched punch 边缘穿孔机
edge purity magnet 边缘色纯化磁铁

edge purlin(e) (屋顶上的)边沿檩条;边檩
edger 磨边(机)器;修边器;修边工具;齐边机;弯曲模膛
edge radius 棱角半径
edge rail 边横条;边横挡;边缘横杆
edger approach table 轧边输入辊道
edger block 小炉侧墙砖
edge reading 边缘读数
edge reduction 侧边压缩
edge reflection 边界反射
edge reinforcement 边缘钢筋
edge resolution 光栅边缘鉴别力;边缘分辨率
edge response 刀口响应
edge response curve 边缘响应曲线
edge restraint 边上约束
edge-restraint condition 边缘约束条件;边界约束条件
edger grain 径切纹理
edge ridge 边饰屋脊
edger impression 滚挤模腔
edge ringing 边缘瞬变
edger mill 立辊轧机;立辊机座;轧边机
edge roll 边卷装饰线脚;边卷筒饰;辊式拉边器
edge rolled 封面卷边的
edge roller 辊式拉边器
edge rolling 滚压工步;滚挤
edge roll machine 辊式拉边机
edge rotation 边上旋转
edge rounding 弄圆边角;边角修圆
edge rounding concrete 圆边混凝土
edger roll 立辊;轧边辊
edger unit 磨边设备
edge runner 轮碾机;碾子;碾磨机;磨轮
edge runner for mixing 混碾机
edge runner mill 碾盘式粉碎机;轮碾机;磨轮式碾机;边碾机;双辊式碾碎机;干式轮碾机
edge runner mixer 湿碾混合机
edge runner pan 轮碾盘;碾盘
edge runner-wet mill 湿碾机
edge run-out 边偏转
edge sander 侧边砂光机
edge scanner 带材边缘位置调整器
edge scarf 竖长嵌接
edge scarfer 火焰清理机;铲疵工
edge scraper 边缘刮刀
edges deviation 跑边
edge seal(ing) 边上密封;封边;刀口密封;边封;边部封接
edge seam 边缘线状裂纹;边缘裂纹;倾斜地层
edge seam welding 端面接头滚焊
edge section 边沿截面
edge segmentation 图像边缘分块
edge sence network 边沿检测网络
edge sense circuit 边沿检测电路
edge sense network 脉冲边沿检测网络
edge sensor of spreader 涂布机导边传感器
edge sequence 边序列
edge shafts 支撑体的柱身(诺曼弟式建设中)
edge shape 边沿形状;边缘装饰线轮廓
edge sharpening 边缘锐化
edge sharpness 边缘清晰度
edge shaver 刨边机
edge-shaving machine 刨版边机
edge shear 边缘剪力
edge shifting 边缘漂移;边缘偏移
edge shot 刨边;修边木板
edge-shot board 修边木板
edge-side compression 侧面压缩
edges imperfection 棱角缺陷
edge skew 斜边砖;单面斜砖
edge slab 边沿平板;边板
edge slump(ing) 边缘坍塌;边缘滑移
edge smoothing machine 磨边机
edge snapping machine 掰边机
edges of plate 钢板的边缘
edge sorter 边选装置
edge spacer 边部隔片
edge spacing 边缘间距
edge spall 边缘剥裂
edge split 边裂(缝)
edge sponging machine 修边机
edge-spread function 刃边扩散函数;边缘扩散函数
edge squeezer 齐边压力机
edge stacking 侧垛法;侧堆法

edge steepness of edge filter 截止滤光片的陡度
edge-stiffened 边沿加强的
edge stiffening 加强边沿;边缘加劲
edgestone 磨石;道路边缘石;边缘石
edge stop roller 挡边轮
edge straightener 边缘碾平器
edge strengthening 边缘加固
edge stress 棱边应力;边缘应力;边缘压力;边界应力
edge stretcher 边拉伸器
edge string 边数据串
edge strip 镶边线条;边缘条;边贴板;围条
edge-strip scanning equipment 带材边缘自动控制器;带材边缘扫描器
edge structure 玻璃带边钩
edge substitution 边缘取代
edge supported 周边支承的;支承边缘
edge-supported slab 边缘支承板
edge-supported interface 边缘支承界面
edge tamping 夯边
edge target 边界目标
edge thickening 边缘加厚
edge thickness 边缘厚度
edge thickness difference 边缘厚度差
edge thickness factor 扼流系数;边缘厚度系数
edge tile 边饰瓦
edge time of migration 偏移边界时间
edge-to-edge 边到边;边靠边
edge-to-edge distance 边沿间的距离;边到边距离
edge-to-edge sharpness 全张像片清晰度
edge toe nailing 斜钉拼板(法);隐钉
edge-to-face flocculation 边对面絮凝
edge tone 边棱音
edgetone amplifier 边振放大器
edgetone effect 尖劈效应
edge-tone element 流振元件
edge tool 修边工具;削边刀;有刃的工具;修饰边缘工具
edge to platform 站台边缘
edge torn 包皮破裂
edge torque moment 边缘扭矩
edge trace 边缘迹线;边缘痕量
edge tracker processor 边沿跟踪器信息处理机
edge tracking 辊边痕迹;边痕
edge trail 边迹
edge tree 林缘树
edge trigger 边缘触发
edge triggered flip-flop 边沿触发的触发器
edge trigger flip flop 边缘触发触发器
edge trim 边缘侧面装饰
edge-trimmed roofing 剪边屋顶
edge trimmer 修缘刨;裁边机;边缘修整器;边缘修剪机;切边机
edge trimming 切边
edge-trimming machine 剪边机
edge-trimming plane 修缘刨
edge-trimming saw 修缘锯
edge trims 切边料
edge type filter 棱角式滤器
edge unit 边沿料
edge value 油水边界异常值;边界值
edge vent 屋盖周边气孔
edge venting 屋顶透气孔
edge view 边视孔
edge warping 边缘翘曲
edge washed 侧面冲洗式
edge water 层边水;边缘水;边水
edge water drive 边缘水冲;边水驱动
edge water drive reservoir 边缘水驱油藏
edge water flood 边缘注水
edge water incursion 边水入侵;边水侵入
edge water limit 边水界限
edge water line 边缘水界限;边水线
edge water pressure 边水压力
edge wave 棱波;边缘波;边沿波
edge wave diffraction 棱波衍射
edge wavelength of edge filter 截止滤光片的截止波长
edgeways 沿边
edgeways instrument 边缘读数仪表
edge weld 边缘焊(接);边缘焊缝;角接焊;对边焊;端面焊缝;端接焊(缝);端焊接
edge welding 边缘焊接
edge welding seam 边缘焊缝

edge well 边井
edge wheel 研磨轮
edge winding 边缘绕组
edge wiping 揩边;擦边
edgewise 沿边(的);竖直的;边对边
edgewise brick paving 沿边砖铺砌
edgewise clay brick paving 竖直砖铺路面
edgewise compression 测面压缩
edgewise conglomerate 竹叶状砾岩
edgewise cube 斜砌方块;(防波堤的)侧置方块
edgewise distortion 沿边变形
edgewise indicator 边转指示器;边缘读数式指示表
edgewise instrument 边转仪器;边缘读数式仪表
edgewise needle 刀刃形针
edgewise placing 竖直位置;沿边铺设
edgewise shear 沿边剪切
edgewise view 边视图
edgewise welding 沿边焊接
edgewise winding 扁绕绕组;扁立绕法;扁带线圈
edgewise wound coil 扁立缠绕线圈
edge with a groove 有槽工具磨边
edgework 边部加工
Edgeworth contract curve 埃奇沃思合同曲线
Edgeworth cycle 埃奇沃思循环
Edgeworth expansion 埃奇沃思展式
Edgeworth index 埃奇沃思指数
edge zone 边缘地带
edge zone of typhoon 台风边缘
edging 卷凸缘;磨边;镶边(线条);轧边;滚压工步;滚压边缘修饰;滚挤;边缘修饰;边条;边饰;去飞边;齐边轧制
edging board 遮挡板;镶边板
edging brush 修坯毛刷
edging grinder 边条切碎机;碎木机
edging lath 边饰板条
edging mill 立轧机架;立辊轧机;轧边机
edging mo(u)ld 边缘线饰;边缘装饰线脚
edging operation 磨边工序
edging pass 立辊孔型;立轧道次;轧边孔型
edging plane 边刨
edging plant 饰边植物;装缘植物;饰缘植物
edging press 轧边压机
edging shears 修边剪刀
edging stand 立辊机座
edging stone 磨边石
edging strip 门边镶条;镶边条;盖檐条;门上镶边木条带
edging trowel (用于新鲜混凝土的)修边镘刀
edgy 带棱的
Ediacaran fauna 埃迪卡拉动物群
edibility 可食用性
edible 适合食用的
edible clay 可食黏土
edible landscaping 食用园林布置
edible oil 食用油
edible oil contamination 食用油污染
edible oil production wastewater 油脂生产废水
edible seaweed 海菜
Edibrac 艾迪布拉斯
edicard 编辑卡
edification 建筑(旧称)
edifice 大型建筑(物);大厦;大建筑物
Edinburgh sink 爱丁堡水槽
edingtonite 钡沸石
Edison accumulator 爱迪生蓄电池
Edison base 螺旋灯座;爱迪生灯座
Edison-base fuse holder 螺旋式熔断器底座
Edison battery 镍铁电池;爱迪生(蓄)电池
Edison bridge 爱迪生电桥
edison cell 爱迪生电池
Edison effect 爱迪生效应;热电放射效应
Edison-Junger accumulator 铁镍蓄电池
Edison lampholder 爱迪生灯座
Edison screw 爱迪生螺纹
Edison screwcap 爱迪生螺帽;爱迪生螺丝灯头
Edison screwholder 爱迪生螺丝插座
Edison screw thread 圆螺纹
Edison socket 爱迪生插座;螺旋式灯口
Edison storage battery 铁镍蓄电池
Edison storage cell 爱迪生蓄电池
Ediswan wiring system 爱迪生瑞布线网
edit 初步整理;编纂;编排;编辑
edit capability 编辑能力
edit check 编辑检查

edit check list 编审意见表
edit code 编辑码
edit command 编辑命令
edit control 编辑控制
edit control character 编辑控制符
edit descriptor 编辑描述符
edit-directed stream 定向编辑流;编排式流
edit-directed transmission 编辑式直接传输;定向编辑传输;编排式传输;编辑制导传输
edit dump 编辑转储
editec 电子编辑器
edited copy 编辑拷贝;编辑复本
edit file 编辑信息
editic acid 乙底酸
edit information file 编辑信息文件
editing 编排;编辑(方法);编集
editing change 编辑性修改
editing character 编辑字符
editing clause 编辑子句
editing console 编辑台
editing controller 编辑控制器
editing function 编辑功能
editing instruction 编辑指令
editing key 编辑键;编辑索引
editing machine 剪辑机;编辑机
editing mode 编辑方式;编辑状态
editing of atlases 地图集编辑
editing operation 编排操作;编辑操作
editing program(me) 编辑程序
editing program(me) center 程序编辑处理中心
editing rack 剪辑吊架
editing routine 编辑例行程序
editing session 编辑会话期;编辑对话期
editing sign control symbol 符号编辑用字符
editing subroutine 编辑子程序
editing symbol 编排符号;编辑符号
editing tape 编辑纸带
editing technique 编辑技术
editing terminal 编辑终端设备
editing type 编辑类型
editing videocorder automatic retrieval device 编辑录像机自动检索装置
edition 版次;版本
editional function 编辑功能
edition binding 特装本
edition de luxe 精装本;精装版(本)
edition note 版次说明
edit item 编辑项目
edit mask 编辑掩码
edit mode command 编辑状态指令
editor 剪辑机;编者;编码程序;编辑者;编辑器
editor command 编辑程序命令
editorial 编辑的
editorial board 编辑部
editorial correction 编辑审校
editorial department 编辑部
editorial document 编辑文件
editorial function 编辑功能
editorial instruction 编辑指令
editorial note 编者按;编辑说明
editorial office 编辑部
editorial panel 编辑小组
editorial plan 编辑计划
editorial policy 编辑方针
editorial processing centre 编辑处理中心
editorial revision 编辑审订
editor in chief 主编(人)
editor program(me) 编辑程序
editor's note 编者按
edit pattern 编辑模式
edit pattern operator 编辑模式操作符
edit plot 编绘原图
edit pulse 剪辑脉冲;编辑脉冲
edit routine 编辑例行程序;编辑程序
edit section 编辑部
edit statement 编辑语句
edit symbol 编辑符号
edit update 编辑修改
edit word 编辑字
Edman degradation technique 埃德曼降解技术
Edmond's balance 爱德蒙天平
edolite 长云角岩
Edridge-Green lantern tests 埃德里奇—格林坦信号灯测验

Edser and Butler's bands 埃泽—巴特勒暗带
EDTA-ammonium salt method 乙二胺四乙酸铵盐速测法
EDTA titration 乙二胺四乙酸滴定法
education 教育;培养
educational administration 教育管理
educational allowance 教育补助
educational appropriations 教育经费
educational area 文教区
educational assessment 教育评估
educational attainment ratio 教育程度比率
educational block 学校大楼;文教建筑群
educational building 教学楼建筑;教育楼;文教建筑物
educational center 文教中心;文化教育中心
educational duty 培养关税
educational ecology 教育生态学
educational environment 教育环境
educational evaluation 教育评价
educational expenses 教育费用
educational facility 教育设施
educational funds 教育经费
educational garden 教育园
educational grant 教育补助金
educational industrialization 教育产业
educational institute 文教机关
educational insurance 教育保险
educational investment 智力投资
educational level of workers 劳动者素质
educational machine 教育机器
educational management 教育管理
educational management information system 教育管理信息系统
educational objective 教育目标
educational opportunities 教育机会
educational or training received 教育程度
educational outreach 教育能达到范围
educational park 教育园
educational plan 教学计划
educational policy 教育政策
educational program(me) 教育程序;教学计划;教育规划;培训计划
educational qualification 学历
educational services 教育机构
educational status 教育程度
educational structure 教育结构;教育建筑物
educational supervision 教育督导
educational tariff 培养关税
educational technology 教育技术(设备)
educational television 科教电视;教育电视
education background 学历
education district 文教区
education of landscape architecture 园林教育
education with electric(al) audio-visual aids 电化教育
educt 离析物;析出物
eduction column 气举柱;气举管;排泄管(道)
eduction gear 排出装置
eduction pipe 放气管;气举管;排泄管(道);排汽管;排气管
eduction port 排泄孔
eduction tube 气举管
eduction valve 泄流阀;放泄阀
eductor 引射器;气举管;喷射器;排泄器;排泄管(道);排放装置;排放管;水射器;射流泵
eductor condenser 排出冷凝器
eductor pump 喷射泵
eductor well point 喷射井点
eductor well point system 喷射井点系统
edudesmin 桉树胶素
edulcorate 精选
Edward balance 爱德华天然气比重天平
Edwardian style 爱德华式建筑(英国)
edwardsite 独居石
Edwards roaster 爱德华型焙烧炉
Edward's tile 爱德华砖(一种注册的窗台陶砖设计)
Edzell shales 埃德泽尔页岩
Eel Brand antifriction metal 伊尔布兰德减摩合金
eelctrotype 电铸版
eel grass 大叶藻
eel ladder 鳗梯
eel migration 鳗鲡迁移
eel oil 鳗鱼油

eernane 羊齿烷
Efco-Northrup furnace 埃费科—诺思拉斯无芯高频感应炉
Efco-Udylite process 埃费科—尤迪莱特光亮镀镍法
Efdolan Green BLN 埃弗多兰绿 BLN
effect 效应;效力;效果;作用;成效
effect an insurance 附加保险;投保
effect a policy 取得保险单;投保
effect assessment 效应评估
effect basecoat 随角异色效应中间涂层
effect by human activity 人类活动影响
effect coating 随角异色效应涂层
effect concentration 效应浓度
effect concentration 50% 半中毒浓度
effect concentration time 50% 半中毒浓度时间积
effect disc 特技插盘
effect-equivalent circle diameter 等效当量圆直径
effect extrapolation model 效应外推模型
effect factors of resources situation 影响资源形势的因素
effect filter 效应滤色器
effect glass 效应玻璃;特技用玻璃
effect in response ratio 响应率的效应
effect instantaneous field-of-view 有效瞬时视场
effect insurance 投保保险
effective 有效的;实际的
effective absorption 有效吸水量(木材等);有效吸收(量)
effective absorption coefficient 有效吸收系数
effective abstraction 有效提取(量)
effective abstractions of precipitation 有效降雨损失量(总雨量与净雨量之差)
effective acceleration 有效加速度
effective acceleration of gravity 有效重力加速度
effective acceptance time interval 有效接收时间间隔
effective accumulated temperature 有效积温
effective accuracy 有效精度
effective acidity 有效酸度
effective acoustic(al) center 有效声源中心
effective acoustic(al) power 有效声功率
effective acoustic(al) pressure 有效声压
effective actuation time 有效激励时间
effective address 有效地址
effective admission 有效进气
effective admittance 有效导纳
effective adsorption 有效吸附
effective adsorption equilibrium constant 有效吸附平衡常数
effective age 有效使用年限(建筑物);有效使用期;有效年限
effective agent 有效药剂
effective aggregate tax rate 实际综合税率
effective air gap 等效气隙
effective air path 有效空中路线
effective algorithm 可行算法
effective alkali 有效碱
effective alkalinity 有效碱度
effective ampere 有效安培
effective amplification 有效放大
effective amplitude 有效振幅
effective analysis 有效分析
effective angle 有效角
effective angle of attack 有效迎角
effective angle of friction 有效摩擦角
effective angle of internal friction 有效内摩擦角
effective angle of obliquity (转子发动机的)有效摆动角
effective angular field 有效视场
effective anisotropy 有效各向异性
effective annual interest rate 实际年利率
effective anomaly 有效异常
effective antenna height 天线有效高度
effective antenna length 有效天线长度
effective aperture 有效口径;有效孔径;实际孔径
effective arch span 有效的拱跨
effective area 有效面积
effective area coefficient 有效面积系数
effective area of concrete 混凝土有效面积
effective area of mice sheet 云母片有效面积
effective area of orifice 孔口有效面积
effective area of reinforcement 钢筋(的)有效面积;有效钢筋面积
effective area of reinforcement in diagonal bends 对角钢筋带的有效面积;弯起钢筋的有效面积
effective area of screen 筛的有效面积
effective area of stain 斑点有效面积
effective area of tidal station 验潮站有效范围
effective array length 有效组合长度
effective aspect ratio 有效长宽比
effective aspect ratio parameter 有效伸长率
effective atmosphere 有效气压;有效大气压;光学有效大气
effective atomic charge 有效原子电荷
effective atomic number 有效原子序数
effective attenuation 有效衰减
effective attenuation factor 有效衰减系数
effective availability 实际利率
effective averaging time 有效平均时间
effective band 有效频带
effective bandwidth 有效频带宽度;有效带宽;谱带宽度
effective base(line) 有效基线
effective beamwidth 等效束宽
effective bearing area 有效支承面积
effective bentonite 有效膨润土
effective berthing energy 有效靠泊能量
effective biological dose 有效生物剂量
effective bit rate 有效位速
effective bit-weight 有效钻压
effective bohr magneton 有效波尔磁子
effective bond (砖墙的)高效砌合;有效砌合
effective brake area 有效制动面积
effective brake drum braking area 有效闸轮制动面积
effective brake length 有效制动长度
effective braking distance 有效制动距离
effective branching factor 有效分支因素
effective bulk weight density of soil 土的有效重度
effective bunching angle 有效聚束角
effective cadmium cut-off 有效镉截止值
effective calcium oxide content test of lime 石灰的有效氧化钙含量试验
effective call 有效呼叫
effective-call meter 有效呼叫计数器
effective CaO and MgO content test of lime 石灰的有效氧化钙和氧化镁含量试验
effective capacitance 有效电容
effective capacity 有效容量;有效能力;有效功率;有效库容
effective capacity of kiln 窑的有效容积
effective capacity of reservoir 水库有效容量;水库有效库容
effective capital recovery rate 有效资金回收率
effective case depth 有效硬化深度
effective cash cost 有效现金成本
effective cathode current 有效阴极电流
effective cavitation number 有效空泡数
effective channel depth 航道有效深度
effective channel width 航道有效宽度
effective charge 有效电荷
effective chart width 有效记录纸宽度
effective check 有效性检查
effective chimney height 有效的烟囱高度
effective chlorine concentration 有效氯浓度
effective circuit 有效电路
effective circulating head 有效流转压头
effective circulation 有效循环;实际流通额
effective cleaning 有效冲洗
effective clearance 有效间隙
effective coefficient 有效作用系数;有效系数
effective coefficient of local resistance 局部阻力有效系数
effective coefficient of permeability 有效透水系数
effective cohesion 有效内聚力;有效黏聚力;有效黏合力;有效凝聚力
effective cohesion intercept 有效黏聚力
effective coil 有效圈
effective collector 有效集电极
effective collision 有效碰撞
effective collision (cross-)section 有效碰撞截面
effective column height 有效柱高;柱的有效高度
effective column length 有效柱长
effective compacted depth 有效夯实深度
effective competition 有效竞争
effective compressibility 有效压缩系数
effective compression ratio 有效压缩比

effective concentration 有效浓度
effective concentration 50% 半有效浓度
effective conclusion 有效结论
effective condition 有效条件
effective conductance 有效电导
effective conductivity 有效电导率
effective confining pressure 有效周围压力;有效围压
effective confusion area 有效干扰面积
effective constant 有效常数
effective constituent 有效分量;有效成分
effective constraint length 有效制约长度
effective contact area 有效接触面积
effective contact area ratio 有效接触面积比
effective contact pressure 有效接触压力
effective contact radius 有效接触半径
effective content 有效容量
effective core penetration 有效岩芯穿透深度
effective core porosity 有效岩芯孔隙度
effective cost 实际成本
effective coverage 可达范围;有效区域
effective credit 有效信贷
effective crimping radius 有效卷曲半径
effective cross current 有效横流
effective crossing coefficient 有效交叉系数
effective crossing over 有效交换
effective cross-section 有效截面;有效横断面;有效断面
effective cross-section(al) area 净截面;有效断面面积;有效截面(面)积
effective cross-section for resonance 有效共振截面
effective (cross-)section of collision 有效碰撞截面
effective cross section of ionization 有效电离截面
effective current 有效电流
effective current area 有效迎流面积
effective current flow 有效水流
effective cut 有效土方
effective cut-off frequency 有效截止频率
effective cycle time 有效循环时间
effective damping 有效阻尼
effective damping constant 有效阻尼常数
effective damping ratio 有效阻尼比
effective data 有效数据
effective data process 有效数据法;有效数据处理
effective data transfer rate 有效数据传输率
effective date 有效日期;有效期;实施期;生效日期
effective date of contract 合同的生效日期
effective date of regulations 条约生效日期;条例生效日期
effective days 实际作业天数;实际工作天数
effective dead-line scheduling 有效期限调度
effective dead time 有效空载时间
effective death rate 实际死亡率
effective debt 实际债务
effective decline rate 有效递减率
effective deficiency 有效亏数
effective deformation 有效体变(强夯);有效沉降量(强夯);有效变形(强夯)
effective degree 有效次数
effective delayed neutron fraction 有效缓发
effective delivery stroke 有效供油行程
effective demand 有效需求;实际需求
effective density 有效密度
effective density of state 有效能态密度
effective depreciation 实际贬值
effective depreciation cause 实际折旧的原因
effective depth 有效水深;有效深度;有效梁高;有效厚度;有效高度;有效板厚
effective depth of beam 有效梁高
effective depth of embedment of sheet pile 板桩的有效嵌固深度
effective depth of grit chamber 沉砂池有效水深
effective depth of section 截面有效深度;截面有效高度
effective depth of sheet pile penetration 板桩有效入土深度
effective depth of slab 有效板高;有效板厚
effective descriptive set theory 有效描述集合论
effective diameter 节径;有效直径;有效粒径;中径;等效值
effective diameter of circulating circle 循环圆有效直径
effective diameter of grain 颗粒有效直径

effective diameter of particle 颗粒有效直径
effective diameter of pipe 有效管径
effective diameter of thread 螺纹中径
effective diaphragm 有效光阑
effective dielectric(al) constant 有效介电常数
effective diffuse radius 有效扩散半径
effective diffusion 有效扩散
effective diffusion coefficient 有效扩散系数
effective diffusion constant 有效扩散常数
effective diffusion velocity 有效扩散速度
effective diffusivity 有效扩散率
effective digit 有效数字(位数)
effective dilution 有效稀释
effective dimension 有效空间;有效尺寸
effective dimensions of lock 船闸有效尺度
effective dipole moment 有效偶极矩
effective directivity factor 有效指向性因数
effective direct radiation 有效直接放射
effective discharge 有效排放(量);有效流量
effective discharge area 有效通过面积
effective discharge time 有效喷射时间
effective disk area 有效圆盘面积
effective dispersion parameter 有效扩散参数;有效分散参数
effective distance 有效距离
effective distortion 有效失真
effective distribution coefficient 有效分布系数
effective dose 有效量;有效剂量
effective dose 50 半数有效量
effective dose equivalent 有效剂量当量
effective draft 有效抽力
effective drainage porosity 有效排水孔隙度;有效孔隙率
effective drawdown 有效降深
effective drawing area 有效绘图面积
effective driving force 有效传动力
effective drying surface 有效干燥面积
effective duration 有效期间;有效历时;有效持续时间
effective dynamic(al) modulus 有效动态模量
effective dynamic permeability 有效动磁导率
effective earthquake force 有效地震力
effective earth radius 有效地球半径;地球有效半径
effective earthwork 有效土方
effective echo area 有效回波区【水文】
effective efficiency 有效效率;有效功能
effective elastic constant 有效弹性常数
effective electromechanical couple factor 有效机电耦合系数
effective electromotive force 有效电动势
effective element 有效元件
effective elongation 有效伸长
effective embedment of sheet pile 桩的有效埋深
effective emission height 有效排放高度
effective emissivity 有效发射率;等效辐射率
effective energy 有效能(量)
effective energy of berthing vessel 靠泊船舶的有效动能
effective entanglement density 有效缠结密度
effective enthalpy 有效焓
effective environment 有效环境
effective equivalent section 有效折算断面
effective error 有效误差
effective evaluation 有效评价
effective evaluation of mine environment 矿山环境影响评价
effective evaporation 有效蒸发(量)
effective evaporation rate 有效蒸发率
effective evapotranspiration 有效总蒸发;有效蒸散(发);有效蒸发蒸腾
effective exchange rate index 实际汇率指数
effective exhaust velocity 有效排气速度
effective exposure 有效接触
effective exposure time 有效曝光时间
effective external fire-fighting 外部有效灭火
effective extinguishing zone 有效灭火区
effective extraction 有效提取(量)
effective facsimile band 有效传真频带
effective factor 有效因数
effective factors of hydrochemical formation 水化学形成的影响因素
effective factors of inundation 充水影响因素
effective fetch length 有效风区长度
effective field 有效场

effective field intensity 有效场强
effective field maintenance 有效野外维护
effective field of fire 有效射界
effective field ratio 有效磁场比
effective figure 有效数字
effective film 有效界膜
effective film thickness 膜的有效厚度
effective filter length 滤水管有效长度
effective filtration pressure 有效滤过压;有效过滤压
effective filtration rate 有效滤率
effective firing range 有效射程
effective flange width 有效翼缘宽度
effective flaw 有效裂隙
effective flaw size 有效裂纹尺寸
effective flow 有效流量
effective flow curve 等效车流曲线
effective flow parameter 有效流动参数
effective flow pitch 有效流距
effective flow resistance 有效流阻
effective fluid pressure 有效流体压力
effective f-number 有效光圈;有效 f 数
effective focal length 有效焦距
effective footage 有效进尺
effective force 有效力;有生力量
effective force vector 有效力向量
effective format 有效幅面
effective formation pressure drop 有效地层压降
effective fractured zone 有效破裂带
effective fragment 有效破片
effective fragmentation 有效爆破
effective full power hours 有效满功率天数
effective functionality 有效官能度
effective gain 有效放大系数
effective gamma 有效灰度系数
effective gap 有效缝隙宽度
effective gap length 有效隙长
effective gasket width 有效垫片宽度
effective gate width 有效选通脉冲宽度
effective ga(u)ge length (强力测试的)有效隔距长度
effective genus 有效亏格
effective grade braking 有效坡道制动
effective gradient 有效坡度
effective gradient of runway 跑道有效坡度
effective grain diameter 有效粒径
effective grain-size 有效粒径
effective grain size of soil 土的有效粒径
effective gravity anomaly 有效重力异常
effective green time 有效绿灯时间
effective gross income 实际总收入
effective gross rent 有效总租金
effective ground contact area 有效接地面积
effective grounded line 有效接地线
effective grounding 有效接地
effective ground pressure 有效接地压力
effective groundwater velocity 地下水有效速度
effective gun bore line 有效炮轴线
effective gust velocity 有效阵风速度
effective half-life 有效半衰期;有效半排出期
effective half period 有效半衰期
effective Hall Parameter 有效霍耳参数
effective head 有效扬程;有效水头;有效落差
effective heat duty 有效热负荷
effective heat exchange area 有效换热面积
effective heating surface 有效受热面
effective heating surface of checker 格子体有效受热面积
effective heating time 有效加热时间
effective height 有效高度
effective height of column 有效柱高
effective height of emission 有效排放高度
effective height of jet 射流有效高度
effective helix angle 有效螺旋角
effective hierarchy access time 有效的层次存取时间
effective hitch point 有效悬挂点
effective hoisting weight 有效提升重量
effective horizon 有效地平线
effective horsepower 有效马力;总有效马力
effective humidity 有效湿度
effective impedance 有效阻抗
effective impulse 有效脉冲
effective impulse width 脉冲有效宽度
effective incidence 有效迎角

effective index 有效折射率
effective index of industrial wastewater emission 有效工业废水排放指标
effective indicated mean pressure 有效平均指示压力
effective inductance 有效电感
effective inertia force 有效贯性力
effective inertia mass 有效惯量
effective influence distance 有效影响距离
effective influence value of water table 有效水位削减值
effective ingredient 有效成分
effective ingredient of pesticide 农药有效成分
effective initial pressure 有效初压力
effective injection 有效注入
effective input 有效输入
effective input admittance 有效输入导纳
effective input resistance 有效输入电阻
effective instruction 有效指令
effective integral 有效积分
effective intensity 有效强度
effective intensity of light 有效光强
effective interaction 互动；有效（交）互作用
effective interest 实际利息
effective interest method 实际利息法
effective interest rate 实际利率
effective interrelated area 共同作用的有效面积
effective investment 有效投资
effective ionic charge 有效离子电荷
effective ionic mobility 有效离子尚度
effective irradiance 有效辐照度
effective isotropic(al) radiated power 全向有效辐射功率
effective jetting distance 有效射距
effective kilogram 有效公斤
effective labo(u)r 有效劳动
effective lateral pressure 有效侧向压力
effective launcher line 有效发射线
effective layer 有用层
effective leakage path 有效漏失面积
effective length 有效长（度）；资用长度
effective length of anchorage 锚定的有效长度
effective length of beam 梁的有效长度
effective length of bolt 螺栓计算长度
effective length of column 有效柱长；柱的有效长度
effective length of delivery 有效输出长度
effective length of filter 过滤器有效长度
effective length of lock 船闸有效长度
effective length of pile 有效桩长
effective length of platform 站台有效长（度）
effective length of station track 站线有效长；车站股道有效长度
effective length of towing rope 拖缆有效长度
effective length of track 线路有效长度；股道有效长；股道有限长度
effective length of turn-back track 折返线有效长度
effective length of vehicle 车辆有效长度
effective lens aperture 透镜有效孔径
effective lethal phase 有效致死期
effective lever arm 有效杠杆比
effective liabilities 实际负债
effective life(time) 有效寿命；有效寿期；有效使用期
effective lighting 有效照明
effective line width 有效谱线宽度
effective liquid water content 有效液态水含量
effective load 有效荷载；有效负载；有效负荷
effective lumped-parameter mass 有效集总参数质量
effectively grounded 有效接地
effective machine format 有效印刷尺寸
effective macroscopic cross-section 有效宏观截面
effective magnetization 有效磁化强度
effective magnetized inclination 有效磁化倾角
effective magneton number 有效磁子数
effective maintenance 有效维护
effective management 有效管理；实际管理机构
effective margin 有效界限；有效富裕
effective marginal tax rates 实际边际税率
effective mass 有效质量
effective mass absorption coefficient 有效质量吸收系数
effective mass approximation 有效质量近似
effective mass density 有效质量密度

effective mean power 有效平均功率
effective mean pressure 有效平均压力；平均有效压力
effective mean temperature difference 有效平均温差
effective measure 有效措施
effective membrane area 有效膜面积
effective microorganism 有效微生物群
effective migration velocity 有效迁移速度
effective migration velocity of particle 有效驱进速度
effective minute ventilation 每分有效通气量
effective mixing length 有效交混长度
effective mobility 有效迁移率
effective model 有效模型
effective modulation 有效调制
effective modulus 有效模量
effective modulus of elasticity 有效弹性模量
effective moisture 有效湿度；有效水分
effective moisture capacity 有效持水量
effective moisture content 有效水量
effective moment of inertia 有效惯性矩
effective monitoring 有效监测
effective monopole radiated power 有效单极（子）辐射功率
effective mortality rate 实际死亡率
effective multiplication constant 有效增殖系数；有效倍增常数
effective multiplication factor 有效增殖因素；有效增殖系数；有效增加系数；有效放大系数；有效倍增系数
effectiveness 可行性；效用；效益；效能；效力；效果；有效（性）；实效
effectiveness factor 效率外推法
effectiveness function 效用函数
effectiveness index 效率指数
effectiveness index of precipitation 降水有效指数
effectiveness of anchor 锚杆效应
effectiveness of compaction 压实效果
effectiveness of excitation 激发效率
effectiveness of heat releasing light method 热释光测量效率
effectiveness of lime treatment 石灰加固效果
effectiveness of sluicing 冲刷效力；冲砂效力
effectiveness of treatment 治理效果
effectiveness test of fire-fighting installation 消防效能试验
effective net head 有效净水头
effective net margin 有效净容限
effective network chain 有效网链
effective neutron cycle time 中子有效寿命时间
effective neutron lifetime 有效中子寿期
effective neutron temperature 有效中子温度
effective nocturnal radiation 有效夜间辐射
effective noise bandwidth 有效噪声带宽
effective noise power 有效噪声功率
effective noise temperature 有效噪声温度
effective normal stress 有效正应力；有效法向应力
effective number 有效数
effective number of plate 有效塔板数
effective numerical aperture 有效数值孔径
ffective observation distance 有效观察距离
effective obstruction free flow area 有效流动截面
effective octahedral stress 有效八面体应力
effective offer 实际报价
effective opening 有效开孔面积；有效孔径
effective operand address 有效操作数地址；操作数有效地址
effective operated hours 有效运转时数
effective operating distance 有效作用距离
effective output 有效输出（功率）；有效生产量；有效功率；有效出力；有效产量
effective output admittance 有效输出导纳
effective output impedance 有效输出阻抗
effective overburden depth 有效覆盖层厚度
effective overburden pressure 有效上覆压力；有效覆盖（层）压力；有效超载压力
effective overlap period 有效重叠期
effective overload output 有效超载功率
effective oxygen 有效氧
effective oxygen content 有效氧含量
effective par 实际票面价值
effective particle density 有效颗粒密度
effective particle diameter 有效粒子直径

effective particle velocity 有效粒子速度
effective pay 有效产层
effective pay factor 生产层有效因素
effective payload 有效营业荷载
effective payment clause 实际付款条款
effective pay rate 有效工资率；实际薪资率；实际工资率
effective pay thickness 产层有效厚度
effective peak acceleration 有效峰值加速度
effective peak number 有效峰数
effective perceived noise decibel 有效感觉噪声分贝
effective perceived noise level 有效感觉噪声级；实际感觉噪声级
effective percentage 有效率
effective percentage modulation 有效调制深度；有效调制率
effective perforation 有效炮眼
effective period 有效周期；有效期
effective periodicity 有效周期性
effective permeability 有效渗透性；有效渗透系数；有效渗透率；有效导磁率；有效磁导率；实际渗透率
effective permeability to oil 原油的有效渗透率
effective permeability to water 水的有效渗透率
effective phase difference 有效相位差
effective picture area 有效图像面积
effective picture signal 有效图像信号
effective picture size 有效幅面
effective pillar length 有效柱长
effective pitch 有效螺距；有效节距；有效铆钉中距
effective pitch angle 有效螺距角
effective pitch ratio 有效螺距比
effective placement 有效充填
effective plate height 有效塔板高度；有效板高
effective plate number 有效塔板数
effective plate volume 有效塔板容积
effective platform area 有效站台面积
effective plume height 有效烟羽高度
effective pneumatic capacitor 膜片气动有效容量
effective population size 有效群体大小
effective pore radius 有效孔径
effective pore space 有效孔隙
effective pore volume 有效孔隙体积；有效孔隙容积
effective porosity 有效疏松度；有效气孔率；有效孔隙率；有效孔隙度；有效空隙率；给水度；实际空隙率
effective porosity of filter 过滤器有效孔隙率
effective possession 有效占有
effective pot life 有效罐存期限
effective power 有效功率；有效动力
effective power head 有效发电水头
effective power input 有效输入功率
effective precipitable water 有效可降水分
effective precipitation 有效降雨量；有效降水（量）；有效沉淀量
effective pressure 有效压强；有效压力
effective pressure head 有效（水）压头
effective prestress 有效预应力；永存预应力
effective prestress in tendon 力筋的有效应力
effective prestress value 有效预应力值
effective price 有效价格；实际价格
effective procedure 有效程序
effective production hour 有效工时
effective profit rate on funds 实际资金利润率
effective propagation velocity 有效传播速度
effective protection 有效防护
effective pull 有效拉力
effective pulse width 脉冲有效宽度
effective pumping speed 有效抽速
effective push 有效推力
effective quantity 有效量
effective radiated power 有效辐射功率
effective radiation 有效辐射
effective radiation constant 有效辐射系数
effective radiation of physiology 生理有效辐射
effective radiation power 有效俘获功率
effective radiation temperature 有效辐射温度
effective radioactive half-life 放射性有效半衰期
effective radium content 有效镭含量
effective radius 有效半径
effective radius of a control rod 控制棒有效半径
effective radius of grouting 灌浆有效半径
effective radius of the earth 有效地球半径；地球

有效半径
effective rain 有效降雨
effective rainfall 净雨量;有效降水(量);有效(降)雨量
effective rainfall hydrograph 净雨过程线;有效雨量过程线
effective rake (起重机的)有效斜度;有效臂长
effective range 有效射程;有效量程;有效距离;有效(测量)范围;有效测程;标度尺的工作部分
effective range of jet 有效射程(水枪)
effective range of spray nozzle 水雾喷头有效射程
effective range of vibrating equipment 振动设备的有效作用范围
effective rate 有效率;实际汇价;实际比价
effective rate of exchange 实际汇价
effective rate of fire 有效射速
effective rate of interest 实际利率
effective rate of protection 实际保护率
effective rate of tax 实际税率
effective ratio 有效度比;有效比
effective reach 有效射程
effective recombination coefficient 有效复合系数
effective recombination velocity 有效复合速度
effective record length 有效记录长
effective red time 有效红灯时间
effective reflectance 有效反射系数
effective reflecting surface 有效反射面
effective reflection 有效反射
effective reflectivity 有效反射率
effective region 有效范围
effective reinforcement 计算钢筋量;有效(配)筋;有效钢筋
effective relative aperture 有效f数
effective relaxation length 有效张弛长度
effective removal cross-section 有效移出截面
effective reproduction amount 有效增殖量
effective residual drag 有效残留阻力
effective resistance 交流电阻;有效阻力;有效电阻;高频电阻
effective resistivity 有效电阻率
effective resonance integral 有效共振积分
effective resource 有效追索
effective resource amount in the near future 近期有效资源量
effective responsive quantum effeciency 有效响应量子效率
effective retarding force (汽车的)有效减速力
effective Reynolds number 有效雷诺数
effective roller length (轴承的)有效滚柱长度
effective rolling radius 有效滚动半径
effective sale 实际销售
effective salinity 有效盐度;有效咸度;有效含盐量
effective sampling area 有效取样面积
effective saturation line 有效浸润线
effective scattering cross-section 有效散射截面
effective scattering mass 有效散射质量
effective screen aperture 有效筛孔
effective screen area 有效筛面积
effective screen cut-point 限制筛分孔;实际筛分粒度
effective screening area 筛面有效筛分面积;筛进工作面积
effective seal depth 有效水封深度
effective sealing material 有效密封材料
effective sealing width of gasket 垫片有效密封宽度
effective section 有效剖面;有效截面;有效断面
effective sectional area 有效截(面)积
effective section modulus 有效剖面模数
effective section recombination 有效复合截面
effective segregation 有效分离;正分离
effective selectivity 有效选择性
effective sensitivity 有效灵敏度
effective separating density 有效分选比重;有效分离密度
effective separating size 有效的分级筛孔尺寸
effective separation factor 有效分离系数
effective separation radius 有效分选半径
effective service area 有效工作区
effective shadow 有效静区
effective shadow zone 有效声影区
effective shear 有效剪力
effective shear area 有效剪切面积
effective shearing rigidity 有效抗剪刚度
effective shear strain energy criterion 有效剪切应变能准则
effective shear strength parameter 有效抗剪强度参数
effective shielding constant 有效屏蔽常数
effective shot depth 有效爆炸深度
effective shrinkage 有效收缩量
effective shrinkage limit 有效缩限
effective shunt impedance 有效并联阻抗
effective sieve aperture size 有效筛眼孔径;有效筛孔度
effective signal 有效信号
effective signal duration 有效信号持续期间
effective signal radiated 有效发射信号
effective simple process factor 单级过程有效系数
effective size 有效粒径;有效尺寸
effective size of aggregate 集料有效尺寸
effective size of grain 有效粒度;有效颗粒度
effective size of sand 砂的有效粒径
effective size of sand grains 砂粒的有效粒径
effective slenderness ratio 有效细长比
effective slip 有效滑距
effective slip ratio 有效滑距比
effective slip velocity 有效滑动速度
effective slit 有效缝隙
effective slit width 有效狭缝宽度
effective snowmelt 有效融雪量
effective solid fire stream 有效密集消防射流
effective sound pressure 有效声压
effective sound pressure level difference 有效声压级差
effective sound velocity 有效声速
effective source area 有效源面积
effective source rock 有效源岩;有效生油岩
effective spacing 有效间距
effective span 计算跨径;计算跨度;有效跨度
effective span/length ratio 有效的跨距/长度比
effective specific gravity 有效比重
effective specific impulse 有效比冲
effective specific surface area 有效比表面积
effective speed 有效速度
effective spot size 有效光点大小;有效斑点尺寸
effective spring length 有效弹簧长
effective stack height 有效烟囱高度;烟囱有效高度
effective stage 效应阶段
effective standard deviation 有效标准偏差
effective steam pressure 有效蒸汽压(力)
effective steepness 有效陡度
effective stiffness 有效劲度;有效刚性;有效刚度
effective stiffness constant 有效劲度常数
effective stiffness matrix 等效刚度矩阵
effective stiffness ratio 有效劲度比
effective stop 有效光阑
effective stopping power 有效阻止本领
effective storage 有效蓄水(量);有效库容
effective storage life 有效储[贮]存期(限)
effective storage space 库场有效面积
effective strain 有效应变;有效菌株
effective strength 有效强度
effective strength parameter 有效强度参数
effective stress 有效应力
effective stress analysis 有效应力分析(法)
effective stress concentration 有效应力集中
effective stress concentration factor 有效应力集中系数
effective stress difference 有效应力差
effective stress law 有效应力定律
effective stress parameter 有效应力参数
effective stress path 有效应力路径
effective stress plot 有效应力图
effective stress strength index 有效应力强度指标
effective stress strength parameter 有效应力强度指标
effective stress theory 有效应力理论
effective stroke 有效冲程
effective suction 有效吸力
effective suction head 有效吸水头
effective superficial porosity 有效表面孔隙率
effective supply 有效供给
effective surface 有效表面;实测面
effective surface area 有效表面积
effective surface retention 有效地面滞留量;有效地面储蓄量
effective susceptibility 有效磁化率
effective sweep width 有效检索宽度
effective swept volume 有效扫除容积
effective switching field 有效开关场
effective target area 有效目标面积;有效靶面积
effective target length 有效目标长度
effective tariff 实际关税税率
effective tax rate 实际税率;实际关税税率
effective temperature 有效温度;感触温度
effective temperature difference 有效温度差;送风温差
effective temperature index 有效温度指数
effective temperature rule 有效温度法则
effective tensile strength 有效抗张强度
effective tension 有效张力;有效拉力
effective terrestrial radiation 陆地有效辐射;有效地球辐射;有效地面辐射;有效大地辐射
effective thermal conductivity 表观导热系数;表观传热系数
effective thermal cross-section 有效热截面
effective thermal efficiency 有效热效率
effective thermal neutron cross-section 有效热中子截面
effective thermal resistance 有效热阻
effective thermal transmittance 有效传热系数
effective thickness 有效厚度
effective thickness of a wall 有效墙宽(用于计算细长比);有效墙厚(用于计算细长比)
effective thickness of gas bed 气层有效厚度
effective thickness of oil bed 油层有效厚度
effective thickness of reservoir 储集层有效厚度
effective thickness of source rock 有效母岩厚度
effective thread 有效螺纹
effective threshold energy 有效阈能
effective throat 有效厚度(焊缝)
effective throat depth 有效焊缝厚度
effective throat thickness 焊缝计算厚度
effective throw (水枪的)有效射程
effective thrust 有效推力
effective time 有效时间
effective time constant 有效时间常数
effective tolerance 有效公差
effective torque 有效转矩;有效扭矩
effective toxicity 有效毒性
effective track length 线路有效长度
effective traction 有效牵引力
effective tractive effort 计算牵引力;有效牵引力
effective tractive force 有效牵引力;有效曳引力
effective tractive power 有效牵引力
effective transfer characteristic 有效传递特性
effective transformation group 有效变换群
effective transmission 有效透光率;有效传输
effective transmission band of channel 通路有效传输频带
effective transmission equivalent 有效传输当量
effective transmission gain 有效传输增益
effective transmission rate 有效传输率
effective transmission rating 有效传输定额
effective transmission speed 有效传输速度
effective transport rate 有效输移率
effective travel 有效行程
effective travel factor 有效行程系数
effective true airspeed 有效真空速
effective tube length 有效管子长度
effective turn 有效匝数
effective unit 统一换算单位;实际单位
effective unit weight 有效重度;有效单位重量
effective unit weight of sand 砂的有效重量
effective up-date rate 有效更新率
effective use time 有效使用时间
effective utilization of funds 资金利用效果
effective valuation 有效赋值
effective value 有效值
effective velocity 有效速度;有效流速;实地流速
effective velocity ratio 有效流速与船速比(螺旋桨处)
effective ventilation 有效换气量
effective ventilatory volume 有效通气量
effective vibration length 有效振动距离
effective virtual address 有效虚拟地址
effective viscosity 有效黏性;有效黏度
effective volatility 有效挥发性;有效挥发度
effective voltage 有效电压
effective voltage gradient 有效电压梯度
effective volt-ampere 有效伏安
effective volume 有效体积;有效容量;有效容积

effective volume capacity 有效容量;有效容积
effective volume content 有效容量
effective volume factor 有效体积因子
effective wake 有效伴流
effective washability curve 可洗性曲线
effective water 有效水
effective water-aluminate ratio 有效水铝酸盐比
effective water-cement ratio 有效水灰比
effective water-holding capacity 有效持水能力;有效持水量
effective water resistivity 水的有效电阻率
effective water resource survey 有效资源勘察
effective water saturation 有效含水饱和度
effective water supply 有效给水量
effective water use 有效用水量
effective watt 有效瓦特
effective wave 有效波(浪)
effective wave height 有效波高
effective wave impedance 有效波阻抗
effective wave length 有效波长
effective wave number 有效卷曲数
effective wave slope 有效波斜度
effective ways 有效方式
effective weight 有效重量
effective weight of soil 土的有效重度
effective well radius 井的有效半径
effective wheel pressure 有效锚定长度;有效轮压(力)
effective width 等效宽度;有效宽度
effective width coefficient 有效宽度系数
effective width of flange 有效翼缘宽度
effective width of lock 船闸有效宽度
effective width of ribbon 有效板宽
effective width of slab 板的有效宽度;有效板宽
effective width of sweep(ing) trains 扫海趟有效宽度
effective wind blocking area 有效风阻面积
effective wind load 有效风载
effective work 有用功,有效功;有效工作
effective work function 有效逸出功
effective working area 有效工作面积
effective working budget 实际执行预算
effective working days 实际作业天数;实际工作天数
effective working hour 有效工作小时;有效工时
effective working time 有效工作时间
effective workload 有效劳动量
effective yield 有效收入;实际收益率
effective-yield amortization 实得摊销
effective yield method 实际报酬率法;实得率法
effective zone 有效范围;有效带
effective zone of partial penetrating well 非完整井有效带
effectivity 有效性
effect lacquer 美饰漆;真空涂漆
effect length of track 到发线有效长
effect lighting 特技照明
effect machine 特技装置;特技机器
effect mechanism 作用机理
effect monitor 效果监察器
effect of acid rain 酸雨影响
effect of aerodynamic(al) downwash 空气动力下沉效应;气动力下沉效应
effect of ag(e)ing 老化作用;老化效应
effect of air pollution on human health 空气污染对人体健康的影响
effect of atmospheric pollution on plants 大气污染对植物的影响
effect of bankruptcy 破产的效力
effect of beam polarization 光束偏振效应
effect of bulking agent 填充剂effect
effect of climatic conditions 气候条件影响
effect of cold bending 冷弯效应
effect of cold work 冷弯效应
effect of confined waters 限制水路效应
effect of construction project 工程经济效益
effect of contraction 约束作用;收缩影响
effect of corrosion inhibition 腐蚀抑制作用
effect of cross-current 横流影响
effect of current 水流影响;水流的影响
effect of degrading 退降效应
effect of demand 需求效果
effect of deposits within the boiler 锅炉内部沉淀的效应

effect of depth 深度影响;深度效应;深度效果
effect of dilution 稀释效应
effect of distortion 变形影响;失真效应
effect of dragging 拖曳作用
effect of earth's curvature 地球曲率影响
effect of embedment 埋置效应
effect of emulsion shrinkage 乳剂收缩效应
effect of errors 误差效应
effect of eutrophication 富营养化效应
effect of experimental variable 实验变数效果
effect of exposure 曝光效果
effect of factor movements 要素移动后果
effect of fault 断层作用;断层影响;断层效应
effect of financial leverage 举债经营效果
effect of get employment 就业效果
effect of gravity 重力效应
effect of grooves 槽效应
effect of holes 孔效应
effect of human activity 人类活动影响
effect of illumination 发光效应;光照作用
effect of impact 着发作用;撞击作用;冲击作用
effect of increased concentration 增浓效应
effect of inertia 惯性作用;惯性效应
effect of inflation 通货膨胀的影响
effect of infrasound to man 次声对人的影响
effect of internal wave 内波效应
effect of investment 投资效果
effect of irradiation 辐照效应
effect of land and sea 海陆效应
effect of lateral friction 侧向摩擦作用
effect of lattice dilatation 晶格膨胀效应
effect of light 光线作用
effect of microwave 微波影响
effect of mode degeneracy 波型退化效应
effect of motion study 动作研究效果
effect of noise to man 噪声对人的影响
effect of odo(u)rs 气味影响
effect of over-consolidation 超固结效应
effect of pesticide 农药影响
effect of pollution 污染影响;污染效应
effect of pooling facility 设备联合经营的效果
effect of progressive failure 渐进破坏效应
effect of radiation on the immune response 辐射对免疫反应的影响
effect of refraction 折射效应
effect of relative project 相关项目效果
effect of relaxation time 张驰时间效应;松弛时间效应
effect of restraint 约束作用;抑制作用
effect of risks 风险的效应;风险的后果
effect of river dilution 河流稀释效应;河流稀释效力
effect of rudder 舵效
effect of salinity 盐度效应
effect of screw and rudder 车舵效应
effect of segregation 分凝效应
effect of self-purification 自净作用;自净效力
effect of sludge deposit 污泥沉积效应
effect of sludge settling 污泥沉积效应
effect of smoothing 修匀效果
effect of SO_2 二氧化硫效应
effect of specimen 试样尺寸效应
effect of spectrochemical smog 光化学烟雾效应
effect of stability 稳定度影响
effect of substitution 替代效应
effect of successive earthquake 连续地震效应
effect of support 支护效果
effect of surcharge 超载影响;附加荷载作用
effect of surroundings 环境影响;周围的影响
effect of the dredging works on the environment and on maritime safety 疏浚工程对环境与海上安全的影响
effect of the escapement 擒纵机构效应
effect of thermal pollution 热污染的影响
effect of torsion 扭转效果
effect of trade 贸易效果
effect of wall potential 壁势效应
effect of wastewater contaminant 废水污染效应
effect of water blocking of mine 矿井堵水效果
effect of water pollution 水污染效应
effect of well-bore storage 井筒储存效应
effect of wind 风力影响;风的影响
effect of yielding 屈服效应
effect on existing adjacent structure 对现有相邻结构物的影响

effector 效应因子;效应部;应变器;格式控制字符;操纵装置;试验器
effect-oriented 面向效率的;重视效率的
effector mechanism 效应机理
effector phase 效率期
effect payment 履行付款
effect pigment 随角异色效应颜料
effect quantity 效应量
effect radius 作用半径
effects 财物
effect filter 效果滤光片
effect microphone 音响效果传声器
effects on decision-making 决策效果
effect sound 特技声
effect to restraint 对约束影响;对抑制影响
effectuation 有效化
effect width of flange 翼缘有效宽度
efferent 输出管
efferent artery 输出动脉
efferentation 输出机能
efferent block 传出阻滞
efferent duct 输出管
efferent echo 出波
efferent fibre 离中纤维
efferent surface 传出曲面
efferent tract 流出道
effervesce 发泡;泡腾
effervescence 急速起泡;沸腾;起泡(沫);泡腾
effervesced steel 沸腾钢
effervescence level 泡沸面
effervescent 起泡的,泡腾的;冒泡的喷泉
effervescent bath tables 泡腾浴台
effervescent granulation 泡腾成粒
effervescent mixture 泡腾混合物;泡腾合剂
effervescent phosphate 泡腾磷酸盐
effervescent salt 泡腾盐
effervescent tablets 泡腾片剂
effervescing clay 碳酸盐黏土;起泡沫的黏土;膨胀黏土;泡沸黏土
effervescing dryer 沸腾式烘干机
effervescing steel 沸腾钢
efficacious opening 有效孔径
efficacy 效应力;效验;效力;功效
efficiency 效益;效能;效率(比);功效
efficiency analysis 效率分析
efficiency apartment 有厨房卫生设备的公寓套房(一至二居室);(有厨房及卫生设备的)小套公寓房间
efficiency apartment or unit 小套公寓房间;一室户公寓
efficiency audit 效率审计;效率检查
efficiency average 平均生产率
efficiency band 效率频带
efficiency-based personnel allocation method 效率定员法
efficiency brought about by information 信息效益
efficiency budgeting 效率预算法
efficiency by inputoutput test 实测效率
efficiency characteristic 效率特性
efficiency chart 效率图
efficiency coefficient method 功效系数法
efficiency contract 效率合约;效率的契约
efficiency cost 效率成本
efficiency curve 效率曲线;功率曲线
efficiency curve of hydraulic coupling 液力耦合器效率曲线
efficiency diagram 效率图
efficiency diode 阻尼二极管
efficiency dip 效率谷点
efficiency dwelling unit 单室居住单元
efficiency earnings 效率收入
efficiency engineer 技术操作工程师;效率专家;效率工程师;工艺工程师
efficiency equivalence 效率等价
efficiency estimate 有效估计(量)
efficiency estimation 效率估计;有效估计(量)
efficiency expert 工效专家;效率专家
efficiency extrapolation method 效率外推法
efficiency factor 效率因子;效率因素;效率因数;效率系数;有效因子
efficiency factor E 筛分效率系数 E
efficiency factor of pile group 群桩效率系数
efficiency formula 效率公式
efficiency frequency characteristic 效率频率特性

efficiency half life 效率半衰期
efficiency hill 等效率曲线图
efficiency hill diagram 综合特性曲线;等效率图
efficiency index 效能指数;效率指数;有效指标
efficiency index of membrane curing media 薄膜养护剂的效率指数
efficiency load curve 有效负荷曲线
efficiency load-range factor 有效负荷范围系数;负荷区效率因数
efficiency locus 效率轨迹
efficiency market 效率市场
efficiency measure 效果的衡量
efficiency modulation 效率调制
efficiency movement 效率增进运动
efficiency of blending 搅拌效率;混合效率
efficiency of buildings 建筑物的使用效果;建筑物的使用价值
efficiency of capital investment 资本投资效果
efficiency of cargo handling 装卸效率
efficiency of cargo-handling per man-shift 装卸工班效率;装卸实际工作工日产量
efficiency of construction 施工效率
efficiency of control 控制效率
efficiency of conversion 变换效率
efficiency of core penetration 岩芯穿透深度
efficiency of counter 计数颇效率
efficiency of covariance 协方差的效率
efficiency of cycle 热循环效率
efficiency of depreciation 折旧效能
efficiency of diagram relative 图表完备程度
efficiency of diesel engine 柴油机效率
efficiency of displacement 驱替效率
efficiency of drawing machine 引上机利用率
efficiency of economic operations 经济运行效率
efficiency of estimator 估计量的有效性
efficiency of free energy capture 自由能获得效率
efficiency of frontier 效率限界
efficiency of fund operations 资金运用效率
efficiency of gain 增重效率
efficiency of generator 发电机效率
efficiency of hammer 落锤效率
efficiency of heat engine cycle 热机械循环效率
efficiency of heat exchanger 换热器的效率
efficiency of heating surface 加热面效率
efficiency of heat utilization 热利用效率
efficiency of imbibition 自吸效率
efficiency of internal combustion engine 内燃机效率
efficiency of irrigation 灌溉效率
efficiency of joint 接合效率
efficiency of labo(u)r 劳动效率;工效;劳动生产率
efficiency of labo(u)r time 劳动时间效率
efficiency of light 光效率
efficiency of loading 装载效率;装货效率
efficiency of making payment 支付能力
efficiency of management 管理效率
efficiency of plant 电站效率
efficiency of preservation 防腐力
efficiency of pulley 滑轮效率
efficiency of pump 泵(的)效率;水泵效率
efficiency of pumping station 泵站效率
efficiency of radiation 辐射效率
efficiency of rectification 整流效率
efficiency of rectifier unit 整流机组效率
efficiency of screening 筛分效率
efficiency of separation 洗煤效率;分选效率
efficiency of shutter 快门效率
efficiency of sieve plate 筛板效率
efficiency of sizing 分粒效率
efficiency of subgrade soil 基土承载能力
efficiency of supply 供料系数;供给效率
efficiency of telescope 望远镜效能
efficiency of the adsorbent 吸附剂效率
efficiency of the furnace 炉效率
efficiency of the shutter contact 闪光接触效率
efficiency of transmission 传输效率;传递效率;输电效率
efficiency of turbine 水轮机效率
efficiency of wastewater treatment 污水处理效率
efficiency of water application 水分利用率
efficiency of water protection measures 水源保护措施实力
efficiency of welded joint 焊缝效率
efficiency-power curve 效率—出力曲线

efficiency rating 效率评价
efficiency ratio 效率比;功效比值
efficiency step up 效率换算(模型效率换算成真机效率)
efficiency system 效率制
efficiency target 效益指标
efficiency test 效率试验;效率测定;生产测定
efficiency testing 效率测定;生产率测定
efficiency testing machine 效率试验机;效率测定机
efficiency tracer method 效率示踪法
efficiency type apartment house 集中式公寓住宅(电梯间、楼梯间布置在住宅的中央)
efficiency unit 效率单位;有小厨房及卫生间设备的居住单元
efficiency value 效率值
efficiency variance 效率差异
efficiency wage 按劳工资;实效工资
efficiency wage system 效率工资制
efficient 效率高的;有效的;因子
efficient algorithm 有效算法
efficient and procuring cause 有效中介条款
efficient apartment 厨房卫生间齐全的公寓
efficient backscattered electron detector 高效反向散射电子探测器
efficient biofilm carrier 有效生物膜载体
efficient biofilm removal method 有效生物膜去除法
efficient blanking 有效消隐
efficient coding 有效编码
efficient coefficient of strengthening 加固效率系数
efficient conversion 有效转换
efficient crew 有效乘员
efficient deck hand 二级水手
efficient delivery 有效供水
efficient departmentalization 有效分权
efficient edge detector 高效边缘检测器
efficient estimator 有效估计(量)
efficient extraction 有效引出
efficient infiltration coefficient of precipitation 有效降水入渗系数
efficient loading 有效装载
efficient management method for all personnel 全员效率管理法
efficient mass density 有效质量密度
efficient microorganism 有效微生物群
efficient nickel plating brightener 高效镀镍光亮剂
efficient of investment 投资效果
efficient of travel(l)ing system 游动系统效率
efficient performance 经营效果
efficient picture coding 有效图像编码
efficient platform length 车站有效站台长度
efficient point 有效点
efficient power-shift transmission 高效(的)动力变换输送
efficient production structure 高效型产业结构
efficient protocol 有效协议
efficient range 有效范围
efficient removal od cuttings 有效地排除岩粉
efficient routine 经济工况
efficient silencer 高效消音器
efficient size 经济尺度
efficient sounding apparatus 有效响器
efficient statistic 有效统计量
efficient system of categories 分类的有效系统
efficient wing span 有效翼展
effigy 画像;肖像;雕像
effloresce 风化
efflorescence 开花期;凝霜;盐华状;风化(物);粉化;粉风;泛霜;泛碱(砖或混凝土表面上泛起的白色盐基类粉末);白花;起霜
efflorescence cleaning 洗刷风化物
efflorescence of salt 盐的起霜;盐的风化
efflorescence on concrete 混凝土表面起霜
efflorescence resistance 抗霜白花性;风化耐久性;抗风化(能力)
efflorescence test 霜白花测试;风化测试
efflorescent 霜状;风化的
efflorescent ice 花状冰
efflorescent-proof 防起霜白花;不风化;抗风化的
efflorescent salt 霜白花盐;风化盐
effluence 流出(物);溢流;溢出
effluent 流出液;流出物;流出的;废水;出液;出水;侧流;排放液;排放物;排出液;污水排放
effluent activity meter 流液放射性测量计

effluent air 排出的气体
effluent angle 出水角
effluent biological oxygen demand 污水生物需氧量
effluent brine 废盐水
effluent cave 出水溶洞;出水洞(穴)
effluent chamber 出水槽
effluent channel 出水渠道;出水槽;排泄道;排水渠
effluent characteristics 出流特征
effluent charge 流出的水流量;废水罚款;排污税;排污(收)费
effluent chemical oxygen demand 污水化学需氧量
effluent clarity 流出液清洁度
effluent collecting trough 出流水收集槽
effluent concentration 流出物浓度;排污浓度
effluent concentration histories 排污浓度曲线
effluent concentration limit 排污浓度限度;排放浓度限制
effluent conduit 出水管道
effluent control 流出物控制;废水(及)废气控制;排污控制;排放控制;排放规定
effluent data 流出物数据;出水数据;排出物数据;排出水数据
effluent design 排污设计
effluent dilution 流水稀释;流出物稀释;出水稀释;废水稀释;污水稀释
effluent discharge 排放废物;排出废液;污染物排放
effluent discharge conduit 排泄水管
effluent disposal 排出水处置;污水处理;废液排出;出水处置;废水处理
effluent disposal in lakes 出水湖泊处置
effluent disposal in rivers 出水河口处置
effluent disposal in the ocean 出水海洋处置
effluent disposal on land 出水陆地处置
effluent disposal standard 污水处理标准
effluent end saturation 流出端饱和度
effluent face 流出面
effluent farm 污水地段;废物地段
effluent fee 排污税;排污费
effluent filtration 出水过滤
effluent flow 污水流
effluent flow diversion work 潜流式分水岭
effluent flux 污水通量
effluent from sewage treatment plant 污水处理厂出水
effluent gas 烟道气;排放气;排出气体;废气
effluent guideline 排污指标
effluent holding reservoir 污水储池
effluent hopper 疏水斗
effluent-impounded body 地下水补给水体
effluent index 排污指标
effluent level 污染物排放量
effluent levy 排污税
effluent limitation 排污限度;排放限制
effluent line 出水管线;出水管道
effluent liquid 流出液
effluent monitor 废水剂量监测计;污水剂量监测计
effluent oil 流出油
effluent oil recovery 污油回收
effluent oil treatment 污油处理
effluent oil treatment equipment 污油处理设备
effluent organic matter 污水有机质
effluent oxygen demand 污水需氧量
effluent per unit length 单位长度出水
effluent pipe 出水管;排气管
effluent plume 出水股流;排污烟缕
effluent point 排污点
effluent polishing 出水终整处理
effluent pump 排污泵;污水排除泵
effluent quality 出水水质
effluent-quality standard 排污标准;排水水质标准;排水质标准;排放质量标准
effluent rate 出流率
effluent reuse 出水回用;出流再用;废水回用;污水再利用
effluent river 赢水河;地下水补给河;出流河(地下河);潜水补给河
effluent sample 试样流出阀
effluent seepage 废水渗漏;污水渗漏;渗漏;渗出
effluent segregation system 废水分流系统;废水分离系统;清污分流系统;排出物分流系统;排出物分离系统;污水分流系统
effluent settling chamber 出水沉淀池;废水沉淀

池;排出物沉降池;排出物沉淀池;污水沉淀池
effluent settling tank 出水沉淀池
effluent sewage (处理后的)污水排泄;流出污水;流出废水;废污水
effluent sewer 下水道
effluent sewerage 下水道污水;排污管道;排出污水
effluent side 出料侧
effluent size of sand 砂的有效直径
effluent specification 排放规范
effluent standard 排污标准;排放标准;出水标准
effluent standard for pollutant from ship 船舶排污标准
effluent stream 地下水逸出水流;赢水河;支流;出流河(地下河);表面水流;潜水补给河;外排流
effluent suspended solid concentration 排污悬浮固体浓度
effluent system 排污系统
effluent tax 废物税;排污税
effluent toxic chemical concentration 污水有机毒化学物浓度
effluent toxicity 污水毒性
effluent toxicity testing 污水毒性试验
effluent treatment 流出物处理;废水处理;排出物处理
effluent treatment plant 污水处理厂;废水处理装置
effluent treatment works 污水处理工程
effluent trough 出水槽;出流槽
effluent turbidity 污水浊度
effluent volume 洗脱液体积
effluent water 流出水
effluent weir 溢流堰;出水堰;出流堰
effluent weir loading in sedimentation tank 沉淀池出水堰负荷
effluent weir of sedimentation tank 沉淀池出水堰口
effluogram 液流图
effluve 高压放电
effluvium 泄出;磁素;排出物;无声放电;散出;[复]effluvia
efflux 流出物;流出量;泄漏;输出通量;时间消逝
efflux aerator 射流曝气器
efflux angle 流出角;排流角;射流角;出气角;出口角(度)
efflux coefficient 流速系数;流出系数;排出系数;射流系数
efflux control 射流控制
efflux cup 流出杯
efflux cup method 流出杯法;流杯法(测粘度方法);黏度杯法
efflux deflector cone 射流偏导锥
efflux door 喷口调节片
efflux equation (孔口的)出流方程
efflux(ion) 射流
efflux method 流出法
efflux nozzle 射流喷口
efflux of time 满期
efflux pump 射流泵
efflux time 流出时间;排出时间
efflux traffic queuing model 流动车队模型
efflux tube 流出管
efflux value 流出值
efflux velocity 排气速度;排出速度;射流速度
efflux visco(si)meter 流出式黏度计;射流黏度计
efforescence 渗斑
effort-controlled cycle 手控周期
effort syndrome 操劳综合症
efforwick test 泛霜试验
effset electrode 偏心式电极
effumability 易挥发性
effuse 倾出;喷出
effuser 扩散器;漫射体;喷管
effuse reflexed 卷边展生的
effusiometer (气体的)扩散计;隙透计;渗速计;气体扩散计
effusion 泻流;隙透;喷发
effusion cell 泄流室
effusion cooling 泻流冷却;隙透冷却;喷射冷却
effusion method 隙透法
effusive 喷发的;喷出的
effusive breccia 喷出角砾岩
effusive eruption 溢流喷发
effusive mass 喷发物
effusive period 喷发期
effusive rock 溢流岩;喷发岩;喷出岩

effusive stage 喷发期
effusive type 喷溢型
effusor 漫射体
eflectography 反射光复印法
Efwardian style 爱德华式建筑(英国)
egagropilus 毛块
Egeria 埃杰里亚
egg 卵形物
egg albumen 白软胶合剂
egg and anchor 卵锚饰;卵镖饰;卵与尖形装饰;卵箭饰【建】;蛋锚相间图案花饰
egg and anchor mo(u)lding (西方古建筑中的)卵与锚形交替的线饰
egg and arrow mo(u)lding (西方古建筑中的)卵与箭形交替的线饰
egg and dart 卵锚饰;卵镖饰;卵与尖形装饰;卵箭饰【建】;蛋矢饰
egg and dart mo(u)lding 蛋形与箭头装饰线脚;(西方古建筑中的)卵与飞标形交替的线饰
egg and spinach glaze 卵青釉
egg and tongue 卵镖饰;卵与尖形装饰;卵箭饰【建】
egg and tongue mo(u)lding (西方古建筑中的)卵与舌形交替的线饰
egg beater flying windmill 直升(飞)机
eggbeater PDC bit 打蛋器型PDC钻头
egg-blue 卵青
egg box foundation 多格基础
egg-breaking plant 打蛋厂
egg cal(l)ipers 卵形卡钳
egg china 薄瓷器
egg coal 蛋级烟煤块
egg coke 小块焦炭
eggcra 天花板嵌灯下的方格架
egg crate 蛋形格栅;花格灯罩;吊顶暗藏荧光灯管反射板
eggcrate canopy 格片挑棚;花格式出檐
eggcrate ceiling 格片顶棚
eggcrate diffuser 格片散光罩;花格散光片
eggcrate louver 方格百叶窗;花格式百页窗
eggcrate model tactical-planning model 战术作业砂盘
eggcrate overhang 格片挑棚
eggcrate type louver 格片(式)百叶窗
eggcrate type shading device 格片遮阳装置
egg cup 蛋杯
egg end 半球形末端板;半球形底板;半球形的底
egg-end 球形封头
egg ended boiler 蛋形端锅炉
Eggert's test 埃杰特快速定碳试验
Eggertz's method 埃格茨法
egg grader 蛋分级器
egg grading machine 蛋分级机
egg green 蛋青色
egg-hole 横梁窝
egg insulator 卵形绝缘子;拉线绝缘子;蛋形绝缘子
egg layer 下蛋式块料摊铺机
egg-laying 产卵
egg-laying area 产卵区
eggletonite 钠辉叶石
egg-mo(u)lding 卵圆饰
egg-paper 绘图纸
egg screen 蛋级筛
egg shape 箱形;蛋形
egg-shaped 卵形的
egg-shaped barrel with base 有底的蛋形圆筒
egg-shaped clothoid 卵形回旋曲线
egg-shaped concrete pipe 蛋形混凝土管
egg-shaped cross-section 卵形截面
egg-shaped cup saucer 蛋形杯碟
egg-shaped drain pipe 蛋形排水管
egg-shaped gallery 蛋形坑道;卵形长廊
egg-shaped kiln 蛋形窑
egg-shaped line 卵形线
egg-shaped ornament 卵形饰
egg-shaped pipe 卵形管(道)
egg-shaped profile 卵形轮廓
egg-shaped section 蛋形横断面;卵形面积
egg-shaped sewer 蛋形排污水管;卵形下水道;卵形污水管;卵形排水管;蛋形污水道
egg-shaped spheroid 卵形椭球
egg-shell china 蛋壳瓷;薄瓷器;薄胎瓷
egg-shell finish 半光面层;蛋光泽面层;半光泽面层;低光泽表面;蛋壳状终饰;蛋壳状加工;蛋

壳光面漆
egg-shell flat 蛋壳样平光
egg-shell flat varnish 蛋壳色泽的平光清漆
egg-shell glaze 蛋壳釉
egg-shell gloss 蛋壳光泽;蛋壳光(泽)
egg-shelling 蛋壳状釉面(釉面缺陷);蛋壳釉面;蛋壳皮;碎裂;鳞片状裂纹
egg-shell meal 蛋壳粉
egg-shell paint 蛋壳色泽的油漆;蛋壳光涂料;蛋壳漆;蛋壳彩画
egg-shell painting 彩绘蛋
egg-shell paper 具有蛋壳光泽的厚绘图纸;易碎纸;光厚绘图纸;蛋壳纸;(具有蛋壳光泽的)厚绘图纸
egg-shell porcelain 薄胎瓷;脱胎瓷;蛋壳瓷
egg-shell pottery 薄胎陶;蛋壳陶
egg sleeker 小型圆角光子;蛋形墁刀;蛋形墁刀
egg spoon 蛋匙
egg tempera 蛋青画
egg tray 蛋盘格
egg white glaze 乳白釉
egg yolk 蛋黄
egg yolk powder 蛋黄粉
eglantine 苯甲酸异丁酯
E glass E 玻璃
eglestonite 氯汞矿;褐氯汞矿
Egnell's law 伊格尼尔定律
ego-defense 自我保护
ego-enhancement 自我提高
egress 终切【天】;出食;出口(指水、气、汽、油等);外溢;外出(权)
egress and ingress 出入
egress aperture 出口孔
egress capacity 出口容量
egress design 出路设计
egress hole 出口孔
egress of heat 传热;放热;散热;热传导
egress opening 出口孔
egress orifice 出口孔
egress pit 出口坑
egress point (高速公路的)出口点
egress port 出口港
egress pressure 出口压力
egress road 出口(指水、气、汽、油等)
egress route 出口通道
egress side 出口侧
egress system 离机系统
egress width 出口宽度
egress window 出路窗
egress window latch 出路窗窗栓
egret 鹭鸶;白鹭
egueite 球磷钙铁矿
eguiphase zone 等相带
egypt blue 硅酸铜蓝
egypt green 硅酸铜绿
Egyptian alabaster 埃及纹大理石
Egyptian architecture 埃及建筑
Egyptian asphalt 埃及地沥青
Egyptian asphaltum 埃及沥青
Egyptian black 埃及黑色无釉玻化器皿
Egyptian blue 埃及蓝
Egyptian capital 埃及式柱头
Egyptian corn 埃及高粱
Egyptian cotton 埃及棉
Egyptian flax 埃及亚麻
Egyptian General Petroleum Corp. 埃及石油总公司
Egyptian gorge 埃及峡谷;埃及隘口
Egyptian green 埃及绿
Egyptian Hall of Vitruvius 维特罗维阿斯的埃及会堂
Egyptianized clay 埃及黏土
Egyptian jasper 埃及碧玉
Egyptian lace 埃及花边
Egyptian minaret 埃及伊斯兰教堂的尖塔
Egyptian prayer-tower 埃及的祈祷塔
Egyptian style 埃及式
Egyptian wind 埃及风
eharge stroke 充液行程(锻压时)
ehemical equipment shop 化工设备车间
EH-PH diagram eH-pH图
Ehrenfest model 埃伦费斯特模型
Ehrenfest's adiabatic law 埃伦费斯绝热定律
Ehrenfest's theorem 埃伦费斯特定理
Ehrenhaft effect 埃伦哈弗特效应
Ehrlich diazo reaction 埃尔利希重氮反应

Ehrlich's side chain theory 侧链学说;埃尔希侧链学说
ehrwaldite 玻基二辉岩
EIA-RS 232 interface 数据终端设备标准接口
eichbergite 艾硫铋铜矿
eiconometer 影像计
eicosadienoic acid 二十碳二烯酸
eicosane 二十烷
eicosanoic acid 花生酸
eicosanol 二十醇
eicosapentaenoic acid 二十碳五烯酸
eicosene dicarboxylic acid 二十碳烯二甲酸
eicosylene 二十碳烯
eiderdown 鸭绒垫
eidetic imagery 鲜明表象
eidograph 绘图用缩放仪;绘图缩放仪;缩放仪;缩放器;缩放绘图仪;伸缩画图器
eidophor 大图像投射器
Eidophor light valve 艾多福光阀
eidophor projector system 大图像投射系统
Eidophor system 艾多福电视投影方式
Eidopohor projector system 艾多福投影系统
Eifelian 艾菲尔阶
Eifelian stage 埃菲尔阶
Eiffel wind tunnel 埃菲尔式风洞
eigen 本征;特征的
eigen displacement method 特征位移法
eigendistribution 特征广义函数
eigen element 本征元素;特征元素
eigenellipse 本征椭圆;特征椭圆
eigen force method 特征力法
eigen frequency 特征频率;本征频率;简正频率
eigen-frequency spectrum 固有频谱
eigenfunction 本征函数;特征函数;特性函数
eigenfunction expansion 特征函数展开式
eigenfunction orbital 本征轨道
eigen magnetic moment 本征磁矩
eigen matrix 本征矩阵;特征矩阵
eigen matrix method 本征矩阵法;特征矩阵法
eigen-mobility 特征迁度
eigenmode 本征模;本征(波)型;正则型;特征模型
eigen-moment 内部禀矩
eigenperiod 本征周期;特征周期;固有周期
eigen polynomial 特征多项式
eigenproblem 本征问题;特征问题
eigenresolution 本征分辨率
eigenroot 本征根
eigenspace 本征空间;特征空间
eigenstate 本征态;特征(状)态;特性态
eigentone 固有振动频率
eigentransformation 本征变换;特征变换
eigenvalue 本征值;本来(的)价值;固有值;特征值
eigenvalue and eigenvector 本征值和本征向量
eigenvalue economizer 本征值节约子
eigenvalue equation 本征值方程
eigenvalue extraction 特征值析取固有振动频率
eigenvalue field problem 特征值域问题
eigenvalue method 本征值法;特征值法
eigenvalue of runoff 径流特征值
eigenvalue problem 本征(值)问题;特征值问题;斯图姆—刘维尔问题
eigenvalue problem of elastic stability 弹性稳定性的本征值问题
eigenvector 特征向量;本征向量;本征矢量;本征矢;特征矢量
eigenvibration 本征振动;特征振动
eigen wavefront 特征波前
eigenwert 特征值;本征值;本来(的)价值
eight and a half brick 八五砖
eight ball 球形全向传声器
eight-bed yard 八个预应力台架场地
eight bit 八位(二进制数)
eight-blade fan 八叶片风扇
eight braided rope 穿插编缆
eight-channel automatic recording viscometer 八线自动记录黏度计
eight-channel cathode-ray oscilloscope 八线阴极射线示波器
eight-channel data processor 八道数据处理器
eight-channel logic analyzer 八通道逻辑分析仪
eight-channel pressure recorder 八道压力记录器
eight-channel recording paper 八导程记录纸;八单位记录纸
eight-code frequency shift-power supply 八信息移频电源
eight colo(u)r plotter 八色绘图仪
eight columned 八柱式的
eight connections and the one level(l)ing assuring that a construction site is connected to water/ power/ roads/ communications/ natural gas/ heat gas/ hot water and sewer/ and that the land is leveled before a building project is begun 八通一平(水、电、道路、通信、天然气、暖气、热水、下水道要通,施工场地要平)
eight convergent points 八全
eight-cut finish 精凿痕面
eight-cyclone dust collector 八回旋式集尘器
eight cylinder engine 八汽缸发动机;八缸发动机
eight-day techograph 八天自记速度计
eight digit binary number 八位二进数
eight digit number 八位数
eight dimensional technique 霜点方法
eight division method of calculating tide 八分算潮法
eighteen-code frequency-shift power supply 十八信息移频电源
eighteenmo 十八开本
eighteen-o'cock statistics 十八点统计
eightfold 八重;八倍(的)
eightfold way 八重法
eight-foot tunnel furance 八英尺隧道加热炉
eight guiding principles 八纲
eighth bend 四十五度弯头
eight-high non-reversing rolling mill 不可逆式八辊轧机
eight-hole brick 八孔砖
eight-hour day 八小时工作制;八小时制
eight-hour day working system 八小时工作制
eight-hour's day 一日八小时劳动制
eight-hour service provided by ship station of the third category 第三类船舶电台工作八小时
eight-hour shift 八小时一班制;八小时工作制;八小时的工作班
eight-hour system of labour 八小时工作制
eight-in-line 直排八汽缸
eight joints 八溪
eight-leaf folding sliding shutter door 八扇折叠式滑动百叶门
eight-level 八单位的
eight level code 八单位码;八电平编码;八级码
eightlings 八连晶
eight-lobe tracery 八叶窗花格
eight-part hammer line 八股锤绳
eight-ply tyre 八层轮胎
eight-pointed 八个尖角的
eight-point mooring system 八点系泊系统
eight-power 八次方
eight queen problem 九宫问题
eights 八叠板;八层叠板
eight sampling intervals per decade 八分度取样
eight-section brocade 八段锦
eight-sided aisle 八角礼拜堂中部通道;八边侧廊
eight-sided bar 八角形条款;八角形钢筋
eight-sided base 八角形基地;八边形基座
eight-sided block 八角形大楼
eight-sided building 八角形建筑物;八边形建筑
eight-sided chapter-house 八角僧侣会堂
eight-sided chimney 八角烟囱
eight-sided cupola 八角半球形屋顶;八边形穹顶
eight-sided dome 八角圆屋顶
eight-sided donjon 八角城堡主塔
eight-sided dungeon 八角地牢
eight-sided foundation 八角基础
eight-sided girder four columns space frame 八角大梁的四柱空间框架
eight-sided ground plan 八角形场地规划
eight-sided keep 八角形要塞;八角形堡垒
eight-sided lantern 八角天窗;八角灯
eight-sided mosaic tile 八角马赛克锦砖;八边形马赛克(瓷)砖
eight-sided pyramid 八角金字塔
eight-sided rod 八角杆
eight-sided spire 八角尖顶
eight-sided steel 八角钢
eight-sided tile 八角砖
eight-sided tower 八角塔
eight-sided turret 八角小塔
eight-sided vault 八角拱顶
eight-sided wire 八角线材
eight soil profiles 八个土壤剖面
eight-speed 八速
eight-speed gearbox 八速齿轮箱;八挡齿轮变速箱
eight-stack base 八垛式炉台
eight-terminal network 八端网络
eight-to-pica leads 薄铅条
eight track strip chart recorder 八通道长图记录仪
eight-wheeler 八轮车
eight-wire traverse 八丝排线器
eight-wood wave channel antenna 八木波道式天线
eight working hour day 八小时工作制
eighty-board 井架工作平台;四立根高处的钻塔工作台
eighty-column card 八十列卡片
eighty-column puncher 八十列穿孔机
eighty-five percentile speed 第百分之八十五位地点车速
eigram 双字母组
eikonal 光程函数
eikonal coefficient 程差系数
eikonal equation 镜像方程;程函方程;程差方程
eikonogen 影源
eikonometer 量影尺;光像测定器
eikonoscope 光电摄像管
Einchluss thermometer 恩克腊斯温度计
Einisil coating 镍硅合金镀覆层
einkanter 单棱石
Einlehner abrasion value 艾因勒纳磨耗值
Einoplas 高密度聚乙烯合成纸
Einstein A coefficient 爱因斯坦自激系数
Einstein B coefficient 爱因斯坦他激系数
Einstein-Bohr equation 爱因斯坦—玻尔方程
Einstein-Bose statistics 爱因斯坦—玻色统计
Einstein coefficient 爱因斯坦系数
Einstein condensation 玻尔—爱因斯坦凝聚
Einstein-de Broglie formula 爱因斯坦—德布罗意公式
Einstein-de Hass effect 爱因斯坦—德哈斯效应
Einstein-de Hass method 爱因斯坦—德哈斯方法
Einstein-de Sitter cosmological model 爱因斯坦—德西特宇宙模型
Einstein-de Sitter model 爱因斯坦—德西特模型
Einstein diffusion equation 爱因斯坦扩散方程
Einstein effect 爱因斯坦效应
Einstein elevator 爱因斯坦升降机
Einstein energy 爱因斯坦能
Einstein equation 爱因斯坦方程(式)
Einstein equation of the field of gravity 爱因斯坦引力场方程
Einstein equations 爱因斯坦方程组
Einstein-Fowler equation 爱因斯坦—否勒方程
Einstein frequency 爱因斯坦频率
Einstein frequency condition 爱因斯坦频率状态
Einstein function 爱因斯坦函数
Einstein fundamental equation 爱因斯坦基本方程式
Einstein law of photochemical equivalence 爱因斯坦光化当量定律
Einstein mass energy formula 爱因斯坦质能公式
Einstein mass energy relation 爱因斯坦质能关系式
Einstein-Maxwell equation 爱因斯坦—麦克斯韦方程
Einstein Observatory 爱因斯坦天文台
Einstein partition function 爱因斯坦配分函数
Einstein photochemical equivalence law 爱因斯坦光化当量定律
Einstein photoelectric(al) equation 爱因斯坦光电方程
Einstein photoelectric(al) law 爱因斯坦光电定律
Einstein-Planck law 爱因斯坦—普朗克定律
Einstein probability coefficient 爱因斯坦概率系数
Einstein-Rosen waves 爱因斯坦—罗森波
Einstein's absorption coefficient 爱因斯坦吸收系数
Einstein's bed-load function 爱因斯坦推移质函数
Einstein's equation for specific heat 爱因斯坦比热方程
Einstein's equation of the field gravity 爱因斯坦引力场方程
Einstein's equivalency principle 等效性原理
Einstein's field equations 爱因斯坦场方程组
Einstein shift 爱因斯坦位移

Einstein's law 爱因斯坦定律
Einstein's law of gravitation 爱因斯坦场方程组
Einstein's law of photochemical equivalence 光化当量的爱因斯坦定律
Einstein's model 爱因斯坦模型
Einstein-Sokes equation 爱因斯坦斯托克方程
Einstein's principle of relativity 爱因斯坦相对性原理
Einstein's relation 爱因斯坦关系式
Einstein's summation convention 爱因斯坦求和约定
Einstein's theory for flow of suspension 爱因斯坦悬浮体流动理论
Einstein's unified field theories 爱因斯坦统一场论
Einstein tensor 爱因斯坦张量
Einstein Tower 爱因斯坦纪念塔(建在德国波茨坦);爱因斯坦塔(建在德国波茨坦)
Einstein transition probability 爱因斯坦跃迁概率
Einstein universe 爱因斯坦宇宙
Einstein viscosity equation 爱因斯坦黏度方程(式)
einstellung 定势
einthoven galvanometer 弦线电流计
Einthoven's galvanometer 艾因托文电流计
Einthoven's triangle 艾因托文三角
einzel lens 单透镜
Eirich mill 艾利奇混砂机
eisenbrucite 铁水镁石
eisengymnite 铁水蛇纹石
Eisenhart model 艾森哈特模型
eisenkiesel 含赤铁石英;铁石英
eisenstassfurtite 铁方硼石
eitelite 碳钠镁石
either directional route 双边线【铁】;双进路
either direction running 双向行车
either-direction signal(l)ing 双向信号
either-or facility 两可融资
either-rotation motor 双向电机
either route 双进路
either-way communication 半双向通信;双向择一通信
either-way market 两可市场
either-way operation 半双工操作;双向运行
ejaculation of flame 射出火焰
ejaculator 射出者
eject 抛出;弹射;弹起;射出
ejecta 喷出物
ejectability 弹射能力
ejecta blanket 溅射覆盖物
ejecta crest height 抛掷堆高度
eject control 抽取控制器;抽出器控制;拆卸器控制
eject drive 弹起驱动
ejected 射出的
ejected beam 引出束;出射束
ejected electron 发射的电子
ejected matter 喷出物质;喷出物
ejected photoelectron 发射的光电子
ejected rock 喷出岩
ejected scoria 喷发火山渣
ejecting 喷射
ejecting action 喷射作用
ejecting force 顶出力
ejecting gear 卸卷机
ejecting mechanism 推出机构
ejecting platform 草捆抛送滑道
ejecting plug 推顶杆
ejecting press 喷射挤压机
ejection 击出;出坯;发射;喷射;喷溅物;喷发;喷出;抛出;排出;脱膜;退壳;推出;射出
ejection air pump 喷吸气泵
ejection capsule 弹射座舱
ejection case mechanism 抛壳机构
ejection click 喷射性滴答声
ejection curve 退出曲线
ejection cylinder 喷油缸
ejection device 弹射装置
ejection during combustion 燃烧时的喷射
ejection efficiency 退出效率
ejection firing device 弹射发射机构
ejection force 弹射力
ejection load 脱模负荷
ejection mechanism 工件自动拆卸机构;弹射机构;抛掷机构
ejection nozzle 喷嘴

ejection of compact 出料
ejection opening 喷嘴;排出孔;喷出口
ejection orbit 弹射轨迹
ejection period 排出期
ejection port 退壳孔
ejection pressure 出坯压力;脱膜压力
ejection pump 喷吸泵
ejection rate 排出率
ejection seat 弹射座椅;弹射位置
ejection sound 喷射音
ejection stroke 喷出冲程;排气冲程
ejection system 顶出系统
ejection test 喷射试验;拍击试验
ejection velocity 抛射速度
ejective fold 隔档褶皱
eject key 抽出键;拆卸键;弹起键
eject lever 弹出控制杆;脱出杆
ejectment 喷射抛出;土地收回诉讼;收回产权诉讼
ejector 集水坑泵;卸土板;引出装置;工件自动拆卸器;顶推器;顶件器;顶杆;拆卸器;起膜器;喷油器;喷吸器;喷(雾)吸器;喷射器;排气辅助器;排除器;排出器;退壳器;推钉机机;推顶器;剔出器;弹射器;水射器
ejector air pump 打气泵;气流式喷射泵;喷射空气泵;喷射抽气泵;喷气(式水)泵;喷气式空气泵;喷射式水泵
ejector arm 推顶臂
ejector automatic speed selector valve 喷射器自动选速阀
ejector beam 顶料杆
ejector booster pump 喷射增压泵
ejector box 顶杆框
ejector brake valve 喷射器制动阀
ejector cable 喷射器绳索
ejector carrier roller 喷射器负重轮
ejector chimney 喷射器烟囱
ejector clutch 喷射器连接阀
ejector condenser 喷射(式)冷凝器;喷射凝汽器
ejector control lever 喷射器控制杆
ejector control valve 喷射器控制阀
ejector device 喷射装置
ejector die 滑动凸模;动型压铸机;凸压模
ejector die half 动型
ejector dredge(r) 射流挖泥船
ejector dryer 喷射干燥器
ejector effect 喷吸作用
ejector exhaust pipe 排气喷管
ejector filler 喷注器;喷射注入器
ejector force 顶杆力;脱模
ejector fore-stage 喷射泵前级
ejector grille 通风导流栅;送风(口)格栅;喷气器通风花格窗
ejector guide roller 喷射器导向滚轴
ejector half 动型
ejector jack 喷射器圆柱
ejector key 推顶键
ejector line 喷射器管道
ejector loop reactor 环状喷射反应器
ejector machine 喷射式机
ejector mark 顶杆压痕
ejector mechanism 工件自动拆卸机构
ejector mixer 喷射式混合器
ejector mixing 喷射式混合
ejector nozzle 喷射嘴;喷口;射流喷嘴
ejector overflow guard 喷射器浅盘溢流防护器
ejector pin 顶销;顶杆;出坯杆;起模杆;推顶杆;推钉
ejector plate 喷射器板;推顶杆板
ejector plug 喷塞
ejector priming 喷射注入法(水泵启动);喷射器;喷射泵启动
ejector pump 抽气泵;喷嘴;喷气引射泵;水抽子;射流泵;喷射(式水)泵
ejector pump for injecting plastic foam 水力泡沫塑料射送器【救】
ejector punch 出坯杆
ejector ram 喷筒活塞;推顶活塞
ejector return pin 复位杆
ejector rod 出坯杆;抛光钩杆;推顶柱
ejector seat 弹射座椅
ejector sequence valve 喷射器系列阀
ejector sleeve 推顶套
ejector speed change valve 喷射器的变速阀
ejector sprinkle head 射流式喷头
ejector stop 喷射器止动装置

ejector strap 喷吸器带条
ejector stroke 顶杆动程;脱模冲程
ejector type agitator 喷射式搅拌器
ejector type dredge(r) 射流式挖泥船
ejector type exhaust manifold 喷射式排气歧管
ejector type gas distributor 环形气相分布板
ejector type through-tubing tool 喷射式过油管下井仪
ejector type ventilator 射流式通风机
ejector type vertical take-off and landing aircraft 射流器式垂直起落飞机
ejector type well-point method 喷射井点降水
ejector vacuum pump 喷射真空泵
ejector valve 喷射器阀
ejector Venturi scrubber 喷射泵文丘里涤气器
ejector water air pump 喷水空气泵;水射空气泵
ejector well point 喷射井点
ejector well point system 喷射井点系统
ejusdem generis rule 同类规则
ekanite 硅钙铀钍矿
eka radium 类镭
eka-silicon 准硅
ekeing 增加;增大;放长
Ekhimi 艾克米乔木(灰褐色,纹理粗疏,用于控制和制普通家具)
ekistical 城市与区域规划的
ekistics 定居学;城市与区域计划学;人类群居学;人类居住学
Ekki 艾基树(产于尼日利亚的紫棕色硬木)
Ekman bottle 埃克曼瓶
Ekman bottom sampler 埃克曼水底采样器;埃克曼底质采样器
Ekman boundary layer 埃克曼边界层
Ekman convergence 埃克曼辐合带
Ekman current 埃克曼海流
Ekman current meter 埃克曼流速仪;埃克曼流速计
Ekman depth 埃克曼深度
Ekman dredge(r) 埃克曼采泥机;埃克曼采泥器
Ekman frictional layer 埃克曼摩擦层
ekmanite 锰叶泥石;锰星泥石
Ekman layer 埃克曼层
Ekman-Merz current meter 埃克曼-梅尔茨流速流向仪
Ekman number 埃克曼数
Ekman reversing water bottle 埃克曼颠倒采样瓶
Ekman rule 埃克曼规则
Ekman spiral 埃克曼螺(旋)线
Ekman square foot bottom sampler 埃克曼采泥器
Ekman transport 埃克曼输送
Ekman(water)bottle 埃克曼水瓶
eksedofacies 风化环境相
Ektar 艾克塔
Ektar lens 艾克塔摄影透镜
ektexis 泌出混合岩化作用
ektexis way 泌出变熔作用方式
ektogenic 外来的
ekzema 盐穹
elaborate 精心制成的;精巧的;阐述
elaborate carved work 细雕
elaborate collaterally 并行加工
elaborate decorative coating 饰面细工
elaborate design 精心设计
elaborated product 加工产物
elaborate embroidery 掺针绣
elaboration 精心装饰;精心制作;精工制造;详尽细节;确立
elaboration of alternatives 精密设计备选方案
elaboration product 精制品;加工产物;精心制作的产品
elaborative sequence 分解顺序
elaeolite 脂光石
elaeolite-syenite 脂光正长石
el(a)eomargaric acid 桐酸
el(a)eometer 油比重计;验油浮计;验油比重计
el(a)eostearic acid 桐酸
elaeostearin 甘油三桐酸酯;桐酸精
elaidic acid 反油酸
elaidic acid test 反油酸试验
elaidin 甘油三反油酸酯
elaidinization oil 反油酸化油
elaidin test 反油酸检验
elaioleucite 油粒
elaiometer 油比重计;验油比重计

elaioplast 油粒
elaiosome 油质体
Elana 埃拉纳聚酯纤维
Elanyl dye 埃拉尼尔染料
elapsed time 已用时间;经过(的)时间;混凝土在搅拌车内的停留时间;消逝时间;占用时间;所经时期;实测时间
elapsed time standard 各阶段最多允许时间标准;经历时间
elapsed time totalizer 使用时间累加器
elapsed years 已过年数
elastance 倒电容(值);弹回性;弹回率
elastane 弹性
elastane fiber 弹性纤维
elastic 橡皮带;有弹性的;弹性的;松紧带
elastic absorption 弹性吸收
elastic abutment 弹性拱座
elastic ach theory 弹性拱理论
elastic acoustic(al) reactance 弹性声(阻)抗
elastic after deformation 弹性后效变形
elastic after effect 滞弹性效应;弹性后效应
elastic aftershock 弹性余震
elastic after-working 弹性后效(应)
elastic air chamber 弹性气室
elastically bedded 弹性搁置的;弹性成层的
elastically bedded plate 弹性基础上安装钢板
elastically built-in 弹性固定的
elastically (em)bedded 弹性埋置的;弹性埋没的
elastically embedded plate 弹性基础上安装钢板
elastically fixed 弹性固定的
elastically isotropic material 弹性各向同性材料
elastically mounted rail 弹性铺设的钢轨
elastically restrained 弹性约束的
elastically restrained edge 弹性约束边
elastically supported beam 弹性支承梁
elastically supported bearing 弹性支承轴承
elastically supported continuous girder method 弹性支承连续梁法
elastically supported girder 弹性支承梁;弹性支承大梁
elastically supported plate 弹性支承板
elastically yielding bearing 弹性支承轴承
elastical metallic expansion ring 弹性金属胀圈
elastic analogy 弹性模拟
elastic analysis 弹性分析
elastic and plastic flow 弹性和塑性流
elastic anisotropy 弹性各向异性
elastic arch 弹性拱
elastic arch method 弹性拱法
elasticated net 弹力网
elasticator 弹性剂
elastic axis 减振轴;缓冲轴;弹性轴
elastic backpull 弹性后张力
elastic bandage 弹性绷带
elastic barometer 变形气压表;弹性气压表
elastic beam 弹性梁
elastic bearing 弹性轴承;弹性支座;弹性支承
elastic bearing pressure distribution 弹性支承压力分布
elastic behaviou(u)r 弹性状态;弹性行动;弹性现象;弹性特性;弹性(行为)
elastic behaviou(u)r of flexible pavement 柔性路面弹性动态;柔性路面弹性作用
elastic bending 弹性弯曲
elastic bitumen 弹性沥青
elastic body 弹性体
elastic bond wheel 弹性胶结磨轮
elastic boundary 弹性边界
elastic boundary condition 弹性边界条件
elastic breakdown 弹性失效;弹性破坏;弹性(变形)断裂
elastic breakdown pressure 弹性失效压力;弹性终止压力
elastic brittle material 弹性脆性材料
elastic buckling 弹性屈曲;弹性压屈;弹性曲屈;弹性屈曲;弹性屈服
elastic buckling load 弹性屈曲负载
elastic buffer 弹性缓冲器
elastic cable 弹性索
elastic calibration device 控制功率计
elastic caoutchouc 橡胶;弹性橡胶
elastic cell 弹性元件
elastic center 弹性中心;弹心
elastic center method 弹性中心法

elastic chamber 弹性室
elastic characteristic 弹性特征;弹性特性
elastic circular sandwich beam 弹性圆形夹合梁
elastic claw (耙的)弹齿
elastic clutch 弹性联轴节;弹性离合器
elastic coating 弹性的涂层;弹性的贴胶
elastic coefficient 弹性系数
elastic coefficient of water 水体积压缩系数
elastic collar 弹性挡圈
elastic collision 弹性碰撞
elastic collision model 弹性碰撞模型
elastic comeback 弹性回复
elastic compaction 弹性挤压作用
elastic compatibility 弹性的相容;弹性相容性
elastic compliance 弹性顺度;弹性顺从
elastic compliance coefficient 弹性顺度系数
elastic compliance constant 弹性顺度常量;弹性柔顺常量
elastic component 弹性构件
elastic composition 弹性组合物
elastic compound 弹性合成物
elastic compressibility 弹性压缩度
elastic compression 弹性压缩(量)
elastic compression energy of gas 天然气弹性压缩能
elastic condition 弹性状态
elastic cone 弹性圆锥;弹力圆锥
elastic connector 弹性连接器;弹性接头
elastic constant 柔顺常数;弹性常数
elastic construction 弹性建筑
elastic construction system 弹性结构体系
elastic contact 弹性接触
elastic container 弹性壳
elastic continuum 弹性连续体;弹性连续介质
elastic control 弹性控制
elastic cord 松紧线
elastic core 弹性核心
elastic core packing 弹性芯子填密物
elastic coupling 弹性联轴器;弹性联轴节;弹性连接
elastic-crack-growth fracture 弹性裂纹增长断裂
elastic cracking strain 弹性破裂应变
elastic creep 弹性蠕动
elastic creep recovery 弹性蠕动复原
elastic critical load 弹性临界负荷;弹性临界荷载
elastic critical shearing stress 弹性临界剪应力
elastic cross-section 弹性截面;弹性反应截面
elastic currency 弹性通货;伸缩性通货
elastic curve 变形曲线;弹性曲线
elastic cushion 弹性垫层
elastic cylinder 弹性柱体;弹性圆柱
elastic damping 弹性阻尼
elastic deflection 弹性弯沉;弹性挠曲;弹性挠度;弹性变位
elastic deflection exponent 弹性挠曲变形指数
elastic deformable aquifer 弹性可变含水量
elastic deformation 弹性形变;弹性挠曲;弹性挠度;弹性变形
elastic deformation curve 弹性变形曲线
elastic deformation limit 弹性形变极限;弹性变形极限
elastic deformation stage 弹性变形阶段
elastic deformation zone 弹缩性变形带
elastic deforming 弹性形变
elastic demand 弹性需求
elastic depression 弹性沉陷
elastic design 按弹性设计;弹性设计
elastic design method 弹性设计方法;弹性阶段设计法
elastic discontinuity 弹性不连续性
elastic dislocation 弹性位错
elastic dislocation theory 弹性位错理论
elastic displacement 弹性移位;弹性位移
elastic distortion 弹性扭曲;弹性畸变
elastic distribution 弹性分布
elastic distribution of bearing pressure 支承压力弹性分布
elastic dolphin 弹性系船柱;弹性靠船墩
elastic draw gear 弹簧车钩
elastic draw hook 弹性拉钩
elastic drift 弹性残余变形;弹性残留变形
elastic drive 橡胶皮带传动;弹性驱动
elastic duck 厚浆黑衬里布
elastic dynamometer 弹性测力计
elastic earth pressure 弹性土压力

elastic effect 弹性效应
elastic element 弹性元件
elastic elongation 弹性伸长
elastic embankment 弹性土堤
elastic end-restraint 弹性终端约束;弹性终端抑制;弹性端部约束
elastic energy 弹性能(量)
elastic energy degradation 弹性能量降级
elastic energy of fluid 流体弹性能
elastic energy of rock 岩石弹性能
elastic entropy 弹性熵
elastic equation 弹性方程(式)
elastic equilibrium 弹性平衡
elastic expandable bobbin 弹性伸缩筒管
elastic expansion joint 弹性膨胀接头
elastic extension 弹性延伸;弹性伸长;弹性膨胀
elastic extenuation 弹性衰减
elastic failure 超过弹性极限的损坏;弹性失效;弹性破坏;弹性断裂
elastic failure criterion 弹性失效准则
elastic fastener 弹性扣件
elastic fastening device 弹性紧固装置
elastic fatigue 弹性疲劳
elastic feedback 弹性反馈
elastic feedback controller 弹性反馈控制器
elastic fender 弹性防撞装置
elastic fendering 弹性碰垫;弹性护舷
elastic fendering device 弹性防撞装置
elastic fiber 弹性纤维;弹力纤维
elastic fiduciary issue system 弹性受托发行制
elastic field 弹性场
elastic filler 弹性填料
elastic finish sheeting 弹性整理稀薄平布
elastic fixing 弹性连接;弹性固接
elastic flow 弹性流量;弹性流(动)
elastic flow creep 弹性流动蠕变
elastic fluid 弹性流体
elastic force 弹性力;弹力
elastic fore-effect 弹性前效
elastic formula 弹性公式
elastic foundation 弹性基础;弹性地基
elastic frame 弹性框架;弹性机座
elastic gain 弹性增益
elastic gel 弹性凝胶
elastic give 弹性变形
elastic glazing compound 弹性腻子(装玻璃用)
elastic gravitational coupled vibration 弹性重力耦联振动
elastic gravitational stiffness 弹性重力刚度
elastic grinding wheel 橡胶砂轮;弹性砂轮
elastic grindstone 弹性磨具
elastic ground 弹性地面
elastic gum 弹性胶
elastic half-space 弹性半空间
elastic half-space spring stiffness 弹性半空间弹簧刚度
elastic half-space theory 弹性半空间理论
elastic hardness 弹性硬度
elastic heave 弹性隆起
elastic helical coil 弹性螺旋圈
elastic heterogeneity 弹性非均匀性
elastic-homogeneous 弹性匀质的
elastic hysteresis 弹性迟滞性;弹性滞后(现象)
elastic hysteresis loop 弹性滞后闭合回线
elastic impact 弹性碰撞;弹性锤击;弹性冲击
elastic impedance 弹性阻抗
elastic imperfection 弹性不完整
elastic initial region 初始弹性区(域)
elastic instability 弹性失稳;弹性不稳定(性);弹性不稳定(度)
elastic interaction 弹性相互作用
elastic intermediate layer 弹性中间垫层
elastic isotropic material 弹性均质材料;弹性各向同性材料
elastic isotropy 弹性各向同性
elasticity 弹性(学);弹力;伸缩性
elasticity after effect 弹性后效
elasticity coefficient 弹性模量;弹性系数;伸缩性系数
elasticity condition 弹性条件
elasticity constant 弹性常数;弹性常量
elasticity correction 弹性修正
elasticity effect 空气密度影响
elasticity equation 弹性方程(式)

elasticity factor 弹性因素
elasticity figure 弹性图形
elasticity form factor 弹性形状因素
elasticity gage 弹性计
elasticity limit 弹性极限
elasticity matrix 弹性矩阵
elasticity measurement 弹性测定
elasticity membrane model 弹性膜模型
elasticity modulus 弹性模数;弹性模量
elasticity number 弹性值;弹性数
elasticity number 1 第一弹性数
elasticity number 2 第二弹性数
elasticity of bending 弯曲弹性
elasticity of bulk 容积弹性;体积弹性
elasticity of compression 压缩弹性
elasticity of demand 需求伸缩性;需求弹性
elasticity of elongation 伸缩弹性;伸长弹性
elasticity of employment 就业弹性
elasticity of export and import 进出口弹性
elasticity of flexure 挠曲弹性;弯曲弹性
elasticity of fluid 流体弹性
elasticity of form 形状弹性
elasticity of gases 气体弹性
elasticity of hose 水带的弹性
elasticity of input substitution 投入替代弹性
elasticity of productivity 生产力弹性
elasticity of response 反应弹性
elasticity of rock 岩石弹性
elasticity of shear(ing) 剪切弹性
elasticity of soil 土壤弹性;土的弹性
elasticity of substitution 替换弹性;替代弹性
elasticity of substitution of factor 要素替代弹性
elasticity of supply 供给弹性
elasticity of supply of factor 要素的供给弹性
elasticity of the production function 生产力弹性
elasticity of torsion 扭转弹性
elasticity of volume 容积弹性;体积弹数;体积弹性模量
elasticity problem 弹性问题
elasticity ratio 弹性比
elasticity supply 供应弹性
elasticity tensor 弹性张量
elasticity test 弹性试验
elasticity test of grease 润滑脂弹性试验
elasticity theory 弹性原理;弹性理论
elasticity volume 弹性容量
elasticized fabric 弹性织物
elasticizer 增塑剂;增韧剂;弹性增进剂;弹性物质;弹性剂;塑化剂
elastic joint 挠性接头;弹性联轴器;弹性联轴节;弹性接头;弹性接合;弹性(接)缝
elastic joint seal(ing) compound 弹性接头密封料
elastic kernel 弹性核
elastic lag 弹性滞后;弹性惯性
elastic lateral bukling 弹性侧向弯曲
elastic layer 弹性层
elastic layered theory 弹性层状体理论
elastic layer of infinite thickness 无限厚的弹性层
elastic limit 弹性限度;弹性权限;弹性极限
elastic limit method 弹性限制法
elastic limit of materials 材料的弹性极限
elastic limit of wood 木材弹性极限
elastic limit strain 弹性极限应变
elastic limit system 弹性限制法;伸缩性发行法
elastic limit under compression 抗压弹性极限
elastic limit value 弹性极限值
elastic line 弯曲中心线;弹性线;弹性(挠)曲线
elastic linear contact seal 弹性线接触密封
elastic line method 弯曲中心线方法;弹性线法
elastic lip 弹性唇片
elastic liquid 弹性液体
elastic load 弹性荷重;弹性荷载
elastic-load method 弹性负载方法
elastic loads method 弹性负载方法
elastic loop dynamometer 弹性环功率计
elastic loss 弹性损失;弹性应力损失
elastic loss factor 弹性损耗因子
elastic loss of prestress 预加应力的弹性损失
elastic mass 弹性物;弹性体
elastic material 弹性物质;弹性(地面)材料
elastic material balance method 弹性物质平衡法
elastic matrix 弹性矩阵
elastic matter 弹性物
elastic mechanics 弹性力学

elastic medium 弹性介质;弹性传导体
elastic member 弹性构件
elastic membrane 弹性膜
elastic membrane analogy 弹性薄膜模拟法
elastic memory 弹性记忆
elastic memory distributor 缓冲存储分配器
elastic memory effect 弹性记忆效应
elastic method 弹性阶段设计法
elastic migration theory 弹性波偏移理论
elastic mineral pitch 弹性矿质硬沥青
elastic model 弹性模型
elastic moderation 弹性缓和
elastic modular ratio 弹性模量比
elastic modulus 弹性计量单位;弹性模量
elastic modulus of rock 岩石的弹性模量;岩石弹性模量
elastic modulus of tensile compression 拉压弹性模数
elastic modulus of the rail foundation 钢轨基础弹性系数
elastic modulus of the rail support 钢轨支点弹性系数
elastic moment 弹性矩
elastic money supply 弹性货币供应
elastic mounting 弹性支座;弹性支承
elastic multiple layered theory 弹性多层体系理论
elastic net 弹性网眼
elastic network 弹性网
elastic neutron 弹性中子
elastic normal stress coefficient 弹性正应力系数
elastic number 弹性值
elastic nut 弹性螺母
elastic nylon 弹性锦纶
elastico-plastic solid 可塑弹性固体
elastic oscillation 弹性振荡
elastico viscosity 弹黏性;弹黏度
elasticoviscous 弹黏性的
elastico viscous liquid 弹黏性液体
elastico viscous solid 弹黏性固体
elastic packing 弹性填料;弹性密封;弹性垫料
elastic packing ring 弹性填密环
elastic parameter 弹性参数
elastic peak 弹性峰值
elastic pendulum 弹性摆
elastic-perfectly plastic 理想弹塑性的
elastic-perfectly plastic body 理想弹塑性体;完全弹塑性体
elastic photoproduction 弹性光生
elastic pile 弹性桩
elastic pinch plate 弹性扣板
elastic pipe 弹性管
elastic plan 弹性计划
elastic-plastic analysis 弹塑性分析
elastic-plastic behavio(u)r 弹塑性行为;弹塑性状态
elastic-plastic bending 弹塑性弯曲
elastic-plastic body 弹塑性体
elastic-plastic-creep analysis 弹塑性蠕变分析
elastic-plastic deformation 弹塑性变形
elastic-plastic drive 弹塑性驱动
elastic-plastic dynamic(al) analysis 弹塑性动力反应分析
elastic-plastic dynamic(al) response analysis 弹性力学反应分析
elastic-plastic flow 弹塑性流
elastic-plastic fracture 弹塑性断裂
elastic-plastic fracture mechanics 弹塑性断裂力学
elastic-plastic interface 弹塑界面
elastic-plasticity 弹塑性
elastic-plasticity of soil 土的弹塑性
elastic-plastic material 弹塑性体;弹塑性材料
elastic-plastic model 弹塑性模型
elastic-plastic of soil 土的弹塑性
elastic-plastic range 弹塑性范围
elastic plate 弹性板
elastic plate method 弹性板法
elastic polymer liquid 弹性聚合物液体
elastic positioning device 弹性定位装置
elastic potential 弹性势
elastic potential energy 弹性势能
elastic potential scattering 势弹性散射
elastic potential theory 弹性势理论
elastic precompression 弹性预压缩
elastic proof stress 弹性极限应力

elastic property 弹性性质;弹性特性
elastic proportional limit 弹性比例极限
elastic quadrant 弹性舵扇;弹簧舵柄弧
elastic quartz 弹性石英
elastic rail fastener 弹性扣件
elastic rail spike 弹性道钉;弹簧道钉
elastic range 弹性区(域);弹性范围
elastic ratio 弹性比值;弹性比
elastic reactance 弹性(阻)抗;弹性反作用力
elastic reaction 弹性作用;弹性后效;弹性反应
elastic rebound 回弹;弹性回跳;回弹;弹性回跳
elastic rebound factor 弹性回弹系数
elastic rebound hypothesis 弹性回跳假说;弹性回弹假说
elastic rebound mechanism 弹性回弹理论
elastic rebound theory 弹性回跳说;弹性回跳理论;弹性回弹理论
elastic rebound theory of earthquake 地震的弹性回跳理论;地震的弹性回弹理论
elastic recoil 弹性回弹;弹性回缩;弹性反冲
elastic recoil analysis 弹性反冲分析
elastic recovery 回弹;弹性回复;弹性恢复(率);弹性复原
elastic reel 弹性伸缩纱框
elastic region 弹性区(域);弹性界限
elastic relaxation time 弹性松弛时间;弹性弛豫时间
elastic release 弹性释放
elastic removal 弹性迁移
elastic reservoir mo(u)lding 弹性体储[贮]罐模塑成型
elastic resilience 回弹性;回弹储能;弹性回能;弹性回复能力;弹性回弹
elastic resin 弹性树脂
elastic resistance 抗弹力;弹(性)阻力;弹性抗力;弹性反力
elastic resistance of ground 地基弹性抗力
elastic resistance-wire ga(u)ge 电阻丝弹性应变计
elastic resisting force 弹性阻力
elastic response 弹性响应
elastic response analysis 弹性反应分析
elastic response line 弹性感应线
elastic restitution 弹性恢复
elastic restoring force 弹性回复力;弹性恢复力
elastic restraint 弹性约束
elastic restraint coefficient 弹性约束系数
elastic return (提花机的)弹力回综
elastic rigidity 弹性刚度
elastic-rim wheel 弹性轮钢车轮
elastic ring 弹性圈;弹性环;弹簧垫圈
elastic rod rail fastening 弹条式扣件
elastic rotor 弹性转子
elastic rubber 弹性橡胶
elastic ruche tape 弹性褶裥带
elastic saddle shaped washer 弹性鞍形垫圈
elastic scaler 弹性尺
elastic scattering 弹性散射
elastic scattering collision 弹性散射碰撞
elastic scattering cross-section 弹性散射截面
elastic scattering electrons 弹性散射电子
elastic scattering resonance 弹性散射共振
elastic scattering wave 弹性散射波
elastic scheme 弹性方案
elastic seal(ing) 弹性密封(材料)
elastic sealing compound 弹性密封料
elastic sealing ring 弹性体封圈
elastic section 弹性横断面
elastic seismic energy 弹性地震能量
elastic seismic response of structure 结构的弹性地震反应
elastic semi-infinite body 弹性半无限体
elastic semi-infinite foundation 弹性半无限地基
elastic sensation 弹性感;弹力感
elastic sensing element 弹性敏感元件
elastic settlement 弹性沉降
elastic shear 弹性剪切
elastic shear deformation 弹性剪切形变
elastic sheet 弹性钢板;弹性模片;弹性薄板
elastic sheet analog 弹性板模拟
elastic shell 弹性壳体
elastic shell equation 弹性薄壳方程
elastic shinguard 弹力护腿
elastic shock 弹性振动;弹性冲击
elastic shortening 弹性压缩量;弹性缩短;弹性收缩(预应力混凝土)

elastic shrinkage 弹性收缩
elastic side wall 弹性边墙
elastic similarity 弹性相似
elastic similarity law 弹性相似定律
elastic slab 弹性板
elastic sleeve 弹性套筒
elastic sleeve bearing 弹性套筒轴承
elastic slip 弹性滑移
elastic soil wedge 弹性土楔
elastic sol 弹性溶胶
elastic solid 弹性体;弹性固体
elastic solution 弹性溶液
elastic space 弹性隔块
elastic spacer 弹性垫条;弹性垫圈
elastic spectral kinetic energy 弹性谱动能
elastic spectrum 弹性谱
elastic sphere 弹性球体
elastic spike 弹性道钉
elastic spindle 弹性锭子
elastic spring 钢丝挡圈;弹簧
elastic spring back 弹性后效
elastic springing 弹性挠曲变形
elastic stabile 弹性稳定
elastic stability 弹性稳定(性);弹性稳定(度)
elastic stage 弹性状态;弹性阶段
elastic state 弹性状态
elastic state of equilibrium 弹性平衡状态
elastic stator 弹性锭子
elastic stiffness 弹性劲度
elastic stiffness coefficient 弹性刚度系数
elastic stiffness constant 弹性劲度常数;弹性劲度常量;弹性刚度常数;弹性刚度常量
elastic stop 弹性制动爪
elastic stop nut 弹性锁紧螺母;弹性防松螺母
elastic storage 弹性储存量
elastic store 缓冲存储器;弹性存储器
elastic stored energy 弹性储[贮]能
elastic strain 弹性应变
elastic strain energy 弹性应变能(量);弹性比功
elastic strain recovery 弹性应变恢复
elastic strength 弹性强度
elastic stress 弹性应力
elastic stress distribution 弹性应力分布
elastic stretch(ing) 弹性伸长;弹性延伸;弹性拉伸
elastic string 弹性弦线
elastic structural arch 弹性结构拱
elastic structural system 弹性结构系统
elastic structure 弹性结构
elastic subgrade 弹性路基;弹性地基
elastic subgrade reaction 弹性基层反力;弹性基床反力
elastic sulfur 弹性硫
elastic supply 弹性供应;伸缩性供应
elastic support 弹性支座;弹性支承;弹性支撑;弹性悬挂架
elastic supported beam method 弹性梁支承法
elastic supported cantilever 弹性支承悬臂梁
elastic supported edge 弹性支承边
elastic supporting stiffness 弹性支承刚度
elastic surface 弹性曲面;弹性表面
elastic surface wave 弹性表面波
elastic suspension 弹性吊架
elastic synthetic 弹性合成物
elastic system 弹性系统;弹性体系
elastic tariff 伸缩关税
elastic tensile strain 弹性拉伸应变
elastic tensor 弹性张量;弹性张力
elastic test 弹性试验
elastic theory 弹性原理;弹性理论
elastic theory computational method 弹性理论计算法
elastic thermal stress 弹性热应力
elastic threads 拉紧线
elastic tire 弹性轮胎
elastic torque 弹性变形力矩
elastic torsion 弹性扭力
elastic traction 弹性牵引
elastic translation 弹性转移;弹性变换
elastic transverse wave 弹性切变波
elastic trip 弹性片
elastic tube 弹性管
elastic tunnel(l)ing effect 弹性隧道效应
elastic turbulence 弹性湍流;弹性扰动
elastic vibrating element 弹性振动元件

elastic vibration 弹性振动
elastic viscoplastic body 弹黏塑性体
elastic viscosity 弹黏性
elastic viscous field 弹黏性场
elastic viscous plasticity 弹黏塑性
elastic wall 弹性墙
elastic washer 弹性垫圈
elastic water drive 弹性水驱
elastic wave 弹性波
elastic wave computerized tomography 弹性波层析成像
elastic wave generator 弹性波发生器
elastic wave holography 弹性波全息摄影术
elastic wave method 弹性波法
elastic wave propagation 弹性波传播
elastic wave theory 弹性波理论
elastic wave velocity 弹性波速(度)
elastic webbing 弹性织物
elastic wedge 弹性槽楔
elastic wedge joint 弹性楔接头
elastic weight 弹性重量;弹性荷重;弹性荷载
elastic weight method 角变位荷载法;弹性荷载法
elastic wheel (研磨的)弹性轮
elastic wire strain meter 弹性钢丝应变计
elastic work schedule system 弹性工作时间制度;弹性工时制
elastic yielding 弹性屈服
elastic zone 弹性范围;弹性区(域)
elastic zoning 弹性分区制
elastification 弹性化
elastivity 介电常数倒数
elastivity coefficient 倒电容系数
elasto 弹塑性
elasto-bending 弹塑性弯曲
elastodiene fibre 二烯类弹性纤维
elastodurometer 弹性硬度计
elastodynamic(al) 弹性动力学的
elastodynamic(al) extruder 弹性动态挤出机
elastodynamic(al) field 弹性动力位移场
elastodynamic(al) model 弹性动力模型
elastodynamic(al) radiation 弹性动力辐射
elastodynamic(al) stress-intensity factor 弹性动力应力强度因数
elastodynamics 弹性体动力学;弹性动力学
elastogel 弹性凝胶
elastohydrodynamic(al) lubrication 弹性流体动力润滑
elastohydrodynamic(al) lubrication theory 弹性流体动力润滑理论
elastohydrodynamics 弹性流体动力学
elastoid 弹性样物质
elastokinetics 弹性动力学
elastomechanics 弹性力学
elastomer 合成胶;高弹性塑料;高弹体;弹性体;弹性灌浆料;弹性高分子物质;弹料;塑料混凝土掺和料
elastomer adduct 弹性体加成物
elastomer adhesive 弹胶体黏合剂
elastomer alloy 弹性体合金
elastomer-based contact solution 高弹性触压胶浆
elastomer connector 橡胶接头
elastomer covering 弹性外套
elastomer dispersion 合成橡胶分散;弹胶分散体
elastomeric 合成橡胶的;橡胶状的;有橡胶特性的
elastomeric adhesives 弹性黏结材料
elastomeric bearing 弹性支座;弹性(体)支承
elastomeric bearing pad 弹性合成橡胶支座
elastomeric bridge bearing 桥梁合成橡胶支座
elastomeric butyl ca(u)lk 丁基橡胶嵌缝
elastomeric coating 弹性敷层
elastomeric coupling 弹性联轴节
elastomeric energy absorber 弹性消能器
elastomeric fender unit 弹性体护舷部件
elastomeric gasket 弹性垫圈
elastomeric joint 弹胶接头
elastomeric joint sealant 合成橡胶嵌缝料;弹胶接缝剂
elastomeric material 高弹性材料
elastomeric meter 弹力计
elastomeric network 弹性网络
elastomeric pad 弹性胶垫
elastomeric pad bearing 板式橡胶支座;弹胶衬垫支承;弹性衬垫承座
elastomeric pad fixed bearing 合成橡胶垫板固定支座
elastomeric parts 弹性材料零件
elastomeric polymer 弹性聚合物
elastomeric property 高弹性能
elastomeric seal 合成橡胶密封
elastomeric sealant 弹性密封料;弹性密封剂
elastomeric sheet 弹性板;弹胶薄片
elastomeric shield material 弹性屏蔽材料
elastomeric state 高弹态
elastomeric strain gate 弹性应变仪
elastomeric tank base joint 弹性箱底接缝;弹胶箱式接头
elastomeric tester 弹性测定器
elastomeric unit 弹性体部件
elastomeric waterproofing 弹性体防水;弹性防水
elastomer material 弹性材料
elastomer paste 弹胶膏体
elastomer plastic 弹性体塑料
elastomer powder 合成橡胶粉;弹胶粉
elastomer seal(ing) 弹胶密封剂;弹性密封
elastomer sheet 弹性体片材
elastomer stator 橡胶定子
elastometer 弹性计;弹力计
elastometry 弹性测定法;弹力测定法
elasto-optic(al) coefficient 弹光系数
elasto-optic(al) effect 弹光效应
elasto-optic(al) material 光弹(性)材料
elastooptics 弹性光学
elastoplast 弹性粘膏;弹性绷带;弹性塑料
elasto-plastic 弹性塑料;弹塑性的
elasto-plastic analysis 弹塑性分析
elasto-plastic area 弹塑性区
elasto-plastic balance 弹塑性平衡
elasto-plastic beam 弹塑性梁
elasto-plastic behavio(u)r 弹塑性状态;弹塑性性质;弹塑性行为;弹塑性现象
elasto-plastic bending 弹塑性弯曲
elasto-plastic body 弹塑性物体
elasto-plastic boundary 弹塑性边界
elasto-plastic buckling 弹塑性屈曲
elasto-plastic coupling 弹性耦合;弹塑性耦合
elasto-plastic deformation 弹塑性变形
elasto-plastic drive 弹塑性驱动
elasto-plastic envelop spectrum 弹塑性包络谱
elasto-plastic failure criterion 弹塑性失效准则
elasto-plastic finite element method 弹塑性有限元法
elasto-plastic fracture 弹塑性断裂
elasto-plastic fracture mechanics 弹塑性断裂力学
elasto-plastic interface 弹塑性界面
elasto-plasticity 弹塑性;弹塑性力学
elasto-plasticity theory 弹塑性理论
elasto-plastic material 弹塑性材料
elasto-plastic matrix 弹塑性矩阵
elasto-plastic matrix displacement analysis 弹塑性矩阵位移分析
elasto-plastic mechanics 弹塑性力学
elasto-plastic medium 弹塑性介质
elasto-plastic model 弹塑性模型
elasto-plastic plate 弹塑性板
elasto-plastic polymer 弹塑性聚合物
elasto-plastic range 弹塑性范围
elasto-plastic region 弹塑性区(域)
elasto-plastic resistance 弹塑性阻力
elasto-plastic response analysis 弹塑性反应分析
elasto-plastic response envelop spectrum 弹塑性反应包络谱
elasto-plastic response spectrum 弹塑性反应谱
elasto-plastics 热塑性弹性材料
elasto-plastic soil 弹塑性土
elasto-plastic solid 弹塑性体;弹塑性固体
elasto-plastic state 弹塑性状态;弹塑态
elasto-plastic stiffness 弹塑性刚度
elasto-plastic structure 弹塑性结构
elasto-plastic system 弹塑性系统;弹塑(性)体系
elasto-plastic type 弹塑性型
elasto-plastic vibration 弹塑性振动
elastopolymer 硫塑料;硫合橡胶;弹性硫塑料;弹性高聚物
elastoprene 二烯橡胶
elastoresistance 弹性电阻
elastoresistance coefficient 抗弹性系数;弹性电阻系数
elastorheodynamic(al) lubricant 弹性变流动力

润滑剂
elastostatics 弹性静力学
elastothiomer 硫化橡胶;弹性硫塑料
elasto-viscometer 弹性黏度计;弹黏计
elasto-viscosity 弹黏性;弹(性)黏度
elasto-viscous 弹黏性的
elasto-viscous behavio(u)r 弹黏性状态
elasto-viscous character 弹黏性特性
elasto-viscous field 弹黏性场
elasto-viscous flow 弹黏性流
elasto-viscous fluid 弹黏性流体
elasto-viscous liquid 弹黏性液体
elasto-viscous solid 弹黏性固体
elasto-viscous system 弹黏性体系;弹黏体系
elaterite 弹性沥青;弹性地沥青
E-layer E 层
elbaite 锂电气石
elbasin 上升盆地
Elbenyl dye 埃尔贝尼尔染料
elbow 急弯;锚链绞花;肘形弯管;关节;拱底石
elbow action handle 肘动把手
elbow amplifier 双弯流型放大器
elbow bend 直角弯管;直角弯管;管子弯头;管子弯曲;弯头;肘形弯管
elbow bend pipe 肘形弯管
elbow board 肘形板;窗台板;镶嵌板
elbow catch 门搭扣;门轧头;肘形锁扣;肘形搭钩;肘形扣
elbow cock 肘形龙头
elbow combustion chamber 肘形燃烧室
elbow connection 弯接头
elbow core box 弯头芯盒
elbow cover opener 弯管开盖器
elbow draft tube 肘形尾水管
elbowed 弯成肘形的
elbowed leader 水落管弯头
elbow equivalent 弯管当量长度
elbow feeder 曲管送料器;弯管给料器
elbow flowmeter 弯管流量计
elbow for nozzle 喷嘴弯头
elbow in the hawse 锚链绞花
elbow jam unt 弯头锁紧螺母
elbow jerk 肘反射
elbow joint 弯头;肘节;肘接;弯头接合;弯管接头
elbow jointed lever 肘节杆
elbow lamp bracket 肘形灯托架
elbow lever 弯肘杆
elbow lining 窗台座头板;窗台镶板
elbow meter 弯管流量计;弯管流速仪
elbow of capture 袭夺(河)弯
elbow on exhaust pipe 废气排放管弯头
elbow operated faucet 肘开式龙头
elbow operated tap 肘开关水嘴;用肘关的水龙头(手术室用)
elbow piece 肘形配件;窗台镶板;弯头(肘形配件)
elbow pipe 直角弯管;弯头;弯管;肘形弯管
elbow rail 肋板;弧形栏
elbow refractor 折轴折射望远镜;折轴反光望远镜;曲折折光计
elbow room 活动余地;回旋余地
elbow-seat height 肘部投影高
elbow separator 弯头脱水器
elbow sight 肘形瞄准具
elbow socket 肘形插口;肘形承口
elbow stay 弯肘连接梁
elbow telescope 肘形望远镜;折轴望远镜
elbow tube 肘管
elbow tubing 肘管
elbow-type girder 肘状梁
elbow type locking mechanism 肘形锁闭器
elbow union 弯头套节;弯头套管节;弯头活接头;弯管接头;直角弯管接头
elbow ventilator 鹅颈式通风筒;弯管通风筒
elbow with socket end 端部带凸角弯头
Elbs reaction 埃尔勃斯反应
Elburz geosyncline 厄尔布尔士地槽
Elcas process 埃尔凯斯(铝表面)两道电泳涂装法
Elcolloy 埃尔科洛依铁镍钴合金;铁镍钴合金
elcolo(u)r plate 场致发光彩色板
elcometer 膜厚测定仪;干膜厚度计
Elconite 埃尔科涅特钨铜合金
elctric mechanical counter 电动机械计数器
elderly housing 老人住房
elderly unit 老年住房单元

Eldred's wire 艾尔德里德镍铁芯镀铜线
elecric instrument testing device 试验电气仪表
elect 选择
election 选择;选举
elective genesis 选择发生
elective share 遗产继承中的选择权
electoral castle 选帝后的宫殿
elector-bed 用测井曲线划分的地层
electornize 电子仪器化
Electorn metal 埃雷克特龙铝镁合金
electose 有填充物的天然树胶
electra 多区无线电导航系统
electraflex 电磁勘探直接找油技术
electret 驻极体
electret microphone 驻极体传声器
electret transducer 驻极体换能器
electric(al) 陡坡电制动闸;电气的;电的;导电的
electric(al) absorption 电的吸收
electric(al) absorption dynamometer 电吸收式测功器
electric(al) accessories 电气附件
electric(al) accident 电气事故;电气故障
electric(al) accommodation ladder winch 电动舷梯绞车
electric(al) accounting machine 电子式会计机;电动记账机
electric(al) accumulator 蓄电池
electric(al) activity coefficient 静电活度系数
electric(al) actuator 电力传动装置
electric(al) actuator valve 电动电调节阀
electric(al) adapter 接插元件
electric(al) adder 电加法器
electric(al) adding machine 电动加法机
electric(al) aerosol analyzer 电气溶胶分析仪
electric(al) aggregate processing 电凝聚处理
electric(al) aggregation method 电凝法
electric(al) air circulator 电动气流旋转器
electric(al) air cleaner 电动空气净化装置
electric(al) (air) compressor 电动空(气)压(缩)机
electric(al) air drier 电热干燥机
electric(al) air drill 电动压气凿岩机
electric(al) air heater 空气电热器
electric(al)-air valve 电动气阀
electric(al) alarm 电气报警器;电力警报器
electric(al) alarm clock 电闹钟
electric(al) alarm system 电气报警系统;电动警报系统
electric(al) altimeter 电测高度计
electric(al) aluminium wire 导电铝线
electric(al) analog indicator 电模拟指示器
electric(al) analog(ue) 电(子)模拟;电气模拟(装置)
electric(al) analog(ue) computer 电(子)模拟计算机
electric(al) analog(ue) for groundwater flow 地下水流电模拟
electric(al) analog(ue) machine 电模拟机
electric(al) analog(ue) method 电模拟法
electric(al) analog(ue) model 电模拟模型
electric(al) analogy 电机类比;电模拟;电比拟
electric(al) analogy test 电模拟试验
electric(al) analogy to streamflow 河川水流的电模拟
electric(al) analogy with electrolyte 电解液电模拟
electric(al) analyzer 电力分析器
electric(al) anchor capstan 电动起锚绞盘
electric(al) anchor winch 电动锚定绞车
electric(al) and electronic equipment fire 电气电子设备火灾
electric(al) and hydraulic dental chair 电动液压牙科椅
electric(al) and mechanical equipement 机电设备
electric(al) and pressure welding 电压力焊接
electric(al) anemometer 电动风速仪;电传风速表;电测风速仪
electric(al) angle 电角(度)
electric(al) annealing furnace 电热退火炉
electric(al) anomaly 电异常
electric(al) antenna 电天线
electric(al) apparatus 电器设备;电气装置;电气设备;电气机械
electric(al) apparatus for explosive atmosphere 防爆电气设备
electric(al) apparatus measuring device 电器试

验测试设备
electric(al) appliance 电气用具;电工仪器;电工设备;耗电器具;电器;电气用品
electric(al) appliance workshop 电器检修间
electric(al) arc 电弧
electric(al) arc air gouging 电弧气割
electric(al) arc cutting 电弧切割(法)
electric(al) arc cutting machine 电弧切割机
electric(al) arc drill 电弧钻
electric(al) arc drilling 电弧钻进
electric(al) arc furnace 电弧炉;电弧炼钢炉
electric(al) arc heating 电弧加热
electric(al) arc ignition 电弧点火
electric(al) arc induction furnace 电弧感应炉
electric(al) arcing 飞弧
electric(al) arc lamp 弧光灯;电弧灯
electric(al) arc lighting 弧光灯
electric(al) arc melting method 电弧熔融法
electric(al) arc process 电弧焊法
electric(al) arc reaction 电弧反应
electric(al) arc resistance 电弧电阻
electric(al) arc spot welding 电弧点焊
electric(al) arc spraying 电弧喷涂;电弧喷镀
electric(al) arc weld 电弧焊
electric(al) arc welder 弧光焊机;电弧焊机
electric(al) arc welding 电弧焊接
electric(al) arc welding electrode 电弧焊条
electric(al) arc welding generator 电弧焊接发电机
electric(al) arc welding machine 电弧焊机
electric(al) arm 电极
electric(al) arraying-puncher 电动式排钉机
electrical ator 电触式测微表
electric(al) attachments 电气附件
electric(al) attraction 电吸引
electric(al) audio-visual instruction 电化教学
electric(al) automatic control system 电力自动控制系统
electric(al) automatic time release 电气自动定时解锁器
electric(al) automatic toaster 电自动面包炉
electric(al) automatization 电气自动化
electric(al) automobile 电动汽车
electric(al) auxiliary machinery 电动辅机
electric(al) auxiliary system 电气辅助系统;厂用电系统
electric(al)-axis 电轴
electric(al) back-to-back test 背靠背电气试验
electric(al) baking 电热烘烤
electric(al) baking apparatus 电气烘烤器具
electric(al) baking oven 电烤炉
electric(al) baking pan 电热饼铛
electric(al) balance 电天平;电平衡
electric(al) ballast tamper 电动道砟夯实机
electric(al) band-saw 电动带锯
electric(al) bar 电热棒
electric(al) baseboard heater 电踢脚板加热器;墙脚板电加热器
electric(al) batching unit 电配料装置
electric(al) bath 电热浴池
electric(al) bathythermograph 电测水深水温记录仪
electric(al) battery 电池(组)
electric(al) battery electrifier 电池充电机
electric(al) battery locomotive 蓄电池式机车;电瓶车
electric(al) battery truck 蓄电池搬运车
electric(al) beacon 灯光信标
electric(al) bedwarmer 电暖床器
electric(al) bell 电铃
electric(al) bell flat push button 电铃平按钮
electric(al) bench drill 台式电钻
electric(al) bench grinder 电动台式磨床
electric(al) bender 电动弯折机
electric(al) bending machine 电动弯管机
electric(al) bias 电偏压
electric(al) bibeam bridge crane 电动双梁桥式起重机
electric(al) birefringence 电场致双折射
electric(al) blanket 电毯;电热毯;电暖毛毯
electric(al) blast furnace 电气高炉
electric(al) blasting 电炮;电(力)爆破;电火花引爆
electric(al) blasting cap 电引信;电力起爆雷管;电雷管
electric(al) blasting circuitry 电力发爆线路

electric(al) blasting machine 电爆机
electric(al) blasting unit 电力放炮机
electric(al) block 电动滑车;电动葫芦
electric(al) block diagram 电路方框图
electric(al) blower 电动鼓风机;电动吹风机
electric(al) blower single-phase 单相吹风机
electric(al) blowpipe 电弧喷焊器
electric(al) blue 钢青色;铁蓝色
electric(al) boat 电力推进快艇
electric(al) boathaulage machine 电动船舶牵引机
electric(al) boat winch 电动起艇机
electric(al) boiler 电热锅炉;电锅炉
electric(al) boiler feed regulator 电热锅炉给水调节器
electric(al) boiling sterilizer 电煮沸消毒器
electric(al) bonding 接触加热金属喷涂
electric(al) book-bundling machine 电动捆书机
electric(al) booster 电升压器;电力升压器
electric(al) booster pump 电动增压泵
electric(al) boosting 电强化加热;辅助电熔;辅助电加热
electric(al) bore comparator ga(u)ge 电测孔径比较仪
electric(al) boring 电气钻探
electric(al) boring machine 电动钻探机
electric(al) bottom hole drilling 井下电动钻井
electric(al) bowl 电饭锅
electric(al) box furnace 箱式电炉
electric(al) brain 电子计算机
electric(al) brain telephone 电脑电话机
electric(al) brake 电制动机;电闸;电(力)制动器;电动制动器
electric(al) braking 电制动(法);电力制动;电空制动
electric(al) braking of motor 电动机电气制动
electric(al) braking power 电制动功率
electric(al) brazing 电热硬钎焊;电热铜焊;电(加)热钎焊
electric(al) bread-cutter 电动切面包机
electric(al) breakdown 电击穿
electric(al) breakdown of P-N junction 结电击穿
electric(al) breakdown strength 击穿电压
electric(al) breakdown strength of gas 气体的抗电击穿强度
electric(al) breakdown voltage 电击穿电压
electric(al) bridge 电桥
electric(al) bridge temperature regulator 电桥式温度调节器
electric(al) broiler 电焙炉
electric(al) brush 电刷
electric(al) brush cutter 电刷切割机
electric(al) brush rough grinder 电刷粗磨床
electric(al) buggy 电动小车
electric(al) builder's pump 电动建筑用泵
electric(al) bulb 电灯泡;灯泡
electric(al) bulb, lamp holder 电灯泡及灯座
electric(al) bulb oven 电灯式干燥箱
electric(al) burn 电灼伤;电烧伤
electric(al) burton 电动辘轳
electric(al) bus 电动公共汽车
electric(al) butterfly valve 电动蝶阀
electric(al) button sewing machine 电动钉钮机
electric(al) buttun 电钮
electric(al) butt welding 电阻接触焊;电对头焊接
electric(al) butt welding apparatus 电弧对接焊器械
electric(al) buzzer 电蜂鸣器
electric(al) cabinet 电气设备室;电控箱;配电箱
electric(al) cable 电线;电缆
electric(al) cable capacitance 电缆电容
electric(al) cable conduit 电缆导管
electric(al) cable connector 电缆接头
electric(al) cable irser 电缆立管
electric(al) cabletowed plough 电力绳索牵引机
electric(al) caging 电力锁定
electric(al) calamine 电异极矿
electric(al) calculating machine 电子计算机
electric(al) caldron 电煮锅
electric(al) calibrating device 电标定装置
electric(al) calibration device 电标定装置
electric(al) cal(l)ipers 电动卡尺
electric(al) calorific installation 电热装置
electric(al) calorifier 电(动)热风机
electric(al) calorimeter 电(气)量热器

electric(al) capacitance ga(u)ge 电容式潮位计
electric(al) capacitor 电容器
electric(al) capacity 电容(量)
electric(al) capacity altimeter 电容高度计;电容测高计
electric(al) capacity analysis 电量分析法
electric(al) capacity moisture meter 电容式水分测定表;电力湿度计;电容湿度计
electric(al) capacity type liquidometer 电容式液位计
electric(al) capstan 电力绞盘;电动起锚(机);电动绞盘
electric(al) car 电车
electric(al) carbon 电刷碳
electric(al) cargo winch 电动起货机
electric(al) car retarder 电动车辆减速器
electric(al) cash-register 电动(现金)出纳机
electric(al)-cast brick 电熔铸砖
electric(al) casting 电熔铸
electric(al) cast mullite brick 电熔(铸)莫来石砖;铁砖
electric(al) cater-pillar crane 电动履带起重机
electric(al) cautery 电灼器;电烙术
electric(al) cautery set 电热烧灼器装置
electric(al) ceiling heating device 天棚电气采暖装置
electric(al) ceiling panel heating 电气天花板供暖
electric(al) cell 光电管;电车
electric(al) cement 导电水泥
electric(al) centering 静电法调整中心
electric(al) central heating 电气中央供热
electric(al) centralized interlocking of shunting area 调车区电气集中联锁
electric(al) central warm-air furnace 电热风炉
electric(al) center 电气中心
electric(al) centrifugal blower 电动离心式吹风机
electric(al) centrifugal pump 电动离心泵
electric(al) centrifuge 电动离心机
electric(al) ceramics 电气陶瓷;电瓷
electric(al) chafing dish 电火锅
electric(al) chain 电路链
electric(al) chain hoist 电动链吊
electric(al) chain hoist with pendant rope control 带吊索操纵器的电动链吊
electric(al) chain saw 电动链锯
electric(al) channel(l)ing machine 电动挖槽机
electric(al) character 电性;电气特性
electric(al) charges 电费;电荷
electric(al) charge density 单位面积电荷数
electric(al) charging 电场电ців
electric(al) charging system 充电系统
electric(al) check valve 电动截止阀;电动防逆阀
electric(al) chemical corrosion 电化学腐蚀
electric(al) chemistry 电化学
electric(al) chlorinator 氯化电炉
electric(al) chopper 电斩波器;电动切碎器
electric(al) chuck 电动卡盘
electric(al) cigar lighter 电气吸烟点火器
electric(al) ciphering machine 电动密码机
electric(al) circuit 电气线路;电路
electric(al) circuit analysis program(me) 电路分析程序
electric(al) circuit shifter 电路转接器
electric(al) circuit theory 电路理论
electric(al) circular saw 电(动)圆锯
electric(al) cleaner (工地用)电力扫除器
electric(al) clippers 电剪刀
electric(al) clock 电钟;电子钟
electric(al) clock receptacle 电钟插座
electric(al) clock synchronizer 电钟同步器
electric(al) clock system 电钟系统
electric(al) cloth cutting machine 电动裁布机
electric(al) clothes dryer 电干衣机
electric(al) clutch 电(动)离合器
electric(al) coagulation 电混凝
electric(al) coagulation and flo(a)tation 电混凝气浮
electric(al) coagulation forceps 电凝钳
electric(al) coal drill 电动煤钻
electric(al) coalescing section 电聚结区
electric(al) code 电气规范;电工规程;电工规范
electric(al) coding machine 电动编码机
electric(al) coffee maker 电咖啡机
electric(al) coffee percolator 电咖啡滤壶

electric(al) coffee pot 电咖啡壶
electric(al) coil 线圈
electric(al) coil tongs 电动钢卷钳
electric(al) coil winder 电动绕线机
electric(al) collaring machine 电动曲线机
electric(al) colo(u)rimeter 电动色度计
electric(al) comb 电梳
electric(al) comductivity rate-of-change detector 电导型温度变化率探测器
electric(al) comfort heating appliance 电热取暖设备
electric(al) communication 电讯;电信;电通信
electric(al) commutator 整流器
electric(al) comparator 电量比较器;电动比较仪
electric(al) component 电气组件;电分量
electric(al) component comprehensive test equipment 电器综合试验装置
electric(al) component dust collector 电器元件除尘装置
electric(al) compression heat pump 电动压缩热泵
electric(al) compressor 电动压缩机
electric(al) compressor governor 电动压缩机调压阀
electric(al) computer 电子计算机
electric(al) concentration 电选(法);电热蒸浓(法)
electric(al) concrete mixer 电动混凝土搅拌机
electric(al) concrete pump 电动混凝土泵;电动混凝土输送泵
electric(al) condenser 电容器
electric(al) condenser oil 电容器油
electric(al) conductance 电导(率);导纳;导电性
electric(al) conductance determination 电导测定
electric(al) conduction 导电(性)
electric(al) conduction field 导电场
electric(al) conduction type flame detector 电导式火焰探测器
electric(al) conductive glass fiber 导电玻璃纤维
electric(al) conductive paint 导电漆
electric(al) conductivity 电导性;导电率;导电性;导电率
electric(al) conductivity alloy 导电合金
electric(al) conductivity analyzer[analyser] 电导率分析器
electric(al) conductivity detector 电导检测器;导电性检测器
electric(al) conductivity fiber 导电性纤维
electric(al) conductivity in a plasma 等离子体导电率
electric(al) conductivity meter 电导仪
electric(al) conductivity method of solution 溶液电导法
electric(al) conductivity of glass 玻璃的电导率;玻璃的导电性
electric(al) conductivity of water 水的电导度
electric(al) conductivity test 电导率试验
electric(al) conductor 电(导)线;电导体;导电体
electric(al) conduit 电线导管;电缆通道;电缆沟;电缆槽
electric(al) conduit line 电导线管路
electric(al) conduit network 电导线管网
electric(al) conduit trench 电力管道沟槽;电缆沟槽
electric(al) cone penetration test 电测静力触探
electric(al) connecting clamp 电连接线夹
electric(al) connecting clamp for busbar 汇流排电连接线夹
electric(al) connection 电连接;电接头
electric(al) connector 电气插件;电路接头;插座;插头;插塞;接线盒;电气接头;电连接器;插座接线盒
electric(al) connector receptacle 电插座
electric(al) consolidation 电气固结法
electric(al) consolidation process 电化固结法(地基加固方法之一)
electric(al) construction 电气施工
electric(al) construction tool 电气施工工具
electric(al) consumer 电力消费者
electric(al) contact 电接点;电接触;电触点;触头
electric(al) contact clock 电接触时钟
electric(al) contact controller 电开关控制器
electric(al) contact detector 电路接触式检测器
electric(al) contact liquid level indicator 电接点液位计;电接触液位指示器
electric(al) contact-making micrometer 电接触

式测微计
electric(al) contact material 电接点材料;电接触器材
electric(al) contactor 电气设备承包者;电气设备承包商;电气设备承包人
electric(al) contact point pressure ga(u)ge 电触点压力表
electric(al) contact(point) thermometer 电接点温度计;电触点温度计
electric(al) contact pressure ga(u)ge 电接触压力表
electric(al) contact receptacle 电插座
electric(al) contact starter 电接触起动器
electric(al) contact sunshine recorder 电接触式日照计
electric(al) continuity 电气连接
electric(al) continuity test 导通试验
electric(al) contractor's pump 电动承包商用泵
electric(al) control 电气控制;电力控制;电动调节;电操纵
electric(al) control box 电控箱
electric(al) control cable 电控电缆
electric(al) control device 电气控制装置
electric(al) control dryer 电控烘干机
electric(al) control equipment 电气控制设备
electric(al) control gear 电气控制装置
electric(al) controlled hoist 电控卷扬机
electric(al) controller 电气控制装置;电控制器
electric(al) control of air brake 电动空气制动机
electric(al) control panel 电气控制屏;电气设备控制板;电控制盘
electric(al) control photo-scattered ceramics 电控光散射陶瓷
electric(al) control room 配电室
electric(al) control room for ventilation and air-condition 通风空调电控室
electric(al) control system 电气控制系统
electric(al) control valve 电力控制阀;电动调节阀
electric(al) control workshop 电管间
electric(al) convection 电热对流
electric(al) convector 电热对流器
electric(al) conversion factorlc reaction 电换算因数
electric(al) conveyer 电动传送设备
electric(al) cooker 电灶;电气炊具;电烤箱
electric(al) cooker range 电灶
electric(al) cooking range 电灶
electric(al) cooking stove 电灶;电橱灶
electric(al) coolant pump 电动冷却液泵
electric(al) copper glazing 电解铜上釉
electric(al) cord 电缆槽
electric(al) cordless steam iron 电池式蒸汽电熨斗
electric(al) core drill 电动岩芯钻
electric(al) coring 电动钻取土样;电钻取芯
electric(al) corona 电晕放电
electric(al) corrosion 电腐蚀
electric(al) corrosion of rail 钢轨电腐蚀
electric(al) cost 电费
electric(al) coupler 电器连接器;车钩电器部分
electric(al) coupling 电器连接;电耦合;电磁联轴节
electric(al) coupling of signal arm 信号臂板选别器
electric(al) coverlet 电床罩
electric(al) crab winch 电动起重绞车
electric(al) crane 电力起重机;电动起重机;电动吊车
electric(al) cranking 电力起动
electric(al) cross 电器混线
electric(al) crystal clock 电晶体钟
electric(al) cup 电热杯
electric(al) curing of concrete 混凝土电热养护
electric(al) current 电流
electric(al) current density 电流密度
electric(al) current failure 电流故障
electric(al) current intensity 电流强度
electric(al) current loop 电流环
electric(al) current meter 电流计;电流表;电测流速仪
electric(al) current relay 电流继电器
electric(al) current supply 供电
electric(al) current tariff 电流税率
electric(al) curtain stabilization 电稳定加固土壤
electric(al) cutter 电气切割机;电动切割机;电动裁剪机
electric(al) cutting machine 电动切割机

electric(al) cyclone furnace 电热旋风炉
electric(al) cylinder 家用电热水器;电加热器
electric(al) damper 电动风阀
electric(al) data collector 电子数据汇集器
electric(al) data exchange and control system 电子数据交换和控制系统
electric(al) decantation 电倾析;电滗析
electric(al) deep freezer 低温箱
electric(al) deep fryer 电炸锅
electric(al) deep well turbine pump 电动深井涡轮泵
electric(al) deflection 电场偏转
electric(al) defroster 电动融冰机;电动除霜器
electric(al) defroster shield 电力除霜器壳
electric(al) degree 电角度
electric(al) dehydration 电脱水
electric(al) dehydrator 电脱水器
electric(al) deicer 电气除冰器
electric(al) delay blasting 延时爆破
electric(al) delay blasting cap 延时电雷管;定时电雷管
electric(al) delay detonator 延发电雷管
electric(al) delay fuse 延期电信管;迟发电雷管
electric(al) delay line 电延迟线
electric(al) density 电荷密度
electric(al) depolarization 通电去极化
electric(al) depth finder 电测深计;电测深仪
electric(al) desalination 电脱盐(作用)
electric(al) desalting 电脱盐;电气脱盐
electric(al) desalting and electrodehydrating 电脱盐和电脱水
electric(al) desalting process 电脱盐法
electric(al) design 电气设计
electric(al) detarring precipitator 电(力)脱焦油沉淀器
electric(al) determination 电测法
electric(al) detonator 电引爆管;电控雷管;电引爆雷管;电力起爆器;电雷管;电爆管
electric(al) detonator firing 电雷管起爆法
electric(al) detonator initiation 电雷管起爆
electric(al) device 电气装置;电力设备
electric(al) dewaterer 电脱水器
electric(al) dew-point hygrometer 电(测)露点湿度计
electric(al) diagnosis 电反应诊断法
electric(al) dial switch 电动拨号开关
electric(al) dialyzator 电渗析器
electric(al) diaphragm emitter 电雾号;电雾雷
electric(al) diaphragm pump 电动隔膜泵
electric(al) differential pressure transmitter 电动差压传感器
electric(al) differentiation 电微分
electric(al) digger 电铲;电掘凿器;电动挖沟机;(手持式)电动掘凿器
electric(al) dip meter 电测浸入式探测计;电测倾斜仪
electric(al) dipole 电偶极(子)
electric(al) dipole moment 电偶极矩
electric(al) dipole radiation 电偶极辐射
electric(al) dipole transition 电偶极跃迁
electric(al) dipper 电(动)铲
electric(al) disappearing stairway 电动扶梯
electric(al) discharge 放电
electric(al) discharge arc 放电电弧
electric(al) discharge convection laser 放电对流式激光器
electric(al) discharge excited gaseous laser 放电激励气体激光器
electric(al) discharge gas-dynamic(al) laser 放电气动激光器
electric(al) discharge lamp 辉光灯;放电管;放电灯
electric(al) discharge laser 放电激光器
electric(al) discharge lighting fixture 放电照明装置
electric(al) discharge machine 放电加工机床
electric(al) discharge machining (电子)放电加工
electric(al) discharge printer 放电式印刷机
electric(al) discharge sawing machine 放电加工锯床
electric(al) discharge synthetical method 放电合成法
electric(al) disc sander 电动打磨砂盘
electric(al) dish washer 电动洗盘机
electric(al) disintegration 电蚀蚀

electric(al) disintegration drilling 电破碎钻进
electric(al) displacement 电位移
electric(al) displacement density 电位移密度
electric(al) displacement meter[metre] 电动式位移计
electric(al) dissipation 电力消散
electric(al) dissipation factor 电损耗因子
electric(al) distance 电距离
electric(al) distance condition 光距条件
electric(al) distance measuring 电子测距法
electric(al) distance unit 光距单位
electric(al) distiller 电热蒸馏水器
electric(al) distribution 电力分配;配电(装置);输电
electric(al) distribution box 配电箱
electric(al) distribution centre 配电中心
electric(al) distribution circuit 配电网
electric(al) distribution circuit conductors 配电网导线
electric(al) distribution network 配电网
electric(al) distribution system 配电系统
electric(al) distribution unit 配电装置;接线盒
electric(al) distributor 电气分配器
electric(al) disturbance 电扰
electric(al) domain wall 电畴壁
electric(al) domestic installation 家用电器装置;家用电器设备
electric(al) domestic water heater 家用电热水器
electric(al) door bell 电门铃
electric(al) door closer 电动关门器
electric(al) door lock 电门锁
electric(al) door opener 电(动)开门器
electric(al) dotter graver 电动刻点仪
electric(al) double-beam bridge crane 电动双梁桥式起重机
electric(al) double layer 偶电层;电双层;电偶层;双电层
electric(al) double refraction 电致折射;电场致双折射
electric(al) doublet 偶极子;电偶极子
electric(al) drainage 电渗;电动排水;电排流;排流器
electric(al) drainager 电排流器
electric(al) draw-bridge lock 开合桥电锁闭器
electric(al) dredge(r) 电动挖泥船
electric(al) drill 电钻;电动钻具;电动凿岩机
electric(al) drill arbor 电钻床架
electric(al) driller's panel 司钻电控盘
electric(al) drill hammer 电气冲击钻
electric(al) drilling machine 电(动)钻机
electric(al) drill stand 电钻架
electric(al) drill with drilling rod 有钎电钻
electric(al) drill with flexible pipe 柔杆电钻
electric(al) drink cooling apparatus 电冷饮料装置
electric(al) drive 电(力)驱动;电力传动(装置);电传动
electric(al) drive fire pump 电动消防泵
electric(al) driven chain block 电动链滑车
electric(al) driven mechanical tube cleaner 电动机械管子清洁器
electric(al) driven welder 电动焊机
electric(al) drive with cascade 串级电气传动
electric(al) driving installation 电气传动装置
electric(al) drying 电加热干燥
electric(al) drying apparatus 电热干燥器
electric(al) drying chamber 电干燥箱
electric(al) drying oven 电烘箱;电干燥炉
electric(al) drying oven with forced convection 电热鼓风干燥箱
electric(al) duct (安装电缆的)通道;(安装电缆的)沟槽;电缆管道;电线导管
electric(al) dumbwaiter 电动货梯
electric(al) dust collection plant 电采尘设备
electric(al) dust collector 电(动)吸尘器;电除尘器
electric(al) dust precipitation 电收尘法
electric(al) dust precipitation chamber 电气除尘室
electric(al) dust precipitator 电气除尘器;电气聚尘器;电气集尘器
electric(al) dynamometer 电气功率表;电(动)测力计;电动测功机;电测功器
electric(al) ear 电耳
electric(al) eddy current dynamometer 电涡流测功机
electric(al) efficiency 电(机)效率
electric(al) element 电池;电力部件

electric(al)-electric(al) 纯电的
electric(al) elevator 电梯;电力升降机;电机驱动电梯
electric(al) emulsion treater 电乳浊液处理器
electric(al) enclosure 电器外罩
electric(al) endosmosis 电渗;电内渗(现象)
electric(al) energy 电能
electric(al) energy consumption 耗电量
electric(al) energy density 电能密度
electric(al) energy loss 电能损耗
electric(al) energy measurement 电能测量
electric(al) energy meter 电能表
electric(al) energy production 发电量
electric(al) energy source 电能震源
electric(al) engine 电机;电火箭发动机
electric(al) engineer 电气工程师;电机员;电机工程师
electric(al) engineering 电气工程;电机工程(学);电工学;电力工程(学)
electric(al) engineering carbon 电工用碳素制品
electric(al) engineering handbook 电工手册
electric(al) engineering installations 电气安装工程
electric(al) engineering material 电工材料
electric(al) engineer's room 电机员室
electric(al) engine oil 电机油
electric(al) engraving machine 电刻机
electric(al) enterprise 电气事业
electric(al) environment 电环境
electric(al) equilibrium 电(荷)平衡
electric(al) equipment 电器;电气设备;电力设备;电动设备
electric(al) equipment fire 电气设备火灾
electric(al) equipment malfunction 电气设备故障
electric(al) equipment noise 电器(设备)噪声
electric(al) equipment potting material 电器灌封材料
electric(al) equipment universal testing stand 高压电路电器设备检查器
electric(al) equivalent 电当量
electric(al) equivalent of calorie 卡路里的电当量;热电当量
electric(al) erosion 电(侵)蚀
electric(al) etcher 电蚀器;电蚀刻器
electric(al) etching 电蚀刻法
electric(al) etching machine 电解蚀刻机
electric(al) excavator 电动挖土机;电铲
electric(al) excitation 电激励
electric(al) expenses 电费
electric(al) experimental equipment 电工实验设备
electric(al) exploder 电雷管;电爆器
electric(al) exploration 电力探查;电法探测;电法勘探
electric(al) exploration instrument 电气勘探仪
electric(al) explosion tested locomotive 防爆电机车
electric(al) explosion valve 电爆阀
electric(al) exposure counter 电动曝光计数器
electric(al) express locomotive 高速电力机车
electric(al) external vibrator 电动外部振捣器
electric(al) eye 光电池;光电管;电眼
electric(al) eye camera 自动摄影机
electric(al) fabric 电绝缘织物
electric(al) facility 电气设备
electric(al) fan 电动通风机;电(风)扇;风扇
electric(al) fan production line 电扇生产线
electric(al) fan timer 电风扇定时器
electric(al) fathometer 电测深计
electric(al) fault 电气故障;电(力)故障
electric(al) feedback 电气反馈
electric(al) feed cooler 电冷却器
electric(al) feeder 馈电线;电源线
electric(al) feed water pump 电动给水泵
electric(al) fence 电围栏;电篱笆;电导墙围
electric(al) fence controller 电牧栏控制器
electric(al) fencing unit 电牧栏设备
electric(al) fidelity 电信号保真度
electric(al) field 电场
electric(al) field gradient 电场梯度
electric(al) field in electric(al) prospecting 电法勘探中的电场
electric(al) field intensity 电场强度
electric(al) field meter 电场强计
electric(al) field mixing tensor 电场混频张量
electric(al) field sensor 电场传感器
electric(al) field strength 电场强度
electric(al) field vector 电场向量;电场矢量
electric(al) filament lamp 白炽电灯
electric(al) filling compound 电缆隔离用混合物
electric(al) filter 滤波器
electric(al) finish 电精饰;电学法表面加工
electric(al) finned strip heater 电加热肋片采暖器
electric(al) fire 电火炉;电气火灾
electric(al) fire alarm 电火灾警报;电气火警报警器
electric(al) fireplace 电壁炉
electric(al) firing 电弧起爆法;电力起爆法;电放炮;电发火;电点火
electric(al) firing circuit 点火电路
electric(al) firing lock 电点火开关
electric(al) firing mechanism 电发火机构
electric(al) fish 发电鱼
electric(al) fish counter 电气数鱼器
electric(al) fishing 电捕鱼
electric(al) fish screen 拦鱼电网
electric(al) fitting 电气安装;电器;电气装修;电气配件
electric(al) fixture 电气设备;电灯固定件;电力设备
electric(al) fixture of a room 室内电气装置
electric(al) flasher 电动闪光器
electric(al) flashing lamp 电闪灯
electric(al) flashlight 电筒
electric(al) flat carriage 电动平车
electric(al) float 电动镘刀
electric(al) floating crane 电动浮式(水上)起重机;电动浮式浮吊
electric(al) flocculation 电(作用)絮凝
electric(al) floccule 电絮凝物
electric(al) floor 电气供暖地板
electric(al) floor panel 电式式地板
electric(al) floor panel warming 地板下电气供热
electric(al) floor polisher 地板擦光机
electric(al) flowmeter 电(动)流量计;电磁流量计
electric(al) fluviograph 电水位计;电测水位计
electric(al) flux 电通量;电通;电焊剂
electric(al) fluxline 电力线(路)
electric(al) food conveyor 电动送餐车
electric(al) food warmer 电热锅
electric(al) force 电力;电场强度
electric(al) force gradient 电场梯度
electric(al) forging machine 电锻机
electric(al) fork lift 叉式电池车;电动叉车
electric(al) fork lift truck 电动叉装卸车
electric(al) fork truck 叉式电动装卸车
electric(al) forming 电冶;电形成;电成型
electric(al) formwork vibrator 电动模壳振动器;电动附着式振动器;电附着模板式振捣器
electric(al) free-point 电流畅通点
electric(al) freezer 电(冰)冻器
electric(al) freight locomotive 货运电力机车
electric(al) freight truck 电动运货汽车;电动运货卡车
electric(al) frog rammer 蛙式夯
electric(al) fryer 电炸锅
electric(al) frying pan 电煎锅
electric(al) fuel ga(u)ge 电动油量计;电测燃料仪;燃料消耗测定仪
electric(al) fuel pump 电动燃料泵
electric(al) fuel spray for preheating air 电气预热设备
electric(al) function generator 电函数发生器
electric(al) furnace 电熔窑;(冶炼用的)电炉
electric(al) furnace brazing 电炉钎焊
electric(al) furnace brick 电炉衬砖
electric(al) furnace heating wire 电炉丝
electric(al) furnace refining 电炉精炼
electric(al) furnace slag 电炉渣;电炉钢渣
electric(al) furnace smelting 电炉熔炼
electric(al) furnace steel 电炉钢
electric(al) furnace steel furnace 电炉钢炉
electric(al) furnace steel heat 电炉钢熔炼
electric(al) furnace tube 电炉炉管
electric(al) fuse 电引燃器;可熔保险丝;电熔丝
electric(al) fusion-welded pipe 电熔焊管
electric(al) fusion weld(ing) 电熔焊
electric(al) gantry crane 电动(龙)门式起重机
electric(al) gas analyzer 电气气体分析器
electric(al) gathering locomotive 集中运输用电力机车
electric(al) ga(u)ge 电力仪表
electric(al) gear 电力机械传动
electric(al) gearshift 电力调挡;电力变速
electric(al) gear shifting 电动变速器
electric(al) generating plant 发电厂
electric(al) generating station 发电站;发电厂
electric(al) generation 电力生产
electric(al) generator 发电机
electric(al) generator set 发电机组
electric(al) geologic(al) survey 电探地基
electric(al) geophysical method 电地球物理法;地球物理电测法
electric(al) gilding press 电动烫金机
electric(al) glass 低碱玻璃;E型玻璃
electric(al) glass fiber 电绝缘玻璃纤维;E型玻璃纤维
electric(al) glow discharge method 电晕放电法
electric(al) goniometer 电测角计
electric(al) governing system 电力调速系统
electric(al) governor 电力调节器
electric(al) grab 电动抓岩机;电动抓斗
electric(al) grease 电力脂
electric(al) grid 供电系统;电栅极
electric(al) grill 电气烤炉
electric(al) grinder 电磨机;电动砂轮机;电动磨床
electric(al) grinding machine 电磨机;电动砂轮机;手提电磨机
electric(al) ground 接地线
electric(al) guillotine shears 电动闸刀式剪切机
electric(al) gun 电子枪
electric(al) hair clipper 电动理发推子
electric(al) hair cutter 电推子
electric(al) haircutting scissors 电动理发剪
electric(al) hair dryer 电吹风机
electric(al) hammer (振捣)电锤
electric(al) hammer drill 电锤钻头
electric(al) hand drill 手(用)电钻;手提电钻
electric(al) hand dryer 电干手器
electric(al) hand generator system 手摇发电机制
electric(al) hand hammer and (rock) drill 手提电锤和电钻
electric(al) hand-held tools 手扶电动工具
electric(al) handlamp 电气手灯
electric(al) hand saw 手提电锯
electric(al) hand shaper 手提电刨
electric(al) hand shears 电动手剪
electric(al) hand tool 手提电动工具
electric(al) hardening 电气淬火
electric(al) harmonic analyser 电谐波分析器
electric(al) harmonic spraying 电谐波喷涂
electric(al) haulage 电机车运输
electric(al) haulage locomotive 运输电机车
electric(al) hazard 电气危害;电气事故
electric(al) heat 电热
electric(al) heat curing 电热养护
electric(al)-heated high pressure sterilizer 电热高压消毒器
electric(al)-heated suit 电热服
electric(al)-heated thermostatic water bath 电热恒温水浴锅
electric(al)-heated underwear 电热服
electric(al) heat energy 电热能量
electric(al) heater 电气采暖器;电炉;电(加)热器
electric(al) heater coil 电热器线圈
electric(al) heater drilling 电热钻进
electric(al) heater for sterilization 消毒用电热器
electric(al) heater section 电加热段
electric(al) heater shovel 电热铲
electric(al) heater unit 电加热装置
electric(al) heat exchanger 电热交换器
electric(al) heat glass 电热玻璃
electric(al) heating 电热(供暖);电(热)采暖;电加热
electric(al) heating air-blowing drier 电热鼓风干燥机
electric(al) heating apparatus 电热装置
electric(al) heating appliance 电热(采暖)设备;电加热器;电暖器具
electric(al) heating blanket 电热毯;电热层;电加热毛毯
electric(al) heating by electromagnetic induction 电磁感应产生的电热
electric(al) heating cable 电热电缆
electric(al) heating ceiling 电热采暖天棚
electric(al) heating ceiling panel 天棚电加热板

electric(al) heating chamber 电预热室
electric(al) heating clothes 电热服
electric(al) heating device 电热器
electric(al) heating drill 电加热钻井
electric(al) heating element 电热元件;电热体
electric(al) heating floor 电热地板
electric(al) heating holding chamber 电均热室
electric(al) heating installation 电加热器安装;电热设备
electric(al) heating method 电加热法
electric(al) heating of high-tensile bars 高拉力钢筋的电加热
electric(al) heating of roads 道路电力加热
electric(al) heating plug 电火花塞
electric(al) heating press 电热压呢机
electric(al) heating safety glass 电加热安全玻璃
electric(al) heating shoes 电热鞋
electric(al) heating shower 电热沐浴器
electric(al) heating system 电热采暖系统
electric(al) heating tube 电热管
electric(al) heating unit 电(加)热元件
electric(al) heat radiator 电热辐射器
electric(al) heat tracing 电伴随加热
electric(al) heat treating furnace 电热处理炉
electric(al) hedge shears 电动修灌木篱墙剪
electric(al) high-speed locomotive 高速电力机车
electric(al) hinge 电控锁铰链
electric(al) hoist 电力提升机;电力起重机;电动升降机;电动卷扬机;电动滑车;电(动)葫芦;电(动)吊车
electric(al) hoist with creep lifting motor 变速电葫芦
electric(al) hole saw mandrel with pilot drill 电锯心轴及导钻
electric(al) hook-up 电力熔化;电接头
electric(al) hopper vibrator 电动料斗振动器
electric(al) horn 电喇叭
electric(al) horn breaker 电喇叭继电器
electric(al) horn relay 电喇叭继电器
electric(al) horsepower 电(动)马力
electric(al) hotbed 电气温床
electric(al) hot compacting press 电热热压机
electric(al) hot plate 电热板;电气平底锅
electric(al) hot tray 电热盘
electric(al) hot water 热水电加热器;电热水器
electric(al) hot water boiler 电热水锅炉;电热锅炉
electric(al) hot water bottle 电热水瓶
electric(al) hot water heater 电热水器
electric(al) humidifier 电加湿器
electric(al) hydraulic controller 电动液压调节器
electric(al) hydrocel 干燥度仪
electric(al) hygrometer 电(阻)湿度计;电测湿度仪
electric(al) igniter 电点火器
electric(al) ignition 电发火;电点火
electric(al) ignition engine 电点火发动机
electric(al) ignition system 电点火系统
electric(al) illumination 电气照明;电力照明
electric(al) illumination information installation 电光式道路情报显示装置
electric(al) illumination sign 电光标志
electric(al) illumination system 电气照明系统;电气照明设施
electric(al) image 电像
electric(al) immersion heater 浸没在水中的电气热水器;浸液电加热器;浸没式电热器
electric(al) impact hammer 电动冲击锤
electric(al) impact wrench 电动冲击扳手
electric(al) impedance 电阻抗
electric(al) impedance meter 电阻测量仪;阻抗计
electric(al) impulse 电脉冲
electric(al) impulse compacting 电冲击成形
electric(al) incinerator 电焚化炉
electric(al) incubator for bacteria 电热细菌培养器
electric(al) indicating controller 电动指示调节器
electric(al) indication lock 表示电锁器
electric(al) indicator 电(动)指示器
electric(al) indicator horizon 电性标志层
electric(al) inductance measuring-testing instrument 电感测试仪
electric(al) inductance strain ga(u)ge 电感应变计
electric(al) inductance strain piece 电感应变片
electric(al) induction 电感(应)

electric(al) induction coil 电感线圈
electric(al) induction furnace 感应电炉;电感应炉;感应炉
electric(al) induction heating 电感应加热
electric(al) induction heating process 电感应加热法
electric(al) induction welding 电感应焊
electric(al) inductivity 介电常数
electric(al) industrial freight trailer 电力工业用运货拖(挂)车;电动货车拖车
electric(al) industrial truck 工业用电动货车;电瓶车
electric(al) industrial truck crane 电力工业用汽车起重机
electric(al) industrial truck trailer 电力工业用卡车拖车
electric(al) industry 电气(制造)工业;电力工业;电机工业
electric(al) inertia 电惯性
electric(al) inertia starter 电力惯性起动器;电动惯性起动机
electric(al) infrared heater 电热红外线采暖器
electric(al) infrared oven 红外(线)电烘箱
electric(al) infrared quartz radiant heater 红外(线)石英辐射电加热器
electric(al) ingot process 电渣法
electric(al) initiating device 电起爆装置
electric(al) initiation 电火花起爆
electric(al) initiation network 电起爆网络
electric(al) instability 电不稳定性
electric(al) installation 电器装置;电器设备;电气装置;电气设备;电气安装
electric(al) installation drawing 电气装置图;电气图纸
electric(al) installation noise 电器设备噪声
electric(al) instrument 电气仪表;电工测量仪表
electric(al) instruments and apparatus 电表仪器
electric(al) insulant material 电气绝缘材料
electric(al) insulating board 电绝缘板
electric(al) insulating compressed asbestos fiber 电绝缘压缩石棉纤维
electric(al) insulating fiber 电绝缘纤维
electric(al) insulating material 电绝缘材料
electric(al) insulating oil 电气绝缘油
electric(al) insulating paint 电绝缘漆
electric(al) insulating paper 电绝缘纸
electric(al) insulating putty 电绝缘油灰;电气绝缘腻子
electric(al) insulating(switch) oil (变压器的)绝缘油
electric(al) insulating treatment 电绝缘处理
electric(al) insulating varnish 电气绝缘清漆;绝缘漆
electric(al) insulation 电气绝缘;电力绝缘;电绝缘(性);电工绝缘
electric(al) insulation oil 变压器油
electric(al) insulation ring 电绝缘环
electric(al) insulation tape 电绝缘带
electric(al) insulation varnish 电绝缘清漆
electric(al) insulativity 电绝缘性
electric(al) insulator 绝缘子;电(气)绝缘子;电绝缘体;电绝缘器
electric(al) integraph 电力积分仪
electric(al) integration 电积分(法)
electric(al) integrator 电积分仪
electric(al) interference 电(气)干扰
electric(al) interlock 电接合
electric(al) interlocker 电动联锁器;电动连锁器
electric(al) interlocking 电气(集中)联锁;电力锁紧
electric(al) interlocking apparatus 电气联锁设备
electric(al) interlocking automatic device 电气联锁式自动化装置
electric(al) interlocking for hump yard 驼峰电气集中
electric(al) interlocking frame 电气联锁机
electric(al) interlocking installation 电气联锁装置
electric(al) interlocking machine 电联动机;电力联锁机;电动联锁机
electric(al) interlocking power rack 电气联锁电源屏
electric(al) internal vibrator 电动插入式振捣器;电动内插式振捣器
electric(al) interrupter 电力断续器
electric(al) interval lubrication installation 电动间歇润滑装置
electric(al) iron 电熨斗;电烙铁
electric(al) isolation 电(工)绝缘
electric(al) isolation fiber 电隔离纤维
electric(al) isolation optic(al) memory 电隔离光存储器
electric(al) jack 电动千斤顶;电动起重机
electric(al) jacket 电热套
electric(al) jib crane 电动单臂起重机
electric(al) jib saw 电动细锯
electric(al) joint 电气接头;电气分断点
electric(al) juice extractor 电动榨汁器
electric(al) jumper lead 跨接线
electric(al) kettle 电热水壶
electric(al) knife 高频电刀;电切刀
electric(al) kymograph 电气脉搏曲线记录器
electric(al) laboratory 电气试验室
electric(al) lamination press 电动层压机
electric(al) lamp 电灯
electric(al) lamp bulb 电灯泡
electric(al) lamp filament 电灯丝
electric(al) lamp holder 电灯座
electric(al) lamp reflector 电灯反射镜
electric(al) lamp socket 电灯座
electric(al) lamp thread 电灯圆螺纹
electric(al) lateral curve log 横向测井
electric(al) layout 电气设计;电气布局
electric(al) lead 电导线
electric(al) lead cover wire 铅皮电线
electric(al) leading screw sounding machine 电动丝杠式触探机
electric(al) leakage 漏电;电漏
electric(al) leakage current breaker 漏电开关
electric(al) leakage tester 漏电试验器
electric(al) length 电长度
electric(al) level 电平
electric(al) level meter 电平表
electric(al) lever 电气握柄
electric(al) life support equipment 电气生命保障设备
electric(al) lift 电梯;电力升降机;电动升降机
electric(al) lift control 电梯控制
electric(al) lift controller 电梯控制器
electric(al) lifter 电动提升机
electric(al) lifting magnet 提升电磁
electric(al) lifting magnet with tines 有叉齿的起重电磁铁
electric(al) lift truck 电力升举载重汽车
electric(al) light 电光;电灯
electric(al) light bulb 电灯泡
electric(al) light charges 电灯费
electric(al) light conduit 电灯线导管
electric(al) lighter 电照明器;电引燃器
electric(al) light fittings 电灯配件;电照明配件
electric(al) lighting 电气照明;电力照明
electric(al) lighting system 电气照明系统;电气照明设施;电光系统
electric(al) linear motor 线性电动机
electric(al) line cutter 电动管道割刀
electric(al) lines of force 电力力线(路)
electric(al) line set packer 电缆坐封式封隔器
electric(al) liquidizer 电液化器
electric(al) liquid level alarm 电气液面报警器
electric(al) load 电气荷载;电力负载;电力负荷
electric(al) loader 电动装载机
electric(al) loading 加感;电枢负载;电气负载;电力负荷
electric(al) local tire spreader 电动局部扩胎机
electric(al) lock 电锁(器);电动锁
electric(al) lock armature 电锁器衔铁
electric(al) lock derail 电锁脱轨器
electric(al) locking 电锁闭;电气锁(闭);电气闭塞
electric(al) lock with handle 带柄电锁器
electric(al) locomotive 电气机车;电力机车;电(动)机车
electric(al) locomotive crab-reel 带绞车的电机车
electric(al) locomotive crane 电动机车起重机
electric(al) locomotive with computer 电脑控制电力机车
electric(al) log 电(动)计程仪;电法测井;电测钻井记录(曲线);电测(探)记录;电测深仪;电测井(曲线图);电测井记录(曲线);电测记录
electric(al) log curve 电测记录曲线
electric(al) logging 电力钻孔测量;电测探法;电

测(井);电测记录
electric(al) logging device 电测仪;电测井装置
electric(al) logging equipment 电测井设备
electric(al) logging meter 电测井仪
electric(al) logging method 电测井法
electric(al) log interpretation 电测解释
electric(al) log interval 电测井段
electric(al) log item 电测项目
electric(al) log resistivity 电测电阻率
electric(al) loss 电损耗
electric(al) low water alarm 电动低水位警报器;低水位电警报
electric(al) luminaire fixture 电气照明装置
electrically actuated convertible top 电动折叠式车顶
electrically alterable ROM 可改写的只读存储器
electrically conducting coating 导电涂料
electrically conducting glass 导电玻璃
electrically conducting rubber floor 导电橡皮地板
electrically conductive adhesive 导电性胶黏剂
electrically conductive concrete 电热养护混凝土;导电混凝土
electrically conductive fiber 导电性纤维
electrically conductive glass 导电玻璃
electrically conductive gum 导电性黏结剂
electrically conductive ink 导电油墨
electrically conductive tile 导电瓦
electrically conductive yarn 导电纱
electrically connected 电气连接的
electrically controlled 电(动)控制的
electrically controlled attenuator 电调衰减器
electrically controlled magazine 电动料斗
electrically controlled mechanical tripdevice 电控机械撞击装置
electrically controlled regulation 电气自动调节
electrically dependent light attenuator 电相关光衰减器
electrically driven 电驱动的;电动的
electrically driven combine 电动联合收获机
electrically driven crab 电动起重绞车
electrically driven duplicator 电驱动复印机
electrically driven fan 电动风机
electrically driven feed pump 电动供给水泵;电动给水泵
electrically driven mixer 电力驱动搅拌机
electrically driven pump 电动泵
electrically driven rotary sidewall sampler 电动钻进式井壁取样器
electrically driven vehicle 电动车辆
electrically energized machine 电气化装置
electrically fired 电燃式
electrically fused magnesite brick 电熔镁砖
electrically heated 电热的
electrically heated backlight 电加热汽车后灯
electrically heated bedding 电热垫褥
electrically heated boiler 电热锅炉;电加热锅炉
electrically heated ceiling 电加热天花板;电加热顶棚
electrically heated concrete 电加热混凝土
electrically heated distilling apparatus 电(热)蒸馏水器
electrically heated flat bed press 电热平板压烫机
electrically heated glass 电加热玻璃
electrically heated glass panel 电加热玻璃板
electrically heated hair-curler 电热卷发器
electrically heated hung ceiling 电加热吊平顶
electrically heated metal suspended ceiling 电加热金属吊平顶
electrically heated mirror 电加热镜子
electrically heated rear window 电加热汽车后窗
electrically heated safety glass 电加热安全玻璃
electrically heated sauna stove 电加热桑拿蒸汽炉
electrically heated tape 电加热筛;电加热带
electrically heated undersuit 电极加热潜水服
electrically heated welding torch 电热焊枪
electrically heated windscreen 电加热挡风玻璃
electrically heated windshield 电加热挡风玻璃
electrically induced adsorption 电诱导吸附(法)
electrically latched 电锁栓的
electrically lighted buoy 电力浮标灯
electrically lit 电气照明
electrically melted 电熔的
electrically melting type Pt substitute furnace 电熔式代铂拉丝炉

electrically non-conductive material 不导电材料
electrically operated 电驱动的;电动操作的
electrically operated brake 电气制动器
electrically operated circular manure grab 电动厩肥圆形抓爪
electrically operated control valve 电动操作控制阀
electrically operated curtain drawing device 电控拉窗帘机
electrically operated flying shears 电动飞剪
electricallyoperated hand scraper 电动手推式铲运机
electrically operated motor car 电动车
electrically operated platform truck 电动装卸车;电池车
electrically operated stop semaphores 电控停止信号机
electrically operated waterflow alarm 电动水流报警器
electrically oriented wave 定向电波
electrically powered 电力驱动的;电动的;电动驱动的
electrically powered press 电动压力机
electrically propelled ship 电力推进船
electrically responsive blasting cap 电敏(起爆)雷管
electrically scanning microwave radiometer 电扫描微波辐射仪;电扫描微波辐射计
electrically scanning radar 电扫描雷达
electrically submersible pump 电动潜水泵
electrically super-conducting alloy 超导电合金
electrically supervised 电气监测
electrically supported gyroscope 静电支承陀螺仪
electrically supported vacuum gyroscope 静电支承真空陀螺仪
electrically suspended gyro accelerometer 电支承式陀螺加速度计
electrically suspended gyro(scope) 电悬式陀螺仪;静电(悬浮)陀螺仪
electrically vibrated chute 电力震动滑槽
electrically welded 电焊的
electrically welded tube 电焊管
electrically welded weave 电焊筛;电焊钢筋网
electrically wound clock 电绕钟
electric(al) machine control 电机控制
electric(al) machine design 电机设计
electric(al) machine operating characteristic 电机运行特性
electric(al) machine 电机
electric(al) machinery 电机;电力机械;电动机械
electric(al) machinery plant 电机厂
electric(al) machine slot-conductor 电机槽内导体
electric(al) machine teaching console 电动机培训仪表
electric(al) magnet 电磁铁
electric(al) magnetic buzzer 电磁蜂鸣器
electric(al) magnetic chuck 电磁吸盘;电磁卡盘
electric(al) magnetic clutch 电磁离合器
electric(al) magnetic engine 电磁力发动机
electric(al)-magnetic forming 电磁成形
electric(al) magnetic iron remover 电磁除铁器
electric(al) magnetic method 电磁法
electric(al) magnetic speed regulation motor 电磁调速电动机
electric(al)-magnetic survey 电磁法测量
electric(al) magnetic variable motor 滑差电机;电磁调速电动机
electric(al) magnetization 电磁激励
electric(al) main 主(输)电线;输电干线
electric(al) maintenance case 电器维修箱
electric(al) mangle 电碾压机
electric(al) manometer 电动压力表;电测压计
electric(al) mantle 电热罩
electric(al) manufacturing 电机制造业
electric(al) mapping 电子作图
electric(al) marker 电标志器
electric(al) marking pen 电标志笔
electric(al) mass filter 电质谱仪;电学滤质器;电学滤质表
electric(al) material 电气器材;电气材料;电工材料
electric(al) materials and appliances 电料
electric(al) measurement 电气测量;电测(学)
electric(al) measuring instrument 电测仪表;电测量仪表
electric(al) mechanical 电动机械的

electric(al) mechanical parts counter 机电式零件计数器
electric(al) mechanical tamper 电动机械路夯
electric(al) medium constant 介电常数;电介质常数
electric(al) megaphone 电动喇叭筒
electric(al) melted refractory tile 电热熔炼的耐火砖
electric(al) melter 电熔窑
electric(al) melting 电熔(炼)
electric(al) melting furnace 电熔炉
electric(al) melting process 电熔炼法
electric(al) membrane demineralization 电膜脱矿化
electric(al) membrane equipment 电气隔膜装置
electric(al) memory glass semiconductor 电存储[贮]玻璃半导体
electric(al) metal 高频金属;电用金属
electric(al) metallic tubing 金属软管
electric(al) metallurgical works 电气冶de厂
electric(al) metal pulverizing device 电力金属喷镀设备
electric(al) metal pulverizator 电力金属喷镀器
electric(al) meter 电气仪表;电表
electric(al) meter enclosure 电表箱
electric(al) meter niche 壁挂式电表箱
electrical method 电动探查方法;电量法;电法
electric(al) method exploration 电阻率探测法
electrical method of sampling 取样电气法
electric(al) microbalance 电动微量天平
electric(al) micrometer 电动量仪;电测微计
electric(al) micrometer tube 电动测微筒
electric(al) migration 电迁移(法)
electric(al) mining shovel 电动矿山铲
electric(al) mini-panel 微型电控盘
electric(al) mixer 电动搅拌机;电动拌和机
electric(al) mixing method 电混合法
electric(al) model 电模型
electric(al) moisture meter 电气测湿计;电动湿度计;电测湿度计
electric(al) molten cement 电熔水泥
electric(al) moment 电矩
electric(al) monopole 电单极
electric(al) monorail crane 电动单轨起重机
electric(al) mosquite driver 电蚊器
electric(al) motive knitting carpet machine 电动织毯机
electric(al) motor 马达;电机;电动马达;电动机
electric(al) motor brush 电动机刷
electric(al) motor car 电动车;电气汽车;电动机车
electric(al) motor coach 电动客车
electric(al) motor controller 电动机控制器
electric(al) motor-drawn channel scraper 电动粪槽刮铲
electric(al) motor drilling 电钻钻进
electric(al) motor drive 电动机传动
electric(al) motor driven 电动机驱动的
electric(al) motor driven butter churn 电动乳脂制作器
electric(al) motor-driven impersonal micrometer 电动超人差测微器
electric(al) motor-driven point mechanism 电动转辙机构
electric(al) motor for rolling way 辊道电动机
electric(al) motor generator 电动发电机
electric(al) motor-generator set 电动发电机组
electric(al) motor movie camera 电动式活动摄影机
electric(al) motor oil 电动机油
electric(al) motor-operated fixed crane 固定式电动起重机
electric(al) motor-operated switch 电机操纵开关
electric(al) motor operation valve 电动阀(门)
electric(al) motor saw 电(力)锯
electric(al) motor short-circuit test instrument 电机短路测试仪
electric(al) motor signal mechanism 电动臂板信号机构
electric(al) motor truck 电气载重车
electric(al) mo(u)lding machine 电动造型机
electric(al) mower 电动割草机
electric(al) mucking machine 电动开挖机
electric(al) mud-slush pump 电动泥浆泵
electric(al) multifunction side-turn nursing bed

电动侧翻多功能护理床
electric(al) multiple set 多机组供电设施
electric(al) multiple unit 电动车组
electric(al) multiple unit railcar 电力多组机动有轨车
electric(al) multiplication 电乘法器
electric(al) multiplier 电动倍电器
electric(al) multipole 电多极
electric(al) multipole field 电多极场
electric(al) multipole radiation 电多极辐射
electric(al) multiunit set 多机组供电设施
electric(al) nailing machine 电动钉箱机
electric(al) network 电网(络);电力网
electric(al) network analyzer 电网络分析器
electric(al) network model 电网络模型
electric(al) network reciprocity theorem 电网络倒易定理
electric(al) neutrality 电荷中和
electric(al) night storage heater 夜间蓄电加热器
electric(al) noise 电噪声
electric(al) null 电零点
electric(al) nuttightening tool 电动上紧螺帽工具
electric(al) octupole moment 电八极矩
electric(al) oil 电绝缘油;电气绝缘用油
electric(al) oil-filled radiator 充油电辐射取暖器
electric(al) oil heater 燃油电加热器
electric(al) oil pump 电动油泵
electric(al) omnibus 电动公共汽车
electric(al) operated control box 电动控制箱
electric(al) operated excavator 电动挖土机
electric(al) operated mucking machine 电动开挖机
electric(al) operation disconnecting switch 电动操作隔离开关
electric(al) operator 电力启闭器;电动执行机构
electric(al) optic(al) converter 电光变换器
electric(al) oscillation 电振荡;电波动
electric(al) oscillator 电振荡器
electric(al) oscillograph 电示波器
electric(al) osmosis 电渗(作用);电渗现象;电渗透(作用)
electric(al)-osmotic stabilization 电渗稳定作用;电渗稳定法;土的电渗加固法
electric(al) outlet 电源插座;电器插口;电气出线口;电气出口线;电力输出
electric(al) output 电输出
electric(al) output signal 输出电信号
electric(al) output storage tube 电输出存储管
electric(al) oven 电炉;电烤箱;电烘箱;电烘炉
electric(al) overburden locomotive 电动超载机车
electric(al) overhead crane 桥式电动起重机
electric(al) overhead travel(l)ing crane 电动桥式起重机
electric(al) overhead travel(l)ing grab crane 桥式电动抓斗起重机
electric(al) ozone installation 电化臭氧装置
electric(al) pad 电热垫;电暖器
electric(al) paint 电化漆
electric(al) paint spray gun 电动喷漆枪
electric(al) pallet truck 电动托盘搬运车
electric(al) panel 配电板
electric(al) panelboard 配电板
electric(al) panel heater 板式加热器
electric(al) panel heating 电热辐射采暖
electric(al) panel room 配电盘室
electric(al) panel-type heater 板式电加热器
electric(al) parameter 电气特性;电(气)参数
electric(al) parameter of layer 地质断面的电性特征;地层的电参数
electric(al) part of semi-automatic coupler 半自动车钩电气部分
electric(al) part of semi-permanent drawbar 半永久车钩电气部分
electric(al) parts 电器零件
electric(al) passenger lift 旅客电梯
electric(al) paving breaker 电动路面破碎机
electric(al) pedestal grinder 支座式电动砂轮机
electric(al) pencil sharpener 电动削铅笔器
electric(al) pendulum 电摆
electric(al) penetration 电气贯穿
electric(al) penetrator 电动穿孔器
electric(al) percussion drill 电动撞钻;电动冲(击)钻
electric(al) percussive drill 电动冲(击)钻

electric(al) perforator 电动打孔机
electric(al) permeater 电测透气性试验仪
electric(al) permissible mine locomotive 矿山防爆机车;防爆式电机车
electric(al) permit to work 电气工作许可证
electric(al) petrol pump 电动汽油泵
electric(al) phase method 电场相位法
electric(al) photoengraver 光电刻板机
electric(al) pick 电镐
electric(al) pickoff 电接触点
electric(al) piezocone penetration test equipment 带孔压电测静力触探装置
electric(al) piezometer 电测压计;电测孔隙水压力仪
electric(al) pig iron 电炉生铁
electric(al) pile driving plant 电动打桩设备
electric(al) piler 电动垛板机;电动堆集机
electric(al) piling winch 电动打桩绞车
electric(al) pilot 电测仪
electric(al) pipe material workshop 电气管材间
electric(al) pipe precipitator 电气管道除尘器;电管除尘器
electric(al) pipe space 电线管井
electric(al) pipe-type precipitator 管式电沉降器
electric(al) plan 电气设备分布图
electric(al) plane 电动刨床;电刨
electric(al) plane characteristic 电场平面内的方向特性
electric(al) planer cutter 电刨刀
electric(al) planer for wood working 木工电刨
electric(al) planimeter 电动求积仪
electric(al) plaster saw 电动石膏锯
electric(al) platen 电热板
electric(al) plate precipitator 电(极)板除尘器;板式电沉降器
electric(al) platform truck 电动平板大卡车;平板式蓄电池搬运车
electric(al) plating 电解沉淀;电镀
electric(al) ploughing 电力耕种
electric(al) plow 电犁
electric(al) plug 电火花塞
electric(al) plug receptacle 电插座
electric(al) plug socket 电插座
electric(al) plunger pump 电动柱塞泵
electric(al) pneumatic temperature regulator 电动气动调温器
electric(al) pneumatic valve 电空阀
electric(al) point detector 岔尖位置电检测器
electric(al) point indicator 岔尖位置电气表示器
electric(al) point machine 电动转辙机
electric(al) point operating mechanism 电动转辙机
electric(al) point pressure type thermometer 电接点压力式温度计
electric(al) point welding 电点焊
electric(al) polarimeter 电子伺服控制光电偏振计
electric(al) polarizability 电极化性
electric(al) polarization 电极化(度)
electric(al) polarization intensity 电极化强度
electric(al) pole 电极;电杆
electric(al) pole centrifugal shaper 电杆离心成型机
electric(al) pole depositing car 电杆灌注车
electric(al) pole mo(u)ld hanger 电杆模型吊架
electric(al) pole predrawn platform 电杆预拉台
electric(al) polisher 电动抛光机
electric(al) polishing machine 电动抛光机
electric(al) porcelain 绝缘瓷(瓶);电工陶瓷;电瓷
electric(al) porcelain insulator 电瓷绝缘子
electric(al) porcelain shearing strength test 电瓷抗剪强度试验
electric(al) porcelain teapot 电瓷茶壶
electric(al) porcelain torsional strength test 电瓷抗扭强度试验
electric(al) pore water pressure cell 电动孔隙水压力盒
electric(al) portable drill 便携式电钻;手提式电钻;手电钻
electric(al) portable fan 便携式电扇
electric(al) portable heater 便携式电热器
electric(al) positive power steer(ing) 电的正向动力控制
electric(al) potential 电位;电势
electric(al) potential difference 电位差;电势差

electric(al) potential energy 电位能;电势能
electric(al) potential gradient 电位梯度;电势梯度
electric(al) potentiometer 电子电位计
electric(al) power 电能;电力;电功率;发电
electric(al) power and electric(al) energy balancing 电力电量平衡
electric(al) power cable 电力电缆
electric(al) power car 电动车
electric(al) power cart 电瓶车
electric(al) power charges 电费
electric(al) power company 电力公司
electric(al) power conductor 供电导线
electric(al) power consumption 耗电量;电力消耗(量)
electric(al) power cost 电力成本
electric(al) power deficiency 缺电
electric(al) power distribution 电力分配
electric(al) power drive 电力联动机
electric(al) power driven 电动的
electric(al) power economy 电气动力经济
electric(al)-powered equipment 电动设备
electric(al) power engineering 电力工程学
electric(al) power expenses 动力电费
electric(al) power failure 电力中断;电力事故
electric(al) power fuse 电雷管;电动火药引信
electric(al) power generating machinery 发电机
electric(al) power generation 发电
electric(al) power industry 电力工业
electric(al) power lifeline 生活供电线路
electric(al) power lift 电力起重机
electric(al) power line 电力线(路)
electric(al) power meter 电功率表
electric(al) power network composition 电网结构
electric(al) power outage 停电
electric(al) power pipe 电力管
electric(al) power plant 动力厂;电厂;发电厂
electric(al) power pool 联合供电网;电力网系统
electric(al) power project 电力工程
electric(al) power pylon 输电铁塔
electric(al) power reactor 发电动力堆
electric(al) power stabilization 电力稳定
electric(al) power station 电力站;电力变电所;发电站
electric(al) power station near coal mines 坑口电站
electric(al) power supply 电源;电力供应;供电
electric(al) power-supply circuit 馈电电路
electric(al) power system 电源系统;电力系统
electric(al) power tool 电动工具;电力工具
electric(al) power transmission 电力输送;输电
electric(al) power transmission lines 电力输送线路
electric(al) power transmission system 电力输送系统
electric(al) power transmission technique 输电技术
electric(al) power unit 电源设备
electric(al) precipitation 电力沉淀;电集尘
electric(al) precipitator 静电除尘装置;静电集尘器;电(气)集尘器;电力沉淀器;电除尘器;电沉降器
electric(al) precision cracker 电动精密粉碎机
electric(al) predictor 电动射击指挥仪
electric(al) preheating furnace 预热电炉
electric(al) press 电动压力机
electric(al) pressboard 电热压纸板
electric(al) pressure 电压
electric(al) pressure cooker 电高压锅
electric(al) pressure cooking saucepan 电压力锅
electric(al) pressure pump 电动压力泵
electric(al) pressure testing pump 电动式试压泵
electric(al) pressure type water heater 电压型热水器
electric(al) prime motor 电气牵引机
electric(al) prime mover 电动牵引机
electric(al) primer 电起爆药包;电起爆管;电雷管;电发火器;电发火极
electric(al) probe 电探针;电试探器
electric(al) probe device 电测装置;电测探计
electric(al) probing device 电测装置
electric(al) process for steel making 电炉炼钢法
electric(al) processing 电气加工
electric(al) process machine 电加工机床
electric(al) product 电气器材

electric(al) production 电力生产
electric(al) profile copyturning equipment 电器型面靠模车削设备
electric(al) profile method 电剖面法
electric(al) profiling 电剖面法；电测剖面(法)
electric(al) profilometer 电接触式轮廓仪
electric(al) property 电性(能)；电特性；电气性能
electric(al) property of minerals 矿物电学性质
electric(al) property of rock and soil 岩土的电学性质
electric(al) propulsion 电(力)推进；电力驱动；电力牵引
electric(al) propulsion system 电火箭发动机
electric(al) prospecting 电阻勘探法；电法找矿；电法探测；电法勘探；电法勘测；电测(勘)探；地层电探法
electric(al) prospecting data interpretation 电法资料解释
electric(al) prospecting data processing 电法资料数据处理
electric(al) prospecting method 电法勘探方法
electric(al) protection 电(器)保护；电气保护(装置)；电防腐
electric(al) protection relay 电力保护继电器
electric(al) protective device 电气保护器件；电力保护装置
electric(al) psychrometer 电测湿度计
electric(al) pugmill 电动捏拌机；电动混料机
electric(al) pulley 电动滑车
electric(al) pulley block 电动滑车
electric(al) pulley tackle 电动辘轳
electric(al) pulsator 电力式脉动器
electric(al) pulse 电脉冲
electric(al) pulse counter 电脉冲计数器
electric(al) pulse electropulse engine 电脉冲发动机
electric(al) pulse frequency 电脉冲频率
electric(al) pulse ignitor 电脉冲点火器
electric(al) pulse motor 电脉冲电动机
electric(al) pulse processing lathe 电脉冲加工车床
electric(al) pulse stimulation 电脉冲刺激
electric(al) pulse stimulator 电脉冲刺激器
electric(al) pump 电(动)泵
electric(al) pump governor 电动泵调节器
electric(al) pyrometer 电测高温计
electric(al) quadrupole 电四极
electric(al) quadrupole lens 四极电透镜
electric(al) quadrupole moment 电四极矩；四极电矩
electric(al) quadrupole shift 电四极移位
electric(al) quadrupole strength 电四极强度
electric(al) quadrupole transition 电四极跃迁
electric(al) quality factor 电学品质因素
electric(al) quantity 电量
electric(al) quick saw 电快锯
electric(al) raceway 电缆管道
electric(al) radiant ceiling system 电缆埋入式顶棚系统
electric(al) radiant heat 电辐射热
electric(al) radiant heater 电辐射加热器
electric(al) radiant heating 电辐射加热；电热辐射采暖
electric(al) radiant stove 辐射电炉
electric(al) radiation 电辐射
electric(al) radiation furnace 电热辐射炉
electric(al) radiator 电气辐射器；电热辐射器；电气散热器
electric(al) railcar 电动有轨车
electric(al) railcar sleeper 电动有轨卧车
electric(al) railcar train 电动有轨列车
electric(al) rail drill 电动钢轨钻(机)
electric(al) rail drilling machine 电动钢轨钻(机)
electric(al) rail grinder 电动钢轨砂轮机
electric(al) rail joint planing machine 电动轨道接头刨平机
electric(al) railroad 电气铁道；电气(化)铁路；电车道
electric(al) rail sawing machine 电动锯轨机
electric(al) railway 电气铁道；电气(化)铁路；电车道
electric(al) railway controller 电离牵引调度员
electric(al) rake-rubber belt conveyer 电扒—胶带输送机
electric(al) rammer 电夯；电动夯(土机)；电动捣固机

electric(al) ramming impact machine 电动夯拍机
electric(al) range 电炉(灶)
electric(al) rate schedule 电力价目表
electric(al) reactor 电抗器
electric(al) readout 电读出
electric(al) receptacle 插座板
electric(al) receptacle plug 电插头
electric(al) reciprocating pump 电动往复泵
electric(al) recorder 电记录器；电动记录仪
electric(al) recording current meter 电动自记流速仪；电动自记海流计
electric(al) reflector radiator 电反射辐射器
electric(al) refrigerator 电制冷库；电(气)冰箱
electric(al) regenerative braking 电力再生制动
electric(al) regulating valve 电动调节阀
electric(al) regulations 电气规程
electric(al) regulator 电调节器
electric(al) relaxation 电驰张
electric(al) relay (电子)继电器
electric(al) relay interlocking 继电式电气集中联锁；继电联锁(装置)
electric(al) remote control 电力遥控；电力遥远控制
electric(al) remote control gear 电力遥控传动装置
electric(al) remote control panel 电动遥控面板
electric(al) remote test pressure ga(u)ge 电气遥测压力表
electric(al) remote test tachometer 电气遥测转速表
electric(al) repair 电气修理
electric(al) repair shop 电气修理车间
electric(al) repair truck 电气修理工程车
electric(al) reproduction 电放声
electric(al) reset relay 电复位继电器
electric(al) resistance 电阻
electric(al) resistance alloy 电阻合金
electric(al) resistance calorimeter 电阻量热器
electric(al) resistance element 电阻元件
electric(al) resistance furnace 电阻炉
electric(al) resistance ga(u)ge 电阻应变仪；电阻式潮位计
electric(al) resistance heater 电阻加热器
electric(al) resistance holding furnace 电阻保温炉
electric(al) resistance humidifer 电阻式加湿器
electric(al) resistance hygrometer 电阻水分测定计；电阻湿度计；电阻湿度表
electric(al) resistance load-ga(u)ge 电阻测力仪
electric(al) resistance manometer 电阻压力计；电阻压力表
electric(al) resistance moisture meter 电阻水分计；电阻式水分测定表；电阻(式)湿度计；电阻式湿度测定仪
electric(al) resistance of rock and soil 岩土(的)电阻率
electric(al) resistance of soil 土的电阻
electric(al) resistance pressure ga(u)ge 电阻压力计
electric(al) resistance pyrometer 电阻高温计
electric(al) resistance seam-welding 电阻缝焊
electric(al) resistance seam-welding machine 电阻缝焊机
electric(al) resistance self-balancing pyrometer 电阻自动平衡高温计
electric(al) resistance sensor 电阻传感器
electric(al) resistance strain ga(u)ge 电阻应变仪；电阻应变器；电阻应变片；电阻应变计；电阻应变规；电阻丝应变片
electric(al) resistance thermometer 电阻温度计；电阻温度表
electric(al) resistance type temperature indicator 电阻式温度指示器
electric(al) resistance type wave height probe 电阻式波高仪；电阻式波高计
electric(al)-resistance-welded steel pipe 电阻焊钢管
electric(al) resistance weld(ing) 电阻焊(接)；触焊(法)；热电阻焊
electric(al) resistance welding tube 电阻焊接管
electric(al) resistance weld mill 电阻焊管机
electric(al) resistance-wire strain ga(u)ge 电阻线弹性应变计
electric(al)-resisting method of prospecting 电阻探测法
electric(al) resistivity 抗电性；电阻率；电弧率

electric(al) resistivity heating of concrete 混凝土的电阻加热
electric(al) resistivity method 电阻探测法；电阻(率)法
electric(al) resistivity method of exploration 电阻率探测法
electric(al) resistivity method of prospecting 电阻率探测法
electric(al) resistivity microscanning image tool 电阻率微扫描成像仪
electric(al) resistivity of soil 土的电阻率
electric(al) resistivity prospecting 电阻率法勘探
electric(al) resistivity survey 电探
electric(al) resistivity technique 电阻率探测技术
electric(al) resistor 电阻(调节)器；电弧器
electric(al) resonance 电谐振；电共振
electric(al) response 电反应
electric(al) response audiometer 电反应听力计
electric(al) revolution indicator 电动转速计；电动转速表
electric(al) revolving shovel 旋转式电铲；电动旋转式挖土机
electric(al) rewind reel 电动重卷机
electric(al) re-wiring 重新布(电)线；电线重布
electric(al) rheostat 电阻箱；变阻器
electric(al) rice cooker 电饭锅
electric(al) riveted joint 电铆接头
electric(al) riveter 电动铆(钉)机；电动铆(接)机
electric(al) riveting 电铆
electric(al) riveting machine 电铆机；电动铆(钉)机
electric(al) roaster(oven) 电烤炉
electric(al) robot 电动机器人
electric(al) rock drill 凿岩电钻；电动钻岩机；电动凿岩机
electric(al) rock drill bit 电动钻岩机钻头
electric(al) rocket 电火箭发动机
electric(al) rock hammer 电动凿岩锤
electric(al) rocking furnace 电弧摆动炉
electric(al) rock loader 电力装石机；电动装岩机
electric(al) rock mucker 电动装岩机
electric(al) rod-curtain precipitator 棒帘式电收尘器；棒帘式电聚尘器；棒帘式电除尘器
electric(al) roll beater 轧辊电热炉
electric(al) room heater 电暖器
electric(al) rot 电腐蚀
electric(al) rotary 电动转盘
electric(al) rotary drilling machine 电动旋转钻机
electric(al) rotary fish screen 电动旋转式鱼栅
electric(al) rotary hammer 电动旋转锤
electric(al) rotating machinery 旋转电机
electric(al) routing analogy 洪水演进电模拟；电流示踪模拟
electric(al) rubber cover wire 胶皮电线
electric(al) rudder angle transmitter 舵角(电传)发送器
electric(al) safety chain 电气安全链
electric(al) safety distance 电气安全距离
electric(al) safety hand lantern (检查油舱用)安全电池灯
electric(al) safety lamp 电矿灯
electric(al) salinity meter 电测盐度计
electric(al) salt bath 电热盐浴
electric(al) sanitary incinerator 电气环境卫生焚化炉
electric(al) saucepan 电锅
electric(al) saw 电锯
electric(al) scaffolding 电动脚手架
electric(al) scalar potential 电标量势；标量电位
electric(al) scanner 电扫描器
electric(al) scanning 电扫描
electric(al) scissors 电剪刀
electric(al) scoreboard 电动记分牌
electric(al) screen 电屏蔽板；电屏
electric(al) screw clay gun 电动螺旋式泥炮
electric(al) screw down gear 电动压下装置
electric(al) screwdriver 电动螺旋钻；电动螺丝刀；电旋凿；电(动)改锥
electric(al) sealing machine 电气熔缝机
electric(al) seasoning 电干燥
electric(al) seawater thermometer 电测海水温度计
electric(al) second meter 电秒表
electric(al) security hinge 电动安全铰链
electric(al) seed bed heater 电热温床
electric(al) seine 电渔网

electric(al) selfpriming centrifugal pump 电自吸离心泵
electric(al) sensing 电传感
electric(al) sensor output 电传感器输出
electric(al) separation 电选;电分离
electric(al) separation method 电力选矿法
electric(al) separation process 电(气)分离法
electric(al) separator 电滤波器;电(力)分离器
electric(al) series 电位序
electric(al) service 电气设备维护和保养
electric(al) service duct 电气公用设施管沟
electric(al) service elevator 电气维修保养电梯
electric(al) service equipment 进户供电设备;电气辅助设备
electric(al) service lift 电气维修保养电梯;保养电梯
electric(al) service pump 电动通用泵
electric(al) service to dry dock 干坞供电
electric(al) servo 电(动)伺服机构
electric(al) servomechanism 电伺服机构
electric(al) servo system 电伺服系统
electric(al) set 电动机组;发电机组
electric(al) setting 电缆坐放
electric(al) sewer 电动缝纫机
electric(al) sewing machine 电动缝纫机
electric(al) shaft 电联动;同速联动;电气竖井
electric(al) shaft furnace 竖井式电炉
electric(al) shaker 电动摇筛器
electric(al) shears 电动剪断机;电剪(刀);电动剪切机
electric(al) sheet 电气钢板;电工钢片
electric(al) sheet or strip 电工钢板或带钢
electric(al) sheet-pile driver 电动板桩打桩机
electric(al) shielding 电屏蔽
electric(al) ship 电力推进船;电动船
electric(al) ship log 电动测程仪【船】
electric(al) shock 电震;电伤;电击;触电
electric(al) shock hazard 触电危险
electric(al) shocking rammer 电动冲击夯
electric(al) shock prevention 触电防护
electric(al) shock tube 电冲击波管
electric(al) shoe polisher 电动擦鞋器
electric(al) shop 电气车间;电机修理车间
electric(al) shovel 电力铲;电动挖土机;电动挖掘机;电(动)铲
electric(al) shutter 电气开关;电气隔离;电气断续器
electric(al) shutter camera 电动快门摄影机
electric(al) shutter door operator 电动卷门操纵器;电动百叶门操纵器;门的电动启闭器
electric(al) sieve 电筛
electric(al) sign 电光标志
electric(al) signal 电信号
electric(al) signal machine 电动信号机
electric(al) signal output 电信号输出
electric(al) signal storage tube 电信号存储管
electric(al) silicification method 电动硅化法
electric(al) silicon carbide 电工碳化硅
electric(al) simulator 电模拟装置;电模拟设备;电模拟器
electric(al)-singeing machine 电板烧毛机
electric(al) single beam bridge crane 电动单梁桥式起重机
electric(al) single beam crane 电动单梁起重机
electric(al) single girder crane 电动单梁起重机
electric(al) single-line systematic diagram 电气单线系统图
electric(al) siren 电警报器;笛笛
electric(al) slag smelting 炉渣电炉熔炼
electric(al) slag welding 电渣焊
electric(al) sleep 电流睡眠
electric(al) slewing crane 电动悬臂起重机;电动旋臂起重机
electric(al) slewing motor 电动转子发动机
electric(al) slip ring 电滑环
electric(al) slipway 电动滑道
electric(al) slot 电选别机
electric(al) slow cooker 电饭锅
electric(al) sludge conveyer worm type 电动螺旋式泥浆输送机
electric(al) smelting 电冶炼;电热熔炼
electric(al) smelting furnace 电气熔炼炉
electric(al) smoke precipitation 电气除尘
electric(al) socket wrench 电动套筒扳手
electric(al) soil moisture meter 电传土壤湿度计;土壤水分电测仪

electric(al) soldering 电气钎焊;电烙;电(加)热钎焊;电焊;钎焊;气焊
electric(al) soldering appliance 电烙焊工具
electric(al) soldering bit 电烙铁头
electric(al) soldering bolt 电烙铁螺杆
electric(al) soldering copper 电烙紫铜烙铁
electric(al) soldering iron 电烙铁
electric(al) soldering pliers 电焊钳
electric(al) soldering wire 电焊锡丝
electric(al) solenoid 螺线管
electric(al) sorting apparatus 电分类器
electric(al) sounder 电测深仪
electric(al) sounding 电测探;电测深
electric(al) sounding machine 电测水深计;电测深计
electric(al) sounding tape 电测卷尺;电声卷尺
electric(al) source 电源
electric(al) source building 供电建设
electric(al) space heater 电暖器;取暖电炉
electric(al) spark (电)火花
electric(al) spark coded source 电火花编码震源
electric(al) spark computer numerial control wire-cut machine 电火花数控线切割机床
electric(al) spark drill 电火花钻井
electric(al) spark drilling 电火花穿孔
electric(al) spark forming 电火花成型加工
electric(al) spark grinding 电火花磨削
electric(al) sparking 电火花加工
electric(al) sparking source 电火花震源
electric(al) spark machinery 电火花加工机床
electric(al) spark machine tool 电火花机床
electric(al) spark machining 电子放电加工;电火花加工
electric(al) spark shaping machine 电火花成型机床
electric(al) spark sintering 电火花烧结
electric(al) spark small hole machine tool 电火花小孔机床
electric(al) specification 电气特性;电气技术规格;电规范说明
electric(al) spectrophotography 光电分光光度学
electric(al) speedgovernor 电调节器;电调速器
electric(al) spot displacement 光点的电位移
electric(al) spot welding 电点焊
electric(al) spot welding machine 电动电焊机
electric(al) squib 动力起爆器;电气导火管;电力导火管;电爆管
electric(al) stabilization 电稳定和固结土壤;土的电法加固
electric(al) stabilization of soil 电处理稳定土壤
electric(al) stacker 电力堆料机;电力堆垛机;电动堆积机
electric(al) staff block system 电气路签闭塞(系统);电气路牌闭塞(系统)
electric(al) staff instrument 电气路签机
electric(al) staff machine 电气路签机
electric(al) standard 电量标准
electric(al) standard equipment 标准电气设备
electric(al) standard equipment for auxiliary power supply 辅助供电标准电气设备
electric(al) standard equipment for lighting 照明标准电气设备
electric(al) standard equipment for monitoring and information 监视和信息标准电气设备
electric(al) standard equipment for train/door control 气动车/门标准电气设备
electric(al) starter 电起动器;电启动器;电力起动器
electric(al) starter motor 电力起动机
electric(al) starter motor battery 发动电动机蓄电池
electric(al) starting 电力发动;电力起动
electric(al) starting system 电起动系统
electric(al) station 电站
electric(al) statistics machine 电统计机
electric(al) steamboiler 电热锅炉
electric(al) steam curing 电热蒸汽养护
electric(al) steamer 电蒸锅;电热蒸器
electric(al) steam ga(u)ge 电动气压机
electric(al) steam generator 电热锅炉
electric(al) steam radiator 电气蒸汽散热器
electric(al) steam sterilizer 电力蒸汽式消毒器
electric(al) steel 电炉钢;电工(用)钢
electric(al) steel furnace 电气炼钢炉
electric(al) steelmaking 电炉炼钢

electric(al) steel making process 电炉炼钢操作法
electric(al) steel sheet 电工硅钢片
electric(al) steering engine 电动舵机
electric(al) steering gear 电动舵机;电(动)操舵装置
electric(al) steering machine 电力控制机械
electric(al) steering system 电舵
electric(al) sterilizer 电消毒器;电气消毒器
electric(al) stirrer 电动搅拌器;电热搅拌器
electric(al) stone ware 电热炉器
electric(al) stop watch 电秒表
electric(al) storage 电存储器
electric(al) storage device 蓄电装置
electric(al) storage heater 电气储存加热器;容积式电加热器
electric(al) storage heating 电荷储存加热
electric(al) storage heating appliance 电气储存加热器
electric(al) storm 雷暴雨;电暴
electric(al) stove 电炉
electric(al) stove wire 电炉丝
electric(al) strain ga(u)ge 电应变计;电(测)应变仪
electric(al) strength 破损低压;耐电强度
electric(al) strength test 介电强度试验;电气强度试验;电绝缘强度试验
electric(al) stress 静电强度;电介质应力
electric(al) stretching 电张拉
electric(al) strip 电工钢片
electric(al) stripping shovel 表层剥离电铲
electric(al) strip tension detector 带材张力电测仪
electric(al) stroboscope 电气频闪观察仪
electric(al) structure 电结构
electric(al) submersible pump 电动潜水泵;电动沉没泵
electric(al) substation 变电站;变电所
electric(al) suction apparatus 电动吸引器
electric(al) suction dredge(r) 电动吸泥机
electric(al) suction pump 电动吸引器;电动吸引泵
electric(al) suction ship 电动吸泥船
electric(al) supply 供电;电源;电力供应
electric(al) supply company 供电公司
electric(al) supply equipment 供电设备;电源(供给)附属设备
electric(al) supply line 供电线(路);供电干线;电源输电线
electric(al) supply of instrument 仪器电源
electric(al) supply shaft 供电井道
electric(al) supply station 供电站
electric(al) surface heater 表面电加热器
electric(al) surface-recording thermometer 地面记录电温度计
electric(al) surface resistivity 表面电阻系数;表面电阻率
electric(al) survey 电力查勘;电力测量;电法勘测
electric(al) surveying 电法探测;电法勘探;电法测量
electric(al) survey-net adjuster 测量网电模拟平差器
electric(al) susceptibility 电极化率
electric(al) swaging 电热型锻
electric(al) switch 电(器)开关;电门
electric(al) switch board 电掣板
electric(al) switch box 接电箱
electric(al) switch circuit controller 转辙电路控制器
electric(al) switcher 电动转辙机
electric(al) switch heater 道岔电热器
electric(al) switching 电气开关
electric(al) switching locomotive 调度电力机车
electric(al) switch lock 电锁
electric(al) switch machine 电动转辙机
electric(al) switch mechanism 电动转辙机
electric(al) switch oil 电键油
electric(al) swivel 电旋转接头;电动水龙头
electric(al) symbol 电气符号
electric(al) synapse 电突触
electric(al) system 电气系统;电力系统
electric(al) systematic drawing 电气系统图
electric(al) system pooling 电力系统联网;电力系统并网
electric(al) table interlocker 台式电联锁器
electric(al) tablet 电路牌
electric(al) tablet block system 电气路签闭塞(系统);电气路牌闭塞(系统)

electric(al) tablet instrument 电气路牌机
electric(al) tablet machine 电动压片机
electric(al) tachometer 电示转速计；电力转速计；电动转速计；电动转速表
electric(al) tachometer indicator 电气式回转速度计指示器
electric(al) tamper 电动夯(土机)；电动打夯机
electric(al) tamping machine 电动捣固机
electric(al) tank 电解槽
electric(al) tape 绝缘胶带
electric(al) tape ga(u)ge 电动水尺；电测水位计；带式电阻水位计
electric(al) tapping machine 电动开孔机
electric(al) tar catcher 电力熔焦油器
electric(al) taximeter 电计程表
electric(al) tea kettle 电茶壶
electric(al) tea pot 电茶壶
electric(al) technical standard 电气技术标准
electric(al) telegraph 电动车钟
electric(al) telemeter 电遥测仪；电遥测计；电力遥测器；电控测距仪；电动遥控测距仪；电测远器；电测距仪
electric(al) telemetering 电遥测术
electric(al) telemetric pressure ga(u)ge 电遥测压力计
electric(al) telemetry 电遥测技术
electric(al) telemotor 电力液压遥控装置；电力遥控传动装置；电动油压舵机
electric(al) teletachometer 电遥测转速计；电遥测转速表
electric(al) telethermometer 电气遥测温度表
electric(al) temperature control 电气温度控制
electric(al) temperature regulator 电动调温器
electric(al) tempering 接触焊机上加热回火；电回火
electric(al) tensimeter 电压力计
electric(al) tension 电压
electric(al) tensioner 电动张紧装置
electric(al) tensioning 电张拉
electric(al) tensioning process 电张法
electric(al) tensometer 电压力计
electric(al) terminal 电气接点
electric(al) terminals of a machine 电机的电气线端
electric(al) test 电栓；电气试验
electric(al) test bench 电工试验台
electric(al) tester 电测试器
electric(al) testing car 电气试验车
electric(al) testing instrument 电气测量仪器；电力试验仪器
electric(al) thermal desiccator 电热干燥器
electric(al) thermal water heater 电热热水器
electric(al) thermal water heating appliance 电热热水设备
electric(al) thermocouple 热电偶
electric(al) thermometer 电(测)温度计；电测温度表
electric(al) thermos flask 电暖水瓶
electric(al) thermostat 恒温开关；电热恒温器
electric(al) thermostat fire alarm system 电热恒温式火灾报警器
electric(al) thickness 电磁测流厚度
electric(al) thickness ga(u)ge 电测厚仪；电测厚计
electric(al) thread 导线铜管螺纹
electric(al) threading machine 电动螺纹机；电动攻丝机
electric(al) ticket date marker 电动票据打日期机
electric(al) tiering machine 电动堆垛机
electric(al) tie tamper 电动砸道机；电动轨枕夯实器
electric(al) timer 电子计时器；电定时器
electric(al) time release 电气定时解锁
electric(al) titling machine 电动烫金机
electric(al) titrimeter 电子滴定计
electric(al) toaster 电烤面包炉
electric(al) -to-fluid transducer 电流体转换
electric(al) token instrument 电气凭证机
electric(al) tool 电动工具
electric(al) tool grinder 电动工具磨床
electric(al) tooltipping machine 电焊刀片机
electric(al) to optic(al) converter 电光转换器
electric(al) to pneumatic transducer 电动气动变换器
electric(al) torch 电筒；电焊枪；手电筒
electric(al) torque meter 电动扭矩表
electric(al) torsiograph 电测扭力仪

electric(al) torsional strength test 电瓷抗剪强度试验
electric(al) torsional vibrometer 电测扭振仪
electric(al) touch-recording apparatus 电触击记录器
electric(al) tough pitch 电韧(火精)铜
electric(al) towing trolley 电气牵引车
electric(al) towing truck 电动曳引卡车；电力牵引车
electric(al) towing winch 电动拖缆绞车
electric(al) toy 电动玩具
electric(al) trace heating 电气式伴随加热法
electric(al) tracer heating system 电伴热系统
electric(al) tracing 电跟踪
electric(al) track jack 电动起道机
electric(al) track tamper 电动轨道夯实机
electric(al) traction 电力牵引
electric(al) traction elemechanical system 电力牵引远动系统
electric(al) traction engineering 电气牵引工程
electric(al) traction feeding system 电力牵引供电系统
electric(al) traction interference 电力牵引干扰
electric(al) traction network 牵引网
electric(al) traction telemechanical system 电力远动系统
electric(al) tractor 电力牵引车；电动拖拉机
electric(al) trailable switch machine 可拉式电动转辙机
electric(al) trailer 电(动)拖车
electric(al) train 电气列车；电动火车
electric(al) train line test 电气列车线试验
electric(al) train staff instrument 电气路签机
electric(al) transducer 电换能器；电传感器
electric(al) transductor 电磁放大器
electric(al) transient 瞬时产生的不稳定电流
electric(al) transmission 电力传动
electric(al) transmission current meter 电力传动流量计
electric(al) transmission line 输电线(路)
electric(al) transmission pole tower 输电塔
electric(al) transmitter 电变送器
electric(al) transporting appliance 电气运输设备
electric(al) transverse weaving machine 电动横织机
electric(al) travel(l)ing crane 移动式电动起重机；移动式电动吊车；电力移动式起重机；电动有轨式起重机；电动移动式起重机
electric(al) travel(l)ing roller 电动移动滚筒
electric(al) trawl cable 拖网电缆
electric(al) treadle 轨道接触器
electric(al) treatment bath 电疗浴缸
electric(al) trench 电缆沟
electric(al) trenching 电剖面法填图
electric(al) troll(e)y 电车
electric(al) troll(e)y hoist 电动移动绞车
electric(al) troll(e)y locomotive 架线式电机车；电动触轮式机车
electric(al) trough 电缆沟槽
electric(al) trowel 电动镘刀；电动抹子
electric(al) truck 电瓶车；电力搬运车；电动卡车；电动搬运车
electric(al) truck crane 电动汽车式起重机
electric(al) tube calipers 管壁厚度电测仪
electric(al) tube-type furnace 管式电炉
electric(al) tubular crane 电动管状起重机；电动管状吊车
electric(al) tubular heater 电的管状加热器
electric(al) tuning 电调谐
electric(al) turntable 电动转盘
electric(al) twinning 电孪生
electric(al) two-beam bridge crane 电动双梁桥式起重机
electric(al) two-speed wiper 电动双速刮水器
electric(al) two-way valve 电动两通阀
electric(al) -type detector 电信号输出探测器
electric(al) tyre crane 电动轮胎起重机
electric(al) tyre groover 电热轮胎切槽器
electric(al) ultra low-temperature incubator 电气超低温细菌培养器
electric(al) underfloor panel heating 地板下电气供暖板
electric(al) underfloor service carried in conduits 地板下安装在管道中的电气线路
electric(al) underfloor warming cable 地板下采暖电缆
electric(al) underfloor warming installation 电气的地板下采暖装置
electric(al) unit 电单位
electric(al) upset forging 电热镦
electric(al) upsetter 电热镦锻机
electric(al) utility industry 电力工业
electric(al) utility supply system 供电系统
electric(al) vacuum ceramics 电真空(陶)瓷
electric(al) vacuum cleaner 电真空吸尘器
electric(al) vacuum drying oven 电真空干燥箱
electric(al) vacuum glass 电真空玻璃
electric(al) vacuum-meter 电真空计
electric(al) valence 电价
electric(al) valve 整流管；电动阀(门)
electric(al) valve actuator 电阀执行机构
electric(al) valve grinder 电动阀磨床
electric(al) valve refacer 电动阀面磨光机
electric(al) valve seat cutter 电动阀座铣刀
electric(al) valve seat grinder 电动阀座研磨机
electric(al) variable speed stirrer 电动变速搅拌机
electric(al) varnish 高度绝缘清漆；电漆
electric(al) vector 电矢；电场矢量
electric(al) vector potential 向量电位
electric(al) vehicle 电动车辆
electric(al) vehicle motor 牵引电动机
electric(al) ventilator 电通风机
electric(al) vibrating beam 电动振动梁
electric(al) vibrating feeder 电振动给料机
electric(al) vibrating plate 电动振动板
electric(al) vibrating screen 电振筛
electric(al) vibrating tamper 电动振动夯；电动振捣板
electric(al) vibration 电振动
electric(al) vibrator 电振(动)子；电(力)振动器；电动振动器；电动振捣器
electric(al) vibratory beam 电动振动梁
electric(al) vibratory plate 电动振动板
electric(al) vibratory ramper 电动振动夯
electric(al) vibratory tamper 电动振捣板
electric(al) visco(si)meter 电黏度计
electric(al) viscosity 电黏性
electric(al) voltage 电压
electric(al) voltage system 电压系统
electric(al) vulcanizer 电力补胎机
electric(al) walking dragline 移动式电动索铲挖土机；步行式电动索铲挖土机
electric(al) wall panel 电热式墙板
electric(al) wall type slewing crane 电动墙式旋臂吊车
electric(al) warmer 电采暖器；电取暖器
electric(al) warming 电热法
electric(al) warming appliance 电采暖器
electric(al) warming device 电采暖器
electric(al) warming oven 电暖炉
electric(al) warming pad 电热垫
electric(al) warming unit 电采暖单元
electric(al) washer 电动洗衣机；电动洗涤机
electric(al) washing machine 电动洗衣机；电动洗涤机
electric(al) water cooler 电(力)冷水器
electric(al) water-filled radiator 冲水电热辐射器
electric(al) water heater 电热温水炉；电热水器
electric(al) water heating appliance 水的电加热装置
electric(al) water heating tap 电暖水器
electric(al) water sounder 电测水深器；电测深仪
electric(al) wave 电波
electric(al) wave filter 电滤波器
electric(al) wave measuring device 电气测波仪
electric(al) wave recorder 电测波浪计
electric(al) wedge valve 电动楔式阀
electric(al) weigher 电秤
electric(al) weighing equipment 电气称重设备
electric(al) weighing system 电子衡重系统；电秤系统
electric(al) welded anchor cable 电焊锚链
electric(al) welded pipe 电焊管
electric(al) welded steel pipe 电焊钢管
electric(al) welder 电焊机；电焊工
electric(al) welder's helmet (head type) 电焊面罩(头带式)
electric(al) welding 电弧焊接；电焊
electric(al) welding assistant's protecting glass

电焊辅助工用护目镜玻璃
electric(al)welding cast copper solder 铸铜电焊条
electric(al)welding equipment 电焊设备
electric(al)welding generator 电焊发电机
electric(al)welding hammer 电焊锤
electric(al)welding machine 电焊机
electric(al)welding plant 电焊厂
electric(al)welding pliers 电焊钳
electric(al)welding rod 电焊条
electric(al)welding set 电焊装置；电焊机
electric(al)welding shield glass 电焊用护目镜玻璃
electric(al)welding strip 电焊片
electric(al)welding transformer 电焊(用)变压器
electric(al)weld-pipe mill 电焊管机
electric(al)well heater 井下电热器
electric(al)well log 电测钻井剖面；电测钻井记录 (曲线)；电测井记录(曲线)
electric(al)well logging 电测井
electric(al)wet and dry bulb hygrometer 电气干湿球湿度计；电动干湿泡湿度计
electric(al)wheel 电动车轮
electric(al)wheel drive 电动轮驱动
electric(al)wheel loader 电动轮式装载机
electric(al)whim 电动辘轳
electric(al)whistle 电笛
electric(al)winch 电力卷扬机；电力绞车；电动起货机；电动卷扬机；电动绞车
electric(al)wind 对流放电
electric(al)winding machine 电动绕线机；电动卷扬机
electric(al)windlass 电动起锚机；电动卷扬机
electric(al)windscreen wiper 电动风挡刮板
electric(al)wind shield wiper 电动风挡刮水器
electric(al)wiper motor 刮水器电动机
electric(al)wire 电线
electric(al)wire braiding machine 电线编织机
electric(al)wire cold header 电动钢丝冷镦机
electric(al)wire conduit 电线套管；电缆套管
electric(al)wire connecting junction 接线器
electric(al)wire hoist 电动葫芦
electric(al)wire plan 电路图；布线图【电】
electric(al)wire resistance 电阻应变仪
electric(al)wire resistance meter 电阻应变仪
electric(al)wire strain ga(u)ge 电线式应变计
electric(al)wire traction instrument 电动钢丝牵引器
electric(al)wiring 电线架设；电气线路的敷设；电气配线；电气布线；电力布线；布电线；安装电线
electric(al)wiring plan 电线布置图；电路图；布线图【电】
electric(al)wiring regulation 装线规则
electric(al)working of points 道岔电操纵
electric(al)works 电气工程；发电站
electric(al)workshop 电修车间；电气车间
Electric(al)World 电世界(刊物)
electric(al)wrench 电动扳手；电动扳钳
electric(al)wrench for railway bolt 铁道螺钉电扳手
electric(al)yarn 电气用绝缘纱
electric(al)zero 电零位；电零点
electric(al)zero signal 电气零点信号
electrician 电气技师；电气工人；电气工程师；电工(技师)
electrician's bag 电工袋
electrician's belt 电工带
electrician's chisel 电工凿
electrician's drill 电工钻
electrician's hammer 电工锤
electrician's knife 电工刀
electrician's level 电工水准仪
electrician's pliers 电工钳
electrician's rubber gloves 电工橡皮手套
electrician's scissors 电剪
electrician's screwdriver 电工螺丝起子；电工螺丝刀
electrician's shop 电工车间
electrician's solder 电工焊料；电工焊锡
electricity 电学；电气；电流
electricity accident 电事故
electricity cable 电缆
electricity cable duct 电力电缆沟槽
electricity charger 充电器
electricity conduit 电导线管
electricity consumer 电用户
electricity consumption 耗电量

electricity cut-off standard 瓦斯煤矿停电标准
electricity expenses 电力费
electricity export 电力输出
electricity failure 电故障
electricity generating station 电厂
electricity generating unit 发电机组；发电单元
electricity heat analogy 电热模拟
electricity imput 电力输入
electricity insulator 电绝缘子
electricity measurer 测电器
electricity meter 电量计
electricity need load 城市用电负荷
electricity production 电力生产
electricity rate 电价；电费率
electricity requirement 电力需求
electricity saving 节约用电
electricity substation 变电站；变电所
electricity supply 供电；电力供应
electricity supply company 供电公司
electricity tariff 电力收费率
electricity utilization 用电；电力利用
electricity works 发电厂
electric-pulse magnet 脉冲电吸铁
electrics 电力机械
electrifiability 可起电性
electrification 电气化；带电；起电；使用电力
electrification by influence 感应起电
electrification detector 带电探测器
electrification ice nucleus 带电冰核
electrification interference 电气化干扰
electrification of steam railway 蒸汽铁路的电化
electrification of terminal 枢纽电气化
electrification plan 电气化计划
electrification scheme 电气化计划
electrification train 电气化列车
electrification work 电气化工程
electrified 电气化的；带电的
electrified body 带电体
electrified fence 电围栏
electrified kilometrage 电气化线路公里里程
electrified line 电化线路
electrified railway 电气化铁路
electrified railway line 电气化铁路线
electrified rate 电化率
electrified section 电气化区段；电气化路段
electrified territory 电气化区段
electrified track mileage 电气化线路英里里程
electrified wire netting 电网
electrify 带电；起电
electrifying 毛绒电光整理
electrino 电微子
electrion 高压放电
electrionised oil 高压放电合成油
electrion oil 高压放电润滑油
electrion process 高压放电法
electrization 电(气)化；电激法
electrize 起电
electrizer 起电盘
electroabrasion 电研磨加工；电腐蚀加工法
electroabsorption 电致吸收
electroabsorption modulator 电吸收调制器
electroacoustic(al) 电声的
electroacoustic(al) analogy 电声类比
electroacoustic(al) apparatus 电音响装置
electroacoustic(al) coupling factor 电声耦合系数
electroacoustic(al) delay line 电声延迟线
electroacoustic(al) dewatering 电声脱水
electroacoustic(al) effect 电声效应
electroacoustic(al) efficiency 电声效率
electroacoustic(al) factor 电声系数
electroacoustic(al) frequency meter 电声频率计
electroacoustic(al) index 电声指数；电声效率
electroacoustic(al) instrument 电声仪器
electroacoustic(al) mill 电声磨机
electroacoustic(al) mill control 磨机的电声控制
electroacoustic(al) modulator 电声调制器
electroacoustic(al) reciprocity principle 电声互易原理
electroacoustic(al) reciprocity theorem 电声互易原理
electroacoustic(al) relay 电声继电器
electroacoustic(al) set up 电声装置
electroacoustic(al) system 电声系统
electroacoustic(al) thermometer 电声温度计

electroacoustic(al) transducer 电声转换器；电声换能器
electroacoustic(al) transformer 电声变压器
electroacoustic(al) wave 电子声波
electroacoustics 电声学
electroacoustomagnetic effect 电声磁效应
electroactive species 电活化粒种
electroactive substance 电活性物质
electroadsorption 电吸附
electroadsorption capacity 电吸附容量
electroadsorption desalination 电吸附脱盐
electroadsorption technique 电吸附技术
electroaerometer 电气体比重计
electroaffinity 电亲合性；电亲合力
electroaluminothermic process 电铝热法
electroanalyser 电分析器
electroanalysis 电(解)分析
electroanalyzer 电分析器
electroanastomosis 电吻合术
electroarc contact machining 接触放电加工
electroarc depositing 放电涂覆
electroassembly 电气安装
electroaudiometer 电测听计
electrobalance 电动天平
electrobath 电镀浴；电镀槽
electrobeam drilling 电子束钻井
electrobeam machining 电子束加工
electrobeam welding 电子束焊(接)
electrobiologic(al) test 生物电试验
electrobiology 电生物学
electrobrightening 电(解)抛光
electrobus 电动公共汽车
electrocaloric coefficient 电热系数
electrocaloric effect 电热效应
electrocalorimeter 电热(量)计；电量热器；电量热计
electrocapacitance altimeter 电容性测高计
electrocapillarity 电毛细管现象
electrocapillary 电毛细现象
electrocapillary adsorption 电毛细吸附
electrocapillary curve 电毛细电毛曲线
electrocapillary effect 电毛细效应
electrocapillary maximum 电毛细极大
electrocapillary phoresis 电毛细泳
electrocar 电动车
electrocarbon 电碳
electrocarbonization 电法炼焦
electrocarbothermic reduction assembly 电热碳还原装置
electrocardiogram 心电图
electrocardiography room 心电图室
electrocarriage 电动拖车
electrocast block 电熔铸铁
electrocast brick 电熔铸耐火砖
electrocasting 电铸法；电熔浇铸
electrocasting machine 电铸机
electrocast refractory 电熔铸耐火材料；电熔耐火材料；电炉熔铸耐火材料
electrocatalysis 电催化(作用)
electrocatalytic oxidation 电催化氧化法
electrocathode 电控阴极
electrocathodoluminescence 场控阴极射线发光
electrocauterization 电烙术
electrocautery 电烙术；电烙器；电灸
electrocautery apparatus 电烙设备
electrocleaning 电清洗
electrocement 电制水泥
electroceramics 电工陶瓷；电瓷
electrochemical 电化学的
electrochemical action 电化(学)作用
electrochemical activity 电化学活性
electrochemical activity coefficient 电化学活度系数
electrochemical actuator 电化学驱动器
electrochemical analysis 电化学分析；电化分析
electrochemical attack 电化学作用；电化侵蚀；电化腐蚀
electrochemical breakdown 电化学击穿
electrochemical capacitor 电解质电容器
electrochemical carbonometer 电化学定碳仪
electrochemical cell 电化学电池；电化还原电池；电化电池
electrochemical chlorine flux monitor 电化学氯量监测仪
electrochemical cleaning 电化学清洗
electrochemical coating 电化学涂层

electrochemical component 电化学分量
electrochemical constant 电化学常数;电化常数
electrochemical corrosion 电化学腐蚀
electrochemical corrosion machining 电解腐蚀加工
electrochemical current 电化电流
electrochemical deburring 电化学去毛刺
electrochemical deburring machine 电化学去毛刺机
electrochemical degradation 电化学降解
electrochemical dehydration 电化学脱水
electrochemical demineralization 电化学脱矿质作用
electrochemical deoxidization 电化学脱氧
electrochemical deposition 电化学沉积;电化学沉淀
electrochemical desalination 电化学除盐
electrochemical desalination process 电化学脱盐法;电化学除盐法
electrochemical desalting 电化学脱盐
electrochemical desalting method 电化学除盐法
electrochemical descaling process 电化学除铁皮法
electrochemical desorption 电化学脱附
electrochemical desulfluration 电化学脱硫(作用)
electrochemical detector 电化学检测器;电化检测器
electrochemical diffused-collector phototransistor 电化学扩散集电极光晶体管
electrochemical discharge machining 电解放电加工
electrochemical disinfection 电化学杀菌
electrochemical disintegration 电化分裂
electrochemical dispersion 电化学分散
electrochemical dissolved oxygen analyzer 电化学溶氧分析仪
electrochemical dissolved oxygen recorder 电化学溶氧记录仪
electrochemical effect 电化(学)效应
electrochemical energy storage 电化学能量储存
electrochemical engineering 电化学工程
electrochemical equation 电化学方程
electrochemical equilibrium 电化平衡
electrochemical equipment 电化设备
electrochemical equivalent 电化(学)当量
electrochemical etching 电抛光;电化蚀刻;电化浸蚀
electrochemical etching and replating process 电化腐蚀和电镀工艺
electrochemical finish 电化学抛光;电化学表面处理;电化学加工
electrochemical forming machine 电化学成形加工机床
electrochemical furnace 电化炉
electrochemical ga(u)ging 电化学测流法
electrochemical grinder 电化学磨床
electrochemical grinding 电解磨削
electrochemical hardening 电化学硬化;电化学加固
electrochemical heterogeneous catalytic process 电化学多相催化工艺
electrochemical incineration 电化学焚化
electrochemical inclinometer 电化学测斜仪
electrochemical induration 电化学固结
electrochemical industry 电化(学)工业
electrochemical injection 电动化学灌浆
electrochemical integrating device 电化积分装置
electrochemical ion exchange 电化学离子交换
electrochemical kinetics 电化学动力学
electrochemical laboratory 电化实验室
electrochemical load 电化负荷
electrochemical luminescence 电化学发光
electrochemical machine 电化学加工机床
electrochemical machining 电解加工;电化学加工
electrochemical measurement 电化学测量
electrochemical method 电化学方法
electrochemical milling 电化学研磨;电化铣削
electrochemical NO_x meter 电化学氮氧化物测定仪
electrochemical oxidation 电化学氧化
electrochemical oxidation index 电化学氧化指数
electrochemical oxidation process 电化学氧化工艺
electrochemical oxidized carbon fiber 电化学氧化碳纤维
electrochemical oxygen demand 电化学需氧量
electrochemical parameter 电化学参数
electrochemical perforator 电化学射孔器

electrochemical phenomenon 电化学现象
electrochemical photolysis 电化学光解
electrochemical plating 电(化学)镀法;电化镀层
electrochemical polish(ing) 电化学抛光
electrochemical polymerization 电化学聚合
electrochemical potential 电化(学)势;电化学电位;电化电(动)势
electrochemical potential energy 电化学势能
electrochemical power source 电化学动力源
electrochemical probe 电化学探针
electrochemical probe method 电化学探头法
electrochemical process 电化(学)过程;电化(学)法;电化学处理
electrochemical processing 电化学加工
electrochemical property 电化学特性
electrochemical protection 电化学保护
electrochemical reaction 电化学反应
electrochemical reactor 电化学反应器
electrochemical recording 电化(学)记录
electrochemical reduction 电化学还原(法);电化还原
electrochemical reduction cell 电化还原电池
electrochemical rotating disc 电化学转盘
electrochemical self-potential 电化学自然电位
electrochemical sensitive recording paper 电化学感应记录纸
electrochemical sensor 电化学传感器
electrochemical separation method 电化学分离法
electrochemical separator 电化学分离仪
electrochemical series 电化学序列;电化序
electrochemical shaping 电化学成形
Electrochemical Society 电化学学会
electrochemical solidification 电化学固结
electrochemical solidification of clay 黏土的电化学固结
electrochemical stabilization 电化学稳定;电化学加固
electrochemical surface grinding machine 电化学平面磨床
electrochemical technique 电化学技术
electrochemical telephotographic paper 电化学传真纸
electrochemical theory 电化学理论
electrochemical thermodynamics 电化学热力学
electrochemical trace analyzer 电化学痕迹分析仪
electrochemical transducer 电化学换能器
electrochemical treater 电化学脱水器
electrochemical treatment 电化学处理
electrochemical valve 电解阀
electrochemical vapo(u)r deposition 电化学气相沉积
electrochemical wastewater 电化学废水
electrochemical wastewater treatment 电化学废水处理
electrochemiluminescence 电化学发光
electrochemistry 电化学
electrochondria 电粒
electrochromatic glass 电致变色玻璃
electrochromatograph 电色谱
electrochromatographic(al) technique 电色谱技术
electrochromatography 电色谱法;电色层分离法
electrochromic 电致变色的
electrochromic behavio(u)r 电变色行为
electrochromic coating 电化着色层
electrochromic device 电变色装置
electrochromic display 电致变色显示技术
electrochromic display device 电致变色显示器
electrochromic property 电变色性能
electrochromics 电变色
electrochromism 电变色
electrochromism spectrometer 电致变色光谱仪
electrochronograph 电子计时仪;电时计;电动精密记时计
electrocision 电切术
electrocladding refractory metal 电镀耐火金属
electroclock 电钟
electrocoagulation 电凝法
electrocoagulation-flo(a)tation process 电混凝气浮法
electrocoagulation process 电混凝法
electrocoagulation with titanium-iron double anodes 钛铁双阳极电混凝法
electrocoating 电泳涂漆;电泳涂装;电涂
electrocoating bath 电泳涂装槽

electrocolo(u)rimeter 电色度计
electrocolo(u)rimetry 电色度学
electrocompressor set 电动压缩机组
electroconductance detector 电导检测器
electroconductive adhesive 导电胶黏剂
electroconductive coating 导电涂层
electroconductive fiber 导电纤维
electroconductive film 导电膜
electroconductive gas analyzer 电导式气体分析仪
electroconductive glass 导电玻璃
electroconductive glaze 电导釉;导电釉
electroconductive paint 导电涂层;导电漆
electroconductive plastics 导电塑料
electroconductive polymer 导电聚合物
electroconductive printing ink 导电印刷油墨
electroconductive rubber 导电橡胶
electroconductive salinometer 电导盐度计
electroconductive synthetic fibre 导电合成纤维
electroconductivity 电导性;导电性
electroconductivity method 电导法
electroconnecting pressure ga(u)ge 电接点压力表
electroconstant 电化常数
electrocontact hardening 电接触淬火
electrocontact type surface roughness tester 电接触式表面粗糙度试验仪
electrocontrol converter 电液转换器
electrocopper 电解铜
electrocopper foil 电镀铜箔
electrocopper glass method 电解玻璃法
electrocopper glazing 防火玻璃窗;电解铜釉;铜条镶嵌玻璃
electrocoppering 电镀铜(法)
electrocoring 电钻取心
electrocorrosion 电(化)腐蚀
electrocorundum 电熔刚玉
electrocoupled oscillator 电子耦合振荡器
electrocoupling 电磁联轴节
electrocracking 电弧裂解
electrocrawler loader 电动履带装载机
electrocrystallization 电结晶
electroculture 电气栽培;电培养
electrocure process 电子束固化法
electrocuring 电子束固化法
electrocution 触电死(亡)
electrocutting machine 电动剪裁机
electrocyc 电环化反应
electrodata center 电数据处理中心
electrodata machine 电动数据处理机
electrode 焊条;电极;电焊条
electrode adjuster 电极调整器
electrode adjusting gear 电极调整装置
electrode admittance 电极导纳
electrode angle 焊条夹角
electrode arm 电极握臂;电板
electrode arrangement 电极系;电极排列
electrode array 电极系;电极排列
electrode assembly 电极装置
electrode bar 电极棒
electrode bias 电极偏压
electrode board 电极夹具
electrode boilder 电热辐射炉
electrode boiler 电极(加热)锅炉;电热锅炉;电极(式)热水器
electrode bushing 电极套;电极夹(头)
electrode cable detector 电缆故障探测器;电缆故障检测器
electrodecantation 电泺析;电倾析
electrode capacitance 电极电容
electrode capacitor 电极电容器
electrode carbon 电极炭
electrode carrier 焊条夹
electrode-carrying superstructure 电极支架
electrode cement 电极胶合剂
electrode chamber 电极室
electrode characteristic 电极特性
electrode circle 电极圆
electrode clamp 电极夹
electrode coalescing area 电极聚结面积
electrode coating 焊条药皮;焊条涂层
electrode coating ingradient 电焊条药皮成分
electrode coating material 电焊条药皮材料
electrode coke 电极焦
electrode collar 电极圈

electrode composition 电解分解(作用)
electrode compound 电极涂料
electrode conductance 电极电导
electrode configuration 电极排列
electrode connection 电极接头
electrode contact 电极接头
electrode contact surface 电极接触面
electrode container 电极筒
electrode contamination 电去污;电净化
electrode control 电极调节
electrode controller 电极控制器
electrode cooling ring 电极冷却圈;电极冷却环
electrode cooling tube 电极水冷管
electrode core wire 电焊条芯材
electrode couple 电极对
electrode current 电极电流
electrode dark current 电极暗电流
electrode density 电极密度
electrode discharge 电极放电
electrode dissipation 电极耗散
electrode distance 极距(指电解极距);电极距
electrode dresser 电极修整器
electrode drop 电极压降;电极电压降
electrode dryer 焊条保温筒
electrode economizer 电极节省器
electrode effect 电极效应
electrode equilibrate 电极平衡
electrode equilibrium potential 电极均衡势
electrode etching 电极蚀刻
electrode extension 焊丝伸出长度;电极外伸长度;电极伸出长度
electrode extrusion press 焊条压涂机
electrode feed 送丝
electrode feeding machine 焊条送进机
electrode flow injection analysis method 电极流动注射分析法
electrode for arc welding 电(弧)焊条
electrode force 电极力
electrode for contact welding 接触焊接电焊条
electrode for electrodialysis 电渗析电极
electrode-forming mechanism 电极成型机械
electrode for underwater cutting 水下割条
electrode for vertical down welding 向下立焊条
electrode furnace 电极炉
electrode gap 电极间隙;电极间距(离)
electrode gap control 电极间隙控制
electrode geometry 电极几何形状
electrode glass 电极玻璃
electrode grid 电极栅
electrode guard 电极防护罩
electrode-hearth arc-furnace 底极电弧炉
electrode holder 焊条(夹)钳;焊钳;电极夹;电焊枪;电焊钳;焊夹
electrode holding and slipping mechanism 电极悬放机组
electrode hole 电极孔
electrode impedance 电极阻抗
electrode insulation 电极间绝缘
electrodeionization 电极电离作用
electrode jaw 电极夹持器
electrode jib 电极支架
electrodeless conductivity system 无电极电导系统
electrodeless discharge 无电极放电
electrodeless discharge lamp 无(电)极放电灯
electrodeless plating 无电极电镀
electrodeless ultraviolet 无电极紫外线
electrode life 电极寿命
electrode-lift regulating device 电极高低调整设施
electrode manipulation 电极处理
electrode material 电极材料
electrode melting rate 焊条熔化率
electrode metal 电极合金
electrode mixer 电极粉料混合器
electrode negative 正接
electrodense 电子致密的
electrode of vertical down welding 立向下焊条
electrode of zeroth kind 第零类电极
electrode oven 电极炉
electrode pad 电极板
electrode paste 电极糊
electrode pickup 电极黏损
electrode plating 电极镀层作用
electrode polarization 电极极化
electrodeposit 电极淀积;电解沉淀;电附着;电解沉积

electrodeposited coating 电镀层
electrodeposited copper 电积铜
electrodeposited nickel 电积镍
electrodepositing 电解沉积
electrodeposition 电泳涂漆(法);电解沉淀;电极沉积;电镀(层);电淀积;电沉着;电积(法);电解沉积
electrodeposition analysis 电沉积分析
electrodeposition cell 电极沉积槽
electrodeposition coating 电沉积涂装;电积涂层
electrodeposition enamelling 电沉积涂搪
electrodeposition flash 电沉积被膜
electrode positioning mechanism 电极升降机
electrodepositor 电镀器;电沉积器
electrode potential 电极电位;电极电势
electrode potential log 电极电位测井
electrode potential log curve 电极电位测井曲线
electrode pressure 电极压力
electrode prong 电极把手架
electrode radiator 电极辐射片
electrode rapping 电极振打
electrode reactance 电极电抗
electrode reaction 电极反应
electrode regulator 电极调节装置;电极调节器
electrode resistance 电极电阻
electrode retainer 电极护管
electrode retaining collar 电极环
electrode rod 电焊条
electrode salt-bath electric(al)furnace 电极盐浴电炉
electrode separation 电极间距离
electrode setback 电极内缩长度
electrodes for spark machining 火花加工焊条
electrode shell 电极筒
electrodesiccation 电干燥法
electrode size 焊芯直径
electrode skid 电极滑移
electrode slip 电极滑移
electrode socket 焊条夹持器;焊条插座
electrode solution 电极溶液
electrodesorption 电脱附
electrode spacing 电极距;电极间距(离)
electrode spread 电极排列
electrode sticking 电极黏着
electrode stroke 电极行程
electrode support 电极支柱
electrode system 电极系
electrode tarrer 静电除焦油器
electrode tip 焊条端部;电极头;电极尖端;电极端部
electrode tip holder 电极座
electrode travel 电极行程
electrode type liquid-level meter 电极式液位计
electrode type salinometer 电极式含盐量测定计
electrode type salt-bath resistance furnace 电极式盐浴电阻炉
electrode vessel 电极室
electrode voltage 电极电位
electrode water heater 电极式热水器
electrode weight 放电极重锤
electrode welding 电弧焊
electrode wire 焊条钢丝
electrode wire reel 焊丝盘
electrode with cast core 铸芯焊条
electrode wrench 电极扳手
electrodiagnosis 电反应诊断法
electrodialyse 电透析
electrodialyser 电渗析器
electrodialysis 电透析;电渗析;电分解
electrodialysis apparatus 电渗析仪器
electrodialysis cell 电渗析槽
electrodialysis demineralizer 电渗析脱矿器
electrodialysis desalination 电渗析脱盐(法)
electrodialysis equipment 电渗析设备
electrodialysis method 电渗析法
electrodialysis plant 电渗析设备
electrodialysis process 电渗析法
electrodialysis reversal 反向电渗析
electrodialysis reversal process 逆电渗析工艺
electrodialysis set 电渗析器
electrodialysis stack 电渗析膜组
electrodialysis wastewater treatment 电渗析法废水处理
electrodialysis watering technology 电渗析脱水技术
electrodialytic cell 电渗析室
electrodialytic demineralization 电渗析除盐
electrodialytic membrance 电渗析膜
electrodialytic process 电渗析过程
electrodialytic treatment 电渗析处理
electrodialyzer 电渗析器
electrodiffusion 电扩散
electrodip process 电浸渍法;电沉积法
electrodischarge machine 电子放电机
electrodischarge machining 放电加工
electrodisintegration 电子核蜕变;电致蜕变
electrodispersion 电分散作用;电分解作用
electrodissociation 电离解(作用);电离(作用);电解作用
electrodissociationary vacuum-ga(u)ge 电离真空计
electrodissolution 电解溶解
electrodissolvent 电解溶剂
electrodissolver 电解溶解器
electrodrainage 电渗排水
electrodressing 电选矿
electrodrill 电钻(具);电动钻具
electroduster 静电喷粉器
electrodusting 静电喷粉
electrodynamic(al) 电动力学的;电动的
electrodynamic(al) actuator 电动执行器
electrodynamic(al) ammeter 电动式安培计
electrodynamic(al) balance 电流秤;电动天平
electrodynamic(al) brake 电气动力制动
electrodynamic(al) bridge 电动式电桥
electrodynamic(al) capacity 自感系数
electrodynamic(al) detector 电动检波器
electrodynamic(al) drift 电动力漂移
electrodynamic(al) force 电动势
electrodynamic(al) galvanometer 电动式电流计
electrodynamic(al) geophone 电动检波器
electrodynamic(al) instrument 力测仪表;电动式仪表;电测力计
electrodynamic(al) levitation 电动磁浮
electrodynamic(al) loudspeaker 电动式扬声器
electrodynamic(al) machine 电动式电机
electrodynamic(al) meter 电功率计;电动式仪表
electrodynamic(al) microphone 动圈式话筒;电动送话器;电动式话筒
electrodynamic(al) multiplier 电动式乘法器
electrodynamic(al) multiwinding brake 电动复绕组制动器
electrodynamic(al) pickup 电动式拾音器;电动拾声器
electrodynamic(al) potential 电动势
electrodynamic(al) power factor meter 电动式功率因素计
electrodynamic(al) pressure ga(u)ge 电测压力计
electrodynamic(al) program(me) controller 电动程序控制器
electrodynamic(al) receiver 电动式受话器;电动式耳机
electrodynamic(al) relay 电动式继电器
electrodynamic(al) retarder 电动缓行器
electrodynamic(al) seismograph 电动式地震仪
electrodynamic(al) seismometer 电动式地震计
electrodynamic(al) separator 电动式分离器
electrodynamic(al) shaker 电动振子;电动振动器
electrodynamic(al) steam engine 电力蒸汽机
electrodynamic(al) suspension 电动悬挂系统
electrodynamic(al) telephone 电动式受话器
electrodynamic(al) timer 电动式定时器
electrodynamic(al) transducer 电动转换器;电动换能器
electrodynamic(al) type high speed impedance relay 电动式高速阻抗继电器
electrodynamic(al) type meter 电动式仪表
electrodynamic(al) type power factor meter 电动功率因素计
electrodynamic(al) type relay 电动式继电器
electrodynamic(al) type reverse-current relay 电动式逆流继电器
electrodynamic(al) type seismometer 电动式地震仪;电动式地震检波器
electrodynamic(al) type single-phase power relay 电动式单相功率继电器
electrodynamic(al) type three phase power relay 电动式三相功率继电器

electrodynamic(al) type three phase wattmeter 电动式三相瓦特计
electrodynamic(al) unit combination instrument 电动单元组合仪表
electrodynamic(al) vibration generator 电动振动发生器；电动式震动发生器
electrodynamic(al) vibration machine 电动振动器；电动式振动台；电动加振器
electrodynamic(al) vibrator 电动振动器
electrodynamic(al) voltmeter 电动式伏特计
electrodynamic(al) wattmeter 电动式瓦特计
electrodynamics 电动力学
electrodynamometer 力测电流计；电力测功计；电功率计；电功率表；电动测功计；电测力器；电测力计；双流作用计
electrodynamometer type impedance meter 电测力计式阻抗表
electrodynamometer type meter 电动式仪表
electrodynamometer type tachometer 电动式转速表
electroendosmosis 电渗；电内渗(现象)
electroengraving 电刻物；电刻术
electroequivalent 电化当量
electroerosion 电浸蚀
electroerosion machine 电腐蚀机
electroerosion machining 电蚀加工
electroerosion process 电腐蚀加工法
electroerosive machining 电子放电加工
electroetching 电刻；电解蚀刻(法)；电解浸蚀
electroexplosive 电起爆炸物；电控引爆器
electroextraction 电解提取(法)；电解萃取
electrofacies 测井相
electrofacies analysis 测井相分析
electrofacies zonation 测井相划分
electrofacing 电镀
electrofax 电子照相(机)；电子摄影机；电子摄影
electrofax paper 电子传真复印纸
electroFenton oxidation method 电芬顿氧化法
electrofilter 电滤(尘)器；电过滤器；电除尘器
electrofiltration 静电聚灰；静电降尘；静电分离；静电沉降；静电沉积；电滤
electrofiltration potential 过滤电位
electrofiltration potential field 过滤位场
electrofining 电精制
electrofishing 电捕鱼
electrofixer 电热蒸化机；电定形器
electroflash for microphotography 显微镜摄影用闪光灯
electroflo(a)tation 电动浮选法
electrofloat process 电浮法
electrofluid 电流体
electrofluid dynamics 电铃动力学
electrofluidized bed 电气硫化床
electroflux refining furnace 电渣精炼炉
electrofocussing 电聚焦
electroforge 电锻
electroformed mo(u)ld 电铸模；电化成型模
electroformed photomask 电镀形成的光掩模
electroformed sieve 电成型筛；电刻筛
electroforming 电铸(造)；电成型
electrofusion 电熔(化)
electrofusion type poly ethylene fitting 电熔聚乙烯管件
electrogalvanised steel-clad fire door 电镀钢包防火门
electrogalvanize 电镀锌
electrogalvanized steel wire 电镀锌钢丝
electrogalvanizing 电解镀锌(法)；电镀锌(法)
electrogas automatic signal 电气体自动信号机
electrogas dynamics 电气体动力学；气电动力学
electrogas enclosed welding 气电立焊
electrogas process 气体保护爆方法
electrogas welding 电气焊；气电焊；气体保护电弧焊
electrogen 光电分子
electrogeneous 发电的
electrogeniometer 电动测相角器
electrogeochemical method 地电化探法
electrogilding 电镀金
electrogoniometer 相位指示器
electrogram 电记录图；大气电势梯度变化图；大气电场自记曲线
electrogrammetry 电重量分析法
electrogranodising 电磷化处理
electrograph 电图；电刻器；电记录器

electrographic analysis 电图分析
electrographic image tube 电子照相管
electrographic paper 电图纸
electrographic pencil 电图铅笔
electrographic recording 电(子)图记录；示波记录
electrographite 人工石墨
electrographite brush (人造)石墨电刷
electrography 电谱法；电刻术；电记录术；电笔
electrogravimetric analysis 电重量分析
electrograving 电蚀剂
electrogyro locomotive 蓄能电机车
electrohardening 电化固结
electrohardening of clay 黏土电化固结
electrohardware 电光硬件
electroheat 电热
electroheated glass 电热玻璃
electroheat equipment 发热装置
electroheating standing-temperature cultivator 电热恒温培养箱
electroheating tensioning method 电热张拉法
electroheterocatalysis 电—多相催化
electrohoretic separation 电屑分离
electrohot regulator 电热调节器
electrohydaulic kneading compactor 电液搓揉压实机
electrohydraulic 电动液压的
electrohydraulic actuator 电动液压执行机构；电动液压促动器；水电传动装置
electrohydraulic actuator package 电动液压作动器组
electrohydraulic bar shears 电动液压式钢筋切断机
electrohydraulic brake 电力液压制动器
electrohydraulic breaker 电动液压轧石机；电动液压破碎机
electrohydraulic cabinet actuator 柜式电(动)液(压)传动装置；电动液压整体型调速控制器
electrohydraulic coil lifter 电动液压式带卷升降台
electrohydraulic control 电动液压控制
electrohydraulic controller 电液控制器；电动液压(式)控制器
electrohydraulic control system 电液调节系统；电动液压控制系统
electrohydraulic control unit 电动液压操纵装置
electrohydraulic control valve 电动液压控制阀
electrohydraulic converter 电液转换器
electrohydraulic crusher 电动液压轧石机；电动液压破碎机
electrohydraulic crushing 电火花水力破碎
electrohydraulic directional valve 电液动换向阀
electrohydraulic dredge(r) 电动液压挖泥船
electrohydraulic drilling 电水锤钻进
electrohydraulic elevator 电动液压升降机
electrohydraulic forming 电液成形；电水锤成型；水中放电成形法
electrohydraulic forming machine 电动液压成形机
electrohydraulic governor 电液调节器；电动液压调节装置；电动液压调节器
electrohydraulic grab 电动液压抓斗
electrohydraulic grab-bucket 电动液压抓斗
electrohydraulic hatch cover 电动液压舱口盖
electrohydraulic lift 电动液压举升器；电动液压电梯
electrohydraulic lifter 电动液压起重机
electrohydraulic notching press 电气液压凹模压床
electrohydraulic pilot system 电动液压操纵系统
electrohydraulic press 电动液压机
electrohydraulic proportional valve 电液比例阀
electrohydraulic regulator 电动液压调节器
electrohydraulic remote-operated valve 电动液压遥控操纵阀；电动液压遥控阀
electrohydraulic servo control 电液伺服控制
electrohydraulic servo control valve 电动液压伺服控制阀
electrohydraulic servo loop 电液伺服回路
electrohydraulic servomechanism 电动液压伺服机构
electrohydraulic servo motor 电动液压伺服电动机；电动液压伺服马达
electrohydraulic servo system 电液伺服系统
electrohydraulic servo testing equipment 电液伺服试验设备
electrohydraulic servo unit 电动液压伺服装置
electrohydraulic servo valve 电液伺服阀；电动液压伺服阀
electrohydraulic shaking table 电动液压振动台

electrohydraulic steering gear 电液式操舵装置
electrohydraulic stepping motor 电液步进马达
electrohydraulic telemotor 电动液压传动装置
electrohydraulic thrustor brake 电力液压推杆制动器
electrohydraulic transducer 电液转换器；电液换能器
electrohydraulic valve 电液控制阀；电动液压阀
electrohydraulic vibrator 电液压激振器
electrohydraulic winch 电动液压绞车
electrohydraulic windlass 电动液压起锚机
electrohydrodynamic(al) analog(ue) 水电比拟；水电模拟
electrohydrometallurgy 电湿法冶炼
electrohydrometer 电湿度计
electrohypress greaser 电动高压润滑装置
electro immersion hot water heating 电沉浸热水器
electroinduction 电感应
electroinductive repulsion 电感应拒斥
electroinjection 电化学加固
electroinsulating property 电绝缘性能
electroinsulating varnish 绝缘电漆
electrointerlocking machine 电机联锁机
electroionization 电离(作用)
electroionization gas 电离气体
electroionization gas laser 电致电离气体激光器
electroionization laser 电致电离激光器
electroionization process 电离合成过程
electroiron 电解铁
electrojet 电离层中的急流
electrokinetic 动电的
electrokinetic component 动电势分量
electrokinetic coupling 电动耦合
electrokinetic effect 动电(学)效应；电动效应
electrokinetic filtration analysis 动电超滤分析
electrokinetic injection 电动注浆
electrokinetic momentum 电动量
electrokinetic phenomenon 电动(力学)现象
electrokinetic potential 动电势；动电(电)位；电动电位
electrokinetics 动电学；电动力学
electrokinetic transducer 电动传感器
electrokinetograph 动电计；电动流速仪；电动测速仪；电动测流计
electrokymogram 电记波照片
electrokymograph 电记波照相仪；电记波(照相)器；电动转筒记录仪
electrokymography 电记波照相术；电记波法
electrolapping process 电解研磨工艺
electroleaching 电沥滤；电浸出
electrolemma 电膜
electroless deposit 化学镀层；化学沉积层；无电沉积
electroless deposition 无电沉积
electroless nickel plated steel 化学镀镍钢；无电镀镍钢
electroless plating 化学(浸)镀；无电(敷)镀
electrolevel 电子水平仪
electrolevel meter 电动测深式料位机
electrolier 枝形吊灯架；枝形电灯架；电烛台；装潢式吊灯架；集合灯
electrolier switch 装潢式开关；装潢灯闪烁器
electrolifting magnet 电磁起重机
electrolines 电力线(路)
electrolinking 电联动
electrolithotripsia 电碎石术
electrolithotrity 电碎石术
electrolock 电锁器；电气锁(闭)
electrolog 电测井
electrolog curve 电测井曲线
electroluminance 场致发光
electroluminescence 场致发光；阴极射线发光；电致发光
electroluminescence junction 场致发光结
electroluminescent 电致荧光
electroluminescent cell 场致发光元件；场致发光盒；场致发光单元；场致发光板；电致发光元件
electroluminescent colo(u)r plate 场致发光彩色板
electroluminescent counting element 场致发光计数元件
electroluminescent diode 场致发光二极管；电荧光二极管；电发光二极管
electroluminescent display 场致发光显示器；电致发光显示器

electroluminescent display filter 场致发光显示过滤器
electroluminescent image display 场致发光显像
electroluminescent lamp 场致发光灯;场致发光板;发光灯
electroluminescent laser 场致发光激光器
electroluminescent material 场致发光材料;电荧光材料
electroluminescent panel 场致发光板;电荧光板
electroluminescent phosphor 电致发光磷光体
electroluminescent-photoconductor circuit 场致发光光电导电路
electroluminescent screen 电发光屏
electroluminescent storage 场致发光存储器
electroluminor 电致发光磷光体
electrolyser 电解器
electrolysing cell 电解槽
electrolysis 电蚀;电解(作用);电分析
electrolysis analysis 电解分析
electrolysis and cell renewal section 电解和修槽工段
electrolysis anti-fouling of seawater 电解海水防污
electrolysis apparatus 电解分析器
electrolysis batch 电解批料
electrolysis bath 电解槽
electrolysis cell 铀电解槽;电解电池
electrolysis condenser 电解电容器
electrolysis copper 电解铜
electrolysis disinfection 电解杀菌
electrolysis field 电解场
electrolysis flo(a)tation 电解浮选(法)
electrolysis inhibiting mastic 电解抑制胶
electrolysis mitigation system 电解减轻设备
electrolysis of solutions 溶液电解
electrolysis of water 水电解
electrolysis process 电解法
electrolysis reduction 电解还原(作用)
electrolysis relay 电解式继电器
electrolysis solution 电解液
electrolysis tube 电解管
electrolysis unit 电解装置
electrolysis with air blowing 鼓风搅拌电解
electrolyte 电解质;电(溶)液
electrolyte activated battery 电解液激活蓄电池
electrolyte balance 电解质平衡
electrolyte circuit 电介质
electrolyte concentration 电解质浓度;电解液浓缩
electrolyte cooler 电解液冷却器
electrolyte dispensing trough 电解液配制槽
electrolyte filter 电解液过滤器
electrolyte hydrometer 电液比重计;电解液比重计
electrolyte impurity 电解质杂质
electrolyte ions 电解质离子
electrolyte osmosis 电解质渗透
electrolyte purification system 电解液净化系统
electrolyte servo valve 电解液伺服阀
electrolyte shell 电解渣壳
electrolyte solution 电解质溶液
electrolyte strength 电解质强度
electrolytic 电解质的;电解的
electrolytic acid cleaning 电解酸洗
electrolytic action 电解作用
electrolytic agent 电解剂
electrolytically deposited black 电解淀积变黑
electrolytically polished slice 电解抛光片
electrolytically zinc-coated sheet 电解锌涂层的钢板
electrolytical refined copper 电解精炼铜
electrolytic aluminium 电解铝
electrolytic aluminium capacitor production line 电解铝电容器生产线
electrolytic analyzer 电解氧气分析仪
electrolytic analysis 电解分析
electrolytic anode 电解阳极
electrolytic apparatus 电解装置;电解仪器
electrolytic arrester 铝电池避雷器;电解放电器;电解避雷器
electrolytic assay 电解检验
electrolytic bath 电解池;电解槽
electrolytic battery charger 电解液蓄电池充电器
electrolytic battery regulator 电解式充电电流调节器
electrolytic beryllium 电解铍
electrolytic bleaching 电解泡白

electrolytic brightening 电解抛光
electrolytic cadmium 电解镉
electrolytic capacitor 电解电容器
electrolytic capacitor paper 电解电容器纸
electrolytic casting 电解注浆(法)
electrolytic caustic soda 电解烧碱
electrolytic cell 电解槽;电解(电)池
electrolytic cell assembly 电解槽装置
electrolytic cell lining 电解槽衬里
electrolytic charger 电解充电器
electrolytic chlorine 电解氯
electrolytic chromium coated steel 电解铬涂层的钢材
electrolytic chromium oxide coateil steel 电解氧化铬涂层的钢材
electrolytic cleaning 电解去垢;电解清洗;电化学清洗
electrolytic coagulation 电解凝聚法;电解结聚法;电解混凝
electrolytic cobalt 电解钴
electrolytic colo(u)ring 电解质着色法
electrolytic condensation method 电解凝聚法
electrolytic condenser 电解(质)电容器
electrolytic conductance 电解电导
electrolytic conduction 电解导电
electrolytic conductivity 电解(液)电导率
electrolytic conductor 电解质导体
electrolytic content 电解反应浓度
electrolytic copper 电解铜
electrolytic copper foil 电解铜箔
electrolytic copper refining 电解铜精炼
electrolytic copper sheet 电解紫铜片
electrolytic copper wire 电解紫铜丝
electrolytic copper wire rods 电解铜盘条
electrolytic corrosion 电蚀;电流腐蚀;电(解)腐蚀;电化腐蚀
electrolytic corrosion test 电(解)腐蚀试验
electrolytic coupled dipolymer 电解偶联二聚物
electrolytic crystal growth 电解晶体生长
electrolytic cyaniding 电解氰化法
electrolytic deburring 电解去毛刺
electrolytic degreasing 电解脱脂
electrolytic degreasing bath 电解脱脂槽
electrolytic deposition 电解淀积;电(解)沉积
electrolytic detector 电解检波器
electrolytic development 电解显影
electrolytic dissociation 电离解(作用);电离(作用);电解质离解;电解解离
electrolytic dissolciation 电解离解
electrolytic dissolution 电解
electrolytic efficiency 电解效率
electrolytic etching 电解蚀刻(法);电解侵蚀;电解腐蚀
electrolytic exchange resin 电解质交换树脂
electrolytic flo(a)tation 电解浮选(法)
electrolytic flo(a)tation unit 电解浮选装置
electrolytic flocculation 电解絮凝
electrolytic formation machining 电解成形加工
electrolytic furnace 电解炉
electrolytic galvanizing 电解镀锌(法)
electrolytic generator 电解用发电机
electrolytic grinder 电解磨床
electrolytic grinding 电解磨削
electrolytic grinding machine 电解磨床
electrolytic grinding wheel 电解磨(削砂)轮
electrolytic growth 电解生长
electrolytic hygrometer 电解湿度计
electrolytic industry 电解工业
electrolytic instrumentectric etcher 电解式仪表
electrolytic interrupter 电解断续器
electrolytic ion 电解离子
electrolytic ionization 电解游离;电解电离
electrolytic iron 电解铁
electrolytic iron powder 电解铁粉
electrolytic lead 电解铅
electrolytic lightening arrester 电解式避雷器
electrolytic loading 电解荷载
electrolytic logging 电解测井
electrolytic machine tool 电解加工机床
electrolytic machining 电解加工
electrolytic machining set 电解加工机
electrolytic magnesium 电解镁
electrolytic manganese 电解锰
electrolytic manganese dioxide 电解二氧化锰

electrolytic manganese metal 电解金属锰
electrolytic mercaptan process 电解液脱硫醇法
electrolytic metal coated steel 电解金属涂层的钢材
electrolytic meter 电解式仪表;电解库仑计
electrolytic method 电解法
electrolytic model 电解模型
electrolytic oxidant detector 电解氧化物检测器
electrolytic oxidation 电解氧化(法)
electrolytic oxidation characteristic 电解氧化特性
electrolytic oxidation method 电解氧化法
electrolytic oxide film 电解氧化膜
electrolytic oxide finish 电解氧化镀层
electrolytic oxide layer 电解产生的氧化层
electrolytic oxygen generator 电解氧气发生器
electrolytic paper 电解纸
electrolytic parting 电解分离
electrolytic photocell 电解光电管;电解光电池
electrolytic photocopying 电解摄影复印
electrolytic pickling 电解酸洗;电解酸蚀;电解浸洗
electrolytic plating 电镀
electrolytic polarization 电解极化
electrolytic polish(ing) 电解抛光
electrolytic potential 电解电势
electrolytic powder 电解粉末
electrolytic process 电解方法;电解影印;电解过程;电解法(处理)
electrolytic protection 电解法防护;阴极保护;电解防蚀法;电极防蚀
electrolytic purification 电解净化
electrolytic quenching 电解(液)淬火
electrolytic reaction 电解反应
electrolytic recording 电解记录
electrolytic recording paper 电解记录纸
electrolytic recovery 电解回收
electrolytic rectifier 电解质整流器;电解整流器
electrolytic reduction 电解还原(作用)
electrolytic refining 电(解)精炼
electrolytic refining bath 电解精炼电槽
electrolytic refining plant 电解精炼车间
electrolytic refining process 电解精炼法
electrolytic refining unit 电解精炼设备
electrolytic regeneration 电解再生
electrolytic resistance 电解质电阻
electrolytic rheostat 电解变阻器
electrolytic route 电解法
electrolytics 电解化学
electrolytic salvage of diamonds 电解法回收金刚石
electrolytic separation 电解分离
electrolytic separation method 电解分离法
electrolytic sewage treatment 电解污水处理
electrolytic soda process 电解制碱法
electrolytic solution 电解溶液
electrolytic solution pressure 电解液压力
electrolytic sorption hygrometer 电解吸附湿度表
electrolytic strip 测湿片
electrolytic surge arrester 电解式电涌放电器
electrolytic tank 电解槽
electrolytic tension 电解液张力
electrolytic time 电解时间
electrolytic tin 电解锡
electrolytic tinning 电解镀锡
electrolytic tinning machine 电镀锡机
electrolytic tin plate 电镀锡薄(钢)板;电镀马口铁
electrolytic tissue 电解用薄纸
electrolytic titration 电解滴定
electrolytic tool grinder 电解工具磨床
electrolytic tough pitch 电解韧铜
electrolytic transducer 电解换能器
electrolytic treatment 电解处理(法)
electrolytic treatment of wastewater 废水电解处理法;废水处理法电解
electrolytic vessel 电解容器
electrolytic voltameter 电解电量计
electrolytic zinc 电解锌
electrolytic zinc plating line 电镀锌作业线
electrolyzable 可以电解的
electrolyzation 电解
electrolyzation time 电解时间
electrolyze 用电解法处理
electrolyzed seawater 电解海水
electrolyzer 电解装置;电解器;电解池;电解槽
electrolyzing cell 电解槽
electromachining 电加工

electromagnet 电磁铁;电磁体
electromagnet armature 电磁铁(的)衔铁
electromagnet assembly 电磁铁组件
electromagnet bearing 电磁轴承
electromagnet core 电磁芯;电磁铁铁芯
electromagnet drive 电磁铁传动
electromagnetic 电磁的
electromagnetic absorption material 电磁吸收材料
electromagnetic accelerometer 电磁加速度计
electromagnetic acoustic(al) instrument 电磁声学仪器
electromagnetic ac relay 电磁交流继电器
electromagnetic adhesive 电磁胶黏剂
electromagnetic adjustable speed motor 滑差电机;电磁调速电动机
electromagnetic agitator 电磁搅拌器
electromagnetic air brake 电磁气闸;电磁空气制动器
electromagnetic air valve 电磁风阀
electromagnetic alternating current relay 电磁式交流电继电器
electromagnetic ammeter 电磁式安培计;电磁式安培表
electromagnetic amplifying lens 电磁放大透镜
electromagnetic anomaly 电磁异常
electromagnetic attraction 电磁引力;电磁吸引
electromagnetic balance 电磁秤
electromagnetic bearing 电磁轴承
electromagnetic bearing wear detector 电磁式轴承磨损探测器
electromagnetic belt pulley 电磁皮带滚筒
electromagnetic blow-out 电磁熄弧;电磁灭弧(器);电磁火花熄灭器
electromagnetic bonding 电磁黏合;电磁焊合
electromagnetic brake 电力制动器;电磁制动(器);电磁闸;电磁刹车
electromagnetic braking 电磁制动
electromagnetic calibration equipment 电磁标定装置
electromagnetic cam-group 电磁凸轮组
electromagnetic capacity 电磁容量
electromagnetic cascade shower 电磁级联簇射
electromagnetic casing-thickness logging tool 电磁套管测厚仪
electromagnetic cathode-ray tube 电磁式阴极射线管
electromagnetic chuck 电磁吸盘;电磁卡盘
electromagnetic clutch 电磁离合器
electromagnetic clutch for dredge pump 泥泵电磁离合器
electromagnetic coil 电磁线圈;电磁铁线圈;电磁感应圈
electromagnetic communication 电磁通信
electromagnetic compatibility 电磁学共存;电磁一致性;电磁适应性;电磁兼容性
electromagnetic compatibility margin 电磁兼容性容量
electromagnetic complex 电磁集合体
electromagnetic component 电磁分量
electromagnetic compression refrigerator 电磁压缩式冰箱
electromagnetic computerized tomography 电磁波层析成像
electromagnetic conductivity 电磁传导率
electromagnetic constant 光速;电磁常数
electromagnetic contactor 电磁开关;电磁接触器
electromagnetic control 电磁控制
electromagnetic control device 电磁控制装置
electromagnetic controlled automatic door 电磁场控制自动门
electromagnetic controller 电磁控制器
electromagnetic control panel 电磁控制盘
electromagnetic core 电磁铁芯
electromagnetic counter 电磁计数器;电磁计量器
electromagnetic countermeasure 电子对抗
electromagnetic coupler 电磁耦合器
electromagnetic coupling 电磁耦合;电磁联轴器;电磁联轴节;电磁接头
electromagnetic coupling effect 电磁耦合效应
electromagnetic coupling meter 电磁耦合测试器
electromagnetic covermeter 电磁层厚仪
electromagnetic crack detector 电磁裂痕探测器;磁力探伤仪
electromagnetic crane 电磁铁起重机

electromagnetic current 电磁流
electromagnetic current meter 电磁流速仪;电磁流速计;电磁海流计
electromagnetic cylinder 螺旋管线圈
electromagnetic damper 电磁阻尼器
electromagnetic damping 电磁阻尼
electromagnetic data 电磁资料
electromagnetic defect detection 电磁探伤
electromagnetic deflection 电磁偏转
electromagnetic deironing 电磁除铁
electromagnetic delay line 电磁延迟线
electromagnetic densitometer 电磁密度计
electromagnetic depth sounder 电磁测深仪
electromagnetic detection 电磁探测
electromagnetic detector 电磁勘探器;电磁检测器;电磁检波器
electromagnetic device 电磁器件
electromagnetic digitizer 电磁数字转换器
electromagnetic directional valve 电磁换向阀
electromagnetic discharge 电磁放电
electromagnetic disc single-track crane 电磁盘式单轨吊
electromagnetic distance measurement 电磁波测距
electromagnetic distance measurement trigonometric level(l)ing 电磁波测距三角高程测量
electromagnetic distance measuring equipment station 电磁波测距站
electromagnetic distance measuring instrument 电磁(波)测距仪
electromagnetic distance meter 电磁波测距仪
electromagnetic distance meter with variable frequency 变频式电磁波测距仪
electromagnetic disturbance 电磁骚扰
electromagnetic door holder 电磁门保持器;电磁定门器
electromagnetic drum 电磁铁滚筒
electromagnetic drum separator 电磁滚筒分离器;电磁滚筒除铁器;电磁鼓形分离器;电磁鼓形除铁器
electromagnetic earthquake 电磁地震
electromagnetic effect 电磁效应
electromagnetic electron lens 电磁电子透镜
electromagnetic electron microscope 电磁式电子显微镜
electromagnetic emission 电磁发射
electromagnetic energy 电磁能(量)
electromagnetic energy absorber coating 电磁能吸收涂层
electromagnetic energy accumulation type welder 电磁蓄能式熔接器;电磁储能式电焊机
electromagnetic environment 电磁环境
electromagnetic environment protection 电磁环境保护
electromagnetic excitor 电磁激振器
electromagnetic exploration 电磁探勘;电磁探查(法)
electromagnetic fathometer 电磁测深仪
electromagnetic feeder 电振给煤机;电磁进料机;电磁进料机;电磁加料器;电磁给料器;电磁给料机
electromagnetic field 电磁场
electromagnetic field equation 麦克斯韦场方程;电磁场方程
electromagnetic field intensity 电磁场强度
electromagnetic field of high frequency 高频电磁场
electromagnetic field tensor 电磁场张量
electromagnetic field theory 电磁场理论
electromagnetic filter 电磁过滤器
electromagnetic fishing tools 电磁打捞工具
electromagnetic flaw detector 电磁探伤器
electromagnetic flow meter 电磁流量计
electromagnetic flow rate meter 电磁流速计
electromagnetic fluid separator 电磁液体分离仪
electromagnetic focal coil 电磁聚焦线圈
electromagnetic focusing 电磁聚焦
electromagnetic force 电磁力;电磁场强度
electromagnetic force welder 电磁加压焊机
electromagnetic forming 电磁成形
electromagnetic frequency spectrum 电磁波频谱
electromagnetic fuel ga(u)ge 电磁式燃油平面指示器
electromagnetic gantry crane 电磁龙门吊
electromagnetic gear 电磁传动装置

electromagnetic ground detector 电磁式接地检测器
electromagnetic guideway 电磁导轨
electromagnetic gun method 电磁枪法
electromagnetic hammer 电磁锤
electromagnetic holder 电磁夹持器
electromagnetic horn 喇叭(形)天线
electromagnetic hydraulic brake 电磁液压制动器
electromagnetic hydraulic valve 电磁液压阀
electromagnetic hydromechanics 电磁流体力学
electromagnetic ignition 电磁点火
electromagnetic impulse counter 电磁脉冲计数器
electromagnetic inductance seismometer 电磁感应地震检波器;电磁感应地震计
electromagnetic induction 电磁感应
electromagnetic induction breaking 电磁感应破碎(法)
electromagnetic induction drying 电磁感应干燥
electromagnetic induction effect 电磁感应效应
electromagnetic inertia 电磁惯量
electromagnetic inspection 电磁检验
electromagnetic instrument 电磁式测试仪表;电磁测试仪表
electromagnetic insurance lock 电磁保险门锁
electromagnetic interaction 电磁相互作用
electromagnetic interference 电磁干扰
electromagnetic interference safety margin 电磁干扰安全系数
electromagnetic interference shielding 电磁干扰屏蔽
electromagnetic intermediate relay 电磁式中间继电器
electromagnetic interrupter 电磁断续器
electromagnetic irrigation 电磁辐射
electromagnetic isotope separation 电磁同位素分离
electromagnetic isotope separation unit 电磁同位素分离设备;电磁同位素分离器
electromagnetic isotope separator 电磁同位素分离器
electromagnetic joint 电磁接合
electromagnetic lens 电磁透镜
electromagnetic levitation 电磁悬浮
electromagnetic lifter 电磁起重机;电磁吊
electromagnetic line source 电磁线源
electromagnetic liquid separation method 电磁液体分离法
electromagnetic lock 电磁锁
electromagnetic log 电磁计程仪;电磁船速仪;电磁测程仪
electromagnetic logging 电磁测井
electromagnetic loudspeaker 电磁扬声器
electromagnetic mass 电磁质量
electromagnetic mass separator 电磁式质量分离器
electromagnetic measurement 电磁测量
electromagnetic meter[metre] 电磁式仪表;电磁流量计
electromagnetic method 电磁(方)法
electromagnetic method of isotope separation 电磁同位素分离法
electromagnetic micrometer 电磁测微计
electromagnetic microphone 电磁微音器;电磁受话器;电磁传声器
electromagnetic millivolt ammeter 电磁式毫伏安培计
electromagnetic mineral 电性矿物
electromagnetic mirror 电磁反射镜;电磁波反射镜
electromagnetic mixing 电磁搅拌
electromagnetic modular guidance 电磁悬浮模块导向
electromagnetic momentum 电磁动量
electromagnetic monitoring 电磁监测
electromagnetic mo(u)lding machine 电磁(式)造型机
electromagnetic multitube water bottle 电磁多筒采水器
electromagnetic noise 电磁噪声
electromagnetic non-thermal effect 电磁非热效应
electromagnetic ore separator 电磁选矿机
electromagnetic orienting tool 电磁定向工具
electromagnetic original field intensity chart 电磁波实测场强值曲线图
electromagnetic oscillation 电磁振荡
electromagnetic oscillation compressor 电磁振

动式压缩机
electromagnetic oscillation refrigerator 电磁振动式冰箱
electromagnetic oscillator 电磁(式)振荡器
electromagnetic oscillograph 电磁(式)示波器
electromagnetic oven 电磁炉
electromagnetic overflow valve 电磁溢流阀
electromagnetic pellet detector 电磁钻粒检测仪
electromagnetic penetration 电磁穿透
electromagnetic percussive welding 电磁冲击焊
electromagnetic photodetector 电磁光电检测器
electromagnetic pickup 电磁(式)拾音器;电磁(式)拾波器;电磁(式)拾音头;电磁(式)拾声器
electromagnetic platen 电磁平台
electromagnetic pollution 电磁污染
electromagnetic pollution source 电磁污染源
electromagnetic positioner 电磁式定位器
electromagnetic position measuring assembly 电磁位置测量装置
electromagnetic potential 电磁势
electromagnetic power press 电磁压力机;电磁成形压力机
electromagnetic pressure ga(u)ge 电磁压力计;电磁压力表
electromagnetic probe 电磁探针
electromagnetic property 电磁性能
electromagnetic propulsion 电磁推进;电磁推动
electromagnetic prospecting 电磁勘探法;电磁(法)勘探
electromagnetic prospecting process 电磁勘探法
electromagnetic protection suit 电磁防护服
electromagnetic pulse 电磁脉冲
electromagnetic pulse hardening 电磁脉冲防护能力
electromagnetic pulse jamming 电磁脉冲干扰
electromagnetic pump 电磁泵
electromagnetic pump wave crest soldering machine 电磁泵波峰焊机
electromagnetic puncher 电磁冲床
electromagnetic radiation 电磁辐射
electromagnetic radiation magnetometer 电磁辐射磁强计
electromagnetic radiation reflector 电磁波反射器
electromagnetic radiation standard 电磁辐射标准
electromagnetic radiation tester 电磁辐射测试器
electromagnetic radiation wave 电磁辐射波
electromagnetic rail brake 电磁钢轨制动器;轨道电磁制动
electromagnetic range 电磁灶
electromagnetic rays propagation log 电磁波传播测井
electromagnetic receiver 电磁(式)受话器
electromagnetic record 电磁记录
electromagnetic rectifier 电磁整流器
electromagnetic registration 电磁记数;电磁记录
electromagnetic relay 电磁式继电器
electromagnetic release 电磁脱机装置
electromagnetic relief valve 电磁溢流阀
electromagnetic reluctance seismometer 电磁阻地震计;电磁(磁阻)地震仪
electromagnetic remote control 电磁遥控
electromagnetic reproducer 电磁复制装置
electromagnetic resolver 电磁分解器
electromagnetic resonance 电磁共振
electromagnetic response of horizontal component 水平分量电磁响应值
electromagnetic response of vertical component 垂直分量电磁响应值
electromagnetic response PPM value 电磁响应PPM值
electromagnetic retarder 电磁缓行器
electromagnetic riveting machine 电磁铆接机
electromagnetic rocket 电磁火箭
electromagnetics 电磁学
electromagnetic safety mechanism 电磁安全机构
electromagnetic scattering 电磁散射
electromagnetic screen 电磁屏蔽装置
electromagnetic seed cleaner 电磁式种子清选机
electromagnetic seismograph 电磁(式)地震仪;电磁(式)地震计
electromagnetic seismometer 电磁地震检波器;电磁地震计
electromagnetic self-force 电磁自力
electromagnetic sensibility 电磁敏感性;电磁敏感度
electromagnetic sensor 电磁传感器
electromagnetic separation 电磁分选;电磁分离(法)
electromagnetic separation of isotope 电磁同位素分离
electromagnetic separation process 电磁分离法
electromagnetic separator 电磁分离器;电磁分离机;电磁除铁器;磁选机
electromagnetic shaker 电磁震动器
electromagnetic shielding 电磁屏蔽
electromagnetic shielding design 电磁屏蔽设计
electromagnetic shielding device 电磁屏蔽装置
electromagnetic shock wave 电磁冲击波
electromagnetic shoe brake 电磁靴式制动器
electromagnetic shut-off 电磁关闭
electromagnetic shutter 电磁快门
electromagnetic sieve 电磁筛
electromagnetic skin depth 电磁感应趋肤深度
electromagnetic solenoid braking 电磁螺线管制动
electromagnetic solenoid switch 电磁螺线管开关
electromagnetic spectrum 电磁(波)谱;电磁振荡频谱;电磁光谱
electromagnetic spectrum analysis 电磁谱分析
electromagnetic speed-adjustable motor 电磁调速电动机
electromagnetic speed regulation 电磁调速
electromagnetic starter 电磁起动器
electromagnetic stirrer 电磁搅拌器
electromagnetic stirring 电磁搅拌
electromagnetic stirring autoclave 电磁搅拌式高压反应器
electromagnetic stored-energy machine 电磁储能机
electromagnetic stored-energy welder 感应焊接机;电磁储能焊接机
electromagnetic stored-energy welding 电磁储能焊
electromagnetic strain ga(u)ge 电磁应变仪
electromagnetic subsurface probing 电磁地下测探
electromagnetic survey(ing) 电磁探测;电磁勘探法
electromagnetic susceptibility 电磁敏感性;电磁敏感度
electromagnetic suspension 电磁悬挂;电磁悬浮
electromagnetic switch 电磁开关
electromagnetic system 电磁式
electromagnetic system of units 电磁单位制
electromagnetic tamper 电动轨道夯实机
electromagnetic teleclinometer 遥控电磁测斜仪;电磁测斜仪
electromagnetic telecommunication 电磁波通信
electromagnetic telemetry 电磁遥测法
electromagnetic telephone 电磁式受话器
electromagnetic testing 电磁检测
electromagnetic theory 电磁理论
electromagnetic theory of light 光的电磁理论
electromagnetic thermal effect 电磁热效应
electromagnetic thickness ga(u)ge 电磁式测厚仪;电磁厚度计
electromagnetic thickness indicator 电磁厚度指示器
electromagnetic thickness meter 电磁测厚仪
electromagnetic thickness tool 电磁测厚仪
electromagnetic tilt angle method 电磁倾角法
electromagnetic time-delay line 电磁延时线
electromagnetic titration instrument 电磁滴定器
electromagnetic transducer 电磁换能器;电磁传感器
electromagnetic transfer device 电磁输送装置
electromagnetic transmission 电磁传输
electromagnetic treadle 电磁式接触器;电磁脚踏板
electromagnetic type disc friction clutch 电磁式圆盘摩擦离合器
electromagnetic type flow meter 电磁式流量计
electromagnetic type relay 继电表;电磁(式)继电器
electromagnetic type retarder 电磁式缓行器
electromagnetic type seismometer 电磁型地震计;电磁式地震检波器;电磁式地震计
electromagnetic type vibrometer 电磁式振动计
electromagnetic unit 电磁(系)单位
electromagnetic units[emu] 电磁单位制
electromagnetic uranium isotope enrichment method 电磁铀同位素浓缩法
electromagnetic valve 电磁阀
electromagnetic velocity meter 电磁流速仪;电磁速度计
electromagnetic velocity profiler 电磁速度廓线仪
electromagnetic vibrating 电磁振动
electromagnetic vibrating feeder 电磁振动加料机
electromagnetic vibrating screen 电磁振动筛
electromagnetic vibration exciter 电磁振动激励器;电磁激振器
electromagnetic vibration generator 电磁式振动台;电磁式振动起振器;电磁式振动发生器
electromagnetic vibration sand screen 电磁振动筛砂机
electromagnetic vibrator 电磁振动器
electromagnetic vibrator feeder 电磁振动喂料机
electromagnetic vibrator knock-out machine 电磁振动落砂机
electromagnetic visco(si)meter 电磁黏度计
electromagnetic water thermometer 电磁式水温表
electromagnetic water treatment 电磁水处理
electromagnetic wave 电磁波
electromagnetic wave filter 电磁波滤波器
electromagnetic wave gyrator 电磁波旋转器
electromagnetic wave phase shift 电磁波相移
electromagnetic wave pollution monitoring 电磁波污染监测
electromagnetic wave probing in borehole 井中电磁波法
electromagnetic wave propagation 电磁波传播
electromagnetic wave propagation log 电磁波传播测井
electromagnetic wave propagation log curve 电磁波传播测井曲线
electromagnetic wave propagation logger 电磁波传播测井仪
electromagnetic wave scattering 电磁波散射
electromagnetic wave shielding coating 电磁波屏蔽涂料
electromagnetic wave shielding door 电磁波屏蔽门
electromagnetic wave spectral band 电磁波谱段
electromagnetic wave spectrum 电磁波谱
electromagnetic well logging 电磁测井
electromagnetic wheel brake 电磁轮闸;轨道电磁式车轮制动
electromagnetic window 电磁窗
electromagnetic wire 电磁线
electromagnetism 电学机械硬件;电磁(学)
electromagnetism dynamic(al) triaxial test system 电磁振动三轴试验装置;电磁振动三轴试验系统
electromagnetization 电磁激励;电磁化
electromagnetometer survey 电磁辐射计测量
electromagnetostatic field 静电磁场
electromagnet screw press 电磁螺旋压力机
electromagnet type detector 电磁式探测器
electromagnet winding 电磁铁绕组
electromalux 玛鲁克丝管
electromanometer 电子液压计;电子压强计
electromanometer transducer 电子压强计换能器
electromatic 电气自动方式;电气自动的;电控自动方式
electromatic drive 电动式自动换档;电磁式自动换挡
electromechanical 电动机械的
electromechanical accelerometer 机电式加速度计
electromechanical actuator 电子机械传动器
electromechanical alarm system 电子机械报警系统
electromechanical amplifier 机电放大器
electromechanical analogy 机电模拟;电机械模拟
electromechanical appliance 机电式器具;机电式调整器
electromechanical arrangement 电力排列器
electromechanical assembly 机电装置
electromechanical brake 机电制动器;机电闸;电闸
electromechanical cab signal(l)ing 接触式机车信号
electromechanical cab signal(l)ing unit 接触式机车信号设备
electromechanical check 机电检查
electromechanical chopper 机电换流器
electromechanical computer 机电式计算机
electromechanical control 电动机械控制
electromechanical counter 机电计数器;机电计量器

electromechanical coupling coefficient 机电偶合系数
electromechanical coupling factor 机电偶合系数
electromechanical coupling system 机电耦合系统
electromechanical data processing 机电式数据处理
electromechanical device 机电装置;电气机械装置
electromechanical dialer 机电拨号器
electromechanical digit 机电数字化器
electromechanical drive 机电传动装置
electromechanical efficiency 机电效率
electromechanical equipment 电力机械设备
electromechanical ga(u)ge 电动机械式压力计
electromechanical hammer 电动机械锤
electromechanical hand time release 机电式人工限时解锁器
electromechanical integrator 电子机械积分仪;电动机械积分仪
electromechanical interlocking 电力机械连锁装置;电机集中联锁
electromechanical interlocking machine 电力机械联动设备;电机联锁机
electromechanical inversor 机电控制器
electromechanical logging cable 机电测井电缆
electromechanically operated fingers 电力机械操作的手爪
electromechanical machining 机电加工
electromechanical oscillograph 机电示波器
electromechanical pick-up 机电传感器;机电(式)拾音器;机电(式)拾波器;机电(式)传感器
electromechanical plotter 机电式绘图机;电子机械绘图机
electromechanical potentiometer 机电电位计;电机电势计
electromechanical power supply 机电电源
electromechanical printer 机电印刷机;电子印刷机
electromechanical property 机电性能
electromechanical record(ing) 机电记录
electromechanical regulator 机电调节器
electromechanical relay 机电继电器
electromechanical robot 电气机械传动机器人
electromechanical scale 电子机械秤
electromechanical scanner 机电扫描装置
electromechanical screw time release 机电螺旋式限时解锁器
electromechanical selector switch 机电选择机键
electromechanical servo 机电伺服机构
electromechanical slot 机电选别器
electromechanical spindle slot 圆盘型机电选别器
electromechanical storage 机电式存储器
electromechanical transducer 机电转换器;机电换能器;机电传感器
electromechanical transformation 机电转换
electromechanical transmission 电力机械传动
electromechanical treadle 机电式接触器;机电式脚踏板
electromechanical vibrator 电动激振器
electromechanical working 电机械加工
electromechanical workshop 机电车间
electromechanics 机电学;电机学;电动机械学
electromechanic sealing device 机电定标装置
electromechanization 电动机械化
electromedia filter 电滤料池
electromelioration 电力土壤改良
electromelting corundum 电熔刚玉
electromembrane process 电隔膜法
electromembrane technology 电隔膜技术
electromerism 气体电离过程
electrometal furnace 电弧熔化炉
electrometal furnace process 电炉炼铁法
electrometallization 电喷镀金属
electrometallurgical method 电冶法
electrometallurgy 电冶金(学);电法冶金
electrometal plant 电冶工厂
electrometer 量电表;静电计;静电测量器;电位计
electrometer amplifier 静电计放大器
electrometer tube 静电计(电子)管
electrometer valve 静电计(电子)管
electrometric 电测(量)的
electrometric analysis 量电分析
electrometric measurement 电(势)测定
electrometric method 电测法
electrometric observation 电子仪器观测
electrometrics 测电学
electrometric stripping analysis 电势溶出分析
electrometric titration 电势滴定;电化学滴定
electrometric titration outfit 电势滴定装置
electrometry 量电法;验电术;测电学;测电术
electromicrography 电子显微摄影
electromicrography plate 电子显微照相干版
electromigration 电迁移(法)
electromobile 电瓶车;电力自动车;电动(汽)车
electromodulation 电调制
electromotance 电动势
electromotion 电力起动;电动;通电
electromotion planimeter method 电动求积仪法
electromotive 电气机车;电力机车;电动力的;电动车
electromotive dental drill 电动牙钻机
electromotive difference potential 电动势
electromotive force 电动势;电动力
electromotive force for interface 界面电动势
electromotive force series 电动势序
electromotive force source 电动势源
electromotive induction 动电感应
electromotive series 置换序列;电位序;电化序;电动(次)序
electromotive spiral hoisting jack 电动螺旋千斤顶
electromotive steam whistle 电动汽笛
electromotive unit[emu] 电动势单位
electromotor 电动马达;电动机
electromotor grease 电动机润滑脂
electromo(u)lding 电铸(铸造)
electromultiphase catalysis 电一多相催化
Electron 埃雷克特龙镁合金
electron absorption 电子吸收
electron accelerating device 电子加速器
electron accelerating voltage 电子加速电压
electron accelerator 电子加速器
electron acceptor 电子(接)受体;电子接受剂
electron activity 电子活动
electron affinity 电子亲合性;电子亲合势;电子亲合力
electron affinity energy 电子亲合能
electron aided chemical vapo(u)r deposition 电子辅助化学气相沉积
Electron alloy 埃雷克特龙镁铝锌合金;镁合金
electron-amplifier 电子放大器
electronating agent 还原剂
electronation 加电子作用;增电子
electronation reaction 还原反应
electron atmosphere 电子云
electron-atom bremsstrahlung 电子原子韧致辐射
electron atomic mass 电子原子质量
electron attachment 电子附着
electron-attracting 吸电子的
electron avalanche 电子雪崩
electron avalanche amplification 电子雪崩倍增
electron avalanche breakdown 电子雪崩击穿
Electron AZF 埃雷克特龙 AZF 铸造铝镁合金
electron balance table 电子平衡表
electron ballast 电子镇流器
electron-beam 阴极射线;电子束
electron-beam-accessed memory 电子束存储器
electron-beam-activated micromachining 电子束显微机械加工
electron-beam addressed memory 电子束寻址存储器
electron beam alignment 电子束校准
electron-beam bonding 电子束压焊
electron-beam counter tube 计数开关管;电子束计数管
electron-beam cure process 电子束固化法
electron-beam curing 电子束固化;电子射线固化
electron-beam current 电子束电流
electron-beam cutter 等离子切割机
electron-beam cutting 等离子切割
electron-beam drafting machine 电子束绘图机
electron-beam drilling 电子束钻孔;电子束钻井
electron-beam drilling machine 电子束钻孔机;电子束钻床
electron-beam efficiency 电子束效率
electron-beam evapo(u)ration 电子束蒸发
electron-beam evapo(u)rator 电子束蒸发器
electron-beam excitation 电子束激发
electron-beam excited laser 电子束激发激光器
electron-beam floating zone furnace 电子束浮区提纯炉
electron-beam furnace 轰击炉;电子束熔炼炉;电子束炉
electron-beam generator 电子束发生器
electron-beam gun 电子束枪
electron-beam heater 电子束加热器
electron-beam image recorder 电子束图像记录器
electron-beam incidence angle 电子束入射角
electron-beam-induced conductivity 电子束感应导电率
electron-beam induced decolo(u)ration 电子束诱导脱色工艺
electron-beam injected transistor amplifier 电子束注入晶体管放大器
electron-beam laser 电子束激光器
electron-beam lithography 电子束平印术
electron-beam machining 电子束加工
electron-beam magnetometer 电子束磁强计;电子束磁力仪;电子式磁强计
electron-beam meltedingot 电子束熔锭
electron-beam melted material 电子束熔炼的材料
electron-beam melting 电子束熔化
electron-beam melting furnace 电子束熔(炼)炉
electron-beam melting purification 电子束精炼
electron-beam melting system 电子束熔炼设备
electron-beam memory 电子束存储器
electron-beam microprobe 电子束显微探头
electron-beam microprobe analysis 电子束探针微量分析
electron-beam parametric amplifier 电子束参量放大器;电子束参数
electron-beam probe 电子束探针;电子束探示器
electron-beam process 电子束熔炼法
electron-beam-pumped free-electron laser 电子束抽运自由电子激光器
electron-beam pumped laser 电子束抽运激光器
electron-beam pumping 电子束抽运
electron-beam recorder 电子束记录器
electron-beam recording 电子束录像;电子束记录(法)
electron-beam refining 电子束精炼
electron-beam regulator 电子束调节器
electron-beam scanner 电子束扫描器
electron-beam scanning 电子束扫描
electron-beam scanning system 电子束扫描系统
electron-beam sintering 电子束烧结
electron-beam spot 电子束光点
electron-beam spot size 电子束斑大小
electron-beam sustainer-pumped laser 电子束持续抽运激光器
electron-beam treatment 电子束加工
electron-beam tube 电子束管;电子衬线管
electron-beam vacuum furnace 电子束真空炉
electron-beam vapo(u)r deposition method 电子束蒸发淀积法
electron-beam welder 电子束焊(接)机
electron-beam welding 电子束焊(接)
electron-beam welding machine 电子束焊(接)机
electron-beam width of cross-section 电子束横截面宽度
electron binding energy 电子结合能
electron bombard 电子轰击
electron bombarded semiconductor amplifier 电子束半导体放大管
electron bombardment 电子轰击
electron bombardment engine 电子轰击式发动机
electron bombardment furnace 电子轰击炉
electron bombardment-induced conductivity 电子轰击感生电导
electron bombardment mass spectrometry 电子轰击质谱法
electron bombardment melting furnace 电子轰击熔炼炉
electron bombardment welding 真空电子束焊
electron bridge 电子桥
electron bunch 电子集束
electron buncher 电子集束器
electron bundle 电子束
electron calculating machine 电子计算器;电子计算机
electron camera 电子照相机;电子摄像机
electron capture 电子俘获
electron capture detection 电子俘获检测
electron capture detector 电子俘获探测器;电子俘获检测器
electron capture gas chromatography 电子俘获

气相色谱
electron capture gas-liquid chromatography 电子俘获气液色谱
electron capture ionization detector 电子捕获电离探测器
electron capture isotope 电子俘获同位素
electron-capture oxygen analyzer 电子捕获氧分析仪
electron capture process 电子俘获过程
electron channel 电子通道
electron channel effect 电子通道效应
electron channelling 电子隧道；电子沟道
electron channel map 电子通道图
electron charge 电荷；电子电荷
electron charge mass ratio 电子荷质比
electron chronometer 电子计时器
electron cloud 电子云
electron-collection counter 电子收集计数管
electron collector 电子集电极
electron collision 电子碰撞
electron colo(u)r scanner 电子分色机
electron commutation 电子整流
electron compound 电子化合物
electron computer 电子计算机
electron concentration 电子浓度
electron conduction 电子传导
electron conductivity 电子电导率
electron configuration 电子构型
electron contributing group 电子供给基团
electron-controlled profile milling 电子控制靠模铣
electron control semi-automatic winder 电子控制半自动绕线机
electron counter 电子计数管
electron coupled frequency converter 电子耦合变频器
electron coupled oscillator 电子耦合振荡器
electron coupler 电子耦合器
electron coupling 电子耦合
electron-coupling control 电子耦合调整
electron coupling oscillator 电子耦合振荡器
electron curing coating 电子射线固化涂料
electron current 电子流
electron cyclotron frequency 电子回旋加速频率
electron cyclotron heating 电子回旋加热
electron cyclotron oscillation 电子回旋振荡
electron cyclotron resonance 电子回旋共振
electron cyclotron resonance heating 电子回旋共振加热
electron cyclotron transverse-wave device 电子回旋横波器件
electron cyclotron wave 电子回旋波
electron defectoscope 电子探伤仪
electron density map 电子密度图
electron density projection 电子密度投射
electron detachment 电子异构体
electron detector 电子探测器
electron device 电子器件
electron diffraction 电子衍射
electron diffraction apparatus 电子衍射装置
electron diffraction camera 电子衍射照相机
electron diffraction contrast effect 电子衍射衬比效应
electron diffraction image 电子衍射图
electron diffractograph 电子衍射仪
electron digital counter 数字型电子读数机
electron discharge machining 电子放电加工
electron discharge machining machines tracer controlled wire 仿形控制的放电加工机床
electron discharge tube 电子放电管
electron displacement 电子位移
electron distribution curve 电子分布曲线
electron donor 电子施主；电子供体；电子给予体；电子给体
electron donor-acceptor 电子给受体
electron donor-acceptor complex 电子给受体复合物
electron dosimetry 电子剂量测定
electron double resonance 电子双共振
electron drift detector 电子漂移检测仪；电子漂移检测器
electron drift velocity 电子漂移速度
electron drift velocity detector 电子漂移速度检测器
electronegative 阴电性的；电负性的；负电性的

electronegative element 阴电性元素；负电性元素
electronegative gas 电负性气体；负电性气体
electronegative gel 阴电性凝胶；负电性凝胶
electronegative ion 阴电性离子
electronegative metal 负电性金属
electronegative potential 阴电势；低电位
electronegative radical 阴电性根
electronegative substituent 电负性取代基
electronegative valency effect 阴电原子价效应
electronegativity 阴电性；电负性；负电性
electron elastic scattering 电子弹性散射
electron-electron bremsstrahlung 电子电子韧致辐射
electron emission 电子发射
electron emission microscope 电子发射显微镜
electron emission regulator 电子发射稳定器
electron emission source 电子源
electron emissivity 电子发射率
electron emitter 电子发射体；电子发射极
electron energy level 电子能(量)级
electroneutrality 电中性
electroneutrality condition 电中性状况
electroneutrality principle 电中和原理
electron evapo(u)ration 电子蒸发
electron event 电子事件
electron exchange 电子交换
electron exchanger 电子(式)交换机
electron exchange resin 电子交换树脂
electron excitation 电子激发
electron feedback 电子反馈
electron feedback loop 电子反馈回路
electron flare 电子耀斑
electron flow 电子流动
electron flux 电子通量
electron flux densitometer 电子通量密度计
electron flux density 电子通量密度
electron flux density indicator 电子通量密度指示器
electron flux density meter 电子通量密度计
electron focusing 电子聚焦
electron fog 电子雾
electron fractograph 电镜断口照片
electronfractography 断口电子显微镜检验
electron gas 电子气
electron gear shifting 电子换挡
electron geometrical optics 电子几何光学
electron gun 电子枪
electron gun density multiplication 电子枪密度倍增量
electron gun vacuum lock 电子枪真空密封装置
electron gyro-frequency 电子回旋频率
electron gyroradiation 电子回旋辐射
electron heating 电子加热
electron hole 电子空穴
electron hole center spectrum 电子空穴心谱
electron hole collision 电子空穴碰撞
electron hole droplets 电子空穴液滴
electron hole pair 电子空穴(碰撞)偶；电子(空)穴对
electronic 电子的
electronic absorption coefficient 电子吸收系数
electronic absorber 电子吸收器；电子吸声器
electronic absorption spectrum 电子吸收光谱
electronic accelerometer 电子加速度计
electronic accounting machine 电子记账机
electronic admittance 电子导纳
electronic aid to navigation 导航电子设备
electronic air cleaner 电子空气清洁器；电子空气净化器
electronic alarm 电子报警器
electronic alignment 电瞄准
electronically agile radar 电子敏捷雷达
electronically conducting glass 电子导电玻璃
electronically conductive glass 电子导电玻璃
electronically controllable coupler 电子控制耦合器
electronically controlled gravity classification yard 电子控制驼峰编组场
electronically controlled screwdriver 电子控制螺丝刀
electronically controlled shower 电子控制的淋浴
electronically controlled telephone exchange 电子控制电话交换机
electronically controlled truck 电子控制的小车
electronically operated mill 自动控制轧机
electronically rotated scanning system 电子旋转

扫描系统
electronically tined pulsator 电子式脉动器
electronically tunable filter 电子可调滤波器
electronically tunable optic(al) filter 电子可调谐光学滤波器
electronically tuned oscillator 电子调谐振荡器
electronic alternate current stabilizer 电子交流稳压器
electronic alternating current switch 交流电子开关
electronical time relay 电子式时间继电器
electronic altimeter 无线电测高计
electronic amplifier 电子(管式)放大器
electronic amplifier-filter 电子放大器滤波器
electronic analog correlator 电子模拟相关器
electronic analog multiplier 电子模拟乘法器
electronic analog(ue) 电子模拟
electronic analog(ue) computer 电子模拟计算机
electronic analysis and analog(ue) equipment 电子分析和模拟设备
electronic and mechanical scanning 电子机械扫描
electronic anemometer 电子风速仪
electronic angular momentum 电子角动量
electronic annunciator 电子信号器；电子信号机
electronic antenna distributor 电子式天线分配器
electronic apparatus 电子器具
electronic arc-control device 电子灭弧器
electronic arc welder 电子弧焊机
electronic asymmetry 电子不对称
electronic attachment 电子附着
electronic attenuator 电子式衰减器
electronic attitude director 电子姿态指引仪
electronic aural responder 电子音响信号应答器
electronic automatic balance instrument 电子式自动平衡仪
electronic automatic balancing recorder 电子式自动平衡记录器
electronic automatic compensation 电子自动补偿
electronic automatic digital computer 电子式自动数字计算机
electronic automatic exchange 电子自动交换机
electronic automatic navigator 电子自动领航仪；电子自动导航仪
electronic automatic switch 电子自动开关
electronic automatic system 电子自动系统
electronic automatic tuning 电子自动调谐
electronic autopilot 电子(式)自动驾驶仪；电子式自动操舵
electronic auto telephone 电子自动电话机
electronic azimuth marker 电子方位标记
electronic balance 电子天平；电子平衡；电子秤
electronic balancer 电子平衡器
electronic ballast 电子镇流器
electronic band spectrum 电子带(光)谱
electronic bank 电子银行
electronic batching counter 电子剂量计数器
electronic batching weigher 定值控制电子秤
electronic battery clock 电池钟
electronic beam 电子束
electronic beam amplifier 电子束放大器
electronic beam centering 电子束定中心
electronic beam controlled discharge 电子束控制放电
electronic beam convergence 电子束汇聚
electronic beam current 电子束电流
electronic beam density 电子束密度
electronic beam generator 电子束发生器
electronic beam irradiation 电子束辐照
electronic beam laser 电子束激光器
electronic beam leveling unit 电子束抄平器
electronic beam machining 电子束加工
electronic beam positioning system 电子束定位系统
electronic beam power 电子束功率
electronic beam power amplifier 电子束功率放大器
electronic beam processing 电子束加工；电子束处理
electronic beam recorder 电子束记录仪；电子束记录器
electronic beam recording micromap system 微型地图电子束记录系统
electronic beam recording system 电子束记录系统
electronic beam scanner 电子束扫描机
electronic beam setting accuracy 电子束稳定精度

electronic beam-steering device 电子束扫描装置
electronic beam welder 电子束对焊机
electronic bearing 电子方位
electronic bearing cursor 电子方位游标
electronic bearing indicator 电子方位指示器
electronic bearing marker 电子方位游标;电子方位标志
electronic belt milling machine 电子磨带机
electronic belt scale 电子皮带秤
electronic belt weigher 电子皮带秤
electronic belt weighing system 电子皮带称量装置
electronic billing computer 开单电子计算机
electronic billing machine 电子会计机
electronic biologic microscope 电子生物显微镜
electronic bottle inspector 电子验瓶机
electronic bottom hole pressure recorder 电子式井下压力计
electronic bottom hole pressure transducer 电子式井下压力传感器
electronic box 电子盒
electronic brain 电脑;电子计算机
electronic brain driven centrifuge 电脑控制离心机
electronic brake control unit 制动微机控制单元
electronic breakdown 电子击穿
electronic bug killer 电子杀虫器
electronic building brick 电子组件
electronic bulk storage 电子大容量存储器
electronic bulletin board 电子通报板
electronic business 电子商务
electronic cabling unit 电子海底电报(发送)装置
electronic calculating punch 电子计算穿孔(机)
electronic calculator 电子计算器;电子计算机
electronic camera 电子摄影机,变像管照相机
electronic camouflage 电子伪装
electronic canister 电子仪器装运箱
electronic cargo winch 电动起货机
electronic carillon 电子钟琴
electronic cartridge 电子盒
electronic cascade 电子级联
electronic cash register 电子现金收入记录机;电子现金出纳机;电子点钞机
electronic casing caliper 套管内壁腐蚀电子检查仪
electronic catcher 电子捕集器
electronic catcher detector 电子捕获鉴定器
electronic central office 电子中心局
electronic ceramics 电子陶瓷
electronic charge 电荷;电子电荷
electronic chart 电子海图
electronic chart and information display system 电子海图显示和信息系统;电子海图显示和数据系统
electronic chart correction 电子海图修改
electronic chart data base 电子海图数据库
electronic chart display 电子海图显示
electronic chart display system 电子海图显示系统
electronic chart scanner 电子扫描读卡仪;电子读卡器
electronic chimes 电子式谐音系统
electronic chopper 电子斩波器
electronic chronometer 电子天文钟;电子时器;电子时计
electronic chronometric tachometer 电子计时式转速计;电子计时式转速表
electronic cipher machine 电子密码机
electronic circuit 电子线路;电子电路;电路
electronic circuit analysis program (me) 电子线路分析程序语言;电子电路分析程序
electronic circuit packaging 电子电路组装
electronic circuitry 电子电路图
electronic clinometer 电子倾斜仪
electronic clock 电子钟
electronic coagulation controller 电子凝聚控制器
electronic coder 电子编码管
electronic coding tube 电子编码管
electronic collection 电子收集
electronic collection time 电子收集时间
electronic colo(u)r analyzer 电子分色机
electronic colo(u)r compositor 电子彩色合成仪
electronic colo(u)r correction 电子分色校正
electronic colo(u)r correction scanner 电子分色校色扫描机;电子校色扫描机
electronic colo(u)r corrector 电子校色器;电子彩色改正器
electronic colo(u)r scanner 电子分色(扫描)机;电子彩色扫描器
electronic colo(u)r separator 光电分离机
electronic colo(u)r splitter 电子分色机
electronic commutator 电子转向器;电子转接器;电子转换器;电子转换开关
electronic compaction meter[metre] 电子压实计
electronic comparator 电子比较仪
electronic compass 电子罗盘;电子罗经
electronic component 电子元件
electronic components laboratory 电子元件实验室
electronic components workshop 电子元件车间
electronic component test bench 电子元件试验台
electronic composition 电子排版
electronic compressor circuit 电子分色扫描头
electronic computation 电子计算
electronic computer 电子计算机
electronic computer control 电子计算机控制
electronic computer-controlled 电子计算机控制的
electronic computer for survey 测地计算机
electronic computer industry 电子计算机工业
electronic computing gunsight 电子计算射击瞄准器
electronic computing machine 电子计算机
electronic computing units 电子计算单元
electronic concentration 电子浓度
electronic condensed-water stirrer 电子冷凝水搅拌器
electronic conduction 电子导电;电子传导
electronic conductive glass 电子导电玻璃
electronic conductivity 电子导率;电子导电性
electronic conductor 电子导(电)体
electronic configuration 电子组态;电子排布;电子构型
electronic confusion area 电子扰乱区
electronic consistency regulator 电子浓度调节器
electronic console 电子仪表台;电子显示器
electronic consolidation test device 电测固结试验仪
electronic contactor 电子接触器
electronic contact printer 电子接触印像机
electronic contact rectifier 电子接触整流器
electronic contour liner 电子等高线仪
electronic control 电子控制;电子调节
electronic control device 电子控制装置
electronic controlled installation 电子控制装置
electronic controlled intermittent tractor 电子控制间断式牵引机
electronic controlled telephone exchange 电子式电话交换机
electronic controlled test rack 电子控制的试验设备
electronic controller 电子控制器;电子调节器
electronic control system 电子控制装置;电子控制系统
electronic cooler 电子冷却装置
electronic cooling 电子冷却
electronic cool(ing) pillow 电子凉枕
electronic coordinate printer 电子坐标打印机
electronic coordinatograph and read-out system 电子坐标仪和读出系统
electronic copier 电子复印机
electronic correlation 电子相关
electronic correlator 电子相关器;电子环形解调电路
electronic cottage 电子化住宅
electronic counter 电子计算器;电子计数器;电子计数管;电子计量电路
electronic counter-countermeasure 电子反干扰措施;电子反对抗
electronic countermeasure 电子对抗;电子干扰措施
electronic countermeasure direction-finder 电子对抗测向仪
electronic counter measure jammer 电子对抗干扰机
electronic counter measure receiver 电子对抗接收机
electronic countermeasure resistant communication system 电子对抗阻力通信系统
electronic counter modulation 电子干扰
electronic counter relay 电子式计数继电器
electronic counter type frequency meter 电子计数式频率计
electronic coupling 电子耦合
electronic crack detector 电子探伤器
electronic crane scale 电子吊秤
electronic cross-section 电子截面
electronic curing 高频固化;高频磁化
electronic current 电子(电)流
electronic curve follower 电子曲线跟踪器
electronic curve tracer 电子曲线示踪器;电子曲线描记仪
electronic data 电子数据
electronic data calculator 电子数据计算器
electronic data display 电子数据显示器
electronic data exchange 电子数据交换
electronic data exchanger 电子数据交换机
electronic data file 电子数据文件
electronic data gathering equipment 电子数据收集设备
electronic data interchange 电子数据交换
electronic data interchange and control system 电子数据交换和控制系统
electronic data interchanger 电子数据交换机
electronic data processing 电子数据处理
electronic data processing center 电子数据处理中心
electronic data processing device 电子数据处理装置
electronic data processing equipment 电子数据处理设备
electronic data processing machine 电子数据处理机
electronic data processing plant 电子数据处理装置
electronic data processing system 电子数据处理系统
electronic data processor 电子数据处理机
electronic data service 电子数据服务
electronic data-switching center 电子数据转接中心;电子数据交换中心
electronic data transmitting set 电子数据传输装置
electronic data transmission communication center 电子数据传输通信中心
electronic decelerometer 电子减速仪
electronic deception 电子欺骗
electronic defense evaluation 电子防务测定
electronic degradation 电子衰减
electronic delocalization 电子非定域位
electronic denier monitor 电子纤度监测仪
electronic density 电子密度
electronic density analyzer 电子密度分析仪
electronic density detector 电子密度探测仪;电子密度检测仪
electronic density distribution 电子密度分布
electronic density index 电子密度指数
electronic density map 电子密度图
electronic density measurement probe 电子密度探测器
electronic desk calculator 台式电子计算器
electronic desk computer 台式电子计算机
electronic desk lamp 电子台灯
electronic detachment 电子分离
electronic detector 电子探测器
electronic detonation meter 电子爆震仪;电子爆震计
electronic device 电子仪器;电子器件
electronic dial exchange 电子式拨号交换机
electronic diamagnetism 电子抗磁性
electronic diary 电子日记
electronic dictionary 电子字典;电子词典
electronic differential analyzer 电子微分分析仪;电子微分分析机
electronic differential rectification 电子微分纠正
electronic differentiator 电子微分器
electronic diffraction 电子绕射
electronic diffraction analysis 电子衍射分析
electronic diffraction camera 电子衍射摄影机;电子衍射摄像机
electronic diffraction diagram 电子衍射图
electronic diffraction instrument 电子衍射仪
electronic diffraction investigation 电子衍射调查
electronic diffraction pattern 电子衍射图样
electronic diffraction photography 电子衍射摄影
electronic diffusion constant 电子扩散常数
electronic digital analyzer 电子数字分析仪
electronic digital calipers 电子数字卡尺
electronic digital computer 电子数字计算机;数字电子计算机
electronic digital density analyzer 电子数字密度分析仪
electronic digital display calipers 电子数显卡尺

electronic digital display combined machine 电子数显组合机
electronic digital display depth calipers 电子数显深度尺
electronic digital display ga(u)ge 电子数显卡规
electronic digital display height scale 电子数显高度尺
electronic digital displaying calipers 电子数显卡尺
electronic digital display machine scale 电子数显机床标尺
electronic digital display micrometer 电子数显千分尺
electronic digital interface 电子数字接口
electronic digital readout micrometer 电子数显千分尺
electronic digital wristwatch 电子数字式手表
electronic dilatometer 电子膨胀仪
electronic directional coupler 电子定向耦合器
electronic discharge 电子放电
electronic discrete variable automatic computer 电子离散变量自动计算机
electronic discriminator 电子鉴别器
electronic displacement 电子位移
electronic displacement polarization 电子移位极化
electronic display 电子显示
electronic display controller 电子显示控制器
electronic distance measurement 激光测距;电子距离测量;电子测距
electronic distance measuring 电子测距法
electronic distance measuring device 电子测距仪(器)
electronic distance measuring equipment 电子测距仪;电子测距设备
electronic distance measuring instrument 电子测距仪(器)
electronic distance measuring trigonometric leveling 电子测距三角高程测量
electronic distancer 电子测距仪
electronic distribution 电子分布
electronic divider 电子分配器;电子除法器
electronic document communication system 电子文档通信系统
electronic document exchange 电子票据交换
electronic document interchange 电子票据交换
electronic dodging device 电子匀光装置
electronic dodging enlarger 电子匀光放大机
electronic dodging printer 电子匀光印像机
electronic dodging system 电子匀光系统
electronic donor 电子受体;电子给予体
electronic door 电子控制的门
electronic drawing board 电子画板
electronic drawing machine 电子绘图机
electronic drift 电子仪器漂移
electronic driftmeter 电子偏航测量仪
electronic dummy 电子半身模型
electronic dynamometer 电子测力计
electronic editor 电子编辑机
electronic efficiency 电子效率
electronic embroidery machine 电子绣花机;电子刺绣机
electronic emission 电子放射;电子发射
electronic emission regulator 电子放射稳定器
electronic emission source 电子发射源
electronic emission spectrum 电子发射光谱
electronic emitter 电子发射器
electronic emitter positioning system 电子发射器定位系统
electronic energy 电子能(量)
electronic energy band 电子能带
electronic energy curve 电子能曲线
electronic energy loss spectroscopy 电子能量损失谱;电子能量损失能谱学
electronic energy-saving lamp 电子节能灯
electronic energy spectrum 电子能谱
electronic engine analyzer 电子式引擎分析器
electronic engineering 电子工程(学)
electronic engraver 电子刻图仪;电子刻版机
electronic engraving 电子刻图;电子刻版;电子雕刻
electronic engraving channel 电子刻版线路
electronic engraving machine 电子刻版机;电子雕刻机
electronic enhancement 电子增强
electronic enhancement equipment 电子增强装置
electronic enhancement viewer 电子增强观测器

electronic equation 电子方程(式)
electronic equilibrium 电子平衡
electronic equipment 电子设备
electronic equipment soldering set 电子设备焊接装置
electronic error correction 电子误差校正
electronic error correlation circuit 电子误差校正线路
electronic etching 电子刻版
electronic exchanger 电子交换器;电子交换机
electronic excitation 电子激发
electronic exposure meter 电子曝光表
electronic eye 电眼
electronic fault indicator 电子故障指示器
electronic feedback 电子反馈
electronic field book 电子外业手簿
electronic field frequency converter 电子场频变换器
electronic file 电子文件
electronic filter 电子滤波器
electronic fingertip control 电子按钮控制
electronic fix 电子定位
electronic flame safeguard 电子火焰防护装置
electronic flash equipment 电子闪光设备
electronic flasher 电子闪光器
electronic flash generator 电子闪光发生器
electronic flash(ing) 电子闪光
electronic flashing mechanism 电子闪光机制
electronic flashlamp 电子闪光灯
electronic flashlight 电子闪光灯
electronic-flash meter 电子闪光计
electronic flash plant 电子闪光设备
electronic flash tube 电子闪光灯
electronic flash unit 电子闪光部件
electronic flood routing machine 电子洪水演算机
electronic flowmeter 电子(式)流量计
electronic flowmeter system 连续式电子流量计
electronic flutter simulator 电子颤振模拟器
electronic follow-up steering apparatus 电子随动操舵仪
electronic formula 电子式;电子公式;电子方程(式)
electronic frequency counter 电子频率计数据
electronic frequency meter 电子(管式)频率计
electronic function generator 电子(式)函数发生器
electronic funds transfer system 电子资金转移系统;电子资金汇兑系统;电子金融传送系统;电子金融传输系统
electronic fuse 电子引信
electronic gas 电子气
electronic ga(u)ge 电子压力计;电子测微仪;电子测微计;电子测量仪
electronic generator 电子(管)振荡器
electronic glass 电子玻璃
electronic glue gun 电子焊枪
electronic governor 电子自动调速器
electronic grade 电子级
electronic grade controller 电子坡度控制器
electronic guidance equipment 电子制导装置;电子测试仪;无线电制导装置
electronic guidance system 电子制导系统
electronic gun 电子枪
electronic gun bombardment 电子枪轰击
electronic gun chamber 电子枪室
electronic gyro compass 电子陀螺罗盘
electronic hairness meter 电子式毛羽测试仪
electronic hanging balance 电子吊秤
electronic heating 电子加热
electronic helicopter 电子直升机
electronic hole opener 电子打孔机
electronic humidistat 电子(式)恒湿器
electronic-hydraulic regulator 电子液压调节
electronic hydrotreater 电子水处理器
electronic ignition 电子点火
electronic ignition system 电子点火系统
electronic image 电子图像
electronic image correlator 电子影像相关器
electronic image enhancement 电子图像增强
electronic image pickup device 电子(式)摄像装置
electronic image storage device 电子录像装置;电子录像设备
electronic image transfer 电子影像转换
electronic image-tube 电子(显)像管
electronic imaging device 电子成像装置

electronic impact 电子撞击
electronic impact furnace 电子撞击炉;电子轰击炉
electronic impact spectrometry 电子撞击光谱法
electronic impedance 电子阻抗
electronic impulse 电子脉冲
electronic impulser 电子脉冲发生器
electronic indicator 电子示功器
electronic industrial stethoscope 电子工业听诊器
Electronic Industries Association 电子工业协会
electronic industry 电子工业
electronic inertia 电子惯性
electronic information exchange system 电子信息交换系统
electronic information network 电子信息网络
electronic installation 电子装置;电子设备
electronic instrument 电子仪表;电子测试仪器
electronic integrating channel 电子积分电路
electronic integrator 电子积分器;电子积分电路
electronic intelligence 电子情报
electronic interchange machine 电子交换机
electronic interference 电子干扰
electronic interlocking 电子联锁
electronic interpreter 电子翻译器
electronic inverter 电子逆变器;电子换流器;电子变流器
electronic ion collection time 电子离子收集时间
electronic ion recombination 电子离子复合
electronic jammer 干扰发射机
electronic jamming 电子干扰;人为干扰
electronic judging and timing equipment 电子裁判计时设备
electronic jump 电子跃迁
electronickelling 电镀镍
electronic keyboard 电子键盘
electronic key generator 电子密钥发生器
electronic keying 电子键控;电子发报
electronic keying unit 电子键控器
electronic key loading device 电子密钥装填器
electronic leak detector 电子检漏仪;电子检漏器
electronic leather area measuring machine 电子皮革面积测量机
electronic lens 电子透镜
electronic level 电子位级;电子水准仪;电子水准器
electronic level(l)ing 电子水准测量
electronic level(l)ing device 电子整平装置
electronic library 自动化图书馆;电子图书馆
electronic library search 电子式图书查找程序
electronic lighter 电子引燃器;电子打火机
electronic lighting 电子照明
electronic linear accelerator 电子直线加速器
electronic line of position 电子位置线;电子设备所定的位置线;电子定位线
electronic line scanner 电子行扫描仪;电子行扫描机
electronic line scanning 电子行扫描
electronic listening device 电子监听装置
electronic loadcell scale 电子秤
electronic lock 电子(门)锁
electronic locking 电子锁定
electronic log 伽马测井曲线图;电子计程仪
electronic logger 电子记录器;辐射测井仪
electronic logging 辐射测井
electronic machinability computer 机械加工性电子计算机
electronic machine 电子仪器
electronic machine test bench 电子机器机器试验台
electronic machine tool control 机床电子控制;电子机床控制
electronic magnetic moment 电子的磁矩
electronic magnetometer 电子(式)磁强计
electronic magneton 电子磁子
electronic mail 电子邮件;电子邮寄;电子信箱
electronic mailbox 电子邮箱
electronic mailbox data entry 电子邮箱数据输入
electronic management system 电子管理系统
electronic map 电子地图
electronic mapping 电子作图
electronic mapping system 电子测绘系统
electronic material 电子材料
electronic measurement 电子(仪器)量测
electronic measuring instrument 电子测量仪表
electronic mechanical plotter 电子机械绘图机
electronic membrane osmometer 电子膜渗透仪

electronic memory 电子记忆;电子存储器
electronic memory apparatus 电子记忆装置
electronic memory device 电子存储设备
electronic memory drum 电子存储磁鼓
electronic memory pressure probe 电子存储压力计
electronic memory recorder 电子存储记录器
electronic message 电子信息
electronic message system 电子信息系统
electronic metal detector 电子金属探测器
electronic meter 电子(仪)表
electronic meter reading 电子仪表读数
electronic metronome 电子节拍器
electronic micro balance 电子微量天平
electronic micro-diffraction 电子显微衍射
electronic microelectrode X-ray analyzer 电子微电极 X 射线分析仪
electronic microga(u)ge 电子厚度计
electronic micrograph 电子显微放大器;电子微写器
electronic micrography 电子显微摄影
electronic micrometer 电子测微仪;电子测微计
electronic microphone 电子扩音器;电子传声器
electronic microprobe 电子(显)微探针
electronic microquartz chronometer 电子微石英钟
electronic microradiography 电子显微射线照相术
electronic microscope 电子显微镜
electronic microscope autoradiography 电子显微放射自显影法
electronic microscope method 电子显微镜法
electronic microscope microanalyzer 电子显微镜微(量)分析器
electronic microscope observation 电子显微镜观察
electronic microscope specimen autoprocessor 电子显微镜标本自动处理机
electronic microscope specimen penetrator 电子显微镜标本渗透器
electronic microscope specimen polymerizer 电子显微镜标本聚合器
electronic microscope spectroscopy 电子显微镜鉴定
electronic microscopic analysis 电子显微镜分析
electronic microscopy 电子显微镜检查;电子显微(检查)法;电子显微检测法;电子显微测定法
electronic microsecond meter 电子微秒表
electronic migration 电子移动;电子迁移
electronic millisecond meter 电子毫秒计;电子毫秒表
electronic mobility 电子迁移率
electronic model 电子模型
electronic module 电子组件
electronic moisture meter 电子湿度计
electronic monitoring device 电子监控装置
electronic mosquito killer 电子杀蚊器;电子灭蚊器
electronic motor control 电动机电子控制
electronic motor controller 电子电动机控制器
electronic multichannel recorder 电子多道记录器
electronic multimeter 电子(式)万用电表
electronic multiple automatic exchange 电子复式自动交换机
electronic multiplier 电子(式)乘法器
electronic multiplier circuit 电子乘法电路
electronic multiplier tube 电子倍增器;电子倍增管
electronic music 电声音乐
electronic navigation 电子导航;无线电导航
electronic navigation(al) aids 电子助航仪器;电子助航设备
electronic navigation(al) chart 电子导航图
electronic navigation(al) equipment 电子导航装置;电子导航设备
electronic navigation(al) method 电子导航法
electronic navigation(al) system 电子导航系统;无线电导航系统
electronic noise jammer 电子噪声干扰器
electronic notebook 电脑记事本
electronic numerical integrator and computer 电子数字积分计算机
electronic operator 电动操作器
electronic ophthalmotonometer 电子眼压计
electronic optic(al) aberration 电子光学像差
electronic optic(al) column 电子光学陈列
electronic optic(al) comparator 光电比较仪
electronic optic(al) control 电子光学控制
electronic optic(al) instrument 电子光学仪器

electronic optic(al) objective 电子光学物镜
electronic optic(al) prism 电子光学棱镜
electronic optic(al) system 电子光学系统
electronic optics 电子光学
electronic optics component 电子光学部件
electronic orbit 电子轨道
electronic oscillations 电子振荡
electronic oscillator 电子振荡器
electronic oscillograph 电子示波器
electronic oven 电子炉
electronic package 电子(仪器)组件;电子器件包
electronic packaging 电子封装;电子(电路)组装
electronic pantograph 电子伸缩绘图器
electronic paramagnetic resonance spectrometer 电子顺磁共振波谱仪
electronic part 电子零件
electronic particle counter 电子颗粒计数器
electronic pen 光笔;电子笔
electronic phase angle meter 电子相位计;电子相角计
electronic photoengraving machine 电子刻版机
electronic photogrammetry 电子摄影测量学
electronic photograph recording 电子照相记录法
electronic photometer 电子光度计
electronic phototype corder 电子照相式直接记录器
electronic phototype-setter 电子照相排字机
electronic photo-voltaic cell 电子光生伏打电池
electronic pickup 电子拾音器
electronic pilot 电子自动驾驶仪
electronic plane-table 电子平板仪
electronic planimeter 电子求积仪
electronic plasma oscillations 电子波
electronic platform scale 电子台秤
electronic plotter 电子绘图仪;电子绘图机
electronic plotting sheet 电子测图板
electronic polarization 电子极化
electronic polisher 电子抛光机
electronic pollution 电子污染
electronic position calculation device 电子定位计算装置
electronic position equipment 电子定位显示器
electronic position fixing system 电子定位系统
electronic position indicator 航迹自记指示器;电子定位器;位置电子指示器
electronic positioning equipment 电子定位仪
electronic positioning unit 电子定位器
electronic position readouts 电子位置读数器
electronic post parcel scale 电子邮寄包裹秤
electronic potentiometer 电子电位(差)计;电子电势仪
electronic power conditioner 电子电源调节器
electronic power converter 电子大功率变流器
electronic power frequency control system 电子电源频率调节装置
electronic presentation 电子显示
electronic presentations of navigational information 航行资料电子显示
electronic pressure controller 电子压力控制器;电子调压器
electronic pressure ga(u)ge 电子压力计
electronic pressure indicator 电压力指示器
electronic pressure transmitter 电子压力变送器
electronic price computing scale 电子计价秤
electronic printer 电子印像机;电子印刷装置;电子印刷机;电子打印机
electronic printing 电子印像;电子印刷
electronic printing control unit 电子晒像控制装置
electronic printing machine 电子印刷机
electronic print reader 电子印刷阅读器
electronic private automatic branch exchange 自动专用交换机
electronic probe 电子探针
electronic probing analysis 电子探针分析
electronic processing 电子处理
electronic product 电子产品
electronic profilometer 电子表面光洁度测量仪
electronic program(me) control 电子程序控制
electronic program(me) decorating spray 电子程序喷涂
electronic programmer 电子程序装置
electronic publishing 电子出版
electronic pulsating stimulator 电子式脉动刺激器
electronic pulsator 电子式脉动器

electronic pulse 电子脉冲
electronic pulse chamber 电子脉冲电离室
electronic pulse counter 电子脉冲计数器
electronic pulse height analyzer 电子脉冲振幅分析器
electronic pump 电子泵
electronic punch 电子凿孔机;电子穿孔机
electronic punch-card machine 电子穿孔卡片机
electronic puncher 电子穿孔器
electronic punching machine 电子穿孔机
electronic rack 电子柜
electronic radiography 电子(辐射)照相术
electronic random number generator 电子随机数发生器
electronic range 雷达试验场;电子设备试验场
electronic range finder 雷达测距仪;雷达测距器;电子测距仪
electronic raster 电子光栅的
electronic raster scanner 电子光栅扫描器
electronic raster scanning 电子扫描
electronic raytube 电子射线管
electronic reader 电子阅读器
electronic reading machine 电子阅读机
electronic readout 电子读出方式
electronic reconnaissance 电子勘测
electronic record annotators 电子记录注记器
electronic recorder 电子记录器
electronic recording 电子记录
electronic recording component 电子记录元件
electronic recording equipment 电子记录装置;电子记录设备
electronic recording spot 电子记录斑点
electronic recording statoscope 电子自动记录高差仪
electronic recording tube 电子记录管
electronic recording unit 电子记录装置
electronic rectifier 电子整流器
electronic refrigerator 电子制冷器;电子冰箱
electronic register control 电子定位控制器
electronic regulator 电子稳压器;电子稳定器;电子调节器
electronic relaxation polarization 电子弛张极化;电子弛豫极化
electronic relay 电子继电器
electronic remote switching 电子遥控转换开关;电子遥控交换设备
electronic repelling 电子排斥
electronic resolver 电子式分解器
electronic resonance spectrometer 电子共振波谱仪
electronic response unit 电子响应装置
electronic rest mass 电子静止质量
electronic reverser 电子反向器
electronic rice cooker 电饭锅
electronic robot 电子自动装置;电子机器人
electronic rotary visco(si)meter 电子旋转黏度计
electronic route storage 电子存储器
electronics 电子学
electronic safety warner 电子安全报警器
electronics aids to navigation 电子助航仪器;电子助航设备
electronic sampler 电子取样器
electronic sampling switch 电子转换器;电子转换开关;电子取样开关
electronic scale 电子秤
electronic scaling 电子计数器
electronic scanned stacked beam radar 电子扫描多波束雷达
electronic scanner 电子扫描仪;电子扫描器;电子扫描机
electronic scanner microscope 电子扫描显微镜
electronic scanning 电子扫描
electronic scanning digitizer 电子扫描数字化器
electronic scanning micrograph 电子扫描显微照片
electronic scanning radar 电子扫描雷达
electronic scanning spectrometer 电子扫描分光仪
electronic scanning system 电子扫描系统
electronic scanning television 电子扫描式电视
electronic scattering 电子散射
electronic schedule telephone 电子调度电话机
electronic scheme 电子原理图
electronic scrambler 电子加扰器
electronic screen 电子显示器荧光屏
electronic screening 电子屏蔽

electronic sculpturing 电子模型法
electronic seam detector 电子裂缝检查仪
electronic sea-wave recorder 电子海波记录仪
electronic sector-scan(ning) sonar 扇形扫描声呐；扇形扫测声呐
electronic security 电子保密措施
electronic seismic check 电子地震检测
electronic seismic detection 电子地震检测
electronic seismic test(ing) 电子地震测试
electronic seismograph 电子(式)地震仪
electronic selector 电子选择器
electronic self-balancing type recorder 电子自动平衡式记录仪
electronic semi-conductor 电子半导体
electronic semiprecision winder 电子半精密卷绕机
electronic sensitive emulsion 电子敏感乳剂
electronic sensitive film 电子敏感胶片
electronic sensor 电子传感器
electronic sentry 电子警戒器
electronic separation 电子分离
electronic separator 电子式分离器；电子分离机
electronic serial digital computer 电子串行数字计算机；串行电子数字计算机
electronic servo 电子随动系统；电子跟踪
electronic sewing machine 电子式绝缘材料焊接机；电子缝纫机
electronics for ocean technology 海洋技术电子学
electronic shaftspeed pickup 电子式轴转速传感器
electronic shell 电子(壳)层
electronic shop 电子车间
electronic shopping 电子购物
electronic shower 电子簇射
electronic shutter 电子快门；电子开闭器
electronic simulation 电子模拟(装置)
electronic simulator 电子模拟装置
electronic sketch master 电子像片绘绘仪
electronic skiving machine 电子削刮机
electronic sky screen equipment 电子空网设备
electronic slide rule 电子计算尺
electronic sorter 电子分类器；电子分类机
electronic sorting 电子分类
electronic sorting machine 电子分拣机
electronic sorting system 电子分类装置
electronic sound reverberator 电子混响器
electronic source 电子源
electronic specific heat 电子比热
electronic spectrograph 电子光谱仪
electronic spectrometer 电子能谱仪
electronic spectrometry 电子谱法
electronic spectroscopy 电子能谱学；电子能谱术；电子能谱法；电子光谱学；电子光谱法
electronic spectrum 电子谱；电子能谱法；电子光谱
electronic speed sensor 电子速度传感器
electronic spin resonance magnetometer 电子自旋共振磁强计；电子自旋共振磁力仪
electronic spin resonance spectrum 电子自旋共振光谱
electronic spin resonator 电子自旋共振器
electronic spot 电子束斑点
electronic spread-sheet 电子表单；电子展开图表
electronic stabilizer 电子稳定器；电子(管)稳压器
electronic state 电子态
electronic statistical machine 电子统计机
electronic stereoviewer 电子立体观察
electronic sterilization 电子消毒；电子杀菌法；电子灭菌法
electronic stimulator 电子刺激器
electronic stopping power 电子阻止本领
electronic storage device 电子存取设备；电子存储设备；电子存储器
electronic storage system 电子存储系统
electronic streak camera 超快扫描摄影机
electronic structure 电子结构
electronic stylus 光笔；电子笔
electronic subassembly 电子组合件
electronic sunflower 电子向日葵
electronic surge arrester 电子电涌放电器
electronic surveillance equipment 电子监视设备
electronic survey-data recorder 电子测量手簿
electronic survey(ing) 电子测量
electronic survey sensor 电子测量传感器
electronics watch 电子手表
electronic switch 电子开关
electronic switch board 电子交换台

electronic switching 电子开关作用；电子交换设备
electronic switching center 电子交换中心
electronic switching system 电子转接系统；电子转换系统；电子交换装置；电子交换系统；电子交换机
electronic switching tube 电子开关管
electronic system simulator 电子系统模拟器
electronic table calculator 台式电子计算器
electronic tabulator control 电子表格控制
electronic tach(e)ometer 电子转速仪；电子视距仪；电子测速仪；电子计数管；(全站型)电子速测仪
electronic tachymeter (全站型)电子速测仪
electronic technology 电子技术
electronic telecontrol 电子遥控
electronic telemeter 电子遥测仪；电子测距仪
electronic telemetering 电子遥测术
electronic telephone exchange 电子电话交换机
electronic teleprinter 电子电传打字电报机
electronic telescope 电子望远镜
electronic television 电子电视
electronic temperature 电子温度
electronic temperature contact controller 接触式电子调温器；电子温度开关控制器
electronic temperature controller 电子温度控制器；电子式温控器
electronic temperature control system 电子温度调节系统
electronic temperature recorder 电子温度计；电子式温度记录器
electronic tensile tester 电子拉伸试验器
electronic tension meter 电子张力测定仪
electronic test equipment 电子测试设备
electronic tester 电子式测试器
electronic testing apparatus 电测仪器
electronic test instrument 电子试验仪器
electronic test pattern generator 电子测试图发生器
electronic test socket 电子测试插座
electronic theodolite 电子经纬仪
electronic theory 电子说
electronic thermometer 电子温度计
electronic thermostat 电子控制的恒温器；电子恒湿器
electronic therm static heat circulator 电子恒温热循环器
electronic thickness ga(u)ge 电子测厚仪
electronic time meter 电子毫秒计；电子毫秒表；电子测时计
electronic timer 电子(式)计时器；电子定时器
electronic timetable 电子时刻表
electronic timing machine 电子校时机
electronic tonometer 电子眼压计
electronic topographic map 电子地形图
electronic torch 电子火炬；电子焊枪
electronic torch generator 电子火炬发生器
electronic torch melting 电子束熔炼
electronic tornado welding 炭极(电)弧焊
electronic torque meter 电子式测扭计
electronic totalizer 电子累积器
electronic total station 电子全站仪
electronic toy 电子玩具
electronic tracer 电子示踪器；电子描图机；电子描绘器；电子故障检寻器；电子跟踪装置
electronic tracing 电子跟踪
electronic traffic-recording machine 电子话务记录机
electronic training device 电子培训设备
electronic transducer 电子换能器
electronic transfer unit 电子转接器
electronic transformer 电子变压器
electronic transient visualizer 电子瞬变观察仪
electronic transition 电子跃迁
electronic translator 电子译码器；电子翻译器
electronic transmission densimeter 电子透射密度计
electronic transmitter 电子发送机
electronic trap 电子捕集器
electronic traverse 电子测距导线
electronic trawl-net 电拖网
electronic treadle 电子式脚踏板
electronic-trilateration 电子三边测量
electronic tube 电子管
electronic tube amplifier 电子管放大器
electronic tube control 电子管控制

electronic tube generator 电子管振荡器；电子管信号发生器
electronic tube glass 电子管玻璃
electronic tube oscillator 电子管振荡器
electronic tube-type amperemeter 电子管式电流计
electronic tuning 电子调谐(的)
electronic tuning hyspteresis 电子调谐滞后
electronic tuning sensitivity 电子调谐灵敏度
electronic tuning unit 电子调谐器
electronic tutor 电子教员；电子教学机
electronic type automatic voltage regulator 电子(管)式自动电压调整器
electronic type regulator 电子式调节器
electronic typewriter 电子打字机
electronic ultra-high frequency tuner 超高频电子调谐器
electronic unit 电子单元
electronic universal thermometer 电子通用温度计
electronic unscrambler 电子辨音器
electronic vacuum dilatometer 电子真空膨胀计
electronic valve 电子管；电子阀
electronic valve action 电子整流作用
electronic valve type power directional relay 电子管(式)功率定向继电器
electronic vehicle scanning equipment 车辆电子扫描设备
electronic ventilator 电子通风机
electronic vertical long-period seismometer 电子立式长周期地震仪
electronic vibration recorder 电子振动记录器
electronic video recorder 电子视频记录装置；电子录像机
electronic video recording 电子录像
electronic viewfinder 电子寻像器；电子取景器
electronic voltage regulator 电子(式)电压调节器
electronic voltmeter 电子(管)伏特计；电子电压表
electronic voltohmmeter 电子伏欧表
electronic vulcanizer 高频硫化装置
electronic walk helper 电子助行器
electronic warning system 电子报警系统
electronic watch 电子表
electronic wave 电子波
electronic wavelength 电子波长
electronic weigher 电子秤
electronic weighing system 电子称量系统
electronic width ga(u)ge 电子测宽仪
electronic window 电子窗
electronic wobbulator 电子摆频振荡器
electronic writing 电子书写
electronic yarn clearer 电子清纱器
electronic yarn selector 电子选纱器
electron illuminating system 电子照明系统
electron image tube 电子图像管
electron impact 电子碰撞
electron impact ionization 电子碰撞电离；电子轰击离子化
electron-induced conductivity vidicon 电子感生导电视像管
electron induction accelerator 电子感应回旋加速器
electron injection 电子注入
electron injector 电子注射器
electron interferometer 电子干涉仪
electron irradiation 电子辐照；电子辐射
electron irradiation damage 电子辐射照伤
electron isomerism 电子同质异能性
electron lattice interaction 电子点阵相互作用
electron lattice theory 电子点阵论
electron lens 电子透镜
electron lepton number 电子轻子数
electron linear accelerator 电子线性加速器
electron magnetic moment 电子磁矩
electron mass 电子质量
electron metal 镁铝合金
electron metallography 电子金相学
electron micrograph 电子显微(镜)照片；电子显微镜相片；电子微写品
electron micropicture 电子显微图像
electron microprobe 电子(显)微探针
electron microprobe analysis 电子微探针分析；微区电子探针分析
electron microprobe analyzer 电子探针分析仪
electron microprobe X-ray analyzer 电子显微探针X射线分析仪
electron microscope 电子显微镜

electron microscope microanalyzer 电镜微量分析仪;电子显微镜微分析器
electron microscope scanning method 电镜扫描法
electron microscope scan(ning) photogrammetry 扫描电子显微测量
electron-microscopical autoradiography 电镜自动放射照相术
electron-microscopical bright field image 电子显微明场象
electron-microscopical dark field image 电子显微暗场象
electron-microscopical examination 电子显微镜检查
electron microscopy 电子显微术
electron mirror 电子镜
electron mirror microscope 电子镜显微镜
electron mobility 电子迁移率;电子流动性
electron mobility detector 电子迁移率检测器
electron multiplicity 电子多重性
electron multiplier 电子倍增器;电子倍增管
electron-multiplier photo tube 电子倍增光电管
electron neutrino 电子中微子
electron neutrino field 电子中微子场
electron-nuclear double resonance helix 电子核磁双共振螺旋管
electron-nuclear double resonance spectroscopy 电子核磁双共振谱学;电子核双共振谱术
electron number 电子数
electron octet 电子八隅体
electronogen 曝光时放出电子的分子
electronogram 电子衍射图(形)
electronograph 电子显像;电子显微镜照片
electronographic camera 电子照相机
electronographic device 电子成像器件
electronographic photometry 电子照相测光
electronographic tube 电子图像传感管
electronography 电子显像术;电子图摄像术
electron optic(al) component 电子光学部件
electron optic(al) image intensifier 电子图像增强
electron optics 电子光学
electron orbit 电子轨道
electron oscillating engine 电子振荡发动机
electron oscillation 电子振荡
electron oscillator 电子振荡器
electron oscillograph 电子示波器
electron pair 电子偶;电子对
electron pair acceptor 电子对受体
electron pair creation 电子偶形成;电子对发生
electron pair donor 电子对给体
electron pair effect 电子对效应
electron pair production 电子对的产生
electron pair production attenuation coefficient 电子对生成衰减系数
electron pair spectrometer 电子偶分光计;电子对谱仪
electron paramagnetic resonance 电子顺磁谐振;电子顺磁共振
electron paramagnetic resonance spectrum 电子顺磁共振波谱
electron paramagnetic resonance spectrum of paramagnetic center 顺磁中心的电子顺磁共振波谱
electron paramagnetism 电子顺磁性
electronphonic music 电声音乐
electron photographic(al) marking-off coating 电子照相画线涂料
electron photographic(al) marking-off method 电子照相画线法
electron-photon scattering 电子光子散射
electron-photon shower 电子光子簇射
electron physics 电子物理学
electron plasma 电子等离子体
electron plasma frequency (电子)等离子体频率
electron plasma wave 电子等离子体波
electron plate 电子板
electronplating wastewater 电镀废水
electron pole 镁铝合金电极;电子极
electron-positron annihilation 电子正电子湮没
electron-positron collision 电子正电子碰撞
electron-positron pair 电子正电子对
electron-positron pair annihilation 电子正电子对湮没
electron-positron pair creation 电子正电子对产生
electron-positron storage ring 电子正电子储存环
electron power tube 电子功率管

electron probe 电子(微)探针
electron probe energy spectrum pattern 电子探针能谱图
electron probe microanalysis 电子探针微量分析
electron probe microanalyzer 电子探针微量分析仪;电子探针微量分析器
electron probe micrograph 电子探针显微照片
electron probe surface mass spectrometry 电子探针表面质谱分析
electron probe X-ray analyzer 电子探针X射线分析器
electron probing analysis 电子探针分析
electron process 高压放电曲线
electron projector 电子投射器
electron-proton spectrometer 电子质子谱仪
electron radiation 电子辐照;电子辐射
electron radiation curing 电子辐射固化
electron radius 电子半径
electron ray 电子射线;电子束
electron ray desulfurization 电子射线脱硫
electron-ray drying 电子射线干燥
electron record(ing) tube 记录用阴极射线管
electron reference level 电子参考水平
electron relay 电子继电器
electron removal 电子异构体
electron repeller 电子反射极
electron resonance breakdown 电子共振击穿
electron resonance magnetometer 电子共振磁强计
electron rest frame 电子静坐标系
electron rest mass 电子(静)质量
electron-rich alloy 富电子合金
electron ring accelerator 电子环加速器
electron scanned complex detector 电子扫描复合检测器
electron scanning microscope 电子扫描显微镜
electron scanning pupillometer 电子扫描瞳孔计
electron scattering continuum 电子散射连续区
electron section 电子截面
electron sharing 电子共有化
electron shell 电子壳层
electron sink 电子冷阱
electrons leaves welding 电子束焊(接)
electron spectrograph 电子摄谱仪
electron spectrometer 电子分光仪
electron spectroscopy for chemical analysis 化学分析用电子光谱(法);化学分析电子谱法
electron spectrum 电子能谱;电子光谱
electron speed regulator 电子调速器
electron spin 电子自旋;电子转动
electron spin density 电子自旋密度
electron spin double resonance 电子自旋双共振
electron spin resonance 电子自旋共振;电子顺磁共振
electron spin resonance dating method 电子自旋共振年代测定法;电子自旋共振测年法
electron spin resonance spectroscopy 顺磁共振光谱学
electron stain 电子着色剂;电子染色;电子染料
electron stimulated desorption 电子致脱附
electron-stream amplifier 电子流放大器
electron-stream potential 电子流电位
electron-stream transmission efficiency 电子流传输效率
electron structure of atom 原子的电子结构
electron switch 电子开关
electron synchrotron 电子同步加速器
electron telescope 电子望远镜
electron temperature 电子温度
electron temperature probe 电子温度探测器
electron theory 电子学说;电子理论
electron time-of-flight spectrometer 电子飞行时间谱仪
electron trajectory 电子轨迹
electron trajectory plotter 电子轨迹标绘器
electron transfer 电子转移;电子迁移;电子传递
electron transfer reaction 电子转移反应
electron transit reactance 电子渡越电抗
electron transit time 电子跃迁时间;电子渡越时间
electron transmission microscope 电子透射显微镜
electron transport chain 电子传递链
electron transport system 电子传递体系
electron trap 电子陷阱
electron tube 真空管;电子管;电真空器件

electron tube amplifier 电子管放大器
electron tube base ga(u)ge 电子管管座量规
electron tube generator 电子管振荡发生器
electron tube instrument 电子管式测试仪器;电子管式测试仪表
electron tube power amplifier 电子管功率放大器
electron tube rack 电子管试验台
electron tube static characteristic 电子管静态特性
electron tube voltmeter 电子管伏特计
electron tunneling 电子贯穿
electronuclear machine 粒子高能加速器
electron vacancy 电子空位
electron vacuum ga(u)ge 电子真空计
electron valve 电子管;电磁阀
electron velocity analyzer 电子速度分析器
electron velocity spectrometer 电子速度谱仪
electron volt 电子伏特
electron voltage 电子伏特
electron water curing 电子水养
electron wave function 电子波函数
electron wavelength 电子波长
electron wave magnetron 电子波磁控管
electron waves 电子波
electron wave tube 电子波管;电子波放大管
electron-withdrawing 吸电子
electron work funcṭon 电子逸出功
electrooptic(al) anisotropy 电学、光学各向异性现象
electrooptic(al) beam splitter 电光分光镜
electrooptic(al) bench 电光具座
electrooptic(al) birefraction effect 电控光双折射效应
electrooptic(al) birefringence 电光克尔效应
electrooptic(al) birefringent effect 电控双折射效应
electrooptic(al) block 电光部件
electrooptic(al) cell 电光元件
electrooptic(al) ceramics 电光陶瓷
electrooptic(al) character recognition 光学字符识别
electrooptic(al) converter 电光变换器
electrooptic(al) countermeasures receiving set 电光对抗接收装置
electrooptic(al) crystal 电光晶体
electrooptic(al) crystal light modulator 电光晶体光调制器;电光晶体调制器
electrooptic(al) deflector 电光偏转器
electrooptic(al) detection 电光探测
electrooptic(al) detector 电光探测器;电光检测器
electrooptic(al) diffraction modulator 电光衍射调制器
electrooptic(al) digital deflector 电光数字式偏转器
electrooptic(al) directional coupler 电光定向耦合器
electrooptic(al) distance measurement 光电测距
electrooptic(al) distance measuring instrument 光电测距仪;光电测距计
electrooptic(al) distance measuring traverse 光电测距导线
electrooptic(al) distancemeter 光电测距仪;光电测距计
electrooptic(al) effect 电光效应;光电效应
electrooptic(al) hardware 电光硬件
electrooptic(al) image intensifier 电光像增强器
electrooptic(al) image tube 光电移像管
electrooptic(al) imaging 电光成像;光电成像
electrooptic(al) input device 电光输入装置
electrooptic(al) Kerr effect 电光克尔效应
electrooptic(al) light-detecting apparatus 光辐射电光探测器;电光光探测器;电光测光器
electrooptic(al) light intensity modulator 电光光强调制器;电光光密度调制器
electrooptic(al) light modulation 光电法光波调制
electrooptic(al) light modulator 电光调节器
electrooptic(al) luminaire 电光源
electrooptically tuned laser 电光可调激光器
electrooptic(al) mark 电光式标志
electrooptic(al) material 电光材料
electrooptic(al) means of communication 电光通信工具
electrooptic(al) memory 电动光学存储器
electrooptic(al) modulating cell 电光调制元件
electrooptic(al) modulation 电光调制
electrooptic(al) modulator 光电(式)调制器
electrooptic(al) multiframe camera 光电多幅照相机

electrooptic(al)news 电光新闻广告
electrooptic(al)of communication means 光电通信工具
electrooptic(al)phase modulator 电光相位调制器
electrooptic(al)polarization converter 电光偏振变换器
electrooptic(al)processor 电光处理机
electrooptic(al)property 电光性质;电光性能
electrooptic(al)range finder 光电测距仪
electrooptic(al)scattering effect 电(控)光散射效应
electrooptic(al)sensor 电光传感器
electrooptic(al)sensor system 光电传感系统
electrooptic(al)short-range surveying 短光电测距
electrooptic(al)shutter 电光快门
electrooptic(al)source 电光源
electrooptic(al)space navigation(al)simulation 电光空间导航模拟
electrooptic(al)spectrograph 电光频谱仪
electrooptic(al)switch(ing) 电光开关
electrooptic(al)system 电光系统
electrooptic(al)telemeter 光电测距仪;光电测距计
electrooptic(al)theodolite 光电经纬仪
electrooptic(al)tracker system 光电跟踪系统
electrooptic(al)transducer 电光变换器
electrooptic(al)transfer characteristic 电光转换特性
electrooptic(al)transit 光电经纬仪
electrooptic(al)transmitter 电光发送机
electrooptic(al)transparent glass-ceramics 电光透明玻璃
electrooptic(al)tube 光电管
electrooptic(al)window 电光窗
electrooptics 电(子)光学;电场光学
electroorganic oxidation 电有机氧化
electroosmose 电渗作用;电渗现象
electroosmosis 电渗作用;电渗现象;电渗(析);电渗透法;电渗透(作用)
electroosmosis installation 电渗装置
electroosmosis method 电渗法
electroosmosis of sludge dewatering 电渗透法污泥脱水
electroosmosis stabilization 电渗加固
electroosmosis transmission coefficient 电渗系数
electroosmosis treatment 电渗(透)处理
electroosmotic consolidation 电渗固结
electroosmotic dewatering 电渗排水;电渗降水;电渗脱水;电渗去水
electroosmotic drainage 电渗排水
electroosmotic driver 电渗激发器
electroosmotic flow 电渗流动
electroosmotic ground stabilization 电渗法地面加固
electroosmotic method 电渗法;电渗透法
electroosmotic permeability coefficient 电渗渗透系数
electroosmotic potential 电渗势
electroosmotic stabilization 电渗稳定作用;电渗稳定法;电渗加固
electroosmotic transmission coefficient 电渗透射系数
electroosmotic transmission of coefficient 电渗传递系数;电渗传导系数
electrooxidation 电氧化;电解氧化(法)
electrooxidation reactor 电氧化反应器
electropaint 电涂层
electropainting 电涂
electropane 导电膜玻璃
electroparting 电分离
electropercussion welding 冲击点焊(法)
electropercussive processor 电冲击处理器
electropercussive welding 电冲击焊;储能焊;冲击焊(法);冲击电焊
electropherography 载体电泳图法
electrophilic 吸电子的
electrophilicity 亲电性
electrophilic reaction 亲电子反应
electrophilic reagent 亲电子试剂
electrophobic 疏电的
electrophone 送受话器;送收话器
electrophonic effect 电响效应
electrophoresis 电泳(现象)
electrophoresis apparatus 电泳装置;电泳仪;电泳器

electrophoresis chromatogram 电泳色谱
electrophoresis effect 电泳效应
electrophoresis experiment 电泳试验
electrophoresis property 电泳性能
electrophoresis scanner 电泳扫描器
electrophoresis separation method 电泳选矿法
electrophoresis tank 电泳槽
electrophoretic analysis 电泳分析
electrophoretic apparatus 电泳装置;电泳仪;电泳器
electrophoretic body-paint line 车身电泳涂漆生产线
electrophoretic buffer 电泳缓冲液
electrophoretic casting 电泳注浆
electrophoretic chromatogram 电泳色谱
electrophoretic clarification 电泳澄清
electrophoretic coating 电镀涂层;电泳涂装;电泳涂漆;电泳涂覆法;电泳涂层;电泳
electrophoretic coating of nickel plus chromium 镍—铬的电镀涂层
electrophoretic column 电泳柱
electrophoretic current 电泳电流
electrophoretic deposition 电泳涂搪;电泳淀积;电泳沉积(法)
electrophoretic display 电泳显示
electrophoretic display device 电泳显示器
electrophoretic effect 电泳效应
electrophoretic examinations 电泳检查
electrophoretic experiment 电泳试验
electrophoretic filtering 电泳过滤
electrophoretic finishing 电泳涂漆
electrophoretic focusing 电泳聚焦
electrophoretic force 电泳力
electrophoretic image display 电泳成像显示器
electrophoretic medium 电泳介质
electrophoretic method 电泳法
electrophoretic mobility 电泳淌度;电泳迁移率
electrophoretic paint 电泳涂料;电泳漆
electrophoretic painting 电泳涂装;电泳涂漆
electrophoretic pattern 电泳图案
electrophoretic powder deposition 电泳粉末沉积法
electrophoretic property 电泳性能
electrophoretic scanner 电泳扫描器
electrophoretic separation 电泳选矿;电泳分离
electrophoretic separation method 电泳选矿法
electrophoretic techniques 电泳技术
electrophoretic variant 电泳变体
electrophorogram 电泳图
electrophorus 静电机摩擦发电;起电盘
electrophotocopy 静电摄影复制品
electrophoto-fluorescence 电控光致发光
electrophotograph 电子摄影
electrophotographic(al)marking off method 电子复印法
electrophotographic(al)paper 静电摄影感光纸
electrophotographic copier 电子照相复印机
electrophotographic copy paper 电子照相复印纸
electrophotographic marking off method 电子复印法
electrophotographic microfilm 电子照相缩微胶片
electrophotographic paper 电子照相感光纸;静电摄影感光纸
electrophotographic printer 电子照相印刷机
electrophotographic printing paper 电子照相印刷纸
electrophotographic projector printer 电子照相投影复印机
electrophotographic recording paper 电子照相记录纸
electrophotography 静电摄影;电(子)照相术;电子摄影术
electrophotoluminescence 电致发光;电控光致发光;场控光致发光
electrophotometer 光电光度计;光电比色计
electrophotometry 光电光度学;光电测光
electrophotonic detector 光电探测器
electrophotosensitive material 电光敏感材料
electrophrenic respiration 电动膈式呼吸
electrophysics 电物理学
electropism 趋电性
electroplane camera 光电透镜摄像机
electroplaque 电板
electroplate 电解沉淀;电镀品;电镀的;电镀板
electroplated chromium finish 电镀铬面

electroplated coating 电镀层
electroplated coating of zinc 电镀锌(法)
electroplated copper back 电镀铜镜背(保护)层
electroplated decoration 电镀装饰
electroplated diamond reaming shell 电镀金刚石扩孔器
electroplated finish 电镀层
electroplated diamond internal circular saw 电镀金刚石圆锯片
electroplated layer 电镀层
electroplated nail 电镀钉
electroplated phototransistor 电镀光电晶体管
electroplated pottery 电镀陶瓷
electroplated scissors 电镀剪
electroplate liquid 电镀液
electroplate zinclead wire 电镀锌铅丝
electroplating 电镀(术)
electroplating bath 电镀浴;电镀槽
electroplating dynamo 电镀用的发电机
electroplating effluent 电镀废液;电镀废水
electroplating factory 电镀厂
electroplating film machine 电镀膜机
electroplating filter 电镀滤器
electroplating industry 电镀工业
electroplating machine 电镀机
electroplating meter 电镀仪
electroplating on plastics 塑料电镀
electroplating operaiton 电镀作业
electroplating rinse water 电镀清洗水
electroplating sludge 电镀污泥
electroplating solution 电镀液
electroplating waste 电镀废料
electroplating(work)shop 电镀车间
electroplax 电板
electropneumatic 电动气动的
electropneumatic actuator 电气气传动;电动气压传动;电动气动装置;电动气动致动器;气压电动装置
electropneumatically controlled 电动气控的;电动风控的
electropneumatic brake 电空制动;电动气动制动器
electropneumatic brake circuit 电空制动电路
electropneumatic braking connector 电动气动制动联结器
electropneumatic contactor 电动气动接触器
electropneumatic control 电力气动控制
electropneumatic controller 电气气动控制器;电动气动控制器;电动气动调节器
electropneumatic convertor 电气气动转换器;电动气动变换器;电气转换器
electropneumatic distributing valve 电动气动分配阀
electropneumatic drilling hammer 电动气动钻锤
electropneumatic emergency brake 电空紧急制动
electropneumatic fire detector 电气联动火情探测器
electropneumatic governor 电气气动调速器;电动气动调速器
electropneumatic hammer 电动气锤
electropneumatic interlocker 电动气压连锁装置
electropneumatic interlocking device 电动气动联锁装置
electropneumatic interlocking machine 电动气动联锁机
electropneumatic loudspeaker 电动气流扬声器
electropneumatic master controller 电动气动制动主控制器
electropneumatic operation 电动气动控制
electropneumatic plant 电动气动设备
electropneumatic point machine 电动气动转辙机
electropneumatic point motor 电动气动转辙机
electropneumatic positioner 电气定位器
electropneumatic pusher 电动气动推料机;电动气动推车器
electropneumatic regulator 电动气动调节器
electropneumatic remote control system 电动气动遥控系统
electropneumatic semaphore 电动气动臂板信号机
electropneumatic service brake 电动常用制动
electropneumatic signal 电动气压信号;电动气动信号机
electropneumatic signaling 电动气动信号装置
electropneumatic signal motor 电动气动信号机
electropneumatic signal valve 信号机电动气动阀

electropneumatic single plate friction clutch 电气单盘摩擦离合器
electropneumatic straight air brake 电动气动直通空气制动器
electropneumatic switch 电动气压开关;电动气动开关
electropneumatic switch apparatus 电动气动转辙机
electropneumatic switch circuit controller 电气气动转辙电路控制器
electropneumatic switch machine 电空气转辙机;电动气动转辙机(器)
electropneumatic switch operating apparatus 电动气动转辙机
electropneumatic transducer 电动气动变换器
electropneumatic transfer switch 电动气动转换开关
electropneumatic transmitting equipment 电动气动传递设备
electropneumatic valve 电动气动(控制)阀
electropneumatic valve positioner 电气阀门定位器
electropolar 电极性的
electropolarization potential 电极化电位
electropolarized relay 极化继电器
electropolishing 电磨光;电(解)抛光;电化学抛光
electropolymer 带电聚合物
electroporcelain 电瓷
electroposition 电淀积
electropositive 阳性的;正电(性)的;电阻性的
electropositive atom 阳电性原子
electropositive element 阳电性元素;正电性元素
electropositive gel 正电性凝胶
electropositive metal 正电性金属
electropositive potential 阳电势
electropositivity 阳电性;正电性
electroprecipitation 电沉淀
electroprecipitator 电力沉淀器;电除尘器
electroprobe 电子探针;电笔;电测头
electroprobe microanalysis 电子探针微分析
electroprobe-X-ray microanalysis 电子探针 X 射线(显)微分析仪
electroprognosis 电预后法
electropsychrometer 电测湿度计;电气湿度计
electropulsograph 脉象仪
electropyrexia 电发热法
electropyrometer 电(阻)高温计;热电偶温度计
electroquartz 电造石英;电熔石英
electroradiography 电放射影
electroradiology 电放射学
electroradiometer (电)放射测量计
electroradioscence 电致辐射
electrorammer 电夯
electrorecording paper 电子记录纸
electrorefine 电解精炼
electrorefining 静电精制;电精炼;电(解)提纯
electrorefining process 电解精炼法
electroreflectance 电反射比
electrorefraction 电限流;电扼流
electroregeneration 电再生
electroregulator 电调节器
electroresection 电切除术
electroresistance alloy 电阻合金
electroresistive effect 电阻压效应
electroresistivity prospecting 电阻探测法
electroresponse 电响应
electroretardation filtration 电阻滞过滤(作用)
electroreversing gear 电磁回动装置
electroscaling 电力除锈
electroscope 静电测量器;验电器;验电盘;试电笔
electroscopy 验电法;气体电离检定法
electroscrubbing 电解洗涤
electrosection 电切除术
electroseismic effect 电振动效应
electrosemaphore 电信号机
electrosensitive metallic paper 电敏金属纸
electrosensitive paper 电敏纸;电感光纸
electrosensitive printer 电灼式打印机
electrosensitive processor 电敏处理器
electrosensitive recording 电灼式记录;电敏(火花刻蚀)记录(法);电火花刻蚀记录法
electroseparation 电解分离
electroseries 电化序列
electroservo 电动伺服机构
electroservo control 电气随动控制;电动伺服控制

electroshape 电力水压成形机
electrosheet copper 电工铜箔
electroshock 触电
electroshop 电镀车间
electrosilicification 电动硅化法
electrosilicification phenomenon 电动硅化现象
electrosilvering 电镀银
electroslag 电渣
electroslag casting 电渣熔铸
electroslag melting 电渣熔炼
electroslag melting process 电渣冶炼法
electroslag refining 电渣精炼
electroslag remelting 电渣重熔
electroslag smelting casting 电渣熔铸
electroslag surfacing 电渣堆焊
electroslag welder 电渣焊机
electroslag weld(ing) 电阻焊;电(磁)渣焊
electroslag welding machine 电渣焊机
electroslag welding with consumable nozzle 熔嘴电渣焊
electroslag welding with plate electrode 板极电渣焊
electroslag welding with wire electrode 丝极电渣焊
electrosmelting 电炉熔炼
electrosmelting bath 电炉熔炼池
electrosmelting of steel 电炉炼钢
electrosmelting plant 电炉熔炼设备
electrosmosis analyzer 电渗析分离仪
electrosol 金属电胶液;电溶胶
electrosolvent method 电解溶剂法
electrosonic and correlation equipment 电声与相关设备
electrosounding level meter 电动探测式料位器
electrospark 电火花
electrospark detector 电火花检测器
electrospark forming 电爆炸成形
electrospark hardening 火花放电硬化法
electrosparking 电火花加工
electrosparking and machinery tool grinding machine 电火花机械复合工具磨床
electrospark machining 电子放电加工
electrospark polishing 电火花抛光
electrospark sintering 放电粉末烧结
electrospark wire-electrode cutting 电火花线切割
electrospectrogram 电光谱图
electrospectrography 电光谱描记术
electrostat 干式氧化锌静电复印机
electrostatic 静电式;静电的
electrostatic accelerator 静电加速器
electrostatic accumulator 静电累加器;电容器
electrostatic acoustic(al)instrument 静电声学仪器
electrostatic adherence 静电吸附;静电附着
electrostatic adhesion 静电吸附;静电附着
electrostatic adsorption 静电吸附作用
electrostatic agent 静电剂
electrostatic air cleaner 静电空气净化器
electrostatic airless spray 静电无空气喷涂
electrostatically augmented baghouse 静电加力袋式除尘器
electrostatically augmented scrubber 静电增强洗涤器
electrostatically facused image intensifier 静电聚焦像增强器
electrostatically focused 静电聚集的
electrostatically focused image tube 静电聚焦像管
electrostatic altimeter 静电式测高计;静电高度计
electrostatic analyzer 静电分析器
electrostatic atomization 静电雾化
electrostatic attraction 电动吸引力;库仑吸引;静电引力;静电吸引;静电吸力
electrostatic balance 静电平衡
electrostatic barrier 静电能垒
electrostatic beam positioner 静电横臂调位器
electrostatic belt 静电带
electrostatic bond 静电键
electrostatic bonding 静电结合
electrostatic calibration device 静电标定装置
electrostatic capacitance 静电(电)容
electrostatic capacitance thickness meter 静电电容(法)测厚仪
electrostatic capacity 静电容(量)
electrostatic cathode-ray tube 静电电子束管
electrostatic charge 静电充电;静(电)电荷

electrostatic charge neutralization 静电中和
electrostatic charge printing tube 静电记录管
electrostatic charging 静电电荷;静电充电
electrostatic cleaner 静电清洁器;静电除尘器
electrostatic coating 静电涂装;静电涂布
electrostatic collection 静电吸尘
electrostatic collector failed element monitor 静电收集破损元件监测器
electrostatic component 静电分量
electrostatic condenser 静电电容器
electrostatic contribution 静电分布
electrostatic control 静电控制
electrostatic convergence 静电会聚
electrostatic copying machine 静电复印机
electrostatic coupling 静电耦合
electrostatic current 静电流
electrostatic danger 静电危险
electrostatic deflection 静电偏转
electrostatic deflection system 静电偏转系统
electrostatic dehydrator 静电脱水器
electrostatic desalting 静电脱盐
electrostatic detearing 静电沥水;静电除漆滴
electrostatic detection 静电探测
electrostatic develop 静电显影
electrostatic discharge 静电释放;静电放电;放静电
electrostatic dosimeter 静电计式剂量计
electrostatic double layer 双电层
electrostatic dry process print 干法静电复印
electrostatic dump 静电打印
electrostatic dust collecting cell 静电集尘单元
electrostatic dust collector 静电集尘器
electrostatic duster 静电式喷粉机
electrostatic dust precipitator 静电吸尘器;静电收尘器;静电降尘器;电集尘器
electrostatic dust sampler 静电粉尘采样器
electrostatic effect 静电影响;静电效应
electrostatic electrify motor 静电起电机
electrostatic electromicroscope 静电式电子显微镜
electrostatic electron lens 静电电子透镜
electrostatic electron optics 静电电子光学
electrostatic elimination 消除静电
electrostatic energy 静电能
electrostatic engine 离子发动机
electrostatic exposure 静电曝光
electrostatic feedback 静电反馈
electrostatic field 静电场
electrostatic field interference 静电场干扰
electrostatic film 静电胶片
electrostatic filter 静电除尘器;静电过滤器
electrostatic filtration 静电过滤
electrostatic fire 静电起火
electrostatic fire detector 静电火花探测器
electrostatic fixing 静电定影
electrostatic flocking 静电植绒
electrostatic flowmeter 静电流量计
electrostatic-fluidized bed 静电流化床
electrostatic-fluidized bed coating 静电流化床涂装;静电沸腾床涂装
electrostatic focusing 静电聚焦
electrostatic focusing lens 静电聚焦透镜
electrostatic focusing system 静电聚焦系统
electrostatic force 静电(作用)力
electrostatic forcing 静电力作用
electrostatic gas cleaner 静电净气器
electrostatic generator 静电发生器;静电发电机
electrostatic glazing 静电喷釉
electrostatic ground detector 静电式接地检测器
electrostatic grounding 静电接地
electrostatic gyro(scope) 静电陀螺仪
electrostatic hand gun 手提式静电喷枪
electrostatic hand spray gun 手提式静电喷枪
electrostatic hazard 静电危险;静电危害
electrostatic heating 静电加热
electrostatic ignition source 静电着火源
electrostatic image 静电图像
electrostatic image section 静电像截面
electrostatic induction 静电感应
electrostatic instrument 静电系仪表;静电式测试仪器
electrostatic interaction 库仑相互作用;静电相互作用
electrostatic interference 静电干扰
electrostatic latent image 静电潜像
electrostatic leakage 静电漏泄

electrostatic lens 静电透镜
electrostatic line copier 静电行式复印机
electrostatic line printer 静电行式打印机
electrostatic loudspeaker 静电(式)扬声器
electrostatic machine 静电起电器;感应起电机
electrostatic machine separation method 静电起电机分离法
electrostatic mark 静电标记
electrostatic measuring instrument 静电式电表
electrostatic memory 静电存储(器)
electrostatic method 静电法
electrostatic microphone 静电式传声器;电容式话筒;电容传声器
electrostatic microscope 静电显微镜
electrostatic motor 静电电动机
electrostatic nebulizer 静电雾化器
electrostatic oscillograph 静电示波器
electrostatic painting 静电涂漆;静电涂镀;静电喷漆
electrostatic particle separator 静电粒子分离器
electrostatic percussing welding 储能焊;冲击焊(法);电能储能焊接
electrostatic percussive welding 静电冲击焊
electrostatic photography 静电摄影(术)
electrostatic platemaking 静电摄影制版
electrostatic plotter 静电(式)绘图仪;静电(式)绘图机
electrostatic positioning 静电置位
electrostatic potential 静电势;静电(电)位
electrostatic potential barrier 静电势垒
electrostatic powder coating 静电粉末涂装;静电粉末喷涂
electrostatic powder coating plant 静电粉末涂装装置
electrostatic powder spraying 粉末静电喷涂
electrostatic precipitation 静电聚灰;静电降尘;静电集尘;静电分离;静电淀积;静电除尘;静电沉降;静电沉积;静电沉淀;电集尘
electrostatic precipitator 静电(式)除尘器;静电滤尘器;静电聚灰器;静电集尘器;静电除尘装置;静电沉积器;静电沉淀器;电气除尘器;电滤尘器;电除尘器
electrostatic precipitator with plate electrodes 阳极板静电感应除尘器
electrostatic precipitator with pulse energization 脉冲激能集尘器
electrostatic pressure 静电压
electrostatic prevention 防静电
electrostatic prevention flooring 防静电地面;防静电地板
electrostatic print 静电复印图
electrostatic printer 静电印刷机;静电复印机;静电打印机
electrostatic printing 静电印刷;静电复制;静电复印图
electrostatic process 静电工艺;静电处理
electrostatic radium 静电半径
electrostatic recorder 静电记录器
electrostatic recording 静电记录(法)
electrostatic recording paper 静电记录纸;静电复印纸
electrostatic relay 静电式继电器
electrostatic repulsion 静电相斥;静电推斥(力);静电斥力
electrostatics 静电学
electrostatic scanning 静电扫描
electrostatic screen 静电屏
electrostatic seismograph 静电地震仪
electrostatic seismometer 静电地震检波器;静电地震计
electrostatic sensitive device 静电敏感器件
electrostatic separation 静电选矿(法);静电分选;静电分离
electrostatic separation method 静电分离法
electrostatic separator 静电分选器;静电分离器
electrostatic shield(ing) 静电屏蔽
electrostatic sorter 电子式分级机
electrostatic space shield stabilization 静电空间屏蔽稳定作用
electrostatic spark 静电火花
electrostatic spinning 静电纺纱
electrostatic spray booth 静电喷釉室
electrostatic spray coating 静电涂漆;静电喷涂(法);静电喷漆
electrostatic spray equipment 静电喷涂设备

electrostatic sprayer 静电喷涂机
electrostatic spray gun 静电喷雾枪
electrostatic spray(ing) 静电喷涂(法)
electrostatic storage 静电存储(器)
electrostatic storage deflection 静电存储偏转
electrostatic stored energy 静电蓄能机
electrostatic strainer 静电除污器
electrostatic strip-oiling apparatus 带材静电涂油机
electrostatic supported gyroscope 静电支承陀螺仪
electrostatic susceptibility 静电敏感性
electrostatic suspended gyroscope 静电悬浮陀螺仪
electrostatic transducer 静电换能器
electrostatic transfer 静电转印
electrostatic treater 静电处理器
electrostatic unit 静电单位
electrostatic viscous filter 静电黏液过滤器;静电油过滤器
electrostatic voltmeter 静电伏特计;静电电压表
electrostatic wattmeter 静电瓦特计
electrostatic wave 静电波
electrostatic welding 静电焊
electrostatic winding insulation 绕组静电绝缘
electrostatic writing head 静电记录头
electrostatographic imaging process 静电摄影成像法
electrosteel 电炉钢
electrostencil master paper 电子油印蜡纸
electrostenolysis 膜孔电积酌;细孔隔膜电解;细孔电解
electrostimulator 电子刺激器
electrostrain effect 电致应变效应;电气应变效应
electrostream drilling 电引流钻孔
electrostriction 电致伸缩
electrostriction ceramics 电致伸缩陶瓷
electrostriction coefficient 电致伸缩系数
electrostriction transducer 电致伸缩换能器
electrostriction vibrator 电致伸缩振动器
electrostrictive 电致伸缩的
electrostrictive compliance 电致伸缩顺度
electrostrictive effect 电致伸缩效应
electrostrictive relay 电致伸缩继电器
electrostrictive transducer 电致伸缩换能器
electrostrictor 电致伸缩体
electrosurgery 电外科
electroswivel 电动旋转接头;电传动水龙头
electrosynthesis 电合成(法)
electrotape 基线电测仪;电子测距装置
electrotaxis 应电作用;移电性;趋电性
electrotechnical 电工学的;电工技术的
electrotechnical carbon 电工炭
electrotechnical ceramics 电工陶瓷
electrotechnical department 电工车间
electrotechnical instrument 电工仪表
electrotechnical measurement 电工测量
electrotechnical porcelain 电瓷
electrotechnical porcelain products 电工陶瓷制品
electrotechnical terminology 电工名词术语
electrotechnics 电工学;电工技术
electrotechnology 电工学;电工技术
electrotellurograph 大地电流测定器
electrothanasia 触电死(亡)
electrotherapy room 电疗室
electrotherm 电热器
electrothermal 电热的
electrothermal alloy 电热合金
electrothermal ammeter 温差电偶安培计
electrothermal atomic absorption spectrometry 电热原子吸收光谱
electrothermal baffle 半导体障板
electrothermal blowing dry box 电热鼓风干燥箱
electrothermal constant-temperature dry box 电热恒温干燥箱
electrothermal curing (混凝土)电热养护
electrothermal effect 电(致)热效应
electrothermal efficiency 电热效率
electrothermal energy conversion 电热能量转化
electrothermal energy 电热能量
electrothermal equivalent 电热当量
electrothermal furnace 电热炉
electrothermal generator 电热发生器
electrothermal hardening (混凝土)电热硬化
electrothermal load 电热负荷
electrothermal maintenance 电热养护

electrothermal mantle 电热罩
electrothermal melting 电熔炼
electrothermal metallurgy 电热冶金
electrothermal meter 电热表;热丝电流表;电热表
electrothermal paraffin vehicle 电热清蜡车
electrothermal pretensioning 电热先张法
electrothermal printer 电热打印机
electrothermal process 电热法
electrothermal propulsion 电热推进
electrothermal recording 电热记录
electrothermal relay 电热继电器
electrothermal smelting 电热熔炼
electrothermal water temperature 电热式水温度
electrothermic 电热的
electrothermic ferroalloy shop 电热铁合金车间
electrothermic ga(u)ge 电热计
electrothermic instrument 电热(式)仪表
electrothermic process 电热法
electrothermic regulator 电热调节器
electrothermics 电热学;电热法
electrothermoluminescence 场控加热发光;电热发光;电控加热发光
electrothermoluminescene 场控热发光
electrothermometer 热电偶温度计
electrothermosensitive recording paper 电热感记录纸
electrothermostat 电(热)恒温箱;电恒温器
electrothermy 电热法
electrotimer 定时继电器;电子定时器
electrotinning 电镀锡
electrotin plate 电镀锡薄钢板
electrotitration 电滴定
electrotitration apparatus 电子滴定仪
electrotome 电刀
electrotomy 电切术
electrotonus 电紧张
electrotreatment of odo(u)r 臭电处理
electrotricycle 电动三轮车
electrotropism 向电性;应电性;趋电性
electrotruck 电动载重车
electrotype metal 英国标准铅字合金
electrotyping 电铸术;电铸法
electrotypograph 电子排字机
electrotypy 电铸术
electroultrafiltration 电超滤(作用)
electrovacuum ceramics 电真空(陶)瓷
electrovacuum engineering 电真空工程学
electrovacuum gear shift 电磁真空变速装置
electrovacuum glass 电真空玻璃
electrovalence 电价
electrovalence difference method 电价差值法
electrovalency 电价
electrovalent 电价的
electrovalent bond 电价键
electrovalent compound 电价化合物
electrovalve 电子阀
electrovibrating feeder 电动振动加料器;电磁振动喂料器;电磁振动喂料机;电磁振动给料机
electrovibrator 电子振荡器;电动振动器
electroviscosity 电黏性;电黏度效应
electroviscous 电滞的
electroviscous effect 电滞效应;电黏效应
electrovoltmeter 静电电压表
electroweak interaction 弱电磁相互作用
electrowelding 电焊
electrowelding machine 电焊机
electrowelding net 电焊网
electrowhite lead 电解铅白
electrowinning 电解冶金法;电解法;电解沉淀;电解沉积;电积金属法
electrozinc coated sheet 电镀锌薄钢板
electr-pneumatic conveyer 电空传送设备
electr-pneumatic valve 电空阀
electrum 琥珀色之金银合金;琥珀金;银金矿;埃勒克特卢姆金银合金
electrum-tarnish method 金银合金硫逸度测定法
electuary 膏剂
elegant art 高尚的艺术
elekron 镁(铜铝)合金
Elektron alloys 埃勒克特龙镁基铝铜轻合金
elelctrochemical power generation 电化发电
Elema 硅碳棒
Elema furnace 埃莱马硅碳棒加热炉
Elema heating resistor 埃莱马加热电阻器

Elemass 电动多尺寸检查仪
element 零件;元素;因素;单质;单元;成分
element abundance of control region 控制区元素丰度
element abundance of prediction region 预测区元素丰度
element abundance patterns 元素丰度模式
element address 单元地址
element interval 组内距
elemental 基本的;本质的
elemental abundance 元素丰度
elemental analysis 元素分析;要素成本
elemental area 像素面积;面面积
elemental array 基本台阵
elemental carbon 元素碳
elemental charge 单电子电荷;元素电荷
elemental chlorine 元素氯
elemental chromatography 元素色谱法
elemental composition 元素成分
elemental constituent 元素组分
elemental content of bitumen 沥青的元素组成
elemental crystal 单质晶体
elemental crystal furnace 单晶炉
elemental floating body 浮体单元
elemental form 基本形(式)
elemental furnace 单元炉缸;单风嘴炉
elemental hydrograph 单元过程线
elemental iron 元素铁
elemental lipid 基本脂
elemental model 基本模型
elemental mole fraction 元素的摩尔数
elemental motion 基本动作
elemental operation 基本操作;要素作业
elemental operational sea training 基本操作海上培训
elemental oxygen 原子氧
elemental particle 基本粒子
elemental phosphorus 元素磷
elemental sulfur 元素硫
elemental sulphur 元素硫
elemental task 基本作业
elemental term 基本项;初等项
elemental user 基本用户
element antenna 振子天线;单元天线
element area 单元面积
element array 元排列
elementary 基本的;初步的
elementary aerodynamics 初级空气动力学
elementary analysis 元素分析;基础分析
elementary antenna 基元天线;单元天线
elementary area 像素面积;像点面积;面面积;单元面积;微分面积;图像单元
elementary arithmetic 基本算术
elementary ash 基元灰分
elementary beam 基本光束;元光束
elementary bearing structure 基本支承结构
elementary bin 基本面元
elementary body 原(粒)体
elementary boundary 基本边界
elementary cable section 单元电缆段
elementary calibration 基本校准
elementary carburetor 简单化油器
elementary cell 基本晶格;基本单元;单元电池;单位栅格;单位晶格;单位晶胞
elementary chain 初级链
elementary charge 单元电荷;基本电荷
elementary colo(u)r filter 三原色滤光片
elementary colo(u)rs 基色;原色
elementary column operation 初等列运算;初等列变换
elementary component 基本成分
elementary composition 基本组成;元素组成;元素成分;组成元素
elementary cone 锥元素
elementary constituent 元素组分
elementary cost accounting 初级成本会计;初步成本会计
elementary current 元电流
elementary cycle 基本周期;基本循环;基本环路
elementary data type 基本数据类型
elementary date of crystal goniometry 晶体测量基本数据
elementary diagram 接线原理图;简图
elementary diffusion equation 基本扩散方程
elementary dipole 元偶极子
elementary directed-tree transformation 初等有向树变换
elementary distance 微距离
elementary error 基元误差;微差
elementary event 基本事件
elementary excitation 元激发
elementary fibril 初级原纤
elementary force 元力
elementary form 元素形成
elementary function 基本函数;基本功能;初等函数
elementary function program(me) 初等函数程序
elementary gas 单质气体;气态元素
elementary gate 基本门
elementary generator 单元发电机
elementary geodesy 初等大地测量学;普通测量学;普通测地学
elementary geometry 初等几何学
elementary granule 基本颗粒;原级粒子
elementary gravity wave 基本重力波
elementary hologram 基元全息图
elementary homomorphism 基本同态
elementary instruction 基本指令
elementary integral 初等积分
elementary item 基本项;初等项
elementary jar 电池瓶
elementary knowledge 基本知识
elementary landscape 单元景观
elementary language 基本语言
elementary layer (应力分析中的)单元层
elementary length 基本长度
elementary line 原线
elementary magnet 单元磁铁
elementary mathematics 初等数学
elementary matrix 初等矩阵
elementary meridianal beam 基本子午光束
elementary microfibril 初级微原纤
elementary microstructure 单元微结构
elementary model 基元模式
elementary move 基本项传送
elementary name 基本元素名
elementary network 基本网(络)
elementary number theory 初等数论
elementary operation 基本操作;初等运算
elementary organic analysis 元素有机分析
elementary organic compound 元素有机化合物
elementary organic polymer 元素有机高聚物
elementary parallelogram 基本平行四边形
elementary particle 基本质点;基本粒子;无质点;元粒子
elementary particle physics 基本粒子物理学
elementary particle reaction 基本粒子反应
elementary path 基本路径
elementary path problem 基本路程问题
elementary period 开始期
elementary plane-wave hologram 基元平面波全息图
elementary primary school 国民小学;公立小学
elementary process 基本过程;元素过程
elementary product 基本积
elementary project 基本项目
elementary proving 基本校准;基本标定
elementary reaction 基元反应;元素反应;微元反应
elementary reflection analysis 基本反射分析
elementary regenerator section 单元再生段
elementary renewal theory 基本更新理论
elementary repeater section 单元中继段
elementary rheology 初等流变学
elementary ring structrue 简单环形构造
elementary row and column operation 初等的行与列运算
elementary row operation 初等行运算;初等行变换
elementary runoff plot 径流(试验)小区
elementary school 初级学校;小学
elementary school building 小学校舍
elementary seismology 地震学基础
elementary sentence 基本命题
elementary solution 基本解
elementary statistical approach 基础统计方法
elementary stream 载流线;水流线
elementary stress 基本应力
elementary strip 基本条带
elementary structure 基本结构;单元构造
elementary sum 基本和
elementary survey(ing) 普通测量(学)
elementary symmetric(al) function 初等对称函数
elementary test 初级检验
elementary theory 一阶理论
elementary theory of numbers 初等数论
elementary time study 基本时间研究
elementary trace 基本道
elementary traffic volume 基本交通量
elementary training 基本训练
elementary transcendent function 初等超越函数
elementary transformation 初等变换
elementary tree transformation 初等树变换
elementary volume 体积单元
elementary wave 基波;元波
element aspect ratio 元素细长比
element association characteristics 元素组合特征
element buckling 构件曲曲;构件弯曲试验
element cell 单元室;单元电池
element center line 构件中心线;构件中心线
element characteristic 单元特征;单元特性
element connection 构件连接
element content 元素含量
element content in the sample 样品中元素含量
element content to be measured 待测元素含量
element control 元件控制
element conveyer 元件运输机
element cross-section 构件的横截面
element design 零件设计;元件设计;构件设计
element displacement 单元位移
element effect 元素效应
element enrichment factor 元素富集因子
element error 元件误差
element error rate 元件故障率;单元出错率;单元差错率
element factor 要素;基本因子
element failure 元件损伤;元部件损坏
element field 构件范围
element file 单元文件
element force 单元力
element geochemistry (元素)地球化学
element housing 元件套
element ID 单元标识符
element identifier 单元标识符
element in array 数组元素
element in seawater 海水元素
element interface 单元交界面
element/isotopic abundance 元素/同位素丰度
element jacket 元件包壳
element joint 节段接头
element length 构件长度
element loading 单位负载
element management system 单元管理系统
element mass matrix 单元质量矩阵
element memory control 单元存储器控制
element migration 元素迁移
element migration experiment 元素迁移实验
element model 箱式模型
element moment 单位力矩
element of absolute orientation 绝对定向元素
element of a cone 锥的母线
element of a cylinder 柱的母线
element of a loaded area 加荷面上的构件;加荷面上的部件
element of arc 弧形元素
element of arc length 弧长元素
element of area 面积元(素)
element of assessment 评价因素
element of centering 归心元素
element of coal seam occurrence 煤层要素
element of confidence 信任要素
element of construction 建筑部件
element of contact 接触元素;切元素
element of cost 成本要素
element of credit 信用要素
element of crystal growth 晶体生长要素
element of current 洋流要素
element of distance 距离元(素)
element of ellipse 变形椭圆元素
element of exterior orientation (相片)外方位元素
element of finite order 有限阶元素
element of fix 定位元素
element of fuel filter 燃油滤芯
element of function 函数元素
element of ground movement 地面移动要素

element of indicatrix 变形椭圆元素
element of inner orientation 内方位元素
element of interior orientation 内方位元素
element of internal orientation 内方位元素
element of matrix 矩阵元(素);矩阵的元素
element of orientation for picture 像片方位元素
element of petroleum 石油元素组成
element of principal diagonal 主对角线元素
element of production 生产要素
element of production expenses 生产费用要素
element of rectification 纠正元素
element of rotation matrix 旋转矩阵元素
element of semi-variable cost 半变动成本的要素
element of solid angle 立体角元素
element of space lattice 空间格子要素
element of structure 构造要素
element of surface 面元素
element of symmetry 对称要素
element of taxonomy 分类要素
element of the earth's field 地磁要素
element of the geomagnetic field 地磁要素
element of the integral 积分元素
element of time 时间要素
element of volume 体积元
element of water power 水力要素
element of water quality 水质要素
element of wave 波浪要素
element of winding 绕组元件
element pair 元对
element ratio 元素比值
element ratio of oil 石油的元素比
elements 元件
element semiconductor 单质半导体
element shape 构件形状
element size 构件尺寸;元素大小
element slope 构件斜度;部件斜度
elements of absolute orientation 绝对方位元素
elements of a fix 定位坐标
elements of an orbit 轨道根数
elements of attitude 岩层产状要素
elements of basic magma 基性岩浆元素
elements of circular curve 圆曲线要素
elements of comparison 比较的要素
elements of difficult migration 难迁移的元素
elements of easiest migration 极易迁移的元素
elements of easy migration 易迁移的元素
elements of estuarine hydrology 河口水文要素
elements of exterior orientation 相片外方位元素
elements of fix 定位要素
elements of interior orientation 相片内方位元素
elements of leading marks 导标要素
elements of migration 可迁移的元素
elements of neutral magma 中性岩浆元素
elements of occurrence 产状要素
elements of orbit 轨道要素;轨道根数
elements of relative orientation 相对方位元素;相对定向元素
elements of statistic(al)theory 统计理论初步
elements of sulfide deposits 硫化矿床元素
elements of the tidal stream 潮流要素
elements of ultra-basic rock 超基性岩元素
elements of variable valence 变价元素
element spacing 元件间距
element stiffness 单元劲度;单元刚度
element stiffness matrix 单元刚度矩阵
element stress 构件应力;单元应力
element string 单元(字符)串
element subdivision 单元剖分
element support 构件支撑
element task 基础工作
element time 基本动作时间
element to be measured 待测元素
element transfer 元素迁移
element transfer capacity 元素迁移能力
element transfer coefficient 元素迁移系数
element transformation 元素转变
element transportation 元素的迁移
element variable 单元变量
element vector 向量元
element welded tight all around 周边密焊元件
element with distributed parameters 分布参数元
element with lumped parameter 集中参数元件

element yield 元素产额
elemi 榄香脂;芳香树脂
elemi balsam 榄香脂;榄香胶;榄香油
elemi gum 榄香胶
elemi resin 榄香树脂
elemol 榄香醇
elemosinaria 救济施舍的房屋
elen shade 毛玻璃灯罩
eleonorite 簇磷铁矿
eleostearate 桐酸酯
eleotrograving 电刻
elephant 大号绘图纸;波纹铁;爱烈芬塔暴风雨;起伏干扰;无游梁抽油设备
elephanta 爱烈芬塔暴风雨
elephant ash 象蜡树
elephant boiler 象形锅炉
elephant dugout 大壕沟;大防空洞
elephanter 爱烈芬塔暴风雨
elephant hunt 大油田勘探
Elephantide pressboard 爱烈芬太压缩板;棉料绝缘板
elephant-legshaped vase 筒瓶
elephant maximus 亚洲象【动】
elephant paper 大画纸
elephant skin 皱纹罩面;皱皮
elephant's peel 皱皮
elephant trunk 溜管;混凝土输送管;象鼻状装置;象鼻管;喷射排泥管
elephant trunk pipe for concrete placing by gravity 混凝土溜管
elephant trunk spout 巨人式混凝土料斗
elerwind 埃勒风
El Escorial 埃斯科里亚尔(16世纪西班牙建筑群,包括宫殿、大学、修道院、陵墓等)
eletric coal drill 煤电钻
eletroconductive resin 导电树脂
eletro-hydraulic steering gear 电动液压舵机
eletrolyzable 可电解的
eletrometric titration 电滴定
eletrostatic colloid 电稳胶体
eleutheromorph 自由形晶
elevate 提升;抬高
elevate above the soil 伸出地面
elevate a turnable ladder 升举
elevated approach 高架(桥)引道
elevated aqueduct 高架渡槽
elevated barrier on pile foundation 桩式防波堤
elevated basin 高位水池;高水池;高架水池
elevated beach 岸边台地;岸边高地;上升滩;上升海滩
elevated bench 上升阶地
elevated bent 高排架
elevated block 上升断块;上升地块
elevated bridge 高架桥
elevated bunker 高位料槽
elevated car 高架电车
elevated chute 高架渡槽
elevated coal track 提升输煤栈桥
elevated coast 上升海岸
elevated conduit 高架导线管
elevated crossing 立体交叉;立交桥;立交道口;高架交叉(道)
elevated delta 上升三角洲
elevated ditch 高架沟渠;填土高渠
elevated duct 增大的波导;高架管道
elevated echo 雨幡回波
elevated emission 高架源排放
elevated expressway 高架快车道;高架高速公路
elevated floor 架空楼板;高架楼板
elevated floor slab 架空楼板
elevated flume 高架水槽;高架渡槽
elevated footing on piles 高承台桩基;高桩承台
elevated footway 高架人行道
elevated freeway 高架快车道;高架超速干道;升高式高速公路;高架高速公路
elevated gallery 高架廊道
elevated goods track 高架货物线
elevated ground counterpoise 架空地网
elevated guideway 高架导轨
elevated heliport 高架的直升机屋顶停机场;高楼屋顶直升飞机停机场
elevated highway 高架公路
elevated highway bridge 高架公路桥
elevated jet condenser 注水冷凝塔

elevated line 高架线(路);高架道路
elevated liquid container 高架储[贮]液容器
elevated loading ramp 高架装料台;高架装料坡道
elevated loop railroad 高架环行铁路
elevated loop railway 环形高架铁路
elevated marble bin 高位玻璃球仓
elevated master stream 举高集水射流;高位集水射流
elevated milking parlor 高架挤奶台;台式牛床挤奶厅
elevated mixer 高位搅拌机
elevated motorway 高架汽车路
elevated overcrossing waiting room 高架跨线候车室
elevated path 递升路线;上升途径
elevated pedestrian crossing 人行天桥
elevated peneplain 上升准平原
elevated pile and beam-slab quay 高桩梁板码头
elevated pile and beam-slab structure 高桩梁板结构
elevated pile footing 高桩承台
elevated pile foundation 高桩承台(格栅);高桩基础;高桩承基
elevated pile grating 高桩(承)台
elevated plain 高原;台地
elevated platform 高架站台
elevated pole 仰极【天文】;上天极
elevated pulpit 抬高的讲坛;高架操纵台
elevated railroad 架空铁路;架空铁道;高架铁路;高架铁道
elevated railway 架空铁路;架空铁道;高架铁路;高架铁道
elevated railway on pier 墩座高架铁路
elevated railway on trestle work 栈道高架铁路
elevated reef 隆起礁;上升礁;上礁
elevated reservoir 高位水库;高水箱;高水塔;高架水柜;高地水库;水塔
elevated road 栈道;高架(道)路
elevated round-about 高架环形交叉
elevated section 高架线路段
elevated shoreface terrace 海岸上升阶地;抬升海岸;升岸阶地;上升水下阶地;上升海岸阶地;上升滨前阶地;上升滨面阶地
elevated shore line 升高海岸线;上升海岸线;上升滨线
elevated sidewalk 高架人行道
elevated silo 高架圆库
elevated single-track trolley bus 高架单轨电车
elevated source 高架源
elevated speed of winch 绞车升降速度
elevated spur track 高架支轨;高架分支轨道;高架岔线
elevated stall 高位式挤奶台
elevated station 高架车站
elevated steel tank 高架钢柜
elevated storage 高位蓄水
elevated storage bin 高架储[贮]料仓;高架储料斗
elevated storage tank 高架蓄水柜;高架储[贮]水池
elevated stream 举高射流
elevated street 高架道路
elevated supply tank 高架供水塔
elevated tank 压力水箱;高位槽;高架水箱;高架水塔;高架水柜;高架(储)罐
elevated temperature 高温带;温升;升高温度
elevated temperature creep 高温蠕变
elevated temperature creep forming 高温蠕变成形
elevated temperature holograph 高温全息照相
elevated temperature impact test 高温冲击试验
elevated temperature inversion 高层逆温
elevated temperature property 高温性能
elevated temperature seal 高温密封
elevated temperature strength 高温强度
elevated temperature tension test 高温拉力试验
elevated temperature test(ing) 高温试验
elevated temperature vessel 高温容器
elevated track 高架线路;(卸货用)高架轨道
elevated track crossing 铁路在上交叉
elevated tramway 高架吊车索道;高架电车道
elevated transportation 高架运输
elevated transportation structure 高架运输结构
elevated type check valve 升降式止回阀
elevated unloading track (直立式低货位)高架卸货线[铁]
elevated water storage tank 高架蓄水箱;高架蓄

水柜
elevated water tank 水塔;高位水箱
elevated wave-cut bench 上升浪蚀台
elevated weighing bunker 高架称料存仓
elevating 升降机构
elevating and conveying machinery 升运和输送机械
elevating angle 高度角
elevating apparatus 举高消防车
elevating appliance 升高设备;高架设备;提升装置
elevating belt 提升运输带
elevating belt conveyer 高架皮带输送机
elevating boom 旋转竖杆的悬臂架;高架吊杆;提升吊杆
elevating capacity 起重能力;提升能力;升举能力
elevating chain 提升链条;输送链;升运链
elevating conveyer 提升输送带;提升输送机
elevating-deck-type mobile drilling rig 升降平台式移动钻井
elevating device 提升装置
elevating dredger 链斗式挖泥机;链斗式挖泥船;提升式挖泥机
elevating drum 升运鼓轮
elevating endless screw 升降蜗杆
elevating fire truck 举高消防车
elevating force 起重力
elevating gear 俯仰装置;提升装置;提升齿轮;升降装置
elevating grader 挖掘式推土机;电铲式平路机;犁扬机;铧式挖土平路机;起土平路机;平土升运机;升运式平土机;升运平路机;升降(式)平路机
elevating handle 高低机转把
elevating handwheel 高低机手轮
elevating hopper feeder 斗式提升送料机
elevating hopper mixer 高架斗式搅拌机
elevating installation 升降装置
elevating jack 升降千斤顶
elevating lever 升降杆;上升杆
elevating loader 链斗装载机;链斗装料机;传送式装载机;提升装载机;提升式装料机
elevating mechanism 高低机;提升机制;升运机构
elevating motor 升降电动机
elevating nozzle 举高喷水炮
elevating piler 升降堆垛机
elevating pinion 高低机齿轮
elevating platform 升降台;升降式平台
elevating platform apparatus 登高平台消防车
elevating platform truck 登高平台消防车;带升降台的载重汽车
elevating roadmixer 高架路拌(拌)和机
elevating rudder 升降舵
elevating scraper 自升式铲运机;提升式铲运机;升运式铲运机;升送式铲运机;升降式刮土机;提升铲
elevating screw 螺旋升运器;螺旋起重器;螺旋起重机;螺旋举重器;升运螺杆;升降丝杆;升降螺旋;上升螺旋
elevating sensation 抬举感
elevating shaft 上升轴
elevating stage 升降台
elevating stops 竖直位移限位器
elevating table 升降台
elevating tackle 高架辘轳;高架滑车
elevating tail gate 升降尾门
elevating unit 高架运输装置;高架运输设备;提升装置
elevating valve 升杆阀
elevating-wagon 自动升降装载车;升运装载车
elevating water device 举高射水装置
elevating wheel 升运鼓轮
elevation 立面;海拔;高程;断面图;地平高度;垂直切面;标高;提高;山顶高度
elevation above sea level 海拔高程
elevation accuracy 仰角准确度;仰角精度;高度准确度
elevation adjustment 高度调节
elevation affection pressure 海拔高度影响压力
elevational design 立面设计
elevational diagram 正视图
elevational drawing 立面图(样);立视图;正视图;正面图;前视图
elevational front 正面图
elevational point 高程点
elevational presentation 立面图像

elevational view 立视图
elevation and latitude 海拔和纬度
elevation and position update 高程与位置修订
elevation and subsidence 升降(作用)
elevation angle 仰角;垂直角;出绳角;倾斜角;竖角;升运角;高度角
elevation angle error 仰角误差
elevation angle of leading marks 导标仰角
elevation anomaly 高程异常
elevation antihunt 天线仰角阻尼
elevation-area curve 高程面积曲线;水位面积曲线
elevation axis 仰角轴;俯仰轴
elevation bearing 仰角方位
elevation cable wrap 仰角缆包
elevation-capacity curve 高程库容(关系)曲线;水位容积曲线;水位库容曲线
elevation-capacity relation 高程容积关系
elevation change 高程变化
elevation circle 垂直度盘
elevation commutator 仰角转换器
elevation computation 高程计算
elevation computer 高程计算机
elevation control 高度调节
elevation control network 高程控制网
elevation control point 高程控制点
elevation control survey 高程控制测量
elevation control system 高程系统
elevation control voltage 仰角控制电压
elevation correction 海拔校正;高程校正;高程改正
elevation correction coefficient 高度校正系数
elevation correction factor 高程校正系数
elevation correction value 高度校正值
elevation counter 高程计数器
elevation coverage 仰角范围
elevation coverage diagram 仰角视野特性曲线;仰角反射特性曲线;垂直切面覆盖图;竖观覆盖图
elevation data 高程数据
elevation data compaction 高程数据压缩
elevation datum 高程水准面
elevation design (城市道路的)竖向设计
elevation detail 立面细部;高程碎部点
elevation determination 高程测定
elevation difference 仰角差;高差;标高差;射角差
elevation difference between riffle and pool 滩槽高差
elevation difference between two adjacent contour lines 等高线间距
elevation-discharge curve 高程泄量关系曲线
elevation-distance curve 高程距离曲线
elevation drawing 正面图;立剖图;立面图
elevation drive 仰角驱动;仰角传动;俯仰传动
elevation drive gear 仰角传动机构
elevation drive motor 仰角驱动电动机;仰角传动电动机
elevation effect 高度影响
elevation effect on atmospheric pressure 海拔高度对大气压力的影响
elevation equation 高程方程
elevation error 仰角误差;高程误差
elevation error signal 仰角误差信号
elevation facing yard 对着庭院的立面
elevation factor 高程系数
elevation figure 高程注记
elevation finder 仰角指示器;仰角探测仪
elevation frequency curve 高程频率曲线
elevation grade 平地升运机
elevation grader 犁扬机
elevation guidance element 仰角引导单元
elevation handwheel 仰角手轮;高低角操纵轮
elevation head 高程水头;位头;位势水头;水力静压头
elevation increment 高程增量
elevation indication 高程指示;高程读数
elevation indicator 仰角显示器;高度指示器;高度显示器;射角指示器
elevation information 仰角信息
elevation knob 俯仰转螺
elevation level 仰角电平
elevation line 高架线路
elevation loss 位能损失
elevation mark 高程标志
elevation meter 仰角计;高程计;测高计
elevation net storage curve (水库的)高程净蓄量曲线

elevation note 高程注记
elevation number 海拔;高程注记;等高线注记
elevation of a building 建筑立面图
elevation of a liquid column 液柱高度
elevation of boiling point 沸点升高
elevation of bore 井口高程;井口标高
elevation of bore hole 孔口标高【岩】
elevation of bridge deck 桥面高程
elevation of building 建筑立面;房屋立视图
elevation of center-line 中线标高
elevation of culvert 涵底标高
elevation of derrick floor 钻塔地板高度
elevation of diffusers 扩散管标高
elevation of end point 端点高程
elevation of floor level 室内地坪标高;室内地面标高
elevation of foundation base 地基开挖深度
elevation of ga(u)ge datum 水尺零点标高
elevation of ga(u)ge zero 水尺零点高程;水尺零点标高
elevation of hole 钻孔标高
elevation of impervious bottom plate 隔水底板高程
elevation of instrument (测量时的)仪器高程
elevation of levelled ground 室外地面标高
elevation of light 灯芯高度
elevation of line of sight 视线标高
elevation of lock bottom 闸底板高程
elevation of lock floor 闸底板高程
elevation of low water 枯水位标高
elevation of natural ground 自然标高
elevation of open pit 露采场标高
elevation of per datum 固定参考点海拔高度
elevation of pole 仰极高度
elevation of rail top 轨面高程
elevation of rubble mound breakwater crest 斜坡堤堤顶高程
elevation of sewer foundation 管道基础标高
elevation of sewer invert 管道沟底标高
elevation of shaft bottom 井底标高
elevation of sight 瞄准器标高;(测量时的)仪器高程;仪器高;视线高
elevation of sight line 视线高程
elevation of soil formation 土层标高
elevation of soil layer 土层标高
elevation of stage 舞台高程
elevation of station 测站高程
elevation of trench bottom 沟槽底标高
elevation of water 水位;水(平)面高度
elevation of water surface 水平面高度;水(平)面高程
elevation of well 井口高程;井口标高;井海拔
elevation of well head 井口标高
elevation of wharf apron 码头前沿高程
elevation of zero point 零点高程
elevation order 高程等级;高程测量等级
elevation perspective 立面透视图
elevation photo control point 像片高程控制点
elevation planning 竖向设计
elevation planning of a building 建筑立面设计
elevation point 高程点
elevation point by independent intersection 独立交会高程点
elevation point with notes 高程注记点
elevation position-finding 仰角位置测定;仰角测位
elevation position-finding antenna 仰角测位天线
elevation position indicator 仰角位置雷达指示器
elevation post 高程标(志)
elevation potentiometer 仰角分压器;仰角电位计
elevation prediction correction 射角的前置修正量
elevation pressure 高程压力
elevation profile 高程断面
elevation radiation pattern 仰角辐射图案
elevation rate 射角速率
elevation resolution 仰角分辨能力
elevation restitution 等高线测绘
elevation rod 避雷针
elevation rotating joint 仰角旋转连接
elevation ruler 高程尺
elevation scale 仰角标度;射角分划
elevation scale projection screen 高程刻度投影屏
elevation scheme 高程布置;(城市道路)竖向设计
elevation selsyn 仰角自动同步机
elevation selsyn drive gear 仰角自动同步传动机构

elevation selsyn transformer 仰角自动同步机变压器
elevation servo power amplifier 仰角伺服功率放大器
elevation setting 仰角定位
elevation slide 高程滑尺
elevation slope line 斜高
elevation stop 俯仰限制器
elevation stowing switch 仰角限制器
elevation surveying 高程测量
elevation synchro gear 仰角同步传动
elevation table 高程表;射角表
elevation tints 高程分层设色
elevation tower 仰角塔
elevation tracker 仰角跟踪器;高低跟踪器
elevation tracking 仰角跟踪
elevation tracking cursor 仰角跟踪指示器
elevation transmission for shaft 竖井高程传递
elevation type 升地式
elevation view 正视图;立视图;立面图
elevation yoke 仰角偏转线圈
elevation zero 高程零点
elevation zone 高程分带
elevator 卸货机;直升电梯;电梯;提引器;提升机;升运机;升降机;升降舵
elevator actuating cylinder 升降动作筒
elevator angle 升降舵偏转角
elevator apartment 高层公寓
elevator apartment house 有电梯的公寓大楼
elevator attendant 电梯司机
elevator automatic dispatching device 电梯自动调度装置
elevator bail 吊卡(提环);提环
elevator balance 升降机平衡
elevator bank 电梯组
elevator bar 防磨凸块
elevator barge 满底泥驳;封底泥驳
elevator belt(ing) 升运机皮带;提升带
elevator bolt 大平头螺栓;方颈埋头螺栓
elevator boom 升降机伸臂
elevator boot 提升机进料箱;升运器底滑脚
elevator bottom 升运器底滑板
elevator bridge 提升(门)开合桥
elevator bucket 吊斗;提升(机勺)斗;升运斗;升降(机屑)斗
elevator cab 电梯桥厢
elevator cable 升降机电缆;升运机用索;升降机供电电缆;升降舵控制索;电梯吊索
elevator cage 笼门;电梯门;电梯车厢;电梯车箱;升降吊笼
elevator canvas 升运器帆布带
elevator capacity 电梯容量
elevator capture 电梯专用
elevator car 升降机轿厢;电梯车厢;梯厢;升降机车厢
elevator car buffer 梯厢减振器
elevator car dimensions 梯厢尺寸
elevator car leveling device 升降机轿厢平层装置;车厢平准装置
elevator car safety 电梯安全装置;升降机安全器
elevator casing 升运机罩;提升机外壳
elevator casing wearing plate 升降机罩衬板
elevator chain 提升链;提升机链条;升运链
elevator chamber 沉浮箱
elevator channel 升运器槽
elevator classifier 提升分级机
elevator component 升降机组成部分
elevator control 升降机控制;升降舵控制
elevator control gear 电梯控制装置
elevator controller 升降机控制器
elevator control lever 升降舵控制杆
elevator conveyer belt 升运机输送带
elevator core 电梯井(筒)
elevator crank 升降机曲柄
elevator crib 升降机箱
elevator cup 升运器斗
elevator digger 升运式挖掘机
elevator digger with transverse delivery 横送升运式挖掘机
elevator door 电梯门
elevator dredge(r) 链索斗挖泥机;升降(式)挖泥机;升降式挖泥船;链斗式挖泥机;链式挖泥船;挖泥提升机
elevator driving rack 升降机主动齿条
elevator dwelling block 有电梯的住宅建筑街区

elevator dwelling house 有电梯的住宅房屋
elevator equipment room 电梯设备房
elevator flight 链耙刮板;提升机刮板
elevator for access to driver's cab 司机室升降台
elevator for high altitude working 高空作业升降车
elevator for root cutter 块根切碎机升运器
elevator frame 链耙骨架;链斗车架
elevator furnace 升降(台)式炉;升降底式炉
elevator gear reducer box 电梯减速箱
elevator groove 吊卡槽
elevator guard 链耙护罩
elevator guide rail 升降机导轨
elevator hall 升降机前厅;电梯(前)厅
elevator hay loader 升运器式装干草机
elevator hoistway 电梯(竖)井;电梯间;升降机井(道)
elevator hoistway door 电梯升降机井门
elevator hopper 提升机料斗;提升机料仓
elevator hopper barge 满底泥驳
elevator horn 升降舵杆
elevator installation 电梯安装;升降机设备(安装)
elevator installation contractor 电梯安装承包者;电梯安装承包商;电梯安装承包人
elevator interlock 升降机联锁装置
elevator jib 升降机臂
elevator kiln 升降窑;升底窑
elevator ladder 挖泥机斗支架;链斗斗架
elevator-ladder dredge(r) 链斗式挖土机;链斗式挖泥船;链斗式挖掘机;链索斗挖土机
elevator landing 电梯平台;升降机平台;电梯走廊;电梯通道
elevator landing entrance 电梯乘客入口
elevator landing stopping device 升降机停层装置
elevator landing zone 升降机平层区
elevator latcher 吊卡栓井架工;(钻探时)塔上工人
elevator-launcher 升降机发射装置
elevator leg 升运器支架
elevator liner 升降机衬垫
elevator link 吊环
elevator loader 升降装载机
elevator loader-excavator 提升式装载挖土机
elevator lobby 电梯(门)厅
elevator machine beam 电梯设备钢梁;升降机梁
elevator machine room 电梯设备房;电梯机房;升降机机房
elevator machinery 起重机械
elevator motor 电梯马达;起重电动机;升运马达;升降机马达;电梯电动机
elevator multiple dwelling 有电梯的多层住宅
elevator nonstop switch 升降机不停层开关
elevator parking device 升降机停车装置
elevator pawl casing 升降机棘爪套
elevator pawl pin 升降机爪销
elevator pawl spring 升降机爪簧
elevator penthouse 屋顶升降机房
elevator periscope 升降式望远镜;升降式潜望镜
elevator pinion 升降机小齿轮
elevator pit 升降机(底)坑;电梯(井)坑;升运器喂料坑
elevator plant 升降机制造厂
elevator platform 升降台
elevator plug (钻杆的)提引塞;提升短节
elevator potential switch 升降机电压开关
elevator power-off protection device 电梯停电保护装置
elevator pump 链斗式提水机;提升泵
elevator quadrant 升降机扇形齿轮
elevator ratchet wheel 升降机棘轮
elevator recess 吊卡凹座
elevator residence block 有电梯的住宅大楼
elevator residence building 高层住宅建筑
elevator residence house 有电梯的房屋
elevator residential 住宅电梯
elevator residential block 有电梯的住宅大楼
elevator residential house 有电梯的居住房屋
elevator room 电梯机房
elevator rope 升降机操作绳;起重机用索;升降缆绳
elevator scoop 提升机料斗;升降斗
elevator screw 升降机螺旋
elevator servo 升降舵伺服系统
elevator shaft 电梯井(筒);升降机井(道)
elevator shaft bush 升降机轴衬
elevator shaft gates 电梯井道栏栅

elevator shaftway door 电梯井门
elevator sheave beam 电梯设备钢梁;升降机滑轮固定梁
elevator slack-cable switch 升降机钢丝绳松弛开关
elevator speed 电梯速度;升降机速度
elevator spider 吊卡—卡盘
elevator split ring 升降机开口环
elevator spout 升运器排出槽
elevator stage 升降舞台
elevator stick 升降舵操纵杆
elevator stirrer 升降机搅动器
elevator tectonics 起落构造
elevator tower 电梯塔;电梯机房;升降塔架;升降机塔(架);升降机井顶楼
elevator tower hoist 升降塔架绞车
elevator travel 电梯行程
elevator truck zone 升降机平层上限距离
elevator-type car park 电梯型停车场
elevator-type load 升运式装载机;带铲斗的升运机
elevator type loader 提升式装载机;升运式装载机
elevator-type loading car 升降式装卸车
elevator-type pick-up loader 升运器式捡拾装载机
elevator well 电梯井;升降机井(道)
elevator wrist pin 升降机关节销
elevator wrist pin keeper 升降机肘节销定位螺钉
elevatory 提起的
eleven points moving average 十一点移动平均
eleven punch 十一行穿孔
elevon 升降副翼
Elexal 铝阳极氧化处理
Elex precipitator 板式电除尘器;平板式电集尘器
elfelite 钠镁大隅石
Elfin 埃尔芬数字管
elfin 高山矮曲林
elfin tree 矮树;高山矮曲林
elfinwood 高山矮曲林
Elfving distribution 埃尔文分布
Elgiloy 埃尔基洛伊耐蚀游丝合金
Elgin extractor 埃尔金萃取器
Elgin sand 埃尔金砂
Elianite 高硅耐蚀铁合金;埃利阿奈特耐蚀高硅铁基合金
elidible hypernotion 可除的超概念
eligibility 合格性;适当性
eligibility for tax credit 符合税收抵免条件
eligible bidder 合格投标者;合格投标人;符合条件的投标者
eligible bill 合法票据
eligible bill for rediscounting 可再贴现票据
eligible cargo 合格货物
eligible country 合格国家
eligible goods 合格货物
eligible goods and service 合格的货物和服务
eligible paper 适当的票据
eligible payment 合格的付款
eligible process 可选进程
eligible product 合格品
eligible source 合格货源
eligible source countries 合格货源国
eligible supplier 合格供货者
eligible tenderer 合格投标者;合格投标人;符合条件的投标者
elimanating pests and preventing disease 除害防治
eliminable 可消除的
eliminable cost 可节省成本
eliminant 消元式;排除剂;排除的
eliminate 消除;排除
eliminate foreign exchange risk 消除外汇风险
eliminate or convert coal-burning boiler 淘汰或改造燃煤锅炉
eliminate smoke and dust 消烟除尘
eliminate unemployment 消除失业
eliminating film 分离膜
eliminating ill weeds 消灭危害性大的杂草
eliminating infection 消除感染
eliminating of outmoded products 淘汰落后产品
eliminating pest 消灭虫害
eliminating trouble some weeds 消灭难以根除的杂草
elimination 铜除;消去;消除;除去;冲消;弃置;排除
elimination action 消去作用
elimination-addition mechanism 消去加成机制
elimination by addition or subtraction 加减消元法

elimination by comparison 比较消元法
elimination by substitution 代入消元法
elimination channeling 减少窜槽
elimination constant 消除常数
elimination entry 冲销分录
elimination error 消除差错
elimination factor 消去系数;消去率;消除率
elimination key 排除法判读样片;排除法判读标志
elimination ledger 抵消账
elimination method 消元法;消去法
elimination of constant 常数消去法
elimination of damage 排出危害
elimination of discrimination treatment 取消歧视待遇
elimination of error 误差消除
elimination of grade crossings 消除平交道口
elimination of group 消去取代基
elimination of heat 排热(法);散热;消除热量
elimination of iron 除铁
elimination of odo(u)r and taste 臭味消除
elimination of parallax 视差消除
elimination of pesticide 农药清除
elimination of static electricity 消除静电
elimination of thermal strain 热应变消除(法)
elimination of unknowns 消元法
elimination of water 消去水;消除水;脱水;除去水
elimination period 剔出期间
elimination process 排除法
elimination rate 消除(速)率
elimination rate constant 消除速率常数;排除速率常数
elimination reaction 消去反应
elimination run 预运转
elimination sulfur 除硫
elimination technique 排除技术
elimination test 消去试验
elimination test of anion 阴离子消去试法
eliminator 空气净化器;空气净化,消除器;限止器;抑制器;液滴分离器;挡水板;分离机;排除器;排出器
eliminator of three-recording errors 三差消除器
eliminator plate 洗涤室挡水板;除水板;洗涤室阻水板
eliminator power 整流电源
eliminator supply 整流电源
elint ship 电子情报船
Elinvar 镍铬恒弹性钢;恒弹性镍铬钢;埃林瓦尔钢;埃林瓦尔(恒弹)合金
elinvar alloy 弹性不变合金
eliquation 液析
elision 沉淀间断作用
elision of cycle stage 沉积缺失
elite market 富有的市场
Elitherm 埃立顿(一种专卖的无缝地面材料)
Elizabethan style 伊丽莎白建筑形式;伊丽莎白式建筑
elk 道砟(的别名)
Elkaloy 埃尔卡洛伊铜合金焊条
elkerite 埃尔克沥青
elkhornite 拉辉正长岩
Elkins formula 1/2/3 艾勒金斯公式
Elkmet 艾尔麦特钨铜合金
Elkonit 艾尔柯尼特合金
Elkonite 埃尔科奈特钨铜烧结合金
Elkonium 埃尔科尼姆接点合金
el(l)侧房;肘形弯管
Elland eave-lead 爱尔兰屋檐铅皮排水槽
Elland stone 爱尔兰石
ell-beam L形梁
elleptic(al) involution 椭性对合
ellestadite 硅磷灰石
ell girder L形大梁
Ellliont type knuckle 埃利奥特变向关节
elliot axle 端叉前轴
Elliot cycle 埃利奥特循环
Elliot sequence control 埃利奥特顺序控制
Elliot test 埃利奥特(闪点测定)试验
Elliot tester 埃利特测定器;埃利奥特(闪点测定)试验器
elliott eye 缆端眼圈
ellipse 椭圆(形)
ellipse area 椭圆面(积)
ellipse element 变形椭圆元素
ellipse head 椭圆封头

ellipse-hyperbolic system 椭圆双曲线系统
ellipse of distortion 底索曲线;变形椭圆
ellipse of elasticity 弹性椭圆
ellipse of errors 误差椭圆
ellipse of inertia 惯性椭圆;惯量椭圆
ellipse of shadow cone 影锥椭圆
ellipse of strain 应变椭圆
ellipse of stress 应力椭圆
ellipse of zero velocity 零速度椭圆
ellipse section 椭圆形断面
ellipse speaker 椭圆形扬声器
ellipse stress 应力椭圆
ellipse template 椭圆样板
ellipse valve diagram (蒸汽机的)椭圆配汽图
ellipsis 省略符号
ellipsograph 椭圆规
ellipsoid 椭圆(粒)体;椭球(体);椭(球)面
ellipsoidal 椭圆形的;椭球状的;椭球体的
ellipsoidal binary 椭球双星
ellipsoidal bound 椭球界
ellipsoidal cavity resonator 椭圆形旋转式空腔谐振器
ellipsoidal chord distance 椭球弦距
ellipsoidal coordinates 椭球(面)坐标
ellipsoidal core antenna 椭圆形铁芯的磁性天线
ellipsoidal curvature 椭球面曲率
ellipsoidal datum 椭球基准
ellipsoidal distance 椭球面距离
ellipsoidal distribution of velocities 速度椭球分布
ellipsoidal drift tube 椭球形漂移管
ellipsoidal excess 椭球面角超
ellipsoidal floodlight 椭球形泛光灯
ellipsoidal function 椭球函数
ellipsoidal geoid 椭球状水准面
ellipsoidal grating 椭球面光栅
ellipsoidal harmonics 椭球(调和)函数;调和函数
ellipsoidal head 椭圆形封头
ellipsoidal head lamp 椭圆形前大灯
ellipsoidal height 椭球体高程;椭球面高
ellipsoidal junction 椭球汇合处
ellipsoidal lava 枕状熔岩
ellipsoidal meridian 椭球子午线
ellipsoidal mirror 椭圆形球面镜;椭球(面)镜;椭变面镜
ellipsoidal model 椭球模型
ellipsoidal nodule 椭球状结核
ellipsoidal norm 椭球模数
ellipsoidal normal 椭球体法线;椭球法线
ellipsoidal of inertia 惯量椭球
ellipsoidal orbit 椭球轨道
ellipsoidal parameter 椭球参数
ellipsoidal polarization 椭球偏振
ellipsoidal reflector 椭球反射镜
ellipsoidal semi-spheric(al) pore 椭圆半圆孔
ellipsoidal shape 椭球形
ellipsoidal shell 椭球面壳
ellipsoidal shell roof 椭圆形壳顶
ellipsoidal spotlight 椭圆聚光灯;椭球体聚光灯
ellipsoidal surface 椭球面
ellipsoidal triangle 椭球面三角形
ellipsoidal trochoidal wave 浅海波;椭圆余摆线波
ellipsoidal tubesheet 椭球管板
ellipsoidal variable 椭球变星
ellipsoidal zenith 椭球天顶;椭球体天顶距
ellipsoid dome 椭圆形穹顶;椭圆形大厦;椭圆形球顶
ellipsoid knob 椭球状柄;蛋形门拉手
ellipsoid method (探伤定位的)椭圆体法;(探伤定位的)椭球体法
ellipsoid of deformation 变形椭球体
ellipsoid of inertia 惯性椭圆体
ellipsoid of revolution 回转椭圆球面;回转椭球;旋转椭球面;旋转(圆)球面
ellipsoid of rotation 旋转椭球体;旋转椭球面
ellipsoid of stress 应力椭圆体;应力椭圆面;应力椭球(面)
ellipsoid of the earth 地球椭圆体
ellipsoid of wave normals 波面法线椭球
ellipsoid particle 椭圆粒
ellipsoid resonator 椭圆形(旋转式)谐振器
ellipsoid vault 椭球状拱顶;椭圆形拱顶
ellipsometer 偏振光椭圆率测量仪;椭圆计;椭率计
ellipsometry 椭圆对称性;椭率测量术
ellipsometry method 椭圆计法

elliptic(al) 椭圆(形)的
elliptic(al) anomaly 椭圆状异常
elliptic(al) antenna 椭圆形天线
elliptic(al) aperture 椭圆孔
elliptic(al) arc 椭圆弧
elliptic(al) arch 椭圆拱
elliptic(al) arch dam 椭圆拱坝
elliptic(al) area 椭圆面(积)
elliptic(al) azimuth 椭圆面方位角
elliptic(al) barrel vault 椭圆形的筒形穹隆
elliptic(al) bearing 椭圆(形)轴承
elliptic(al) blade 椭圆叶片(推进器)
elliptic(al) blade section 椭圆叶片截面
elliptic(al) bolt hole 椭圆螺栓孔
elliptic(al) build package 椭圆形卷装
elliptic(al) cam 椭圆凸轮
elliptic(al) catenary 椭圆悬链线
elliptic(al) chuck 椭圆卡盘
elliptic(al) coil 椭圆线圈
elliptic(al) coiling (管子的)椭圆盘绕
elliptic(al) collineation 椭性直射
elliptic(al) column 椭圆柱
elliptic(al) compasses 椭圆(画)规
elliptic(al) complex 椭圆复形
elliptic(al) conchoid 椭圆蚌线
elliptic(al) concrete pipe 椭圆混凝土管
elliptic(al) conducting cylinder 椭圆导电柱
elliptic(al) conduit 椭圆形通道;椭圆形涵洞
elliptic(al) cone 椭圆锥面
elliptic(al) congruence 椭性线汇
elliptic(al)-conic(al) curved surface 椭圆锥曲面
elliptic(al) conoid 椭圆锥体
elliptic(al) coordinates 椭圆坐标;椭球坐标
elliptic(al) cosine wave 椭圆余弦波
elliptic(al) (cross-)section 椭圆截面;椭圆横断面
elliptic(al) cupola 椭圆屋顶建筑
elliptic(al) curtain in the chain system 椭圆花环形链幕
elliptic(al) curvature 椭圆的曲率
elliptic(al) curve 椭圆曲线
elliptic(al) cutwater 椭圆分水尖
elliptic(al) cyclic group 椭圆循环群
elliptic(al) cylinder 椭圆柱(面)
elliptic(al) cylinder coordinates 椭圆柱坐标
elliptic(al) cylinder function 椭圆柱函数
elliptic(al) cylindric(al) coordinates 椭圆柱坐标
elliptic(al) cylindric(al) surface 椭圆柱面
elliptic(al) cylindric(al) wave function 椭圆柱面波函数
elliptic(al) differential equation 椭圆(型)微分方程
elliptic(al) differential operator 椭圆型微分算子
elliptic(al) dispersion 椭圆色散
elliptic(al) distortion 筒体弯形
elliptic(al) distribution 椭圆分布
elliptic(al) dome 椭圆形穹顶;椭圆屋顶
elliptic(al) dot screen 链形网屏
elliptic(al) edges 椭圆边
elliptic(al) equal-area projection 等积椭圆投影
elliptic(al) equation 椭圆型方程;椭圆时差
elliptic(al) error 椭圆误差
elliptic(al) field 椭圆场
elliptic(al) file 椭圆形锉
elliptic(al) fin tube 椭圆管式翅片管
elliptic(al) flat-plate 椭圆形板
elliptic(al) function 椭圆函数
elliptic(al) function field 椭圆函数域
elliptic(al) gallery 椭圆形围栏;椭圆形坑道
elliptic(al) gear 椭圆齿轮
elliptic(al) geometry 椭圆几何
elliptic(al) glass reflector 双曲线反射器
elliptic(al) grate 椭圆形炉栅
elliptic(al) guide 椭圆波导
elliptic(al) harmonic motion 椭圆和谐运动
elliptic(al) head 椭圆封头
elliptic(al) hole 椭圆形井眼;椭圆孔
elliptic(al) homology 椭性透射
elliptic(al) horn 椭圆形喇叭
elliptic(al) integral 椭圆积分
elliptic(al) involution 椭圆对合;椭性对合
elliptic(al) iris waveguide 椭圆形光圈式波导管
elliptic(al) irrational function 椭圆无理函数
elliptic(al) isoline 椭圆形等值线
elliptic(al) liferaft 椭圆形救生筏
elliptic(al) loading 按椭圆分布的荷载;按椭圆分

布的负荷
elliptic(al) loudspeaker 椭圆形扬声器
elliptically polarized light 椭圆偏振光
elliptically polarized radiation 椭圆偏振辐射
elliptically polarized wave 椭圆偏振波;椭圆极化波
elliptically symmetric distribution 椭圆对称分布
elliptic(al) magnetic axis 椭圆形磁轴
elliptic(al) meridian 椭圆经线
elliptic(al) mirror 椭圆镜
elliptic(al) model 椭圆模型
elliptic(al) modular function 椭圆模函数
elliptic(al) modular function field 椭圆模函数域
elliptic(al) modular group 椭圆模群
elliptic(al) motion 椭圆运动
elliptic(al) motion vibrating screen 椭圆运动振动筛
ellipticalness 椭圆形
elliptic(al) non-Euclidean geometry 椭圆形非欧几何学
elliptic(al) norm 椭圆形模
elliptic(al) nosed cylinder 椭圆形头部汽缸
elliptic(al) operator 椭圆算子
elliptic(al) orbit 椭圆轨迹;椭圆轨道
elliptic(al) orbiting capture 椭圆轨道捕获
elliptic(al) orbiting velocity 椭圆轨道速度
elliptic(al) orbit life 椭圆轨道寿命
elliptic(al) orifice 椭圆孔口
elliptic(al) paraboloid 椭圆抛物面
elliptic(al) paraboloidal roof 椭圆抛物面壳顶
elliptic(al) paraboloid shell 椭圆抛物面壳(体)
elliptic(al) parallel 椭圆纬线
elliptic(al) partial differential equation 椭圆型(偏)微分方程;椭圆型差分方程
elliptic(al) partial differential operators 椭圆型偏微分算子
elliptic(al) pencil of circle 椭性圆束
elliptic(al) pin 椭圆形销
elliptic(al) plane geometry 椭圆平面几何学
elliptic(al) plate 椭圆板
elliptic(al) point 椭圆点
elliptic(al) pointed arch 椭圆尖(头)拱
elliptic(al) polarization 椭振;椭圆偏振;椭圆极化
elliptic(al) polarized antenna 椭圆极化天线
elliptic(al) polarized light 椭圆偏光
elliptic(al) problem 椭圆形问题
elliptic(al) projection 椭圆投影;椭球投影
elliptic(al) projectivity 椭性直射
elliptic(al) pseudo-differential operator 椭圆型伪微分算子
elliptic(al) quadratic surface 椭圆型二次曲面
elliptic(al) quadric hypersurface 椭圆型二次超曲面
elliptic(al) rate 椭圆率
elliptic(al) reflector 椭圆反射镜
elliptic(al) region 椭圆区域
elliptic(al) representation 椭圆表示法
elliptic(al) restricted problem 椭圆型限制性问题
elliptic(al) Riemann surface 椭圆型黎曼曲面
elliptic(al) ring 椭圆环
elliptic(al) ring structure 椭圆环形构造
elliptic(al) roof 椭圆形屋顶
elliptic(al) rotating field 椭圆旋转场;椭圆(形)旋转磁场
elliptic(al) rotational curved surface 椭圆回转曲面
elliptic(al) scanning 椭圆扫描
elliptic(al) section 椭圆形截面
elliptic(al) sector 椭圆扇形
elliptic(al) segment trajectory 椭圆弧轨迹
elliptic(al) semidiurnal constituent 椭圆半日分潮
elliptic(al) semiflexible waveguide 椭圆半软式波导
elliptic(al) shaft 椭圆形井筒
elliptic(al) shape 椭圆形
elliptic(al) shell cupola 椭圆壳体圆屋顶
elliptic(al) sine function 椭圆正弦函数
elliptic(al) singular point 椭圆奇点
elliptic(al) skirted piston 椭圆裙活塞
elliptic(al) slab 椭圆平板
elliptic(al) slot 椭圆形槽
elliptic(al) space 椭圆空间
elliptic(al) spike 椭圆形穗
elliptic(al) spot 椭圆形斑
elliptic(al) spring 双弓板弹簧;椭圆形叠板弹簧;椭圆(形)弹簧;椭圆板板(弹)簧
elliptic(al) spring bolster 椭圆形板弹簧座

elliptic(al) stair(case) 椭圆(楼)梯;盘旋楼梯
elliptic(al) stern 椭圆(形)船尾
elliptic(al) stone arch bridge 椭圆形石拱桥
elliptic(al) strand rope 椭圆形股钢丝绳
elliptic(al) subcarrier 椭圆色度副载波
elliptic(al) subsystem 椭圆次系
elliptic(al) surface 椭圆曲面
elliptic(al) sweep 椭圆扫指
elliptic(al) system 椭圆系统
elliptic(al) tide 椭圆潮
elliptic(al) time base 椭圆时基(扫描)线
elliptic(al) tire 椭圆轮胎
elliptic(al) trace 椭圆扫描
elliptic(al) trajectory 椭圆轨道
elliptic(al) trammel 椭圆仪;椭圆规
elliptic(al) transformation 椭圆交换;椭圆变换
elliptic(al) trochoid 椭圆摆线
elliptic(al) trochoidal wave 椭圆余摆线波;浅海波
elliptic(al) tube 椭圆管
elliptic(al) tubesheet 椭圆管板
elliptic(al) type 椭圆型
elliptic(al) type of Riemann surface 椭圆型黎曼曲面
elliptic(al) type weight 铅鱼;椭圆形测深(重)锤
elliptic(al) valve 椭圆形阀
elliptic(al) vault 椭圆穹隆;椭圆拱顶
elliptic(al) vessel 椭圆形容器
elliptic(al) vibrating screen 椭圆形振动筛
elliptic(al) vibration 椭圆振动
elliptic(al) wave 椭圆波
elliptic(al) waveguide 椭圆形波导管
elliptic(al) wheel 椭圆(车)轮
elliptic(al) winch 椭圆形绞盘
elliptic(al) wire 椭圆形截面织针
ellipticity 长短轴比;不圆性;不圆度;扁率;椭(圆)率;椭圆度
ellipticity coefficient 扁率系数
ellipticity condition 椭圆率条件
ellipticity effect 椭球状效应
ellipticity of cement pipe 水泥管椭圆率
ellipticity of ellipse 椭圆扁率
ellipticity of ellipsoid 椭球扁率
ellipticity of light ellipse 光椭圆扁率
ellipticity of spheroid 椭圆扁率;椭球体扁率
ellipticity of the earth 地球椭率;地球扁率
ellipticity parameter 椭圆率参数
ellipticity ratio 椭圆度比
ellipticity tables 椭圆表
elliptic-nosed cylinder 椭圆形头部汽缸
elliptic(al) one of strain ellipsoid 应变椭球圆锥面
elliptic one of strain ellipsoid 应变椭球圆锥面
Ellis fluid 埃利斯流体
ellisite 硫砷铊矿
Ellis model for flow 埃利斯流动模型
Ellis mortar 埃利斯钵
ellitoral 远岸线海底的
ellitoral zone 远岸浅海底带
Ellwood technique 埃利伍德法(用于投资分析)
elm 榆树;榆木
Elmarit 埃马里特硬质合金
elm bark 榆树皮
elm bast 榆树皮纤维
elmendorf machine 单板片削机
Elmendorf tearing tester 爱尔门道夫式撕裂度测定仪;埃乐曼多夫扯裂试验仪
Elmillimess 电动测微仪
elmination of burst noise 去野值
elminator 滤波电路
Elmo pump 爱尔莫真空泵
El Nino 厄尔尼诺现象(又称圣婴现象);厄尔尼诺海流
Elnino current 厄尔尼诺海流
El Nino/Southern oscillation 厄尔尼诺/南方涛动
elobw type lock 肘形锁闭器
e-lgP curve 压缩曲线【岩】
eloignment 发还抵债物品
elomag process 镁合金阳极氧化法
elongate 拉长;细长形;伸长
elongate along the axis of the root 沿根轴伸长
elongated aggregate 细长骨料;细长集料;针状骨料;细长粒料
elongated anticline 狭长背斜
elongated anticline trap 狭长背斜圈闭
elongated apical system 延伸顶系

elongated axis 伸长轴
elongated blister 小长气泡
elongated bubble 小长气泡
elongated charge 长条状爆炸震源;长条形炸药
elongated cocoon 长茧
elongate delta 纵长三角洲
elongated focusing electrode 加长聚焦电极
elongated fold 纵长褶皱
elongated four-spot pattern 伸长四点井网
elongated grain 拉长的晶粒
elongated hole 延伸孔
elongated lens 扁长矿体;延长镜头;长条形透镜体
elongated material 长粒(砂石)材料
elongated particle 细长粒子;细长颗粒;针状颗粒
elongated piece 长形件;长粒料;长骨料
elongated shoot 延长枝
elongated slot 长槽
elongated structure 纵长构造
elongated tent shaped stockpile 长角柱形料堆
elongated trough 延长凹槽
elongate with long axis 沿长轴伸展
elongating mill 延伸轧机
elongation 拉长;延展率;延伸;天体距角;伸展;伸长(度)
elongational flow 拉伸流动;延伸流动;伸长流动
elongational rate 伸长应变率
elongational viscosity 延伸黏度;伸长黏性;伸长黏度
elongation areas 伸长区
elongation at break 裂断伸长;极限延伸率;致断伸长(度);断裂延伸率;断裂伸长度;断裂伸长(率);扯断伸长率
elongation at constant load 定载伸长
elongation at failure 扯断伸长率;破坏伸长率;破坏伸长度;衰坏伸长
elongation at fracture 断裂时延伸
elongation at rupture 断裂时试样伸长量;断裂时试样的拉伸;断裂伸长度;断裂伸长(率);断裂拉伸;扯断伸长率
elongation at yield 屈服伸长(率);屈服伸长度
elongation between ga(u)ges 标距伸长(率)
elongation coefficient 伸长系数
elongation due to tension 拉伸;受拉伸长
elongation factor 延伸因子;延长因子;伸长系数
elongation for each pass 每道次延伸量
elongation growth 伸长生长
elongation in deep water 在深水中的生长力
elongation index 细长指数;伸长指数
elongation in tension 受拉伸长
elongation line 延长线(路)
elongation load 延伸试验负载
elongation modulus 延伸系数
elongation of circumpolar star 拱极星大距
elongation of internode 节间延长
elongation of rupture 断裂处延长
elongation of valve stem 阀杆伸长
elongation pad 延长器
elongation pass 延伸孔型
elongation percentage 延伸(百分)率
elongation per unit length 相对伸长;单位(长度)伸长;伸长率
elongation ratio 延伸率;延迟率;延长率;伸长比(值)
elongation recorder 伸长记录仪
elongation scale 伸长标度
elongation set 伸长变定
elongation sign 延性符号;延长符号
elongation stage 伸长期
elongation strain 拉伸应变;延伸应变;伸长应变;伸长形变
elongation stress 延伸应力
elongation tensor 伸长张量
elongation test 拉伸试验;拉长试验;延伸(率)试验
elongation zone 伸长区
elongator 延伸轧机;延伸机
ELO-Vac process 电炉真空脱碳脱气法
eloxal 铝阳极氧化处理
eloxal coating 铝阳极涂层处理
eloxal process 铝表面阳极氧化法
elpasolite 钾冰晶石
El Paso's smut 埃尔帕索烟尘(美国)
Elphal 电泳涂铝;埃尔法尔电泳涂铝钢带
elpidite 斜钠锆石;纤硅锆钠石
Elrod-Maron-Krieger equation for shear rate 埃

尔罗德—马伦—克里格剪切速率方程
Elsan chemical closet 埃桑化学马桶(一种专利产品)
Elsasser band model 艾尔萨沙带模式
Elsasser's radiation chart 艾尔萨沙辐射图
Elster-Geitel effect 埃尔斯特—盖特尔效应
Elsterian glacial epoch 埃尔斯特冰期
Elsterian glacial stage 埃尔斯特冰期
Eltanin fracture zone 埃尔塔宁断裂带
Eltermit 爱特尼特石棉水泥板
Elton's ring test 埃尔顿氏环试验
el-train 高架铁路电气车
eluant 洗提液;洗提剂
eluant applied in sequence 按顺序加的洗提液
eluant component 洗脱液组分;洗提液组分
eluant composition 洗提液成分
eluant gas 洗脱用气体
eluant strength 洗脱液强度;洗提液强度
eluant strength gradient 洗脱液强度梯度;洗提液强度梯度
eluant stripper column 洗脱液清洗柱;洗提液清洗柱
eluate 洗脱物;洗出液
eluate grade 洗出液浓度
elucidate 阐明
eluent 洗脱液;洗脱剂;洗出母液
eluent composition 洗脱液组成
eluent concentration 洗脱浓度
eluent gas 洗脱气体
eluent mixture 洗脱混合液
eluent property 洗脱性质
eluent strength 洗脱液浓度
eluotropic 洗脱的
eluotropic series 洗脱序
eluotropic strength 洗脱强度
elusion 避免
elute 洗脱;洗提
eluted resin 洗净树脂
eluting 淋洗;洗脱
eluting agent 洗脱剂
eluting effect 洗脱效应;洗提效应
eluting order 洗脱顺序;洗提顺序
eluting peak 流出峰;洗脱峰;洗提峰
eluting power 洗脱能力;洗提能力
eluting reagent 洗提剂
eluting sequence 洗脱顺序;洗提顺序
eluting temperature 洗脱温度;洗提温度
elution 洗脱;洗提(法);洗净
elution analysis 洗脱分析
elution band 洗脱带
elution chromatography 洗脱色谱法
elution column 洗提塔
elution constant 洗提常数
elution curve 流出曲线;洗提曲线
elution cycle 洗提循环
elution development 洗提展开
elution efficiency 洗脱效率
elution fractionation 洗提分级
elution gas chromatography 洗提气体色层分离法;气体色谱法;淘析气体色层分离法
elution method 淋洗法;洗脱法;洗提法;冲洗法
elution mode 洗脱方式
elution order 流出顺序;洗提顺序
elution period 洗提周期
elution process 洗脱过程
elution program(me) 洗脱程序
elution requirement 洗脱条件;洗提条件
elution section 洗提区
elution system 洗脱系统
elution time 洗脱时间;洗提时间
elution volume 洗脱体积;洗提容积
elutive power 洗脱能力
elutriant 洗脱液;洗脱剂;洗提液;洗出母液
elutriate 洗矿;淘洗;淘析;水选
elutriated product 淘析精矿
elutriated sludge 澄析污泥
elutriated water 澄析水
elutriate test 淘选试验
elutriating 淘析
elutriating apparatus 分粒装置;淘析器
elutriating flask 淘析瓶
elutriating funnel 淘析漏斗
elutriating glass 洗净玻璃
elutriation 空气分选;颗粒水析分级;洗选;洗净;冲洗;沉淀分析法;淘选;淘洗;淘析;水选;水析
elutriation analysis 淘析法;水簸分析
elutriation apparatus 淘洗设备
elutriation by air 空气淘析
elutriation by water 水力冲洗
elutriation leg 析出段
elutriation method 冲洗法;分级离析法;淘选法;淘洗法;水析分级法
elutriation of sludge 污泥淘析;污泥(的)淘洗
elutriation process 淘洗过程
elutriation ring 淘析环
elutriation separator 淘洗分选器
elutriation sludge 淘洗污泥
elutriation test 淘洗试验;淘析试验;淘分试验
elutriation water 淘洗水
elutriation wet analysis 淘析
elutriator 洗提器;淘洗器;淘析器;淘分机
elutriator-centrifugal apparatus 离心式淘析器
elutriator-centrifuge 淘洗离心机;淘析离心机
eluvial 残积的
eluvial-accumulative landscape 残积堆积景观
eluvial breccia 残积角砾岩
eluvial deposit 淋溶沉积;残积物;残积床
eluvial facies 残积相
eluvial gravel 残积砾石
eluvial horizon 溶提层;淋溶层;淋滤带;残积层;B层土壤
eluvial landscape 残积景观
eluvial laterite 残积红土
eluvial material 残积物
eluvial ore deposit 残积矿床
eluvial placer 残积砂矿
eluvial sand 残积砂
eluvial soil 沉积土;淋溶土;淋滤土;残积土
eluviation 淋溶(作用);淋滤(作用);残积作用
eluvium 风化细砂土;淋溶层;残积层
eluvium gravel 残积砾石
elvan 白色英斑岩;淡英斑岩
elvegust 埃尔维阵风
Elverite 埃尔韦莱特耐蚀铸铁
Elwotite (埃尔沃太特)硬钨合金
elyite 铜铅矾
Ely's test 艾利试验
Elzholz's mixture 埃尔兹霍兹混合剂
emagram 埃玛图(气温气压图);气压温度图;温度对数压力图
E-mail 电子邮件;电子信箱
e-mail address 电子邮件地址
eman 埃曼;唉曼
emanating power 发射气的能力;射气率
emanation 射气;(放射性)射气
emanation chamber 射气箱
emanation coefficient 放射系数;射气系数
emanation coefficient value 射气系数值
emanation concentration value in soil 土壤射气浓度值
emanation concentration value in water 水中射气浓度值
emanation electrometer 放射性静电计;射气静电计
emanation electroscope 射气验电器
emanation exhalation 喷气
emanation flux density 射气流密度
emanation prospecting 射气勘探;射气测量
emanation rate 射气率
emanation security 放射保密措施
emanation strength 射气强度
emanation technique 射气技术
emanation test 射气试验氡
emanation thermal analysis 射气热分析
emanation thermal analyzer 放射热分析仪
emanation total consistence 射气总浓度
emanative 放射性的
emanator 测氡仪;放射性气体仪;射气测量计
emanatorium 射气治疗院
emanator type 射气仪类型
emanium 射气
emanometer 测氡仪;射气仪;射气计
emanometry 射气测量法
Emanual style (16世纪初葡萄牙的)埃曼努尔式建筑
emarginate 边缘有凹痕的;凹缘的
emarginated tail 凹尾
emasculator 去势器
embacle 堆积冰块;冰群;冰堆

embank 筑堤防堵
embanked area 筑有围堤的地区;堤坝区域;围垦区;围堤区域;围堤面积
embanked ground 堤内(土)地;堤后土地
embanked reach 有堤河段;堤防段;设堤河段
embanking 筑路堤;筑堤挡水;筑堤
embanking materials 筑堤材料
embankment 路堤;河堤;河岸堤防;筑堤;堤防;堤(岸);防护堤;(指土石坝的)坝体;人工堤防;填方挡土墙;塘堤岸
embankment altitude 路堤高度
embankment composite dam 土石坝
embankment construction 路堤施工;堤坝工程
embankment core 路堤心
embankment crown retainer 路堤拱度保持器
embankment dam 堤坝;土石坝;填筑坝
embankment dam construction 堤坝施工
embankment dam stabilization 土石坝加固
embankment dam subsoil 坝堤下层土
embankment dam width 堤坝宽度
embankment failure (指土石坝)溃坝;堤溃破;坝;土堤破坏
embankment fill 筑堤;路堤填土;路堤填料;填土掩埋
embankment foundation 路堤基础;堤基;坝基
embankment ground 堤内土地
embankment line 堤岸线
embankment material 筑堤材料;筑坝材料;堆筑材料;坝体土石料;填筑材料
embankment mower 堤埂割草机
embankment pile 护堤桩
embankment project 堤防工程
embankment protection 路基防护;堤堤护坡;路堤防护;护堤;堤坝防护;堤岸防护
embankment protection from wind 路堤防风吹蚀
embankment repair 培修堤防
embankment settlement 路堤沉陷;路堤沉降
embankment slope 路堤坡度;路堤边坡;堆筑体边坡
embankment slope angle 路堤(边)坡角
embankment stabilization 堤坡加固
embankment type dam 路堤式坝
embankment type dam construction 路堤式坝施工
embankment wall (护)堤墙
embankment widening 加宽路堤
embankment width 路堤宽度
embargo 扣留;禁止通商;禁止贸易;禁止(船舶)进出口;禁止出口;禁运;封锁;封港(令)
embargo list 禁运货单
embargo on ships 禁止船舶出口
embargo on the export of gold 禁止黄金出口
embargo on wagon loading 停止装车
embark 冒险投资;乘船;上船
embarkation 登船;搭载物;乘船
embarkation area 登乘区域
embarkation card 出境卡
embarkation deck 登艇甲板
embarkation entrance 上船入口
embarkation ladder 搭载梯
embarkation lamp 登艇灯
embarkation notice 登船通知书
embarkation point 堆货场(地);登船地点
embarkation quay 装船码头;登船码头;上船载货的(横)码头
embarking 上船
embarkment 装船;上船
embark money in an enterprise 投资于某企业
embassy 大使馆;使馆
embata 恩巴塔风
embattement 城垛
embattle 筑以城垛
embattled 有城垛的;饰以锯齿形的
embattled bridge parapet 饰以锯齿形的桥上护墙;城垛的桥上护墙;锯齿形桥栏
embattled mo(u)lding 锯齿形线饰
embattled parapet wall 饰以锯齿形的女儿墙;城垛的女儿墙;锯齿形胸墙
embattlement 凸状壁;雉堞;城垛
embattlemented 凸状壁的;城墙状的;城垛的;齿状墙顶
embay 驶进港湾
embayed 多湾的;港湾状
embayed coast 多湾海岸;湾形海岸
embayed coastal plain 多湾海岸平原

embayed mountain 多湾山
embayed shore 多湾海滨
embayment 横越地槽；河湾；海湾形成；海湾(构造)；形成港湾；支地槽；造山带内凹；构造凹处；成弯；岸湾；形成弯状；弯入
embed 埋置；装入；灌封；放入；包埋；嵌入
embedability 挤入性；压入性；压入能力
embedded 夹在层间的；埋置的；埋入的；装入的
embedded abutment 埋置式桥台
embedded application 嵌入式系统
embedded arc welding 埋弧焊
embedded bar 预埋钢筋
embedded blank 嵌入式空白
embedded block 预埋砌块
embedded bolt 埋置螺栓；预埋螺栓；预埋螺杆
embedded chain 嵌入链
embedded chips 埋入的石屑
embedded code 嵌入码
embedded column 埋入墙内的柱；暗柱
embedded command 嵌入命令
embedded component 嵌入分量
embedded computer 嵌入式计算机
embedded conductor 嵌入导体
embedded conduit 埋置管道；埋入式管道；暗管
embedded conduits 埋管
embedded construction 隐蔽工程
embedded control channel 嵌入控制通路
embedded controller 嵌入式控制器
embedded core 补砂芯
embedded cylindric(al) prestressed concrete pipe 带钢管预应力混凝土管
embedded cylindric(al) shell 埋入式圆柱壳
embedded depth 埋置深度；埋深
embedded depth of foundation 基础埋置深度；基础埋深
embedded detector 埋入式探测器
embedded electric heating cable 埋置的供热电缆
embedded element 预埋构件
embedded footing 埋入式基脚；嵌入式基脚；嵌固基础
embedded footing on piles 低承台桩基；低桩承台
embedded grit 型砂斑点
embedded hanger 预埋吊件
embedded heating element 嵌入式加热元件
embedded heating panel 内埋式采暖板
embedded heterostructure 嵌入式异质结构
embedded in concrete 埋置入混凝土中；埋置在混凝土中；埋入混凝土内；预埋在混凝土中
embedded in ground installment 埋地敷设
embedded in masonry 砌入墙体中的
embedded inserts (管子或电线等)预埋件
embedded installation 暗装；暗设
embedded instrument 埋设仪器；埋入仪器
embedded key 装填键
embedded length 锚固长度；埋置长度；埋入长度；(桩的)入土长度；嵌入长度
embedded length of bar 钢筋埋入长度
embedded Markov chain 嵌入马尔可夫链
embedded matrix 嵌入矩阵
embedded mesh 预埋网
embedded metal work 金属埋件
embedded object 埋入物；嵌入对象
embedded panel 埋置板
embedded panel system 壁板式供暖系统
embedded parts (管子或电线等)预埋件；嵌入件；嵌入部分；埋设件；埋置部分；埋入部分
embedded penstock 埋入式压力水管；埋入式压力管道；埋入式压力钢管；埋藏式压力水管
embedded pipe 埋置管道；埋入式管道；埋藏管道；暗装管(道)；暗管；埋管
embedded pipe cooling 埋管式冷却(法)；埋管冷却
embedded prime divisor 嵌入素除子
embedded reactor shell 埋藏式反应堆外壳
embedded reinforcement 配置钢筋；埋置钢筋；预埋钢筋
embedded sensor 埋置式传感器
embedded software 嵌入式软件
embedded spiral case 埋置蜗壳；埋藏蜗壳；预埋式蜗壳
embedded steel 埋置钢筋；埋入钢筋
embedded stone pitchin 嵌砌式砌石护坡
embedded stone pitching 埋砌石护坡；埋藏式砌石护坡
embedded strain ga(u)ge 埋入应变片；埋入式应

变计
embedded strap 预埋条状物
embedded stripe double-heterostructure 嵌入式条状双异质结构
embedded structure 埋设结构
embedded suction hose 埋吸胶管
embedded switch 预埋开关
embedded system 预埋系统；嵌入式系统
embedded tank 埋入式储[贮]水池；埋入式水池
embedded tap 暗装旋塞
embedded teeth 包埋牙
embedded temperature detector 埋置温度探测；埋置温度传感器；埋置式温度探测器；埋置式温度计；埋入式温度探测器；埋入式温度计；预埋式温度计；预埋式温度探测器；预埋式温度计；预埋式温度传感器；嵌入温度传感器；嵌入温度探测器；嵌入(式)温度计；嵌入式检温计；嵌入式测温计
embedded tie 埋置拉杆
embedded tube 暗装管(道)
embedded type 嵌入式
embedded wavelet 嵌入子波
embedded wire acid and alkali suction hose 埋线吸酸碱胶管
embedded wire oil suction hose 埋线吸油胶管
embedded wire water suction hose 埋线吸水胶管
embedded wiring 暗敷线
embedded wood brick 预埋木砖
embedded works 隐蔽工程
embedding 浇铸封闭；埋入；埋封；包埋
embedding compound 埋封胶
embedding mapping 嵌套映射
embedding medium 嵌入介质
embedding technique 嵌入法
embedding theorem 嵌入定理
embedment 浇铸封闭；埋置；埋入；埋封；包埋；嵌入(件)
embedment anchor 埋入锚
embedment depth (隧道)埋置深度；埋深
embedment depth of foundation 基础埋置深度
embedment inserts 埋入件
embedment length 埋置长度；埋入长度；(桩的)入土长度；嵌入长度；钢筋埋置长度
embedment length equivalent 钢筋等效埋入长度
embedment method 埋入法
embedment of footing 基础埋入法
embedment of holding foundation for ground tackle 埋地窟【救】
embedment of reinforcement 钢筋埋入
embedment pressure 埋入压力
embedment strength 嵌入强度
embeliate 恩贝酸盐
embelic acid 恩贝酸
embellish 装饰；修饰
embellishment 润色；修饰；艺术加工；装饰(品)
embellishment work 装饰工程
ember 灰烬；余烬
embezzlement 盗用公款
embezzlement of public funds 挪用公款
emblazon 盛装【建】
emblem 徽章；厂标；标志
emblem mark 厂标
emblem oil 巴林油
embodied technical progress 被体现的技术进步
embodiment 具体化；化身；体现
embodiment of process 工艺的实施方式
embody 具体化；包括有；包含(物)
emboldening 加强
embolism 栓塞
embolismic year 闰年
embolium 绿片
embolus 插入物；栓子
emboly 陷入
embosom 遮掩；围护
emboss 作浮雕；浮花雕刻；浮雕；压纹；压花
embossed 隆起的；压纹的
embossed acoustic(al) tile 浮雕(花纹)天花板
embossed alumin(i)um foil 凹凸压铸铝箔；浮雕铝箔
embossed calico 压花装订布
embossed candle 浮雕蜡
embossed cloth 压花布
embossed coating surface 浮雕涂覆表面
embossed decoration 压印花纹；堆贴；浮花雕饰

embossed design 凹凸浮雕设计；压花设计；浮雕设计
embossed effect 压花效应
embossed film 压花薄膜
embossed finish 压花涂饰剂
embossed glass 单面压花玻璃；浮雕玻璃
embossed hardboard 浮雕硬木板；压花硬质纤维板；压花硬木板
embossed integrated optics 浮雕集成光学
embossed lacquer 堆漆
embossed linoleum 压花漆布
embossed metal ornamentation 金属浮雕饰
embossed ornament 雕刻凸饰
embossed painting 沥粉
embossed panel 浮雕面板
embossed paper 凹凸纸；压花墙纸；轧花墙纸；绸纹纸；浮雕墙纸；凸凹纸
embossed pattern 凹凸浮雕模型；压花图案；浮雕纹样；浮雕花式；凸纹图案(立体图案)
embossed plastic material 压花塑料
embossed plastic relief map 塑料立体地图
embossed plastics 压花塑料
embossed plate 凸板
embossed plate evapo(u)rator 凹凸板式蒸发器
embossed plate print 凸板打印机
embossed plywood 浮雕胶合板
embossed printing 浮凸印刷
embossed sheet 压纹板；压花金属薄板；压花板；模压板
embossed sheeting 压花片材
embossed sheet metal panel 压制金属薄板面板
embossed texture 凹凸织物；压花织物
embossed tile 雕纹花砖；浮雕面砖；浮雕瓷砖
embossed tube 压印管
embossed wall paper 凹凸浮雕墙纸；凸纹墙纸
embossed work 刻花加工；模压加工；压印加工；浮雕细工
embosser 轧花机；印纹轧光机；压纹机；压花机
embossing 压制成波浪形；压纹；模压加工；浮雕；压型；压花(法)；雕刻凸饰；凹凸轧花；凹凸印(刷)；起伏成形
embossing black 防蚀蜡
embossing calender 压纹机；浮花辊压机；凹凸轧花机
embossing die 压凸冲模；压筋模；压花头；压花模
embossing flask 浮雕瓶
embossing iron 压花铁
embossing machine 压花机
embossing pad 压花垫
embossing plate 压花板材料
embossing point 搭乘公共汽车地点
embossing press 肘杆式压力机
embossing roll 压花辊
embossing roller 螺旋辊子；压纹辊；压花辊子
embossing seal 钢印
embossment 凸起；浮花装饰；模压加工；压花；浮(花)雕(饰)；凸雕
emboss roller 压花滚筒
embouchure 河口；峡门；峡谷宽阔河段；炮口
embow 弯成弧形
embowed 弯曲的；弧形的
embowment 弯成弧形
embrace reflex 紧抱反射
embracing grammar 包围文法
embranchment 支脉；支流；分支(机构)
embrasure 内宽外窄的开口漏斗状碱面墙；楔状隙；堡眼；八字门窗口；枪眼；斜面窗洞
embrasured watchtower 箭楼
embrechite 残层混合岩
embreyite 磷铬铅矿
embrittle 使脆化
embrittlement 脆裂；脆化(性)；脆变；变脆
embrittlement by aging 时效脆性
embrittlement detector 脆度检测器
embrittlement temperature 脆裂温度；延性—脆性转变温度；脆化点
embrittlement tendency 脆裂趋向
embrittling 脆裂
embrocation 擦剂
embroidered carpet 绣花地毯；刺绣地毯
embroidered painting 刺绣画
embroidered picture 刺绣画片
embroidered portrait 刺绣肖像
embroidery 刺绣品；(自记仪器的)曲线弯曲度

embroidery clipping machine 刺绣挖剪机
embroidery cotton 刺绣纱线
embroidery frame 刺绣绷子
embroidery machine 刺绣机
embryo 胚芽
embryo deposit 初始沉积
embryo fisheries 初始渔业
embryogeosyncline 萌地槽
embryogeosyncline stage 萌地槽阶段
embryonic folding 雏形褶皱
embryonic ore formation 雏形成矿
embryonic stage 萌芽阶段；酝酿阶段；雏形期；胚胎期
embryonic volcano 雏(形)火山
embryoplatform 萌地台
embryoplatform stage 萌地台阶段
embryo volcano 胎火山
embryulcus 牵胎钩
emcee (电台节目的)主持人
Emden equation 艾姆登方程
Emden function 艾姆登函数
emeleusite 高铁锂大隅石
emend 校勘
emendation 校勘
emerada 人造尖晶石
emerald 绿宝石；艳绿色；祖母绿；翠绿；纯绿柱石；翡翠(色)
emerald alloyed abrasive 钒刚玉
emerald brass alloy 翡翠黄铜合金
emerald feather 文竹
emerald green 碱式醋酸铜绿；鲜绿色；翠绿；翡翠绿；巴黎绿
emeraldine 翠绿亚胺
emerald nickel 翠镍矿
emerald oxide of chromium 鲜绿氧化铬；水合氧化铬
emerald pearl (产于挪威的)绿珠花岗岩
emeraudine 透视石
emerge 露出；呈现；浮现；发生
emerged hydrophytes 出水水生植物
emerged plant 出水水生植物
emerged shoreline 上升(海)岸线
emerged wedge 露出楔；出水楔形部分
emergence 显露；出现；显露；突出体
emergence angle 出射角
emergence nodal point 后节点
emergence of land 陆地上升
emergency 紧急；急救；应急(的)；意外；出现；安全停车；事故
emergency access 紧急通道；救灾通道
emergency access for vehicles 车辆紧急通道
emergency accommodation 应急设备；应急供应；紧急寄宿处；应急住所
emergency acoustic(al) system 紧急声控系统
emergency action 紧急行动；紧急措施；紧急处置；应急行动；应急措施；安全措施
emergency action plan 应急活动计划
emergency action team 紧急行动队
emergency aid center 急救中心；急救站
emergency air compresser 应急空气压缩机
emergency airdrome equipment 机场抢救设备
emergency air exhaust 事故排风
emergency air inlet 安全气孔
emergency air system 应急空气系统
emergency alarm bell 紧急报警警铃；报警铃
emergency alarm communication equipment 紧急报警通信设备
emergency alarm signal 应急警铃信号；事故报警信号
emergency altitude (飞行)极限高度
emergency ambulance 急救车
emergency ambulance service 急救车服务
emergency aminonia relief device 紧急泄氨装置
emergency amortization 紧急偿还；非常情况下的加速折旧
emergency analysis 应急情况分析
emergency anchorage 临时锚地；应急锚地
emergency and accident department 紧急抢救收留站
emergency and rescue equipment 应急救援设备
emergency antenna 应急天线；备用天线
emergency apparatus 应急装置；应急设备；备用设备；保险装置；安全装置
emergency application 紧急制动；紧急应用

emergency application position 紧急制动位(置)
emergency application valve 紧急作用阀
emergency area 紧急地区；灾区；事故区
emergency arrangement 应变部署
emergency ascent 紧急上升
emergency astern test 紧急倒车试验【船】
emergency auxiliary power 事故备用电源
emergency back-up fuel 应急燃料
emergency ballast 紧急压载
emergency bank repair 河岸抢修；堤岸抢修
emergency basin 备用池
emergency battery 紧急备用电池；应急(用)电池；备用电池组
emergency battery room 事故备用蓄电池室
emergency beacon 紧急救援无线电信标
emergency bearing oil pump 备用(润滑)油泵
emergency bilge pump 紧急舱底污水泵；应急污水泵；应急舱底水泵；应急舭水泵
emergency bin discharge gate 料仓应急卸料门
emergency blow-down 紧急放空
emergency blowdown valve 紧急排污阀
emergency blowout preventer 应急防喷器
emergency boat 应急船
emergency brake 紧急制动器；紧急闸；紧急刹车；应急刹车；保险制动器；保险闸；安全制动器；安全闸；事故制动器
emergency brake application 紧急制动位；紧急加闸
emergency brake application position 紧急制动作用
emergency brake band 紧急闸带
emergency brake carrier plate 紧急闸座板
emergency brake distance 紧急制动距离
emergency brake gear 紧急制动装置；紧急闸装置；应急制动装置；保险制动装置；安全制动装置
emergency brake lever 紧急闸杆
emergency brake lever spring 紧急闸杆弹簧
emergency brake of ship chamber 承船厢事故制动器
emergency brake switch 紧急闸开关；紧急刹车开关
emergency brake system 紧急备用制动系统
emergency brake valve 紧急制动阀
emergency braking 紧急制动(位)；紧急刹车
emergency braking control relay 应急制动控制继电器
emergency braking deceleration 紧急制动减速
emergency braking deceleration device 紧急制动减速装置
emergency breakaway valve 紧急断气阀
emergency breakdown repairs 紧急破损修理
emergency breaker 紧急备用破碎机
emergency break-off 紧急停车(装置)
emergency bridge 临时桥；应急(用)桥；便桥；备用桥；战备桥
emergency broadcasting system 紧急广播系统
emergency broadcast system 应急用广播系统
emergency bulkhead 安全闸(门)；安全堵头
emergency buoy 应急救生浮标
emergency buoyancy for operating depth 工作深度应急浮力
emergency button 应急(按)钮；备用按钮；事故按钮
emergency by-pass 安全旁通(管)；安全旁路；事故旁通；事故旁路
emergency cable 应急电缆
emergency call 紧急情况报警；紧急集合；紧急呼叫；紧急电话
emergency call box system 紧急呼叫系统
emergency call key 紧急呼叫键
emergency call-on button 事故引导按钮
emergency call system 紧急传呼系统
emergency capacity 应急储备；安全储备；事故备用容量
emergency car 救急车
emergency carrier generator 备用载频振荡器
emergency cart 抢救推车
emergency case 紧急情况
emergency caterpillar gate cut-off 紧急履带闸门关闭
emergency cell 应急(用)电池；备用电池
emergency center 应急基地
emergency check valve 紧急止回阀
emergency church 应急教堂
emergency chutdown circuit 紧急关闭线路
emergency chute 飞机发生事故时

emergency circumstance 紧急状态
emergency closing device 应急关闭装置；事故关闭装置
emergency closing valve 应急阀(门)
emergency closure 紧急闸(门)；紧急截流；备用闸(门)；保险闸(门)；事故闸(门)；事故关闭
emergency closure device 事故关闭装置；事故关闭设备
emergency closure gate 应急关闭装置；应急闸(门)
emergency cock 紧急旋塞；应急旋塞；应急开关
emergency command center 事故指挥中心
emergency communication 紧急通信；应急通信；事故通信
emergency communication net 应急通信网
emergency communication terminal 应急通信终端站
emergency compass 应急罗盘
emergency condition 紧急情况；应急状态；临时情况
emergency connection 紧急连接；应急接线；事故接线
emergency consent 事故插座
emergency construction 紧急防险建筑；防险建筑物
emergency control 紧急控制；安全控制；事故控制；事故操纵
emergency control measure 紧急防治措施
emergency control station 应急控制站
emergency control switch 故障控制开关
emergency coolant injection 应急冷却剂注入
emergency coolant pump 紧急冷却液泵
emergency coolant recirculation 应急冷却剂再循环
emergency coolant system 应急冷却系统
emergency cooling 紧急冷却；骤冷
emergency cooling system 应急冷却系统
emergency copy 紧急复印
emergency core cooling 紧急堆芯冷却
emergency core cooling system 堆芯事故冷却系统
emergency corridor (防火灾用)应急走廊
emergency coupler 紧急车钩装置
emergency coupling 紧急连接器
emergency coupling device 紧急连接器
emergency crane 抢修吊车
emergency crash work 飞机失事救援作业
emergency credit 紧急信贷
emergency credit assistance 紧急信贷援助
emergency crop 短生长期植物
emergency crossover 紧急转车道；临时渡线
emergency cut-off 紧急停车；应急切断装置
emergency cut-out 应急关断；紧急断流
emergency cylinder 应急气瓶
emergency dam 临时应急坝；防险坝
emergency decontamination measure 应急消毒措施
emergency decree 安全技术规程
emergency department 急诊部
emergency detection system 应急检测系统
emergency device 应急装置
emergency diesel-electric(al) generator set 备用柴油发电机组
emergency diesel generator 备用柴油发电机
emergency disconnection 应急断开
emergency disengage switch 应急断开电门
emergency disinfection apparatus 应急消毒设备
emergency disposal 应急处置
emergency dispute 紧急争端
emergency district 禁伐林区；应急林
emergency dive 紧急下潜
emergency docking facility 应急靠泊设备
emergency door 紧急出口(门)；防火门；防爆门；安全门；太平门；事故出口
emergency door handle 太平门门把手
emergency door stop 双向门挡；安全门轧头；安全门挡
emergency dose 应急剂量
emergency drain 紧急排放
emergency drill 应急演习
emergency drive 备用传动装置
emergency drop door 紧急落板装置；紧急落板口
emergency dump steam 事故排汽
emergency duty 紧急关税；事故备用
emergency dwelling 应急住宅；避难所
emergency dwelling unit 应急住所
emergency egress 紧急出口
emergency egress opening 应急出口孔洞

emergency egress window 紧急出口窗户
emergency ejection 应急弹射
emergency ejectioning device 应急抛载装置
emergency ejector seat 应急弹射座椅
emergency electric(al) current 应急电流
emergency electric(al) supply 紧急供电
emergency electric(al) supply unit 备用供电设备
emergency electric(al) system 紧急电气系统；紧急电力系统
emergency elevator 备用提升机；消防电梯
emergency emptying of reservoir 水库紧急放空
emergency engine 救急机车；紧急备用机；应急发动机；备用机车
emergency engineering project 紧急工程计划
emergency engine kill 紧急停机
emergency entrance 紧急入口
emergency entrance marking 紧急入口标志
emergency environment pollution accident 突发性环境污染事故
emergency episode 严重事件
emergency equipment 应急设备
emergency escape 紧急避难
emergency escape cable 应急脱离索
emergency escape capsule 应急脱离舱
emergency escape chamber 应急脱险舱
emergency escape control 应急离机操纵
emergency escape equipment 应急离机设备
emergency escape ladder 应急梯
emergency escape lane 紧急安全车道
emergency escape procedure 应急离机程序
emergency escape provision 应急离机设备
emergency escape system 应急离机系统
emergency evacuation 紧急疏散；紧急撤退；应急撤离程序
emergency evacuation kit 紧急疏散器具箱
emergency evacuation plan 紧急撤离计划
emergency evacuation route 紧急疏散路线
emergency evacuation smoke hood 紧急疏散防烟头罩
emergency exchange 应急交换
emergency exhaust 紧急排汽；紧急排气；紧急排风；应急排汽；应急排气；应急排风；事故排气；事故排气
emergency exhaust fan 应急排气风扇；辅助排气风扇
emergency exhaust valve 应急排气阀
emergency exhaust valve head connection 紧急排汽阀盖接头
emergency exit 紧急出口；应急出口；防火门；防爆门；备用引出端；安全门；安全出口；太平门；事故出口；疏散出口；人流疏散道路
emergency exit hatch 应急出口舱盖
emergency exit ladder 应急出口梯
emergency exit light 安全出口灯
emergency exit lighting 紧急出口照明
emergency exit placard 应急出口标牌
emergency exit sign 紧急出口标志
emergency exit spoiler 应急出口挡板
emergency exit structure 应急太平门建筑物；应急出口建筑物
emergency exit window 紧急出口窗(户)
emergency exposure 紧急照射；应急照射
emergency extension 应急放下
emergency extension lever 应急放起落架手柄
emergency extension system 应急放下系统
emergency facility 紧急设备；应急设施；应急措施设备
emergency fan 事故风扇
emergency feature 应急用装置
emergency feed boron detector 应急送硼探测器
emergency feeder 应急喂料机；备用馈电线
emergency feed line 事故电源线
emergency ferry 临时渡口
emergency field 备耕地
emergency filter 事故备用过滤器
emergency fire pump 紧急消防泵
emergency first-aid kit 紧急救护用具；紧急救护箱；紧急救护器材
emergency first aid treatment guide 紧急急救处理指南
emergency flo(a)tation gear 紧急浮水装置
emergency floating caisson 安全浮箱闸门；安全浮箱(船坞用保险门)
emergency flood fighting 防汛抢险

emergency flood flow 紧急洪流；意外洪水流量
emergency food 应急食品
emergency forest 禁伐区
emergency free ascent 紧急自由上升
emergency frequency 应急频率
emergency fuel cut-off switch 应急燃料切断开关
emergency fuel regulator 应急燃料调节器
emergency fuel tank 应急燃料箱
emergency fuel trip 紧急停料装置；燃料紧急自动切断器
emergency fund 应急基金；应急费用；准备金
emergency gallery 应急安全坑道
emergency gasoline tank 应急汽油箱
emergency gate 紧急排放口闸门；检修闸门；应急闸门；备用闸门；备修船用；保险闸门；事故闸门
emergency gate cut-off 事故闸门关闭
emergency gate slot 事故闸门槽
emergency gear 应急齿轮；安全齿轮
emergency generator 紧急发电机；应急发电装置；应急发电机；备用发动机；备用电机；备急发电机；事故(备)用发电机
emergency generator room 应急发电机室
emergency gliding 紧急滑行
emergency governor 紧急调速器；急流调节器；限速器；应急调速器；防流调速器
emergency grounding 应急接地
emergency guard gate 应急闸门；临时备用闸门；保险闸门
emergency hand control 手动应急控制
emergency hand-drive 临时手动装置；应急手动装置；事故手动装置
emergency handle 紧急手柄
emergency hand pump connector 应急手泵连接器
emergency hand trip 危急手动脱扣
emergency hatch 应急舱
emergency head backup connection 自动连接器助接
emergency high exposure 事故时高照射
emergency hospital 急救医院
emergency house 应急临时住宅
emergency housing 应急住房
emergency housing scheme 应急居住计划
emergency identification signal 应急识别信号
emergency illuminance 应急照明
emergency illumination 应急照明
emergency import 紧急输入
emergency import duties 紧急进口税
emergency incident 紧急事件
emergency information 紧急信息；紧急通告
emergency installation 应急装置；应急设备；备用装置
emergency instruction 紧急指令
emergency interconnection 紧急联络；应急互接
emergency issue 临时版
emergency joint bar 救险鱼尾板；应急接头夹板
emergency junction 紧急连接
emergency key 应急钥匙；应急电键；应急按钮
emergency keyer panel 遇险信号控制板
emergency kit 应急包；应急工具
emergency ladder 应急梯子
emergency lagoon 应急池
emergency lamp 应急灯；保险灯；事故备用灯
emergency lamp signal telegraph 应急信号灯(传令钟)
emergency landing 紧急着陆；紧急降落
emergency landing field 迫降场
emergency landing flare 紧急着陆照明弹
emergency landing gear 应急起落架
emergency landing ground 应急着陆场
emergency landing procedure 应急着陆程序
emergency landing signal 紧急着陆信号
emergency landing site 迫降场
emergency landing strip 应急着陆地带
emergency landing survival accident 紧急着陆救生事故
emergency level 紧急状态；紧急等级；紧急标准
emergency life boat 应急救生艇
emergency life support system 应急生命保障系统
emergency lift 事故电梯
emergency light 紧急灯；应急灯；故障信号灯；事故照明器；事故表示灯
emergency lighting 紧急照明；应急照明；备用照明；安全照明；事故照明
emergency lighting battery 应急照明电池

emergency lighting distribution box 事故照明配电箱
emergency lighting distribution equipment 事故照明配电设备
emergency lighting set 应急照明电机
emergency lighting system 紧急照明系统
emergency lighting wiring 事故照明线路
emergency light reflex 强光反射
emergency light switch 应急照明电门
emergency line 事故备用线路
emergency lining support 暂时性衬砌
emergency load dump 事故甩负荷
emergency loading point 紧急装油点
emergency loan 紧急贷款
emergency location transponder 应急定位应答器
emergency locator transmitter 紧急定位信标；紧急定位器发射机；应急定位发射机；事故地点发射机
emergency lock 紧急闸(门)；检修门；备用闸；备用门；安全闸(门)；太平闸；事故闸(门)
emergency locomotive 抢险机车
emergency lubrication 应急润滑
emergency luminaire 应急灯具
emergency maintenance 紧急维修；应急维修；事故维修
emergency maintenance time 应急维修时间
emergency man 临时雇员
emergency management 紧急处理；应急措施
emergency maneuver 应急操纵
emergency manual release 紧急手动解锁装置
emergency material 应急(用)备料
emergency measures 紧急措施；急救方法；急救措施；应急措施；抢救措施
emergency measures of environment(al) pollution 环境污染应急措施
emergency medical service 紧急救援医疗服务
emergency medical service communication 紧急医疗服务通信
emergency medical services 急救医疗服务
emergency medical services and technical support services program(me) 急救医疗和技术援助服务计划
emergency medical system 急救医疗系统
emergency message 紧急呼救信号
emergency message handling 应急信息处理
emergency milling-out 事故磨铣
emergency milling-up 事故磨铣
emergency mobile communication 应急移动通信
emergency mode 应急(运行)方式
emergency monitor(ing) 应急监测
emergency network 备用电网；事故供电网
emergency network of the situation of a disaster 紧急灾情通信网
emergency number 紧急电话号码
emergency O₂ valve 氧气快速开关
emergency-off 应急断电
emergency office 失事紧急救援处
emergency off local 现场事故断开
emergency off-take point 紧急排空点
emergency oil storage installation 事故储[贮]油设施
emergency open 紧急打开
emergency opening 紧急出口
emergency operating level 非常运行水位；非常操作水位
emergency operating order 紧急操作规程
emergency operating room 急救手术室
emergency operation 紧急操作；应急运转；应急工作；应急操作
emergency operation center 紧急行动中心
emergency operations system 应急操作系统
emergency organization 应急组织
emergency outage 紧急停运；事故停机
emergency outfit 应急设备
emergency outlet 应急泄洪口；安全出(水)口；事故排放口；事故(排)出口
emergency outlet of inverted siphon 倒虹管紧急出口
emergency overflow 紧急溢流；应急溢流
emergency overflow pipe 应急溢流管
emergency override switch 应急超控电门
emergency over-speed trip 超速紧急脱扣；超速保安器

emergency oxygen 应急氧气
emergency oxygen apparatus 应急氧气设备
emergency oxygen equipment 应急供氧设备
emergency oxygen pack 急救氧气袋
emergency pack 应急电源
emergency packer 紧急用封隔器;安全封隔器
emergency panel 备用配电盘;事故配电盘
emergency panel lights switch 仪表板应急照明电门
emergency parking 应急停车场
emergency parking area 紧急停车带
emergency parking bay 港湾应急停车场
emergency parking strip 紧急停车带
emergency passage(way) 避难通道
emergency period 非常时期
emergency personnel 应急人员
emergency phase 紧急阶段
emergency phone 应急电话
emergency phone cable 应急电话电缆
emergency phone system 应急电话系统
emergency pipe line 应急管线
emergency piston 紧急活塞
emergency plan 应急计划;应急方案
emergency plant 应急(备用)设备
emergency plant cool-down system 电站事故冷却系统
emergency point-to-point communication facility 点间备用通信设备
emergency position 紧急制动位(置)
emergency position-indicating radio beacon 应急无线电示位标;紧急示位无线电信标;紧急示位无线电示位标
emergency position-indicating radiobeacon station 紧急示位无线电信标电台;紧急示位无线电示位标电台
emergency power 应急电源;紧急备用动力;应急动力;事故电源;备用电源
emergency power controlling set 应急电源控制台
emergency power cut-off 紧急能源切断器
emergency power generating set 备用发电机组
emergency power generating system 紧急发电系统
emergency power handle 应急电源手柄
emergency power installation 应急电力装置
emergency power load 紧急电力荷载
emergency power-off control 紧急断电控制
emergency power package 应急电源组
emergency power plant 事故备用电站
emergency power shut-off 紧急停电;应急停电
emergency power source 紧急电源;应急电源;事故电源
emergency power station 应急电站
emergency power supply 紧急电力供应;应急动力(电)源;备用电源;保安电源;事故电源
emergency power supply unit 事故电源装置
emergency preparedness 紧急状况的准备工作
emergency pressure increasing valve 紧急增压阀
emergency pressure reducing valve 紧急减压阀
emergency procedure 应急(离机)程序;应急操作步骤
emergency program(me) 应急预案;应急计划
emergency project 紧急(工程)计划;紧急工程(项目);应急工程项目;备耕方案
emergency protection 故障防护
emergency protection against leakage 漏洞抢险
emergency protection against loophole 漏洞抢险
emergency provision 应急预防措施
emergency pulse 呼救信号;呼救脉冲
emergency pump 应急泵;备用泵;危急排水泵;事故应急泵;事故备用泵
emergency push button 紧急按钮
emergency pushbutton switch 备用按钮开关
emergency quick-release device 应急快速解脱装置
emergency radio 无线电设备
emergency radio alarming, protection equipment 无线防护报警装置
emergency radio channel 呼救信号波道;应急无线电信道
emergency radio dispatching system 无线调度防灾系统【铁】
emergency rail 应急轨;备用轨
emergency rail clamp 应急轨夹
emergency rail-joint 断轨救急接头
emergency rate 紧急费率

emergency ration 应急干粮
emergency ray 出射线
emergency recall 紧急召回
emergency receiver 应急收讯机;应急收报机;应急接收机
emergency reclaim hopper 应急取料斗
emergency recorder plot 应急记录标绘
emergency red aspect 紧急停车显示
emergency reference level 应急参考水平
emergency refuge center 紧急避难中心
emergency relay valve 紧急继动阀
emergency release 紧急泄放;紧急缓解;应急安全装置
emergency release feature 紧急解锁功能
emergency release notification 紧急释放通知
emergency release push 应急释放开关
emergency relief 紧急救援;事故放水口
emergency relief measures 应急减灾措施
emergency relief valve 安全阀;紧急安全阀
emergency repair 紧急修理;紧急抢修;计划外修理;急修;应急修理;抢修;事故修理;事故检修;紧急补修
emergency repair car 抢修车
emergency repair ship 应急修理船
emergency repairs to buildings 房屋抢修
emergency repair truck 抢修车;紧急修理车;应急修理车
emergency reporting system 应急报告系统
emergency request 紧急申请
emergency requirement 紧急要求;应急需要
emergency rescue 紧急救援
emergency rescue mission 紧急救援任务
emergency rescue operation 应急救援操作
emergency reserve 紧急预备费;紧急储备(金)
emergency reserve fund 应急备用基金
emergency reserve(power) station 应急电站
emergency reserve station 应急备用站
emergency reservoir 紧急风缸
emergency response 紧急响应;对紧急状况的反应
emergency response management 紧急响应的组织管理
emergency response plan 应急响应计划
emergency response procedure 紧急响应程序
emergency restart 紧急重新启动;紧急再启动;应急再启动
emergency restoration 紧急修复
emergency retreat 紧急躲避处
emergency risks 紧急危险
emergency road 备用道路
emergency rod (原子堆)事故棒
emergency roof walkway (防火灾用)应急屋顶走道
emergency room 急诊室;急救室;紧急情况室
emergency rope 救急(绳)索;应急绳索
emergency rotary converter 应急旋转换流机
emergency route 紧急(逃逸)路线;紧急逃离路线;应急通路;应急路由;备用迂回路线;安全路线
emergency run 紧急复印
emergency-run characteristic 应急运转特性
emergency safe switch 应急保险电门
emergency safety evaluation 应急安全评价
emergency safety switch 应急安全开关
emergency safing circuit 应急保险电路
emergency satellite communications system 应急卫星通信系统
emergency satellite position indicating radio beacon station 卫星紧急无线电示位电台
emergency scene 事故现场
emergency seat 安全座
emergency security operation 紧急保卫活动
emergency sensitiveness 紧急制动灵敏度
emergency separation valve 紧急隔断阀
emergency service 救援(站);紧急供电;应急通信业务;应急供电;应急服务(机构);紧急服务机构
emergency service line 应急服务线
emergency service vehicle 抢救车
emergency set 应急装置;应急设备;应急电台;备用(应急)装置;备用台;备用机组;抢救包
emergency shaft 应急井;备用井
emergency shelter 紧急掩蔽所;临时庇护所;紧急避难所;应急掩蔽部;应急避难所
emergency ship salvage materials system 船舶应急救捞器材系统
emergency shower 事故喷淋器;紧急喷淋;急救莲蓬头

emergency shutdown 紧急开关;紧急停运;紧急停堆;紧急停车;紧急关闭;事故停机;安全开关;事故停车
emergency shutdown assembly 紧急停堆安全装置
emergency shutdown limit 紧急停堆极限
emergency shutdown line 紧急关闭线路
emergency shutdown member 事故停堆棒
emergency shutdown switch 紧急停堆开关
emergency shutdown valve 紧急停堆阀门
emergency shut-in valve 紧急关闭阀
emergency shut-off 紧急关断;紧急停机;紧急停堆;事故(紧急)关闭
emergency shut-off device 紧急停机装置;紧急切断装置
emergency shut-off rod 安全棒
emergency shut-off valve 紧急关闭阀;紧急切断阀
emergency signal 紧急信号;危急信号;事故信号
emergency signal and communication 紧急信号与通信
emergency signal(l)ing device 事故讯号装置;事故信号装置
emergency signal(l)ing mark 紧急讯号标志
emergency signal service message 紧急公文通信电文
emergency signal transmitter 事故信号发送器
emergency situation 紧急情况
emergency slide gate 检修插板门;事故滑动门
emergency snow clearing 紧急除雪
emergency special tax 非常特别税
emergency speed 紧急航速;应急速度;最高航速
emergency speed governor 紧急调速器
emergency spilling 非常溢洪
emergency spillway 紧急溢洪道;应急溢洪道;非常溢洪道;保险溢洪道;安全溢洪道;事故溢洪道
emergency squad 应变小组;应变突击队;抢险队
emergency stair(case) 太平梯;应急梯;安全(楼)梯;疏散楼梯
emergency standby power 应急备用电源
emergency standby system 紧急备用系统
emergency state 紧急状态
emergency station service transformer 厂用事故变压器
emergency steering gear 应急舵机
emergency tire chain 应急胎链
emergency tire sleeve 紧急胎套
emergency stock 紧急储备;应急储备
emergency stockpile 备用堆料
emergency stop 紧急停机;紧急停车;急刹车;安全掣
emergency stop apparatus 紧急停车装置
emergency stop button box 站台紧急按钮箱
emergency stop cock 紧急制动旋塞;应急开关;应急管栓;应急关断开关
emergency stop indicator 紧急停止指示器
emergency stop lever 紧急制动杆
emergency stoplog 应急叠梁闸门;备用叠梁闸门;事故叠梁闸门
emergency stopping 应急隔墙
emergency stopping lane 临时停车道;紧急停车车道;应急停车(车)道;事故停车道
emergency stop protection 紧急停止保护装置;紧急停车保护装置
emergency stop push button (击打式)紧急停车按钮
emergency stop signal 紧急停机信号
emergency stop switch 紧急停机开关;紧急停车开关;电梯安全电闸
emergency stop valve 防火开关;危急遮断阀
emergency storage 应急储备;备用仓库
emergency strap 应急牵夹
emergency stretcher 应急担架
emergency supply (紧急)备用电源;应急电源
emergency support 车底制动拉杆安全托
emergency support vessel 紧急支援船
emergency surcharge 紧急附加费
emergency surfacing 紧急出水;(潜水时)应急出水面
emergency surfacing decompression 紧急出水减压
emergency survival 紧急救生
emergency switch 紧急停车事故开关;紧急(保险)开关;应急开关;保险开关;事故开关
emergency switchboard 应急配电盘;应急配电板

emergency switching-off 应急切断;事故切断
emergency system 应急系统
emergency tank 应急油箱;事故储罐
emergency tapping tank 放玻璃液池
emergency tariff 紧急关税
emergency team 抢救队
emergency telemetry 应急遥测术
emergency telephone 应急电话;紧急电话
emergency telephone number 紧急电话号码
emergency telephone place 紧急电话亭
emergency telephone post 紧急电话亭
emergency telephone selector 应急电话选择器
emergency telephone station 紧急电话台
emergency telephone system 应急电话系统;紧急电话系统
emergency temporary construction 应急临时建筑
emergency tender 抢险救援消防车
emergency tender with crane 带起重机的抢险救援车
emergency terminal switch 终端安全开关
emergency theory 应急说
emergency through-water communication 应急水下通信
emergency through-water speech 应急水下通话
emergency throw-over equipment 事故转换设备
emergency tiller 应急舵柄
emergency tire chain 应急胎链
emergency tire sleeve 紧急胎套
emergency tool box 紧急工具箱
emergency torque release 紧急倒扣释放
emergency tow rope 应急拖缆
emergency traffic 紧急交通
emergency train 抢险列车
emergency transfer capability 紧急转换能力;备用转换功率;备用交换功率
emergency transmitter 应急发射机
emergency transmitter beacon 应急发射指向标
emergency treatment 急救措施;应急处理;抢救
emergency treatment after failure 故障处理;故障办理
emergency treatment person 急救护理员
emergency trip 紧急脱扣;紧急停堆;紧急跳闸;紧急断开机构;应急脱扣
emergency trip magnet valve 紧急止流电磁阀;紧急脱钩电磁阀
emergency trip mechanism 应急跳闸机构
emergency trip speed 紧急跳闸转速
emergency trip switch 紧急脱钩开关
emergency trip valve 紧急截止阀
emergency trip wire system 事故跳闸电气系统
emergency tritium containment system 氚事故控制系统
emergency truck 抢险救援消防车
emergency truck garage 卡车急救车库
emergency turbine generator 应急涡轮发电机
emergency under-water work 应急水下作业
emergency unit 应急装置;备用装置;备用元件;备用设备;备用机件;保险装置
emergency usage 紧急使用
emergency use 紧急备用
emergency valve 应急阀(备用)阀;应急动阀;安全排气阀;事故阀;安全阀
emergency valve gallery 紧急启闭坑道
emergency valve nut 安全阀螺母
emergency valve piston 安全阀活塞
emergency valve seat 安全阀座
emergency vehicle 紧急备用车辆;应急车辆
emergency vehicle automatic detector 紧急车辆自动检测
emergency vehicle priority 紧急车辆优先权
emergency vent 应急通风;辅助通风;应急排气口;应急排气孔;安全排气口
emergency ventilation 紧急通风;应急通风;事故通风
emergency ventilation shaft 事故通风井
emergency ventilation system 事故通风系统
emergency venting 应急通风口
emergency vent valve 紧急放风阀
emergency voice alarm communication system 应急话音报警通信系统
emergency voice communication system 应急话音通信系统
emergency ward 急诊室
emergency warning circuit 应急报警电路
emergency water stop 紧急停水装置

emergency water supply 事故给水;备用水源;紧急供水
emergency water tank 备用水箱
emergency water valve 应急水阀门;备用水阀门
emergency weir (临时)应急堰
emergency wheelsets 紧急备用轮对
emergency window 应急窗;安全窗
emergency withdraw 紧急撤退
emergency works 急救工作;抢险工程;应急工程
emergent 紧急(的);自然发生的;出射;发出的;突出的;射出的
emergent aquatic 临危水生生物;临绝水生生物;濒于危绝的水生生物
emergent aquatic plant 挺水植物
emergent coast 上升海岸
emergent country 发展中国家
emergent face 出射面
emergent gas 逸出气体
emergent light 出射光
emergent light beam 出射光束
emergent medium 出射介质
emergent node 后方节点
emergent particle 出射粒子
emergent pollution event 突发性污染事件
emergent power 出现功率;出射功率
emergent pressure 临界压力
emergent pupil 出射光瞳
emergent radiation 出射辐射
emergent reef 出水礁
emergent repair 临时性修理
emergent structure 紧急结构
emergent surface 出射面
emergent thrust front 上升的冲断层前缘
emergent treatment 应急处理
emergent wave 出射波
emerging and high-tech industries 新兴产业和高技术产业
emerging beam 呈现射束
emerging bubble 露头气泡;浮起气泡
emerging countries 新兴国家
emerging nation 新兴国家
emerging particle 出发粒子
emerging period of maximum discharge of spring 泉最大流量出现时间
emerging period of minimum discharge of spring 泉最小流量出现时间
emerging pollutant 新型污染物
emerging ray 出射光线
emerging wavefront 出射波前
emergy sand 刚玉砂层
emeritus professor 名誉教授
emersiherbosa 沼泽草本群落
emersion 浮出;脱出
emersion angle 出水角
emersion time 复现时刻
Emerson beater 爱默生打浆机
Emerson Plan 爱默生计划
emery 金刚砂;刚(玉)砂;刚石粉;宝砂;天然刚玉
emery abrasive 金刚砂磨料;刚玉砂研磨剂
emery bar 棒状抛光膏
emery belt 金刚砂(布)带;金刚砂抛光带
emery bit 金刚砂钻头
emery brick 金刚砂砖
emery buff 金刚砂磨光轮;刚玉磨光轮;布轮砂轮;皮砂轮
emery cake 金刚砂磨块;研磨用磨块;刚砂饼
Emery cell 艾麦黎型浮选机
emery circular saw 金刚砂圆锯
emery cloth 金刚砂布;研磨布布;布布
emery-coated beam 金刚砂卷布辊
emery construction aggregate 金刚砂建筑骨料
emery-covered roller 金刚砂卷布辊;金刚砂导布辊
emery cutter 砂轮
emery cylinder 砂轮鼓
Emery-Dietz gravity corer 埃默里—迪茨重力取样器
emery disk 刚砂磨盘
emery file 金刚砂锉
emery fillet 金刚砂布带;刚玉砂布带
emery flour 细金刚砂粉
emery grinder 金刚砂磨石;金刚砂磨床;刚玉磨床
emery grinding grease 金刚砂研磨膏
emery grinding machine 金刚砂磨床;刚玉砂磨床
emery grit 金刚砂粒
emery-ground concrete 金刚砂磨光混凝土
emery paper 金刚砂纸;刚玉砂纸;砂纸

emery papering machine 砂纸抛光机
emery paste 金刚砂糊;研磨膏
emery powder 金刚砂(粉);刚玉粉
emery rock 刚玉岩
emery sand cloth 刚砂布
emery sand paper 刚砂纸
emery sharpener 金刚砂磨床
emery stick 薄的磨光锉
emery stone 金刚石;金刚砂(磨)石;刚砂石
emery tape 金刚砂卷带;砂带
emery wheel (金刚)砂轮
emery wheel dresser 修整砂轮机;金刚砂轮修整器
E-metal 锌铝合金
emetine bismuth iodide 碘化铋吐根碱
emigrant(out) 移居者
emigrant ship (向国外移居的)移民船
emigrate 移居国外
emigration 移民;移居(国外)
emigration anchorage (从国外移入)移民锚地
emigration migration 迁出洄游
emigration policy 移民政策
emigration ship (从国外移入)移民船
Emila rotary viscometer 埃米拉旋转黏度计
emimorphic form 异极形
eminence 高地处;高处;高丘;高地
eminent domain 征用(土地)权;土地征用权;铁路土地使用权
eminentia arcuata 弓状隆起
eminently hydraulic lime 天然水泥;罗马水泥;强水硬性石灰;优质水硬石灰
eminently hydraulic lime mortar 显著的水硬性石灰浆
eminent period 卓越周期
emissarium 地下溢洪渠;地下水道;地下排水渠
emissary 排水道;分水道
emissary sky 预兆性天空【气】
emissing ability 发射率
emission 放射;发射;发散;排污物;排放(物);排出污染物
emission amount of sludge 排污泥量
emission analyzer 排气污染成分分析仪
emission angle 光投射角;发射角
emission antenna 发射天线
emission area 发射面积
emission at par 平价发行
emission band 辐射频带;发射带
emission beam angle between half power points 半功率点间的发射束角
emission behavio(u)r 发射性能;排放性能;排放情形
emission by high chimney 高烟囱排放
emission cathode 发射阴极
emission cell 发射式光电元件;发射光电管
emission certification (汽车发动机的)排气污染鉴定;排放许可证
emission characteristic 发射特征;发射特性;排放特性
emission coating 放射层
emission coefficient 发射系数
emission component 发射子线
emission concentration 排气污染物浓度;排放浓度
emission concentration limit 排污浓度限度
emission control 排污控制;排放控制
emission control deadline 排放控制极限
emission control device 排气污染控制装置
emission control equipment 排污控制设备;排放控制设备;排气净化装置
emission control standard 排污控制标准;排放控制标准
emission control system 排气净化系统
emission criterion 排放标准
emission current 放电电流;发射电流
emission decay 电流衰变
emission delay 发射延时;发射延迟
emission detector 发射型探测器;发射检测器
emission duration 发射持续时间
emission efficiency 发射效率
emission electrode 发射电极
emission electron microscope 放射电子显微镜;发射(式)电子显微镜
emission factor 发射因素;排污因子;排放因子;排放因素;排放系数
emission flame photometer 发射光学火焰光度计
emission flame photometry 发射火焰光度法

emission flocculi 发射谱斑
emission formation 排气污染物的形成
emission for sulfur dioxide 二氧化硫排放
emission frequency 发射频率
emission from a stationary source 固定源空气排放
emission from diesel vehicle 柴油机车排气
emission interference 发射干扰
emission inventory 日放逸量；排放数据库；排放记录；排放表
emission inventory of greenhouse gases 温室气体排放数据库
emission law 发射定律
emission level 排污水平；排气污染程度；排放水平；排放标准
emission limit 排放限度；排放限度
emission limited diode 发射限制二极管
emission line 发射谱线
emission-line galaxy 发射线星系
emission-line object 发射线天体
emission-line star 发射线星
emission-line variable 发射线变量
emission load 排放负荷
emission measure 辐射量度；发射量
emission measurement 放出物的计量；排放物测量；排放计量
emission mechanism 发射机制
emission microscopy 发射显微术
emission mitigation 排污消减
emission modulation 放射调制
emission monitoring 排污监测；排放监测；散发监视
emission monochromator 发射光单色仪
emission nebula 发射星云
emission of gas 瓦斯涌出
emission of heat 热能发散；热辐射
emission of lava 熔岩放热；岩浆发散
emission of light 光线发射；发光
emission of neutrons 中子发射
emission of radiant energy 辐射能发射
emission of radiation 辐射
emission of smoke 烟尘扩散；烟雾发散
emission pattern 发射图样
emission peak 发射峰
emission performance 排放特性
emission per unit length 单位长度排放量
emission phototube 放射光电管；发射光电管
emission plate 放射性硬片
emission point 发射点；排放点
emission polarization tensor 辐射偏振张量
emission power 发射强度；发射功率
emission probability 发射概率
emission rate（污染物）放出率；发射（速）率；发射强度；喷出率；排污率；排放（速）率；喷洒率
emission rate constant 排污速率常数
emission readiness monitoring operation 发射准备情况检查操作
emission reduction right 排放消减权
emission reduction unit 排放消减单位
emission regulation 发射调节；排气污染法规
emission regulator 电子放射稳定器；电子发射调节器
emission requirement 排污要求；污染排放要求
emission resonance 发射共振
emission rights 排污权；排放权
emission sampling 排放物取样
emissions monitoring system 废气监测系统
emission source 发射源；排污源；排放源
emission source sampling 排污源采样；排放源采样
emission spectrochemcial analysis 发射光谱化学分析
emission spectrograph 发射摄谱仪
emission spectrographic analysis 发射光谱分析
emission spectrographic detector 发射光谱检测器
emission spectrographic method 发射摄谱仪分析法
emission spectrography 发射摄谱法
emission spectrometer 发射光谱仪；发射分光计
emission spectrometric analyzer 发射光谱分析仪
emission spectrometric method 发射光谱测定法
emission spectrometry 发射光谱测量学
emission spectrophotometer 发射分光光度计
emission spectrophotometry 发射分光光度测定法
emission spectroscope 发射分光镜
emission spectroscopy 发射光谱学；发射分光法

emission spectrum 发射频谱；发射（光）谱
emission spectrum instrument 发射光谱仪器
emission standard 扩散标准；挥发物散发标准；辐射标准；发射标准；排污标准；排放标准
emission standard of smoke and soot 烟灰排放标准
emission standards for sulfur dioxide 二氧化硫排放标准
emission status 排放状况
emission strength 发射强度；排放浓度
emission substance 排放物质
emission surface 发射面
emission tax 排污税；排出物税收
emission test 排气污染试验
emission testing 排放试验
emission theory 放射理论；微粒说；微粒论
emission threshold 发射阈
emission-time pattern 排放—时间特性
emission trading 排污交易；排放交易
emission type electron microscope 发射式电子显微镜
emission type tube tester 发射式电子管测试仪
emission uniformity 喷洒均匀度
emission vacuum spectrometer 发射真空分光计
emission vacuum spectroscopy 发射真空光谱学
emissive 发射的
emissive ability 发射能力
emissive frequency 发射频率
emissive power 辐射本领；发射强度；发射功率；发射本领
emissive power level 发射功率电平
emissivity 表面辐射率；热辐射系数；发射率
emissivity coefficient 发射系数
emissivity factor 散射率因素；发射因素；发射系数；发射率
emissivity of a surface 表面发射系数
emit data 发射数据
emit field 常数字段；发射域
emit output 发射输出
emitron 光电摄像管
emitron camera 光电（摄像）管摄像机
emittance 黑度；出射度；辐射本领；发射能力；发射率；发射度
emitted dose 喷射量
emitted electron energy 发射电子能量
emitted energy 发射能量
emitted particle 发射粒子
emitted power 发射功率
emitted power level 发射功率电平
emitted radiation 发射辐射
emitted wave 发射波
emitter 放射器；发射源；发射体；发射极；滴水喷头
emitter amplifier 射极输出器
emitter and receiver 发射器和接收器
emitter ballast resistance 发射极镇流电阻
emitter barrier 发射区势垒
emitter base diffusion 发射极基极扩散法
emitter base diffusion transistor 发射极基极扩散晶体管
emitter beam lead 发射极梁式引线
emitter bias 发射极偏压
emitter characteristic 射极特性
emitter-collector separation 射极集极间距
emitter control 发射极控制
emitter-controlled oscillator 发射极控制振荡器
emitter-coupled 射极耦合（的）
emitter-coupled cell 射极耦合单元
emitter-coupled logic 发射极耦合逻辑
emitter-coupled logic gate 发射极耦合逻辑门
emitter-coupled-pair modulator 发射极耦合对调制器
emitter-coupled transistor logic 发射极耦合晶体管逻辑
emitter-coupled trigger 射极耦合触发器
emitter current （发）射极电流
emitter cut-off current （发）射极截止电流
emitter degeneration 射极负反馈
emitter diffusion 发射极扩散
emitter diffusion capacitance 射极扩散容量
emitter diffusion layer 射极扩散层
emitter electrode 发射极
emitter follower 发射极输出器；发射极输出放大器；发射极跟随器；射极输出器

emitter follower amplifier 射极跟随放大器
emitter follower circuit 发射极跟随器电路
emitter follower logic 射极跟随器逻辑
emitter function logic 发射极功能逻辑
emitter function logic circuit 发射极功能逻辑电路
emitter junction 发射（极）结
emitter layer 发射极层
emitter leak current 发射极漏电流
emitter leg 射极引线
emitter pattern 发射极图形
emitter power 射极功率
emitter pulse 发射器脉冲；射极脉冲
emitter region 发射区；发射极（区）；射极区
emitter resistance 发射（极）电阻
emitter terminal 发射极端子
emitter time constant 发射极时间常数
emitter timing monostable circuit 射极定时单稳电路
emitter timing multivibrator 射极定时多谐振荡器
emitter-to-base voltage 基极发射极间电压
emit(ting) 逸出
emitting area 发射区；发射面积
emitting atmosphere 发光大气
emitting diode 发射二极管
emitting diode coupler 发射二极管耦合器
emitting electrode 发射（电）极
emitting electron 发光电子
emitting layer 发光层
emitting molecule 发射分子
emitting probe 发射探针
emitting sole tube 底极发射管
emitting stage 发射级
emitting substance 发射物
emitting surface 放射面；发射（表）面
Emley plastometer 恩莱（氏可）塑性仪
Emley's plasticimeter 恩莱塑性试验机
emlgator 乳化器
Emma 声频信号雷达站
Emmel cast iron 埃姆尔高级铸铁
Emmel process 埃姆尔铸铁炼制法
emmetropia 正视眼；正常视觉；折光正常
emmetropic eye 正视眼
Emmi 埃米（一种有机录杀菌剂）
emmonsite 碲铁石
E mode 横磁模式
emoline oil 艾摩贝油；艾摩林低黏度油
emollescence 软化作用
emollient 软化剂
emoluments 酬劳薪金及津贴；报酬
emoticon 情感图标
emotional firesetter 报复放火者
Emperor seamount 皇帝海山
Emperor seamount chain 皇帝海山链
emphasis 加重
emphasis circuit 校正电路
emphasis network 预斜网络
emphasis train 重点列车
emphasis transmission 斜发送
emphasize 强调
emphasize application season 注重喷洒时间
emphasized second marker 加重秒信号
emphasized second marks 加重秒信号
emphasized second signal 加重秒信号
emphasizer 加重器；加重电路；预加重网络
emphyteutic lease 永租权
empire 电绝缘漆；（十八世纪）法国建筑风格
empire cloth 涂布；胶布；绝缘油布；绝缘（胶）布；油基绝缘漆布
empire drill 砂矿钻机
Empire drilling 恩派尔型旋转冲击钻探
empire paper 绝缘油纸
Empire State Building （纽约）帝国大厦
Empire style 家具的设计样式；（拿破仑称帝后）法国建筑
empire tube 绝缘套管
empiric(al) 经验的；实验的
empiric(al) analysis 经验分析
empiric(al) assumption 经验假设；经验假定
empiric(al) average 经验平均值
empiric(al) Bayes' estimator 经验贝叶斯估计量
empiric(al) Bayes procedure 经验贝叶斯过程
empiric(al) coefficient 经验系数
empiric(al) comparison 经验比较
empiric(al) constant 经验常数；经验常量；实验常数

empiric(al) correction 经济改正值
empiric(al) correction factor 经验修正因数;经验修正系数
empiric(al) correlating constant 经验对比常数
empiric(al) cumulative distribution function 经验累积分布函数
empiric(al) curve 经验曲线
empiric(al) data 经验数据
empiric(al) demand function 经验需求函数
empiric(al) design 经验设计
empiric(al) discriminate function 经验判别函数
empiric(al) distribution 经验分布
empiric(al) distribution function 经验分布函数
empiric(al) documentation 经验(性)资料
empiric(al) drying equation 经验干燥方程
empiric(al) econometrics 经验经济计量学
empiric(al) equation 实验公式;经验方程(式)
empiric(al) estimation method of tritium 氚经验估算法
empiric(al) evaluation 经验评价
empiric(al) experiment 经验试验
empiric(al) exponential decline curve 经验指数递减曲线
empiric(al) factor 经验系数
empiric(al) fit 经验拟合;经验符合
empiric(al) fit constant 经验拟合常数
empiric(al) formula 经验(公)式;成分式;实验式
empiric(al) formula by method of average 用平均法求经验公式;平均经验公式
empiric(al) formula by plotting 用图解法求经验公式
empiric(al) formula in smoothing 修匀的经验公式
empiric(al) formula of body 坯式;坯的实验式
empiric(al) formula of glaze 釉的实验式
empiric(al) formula of rainfall 降雨经验公式
empiric(al) formulas by plotting 图解法经验公式
empiric(al) frequency 经验频率
empiric(al) function generator 经验函数发生器;实验函数生成程序;实验函数发生器
empiric(al) generalization 经验概括化
empiric(al) grid residual system 经验网格剩余系统
empiric(al) hypothesis 经验假设
empiric(al) index number 经验指数
empiric(al) information 经验知识
empiric(al) isotherm 经验吸附等温线
empiric(al) law 经验定律;经验性规律
empirically based operations research 经验运筹学
empirically derived model 经验推论模型
empiric(al) management school 经验主义管理学派
empiric(al) mass formula 经验质量公式
empiric(al) materials 经验材料
empiric(al) mean 经验平均值
empiric(al) method 经验(方)法
empiric(al) mode 经验众数;经验模式
empiric(al) model 经验模型;实验用模型
empiric(al) norm 经验定额
empiric(al) observation 经验观测值
empiric(al) orientation 经验定向
empiric(al) orthogonal function 经验正交函数
empiric(al) parameter 经验参数
empiric(al) point data 经验点据
empiric(al) prediction method 经验预测法
empiric(al) probability 经验概率
empiric(al) production rate 经验产量
empiric(al) proportion 经验配合比
empiric(al) proportioning 经验配料比;经验配合(法);经验配合比法;按经验配合
empiric(al) quantitative approach 经验定量方法
empiric(al) rate-time equation 产量时间经验公式
empiric(al)-rational change strategy 经验理性的变革策略
empiric(al) regression relationship 经验回归关系
empiric(al) regularity 经验性规律;经验规律性
empiric(al) relation 经验关系式
empiric(al) relation curve 经验关系曲线
empiric(al) relationship 经验关系(式);经验关系曲线
empiric(al) research 实证研究
empiric(al) rule 经验定则
empiric(al) sample mean 经验样本(平)均值
empiric(al) school 经验学派
empiric(al) series 经验数列

empiric(al) significance level 经验显著性水平
empiric(al) solution 经验解
empiric(al) standard deviation 经验标准偏差
empiric(al) statistic(al) rate 经验统计定额
empiric(al) statistics 经验统计
empiric(al) statistics analysis method 经验统计分析法
empiric(al) stress distribution method 实验应力分布法
empiric(al) study 实证研究
empiric(al) temperature profile 经验温度剖面
empiric(al) test 经验检验
empiric(al) validity 经验的正确性
empiric(al) value 经验值;经验数值;实验数值
empiric(al) variance 样本方差
empiric(al) water flood prediction method 注水效果的经验预测法
empiric(al) yield table 经验收获表
empiricism 经验主义
emplacement 指定一定位置;放置;放列动作【铁】;安置;安放;侵位
emplacement age 就位年龄;侵位年龄
emplacement hole 放置孔穴
emplacement mechanism 成矿机制;成矿机理;成矿方式;侵位机制
emplacement of blast-hole 炮孔定位
emplacement of borehole 炮孔定位
emplacement stage of intrusion 侵入体侵入期次
emplaster 灰膏
emplectite 硫铜铋矿;恩硫铋铜矿
emplecton 空斗石墙
emplectum 空斗石墙;一丁一顺石砌体(片石填心)
employ 占用;采用;聘请;雇佣
employability 受雇就业能力
employable 有使用价值的;可使用的
employed labor force 雇佣劳动力;就业劳动力
employed persons 就业人员数
employed population 就业人口
employee 职工;雇员;员工
employee benefit 员工津贴;职工福利
employee benefit fund 员工福利基金
employee bonus 员工酬劳金;职工酬劳金
employee comfort and services in layout 员工福利服务设施安排
employee compensation 员工补偿金
employee counseling 员工咨询
employee discount 雇员折扣
employee dormitory (职工)单身宿舍
employee-employer relations 劳资关系
employee hours 雇员工时数
employee identification 职工身份证
employee incentive 职工积极性
employee incentive fund source 职工奖励基金
employee insurance 雇员保险
employee insurance fund 职工保险基金
employee loyalty insurance 雇员忠诚险
employee orientation 职工导向
employee pass 员工票
employee payroll 员工薪金册
employee pension fund 职工抚恤金
employee questionnaire method 职工调查法
employee rating form 职工考核表
employee relations 职工关系
employee relief fund 员工救济金;职工救济基金
employee retirement allowance 员工退休津贴
employee saving fund 职工储蓄金
employee's benefit fund 职工福利基金
employee's business expenses 职工业务支出
employee's canteen 雇员小卖部;雇员食堂
employee's card 工作证
employee's educational expenses 职工教育经费
employee services and benefit 职工福利费
employee's flo(a)tation expenses 职工的筹款费用
employee's investment fund 职工投资款
employee's pension fund 职工退休基金
employee's retirement fund 职工退休基金
employee stock option 职工股票购买权
employee training 雇员/员工培训
employee-training expenses 职工培训费
employee-training plan 职工培训计划
employee turnover 雇员调动;职工流动(率)
employee turnover ratio 职工周转率

employee welfare 员工福利;职工生活福利
employee welfare facility 员工福利设施
employer 业主;雇主;发包人
employer's agent 雇主代理人
employer's default 雇主
employer's drawings 雇主的图纸
employer's equipment 业主设备
employer's estimate 标底
employer/server relationship 雇主/服务者关系
employer's facility 业主设施
employer's liability 雇主(的)责任
employer's liability and worker's compensation insurance 雇主责任和工人赔偿保险
employer's liability insurance 业主义务保险;雇主责任保险;雇主义务保险
employer's risks 业主风险;雇主风险
employment 就业;职业;职务;雇用;人事任用;求业者
Employment Act 就业法
employment advertisement 招牌广告
employment agency 职业介绍所
employment agent 经营职业介绍所
employment agreement 雇佣合同
employment and suitability test 适用性试验
employment and unemployment 就业和失业
employment base 就业基数
employment center 就业中心
employment chance 就业机会
employment clause 雇佣条款
employment coefficient 就业系数
employment contract 雇用合同;雇佣合同
employment creation 创造就业机会;提供就业
employment department 就业部门;人事部门
employment department expenses 劳动部门费用
employment disease 职业病
employment effects 就业效果
employment exchange 劳务介绍所
employment experience 就业经历
employment forecast 就业预测
employment function 就业函数
employment generation 提供就业
employment income deduction 发薪扣除所得税
employment index 就业指数
employment insurance 就业保险
employment interest 就业利息
employment letter 就业情况说明
employment multiplier 就业乘数
employment of capital 资本(的)运用
employment office 职业介绍所
employment of plant 设备的使用
employment opportunity 就业机会
employment potential 就业潜力
employment rate 就业率
employment ratio 就业率
employment scheme 就业计划
employment security 就业保障
employment stabilization fund 就业稳定基金
employment statistic 就业统计
employment status 就业状况;就业情况
employment structure 就业结构
employment structure of urban population 城市人口职业构成;城市人口业结构
employment survey 就业调查
employment system 就业制度
employment tax 工资税
employment test 就业测验
empolder 围垦;圩
emporium 商业中心;大百货商店;大商场;百货公司;商场
empressite 粒碲银矿
empress slate 大石板瓦
empretron 手弧离子管
emptied floating dry dock 排空压舱水的浮船坞
emptier 卸载器;倒空装置;清洁车;退火炉出炉工
emptiness problem 空集问题
empty 空的;抛空;使空出来
empty and load brake equipment 空重车制动装置
empty and load change-over cock 空重车转换塞门
empty and load valve 空重车转换阀
empty argument 空变元
empty band 空带
empty barge anchorage 空驳停泊区
empty barge harbo(u)r 空驳停泊区
empty barge storage 空驳停泊区

empty bed contact time 空床接触时间
empty bottle store 空瓶子堆场
empty by gravity 重力卸车
empty car 空车
empty car accumulating track 空车集结线
empty car flow 空车流
empty case ejector mechanism 排筒机构
empty-cell pressure process 空细胞加压不完全浸注法(木材防腐处理)
empty-cell process 空细胞法(木材防腐的压力处理法);空格舱法;定量浸注法;不完全浸注法
empty-cell treatment 空格舱法
empty container 空容器
empty container gate-out advise (集装箱)空箱出场通知
empty container inventory report (集装箱)空箱存场报告;空集装箱清点报告(表)
empty descriptor segment 空描述段
empty device 卸料装置
empty emergency ditch 备用排水渠
empty fissure 空裂隙
empty form 空白表格
empty graph 空图
empty hole 空炮眼
emptying 制动放空;(管道)吹扫;放空;排空
emptying and filling conduit 排水冲水廊道
emptying and filling culvert 排水冲水廊道
emptying and filling gallery 排水冲水廊道
emptying auger 出料搅龙
emptying by fluidization 流态化卸料;充气卸料
emptying chute 卸料滑槽
emptying cock 排空旋塞
emptying culvert 泄水涵道;泄水涵洞;放空涵洞
emptying device 卸载装置;泄水设备;放水设备;放空设备;排水设备
emptying for inspection 检查用放空
emptying gate 泄水闸;放水闸(门);排泄门
emptying hose 放空胶管;排放胶管
emptying lock 船闸放水
emptying of reservoir 水库放空
emptying of throats 喉道排空
emptying point 卸料点
emptying pump 排放泵
emptying system 卸载系统;泄水系统;放水系统;放空系统;排油系统;排水系统;排气系统
emptying time 卸载时间;泄水时间;泄空时间;放空时间;排水时间;排空时间
emptying valve 放空阀;卸料阀;泄水阀;泄空阀;排油阀;排水阀(门);排气阀
empty land 空地;未占用土地
empty line 空闲线路
empty-loaded braking system 空重车可调制系统;有空重车的制动系统
empty-loaded change-over brake 空重车转换制动机;空重车可调制动机
empty-loaded change-over device 空重车转换装置
empty medium 空的媒体;空白媒体
empty nest 陋巷
empty nester 身边无子女的夫妇;单身
empty queue 空队列
empty reel 空盘
empty return speed 空车回程速度
empty rolling stock amount of discharging 卸空车数
empty run 空行程
empty set 空系;空集(合)
empty ship 空船
empty siding 空车停留线
empty sinking 空拔
empty space 空域;真空
empty store 空存储
empty string 空行;空串
empty substitution 空置换
empty symbol 空符号
empty tape 空带
empty track 空闲线路
empty track for shunting 岔道空轨道
empty trip 空车行程;空程
empty tub 空矿车
empty tube column 空管柱
empty upper defect level 空缺陷上能级
empty van (集装箱)空箱
empty visual field 无参考目标视野
empty wagon 空车

empty wagon kilometres 空车走行公里
empty wagon regulation diagram 空车调整图
empty wagon regulation plan 空车调整计划
empty wagon safety valve 空车安全阀
empty wagons flow 空车流
empty wagon turnround time 空车周转时间
empty weight 无载重量;空机重量;净重;空重;空船重;皮重
empty weight of equipment 设备空重
empyema 积液
Emscher filter 埃木舍过滤器;稳压池;双层沉淀池
Emscherian 埃木舍阶
Emscher well 埃木舍井
Emsian 艾姆斯阶
emssivity 放射率
Emsworth water sampler 爱姆斯沃司水样采集器
emulate 赶上
emulated data 仿真数据
emulate system architecture 仿真系统结构
emulation 模拟;仿真
emulation among workers of the same type of work 同工种竞赛
emulation bus 模拟总线;仿真总线
emulation bus trace 仿真总线跟踪
emulation command 仿真命令
emulation job 仿真作业
emulation mode 仿真方式
emulation program(me) 仿效程序
emulation terminal system 仿真终端系统
emulator 竞争者;模拟器;模快器;仿真器;乳化剂
emulator control 仿真器控制
emulator debug 仿真器调试
emulator generation 仿真器形成
emulator mode 仿真方式
emulator package 仿真数据包;仿真软件包
emulator section 仿真器部分
emulgator 乳化剂
emulgent 泄出的
emulphor 乳化剂
emulsibility 乳化性
emulsible oil 乳化油
emulsifiability 可乳化性;可乳化本领;乳化性;乳化度
emulsifiability test 乳化试验
emulsifiable 可乳化的
emulsifiable base 可乳化基质
emulsifiable concentrate containing 含有效成分
emulsifiable material 乳化材料
emulsifiable oil 无水分的油;(可)乳化油
emulsifiable paint (可)乳化漆
emulsifiable paste 乳化膏
emulsification 乳化(作用);乳化的过程
emulsification apparatus 乳化(试验)器
emulsification polymerrization 乳液聚合;乳化聚合(作用)
emulsification property 乳化特性
emulsification test 乳化试验
emulsified acid 乳化酸
emulsified ammonium nitrate 乳化硝酸铵
emulsified asphalt 沥青乳胶;乳状沥青;乳化(地)沥青
emulsified asphalt varnish 乳化沥青漆
emulsified asphalt with lime paste 石灰膏乳化沥青
emulsified base 乳化基质
emulsified binder 乳化黏合剂
emulsified bitumen 沥青乳胶;沥青乳化液;沥青乳(浊)液;乳化(地)沥青
emulsified bituminous materials 乳化沥青材料
emulsified coolants 乳化冷却剂
emulsified crude oil 乳化原油
emulsified epoxide 可乳化环氧化物
emulsified explosive 乳化炸药
emulsified grease 乳化脂
emulsified hydrocarbon fuel 乳化烃燃料
emulsified mineral oil 可乳化矿物油
emulsified oil 乳化油
emulsified oil droplet 乳化油滴
emulsified oil quenching 乳化油淬火
emulsified oil wastewater 乳化油废水
emulsified paint 乳化涂料
emulsified particles 乳化颗粒
emulsified pearlite 乳状化珠光体
emulsified reclaimed-rubber coating 回收橡胶乳化涂层

emulsified refinery wastewater 炼油厂乳化废水
emulsified resin 乳化树脂
emulsified rubber 乳化橡胶
emulsified silicone oil 乳化硅油
emulsified solvent 乳化溶剂
emulsified system 乳化剂灭火法(一种油面灭火法);乳化剂灭火系统
emulsified water 乳状水
emulsifier 多效乳化剂;乳化物质;乳化剂
emulsifier-free latex 无乳化剂的乳胶
emulsifier injector 乳化器
emulsifier layer 乳化剂层
emulsifier viscosity 乳化剂黏度
emulsify 使乳化;使成乳剂;乳化
emulsifying 乳化
emulsifying ability 乳化性
emulsifying additive 乳化添加剂
emulsifying agent 乳化剂;乳化剂
emulsifying agitator 乳化搅拌器
emulsifying capacity 乳化能力
emulsifying colloid 乳化胶体
emulsifying efficiency 乳化效率
emulsifying hydrocarbon 乳化烃
emulsifying lubricants additive 乳化润滑油添加剂
emulsifying machine 乳化机
emulsifying mill 乳化磨
emulsifying mixer 乳化混合器
emulsifying oil 可溶油;乳化油
emulsifying power 乳化力;乳化本领
emulsifying process 乳化方法;乳化过程
emulsifying property 乳化性;乳化特性
emulsifying salt 乳化用盐
emulsifying tank 乳化槽
emulsifying tower 乳化塔
emulsifying wax 乳化蜡
emulsion 乳浊液;乳胶;乳剂
emulsion adhesive 乳胶黏结
emulsion aeration 乳化液充气
emulsion alkyd 醇酸树脂乳化液
emulsion asbestos sheet 石棉乳胶板
emulsion asphalt 乳化沥青
emulsion band 乳化带
emulsion base 乳胶基质;乳剂基质
emulsion batch number 乳剂批号
emulsion binder 乳液黏结剂;乳液基料;乳胶黏结剂;乳胶黏合剂
emulsion binding medium 乳胶黏结剂
emulsion blocking agent 乳状液封堵剂
emulsion bonded mat 乳液黏结毡
emulsion bonding adhesive 乳胶黏结剂
emulsion bonding agent 乳胶黏结剂
emulsion bonding medium 乳胶黏合介质
emulsion breaker 乳胶分解剂;破乳器;破乳剂
emulsion break(ing) 乳胶分解;破乳
emulsion-breaking voltage 破乳电压
emulsion calibration curve 乳胶校正曲线;乳剂特性曲线;乳化校正曲线
emulsion carrier 载胶体;乳胶载体
emulsion-causing 产生乳化
emulsion cement(-ing agent) 乳胶胶凝剂;乳胶黏结剂
emulsion chamber 乳化室
emulsion chemistry 乳胶化学;胶体化学
emulsion cleaner 乳状洗涤剂;乳胶洗净剂
emulsion cleaning 乳化液清洗
emulsion cleaning agent 乳液清洁剂
emulsion-coated film 感光胶片
emulsion-coated material 涂膜材料
emulsion coating 乳液表面涂层;乳液表面处治;乳胶涂料;(混凝土养护用)乳胶涂层;乳化涂料
emulsion collector 乳胶收集体
emulsion colloid 乳(状)胶体
emulsion contrast 乳胶对比度
emulsion copolymerization 乳液共聚
emulsion counting 乳胶计数
emulsion cracking 乳胶裂化
emulsion creaming 乳剂分层
emulsion cutting oil 乳化切削油
emulsion dampening 乳剂湿润
emulsion degreasing 乳化脱脂
emulsion density 乳胶光密度
emulsion dispersion 乳剂分散(性)

emulsion dy(e)ing 乳化液染色
emulsion explosive 乳化炸药;乳化炸药
emulsion flo(a)tation 乳化浮选
emulsion flow property 乳剂流动性
emulsion flow regime 乳状流型
emulsion fluid 乳状液;乳化液
emulsion fog 乳胶灰雾
emulsion for artificial light source 人工光源乳剂
emulsion for road construction 公路建筑用乳胶
emulsion fuel 乳化燃料
emulsion-glass interface 乳剂—玻璃板界面
emulsion glue 乳胶黏合剂;乳化胶
emulsion granularity 乳剂颗粒度
emulsion grouting 乳油灌浆
emulsion hand sprayer 手动乳液喷洒器
emulsion inhibitor 抗乳化剂;乳胶阻化剂;乳化抑制剂;脱乳胶剂
emulsion injection 乳液注入
emulsion interference filter 乳胶干涉滤光片
emulsion inverse 乳液转相;乳液逆转
emulsion inversion 乳剂变型
emulsionize 乳化
emulsion-laser storage 乳胶激光存储器
emulsion layer 乳胶层;乳剂层;乳化层
emulsion machine 乳化机
emulsion mask 乳胶掩模
emulsion medium 乳胶介质
emulsion membrane 乳胶薄膜;乳化液层
emulsion mixer 乳化混合器
emulsion mud 乳化泥浆
emulsion numbers 乳剂号
emulsion of chloroform 氯仿乳剂
emulsion of road tar 道路焦油乳胶
emulsion of tar 乳化沥青焦油
emulsion of tar/asphalt mix(ture) 焦油沥青混合乳胶
emulsion of tar/bitumen mix(ture) 焦油沥青混合乳胶
emulsion of water-gas tar 水煤气沥青乳胶
emulsion oil in water 水包油乳化液
emulsion ointment base 乳剂软膏基质
emulsion opal glasses 乳光玻璃
emulsion paint 乳液涂料;乳胶(状)漆;乳剂颜料;乳化(油)漆
emulsion paint with aggregate 加集料乳胶漆;加骨料乳胶漆
emulsion particle 乳胶粒;乳胶(微)粒子
emulsion paste 乳化浆
emulsion photometer 乳胶光度计
emulsion polish 乳胶蜡;乳胶抛光剂
emulsion polymer 乳液聚合物;乳胶聚合物
emulsion polymer adhesive 乳液聚合物胶黏剂;乳胶聚合物胶黏剂
emulsion polymerization 乳液聚合(作用);乳胶聚合;乳胶聚合作用
emulsion polymerizing 乳胶聚合作用
emulsion preventative surfactant 防乳剂
emulsion preventer 防乳剂
emulsion printing 乳化浆印花
emulsion process 乳胶法;乳化过程;乳化法
emulsion product 乳胶生产
emulsion pump 乳化液泵
emulsion putty 乳液腻子
emulsion reaction 乳化反应
emulsion resistance 抗乳化(性)
emulsion resolving 乳胶分解;脱乳化
emulsion response 乳胶响应
emulsion sealant 乳液密封膏
emulsion sensitivity 乳胶灵敏度
emulsion separation 乳胶分离
emulsion shelf life 乳剂搁置寿命
emulsion shift 乳胶位移;乳剂漂移
emulsion shrinkage 乳胶收缩
emulsion side 乳剂面
emulsion sintering method of spinning 乳液烧结纺丝法
emulsion sludge 乳胶淤渣沉液;乳化淤渣
emulsion slurry 乳化砂浆;乳液稀砂浆;乳浆
emulsion speed 乳胶感光速率;乳剂感光度
emulsion splitter 破乳剂
emulsion sprayer 乳液喷射器
emulsion spray nozzle 乳液喷嘴
emulsion stability 乳状液稳定度;乳液稳定性;乳胶稳定度;乳剂稳定性;乳化(液)稳定性

emulsion stability test 乳胶安定性试验
emulsion stabilizer 乳胶稳定剂;乳(状)液稳定剂;乳化稳定剂
emulsion stabilizing agent 乳胶稳定剂
emulsion strength 乳化强度
emulsion surface 乳化面;乳剂面
emulsion tack coat 乳胶黏膜
emulsion technique 乳化技术
emulsion technology 乳剂技术;乳剂工艺
emulsion test 乳剂试验;乳化试验
emulsion testing 乳剂检验
emulsion texture 乳滴状结构
emulsion theory 乳化理论
emulsion thickener 乳化糊
emulsion thickness 乳厚度
emulsion tower 乳化塔
emulsion track 乳胶径迹
emulsion treater 乳状液处理器;乳化处理器
emulsion treatment 沥青乳液处治;乳液处理
emulsion tube 乳液管
emulsion type 乳状液类型
emulsion type acrylic resin 乳液型丙烯酸树脂漆
emulsion type alkyd 乳液型醇酸
emulsion type binder 乳液型胶合剂
emulsion type cleaner 乳液型洗净剂
emulsion type lubricant 乳液型润滑油
emulsion-type metalworking liquid 乳化型金属加工液
emulsion-type water-carried paint 乳胶型水性涂料
emulsion varnish 乳液清漆;乳胶清漆
emulsion velocity 显影速度
emulsion viscosity 乳化剂稠黏度
emulsion waste oil 乳状废油
emulsion water 乳胶水;水乳液
emulsion wax 乳化石蜡;乳状蜡
emulsity agent 乳化剂
emulsive 乳化的;乳状的
emulsive waterproof paint 水乳型防水涂料
emulsoid 乳胶(液);乳胶体;乳浊液
emulsol (浮选用)油显剂
emulsor 乳化器
emulsum 乳剂
emulsus 乳状
Emy's roof 爱米式屋顶
enable 恢复操作;撤消禁止门的禁止信号;使能够
enable buffer 使能缓冲器
enabled 使能够
enabled condition 允许条件
enabled instruction 启动指令
enabled interruption 允许中断
enabled operator 启动算符
enabled state 允许状态
enable input 起动输入;启动输入
enable level 允许电平
enable line 起动信号线
enable position 开启位置;起动位置;启动位置
enable pulse 起动脉冲;使通脉冲
enable signal 允许信号;使能信号
enable switch 启动开关
enabling act 授权法案
enabling address 启动地址
enabling assets 辅助资产
enabling counting 启动计数
enabling count memory 启动计数存储器
enabling declaration 行使权利文书;授权申明
enabling gate 启动门
enabling legislation 授权法案
enabling pulse 起动脉冲;使通脉冲
enabling signal 恢复操作信号;许可信号;起动信号;启动信号;使能信号
enalite 水硅钍铀矿
enamel 釉质;釉瓷;瓷漆;珐琅;搪玻璃
enamel back tubing 搪瓷敷层管
enamel badge 搪瓷釉标
enamel bead 搪瓷珠
enamel blue 大青色;藤青;搪瓷青
enamel brush 瓷漆刷
enamel building 搪瓷建筑物
enamel clay 釉瓷黏土;涂搪瓷用黏土
enamel coating 搪瓷涂层
enamel colo(u)r 釉上彩色料;珐琅彩;搪瓷颜料;釉瓷色
enamel cooler 搪瓷冷却器
enamel counter 搪瓷计量器

enamel covering 漆皮;漆包皮
enamel defect 搪瓷缺陷
enamel drop 釉珠
enamel dye 搪瓷颜料
enamel evapo(u)rator 搪瓷蒸发器
enamel eye 漆孔
enamel finish 末道瓷漆;表层瓷漆
enamel firing 烤花;搪烧
enamel firing furnace 搪烧炉
enamel frit 珐琅粉;搪瓷熔块
enamel furnace 搪瓷窑
enamel furniture 装配搪瓷
enamel gauging tank 搪瓷计量罐
enamel-germ 釉胚
enamel glass 釉彩玻璃;搪瓷玻璃
enamel glaze 搪瓷釉
enamel glazed coating 搪瓷上釉涂层;瓷釉涂层
enamel heat exchanger 搪玻璃换热器
enamel high pressure still 搪瓷高压釜
enamel impregnation 上釉
enamel-insulated alumin(i)um wire 漆包铝线
enamel-insulated cable 漆包电缆
enamel-insulated wire 漆绝缘线;漆包线
enamel kiln 锦窑;釉窑;搪瓷窑;搪瓷(烤花)窑
enamel label 搪瓷釉商标
enamel lacquer (纤维素)瓷
enamel lamella 釉板
enamel lampshade 搪瓷灯罩
enamel layer 瓷层;搪瓷层
enamel layer stress test 瓷层应力检验
enamel leather 漆革;漆皮
enamel(l)ed 涂珐琅的;上釉的;漆包的;搪瓷的;涂漆的
enamel(l)ed and cotton covered wire 纱包漆包线
enamel(l)ed asbestos-cement 涂层石棉水泥
enamel(l)ed blade 搪瓷臂板
enamel(l)ed brick 釉(面)砖;琉璃瓦;釉瓷砖
enamel(l)ed bulb 搪瓷球形物;加釉玻璃壳
enamel(l)ed cable 漆包线;绝缘线;漆包电缆
enamel(l)ed cast iron 搪瓷铸铁
enamel(l)ed cast-iron(bath)tub 搪瓷铁浴盆;搪瓷铸铁浴盆;铁胎搪瓷
enamel(l)ed cloth 漆皮布
enamel(l)ed coating 漆层
enamel(l)ed concrete 磨光混凝土
enamel(l)ed cooking utensil 搪瓷烧器
enamel(l)ed copper wire 漆包铜线
enamel(l)ed duck 涂层帆布
enamel(l)ed glass 釉面玻璃;釉瓷玻璃;搪瓷玻璃
enamel(l)ed hardboard 搪瓷硬板;搪瓷面硬质纤维板;釉瓷面硬质纤维板
enamel(l)ed hotplate 搪瓷灶
enamel(l)ed ironware 钢胎搪瓷制品;搪瓷铁器
enamel(l)ed leather 漆皮
enamel(l)ed material 搪瓷材料
enamel(l)ed paint 搪瓷漆;搪瓷画
enamel(l)ed painting 搪瓷画
enamel(l)ed paper 印图纸;图版纸;铜版纸
enamel(l)ed pressed steel 搪瓷钢板
enamel(l)ed product 搪瓷制品
enamel(l)ed reflector 搪瓷反射罩;涂瓷漆反射器
enamel(l)ed silo 搪瓷粮仓
enamel(l)ed slate 上了漆的板;上珐琅的板
enamel(l)ed steel 搪瓷钢
enamel(l)ed strip 涂漆带钢;涂珐琅钢带
enamel(l)ed tile 釉面砖;上釉精加工;琉璃瓦;釉瓷瓦
enamel(l)ed vessel 搪瓷容器;搪玻璃容器
enamel(l)ed wire (瓷)漆包线
enamel(l)ed wire cable 漆包线电缆
enamel(l)ed wire wound resistor 漆包线绕电阻
enameller 搪瓷工匠
enamel lined 搪瓷的
enamel lined pump 搪瓷泵
enamel(l)ing 上涂料;涂搪;涂瓷漆;搪瓷;搪玻璃;上釉
enamel(l)ing by pouring 注浆涂搪
enamel(l)ing clay 涂釉土坯;涂搪瓷用黏土
enamel(l)ing firing 搪烧
enamel(l)ing furnace 烘焙炉
enamel(l)ing iron 搪瓷钢板
enamel(l)ing metals 施珐琅用金属
enamel(l)ing sheet steel 钢板搪瓷;搪瓷用钢板

enamel(l)ing stove 上釉(火)炉
enamel(l)ized variable resistor 珐琅可变电阻器
enamel manganin 漆包锰铜线
enamel melting furnace 搪瓷熔炉
enamel organ 釉器;造釉器
enamel oxide 搪瓷氧化物
enamel paint 瓷漆;搪瓷漆;亮漆
enamel painter 搪瓷画家
enamel painting 搪瓷涂敷
enamel painting on porcelain 瓷胎画珐琅
enamel pan 搪瓷锅
enamel paper 蜡图纸;单面上光纸;涂布纸;印图纸;铜版纸
enamel paraffin(e) wire 蜡浸漆包线
enamel pearl 釉珠
enamel pipe 搪瓷管
enamel polymerization still 搪瓷聚合釜
enamel powder 珐琅粉
enamel prism 釉棱柱
enamel reactor 搪瓷反应罐
enamel reflector 搪瓷罩
enamel rod 釉(棱)柱
enamel sanitary wares 卫生搪瓷
enamel seal 搪瓷封接
enamel sheet 搪瓷(用)钢板
enamel silk-covered wire 丝包漆包线
enamel slip 釉浆
enamel spindle 釉梭
enamel still 搪瓷釜
enamel storage tank 搪瓷储[贮]罐
enamel stove 烧珐琅质的炉;烧釉炉
enamel strip 涂塑料钢片;涂塑钢片
enamel tank furnace 彩烧池炉
enamel thickness test 瓷层厚度测定
enamel tile 瓷砖
enamel tufts 釉丛
enamelum 釉质
enamel varnish 瓷漆
enamel ware 搪瓷器(皿);搪瓷制品
enamel wash basin 搪瓷脸盆
enamel white 锌钡白
enamel wire 漆包线
enanthal 庚醛
enanthic acid 庚酸
enanthic aldehyde 庚醛
enanthine 庚炔
enanthylic acid 庚酸
enantiomer 对映体
enantiomorph 对映体;对映结构体
enantiomorphism 对映形态
enantiomorphous 对映结构的
enantiomorphous class 对映对称型
enantiotropic 互变性的;对映性的
enantiotropic inversion 双变性转变
enantiotropic transformation 双向转变
enantiotropy 互变性;互变(现象);对映(异构)现象
enargite 硫砷铜矿
enathaldehyde 庚醛
enathol 庚醇
enaulium 固沙群落
en axe 在轴线上;轴对称的
en-block 单体;单块
en-block cast 整体铸造
en-block construction 整体结构;整块结构;单块结构
en bloc resection 整块切除
encallow 制砖取土场表层土;型砖;坯砖
encallowing 移走型砖(由制砖黏土场顶面层处移去型砖)
en-camp 建立营地
encampment 营帐;营区;营房
encapsulant 密封物;密封剂;密封材料;包封剂
encapsulate 用胶囊包;封装;包胶
encapsulated 压缩的;密封的;封装的;用胶囊包起来的;包胶的;包裹的
encapsulated circuit 封装电路
encapsulated demodulator 密封解调器
encapsulated fertilizer 胶模肥料
encapsulated fuel 加密封装的燃料元件
encapsulated Ge detector 密封锗探测器
encapsulated pitting 囊状点蚀
encapsulated product 密封产品
encapsulated strain ga(u)ge 屏蔽应变片;屏蔽应变计

encapsulated tree 隔水采油树
encapsulated winding 绝缘绕阻;浇注绕组;密封绕组
encapsulated window 密封窗;气囊窗
encapsulating 包封
encapsulating compound 浇注胶;封铸用混合料
encapsulating glass 封接玻璃
encapsulating mud 囊护泥浆
encapsulating suit 密封服
encapsulation 胶囊化;内嵌;密封(设计);密封;用胶囊包;封入胶内;封闭止漏;包围;包囊涂封;包胶;包裹;屏罩;陶瓷封料
encapsulation technique 成囊技术
encapsulization 包囊化
encarpa 华果形装饰;垂花饰
encarpus 垂花饰【建】;垂花装饰;[复]encarpi
encased 包装
encased back gear 封闭式背齿轮
encased bath tube 包装起来的浴缸管子
encased beam 包镶梁;包(钢)梁;嵌入梁
encased column 混成柱
encased conveying 集装箱运输
encased gear 封闭式齿轮传动装置
encased knot 枯木节疤;暗木节;皮包节;死节
encased magnet 铠装磁铁
encased sheet pile 壳桩;有(外)壳(的)桩
encased stanchion 混成柱
encased steel beam 包壳钢梁
encased steelwork 埋置钢结构
encased steelwork for fire-proofing 钢结构防火包壳
encased structure 有外壳(的钢)结构
encased turbine 有壳轮机
encase knot 死结
encasement 装箱;套子
encasement concrete 包裹混凝土
encasement medium 包封介质
encasement of individual fiber 单丝隔离
encashment 兑现;取现
encasing 护壁【岩】;罩子;嵌入;砌面;套子
encasing in concrete 混凝土外包
encasing of girders 大梁外壳
encasing rock 埋着的岩石
encasing steelwork for fire-protection 钢结构防火包壳
encastre 嵌入梁;固端梁;端部固定;预置梁;端头嵌固
encastre bending moment 端部固定的弯矩
encastre condition 端部固定条件
encastre(d) beam 端固结的梁;固端梁
encastre moment 端部固定的力矩
encasure 水泥包装;包装
encaustic 蜡画;上釉烧的
encaustic brick 琉璃砖;釉面砖;彩砖;彩色釉砖
encaustic decoration 釉彩饰;烧蜡彩画釉烧成装饰处理的砖、瓷砖、玻璃、陶器等
encaustic painting 釉烧画;蜡绘;蜡画(法);瓷画
encaustic paving 瓷砖面层
encaustic tile 琉璃瓦;彩(色)瓦;彩花陶瓷面砖;烧彩砖瓦;釉面瓦
enceinte 围墙;围地;围栏
enceinte wall 庙宇围墙;建筑群围墙
encephalitis B 流行性乙型脑炎
enchain 锁住
enchancement factor 随机因素
enchase 嵌花;镶花;雕镂
enchased 镶嵌的;镶边的;镂刻的;镂刻的;雕镂的
enchased decoration 嵌花装饰
enchasing 镂刻
enchylema 细胞液;透明质
encipher 用代号【测】;编码
enciphered data 加密数据;编码资料
enciphered facsimile communication 机密传真通信
enciphered message 加密信息
enciphering 译成密码
encipherment protection 加密保护
encipheror 编码器
encircle 环绕;包围
encircled by a palisade 用木栅栏围住的
encirclement 圈闭
encircling 环礁
encircling arcuate faulting 旋回弧形断裂
encircling chapel 围住的小教堂

encircling groove 圆形槽
encircling reef 堤礁;堡礁(指与海岸平行的珊瑚礁)
encircling ring 环形垫
Encke division 恩克环缝
Encke's method 恩克方法
Encke's roots 恩克根
enclasp 紧握
enclave 残遗群落;包含物;残遗单优种群落;飞地;被围物
enclave allomorphe 同源异构包体
enclave antilogene 异质包体
enclave economy 飞地式经济
enclave enallogene 异源包体
enclave homoeogene 同源包体
enclave polygene 多源包体
enclave swarm 多体群
enclose 内含节;附入;包装;包围
enclose by masonry wall 用砖石墙围住
enclose circuit 编码电路
enclose cutter head 带罩切削盘
enclosed 被围的
enclosed accommodation space 围蔽起居处所
enclosed accumulator 封闭式蓄电池
enclosed aerated filter 封闭曝气过滤池;封闭冲气滤池
enclosed air-side conveyer 封闭式输送斜槽
enclosed arcade 环围式拱廊
enclosed arc lamp 封闭式弧光灯
enclosed area 围住的面积;有界区域;封闭区(域)
enclosed auger 闭式螺旋
enclosed basin 内陆盆地;内陆流域;封闭(式)盆地;封闭式流域
enclosed bay 有屏障海湾;隐蔽海湾;封闭港湾
enclosed bearing 封闭式轴承
enclosed belt conveyer 封闭式皮带运输机;封闭式皮带输送机
enclosed berth 闭合式泊位
enclosed block 围住的大楼;封闭式滑轮
enclosed bodies of waters 孤立水域;孤立水区;封闭水域;闭合水域;闭合水区
enclosed building 封闭式建筑
enclosed bus(bar) 封闭母线
enclosed cabinet for drying 烘干间
enclosed cabinet for drying washing 密闭的洗衣干燥小室
enclosed cam shaft 闭式凸轮轴
enclosed carbon-arc exposure apparatus 封闭碳弧灯照射老化仪
enclosed carbon arc lamp 封闭式碳极弧灯
enclosed carbonate light 封闭式碳精弧光灯
enclosed cell 封闭式电池
enclosed circulating air cooling 密闭循环空气冷却
enclosed clinker storage 封闭式熟料储库
enclosed coal(conveyer) trestle 运煤廊
enclosed coded source 闭合式编码震源
enclosed combustion burner 封闭燃烧器;对流燃烧器
enclosed compartment 围蔽舱室
enclosed compressor 封闭式压缩机
enclosed construction 封闭结构
enclosed conveyer conduit 封闭式输送管道
enclosed conveyer 封闭式输送机
enclosed cooling system 封闭式冷却系统
enclosed court 封闭式球场;封闭式场地;封闭的院子;四周有墙或建筑物围绕的庭院
enclosed courtyard 封闭式庭院
enclosed cutter 带罩切削器;带罩绞刀;封闭式绞刀;闭式绞刀
enclosed deck 围蔽甲板
enclosed dock 闭合(式)港池;闭合船坞;湿(船)坞;带闸港池
enclosed door bell 封闭式门铃
enclosed(drilling) derrick 封闭式钻塔;遮蔽的钻井架
enclosed drive 闭式传动
enclosed dyeing machine 封闭式染色机
enclosed electric(al) machine 封闭式电(动)机
enclosed end bay 暗间
enclosed environment 封闭环境
enclosed excavation method 封闭开挖法
enclosed exit stairwell 密闭的安全楼梯井
enclosed expansion joint 隐蔽伸缩缝
enclosed feed gear 封闭式进刀齿轮

enclosed figure 封闭的图形
enclosed fiord 围蔽海峡
enclosed fitting 密封罩;封闭罩
enclosed flow 封冻径流;冰下水流
enclosed furnace 封闭(式)炉
enclosed fuse 管形熔断片;封闭式保险丝;封闭式熔断器
enclosed gear 封闭式齿轮传动装置
enclosed geometric figure 封闭的几何图形
enclosed ground 隐蔽地区
enclosed guide rod 闭式导杆
enclosed harbo(u)r 闭合式港口
enclosed hood 密闭罩
enclosed housing 封闭箱体
enclosed impeller (封)闭式叶轮
enclosed isolated phase bus 全封闭式分相母线
enclosed jointly by higher potential plane and impermeable barrier 由高势面和非渗遮挡共同封闭
enclosed knife switch 金属盒开关;密闭闸刀开关;封闭式闸刀开关;铁壳开关
enclosed knot (木材)暗节
enclosed lake 内陆湖泊;内湖
enclosed laser device 封闭型激光装置
enclosed launder 闭式流槽
enclosed lifeboat 封闭式救生艇
enclosed low potential area 闭合的低势区
enclosed mall 封闭式购物中心
enclosed mall shopping center 覆盖的具有走道的商业区
enclosed mass 包块
enclosed meander 环形河湾
enclosed motor 密封式电动机;封闭电动机
enclosed paddle worm 封闭的桨叶式推进器
enclosed parking garage 封闭式停车库
enclosed pasture 围栏草地
enclosed phase 内封相;被围相
enclosed plaza 封闭式广场
enclosed plunger conveyer system 封闭柱塞式输送体系
enclosed porch 封闭式门廊
enclosed press 封闭式压力机
enclosed promenade deck 围蔽散步甲板
enclosed propeller shaft 闭式螺旋桨轴
enclosed radiator 暗装散热器
enclosed refrigerator 封闭式制冷机
enclosed relay 封装式继电器
enclosed scale thermometer 内标尺式温度计;双套管温度计
enclosed sea 内海;封闭海
enclosed seal 封闭装置
enclosed section 盖封区段
enclosed separately ventilation 闭路单独通风
enclosed shaft 封闭竖井
enclosed single point press 闭式单点压力机
enclosed single-shell drier 封口单壳干燥器
enclosed slag 夹渣
enclosed slot 封口槽
enclosed small container 封闭型小型集装箱;封闭式小型集装箱
enclosed space 围住的空间;公侯的包厢;封闭位置;封闭空间;封闭处所;封闭舱位;封起来的场地;围蔽处所;建筑空间
enclosed space tonnage 封闭舱室吨位
enclosed spiral conveyer 封闭式螺旋输送机
enclosed spray type cooler 密封喷雾式冷却器
enclosed stair(case) 外包式楼梯;封闭式楼梯;隔绝楼梯(间)
enclosed stairway 封闭楼梯间
enclosed stairwell 封闭楼梯间
enclosed starter 封闭型起动器;封闭式起动器
enclosed style enamel 嵌线珐琅
enclosed switch 封闭型开关;封闭式低压开关
enclosed switch board 封闭式配电盘
enclosed trickling filter 封闭式滴滤池
enclosed tubular bearing 密封套管轴承
enclosed tubular cut-off 封闭管式熔断器
enclosed type 密闭式;封闭型;封闭式;闭锁式
enclosed type breast drill 封闭式胸摇钻
enclosed type bus bar 封闭式母线
enclosed type combination switch 密闭组合式开关
enclosed type combined electric(al) apparatus 封闭式组合电器

enclosed type constitutional electric(al) apparatus 封闭式组合电器
enclosed type construction 封闭式构造
enclosed type duplex lantern 闭式双航标灯
enclosed type electric(al) equipment 封闭型电气设备
enclosed type fuse 封闭式熔断器
enclosed type grab 密闭式抓斗;封闭式抓斗
enclosed type impeller 闭式叶轮
enclosed type induction motor 封闭式感应电动机
enclosed type layout 封闭式布局
enclosed type motor 封闭式电(动)机;封闭型电动机
enclosed type switch 封闭型开关;封闭式开关
enclosed type turbine 封闭式水轮机
enclosed ventilated machine 封闭式风冷电机
enclosed ventilated motor 封闭通风式电动机
enclosed ventilated type motor 封闭通风型电机
enclosed walkway 封闭人行道
enclosed watch 密封表
enclosed waters 孤立水域;孤立水区;封闭水域;闭合水域;闭合水区
enclosed welding 强制成形焊接
enclosed well 封闭井
encloser 罩壳;外罩;外壳
enclose tideland for cultivation 围垦
enclosing beach for land reclamation 围垦滩涂
enclosing by masonry wall 用墙围住
enclosing cage 紧闭的笼;封闭笼
enclosing cover 轮盖;紧密盖
enclosing cover for impeller 叶轮轮盖
enclosing design function 闭合设计函数
enclosing dike 圩垸
enclosing lake for land reclamation 围湖造田
enclosing masonry wall 围墙;外围砖(石)砌墙
enclosing masonry wall column 围墙柱
enclosing masonry wall facing 围墙墙面
enclosing masonry wall lining 围墙墙衬
enclosing masonry wall(sur)facing 围墙表面
enclosing masonry work 围墙工
enclosing rock 围岩
enclosing safety cage 封闭安全笼
enclosing sheeting 围住的板桩
enclosing stratum 围岩层
enclosing structure system 密闭结构系统
enclosing the sharp bend and opening a more straight channel 封弯走直
enclosing wall 小墙;围岩;围墙;外围墙
enclosing wall building component 围墙建筑组成部分;外墙建筑组成部分
enclosing wall building member 围墙建筑构件;外墙建筑构件
enclosing wall building unit 围墙建筑单元;外墙建筑单元
enclosing wall construction 围墙建筑施工;外墙建筑施工
enclosing wall facing 围墙饰面;外墙饰面
enclosing wall finish 围墙修饰;外墙修饰
enclosing wall lining 围墙衬砌;外墙衬砌
enclosure 信中附件;吹填围埝;附寄物;封装;封入;保护箱罩;包围;排烟罩;围绕;围栏;围场;外壳;设备外壳
enclosure and pressure proportion 气体密封腔与压力比
enclosure bulkhead 围密舱壁;围(封)舱壁
enclosure dock 闭合式港池
enclosure ecosystem 围隔生态系统
enclosure effect 封闭效应
enclosure exhaust ventilation 密闭排气通风
enclosure fire 封闭空间火灾
enclosure fire model 封闭空间火灾模型
enclosure framing 围墙结构;围栏框架;围护构架
enclosure method 封闭方法
enclosure movement 占地运动
enclosure of noise 噪声围蔽
enclosure of oil (石油)圈闭油
enclosure of radiator 散热器外罩;散流器外罩
enclosure of radiators 汽炉围栏
enclosure of space 空间的围闭
enclosure plan 周边式布置
enclosure planting 围植
enclosure road 闭合式锚泊地
enclosures 维护结构
enclosure space 封闭空间

enclosure theorem 界限定理
enclosure wall 大围墙;围(护)墙
encode 密码;代码化;编码
encode address 编码通信处;编码地址
encode block 码组
encode by group 按组编码
encode combination 码组合
encoded abstract 编码(式)文摘
encode data 编码数据
encoded control 编码控制
encoded image 编码图像
encoded keyboard 编码键盘
encoded mode 编码方式
encoded multiplex spectrometer 编码多路分光计
encoded photogrammetric plotter 编码摄影绘图仪;编码立体绘图仪;编码立体绘图机
encoded point 编码点
encoded question 编码问题;编码的题目
encoded sweep technique 编码扫描技术
encoded Turing machine 编码图灵机
encode mode 编码模式
encode processor 编码处理器
encoder 编码装置;编码员;编码器
encoder block diagram 编码器方框图
encoder digital electronic balance 码盘数字电子秤
encoder group 编码器组
encoder matrix 编译矩阵;编码矩阵
encoder output 编码器输出
encoder relay 编码继电器
encoder room 分拣室
encode table 编码表
encode video information 编码视频信息
encoding altimeter 编码高度计
encoding board 编码盘
encoding by bit 按位编码
encoding constraint length 编码约束长度
encoding impairment 编码缺损
encoding key 编码键
encoding law 编码(规)律
encoding mask 编码掩膜
encoding method 编码方法
encoding operation 编码操作
encoding pack 编码组件
encoding rate 编码率
encoding strip 编码条
encoding technique 编码技术
encoignure 墙角家具(法国)
encompass 围绕;包含
encompassing block 包括分程序
encompass(ment) 环绕
encorbel(l)ment 悬挑(件)
encounter 会遇;遇到;冲击
encounter danger 遇到危险
encounter frequency 遭遇频率;冲击频率;碰撞频率
encounter hypothesis 偶遇假说
encounter radius 会遇半径
encounter rate 会遇率
encounter sequence 相互作用序列
encounter superficial 冲击面
encouragement policy 鼓励政策
encourage native products 提倡国货
encouraging a high level of consumption 拼高消费水平
encrimson 红色油漆;红色涂料
encrinal limestone 海百合灰岩
encrinite 海百合灰岩
encroach 侵占;侵蚀;侵入
encroaching water advance 水侵前缘推进
encroachment 阻滞作用;入侵;侵占(邻地);侵蚀(地);侵入;侵界;侵害
encroachment area 侵入面积
encroachment by sand 砂的侵害
encroachment line 边水前侵线;侵蚀线;侵入线
encroachment of seawater 海水入侵;海水倒灌
encroachment of water 水的入侵
encroach on ecologic(al) balance 侵害生态平衡
encroach on standars of pollution control 侵犯污染控制法
encrust 结皮;镶饰;包以外壳的
encrustation 结(硬)壳;形成层理;镶贴;匣钵上集附的渣;硬壳化;饰面细工
encrustation of pipes 管道结垢
encrusted piping 积垢的管道
encrusted tubing 积垢的管道

encrusting matter 结壳物质;甲壳;硬壳;包体;包皮材料
encrusting substance 包被物
encrustment 结壳作用;结垢
encrypt 译成密码;编码
encrypting 译成密码
encryption 加密;编(译)密码
encryption algorithm 编码算法
encryption-based protection protocol 加密保护协议
encryption description 编码术语
encryption technique 加密技术
encumbered with debts 负债的
encumbrance 阻碍物;财产的留置权;负担;妨碍;保留数;债权;不动产上的负担
encumbrance accounting 约定付款会计
encyclop(a)edia 丛书;百科全书
encystation 成囊
encysted 包绕的;包裹的
end 截止;截土;衣料段;后倾自卸车;末尾;末端;终止
end absorption 终端吸收
end acroterion 顶端山墙饰物;正门三角墙端点上的饰物(希腊建筑)
end activity 终端活动
end address 结束地址;终点地址
end adjuster 端调整器
end adjustment 端面调整
end-all 最终目标
endamage 使损坏
end anchor 端锚
end anchorage 钢筋弯钩;端锚(墩);端部锚固
end anchorage failure 终端锚固毁坏
end anchorage of bars 钢筋弯钩;钢筋端部锚固;曲拐弯头
end anchorage type bolt 端头锚固型锚杆
end anchor block 端镇墩(锚墩、地锚、锚枕)
end-anchored 端部锚固的
end-anchored reinforcement 端部锚固的钢筋;端锚钢筋;钢筋端部锚固;带弯钩钢筋
end-anchored tendon 两端锚固的预应力钢筋
end anchoring system 端部锚固系统
end and shear legs lifting 顶吊打捞法
endanger 危及
endangered animal 濒于灭绝的动物
endangered animal species 濒危动物物种
endangered area 危险区(域)
endangered plant 濒于灭绝的植物
endangered plant species 濒危植物物种
endangered species 临绝物种;濒于灭绝的物种;濒危(物)种
endangered wildlife 濒危野生动物
endanger the safety of people and properties 危害人民群众生命财产的安全
end angle 终止角
end arch 拱头拱砖;拱拱;内始式
end arch brick 拱端砖;拱拱砖
end area method 端面积法
end-around 首尾循环(的)
end-around borrow 循环借位;端回借位
end-around carry 循环进位;端回进位;舍入进位
end-around carry shift 循环进位移位
end-around shift 循环移位
end arrangement 纵向排列
end assemblage point 端接合点;端点装配
end axle 轴端;端轴
end baffle 消力槛;尾槛
end band 端带;边瓦;山墙辅助瓦
end bank 端部炉坡
end bar 端钢筋;端部钢筋
end bay 尽间;末端开间;端跨(度);端架间;端部闸孔;端部桥孔;(靠岸的)边段;梢间【建】;末端跨;端闸孔;端桥孔
end beam 端梁
end bearing 端(支)承座;末端支承;端轴承;端支座;端部支承
end bearing capacity of pile 桩端承载能力;桩承载量;桩端承(重能)力
end bearing cover 端轴承盖
end bearing flange 端承凸缘;端承梁翼;端承法兰
end bearing of pile 桩尖承载力;支承桩;端承桩
end bearing steel H-pile 端承式工字钢桩;端部承载的工字钢桩
end bell 钟形端板;(电动机)端盖;端口;电缆(锥形)尾端;外罩;(仪器的)外壳

end bending moment 终端弯矩;固定端弯矩
end bent 端部弯曲
end bevel 坡口
end bit 结束位
end block 引线端子;墩头;墩粗端;端块;旋开桥平衡锤;煤气发生炉隔墙;端锚区;端基封闭
end blocking 后翻板
end block of swing bridge 旋开桥平衡锤
end block reinforcement 端头较粗的钢筋
end board 端板
end-body 终体
end body brace 车体端部撑条
end boiling point 终沸点
end bond 终端棒;端键
end-bound track 两端捣实轨道
end box 终端盒
end box sealing 终端盒密封
end bracing 端部斜撑;端(部)横撑
end bracket 结束括号;轴承座;端肘板;端支架;端部牛腿;尾架
end brake 终点制动;端闸
end branch 末端分支
end built-in 末端砌入;末端砌固;端末固定
end bulb (抹灰形成的)边缘凸包;终球;棒头
end bulkhead 端隔墙;端(部)舱壁
end bulkhead for air duct 风渠挡头板
end bull plant 终点站
end bull station 终点站
end burning 端面燃烧
end butt 端头
end butt joint 端头对接;端接缝
end cam 端面凸轮
end cap 管子堵头;管端盖帽;管端盖板;管底盖;端帽;端盖;堵头;管头塞
end capacity 极限容量
end cap orifice 带帽管口
end capped 封端的
end capping 封端
end cast method 密实浇注法
end cavity 终端谐振腔;端部空腔谐振器
end cell 末端(调压)电池;端电池;附加电池
end cell rectifier 末端电池整流器
end cell switch 末端电池转换开关
end cell system 末端电池式
end center arrangement 端中布置
end-centered orthorhombic 底心正交
end channel 端部槽钢
end charging 端面加料
end check 木材辐裂;木材裂;端纹;端面裂纹;端面裂缝;端裂;(木材)端部裂缝
end chock 端面挡板
end clamp plate 端压板
end clean out 管端清扫口
end clearance 轴向间隙;端隙;端面间隙
end clearance angle 端刃后角
end cleats 端节理;次要内生裂隙组;次生解理
end clip 端卡
end closure 端盖
end coaming 舱口端围板
end-coating 端面涂刷;端面涂层
end coil 末端线圈;终端线圈;端部线圈;无效圈;尾圈
end-coincidence method 端点符合法
end collar 端(辊)环;边辊环
end column 终端柱;结束列;结束栏;端柱
end comb 端栉
end composition 最终成分
end compression 端部压力;端部受压
end conclusion 终结
end condition 末端条件;端部条件;终点条件
end conditioning (管材)管头预加工
end conductor 导体端部
end cone of pile 料堆锥
end connecting system 终端连接系统
end connection 端连接;端(部连)接
end connection angle 端连接角钢
end-connection reactance 绕组端部漏抗
end connector 终接器;终端连接器;端接头;索头连接器
end constraint 终端约束
end construction 堵头堰;堵头建筑
end construction tile 端部结构空心砖;竖孔砖;尾端结构砖
end contact method 通电磁化法
end contraction 末端收缩;堰端收缩;堰壁收缩;端部收缩
end cooling 末端冷却
end cornice girder 端楣板撑柱
end correction 末端改正;终端校正;端面校正;端部校正
end count 单丝支数
end coupling 端连接
end cover 端盖
end crack 端部裂缝
end crest 端山脊
end crib 末端井框
end cross frame 端交叉联架;端叉架
end cut 端切面;帮槽
end-cut brick 砍头砖;端切砖
end cuts 最后馏分;尾馏分
end cutter nippers 顶切钳
end cutting edge 副切削刃
end cutting edge angle 离角;(刀具的)副偏角
end cutting nippers 中心剪切钳;顶切剪钳
end cutting of green pipe 管端湿切
end cutting pliers 钢丝钳;顶切钳
end cutting reamer 端切削铰刀
end cutting stripper 顶切剥皮器
end cylinder 末端汽缸
end deck 端部甲板
end delivery date 最后交货日期
end dep 终点站
end depot 终点站
end depth of drilling coal 止煤深度
end device 终端装置
end diagonal 终端对角撑;端支撑杆;端部斜撑
end diaphragm 终端横端墙;端部隔板
end digging 纵向挖掘
end disc 端圆盘
end discard 最终尾矿
end discharge 端部卸载;后倾自卸车
end discharge door 尾卸门
end discharge mill 尾卸磨
end discharge tipper 尾卸车
end discharge truck 尾卸式自卸(汽)车
end discharge truck mixer 后卸式拌和车;尾卸式搅拌车;尾式拌和车
end distance 端距;端距离
end distortion 末端畸变;终端失真
end door 终端门;拉引室门;端门;尾门
end-door car 端卸式矿车
end-door window 拉引室窗
end double bonds 末端双键
end down 断头
end drawn sprinkler 端拖式喷灌机
end driver 端齿状驱动顶尖
end dump 尾卸的
end dump body 后倾卸车身;尾卸式车身
end dump car 后卸车;端卸式矿车;尾卸(式卡)车;尾部卸料车
end-dumped 尾卸的;从尾部堆填的
end-dumped wagon 尾卸式(运)货车
end dumping 尾卸式
end dumping method 立堵法;端卸法;端进法(施工);尾卸法
end-dump truck 尾卸式汽车;后卸式卡车;后倾自卸车;自卸卡车
endeavour 试图
endectoplastic 内外共同形成的
end edging machine 端折机
endedness 零碎物;边角
end effect 末端效应;端部效应
end-effect simulation 端部效应模拟
end effect zone 末端效应带
end eigenvalue 末端本征值
end elevation 端视;端立面;侧视图
endellite 多水高岭石;多水埃洛石;水埃洛石
Endell's method 恩德尔耐碱试验法
endemia 地方(性流行)病
endemic 地区性的;地方性;地方的;土生的;特有的
endemic bone fluorosis 地方性氟骨病
endemic center 地方中心
endemic cretinism 地方性克汀病
endemic dental fluorosis 地方性氟牙病
endemic disease 地方性疾病;地方(性流行)病
endemic fluorine 地方流行性氟中毒(病)
endemic fluorosis 地方性氟中毒
endemic genus 特有种属
endemic goiter 地方性甲状腺肿

endemic infection disease 地方性传染病
endemic influenza 地方性流感
endemicity 地方性
endemic plant 当地植物
endemic species 地方(品)种;当地种;本地种;土著种;特有种
endemism 特有分布现象
endemy 地方病
end equipment 末端设备
enderbite 紫苏花岗闪长岩
end error 终端误差
endexine 外壁内层
endexinium 外壁内层
endexoteric 内外因の
end facade 终端立面
end face 端面;丁面砖
end face run-out 端面跳动
end-face seal 端面密封;端面式机械密封
end fain time 终了增able时间
end-fatigue 边缘疲劳
end fed 尾端进料
end-fed antenna 底端馈电天线
end-fed knife 尾端进料刀
end-feed 纵向定程进刀;端部馈电;端加料;侧端加料
end-feed centerless grinding 纵向进给无心磨
end-feed cross-flow zone refiner 侧端加料纵横流动区域精炼炉
end-feed grinding 纵向定程进刀磨削
end-feed magazine 端部进料料斗
end-feet 终钮
end fender 端部护舷
end file 结束文件
end file label 结束文件标号
end file record 结束文件记录
end filler (檐沟的)端盖
end filling and emptying system 闸首灌泄水系统;闸首泄水系统;端部充泄水系统
end filling conduit 闸首灌水管道;闸首充水管道
end filling conduit system 闸首灌水管道系统;闸首充水管道系统
end filling system 闸首灌水系统;闸首充水系统;端部充水系统
end finish 管端加工要求
end-fire aerial array 端射天线阵
end-fire antenna 顺射天线阵
end-fire antenna array 端射天线阵
end-fire array 轴向辐射天线阵;顺射天线阵
end-fire array antenna 端射天线阵
end-fired furnace 马蹄焰炉;端焰炉
end-fire magnetic dipole 端射磁偶极子
end fixing 端部固定;固定端
end fixity 终端固定性;端部固定
end fixity constant (梁、柱等的)端固常数
end-flake 终板
end float 轴向端游隙;轴向浮动
end floor beam 楼盖端梁;端横梁
end force 端面压力
end frame 支腿;端框(架);终端架
end framing 终端框架
end freight platform 尽端式货物站台【铁】
end frogs 两端锐角辙叉
end gain 绿灯后利用时间
end gap 端隙
end gas 终结气体;尾气;废气
end gate distributor 车尾撒布器;安装在车尾的撒肥装置
end gate seeder 车尾悬挂的撒播机
end gate spreader 车厢后部撒肥器;车尾撒布器
end gate stanchion 闸机端挡
end ga(u)ge 端面规块;端点量规;端(测)规;挡料装置
end-ga(u)ge interferometer 干涉检定仪
end ga(u)ge pin 端部定位销
end geophone 端点检波器
end girder 端(大)梁;端部大梁
end glow fiber 末端发光光纤
end grain 端纹;端切面;端面纹路;端面纹理;端面晶粒;端纹向外的(木材)
end-grain cutting 端纹切削
end-grain nailing 横切面固定法
end-grain wood 横纹木材
end group 末端基
end-group analysis 端基分析(法);末端分析
end guide 端导规

end hardening 顶端淬火法
end hardness 端面硬度
end hauling 一端搬运
end haul slipway 直式滑道
end head 镜头端部
end hinge 端铰
end hole 边缘孔;边井
end hook 终端吊钩;端钩
end housing 端盖
ending 端接法;末尾;末端
ending activity 结束活动
ending balance 期末余额
ending beat 端面跳动
ending inventory 期末存货
ending inventory valuation 期末存货估计
ending of ebb 落潮末
ending of flood 洪水消落;洪水退落;汛(期)末;涨潮末;潮水消退
ending of ice sheet 冰层消失
ending of snow cover 积雪融化
ending oil-water segregator 终端油水分离器
ending place 终止地点
ending pulsation 端面跳动
ending sign 终了信号
ending wave 终末波
end installation 终端设备
end instrument 末端设备;终端仪表;端点仪器
end insulator 终端绝缘子
end interval 端部间隙
end item 最终产品;个别项目;成品
end joint 端接(头);端部接缝;端部接头
end journal 端轴颈;端框
end journal bearing 端枢轴承
end key 结束键
end knee brace 端斜撑
end knot 端结
end lamp 车尾灯
end land 刀尖
end lap 末端搭接;端搭接;后航向重叠;端接接;板端接法;前重叠
end lap joint 端搭接(头);端部搭接
end lap sealer 端部搭接密封剂
end lap weld 搭接焊
end lap welding 搭合焊;端搭焊接
end lathe 卡盘车床
end launching 纵向下水
end leakage 端部泄漏
endless 循环不止的;无(穷)尽的;无接头环带
endless apron 循环输送带
endless apron hay loader 输送带式装干草机
endless band 环(形)带;无端带
endless band elevator 链斗式提升机;环带式提升机;环链式升降机;带式升降机
endless band lift 带式升降机
endless bed drum sander 带式进给辊筒砂光机
endless belt 凝固皮带;环形皮带;环(形)带;无极带
endless-belt conveyer 环(形)带式运输机;环带运输器;环带运输机;环形皮带运输机;环带传送器;循环(式)输送机;皮带循环运输机;皮带输送机;皮带环运输机;皮带运输机;皮带送料机;胶带输送机
endless belt elevator 链斗式提升机;带式升降机
endless belt peener 环带式喷丸机
endless belt trencher 带式挖槽机
endless-bucket trencher 环斗式挖槽机;多斗挖沟机
endless cable 无极钢缆
endless-cable system 环索系统
endless chain 轮链;环状链;循环链;无极链;无端链
endless chain coalcutter 无极链式截煤机
endless chain conveyer 循环链式输送机
endless chain dredge(r) 环链式挖泥机;轮链式污泥机
endless chain feeder 无极链式给矿机;无极链给料机
endless chain for fastening wire rope 铸子链
endless chain grate 无极链炉箅
endless chain haulage 链带移运
endless chain haulage system 环链式运输装置
endless chain log haul-up 无端链条拉木机
endless chain trencher 环链式多斗挖槽机
endless chain trench excavator 链式挖沟机;链斗式挖沟机;环链式多斗挖土机;环链式多斗挖沟机;循环式多斗挖沟机;循环多斗式挖沟机;无极多斗式挖掘机

endless conveyer 环带运输机;无极运输机
endless conveyer chains 循环输送链
endless crew 蜗杆
endless film 环状胶带;循环胶片;无端胶片
endless grate 链条炉排
endless groove 环状槽
endless horizontal drive 环状水平传动
endless joint 环状接头;环状接合
endless long splicing of wire rope 钢丝绳镶圈
endless loop 循环带;无限循环;无接头循环;无端(循)环;无端回线
endlessly rooted tree 无端有根树
endless main 无极绳运输主要平巷
endless mortising chain 环形开榫链
endlessness 无限
endless paper 卷筒纸
endless pocket conveyer 带式连续输送机
endless punched-pattern control-paper 冲孔纹样控制循环纸带
endless raster 无尽光栅
endless reeving 无端穿绕
endless-ring lock 钢环式
endless rolling 无头轧制
endless rope 环(吊)索;无极绳
endless rope car-haul 无极钢绳矿车运输
endless rope haulage 无极绳运输(机)
endless rope haulage system 环索运输系统
endless rope scraper haulage 无极绳耙斗搬运
endless rope system 无极绳运输系统
endless rope traction 无极绳牵引
endless saw 环形(带)锯;带锯;带锯(机)
endless saw blade 环形带锯条;环形带锯片
endless screw 无限螺旋;输送螺旋
endless sling 循环吊链
endless staple fiber 未切断丝束
endless strap 环带;无端皮带
endless tangent screw sextant 微调螺杆六分仪
endless tape 环形带;无端带
endless thrower 无极带式抛掷充填机
endless tow 长丝束
endless track 环(形)道
endless tubing 无接头油管
endless wire 无端铜网
endlichite 砷钒铅矿
end lift 端吊作业
end lifting 端台升打捞
end-lifting apparatus 端起重装置
end lighting 终端照明;终端采光
end line 结东行;终点行;端边线
end liner 底衬;侧衬板
end lining 磨头衬板;(集装箱)端壁内衬板
end link 锚链末端链环;终端连杆;端环
end load 末端荷载;端(部)荷载
end loaden 后端装载机
end loading 端部装车;端部加荷;船尾门装卸;船首门装卸;车尾装卸
end loading berth 轮渡式泊位;顶靠泊位
end loading dock 端式装车停车处
end loading platform 端式装车站台
end loading ramp 端式装车斜坡台
end-loading siding 端部装车线
end loop 终端小窗;终端墙洞;终端环
end machining tool 端加工机床
end manhole 端部人孔;尾端窨井;尾端人孔
end mark 结束符;结束标志;末端标记;终止标记;断面记号
end marker 端点标志
end masonry wall 边墙;侧圬墙
end match 端头拼接;端头企口榫接
end matched 企口的;端头榫结合的
end matched board 端接企口板
end matched lumber 端头企口板;端头拼接板;端边企口板
end matcher 多轴制榫机
end matching 端拼
end measure 端面量具
end measuring ga(u)ge 内径杆规
end measuring instrument 触模型测量仪
end measuring machine 测长机
end measuring rod 内径杆规
end member 端员;端元【地】
end-member component 端员组分
end-member delta type 端员三角洲类型
end-member molecular matching procedure 端

员分子配比法
end-member rock 端员岩
end mill 立铣刀;立铣床
end-milled keyway 成形键槽
end mill gashing and radius grinder 端铣刀和侧面铣刀磨床
end mill head 立铣头
end milling 端铣(削);立铣
end milling attachment 端铣附件
end milling cutter 端铣刀;立铣刀
end milling machine 端铣机
end mill with straight teeth 直齿端铣刀
end mirror 端面镜
end moment 端(部)力矩
end money 备作超过预算之用的款项
end moraine 终碛;终碛石;终端冰碛;端碛;边堆石;尾碛
end-moraine bar 终碛堤
end moraine lake 终碛湖
end motion 端运动
end mo(u)ld 端模
end-mounted spindle 研磨芯轴
end movement 端移动;端位移
end nailing 钉住末端
end node 结束结点;末端结点;末端节点;端节点
end nozzle (檐沟的)端部下水口
end nut 端螺母
endo 把管子头对头地排放
endo-adaptation 内适性
endobatholite zone 内岩基带
endobatholithic 内岩基的
endobenthos 内在底栖生物
endobiont 内栖生物
endobiophyta 内生寄生植物
endobiosis 内在底栖生物
endobiphyta 内生寄生生物
endoblast 内变晶
endocapillary layer 毛细管内层
endoceratite 内角石属的化石
endochronic approach 自连续法
endochronic theory 内时理论;延迟理论;自连续理论
endocone 内锥
endo-configuration 内向构型
endocrator 巨坑
endocyclic 内环的
endocyclic compound 桥环化合物
endocyclio 桥环的
endocytosis 内吞作用
endodyke 内成岩墙
endodynamic(al) soil 内动力型土(壤)
endodynamic(al) succession 内因动态演替
endodynamomorphic soil 内动力型土(壤);岩成土壤
endodynamorphic 内动力型的
endodyne 自差
endoenergic 吸能的
endoergic 吸能的;收能的
endoergic collision 第一种碰撞
endoergic reaction 收能反应
endo-exo configuration 内外构型
endo-exoenetic isomerization 内外异构化
endo-exogenetic succession 内外因演替
end of address 地址终端;地址结束(符)
end of address code 地址结束代码
end of arm speed 机械臂末端速度
end of astronomic(al) evening twilight 天文昏影终
end of backwater 回水末端
end of block 块结束;信息组结束;信息块结束
end of block character 块结束符
end of block signal 信息组结束信号;数据组末端信号
end of brick 砖端
end of calling code 呼叫结束码;调用结束码
end of chain 链结束(符)
end-of-charge voltage 充电终止电压
end of circular curve 圆曲线终点
end of contract 合同终止
end of conversion 换算结束;转换端
end-of-conversion pulse 变换终止
end-of-copy signal 图像终了信号
end of curve 曲线端(点);(平)曲线终点
end of curve sign 曲线终点标
end of data 数据结尾
end of data mark(er) 数据结束标志
end of data set 数据集结束
end-of-data-set exit routine 数据集结尾出口程序
end of date 截止日期
end-of-day glass 混色玻璃;混合玻璃
end-of-day joint 日施工缝
end-of-day surcharge 日工作末的超载
end of design 设计终点
end of ebb 落末
end-of-encode 编码结束
end of extent 区结束(符)
end off 结束;尖灭;包端松散;终止
end office 终端(使用)办公室;终端局;端局
end of file 文件结束;外存结束
end-of-file character 文件结束符
end of flame 火梢
end of flood 涨末
end of flood current 洪水流终止
end-off shift 舍尾移位;舍入移位
end of identify 鉴别终止
end of image 图像结束
end of input 输入结束
end of job 作业结束;工作结束
end-of-job routine 作业结束程序
end of life 寿命终止(可靠性术语);寿命末期
end-of-life salvage value 使用期满时残值
end-of-line loop turnaround 路端掉头环道
end of list 表结束
end of list control block 表结束控制块
end of longwall face 纵向端墙墙面
end of map(ping) 绘图结束
end of medium 介质末端
end of medium character 记录媒体终端符
end of message 信息结束;电报结束;通报终了
end-of-message code 信息结束代码
end-of-message character 信息结束(字)符
end of mid-span intercepted 截断中跨端
end of month 月底
end of nautical evening twilight 航海昏影终
end of oil production 采油末期
end of passage 航程终了
end-of-period adjusting entry 期末调整分录
end-of-period cutoff 期末截账
end-of-period statements 期末报表
end-of-pipe approach 集水处理
end-of-pipe technology 最后技术
end-of-pipe test 终端检验
end-of-point technology 最后封尾技术
end of procedure division 过程结尾部分;过程部分结尾
end of program(me) 程序结束
end of pulsing 脉冲发送结束信号
end of railway 路线终点【铁】;铁路终点(站)
end of record 记录终止;记录尾部;记录结束
end-of-record(ing) delimiter 记录结束符
end-of-record(ing) gap 记录结束间隙
end-of-record(ing) word 记录结束字;记录终止字
end of reel 卷尾(的);带卷结束;带卷尾
end-of-reel mark 带终解标志
end-of-reel marker (带)卷尾标志
end-of-reel routine 卷尾例行程序;卷尾处理程序
end of route 路线终点
end of run 运行终点
end-of-run routine 运行结束(例行)程序
end-of-run temperature 运转末期温度
end of scale 标度终点
end of scan 扫描结束
end-of-selection signal 选择完成信号
end of sequence 序列结束
end of set(ting) 凝结终止
end of sheet pile 板桩尾部
end of shift 移位终止
end of side span intercepted 截断边跨端
end-of-study report 研究总结报告
end of submerged pipeline 潜水管线末端
end of subroutine 子程序结束
end of sun screen 遮光栅终点
end of switch 道岔端点
end-of-tape 带结束(标志);带末端
end-of-tape label 带尾标记
end-of-tape marker 带结束标记
end-of-tape warning 带结束警告;带末期预告
end of text character 文末字符;文本结束符
end of the chain 链端
end of thread 螺纹退刀扣;螺纹端
end of track 线路终点;磁道结束;铁路终点(站)
end of transition curve 缓和曲线终点
end-of-transmission 传输结束
end-of-transmission block 信息块传送结束
end-of-transmission block character 信息组传输终止符;信息组传输结束符;字组传输结束符;传输区终了字符;数据块传送结束符;传输块结束符
end-of-transmission card 传送终止卡;传输终止卡;报终卡
end-of-transmission character 传送终止符;传输结束(字)符;传输终止符
end-of-transmission code 传送结束码;传输结束码
end-of-transmission recognition 传送结束识别
end of volume 卷终;卷结束(符)
end-of-volume label 卷尾标号;卷结束标号
end of warning area 警告区域末端
end-of-work signal 作业终止信号
end of year method 年末法
endogamy 同系配合
endogas unit 裂化气体设备;吸热型气体发生设备
endogen 内生
endogenesis 内源;内成
endogenetic 内源(性)的;内营力的;内生的;内成的
endogenetic action 内成作用
endogenetic deformation 内成形变;内成变形
endogenetic deposit 内生矿床
endogenetic energy 内生能
endogenetic fissure 内成裂隙;内成裂缝
endogenetic force 内(营)力
endogenetic metallogenic formation 内生成矿建造
endogenetic process 内力作用
endogenetic rock 内成岩
endogenetic rocks 内成岩类
endogenetic sand 内生砂
endogenetic sediment 内成沉积
endogenetic subsidence 内因沉降
endogenetic succession 内因演替
endogenetic texture 内成结构
endogen force 内力
endogenic 内原(性)的;内营力的
endogenic action 内生作用;内力作用;内成作用
endogenic anomaly 内生异常
endogenic energy 内生能
endogenic force of geologic(al) function 内力地质作用
endogenic geologic(al) process 内动力地质作用
endogenic halo 内生晕
endogenic inclusion 内源包体
endogenic mineralization 内生成矿作用
endogenic movement 内生运动
endogenic process 内生作用过程;内力作用过程;内成作用过程
endogenic reaction 吸能代谢反应
endogenic rock 内生岩;内成岩
endogenic subsidence 内因沉降
endogenous 内原(性)的;内营力的
endogenous activity 内生活动
endogenous cave 内生洞穴
endogenous change 内生变动
endogenous cycle 内生环
endogenous dome 内成穹隆;内成隆起
endogenous ejecta 内成喷出物
endogenous enclosure 内在夹杂物;内源包体;内生包体
endogenous explanatory variable 内生解释变量
endogenous fire 自发火
endogenous fluid 内生流体
endogenous force 内(营)力
endogenous growth 内生生长
endogenous halo 内生晕
endogenesis inclusion 内在夹杂物;内源包体;内生包体
endogenous income hypothesis 内生收入假设
endogenous investment 内生投资
endogenous lagged variable 内生滞后变量
endogenous metabolism 内原代谢(作用)
endogenous modulator 内源性调节器
endogenous movement 内因运动
endogenous origin 内生源
endogenous periodicity 内源周期性
endogenous phase 内源相
endogenous phase of growth 内源生长期

endogenous pollutant mass 内源污染物质量
endogenous process 内生过程
endogenous pulsed source 内生脉冲源
endogenous pyrogen 内热源
endogenous rhythm 内源节律
endogenous spore 内生包体
endogenous susceptibility 内在敏感性
endogenous uranium ore 内成铀矿
endogenous variable 内源变量;内生变量
endogenous variate 内生变量
endogen tree 无年轮树;内生树
endogeny 内生;内成
endoglyph 层内痕
endokinetic fissure 内生裂隙;内生裂缝;内成裂隙;内成裂缝;内藏裂纹;自裂缝
endokinetic joint 内成节理
endolithic brecciation 内成角砾作用
endolotihic breccia 内成角砾岩
endolytic compound 内分解化合物
endomagmatic hydrothermal differentiation 内成岩浆热液分导作用
endometamorphism 内变质(作用)
endometry 内腔容积测定法
endomigmatization 内质混合岩化作用
endomigmatization way 内混合岩化方式
endomomental 脉冲吸收的;瞬时吸收
endomorph 内容矿物
endomorphic system 符号序列变换系统
endomorphic zone 内变质带
endomorphism 内变质(作用);自同态
endomose 内渗透现象;内渗
end-on 末端对接;对遇;端对;主解理直角工作面;端头向前的
end-on aerial array 端射天线阵
end-on barrel steering gear 套筒拉链舵机
end-on collision 尾追撞车
end-on coupling 终端耦合;端键
end-on courses 接续学程
end-on directional array 轴向辐射天线阵
endonexine 内里层
end-on fire antenna 轴向辐射天线;端射天线
end-on object 端向天体
end-on slipway 纵向滑道
end-on spread 端点放炮排列
end-on system 端进法;递进制;外伸施工法;推进法
endonucleolus 内核仁
end-on view 端视图
endoparasite 内寄生物
endoparticle 内颗粒
end opening 端跨度;端孔
endoperistome 内齿层
endophyte 内生生物;内部寄生菌
endophytic algae 内生藻类
endopleural ridge 侧内脊
endoploidy 内倍性
endopolyploid 内多倍体
endopolyploidization 内多倍体化
endopolyploidy 内多倍性
endoradiosonde 体内无线电探头
endoreduplication 内重复
end organ 灵敏元件;终器
endor(h)eic basin 内流盆地;内陆盆地;内陆流域
endor(h)eic drainage 无排水地区;内陆河流流域;内陆水系;无泄水区
endor(h)eic drainage basin 内陆河流流域
endor(h)eic lake 内陆湖;内流湖
endor(h)eic region 内流区
endor(h)eic system 内流水系
endorheism 内陆流域
end ornament 顶饰;墙墙上的装饰
endorsable cheque 可背书支票
endorse a bill 背签票据
endorse a passport 签护照
endorse clause 背书条款
endorsed bond 有担保债券;背书债券
endorsed in blank 空白背书
endorsed notes payable a-c 背书的应付票据账户
endorsee 被转让人;票据承受人;受让者;受让人
endorse in blank 空白背书
endorsement 背书;签署;票据等后面的签名;票背签字;批改条款;批单
endorsement commission 背书手续费
endorsement confirmed 已确认的背书;背书证实
endorsement for collection 委托取款背书;托收背书
endorsement guaranteed 已担保的背书
endorsement in blank 空白背书,无记名式背书
endorsement in full 详细背书;完整背书;完全背书;特别背书
endorsement in pledge 抵押背书
endorsement in security 保证背书
endorsement irregular 背书不符
endorsement of bill 单据的背书
endorsement of policy 保险单背书
endorsement to order 指派式背书;指定式背书
endorsement without recourse 免除担保付款的背书;免除偿还的背书;无追索权背书
endorse of ticket 签证
endorse over a bill to another one 背书票据转让他人
endorser 转让人;背书人
endorser for accommodation 通融资金背书人
end or side rolling hatchcover 平滚式舱盖装置
endoscope 内腔镜;内窥镜;铸件内表面检查仪;管状仪器
endoscopic film projector 内腔镜摄片投影仪
endoscopic picture 内窥图像
endoscopy 内腔镜检查;内窥镜检查;铸件内表面检查
endoseptum 内隔壁
endoser 接触内视镜
endosharn 内矽卡岩
endosite 内寄生物
endosmometer 内渗仪;内渗压测定器;内渗(透)计
endosmose 内渗现象
endosmosis 内渗透现象;内渗;向内渗透
endosmosis-mose 内渗透现象
endosmotic equivalent 内渗化当量
endosome 核内体
endosphere 内球
endostoma 内口板
endostracum 内介壳
endostratic breccia 层内角砾岩
endosulfan 六氯硫丹;硫丹;有机氯杀虫剂
endosulfan sulfate 硫酸硫丹
endosymbiosis 内共生
endotectonic 内力构造的
endotergite 背内骨
endothal 草多素
endothall 草藻灭
endothelium 内皮
endotherm 吸热(线)
endothermal 吸热的
endothermal change 吸热变化
endothermal reaction 吸热反应
endothermic 吸热的
endothermic animal 内热动物
endothermic atmosphere 吸热气氛
endothermic change 吸热变化
endothermic character 吸热特性
endothermic compound 吸热化合物
endothermic conversion 吸热转化
endothermic degradation reaction 吸热降解反应
endothermic disintegration 吸热转化
endothermic effect 吸热作用;吸热效应
endothermic fuel 吸热燃料
endothermic gas 吸热型气体
endothermic peak 吸热峰
endothermic process 吸热过程
endothermic reaction 细热反应;吸热反应
endothermic transition 吸热转换
endothermic valley 吸热谷
endotherm knife 高频电刀
endothiobacteria 内生硫细菌
end out 尖灭;断头;缺径
end-over-end mixer 立式圆筒混合机;直立圆筒混合机
end overlap 航向重叠
endovolcanic structure 内火山构造
endow 资助
endowment 基金
endowment annuity 养老年金
endowment assurance 养老保险;人寿定期保险
endowment fund 留本基金;捐赠基金
endowment insurance 储蓄保险;养老保险;人寿(定期)保险
endowment of fixed assets 固定资产捐赠
endowment of human capital 人力资本赋予量
endowment policy 养老保险单
endow them with new use values 赋予新的使用价值
endow with subname 赋予副名
end packing 外包装
end panel (连结板、楼盖板的)端跨;终端屏;终端配电板;(集装箱)端壁板;(集装箱)端面
end panel columns resisting wind load 抗风端跨柱;端墙抗风柱
endpaper map 衬页地图
end paragraph name 结束段名
end part 尾部
end partition wall 挡头墙
end patch 补强衬小角
end peak 尾峰
end peg 端点桩【测】
end peneplain 终期准平原
end period 终结期
end peripheral discharge 尾端周边卸料
end phase shifting 端点相位漂移
end piece 尾端件;(房屋屋脊的)端部构件;终端片;出口端;末端片;挡砖;尾段
end pier 近岸桥墩;终端杆;边墩
end-piled loading 纵向堆垛法
end pillar 端部支柱
end pin 尾销
end plane 端面
end plank washer plate 端板压板
end plate 磨头盖板;后端板;终板;端(面)板;端盖;底型;封头
end-plate fin 端板式垂直尾翼
end plate grid 终板栅
end plate hanging hook 后端板吊钩
end plate nozzle 后端板接管
end plate oiler 末端板加油器
end plate side 后端板侧
end platform 尽端式站台;端末月台
end play 轴向窜动;轴向间隙;轴端余隙;轴间间隙;端隙;径向游隙;轴向游隙
end play device 轴端余隙器件;轴间间隙装置;摇杆机构
end plinth tile 端墙勒脚砖
end plug 端塞;堵头
end point 终端;终点;端点;边界点;弯倒温度
end point analysis 最终结果分析;最后结果分析
end point condition 端点条件
end point control 终点控制;干点控制;端点控制
end point control problem 终端控制问题
end point data 端点数据
end point detector 端点检测器
end point energy 极点能量;终点能量;极限能量
end point gasoline 终点汽油
end point method 终点法
end point mobility ratio 端点迁移率;端点流度比
end point mutation 终点突变
end point node 末端节点;端部节点
end point number of migration 偏移终了点号
end point of alcoholysis 醇解终点
end point of fraction 终点馏分
end point of fractionation 分馏终点
end point of seismic fault 地震断裂端点
end point of titration 滴定终点
end point recovery 端点采收率
end point relative permeability 端点相对渗透率
end point saturation 端点饱和度
end point support 端点支承
end point temperature 终点温度
end-point voltage 截止电压;终点电压
end pole 末端杆;末端电线杆;终端杆
end portion 末段;尾相
end portion earthquake 地震尾相
end portion of painted beam 箍头
end position 终点位置
end position of segment 分段终点位置
end post 端杆;端压杆
end pot 终端块件;铺地空心块
end pressure 最后压力;轴向压力;终压(力);端压(力);端头压力
end-process 终加工
end product 最终结果;最终产物;最终产品;最后制品;最后成果;最后产物;最后产品;终产物;终产品;成品;制成品
end product interpretation 最终成果解释

end products of weathering 风化的终极产物
end product specifications 最终产品规格;最终(工程)质量规定
end projecting firebox 外伸引火箱
end property 成品的性能
end prow 突出的前端
end pulley 回程鼓;导向滑轮;尾滑车;(皮带输送机)端部滑轮
end quenched bar 端淬试棒
end quench hardenability test 顶端淬火试验
end quench hardening test 端面淬火试验
end quench(ing) 顶端淬火;端淬
end quenching test 端淬试验
end-racking 叉堆法
end ratchet 端面棘轮
end reaction 终结反应;支座反力;端(部)反力
end reaction of beam 梁端反力
end reading 最终读出
end readout 最终读出;最后读出;终读数
end reamer 不通孔铰刀;前锋铰刀
end record 结束记录
end reflection loss 末端反射损失
end-region magnetic field of electric(al) machine 电机端部磁场
end reinforcing rib 端部加强筋
end release 终端安全释放机构
end relief 端形;齿端倒角
end relief angle 副后角
end repeater 终端中继器
end-reply packet 终点回答包
end resistance 端阻力;桩尖阻力;尾端阻力
end-resistance coefficient 桩尖阻力系数
end-resistance material 末端电阻材料;终端电阻材料
end restraint 终端约束;嵌固端;固端约束;杆端约束;端束缚;端(部)约束
end restraint effect 端限制效应;端部约束效应
end restraint moment 固端弯矩
end result 成品鉴定
end rib 端肋
end ridge 端脊
end ridge tile 端脊瓦
end ring 锁紧环;锚链末端环;轴承挡圈;(终)端环
end-ring rotor 带短路环转子;带端环转子
end-road shift 循环位移
end roller 端部滚轮
end-rolling hatch cover 纵向滚动舱口盖
end rotation 终端旋转;端转角;端旋
end rotational stiffness 端旋劲度
end round 齿端修圆
end-rounding 路头调向环道;路端掉头环道
end-runner mill 双辊研磨机;机动研钵;端辊研磨机
ends 零星物件;零星木材
end saw 截头锯;端锯
end scale 尺端刻度
end scale value (仪表)满表值;满(标)值;(仪表)满标值
end scarf 互相缺接法;端头嵌接
end screw brake 端部螺旋制动器
end seal 端封;端部密封
end sealing (电缆的)终端扎结
end section (涵管喇叭形的)进出口节;防波堤的端面;端末管节;末段;终端段;端截面;尾段
end sensing 终端读出
ends fixed 电缆端固定
ends free 电缆端不固定
end shackle 联锚卸扣;联锚卸机;接锚卸扣;末端卡扣;锚端卸扣
end shake 木材辐裂;木材端裂
end-shake vanner 端摇床
end shear 端剪力
end shears 剪边机
end sheathing 端板
end sheathing brace 端板横铁
end sheathing diagonal stay 端板对角撑
end sheet (集装箱)端壁板;端板
end shield 末端护罩;端部罩盖;端(护)罩;端护板
end shield bearing 机油滤清器内盖头轴承
end shielding 末端屏蔽
end shot 末端链节
end side arrangement 端侧布置
end sill 消能坎;消力槛;端系杆;端梁;尾槛
end sill bracket 端梁角补强铁
end sill corner plate 端梁角补强铁

end sill cover plate 端梁盖板
end sill gusset plate 端梁角补强铁
end sill plate 端梁铁板
end sizing 端径校准;顶端直径调整
end skew brick 斜端砖;端斜砖
end skew on edge 侧斜端砖;条面斜砖
end skew on flat 大面斜砖;平斜端砖
end sleeve 末端套管;终端筒形轴承;端部套筒
end slip of bars 钢筋端部滑移;钢筋端头滑动
end slope 端部斜面板
end slope of groin 丁坝头部坡度
end slope of groyne 丁坝头部坡度
end slot 端槽;端部切槽
end slot piston pin 端槽活塞销
ends-means analysis 目的一手段分析
end socket 端套筒;端头
ends of piston ring 活塞圈末端
end span 端跨(度);边跨
end speed 末速(度)
end speed zone sign 车速限制区终点标志
end split 木材辐裂;木材端裂
ends protected only 只包扎端部
end stacking 竖堆法
end stair(case) 终端楼梯
end stake number of line 测线终止桩号
end stake number of the first receiver line 首检波点线终了桩号
end stake number of the last receiver line 末检波点线终了桩号
end stake socket 端柱插口
end standard 顶端标准
end standard rod 端点式标准尺;标准端测规
end standard staff 端点式标准尺
end stiffener 端部加劲物;端部加劲条;端部加劲肋;端部加劲杆
end stiffener angle 端部加劲角钢
end stiffening angle 端部加劲角钢
end stiffness 端点硬度
end stiffness of member 杆端刚度
end stile 端侧柱;边端门梃;边端窗梃
end stirrup 端箍筋
end stone 终端宝石;托钻;止推宝石轴承
end stop 行程限位器;终点止动装置;终点挡板;端头止挡
end stopper of chain 链端止剂
end stopping of chain 链的终止
end stress of member 杆端应力
end suction centrifugal pump 端吸离心泵
end suction fire pump 末端吸入消防泵
end suction pump 末端吸入泵;单侧吸入泵
end support 端部支撑;端(部)支承;端支座;端承(桩)
end support column 支端柱
end surface 端面
end swivel 尺钩
end tab 引弧板;引出板;(焊缝)端部附片
end tab assembly 引弧板
end table (沙发旁的)茶几
end tank farm 终点站
end tape 端贴尺;端点尺
end tapering 减薄收尾
end tenoner 端部开榫机
end thread 端螺纹
end thrust 轴向推力;轴端推力;端推力
end thrust ball bearing 止推滚珠轴承
end thrust bearing 止推轴承
end tie-bar 端系行;端拉杆;末端拉杆
end tie-plate 端部系紧板
end tightening 端部紧固
end tile 端盖瓦
end tilt ingot car 端倾式送锭车
end time 结束时间;终止时间;终了时间
end time of migration 偏移终了时间
end tipper 后卸式自卸车;后倾(式)自卸车;翻斗车;前翻斗手推车;尾卸式自卸(汽)车
end tipping barrow 翻斗手推车;尾卸式手推车
end tipping lorry 尾卸(式)汽车;尾卸式卡车
end-tipping motor lorry 自动尾卸载重汽车;自动尾卸载重卡车;自动尾卸载重汽车;尾卸载重卡车
end-tipping vehicle (自动)尾卸车辆
end-tip trailer 后倾卸式拖车
end-to-end 衔接;终端到终端;首尾衔接地;首尾相连;端对端;端对端;不断(的);(首尾相连(的)
end-to-end anastomosis 对端吻合术

end-to-end arrangement 纵向排列
end-to-end circuit 端到端电路
end-to-end communication 终端站间通信
end-to-end discharge 双端卸载
end-to-end distance 末端距
end-to-end flow control 端到端流量控制;点到点流量控制
end-to-end inspection 端到端检查
end-to-end joint 对接;端(点连)接;对抵接头;(管道)平接;管道对接
end-to-end measurement 直通测量线路;对端测量
end-to-end placed cells 纵向排列电解槽
end-to-end protocol 端到端(传输)协议;点到点传输协议
end to end service 端对端业务
end-to-end session 端到端会话期;端到端对话
end-to-end setup 纵向排列
end-to-end signaling 端到端信号方式
end-to-end test 端到端双向测试;点到点双向测试
end to grafting 对头接
end to joint 对接
end tooth 端齿
end-to-reel marker 带卷结束标记
end transom and pole piece 端横梁及磁极块
end trap 管端凝汽阀
end travel 轴端移动
end trimming 端截头
end trimming machine 端部修饰机
end trimming shears 切边剪
end truck 端车
end truck controller 端车控制器
end turn 线圈端部
end turning 绕焊
end twist 端部扭曲;端扭
end upsetting 管端加厚
endurable pressure 持久压力
endurance 抗磨度;耐用性;耐用度;耐疲劳度;耐久性;耐久(力);续航力;持续;持久(力)
endurance action 疲劳作用
endurance behavio(u)r 疲劳性质;耐久性能
endurance bending failure 疲劳弯曲破裂;疲劳弯曲毁坏
endurance bending strength 抗弯疲劳强度
endurance bending test 疲劳弯折试验;弯曲疲劳试验
endurance character 疲劳特性
endurance characteristic 疲劳特征
endurance coefficient 疲劳系数;耐久系数
endurance crack 疲劳裂纹;疲劳裂缝
endurance curve 持久曲线
endurance degree 疲劳度;耐久程度
endurance expectation 估计使用寿命;估计使用期限;估计使用年限;估计耐用年限
endurance failure 疲劳破坏;耐久破坏;疲劳破损
endurance fatigue 疲劳破坏
endurance fatigue limit 耐劳极限
endurance fracture 疲劳断裂
endurance impact test 冲击疲劳试验
endurance in bending 弯折疲劳
endurance in compression 压缩疲劳
endurance in flexure 挠曲疲劳
endurance life 耐用年限;耐久寿命;疲劳寿命
endurance life characteristic 耐用寿命特性
endurance life test 持久性试验
endurance limit 持久极限;持久强度;抗疲限度;抗磨极限;耐磨限度;耐磨极限;耐力限度;耐久限度;耐久极限;持久限度;持久极限;疲劳限度;疲劳耐限;疲劳界限;疲劳极限
endurance limit at pulsating stress 脉冲应力疲劳极限
endurance limit on base 对称的应力换向循环下的疲劳极限
endurance limit stress 耐久极限应力
endurance limit under pulsating tension 脉冲拉伸疲劳极限
endurance limit under pulsating torsion 脉冲扭转疲劳极限
endurance load 疲劳负载
endurance of bond 黏合疲劳;黏合耐久性
endurance of the concrete 混凝土疲劳;混凝土耐久性
endurance period 持续时间
endurance phenomenon 疲劳现象;持久性现象;耐久现象

endurance range 应力极限;持久限;持久范围
endurance ratio 耐久系数;耐久比;持久比;疲劳系数;疲劳强度极限与抗断强度极限之比;疲劳(强度)比
endurance road test (汽车的)耐久性路线;道路耐久性试验
endurance road testing 耐久性道路试验
endurance run 续航力测定航次
endurance running 耐久性运转
endurance rupture 疲劳破裂
endurance state 持久状态
endurance strength 耐久限度;耐久强度;持久强度;疲劳强度
endurance tension test 持久张力试验
endurance test 耐用性试验;耐疲劳度试验;耐力试验;耐久(性)试验;续航力试验;持久(性)试验
endurance test bed 耐久性试验台
endurance testing machine 持久试验机
endurance testing machine for torsion repeated 重复扭转疲劳试验机
endurance to alternate wetting and drying 耐干湿循环性
endurance transverse stress test 横向弯曲疲劳试验
endurance trial 耐久试验;续航力试验
endurancing 耐久性试验
enduration period 持续运转时间
enduring water-proof membrane 耐久水膜
end use 最终用途;产品用途
end-use competition 最终用途竞争
end-use device 端点使用设备
end-use performance 产品性能;使用期性能
end-user 用户;末端用户;直接用户;终端用户;最终用户;最终消费者;最后用户
end-use tariffs 终端用户费率表
end-use temperature 最终使用温度
end-use test(ing) 使用期试验
end vacuum 极限真空
end value 结果值;最终值;端值
end valve 端阀;尾阀
end velocity 末速(度)
end velocity model 最终速度模型
end view 端视;侧面图;侧端图;背面图;(端)侧视图
end view drawing 侧视图;侧面图;端视图;端面图
end-voltage 终止电压
end wall 磨头;后墙;端墙;端壁;边墙;投料端墙;尽端墙
end wall balcony 端墙阳台
end wall boundary layer 端壁边界层
end wall bracket 端墙托架
end wall car end No.1 一号车底架端头
end wall liner 磨头衬板
end wall loss 端墙损失
end wall paneling 端墙镶面板
end wall strength test 端壁强度试验
end warning area 结束警告区
end washer 端部垫圈;端部垫饼
end-wastage 残头废料
endway 末端向前的;末端朝上的;竖的
end wear 端部损耗;端部磨损
end web 端腹板
end wheel press 轮端成形压力机
end winding 端部绕组
end winding region 绕组端部区域
end window 端(面)窗;端窗式的
end window counter 钟罩形计数管;端窗计数器
end window counter tube 端窗计数管
end window X-ray tube 端窗X射线管
end wire strippers 顶切剥线钳
endwise 末端向前(的);末端朝上(的)
endwise feed 终向馈给
endwise mismatch 纵向错移
endwise slip 轴向滑动
end wobble 端面震摆
end wrench 平扳手;单头扳手
endyked 用堤围住的;设围堤的
end yoke 端叉
endy veneer 斜纹刨切单板
end zone 结束区
en echelon 雁列状;梯形状
en echelon angles 雁列角
en echelon arrangement 雁行排列
en echelon cracks 雁列裂纹
en echelon fault blocks 雁列断块

en echelon faults 雁列(状)断层
en echelon fissure 雁行裂缝
en echelon folding 雁行褶皱作用
en echelon folds 雁行褶皱;雁列褶皱
en echelon joints 斜列节理;雁行节理;雁列节理
en echelon lenses 雁行透镜体
en echelon lines 雁列线
en echelon magnetic anomaly 雁行磁异常
en echelon offset 雁列错断
en echelon structures 雁行状构造;雁行式构造;雁列(状)构造;斜列式构造;梯阵(式)构造
enemel prism 釉柱
enen mode 偶数模
emergency transmitter 应急发送机
energetic 有能力的;高能的
energetic atom 高能原子
energetic decay 高能衰变
energetic developer 高效显影液
energetic disturbance 动力不正常现象
energetic emission 高能发射
energetic encounter 高能碰撞
energetic gamma-ray 高能伽马射线
energetic meson 高能介子
energetic particle 高能粒子
energetic particle burst 高能粒子爆
energetic particle event 高能粒子事件
energetic particle instrument 高能粒子仪器
energetic plasma 高能等离子体
energetic recoil atom 高能反冲原子
energetic recoil nucleus 高能反冲核
energetic recoil particle 高能反冲粒子
energetic reflection 高能反射
energetics 力能学;能量学;动能学
energetic shower 高能簇射
energetic solar particles 高能太阳粒子
energetic solar proton 高能太阳质子
energetic storm particle 高能暴粒子
energise = energize
energised 已通电的
energization 增能;供能
energize 供给能量;通电流;送电
energized 已通电的
energized acid 增能酸
energized circuit 带电电路
energized conformation 高能构象
energized force by pressure 压力自紧力
energized line 带电电路
energized liquid 增能液
energized network 赋电网络;被激励网络
energized packing element 增能密封元件
energized part 通电部件
energizer 激发器;增能器;增能剂;催化剂;渗碳加速剂
energizing 供给能量
energizing circuit 励磁电路;激励电路
energizing circuit relay 激励电路继电器
energizing gas 激发气体
energizing loop 激磁回路
energizing lug 驱动用凸铁
energizing requisition 励磁条件
energizor 增能器;增能剂
energometer 脉能测量器
energometry 能量测定器
energonic reaction 收能反应
energy 能量;活力;放射能
energy absorbed in compression 压变形吸收能
energy absorbed in fracture 断裂吸收能;断裂所吸收的能量
energy absorber 减震器;减能器;能量转换器;能量吸收器;消能设备;消能器;消能工;消力槛
energy absorbing bumper system 能量吸收式保险杠系统
energy absorbing capacity 消能能力;吸收能量
energy absorbing element 吸能构件
energy absorbing frame 能量吸收式车架
energy absorbing front end (汽车的)能量吸收前端
energy absorbing hitch 减振挂钩
energy absorbing material 能量吸收材料;吸能材料
energy absorbing method 能量吸收法
energy absorbing roadside crash barrier 能量吸收式路边防撞护栏
energy absorbing steering assembly 能量吸收型转向装置

energy absorbing steering column 能量吸收式转向柱
energy absorbing system 耗能系统;能量吸收系统
energy absorbing tube 能量吸收管
energy absorption 能量吸收
energy absorption capability 能量吸收能力
energy absorption capacity 能量吸收能力
energy absorption coefficient 能量吸收系数
energy absorption cross-section 能量吸收截面
energy absorption process 能量吸收过程
energy acceptance 能量接收度
energy accumulation 能量累积;能量积累
energy accumulation factor 能量积累因子
energy adsorption potential 吸能潜力
energy albedo 能量反照率
energy alternative 替代能源
energy amplification 能量放大;功率放大
energy analysis 能量分析
energy analysis method 能量分析法
energy analyzing magnet 能量分析磁铁
energy analyzing microscope 能量分析显微镜
energy and natural driving type 能量和天然驱动类型
energy and pressure of liquid 液体的能量和压强
energy and transportation fund 能源交通基金
energy approach 能量法
energy attenuation 能量衰减
energy auditing 能源审计
energy balance 能量平衡;能量结算
energy balance equation 能量平衡方程
energy balance method 能量平衡法
energy balance sheet 能量平衡表
energy band 能量带;能带
energy band diagram 能带图
energy band gap 能带间隙
energy band model 能量带模型
energy band pass 能量带通
energy band structure 能带结构
energy band theory 能带学说
energy barrier 能峰;能障;能量(位)垒;能垒
energy based index 基本能量指标
energy beam 能量射束
energy beaming 能量辐射
energy bite 能量间隔
energy bound 能量极限值
energy brush 供能电刷
energy brushes 主电刷
energy budget 能源预算;能量预算;能量平衡;能量核算;能量估算
energy budget method 能量预算法;能量平衡法;能量估算法
energy budget of wave 波浪能量估计
energy built-up factor 能量积累因数;能量积累因子
energy by ocean current 海流能
energy by wave motion 波浪运动能
energy calibration 能量刻度
energy calibration standard 能量校准标准
energy capacity 能量
energy cell (柴油机)辅助燃烧室;预燃室
energy chain 能链
energy charges 用电费
energy circuit 能路
energy coefficient 能量系数
energy coincidence 能量一致
energy component 有效部分;有功分量;有功部分;电阻部分;实数部分
energy concentration 能量集中
energy confinement time 能量约束时间
energy conservation 节约能源;节省能量;节能;能源节约;能源合理利用;能源保护;能量守恒;能量不灭;能量不变;能量保持
energy conservation equation 能量守恒方程
energy conservation law 能量守恒定律
energy conservation principle 能量守恒定则;能量守恒定理
energy conservation program(me) 节能计划;能源保护计划
energy conservation technique 节能技术
energy-conserving 节约能源的
energy constant 能量常数
energy consumption 能源消耗(量);能源消费;能量消耗;能耗
energy consumption brake 能量吸收制动器
energy consumption elasticity coefficient 能源

消费弹性系数
energy consumption growth coefficient 能源消耗增长系数
energy consumption per unit of output value 单位产值能耗
energy consumption per unit working 单位工作能耗
energy consumption structure 能源消费结构
energy containing eddy 载能涡漩
energy content 能量值;能的储量;含能量
energy content of road-building materials 筑路材料能量含量
energy continuum 连续能区;能量连续区
energy control 能源管理
energy converge 能量聚集
energy conversion 能源转换;能量转换;能量转变;能量变换
energy conversion device 能量转换装置;换能器
energy conversion efficiency 能量转换效率
energy conversion engineering 能量转换工程
energy conversion factor 能量转换系数
energy conversion rate 能量转换率
energy conversion system 能量转换系统
energy conversion technique 能量转换技术
energy converter 能源转换器;能量转换器;换能器
energy correction factor 能量校正系数
energy cost 能源成本;能量消耗;能量价格
energy coupling 能量耦合
energy coversion efficiency of a scintillator 闪烁体能量转换效率
energy crisis 能源危机;能量危机
energy criterion 能量判据
energy crop 能源作物
energy current 能流;有效电流;有功电流
energy curve 能量曲线
energy cut-off 能量阈
energy cycle 能量循环
energy damper 能量阻尼器
energy datum 能量数据
energy decay 能量衰减
energy decay function 能量衰减函数
energy decrement 能量减量
energy deficiency 能量缺乏;能量不足
energy deficit 能量缺乏
energy-deflection curve 能量—变形曲线;(码头护弦木的)能量—挠度关系曲线
energy degradation 能量损失;能量降级;能量递降
energy deliver by hammer per blow 打桩每次传送功能
energy-delivering 能量传送的;传送能量的
energy demand 能源需求;能量需要
energy demand trend 能源需求趋势
energy density 能量密度
energy density function 能量密度函数
energy density of electromagnetic field 电磁场的能量密度
energy density of radiation 辐射能密度
energy density of sound 声的能量密度
energy density spectrum 能量密度谱
energy dependence 能量依赖(性);能量相关性
energy-dependent flux 能量相关通量
energy deposition 能量积存;能量沉积
energy deposition event 能量沉积事件
energy depot 能源库
energy development 能源开发
energy diagram 能量图(解);热力图
energy difference 能量差
energy diffusion 能量扩散
energy dilemma 能源危机
energy discrimination method 能量甄别法
energy dispersal 能量扩散
energy dispersal characteristics 能量扩散特性
energy dispersal signal 能量扩散信号
energy disperser 减震器;减振器;能量扩散器;缓冲器;消能装置
energy dispersion 能量色散;能量分散
energy dispersion analysis 能量色散分析
energy-dispersion baffle 消能墩;消力墩
energy dispersion block 能量块;消能墩;消力墩
energy dispersion method 能量色散法
energy dispersion spectrometer 散能量分光计
energy dispersion X-ray analysis 能量色散X射线分析
energy-dispersive analysis 能散分析

energy-dispersive microanalysis 能量消散微观分析;能量散逸微观分析;能量分散微观分析
energy-dispersive spectroscopy 能量分散光谱法
energy-dispersive X-ray analysis 能量分散X射线分析
energy-dispersive X-ray detector 能量色散X射线探测仪
energy-dispersive X-ray fluorescence 能量色散X射线荧光
energy-dissipating bucket 消力戽
energy-dissipating concrete block 消能混凝土块体
energy-dissipating dents 消力齿;消能齿
energy-dissipating device 消能设备
energy-dissipating element 能量耗散部件
energy-dissipating pile 消能桩;防撞桩
energy-dissipating sill 消能坎;消力槛;消能槛
energy-dissipating wedge 能量楔
energy dissipation 能源浪费;能量消散;能量损耗;能量散逸;能量浪费;能量耗散;能量分散;耗散;消能
energy dissipation below spillway 溢洪道下消能
energy dissipation by hydraulic jump 水跃消能
energy dissipation of roller bucket 消力戽消能
energy dissipation of ski jump type 挑流消能
energy dissipation of surface regime 面流消能
energy dissipation revetment 消能护岸
energy dissipation structure 消能建筑物
energy dissipator 能量消散器;能量消散术;消能装置;消能器;消能结构;消能构筑物;消能工
energy distribution 能源分布;能量分配;能量分布
energy distribution curve 能量分配曲线;能量分布曲线
energy distribution function 能量分布函数
energy distribution system 能源分配系统
energy divergence 能量发散;能散度
energy doubler 能量倍增器
energy economics 能源经济学
energy-economic system 动能经济系统
energy economist 能源经济学家
energy economy 动能经济
energy effect 能量效应
energy efficiency 能源效应;能源效率;能效(率);能量效率;能量系数
energy-efficiency ratio 能效率;能(量)效(率)比;能源效率比
energy-efficient 高能效的
energy-efficient building 节能建筑
energy-efficient equipment 节能设备
energy-efficient policy 节能政策
energy-efficient precast wall panel 低能耗预制墙板
energy-efficient space 节能空间
energy-efficient treatment 高能敏处理
energy-efficient wall panel 节能预制墙板
energy eigenstate 能态
energy elasticity 能源弹性值;能量弹性
energy elasticity coefficient 能源弹性系数
energy emission 能量发射
energy environment 能源环境
energy environmental policy 能源环境政策
energy equation 能量方程(式);伯努里方程式
energy equilibrium 能量平衡
energy equipartition 能量均分
energy equipartition time 能量均分时间
energy equivalence 能量当量能谱
energy equivalent 能量(源)当量
energy error 能量误差
energy exchange 能量交换
energy exchange flux 能量交换通量
energy exchange time 能量交换时间
energy expenditures 能量消费;能量消耗
energy exploration 能源勘探
energy facility 能源设施
energy factor 能量因子;能量因素;能量因数
energy feedback 能量反馈
energy flow 能通量;能流;能量流(动)
energy flow chart 能流图
energy flow diagram 能流图
energy flow in ecosystem 生态系内能量流动
energy flow mode 能流模式
energy flow model 能流模型
energy flow optimization model 能源流优化模型
energy fluence 能通量;能量注量;能量流量
energy fluence rate 能通量密度;能量流量率

energy flux 能通量;能流
energy flux density 能通量密度;能流密度
energy flux rate 能量流率
energy focusing 能量聚焦
energy forecasting 能源预测
energy from refuse 垃圾能
energy front 能量峰面
energy function 能量函数
energy future 能源未来;能源期货
energy gain 能量增益
energy gap 禁带宽度;能域;能隙;能带宽度
energy gap analysis 能域分析
energy-gathered shooting 聚能爆炸
energy grade line 能量坡线;能量梯度线;能量坡降线;能量均衡线
energy gradient 水能线;能坡;能量梯度;能量变化率;水能梯度
energy gradient line 能坡线;能量梯度线
energy-gram 能量图(解)
energy ground feedback 能量面土反馈
energy group 能量群
energy growth rate 能源增长率
energy guideline factor 能量基准因素
energy-handling capability 能量处理能力
energy head 能(量)头;能量水头;能量差;能级差;能高【铁】
energy head line 能高线
energy head of retarder location 制动能高【铁】
energy head per unit length 单位长度的能量水头
energy hill 能量高点
energy-hungry advanced country 缺能源的工业先进国
energy hunting 能量波动
energy imparted to matter 物质授予能量
energy import 能量输入
energy impulse 能脉冲
energy-impulse matrix 能量动量矩阵
energy increment 能量增量
energy index 能量指数
energy induced global climate change 能源诱发的全球气候变化
energy inequality 能量不等式
Energy Information Administration 能源信息管理局
energy input 能量输入;输入能
energy-insensitive detector 能量不灵敏探测器
energy integral 能量积分
Energy Intelligence Agency 能源情报局(美国)
energy intensive 能源紧张的;能量密集的;耗能(多)的;耗电量大的;能源密集的
energy intensive industry 能量密集型工业
energy intensive project 大量消耗能源的工程项目
energy intensive technology 能源密集的技术
energy interval 能量间隔
energy jitter 能量漂移
energy jump 能量跃迁;能量跳变
energy kernel 能量核
energy killer 能量消杀率;消能装置
energy law 能量定律
energy level 能(量)级
energy level-crossing 能级交叉
energy level crossing method 能级交叉法
energy level depletion 能级耗尽
energy level depopulation 能级粒减少
energy(level)diagram 能级图;能级分配图
energy level distribution 能级分布
energy level gap 能级距离
energy level multiplicity 能级多重性
energy level parameter 能级参数
energy level population 能级粒子数
energy level recorder 能级记录仪
energy level scheme 能级图
energy level spacing 能级间隔
energy level splitting 能级分裂
energy level structure 能级结构
energy level transition 能级跃迁
energy level width 能级宽度
energy liberation 能量释放
energy life time 能量寿命
energy limit 能量极限
energy line 能线;能坡线;能量线
energy load 动力荷载
energy loss 能量损失;能量损耗
energy loss by radiation 能量辐射损失

energy loss control 能量损失控制
energy loss factor 能量损失系数
energy loss in hydraulic jump 水跃能头损失；水跃能量消耗
energy loss mechanism 能头损失机理；能量损失机理
energy loss per atom 每原子能量损失
energy loss per ion pair 每离子对能量损失
energy loss spectroscopy 能量损失谱
energy-loss-time 能量置换时间
energy-magnitude 能量—震级
energy-magnitude relationship 能量—震级关系
energy management 能源管理；能量消耗监控；能量控制；能量管理
energy management contract 能源管理合同
energy mass equivalence 能量质量当量
energy matching 能量匹配
energy measurement 能量测量
energy metabolism 能量代谢
energy meter 累积式瓦特计；能量表；火表
energy method 能量法
energy migration 能量迁移
energy minimization 能量减到最小法
energy mix 混合能源；不同来源的能源混合使用
energy model 能量模型
energy-modulated beams 能量已调制束
energy modulation 能量调制
energy-momentum conservation 能量动量守恒
energy-momentum pseudotensor 能量动量赝张量
energy-momentum relationship 能量动量关系
energy-momentum tensor 能量动量张量
energy monochromaticity 能量单色性
energy multiplication factor 能量倍增因子
energy need 能源需求；能源需要
energy noise immunity 能量—噪声抗扰度；能量抗扰度
energy of absolute zero 绝对零点能量
energy of activation 活化能
energy of adhesion 黏合能
energy of blow 冲击能(量)
energy of deformation 应变能(量)；变形能(量)；变形功
energy of desorption 解吸能
energy of disloation 位错能
energy of dissipation 消能
energy of ensemble member 集元能量
energy of flow 流动能(量)
energy of hardening 硬化能
energy of initial set(ting) 初凝能
energy of initiation 起爆能
energy of light 光能
energy of mixing 混合能
energy of molecular dissociation 分子离解能
energy of motion 运动能
energy of nature 天然能量
energy of photoelectron 光电子能量
energy of pile driving 打桩能量
energy of position 电势能；位能；势能
energy of resonance absorption 共振吸收能量
energy of rotation 旋转能；转动能(量)
energy of rupture 断裂能(量)；破裂能量；破坏能(量)
energy of sea current 海流能
energy of set(ting) 凝结能
energy of solvation 溶剂化能
energy of storm 风暴能量
energy of strain 应变能(量)
energy of stream flow 流水能量
energy of thermal motion 热运动能
energy of the sun 日光能
energy of turbulence 涡动能；紊动能；湍流能(量)
energy of vibration 振动能(量)
energy of voluntary activity 随意活动用能
energy of zero point 零点能量
energy operator 能量算符
energy option 能源交易选择权
energy oscillation 能量振荡
energy output 能量输出；电能生产量；发动机有效功率；发电量；输出能(量)
energy path(way) 能量途径
energy performance standard 能源效率标准
energy per mole 摩尔能量
energy per unit mass 每单位质量能量

energy pipelines and systems 能源管道与系统
energy plan(ning) 能源计划；能源规划
energy plus wastewater treatment 产能污水处理
energy policy 能源政策
energy poor 能源贫乏(国家)
energy preservation 能量保护
energy pricing study 能源价格研究
energy principle 能量原理
energy-probe 能量探测器
energy problem 能源问题
energy producing 产生能量的
energy product 能积
energy-product curve 能量(乘)积曲线；能积曲线
energy production 能量产生
energy production and consumption 能源生产与消耗量
energy production capacity 能源生产能力
energy production elasticity 能源生产弹性系数
energy production structure 能源生产结构
energy profile 能线图
energy program(me) 能源计划
energy projection operators 能量投影算符
energy pyramid 能量(金字)塔
energy quantum 能量子
energy quay 能源码头
energy radiation 能量辐射
energy radiation rate 能量辐射率
energy radiator 散热器
energy range 能量范围
energy-range relation 射程能量关系
energy rate 能率
energy ratio 能量比
energy reaction 能流反应
energy recovery 能量回收；能量恢复
energy recovery from sludge 污泥能源回收
energy recovery from solid wastes 固体废物能源
energy recovery from waste 从废料回收能源
energy recovery power 能流回收
energy recovery system 能量恢复系统
energy reflection build-up factor 能量反射积累因数
energy reflection coefficient 能量反射系数
energy reflectivity 能量反射系数；能量反射率
energy regeneration 能量再生
energy region 能(量)区；能量范围
energy regulation 能量调节
energy regulator 能量调节器
energy release 能量释放
energy release rate 能量释放(速)率
energy replacement 能量置换
energy-replacement time 能量置换时间
energy requirement 能量需要；需能量
Energy Research and Development Administration 能源研究和发展管理局
energy reserve 能量储备；能源储量；能源储备
energy residual curve for low-frequency band-pass 低通能量剩余曲线
energy resolution 能量分解；能量分辨(率)
energy resolving power 能量分辨本领
energy resources 能源(资源)
Energy Resources Conservation Board 节能局
energy resources data system 能源数据系统
energy resource survey 能源勘探；能源勘察
energy response 能量响应；能量特性
energy response function 能量响应函数
energy retention 能量存留
energy revolution 能源革命
energy rich 能源丰富国家
energy-rich bond 高能键
energy-rich chemical fuel 高能化学燃料
energy-rich compound 高能化合物
energy-rich phosphate 高能磷酸代物
energy-rich thioester bond 高能硫酯键
energy-saving 能源节约；节约能源的；节省能源的；节能
energy-saving building 节能建筑
energy-saving checkout automobile 节能检测车
energy saving design 节能设计
energy-saving investment 节能投资
energy-saving lighting 节能照明
energy-saving operation 节能营运
energy-saving policy 节能政策
energy-saving product structure 节能型产品结构
energy-saving space 节能空间

energy-saving technology 节能技术；节能工艺
energy scale 能量标度
energy scale calibration 能流标度校准
energy scaling factor 能量定标因数
energy scheme 能量图(解)
energy security 能源安全
energy selection 能量选择
energy-selective slit 能量选择缝
energy-sensitive 能量变化灵敏的
energy sensitivity 能量灵敏度
energy separation 能量间隔
energy shortage 能源短缺
energy simulation and optimization 电能模拟及优化
energy sink 能阱；能汇
energy slope 能力比降；能坡；能量梯度；能量坡降；能量比降；水能梯度
energy source 能源；资源
energy source assessment 能源评议
energy source boat 震源船
energy source building 能源建设
energy source controller 能源控制装置
energy source feed 能源供应
energy source industry 能源工业
energy source of exogenic process 外营力能量来源
energy sources policy 能源政策
energy source supply 能源供应
energy source synchronizer 震源同步装置
energy spacing 能量间隔
energy spectral density 能谱密度
energy spectrometer 能谱仪
energy spectrometry 能谱分析法
energy spectrum 能(量)谱；能量分布
energy spectrum analysis 能谱分析
energy spectrum density 能量谱密度
energy spectrum function 能谱函数
energy spectrum width 能谱宽度
energy spread 能量扩展度；能量分散；能量发散
energy spreading waveform 能量展布波形
energy spread of ion beam 离子束的能量发散
energy stability 能量稳定度
energy stabilization 能量稳定
energy stage 能量级
energy state 能态
energy state density 能态密度
energy statistics 能源统计
energy storage 能量存储；能量储存；能量储备
energy storage accumulation 蓄能
energy storage braking 能量积聚制动
energy storage capacitor 蓄能电容器；储能电容器
energy storage capacity 储能量
energy storage device 储能装置
energy storage material 能量储存材料；储能材料
energy storage system 储能系统
energy storage unit 储[贮]能元件
energy storage welding 储[贮]能焊；脉冲焊接
energy straggling 能量歧离；能量分散
energy stream 能流
energy structure 能源结构
energy supply 能量供应；供能
energy surface 能面
energy system 能量系统；能量体系
energy systems laboratory 能源设备实验
energy technician 能源管理技术员
energy-temperature curve 能量—温度曲线
energy tensor 能(量)张量
energy term 能量项
energy terminal 能源码头
energy theorem 能量法则；能量定律
energy thickness 能量损失强度
energy threshold 能阀
energy to failure 断裂能(量)
energy to fracture 断裂能(量)；冲击韧性
energy to initiate crack 初始裂纹能量
energy transducer 换能器
energy transfer 能源转换；能量转移；能量转换；能量迁移；能量传输；能量传输；能量传递
energy transfer coefficient 能量传递系数
energy transfer function 能量转移函数
energy transfer matrix 能量转移矩阵
energy transfer mechanism 能量转换机理
energy transfer process 能量转移过程
energy transfer time 能量转移时间

energy transformation 能量转换;能量转化
energy transformer 能量转化者
energy transition temperature 能量—脆性转变温度
energy transmission 能量转换;能量传递
energy transmission build-up factor 能量透射积累因子
energy-transmitting boundary 能量传递边界
energy transport 能量输运;能量输送
energy trap(ping) 能量陷阱;能阱
energy trap vibration mode 能陷振动模式
energy-type industry 耗能型工业
energy uniformity 能量均匀性
energy unit 能量单位
energy use 能源消费;能源利用;能量的利用
energy used in driving 打桩能量(消耗)
energy user 能源利用者
energy utility management 能源使用管理
energy utilization 能源利用;能量利用
energy utilization efficiency 能源利用效率
energy utilization rate 能源利用率
energy utilization technology 能源利用技术
energy-valley 能谷
energy value 能量值
energy-variant sequential detection 能量变化顺序检测
energy waste 能量损耗
energy wharf 能源码头
energy window 能量窗
energy without pollution 无污染能源
energy wood 能源材;薪材
energy yield 能量输出;能量产量
enerjet gun 半销毁式过油管射孔器
enestella 窗扇壁龛
eneyne 烯炔
enfaced paper 具名支票
enfacial junction 贴面接合
enfacial junction cement texture 贴面结合胶结物结构
enfilade 门轴线;直路;正对(指相对的物体)
enfoldig batch charger 包料式加料机
enfolding charger 裹入式加料机
enforce 强制
enforceability 法律的可执行性;强制性
enforceable contract 有效合同
enforce a block 实行封锁
enforce a rule 实施规则
enforced diversion 强制驶出
enforced liquidation 强制清算
enforced meander 强迫蛇曲;强化曲流
enforced settlement 强迫下沉
enforcement 实施;执行;强制
enforcement of rights 执行权利
enforcement rules 施行细则
enforce payment 强制性支付
enforce traffic flow 强制车流
enforcing authority 制服官员
enframe 作框架用;装在框内;配框架
enframed 已装框架的
enframed scenery 框景
enframement 框景
enframing arcades 拱廊装配;拱廊架设
engadinite 少英细晶岩
engage 经营;预约;聘请
engage a clutch 挂离合器
engage a gear 挂挡
engage angle (铣刀的)接触角
engaged angle 啮合角
engaged bollard 标柱
engaged chimney 内装壁炉;罗马式烟囱
engaged column 附墙(圆)柱;壁柱;半身柱;嵌墙柱;贴墙柱
engaged condition 占线状态
engaged Corinthianesque column 部分在墙内的考林斯式圆柱
engaged fire-place 内装壁炉;罗马式烟囱
engaged gear 从动轮
engaged lamp 占线信号灯
engaged line 忙线
engaged pier 附墙方柱;嵌墙墩
engaged pier capital 附墙墩帽;部分砌入墙内的墩帽
engaged shaft 附墙柱子;附墙竖片;部分砌入墙内的柱子;部分砌入墙内的竖井
engaged signal 占线信号

engaged switch 接通开关
engaged test 在线测试
engaged tone 忙音
engaged wheel 联动轮;从动轮;相啮合齿轮;相吻合齿轮
engage in 参与
engage in all kinds of foreign exchange operations 经营外汇业务
engage in labo(u)r to a very small extent 附带劳动
engage in speculation 投机倒把
engage in speculation and profiteering 投机倒把
engagement clause 保证条款
engagement factor 接触比;重合度
engagement letter 聘书
engagement memorandum 业务备忘录;工作备忘录
engagement of labo(u)r 劳务的雇佣
engagement of shipping list 承受托运单
engagement of staff 职员的雇用
engagement process 啮合过程
engagement range 接合范围;啮合范围;捕获距离
engagements 新雇用者(指在任何一个时期中进入就业的总人数)
engagement screw 衔接丝杆
engagement sleeve 齿轮离合器
engagement system 啮合方式
engagement with the fish 抓住落鱼;打捞孔内脱落钻具
engager 衔接器
engage relay 接通继电器
engage spring 啮合弹簧
engage switch 啮合器
engage test 忙碌试验
engaging and disengaging gear 离合装置
engaging angle 啮合角
engaging arm 衔接臂
engaging contact 进啮
engaging fish 捞住落鱼
engaging friction 进入摩擦;啮合摩擦
engaging friction cone 同步摩擦锥轮
engaging lever 离合杆;开动杆;接合杆
engaging lug 连接耳
engaging means 接通装置;接通机构
engaging mechanism 接合机构;啮合机构
engaging reverse gear 接倒挡齿轮
engaging side of gears 齿轮啮合边
Engel coefficient 恩格尔系数
Engel curve 恩格尔曲线
Engel curve analysis 恩格尔曲线分析
Engel elasticity 恩格尔弹性
Engelhardt testing 恩格尔哈特实验
Engelhardt value 恩格尔哈特值
Engelian coefficient 恩格尔系数
Engel law 恩格尔定律
engender 产生
engine 机器;机车【铁】;发动机
engine accessory 发动机辅助装置
engine afterburner system 发动机加力燃烧系统
engine air capacity 发动机的空气流量
engine air intake precleaner 发动机进气粗滤器
engine alternator 机动同步发电机
engine alternator set 发动机交流发电机组
engine altimeter 测量发动机增压的高度计
engine altitude chamber 发动机高空低压实验室
engine analyzer 引擎分析器;发动机综合试验机
engine and boiler room 机炉舱
engine and gearbox unit 发动机和齿轮箱组
engine and pump assembly 泵机组
engine anti-freeze 发动机防冻剂
engine anti-icing system 发动机防冰系统
engine arm 发动机起动摇把
engine ash 炉渣;机车炉灰
engine assembly 发动机装配
engine attendant 机工;司机
engine balance 发动机平衡;发动机飞轮上的平衡重
engine base 机座;引擎基座;发动机座
engine baseplate 发动机支承板;发动机底板
engine bay 机车停留线;机车间
engine bay cooling system 发动机舱冷却系统
engine bearer (发动)机座
engine bearer foot 发动机支座支撑脚
engine bed (发动)机座
engine block 发动机组;发动机缸体;汽缸体

engine blower 发动机风扇
engine body (发动机)机身;发动机身;发动机机体
engine book 发动机记录簿
engine boss 消防车领班
engine bracket 发动机支架;发动机托架
engine brake 发动机制动;发动机闸
engine braking 发动机制动
engine break-in 机器轧碎;机器碾平;发动机试车;发动机启动
engine breathing ability 发动机通风能力
engine builder 主机制造商
engine-building 发动机制造
engine bulkhead 发动机隔壁
engine burn 钢轨烧伤;发动机燃烧过程
engine capacity 发动机容量;发动机功率
engine car 动力车;发动机吊舱
engine case (发动)机箱
engine casing 机器罩;机壳;机舱棚;电动机罩;发动机罩
engine cast pad 发动机壳垫
engine chart 发动机图表
engine cinder 炉渣
engine cleaner 发动机清洗器;发动机清洁器
engine cleaning agent 发动机洗涤剂
engine cleaning gun 机械清洁枪
engine cleansing agent 发动机清洗剂
engine clutch 发动机离合器
engine cold running test installation 发动机冷试转设备
engine company 消防分队
engine compartment (发动机)机舱
engine compartment drain valve 机舱放泄阀
engine compartment heater 机舱加热器
engine compartment heater duct 机舱加热器导管
engine compartment terminal box 机舱接线盒
engine components 发动机部件
engine compound 发动机并车
engine conditioning 发动机调节
engine conditioning oil 发动机清洗油
engine condition monitoring 发动机状态监控
engine configuration 发动机布置;发电机布置
engine conk 发动机运转中断
engine continuous output 发动机持续功率
engine control characteristic 发动机控制特性
engine control gear 发动机控制机构
engine controller 发动机调节器
engine control lever 发动机操纵杆
engine control panel 发动机控制盘
engine controls 发动机控制机构
engine control stand 发动机操纵台
engine control system 发动机控制系统;发动机操纵系统
engine coolant 发动机冷却液
engine coolant heater 发动机冷却液加温器
engine cooling 发动机冷却
engine cooling thermometer 发动机冷却系温度计
engine cooling water 发动机冷却水
engine counter 机轴转数表;发动机转数表
engine cover 机罩;发动机(机)罩
engine cradle 发动机架
engine crank 发动机曲柄
engine crankcase 发动机曲柄箱
engine crankcast 发动机机匣
engine cranking motor 发动机起动马达
engine crankshaft 发动机曲轴
engine crank shaft loading 发动机曲轴负载
engine crew 轮机组
engine crossdrive 发动机横向传动
engine cross-drive casing 发动机横向传动箱
engine crossfeed 发动机交叉供油
engine cut-off 发动机自动停机装置;停机;关闭发动机;发动机停车
engine cut-off timer 发动机停车计时器
engine cut-off velocity 发动机停车瞬时速度
engine cut-out 停车
engine cycle 发动机循环;热机循环
engine cycle pressure ratio 发动机循环增压比
engine cylinder (汽)缸
engine cylinder displacement 发动机汽缸排量
engine cylinder oil tank 发动机汽缸油箱
engine cylinders in line arrangement 发动机汽缸直线单行排列
engine cylinders opposed arrangement 发动机双排汽缸相对平放

engine cylinders radial arrangement 发动机双排汽缸星形排列
engine degasser 电动脱气器
engine department 轮机部
engine deposits 发动机沉积物
engine design 电机设计
engine designer 主机设计者
engine detonation indicator 发动机爆燃指示器
engine dies 发动机熄火停转
engine disintegration 发动机碎裂
engine displacement 发动机排量
engine distillate 发动机用轻油
engine district 机车段
engine drag 发动机阻力
engine-drive 发动机传动
engine-driven 机动的;发动机驱动的
engine-driven air compressor 机动空气压缩机
engine-driven generator 机动发电机;发动机驱动的发电机
engine-driven generator power system 机动发电机动力系统
engine-driven hydraulic pump 机动液压泵
engine-driven knapsack sprayer 背负式机动喷雾机
engine-driven mastic asphalt mixer 发动机驱动沥青玛琋脂搅拌机
engine-driven pump 发动机传动泵
engine-driven rotary sweeper 发动机驱动旋转式扫路机;机动旋转式扫路机
engine-driven vacuum pump 机动真空泵
engine-driven winch 发动机驱动绞车;机动绞车
engine driver 司机
engine dry weight 发动机自重
engine durability 发动机的耐用性
engine dust pan 发动机防尘盘
engine efficiency 发动机效率
engine endurance test 发动机耐久试验;发电机耐久性试验
engineer 工程师
engineer cadet 轮机实习员;实习轮机员
engineer-construction firm 建筑工程公司
engineer department 机舱部门
Engineer District 美国陆军工程师团分局
Engineer Division 美国陆军工程师团区局
engineered 设计的;工程监督的
engineered aquatic system 工程水系统
engineered brick (特定)工程用砖;高强度砖;半轴砖
engineered fill 控制回填土;质控(回)填土
engineered performance standard 设计绩效标准
engineered pressure 设计压力
engineered safety feature 外设安全措施
engineered safety system 外设安全系统
engineered water system 工程水系统
engineered water treatment system 工程水处理系统
engineer for erection(work) 装配工程师;架设工程师
engineer-hydrologist 工程师兼水文学家
engineer in charge 主任工程师;总工程师;值班工程师;主管工程师
engineer in chief 总工程师
engineering 工程(学);工程技术;设计
engineering acceptance 工程验收
engineering achievement 工程成就
engineering acoustics 工程声学
engineering act 工程法规
engineering activities of human 人类工程活动
engineering administration manual 工程管理手册
engineering advancement (工程)技术发展
engineering adviser 工程顾问
engineering aerodynamics 工程空气动力学
engineering agreement 工程协议
engineering alternate 工程方案
engineering amount per unit ore reserve 单位储量工程量
engineering analysis 工艺分析;工程(经济)分析
engineering analysis of cost 成本技术分析
engineering analysis report 工程分析报告
engineering and administration data-acquisition system 工程管理数据采集系统
engineering and construction 设计与施工
engineering and design department 设计处
engineering and manufacturing 技术与制造过程

规格手册
engineering and technological research 工程技术研究
engineering approximate estimate 工程概算
engineering approximation 工程近似(法)
engineering architecture 工程建筑
engineering arrangement map of exploration 勘探区工程布置图
engineering assembly parts list 工程装配零件单
engineering association 工程协会
engineering atmosphere 工程大气压
engineering bacteria 工程菌
engineering bacteria construction 工程菌构建
engineering bacteria species 工程菌菌种
engineering-biological method of construction 工程生物学施工方法
engineering biology 工程生物学
engineering boat 工程船
engineering boring 工程地质钻探
engineering brick 工程(用)砖;高强抗蚀砖;高强(度)砖;半釉砖
engineering brick facing 贴面砖工程
engineering brick floor(ing) 贴地面砖工程;铺地板用工程砖
engineering brick for manholes 筑人孔用工程砖
engineering brick masonry 工程砖圬工墙
engineering brick paving 工程砖铺砌(路面)
engineering brief 工程简讯
engineering calculation 工程计算
engineering capability 工程能力
engineering capacity 设计能力
engineering car 工程车
engineering case 工程实例
engineering casting 工程铸件
engineering cast iron 工程铸件
engineering ceramics 工程陶瓷
engineering change 工程更改;工程改变
engineering change capability 工程改变能力
engineering change order 工程更改指令;技术更改指令;设计变动命令书
engineering change plane 工程修改层
engineering change procedure 技术更改程序;工程更改程序
engineering change proposal 技术更改建议;修改设计建议;工程更改建议书
engineering change proposal work statement 技术更改建议说明书
engineering change request 工程更改申请;工程变更申请
engineering change schedule 工程更改日程
engineering change statement 工程更改说明书
engineering characteristic 工程特征
engineering characteristics of rock 岩石的工程特性
engineering classification 工程分类
engineering classification of soil 土的工程分类
engineering college 工程学院
Engineering Committee on Oceanic Resources 海洋资源工程委员会
Engineering Committee on Ocean Research 海洋研究技术委员会
engineering community 工程界
engineering company 工程公司
engineering component 机械部件
engineering component list 机械部件清单
engineering component parts list 工程(零)部件清单
engineering compromise 工程综合考虑;工程折中方案
engineering concept 工程概念
engineering condition 工程条件
engineering condition of rock and soil 岩土工程条件
engineering construction 工程建设;工程施工
engineering construction(bore)hole 工程施工钻孔
engineering construction drill 工程施工钻机
engineering construction standard 工程建设标准
engineering consultancy firm 工程咨询公司
engineering consultation 工程咨询
engineering consultation service 工程咨询服务
engineering consulting contract 工程咨询合同
engineering consulting firm 工程咨询公司
engineering contract 建筑工程承包合同;工程(承

包)合同
engineering contraction audit 工程承发包审计
engineering control 工程控制
engineering control inspection 工程控制与检查
engineering control manager 工程控制经理
engineering control network 工程控制网
engineering control of air pollution 空气污染的工程控制
engineering control room 工程控制室
engineering conveyer 工程输送机
engineering corporation 工程公司
engineering cost 工程技术费;工程成本;施工费(用);工程费(用)
engineering creep 工程蠕变
engineering cybernetics 工程控制论
engineering data 技术资料;技术数据;工程(技术)资料;工程(技术)数据;设计资料
engineering database 工程数据库
engineering data collection 技术资料收集
engineering data gathering 设计资料收集
engineering data information system 工程数据信息系统
engineering data processing 工程数据处理
engineering data requirement 工程数据(的)要求
engineering data search 收集设计资料;收集技术资料
engineering data treatment 工程数据处理
engineering decision 工程决策
engineering demonstrated inspection 有技术根据的检查
engineering department 技术处;技术部门;工程系;工程局;工程处;工程部门
engineering departmental instructions 工程部门说明书
engineering department instructions 工程部门说明书
engineering department notice 工程部门通知书
engineering design 工程设计
engineering design alternate 工程设计方案
engineering design change 工程设计(的)改变;工程设计(的)变更;技术设计改变;工程设计更改
engineering design data 工程设计数据
engineering design data package 工程设计资料包
engineering design department 工程设计部门
engineering design division 工程设计部门
engineering design firm 工程设计公司
engineering design map of exploration 勘探区工程设计图
engineering design modification 工程设计(的)修改
engineering design plan 技术设计方案;工程设计方案
engineering design review 技术设计审查;工程设计;审查工程设计
Engineering Design Review Board 工程设计评审委员会
engineering design standards 工程设计标准
engineering details 工程细目
engineering detail schedule 工程的详细时间表
engineering development 技术发展;工程研制;工程技术开发;工程技术发展
engineering diploma 工程执照;工程证书
engineering discipline 工程学科;工程规范;工程规定
engineering district 工务(分)段;工程总段
engineering documentation 工程编录
engineering drafting 工程制图
engineering drawing 工程(制)图;工程图纸;工程图样;工程画
engineering drawing change 工程图纸的改变;工程图纸的变更
engineering drawing list 工程图纸清单
engineering drawing number 工程图编号
engineering drill bit 工程钻钻头
engineering driller 工程钻机
engineering drilling 工程地质钻探
engineering dynamic(al)geology 工程动力地质学
engineering-economic analysis 工程经济分析
engineering economics 技术经济学;工程经济学
engineering economist 工程经济学家
engineering economy 工程经济(学);工程分析;设计方案经济比较

engineering economy study 技术经济分析;工程经济研究
engineering education 工程教育
engineering effective target 工程效率指标
engineering effort 技术工作
engineering elasticity 工程弹性力学
engineering electromagnetics 工程电磁学
engineering electrophysics 工程电物理学
engineering element 工程要素
engineering environmental hydrogeology 工程环境水文地质学
engineering estimate 设计估算
engineering evaluation 工程评价
engineering evaluation test 工程评估试验;工程估价试验
engineering examination 工程验证
engineering example 工程实例
engineering facility 工程设备
engineering factor 技术条件;工程原理;工程因素
engineering feasibility 技术可能性;技术(上的)可行性;工程现实性;工程可行性
engineering feasibility analysis 技术可行性分析
engineering feasibility study 工程技术可行性研究
engineering feat 杰出的工程技术成就;工程技术业绩
engineering feature 工程特征
engineering firm 工程公司
engineering flow sheet 工艺流程图
engineering fluid mechanics 工程流体力学
engineering for nuclear fuel reprocessing 核燃料后处理工程
engineering for pollution control 污染控制技术
engineering function 技术职能;技术功能;工程(上的)职能
engineering gap 技术差距
engineering geography 工程地理(学)
engineering geologic(al) analogy 工程地质类比(法)
engineering geologic(al) analysis 工程地质分析
engineering geologic(al) analysis map 工程地质分析图
engineering geologic(al) appraisal 工程地质鉴定
engineering geologic(al) calculation 工程地质计算
engineering geologic(al) characteristic 工程地质特征
engineering geologic(al) classification for rock and soil 岩土工程地质分类
engineering geologic(al) classification of rock 岩石的工程地质分类
engineering geologic(al) classification of rock mass 岩体的工程地质分类
engineering geologic(al) classification of soil 土的工程地质分类
engineering geologic(al) columnar section 工程地质柱状图
engineering geologic(al) conclusion 工程地质结论
engineering geologic(al) condition 工程地质条件
engineering geologic(al) condition analysis 工程地质条件分析
engineering geologic(al) condition assessment for urban architecture based on remote sensing 遥感城市建筑工程地质条件评估
engineering geologic(al) condition evaluation for large-scale building foundation 大型建筑基础工程地质条件评价
engineering geologic(al) condition of ore district 矿区工程地质条件
engineering geologic(al) demonstration 工程地质论证
engineering geologic(al) detail map 工程地质详图
engineering geologic(al) division 工程地质学分科
engineering geologic(al) divisional map 工程地质分区图
engineering geologic(al) drilling 工程地质钻探;工程地质钻进
engineering geologic(al) drilling rig 工程地质钻机
engineering geologic(al) estimate 工程地质评价
engineering geologic(al) evaluation 工程地质评价
engineering geologic(al) exploration 工程地质勘探
engineering geologic(al) exploration of air field 机场工程地质勘察
engineering geologic(al) exploration of city 城市工程地质勘察

engineering geologic(al) exploration of grade separation 立交桥工程地质勘察
engineering geologic(al) exploration of theatre ground 剧场工程地质勘察
engineering geologic(al) exploration of tunnel 隧道工程地质勘察
engineering geologic(al) exploration types of shaft 矿井工程地质勘探类型
engineering geologic(al) exploratory drill 工程地质勘察钻机
engineering geologic(al) exploring mining 工程地质坑探
engineering geologic(al) geophysical exploration 工程地质物探
engineering geologic(al) hole 工程地质钻孔
engineering geologic(al) improvement 工程地质改造
engineering geologic(al) investigation 工程地质勘察;工程地质调查
engineering geologic(al) investigation of bridge and road 道路与桥梁工程地质勘察
engineering geologic(al) investigation of canal 渠道工程地质勘察
engineering geologic(al) investigation of damsite 坝址工程地质勘察
engineering geologic(al) investigation of dock 码头工程地质勘察
engineering geologic(al) investigation of draining flooded fields 排洪区工程地质勘察
engineering geologic(al) investigation of electric(al) station buildings 电站厂房工程地质勘察
engineering geologic(al) investigation of goods yard 货场工程地质勘察
engineering geologic(al) investigation of gulf 海湾工程地质勘察
engineering geologic(al) investigation of harbo(u)r area 港区工程地质勘察
engineering geologic(al) investigation of harbo(u)r 港湾工程地质勘察
engineering geologic(al) investigation of irrigation region 洪区工程地质勘察
engineering geologic(al) investigation of natural building materials 天然建筑材料勘察
engineering geologic(al) investigation of nuclear power station 核电站工程地质勘察
engineering geologic(al) investigation of outer harbo(u)r 港外工程地质勘察
engineering geologic(al) investigation of pumping station site 扬水站址工程地质勘察
engineering geologic(al) investigation of reservoir 水库工程地质勘察
engineering geologic(al) investigation of shipyard 船坞工程地质勘察
engineering geologic(al) investigation of sluice gate site 闸址工程地质勘察
engineering geologic(al) investigation of spillway 溢洪道工程地质勘察
engineering geologic(al) investigation of tunnel 隧道工程地质勘察
engineering geologic(al) investigation of underground construction 地下建筑工程地质勘察
engineering geologic(al) investigation of water conservancy facility and hydroelectric(al) station 水利水电工程地质勘察
engineering geologic(al) investigation of water tunnel 输水洞工程地质勘察
engineering geologic(al) investigation report 工程地质勘察报告
engineering geologic(al) location of line 工程地质选线
engineering geologic(al) long-term observation 工程地质长期观测
engineering geologic(al) map 工程地质图
engineering geologic(al) mapping 工程地质调绘;工程地质测绘
engineering geologic(al) measuring 工程地质测试
engineering geologic(al) of rock mass 岩体工程地质(力学)
engineering geologic(al) of soil mass 土体工程地质
engineering geologic(al) predicting 工程地质预测
engineering geologic(al) predicting map 工程地质预测图
engineering geologic(al) problem 工程地质问题

engineering geologic(al) process 工程地质作用
engineering geologic(al) profile 工程地质剖面图
engineering geologic(al) property 工程地质性质
engineering geologic(al) property of frozen soil 冻土的工程地质性质
engineering geologic(al) property of rock and soil 岩土的工程地质性质
engineering geologic(al) prospecting 工程地质勘探;工程地质勘察
engineering geologic(al) recommendation 工程地质建议
engineering geologic(al) scheme map 工程地质略图
engineering geologic(al) section 工程地质剖面图;工程地质断面图
engineering geologic(al) test 工程地质试验;工程地质测试
engineering geologic(al) test instrument 工程地质测试
engineering geologic(al) unit 工程地质单元
engineering geologic(al) zoning 工程地质区划;工程地质分区
engineering geologic(al) zoning map 工程地质分区图
engineering geologist 工程地质师
engineering geology 工程地质(学)
engineering geology investigation 工程地质勘察
engineering geology mapping of harbo(u)r 海港工程地质测绘
engineering geology mapping of irrigation area 灌区工程地质测绘
engineering geology mapping of road 线路工程地质测绘
engineering geology mapping of urban 城市工程地质测绘
engineering geology of mine environment 矿山环境工程地质
engineering geology of underground structures 地下建筑工程地质
engineering geology petrofabric 工程地质岩组
engineering geomechanics 工程地质力学
engineering geomorphology 工程地貌学
engineering geophysical prospecting 工程地球物理勘探
engineering geophysical sounding 工程地球物理测探
engineering geophysics 工程物探;工程地球物理学
engineering geotechnic(al) test 土工试验
engineering goods 工程器材
engineering graduate 工科大学毕业生
engineering graphics 工程图(学);工程图示学;工程图画学
engineering hydraulics 工程水力学
engineering hydrology 工程水文学;工程水力学
engineering improvement 工程改建
engineering improvement of soil and rock 岩土工程改良方法
engineering improvement time 工程改进时间
engineering index 工程(技术文献)索引
engineering index of soft soil 软土工程指标
engineering industry 机械工业
engineering industry association 工程工业协会
engineering information control system 工程技术情报管理系统
engineering inspection and acceptance specification 工程验收规范
engineering inspection specification 工程检查标准
engineering institute 工程研究院
engineering instruction 技术细则;工程说明书;技术说明书;技术(操作)指南;工程细则
engineering insurance 工程保险
engineering intensity 工程烈度
engineering intensity scale 工程烈度表
engineering investigation 工程勘察
engineering investigation (and survey) 工程勘察
engineering investment 工程投资
engineering item description 工程项目说明(书)
engineering job analysis 工程(任务)分析
engineering job order 工程任务单
engineering judgment 工程评价
engineering laboratory 技术试验室;技术实验室;工程(技术)试验室;工程(技术)实验室
engineering landslide 工程滑坡
engineering legislation 工程技术规范;工程法制;

工程法规;工程法则
engineering level 技术水平;工程水准仪;工程水平(仪)
engineering level(l)ing 工程水准测量
engineering level or technical level 工程水准测量
engineering lofting 按样板裁切板状材料
engineering maintenance fee 工程维持费
engineering management 工程管理
engineering manager 工程设计经理
engineering manual 工程手册
engineering map 工程(地)图
engineering mapping 工程地质制图
engineering material 工程材料;土木材料
engineering material specification 工程材料规范
engineering materials research laboratory 工程材料研究实验室
engineering mathematics 工程数学
engineering measure 工程度量;工程措施
engineering mechanics 工程力学
engineering mechanics of material 材料工程力学
engineering metallurgy 工程冶金学
engineering meteorology 工程气象学
engineering metrology 工程计量学
engineering model 工程结构模型
engineering modification of rock and soil 岩土工程改良
engineering news formula 工程新闻公式
engineering news pile driving formula 工程新闻打桩公式
engineering news record 工程新闻记录
engineering norm 技术定额
engineering number 工程代号;工程编号
engineering oceanography 工程海洋学
engineering of artificial recharge 人工补给工程
engineering of exploration 勘探工程
engineering officer 责任工程师;工程管理员
engineering of leaky riverbed 河床滞流工程
engineering of mineral deposit dewatering 矿床疏干工程
engineering of nuclear power 核动力工程(学)
engineering of survey 工程安全
engineering of waste disposal 排污工程
engineering operation 工程设施
engineering optics 工程光学
engineering order 技术指令;工程等级
engineering-orientated 工程技术方面
engineering-oriented 工程(技术)有关的
engineering painting 工程画
engineering parameter 技术参数
engineering parameter of well arrangement 水井布局的工程参数
engineering performance standard 技术性能标准;工程性能标准
engineering philosophy 技术原则;工程原理;设计原则;设计思想
engineering photogrammetry 工程摄影测量
engineering photography 工程摄影学
engineering picture 工程画
engineering plan 工程图
engineering planner 工程计划人员;工程规划人员
engineering planning 工程计划
engineering planning and analysis system 工程计划分析系统
engineering plant 工程器材
engineering plastics 工程塑料
engineering plot 工程图
engineering possibility 技术可靠性
engineering practice 技术实践;工程实践
engineering practice amendment 工程实践修正
engineering precaution 工程预防
engineering preliminaries (勘察、设计、预算等)工程准备事项;工程前期工作
engineering preparation (勘察、设计、预算等)工程准备事项
engineering problem 技术问题
engineering process 工程处理方法;工程程序
engineering procurement 工程采购
engineering-procurement-construction 设计—成交—建造
engineering project 工程项目;工程设计;工程计划
engineering property 机械性质;工程性质;工程性能;工程特性
engineering property of soil 土的工程性质
engineering property of soil mass 土体的工程性质

engineering proposal 工程方案
engineering psychology 工程心理学
engineering purchasing specification 工程采购规格
engineering purchasing specification manual 工程采购规格手册
engineering quality 工程质量
engineering quality index of rock mass 岩体工程质量指标
engineering query note 设计询问通知
engineering quota 工程定额
engineering receiving 工程验收
engineering receiving inspection 工程验收检查
engineering record 工程记录
engineering regulation 工程规则;工程条例;工程法规
engineering reinforcement 加固工程
engineering rejection well 工程报废井
engineering reliability 技术可靠性;工程可靠性
engineering reliability and quality control 工程可靠性与质量控制
engineering reliability review 工程安全性检查
engineering remote sensing 工程遥感
engineering repair truck 工程修理车
engineering repair works 维修工厂
engineering report 技术报告;工程(技术)报告
engineering requirement 工程要求
engineering research 技术研究;工程(学术)研究
engineering research report 工程研究报告
engineering road-blocks 工程难题
engineering rock and soil 岩土工程学
engineering rock mass 工程岩体
engineering scale 工程(用)比例尺
engineering scale of water conservancy facility and hydroelectric(al)station 水力水电工程规模
engineering schedule 工程计划(进度)
engineering schedule plan 工程进度计划
engineering scheme hydro-electric (al) engineering 水电工程
engineering science 技术科学;工程科学
Engineering Science Database of United Kingdom 英国工程科学数据库
engineering seismics 工程地震
engineering seismograph 工程地震仪
engineering seismology 抗震工程学;工程地震学;地震工程学
engineering service group 工程维修组
engineering service loan 工程服务贷款
engineering service memo 工程维修备忘录
engineering service publication 工程技术业务出版物
engineering services 工程设计服务;工程技术服务
engineering ship 工程船
engineering shop memo 工程技术工场备忘录
engineering shop(work) 机械加工车间
engineering simulator 工程模拟器
engineering simulator for consolidation well 固井用工程模拟器
engineering simulator for drilling 钻井用工程模拟器
engineering simulator for offshore drilling 海上钻井用工程模拟器
engineering site 工程位置
engineering soil condition 工程土壤状态
engineering soil map 土分布图
engineering soils' classification system 工程土的分类体系
engineering soil survey 工程土壤测量
engineering source data 工程技术来源资料
engineering special test equipment 工程特殊试验设备
engineering specification 工程(技术)说明书;工程技术规范;工程规格(书);工程标准
engineering staff 工程技术人员
engineering standard 技术标准;工程(技术)标准
engineering standard department 工程标准部门
engineering standards manual 工程技术标准手册
engineering strain 工程应变
engineering structure 工程结构;工程建筑物;工程构筑物
engineering study 工程研究
engineering sub-contract 工程分包
engineering summary 工程总结
engineering summary report 工程总结报告

engineering supervision 技术监督;工程视察;工程检查;工程监理;工程监督;工程管理
engineering supervision company 工程监理公司
engineering supervision personnel 工程监理人员
engineering support group 工程支援组
engineering survey(ing) 工程测量(学)
engineering survey of hydraulic hinge 水利枢纽工程测量
engineering survey to navigation 港口
engineering system 工程系统
engineering system of units 工程单位制
engineering target 工程指标
engineering task assignment 工程任务分配
engineering technical design specification 工程技术设计规范;工程技术设计标准
engineering technical latter 工程技术许可证
engineering technical personnel 工程技术人员
engineering technique 工程技术;工程技巧;施工技术
engineering technique files system 工程技术档案制度
engineering test 技术检验;工程试验
engineering test evaluation 工程试验鉴定
engineering test facility 工程试验设备
engineering test laboratory 工程试验实验室
engineering test order 工程试验指令
engineering test procedure 工程试验程序
engineering test program(me) 工程试验规划
engineering test reactor 工程试验反应堆
engineering test reactor critical assembly 工程试验堆临界装置
engineering test report 工程试验报告
engineering test request 工程试验要求
engineering test unit 工程试验装置
engineering theodolite 工程经纬仪
engineering thermodynamics 工程热力学
engineering thermoplastic elastomer 工程热塑性弹性体
engineering time 检修时间;维(护)修时间
engineering train 工程(列)车;【铁】
engineering transit 工程经纬仪
engineering truck 轨道维修工作车;工程车
engineering turnover 工程维修费
engineering undertaking 工程业务
engineering unit 工程单元;工程单位
engineering unit conversion factor 工程单位转换系数
engineering unit system 工程单位制
engineering valid period 工程有效期限
engineering van 工程车
engineering verification 技术鉴定
engineering vessel 工程船
engineering viewpoint 工程观点
engineering well 工程井
engineering work 机械制造厂;工程建筑
engineering worker 技工
engineering workload 工程量
engineering works 工程设施;工程建筑物
engineering workshop wastes 机械工厂废水
engineering work station 工程设计工作站;设计工作站
engineer-in-training 助理工程师;见习工程师;实习工程师
engineer of standard 技术标准工程师
engineer of tracks 轨道工程师
engineer of way 养路工程师
engineer on duty 值班轮机员
engineer's alarm panel 机舱警报盘
engineers and technicians 工程技术人员
engineer's approval 工程师认可证明
engineer's brake valve 司机制动阀
engineer's brake valve spindle 司机制动阀杆
engineer's chain 工程师用链;工程(测)链;测链(长66英尺,1英尺=0.3048米)
engineer's cross pein hammer 扁尾锤
engineer's hammer 钳工锤;工程锤
engineer's indicator 机车信号机
Engineers Joint Council 工程师联合理事会
engineer's kit 工具包
engineer's level 水准仪;活镜水准仪;工程(师用)水准仪
engineer's logbook 机舱日志
engineer's pliers 万能钳(子);通用钳
Engineers Registration Board 工程师注册局;工

程师登记局
engineer's representative 工程师代表
engineer's scale 工程用比例尺
engineer's site office 工程师的工地办公室
engineer's square 丁字把尺
engineer's store rooms and workshops 机舱储[贮]物间和工作间
engineer's theodolite 工程经纬仪
engineer stores depot 工程兵存储[贮]仓库
engineer's transit 工程经纬仪
engineer superintendent 总轮机长
engineer surveyor 轮机验船师
engineer testing 工艺试验
engine exhaust(gas) 发动机排气侧
engine exhaust manifold 发动机排气歧管
engine exhaust pipe 发动机排气管
engine failure 机损；发动机故障
engine failure sensing and shutdown system 发动机故障显示与停车装置
engine fan pulley 风扇主动皮带轮
engine feed pump 发动机供料泵
engine filter 发动机滤清器
engine fire 发动机着火
engine fire-extinguisher push-bottom 发动机灭火按钮
engine fire handle 发动机灭火系统手柄
engine fire panel 发动机消防系统控制板
engine fire-protection 发动机防火钢
engine fire selector switch 发动机灭火系统选择门
engine fire shut-off cock 发动机灭火开关
engine firewall 隔火墙；发电机防火壁
engine fire warning light 发电机火警信号灯
engine fire warning system 发电机火警系统
engine firing order 发动机发火次序
engine fitter 机器装配工；发动机装配工
engine fitting 机械属具
engine fitting-out shop 轮机车间
engine fluid 发动机防冻液
engine for boat 船舶引擎
engine fork lift 内燃叉车起重机
engine fouling 发动机污脏
engine frame 发动机机架
engine-frame mounting 发动机支撑框架
engine friction 发动机摩擦力
engine front 发动机前面面积
engine front support 发动机前支架
engine front support bracket 发动机前支架
engine fuel 机车燃料；发动机燃料
engine fuel additive 发动机燃料添加剂
engine functioning 发动机工作
engine gasket 发动机密封垫
engine ga(u)ge 发动机指示器
engine ga(u)ge unit 发动机仪表组
engine gearbox 发动机齿轮箱
engine gearbox unit 发动机齿轮箱组
engine-generator 引擎发电机
engine governed speed 发动机限速；发动机额定转速；发动机调速
engine governor 发动机调节器；引擎调速器；发动机调速器
engine governor testing stand 柴油机调速器试验台
engine grease 发动机润滑剂
engine ground strap 发动机接地片
engine guard 发动机护板
engine gudgeon pin 发动机活塞销
engine gum 发动机胶状沉积；发动机工作中形成的胶状物质
engine hand piece 发动机手柄
engine hands 机舱人员
engine hatch 发动机舱口
engine hatch way 机舱(舱)口
engine health monitoring 发动机安全检查；发动机安全检查
engine heater 发动机保温箱
engine heat indicator 发动机温度计
engine heating rate 发动机发热速率
engine high temperature alarm switch 发动机高温告警开关
engine hoist 发动机起重机
engine hoisting sling 发动机吊索
engine hood 发动机盖；发动机(护)罩
engine hood panel 发动机盖板
engine hood upper side panel 发动机盖上侧板

engine hour meter 发动机(运转)小时计；发动机计时器
engine house 机器房；机车库【铁】；发动机房
engine house lead track 机车库引出线
engine housing 发动机房
engine idling speed adjusting screw 发动机怠速调整螺丝
engine indicator 示功器
engine in frame aligement device 发动机在车架上定中心设备
engine inlet ram 发动机进气口冲压
engine in situ 原位发动机；底盘发动机
engine installation 动力装置
engine intake 发动机进气口
engine jacket 发动机套
engine jaw 发动机曲轴爪
engine jet pipe 发动机喷管
engine keelson 承受发动机的内龙骨
engine kill 发动机熄火
engine-kilometer 机车·公里
engine knock 发动机爆震；发动机爆声；发动机爆燃；敲缸
engine knocking 发动机敲缸；发动机爆震
engine lacquer 发动机用漆；发动机漆膜沉积
engine lathe 普通车床
engine licence 主机执照
engine life 发动机寿命
engine lifting bracket 发动机吊架
engine lifting fixture 发动机吊架
engine lifting hook 发动机吊钩
engine line 机车(走行)线
engine load 发动机荷载
engine location 发动机位置
engine logbook 机舱日志
engine long stroke 长冲程发动机
engine louver 发动机放气孔
engine lubricating oil 发动机润滑油
engine lubrication system 发动机润滑系统
engine lug 发动机凸耳
engine lugging 发动机振动
engine maintenance area 发动机保养范围
engine maintenance man 机舱维修工人
engine man 司机
engineman's automatic brake valve 司机自动制动阀
engineman's brake valve 司机制动阀
engineman's indicator 机车信号机
engine map(ping) 发动机特性测定；发动机台试
engine mechanic 发动机技术员
engine misfiring 发动机不发火
engine misses 发动机熄火
engine mission 发动机变速箱
engine mission ga(u)ge 发动机变速箱转数表
engine mission meter 发动机变速箱转数计
engine moment 发动机力矩
engine monitor 发动机监视器；发动机监测器
engine mount 发动机支架
engine mounted longitudinally 纵向布置发动机
engine mounted transversally 横向布置发动机
engine mounting 发动机装置；发动机架；发动机固定方式
engine mounting bracket 发动机安装支架
engine mounting construction 发动机机架结构
engine mounting ring 发动机环架
engine mounting structure 发动机机架结构
engine mud sill 发动机地梁
engine muffler 引擎消音器；引擎消声器
engine nacelle 发动机(短)舱
engine nameplate 发动机标牌
engine noise 发动机噪声
engine number 发动机号数
engine off 发动机停车
engine oil 机油；机器(润滑)油；发动机(润滑)油
engine oil capacity 发动机油容量
engine oil conditioner 机油脱水装置
engine oil cooler 机油冷却器
engine oil dilution system 发动机润滑油稀释系统
engine oil filling cap 发动机加油口盖
engine oil filter 发动机润滑油滤清器
engine oil gallery 发动机润滑油路
engine oil heater 机车燃油加热器
engine oil level indicator 发动机油面高度指示器
engine oil pan 发动机油盘
engine oil pressure ga(u)ge 机油压力计；(发动机)机油压力表

engine oil sludge 发动机油渣
engine oil sump 发动机油槽
engine oil supply passage 发动机输油路
engine oil system 发动机滑油系统
engine oil tank 发动机油箱
engine-on landing (飞机)动力降落
engine opening 机舱(舱)口
engine operating sequence 发动机试验操作序列
engine operating temperature 发动机操作温度
engine operation command 发动机操作指令
engine order 车令
engine out emission 发动机排出物
engine output 发动机输出量；发动机有效功率；发动机输出功率
engine overhaul 发动机修理
engine overheat 发动机过热
engine overspeed switch 发动机超速开关
engine pack 发动机组件
engine pan 发动机底盘
engine panel 发动机配电板
engine parameter 发动机参数
engine parts 发动机零件
engine pedestal 发动机台架
engine performance 发动机性能；发动机特性
engine performance chart 发动机性能图表
engine performance curve 发动机特性曲线
engine performance test 发动机性能试验
engine pickling 发动机封藏
engine pickups 发动机支台
engine piston 发动机活塞
engine pit 机车修理坑；修(机)车坑；飞轮槽；发动机检修坑；曲柄箱油槽；汽车检修坑
engine plane 斜坡道；斜吊斗
engine plant 发动机车队
engine platform 油发动机台
engine plow 机动犁
engine power 发动机马力；发动机功率
engine-powered 用发动机作动力的
engine power rating 发动机功率定额；发动机额定功率
engine preheater 发动机预热器
engine pressure 发动机压力；发动机爆燃压力
engine preventive maintenance 发动机预防性保养
engine prewarming 发动机预热
engine primer 引擎起动油泵；发动机起动注油器；发动机起动器
engine priming 发动机起动注油
engine priming fuel 发动机起动燃料
engine programmer 发动机程序设计
engine propeller unit 发动机螺旋桨组
engine pulley 发动机皮带轮
engine quit 发动机突然停转
engine racing 发动机的高速运转；发动机超速转
engine radiation 发动机热辐射
engine radiator 引擎散热器；发动机散热器
engine radius rod bracket 发动机半径杆托架
engine ram 发动机进气口冲压
engine rear support 发动机后支架
engine reconditioning 发动机修理
engine regulator 机器调速器
engine release line 机车出入库线
engine removal 发动机卸下
engine renovation 发动机翻新；发动机大修
engine repair section 发动机修理组
engine repair stand 发动机修理台
engine resistance 机器阻力
engine restart 发动机重新启动
engine revolution 发动机转数
engine revolution counter 主机转数记录器；发动机转数计；发动机转数表
engine revolutions per minute 发动机每分钟转数
engine rig test 发动机试验台试验
engine road 机车走行线
engine room 机器房；机舱；发动机房
engine room annunciator 机舱传令钟
engine room arrangement 机舱布置
engine room auxiliaries 机舱辅机
engine room bulkhead 机舱舱壁
engine room canvas cowl 机舱帆布通风筒
engine room casing 机舱棚；机舱(口)围壁
engine room compartmentation 发电机房的分隔
engine room control center 机舱控制中心
engine room double bottom 机舱双层底

engine room flat 机舱平台
engine room hand 机舱加油
engine room hatch 机舱口
engine room log 机舱日志
engine room platform 机舱平台
engine room remote control 机舱遥控
engine room skylight 机舱天窗
engine room space 机舱容积
engine room space deduction 机舱容积减除额
engine room ventilation 机舱通风机间
engine rough 机舱转动不稳
engine roughness 发动机工作不平稳
engine run 发动机试车
engine runner 司机
engine running track 机车行驶线
engine run-round track 机回线;机车迁回线
engine runs hot 发动机轴承发热
enginery 机械类;机能;制造或使用机械的技术
engine saddle 发动机座板
engine salvage machine 发动机补修机
engine-scope 发动机电子显示器
engine scuffing 发动机磨损
engine seat(ing) 机座;发动机座
engine section 机器部件;发动机部分
engine sequence panel 发动机次序操纵台
engine service track 机车整备线
engine serving track 机车整备线
engine set protecting relay 机组保护继电器
engine shaft 发动机轴
engine shed 机车棚;机器棚;机车库【铁】;发动机棚;发动机房
engine shield 机器罩
engine shop 发动(机)制造厂
engine shorting-out 发动机停火
engine shroud 发动机挡板
engine shutdown device 发动机停机装置
engine side plates 发动机侧板
engine-sized paper 机内施胶纸
engine skid 橇装引擎
engine sleepers 机座
engine sludge 发动机淤渣
engine snubber 发动机减振器
engine solar oil 粗柴油;柴油机燃料;发动机瓦斯油;发动机太阳油
engine speed 发动机转数;发动机转速
engine speed indicator 机车速度指示器;发动机转速表
engine speed recorder 发动机转速表
engine speed reducer 发动机减速装置;发动机减速器
engine speed regulator 发动机调速器
engine spigot 发动机塞
engine splutter 发动机敲击
engine stabilizer arm 发动机安全臂
engine stabilizer bar 发动机稳定器杆
engine stabilizer bracket 发动机稳定架
engine stabilizer pad 发动机安定垫
engine stabilizer spring 发动机稳定器弹簧
engine stalling 发动机自动停车
engine stand 发动机座;发动机(支)架
engine start 发动机起动
engine starter 发动机起动机
engine starting contactor 发动机起动接触器
engine starting fuel 发动机起动燃料
engine starting gear 发动机起动装置
engine starting handle 发动机起动手柄
engine starting mechanism 发动机起动机构
engine starting system 发动机起动系统
engine start switch 发动机起动开关
engine sticking 发动机活塞卡住
engine stop 发动机停止
engine stop valve 停车阀
engine stores 轮机部储[贮]藏品
engine stripping 发动机拆卸
engine stroke 发动机活塞冲程
engine strutting 发动机支撑
engine sump 发动机曲轴箱;发动机底壳
engine sump well 发动机滑油沉淀器
engine supercharger 发动机增压器;发动机增压机
engine support 发动机支架;发动机底座
engine support bar 发动机支杆
engine support cushion 发动机支架垫
engine supporting arm 发动机曲轴箱支座臂
engine supporting plate 发动机座板

engine support insulator 发动机支架软座
engine support oil shield 发动机支架防油罩
engine support spring 发动机支架弹簧
engine sweeper 路面清扫机;机动路面清扫机
engine system 发动机系统
engine tach(e)ometer 发动机转速计;发动机转速表
engine telegraph 机舱传令钟;车钟
engine temperature ga(u)ge 发动机温度计
engine temperature indicator 发动机温度计
engine temperature sensing unit 发动机温度传感器
engine terminal capacity 机务段通过能力
engine test 主机测试;发动机试验
engine test bed 发动机试验台
engine test bench characteristic 发动机试验台特性
engine test cell 发动机试(验)车间
engine tester 发动机试验器
engine test facility 发动机试验设备
engine testing equipment 发动机试验设备
engine testing of lube oil 润滑油发动机
engine testing room 发动机试验室
engine test method 发动机试验方法
engine test plant 发动机试验装置
engine test stand 发动机试验台
engine test tool kit 发动机测试工具箱
engine thermostat housing drain valve 发动机恒温器壳放泄阀
engine three point suspension 发动机三点固定
engine throttle 发动机油门;发动机节流阀
engine throttle bellcrank 发动机节气门操纵臂
engine throttled back 发动机节气门关闭
engine thrust measuring 发动机推力测定
engine timing case 发动机定时齿轮壳
engine timing pointer 发动机定时指针
engine tool 机修工具
engine tool kit 发动机成套工具
engine torque 发动机转矩;发动机扭矩
engine-transmission unit variable coupling 发动机传动装置的可变连接
engine trends 发动机发展趋向
engine trouble 发动机故障
engine truck side bearing 机车转向架旁轴承
engine truck swing frame 机车转向架摆架
engine truck swing link 机车转向架摆杆
engine tuneup 发动机调整
engine turnable 机车转车台
engine-turned ware 车旋陶瓷制品
engine turning at peak revolution 最大转数旋转的发动机
engine turning lathe 陶瓷坯刻花车床
engine turntable 转车台
engine type 发动机型号
engine-type plate 发动机标牌
engine underpan 发动机底盘
engine under seat 座下发动机
engine unit 发动机
engine unit sender 发动机组传感器
engine unloading pressure regulator 发动机卸荷调压器
engine upper rubber mounting assy 发动机上软垫总成
engine vacuum checking ga(u)ge 发动机真空测试计
engine valve 发动机阀
engine varnish 发动机清漆
engine vibration 发动机振动
engine volume coefficiency 发动机容积效率
engine volume efficiency 发动机容积效率
engine warm-up 发动机加温
engine wash 发动机清洗
engine water inlet 发动机进水口
engine water outlet hose connection 发动机出水软管接头
engine wear 发动机磨损
engine weight per horsepower 每马力发动机重(量)(1马力=735.49875W)
engine winterization system 发动机防冻系统
engine with opposed cranks 对置曲柄发动机
engine with opposing cylinders 对置汽缸发动机
engine with outside guide 十字头发动机
engine with supercharger 增压发动机
engine wobble 发动机摆动
engine working shop 轮机车间

engine works 发动机厂
engine workshop 发动机车间
englacial 冰川内的
englacial debris 冰内碎屑物;冰川碎屑物
englacial drainage 冰(川)内水系
englacial drift 冰川内碛
englacial environment 冰内环境
englacial lake 冰内湖
englacial load 冰川内碛
englacial melt(ing) 冰川内(部)融(解)
englacial moraine 内冰碛;冰川内碛
englacial river 冰川内(部)河流
englacial stream 冰内河(流);冰川内(部)河流
englacial till 内冰碛;冰内冰碛
englacial zone 冰内带
England Epsilon structural system 英格兰山字形构造体系
Engler degree 恩格勒黏(滞)度;恩格勒度
Engler's curve 恩格勒蒸馏曲线
Engler's degree 恩格勒黏(滞)度;恩格勒度
Engler's distillation flask 恩格勒蒸馏烧瓶
Engler's distillation test 恩格勒蒸馏试验
Engler's flask 恩格勒(烧)瓶
Engler's number 恩格勒黏度值;恩格勒黏度数
Engler's orifice visco(si)meter 恩格勒锐孔黏度计
Engler's seconds 恩格勒(黏度)秒数
Engler's specific viscosity 恩格勒比黏度
Engler's unit 恩格勒黏度单位
Engler's visco(si)meter 恩格勒黏度计
Engler's viscosity 恩格勒黏(滞)度
Engler-type visco(si)meter 恩格勒黏度计
Engler viscosity 恩格勒黏(滞)度
English antique lace 英式仿古机制花边
English architecture 英(国)式建筑;英国建筑
English basement 英式地下室;英国式房屋底层;半地下室
English bond 英式砌法;英(国)式砌砖法(顶砖层与顺砖层交错);英国式砌合;英国十字缝砌砖法(即荷兰式砌法);丁顺隔皮砌法
English brown oak 英国褐栎属
English calibration 英制计量单位
English candle 英制烛光;英国烛光
English-Chinese garden (十世纪英国流行的)中英混合式庭园
English Cotswold architecture 英国科茨寓德式建筑
English count 英制支数
English cross bond 英式十字缝砌法;英式(十字)砌法;英国式砌合;荷兰式砌合;荷兰式砌法
English crystal 铅玻璃
English degree 英制硬度
English dinas 英国砂石
English edging 英式滚边
English Elizabethan 英国伊丽莎白式建筑
English engineering unit 英制工程单位
English fathom 英寸(1英寸=2.54厘米)
English garden-wall bond 英国式园墙砌砖法;英国园墙式砌合法;三顺一丁砌墙法
English ga(u)ge 英制机号
English Gothic architecture 英国哥特式建筑
English half-timbered architecture 英国半木式建筑
English hardness (水的)英制硬度
English heating test 润滑脂加热试验
englishite 水磷铝钾石;水磷钙钾石
English ivy 常春藤
English knot net making machine 英式有结构网机
English landscape style garden 英国风景式庭园
English lever 尖齿式擒纵叉;英式擒纵叉
English melting point 英国法蜡熔点
English metal 英国合金
English method of timbering 英(国)式隧道支撑法
English mile 英里
English oak 英国栎
English pink 荷兰粉红
English red 英国红
English reel 英式卷纸机
English Renaissance architecture 英国文艺复兴建筑
English rib 英式罗纹组织
English Romanesque 英国罗马式建筑艺术;英国罗马式(建筑)
English roof(ing) tile 英(国)式屋面瓦
English roof truss 英(国)式屋架
English screw 英制螺纹

English screw pitch ga(u)ge 英制螺距规
English sea mile 英制海里(1 英制海里=6080 英尺,折合 1853.24 米)
English sennit 索辫
English shingle 英国式屋面瓦
English spanner (英式)活动扳手
English standard thread 英制标准螺纹
English style garden 英国式园林
English system 英制
English system of timbering 木支撑的英式体系;木支撑的英式方法
English system of wet twisting 英式湿捻
English thread 英国标准螺纹
English tile 英式屋面瓦
English time signals 英式无线电信号
English ton 英吨
English translation 英译
English translucent china 英国半透明瓷
English truss 英(国)式桁架(三角桁架)
English tubing 英式敷设管道方法
English Tudor architecture 英国都铎式建筑
English twist multiplier 英制捻系数
English-type tubbing 英式预制弧形块井壁
English unit 英制单位
English velvet 英国丝绒
English vermilion 英国银朱;亮朱红
English version 英译本
English walnut 胡桃
English white 英国白;白垩
English Wrenaissance 英国雷恩文艺复兴时期
English yellow 英国黄
engobe 釉底料
engobe coating 化妆土;上底釉
engobing 上釉底料
engorgement 拥塞;充满;装料口;舱口
engraft 移入(物)
engrafted river 接枝河
engrafted stream 接枝河
engrail 镶双锯齿形花边
engrailed (刻成)齿纹的;波形的
engrailment 刻成齿纹状
engrain 深染;染成木纹色
engrain lining paper 深染的衬里纸;染印有木纹的衬里纸;木纹色衬
engrain wallpaper 深染的墙纸;染印有木纹的墙纸;木纹色墙纸
engrain wallpaper coat 深染的墙纸涂层
engrave 雕刻(制版)
engraved block 木刻印版
engraved coating 槽辊涂装
engraved copper plate 雕刻凹铜板
engraved decoration 装饰刻花
engraved glass 刻花玻璃;雕刻玻璃;雕花玻璃
engraved glass screen 雕刻网屏
engraved halftone screen 雕刻网屏
engraved letters 刻字
engraved-on-stem thermometer 刻标棒状温度计
engraved outline 刻绘轮廓线
engraved point 铭记点
engraved porcelain 雕瓷
engraved roll 刻花辊;雕刻辊
engraved roll coater 凹辊涂布机
engraved roll coating 凹辊涂布
engraved roller 螺旋辊子;凹辊
engraved sign 雕刻标志
engraver 刻图仪;刻点仪;刻版机;雕刻师
engraver beetle 雕刻小蠹
Engravers brass 恩格雷弗斯含铅黄铜;雕刻黄铜
engraver's needle 雕刻针
engraver subdivider 点圆刻图仪
engraving 镂刻;雕版;刻模;磨刻;雕刻艺术品;雕刻术;版画
engraving cutter 刻模铣刀
engraving diamond 刻图钻石
engraving disc revolver 轮形旋刻刀
engraving head 刻图头
engraving ink 雕刻油墨
engraving in relief 浮雕压花
engraving knife 雕刻刀
engraving machine 刻模机;刻版机;雕纹机;雕刻机
engraving needle 刻图针;雕刻(钢)针
engraving plate 雕刻板
engraving process 雕刻方法
engraving sheet 刻图版

engraving stylus 雕刻刀
engraving tool 刻刀
engrossing 绘制大型花饰字
engross the market 垄断市场
Engset distribution 恩格塞特分布
engulf 席卷;吞没
engysseismology 近源地震学;工程地震学
enhance 增进;增强
enhanced accuracy 提高准确度
enhanced activated sludge process 强化活性污泥法
enhanced activated sludge system 强化活性污泥法
enhanced ammonia nitrogen removal 强化除氨氮
enhanced beam 增强束
enhanced biodegradation 强化生物降解
enhanced biological dephosphorization 强化生物除磷
enhanced biological phosphorus removal 强化生物除磷
enhanced biological phosphorus removal biomass 强化生物除磷生物量
enhanced bio-treatment 强化生物处理
enhanced carrier demodulation 增强载波解调;加强载波解调
enhanced coagulation 强化混凝
enhanced coagulation-bio-ferric process 强化混凝生物铁法
enhanced contrast 增强反差
enhanced conventional water treatment process 强化常规水处理工艺
enhanced diffusion 反常扩散
enhanced emission 增强发射
enhanced-enhanced chemical vapo(u)r deposition 等离子强化化学汽相沉积
enhanced field 增强场
enhanced filtration 强化过滤
enhanced filtration efficiency 强化过滤效果
enhanced flocculation deposition 强化絮凝沉淀
enhanced gas injection 注气
enhanced group call 增强群呼【无】;强化群呼
enhanced image 增强图像
enhanced molecular ozone reaction method 强化分子臭氧反应法
enhanced oil recovery 提高采收率法采油
enhanced phosphorus removal 强化除磷
enhanced prop treatment 高浓度支撑剂压裂
enhanced rad 增强辐射
enhanced rapid gravity filtration 强化快速重力渗滤
enhanced recovery 强化开采
enhanced recovery factor technique 提高采收率的措施
enhanced recovery pump 强化采液泵
enhanced reductive degradation 强化还原降解
enhanced reflecting safety glazing material 增发射安全玻璃
enhanced seismic profile 加强的地震剖面
enhanced sensitivity 增敏
enhanced sensitivity manostat 增高灵敏度的冲压器
enhanced solubility 增溶溶解
enhanced spectral line 增强(光)谱线
enhanced spectrum 增强光谱
enhanced surface water treatment rule 强化地表水处理法规
enhanced swirling clarifier 强化水旋澄清池
enhanced tube 强化管
enhanced type 增强型
enhancement 增强;加强
enhancement action 增强作用
enhancement coefficient 增大系数
enhancement effect 增强效应;增感效应;提高效应
enhancement factor 增强因子
enhancement load 增强型负载
enhancement mode 增强(型模)式;增强型
enhancement of anomaly 异常强化
enhancement of monitoring 加强监测
enhancement type 增强型
enhancement type channel 增强型沟道
enhancement type curve 增强型曲线
enhancer 增强因子;增强ീ;增强剂;放大镜
enhancing toxic action 增毒作用
enhydrous 结晶含水;内部含有水分的;含水的

eniclobrate 恩尼贝特
enigmatite 三斜闪石
enjoyment of right 行使权利
Enka nylon 恩卡尼纶
Enkaswing 恩卡斯温聚氨基甲酸酯弹性纤维(荷兰)
Enkather fiber 恩卡特姆纤维
enkindle 点燃
enlarge 扩张;增大
enlarged-and-constricted section 交替放大和收缩的截面
enlarged base (桩工)扩底
enlarged-base pile 扩底桩;底部扩大桩
enlarged culvert inlet 流线型洞口
enlarged detail 大样;放大详图
enlarged discharging area 扩大的过水地面
enlarged drawing 放大图
enlarged image 放大(图)像
enlarged image with partial coherent light 部分相干光放大图像
enlarged intersection 加宽式交叉口;展宽式交叉(路口)
enlarged link 加大链环
enlarged model 放大模型
enlarged mosaic 放大镶嵌图
enlarged period 扩大时期
enlarged photo 放大相片
enlarged picture 放大相片
enlarged pile 扩底(脚)桩
enlarged preliminary design 扩大初步设计
enlarged print 放大相片
enlarged relay 放大继电器
enlarged reproduction 扩大再生产
enlarged sample 扩大(的)样本
enlarged scale 放大比例(尺)
enlarged section 扩大断面;放大截面;放大断面
enlarged side wall 边墙加厚
enlarged sintering zone 直径扩大的烧成带
enlarged sketch 放大图样
enlarged toe pile 大头桩;扩脚桩;扩底桩
enlarged unit line connection 扩大单元结线
enlarged view 放大图
enlarged viewing screen 放大的目视荧光屏
enlarged water-carrying cross-section 扩大的过水地面
enlargement 扩建;扩大;扩充;加宽;增大;增补;放大;膨大
enlargement diagram 放大复制图
enlargement factor 扩大因数;放大因素;放大因数;放大系数;放大倍数
enlargement in excavation 扩大开挖
enlargement in section 截面扩大
enlargement loss 扩张损失;扩展能量损失;扩大损失;放大损失
enlargement mode 增强型
enlargement of ring grooves 环槽扩大
enlargement print 放大相片
enlargement radiography 放大射线照相法
enlargement range 放大幅度
enlargement ratio 放大率;放大比
enlargement ratio of telescope 望远镜放大率
enlargement scale 放大比例(尺)
enlargement test 扩管试验
enlargement to a building 建筑的扩建部分
enlarge on 详其
enlarger 扩大机;光电倍增器;放像机;放大器;放大机
enlarger lamp 放大机灯泡
enlarger lens 放大机镜头
enlarger printer 放印机;放大印刷机
enlarger zone 扩大带
enlarging 外围切土
enlarging bit 扩眼钻头;扩孔钻
enlarging bulb 放大灯泡
enlarging camera 放样照相机
enlarging filter 放大滤光镜
enlarging hammer 扁锤
enlarging hole 扩孔
enlarging lens 放大镜
enlarging machine 放影机;放大机
enlarging meter 放大光度计
enlarging mo(u)lding 线脚的放大;放大线脚【建】
enlarging objective 放大物镜
enlarging of harbo(u)r 港口扩建(工程)

enlarging paper 放大纸
enlarging photometer 放大光度计
enlarging photopaper 放大相纸
enlarging plant 工厂扩建
enlarging rod 放大机竖杆
enlarging roller bit 扩大滚柱钻头
enlarging spring construction 扩泉工程
enlarging technique 放大技术
enlarging timer 放大曝光定时器
enlistment age 役龄
enlivening the markets 启动市场
Enlund method 快速测碳法
en masse elevator 密斗提升机；垂直链槽式提升机
en masse conveyer 埋刮链输送机；连续流动输送机；埋刮板(链槽式)输送机
ennead 九个一组
enneagon 九角形；九边形
enneahedron 九面体
enneastyle 九柱式
enneastyle building 九柱式建筑
enneastylos 九柱式建筑
enneri 干河谷
ennuple 标形
Enochkin series 埃诺奇肯统【地】
enocuproine 新亚铜试剂
enol 烯醇
enolate 烯醇化物
enol form 烯醇型；烯醇式
enolic ester type 烯醇型脂
enolizability 烯醇化程度
enolization 烯醇化(作用)
enol phosphate 烯醇磷酸酯
enol pyruvic acid 烯醇式丙酮酸
Enor motor 埃诺罗式叶片液压马达
enormous 巨大的
enormous quantities of sediment 巨量泥沙
enoscope 视车镜；折光镜
enough sunshine duration 日照时间充足
enphytotic 地区性植物发病
enplaning road 上飞机道路
enquiry 询价；查询；探价
enquiry agent 征询机构
enquiry character 询问字符；查询字符
enquiry character sequence 询问符次序
enquiry circuit 查询电路
enquiry data 调查数据
enquiry desk 问讯处柜台
enquiry signal 询问信号
enquiry station 询问台
enrich 增浓；加工装饰；富化
enriched 装修过的
enriched blast 富氧鼓风
enriched boron trifluoride neutron detector 富化的三氟化硼中子探测器
enriched buckling 浓缩燃料堆芯曲率
enriched chamber 浓缩室
enriched compound 浓缩化合物
enriched feed 浓集产物供料
enriched fraction 浓缩馏分
enriched fuel 浓缩燃料
enriched-fuel reactor 浓缩燃料堆
enriched gas 浓缩汽油；富化气
enriched gas drive 富化气区
enriched gas injection 加浓瓦斯喷射
enriched gas oil 浓缩粗柴油
enriched lake 富营养湖(泊)
enriched layer 浓缩层
enriched material 浓缩物(质)；浓缩材料
enriched medium 增菌培养基
enriched mixture 浓缩混合物；富油混合气
enriched nuclear fuel 浓缩核燃料
enriched oil 浓缩油；增加了轻馏分的原油；富化油
enriched oxidation zone 氧化富集带；富集氧化带；富集法
enriched oxygen aeration 富氧曝气法
enriched pulverized refuse 浓缩粉状垃圾
enriched reactant gas 浓集反应气体
enriched reactor 浓缩反应堆
enriched seawater 浓缩海水
enriched target 浓缩靶
enriched uranium 浓缩铀
enriched uranium aqueous homogeneous reactor 浓缩铀水均匀反应堆
enriched uranium booster 浓缩铀点燃棒
enriched uranium carbide 浓缩碳化铀
enriched uranium carbide particle 浓缩碳化铀粒子
enriched uranium fission chamber 浓缩铀裂变室
enriched uranium graphite moderated reactor 浓缩铀石墨慢化反应堆
enriched uranium heavy water moderated reactor 浓缩铀重水慢化反应堆
enriched uranium light water moderated reactor 浓缩铀轻水慢化反应堆
enriched uranium oxide 浓缩氧化铀
enriched uranium reactor 浓缩铀(反应)堆
enriched uranium solid homogeneous reactor 浓缩铀固体均匀反应堆
enriched uranium swimming pool reactor 浓缩铀水池式反应堆
enriched water 加浓水
enriching 选集
enriching agent 富化剂
enriching column 富集柱
enriching device 多加燃油器
enriching needle 加浓针
enriching recovery 浓缩回收
enriching section (提浓)精馏段；(塔中)浓缩段
enrichment 加浓；浓缩；增富法；富有；富集(作用)；富化
enrichment coefficient 浓缩系数
enrichment control 浓缩度控制
enrichment culture 加富培养；富集培养
enrichment facility 浓缩装置
enrichment factor 浓缩因子；浓缩系数；浓缩率；富集系数；富化系数
enrichment horizon 富积层
enrichment layer 浓缩层
enrichment medium 加富培养基
enrichment meter 浓缩度计
enrichment mode 浓缩模
enrichment of elements 元素富集作用
enrichment of lakes 湖泊富集
enrichment of mixture 混合物浓缩
enrichment of soil 土壤加肥
enrichment plant 浓缩工厂；选矿车间
enrichment process 浓缩法
enrichment ratio 富集比
enrichment-system 加浓装置
enrichment zone 富集带
enrich the top-soil layer 使表土层肥沃
enrico femit atomic power plant 恩里科费米原子能电站
enrico femit fast breeder reactor 恩里科费米快中子增殖反应堆
enrockment (基底)填石；基底抛石；堆石；抛石(体)；抛石基床
enrol(l) 登录
enrollment 记录处；注册
en route 在途中
enroute chart 航线图；空中航道导航设备图；无线电导航图
enroute surveillance radar 航路监视雷达
enroute turning area 航途转弯区
ensemble 集合机；集合化；信号群；系综；总效果
ensemble aggregate 集集合
ensemble average (数学期望的)集平均值；集平均；系综平均(值)；总体(平)均值；总体平均(数)；总集平均；统计平均值
ensemble correlation function 总体相关函数
ensemble mean 总体平均
ensemble of communication 信息集合
ensemble of electrons 电子集
ensemble spectral density 总体谱密度
ensialic 硅铝层上的
ensialic basin 硅铝壳盆地；硅铝层盆地
ensialic geosyncline 内硅铝地槽【地】；硅铝壳(上)地槽；硅铝层地槽
ensign 舰旗；国家旗帜；标志；旗
ensign halyard 船尾旗索
ensign staff 船尾旗杆
ensilage 青储[贮]饲料
ensilage crop 青储[贮]作物；青饲料作物
ensilage dump blower 强力鼓风机
ensilage technique 青储[贮]饲料技术
ensimatic 硅镁层上的
ensimatic geosyncline 内硅镁地槽；硅镁壳上地槽；硅镁层地槽
Enslin apparatus 恩斯林装置；恩斯林黏土吸水量装置
ensonified area 水声仪器监听海区
ensonified zone 水下声音传播区
enstatite 顽(火)辉石
enstatite chondrite 顽火辉石球粒陨石
enstatite peridotite 顽火辉石橄榄岩
enstatite picrite 顽火辉石苦橄岩
En steel 工程用低碳钢(英国标准)
enstenite 斜方辉石(类)
ensuing earthquake 续发地震
ensurance period 保用期
ensuring moment 引起的弯矩
ensuring year 次年；翌年
entablature 檐部；柱上楣构；柱顶盘；支柱层；古典柱式顶部；发动机机架；墙上楣构；上柱列
entablement 柱上楣构；柱顶线盘；像台；像座
entad 向内
entagenic acid 过岗龙酸
entail 需要；限定继承权；细雕；线脚装饰处理
entamoeba 内阿米巴
entangled chain 缠结链
entangled molecule 缠结分子
entanglement 缠结；铁丝网
entanglement cross link 缠结交联
entanglement network 缠结网络；缠结(分子)网
entasis 柱收分；收分(指柱)；圆柱收分线；柱中微凸线【建】；(西方古典式柱身的)收分曲线微凸线；[复]entases
entasis of a column 柱身的收分
entasis of column 柱收分
entasis reverse 凸肚线型板；圆柱中微凸线样板
entasis treatment 卷杀
entended stream 延长河
enter 进入；送入；参加
enter a bid 投标
enter a caveat 申请停办手续
enter a judgement 作出判断
enter an item in an account 入账
enter an item of expenditures in the accounts 出账
enter bid 投标
enterclose 通道；隔墙；穿堂；间壁；过道；格廊
enter contract 缔结契约
enter duty free 免税进口
entered value 报关价值
enter guide 导口
enteric bacilli 肠杆菌
enteric fever 伤寒病
enter in 进口申报；向海关申报进口
entering and leaving signals 进出港信号；驶入和驶出信号
entering angle 咬入角
entering book 登账
entering catch 进入锁环
entering edge 前缘
entering end 进口
entering gas quantity 进入气体量
entering into occupied track 进入占用线
entering into section 进入区间
entering into unexpected track 进入异线
entering into wrong track 进入异线
entering/leaving lock in a rectilinear way 直线进/出闸
entering lock in a curvilinear way 曲线进闸
entering momentum 输入动量
entering of flood 洪水进入
entering of tidal wave 海浪进入；潮浪进入；潮波涌入
entering port 入港
entering speed 进口速度
entering surface 入射面
entering temperature 进口温度
entering time 进闸时间
entering variable 进入变量；插入变量
entering warehouse 入库(存货)
enter into 参加；参与；签订
enter into a contract with 与……订(立)合同
enter into contract 订合同
enter into force (开始)生效
enter into negotiations 参加谈判
enter into the account book 入账
enter inward 进口申请；进口报关
enter key 进入键；输入键
enter key request 输入键请求

enter mark 进口轧痕
enter mode 进入形式
enter new order 接受新订单
enterococci 肠道球菌
enterokinetic 蠕动的
enterolithic 内变形的【地】
enterolithic structure 盘肠构造
enter out 向海关申报出口;报关出口
enter outwards 出口申请;出口报关
enterphone 看门电话
enter point 输入点
enter port 进港
enterprise 企业(单位);事业
enterprise accounting 企业会计
enterprise administration expenses 企业管理费(用)
enterprise bankruptcy 企业破产
enterprise circle 企业界
enterprise conglomerate 企业集团
enterprise cost 企业成本
enterprise debt 企业债务
enterprise developing fund 企业发展基金
enterprise expansion fund 企业发展基金
enterprise expenses 事业支出
enterprise failure 企业倒闭
enterprise financial benefit 企业财务效益
enterprise fund 企业基金
enterprise funds system 企业基金制
enterprise group 企业集团
enterprise income tax 企业所得税
enterprise incurring long run losses 长期亏损的企业
enterprise investment 企业投资
enterprise law 企业法
enterprise management 企业管理
enterprise management of coal industry 煤炭企业管理
enterprise management of dye-stuffs industry 染料企业管理
enterprise management of paints industry 涂料企业管理
enterprise management system 企业管理体系
enterprise objective 企业任务;企业目的
enterprise owned by a sole investor 独资企业
enterprise profit drawing 企业利润提成;企业利润提取
enterprise profit partly reserved 企业利润留成
enterpriser 企业家
enterprise relocation expenses 企业搬迁费
enterprise repayable ability analysis 企业偿债能力分析
enterprise reserve fund 企业储备基金
enterprise's ability of self-financing 企业自筹资金能力
enterprises and establishments 企事业
enterprise's economic appraisal of a project 项目企业经济评价
enterprise self-raised fund 自筹资金企业
enterprises in townships(and towns) 乡镇企业
enterprises owned by foreign capitalists 外资产业
enterprise's ownership resources 企业自有资金
enterprise standard 企业标准
enterprise station 企业站
enterprises using advanced technology 技术先进型企业
enterprise termination and liquidation 企业结业清算
enterprise with advance technology 先进技术企业
enterprise with exportable products 产品出口企业
enterprise zone 企业区;兴业区
enterprising 有事业心的
enter selection 进入选择项
enter speed at retarder 缓行器进口速度
entertain claim 接受索赔
entertaining expenses 交际费;招待费
entertainment 招待会;娱乐场;表演会;交际
entertainment center 娱乐中心
entertainment expenses 娱乐费;交际(接待)费用;招待费
entertainment fee 招待费
entertainment on the expense account 由企业开支的招待费
entertain offer 考虑发盘;接受报盘
entertain order 接受订货

enter the country 入境
enter up an accounts 入账
entexis 注入混合作用
enthalpimetry 峰值热焓分析法
enthalpy 热焓;热含量
enthalpy balance 焓平衡
enthalpy-concentration diagram 焓浓图;热焓浓度图
enthalpy-controlled flow 受控焓流
enthalpy control system 焓值控制系统
enthalpy diagram 焓图表
enthalpy difference 焓差
enthalpy-entropy chart 焓湿图
enthalpy-entropy diagram 焓—熵图
enthalpy gain 热焓增加
enthalpy-humidity diagram 湿空气焓湿图
enthalpy-humidity difference ratio 焓湿差比
enthalpy-moisture chart 焓湿图
enthalpy-moisture ratio 焓湿比
enthalpy of adsorption 吸附焓
enthalpy of fuel 燃料热含量;燃料比热
enthalpy of moist air 湿空气量
enthalpy of reaction 反应热焓
enthalpy of transition 转变热焓
enthalpy potential 焓电势;热焓电位;焓差
enthalpy potential method 焓差法
enthalpy-temperature diagram 焓温图
enthalpy titration 温度滴定
enthrakometer 超高频功率(测量)计
enthusiasm 积极性
entiometer 微电位计
entire 全体;全部(的);完全的
entire agreement clause 完整条款
entire allocation 总体布局
entire aperture 全壳口
entire area 全部面积
entire array 整数组
entire brake time 全制动时间
entire car 整车
entire contract 不可分合同
entire distribution 总体布局
entire ecologic(al) system 全生态系统
entire function 整函数
entire injection interval 总注入层段
entire life 总寿命
entirelt ventilation 全面通风
entirely accurate 完全准确
entirely in spheric(al) portion 全部在球体部分
entirely oil-wet 完全亲油
entirely shut 完全闭塞
entire model 整体模型
entire pay calculated by piecework 全计件工资
entire rational function 整有理函数
entire release time 全缓解时间
entire social income tax 全额社会所得税
entire split system 全分流系统
entire tenancy 全权拥有;完整租借
entire thermal resistance 总热阻
entirety 整体(性);全部
entire wavelet 完整子波
entisol 新成土
entitle 有权;赋予权利
entitlement 权利
entitlement city 受资助城市
entitlement funding 按指定用途提供资金
entitlement to housing 住房申请权
entity 机构;法人;本体;实物;实体
entity relationship 实体关系
entity relationship analysis 实体关系分析
entity relationship diagram 实体关系图
entity relationship model 实体关系模型
entity set model 实体集合模型
entocodon 内钟
entodorsum 内背板
ento-ectad 由内向外
entogenous 内原(性)的
entognathous type 内口式
entombment 埋葬;埋没
entomogenous 虫上寄生的
entomogenous fungi 虫寄生真菌
Entomological Society of America 美国昆虫协会
entomophilous flower 虫媒花
entomotoxicity 虫毒性
entoolitic 内呈鲕粒状的

entopic 正常位置的
entoplasm 内浆
entorganism 内寄生物
entosarc 内浆
entotergum 内背板
entourage 周围环境;周围的人;配景;四周
entourage of building 建筑配景
entrails 机内
entrain (悬浮)夹带;引入空气
entrained air (有意在混凝土或砂浆中产生的)气泡;漏入空气;夹杂空气;夹带(的)空气;混入空气(液压系统中混入的小气泡);含气;输入空气
entrained air content 漏气量
entrained air depth 掺气水深
entrained air indicator (在混凝土或砂浆中)引入空气量的指示器
entrained air void 夹进空气的孔隙
entrained drip 夹带液滴
entrained fluid 曳出流体
entrained oil 夹带(走)的油;带走的油
entrained particle gasification 气流床气化
entrained sand particle 夹带砂粒
entrained sediment 夹带泥沙;挟带的泥沙
entrained solid 夹带的固体
entrained steam 残留蒸汽
entrained stone sand mortar 混入石块的砂浆;混入砂石的灰浆
entrained water 附连水
entrainer 夹带物;夹带剂;共沸剂
entraining 吸入的
entraining agent 夹带剂
entraining method of foaming 泡沫引气法
entraining velocity 夹带速度
entrainment 流水攻砂;卷吸(作用);夹卷;夹带;挟带;诱导作用;掺气;雾沫
entrainment coefficient 携带系数
entrainment eliminator 除沫器
entrainment filter 夹带物过滤器;收集过滤器
entrainment force 夹带力
entrainment mechanism 携带机理
entrainment meter 含气量测定仪;引气量测定仪
entrainment of air 卷入空气;夹入空气;引气;渗入空气
entrainment of frequency 频率诱导
entrainment of sut 带走粉尘量
entrainment rate 夹带速率
entrainment ratio 含气率;诱导比;含量比
entrainment separation 雾沫分离器
entrainment separator 液滴分离器;雾沫分离器
entrainment trap 液滴分离装置;液滴捕集;雾沫分离器
entrainment velocity 携带速度
entrance (港池的)口门;进入;进流段;进流端(船首受水部分);进口;入射(港口的)入口;入境;入海口;入港(手续);起始
entrance and egress 进出口门
entrance and exit 进出口门
entrance angle 口门角度;进水角;进入角;进口角(度);入射角;入口角
entrance arch 入口拱;进口门拱
entrance area 入口区域;口门区【港】
entrance bar 拦门(栓)沙
entrance bay 进口河湾
entrance boundary 入口边界
entrance boundary point 入口边界点
entrance branch 引入歧管;入口分支
entrance bucket 导叶
entrance bundle of rays 入射光束
entrance buoy 进口浮标;入口浮标
entrance bushing 进线套筒;引线导管;引入线绝缘套
entrance cable 引入电缆
entrance caisson 坞门
entrance canal 进口渠道
entrance canopy 入口雨篷
entrance capacity 进口能力;容水量
entrance capillary pressure 入口毛细管压力
entrance channel 进口水道;进口河道;进口航道;进(出)港航道;引航道;入射道;入港航道;进口航槽
entrance channel chart 进港航道图
entrance condition (港口航道等的)进口条件;入口条件
entrance corner 进口拐角

entrance corridor 入口走廊;进口走廊
entrance court 前院
entrance culvert 进水涵管;进水涵洞
entrance curve 进口曲线;引入曲线;入口曲线
entrance designator 进口指名器
entrance design radius 入口设计半径
entrance diameter (入口的)进口直径
entrance direction 入射方向
entrance discharge curve 进口流量曲线
entrance door 进口大门;大门;入口门;外门
entrance door location 入口车门的布置
entrance dose 入口剂量
entrance driveway 入口车行道;进口车道
entrance effect 入口(段)效应
entrance-exit button 进出选路式按钮
entrance-exit console 进出选路式操纵台
entrance-exit interlocking 出入口联锁
entrance-exit line (机车的)出入库线
entrance-exit line from depot 出入段线
entrance-exit panel 进出集中装置控制板
entrance facade 入口处立面;进口门面
entrance face 入射面
entrance fee 入场费
entrance flow 入口流量
entrance flume 进水渡槽
entrance foyer 剧场休息室入口;进口门厅
entrance frame 电梯进(口)门框
entrance friction 进口阻力
entrance front 入口的正面
entrance gallery 进口廊道;进出廊道;进厂廊道
entrance gate 进水闸(门);进水门;进口大门;进口(闸)门;进港大门;进船闸;厂区大门;入口(闸)门
entrance geometry 入口几何形状
entrance half-angle 入射半角
entrance hall 入口大厅;门厅;入口大厅
entrance harbo(u)r 进口港
entrance head 流入水头;进口水头
entrance heating 入口供暖
entrance hole 进人孔
entrance illumination 入口照明
entrance installation 安装进门口
entrance jetty 入口导堤
entrance knob 进口旋钮
entrance lamp 门灯
entrance lane 进入(快速干道的)车道
entrance lattice gate 入口花格大门
entrance level 入口平面;入口水位
entrance light beam 入射光束
entrance lighting 进站照明;入口照明
entrance line 进站线(路);进库线;入口线
entrance lip (引)进刃
entrance lobby 休息室入口;门廊入口;前厅
entrance lock 进(口)船闸;进料密封装置;进口闸门;闸门;过渡舱;港池闸门;船坞进口船闸;潮汐船闸;入口船闸
entrance loss 进(水)口(水头)损失;进口能量损失;入口损耗;入口(水头)损失
entrance masonry wall 砖石墙入口
entrance notice 进港通知;准许进口通知
entrance of branch channel 汊道进口
entrance of dry dock 干船坞入口;坞首;干船坞进口
entrance of feed liquid 料液进(入)口
entrance of harbo(u)r 港的入口
entrance of intersection 交叉口出口
entrance of port 港口口;港的进口
entrance of stair(case) 楼梯口
entrance of the pond 水塘入口处
entrance of tunnel 洞口
entrance of washing water 洗涤水入口
entrance orifice 进水孔口;入口节流(圈)
entrance packing 洞门密封
entrance permit flag 准许入港旗
entrance piazza 广场入口;步廊入口;进口长廊
entrance pier 进港导堤;(进港)导堤
entrance point 进口端;入射点
entrance porch 正门;入口门;大门
entrance port 舷门;入射口
entrance portal 隧道门入口;桥门入口;大门入口;进口桥架
entrance pressure 进压;入口压力
entrance pressure drop 入口压降
entrance prism 入射棱镜
entrance pupil 入射光瞳孔

entrance pylon 进口塔门
entrance ramp 入口匝道;(道路交叉处)入口坡道
entrance ramp closure 关闭入口匝道
entrance ramp control 入口匝道控制
entrance ramp marking 入口坡道路面标示
entrance ramp metering 入口匝道限流
entrance ray angle 入射光线角
entrance region 进口区(段);助流区;入口区
entrance reinforcement (of tunnel) (隧道的)加强照明;隧道加强照明
entrance relay 进口继电器
entrance restriction 限制驶入;限制进入
entrance roadway 入口车行道
entrance roof overhang 屋顶小房间的入口;披屋入口
entrance route 接车进路
entrance salary 就职薪金
entrance separator 进线隔离
entrance side 入口侧
entrance signal 进站信号(机)
entrance slit 入射狭缝;入口狭缝
entrance slit height 入射狭缝高度
entrance speed at a retarder 减速器入口速度
entrance stair(case) 楼梯(间)入口;进口楼梯
entrance steps 入口梯级;入口台阶
entrance switch 入户线接线盒
entrance taper 进口锥形岛;入口锥形岛
entrance terminal 进口终点;入口终点
entrance terrace 入口平台;进口平台
entrance test ratio 入口检验比
entrance ticket 门票
entrance to a basement 地下层入口
entrance to a cellar 地下水入口
entrance to arcade 拱廊入口
entrance to haro(u) or harbor 港湾进口;港口门
entrance to river 河道入口
entrance track 列车进站线;入口轨道
entrance transition 进口渐变段
entrance tunnel 进站地道
entrance turn 进口回车道;进口回车场;转弯驶入
entrance velocity 进口流速;入口速度
entrance velocity of whirl 入口涡流速度
entrance visa 入境签证
entrance wall 进料护板;进口墙板
entrance water head 入口水头
entrance wave 入口波
entrance way 入口道路
entrance well 进水口;进水井
entrance width 口口宽度;港口宽度;入口宽度
entrance window 入射窗
entrance zone 进口段;入口(区)段
entrant sound 透声
entrap 陷阱;圈闭
entrapment 截流;夹杂;捕集;圈闭
entrapment vacuum pump 捕集真空泵
entrapped air 截留空气;截流空气;夹带空气;偶成气泡;混凝土含气泡;意外气泡;带入的空气;残余气体;残存气体;封闭气泡;封闭空气;封闭空气;潜在空气
entrapped air content 夹入气含量
entrapped air void 封好的气孔;空隙;(非故意掺入的)截留的空气孔隙
entrapped basin 圈闭海盆
entrapped dirt 卷入杂质
entrapped gas 夹气;滞留气;带走的气体
entrapped humidity 吸入水分;封入的湿气
entrapped moisture 截留水分;吸着水分
entrapped phase 圈闭相
entrapped slag 夹渣
entrapped water 截留的水分
entree 入场权;进入权
entrefer (电机的)铁间空隙
entrench 用壕沟围绕;挖壕;以壕围绕
entrenched belt 槽形胶带运送机;槽形胶带输送机
entrenched fan 嵌入式洪积扇
entrenched foundation bed 暗基床
entrenched meander 嵌入曲流;嵌入河曲;深切弯曲河段;深切曲流;深切河湾
entrenched provision 特别维护条款
entrenched river 嵌入河;深切河流
entrenched stream 嵌入河;深切河流
entrenched terrace 嵌入阶地
entrenched valley 深切(河)谷
entrenching shovel 挖壕锹

entrenchment 修壕沟防护;防御设施;堑壕;挖壕沟;挖槽
entrepot 贸易中心;货物集散地;转运口岸;转口;仓库
entrepot duty 转口税
entrepot port 转口货港;中转港(口)
entrepot storage 中转库
entrepot trade 中转贸易;转口贸易
entrepot warehouse 转口货仓库
entrepreneur 主办者;主办人;创业家;承包者;承包商;承包人;企业家;实业家
entrepreneurial capitalist 职能资本家
entrepreneurial decision 企业家决策
entrepreneurial factor 企业管理因素
entrepreneurial income 业主收入
entrepreneurship 企业精神
entresol 半层;夹层(结构);半楼
entresol floor (楼层间阁楼的)夹层地板
entripsis 敷擦法
entrochal limestone 含角质岩的石灰石;含化石睡莲茎的石灰岩
entrophication of surface waters 地表水体富营养化
entroport 转口港
entropy 岩相混合程度;平均信息量
entropy analysis 熵分析
entropy coding 熵编码
entropy contribution 熵分布
entropy crisis 熵单位
entropy elasticity 熵弹性
entropy flow 熵流
entropy function 熵函数
entropy minimization 熵最小化
entropy of activation 活化熵
entropy of a partition 划分的熵
entropy of a transformation 柯尔莫哥洛夫—希奈不变量
entropy of binary source 二元信源熵
entropy of dilution 稀释熵
entropy of evapo(u)ration 蒸气熵;汽化熵
entropy of fusion 熔解熵
entropy of hydration 水化熵
entropy of information 信息熵
entropy of joint event 联合事件熵
entropy of mixing 混合熵
entropy of network formation 网络形成熵
entropy of superheating 过热熵
entropy of the source 源熵
entropy of transition 转变熵
entropy pollution 熵污染
entropy principle 熵原理
entropy ratio map 熵比图
entropy spring 熵跃
entropy-temperature curve 熵—温曲线
entropy trapping 熵捕获
entropy weight 熵权
entrucking (载重汽车的)装车
entrust 信托;托管
entrusted person 受委托者;受委托人
entrusting party 委托方;委托人
entrusting party's account 委托人的账
entrust ment 委托
entry 款目;进入;进路;记入(项目);项目;登录;登记(项);词条;分录;表目;报关手续;入口通道;入口路;入境;输入(项目)
entry account 登账
entry address 入口地址
entry angle 口门角度;进口角(度)
entry association 入口结合
entry attribute 表目属性;入口属性
entry block 项目块;表目块;入口块
entry bridle 入口张紧装置
entry cell 驳格导口【船】
entry clothing 避火服
entry cluster 登记项簇
entry coil conveyer 供卷运输机
entry condition 进入条件;记入条件;入口条件;启动条件;数据输入条件
entry cost 参加成本
entry data 入口数据
entry date 进入日期
entry declaration 进口申请;入港申报(单)
entry-driving machine 掘进机
entry edge 入口缘

entry end 进口端;入口端
entry equipment 分录设备
entry-exist visa valid for a single journey 一次(性)有效出入境签证
entry-exit gate 进出闸机
entry-exit permit 进出许可
entry fire fighting 深入火场灭火战斗
entry for consumption 进口货物报单
entry for free goods 免税品进口报单;免税货物(进口)报单
entry for home use 完税进口报关单
entry formality 入港申报手续
entry forms 入境调查表
entry gap 啮入间隙;罗隙进口
entry gate 进闸机;进口大门;进港大门
entry guards 进口护板
entry guide 集装箱导器;导口
entry hatch 入舱口
entry head 进口水头
entry hole 入口直径
entry(ing)time 进闸时间
entry in log 载入航海日志
entry instruction 进入指令;入口指令
entry into power 逐渐加荷
entry into the environment of a toxic material 有毒物质进入环境
entry inward 进口报关;报进口
entry joint inspection 进口联检
entry length 进水口长度
entry loader 平巷装车机
entry lock 船闸进入口;进口闸门;入口锁
entry locker 过渡舱
entry loss 进入损失;进气降压;进口损失;入口损失;入口损耗
entry mask 入口屏蔽
entry mode program(me) 入口态程序
entry name 入口名
entry name attribute 登记项名字属性
entry name required for procedure statement 过程语句所要求的输入名
entry neck 入口孔颈
entry of a matrix 矩阵的元
entry of appeal 提出上诉
entry of cell guide 箱格导柱口
entry of contract into force 合同生效
entry of direct sunshine 直射阳光的进入
entry of notes dishonored 拒付票据分录
entry of stab wound 刺入口
entry of train into station 列车进站
entry outward 报出口
entry permit 进舱许可
entry point 进入点;吸液点;子程序进入点;(指令)转移点;入口点;指令变换点;输入点
entry pore 入口孔隙
entry port 进口港
entry portal 射入口
entry position 登记项位置;入口位置
entry pressure 挤入压力
entry price 入账价格
entry procedures 入境手续
entry profile 注入剖面
entry program(me) 入口程序
entry protection 进路保护装置
entry radius 入口半径
entry ramp 驶入匝道
entry region 进出区(段)
entry section 带卷送进区段
entry sequenced data set 顺序输入数据集
entry sequenced file 入口顺序文件
entry shape 进口形状
entry side 进口端
entry slot 进口缺槽
entry speed 进入速度;进港航速;入口速度
entry spin 进气预旋
entry statement 入口语句
entry symbol 入口符号
entry tension roll 进口张紧辊
entry terminal 输入终端
entry time 进舱时间;进入时间;报关时间;驶入时间
entry tool 破拆工具
entry variable 入口变量
entry visa 入境签证
entry way 入口(道路);通路
entry zone 进出区(段);入口区

Entwistle gearing 恩氏齿轮传动
enucleator 摘除器;摘出器
enumerability 可枚举性
enumerable 可数的
enumerable set 可数集;可枚举集
enumerate 计算;列举;枚举
enumerated data 不连续值
enumerated population (人口普查区内的)所有人口
enumerating all paths 列举所有路径
enumerating function 枚举函数
enumeration 列举;计点;枚举;细目
enumeration algorithm 枚举算法
enumeration correlation 计数相关
enumeration data 计数资料;计数数据
enumeration file 枚举文件
enumeration method 枚举法
enumeration of bacteria 细菌计数
enumeration of binary tree 二叉树的枚举
enumeration of labelled structure 带标号结构的枚举
enumeration of labelled trees 带标号树的枚举
enumeration theorem 枚举定理
enumeration tree 枚举树
enumeration type 枚举类型
enumeration unit 区划单位
enumerator 计数器;计数机
enumerator's schedule 普查员用表
enunciator 信号器
enunciator jack 信号机塞孔
enunciator relay 信号继电器
envelop curve 包络曲线
envelop delay distortion 包线延迟畸变
envelope 蒙皮;封皮;被膜;包围;包面;包络(层);包壳;包晶;包迹;包裹物;包封(位);气囊
envelope amplitude 包络振幅
envelope barrier 包面栅栏
envelope cancellation 包线补偿;包迹对消
envelope card 包络
envelope cavity 围护结构空腔
envelope curve 包络(曲)线
envelope cut-off 包迹截止波长
envelope delay 包线(延迟);包络延时线;包(线)延迟;包络时延;群时延
envelope delay data 包络线延迟数据
envelope delay distortion 包络线延迟畸变
envelope delay distortion meter 包络延迟失真测量仪
envelope delay scanner 包线延迟特性扫描仪;包络延迟特性扫描器;包络延迟特性测定器
envelope delay time meter 包络延时测量器
envelope delay tracer 包络延迟显迹器
envelope demodulation circuit 包络解调电路
envelope detection 包络检波
envelope detector 检波器;包线检波器;包络探测器;包络检波器
envelope diagram 包络图
envelope distortion 包络失真;群时延失真
envelope distribution 包络分布
enveloped pavement 封闭式路面结构
enveloped thermistor 密封型热敏电阻器
envelope factor 包迹因数
envelope fluctuation 包络波动
envelope function 包络函数
envelope glissette 推成包络
envelope grouting 封套灌浆
envelope halo 包裹晕
envelope kiln 封套窑
envelope line 包络线;外包线
envelope machine 制信封机
envelope match 匹配封装
envelope matching 包络线重合
envelope material 包装材料
envelope method 包络法;包迹法
envelope modulation 包络调制
envelope multiplex 包络线多路传送
envelope noise 包络噪声
envelope of curves 曲线(的)包络
envelope of failure 破坏包(络)线
envelope of grading 土地平整的包络线;级配的包络线
envelope of hood 罩套
envelope of rupture 破裂包线
envelope of surfaces 曲面的包络面

envelope oscilloscope 包线示波器;包络示波器;包迹示波器
envelope pattern 包络方向图
envelope pavement 密封式路面结构
envelope phase shift 包络相移
envelope principle 包容原则;包络原理
envelope reaction 被膜反应
envelope recording 包络记录
envelope rock 封套岩石
envelope select 包络选择
envelope surface 包络(曲)面
envelope synchronization 包迹同步
envelope table 折页桌
envelope test 静态试验
envelope threshold detection 包络阈检测
envelope threshold detector 包络阈检定器
envelope to cycle difference 包周差
envelope to cycle discrepancy 包周差
envelope type filter 袋式集尘器
envelope type forming hood 信封式成形帘
envelope velocity 包络速度;群速(度)
envelope viewer (脉冲)包线指示器;包络显示器;包络观察器
envelope widescope 包迹宽带示波器;包络宽带示波器
enveloping algebra 封裹代数
enveloping asbestos 花包棉
enveloping curvature 包络曲率
enveloping curve 包线【电】;包络(曲)线
enveloping curve of internal force 内力包络图
enveloping glass 封接玻璃
enveloping line of material 材料包络图
enveloping plane 包络面
enveloping scheme 环绕式布置;环流式布置
enveloping solid 包络体
enveloping space 包络空间
enveloping surface 包络面
enveloping surface fold 包络面褶皱
enveloping worm 包络蜗杆
enveloping worm-gear reducer 球面蜗轮减速箱
enveloping worm wheel 包络蜗轮
envelopment 封套;包封;封皮
envelop of stochastic process 随机过程包络
envelop plane 包络面
envelop to circles 圆的包迹
envenomation 表面变质
environ-geology 环境地质学
environics 环境学
environment 环境;四周
environment, health, safety 环境健康安全
environment act 环境法令
environmental 环境的;周围的
environmental abnormality 环境异常
environmental absolute capacity 环境绝对容量
environmental accident 环境事故
environmental accounting 环境核算
environmental acoustics 环境声学
environmental action 环境作用;环境行动
environmental adaptability 环境适应性
environmental adaptation 环境适应;适应环境
environmental administration 环境管理
environmental administration charges 环境管理收费
environmental administration liability 环境行政责任
environmental administrator 环境管理人员
environmental adviser 环境顾问
environmental aerodynamics 环境空气动力学
environmental aerology 环境大气学
environmental aesthetics 环境美学
environmental aesthetics assessment 环境美学评价
environmental agent 环境作用;环境介质
environmental alternation 环境改造;环境改变
environmental amelioration 环境改善;环境改良
environmental amenity 环境舒适;环境适宜(度)
environmental analysis 环境分析
environmental analytical chemistry 环境分析化学
environmental and architectural treatment 环境和建筑处理
environmental annual capacity 环境年容量
environmental anomaly 环境异常
environmental appraisal 环境评价;环境鉴定
environmental appraisal of a project 项目生态环

境评价;项目环境(生态)评价
environmental aquatic chemistry 环境水化学
environmental arbitration 环境仲裁
environmental architecture 环境建筑(学)
environmental area 环境地域
environmental art 环境艺术
environmental aspect 环境状况;环境因素;环境情况;环境概貌
environmental aspect of human settlement 人类住区的环境方面
environmental aspects of dredging 疏浚的环境方面
environmental assessment 环境评价;环境评估
environmental assessment and review process 环境评审法
environmental assessment of multiparameter 多项参数环境评价
environmental assessment of single element 环境单要素评价;单要素环境评价;单项参数环境评价
environmental assets 环境资产;环境财富
environmental assimilating ability 环境同化能力
environmental assimilating capacity 环境容量;环境同化能力
environmental assimilation capacity 环境同化容量
environmental association 生态组合;生态关联
environmental assurance system 环境保险体系
environmental attribute 环境属性;环境质量分类;环境特征
environmental audio extension 环境音效扩展
environmental auditing 环境审计;环境稽核
environmental auditor 环境审核团
environmental awareness 环境意识
environmental background 环境背景
environmental background value 环境本底值;环境背景值
environmental barrier 环境壁垒
environmental baseline 环境基线
environmental basic policy 环境基本政策
environmental bearing capacity 环境承受力
environmental beauty 环境美
environmental behavior mentality 环境行为心理
environmental behavoir 环境行为
environmental benefit 环境效益
environmental bioengineering 环境生物工程
environmental biogeochemistry 环境生物地球化学
environmental biological assessment 环境生物效应
environmental biological factor 环境生物因子
environmental biological impact 环境生物影响
environmental biological preparation 环境生物制剂
environmental biology 环境生物学
environmental biophysics 环境生物物理学
environmental biotechnology 环境生物技术;环境生物工艺学
environmental birth defect 环境性出生缺陷
environmental botany 环境植物学
Environmental Brief 环境简报;环境期刊
environmental budget 环境预算
environmental cab 推土机驾驶员室
environmental cabinet 环境箱;人造环境室
environmental calibrator 环境刻度器
environmental cancer 环境性癌
environmental cancer attack rate 环境性癌发生率
environmental capability 环境容量
environmental capacity 环境容量
environmental capacity mode 环境容量模式
environmental capacity model 环境容量模型
environmental capacity of air 大气环境容量
environmental capacity of groundwater 地下水环境容量
environmental capacity of population 人口环境容量
environmental capacity of soil 土壤环境容量
environmental capacity of urban atmosphere 城市大气环境容量
environmental capacity of water 水环境容量
environmental carcinogen 环境致癌物
environmental carcinogenesis 环境致癌作用
environmental carrying capacity 环境负载容量;环境承载容量
environmental challenge 环境要求;环境(的)挑战
environmental chamber 环境试验室;环境控制

(试验)室;环境舱;人工环境室
environmental change 环境变化
environmental characteristic 环境特征;环境特性
environmental chemical 环境化学的
environmental chemical carcinogen 环境化学致癌物
environmental chemical condition 环境化学条件
environmental chemical data and information network 环境化学品数据与资料网络
environmental chemical effect 环境化学效应
environmental chemical model 环境化学模式
environmental chemical pollutants 环境化学污染物
environmental chemicals 环境化学物
environmental chemistry 环境化学
environmental civil liability 环境民事责任
environmental class 环境分级
environmental clean up 环境净化
environmental climate 环境气候
environmental climatic map 环境气候图
environmental colo(u)rs 环境色
environmental compatibility 环境兼容性
environmental complaint 环境问题投诉
environmental complex 环境总体;环境综合体;环境体系;环境合成物
environmental compliance certificate 环境合格证(书)
environmental concentration 环境浓度
environmental concept 环境概念
environmental concern 环境关心
environmental condition 环境状态;环境状况;环境条件;外界条件
environmental conditioning 环境改善;环境改良;环境调节
environmental condition qualification specification 环境条件限制规范
environmental conditions at time of maturity 成熟期的环境条件
environmental conditions determination 环境条件测定
environmental consciousness 环境意识
environmental consequence 环境影响;环境后果
environmental conservation 环境保护
environmental consideration 环境研究;环境顾虑
environmental constraining capacity 环境约束(能)力
environmental constraint 环境约束;环境限制
environmental contaminant 环境污染物
environmental contamination 环境污染
environmental contamination burden 环境污染负荷
environmental contamination from reactor 反应堆的环境污染
environmental contamination of groundwater 环境地下水污染
environmental contingency planning 环境应急计划;环境应急规划
environmental control 环境控制;环境防治
environmental control building 环保治理建设
environmental control engineering 环境控制工程
environmental controller 环调
environmental controlling equipment room 环控机房
environmental control measure 环境治理措施
environmental control of groundwater quality 环境地下水水质控制
environmental control of underground station 地下车站环(境)控(制)
environmental control organization 环境控制组织
environmental control system 环境控制系统;环(境监)控系统;环境保护系统;防污染装置;防污染系统
environmental control technology activities 环境控制技术活力
environmental control unit 环境控制装置;环境调节装置
Environmental Coordination Board 环境协调委员会
environmental core policy 环境核心政策
environmental correlation 环境相互关系;环境关联(作用)
environmental cost 环境费用;环境成本
environmental cost benefit analysis 环境费用效益分析

environmental countermeasure 环境对策
environmental cracking 环境致裂;环境破裂
environmental crime 环境犯罪
environmental criminal liability 环境刑事责任
environmental crisis 环境危机
environmental criteria 环境指标;环境(判定)标准;环境基准
environmental cure 环境对策
environmental current 环境水流
environmental cycle 环境循环
environmental damage 环境损失;环境损害;环境破坏
environmental data 环境资料;环境数据
environmental database 环境数据库
environmental data buoy 环境数据浮标
environmental data compendium 环境资料概要
environmental data dictionary 环境数据字典
environmental data on chemicals 关于化学品的环境资料
environmental data service 环境资料服务
environmental decay 环境衰退;环境破坏
environmental decision analysis 环境决定分析;环境决策分析
environmental decision-making 环境决策
environmental decision model 环境决策模式
environmental decision process 环境决策过程
environmental decision support system 环境决策支持系统
environmental decision technology 环境决策技术
environmental defence fund 环境保护基金
environmental degradation 环境质量下降;环境退化;环境劣化;环境变坏
environmental dense fund 保卫环境基金
environmental density 环境密度
environmental de-pollution 环境去污
environmental design 环境设计
environmental design profession 环境设计专业;环境商品和服务业
environmental despoliation 环境掠夺
environmental destruction by traffic 交通公害
environmental detection 环境检测
environmental detection control center 环境监测控制中心
environmental detection set 环境检测装置
environmental detector 环境检测器
environmental deterioration 环境退化;环境恶化;环境变坏
environmental determinism 环境决定(论);环境决策论
environmental deviation 环境差异
environmental dilemma 环境困境
environmental diplomacy 环境外交
environmental disaster 环境灾难;环境灾害;生态灾难
environmental disaster control 环境灾害控制
environmental disaster events 环境灾难事件
environmental disaster monitoring 环境灾害监测
environmental diseases 环境(性)病
environmental diseconomy 环境不经济
environmental disfunction 环境功能丧失
environmental disorder 环境混乱
environmental dispute 环境纠纷
environmental dispute resolution 环境纠纷的调解
environmental disruption 环境污染;环境失调;环境破坏
environmental distribution 环境分配
environmental disturbance 环境失调;环境扰乱
environmental disutility 环境失效
environmental diversity 环境多样性
environmental diving 环境潜水
environmental division 环境区划
environmental due diligence 环境尽责
environmental ecological evaluation 环境生态评价
environmental ecological function 环境生态功能
environmental ecological hazard 危害生态
environmental ecological planning 环境生态规划
environmental ecological revolution 环境生态变革
environmental ecology 环境生态学
environmental ecology function 环境生态功能
environmental economic input-output table 环境经济投入产出表
environmental economic issues 环境经济问题
environmental economic planning 环境经济规划
environmental economic policy 环境经济政策

environmental economics 环境经济(学)
environmental economic system decision-making 环境经济系统决策
environmental economic system forecasting 环境经济系统预测
environmental economic system planning 环境经济系统规划
environmental economic target system 环境经济指标体系;环境经济目标体系
environmental economist 环境经济学家
environmental economy 环境经济学
environmental edaphology 环境土壤学
environmental education 环境教育
environmental education for the whole people 全民环境教育
environmental education in college 大学环境教育
environmental education in primary school 小学环境教育
environmental education on the job 在职环境教育
environmental education system 环境教育体系
environmental effect 环境影响;环境效应
environmental efficiency 环境效率
environmental electricity 环境电学
environmental electrochemistry 环境电化学
environmental electromagnetism 环境电磁学
environmental element 环境因素;环境要素
environmental emergency 环境紧急事故;环境紧急情况
environmental emergency response 环境应急响应
environmental emission 废气废物排入环境
environmental engineer 环境工程师
environmental engineering 环境工程(学);环境(保护)技术;环境保护工程;环境模拟工程;发展中的现代技术;模拟运转条件的技术
environmental engineering activities 环境工程活动范围
environmental engineering computer aided design 环境工程计算机辅助设计
environmental engineering consulting 环境工程咨询
environmental engineering geological map 环境工程地质图
environmental engineering geology 环境工程地质学
environmental engineering laboratory 环境工程实验室
environmental engineering system 环境工程系统
environmental engineering technologist 环境工程技术家
environmental entity 环境实体;环境本质
environmental entomology 环境昆虫学
environmental epidemic 环境流行病;周围的流行性传染病;外界传染病
environmental epidemiological survey 环境流行病调查
environmental epidemiology 环境流行病学
environmental error 环境误差
environmental estrogen 环境雌激素
environmental ethics 环境伦理学;环境道德
environmental evaluation 环境评价
environmental evaluation system 环境评价系统
environmental evidence 环境标志
environmental expert system 环境专家系统
environmental exposure 环境接触;环境风险;环境暴露
environmental exposure concentration 环境暴露浓度
environmental exposure indicator 环境暴露指标
environmental exposure level 环境暴露量
environmental externalisties of economic activities 经济活动对环境的影响
environmental extremes 极端环境;恶劣环境
environmental facies 环境相
environmental factor 环境因子;环境因素;环境要素;周围因素;外界因素;周围介质;外界(条件)因素;外部介质条件
environmental factor class 环境因素等级
environmental false alarm 环境误报
environmental fate 环境灾难;环境归宿
environmental feasibility 环境可行性
environmental field 环境范围;环境场
environmental field test 环境野外测试
environmental flow 环境流量
environmental flow assessment 环境流量估计

environmental fluctuation 环境变动
environmental fluid dynamics code 环境流体动力学代码
environmental fluid mechanics 环境流体力学
environmental follow-up 有关环境的后续行动
environmental force 环境力
environmental forecasting 环境预测;环境预报
environmental forecasting assessment 环境预测评价
environmental friendly 环境友好
environmental friendly products 环保产品
environmental function 环境功能
environmental functional coefficient 环境功能系数
environmental functional district planning 环境功能区划
environmental fund 环境基金
environmental gain 环境意识上升;环境认识提高
environmental gap 环境空隙;环境间隔;环境差距
environmental geochemistry 环境地球化学
environmental geography 环境地理学
environmental geological condition of ore district 矿区环境地质条件
environmental geological index 环境地质标志
environmental geological map 环境地质图
environmental geological problem 环境地质问题
environmental geology 环境地质(学)
environmental geology data base 环境地质数据库
environmental geomorphological map 环境地貌图
environmental geomorphology 环境地貌学
environmental geoprocess 环境地质作用
environmental geoscience 环境地学;环境地(球)科学
environmental geotechnics 环境岩土工程学
environmental geotechnology 环境岩土工程学
environmental goods and services 环境商品和服务业
environmental gradient 环境梯度
environmental greening 环境绿化
environmental green space 环境绿地
environmental greybox model 环境灰箱类型
environmental guideline 环境准则;环境方针
environmental harassment 环境困扰
environmental hazard 环境危害;环境公害
environmental hazard geology 环境灾害地质学
environmental hazard indicator 环境危害指示物
environmental health 环境卫生
environmental health bureau 环境卫生局
environmental health criteria 环境卫生基准
environmental health directorate 环境卫生局
environmental health engineering 环境卫生工程
environmental health evaluation 环境卫生评价
environmental health hazard 环境健康危害
environmental health impact assessment 环境健康影响评价
environmental health measure 环境卫生措施
environmental health monitoring(and surveillance) 环境卫生监督
environmental health planning 环境卫生规划
environmental health risk 环境健康风险
environmental health standard 环境卫生标准
environmental heat 环境热学
environmental heritage 环境遗产
environmental hierarchical statistical map 环境分级统计图
environmental hormone 环境激素
environmental hydraulics 环境水力学
environmental hydrobiology 环境水生物学
environmental hydrochemistry 环境水化学
environmental hydro economics 环境水经济学;环境水利学
environmental hydrogeochemistry 环境水文地球化学
environmental hydrogeochemistry analog(ue) 环境水文地球化学模拟
environmental hydrogeological map 环境水文地质图
environmental hydrogeology 环境水文地质(学)
environmental hydrogeology analog(ue) 环境水文地质模拟
environmental hydrogeology survey 环境水文地质调查
environmental hydrological effect 环境水文效应
environmental hydrological regime 环境水文效应
environmental hydrology 环境水文学

environmental hygiene 环境卫生学
environmental hypoxia 环境缺氧;环境低氧
environmental ill 公害
environmental image 环境图像
environmental impact 环境影响;环境冲击
environmental impact analysis 环境影响分析
environmental impact assessment 环境影响评价;环境影响估价;环境影响报告(书);环境冲击评价
environmental impact assessment follow-up 环境影响事后评价
environmental impact assessment groundwater 地下水环境影响评价
environmental impact assessment method 环境影响评价法
environmental impact assessment of harbo(u)r construction 港口工程施工环境影响评价
environmental impact assessment of solid waste disposal 固体废物处置环境影响评价
environmental impact assessment process 环境影响评价法
environmental impact assessment report 环境影响评价报告
environmental impact assessment statement 环境影响评价声明
environmental impact evaluation follow-up 后继环境影响评价
environmental impact factor 环境影响因素
environmental impact identification 环境影响识别
environmental impact index 环境影响指数
environmental impact list 环境影响报告表
environmental impact of construction 施工环境影响
environmental impact potential index 环境影响潜在指数;潜在环境影响指数
environmental impact procedure 环境影响通信;环境冲击过程
environmental impact report 环境影响报告(书)
environmental impact report format 环境影响报告格式
environmental impact scoping 环境影响界定
environmental impact statement 环境影响声明;环境影响鉴定;环境影响报告(书);环境变化报告书
environmental impact statement and review process 环境影响鉴定评审法
environmental impact statement of port 港口环境影响报告书;环境影响鉴定报告
environmental impact study 环境影响研究
environmental impact system 环境影响系统
environmental impact units 环境冲击单位
environmental impairing activity 环境损害行为;环境损害范围
environmental impairing liability 环境损害责任
environmental implication 环境关联;铅直的环境问题
environmental improvement 环境改善;环境改良;环境改进
environmental inadequacy 环境不适应
environmental incentives 环境激励措施
environmental incident 环境事件
environmental index 环境指数;环境指标
environmental indicator 环境指示物
environmental industry 环境(行)业
environmental influence 环境影响;环境效应
environmental information 环境资料;环境信息
environmental information center 环境信息中心
environmental information collection center 环境信息采集中心
environmental information collection system 环境信息采集系统
environmental information demand analysis 环境信息系统的需求分析
environmental information source 环境信息源
environmental information system 环境信息系统
environmental innocuity management of inland waters 内陆水域环境的无害管理
environmental instrument 环境仪器
environmental insult 环境对人体的危害
environmental interfacial action 环境界面作用
environmental intrusion 环境干扰
environmentalism 环境决定(论);环境(保护)论
environmental isopleth map 环境等值线图
environmental isotope 环境同位素

environmental isotope geochemistry 环境同位素地球化学
environmental isotopes test 环境同位素测试法
environmental isotopic hydrogeology 环境同位素水文地质学
environmental issue 环境问题
environmental issue in priority 环境优先论
environmentalist 环境(专)家;环境学家;环境问题专家;环境工作者;环境保护(工作)者;研究环境问题的专家
environmental items 环境项目
environmental judicature 环境司法
environmental Kuznets curve 环境库兹涅茨曲线
environmental label 环境标志
environmental labeled method of artificial compound 人工合成化合物环境标记法
environmental label(l)ing 环境标记;标明对环境的影响
environmental lapse 环境演化
environmental lapse rate 环境直减率;环境推移率;环境递减率
environmental law 环境法(律);环境保护法
environmental law enforcement 环境执法
environmental law of France 法国环境法
Environmental Law of Great Britain 英国环境法
Environmental Law of Japan 日本环境法
Environmental Law of the United States 美国环境法
environmental legal liability 环境法律责任
environmental legislation 环境立法;环境法规
environmental level 环境水准;环境水平
environmental levy 环境(征)税
environmental liability 环境责任
environmental lighting 环境照明
environmental limit 环境限制;环境限度;环境极限
environmental limit concentration 环境浓度极限
environmental limit condition 环境限制条件
environmental link board 环境联络委员会
environmental literature 环境文学
environmental load 环境荷载;环境负载;自然条件荷载
environmental load capacity 环境负荷量;环境承载力
environmental lubrication 贯流式润滑
environmentally acceptable biodegradation 环境允许生物降解
environmentally acceptable substitute 环境可接受的代用品;合乎环境要求的代用品
environmentally aligned channel management 环境均衡的河槽管理
environmentally aligned channel management measure 环境均衡的河槽管理措施
environmentally aligned channel management scheme 环境均衡的河槽管理方案
environmentally compatible paint 符合环保要求的漆
environmentally conscious 有环境意识的
environmentally correct 环保意识强烈的;环保观念正确的
environmentally damaging 环境破坏的
environmentally damaging activities 破坏环境的经济活动
environmentally damaging liability 破坏环境责任
environmentally determined economy 考虑到环境因素的经济
environmentally favo(u)rable energy option 对环境有利的能源选择
environmentally favo(u)rable energy strategy 对环境有利的能源战略
environmentally friendly 不损害环境的
environmentally friendly agriculture 环境友好的农业;生态农业
environmentally friendly binder 不污染环境的漆基
environmentally friendly dredge(r) 环保型挖泥船
environmentally friendly energy option 对环境有利的能源选择
environmentally friendly enterprise 环境友好企业
environmentally friendly pigment 不污染环境的颜料
environmentally friendly polymer 不污染环境的聚合物
environmentally friendly technology 环境友好技术
environmentally harmful development 对环境有害的开发

environmentally hazardous 危害环境的
environmentally hazardous development 对环境有害的开发
environmentally important redox half-cell reaction 环境优势氧化还原半电池反应器
environmentally oriented 环境工程有关的
environmentally protected railway 环保铁路
environmentally related disease 与环境有关的疾病
environmentally safe coating 对环境安全的涂料
environmentally safe product 对环境安全的产品
environmentally sealed 对环境封闭的;密封的
environmentally sensitive area 环境敏感(地)区
environmentally sensitive policy 环境敏感性政策
environmentally sound 环境无害;合乎环境要求的
environmentally sound characteristic 环境无害特性
environmentally sound management 环境有效管理
environmentally sound product 环境无害产品;合乎环境要求的产品
environmentally sound sustainable development 合乎环境要求的可持续发展
environmentally sound technological process 环境无害的工艺过程
environmentally sound technology 环境无害技术;合乎环境要求的工艺
environmentally stable 在环境中稳定的
environmental management 环境管理
Environmental Management Association 环境管理协会(美国)
environmental management for enterprise 企业环境管理
environmental management for enterprise of village and town 乡镇企业环境管理
environmental management indicator 环境管理指标
environmental management information system 环境管理信息系统
environmental management model 环境管理模式
environmental management of industrial enterprise 工业企业环境管理
environmental management of nuclear energy 核能的环境管理
environmental management program(me) 环境管理方案
environmental management review 环境管理评审
environmental management science 环境管理学
environmental management system 环境管理制度;环境管理系统;环境管理体制;环境管理体系
environmental map 环境感知图;环境地图
environmental mapping 环境制图
environmental mathematical model 环境数学模式
environmental mathematics 环境数学
environmental matrices 环境基质
environmental measure 环境措施
environmental measurements laboratory 环境测量实验室
environmental media contact corrosion 周围介质接触腐蚀
environmental mediation 环境调解
environmental medical monitoring 环境医学监测
environmental medical science 环境医学
environmental medical surveillance 环境医学监测
environmental medicine 环境医学
environmental medium 环境介质
environmental mercury contamination 环境汞污染
environmental mess 环境困境
environmental message 环境信息
environmental meteorology 环境气象学;环境的气象状态
environmental microbial technology 环境微生物技术
environmental micrology 环境微生物学
environmental mineralogy 环境矿物学
environmental mineral resources exploration based on remote sensing 遥感环境矿产资源调查
environmental misconduct 环境过失;环境处置不当;环境处理不当
environmental mobility 环境流动性
environmental modification 环境改变;环境变异
environmental monitoring 环境监控;环境监测
environmental monitoring analysis 环境监测分析
environmental monitoring and control 环境监测及控制
environmental monitoring and surveillance 环境监测及监督
environmental monitoring for background value 环境背景值监测
environmental monitoring information network 环境监测情报网
environmental monitoring installation 环境监控制装置
environmental monitoring laboratory 环境监测实验室
environmental monitoring management 环境监测管理
environmental monitoring network 环境监测网络化
environmental monitoring program(me) 环境监测计划;环境监测程序
environmental monitoring project 环境监测项目
environmental monitoring quality assurance 环境监测质量保证
environmental monitoring system 环境监控系统;环境监测系统
environmental moral beneficiary 环境道德收益者
environmental moral community 环境道德共同体
environmental moral consciousness 环境道德意识
environmental moral education 环境道德教育
environmental moral litigant 环境道德当事人
environmental moral norm 环境道德规范
environmental moral standard 环境道德标准
environmental mutagenic factor 环境诱变因素
environmental mutagen information center 环境诱变源情报中心
environmental need 环境要求
environmental network map 环境网格图
environmental niche 环境性境;环境龛;环境小生境
environmental noise 环境噪声
environmental noise criteria 环境噪声基准;环境噪声标准
environmental noise legislation 环境噪声法规
environmental noise level 环境噪声级
environmental non-biological factor 环境非生物因子
environmental nuisance 环境损害;环境公害
environmental objective 环境目标
environmental objective management 环境目标管理
environmental oceanography 环境海洋学
environmental optics 环境光学
environmental optimal decision 环境最优决策
environmental optimization 环境最优化
environmental optimum condition 环境最适条件
environmental organic chemistry 环境有机化学
environmental organic geochemistry 环境有机地球化学
environmental outage 环境系统运行中断
environmental parameter 环境参数
environmental particulate concentration 环境微粒浓度
environmental pathology 环境病状;环境病理学
environmental pathway 环境途径;进入环境途径
environmental pattern 环境模式;环境格局
environmental perception 环境认识;环境感观;环境概念
environmental performance 环境表现;环保绩效
environmental periodicity 环境周期性
environmental permit 环境许可证
environmental pesticide control act 环境农药控制法令
environmental philosophy 环境哲学
environmental physics 环境物理学
environmental physiology 环境生理学
environmental picture of product 产品的环境概貌
environmental planning 环境规划
environmental planting 环境绿化
environmental plasma 外围等离子体
environmental plotting 环境制图
environmental police 环境警察
environmental policy 环境政策
environmental policy act 环境政策法规
environmental policy evaluation 环境政策评估;环境政策评定
environmental policy for product 产品环境政策
environmental policy risk assessment 环境政策风险评价

environmental policy science 环境政策学
environmental policy system 环境政策体系
environmental pollutant 环境污染物
environmental polluter 环境污染物
environmental pollution 环境污染
environmental pollution assessment 环境污染评价
environmental pollution burden 环境污染负荷
environmental pollution chemistry 环境污染化学
environmental pollution control 环境污染控制
environmental pollution diseases 环境污染性疾病
environmental pollution forecasting 环境污染预测
environmental pollution index 环境污染指数；环境污染指标
environmental pollution indicator 环境污染指示剂
environmental pollution loss 环境污染损失
environmental pollution monitor 环境污染监测仪
environmental pollution monitoring 环境污染监测
environmental pollution monitoring and resource program(me) 环境污染监测对策计划
environmental pollution monitoring ship 环境污染监测船
environmental pollution processor 环境污染处理设备
environmental pollution psychology 环境污染心理
environmental pollution sources 环境污染源
environmental potential 环境势
environmental prediction 环境预测
environmental presence indicator 环境指示生物
environmental pressure 环境压力
environmental priority 环境优先项目
environmental priority pollutant 环境优先污染物
environmental probabilism 环境可能论
environmental problem 环境问题
environmental product profile 环境产品的概貌
environmental professional 环境工作专业人员
environmental profile 环境概貌
environmental program(me) 环境规划
environmental program(ming) 环境规划
environmental project 环境规划
environmental property 环境性质；环境特性
environmental protection 环境保护
environmental protection act 环境保护条例；环境保护法令
environmental protection agency 环(境)保(护)机构；环境保护署
environmental protection agency progress report 环境保护局进度报告
environmental protection assessment 环境保护评议
environmental protection collaboration 环境保护合作
environmental protection committee 环境保护工作委员会
environmental protection convention 环境保护协定
environmental protection costs 环境保护费用
environmental protection equipment 环境保护设备
environmental protection expenditures 环境保护支出
environmental protection expenses 环境保护费用
environmental protection guide 环境保护指南
environmental protection handbook 环保手册
environmental protection investment 环境保护投资
environmental protection law 环境保护法
environmental protection law enforcement inspection 环保执法检查
Environmental Protection Law of the People's Republic of China 中华人民共和国环境保护法
environmental protection limit 环境保护界限
environmental protection map 环保地图
environmental protection matters 环境保护要素
environmental protection objective 环境保护目标
environmental protection of mine 矿区环境保护
environmental protection of plant 工厂环境保护
environmental protection planning 环境保护计划
environmental protection policy 环境保护政策
environmental protection program(me) 环境保护规划
environmental protection standard 环境保护标准
environmental protection subsidy funds 环境保护补助资金
environmental protection swap 环境保护交换

environmental psychology 环境心理学
environmental purification 环境净化
environmental purification mechanism 环境净化机理
environmental qualification 环境鉴定；环境合格
environmental quality 环境质量
Environmental Quality Act 环境质量法令
environmental quality assessment 环境质量评价
environmental quality assessment method 环境质量评价方法
environmental quality assessment of groundwater 地下水环境质量评价
environmental quality assessment procedure 环境质量评价程序
environmental quality atlas 环境质量图集
environmental quality comprehensive assessment 环境质量综合评价
environmental quality comprehensive index 环境质量综合指数
environmental quality control 环境质量控制
environmental quality criterion 环境质量基准；环境质量准则
environmental quality evaluation 环境质量评价
environmental quality evaluation procedure 环境质量评价程序
environmental quality forecast 环境质量预报
Environmental Quality Improvement Act 环境质量改善法案
environmental quality index 环境质量指数；环境质量指标
environmental quality index of single element 环境质量单要素指数；单要素环境质量指数
environmental quality management 环境质量管理
environmental quality map 环境质量图
environmental quality map of river basin 流域环境质量图
environmental quality modeling 环境质量模拟
environmental quality monitoring 环境质量监测
environmental quality objective 环境质量目标
environmental quality parameter 环境质量参数
environmental quality pattern 环境质量模式
environmental quality recovery 环境质量恢复
environmental quality retrospective evaluation 环境质量回顾评价
environmental quality review 环境质量评审
Environmental Quality Review Act 环境质量评审法令
environmental quality sensor 环境质量传感器；环境大气质量检测传感器
environmental quality single index 环境质量单一指数
environmental quality standard 环境质量标准
environmental quality standard for groundwater 环境地下水水质标准
environmental quality standard for surface water 地面水环境质量标准；环境地表水水质标准
environmental quality standard maximum acceptable 环境质量标准可接受最大值
environmental quality statement 环境质量报告书
environmental quality variation 环境质量变化
environmental racism 环境种族主义
environmental radiation 环境辐射；环境放射(性)
environmental radiation monitor 环境辐射监测仪
environmental radiation surveillance activity 环境辐射监视工作；环境放射性监视工作
environmental radioactivity 环境放射性；环境辐射性
environmental radioactivity background 环境放射性本底
environmental radioactivity investigation 环境放射性调查
environmental range 环境范围
environmental realism 环境的真实性
Environmental Realm Plan of the European Communities 欧洲共同体委员会环境领域计划
environmental receptivity 环境受纳性；环境受纳能力；环境承受能力
environmental reconstruction 环境重建
environmental recovery 环境恢复
environmental reform 环境改造
environmental refugees 环境难民
environmental regional planning 环境区域规划
environmental registration map 环境定位图
environmental release 倾弃于环境；排入环境

environmental remediation 环境修复
environmental remote sensing 环境遥感
environmental renovation 环境复原
environmental repair fund 环境治理基金
environmental report 环境报告
environmental requirement 环境要求
environmental research 环境研究
environmental research and development program(me) and policy 环境研究和发展规划与政策
environmental research satellite 环境研究卫星
environmental reservoir capacity 环境库容
environmental resistance 环境阻力；环境抗性；环境抵抗力
environmental resistance hypothesis 环境阻力说
environmental resource geology 环境资源地质学
environmental resource information system 环境资源系统
environmental resources 环境资源
environmental resource satellites 环境资源卫星
environmental resource scarcity 环境资源稀缺性
environmental response team 环境事故反应队伍
environmental responsibility 环境责任感
environmental restraint 环境限制
environmental restriction 环境效益；环境限制
environmental return 环境回复；环境恢复
Environmental Reviewing Council 环境评审委员会
environmental right 环境权(利)
environmental risk 环境危险；环境风险
environmental risk assessment 环境危险评价；环境风险评价
environmental risk policy 环境风险政策
environmental river flow hydraulic 环境河道水流水力学
environmental route 进入环境途经
environmental safety 环境安全
environmental safety assessment 环境安全评价
environmental safety standard 环境安全标准
environmental sample 环境样品；环境样本；环境样板
environmental sampler 环境取样器
environmental sampling media 环境采样介质
environmental sanitation 环境卫生
environmental satellite 环境卫星
environmental satellite system 环境卫星系统
environmental science 环境(科)学
environmental science and technology 环境科技
environmental sciences as a whole 环境科学整体化
environmental sciences methodology 环境科学方法论
environmental scientist 环境科学家
environmental seal of approval 环境合格标记；合乎环境要求标记
environmental selection 环境选择
environmental self-cleaning capacity 环境自净能力
environmental self-purification 环境自净(作用)；环境自净(化)
environmental selling 模拟环境销售
environmental sensitivity 环境敏感性
environmental sensor 环境传感器
environmental service project 环保服务计划
environmental setting 环境背景
environmental severity factor 环境的劣度系数
environmental signalling phenomenon 环境信号现象
environmental simulation 环境模拟
environmental simulation equipment 环境模拟装置；环境模拟设备
environmental simulation test 环境模拟试验
environmental social policy 环境社会政策
environmental society economic index 环境社会经济指标
environmental socioeconomic indicator 环境社会经济指标
environmental sociology 环境社会学
environmental soil map 环境土壤图
environmental solution 环境溶液
environmental space 环境空间
environmental specification 环境规格
environmental spectrum 环境范围
environmental spillover 因考虑环境问题带来的其他问题

environmental stability 环境稳定性
environmental stability assessment based on remote sensing 遥感环境稳定性评价
environmental standard 环境标准
environmental standard of pollutant 环境污染物标准;污染物环境标准
environmental standard system 环境标准体系
environmental state 环境状态
environmental statement 环境报告(书)
environmental state of water body 环境水体状况
environmental statistical data 环境统计资料
environmental statistical map 环境统计地图
environmental statistical table 环境统计表
environmental statistics 环境统计(学)
environmental statistics yearbook 环境统计年报
environmental status 环境状况
environmental stock exchange 环境股票交易
environmental strategy 环境策略
environmental strategy objective 环境战略目标
environmental stratigraphy 环境地层学
environmental stress 环境重点;环境应力;环境胁迫
environmental stress cracking 环境致裂;环境应力诱发裂缝;环境应力裂纹;环境应力开裂;环境应力龟裂;外界应力龟裂
environmental stress crack resistance 耐环境应力抗裂性
environmental stress index 环境胁迫指数
environmental stressor 带来环境压力的因素
environmental structure 环境结构
environmental study 环境研究
environmental study area 环境调研区
environmental subsidy 环境补贴
environmental suit 环境诉讼
environmental supervision and management 环境监察管理
environmental surveillance 环境监视
environmental surveillance satellite 环境监视卫星
environmental surveillance system 环境监视系统
environmental survey 环境观测;环境调查
environmental surveying 环境图测量
environmental survey satellite[ESSA]环境调查卫星;环境探查卫星;环境勘测卫星;环境监测卫星;环境观测卫星;艾萨测量卫星
environmental system 环境系统;环境体系
environmental system of future population 未来人口环境系统
environmental systems engineering 环境系统工程(学)
environmental systems monitor 环境系统监控器
environmental tax 环境税
environmental taxes and fee 环境税费
environmental technical specification 环境技术规范
environmental technology 环境技术;环境工艺
environmental technology profile 环境技术概貌
environmental temperature 环境温度;周围温度
environmental teratogen 环境致畸原
environmental terminology 环境术语
environmental test control center 环境试验管理中心
environmental test(ing)环境试验;运行条件的模拟试验
environmental test planning 环境检验计划;环境检验规划
environmental test report 环境试验报告
environmental theory 环境理论
environmental therapy 环境疗法
environmental threat 环境威胁
environmental tolerance 环境容忍性;环境容限;环境忍耐力;环境耐受力
environmental tort 环境侵权行为
environmental toxic chemicals 环境有毒化学物质
environmental toxicity 环境毒性
environmental toxicology 环境毒物学;环境毒理学
environmental trace analysis 环境质量分析;环境痕量分析
environmental tracer 环境示踪剂;天然示踪剂
environmental training 环境训练;环境培训
environmental training program(me)环保训练计划
environmental training seminar 环保培训研讨会
environmental training workshop 环保培训研习班
environmental transition 环境变迁
environmental trend 环境趋势

environmental typical policy 环境典型政策
environmental valuation 环境评价
environmental value of water 水环境值
environmental vandalism 环境破坏行为
environmental variable 环境变量;环境(引起的)变数
environmental variance 环境方差
environmental variation 环境变异
environmental vegetation map 环境植被图
environmental velocity 环境流速
environmental vibration 环境振动
environmental warfare 环境冲突
environmental warning services 环境警报服务
environmental wastewater treatment 环境废水处理
environmental water 环境水
environmental water conservancy 环境水利
environmental waterfront 环境沿水区
environmental water pressure 环境水压
environmental water quality 环境水质
environmental water quality criteria 环境水质基准
environmental water quality monitoring network 环境水质监测网
environmental water quality standard 环境水质标准
environmental water sample 环境水样
environmental water sampler 环境水样器
environmental wind 环境风
environmental zoning 环境区划
environment and habitat 环境和生境
environment and natural resources 环境与自然资源
environment and sedimentation 环境与沉积
environment assessment of project 项目环境评价
environment at station site 站址环境
environment-based learning 以环境为基础的学习
environment capacity of tourism 旅游环境容量
environment containment standard 环境保护标准
environment control of main substation 主变电站环控
environment correction 环境校正
environment disturbance 环境干扰
environment division 环境部分;设备部分
environment fitness 环境适宜性
environment-friendly region 生态示范区
environment-geared project 面向环境的项目;围绕着环境的项目
environment-hydrogeological survey 环境水文地质调查
environment inquiry 设备询问
environment mapping 环境图测量
Environment Ministry of United Kingdom 英国环境部
environment model 环境模型
environment modification 环境改造
environment of a label constant 标号常数环境
environment of competition 竞争环境
environment of eruption 喷发环境
environment of sedimentation 沉积环境
environment of work 作业环境;工作环境
environment parameter observation 环境参数观测
environment pollution 公害
environment protecting plant 环保植物
environment protection agency 环境保护局
environment protection engineering 环境保护工程
environment quality evaluation of mine 矿山环境质量评价
environment simulator 设备模拟程序
environment subjected to misuse 易于被糟蹋的环境
environ-politics 环境政治学
environs 近郊;附近;周围地区
envisage 设想
envisagement 展望
envitrinite 非结构镜质体
enwind 缠绕
enyne 烯炔烃;烯炔化合物
enzootic 动物地方病
enzymology 酵素化学
Eoarticulate 始铰纲
eobiontic 原始生物的
Eocambrian 始寒武统【地】
Eocambrian glaciation 始寒武纪冰川作用

Eocambrian period 始寒武纪【地】
Eocene 始新世【地】
Eocene clay 始新世黏土
Eocene climate 始新期气候
Eocene epoch 始新世【地】
Eocene era oils 始新纪石油
Eocene period 下第三纪【地】
Eocene series 始新统【地】
Eocene system 下第三系【地】;始新系【地】
Eocene transgression 始新世海浸
Eocretaceous 始白垩层【地】
eodiscid 始盘类
Eogene climatic zonation 老第三纪气候分带
Eogene period 老第三纪【地】;早第三纪【地】
Eogene planation surface 老第三纪夷平面
Eogene system 老第三系【地】;早第三系【地】
eogenetic 早期成岩的
eogenetic carbonate cement 成岩初期碳酸盐胶结物
eogenetic carbonatization 早期碳酸盐化作用
eogenetic glauconite 成岩早期海绿石
eogenetic porosity 成岩早期孔隙
eogenetic stage 始成岩阶段
eogenetic syntaxial quartz overgrowth 成岩早期共生石英增生物
eogenetic zone 早期成岩带
eohypse 古地面等高线
eolation 风蚀作用;风成作用
eolian 风积的
eolian abrasion 风磨耗(作用);风蚀
eolian anemometer 风琴式风速计
eolian basin 风成盆地
eolian constituent 风成组分
eolian deposit 风积物;风积层;风成沉积
eolian dune 风成丘
eolian environment 风成环境
eolianite 风成岩
eolian mantle rock 风积表皮岩;风成覆盖岩
eolian placer 风成砂矿
eolian plain 风成平原
eolian reservoir 风积储集层
eolian rock 风成岩
eolian sand 风积沙;风成沙
eolian sand ripple 风积沙波纹
eolian soil 风成土
eolian sounds 风吹声
eolian spot 风成斑点
eolian suspension 风悬浮作用
eolian transportation 风力搬运
eolith 始石器;风成岩
Eolithic Age 始石器时代
Eolithic era 原始石器时代
Eolithic Period 始石器时代
eolotropic 各向异性的
eometamorphism 始变质作用
eon 世;时代;年纪;极长期
eoorogenic phase 潜伏造山幕
Eopaleozoic 早古生代【地】
eoplatform 始地台
Eopleistocene 早更新统
eoposition 风蚀沉积作用
eosin 溴化荧光素钠盐;伊红
eosin blue 荧光蓝
eosine 曙红
Eosine bluish 带蓝曙红
eosine lake 曙红色淀
eosin stain 伊红染液
eosin Y 黄色曙红
eosin yellow 荧光黄
eosiohypse 古地面等高线
eosphorite 磷铝锰矿
eotic animal 流水动物
Eotvos correction 厄缶校正;厄缶改正
Eotvos correction value 厄缶校正值
Eotvos effect 厄缶效应
Eotvos equation 厄缶方程
Eotvos torsion balance 厄缶扭秤
Eotvos unit 厄缶单位
eoumarone-indene resin 氧茚树脂
eoving cloth 粗格布
eovolcanic 原始火山的
Eozoic era 始生代
Eozoic group 始生界
epact 闰余;岁首月龄

Eparchean 后太古代的
epaul(e)ment (军工筑城)肩墙
epaulette-tree 白辛树
epaxial 轴上的
EPBS [Earth Pressure Balance Shield] 土压平衡盾构
Epcdauros theatre 埃比道拉斯剧场
epeiric 陆表的;大陆边缘的
epeiric-pelagic facies 陆缘—远洋相
epeiric sea 陆缘海;陆表海;浅海
epeiroclase 地台裂缝
epeirocratic 陆地克拉通的【地】
epeirocratic condition 造陆优势期
epeirocratic craton 陆地克拉通(稳定地块)
epeirocratic period 造陆通期
epeirocraton 陆地克拉通(稳定地块)
epeirogenesis 造陆作用;造陆运动
epeirogenetic 造陆的
epeirogenetic movement 造陆作用
epeirogenetic uplift 造陆隆起
epeirogenic 造陆运动的
epeirogenic facies 造陆相
epeirogenic folds 造陆褶皱
epeirogenic movement 造陆运动
epeirogenic phase 造陆相
epeirogenic sedimentation 造陆沉积作用
epeirogenic taphrogenesis 造陆地裂运动
epeirogenic uplift 造陆上升
epeirogeny 造陆作用;造陆运动
epeirophoresis theory 大陆漂移(学)说
epergne 中心饰;餐桌饰架
epersalgia 过劳病
epexegetical 解释性
ephaptic transmission 旁触传递
epharmonic convergence 适应趋同
ephebeion (古希腊、古罗马建筑中的)运动场所
Ephedra sinica 草麻黄
ephedroid 麻黄型穿孔板
ephedroid perforation plate 麻黄式穿孔板
ephemera 瞬息
ephemeral channel 季节性河槽;雨源河床
ephemeral data 瞬息数据
ephemeral fever 短暂热;三日热
ephemeral flow 季节性水流
ephemeral lake 季节性湖泊
ephemeral plant 短生植物;短命植物
ephemeral plant label 短生植物标记
ephemeral plant region 短生植物区
ephemeral region 瞬现区
ephemeral stream 季节性流水;季节性河流
ephemeral system 暂时性水系
ephemeris 历表;星历表;天文历表
ephemeris data 星历表数据
ephemeris day 历书日;星历日
ephemeris error 星表误差
ephemeris hour angle 历书时角
ephemeris longitude 历书经度
ephemeris meridian 历书子午线
ephemeris of the moon 月历表
ephemeris second 历书秒;星历秒
ephemeris sidereal-time 历书恒星时
ephemeris table computation 星历表计算
ephemeris table computation synchronous satellite 星历表计算同步卫星
ephemeris time 历书时(间);星历时;星历表时间
ephemeris transit 历书中天
ephepbeion (古希腊、古罗马建筑中的)运动场所
ephesite 钠珠云母;钡珠云母
ephippium 鞍状壳
epi 离岸沙坝;尖屋顶端部的覆盖物;尖;直沙嘴;覆盖物
epi-Algonkian orogenic period 后阿尔冈纪造山山期
epi-anticlinal fault 背斜边缘断层
epibatholite zone 浅岩基带
epibelt 浅成带
epibenthic 浅水底的
epibenthic environment 浅海底环境
epibenthic fauna 浅水底栖动物系
epibenthic organisms 浅海底栖生物
epibenthic population 浅海底栖生物种群
epibenthile 浅海底的
epibenthos 浅水底栖生物;浅海底栖生物
epibiont 附生

epibiota 海底生物
epibiotic 底层上生的
epibiotic species 残遗种
epiblem 根被皮
epibole 极盛带
epibolic invagination 外包内陷
epibolic stress 弛失应力
epibolite 间层状混合岩;顺层混合岩
epibond 环氧树脂类黏结剂;环氧树脂类黏合剂
epiborneol 表冰片
epicadmium 超镉
epicadmium energy 超镉能
epicadmium energy region 超镉能区
epicadmium fission 超镉(中子)裂变
epicadmium flux 超镉通量
epicadmium neutron 超镉中子
epicadmium resonance integral 超镉共振积分
epicap 功率二极管
epicenter 集中点;种源;中心;震中
epicenter azimuth 震中方位角
epicenter bias 震中偏差
epicenter concentration 震中集中
epicenter determination 震中测定
epicenter location routine 震中定位程序
epicenter longitude and latitude 震中经纬度
epicenter migration 震中迁移
epicenter of earthquake 地震震中
epicenter scale 震中表
epicenter shift 震中迁移
epicentral 震中的
epicentral acceleration 震中加速度
epicentral area 震中区;震中面积
epicentral bias 震中偏差
epicentral distance 震中距
epicentral intensity 震中烈度
epicentral location 震中位置
epicentral of earthquake 震中对点
epicentral region 震中区
epicentral zone 震中区;震中带
epicentre 震中;中心; = epicenter
epicentre coordinates 震中坐标
epicentre location 震中位置
epicentre map 震中图
epicentric distribution 震中分布
epicentrum 震中
epichlorite 次绿泥石
epichlorohydrin 氯甲代氧丙环;表氯醇
epichlorohydrin elastomer 环氧氯丙烷弹性体
epichlorohydrin poisoning 环氧氯丙烷中毒
epichlorohydrin polymer 环氧氯丙烷聚合物
epichmia 表迹;(遗迹化石的)上迹
epicholestanol 表胆甾醇
epiclast 表生碎屑
epiclastic 外力碎屑的
epiclastic conglomerate 表生碎屑砾岩;外力碎屑砾岩
epiclastic debris 表生碎屑
epiclastic fragmental facies 外力碎屑岩相
epiclastic rock 表生碎屑岩;外力碎屑岩
epiclastic sand 表生砂
epiclastic sedimentation 外力沉积作用
epiclastic volcanic fragment 外力火山碎屑
epicon 外延硅靶摄像管
epicontinental 陆缘的
epicontinental basin 陆缘盆地
epicontinental deposit 陆表海沉积
epicontinental facies 陆表相
epicontinental geosyncline 陆缘地槽
epicontinental marginal sea 陆缘浅海
epicontinental neritic facies 陆表浅海相
epicontinental platform 陆缘地台
epicontinental sea 陆缘海;陆架海;内陆海;边缘海;浅海;大陆架海区
epicontinental sediment 陆缘沉积(物)
epicontinental sedimentation 陆缘沉积(作用);陆架沉积作用
epicontinental warp 陆缘翘曲
epictesis 重复吸收
epicycle 小旋回;周转圆【数】;周转圈;本轮
epicycle motor 行星减速电动机
epicycle theory 本轮说
epicyclic(al) 周转圆的
epicyclic(al) arm 周转臂
epicyclic(al) clutch brake 周转离合器闸

epicyclic(al) drive 周转(齿轮)传动
epicyclic(al) gear 行星齿轮;周转齿轮
epicyclic(al) gearbox 行星齿轮箱
epicyclic(al) gear drive 周转传动
epicyclic(al) gearing 行星传动;周转轮系;行星齿轮(组)
epicyclic(al) gear train 周转轮系
epicyclic(al) gear transmission 行星齿轮式变速器;行星齿轮传动
epicyclic(al) motion 行星运动;周转圆运动;本轮运动;外摆线运动
epicyclic(al) reduction gear 周转减速装置
epicyclic(al) reduction gear of planetary type 行星式齿轮减速
epicyclic(al) reduction gear of solar type 恒星式齿轮减速
epicyclic(al) reduction gear of star type 定星式齿轮减速
epicyclic(al) reduction gear solar type 恒星周转齿轮减速
epicyclic(al) reduction gear star type 定星式周转齿轮减速
epicyclic(al) reduction gear unit 行星齿轮减速装置
epicyclic(al) singleplanet helical gear 单排行星斜齿轮
epicyclic(al) train 行星齿轮系;周转轮系;外摆线轮系
epicyclic(al) train of gears 周转齿轮系
epicyclic(al) transmission 行星传动
epicycloid 圆外旋轮线;外摆线
epicycloidal 外摆线的
epicycloidal and hypocycloidal gear tooth 内外摆线轮齿
epicycloidal gear 外摆线齿轮
epicycloidal gear cutter 外摆线齿轮铣刀
epicycloidal or planetary gear train 周转轮系
epicycloidal tooth 外摆线齿
epicycloidal wheel 外摆(线)轮
epicycloidal wheel tooth 外摆线齿轮
epicycloids 外圆滚线
epicylic reduction gear unit 行星减速齿轮装置
epidemic 流行的;流行病流行
epidemic catarrhal fever 流行性卡他热;流行性感冒;流行性疱疹热
epidemic curves 流行曲线
epidemic degree of disease 疾病流行强度
epidemic diarrhea 流行性腹泻
epidemic disease 流行病
epidemic factor 流行因素
epidemic focus 疫源地
epidemic hepatitis 流行性肝炎;传染性肝炎
epidemic index 流行指数
epidemic meningitis 流行性脑膜炎
epidemic prevention 防疫
epidemic prevention of warships 舰艇防疫
epidemic prevention station 防疫站
epidemic process 流行过程
epidemic spread 流行扩散
epidemic strength 流行强度
epidemic virus 流行性病毒
epidemiogenesis 流行病发生
epidemiologic(al) investigation 流行病学调查
epidemiologic(al) process 流行过程
epidemiologic(al) surveillance 流行病学监视
epidemiologic(al) survey 流行病学调查
epidemiologic data 流行病学资料
epidemiologic observation 流行病学观察
epidemy 流行病
epiderm 浅硅铝层
epidermal 表皮的
epidermal area 表皮层
epidermal deformation 硅铝壳表层变形
epidermal desquamation 表皮剥脱
epidermal glide tectonics 表皮滑动构造;表层滑动构造
epidermal layer 表皮层
epidermal type of gravitate tectonics 表皮型重力构造
epidermic 地球表层的
epidermic fold 表层褶皱
epidermic method 表皮法
epidermis 壳;地壳的沉积层;表皮(层)
epidermization 表皮形成
epidiabase 变辉绿岩

epidiascope 两用幻灯机；两射投影灯；两射放映机；幻灯(放映)机；反射透射两用投影仪；透反射两用幻灯机；实物幻灯机
epididymite 板晶石
epidiorite 变闪长岩
epidolerite 变粒玄岩
epidosite 绿帘(石)岩
epidote 绿帘石
epidote actinolite amphibolite 绿帘阳起角闪岩
epidote actinolite schist 绿帘阳起片岩
epidote amphibolite 绿帘角闪岩
epidote-amphibolite facies 绿帘石角闪岩相；绿帘闪岩相
epidote chlorite schist 绿帘绿泥片岩
epidote cummingtonite schist 绿帘石镁铁闪石片芒
epidote gneiss 绿帘石片麻岩
epidote-hornblende hornfels facies 绿帘石角闪石角岩相
epidote hornblende plagioclase gneiss 绿帘角闪斜长片麻岩
epidote hornblende schist 绿帘角闪片岩
epidote hornfels 绿帘石角岩
epidote-mica schist 绿帘石云母片岩
epidote rock 绿帘石岩
epidote sharn 绿帘石矽卡岩
epidote-tremolite schist 绿帘透闪片岩；绿帘石—透闪石片岩
epidotization 绿帘石化(作用)
epieugeosyncline 后成优地槽；次生优地槽
epifauna 表栖动物群
epifaunal organism 表栖动物有机体
epifluorescence microscopic count 表射荧光显微镜计数
epifluorescent microscope 落射荧光显微镜
epifocal 震中的
epifocus 震中
epigene 表成的
epigene action 外生作用；外力作用；外成作用
epigenesis 渐成论；后生作用；后成说
epigenetic 后生的；后成的
epigenetic action 外力作用
epigenetic anomalous inclusion 后生异常包裹体
epigenetic anomaly 后生异常；后成异常
epigenetic breccia 后生角砾岩
epigenetic concentration 后成富集作用
epigenetic concretion 后生结核
epigenetic deposit 后生矿床；后成矿床
epigenetic dispersion 后生分散
epigenetic dispersion pattern 后生分散模式
epigenetic dolomite 后生白云岩
epigenetic dolostone 后成白云岩
epigenetic drainage 叠置水系；上遗水系
epigenetic fabric 早成组构
epigenetic interstice 后生间隙
epigenetic interstitial water 后生间隙水
epigenetic mineral 后生矿物
epigenetic mineralization 后生成矿作用；后成矿化作用
epigenetic phase 后生相
epigenetic river 遗传河；叠置河；上遗河
epigenetic rock 后生岩
epigenetic stratabound deposit 后生层控矿床
epigenetic stream 上遗河
epigenetic structure 表皮构造；浅层构造【地】
epigenetic valley 上遗谷
epigenetic volcanism 后期火山作用
epigenetic volcano 后期火山
epigenetic water 后生水
epigenic 外成的
epigenic dolomite 后生白云岩
epigenic sediment 外表沉积物
epigenite 砷硫铜铁矿；砷硫铁铜矿
epigensis 外生作用
epigeosyncline 陆表地槽
epiglyph 层顶痕
epigneiss 浅带片麻岩
epigranite 浅成花岗岩
epigranular 粒度均匀；等粒的
epigraph 铭文；碑文
epigraphy 铭文艺术；铭文学；金石学
epigynous flower 上位花
epigyny 上位式
epi-Hercynian 海西期后的
epi-Huronian orogenic period 后休伦造山期【地】

epihydric alcohol 缩三甘油
epihymenium 上子实层
epi-illuminator 反射照明器
epi-impsonite 浅脆沥青
epikote 环氧类树脂
epikote resin-based adhesive compound 环氧树脂胶黏附混合物
epileucite 变白榴石
epilimnion (layer)暖等温水层；湖面温水层；变温层；上部(湖面)温水层
epilittoral zone 海岸带上的
epilog 最终测井解释成果图；收尾(过程)
epilogue 收尾程序
epimagma 后期岩浆残液；外岩浆
epimagmatic 浅岩浆的
epimatrix 外杂基
epimer 差向(立体)异构体
epimeride 差向异构体
epimerism 差向异构
epimerization 差向异构作用；差向(立体)异构化
epimetsmorphism 浅变质作用
epimigmatization 浅混溶作用
epimigmatization way 浅部混合岩化方式
epimirror 显微镜外镜
epimorph 晶体的天然色痕；外形模；外附(同态)体
epimorphism 满同态；满射
epinaos 内殿后部房间
epinasty 偏上生长
epineritic environment 半浅海环境
epiorganism 附生生物
epipedon 表土特性
epipelagic 光合作用带的
epipelagic fauna 深海上层动物区系
epipelagic fish 上层鱼
epipelagic fishes 上层鱼类
epipelagic region 海洋上层区；海洋上层带；海洋光合作用带；深海变层区
epipelagic zone 海洋光合作用带
epiphenomenonlism 副现象论
epiphenomenon 副现象；附带现象
epiphta arboricosa 树上附生植物
epiphyte 附生植物
epiphytic microorganism 附生微生物
epiphytic plant 附植植物；附生植物
epiphytic vegetation 附生植被
epiphytism 附着生活
epi-planar integrated circuit 表面集成电路
epiplankton 上层浮游生物
epiplasma 超等离子体
epiplatform 地台浅部；边缘地台
Epipleistocene 晚更新世【地】；上更新世【地】
epipodium 高环
epipolar 核轴
epipolar beam of ray 核心光束
epipolar correlation 核线相关
epipolar geometric(al) principle 核面几何原理
epipolar geometry 核面几何学
epipolarized light 表射偏振光；外表偏振光
epipolar line 核线
epipolar pencil of ray 核心光束
epipolar plane 核面
epipolar point 核点
epipolar ray 核线
epipolar-scan correlation 核线扫描相关
epipolar scanning 核线扫描
epipolic 荧光(性)的
epipolic dispersion 荧光色散
epi-position 表位
Epiproterozoic 上元古代【地】
Epiprotozoic 表原生代【地】
epipsammon 砂底动物
epirelief (层面的)上凸
epiric sea 边缘海
epirock 浅带变质岩
epicraton 陆缘克拉通【地】
epirogenic 造陆的
epirubicin 表柔比星
episcleral space 环球间隙
episcolecite 钙沸石
Episcopal church 英国国教
Episcopal palace 英国国教的宫殿
episcope 反射映画器；反射投影器；反射幻灯机；不透明物投影放大器

episcopic illumination device 反射照明装置
episcopic projector 反射投影仪
episcotister 截光器；减光转盘
episemantic 信息载体产物
episericite 绢云母
episode 一个事件
episodic erosion 侵蚀幕
episodic movement 短期地壳运动
episodic subsidence 阶段性沉降
epistatic deviation 上位离差
epistatic interaction 上位相互作用
epistatic parameter 上位参量
epistatic variance 上位方差
epistilbite 柱沸石
epistle side 教堂的右侧(面对教堂)；教堂祭合右侧
epistolite 硅铌钛矿；水硅钠铌石
epistomium 水管喷口
epistratal 后成底层的
epistrophe 趋表分布
epistrophic movement 表层地壳运动
epistyle 框缘；下楣；柱顶过梁；门窗头条板；额枋
epistylium 框缘；柱顶过梁；门框楹柱线条
episutural basin 接合带盆地
epitaph 墓志铭；碑铭
epitaphy 碑铭学
epitaxial 外延的
epitaxial alloyed device 外延合金器件
epitaxial autodoping 外延自动掺杂
epitaxial backfill 外延后填
epitaxial cell 外延元件
epitaxial chemical vapo(u)r deposition growth 外延化学气相沉积生长
epitaxial deposited silicon 外延淀积硅
epitaxial deposition 外延沉积
epitaxial diffuse 外延扩散
epitaxial diffused method 外延扩散法
epitaxial diffused phototransistor 外延扩散型光电二极管
epitaxial diposition 外延淀积
epitaxial effect 外延生长作用
epitaxial film 外延膜；外延层
epitaxial furnace 外延炉
epitaxial garnet film 外延榴石膜
epitaxial growth 外延生长
epitaxial interface 外延界面
epitaxial isolation method 外延隔离法
epitaxial junction 外延结
epitaxial layer 外延层
epitaxial ledge 外延台阶
epitaxially grown silicon crystal 外延生长硅晶体
epitaxial mesa transistor 外延台式晶体管
epitaxial method 外延法
epitaxial overgrowth 异轴增生
epitaxial planar technique 外延平面技术
epitaxial planar transistor 外延平面晶体管
epitaxial polycrystalline film 外延多晶膜
epitaxial process 外延过程
epitaxial region 外延区
epitaxial silicon 外延硅
epitaxial substrate 外延生长衬底
epitaxial susceptor 外延用衬托器
epitaxial vapor growth 外延气相生长
epitaxial wafer 外延片
epitaxis 外延生长
epitaxy 晶体取向接长；浮生；取向附生；外延(性)
epitaxy in closed device 闭管外延
epi-tectonic 浅层构造【地】
epitetracyclin 差向四环素
epithalassa 海洋表温层
epitheca 上壳
epithelial attachment 附着上皮
epithermal 超热(的)
epithermal absorption 超热中子吸收
epithermal absorption limit 超热中子吸收极限
epithermal activation 超热中子激活
epithermal activity 超热中子放射性；超热活性
epithermal capture 超热(中子)俘获
epithermal deposit 浅成热沉积；浅成低温热液矿床
epithermal energy 超热能
epithermal fission 超热(中子)裂变
epithermal leakage 超热中子漏失
epithermal neutron 超热(能)中子
epithermal neutron activation 超热中子活化
epithermal neutron activation analysis 超热中子

活化分析
epithermal neutron log 超热中子测井
epithermal process 浅成热液作用
epithermal reactor 超热中子反应堆
epithermal region 超热区(域);超热能区
epithermal thorium reactor 超热中子钍反应堆
epitome 节录;梗概;缩影
epitomize 概括
epitope 表位
epitrochoid (长短辐圆)外旋轮线
Epitropic(al) fiber[fibre] 埃比特罗比克纤维
Epitropic(al) filament yarn 埃比特罗比克长丝
epivalve 上壳瓣
epizoan 附生动物
epizonal 浅变质带的
epizonal dynamometamorphism 浅成动力变质作用
epizonal metamorphism 浅(带)变质作用
epizonal rock 浅成岩
epizone 浅(成)带;浅变质带
epizoon 外套生物
epoch 历元;恒定相位延迟;新纪元;潮相;时代
epoch-making 开新纪元的;划时代的
epoch of mineralization 成矿期
epoch of neogenticum 新地时期
epoch of neutralization 中和纪元
epoch of observation 观测历元
epoch of orientation 定向历元
epoch of partial tide 分潮迟角
epoch of perigee passage 近地点通过时刻
epoch of place 位置历元
epoch of style 新时代风格
Epolene 埃波纶低分子量聚乙烯
Epon 环氧类树脂(商品名)
epoxidation 环氧化作用
epoxide 环氧化物
epoxide alloy 环氧(树脂)金属;环氧树脂合金
epoxide cement 环氧黏接剂;环氧结合剂;环氧胶结剂
epoxide equivalent 环氧基当量
epoxide group 环氧基团
epoxide powder 环氧粉末
epoxide powder coating 环氧粉末涂料
epoxide resin 环氧树脂
epoxide-resin glue 环氧树脂胶
epoxide resin paint 环氧树脂涂料
epoxides 环氧衍生物
epoxide stabilizer 环氧类稳定剂
epoxiding agent 环氧化剂
epoxidised cyclic(al) acetal 环氧化环状缩醛
epoxidised linseed oil 环氧化亚麻仁油
epoxidized ester 环氧化酯
epoxidized oil 环氧化油
epoxidized plasticizer 环氧型增塑剂;环氧化增塑剂
epoxidized polybutadiene 环氧化聚丁烯
epoxidized soybean oil 环氧化大豆油
epoxidized vegetable oil 环氧化植物油
epoxidizing agent 环氧化剂
epoxirane formation 环氧化作用
epoxy acrylate 环氧丙烯酸酯
epoxy-acrylic coating 环氧丙烯酸树脂涂层
epoxy adduct 环氧加成物
epoxy adhesive 环氧黏结剂;环氧(树脂)黏合剂;环氧胶黏剂
epoxy adhesive for metals 环氧金属胶黏剂
epoxy alkyd varnish 环氧醇酸清漆
epoxy alloy 环氧树脂合金
epoxy amine 胺固化环氧树脂
epoxy anti-corrosive powder coating 防腐型粉末涂料
epoxy asphalt 环氧沥青
epoxy-asphalt concrete 环氧沥青混凝土
epoxy-asphalt material 环氧沥青材料
epoxy-asphalt paint 环氧沥青漆
epoxy baking enamel 环氧系烘烤瓷漆
epoxy-based bonding agent 环氧基黏结剂
epoxy-based cement(ing agent) 环氧基水泥胶凝剂
epoxy-based crack filler 环氧基裂缝充填剂
epoxy-based enamel 环氧基珐琅
epoxy-based hard gloss paint 环氧基坚硬有光油漆
epoxy-based mortar 环氧基砂浆
epoxy-based paint 环氧基油漆
epoxy-based paste 环氧基软胶
epoxy-based resin 环氧基树脂
epoxy-based vinyl ester resin 环氧基乙烯基酯树脂
epoxy binder 环氧(树脂)胶结剂;环氧结合剂
epoxy bitumen 环氧沥青
epoxy-bitumen material 环氧沥青材料
epoxy bituminous paint 环氧沥青漆
epoxy bonded (用)环氧树脂黏合的
epoxy bonded fiber-glass board 环氧树脂纤维玻璃板
epoxy bonded joint 环氧黏合缝
epoxy bonding 环氧基黏合
epoxy bonding adhesive 环氧黏结剂;环氧胶黏剂
epoxy brombenzene 环氧溴苯
epoxy butane 环氧丁烷;氧化丁烯
epoxy cast dry type coupled transformer 环氧浇铸干式耦合变压器
epoxy cement 环氧树脂黏结剂
epoxy chloroparaffin 环氧氯烃
epoxy chloropropane 环氧氯丙烷
epoxy coal coating resin 环氧—煤涂层树脂
epoxy coal floor cover(ing) 环氧—煤地板涂层
epoxy-coal-hydrocarbon binder material 环氧—煤—碳氢化合物黏结材料
epoxy-coal-hydrocarbon powder coating 环氧—煤—碳氢化合物粉状涂层
epoxy-coal-hydrocarbon resin 环氧—煤—碳氢化合物树脂
epoxy-coal-hydrocarbon resin paste 环氧—煤—碳氢化合物浆膏
epoxy coal tar 环氧煤焦油
epoxy coal tar material 环氧—煤焦油材料
epoxy coal tar paint 环氧煤焦油漆
epoxy coal tar varnish 环氧沥青清漆
epoxy-coated rebar 有环氧树脂涂层的钢筋
epoxy coating 环氧(树脂)涂料;环氧(树脂)涂层
epoxy coating resin 环氧树脂涂层
epoxy compound 环氧化合物
epoxy concrete 环氧树脂混凝土
epoxy conductive terrazzo 环氧树脂防静电水磨石
epoxy-containing concrete 掺和环氧树脂的混凝土
epoxy cotton paper 环氧棉纸
epoxy crack filler 环氧填缝料
epoxy cure-seal-harden compound 环氧处理密封凝固剂
epoxyde 环氧化物
epoxy derivative 环氧衍生物
epoxy dicarboxylic acid 环氧二羧酸
epoxy enamel 环氧瓷漆
epoxy encapsulated transformer 环氧树脂外包的变压器
epoxy equivalent 环氧当量
epoxy ester 环氧酯(类)
epoxy ester primer 环氧酯底漆
epoxy ester undercoat 环氧酯中间涂层
epoxy fiberglass 环氧玻璃纤维
epoxy film 环氧薄膜
epoxy film adhesive 环氧树脂薄膜黏合剂
epoxy foam 环氧泡沫
epoxy foam plastics 环氧泡沫塑料
epoxy foams 环氧泡沫体
epoxy formulation 环氧组成
epoxy-gel coating 环氧树脂胶衣
epoxy-gel mortar 环氧树脂胶泥
epoxy glass 环氧玻璃
epoxy glyceride 环氧甘油酯
epoxy group 环氧基;表氧基
epoxy grout 环氧薄浆
epoxy hardener 环氧固化剂
epoxy injection 环氧注射
epoxy injection adhesive 环氧注入胶黏剂
epoxy insulation 环氧树脂绝缘
epoxy insulation varnish 环氧绝缘漆
epoxy isocyanate paint 环氧异氰酸漆
epoxy joint 环氧(树脂)连接
epoxy laminate 环氧薄片
epoxylite 环氧(类)树脂
epoxy mastic 环氧胶泥
epoxy membrane 环氧膜层;环氧薄膜涂层
epoxy mix(ture) 环氧混合料
epoxy-modified cement 环氧树脂改性水泥
epoxy mortar 环氧树脂胶泥;环氧水泥砂浆;环氧(树脂)砂浆;环氧灰浆
epoxyn 环氧树脂黏结剂;环氧树脂黏结剂;环氧树脂类黏合剂
epoxy novolac adhesive 线型酚醛环氧黏合剂
epoxy novolac resin 环氧酚醛清漆树脂
epoxyoleic acid 环氧化油酸
epoxy oxygen 环氧态氧;环氧基氧
epoxy paint 环氧涂料;环氧漆
epoxy paint on screen 砂浆面层涂环氧漆
epoxy paste 环氧膏
epoxy-phenol aldehyde glass cloth laminated board 环氧酚醛玻璃布板
epoxy-phenol aldehyde glass cloth laminated rod 环氧酚醛玻璃布棒
epoxy-phenolic binder 环氧酚醛胶结料;环氧酚醛胶结剂
epoxy-phenolic cotton paper 环氧酚醛棉纸
epoxy-phenolic resin 环氧酚醛树脂
epoxy-phenolic resin paint 环氧酚醛(树脂)漆;环氧酚醛树脂涂料
epoxy plasticizer 环氧增塑剂
epoxy plastics 环氧(树脂)塑料
epoxy polyamide paint 环氧聚酰胺漆
epoxy polyester powder coating 环氧聚酯树脂粉末涂料
epoxy-polyethylene coating 环氧聚乙烯涂层
epoxy polymer 环氧聚合物
epoxy powder 环氧粉末
epoxy powder coating 环氧(树脂)粉末涂料
epoxy powder coating material 环氧树脂塑封材料
epoxy primer 环氧底子油;环氧底漆
epoxy propane 环氧丙烷
epoxy propane rubber 环氧丙烷橡胶
epoxypropyl phenylether 环氧丙基苯醚
epoxy putty 环氧树脂腻子
epoxy radical 环氧基
epoxy repair of crack 用环氧树脂补裂缝
epoxy resin 环氧树脂
epoxy resin adhesive 环氧树脂黏合剂;环氧树脂胶合剂
epoxy resin binder 环氧树脂结合料
epoxy resin concrete 环氧树脂混凝土
epoxy resin dipping paint 环氧树脂浸渍漆
epoxy resin electrophoresis membrane 环氧树脂电泳膜
epoxy resin enamel 环氧树脂瓷漆
epoxy resin ester 环氧树脂酯
epoxy resin filling equipment 环氧树脂灌注设备
epoxy resin finish 环氧树脂面漆
epoxy resin flooring 环氧树脂楼地面
epoxy resin floor paint 环氧树脂地板漆
epoxy resin glue 环氧树脂胶
epoxy resin grouting 环氧树脂灌浆
epoxy resin hardened at room temperature 室温硬化环氧树脂
epoxy resin joint 环氧树脂接缝
epoxy resin lining 环氧树脂衬
epoxy resin mortar 环氧树脂胶泥;环氧树脂砂浆
epoxy resin paint 环氧树脂涂料;环氧树脂漆
epoxy resin pattern 环氧树脂铸模;环氧树脂模型
epoxy resin poisoning 环氧树脂中毒
epoxy resin powder 环氧树脂粉末
epoxy resin protective coating 环氧树脂防护层
epoxy-resin-sand mortar 环氧树脂砂浆
epoxy resin sealant 环氧树脂嵌缝膏
epoxy resin silver paint 环氧树脂银漆
epoxy resin varnish 环氧树脂清漆
epoxy ring 环氧环
epoxy siloxane 环氧硅氧烷
epoxy splicer 环氧树脂接头机
epoxy stabilizer 环氧稳定剂
epoxy stearic acid 环氧硬脂酸
epoxy styrene 环氧苯乙烯
epoxy substrate 环氧基质
epoxy tar 环氧柏油
epoxy tar coating 环氧焦油涂层
epoxy tar paint 环氧焦油漆
epoxy teflon 环氧聚四氟乙烯
epoxy terrazzo 环氧水磨石
epoxy topping 环氧地面层
epoxy value 环氧值
epoxy varnish 环氧清漆
epoxy wearing surface 环氧磨耗层
epoxy zinc rich primer 环氧富锌底漆

Eppenbach homomixer 埃彭巴赫高速搅拌器
Eppenbach mixer 埃彭巴赫混合机
Eppenstein's principle 爱本斯坦原理
Eppley pyrheliometer 埃普利太阳热量计;埃普利日射强度计
Epprecht viscometer 埃普雷切特黏度计
eprazinonum 苯丙哌酮
Eprolithus 埃普罗颗石
epsilon-type structural system 山字形构造体系【地】
epsilon-type structure 山字形构造【地】
epsilon-type tectonic system 山字形构造体系【地】
epsomite 泻盐矿;泻盐
Epsom salt 泻盐(硫酸镁结晶)
Epstein hysteresis tester 爱泼斯坦磁滞仪
Epstein method 爱泼斯坦法
Epstein square 爱泼斯坦方圈仪
Epstein test 铁损试验
Epstein tester 爱泼斯坦测试仪
epure 精制地沥青;原尺寸图案;足尺图案
equability 均等;平静
equable moderate 温度均衡而适中
equal 均一的;相等的;等于;等同;等号;同样的人
equal-addendum tooth 等齿顶高齿
equal advantage 同等利益
equal-alternation wave 等交变波
equal altitude 相等高度
equal-altitude circle 等高圈
equal-altitude method 等高法;等高度法(求时钟差)
equal-altitude method of multistar 多星等高法
equal-altitude method with east and west stars 东西星等高法
equal-altitude observation 等高观测
equal amount installment payment 等额分期付款
equal-angle bar 等边角铁
equal-angleiron 等边角钢
equal-angle map 等角投影地图
equal-angle point 等角点
equal-angle projection 等角投影图;等角(度)投影
equal-angle steel 等边角钢
equal-angle transverse projection 等角横切椭圆柱投影
equal angular distortion 等角变形
equal annual accrual method 每年等额应计法
equal annual depreciation payment method 平均年金折旧法
equal-area chart 等面积图
equal-area conic(al) projection 等积圆锥投影
equal-area elliptic(al) projection 等积椭圆投影
equal-area latitude 等积投影纬度
equal-area map 等(面)积投影图
equal-area map projection 等积地图投影
equal-area net 等面积投影网;施密特网
equal-area plot 等面积布置
equal-area projection 等(面)积投影
equal-arm balance 等臂天平
equal arm bridge 等臂电桥
equal authenticity 具(有)同等效力;同等效力
equal azimuth 等方位角
equal azimuth projection 等方位投影
equal brightness photometer 等亮度光度计
equal chance 均等机会;同等机会
equal circumference 等圆周
equal clearances 等余隙
equal colo(u)r band (of water) 等水色带
equal competition 平等竞争
equal-compost distance classification 等混合距离法分类
equal concentration reduction 等浓度削减
equal corner iron 等边角铁;等边角钢
equal cost contour 等成本曲线
equal cost curve 等成本曲线
equal cost line 等价线
equal coverage 同等保证条款
equal cracking 等分裂化
equal current 均流
equal current cable 均流电缆
equal-deflection method 等偏转法
equal-degree-of-saturation method 等饱和度法
equal delay angle control 等滞后角控制
equal depression 等量沉陷
equal depth 等深度

equal depth map 等深度图
equal diameter 等直径
equal dignities rule 平等效力条例(当事人对代办人的书面委托)
equal displacement 等位移
equal distance 等距(离)
equal distribution 均等分布;平均分配
equal distribution of load 负载均匀分布
equal distribution of wealth 财富的平均分配
equal division 平均分配
equal drawdown map groundwater 地下水位等降深图
equaled time of payment 平均付款期
equal effect 同等效力
equal effect solution 等效解法
equal employment opportunity 同等就业机会
Equal Employment Opportunity Commission 同等就业机会委员会(美国)
equal-energy depth 等能量水深
equal-energy gradient 等能量梯度
equal-energy phosphor 等能荧光体
equal-energy source 等能源
equal-energy spectrum 等能量(光)谱;等能光谱
equal-energy stimulus 等能量刺激
equal-energy theory 等能量理论
equal-energy white 等能白色
equal exchange 等价交换
equal exposure 等量曝光
equal falling particles 等速下沉颗粒;等降粒子
equal force line 等力线
equal friction method 等摩阻法;(送风管道的)均等摩阻计算法
equal gain combiner 等增益合并器
equal gain diversity combining 等增益分集组合
equal hardness value 等硬度值
equal height tooth cutting method 等高齿切削法
equal heterodyne 等幅外差法;等幅差拍
equal horsepower curve 等马力曲线
equal housing opportunity 均等住房机会
equal inclination frings 等倾条纹
equal inclination interferometer 等倾干涉仪
equal inclination method 等倾角法
equal-increment solution 等增长解
equaling file 细平齐头锉;扁锉
equaling needle-handle file 平齐头针锉
equaling tube 均衡管
equal in quality 相同质量
equal instalment system 等额分期付款制;均等分摊法
equal in strength 强度相等
equal intercept restrictions 相等截距约束
equal interval flood peak 等时距洪峰
equal interval light 等明暗光
equal interval peak 等时距(洪)峰
equal interval quantizing 线性量化;等间隔量化
equalising bar 均衡杆
equalising pond 均衡蓄水池
equalising reservoir 均衡蓄水池;均衡风缸;平衡水库;平衡水柜;平衡储槽;调节水库
equalitarian distribution method 平均分配方式
equalite-layer-thickness 等时间层厚
equality 均一性;相等性;等式
equality and mutual benefit 平等互利
equality circuit 相等电路
equality constraint 等式约束
equality constraint condition 等式约束条件
equality gate 同门
equality of brightness photometer 等亮度(场)光度计
equality of contrast 反差相等
equality of contrast photometer 等对比(场)光度计;等反衬光度计
equality of fuzzy sets 模糊集相等
equality of luminosity 发光相等
equality of luminosity photometer 等亮度场光度计
equality of luminous photometer 等亮度光度计
equality of mass 质量相等
equality of pressure 压力等式
equality of rate of profit 利润率均等化
equality of taxation 纳税平等
equality system 等式组
equality unit 等同装置
equalization 均值化;均涂作用;均衡(调节);均衡(化);校平;相等(性);废水均化;平衡;均衡

equalization allowance 均衡准备
equalization basin 均衡池
Equalization Board 平公局(负责审核财产税税收及投诉地方政府评估房地产税的机构)
equalization box 平衡箱
equalization brake gear 平衡制动装置;平衡闸装置
equalization characteristics of bracking 平均制动性能
equalization claim 统一债权
equalization effect 均平效应
equalization fund 平准基金
equalization grants 平衡补助金
equalization network 均衡网络
equalization neutralization 均衡中和
equalization of assessments 确定税款平等
equalization of boundaries (计算不规则面积的)边界取舍
equalization of burden 负担均等化
equalization of discharge 均衡出水;平衡出水
equalization of earthwork 土方平衡
equalization of embankments and cuttings 填挖平衡
equalization of incidence 平均负担
equalization of land rent 平均地租
equalization of pressure 压力的均衡;均衡压力;压力平衡
equalization of profit 利润平均化
equalization of profit rates 利润率平均化
equalization of profit rates on capital 资本利润率平均化
equalization of tax 赋税均等化
equalization of tax payment 纳税平等
equalization payment 均衡支付
equalization point 均等点
equalization pond 均匀滞流池;稳流池;调节池
equalization reserve 均衡补偿准备
equalization reserve account 平均储备金账户;平均准备账户
equalization stage 均等期
equalization tank 均衡槽;调节池
equalization treatment 平衡处理
equalize 平衡;使均衡;使均等
equalized clamp 平衡卡头
equalized delay line 补偿延迟线
equalized heat distribution (method) 等热分配法;均等热分配(法)
equalized reservoir 反调节水库;反调节库容;调压水库;调蓄水库
equalized side lobe antenna 均衡旁瓣天线
equalized tank 调压水箱;调压水塔
equalized tip 均压焊嘴;均压电极头
equalized tip holder 均压焊嘴夹头
equalized vector 均等向量
equalizer 均值器;均压器;均压母线;均压管;均衡器;补偿器;起模同步机构;平衡器;平衡池;天平梁
equalizer amplifier 均衡放大器
equalizer-analyzer combination 平衡器—分析器组合
equalizer bar 平衡梁;平衡架;平衡杆
equalizer bar pivot 平衡杆枢
equalizer bar suspension 平衡梁悬架
equalizer bay 均衡器架
equalizer beam 平衡梁
equalizer bearing 平衡轴承
equalizer block 平衡器功能块
equalizer circuit 均压电路
equalizer coil 均压线圈
equalizer connection 均压线连接
equalizer curve 均衡特性曲线
equalizer disc 调整块
equalizer frame 平衡器架
equalizer fulcrum 均衡梁支点
equalizer network 均衡网络;校正四端网络
equalizer pipe 平衡管
equalizer ring 均压环;均衡环
equalizer rod 均涂棒;刮涂棒
equalizer saddle 平衡梁鞍座
equalizer safety strap 平衡器安全环
equalizer set 均衡器组
equalizer spring 平衡器弹簧;补偿弹簧
equalizer spring suspension 平衡弹簧悬架
equalizer support 均衡梁支架
equalizer supporting bar (车辆)平衡器支杆
equalizer switch 均压开关;均衡开关

equalizer valve 平衡器阀
equalizing 均衡的;匀涂
equalizing abrasion 均匀磨损;均匀磨耗
equalizing amplifier 平衡放大器
equalizing bar 均衡杆;平衡架;平衡杆
equalizing basin 均衡池;反调节池
equalizing beam 平衡梁
equalizing beam for counterweight 均衡重平衡梁
equalizing bed 管沟垫层;垫平整的基底;(铺管道时用的)垫床;铺管垫床;稳管垫床;碎石底座;沟渠管道混凝土底座;垫床找平基床
equalizing bolt 均衡螺栓;平衡螺栓
equalizing buffer 平衡缓冲器
equalizing buffer spring 均衡缓冲弹簧
equalizing busbar 均压母线;均压汇流条
equalizing cable 均压电缆;均压电缆
equalizing capacitor 均衡电容器
equalizing charge 均衡充电
equalizing compartment 平衡舱【船】
equalizing components 平衡元件
equalizing conductor 平衡(导)线
equalizing connection 均压连接;等位连接;平衡连接
equalizing culvert 平压涵洞;平水涵洞
equalizing current 均压电流
equalizing device 均衡装置;平衡装置
equalizing differences 补偿差别
equalizing discharge valve 均衡放风阀
equalizing discount rate 平衡贴现率
equalizing dynamo 平衡发电机
equalizing effect 调节作用
equalizing effect of lake 湖泊调节作用
equalizing file 平衡锉;细平头锉
equalizing frame 平衡架
equalizing gear 差动齿轮;补偿装置;平衡装置
equalizing graduating valve 均衡递动阀
equalizing hole 卸负荷孔;平衡孔
equalizing influence 互相抵消的影响
equalizing joint 平衡接头
equalizing lapping 配合研磨法
equalizing lever 平衡杆
equalizing main 平衡管道
equalizing mechanism 平衡机构
equalizing method 等量法
equalizing network 补偿回路
equalizing pipe 压力平衡管;平压管
equalizing piston packing ring 平衡活塞密封圈
equalizing pond 均衡蓄水池;平水池;调节水池
equalizing pool 平水池
equalizing port for needle 指针平衡口
equalizing pressure 平衡压力
equalizing projector 平衡投影器
equalizing prong 均衡支架
equalizing pulley 平衡轮
equalizing pulse 均衡脉冲
equalizing pulse interval 均衡脉冲间隔
equalizing pulse signal 均衡脉冲信号
equalizing reactor 均流电抗器
equalizing reservoir 均衡蓄水池;均衡风缸;平衡水库;平衡水柜;平衡水池;平衡储槽;调压水库;调节水库;调节(水)池
equalizing resistance 均压电阻
equalizing rolling mill 平整辊机
equalizing rudder 平衡舵【船】
equalizing section 池窑冷却部;温度均化区;(窑池的)温度调节区
equalizing slide valve 均衡滑阀
equalizing spring 均衡弹簧
equalizing stage 均衡期
equalizing sub 平衡接头
equalizing tank 均衡水池;均衡槽;平水塔;调压水箱;调压水塔
equalizing the dividend 均期股利
equalizing value 均值
equalizing valve 平衡阀;均衡阀;减压阀
equalizing washer 平衡垫圈
equalizing yoke guide 平衡轭架导承
equal labo(u)r values system 等劳制
equal laid wire rope 平行捻钢丝绳
equal lane utilization 车道等利用度
equal lay 平行敷设;等敷设距离;(缆索的)均绞
equal lay rope 等捻绞绳索
equal leg 对称焊脚
equal-leg angle 等肢角钢;等边角钢

equal-leg angle iron 等边角钢
equal-leg angle section 等边角剖面
equal-leg angle steel 等边角钢
equal-leg bulb-angle 等边圆头角钢;等边球头角钢
equal-leg frame 等腿框架;等边框架
equal length code 等长码;等长度编码
equal-level patching bay 等值接线架
equal life 相同使用期;等寿命
equal-limb line method 等月缘线法
equalling 均衡
equalling needlehandle file 平齐头针锉
equal listener response scale 等闻响应标度
equal load increment 等增加载
equal load increment method 等量加载法
equal-loudness 等响线
equal loudness contours 音量等响线
equal loudness curve 等声响曲线
equally authentic 具有同等效力
equally continuous 同等连续的
equally distributed sequence 等分布序列
equally likelihood principle 等似然率原理
equally likely event 等可能事件
equally loaded 均衡负载;等载的
equally probable 等概率的
equally probable contour 等概率线
equally spaced 等距(离)的;等间隔的;等齿距的
equally spaced data 等间距数据
equally spaced fringes 等间距条纹
equally spaced line 等距线
equally spaced points 等距(离)点;等间距点
equally spaced reference 等间距基准点
equally split 均等划分;等分绿(灯)信(号)比
equally tempered scale 等音程阶
equally tilted photography 等斜摄影;等倾摄影
equal matrix language 等矩阵语言
equal measurement 等精度测量
equal moment theory 等弯矩理论
equal nipple 等径螺纹接套
equal-noisiness contour 等噪线
equal number 相等数
equal observation 等精度观测
equal offset traces 等炮检距道
equal opportunity 均等机会
equal panel 等节间
equal pay for equal work 同工同酬
equal payment 等额支付
equal-payment series 资本回收系数
equal-payment-series-capital recovery-computation 按期定额支付资金回收计算法
equal-payment-series compound amount-computation 按期定额支付复利计算法
equal-payment-series-present value-computation 按期定额支付现值计算法
equal-payment-series-sinking fund-computation 按期定额支付偿债基金计算法
equal percentage characteristic 等百分比特性
equal percentage flow characteristic 等百分比流量特性
equal performance of labo(u)r 同等的劳动
equal-pitch contour 等音调线
equal planar unit 同平面单元
equal pore-pressure ratio 等孔隙压力比
equal potential 均压
equal power sampling 等动力采样(法)
equal precision curve 等精度曲线
equal pressure 等压(式);等压力
equal pressure curve 等压线
equal pressure method 等压法
equal probability contour 等概率线
equal probability curve 等概率曲线
equal proportions 同等数量
equal pulse 均脉
equal quantity monthly 每月等量装运
equal range map of groundwater level change 地下水等变幅图
equal ratio 等比
equal-reaction approach 等反力近似法
equal-ripple filter 等波纹滤波器
equal-ripple passband filter 等波纹通带滤波器
equal-ripple property 相等波纹性质
equal-ripple stopband filter 等波纹阻带滤波器
equal root 等根【数】
equal sacrifice 平均负担租;平均负担(税)
equal scale distortion 等尺度变形

equal section charge 等分装炸药
equal sensibility array 等灵敏度组合
equal set 相等集
equal settlement 均匀沉降;等量沉陷;等量沉降(量)
equal settlement plane 等沉陷面;等沉降面
equal settling factor 等落比;等沉比
equal settling particle 等沉粒
equal settling velocity 等沉降速度
equal-sided angle iron 等边角钢
equal sides 等边
equal sign 等号
equal signal system 等信号无线电导航系统
equal-signal white 等信号白色
equal-sized meshes 尺寸相同的网眼;尺寸相同的筛孔
equal-spaced 等间隔的;等距离布置的
equal span 等跨
equal speed drying 等速干燥
equal sphere 等球体
equal status of legal persons 平等的法人身份
equal step increment 等仓长增量;等步增长量
equal stop 对称触止
equal strain method 等应变法
equal strength 等强度
equal-strength explosive 等效炸药
equal stress 等应力
equal subsidence 等量沉陷
equal tails test 等尾检验
equal thick fringes 等厚条纹
equal-thickness fringe 等厚边缘
equal thickness interferometer 等厚干涉仪
equal time 均等机会
equal travel time 等运行时间
equal travel time surface 等运行时面
equal treatment 平等待遇
equal treatment of bidders 平等对待投标者
equal turbidity method 等浊滴定法
equal-utility contour 效用等高线
equal utility curve 等效用曲线
equal validity 同等效力
equal-value map 等值图
equal velocity length 等速核长度
equal vertical strain 相同竖向应变
equal vibration feeling contour 振动等响曲线
equal wear criterion 等磨损准则
equal wear theory 等磨损理论
equal weight 等重;等权
equal weighting 同量加权
equal weight value 等权值
equal-zero indicator 等于零指示器;等零指示位
equant 等轴的;等径的;等分(的)
equant element 等维要素
equantequation 均衡点
equant grains 等轴颗粒
equant micrite crust 等厚微晶外壳
equant-pore porosity 等径孔隙度
equant-shaped 等形的
equant-shaped basin 形状相同的盆地
equate 立方程式;等同;使相等;使平均
equated distance 换算距离
equate directive 使相等命令
equated maturity 平均到期日
equated ton kilometre 换算吨公理
equating 整平
equating coefficient 使系数相等
equating expections 相同期望
equation 相等(性);等式;等分;方程(式);反应式
equational reduction 均等减数分裂
equation clock 均时钟差
equation discharge-drawdown curve 涌水量水位降深曲线方程
equation factor 换算率
equation for steady incompressible flow 不可压式稳定流动方程;不可压的稳定流动方程
equation governing the motion 运动方程
equation in one unknown 一元方程
equation of a fuzzy relation 模糊关系方程
equation of angular motion 角运动方程
equation of center[centre] 偏心差;中心差
equation of compatibility 协调方程;相容性方程
equation of compatibility of deflection 挠度协调方程
equation of compatibility of displacement 位移

协调方程
equation of compatibility of rotation 转角协调方程
equation of compatibility of warping 翘曲协调方程
equation of compound 复利方程
equation of compressible flow 可压缩流体流动方程
equation of condition 条件方程(式)
equation of conservation 守恒方程
equation of constraint 约束方程(式)
equation of continuity 连续性方程;连续方程(式)
equation of continuous flow 流体连续方程
equation of correlation 相关方程
equation of diffusion 扩散议程;扩散方程
equation of displacement transformation 位移变化方程
equation of drilling rate 钻速方程
equation of dynamics 动力方程
equation of energy balance 能量平衡方程
equation of equal altitude 近午等高度改正量;等高差
equation of equilibrium 平衡方程(式)
equation of equinoxes 分点差;二分差方程
equation of evolution 发展方程
equation of exchange 交易方程式;费歇尔氏交换方程式
equation of first variation 一次变分方程;第一变分方程
equation of gas state 气体状态方程
equation of heat conduction 热传导方程
equation of higher degree 高次方程
equation of higher order 高阶方程
equation of hydrologic equilibrium 水均衡方程
equation of intersector equilibrium 门部间流量均衡方程
equation of kinetic and potential energy 动能与势能方程
equation of law and regulation 法规方程
equation of light 光行时差;光差
equation of linear regression 线性回归方程(式)
equation of locus 轨迹的方程
equation of mass conservation 质量守恒定律方程
equation of material balance 物质平衡方程(式)
equation of mathematic(al) physics 数学物理方程
equation of moments 力矩方程(式)
equation of momentum 动量方程(式)
equation of motion 运动方程
equation of n th degree n 次方程
equation of n th order n 次方程
equation of payments 平均分期付款
equation of piezotropy 压性方程
equation of radiative transfer 辐射转移方程;辐射传递方程
equation of recession 退水方程
equation of regression 回归方程(式)
equation of regression line 回归线方程;退水曲线方程
equation of state 状态方程(式);条件方程(式);物态方程(式)
equation of state of real gas 实际气体状态方程
equation of static equilibrium 静态平衡方程
equation of structure 结构方程
equation of tape length 尺长方程式
equation of the center 中心差
equation of the equinoxes 赤经章动
equation of the first degree 一次方程(式)
equation of three-moments 三弯矩方程
equation of time 时间方程;时差(率)
equation of transfer 转移方程
equation of transformation 变换方程
equation of trend surface 趋势面方程
equation of variation 变分方程
equation of varied flow 变流方程
equation of volumetric(al) balance 体积平衡方程式
equation root 方程的根
equation set 方程组
equations of compatibility of strains 应变协调方程
equation solver 解算装置;解算机;方程求解器;方程解算装置;方程解算器;方程解算机;求解器
equations system method 方程解法
equation statement 方程语句

equation technique 方程式法
equation type 方程类型
equator 昼夜平分线;中纬线;中规;大圆;赤道(线)
equator correction 赤道改正
equator earth terminal 赤道地面站
equatorial 赤道装置
equatorial acceleration 赤道加速现象;赤道加速度
equatorial aeronomy 赤道高层大气物理学
equatorial air 赤道气团;赤道空气
equatorial air mass 赤道气团;赤道空气团
equatorial angle 赤道角
equatorial angle value 赤道角值
equatorial anti-cyclone 赤道反气旋
equatorial aperture 赤道萌发孔
equatorial armilla 赤道经纬仪
equatorial armillary sphere 赤道经纬仪
equatorial array 赤道电极排列
equatorial astronomic(al) coordinate system 天文赤道坐标系
equatorial atmospheric dynamics 赤道大气动力学
equatorial axis 赤道轴;赤道直径
equatorial band 赤道带
equatorial belt 赤道带
equatorial belt of convergence 赤道辐合带
equatorial bond 赤道键;平伏键
equatorial bulge 赤道隆起(部分);赤道区鼓起
equatorial calm 赤道无风带
equatorial camera 赤道式摄影机
equatorial chart 赤切投影海图;赤道区域图;赤道地区海图
equatorial circle 赤道圆;赤道圈
equatorial circumference 赤道圆;赤道圈
equatorial climate 赤道气候
equatorial climate zone 赤道气候带
equatorial continental air 赤道大陆气团;赤道大陆空气
equatorial contour 赤道轮廓
equatorial convergence zone 赤道辐合区
equatorial convergence zone meteorologic(al) equator 热带辐合区
equatorial coordinates 赤道坐标
equatorial coordinate system 赤道坐标系(统)
equatorial correction 赤道改正
equatorial countercurrent 赤道逆流
equatorial current 赤道潮流;南赤道海流;赤道(洋)流
equatorial cylindric(al) orthmorphic chart 赤切圆柱形投影海图
equatorial cylindric(al) orthomorphic chart 渐长海图;墨卡托投影海图
equatorial day 赤道日
equatorial depression 赤道低压
equatorial diameter 赤道直径
equatorial doldrums 赤道无风带
equatorial drift 赤道暖流
equatorial drift current 赤道洋流
equatorial dry zone 赤道干旱带
equatorial easterlies 赤道东风带
equatorial ejection 赤道抛射
equatorial electrojet 赤道电射流;赤道电急流
equatorial flattening 赤道扁率
equatorial flow 赤道流
equatorial flow region 赤道低压区
equatorial fringe 赤道边饰
equatorial front 赤道锋;热带锋
equatorial gnomonic chart 赤切心射海图
equatorial gnomonic graticules 赤切心射标线
equatorial gravity 赤道重力
equatorial gravity value 赤道重力值
Equatorial Guinea 赤道几内亚
equatorial gully 赤道沟
equatorial gyre 赤道环流
equatorial high 赤道高压
equatorial horizontal parallax 赤道地平视差
equatorial instrument 赤道仪
equatorial interval 赤道星距
equatorial latitude 近赤道纬度
equatorial limb 赤道轮廓
equatorial line 赤道线
equatorial low 赤道低压
equatorial low belt 赤道低压带
equatorial low pressure trough 赤道低压槽
equatorial maritime air 赤道气团;赤道(海洋)空气
equatorial moment of inertia 中线惯性矩

equatorial monsoon 赤道季风
equatorial mounting 赤道仪座;赤道(式)装置
equatorial net 赤道网
equatorial orbit 赤道轨道
equatorial parallax 赤道视差
equatorial plane 赤道面
equatorial plante 赤道平面
equatorial precipitation belt 赤道多雨带
equatorial profile 球环带外形
equatorial projection 赤切投影;赤道凸出物;赤道投影(法);二维平面投影;二度平面投影
equatorial prominence 赤道日珥
equatorial quantum number 赤道量子数
equatorial radius 赤道半径
equatorial rain forest 热带雨林
equatorial rainforest climate 赤道雨林气候
equatorial rectangular coordinates 赤道直角坐标
equatorial reflection (X 线衍射图中的)赤道反射
equatorial region 赤道海区
equatorial satellite 赤道卫星
equatorial scale 赤道尺度
equatorial second 赤道秒
equatorial section 中纬切面
equatorial sediment bulge 赤道沉积物增厚
equatorial sundial 赤道日晷仪;赤道日规
equatorial symmetry 赤道对称
equatorial system 赤道系统
equatorial system of coordinates 赤道坐标系(统)
equatorial telescope 赤道仪;赤道望远镜
equatorial tidal current 赤道潮流
equatorial tide 赤道潮(汐)
equatorial torquetum 简仪
equatorial trapped wave 赤道拦获波
equatorial tropical zone 赤道热带
equatorial trough 赤道槽
equatorial undercurrent 赤道潜流
equatorial upwelling 赤道上升流
equatorial vortex 赤道涡流
equatorial warming 赤道增温
equatorial water 赤道水
equatorial wave 赤道波
equatorial wave theory 赤道波理论
equatorial westerlies 赤道西风带
equatorial wind 赤道风
equatorial zone 赤道海区;赤道带
equator latitudinal tectonic belt 赤道纬向构造带
equator of date 瞬时赤道
equator of epoch 历元赤道
equator of lens 中纬线
equator side 赤道侧
equator tide 赤道潮(汐)
equestrain tourism 骑马旅游
equestrian statue 骑士的塑像
equiaccuracy chart 等精度(曲线)图
equiaccuracy curve 等精度曲线
equiaffine 等仿射
equi-amplitude 等幅
equiamplitude signal 等幅信号
equiangular 等角的
equiangular circular arc network(lattice) for locating 定角圆弧格网
equiangular figure 等角形
equiangular hyperbola 等角双曲线
equiangular lattice 等角格网
equiangular polygon 等角多角形;等角多边形
equiangular positioning grid 等角定位格网
equiangular spiral 恒向螺旋线;等角螺线
equiangular spiral antenna 等角螺线天线
equiangular transformation 等角变换
equiangular triangle 等角三角形
equiangulator 观象仪;等高仪
equianharmonic curve 等交比曲线
equiareal mapping 等积映射;保积映射
equi-areal projection 等(面)积投影
equi-area mapping 等积投影制图
equi-area method of reinforcement 等面积补强法
equi-arm 等臂的
equiasymptotical 等度渐近的
equiasymptotical stability 等度渐近稳定性
equiaxed 等轴的
equiaxed crystal 等轴晶(体)
equiaxed dendritic structure 等轴树枝状结构
equiaxed ferritic grain 等轴铁素体晶粒

equiaxed grain 等轴晶粒
equiaxed grain structure 等轴结晶构造
equiaxed structure 等轴组织
equiaxed system 等轴晶系
equiaxial 等轴的
equiaxial crystal 等轴晶
equiaxial grain 等轴粒子
equiaxis 等轴
equiband 等带宽;等边带
equiband coder 等带宽编码器
equiband demodulator 等带宽解调器
equiband receiver 等带宽接收机
equibinary polymer 平衡二元聚合物
equiblast 等风
equiblast cupola 等风冲天炉
equicenter 等心
equicohesive 等强度的;等内聚的
equicohesive temperature 等凝集温度;等内聚温度
equi-composition method 等组成法
equi-composition standard 合成校正试样;合成标样
equiconjugate 等共轭的
equiconjugate diameters 等共轭直径
equicontinuous family of function 等度连续函数族
equicontrollability 等控性
equiconvergent 同等收敛
equicrescent variable 等半圆变数
equicrural 等腰的
equidensitometry 等密度测量
equidensity 等密度
equidensity curve 等密度曲线
equidensity film 等密度(线)胶片
equidensity image 等密度影像
equidensity pseudocolo(u)r encoding 等密度假彩色编码
equideparture 等距平
equidep gear 等高齿圆锥齿轮
equidepth section 等厚断面
equidifferent 等差(的)
equidimension 等尺寸;同大小
equidimensional 等轴的;等因次的;等维的;等量纲的;等尺寸的
equidimensional grains 等径颗粒
equidimensional grid 等维网络
equidimensional halo 等量度晕
equidirectional 同向
equidisplacement chart 等位移量曲线图
equidistance 等距(离)
equidistance curve 等距(离)曲线
equidistance line 等距(离)线
equidistance motion 等距(离)运动
equidistance principle 等距(离)原则
equidistance projection 等距(离)投影(法)
equidistant 等距(离)的;等间距的
equidistant curve 等距(离)曲线
equidistant equatorial azimuthal projection 等距(离)赤道方位投影
equidistant fringes 等间距条纹
equidistant grid 等距(离)网格
equidistant isometric(al) latitude 等距(离)纬度
equidistant lattice 距离格网
equidistant line 等距直线;等距(离)线
equidistant locking 等距(离)锁面
equidistant locking pallets 等距(离)螯尖
equidistant node 等距(离)节点
equidistant polar projection 等距(离)极方位投影
equidistant projection 等距(离)投影(法)
equidistant pulse 等距(离)脉冲
equidistant spacing 等距(离)
equidistant structural control of ore deposits 构造等间距控矿
equidistant surface 等距(离)曲面
equidistributed 等分布的
equidistribution 均匀分布;均匀分布;等分布
equidiurnal effect 等日效应
equi-energy spectrum 等能量(光)谱
equi-error contour 等误差线
equifield intensity 等场强
equifield intensity curve 等场强曲线
equifinal 等效的
equiflux heater 均匀加热炉;双面辐射式加热炉
equiformal 相似的
equiform geometry 相似几何
equiform group 相似变换群

equiform transformation 相似变换
equigeopotential surface 地球等位面
equiglacial line 等冰况线
equigranular 等粒状(结构)的;等粒度的;同样大小(颗粒)的;等颗粒的
equigranular aerated concrete 等粒度的加气混凝土
equigranular blastic texture 等粒变晶结构
equigranular concrete 等粒度混凝土
equigranular mortar 等粒度砂浆
equigranular psamnitic texture 等粒砂状结构
equigranular texture 均匀粒状结构;花岗岩状结构;等粒组织;等粒(状)结构
equi-inclination Weissenberg method 等倾魏森堡法
equi-inclination Weissenberg photograph a-symmetry 等倾魏森堡图的对称
equi-inertial filling and emptying system 等惯性输水系统
equi-interference method 等干涉法
equi join 等结合;等接
equilater(al) 等边形;等轴的;等面(的);两侧对称(的);等边(的)
equilateral angle steel 等边角钢
equilateral arch 尖拱;等角拱;等边(二心)拱;二边心桃尖拱
equilateral cross curve 等边十字线
equilateral hyperbola 等轴(直角)双曲线;等边双曲线
equilateral point arch 二心边心桃尖拱
equilateral points 等边点
equilateral polygon 正多边形【数】;等边多边形
equilateral roof 等角屋顶(其跨度和椽子形成一等边三角形)
equilateral spheric(al) triangle 等腰球面三角形
equilateral state of stress 多面等应力状态
equilateral triangle 正三角形;等边三角形
equilateral triangle point 等边三角形点
equilateral triangular 等边三角形的
equilateral turnout 单式对称道岔【铁】
equilibrant 均衡力;平衡物;平衡力
equilibrant system 平衡力系
equilibrated valve 预启阀;压力平衡阀
equilibrate process 平衡过程
equilibrating action 平衡作用
equilibrating movement of short-term capital 短期资金平衡移动
equilibration 平衡作用;平衡(化)
equilibration time 到达平衡时间;平衡稳定时间
equilibrator 平衡装置;平衡物;平衡机
equilibrator spring 平衡器弹簧
equilibrious 平衡的
equilibristat 平衡计
equilibrium 均衡;平衡
equilibrium about rolling axis 滚动力矩平衡
equilibrium absorption 平衡吸收;平衡吸附
equilibrium absorption model 平衡吸附模型
equilibrium accelerating voltage 平衡加速电压
equilibrium activity 平衡作用;平衡活度;平衡放射性
equilibrium adsorption 平衡吸收;平衡吸附
equilibrium adsorption amount 平衡吸附量
equilibrium adsorption relationship 平衡吸附关系
equilibrium air distillation 平衡空气蒸馏
equilibrium amount 均衡数量;平衡量
equilibrium amplitude 平衡振幅
equilibrium analysis 均衡分析;平衡分析
equilibrium angle 平衡角
equilibrium anomaly 均衡异常
equilibrium approach 平衡解题法;平衡(近似)计算法
equilibrium argument 平衡潮汐论
equilibrium at rest 静态平衡;休止平衡
equilibrium ball charge 平衡装球量;平衡磨球装量
equilibrium ball load 平衡装球量
equilibrium ball valve 平衡球阀
equilibrium bank 平衡砂堤
equilibrium beam 平衡束
equilibrium beam dimensions 平衡束流尺寸
equilibrium beam size 平衡束流大小
equilibrium between scour and deposition 冲淤平衡
equilibrium between transport and deposition 冲淤平衡

equilibrium binding constant 平衡接合常数
equilibrium blast cupola 等风冲天炉
equilibrium blast furnace 等风冲天炉
equilibrium boiling point 平衡沸点
equilibrium bomb 平衡弹
equilibrium boundary layer 平衡边界层
equilibrium box 平衡箱
equilibrium brightness 平衡亮度
equilibrium capillary pressure 平衡毛细管压力
equilibrium carbon dioxide 平衡二氧化碳
equilibrium carriers 平衡载流子
equilibrium catalyst 平衡催化剂
equilibrium chemical specification 平衡化学表征
equilibrium chemistry 平衡化学
equilibrium coefficient 平衡系数
equilibrium compaction 平衡压实
equilibrium compliance 平衡柔量
equilibrium composition 平衡组分;平衡成分
equilibrium compression 平衡压缩
equilibrium concentration 平衡浓度
equilibrium concentration of vacancy 空位平衡浓度
equilibrium condensation 平衡冷凝
equilibrium condensation curve 平衡冷凝曲线
equilibrium condition 均衡条件;平衡状况;平衡条件
equilibrium configuration 平衡形状
equilibrium constant 平衡常数;平衡常量
equilibrium constant at constant pressure 恒压平衡常数
equilibrium constant at constant volume 定容平衡常数;定容平衡常数
equilibrium contact angle 平衡接触角
equilibrium contact stage 平衡接触级
equilibrium-controlled reaction 平衡控制反应
equilibrium conversion 平衡转化
equilibrium copolymerization 平衡共聚合(作用)
equilibrium core 平衡堆芯
equilibrium corrected value 平衡修正值
equilibrium criterion 平衡准则;平衡数据;平衡判据
equilibrium crystallization 平衡结晶作用
equilibrium curve 均压曲线;平衡曲线
equilibrium curve of adsorption process 吸收平衡线
equilibrium cycle 平衡循环
equilibrium delay 平衡延迟
equilibrium density 平衡密度
equilibrium depth 平衡水深;平衡深度
equilibrium dew point 平衡露点
equilibrium diagram 相图;状态图;平衡图;金相图
equilibrium dialysis 平衡透析;平衡渗析法
equilibrium dissolved oxalate concentration 平衡溶解草酸盐浓度
equilibrium dissolved oxygen concentration 平衡溶解氧浓度
equilibrium distance 平移分布距离
equilibrium distillation 平衡蒸馏
equilibrium distillation curve 平衡蒸馏曲线
equilibrium distribution 均衡分布;平衡分布
equilibrium distribution coefficient 平衡分布系数
equilibrium division 平衡分割
equilibrium dose constant 平衡剂量常数
equilibrium draft 平衡通风
equilibrium drainage 均衡排泄
equilibrium economics 均衡经济学
equilibrium enrichment coefficient 平衡浓缩系数
equilibrium enrichment factor 平衡浓缩系数
equilibrium equation 均衡方程(式);平衡方程(式)
equilibrium eutectic 平衡低共熔物
equilibrium evapo(u)ration 平衡汽化
equilibrium experiment 平衡实验
equilibrium failure rate 平衡故障率
equilibrium figure 平衡形态;均衡图
equilibrium film thickness 平衡膜厚度
equilibrium filtration rate 平衡滤失速度
equilibrium firm 均衡厂商
equilibrium flash curve 平衡闪蒸曲线
equilibrium flash still 平衡闪蒸锅
equilibrium flash vapo(u)rization 平衡闪蒸(法)
equilibrium flash vapo(u)rization still 平衡汽化釜
equilibrium flash vapo(u)rizer 平衡闪蒸设备;平衡闪蒸器
equilibrium flow 平衡流

equilibrium flux 平衡通量
equilibrium force 平衡力
equilibrium formula 平衡公式
equilibrium freezing 平衡凝固
equilibrium freezing-in 平衡凝入
equilibrium gas 平衡气
equilibrium gas drive 平衡气驱
equilibrium gas saturation 平衡气体饱和率
equilibrium grade of reserve by mining 储采均衡程度
equilibrium gradient 平衡坡降
equilibrium hole inclination 平衡井斜
equilibrium humidity 平衡湿度
equilibrium hypothesis 平衡假设
equilibrium income 均衡收入
equilibrium index 平衡指数
equilibrium in pure strategy 纯策略中的平衡
equilibrium interest rate 均衡利率
equilibrium interfacial tension 平衡界面张力
equilibrium internuclear distance 平衡核间距
equilibrium ion 平衡离子
equilibrium kinetics 平衡动力学
equilibrium lattice position 晶格平衡位置
equilibrium level 平衡水平;平衡水准;平衡水平
equilibrium level of income 均衡收入水平
equilibrium level of national income 均衡国民收入水平
equilibrium line 均衡线;平衡曲线
equilibrium liquid 平衡液
equilibrium load 平衡沙量;平衡荷载;平衡负载
equilibrium mass distribution 平衡质量分布
equilibrium meniscus 平衡弯液面
equilibrium method 平衡(方)法
equilibrium mixture 平衡混合物
equilibrium mode 平衡模式
equilibrium model 均衡模型;平衡模型
equilibrium modulus 平衡模量
equilibrium moisture content 干燥平衡水;平衡水汽含量;平衡水分;平衡湿含量;平衡含水率;平衡(状态)含水量;土壤平衡含水率;木材平衡含水量
equilibrium moisture content of soil 土壤平衡含水量
equilibrium moisture content of timber 木材平衡含水率
equilibrium moisture content of wood 木材平衡含水率
equilibrium moisture regain 平衡回潮率
equilibrium moment 平衡力矩
equilibrium momentum 平衡动量
equilibrium multiplier 均衡乘数
equilibrium number 平衡数
equilibrium of angles of rotation 转角平衡
equilibrium of firm 厂商平衡
equilibrium of floating bodies 浮体平衡
equilibrium of forces 力的平衡
equilibrium of grains 颗粒平衡
equilibrium of moments 力矩平衡
equilibrium of motion 运动平衡
equilibrium of pressure 平衡压力
equilibrium of sediment transportation 输沙平衡
equilibrium of stock 资源平衡
equilibrium of stress state 应力平衡状态
equilibrium of supply and demand 供устройства需平衡;供求平衡
equilibrium of the earth's crust 地壳均衡
equilibrium oil 平衡油
equilibrium operating 均衡运行
equilibrium operating temperature 平衡(工作)温度
equilibrium orbit 平衡轨道
equilibrium orbit radius 平衡轨道半径
equilibrium ozone concentration 臭氧的平衡浓度
equilibrium parity zone 均衡平价区
equilibrium partial pressure 等分压力
equilibrium pH 平衡pH
equilibrium phase 平衡相
equilibrium phase angle 平衡相角
equilibrium phase behavio(u)r 平衡相态
equilibrium phase change 平衡相变
equilibrium phase diagram 平衡相图
equilibrium phosphorus concentration 平衡磷浓度
equilibrium plasma 平衡等离子体
equilibrium plasma temperature 平衡等离子体温度
equilibrium point 均衡点;平衡点
equilibrium point algorithm for computing 计算中的平衡点算法
equilibrium point detector 平衡点检测器
equilibrium point measurement 平衡点测量
equilibrium polygon 平衡多边形
equilibrium population 平衡群落
equilibrium porosity 平衡孔隙度
equilibrium position 均衡部位;平衡位置
equilibrium potential 平衡电位;平衡电势
equilibrium power level 平衡功率级
equilibrium pressure 平衡压力
equilibrium price 均衡价格;(供求)平衡价格
equilibrium principle 平衡原理
equilibrium prism 平衡棱镜
equilibrium probability 平衡概率
equilibrium problem 平衡问题
equilibrium process 平衡过程
equilibrium production 均衡生产
equilibrium profile 平衡纵剖面;平衡断面;天然纵剖面
equilibrium profile of beach 海滩平衡剖面
equilibrium profile of coast 海岸均衡剖面
equilibrium proton 平衡质子
equilibrium purchase point 平衡购入点
equilibrium quantity 均衡数量;均衡产量
equilibrium radiation 平衡辐射
equilibrium radius 平衡半径
equilibrium range 平衡滑翔距离
equilibrium rate 平衡汇率
equilibrium rate of exchange 均衡汇率
equilibrium rate of growth 均衡增长率;平衡成长率
equilibrium rate of inflation 均衡通货膨胀率
equilibrium rate of interest 平衡利率
equilibrium ratio 平衡(汽化)比
equilibrium-reaction temperature 平衡反应温度
equilibrium redox specification 平衡氧化还原表征
equilibrium reflux 平衡回流
equilibrium reflux boiling point 平衡回流沸点
equilibrium relationship 均衡关系;平衡关系
equilibrium relative humidity 平衡相对湿度
equilibrium requirement 平衡要求;均衡需求
equilibrium reservoir pressure 均衡水库压力;油层平衡压力
equilibrium response 平衡特性
equilibrium return 均衡报酬
equilibrium ring 平衡环
equilibrium rudder angle 平衡舵角
equilibrium running 平衡运转
equilibrium saturation 平衡饱和(度)
equilibrium scour 平衡冲刷
equilibrium segregation coefficient 平衡分凝系数
equilibrium sense 平衡感
equilibrium separation coefficient 平衡分离系数
equilibrium shear compliance 平衡剪切柔量
equilibrium shear modulus 平衡剪切模量
equilibrium slide valve 平衡滑阀
equilibrium slope 自然坡度;平衡坡度;平衡比降
equilibrium solar tide 平衡太阳潮
equilibrium solid solubility 平衡固溶度
equilibrium solubility 平衡溶解度
equilibrium solution 均衡解决;平衡溶液;平衡解
equilibrium solution composition 平衡溶液成分
equilibrium species 稳态种
equilibrium specification 平衡表征
equilibrium spectrum 平衡谱
equilibrium speed 平衡速率;平衡速度
equilibrium spheroid 平衡球状体;平衡球面
equilibrium stage 平衡级
equilibrium state 平衡状态;平衡(状)态
equilibrium still 平衡蒸馏锅
equilibrium strategy 平衡策略
equilibrium structure 平衡结构
equilibrium study 平衡研究
equilibrium superelevation 平衡超高(度)
equilibrium surface structure 平衡表面结构
equilibrium surface tension 平衡表面张力
equilibrium swelling 平衡溶胀;平衡膨胀
equilibrium system 平衡系统
equilibrium temperature 平衡温度;稳定温度
equilibrium tendency 平衡趋势
equilibrium test 平衡检查
equilibrium theory 均衡(学)说;均衡理论;平衡理论
equilibrium theory of tide 潮汐平衡理论;潮汐均衡理论;平衡潮学说;平衡潮理论
equilibrium thickness 平衡厚度
equilibrium tide 平衡潮(汐);天文潮
equilibrium tide level 天文潮位
equilibrium time 平衡时间
equilibrium total concentration 总平衡浓度
equilibrium torsion 平衡扭转
equilibrium unit 平衡装置
equilibrium value 均衡值;平衡值
equilibrium vapo(u)rization 平衡蒸发
equilibrium vapo(u)rization ratio 平衡汽化比
equilibrium vapo(u)r phase 平衡汽相
equilibrium vapo(u)r pressure 平衡蒸气压
equilibrium viscosity 平衡黏滞度
equilibrium water 均衡水;平衡水;平衡含水率;平衡含水量
equilibrium water content 平衡水分
equilibrium water flux 平衡水通量
equilibrium water surface 平衡水面
equilibrium wetting 平衡润湿
equilibrium writing 均衡记录
equilong 等距(离)的;等长的
equilong circle arc grid 等距(离)圆弧格网
equilong transformation 等距变换
equiluminous 等照度
equiluminous curve 等照度曲线
equiluminous sphere 等照度球面
equiluminous surface 等照度面
equiluminous transfrom 等照度变换
equilux 等照度
equilux sphere 等勒克斯球面
equi-magnetic 等磁(的)
equi-marginal principle 边际均等原理
equi-marginal utility 边际效用均等
equimolal 克分子数相等;当量克分子的
equimolal diffusion 等分子扩散
equimolar 当量分子的
equimolar counter-diffusion 等摩尔对向扩散
equimolar series method 连续浓度变更法;等摩尔系列法
equimolar solution 等摩尔溶液
equimolecular diffusion 等分子扩散
equimolecular solution 等分子溶液
equimorphism 全同态映射
equimultiple 等倍数;等倍量
equinoctial 昼夜平分线的;分点的;天球赤道;天赤道
equinoctial circle 昼夜平分线;二分圆
equinoctial circle plane 二分圆面
equinoctial closure 二分圆
equinoctial closure plane 二分圆面
equinoctial colure 分点圈
equinoctial flower 定时开闭花
equinoctial gale 分点风暴;二分点风暴;春、秋分时暴风雨
equinoctial high water 分时高潮;春、秋分高潮
equinoctial line 昼夜平分线
equinoctial low water 分时低潮;分点低潮
equinoctial neap tide 昼夜平分点小潮
equinoctial point 春分或秋分点;分点;二分点
equinoctial rain 二分雨
equinoctial spring tide 分点大潮;特大潮
equinoctial storm 分点风暴;二分点风暴
equinoctial system of coordinates 赤道系统;分至坐标系
equinoctial tide 分点潮;二分潮
equinoctial time 分至时
equinoctial year 民用年;回归年;分至年
equinox 昼夜平分点;春分或秋分点;分点;二分时刻;二分点
equinox correction 春分点改正
equinoxial colure 二分圈
equinoxial tide 二分点潮;分点潮(春、秋分分期的潮汐)
equi-orthometric surface 等正高面
equip 准备
equipage 装置;装备;船具
equiparte 伊奎帕特雨
equipartition 均分;匀配;匀隔;平均分隔
equipartition law 均匀分布定律
equipartition of energy 能量均分
equipatos 伊奎帕特雨
equiphase 相等的【物】;等相(位)的;同相的
equiphase plane 等相平面

equiphase surface 等相(位)面
equiphase zone 等相位区
equiplanation 高纬度均夷作用;冰冻夷平作用
equipluve 等雨量线
equipment 舾装;舣装品;仪器;装置;器械;设备
equipment acceptance requirements and inspections 设备接收的要求与检验
equipment acquisition 设置购买费
equipment airlock 设备运输气闸
equipment and facility console 装备与设施控制台
equipment and materials 器材;设备材料
equipment and process research 设备与工艺研究
equipment and repair quay 舣装码头
equipment and spare parts 设备与零备件
equipment and supplies 设备和配件
equipment arrangement 设备布置;设备安装
equipment assembly 设备组装;设备安装
equipment augmentation 设备扩充
equipment availability 设备完好率
equipment average outage time 设备平均停运时间
equipment balk diagnosis 设备故障调查
equipment box 设备箱
equipment breakdown 设备破损;设备故障;设备事故
equipment calibration procedure 设备校准程序
equipment capacity 设备能力
equipment capacity factor 设备利用率;设备负载因数
equipment center[centre] 设备中心
equipment change-over time 设备调换时间;设备更换时间
equipment change request 设备更改申请
equipment characteristic 设备性能
equipment characteristic distortion 设备特性失真
equipment charge 设备加油;设备充电
equipment charge out rates 施工机械台时费
equipment code 设备规范
equipment combination 仪器组合
equipment compatibility 设备相容性;设备兼容(性);设备互换性
equipment complex 整套设备;复合设备;设备综合
equipment component 设备元件;设备部件
equipment component list 全套设备部件零件表;设备元件明细表;设备部件零件表
equipment configuration control 设备外形检查
equipment control 设备控制;设备管理
equipment control switch 设备控制开关
equipment cooling 设备的冷却
equipment cost 设备购置费;设备费(用);设备成本
equipment cost estimate 施工机械费预算
equipment credit for small business 对小企业设备贷款
equipment cupboard 仪器储柜
equipment dedicated to project 工程专用设备
equipment delay-output 设备延迟输出
equipment dependability 设备可靠性
equipment depot 设备仓库
equipment depreciation 设备折旧
equipment depreciation charge apportion and over haul charges 设备折旧费摊销费及大修费
equipment design 设备设计
equipment dispatch order (集装箱)空箱提交单
equipment division 设备科
equipment drawer 设备柜
equipment drawing 设备图(纸);装置图
equipment economic life 设备经济使用寿命
equipment engineering 设备工程
equipment error 设备误差
equipment facility 现有生产设备
equipment factor 仪器因素
equipment failure 设备故障
equipment failure rate 设备故障率
equipment fault diagnosis 设备故障诊断
equipment floor 放置设备的楼层
equipment flow chart of information 施工机械使用核算流程图
equipment flowsheet 设备流程图
equipment for controlled station 被控制设备
equipment for cutter positioning 刀具定位装置
equipment for electronic anti-interference 电子反干扰设备
equipment for excavation 采掘设备
equipment for flue curing 烟熏干燥装备
equipment for illumination 照明装置

equipment for maintenance shop 修配车间设备
equipment formats 施工机械表格式
equipment for mechanics of materials 材料力学试验设备
equipment for terminal station 终端站设备
equipment for water supply works 水源工程设备
equipment foundation 设备基础
equipment funds 设备资金;设备基金
equipment ga(u)ge 设备限界
equipment gear 辅助齿轮;备用齿轮
equipment ground 设备(外壳)接地;接地连接
equipment grounding 接地连接;设备接地母线
equipment hatch 设备舱口
equipment-hour 台时
equipment identification code 设备识别符号
equipment information 设备资料
equipment in proximity to contact net system 临近接触网系统的设备
equipment input noise sound level 设备输入噪声级
equipment inside the tunnel 洞内设备
equipment inspection 设备检查
equipment inspection expenses 设备检验费
equipment installation 设备安装
equipment installation and checkout 设备安装与检修
equipment installation cost 设备安装费
equipment installation notice 设备安装通知
equipment interchange 设备交换
equipment interchange receipt (集装箱码头的)设备交接单
equipment interface 设备接口
equipment in tunnel 洞内设备
equipment in value 有价值的设备
equipment inventory 设备库存量
equipment investment 设备投资
equipment investment function 设备投资函数
equipment investment plan 设备投资计划
equipment isolation 设备隔振
equipment item 设备项目;单项设备;设备细目
equipment layout 设备布置
equipment lease 设备租赁
equipment leasing 设备租赁
equipment leasing by financing 融资租赁
equipment level 设备层
equipment life 设备寿命
equipment list 设备(一览)表;设备清单
equipment lock 设备运输气闸
equipment locker 设备柜
equipment locker frame 设备柜支架
equipment locker lighting 设备柜照明
equipment maintenance 设备维修;设备保养
equipment malfunction 设备失灵
equipment manufacture 设备制造
equipment manufacturing industry 装备工业
equipment modification 改进设备;设备改装;设备改造;设备改良
equipment modification list 设备变更明细表
equipment name 设备名(称)
equipment number 舣装数;设备号
equipment of a ship 船舶装备
equipment of balance test 均衡试验设备
equipment of boring flow measurement 钻孔流量测定工具
equipment of dispersion test 弥散试验设备
equipment officer 消防设备管理员
equipment of groundwater modeling test 地下水模拟试验设备
equipment of hydrogeologic(al) exploration test 水文地质勘探试验设备
equipment of lifeboat 救生艇属具
equipment of life float 救生浮属具
equipment of manual lifeboat 人力救生艇属具
equipment of measuring the groundwater level 地下水位测量工具
equipment of pumping test 抽水试验设备
equipment operational procedure 设备操作顺序
equipment optic(al) fiber cable 设备用光缆
equipment optimization 设备最优化
equipment outlay 装备费用;设备费(用)
equipment out-of-use 设备停用
equipment output port 设备输出口
equipment outside the tunnel 洞外设备
equipment overhaul 设备检修

equipment package 整套装置;整套设备
equipment patch bay 设备接线架
equipment performance 设备性能
equipment performance log 设备运转状况记录簿;设备性能表;设备工作日志
equipment periodic maintenance 设备定期维护
equipment planning 设备计划
equipment pool 设备修理场;设备库;设备供应源
equipment practice 设备运行
equipment precinct 设备管理区
equipment production study 设备生产研究;设备生产调查
equipment product line 设备生产线
equipment protection system 机组备用制;机组备用方式
equipment quality cost 设备质量成本
equipment quay 舣装码头
equipment rack alarm 设备机架报警
equipment rebuilding 设备大修
equipment receipt 设备收据;(集装箱码头的)设备交接单
equipment record card 设备记录卡片
equipment recordkeeping 设备记录簿
equipment records 施工机械的记录
equipment registration 产品注册;产品定型
equipment regulations 设备操作规程
equipment remove 设备拆迁
equipment renewal 设备更新
equipment renovation 设备更新
equipment rental 施工机械的租金;设备租用;设备租金
equipment rental charges 设备租赁费;设备租借费
equipment rental compilation 出租设备汇编
equipment rental contract 设备出租合同
equipment rental cost 设备租赁费;设备租借费
equipment rental rate 设备租费
equipment renting 设备出租
equipment repaired method 设备修理方法
equipment repair yard 设备修理场;设备维修场地
equipment replacement 设备更新
equipment replacement model 设备更新模型;设备更新模式
equipment replacement policy 设备更新政策
equipment replacement problem 设备更新问题
equipment requirement 设备要求
equipment requirement specification 设备要求规格
equipment restoration 设备修复
equipment room 设备室;设备间;设备房
equipment room at station end 站台端部设备用房
equipment schedule 设备(一览)表
equipment selection 设备选择;设备选型
equipment selection model 选购设备模式
equipment service 设备维修
equipment set 整套装置;整套设备
equipment shakedown 设备试运转
equipment shanty 设备小屋;存放设备的小屋
equipment shed 设备库
equipment siting 设备定位
equipment sparing 设备备用
equipment specification 设备说明书;设备技术规范;设备技术规定
equipment standard 设备标准
equipment station 器材站
equipment status monitoring 设备状况监视
equipment storage hut 存放设备小屋;存放设备棚屋;机械库
equipment stor(e)y 放置设备的楼层
equipment subsidy 征用船舶的设备补贴
equipment supplier 设备供应商
equipment supply depot 设备供给仓库
equipment supporting deck 设备支承构架
equipment system 设备系统
equipment test facility 设备试验装置;设备检验装置
equipment tonnage 舣装吨位;船舶属具吨位
equipment train 设备系列
equipment trouble 设备故障
equipment trust 成套设备托拉斯
equipment trust bond 设备信托(公司)债券
equipment unit 硬设备;单项设备;设备部件
equipment upgrading 设备改造
equipment utility constant 设备利用常数
equipment utilization 设备利用
equipment utilization quota 设备利用定额

equipment work 设备工程
equipment workshop 设备车间
equipment yard 设备维修场地
equipoise 均势;均衡;静平衡位置;平衡物;平衡力;使相称;使平衡
equipoise rudder 平衡舵
equipolarization 等配极变换
equipolar loudspeaker 等极式扬声器
equipollence 等力
equipollent 均等的;等力的;等价
equipollent load 等力荷载
equipollent press forming 均衡压制成型
equiponderance 等重;等力;等功
equiponderant 均衡的;等重物
equiponderant state 均衡状态
equiponderate 均衡;等重的;补整;平衡;使平衡
equipotent 等力的
equipotential 等势(的);等(电)位的;等电势;潜力上均等的
equipotential boundary 等位边界
equipotential cathode 等电位阴极;旁热式阴极
equipotential connection 等电位连接
equipotential contour 等势同高线;等电位线图
equipotential curve of velocity 等速度势线
equipotential layer 等位层;等势层
equipotential lens 等电位透镜
equipotential line 等水头线;等势线;等(电)位线
equipotential metallization 等电位金属化
equipotential method 探查法;等位法;等势法
equipotential pitch 等位距
equipotential plane 等势面;等(电)位面
equipotential plate 等位片;等势片
equipotential point 等电位点
equipotential ring 等位环;等势环
equipotential slope 等电位倾斜面
equipotential surface 同位面;等位面;等势面;同势面
equipotential surface of gravity 重力等势面
equipotential surface of velocity 等速度势面
equipotential survey 等势测量
equipotential system 等电势系统
equipotential temperature 等位温
equipotential winding 均压绕组
equipped 装备的
equipped capacitor 装机容量;配备容量
equipped with 装备有
equipping 装备的;配备的
equipping for land/marine work 为了适用陆上/海上作业安装
equipressure boiler 等压锅炉
equipressure cycle 等压循环
equipressure line 等压线
equipressure surface 等压面
equiprobability 机率相等;等概率
equiprobability curve 等概率曲线
equiprobable choice 等可能选择
equipt 装备;配备
equirectangular projection 等矩形投影
equiripple response 等连波响应
equisaturation 等饱和度
equisaturation program(me) 等饱和度程序
equiscalar 等纯量曲面;等标量
equisetum peat 木贼泥炭
equi-shaliness line 等泥质含量线
equisignal 等信号
equisignal effect 等信号效应
equisignal glide path 等信号滑翔道
equisignal localizer 等信号着陆指标;等信号式定位器
equisignal navigator 等信号导航台
equisignal radio-range beacon 等信号着陆信标
equisignal radio station 等信号无线电台
equisignal sector 等信号区
equisignal system 等强信号制
equisignal type 等信号型
equi-spaced 等间距的;等步长的;均布;等间隔的;平均间隔
equistability 等稳定性
equi-stepped 等步长的
equisubstantial 等质
equitable 公正的
equitable apportionment doctrine 衡平法分配原则;按比例分配
equitable assets 平衡法上遗产

equitable assignment of debt 平衡法上债务让与
equitable distribution 合理分配
equitable distribution of burden 合理负担
equitable interests 平衡法上利益
equitable mortgage 平衡法抵押
equitable redemption 回赎
equitable right of redemption 回赎权
equitable risk 合理危险性
equitable servitude 土地无限使用权
equitable settlement 公平解决
equitable share of world export trade 世界出口贸易公平比率
equitable sharing 均等共享
equitable tax 纳税公平
equitable waste 平衡法上毁坏
equi-temperature 等温(度)
equi-temperature metamorphism 等温变质(作用)
equitime 等时
equitransference 等传输
equity 资产净值;主权;公认之事;公平;公道;产权
equity build-up 资本增益;权益增加
equity capital 自有资本;主权资本;股东资本;股本资本;产权资本;权益资本;投资于新企业的资本;投入资本
equity capital turnover 投资总额周转率
equity convertible Eurobonds 可转换股权欧元债券
equity cost 权益成本
equity crude oil 股本油
equity-debt ratio 资产负债比(率)
equity dividend 股息
equity dividend rate 股息率
equity earnings 参股收益
equity financing 增股筹资;股本筹措;权益融资
equity investment 直接投资;股票投资;股本投资
equity investment audit 股票投资审计
equity investor 权益投资者
equity joint venture 合资(经营);股权式合资经营;股份合营(企业)
equity line of credit 资产信贷额度
equity loan 资产抵押贷款
equity margin 权益保障金
equity method 吸收法(即产权净值法)
equity method of valuation 净值估价法
equity mortgage 资产抵押贷款
equity of redemption 回赎权;平衡法上赎回权;赎回抵押权
equity ownership 业主权益;资产所有权;产权所有权;产权衡平法
equity participant 参股者
equity participation 资本参与;资本分担;出资方式;参股
equity principle of taxation 纳税公平原则
equity purchaser 权益购买人;产权购买人
equity ratio 股权比率;产权率;主权比率;产权比率;权益比率
equity receiver 产权清算管理人;清理财产接管人
equity security 股票
equity stake mortgage 收益分成抵押
equity tax 公平税
equity trading 举债经营
equity transaction 股票交易;净值交易事项;权益交易
equity treatment of commerce 通商待遇公平
equity-type investment 资本性的投资
equity value 权益值;产权值
equity yield rate 股本回报率
equivalence 对等物;等值(性);等效;等价;等积;当量的
equivalence algorithm 等价算法
equivalence between list 列表间的等价性
equivalence between two algorithm 两个算法间的等价性
equivalence class 等价类
equivalence declaration 等价性说明
equivalence distribution 当量分布
equivalence element 符合元件
equivalence factor 换算系数;等值系数;当量系数
equivalence group 等价组;等价群
equivalence isomorphism 等价类质同像
equivalence of dependency 依赖的等价
equivalence of expression 表达式的等价
equivalence of mixed layer 混层等价
equivalence of query 查询等价

equivalence of same layer 同层等价
equivalence of standards 标准的等效性
equivalence operation 等价操作
equivalence point 等量点;等当(量)点
equivalence principle 等效原理
equivalence problem 等价问题
equivalence problem of flowchart schema 框图模式的等价问题
equivalence proposition 等值命题
equivalence radius of pool 油藏等效半径
equivalence radius of supply area 供给区等效半径
equivalence relation 等(价)值关系
equivalence statement 等值语句;等价语句
equivalence static load 等效静荷载
equivalence transformation 等价变换
equivalence zone 最适比例带
equivalency 等值
equivalent 相等物;相等的;相当的;对应词;对等;等值的;等效的;等量的;等价的;当量的;同价品
equivalent absolute nozzle flow 喷嘴绝对等值流
equivalent absorption 等效吸收
equivalent absorption area 吸声量;等效吸收面积;等效吸声面积;等效吸声量
equivalent absorption surface 等效吸收表面
equivalent acidity 当量酸度
equivalent activity 等效性;等量活动
equivalent address 等效地址
equivalent admittance 等效导纳
equivalent airspeed 等量气速;当量空速
equivalent alternative 等效行动方案
equivalent altitude 等效高度
equivalent amplification area 等效放大面积
equivalent amplification function 等效放大函数
equivalent angle of internal friction 等效内摩擦角
equivalent annual cost 每年成本等值;等值年成本
equivalent annual cost method 等值年成本法
equivalent annual worth comparisons 年等值比较法
equivalent anomaly value 当量异常值
equivalent antenna 等效天线
equivalent aperture 等效孔径
equivalent aquifer 等效含水层
equivalent area 换算面积;等效面积;当量面积
equivalent articulation loss 等效清晰度衰减
equivalent aspect 等效示像
equivalent automaton 等价自动机
equivalent axis coil 等效轴线圈
equivalent background irradiance 等效背景辐照度
equivalent bandwidth 等效频带宽;等价带宽
equivalent base 等价基
equivalent base shear method 底部剪力法
equivalent basicity 当量碱度
equivalent beam 等效梁
equivalent beam method 等效梁(方)法;等值梁法
equivalent beamwidth 等效束宽
equivalent bed 同位(地)层
equivalent bedrock 等效基岩
equivalent bending moment 等效弯曲力矩;等效弯距;等量弯矩
equivalent bending rigidity 等代抗弯刚度
equivalent binary digit 等值二进数;等效二进位数;等价的二进制数字;等效二进位;等价二进位;等价二进数
equivalent binary digit factor 等价二进制数位因子;等价的二进制因子
equivalent biochemical oxygen demand 等效生化需氧量
equivalent bit rate 等效比特率
equivalent blackbody temperature 等效黑体温度
equivalent breadth 等值宽度
equivalent capacitance 等效电容
equivalent car 当量小汽车
equivalent carbon content 当量碳含量
equivalent change 等效变化
equivalent characteristic 等效特性曲线
equivalent characteristic curve of hydraulic coupling 液力耦合器等效特性曲线
equivalent charge 当量电荷
equivalent charge density 当量电荷密度
equivalent circle diameter 当量圆直径
equivalent circle diameter of tyre imprint 轮迹当量圆直径
equivalent circuit 等值电路
equivalent circuit of piezoelectric(al) resonator

压电振子等效电路
equivalent circuit of transducer 换能器等效电路
equivalent circular section 相当圆截面
equivalent circulating density 当量循环密度
equivalent circulating mud weight 循环当量比重
equivalent clothoid 当量缓和曲线
equivalent coal 当量煤
equivalent code 等码
equivalent coefficient 当量系数
equivalent coefficient of local resistance 局部阻力等值系数
equivalent coefficient of mass 质量等效系数
equivalent coefficient of rock 岩石等效系数
equivalent cohesiometer value 当量黏聚值
equivalent collecting area 等效面积
equivalent combination 等效配合法
equivalent combustible weight 当量可燃物重量
equivalent compaction pressure 等效击实压力;当量击实压力;等效夯实压力
equivalent compressive force 等效压力
equivalent compressive stress 当量压应力
equivalent concentrated reaction 等价集中反力
equivalent concentration 等效集中;当量浓度
equivalent concentration load 等效集中荷载
equivalent concept 等价方程
equivalent condition 等面积条件
equivalent conductance 等效电导;当量导电(率)
equivalent conductivity 等值电导率;等效电导率;等效导流能力;等效传导性;当量电导率
equivalent conic(al) projection 等积圆锥投影
equivalent consolidation pressure 等效固结压力
equivalent constant 等效常数
equivalent constant wave 当量固定波
equivalent contact angle 等效接触角
equivalent continuous A sound level 等效连续A声级
equivalent continuous A-weighted sound level 等效连续A声级
equivalent continuous noise level 等效连续噪声级
equivalent continuous sound pressure level 等效连续声压级
equivalent correction 等效校正
equivalent cost 相等成本
equivalent crack length 等效裂纹长度
equivalent cross-section 换算断面;等值截面;等效截面;等效断面
equivalent cross-section area 等效断面面积
equivalent cube 等效立方体
equivalent cube method 等效立方体方法
equivalent cube test 等效立方体试验
equivalent current 当量电流
equivalent curve 等值曲线;等效曲线
equivalent cylindric(al) projection 等积圆柱投影
equivalent damping constant 等效阻尼常数
equivalent damping ratio 等效阻尼比
equivalent deepwater wave height 换算深水波高;等效水深波高;等效深水波(波)高;等效换算深水波高
equivalent defect 等代缺陷
equivalent depth 等效水深;等效水深;等效深度;等量深度;当量深度;当量厚度
equivalent depth wave height 换算水深波高
equivalent design flow 当量设计流量
equivalent design method (路面的)等效设计法
equivalent detector 等效检波器
equivalent deviation 等价偏差
equivalent device 等效装置
equivalent diagonal strut 等效斜撑
equivalent diameter 换算直径;相当直径;等效直径;等效粒径;当量直径;标称直径
equivalent difference 当量差
equivalent diffusion distance 当量扩散距
equivalent direct radiation 等效散热
equivalent discharge 换算流量;等效流量
equivalent distribution 等效分布
equivalent dose rate 等效剂量率
equivalent ductility factor 等效延性系数
equivalent duration 等效持续时间
equivalent duration sound level 等效持续声级
equivalent earth's radius 等效地球半径
equivalent eccentric wheel 等效偏心轮
equivalent education 同等学历
equivalent effective radius 当量有效半径
equivalent effective strain 等效应变

equivalent effective stress 等效应力
equivalent elastic analysis 等效弹性分析
equivalent elastic method 等效弹性法
equivalent elastic set 弹性沉降当量;弹性变定当量
equivalent electric(al) circuit 等效电路
equivalent electron density 等效电子密度
equivalent electrons 等效电子
equivalent embedment length 等效埋入长度
equivalent energy 能当量
equivalent equation 等价方程
equivalent evapo(u)ration (锅炉)等温蒸发量;蒸发当量;等效蒸发量;等量蒸发;当量蒸发(量)
equivalent exchange 等价交换
equivalent excited effective mass 受激有效质量
equivalent explosive 等效炸药
equivalent exposure 等值暴露面
equivalent factor 换算系数
equivalent fire endurance 等效耐火性
equivalent fire severity 相当的火灾烈度
equivalent flat plate area 当量平板面积
equivalent flaw diameter 当量缺陷直径
equivalent flocculation 当量絮凝作用
equivalent fluid 等效流体
equivalent fluid method 等流体法
equivalent fluid pressure 当量液压
equivalent focal distance 等效焦距
equivalent focal length 等值焦距;等效焦距
equivalent focus 等值焦点;等效焦点
equivalent footing analogy 等效基础模拟
equivalent force 等效力;当量力
equivalent forced outage rate 当量强迫停机率
equivalent formula 当量式
equivalent four-wire system 等效四线制
equivalent fracturing 等效压裂
equivalent frame 等效框架
equivalent frame method 等效框架法;等代框架法
equivalent free-falling diameter 等效自由沉降直径
equivalent fresh air supply 等量送风
equivalent friction loss 当量摩擦损失
equivalent frontal area 最大截面的等效面积
equivalent full-power days 等效满功率天数
equivalent future 等价远景
equivalent gap 等效缝隙
equivalent gate current 等效栅极电流;等效门电流
equivalent ga(u)ge reading 换算水位读数;相应水位读数;等值水位读数;等效水位读数
equivalent generator 等效发生器;等效发电机
equivalent generator law 等效发生器法则
equivalent generator theorem 等效发电机定理
equivalent girder 相当桁;等强桁
equivalent grade 换算坡度;换算坡道;等值坡度;等效坡度;等效粒级;当量粒度
equivalent gradient 当量坡度
equivalent gradient of additional resistance 附加阻力换算坡度
equivalent gradient of additional resistance on curve 曲线附加阻力换算坡度
equivalent gradient of additional resistance on tunnel 隧道附加阻力换算坡度
equivalent gradient resistance 计算坡道阻力
equivalent grain diameter 等效颗粒直径
equivalent grain size 等效粒径;等价粒径;当量粒径
equivalent ground dimension 相应实地尺寸
equivalent head wind 等效逆风
equivalent heat 当量热
equivalent heat conductivity 等效导热系数
equivalent height 虚高;等值高度;等效高度
equivalent height of surcharge 超载的换算(等效)高度
equivalent hiding power 等效遮光力
equivalent homogeneous system 等效均质系统
equivalent horsepower 当量功率
equivalent hydrostatic(al) pressure 当量静水压力;当量水压
equivalent hysteretic damping ratio 等效滞变阻尼比
equivalent impedance 等值阻抗;等效阻抗
equivalent index of thin-film 薄膜等效折射率
equivalent inductance 等效电感
equivalent information content 等效信息量
equivalent injection time 等效注入时间
equivalent intensity 等效强度
equivalent isentropic velocity ratio 等熵速度比当量

equivalent isometric(al) latitude 等容积纬度
equivalent isotropic radiated power 等效各向同性辐射功率
equivalent law 互等定律;等价定律
equivalent law of shear(ing) stress 剪应力互等定律
equivalent layer 等值层;同位层;等效层
equivalent layer depth 当量层厚
equivalent length 等值长度;等效长度;当量长度
equivalent length of pendulum 等值单摆长
equivalent length of pipe 管道当量长度
equivalent length of straight 等效直管长度
equivalent length of tracks 铁路线路折合长度
equivalent length on penetrated crack 当量穿透裂纹尺寸
equivalent lens 等焦透镜
equivalent level 等效电平
equivalent linearization 等效线性化
equivalent linear method 等值线性法
equivalent linear model 等效线性模型
equivalent line capacitance 线路等值电容
equivalent liquid pressure 当量液压
equivalent liquid weight 当量液重
equivalent liter 当量升
equivalent live load 均布活荷载;等效活荷载
equivalent load(ing) 等效静载;等效荷载;等效负载;等效负荷;等代荷载;等值荷载;等值负载;当量荷载
equivalent load method 等效荷载法
equivalent loudness 等(效)响度
equivalent loudness level 等效响度级
equivalent lumped system 等效集总体系
equivalent main 当量管道
equivalent map 等(面)积投影地图
equivalent map projection 等积地图投影
equivalent mass 等效质量
equivalent mass density 等效质量密度
equivalent mass of structure 结构的等效质量
equivalent material 等效材料;等代材料;替代材料
equivalent matrice 等价矩阵
equivalent matrix 当量矩阵
equivalent mean diameter 等效平均粒径
equivalent mean diameter of aggregate 骨料等效平均粒径
equivalent mean shearing stress 等效平均剪应力
equivalent member 等效杆件
equivalent mesh current 等效回路电流
equivalent method 等值法
equivalent mixture 当量混合物;当量混合料
equivalent modulus of deformation 当量形变模量
equivalent modulus of elasticity 等价弹性模量;当量形变模量;当量弹性模量
equivalent moisture 相当湿度;等价含水率
equivalent moment 等量力矩
equivalent moment of force 等量力矩
equivalent moment of inertia 当量惯性矩
equivalent mud density 当量泥浆密度
equivalent name 等价名称
equivalent Na standard deviation map 当量钠标准偏差图
equivalent network 等值网络;等效网络
equivalent neutral density 等效中性密度;当灰密度
equivalent nitrogen pressure 等效氮气瓶压力
equivalent nodal force 等效节点力;等价节点力
equivalent nodal load 等效节点荷载
equivalent noise 噪声等效声压;等效噪声
equivalent noise conductance 等效噪声电导
equivalent noise level 等效噪声级
equivalent noise method 等值噪声法
equivalent noise pressure level 等效噪声压级
equivalent noise resistance 等效噪声电阻
equivalent noise temperature 等效噪声温度
equivalent normal form method 等价范式法
equivalent nozzle 等效喷管
equivalent nuclei 等效核
equivalent number cycles 等循环作用次数
equivalent number of pulse 等效脉冲数
equivalent object program(me) 等价目标程序
equivalent observation 等效观测;等价观测
equivalent observation equation 等价观测方程
equivalent of binary tree 等价的二叉树
equivalent of coal 标准煤
equivalent of heat 热当量
equivalent of work 功当量

equivalent opacity 等效不透明度;当量不透明性
equivalent opening size 当量筛孔尺寸
equivalent optical thickness of thin-film 薄膜等效厚度
equivalent orifice area 等效孔口面积
equivalent outdoor temperature 等效室外温度
equivalent parallel resistance 等效并联电阻
equivalent parameter 等效参数
equivalent particle diameter 颗粒当量直径;等效粒径;等容粒径;当量粒径
equivalent particle size 粒子当量尺度
equivalent pendulum 等值摆;等效摆
equivalent performance 等效性能;实绩统一换算
equivalent period 等效周期
equivalent periodic(al) line 等周期线;等效周期线路
equivalent permeability 等效渗透率
equivalent permeability coefficient 等效渗透系数
equivalent per million 百万分之一当量
equivalent phenomenon of electric(al) sounding curve 电测深曲线等价现象
equivalent photograph 等效照片;等效相片
equivalent pipe 等效管路;等似管;等流管道;当量管(道)
equivalent piston 等效活塞
equivalent plane frame 等效平面框架
equivalent plane wave 等效平面波
equivalent plane wave reverberation level 等效平面波混响级
equivalent point 等价点;等当(量)点;当量点
equivalent point group 等效点系
equivalent-point potential 等价点电势
equivalent point source 等效点源
equivalent polygon 等积多边形
equivalent pore pressure 当量孔隙压力
equivalent porosity 等效孔隙度;当量孔隙度
equivalent position group 等效位置组
equivalent potential source 等效电势源
equivalent potential temperature 相当位温;等效位温
equivalent power 等效功率
equivalent pressure 等效压力
equivalent principal distance 等效主距
equivalent principle 等效原理;等价原理
equivalent production time 等效生产时间
equivalent profile 等效外形;等效断面
equivalent projection 等(面)积投影
equivalent proportion 当量比(定律)
equivalent radium content 等效镭含量
equivalent radius 等效半径;等积半径;当量半径
equivalent rainfall 等效降雨量
equivalent reactance 等效电抗
equivalent rectangular stress block 等效矩形应力图
equivalent rectangular stress distribution 等效矩形应力分布;等值矩形应力分布
equivalent reflectance 等效反射系数
equivalent relation 等价关系
equivalent relative density 当量相对密度
equivalent relative roughness 当量相对粗糙度
equivalent replacement method 等量置换法
equivalent resilient modulus 当量回弹模量
equivalent resistance 等效电阻
equivalent resistivity 等效电阻率
equivalent resistivity of formation water 地层水等效电阻率
equivalent roughness 等值糙率;等值糙度;等效糙率;当量粗糙度
equivalent roughness coefficient 等值糙率系数;当量粗糙系数
equivalent round 换算直径;等效直径;等效圆
equivalent sand grain roughness 当量砂径糙率
equivalent saturation 等效饱和度
equivalent scale 当量比例尺
equivalent section 等圆断面;等效截面;等效断面;等代断面;当量截面
equivalent sectional radius 等效截面半径
equivalent section area 等效断面面积
equivalent seismic force 等效地震力
equivalent seismic load 等效地震荷载
equivalent separation 等效间距
equivalent sequence 等价序列
equivalent settlement coefficient for calculating settlement of piled 桩基等效沉降系数

equivalent shaft 等效轴(杆)
equivalent shear modulus 等效剪切模量
equivalent sheathed explosive 当量套筒炸药
equivalent simple oscillator 等效简单振子
equivalent sine wave 等效正弦波
equivalent sine wave intensity 等效正弦波强度
equivalent single conductor 等值单根导线
equivalent single course 直航向
equivalent single wheel load 当量单轮荷载
equivalent site 等同位置
equivalent size 当量粒度;等值粒度;等效尺寸;当量尺寸
equivalent(s) of charge 资费等值
equivalent soil layer method 等值层法
equivalent sound absorption 等效吸声量
equivalent sound level 等值声级;等效声(压)级
equivalent sound level difference 等效声压级差
equivalent source 等效电源
equivalent space charge 等效空间电荷
equivalent span length 当量跨距
equivalent specific fuel consumption 当量比油耗
equivalent specific heat 等效比热
equivalent specific weight 当量比重
equivalent spectrum curve 等效谱曲线
equivalent sphere 等效球;当量球体
equivalent sphere diameter 等球粒直径
equivalent sphere illuminance 等效球照度
equivalent sphere illumination 等效球面照明
equivalent spheric(al) diameter 当量球径
equivalent spring 当量弹簧
equivalent spur gear 当量直齿轮
equivalent star capacitance 等效星形电容
equivalent state 等效状态;等价状态
equivalent statement 等价说明
equivalent static analysis 等效静力分析
equivalent static criterion 等效静力准则
equivalent static load 等代静载;等效静荷载
equivalent static method 等效静力法
equivalent stationary motion 等效稳态运动
equivalent stationary response 等效稳态反应
equivalent steam 当量蒸汽;标准蒸汽
equivalent stiffness 等效刚度;等效刚度;当量刚度
equivalent stiffness ratio 等效刚度比
equivalent stopping power 等效阻止本领
equivalent strain 等效应变
equivalent stress 换算应力;相当应力;折算应力;等效应力
equivalent stress yield criterion 等效应力屈服条件
equivalent string 等效数字符串
equivalent structure 等效结构
equivalent surcharge 等值超载
equivalent surface 等效面;等位面
equivalent surface diameter 等效表面直径
equivalent system 等效系统;等价系统
equivalent system for qualitative investigation 当量质量集中系
equivalent system of equations 等价方程组
equivalent table 换算表
equivalent tail wind 等效顺风
equivalent taxable yield 相等课税收益率
equivalent temperature 相当温度;等效温度;当量温度
equivalent tensile force 等效拉力;当量张力
equivalent test 等效试验法;等效实验
equivalent Th colo(u)red map 当量钍彩色图
equivalent theorem 等效定理
equivalent thermal conductivity 表观导热系数;表观传热系数
equivalent thickness 等值厚度;等效厚度
equivalent thickness concrete slab 等效厚度混凝土板
equivalent Th standard deviation map 当量钍标准偏差图
equivalent time 等值时;等较时间
equivalent time horizon 等时面
equivalent to 折合;等于
equivalent-to-element 符合元件
equivalent torsion stiffness 等代扭刚度
equivalent track kilometers 铁路线路折合长度;换算轨道公里
equivalent traffic(volume) 当量交通量
equivalent traffic volume of grade crossing 道口折算交通量
equivalent transfer function 等效传递函数

equivalent transformation 等价变换
equivalent transform of block diagram 方框图等效变换
equivalent transverse beam method 等代横梁法
equivalent tree 等价树
equivalent trend line 当量趋势线
equivalent triangle 等积三角形
equivalent triangular burst 等效三角形脉冲群
equivalent tube 等似管;当量管(道)
equivalent twenty-four hours pulse rectify 等效24小时脉波整流
equivalent twisting moment 等量扭矩
equivalent U colo(u)red map 当量铀轴彩色图
equivalent uniform annual cost 每年相等的费用
equivalent uniform distributed load 等效均布荷载
equivalent uniform live load 等效均布活载
equivalent uniform load 换算均布荷载;等效均布荷载;等代均布荷载
equivalent uniform solid 当量均质体
equivalent unit 换算单位;约当产量;等效部件;当量单位;统一换算单位;实际单位
equivalent utility curve 等效用曲线
equivalent value 换算值;等效值;当量值
equivalent valve 等效电子管
equivalent vapo(u)r volume 当量蒸汽体积
equivalent veiling luminance 等效罩纱亮度
equivalent velocity 等效速度
equivalent viscosity 等效黏度
equivalent viscous damping 等效黏滞(性)阻尼
equivalent volt 电子伏特
equivalent volume 等效体积
equivalent volume diameter 等效体积直径
equivalent water level 相应水位;相当水位;等值水位;等(效)水位
equivalent water stage 等效水位
equivalent wavelet 等效子波
equivalent weight 换算重量;化合当量;等效重量;当量(重量)
equivalent weight of acid and base 酸碱当量
equivalent weight per million method 当量百分法
equivalent weight ratio of chloride ion-sodium ion to magnesium ion 氯离子钠离子/镁离子当量比
equivalent weight ratio of sodium ion-chloride ion to sulfate ion 钠离子氯离子/硫酸根离子当量比
equivalent weight ratio of sodium ion to chloride ion 钠离子/氯离子当量比
equivalent weight replacement method 等重量代替法
equivalent wet bulk temperature 等效湿球温度
equivalent wheel load 等效轮载;等代轮载
equivalent width 换算宽度;等值宽度;等效宽度;等宽
equivalve 等壳瓣;等瓣的
equivelocity contour 等速恒值线;等流速线
equiviscous 等黏滞性的
equiviscous temperature 等黏(滞)温度
equiviscous temperature application range 等黏度温度涂布段
equivocal data 不可靠数据
equivocal surface 模糊曲面
equivocation 模糊(度);多义;不明确;条件信息量总平均值
equivolume inclination 等体积倾斜
equivoluminal 等体积的
equivoluminal wave 等体积波
equi-wavelength 等波长
equi-wavelength pattern 等波长图
era 纪元;代【地】;时代
Era 耐蚀耐热合金锯
eradiation 地面辐射
eradicant 铲除剂;铲除的
eradicant action 铲除作用
eradicant eating 铲除法
eradicant fungicide 直接杀菌剂;铲除性杀菌剂
eradication of disease 消除疾病
eradication of pests and elimination of diseases 除害灭病
eradicator 消色灵;除草机;褪色灵;根除器
eraducation 根除
Erap Agreement 埃拉普协议
erasability 记录可消除性;可擦性;可擦度;消磁程度
erasability of storage 存储器的可擦除性

erasable 可清除的;可擦(除)的
erasable area 可清除区
erasable memory 可擦存储(器)
erasable optic(al) disk 可擦写光盘
erasable programmable 可擦除编程
erasable read-only memory 可擦只读存储器
erasable storage 可清存储器;可擦存储(器)
erase 抹去;抹迹;抹掉;抹除;消除;擦去;擦掉;擦除;清除
erase amplifier 消音放大器;消抹放大器;消除放大器
erase character 抹去字符;擦除符;取消符;删去符;删除字符
erase circuit 抹迹电路
erase command 清除指令
erase error 擦除错
erase frequency 消磁频率
erase head 抹音(磁)头;抹去磁头;消磁头;擦除磁头
erase input 消去信号输入
erase interlock 抹音联锁装置
erase key 消磁按钮
erase list 清除表格
erasement 消磁
erase oscillator 抹除振荡器;消迹振荡器;消磁振荡器
eraser 消磁头;消磁器;消除用具;消除器;橡皮(擦);撤该消除器;擦去器;擦抹用具;擦除者;擦除器;清洗器
erase residual 清洗残余
eraser gate 消除装置开关;擦去装置门
eraser knife 刮具;刮刀
eraser shield 擦图(挡)片
eraser system 消磁系统
erasing ability 可抹性;抹去能力
erasing factor 消声因数;消声系数;消磁因素
erasing head 抹音(磁)头;消磁头
erasing knife 刮刀
erasing move 擦除动作
erasing of binary tree 二叉树的擦去
erasing of information 信息清除
erasing of tree 抹去树
erasing production 删符产生式
erasing pulse 擦除脉冲
erasing record 消去记录
erasing rubber 擦字橡皮
erasing shield 擦图孔板;擦图(挡)片
erasing speed 抹去速度;抹迹速率;擦除速度
erasing time 抹去时间;消磁时间;擦除时间
erasing tool 清除工具
erasion 抹掉;消迹;刮术
erasure 消除(物);涂擦处;删去处
erasure channel 删除信道;删除通道【计】
erasure effect 消音效果;消磁效果
erasure locator 疑迹定位器;删除定位器
erasure signal 消除信号;清除信号
Eratosthenes sieve method 厄拉多塞筛法
eratosthenian 爱拉托辛纪
e-ray 非常光线
Erba deep 埃尔巴海渊
Erben's reflex 埃尔本反射
erbium 铒激光器
erbium-doped optic(al) fiber amplifier 掺铒光纤放大器
erbium-doped silica fiber amplifier 掺铒二氧化硅光纤放大器
erbium-doped single-mode fiber 掺铒单模光纤
erbiumdoped yttrium aluminium garnet crystal 掺铒钇铝石榴石晶体
erbium-doped yttrium lithium fluoride laser 掺铒氟化钇锂激光器
erbium ores 铒矿
erbium oxalate 草酸铒
Erb's point 欧勃点
Erb's wave 欧勃波
ercapsulate fertilizer 包被肥料
erdite 水硫铁钠石
Erechtheion 伊瑞克提翁神庙(古希腊)
erect 架立;直立(的)
erectable 可装配的
erectable whipstock 活动式造斜器
erect anticline 直立背斜
erect beacon 树立瓴标
erect boundary markers 立界柱
erect branch 直立枝

erect centering 竖立拱模
erect cover 直立覆盖植物
erected container 垂直安装的容器
erect embryo 直立胚
erect falsework 架立脚手架
erect film 立位片
erect fold 直立褶皱
erect framework 安装模板
erect head 直立穗
erectility 安装能力
erect image 正像
erect-image telescope 正像望远镜
erect-image view finder 正像取景器
erecting 装配
erecting act 安装证书
erecting arm 管片拼装架
erecting bay 装配台;装配(车)间
erecting bed 安装(现)场;安装台
erecting bill 构件明细表;安装顺序清单;安装材料单;工程安装顺序清单(指在常温湿度下)
erecting by floating 浮运架设(法);浮运架桥法;浮运法架设
erecting by floating into position 浮运就位法
erecting by overhang 吊挂架设;悬伸架设;吊挂装置
erecting crane 装配吊车;安装(用)起重机
erecting deck 架设桥台;装配平台
erecting derrick crane 安装用转臂起重机;安装用转臂吊机
erecting device 架设机构;安装装置
erecting diagram 装配图(样);安装图
erecting drawing 安装图
erecting equipment 安装设备;起重设备
erecting eyepiece 正像目镜
erecting fixture 安装架
erecting floor 装配场地;安装台
erecting frame 脚手架;装配用构架;安装架
erecting horn 安装喇叭
erecting image 直立图像
erecting jib 安装用起重杆;安装(用)吊杆;装配吊杆
erecting lens 正(像)透镜
erecting lens system 正像透镜系统
erecting machinist 装配工
erecting mast 安装用起重杆;装配桅杆
erecting overhang 吊挂架设
erecting platform 装配舞台;架设平台
erecting pontoon scaffolding 浮运脚手架
erecting pontoon stage 浮运脚手架;浮式脚手架
erecting prism 转正棱镜;正像棱镜
erecting scaffold 脚手架
erecting shop 装配(车)间;露天组装场;装配厂
erecting stage 脚手架;装配期;装配(平)台;安装现场;安装期;安装阶段
erecting system 正像系统
erecting telescope 正像望远镜
erecting tools 架设工具;安装工具
erecting tower 安装(用)塔架;吊装塔
erecting welding 装配焊接;安装焊接
erecting work 安装工作
erecting yard 装配(现)场;安装场
erection 架设;组装;装配;吊装;船楼;安装;竖立;设备安装;上层建筑;直立
erection aid 架设工具;安装工具
erection allowance 装配容许偏差
erection all risks 安装一切险
erection-all-risks insurance 安装工程一切保险
erection and commissioning 安装与调试
erection and construction plot 安装施工图表
erection and dismantling bay 装配(及)拆卸车间
erection area 装配间;安装场
erection bar 架立(钢)筋;吊装钢筋;安装杆
erection bay 装配间;安装间
erection bolt 组装螺栓;装配螺栓;安装螺栓
erection brace 安装支撑
erection bracing 安装支撑;架设支撑
erection by crane or derrick 吊装架设法
erection by dragging 拖拉施工
erection by falsework 脚手架设法
erection by floating 浮运架桥法;浮运落梁法;浮运法安装
erection by lateral rolling 横向移架设法
erection by launching 拖拉法;推进式架设;伸展架设法

erection by launching girder 导梁(架设)法
erection by longitudinal pulling 纵向拖拉法
erection by overhang 吊挂装配
erection by protrusion 悬臂拼装法;悬臂架设;伸展架设法;伸臂架设法
erection by protrusion and floating 伸臂浮运架设安装
erection by staging 用脚手架进行安装
erection by swing 旋转法安装
erection by swing method 旋转法施工;旋转法架设
erection by the use of a mobile gantry 活动脚手架设法
erection by working cable 空中吊索架设法
erection cable 安装缆索;安装绳索
erection clearance 安装净空
erection code 安装规程
erection column 竖立柱子;安装柱
erection contract 架设合同;安装合同
erection cost 安装费(用)
erection crane 安装(用)吊车;装配用吊;安装(用)起重机
erection crew 架设工作;安装队
erection deck 船楼甲板
erection device 架设设备;安装方法
erection diagram 架设图;装配图(样);安装图
erection dip 安装垂度
erection drawing 安装图样;装置图(样);装配图(样)
erection elevation 安装标高
erection end 船楼端
erection equipment 安装设备;安装润滑剂
erection error 安装误差
erection floating method 浮运架设(法)
erection force 安装队
erection frame 安装支架
erection gang 安装工作队
erection hinge 铰链装置;装置铰链
erection information 安装资料
erection insurance 安装保险
erection jack 安装用千斤顶
erection jig 安装用夹具
erection joint 装配接头;安装连接;安装接头
erection load 施工荷载;架设荷载;装配荷载;装配负载;安装荷载
erection loop 吊环;安装环
erection mark 装配证号;安装标高
erection material 安装材料
erection measurement 安装测量
erection method 安装方法
erection nut 安装螺帽
erection of bridge 架设桥梁;桥梁架设
erection of derrick 立井架
erection of engine 发动机安装
erection of forms 立模板;架设模板
erection of formwork 立模
erection of framing 构架安装
erection of machinery 机器安装
erection of shuttering 支模
erection of steel work 钢结构安装;钢架装置
erection of tank 储罐的安装
erection on site 现场安装
erection on the site 安装在工地上
erection party 安装工作队
erection pass 安装通道
erection person 安装人员
erection plant 装配设备
erection platform 架设平台;装配平台;安装平台
erection pressure 安装压力
erection procedure 安装步骤;安装程序
erection rate 安装速度
erection reference plane 安装基准面
erection reinforcement 脚手架安装
erection repair shed 架修库
erection rig 安装专用设备
erection rope 安装绳
erection sag 安装垂度
erection scaffold(ing) 安装(用)脚手架
erection schedule 安装进度;安装进程;安装计划
erection scheme 建造计划
erection screw 安装螺旋;安装螺丝
erection seat 装置座;安装支座
erection sequence 安装顺序;安装程序
erection service 安装服务
erection shop 装配(车)间

erection space 装配面积;安装面积
erection speed 安装速度
erection stage 舞台安装
erection strength 安装强度
erection stress 架设应力;装配应力;安装应力;施工应力
erection supervision 安装监督
erection support 安装支承;安装支座
erection survey 安装测量
erection system 垂直安装系统
erection team 安装(工作)队
erection time 安装期限
erection tolerance 安装允许误差;安装容许误差;安装公差
erection torque motor 架设转矩电动机;竖起力矩电动机
erection tower 吊装塔架;安装塔架;施工起重塔架
erection truss 安装桁架
erection unit 安装设备
erection weight 安装重量
erection weld 安装焊缝
erection welding 装配焊接;安装焊接
erection with cableway 缆索吊装(法);无支架吊装法
erection without scaffolding 无支架施工;不设脚手架装置;无支架安装(法);无脚手架安装
erection with pontoon 浮囷架设
erection with spacing 安装间距
erection work 装配工作;装配工种;安装工作;安装工程
erect number 正像数字
erect of signal 造标
erector 举重臂;建立者;架设器;装配者;装配机具,装配工;安装器;安装机;安装工(人);索具工人;升降架
erector arm 举重臂;起重臂
erector arm of tunnel(l)ing 隧道掘进机起重臂
erector beam 千斤顶;起重臂
erector jack 安装机千斤顶
erector of shield 盾构举重臂
erector track 拼装轨道
erect position 立位;站位
erect scaffold 竖立脚手架
erect stem 直立茎
erect telescope image 望远镜正像
E-region E区
eremacausis 木材在大气中慢性腐烂;慢性氧化;缓慢氧化
eremophyte 荒原植物;荒漠植物;旱生植物
Erfle eyepiece 埃弗利目镜
Erfurt faience 埃尔福锡釉陶器
erg 纯砂沙漠;尔格;砂质沙漠;砂质荒漠;沙质沙漠
Ergal 铝镁锌系合金;厄盖尔铝镁锌系合金
ergastic material 后含物
ergastic matter 后含物
ergastic substance 后含物
ergastulum (古罗马囚禁奴隶的)私有监狱
ergatandromorph 工雄蚁
erg desert 纯沙漠
ergh 纯砂沙漠
ergodic 遍历各态历经
ergodic condition 遍历条件
ergodic hypotheses 各态历程假说
ergodic hypothesis 遍历性假说;遍历假设
ergodicity 各态历经性;各态历程(性);遍历性
ergodicity law 各态历经定律;遍历定律
ergodic motion 各态历程运动
ergodic noise 遍历噪声
ergodic process 各态历程过程;遍历过程
ergodic property 各态历程性质;遍历性
ergodic random process 各态历程随机过程;遍历性随机过程
ergodic state 各态历程状态;遍历状态
ergodic system 埃尔过德系统
ergodic theorem 各态历程定理;遍历定理
ergodic theory 各态历程理论;遍历理论
ergogram 测力图;尔格图;示功图
ergograph 测力器;测功器;测功计;疲劳记录计;示功器
ergography 测力法
ergometer 功率计;测力计;测功计;尔格计
ergometric 测力的
ergomomic study 工作条件研究
ergon 尔格子;尔刚(光子能量单位)

ergonometrics 工效学;人类工程计量学
ergonomical 人体功率学的
ergonomics 工效学;工程生理学;人与机械控制;人因工程(学);人体功率学;人类工程学;人机(工程)学
ergonomic study 劳动强度研究;工程生理研究
ergonomic survey 人类工程学调查
ergonomtrics 人类工效学
ergoregion 能区
ergosphere 能层
ergosphere effect 能层效应
ergostane 麦角烷
ergosterol 麦角甾醇
ergotropic system 增进抵抗力系统
Ergun-Altun tectonic zone 额尔古纳—阿尔金构造带【地】
Ergun fold system 额尔古纳褶皱系【地】
Ergun geosyncline 额尔古纳地槽【地】
Ergun He tectonic segment 额尔古纳河构造段【地】
Ergun old land 额尔古纳古陆【地】
Erian 伊里亚统【地】
Erian orogeny 伊里亚造山运动【地】
erianthus gianteus 芒草
ericaite 铁方硼石
Erical mo(u)ld 可压锭模;挤压锭模
ericelal 沼泽植物
Erichsen cupping machine 埃氏冲盂试验机
Erichsen deep drawing cup test 埃氏拉延试验
Erichsen depth index 埃氏深度指标
Erichsen ductility machine 埃氏金属板延展性试验机;埃里克森深拉试验机
Erichsen indentation test 杯突试验;埃里克森压痕试验
Erichsen number 杯突(深度)值;埃里克森杯突深度值
Erichsen test 拉伸性能试验;杯突试验;埃氏杯突试验;埃里克森试验
Erichsen test coupon 拉伸试件
Erichsen tester 杯突试验机;埃里克森试验机
Erichsen value 埃氏(杯突深度)值;埃里克森氏杯突深度值
Ericsonia 埃氏颗石
Ericsson cycle 埃里克森循环
ericssonite 钡锰闪叶石
Ericsson screw 埃里克森螺纹
Ericsson selector 埃里克森选择器
Ericsson separating filter 埃里克森分隔滤波器
Ericsson system 埃里克森制
Eridite 电镀中间抛光液
Erienmeyer flask 爱伦美瓶
Erie Railroad Company 伊利铁路公司(美国)
erigeron oil 飞蓬油
erikite 磷铈石
erinite 墨绿砷铜石;玄武岩蚀变红黏土;翠绿砷铜矿
Eriocheir sinensis 毛蟹;中华绒螯蟹
eriochrome black 依来铬黑
eriochrome blue black (依来)铬蓝黑
eriochrome red 依来铬红
eriogonum inflatum 紫葳花
eriogreen B 黄光酸性绿
eriometer 衍射测微器;衍射测微计;绕射测微计
erionite 毛沸石
eriophorum peat 苔藓泥炭
eriophorum peat moss 苔藓泥炭沼
erisma 墙壁扶垛;扶壁;临时支撑扶架
Eritrea 厄立特里亚
erkensator 立式离心除砂机
Erlang 厄尔兰
Erlang density function 到达时间分布密度函数(排队论)
Erlang distribution 厄尔兰分布
Erlanger blue 厄尔兰格蓝;铁蓝
Erlanger loan 厄尔兰贷款
Erlangian arrival 厄尔兰到达
Erlangian distribution 厄尔兰分布
Erlang's loss formula 厄尔兰损失公式
Erlang's density formula 厄尔兰密度分布公式
Erlang's distribution 厄尔兰分布
Erlang's formula 厄尔兰(损失)公式
Erlang's loss formula 厄尔兰公式
Erlenmeyer flask 厄伦美尔(烧)瓶;锥(形烧)瓶;三角烧瓶

erlichmanite 硫锇矿
ermakite 褐蜡土
Ermalite 厄马拉依特(重载高级)铸铁
ernstite 羟磷钼锰石
erodability 可侵蚀性;可侵蚀度
erode 腐蚀
eroded 有蚀痕的;被浸蚀的;被腐蚀的;缺刻状的;侵蚀的
eroded canyon 侵蚀峡谷
eroded crater 侵蚀坑
eroded field 冲刷土地;侵蚀土地
eroded formation 冲刷形成的地层;侵蚀形成的地层
eroded hole 冲刷形成的洞穴;冲刷孔;冲刷坑
eroded limestone 太湖石
eroded material 冲刷下来的泥沙;冲刷物
eroded sediment 冲刷物
eroded slope 被冲刷边坡;侵蚀性边坡
eroded soil 侵蚀土(壤)
eroded spot 冲蚀点
eroded valley 侵蚀谷
eroded volcano 侵蚀火山
erode material 冲刷物
erodent 侵蚀物;侵蚀剂;侵蚀的
erodibility 冲刷性;易(侵)蚀性;冲刷性;侵蚀性;侵蚀度;受浸蚀性
erodibility constant 冲刷常数
erodibility factor 冲刷系数
erodibility index 侵蚀程度指标
erodible 可冲蚀的;有侵蚀性;易受侵蚀的;易受腐蚀的;易(侵)蚀的;易冲蚀的河岸
erodible bank 冲刷性岸坡
erodible bed 易冲刷河床;动床;冲刷性河床
erodible bed channel 活动河床;易侵蚀槽;易冲刷河槽;不稳定河床
erodible bed engineer model 动床河工模型
erodible bed model 动床模型
erodible bed stream 可冲床底河川
erodible channel 可侵蚀槽;可冲刷河槽;冲刷(性)河槽
erodible material 易受蚀材料
erodible river 冲刷性河流
erodible soil 易受(侵)蚀的土壤;易受冲蚀的土壤;易蚀土壤
erodible stream 冲刷性河流;砂床河川
eroding agent 冲刷的原动力;侵蚀原动力
eroding bank 冲刷岸坡;崩岸;侵蚀(河)岸;塌岸
eroding channel 冲刷河槽;侵蚀河槽
eroding force 侵蚀力
eroding process 冲刷过程
eroding rate 冲刷速率
eroding-reworking current 侵蚀改造水流
eroding shore 冲刷海岸
eroding velocity 冲刷速度;侵蚀速度;起动速度
eroding water 侵蚀水
erogonum ovalioium 乔麦
erohbergite 斜方碲铁矿
eropoglyph 虫痕
Eros 爱神星
EROS data center of geologic(al) survey U S A. 地质调查所埃罗斯资料中心
erose 啮齿状的;不整齐齿状的;缺刻状的
erosibility 可侵蚀性;可侵蚀度
erosion 侵蚀(作用);糜烂;冲刷;冲蚀;腐蚀;剥蚀(作用);溶蚀;酸蚀;水点侵蚀;烧蚀
erosional 侵蚀的
erosional action 侵蚀作用
erosional agent 侵蚀原动力;侵蚀营力;侵蚀因素
erosional and depositional behavio(u)r 冲淤状况;冲淤特性
erosional base 侵蚀基面
erosional basin 侵蚀盆地;冲刷盆地
erosional basis 侵蚀基面
erosional bedding contact 侵蚀层面接触
erosional bench 侵蚀阶地
erosional boundary 侵蚀边界
erosional by thermal water 热水侵蚀
erosional caldera 侵蚀破火口
erosional carbon dioxide 侵蚀性二氧化碳
erosional channel 侵蚀河槽
erosional characteristic 冲刷特性
erosional characteristics 侵蚀特性
erosional class 侵蚀等级
erosional coast 侵蚀海岸
erosional contact 侵蚀接触

erosional control 侵蚀治理;侵蚀控制;侵蚀防治
erosional crater 侵蚀坑;侵蚀火山口
erosional cycle 侵蚀循环;侵蚀旋回
erosional delta 侵蚀三角洲
erosional depth of intrusive body 侵入体剥蚀深度
erosional drilling 冲蚀钻井
erosional effect 侵蚀作用
erosional exposed surface 已风化侵蚀
erosional factor 侵蚀因子;侵蚀因素
erosional feature 侵蚀特征;冲刷特征
erosional flood plain 侵蚀河漫滩
erosional force 侵蚀力;冲蚀力
erosional form 侵蚀形式
erosional gap 侵蚀山口;侵蚀间断
erosional groove 侵蚀沟(槽)
erosional gully 侵蚀沟(槽)
erosional hiatus 侵蚀间断
erosional index 侵蚀指数;侵蚀率;侵蚀度
erosional intensity 侵蚀强度
erosional investigation 侵蚀调查
erosional lake 侵蚀湖
erosional landform 侵蚀地形
erosional level 侵蚀剥削面;侵蚀剥离面
erosional life 侵蚀寿命
erosional loss 侵蚀流失
erosional mark 侵蚀痕
erosional modulus 侵蚀模数;侵蚀模量
erosional nick-point 侵蚀交叉点
erosional pattern 侵蚀形态
erosional pavement 侵蚀岩屑积层
erosional pillar 侵蚀墩
erosional pit 侵蚀坑
erosional plain 侵蚀平原
erosional plateau 侵蚀高原
erosional platform 侵蚀台地
erosional process 侵蚀作用;侵蚀过程
erosional protection 侵蚀防护
erosional ramp 侵蚀礁坡
erosional rate 侵蚀速率
erosional ratio 侵蚀率;侵蚀比
erosional ravaged area 侵蚀破坏(地)区
erosional retreat 侵蚀后退
erosional scar 侵蚀痕
erosional shore 侵蚀海岸
erosional slope 侵蚀斜坡
erosional soil 侵蚀土(壤)
erosional spring 侵蚀泉
erosional surface 侵蚀面
erosional survey 侵蚀调查
erosional terrace 侵蚀阶地
erosional thrust 侵蚀冲断层
erosional transgression 侵蚀海进
erosional truncation 削蚀作用
erosional unconformity 侵蚀不整合
erosional valley 侵蚀壳;侵蚀谷
erosional velocity 侵蚀速度
erosion and deposition 冲淤
erosion attack 冲蚀破坏
erosion bank 冲刷河岸
erosion basin 冲刷池;冲蚀盆地
erosion by current 水流冲刷
erosion by water 水蚀;水侵蚀
erosion by water action 水的冲刷(作用);水的侵蚀(作用)
erosion by wind action 风的侵蚀(作用);风蚀
erosion caused by 磨蚀原因
erosion cave forms 洞穴蚀成形态
erosion cave shapes 洞穴蚀成形态
erosion control 侵蚀控制;冲蚀控制;冲刷防治;冲刷控制;冲刷防护;防冲;水土保持措施
erosion control below dam 坝下冲刷防治;坝头冲刷防治
erosion control dam 防冲坝
erosion control fillet 防冲层
erosion control mattress 防冲沉排
erosion control measure 防冲措施
erosion control of land 陆地水土保持
erosion control plastic net 水土保持塑料网
erosion control works 防冲设施;防冲建筑物;防冲构筑物
erosion corrosion 浸蚀腐蚀;风化腐蚀;磨蚀腐蚀;冲蚀腐蚀
erosion-corrosion by thermal water 热水侵蚀腐蚀联合作用

erosion cycle 冲刷循环;冲蚀循环;冲积循环
erosion datum 冲刷基面
erosion depth 冲刷深度
erosion depth of wave 波浪冲刷深度
erosion device 冲蚀探测装置
erosion drilling 高速喷蚀钻进
erosion durability of soil 土壤抗蚀性
erosion form 侵蚀形态
erosion gully 冲(刷)沟
erosion history 冲刷过程
erosion hollow 冲刷坑
erosion index 冲刷指数
erosion in the bore 内腔腐蚀
erosion lake 侵蚀(形成的)湖
erosion landform of loessland 黄土侵蚀地貌
erosion loss 冲蚀流失;水土流失
erosion marine cycle 海蚀旋回
erosion mark 浸蚀痕;冲蚀痕;酸蚀痕
erosion material 泥沙;冲刷物
erosion of a bank 河岸侵蚀;河岸冲刷
erosion of electric(al) contacts 电接触烧蚀
erosion of glacier 冰川侵蚀
erosion of groundwater 地下水侵蚀性
erosion of levee slope 堤坡冲刷
erosion of refractory 耐火材料的侵蚀现象
erosion of river 河水侵蚀
erosion of rivers 河流侵蚀作用
erosion of shoulder 路肩坍塌
erosion of slope 边坡侵蚀;边坡溜坍
erosion of soil 土壤侵蚀
erosion of streams 河水侵蚀;河泥侵蚀
erosion of thermal 热腐蚀
erosion of vulva 阴蚀
erosion pavement 浸蚀砾幕;护坡;护堆;防侵蚀砾层;防冲(刷)铺砌层;防冲护坡
erosion pillar 侵蚀的支柱;腐蚀的支柱
erosion pilot 冲蚀控制器;冲蚀监测器
erosion pit 冲刷坑
erosion plain 冲刷平原
erosion platform 浪蚀台地;海蚀台地
erosion probe 冲刷探测器
erosion problem 腐蚀问题
erosion process 冲刷过程
erosion-proof 防冲的
erosion protection 冲刷防护;冲刷保护;防冲(刷)设施)
erosion protection shield 防蚀片
erosion quantity 冲刷量
erosion rate 冲刷(速)率;冲刷速度;冲蚀速率
erosion resistance 抗侵蚀性;抗侵蚀能力;抗冲刷性;抗冲刷能力;耐侵蚀性;耐磨耗性;耐腐蚀性;耐冲刷性
erosion resistant 抗蚀材料;耐蚀性的;耐冲蚀的
erosion resistant channel 抗侵蚀河槽;抗冲刷河槽
erosion resistant coating 防侵蚀涂层;防腐层
erosion resisting 耐受冲蚀
erosion-resisting characteristic 抗侵蚀性能;抗冲(刷)性能
erosion-resisting insulation 耐腐蚀绝缘
erosion ridge 风蚀雪波
erosion sand plug 冲蚀砂柱
erosion sand probe 砂粒冲蚀探头
erosion scab 冲蚀结疤;侵蚀铸瘤
erosion slope 侵蚀坡度;冲刷坡度
erosion spot 冲蚀点
erosion surface 冲刷面;冲蚀面;风化面(蚀面);风化剥蚀面;侵蚀面
erosion target 冲蚀靶
erosion terrace 浪蚀阶地
erosion test 侵蚀试验
erosion torrent 冲蚀急流
erosion valley 冲沟
erosive 冲刷的;腐蚀的;侵蚀的
erosive action 侵蚀作用;冲刷作用
erosive action of sand 砂的侵蚀作用
erosive agent 侵蚀力
erosive attack 腐蚀作用
erosive bed 冲刷性河床
erosive capacity 冲刷能力;侵蚀能力
erosive channel 冲刷性河槽
erosive collision 侵蚀性碰撞
erosive current 冲蚀水流
erosive force 冲刷力;侵蚀力
erosiveness 腐蚀度;侵蚀性;侵蚀能力;侵蚀度

erosiveness evaluation grade 侵蚀性评价等级
erosive power 冲刷能力;冲刷力;侵蚀能力
erosive river 易冲刷的河流;冲蚀性河流
erosive stream 冲刷性河流
erosive terrace 侵蚀阶地
erosive trace of water on fissure surface 裂隙面水蚀痕迹
erosive velocity 冲刷流速;侵蚀速度;侵蚀流速
erosivity 侵蚀性;侵蚀系数;侵蚀程度;磨蚀性;冲刷能力;侵蚀度
erpoglyph 蠕虫铸型
erps cupola 螺旋风口式化铁炉
errant algae 漫游藻类
errantia 漫游生物;浮游植物群落;漂浮植被
errata 正误表;改正表;订正表
erratic 移动的;分散的;不定的;变化无常的;漂游的;漂移物;无规律的
erratic behavio(u)r 不规律性
erratic block (of rock) 漂石;漂砾;孤石;巨漂砾;转石;漂块
erratic boulder 孤石;漂石;漂砾;漂块
erratic cell structure 不均匀孔眼结构
erratic current 不稳定电流
erratic curve 漂移特征曲线;无规律曲线
erratic deposit 漂砾沉积
erratic deposition 不规则沉积
erratic drift 不规则掉格
erratic error 偶然误差;不规则误差
erratic feed 错误给料
erratic feeding 不均匀给进
erratic firing 不规则发火
erratic flow 扰动流;涡流;紊流
erratic fluctuations 不稳定波动
erratic form 漂砾层;漂积物形状;漂积层
erratic function 不规则函数
erratic germination 不稳定发芽
erratic high-grade 特高品位
erratic index 漂砾指数
erratic load 不规则荷载
erratic loading 不规则荷载
erratic operation 不稳定工作
erratic orebody 漂积矿体
erratic process 漂游过程
erratics 漂石
erratic soil 冲积土(壤);漂砾土;无规律的土壤
erratic soil profile 不规则土层剖面
erratic soil structure 漂砾土结构;土的不规则结构
erratic stitch pattern 不规则变化花纹
erratic subsoil 杂乱土层;不均一土;不规则土层;不规则基土;漂砾质基(底层)土
erratum 勘误(表);书写错误
erroneous 有误差的
erroneous analysis 误差分析
erroneous conceptions 错误概念
erroneous correlation 不正确的对比
erroneous picture 错误概念
error 过失;错误;出错;差错;误差
error accumulation 误差累积
error actuated system 误差控制系统
error adjustment 误差调整
error alarm list 故障信号表
error allowance 误差容许限
error ambiguity 歧义性误差
error amplifier 误差放大器
error analysis 误差分析
error and omission excepted 遗误除外;保留更改差错权
error angle 误差角;失调角
error band 误差范围;误差带
error bar 误差条线;误差棒
error between flight line and cross line 测线切割线交点误差
error bound 误差界(限);误差角
error box 偏位框;误差框
error budget 误差预算
error burst 错误群;错误猝发;错误串;错码群;差错区间;差错脉冲群;差错猝发【计】;误差区间;突发差错
error by good faith 善意过失
error calculation 误差计算
error calculus 计算误差
error call 错误调用;误差呼叫
error cancelling 误差消除
error cause elimination 消除错误原因

error cause removal 消除错误原因
error character 错误字符;误差记号
error characteristic 误差特性
error check 错误检查;误差检查;误差校验
error check and correction 错误检查和校正
error check character 误差校验符
error checking 验错;错误校验;检验误差的;检验错;误差校验
error checking and correction 检错与纠错;错误检验和纠正;错误检测与校正;误差检验与校正
error checking arrangement 误差校验装置
error checking code 误差校验码;误差检验码;自检验码
error checking information 误差校正信息
error checking protocol 错误检查协议
error check program(me) 错误校验程序
error circle 误差圆
error circular radius 误差圆半径
error code 错误(指示)码;误码
error code diagnostic message 错误代码诊断信息
error code word 误码字
error coefficient 误差系数
error comment 出错注释
error compensation 平差;误差补偿
error compensator 误差补偿器
error condition 出错条件;差错状态;误差条件
error constant 误差系数;误差常数
error control 错误控制;差错控制;误差控制
error control character 误差控制记号
error control code 误差控制码
error control coding 差错控制编码
error control equipment 差错控制设备
error control system 错误控制系统
error control technique 出错控制技术;误差控制技术
error correcting 改正错误的;差错改正;误差校正;误差改正的;纠错的
error correcting capability 纠错能力
error correcting code 纠错码【计】;误差纠错码组
error correcting code 错误校正码
error correcting decoder 误差校正译码器
error correcting parser 误差修正剖析器
error correcting parsing 误差修正剖析
error correcting program(me) 误差校正程序
error correcting routine 错误校正程序;纠错程序
error correcting system 误差校正系统;纠错系统
error correction 错误校正;误差修正;误差校正
error correction circuit 误差改正电路
error correction code 纠错码【计】;误差校正符号;误差校正(代)码
error correction console 误差校正操纵台
error correction logic 误差校正逻辑
error correction operation 纠错操作
error correction procedure 纠错过程;纠错程序
error correction profile 误差校正剖析
error correction save point 纠错保留点
error correction servo 误差更正的伺服机构
error correction training procedure 误差校正训练步骤
error corrector device 误差校准器
error covariance 误差协方差
error criterion 误差准则;误差判据
error curve 误差曲线
error curve of mean 平均值的误差曲线
error deformation string 误差畸变链
error detecting 错误检测;误差检查;检测误差
error detecting and correcting code 误差检测及校正码
error detecting and feedback system 错误检测和反馈系统
error detecting capability 检错能力
error detecting code 检错码;检测码;错误检测码;误差检测电码;误差检测代码;自检验码
error detecting code system 错误检测码方式
error detecting facility 误差检测设备
error detecting program(me) 错误检测程序
error detecting routine 错误检测(例行)程序
error detecting system 错误检测系统;误差检出方式;检错系统
error detection 检错;错误检测;缺陷检测;误差检测
error detection and correction 误差检测与校正
error detection and recovery 错误检测和补救
error detection code 差错检测码;误差检测码
error detection signal 错误检测信号

error detection system 误差检测系统
error detector 检错器;错误检测器;误差检测器
error detector circuit 误差检测电路
error detector element 错误检测装置
error determine 误差测定
error diagnostics 错误诊断;出错诊断;误差诊断
error distance 误差距离
error distributing code 误差分配码
error distribution 误差分配;误差分布
error distribution principle 误差分布原理
error due to bias 偏倚误差
error due to curvature of earth 地球弯曲误差
error due to curvature of the earth 地球曲率误差
error due to emulsion displacement 乳剂位移误差
error due to inclination of horizontal axis 水平轴倾斜误差
error due to inclination of vertical axis 竖直轴倾斜误差
error due to non-uniformity of space phase 空间相位不均匀误差
error dump(ing) 误差清除;检错转储[贮]
errored second 误码秒
errored second ratio 误块秒比
error effects 误差效应
error ellipse 误差椭圆
error ellipse element 误差椭圆单元
error ellipsoid 误差椭球
error energy 误差能量
error equation 误差方程(式)
error equation coefficient 误差方程系数
error equation for interpolating 内插误差方程式
error estimate 误差估计
error evaluation 误差估计
error evaluator polynomial 误差计值多项式
error excepted 允许误差;错误除外
error exit 错误出口
error field 误差场
error figure 误差图形
error file 出错(登记)文件;养错文件
error flag 错误标志;错误标记;误差标记
error free 无误差;无误差的;无差错的
error free channel 无错误信道
error free coding 无错误编码
error free encoding 无错误编码
error free information 无错误信道
error free operation 无误差操作;无错操作;正常操作
error free running 无误差运行
error free running period 无差错运转周期;正常运行期;无误差运转期
error free simulated photograph 无误差模拟相片
error free transmission 无误差传输
error frequency 差错频率
error frequency limit 错误频率极限;出错频率界限
error function 误差函数
error function complement 补余误差函数
error function integral 误差函数积分
error function test 误差函数试验
error given in percent 以百分比表示的误差
error graphic model 错误图形模型
error handler 错误处理程序
error handling 出错处理;错误处理
error handling routine 错误处理程序
error in address 地址误差;地址错误
error in area 面积误差
error in bearing 方位误差
error in calculation 计算误差
error in circle graduation 度盘刻度误差
error in classification 分类误差
error in data 数据误差
error in depth 景深误差
error in depth measurement 深度测量误差
error indicating circuit 误差指示电路
error indicating system 误差指示系统
error indication 出错指示;误差指示
error indication facility 误差指示设备
error indicator 出错指示符;误差指示器
error induced grammar 误差导出文法
error in economic analysis 经济分析的误差
error in electromagnetic distance measurement 电磁波测距误差
error in equation 方程(内)误差
error-in-equation model 方程误差模型
error in exactness 精确度误差

error in fix 船位推算误差
error in focusing 调焦误差
error information 误差信息;误差数据
error in geometric(al) form 几何形状误差
error in heighting 测高误差
error in judgement 判断误差
error in label 符号错误;符号差错;符号(部分)出错
error in lable 符号误差
error in latitude 纬度误差
error in level(l)ing 水平测量误差
error in line 方向误差
error in longitude 经度误差
error in measurement 量测误差;测量误差
error in navigation 驾驶过失;导航误差
error in numeric(al) calculation 数值计算误差
error in observation 观察误差;观测误差
error in observed altitude 观测高度误差
error in omission 漏计误差
error in operation 运算误差;操作误差;操作错误;操作(码)差错
error in operation code 操作码错误
error in pointing 指向误差
error in point measurement 点位测定误差
error in reading 连续误差;读数误差;读数错误
error in regressor 回归因子误差
error in rounding number 数字舍入误差
error inspection 误差检验;误差检测
error in spectrochemical analysis 光谱分析误差
error integral 误差积分
error integrator 误差积分器
error interrupt 错误中断;出错中断;差错中断;误差中断
error interval 误差区间
error in variable 变量误差
error in variable bias 变量误差偏倚
error in variable model 变量误差模型
error in viewing 视差
error latence 错误潜伏期
error law 误差(定)律
errorless 无误差的;无错误的
error limit 误差限度;误差极限;误差范围
error list 故障表;错误表
error listing 故障列表;出错列表
error locating 错误定位
error locating code 误差定位码
error-locator polynomial 错误定位多项式
error lock 错误封锁;出错封锁
error log 出错记录
error logger 错误记录程序;出错登记程序;误差记录程序
error logging 错误记录
error log manager 出错登记管理程序;错误记录管理;出错记录管理程序
error macro 查错宏指令
error magnitude 误差大小
error map 有误差的地图数据
error margin 误差容限
error mark 错误标记
error matrice 误差矩阵
error matrix 误差矩阵
error mean 平均误差
error measuring element 误差测量环节
error measuring function 误差测量函数
error measuring means 偏差检测装置;误差测量装置
error measuring system 误差测量系统
error memory 差错存储器
error message 错误信息;错误报文;出错信息;查错信息;误差信息
error meter 误差测量计
error method 误差法
error model 误差模型
error monitor 误差监督程序
error multiplication 误码增殖
error multiplication factor 误码增殖因子
error multiplier 误差乘法器
error norm 误差准则;误差范围
error of adjustment 校准误差
error of alignment 定线误差
error of allocation 分配误差;配置误差
error of approximate 近似误差
error of assessment 估算误差
error of axis of tilt 倾轴误差
error of back lash in three-arm protractor 三杆

分度仪隙动差;三杆分度器隙动差
error of bend in three-arm protractor 三杆分度仪弯曲差;三杆分度器弯曲差
error of calculation 计算误差;计算错误
error of calibrated focal length 检定焦距误差
error of calibration 校检误差
error of centering 对心误差
error of centration 共轴性误差
error of centric position 对心误差
error of circle graduation 度盘分划误差
error of closure 闭塞差;闭合差【测】
error of closure in azimuth 方位角闭合差
error of closure level circuit 水准测量闭合差
error of closure of angles 角度闭合差
error of closure of horizon 圆周闭合差
error of closure of traverse 导线闭合差
error of closure of triangle 三角形闭合(误)差
error of collimation 准直差;透视差;视准(误)差
error of compass 罗盘差
error of compensation 掺入误差
error of computation 计算错误
error of coordinates 坐标误差
error of direction 方向误差
error of division 分度误差
error of due to curvature of position circle 船位线曲率误差
error of eccentricity 偏心差
error of eccentricity in three-arm protractor 三杆分度仪偏差;三杆分度器偏差
error of echo sounder 测深仪误差
error of enclosure 闭合误差
error of estimate 估计(量的)误差;估读误差
error of estimation 估算误差;估计误差;估读误差
error of first kind 第一类错误
error of fitting 拟合误差
error of flight height 航高差
error of focusing 调焦差误
error of geologic(al) observation point 地质点定位误差
error of graduation 刻度误差;分划误差
error of gyrocompass 陀螺罗经误差
error of gyrocompass heading 陀螺罗经舰向误差
error of heading 舰向误差
error of heading of gyrocompass 陀螺罗经舰向误差
error of height 高度误差
error of horizontal axis 水平轴误差
error of indication 指示错误;指示误差
error of inner orientation 内方位误差
error of input data 输入数据误差
error of interior orientation 内方位误差
error of internal orientation 内方位误差
error of interpolation 插值误差
error of irradiation 光渗差
error of judgment 判断错误
error of length 长度误差
error of lens distortion 透镜畸变误差
error of mean square 均方(误)差;中误差;平均误差
error of measurement 测量误差
error of observation 观察误差;观测误差
error of observation data 观测数据误差
error of omission 遗漏误差;省略误差
error of perpendicularity 垂直度误差;垂直差
error of piercing (隧道的)贯通误差
error of pivot 轴颈误差
error of plotting 绘图误差
error of plotting points 展点误差
error of position 位置误差
error of position line 船位线误差
error of radio direction finder 无线电测向仪误差
error of reading 观读误差;读数误差
error of re-check observation 测回差
error of reckoning 船位推算误差
error of regression 回归误差
error of relative orientation 相对定向误差
error of repetition 重复误差
error of replication 重复误差
error of representation 表征误差
error of safe side 安全误差
error of satellite orbital elements 卫星轨道根数误差
error of scale 刻度误差;比例误差;缩尺误差
error of service 业务差错

error of sextant 六分仪误差
error of sighting 瞄准误差;照准误差
error of star catalog 恒星星表误差
error of survey 测量误差
error of swing 转角误差
error of the estimated ore reserves 估算矿石储量误差
error of the first kind 第一种错误
error of the second kind 第二类型错误
error of the vertical axis 垂直轴误差
error of tilt 倾斜误差;倾角误差
error of transferring 位置线移位误差
error of transmission 传输错误
error of traverse 导线闭合误差
error of truncation 截断误差
error of vertical collimation 垂直度盘瞄准差
error of vertical index 垂直度盘指示差
error of zero 零点误差
error on safe side 安全误差
error on the safe side 安全误差
error on the side of safety 安全性的计算误差
error option 误差选择
error parallelogram 误差平行四边形
error pattern 错误型;误差图样;误差方向图
error pattern generator 错误型发生器
error percentage 误差百分数;误差百分率
error per digit 每位误差;每个数位误差
error performance 误差特性
error phasing 误差调相
error pick-up 误差传感器;失配信号传感器
error print 错误打印
error probability 错误概率;误差概率
error probability diagram 误差概率图
error probability function 误差概率函数
error processing 错误处理;误差处理
error process(ing) procedure 错误处理过程
error process(ing) subsystem 错报处理子系统
error production 误差产生
error-prone 易于出错的;容易产生误差的
error-prone repair 易误修复
error propagation 错误传播;误差传播
error propagation limiting code 错误传播受限码
error protected transmission system 防误传送方式
error protection 故障保护;错误保护;误差防护
error pulse 误差脉冲
error queue 错误队列
error range 误差范围
error rate 故障率;差错率;出错率;误码率;误差率
error rate for human incurring fault 人为故障发生率
error rate measuring equipment 误差率测量设备
error rate monitor 差错率监视器
error rate of drilling depth 孔深误差
error rate of keying 按键误差率
error rate of translation 翻译误差率
error ratio 错误率;误码率;误码比;误差率;误差比
error record 错误记录
error recording 故障记录
error recording data set 故障记录数据集
error recovery 错误校正;错误恢复;出错恢复;误差恢复
error recovery manager 错误恢复管理程序
error recovery procedure 错误恢复过程;误差校正过程
error recovery routine 错误校正程序
error register 误差记录器
error registration ATO on-board 车载 ATO 故障记录【铁】
error regression 机误回归;回归误差
error resonance 误差共振
error retry 误差重算
error return 出错返回
error return address 错误返回地址;出错返回地址
error routine 差错程序;查错程序;查错程序
error sampled control system 误差抽样控制系统
errors and omissions excepted 错误遗漏不在此限;错漏除外;差错查查;如有错漏可予更改
errors and omissions insurance 过失保险
error search program(me) 诊断程序
error sensing device 误差传感装置
error sensitivity 误差灵敏度
error sensitivity number 误差敏感度数
error sensor 误差敏感元件

error sequence 误差序列
error severity code 标志错误严重性的代码
errors expected 错误不在此限
error signal 出错信号;查错信号;误差信号
error signal decoder 误差信号解码器
error signal detection 误差信号检测
error signal detector 误差信号检测器
error signal encoder 误差信号编码器
error signal generation 误差信号产生
error signal generator 误差信号发生器
error signal transmitter 失调传感器
errors in the invoice 发票金额错误
errors of account 错账
errors of book entry 记账错误
errors of compensation 相抵错误
errors of omission 遗漏错误
errors of pessimism 悲观误差
errors of sampling 抽样误差
error source 误差(来)源
error span 误差距离
error spectral density 误差谱密度
error spectrum density 误差谱密度
error square 误差平方
error squared criterion 误差平方准则
error state 异常状态;误差状态
error statistics by volume 卷错误统计
error status 错状态
error status register 错误状态寄存器
error storage 差错存储器
error sum of squares 误差平方和
error symbol 错误符号;误差符号
error synthesis(method) 误差综合(法)
error system 偏差查寻系统
error tape 差错带
error term 残项;误差项
error test(ed) 误差检验
error theory 误差理论
error threshold 误差界限
error tolerance 误差容限
error transfer function 误差传递函数
error transformation 误差转换
error transformer 误差信号变换器
error trapping 错误捕捉
error treatment 误差处理
error triangle (船位)误差三角形
error type 误差类型
error variance 机误变量;误差偏差;误差离散;误差方差;误差变量
error vector 误差向量
error voltage 误差电压
error voltage polarity 误差电压极性
error volume analysis 卷错误分析
errow detector 误差检测器
ersatz 代用的
ersatz material 代用材料
Ertix diwa system 额尔齐斯地注系
Ertix fold-fault belt 额尔齐斯褶皱断裂带
Ertixiite 额尔齐斯石
Ertix tectonic knot 额尔齐斯构造结
erubescite 斑铜矿
Eruc 埃勒克纤维
erucic acid 芥酸
erucic acid oil 芥酸油
eructation 嗳气
eructation with fetid odo(u)r 嗳腐
erucyl amide 芥酸酰胺
erupt 喷发
erupt completely 出齐
erupting wave 突发波
eruption 萌出;爆发;喷发;喷出物
eruption canal 喷发管道
eruption fissure 喷发裂隙;喷发裂缝
eruption intensity 喷发类型
eruption laccolith 喷发岩盖
eruption modes 喷发强度
eruption period 萌出期
eruption pipe 喷发管道
eruption rhythm 喷发韵律
eruption symptom 喷发征兆;喷发前兆
eruption type 喷发形式;喷发方式
eruptive 喷出的
eruptive activity of volcano 火山喷发活动
eruptive arch 爆发拱
eruptive breccia 火山角砾岩;喷发角砾岩

eruptive deposit 喷发沉积
eruptive facies 喷发(岩)相
eruptive fountain 喷(射)泉
eruptive laccolith 喷发岩盖
eruptive material (火山)喷发物;火山喷出物
eruptive mineral deposit 喷发矿床
eruptive process 喷发作用
eruptive prominence 爆发日珥
eruptive rock (火山)喷发岩;火成岩;喷出岩
eruptive sheet 喷出层
eruptive spring 喷泉
eruptive stock 喷出的堆积物
eruptive tuff 凝灰岩;泥流;喷发凝灰岩
eruptive variable 爆发变星
erysodine 刺桐定碱
erysonine 刺桐宁
erysothiopine 刺桐硫平
erysovine 刺桐碱
erythema caloricum 热力红斑
erythema solare 日晒红斑
erythraline 刺桐灵
erythramine 刺桐胺
erythratine 刺桐亭
Erythrean 厄立特里亚古海
erythrene 间丁二烯;刺桐烯
erythrina 刺桐
erythrinan 刺桐烷
erythrine 钴华
erythrite 赤丁四醇;钴华
erythrite peachblossom ore 钴华
erythritol 丁四醇;赤藓醇
erythrityl tetranitrate 丁四硝酯;赤藓醇四硝酸酯
erythro configuration 赤系构型
erythrocyte sedimentation rate 沉降反应
erythroglaucin 毛罂红
erythroidine 刺桐定
erythrol 赤醇
erythronic acid 赤酮酸
erythronolactone 赤酮酸内酯
erythrosiderite 红钾铁盐
erythrosine 赤藓红
erythrosine sodium 赤藓红钠
erythrosine sodium salt 赤藓红钠盐
erythroskyrin 红天精
erzbergite 文石—方解石互层
Erz cement 铁水泥
Erzgebirgian orogeny 埃尔茨造山运动【地】
Esbach's reagent 埃斯巴赫试剂
Esbach's test 埃斯巴赫试验
esboite 奥球闪长岩
escalade 活动人行道;用梯攀登;爬云梯
escaladed bid 提供标价
escalating coefficient of material price 材料费上涨系数
escalating prices 上涨价格
escalating rate 上升率
escalating wage 物价补贴工资
escalation 价格调整;应变浮升;逐步上升
escalation clause 合同滑动条款;物价变动条款;调整条款;收费自动调整条款;升级条款;伸缩条款
escalation contingency 提升的不可预见费
escalation during construction 施工期涨价
escalator 连续提升机;阶梯式自动电梯;自动行人传送带;自动扶梯;自动电梯;增减手段;升降梯
escalator arrange 自动楼梯布置
escalator bond 调整债券;伸缩债券
escalator clause 价格调整条款;伸缩条款;调整(价格伸缩)条款
escalator drive 自动梯传动装置
escalator method 迭代法;梯降法
escalator pit 自动扶梯坑
escalator shutter 自动楼梯百叶门
escalator tunnel 自动楼梯隧道
escalloped 片状作边缘曲线饰
escapable cost 可免成本;可节省成本
escapage 泄水量
escape 漏出;扩充功能符;空刀槽;逸水门【给】;逸出;放出;疏散机轮;退水闸,退刀槽;逃脱;司行轮;疏散(口)
escape and evasion graph 侦察行动图
escape angle 擒纵角
escape behaviou(u)r 逃避行为
escape booms (海洋钻探的)救生浮筒
escape canal 泄水渠(道);排水渠(道);排水沟

(渠);退水渠
escape capability 逃逸能力
escape character 换码符;信息漏失符号;逸出字符;转义字符
escape chute 救生滑槽
escape clause 免罚条款;例外条款;免责条款;免除条款;回避条款;逃避条款
escape cock 放水旋塞;放水龙头;放水阀;放气旋塞;安全塞;排泄旋塞;排泄阀;排放阀
escape compartment 脱险舱;逃生舱
escape conditioning 逃走条件作用
escape cone 逃逸锥(面)
escape corridor 安全防火廊;安全通路;安全通道;疏散走廊
escape covert 逃避隐蔽植被
escape crank 操纵柄
escape depth 逸出深度
escaped fire 超过初期灭火能力的火灾
escape door 紧急离机门;安全出口;太平门
escaped product 漏失的油品
escape emergency 应急离机事故
escape engine 分离发动机
escape equipment 应急离机设备
escape exit 应急(离机)出口;安全出口
escape factor 逃逸因子;逃脱因子
escape from stage 舞台疏散口
escape gallery 应急安全坑
escape gate 放水闸(门);安全闸(门)
escape gradient 逸出坡降
escape grid 凹板筛
escape handle 应急离机手柄
escape hatch 应急出口;太平门;应急通道;应急舱口;舱口;安全出口;安全舱口;脱险舱口;逃生舱口;太平舱口;紧急出口
escape hatch opening 应急离机舱口
escape hole 应急舱口;放水孔;排水口;排水孔;排气口;排出口;太平舱口
escape instruction 流出指令
escape kit 应急离机救生包(内装地图)
escape ladder 太平梯;安全梯;应急梯;疏散爬梯
escape level 脱逸能级;逃逸能级
escape lighting 疏散照明
escape lighting system 疏散照明系统
escape lighting unit 疏散照明装置
escape line 固定在摇篮或伸缩梯顶级梯档的绳子
escape lock 备用闸;安全闸(门);太平闸;事故闸(门)
escape lung 单人呼吸器具
escape margin 逃生安全系数
escapement 间隔器;棘轮装置;制动,闭锁;擒纵机;牵纵拐肘;排出口;司行轮
escapement crank 擒纵曲柄
escapement device 擒纵装置
escapement error 擒纵误差
escapement faults 擒纵机构误差
escapement file 带方形断面尾的锉;擒纵机锉
escapement impulse 擒纵机构冲击
escapement lever 擒纵叉
escapement mechanism 擒纵机构
escapement motion 擒纵器
escapement noise 擒纵机构噪声
escapement plate 擒纵机构夹板
escapement repeater 出口围盘
escapement stop 杠杆式销子挡器;擒纵式挡器;擒纵机挡器
escapement system 擒纵系统
escapement timing 带制动的时滞装置
escapement wheel 摆轮;擒纵(机)轮
escape of chips 排屑
escape of dust 扬尘;粉尘逸出;(路面磨损后造成的)起尘
escape of gas 漏气
escape of neutron 中子漏泄
escape of oil 漏油
escape of radioactivity 放射性泄漏;放射性漏出
escape opening 安全门;安全口;排气孔
escape orifice 泄放孔;排泄孔
escape path lighting system 疏散通道照明装置
escape peak 泄漏峰;(探测器的)逃逸峰
escape pinion 擒纵齿轮
escape pipe 泄水管(道);溢水管;出气管;放气管;放气管;安全管;排气管;放出管
escape pit 安全井
escape plan 避难计划

escape probability 漏失概率;逸出概率;逃脱几率
escape pump 泵浦摇梯消防车
escaper 排放器
escape ramp 疏散坡道
escape rate 逃逸速度
escape rate coefficient 逃逸率系数
escape react test 回避反应试验
escape regulator 排泄调节器
escape road 机车回车线
escape rocket 救生火箭;应急分离火箭
escape roof(walk)way 安全屋顶走道
escape rope 疏散绳
escape route 排泄途径;漏失途径;迂回进路;安全撤离路线;排出通道;排出管路;疏散路线
escape route lighting 疏散路线照明
escape route sign 疏散路线标志
escapes apparatus 自救器
escape scuttle 应急出口;脱险口
escape sequence 应急离机程序;代码换向序列
escape shaft 安全竖井;太平竖井
escape sign 避难标志
escape sluice 泄水闸;安全泄水闸
escape speed 逃逸速度
escape stair(case) 太平(楼)梯;安全梯;防火楼梯;疏散楼梯
escape staircase well 太平梯井口
escape stair way 太平梯
escape system 避难系统
escape test 逃逸试验
escape tower 应急脱塔;逃离塔
escape trace 逃逸迹
escape trajectory 逸离轨道
escape trunk 应急出口围阱;应急出口(围壁);潜水艇逃生舱;逃生孔
escape tube 放汽管;安全管
escape tunnel 卸载地道
escape valve 泄气阀;溢流阀;逸泄阀;放泄阀;放汽阀;保险阀;安全阀
escape velocity 出流速度;脱离速度;逃逸速度
escape vent 逸气口;应急出口
escape way 第二条出口;应急出口;安全通路;安全通道
escape weir 泄水堰
escape wheel 擒纵轮
escape wheel and pinion 擒纵轮
escape wheel bridge 擒纵轮夹板
escape wheel cock jewel 擒上钻
escape wheel foot jewel 擒下钻
escape wheel grinding machine 擒纵轮磨床
escape wheel guard 擒纵轮护板
escape wheel pinion 擒纵轮齿轴
escape wheel teeth 擒纵轮齿
escape window 消防窗户;太平窗(户)
escape work 泄水构筑(物);泄水建筑(物)
escape zone 避难区(域)
escaping gas 挥发性气体
escaping structure 逃逸构造
escaping tendency 逃逸趋势
escar 蛇(形)丘
escarp 内壕;壕沟内壁;筑陡坡
escarping 筑陡坡
escarpment 鬣丘;马头丘;削壁;陡崖;单斜山
escenter 旁心;旁切圆心
escenter of a triangle 三角形(的)旁心
eschar 蛇形丘
escharotic 苛性剂;腐蚀性的
escheat 产业归公(无继承);没收地;没收财物;土地复归
Escherichia 埃希菌属
Escherichia coli 大肠埃希氏菌
Escherich's reflex 埃舍利希氏反射
Eschka method 埃斯卡测硫法
Eschka mixture 埃斯卡混合溶剂
Eschka mixture for sulfur determination 埃斯卡测硫混合剂
eschshltzia mexicana 罂粟
Eschweiler-Clarke modification 埃施魏勒—克拉克改进反应
eschynite 易解石
Esclangon effect 埃斯克朗贡效应
Escobillo fiber 埃斯科比洛纤维
esconson 窗框(边框)内侧
escorial (炉)渣堆
escort 护舰部队;护航(飞机)

escort carrier 小型航空母舰
escort fee 押运费
escort goods in transportation 押运
escorting of parcel 押运
escort minesweeper 护航扫雷艇
escort ship 护航船
escort tanker 护航的油船
escort vessel 护送船;护航舰;护航船
escribe 旁切
escribed circle 旁切圆
escribed circle of a triangle 三角形的旁切圆
escribed sphere 旁切球
escritoire 具有一排抽屉的书桌
escrow 寄存担保品;有条件的契约;暂管;待完成的担保证书;代管;附条件交付契据;附履行条件交付;契据
escrow account 记账账户;第三者保管账户;条件交付账户
escrow agreement 有条件转让协议;有条件的转让契约
escrow barter 托付易货贸易
escrow closing 中介结束
escrow contract 契约法
escrow cost 契约费用
escrow credit 寄托信用证;托付信用证
escrow fee 契约手续费
escrow holder 契约持有人
escrow instruction 中介委托书
escrow letter of credit 托付信用证;条件交付信用证
escrow officer 中介人
escutcheon 孔罩;刻度盘饰框;金属框子;门锁板;钥匙孔盖;钥匙孔板;遮护板;管子箍;盾饰;船名板;复板;锁眼盖;饰框
escutcheon pin 钥匙孔盖销;盾形针;盖板钉
escutcheon plate 门锁板;门把手垫板;饰框
Esequibo cotton 埃塞奎博棉
Eshka method 埃史卡测硫法
eskar 冰河砂堆;蛇形丘
eskebornite 硒黄铜矿;铁硒铜矿
esker 冰河砂堆;蛇(形)丘
esker delta 冰河口砂质沉积;蛇丘三角洲
esker fan 蛇丘扇形丘
eskerine 蛇丘的
esker knobs 蛇丘丛
esker ridge 蛇丘脊
esker trough 蛇丘谷
Eskimo dog 北极犬
Eskimo fabric 爱斯基摩缎纹呢
eskimoite 埃硫铋铅银矿
eskolanite 绿铬矿
Eslon 埃斯纶聚酯短纤维
esmeraldaite 杂褐铁矿
esmeraldite 云英花岗岩;多英白云母岩
Espa 埃斯帕弹性纤维长丝纱
espacement 间距;植距
espagnolette 通天长插销开关;长窗钩;长钉钩
espagnolette bolt 长插销;通天插销
espalier 花木架;树墙
espalier drainage 格水系
espalier drainage pattern 方格水系
espalier lath 花棚板条
espalier training 篱形整枝
esparto 西班牙草;北非芦苇草
esparto fabric 编织用禾本科草织物
esparto grass fiber 北非茅草纤维
esparto pulp 茅草纸浆
esparto wax 茅草蜡
especial 特别的
esperite 硅钙铅锌矿
espews 无定形扫描信号
espichellite 橄闪煌斑岩;橄闪粗玄斑岩
espionage 间谍活动
esplanade 空地;码头面;斜堤;峡谷邻近的宽阶地;岩性阶地;岩石阶地;闸顶面;草地;填筑面;散步路;广场
Espoir 爱斯派义耳黏胶长丝织物
esquisse 快速设计草图;快速设计草案;草拟图稿
esquisse-esquisse 快速设计草图;快速设计草案
essange oil 埃散油
essaying 取样
essayist 论文著者
essence 精华;本质
essence exhaustion 精极
essence of an invention 发明的实质

essence of the contract 契约要素
Essen coefficient 埃森系数
essential 精制的;基本的;要素;要件;根本的,本质的;实质的
essential analysis 基本分析
essential association 基本结合
essential boundary condition 本质边界条件;实在边界条件
essential characteristic 本质特征
essential cofactor 必需的辅助因素
essential colo(u)r 主色;本色
essential component 主要部件;主要分量;主要成分;必要成分
essential condition 基本条件;必要条件;必备条件
essential constant 本质常数
essential convergence 本质收敛
essential cooperative game 本质合作对策
essential data 基本资料;基本数据
essential difference 本质区别
essential ejecta 本源抛出物
essential element 主要元素;必需元素;必要元素
essential element of analysis 分析的基本元素
essential elements of drill hole 钻孔要素
essential environment(al)factors 主要环境因素
essential facility 重要设备;重型设备
essential factor 要素
essential fatty acid 必需脂肪酸
essential fatty acid index 必需脂肪酸指数
essential goods 必需品
essential group 必需基团
essential hazard 本质冒险;实质险态
essential information 基本信息
essential ingredient 主要成分
essentialities of design 设计要点
essentiality 本性;实质性
essentially 本质上;实质上
essentially bounded functions 本质有界函数
essentially complementary partition 本质余划分
essentially complete class of decision rule 决策规则的本质完全类
essentially equivalent 实质等价
essentially general game 本质一般对策
essentially non-negative matrix 本性非负矩阵
essentially periodic sequence 本性周期序列
essentially positive matrix 本性正矩阵
essential map element 地图基要素
essential maximum 本质最大值
essential microelement 必需微元素
essential mineral 基本矿物;主要矿物;必需矿物质
essential mineral element 必需矿物质元素
essential multiple output prime implicant 实质多重输出本原
essential of curve 曲线要素
essential oil 香精油;芳香油;天然香料货
essential parameter 基本参数
essential parameter of non-hydrocarbon compound 非烃化合物的主要参数
essential part 基本部分
essential point 本质点
essential prime implicant 基本素蕴涵;实质本原蕴涵
essential provision fee 基本预备费
essential quality of damages 实质上的损害
essential singularity 本性奇点
essentials of preliminary design 初步设计要点
essential species 主要种
essential strategy 本质策略
essential supremum 本质上确界;本性上确界
essential technique 基本功
essential term 本质项
essential tone 基本色调
essential useful component 主要有用组分
essential variables of tube-to-tube sheet welds 管子与管板焊接的主要参数
essential water 组成水分;必要水分
esserbetol 聚醚树脂
Essevi 埃塞维黏胶长丝
Essex board 厄塞纤维板
Essex board measure 厄塞板尺计算法
essexite 厄塞岩;碱性辉长岩
Esson coefficient 比转矩系数;比力矩系数
ESSO Oil 埃索石油公司
establish 开业;开设;开办;设置;设立
establish a business 创业

establish an account 开设账户;开户
established angle 安装角
established business house 老铺
established channel 老航道;原有航道
established coenosium 建成群落
established connection 确立连接
established customs 确定的惯例;既定惯例;常规
established data 确定的数据
established fact 既成事实
established flow 定型流动
established harbo(u)r 老港(口);原有港口;已有港口
established harbo(u)r basin 原有港池
established international practice 公认的国际惯例
established investment 确定的投资量;既定投资
established line bearing 非标准轴承
established policy 既定方针;一贯的政策
established port 老港(口)
established port area 老港区
established post 常设员额
established practice 惯例;成规
established principle 既定原则
established procedure 规定的操作法;规定的(操作)程序
established programme 既定方案
established reliability 确立可靠性
established reserves 确定的储量
established right 确定的权利
established scale 规定比例尺
established sector 原有港区;老港区
established system 建成的(道路)系统
established time of name of tectonic movement 构造运动名称创立的时间
establishing sales territories 开辟推销地区
establishment 建立;机构;形成过程;成林过程;潮候时;产业单位;编制;确立;企业机构
establishment and control of inventory 存货的设置和管理
establishment and supplementary estimates 概算及补充概算
establishment burden rate 确定的负荷率
establishment charges 管理费;开办费;组织费;创办费;筹建费(用);设置费用;设施费
establishment of aids-to-navigation 航标设置
establishment of a port 港口建立
establishment of base line 测定基线
establishment of car detention standards 车辆停留时间规定标准
establishment of credit 开立透支账户
establishment of datum 基准建立
establishment of fund 基金设置
establishment of letter of certificate 开证
establishment of letter of credit 开立信用证;信用证的开立
establishment of port 港口(平均)朔望高潮间隙;标准潮汛
establishment of standard 标准的建立
establishment of the port 潮候时差;标准潮汛
establishment of unit 单位编制
establishment period 成苗期;成林期
establish the cause of a fire 调查火灾原因
establish vertical control net 测设高程控制网
estaminet 小餐馆;小咖啡馆
estancia 大庄园;大牧场(南美)
Estane 埃斯坦聚氨基甲酸乙酯弹性纤维
estate 遗产;庄园;产业;财产;房地产;不动产
estate administrator 遗产管理人
estate agent 地产经纪人;房地产中间商;土地管理人
estate area 地产面积
estate at sufferance 容许资产
estate at will 遗嘱上的地产
estate battery garage 成排的汽车库房地产
estate car 旅行车;客货两用轿车(英国);轻便旅行车
estate car tourist coach 旅行车
estate contract 不动产契约
estate corpus 财产本金
estate development 地产开发
Estate Duties Investment Trust 遗产税投资信托公司
estate duty 遗产税;财产税
estate encumbered with mortgages 已抵押出去的地产
estate for life 终身属有的地产
estate for years 有年限的地产

estate from year to year 自动逐年延期租赁权
estate income 财产收益
estate in fee 无条件继承的不动产
estate in ga(u)ge 作抵押品的地产
estate in remainder 地产的他人继承权
estate less than freehold 非自由拥有地产
estate management 物业管理
estate of inheritance 世袭财产
estate planning 住宅群规划;住宅群公用地综合
estate road 种植园道路;庄园(内)道路
estate sprayer 场内建筑物及树木消毒用喷雾机
estate tax (房地产)遗产税
estate taxation 租赁的地产
estate unit 房地产单位
estator 遗嘱人
estavel 落水洞泉;间歇性出露;消盖水洞;涌泉;地下河(流);反复泉
esteem value 品位价值
Este porcelain 依斯特瓷器(软瓷,意大利)
ester 酯
Estera 埃斯特拉醋酯短纤维
ester acid 酯酸
ester alkyl 酯烷基
ester condensation 酯缩合(作用)
estercrete 酯强混凝土
esterdiol 酯二醇
esterellite 英闪玢岩
ester gum 酯(树)胶;松香甘油酯
ester gum varnish 酯树胶清漆
esterification 酯化(作用)
esterification catalyst 酯化催化剂
esterification equivalent weight 酯化当量
esterification number 酯化值
esterification rate 酯化速度
esterified 酯化的
esterified natural resin 酯化天然树脂
esterified natural resin varnish 酯化天然树脂清漆
esterified oil 酯化油
esterified resin 酯化树脂
esterify(ing) 酯化
ester interchange 酯交换(作用)
ester linkage 酯键
ester number 酯值
estero 河口湾旁沼泽;河口湾旁滩地
ester of fatty acid 脂肪酸酯
ester oil 合成酯类润滑油;由酯组成的合成油
Esterophile dye 埃斯特罗菲染料
Esteroquinone dye 埃斯特罗基农染料
ester plasticizer 酯增塑剂
ester solvent 酯系溶剂
ester tank wagon 水罐车
ester type synthetic(al) drying oil 酯式合成干性油
ester value 酯(化)值
ester wax 酯蜡
Esterweld 埃斯特韦尔德聚酯长丝
esthete 美学家
esthetically pleasing 赏心悦目的;美而愉快的
esthetic appeal 美学要求
esthetic area 游览区
esthetic aspect 美观方面
esthetic charm 美的吸引力
esthetic effect 美的效果
esthetic feeling 美的感觉
esthetic idea 美的思想;美的观念
estheticism 唯美主义
esthetic pollution 感官污染
esthetic quality 感官水质
esthetics 美学
esthetic sense 美学的感觉
Esthonian oil shale 爱沙尼亚油页岩
estimability of coefficient 系数的可估性
estimability of parameters 参数的可估性
estimable function 可估函数
estimate 预算;预计;估算;估量;估计(量);评估
estimate accuracy 估算精度
estimate acoustic(al) impedance function 估计声阻抗函数
estimate and budget 概预算
estimate basis 计价依据
estimate breakdown 投资估算分项金额
estimate burden rate 预算分配比例
estimate by eye 目测
estimate by interval 区间估计
estimate by ratio 按比例估计

estimate clerk 预算员
estimated 预计的;估计的
estimated accuracy 估计精度
estimated actual radium analyzer[analyser] 估计真实镭含量分析仪
estimated actual radium content 估算实际镭含量
estimated additional income 增加收入估计数
estimated additional requirements 预计增加经费;估计所需增加费用
estimated amount 预算造价;预算数量;预算工程量;估计数量
estimated amount of damage 损失数额估计
estimated appropriation 估定拨款数额
estimated balance sheet 预计资产负债表;资产负债估算表;估计资产负债表
estimated bearing by eye 目测方位
estimated burden rate 预计分摊率
estimated capacity 估计能力
estimated capacity per articulated unit 每节铰接车厢载容量
estimated capital 预算成本;估计成本
estimated cash requirement 预计现金需要量
estimated complete date 预计完工日期
estimated completion date 估计完工日期;估计完成日期
estimated completion time 预计完工时间;预计完成时间;预定完成时间
estimated contract coverage 预计合同范围
estimated contract payments 付款计划表
estimated cost 预算造价;预算价(值);预算费用;预算成本;估算造价;估计费用;估计成本;概算造价;概算费用
estimated cost accounting 估计成本会计
estimated cost calculating 估计成本计算
estimated cost of construction 预计造价
estimated cost of construction work 工程成本估计
estimated cost system 估计成本制度
estimated course 预计航向
estimated damage ratio 估算损害比
estimated data 估算数据
estimated date of arrival 估计到达日期
estimated date of availability 估计获得日期
estimated date of completion 估计的完工日期;预计完工日期;估计完成日期
estimated date of departure 估计开航日期
estimated decrease in income 收入估计减少数
estimated delivery date 估计交货(日)期
estimated delivery time 估计交货时间
estimated design load 估计设计负荷;估计的设计荷载
estimated disbursement 预计使费;预计费用
estimated distance 估算距离
estimated distance by eye 目测距离
estimated economic life 估计的经济年限
estimated erosion depth 预估冲刷深度
estimated error 估计误差
estimated expenditures 估计支出
estimated expenditures and revenue 收支概算
estimated expenditures report 支出概算书
estimated expenses 预计费用;费用估计数
estimated flow 估计流量
estimated for lump-sum appropriations 整付经费概算书
estimated frequency 估计频率
estimated fund statement 估计资金表
estimated future traffic 估计远景运量;估计未来交通量
estimated horse power 估计马力
estimate differential settlement 预估差异沉降
estimated income 估计收益;估计收入
estimated income tax payable 估计应付所得税
estimated inventory 估计存货
estimated investment 估计投资量
estimated latitude 估计纬度
estimated length of weld 换算焊接长度
estimated liabilities 预计负债
estimated life 估计寿命;估计年限
estimated life of plant 工程设备的估计使用期限
estimated location of epicenter 估测震中位置
estimated longitude 估计经度
estimated loss 预计损失
estimated luminance 估定亮度
estimated market value 估计市价
estimated maximum load 估计最大负荷

estimated net energy 估计净能
estimated net energy value 估计净能值
estimate documentation 预算文件;预算书
estimated of financial returns 财务收入概算
estimated performance 预测性能;估算性能;估计性能;估计使用期限;估测动态
estimated physical life 估计实际年限
estimated pore pressure 估计孔隙压力
estimated position 估算位置;估计位置;估计船位
estimated position plot 航迹绘画
estimated premium 估计保险费
estimated price 估算价格;估计价格
estimated price method 预定价格法
estimated profit 预算利润;估计利润
estimated profit and loss statement 估计损益表
estimated receipts 收入概算
estimated residual or scrap value 估定残值
estimated residual value 估计残值
estimated scrap value 估计残值
estimated seismic intensity 估测地震烈度
estimated service life 估计使用寿命;估计使用年限
estimated settlement rate 预计沉降速率
estimated shipping date 预计装运日期
estimated ship's position 估计船位
estimated square feet 估计平方尺
estimated standard deviation 标准估计差
estimated statement 估计报表
estimated steam rate 估计耗气率
estimated stress 预计应力
estimated supply day 预计供应日
estimated theoretic(al) depreciation 估计折旧额
estimated theory 估计理论
estimated time of arrival 预算到港时间;预抵期;估计到达时间;船舶预抵港时间;预计到达时间
estimated time of completion 估计的完工时间
estimated time of delivery 预计交货时间
estimated time of departure 预计离港时间;预算开航时间;预离期;估计起航时间;预计离岸时间;预计开航时间;估计出发时间;船舶预计离港时间
estimated time of enroute 预计途中时间;估计途中时间
estimated time of finishing discharging 预计卸完时间;预计卸毕时间
estimated time of finishing loading 预计装完时间
estimated time of return 预计返回时间
estimated total loss 估计总损失
estimated toxic threshold 毒性阈值
estimated ultimate reserves 估算最终储量
estimated uncollectible current taxes 本期税收损失估计
estimated usable period 估定使用期
estimated useful life 估计使用年限
estimated value 预算值;估计值;估定价值;测定值
estimated value of universal Kriging 泛克立格估计值
estimated variance 估计方差;方差估计
estimated viscosity 估定黏度;测定黏度
estimated wage rate 预计工资率
estimated water supply 计划供水
estimated water supply district 计划供水区
estimated weight 估计重量
estimated yearly operating cost 估计每年营业费用
estimate Lagrange multipliers 估算拉格郎日乘子
estimate lower limit on bids 标底
estimate method of primary radioactive carbon-14 估算碳14原始放射性的模式
estimate method of work 工程估算方法
estimate of car requirements 车辆需要推算
estimate of construction 工程概算
estimate of construction cost 工程建筑费用详细估价表
estimate of correlation 相关估计量
estimate of cost 预算;概算;成本估算;成本估计;费用估算
estimate of discharge 流量估算
estimate of error 误差的估计量
estimate of expenditures 支出概算
estimate of heat loss 热损耗估算
estimate of output capacity 生产能力估计
estimate of production capability 生产能力估算
estimate of productivity 生产能力估计
estimate of regression coefficient 回归系数估计
estimate of revenue 岁入概算书

estimate of seismic intensity 地震烈度估计
estimate of standard deviation 标准差估计
estimate of standard error 标准差估计值
estimate of traffic 交通量估算;交通量估计
estimate of variance 方差估计
estimate of wagon flow 车流预测
estimate on cubic(al) meter basis 立方米预算
estimate on square meter basis 平方米预算价
estimate premium 估计保险费
estimate price 估价
estimate revenue 岁入预算
estimate separation 估计距离
estimate sheet 预算单;预算表;估价单
estimate slip 估计滑距
estimates of construction 施工预算
estimates of income 收入概算
estimates of optimum plot size 适宜面积的确定
estimates of petroleum reserves 石油储(藏)量估计
estimates system 概算填补法
estimate summary 预算总表;预算的汇总
estimate survey 估测调查;粗测;初步调查
estimate time and amount of the flow of funds 估计资金的流转时间和数量
estimate total errors 估计总误差
estimate value 估值
estimating 估价;估计;编制预算;编写预算
estimating cost recoverability 估计成本收回的可能性;成本收回的可能性估计
estimating engineer 预算工程师
estimating equation 估算方程;估计方程
estimating floods at bridge site 估计桥址洪水
estimating forms 预算用表
estimating index 估算指标
estimating leaf water potential 叶片水热能的估计
estimating method of construction coefficient 工程系数估算法
estimating method of index 指数估算法
estimating microscope 刻度放大镜;读数放大镜
estimating neighbo(u)rhood method 估计邻域法
estimating of cost 成本估计
estimating of molecular weight 分子量的计算
estimating of viscosity 黏度的估算
estimating pile bearing capacity by Engineering News Formula 工程新闻公式估算单桩承载力
estimating pile bearing capacity by Hiley formula 希列公式估算单桩承载力
estimating procedure 预算程序
estimating saturated hydraulic conductivity 估计饱和水的导电性
estimating techniques 估算技术
estimating water potential 水势能的估计
estimation 预算;估算;估计;估定;评定;推定
estimation block 估计块段
estimation by experience 经验估计
estimation by rule of thumb 经验估计
estimation by the method of moment 矩法估计
estimation error 估算误差;估测误差
estimation error of surface 面积估计误差
estimation error of volumes 体积估计误差
estimation for population mean 总体均数估计
estimation for population rate 总体率估计
estimation for population variance 总体方差估计
estimation for salty foundation soil 岩渍土的地基评价
estimation interval 估计区间
estimation mean 估计均值
estimation method of manhours by experience 经验估工法
estimation method of reserve 储量估算方法
estimation neighbo(u)rhood method 估计邻域法
estimation of accuracy 精度的估计
estimation of approximate quantities 工程量估算
estimation of burial time 埋葬时间测定
estimation of capital cost 估算基建投资
estimation of cost 估价
estimation of dilution 稀释度估算
estimation of elasticity 弹性估计
estimation of error 误差估计
estimation of formation mean direction 岩层平均方向的估计
estimation of intensity 估计地震烈变
estimation of interval 区间估计
estimation of maximum value of ground surface movement in mined-up region 采空区地表最大变形值的估算
estimation of minerogenetic probability 成矿概率估计
estimation of missing data 漏失数据估计;缺项估计
estimation of missing value 漏失数据估计;缺项估计
estimation of parameters 参量估计;参数估值;参数估计
estimation of point 点值估计
estimation of pollution load influence 污染负荷影响估算
estimation of population 人口推算
estimation of population parameters 总体参数估计
estimation of position of epicentre 估测震中位置
estimation of railway internal control system 铁路内部控制制度评审
estimation of ratio 比率估计
estimation of reserves 储量估计
estimation of sales volume 销售量估计
estimation of sample size 样本含量估计
estimation of sample size with proportions 比例抽样样本容量估计
estimation of seismic intensity 地震烈度鉴定
estimation of through-error 贯通误差预计
estimation of truncation 截断误差估计
estimation population 估计人口
estimation range 预测区域
estimation standard deviation 估计标准差
estimation theory 估计理论
estimation variance 估计方差
estimation variance of surface 面积估计方差
estimation variance of volumes 体积估计方差
estimative figure 估计的数字
estimative size of meteorite 估计陨石大小
estimator 预算员;估算员;估价人员;估量值;估价者;估价师;估计量;推算子;设计员
estoppel by deed 禁止推翻契据
estoppel of water rights 水权翻案禁条
estover 佃户使用租地林木权
estrade 室内矮平台;讲台
estradiol 雌甾二醇
estrane 雌甾烷
estrane diol 雌甾烷二醇
estriol 雌甾三醇
Estrol dye 埃斯特罗染料
Estron 埃斯特纶醋酯长丝;醋酸纤维素
estrone 雌甾酮
Estron spunyarn 埃斯特纶醋酯短纤纱
estuarial 河口的
estuarial circulation 河口环流
estuarial crossing 港口立交桥
estuarial dredging 河口疏浚;河口挖泥
estuarial model 河口模型
estuarial sediment 河口泥沙(沉积)
estuarine 河口的;感潮河口的
estuarine, continental shelf and marine model 河口、陆架和海洋模型
estuarine animal 河口动物
estuarine berth 河口泊位
estuarine biology 河口生物学
estuarine chemistry 河口化学;港湾化学
estuarine circulation 河口环流
estuarine clay 河流三角洲沉积黏土;海湾黏土
estuarine communities 河口生物群落;港湾生物群落
estuarine cycle 河口循环
estuarine delta 河口(湾)三角洲
estuarine deposit 河口淤积;河口湾沉积(物);河口沉积物;河口沉积(层);海湾沉积
estuarine deposition 港湾沉积作用
estuarine ecology 河口生态学;港湾生态学
estuarine ecosystem 河口生态系统;港湾生态系统
estuarine environment 河口(湾)环境;海湾环境
estuarine facies 河口(湾)相;海湾相;港湾相【地】
estuarine fauna 河口动物区系
estuarine fishery 河口渔业
estuarine fishes 河口鱼类
estuarine flat 河口潮漫滩;海湾低洼地;港湾低洼地
estuarine flood 河口潮洪
estuarine flow 河口水流;海湾水流
estuarine harbo(u)r 河口港;港湾
estuarine hydrology 河口水文学
estuarine invertebrate 河口无脊椎动物
estuarine lake 河口湖;海湾湖
estuarine lake of intermittent stream 间歇性河口湖
estuarine marine science 港湾海洋科学
estuarine microbiota 河口微生物
estuarine mixing 河口混合
estuarine mixture 河口水流混合
estuarine mud 河口软泥;港湾泥
estuarine navigation channel 河口航道
estuarine oceanography 河口海洋学
estuarine ooze 河口软泥
estuarine organism 河口生物
estuarine pollution 河口污染
estuarine process 河口演变过程
estuarine project 河口工程
estuarine river-flow reach 河口河流段
estuarine river net 河口河网
estuarine sand 河口沙(也称河口砂)
estuarine sediment 河口沉积(物);河口泥沙(沉积);河口沉积物;港湾沉积物
estuarine series 港湾统【地】
estuarine species 河口物种
estuarine surface sediment 河口表层沉积物
estuarine tidal current reach 河口潮流段
estuarine waste load allocation 河口废负荷分配
estuarine water 海湾水;河水
estuarine water quality assessment 河口水质评价
estuarine waters 河口水域;河口水体
estuary 江口段;江河入海口;河流入海口;河口;港湾;潮区;入海(河)口
estuary area 河口区(域)
estuary branch 河口汊道
estuary channel 河口水道;河口航槽
estuary clay 河口黏土
estuary closure 河口闭塞
estuary coast 港湾海岸;湾形海岸
estuary dam 河口坝
estuary delta 河口三角洲
estuary deposit 河口淤积;河口泥沙(沉积);河口沉积物;河口沉积(层);海湾沉积;港湾淤积;港湾沉积
estuary district 河口区(域);河口地区
estuary engineering 河口工程学
estuary facies 港湾相【地】
estuary firth 河口湾
estuary geography 河口学
estuary guide jetty 河口导堤
estuary harbo(u)r 河口港
estuary hydraulic condition 河口水力条件
estuary hydrology 河口水文学
estuary model 河口模型
estuary mud bar 河口泥滩
estuary pollutant 河口湾污染物
estuary pollution 河口(湾)污染;港湾污染
estuary port 河口港
estuary process 河口演变过程
estuary project 河口工程
estuary region 河口区(域)
estuary regulation 河口整治
estuary sand trap 河口沙体圈闭
estuary sediment 河口沉积物
estuary sedimentation 河口淤积
estuary station 河口测站
estuary tide 河口潮(汐)
estuary tug 河口拖轮
estuary water quality 河口水质
estuary water quality model 河口水质模型
estuary weir 河口堰
estuary with large-scale island 大型岛屿河口
estutaine conservation area 河口保护区
etalon 波长测定仪;标准(量)具;校准器;标准件
etalon clock 标准钟
etalon optical power 标准光强度
etalons in tandem 串列标准具
etalon source constant 标准源常数
etalon source content 标准源含量
etalon time 标准时(间)
etamsylate 酚磺乙胺
etang 泻湖;浅水海湾;滩积内陆湖
eta patch 加强补片
Etard's reaction 埃塔德反应
etch 胶印药水;侵蚀;镂蚀;浸蚀;蚀刻

etch angle 蚀痕角
etchant 腐蚀刻;镂蚀剂;蚀刻剂
etchant resist 抗蚀膜
etch back 内腐蚀;深腐蚀
etch-bleach 漂白反像法
etch channel 腐蚀隧道
etch copy 槽纹图板
etch cut 蚀割
etch cut method 腐蚀截割法
etch cutting 腐蚀切割
etch cutting method 腐蚀切割法
etched alumin(i)um plate 腐蚀铝板
etched cavity 蚀刻腔
etched channel 酸蚀孔道
etched circuit 蚀刻电路
etched circuit board 蚀刻线路板
etched concrete 腐蚀混凝土;酸蚀混凝土
etched decoration 蚀刻装饰
etched dimple 浸蚀陷斑;侵蚀陷斑
etched ellipse 刻蚀椭圆痕
etched figure 侵蚀像;蚀像
etched foil process 腐蚀法
etched glass 刻花玻璃;磨砂玻璃;毛玻璃;无光玻璃;酸蚀毛玻璃
etched grating 蚀刻光栅
etched groove 蚀沟
etched hill 蚀丘
etched image 蚀刻图像
etched like finish 浸蚀型光洁度
etched line 蚀刻槽纹
etched mirror 蚀刻反射镜
etched plate 腐蚀板
etched reflector 蚀刻反射镜
etched resin cast 蚀刻的树脂铸模
etched specimen 浸蚀试件
etched surface 腐蚀面
etched wiring 腐蚀法印刷电路
etched zinc plate 腐蚀锌板
Etcheminian 埃奇米尼阶
etcher 标记与图案酸洗器;蚀刻师;蚀刻器
etcher roll 蚀刻辊
etcher's needle 刻图针;蚀刻用针
etcher's wax 蚀刻用石蜡
etch figure 浸蚀图;蚀刻图
etch hillock 蚀丘;溶解小丘
etch hole 腐蚀孔点;蚀刻孔
etching 浸润;腐蚀法;酸洗;蚀刻(法);镂蚀;浸蚀
etching acid 腐蚀酸
etching agent 镂蚀剂;浸蚀剂;侵蚀剂;蚀刻剂
etching bath 蚀刻槽
etching device 蚀刻装置
etching dye 蚀刻染色
etching figure 浸蚀像
etching glass 蚀刻玻璃
etching ink 耐酸漆;腐蚀墨
etching-interfering method 腐蚀干涉法
etching knife (蚀)刻刀
etching machine 腐蚀制版机;蚀刻机
etching mark 蚀刻痕迹
etching mask 腐蚀(性)掩模;蚀刻掩模
etching material 蚀刻材料
etching method 刻蚀法;箔腐蚀法;蚀刻法
etching of slope toe 冲蚀坡脚
etching orientation 腐蚀方向
etching paper 腐蚀版印刷用纸
etching pattern 侵蚀形态;蚀刻图像
etching period 蚀刻时间;浸蚀期
etching pit 蚀刻法
etching plate 腐蚀凹板
etching powder 蚀刻粉
etching power 刻蚀能力
etching primer 金属底层处理用漆;洗涤底漆;腐蚀性涂料;防护涂料;防护涂层;防腐蚀涂料;磷化底漆
etching process 刻蚀过程
etching rate 浸蚀速度;侵蚀速率
etching reaction 侵蚀反应
etching reagent 浸蚀剂;腐刻玻璃试剂;侵蚀剂;蚀刻剂
etching residue 腐蚀残渣
etching resist 抗蚀层;浸蚀涂层
etching ring 刻蚀环痕
etching room 腐蚀制版车间;酸洗间
etching solution 浸蚀溶液;蚀刻溶液
etching solution concentration 蚀刻溶液浓度
etching solutions for examination of materials 检验材料的侵蚀溶液
etching solution temperature 蚀刻溶液温度
etching stamp 酸蚀印记
etching structure 侵蚀构造
etching tank 浸蚀槽
etching technique 蚀刻技术
etching test 浸蚀试验;浸蚀试法;腐蚀试验;蚀刻试验
etching time 蚀刻时间;刻蚀时间
etching to frequency 腐蚀到所需频率
etching treatment 蚀刻处理
etching trough 腐蚀槽
etching varnish 蚀刻清漆
etching waste liquor 蚀刻废液
etch into relief 蚀刻成凹凸
etch line 蚀刻线
etch mark 刻痕
etch method 蚀刻测斜法
etch off 腐蚀掉
etch pattern 腐蚀图形;蚀刻型;腐蚀图案
etch period 刻蚀时间
etch pit 蚀痕;侵蚀陷斑;腐蚀陷斑;浸蚀坑;腐蚀坑;蚀刻坑
etch-pit counting method 蚀刻坑计数法
etch pit method 浸饰坑法;腐蚀坑法
etch plain 刻蚀平原
etch-polish(ing) 蚀刻抛光
etch primer 反应性底漆;磷化底漆;防蚀底漆
etch printing 蚀刻印刷
etch-proof 抗腐蚀;防腐蚀的
etch-proof resist 防蚀剂
etch reaction 浸蚀反应
etch reagent 酸洗剂
etch-resistant coating 抗蚀(刻)涂层;防浸蚀涂层
etch-resistant layer 抗腐蚀层
etch-resistant resin 光致抗蚀剂树脂
etch ring 蚀刻环线
etch slide 蚀刻片
etch slip 擦板条
etch solution test 颜料耐蚀试验
etch sternway 后退
etch test 酸蚀试验
etch tube 蚀刻管
etch virus 蚀刻病毒
etendue 集光率
eternal equation 永恒方程式
eternal frost 永冻层;多年冻土
eternal frost climate 永冻气候;冰冻气候
eternal frozen ground 永冻土;永冻地
eternal strain 永久应变
Eternit 石棉水泥;爱特尼石棉水泥屋顶材
eternit pipe 永久性管;不朽管;石棉水泥管
eternit roofing 石棉水泥屋面
eternit roof sheathing 石棉水泥屋顶防水层
Eternit slab 石棉水泥板
Eternit slate 石棉水泥板岩
eternity corrugated sheet 石棉水泥波纹板
Etesian climate 地中海气候
etesiens 地中海季风
ethanal 乙醛
ethanamide 乙酰胺
ethane 乙烷
ethane bacteria 产乙烷菌
ethane diacid 草酸
ethane dicarboxylic acid 乙烷二甲酸
ethanedioic acid 乙二酸;草酸
ethanedioyl chloride 乙二酰氯
ethane disulfonic acid 乙烷二磺酸
ethanedisulphonate 乙二磺酸盐
ethane peroxide 过氧乙烷
ethane tetracarboxylic acid 乙烷四甲酸
ethanol acid 乙醛酸
ethanolamine 乙醇胺;氨基乙醇
ethanolamine methyl ether 氨基乙醇甲醚
ethanol amine process 乙醇胺法
ethanol gel test 乙醇胶试验
ethanolic extract 乙醇萃取物
ethanoyl(acetye) 乙酰
ethanthiol 乙二硫醇
ethargy group 勒群
ethenoid group 乙烯型基
ethenoid plastics 乙烯型塑料
ethenoid polymer 乙烯型聚合物
ethenoid resin 乙烯(型)树脂
ethenol 乙烯醇
ethenone 乙烯酮
ethenyl 次乙基
ether 乙醚
ether acid 醚酸
ether alcohol 醚醇
etherate 醚合物
Ether board 埃特板
ether capsule 乙醚胶囊
ether cleavage 醚裂开
ether discharger 乙醚排水装置
ether drag 以太曳引
ether drift 以太漂移
ethereal 醚制的
ethereal blue 纯蓝
ethereal essence 醚香料
ethereal liquid 醚状液
ethereal oil 醚油
ethereal sulfate 硫酸酚酯
ethereal sulphate 硫酸酚酯
ether ester 醚酯
ether extraction 用乙醚萃取
etheric acid 乙酰乙酸
etherification 醚化(作用)
etherified resin 醚化了的尿素树脂
etherify 醚化
etherifying agent 醚化剂
ether-insoluble resin 醚不溶树脂
ether link 醚键
Ethernet 以太网(络);以太计算机网
ether oil 醚油
ethero-sulfuric acid 硫酸氢乙酯
ether peroxide 过氧化二乙醚
ether ring 醚环
ether scanner 全景搜索接收机;全景接收机
ether-soluble extractive 醚溶性浸出物
ether starting aid 乙醚启动器
ether wave 以太波;电磁波
ethical sales 公平销售
ethics 规矩
ethics of profession 职业道德
Ethide 二氯硝基乙烷(一种熏蒸杀虫剂)
ethinylation 乙炔化(作用)
ethionic acid 乙二磺酸
Ethiopian architecture 埃塞俄比亚建筑
Ethiopian zoogeographic region 埃塞俄比亚动物地理区
ethmolith 漏斗(状)岩盘;岩漏斗
ethnic group 种族集团
ethnic minorities 少数民族
ethnic tour 民俗旅游
ethnic tourism 民俗旅游
ethnoecology 民族生态学
ethnography 民族地理学
ethnonymics 民族学
ethnozoology 人文动物学
ethogram 习性谱
ethoxal 乙草酸
ethoxide 乙氧基金属
ethoxyacetic acid 乙氧基乙酸
ethoxyaniline 乙氧基苯胺
ethoxy benzene 乙氧基苯
ethoxy benzoic acid 乙氧基苯(甲)酸
ethoxy-benzoin 乙氧基苯偶姻
ethoxybenzoyl 安纳晶
ethoxy castor oil 乙氧基蓖麻油
ethoxy cellulose 乙氧基纤维素
ethoxy determination 乙氧基测定
ethoxy-ethanol 乙氧基乙醇
ethoxy ethanol acetate 乙氧基乙醇乙酸酯;乙二醇乙醚乙酸酯
ethoxyethyl laurate 月桂酸乙氧基乙酯
ethoxyethyl phthalate 乙氧基乙基邻苯二酸盐
ethoxyethyl ricinoleate 乙二醇乙醚蓖麻酸酯;蓖麻酸乙氧基乙酯
ethoxy hexane 乙氧基己烷
ethoxylate 乙氧基化物
ethoxyline resin 环氧树脂
ethoxy methane 乙氧基甲烷
ethoxy octane 乙氧基辛烷
ethoxy pentane 乙氧基戊烷

ethoxy propane 乙氧基丙烷
ethoxy triethyl silane 乙氧基三乙基硅烷
ethoxytrimethylsilane 乙氧基三甲基(甲)硅
ethvlidene ether 醛缩醇
ethyl 乙基
ethylabietate 松香酸乙酯
ethyl acetate 乙酸乙酯;醋酸乙酯
ethyl acetic acid 丁酸
ethyl acetoacetate 乙酰乙酸乙酯
ethyl acetylene 乙基乙炔
ethyl acetyl glycola(e)乙酰基乙醇酸乙酯
ethyl acetyl ricinoleate ether 乙酰蓖麻酸乙酯
ethyl acrylate 丙烯酸乙酯
ethyl alcohol 酒精;乙醇
ethyl alcohol-benzene bitumen A content 乙醇苯沥青A含量
ethyl aluminum dichloride 二氯乙基铝
ethyl amine 乙胺;氨基乙烷;胺基乙烷
ethyl aminobenzoate 氨基苯甲酸乙酯
ethyl amyl ketone 乙戊酮;乙基戊基甲酮
ethyl arsine sulfide 乙胂化硫
ethylation 乙基化(作用)
ethyl benzene 苯乙烷;乙(基)苯;苯基乙烷
ethylbenzene sulfate 硫酸乙苯
ethyl benzenesulfonate 苯磺酸乙酯
ethyl benzoate 苯甲酸乙酯
ethyl benzoylacetate 苯酰乙酸乙酯
ethyl benzoylformate 苯甲酰乙酸乙酯
ethyl benzyl cellulose 乙基苄基纤维素;乙苄纤维素
ethyl benzyl ether 乙基苄基醚
ethyl borate 硼酸乙酯
ethyl bromide 溴化乙
ethyl butyl carbonate 碳酸乙丁酯
ethyl butyrate 丁酸乙酯
ethyl caprilate 辛酸乙酯
ethyl caproate 己酸乙酯
ethyl carbamate 氨基甲酸乙酯;乌拉坦
ethyl carbonate 碳酸盐乙酯;碳酸二乙酯
ethyl cellosolve 乙基溶纤剂
ethyl cellulose 乙基纤维素
ethyl cellulose lacquer 乙基纤维素漆
ethyl cellulose plastics 乙基纤维塑料
ethyl chaulmoograte 晁模酸乙酯
ethyl chloride 氯乙烷
ethyl chloroacetate 氯醋酸乙酯
ethylcyclohexane 乙基环己烷
ethylcyclopentane 乙基环戊烷
ethyl decylate 癸酸乙酯
ethyl diacetoacetate 二乙酰乙酸乙酯
ethyl diazoacetate 重氮基醋酸乙酯
ethyl dibenzyl-glycolate 二苄基乙醇酸乙酯
ethyl dibromoacetate 二溴乙酸乙酯
ethyl dibunate 地布酸乙酯
ethyl dibutylcarbamate 二丁氨基甲酸乙酯
ethyl dichloroacetate 二氯乙酸乙酯
ethyl dinitrobenzoate 二硝基苯甲酸乙酯
ethyl enanthate 庚酸乙酯
ethylene 乙烯;次乙基
ethylene acrylate copolymer 乙烯丙烯酸酯共聚物
ethylene acrylic acid 乙烯丙烯酸
ethylene butadiene copolymer 乙烯丁二烯共聚物
ethylene-butylene copolymer 乙烯丁烯共聚物
ethylene carbonate 碳酸盐乙酯;碳酸次乙酯
ethylene chloride 氯化乙烯
ethylene compressor 乙烯压缩机
ethylene copolymer 乙烯共聚物
ethylene copolymer bitumen 乙烯共聚沥青
ethylene dialdehyde 乙二醛
ethylene-diamine 乙二胺
ethylene diamine dinitrate 二硝酸化乙二胺
ethylene-diamine tetracetic acid 乙二胺四乙酸
ethylene dibromide 二溴化乙烯
ethylene dichloride 二氯化乙烯
ethylene difluoride 二氟化乙烯
ethylene dihalide 二卤化乙烯
ethylene fluoride 氟化乙烯
ethylene glycol 亚基乙二醇;乙二醇;甘醇
ethylene glycol butyl ether 乙二醇一丁醚
ethylene glycol dimethacrylate 乙二醇二甲基丙烯酸酯
ethylene glycol dimethyl ether 乙二醇二甲醚
ethylene glycol monoacetate 乙二醇-乙酸酯
ethylene glycol monobutyl ether 丁基溶纤剂

ethylene glycolmonoethyl ether acetate 醋酸酯溶纤剂
ethylene glycol monoethyl ether laurate 乙二醇单乙醚月桂酸酯
ethylene glycol monomethyl ether 乙二醇一甲醚
ethylene glycol monomethyl ether acetate 乙二醇单甲醚乙酸酯
ethylene glycol monomethyl ether acetyl ricinoleate 乙酰蓖麻酸甲氧基乙酯
ethylene glycol monomethyl ether acetyl ridnolcatelene 乙酰蓖麻酸乙二醇单甲醚酯
ethyleneimine 吖丙啶;乙亚胺
ethylene naphthalene 苊
ethylene oxide 环氧乙烷;氧(化)乙烯;氧丙环
ethylene oxide addition polymer 环氧乙烷加成聚合物
ethylene oxide polymer 聚环氧乙烷;环氧乙烷聚合物
ethylene oxide propylene glycol condensate 环氧乙烯丙二醇缩聚物
ethylene perchloride anti-corrosive enamel paint 过氯乙烯防腐瓷漆
ethylene perchloride enamel paint 过氯乙烯瓷漆
ethylene perchloride primer 过氯乙烯底漆
ethylene perchloride white varnish 过氯乙烯白漆
ethylene plant 乙烯厂
ethylene plastics 乙烯塑料
ethylene polymer 乙烯聚合物
ethylene polymerised oil 乙烯聚合油
ethylene product 乙烯产品
ethylene project 乙烯工程
ethylene-propylene copolymer 乙烯丙烯共聚物;乙丙共聚物
ethylene-propylene diene monomer 乙烯丙烯二烯系单体;乙丙二烯单体
ethylene-propylene diene tripolymer 乙烯丙烯二烯三元聚物
ethylene-propylene elastomer 乙烯丙烯弹性体
ethylene-propylene methylene lintage 乙烯丙烯橡胶
ethylene-propylene packing 乙丙胶垫
ethylene-propylene rubber 乙烯丙烯橡胶;乙丙橡胶;三元丙橡胶
ethylene-propylene rubber gasket 三元乙丙橡胶密封垫
ethylene-propylene terpolymer 三元乙丙胶
ethylene-rich gas 富乙烯气体
ethylene tanker 乙烯船
ethylene terephthalate polymer 对苯二酸乙二醇缩聚物
ethylene tetrachloride 四氯乙烷;四氯化乙烷
ethylene-tetrafluoethylene copolymer 乙烯四氟乙烯共聚物
ethylene tube 乙烯管
ethylene unit 乙烯装置
ethylene-vinyl acetate 乙烯乙酸乙烯酯
ethylene-vinyl acetate copolymer 乙烯醋酸乙烯共聚物;乙烯乙酸乙烯(酯)共聚物
ethylene-vinyl acetate copolymer hot-melt adhesive 乙烯乙酸乙烯热熔胶
ethylene-vinyl acetate rubber 乙烯乙酸乙烯酯橡胶
ethyleric glycol butyrate 乙二醇丁酸酯
ethyl ester 乙酯
ethyl ether 醚;乙基醚;二乙醚;乙醚
ethyl ethoxyl propionate 乙氧基丙酸乙酯
ethylethylene 丁烯-1
ethyl flooring 乙烯地板
ethyl fluid 乙基液
ethyl fluoroacetate 氟乙酸乙酯
ethyl-formamide 乙替甲酰胺
ethyl gasoline (含四乙铅的)乙基汽油
ethyl glycol terephthalate 对苯二甲酸乙二酯
ethyl gravitometer 乙种比重计
ethyl heptanoate 庚酸乙酯
ethyl hexanoate 己酸乙酯
ethyl hexanol 异辛醇;乙基己醇
ethyl hexoate 己酸乙酯
ethylhexoate drier 异辛酸盐催干剂
ethyl hexyl acrylate 丙烯酸异辛酯
ethyl hydride 乙烷
ethyl hydrogen sulfate 硫酸氢乙酯;乙基硫酸
ethyl hydroxyethyl cellulose 乙基羟乙基纤维素
ethyl hypochlorite 次氯酸乙酯
ethylidene biuret 亚乙基缩二脲

ethylidene chloride 不对称二氯乙烷
ethylidene dichloride 亚乙基二氯
ethylidene diurethane 亚乙基二氨基甲酸酯
ethylidene ether 乙醛缩二乙醇;缩醛
ethylidene fluoride 不对称二氟乙烷
ethylidyne 次乙基
ethyliminum 乙亚胺
ethyl iodide 乙基碘
ethyl isocyanate 异氰酸乙酯
ethylized fuel 乙基燃料
ethyl lactate 乳酸乙酯
ethyl liquor 乙基液
ethyl malonate 丙二酸二乙酯
ethyl mercury acetate 乙酸乙汞
ethyl methacrylate 甲基丙烯酸乙酯
ethyl methyl ether 乙基甲基醚
ethyl methyl ketone 甲乙酮
ethyl mobutyl ketone 乙基异丁基甲酮
ethyl n-butyl ketone 乙基正丁基甲酮
ethyl oenanthate 庚酸乙酯
ethylogen 乙烯原
ethyl oleate 油酸乙酯
Ethylon 埃塞纶聚乙烯鬃丝
ethyl orange 乙基橙;乙橙
ethyl orthacetate 原乙酸乙酯
ethyl orthosilicate 硅酸乙酯
ethyl oxalacetate 草醋酸乙酯
ethyl oxalate 草酸乙酯
ethyl-para aminobenzoate 对氨基苯甲酸乙酯
ethylparaben 对羟基苯甲酸乙酯
ethyl para-chlorophenoxyisobutyrate 对氯苯氧异丁酸乙酯
ethyl-para-hydroxybenzoate 对羟基苯甲酸乙酯
ethyl-para-hydroxyphenyl ketone 对羟基苯丙酮
ethyl peroxide 过氧乙醚;过氧化二乙基
ethyl persulfide 二乙基化二硫
ethyl petrol 含四乙基铅汽油;(含四乙铅的)乙基汽油
ethyl phenacemide 苯丁酰脲
ethyl phenolate 苯乙醚
ethyl phenylacetate 苯乙酸乙酯
ethyl phthalyl ethyl glycolate 邻苯二甲酰(乙酯)乙醇酸乙酯
ethyl p-methyl benzoate 对甲基苯甲酸乙酯
ethyl p-nitrobenzoate 对硝基苯甲酸乙酯
ethyl polysilicate 聚硅酸乙酯
ethyl propiolate 丙炔酸乙酯
ethyl propionate 丙酸乙酯
ethyl propyl-acetoacetate 丙乙酰乙酸乙酯
ethyl p-toluenesulfonate 对甲苯磺酸乙酯
ethyl pyruvate 丙酮酸乙酯
ethylquinolinium iodide 碘化N—乙基喹啉盐
ethyl red 乙基红
ethyl rubber 乙基橡胶
ethyl salicylate 水杨酸乙酯
ethyl silicate 硅酸乙酯
ethyl silicate film 硅酸乙酯薄膜
ethyl silicone resin 乙基硅酮树脂;乙硅酮树脂
ethyl silicon oil 乙基硅油
ethyl succinate 琥珀酸乙酯;丁二酸乙酯
ethyl sulfate 硫酸二乙酯;硫酸乙酯
ethyl sulfide 二乙硫
ethyl sulfuric acid 乙基硫酸
ethyl-t-butyl peroxyoxalate 过氧草酸乙基特丁酯
ethyltetrahydronaphthalene 乙基四氢化萘
ethylthioethane 二乙硫
ethylthiophene 乙基噻吩
ethyl trifluorosilane 乙基三氟(甲)硅烷
ethyl trimethyl silane 乙基三甲基(甲)硅烷
ethyl urethane 氨基甲酸乙酯
ethyl vinyl ether 乙基乙烯基醚;乙氧基乙烯
ethyl xanthate 黄原酸乙酯
ethyl xanthogenate 黄原酸乙酯
ethyl xanthogen disulfide 草编织品
ethynyl 乙炔基
ethynylbenzene 乙炔苯
ethynyl carbinol 乙炔基甲醇
ethyudene diacetate 乙醛二乙酸酯
etilogic solid waste 致病原体固体废物
etintidine 丙炔替丁
etioblast 白化囊泡
etiocholane 本胆烷
etiocholanedione 本胆烷二酮
etiocholanolone 本胆烷醇酮;

etiolate 黄化
etiolation 褪色;黄化现象
etiological examination 病原学检查
etiologic solid waste 致病性固体废物
etiology 病原学
etioplast membranes 白化体膜
etioporphyrin 初卟啉
etiopyllin 初卟啉合镁盐
Etna volcano 埃特纳火山
Etroeungtian 艾特隆
Etruria Ma 1s 埃特鲁利亚泥灰岩
Etruscan architecture 伊特鲁斯坎建筑(古意大利)
Etruscan column 伊特鲁斯坎建筑的柱
Etruscan style 伊特鲁斯坎风格
Etruscan temple 伊特鲁斯坎庙宇
Ettinghausen effect 艾廷豪森效应
ettle 废矿石
ettringite 钙铝矾;钙矾石;水泥杆菌;三硫型水化硫铝酸钙
E-twin lamellae E 双晶纹
Etype graphite 晶间石墨
euabyssite 放射虫岩
euaster 真星型
euautochthonous coal 纯原地生成煤
eubacteria 真细菌
eubacterium 真细菌
eu-bitumen 液体沥青
eucairite 硒铜银矿
eucalypt ecology 桉树生态
eucalyptene 桉树烯
eucalyptol 桉油醇;桉叶油素;桉叶醇;桉树脑
eucalyptole 桉叶脑
eucalyptolene 桉树脑烯
eucalypt species 桉树品种
eucalyptus 桉树属植物;桉树(木材);桉属
Eucalyptus citriodora 柠檬桉
eucalyptus gum 桉树胶
eucalyptus kino 桉树胶
eucalyptus leaves 桉树叶
eucalyptus oil 桉油;桉叶油;桉树油
eucalyptus resin oil 桉树焦油
eucalyptus tar 桉树焦油
eucazulene 桉蓝烃
eucelyptol 桉油精
euchlorin(e) 碱铜矾
euchlorite 绿云母
euchroite 翠砷铜石;翠砷铜矿
euclase 蓝柱石
Euclidean algorithm 欧几里得辗转相除法;欧几里得算法
Euclidean body 欧几里得体
Euclidean complex 欧几里得复形
Euclidean connection 欧几里得联络
Euclidean distance 欧几里得距离
Euclidean domain 欧几里得整环;欧几里得区
Euclidean field 欧几里得域
Euclidean geometry 欧几里得几何
Euclidean geometry code 欧几里得几何码
Euclidean group of motions 欧几里得运动群
Euclidean integral domain 欧几里得整环
Euclidean measure 欧几里得测度
Euclidean motion 欧几里得运动
Euclidean norm 欧几里得模量;欧几里得范数
Euclidean plane 欧几里得平面
Euclidean polyhedron 欧几里得多面体
Euclidean postulate 欧几里得公设
Euclidean rigid body 欧几里得刚体
Euclidean ring 欧几里得环
Euclidean simplex 欧几里得单形
Euclidean simplicial complex 欧几里得单纯复形
Euclidean solid 欧几里得固体
Euclidean space 欧几里得空间
Euclidean straight line 欧几里得直线
Euclidean vector space 欧几里得向量空间
Euclid norm 欧几里得范数
Euclid's algorithm 欧几里得算法
Euclid's geometry 欧几里得几何
Euclid-solid 欧几里得体
Euclid's postulate 欧几里得公设
eucolite 负异性石
eucolloid 真胶体
eucrite 钙长易变辉石无球粒陨石;钙长石陨石;钙长辉长岩

eucryptite 锂霞石
eucrystalline 良晶(质)的;优晶质的
eucrystalline texture 全晶质结构
eudesmane 桉烷;桉烷
eudesmene 桉叶烯
eudesmin 桉素
eudesmol 桉叶油醇;桉醇
eudialite 异性岩;异性石
eudidymite 双晶石
eudiometer 量气计;量气(纯度)管;空气纯度测定管;测气管;容积变化测定管;气体燃化计;气体纯度测定器
eudiometry 空气纯度测定(法);气体分析法;气体测定法
eufroe 紧绳器;天幕吊板
eugelinite 充分分解凝胶体;致密凝胶体
eugenol 丁香酚
eugenol type basil oil 丁香罗勒油
eugeocline 活动正地槽;优地斜
eugeogenous 易风化成碎屑的;风化岩屑的;易风化的
eugeogenous rock 易风化岩
eugeosycline 优地槽【地】
eugeosyncline 活动正地槽
eugeosyncline type formation 优地槽型沉积建造
Euglena deses 静裸藻
Euglypha 鳞壳虫属
eugranitic 花岗岩状的
eugsterite 尤钠钙矾
euhedral 自形的;全形的
euhedral crystal 自形(结)晶
euhedral-granular 自形粒状的
euhedral-granular texture 自形(晶)粒状结构
euhyponeuston 真水面下漂浮生物
eukamptite 水黑蛭石
eukarytic 真核微型浮游生物
Eular head 欧拉压头
Euler-Bernouli beam theory 欧拉—佰务利梁理论
Euler-Bernouli theory 欧拉—佰务利理论
Euler-Cauchy method 欧拉柯西法
Euler's angle 欧拉角
Eulerian buckling stress 欧拉屈曲应力
Eulerian change 欧拉变换
Eulerian circuit 欧拉线路
Eulerian column formula 欧拉柱公式
Eulerian coordinates 欧拉坐标
Eulerian correlation 欧拉相关
Eulerian crippling stress 欧拉撕裂应力
Eulerian equation 欧拉方程
Eulerian equation of motion 欧拉运动方程
Eulerian free period 欧拉自由周期
Eulerian index 欧拉指数
Eulerian intrinsic(al) equation 欧拉裹性方程
Eulerian load 欧拉荷载
Eulerian matroid 欧拉拟阵
Eulerian mean flow 欧拉平均流
Eulerian method 欧拉法
Eulerian motion 欧拉运动
Eulerian number 欧拉数
Eulerian numerical method 欧拉数值法
Eulerian path 欧拉路径
Eulerian period 欧拉周期
Eulerian pole 欧氏极
Eulerian square 欧拉方
Eulerian tour 欧拉游程
Eulerian trace 欧拉迹
Eulerian trajectory 欧拉轨道
Eulerian wind 欧拉风
Euler-Lagrange equation 欧拉—拉格郎日方程
Euler number 1 第一欧拉数
Euler number 2 第二欧拉数
Euler-Poincare equation 欧拉—庞加莱方程
Euler-Rodriguse parameter 欧拉—罗德里格参数
Euler's angle method 欧拉角法
Euler's approach 欧拉求解法
Euler's buckling 欧拉屈曲
Euler's buckling formula 欧拉屈曲公式
Euler's buckling length 欧拉屈曲长度
Euler's buckling load 欧拉屈曲荷载
Euler's buckling stress 欧拉弯曲应力
Euler's characteristic 欧拉示性数
Euler's circle 欧拉圆
Euler's column formula 欧拉柱公式
Euler's constant 欧拉常数

Euler's coordinates 欧拉坐标
Euler's crippling stress (纵向弯曲)欧拉临界应力;欧拉临界(断裂)应力
Euler's criterion 欧拉判别准则
Euler's criterion for residues 欧拉剩余判别准则
Euler's curve 欧拉曲线
Euler's cycle 欧拉循环
Euler's diagram 欧拉图
Euler's differential equation 欧拉微分方程
Euler's equation 欧拉方程(式)
Euler's expansion 欧拉展开
Euler's expansion formula 欧拉展开式
Euler's force 欧拉力
Euler's formula 欧拉公式;欧拉方程式
Euler's formula for long columns 欧拉长柱公式
Euler's fracture stress 欧拉断裂应力
Euler's function 欧拉函数
Euler's graph 欧拉图
Euler's hydrodynamical law 欧拉水动力定律
Euler's hydrostatical law 欧拉静水力学定律
Euler's hyperbola 欧拉双曲线
Euler's identity 欧拉恒等式
Euler's infinite product representation 欧拉无穷积表示
Euler's integral 欧拉积分
Euler's law 欧拉定律
Euler's line 欧拉线
Euler('s) load 欧拉荷载
Euler's loop 欧拉循环
Euler's mesh formula 欧拉网孔数公式
Euler's method 欧拉(方)法
Euler's method of elimination 欧拉消去法
Euler's number 欧拉(指)数
Euler's period 欧拉周期
Euler's polyhedron formula 欧拉可分割空间公式
Euler's polynomial 欧拉多项式
Euler's product 欧拉积
Euler's square 欧拉方
Euler's stress 欧拉应力
Euler's summation formula 欧拉求和公式
Euler's theorem 欧拉定理
Euler's theory 欧拉理论
Euler's transformation 欧拉变换
Euler's triangle 欧拉三角形
Euler's turbine equation 欧拉涡轮方程
eulimnetic 湖心的;湖泊浮游生物;湖泊的
eulimnoplankton 湖心浮游生物;湖泊浮游生物
eulite 钙莱辉石;易熔辉石
eulittoral belt 湖沿岸带
eulittoral zone 湖沿岸区;湖潮间带
eulysite 榴辉铁橄岩
eulytite 门铋矿;硅铋石
eumetric fishing 适合捕捞
eumetric yield 适合产量
eumorphic 正形的
eumycetes 真菌类
eupatheoscope 温度自记仪;热损仪;热损失估测仪
eupelagic clay 远洋黏土
eupelagic deposit 远洋沉积;纯远海沉积;纯大洋沉积
eupelagic facies 远洋相;泛光海相
eupelagic plankton 远洋浮游生物
eupelagic sediment 远洋沉积(物);纯远洋沉积;纯大洋沉积
eupholite 滑石辉长岩
euphonic zone 光亮区
euphorin 苯基氨基甲酸乙酯
euphotic 透光的;透光层的
euphotic layer 真光层;透光层
euphotic zone 透光带
euphotide 槽化辉长岩
euphroe 紧绳器;天幕吊板
Euphrosyne 美乐女神星
euphyllite 钠钾云母
euplankton 真浮游
eupolymer 优聚合物
euporphyric texture 全斑结构
euprofundal 深湖底的
eupyrexia 微热
Eurafrica 欧非共同体
Euramerica 欧美大陆
Euramerican bivalve region 欧美双壳类地理大区
Euramerican floral realm 欧美植物地理区系
euraquilo 尤拉奎洛风

Eurasia 欧亚大陆
Eurasia-Arctic foraminifera realm 欧西—北极有孔虫地理区系
Eurasia-Asian coral realm 欧亚珊瑚地理区系
Eurasia basin 欧亚海盆
Eurasian continental bridge 欧亚大陆桥
Eurasian epsilon structural system 欧亚山字形构造体系【地】
Eurasian paleocontinent 欧亚古陆
Eurasian plate 欧亚板块
Eurasia realm 欧亚区系
Eureka wire 铜镍合金丝
eurhythmy 房屋建筑配置协调一致性
eurite 致密封烷质岩石；霏细岩
Euroacryl 欧罗阿克里尔（聚丙烯腈系纤维）
Euro-African roads 欧—非道路
euroa-quilo 尤拉奎洛风
Eurobank 欧洲银行
Eurobitum 欧洲沥青固化设备
Eurobond 欧洲债券
euroclydon 尤拉奎洛风
Eurocurrency 欧洲货币
Eurocurrency market 欧洲货币市场
Euroflock abrasion tester 欧罗磨损试验仪
Euroflock shearing tester 欧罗剪切牢度试验仪
Eurofranc 欧洲法郎
Euromarket 欧洲共同市场
Euromoney 欧洲货币
Euronet 欧洲共同市场科技数据通信网络
Europe 涂树脂的缆索
European alder 欧洲桤木
European Article Number 欧洲商品编号
European ash 欧洲白蜡树
European aspen 欧洲山杨
European Association of Exploration Geophysicists 欧洲地球物理勘探工作者协会
European Association of Petroleum Geologists 欧洲石油地质家协会
European Atlantic province 欧洲大西洋部
European Banks of International Company 欧洲国际银行
European barge carrier system 欧洲驳船运输方式；欧洲驳船货运系统
European beech 欧洲山毛榉
European birch 欧洲桦
European boreal faunal region 北欧北大西洋动物区
European car 欧洲汽车
European chestnut 欧洲栗木
European coal and steel industry 欧洲煤炭钢铁工业
European Coil Coating Association 欧洲卷材涂装工业协会
European Commission 欧洲委员会
European Committee for Standardization 欧洲标准化委员会
European Common Market 欧洲共同市场；西欧共同市场
European Communities 欧洲共同体
European community information services 欧洲共同体情报服务机构
European Computer Manufacturer's Association 欧洲计算机制造商学会；欧洲计算机厂家协会
European Computer Network 欧洲计算机网络
European conference of Minister's of Transport 欧洲运输部长会议
European conventional style sprinkler 欧洲传统式洒水喷头
European currency unit 欧洲货币单位
European Economic Communities 欧洲经济共同体
European Environmental Agency 欧洲环境机构
European Environmental Festival 欧洲环境节
European Environmental Monitoring and Information Network 欧洲环境监测和资料网络
European Federation 欧洲联盟
European fir 欧洲白冷杉
European foraminifera region 欧洲有孔虫地理区系
European Forestry Commission 欧洲林业委员会
European Free Trade Association 欧洲自由贸易联盟
European Geophysics Society 欧洲地球物理学会
European group on ocean stations 海洋观测站欧洲小组
European high-rate aeration 欧洲式高效曝气(法)
European hophorn beam 欧洲铁木
European horsechestnut 欧洲七叶树
European Information Center for Nature Conservation 欧洲自然保护资料中心
European Information Network 欧洲信息网；欧洲情报网
European International Contractors 欧洲国际承包商协会
European larch 欧洲落叶松
European lime 欧洲椴木
European main ports 欧洲主要港口
European Marine Pilots Association 欧洲引水员协会
European Mediterranean province 欧洲地中海部
European Monetary System 欧洲货币体系
European Monetary Union 欧洲货币联盟
European nautiloid region 欧洲鹦鹉螺地理大区
European network 欧洲网络
European-north American nautiloid region 欧洲—北美鹦鹉螺地理大区
European oak 欧洲栎
European option 期货期权
European oyster 欧洲牡蛎
European porcelain 欧罗巴瓷
European port 欧洲港口
European railway research institute 欧洲铁路研究所
European red elder 接骨木
European restaurant 西餐馆
European Science Fund 欧洲科学基金会
European Seismological Commission 欧洲地震学委员会
European-Siberian floristic subregion 欧洲西伯利亚植物亚区
European silver fir 欧洲冷杉
European space administration remote sensing satellite 欧空局遥感卫星
European Space Agency 欧洲空间局
European Space Agency Information Retrieval Service 欧洲空间组织情报检索服务中心
European Space Data Center 欧洲空间数据中心
European spindle-tree 欧洲卫矛
European spruce 欧洲云杉
European standard 欧洲标准
European switching network 欧洲交换网络
European system 欧洲系统
European Translation Center 欧洲文献翻译中心
European Transport Organization 欧洲运输组织
European Tugowners Association 欧洲拖轮船东协会
European vegetation map 欧洲植被图
European walnut 欧洲核桃木
European white birch 欧洲白桦
European white poplar 欧洲白杨
European yew 欧洲紫杉
Europe-Australia route 欧洲—澳大利亚航线
Europe barge carrier 欧洲载驳船
Europe chestnut 欧洲栗木
Europe class ship 欧洲标准船舶
Europe paleocontinent 欧洲古陆
Europe self-propelled barge 欧洲型自航驳船
Europe self-propelled ship 欧洲型自航船
Europe ship 欧洲型自航驳船
Europe type 欧洲式型号
europium halide 卤化铕
europium ores 铕矿
Europlug 欧洲国家通用插头
europous chloride 二氯化铕
Euro-press 辊压挤泥机
Euro-Signal Network 欧洲信号网
Eurosterling 欧洲英镑
eurotia ceratoides 优若藜
Eurovision 欧洲电视网
eurybathic 广深性的；广阔深的
eurychoric plant 广域分布植物
eurychoric species 广域分布物种
eurygamous 敞配型
euryhagy animal 广食性动物
euryhaline 广盐性种；广盐性的(海洋生物)
euryhaline animal 广盐性动物
euryhaline fishes 广盐性鱼类
euryhaline marine animal 广盐性海洋动物
euryhaline organism 广盐性生物；广盐性动物
euryhalinous 广盐性的(海洋生物)
euryhygric animal 广湿性动物
eurymeric 正型
euryoecic 广栖性的
euryoecious 广栖性的
euryoxybiont 广氧性生物
euryoxybiotic animal 广酸性动物
euryphotic 广光性的
euryphotic algae 广光性藻类
euryphotic animal 广光性动物
eurytherm 广温性；广温生物
eurythermal 广温性的；广温生的
eurythermal organism 广氧性生物；广盐性生物；广温生物
eurythermic 广温的
eurythermic animal 广温动物
eurytope 广生境
eurytropic 广适性的；广幅营养性的；广分布的
euscope 显微扩视镜
eustasism 海面升降；海面进退
eustasy 海面升降；冰河变异
eustatic 海面升降的
eustatic change 海面升降变化
eustatic change of sea level 全球性海平面变化
eustatic event 海面升降事件
eustatic fluctuation 海面(升降)变动
eustatic lake 变换湖
eustatic movement 海面升降运动；海面进退运动
eustatic observation 海平面变化观测
eustatic rejuvenation 海面升降回春；海面升降更新
eustatic sea level change 海平面升降变化
eustatism 海面升降；水动型海面升降
Eustone process 奥斯顿法(制铅白粉的分室炉法)
eustratite 透长辉煌岩
eustylos (古希腊、古罗马神庙的)柱式
eutasis reverse 柱上微凸线样板
eutasis rule 柱上微凸线样板
eutaxic deposit 成层矿床
eutaxite 条纹杂岩
eutaxitic 条纹斑状的
eutaxitic structure 条纹斑状构造
eutecrod 易熔焊条；共晶焊条
eutectic 易熔质；共晶；低共熔体的；低共熔(点的)
eutectic alloy 易熔合金；共晶合金；低共熔合金
eutectic austenite 共晶奥氏体
eutectic bonding 易熔焊接
eutectic boundary curve 共熔晶界线
eutectic carbide 共晶碳化物
eutectic cast iron 共晶铸铁；低熔生铁
eutectic cell 共晶团
eutectic cell structure 共晶团组织
eutectic cementite 共晶渗碳体
eutectic change 固液两态变化；共晶转变
eutectic colony 共晶团
eutectic composition 共晶成分；低共熔组成；低共熔成分
eutectic crystal 共晶体；低共熔晶体
eutectic crystallization 共晶结晶
eutectic deformation 共晶变形
eutectic diagram 低共熔图
eutecticevaporate 低共熔
eutectic freezing 共晶凝固
eutectic graphite 共晶石墨
eutectic horizontal 共晶水平线；低共熔温度线
eutectic ice 冻结的低共熔溶液；低共熔冰
eutectic melting 共晶熔化
eutectic metal 低共熔合金
eutectic mixture 共晶混合物；共结混合物；低共熔混合物
eutectic pig iron 共晶生铁
eutectic plate 低共熔片
eutectic point 易熔点；共熔点；共晶点；共结点；低共熔点
eutectic point method 共结点法
eutectic powder 共晶粉末
eutectic ratio 共晶比
eutectic reaction 共晶反应
eutectic state 低共熔态
eutectic steel 共晶钢
eutectic structure 共晶组织；共结构造
eutectic substance 低共熔物
eutectic system 共晶系统；低共熔体系
eutectic temperature 共晶温度；低共熔温度；低共

熔点
eutectic texture 共晶结构;共结结构
eutectic transformation 共晶转变
eutecticum 共晶
eutectic welder 低温焊机
eutectic welding 低温焊(接)
eutectiferous 共晶体的
eutectiform 共晶状
eutectofelsite 共结霏细岩
eutectoid 类低共熔体;共析(体);共析熔融物;共析混合物;低共熔体;不均匀共熔体
eutectoid alloy 共析合金
eutectoid cementite 共析渗碳体
eutectoid composition 共析成分
eutectoid ferrite 共析铁素体
eutectoid graphitization 共析石墨化
eutectoid horizontal 共析水平线
eutectoid interval 共析转变温度范围
eutectoid point 易熔点;固态熔液低共熔点;共析点
eutectoid reaction 共析转变;共析反应
eutectoid steel 共析钢
eutectoid structure 共析结构
eutectoid temperature 共析温度
eutectoid transformation 共析蜕变
eutectometer 凝点记录仪;铸铁碳当量测定仪
eutectoperthite 共析条纹长石
eutectophyre 共结斑岩
eutectrol process 共析法
euthenics 优境学
euthynteria 庙宇建筑(希腊)
eutopic reduction 正复
eutopium chelate laser 铕螯合物激光器
eutrification of water body 水体富营养化
eutrification-related factor of water body 水体富营养化相关因素
eutrophia 富营养化
eutrophic 富营养的;富营养分的
eutrophic aquaculture waters 富营养养殖水体
eutrophicated river water 富营养化河水
eutrophicated water 富营养化水体
eutrophication 加富过程;海藻污染;超养化;富营养化(过程)
eutrophication analysis 富营养化分析
eutrophication assessment criteria 富营养化评价基准
eutrophication cause 富营养化成因
eutrophication condition 富营养化状况
eutrophication control area of specialization 限控富营养化区
eutrophication course 富营养化进程
eutrophication effect 富营养化效应
eutrophication factor 富营养化因素
eutrophication index 富营养化指数
eutrophication management optimization model 富营养化管理优化模型
eutrophication mathematics model 富营养化数学模型
eutrophication mechanism 富营养化机理
eutrophication monitoring 富营养化监测
eutrophication of inland waters 内陆水体富营养化
eutrophication of lake 湖泊富营养化
eutrophication of lake ecosystem 湖泊生态系统富营养化
eutrophication of water body 水体富营养化
eutrophication of waters 水体富营养化
eutrophication rate 富营养化速率
eutrophication survey 富营养化调查
eutrophic condition 富营养状况
eutrophic criterion 富营养标准
eutrophic criteria of lakes 湖泊富营养标准;湖泊富营养化指标
eutrophic environment 富营养环境
eutrophic kettle lake 富营养锅穴湖
eutrophic lake 滋育湖;富营养湖(泊);滨海湖
eutrophic mire 过滋育沼泽;富营养沼泽;富营养泥沼
eutrophic monomictic lake 富营养单对流湖
eutrophic peat 滋育泥炭
eutrophic raw water 富营养原水
eutrophic reservoir 富营养水库
eutrophic river 富营养河流;富营养河道
eutrophic river system 富营养河系
eutrophic state 富营养状态;富营养状况
eutrophic status 富营养状况

eutrophic stream 富营养河流;富营养河道
eutrophic swamp 滋育沼泽;滋育木本沼泽
eutrophic wastewater 富营养废水
eutrophic water 过肥水
eutrophic water body 富营养水体
eutrophic water supply reservoir 富营养供水水库
eutrophied softwater 富营养软水
eutroph lake 滨海湖
eutrophy 营养丰富;富营养性;滋养湖;富营养化
eutropic crystal 共晶
eutropic series 异序同晶系
eutropy 异序同晶现象
eu-ulminite 充分分解腐木质体
eu-vitrain (充分分解)无结构镜煤;纯镜煤
eu-vitrinite (充分分解)无结构镜质体
euxenite 黑稀金矿
euxenite ore 黑稀金矿矿石
euximic facies 静海相
euxinic 静海的
euxinic basin 滞海盆地
euxinic deposit(ion) 静海沉积;幽深沉积;滞海沉积;富有机质沉积;闭流海沉积
euxinic facies 滞海相
evactor cooling system 高真空冷却系统;抽空式冷却系统
evacuant 排除的
evacuate 除清;抽空(排泄);搬空;排空
evacuated air 排泄空气
evacuated area (机场)疏散区
evacuated bulb 抽空灯泡
evacuated chamber 真空室;真空容器
evacuated container 真空箱
evacuated die-casting process 真空压铸
evacuated housing 抽真空罩
evacuated insulation 真空隔热
evacuated mo(u)ld 空腔模
evacuated pore 真空孔隙
evacuated route 疏散路线
evacuated space 搬空地方
evacuated tube 真空管;电子管
evacuating equipment 抽气设备
evacuating machine 排气机
evacuating the flood plain 洪泛区拆迁
evacuation 泄出物;抽(真)空;避车台;排空;疏散
evacuation alarm 疏散警报
evacuation area 撤退地区
evacuation capability 疏散能力
evacuation capacity of harbo(u)r (港口的)疏运能力
evacuation capacity of port (港口的)疏运能力
evacuation chute 疏散滑槽
evacuation drill 疏散演习
evacuation equipment kit 疏散器材
evacuation hospital 转运医院
evacuation line 排气管道
evacuation network computer model 疏散网络计算机模型
evacuation of capsule 包套抽空
evacuation of cargo from port 疏港
evacuation of sediment 泥沙清除;水库清淤
evacuation passageway 疏散通道
evacuation plan 疏散计划
evacuation procedure 疏散程序
evacuation route 疏散路线
evacuation signal 疏散信号
evacuation slide 紧急出口;疏散滑门
evacuation strobe light 疏散指示灯
evacuation time 排气时间;疏散时间
evacuation time up to operating pressure 抽空时间
evacuation valve 抽空阀
evacuator 排气器;抽气器;抽空装置;排出器
evadable 可逃避的
evade a tax 偷税
evade declaration of dutiable goods 漏报应交税货品
evade duty 漏税;逃税
evade foreign exchange 逃汇
evade paying taxes 偷税漏税
evade payment of duty 避税
evade tax 逃避税收;偷税(漏税)
evading 避免
evagination 翻出
evaluate 估定;求值;评定

evaluate a deposit 鉴定矿床;评价矿床
evaluated bid price 评标价
evaluated data 评价过的数据
evaluated error 估计误差;测定误差;评定误差
evaluated maintenance programming 评价维修规划
evaluated nuclear data file 评价过的核数据汇编
evaluate method for minerogenetic prospect 成矿远景评价方法
evaluate photochemical method 评价光化学法
evaluate the fuels 评定燃料
evaluate tree function 计算树函数
evaluating 估定的
evaluating alterative course 替代方案评价
evaluating alternative project 代替方案评价
evaluating for regional steady degree 区域稳定性程度评价
evaluating imported projects 审查进口项目
evaluating index 评价指标
evaluating indicator 评价指标
evaluating material 材料作价
evaluating method 评价方法;估价方法
evaluating method for coal-field prospect 煤田远景评价方法
evaluating method for geothermal field prospects 地热田远景评价方法
evaluating method for oil-gas-bearing prospect 油气远景评价方法
evaluating parameter model 求参模型
evaluating road accident data 道路交通事故评估数据
evaluating technology 评价技术
evaluating water quality 评价水质
evaluation 计值;估价;估计;估定;求值;评价(总结);评估;评定
evaluation alternatives 方案评价
evaluation and review technique 图示评审技术
evaluation by ranking 分等估价法
evaluation calculating method 评价计算方法
evaluation card 求值卡
evaluation chart 评质图;评估图
evaluation clause 估价条款
evaluation criterion 评价标准;评估准则;评估标准
evaluation for ambient environment 背景环境评价
evaluation for customs purpose 关税的估定
evaluation formula of water environment quality 地下水环境质量评价公式
evaluation function 评价函数
evaluation grade 评价等级
evaluation grade of boiler water quality 锅炉用水评价等级
evaluation grade of nutritive water 肥水评价等级
evaluation group 评审小组;审评小组
evaluation index 评价指标
evaluation map 评价图
evaluation map of mine environment 矿山环境质量评价图
evaluation method 评价方法
evaluation method of soil slope stability 土坡稳定性评价方法
evaluation model 评价模型
evaluation module 评价模件
evaluation module interface 评价模件接口
evaluation number 评价指数
evaluation of a primary 一个初等量的求值
evaluation of argument subscript 变元下标求值
evaluation of average attenuation coefficients 平均衰减系数求取
evaluation of berthing energy 靠船能量计算
evaluation of berthing load 靠船力计算
evaluation of bids 投标评估;评标
evaluation of construction design 施工图设计审查
evaluation of cumulative reservoir water quality impact 水库累积水质效果评价
evaluation of damage 损坏估价
evaluation of determinants 行列式求值
evaluation of development program(me) of new products 新产品开发方案评价
evaluation of employee performance 职工考核
evaluation of estimate 评价估算
evaluation of exploration precision 勘探精度评价
evaluation of expression 表达式求值
evaluation of fixed assets 固定资产的估计
evaluation of goods in the perils of sea 海损货

物估价
evaluation of groundwater quality 地下水水质评价
evaluation of hardware 硬件的估算
evaluation of imaging quality 成像质量评价
evaluation of inventory 库存估价
evaluation of iron deficiency 评定铁的不足
evaluation of land 土地估价
evaluation of long-term reservoir water quality impact 长期水库水质效果评价
evaluation of mooring load 系泊荷载计算
evaluation of new materials 新材料评价
evaluation of new technology 新技术评价
evaluation of opening stability 洞室稳定性评价
evaluation of product planning 生产计划的评价
evaluation of quality 质量评估
evaluation of reliability program(me) 可靠性规划评价
evaluation of result 结果评价
evaluation of simulation model 仿真模型评价
evaluation of source rock 烃源岩评价
evaluation of spline function 样条函数计算
evaluation of stability of regional crust 区域地壳稳定性评价
evaluation of tender 评标;投标评估
evaluation of thermal environment 热环境评价
evaluation of trap 圈闭评价
evaluation of water corrosivity 环境水侵蚀性评价
evaluation of work and allotment of points 评工记分
evaluation period 计算期;评价期
evaluation procedure 评价步骤
evaluation process 估价程序;评价程序
evaluation program(me) 求值程序;评价程序
evaluation report 鉴定报告;评价报告
evaluation rule 评价法则
evaluation sheet 评估表
evaluation standard of railway economic efficiency audit 铁路经济效益审计的评价标准
evaluation study 评核研究
evaluation system 评价系统;评估方式
evaluation test 鉴定试验;评价试验
evaluation test of power plant on board 动力装置船上鉴定试验
evaluation test program(me) 鉴定试验规划
evaluation test specification 鉴定试验说明书;鉴定试验规范
evaluation trial 鉴定性试验
evaluative model for multiple criteria 多准则评价模型
evaluator 鉴别器
evanescent field 渐逝场;瞬逝场
evanescent-field coupling 渐逝场耦合
evanescent lake 易消失的湖;短暂的湖;干涸湖
evanescent light wave 瞬逝光波
evanescent region 渐消失区
evanescent voltage 衰减电压
evanescent wave 渐逝波;消失波;消散波;损耗波;瞬逝波
Evanohm 埃弗诺姆镍铬系电阻合金
Evans-Burr unit 伊万斯—伯尔单位
Evans classifier 伊万斯分级机
evansite 核磷铝石
Evans launder classifier 流槽分级机
Evans mill 伊万斯带钢轧机
evapolensic texture 蒸发纹层结构
evapoporphyrocrystic texture 蒸发斑状结构
evaporate 使蒸发
evaporated product 挥发的油品
evaporate ratio 蒸发比
evaporation cooling with Freon 用氟利昂升华冷却
evaporitic-solution breccia 盐溶角砾岩
evaporitic environment 蒸发环境
evapotranspiration 蒸散作用;蒸散(发);蒸发蒸腾(作用);蒸发散发;土壤水分蒸发蒸腾的损失总量
evapotranspiration and water use efficiency 蒸发蒸腾作用与水分利用率
evapotranspiration loss 蒸散损失;地面和植物总蒸发量
evapotranspiration potential 蒸散潜能
evapotranspiration ratio 蒸发蒸腾比
evapotranspiration tank 蒸散箱
evapotranspire 蒸散
evapotranspirometer 蒸散计;(土壤)蒸发器

evapotron 涡动通量仪
evapo(u)rability 可蒸发性;汽化性;挥发性
evapo(u)rable 易蒸发的
evapo(u)rable water (可)挥发水;(可)蒸发水
evapo(u)rant 蒸发物;蒸发剂
evapo(u)rant ion source 蒸发物离子源
evapo(u)rate 蒸发
evapo(u)rate barrier 蒸发障
evapo(u)rate basin model 蒸发盆地模式
evapo(u)rate combustion 蒸发燃烧
evapo(u)rated alloying technology 蒸发合金工艺
evapo(u)rated black 蒸镀发黑处理;蒸镀变黑
evapo(u)rated crystallizer 蒸发结晶器;蒸发结晶机
evapo(u)rated deposit 蒸发沉积物
evapo(u)rate deposit 蒸发沉积矿床
evapo(u)rate deposition 蒸发沉积作用
evapo(u)rated film 蒸镀薄膜
evapo(u)rated film of alumin(i)um 蒸镀铅膜
evapo(u)rated gold antimony contact 蒸发的金锑接触
evapo(u)rated latex 蒸发橡浆
evapo(u)rated liquid activity meter 蒸发液体式试样放射性测量计
evapo(u)rated liquor 蒸发液
evapo(u)rated make up 蒸发补给水
evapo(u)rated nitrogen inlet 氮蒸气入口
evapo(u)rated product 挥发性产品
evapo(u)rated vegetables 脱水蔬菜
evapo(u)rated volume 蒸发体积
evapo(u)rated waste 蒸浓废物
evapo(u)rate station 蒸发站
evapo(u)rating 蒸发
evapo(u)rating basin (内陆湖流域的)蒸发盆地
evapo(u)rating capacity 蒸发能力;蒸发量
evapo(u)rating capacity of chain system 链条系统蒸发强度
evapo(u)rating carburetor 蒸发汽化器
evapo(u)rating chamber 蒸发室
evapo(u)rating chilling 蒸发冷硬
evapo(u)rating coil 蒸发蛇管;蒸发盘管
evapo(u)rating column 蒸浓柱;蒸发塔
evapo(u)rating dish 蒸发皿
evapo(u)rating excretion 蒸泄
evapo(u)rating field 蒸发场
evapo(u)rating ga(u)ge 蒸发量测定计
evapo(u)rating halo method 蒸发晕法
evapo(u)rating heat 蒸发热
evapo(u)rating installation 蒸馏设备
evapo(u)rating intensity of water 水分蒸发强度
evapo(u)rating method 汽化法
evapo(u)rating pan 蒸发锅
evapo(u)rating pipe 浓缩管;蒸发管
evapo(u)rating plant 蒸发装置
evapo(u)rating point 蒸发点
evapo(u)rating pot 蒸发罐
evapo(u)rating pressure 蒸发压力;汽化压力
evapo(u)rating pressure regulating valve 蒸发压力调节阀
evapo(u)rating surface 蒸发表面
evapo(u)rating temperature 蒸发温度;汽化温度
evapo(u)rating tower 蒸浓塔
evapo(u)rating unit 集装箱送冷气装置;蒸发设备
evapo(u)rating water 蒸发水
evapo(u)rating zone 干燥带
evapo(u)ration 蒸气;蒸发作用;蒸发(量);脱水法;升华逸散作用
evapo(u)ration and concentration 蒸发浓缩作用
evapo(u)ration area 蒸发区;蒸发面积
evapo(u)ration basin 蒸发池;大型蒸发器
evapo(u)ration boiler 蒸发锅
evapo(u)ration burner 蒸发燃烧器
evapo(u)ration capacity 最大蒸发量;蒸发(容)量;蒸发能力
evapo(u)ration capacity observation 蒸发量观测
evapo(u)ration carburettor 蒸发式化油器
evapo(u)ration cathode 蒸发阴极
evapo(u)ration chamber 蒸发室
evapo(u)ration characteristic 蒸发特性
evapo(u)ration coating 蒸发涂敷;真空镀膜
evapo(u)ration coefficient 蒸发系数
evapo(u)ration coil 蒸发旋管;蒸发器
evapo(u)ration condensation 蒸发凝聚
evapo(u)ration condensation material mechanism 蒸发凝聚机理

evapo(u)ration condensation material transfer mechanism 蒸发凝聚传质机理
evapo(u)ration condensation mechanism 蒸发冷凝机理
evapo(u)ration control 蒸发(率)控制;蒸发调节
evapo(u)ration controller 蒸发控制器
evapo(u)ration cooled tube 蒸发冷却管
evapo(u)ration cooler 增湿塔
evapo(u)ration cooling 蒸发冷却;汽化冷却
evapo(u)ration current 蒸发流
evapo(u)ration curve 蒸发曲线
evapo(u)ration decomposition rate 蒸气分解率
evapo(u)ration discharge 蒸发消耗(量);蒸发排泄;蒸发量;汽化度
evapo(u)ration discharge from soil surface 土面蒸发量
evapo(u)ration dish 蒸发皿
evapo(u)ration dissipation 蒸发消散
evapo(u)ration drying 蒸发干燥
evapo(u)ration efficiency 蒸发效率
evapo(u)ration factor 蒸发系数
evapo(u)ration feed liquor 蒸发器进料
evapo(u)ration fog 蒸发雾
evapo(u)ration free curing 无蒸发养护
evapo(u)ration from ice 冰面蒸发
evapo(u)ration from land 地面蒸发
evapo(u)ration from land area 陆地蒸发
evapo(u)ration from land surface 陆面蒸发
evapo(u)ration from plant 植物蒸发
evapo(u)ration from snow 雪面蒸发
evapo(u)ration from soil 土壤蒸发
evapo(u)ration from swamp and wetted area 沼泽湿地蒸发
evapo(u)ration from vegetation 植物蒸发
evapo(u)ration from water surface 水面蒸发(量)
evapo(u)ration ga(u)ge 蒸发量测定计;蒸发计
evapo(u)ration getter-ion pump 蒸发吸气离子泵
evapo(u)ration gum test 蒸发胶质试验
evapo(u)ration heat 蒸发热(量);汽化热
evapo(u)ration hook ga(u)ger 蒸发器
evapo(u)ration installation 蒸发装置
evapo(u)ration intensity 蒸发强度
evapo(u)ration in the field 田间蒸发
evapo(u)ration ion pump 蒸发离子泵
evapo(u)ration jig 蒸发架
evapo(u)ration layer 蒸涂层;蒸发层
evapo(u)ration load 蒸发负荷
evapo(u)ration loss 蒸发损失;蒸发损耗
evapo(u)ration loss for asphalt 沥青蒸发(减)量试验
evapo(u)ration loss test 蒸发损失试验
evapo(u)ration loss test for asphalt 沥青挥发分试验
evapo(u)ration mask 蒸发罩
evapo(u)ration meter 蒸发计;蒸发表
evapo(u)ration method 蒸发法
evapo(u)ration model 蒸发模型
evapo(u)ration neutron 蒸发中子
evapo(u)ration nuclear model 蒸发核模型
evapo(u)ration nucleon 蒸发核子
evapo(u)ration number 汽化值
evapo(u)ration observation 蒸发观测
evapo(u)ration of electron 电子放射
evapo(u)ration of groundwater 地下水蒸发
evapo(u)ration of land surface 地面蒸发
evapo(u)ration of moisture 水分蒸发
evapo(u)ration of terrain 地面蒸发
evapo(u)ration of water 水分蒸发
evapo(u)ration opportunity 可能蒸发率;蒸发可能性;蒸发可能率;蒸发机率
evapo(u)ration pan 蒸发器;蒸发锅;蒸发盘;蒸发皿
evapo(u)ration particle 蒸发粒子
evapo(u)ration pilot 蒸发试验小区
evapo(u)ration pit 蒸发坑
evapo(u)ration plant 蒸发装置
evapo(u)ration pond 蒸发塘;蒸发池
evapo(u)ration potential 蒸发势
evapo(u)ration power 可能蒸发(量);蒸发能力
evapo(u)ration pressure 蒸发压力
evapo(u)ration process 蒸发过程
evapo(u)ration pump 蒸发泵
evapo(u)ration quantity of phreatic water 潜水

蒸发量
evapo(u)ration-rainfall ratio 蒸发雨量率;蒸发雨量比
evapo(u)ration rate 挥发率;蒸发速率;蒸发速度;蒸发(比)率
evapo(u)ration rate analysis 蒸发速度解析法
evapo(u)ration rate determination 蒸发率测定
evapo(u)ration ratemeter 蒸发率计
evapo(u)ration rate of groundwater 地下水蒸发量
evapo(u)ration ratio 蒸发倍率
evapo(u)ration reduction 蒸发抑制
evapo(u)ration regulation area 蒸发调节区
evapo(u)ration residue 蒸发残渣;蒸发残余物
evapo(u)ration retardant 缓蒸发剂;蒸发阻滞
evapo(u)ration salt 蒸发盐
evapo(u)ration section 蒸发(工)段
evapo(u)ration shield 蒸发挡板
evapo(u)ration source 蒸发源
evapo(u)ration source shutter 蒸发源活门
evapo(u)ration source turret 蒸发源转盘
evapo(u)ration stage 蒸发级
evapo(u)ration station 蒸发站
evapo(u)ration station on water surface 漂浮蒸发站
evapo(u)ration suppressant 蒸发抑制剂
evapo(u)ration suppression 抑制蒸发
evapo(u)ration suppressor 蒸发抑制剂
evapo(u)ration surface 蒸发面;蒸发表面
evapo(u)ration synthesis 蒸发合成
evapo(u)ration tank 蒸发箱;蒸发池;大型蒸发器;(大型)蒸发皿
evapo(u)ration technology 蒸发工艺
evapo(u)ration temperature 蒸发温度
evapo(u)ration test 蒸发试验;汽化度测定
evapo(u)ration to dryness 蒸干
evapo(u)ration trail 蒸发尾迹
evapo(u)ration type 蒸发型
evapo(u)ration velocity 蒸发速度
evapo(u)ration zone 干燥带;蒸发段
evapo(u)rative 蒸发的
evapo(u)rative apparatus 蒸发设备
evapo(u)rative capacity 蒸发(容)量;蒸发能力;蒸发本领
evapo(u)rative centrifuge 蒸发(式)离心机
evapo(u)rative coating 蒸发涂敷
evapo(u)rative concentration-flame atomic adsorption spectrometry 蒸发浓缩—火焰原子吸收光谱法
evapo(u)rative condenser 蒸发式冷凝器;蒸发凝汽器;蒸发冷凝器
evapo(u)rative cooled device 蒸发冷却装置
evapo(u)rative cooler 蒸发冷却器
evapo(u)rative cooling 蒸发冷却
evapo(u)rative cooling equipment 蒸发制冷设备
evapo(u)rative cooling generator 蒸发冷却发电机
evapo(u)rative cooling system 蒸发冷却系统
evapo(u)rative cooling tower 蒸发式冷却塔
evapo(u)rative crystallizer 蒸发结晶器;蒸发结晶机
evapo(u)rative dolomite 蒸发白云岩
evapo(u)rative drying 蒸发式烘燥;蒸发式干燥
evapo(u)rative duty 蒸发率
evapo(u)rative efficiency 蒸发效率
evapo(u)rative emission 蒸发性排放
evapo(u)rative emission control 蒸发性排放控制
evapo(u)rative emission controller 蒸发性排放控制器
evapo(u)rative factor 蒸发系数
evapo(u)rative freezing 蒸发冷冻
evapo(u)rative heat loss 蒸发散热
evapo(u)rative heat regulation 蒸发热调节
evapo(u)rative heat transfer coefficient 蒸发换热系数
evapo(u)rative loss control device 蒸发损失控制设施
evapo(u)rative power 蒸发力;蒸发本领
evapo(u)rative pumping 蒸发泵作用
evapo(u)rative rate 蒸发率
evapo(u)rative reflux 蒸发回流作用
evapo(u)rative refrigerant condenser 蒸发式制冷剂冷凝器
evapo(u)rative surface condenser 蒸发表面冷凝器

evapo(u)rative tower 蒸发塔
evapo(u)rative type cooler 蒸发式冷却器
evapo(u)rative type cooling tower 蒸发式冷却塔
evapo(u)rative water 蒸发水分
evapo(u)rativity 蒸发能力;蒸发率;蒸发度
evapo(u)rator 浓缩器;蒸发器;蒸发皿
evapo(u)rator blower 蒸发器吹风机
evapo(u)rator boat 蒸发器皿
evapo(u)rator body 蒸发器身
evapo(u)rator boiler 蒸发器的沸腾器
evapo(u)rator coil 蒸发蛇形管;蒸发器旋管;蒸发器盘管
evapo(u)rator condenser 蒸发凝气机;蒸发冷凝器
evapo(u)rator-condenser shell 蒸发器冷凝器壳体
evapo(u)rator core assembly 蒸发芯子总成
evapo(u)rator cryostat 蒸发器恒温器
evapo(u)rator current 蒸发器电流
evapo(u)rator element 蒸发器元件
evapo(u)rator feed pump 蒸发器给水泵
evapo(u)rator filament 蒸发器灯丝
evapo(u)rator filament support 蒸发器灯丝支架
evapo(u)rator ga(u)ge 蒸发计
evapo(u)rator getter pump 蒸发吸气泵
evapo(u)rator heating surface 蒸发器加热表面
evapo(u)rator ion pump 蒸发离子泵
evapo(u)rator material 蒸发器材料
evapo(u)rator metal 蒸发器金属
evapo(u)rator potential 蒸发器电势
evapo(u)rator pressure control valve 蒸发器压力控制阀
evapo(u)rator regulating valve 蒸发压力调节阀
evapo(u)rator room 蒸发室
evapo(u)rator scale 蒸发器垢层
evapo(u)rator source 蒸发源
evapo(u)rator source carrier 蒸发源支持器
evapo(u)rator station 蒸发站
evapo(u)rator strip 蒸发带
evapo(u)rator strip holder 蒸发带夹持器
evapo(u)rator support 蒸发器支架
evapo(u)rator surface 蒸发表面
evapo(u)rator tank 蒸发槽
evapo(u)rator tower 蒸发塔
evapo(u)rator tube 蒸发器管
evapo(u)rator vapo(u)r 蒸发器的二次蒸汽
evapo(u)rator wire 蒸发丝
evapo(u)rator with heat pump 热泵蒸发器
evapo(u)rator with rotating brush 带回转刷的蒸发器
evapo(u)rigraph 蒸发记录仪
evapo(u)rimeter 蒸发仪;蒸发计;蒸发测定器;蒸发表;气化计
evapo(u)rimetry 蒸发测定法
evapo(u)ri(o)meter 蒸发计
evapo(u)rion pump 蒸发离子泵
evapo(u)rite 蒸发盐;蒸发岩
evapo(u)rite basin 蒸发盐盆地
evapo(u)rite deposit 蒸发沉积矿床
evapo(u)rite deposition 蒸发盐沉积
evapo(u)rite facies 蒸发岩相
evapo(u)rite mineral 蒸发(岩)矿物
evapo(u)rite platform facies 蒸发岩台地相
evapo(u)rites 蒸发后剩余残渣
evapo(u)rite sediment 蒸发岩沉积
evapo(u)rite sequence 蒸发岩序列
evapo(u)rization front 汽化前缘
evapo(u)rizer 汽化室;汽化器
evapo(u)r method 汽化降温法
evapo(u)rogram 自记蒸发量记录(曲线);蒸发量曲线
evapo(u)rograph 蒸发成像仪
evapo(u)rograph cell 蒸发计
evapo(u)rography 蒸发成像术
evapo(u)rometer 挥发速度计;蒸发仪;蒸发计;蒸发测定器;蒸发表;气化计
evapo(u)roscope 蒸发镜
evapo(u)rous water 汽化水
evapo(u)r pressure 蒸发压力
evapo(u)rtion degree 蒸发度
evase 扇风机出风扩散道;出风扩散道
evasion and avoidance 偷税漏税;逃税(与)避税
evasion of exchange control 逃汇
evasion of tax 逃避税收
evasion of taxation 偷漏税;逃税
evasive routing 避让航线

evatron 自动控制用热离子变阻器;电子变阻器
evection 出差
evectional tide 出差潮
evection in latitude 黄纬出差
eveite 羟砷锰矿
evel(l)ing process 水准测量
even 均匀的;均等的;偶数的;平稳;平坦;平滑;平的;双数
even-aged 同龄的
even-aged forest 同龄林
even-aged stand 同龄林分
even-A isotope 偶A同位素
even and odd functions 偶函数与奇函数
even angle method 定角测量法(视距测量)
even bargain 公平交易
even bearing 均匀支承;表面压力均布的轴承
even bedded 均匀分层的;基床平整的;平整成层的;平整层状的;路床平整的
even bedded subgrade 均匀型地基
even bed of mortar 灰浆平整(垫)层
even bit 偶数比特
even bladed propeller 偶数桨叶螺旋桨
even blend 均匀掺混
even break 均等机会
even carbon number predominance 偶碳数优势
even channel 偶数通路
even check 偶数校验
even colo(u)r 均匀颜色
even component 偶支
even coupling 偶耦合
even-crested ridge 平顶山
even decompression 均匀减压
even depth 相等深度;等深度
even-diameter hole 等径钻孔
even distribution 均匀分布;平均分布
even-duty bit 均匀负载钻头;均匀负荷钻头;等负荷刃钻头
even dye 均匀染料
even elements 偶数元素
evener 整平器;整平机;牵引均衡横木;平衡器
even-even 偶数对
even-even element 偶偶元素
even-even nucleus 偶偶核
even field 偶数场
even flow 平稳流动
even flowing 等传递动力
even flowing power 匀流功率;等功率传动
even focusing 偶聚焦
even fracture 均匀断口;平坦断口;平(滑)断口
even front 均匀前缘
even function 偶函数
even generalized function 偶广义函数
even grain 均匀纹理;均匀木纹
even-grained 等粒状(结构)的;颗粒均匀的
even-grained aerated concrete 颗粒均匀的充气混凝土
even-grained concrete 颗粒均匀的混凝土
even-grained mortar 颗粒均匀的砂浆;颗粒均匀的灰泥
evengranular 颗粒均匀的
even ground 平坦地区
even harmonic 偶次谐波
even harmonic operation 偶次谐波工作
even illumination 均匀照明
evening college 夜大学
evening glow 暮辉
evening observation 黄昏观测
evening peak 夜间峰值(指电力负荷);晚高峰
evening peak hours 晚高峰小时;晚高峰时间
evening peak traffic 傍晚高峰交通
evening rush hours 晚高峰小时
evening school 夜校
evening sight 黄昏观测
evening tide 潮汐;潮水;汐;晚潮
evening twilight 暮光
evening university 夜大学
even joint 平头接头;平头接合;平接
even keel 等吃水【船】;(船)首尾吃水相等;平载;平浮;平吃水
even-keel objective 平稳目标
even-keel ship 等吃水船
evenkite 鳞石蜡
even level 偶数级;偶数层
even line 光滑线

even load(ing) 均匀荷载;均布荷载;均布负载
even location 偶数存储单元
even-lock 不调梭缝迹
even lot 整批
evenly bedding 平行层理
evenly bedding structure 平行层理构造
evenly distributed 均匀分配的;均匀分布的
evenly distributed load 均匀分布荷载;均布荷载
evenly distributed throughout the matrix (金刚石)均匀分布在胎体上
evenly distribution joint gaps 调整轨缝
evenly divided 等分刻度
evenly graded 颗粒均匀的;级配均匀的
evenly spaced energy level 均匀间隔能级
even-mass 偶质量
even-mass isotope 偶质量数同位素
even multiple 偶倍数
evenness 匀度;平直度;平坦度;平坡;平面度;平滑度
evenness calender 均匀轧车
evenness degree 均匀度
evenness index 均匀性指数;均匀度指数
evenness of weld 焊缝均一性;焊缝的匀整性
evenness test device 平整度试验设备
evenness transducer 均匀度传感器
even-N isotope 偶N同位素
even number 偶数;双数
even-numbered channels 偶数信道
even-numbered days 双日
even-numbered error 偶数差错
even-numbered line 偶数行
even-numbered pass 偶数道次
even numbered of lines 偶数行数
even number of teeth 偶数齿数
even number of threads 偶数螺纹
even number relay 偶数继电器
even number step 偶数步
even-odd check 奇偶校验
even-odd element 偶奇元素
even-odd generator 奇偶发生器
even-odd nuclei 偶奇核
even-odd nucleus 偶奇核
even-odd rule 偶奇定则
even-odds 成败机会相等
evenometer 光电式均匀度测定仪
even-order 偶数阶;偶次的
even-order harmonic 偶次谐波
even-order harmonic distortion 偶次谐波失真
even out the distribution of industry 平衡工业发展布局
even parity 偶同位;偶数奇偶性
even-parity check 偶校验;偶数同位校验;偶数奇偶校验
even-parity state 偶宇称态
even-parity vibration 偶宇称振动
even partition 偶分拆
even perfect number 偶完全数
even permutation 偶置换;偶排列
even pitch 均一坡度;偶数螺距;整螺距;平坦坡度
even power 偶数幂
even pressure 均匀压力
even pulse 均脉;偶脉冲
even pulse response 偶脉冲响应值
even quantum number 偶量子数
even running 均衡运行;平稳运行
even running surface 平整的路面
even slope 均匀坡
even-span greenhouse 两坡顶温室
even speed 均匀速度
even spray nozzle 等宽式喷嘴
even state 偶态
even subgraph 偶子图
even substitution 偶代换
even summation 偶和
even-N surface 平整的表面;平滑面
even symmetry 偶对称
event 结局(指网络模型);事件
event algebra 事件代数
event attribute 事件属性
event-based simulation 根据事件模拟法
event built-in function 事件内函数
event chain 事件链
event commutativity 事件的交换性
event control block 事件控制块

event controller 事件控制器
event counter 信号计数器;事件计数器
event data 事件数据
event delay 事件延迟
event detector 事件探测器
event-directed simulation 面向事件的模拟;指向事件模拟
event-driven 事件驱动;从动事件
event driven executive 事件驱动执行程序
event driven system 事件驱动系统;实时系统
even-temperature condition 等温条件
even term 偶项
even term of atom 原子的偶数项
event establishment 事件建立
even texture 均匀质地;均匀结构;均匀构造
even-textured 均匀纹理的;均一构造的
event flag 事件标志;事件标记
event flag cluster 事件标志束;事件标记组
event frequency 事件频率;事件频度
event-generating process 事件产生过程
event horizon 事界;事件水平线
even thread 偶螺纹
event identification time 事件识别时间
even time 对等工休制
event in operation 作业项目
event list 事件表
event location 事件位置
event magnet 步调磁铁
event mark 记事符号
event marker 结果标示器;事件标记笔
event mean concentration 事件平均浓度
event monitoring 事件监督
event name 事件名
event node network 事件节点网
event number 结点号码;事件号码
event numbering 事件编写
event of default 违约事项
event of magnitude zero 零级地震
event option 事件任选(项)
event-oriented 面向事件的
event-oriented simulation 离散模拟;后据事件模拟法
event posting 事件配置;事件记入;事件登记
event power failure 电源故障
event probability regression 事件概率回归
event pseudo-variable 事件伪变量
event queue 事件队列
event record 事件记录
event recorder 动作记录器;事件记录器
event recording 行为事件记录
event recording system 事件记录系统
event report 事件报告
event reporting system 事件自报式系统;事件报告系统
event representation 事件表达
event resolution 同相轴分辨率
event routine 事件程序
event scan(ning) 事件扫描
event-scan(ning) mechanism 事件扫描机构
event schedule 事项进度表;事件调度程序
event-sensing card 事件感知卡
event's failing 事件不出现
event's happening 事件出现
event size 事件规模;事件大小
event slack 节点迟缓
event slack time 事件松弛时间
events of termination 终止协议事件
events subsequent to balance sheet 决算后发生事件
event storage and distribution unit 事件存储与分配部件
event stratification 事件层理
event stratigraphy 事件地层学
events which are not mutually exclusive 非互斥事件
event synchronization 事件同时性;事件同步
event tracer 事件跟踪程序
event tree 事件树;事故树
eventual deviation 偶然偏差;稳态偏差
eventual failure 最后断裂;完全破坏
eventual flood 可能(发生的)洪水;特大洪水
eventuality 偶然性;非常时期;不测事件
eventual purification 最终净化作用;最后净化
event variable 事件变量

even twisting moment 固定扭矩
even up accounts with 结清账目
even wear 磨损均匀性
even windrow 整平料堆
even working 整帖装订
ever-bearer 常果植物
ever-bloomer 常果植物
Ever-brass 埃弗无缝黄铜管
Everbrite metal 埃瓦布赖特耐蚀铜镍合金
ever changing braided river 经常变化的多汊河流
ever changing braided stream 经常变化的多汊河流
Everdur 埃维杜尔铜合金;铜硅锰合金
Everdur alloy 埃维杜尔合金
Everdur copper 埃维尔耐蚀硅青铜
Everest metal 埃瓦雷斯特铅基轴承合金
Everest theodolite 埃瓦雷斯特经纬仪
ever-expanding 不断膨胀的;不断扩大的
ever flowering rose 月季花
ever frost 多年冻土;永冻(土)
ever-frost layer 永冻层
ever frozen 永冻层
ever-frozen layer 多年冻层;永冻层
ever-frozen soil 永冻土
ever-frozen subsoil 永冻基土
everglade 开阔沼泽;沼泽地;沼泽草地;丘陵沼泽;轻度沼泽化低地;湿地
ever greater demands 得寸进尺
evergreen 冬青;常绿树;常绿植物;常绿的;万年青
evergreen broad-leaved sclerophyllous forest 常绿硬叶阔叶林
evergreen bushland 常绿矮灌丛
evergreen chinquapin 栲(树)属;栲林
evergreen community 常绿群落
evergreen cutting 常绿果树
evergreen evonymus 大叶黄杨
evergreen firebreak 常绿防火林
evergreen forest 常绿林
evergreen forest on coral island 珊瑚岛常绿林
evergreen fruit tree 常绿果树
evergreen hedge 常绿绿篱
evergreen herbage 常绿草本植物
evergreenite 流英正长岩
evergreenness 常绿性;常绿现象
evergreen plant 常青植物;常绿植物
evergreen seasonal forest 常绿季雨林
evergreen shrub 常绿灌木
evergreen silva 常绿乔木
evergreen silvae 常绿木本群落;常绿林
evergreen thicket 常绿灌木
evergreen tree 常绿树;长绿树
evergreen undershrub 常绿小灌木;常绿下层灌木;常绿短灌木
evergreen woodland 常绿树林带;常绿森林
ever-growing 不断增长的
everhanging eaves 飞檐
Everhard effect 埃弗哈特效应
ever-increasing 不断增强的;不断增加的;不断增长的
ever increasing encroachment 得寸进尺
Everite 爱弗来组合材料
Everlube 耐寒性润滑油
ever present inertia force 永存惯性力
ever-reduced 不断减小的
everseal 沥青防水剂(一种专利产品)
evershed ducter 小电阻测量表
Evershed effect 埃弗谢德效应
e versus lgP curve 压缩关系曲线
evert corner 阳条(瓷砖)
everted rimmed shallow bowl 翻唇浅碗
everyaspect of forestry 林业的各个方面
everyday 日常的
everyday architecture 普通建筑风格
everyday life 日常生活
everyday routine 日常工作;每日的例行公事
everyday wave base 正常浪基面;正常波基面
every definite quantity of a commodity 每一定量某种商品
every industry or trade 各行各业
every monthly penetration rate 平均月机械钻
every other day 每隔一天;每隔一日
every other month 每隔一月;隔月
every other run 分班

every person 每人
every second well 每隔一口井
every stitch set 张所有的帆
everything on the earth 地面上的每样东西
every two days 每隔一天
everywhere convergent 处处收敛
everywhere dense 处处稠密
everywhere-dense manifold 遍密集簇
every year's investment 分年度投资
eviction 退租;收回租地;收回财产
evict property 收回财产
evidence 证据;凭证
evidence at assessment 课税依据;课税根据
evidence collateral 旁证
evidence direct 直接证据
evidence documentary 书面证据
evidence for the defence 被告证人
evidence incomplete 证据不全
evidence insufficient 证据不足
evidence of damage 损坏征象
evidence of debt 借据
evidence of glaciation 冰川踪迹;冰川(作用)遗迹
evidence of payment 付款凭证
evidence of title 所有权证据
evidence record 凭证记录
evidences of title 所有权证据
evidence title 所有权证明
evidential burden 举证责任
evidential document 凭证文件
evidentiary hearing 听证会
evil consequence 恶果
evisceration 除脏术
Evison wave 埃弗森波
evocon 电视发射管
evoked potential 诱发(性)电位;诱发电位
evoke module 移送模块;制订模块;订制模块
evolute 渐屈线;渐开线;展开线;法包线;外旋式(渐曲线)
evolute line 法线
evolute of a surface 渐屈面
evolution 开方;进化;渐进;演化;演变;发展
evolutional unit 演化单位
evolutionary convergence 进化趋同
evolutionary divergence 进化趋异
evolutionary ecotoxicology 进化生态毒理学
evolutionary operation 渐近操作;改良操作法;调优运算
evolutionary path 发展途径
evolutionary pattern 演化模式
evolutionary phase 演化阶段
evolutionary pollution ecology 进化污染生态学
evolutionary process 演化过程;演变过程
evolutionary regularity 演变规律
evolutionary series 演化序列
evolutionary species 进化种
evolutionary stochastic process 发展随机过程
evolutionary taxonomy 演化分类
evolutionary track 演化程
evolutionary trend 演化趋向
evolutionary zone 演化带
evolution criterion 发展判据
evolution map of morphology 地貌演化图
evolution of environment 环境演化
evolution of gas 放出气体
evolution of geosyncline-platform 地槽台的演化
evolution of heat 热的放出
evolution of meandering channel 蜿蜒水道的演变
evolution of model(l)ing 模型化(的)进展
evolution of objective 目标的演变;目标的发展
evolution of organic matter 有机质演化
evolution of pollution 污染演化
evolution of reservoir bank 库岸演变
evolution of riverbed 河床演变
evolution of river course 河道演变
evolution of seismicity 地震活动性演变
evolution of shoal 浅滩演变
evolution of the earth 地球演化
evolution pathway of kerogen 干酪根演化途径
evolution stage of organic matter 有机质演化阶段
evolution test 评定质量试验
evolution type of mountain 山地演化类型
evolutoid 广渐屈线
evolved gas analysis method 逸出气分析法
evolved gas curve 逸出气曲线

evolved gas detection 逸出气检测法
evolvement 演化
evolvent 渐伸(开)线;渐开线
evolving fault 进展性故障
evolving magnetic feature 演变磁结构
evorsion 涡流侵蚀
evorsion hollow 瓯穴;河床;锅穴;涡蚀穴
Evos door 装箱门件总成(一种建筑材料)
ewaldite 碳铈钙钡石
Ewald-Kornfeld method 埃瓦耳德—科恩菲尔德法
Ewald method 埃瓦耳德法
Ewald sphere 埃瓦耳德球
E wave 横磁波
ewcentralize 恢复到中心位置
ewer 大口水壶;水罐
ewer in human form 人形壶
ewer with inner tube 内管壶(瓷器名)
ewer with raised lotus petals 凸莲瓣茶壶(瓷器名)
ewer with small mouth and double handles 小口双耳壶(瓷器名)
ewery (存放大口水罐、桌布、餐巾的)房间
exacerbation 加剧;再燃
exacitude 正确性
exact 精确;准确的;正确的
exact additive functor 正合加性函子
exact analysis 精确分析
exact approximation 正合逼近
exact category 正确范围;正合范畴
exact coefficient of deviation 准确自差系数
exact cohomology sequence 正合上同调序列
exact coproducts 正合上积
exact couple 正合偶
exact curve 精确曲线
exact details 底细
exact differential 正合微分;恰当微分;完整微分
exact distribution 精确分布
exact duplicate 复印本
exact endomorphism 正合自同态
exact equation 恰当方程
exact fit 精确配合
exact focus 准确焦点
exact focusing 精确调焦
exact functor 正合函子
exact height 准确高度
exact in every particular 正确到毫厘不差
exact information 可靠情报
exacting limit 精密极限
exacting terms 苛刻的条件
exact instrument 精密仪器;精密仪表
exact interest 实利息;实计利息
exactitude 精确
exact length 准确长度
exact linear function 精确线性函数
exact linear relation 精确线性关系
exact localizing system 正合局部化系统
exact local time 准确地方时
exact location 精确位置;精确定位
exact location finder 精确定位仪
exactly divisible 整除;除尽
exact method 确切法
exactness 精确(度);精密(度);正合性;确切性
exactness axiom 正合性公理
exactor 激发机
exact plastic theory 精确塑性理论
exact point 恰当点
exact position 精确位置;精确定位
exact position finder 精确定位仪
exact quantity 确数
exact register 精确套合
exact relation(ship) 精确关系
exact scale 精确刻度
exact science 精确科学;精密科学
exact sequence 正合序列
exact solution 精确解
exact solution condition 恰当溶液条件
exact square 正合平方
exact statistical method 精确统计法
exact unit 正确单位
exact value 精确值
exact width 准确宽度
exaggerated claim 夸大的索赔
exaggerated grain growth 异常晶粒生长
exaggerated model 夸大地形模型

exaggerated ratio 夸大率
exaggerated relief model 夸大地形模型
exaggerated scale 扩大比例;逾常比例;超常比例;放大比例(尺);变态比例(尺);特别放大比例
exaggerated stereoview 夸大立体观测
exaggerated test 过度条件下检验;超定额试验;(特别不利条件下进行的)超常试验
exaggerated vertical scale 放大垂直比例
exaggeration model 放大模型
exaggeration of damages 夸大损失
ex airport 机场交货
ex all 不附权利(证券);保留全部权利
exaltation 超加折射
exaltation phase 超常相;超常期
exalted carrier 恢复载波
examinant 检查人
examinatin of wear 磨损检验
examination 检验;检查;测验;审查
examination anchorage 候潮锚地
examination and acceptance of materials 材料验收
examination and approval 审批
examination and approval authority 审批单位
examination and repair 检修
examination and repair of passenger train stock 车底检修
examination and verification 审核
examination(bore)hole 检查钻孔
examination by sectioning 剖开检验
examination by sensory organs 感官检查
examination copy 审校样
examination cubicle (医院中的)检查室
examination finding 检查所见
examination into the why and how of an accident 对事故的起因和经过作周密调查
examination meeting of standard 标准审会会
examination of account 审查账目
examination of auditor 审计检查
examination of bids 验标;投标审核;鉴定投标
examination of budget 预算审定
examination of conscience 反省
examination of construction design 施工图设计审查
examination of discovery 矿点检查
examination of fibers 纤维检验
examination of materials 材料检验
examination of mixed stain 混合斑检验
examination of municipality water supply 城市给水检验
examination of pollution index 污染度检测
examination of ropes 绳索检验
examination of sanitary water 上水检验
examination of sectioned specimens 截面试样的检验
examination of the building ground 建筑场地土质调查
examination of wastewater 废水检验;污水检测
examination of water 水质检查
examination of water quality 水质检验;水质检测;水质分析
examination of water supply hygiene 给水卫生审查
examination of wound 相度损伤
examination on-the-spot 就地诊勘
examination-oriented education system 应试教育
examination period 检查周期
examination pit 检验坑
examination rack fitted on wagon for checking out-of-ga(u)ge goods 超限货物检查架
examination report of anomaly 异常检查报告
examination report of ore discovery 矿点检查报告
examination table 检验结果表;检测成果表;调查表
examination under ultraviolet light 紫外线照射检查法
examine and approve 审定
examine and approve budget 审核预算
examine and receive 验收
examine and verify 审核
examine as to substance 实质审查
examined 验讫
examined and verified by 审核
examined cross-section of goods 货物的检定断面
examined in depth 深入研究
examine documentary value 检查编录值

examined section 检算截面;验算截面
examine environmental sample 检定环境样本
examine for wear and tear 磨损检验
examiner 量隙规;验收员;验检员;复验工
examine suggestions of design 设计审查意见
examine the account 盘账
examine the project proposal 审查项目建议书
examine through(out) 彻底检查
examining amount of work 检查工作量
examining endoscope 检查窥镜
examining officer 检查员;海关验关员;海关检查人员
examining operational policy 审查经营方针
examining rack for out-of-ga(u)ge freight 超限货物检查架
examining the records of staff and workers 人事考核
example based system 基于事例的系统
example derived from nature 自然模式
example of calculation 计算实例
example of case 案例
example program(me)graph 实例程序图
example ship 样船
examples of application 应用实例
examples of geologic(al) expert system 地质专家系统实例
ex ante analysis 事先分析
ex ante forecast(ing) 事前预测
ex ante measurement of benefit 效益预测;工程竣工前的效益估算
ex ante payoff 事先偿付
exaration 冰川剥蚀
ex barge 驳船交货
ex barge price 驳船价
ex bond (纳税后)关栈交货;(纳税后)开栈交货
ex bonus 无奖金
ex buyer's bonded warehouse duty paid 完税后买方关栈交货价格
ex buyer's godown 买方仓库交货价格
escape route 运移通道
excavate 采掘;挖掘
excavated and cast-in-place pile 挖孔灌注桩
excavated area 开挖面积;开挖区
excavated artificial harbo(u)r 开挖成的人工港;挖入式人工港
excavated artificial trench 人工开挖沟
excavated bottom slope 开挖坡度;开挖底坡
excavated cavity 挖穴法
excavated diameter 挖掘直径
excavated dock basin 挖入式港池
excavated earth 挖出的土;开挖(出的)土
excavated face 开挖面
excavated-in harbo(u)r basin 挖入式港池
excavated-in port 挖入式港口
excavated material 开挖(土)料;挖掘料;挖出物(质);挖出土料
excavated pit 挖坑
excavated profile grade 开挖纵坡
excavated section 开挖断面;掘进断面
excavated side slope 开挖边坡
excavated spoil 开挖废渣;开挖的废土
excavated substance 挖出物(质)
excavated surface 开挖面
excavated tank 开挖的蓄水池
excavated volume 开挖体积;开挖(方)量;挖方体积;挖方(量);土石方量
excavate holes 打千斤顶洞
excavate in dry 干开挖
excavate underwater 水下开挖
excavate volume 土石方量;开挖量
excavate without timbering 无支撑挖掘
excavate with timbering 加撑开挖;支撑挖掘
excavating 挖掘
excavating area 挖掘区
excavating attachment 挖掘工具附件
excavating base depth 清基深度
excavating bucket 挖斗;铲斗
excavating cableway 开挖(用)索道;挖土用架空索道
excavating chute 挖土滑运槽
excavating clamshell 挖掘抓斗
excavating cycle 掘进循环
excavating depth 开挖深度;挖掘深度
excavating diameter 开挖直径

excavating drain trench 开挖的排水沟
excavating engineering 开挖工程
excavating equipment 挖掘设备
excavating face 开挖面
excavating gear 挖掘装置
excavating grab 挖掘抓斗
excavating hand conveyer 挖土皮带输送机
excavating leader 运渣机
excavating loader 开挖装载机;开挖运渣机
excavating machine (履带式)挖土机;开挖机(械);挖泥机
excavating machinery 露天采掘机械;挖掘机械
excavating method 掘井方法
excavating plant 开挖设备;挖土机械设备
excavating position 挖掘位置
excavating productivity 掘进生产率
excavating pump 泥浆泵;吸泥泵;排出泵;挖泥泵
excavating radius 挖掘半径
excavating range 挖掘范围
excavating shaft 开挖竖井
excavating wheel 挖掘轮;挖土轮
excavation 开窖;开挖;掘蚀;采掘;凹陷;挖掘;挖方;挖除
excavation and cart-away 挖方(运土)工程
excavation and carting away 挖方和运土
excavation and filling 挖填(方)
excavation base 开挖基线
excavation below arch 拱部的下开挖
excavation blasting 爆破挖掘
excavation by blasting 爆破开挖
excavation cost 开挖费用
excavation cycle 掘进循环
excavation deformation 挖掘错动
excavation depth 挖掘深度;开挖深度
excavation effect 挖掘效应
excavation equipment 开挖设备;挖泥设备;挖掘设备
excavation face 开挖面
excavation for foundation 基础开挖;基坑开挖
excavation from one end 单向开挖
excavation heap 挖掘堆积
excavation hoisting method 掘进提升方法
excavation in cohesive soil 黏性土开挖
excavation in dry 干挖
excavation in dry ground 无水地层开挖
excavation in foundation 地基开挖
excavation in open cut 明挖方量
excavation level 挖方高程
excavation limit 开挖范围(界)线;开挖范围
excavation line 开挖(轮廓)线;基槽开挖线
excavation line for regulation 整治开挖线
excavation machinery 开挖机械;挖掘机械
excavation method 开挖方法
excavation of cutting 路堑开挖
excavation of earth 挖土
excavation of parallel trench 开挖平行槽
excavation of side-wall at interval 间隔式边墙开挖
excavation of underwater foundation trench 水下基槽开挖
excavation of underwater pipe ditch 水下管沟开挖
excavation plant 挖泥设备
excavation progress 采掘进度
excavation protection 开挖防护
excavation quantity 挖方(数)量;开挖方量;挖土数量;挖方量
excavation scraper 铲运机
excavation section 开挖断面
excavation side 开挖侧面;开挖边
excavation sinkage (原地打滑引起的)挖掘性下陷量
excavation slope 挖方边坡;开挖边;基坑边坡;挖方坡度
excavation-stability condition 开挖稳定条件;挖土稳定条件;挖掘稳定条件
excavation subsidence 采掘塌陷
excavation support 巷道支架;挖掘支架
excavation unit 挖泥设备
excavation waste 弃渣
excavation waste dump 弃渣场
excavation widening 加宽路堑
excavation without timbering 无支撑挖掘;无支撑开挖
excavation with sheeting piling 打板桩挖掘
excavation with timbering 有支撑开挖;支撑挖掘

excavation work 开挖工作;开挖工程;挖方工程;土方工程;挖土工程;挖掘工程
excavation work without timbering 无支撑挖土工程
excavation zone 挖泥区域
excavator 开挖机;掘凿机;掘土机;打洞机;挖土机;器;挖掘机;挖掘工人
excavator base machine 挖掘机机身
excavator bucket 电铲勺斗;铲斗;挖土(机)斗;挖掘(勺)斗;挖斗
excavator bucket capacity 挖土机铲斗容量
excavator cab(in) 挖土机驾驶员室
excavator cable 挖土机缆索
excavator deck 挖土机平台
excavator digging below ground level 挖土机在地面以下挖掘
excavator drive 挖土机推进
excavator driver 挖土机驾驶人
excavator drum 挖土机鼓
excavator engine 挖土机发动机
excavator equipment (挖土机的)挖铲装置
excavator factor 挖掘系数
excavator grab 挖泥抓斗;挖掘机抓斗;挖掘机铲斗;挖斗
excavator grease 挖土机润滑脂
excavator-loader 装载机;装料机;挖掘装载机
excavator mat 挖土机支垫;挖土机垫物
excavator motor 挖土机发动机;挖土机电动机
excavator-operated stamper 挖土机操作的捣实机
excavator pass 挖掘剥离循环
excavator primer 挖土机启动注油器
excavator rope 挖土机钢丝索
excavator spoon 挖匙
excavator supporting mat 挖土机支承底板
excavator tooth 挖土机牙齿
excavator track 电铲履带
excavator type bulldozer 畚箕式刀片推土机;铲斗式推土机;簸箕式刀片推土机;挖掘机式推土机
excavator type shovel 铲斗式挖掘机
excavator wearing part 挖土机磨损部件
excavator work 挖土机工作;挖土作业
execution at the close 收市执行
exceed 超过;超出
exceedance 超过数
exceedance frequency 超过某一特定值的频率
exceedance interval 重现期;超现期
exceedance probability 相等或超过概率;超过概率
exceedance test 超过数检验
exceed capacity 超过范围
exceed capacity check 溢出校验;(加法器的)过容量检查
exceed credit 超过贷款
exceed drum capacity (程序)超过磁鼓容量
exceeding length 超长尺寸
exceedingly 过度地
exceedingly rich to water supply 异常丰富的水资源
exceeding the permitted weight per axle 超重轴;超过轴重
exceeding the pressure 超过容许压力
exceed label capacity 超过标号容量
exceed one's competence 超越某人权限
exceed the budget 超出预算
exceed the high limit of flight 飞行超高限
exceed the price 超过价格
exceed the speed limit 超速行驶
Excelate 天然石板装饰件(一种专利建筑材料,主要用于外墙装饰)
excellence 优点;卓越
excellent 高超
excellent bond 强键
excellent clear displayed 显示极明显
excellent design 优秀设计
excellent diamond 优等金刚石;高质量金刚石
excellent doctor 良工
excellent growing condition 良好生长条件
excellent harvest 大丰收
excellent picture 优质图像
excellent river stretch 优良河段
excellent-stretch simulation method 优良河段模拟法
excellent visibility 极好能见度;优异能见度;优良能见度;最佳能见度;最好能见度
excellent water 优质水

excelsior 细刨花;刨花(填料);上等木屑;锯屑
excelsior absorbent board 木丝吸收板
excelsior acoustic(al)board 木丝吸声板
excelsior board 木丝(板);细刨花板
excelsior building slab 木丝建筑平板
excelsior chutting machine 木丝机
excelsior concrete 木丝混凝土
excelsior concrete slab 木丝混凝土平板
Excelsior cotton 埃克斯塞尔西棉
excelsior covering rope 木丝保护绳
excelsior cutting machine 锯屑机
excelsior filter 木丝滤器
excelsior hollow filler 木丝空隙填充料
excelsior insulation 木丝绝缘
excelsior slab 木丝平板
excelsior slab partition 木丝平板隔墙
excelsior sound-control board 木丝声控板;木丝调音板
excelsior tissue 高级薄纸;刨花纸
excel tester 蓄电池电压表;电池测试器
excenric 偏心的
excenter 旁心;外心
excentralizer 偏心器
excentricity 不共心性
excentric shaft 偏心轴
except 除外
except circuit 禁止电路
excepted perils 例外危险;除外危险
excepted perils clause 例外危险条款;除外危险条款
excepted risk 意外风险
exceptional advantages 特别利益
exceptional case 例外事件;例外情况
exceptional circumstance 特殊情况
exceptional condition 异常条件;例外条件
exceptional dimension freight 阔大货物
exceptional dimension goods 阔大货物
exceptional direction 例外方向;特殊方向
exceptional emergency 非常紧急情况
exceptional exposure 特殊暴露潜水
exceptional exposure dive schedule 特殊暴露潜水方案
exceptional exposure excursion time 特殊暴露巡潜时间
exceptional flood 特大洪水
exceptional flood level 非常洪水位;特大洪水位
exceptional ground pressure 地层异常压力
exceptional hardness 超硬度
exceptional importance 异常重要
exceptional integers 例外整数
exceptional Jordan algebra 例外约当代数
exceptional length freight 超长货物
exceptional length goods 超长货物
exceptional list 货损清单
exceptional load 特殊载重
exceptionally big bursting water 特大突水
exceptionally bulky goods 特大货物
exceptionally heavy goods 特大重件
exceptionally high flood 非常洪水;特大洪水
exceptional measures 非常措施
exceptional overload 额外超载
exceptional service 额外服务
exceptional shower 特大暴雨
exceptional urgency 特殊紧急事态;例外紧急事件
exceptional use 特殊用法
exceptional value 例外值;特小值;特大值;非常(价)值
exceptional visibility 九级能见度;特佳能见度
exceptional water level 异常水位;非常(洪)水位;特定水位;特大水位
exceptional well 特殊井
exception clause 免罚条款;例外条款;免责条款;除外条款
exception code 异常码
exception concept 例外概念
exception dispatcher 异常分派程序
exception enable 异常允许
exception exit 异常出口
exception from liability 免责事项
exception handler 异常处理程序
exception handling 例外处理;异常情况处理;异常处理
exception income limits 住房资助的例外收入极限
exception inquiry 例外咨询
exception item 异常项

exception-item encoding 异常项编码
exception list 例外清单;异常情况单
exception message 事故消息
exception paragraph 例外段落
exception principle 例外原则;例外法则
exception principle system 例外原则系统;异常原则系统
exception rate 特别减价运费
exception report(ing) 异常报告;例外报告;例外报表;异议报告
exception request 异常请求
exception response 异常应答;异常响应;异常回答
exceptions 免责条款
exception scheduling routine 例外调度程序;异常调度(例行)程序
exception service routine 异常服务例行程序
exceptions form liability 赔偿责任的免责事项;免赔事项
exceptions noted on the bill of lading 提单批注(的免责事项)
exception vector 异常向量
exceptive warranty 例外保证;除外担保
except operation 禁止操作
except otherwise herein provided 除另有规定外
except risks 例外风险
Excer process 埃克萨法
excerpt 选录;摘录
excerption 摘录
excess 绝对免赔限度;余物;过量;过多;过度;超(过)额;超(出)量;额外
excess absorption 过剩吸收
excess accelerated depreciation 超额加速折旧
excess activated sludge 过剩活性污泥;剩余活性污泥
excess air 剩余空气
excess air burner 过量空气燃烧器
excess air chimney 过剩空气烟囱
excess air coefficient 过剩空气系数;过量空气系数
excess air factor 过剩空气因数;过剩空气系数;过量空气系数
excess air inleakage 漏入的过剩空气
excess air pressure 超气压
excess air ratio 过剩空气系数
excess air stack 过剩空气排放管
excess amount of dust 过量尘埃
excess and defficiency 超额或不足
excess and total meter 总额和超额积算表
excess area 剩余面积
excess argon 过剩氩
excess attenuation 逾量衰减
excess baggage 超重行李
excess baggage ticket 超重行李票
excess bank reserves 超额银行准备金
excess bleed 过量流失
excess budget 预算超额;超出预算
excess cant 过超高
excess capacity 超额生产能力;超额容量;超额(能)力)
excess carburizing 过度渗碳
excess carrier ratio 过剩传送比
excess cash 超额现金
excess cement 多余的水泥
excess charge 过剩电荷
excess charge carrier 多余载流子
excess-chlorinating 过量氯化
excess chlorination 过量氯化(法)
excess clause 船价超met保价条款
excess clearance 起拱度
excess code 加编码;余码
excess coefficient 过量系数
excess combustion air 过量燃烧空气
excess condemnation 超额征用土地;超额征收(土地);额外征用土地
excess conduction 过剩(型)导电
excess conductivity 过剩型导电率
excess consumption 超额消费
excess convexity (焊接的)过凸
excess cooler air 冷却机过剩空气
excess coverage 超额保险
excess current 过载电流;过(剩)电流;多余流量;多余电流
excess-current protection 过载电流保护
excess-current protective breaker 过载电流保护断路器

excess death 过量死亡
excess death number 超死亡数
excess death rate (过)高死亡率
excess deforestation 过量采伐
excess demand 过量需求;过度需求;超额需要
excess demand correspondence 过量需求对应
excess demand function 超额需求函数
excess demand tariff 超需量收费制
excess disbursement 超额支付
excess discharge fee 超标排污费
excess disinfectant concentration index value 剩余消毒剂浓度指数值
excess draft 过量取水;过量气流(船舶);过量吃水;超量取(地下)水;富裕水深
excess draught 过量气流;(船舶)过量吃水
excess dredging 过量疏浚
excess earnings 超额盈利;超额收益
excess electron 多余电子
excess eletron 过剩电子
excess elevation 过超高
excess elongation 过度伸长
excess energy 剩余能量;过剩能量;多余能量
excess enthalpy 过剩焓
excess entropy 超额熵
excess excavation 过量开挖
excess excavation lining 超挖衬砌
excess fall 过大落差
excess fare 补票费
excess flow check valve 溢流止回阀
excess flow test 过流量试验
excess flow valve 流量限制阀;溢流阀
excess fluid head 过剩流体压头
excess force 剩余力
excess franchise 绝对免赔限度
excess function 超额函数
excess funds 超额资金
excess general average 超过共同海损
excess geometric(al)head 过剩几何压头
excess glass 飞边
excess glaze 釉缕
excess gravel 超量砾石
excess head 超压位差;剩余水头
excess heat 余热
excess horsepower 剩余马力
excess humidity 过剩湿度
excess hydraulic pressure 超水力压力
excess hydrostatic(al)pressure 超静水压力;剩余静水压力
excess income 超额收益
excess income tax 超额所得税
excess in deviation 过偏差
excess inflow 超额流入
excess insurance 超过损失保险;超额保险
excessive 过量的;过分的;过多的;过度的
excessive air 过剩空气(量);过量空气
excessive air pressure 过剩气压
excessive allowance 超越宽度
excessive amortization 超额偿还
excessive amount of dust 煤尘超限
excessive approximate value 偏大近似值
excessive camber 过高上拱度;过高路拱;过大路拱
excessive carrying capacity 富余挟沙能力;富余挟带能力
excessive chlorine content 含氯度超标
excessive clay 过黏土
excessive clearance 过量余隙
excessive comminution 过分粉碎
excessive concentration of population and industry 人口和工业过度集中
excessive condemnation 多余土地征用;超额征用
excessive cost 超额成本
excessive current 过电流
excessive damages 超额损害赔偿
excessive deficit 超额赤字
excessive deformation 过度变形
excessive demand 需求过旺
excessive demand inflation 需求过多引起的通货膨胀
excessive depreciation deduction 折旧超额扣除
excessive detail 负载过多
excessive displacement 过大位移
excessive dogleg 过度的狗腿
excessive effects 重叠效果
excessive electric(al)current protection 过电流

保护
excessive evapo(u)ration 过量蒸发
excessive expenditures 超额开支
excessive fall 过大坡浆;过大落差
excessive fill 多余填土
excessive friction 过分摩擦
excessive function 超过函数
excessive grade 过大坡度
excessive gradient 过大坡度
excessive heating 过热
excessive idling 空转
excessive inflation 过度通货膨胀
excessive issue 超额发行
excessive lime 过量石灰
excessive lime process 过量石灰法
excessive load 超载
excessive loss 过大损耗
excessively drained 排水良好的
excessively drained field 排水良好的地块
excessively high level of phosphorus 过高的磷水平
excessively high salinity 过高盐渍度水
excessively long-distance traffic 过远运输
excessive moistening 过度湿润
excessive moisture 过量水分;过度湿润;多余水分
excessive multiplication 过量繁殖
excessive noise 过量噪声
excessive of groundwater 地下水过量开采
excessive overhang of tool 刀具悬臂过长
excessive overtime 过量加班
excessive pantograph lift 受电弓提升过多
excessive penetration (焊接时)塌陷
excessive precipitation 过量降水;非常(多的)降水量
excessive pressure 剩余压力;过大压力;超压(力);超量压力
excessive production 过度生产;过度开采
excessive production of 强化开采
excessive profit 超额利润
excessive profit tax 超额利税
excessive pumping 过分抽水
excessive rain(fall of long duration) 雨量过大;多余雨量;过多的雨水;特大雨量;雨水过多;霪雨;久雨;非常降雨
excessive rainfall of short duration 短时暴雨
excessive reserve fund 超额储备金
excessive rolling 过压缩
excessive sensitivity 过高灵敏度;多余灵敏度
excessive settlement 过多沉降;过度沉陷;过度沉降;过大沉降
excessive shaft runout 轴摆度过大
excessive shock 过度振动
excessive shrinking 过量收缩
excessive slippage 过量滑移
excessive sludge 剩余污泥
excessive snowfall 过量降雪
excessive spatter 严重飞溅
excessive star 超恒星
excessive steam pressure switch 蒸汽超压开关
excessive stiffness 过度刚性
excessive stock 过多存货
excessive storms 过量暴雨
excessive strain 超限应变
excessive stress 过大应力;超限应力
excessive substitution 超量替代
excessive sulfur content 过高硫含量
excessive sulphur content 过高含硫量
excessive superelevation 余超高
excessive supply 超额供给
excessive tapping 放过剩的玻璃料
excessive temperature differentials 温差过大
excessive thermal shock 过量热振荡
excessive urbanization 过度都市化;过度城市化
excessive use of fertilizer 施肥过多
excessive use of water 过量用水
excessive velocity 过高流速;超速
excessive vibration 剧烈振动;过大振动
excessive voltage 超额电压
excessive water temperature switch 水温超限开关
excessive wear 过度磨损;过度磨耗
excessive weight 过重
excessive yield 超额产量
excessive yielding 超屈服
excess lactate 过量乳酸盐

excess land 多余土地
excess length or weight train 超长或超重列车
excess-lime softening 过量石灰软化法
excess liquidity 流动性过剩;超额换现能力
excess load 净增荷载;过载;过负载;超(限)荷载;超负荷;超载
excess loan 超过法定限额的银行贷款
excess loss 额外损失;额外损耗
excess luggage 超重行李
excess luggage voucher 超重行李运货收据
excess mass 多余质量
excess material 过剩材料
excess materials ledger 材料余额簿
excess materials requisition 超额领料单
excess metabolism 过量代谢
excess metal 余块;多余金属
excess meter 积算超量功率表;积算超量电度表;过量计;超量电度表
excess moisture 过剩湿分;过剩水分;过多水分;多余水分;水分过剩;水分过多
excess mortality 超死亡
excess multiplication constant 过剩倍增常数
excess multiplication factor 增减因数;过剩倍增系数;超过倍率
excess neutron 过剩中子
excess neutron flux shutdown 中子通量过高停堆
excess-n notation 余n表达式
excess noise 过量噪声;超噪声;闪变噪声
excess noise contribution 过量的噪声影响
excess noise power 超准噪声功率
excess noise ratio 超准噪声比;超噪比
excess noise temperature 超准噪声温度
excess of arc 弧尺超量部分
excess of binder 过量结合料
excess of concrete 过量混凝土
excess of elements 元素过剩
excess of exports 出超;顺差
excess of hatch(way) 舱口超额吨位
excess of heat 热出大;富裕热
excess of imports 逆差
excess of jurisdiction 超越管辖权
excess of lime 石灰过剩
excess-of-loss 超额赔款(保障)
excess-of-loss insurance 超额损失保险
excess-of-loss ratio cover 超额损失率保障
excess-of-loss ratio reinsurance 超额损失率分保
excess-of-loss ratio reinsurance treaty 超额损失率分保合同;超额赔款率合约再保险
excess-of-loss ratio treaty reinsurance 超额赔款率合约再保险
excess-of-loss reinsurance 超额损失分保
excess-of-loss reinsurance treaty 超额损失分保合同;超额赔款分保合同
excess of materials 原材料积压
excess of rates 超额收费
excess of stroke 超程
excess of triangle 三角形角盈;三角形角超
excess of water 剩余水;过量水分
excess oil 余油
excess-oil return line 剩油回输管线
excess oxygen 氧量过剩
excess oxygen level 过剩氧含量
excess payment 超额支付
excess photon noise 过量光子噪声
excess pigment dispersion 过量颜料的弥散
excess planting 超量播种
excess population 过剩粒子数
excess pore pressure 超孔隙压力
excess pore water 超孔隙水
excess pore water pressure 超孔隙水压力;超静水压力;剩余孔隙水压力
excess porewater pressure 超孔隙水压力
excess power 剩余动力
excess precipitation 超渗降水量
excess present value 超(过)现值数
excess present value index 超现值指数;超收现值指数;超过现值指数
excess pressure 余压;过度压力;超压(力)
excess pressure cut-out 超压切断
excess pressure gradient 过压梯度
excess pressure head 过剩静压头
excess-pressure relief by-pass 过压保护旁通管
excess pressure simulator 超压压力模拟器;超压(力)模拟器

excess pressure valve 余压阀;过压阀
excess production 超额生产
excess production capacity 生产能力过剩
excess productive capacity 超额生产能力
excess profit 超额利润;超额收益
excess profit credit 超额利润扣除
excess profit duty 超额利润税
excess profit levy 超额利润附加税
excess profits tax 超额利润税;超额利得税
excess-profit-tax credit 超额利润税贷款
excess purchasing power 过剩购买力
excess quantity 过剩量
excess rain 超渗雨
excess rainfall 径流雨量;过量雨量;超渗雨量;多余雨量
excess reactivity 后备反应性;过剩反应性;过量反应性;剩余反应性
excess reflux 过量回流
excess reinsurance 超(过)额再保险
excess reserve 超额准备(金)
excess resin 过量树脂
excess resonance integral 过剩共振积分
excess revolution 超转(速)
excess risk 过分危险
excess rolling 过度碾压
excess share 超额股
excess sludge 过剩污泥;剩余污泥
excess sludge concentration 剩余污泥浓度
excess sludge production 剩余污泥产量
excess sludge reduction 剩余污泥减量
excess sludge treatment 剩余污泥处理
excess slurry volume 水泥浆附加量
excess sound pressure 峰值声压
excess stock 余料;存货过多
excess supply capacity 过剩供应能力
excess supply curve 超额供应曲线
excess surface water 地面积水;多余地表水
excess surplus value 超额剩余价值
excess system loss 额外系统损失
excess tax credit 超额税收抵免
excess temperature 过温度;超温
excess temperature limit 超温极限
excess temperature shutdown 温度过高停堆
excess thermodynamic(al) function 超热力学函数;超额函数
excess thermodynamics 超常热动力学
excess thickness 过大厚度
excess three code 余三码;超三代码
excess time 剩余时间
excess time fare 超时费
excess torque 剩余力矩
excess value 超过价值
excess viscous dissipation 超黏滞扩散
excess voltage 过电压;超过正常的电压
excess voltage protection 过电压保护
excess voltage relay 过电压继电器
excess volume 过剩体积
excess water 过剩水(量);过量水分;多余水分;水分过多
excess water removal 去除为增加混凝土和易性而超用的水
excess weight 过重;超重量;附加重量
excess weld metal 焊缝补强金属;(焊缝的)余高;补强金属;补强(焊料)
excess windfall profit 超额暴利
excess workers 过剩人员
exchange 流转;交易所;交换局;换汇;互换;兑换;外汇交易
exchangeability 可交易性;可交换性;互换性
exchangeable 可交换的;可兑换的
exchangeable acidity 代换性酸度
exchangeable bases 可互换(的)基础;换算单位
exchangeable bond 可兑换债券
exchangeable cation 可交换阳离子;交换性阳离子
exchangeable events 可交换事件
exchangeable form 可交换态
exchangeable for same zone 同带可以互换
exchangeable fraction ratio 可交换百分比
exchangeable ion 可交换离子;代替性离子
exchangeable ion energy 可交换电子能量
exchangeable object 交换对象
exchangeable paper currency 可兑换纸币
exchangeable part 可换零件
exchangeable power 可交换功率

exchangeable sodium 可交换性钠
exchangeable sodium percentage 可交换钠百分数
exchangeable storage 可换储存器
exchangeable value （可）交换价值；交换价格
exchange acidity 交换（性）酸度
exchange address register 更换地址寄存器
exchange adjustment 汇率调整；调整汇率
exchange adsorption 交换吸附
exchange algorithm 交换算法
exchange alumin(i)um-ion capacity 交换性铝离子含量
exchange arbitrage 套汇
exchange area （电话）交换区
exchange area cable 市内电缆
exchange as per endorsement 背书汇兑条款
exchange at equal value 等价变换
exchange at frontier 边境兑换
exchange at par 平价兑换
exchange base 交换基
exchange-based cost for the export of products 出口商品换汇成本
exchange board 交换机
exchange broker 外汇经纪人
exchange budget 外汇预算
exchange buffering 交换中间寄存；交换缓冲；更换缓冲
exchange-Ca ion capacity 交换性钙离子含量
exchange call 市内呼叫
exchange capacity 交换（容）量；交换能力
exchange capacity of soil 土壤交换量
exchange cation 交换阳离子
exchange certificate 兑换券
exchange charge 交换电荷
exchange check 交换支票
exchange chromatography 交换色谱法；交换色层法
exchange class 交换等级
exchange clause 货币条款；汇款条款；汇兑条款
exchange clearing 外汇结算
exchange clearing agreement 外汇结算协定
exchange coefficient 交流系数；交会系数；交换系数
exchange collision 交换碰撞
exchange column 交换柱
exchange complex 交换综合体
exchange concentrator 交换局集中器
exchange connection 交互连接
exchange constant 交换常数
exchange contract 兑汇合同
exchange contract confirmation 汇兑合同确认书；外汇合同确认书
exchange control 外汇管制；外汇管理
exchange control authorization 外汇管制当局
exchange control department 外汇管理局
exchange correlation 交换相关
exchange counter 外汇兑换处
exchange cover 外汇储备；补偿交易
exchange current 交换流；交变电流；互换流
exchange degeneracy 交换衰减；交换简并性
exchanged heat 转化热；传递热
exchange diffusion 交换扩散
exchange discount 外汇贴现；外汇贴水
exchange discount rate 外汇贴水率
exchange distillation process 交换蒸馏法
exchange draft 兑汇票据
exchange dumping 外汇倾销
exchange economy 市场经济；交易经济
exchange effect 交换效应
exchange efficiency 交换效率
exchange equation 交换方程
exchange experience 交流经验
exchange facility 交换设施
exchange factor 交换因数
exchange fault 交换机故障
exchange flow 外汇流
exchange flow rate 交换流速
exchange force 交换力
exchange for collection 托收汇兑款项
exchange for physical 转现货交易
exchange for telephone general office 电话总局交换机
exchange freedom 外汇自由
exchange frequency 交换频率
exchange fund 汇兑基金
exchange gain 汇兑盈余

exchange gap between 交换差价
exchange hydrogen-ion capacity 交换性氢离子含量
exchange idea 交换意见
exchange instruction 交换指令
exchange interaction 交换作用
exchange interaction energy 交换相互酌能
exchange inversion 交换反演
exchange labo(u)r 变工
exchange line 交换线；用户线
exchange locality 交货地点
exchange loss 汇兑损失
exchange margin 外汇差额
exchange market 外汇市场
exchange membrane 交换膜
exchange memo 外币兑换水单
exchange message 交换信息
exchange-Mg ion capacity 交换性镁离子含量
exchange movement 汇价浮动
exchange narrowing 交换窄化
exchange network 交流网；交换网（络）
exchange network facility of interstate access 州际通路交换网络设备
exchange network service 交换网络服务
exchange object 交换对象
exchange of abilities 能力交换
exchange of air 空气交换
exchange of communication 交换函电
exchange of complete units 成套机组交换
exchange of elements 元素交换作用
exchange of equipment 设备交换
exchange of experience 经验交流
exchange offerings 交换发行
exchange of foreign currency 外币兑换
exchange of future for cash 以期货现货
exchange of half time 半交换期
exchange of heat 热交换
exchange of information 资料交换；情报交换
exchange of kinetic energy 动能交换
exchange of know-how 技术交流
exchange of letter 换文
exchange of liability on the bill 免除汇票上的责任
exchange of members 构件交换
exchange of momentum 动量交换
exchange of reinsurance 交换分保
exchange of residential property 居住产业的交换
exchange of skill 技术交流
exchange of technical know-how 技术交流
exchange of unequal values 不等值交换
exchange operator 交换（运）算符
exchange package 交换程序包；更换包
exchange parity 外汇评价
exchange pattern 交换模式
exchange payment 外币付款
exchange phase connection 换相连接
exchange pile 交换柱
exchange pipe 流道管
exchange position 现有外汇额
exchange post 交换柱
exchange potassium-ion capacity 交换性钾离子含量
exchange premium 外汇贴现升值；外汇升水；贴进
exchange process 交换法
exchange profit 外汇利润
exchange proviso clause 外汇保留条款
exchange quota 外汇限额
exchange quota system 外汇限额制度；外汇配额制度
exchange quotation 外汇行情；汇兑牌价
exchanger 交换器；交换剂；换汇机；交换程序
exchange rate 交换（速）率；汇价；汇兑（牌）价；汇（兑）率；汇兑换算表；外汇（汇）率；外汇兑换率
exchange rate adjustment 调整外汇汇率
exchange rate fluctuation 外汇汇率波动
exchange rate quotation 外汇牌价
exchange reaction 交换反应
exchange register 交换寄存器
exchange resin 交换树脂
exchange restriction 外汇限制
exchange risk 汇兑风险；外汇风险
exchanger-type subcooler 热交换式低温冷却器
exchange sector 经济交换部门
exchange separation process 交换分离法
exchange service 交换业务；交换服务

exchange settlement 外汇结算
exchange shop 钱庄
exchange site 交换点
exchange slip 汇兑单
exchange sodium-ion capacity 交换性钠离子含量
exchanges of plant and equipment 厂房和设备的交换
exchange sort 交换分类
exchange splitting 交换分裂
exchange stabilization fund 外汇稳定资金；外汇平准基金
exchange state 更换状态
exchange station 交换站；车辆交接站
exchange surrender certificate 外汇转移证；外汇缴售凭证
exchange system 交换制；交换系统
exchange table 外汇换算表
exchange tax 外汇税
exchange terminal 交换终端
exchange termination 交换终端
exchange theory 交换理论
exchange track 交换车停留线；车辆交接线
exchange train 小运转列车
exchange transactions 交换往来
exchange trunk carrier system 局间中继线载波系统
exchange unit 互换件
exchange view 交换意见
exchange volume 交换量
exchange voucher 外汇单据
exchange work 变工
exchange yard 交换场；车辆交接车场
exchanging adsorption coefficient 交替吸附系数
exchanging air sampling method 空气置换采样法
exchanging land for housing 以地要房
exchequer 国库
exchequer equalization account 财政收支平衡账户
exchequer equalization grant 财政收支平衡补贴
excide 切开
excide battery 糊制极板蓄电池
excimer 激元；激基体；激发原；激发物；激发基态复合物
excimer laser 激发物激光器；准分子激光器
excipient 赋形剂
exciplex 复合受激态
exciplex dye laser 复合受激态染料激光器
excircle 旁切圆；外圆；外切圆
excise （领）许可证税；货物税；执照税；收货物税
excise duty 消费税；土产税
excise tax 营业税；特种消费行为税；消费税
excising forceps 切断钳
excision 被切去部分；切除
excision repair 切割修补；切补修复
excision theorem 分割定理
excitability 灵敏性；可激发性
excitant 励磁溶液；激发剂；兴奋剂；刺激性的；刺激物
excitation 激振；激励；激发
excitation anode 激励阳极
excitation band 激发带
excitation capacity 励磁容量
excitation center 激发中心
excitation characteristic 激发特性
excitation circuit 激励电路；触发电路
excitation coefficient 激振系数
excitation colo(u)r purity 彩色纯度
excitation condition 励磁条件；激发因素；激发条件
excitation contactor 励磁接触器
excitation control 励磁控制
excitation control relay 励磁控制继电器
excitation cross-section 激发截面
excitation cubicle 励磁柜
excitation current 激励电流
excitation curve 励磁曲线；激发曲线
excitation device 激发装置
excitation earthquake 激发地震
excitation electron 激发电子
excitation emission matrix fluoresence spectrometry 激射基体荧光光谱法
excitation energy 激发能（量）
excitation energy of compound nucleus 复合核激发能
excitation energy transfer 激发能传递
excitation factor 激励因子
excitation flux 励磁通量；励磁磁通

excitation force 激发力
excitation forcing 强迫励磁
excitation forcing limiter 强磁限制器
excitation frequency 激振频率;激励频率;激发频率;激动频率
excitation function 激励函数;激发函数
excitation function generator 激振函数发生器
excitation generator 励磁发电机
excitation impedance 激磁阻抗
excitation index 激发指标
excitation intensity 激发强度
excitation lag 激发滞后
excitation level 激发能级
excitation limiting relay 励磁限制继电器
excitation loss 励磁损失;激磁损耗;磁芯损耗
excitation-loss relay 失磁继电器
excitation matrix 激励矩阵
excitation mechanism 励磁机理;激励机制;激发机理;激发机制;激发机理;激发机构
excitation monochrometer 激发单色仪
excitation of spectra 光谱激发
excitation of turbulence 湍动激发
excitation parameter 扰动参数
excitation potential 激励电位;激励电势;激发电位;激发(电)势
excitation power supply 激励电源
excitation probability 激发概率
excitation pulse 激发脉冲
excitation purity 激发纯度;兴奋纯度;色纯(度)
excitation rate 激发速率
excitation reduction 减励磁
excitation-response relation 激振反应关系
excitation source 激发源
excitation spectrum 激发光谱
excitation stage 激发能级
excitation state 激发态
excitation suppression 减励磁;灭磁
excitation surge current 励磁涌流;激磁冲击电流
excitation system 励磁系统;激励系统
excitation temperature 激发温度
excitation thermal conductivity 激子热导率
excitation transfer 激发转移;激发迁移
excitation variable 激发变量
excitation voltage 励磁电压;激励电压
excitation volume 激发体积
excitation wave 激振波;激发波
excitation wavelength 激发波长
excitation winding 励磁绕组;激励线圈;激励绕组;激发绕组;激磁绕组
excitative peak 激发峰
excited atom 受激原子
excited band 受激能带
excited cavity 受激共振腔
excited center 受激中心
excited complex 受激络合物
excited condition 激发条件
excited electron 受激电子
excited electronic state 电子激发态
excited energy level 受激能级
excited ethylene 受激乙烯
excited-field 受激场
excited frequency 激振频率
excited in phase 同相激励
excited ion 受激离子
excited level 受激电平
excited magnet 受激磁铁
excited phosphor 受激磷光粉
excited resonance 受激共振
excited singlet state 受激单重态
excited state 激发(状)态;受激(状)态
excited state spectroscopy 激发态光谱学
excited state well 激发态阱
excited transition 受激跃迁
excited vibration 激发振动
excited wave 受激波
excite heat by friction 摩擦生热
exciter 起振器;励磁机;激振器;激励器;激励灯;激发器;激磁机;主控振荡器;主控振荡槽路;振动试验台;发器
exciter alternator 交流励磁机
exciter brush 励磁机电刷
exciter dome 励磁机上盖端
exciter field breaker 磁场电流断路器
exciter field rheostat 激磁式变阻器

exciter filter 激发滤色镜
exciter lamp 激发管;激励灯;激发灯;照明灯
exciter motor-generator 励磁电动发电机
exciter rectifier 激励整流器
exciter selsyn 励磁自动同步机;激磁自动同步机
exciter set 励磁机组
exciter tube 励磁管;激励管;主振管
exciter turbine 励磁透平;励磁水轮机;激磁用透平
exciter voltage 励磁机电压
exciting agent 激发剂
exciting bank 激励装置
exciting beam 激励光束
exciting circuit 励磁电路;激励电路
exciting circuit time constant 励磁回路时间常数
exciting coil 励磁线圈;激励线圈
exciting collision 激发碰撞
exciting contact 励磁接点
exciting current 激磁电流;磁化电流
exciting discharge 励磁放电
exciting dynamo 励磁电机;激磁机
exciting earthquake 激发地震
exciting electrode 激励电极;激发电极
exciting energy 激能
exciting field 励磁场;激发场
exciting force 激振力;激发力;扰动力【船】
exciting frequency 激发频率
exciting harmonic 激振谐率
exciting irradiation 激发照射
exciting light 激励光;激活光;激发光
exciting line 激发(谱)线
exciting loop 激发回路
exciting magnet 励磁磁铁
exciting mechanism 激振机制;激振机构
exciting nozzle 激振用喷嘴
exciting power 励磁功率;激振功率;激励能力;激励功率;激发功率
exciting radiation 激发辐射
exciting rectifier 激发整流器
exciting star 激发星
exciting steam 激振用蒸汽
exciting transducer 激磁换能器
exciting transformer 激励变压器
exciting voltage 励磁电压
exciting voltage pulse shape 激发电压脉冲形状
exciting watt 励磁功率
exciting winding 励磁绕组;激励线圈;激励绕组;激发绕组;激磁绕组
excitomotion 反射运动;反射机能
exciton band 激子(能)带
exciton condensation 激子凝聚
exciton density 激子密度
exciton diffusion 激子漫射
excitonics 激子学
exciton level 激子能级
exciton recombination radiation 激子复合辐射
exciton spectrum 激子光谱
excitor 激励器;主控振荡槽路
excitor control 激励调整
excitor rheostat 激磁变阻器
excitory input 激励输入
excitron 激励管;激弧管
excitron locomotive 激励管电力机车
exclave 飞地【地】
exclosure 禁牧区
exclude 除外
excluded cargo 不准装货物
excluded corrosion allowance 不包括腐蚀裕度
excluded ports 除外港口
excluded region 禁区
excluded space 占住空间
excluded volume 已占容积;已占空间
excluded volume effect 体积排除效应
excluder 封隔器;渠首排沙设施;排除器;排除剂
excluder pigment 防锈颜料
exclusion 隔断;除外责任;分离术;表面排斥;屏除;排阻;排除
exclusion and eradication versus 排除和根除对比
exclusion area 禁止区;禁入区;禁区;非居住区
exclusionary rule 不采纳规定
exclusion chromatography 凝胶色谱(法);排阻色谱法;排除色谱法;筛析色谱法
exclusion clause 限外条款;不属保险事项的条款
exclusion effect 排斥效应

exclusion energy 不相容能量
exclusion from income 不计列所得额
exclusion from taxable income 不计列应缴税所得额
exclusion gate 禁(止)门【计】
exclusion limit 排阻(极)限
exclusion method 排他法
exclusion of dividend received from profit 利润分红除外
exclusion of draught 气封;排除通风
exclusion of interbedded water 封堵夹层水
exclusion of water 堵水
exclusion operation 排他运算
exclusion principle 不相容原理;排他原则
exclusion under direct vision 直视分离术
exclusive 稀缺商品;专一的;独占的;除外的
exclusive agency 唯一代理
exclusive agency agreement 独家代理协议书
exclusive agency listing 房地产独家代理
exclusive agency policy 专卖政策
exclusive agent 独家经销商
exclusive attribute 排他属性;排斥属性
exclusive branch 互斥转移;排斥转移
exclusive bus lane 公共汽车专用车道
exclusive busway 公共汽车专用快速(车)道
exclusive call 排他性调用
exclusive circuit 专用电路;闭锁电路
exclusive contract 排他性合同
exclusive data 专用数据
exclusive dealing 专卖;独家经销
exclusive disjunction 非兼析取
exclusive dispatching facility 排他调度方式
exclusive distribution 总经销;独家经销
exclusive distribution agreement 排他性销售协议
exclusive distributor 包销人
exclusive district 专用区
exclusive economic zone 专属经济区;专管经济区(二百海里内)
exclusive economic zone concept 专属经济区概念
exclusive events 互斥事件;不相容条件;不相容事件
exclusive filter 窄频滤波器
exclusive fisheries zone 专属渔区
exclusive fishery limits 专属渔区
exclusive fishery zone 专属渔业区;专属捕鱼区
exclusive fishing jurisdiction 专属捕鱼管辖权
exclusive fishing zone 专属渔区
exclusive industrial district 专属工业区;工业专用(地)区
exclusive industrial water supply 自备工业给水工程
exclusive jurisdiction 专属管辖权;排他性的管辖条款
exclusive lane 专用车道
exclusive lanes upstream of bottlenecks 瓶颈上游专用车道
exclusive licence 专营许可证
exclusive licence contract 独家许可证合同
exclusive license 专营许可证;独占许可证;独立专利权
exclusive lines 专卖品
exclusive lock 互斥型锁
exclusively owned technology 专有技术
exclusive method 排除法
exclusive of loading and unloading 装卸除外
exclusive operator 异常算子
exclusive option 排斥任选
exclusive or 按位加(无进位加法)
exclusive or gate 异或门【计】
exclusive or operator 按位加算符
exclusive pedestrian phase 行人专用相
exclusive privilege 专有特权
exclusive problem 排斥问题
exclusive railway 专用铁路;专用铁道
exclusive reaction 专一反应
exclusive reference to standards 排他性引用标准
exclusive relationship theory 生物排斥说
exclusive remedies 排他的补救
exclusive requirement 必要要求
exclusive residential district 居住专用地区;专属住宅区;住宅专用区;高级住宅区
exclusive right 专有权;专营权;专利权;独家代理权
exclusive right to sale 专卖权
exclusive right to sale listing 独家销售代理权
exclusive right to sell 独家经销权

exclusive right-turn lane 右转专用车道
exclusive road 专用道路
exclusive safety device 紧急保险装置;专用安全装置
exclusive sale agreement 包销协议
exclusive sales 包销
exclusive sales contract 包销合同
exclusive segment 互斥(程序)段;排除段;排除部分
exclusive shear clutch 断销式分离合器
exclusive species 专见种;确限种
exclusive stress on material benefits 福利主义
exclusive supply 独家供货
exclusive surveyor 专任验船员
exclusive turning lane 转弯专用车道
exclusive use 专有使用;专用;单独使用
exclusive use berth 专用泊位
exclusive use district 专用区
exclusive use for the works 工程专用的
exclusive use zoning 专用地区划;专用区域规划
exclusivity agreement 独家经营协议
exclusivity clause 排他性条款
excogitation 计划;方案;计策
ex-contractual claims 非合同规定的索赔
ex coupon 不附息票;不带息票;无息票
excrement 粪便;排泄物
excrement filter 粪便过滤器
excrement of animals 牲畜粪便
excrement use 粪便利用
excrescence 突出体
excreta 分泌物;排泄物
excreta pit 粪坑
excrete 排泄【生】
excretion 排泄【生】
excretion concretion 内增结核
excretion monitoring 排泄物监测
excretion of toxicants 毒物排泄
excretion product 排泄物
excretion rate 排出率
excretion test of fish 鱼类排泄试验
excretory product 排泄(产)物
excretory system 排泄系统
excubitorium (古罗马)守夜人的职位
exculpatory 开脱罪责的;无责任的
exculpatory clause 协议中有关开脱责任的条款
excurb 市郊富人住宅区
excursion 功率骤增;短途旅行;变化范围;摆幅;漂游;偏振
excursion area 游览区
excursion boat 游艇;游览船
excursion bus 游览公共汽车
excursion center 游览胜地
excursion diving 短时间潜水
excursion diving operation 巡回潜水作业
excursion factor 超限系数
excursion fare 旅游票价
excursion ferry 游览渡船
excursionist 旅游者
excursion steamer 游览轮
excursion time 上浮下潜时间
excursion train 旅游列车;游览列车
excursion vessel 旅游船
excursus 附记
excurvature 外弯
excusable compensable delay 可原谅并应给予补偿的拖期
excusable delay 可谅解的延迟;可容许迟延;可原谅拖期
excusable non-compensable delay 可原谅但不给予补偿的拖期
ex customs 关税已付价
ex customs compound 海关交货价
exd 试验过的
ex dividend 不带股息;无红利;无息股
ex dock 码头交货价(格);码头交货;码头交换价(格);目的地码头交货
exdogenic force of geologic(al) function 外力地质作用
exducer 出口导流器;出口导风轮
exector 遗嘱执行人
execuor 操纵器
executable 可执行态;可执行的
executable array statement 可执行数组语句
executable file 可执行文件

executable image 可执行映像
executable instruction 可执行指令
executable module 可执行模块
executable program 可(可)执行程序
executable program(me) unit 可执行程序单元
executable state 可执行状态
executable statement 可执行语句
executable unit 执行单位
executant 执行者;实行者
execution road network 实施线网
execute 执行;编制;让渡财产;实行;施行
execute, complete and remedy defects 施工、完工和修补缺陷
execute a contract 在合同上签字
execute a deed 签名使契约生效;签名使契据生效
execute an estate 让渡财产
execute an order 接受订货;接受订单;交付订货
execute a pincer movement 执行钳形运动
execute a plan 实现计划
execute card 执行卡(片)
execute contract 执行契约;签署合同
execute cycle 执行周期
executed amount (已)完成工作量
executed consideration 已履行的对价
executed contract 已执行完毕的合同
executed contract of sale 已完成销售合同
executed copy 签定生效的副本
executed sale 已完成的销售
executed work amount 已完成工作量
execute exception 执行异常
execute instruction 执行指令
execute phase 执行阶段
execute register content 执行寄存器内容
execute statement 执行语句
execute store 执行存储器
execute treaty 执行条款
execute truncate 执行中断
execute vector 执行向量
executing accounting 执行会计
executing routine support 执行例行程序支援
executing state 执行状态
execution 制作;实行;实施;施工
execution area 执行区域
execution behavio(u)r 执行状态
execution capacity 施工能力
execution circuit 执行电路
execution command 执行指令
execution control 施工管理
execution control function 执行控制功能
execution control program(me) 执行控制程序
execution cycle 执行周期;完成周期
execution dead time 执行迟延时间
execution design 施工图设计
execution end 执行端
execution error 执行错误
execution error detection 执行错误检测
execution function 执行控制功能
execution in alignment 对线施工
execution in stages 分阶段施工
execution instruction 执行指令
execution interruption 执行中断
execution method 施工方法
execution mode 执行方式
execution module 执行模块
execution of construction work 工程实施
execution of contract 履约;履行合同;合同生效;执行合同;契约生效
execution of order 执行订单
execution of the budget 预算执行
execution of the order 执行订单
execution of the works 工程施工
execution of work 施工
execution order of node 节点执行次序
execution path 执行通路
execution period 完成周期
execution phase (指)令执行阶段;执行状态
execution pipeline 执行流水线
execution plan 工程实施计划
execution priority 执行优先权;执行优先级
execution process 行政措施
execution program(me) for works 施工程序;工序;工程进度表
execution route 执行路径
execution sale 强制拍卖

execution scheme drawing 施工计划图
execution step 执行步骤
execution time 执行时间;施工时间
execution unit 执行部件
executive 经理;执行者;执行管理人
executive agreement 行政协定
executive authorities 行政当局
executive board 执行委员会;执行局
executive branch 行政部门;执行部门
executive chairman 执行总裁
executive circuit 执行电路
executive command 执行命令
executive committee 执行委员会;常务委员会
executive communication 执行通信
executive component 执行部件;操作元件
executive construction organization design of long-term large-size and key project 长、大、重工程实施性施工组织设计
executive construction organization design of unit project 单元工程实施性施工组织设计
executive control function 执行控制功能;执行控制操作
executive control multiprogramming 行政管理多道程序设计;执行控制多道程序设计
executive control program(me) 执行控制程序
executive control routine 执行控制程序
executive control system 执行控制系统
executive control utility routine 执行控制实用程序
executive cost 施工费(用)
executive council 执行理事会
executive cycle 执行周期
executive decision 行政决策
executive deck 执行(程序)卡片组
executive diagnostic system 执行诊断系统
executive directors 执行董事会
executive dumping 执行转储
executive employment system 干部聘用制
executive facility assignment 执行设备分配
executive file-control system 执行文件控制系统
executive function 执行控制操作
executive guard mode 执行防护方式;执行保护方式
executive instruction 执行指令;管理指令
executive language control 执行语言控制
executive level 领导层;高管层
executive logging 执行记录
executive management 行政管理
executive mode 执行状态;执行方式
executive module 执行组件
executive monitor 执行监控器
executive officer 执行官员;副舰长
executive operating system 执行操作系统
executive order 执行指令
executive organ 执行机关
executive overhead 执行总开销
executive park (远离市中心的)市外商业机构办公区
executive phase 执行阶段
executive plan 执行计划
executive preparator 行政编制
executive program(me) 执行程序;管理指令;管理程序
executive program(me) component 执行程序成分;管理程序的组成
executive program(me) control 执行程序控制
executive pulse 执行脉冲
executive real-time system 执行实时系统
executive report 行政报告;执行报告
executive request 执行请求
executive resident segment 执行驻留段
executive routine 检验程序;执行程序;管理(例行)程序
executive salaries and bonuses 行政的薪金和津贴
executive schedule maintenance 执行调度保持
executive secretary 执行秘书;常务秘书
executive staff 行政职员;主管人员;高级职员
executive statement 执行语句
executive subsystem program(me) 执行子系统程序
executive summary 经营综合报告
executive supervisor 执行监控器;执行管理程序
executive system 执行系统;操作系统
executive system concurrence 执行系统并行性
executive system concurrency 执行系统并行能力

executive system control 执行系统控制
executive system function 执行系统功能
executive system multiprocessing 执行系统多重处理
executive system multiprogramming 管理系统多道程序设计
executive system of multiprogramming 多道程序的执行系统
executive system routine 执行系统程序
executive system utility 执行系统应用程序
executive termination 执行终止
executive time 执行时间
executive unit 执行单位
executive utility 执行应用程序;执行实用程序
executive veto 行政否决权
executive worker 行政人员
executor 指定遗嘱执行人;执行元件;执行程序
executory contract 待执行的契约;待执行的合同;尚待履行的合同
executory sale 未完成的销售
exedra 有凳门廊;半圆式露天建筑;开敞式座谈间
exedra arch 教堂前廊的拱
exedra arch impost 教堂前廊拱墩
exedra window 教堂前廊窗
exehange register 存储寄存器
exemplar 样件;样品;试样
exemplary damages 惩罚性赔偿;超过实际损失的赔偿
exemplary documentation 样板文件管理
exemplification of probate 经盖公章
exemplify 举例说明
exempt 免税人
exempt allowables 允许免税限度
exempt bonds 免除债券
exempt coastal zone 沿海豁免区
exempt concentration 最大允许浓度
exempt corporation 非课税法人
exempted discharge 免税排水
exempted exchange 免税交易
exempted from charges 免费
exempted from inspection 免检的
exempted goods 免税(进口)货物
exempted space 豁免吨位
exempt employees 免受限制的职工
exempt form customs examination 免于验关
exempt from commercial and industrial taxes 免除工商税
exempt from consumption tax 免征消费税
exempt from customs duty 免关税
exempt from customs examination 免验;免给
exempt from filing annual income tax returns 免办年度所得税申报
exempt from petroleum revenue tax 免缴石油收益税
exempt from the levying of tariff 免征关税
exempt goods 免税货物
exempt income 免于纳税的收入;免税所得额;免税收入;免除所得
exemption 免责;免税(法);免检;免除;豁免
exemption certificate 免税证(明)书;免税单;免除证书
exemption certification 免税证书
exemption clause 免罚条款;免税条款;免除条件;豁免条款
exemption form tax(ation) 免税
exemption for reinvested earnings 投资收益的免税
exemption from custom duty 关税减免
exemption from customs examination 免拆验
exempt(ion)from income tax 免除所得税
exemption from prosecution 免于起诉
exemption from the local income tax 免征地方所得税
exemption limit 免缴所得税界限
exemption mechanism 免税机制
exemption method 免税方法
exemption of debt 免偿债务
exemption of duty 免税放行
exemption of income 免税收益
exemption period 免税期
exemption point 免税点
exemptions of diameter and volume 直径和容积的免查
exemption system 免税制度

exemptions from safeguards 免于核监督
exempt pilot 非强制领航
exempt property 免税资产;免税地产;免税财产
exempt securities 免办登记证券
exempt solvent 豁免溶剂;符合环保法规要求的溶剂
exempt zone 免税区
exenteration 除脏术
exequatur (驻在国发给商务人员的)许可证书
exercise 练习;行使;实习
exercise area 演习区
exercise chart 练习海图
exercise class 实习课
exercise control 进行控制;进行操纵
exercise due diligence 谨慎处理
exercise en loge 自由设计图纸(法国)
exercise equipment 训练设备
exercise judgment 作出判断
exercise of authority 行使职权
exercise of boat's crew 救生艇水手操练
exercise option 行使期权
exercise power 行使权利
exerciser 练习程序;行驶职权的人;运动器械;演算器;操作器;受训练者
exercise the option 兑现选择权
exercise yard 活动场地
exergy 放射本领
exergy analysis 熵析
exeric reaction 放能反应
exernal layer 外层
exertion 行使职权
exert pressure 加压;施加压力
ex-factory 工厂交货
ex-factory inspection 出厂检验
ex factory pass rate 出厂合格率
ex-factory price 工厂交货价(格);出厂价(格)
exfiltrate 漏出
exfiltration 逐渐渗出;渗出
exfiltration pipe 渗水管
exfoliation mountain 页片剥离山
exfocal 焦外的
exfoliate 页状剥落;层离
exfoliated vermiculite 层状蛭石;膨胀蛭石
exfoliated vermiculite aggregate 膨胀蛭石集料;膨胀蛭石骨料
exfoliated vermiculite concrete 片状剥落的蛭石混凝土
exfoliated vermiculite-gypsum plaster 片状剥落的蛭石石膏灰泥
exfoliated vermiculite mortar 膨胀云母灰泥
exfoliated vermiculite plaster 膨胀蛭石灰浆
exfoliation 落屑;鳞片样脱落;页状剥落;叶状剥落;剥脱;剥离作用;表皮脱落;表皮剥脱;片状剥落
exfoliation corrosion 层蚀;剥落腐蚀;剥离腐蚀;片状剥落腐蚀
exfoliation dome 页状剥落丘
exfoliation joint 页状剥落节理
exfoliation phenomenon 剥离现象
exfoliative 剥落的
exfractor 顶架
ex-godown 卖方仓库交货价(格);仓库交货
ex-godown price 仓库交货价(格)
exgratia claims 道义索赔;通融索赔
exgratia payment 通融赔款
exhalant 蒸发性的;出管
exhalation 呼气;发散
exhalation deposit 喷气矿床
exhalation from oceanic ridge 大洋中脊喷出气
exhalation-sedimentary mineral deposit 喷气沉积矿床
exhalative process 喷气作用
exhale 蒸发管;放出;发散管
exhalent 蒸发性的
exhalite 喷气岩
exhanst gas outlet 烟气出口
exhaust 抽空;排气式;排风
exhaust advance 排气提前
exhaust afterburner 排气后燃(净化)器
exhaust after treatment device 废气后处理净化装置
exhaust air 余风;废气;排气;排净空气;排出空气
exhaust air box 排气室
exhaust air decontamination system 排气去污系统

exhaust air duct 排风(管)道;排气管道;废气管道
exhaust air fan 排气风机;排气风扇
exhaust air filter 排风过滤器
exhaust air grill(e) 排气格栅
exhaust-air monitor 排气监测器
exhaust air plenum 排风静压室
exhaust air pollution standard 排气污染标准
exhaust air port 排气闸门
exhaust air pre-filter 排气前置过滤器
exhaust air utilization 废气利用
exhaust air volume 排风量
exhaust airway 排风(管)道
exhaust alarm 排气警报器
exhaust algorithm 穷举法
exhaust analyzer 排气分析器
exhaust and intake manifold 排气及进气歧管
exhaust back pressure 排气反压力
exhaust bake-out 抽气烘干;排气烘烤
exhaust beater 排气式打手
exhaust belt 排气总管
exhaust blading 排气叶片
exhaust blanking plate 排气管扩板;排气管护板
exhaust blast 废气喷射
exhaust blow down loss 排气喷泄损失
exhaust blower 抽气机;抽风机;排气机;排气风箱;排风机
exhaust boiler 废热锅炉;废气锅炉
exhaust box 减音器;排气箱
exhaust brake 排气制动;排气闸
exhaust brake indicator lamp 排气制动指示灯
exhaust braking 排气受阻
exhaust branch pipe 排气支管
exhaust by-pass valve 排气旁通阀
exhaust by-pass valve control wire switch 排气旁通阀控制线转环
exhaust cam 排气凸轮
exhaust cam shaft 排气凸轮轴
exhaust canopy 排气罩
exhaust capacity 抽气能力
exhaust casing 排汽缸
exhaust cavity 排气穴
exhaust center 中立泄流
exhaust chamber 流出腔;抽气箱;抽风箱;排液腔;排气箱;排出室
exhaust chest 排气箱
exhaust chimney 排气烟囱;排气塔
exhaust choke 排出节流
exhaust clack 排气活门;排气瓣
exhaust close 排气停止;排气开关;停止排气(阀门)
exhaust coefficient 排气系数
exhaust collector 排气总管;排气收集器
exhaust collector pipe 排气歧管
exhaust collector ring 排气集合环
exhaust column 抽风管
exhaust condition 排气状态
exhaust conditioning 废气调节
exhaust conditioning box 排气净化箱
exhaust cone 喷口调节锥;调整针案
exhaust connecting branch 排水接户支管
exhaust contaminant 排气污染物
exhaust cooling method 排气冷却法
exhaust cover 排气口罩
exhaust curve (示功图的)排气曲线
exhaust cut-out 排气停截阀;排气闭路阀;排气(消声器)开关
exhaust cycle 排气循环
exhaust damper 排气风门;排尘浮动门
exhaust deflecting ring 排气偏转环
exhaust deflection control 排气偏转控制
exhaust deflector 排气偏导装置
exhaust detonation 排气爆震
exhaust diffuser 排气扩压器
exhaust draught fan 排气风扇
exhaust-driven 排气式;排气传动
exhaust-driven gas turbine 排气涡轮;废气涡轮;废气驱动式汽轮机
exhaust-driven supercharger 排气驱动增压器;废气驱动增压器
exhaust-driven turbo-charger 废气涡轮增压器;废气涡轮增压机
exhaust duct 排气通道;排风管;排气风筒
exhaust duct ventilation 排气通风(法)
exhaust dust device with bag 袋式除尘装置

exhausted 耗尽;用过的;无用的
exhausted air quantity 废气量
exhausted area 采竭地区
exhausted bath 废液
exhausted borrow pit 废取土坑
exhausted cell 放完电的电池
exhausted edition 绝版
exhausted enclosure 密闭罩;废外壳
exhaust edge 排气缘
exhausted liquid 废液
exhausted liquor 废液
exhausted lye 废碱液
exhausted soil 耗尽土壤;耗竭土(壤)
exhausted solution 废溶液
exhausted water 废液;废水
exhausted well 已枯竭的井;枯竭井
exhaust ejector primer pump 排气喷射引水泵
exhaust elbow 排气弯管
exhaust emission 废气污染(物);废气排放
exhaust emission analyzer 排气分析器
exhaust emission characteristics 废气污染特征
exhaust emission limit 废气污染限值
exhaust emission target 排气净化目标
exhaust end 排料端
exhaust equipment 排风设备
exhauster 燃气排送机;引风机;抽气装置;抽风机;排送机;排气器;排气机;排尘机;水力吹风管
exhauster chamber 排气箱
exhauster draft fan 排气器通风风扇
exhauster draught fan 排气风扇
exhauster governor 真空泵调压器
exhauster regulating valve 真空泵调节阀
exhaust expansion 排气膨胀
exhaust explosion 排气管爆燃
exhaust extension 排气管延伸
exhaust fan 抽风机;排气(风)扇;排气扇;排风机
exhaust fan room 排风机室
exhaust feed heater 废热加热器;废气给水加热器;排气加热器
exhaust filter 排气滤器
exhaust-fired-boiler combined cycle 排气补燃锅炉联合循环
exhaust fitting 排气管接头
exhaust flame 排气火焰
exhaust flame damper 排气管阻焰器
exhaust flaming 排气着火
exhaust flap 排气瓣;排气阀
exhaust flow 废气流
exhaust flue 排气风管
exhaust from cooler 冷却余风
exhaust from mill 磨机排出的余风
exhaust fume 排烟;排出的废气
exhaust fume hood 毒气柜;通风橱;去烟橱
exhaust gas 排出的气体;废气;排(出的废)气
exhaust gas afterburner 排气补燃器
exhaust gas analysis 排气分析
exhaust gas analysis system 排气分析系统
exhaust gas analyzer 排气分析器;排气分析仪;废气分析器
exhaust-gas boiler 排气锅炉
exhaust gas cabinet 排气柜
exhaust gas catalytic converter 废气催化转化器;排气催化转化器
exhaust gas catalytic reactor 排气催化反应器
exhaust gas catalytic system 排气催化系统
exhaust gas composition 排气成分
exhaust gas converter 排气转化器
exhaust gas counter pressure 排气背压力
exhaust gas desulfurization 排烟脱硫;排气脱硫
exhaust gas disposal 排气处理
exhaust gas drilling 废气洗井钻进
exhaust gas duct 排气风箱;排气风筒
exhaust gas economizer 废气预热器
exhaust gas filter 废气过滤器
exhaust gas from car 汽车排气
exhaust gas from incinerator 焚烧炉排气
exhaust gas heat-recovery equipment 废气热回收设备
exhaust gas instrumentation 排气分析仪表
exhaust gas intake port (气波增压器)排气进口
exhaust gas muffler 排气消音器
exhaust gas plant 废气处理装置
exhaust gas purification 废气净化
exhaust gas purifier 排气净化器;废气滤清器

exhaust gas purifying equipment 废气净化装置
exhaust gas recirculation 废气(再)循环;排气(再)循环
exhaust gas recirculation controller 排气(再)循环控制器
exhaust gas recirculation device 废气再循环装置
exhaust gas recirculation valve 废气再循环阀
exhaust gas recirculation 废气环流
exhaust gas recirculation pipe 汽车漏气回流管
exhaust gas recycling 排气循环
exhaust gas recycling system 排气再循环系统
exhaust gas sampling 排气取样
exhaust gas scrubber 废气洗涤器
exhaust gas silencer 废气消音器
exhaust gas smoke cleaner 排气黑烟滤清器
exhaust gas suction fan 抽气式吸风机
exhaust gas system 排气系统
exhaust gas temperature 排气温度;废气温度
exhaust gas turbine 废气涡轮机;废气透平;废气汽轮机;废气燃气轮机
exhaust gas turbine supercharger 排气涡轮式增压器
exhaust gas turbocharger 废气涡轮增压器;废气涡轮增压机
exhaust gas water heater 废气热水器
exhaust gear 排气装置
exhaust grill(e) 排气算子;排气(格)孔;排气(格)眼;排气栅格;排气窗
exhaust guidance 排气导管
exhaust head (蒸汽机的)排气头;排气头
exhaust header 排气集管
exhaust heat 废热
exhaust heat boiler 余热锅炉
exhaust-heated atomizer 排气加热喷雾器
exhaust-heated cycle 废气加热循环;废气回热循环;废气加热循环
exhaust-heated open cycle 排气供热开口循环
exhaust heater 排气加热器
exhaust heat exchanger 废气热交换器
exhaust heating 排气加热
exhaust heat loss 排气热损失
exhaust heat recovery boiler 废热锅炉
exhaust hole 排气孔
exhaust hole in gable 山墙排气孔
exhaust hood 吸气罩;拔气罩;排汽室;排汽缸;排气;排风罩;排烟罩
exhaust horn 排气角管
exhaust hose 排气软管;排气胶管
exhaustible resources 可耗竭资源;可耗尽资源;有限资源
exhausting and blowing combined ventilation system 吸压混合式通风
exhausting ductwork 排气管网
exhausting exit 放气孔
exhausting labo(u)r 繁重的体力劳动
exhausting machine 抽气机
exhausting section of column 塔的汽提段【化】
exhausting system 抽风系统
exhausting the air in brake cylinders 放尽制动汽缸余气
exhausting tower 抽气塔
exhausting tube sealing machine 玻壳接管机
exhausting-type ventilation 吸出式通风
exhausting with excessive pressure 超压排风
exhaust inlet casing 废气进气壳
exhaust installation 排气装置
exhaustion 耗尽;耗竭;用尽;抽空;彻底研究;衰竭
exhaustion degree 抽空度
exhaustion expenditures 消耗性支出
exhaustion hypothesis 耗竭学说
exhaustion of effect 有效期满
exhaustion of groundwater resources 地下水源枯竭
exhaustion of soil 地力衰竭;土壤衰竭;土壤耗竭;土地肥力下降
exhaustion period 亏耗期
exhaustion plate 排气板
exhaustion range 耗损范围
exhaustion region 耗尽区
exhaustive expenditures 消耗性经费;消耗性支出
exhaustive exploitation 过度开采
exhaustive farming 滥耕
exhaustive fluorinated ethylene propylene 聚全氟乙丙烯树脂

exhaustive grouping 穷举归组
exhaustive index 穷举索引
exhaustive list 详细清单
exhaustive methylation 摧毁性甲基化作用;彻底甲基化
exhaustive search 穷竭搜索
exhaustive test 抽气试验;全面性试验;穷竭检查
exhaustive testing 穷举调试
exhaustive work 详细工作清单
exhaustive yield 枯竭性抽水量
exhaustivity 穷举
exhaust jacket 废气套;排气加热器
exhaust jacketed carbureter 废气套预热汽化器
exhaust jet stream 排气射流
exhaust lag 排气滞后
exhaust lap 排气余面;排气(侧)余面
exhaust lead 排气导程;提前排气
exhaust line 耗损线;排泄管(道);排汽管路;排气线;排废线;排出(曲)线
exhaust loss 排气损失
exhaust main 排气总管
exhaust manifold 排气歧管;排出管汇
exhaust manifold companion flange 排气歧管连合凸缘
exhaust manifold gasket 排气歧管密封片
exhaust manifold heat valve 排气歧管加热阀
exhaust manifold jacket 排气歧管套
exhaust manifold reactor 排气歧管反应器
exhaust muffler 排气消音器;排气消声器
exhaust muffler cut-out 排气消声器开关
exhaust noise 排(废)气噪声
exhaust note 排气音调
exhaust nozzle 排气(喷)嘴;排气喷管
exhaust nozzle control 排气口控制
exhaust of engine 发动机排气侧
exhaust open 开放排气;排气打开
exhaust opening 吸风口;排气口;排气孔;排气开度
exhaustor 水力吹风管
exhaust orifice 排气口;排气孔
exhaust outlet 排气孔;放油孔;废气口;排水孔;排气出口;排风口;排出支管
exhaust outlet casing 废气出口壳
exhaust partition wall 排气隔板
exhaust passage 出风口;排气通路;排气(通)道
exhaust period 排气周期
exhaust phase 排气相
exhaust pipe 排水管;放空管;出气管;排气管;排气阀
exhaust pipe blower connection 废气管送风器接头
exhaust pipe blower nozzle 废气管送风器喷嘴
exhaust pipe bracket 排气管托架
exhaust pipe clamp 排气管夹
exhaust pipe combination 废气管接头
exhaust pipe extension 排气管延伸
exhaust pipe flange gasket 排气管凸缘密封片
exhaust pipe flange seal 排气管凸缘气封
exhaust pipe nozzle 排气管喷嘴
exhaust pipe packing 排气管填密物
exhaust pipe shield 排气管罩
exhaust pipe support 排气管支架
exhaust pipe with hinged flaps 铰接摺瓣排气管
exhaust pipe with intermediate nozzles 中间带喷嘴排气管
exhaust piping 废气管道
exhaust piston 排气活塞
exhaust pit 排气坑
exhaust plug 排气口堵
exhaust plumbing 排气管道装置
exhaust point 抽汽点;排汽点
exhaust pollutant 排气污染物
exhaust pollution control 排气污染控制
exhaust port 排出孔;排气小炉;排烟口;排气门;排气口;排出口
exhaust port deposition 排气口积炭
exhaust port ignition cleaner 排气管点火废气净化系统
exhaust power fluid 废动力液
exhaust pressure 排气压力;出口压力;排出压力
exhaust pressure modulated exhaust gas recirculation valve 排气压力控制型废气再循环阀
exhaust pressure ratio passage 排气压比
exhaust price 押金缺少时的价格
exhaust primer 废气引水装置

exhaust process 排气过程
exhaust product 排出的燃烧产物
exhaust pulse air injection (抽吸式)排气脉冲空气喷射
exhaust pulse pressure-charging 废气脉冲增压
exhaust pump 排气泵
exhaust purifier 废气滤清器;排气净化器
exhaust pyrometer 排气高温计
exhaust-recirculation method 排气循环法
exhaust regulating ring 排气调节环
exhaust regulator 排气调节器
exhaust reheat(ing) 排气再热
exhaust reservoir 排气储筒
exhaust resistance 排气阻力
exhaust ring 排气圈
exhaust riser 排气口
exhaust roar 排气噪声
exhaust rocker 排气摇杆
exhaust rod 排汽阀拉杆
exhaust room 排气室
exhaust schedule 排气规范
exhaust screen 排气滤网
exhaust scrubber 废气洗涤器;废气滤清器
exhaust shaft 排气(竖)管;排气井;排风竖井
exhaust shroud 排气管套
exhaust side 熔窑放液侧;排气侧
exhaust side of engine 发动机排气侧
exhaust side trap 排出侧阱
exhaust silencer 排气消音器;排气消声器
exhaust slot 排出口
exhaust smoke 排出烟雾;排(出的)烟
exhaust smoke denitrogeneration 排烟脱氮
exhaust smoke density 排烟浓度
exhaust smoke ejector 排烟喷射器
exhaust smoke level 排(出的)烟量
exhaust smoke window 排烟窗
exhaust snubber 排气消声器
exhaust sound 排气声音
exhaust space 排气室
exhaust stack 排气烟囱;排气器;排风塔
exhaust stacking 排气管;抽风管
exhaust stage 排汽级
exhaust stage blade 末级叶片
exhaust stator blade 排气导流叶片
exhaust steam 废(水)蒸汽;废汽;排汽;排出蒸汽
exhaust steam casing 排汽缸
exhaust steam end 排汽端
exhaust steam from steam hammer 汽锤废汽
exhaust steam heating 废汽供暖;废气供暖;废气采暖;排气供热
exhaust steam injector 乏汽注水器;排汽喷射器;排气喷射器
exhaust steam inlet 排汽进口
exhaust steam jet 废汽喷口
exhaust steam main 总排汽管;排汽总管
exhaust steam moisture 排汽湿度
exhaust steam nozzle (机车的)排汽喷嘴
exhaust steam passage 废汽通道
exhaust steam pipe 排(蒸)汽管;废(蒸)汽管
exhaust steam port 排汽孔
exhaust steam preheater 排汽预热器;乏汽预热器
exhaust steam pressure 排汽压力
exhaust steam separator 废汽分离器
exhaust steam supply pipe 排汽供给管
exhaust steam turbine 废汽汽轮机;废汽涡轮机;废汽透平
exhaust steam valve 排汽阀
exhaust strain 废气滤清器
exhaust stream 排出气流
exhaust stroke 排气行程;排气冲程
exhaust suction pipe 排气喷射器
exhaust suction stroke 排气吸气冲程
exhaust system 排气系统;放气装置
exhaust system of ventilation 吸出式通风;通风排气系统
exhaust tank 排气罐
exhaust technique 排气技术
exhaust temperature ga(u)ge 排气温度计
exhaust thrust 排气推力
exhaust time 排气时间
exhaust tower 排气塔
exhaust trumpet 喇叭形排气管
exhaust trunk 排气总管
exhaust tube 抽风管

exhaust tumbling barrel 抽气摇摆滚磨筒;排气滚净筒
exhaust tumbling mill 排气滚净筒
exhaust turbine driven supercharge 排气涡轮传动增压器
exhaust turbo-blower 排气轮机增压器
exhaust turbo-compressor 排气汽轮式压缩机
exhaust type steam turbine 排汽式涡轮
exhaust unit 排气装置
exhaust valve 放空阀;排气活门;排气阀;排放阀;排出阀
exhaust valve cam 排气阀凸轮
exhaust valve cap 排气阀盖
exhaust valve clearance 排气阀间隙
exhaust valve guide 排气阀导管
exhaust valve head 排气阀头部
exhaust valve insert 排气阀镶座
exhaust valve lifter 排气阀挺杆
exhaust valve lifting cam 排气阀凸轮
exhaust valve lifting gear 排气阀升降装置
exhaust valve mechanism 排气阀机构
exhaust valve pin 排气阀销
exhaust valve regulation 排气阀调整
exhaust valve regulator 排气门调节器
exhaust valve spindle 排气阀轴
exhaust valve spring 排气阀弹簧
exhaust valve stem 排气阀杆
exhaust valve tappet 排气阀挺杆
exhaust velocity 排气速度
exhaust vent 出风孔;排气孔
exhaust ventilating duct 排气通风管
exhaust ventilation 抽(气通)风;抽出式通风;排气通风
exhaust ventilation system 抽出式通风系统;排气通风系统
exhaust ventilator 排气通风装置;排气通风筒;排气器
exhaust volume 排气容积;排气量
exhaust water pipe 放水管
exhaust wetness 排气湿度
exhaust whistle (排)气笛
exhibit 展览品;展出;陈列品
exhibition 奖学金;显示;展览(会);呈现;陈列
exhibition architecture 展览建筑
exhibition area 展览区
exhibition block 展览区段;展览大楼
exhibition building 展览馆;陈列馆
exhibition car 展览车
exhibition center 展览中心
exhibition drawing 展览(的)图片
exhibition gallery 展览廊
exhibition garden 展览花园
exhibition goods 展览品
exhibition ground 展览场地
exhibition hall 展览厅;展览馆;陈列馆
exhibition model 展览模型
exhibition of applied art 应用艺术展览
exhibition palace 展览宫;展览大厦
exhibition park 展销会场
exhibition pavilion 展览大帐篷
exhibition place 展览场所
exhibition print 展览图
exhibition room 展览室;陈列室
exhibition sale 展销
exhibition specimen 陈列标本
exhibition stand 展览看台
exhibition vessel 展览船
exhibitor 展出单位;提出人
exhilaration 经济兴旺
exhuast gas desulfurization 废气脱硫
exhumation 发掘;剥露作用
exhume 掘墓;掘出;发掘
exhumed landscape 剥露景观
exhumed monadnock 裸露残丘
exhumed topography 剥落地形;剥露地形
exichnia 外迹
exide accumulator (一种牵引设备用的)蓄电池
Exide ironclad battery 铠装铅锑蓄电池
exigence 急事
exigent machine check 紧急机器检验
exiguous detail 负载过少
ex importation dock price 入口港栈货价
exinite 壳质组;壳质煤素质
exinoid 类壳质的

exinoid group 壳质组
exinonigritite 壳质煤化沥青
ex interest 不带利息;无利息
existence doubtful (指图示水下障碍物)疑存
existence mode of dissolved element 水中元素存在形式
existence of solution 解的存在性
existence theorem 存在定理
existent corrosion 原在腐蚀
existent gum 实际胶质
existential closure 存在闭包
existentialism 自觉存在论
existentially quantified variable 存在量词化变元
existential quantifier 存在量词
existential specification 存在规范
existent state of moisture 水分的存在状态
existing 已建的
existing amount of budgetary expenditures 预算支出限额
existing assets 现存财产
existing ballast 现有道砟
existing bed 原有河底
existing berth 现有泊位
existing building 现有建筑(物);原有房屋;已建的建筑
existing capital goods 现存资本商品
existing circumstance 现状;现况;实际状态;实际状况
existing code 现行规范
existing combined sewerage system 现有合流管网系统
existing company 现存企业;现存公司
existing condition 现状;现有条件;生活情况;生存条件
existing condition assessment of river environmental quality 河流环境质量现状评价
existing construction 现有建筑
existing construction map 现有结构图
existing corrosion 现存腐蚀
existing data 现有资料
existing data sources 现有资料来源
existing design 现有设计
existing double-track line 既有双线铁路
existing enterprise 现存企业
existing environment 环境现状
existing equipment 现有(生产)设备;现有设施;原有设备
existing facility 现有设施;现有设备;原有设备
existing flow 现有流量
existing gallery 既有巷道
existing grade 现有(地面)高程
existing ground 原地面
existing ground level 原地面高程;原地面标高
existing ground line 原地面线
existing ground surface 原地表面
existing harbo(u)r 已有港口
existing house 现有房屋;原有房屋
existing housing 现有住房
existing housing stock 现有住房量
existing land-use map 土地使用现状图;土地利用现状图
existing land-use pattern 土地使用现状
existing law 现行法律
Existing Legislation Clause 现行法律条款
existing line 既有线
existing locomotive amount statistics 机车现有数统计
existing machinery 现有机械
existing main line network 既有干线网
existing masonry 现有圬工
existing material condition of production 现成的物质生产条件
existing materials 现有材料
existing mortgage 现行抵押(契据)
existing multifamily rental housing 现有供多户居住的出租房屋
existing perforation 老炮眼
existing pipework 现有管网
existing plan 既定方案
existing plant force 现有工厂能力
existing preferential duties 现行特惠关税
existing preferential tariff 现存特惠关税
existing price 现行价格
existing product 现有产品

existing production capacity 现有生产能力
existing railroad 既有铁路;现有铁路
existing railway 既有铁路
existing rain gauge networks 现有雨量计网
existing rate 目前的工资率
existing residential building-permanent 现存永久性住宅
existing residential building-temporary 现存临时性住宅
existing riverbed 原有河底
existing road 现状道路;现有道路
existing route 现有路线
existing seabed depth 原海底深度
existing seismic-resistant capacity 现有的抗震能力
existing seismic-resistant design provision 现有抗震设计规定
existing settlement 原有沉降
existing ship 现有船
existing stock (现期)存货
existing structure 既有建筑物;现有建筑结构
existing submersible 现有潜水器
existing track 既有线
existing traffic 目前运量;现有运量;现有交通
existing traffic system 现有交通系统
existing traffic volume 现有交通量
existing tree 原有树木
existing type 现有形式
existing use 现行用途
existing utility 现有设施;现有公用(事业)设施
existing value 现存值
existing wastewater treatment plant 现有污水处理厂
existing water-course 原有水道
existing water table 现有水位
existing waterway 现有河道;原有水道
existing well 现有井
exist-voice model 退出呼声模型
exit 引出端;延迟出口;出射;出口(指水、气、汽、油等);安全出口;疏散口;安全门
exit access 出口引道
exit age distribution 出口龄期分布
exit air 排出空气;空气排出口
exit air chimney 排气烟囱
exit air duct 排气管道
exit air fan 排气风扇
exit air opening 排气出口
exit air shaft 出气竖井
exit air station 排气站
exit analysis 出口分析;废气分析
exitance 出射率;出射度
exit and entrance control 控制进出
exit angle 出射角;出口角(度)
exit aperture 出口光阑
exit area 喷管出口截面(面)积
exit area of nozzle 喷嘴出口面积
exit area ratio 喷管面积比
exit barrier 出口检票栅门
exit beam 引出束
exit beam deflection 出射光束的偏转
exit blade angle (螺旋桨)叶片出口角
exit branch 出水支管;出口管;出口支路
exit bridle 出口张紧装置
exit button 出口按钮
exit button relay 终端按钮继电器
exit capacity 出口能力;安全门容量
exit channel 出水渠
exit cleaning 废气净化;废气除尘
exit coefficient 出口流量系数;排气系数
exit condition 出口条件
exit cone 出气整流锥;出口锥管
exit connection 出口连接
exit control alarm 出口控制警钟;出口警报器
exit corridor 出口门廊;出口廊道;疏散走廊
exit curve 引出曲线
exit design 出口设计
exit designator 出口指示器
exit device 出口设施;(逃生门)安全(门)装置;紧急出口装置
exit diameter 出口截面直径
exit direction 出射方向
exit direction sign 出口方向标志
exit discharge 出口疏散;出口场地
exit distance 安全疏散距离
exit door 出口门;安全出口门;太平门

exit dose 出射剂量;出端剂量
exit drills in the home 撤离住宅训练
exit dust 排出粉尘
exit emergency 紧急出口
exit end 出射端
exit-end water saturation 出口端含水饱和度
exit face 出口面
exit facility (道路的)引出设备;出口设施;安全门设施
exit flue 排烟(烟)道;排烟管(道)
exit from a story 楼层出口
exit gap 翼隙出口
exit gas 出气;排出气
exit gas installation 排气烟囱
exit gas line 排气线路
exit gas pipe 排气管
exit gas pipeline 排气管路
exit gas recycling 废气循环
exit gas socket 排气管座;排气管套
exit gas system 排气系统
exit gas temperature 排气温度
exit gate 出闸机;出口大门
exit gradient 出逸梯度;出逸坡降;出口坡降;出口坡度
exit guide 出口导板
exit guide blade 出口导叶
exit guide vane 出口导叶
exit handle 退出处理程序
exit handler 出口处理程序
exit hardware 出口金属构件
exit head loss 出口水头损失
exit hole 出口孔
exit identification 出口标志
exit illumination 出口照明;太平门照明
exit indicator 出口指示器;出口标示牌
exitine 外壁内层
exit instruction 引出指令;出口指令
exit knob 出口旋钮
exit label 出口标号
exit lamp 太平灯
exit lane 出口走廊;驶出车道
exit light 出口灯;安全出口灯
exit lighting 出口指示灯;出口照明;太平门指示灯
exit light outlet 安全出口灯出线口
exit lip 引出刃
exit list 出口表
exit lock 太平门锁止装置
exit loss 出口(水头)损失;输出端损失;输出端损耗
exit macroinstruction 出口宏指令
exit marking 出口标记
exit mine air 矿坑空气出口
exit nozzle throat 喷管临界截面
exit of intersection 交叉口出口
exit of stab wound 刺出口
exit opening 安全窗;太平门;屋顶窗;安全门
exit orifice 出口孔
exit passageway 出口通路;出口通道;安全通路;安全通道;疏散通道
exit piler 出口堆板机
exit pit 出口坑
exit plan 出口平面图
exit plane 出口面
exit point 出站口岔;出口端;引出点;逸出点;出口点
exit port(al) 出射口;出口宽度;出口孔
exit position 出口位置
exit pressure 出口压力
exit pressure drop 出口压降
exit probability 出射机率
exit procedure 出口过程;出口程序
exit program(me) 出口程序
exit ramp 下管滑道;出口坡道;驶出匝道
exit ramp closure 关闭出口匝道
exit ramp control 出口匝道控制
exit ramp marking 出口匝道标示
exit ramp of highway 出公路的匝道
exit rated ingredient ore 出矿配矿
exit ray 出射光线
exit recommendation 规劝驶出
exit region 急流出口区
exit relay 终端继电器;出口继电器
exit ring 出口圈
exit river 泄流河
exit road 驶出道路;出口道路
exit roadway 出口车行道

exitron 激励管
exit route 出口路线;发车进路
exit routine 出口(例行)程序
exit section 出洞段
exit separation 出口间距
exit side 出料侧;出口侧
exit sign 出口标志
exit signal 出站信号【铁】
exit signal(l)er 出站信号机
exit skirt 出口扩张部;出口的扩张部分
exit sliproad 出口岔道
exit slit 出射狭缝
exit speed 逸出速度;出闸速度;出水速度;驶出速度
exit speed at retarder 缓行器出口速度
exit stairway 出口楼梯
exit statement 出口语句
exit stream 泄流河
exit switch actuator 出纸口开关致动器
exit taxiing (飞机)起飞前滑行
exit taxiway (飞机)出口滑行道
exit temperature 出射温度;出口温度
exit temperature of clinker 熟料排出温度
exit terminal 出口终点
exit time 逸出时间;出闸时间;出水时间;驶出时间
exit track 出站线;出发线【铁】
exit travel distance 出口疏散距离
exit tray 出纸口接盘
exit tube 泄管;出管
exit tunnel 出口隧道
exit turn 转弯驶出
exitus papilla 出口乳头状突起
exit velocity 逸出流速;出口流速;出口(处)速度;输出速度;驶出速度
exit velocity triangle 出口速度三角形
exit-visa 出境签证
exit visibility 出口可视
exit wall 磨尾;出料端墙板
exit warning sign 出口警告标志
exitway 出口通路
exit weir 出口堰
exit width 安全出口宽度
exit window 初射窗;出射窗
exit zone 出口(区)段
ex lighter 驳船交货
ex lighter port of arrival 目的港船上交货价
exline correction 偏流线修正
ex meridian 近子午线的;近中天;傍午;近午
ex-meridian altitude 近子午圈高度;近中天高度
ex-meridian below pole 下近中天
ex-meridian observation 近子午圈观测
ex-meridian zenith distance 近子午圈天顶距
Exmet 金属网牌号(一种用于加强砖工的)
ex-mill 工厂交货
ex mine 矿山交货(价);矿场交货
exoatmosphere 外大气层
exoatmospheric radiation 外大气层辐射
exoatmospheric rays 外大气层射线
exobiology 地球外生物学
exobiophase 外生物相
exochomophyte 石面植物
exo-cis-addition 外向-顺式加成(反应)
exocondensation 支链缩合;外缩作用;外缩成环
exoconfiguration 外向构型
exocontact 外部接触
exodeviation 外转
exodiagenesis 外生成岩作用;表生成岩作用
exodyke 沉积岩墙
exoelectric(al) 放电的
exoelectric(al) reaction 放电反应
exo-electron 外电子
exo-electron dosimeter 外逸电子剂量计
exoenergic 放能的
exoenergic nuclear reaction 放能核反应
exoenergic process 放能过程
exoenergic reaction 放能反应
exoergic 放能的
exoergic collision 第二种碰撞
exoergic process 释能过程
exoeric 放热的
exoeric nuclear reaction 放能核反应
exoexine 外表层
exogas 放热性气体;放热型气体
exogeneous ore deposit 外生矿床

exogenestic movement 外生运动
exogenetic 外生的;外成的
exogenetic action 外成作用
exogenetic deformation 外成变形
exogenetic force 外(营)力
exogenetic inclusion 外来包体【地】;外成包体
exogenetic joint 外成节理
exogenetic mesomixis 消耗过高
exogenetic metallization 外生成矿作用
exogenetic metallogenic formation 外生成矿建造
exogenetic ore deposit 外生矿床
exogenetic process 外成作用
exogenetic rock 外成岩
exogenetic sediment 外成沉积
exogenic 非同缘的;外生的;外成的
exogenic action 外力作用
exogenic arenitic lava 外生砂屑熔岩
exogenic boulder lava 外生巨角砾熔岩
exogenic breccia lava 外生角砾熔岩
exogenic colo(u)r 外色
exogenic force 外(营)力
exogenic geologic(al) process 外动力地质作用
exogenic inclusion 外源包体
exogenic ores 外生矿床
exogenic process simulating 表生作用模拟实验
exogenic subsidence 外因沉降
exogenic toxicosis 外因性中毒
exogenous 外源的;外生的;外来的;外成的
exogenous action 外生作用
exogenous branching 外长分枝
exogenous cave 外生洞穴
exogenous cycle 外生环
exogenous dome 外成穹丘;外成穹隆
exogenous ejecta 外成变形
exogenous electrification 外源起电
exogenous environmental factor 外生环境因素
exogenous factor 外界因素
exogenous failure 外因故障
exogenous force 外营力
exogenous formation 外生形成
exogenous impurity 外来杂质
exogenous inclusion 捕虏体【地】;外源包体
exogenous infection 外源传染
exogenous metabolism 外源代谢
exogenous metallic inclusion 外来金属夹杂物
exogenous microognism 外源微生物
exogenous mine fire 井下外因火灾
exogenous non-metallic inclusion 外来非金属夹杂物
exogenous organic matter 外源有机物
exogenous origin 外生源
exogenous plant 外长植物
exogenous pollutant 外源污染物;外生污染物
exogenous process 外因作用【地】;外力作用
exogenous rocks 外成岩类
exogenous solidification 外生长凝固
exogenous subsidence 外因沉降
exogenous toxin 外生素
exogenous tree 外长树
exogenous uranium ore 外生铀矿
exogenous variable 局外变量;外生变数;外变量
exogenous zeitgeber 外源同步因子
exogens 外生性植物
exogen tree 外长树;有年轮树
exogeosynclinal 外地槽的
exogeosyncline 外枝准地槽;外(枝副)地槽;外源准地槽;外坡准地槽
exoglyph 层内痕
exoil field cost 油田成本
exokinematic fissure 外生裂隙
exokinetic 外成的
exokinetic fissure 外动力裂缝;外成裂缝
exokinetic joint 外成节理
exo-mantle 外地幔
exometamorphism 外变质(作用)
exomigmatization 外质混合岩化作用
exomigmatization way 外源混合岩化方式
exomomental 发射脉冲的
exomorphic 外变质的
exomorphic zone 外(接触)变质带
exomorphism 外变质(作用)
exomorphosed 外变质的
exonarthex 外门厅
exoneration 解除负债

exoneration clause 免罚条款;例外条款
exonym 外来惯用名
exoolitic 外生鲕状的
exo-Pacific tectonic zone 环太构造外带
exoperistome 外齿层
exopexy 外固定术
exopolygene 外生多源包体
exorbitant expenditures 消耗过高
exorbitant price 过高的价格;非法价格
exorbitant profit 超额利润;非法利润
exordium 序言
exorheic basin 外流盆地;外流流域
exorheic lake 排水湖;外流湖
exorheic region 河流通海地区;外流区
exorheism 外洋流域
exoskeletal plate 骨片
exoskeleton 外骨骼
exosketetal construction 骨架式建造术
exosmometer 外渗测定器
exosmose 外渗(现象)
exosmosis 外渗(透)
exosphere 外逸层;外大气圈;外(大)气层
exospheric phenomenon 外大气层现象
exostructure 外壳载的结构
exotemp 发热回火
exotherm 放热曲线
exothermal 放热的
exothermal change 放热变化
exothermal reaction 放热反应
exothermic 放热的
exothermic animal 外热动物
exothermic atmosphere 放热型保护气氛;放热式气氛
exothermic auxiliary reaction 放热副反应
exothermic change 放热变化
exothermic character 放热特性
exothermic clinkering process 形成熟料的放热过程
exothermic compound 放热化合物;发热剂
exothermic curve 放热曲线
exothermic decomposition 放热分解
exothermic effect 放热效应
exothermic ferro alloy 铝热法炼制铁合金
exothermic gas 放热型气体;发热气体
exothermic heating 放热加热
exothermicity 放热性
exothermicity gas 放热性气体
exothermic mixture 发热混合物;发热混合料
exothermic nuclear reaction 放热核反应
exothermic padding 发热贴片
exothermic paint 发热涂料
exothermic peak 放热峰
exothermic process 放热过程;散热过程
exothermic reaction 放热反应
exothermic reducing reaction 放热还原反应
exothermic riser sleeve 放热冒口套
exothermic smelter shop 金属热熔炼车间
exothermic welding 热熔焊;铝热焊(接)
exotic 外来货币;外来的
exotic atmosphere 异国情调
exotic block 浮动岛;外来岩块
exotic breed 外来品种
exotic composition 特殊高能燃料
exotic disease 外来病
exotic environment 外部环境
exotic fashions 异国风尚
exotic fish 外来鱼
exotic fuel 进口燃料;稀有燃料;高热值燃料
exotic insect 侵入害虫
exoti(ci)sm 异国情调;外来语
exotic mantle materials 外来地幔岩质
exotic material 特殊材料
exotic plant 引入的植物;外地植物
exotic product 外国产品
exotic river 外源河
exotic solvent 特种溶剂
exotic species 外来(物)种
exotic stream 外源河
exotic terrane 异源地体
exotic tree 引进树种
exotic tree species 外来树种
exotoxin 外毒素
exotrophy 外向性
expand 扩张;胀形;胀管口;展宽

expandability 可扩充性;膨胀性;膨胀度
expandable bathythermograph 吊挂式温深仪
expandable clay 膨胀黏土
expandable form 展开式模板
expandable fund 扩充基金
expandable instruction set 可扩充指令系统
expandable material 膨胀性材料
expandable plastics 可膨胀塑料;可发性塑料;充气塑料
expandable polystyrene 膨胀聚苯乙烯;可发(泡)性聚乙烯;发泡聚苯乙烯
expandable polystyrene bead 可发性聚苯乙烯珠粒
expandable polystyrene board 发泡聚苯乙烯板
expandable polystyrene pellet 可发性聚苯乙烯颗粒
expandable ring support 可展式环形支护
expandable shell (杆柱)膨胀套壳
expandable spreader 扩张式吊具;集装箱可伸缩吊具;吊缩式吊架
expandable starter system 可扩充起动器系统
expandable thermoplastic microsphere 可膨胀热塑性微球体
expandable tip pile 扩端桩
expanded aggregate 膨胀集料;膨胀骨料
expanded aggregate concrete 饱和性集料混凝土;饱和性骨料混凝土;膨胀(性)集料混凝土;膨胀(性)骨料混凝土
expanded alumin(i)um grating 网眼铝栅板
expanded alumin(i)um sheet 铝板网
expanded anchor 膨胀螺栓
expanded and welded tube joint 胀焊
expanded area ratio 桨叶展开面积比
expanded bed 多层孔气垫;膨胀床
expanded-bed activated carbon contactor 膨胀床活性炭接触器
expanded-bed adsorption 膨胀床吸附;升流床吸附(法)
expanded-bed height 膨胀床高度
expanded-bed reactor 膨胀床反应器
expanded blast-furnace slag 膨胀性高炉炉渣;膨胀高炉矿渣;多孔高炉矿渣
expanded blast-furnace slag concrete 膨胀高炉矿渣混凝土
expanded-bolted segment 扩径式螺栓连接管片
expanded bore pile 膨胀钻孔桩
expanded brick 膨胀砖
expanded canal 扩展渠道
expanded cement 膨胀水泥
expanded-center display 空心扫描显示器
expanded channel 扩展渠道;扩散型管道
expanded cinder 膨胀炉渣;多孔炉渣
expanded cinder aggregate 膨胀炉渣集料;膨胀炉渣骨料
expanded cinder concrete 膨胀炉渣混凝土
expanded cinder concrete cavity block 膨胀炉渣混凝土空心砌块
expanded cinder concrete pot 膨胀炉渣混凝土罐
expanded cinder concrete slab 膨胀炉渣混凝土平板
expanded cinder concrete wall slab 膨胀炉渣混凝土墙板;多孔炉渣混凝土墙板
expanded cinder slab 膨胀炉渣板材;多孔炉渣板材
expanded city 卫星城(市)
expanded clay 黏土陶粒;膨胀黏土
expanded clay aggregate 膨胀黏土集料;膨胀黏土骨料;多孔黏土集料;多孔黏土骨料;烧胀黏土骨料
expanded clay aggregate filter 膨胀黏土团粒滤池
expanded clay concrete block 膨胀黏土混凝土砌块;多孔黏土混凝土砌块
expanded clay concrete wall slab 膨胀黏土混凝土墙板;多孔黏土混凝土墙板
expanded column head 扩展式柱头
expanded composition cork 膨胀的合成软木;多孔的合成软木
expanded concrete 加气混凝土;膨胀混凝土
expanded concrete aggregate 膨胀混凝土集料;膨胀混凝土骨料;多孔混凝土集料;多孔混凝土骨料
expanded concrete block 膨胀混凝土砌块;多孔混凝土砌块
expanded concrete cavity block 膨胀混凝土空心砌块;多孔混凝土空心砌块
expanded concrete hollow block 膨胀混凝土空

心砌块;多孔混凝土空心砌块
expanded concrete insulating brick 膨胀混凝土绝热砖;多孔混凝土绝热砖
expanded concrete insulation grade slab 膨胀混凝土绝热板材;多孔混凝土绝热板材
expanded concrete plant 膨胀混凝土工场;多孔混凝土工场
expanded concrete pot 膨胀混凝土罐筒;多孔混凝土罐筒
expanded concrete product 膨胀混凝土产品;多孔混凝土产品
expanded concrete purpose-made block 膨胀混凝土专用砌块;多孔混凝土专用砌块
expanded concrete purpose-made brick 膨胀混凝土专用砖;多孔混凝土专用砖
expanded concrete slab 膨胀混凝土板材;多孔混凝土板材
expanded concrete special block 膨胀混凝土特殊砌块;多孔混凝土特殊砌块
expanded concrete tile 膨胀混凝土瓦;多孔混凝土瓦
expanded contrast 扩展衬度;扩大对比度
expanded cork 膨胀塞子;膨胀软木
expanded cork plate 膨胀软木片
expanded cork sheet 膨胀软木片
expanded cork slab 膨胀软木板;多孔软木板
expanded cushion 膨胀垫
expanded diameter conductor 扩径导线
expanded display 扩大显示
expanded ebonite 膨胀硬橡胶;多孔硬橡胶
expanded fiber strand 多孔纤维竹节纱
expanded film 扩张薄膜;多孔薄膜
expanded flange 扩张法兰;碾制凸缘
expanded flow 膨胀流(动)
expanded fly ash 粉煤灰陶粒
expanded foam 多孔泡沫
expanded foam plastics 多孔泡沫塑料
expanded-foot glacier 宽尾冰川
expanded form 展开式
expanded form of value 扩大的价值形式
expanded formula 展开式
expanded function 扩展函数
expanded function operator panel 扩充功能操作板
expanded glass 泡沫玻璃(板)
expanded glass sphere 膨胀玻璃球
expanded granulated cork 膨胀粒化软木;多孔粒化软木
expanded granulated cork slab 膨胀粒化软木板;多孔粒化软木板
expanded graphite 膨胀石墨
expanded hard rubber 膨胀硬橡皮;多孔硬橡皮
expanded human sources 膨胀劳动力资源
expanded in-situ capability 就地气泡性能
Expanded International Monetary Fund 扩大的国际货币基金组织
expanded joint 扩口接头;扩口接合;胀连;伸缩缝
expanded joist 伸展格栅
expanded lending 扩大放款
expanded lime concrete 膨胀石灰混凝土;多孔石灰混凝土
expanded lime concrete tile 膨胀石灰混凝土砖瓦
expanded line scale 放线比例尺
expanded lining 胀开式衬砌
expanded loss 扩展损失
expanded material 发泡材料
expanded memory 扩展内存
expanded memory specification 扩充内存
expanded memory standard 扩展内存规范
expanded mesh 膨胀网;金属网
expanded mesh screen 金属网眼屏(反眩光)
expanded metal 拉展金属件;多孔金属网;冷胀合金;灰慢钢板;金属网;多孔拉制金属网;多孔金属板;网形铁;板网
expanded metal and plaster ceiling 钢板网抹灰吊顶
expanded metal fabric 金属板网
expanded metal fabric reinforcement 网眼钢筋
expanded metal fence 金属网栅栏
expanded metal grate 金属网炉条;金属网护栅
expanded metal grid 金属网格子
expanded metal guard 金属网护罩
expanded metal integral lath(ing) 整块网眼钢皮板条
expanded metal lath ceiling 钢板网顶棚

expanded metal lath(ing) 拉展金属网;金属网板条;钢丝伸展网;金属拉网;钢网;网眼钢皮
expanded metal lath partition 金属拉网抹灰隔墙
expanded metal mesh 金属网
expanded metal net 金属板网
expanded metal partition 金属拉网(抹灰)隔墙;钢板网隔断
expanded metal reinforcement 金属网加强(物);钢板网配筋;网眼网配筋
expanded metal sheet 网眼钢板;金属网
expanded metal strip 金属网带条
expanded mineral 膨胀矿石
expanded natural rubber 膨胀的天然橡胶
expanded neoprene 膨胀氯丁橡胶
expanded nozzle 扩散喷嘴
expanded order 扩展命令
expanded packer 膨胀式封隔器
expanded partial-indication display 局部扩展显示(器)
expanded pe(a)rlite 膨胀珍珠岩;多孔珍珠岩;膨胀珠光体
expanded pe(a)rlite acoustic(al) tile 膨胀珍珠岩吸声板
expanded pe(a)rlite aggregate 膨胀珍珠岩集料;膨胀珍珠岩骨料
expanded pe(a)rlite cement product 膨胀珍珠岩水泥制品;水泥膨胀珍珠岩制品
expanded pe(a)rlite coarse aggregate 膨胀珍珠岩碎石
expanded pe(a)rlite concrete 膨胀珍珠岩混凝土
expanded pe(a)rlite core 膨胀珍珠岩芯板
expanded pe(a)rlite mortar 膨胀珍珠岩砂浆
expanded pe(a)rlite product 膨胀珍珠岩制品
expanded pe(a)rlite sand 膨胀珍珠岩砂
expanded perlite 多孔珍珠岩;膨胀珠光体
expanded pipe network 扩展管网
expanded plan position indicator 扩大平面指示器
expanded plastic board 膨胀塑料板
expanded plastic box 多孔塑料箱
expanded plastic concrete 膨胀塑料混凝土;多孔塑料混凝土
expanded plastic cupola 膨胀塑料圆屋顶;多孔塑料圆屋顶
expanded plastic dome 膨胀塑料穹窿;多孔塑料穹窿
expanded plastic insulation 膨胀塑料绝缘;泡沫塑料绝缘
expanded plastic lightweight concrete 膨胀塑料轻质混凝土
expanded plastic plaster baseboard 膨胀塑性石膏踢脚板
expanded plastics 多孔塑料;充气塑料;膨胀塑料;泡沫塑料;发泡塑料
expanded plastic seal(ing) 泡沫塑料密封
expanded plastic sheet 泡沫塑料薄板
expanded polyethylene draught seal 膨胀聚乙烯缝隙密封
expanded polypropylene 泡沫聚丙烯
expanded polystyrene 多孔聚苯乙烯;发性聚苯乙烯;膨胀(性)聚苯乙烯;泡沫聚苯乙烯;聚苯乙烯泡沫塑料
expanded polystyrene aggregate 聚苯乙烯膨胀集料
expanded polystyrene beads 聚苯乙烯膨胀微珠
expanded polystyrene block 聚苯乙烯泡沫块
expanded polystyrene board 聚苯乙烯泡沫板
expanded polystyrene pattern 泡沫聚苯乙烯模
expanded polystyrene sheet 聚苯乙烯泡沫片材
expanded polystyrene tile 泡沫聚苯乙烯砖瓦
expanded polyurethane 聚氨酯泡沫塑料;膨胀性聚氨酯
expanded polyurethane board 膨胀聚氨基甲酸酯板;多孔聚氨基甲酸酯板
expanded polyurethane sheet 膨胀聚氨基甲酸酯薄板;多孔聚氨基甲酸酯薄板
expanded polyurethane strip 膨胀聚氨基甲酸酯条带;多孔聚氨基甲酸酯条带
expanded polyvinyl chloride 膨胀聚氯乙烯
expanded preliminary design 扩大初步设计
expanded production 扩大生产
expanded pumice concrete 浮石混凝土;多孔混凝土
expanded pure agglomerated cork 膨胀块状纯软木

expanded pure baked cork 多孔烘干纯软木
expanded PVC 膨胀聚氯乙烯
expanded range 延伸范围
expanded ratio 膨胀倍数
expanded reinforcement 拉制钢筋
expanded relative form of value 扩大的相对价值形式
expanded relative value-form 扩大的相对价值形式
expanded reproduction 扩大再生产
expanded rigid polyurethane 多孔刚性的聚亚胺酯
expanded rigid polyvinyl chloride 膨胀刚性聚氯乙烯;多孔刚性聚氯乙烯
expanded rubber 多孔橡胶;膨胀橡胶;橡胶海绵
expanded rubble 泡沫循环
expanded sample 扩大的样本
expanded scale 扩展刻度
expanded scale spectroscopy 扩展量程光度法
expanded scan 扩展扫描
expanded shale 多孔页岩;膨胀(油)页岩;烧胀页岩;膨胀泥板岩
expanded shale aggregate 膨胀页岩集料;膨胀页岩骨料
expanded shale clay 膨胀页岩黏土
expanded shale clay concrete 膨胀页岩黏土混凝土
expanded shale coarse aggregate 膨胀页岩粗集料;膨胀页岩粗骨料
expanded shale fine 膨胀性页岩粉屑
expanded sheet metal 拉展金属网;金属拉网;钢丝网板;网眼薄钢皮;网眼薄板
expanded slag 多孔熔渣;多孔矿渣;膨胀(性)矿渣
expanded slag aggregate 膨胀矿渣集料;多孔矿渣骨料;多孔矿渣骨料
expanded slag beads 矿渣膨珠
expanded slag block 膨胀矿渣砌块
expanded slag building tile 膨胀矿渣建筑砖瓦
expanded slag concrete 膨胀矿渣混凝土
expanded slag concrete block 膨胀矿渣混凝土块;多孔矿渣混凝土砌块
expanded slag concrete plank 膨胀矿渣混凝土板;多孔矿渣混凝土板
expanded slag concrete tile 膨胀矿渣混凝土砖瓦;多孔矿渣混凝土砖瓦
expanded slag pellet 膨珠
expanded slag powder 膨胀矿渣粉末;多孔矿渣粉末
expanded slag tile 膨胀矿渣瓦;多孔矿渣瓦
expanded slate 烧胀板岩;膨胀板岩
expanded slate aggregate 多孔页岩骨料;膨胀板岩集料;膨胀板岩骨料
expanded slate concrete 膨胀板岩混凝土
expanded slate factory 烧胀板岩工厂
expanded sludge 膨胀污泥
expanded steel diamond mesh 菱形钢板网
expanded steel packing element 钢板网蜂窝填料
expanded succession 加厚序列
expanded sweep 扩展扫描
expanded test 扩张试验(管口)
expanded tube hole groove 胀管槽
expanded tube joint 胀管连接;胀管接头
expanded tube method 管材扩张制服法
expanded type air filter 楔形空气过滤器
expanded urea-formaldehyde 膨胀尿素甲醛
expanded value form 扩大的价值形式
expanded vermiculite 膨胀蛭石
expanded vermiculite aggregate 膨胀蛭石集料;膨胀蛭石骨料
expanded vermiculite mortar 膨胀蛭石砂浆
expanded vermiculite plaster 膨胀蛭石灰浆
expanded vermiculite product 膨胀蛭石制品
expanded view 展开图
expanded volcanic glass 膨胀火山玻璃体
expanded volume 膨胀容积
expander 扩张器;扩展器;扩孔器;扩大器;开幅辊;骤冷器;增浓剂;撑模器;膨胀器;膨胀机
expander amplifier 频带伸展放大器
expander and remover stabilizer 扩大器和移动器的稳定器
expander board kit 成套扩展电路板
expander-booster compressor 膨胀机增压压缩机
expander punch 扩管器冲头
expander ring 胀圈
expander roll 胀管器滚子;胀杆

expander system 扩束系统
expander tube 扩张器管
expander tube wheel brake 软管式机轮制动
expander type ring 撑胀器活塞环
expand escapement 扩充走ેલ;扩充的棘轮装置
expand exploration of mine district 矿区扩勘
expand extreme dark 增大最暗部分的黑度
expandibility 发泡性
expanding action 膨胀作用;扩大作用
expanding agent 发泡剂;膨胀剂
expanding anchor 膨胀锚固螺栓
expanding and non-contracting cement 膨胀性无收缩水泥
expanding arbor 胀轴;扩管器;胀开心轴;胀杆
expanding area 扩张区
expanding arm 膨胀臂
expanding auger 扩孔钻
expanding band brake 胀带闸;膨胀带闸;内带式刹车
expanding band clutch 胀带(式)离合器
expanding beach 波浪消能滩(堆石)
expanding bed 增厚地层;变厚的矿层;膨胀床层
expanding bench 扩孔机床胀带闸
expanding bit 伸缩钻头;扩孔钻头
expanding bolt 膨胀螺栓
expanding brake 胀闸;外胀式制动器;胀式制动器
expanding braker 胀闸
expanding bubble 膨胀气泡
expanding bushing 扩张式衬套
expanding carburet(t)or 膨胀式化油器
expanding cement 膨胀水泥
expanding cement concrete 膨胀水泥混凝土
expanding cement mortar 膨胀(性)水泥砂浆
expanding chemical 膨胀化学剂
expanding chuck (胀开式)弹簧筒夹
expanding clay 膨胀性黏土
expanding cone 胀式圆锥
expanding current 扩展水流
expanding cutter 扩张式管子割刀;伸出式刮刀
expanding cutting arm 伸缩式切削臂
expanding device 扩张装置;扩张设备;膨胀装置;伸缩调节器
expanding die 胀形模
expanding disc 膨胀盘
expanding dowel 张开销钉
expanding drill 扩孔钻
expanding drilling tool 扩孔钻井工具
expanding drum coiler 胀缩卷筒式卷取机
expanding drum uncoiler 胀缩卷筒式开卷机
expanding economy 扩张型经济
expanding equilibrium 扩大平衡
expanding flue 弹性火焰管
expanding gas 膨胀气
expanding grade 胀管度
expanding grout 膨胀性灌浆;膨胀水泥灌浆
expanding harrow 能伸缩耙
expanding inner jaw 胀形内爪
expanding inside brake 内胀式制动器
expanding in situ 现场膨胀
expanding investment 投资扩张
expanding latch segment 伸展式锁块
expanding lattice clay 膨胀性晶格黏粒
expanding lid 膨胀盖
expanding machine 扩径机
expanding mandrel 胀开(式)心轴;泥芯件;可胀式心轴;可调(式)心轴
expanding mandrel method 膨胀心轴法
expanding material 膨胀材料
expanding metal 膨胀金属;膨胀合金
expanding mill 扩孔碾磨机;扩径机
expanding mortar 膨胀(性)砂浆
expanding node 扩展节点;扩充节点
expanding nozzle 扩张型喷管;膨胀喷嘴
expanding or welding of tube with tube sheet 管子与管板的胀接或焊接
expanding pilot 自动扩展式点火器;扩张引导;膨胀引导
expanding pipe 膨胀管
expanding plant 膨胀设备;发泡工厂
expanding pliers 扩边钳
expanding plug 可扩张的管塞;胀(形)塞;胀开式封隔器;膨胀塞
expanding press 扩孔冲床
expanding pulley 伸缩轮

expanding reach 扩张河段;扩大段;伸展范围
expanding reamer 扩张(式)铰刀;可调铰刀
expanding reflection spread 展开反射排列
expanding ridged tractor roller 可扩张拖拉式凸纹滚柱
expanding ring 皮碗弹簧圈;伸缩圈
expanding ring clutch 胀环离合器
expanding rock 膨胀性围岩
expanding roller 展幅机
expanding roller sizer 带张开辊的分级机
expanding section 扩散段;扩大段
expanding shale 膨胀页岩
expanding shell 膨胀壳(体)
expanding-slot 可调缝隙的
expanding-slot atomizer 可调缝隙式喷油嘴;可调缝隙喷雾器
expanding space 可伸缩隔离衬垫
expanding spread 展开排列
expanding spread profile 展开排列剖面
expanding spread vertical-loop technique 展开排列垂直回路技术
expanding square search 正方形式搜索;展开方形搜索
expanding square search pattern 方形扩展搜寻方式
expanding stopper 膨胀塞子
expanding stopple 膨胀塞
expanding subspace 扩张子空间
expanding support 伸展支撑
expanding tap 可配丝锥;可调丝锥
expanding telescope 束扩展望远镜
expanding tensional motion 膨胀拉力运动
expanding test 扩胀试验;扩管试验;扩大试验;胀管试验;管子扩口试验;膨胀试验
expanding town 扩大城镇
expanding tube 膨胀管
expanding type tubing packer 膨胀式管形封隔器
expanding value 膨胀值
expanding valve 膨胀阀
expanding volume 膨胀体积
expanding waterway 内窄外宽的水槽;(在钻头底唇上的)逐渐深宽的水槽
expanding-wedge brake 张楔式制动器
expanding with wedge block 楔块扩孔
expanding workload 日益增多的工作量
expand market 扩展市场
expand middle tones 扩大中间调
expand production capacity 扩大生产能力;扩大生产规模
expand test 胀管试验;钢管扩口试验
expand text 扩充文本【计】
expand the line of credit 扩大信贷
expand tube 胀管
expand tube joint with tubesheet 胀管连接
expand with heat and contract with cold 热胀冷缩
expaned alloy 膨胀合金
expaned granular sludge bed reactor 膨胀颗粒污泥床反应器
expaned metal 膨胀合金
expanse of beach 宽广的海域
expanse of water 汪洋大海
expansibility 扩展性;可延伸性;可膨胀性;可扩展性;胀度;膨胀性
expansible bend 膨胀弯管
expansible force 膨胀力
expansible house 扩建住宅
expansible joint 膨胀接合
expansible polystyrene 发泡级聚苯乙烯
expansile 膨胀性的
expansile granular sludge bed reactor 膨胀颗粒污泥床反应器
expansimeter 膨胀计;膨胀度计
expansing crack 膨胀裂缝
expansion 扩张;扩展;扩建;扩大;扩充;均整;碾轧;胀大;展平;胀(余);伸缩缝
expansion ability 膨胀能力
expansion adjusting rail 伸缩调整轨
expansion admixture 膨胀附加剂;膨胀掺和剂
expansion agent 膨胀剂
expansion(al)cooling 膨胀冷却
expansion allowance 预留膨胀缝;中胀容许量;伸缩留量
expansion alloy 膨胀合金

expansional phase 膨胀状态
expansional state 膨胀状态
expansion anchor 扩大式壁锚;膨胀锚固;膨胀锚杆;扩胀锚杆
expansion and contraction 伸缩
expansion and contraction bearing 伸缩支座
expansion and contraction joint 伸缩缝
expansion apparatus 膨胀仪
expansion appliance 膨胀补偿器
expansion arc suppressing 膨胀消弧
expansion area 膨胀区
expansionary budget 扩张性预算
expansionary force 扩张力
expansionary monetary policy 扩展性货币政策
expansionary phase 扩张阶段
expansionary policy 经济扩张政策
expansion attic 备用阁楼
expansion ball joint 球形补偿器
expansion bath 膨胀槽
expansion bath treatment 膨胀浴处理
expansion bay (供弯管伸缩的)膨胀凹槽
expansion bead 伸缩垫圈
expansion bearing 活动支座;膨胀轴承;膨胀支座;伸缩支座;伸缩支承;伸缩承座
expansion bearing for bridge 桥梁活动支座
expansion bellows 膨胀伸缩式波纹管
expansion bend 胀缩弯头;胀缩弯管;膨胀弯管;膨胀补偿器;伸缩弯管;伸缩器
expansion bent 伸缩(补偿)弯管
expansion bit 扩孔钻头;扩孔器;活络螺旋钻头;变径钻头;伸缩钻头
expansion block 膨胀垫
expansion boiler 膨胀式沸水器
expansion bolt 扩张式地脚螺栓;扩开螺栓;支撑螺栓;胀管螺栓;膨胀螺栓;伸缩(螺)栓
expansion box 储气箱;膨胀箱
expansion brace 伸缩拉条
expansion bracket 伸缩架
expansion braker 胀闸
expansion buoyancy 膨胀浮力
expansion bushing 膨胀衬套
expansion cam 膨胀凸轮
expansion cap (混凝土路面的)胀缝传力杆套;传力杆帽;膨胀帽
expansion capacity 扩展能力;扩充能力
expansion card 扩展卡;扩充插件板
expansion cascading 扩展级联
expansion caused by heat and contraction caused by cold 热胀冷缩
expansion cement 膨胀水泥
expansion chamber 扩大室;减压卸料室;云室;膨胀室;膨胀盒
expansion chamber collector 膨胀室集尘器
expansion chamber micropore muffler 小孔膨胀消声器
expansion chamber muffler 膨胀室消声器
expansion-chamber type absorber 扩大空洞型消音器
expansion character 膨胀性能
expansion chuck 夹紧卡盘
expansion chucking reamer 机用扩张式铰刀
expansion circuit breaker 膨胀断路器
expansion clamp 膨胀卡头
expansion clearance 膨胀余隙;膨胀间隙
expansion cloud chamber 膨胀云室
expansion clutch 内涨式摩擦离合器
expansion cock 节流旋塞
expansion coefficient 展开系数;膨胀系数
expansion coefficient of tape(wire) 基线尺膨胀系数
expansion coil 冷凝盘管;蒸发盘管;膨胀盘管
expansion compensation 膨胀补偿
expansion compensator 膨胀补偿器
expansion component 膨胀组分
expansion concrete 膨胀混凝土
expansion coupling 膨胀连接器;胀缩联轴节;补偿联轴节;膨胀接头;伸缩接头
expansion crack(ing) 膨胀(裂)缝;膨胀裂纹;伸缩裂缝;胀裂
expansion crosstie 腰板横梁
expansion crown stay bolt 活节顶撑螺栓
expansion cup 胀杯
expansion curve 膨胀曲线
expansion cushion 膨胀补偿器

expansion cutter 扩张式管子割刀;伸出式刮刀
expansion cylinder 滚筒;膨胀筒
expansion deaerator 膨胀式除氧器
expansion deflection 膨胀变位
expansion demand 扩充需求
expansion device 膨胀装置;伸缩调整器
expansion device for continuous length of rails 焊接长钢轨伸缩调整器
expansion dome 膨胀穹
expansion drawing 放样图
expansion drill 活络螺旋钻
expansion drive 膨胀驱
expansion drum 膨胀罐
expansion due to free lime 游离石灰膨胀
expansion due to gypsum 石膏膨胀
expansion due to magnesia 氧化镁膨胀
expansion due to shear 剪切膨胀
expansion eccentric 膨胀偏心轮
expansion effort 扩建工作
expansion end (可)伸缩端;膨胀端;构件自由端
expansion end bearing 活动支座
expansion energy 膨胀能
expansion engine 膨胀式致冷发动机;膨胀机
expansion enterprise 扩大订货
expansion equation 展开(方程)式
expansion [expanded] tube joint 胀接
expansion factor 膨胀因素;膨胀系数
expansion fan 扩散线族;稀散波
expansion fastener 扩张锚杆
expansion fissure 膨胀裂隙;膨胀裂缝;膨胀节理【地】
expansion fit 膨胀配合
expansion flow 扩张(水)流
expansion fog 膨胀雾
expansion fold 膨胀褶皱
expansion force 膨胀力;伸张力
expansion fork 膨胀叉
expansion formula 展开(公)式
expansion fund 发展基金
expansion furnace 前置炉膛
expansion gallery 廊道扩大段
expansion gap 膨胀缝;伸缩间隙
expansion gas 膨胀气
expansion gas drive 膨胀气驱
expansion ga(u)ge 测膨胀器;膨胀计
expansion gear 膨胀装置
expansion gland 膨胀接头的填料压盖;填密胀圈
expansiong mandrel 可胀式心轴
expansion guide 膨胀导承
expansion gun 胀接式点焊枪
expansion hatch(way) 膨胀井舱口;膨胀口
expansion hinge 伸缩铰链
expansion hoop 膨胀环箍
expansion in autoclave test 压蒸膨胀(率)
expansion in channel 水道扩大段
expansion index 回弹指数;膨胀指数
expansion indicator 膨胀指示器
expansion in directed path 展成有向路
expansion in negative powers 按负幂展开
expansion in partial fractions 展成部分分数
expansion in powers 按幂展开
expansion in series 级数展开;展开成级数;展成级数
expansion insert 膨胀衬垫
expansion in Taylor series 展开为泰勒级数
expansion into continued fraction 连分式展开式
expansion into powerseries 幂级数展开
expansion investment 扩建投资
expansionist measures 扩张措施
expansionist policy 经济扩张政策
expansion joint 张缩接合;胀缝;张缩接头;钢轨伸缩调节器;补偿节;变形缝;膨胀结合;膨胀节(理)【地】;膨胀接头;伸缩节;伸缩接头;伸缩缝
expansion joint assembly 伸缩缝装配
expansion joint cap strip 伸缩缝金属套筒
expansion joint cover 伸缩缝盖板
expansion joint filler 伸缩缝填料
expansion joint for bridges 桥梁胀缝;桥梁伸缩缝
expansion jointing 设置伸缩缝;设置伸缩缝
expansion joint iron 伸缩接头铁;伸缩接缝铁
expansion joint mastic(sealer) 伸缩接头黏胶;伸缩接头密封剂
expansion joint material 伸缩缝材料;弹性填缝料
expansion joint of bridge deck 桥面伸缩缝

expansion joint profile 伸缩接头外形
expansion joint ridge capping 膨胀接合屋脊顶梁
expansion joint seal 伸缩缝密封体
expansion joint sealant 伸缩缝止水层;伸缩缝密封层
expansion joint sealing 膨胀接合的密封口;伸缩缝盖缝条;伸缩缝盖缝料
expansion joint section 伸缩接头横断面
expansion joint tape 伸缩缝狭带
expansion joint trim 伸缩接头的修饰
expansion joint unit 伸缩接头单元
expansion joint waterstop 伸缩接头水密封;膨胀节水密封;膨胀接合止水带
expansion knot 伸缩节
expansion line 引出线;气体膨胀的压力变化曲线;膨胀线;膨胀过程
expansion link 膨胀杆;伸缩杆
expansion link guide 膨胀杆导承
expansion link pin 伸缩杆销
expansion loop U形补偿器;胀缩器;胀圈;膨胀圈;膨胀环;盘形管膨胀补偿器;伸缩圈;膨胀器;伸缩节;伸缩(补偿)弯管
expansion loss 扩张损失
expansion machine 膨胀机
expansion mandrel 可调式心轴;胀开心轴
expansion mechanism 膨胀机理
expansion metal sheet 金属伸缩片
expansion method 展开法;膨胀法
expansion multiple 膨胀倍数
expansion multiple after water absorption 吸水膨胀倍数
expansion nozzle 扩散形喷管
expansion of a function 函数展开
expansion of area 面积扩大
expansion of brickwork 砌体膨胀
expansion of business 扩大营业
expansion of capital value 资本的价值增值
expansion of determinants 行列式展开(式);展开行列式
expansion of established enterprises 现有企业的扩充
expansion of existing sewerage system 现有污水系统扩建
expansion of gravitational into spheric(al) harmonics 引力球谐函数展开式
expansion of node 节点扩展
expansion of packing medium 填料膨胀
expansion of powers and transfer of profits 扩权让利
expansion of rail 铁轨膨胀
expansion of state financial resources 扩大国家财力
expansion of the ozone hole 臭氧层空洞的扩大
expansion of vapo(u)r 汽膨胀;膨胀汽
expansion of wave 波浪扩散
expansion opening 膨胀室
expansion orbit 扩张轨道
expansion or contraction stress 伸缩应力
expansion-oriented system 面向扩充系统
expansion pad 膨胀垫;伸缩垫
expansion pad cap 伸缩垫盖
expansion pedestal 膨胀支;伸缩(端)支座
expansion period 膨胀周期
expansion phase 膨胀相
expansion piece 补偿器;膨胀片;膨胀件;伸缩器;伸缩节;伸缩调整件;补充器
expansion pipe 泄流管;补偿器;膨胀节;膨胀管;伸缩管
expansion pipe bend 伸缩管弯头;伸缩弯管
expansion pipe joint 输油管伸缩接头
expansion plan 扩展计划;扩建计划;展开图
expansion plate 膨胀板;伸缩板
expansion plug 膨胀栓;膨胀塞
expansion plug snap ring 膨胀塞卡环
expansion post 膨胀式支柱;伸长式支柱
expansion pressure device 膨胀压力试验装置
expansion pressure measurement 膨胀压力测定
expansion pressure of surroundings 膨胀围岩压力
expansion pressure ratio 膨胀压比;膨胀比(率)
expansion pressure test 膨胀压力试验
expansion pressure testing apparatus 膨胀试验仪
expansion process 膨胀过程
expansion producing admixture 膨胀性掺加料;膨胀性掺和料;膨胀剂

expansion project 扩建项目;扩建企业;扩建计划;扩建工程;扩大方案
expansion proposal 扩建方案
expansion pulsation 膨胀脉动
expansion pyrometer 膨胀高温计
expansion rate 扩充率;膨胀率
expansion rate of foam 发泡倍数
expansion ratio 扩张度;展缩比;发泡率;发泡倍数;膨胀率;膨胀比(率)
expansion ratio of fluidized bed 流化床膨胀比
expansion reaction 膨胀反应
expansion reamer 扩展式扩大器;扩张式铰刀;扩展式铰刀;可调(径)铰刀
expansion refrigeration 膨胀致冷
expansion regulating valve 伸缩调节阀
expansion regulator 膨胀调节器
expansion reinforcement 伸展钢筋;伸缩钢筋;膨胀钢筋
expansion relief valve 膨胀安全阀
expansion reproduction 扩大再生产
expansion rivet 膨胀铆钉
expansion ring (膨)胀圈;伸缩圈;伸缩环
expansion rock bolt 胀壳式锚杆
expansion roller 带滚柱的支承;伸缩滚轴
expansion roof 膨胀顶
expansion roof tanks 膨胀顶储罐
expansion room 扩散室
expansion rule 胀尺
expansion scab 膨胀结疤
expansion screw 扩开螺钉;可调螺丝;可调螺钉
expansion section 扩散段;扩大段
expansion sheet 膨胀板
expansion shell bolt 胀销螺栓;胀壳式锚杆;膨壳式锚杆;胀毂式锚杆;涨壳式锚杆
expansion shell type heater 膨胀壳式加热器
expansion shield 扩展护罩;膨胀螺栓
expansion shim 填缝片
expansion shock 稀散波;膨胀激波
expansion shoe 伸缩桥靴
expansion sleeve 胀紧套筒;膨胀套筒;膨胀套管;(传力杆的)伸缩套管;伸缩套筒
expansion sleeve bolt 胀壳式锚杆;撑帽式锚栓;撑帽式螺栓;撑帽式杆柱
expansion slide valve 膨胀导阀
expansion sliding block 可伸缩滑块
expansion slot 扩展(插件)槽;扩充插槽;胀槽;伸胀孔
expansion soil 膨胀土
expansion space 膨胀余量;膨胀空隙;膨胀空间;膨胀缝
expansion spacing 胀缝隙
expansion spending 扩建支出
expansion spring 膨胀弹簧
expansion stage 膨胀阶段
expansion stay 伸缩拉撑
expansion staybolt 伸缩撑螺栓
expansion steam trap 恒温式疏水器
expansion stopple 膨胀塞
expansion strain 膨胀应变
expansion strength 真空波强度;膨胀强度
expansion strip 伸缩嵌条;弹性绝缘填料
expansion stroke 膨胀冲程
expansion structure 膨胀构造
expansion stuffing box 膨胀填料盒
expansion styropor 膨胀型聚苯乙烯材料
expansion suspender 伸胀悬杆
expansion tank 扩容器;缓冲罐;膨胀箱;膨胀柜;膨胀舱;膨胀水箱
expansion tap 可胀丝锥;可调丝锥
expansion temperature 膨胀冲程终了时的温度
expansion tendency due to magnesia 氧化镁的膨胀作用
expansion test 膨胀试验
expansion test using Le Chatelier needles 雷氏夹法水泥膨胀试验
expansion texture 膨胀结构
expansion the network of banking of business 扩大银行机构网
expansion theorem 展开定理
expansion thermometer 膨胀式温度计
expansion time 膨胀时间
expansion tissue 伸展组织
expansion trap 膨胀式除水器

expansion trough 膨胀槽
expansion trunk 膨胀围壁;膨胀筒;膨胀井;膨胀舱
expansion trunk hatchway 膨胀井舱口
expansion tube 补偿管;膨胀管;伸缩管
expansion turbine 膨胀涡轮
expansion-type bolt 膨胀型锚杆;膨胀式锚杆
expansion-type rock bolt 胀壳式锚杆
expansion type reservoir 扩展式蓄水池
expansion U-bend U 形伸缩管;膨胀 U 形管;U 形胀缩器;U 形膨胀接头
expansion use of mineral commodities 扩大矿产品用途
expansion value 膨胀值
expansion valve 截流滑阀;节流阀;安全阀;膨胀(调节)阀
expansion valve gear 膨胀阀动装置
expansion vessel 膨胀容器;膨胀箱
expansion wall scraper 扩张式井壁刮刀;伸缩井壁刮刀
expansion washer 伸缩垫圈;膨胀垫圈;弹簧垫圈;弹簧垫片;伸缩垫片
expansion water pipe 膨胀水管
expansion water tank 膨胀水箱;补水箱
expansion wave 扩散波;稀疏波;膨胀波;伸缩波
expansion work 扩建工程;膨胀功
expansive 膨胀性的
expansive action 膨胀作用
expansive agent 膨胀剂
expansive bar test 膨胀杆试验法
expansive bit 伸缩式钻头
expansive cement 自应力水泥;膨胀水泥
expansive cement concrete 膨胀(水泥)混凝土
expansive cement general 普通膨胀水泥
expansive cement grout 膨胀性水泥砂浆
expansive cement grouting compound 用膨胀水泥灌浆混合物
expansive cement mortar 膨胀水泥灰浆
expansive center bit 伸缩式中心钻头
expansive classification 扩展式分类法
expansive clay 膨胀性黏土
expansive colo(u)r 放大色;似胀色
expansive component 膨胀性成分;膨胀组分
expansive concrete 膨胀性混凝土;膨胀混凝土
expansive constituent 膨胀组分
expansive cracking 膨胀开裂
expansive curvature 膨胀曲率
expansive force 胀力;膨胀力
expansive formation 膨胀性地层
expansive foundation 膨胀性地基
expansive fusion caking 膨胀熔融黏结
expansive ground 膨胀性地基;膨胀性地层
expansive grout 膨胀性砂浆
expansive growth 膨胀性生长
expansive keynote 扩充要旨
expansive material 膨胀性材料
expansive metal 膨胀金属
expansiveness 夸张性
expansive oil well cement 膨胀油井水泥
expansive power 膨胀力
expansive pressure 膨胀(压)力
expansive rivet 膨胀铆钉
expansive rock 膨胀性岩石
expansive rubber 膨胀橡胶
expansive soil 膨胀性土(壤);膨胀土
expansive working 膨胀操作
expansivity 延伸性;膨胀性;膨胀能力
ex-parte award 一方当事人在场所作出的仲裁裁决
ex-parte evidence 一方所提出的证据
expatriate allowance 出国津贴
expatriate employee 驻外职员
expatriate labo(u)r 外籍工人;外籍劳工;外国劳动力
expatriate personnel 派出人员
expectance 预期;期望;期待
expectancy = expectance
expectancy hypothesis 期望假说
expectancy life 概率寿命
expectancy model of motivation 激励期望模型
expectancy theory 期望理论
expectation 预料;期望;期待
expectation criterion 期望值准则
expectation curve 期望曲线
expectation of death 预期死亡
expectation of life 预期寿命;估计寿命;期望寿命;平均(概率)寿命
expectation payment 期望支付
expectation value 预期值;期望值
expect cost 期望费用
expect departure time 计划开航时间
expected 预计的
expected accuracy 预期精(确)度
expected angle 期望角
expected approach clearance time 预料准许进场时间
expected approach time 预计接近时间;期望到达时间
expected average life 预期平均寿命
expected balance sheet 预计资产负债表
expected behavio(u)r 期望行为
expected benefit 预期效益
expected benefit-cost method 预期成本效益法
expected breaks 期望破损数
expected completion time 预计完工时间
expected conclusion 意料结果
expected consumption 期望消费
expected cost 预期造价;预期成本
expected cost of building 预计造价
expected cost of production 期望生产费用
expected data 预定数据;期望数据
expected data of arrival 预计到达日期
expected date of completion 预计完工日期
expected death rate 预期死亡率
expected delay 期望延迟
expected distance of sliding 预计滑动距离
expected duration 预期持续时间
expected earning 预期收益
expected earthquake 期望地震
expected elapsed time 期望消逝时间;期望间隔时间
expected energy not served 应发未发电量
expected environmental concentration 期望环境浓度
expected fatigue life 预期疲劳寿命
expected financial conditions changing sheet 预计财务状况变动表
expected gradation curve 预计粒度曲线
expected gradient of river bed 预期坡度
expected income 预期收入
expected internal rate of return 期望值内部收益率
expected length of the waiting line 期望排队长度
expected life 预期寿命;预计寿命;预计使用年限;推测使用期
expected life span 期望寿命
expected life time 预期使用期
expected line 期望线
expected loglog 期望重对数
expected loss 预期损失
expected marginal loss 期望边际损失
expected marginal profit 期望边际利润
expected monetary value 期望货币值
expected morbidity rate 预期患病率
expected mortality 预期死亡率
expected normal frequency 期望正态频数
expected normal price 期望正常价格
expected number 期待数
expected number in the queue 排队期望人数
expected number in the system(s) 系统内期望人数
expected number of failure 期望故障数
expected opportunity 期望机率损失
expected outcome 期待结果
expected payoff 期望支付
expected performance time 期望完成时间
expected price 预期价格
expected principle 期望原理
expected probit 期望概率值
expected profit 预期利润
expected profit and loss sheet 预计损益表
expected profit with perfect information 具有完全信息的期望利润
expected progress from selection for single traits 单独性状选择的预期
expected quality level 预期质量水平;预期品质水准;期望质量水准
expected quantities short 预期数量短缺
expected quantity of metal reserves 期望金属量
expected quantity of reserves 期望矿石量
expected rate of overtopping 预计满顶流量
expected rate of profit 预期利润率
expected rate of return 预期报酬率
expected rate of return on investment 预期投资回收率
expected regret value 期望后悔值
expected reliability 可靠性期望值
expected reserves 预计储量;推测储量
expected response 期望反应
expected result 期望结果
expected return 预期收益;期望报酬
expected revenue 期望税收;期望收益
expected reward 预期报酬
expected risk 期望风险
expected road surface performance 预期路面性能
expected sailing time 预计启航时间;预计开航时间
expected service life 预期使用寿命
expected service time 期望服务时间
expected shortage 预期缺货
expected time 预计执行时间;期望时间
expected time of arrivals 预计到达时间
expected time of delivery 预计交货期;预计交船期
expected time of departure 预计起航时间;预计开航时间
expected time of finishing discharging 预计卸完时间;预计卸毕时间
expected time of finishing loading 预计装完时间
expected time of return 预计返回时间
expected total reward 期望总报酬
expected utility 期望效用;期望效益
expected utility hypothesis 期望效用假设
expected utility value 期望效用值
expected value 期望值;预计值;期望值;期待值
expected value criterion 期望值标准
expected value of a probability distribution 一种概率分布的期望值
expected value of imperfect information 不完全信息预期价值
expected value of perfect information 完全信息期望值;完全情报期望值
expected value of sample information 取样信息的期望值
expected value of stock-out 预期缺货量
expected variance 期望方差
expected waiting time 预期等候时间
expected wealth value 预期资产价值
expected working period 预计施工期限
expected yield 预期产量;预期产额
expecting simple variable 期望简单变量
expecting variable 期望变量
expect ready to load 计划装货准备就绪日期
expedience 便利;权宜之计;权宜措施
expediency = expedience
expedient measure 简便措施;高效措施;权宜之计;权宜措施
expedient measure to meet an emergency 应急权宜措施
expedite 加快速度
expedited data unit 快速数据块
expedited flow 加急(数据)流
expedited message handling 快速信息处理;加速信息处理;加速消息处理
expediter 稽查员;督工员;材料检查员;器材调度员;催料员;材料进度管理员
expediting cost 加速费用
expedition 考察;远征;探险
expeditionary observation 考察
expeditionary research 考察性调查
expeditionary ship 调查船
expeditionary topography 勘测
expedition party 考察队
expedition pump 应急泵
expeditions military soil stabilization 军事工程土壤快速稳定法
expedition team 考察队
expedition theodolite 勘测经纬仪
expedition transit 勘测经纬仪
expeditor 器材调度员
expedor phase advancer 进相感受器
expel 追出
expellant bag 弹性箱
expellant gas 排出的气体
expellent 排除的
expeller 螺旋式压榨器;排除器;推出器
expend 扩展;花费;耗费;消费

expendable 消耗物;消耗的;不可恢复的
expendable ballast 可弃压载
expendable bathythermograph 船用抛弃式温深仪;船用抛弃式深度温度计;投弃式温深计;抛弃式海水温度计;投放式海水温度计
expendable bathythermograph data 投弃式温深计观测资料
expendable bathythermograph section 投弃式温深计观测断面
expendable bathythermography 抛弃式海水温度法;投放式海水温度法
expendable box 消费箱
expendable buoy 不回收的浮标
expendable carrier 消耗式载体
expendable conductivity temperature depth-sonde 抛弃式电导海水温度计;投放式电导海水温度计
expendable container 简易集装箱
expendable drive point 可耗式触探头
expendable equipment 消耗性装置;一次使用装置;不可回收装置
expendable fund 备用基金
expendable gun 销毁式射孔器
expendable instrument 不回收式仪器;投弃式仪器
expendable mo(u)ld 一次铸模
expendable packaging 消耗性包装;不回收的包装
expendable pallet 简易托盘
expendable pattern casting 一次模铸造
expendable pattern material 一次模型材料;熔模材料
expendable plug 一次性桥塞
expendables 消耗品
expendable supply 消耗性用品;消耗品;低值易耗品
expendable through-tubing perforator 销毁式过油管射孔器
expendable water supply 消耗性用水
expendable weight 耗重
expendable zinc core 可熔消锌芯
expended appropriation 已用拨款
expending beach 消能(海)滩;消浪海滩;波浪消能海滩
expending equilibrium 消费平衡;消费均衡
expenditure 开支;经费;支出(成本);财政支出;费用(开支)
expenditure accounting 成本费用核算
expenditure capability 支付能力
expenditure category 开支类别
expenditure center 费用中心
expenditure control 支出控制
expenditure cut 支出削减
expenditure expansion 支出扩大
expenditure for acquisition and use of fixed assets 因购置及使用固定资产的费用
expenditure for not buying commodities 非商品支出
expenditure for procurement 采购支出
expenditure for public works 公共工程支出
expenditure for unexplained purpose 未说明用途的支出
expenditure fund deposit 经费存款
expenditure-income relation 收支关系
expenditure incurred 发生费用
expenditure is determined by revenue 以收定支
expenditure items 开支项目;消耗品
expenditure not deductible 纳税时不能扣除的支出项目
expenditure not related to volume of traffic 与运输量无关的支出
expenditure of capital 投资费(用)
expenditure of construction 建造费(用)
expenditure of economic construction 经济建设费
expenditure of energy 能源消耗;能量消耗;能量损耗
expenditure of outlay cost 各项支出费用
expenditure of protecting safety for workers 劳动安全保护费
expenditure of protecting security for workers 劳动安全保护费
expenditure of scientific research 科学研究费
expenditure of time 工时消耗
expenditure of water 水的消耗
expenditure of work 工程费用
expenditure on construction 建筑费(用);施工费用指数;施工费(用);建造费用

expenditure on national defence 国防支出
expenditure on new plant and equipment 新厂房和新设备投资额
expenditure on power 动力费
expenditure on public debt 公债支出
expenditure on rolling stock 机车车辆购置费
expenditure out of budget 预算外支出;预算外开支
expenditure pattern (支出)费用构成
expenditure plan 用款计划
expenditure rate 支出比例
expenditure rate method 支出率法
expenditure reduction 经费节减
expenditure related to volume of traffic 与运输量有关的支出
expenditure subsequent to acquisition and use of fixed assets 购置及使用固定资产而产生的费用
expenditure-switching policies 支出转移政策
expenditure taxes 消费支出税;开支税
expensating ga(u)ge 补偿片
expensation trade 补偿贸易
expense 开支;消费
expense account 报销账单;开支账(户);费用账户
expense accrued 应付未付费用
expense adjunct account 费用附加账户
expense adjust account 费用调整账户
expense allocation 费用分摊;费用分配
expense allowance 交际接待费
expense analysis book 费用分析账
expense and receipts 收支
expense and receipts in balance 收支相抵
expense arising from outside manufacture 外加工费用
expense assets 消费资产;费用资产
expense associated with the earning of interest 与利息收入有关的费用
expense attributable to the investment income 因投资收入而发生的费用
expense audit 费用审计;费用审查
expense balance receipts 收支相抵
expense belonging to the preceding financial year 上财政年度费用支出
expense budget 费用预算
expense burden 费用负担;费用负荷
expense category 费用类别
expense ceiling 费用开支标准
expense center 费用中心
expense classification 费用分类
expense constant 费用常数
expense contra account 费用抵销账户
expense control 费用管理;费用控制
expense cost 费用成本
expense deducted 缩减开支
expense distribution sheet 费用分配表
expense flow 费用流量
expense for continued project 续建项目费用
expense for protecting mines 矿山维护费
expense for subsiding subordinary unit 对附属单位补助费用
expense for urban development and maintenance 城市建设和维护费
expense fund deposit 费用存款;经费存款
expense increase coefficient 费用递增系数
expense incurred for opening of business 开业费
expense in local currency 用本国货币支付的费用
expense in the trial manufacture of new products 新产品试制费
expense invoices 费用清单;费用发票
expense item 费用项目
expense journal 费用日记账
expense ledger 费用分类账;经费总账
expense liquidated 费用付现
expense loading 费用附加;费用负荷
expense management 费用管理
expense method of apportionment of research cost 研究成本分配的费用法
expense not allocated 未分配费用
expense of agency business 代办业务支出
expense of circulation 流通费用
expense of communication 交通费用
expense of developing 开发费用;发展费用
expense of idleness 闲置费(用)(指设备);窝工费用
expense of overall testing with load 负荷联合试车费

expense of production 生产费(用)
expense of repair materials 检修材料费(用)
expense of reproduction 再生产费用
expense of supervision 监督管理费用
expense of taxation 税费
expense of technical innovation and examination 技术革新及试验费
expense of workshop 车间经费
expense on design works 设计费
expense paid in advance 预付费用
expense payable 应付费用
expense preference 费用偏好
expense quota 费用限额;费用开支定额;经费限额
expense ratio 费用比率
expense standard 费用定额;费用标准
expense statement 费用表
expense stop 费用额度
expense surveillance 费用支出监督
expense variance 费用差异
expensive 贵的
expensive goods 贵重货(物);高价货物
expensive roof of tank 气柜浮动顶盖
experience 经验;经历;实践
experience-based estimation of workhours 经验估工法
experienced 老练的;经验丰富的;经历过的;有(实践)经验的;熟练的
experienced contractor 有经验承包商
experienced engineer 有经验工程师
experienced expert 经验专家
experienced worker 熟练工(人)
experience factor 经验系数
experience gained in operation 操作经验
experience mortality 经验死亡率
experience object interpretation 经验直观解释法
experience rating 经验税率;经验费率;经验定额
experience school of management thought 经验管理学派
experience survey 经验调查
experience table 人寿保险公司的死亡率统计表
experiment 试验;实验
experimentable line 实验铁路线
experimental 经验的;实验的
experimental analogic method 模拟实验法;类比试验法;实验模拟法
experimental analogy theory 实验模拟理论
experimental analysis 实验分析
experimental animal 实验动物
experimental apparatus 实验装置;实验仪器
experimental area 实验区
experimental arrangement 试验排列;实验计划;实验布置
experimental banking 试验性填土
experimental basin 径流实验站;实验水池;实验流域
experimental beams 实验束流
experimental beater 实验用打浆机;实验室打浆机
experimental bench 试验台;实验台
experimental beryllium oxide reactor 氧化铍实验反应堆
experimental biology 实验生物学
experimental boiling water reactor 实验性沸水反应堆
experimental breeder reactor 实验(性)增殖反应堆
experimental building 实验性建筑
experimental cartographic facility 实验制图设备
experimental catchment 径流实验站;试验流域
experimental cell 试验电解槽;实验回路栅元
experimental channel 实验管道
experimental check 实验校验;实验检查
experimental chemistry 实验化学
experimental circuit 实验电路
experimental city 实验城市
experimental communications satellite 实验通信卫星
experimental concrete 试验性混凝土
experimental concrete surface 试验性混凝土路面
experimental condition 试验条件;实验条件
experimental considerations 实验思路
experimental constant 实验常数
experimental converter 试验转炉;实验转炉
experimental correlatogram 实验相关图
experimental cost 试验成本
experimental data 经验数据;试验数据;实验资

料;实验值;实验数据
experimental data base 实验数据库
experimental data processing system 实验数据处理系统
experimental department 试制车间
experimental design 试验(性)设计;经验设计;实验设计
experimental design work 实验设计
experimental desk 试验桌
experimental detail 实验细节
experimental determination 实验测定
experimental device 实验装置
experimental distribution 实验分布
experimental diving 实验潜水
experimental dredge(r) 试验性挖泥船
experimental duties 实验任务
experimental elasticity 实验弹性力学
experimental engineer 技术研究工程师
experimental engineering 技术研究工作
experimental equation model 经验方程模型
experimental error 试验误差;实验误差
experimental establishment 实验站
experimental estate 实验性的房地产
experimental evidence 试验证据;试验根据;实验证据
experimental expenses 试验费;实验费用
experimental exploitation 试验性开发
experimental extinction 实验性消退
experimental face 实验工作面
experimental facility 实验装置;实验设备
experimental farm 试验农场;实验农场
experimental feature 实验特性
experimental field 试验现场;试验地
experimental findings 实验性的调查结果;实验数据
experimental flight 飞行试验
experimental fluid mechanics 实验流体力学
experimental flume 实验水槽
experimental forecast 经验预报
experimental formula 经验公式;实验公式
experimental frequency curve 经验频率曲线
experimental fusion device of magnetic mirror type 磁镜型实验用核聚变装置
experimental fusion power reactor 实验聚变动力反应堆
experimental gas cooled reactor 实验性气冷反应堆
experimental geochemistry 实验地球化学
experimental geochemistry diagram 实验地球化学图
experimental geology 实验地质学
experimental geomorphology 实验地貌学
experimental highway 试验性公路
experimental hole 实验孔道
experimental housing mortgage insurance 实验性住房抵押贷款保险
experimental installation 实验装置;实验设备
experimental investigation 试验性调查;试验研究;实验研究法
experimental irrigation station 灌溉试验站
experimentalist 科学实验人员
experimental joint 试验接缝
experimental kiln 试验窑
experimental laboratory 实验室
experimental labour 试验工人
experimental law 实验规则;实验(规)律;实验定则;实验定律
experimental layout 实验仪器配置
experimental light 试用灯标
experimental loop 实验回路
experimental mass-transfer coefficient 实验传质系数
experimental mean pitch 实验平均螺距;实际平均螺距
experimental mechanics 实验力学
experimental memo 实验备忘录
experimental meson spectroscopy 实验介子谱学
experimental meteorology 实验气象学
experimental method 试验方法;实验(方)法
experimental methodology 实验研究法
experimental middle school 实验中学
experimental mine 实验矿井
experimental mineralogy 实验矿物学
experimental mining country 试采国
experimental mining method and device 试采方法和设备
experimental mining ship 试采船
experimental mining site 试采地点
experimental mining time 试采时间
experimental mistake 试验错误
experimental model 试验模型;实验(性)模型
experimental model analysis 实验模型分析
experimental model basin 船模试验池
experimental model net 模型试验网
experimental model test 模拟实验
experimental morphology 实验形态学
experimental mounting 实验装置
experimental multispectral scanner 实验性多谱段扫描仪
experimental network 试验网
experimental nuclear physics 实验核物理学
experimental observation 实验观察
experimental organic cooled reactor 实验性有机冷却反应堆
experimental packet switching service 包交换实验网络;实验包交换业务
experimental parameter 试验参数
experimental performance 实验性能
experimental period 实验周期;实验时间
experimental petrology 实验岩石学
experimental phase 实验阶段
experimental physics 实验物理学
experimental pit 实验坑
experimental pitch 实验螺距
experimental plat 试验场
experimental plot 试验小区;实验小区
experimental point 实验点
experimental port 实验孔
experimental power reactor 实验动力反应堆
experimental preparation 实验准备
experimental prestressed concrete road 试验性预应力混凝土路
experimental pricing 实验定价法
experimental probe 试探器
experimental procedure 实验程序;实验步骤
experimental process simulation 实验(性)过程模拟
experimental program(me) 实验方案
experimental project 试验项目;实验性项目
experimental proof 实验证据
experimental prototype 试验样机;实验模型
experimental provision 实验装置
experimental radar equipment 实验性雷达设备
experimental radio receiver 实验型无线电接收机
experimental reactor 实验(性)反应堆
experimental reforms on a local basis 局部的试验性改革
experimental regulation 试行章程
experimental reoccurrence period 经验重现期
experimental research satellite 实验研究卫星
experimental reserch program(me) 实验研究计划
experimental resorption effect 实验吸回效应
experimental result 试验成果;实验结果
experimental results reduction 实验结果归纳
experimental road 试验路;实验路
experimental rock mechanics 实验岩石力学
experimental room 实验厅
experimental run 试验运转
experimental satellite communication earth station 实验卫星通信地面站
experimental section 试验段
experimental seismology 实验地震学
experimental sequence 实验程序
experimental set-up 实验准备;实验计划;实验布置
experimental sex seversal 人工变性
experimental ship 试验船
experimental shop 试制间;实验车间
experimental simulation method 模拟实验法
experimental site 试验现场
experimental soil engineering 试验土工学
experimental soil mechanics 试验土力学
experimental stage 实验阶段;实验阶段
experimental starting 试起动
experimental station 试验站;实验站;实验电台;实验处;实验场
experimental station system 试验站网
experimental strategy 实验策略
experimental stress analysis 实验应力分析
experimental structural geology 实验构造地质学
experimental study 试验研究;实验(性)研究
experimental subsea station 海下试验站
experimental superheat reactor 实验性过热反应堆
experimental surface 试验路面
experimental survey 实验调查法(市场调查的方法之一)
experimental synchronous satellite 实验同步卫星
experimental table 实验台
experimental tank 试验(水)池;试验(水)槽;(水)槽
experimental technique 实验技术
experimental tectono-geochemistry 实验构造地球化学
experimental telecommunications satellite 实验电信卫星
experimental temperature 实验温度
experimental test 实验性检验;试点试验
experimental test cell 实验浮选槽
experimental theatre 实验剧院
experimental tidal power plant 潮汐实验电站
experimental time 实验时间
experimental uncertainty 实验的不确定性
experimental unit 试验单位;实验单位
experimental value 经验数值
experimental variation 试验偏差;试验变差;实验变数
experimental variogram 实验变差函数
experimental variogram value 实验变差函数值
experimental verification 实验验证;实验性鉴定
experimental voltage 实验电压
experimental wall 试验墙
experimental watershed 径流实验站;试验流域;实验性集水区;实验流域
experimental well 试验井
experimental work 实验工作
experimental work in selected point 试点
experiment and research expenses 试验研究费
experimentary safety vehicle 汽车安全试验
experimentation 试验方法;实验法
experimentation cost 试验成本
experimentation expenses 已耗费用
experimentation expired appropriation 过期拨款
experimentator 实验员
experiment block 试验区
experiment building 试验房屋
experiment by photoelastic method 光弹实验
experimenter 实验者
experiment facility 试验设备
experiment field 试验场地
experiment in cultivation 栽培试验
experiment investigation 实验研究
experiment manufacturing cost 试制费
experiment measuring device 试验量测装置
experiment of tectonic simulation 构造模拟实验
experiment of weighing 称重试验
experiment on self aerated flow 水流自动掺气试验
experiment package 实验组件
experiment packet 实验仪器组
experiment plot 试验区
experiment research and development expenses 试验研究及开发费用
experiment rig 实验台
experiment set-up 试验装置
experiment station 实验站;实验场
experiment station of water and soil conservation 水上保持实验站
experiment thimble 实验套管
experiment uncertainty 试验不确定性
experiment value 实验值
experiment with clay model 泥巴试验
experiment with clay on shear fractures 扭裂缝泥巴实验
experiment with skips in the row 垄上留空试验
expert 行家;专家
Expert Advisory Panel on Emergency Relief Operations 紧急救济行动专家咨询小组
expert committee 专家委员会
expert consultancy services system 专家咨询系统
expert consultant 咨询专家
expert consulting system 专家咨询系统
expert decision-making 专家决策
expert engineer 专业工程师
expert evidence 鉴定(人证明)
expert fee 专家酬金

expertise 鉴定;鉴别;专业知识;专业人员;专门知识;专门技能;专家评价;专家经验;专家鉴定;专长
expertise report 鉴定书
expertization 技术鉴定
expertize 专业鉴定;专家鉴定
expert knowledge 专业知识;专门知识
expert manager 熟练的管理人员
expert meeting 专家会议
expert opinion 专家主张;专家意见;专家评价
expert panel 专门小组委员会;专家(小)组;专家团
expert program(me)s 专家程序系统
expert report 专家报告
expert('s) appraisement 专家鉴定
expert('s) examination 专家鉴定
expert skill 专门技术;专门技能
expert's report 鉴定报告
expert('s) statement 专家鉴定
expert's survey 专家报告
expert sub-group 专家小组
expert system 专家系统
expert system module 专家系统模块
expert system of image interpretation 图像判读专家系统
expert testimony 专家证明;专家证据
expert witness 鉴定员;鉴定人;专家证人
expert worksmanship 熟练的手艺
expextation variance 期望值方差
ex pier 码头交货
exploration of coal mines 煤矿勘探
expiration 截止日期;满期;呼气;终了;到期;期满
expiration date of letter of credit 信用证(截止)有效日期
expiration notice 期满通知(书)
expiration of contract period 合同期满
expiration of effect 有效期满
expiration of license 执照期满
expiration of limitation period 有效期(限)终止;有效期(限)终了;有效期(限)截止
expiration of patent 专利权期满
expiration of policy 保险单期满;保单期满
expiration of the contract 合同期满
expiration of the contract period 合同期满
expiration of the service value 超过租赁资产已服务的价值
expiratory reserve 呼气储备
expiratory reserve volume 补呼气量
expire 满期;期满
expired cost 过期成本;已消逝成本;已耗成本
expired expense 已消逝费用;已耗费用
expired patent 过期专利
expired patent value 已消逝专利权价值;已耗专利权价值
expired utility 已耗效用
expiry 满期
expiry date 有效日期;终止日期;到期日;失效日期
expiry date for interest 结息日
expiry date of L/C 信用证有效性
expiry of a contract term 合同有效期满
expiry of contract 合同终止;合同期满
expiry of tenancy 租约期满
expitaxial growth by melting 熔融外延法
ex-pit transaction 场外交易
explained sum of squares 解释了的平方和
explained variable 被说明变数;被解释变量
explained variance 已说明方差
explained variation 可解释变差
explaining variable 解释变量
explanade 旷地
explanation 解释;说明
explanation facility 解释机制
explanation of numeric(al) method 数值法说明
explanation of symbols used on core logs 岩芯记录标志的解释
explanation on calculation method 计算方法说明
explanation subsystem 解释子系统
explanation variable 说明变量
explanatory 说明的
explanatory ability 解释能力
explanatory comment 说明注解
explanatory drawing 说明图(纸)
explanatory interface module 解释接口模块
explanatory label 图例表
explanatory legend 图例(说明);说明图例
explanatory note 注释;补充说明;(说明)附注

explanatory notes 注释说明
explanatory notes of drawings 图纸说明
explanatory notes to the financial statement 财务报表注释说明
explanatory paragraph 审计解释部分
explanatory power 解释功效
explanatory statement 叙述式报表
explanatory text 解说地图;图注;说明注记
explanatory variable 解释变量
ex plane 空运机交货;输出空运港机上交货价
explant 移出物
explantation 农场交货;移出
explementary 共轭
explementary angles 共辅角
explement of angle 周余角
expletive 多余的;填塞的;填空石;填补物
expletive stone 填空石
explication 精确定义
explicit 显
explicit address 显(式)地址
explicit constraint 显约束
explicit cost 明显成本;显含成本;直接以货币支付的成本;外现成本
explicit debt 明显债务
explicit declaration 显式说明
explicit definition 显定义
explicit difference formula 显式差分公式
explicit difference scheme 显式差分格式
explicit estimation technique 显式估计法
explicit expression 显式(表达式);显式表示法
explicit finite difference 显式有限差分(法)
explicit finite difference calculation 显式有限差分计算
explicit finite difference scheme 显式有限差分格式
explicit formula 显式(公式)
explicit formulation 显式公式化;显式(表达式)
explicit function 显函数
explicit identification 显式辨识
explicit interest 明息;(直接)以货币支付的利息
explicit interest rate 名义利率
explicit library 显式程序库
explicit mathematical model 显数学模型
explicit method 显式(方)法
explicit model 明显的模型
explicit occurrence 显出现
explicit parameter 显(式)参数
explicit program(me) 显程序;明细计划
explicit quadratic function 二次显函数
explicit relation 显关系
explicit relaxation 显式松弛
explicit representation 显式(表示)法
explicit saturation calculation 显式饱和度计算
explicit scheme 已式格式
explicit solution 显(式)解
explicit specific 明确的
explicit type 显式类型
explicit type association 显式类型结合
explicit unary operator 显式一元算符
explicit value 确切值
explicit weighting 显式加权
explode 剖析
explode a bombshell 提出惊人的意见
exploded channel 炸槽
exploded drawing 揭示立体图(去顶去侧,可见内部)
exploded outcrop effect 爆破露头效应
exploded pile 爆扩桩
exploded view 解释图;零件分解图;立体图;展示图;分解图;部件分解图;剖视图;立体影像
exploder 雷管;引信;打炮器;爆炸装置;爆炸物;爆炸剂;爆炸工;爆破工;起爆器
exploder tester 引信检验器;放炮检验器
exploding 爆发
exploding bridge wire initiator 桥式爆炸发生器
exploding composition 爆炸混合物
exploding film Q-switching 爆炸膜 Q 开关
exploding mixture 爆炸混合物
exploding population 激增的人口
exploding reflector 爆炸反射面
exploding speed 爆速
exploding wire 引爆线;爆炸导线
exploding wire source 爆炸线闪光光源
exploer-typer satellite 勘探卫星
exploit 开采

exploitability 可利用性;可利用率;可利用度;可开发性;开发价值
exploitable girth 可利用干围级
exploitable mine pay 可采矿石
exploitable phase 可捕阶段
exploitable thickness 可采层厚度
exploitable tidal resources 可开发潮汐资源
exploitation 矿床开采;开发;开采;捕获利用
exploitation and development cost 开发和发展费用
exploitation cutting 单纯利用采伐
exploitation engineer 采掘工程师
exploitation engineering 开采工程
exploitation felling 单纯利用伐
exploitation field 开垦地
exploitation losses 开采损失
exploitation of an invention 发明利用
exploitation of marine food product 海洋食物资源开发
exploitation of ocean 海洋开发
exploitation of resources 资源开发
exploitation percent 采伐率
exploitation plan 开采计划
exploitation value 利用价
exploitation velocity 开采速度
exploited stock 开发鱼群
exploit(ing) 开拓(采矿)
exploiting and using of natural resources 资源的开发和利用
exploit the natural resources 开发资源
exploit the particular advantages of each region 发挥各地优势
exploit the virgin land 开垦荒地
exploration 考察;勘探;勘察;勘查;查勘;勘探;探索;探究;探查;钻探
exploration accuracy 勘探精度
exploration activities 勘探活动(包括航测、地质研究、物探及钻探等)
exploration activity 油勘探
exploration adit 勘探平洞;勘探坑道;勘探洞;勘矿平洞
exploration and developing(bore) hole 探采结合孔
exploration and developing drill-hole 探采结合孔
exploration and development cost 开发和发展费用
exploration and exploitation of seabed 海底探测与开发;海床的勘探开发
exploration and production 勘探与生产
exploration area 勘探区
exploration area structure 勘探区构造
exploration bit 勘探钻头
exploration boring 钻探;勘探钻进
exploration budget 勘探预算
exploration by echoes 回声探测(法)
exploration-classification 探查分类
exploration cost 勘探成本
exploration cost in extractive industries 采掘工业中的勘探成本
exploration crew 勘探队
exploration cycle 勘探周期
exploration data 勘探资料
exploration data interpretation system 勘探资料解释系统
exploration data processing 勘探数据处理
exploration department 勘探处;勘探部门
exploration depth 勘探深度;探测深度
exploration diving 勘察潜水
exploration drift 勘探平洞;勘探孔;探矿坑道
exploration drift cost 坑探成本
exploration drill 地质钻机
exploration drill hole for hydrogeology 水文地质勘探孔
exploration drilling 钻探;勘探钻井;勘探钻进
exploration economics 勘探经济学
exploration efficiency 勘探效率
exploration engineering 勘探工程;探矿工程
exploration equipment 探测设备
exploration error 勘探误差
exploration evaluation 地质勘查;地质勘测
exploration expenditures 勘探费用
exploration expenditure obligation 勘探费用义务
exploration expense economizing charges 地勘费用节约额

exploration for construction 施工勘察
exploration funds 勘探资金
exploration gallery 探查坑道
exploration geochemistry 勘探地球化学
exploration geologist 勘探地质学家
exploration geology 勘探地质学
exploration geophysics 勘探地球物理学;探测地球物理学
exploration geophysist 勘探地球物理学家
exploration grid method 勘探网法
exploration higher technology 探矿技术高
exploration hole 勘探孔;探测孔
exploration information system 勘探信息系统
exploration instrument for geochemistry 地球化学勘探仪器
exploration instrument for geophysics 地球物理勘探仪器
exploration instrument in geology 地质勘探仪器
exploration investigation 探索性研究
exploration investment 勘探投资
exploration island 勘探人工岛
explorationist 勘探家
exploration line method 勘探线法
exploration logging 勘探测井
exploration lower technology 探矿技术低
exploration manager 勘探经理
exploration map 勘探图
exploration method 勘探方法
exploration method by anologic(al)evaluation 类比评价法
exploration method by combined drilling-geophysical engineering 钻探与物探结合法
exploration method by combined drilling-opening engineering 钻坑探结合法
exploration method by combined exploratory-exploiting engineering 边采边探法
exploration method by combined opening-geophysical chemical engineering 坑探与物化探结合法
exploration method by drilling engineering 钻探工程法
exploration method by opening engineering 坑探工程法
exploration method by profile 剖面法
exploration method by sampling 取样法
exploration of chimney foundation 烟囱地基勘察
exploration of coal resources 煤炭资源勘探
exploration of factory foundation and vibrating ground 振动基础厂房地基勘察
exploration of launch tower site 发射塔址勘察
exploration of mineral deposits 矿床勘探
exploration of ocean 海洋开发
exploration of seabed 海底勘探
exploration of water tower foundation 水塔地基勘察
exploration operation 勘探作业
exploration permit 勘探许可
exploration plan 勘探计划
exploration planning 勘探设计
exploration point 探查点
exploration point of a deposit 矿床勘探点
exploration potential 勘探远景
exploration procedure 勘探程序;查勘程序
exploration-production hole 勘探生产孔
exploration-production well 勘探生产井
exploration program(me) 勘探计划
exploration project 勘探设计;勘探计划;勘探项目
exploration prospect evaluation 勘探远景评价
exploration prospecting instrument 勘探器具
exploration reserve proportion 勘探储量比例
exploration risk 勘探风险
exploration seismology 勘探地震学
exploration shaft 探查竖井
exploration ship 勘探船;探险船
exploration stage 勘探阶段
exploration strategy 勘探策略
exploration survey 勘探(测量);勘测;探测;踏勘测量
exploration target 勘探目的层
exploration time 探测时间
exploration trench 验探槽;探坑;探槽
exploration tunnel 勘探隧洞;勘探坑道;勘探洞;勘探导坑;勘矿坑道
exploration type 勘探工程类型

exploration type of ore deposits 矿床勘探类型
exploration types of opencast slope 露天矿边坡勘探类型
exploration vessel 科学调查船;勘探船
exploration well 勘探井
exploration well cost 探井费用
exploration work 勘探工作
explorative technologic(al)research 探矿技术研究
explorator 靠模
exploratory 探测的
exploratory adit 勘探平洞;勘测导洞
exploratory analysis 探索性分析
exploratory behavio(u)r 探究行为
exploratory(bore)hole 超前钻孔;试钻钻孔;勘探钻孔
exploratory boring 勘探钻孔;勘探钻井;勘探钻进;钻探;工程地质钻进;试钻
exploratory consultation 探讨性磋商
exploratory data analysis 探索性数据分析
exploratory decision 勘探决策
exploratory development 勘探开发;应用研究;探索研究;探索性研制;探索性开发;探索性发展
exploratory drift 勘探平巷;探坑;探洞
exploratory drilling 勘探钻井;勘探钻进;勘测钻探
exploratory drilling engineering 钻探工程
exploratory excavation 勘察挖坑;勘探(性)开挖
exploratory experiment 探索试验
exploratory fleet 勘探船队
exploratory forecast 探测性技术预测
exploratory forecasting 探索性预测
exploratory forecasting technique 探索性预测法
exploratory grid 勘探网;钻探网格
exploratory heading 勘探导洞;探洞
exploratory heading excavation 探洞开挖
exploratory hole 勘探孔
exploratory holes in face 开挖面的勘测孔
exploratory intensity 勘查工作程度
exploratory investigation 调查研究;初步勘探;初步勘察;探讨
exploratory level 勘探中段
exploratory line 勘探线
exploratory line azimuth 勘探线方位
exploratory line interval 勘探线间距
exploratory line number 勘探线号
exploratory line profile map 勘探线剖面图
exploratory line survey 勘探线测量
exploratory map 勘测图
exploratory method 探索法
exploratory mission 考察团;考察任务
exploratory move 勘探移动
exploratory multiprocessor 勘探多用处理机
exploratory of adit 勘察坑道
exploratory operation 探查术
exploratory orbitotomy 开眶探查术
exploratory period 勘探期
exploratory pit 探坑;探井
exploratory point 勘探点
exploratory puncture 试探穿刺
exploratory raise 探矿天井
exploratory report 勘探报告
exploratory research 探索(性)研究
exploratory sampling 勘探采样;探查抽样
exploratory scenario 调查提纲
exploratory search 考察;调查
exploratory shaft 勘探竖井;探坑;探井
exploratory spot 勘探点
exploratory study 探测性研究
exploratory survey 探测;勘测;探索调查;踏勘测量
exploratory talk 探讨性会谈
exploratory target 勘探对象
exploratory test 探查试验
exploratory test well 预探井
exploratory trench 探槽
exploratory trephination 钻孔探查术
exploratory tunnel 勘探隧道;探洞;探测隧道
exploratory tunnel excavation 探洞开挖
exploratory tunneling engineering 坑探工程
exploratory types of coal mine 煤矿勘探
exploratory well 探井
exploratory work 勘探工程;探索(性)研究
explored 勘探过的
explored depth 勘探深度
explored range 勘探范围
explored reserves 探明储量

explore for oil 勘探石油
explorer 勘探人员;地质勘探工作者;测试线圈;探针;探索器;探测员;探测线圈;探测器;探测机
exploring antenna 探测天线
exploring block 探测部件
exploring brush 测试(弧)电刷
exploring coil 测试线圈;探索线圈;探测线圈
exploring drift 调查导洞
exploring drill 探查孔
exploring electrode 探查电极
exploring element 探测元件
exploring forecasting technique 探索预测技术
exploring mining 坑探;勘探(性)开采;边探边采
exploring mining system 坑道工程系统
exploring shallow system 浅井工程系统
exploring spot 亮点;探索光点;搜索光点
exploring trench 槽探
exploring tube 探测管
exploritation oreblock method 开采块段法
explosibility 可爆性;爆炸性
explosibility curve 爆炸性曲线
explosibility index 爆炸性指数
explosibility limit 爆炸性极限
explosibility of coal dust 煤尘爆炸性
explosibility test 爆炸性试验
explosible 可爆炸的
explosimeter 可燃气体浓度测定仪;测爆仪;测爆计;爆炸计;爆炸测量器;气体可爆性测定仪
explosing-proof lift 防爆升降机
explosion 爆炸;爆破;爆发
explosion accident 爆炸事故
explosion action 爆炸作用
explosion along tunnel roof (隧洞)顶部爆破
explosion breccia 爆发角砾岩
explosion bubble 爆破气泡
explosion bulge test 爆破试验
explosion by influence 感应爆炸
explosion carter fumarole 爆炸坑里的喷汽孔
explosion center 爆炸中心;爆心
explosion chamber 灭弧箱;消弧室;爆发室;燃烧罐
explosion cladding 爆炸复合
explosion cloud 爆炸云
explosion collapse 爆炸崩坍;爆炸崩塌
explosion command 爆炸命令
explosion compactor 爆炸式夯实机
explosion concentration 爆炸浓度
explosion control 爆炸控制
explosion crater 爆裂火山口;爆发火山口
explosion depth 爆炸深度
explosion detector 防爆检测器;爆炸探测器
explosion diaphragm 防爆膜板
explosion disc 防爆盘
explosion disk 防爆盘
explosion ditching method 爆破成沟法
explosion door 防爆阀;爆炸安全门;防爆门
explosion drilling 爆破钻进
explosion duct 防爆管
explosion earthquake 爆破地震;爆裂地震
explosion energy 爆炸能量
explosion engine 爆燃式内燃机;爆发内燃机
explosion equipment 放炮设备
explosion-expanded piling 爆扩桩
explosion fissure 爆炸裂隙;爆炸裂缝
explosion flag 防爆物
explosion flap 防爆门;防爆阀
explosion focal point 爆炸焦点
explosion focusing peak magnitude 爆聚峰值
explosion forming 爆炸成型
explosion gas pipette 爆炸气管
explosion gas turbine 爆炸式燃气轮机;爆燃式燃气轮机
explosion generated ground motion 爆破引起地面运动
explosion grading 爆炸猛度测定
explosion hardening 爆炸硬化
explosion hardware 防爆五金件
explosion hatch (储罐)活动保险盖;防爆口
explosion hazard 易爆性;爆炸危险;爆炸事故
explosion hazard gas 易爆气(体)
explosion hazard protection 爆炸危险防护
explosion hazards of digester gas 消化池气(体)爆炸危险
explosion heat 爆炸热
explosion hole on lead plate 铅板炸孔

explosion imprint 爆炸痕迹
explosion in air 空爆
explosion index 爆炸指数
explosion-induced earthquake 爆炸诱发的地震；爆炸激发的地震
explosion in ground 地爆
explosion injury 爆炸伤
explosion insurance 爆炸保险
explosion interval 爆发间隔
explosion laser 爆发激光器
explosion-like impact 爆炸式冲击
explosion limit 爆炸极限
explosion load(ing) 爆炸荷载；爆炸荷载
explosion malfunction 爆炸故障
explosion mechanics 爆破力学
explosion meter 爆炸表
explosion method 爆炸方法；爆炸法；速燃法
explosion motor 爆燃式发动机
explosion notebook 爆破记录
explosion of coal mines 煤矿爆炸
explosion of firedamp 煤气爆炸
explosion of manhole 检查井爆炸
explosion of monetary crisis 爆发货币危机
explosion of sewer 管道爆炸
explosion period 爆燃期
explosion-pipe eruption 爆炸管爆发
explosion pipette 爆炸球管
explosion pot 起爆灭弧室
explosion pressure 爆炸压力
explosion pressure relief 爆破降压
explosion produced high-frequency background 爆炸产生的高频背景
explosion product 爆炸产物
explosion-proof 隔爆的；防爆(作用)；防爆炸的
explosion-proof and electric(al) heating drying box 电热防爆干燥箱
explosion-proof apparatus 防爆装置
explosion-proof armo(u)red door 防爆装甲门
explosion-proof box 防爆外壳
explosion-proof centrifugal blower 防爆离心鼓风机
explosion-proof class 防爆等级
explosion-proof compressor 防爆式压缩机
explosion-proof connection box 隔爆接线盒
explosion-proof construction 防爆设施；防爆结构
explosion-proof criteria 防爆标准
explosion-proof detector 防爆探测器
explosion-proof door 防爆门
explosion-proof electric(al) machine 防爆型电机
explosion-proof equipment 防爆装置
explosion-proof gas detector 防爆瓦斯探测器
explosion-proof generating line box 隔爆母线盒
explosion-proof glass 防爆玻璃
explosion-proof glazing 防爆窗玻璃
explosion-proof housing 防爆外壳
explosion-proof lamp 防爆灯
explosion-proof lighting 防爆(式)照明
explosion-proof luminaire 防爆照明器；防爆灯具
explosion-proof machine 防爆式电机；防爆机器
explosion-proof machinery 防爆机械
explosion-proof material 防爆材料
explosion-proof motor 防爆型电机；防爆马达；防爆电(动)机
explosion-proof moving substation 隔爆移动式变电站
explosion-proof packaging 防爆包装
explosion-proof safety fuse box 隔爆熔断器盒
explosion-proof safety hook 防爆安全钩
explosion-proof stopping 防爆隔墙
explosion-proof switch 防爆(式)开关
explosion-proof tank 防爆槽
explosion-proof transformer 防爆(式)变压器
explosion-proof tube 安全气道
explosion-proof type 防爆型
explosion-proof type detector 防爆式探测器
explosion-proof type induction motor 防爆式感应电动机
explosion-proof ventilated synchronous motor 防爆通风型同步电动机
explosion-proof wall 防爆墙
explosion protection 防爆装置；爆炸防护
explosion protection door 防爆门
explosion-protection measures 防爆措施
explosion protection system 防爆系统

explosion punching 爆炸冲压
explosion ram 内燃(机式)打夯机
explosion rammer 爆炸式夯实机；火力夯；爆炸夯
explosion ratio 爆炸比(空气和燃料)；爆燃比
explosion relief cover 防爆盖
explosion relief device 爆炸排气装置
explosion relief flap 防爆阀门
explosion relief membrane 防爆隔膜；爆炸安全隔膜
explosion relief panel 爆破通风板；爆破减压板
explosion relief port 泄爆门
explosion relief provision 泄爆装置
explosion relief valve 防爆安全阀；爆炸安全阀
explosion relief venting 爆炸排气装置
explosion-resistant enclosure 防爆围墙；防爆外壳
explosion-resistant structure 抗爆结构物
explosion risk 爆炸危险；爆炸风险；爆炸安全性
explosion riveting 爆炸铆
explosion rupture disc device 防爆片装置
explosion-safe 防爆的
explosion safety distance 爆破安全距离
explosion safety glazing 爆炸防爆窗玻璃
explosion scene 爆炸景象
explosion seam 防爆缝
explosion seismic effect 爆破地震效应
explosion seismic observation 爆破地震观测
explosion seismology 爆炸地震学；爆破地震学
explosion sensor 爆炸传感器
explosion shaping 爆炸成型
explosion shock wave 爆炸冲击波
explosion simulating earthquake 爆炸模拟地震
explosion site 爆炸位置；爆炸地点
explosions of digestion tank 消化池爆炸
explosion sound 爆炸声
explosion spread 爆炸范围；爆破展开
explosion stack 防爆管
explosion state 爆炸态
explosion-stimulated earthquake 爆炸激发的地震
explosion stroke 爆发冲程
explosion suppressing peak value 抑爆峰值
explosion suppression 爆炸遏制
explosion suppression control unit 抑爆控制器
explosion suppression system 爆炸遏制系统
explosion tamper 爆燃式夯具
explosion technique 爆破技术
explosion temperature 爆炸温度
explosion-tested equipment 防爆设备
explosion test ground 爆炸试验场
explosion test method 爆炸试验法
explosion to pieces 爆碎
explosion track 爆炸痕迹
explosion trap 爆炸阱
explosion trench 爆破路槽；爆破沟槽
explosion-triggering earthquake 爆破诱发地震
explosion tube 爆炸管
explosion tuff 爆发凝灰岩
explosion turbine 爆燃式透平机
explosion type core blower 射芯机
explosion type rammer 爆燃式夯具
explosion velocity of detonating cord 导爆索爆速
explosion velocity of explosive 炸药爆速
explosion vent 爆裂口；防爆门；爆破卸压孔；安全气道
explosion venting 防爆通风
explosion wave (爆炸)冲击波；爆炸波；爆破波
explosion wave pressure 爆炸波压力
explosion welding 爆炸焊(接)
explosion zone 爆炸区(域)
explosive 火药；(不熔)炸药；爆炸的；爆炸剂；爆炸物
explosive accident 爆炸事故
explosive action 爆破作用
explosive action circle 爆破作用圈
explosive action index 爆破作用指数
explosive-actuated device 爆炸致动装置；炸药驱动装置
explosive-actuated gun 射钉枪；爆炸驱动枪
explosive agent 炸药；爆炸剂；爆炸物
explosive anchorage 易爆货物锚地；装卸爆炸性货物的锚地；爆炸品锚地
explosive anchorage area 易爆炸品锚地
explosive area 爆炸区(域)；危险区(域)
explosive assembly 爆炸装置
explosive atmosphere 有爆炸危险的空气；有爆炸危险的环境；易爆气体环境；爆燃性空气

explosive attack 爆破作用
explosive blast 爆炸冲击波
explosive boiling 暴沸
explosive bolt 分裂螺栓；分两螺栓
explosive bonding 爆炸黏合；爆炸熔黏；爆炸焊(接)
explosive breccia 爆发角砾岩
explosive brisance (炸药)猛度
explosive bulge test 爆炸膨胀试验
explosive burning 爆燃
explosive burst 爆炸性破裂
explosive cable 引爆电缆
explosive capapcity 爆炸威力
explosive cargo 爆炸性货物
explosive carriage 炸药运输车
explosive cartridge 药包筒；药包卷；炸药卷；爆破筒
explosive cartridge density 药包装药密度
explosive cartridge length 药包长度
explosive cartridge type 药包类型
explosive cartridge weight 药包重量
explosive casting 爆破抛掷；抛掷爆破
explosive characteristics of fuel 燃料的爆炸特性
explosive charge 射孔弹；炸药装填量；炸药(费)；爆炸药量
explosive charge method 装(炸)药方法
explosive cladding 爆炸复层法
explosive combustion 爆炸性燃烧
explosive compacting 爆炸密实
explosive compaction 爆炸挤实；爆炸挤密
explosive compaction method 爆炸挤密法；爆炸压密法
explosive component 爆炸组分
explosive compound 炸药；爆炸(性)化合物；爆炸物
explosive consolidation 爆破固结；爆炸压密
explosive consumption 耗药量；炸药消耗量
explosive consumption per cubic meter 每立方米坑道炸药消耗量
explosive consumption per meter driftage 每米坑道炸药消耗量
explosive contaminated wastewater 爆发性污染废水
explosive control 爆炸控制
explosive cord 导炸索
explosive core 炸药药芯
explosive coupling distance 殉爆距离
explosive cutter 爆炸切割器
explosive cycle 爆发性周期
explosive D 苦味酸铵炸药；D 型炸药
explosive damage 爆炸破坏
explosive decompression 暴发性减压
explosive density 爆炸密实度(按每立方厘米克计)
explosive detector 爆炸探测器
explosive device 引爆装置
explosive dispersion 爆炸分散(法)
explosive displacement method 爆夯挤淤法
explosive disposal 未爆弹处置
explosive disruption 炸散；炸裂
explosive distance 爆距
explosive distance indicator 爆炸距离指示器
explosive door 防爆门
explosive drilling 爆炸钻井；爆破钻进；爆力钻进
explosive-driven 爆炸驱动的
explosive-driven anchor 爆抓锚
explosive-driven fastening 爆力打入式紧固件
explosive dust 爆炸性煤尘；爆炸性矿尘；爆炸性粉尘
explosive dust atmosphere 爆炸性粉尘大气
explosive dust mixture 爆炸粉尘混合物
explosive earthquake 爆炸地震
explosive echo ranging 爆炸回声定位；爆炸回声测距
explosive echo-ranging sonobuoy 爆炸回声测距声呐浮标
explosive effect 爆炸效力；爆炸效果；爆炸威力；爆破效果
explosive efficiency 爆炸效率
explosive element 爆炸装药
explosive energy 爆炸能量
explosive energy source 炸药震源
explosive engine 爆发内燃机
explosive evapo(u)ration 沸腾蒸发

explosive expanded pile 爆扩桩
explosive expanding 爆炸胀管
explosive expansion joint 爆炸胀接
explosive explosion 炸药爆破
explosive factor 炸药消耗率;炸药爆破系数;单位炸药消耗量
explosive fission 爆炸裂变
explosive fixing 安装爆炸物
explosive flare 爆发耀斑
explosive flux compression 爆炸磁通量压缩
explosive fog signal 爆炸式音响雾号;爆响雾号
explosive force 爆(炸)力;爆破力
explosive forging 爆炸锻造
explosive forming 爆炸成型
explosive fracturing 爆炸压力
explosive fringe 爆燃区
explosive fuel 易爆燃料
explosive fuel vapo(u)r 爆炸燃料蒸汽
explosive funnel 爆破漏斗
explosive gas 易爆气(体);爆炸(性)气体
explosive gas analyzer 爆炸性气体分析器
explosive gas detector 爆炸气体检测器
explosive gas indicator 爆炸性气体指示器
explosive gas mixture 爆炸气体混合物
explosive gelatin(e) 胶棉炸药;胶质炸药;硝化甘油
explosive gelating 胶棉炸药;胶质炸药
explosive generated ground motion 爆炸引起的地面运动
explosive goods 易爆炸品
explosive grading (机构)猛度测定
explosive grinding 爆炸粉磨
explosive hazard 爆炸危险
explosive hazard indicator 爆炸危险指示器
explosive hazard protection 爆炸危险防护
explosive heat 爆炸热
explosive high-speed shutter 爆炸高速快门
explosive holding depot 爆炸物仓库
explosive impulsion 爆破冲动
explosive increase 迅速增长
explosive index 爆炸指数;爆发指数
explosive index of coal dust 煤尘爆炸指数
explosive initiation 起爆
explosive injury 爆震伤
explosive isostatic pressing 爆炸等静压形成法;爆炸等静压成型法
explosive jar 爆炸振击器
explosive joining 爆炸连接;爆炸接合
explosive laser 爆炸激光器
explosive limit 着火浓度极限;爆炸(性)极限;爆炸范围
explosive limits of hydrocarbon 烃类爆炸极限
explosive load 装药量
explosively anchored 炸药爆破锚固锚杆法
explosively anchored rockbolt 爆固式锚杆
explosively calibrated seismic model 爆破校准地震模型
explosively driven pulsed chemical laser 爆炸驱动脉冲化学激光器
explosively embedded anchor 爆抓锚
explosive magazine 炸药库
explosive material 易爆材料;爆炸物(质);爆破器材;易爆物料
explosive metal 爆炸金属
explosive metal forming 金属爆炸成形
explosive meter 爆炸计
explosive methane 爆炸性沼气
explosive mixer 炸药混合器
explosive mixture 混合炸药;混合炸物;爆炸混合体
explosive motor 内燃发动机
explosiveness 可爆性
explosive noise 爆炸噪声
explosive oil 甘油三硝酸酯;硝化甘油;爆炸油
explosive ordinance disposal control center 爆炸物处理控制中心
explosive ordinance disposal squadron 爆炸物处理中队
explosive ordinance disposal unit 爆炸物处理小队
explosive ordinance reconnaissance 爆炸物侦察
explosive oscillation 爆发性振荡
explosive oxidizer 炸药的氧化剂
explosive perforation 爆炸打孔
explosive plant waste 炸药厂废水

explosive point 爆炸点
explosive powder 爆炸火药
explosive power 炸药威力;爆炸(能)力
explosive powerload 爆炸力载体
explosive pressing 爆炸压制;爆炸加压形成法
explosive process 爆发性过程
explosive processing 爆炸等静压形成法;爆炸等静压成型法
explosive-produced air shock 空中爆炸的冲击波
explosive-proof grade 防爆等级
explosive pulse 爆炸冲量;爆炸冲动
explosive-pumped gas dynamic(al) laser 爆炸抽运气动激光器
explosive punching 爆炸冲压
explosive ramming 爆炸紧实
explosive range 爆炸范围
explosive reaction 爆炸反应
explosive release 爆炸装置
explosive replacement method 爆破置换法
explosive ring 爆炸破坏
explosive risk 爆炸危险(性);爆炸安全性
explosive risk indicator 爆炸危险指示器
explosive rivet 带炸药铆钉;爆炸铆钉
explosive rock 喷发岩
explosives (易)爆炸品
explosives and dangerous goods anchorage 易爆及危险货物锚地
explosives area 爆破器材堆场
explosive seismic observation 爆破地震观测
explosive seismic origin 爆炸震源
explosive separation 机构脱落;爆炸分离
explosive-set anchor 爆炸安装锚
explosive severity 爆炸烈度
explosives for construction 施工炸药
explosives handling 爆炸物装卸;爆炸物处理
explosive signal 爆炸信号
explosive sintering 爆炸烧结
explosive site 爆炸位置
explosives-loading factor of hole 炮眼装药系数
explosives-loading length 装药长度
explosives-loading weight of hole 炮眼装药量
explosives magazine 炸药库;炸弹库
explosive snuffing 爆炸灭火
explosive solid waste 爆炸性固体废物
explosive sound source 爆炸声源
explosive source 炸药式震源;爆源
explosive speciation 爆破式物种形成
explosive specific charge 单位药耗;单位炸药消耗量
explosive speech 爆炸式语言
explosive state 爆炸态
explosive-stimulated earthquake 爆破激发地震
explosive strength 炸药威力
explosive stripping 爆炸剥离
explosive substance 炸药;爆炸物;爆炸剂;爆炸性物质
explosive substance in sewage 污水中易爆物质
explosive swaging 爆炸成型
explosive technique 爆炸工艺;爆破技术
explosive train 火药系;导火药;导爆索;传爆系统;分段装药
explosive type 爆发型
explosive-type engine 点火爆发型内燃机
explosive-type malfunction 爆炸型故障
explosive valve 爆发活门
explosive vapo(u)r 可爆蒸汽
explosive vapo(u)r indicator 爆炸性气体指示器
explosive vent 爆裂口
explosive venting 泄爆
explosive vibration 爆破振动
explosive vibration observation 爆破地震观测
explosive view 爆炸图
explosive volcano 爆裂式火山
explosive warming 爆炸性增温
explosive waste 爆炸性废物
explosive waste water 爆炸性废水
explosive wave 冲击波;爆炸波
explosive weight 爆炸量
explosive weld(ing) 爆炸焊(接)
explosive wound 爆炸伤
explosive yield 爆炸当量
explosivity 爆炸性
explosivity limit 爆炸性极限;爆炸限度
expneses of harvesting and of collection of revenue 林木采收费
expo 展览会;市集
ex point of origin 现场交货价格;由发货点算起
expometer 露光计;曝光计
exponent 幂数;指数;代表者;说明者
exponent arithmetic 阶运算
exponent character 阶符
exponent code 阶码
exponent counter 阶计数器
exponent curve 指数曲线
exponent distribution 指数分布
exponent equation 指数方程
exponent factor 指数因子
exponent frame 阶框架
exponential 指数的
exponential absorption 指数吸收
exponential absorption law 指数吸收定律
exponential approximation 指数近似(法);指数逼近
exponential assembly (反应堆的)指数(实验)装置
exponential asymptotical stability 指数式渐近稳定
exponential atmosphere 指数大气
exponential attenuation 指数减弱
exponential attenuation method 指数衰减法
exponential average 指数平均(数)
exponential convergence 指数收敛
exponential counter 指数计数器
exponential curve 指数曲线
exponential curve fitting 指数曲线拟合
exponential curve forecasting 指数曲线预测法
exponential curve of fit 拟合指数曲线
exponential damping 指数衰减;指数式阻尼;指数阻尼
exponential damping factor 指数衰减因素
exponential decay 指数(性)衰减;指数式衰减
exponential decay curve 指数衰减曲线
exponential decay law 指数衰减律;指数衰变律
exponential decay time constant 指数衰减时间常数
exponential decline 指数递减
exponential decline equation 指数递减方程
exponential decrease 指数递减
exponential density function 指数密度函数
exponential dilution method 指数稀释法
exponential discharge 指数衰减;指数律放电
exponential disintegration 指数衰变
exponential distribution 指数分布
exponential doubling time 指数倍增时间
exponential equation 指数方程
exponential experiment 指数实验
exponential extrapolation 指数外推法
exponential family of distributions 指数分布族
exponential filter 指数滤波器
exponential filtering 指数滤波
exponential flow 指数(式)流动
exponential form 指数形式
exponential formula 指数公式
exponential function 幂函数;指数函数
exponential function approximation 指数函数近似
exponential function model 指数函数模型
exponential functor 指数函子
exponential gain correction 指数增益校正
exponential generating function 指数母函数
exponential gradient 指数梯度
exponential gradient device 指数梯度装置
exponential growth 指数增长;指数式成长;指数生长
exponential growth curve 指数生长曲线
exponential growth level 指数增长级
exponential holder 指数式保持器
exponential horn 指数曲线形振幅扩大棒;指数曲线形喇叭
exponential increase 指数增长;指数式增加
exponential integral 幂积分;指数积分
exponential integral function 指数积分函数
exponential interpolation 指数内插(法);指数插值
exponential lag 指数滞后
exponential law 指数定律
exponential law of attenuation 衰减指数定律
exponential law of error 误差指数律
exponential line 指数传输线
exponential loss of pollutant mass 指数污染物损失
exponential loudspeaker 指数式扬声器

exponentially damped quantity 指数减幅量
exponentially damped sinusoid 指数阻尼正弦曲线
exponentially damping quantity 指数阻尼量
exponentially decaying rate 指数衰减率
exponentially stable 指数式稳定的
exponentially tapered line 指数锥削形传输线;指数衰减线路
exponentially weighted moving average 指数型加权移动平均
exponential matching 指数匹配
exponential method 指数法
exponential motion 按指数规律运动
exponential notation 指数记数法;指数记号;指数计数制
exponential operator 指数算子
exponential order 指数阶
exponential oscillation 指数振荡
exponential part 指数部分
exponential part format 指数部分格式
exponential part of number 数的指数部分;数的阶部分
exponential phase 指数期
exponential phase of growth 指数生长期
exponential pile 指数(反应)堆
exponential profiled hopper 仓壁指数曲线形斗仓
exponential pulse 指数形脉冲;指数式脉冲
exponential quantity 指数值;指数量
exponential ramp 指数倾斜
exponential random variable 指数随机变量
exponential reactor 指数(反应)堆
exponential region 指数(分布)区
exponential relationship 指数关系
exponential representation 指数表示
exponential rising 指数律上升
exponential scale 指数比例尺
exponential series 指数级数
exponential service 指数分布服务
exponential smoothing 指数修匀(法);指数平滑;指数调整;指数平滑
exponential smoothing (method) 指数平滑法
exponential solution 指数解
exponential subroutine 指数子程序
exponential sum 指数和
exponential survival curve 指数存活曲线
exponential sweep 指数(式)扫描
exponential sweep generator 指数扫描振荡器
exponential taper 指数递减
exponential tapered line 指数锥削线
exponential term 指数项
exponential theorem 指数定理
exponential time base 指数式时基
exponential time delay 指数特性时间延迟
exponential tool table 指数曲线形工具架
exponential transformation 指数变换
exponential transient 指数型过渡历程;指数瞬变过程
exponential transmission line 指数传输线
exponential trend 指数趋势法
exponential trend curve 指数型趋势曲线
exponential tube 指数(特性曲线)管
exponential type 指数型
exponential type curve 指数型曲线
exponential type distribution 指数型分布
exponential valuation 指数赋值
exponential value 指数值
exponential voltage change 指数变化电压
exponential wave 指数波
exponential waveform 指数曲线波形
exponential weighting 指数加权
exponential well 指数势阱
exponentiate 指数化
exponentiation 指数表示;阶的表示;取幂
exponentiation operator 取幂运算
exponentiation sign 取幂符号
exponent-marker 阶码标记符
exponent number 阶数
exponent of expansion 膨胀指数
exponent overflow 阶上溢;阶码溢出
exponent overflow exception 阶上溢异常
exponent part 阶部分;指数部分
exponent part format 阶部分格式;指数部分格式
exponent part of number 数字阶部分
exponent picture character 指数图像字符
exponent range 阶码范围

exponent sign 幂指数(符号);指数符号
exponent smoothing 指数平滑法
exponent specifier 指数区分符
exponent uncorrelation 不相关指数
exponent underflow 阶下溢
exponent underflow exception 阶下溢异常
exponent underflow mask 阶下溢屏蔽
exponometer 曝光时计
export 码头交货;出口(指货物等);输出(品)
exportable surplus 可出口盈余;可出口剩余量
export account 出口往来账户
Export Administration Act 出口管理法(美国)
export advance 出口预付款
export advanced system 出口预付制
export agent 出口代理商
export and import 进出口(指外贸方面)
export and import licensing system 出进口许可制度
export and import price index 进出口物价指数
export and import records or performance 进出口实绩
export and import revolving fund 进出口周转基金
exportation of public nuissance 公害输出
export authorization 出口授权书;出口核准制
export availability 可出口量
export bank 出口银行
export bill 出口(清)单;出口汇票
export bill for collection 出口汇票托收
export bounties 出口奖励金
export bounty 出口津贴;出口补贴;输出奖励金
export by deferred payment 延期收款出口
export calculation sheet 出口价格计算表
export capability 出口能力
export cargo 出口货(物)
export cargo at all harbo(u)rs 海港出口货物量
export cargoes 吐量
export cartels 出口卡特尔
export-centered enterprises 出口主导型企业
export certificate 出口证明书
export certification 出口许可证
export charges 出口费用
export collection 出口托收
export commercial sample carnet 商业样品进口证
export commission 出口代理佣金
export commission agent 出口代理商
export commission house 出口代理商;出口代办行
export commoditiy 出口商品
export container 出口集装箱
export contract 出口合同
export control 出口管制;出口管理
export cost insurance 出口费用保险
export cost in terms of foreign exchange 出口换汇成本
export credit 出口信用;出口信贷
export credit guarantee 出口信贷保函;出口信贷保证;出口货信用担保
Export Credit Guarantee Department 出口信贷保证局(英国)
export credit guarantee facility 出口信贷担保办法
export credit guarantee scheme 出口信用担保制度
export credit insurance 出口信用保险;出口信用保险;输出信用保险
export credit insurance facility 出口信贷保险业务
export credit loan 出口信贷性贷款
export credit on deferred payment 延期收款出口信贷
export credit rate 出口信贷利率
export credit scheme 出口信贷计划
export data 出口数据
export debit a-c 出口支出账户
export declaration 出口申请书;出口申报(单);出口报关单
Export Development Corp. 出口发展公司
export discount 出口折扣
export distribution agreement 出口经销协议
export document 出口文件;出口单据
export drawback 海关退税
export drop shipper 出口中间商
export dumping 出口倾销;输出倾销
export duty 出口税
export enhancement program(me) 增强出口计划
export enterprise 出口企业
export entry 出口报单
exporter 出口商;输出者

exporter's advance 出口商货款
exporter's currency 出口国货币
export exchange 出口外汇
export expansion 扩大出口;出口扩展
export factoring 出口货款保收业务
export finance house 出口金融公司
export finance insurance 出口信贷保险;出口融资保险
Export-Financiering-Maatschappij 出口信贷公司(荷兰)
export financing 出口投资;出口融资;出口资金融通
export flow function 出口流量函数;出口流动函数
export goods 外销货
export goods withdrawn for sale on home market 出口转内销商品
export harbor 输出港
export house 出口行;出口商号
Export-Import Bank 进出口银行(美国)
Export-Import Bank of Japan 日本进出口银行
export-import cover ratio 进出口比率;出进口比率
export-import financing 进出口资金融通
export import firm 进出口商
export-import operation 进出口营业
export incentives 出口刺激
export indemnification 出口补偿
exporting unemployment 输出失业
export inspection system 出口检验制度;出口检查制度
export installment 分期付款出口
export insurance 出口保险;输出保险
export insurance covers 80% of commercial risk 出口保险抵偿商业风险的80%
export-intensive industry 出口密集型产业
export investment 出口投资
export invoice 出口(货物)发票
export item 出口项目
export jetty 出口码头
export kerosene 出口煤油
export labo(u)r power 输出劳动力
export-led growth 出口增加带动的经济增长;出口带动的增长
export-led industrialization 面向出口的工业化
export-led industrial structure 出口导向型产业结构
export letter of credit 出口信用证
export letter of credit received 开来出口信用证;收到出口信用证
export license 出口许可证;输出许可证
export licensing regulation 出口许可条例
export licensing system 出口许可证制度
export list 海关出口货物分类表
export loan 输出贷款
export management company 出口代理公司
export manifest 出口舱单
export mark 出口标志(指商品出口);产地标志
export market entry guarantee scheme 出口市场开发保险制
export merchant 出口商
export multiplier 出口乘数
export of capital 资本输出
export of hazardous waste 危险废物的出口
export order 出口订单
export organization 外销机构
export orientation 出口导向
export-oriented economy 出口(主导)型经济;外向型经济
export-oriented enterprise 出口型企业;出口基地企业
export-oriented industry 面向出口的工业;出口导向工业
export-oriented refinery 出口型炼油厂
export package 出口包装
export packaging 出口包装
export packing for buyer's account 出口包装费由买方负担
export permit 出口许可证;输出许可证
export pier 出口码头
export point 输出地点
export prepayment 出口预付款
export price 出口价格
export price index 出口价格指数
export priority 出口优先权
export process(ing) free zone 出口加工免税区
export process(ing) zone 出口加工区;加工出口区
export production 输出生产

export-production-first policy 出口产业优先政策
export products 出口产品
export promotion charges 出口推广费用
export promotion department 出口促进部
export promotion loan 出口融资
export quantum 出口量
export quarantine 出口检疫
export quota 限额输出;出口限额;输出限额;出口配额
export quota certificate 出口配额证明书
export quota system 出口配额制
export quotation 出口报价
export rate 出口汇率;出口航运费率
export ratio 输出比例
export rebate 出口回扣;出口减税
export refinery 出口炼(油)厂
export regulations 出口条例
export-related loan 出口有关贷款
export restitution 出口补偿
export restraint 出口限制
export restriction 出口限制;输出限制
export retention quota 出口外汇保留额
export riser 输出立管
exports 出口物;出口商品;出口货
export sale 外销
export sales department 出口销售部门
Export Services Division 出口服务处
exports of goods and material services 货物和物质性服务出口
export structure 出口结构;输出结构
export subsidy 出口津贴;出口补贴;出口贴补
export substitution 出口替代
export superiority 出口优势
export supply 出口供应
export surplus 出超
export target 输出目标制
export target system 出口指标制度
export tax 出口税
export tax rebates 出口退税
export-tax relief 出口税救济
export technique 输出技术
export tender 出口招标
export tender risk avoidance 出口投标防止风险期权
export terminal 出口油库;出口码头
export trade 出口贸易;外贸出口
Export Trade Act 出口贸易法(案)
export trade bill 出口贸易票据
export trader 出口商
export trading company 出口贸易公司
export unit value index 出口商品单位价值指数
export usance bill 出口远期汇票
export value 出口价格
export volume 出口量
export volume index 出口量指数
export ware 洋器
export weight 出口重量
export wharf 出口(货)码头
expose 露出;辐照;暴露;曝露;曝光
exposed 被辐照的;外露的
exposed aggregate 露面集料;浮露骨料;外露骨料;洗石面(混凝土表面);洗出骨料;水刷石
exposed aggregate concrete 露石混凝土;外露骨料混凝土
exposed aggregate finish 外露骨料饰面;水刷石饰面;混凝土裸露骨料处理
exposed aggregate panel 外露骨料墙板
exposed aggregate surface 露明集料面;露明骨料面
exposed aggregate texture 面露集料纹理;面露骨料纹理;外露骨料饰面
exposed anchorage 暴露的锚地;未掩护的锚地
exposed anticline 外露背斜层
exposed application 外露使用
exposed area 外露区域;暴露面积;无屏障地带;一般外露面积;屋面材料外露面积
exposed berth 无屏障泊位
exposed block 裸露的砌块
exposed body 露胎;缺釉;脱釉
exposed bone 露骨
exposed brickwork 裸露的砌砖工程
exposed building 暴露建筑物
exposed building face 建筑外露表面
exposed burning area 暴露燃烧面积

exposed carcassing 明线;明管
exposed cast component 裸露的烧制成分
exposed cast member 裸露烧制构件
exposed ceiling 露天顶棚
exposed ceiling grid system 明龙骨吊顶系统
exposed coalfield 出露煤田;暴露式煤田
exposed coast 裸露海岸;暴露海岸
exposed concrete 清水混凝土
exposed concrete beam 裸露的混凝土梁
exposed concrete casting 裸露的混凝土浇制件
exposed concrete column 裸露的混凝土柱
exposed concrete facade building component 裸露的混凝土立面建筑构件
exposed concrete finish 裸露混凝土的整修
exposed concrete form 裸露混凝土形状;可以看到的混凝土形状
exposed concrete form work 裸露混凝土模板工程
exposed concrete panel 裸露混凝土面板
exposed concrete shuttering 裸露混凝土模板
exposed concrete stair(case) 裸露混凝土楼梯间
exposed concrete texture 裸露混凝土结构
exposed concrete wall 混凝土外墙
exposed conduit 明管
exposed core (焊条的)夹持端
exposed cutting surface (金刚石)出刃;外露切削刃
exposed deck 露天甲板;敞篷甲板;无铺板甲板
exposed deep water 深海
exposed deposit 浮露矿床
exposed downpipe 明雨水管
exposed drive 开式传动
exposed edge 外露面;外露棱边
exposed electric(al)wire 外露电线;外露导线
exposed end 裸露端
exposed explosion 裸露爆破
exposed exterior sealer 外露密封剂
exposed face 外露面
exposed facing block 外露预制(砌)块
exposed field concrete 现拌外露混凝土
exposed finish 贴面修饰;暴露施工
exposed finish tile 外饰面砖;外露装饰瓷砖;外露装修瓷砖
exposed flashing(piece) 裸露的防雨板;裸露的防水板
exposed framing glass curtain wall 明框玻璃幕墙
exposed gear 开式齿轮
exposed glass area 采光玻璃面积
exposed granite aggregate finish 花岗岩集料面;花岗岩骨料面
exposed gravel aggregate panel 露明砾石板
exposed installation 露明式装置(法);明设;明管布设法
exposed intake 明流进水(口)
exposed joint 明接头;明接(合);明缝
exposed joist (露)明格栅
exposed liabilities 暴露负债
exposed lightweight concrete block 裸露轻质混凝土砌块
exposed live part 外露带电部件
exposed location 暴露的地方;无屏障地带
exposed location single buoy mooring 开敞水域单点系泊;外港单浮筒系泊
exposed masonry(work) 裸露砖石砌筑;清水墙;露面圬工;清水砌体建筑
exposed nailing 明钉法;外露钉(钉法)
expose dose 照射剂量
exposed overlying layer 地窗(暴露的覆盖层)
exposed pallets 外露擒纵叉
exposed part 裸露部分
exposed penstock 露天式压力(水)管
exposed pipe 露天式管道;明装管道;明管
exposed piping 明管
exposed population group 暴露人群
exposed portion 外露部位
exposed position 裸露位置
exposed post 明梁(承重)柱
exposed precast concrete component 裸露预浇混凝土构件
exposed radiator 明散热器
exposed reinforced concrete 裸露的钢筋混凝土
exposed reinforcement 露筋
exposed rib 裸露的肋
exposed road 空旷地区的道路;野外道路
exposed rock 明礁
exposed rock formation 裸露岩层

exposed rock surface 露头岩面;岩石出露面;浮露岩面
exposed side 向风面;暴露面
exposed site-placed concrete 裸露在工地上的混凝土
exposed situation 暴露的场所;无掩遮物的地带
exposed soil 裸土
exposed soil and rock formation 出露土层
exposed spiral case 外露式蜗壳
exposed station 露天台站;露天观测场;露天场地
exposed steel construction 外露式钢结构
exposed steel framing 裸露钢框架
exposed subsoil 浮露底土
exposed surface 裸露表面;冲刷面;暴露(表)面
exposed suspension system 外露悬挂法;露龙骨吊顶系统
exposed tank 暴露油罐
exposed tile 裸露花砖
exposed to air 曝露在空气中
exposed to contact 容易接触的
exposed to water 与水接触;受水;受潮
exposed to weather 露天;放在露天;曝露在大气中
exposed types of coalfield 煤田暴露(类)型
exposed unit 明露元件
exposed wall 外墙;外墙(围)墙
exposed waters 开敞水域;无屏障水域
exposed wiring 明线布置;明线(布线);明敷线
exposed zone 暴露区
exposing 外露
exposing building 有造成蔓延火灾危险的建筑物
exposing chart 曝光表
exposing engine map of survey area 测区揭露工程图
exposing lamp 曝光灯
exposing meter 曝光表
exposing system 曝光系统
exposition 展览会;显露;暴露;曝光
exposition architecture 展览会建筑
exposition area 展览会面积;陈列面积
exposition block 展览会大楼
exposition garden 展览会花园
exposition ground 展览场地
exposition hall 展览大厅
exposition on human settlement 人类住区展览
exposition palace 展览宫
exposition pavilion 展览馆
exposition stand 展览看台
expositor 解说员;评注者
exposive wedge 爆破楔
ex-post analysis 事后分析
ex post calculation 事后计算
ex post forecast 事后预测
ex post forecast error 事后预测误差
ex post investment 事后投资
ex post measurement of benefit 工程竣工后的效益估算;实际效益估算
ex post payoff 事后偿付
ex post simulation 事后模拟
exposure 露头;开敞程度;显露;照射量;从事危险工作的时间;出露;陈列;辐照量;辐射曝露量;暴露(风险);曝射量
exposure albedo 照射反照率
exposure area 开架区
exposure assessment 接触评定;暴露评价
exposure axis 曝光轴
exposure buildup factor 照射积累因子
exposure button 曝光按钮
exposure cage 照射栅格;照射容器
exposure calculating chart 曝光计算表
exposure calculator 曝光计算表
exposure cavity 照射腔
exposure cell 照射室
exposure chamber 接触室
exposure characteristic curve 曝光特性曲线
exposure chart 露光表;照射(时间)表;曝光图表
exposure clock 曝光钟;曝光计
exposure coefficient 曝光系数
exposure compensation 曝光补偿
exposure container 照射容器
exposure control 曝光控制
exposure control band 曝光控制带
exposure controller 曝光控制器;曝光控制计
exposure control stripe 曝光控制带
exposure counter 曝光计数器

exposure cracking 曝露龟裂;天候龟裂
exposure cycle 曝光周期
exposure cycling rate 曝光周率
exposure determination 曝光测定
exposure dimmer 照射衰减器
exposure distance 具有火灾危险的距离;曝露距离;危险距离
exposure dose 接触剂量;照射剂量;辐射剂量;暴露剂量;曝射剂量
exposure dose rate 照射剂量率
exposure duration 暴露时间;曝光时间
exposure effect 大气影响效应;大气暴露效应;曝光效果
exposure-effect relationship 暴露效应关系
exposure error 曝光误差
exposure facility 曝晒设施;曝晒设备
exposure factor 照射量因子;暴露因子;曝光系数
exposure field 照射场
exposure fire 暴露火灾
exposure gate 曝光口
exposure guide 曝光一览表;曝光手册
exposure hazard 着火危险性;暴露危险
exposure head 曝光头
exposure hole 照射孔道
exposure hours 全体雇员的工时数;曝光时间
exposure index 曝光指数
exposure indicator 照射量指示器;曝光指示器
exposure integrator 曝光积分器
exposure intensity 照射强度;曝光强度
exposure interval 曝光间隔
exposure in two parts 两次曝光
exposure label(l)ing 照射标记
exposure lamp 曝光灯
exposure latitude 露光宽裕度;曝光时限;曝光宽容度;曝光范围
exposure level 接触程度;暴露水平;曝光量级
exposure limit 暴露极限
exposure line of coal seam 煤层露头线
exposure logarithm 曝光量对数
exposure machine 曝光机
exposure maximum temperature of source rock 烃源岩经历的最高温度
exposure measurement 曝光测定
exposure meter 露光计;照射量计;暴露表;曝光计;曝光表
exposure meter reading 曝光表读数
exposure needle 曝光指针
exposure number 航空相片编号
exposure of aggregate 集料外露;骨料外露;骨料浮露;粒料外露
exposure of lining 露砖
exposure of rain ga(u)ge 雨量器的承雨面;雨量计的承雨面
exposure panel 曝晒(试验)样板
exposure parameter 接触参数
exposure pathway 照射径
exposure pattern 暴露方式
exposure period 曝光周期
exposure phase 接触相
exposure point 曝光点
exposure protection 暴露保护设备
exposure rack 曝晒架
exposure range 曝光(量)范围
exposure rate 接触速率;照射(量)率
exposure rate constant 照射量率常数
exposure ratemeter 照射(量)率计
exposure recording paper 曝光记录纸
exposure regulator 曝光节制器
exposure scale 照相机曝光量范围;曝光标(度)
exposure shutter 曝光快门
exposure site 露头位置;暴露地点;曝置场;曝晒场
exposure slit 曝光缝
exposure speed 曝光速度
exposure station 曝光站;曝光点;摄站
exposure step 曝光梯度
exposure suit 防寒衣
exposure system 曝光系统
exposure table 曝光用表
exposure test 接触检验;暴露试验;曝晒试验;曝露试验;曝光试验
exposure test fence 曝露试验台
exposure test frame 曝晒试验架
exposure test strip 曝光试验条
exposure time 接触时间;照射时间;辐照时间;曝光时间
exposure timer 曝光定时器
exposure time range 照射时间范围
exposure time table 照射时间表
exposure to adverse condition 暴露于不利条件
exposure to atmosphere 大气暴露
exposure to exchange risk 冒受汇兑风险
exposure to fire 曝热起火
exposure tolerance 曝光限
exposure to pollutant 曝露于污染物
exposure to sunlight 曝光;曝露于日光下
exposure to translation risk 折算风险暴露
exposure to weather 露天
exposure tube 照射管
exposure unit 曝光装置
exposure value 曝光值
exposure velocity 接触速度;曝光速度
expound 阐述;阐明
express 快运;快汇;快递;急件;表达
express abrogation 明确废除
express acceptance 明示承诺
expressage 快运(费);捷运费
express agency 快件运输公司;快运公司;捷运行
express alternate program(me) block 加速交替程序块
express analysis 快递分析
express arrangement 明示和解
express artery 快速干道;高速干道
express assignment 明示转让
express authority 明示授权
express bailment 明示保释
express boat 快艇;快船
express boiler 快升温锅炉
express bus 快速公共汽车
express-bus on freeway 高速公路快速公共汽车
express car 快运包囊车
express cargo boat 快速货船
express coach 特快车
express company 快运公司;快件运输公司;快递公司
express condition 明示条件
express consent 明示同意
express consideration 明示约因
express consignment 快递货;快件货;快达货
express container ship 快速集装箱
express contract 明约;明示合同
express deed 明示契约
express delivery 快递;明示交付
expressed covenant 明文条款
expressed folio 对开纸
expressed oi 不挥发油
expressed scale 理论比例尺;规定比例尺
expressed warranty 明示保证
expressed water 挤出水
express elevator 快速电梯;高速电梯
express engine 特别快车机车
expresser 压榨器
express exit ramp 快速出口坡道
express extradition 明示引渡
express fee 快递费;快办费;捷运费
express freight train 快速货车;特快货物列车
express goods 快运货(物);快运的货物;快件货物;快递货
express goods elevator 快速货运电梯
express goods lift 快速货运电梯
express guarantee(ship) 明示担保
express highway 快速公路;高速公路
expressible method of salinity 盐度表示法
expression 压榨(法);榨出;表示式;表达(式)
expression art 表现艺术
expression as bound or length 界或长度表达式
expression field 表达式区段
expression forceps 压榨镊(钳)
expressionism 表现主义
expression method of heavy placer prospecting result 重砂测量成果表示法
expression of interest 显示其重要性
expression of result 结果的表达
expression of toxicity 毒性符号
expression parsing 表达式分析
expression precision 表达式精度
expression roll 压辊
expressive method of infiltration 渗流的表示方法
expressiveness 可表达性
expressivity 表现度
express laboratory 快速化验室
express letter 快信
express lift 快速吊机;快速电梯;高速电梯
express line 快运线路
express liner 快速定期船;快速班船;定期快轮;特快班轮
express locomotive 高速机车
expressly agreed terms 明文规定
express mail 快件;快递邮件
express motorway 高速国家公路
express objection 明示异议
express offer and acceptance 明示要约与承诺
express or implied 明示或默示
express paid 快递费已付
express paid by post 加快费已邮付
express paid by telegraph 电汇付讫
express passenger elevator 快速客运电梯
express passenger lift 快速客运电梯
express passenger train 快速旅客列车;特快旅客列车;特快客运列车
express pile 大头桩
express post 快递邮件
express proclamation 明文公告
express promise 明示诺言
express provision 明文条款;明文规定;阐明条款
express pump 高速泵
express recognition 明示承认
express renunciation 明示拒绝履行
express repeal 明显废止
express revenue 快件运输收入
express road 高速公路
express service 快速服务
express statistics 快速统计
express summary of operation 生产快报
express telegram 急电
express terms 明文规定;明示条款
express terms of contract 合同的明文规定
express ticket 快车票
express track 快运线路(美国)
express traffic 快速交通
express train 快车;特快列车;特别(旅客)快车
express train locomotive 快车机车
express transportation 快运;急件运输
express truck 运送快件的列车
express trust 明示信托;书面信托
express type Francis turbine 特快型混流式水轮机;特快型法兰西斯式水轮机
express wagon 快车
express waiver 明示放弃
express warranty 明示担保;载明保证
expressway 快行道;快车路;快速(公)路;快速干道;快速道路;高速公路;高速干道
ex privileges 不予优待;无优惠
expromissor 替代债务人
expropriate 没收
expropriation 没收财产;征用(土地);让渡
expropriation contingency 紧急征用
expropriation of building land 建筑土地征用
expropriation of land 征地;征用土地;土地(的)征用
expropriation of property 财产没收
expropriation proceedings 征地手续
exptended sample 扩充样本
expulsion 喷溅
expulsion efficiency 排驱效率
expulsion element 熄弧室
expulsion force 排出力
expulsion from source rock 从母岩系统驱出
expulsion fuse 冲出式熔丝;冲出式熔断器
expulsion fuse unit 冲击式熔断丝;冲出式熔断器
expulsion gap 冲出式熔丝保护放电器
expulsion mechanism 驱动机理
expulsion mechanism of gravitational compaction 重力压实排驱机理
expulsion of arc 弧的吹熄
expulsion of water 去水;脱水
expulsion pressure 喷射压力
expulsion protective gap 冲出式保护放电器
expulsion type fuse 冲出式熔断器
expulsion type surge arrestor 吹弧式避雷器
expulsive efficiency of hydrocarbon 排烃效率
expulsive force 推动力
expulsive hydrocarbon coefficient 排烃系数

expulsive stage 排出期
expunction 抹掉
expunge the class 注销船级
expurgated bound 修正限
expurgated code 删信码
expurgated edition 修订版
ex quay 码头交货价(格);码头交货;码头交换价(格);码头完税价;目的港码头交货
ex quay duty paid 码头交货价(格)
ex quay term 码头交换条件
exquisite 精致的
ex rail 铁路旁交货;铁路交货(价)
ex-refinery price 出炼厂价
exrographic(al) printer 静电印刷机
exrophreatophyte 适旱性植物
exroradiography 干放射性照相术
exsecant 外正割
ex-seller's godown 卖方仓库交货价(格)
ex-seller's warehouse 卖方仓库交货价(格)
ex-ship 船边交货;目的港船上交货(价);船上卸下的;船上交货(价格)
exship's hold 舱底交货的到岸价格
ex-ship to rail 卸船装火车
ex-showroom 展销交货(价格);展室交货价
exsiccate 使干燥
exsiccated alum 干燥明矾;干燥铝矾
exsiccated calcium sulfate 干燥硫酸钙
exsiccation 干燥法;除湿作用;土地变干
exsiccation fever 缺水热
exsiccative 干燥的;使干燥的
exsiccator 干燥器;除湿器;保干器
ex-situ conservation 迁地保护
ex-situ treatment 易地处理
exsolution 离溶;矿物析去;固相分离;出溶(作用)
exsolution lamellae 初溶层;出溶纹(层)
exsolution of solid-solution 固溶体离溶
exsolution paragenesis 离溶共生
exsolution substance 出溶物
exsolution texture 固溶体分解结构;出溶结构
exsorption 外吸渗
exstock anchor 不计重的有锚杆
ex-store 出厂;仓库交货(交货)
exsuccation 吸出术
exsuction 吸出术
exsudatinite 热析沥青体;渗出沥青体
exsufflator 排气器
exsurgence 喀斯特泉;溶洞泉
extanded source 展源
Extar 埃克斯塔阻燃聚酯纤维
extemporaneous explosive 临时炸药
extempore 即席;当场
extend 扩张;扩充;蔓延
extendability 可延伸性
extendable 可延伸的
extend all over 遍及
extend a loan 发放贷款
extend a time limit 放宽期限
extend depth of fault 断层延伸深度
extended address 扩充地址
extended addressing 扩展寻址【计】;扩充访问;扩充编址
extended aeration 延时曝气
extended aeration activated sludge process 延时曝气活性污泥法
extended aeration process 延时曝气法
extended aerator 延时曝气器
extended antenna 加长天线
extended application 扩大使用
extended application rate 延长应用率
extended architecture 扩充体系结构
extended area 扩张面积;扩允(存储)区
extended area service 扩展区域服务;扩展区服务;区域性扩充设施;市内价郊区通信业务
extended arithmetic element 扩展的运算单元;扩充运算元件;扩充运算器;扩充运算单元;扩充运算部件
extended atmosphere 厚大气;延伸大气
extended background 展宽背景;大面积背景
extended bar rotor 深槽鼠笼转子
extended binary tree 扩展的二叉树;扩充的二叉树
extended bonds 延期(偿付)债券;延付债券
extended branch 加长水枪
extended broad range regulator 超广限式调整器

extended card kit 配套扩充插件
extended charge 间隔药包;柱状药包;直列式填装炸药;分段装药
extended clamp for suspension 长吊环
extended code 扩展码
extended colo(u)r 破图廓图
extended column concept 延长墩柱概念
extended consequent river 延长顺向河
extended consequent stream 延伸顺向河;延长顺向河
extended control mode 扩充控制方式
extended core storage 扩展磁芯存储器
extended cover 扩展责任;扩展保险;延期承保
extended coverage 延伸保险范围;扩大保险范围
extended coverage endorsement 扩展保险责任批单;扩展保险所加条款;展延保险批单
extended coverage insurance 其他产业保险
extended coverage sidewall sprinkler 扩大覆盖面边墙式洒水喷头
extended coverage sprinkler 扩大覆盖面洒水喷头
extended cover clause 延期承保条款
extended credit 延期信用证;延期信用证;附带预支条款的循环信用证
extended data out random access memory 扩展数据输出动态存储器
extended Debye-Huckel equation 扩展德拜—休克尔方程
extended delivery 伸展式收纸装置
extended delta connection 延长三角形接法
extended digestion effect 延时消化效应
extended discharge 延长喷射
extended discharge system 延长排放系统
extended discharge time 延长喷放时间
extended dislocation 扩展位错;扩展错位;扩散位错;延伸位错
extended dominance 扩充支配
extended DR equation 扩展杜比宁—拉杜施凯维奇方程
extended drill stem test 长期钻柱测试
extended-duration principle 延时原则
extended entity data 扩充实体数据
extended explanation 详细说明
extended facility 中期贷款
extended family 大家庭
extended filamentary radio nebula 延伸纤维射电星云
extended filtration 延时过滤
extended filtration process 延时(过)滤法
extended fire box 下涨式火箱
extended floating point 扩充浮点
extended floating point number 扩充浮点数
extended floating point operation 扩充浮点运算
extended-foil construction 箔片延伸结构
extended forecast(ing) 中期预报
extended foundation 开展基础;扩式基础;扩大基础
extended fracture 延长断裂
extended function 扩充功能
extended function space 扩展的函数空间
extended furnace 前置燃烧室;前置炉膛
extended ga(u)ge bit 加长保径钻头
extended grammar 扩充文法
extended group coded recording 扩展组编码记录;扩充组编码记录
extended high rate 长期高产
extended industry standard architecture 扩充工业标准结构总线
extended insurance 延期保险;展延保险
extended interaction klystron 分布(互)作用速调管
extended interaction oscillator 分布作用振荡器
extended interaction resonator 分布互作用谐振腔
extended interaction tube 长作用区管
extended interface 扩充接口
extended joint 延长节理
extended kilometer (车站)延展公里
extended Langmuir equation 扩展朗格缪尔方程
extended lateral range electric(al) conductivity 扩大侧向电导率
extended launder furnace 带延伸流槽的熔炉
extended length of line 线路延展长度
extended length of main line 正线延展长度
extended length of railway 铁路延展里程;铁路延展长度
extended length of siding 岔线延展长度

extended length of special using siding 特别用途线延展长度【铁】
extended length of station track 站线延展长度
extended length of the line used by section 段管线延展长度
extended letter of credit 延期信用证
extended line 延伸线(路);延长线(路);纵测线
extended lock mode 扩展锁定方式;扩充锁定方式
extended logout 扩充记录输出
extended lot 增批地段
extended low-surface brightness source 低面亮度展源
extended mean value 广义均值
extended memory 扩充存储器
extended mode 扩展状态;扩充方式
extended model 扩展模型
extended monitored mode 扩展监督方式
extended Muskat method 扩展的麦斯卡特分析法
extended network 扩展网络;扩充网络
extended networking 扩展联网;扩充组网性能
extended nozzle 长水口砖
extended nozzle bit 加长喷嘴钻头
extended object 延伸目标
extended operation 扩展操作;扩充操作
extended operation code 可扩充的操作码
extended operation instruction 扩展操作指令
extended operator control station 扩充操作员控制站
extended orthogonal group 扩张的正交群
extended partition specification table 扩充(分)区说明表
extended pasture spelling 草地长期休闲
extended period 延长时期;持续期
extended period simulation 延续时模拟
extended photosphere 延伸光球
extended pigment 增容颜料;填充性颜料
extended pipe network 扩展管网
extended pit aperture 外表纹孔口
extended pitch chain 加长节距链
extended point transformation 开拓点变换
extended precedence grammar 扩充优先文法
extended precision 扩充精度
extended precision floating point 扩(充)精度浮点
extended precision word 扩充精度学
extended preliminary design 扩大初步设计
extended price 评估价;扩充费用;预算价格
extended primary operator control station 扩充主操作员控制站;扩充的基本操作员控制站
extended problem 扩充问题
extended property line 延伸的用地线;延伸的建筑红线;延伸的地界线
extended protect 延期拒付证书
extended protest 海难补充报告;(船长呈交公证行的)延期海难抗辩书
extended punched card code 扩充穿孔卡码
extended quay dock 顺岸式码头港池
extended range 扩展范围;扩大域;延长范围
extended range reflection 延程反射
extended range Shoran 高灵敏度肖兰系统
extended range turbidimeter 广量程浊度计
extended reach drilling 延伸钻井
extended reach well 延伸井
extended real number 广义实数
extended reals 扩充实数集
extended reproduction funds 扩大再生产基金
extended requirement space 扩充要求空间
extended resolution 扩展分辨率
extended result output 扩充结果输出
extended ring fitting 长定位环
extended risk guarantee programme 放款风险担保计划
extended risk guarantees 扩大风险保险
extended risk insurance 扩大风险保险
extended river 延伸河(流);延长河
extended route 扩充路径
extended scattering medium 延伸散射介质
extended secondary operator control station 扩充辅助操作员控制站;扩充的辅助操作员控制台
extended service 长期运行
extended-service lamp 长命灯
extended shank bib 长把旋塞;长把龙头
extended shock source 延伸振源
extended slide valve rod 伸长滑阀杆
extended slot 扩散槽

extended slurry 缓凝水泥浆
extended source 扩展源
extended splitting 扩张式分岔
extended spread 纵排列
extended state 广延态
extended stem 伸缩杆
extended Stokes formula 广义司托克斯公式
extended storage 扩充存储器
extended storage test 延长储层测试
extended stream 延伸河(流)
extended structure 扩建建筑物
extended succession 延续层序
extended surface 延伸面;展开面;添加受热面;有肋面
extended surface elements 带有加热表面的部件
extended surface tube 鳍片管
extended switch 延时开关
extended symbol processing 扩充符号处理
extended system contents directory 扩充系统内容目录
extended target 展开目标
extended term amortization 延期分期偿付
extended terms insurance 延期保险
extended theorem of mean value 广义中值定理
extended threshold demodulator 门限扩展解调器
extended time scale 扩展时间量程;扩展时标;慢速时间比例
extended valley 延长谷
extended value on assignment 赋值时的展开值
extended vector 扩充向量
extended Von Mises failure criteria 扩展的冯·米塞斯准则
extended X-ray absorption fine structure 扩展X射线吸收精细结构
extended X-ray energy absorption fine structure 外延X射线能量吸收精细结构
extended X-ray energy loss fine structure 外延X射线能量损失精细结构
extended zero-sided Lindenmayer system 广义林氏无关系统
extender 扩充器;混合剂;延伸材料;延长器;增容剂;增量剂;增充剂;充填剂;补充料;补充剂;膨胀器;通信距离延长系统;填料;填充剂
extender lens 扩束透镜
extender pigment 油漆调和颜料;增量剂型颜料;填充用原料;填充性颜料;体质颜料
extender plasticizer 增量增塑剂;辅助增塑剂
extender plate 延伸板
extender polymer 增量聚合物
extender resin 增容树脂
extender rool module 下井仪模块开展器
extend ga(u)ge 增大金刚石钻探侧刃
extendible 可延伸的
extendible and retractable dog 伸缩爪
extendible compiler 可扩充编译程序
extendible derrick 伸开井架
extendible portion 延伸部分
extendibles 可延长债卷
extend information 扩展信息;扩充信息
extending 伸展作用
extending assets 递延资产
extending bracing 直撑
extending ladder 消防云梯;伸展梯;伸缩梯;活动云梯
extending-leg tripod 活腿三脚架
extending map of shallow mine 浅井展开图
extending map of slop mine 斜井展开图
extending map of small cylindrical pit 小圆井展开图
extending of rating curve 展延水位—流量关系曲线;水位—流量关系曲线的延长
extending of short-term records 根据短期记录推算;展延短期系列
extending-retracting speed of shuttle 货叉伸缩速度
extending shoot 长新梢
extending stroke 活塞推程
extending table 伸展台;伸缩案面台
extending tower 可伸展天线杆
extending vector 扩充向量
extend letter of credit 延长信用证有效期
extend neck 接筒
extend offer 延长发盘
extend one's arms horizontally 两臂平伸

extend the expiration date 延展有效期;延长有效期
extend the life 延长使用寿命
extend the terms 放宽条件
extend the time for filing claims 延长索赔期
extend the time limit for filing clause 延长索赔期限
extend trade and commerce 扩大贸易及商务;推广贸易及商务
extend type 扩充形式
extend working spring 牵引簧
extened instruction 扩展指令
extensibility 拉伸性;可(延)展性;延展性;延展度;延伸性;延伸率;延伸度;膨胀性;伸展性;伸长性;伸长率;伸长度
extensibility of concrete 混凝土拉伸性
extensible 可延伸的
extensible arm 伸缩式支臂
extensible belt conveyer 可伸缩的带式输送机
extensible boom 可伸支臂;伸缩式臂杆;伸缩吊臂
extensible boom platform 伸缩臂架载人平台
extensible chute 可伸缩式溜槽
extensible conveying pipe 伸缩式输送管
extensible conveyer 伸缩式运输机
extensible cover 伸缩护罩
extensible discharge trough 伸缩式卸料溜槽
extensible exercise head 伸长式操寶头
extensible fork truck 货叉可延长的叉车
extensible language 可扩充语言
extensible link servo 可伸缩拉杆伺服机构
extensible pan 伸缩式溜槽
extensible portion 延伸时间
extens(i)ometer 延伸仪;伸长计;伸张仪;伸缩仪;扩展仪
extension 扩展名;扩大;扩充;宽展期;延伸;延伸;延期还债认可书;延期;延长(水带线);展期;广度;电话分机;分机(站);范围;牵伸;外延;外伸长度;伸展;伸出部分;伸长
extension agreement 延期协议;延期契约
extensional beam 外伸梁
extensional deformation 延伸变形
extensional fault 张性断层;伸展构造
extensional faulting 张性断裂
extensional flow 拉伸流动;伸长流动
extensional fracture belt 张裂带
extensional fracture region 张裂区
Extensional fracture region of central south China Sea basin 中国南海海盆张裂区
extensionality 外延性
extensional marginal basin 扩张边缘盆地
extensional motion 拉伸运动
extensional movement 拉张运动
extensional rheology 延伸流变学
extensional rigidity 拉伸刚度;抗拉刚度
extensional stiffness 延伸劲度
extensional strain 伸长应变
extensional structure 张性构造
extensional tectonics 伸展断层
extensional tectonic type 伸展构造型
extensional test 拉伸试验
extensional vibration 扩张振动;纵向振动
extensional viscosity 拉伸黏度
extension and renewal clause 延期和续定条款
extension apparatus 牵引器;牵伸器
extension arbor for Morse taper bush with tang 带扁尾莫氏锥套接长杆
extension arbor for roughing boring tool 粗镗刀接长杆
extension area 扩展区
extension arm 加长(钻)杆;加长臂;延伸臂;伸缩杆;伸出臂;伸长臂
extension at break 断裂伸长(率)
extension bar 接长杆;加长杆;伸出杆
extension barrel 延伸筒
extension base 延伸机座;延长机座
extension beam trammel dividers 长梁分规
extension bell 分设钟;分铃
extension bell call 分铃呼叫
extension bellows 伸缩皮腔
extension blind stop 百叶窗边框压条
extension bolt 附加门栓;长(柄)插销;拧接螺栓;插销;门插销
extension bond 扩充公司债
extension boom (起重机的)延长伸臂;伸长臂
extension boring and turning mill 大型立式铣床

extension cable 延伸缆索;延伸电缆;伸长电缆
extension casement hinge 阔隙窗铰;伸张架式铰链
extension chute 延伸滑槽;伸缩滑槽
extension circuit 扩充电路;展接电路;增设电路
extension clerk 助理收货员
extension coefficient 移距系数;延伸系数;延长系数;增长系数;伸长系数
extension coil 加长线圈;延长线圈
extension commission 延期手续费
extension-compression cycle 拉伸—压缩循环
extension contract 延长合同
extension cord 软线;软缆;接长线;延长绳路;延长(软)线;伸展绳
extension core barrel 可接长岩芯管;可接长取芯筒;加长岩芯管;伸缩岩芯管
extension coupling (带导向环的)加长联轴节;(带导向环的)加长接头;加长短节;伸缩联轴器
extension crystallization 伸展结晶
extension cycles 伸张次数
extension delivery 输纸装置
extension device 活动塔式脚手架;竖向调高装置;竖向调高器
extension die 延伸模
extension dividers 加长分线规
extension drill 接长钻
extension drilling 扩边钻井
extension drill steel equipment 接杆凿岩工具
extension drive shaft 延伸传动轴
extension elongation 拉伸延长;受拉伸长;伸出长度
extension fault 拉伸断层;伸展断层;张(性)断层
extension fee 延期费;延期费(用);展期费
extension field 扩展域
extension filter 辅助滤波器
extension fire alarm box 区域火灾报警箱
extension fitting 伸缩套筒式管接头
extension fixture 延长装置;(电杆)展接装置
extension flap 后伸式襟翼
extension flex 延长线;软线;软缆;接长线
extension flight conveyer 延伸链动输送机
extension flush bolt 长柄插销;伸缩式埋头螺栓;门插销;附加门栓
extension fork 长臂叉车
extension fracture 延伸断裂;延长断裂;张性破裂;伸长断裂
extension girder 伸缩大梁
extension hand wheel block 滑车组
extension hanger 吊挂尾管装置;尾管悬挂器
extension head 伸缩式机罩
extension hinge 阔隙窗铰;长脚铰链;长脚合页
extension hunting 扩充寻找;扩充搜索;扩充查找
extension indicator 伸长指示器
extension instrument 附加仪表;外接仪表
extension jamb 延伸边框
extension jib 加长臂
extension joint 延(长)节理;张(性)节理
extension key 伸缩钥匙;伸出键
extension ladder 消防云梯;伸缩梯
extension lamp 携带式灯
extension lathe 伸出座车床
extension lead 伸长引线
extension leader 主要延长枝
extension length of ladder 梯子拉出长度
extension limit 延长限度
extension line 延伸线(路);延长线(路);分线机;分机线
extension line for dimension 尺寸标注(引出)线
extension link 伸缩连件件;伸缩连接器
extension machine 伸出座车床
extension memory 扩展存储器
extension meter 延伸仪;伸长计
extension micrometer 伸长式千分尺;伸长式测微计
extension modulus 张拉模量;伸长模量
extension nipple 加长公接头;加长短节
extension number 分机号码
extension of a field 域的扩张
extension of algebra 代数扩充
extension of availability 有效期延长
extension of bracket 出跳
extension of building 扩建
extension of control 控制网加密
extension of defects liability period 缺陷责任期的延长
extension of excavation 掘进延长
extension of factory 工厂扩建

extension of field 矿区扩展;场地扩建
extension of foundation 基础的延伸;基础的扩展;基底扩大
extension of game with exchange of information 交换信息对策的扩充
extension of game without exchange of information 不交换信息对策的扩充
extension of ladder 拉梯
extension of line 线路延长;展线
extension of loan 借款的延期偿还;贷款偿还期的延长
extension of maturity 延长期限
extension of mortgage 抵押延期
extension of ore into wall 深入围岩的矿脉
extension of photo 胶片伸长
extension of plastic zone 塑性区扩大
extension of rating curve 关系曲线延长;水位流量关系曲线的延长
extension of real function 实函数扩充
extension of records 延展实测系列;延长记录;展延实测系列
extension of relation curve 关系曲线延长
extension of service 扩充服务;增加班次(或线路)
extension of short-term record 短系列记录展延
extension of the shaft 井筒的延伸;轴的延长端
extension of the work cycle 延长加工周期
extension of the works 工程延期
extension of time 延长期
extension of time for completion 延期;竣工期限的延长
extension of time for filing 延长报税期限
extension of time for payment 延长付款期限
extension of track 线路延长
extension of validity of bids 延长投标有效期
extension of work 工厂扩建
extension ore 外延储量
extension percent 伸长率
extension period 伸展期
extension piece 延伸管件;伸伸部件;连接件;接头件;内外螺纹管接头;引伸接头;伸长片
extension pipe 接长管;加长管;延伸管;延长管;伸长管
extension plate 延长板
extension principle 扩展原理
extension producer 扩边井;附加生产装置
extension project 扩建工程
extension proposal 扩建意见;扩建方案
extension rate 伸长速率;伸长速度
extension ratio 牵伸比;伸张比
extension register 扩充寄存器
extension ringer 分铃;分机振铃器
extension rod 伸缩杆;接杆;加长杆;拧接钻杆;延伸杆;塔式水准尺;塔尺【测】;伸缩(标)尺;伸长杆
extension rule 伸缩尺;伸长尺
extensions 扩充费用
extension scale 扩展刻度;延长刻度;伸缩尺
extension screw 螺旋延长部
extension services 附设服务
extension set 增设装置;分机
extension shaft 套管伸缩轴
extension side 扩大边
extension speed 伸长速率;伸长速度
extension spring 拉伸弹簧;拉簧;牵(引)簧;牵伸簧
extension station 扩充站;辅站
extension steel 接杆钎子;长钎子
extension stem 接长钻杆;延伸柄;伸缩柄;伸缩棒
extension strain 伸长变形
extension sub 延伸接头
extension subscriber 分机用户
extension table 可伸长式桌;伸缩桌;伸缩案面桌
extension tap 长柄丝锥
extension taper 伸出锥度
extension telephone set 电话分机
extension test 扩探;甩出预探井;伸长试验
extension thermometer 长管温度计
extension tongs 钻杆钳;长臂管钳;伸缩钳
extension trench braces 活络支撑
extension trestle ladder 伸缩式高架梯
extension tripod 伸缩(式)三脚架
extension tube 加长管;延伸管;附加遮光罩;外接镜管
extension tube of lens 镜头延伸管

extension-type 延伸型的;可伸缩的;套管式;伸缩式的
extension valve case 伸出阀箱
extension warping barrel 伸缩式带缆卷筒
extension well 扩边井;开发井;蔓延井;附加井;探边井
extension wire 延长线(路);附加线路;分接线路;备用线路
extensive 广阔的;广泛的;大规模
extensive accumulation 大面积堆载
extensive agriculture 粗放栽培;粗放衣作;粗放农业;粗放经营
extensive air shower 广延空气簇射
extensive arched structure 大面积拱形构造
extensive authority 扩大权限
extensive burn 大面积烧伤
extensive city planning 扩大区域城市规划
extensive effort 大量工作
extensive employment 广泛雇工
extensive farm 非集约牧场
extensive farming 广耕法;粗放衣作;粗放农业(经营);粗放经营
extensive form 扩展形
extensive grassland 大草原
extensive laceration 大面积裂伤
extensive lines 扩大产品
extensive magnitude 扩大量
extensiveness input, output control 扩展输入输出控制装置
extensive network 扩大台网
extensive observation 扩大观察
extensive operating 扩大经营
extensive operating range 扩大经营范围
extensive order 大批定货
extensive pasture 非集约牧场
extensive product strategy 开放性产品策略
extensive property 广延性(质)
extensive quantity 扩大量;广延量;外延量
extensive repair 大修理
extensive reproduction of fixed assets 固定资产扩大再生产
extensive root system 延伸根系
extensive sampling 扩大取样;扩大抽样
extensive scale model 大比例尺模型
extensive survey 粗放调查;普查
extensive system 分支系统
extensive territorial waters 延伸领海
extensive town planning 大区域城镇规划
extensive transaction 巨额贸易;巨额交易
extensive use 广泛应用
extensive use of chemical fertilizer 化肥化
extensive vertical movement 大面积升降运动
extensometer 应变计;引伸仪;延度仪;延度计;变形测定器;伸缩仪;伸缩计;伸长计
extensometer arrangement 伸长仪装置
extensometer balance 应变秤;变形秤
extensometer bridge 应变电桥
extensometry 应变测定
extensor 延展器
extensor reflex 展反射
extent 一大片;广度;程度;范围;范例
extention(al) strain 拉伸应变
extent of agreement 协议范围
extent of armo(u)r layer 护面层范围
extent of authority 权限
extent of compensation 赔偿范围
extent of contract 承包范围
extent of damage 损坏范围;(受)损失程度
extent of damping 阻尼度;衰减度
extent of delivery 交货范围
extent of error 误差范围
extent of fold 褶皱展布面积
extent of ground subsidence 地面沉降范围
extent of housing overcrowding 住房拥挤程度
extent of hydration 水化程度
extent of jurisdiction 管辖范围
extent of liability 责任范围
extent of liquid penetrant testing 液体渗透探伤范围
extent of magnetic inspection 磁力探伤范围
extent of polymerization 聚合程度
extent of porosity 孔隙度
extent of power 权限
extent of reaction 反应程度

extent of resettlement 移民安置范围
extent of supply 供应范围
extent of territorial sea 领海宽度
extent of territorial waters 领航范围;领海范围
extent of the error 偏差范围;误差量
extent of the root system 根系的长度
exterior 外用的;外观;外部(的);外表的
exterior aerial 室外天线
exterior air 外界空气;室外空气
exterior algebra 外代数
exterior algebra bundle 外代数丛
exterior ambient temperature 外部周围温度
exterior and interior 表里
exterior angle 外角
exterior antenna 室外天线
exterior apartment 公寓外面(阳台等)的面积
exterior application 外部应用
exterior architecture 外部建筑艺术;外部建筑风格
exterior attack 外攻
exterior automotive trim 汽车外表装饰
exterior balcony 外阳台;凸阳台
exterior beam 外纵梁
exterior bending moment 外部弯矩
exterior black gloss paint 外用黑色有光漆
exterior blind 外部遮阳;外部遮帘
exterior border 外图廓
exterior boundary 外(边)界
exterior brick 外部砖
exterior brick wall 外部砖墙
exterior bridge support 桥梁外部支承
exterior building panel 外部建筑(面)板
exterior capacity 外容量
exterior cheek brake 外部夹板制动器
exterior chimney 外部烟囱
exterior chlorinated rubber paint 外部用氯气处理的橡胶涂料
exterior chord 外弦杆
exterior circulation upflow anaerobic sludge bed reactor 外循环升流式厌氧污泥床反应器
exterior cladding 构架覆盖;外墙饰面;外部涂装;外部覆面;外包
exterior clamp 外对管器
exterior climate 外部气候
exterior coat 外部涂层;外部粉刷
exterior coating 表面涂层;外用涂料;室外用涂料
exterior coating of pipe line 输送管的外涂层
exterior colo(u)r 外部色彩
exterior column 外柱
exterior common area 户外公用面积
exterior component 室外工程
exterior concrete form 外部混凝土模板
exterior conctractor's hoist 外用的承包商绞车;外用的承包商吊车
exterior condition 外部条件
exterior contact 外切
exterior content 外容度
exterior core 外部中心
exterior corner 外部转角处
exterior corner reinforcement 外墙护角
exterior corridor 外走廊
exterior corridor apartment 外廊式公寓
exterior corrosion 外腐蚀;表面腐蚀
exterior crack 外表接缝
exterior crest 外顶线
exterior cubic(al) marble 砌外墙用大理石
exterior cupola 外部圆屋顶
exterior decoration 外部装修;外部装饰
exterior derivative 外导数
exterior differential 外微分
exterior differential form 外微分形式
exterior division 外分
exterior dome 外部穹隆
exterior door 外门
exterior drainage 外落水;外流水系
exterior drainage system 外部排水系统
exterior durability 户外耐久性;外用耐久性;外露耐久性
exterior emulsion paint 外用乳胶漆
exterior escape stairway 室外太平梯;室外防火梯
exterior exposure 外部暴露
exterior extent 外广延(度)
exterior face 外面
exterior facing 外部饰面;外饰面;外墙饰面
exterior fan 室外抽风机

exterior fiber 外部纤维(层)
exterior final rendering 外部的最后粉刷
exterior final rendering mix(ture) 外部的最后粉刷混合物
exterior final rendering stuff 外部的最后粉刷材料
exterior finish 外装修;外用罩面漆;外饰面;室外装修
exterior finishing 外表面涂饰;(胶合板)板边密封
exterior finish work 外檐装修
exterior fire 外部火
exterior fire escape 屋外太平梯
exterior fire hydrant system 外部消火栓系统
exterior floor slab 外部地板
exterior focusing 外聚焦
exterior focusing telescope 外对光望远镜;外调焦望远镜
exterior force 外力
exterior form 外形式;外模板
exterior frame 外图廓
exterior function 外罚函数
exterior gallery 外部长廊
exterior girder 外部大梁
exterior glass wall 外部玻璃墙
exterior glazed 外部镶嵌玻璃的;外部涂釉的
exterior glazing 外装玻璃(法);外墙贴釉法
exterior gloss paint 外部光泽油漆;外部光泽涂料
exterior-grade plywood 室外用胶合板;防水胶合板
exterior gypsum plaster 外用抹灰石膏
exterior handrail 外部扶手栏杆
exterior heat syndrome 表热
exterior home decoration 外部住宅装修;外部住宅装潢
exterior house paint 屋外(用)油漆;室外用建筑漆
exterior illumination 室外照明
exterior ingress 外初切
exterior installation 室外安装
exterior insulation 外部保温;外部隔绝
exterior insulation and finish system 外保温饰面系统
exterior-interior angle 同位角
exteriority 外表面
exteriorization 外置术;外向化
exteriorize 外向化
exterior label 外部标号
exterior lamp post 室外灯柱
exterior latex paint 室外用乳胶涂料;外用胶乳涂料
exterior layer 外部敷设层
exterior leaf 外部门扇;外部窗扇
exterior liabilities 对外负债
exterior lighting 户外照明;屋外照明;外部照明;室外照明
exterior lighting system 外部照明系统
exterior lighting unit 外部照明装置;室外照明装置
exterior lineup clamp 外对管器
exterior lining 外部衬砌
exterior loading 外部加载
exterior marble 外用大理石
exterior margin 外缘;外图廓
exterior mask 外图廓蒙片
exterior masonry 外部砖石圬工
exterior masonry wall 外部砖石墙;外部圬工墙
exterior masonry wall column 外墙支柱
exterior masonry wall lining 外墙装饰;外墙嵌饰
exterior masonry work 外表圬工
exterior material 外覆盖层材料
exterior materials lift 外部材料升降机
exterior measure 外测度
exterior mixing blade 外部搅拌叶片
exterior mosaic finish 外部马赛克终饰;外部马赛克精加工;外部锦砖终饰;外部锦砖精加工
exterior multiplication 外乘(法)
exterior nailing 外用钉
exterior node 外节点
exterior noise 外部噪声
exterior noise insulation 室外隔声
exterior non-linear resonance 外部非线性共振
exterior normal 外法线
exterior normal derivative 外法向导数
exterior of a set 集合的外部
exterior of surface 外表面
exterior oil paint 室外油质油漆;室外油质涂料
exterior orientation 外方位;外部定向
exterior orientation elements 相片外方位元素

exterior outline 轮廓线
exterior paint 面层油漆;外用(油)漆;室外用油漆;户外油漆
exterior painting 外涂漆
exterior painting work 外部涂刷油漆工程
exterior panel 端隔间板;边区格;外部镶板;外部嵌板;外部节间;外部护面板;外部壁板;端跨部分(无梁楼板结构的)
exterior part of ingot 钢锭外皮;钢锭头
exterior passenger elevator 外部旅客电梯
exterior penalty function 外部补偿函数
exterior perspective center 外透视中心
exterior pipe system 外部管网;外部管道系统
exterior plaster 外墙泥层;外粉刷
exterior plaster aggregate 室外灰泥集料;室外灰膏集料
exterior plaster base 外墙抹灰基底
exterior plastering 外部抹灰(工作)
exterior plastering coat 外部抹灰层
exterior plastering mix(ture) 外部抹灰混合物
exterior plastering practice 外部抹灰实践
exterior plastering scheme 外部抹灰方案
exterior plastering stuff 外部抹灰材料
exterior plastering system 外部抹灰系统
exterior plastering technique 外部抹灰技术;外部抹灰工艺
exterior plywood 防水胶合板;(室)外用胶合板
exterior point 外点
exterior pole position 外极位置
exterior post type container 外柱式集装箱
exterior power 外幂
exterior power sheaf 外幂层
exterior precast block 外部预制砌块;外部预浇砌块
exterior precast brick 外部预制砖
exterior prestressed column 外部预应力柱
exterior prestressed concrete column 外部预应力混凝土柱
exterior primer 外部底漆
exterior product 外积
exterior product algebra 外积代数
exterior-protected construction 围护结构
exterior-protected structure 围护结构
exterior pulpit 外部布道讲坛
exterior rendering aggregate 外部抹灰集料;外部抹灰骨料;外部打底集料;外部打底骨料
exterior rendering coat 外部底涂层;外部打底灰
exterior rendering practice 外部抹灰习俗;外部抹灰方案;外部打底习俗;外部打底方案
exterior rendering stuff 外部抹灰材料;外部打底材料
exterior rendering technique 外部抹灰技术;外部抹灰工艺;外部打底技术;外部打底工艺
exterior resident(ial) area 外居住面积(等于公寓外面的面积加户外公用面积)
exterior resonance 外部共振
exterior reveal 门窗外侧面墙
exterior scaffold(ing) 外部脚手架
exterior scaling 外部密封垫
exterior semi-degree 外半次
exterior sewer system 外部排水系统
exterior shaft 外柱身
exterior sheathing 外模板
exterior sheet 外装饰用薄板
exterior shell 外部薄壳
exterior shell component 外部薄壳构件
exterior shell member 外部薄壳构件
exterior ship-hoist 室外吊斗提升机
exterior side yard 临街侧院
exterior siding 护墙板;外壁板
exterior signal lighting 车外信号照明灯
exterior skin 外壳(层);外表面
exterior slatted blind 外页岩百叶窗
exterior space 外部空间;室外空间
exterior span 外跨
exterior stain 外部污斑;外用着色剂
exterior stair case 防火楼梯;户外楼梯;外部阶梯;室外楼梯
exterior stairrail 外楼梯围栏
exterior stair way 外楼梯
exterior steps 室外台阶
exterior stop (玻璃)外压条;外定位条
exterior strain 内应变
exterior stress 外应力

exterior stretched concrete column 外部延伸的混凝土柱
exterior string 外楼梯斜梁
exterior stringer 外纵梁
exterior structural system 外部结构系统
exterior structure 外部结构
exterior stucco 外墙拉毛粉刷
exterior stud 外部间柱;外部灰板墙柱
exterior supply water 外水源
exterior support 外部支承
exterior surface 外表(面);外面(墙面)
exterior surfacing 外表铺面
exterior-teeth gear 外齿轮
exterior-teeth spur gearing 外正齿轮
exterior temperature 室外温度;外界温度
exterior temperature sensing device 室外温度传感器
exterior template 外廓样板
exterior tendon 外部钢筋束
exterior tensioned concrete column 外部张力混凝土柱
exterior territorial waters 外部领海
exterior thread 外部螺纹
exterior tieback 牵索;锚定装置;拉杆;窗帘钩
exterior tier 外部一排;外部一层
exterior tile 外部瓷面砖
exterior to curve 曲线外部区域
exterior torque 外部转矩
exterior torsion(al) moment 外部扭转力矩
exterior traffic paint 马路划线漆;外用路标漆
exterior trim 外部贴脸;屋顶线脚;外饰线材;外部(门头)装修;室外装修
exterior trim paint 外装饰物用漆;外用装饰漆
exterior trim part 外装品;外部装饰品
exterior twist(ing) moment 外部扭转力矩
exterior type plywood 室外用胶合板;外(装)用胶合板
exterior undercoat(er) 外部底涂层
exterior undercoat plaster 外部底涂层灰浆
exterior upset 外部镦锻
exterior upset drill pipe 外表镦锻钻管
exterior use 外部用途;外部应用
exterior varnish 外用清漆
exterior varnishing 外部加粉刷
exterior veneer 外部饰面
exterior vibration 外部振动
exterior vibrator 表面振捣器
exterior view 表面图;外视图
exterior vinyl paint 外部乙烯基颜料;外部乙烯基漆
exterior wall 围墙;外墙
exterior wall block 外墙砌块
exterior wall building component 外墙建筑构件
exterior wall cladding 外墙挂板
exterior wall column 围墙柱;外墙柱
exterior wall construction 外墙建筑
exterior wall facing 外墙饰面
exterior wall finish 外墙终饰;外墙饰面
exterior wall lining 外墙涂层;外墙衬
exterior wall member 外墙构件
exterior wall panel 外墙镶板;外墙嵌板
exterior wall slab 外墙板
exterior wall(sur)facing 外墙饰面(材料)
exterior wall tile 外墙瓷面砖
exterior wall unit 外墙构件
exterior waterproofing 外防水
exterior wear 表面磨耗
exterior window 外窗
exterior window check 外窗边槽口
exterior window frame 外窗框
exterior window lining 外窗涂层
exterior window rabbet 外窗边槽口
exterior window sill 外窗台;外窗盘
exterior window trim 窗套
exterior wood stain 外部木材污斑
exterior work 户外工作
exterior yard 外面庭院;室外庭院
ex-terminal 目的地码头交货
exterminate the four pests 除四害
exterminatory earthquake 毁灭性地震
externa bark 外树皮
external 对外的;外面的;外部的;外表的
external ablation 外部烧蚀
external absorbent method 外吸收法
external accelerator 外部催化剂

external access 对外交通;外部沟
external accounts 境外账户
external accuracy 外部精度
external action 外部影响
external activate 外部起动
external activator 外部催化剂
external activity 外在活动
external address in 外部地址输入
external agent 外营力
external aileron 外加副翼
external air 户外空气
external air injection manifold 外装式空气喷射总管
external algebra 外代数
external ambient temperature 外部周围温度
external and internal thickening (钻杆)内外加厚
external angle (凸墙)外角
external angle head 阳角砖
external angle to cove skirting 阳角壁脚弯砖
external antistatic agent 外用抗静电剂
external antistatics 外用抗静电剂
external aperture of cochlear aqueduct 蜗水管外口
external applied load 外荷载
external architecture 外部建筑风格
external area 外域
external arithmetic 机外运算;外部运算
external armature 外电枢
external armature alternator 外电枢式同步发电机
external assistance 外部援助
external attribute 外部属性
external auxiliary column 外部辅助柱
external balcony slab 外阳台板
external band brake 外带内紧式制动器
external band clutch 外带离合器
external-band-type brake 外带式制动器
external basin 外流盆地;外流流域
external beam 外束
external beam current 外束流
external beam path 外束路线
external behavio(u)r 外部特性
external bending moment 外部弯矩
external benefit 外部效益;外部利益
external bird's beak 阳角尖嘴砖
external bisector 外分角线
external blind 外部遮阳;外部遮帘
external block brake 外闸瓦制动器
external border 外图廓
external borrowing 对外借款
external boundary 外边界(线)
external boundary condition 外部边界条件
external boundary region 外边界区
external bound block 外带滑车
external bracing 外拉;外部拉条
external brake 外闸(制动);外抱制动器
external branch 外支
external breakwater 外堤;外(部)防波堤
external breeding ratio 外增殖比
external bremsstrahlung 外韧致辐射
external brick 外部用砖
external brick wall 外砖墙
external broach 外拉刀
external building material 外覆盖层建筑材料
external building panel 外部建筑(面)板
external burner lip 油枪口;烧嘴
external burning 外部燃烧
external-burning engine 外燃发动机
external-burning ram 外燃冲压发动机
external burst flag 外接色同步
external bursting pressure 外部破坏压力
external bus 外汇流条
external bus instruction 外总线指令
external buzzer 外接蜂鸣器
external cable 舱外电缆;外部电缆;室外电缆
external calibration 外部校准
external call 外部呼叫;外部调用
external calliper ga(u)ge 外卡规
external can coating 罐头外壁涂料
external carbon dosage 外碳源投加量
external casing corrosion 套管外部腐蚀
external casing packer 套管外(部)封隔器
external casing perforation gravel pack 管外砾石充填
external catalyst 外部催化剂

external catode counter 外阴极计数管
external catode inverted magnetron 外阴极磁控管
external cause 外因;外来事故
external cavity klystron 外空腔调速管
external cellar wall 室外地下室墙
external center 尖端
external centerless grinding 外圆无心磨
external channel program(me) 外部通道程序
external characteristic 外部性特征;外(部)特性
external characteristic curve 外特性曲线
external cheek brake 外缩制动器
external chlorinated rubber paint 外部用氯气处理的橡胶涂料
external circle trimming 外圆修正
external circuit 外电路;外(部)回路
external circulating system 外部循环系统
external cladding 外护板;外覆盖层
external cladding tile 外部覆面瓷砖
external clamp 外对管器
external classification 外部分级
external cleaning 外部清洗
external cleaning pipe machine 外部清管器
external climate 室外气候
external clock 外部时钟
external clock divider 外时钟分频器
external clocking 外部定时
external clock signal 外部时钟脉冲信号
external clutch gear 离合器外齿轮
external coagulant 外施凝固剂
external coat(ing) 外(表面)涂层;外部涂刷
external coat of paint 外表面油漆涂层
external coconut piece (配件砖)阳三角
external coherence system 外相干
external coherent 外相干的
external coil 外蛇形管
external column 外柱;外部管柱
external combustion chamber 外燃室
external combustion cycle 外燃循环
external combustion engine 外燃机
external combustion turbine 外燃式透平
external command 外命令
external common tangent 外公切线
external common tangent plane 外公切面
external communication 外部通信
external composition 外部合成
external compressor 外部压气机
external concavity 外部凹陷
external concrete vibrators with motor 带电动机的混凝土振动器
external conctractor's hoist 外用的承包商绞车;外用的承包商吊车
external condensation process 外冷凝法
external condenser 外冷凝器
external condition 外界条件;外部条件
external condition of packing 包装外貌
external conditions for design 设计的外部条件
external conductive casing 导电外壳
external cone 外锥
external confining pressure 外关闭压力
external conic(al) refraction 外锥形折射
external conjugate diameter 外直径;外前后径
external connection 外接(头)
external consistency 外部符合
external construction 外部结构;外部构造
external contact base 边接触的电子管底
external contamination 外部污染
external contour saw 外轮廓锯法
external contracting 外缩
external contracting brake 外缩式制动器;外抱闸
external contracting brake band 外闸带
external control 外部控制
external control analysis 外检(分析)
external control point 外控点
external control point of route 路线外控点
external conversion 外转换
external conversion coefficient 外转换系数
external conversion ratio 外转换比
external convertibility 境外兑换性
external cooler 外部冷却器
external cooling 外冷;外部冷却
external cooling circuit 外部冷却回路
external cooling reactor 外部冷却反应堆
external cordon (城市的)外围线
external-cordon survey 外围线交通调查

external core 外部心板
external corner 外墙转角;阳角
external corner reinforcement 外墙护角
external corona 外电晕
external corridor 外部走廊;外部廊沿
external corrosion 外部锈蚀;外(部)腐蚀;外蚀
external cost 外部费用
external covering 外护板;外覆盖层
external crack 表面裂纹;表面裂缝;外部开裂;外表裂纹
external critical damping resistance 外部临界阻尼电阻
external cross-region 外十字区
external cupola 外部圆屋顶
external current 外电路电流
external cut 外部切除
external cutter 外割刀
external cutting 外圆车削
external cutting off 外切断
external cyclone 外置式旋风筒
external cylinder ga(u)ge 外径圆柱规
external cylindrical ga(u)ge 环规
external damage 外部损害
external database 外部数据库
external data modem 外接数据调制
external data structure 外部数据结构
external debt 外债
external debt servicing 外债清偿;外债偿付
external decimal digit 外部十进制数
external decimal item 外部十进制项目
external declaration 外(部)说明
external decoration 外部装潢
external decorative feature 外部装潢面貌
external decorative finish 外部装潢最后装修
external definition 外部定义
external definition symbol 外部定义符号
external delay 外因延迟;外部延迟
external delay time 外延迟时间
external delivery point (液压系统)外接输出端
external delustering 外部消光
external depreciation 外在折旧
external detector 外检测器
external device 外置器件;外部装置;外部设备
external device address 外部设备地址
external device code 外部设备码
external device communication 外部设备通信
external device control 外部设备控制
external device data flow 外部设备数据流
external device instruction 外部设备指令
external device interrupt 外部设备中断
external device operand 外部设备操作数
external device operation code 外部设备操作码
external device response 外部设备响应
external device start 外部设备起动
external device status 外部设备状态
external diameter 外径
external diameter of disc wheel 工作轮外径
external diameter of pipe 管外径
external diameter of screen tube 滤水管外径
external diameter of tube 管的外径
external diameter of well pipe 井管外径
external diaphragm 外光阑
external differential compensating pinion 外差动补偿齿轮
external diffraction 外衍射
external diffusion 外扩散
external dike 外堤
external dimension 外形尺寸;轮廓尺寸;外部尺寸
external direct sum 外直和
external discontinuity lattice 外部不整合晶格
external diseconomics 外部不利条件;不利外部条件
external diseconomy 外部经济负效果;外部不经济性
external distance 外矢距;外距
external diversion dike 外导流堤
external diversion sand dike 外导流砂堤
external dollar 境外美元
external dollar market 境外美元市场
external dome 外部穹隆
external door 外门
external dose 外剂量
external downpipe 外部落水管
external drag 外部阻力

external drainage 入海水系;外流水系;外部排水;通海水系
external drainage system 外排水系统
external drencher 外部水幕喷头;外部防火淋水器
external drencher system 外部水幕系统
external dressing 外部整修;外部涂油;外部加油
external drip feed 滴油润滑器
external drive 外部驱动;外部激励
external drum 外鼓
external dump 外部排土场
external economics 对外经济学;外部经济;外部(的)有利条件;有利外部条件
external economizer 外置式省煤器
external economy 外在节约因素;外部经济(性)
external effect 外部因素
external efficiency 外在效率
external electric(al) potential 外电势
external electric(al) supply socket 外接电源插座
external electromotive force 外电动势
external element 外界元件
external enamel 外用瓷漆
external enamel epithelium 外釉(质)上皮
external energy input 外部能量输入
external engine 外置发动机
external engine control 发动机外部控制
external enrichment 外部增添装饰
external entry point 外部入口点
external environment 外(界)环境;外部环境
external environment(al) condition 外界环境条件
external eonomy 外部经济效果
external equalizer 外平衡装置
external equipment 外围设备;外部设备
external error 外在作用误差;外因误差;外部错误
external eruption 外喷溢
external escape route 室外疏散路线
external escape stairway 外部疏散楼梯
external evacuation route 室外疏散路线
external event processor module 外部事件处理机模件
external examination 外部检验
external examining errors of thorium linear reserves 钍线储量外检误差
external examining errors of uranium linear reserves 铀线储量外检误差
external excitation 外部激发
external exposure 外照射;外体暴露
external exposure dose 外照射剂量
external extendibility 外部扩充性
external factor 外因;外界因素
external failsafe 外部保险
external failsafe timer 外部保险计时器
external faired pipe 外形流线化与导管
external fan 外风扇
external fan kiln 外装风扇干燥窑
external fault 外部故障
external feedback 外反馈
external feedback magnetic amplifier 外反馈式磁放大器
external feedback signal 外部反馈信号
external feedback type 外反馈式
external feedback type controller 外反馈控制器
external feedback type oscillator circuit 外反馈式振荡电路
external feed line 外部给水管路
external fiber stress 外缘纤维应力
external field 外磁场
external final rendering 室外最后一层粉刷
external finance 外来资金;外部财务
external financial report 对外财务报告
external financing 外部筹资
external fire 外燃;外部火
external fire exposure 外部暴露于火
external fire hazard 外部火灾危险;外部火灾事故;外部火灾公害
external firing 外部燃烧
external firing system 外砌式燃烧炉;外部燃烧系统
external flame 外部火焰
external flashing 外部闪光
external floor slab 室外地板
external flow 外流
external flow vehicle 形成周边流幕;沿艇外流动;外流气垫艇
external fluid phase 外相流体
external flush-jointed coupling 平口接箍;外平接头

external flux 外流量
external focusing 外对光
external force 外力
external force drift 外力漂移
external force feed 外槽轮排种器
external force retarder 外力式缓行器
external form 模套;外形;外部形式
external freight volume 对外货运量
external friction 外摩擦(力)
external fumarole field 火口外围的喷气孔田
external function 外罚函数;外部函数
external function name 外函数名
external function reference 外函数引用
external function subroutine 外函数子程序
external fund 外来资金
external furnace 炉外燃烧室;外置炉膛
external gallery 外部长廊
external galvanized iron sheet sill 镀锌铁皮外窗台
external gamma ray detector 管外伽马射线探测仪
external gas cap 外部气顶
external gas injection 外部注气
external ga(u)ge 外径规
external gear 外(啮合)齿轮
external gearing 外啮合传动装置;外齿轮传动装置
external gear oil pump 外齿轮油泵
external gear rotary pump 外啮合齿轮泵
external gear-type oil pump 外啮合齿轮式油泵
external geomagnetic field 外地磁场
external gland 端部汽封
external glazing 装配外墙玻璃
external gloss paint 外部发光漆
external gravity field 外部重力场
external grid 外格
external grinding 外圆磨削
external groove sidevall 外槽壁
external ground 外接地
external guidance 遥控制导
external handrail 室外扶手
external hardware 外部硬件
external harmonics 外调和
external hazard 外来火险
external haze 表面混浊
external header 外联箱
external heat 外热
external heater 外加热器
external heat exchange medium 外部热交换介质
external heat exchanger 外换热器
external heating 外(部)加热
external heating medium 外部加热介质
external heating surface 外受热面
external home decoration 房屋外部装饰
external honing machine 外圆珩床
external hydraulic pressure 外周水压;外水压力
external hydrostatic(al) pressure 外静水压力
external idle time 外因造成的停工时间;外部停机时间
external illumination 外部照明
external illumination sign 外部照明式标志
external impedance 外阻抗
external impressed current 外接电流
external inconsistency 外不相容性
external indication 外标记
external indicator 外指示器;外指示剂
external influence 外界影响
external inhibit interrupt 外部禁止中断
external injection 外注入
external input 外来投入;外部输入
external institution 境外机构
external instrumentation system 外部仪表测量系统
external insulation 外部(热)绝缘;外部隔热;外绝缘;外部保温
external integrating capacitor 外积分电容器
external interest rate 境外货适用的利率
external interference 外界干扰;外干涉
external interference effect 外干涉效应
external interheater 外中间加热器
external-internal-cordon 城市外围内围警戒线
external-internal-cordon survey 外围内围线交通调查
external interrupt 外部中断
external interrupt enable 允许外部中断
external interrupt processor 外部中断处理器
external interrupt symbol 外部中断符号

external investment 国外投资
external investment decision 对外投资决策
external irradiation 外辐照;外部照射
external irradiation dose 外照射剂量
externalir radiation exposure 外照射
external irradiation hazard 外照射危害
external irradiation protection 外照射防护
externality 外在因素;外部影响;外部性;外部(经济)效果
externality of transportation 运输外部性
externalization 外表化;体现
external joint 外部节点
external kinetic energy 外部动能
external knurling 外压花
external label 外部标记;外部标号
external label(l)ing 外标记
external lantern 表面灯
external latent heat 外潜热
external layer 外层
external leaf 外部页扇
external leakage 外部漏泄
external lender 境外货款人
external light 外界光
external lighting 外部照明设备
external light intensity 外部光强度
external light path 外光路
external limit ga(u)ge 外径极限规
external limiting layer 外界层
external limiting membrane 外界膜
external line alternation 外接行交替
external lineup clamp 外对管器
external lining 外衬垫
external linking 外部链接
external liquidity 对外清偿能力
external load 外(加)荷载;外负载;外部负荷
external loading 外部加载;外部挤压作用
external loading test for drain pipe 排水管外压试验
external load of pipe 管道外载
external loan 外债
external lock signal 外锁信号
external logic 外部逻辑
external loop three phase fluidized bed 外循环三相流化床
external loop three phase fluidized bed reactor 外循环三相流化床反应器
external losses 外部损失
external lubrication 外部润滑
external lug 外部接头
externally acting brake 外作用闸
externally activated system 外部活化体系
externally applied force (施加的)外力
externally applied load 外加荷载
externally applied tanking 外贴地下防水层
externally braced aeroplane 外拉撑飞机
externally braced monoplane 外张单翼机
externally catalyzed resin 外部催化剂树脂
externally coherent pulse doppler radar 脉冲外相干多普勒雷达
externally defined symbol 外部规定的符号
externally determinate system 外静定体系
externally fired boiler 外燃锅炉
externally fired fire tube boiler 外燃火管锅炉;外燃火管锅炉
externally fired superheater 外燃式过热器
externally fired water tube boiler 外燃水管锅炉
externally flush-coupled rods 外部连接钻杆
externally heated arc 外部加热弧
externally heated oven 外热式烤箱;间接式烤箱
externally heated pressure vessel 外加热高压釜
externally heating 外面加热
externally irregular point 外非正则点
externally maintained field 外保持场
externally mounted fuel tank 外油箱
externally operable switch 外面操作开关
externally plastered 外部抹灰过的;外部粉刷过的
externally pressurizde gas lubrication 外部供压气体润滑
externally programmed computer 外部程序式计算机
externally pulsed system 外脉冲同步系统;外同步脉冲装置
externally quenched counter tube 外猝灭计数管
externally reflected component 外面的反射部分

externally rendered 外表面粉刷过的
externally reversible motor 双向启动可逆电动机
externally specified index 外部专用索引
externally specified index operation 外定索引操作
externally stable block 外力作用下稳定的块体
externally stored program(me) 配线程序；外部存储程序
externally tangent circle 外切圆
externally temperature influence 外界温度影响
externally unstable block 外力作用下不稳定的块体
external magnet 外磁铁
external magnetic field 外在磁场；外部磁场
external marble 外部大理石
external market 境外市场
external marks 外部标志
external masonry 外部砖石圬工
external masonry wall column 圬工外墙柱
external masonry wall facing 圬工外墙饰面
external masonry wall lining 圬工外墙衬里
external masonry wall of double-leaf cavity construction 外部圬工空心砖砌筑；双壳外墙
external material 外覆盖层材料；杂质
external material lift 外部材料升降机
external measurement 外部丈量【船】；外部尺度
external measuring instrument 外径测量仪
external mechanical seal 外装式机械密封
external medium 外存储
external member 包容件；外部构件；套件
external memory 外存储器；外部记忆装置
external memory computer 外存储计算机
external memory(storage) 外部存储器
external micrometer 外部测微计
external migration 移居国外；外迁(移)
external miogeosynclinal zone 外冒地槽带
external mirror 外反射镜
external mitre 外隅角
external mix 外部混合
external mixing blade 外部搅拌叶片
external mixing type atomizer 外混式雾化喷嘴
external mix oil burner 外混式油喷燃器
external mix type gun 外部混合型喷枪
external mobility 外观迁移率
external model 外部模型
external moderated reactor 外部慢化堆
external modulation 外调制
external modulator 外调制器
external module 外部模件
external moment 外(部)力矩
external mosaic finish 外部马赛克精加工；外部锦砖精加工
external motion resistance 外部运动阻力
external mo(u)ld 外模
external mo(u)ld lubricant 外施脱模剂
external multitubular heater 外多管加热器
external mute 外部切除
external name 外部名
external nodal point 外结点；外节点
external nodal point of first kind boundary 一类边界外节点
external nodal point of secondary kind boundary 二类边界外节点
external noise 外界杂音；外(部)噪声
external noise factor 外部噪声因数
external noise level 外部噪声电平
external noise penetration 外部噪声穿透
external number 外部数
external object 外部目标
external observation 外部观察；外部观测
external octagon brick 外八角砖
external of time for payment 推迟付款时间
external oil cooler 外部油冷却器
external oil paint 室外油漆
external oil supply 外面供给润滑油
external opening 外孔
external operating ratio 外部操作时间比
external operating valve 外操纵阀
external operation 外表加工
external operation ratio 运行率【计】
external optic(al) density 外光密度
external ordering 外次序
external ornamental feature 外部装饰特征；外部装饰面貌
external ornamental finish 外部装饰最后修整；外部装饰精加工

external packer 裸眼封隔器
external packing 外包装
external packing leakage 端部汽封漏泄
external paint 外部发光颜料；外用涂料
external painting 外表涂漆
external painting(work) 室外油漆工程
external panel 室外镶板；室外面板
external parameter 外部参数
external payment 国外支付；国外付款；国际支付(款额)
external pelivimetry 外测量
external periodic(al)force 周期性外力
external perspective center 外透视中心
external phase 外相
external phase of an emulsion 乳胶体外相
external phasing (天线)外部定相
external phasing gears (单纯旋转发动机的)外轴式相位机构
external phosphorus loading 外磷负荷
external photoeffect 外光电效应
external photoelectric(al) effect 外(部)光电效应
external physical characteristic 外型特征
external piloting 外部先导控制
external piloting control 外部领示控制
external pipe cleaning machine 外清管器；外清管机
external pipe coating 管子外涂层
external pipe system 外部管网；外部管道系统
external pipe thread 管(子)外螺纹；外管螺纹
external piping 室外管道
external planetary arrangement 外啮双排行星齿轮机构
external plasma 外部等离子体
external plaster 外墙灰泥层；外粉刷
external plaster aggregate 外墙灰泥集料；外墙灰泥骨料
external plaster base 外墙抹灰基底
external plaster facade 抹灰泥的外墙正面
external plastering 外部抹灰泥；外部加粉刷
external plastering coat 外部抹灰层
external plastering mix(ture) 外部抹灰拌和物
external plastering practice 外部抹灰实践
external plastering scheme 外部抹灰示意图
external plasticization 外增塑(作用)
external plasticizer 外(用)增塑剂
external plate impedance 板极电路阻抗；外屏极阻抗
external plexiform layer 外丛状层
external pliers 轴用卡簧钳
external plumber 屋外管子工
external point 尖端；外点
external point of division 外分点
external pontoon floating roof tank 浮船式外浮顶油罐
external porosity 外部孔隙率
external potential 外电势
external potential energy 外部位能
external power 外电源
external power grid 外电网
external power receptacle 外电源插座
external power source 外接电源；外部电源
external power supply 外部电源
external precast block 外部预制砌块
external pressure 外压强；外压(力)；外部压力
external pressure cylinder 外压圆筒
external pressure flange 外压法兰
external pressure heads and shells 外压封头和壳体
external pressure region 外压区
external pressure resistance 耐外延性
external pressure sphere 外压球壳
external pressure strength 外压强度
external pressure test 外压试验
external pressure tester 外压试验器
external pressure tube 外压管
external pressure vessel 外压容器
external prestress 表面预应力；外加预应力
external prestressed column 外部预应力柱
external prestressed concrete column 外部预应力混凝土柱
external prestressing 体外预应力
external prestressing tendon 体外预应力索
external presure confinement 外压力约束
external primer 外部底漆

external principal column 老檐柱
external procedure 外部过程
external profile diameter 外缘直径
external program(me)interrupt 外部程序中断
external programming 外部程序设计
external projection 外心影；外投影法
external protection works 外部防护工程
external pulpit 室外讲台
external pyramidal layer 外锥体层
external Q 外Q值
external quality factor 外部品质因素
external quenching 外淬灭
external radial boundary 外部径向边界
external radiation 外(部)辐射
external radiation dose 外部辐射剂量
external radiation exposure 外部辐射暴露
external radius 外半径
external ram 外置油缸
external ram control valve 外置油缸控制阀
external ratio 外分比
external reactance 外电抗
external reality 客观现实
external receptor 外部感受器
external recessing 外开槽
external reference 外部引用；外部基标；外部调用
external reference coil 参比线圈
external reference item 外部引用项
external reference method 外标法
external reference record 外部引用记录
external reference symbol 外部引用符号
external reflection 外(部)反射
external reflux 外部回流
external reflux of column 分馏柱外回流
external register 外(部)寄存器
external reinforced column 外部加强支柱
external release agent 外脱模剂
external rendering 外部粉刷；外部抹灰集料
external rendering aggregate 外墙粉刷；外部抹灰集料；外部抹灰骨料
external representation 外部表示
external request 外部请求
external rescue 外部救援
external reset 外部复位
external resistance 外力阻抗；外(电路)电阻；外(部)阻力
external resistor 外电路电阻器
external resonance 外共振
external reversing mechanism 外部换向机构
external ring 外环
external rotation 外旋转；外回旋
external rotatory position 外旋位
external-rotor motor 外转子式电动机
external rubber (水轮发电机的)外摩擦橡皮
external saddle 外鞍
external safety feature 外部安全特征
external sand barrier 外拦门砂
external scaffold(ing) 外部脚手架；外脚手架
external schema 外部模式
external scrap 外来废料
external screw 螺钉
external screw cutting tool 外螺纹车刀
external screw ga(u)ge 外螺纹检查规
external screw thread 外螺纹
external screw thread finish 外螺纹瓶口
external seal 外装式密封
external seal flush 外部外密封冲洗
external secant 外矢距；外割
external selection 外部选择
external self-locking flag 外部自锁表征
external sensation 外部感觉
external sense and control line 外部传感及控制线
external sewer network 外部排水管网
external sewer system 外部排水系统
external shaving 外缘整修
external sheath 外层包覆
external sheet-metal covering 外部金属板包皮
external shield 外屏蔽
external shield plastics 外屏蔽耐火塑料；外屏蔽防火塑料
external shoe 外蹄片
external shoe brake 外蹄式制动器
external short-term liabilities 对外短期负债
external shoulder angle 阳角扇形砖
external shoulder turning 外轴肩车削

external shutter 外板窗;外百叶窗
external signal 外部信号
external signal interrupt 外信号中断
external signal line 外(部)信号线
external sine-wave oscillator 长线槽路正弦波振荡器
external skin 砖墙外皮(外层墙);外覆盖层
external skip-hoist 外用吊斗提升机
external sleeve gun 外套筒式空气枪
external socket 外电源插座
external solvency 对外偿付能力
external sort 外排序;外部分类
external source 外部信源
external space 外空间
external spectrum 外光谱
external spiral 外螺旋
external spline 外花键
external spur 外距
external square 外方头
external stability 外部稳定度;外部固定
external stability number 外固数
external stability of a solution 解的外部稳定性
external stair(case) 外梯级;外台阶
external standard 外标
external standard source 外标准源
external stop 外部光阑
external storage 外存储器;外储存
external store 外(部)存储器
external straight turning 外直线车削
external strain ga(u)ge 表面应变仪;表面应变计
external stress 外缘应力;外(部)应力
external stress corrosion cracking 外应力腐蚀裂纹
external structure 外(部)结构;外部构造
external study 越界调查
external styling 外部造型
external subcarrier 外接副载波
external subroutine 外部子程序
external supercharger 外(装)增压器
external superheater 外置(式)过热器
external surface 外表(面)
external surface area 外表面积
external surfacing 外表整修面
external suture line 外缝合线
external symbol 外部符号
external symbolic reference 外部符号引用
external symmetry 外对称
external synchronization 外同步
external synchronizer 外同步器
external system design 外部系统设计
external tank-type heater 外置式冷却水加热箱
external taper 外锥度
external taper turning 外锥度车削
external tapping 外接口
external target 外靶
external-teeth gear 外齿轮
external-teeth spur gearing 外正齿轮
external telephone 外线电话
external temperature 周围介质温度;外界温度
external temperature influence (仪表的)周围温度影响
external temperature sensing device 周围温度传感器;外部温度传感器
external tendon 外部钢筋束
external terminal 外部接头
external terminal thread 外终丝
external thermocouple 外热电偶
external thread 外(部)螺纹;阳螺纹
external thread cutting 外螺纹切削
external threaded drill rod 外螺纹钻杆
external thread ga(u)ge 螺纹量规
external thread grinder 外螺纹磨床
external threading 外螺纹车削
external thread valve 外螺纹阀
external tieback 外牵索;外拉索;外部锚定装置
external tier 外部一排;外部一层;单层砖墙
external tile 外墙砖
external tooth 外齿
external toothing 外啮合
external topography 外部地形
external torque 外部转矩
external torsion(al) moment 外部扭力矩
external tracing table 外解绘图桌
external trade 对外贸易;外贸
external trade by regions 分地区的外贸额

external transaction 对外贸易
external transportation 对外运输
external transport instruction 外部输送指令
external transport(ion) 厂外运输
external trigger 外触发器
external trigger input 外触发输入
external trigger pulse 外触发脉冲
external trimmer 外微调电容
external tube cleaning machine 外清管器;外清管机
external turning tool of lever design 杠杆式外圆车刀
external twist(ing) moment 外部扭转力矩
external-type idle limiter 外装式怠速浓度调程限制器
external undercoat(er) 外部底涂层
external undercut 焊缝外边缘咬边
external unit 外围设备;外部设备
external unwinding 外退解
external upset 外加厚
external upset casing 外加厚套管
external upset drill pipe 外加厚钻管;外加厚钻杆
external upset drill rod 外加厚钻钎;外加厚钻杆
external upset end 外加厚端
external upset tubing 外加厚油管;(钻杆端部的)外加厚管
external valuation 外部估价
external variance 外部方差
external varnish 外用清漆
external varnishing 外部上(清)漆
external veneer 外部饰面
external version 外倒转术
external vibration 附着式振捣;表面振捣;外来振动;外部振动;外部振捣(适用于模板小配筋密的构件)
external vibrator 模板振捣器;附着式振捣器;表面振捣器;(表面)附着式振捣器;外部振动器;外部振捣器
external view 外貌;外形图
external vinyl paint 外部乙烯基颜料;外部乙烯基漆
external void fraction 外部空隙分率
external voltage 外加电压
external wage structure 企业间工资结构
external wall 围墙;外墙;外堤
external wall construction 外墙结构
external walling 围墙砌筑;外墙砌筑
external walling column 围墙柱;外墙柱
external walling lining 围墙衬砌;外墙衬砌
external wall panel 外墙板
external washing 外部洗刷
external waste 围岩
external water circulation 外水分循环
external water drive 外部水驱
external water-feed machine 旁侧供水凿岩机;外部供水机
external water pressure 外水压力
external water separator 外水分离器
external wave 外波
external wear 表面磨耗
external weight 外部权
external window 外窗
external window shutter 外板窗;外百叶窗
external window sill 外窗台
external wiring diagram 外部接线图
external work 外作业;外力所做的功;外功;外部修饰工作;外表面加工
external world 客观世界
external X-cross 外交叉
external zone 外围地区
externide 外构造带
externo-dorsal ray 外背辐肋
externo-lateral ray 外侧辐肋
exterofection 外反应作用
extinct 耗尽的;熄灭
extinct animal 灭绝动物
extinct bed 干涸湖
extinct block 死亡裂谷
extinct elements 灭迹元素
extinct geothermal field 已消亡的地热田
extinct group 绝灭类群
extinct hydrothermal area 已消失的水热活动区
extinction 灭绝;消亡;消退;消失作用;消灭;消光;吸光度;淬灭;生物绝灭

extinction angle 消弧角;消光角
extinction coefficient 灭亡系数;消声系数;消弱系数;消光系数;吸光系数
extinction coefficient of allowable smoke concentration 容许消光系数
extinction coefficient of exhaust gas 废气消光系数
extinction coil 消弧线圈
extinction contour 消光轮廓
extinction contrast 消光衬比
extinction direction 消光方向
extinction distance 消光距离
extinction effect 消灭效果
extinction factor 消光因数;消声系数
extinction limit 熄灭极限
extinction meter 消光法曝光表
extinction moisture content 熄灭含水量
extinction of a flame 熄火
extinction of arc 消弧
extinction of debt 还清债务
extinction of fuel 燃料的熄灭
extinction of spark 火花消除
extinction of species 品种灭绝
extinction position 消光位置
extinction potential 消电离电位;熄灭电势;熄火电位
extinction probability 消灭概率
extinction ratio 消声系数;消声比
extinction recycle operation 全循环操作
extinction rule (结晶面的)消光规则
extinction test 消光试验
extinction time 消灭时间
extinction turbidimeter 消光浊度计
extinction using gas 气体灭火
extinction using vapo(u)r 蒸汽灭火
extinction value 消光值
extinction voltage 灭弧电压;熄灭电压
extinction voltage of arc 消弧电压
extinction zone 消失带;消减带
extinctive 消退的
extinctive atmosphere 灭火气氛
extinctive inhibition 消退性抑制
extinctive prescription 消灭时效
extinctivity 消光率;熄灭率
extinct lake 消亡湖;干涸湖;死湖
extinct natural radionucleus 消失的天然放射性核
extinct natural radionuclides 消失的天然放射性核素
extinct rift valley 死裂谷
extinct species 绝灭种
extinct volcano 死火山
extinct volcano crater 死火山口
extine 外壁
extingiushment 灭火
extinguish 消灭
extinguish a claim 使索赔无效
extinguishant 灭火剂
extinguish claim 使索赔无效
extinguished 灯标失明
extinguished light 已熄灭的灯标
extinguished volcano 熄火山
extinguished waterfall 已灭瀑布
extinguisher 灭火器;灭弧器;消火者;消火器;消弧圈;消弧器;消防器材;消除器;熄灭者;熄灭器;熄灯器
extinguisher with high expansion of foam 高倍数泡沫灭火器
extinguish gun 灭火枪
extinguishing ability 灭火能力
extinguishing agent 灭火药剂
extinguishing bomber 灭火手雷
extinguishing by cooling 冷却灭火
extinguishing by smothering 窒息灭火
extinguishing coefficient 消光系数;衰减系数
extinguishing coil 消弧线圈
extinguishing device 灭火器件
extinguishing device without piping system 无管网气体灭火装置
extinguishing engineering 灭火工程
extinguishing equipment 灭火设备
extinguishing foam 灭火用泡沫材料;灭火泡沫
extinguishing grenade 灭火手雷
extinguishing installation 灭火装置;灭火设备
extinguishing jet 灭火射流
extinguishing line 灭火路线

extinguishing medium 灭火药剂
extinguishing method 灭火(方)法
extinguishing pipe 消防管
extinguishing product 灭火药剂
extinguishing pulse 消除脉冲;熄火脉冲;擦除脉冲
extinguishing requirement 灭火要求
extinguishing run 灭火试验
extinguishing system 灭火系统
extinguishing voltage 灭弧电压
extinguishmaterial 灭火料
extinguishment 熄灭
extinguishment by dilution 稀释灭火
extinguishment by emulsification 乳化灭火
extinguishment characteristic 灭火性能
extinguishment fund 偿债基金
extinguishment material 灭火材料
extinguishment method 灭火(方)法
extinguishment reserve 债务清偿准备金;销除债务准备;偿债准备
extinguishments by smothering 窒息灭火法
extinguishments equipment 消防设施
extinguishment stream 灭火射流
extinguish steam 灭火蒸汽
extinguish the fire 灭火
extinguishing forest fire 消灭林火
extinguishing grass fire 消灭地表火
extirpation 摘除
extirpator 中耕除草机;摘除器
extoparasite 外寄生物
extorsion 外旋
extra 附加的;额外(物);额外附加物;额外的;特级品
extra-abdominal monitoring 腹壁外监护
extra absorption 额外吸收层
extra accuracy 超精度
extra acid 额外酸
extra address 附加地址
extra-air inlet 副进气口
extra allowance 额外津贴;特别津贴
extra amount 多余部分
extra area effect 地区外影响
extra arid desert 超干旱荒漠
extra band 额外带
extra-banking 额外填土;超高填土;附加填土;填土抛高;外加填土
extrabasinal river 盆外河流
extrabasinal rock 盆外岩
extra belt vegetation 超带植被
extra best best 最高质量
extrabold 粗黑体
extra bound 手工精装本
extrabowline 船首增索
extra-bright 特高亮度
extra-budgetary 预算外
extra-budgetary fund 预算外经费;预算外资金
extra-budgetary program(me) 预算外计划
extra-budgetary source 预算外资源;预算外来源
extra burner 临时燃烧器
extra burning period 额外快速燃烧期
extra catch 额外捕获物
extra cell 附加单元
extra central telescope 附加对中望远镜;偏心望远镜
extra charges 附加费(用);额外费用;附加负载;附加负荷
extracharge of educational funds 教育费附加
extracharge on long length 超长货物附加费
extra check 特别对照
extraclast 外(来)碎屑
extra clearance period 附加清车时间
extracode 附(加)码;加码;附加代码;附加程序
extracode control register 附加码控制寄存器
extracode routine 附加码程序
extra coke charge 接力焦炭
extra conductive furnace black 超导电炉黑
extracontinental basin 陆外盆地;外陆盆地
extracontinental geosyncline 陆外地槽
extra contour 助曲线;辅助等高线
extra contrasty 特大反差的
extra control 附加控制;外部控制
extra controller 附加控制器;外部控制器
extra cooling air 补充冷却空气;外加冷却空气
extracoronal retainer 冠外固位体
extra cost 新增成本;追加成本;额外费用;额外成本
extract 开采;浸膏;浸出液;浸出物;挤干;蒸馏出;摘录;萃取物;抽提;抽取;抽出物;拔出;片段;提取液;提取(物);提炼
extractability 可萃取性
extractable 可提取的
extractable organic carbon 可萃取有机碳
extractable organic chlorine compound 可萃取有机氯化物
extractable organic halide 可萃取有机卤化物
extractable organic halogen 可萃取有机卤素
extractable organic matter 可萃取有机物;抽提有机质
extractable softener 抽出性软化剂
extract air chamber 低气压室;抽空气室
extract and oxidation sweetening 抽提物氧化脱硫
extractant 浸媒;萃取物;萃取剂;提取剂
extractant reagent 萃取剂
extract a root 开方;求根
extract atmospheric flash tower 抽出液常压闪蒸塔
extract duct 排气管;抽气管;拔风管
extracted activated sludge 萃取活性污泥
extracted asphalt 提炼过的沥青;抽提沥青
extracted beam 引出束
extracted bitumen 提取的沥青
extracted core 抽洗过的岩心
extracted digit 抽出数位
extracted-head wire wound potentiator 抽式线绕电位器
extracted ore tonnage 出矿量
extracted solid 萃取残渣
extracted solid discharge 萃取残渣出口
extracter 析取字;抽提器
extract exchanger 提取热交换器
extract fan 抽风机
extract from log 航海日志摘要;航海日记摘要
extract furnace 提取加热炉
extracting 提取
extracting agent 萃取剂;提取剂
extracting column 提取塔
extracting jaw 取出爪
extracting level 开采水平
extracting machine 绞榨机;轧液机;萃取机;提液机
extracting method 抽出法
extracting power 萃取能力
extracting screw 卸料螺旋输送机;卸料绞刀
extracting screw conveyer 卸料螺旋输送机
extracting solution 提取液;蒸馏液;浸提液
extracting solvent 萃取溶剂
extracting stage 萃取级
extracting tower 萃取塔
extracting water pattern from casing well 管井的取水方式
extract instruction 开方指令;析取指令;抽取指令;抽出指令;提取指令
extraction 林产集运;浸出;析取;摘录;萃取;抽出(物);分出;拔除;拔出;取出;求根;牵引出术;提取(法);提炼;提出(物)
extraction agent 萃取剂;提炼剂
extraction air sampling 抽气采样
extraction and purification 提炼
extraction and stripping apparatus 萃取分离设备
extraction apparatus 萃取设备;萃取器;抽提器;提取器
extraction back pressure turbine 背压抽汽式汽轮机
extraction battery 萃取器组;抽提器组
extraction cascade 级联萃取设备
extraction chamber 提取室
extraction check valve 抽汽逆止阀
extraction chromatography 提取色谱(法)
extraction coefficient 萃取系数;提取系数
extraction column 萃取柱;抽提柱;抽提柱;提取塔
extraction condensing turbine 抽汽式凝汽汽轮机
extraction conduit 排泥管
extraction constant 萃取常数;提取常数
extraction control valve 抽汽控制阀;抽汽调节阀
extraction cycle 回热循环;抽提循环
extraction data 提取数据
extraction device for pipe 脱管机
extraction efficiency 抽提效率;提取效率
extraction efficiency of geothermal energy 地热能吸出效率
extraction equipment 出料设备
extraction factor 抽提因数;提取因数
extraction fan 吸气式抽风机;抽吸卸料用抽风机;抽气风扇;抽风机;排气风扇
extraction feedwater feating 给水回热加热
extraction feedwater heating 抽汽给水加热
extraction flame atomic absorption method 萃取火焰原子吸收法
extraction flask 萃取瓶;浸出瓶;抽提烧瓶;提取瓶
extraction flow 萃取流量
extraction fulorimetric analysis 提取荧光分析
extraction gas turbine 抽汽式燃气轮机
extraction graphite furnace method 萃取石墨炉法
extraction gravimetric method 萃取重量法
extraction heater 抽汽加热器
extraction indicator 提取指示剂
extraction instruction 抽取指令;取出指令
extraction into solvent 溶剂萃取
extraction isotherm 萃取等温线
extraction jack 拔桩千斤顶;拔桩机
extraction kinetics 萃取动力学
extraction line 抽汽管道
extraction matter 提取物质
extraction mechanism 萃取机理
extraction metallurgy 提取冶金(学)
extraction method 萃取法;抽提法;抽取法;提取法
extraction non-return valve 抽汽逆止阀
extraction of absorption coefficients 吸收系数的提取
extraction of a root 开方(法)
extraction of oil 油的提取
extraction of pile 拔桩
extraction of root 求根
extraction of signal 分出信号
extraction of sleeper screw 拔道钉
extraction of soil solution 提取土壤溶液
extraction of soluble organic matter 可溶有机质抽提
extraction of square root 开平方;开方法
extraction of steam 抽汽;排汽
extraction of the cubic(al) root 开立方
extraction of the profit 取得利润
extraction of the square root 开方
extraction of water from slurry 料浆脱水
extraction of wavelet 子波提取
extraction opening 抽汽口;抽气口
extraction ore supervision 出矿管理
extraction packed column 填充抽提塔
extraction percentage 提取率
extraction-photometric method 提取光度法
extraction pipe 抽气管
extraction plant 萃取装置;萃取设备;抽提装置
extraction point 抽汽口
extraction potential 引出电位
extraction pressure 抽汽压力
extraction pressure governor 抽汽压力调节器
extraction procedure 提炼工序;提炼方法
extraction process 萃取过程;萃取法(处理);提取工艺
extraction process in wastewater treatment 废水处理萃取法
extraction pump 抽气泵;抽出泵;排气泵
extraction rate 提取率;萃取率;浸提率
extraction ratio 提取率
extraction regulating valve 抽汽调节阀
extraction reserves 备采矿量
extraction roll 压出辊
extraction scale 基本编图资料比例尺
extraction-scrubbing column 提取—洗涤柱
extraction separation 提取分离(法)
extraction shell 提取筒壳
extraction solvent 萃取剂;萃取(用)溶剂;抽提溶剂;提取剂
extraction spectrophotometric method 萃取光度法
extraction spectrophotometry 萃取分光光度法
extraction stage 抽汽级
extraction stage heater 抽汽加热器
extraction steam 抽汽
extraction steam engine 抽汽式蒸汽机
extraction steam for factories 工业抽汽
extraction steam heating 抽汽加热
extraction steam pipe 抽汽管
extraction steam regulation 抽汽调节
extraction steam turbine 抽汽式汽轮机
extraction-step 提取段
extraction stop and non-return valve 抽汽截止

extraction stop valve 逆止阀
extraction stop valve 抽汽截止阀
extraction substance 提取物质
extraction suppression 萃取抑制
extraction technique 抽出方法;提取技术
extraction temperature 萃取温度;抽汽温度
extraction test 浸析试验;萃取试验;抽提试验;抽取试验
extraction thimble 浸出用滤纸筒;抽提套管;提取壳筒
extraction time 萃取时间
extraction titration 萃取滴定(法)
extraction tool 拔插件工具
extraction tower 抽提塔
extraction treatment 萃取处理
extraction treatment of wastewater 废水萃取处理法
extraction tube 提取管
extraction turbine 抽汽式涡轮机;抽汽式汽轮机;抽汽式透平;抽(汽)式汽轮机
extraction turbo-generator unit 抽气式汽轮发电机组
extraction unit 萃取设备;抽气设备
extraction valve 抽汽阀
extraction ventilation system 抽气(式)通风系统
extraction well 抽水井
extraction yield 提取率;实收率
extractive 可提炼的;浸出物;耗竭自然资源的;抽出的;提取物;提取的;提出物
extractive agent 提取剂
extractive crystallization 抽提结晶
extractive distillation 萃取蒸馏;抽提蒸馏;提取蒸馏
extractive effluent 可提取的流出物
extractive-free wood 无提取物木材
extractive industries 天然生产业(矿业、农业、渔业等)
extractive industry 采掘工业
extractive material 提炼物
extractive matter 提炼物
extractive membrane bioreactor 萃取膜生物反应器
extractive metallurgy 精practical冶金(学);提取冶金(学)
extractive process 纯油提过程;提取制造程序;提炼(制造)过程
extractive substance 提取物质
extractive technique 提取技术
extractive test 萃取试验
extractive titration 提取滴定(法);萃取滴定(法)
extract layer 萃取层;抽提层;提取层
extract of a letter 函件摘录
extract of an account 账目摘要
extract oil 抽提油;提取油
extract operation 取操作
extractor 检出器;选拔器;析取字;引出装置;摘出器;萃取器;出料装置;抽筒子钩;抽提塔;抽出器;分离机;拔桩器;拔出器;起钉机;抛料机;抛壳机构;排气辅助器;爬го钩;脱模工具;退子钩;退壳器;退弹簧;提取者;提取设备;提取器;提炼器
extractor basket 提取器筐
extractor claw 退壳器爪
extractor fan 排气风扇;抽风机;抽风(风)扇
extractor fork 带卷推出机
extractor ga(u)ge 分离型电离真空计
extractor groove 退壳槽
extractor hook 抽筒子钩
extractor ionization ga(u)ge 超高真空分离规
extractor of pile 拔桩机
extractor pivot 退壳器支框
extractor plunger 退壳器柱塞
extractors 采掘性企业
extractor sub 抽取器接头;提取器接头
extractor transport unit 御运装置
extractor-weigher 给料称量器
extract phase 提取相
extract pressure flash tower 抽取加压闪蒸塔;提取加压蒸发塔
extract residues of cotton seeds 堵漏用棉花籽壳
extract separation 抽(提)物分离
extract sand 附加充填砂
extract spatula 浸膏调刀
extract storage 提出物储器
extract stripper 抽出液气提搭
extract sweetening 抽提物脱硫(法)
extract system 排风系统;抽气系统;抽风系统
extract tower 提取塔

extractum 浸膏;[复]extracta
extract ventilation 抽气(通风);通风
extract ventilation block 抽气式通风机组
extract ventilation chimney 抽气式通风烟囱
extract ventilation concrete block 吹气式通风混凝土砌块
extract ventilation duct 抽气式通风通道
extract ventilation line 抽气式通风管道
extract ventilation pipe 抽气式通风管
extract ventilation shaft 抽气式通风竖井;抽气式通风井道
extract ventilation tile 通气砖;抽气式通风砌块
extract ventilation unit 排风机组;抽气式通风机
extract ventilator 排气通道
extract venting block 排气通道砌块
extract venting chimney 排气烟囱
extract venting duct 排气风道
extract venting line 排气路线
extract venting pipe 排(气)式汽轮机
extract venting shaft 排气竖井;排气井道
extract venting tile 排气道用砖
extract wood 萃取用材
extra current 暂时电流;额外电流
extracurricular activity 课外活动
extra cutter head 附加进刀架
extra cutting 额外采伐
extra deep drawing 极深冲
extra deep-drawing steel 极深冲钢;优质深冲钢;超深冲钢
extradeep well 超深井
extra-dense barium crown glass 特重火石;特重钡冕玻璃
extra dense flint 特种火石玻璃;特重火石玻璃
extra dense flint glass 超重火石玻璃;特重火石玻璃
extra-departmental data 辅助编图资料
extra depreciation 额外折旧
extra depth 超挖深度[疏];超深;(航道等的)富余水深;附加水深
extra depth for siltation 预留备淤深度
extra depth of dredging 挖泥超深
extra-digging 超挖(量)
extra discount 额外折扣
extradition 引渡
extra dividend 红利;附加股利;额外股息;额外股利;特别红利
extrado 拱背
extrado of arch 拱背
extrados 拱弧的外曲线;拱背线;外弧面;外拱圈
extrados curvature 外弧曲率
extrados of arch 外拱圈;拱背
extrados of cast-in-situ concrete lining 现浇混凝土衬砌外轮廓周边
extrados radius 拱外弧半径
extrados springing 拱背起拱(线)点
extrados springing line 拱外弧起拱线;外拱圈起拱线
extra dot 补充点
extra-dredging 超挖量【疏】;超挖
extra drinking water 特优饮用水
extra duty 特别税;超负荷;额外负荷
extra-duty glaze 特种釉
extra-duty glazed tile 特种釉面砖
extra dynamite 特别黄色炸药
extraeconomic coercion 超经济强制
extra entropy production 额外熵产生
extra equipment 附加设备;特殊设备
extra error 附加误差
extra excavation 额外挖方
extra excitation 非常激发;额外激发
extra expenditures 临时费用
extra expenses 独立费(工程成本中除去直接费和间接费以外部分);额外开支;额外费用
extra exposure 额外曝光
extra facility 外部设备
extra factor 额外因素
extra-fill 超填;超高填方;附加填土;填土抛高
extra fill sand 附加充填砂
extra fine 特细号;特精密加工;超等
extra-fine black stone 超细黑石
extra-fine fit 特级精度配合;一级精度配合
extra-fine glass wool 超细玻璃棉
extra-fine glass-wool felt 超细玻璃棉毡
extra-fine grade 一级精度

extra-fine grinding 超细研磨
extra-fine screw 特细牙螺纹
extra-fine steel 超细钢;优质钢
extra fine thread 超细螺纹;超细牙螺纹
extrafocal image 焦外像
extrafocal photometer 焦外光度计
extrafocal photometry 焦外光度测量;焦外测光
extra-foresight 附加前视
extra freight 额外运费;附加运费
extra gain 额外收益
extra gang 特别工班;辅助工作队
extra ga(u)ge 分离规
extra gift 赠品
extraglacial deposit 远冰川沉积;冰川外沉积;外冰川沉积
extra good quality 超等质量
extra grade 特级
extra half of atoms 额外半面
extra hand 临时工;加长把
extra hard steel 超硬钢(钎);特硬钢
extra hard twist 高强捻
extra hazard 特别火险
extra-hazard occupancy 高火险用房
extrahead 反常带头
extra heavy 超重的;特强;特大功率的;特重级;加厚的
extra heavy concentration 特高浓度
extra heavy contact 特强接点
extra heavy crown glass 特重冕玻璃
extra heavy crude oil 超重原油
extra heavy drill rod 特重钻杆;加重钻杆
extra heavy duty 特重载
extra heavy duty crane 特重型起重机
extra heavy duty truck 特重载重车
extra-heavy duty type 特重型
extra-heavy duty tyre 特重轮胎
extra heavy duty winch 特重型绞车
extra heavy frame 特重车架
extra heavy fuel oil 超重燃料油
extra-heavy loading 特重荷载
extra-heavy oil 超浓缩油
extra heavy pipe 特强管;加重管;粗管;特重管(子);特厚管;加厚管
extra heavy sheet 特重厚玻璃板;特厚玻璃板
extra heavy sheet glass 超重玻璃板
extra heavy spring 特重型弹簧
extra heavy strength concrete pipe 特强混凝土管
extra heavy tube 加厚管
extra heavy type grab 特重型抓斗
extra heavy type grab bucket 特重型抓斗
extra heavy valves and fittings 加重型阀门及管件;特重阀和附件
extra heavy wash rod 特重冲洗管
extra high build epoxy bituminous paint 超厚浆环氧沥青漆
extra-high compression engine 超高压缩比发动机
extra-high frequency 极高频
extra-high grade 特高品位
extra-high-leaded brass 超高铅黄铜
extra high-leaded brass alloy 超高铅黄铜合金
extra-high pressure 超高(电)压
extra-high pressure mercury(vapor) lamp 超高压水银灯
extra-high refractive index optic(al) glass 超高折射光学玻璃
extra-high speed copier 超高速复印机
extra-high tensile steel 特高强度钢
extra-high tension 极高压;超高压【电】;特高压
extra-high tension cable 超高压电缆;特高压电缆
extra-high tension cable paper 超高压电缆纸
extra-high tension cage 极高压罩
extra-high tension circuit 超高压电路
extra-high tension insulator 超高压绝缘子
extra-high tension pulse generator 超高压脉冲发生器
extra-high tension rectifier 超高压整流器;特高压整流器
extra-high tension supply 超高压电源
extra-hightension unit 超高压设备;特高压部件
extra-high tide 特大高潮
extra-high voltage 极高压;超高压(电);特高压
extra-high voltage bushing 超高压套管
extra-high voltage electric(al) field 超高压电场
extra-high voltage generator 特高压发生器;超高

压发生器
extra income 额外所得
extra-instruction 附加指令;宏指令;广义指令【计】
extra interpolation 超插入法
extra investment 额外投资
extra item 额外项;额外项目
extrajection 外向投射
extra job 额外工作
extrajudicial document 非法律文件
extra large 特大号;特大的
extra-large orifice sprinkler system 特大喷口喷水灭火系统
extra latten 热轧特薄板
extra-lemon pale 超柠檬白
extra-light drive fit 轻压配合
extra-light duty crane 特轻型抓斗;特轻型起重机
extra-light flint glass 超级火石玻璃;特轻火石玻璃
extra-light loaded 特别轻载
extra-light-loaded circuit 特轻加载线路
extra-light loading 特轻加载
extra-light reflector 超光源反射器
extra limit 外加限制的
extra limiter 外加限止器
extra-limit issue 额外发行;限外发行
extra-limy slag 高石灰含量矿渣;特强石灰炉渣
extra-load 额外荷载
extra-load bearing capacity 额外荷载量;额外负载量;额外承载(能)力
extra-load carrying capacity 额外荷载量
extra-long flat car 特长平车
extra-long insertion forceps 超长插入镊
extra-long near-bit stabilizer 超长近钻头稳定器
extra-long oil 超长油度
extra-long perforation tunnel 特长射孔孔道
extra-long range forecast 超长期预报
extra-long staple 超级长绒
extra-long wheelbase 超长轴距
extra-loss 附加损失;附加损耗;额外损失;额外损耗
extra-low carbon stain less steel 超低碳不锈钢
extra-low carbon steel 极低碳素钢
extra-low frequency 超低频
extra-low potential 超低电压(低于30V)
extra-low pressure tyre 超低压轮胎
extra-low refractive index optic(al)glass 特低折射率光学玻璃
extra-low voltage lighting 特低压照明
extramagmatic 外岩浆的
extraman 机械手
extra marker 外加指标点
extramaster 远航船长;资深船长
extra memory 外加存储器
extra-meridian observation 近子午圈观测
extra mild steel 特软钢
extra-milled fiber 特制磨切纤维
extramolecular 分子外的
extra-moraine lake 冰堰湖
extra mortality rate 超额死亡率
extramural 壁外的;市外的;校外的;墙外的
extramural absorption 光纤外部吸收
extramural adsorption 界外吸收
extramural cladding 光纤外包层
extra narrow pillar file 特窄平齐头锉
extraneous 外附物
extraneous action 外部作用
extraneous air 外部空气
extraneous air separator 外部空气离析器
extraneous argon 外来氩
extraneous ash 外来灰分
extraneous cracking 外部裂化
extraneous current 额外中流
extraneous dirt 外来杂质
extraneous earnings 额外收益
extraneous electricity 外来电流
extraneous element 外来元素
extraneous estimate 额外估计(值)
extraneous estimator 额外估计量
extraneous expenses 额外费用
extraneous factor 额外因子;外界因素
extraneous field 外界场
extraneous force 外力
extraneous gas 外来气
extraneous high frequency 外路高频
extraneous income 额外收益;额外收入
extraneous information 额外信息;外部信息
extraneous interference 外来干扰
extraneous light 外来光;外部光线
extraneous load 外来负荷
extraneous locus 额外轨迹
extraneous loss 额外损失;额外损耗
extraneous magnetic effect 外磁效应
extraneous material 外来杂质
extraneous matter 杂质
extraneous mineral mater 外来矿物质
extraneous modulation 寄生调制
extraneous moisture 外在水分
extraneous noise 外来噪声
extraneous observation 额外观测(值)
extraneous peak 假峰;外来峰
extraneous pressure 外来压力
extraneous quantitative information 额外数量信息
extraneous regressor 额外回归自变量;额外回归因子
extraneous response 外来响应
extraneous risks 附加(风)险;附加危险;额外风险;特殊危险
extraneous rust 外部锈蚀
extraneous signal 局外信号
extraneous solution 额外解
extraneous sources 外源
extraneous term 外加项
extraneous thermal effect 外界温度影响
extraneous variable 客变量
extraneous waste 外来废石
extraneous water 外部水
extraneous water production 外来水产量
extraneous wave 局外波;寄生波;寄生信号
extranet 企业外部网【计】;外延网
extra-network ion 网络外离子
extra-normal soil 异常土(壤)
extranuclear process 核外过程
extranuclear structure 核外电子结构
extra-nutrition 补充营养
extra-oil production 增产油量
extra-orange pale 超橙黄白
extra order 附加位;外加指令
extraordinarily high temperature 超高温
extraordinarily serious natural calamities 特大的自然灾害
extraordinary 非常折射线
extraordinary budget 特殊预算;临时预算;非常预算
extraordinary case 非常情况
extraordinary component 非寻常光波
extraordinary cost 特别费用
extraordinary decision 非常性决策
extraordinary depreciation 非常折旧;特别折旧
extraordinary disbursement 非常支出
extraordinary expenditures 特殊支出;临时费用;临时支出;非常支出;特殊费用
extraordinary expenses 临时经费;非常费用
extraordinary flood 非常洪水;特大洪水
extraordinary gains 非常收益
extraordinary gains and losses 非常损益
extraordinary high tide 异常高潮;特大高潮
extraordinary high tide level 异常高水位;异常高潮(位)
extraordinary image 异常像;非常像
extraordinary inclusion 异常包裹体
extraordinary income 非常收入
extraordinary index 非寻常波折射率
extraordinary index of refraction 非常折射率
extraordinary item 特殊项目;临时项目;例外项目;非常项目
extraordinary light 非常光(线)
extraordinary light ray 异常光线
extraordinary loss 非常损失;特别损失
extraordinary maintenance 临时养护;非常养护;特别养护
extraordinary mode 非常模式
extraordinary obsolescence 非常陈旧
extraordinary payment 临时支出;意外支出;特别费用
extraordinary precaution 特殊预防措施
extraordinary profit 超额利润;非常利润
extraordinary profit and loss 额外损益
extraordinary property loss 财产的异常损失
extraordinary rainfall 特大降水
extraordinary ray 非常射线;非寻常光;非常光(线)
extraordinary ray spectrum 非常光线光谱
extraordinary(ray)wave 非寻常光波
extraordinary receipts 非常收入;额外收入;特殊(部门)收入
extraordinary refractive index 非常折射率
extraordinary repairs 非常修理;特别修理费
extraordinary resolution 非常决议
extraordinary revenue 非常收入
extraordinary scour 异常冲刷;强烈冲刷;特大冲刷
extraordinary sea 特大浪
extraordinary serious natural calamity 非常严重的自然灾害
extraordinary spoilage 非常损坏
extraordinary spring tide 特大高潮
extraordinary storm 非常暴雨;强烈暴雨;特大暴雨
extraordinary tax 特别税;非常税
extraordinary-tonnage mine 特大型矿山
extraordinary traffic 异常的交通运输
extraordinary wave 异常波;非常波;特大波
extraordinary weather 异常天气;反常气候;反常(的)天气
extra orientation element 外定向元素
extra output order 附加输出指令;补充输出指令
extra over 额外附加费
extra over price 高价;额外加价
extrapacking 特殊包装
extra pair 备用线对
extra-pale 超苍白;特级浅色
extra passenger train 临时旅客列车
extra payment 附加工资;额外报酬
extra-period fire 超时限火;超时火灾
extra plane of atoms 额外原子平面
extra plan profit 超计划利润
extra ply 附加层
extra point 附加点
extrapolability 推断力
extrapolate 外推;外插
extrapolated boundary 外推边界
extrapolated building-up pressure 外推压力恢复值
extrapolated Curie temperature 外推居里温度
extrapolated curve 外推曲线;外插曲线
extrapolated cutoff 外推截止电压
extrapolated end point 外插端点
extrapolated height 外推高度
extrapolated insurance data 外推保险数据
extrapolated intercept 外推截距
extrapolated intersection time 外推交点对应时间
extrapolated ionization range 外推电离程
extrapolated length 外推长度
extrapolated mean 外推平均
extrapolated method 外推法
extrapolated method of lateral pressure curve 旁压曲线外推法
extrapolated octane number 外延辛烷值;外推辛烷值
extrapolated onset point 外推起始点
extrapolated onset temperature 外推起始温度
extrapolated pressure 外推压力
extrapolated range 外推射程
extrapolated side 外推面
extrapolated state 外推状态
extrapolated value 外推值;外差值;推定数值
extrapolating 外插
extrapolating system 外推制
extrapolation 域外评价法;归纳;外推(法);外插;推测
extrapolation chamber 外推电离箱
extrapolation coefficient 外推系数
extrapolation design 外推法设计
extrapolation distance 外推距离
extrapolation error 外延误差;外推误差
extrapolation formula 外推公式
extrapolation function 外推函数
extrapolation ionization chamber 外推电离室
extrapolation method 外插法;外延法;外推法
extrapolation method with equation discharge drawdown curve 流量降深曲线方程下推法
extrapolation number 外推数
extrapolation of gravity 重力外推
extrapolation of historic(al)records 历史记录推断
extrapolation of rating curve 展延水位—流量关系曲线;水位—流量关系曲线外延
extrapolation operator 外推算子
extrapolation procedure 外推程序

extrapolation process 外推过程
extrapolation type satellite's guidance system 外推法卫星制导系统
extrapolation value 外推值
extrapolation viscosity 外推黏度
extra power 外部能源
extra precision 超精度
extra premium 额外奖励;额外保险费;附加保险费
extra pressure drop 额外压力降
extra pressurization 特别增压
extra price 附加价格
extra profit 超额利润;额外利润
extra project 额外工程
extra pseudo order 附加假指令;补充假指令;外加伪指令
extra pulse 过脉冲
extra-pure grade 超纯级
extrapyramidal system 锥体外系统
extrapyramidal tract 锥体外束
extra quality 特优质量
extra quantity 额外数量
extra-railway 铁路外的
extra rapid hardening cement 特快硬水泥
extra rapid hardening Portland cement 特快硬硅酸盐水泥;特快硬波特兰水泥
extra-rapid stitching machine 高速缝合机
extra reflection 额外反射
extra regressor 附加回归因子
extra reinforced bar 加设护顶梁
extra reinforcement 额外增强
extra risks 特约险;特别险
extrasample performance 超样本特性;附加样本特性;额外样本方法
extraseasonal fruiting 季节外结果
extra-sensitive 超灵敏度的;高敏(感)性的
extra-sensitive clay 高度灵敏黏土;高塑性黏土;超敏感黏土;过敏性黏土;超灵敏黏土
extra-sensitivity 超灵敏度;高敏性;特高灵敏度
extra service 额外服务
extra shallow wave 越浅水波;超浅水波
extra shift 加班
extra ship cargo 额外船货
extra short oil 超短油度
extra-sintering phenomenon 额外烧结现象;同化烧结现象
extra-size folio 大四开图纸
extra slack running fit 特松动配合;松动配合
extra sliding member 特别滑动构件
extra slim hole 超狭孔
extra slim taper file 细尖三角锉
extra slow lane 外加车道
extra small bearing 特小型轴承
extra soft 机软的
extra soft steel 特软钢
extra special improved plough steel wire 超级制绳钢丝
extra-spectrum colo(u)r 谱外色
extra stack section 超级叠加剖面
extra statutory concession 法外宽免
extra statutory expenditures 非法定支出
extra steam 额外蒸汽
extra step 附加步
extra stern line 船尾增线
extra still 减压重蒸馏;减压拔顶蒸馏
extrastratal silica cement 外源氧化硅胶结物
extrastratal source 层外来源
extra-strength pipe 超强度管;超强钢管
extra-stress 额外应力
extra strong 加强的
extra strong coupling 高强接头
extra strong paper 超强纸
extra strong pipe 特强管;特厚壁管;粗管;超强钢管;超厚壁管
extra strong steel pipe 特强钢管
extra-super-duralumin 超高强度硬铝
extra-super-duralumin alloy 超(级)硬铝合金
extra-super long dash 特长虚线
extra surplus value 超额剩余价值
extra systole 期外收缩;过早收缩
extra task 额外工作
extra tax on corporate profit 超额公司利得税
extra tax on profits 超额利得税
extra tendon 外加力筋
extraterrestrial 地球外的;管辖地区以外的

extraterrestrial γ-ray 地外伽马射线
extraterrestrial civilization 地外文明
extraterrestrial environment 行星际环境
extraterrestrial feature 星体地物
extraterrestrial geodesy 宇宙大地测量学
extraterrestrial inteligence 地外智慧
extraterrestrial light 行星际光
extraterrestrial noise 地(球)外噪声
extraterrestrial photograph 从外层空间拍得的地面相片;地外相片
extraterrestrial photography 地外摄影
extraterrestrial radiation 大气顶ди太阳辐射;地球外辐射;大气圈外太阳辐射;行星际辐射;地外辐射
extraterrestrial signal 宇宙信号
extraterrestrial soil 地球外的土
extraterrestrial source 地外源
extraterrestrial trajecory 行星际轨迹
extraterritoriality 治外法权
extraterritorial sales clause 境外销售条款
extrathermal ionization 超热电离
extrathermodynamic(al) method 超热力学法
extrathermodynamics 超热力学
extra-thick sheet glass 超厚玻璃板;特厚玻璃板
extra thick steel wire 特粗钢丝
extra thin 超薄型的
extra thin fine thread small hexagonal nut 特薄细牙六角小螺母
extra thin sheet glass 超薄(型)玻璃板;特薄玻璃板
extra thin steel wire 特细钢丝
extra thin-walled bit 特薄壁钻头
extra thin-walled casing 特薄壁套管
extra thirds 特三裁
extra traction 附加牵引力
extra-tropic(al) 热带以外的
extra-tropic(al) belt 热带外地区;温带;中纬度地带
extra-tropic(al) cyclone 温带气旋;中纬度气旋
extra twist 外加捻度
extraultra-viblet radiation 超紫外辐射
extra-urgent 特别紧迫的
extravagance and waste 铺张浪费
extravagant 浪费
extra variable 附加变数
extravasation 溢出;熔岩外喷
extravaste 熔岩外喷
extravehicular activity 舱外活动
extraversion 外向
extra voting rights 附加表决权
extra wage 附加工资;额外工资
extra water 附加水;补给水
extra water pressure 附加水压力
extra weft figuring 特纬纱形成的花纹
extra-weight drill pipe 加重钻杆
extra wheel 预备轮;备用轮(胎)
extra white 特白
extra-wide base tire 特宽轮胎
extra-wide cloth 超阔幅布
extra-wide field stereo-microscope 超宽视场立体显微镜
extra work 额外工作量;加班;新添工程;附加工程;额外工作
extrema 极值
extremal 致极函数
extremal curve 极值曲线
extremal distance 极值距离
extremal distribution 极值分布
extremal field 极值场
extremal function 极值函数
extremal graph 极值图;极图
extremalization 极端化
extremal probability paper 极值概率纸;极值(分布)机率格纸
extremal problem 极值问题
extremals 极值曲线
extremal vector 极值向量;极大向量;极描向量
extreme 尽端的;末尾的;最终的;最大程度;外项
extreme and mean ratio 黄金分割;中末比;外内比
extreme arch 终端拱
extreme arid 极干旱
extreme arid area 极干旱地区
extreme boundary lubrication 极压润滑;极限边界润滑
extreme breadth 最大宽度;全宽(度)
extreme case 终端箱;特例
extreme climate 极端气候

extreme close-up 大特写镜头
extreme close-up photograph 超近景摄影
extreme components 极端组分
extreme compression fiber 受压最强面
extreme condition 极端条件;最恶劣的条件
extreme control 极值控制;极值调节
extreme cross girder 终端横大梁;终端横撑杆
extreme crosswind 强侧风
extreme deep diving 极限深度潜水
extreme deformation 极端变形
extreme depth 最大深度
extreme descent 急剧下降
extreme design pressure 最高设计压力
extreme deviation 极差
extreme difference 极差
extreme dimension 极限尺寸
extreme discharge 极限流量;特大流量
extreme disequilibrium 极不平衡
extreme distant earthquake 极远震
extreme draft 最大吃水(深度)
extreme draught 最大吃水(深度)
extreme earthquake event 强烈地震事件
extreme earthquake motion 强烈地震运动
extreme end 最末端
extreme end of chassis 底盘末端
extreme environment 极端环境
extreme event 极端事件
extreme face 极面
extreme feasible vector 极可行向量
extreme fiber 最外边(的)纤维
extreme fiber stress 最外(缘)纤维应力;外缘纤维应力
extreme fire behavio(u)r 极端火灾行为
extreme fire danger 极端火险
extreme flood 极限洪水;特大洪水
extreme gear 终端齿轮
extreme higher position 最高位置
extreme highlight 最明部分
extreme high reliability 非常高的可靠性
extreme high-speed gas dynamics 高超音速气体动力学
extreme high vacuum 极高真空
extreme high voltage 特高电压
extreme high water 极高水位;极高潮位;最高洪水位;最高高潮(位)
extreme hillside combine 陡坡用联合收割机
extreme infrared 超红外
extreme injury 极严重的损伤
extreme in position 进入报端位置
extreme length 最大长度;船舶全长
extreme light weight hide 最轻磅皮
extreme limit 极限值
extreme line 极线
extreme line casing 管端成平坦线的套管;公母扣直接连接的套管;外加厚套管
extreme-line thread 极线扣套管
extreme-line thread coupling 极线扣接箍
extreme line tubing 流线型油管
extreme load 极端负载;特殊荷载
extreme load combination 极端荷载组合
extreme loading condition 特殊荷载条件
extreme long-shot 大全景镜头
extreme low carbon steel 超低碳钢
extreme low discharge 极限枯水流量;极限低流量
extreme lower position 最低位置
extreme low water 最低枯水位;最低低潮(位)
extremely abrasive ground 强摩擦性岩层
extremely arduous condition 极端状态;最艰难的情况
extremely arid 极端干燥的;极端干旱的
extremely coarse 极粗的
extremely coarsely crystalline 极粗晶质
extremely complex morphologic(al) 形态很复杂的
extremely crushed 极破碎
extremely dangerous material 特别危险品
extremely difficultly falling roof 极难冒落顶板
extremely dried mix 特干混合料
extremely finely crystalline 极细晶质;隐晶质
extremely hard 特硬
extremely hard condition 最恶劣的条件
extremely hazardous substance 极危险物质;特别危险品
extremely high frequency 极高频

extremely high mountain 极高山(绝对高度五千米)
extremely high-speed oscilloscope 极高速记录示波器
extremely high strength rock 强度极高的岩石
extremely high tension generator 超高压发生器
extremely high water level spring tide 大潮极高水位
extremely hydrographic(al) year 非常水文年
extremely important test 非常重要的试验
extremely inverse current relay 反时限电流继电保护装置
extremely inverse relay 极端反时限继电器
extremely low frequency 极低频
extremely low frequency antenna 极长波天线
extremely low water 极低水位;极低潮
extremely low-water level spring tide 大潮极低水位
extremely quick 非常快的
extremely selfcombustion 很易自燃
extremely sensitive 高度敏感的
extremely short pulse 极短脉冲
extremely slow 非常慢的
extremely small orebody 极小矿体
extremely strong complex 极强络合物
extremely thin-bedded orebody 极薄层矿体
extremely thin-walled bit 特薄壁钻头
extremely toxic substance 剧毒物质
extremely turbulence 极动荡
extremely viscous 非常黏的;特黏的
extreme mean 最大最小平均值;极限平均值
extreme method 极值法
extreme minimum temperature 极端最低温度
extreme narrowing approximation 极窄近似法
extreme operating condition 极端操作条件;最恶劣使用条件
extreme optimal strategy 极最优策略
extreme optimal vector 极最优向量
extreme parallel 极限纬圈
extreme path 极端光程
extreme point 极值点;极限点;满点;端点
extreme pollution 极度污染
extreme position 极限位置
extreme pressure 极限压力;极端压力;最高压力;特压
extreme pressure additive 极压添加剂;特压添加剂;耐特压添加剂;超高压添加剂
extreme pressure film 极压润滑油膜
extreme pressure gear oil 极压齿轮油
extreme pressure lubricant 极压润滑剂;耐(热)高压润滑剂
extreme pressure soap 特压皂
extreme pressure turbine oil 极压汽轮机油
extreme range 最大射程
extreme ray 极射线
extreme relativistic limit 相对论极限
extreme response 极值反应
extreme scenario 最后方案;特殊情况
extreme service pressure 最高使用压力
extreme service shoe 最重工作鞋
extreme small discharge 最小流量
extreme span 极限跨度;全翼展
extreme spread 最大散布
extreme stray of concrete quality 混凝土质量的最大离散性
extreme stress 极大应力;外缘应力
extreme temperature 极端温度;最高温度
extreme temperature refractory 超高温耐火材料
extreme tension region 构造不稳定地区
extreme term 外项
extreme terms of proportion 比例外项
extreme theory 极值理论
extreme tidal range 极端潮幅;极潮差;最大潮差
extreme tide 非常潮位
extreme tide range 最大潮差
extreme trace 末道;超痕量
extreme trace analysis 极痕量分析
extreme ultraviolet 超紫外
extreme ultraviolet laser 极(远)紫外激光器
extreme ultraviolet photometer 极远紫外光度计
extreme ultraviolet radiation 远紫外辐射
extreme ultraviolet ray 远紫外线
extreme ultraviolet region 远紫外区
extreme upsurge 最高(上)涌浪

extreme value 极量值;极大值;极(限)值;极端值
extreme value criterion 极值准则;极值判据
extreme value distribution 极值分布
extreme value index 极值指数
extreme value problem 极值问题
extreme value response 极值响应
extreme vector 极向量
extreme vessel breadth 最大船宽
extreme vessel length 最大船长
extreme wave condition 极大波况
extreme wind velocity 最大风速
extremital 末端的
extremities of fenders 保护板末端
extremity 绝境;极限;末端;非常手段
extremity of an interval 区间的末端
extremity piece 端件
extremity routine 末端程序
extremity value 极值;极量值;极大值
extremum 极值
extremum control system 极值控制系统;极差控制系统
extremum method 极值法
extremum of a fuzzy functions 模糊函数(的)极值
extremum principle 极值原理
extremum problem 极值问题
extremum property 极值性
extremum seeking method 极值寻找法
extricate 游离
extrication 解救;摆脱
extrinsic 非固有的;非本征的;外来的;外赋的
extrinsic absorption 非固有吸收
extrinsic acceptance 票外承兑
extrinsic conduction 非本征导电
extrinsic conductivity 外赋传导率
extrinsic current 外延电流
extrinsic curvature 外在曲率
extrinsic cycle 外循环
extrinsic detedctor 非本征激发的探测器
extrinsic diffusion 非本征扩散
extrinsic evidence 外部证据
extrinsic factor 外源因素;外部因子
extrinsic feature 外观
extrinsic incubation period 外潜伏期
extrinsic insoluble 外来(的)不溶物
extrinsic instability 非本征不稳定性
extrinsic internal photoeffect 非本征内光电效应
extrinsic junction loss 非本征连接损耗
extrinsic luminescence 非本征发光
extrinsic nucleation 非本征成核
extrinsic photo conductor 非本征光敏电阻
extrinsic photo-effect 非本征光电效应
extrinsic photoemission 非本征光电发射
extrinsic problem 外在问题
extrinsic property 非固有性质;非本征性质;非本能性质
extrinsic region 外赋区
extrinsic rewards 额外报酬;外在的酬赏
extrinsic rewards and punishments 附带的奖惩
extrinsic semiconductor 非本征半导体
extrinsic silicon detector 非本征硅探测器
extrinsic sillca glass 非本征石英玻璃
extrinsic sol 外来溶胶
extrinsic stability 非本征稳定性
extrinsic value 非固有的价值
extroversion 外向性;外倾
extrudability 可挤压性;可挤出性;压出可能性
extrudable insulation 压挤绝缘;注射式绝缘
extrudant 压出半成品
extrudate 挤出物;压出物;压出型材
extrudation 压出
extrude 压制;压出;冲压
extrude aluminium bronze tube 挤压铝青铜管
extrude angle 挤压(的)角钢
extrude bar 挤压条
extruded 挤出型的
extruded alumin(i)um alloy 压挤的铝合金
extruded alumin(i)um mounting 挤压铝线脚
extruded alumin(i)um sill 挤压铝窗台
extruded angle section 挤出角材
extruded article 挤压制品
extruded asbestos cement 挤制石棉水泥
extruded bar 挤出棒材;挤压棒材
extruded board 挤压的板
extruded brass 压延黄铜制品

extruded brass section 压延黄铜型材
extruded channel 挤压的槽钢
extruded chipboard 压延的碎胶合板;压延的刨花板
extruded clay roofing tile 挤制陶土屋顶瓦
extruded coating 挤塑涂层;挤塑贴面
extruded concrete 计量混凝土;挤压混凝土
extruded concrete lining 挤压混凝土衬砌
extruded concrete precast unit 挤压混凝土预制构件
extruded concrete profile 压制混凝土型材;模压混凝土型材;挤压混凝土型材
extruded concrete section 压制混凝土型材;模压混凝土型材;挤压混凝土型材
extruded concrete shape 压制混凝土型材;模压混凝土型材;挤压混凝土型材
extruded concrete trim 挤压成的混凝土装饰物
extruded concrete unit 挤压成的混凝土构件
extruded electrode 挤压成型焊条;压涂焊条
extruded floor(ing) tile 压挤的地面砖
extruded fuel 挤压燃料
extruded glass 压型玻璃
extruded glazed tile 压出釉面砖;压出釉面瓦;压出玻璃瓦
extruded graphite 挤压成型石墨
extruded hole 挤压翻孔;挤压成带法兰的孔
extruded interlocking clay roof(ing) tile 压制陶土咬接瓦
extruded liner 模压橡胶内套
extruded metal 压延金属
extruded particle 挤压碎料
extruded particle board 挤压碎料板;挤压刨花板;挤压刨花板
extruded pipe 挤压法制出的管子;压制管;冲制管
extruded plastic coating 挤压塑料涂层
extruded polystyrene 挤压聚苯乙烯
extruded porous sections 挤压多孔零件
extruded product 挤压制品;挤出制品
extruded railing 挤压成型的栏杆
extruded rib section 轧制带肋型材
extruded rib shape 挤压成形的肋条型材
extruded rib trim 栅螺纹镶边
extruded rib unit 挤压成形的肋形(建筑)部件
extruded rigid corrugated PVC sheet 挤压硬质聚氯乙烯波纹板
extruded rock 侵出岩
extruded rod 挤压条材
extruded rubber parts 挤制的橡胶零件
extruded sample 挤压样
extruded section 压制型材;挤压型材;压制叶型;挤压型钢;压挤截面;挤压(零)件
extruded shape 压制型材;挤压型材;挤压成型;推挤型
extruded sheet 挤塑板;挤板材;喷出岩席
extruded spar 挤压成型翼梁
extruded stock 挤压坯
extruded structural pipe 挤压结构管材;挤压结构钢管
extruded structural shape 挤压结构型材
extruded structural tube 挤压结构管材
extruded tile 挤制砖;挤制瓦
extruded tube 挤压管材
extruded unit 挤压成形的建筑构件
extruded vinyl fitting 挤压成形的乙烯配件
extruded vinyl furniture 挤压制的乙烯家具
extruded vinyl hardware 压挤的乙烯小五金
extruded vinyl section 挤压制的乙烯型材
extrude lining 挤压衬砌
extruder 塑孔器;挤压机;挤泥机;挤出机;压挤机;顶样器;顶挤器
extruder and cooling train 压出冷却联动装置
extruder barrel 压出机料筒
extruder base 压出机机座
extruder blan 压出坯料
extruder blanks 压出坯料
extruder bore 挤压机内径;压出机机膛;压出机机膛
extruder-calender 压出机延机
extruder core 挤压机芯型;压出机芯型
extruder cylinder liner 挤压机机筒内衬;压出机机筒内衬
extruder die 挤压机口型;压出机口型
extruder head 挤压机机头;压出机机头
extruder-head core 压出机机头芯型
extruder output 压出量

extruder-pelletizer 螺杆挤压造粒机
extruder pressure 挤压压力;压出压力
extruder rate 挤压速度;压出速度
extruder screen pack 挤出机过滤网组
extruder screw 螺旋挤压机;挤压机螺杆;挤出机螺杆;压出机螺杆
extruder size 压出机型号
extruder stand 压出机机座
extruder type compactor 挤压式压实机
extruder type spinning machine 螺杆式纺丝机;挤压式纺丝机
extruder with side delivery head 直角(机头)挤压机
extruder with straight delivery head 直通(机头)挤压机
extrude spinning method 挤压纺丝法
extrude tubes 挤压管材
extruding 挤压(成型);挤压;压出
extruding aids 挤压助剂;压出助剂
extruding-desiccation machine 挤压脱水机
extruding device 推撑器
extruding-head 压出机机头
extruding machine 挤压机;挤芯(棒)机;挤棒机;压压机;压出机
extruding press 挤压机
extruding process die 挤压模
extruding screw 挤压螺杆
extruding test 挤压试验
extruding tool 冲孔翻边模
extruding trauma 挤压伤
extruding wire coating 挤塑电线涂层
extruser 螺旋挤出机
extrusion 挤压(件);压出;冲塞
extrusion and press process 滴料热压成型法
extrusion auger 螺旋式挤压机
extrusion billet 挤压坯段;挤压坯料;挤压方坯
extrusion-blow mo(u)lding 挤压吹制成型
extrusion casting 模压铸造
extrusion chipboard 挤压(成型)刨花板;挤压碎料板
extrusion coater 挤涂机
extrusion coating 挤压涂装;挤压涂层;挤出贴面;喷涂
extrusion colo(u)ring 挤压着色
extrusion compound 挤塑用混合料
extrusion concrete 挤压混凝土
extrusion concrete girder 挤压制成的混凝土大梁
extrusion concrete section 挤压制成的混凝土型材
extrusion contour 展宽轮廓
extrusion cooking 挤压蒸煮
extrusion data 展宽数据
extrusion defect 挤压缺陷
extrusion die 挤型模;压挤钢模
extrusion die swell 模口挤出膨胀
extrusion-dried rubber 压出干燥法橡胶
extrusion dryer 螺压脱水机
extrusion effect 挤压效应
extrusion flow 挤压流
extrusion force 挤压力
extrusion forging 挤压锻造
extrusion forming 加压成型;挤压成型;挤出成型
extrusion gear pump 挤出齿轮泵
extrusion grade 挤出级
extrusion graph 挤压曲线
extrusion horst 挤出型地垒
extrusion index 压出指数
extrusion ingot 挤压(坯)锭
extrusion interlocking clay tile roof(ing) cover(ing) 压制陶土咬合瓦屋面覆盖层
extrusion jack 顶样器
extrusion laminating 挤出层压
extrusion line 压出流水线
extrusion load 挤压荷载
extrusion machine 挤压机;挤坯机;挤管机
extrusion mark 挤压痕迹;挤出纹路
extrusion metal 挤压金属
extrusion method 挤压法;压涂法
extrusion method compaction by explosion 爆破挤密法
extrusion method of pile driving 打桩挤密法
extrusion method of sand pile 砂桩挤密法
extrusion method of toss a rock 抛石挤密法
extrusion moisture 挤压水分
extrusion mo(u)lding 挤压模塑法;挤压成型;挤塑;压挤成型

extrusion nozzle 喷丝板
extrusion of metal 金属的挤压成型;挤出金属
extrusion of pore water 孔隙水逸出
extrusion plant 挤压设备;挤压工厂
extrusion plastometer 挤压式塑性计
extrusion press 挤压机;热模压机
extrusion pressing 热模压加工
extrusion pressure 挤压(压)力;压出力;单位挤压力
extrusion process 挤压加工;挤压过程;挤压(成型)法;挤塑法;挤出成型法;压铸工艺
extrusion product 挤压制品
extrusion quality 压出质量
extrusion ratio 挤压比
extrusion rheometer 挤出式流变仪
extrusion rheometry 挤出式流变测定法;挤出流变测定法
extrusion section 压挤断面;挤压型材
extrusion shape 压挤型材
extrusion shroud 挤压防护罩
extrusion stamp 捣锤;风冲子
extrusion stress 挤压应力
extrusion system 挤压系统
extrusion test(for joint filler)(嵌缝板的)挤出试验
extrusion type capillary visco(si)meter 熔体指数测定仪
extrusion type injection machine 压出式注压机
extrusion type test 挤出型测试法
extrusion under vacuum 真空挤压成型
extrusion velocity 挤压速度
extrusion viscometer 挤出式黏度计
extrusion weldrod coating hydraulic press 焊条挤压镀层机
extrusive 喷出的
extrusive body 喷出(岩)体
extrusive facies 喷出相
extrusive ingeous rock 喷出火成岩
extrusive rock 喷出岩
extrusive window sash plastics 挤出的窗框塑料
extrustion of metals 金属件冲压
exttaction water 提取水
exudant 渗出物
exudate 渗出物;分泌物;渗出液
exudating temperature 渗出温度
exudatinite 沥青侵入体
exudation 流出物;溢液;胀箱;跑火;渗出作用;渗出液;渗出(物)
exudation pressure 土中水分渗出压力;渗泥压力;渗(流)压力
exudation pressure test 渗水压力试验
exudation theory 挤出学说
exudation vein 分凝脉
exude 渗出;流出
exurb 城市远郊富裕阶层居住区;富人市郊住宅区
exurban 城市远郊的;市远郊的
exurbanite 市郊居民
exurbia 城市远郊
exutory 诱导剂
exuviate 脱壳
exvacuated time 抽空时间
ex voto church 还愿教堂
ex-warehouse 卖方仓库交货价(格);仓库交货
ex-warehouse price 仓库交货价(格)
ex warrants 无执照
ex-wharf 码头交货价(格);码头交货;码头交换价(格);目的港码头交货(价);目的地码头交货
ex-works 现场交货;工厂交货
ex-works price 工厂交货价(格)
ex-works test 出厂试验
exydisulfoton 砜拌磷
eye 孔(环);环首;信管口;中心进气通道;洞眼;穹隆顶窗;喷火孔
eye-and-ear method 耳目法
eye-and-eyebrow structure 眼屑状构造
eye and eye turnbuckle 双环伸缩螺丝
eye and hook turnbukle 环钩伸缩螺丝
eye and key method 目捷法
eye appeal 具有动人吸引力的
eye assay 用眼评定;肉眼鉴定;肉眼鉴别;肉眼检查
eye axis 视轴
eyeball 实测点
eyeball assay 肉眼检查
eyeballs in 凹眼球

eyeballs out 凸眼球
eye band 桅顶系绳籀
eye bar 环头铁杆;有孔杆;羊眼杆;眼杆;带环(拉)杆;孔杆
eye-bar head 梢眼放大端头;孔眼杆件端(穿销钉用)
eye-bar hook 眼杆钩
eye bar packing 眼杆填料;眼杆垫板
eyebar suspension bridge 铰链(眼杆)悬索桥;眼杆悬索桥
eye base 眼距;眼基线
eyebase for steel mast 钢柱拉杆底座
eye base line 眼基线
eyebase setting 眼基线调节【测】
eye bender 弯套管机
eye bit 钩环扁钻
eye block 带眼滑车
eye bolt 环首螺钉;有眼螺栓;羊眼螺栓;眼螺栓;吊环螺栓;带眼螺栓;带孔螺栓;带环螺栓;凹头螺栓;单耳连接螺栓
eyebolt and key 环首螺栓;(带)插销螺栓
eyebolt for bilge block 侧垫木穿钉(干坞坞底埋件)
eyebolt shaft 加长活塞杆
eye bracket on derrick 吊杆上眼肘板
eye brow 前缘翼缝;遮水楣;窗头线饰;窗楣;屋顶窗;滴水;波形老虎窗;眉状结construction
eyebrow dormer 低(矮)老虎窗;波形老虎窗;矮老虎窗;低屋顶窗
eyebrow of side scuttle 舷窗遮水楣
eyebrow window 墙顶窗
eyecap 目镜转向棱镜
eye-catcher 醒目广告
eye-circle 出射光瞳
eye coal 鸟眼状煤;眼球状煤
eyecup 目镜环
eye diagram 可见图形
eye diameter 孔眼直径;入口直径
eye diaphragm of caisson 沉箱的眼窗隔板
eye draft 目测草图
eye end 有眼端
eye end clamp for strands 定位环线夹
eye end rod 耳环杆
eye estimate 目视估计;目估;目测估计值
eye estimate method 目估法
eye estimation 目视估计;目估
eye examination 目测
eye failure 孔眼引起的失效
eye fidelity 映像保真度
eye field 眼状油气田
eye filter 目镜滤光片
eye for towing 拖曳用孔眼
eye ga(u)ge 目测放大镜
eyeglass 目镜;观察镜;表玻璃;镜片
eye gneiss 眼球片麻岩
eye goggles 眼护
eye guard 目镜装置;护目;目镜板;防蚀目罩;目镜安全罩
eye height 视线高度
eye hitch 牵引环
eye hole 窥视孔;检视孔;麻坑
eye hook 链钩;眼钩
eye hook with safety latch 带安全阀的耳钩
eye joint 链连接;活节;环孔接合;眼榫接头;眼圈接合
eye lens 目镜
eye lens of the eyepiece 目镜的接目镜
eye lens slide 目镜滑轨
eyelet 窥视孔;眼;金属圈;小眼孔;小窥视孔;眼孔;打小孔;穿绳孔;锁缝
eyelet bolt 活节螺栓
eyelet grommet 卷索眼;索耳
eyelet hole 锁眼
eyeleting 镶唇
eyelet machine 冲孔机
eyelet nozzle 带固紧环喷嘴
eyelet pincers 孔眼钳;环钳
eyelet pliers 孔眼钳;打眼钳
eyelet punch 孔眼钳;环钳;打眼冲头
eyeletter 打小眼机;打鸡眼机
eyeletting 用空心铆钉接合板材;在孔口中形成唇;打出小孔
eyeletting and riveting machine 打眼和铆接机
eyelet welding 孔焊;小孔熔焊

eyelet wire 带环线
eyelet work 衡孔;打孔眼;冲孔
eye-level 眼高度;视高度;视高程
eye-level rise 视线升高
eye-level shelf location 开架商品陈列
eye-level viewfinder 目视水准瞄准器
eyelid (半圆形调节片的)可调节喷口
eyelid retractor 支撑器
eyelid shutter 眼脸式快门
eyelid-type nozzle 可调节双瓣式喷嘴
eye light (摄影用)眼照明;眼神灯
eye line (透视图上的)视干线
eye lotion dropper 点眼器
eyemark 目标
eye measure(ment) 目测
eye-nose mask 眼鼻面罩
eye nut 环首螺母;环道螺母;有眼螺母;吊环螺母
eye observation 目测
eye-observation optic(al) decrepitation record method 目测光学爆裂记录法
eye of a dome 穹隆孔眼;穹隆顶孔
eye of a volute 涡卷心
eye of axe 斧柄眼
eye of caisson 沉箱眼窗
eye of cyclone 气旋眼
eye of dome 穹隆顶窗;穹顶孔眼
eye of hatchet 斧柄眼
eye of hurricane 飓风眼
eye of storm 风眼;风暴中心(区);风暴眼
eye of tropical cyclone (热带气旋)风眼
eye of typhoon 台风中心;台风眼
eye of wind 风眼;风穴
eye pattern 眼状显示图;目视图;出射图;目视模型;眼图
eye peruse eye survey 目测
eyepice aberration 目镜的像差
eyepiece 镜头;目镜;眼罩;耳承
eyepiece assembly 目镜总成;目镜管
eyepiece barrel 目镜筒
eyepiece cap 目镜帽;目镜盖
eyepiece collective 目镜聚光透镜
eyepiece cross-hair 目镜十字线
eyepiece diaphragm 目镜光阑
eyepiece end 接目端
eyepiece filter 目镜滤光片
eyepiece focal length 目镜焦距
eyepiece focusing 目镜调焦
eyepiece focusing ring 目镜调焦环
eyepiece frame 目镜框
eyepiece goniometer 目镜量角器
eyepiece graticule 目镜分划板
eyepiece grid 目视网格
eyepiece head 目镜头
eyepiece housing 目镜箱
eyepiece lamp 目镜灯
eyepiece lens 目镜透镜
eyepiece magnification 目镜放大率
eyepiece mark 目镜标记
eyepiece micrometer 接目镜测微计;目镜千分尺;目镜测微器;目镜测微计
eyepiece micrometer integrated measurement method 目镜微尺累积测量法
eyepiece micrometer measure method 目镜微尺测量法
eyepiece mount 目镜座
eyepiece network measure grain areal method 目镜网格测量颗粒面移法
eyepiece of level tube 水准管目镜
eyepiece prism 目镜棱镜
eyepiece reading 用目镜读数;目镜读数
eyepiece ring 目镜环
eye pieces 接目镜
eyepiece screw 目镜螺旋;目镜螺丝
eyepiece screw-micrometer 螺杆式目镜测微计
eyepiece selector lever 目镜选择杆
eyepiece sleeve 目镜筒
eyepiece slide 目视筒;目镜(滑)筒
eyepiece spectroscope 目镜分光镜
eyepiece stalk 目镜筒
eyepiece stop 目镜光栏
eyepiece tube 目镜筒;目镜管
eyepiece with micrometer 测微目镜
eye pit 视孔
eye plate 眼板;闸缸推杆头;三眼环;三角(眼)铁;三角眼板
eyeplate for mooring 系缆眼板
eye plate on deck 甲板眼板
eye-pleasing 悦目
eye point 眼点;出射点
eye-protecting material 护目材料
eye-protection glasses 护目镜
eye protector 护目设备;防护眼镜;护目(镜)
eye pupil 眼瞳孔
eye reach 视野
eye reading 用视读数;目读
eye-refraction difference 眼折射差
eye-rest colo(u)r 悦目色
eye ring 吊环;吊耳;耳环;耳柄;提引环;提手;套环
eye screw 环首螺栓;环首螺钉;羊眼螺丝;眼圈螺钉;吊螺栓;有眼螺钉
eye screw shackle 环眼螺栓卸扣
eye seizing 绳圈合扎
eye sensitivity curve 可见度曲线
eye separation 眼距
eye shade 目镜色片;目镜挡光片;遮光眼罩
eye shield 目镜色片;目镜挡光片;护目罩
eyeshot 眼界;视野
eyesight 眼界;观察;目视孔;目力;视野(观察);视力
eyesight acuity 视力敏锐度
eyesight adjustment 调焦
eyesight elbow 风口弯头
eyesight navigation 目测航行;助导航
eyesight tester 视力检验星
eye sketch 略图;目测草图;草测图
eye slit 观察孔
eyes of her 船的最前端
eye-sore 碍眼物
eye splice 接环;环接索眼;琵琶头(绳结);索眼(插接);索环;索端结扣眼圈;环接合;插扣
eye split 观察缝
eyestone 眼石
eye-stop 路标
eye strain 视觉疲劳;眼疲劳
eye structure 眼球状构造
eye survey by microscope 显微镜目测法
eye-survey method 目测法
eye tackle 带眼绞辘
eye thimble 眼环;眼圈
eye tunnel 双孔隧道
eye visible crack 肉眼可见裂缝
eye-wall chimney 眼壁柱
eye washing bath (劳动保护用)洗眼设备
eye window 眼形窗
eye-witness 见证人;目击者
eye work 目测
Eykman formula 艾克曼公式
eykometer 泥浆凝胶强度和剪切力测定仪;埃柯仪(泥浆和凝胶强度和剪切力测定仪)
eylettersite 磷钍铝矿
eyot 湖心岛;湖泊中小岛;河洲;河心(小)岛;河或湖中小岛
Eyring equation 艾林方程(式)
Eyring formula 艾林公式
Eyring molecular system 艾林分子体系
Eyring theory 艾林理论
Eyring viscosity 艾林黏度
Eyring viscosity formula 艾林黏度公式
Eytelwein's formula 埃特温公式(计算排水管流速公式)
eytlandite 钶乙矿
ezcurrite 意水硼钠石
eztlite 红碲铅铁石
E-Z tree 简易树

F

FAA [Federal Aviation Administration] System 岩土美国联邦航空局分类系统
Faber flaw 法伯瑕
Faber visco(si)meter 法贝尔黏度计
fabianite 法硼钙石
Fablan system 冲击钻进法;费布伦钻眼法
fabric 纤维品;组构;织造品;织物;编织物;编织品
fabric air permeability test 织物透气性试验
fabric analysing glass 织物分析镜
fabric analysis 岩组分析;组构分析;织物分析
fabric and metal lined plate type 纤维和金属衬里板式
fabric anisotropy 组构各向异性
fabricant 制作者;制造人;厂商
fabricate 制造;制备;结构加工
fabricated bar 钢筋网;网格钢筋
fabricated brick 机制红砖
fabricated bridge 装配式桥(梁)
fabricated building 装配式结构;装配式房屋
fabricated by the buyer 由买方制造
fabricated claim 虚构的索赔
fabricated construction 装配式结构;装配式建筑
fabricated cost 造价
fabricated crossing 装配式辙叉
fabricated dam 装配式坝
fabricated frame 装配式机架
fabricated materials 已加工原材料
fabricated metal 金属制品
fabricated metal industry 金属制品工业
fabricated on site 现场安装;就地安装
fabricated parts 现成构件;互换配件
fabricated product 装配式成品;加工制品
fabricated runner 组合式转轮;焊接式转轮
fabricated section 拼合截面;安装件;组合截面(铆接式焊接)
fabricated shaft 组合式主轴;焊接轴
fabricated shaper 成型机
fabricated shapes 加工型材
fabricated ship 组装船;分段装配船
fabricated short rail 缩短轨
fabricated single-post shore 装配式单杆支柱
fabricated steel 成型钢
fabricated steel block 装配式钢坞墩
fabricated steel body 装配式钢体
fabricated steel bridge 装配式钢桥;装拆式钢桥
fabricated steel liner 预制(钢筋)网衬砌
fabricated stem 钢板装配船首材
fabricated stern frame 组合船尾肋骨
fabricated structure 装配式结构;装配式建筑物
fabricated terrazzo 预制水磨石
fabricated textile 仿织造织物
fabricated track work 组装轨道工作
fabricated universal steel members 万能钢杆件;万能杆件
fabricated vessel 组合船;分段装配的船舶;分段建造的船舶
fabricating bay 生产车间
fabricating cost 制造费(用);建筑造价;建造费(用);安装费(用)
fabricating cycle 生产周期;建造周期
fabricating defect 制造缺陷;生产故障
fabricating dimension 制造尺寸
fabricating dispatch 制造工艺卡
fabricating drawing 生产图纸;加工图
fabricating industry 装配工业
fabricating method 制作法;生产方法
fabricating process 生产过程;制造过程;变形制造程
fabricating program(me) 生产计划
fabricating shop 装配(车)间;分段建造车间
fabricating technique 生产技术
fabricating technology 制造技术;制造工艺
fabricating yard 制作场;施工现场;安装场地;组装场
fabrication 装配;制作;制造加工;建造;生产;二次加料
fabrication cost 制造费用;造价;生产成本;建造费用
fabrication craft 制造工艺
fabrication drawing 加工图

fabrication facility 装配设施
fabrication of structural steel 建造钢的装配
fabrication on site 现场制作
fabrication operation 生产操作
fabrication order 生产凭单;生产命令
fabrication plant 加工软化
fabrication plant of timber 木材加工厂
fabrication platform 装配平台
fabrication procedure 制造过程
fabrication process 制造过程;制造工艺
fabrication schedule 制造程序
fabrication sequence 制造程序
fabrication shop 装配(车)间
fabrication specification 制造说明书;制造技术规范
fabrication technology 制造工艺
fabrication tolerance 制造公差
fabrication weldability 工艺焊接性
fabrication yard 装配场;制作场
fabricator 修整工;装配工;制作者;制造者;制造公司;金属加工厂;加工者
fabric backing 织物背衬;织物反面
fabric base 织物垫底
fabric belt 纤维带;纤维皮带
fabric board mill wastewater 纤维板厂废水
fabric breathability 织物透气性
fabric bursting test 织物顶破强力试验
fabric carcass 织物胎壳
fabric cling-testing 织物黏附性试验
fabric coating 织物涂料;织物涂胶
fabric coating unit 织物覆胶装置
fabric collector 袋式除尘器
fabric colo(u)r 金属丝颜色
fabric composition 织物的组成
fabric construction 织物结构
fabric cotton 棉织物;棉布
fabric count 织物经纬密度
fabric covering of wall 壁衣
fabric cutting machine 裁布机
fabric data 组构数据
fabric defects 织物疵点
fabric design 织物设计
fabric diagram 岩组图;组构图
fabric dipping unit 织物浸胶装置
fabric distortion 织物变形
fabric domain 组构域
fabric draft 织物设计草图
fabric drying machine 织物烘燥机
fabric dust 纤维性粉尘;纤维性尘埃
fabric dust collector 织物集尘器
fabric dust filter 布袋集尘器
fabric elasticity test 织物弹性试验
fabric element 组构要素;织物滤芯
fabric-enclosed sand drain 袋装砂井
fabric extension 织物伸长
fabric external appearance inspection 织物外观检验
fabric fall 织物悬垂性
fabric feature of cleavage 劈理组构特征
fabric fiber 织物纤维;屋面油毡
fabric filler 织物填料
fabric filter 纤维织物除尘器;纤维过滤器;织物吸尘器;织物过滤器;袋滤器;布滤器
fabric filter collector 织物除尘器;织物滤尘器
fabric filter dust collector 布袋过滤式集尘器
fabric filtration 布滤
fabric finish 织物整理
fabric flamability 织物可燃性
fabric flashing 油毡防雨层
fabric form 织物模板
fabric for roofing 屋面毡
fabric fuel tank 软燃料箱
fabric geometry 织物几何学
fabric growth 织物增长
fabric heat conductivity test 织物传热性试验
fabric hose 夹布胶管
fabric impact test 织物冲击强力试验
fabric index 定向程度指数
fabric inflating lead 布制充气管

fabric insert(ing) 织物垫片
fabric inspection 织物检验
fabric inspection of penalty system 织物扣分式检验法
fabric joint 织物挠性万向节
fabric laminate 织物层压材料
fabric laying jumbo 钢筋网铺设大型设备
fabric light 织物照明
fabric lineation 组构线理
fabric matrix 织物结合料
fabric metal 网用金属丝;编织用金属丝
fabric of aggregate 团集体组构
fabric ply 织物层
fabric porosity 织物透孔性
fabric proofing 织物覆胶
fabric pull 织物牵引力
fabric-reinforced 织物增强的
fabric-reinforced seal 织物加强密封件
fabric reinforcement 织物加强件;钢丝网配筋;钢筋网;焊接钢架
fabric resilience 织物回弹性
fabric retaining wall 土工布挡土墙
fabric ribbon 编织色带
fabric rubber lined hose 有橡胶衬里纤维水带
fabric scrim 织物料
fabric-seal bearing 织料密封轴承
fabric separator 布袋集尘器
fabric sewability tester 织物缝纫性能试验仪
fabric sheen 织物光泽
fabric sheet(ing) 织物片;纤维织物(片材)
fabric shift tester 织物经纬纱线滑移性试验仪
fabric shock test 织物冲击强力试验
fabric shrinkage 织物缩水率
fabric shrinkage test 织物缩水试验
fabric silket machine 织物丝光整理机
fabric slitting 织物剖幅
fabric specification 织物规格
fabric stiffness test 织物硬挺度试验
fabric strength test 织物强力试验
fabric structure 织物结构
fabric take-off 织物卷出
fabric take-up 织物卷取
fabric take-up and batching device 织物卷取和打卷装置
fabric tank 可折叠罐;软桶
fabric tape 织带;布质卷尺
fabric texture 织物组织
fabric tire 织物胶轮胎
fabric-to-fabric bonding 织物与织物黏合
fabric-to-fabric lamination 织物与织物层压黏合
fabric-to-foam laminating 织物与泡沫塑料层压黏合
fabric treating unit 织物处理装置
fabric-type dust collector 织物过滤集尘器
fabric underlay 织物衬底
fabric universal joint 软性万向节
fabric wall covering 壁布;贴墙织物;贴墙布
fabric web 织物幅
fabric width 网幅;金属网幅;织物幅宽
fabric wire 钢丝网
fabric wire ga(u)ge 金属网线规;网丝规;金属网丝规
fabric work 建造房屋构架(工作)
fabridam 尼龙坝;合成橡胶坝
fabriform 焊制结构
fabriform concrete 模袋混凝土
fabriform protection 尼龙织物护岸
Fabrikoid 伪造布革;法布里德(一种防水织物)
Fabritius correction 法布里蒂乌斯改正
fabroil 纤维胶木;夹布胶木
fabroil bearing 夹布胶木轴承
faroil gear 纤维胶木齿轮;夹布胶木齿轮
Fabry-Barot method 法布里-巴罗法
Fabry lens 法布里透镜
Fabry-Perot 法布里-珀罗
Fabry-Perot amplifier 法布里-珀罗放大器
Fabry-Perot cavity 法布里-珀罗谐振腔
Fabry-Perot etalon 法布里-珀罗标准具;标准具

Fabry-Perot filter 法布里—珀罗滤光器;法布里—珀罗滤光片
Fabry-Perot fringes 法布里—珀罗条纹;法布里—珀罗干涉条纹
Fabry-Perot injection laser 法布里—珀罗注入式激光器;法布里—珀罗半导体注入式激光器
Fabry-Perot interference spectroscope 法布里—珀罗干涉分光镜
Fabry-Perot interferometer 法布里—珀罗干涉仪;法布里—珀罗干涉计
Fabry-Perot laser cavity 法布里—珀罗激光腔
Fabry-Perot modulator 法布里—珀罗调制器
Fabry-Perot photograph 法布里—珀罗照片
Fabry-Perot recycling spectrometer 法布里—珀罗重复分光计
Fabry-Perot reflector 法布里—珀罗反射器
Fabry-Perot resonator 法布里—珀罗谐振器
Fabry-Perot spectrogram 法布里—珀罗光谱图
Fabry-Perot spectrometer 法布里—珀罗分光计
Fabry-Perot type cavity 法布里—珀罗腔
Fabry-Perot type laser 法布里—珀罗型激光器
facade 正(立)面;立面;建筑(物)立面
facade articulation 房屋正面连接
facade beam 房屋正面梁
facade brick panel 房屋正面砖板
facade building component 房屋正面构件
facade building member 房屋正面建筑构件
facade building unit 房屋正面建筑构件
facade cleaning 房屋正面清洗
facade cleansing 房屋正面清洗
facade cleansing agent 房屋正面清洗剂
facade coat 房屋正面涂层
facade column 房屋正面立柱
facade construction 房屋正面构造
facade covering 房屋正面覆盖层
facade density 外部装饰设计
facade design 房屋正面设计
facade development 房屋正面展开
facade division 房屋正面划分
facade double T frame 房屋正面双T形框架
facade easement 保持建筑外貌的守约
facade elevator 屋正面清洁用吊车
facade facing 房屋正面饰面
facade girder 房屋正面大梁
facade H frame 房屋正面H形框架
facade joint 房屋正面接合
facade lining 房屋正面加衬
facade masonry wall 房屋正面砖石墙
facade of building 房屋正面
facade of station 车站正面
facade painting 房屋正面油漆
facade panel 房屋正面预制板;外墙板
facade pavilion 建筑物前面突出部分
facade polyvinyl chloride coat 房屋正面聚氯乙烯涂层
facade protection 房屋正面保护
facade protection agent 房屋正面保护剂
facade PVC coat 房屋正面聚氯乙烯涂层
facade rendering 房屋正面抹灰
facade scaffold(ing) 屋正面脚手架
facade slab 房屋正面石板
facade stucco 房屋正面拉毛粉刷
facade system 房屋正面建筑
facade tower 建筑物前面构筑的塔楼
facade treatment 外部装饰
facade unit 房屋正面(建筑)构件
facade ventilation grating 房屋正面通风栅
facade wall 正面墙
facade with air circulation 通风的房屋正面
facctte 柱槽突面
face 支撑面;割面;膜面;面向;面临;面对;面部;乳剂面;前面;使用面;上面;端面;地貌面;采脂面;采矿工作面;采掘面;材面;表面;版面;板面
face a crisis 面临危机
face advance 工作面推进;工作面进度;齿面接触提前量
face airing 正面通风;工作面通风(开凿隧道)
face amount 面值;面额;票面金额
face-amount certificate 面值证书
face and fill 面上和装满
face-and-side cutter 三面刃铣刀
face-and-slab plan 分段回采程序
face angle 面角;齿面角
face angle of a polyhedral angle 多面角的面角

face angle of a polyhedron 多面体的面角
face arch 前拱
face area of sample 样品表面积
face ashlar 表面琢石
face bar 面材
face-bedded 竖向石砌筑;(沉积岩类等层状石材的)竖向纹理砌筑
face-bedded stone 竖向石砌体
face belt 工作面皮带运输机
face belt conveyer 工作面胶带运输机
face belt track 工作面皮带运输机道
face-bend specimen 表面弯曲试样
face-bend test 正面弯曲试验;表面弯曲试验;(焊缝的)正弯(曲)试验;面弯试验
face-bend test specimen 面弯试样
face block 饰面砌块;端面大砖
face board 贴面胶合板;贴面板
face bond 面接
face bonding 正面焊(接);面接合法;平面焊接
face-bow 面弓
face-breasting jack (盾构的)开挖面千斤顶
face brick 面砖;护面砖;外墙用砖;饰面砖
face brick bond 面砖砌合
face bucket ladder excavator 工作面多斗式挖土机
face building component 饰面建筑构件
face building member 饰面建筑构件
face building unit 饰面建筑构件
face cam 平面凸轮
face capacity 掌子面工效
face-centered 面心的
face-centered closest packing 面心最紧密填充
face-centered crystal 面心晶体
face-centered cube 面心立方体
face-centered cubic(al) 面心立方的
face-centered cubic(al) copper-gold alloy 面心立方铜金合金
face-centered cubic(al) packing 面心立方堆积
face-centered cubic(al) structure 面心立方结构
face-centered cubic(al) lattice 面心立方晶格;面心立方格子
face-centered grating 面心网格
face-centered lattice 面心晶格;面心格子;面心点阵
face-centered monocline system 面心单斜晶系
face-centered orthorhombic 面心正交
face-centered orthorhombic lattice 面心正交晶格
face-central cubic(al) array 立方面心排列
face checking 面板离缝;表面开裂
face chuck 平面卡盘
face clay brick 饰面黏土砖
face clearance 铣刀端面后角;表面留隙
face cleat 面割理;主要劈理;主解理面
face cloth 方巾
face coat(ing) 表面涂层;面漆
face colo(u)r 正视颜色;正面颜色;面色
face component 饰面构件
face concrete 饰面混凝土
face cone 面锥
face contact 面接触;按压接触;按钮开关
face conveyer 工作面输送机
face count 材面检量
face cut 基准材面
face cutter 端铣刀
face cutting 端面车削;车平面
face cutting edge 端面切削刃
face cutting radius 唇部切削半径(钻头)
faced and drilled 表面加工及钻孔
faced block 饰面砌块;面饰砌块
faced building mat 贴面卷材
faced building quilt 贴面卷材
faced concrete panel 饰面混凝土板
faced disk 门锁垫盘
face defect 面缺陷
faced facade 房屋装饰正面
faced glass wool product 贴面玻璃棉制品
faced hammer 琢面(石)锤;平锤
face diagonal 面对角线
faced-in 双列内向牛舌
faced insulation(material) 贴面绝热材料
face discharge 底面排水(钻头)
face discharge bit 唇面排水眼钻头;唇面排水孔钻头;底冲式钻头
faced joint 光面接缝;光面接缝
faced masonry 镶面圬工
faced on both sides 两边饰面

face down 面向下
face-down bonder 倒装焊接机
face-down bonding 倒装焊接;侧焊;面朝下焊接(法);倒焊(法)
face-down feed 背面馈送
faced plywood 覆面胶合板;贴面胶合板;装饰夹芯板
face drain 表面排水
faced rigid cellular phenolic thermal insulation 贴面硬质酚醛泡沫塑料保温板
face drill 工作面用钻机
face driver center 端面拨动顶尖
faced surface 削光面
faced wall 光面墙;饰面墙;正面墙;镶面墙
faced washer 平面垫圈
faced with danger 面临危险
faced with granular cork 粒状软木镶面
faced with granular material 颗粒材料镶面
faced with limited resources 面临有限资源
face east 朝东
face edge 规准面边;基面棱边
face ejection 唇面排水
face ejection bit 唇面排水眼钻头;唇面排水孔钻头;底冲式钻头
face end 工作面终点;工作面端头
face equipment 面板设备
face excavated by shovel 用铲开挖竖直面
face excavator 正铲挖土机
face-face contact 面面接触
face fall 工作面塌跨
face fastening 板面固定
face-fed brazing 外加钎料的钎焊
face finish 表面涂饰剂
face flange 平面法兰;平面凸缘
face flushing 工作面冲洗;工作面通风
face formwork 饰面模板
face gear 平面齿轮;端面齿轮
face glass 前దిష్ట;玻璃面板;槽面镶嵌玻璃
face glazing 槽面镶嵌玻璃;装窗玻璃;表面釉饰
face grain 表面纹理
face grinder 平面磨床;表面磨光机
face grinding 磨端面;平面研磨;表面磨削
face grinding device 端面磨削装置
face grinding machine 平面磨床;端面磨床
face grinding wheel 平面磨轮
face guard 面罩;面具;护面罩;防护面具
face hammer 琢石锤;琢面锤;平锤
face-hardened 表面变硬;使表面变硬;表面淬硬的
face hardening 表面硬化;表面淬火
face haulage 回采工作面运输
face head attachment 端面刀架附件
face height 工作面高度
face in frames 框架中的侧面
face infusion pressure pipe 表面灌注压力
face jack (盾构的)开挖面千斤顶;挡土千斤顶
face joint 明缝接头;明缝;端面接合;出面接缝;材面接合;表面接缝;露面接缝
face knot 外皮节子;材面节疤;表面节疤
face lapping mill 平面研磨机
face lathe 落地(式)车床
face layer 面层
face left(position) 盘左;正镜【测】
face length 面长
face-lift 改建;整容;刷新;表面修整;表面修饰;整顿刷新
face-lifting 建筑物改建;翻新
face lighting 工作面照明
face-line 码头前沿线;(码头的)前沿线
face-line of teeth 齿顶线
face-line of wharf 码头前沿线;码头岸线
facellite 钾霞石
face-loading equipment 工作面机械化装载设备
face machined flat 磨平的端面
face man 工作面的工人
face mark 面层符号;材面标记;表面标记;刨面标记
face mask 面罩;面具;防毒面具
face-masked pressure respiration 面罩加压呼吸
face measure 正面宽度(建筑物);立面宽度(建筑物);木板面积;单面量度;材面量度;表面量度
face measuring instrument 平面测量仪;表面测量仪
face mechanization 工作面机械化
face member 饰面构件
face mill 平面铣刀
face miller 端面铣床

face milling 表面铣削;端铣;端面铣削;表面研磨
face-milling cutter grinder 端面铣刀刃磨机
face-milling machine 磨墙面机;磨机
face mix 特制饰面混和料;水泥石屑拌和料;饰面混合(材)料;筛面混合料
face mixture 面层混合料(混凝土)
face moment 端部节点弯距
face mo(u)ld 面样模板;划线模板;表面形状检验仪;样板;面模
face nail 平头小钉
face nailing 表面钉住;露头钉面;面钉
face of a block 街段的一面
face of a dihedral angle 二面角的面
face of an arch 拱面
face of brick 砖面
face of coupling 接头连接面
face of crystal 晶面
face of drawing 图纸的正面
face of existing reclamation bund 原吹填围堰面
face off 倒角;侧角
face of form 模板内面
face of gear 齿轮面
face of hole 孔底
face of map 图面
face of oiling 模板面涂油
face of polyhedron 多面体的面
face of propeller 压力面
face of pulley 滑轮端面;滑轮侧板
face of sand 砂纸表面;砂堆
face of slope 坡面
face of the channel 探槽工作面
face of theodolite 经纬仪望远镜位置;望远镜位置
face of the screen 屏面;屏蔽面
face of the value 支票票面值
face of the well bore 孔壁;井壁
face of tool 钻头前刃面
face of tooth 齿的侧面
face of weld 焊接面;焊缝表面
face of well 孔壁;井壁
face of wheel 轮面
face of work area 掌子面
faceometer 面直径测量器
face-on object 正向天体
face panel 面板
face paper 贴面纸
face par 票面价格
face parallel cut 平行面切割
face-passing point 对向道岔
face perpendicular cut 垂直面切割
face-piece 面壳;面具;面罩
face pin spanner 端面扳手
face piping (滤池系统的)外管路
face-plate 面板;划线用平板;划线平台;平台基准面;荧光屏(阴极射线管);面盘;花盘;平面卡盘
face-plate rheostat 平板式变阻器
face-plate screw 面板螺钉
face plywood 复面胶合板
face pofiling 端面仿形车削
face point 表面接缝
face point switch 对向道岔
face position of telescope 外业计算的点位
face powder 擦面粉
face preparation 切割工作面
face profile 平面轮廓
face profiling 车面仿形车削
face protection 护面
face protector 面部保护用具
face-pumping 面泵浦
face putty 正面腻子;(嵌门窗玻璃的)面油灰;外用腻子;外露油灰;饰面用油灰;(装窗玻璃的)露面油灰
face puttying (嵌门窗玻璃的)面灰;抹刮油灰
face rate of interest 名义利率
facer block 端面大砖
face recovery coefficient 工作面的生产率
face relief angle 端面后角
face right 盘右;倒镜
face right position 盘右
face roller 转子;叶轮;滚轮;导论
face runout 端面跳动
faces 斧刃面
face sampling 暴露面取样
face saver 面层保护罩;护面罩
face seal 面密封

face seam 表面黏接;表面接缝;表面黏结(地毯)
face seating pressure 端面压力
face-sheet (夹层结构的)面板
face-shell (混凝土空心砌块的)边壁,空心砌块侧面
face shield 脸护;面罩;护面罩;电焊面罩
face shovel 正(向)铲;正铲挖土机;正铲挖掘机;面铲
face shovel attachment 正铲挖土机附件
face shovel bucket 正铲挖土机斗
face shovel fitting 正铲挖土机装置
face shovel position 正铲挖土位置
face side 材料正面;正边;工作面;木(作表)面
face signal 回采工作面信号
face slab 镶面板;面板;饰面板
faces occlusalis 咬合面
face spanner 水平调节扳手;半圆头扳手;端面扳手;叉形螺母扳手
face spanner for slotted lock rings 开槽锁环扳手
face sprag 工作面斜支柱;工作面支柱
face-sprigging 插型钉
face stability 开挖面稳定
face stock 露面镶板
face stone 面石;饰面石;底面金刚石(钻头)
face string 露明楼梯斜梁;外露楼梯基;露面楼梯斜梁
face stringer 露明楼梯小梁;出面(的)楼梯小梁;露面楼梯小梁
face structure 面结构
face support 工作面支护
face supporting 开挖面支护
face surface 工作面
face symbol 面状符号
facet analysis 面分析法
facet-classification 小平面分类
faceted 翻光面
faceted boulder 磨面巨砾
faceted classification 逐面分类法;分面分类法
faceted glass (有刻凿面的)彩色玻璃
faceted lens 多面体透镜
faceted pebble 多面体卵石
facet effect 刻面效应
face template 划线样板
face terrace 海岸阶地
face the challenge 面临挑战
face tile 贴面砖;饰面砖
face timbering 工作面支护;工作面支撑
face tissue production line 面巾纸生产线
facet method 小块平面法
facet mirror 分段镜
face-to-face 面对面地;面对面
face-to-face coupling 面对面耦合
face-to-face distance 平行面间距
face-to-face duplex ball bearing 成对双联向心推力球轴承
face-to-face duplex bearing 成对双联轴承
face-to-face duplex tapered roller bearing 成对双联圆锥滚子轴承
face-to-face groin 对口丁坝
face-to-face negotiation 面洽;当面谈判
face-to-face picturephone 图像电话
face-to-face pile carpet 双层绒头地毯
face-to-face spur dike 对口丁坝
face tooth 端面铣刀齿
face towel 毛巾
facet(te) 凸线;平圆面;柱槽筋;(多面体的)小平面;小(晶)面;镶面;刻面;盘刻;板刻
facetted mirrors 组合反射镜
facetted pebble 棱石
face tube 皮托管
face type 连接面形式
face unit 饰面构件
face unit formwork 工作面单元模壳
face unit mo(u)ld 母模;面模;划线样板
face up 面向上;面朝上;滑配合;配研;涂色;对研
face-up bonding 正面焊接(法);面朝上焊接(法)
face-up feed 正面馈送
face value 面值;面额;票面价值
face velocity 迎面风速;罩口风速;(空气流入设备的)面速度
face veneer 露面镶板;面层单板;饰面板;表面装饰薄板;表层装饰薄板
face vener 表层薄板
face voltage 工作面电压
face waling 工作面横撑;工作表面横撑
face wall 胸墙;镶面墙;面墙;齐胸高的女儿墙;挡土墙

face wheel 平面轮
face width 齿宽;表面宽度;刨光板面净宽度
face width of tooth 齿面宽
face-window 面窗
face work 镶石工作;镶面工作;抹面(工作);饰面工(程);车端面;涂面工;贴面工
faceworker 工作面的工人
facework material 镶面材料;饰面材料
facia 招牌;挑出牌;封檐板
facia board 檐口板;(柱顶横梁的)横带;封檐板;牌匾
facial 正面上的
facial angle 面角
facial carina 颜隆线
facial cleft 面裂
facial-contracted wire rope 面接触钢丝绳
facial deformity 面部畸形
facial density of ore layer 矿层面密度
facial density of surrounding rock 围岩面密度
facial dimension 面部距离
facial force 面积力
facial index 面指数
facial plane 面部平面
facial restoration 面修复术
facial sign 面征
facial tissue 搽面纸
facial tissue holders 表面式餐巾纸
facial triangle 面部三角区
facia panel 仪器盘
facia sign 门面招牌
facient 因子;因数;乘数
facieology 岩相学
facies 相【地】;外观
facies association 相组合
facies change 相的变化;相变
facies controlled source rock 烃源岩相分布
facies dorsales 背侧面
facies fossil 指相化石
facies group 相组;相阻
facies inferolateralis 下外侧面
facies map 相图
facies medialis 内面;内侧面
facies micrometer 目镜测微计
facies model 相模式
facies of igneous rocks 火成岩岩相
facies posterior 后面
facies suite 相套
facies-unconformity trap 岩相不整合圈闭
facies ventralis 腹侧面
facies zone 相带
facies zone within intrusive body 侵入体岩相分带
facilitate 促进
facilitated area 易化区
facilitated diffusion 易化扩散
facilitating agency 服务代理机构
facilitating functions 便利功能
facilitation 易化
facilitator 服务商;服务性企业
facility plan 处理设施初步设计(废水)
facility purchase order 设备购买订货单
facility 设施;装置;设备
facility administration control and time schedule 设备管理控制和时间调度程序
facility and equipment 工厂和设备
facility assignment 设备分配
facility charges 设备费(用)
facility comparison 设备比较
facility control console 设备控制台
facility cost 设备成本
facility design 工具设计;设备设计
facility design criteria document 设备设计标准文件
facility detail 详细设施
facility dispersion 设施分散
facility engineering change proposal 设备工程更改建议
facility for accelerated service testing 快速试验环线(美国)
facility for berthing 靠船设备
facility for cooperative effort 并行工作的方便性
facility for dehydration treatment 脱水处理设施
facility for dispatch 装卸疏运设备;疏运设备
facility for drawing 绘图设备
facility for other contractors 为其他承包商提供方便

facility for plotting 绘图设备
facility for repair 维修设备
facility for scribing 刻图工具
facility for simultaneous interpretation 同声传译设备
facility for testing 为检验提供设备
facility for transportation 运输设备
facility for water intake and drainage 取排水设备
facility installation review 设备安装检查
facility location 设备布置；厂址选择
facility management 设备管理
facility modernization 装备现代化
facility of access 有效性
facility of payment clause 支付约定条款
facility operation department 设施部门
facility planning 设备计划
facility power panel 设备电源板
facility prosperity 福利设施
facility purchase order 设备购买订货单
facility record 设施记录
facility replacement 设施更新
facility replacement decision 设备更新决策
facility request 设备要求；设备请求
facility request separator 技术能力请求分解器
facility terminal cabinet 设备接线盒
facility usage ratio 设施利用率
facility visit 现场访问
facing 镶面(工作)；罩面；刮削；盖面；加工表面；面向；面料；面对；面层；面板；密封面；护面；砌面；铺面；平面加工；贴面；饰面；端面车削；衬片；衬面；敷面
facing aggregate 饰面(混凝土)集料；饰面(混凝土)骨料
facing alloy 表面加硬用硬合金；表面堆焊硬合金
facing arm 横旋转刀架；横刀架
facing bar 支撑板
facing between centers 顶尖间车削端面
facing block 外露预制(砌)块；饰面砌块
facing block partition 饰面砖砌筑的隔墙
facing board 饰面板
facing bond 砌面；顺砌
facing brick bond 饰面砖砌筑
facing brickwork 饰面砖工
facing (clay) brick 贴面砖；面砖；饰面砌筑砖
facing composition 饰面材料成分
facing compound 饰面复合材料
facing concrete 装饰混凝土；面层混凝土；护面混凝土
facing concrete aggregate 护面混凝土集料
facing concrete block 护面混凝土块体
facing concrete slab 镶面混凝土板
facing cows in 双列内向牛舍；对头式栓头
facing cut 面铣；端面铣
facing cutter 铣刀盘；平面铣刀
facing element concrete 面板混凝土
facing engineering brick 房屋建筑饰面工程用砖
facing expansion joint 面层伸缩缝；面层接缝
facing foil 饰面金属箔；饰面箔
facing ga(u)ge 总length头测定仪；对向轨距
facing grouting 饰面喷浆
facing hammer 修整锤；琢面锤
facing head 回转刀架；平旋盘
facing joint 面层接缝；护面接缝
facing jointer 表面刨光机
facing kit 饰面工具
facing layer 饰面层
facing machine 镶面机；刨床
facing masonry wall 护面圬工墙
facing masonry work 护面圬工砌作
facing mass 饰面材料
facing material 覆面材料；罩面材料；面料；面材；护面材料；涂料；贴面材料；饰面(材)料；表面装饰料；表材
facing member 饰面构件
facing method 涂面方法
facing mix(ture) 外露面混料；饰面混合(材)料；涂面混合料
facing movement 逆向行驶
facing of embankment 路堤边坡；块石护坡
facing of pitching 铺砌护面
facing of slope 地面坡向
facing operation 端面加工
facing panel 饰面板
facing pavio(u)r 高温焙烧砖；面层铺路机；面层铺路工；过火面砖
facing plywood 饰面胶合板
facing point4 对向道岔；敌对道岔
facing point crossover 双向渡线；对向渡线
facing point lock 对向道岔锁闭器
facing porcelain 瓷身面
facing punch 精整冲头；端面冲头
facing reverse pin porcelain 反订瓷面
facing ring 衬片环
facing rivet 覆面铆钉
facing route 对向进路
facing sand 覆面砂
facing set 平面刮刀
facing sheet 镶面薄板
facing slab 饰面板；镶面板；面板
facing slide 端面切削滑板
facing slip 砂面砖
facing stone 面石；护面石料；护面石；贴面石；饰面石
facing stone slab 饰面石板
facing stone ware 饰面用石制件
facing stop 端面挡块
facing surface 饰面；密封面层；密封面
facing system 镶面方法
facing the limited resources 面临有限资源
facing tile 饰面砖；墙面砖；外墙面砖；贴面砖；面砖
facing tile partition(wall) 饰面砖隔墙
facing tool 端面车刀
facing turnout 对向道岔
facing-type cutter 套式面铣刀
facing up 滑配合；配研；对研
facing wall 护面墙
facing with marble 大理石饰面
facing with pebbles 卵石饰面
facing work 饰面工作
facis board 封檐板
Fa colo(u)r 法华(指陶瓷)
facs 设施；设备
facsimile 传真；精确复制；模写；摹写
facsimile communication equipment 传真通信设备
facsimile communication system 传真通信系统
facsimile (copying) telegraph 传真电报
facsimile edition 传真版
facsimile equipment 传真机
facsimile map 摹绘图
facsimile optic-fibre 传像光学纤维
facsimile paper 传真感光纸
facsimile posting 摹写登记
facsimile printing 摹真印花
facsimile radio 无线电传真
facsimile receiver 传真接收机
facsimile recording 传真记录
facsimile room 无线电传真室
facsimile service 传真业务
facsimile signal 传真信号
facsimile signature 印鉴样本；复制签字
facsimile system 传真系统
facsimile telegram 真极电报
facsimile telegraph 传真电报
facsimile terminal 传真终端
facsimile transmission 传真输送；传真发送
facsimile weather chart 传真气象图
facsimile weather map 传真气象图
factable 墩帽(的一种)
fact finder 调查者
fact finding 实况调查；实地调查的；进行实地调查；调查的
fact-finding meeting 调查会
fact-finding technique 调查研究技术
factice 油胶；油膏
factional sampling 分级取样
factis 亚麻油橡胶
factitial 人工的
factitious 人工的
factitious ultramarine 合成群青(蓝)；人造群青(蓝)
factor 系数；因子；因素；因数；因式；中间商人代理商；指数；融资商；土地经管人(苏格兰)；代收公司；保收人
factorable motion 可析因运动
factorable polynomial 可分解的多项式
factor affecting 影响因素
factorage 佣金；融资商佣金；手续费；代理商佣金；代理厂商；代理业

factor algebra 因子代数
factor analysis 因子分析；因子分解；因素分析(法)；因数分析；因式分解；要素分析
factor analysis method 因子分析法
factor analysis method of data 数据因子分析法
factor analysis model 因子分析模型
factor analysis of soil slope stability 土坡稳定性因素分析
factor augmenting technical progress 要素增广的技术进步
factor combination 要素组合
factor comparison 因素比较
factor comparison system 因素比较法
factor compensation 因子补偿作用
factor cost 因素成本；要素成本
factor-cost value 要素成本价值
factored load 设计极限荷载；乘上系数的荷载；极限设计荷载
factored moment 计算力矩
factor-factor relationship 要素间关系
factor for impact 冲击系数
factor for overcapacity 过载因数
factor group 因子群；商群
factor group analysis 因子群分析
factorial 阶乘积；级乘；代理厂商的
factorial analysis 因子分析
factorial arrangement 因子排列
factorial design 析因设计；因子设计
factorial development 因子显影
factorial discriminant analysis 因子判别式分析
factorial distribution 阶乘分布
factorial ecology 分解生态学
factorial effect 因子效应
factorial experiment 析因试验；因子试验；阶乘试验
factorial experiment design 析因实验设计
factorial function 阶乘函数
factorial information retrieval system 事项情报检索系统
factorial jump function 阶乘跳跃函数
factorial lumber 加工用材
factorial moment 阶乘矩
factorial moment generating function 阶乘矩母函数
factorial multinomial distribution 阶乘多项式分布
factorial operator 阶乘运算符
factorial polynomial 阶乘多项式
factorial power 阶乘的幂
factorial sampling 均匀分格抽样
factorial series 阶乘级数
factorial ship 加工船
factorial test 析因试验；因子试验
factorial trials 析因试验
factories act 工厂法
factories and mines road 厂矿道路
factories-mines road 厂矿道路
factoring 因子选择；因子分解；因式分解；托收信贷行；代理经营；出售应收账款；财务代理
factoring commission 经纪人佣金
factoring correlation matrix 析因相关矩阵
factoring firm 经纪人商行；客账经纪商号
factoring method 因数法
factoring of receivables 应收款让售
factoring service 代收服务
factor in precipitation 沉淀系数
factor input coefficient 要素投入系数
factor interaction 因子相互作用
factorist 因素论者
factorization 因子分解；因式分解(法)；编制计算程序
factorization method 因子分解(方)法
factorization of algebraic(al) equations 代数方程的因式分解
factorization of a transformation 变换(的)因子分解；变换因式分解
factorization technique 因子分解技巧
factorize 因式分解
factor level 因素水平
factor loading 因子载荷；因子负荷量
factor lytic 裂解因子
factor matrix 因子矩阵
factor model 因子模型
factor modulus 因子模
factor of adhesion 黏着系数；附着系数
factor of a model 模型因子

factor of an integer 整数的因子
factor of a polynomial 多项式的因子
factor of assurance 安全系数
factor of cargo permeability 货物渗透因数
factor of circuit 流程系数
factor of cracking 裂化系数
factor of cubicity (碎石的)立方体系数
factor of dilution of radiation 辐射稀化因子
factor of erosion 侵蚀因素
factor of evaluation 评价因素
factor of expansion 膨胀系数;热膨胀因数;热膨胀系数
factor of foundation bearing capacity 地基承载系数
factor of fuel utilzation 燃料利用系数
factor of hatches 舱口系数
factor of ignorance 无知因数;安全系数
factor of importance 重要性系数
factor of influence 影响因子
factor of karst 岩溶率
factor of life scatter 寿命分散系数
factor of location 区位因素;布局因素
factor of merit 优质率;质量因子;灵敏值;品质系数
factor of mineralizing control 成矿控制因素
factor of over-capacity 超荷载因素
factor of pavement width 铺面宽度系数
factor of plant location 厂址选择因素
factor of porosity 孔隙率
factor of probability 几率
factor of production 生产要素;生产的因素
factor of proportionality 比率;比例因子;比例因数
factor of quality 质量因素;质量因数
factor of reduction 折减因子;折减因素;折减因数;折减系数
factor of rigidity 刚性系数
factor of runoff 径流因素;径流因数;径流率;径流率
factor of safety 安全因数;安全因子;安全因素;安全系数;安全率;安全度;保险系数
factor of safety against cracking 抗裂安全系数
factor of safety against failure 抗破坏安全系数
factor of safety against flo(a)tation 抗浮安全系数
factor of safety against overturning 抗倾覆安全系数
factor of safety against sliding 抗滑安全系数
factor of safety against sliding failure 抗滑动安全系数
factor of safety of dry dock against flo(a)tation 干坞抗浮安全系数
factor of safety of slope 边坡稳定系数
factor of saturation 饱和系数;饱和率
factor of stowage 货物的积载系数;货物积载因数
factor of stress concentration 应力集中系数;集中应力系数
factor of stress-concentration in surrounding rock 围岩应力集中系数
factor of stress utilization 应力利用系数
factor of subdivision 分舱因数;分舱系数
factor of synchronous operation 同步运行系数
factor of thaw compression 融化压缩因子;融化压缩因数
factor of the habitat 生境因子;生境因素
factor of tooth wear 牙齿磨损系数;钻头类型系数
factor of transmission 传播因素
factor of utilization 公用因数
factor of variability 变动性因素
factor of vibration insulation 隔振系数
factor pair 因子对
factor pattern 因子模型
factor plane map 因子平面图
factor-price differentials 要素价格差别
factor-price equalization 要素价格均等化
factor-price frontier 因子价格界限;要素价格边界
factor productivity 要素生产力
factor proportion 因素比例;要素比例
factor rate 要素比率
factor representation 因子表示
factor reversals 要素反转
factor reversal test 因子互换检验
factors affecting response of testing 影响试验反应的各种因素
factor-saving innovation 节约生产要素的技术革新
factors controlling rate of dissolution 控制溶解速度的因素

factor score 因子得分
factor service 要素服务
factor set 因子组;因子集
factors influencing exogenic process 影响外力地质作用的因素
factors of mineral deposition and geologic(al) condition 矿产地质因素
factors of politics and society 政治社会因素
factors of production 生产要素;生产因素
factors of prospecting and exploration ex-post analysis 地质勘查因素的事后分析
factors of regional engineering geology 区域工程地质因素
factors of safety against buoyancy 抗浮安全系数
factors of slope shape 边坡形态要素
factors resulting from policies 政策性因素
factors that modify toxicity 毒性影响因素
factor structure 因子结构
factor substitution 因素代替
factor system 因子组
factor theorem 因子定理
factor theorem of algebra 代数学因式定理
factor total 因子总计
factor transformation 因子变换
factor utilization 要素利用
factor value 要素价值
factor weight 因数重量
factory 制造厂;工厂;商行在国外的代理处
factory acceptance test 工厂验收试验
factory act 工厂法;工厂管理条例
factory-adjusted 出厂调整
factory adjusted control 出厂调整控制
factory administration 企业管理
factory administrative expenses 厂部经费
factory and shop lumber 门窗装修木料;工厂用和再加工用木料
factory-applied coating 在工厂上的涂层
factory assembled system of refrigeration machine 制冷机组
factory assembly 工厂装配
factory atmosphere 厂区气氛
factory attached to school 实习工厂
factory automation 工厂自动化
factory bend 预制弯头
factory block 厂房
factory blower 工厂鼓风机
factory bonding 工厂接合
factory builder 工厂预制房屋制造商
factory building 工厂建筑;厂房
factory building construction 厂房建筑
factory built 工厂化建筑
factory-built building 建房工厂化
factory-built chimney 工厂预制烟囱
factory-built door 工厂预制的门
factory-built door element 工厂预制的门构件
factory-built girder 工厂预制梁
factory-built house 工厂预制房屋
factory-built partition(wall) 工厂预制隔墙
factory-built roof truss 工厂制造屋架
factory-built stair(case) 工厂预制楼梯
factory burden 制造间接费;工厂间接费用;工厂负荷
factory calibration 工厂校准
factory casting 工厂浇制
factory chimney 工厂烟囱
factory-chimney builder 工业烟囱修建工
factory construction 工厂施工;工厂建筑
factory construction abroad 海外工厂建设
factory content 预制百分率
factory contract 工厂合同
factory contractor 承包厂商
factory cost 制造成本
factory cost report 制造成本报告书
factory cost sheet 制造成本单
factory crane 厂房起重机
factory cullet 工厂(回炉)废碎玻璃;工厂碎玻璃
factory-data collection 工业数据收集
factory director 厂长
factory direct service 工厂直接服务
factory district 工厂区
factory-dock 厂用码头
factory effluent 工厂排放液;工厂排放物;工厂废水;生产废水
factory ell 预制弯头
factory engineer 厂家工程师

factory engineering 工厂工程(学)
factory engineering inspection 工厂工程检验
factory entrance 工厂入口;工厂大门
factory environment 工厂环境
factory equipment 工厂设备
factory expense analysis 制造费用分析
factory expense ledger 制造费用分类账
factory expense variance 制造费用差异
factory external transportation line (工厂的)对外道路
factory fabricated 工厂装配
factory fabricated filtration system 工厂制造的过滤系统
factory farm 工厂化农场
factory-filled particle board 厂制镶边木屑板
factory finished 厂制装修品
factory-finished board 工厂饰面板
factory finishing 工场施工
factory-fitting 工厂照明装置;工厂另配件
factory formula 生产配方
factory for precast concrete buildings 预制混凝土房屋的工厂
factory for prefabrication 预制构件厂;预制厂
factory fraction 预制百分率
factory gardening 工厂氯化
factory gate price 出厂价(格)
factory general expenses 工厂管理费用
factory girl 女工
factory ground 工厂场地
factory hand 工人
factory hangar 工厂棚屋
factory hooter 工厂汽笛
factory hygiene 工厂保健
factory illumination 工厂照明
factory-in road 厂内道路
factory inspection 工厂检验;厂内检查
factory installed 工厂装配
factory insurance association 工厂保险协会
factory invoice 厂商发票
factory laboratory 工厂试验室
factory land-use coefficient 厂区土地利用系数
factory layout 厂区配置
factory-level accounting 厂级经济核算
factory lighting 工厂照明
factory-like production in livestock husbandry 畜牧业生产工厂化
factory load 工厂负荷
factory lumber 加工木材;粗加工木材;粗加工木材;加工用材;工厂定级木材;厂制木材
factory-made block 工厂预制房屋
factory-made building 工厂预制建筑物
factory-made component 工厂制造的构件;工厂预制(的)构件
factory-made door 成品门
factory-made house 工厂预制房屋
factory-made housing 工厂预制住房
factory-made terrazzo 预制水磨石
factory-made trim 预制门窗贴脸
factory main road 工厂干道;厂内大路
factory management 工厂管理;生产组织
factory manager 厂长
factory-mounted 在工厂安装的
factory mutual company 工厂相互保险公司
factory mutual laboratory 工厂相互保险研究所
factory mutual system 工厂相互保险公司
factory noise 工厂噪声
factory operation 工厂作业
factory order 生产通知单
factory-out road 厂外道路
factory overhead 厂务费;制造间接费用;工厂杂项支出;工厂费用;(工厂的)生产费用
factory pack 随车供给的涂料
factory packed 出厂包装的
factory-painted 工厂油漆的
factory percentage 预制百分率
factory pipe bend 工厂制造的弯头;工厂预制弯头
factory planks 制门窗板材;工厂加工板材;加工木板
factory plant 工厂设备
factory preassembly 工厂预装配
factory precasting 工厂预制;工厂浇制
factory precast unit 工厂预浇单元
factory prefabricated filtration system 工厂预制的过滤系统

factory prefabrication 工厂预制
factory price 出厂价(格)
factory primed 工厂涂底漆的;厂内预涂
factory-primed board 工厂打底板
factory priming 工厂涂底漆
factory processing 工厂条件下加工
factory-produced component 工厂预制构件
factory production 工厂预制;工厂生产
factory programming 工厂编制程序
factory quay 厂用码头
factory railway 工厂专用线;工厂(专用)铁路;工厂铁道
factory record 写实记录
factory regulation 工厂规章;工厂管理条例;工厂法规
factory repair shed 厂修库
factory report 写实报告
factory requirement 工厂需用量
factory research 工厂研究
factory-sealed double-glazing unit 密封双层玻璃窗
factory seam strength 工厂接缝强度
factory-set bit 厂镶金刚石钻头
factory-set price 厂定价格
factory-set value 出厂调整值
factory sewage 工厂污水;工厂废水;生产废水
factory shed 工厂棚屋
factory ship 加工船;水产加工船
factory siding 工厂专用线
factory site 工厂用地;厂址
factory site decision 建厂厂址的决定
factory stack 工厂烟囱
factory tailings/rejects 工厂废石/垃圾
factory tax 出厂税
factory terminal 厂用码头
factory test 工场试验;工厂试验
factory test and inspection plan 工厂试验与检查计划
factory test equipment 工厂试验设备
factory testing 生产性试验;工厂条件下试验;工厂测试
factory test pressure 工厂试验压力;工厂测试压力
factory timber 粗加工木料;粗加工木材
factory timber industrial lumber 加工用材
factory-trawler 拖网加工渔船
factory trial 厂内试车法;厂内试运转
factory vessel 修理船;水产加工船
factory waste 工业废料;工厂废料
factory wastes per shift 工厂每班废水
factory waste water 工业废水
factory wharf 厂用码头
factory work 工厂作业
factotum 杂工
fact processing 处理事实
fact retrieval 事项检索;事实检索;事件检索
fact retrieval system 事实检索系统
fact sheet 情况说明书
facts survey 初测
factual evidence 事实的证据
factual material map of ore deposits 矿床实际材料图
factual premise 事实前提
factual proposal 实际建议
factual report 事实报告
factum probantia 举证事实
facula 光斑
facular area 光斑区
facular granule 光斑米粒
facular point 光斑亮点
facular region 光斑区
faculous region 光斑区
facultative 临时的;任意的;授权的
facultative aerated lagoon 兼性曝气氧化塘
facultative aerobe 兼性需氧生物;兼性需氧菌;兼性好氧菌
facultative anaerobe 兼性厌氧生物;兼性厌氧菌;兼性厌气菌
facultative anaerobic autotrophic bacteria 兼性厌氧自养菌
facultative anaerobic bacteria 兼性厌氧菌;兼性厌气菌
facultative anaerobic microorganism 兼性厌氧微生物
facultative anaerobiont 兼性嫌气生物
facultative autotroph 兼性自养生物

facultative bacteria 兼性细菌;兼性菌
facultative bioattached hydrolysis acidification 兼性生物附着水解酸化
facultative biological process 兼性生物工艺
facultative biotrophic parasite 兼性生体营养寄生物
facultative chemoautotroph 兼性化能自养生物;兼性光能自养生物;化能自养生物
facultative chemoorganotrophy 兼性化能有机营养;兼性有机能营养
facultative dentro-aeration process 兼性脱硝基曝气工艺
facultative endorsement 任意背书
facultative heterotrophic bacteria 兼性异氧菌
facultative microorganism 兼性微生物
facultative oxidation pond 兼性氧化塘
facultative parasite 兼性寄生物;兼寄生物
facultative photoautotroph 光能自养生物;兼性光能自养生物
facultative photoheterotroph 兼性光能异养
facultative plankton 兼性浮游生物
facultative plant 不定型植物
facultative pond 兼性池;兼性塘;稳定塘
facultative process 兼性处理过程
facultative reinsurance 临时分保
facultative saprophyte 兼性污水生物;兼性腐生植物;兼性腐生菌;兼腐生物
facultative sludge basin 兼性污泥池
facultative treaty 临时分保合同
facultative wetland 兼性湿地
faculty 本领
faculty block 院系大楼
faculty principle of taxation 课税能力原则
faculty reading room 专业阅览室
fadding 揩漆(用软包上漆);虫胶漆打底
fade 减弱;褪色
fade and dissolve 渐现渐隐
fade and lap-dissolve circuit 淡变和叠化电路
fade-and-recovery test 衰退和恢复试验
fade area 消失区;衰落区
fade-away 逐渐消失
fade characteristics 衰退特性
fade chart 盲区图;衰落区图
faded colo(u)r 褪色的颜色
fade down 逐渐渐隐;图像消失;图像渐灭;衰减
fade in 渐现;渐显;图像渐显;淡入
fade indicator 衰落指示器;衰减指示器
fade-in fade-out 淡入淡出
faden reaction 丝状反应
fadeometer 减弱控制器;褪色计;耐晒试验器
fade out 渐隐;图像渐隐;淡出;衰落现象
fade-out system 渐隐系统;淡出系统
fade over 淡入淡出;淡出淡入
fadeproof 不褪色的
fader 音量控制器;增益调整器;光量调节器;渐减器(音量、照明);混频电位器
fader control 照明渐减调整;光量控制;衰退控制
fader potentiometer 信号电平比调整电位计;双路混合器
fade up 增亮;图像增亮
fade zone 消失区;盲区;衰落区
fadgenising 锌基模铸件表面机械抛光;电镀前铸件表面机械抛光
fading 用纱团擦涂;枯萎病;揉轧涂装;褪色(作用);涂面变色;衰落;衰减作用
fading amplifier 抗衰落放大器;淡变放大器
fading area 衰落区
fading bandwidth 衰落带宽
fading channel 衰落信道
fading characteristic 衰减特性
fading control 衰落控制
fading depth 衰落度
fading distribution 衰落分布
fading effect 衰退效应
fading loss 衰落损失
fading machine 衰落试验器;衰减试验器
fading margin 衰落余量;衰落储备;衰落边际;衰减极限
fading memory 衰退记忆
fading memory filter 存储衰落筛选程序
fading of brake 制动衰减
fading of settlement 沉降衰减
fading of the settlements 沉降消失;沉降减小
fading on tropopshericscatter path 对流层散射电路衰落

fading peak 衰落峰值
fading period 衰落周期
fading range 衰退范围
fading rate 衰退率
fading region 衰减区
fading rise 衰落增长
fading safety factor 衰减安全系数
fading signal 衰落信号;衰减信号
fading spectrum 衰落频谱
fading unit 衰落装置;输入功率调节装置
faecal coliform bacteria 粪便大肠杆菌群
faecal pellet 粪球粒
faeces pit 粪便坑
faeces seepage 排泄物渗漏;粪便渗漏
faeeingtonite 磷镁石
faffinate 萃余液
fag-end 散端;散边;末尼;残浆;废渣;末端
Fager green cell 费戈格林直流槽型浮选机
Fagergren flo(a)tation cell 法格伦浮选池
Fagergren flo(a)tation machine 法格伦式回转子浮选机
Fagersta cut 费格斯塔超前打眼法
fagging 撑板
fag(g)ot 柴捆;柴草;束铁;成束熟铁块;成捆熟铁板条
fag(g)ot dam 柴堤;柴捆坝
fag(g)ot dike 柴堤
fag(g)ot drain 柴排水沟
fag(g)oted iron 锻焊的层状熟铁;束铁
fag(g)ot fender 捆条碰垫
fag(g)ot filling 沉排填石
fag(g)ot steel 束钢;成束钢条
fag(g)ot wood 薪炭材;木块(路面用);柴排;柴捆
fag station 吸烟站
Fagus sylvatica 欧洲水青冈;欧洲山毛榉
faheyite 磷铍锰铁石
fahlband 黝矿带
fahl-erz 黝矿
fahlum metal 法拉姆铅锡合金
Fahlun brilliant alloy 法伦银亮合金
fahlunite 褐块云母
Fahnestock clip 法耐斯托克线夹
Fahralloy 耐热铁铬镍铝合金;法拉洛伊耐热铁镍铬铝合金
Fahrenheit 华氏
Fahrenheit's degree 华氏(温)度
Fahrenheit's hydrometer 华氏比重计
Fahrenheit's scale 华氏温度刻度
Fahrenheit's temperature scale 华氏温标;华氏表刻度
Fahrenheit's thermometer 华氏温度计
Fahrenheit's thermometric(al) scale 华氏温标
Fahrenwald flo(a)tation cell 法兰瓦尔特型浮选机;法兰瓦尔特分级机
Fahrig metal 锡铜轴承合金
Fahrite 法里特耐热耐蚀高镍合金
Fahry alloy 锡铜轴承合金
fahuacai 法华彩
fahua colo(u)r 法华彩
fahua green 法翠
faience 锡釉陶器;彩釉陶器;彩釉蓝;彩陶(器);彩色瓷器
faience beads 多晶石英珠
faience grille 釉陶栅栏
faience mosaic 嵌花地砖;釉陶面砖;釉陶马赛克;釉陶锦砖;嵌花地砖;彩釉马赛克
faience mosaic tile 釉陶马赛克地砖
faience pottery 彩陶
faience tile 陶器面砖;彩釉瓷砖;釉陶面砖;釉陶锦砖
faience ware 上釉的陶器;彩釉陶器
faikes 裂纹金刚石钻头
fail 衰退
fail-active 固着积极防护的
fail address 故障地址;失效地址
fail category 失效类别
fail closed 出故障时自动关闭的
fail data 失效数据
failed area 破坏面积;破坏范围;造林失败地
failed arm 衰退支
failed bank 破产银行
failed fuel element 破损燃料元件
failed hole 不爆破炮眼;瞎炮;失效炮眼
failed place 造林失败地
failed rift 衰亡裂谷

failed surface 破坏表面;衰坏面
failed test piece 不合格试件
failed test sample 不合(规)格样品
fail in bending 弯曲破坏;受弯坡坏;弯曲损坏
fail in bond 黏着破坏;黏接失效
fail in compression 压缩破坏;压力破坏;受压损坏;受压破坏
failing load 破坏荷重;破坏荷载;破坏负荷
failing moment 破坏弯矩
failing out 尖灭
failing stress 破坏应力;破坏(性)压力
failing vision 视力衰退
fail in negotiations 谈判失败
fail in shear 剪切损坏;剪切破坏;剪力破坏;受剪破坏
fail in tension 拉力破坏;受拉破坏
faille 斜纹绸;菲尔绸;罗缎
fail-obvious condition 明显故障状态
fail open 出故障时自动打开
fail passive 无大害故障;工作可靠但性能有下降
failpoint 破坏点;失效点
fail safe 失效保险;失效保护;差错防止;防故障设备;故障自动矫正;故障(自动)保险;故障自动保护的;故障—安全;可靠的;绝对可靠的;防错;破损安全
fail-safe analysis 可靠性分析
fail-safe concept 故障保险概念
fail-safe control 无故障控制;保安控制;失效保险控制
fail-safe design 可靠设计
fail-safe device 防止事故装置;安全保障装置
fail-safe disconnect 故障保险切断
fail-safe equipment 破损安全装置
fail-safe facility 故障保险设施
fail-safe mechanism 防障装置
fail-safe operation 安全运作;安全运行;防障装置操作
fail-safe program(me) 不安全的程序
fail-safe safety interlock 失效保护安全联锁;故障保险防护装置
fail-safe stop accuracy 无故障停车精度
fail-safe system 安全系统;故障自动防护系统;保险系统;故障保险系统;安全装置
fail-safe test 破损安全试验
fail-safety 破损安全;故障可靠性;绝对可靠性
fail-safe valve 故障安全阀
fail soft 有限可靠性;故障自动缓和的;故障弱化;降级进行(系统出故障后);工作可靠的;工作可靠但性能下降;失效弱化;无大害故障
fail-soft behaviour 故障软化特性
fail-soft capability 故障弱化能力
fail softly (程序的)软化故障
fail-soft system 故障弱化系统
fail system 失效保险系统
fail temperature 失效温度
fail test 未通过试验
fail to be elected 落选
fail to fire 不发火(发动机或者炮)
fail to go 故障
fail to heal 不愈合
fail to submit a true report 不据实申报
failure 失灵;折断;故障;溃决;破损;破坏;断裂;破产;违约;损坏;衰竭;失效;断裂;变钝
failure accumulation 故障积累
failure alarm 故障报警
failure analysis 破损分析;破坏分析;失效分析;断裂分析;故障分析
failure analysis report 故障分析报告;事故分析报告
failure arc 毁坏弧;滑动弧;破坏弧
failure arc of earth slope 土坡滑动弧
failure area 破坏区
failure bearing stress 破坏承载应力
failure bending angle 破坏弯角
failure bending moment 破裂弯矩
failure bit 故障位
failure breaking 断裂
failure by buckling 压曲破坏
failure by bulging 膨胀破坏
failure by heave 冻胀破坏
failure by heaving 隆起破坏;冻胀破坏
failure by piping 管涌破坏;管涌
failure by plastic flow 塑流破坏
failure by pull 拉曳破坏;拉坏
failure by rupture 破裂破坏;破裂变形
failure by shear 剪切破坏;剪力破坏

failure by sinking 沉陷破坏
failure by spreading 铺展破坏
failure by subsurface erosion 下层土冲刷破坏;地下侵蚀破坏
failure by tilting 倾斜破坏
failure capacity 破坏能力
failure cause 故障原因
failure cause data report 故障原因数据报告
failure caused by bending 弯曲损坏
failure circle 破坏弧
failure classes 故障类
failure condition 断裂条件;破坏条件
failure condition in saturated soil 饱和土的损坏状况
failure cost 故障费用;损失费用
failure cost function 故障费用函数
failure crack 裂断;断裂纹
failure criteria 破坏判断
failure criterion 破坏判据;故障判定标准;破坏评定标准;破坏(判别)准则;失效准则;失效标准;失效判据;断裂判据
failure criticality analysis 故障致命性分析
failure cross-section 破坏断面
failure current 故障电流;故障电极电流
failure data analysis 故障数据的分析
failure deformation 破坏变形
failure density 失效密度
failure density function 破裂密度函数
failure detection 故障检测
failure detection system 破坏监测系统
failure diagnosis time 故障诊断时间
failure distribution 故障分布;缺陷分布;破坏分布;失效分布
failure due to fatigue 疲劳破坏
failure due to instability 失稳破坏
failure due to passive pressure 被动土压力引起的破坏
failure due to scouring 冲刷破坏;冲毁
failure effect analysis 故障影响分析
failure effect of earthquake 震动破坏效应
failure effect of ground surface 地面破坏效应
failure end point 破坏终点
failure energy 破坏能(量)
failure envelope (土的)强度线;破坏包(络)线
failure exception mode 失败异常方式
failure-free operation 正常运转;正常运行;无故障运行
failure function 故障函数;破坏函数
failure go-to field 失败区域
failure hydraulic gradient 破坏水力梯度
failure hypothesis 破坏假说
failure identification 故障标识
failure in buckling 压曲破坏
failure index 破坏指数
failure indication 失事指示信号
failure in service 使用中的故障
failure in shear 剪切损坏
failure in tension 拉伸损坏
failure isolation 故障隔离
failure latency 故障潜伏
failure law 破坏规律
failure level 失效级
failure life 失效寿命
failure limit 破坏极限
failure limit state 破坏极限状态
failure line 破坏线
failure load 破坏荷载
failure load by buckling 屈服破坏荷载
failure loading 极限荷载
failure logging 故障记录;故障登记;失效记录
failure mechanism 故障机理;破损机理;破坏机制;破坏机理;破坏机构;失效机理;失效机构
failure mode 故障种类;故障模型;破裂方式;破坏形式;破坏方式;失效模式
failure mode analysis 故障模式分析;失效模式分析
failure mode and effect analysis 故障模式和影响分析
failure mode effect and criticality analysis 失败模式的效应和鉴定分析
failure model 失效模型;失效模型
failure model and effect analysis 失效模式与影响分析
failure mode of slope 斜坡破坏形式
failure mode of stress rock mass 应力岩体破坏

方式
failure modes and effect analysis 失败模式和效应分析
failure moment 破坏力矩
failure monitor 故障监控仪
failure monitoring system 破坏监测系统
failure node 失效节点
failure of air supply 供气中断
failure of beam 梁的破坏
failure of concrete 混凝土破坏
failure of dam 水坝失事;坝的失事
failure of earth slope 滑坡;土坡坍塌;土坡坍毁
failure of fusion 闭合不全
failure of light 灯光失常
failure of oscillations 停止振荡
failure of performance 未履行合同
failure of rock 岩石破坏
failure of shot 拒爆
failure of surrounding rock 围岩破坏
failure of the soil 土的破坏
failure pattern 破坏样式;破坏模式
failure percentage 装箱破损率
failure physics 失效物理
failure plane 临界面;破裂面;破坏(平)面
failure point 破裂点;破坏点
failure prediction 故障预测;缺陷预测;破损预测;破坏预测;失效预测
failure probability 破损概率;破坏概率;失效概率
failure rank 破坏级别
failure rate 故障率;破坏率;事故率;失效率;失败率
failure rate acceleration 失效率加速度
failure rate acceleration factor 故障率加速系数;失效率加速系数
failure rate allocation 故障率分配
failure rate average function 失效率平均函数
failure rate characteristic 故障率特性
failure rate data 故障率数据
failure rate function 故障率函数
failure rate level 故障率水平;失效率等级
failure rate of lot tolerance 批量产品容许失效率
failure rate sampling plan 故障率采样计划
failure rate test 故障率试验
failure ratio 破坏比
failure record 故障记录;失效记录
failure recovery 故障排除;故障恢复;故障后的恢复
failure repair 故障修理
failure report 故障报告
failure section 破坏截面;损坏段落
failure sign 破损迹象
failure slope 坍坡
failure stage 破坏阶段
failure state 失效状态
failure strain 破坏应变;破坏变形
failure strength 破坏强度
failure stress 裂断应力;破坏应力
failure stress and strain 破坏应力与应变
failure stress condition 破坏应力状态
failure surface 破毁面;破坏面;失效面
failure terms 故障测定
failure test 可靠性试验;可靠度试验;破坏(性)试验
failure testing 故障检测
failure theory 破坏理论
failure to actuate 不动作
failure to breech 关不上闩
failure to complete required paper work 没有完成必要的凭证手续
failure to deliver clause 未交货条款
failure to deliver the goods 未交货
failure to follow instruction 违反操作程序
failure to inspect 疏于检查
failure to maintain 疏于保养;疏于维护
failure to perform 未履行合同
failure to take delivery 未提货物
failure tree analysis 失效树分析;故障树分析(法)
failure type of rock 岩石破坏形式
failure unit 失效单位
failure voltage 破坏电压
failure warning 故障警告;故障警报
failure warning indicator 故障告警指示器
failure warning relay 故障警告继电器
failure wedge 破裂楔体;破裂楔件;破坏楔体;破坏楔体
failure zone 断损范围;破坏区;破坏范围;断损区域
fail year 歉收年

faint 弱的;衰弱的;暗淡的
faint colo(u)r 暗淡色
faint companion 暗伴星
faint contour 不清晰等高线;不明显等高线
faint difference 细微差别
faint galaxy 暗星系
faint haze 薄雾
faint impression 压印图案模糊;花纹模糊
faint in place 安装原处
faint light 弱光;微弱灯光
faint light source 弱光源
faintly colo(u)red 暗淡色的
faintly curved reach 微弯河段
faint map 难读地图
faint meteor 暗流星
faint negative 弱反差
faintness 微弱
faint odo(u)r 微弱恶臭
faint red 淡红色
faint satellite 暗卫星
faint star 暗星
faint voice 微弱声音
faint yellow 淡黄色
fair 展销会;公正的;交易会;平顺;商品交易会;定期集市;博览会
fair and reasonable 公平合理
fair and square 公平合理
fair average 适当平均数
fair average quality 中等质量;中等品(质);质量一般良好;良好中等质量;良好平均质量;良好平均品质
fair batten 顺光线木条
fair building 展览厅
fair buoy 进口浮标;航道口浮标
fair chart 印刷图
fairchildite 碳酸钾钙石;碳钾钙石
Fairchild photographic(al) flight analyzer 费尔柴耳德型摄影飞行分析器
fair copy 清样;清稿
faircrete 纤维加气混凝土;装饰用混凝土板;装饰混凝土;混凝土装饰板
fair current 顺流
fair curve 修正曲线;整形曲线;光顺曲线;平面曲线
fair cutting 修整切削;细砍砖
fair deal 公平交易
fair draft 清绘
fair draught 清绘
fair drawing 清绘(原图);地图清绘
faired 流线型的
faired cable 流线型绳缆
faired contour balance 流线型减阻平衡
faired curve 整平曲线;展平曲线
fair ends 琢舂露头
fair face 外露面;(从驾驶员位置上观察的)路面可见部分
fair-faced aggregate 饰面混凝土集料
fair-faced block 饰面砌块
fair-faced brickwork 清水砖砌体;平滑面砌砖工程;清水(勾缝)砖砌体
fair-faced cast compound unit 饰面用的混凝土组合件
fair-faced cast-concrete component 饰面用的混预制凝土构件
fair-faced cast-concrete compound unit 饰面用的预制混凝土组合件
fair-faced cast-in-place concrete 现浇饰面混凝土
fair-faced cast-in-situ concrete 现浇饰面混凝土
fair-faced clay brickwork 饰面黏土砖砌体
fair-faced concrete 饰面混凝土;装饰混凝土;光面混凝土;清水(面)混凝土
fair-faced concrete aggregate 饰面混凝土集料;饰面混凝土骨料
fair-faced concrete beam 饰面混凝土梁
fair-faced concrete cast(ing) 饰面混凝土浇注
fair-faced concrete column 饰面混凝土柱
fair-faced concrete facade building unit 饰面混凝土房屋正面部件
fair-faced concrete finish 混凝土饰面
fair-faced concrete forms 饰面混凝土模板
fair-faced concrete panel 饰面混凝土板
fair-faced concrete shuttering 饰面混凝土模板
fair-faced concrete slab 饰面混凝土板
fair-faced concrete stair(case) 饰面混凝土楼梯

fair-faced concrete texture 饰面混凝土结构
fair-faced concrete wall 饰面混凝土墙
fair-faced field concrete 现浇饰面混凝土
fair-faced finish 饰面
fair-faced granite aggregate finish 水刷石饰面
fair-faced lightweight concrete 饰面轻质混凝土
fair-faced lightweight concrete block 饰面轻质混凝土砌块
fair-faced lightweight concrete tile 饰面轻质混凝土砖
fair-faced masonry(work) 饰面砖石砌筑
fair-faced poured in-place concrete 就地浇注饰面混凝土
fair-faced precast compound unit 预制饰面混凝土组合件
fair-faced precast concrete component 预制混凝土饰面构件
fair-faced precast concrete member 预制混凝土饰面构件
fair-faced precast concrete unit 预制混凝土饰面件
fair-faced reinforced concrete 饰面的钢筋混凝土
fair-faced site-placed concrete 现浇饰面混凝土
fair-faced wall 清水墙
fair face of concrete 混凝土清水面
fairfieldite 磷钙锰石
fair game 公平博弈
fair ground 集市场所;博览会场;露天交易会场
fair housing 公平住房政策
fair housing program(me) 公平住房计划
fairing 风嘴;整流装置;整流罩;整流片;光顺;流线型罩;流线型外壳;流线(型)体;减阻装置;导流罩
fairing cap 推进器导流槽
fairing of curves 曲线光顺
fairing spoke 整流形辐条
fair in place 状况良好
fairlead 绳索滑车;系船缆导轮;引线孔;引线管;转动系缆柱;滚柱导缆钳;滚柱导缆器;导片;导向套;导向滑轮;导线管;导索器;导索架;导缆板;导缆轮;导缆孔;导缆钩
fair leader 导向器;导索器;卷扬机械;导缆器;导缆滚筒;导缆滚轮;碾子
fairleader rack 导缆孔框
fairleader with horizontal roller 水平滚轮导缆器
fairlead for rudder chain 舵链导轮
fairlead for side ropes 边缆导缆钳
fairlead logging arch 导索集材拱架
fairlead of rudder chain 舵链导轮
fairlead roller 支承滚轮;支轮
fairlead sheave 导索滑车
fairlead sheave shroud 导索滑轮罩
fairlead truck 导索滚轮
fairlight 气窗(英国);门顶窗
Fairlight clay 费尔莱特黏土
fair line 光顺线
Fair-Mannington type elevator 带自动挡销的筒状铰链式提引器
fair market 公平市价
fair market price 公平市场价格
fair market value 公平市场价值
fairness limit 允许变形量;容许变形量
fairness of financial statements 财务报表公允性
fair-plastic clay soil 中塑性黏质土
fair-plastic silt 中塑性粉土
fair-price shop 平价商店
fair raking cutting 修整切削
fair rate of return 公允收益率
fair rent 公平地租
fair rental 公平租价
fair returns 合理报酬
Fair's graticules 费尔分度镜
fairshaped 流线型的
fair sheet 测深报告表
fair source rock 较好烃源岩
fair stand 博览会的陈列柜
fair tide 顺航潮流;顺潮
fair tide riding 乘潮航行
fair trade 公平贸易
fair-trade agreement 不擅自降低零售价格的协定
fair valuation 合理估价;合理的估价
fair value 合理价格;公平价值
fair visibility 中上能见度
fair wage 合理工资
fair water 导流;指section台围壳
fair water cap 导流帽

fair water cone 导流帽;流向袋;水锥(流向指示器)
fair water fin 导流鳍;导流板
fair water piece 导流板
fair water sleeves 导流套筒
fairway 主航道;航道;通路;安全通航区;安全航路
fairway arch 通航孔
fairway arch span 通航拱跨
fairway buoy 航道(进口)浮标;航道灯标;航标
fairway buoy engineering vessel 航道浮标工程船
fairway line 主航道线
fairway speed 航道航速
fairway transferring 改槽
fair wear and tear 正常磨损;正常磨耗
fair weather 好天气;晴天(积云)
fair weather cirrus 晴天卷云
fair weather cumulus 淡积云
fair weather landing strip 好天气着陆地带(机场)
fair weather road 晴天通车道路
fair weather runoff 晴天径流
fair weather sailor 无航海经验的水手
fair weather ship 经受不住大风浪的船
fair wind 惠风;顺风
fairy lamp 彩色小灯
fairyland 幻境
fait accompli 既成事实
faithful functor 一一的函子
faithful linear representation 一一线性表示
faithfully flat morphism 一一平坦射
faithfully full functor 一一全函子
faithful module 一一的模
faithful permutation representation 一一置换表示
faithful representation module 一一表示模
faith money 信用货币
fake 盘索;绳圈
fake and poor product 伪劣品
fakement 赝品;伪造品
fakes 云母板状岩;砂质页岩;板层建造;云母质砂岩
FAK rate 不分级运费率
falaise 古海崖;古海滩脊
falcate 弯月形
falcated teal 罗纹鸭;镰鸭
falcondite 镍海泡石
Falcon engine rust test 福尔肯发动机锈蚀试验
faldstool 折叠椅;(祈祷用的)跪几
fale itemization of accounts 账目不实
Fales grain 闭砂
Fales-Stuart windmill 双叶农用风车
Falex friction machine 法列克司摩擦试验机
Falex tester 润滑剂耐热耐压试验机
Falk flexible coupling 蛇形弹簧联轴器
Falkland Current 福克兰海流
Falk rail-joint 福尔克钢轨接头
fall 减水;落潮;辘绳;降落;见落陨星;秋季;起重机索;坡降;坍塞;倒闭
fallacious derivation 谬误推理;谬误推导
fallacious inference 谬误推理
fallacy 假饰;谬误
fall apart 分离
fall away 渐远;分开
fallaway section 脱落部分;分离段;分离度
fall back 退却;退后;回落;不履行(契约);引退;降落原地;低效运行;低效率运行;待援;撤退
fallback function 后退功能
fallback mode 低效方式运行;后退方式
fallback procedure 撤退过程
fallback recovery 撤退校正
fallback state 低效运行状态;后退状态
fall bar 轴转关门闩;轴转关门扣;门插销
fall behind 落后
fall block 活动滑车;受力滑车;动索滑车;动滑轮;带绳滑轮;卷帆滑车
fall calm 突然无风
fall characteristic 下降特性
fall cloud 下降云
fall colo(u)r 秋色
fall-cone test 落锥试验
fall cover 吊艇罩
fall-dart test 落锥冲击试验
fall delay 下降延迟
fall diameter 沉降球直经
fall down 流下去;漂下来;倒下;装船误期;次级材
fall due 满期
fall dust 降尘

faller 落体;练条机;砍伐树木的人
faller screw 针板螺杆
fall flat 没有达到预期效果
fall flood 秋汛;秋季洪水;秋洪
fall foul of a ship 碰船
fall from the plant 脱离
fall head(of water) 落差;下降水头;降落水头;压头高度;压头差;跌落水头
fall home 内倾
fall-in 崩陷;一致;进入同步;坍方;塌陷;凹进去;内倾
fall increaser 落差增流器
falling 落像;塌落;凹陷
falling angle 落角
falling apron 允许坍塌的护脚棱体;水下堆石护坡;堆石防冲护坦
falling apron principle 堆石下落护坡法
falling back 可放下的
falling-ball impact test 落球冲击试验
falling-ball method 落球法
falling-ball strength tester 落球式强度测定仪
falling ball test 落球冲击试验;落球试验
falling-ball viscometer 沉球式黏度计;落球黏度计
falling-ball visco(si)metry 落球黏度法
falling body 落体
falling body visco(si)meter 落球黏(滞)度仪;落体黏度计
falling characteristic 降落特性
falling coaxial cylinder viscometer 同轴筒下落式黏度计
falling cone method 沉锥法
falling count 落下次数
falling current of air 气流下降
falling curve 落洪段;退水曲线;退水段
falling cylinder viscometer 圆筒下落式黏度计;落筒式黏度计
falling dart 落锥
falling dart test 落箭试验
falling debris 倒塌房屋的瓦砾
falling diameter 降落直径
falling door 降落式门;吊门
falling-drop method 落滴法
falling due 到期
falling film 降落液膜
falling-film evapo(u)rator 降膜蒸发器;降膜浓缩器
falling film heat exchanger 降膜热交换器
falling film molecular still 降膜分子蒸发设备
falling film photocatalytic reactor 降膜式光催化反应器
falling film reactor 降膜式反应器
falling film type absorber 湿壁降膜吸收塔
falling flood stage 落洪期;退水期;降水期
falling gate 卧式闸门;卧式坞门;卧倒门
falling glass 气压柱下降
falling gradient 下坡度;下坡;下降坡度;降坡
falling ground 陷落地层
falling head 下降高度;水头落差
falling-head permeability test 变水头渗透试验
falling-head permeability test in laboratory 室内变水头渗透性试验
falling head permeameter 降水(头)渗透仪;变水头渗透仪
falling-head permeameter test 降水头渗透试验
falling height 下降高度
falling hinge 水平铰链
falling-in 陷落;坍方;滑坍;啮合;内倾
falling-in of bank 路堤坍塌
falling latch 下落式(棘轮)掣子
falling lead track 溜放线
falling-leaf gate 卧倒门
falling level 下降水位
falling limb 消落段;过线下降段;落段;退水段
falling limb of hydrograph 水文过程线的退水段
falling losses 采下损失
falling losses ratio 采下损失率
falling main 下行主管(下水道);下行竖管
falling market 市场跌落
falling mass 落体
falling mo(u)ld 深度模式(楼梯扶手弯曲部分);曲型线条;弯曲形式
falling needle visco(si)meter 落针黏度计
falling objects protective structure 落下物防护装置
falling of dike 决堤
falling of embankment 决堤
falling off 转向下风

falling-off illumination 光损耗
falling-off of contrast 反差消减
falling of water 水退;水降
falling of water table 水位降落;地下水位降落
falling over 极度倒伏;倒伏
falling pawl 下落式(棘轮)掣子;下落卡子
falling period 消落期;下降周期
falling plaster 抹灰脱落
falling portion 下坡段
falling portion of waves 波浪下坡段
falling pressure 下降压力
falling price 跌价
falling rate 下沉速率
falling rate of profit 利润率下降
falling rate period of drying 降速干燥阶段
falling rock 落石
falling rock alarm 落石报警
falling sand abrasion test 落砂磨损试验
falling sand test 落砂试验;耐磨性测定;喷砂试验
falling segment 消落段
falling slag 高钙高炉渣
falling slope 下坡道
falling sluice 自动泄水闸;降落式泄水闸;跌落式泄水闸
falling soil 崩落土
falling speed 下落速度;下降速度
falling sphere 落球
falling sphere test 落球试验
falling sphere visco(si)meter 落球黏度计;沉球式黏度计
falling stage 消落段;下降水位;水位下降期
falling stage-flow stage relation 落水期水位流量关系
falling stile 垂落门梃
falling strain 破裂应变
falling stress 破裂应力
falling surface curve 下降水面曲线
falling temperature technique 降温法
falling test 落锤试验
falling tide 落潮;退潮;低潮
falling top 折叠式车顶
falling torque 下降转矩
falling track 溜放线
falling tup machine 落锤机
falling type safety fuse filament 跌落式保险熔断丝
falling velocity 下落速度;降落速度;沉降速度;沉速
falling velocity scale 沉速比例尺
falling volt-ampere characteristic 下降伏安特性
falling water (1936年赖特设计的著名美国宾夕法尼亚洲的)瀑布别墅
falling weather 雪季;雨季;坏天气
falling wedge 伐木楔子
falling weight 下落重量;落下重量;落锤(打桩用)
falling weight deflectometer 落锤式弯沉仪
falling weight impact test 坠重冲击试验
falling weight test 落锤试验;冲击试验
falling weight type butterfly valve 落锤式蝶形风门;落锤式蝶形阀
fall in price 落价
fall in temperature 温度下降
fall in the bank rate 降低贴现率
fall into step 进入同步
fall in water-level 水位降落
fall irrigation 秋季灌溉;秋灌
fall lever 补偿杆
fall line 起重索;瀑布线;跌水线;吊索
fall line in terrain 示坡线
fall loss 水头损耗
fall measurement 落差测量;倾斜测量
fall meteorite 降落陨石
fall of 压向下风
fall of a lock 船闸下游水位差
fall of channel 沟道坡度
fall of current 电流下降
fall off 下降;落下;减退;偏振;偏出航向;脱落;衰减;侧降
fall-off curve 压降曲线;降压曲线
fall-off in illumination 照度衰减
fall-off law 衰减定律
fall-off meter 偏振测量仪
fall of ground 岩石冒落;冒顶;地层陷落
fall of pipe 管道坡度
fall of potential 电压降;电位降;电势降

fall-of-potential method 电位降法
fall-of-potential test 电压降试验;电压降测试法
fall of pressure 压降
fall of rain 降雨
fall of ram 落锤高度
fall of river 河流落差;河流比降
fall of river reach 河段落差
fall of rock 岩石滚落
fall of roof 屋顶坡度
fall of snow 降雪量;降雪
fall of stage 水位下降;水位降落
fall of stream 河流落差;河流比降;水流坡降;水流落差
fall of temperature 温降
fall of tide 落潮;退潮
fall of viscosity 黏度降落
fall of water 跌水;水流落差;水降
fall of water-level 水位下降;水位降落
fall of water surface 水面落差
fallout 外倾;附带成果;落下灰;落尘;散落;撒落;沉降物;沉降灰;放射尘;放射性微尘
fallout area 放射性沉降区
fallout collector 沉降物收集器
fallout contour 微粒回降等强线
fallout damage 落下灰损伤
fallout dose rate 散落剂量率
fallout from nuclear explosion 核爆炸散落物
fallout front 降水垂界;降水前沿
fallout injury 落下灰损伤
fallout measure 错检率
fallout measurement 散落物测量;散落物测定
fallout meter 沉降物测定仪;辐射量测试器
fallout monitoring 沉降物监测
fallout of pesticide 农药的沉降;农药沉降
fallout of radioactive material 放射性散落物
fall out of step 失去同步
fallout particles 落下灰粒子
fallout pattern 微粒回降等强度分布
fallout predictor 沉降物预测器
fallout protection 放射性微粒保护
fallout ratio 错检率
fallout shelter (放射性的)坠尘掩蔽处;沉降物掩蔽所;放射性物体掩蔽室;放射性微粒(回降)掩蔽所;放射性微粒(回降)掩蔽处;防核尘地下室
fallout tester 沉降物测试器
fallout wind 沉降风
fall over 下倾
fall overboard 落水
fall overturn 秋季翻转;秋季对流
fallow 休闲的;未开垦的;淡黄色的
fallow crop 休闲作物;休闲地作物
fallowed and non-fallowed land 休耕与不休耕地
fallow field 休闲田;休闲地;休耕地
fallow ground 休耕地
fallowing 休闲制
fallowing of soil 土壤的休耕
fallow land 未开垦土地;休闲地;休耕地
fallowness 休闲状态;休闲(土地);尚未被利用
fallow paddy field 休闲水田;休闲稻田;休耕水田
fallow soil 休闲土壤;土地休耕
fallow system 休闲制度
fallow time 休耕时间
fall penetrating well 全透水井
fall pipe 水落管
fall-pipe filter 水落管滤池
fall plowing 秋耕
fall plowing time 秋耕日
fall precipitation 秋季降雨(量)
fall pruning 秋剪
fall rate 下沉速率
fall rope 滑车索;吊索;吊绳
fall short 不足
fall short of the mark 不合格
fall short of the specifications 不合规格
falls of stone 坠石;石块坠落
fall-stage-discharge relation 落差水位流量关系;坡降水位流量关系
fall-stage-discharge relation curve 坡降水位流量关系曲线
fall-stage-discharge relationship 坡降水位流量关系
fall test 冲击试验
fall through 归于;导向
fall through machine 机内物料的落差
fall time 下落时间;下降时间;熄灭时间;降落时间

fall to arrears 迟付
fall to leeward 向下风偏转
fall trap 陷阱；带活门的陷阱
fall tube 排水管；落水管
fall turnover 秋季翻转
fall velocity 下降速度；落速
fall velocity of sediment 泥沙沉速；泥沙沉降速度
fallway 运货竖道；楼面井；升降道；吊物竖井；吊物竖道；运货洞；竖井
fall wheat 冬小麦
fall wind 下坡风；下降风；下吹风
false acacia 洋槐；刺槐
false account 假账
false accounting 伪造帐账目
false add 假加；无进位加法；半加
false air 漏入空气；漏风
false air coefficient 漏风系数
false alarm 虚警；误发警报；错误警告；错误报警
false alarm of fire 误报的火警
false alarm probability 虚警概率
false alarm rate 虚率；火灾误报率
false alarm time 虚警时间
false annual ring 假年轮
false anomaly 假异常
false appearance 假像
false arcade 假拱廊
false arch 虚拱；假拱
false arched girder 假拱梁
false assumption 错误假设
false attic 假顶楼；假屋（顶）层；假顶层
false backe line 假贝克线
false balustrade 假扶手；假栏杆
false base 活底座
false beach 假滩；岸外坝
false beam 假梁；不受载梁；不承压梁
false bearing 虚假定位；正反像限角之差【测】；间接承受；假支承
false bedding 交错层；假层理
false bedding structure 假层理构造
false bellies 假腰【船】；防擦护板
false bill 假票据
false billing 虚假填报（货物运单）
false blossom 假花
false bob 假摆；模拟摆
false body 假黏度；假稠性；假稠度；触变体
false bottom 假基岩；假湖底；假海底；（砂矿床的）假底板；假底；活动底板；活底（的）；幻底；深散射层
false bottom bucket 活底料斗；活底铲斗；活底戽斗
false bottom retort 假宽甑
false branching 假分枝
false brick 托座砖
false brinelling 虚假硬度；摩擦腐蚀压痕
false burning 生烧
false call 误调入
false cantilever principal rafter 昂（指古建筑）
false ceiling 假天花板；假平顶；设备吊顶；假吊顶
false ceiling of lightweight blocks 轻质砌块假平顶
false ceiling of lightweight slabs 轻质砌块假平顶
false ceiling slab 假平顶板
false ceiling with slanting joints 斜缝假平顶
false ceiling with straight joints 直缝假平顶
false center 假顶尖
false channel 辅航道
false characteristic curve 假特性曲线
false cheek 箱圈
false cirrus 伪卷云
false claim 虚假索赔
false cleavage 假劈理
false code 伪代码
false code check 假代码检验；非法代码校验
false colo(u)r 假色；虚色；假彩色；伪彩色；不坚牢色
false colo(u)r composite 假彩色合成
false colo(u)r composite image 伪彩色合成影像；假彩色合成图像
false colo(u)r composite picture 假彩色合成图像；假彩色合成片
false colo(u)r film 假彩色胶片
false colo(u)r photographic(al) material 假彩色片
false colo(u)r photography 假彩色摄影
false colo(u)r reversal film 假彩色反转胶片
false colo(u)r satellite photomap 假彩色卫星像片图
false combination 假组合

false command 错误指令
false condition 假设条件
false conductance 假电导
false contouring 虚假轮廓线
false coordinates 假设坐标
false core 假台心
false course 假航向；模拟航向
false cupola 假圆顶
false curved girder 假曲梁
false data 假数据
false deck 临时平台；假板（码头结构的减压板、减压平台等）；埋入式面板；覆盖式面板
false declaration 虚假声明；虚报
false dip 假倾斜；假倾角；假顶角
false dip projection method 视倾斜投影法
false dismissal probability 漏检概率
false dismissal time 漏警时间
false distance 虚距
false doem 假圆顶
false dome 假穹顶
false door 假门
false double-winged temple 带假双壁的庙宇
false drop 误选；误检
false dynamite 含硝化甘油少的炸药；低硝甘炸药
false easting 东移假定值（指横坐标，用来避免横坐标出现负值）
false echo 假回波
false ellipse 五心拱券；近似椭圆；圆弧椭圆
false ellipse arch 五心装饰拱；装饰成的三心拱；近似椭圆拱；三心装饰拱【建】
false entry 伪造记录
false equilibrium 假平衡
false equilibrium state 假平衡状态
false error 假错
false exclusion 错误排除
false exit 假出口
false fenestration 假开窗
false filter bottom 滤质假底
false fire warning light 假火警信号灯光
false firing 误燃；误触发
false flax 亚麻芥
false floor 假地板；活地板；空格底板；架空地板；楼层；假楼板；假底；夹层地板
false folding 假褶皱
false form 临时性模板；假像（结构）；假模；假晶
false front 装饰正面；假门面；假锋【气】；伪锋【气】；散热器前护栅；房屋假立面
false galena 闪锌矿
false glide path 模拟滑翔道
false golden cogongrass 拟金茅
false ground water table 栖留水水位；栖留水水面；静滞地下水位
false header 假一丁砖；假丁砖；半砖
false-heart 伪心材
false heart wood 假心材
false hinge 假铰
false hood 虚假
false horizon 假水平线；假地平（线）
false host 假主机
false identification 错误辨识
false image 虚像；幻影；误像
false impression 假像
false indication 乱显示
false indication of a switch 道岔错误表示
false jaw 虎钳口
false joint 假接合；假(接)缝；假灰缝
false keel 龙骨护板；假龙骨；副龙骨；防擦龙骨；保护龙骨
false keelson 冠内龙骨(木船)；内龙骨护板
false key 活键
false lapis 天蓝石
false leader 零杆；索桅式打桩架；导桩架
false leak 虚漏
false light 假光；(灯塔的)虚假光
false loading method 假负载法
false locking 错误锁闭
false longeron 辅助纵梁
falsely accused person 被诬告的人
falsely peripteral temple 假列柱式庙宇
falsely prostyle temple 假柱廊式庙宇
false maneuver 假操纵
false masonry(work) 假圬工
false member 零杆；空转构件；(桁架中的)伪杆
false membrane 假膜

false Mooneys 假门尼黏度
false mounting 假爬跨
false mud crack 假泥裂
false neutron 不稳定性中子
false northing 北移假定值(指纵坐标，用来避免纵坐标出现负值)
false ogive 整流罩；风帽
false operation 虚假动作
false origin 假定原点
false panel 墙壁屋顶粉面；护壁镶板
false papers 假证件；伪造文件
false parallax technique 假视技术
false pass 空走轧道
false path 假路径
false peripteral 假列柱式（建筑）
false pile 桩帽；送桩；假桩
false point 临时性沙嘴；三字点磁罗经
false position 试值法；试位法
false profit 虚盈实亏
false proscenium 假台口；活动台口
false pyroelectricity 第三热电性
false quay 透空式顺岸码头；辅助码头
false rafter 飞椽；假椽
false range 假距离
false rejection 误拒绝
false release 错误解锁
false release by itself 自动缓解
false relief effect 假立体效应
false reporting 虚报
false rib 假肋
false roof 假屋顶；伪顶；吊顶；吊悬屋顶
false ruby 假红宝石
false salmon 类鲑鱼；假鲑鱼
false selvedge 假边
false set 过早硬化；过早凝结；假终贯桩；假凝现象；假凝（结）；轻型临时支架
false set cement 假凝水泥
false set of cement paste 水泥的假凝
false setting 轻型临时支架；(水泥的)过凝结
false shaft 虚轴
false shelter 微粒掩蔽所
false signal 虚假信号；假信号；错误信号
false smoke 假烟
false social value theory 虚价社会价值理论
false sorts 假类
false start 假起动；不正确开动
false statement 虚假的陈述
false station 偏心站；测站偏心校正
false stern post 船尾柱贴片
false stopping of a signal 错误关闭信号
false stull 临时横撑
false surface 假面
false symmetry 假对称
false synchronization 虚同步；假同步
false target 假目标
false target generator 假目标发生器
false tax 虚税
false tenon 紧固榫；假榫；嵌入榫
false timber root 木板假顶
false tongue 假榫(舌)；对接销键；穿条
false topaz 黄晶；仿造黄晶；假黄玉
false trading 虚假贸易
false tube 假管
false twist 虚捻；假捻
false vault 假拱顶
false warm sector 假暖区
false white rainbow 雾虹
false window 百叶窗；假窗；盲窗
falsework 鹰架；工作架；临时支架；临时支撑；模板支架；脚手架；拱模
false zero 虚零
false zero method 虚零点法
false zero test 零值法测试；虚零检验
false zodiacal light 假黄道光
falsification 假造；伪造；反证
falsification of colo(u)rs 颜色失真
falsification of document 伪造文件
falsification of econometric model 经济计量模型的误用
falsification of tone values 色值失真
falsification the account 虚报亏损
falsify accounts 造假账；伪造账目
falsifying the account 虚报亏损
falsify production cost 谎报成本

falsing rock formation 假基岩
falsity 不真确
faltboat 橡胶帆布艇;可折叠帆布艇
Falten apparatus 福尔特软化点测定仪
faltering demand 需求下降
falt fillister screw 凹槽扁头螺杆
faltung 褶积;褶合
faltung integral 褶合积分
faltung theorem 褶合定理
falut detector 故障指示器
falx straw 亚麻杆
famatinite 块硫锑铜矿;脆硫锑铜矿
fambe 窑变
Famennian 法门阶【地】
familiarise 使通晓;使通俗化;使熟悉;使精通
familiarization cost of workers and staff 职工专业训练费用
familiarize 使通晓;使通俗化;使熟悉;使精通
familiar with 精通
famille rose 粉红色系列色料
famille-rose porcelain 粉彩瓷
family boarding home (供食宿的)出租房子
family brand 统一品牌
family card 分类卡
family census 家庭普查
family chip 系列片
family circle 包厢;(剧场的)小包厢;(剧场的)家庭座位
family composition 家庭构成
family curves 曲线组;曲线族;曲线簇
family day care home 家庭日托儿所
family disaster plan 家庭防灾计划
family dwelling unit 住宅;寓所
family game center 家庭游戏中心
family hall for worshipping Buddha 佛堂
family household 家庭住户
family housing 家属宿舍
family land plot 宅基地;自留地
family life cycle 家庭人口变化周期
family make-up 家庭人员组成
family mansion 住宅;宅第;(罗马人的)私人住宅
family mo(u)ld 一模多铸铸型;多腔铸型
family of bridges 桥簇
family of cables 缆索系列;电缆系列
family of characteristic curves 特征曲线族;特性曲线族
family of characteristics 特征曲线族;特性曲线族
family of circle 圆族
family of curves 曲线组;曲线族
family of ellipse 椭圆族
family of family 缆索簇
family of functions 函数族
family of half-curves 半曲线族
family of high cloud 高云族
family of lines 线族
family of maps 图组
family of ropes 缆索系列
family of sets 集族
family of spirals 螺线族
family of stationary curves 平稳曲线族
family planning 家庭计划
family room 多用途房间;娱乐室;家庭活动室
family status life cycle 家庭状况生命周期
family structure 家庭结构
family tree 谱系;家谱;系谱(树)
family unit 房屋
family unit price 家庭单元价格
family-use grinder 家用粉碎机
famine price 严重缺货时的市价;缺货造成的高价
faming price 缺货行市
famotidine 法莫替丁
famous and precious 名贵
famous brand 名牌货;名牌
famous brand high-quality products 名优产品
famous brand products 名牌产品
famous cultural city 文化名城
famous historical city 历史名城
famous historic sites and scenic spots 风景名胜区
famous painting 名画
famous product 名产
famous scenery 风景名胜
famous scenic spot 胜地
famphur 伐灭磷(杀虫剂);氨磺磷
fan 峡谷滚石堆;排风机;排风扇;通风器;扇状物;扇状流;扇形地;风选机;风扇;风机
fan adjusting pulley 风扇调节皮带轮
fan advance 扇形推进
fan air heating 通风机供暖
fan air intake 风扇进气口
Fanal blue 法纳尔蓝
fan anchorage 扇形锚固
fan and its incorporated equipment 风机及其配套设备
fan antenna 扇形天线
fan-assisted burner 低压鼓风机燃烧器
fan-assisted heater 鼓风机燃烧采暖炉
fan-assisted ventilation 通风机通风
fan-assisted warm-air central heating 通风机送热集中供暖
fan-assisted warm air heating 风扇辅助热空气供暖
fan atomizer 风扇(式)喷雾器;风扇式燃烧器
fan baffle 风扇导流板
fan base of alluvial fan 冲积扇扇尾
fan beam 扇形射束;扇形波(射)束;扇束
fan beam beacon 扇形波束信标
fan belt 风扇(皮)带;风扇传动皮带
fan blade 风扇叶子板;风扇叶片;风机叶片
fan blade arm 风扇叶片支架
fan blast 通风机的鼓风
fan blast deflector 排风扇气流折射板
fan blower 鼓轮风箱;鼓风机;送风机;风扇;排风扇
fan board 风扇架
fan bolt adjusting pulley bracket 通风器螺栓调整滑轮支架
fan bolt tightener 通风器螺栓紧固器
fan boring 扩孔
fan box 风扇盒
fan brake 叶片式空气制动器
fan building 风机房
fan burden 扇形炮孔间距
fan cameras 扇形排列摄影机组
fan cardan shaft 风扇万向轴
fan casing 通风机罩
fan chamber 扇风机房
fan characteristic 风扇特性曲线
fan chart 扇形图
fan chimney 扇风机出风筒
fancier 设计师
fanciful ornamentation 奇形装饰
fan cleavage 扇状劈理;扇形劈理
fan coil 风机盘管
fan coil air-conditioning system 风机盘管空气调节系统
fan coil cooling unit 风机盘管空调系统
fan coil system 风机盘管系统
fan coil unit 冷暖气送风机;风机盘管机组
fan collector 风扇除尘器
fan complex fold 扇状复式褶皱
fan compression ratio 风扇压缩比
fan connector 扇形连接器
fan construction 风机施工
fan control 风扇调节器
fan control word 扇控制字
fan convector 对流散热器
fan convector heater 鼓风式对流加热器;扇式对流加热器;扇动对流加热器
fan conveyer 转动式输送机;旋转式运送机;轴承式输送机;旋转式输送机
fan-cooled 风扇冷却(的)
fan-cooled cooling 鼓风冷却;风扇冷却
fan-cooled machine 风冷电机
fan-cooled motor 全封闭风冷式电动机
fan cooler 风扇冷却器;风机冷却器
fan cooling 电扇冷却;风扇冷却
fan coupler 风扇耦合器
fan cut 扇状展开;扇形掏槽;扇形掘进
fancy accounting 粉饰结算
fancy alloys 装饰合金
fancy breed 观赏品种
fancy butt 花边墙面板
fancy carton 彩色纸盒
fancy coal 上等煤
fancy door 双开门;双道门
fancy embossing card 美术纸板
fancy fabric 变化组织织物
fancy fair 杂货市集
fancy finish 美术涂饰剂
fancy glass ware 艺术玻璃器皿
fancy glaze 花釉;复色釉
fancy goods 新奇商品;杂货;小工艺品
fancy lantern 彩灯
fancy leather 装饰革
fancy lump coal 上等大块煤
fancy packing box 彩印包装盒
fancy paint 装饰漆;美术装饰漆
fancy paper 彩色纸
fancy plywood 花色胶合板;装饰性胶合板;饰面胶合板
fancy points 观赏性状
fancy price 特昂价格;高价
fancy sheet metal 装饰金属板;花色金属板
fancy soap 香皂
fancy stocks 热门股票
fancy veneer 装饰镶面板;装饰单板;装饰薄板;上等单板
fancy wood 精选木材;花色木材;贵重木材
fancy work 钩编织品;刺绣品
fancy yarns 花色纱
fan delivery 风扇功率;风机风量
fan delta 扇状三角洲;扇形三角洲;三角形冲积扇
fan delta deposit 扇三角洲沉积
fan delta facies 扇三角洲相
fan delta sequence 扇三角洲层序
fan diagram 扇形图
fan dial 扇形刻度盘;扇形度盘;半圆形刻度盘
fan diffuser 扇风机扩散器
fan discharge 扇风机出风口
fan drain system 扇形排水系统
fan-draught cooler 风扇冷却器
fan drift 引风道引风平巷;扇形堆积;风扇机引风道;风扇机通风道;通风道
fan-drill 扇形孔凿岩机
fan drilling 扇形钻孔;扇形孔凿岩;扇形布孔钻进
fan drill rig 扇形孔凿岩台车
fan drive 通风机传动装置;风机传动装置;风机驱动
fan-driven generator 风力发电机
fan driven turbine 驱动风扇涡轮
fan drum 风扇外壳
fan-drying 排风干燥
fan duster 翼式除尘机
fan economizer 风扇机省功器
fan effect 扇形效应
fan efficiency 风扇效率;风机效率
fan electromotor 风扇电机
fan end thrust ball bearing 风扇端止推滚球轴承
fan engine 鼓风机;风扇机
fan evase 风扇机出风筒
fan evasion stack 扇风机出风筒
fan exhaust 风扇排气
fan failure device 风机故障切断装置
fan filter 扇形滤光器;扇形滤波
fan Fink truss 扇形芬克桁架
fan flywheel 带扇飞轮
fan fold 扇状褶皱;扇状褶曲;扇形褶皱
fang 燕尾开脚;(锚的)鱼尾
fan gate 扇形闸门;扇形浇口
fang bolt 棘螺栓;锚栓;锚定螺栓;底脚螺栓;板座(螺)栓
fan gear housing 风扇齿轮箱
fan-generator 风扇发电机
Fanger comfort equation 范格热舒适方程
Fangeron kiln 法格仓窑
fanging nozzle 接收喷嘴
fan glacier 扇形冰川
fanglomerate 扇岩砾;扇砾岩;扇积砾
fango 矿泥
fan governor 风扇调节器
fan groining 扇形穹顶
Fangshan stone 房山石
fan guard 拉力支架;脚手架上防止杂物落下的挡板;风扇罩;脚手架上护网
fan head 通风机压头;扇顶区
fanhead nozzle 扇形头喷嘴
fanhead of alluvial fan 冲积扇扇顶
fan heater 通风机式加热器;风扇式空气加热器
fan heating 激动式热风供暖;风扇加热
fan heating convector 扇式加热器对流器
fan holes 扇形布置的钻孔;扇形布置的炮眼;扇形炮孔
fan horsepower 风机功率
fan house 通风机室;通风机房;扇风机房
fan housing 风扇罩

fan hub 扇风机工作轮轮壳
fan idler 鼓风机惰轮
fan idler bellcrank 风机惰轮摇臂
fan impeller 扇风机叶轮；风扇叶轮
fan-in 输入端数；输入；扇入
fan-in argument 扇入变元
fan inlet 通风(机)进气口；风扇进风口
fan-in network 扇入网络
fan inversion 风扇反转
fanion 小旗；测旗；测量旗
fan jet 扇形喷流；鼓风式喷气发动机
fan joint 风扇接头
fan jumbo 扇形孔凿岩台车
fan-key cock 阔把旋塞；阔把龙头
fan ladders 扇形梯线
fan layout 扇形观测系统；风扇机布置
fanlight 腰头窗；气窗；上亮子；扇形(气)窗
fanlight catch 楣窗；(门上的)亮子；扇形窗上的扇形档；扇形窗窗钩
fanlight opener 楣窗开关
fanlight opening gear 扇形窗开闭传动装置
fanlight quadrant 腰窗档；扇形窗上的扇形档
fanlight transom(e) window 天窗；扇形窗；气窗；楣窗
fan-like 扇形的
fan-like distributory 扇状分流
fan-like diversion 扇状分流
fan-like drainage 扇状水系
fan-like pattern 扇面形
fan-like structure 扇形构造
fan line 风扇线
fan longitudinal cable arrangement of cable-stayed bridge 斜拉桥扇形纵向布置
fan maker beacon 扇形标志
fan mark 扇形标志
fan marker 扇形指点标
fan marker beacon 扇形指点标
fan motor 风扇马达；风扇电动机
fanned 扇形的
fanned-beam antenna 扇形射束天线；扇形波束天线；扇形定向天线
fanner 通风机；鼓风机操作工；鼓风机；风扇
fanning 用通风机吸尘；呈扇形展开；扇形展开；扇形编组；通风
fanning beam 扇形射束
fanning bottom 降低孔底的钻压
fanning friction factor 通风管摩擦系数
fanning machine 通风器；风选机
fanning mill 风车
fanning mill chute 风选机喂入槽
fanning mill shutter adjuster 风选机进风门调节器；分选机进风门调节器
fanning plume 扇形烟羽；成扇形烟缕
fanning pollutant 弱状污染物
fanning strip 扇形片；扇形(端子)板
fanning the bottom of the borehole 孔底减压钻进
Fanno flow 法诺流动
fan noise 风扇噪声；风机噪声
fan nozzle 扇形喷嘴
Fann viscosity 范氏黏度计
fan of filaments 纤维扇面
fan of outwash sediment 冰水沉积扇
fan operation 风机运行
fan out 散开；展开；输出；扇形展开；负载数；分开电缆芯；逻辑输出；输出端(数)；扇出
fan-out argument 扇出变元
fan-out branch 扇出转移
fan-out capability 输出能力
fan-out feature 扇出特点
fan-outlet 扇风机出风道；电扇出线口
fan-out limitation 扇出极限
fan-out logic function 扇出逻辑函数
fan-out modem 扇出调制解调器
fan-out network 扇出网络
fan-out node 扇出节点
fan-output 风机出力
fan-out screw 调整螺栓以增加钢丝绳的给进长度
fan-out stem 扇出干线
fan-outwash 冰水扇形地
fan palm 扇形棕榈
fan pattern 扇形布置
fan-pattren holes 扇形炮孔组
fan propeller 风扇转子
fan pulley 风扇皮带轮

fan rating 扇风机额定能力；风机额定性能
fan ratio 扇出率
fan ray 扇形射线
fan reflector 扇形反射器
fan reversal 扇风机反转
fan ring 风扇外环
fan roof 扇形花格交叉屋顶；扇形花格交叉拱顶
fan room 鼓风机室；通风机室；通风机间；风机房
fan-room yard 风机房场坪
fan round 扇形炮孔组
fan runner 扇风机操作工
fan section 扇形区；风机段
fan selection 风机选择
fan sewer system 扇形排水管系统；扇形断面下水道系统
fan shaft 扇风机轴；风扇轴
fan shaft bearing 风扇轴轴承
fan shaft gear 风扇轴齿轮
fan-shaped 扇状；扇形的；扁形的
fan-shaped allurium 冲积扇
fan-shaped alluvial tract 扇形冲积地带
fan-shaped aquifer 扇形含水层
fan-shaped arc 扇形弧
fan-shaped basin 扇形港池
fan-shaped beam 扇形射束；扇形光束
fan-shaped blastic texture 扇状变晶结构
fan-shaped brick 扇形砖
fan-shaped centering 扇形拱架
fan-shaped crack 扇形裂缝
fan-shaped delta 弧形三角洲；扇形三角洲
fan-shaped dock 扇形港池；扇形码头
fan-shaped door 扇形门
fan-shaped drainage 扇形水系
fan-shaped earth pulling process 扇形拉土法
fan-shaped falsework 扇形支架
fan-shaped floor tile 扇形铺地砖
fan-shaped fold 扇状褶曲
fan-shaped gate 扇形闸门；扇形门
fan-shaped grating 扇形光栅
fan-shaped laser beam 扇形激光束
fan-shaped manhole 扇形人孔
fan-shaped pattern 扇形图案
fan-shaped paving 扇形铺砌(路面)
fan-shaped plan 扇形平面
fan-shaped plate feeder 扇形板给料器
fan-shaped round 扇形掏槽；扇形布置的钻孔；扇形布置的炮眼；扇形炮孔组
fan-shaped shingle bank 扇形砾石滩
fan-shaped structure 扇状构造
fan-shaped system 扇形体系
fan-shaped tail 扇状尾
fan-shaped tenon 扇形榫头
fan-shaped tracery 扇形花格装饰
fan-shaped training 扇形整枝
fan-shaped transverse dredging method 扇形横向挖泥法
fan-shaped tree 扇形树
fan-shaped weir 扇形闸门堰
fan-shaped window 扇形窗
fan shooting 扇形炮眼爆破；扇形放射；(地震勘探法的)扇形爆炸
fan shroud 风机护罩
fan shutter 通风机闸门；通风调节板；风扇机节风闸门
fan shutter adjustment lever 风扇进风门调节杆
fan shutter control 风扇进风门调节器
fan slip 风压损失；风扇滑流
fan sound 风机噪声
fan spider 扇风机星形轮
fan spindle bearing 风扇轴承
fan spoked wheel 风扇辐轮
fan-spray nozzle 雾化膜喷嘴
fan stack 风扇排气管
fan static head 扇风机静压头
fan static pressure 扇风机静压力；风机静压力
fan stern 扇形船尾
fan system 扇形式系统(排水系统布置)；扇形系统
fantail 鸭尾片；扇形尾
fantail arch 扇形拱；(马丁炉的)炉门拱
fantail burner 鸠尾燃烧器；扇尾形火焰喷燃器
fantail deck 船尾甲板
fantail joint 鸠尾接合；鸠尾连接；鸠尾结合
fantail mullet 扇尾鲻
fantail roof 扇形屋顶

fantail stern 扇形船尾
fantail suction hood 扇形吸入罩
fan talus 扇状岩(屑)堆；扇形岩堆
fantascope 幻视器
fantasound 立体声
fantastic architecture 奇趣建筑
fantastron multivibrator 幻像多谐振荡器
fan terrace 扇状阶地；扇形(冲积)阶地；(冲积的)扇阶地；冲积扇阶地
fan test 扇风机试验；风机试验
fan test object 扇形测试物
fan thrust reduction 风机推力减损
fan thrust reverser 风扇推力反向器
fan total head 风机总压头
fan total pressure 风机总压力
fan tracery 扇形花格架；扇形格架
fan tracery vault 扇形格式穹顶；倒伞形格式穹顶
fan trained cordon 扇状单干形
fan trained tree 扇状整枝树
fan truss 扇形桁架
fan-type aeroengine 扇形航空发动机
fan-type anchorage 扇形锚碇
fan-type atomizer 风扇式喷雾器
fan-type cable-stayed bridge 扇形斜拉桥
fan-type coal mill 风扇煤磨
fan-type dynamometer 风扇式测力计
fan-type engine 扇形发动机；风扇式发动机
fan-type floor furnace 风扇式地板炉
fan-type heater 风扇式空气加热器
fan-type heating 扇风式热风供暖
fan-type nozzle 风扇式喷嘴
fan-type pump 风扇式泵
fan-type stay cable 扇形拉索
fan-type support 扇形支架
fan-type vented wall furnace 风机式壁炉
fan-type ventilator 风扇式通风机
fan unit 扇风设备
fan valve 扇形阀(门)
fan vault(ing) 扇形穹顶；扇状展开的拱顶构造；扇形穹拱；扇形拱顶；扇形拱(顶)
fan veined 扇状脉的
fan-ventilated moter 风扇冷却型电动机
fan ventilation 风机通风；风扇机通风
fan ventilator 风扇通风器；风扇通风机
fan-volute 扇风机出风扩散螺道
fan wafter 风扇
fan wheel 扇风机风轮
fan window 扇形窗；通风窗
fanwise paving 扇形铺砌(路面)
fanwork 扇形花格架
faq 码头交货
farad 法拉
Faradaic impedance 法拉第阻抗
Faraday 法拉第
Faraday-ice bucket experiment 法拉第冰桶试验
Faraday's balance 法拉第天平
Faraday's birefringence 法拉第双折射
Faraday's cage 法拉第罩；法拉第屏蔽；法拉第笼
Faraday's constant 法拉第常数
Faraday's cup 法拉第杯
Faraday's current 法拉第电流
Faraday's dark space 法拉第暗区
Faraday's depolarization 法拉第消偏振
Faraday's disc dynamo 法拉第圆盘发电机
Faraday's disk machine 法拉第圆盘机
Faraday's driver 法拉第驱动器
Faraday's effect 法拉第效应
Faraday's equivalent 法拉第当量
Faraday's generator 法拉第圆盘机
Faraday's glass 法拉第玻璃
Faraday's impedance 法拉第阻抗
Faraday's isolator 法拉第绝缘体
Faraday's law 法拉第定律
Faraday's law of electrolysis 法拉第电解定律
Faraday's law of electromagnetic induction 法拉第电磁感应定律
Faraday's law of induction 法拉第感应定律
Faraday's magnetooptic(al) effect 法拉第磁光效应
Faraday's pulsation 法拉第脉动
Faraday's rotation 法拉第旋转
Faraday's rotation effect 法拉第旋转效应
Faraday's rotation glass 法拉第磁光玻璃
Faraday's rotation isolator 法拉第旋转式隔离器
Faraday's rotator 法拉第旋转器

Faraday's screen 法拉第屏蔽
Faraday's shield 法拉第屏蔽
Faraday's shutter 法拉第快门
Faraday's tube 法拉第管
farad bridge 电容电桥
faradic current 感应电流
faradic electricity 感应电
faradimeter 感应电流计
Farad/meter 法拉/米(电容率单位)
farallonite 硅钨镁矿
Farallon plate 法拉郎板块
farameter 法拉计
far and away 远远
far-and-near combination 远近结合
faratron 液面控制器
faratsihite 杂绿脱高岭石;铁高岭石
far bank 远外堤岸
far bar 远沙坝
far beam 远光
fardage 货垫;垫衬材;承载垫层
fare 供使用的设备;乘船费;车船费
far earthquake 远震
far east floral region 远东植物地理大区
fare free age 免费年龄
Fareham reds 法列哈姆式红面砖
far end 远端
far-end cross talk 远端串扰
far-end field 远端场
far-end infrared frequency 超级高频
far-end interference 远端干扰
far-end operated terminal echo suppressor 远端操作的终端回波抑制器
fare register 计费器
fare subsidy 票价补贴
fare table 票价表
farewell buoy 告别浮标(出港航路最后一个浮标);海口浮标;海界浮标
Farewell rock 磨刀砂岩
farewell whistle 告别汽笛
fare zone 车费票价区
far face 远视面
far field 远场;夫琅和费区
far-field analyser 远场分析器
far-field approximation 远场近似
far-field Cassegrainian antenna 远场卡塞格伦天线
far-field displacement 远场位移
far-field distribution 远场分布
far-field earthquake motion 远场地震地面运动
far-field expressions 远场表示式
far-field fringes 远场条纹
far-field holography 远场全息术
far-field interference pattern 远场干涉图
far-field noise 远场噪声
far-field pattern 远场图形
far field pattern of laser 激光远场图
far-field polar radiation 远场极辐射
far-field region 远场区域
far-field seismic signal 远场地震信号
far-field sensitivity 远场灵敏度
far-field spectrum 远场谱
far-field stress 远场应力
far field testing 远场测试
far from surface water 远离地表水体
far future 远景
far-future stage 远期
fargite 钠沸石
farina 谷粉
far infrared 远红外
far infrared absorption spectrum 远红外吸收光谱
far infrared band 远红外区;远红外波段
far infrared communication 远红外通信
far infrared controller 远红外控制器
far infrared detector 远红外探测器
far infrared drying 远红外线干燥;远红外干燥
far infrared gear 远红外仪器
far infrared heater 远红外加热器
far infrared image 远红外成像
far infrared imagery 远红外成像
far infrared interference 远红外干涉
far infrared interferometer 远红外干涉仪
far infrared laser 远红外线激光器
far infrared photoconductivity 远红外光电导
far infrared photoconductor 远红外探测器
far infrared radiation 远红外线辐射

far infrared ray 远红外线
far infrared region 远红外区
far infrared reststrahlen spectrum 远红外余辉带光谱
far infrared scanning 远红外扫描
far infrared spectrum 远红外光谱
far infrared transmission filter 远红外透射滤光片
far infrared window 远红外窗口
Faringdon sponge bed 法林登海绵层
farinose 被粉的
Farkas lemma 法卡斯引理
Farleigh Down stone 法雷当石
farm 临时堆货场;农庄;田间;饲养场
Farm and Land Institute 农田土地协会
farm bridge 农村桥梁
farm broker 田地掮客;田地经纪人
farm building 农业建筑;农村建筑;农场建筑(物)
farm building construction 农村建筑工程
farm capital 农场固定资产
farm cart 农用拖车
farm consumption use 田间耗水量
farm consumptive use 田间灌水定额
farm crane 农用起重机
farm crop 农作物
farm dam 小型灌溉坝;小土坝;农用小土坝
farm delivery requirement 净需水量;农田净灌溉需水量;农田供水量;田间放水需要量
farm distribution system 田间配水系统
farm ditch 农渠;农沟;毛渠
farm drainage 农田排水
farm drain tile 农田用瓦管;农田排水瓦管
farm duty (of water) 田间用水率;田间灌水定额;农田灌溉率;农场灌溉率;灌溉(用水)率
farm dwelling 农村房屋
farm electrification 农业电气化
farm employee housing 农村雇工的房屋
farm engine 农用发动机
farm equipment factory 农机制造厂
farmer 农场主;承包者
farmer's family labour return 农户劳动收入
farmer's law 农业法
farmer's lung 农民肺
Farmers reducer 法梅减薄液
farmer's year 农事年
farm fence 农田栅栏
farm field 农田水利工程;农田
farm forestry 农场式林业
farm garden 园艺场
farm gate price 农场出售价
farmhand 农业工人;农场工人
farm head gate 农用斗门
farmholding 农家;农户
farm house 农舍
farm implement 农用机具;农具
farming 耕作;耕种
farming and forestry badge 农林徽
farming crossing 农用道口
farming efficiency 农耕效率
farming industry 农产品加工(工)业
farming landscape 农业景观
farming method 耕作方法;耕作法
farming of bivalve 牡蛎养殖
farming practice 农耕方式
farming research 农业研究
farming settlement 农业聚落
farming system 耕作设计;农作制度
farm insecticide 农药
farm investment 农业投资;农场投资
farm irrigation layout 农田灌溉布置
farm labo(u)r camp 农业工人宿舍区
farm labo(u)rer's quarters 农业工人宿舍
farmland 农业用地;农业土地;耕地;农田;田地
farmland ecosystem 农田生态系统
farmland management 农业土地管理
farmland running to weeds 草荒
farmland shelter forest 农田防护林
farmland water 农业土壤用水
farm lateral 田间支渠
farm leveller 农用平地机
farm loading equipment 农用装载设备
farm machine 农场用机器
farm machinery 农用机械;农业机械
farm machinery and implements 农业机械及农具
farm machinery industry 农业机械工业

farm machinery production 农业机械生产
farm management 农场管理;田间管理
farm manure 农家肥(料)
farm mechanization 农业机械化
farm mortgage 农场抵押
farm motor 农用电动机
farm operator 农场经营人员
farm out 佃出;出租;包出工件;把收入、工作、活计等包出去;合同分包
farm out agreement 油田租出协议
farm percolation loss 农田渗透损失;农场渗透损失;田间渗漏损失
farm planning 农田规划
farm plot 农田水利工程
farm pollution 农田污染
farm pond 小型灌溉(水)塘;养鱼池;养鱼场;农业用储[贮]水池;农田池塘;农水水塘;田间蓄水坑塘
farm posthole auger 合抱式取土螺钻;合抱式取土钻
farm power 农用动力;农业动力
farm product 农产品
farm production cost 农场生产费用
farm product market 农贸市场
farm products processing waste 农产品加工废物
farm research station 农业试验站
farm reservoir 农用水库;农场蓄水池
farm road 乡村路;农村道路
farm roller 农用压路机
farm runoff 农田径流
farm service building 农村辅助建筑物
farm shelter belt 护田林带
farm sprinkler system 农场喷灌系统
farms producing good poultry and animal strains 畜禽良种场
farmstead 农庄;农场建筑(物);农场及其建筑物
farmstead engineering 农场工程
farm structure 农村构筑物
farm survey 农田丈量;农田调查;农田测量
farm tools showroom 陈列农具室
farm tractor 农用拖拉机
farm tractor fuel 农业拖拉机燃料
farm tractor tyre 农用拖拉机轮胎
farm truck 农场载货汽车
farm wagon 农用拖车;农用挂车;农村运货大车;农场用带篷挂车;四轮大车
farm waste 农田废物;农场废物
farm water requirement 田间需水量
farm water supply engineering 农场给水工程
farm waterway 垄沟;农场内部水道
farm woodland 农场式林地
farm woodlot 农场式林地
farm woods 农场式森林
farm yard 农家庄院;农家庭院;农场空地;农场空地;场院;农业场地
farmyard manure 农家肥(料);农场肥料
farmyard waste 农家场院废物
farmyard wastewater 农场废水
farnesene 法呢烯
farnesol 双苯醇
faro 小珊瑚礁
faroelite 杆沸石
far off 遥远
faro reef 小环礁
far point 远点;明视远点
far-reaching 有深远影响的;深远的
far-reaching design 远景设计;长远的计划
far-reaching headlamp 远光头灯;探照头灯
far-reaching impact 深远影响
far-reaching plan 远景规划
far-red light 远红光
far region 夫琅和费区
Farrel system 法雷尔方式
farringtonite 磷镁石
farrisite 闪辉黄煌岩
Farromastic (一种专卖的用于金属窗的)上光剂
Farro's process 费洛法
farrowing cage 产仔栏
farrowing house 产仔猪舍
farrowing hut 产仔临时猪舍
farrowing stall 产仔栏
far-seeing plan 远景规划
far shot-point trace number 远泡点道号
far-side bus stop 越过交叉口的公共汽车站
far side traffic lane 超车(车)道

far-sight 远见(力)
far-sighted 远视的
farsightedness 远视;远见
far-super heavy nuclei 远超重核
farther state 延伸声明
farthest range 最大测距
far-ultraviolet radiation 远紫外辐射
far ultraviolet ray 远紫外线
far ultraviolet region 远紫外区
far vane 接物端
farvitron 线振质谱计;分压指示计
far zone 远区;夫琅和费区;边远区
far zone condition 远区条件
fascia 汽车仪表板;盘口面;托板;挑口饰;窗过梁饰带;封檐(底)板
fascia beam 装饰梁;前沿横护梁
fascia board 牌匾;仪表板;挑口板(檐槽);封檐板
fascia brachii 臂筋膜
fascia bracket 檐沟托座
fascia plate 封口板
fascia propria 固有筋膜
fasciated yarn 包缠纱
fasciation 带化现象
fascia transversalis 横筋膜
fascia wall 胸墙;面墙
fascicled leaves 簇生叶
fascicular 束状
fascicular tree 丛生树
fasciculate 束状
fasciculate root 丛状根
fasciculation potential 束电位
fasciculite 角闪石
fascine 梢捆;埽料;柴笼
fascine barge 梢料船;柴捆船
fascine building 柴木房屋
fascine bundle 束柴;梢捆;柴束栏(册);柴捆
fascine cradle 束木支架
fascine dam 梢捆坝;沉捆坝;柴捆坝;草坝
fascine dike 木排护堤;柴笼堤坝;梢护堤;捆护堤
fascine filling 沉排填石
fascine foundation 梢料基础;柴排基础;柴捆护工基础;束木基础
fascine groin 梢捆丁坝
fascine groyne 梢捆丁坝
fascine hurdle 柴篱;柴束栏栅;柴束栏
fascine layer 梢料填层
fascine matters 梢蓐
fascine mattress 梢枝排;梢蓐料沉褥;沉排;梢排席;柴排
fascine mattress bedding 柴排垫层
fascine pole 梢枝束;柴束
fascine project 梢捆工程
fascine revetment 木排护岸;梢蓐护岸;埽工护岸;沉排护岸;柴笼护岸;梢捆护岸
fascine revetment works 柴束护岸工程
fascine road 柴束路
fascine roll 梢笼;梢棍;沉辊;柴笼
fascine roll closure 柴笼截流;沉辊截流
fascine whip 梢枝束;梢笼;梢鞭
fascine wood 柴木树;梢料
fascine work 堆梁柴捆;梢捆建物;梢捆工程;梢工;埽工;沉排工作;沉排护岸工事;沉柴排工作;柴束法
fasciole 带线
fasciculation 缩聚
fase corrdinate 假坐标
fashion 样式;流行的;款式;风格;方式
fashionable 流行式;流行的
fashionable architect 时髦建筑师;流行建筑师
fashion colo(u)r 流行色
fashion count 流行计数
fashioned 成型的
fashioned iron 型铁;型钢;异型钢
fashioned vehicle 花车
fashioner 设计者
fashioning flat knitting machine 成型扁平针织机
fashion of data record 数据记录方式
fashion of depth record 深度记录方式
fashion parts 异形配件;定型部件
fashion photography 样式摄影术
fashion plate 船首线型板
fashion-plate stem 钢板装配船首材
fashion style 时样
fassaite 深绿辉石

fast 牢固;紧结;不褪色的
fast-access information retrieval 快速访问信息检索;快速存取信息检索
fast-access memory 快速存储器
fast-access memory chip 快速存取存储器芯片
fast-access retrieval system 快速访问信息检索系统;快速存取信息检索系统
fast-access storage 快速存储器
fast-acting 快速作用
fast-acting closing valve 快动闭阀
fast-acting hydraulic controlled self centrical chuck 液压控制快速定心卡盘
fast-acting insecticide 速效杀虫剂
fast-acting relay 高速继电器;快速用继电器;快动继电器
fast-acting tracking device 快动跟踪系统
fast action cramp 快速夹紧;高速固定
fast adaptation 快速适应
fast address 快速地址
fast advance cam 快拨凸轮
fast amplifier 快速放大器
fast and loose pulley 固定轮和游滑轮;固定轮和动滑轮
fast as can 尽速交运
fast automatic gain control 快速自动增益控制
fast axis 快速轴;快光轴
fast-back 有长坡度顶的汽车
fast-back algorithm 快返回算法
fast-back parser 快返回分析程序
fast-back top-down analysis 快返回自顶向下分析
fast blink 快速闪烁
fast body waves 快波群
fast bolometer 快速辐射热测量计;快速测辐射热计
fast brake 快速制动器
fast braking 快速制动
fast break 快速断电
fast breeder reactor 快中子增殖反应堆;快速增殖反应堆
fast burn 快速燃烧
fast burn engine 高速燃烧发动机;快速燃烧发动机
fast burner 高速燃烧器
fast-burning 快速烧成
fast-burning charge 易燃药柱
fast-burning coke 易燃焦
fast-burning composition 速燃成分
fast-burning fuse 速燃导火线;速燃导火索
fast-burning powder 速燃火药
fast-burst reactor 快中子脉冲反应堆;快脉冲反应堆
fast cableway 快速索道;快速缆道
fast-carry lookahead 快速超前进位
fast-carry look-ahead arithmetic 快速超前进位算术运算
fast-cash 快速付现法
fast ceramic reactor 快中子陶瓷反应堆
fast channel 快通道
fast chemical reaction 快速化学反应
fast chopper 快速断路器
fast circuit 快速电路
fast coagulating cement 速凝水泥
fast coding 快速编码
fast coincidence circuit 快速符合线路
fast colo(u)r 坚牢的颜料;坚牢的染料;不易褪色的颜色;不褪(的)颜色
fast compact colo(u)r printer 彩色小型快速印片机
fast compression process 快速压缩试验法
fast compression test 快速压缩试验
fast connect 快速连接
fast control memory 快速控制存储器
fast conversion 快速转换
fast cool zone 急冷工段
fast core memory 快速磁芯存储器
fast cornering 快速回转
fast coupling 硬性联轴节;刚性联轴节;可移式刚性联轴器;死刚轴节
fast crosstalk 快速交错
fast-curing resin 快速硫化树脂
fast-curing system 快速硫化体系
fast-delay cap 微差延发雷管;毫秒延时雷管;毫秒延发雷管;毫秒延迟雷管
fast delay detonation 微差爆破
fast deployment logistic ship 快速支援船
fast detector 快速探测器
fast digital processor 快速数字处理机
fast discrete cosine transform 快速离散余弦变换

fast drift burst 速漂爆发
fast-driving road 快车道
fast-drying 快干的
fast-drying ink 快干油墨
fast-drying material 快干剂;快干材料
fast drying paint 快干漆
fast dye 不褪色染料
fast dyed yarn 不褪色纱线
fast effect 快效应
fasten 扎牢;扣牢;扣紧;夹紧
fasten alumin(i)um 铝连接件
fastened cardboard 黏贴纸板
fastened steel sheet 黏贴钢板
fastener 线夹;系固物;拉链;U形铁箍;固定器;钩扣;连接材料;扣件;紧固(零)件;接线柱【电】;接合件;夹持器;卡子;水落铁卡;闭锁(器)
fastener tape 搭扣带
fastener treatment 紧固件遮盖处理
fastening 支撑设备;扎牢;固紧件;连接(法);扣件;紧结物(如锁、闩、钩等);紧固(接头)
fastening angle 固定角铁;联系角铁;连系角铁
fastening arrangement 固紧装置
fastening bar 连接杆
fastening bolt 扣紧螺杆;桩栓;螺栓标志;扣轨螺栓;紧固螺栓;连接螺栓
fastening device 系泊装置;紧固方法;紧固装置
fastening-down temperature of rail 锁定轨温
fastening element 紧固件
fastening gripping device 锚固夹具
fastening lug 紧固凸耳
fastening nail 紧固用钉
fastening nut 紧固螺母
fastening of the tyre by spring clip 用弹簧卡子紧固轮箍
fastening of the tyre by spring ring 用弹簧圈紧固轮箍
fastening part 紧固件
fastening piece 连接件;紧固件
fastening pile 系固桩
fastening point 紧固插座
fastening resistance 扣件阻力
fastening ring 固定圈
fastenings 连接件
fastening screw 紧固螺钉
fastening socket 紧固套筒
fastening thread 紧固螺纹
fastening tool 紧固工具
fastening torque 夹紧力矩
fastening wire 轧用钢丝;扎用钢丝;紧固用钢丝;绑扎用钢丝
fastening with transverse cables anchored to ground 用横缆锚地紧固
fasten with bolt 用螺丝扣紧;用螺栓固定
faster handler 快速处理机
faster processor 快速处理机
faster than real time simulation 超实时仿真
faster welding 快速焊接
fastest mile 最大风速英里
fastest ratio 最小传动比
fastest sweep 最快扫描
fast ethernet 快速以太网
fast-exit taxiway 高速脱离滑行道
fast exponential experiment 快中子指数实验装置
fast extrusion furnace black 快速压出炉黑
fast fading 快衰落
fast fading margin 快衰落余量
fast-falling level 猛烈水位
fast-fast concrete 耐酸混凝土
fast feathering 出羽快
fast feed 快速走带;快速进给;快速给进
fast feed gear 快速给进齿轮
fast film 快速感光片
fast-fired coating 快速氧化膜
fast fired colo(u)r 快烧颜料
fast firing 快烧;快速烧成
fast firing colo(u)r 快烧颜料
fast firing glaze 快速烧成釉
fast firing kiln 快速烧成窑
fast firing pusher-type kiln 快烧推板式窑
fast firing walking beam kiln 快烧步进式窑
fast fission factor 快中子增殖系数
fast flange 固定法兰(盘);固定凸缘(盘)
fast-flowing 水流湍急的
fast food restaurant 快餐餐厅

fast foods 快餐;现成食品
fast food service installation 快餐服务设备
fast footage 快速进尺
fast forward 快速进带
fast forward control 快速正向控制
fast forward controller 快速正向控制器
fast Fourier analysis 快速傅立叶分析
fast Fourier inverse transform 快速傅立叶反变换
fast Fourier transform 快速傅立叶变换
fast Fourier transformation algorithm 快速傅立叶算法
fast Fourier transformation analysis system 快速傅立叶变换分析装置
fast fracture 快速断裂
fast-freezing slag layer 快凝渣层
fast freight 快运货物;快件
fast freight line 快运货物直达运输线
fast freight traffic 快运货物运输;快速货物运输
fast freight train 快运货物列车
fast frequency 快速频率
fast frequency shift key(ing) 快速频移键控
fast gaining 增重快
fast gear 快速齿轮
fast goods 快运货物;快件
fast goods transport 快速货物运输
fast governor 快速调速器;快调速器
fast groove 稀纹
fast grower 速生植物
fast-growing plant 速生植物
fast-growing species of trees 速生树种
fast-growing trees 速生林
fast-growing tree species 速生树种
fast-hardening cement 快硬水泥
fast-hardening concrete 快硬混凝土;快凝混凝土
fast head 固定式车床头
fast headstock 固定式前顶针座
fast helium analysis 快速氦分析仪
fast-high-temperature cure 快速高温硫化;快速高温固化
fast hitch mechanism 快速连接装置
fast-hitch plow 快速挂结犁
fast ice 固定冰;紧密冰;坚固冰;贴岸冰
fast ice boundary 固定冰边界
fast ice edge 固定冰边缘
fast idle 高速空转;快速空转
fast idle cam 高速空行程用凸轮
Fastie-Ebert monochromator 法斯梯—艾伯特单色仪
Fastie-Ebert spectrometer 法斯梯—艾伯特分光计
fastigium (门窗的)三角饰顶点;尖顶;高峰期(间);屋脊;顶尖(屋脊、人字山头、山墙顶端等)
fastigium of house 房顶尖
fast image intensifier 快像增强器
fast-interface-state capture 快速界面态俘获
fast-interface-state loss 快速界面态损失
fast intraurban transit link 城市内部快速交通联系
fast ion conductor 快离子导体
fast joint butt 固轴铰链
fast land 高潮面以上陆地
fast lane 快车道;内侧车道
fast laser pulse 激光短脉冲
fast lens 快镜;强光透镜
fast light 耐光
fast line-end 快绳端
fast line stresses 快绳拉力
fast load 快速装入
fast lock 快同步
fast mass storage 快速大容量存储器
fast mill 高速轧机
fast mixing 快混合
fast molten-steel temperature ga(u)ge 钢液快速测温仪
fast motion 快速移动
fast motion gear 速动齿轮
fast-moving 快速移动的
fast-moving depression 快速移动的低气压
fast-movingtraffic 快速交通
fast-moving vehicle lane 高速车道;快车道
fast multiplier 高速乘法器;快速乘法器
fastness 坚牢(性);坚固度;急速
fastness against damage 抗损坏的坚牢度
fastness degree 坚牢程度
fastness property 牢固性
fastness rate 耐光度

fastness test 坚牢度试验
fastness to abrasion 耐磨损牢度
fastness to acid spotting 耐淡碱渍牢度
fastness to air pollution 耐空气污染牢度
fastness to alcohol 耐醇性
fastness to alkali 耐碱性;耐碱度
fastness to alkali spotting 耐淡碱渍牢度
fastness to anti-crease processing 耐皱加工牢度
fastness to artificial light 耐人造光牢度
fastness to atmospheric gases 耐大气牢度
fastness to bleaching 耐漂白牢度
fastness to bleeding 耐渗漏度;耐泌水度
fastness to blooming 耐霜性;防起雾性
fastness to boiling 耐煮度
fastness to brushing 耐刷洗牢度
fastness to carbonizing 耐炭化牢度
fastness to cement 耐水泥胶结度
fastness to chemical washing 耐化学洗涤牢度
fastness to chlorinated water 耐氯水牢度
fastness to chlorination 耐氯化牢度
fastness to chlorine 耐氯牢度
fastness to chlorine-bleaching 耐氯漂牢度
fastness to cold 耐冷度
fastness to crocking 耐摩擦脱色牢度
fastness to crossdy(e)ing 耐交染牢度
fastness to daylight 耐天然光牢度
fastness to decatizing 耐蒸度
fastness to degumming 耐脱胶牢度
fastness to dryheat 耐干热牢度
fastness to dust 耐尘度
fastness to gas fumes 耐烟气牢度
fastness to heat 耐热度;耐热牢度
fastness to hot pressing 耐热压烫牢度
fastness to hot water 耐热水牢度
fastness to light 不褪色性;耐日光牢度;耐光度;不透光性
fastness to lime 耐石灰性
fastness to mercerizing 耐丝光牢度
fastness to nitrogen oxides 耐氧化氮牢度
fastness to oil 耐油性
fastness to organic solvent 耐有机溶剂牢度
fastness to overpainting 底面漆配套稳定性
fastness to overvanishing 耐再涂性
fastness to perspiration 耐汗度
fastness to planking 耐酸缩绒牢度
fastness to rain 耐雨淋牢度
fastness to recoating 底面漆配套稳定性
fastness to rubbing 耐磨度;耐摩擦牢度
fastness to sea water 耐海水牢度
fastness to soaping 耐皂洗牢度
fastness to soda boiling 耐碱煮牢度
fastness to soil burial 耐土埋牢度
fastness to soluble glass 耐水玻璃性
fastness to spirit 耐酒精性
fastness to steaming 耐汽蒸牢度
fastness to stoving 耐二氧化硫气体牢度
fastness to sublimation 耐升华牢度
fastness to swelling 耐膨胀性
fastness to trubenizing 耐胶合牢度
fastness to washing 耐洗性;耐洗牢度
fastness to water 水浸牢度
fastness to water glass 耐水玻璃性
fastness to water spotting 耐水渍牢度
fastness to wear 耐水浸牢度;耐穿着牢度
fastness to weather 耐气候性
fastness to weathering 耐气候牢度;耐候性
fast neutron 快中子
fast neutron activation analysis 快速中子活化分析
fast neutron reactor 快中子反应堆
fast neutron source 快速中子源
fast neutron spectrometry 快中子测谱学
fastopake inks 打蜡表面上印刷用油墨
fast-operating relay 快动作继电器
fast operation 快速运行;快速操作;快动作
fast operation on station 车站快速作业
fast overcurrent protection 快速过流保护(装置)
fast packet 快速分组
fast paint 速干油漆
fast parallel arithmetic 快速平行运算器;快速并行运算器
fast parallel arithmetic unit 快速平行运算装置
fast particle 快粒子
fast pass 最终传送
fast passenger train 旅客快车

fast path dependent region 快速路径相关区
fast path exclusive transaction 快速路径专有事项处理
fast path feature 快速路径性能
fast path potential transaction 快速路径潜在事项处理
fast patrol 快艇
fast period 快周期
fast pin butt 固定铰链
fast pin butt hinge 固杆铰链
fast pin hinge 固定销铰(链);紧(固)销铰链
fast plutonium reactor 快中子钚反应堆
fast powder 高速炸药;速爆炸药
fast pull down system 快速关闭方式
fast pulley 主动皮带轮;固定皮带轮;固定(滑)轮;紧轮;定滑轮;定滑车
fast pumping installation 高速抽水装置
Fastraverse press 快速双效水压机
fast rays 快光
fast reaction 快速反应
fast reactor 高速中子炉;快速反应堆
fast red 坚牢红
fast red A 黄光酸性红
fast red pigment 坚牢红颜料
fast refractive infrared optic(al) system 快速折射红外光学系统
fast register 快速寄存器
fast registration 计数偏快
fast relay 高速继电器;快速继电器
fast repair method 快速修理法
fast repetitive analog(ue) computer 快速重复模拟计算机
fast response 快速响应;反应快;快响应
fast-response infrared device 快速响应红外装置
fast-response instrument 灵敏仪器
fast-response photomultiplier 快速光电倍增器
fast-response probe 灵敏传感器
fast-response recorder 快记录器
fast response sprinkler 快速反应洒水喷头
fast-response time 快速反应时间
fast-response transducer 小惯性传感器;快速传感器
fast return 快速返回
fast return control 快速返回控制(器)
fast reverse 快速换向开关
fast rewind 快速倒带
fast rewind control 快速反绕控制器;快速反绕控制
fast-rising level 猛涨水位
fast road towing 快速道路牵引车
fast runner 快行车(在编组场)
fast running car 快行车(在编组场)
fast running mill 高速磨机
fast scan 快速扫描
fast screen 短余辉荧光屏
fast screen tube 短余辉电子管
fast select 快速选择
fast-selling products 热门货(品)
fast-selling stock report 快货销售库存报告
fast-setting 快凝;快凝的;(混凝土的)急凝;快干性
fast-setting cement 快凝水泥
fast-setting concrete 快凝混凝土
fast-setting glass 速凝玻璃
fast-setting ink 快干油墨
fast-setting patching compound 快凝胶合剂
fast-setting sheet 快凝片材
fast-setting to cement 水泥快凝的
fast-setting to lime 石灰快凝的
fast shaded graphics 快速浓淡图
fast sheave (天车的)快绳滑轮
fast sheet 无框固定玻璃;固定(玻璃)窗
fast shot 瞎炮
fast shutter 快速快门
fast signal 快信号
fast sintering 快速烧成
fast sky blue 坚固天蓝
fast solvent 快速挥发溶剂
fast speed 快速;高速的
fast-speed tramway 快速有轨电车线路
fast-spiral drill 高速螺旋钻
fast-spreading 迅速传播的
fast steamship 快速汽船
fast sweep 快速扫描
fast switch cabinet with direct current infeeder 直流进线快速开关柜
fast switching arrangement 快速开关装置

fast switching channel 快切换通道
fast switch over 快速转接
fast synchronisation 快速闪光同步
fast telescope 强光力望远镜
fast teletype 快速电传打字机
fast tender 高速交通艇
fast thermal coupled reactor 快热耦合反应堆
fast thermocouple 小惯性热电偶；快速热电偶
fast time 短时的
fast time constant 快时间常数；短时间常数
fast time constant circuit 短时间常数电路
fast time control 快时间控制
fast time scale 快速时间比例；快速时标；快时标
fast time simulation model 快速模拟模型
fast timing 快定时
fast to light 不变色的
fast-to-light paint 受光不褪色的漆
fast towing chassis 快速牵引车底盘
fast track construction 快速施工法；边设计边施工（法）
fast track(ing) 边设计边施工（法）
fast track procedure 快速行动程序
fast traffic 快速交通
fast traffic lane 快行道
fast traffic path 快行道
fast train 快车
fast travel 快速行程
fast-travel crane 快速移动起重机；快速移动吊车
fast-travel excavator 快速移动挖土机
fast-travel hydraulic crane 快速移动液压起重机；快速移动液压吊车
fast-travel hydraulic loader 快速移动液压装载机
fast traverse 快速横动
fast traverse forwards 快速接近
fast trawler 快速拖网渔船
fast turnaround 快速转换；快速周转；快速运转；快速交接；快速换向
fast valve 高速阀；快关阀
fast valving 快动
fast vehicle 快速车辆
fast vehicular traffic 快速交通
fast-vibration direction 快速振动方向
fast way 快行道
fast window 稳固窗；固定窗
fast-working development 快速显影液
fast yellow 坚牢黄
fast zoom mode 快速缩放方式
fat 黄油
fatal 致死的
fatal accident 严重事故；人身事故；死亡事故；失事；伤亡事故
fatal-accident rate 死亡事故率
fatal concentration 致死浓度
fatal consistency of killing half of the tested animals 半数致死浓度
fatal construction mishap 施工死亡事故
fatal crash 人身事故
fatal dose 致死剂量
fatal dose of killing half of the tested animals 半数致死量
fatal dryness 致死干（燥）度
fatal error 致命性错误
fatal error message 严重错误信息
fatal high temperature 最高致死温度；致死高温
fatal humidity 致死湿度
fatalities 死亡人数
fatality 灾难；死亡事故
fatality rate 行车事故率；致死（速）率；病死率
fatal low temperature 最低致死温度；致死低温
fatal rate 致死率；死亡率
fatal road accident 道路死亡事故
fatal temperature 致死温度
Fata Morgana 复杂蜃景
fat and oil 油和脂
fat area 厚斑
fat asphalt 肥沥青
fat asphalt mixture 多（地）沥青混合物；多（地）沥青混合料；富沥青混合料
fat beams and columns 肥梁胖柱
fat board 托灰板；砂浆板（瓦工用）；灰浆（托）板
fat box 油脂匣
fat brown 脂溶棕
fat clay 油性黏土；高塑性黏土；可塑性黏土；富黏土；肥黏土

fat clay of high plasticity 高塑性肥黏土
fat coal 烟煤；沥青质煤；肥煤
fat collector 撇油池；储油池
fat collector separator 油脂分离器
fat colo(u)r 油溶染料
fat concrete 水泥用量多的混凝土；多灰混凝土；富混凝土；肥混凝土
fat containing waste 含脂肪废料
fat content 脂肪含量；含脂量
fat dye 油脂染料
fat edge 堆集边缘（涂料）；底边漆梗；淤边；厚边
fat-extracted 脱脂的
fat farm 减肥中心
fat-free 无脂的
father 创造者
father of node 节点上层
fathogram 回声测深计；水深图；测深图
fathogram record 测深记录簿；测深记录
fathom 英寻；拓；材积名
fathomable 深度可测的
fathom curve 等深线（以英制单位寻计）；等深曲线；等海深线
fathomed accuracy 测深精度
fathometer 回响测深仪；回声测深仪；回波测深仪；探测仪（器）；水深仪；水深计；超声波测深器；测深仪；测深计
fathometer sounding 测深仪测深
fathom line 等深线（以英制单位寻计）
fathom scale 测深比例尺；标深尺
fatigue 疲劳度；疲劳
fatigue action 疲劳作用
fatigue allowance 耐劳津贴；疲劳限度；疲劳补偿
fatigue analysis 疲劳分析
fatigue analysis of attachment 连接的疲劳分析
fatigue area 疲劳破坏区域
fatigue at high temperature 高温疲劳
fatigue auditory 听觉疲劳
fatigue behavio(u)r 持久特性；疲劳特性
fatigue bending failure 弯曲疲劳破坏
fatigue bending test 弯曲疲劳试验
fatigue bend test 疲劳弯曲试验
fatigue bond strength 疲劳粘结强度
fatigue break 疲劳裂口；疲劳断裂
fatigue breakdown 疲劳破坏
fatigue capability 疲劳能力
fatigue characteristic 疲劳特性
fatigue coefficient 耐久系数；疲劳系数
fatigue corrosion 疲劳腐蚀
fatigue crack 疲乏裂缝；疲劳裂纹；疲劳断裂
fatigue crack growth 疲劳裂纹增长
fatigue crack(ing) 疲劳裂缝；疲劳开裂；疲劳龟裂
fatigue crack propagation 疲劳裂纹扩展
fatigue crack propagation test 疲劳裂纹扩展试验
fatigue crack tip 疲劳裂纹尖端
fatigue curve 疲劳曲线
fatigue damage 疲劳损失；疲劳损坏
fatigue damage accumulation 疲劳损失累加
fatigue damage indicator 疲劳损伤指示器
fatigue data 疲劳（试验）数据
fatigue decreased proficiency boundary 疲劳效率减退界限
fatigue deformation 疲劳变形
fatigue degree 疲劳程度
fatigue design 疲劳设计
fatigue dress 工作服
fatigue durability 耐疲劳性
fatigue effect 疲劳效应
fatigue endurance 疲劳寿命
fatigue endurance limit 耐疲劳极限；疲劳强度极限；疲劳持久极限
fatigue endurance test 疲劳耐久试验
fatigue factor 疲劳因素；疲劳系数
fatigue failure 疲劳失效；疲劳破坏；疲劳毁损（金属）；疲劳故障；疲劳断裂
fatigue flake 疲劳剥落
fatigue fracture 疲乏裂缝；疲劳破裂（材料）；裂缝；疲劳断裂
fatigue fracture mechanism 疲劳断裂机理
fatigue fragmentation of rock 岩石疲劳破碎
fatigue hardening 疲劳硬化
fatigue impact strength 疲劳冲击强度
fatigue in bending 弯曲疲劳
fatigue in compression 压缩疲劳

fatigue indicator 疲劳度指示器
fatigue in flexure 挠曲疲劳
fatigue initiation 疲劳起始
fatigue in tension 拉伸疲劳；张拉疲劳
fatigue life 疲劳期；疲劳负荷寿命
fatigue life capability 耐疲劳能力
fatigue life ga(u)ge 疲劳寿命片
fatigue life time 疲劳寿命
fatigue limit 疲劳限度；疲劳强度；疲劳耐限；疲劳界限；疲劳极限；持久极限
fatigue limit for torsion 扭转疲劳极限
fatigue limit of material 材料疲劳极限
fatigue limit test 耐久极限试验
fatigue load(ing) 交变荷载；交变负载；疲劳荷载；疲劳加载
fatigue loss 疲劳损耗
fatigue measurement 疲劳测定
fatigue mechanism 疲劳机理
fatigue metal 金属疲劳
fatigue meter 疲劳强度计；疲劳度测定仪
fatigue model of soil cyclic(al) mobility 土（壤）循环流动疲劳模型
fatigue nucleus 疲劳核
fatigue of bond 黏结疲劳（混凝土）
fatigue of concrete 混凝土疲劳
fatigue of materials 材料疲劳
fatigue of metal 金属疲劳
fatigue of soil 土壤疲乏
fatigue of subgrade 路基的疲劳
fatigue of wood 木材抗疲劳性
fatigue phenomenon 疲劳现象
fatigue pitting 疲劳剥蚀
fatigue point 疲劳界限；疲劳极限点；疲劳极限；疲劳点
fatigue prediction 疲劳预测（值）
fatigue product 疲劳产物
fatigue-proof rubber 耐劳橡胶
fatigue property 疲劳性能
fatigue range 疲劳限度；疲劳范围
fatigue ratio 疲劳系数；疲劳强度比；疲劳率；疲劳（应力）比
fatigue reduction factor 疲劳折减系数；疲劳折减因子
fatigue reliability 疲劳可靠度
fatigue residual life of bridge 桥梁疲劳剩余寿命
fatigue resistance 抗疲劳性；耐疲劳性；疲劳强度；疲劳抗力
fatigue rig 疲劳试验设备
fatigue rig of materials 材料疲劳试验设备
fatigue rupture 疲劳破坏；疲劳断裂
fatigue shear apparatus 疲劳剪切仪
fatigue specimen 疲劳试样
fatigue strength 疲劳强度
fatigue strength coefficient 疲劳强度系数
fatigue strength for finite life 有限寿命疲劳强度
fatigue strength reduction factor 应力集中因子；疲劳强度折减因子；疲劳强度缩小系数
fatigue strength under oscillation stresses 振动疲劳强度
fatigue strength under pulsating tensile stresses 脉动张应力疲劳强度
fatigue strength under reversed bending stress 反复弯曲应力疲劳强度
fatigue stress 交变应力；疲劳应力
fatigue stress concentration factor 疲劳应力集中系数
fatigue stress of steel members 钢材疲劳应力
fatigue stress ratio 疲劳应力比（值）
fatigue tension test 张力疲劳试验；拉伸疲劳试验；拉力疲劳试验；抗拉疲劳试验
fatigue tester 疲劳试验机
fatigue test(ing) 耐久试验；疲劳试验；疲劳检查
fatigue testing at low-temperature 低温疲劳试验
fatigue testing machine 疲劳试验机
fatigue testing rig 疲劳试验设备
fatigue testing specimen 疲劳试样
fatigue testing types of stress 疲劳试验应力类型
fatigue testing under actual service condition 真实工作条件下疲劳试验
fatigue testing under axial loading 轴向荷载疲劳试验
fatigue testing with several load steps 多级加载疲劳试验
fatigue test of materials 材料疲劳试验

fatigue test specimen 疲劳试样
fatigue threshold 疲劳阈值
fatigue toxin 疲劳毒素
fatigue ultimate moment 疲劳极限力矩
fatigue under flexing 弯曲疲劳
fatigue uniform 工作服
fatigue value 疲劳极限
fatigue wear 疲劳磨损
fatigue testing at elevated temperature 高温疲劳试验
fat lime 浓石灰;肥石灰;纯质石灰;高钙石灰;富石灰
fat link 多链传输组;粗链路
fat liquor 油乳液;皮革上光油
fat liquoring 涂油
fat liquoring complex agent 复合加脂剂
fat-lub test 液体中含油量的测定
fat mix(ture) 富混合料;浓拌和;富拌和料;(水泥、石灰等用量多的)富灰混合比
fat mortar 浓砂浆;浓灰浆;黏性砂浆;强黏砂浆;强黏胶泥;富砂浆
fat oil 饱和油;饱和的油
fat orange 脂橙
Fatou-Lebesgue lemma 法都—勒贝格引理【数】
Fatou's lemma 法都引理【数】
fat paint 厚漆
fat pine 长叶松
fat pitch 富脂肪酸沥青
fat purse 富裕
fat red 脂溶红
fat sand 黏砂;含黏土型砂;肥砂
fat-sided ship 直舷船
fat soil 沃土;肥土
fat solubility 高脂溶性
fat-solubility of pesticide 农药脂溶度
fat solvent 油脂溶剂
fat spark 强火花
fat splitting 油脂分解
fat spot 油斑;沥青过多地点;沥青过多处
fat surface 多油面层
fatten(ing) 催肥;增稠;稠化;肥育
fattening up 熟化;发透石灰膏(石灰膏经水泡约三十天的)
fatting up 泛油(路面泛出多余沥青)
fat turpentine 稠化松节油
fatty acid 油脂酸;脂肪酸
fatty acid amide 脂肪酸酰胺
fatty acid anhydride 脂肪酸酐
fatty acid dehydrogenase 脂肪酸脱氢酶
fatty acid inositol ester 脂肪酸环己六醇酯
fatty acid lake 脂肪酸色淀
fatty acid mannitol ester 脂肪酸甘露糖醇酯
fatty acid of linseed oil 亚麻仁油脂肪酸
fatty acid pentaerythritol ester 脂肪酸季戊四醇酯
fatty acid pitch 脂肪酸沥青
fatty acid polyvinyl alcohol ester 脂肪酸聚乙烯醇酯
fatty acid process 脂肪酸法
fatty acid sorbitol ester 脂肪酸山梨醇酯
fatty acid vinylester 脂肪酸乙烯酯
fatty alcohol 脂肪族醇
fatty compound 脂肪族化合物
fatty cutting oil 脂肪切削油;切削用乳液
fatty edge 溜边
fatty group 脂肪族基;脂肪族
fatty oil 脂肪油;油脂
fatty oil processing factory 油脂加工厂
fatty paint 变稠的漆
fatty pitch 软柏油脂
fatty surface 脂状面
fat wood 松木;轻木;多脂松
fat yellow 脂溶黄
faubourg 市郊;郊区
faucal 壳口
Faucault mercury interrupter 弗科水银断续器
fauces 咽口;通道(罗马建筑中的一种)
faucet 旋塞;龙头;开关;水龙头承插口;水龙头;承(插)口(管子)
faucet aeration 龙头曝气
faucet aerator 龙头加气器
faucet-attached unit 接口龙头装置
faucet control 由水龙头控制;龙头控制
faucet ear 管子吊环
faucet end 龙头管端;龙头端管
faucet joint 套筒接合;龙头接嘴;龙头接头;套接接头;水龙头接头
faucet-mounted filter 有龙头过滤器
faucet pipe 套接管子;缩节(即套接管);承接管;水龙头接管
faucet seat 龙头座
faucet seat dresser 龙头座修整器
Faugeron kiln 福杰隆窑
faujasite 八面沸石
Fauld's tool (一种可以换刃片的)弗尔德圬工凿子
faulf-free 无故障
faul notch 断层鞍部
faulschlamm 骸泥
fault 过失;缺陷;堆垛层错;断裂【地】;断层;错误;层错;分枝断层
fault action 断层作用
fault alarm signal 故障报警信号
fault analysis 故障分析;事故分析
fault-angle basin 断层角盆地
fault anticline structure zone 断裂背斜带
fault apron 断层群;断层冲积扇
fault-avoidance 故障免除;冲故障;免错;避免错误
fault barrier 断层遮挡
fault basin 断裂盆地;断层盆地
fault basin lake 断层盆地湖
fault belt 断层带
fault bench 断层阶地
fault bit 故障位
fault block 断裂地块;断块
fault-block basin 断块盆地
fault-block buried hill zone 断块潜山带
fault-block landform 断块地貌
fault-block mountain 断块山
fault-block movement 断块运动
fault-block oil-gas field 断块油气田
fault-block pool 断块油藏
fault-block region 断块区
fault-block structure 断块构造
fault-block tectonics theory 断块构造学说
fault-block type 断块型
fault body 断层体
fault boundary 断层边界
fault break 断层裂缝
fault breakout 断层崩落
fault breccia 断层角砾岩;断层角砾岩
fault breccia zone 断层角砾岩带
fault bundle 断层束;断层群
fault clay 断层黏土;断层泥
fault clearance 故障清除
fault clearing 排除故障
fault clearing time 故障清除时间
fault clerk's desk 障碍控制台
fault cliff 断层崖
fault closure 断层闭合
fault coast 断层海岸;断层岸
fault collapsing 故障收缩
fault concrete 劣质混凝土
fault condition 故障条件;故障状态
fault connection 错误连接
fault control 故障控制;事故控制;事故监督
fault-controlled memory 故障控制存储器
fault controlled stream 断层控制的河流
fault-controlled system 断裂控制系统
fault correct 差错纠正
fault correction time 故障修复时间
fault correlation 断层对比
fault coverage 故障覆盖率
fault creep 断层蠕动
fault crevice 断隙;断层裂隙
fault crossing design 跨越断层设计
fault crush zone 断层破碎带
fault current 接地电流;故障电流;漏电;事故电流
fault current circuit breaker 故障电流断路器
fault dam 断层坝
fault dam spring 断层阻截泉;断层堰塞泉;断层壅水泉;断层堤泉
fault debris 断层岩屑
fault deposit 断层矿床
fault depression 断陷
fault depth 断层深度
fault detect(ion) 故障探测;故障检查;缺陷检验;故障检测
fault detection system 故障检测系统
fault detector 故障探测器;毁损指示器;探伤器;断层探测器;探伤仪
fault diagnosis 故障诊断
fault diagnostic system 故障诊断系统
fault dictionary 故障字典;故障辞典
fault dimension 断层大小
fault dip 断斜;断层倾斜;断层倾向;断层倾角
fault dipping against the beds 与地层倾斜不一致的断层
fault displacement 断距;断层移距;断层位移
fault displacement measurement 断层位移测量
fault distinguish 故障区分
fault dominance 故障支配
fault-downwarping belt 断陷带
fault-downwarping region 断陷区
fault drag 损坏的路刮;损坏的刮路机;失效的路刮;失效的刮路机;断层牵连
fault-drag mode 断层拖曳模式
fault drop 断层落距
fault earthquake 断层地震
faulted bedding plane 错断层面;错动层面
faulted contact 断层接触;侵入接触
faulted island 层错岛
faulted joint 断层缝;断裂缝;错缝;(路面的)错断接缝
faulted mountain 断块山;断层山岭;断层山(地)
faulted outer edge bar 断层外缘坝
faulted overfold 断裂倒转褶皱
faulted seam 断裂矿层
faulted segment 层错断片
faulted slabs 错台【道】
faulted strata 断层切割地层
faulted structure 断裂结构
faulted zone 断裂带;断层带
fault effect 断层作用
fault electrode current 故障电极电流
fault equivalence 故障等效
fault escarpment 断层崖
fault facet 断层三角面
fault finder 障碍检查器;故障指示器;故障寻迹器;故障探测器;故障查找器;裂缝观察器;探伤器;探伤仪;故障检查设备
fault finding 故障寻找;故障探测;检修;查找故障
fault fissure 断层裂纹;断层裂隙;断层裂缝
fault flag 故障标记
fault fold 断褶;断层褶皱
fault-fold belt 断裂褶皱带
fault-fold buried hill zone 断褶潜山带
fault-folded belt 断褶带
fault-folded region 断褶区
fault-folding 断褶活动;断层褶皱
fault-folding activity 断褶褶皱活动
fault-folding movement 断褶运动
fault-folding structure 断褶构造
fault free 无故障的
fault frequency 故障频率
fault gap 断层峡谷;断层隘口
fault ga(u)ge 断层仪
fault gouge 断层泥
fault gouge zone 断层泥带
fault ground bus 故障接地母线
fault group 断层群
fault hade 断层余角
fault holding 障碍保持
fault image 假图像;失真影像;失真图像
fault indicating lamp 故障指示灯
fault indicator 故障指示器;故障探测器;损伤指示器
fault in enlargement 放大误差
fault infeed 带故障馈电
fault information 故障信息
faulting 断裂作用;断裂(运动);断层作用;断层运动;断层活动;断层错动
faulting activity 断层活动性
faulting integral 卷积积分
faulting length 断层作用长度
faulting of slab ends 板端错台;错台【道】;路面板端相对垂直错动;路面板边的垂直错位
faulting of transverse joints 横向接缝错位
faulting rupture crack 断层破裂缝
fault in management or navigation 管理或驾驶的过失;在管理或导航的过失
fault in material 材料缺陷
fault in seam 接缝缺陷
fault-intolerance approach 不容错方法
fault isolation mode 错误隔离方式
fault landform 断层地貌

fault-latency period 故障潜伏期
fault ledge 断层崖
fault length 断层长度
faultless 无故障的
faultless gear change 无冲击变速
faultless responsibility 无过失责任制
fault liability 过失赔偿责任
fault line 裂纹线;断层线
fault lineament 断层线性体【地】
fault-line scarp 断层线崖;断线崖;断线岸
fault-line scarp shoreline 断线崖滨线
fault-line shoreline 断线谷滨线
fault-line valley 断层线谷;断线谷
fault localization 查找故障
fault localizer 障碍位置测定器;故障点探测器
fault locating 故障定位
fault locating instrument 故障定位仪
fault locating program(me) 故障定位程序
fault locating technology 故障定位技术
fault locating test 故障定位测试
fault location 障碍勘测;故障位置;故障定位;故障部位测定;断层位置
fault location determination 故障位置测定
fault location monitor 故障位置监测器
fault location problem 故障定位问题;故障查找问题
fault location program(me) 故障定位程序
fault location test 故障定位测试;错误定位测试
fault location test method 故障定位测试法
fault location time 故障探测时间;故障部位寻找时间
fault locator 故障位置测定器;故障探测器;故障定位器
fault map 断层图
fault masking 故障屏蔽
fault matrix 故障矩阵
fault mechanics 断层力学
fault mechanism 断层机制
fault misclosure 断层闭合差
fault model 故障模型
fault monitoring system 故障监视系统
fault motion 断层运动
fault movement 断裂运动;断层运动;断层错动
fault nature 断层性质
fault number 断层编号
fault of extrados 外弧断层
fault of intrados 内弧断层
fault of slab ends (路面的)板端相对垂直错动
fault orientation 断层方位
fault oscillograph 故障示波器
fault outcrop 断层露头
fault parameter 断层参数
fault pattern 断层型式;断层类型
fault permeability 断层渗透率
fault pit 断层坑
fault plane 断层面
fault plate 带疵板
fault plateau 断层高原
fault point 故障点
fault polish 断层磨光面;断层镜面
fault pool 断层油藏
fault power 故障情况下的功率;电源故障
fault prediction device 故障预测装置
fault prevention 故障预防;防止事故
fault prevention analysis 故障预防分析
fault probability 故障概率
fault propagation 断层扩展
fault radius 断层半径
fault range 断层规模
fault rate 故障率
fault rate threshold 故障率阈值;故障率门限
fault recognition 故障识别
fault-recognition program(me) 故障识别程序
fault recorder 故障记录器
fault recording 障碍记录;故障记录
fault redundancy 故障冗长度
fault region 断层区
fault-related cleavage 邻断裂劈理
fault-related flour 断层岩粉
fault relay 故障继电器;事故继电器
fault report 故障报告
fault resistance 故障地点电阻
fault ribbon 层错带
fault ridge 断层脊
fault rift 断层裂谷
fault rock 断层岩

fault rupture 断层破裂;断层断裂
fault rupture length 断裂长度;断层破裂长度
fault saddle 断层鞍状构造;断层鞍部
fault safe 故障—安全
fault scale 断层规模
fault scarp 断层悬崖;断层崖
fault-scarp shoreline 断崖滨线
fault secure 故障保险
fault secure circuit 故障保护电路;故障安全电路
fault sensing circuit 故障敏感电路
fault sensing system 故障敏感系统
fault sensor 故障传感器
fault set 断层组
fault shoreline 断层岸线
fault signal(l)ing 事故信号;故障信号
fault signature 故障表征
fault simulation 故障模拟
fault size 断层规模;断层大小
fault slickenside 断层擦痕面
fault slip(page) 断层滑动
fault space 断层间隔
fault splinter 连接两断层末端斜坡
fault spot 故障点
fault spring 断层泉
fault state 故障状态
fault step zone 断阶带
fault striae 断层擦痕
fault striation 滑动镜面;断层擦痕
fault strike 断层走向
fault-strike slip basin 走向滑移盆地;走向滑动盆地
fault structure 断裂构造
fault subsidence basin 断陷盆地
fault surface 断层面
fault system 断层系
fault terrace 断层阶地
fault test generation 故障测试生成
fault testing 故障测试
fault threshold 故障门限
fault throw 断距;断错;断层位移;断层落差;断距
fault throwing 人工短路跳闸
fault time 故障(停机维修)时间
fault-tolerance 容错
fault tolerant 失效容限
fault tolerant approach 容错法
fault tolerant computer 容错计算机
fault tolerant computing 容错计算
fault tolerant system 容错系统
fault tolerant technique 容错技术
fault toleration 容错
fault trace 故障追踪;故障追索;故障跟踪;裂纹迹;断层线;断层迹
fault tracing 故障追踪;故障追索
fault train 故障列车
fault trap 断层圈闭
fault treatment 故障处理
fault tree 故障树;事故树
fault tree analysis 故障树分析(法);事故树分析;失效树分析
fault-trellised drainage 不规则格状水系
fault trench 断层沟
fault trend 断层方向
fault-trough 堑沟;断层裂谷;断(层)槽;地堑
fault-trough coast 地堑海岸
fault trough lake(sag) 断陷湖;断层槽湖;地堑湖
fault trough submarine valley 断槽海底谷
fault type 断层类型
fault-unconformity combination trap 断层不整合复合圈闭【地】
fault-up-dip edge out combination trap 断层上倾尖灭复合圈闭【地】
fault-upwarping belt 断隆带
fault-upwarping region 断隆区
fault valley 断裂谷;断层谷
fault-valley-side trap 断谷侧圈闭
fault vector 层错矢量
fault vein 断层脉;断层矿脉
fault wall 断盘;断层壁;断壁
fault warning 故障警报
fault warning receiving station 故障警报接收站
fault warning routing equipment 故障警报发送设备
fault water 裂隙地下水;断层水;断层带地下水
faulty act 错误行为

faulty burning 生烧或过烧
faulty casting 废铸件
faulty circuit 有故障(的)电路;故障电路
faulty coal 劣质煤
faulty colo(u)r vision 错误色观察
faulty component 故障部分
faulty concrete 不合格混凝土;劣质混凝土
faulty exposure 不正确曝光
faulty fusion 未熔合;未焊透
faulty goods 劣货
faulty indicator 探伤器
faulty insulator 故障绝缘子;不合格绝缘子
faulty item 报废项目
faulty line 故障线(路)
faulty lubrication 润滑不当;不合规定的润滑
faulty machining 机加工不当
faulty operation 错误操作
faulty operation and management 经营管理不善
faulty packing 包装不完备
faulty packing of cargo 货物包装不良
faulty parts 报废零件
faulty prosthesis 不良假体
faulty restoration 不良修复
faulty separation 错误的区分;不正确的分类
faulty storage 储[贮]藏不良;保管不良
faulty technology 探伤技术
faulty vision 错误视觉【测】
fault zone 断层区;断层地带
fault zone valley 断层地带山谷;断层地带流域
fauna 区系;动物志(某地区或时期的);动物群;动物区系
fauna and flora 动植物
faunal extinction 动物区系消亡
faunal province 动物群区;动物区系
faunal region 动物区系
faunal stage 动物群阶
faunal zone 动物区系带;动物分布带
faunal zoogeography 区系动物地理学
faunizone 动物群岩层带
faunule 小动物群
faure-type plate 涂浆型极板
Fauser process 佛瑟法
faustite 锌绿松石
Faust jig 福斯特型跳汰机
Faust's test 福斯特试验
fauteuil 扶手椅
fauton (埋于混凝土中的)金属杆
faux hois 仿纹涂装
faux jour 隔墙窗;隔断窗
favas 六角形构件;蜂窝式
favela 木屋棚户区(巴西);贫民窟(葡萄牙)
favi 六角砖;六角瓦
favo(u)r 有利的;优惠;赞成
favo(u)rable array location 有利台阵场地
favo(u)rable atmosphere condition 有利大气条件
favo(u)rable balance 顺差
favo(u)rable balance of foreign exchange and outlays 外汇收支顺差
favo(u)rable balance of trade 贸易顺差;出超;贸易出超
favo(u)rable beds 有利岩层
favo(u)rable case 有利情形
favo(u)rable condition 有利条件
favo(u)rable condition for car rolling 溜车有利条件
favo(u)rable condition 优惠条件
favo(u)rable current 顺流;顺气流
favo(u)rable economic condition 经济优势
favo(u)rable event 有利事件
favo(u)rable exchange 有利汇兑;顺汇
favo(u)rable export terms 具有优越性的出口产品
favo(u)rable geometry 有利几何条件
favo(u)rable grade 有利的坡度
favo(u)rable interference 有效干扰
favo(u)rable location 有利地点
favo(u)rable numerical interval of ore-control factors and indexes 控矿因素和标志的有利数值区间
favo(u)rable observation condition 有利观测条件
favo(u)rable opposition 大冲
favo(u)rable phase 有利相位
favo(u)rable position 有利地位
favo(u)rable price 优惠价(格)
favo(u)rable result 有利结果

favo(u)rable spiral 良性循环
favo(u)rable stream crossing 有利跨河地点
favo(u)rable target 有利目标
favo(u)rable terms 优惠条款;优惠条件
favo(u)rable trade balance 贸易顺差
favo(u)rable trade payment 贸易顺差
favo(u)rable variation 有利变异
favo(u)rable weather 有利天气
favo(u)rable wind 顺风
favo(u)red nation 优惠国
favo(u)red state 有利状态
favo(u)red transition 有利跃迁
favus 蜂窝型;[复]favi
fawcettine 佛石松碱
Fawcett's floor 陶土槽形板楼盖
Fawcett's lintel 空心耐火砖过梁
fawn brown 淡褐色;淡黄褐色;小鹿毛色
fawn foot 鹿脚形斧柄
fax 摹写;传真
faxcasting 电视传真广播;电视广播
Faxen drag factor 曳力系数
fax message 传真电文
fax service 摹写传输业务
fay 紧镶;紧密连接;密配合
fayalite (硅酸铁形成的)砖面黑斑;正硅酸铁;镁橄榄石;铁橄榄石
Fay calliper 外径内径组合卡钳
Faye anomaly 法耶异常
Faye correction 法耶校正;法耶改正
fayence tile 釉陶面砖;釉陶锦砖
faying face 钎焊面
faying flange 接合边
faying surface 接触面;吻合面;贴合面;搭接面;重叠面;连接面;接合面
fazenda 大农场(巴西)
F-band absorption F 波段吸收
Fce clause 摆平投标
F distribution F 分布
fear of unemployment 失业恐慌
fease 解散绳股
feasibility 现实性;可行性;可能性
feasibility analysis 可行性分析
feasibility and reconnaissance survey 可行性研究和选点调查
feasibility assessment 可行性评价;可行性的估计
feasibility assessment of building 投建可行性评议
feasibility assessment of mine building 投建可行性评议【岩】
feasibility assessment of ore exploitation 矿产开发可行性评价
feasibility criterion 可行性标准
feasibility design 可行性设计
feasibility evaluation 可行性评估
feasibility exploration 可行性勘探
feasibility group 可行性研究小组
feasibility investigation 可行性(调查)研究
feasibility of imputation 分配可行性;分配的可行性
feasibility of linear programming problem 线性规划问题的可行解区域
feasibility of momentum grades 动能坡度的可行性
feasibility report 可行性报告
feasibility stage 实施阶段
feasibility study 可行性(调查)研究;可能性研究;技术经济论证
feasibility study of port project 港口工程的可行性研究
feasibility study of railway construction 铁路建设项目可行性研究
feasibility study report 可行性研究报告
feasibility survey 可行性调研
feasibility survey stage 可行性研究阶段
feasibility test 可行性试验;可行性检验;可能性检验
feasibility testing 可行性测定
feasible arc 可行弧
feasible basic solution 可行基本解
feasible computability 可行可计算性
feasible constraint 可行约束
feasible correction 可行校正
feasible decent direction 可行下降方向
feasible decomposition 可行分解
feasible direction 可行方向
feasible domain 可能领域
feasible limit of increase in financial revenue 财政收入增长可行限量

feasible method 可行法
feasible normal plant capacity 可达到的正常工厂生产能力
feasible path 可行通道
feasible plan 可行的计划
feasible point 可行点
feasible program(me) 可行方案
feasible region 可行域;可行区(域)
feasible report 可行性报告
feasible schedule 可行的计划
feasible sequence 可行序列
feasible set 可行集(合)
feasible solution 可行解;容许解(法)
feasible vector 可行向量;可行矢量
feasible zone 可行域
feather 旋翼周期变距;羽纹;铸造披缝;铸造毛刺;改变车叶距;桨叶水平运动;潜望镜引起的微波;网格变形;凸出部分;塞缝片;风羽;变桨距;横舌榫
feather alum 毛矾石;铁明矾
Feather analysis 费瑟分析法
feather an oar 平掠回桨
feather bar 钢钎
feather bed 羽毛状池底
featherbedding (超过需要的)工人雇佣;限产超雇;强迫雇用;超过工作需要的人员雇佣;超额雇工
feather board 薄边板
feather brick 斜切楔形砖;翅砖
feather check(ing) 细裂纹;产生裂缝;发丝(状)裂缝;羽毛状裂缝
feather chenille 羽毛雪尼尔线
feather-cloth figure (木材的)羽毛花纹
feather cracking 发丝裂缝;发裂;细裂纹;发丝状裂缝
feather crotch 羽状纹板
feather crystal 羽毛状晶体
feather cutter 企口凿
feather duster 鸡毛尘掸
feather duster machine 羽毛清洁机
feathered 导键连接的
feather-edge 羽状薄边;斜割边;薄边;阴角抹子;楔形边(一边厚、一边薄);削边(尺)
feather-edged 羽毛状边饰;薄缘的;薄边的;刀刃状的
feather-edged board 薄边板;楔边板;楔形板;风雨板;斜削板
feather-edged boarding 薄边板;楔边板(安装)
feather-edged brick 斜削砖;楔形砖;弧形砖;削边砖
feather-edged coping 单坡屋顶;薄边式盖顶;薄边盖顶石;斜面墙帽
feather-edged cross-section 薄边(式)断面
feather-edged file 剑形锉;薄口锉;刀形锉
feather-edged section 薄边(式)断面
feather edging 斜削边;坡口切边;磨薄边缘;毛状饰;薄边;渐细等高线
feathered tin 羽状锡
feathered washboard 有羽毛状的;银色片状玻璃,磨砂的
feather end 一端楔形砖;宽端弯头楔形砖
feather end on edge 窄端头楔形砖
feather fault 羽状断层
feather figure 羽状花纹;鹑羽纹
feather fractures 羽状结构
feather grain 羽状纹理
featheriness 羽毛状
feathering 絮羽化;羽状物;羽状边缘;羽毛样子;泅色;叶瓣饰;桨叶水平旋转;渐细等高线;顺桨;粗糙边缘;出羽
feathering airscrew 顺桨螺旋桨
feathering auger 带偏心扒杆的螺旋
feathering blade runner 顺桨叶轮
feathering compound 整平合成物;找平合成物
feathering Francis runner 斜翼式法兰西斯转轮;斜翼式水轮机转轮
feathering hinge 顺桨铰链
feathering out 薄边式铺开;尖灭
feathering paddle 活桨叶
feathering paddle wheel 动叶明轮
feathering pitch 顺桨桨距
feathering propeller(runner) 顺桨螺旋桨
feathering propeller turbine 顺桨螺旋桨涡轮机;顺桨螺旋桨透平;螺旋桨式水轮机
feathering pump 螺旋桨式泵;顺桨泵
feathering step 常平踏板
feathering stroke 掠水面划船法
feathering vane 桨叶

feathering vane runner 转桨式转轮
feathering vane wheel 转桨式水轮机
feather joint 插榫结合;铰链接合;滑键接合;嵌榫拼接;嵌槽接合;榫连接;插楔接合;羽状节理
feather key 滑键;导向键
feather length 羽毛边长度
feather one's nest 私吞
feather ore 硫锑铁铅矿;羽毛矿
feather out 羽灭
feather piece 榫舌;暗销
feather pitch 顺桨螺距
feather rot 羽毛状腐病
feathers 羽状裂缝
feathers and downs 羽毛羽绒类货
feather sander 羽毛抛光机;羽毛打磨机
feathers and plug 石楔及垫片
feather shape 羽毛状
feather side 侧面楔形砖
feather texture 羽状影纹
feather tip 羽状端部;羽状接头;羽状触点;薄边板
feather tongue 横舌榫;销板;斜削销;导销
feather twill 山形斜纹
feather valve 卸载阀;滑阀;弹性阀;放气阀
feather way 滑键槽
featherweight paper 轻磅纸
featherweight precast concrete channel slab 预制轻质混凝土槽形板
feathery 羽毛状
feathery crystal 羽状晶体
feathery crystalline material 羽毛状结晶物质
feathery needles 羽毛针状体
feature 形迹;面貌;特色;地物,地图项
feature article 专文;紧急事项
feature attribute 特征码
feature based matching 基于特征影像匹配
feature code 地面要素编号
feature code list 特征码表
feature codes menu 特征码清单
feature control symbol 特征控制符号
featured edge 特形边
feature detection 特征监测
feature extraction 特征提取【测】
feature extraction program(me) 特征抽出程序
feature extractor 要素检出器
feature for attaching communication 加接通信的特性;附属通信性能
feature monitoring 特征监测
feature of boudinage 香肠构造特征【地】
feature of breakers 碎波特征;碎浪特征
feature of diapiric structure 底辟构造特征
feature of each movement 每次活动性质
feature of en echelon vein 雁列脉特征
feature of exposures 野外露头
feature of fault scarp 断层崖特征
feature of fold shape 褶曲形态特征
feature of fracture zone 断裂带特征
feature of gravitative gliding structure 重力滑动构造特征
feature of impact crater 撞击构造特征
feature of inclusion 包裹体镜下特征
feature of joint surface 节理面特征
feature of lineation 线理特征
feature of lithofacies variation 岩相变化特征
feature of military value 军事目标
feature of polygenetic compound ore deposit 多因复成矿床特征
feature of project 工程特征
feature of recent movement 断层现代活动特征
feature of regional geologic(al) survey 区域地质调查性质
feature of regional tectonics 区域构造特征
feature of rim syncline 周缘向斜特征
feature of river environmental system 河流环境系统特征
feature of rock type 岩性特征
feature of scarp surface 断层面特征
feature of secondary magnetic field 二次磁场特性
feature of seismic activity 地震活动特征
feature of stratigraphic(al) development 地层发育特征
feature of style 建筑风格;风格
feature of superficial cracks 地裂缝性质
feature record 要素记录
feature selection 特征选择;特征分类

feature space 特征空间
feature strip 特制缝条;特制缝带
feature window 有特色的窗
featurization 特制片
feaze 解散绳股
feazings 解散绳股
febetron 冷阴极脉冲
febrile crisis 热骤退
febrile lysis 热渐退
February 二月
fecal bacteria 粪便细菌
fecal borne 粪便传播的
fecal coliform bacteria 粪大肠菌
fecal contaminant 粪便污染物
fecal harmless treatment 粪便无害化处理
fecal material 粪物质
fecal microorganism 粪便微生物
fecal pellet 粪粒
fecal pollutant 粪污染物;粪便污染物
fecal pollution 粪污染;粪便污染
fecal sewage 粪便污水
fecal treatment 粪便处理方法;粪便处理
fecal waste 粪便排泄物
feces 粪便
feces sewage 粪便污水
Fe-chain coagulant 铁系混凝剂
Fechner colo(u)r 费克纳色
Fechner fraction 费克纳分数
Fechner law 费克纳定律
Fe-C micro-electrolysis-biochemical process 铁炭微电解生化法
Fecraloy alloy 费克拉洛伊合金
fect 电场效应
fecula 污物;排泄物;粪便;渣滓
feculence[feculency] 混浊
feculent 混浊的
feculose 淀粉羧酸酯
fecundate 使丰饶;使肥沃;使多产
fecundity 繁殖力
fecundity of the sea 海洋生殖力;海洋生产力
fecund soil 沃土;肥沃的土壤
Fe-Cu ore 铁铜矿石
feddan 费丹(埃及面积单位,1 费丹 = 4200 平方米)
feddback relay amplifier 反馈继电放大器
Fedepth meter 钢筋位置测定仪
Federal 联邦政府的
Federal Advisory Council 联邦咨询委员会
federal agency 联邦机构
Federal aid 联邦政府补助
Federal aid road 联邦政府资助公路
Federal arc converter 费特尔电弧换流机
Federal Aviation Administration 联邦航空局
Federal Aviation Agency 联邦航空署
Federal Aviation Agency Classification 联邦航空署土分类法(美国)
Federal Aviation Regulation 联邦航空条例
Federal budget 联邦预算
Federal Clean Water Act 联邦清洁水法令(美国)
Federal Communications Committee 联邦通信委员会
federal control 中央控制
Federal debt subject to limitation 限额内联邦债务
Federal Deposit Insurance Corporation 联邦储备保险公司
Federal Emergency Management Agency 联邦应急管理局;联邦紧急情况管理署
Federal Emergency Relief Law 联邦紧急救济法(美国)
Federal Energy Adminstration 联邦能源管理局(美国)
Federal Environmental Agency 联邦环境局(美国)
Federal Environmental Pesticide Control Regulation 联邦环境农药控制条例(美国)
Federal Fair Housing Law 联邦平等居住法(美国)
Federal Fire Council 联邦消防委员会(美国)
Federal Fund Rate 联邦资金利率(美国)
Federal Funds 联邦资金;联邦储备基金(美国)
Federal Geodetic Coordinating Committee 联邦大地测量协调委员会(美国)
federal highway 联邦道路;国道
Federal Highway Administration 联邦公路管理局(美国)
Federal Home Loan Bank 联邦居民贷款银行(美国)
Federal home loan bank system 联邦住宅贷款银行制度(美国)
Federal Home Loan Mortgage Corp. 联邦住房贷款抵押公司(美国)
Federal Housing Act 联邦住宅法(美国)
Federal Housing Administration 联邦住宅管理局(美国);联邦住宅局(美国)
Federal Housing Authority 联邦住宅管理局(美国)
Federal Hydro Electric(al)Board 联邦水电局(美国)
Federal income tax 联邦所得税
federal land 联邦所有的土地;国有土地
Federal land bank system 联邦土地银行系统
federal law 联邦法律
Federal Maritime Commission 联邦航运委员会(美国)
Federal motorway 联邦公路
Federal motorway viaduct 联邦公路跨线桥;联邦高架道路
Federal National Mortgage Association 联邦全国抵押协会(美国)
Federal Pollution Control Act 联邦污染控制法令(美国)
Federal Radiation Council 联邦辐射委员会(美国)
Federal Radiological Monitoring and Assistment Plan 联邦放射性监测和援助计划(美国)
Federal Railway Administration 联邦铁路管理局(美国)
Federal Reclamation Laws 联邦垦务条例(美国)
federal regulations compliance 遵照联邦规则
Federal Reserve Bank 联邦储备银行(美国)
Federal Reserve Note 美钞
Federal Reserve System 联邦储备系统(美国)
Federal Revenue Stamps 联邦印花税票(美国)
Federal Safe Drinking Water Standards 联邦安全饮用水标准(美国)
Federal specification 联邦技术规范(美国)
Federal Specification paint 符合联邦规范的漆(美国)
Federal Specifications Executive Committee 联邦规范执行委员会(美国)
Federal Standard method 联邦标准方法(美国)
Federal static boundary 联邦界
Federal style 联邦古典复兴式风格
Federal Telecommunications System 联邦远程通信系统;联邦电信系统
Federal Test Method Standards 联邦试验方法标准(美国)
Federal Water Pollution Control Act 联邦水污染控制法令(美国)
Federal Water Pollution Control Act Amendments 联邦水污染控制修正法令(美国)
Federal Water Pollution Control Administration 联邦水污染管理局(美国)
Federal Water Quality Administration 联邦水质管理局(美国)
Federal waterway 联邦水道(美国);联邦河道(美国)
Federal Wire System 联邦有线系统(美国)
federated computer system 联合计算机系统
federation 联合会
Federation for Cooperative Housing 合作住房会(英国)
Federation International Des Ingenieurs Conseils[FIDIC] 国际咨询工程师联合会
Federation of British Industries 英国工联
Federation of Building Block Manufacturers 建筑砌块制造联合会(英国)
Federation of Civil Engineering Contractors 土木工程承包商联合会(英国)
Federation of Concrete Specialists 混凝土技术专家联合会(英国)
Federation of International prestressing concrete 国际预应力混凝土联合会
Federation of Societies for Coatings Technology 美国油漆工艺学会联合会(美国);涂料工艺学会联合会(美国)
Federation of Swedish Industries 瑞典工业联合会
fedorite 硅钠钙石

Fedorov stage 万用旋转台;费多罗夫旋转台
Fedorov universal microscope 费多罗夫台
fed waters 供水水域
Fedwire 联储电信
fee 收费;费用
fee appraiser 收费评估人;收费鉴定人
feeble 微弱的;薄弱的
feeble current 弱电流
feeble field 弱场
feeble magnetism 弱磁性
feeblemindedness 低能
feeble shock 微弱地震
feebly cohesive soil 弱黏性土;弱黏(聚)性
feebly hydraulic lime 弱性水硬石灰;弱水硬性石灰
fee charged for the use of 占用费用
fee checking account 收费支票户
fee claim 费用索赔
fee curve 工程预算曲线;预算百分比(曲线)
feed 走刀;供片;给料;馈送;馈给;馈电;进刀;加载;送料;送进量;饲料;供料;进料;加料
feed adjustment 馈电调整;进刀调整
feed agitator 饲料搅拌器
feed air nozzle 输气喷管
feed alley 饲喂过道;饲料通道
feed and filter tank 进给过滤箱
feed and pull-out 给进和提升
feed apparatus 供料装置;给水器;给料装置
feed apron 进料(挡)板;裙板式进料机;裙板式供料器;皮带式供料器;输送机进料板;带式供料器;板式喂料机;板式给料机
feed arm 馈电臂
feedback 应答;转手资料;回授;提供的成果;反馈
feedback accumulator 回授储蓄器
feedback action 反馈作用
feedback adjustment 反馈调整
feedback admittance 回授导纳;反馈导纳
feedback amplification 反馈放大
feedback amplifier 反馈放大器
feedback amplifier characteristic 反馈放大器特性
feedback automation 反馈自动化
feedback bellows 反馈波纹管;反馈膜盒
feedback bias 回授偏压
feedback branch 反馈支路
feedback bridging fault 反馈桥接故障
feedback channel 反馈信道;反馈通道
feedback characteristic 反馈特性
feedback circuit 回授电路;反馈路线;反馈电路
feedback clamp circuit 反馈箝位电路
feedback classroom 反馈教室
feedback coefficient 反馈系数
feedback coil 反馈线圈
feedback communication 反馈通信
feedback compensating winding 反馈补偿绕组
feedback compensation 反馈补偿
feedback compensation technique 反馈校正技术
feedback control 回授控制;回授调整;反馈抑制;反馈控制;反馈调整;反馈调节
feedback control based on dosed water quality 基于剂量水质的反馈控制
feedback-controlled optics 反馈控制光学
feedback controller 反馈控制器
feedback control loop 反馈控制回路;反馈环路
feedback control of cost 成本反馈控制
feedback control signal 反馈控制信号
feedback control system 反馈控制系统
feedback counter 反馈计数器
feedback couple 回授耦合
feedback coupling 回授耦合;反馈耦合
feedback current 反馈电流
feedback cutter 反馈切割器;反馈刻纹头
feedback cutter head 反馈式刻纹头
feedback decoder 反馈译码器
feedback decoding 回授译码;反馈解码
feedback device 反馈装置
feedback differentiator 反馈微分器
feedback divider 反馈分频器
feedback dynamics 反馈动力学
feedback effect 反馈效应
feedback element 反馈元件;反馈环节;反馈部件
feedback encoding 反馈编码
feedback envelope 反馈包迹
feedback equalizer 反馈均衡器
feedback excitation 反馈励磁;反馈激励

feedback factor 反馈因数;反馈系数
feedback filter 反馈滤波器
feedback force 反馈力
feedback function 反馈函数
feedback gain 反馈增益
feedback generator 回授振荡器;反馈振荡器;反馈发生器
feedback grinding 圈流粉磨;闭路循环粉磨
feedback heater 反馈加热器
feedback impedance 反馈阻抗
feedback index 反馈指数
feedback influence 反馈影响
feedback information 重整资料;反馈信息
feedback inhibition 反馈抑制
feedback integrator 反馈积分器
feedback laser 反馈激光器
feedback limiter 反馈限制器;反馈限幅器
feedback line 回授线圈;回授线;反馈线路
feedback link 反馈链
feedback linkage 反馈传动杠杆
feedback loop 回授电路;反馈回路;反馈环(路);反馈电路
feedback loop system 反馈回路系统
feedback mechanism 反馈作用;反馈机制
feedback method 反馈法
feedback modulator 反馈调制器;反馈调节器
feedback module 反馈模块
feedback of environment(al)policy 环境政策反馈
feedback oscillation 反馈振荡
feedback oscillator 反馈振荡器
feedback parameter 反馈参数
feedback path 反馈通路;反馈通道;反馈路线;反馈路径
feedback potentiometer 反馈电位计;反馈电势计
feedback process 反馈程序
feedback quantity 反馈值;反馈量
feedback ratio 回授比;反馈比
feedback recorder 回授记录装置
feedback regulation 反馈调节
feedback regulator 反馈调节器
feedback relationship 反馈关系
feedback report 反馈报告
feedback repression 反馈阻遏
feedback resistance 反馈电阻
feedback resistor 反馈电阻器
feedback response 反馈响应
feedback sawtooth generator 反馈式锯齿波发生器
feedback searching 反馈检索
feedback servomechanism 反馈伺服机构
feedback shift register 反馈移位寄存器
feedback signal 回授信号;反馈信号
feedback solution 反馈解
feedback stabilisation 反馈稳定作用
feedback stabilization 反馈稳定
feedback-stabilized amplifier 反馈稳定放大器
feedback suppressor 反馈抑制器
feedback system 考核制度;反馈系统;反馈方式
feedback system automation 反馈系统自动化
feedback test 反馈试验
feedback time 反馈时间
feedback-time lag effects 反馈时间差距影响
feedback to all levels 各级的反馈
feedback transducer 反馈转换器;反馈换能器;反馈传感器
feedback transfer function 反馈传递函数
feedback transformer 反馈变压器
feedback transmitter 反馈发射机
feedback type 反馈型
feedback type current transformer 反馈式电流互感器
feedback variable 反馈变量
feedback voltage 反馈电压
feedback voltmeter 反馈式伏特计
feedback winding 反馈绕组
feedback with evaluation 反馈评估
feed bar 供料器杆;给料(机)杆;给料机;进给杆;供料机(杆)
feed barn 饲喂场
feed base (天栈的)馈电边
feed batching 进料投配
feed bed 料层;熔烧料层
feed belt 供料皮带;给料输送带;给料输送带;进给皮带;加料皮带;皮带给料机;送料带;输送带;传送皮带

feed belt conveyer 进料皮带输送机
feed bin 进料仓;原料仓;装料斗;给料斗;料仓;喂料仓;饲料仓
feed bin of mill 磨机喂料仓
feed block 给油导管;夹钳式送料机构
feed blower 饲料吹送机
feedboard 供纸板;馈送盘
feedboard raising and lowering control 供纸板升降控制;馈送盘升降控制
feedboard raising and lowering controller 供纸板升降控制器
feed boot 喂料仓
feed box 给料箱;料箱;进料箱;进料盒;浇口箱;饲料箱;供料箱;走刀箱;进给箱
feed bridge 送料搭边;送进搭边
feed bucket elevator 进料斗升运机;喂料用斗式提升机
feed bunk 饲喂槽
feed by the brake 用制动器给进
feed cable 送电线;电源电缆;馈电电缆;馈电线
feed cam 进给凸轮
feed canal 给水渠;传动塔轴;灌溉引水渠
feed cap 进料盖
feed cart 饲料分送车
feed change 进给变速;进刀箱
feed change gears 进刀机构变速齿轮
feed change lever 给进变速把手;进刀变速手柄
feed change unit 给进变速装置
feed charge 装料
feed charge-in 进料
feed-check valve 供水止回阀
feed chute 给料(溜)槽;进料斜槽;进刀槽;喂料(溜)槽;送料(斜)槽;输送斜槽
feed circuit 供电电路;补油回路
feed cistern 给水箱;蓄水供水箱;供水箱;供水池
feed cleaner 给水滤器;饲料清选机
feed clutch 自动进退刀离合器
feed cock 给水旋塞;给水龙头;进水龙头;进气龙头
feedcode lock 进给代码闭锁
feed collector 饲料收集器;饲料集存器
feed collet 料夹
feed cone 进给锥轮
feed cone pulley 进刀锥形轮
feed control 自动加料;供纸控制;馈送控制;进料控制;进给控制;进刀控制;给水控制(锅炉)
feed controller 供纸控制器;给料调节器
feed control lever 进给控制手柄
feed control valve 给进控制阀;进给控制阀
feed conveyer 进料输送机;给料输送机;给料运输机;饲料传送机;送料输送机
feed conveyor trough 自动送料饲槽
feed cord 馈电软线
feed cutoff 阻塞套
feed cut-off gate 喂料截流阀
feed cutter 饲料切割机
feed cylinder 给进油缸
feed degree of domestic resources 国内资源保证程度
feed device 供料装置;喂料装置
feed dial 进料刻度盘
feed disc 加料盘;盘式喂料器;盘式喂料机;排料盘;喂料盘
feed dispenser 饲料分送器
feed distribution pump 给水分配泵
feed distribution system 进给分配系统
feed distributor 饲料分配器
feed ditch 引水沟;灌(水)沟;灌溉渠(道);灌溉沟
feed ditch for irrigation 灌溉引水渠
feed dog 传动卡爪
feed door 加料炉门
feed downpipe 下料管
feed drive gearing 进给传动装置
feed drum 筒式喂料机;筒式送料机;送料筒
feed elevator 进料升运机;喂料用斗式提升机;饲料提升机
feed end 进料端;加料和卸料位置;加料端;喂料端;送电端
feed end head 喂料端磨头
feed end hood 冷烟室;冷却端;除尘室
feed end housing 装料端护罩;冷烟室;冷却机热烟室;降尘室
feed end refuse extraction chamber 给料端排渣室
feed end ring 料浆结圈;挡料圈
feed end trunnion 进料端轴颈;进料端空心轴

feed end wall 投料端墙
feed engagement 每齿进给量
feed equipment 供料设备
feeder 装料装置;支矿脉;谷舱灌补器;供纸器;供料器;给水器;给料器;给料机;漏斗;馈线;馈送器;进料机;进给器;进刀装置;浇道;加料装置;加料器;加机;区间集散;喂料机;送料装置;送料器;输送器;反馈线
feeder airport 支航线机场
feeder and distribution line 集散运输线
feeder antenna 馈线式天线
feeder apparatus 送料装置
feeder assembly 馈线组件
feeder beach 给养海滩;(人工拓宽海滩,借沿岸水流补给下行海岸的)补给性海滩;补给沙滩;补给海滩
feeder belt 进料器皮带;给料筛带
feeder boom 拦河埂
feeder boot 供料罩
feeder bowl 供料机料盆;供料机料盒
feeder box 馈电箱;电缆汇接室;分线箱
feeder-breaker 装料破碎机
feeder bridge 支线桥梁
feeder brush 馈电刷
feeder bus-bar 馈路母线;馈电线
feeder bush 冒口圈
feeder bus system 公共汽车支线系统
feeder cable 馈电电缆;馈电线;用户馈电电缆;馈线电缆;输电电缆;电力干线
feeder canal 引水渠;补给渠
feeder canvas 进料帆布输送带
feeder-category airport 地方支线机场
feeder cattle 架子牛
feeder chain 进料器链
feeder channel 供水渠道;供料通路;料道
feeder clamp 馈线(线)夹;馈电线端子;馈电接线柱;双线等距线夹;电连接线夹
feeder clip 馈线夹
feeder compartment 给料室
feeder compression clamp 馈线压接头
feeder conductor 配出线
feeder connection 供料机通道口
feeder container ship 支线集装箱船;集装箱集散(货)船
feeder conveyer 装料传送机;供料皮带;送料输送机;进料机;传送进料器
feeder cord 馈电软线
feeder current 支流;补流;补给流
feeder distribution center 馈线分配中心;馈电配线盘;馈电分配中心
feeder drain 分支排水沟
feeder drive eccentric 给料偏心轮
feeder drive link 进刀传动杆
feeder drop 馈线电压降;馈路电压降
feeder ear 供电端;馈线耳;馈电线夹
feeder entrance brick 料道入口砖
feeder equipment 馈电装置;送料设备
feeder floor 给料台
feeder gas 流出瓦斯
feeder gate 供料闸板;供料机闸砖
feeder green 青饲料
feeder head 钢锭收缩头;收缩头
feeder height 冒口高度
feeder highway 公路支路;支线公路;公路支线
feeder hole 探查孔
feeder hopper 装料箱;进料斗
feeder house 进给室
feeder hundreds dial 进给器百分度盘
feeder industry 补给性工业
feeder insulator 馈线绝缘子
feeder interface unit 馈线接口单元
feeder lighter above ship vessel 短途载驳货轮
feeder line 供气干线;支线;馈线;进水线;进料管线;短途运输线
feeder line and earth wire support clamp 馈线合地线支撑线夹
feeder liner 支线运输机
feeder link 馈电链路
feeder loss 馈电损耗;馈电耗损
feeder machine 供料机;给料机
feeder main 供气干线;给水总管;馈电干线
feeder messenger wire 馈线吊线
feeder modulus 冒口模数
feeder neck 冒口颈

feeder needle 供料针冲头
feeder network 馈线网
feeder nose 供料盆;投料口
feeder opening 供料机通道口;供料机开口
feeder panel 馈电线控制板;馈电盘;馈电(控制)板
feeder pillar 馈线柱;馈电柱;分支箱
feeder plug 供料机闸砖
feeder port 子港;支线港;集散港;集散点;区间集散港
feeder process 供料过程
feeder protection equipment 馈线保护装置;馈电保护装置
feeder railway 铁路支线
feeder ratchet 进料器棘轮机构
feeder ratchet stop link 进给棘轮停止杆
feeder ratchet stop pin 进给棘轮制动销
feeder ratio 供料机传动比;供料传动比
feeder reactor 线路电抗器;馈路电抗器;馈电扼流圈
feeder removal 去除冒口
feeder reservoir 支流水库;饲料储藏
feeder road 专用线(路);支线公路;支路;公路支线;分路支线
feeder rod 捣冒口棒
feeder screen 给料筛;进料筛
feeder section 馈路区间;馈电区域
feeder service 集散支线;集散运输;支线运输;区间疏远业务;短途运输服务
feeder service port 支线港
feeder ship 支线船舶;短途运输船;分运
feeder skip hoist 进料吊斗提升机
feeder skip hoist for silos 筒仓进料吊斗提升机
feeder sleeve 供料机勺料筒;料筒;冲头套管
feeder spout 供料机盆;供料机料盒
feeder street 分支街道;支线街道
feeder support clamp 馈线支架线夹
feeder support insulator 馈线支撑绝缘子
feeder suspension 馈线悬挂
feeder switch 馈路开关
feeder system 供气干线;供水系统;供料系统;供电系统;短途运输系统
feeder tens dial 进给器十分度盘
feeder termination 馈线下锚
feeder trough 给料槽
feeder tube 供料机勺料筒;料筒;冲头套管
feeder-type airport 馈线式航空港
feeder unit 续纸装置
feeder vessel 支线船舶
feeder voltage regulator 馈路电压调整器
feeder weigher 称给料器
feeder well block 供料机料盆
feeder wire clamp 馈线连接线夹
feeder wire support 馈线支架
feeder with elliptic(al) movement 椭圆形运转的进料器
feed expansion honing 定进给量珩磨
feed filter 给水滤器
feed finger 进料指
feed flow adjustment 流送量调节
feed force 进刀力
feed for retreatment 再选给料
feed forward 正向输送;前馈;正向馈电传送
feed forward control 前馈控制
feed forward of information 信息前馈
feed forward optimalizing 馈情最优化
feed forward principle 前馈原理
feed forward system 前馈系统
feed-fraction 给料粒度级
feed frame 饲料框架
feed friction 进给摩擦
feed function 进给功能
feed gallery 馈电抗道
feed gas 原料气
feed gate 进料闸(门)
feed gear 给进齿轮;进刀装置;送刀装置
feed gear bonnet 进给齿轮罩
feed gearing 进刀传动装置
feed gearing mechanism 齿轮进给机构
feed gear wheel 进给齿轮
feed glass 给油指示器;控制玻璃
feed glass washer 玻璃眼垫圈
feed governor 给水调节器
feed grinder 饲料粉碎机
feed grinding 进给研磨法;横向进磨法
feed gun 进料枪;加料枪

feed handling equipment 饲料装卸分送设备
feed head 馈给压头;进料头;进料口;(补缩的)冒口;气锅炉上给水箱
feed head of steam turbine 蒸汽涡轮给水箱
feed heater 进料加热器
feed heat exchanger 喂料热交换器
feed hold 进给保持
feed hole 馈送孔;导孔;传动导孔;输送孔;中导孔
feed hopper 装料斗;料斗;给料漏斗;进料(漏)斗;送料斗
feed hopper liner 料斗衬板
feed horn 馈电喇叭;号角形馈电器;喇叭天线
feed hose 供给管
feed-in 给进;馈入;送进;淡入
feeding 装料;馈送;进油;进给;喂料;投料;补缩;供料;馈电;进料;加料;发满
feeding a casting 铸件补缩
feeding apparatus 进料装置;给水装置;喂料装置
feeding area 饲喂区
feeding area of geothermal reservoir 热储补给区
feeding bank 送油槽
feeding belt 进料(皮)带
feeding belt conveyer 皮带输送机;皮带送料机;皮带送料带;皮带喂料机;喂料皮带机
feeding box 进给箱
feeding bucket 进料斗
feeding bucket elevator 进料斗升运机;喂料斗式提升机
feeding burrow 觅食潜穴
feeding by weight 按重量喂料
feeding canal 供水渠;进料道;进料槽;给水沟
feeding capacity 供给容量;补缩能力
feeding carriage 进给托架
feeding cell 进料室
feeding center 供食中心;馈电中心;牲畜肥育站
feeding channel 送料道;馈应孔道;给料道
feeding channel colo(u)ration 料道着色
feeding chute 进料管;进料(溜)槽;装料槽;供料溜子;进料滑槽
feeding compartment 进料室
feeding condition 补给条件
feeding conduit 供水管道;供水道;输送管(道)
feeding conveyer 进料输送机
feeding crane 装料起重机
feeding crop 饲料作物
feeding current 馈电电流
feeding device 进料装置;进给装置;给纸装置;加料装置;加料器
feeding dial 进料盘
feeding disc 进料盘
feeding distance 进给距离;补缩距离
feeding ditch 引水沟;灌溉渠;供水沟
feeding drawer 喂料耙
feeding elevator 进料升运机
feeding end 投料端;池窑投料端;池窑加料端
feeding equipment 给料设备;加料装置;供料设备
feeding facility 饲养设备
feeding finger 送料叉
feeding fissure 补给裂缝
feeding floor 饲喂间
feeding force 进给力
feeding funnel 进料斗;送料斗;上料斗
feeding gate 进料门
feeding grid 饲料筛
feeding habit 索饵习性;食性;摄食习性
feeding head 绕丝头;补缩冒口
feeding hole 进料口
feeding hopper 进料斗;进料仓
feeding house 饲畜舍
feeding industry 牲畜肥育业
feeding leg 溜送槽
feeding length (钻进的)进给长度
feeding line 进料管线
feeding machine 供料机;给料机;进料机;加料机械;喂料机;饲料机;饲料分送机
feeding mechanism (钻机的)给进机构;进给机构;供料机构;进料机构;索饵机制
feeding method 馈电(方)法
feeding migration 索饵洄游;摄食洄游
feeding neck 冒口颈
feeding of a river 河水补给;河流补给
feeding opening 进料门
feeding pack 馈电部件
feeding pan 进料槽

feeding passage 进料通道
feeding pipe 进料管
feeding pitch 送料节距
feeding platform 投料平台
feeding point 馈电点;投料点
feeding pot 供液壶
feeding pressure 给进压力;进刀压力
feeding process 供料方法
feeding pump 给料泵;进料泵;送料泵
feeding quantity 进料数量
feeding rate 加料速度
feeding reservoir 蓄料池;供水水库;给水池
feeding rod 进给杆;捣冒口棒;补缩捣杆
feeding roll 进料滚轴;进料辊;喂料辊
feeding roller table 摆抖机进袋辊道
feeding screw 螺旋喂料机;螺旋进料器;喂料绞刀
feeding season 放牧季节
feeding section 牵引供电臂
feeding sequence 装料程序
feeding shoot 进料滑槽
feeding side 投料端
feeding silo 进料筒仓
feeding skip 供料斗;进料斗
feeding slip 进浆
feeding space 进料室
feeding speed 进刀速度
feeding spool 供片卷轴
feeding spout 进料斜槽
feeding store 进料储[贮]仓
feeding structure 觅食构造
feeding stuff cuber 饲料制粒机
feeding system 供料系统;冒口补缩系统;喂料系统;补给系统
feeding tank 进料罐;进料槽;送料槽
feeding thrust 给进力
feeding time 供片时间
feeding tipple conveyer 干燥机进板分配桥
feeding trace 觅食迹
feeding transformer 电源变压器
feeding trough 进料槽;饲料槽
feeding unit 进给装置;进料装置;供料设备;给纸装置;加料装置
feeding-up 过稠;硬化;加料;稠化;发满
feeding voltage 馈电电压
feeding water 供水;补给水
feeding web 进料输送带
feeding zone 冒口补缩区
feed inlet 原料入口;进料口;加料口
feed-in-pull-out technique 馈入—拉出技术
feed intake 采食量
feed jumper 馈线跳接板
feed launder 加料流槽
feed leg 动力支架;风动钻架
feed length 给进长度(钻机);推进长度
feed lever 给进手把
feed lime 饲用石灰
feed line 整备线;供料(液)管线;给水管(路);进给线;饲料输送管道;动力线;供给线;给水管
feed liquid 供液;料液
feed liquid inlet 供液口
feed liquid pump 供液泵
feed liquor 料液
feed liquor inlet 滤浆入口
feed loader 牵引式饲料装运车
feedlot 露天畜栏;饲养场;饲养圈;牲畜饲养场;肥育地
feedlot lagoon 牲畜饲喂场饮水池
feedlot runoff 牲畜饲喂场排水
feedlot wastewater 牲畜饲喂场废水
feed lump size 喂料块度
feed machine 馈送机
feed magazine 进料斗
feed main 馈电干线;进料总管;给水干管
feed make-up boiler 配料蒸锅
feed mark 进刀痕迹
feed material filling ratio 喂入物料的喂料比
feed measuring 进料测定
feed mechanism 馈送机构;进料机件;进料机构;进刀机制;输入机制;输入机构;上料机构
feed metal 原料金属
feed metering 饲料计量
feed mill 送料集;饲料粉碎机
feed mixer 饲料拌和机
feed mixer-and-elevator combination wagon 饲

料拌和分送小车
feed mixing machine 饲料混合机
feed mixing pump 进料搅拌泵
feed motion 进给动作
feed motor 进料发动机;进料电动机;进给电动机
feed mouth 进料口
feed movement 进给运动
feed nozzle 进料嘴
feed nut 进给螺母
feed of drill 钻头进程;进尺【岩】
feed of fuel 燃料输送
feed oil pressure 给油压力
feedome 馈线罩
feed on rush 失控料流
feed opening 进料口;进料孔;加料孔
feed operating link 进刀连杆
feed-out 馈出
feed particle 给料颗粒
feed path 进给路
feed per minute 每分钟走刀量;每分钟进刀量;每分钟进尺量
feed per revolution 每转走刀量
feed per stroke 每行程走刀量;每次行程走刀量
feed pipe 供液管;给水管;给料管;料管;加料管;送料管;送料管;供水管
feed pitch 馈送孔间距;进给传送孔距;输送孔距;导孔间距
feed plant 供料设备
feed plunger 供料机冲头
feed point 供应点;馈电点;补给点
feed point impedance 馈电点阻抗
feed port 进料口;喂料口
feed preheater 原料预热器;进料预热器;喂料预热器
feed preparation 备料
feed pressure 供油压力;供水压力;进给压力;输送压力
feed processing building 饲料加工间
feed processor 饲料加工机
feed proportioning 进料配(合)比;配料
feed pull 提升力
feed pulley 主动滑轮;驱动进料三角皮带轮
feed pull maximum 最大提升力(指钻机立轴)
feed pump 供水泵;供给泵;料液泵;料浆泵;进给泵;送料泵;给水泵
feed pump regulator 给水泵调节器
feed punch 馈送穿孔;送卡穿孔机;导孔穿孔
feed quantity 进料数量
feed quill 给进套筒
feed rack 给进齿条;进给齿条
feed range 给进范围;进给范围
feed rate 给料速度;供料速度;馈送率;进料速度;进料率;进给速度;(焊丝的)送进速度;进料速率
feed rate indicator 给进速度指示器;进给率指示器
feed rate number 馈送率数
feed rate of welding wire 焊丝送进速度
feed rate override 进给速度修调
feed ratio 给进率;馈电比;进给比
feed reactor 馈电电抗器;线路电抗器
feed reel 折卷机;供带盘;进料卷取机;进给卷盘
feed regulating lever 进刀调节手柄
feed regulating valve 给水调节阀
feed regulator 供料调节器;供给调整器;给料调节器;加料控制器
feed reservoir 补给水池
feed reverse 进给换向
feed reverse lever 进刀反向手柄
feed reversing 反向进给
feed reversing gear 反向给装置
feed riser 供水竖管
feed rod bracket 走刀杆托架
feed roll 给纸辊;馈馈辊;送料辊;输送滚筒
feed roller 供料辊子;进给辊;喂料辊
feed-roll granulator 供料辊成粒机
feed sampler 给料取样机
feed scale 喂料计量器
feed screw 供料螺杆;给进螺杆;螺旋送料器;螺旋送料机;螺旋给料器;进给螺杆
feed screw collar 进给螺杆环
feed screw mechanism 进料螺杆机构
feed screw nut 丝杆螺母
feed selection lever 走刀箱变速杆;进给变速杆
feed service 集散支线;集散运输
feed shaft 给进轴;进给轴;垂直火道
feed shelf 馈送带

feed shelter 进给盘
feed-shifter 给进变向把
feed shoe 料槽;加料刮板;给料(刮)板
feed shoot 进料槽
feed side 进料侧
feed silo 进料筒仓
feed size 喂料粒度
feed skip 给料斗
feed slide 进给滑板
feed slug 滞料
feed slurry 滤浆
feed slurry inlet 滤浆入口;料浆入口;料浆进口
feed solids 给料固体
feed speed 给进速度
feed spindle 给进立轴
feed split ratio 进料碎裂比(粗料/细料)
feed-sponge 海绵金属料
feed spool 供片轴;供片卷轴;输带轴
feed spout 进料斜槽;给料溜管;喂料斜槽;喂料溜子
feed station 供应点
feed status 馈送状态;输送状态
feed steam 进汽
feed stock 原材料;原料;进料
feed stopper 进给停止器
feed stop valve 给料截止阀;进给安全阀
feed storage 饲料库
feed store 饲料库
feed strainer 进料滤器
feed stream 原料流;料流;供入液流
feed stroke 给进行程
feed stuff 饲料
feed surge 失控料流
feed synchronization 喂料同步化;同步喂料机
feed system 供给系统;给油系统;给进系统;馈电系统
feed table 盘式给料器;盘式给料机;送料盘
feed tank 储[贮]水箱;储料容器;备用油箱;进料箱;供料槽;给水箱;进给箱;喂料池;进刀箱
feed-tape 供带机
feed throat 喂料喉
feed-through 直通;馈通;馈电导体;导孔;穿通线
feed-through capacitor 穿心式电容器
feed-through collar 引线环
feed-through conductance 馈通电导
feed-through connector 直连插头座;传输(用的)接插件
feed-through insulator 穿通端子
feed-through power meter 馈通功率测量仪;通过功率测量仪
feed-through signal 馈通信号
feed-through spool 转动送进筒;转动送盘管
feed-through switch 穿线开关
feed-through system 馈通式
feed-through terminal 穿通送进管;穿通接线柱;穿通端子
feed-through voltage 馈通电压
feed-through wire 穿通线
feed-thru connection 正反面连接
feed thrust maximum 立轴最大给进压力
feed track 整备线;馈送道;送车线;输送(轨)道
feed travel 钻进深度;进给行程;进料运程;进给行程;推进器行程
feed tray 给料盘;馈送支架;馈送托盘;进料塔盘;送料盘
feed tube 料管;虹吸浇口
feed unit 进料装置;进料设备;进刀机构
feed valve 供气阀;供水阀;供料阀;供给阀;给水阀;给气阀;送料阀
feed valve block 给水阀组
feed valve gasket 进给阀垫密片
feed valve stem nut 进给阀杆螺母
feed velocity 进料速度;进刀速度
feed water 给水(的);补给水;给水管路
feed-water booster pump 给水升压泵
feed-water connection (锅炉的)给水入口管;给水管接头
feed-water control 给水控制;供水控制
feed-water cycle system 给水循环系统
feed-water duct 供水管道
feed-water ejector 给水喷射器
feed-water equipment 给水设备
feed-water filter 给水过滤器;补给水过滤器
feed-water flow 给水流量
feed-water heater 给水加热器;供水加热器

feed-water heating apparatus 给水加热器
feed-water heating loop 给水加热回路
feed-water injector 给水注入器;喷水器
feed-water inlet 供水口
feed-water joint 给水管接头
feed-water line 给水管线
feed-water main 供水总管;给水干管
feed-water make-up 补给水
feed-water pipe 给水管道
feed-water(pipe)coupling 给水管接头
feed-water piping 给水管道;给水管网
feed-water plug 注水口旋塞
feed-water preheater 给水预热器
feed-water preparation 给水准备
feed-water pump 给水泵
feed-water purifier 给水净化器
feed-water quality 给水品质
feed-water regulator 供水调节装置;供水调节器;给水调节器
feed-water softening 给水软化
feed-water space 给水柜
feed-water strainer 给水过滤器
feed-water system 供水系统;给水系统
feed-water tank 给水箱
feed-water treatment 给水处理
feed-water treatment plant 补给水处理设备
feed-water unit 给水装置
feedway 供给装置;饲喂过道
feed well 给水井;给料孔;供水口
feed wheel 推进轮;送料轮;供带轮;给进轮;进给轮
feed wire 馈线;馈电线
feed worm 给进蜗杆;进给蜗杆
feed zone 加料段
feee 解散绳股
fee estimate 收费估计
fee for analysis certificate 化验证明书费
fee for arbitration 仲裁费(用)
fee for certificate of origin 产地证明书费
fee for construction site facility 临时设施费
fee for expert opinion 专家咨询费
fee for land use 土地使用费
fee for permit 牌照税
fee for possession and use of 占用费用
fee for refund 退赔费
fee for technology 技术费用
fee for trial production of new products 新产品试制费
fee for using a site 场地使用费
fee influent 料液
feel 感觉;手感
feel dizzy 发晕
feeler 隙片;原薄规;灵敏元件;厚薄规;探(纬)针;探头;探深杆;探钩;探测杆;塞尺;触角;触点;测深竿;测深杆;测厚规;复制工作
feeler blade 隙片;测隙片
feeler block 对刀块
feeler control 仿形控制器
feeler dilatometer 探针膨胀仪
feeler filling changing device 探针式换纬装置
feeler ga(u)ge 千分尺;厚薄规;千分垫;塞规;塞尺;测隙规;测厚规
feeler head 测隙装置;测隙头
feeler inspection 探针检验;触探
feeler leaf 塞尺片
feeler lever 探纬针杆;探杆;触杆
feeler lever ga(u)ge 触杆规
feeler mechanism 检测机构;仿形机构
feeler motion 探纬运动
feeler pin 触针
feeler plate 探纬板
feeler plug 测孔规
feelers 测隙规
feeler stock 量隙规;千分垫
feeler switch 测试键
feeler wheel 仿形轮
feeler wire 探触器
feel giddy 发晕;发昏
feeling 触头
feeling bottom 开始受水底的影响
feeling elongation meter 感触延伸仪;触头膨胀仪
feeling of being controlled 被支配感
feeling of being revealed 被泄露感;被洞悉感
feeling of comfort 舒适感

feeling of passivity 被控制感;被动体验;被动感
feeling of quantity 量感
feeling of repose 安静感
feeling of space 室内感触;空间感觉
feeling of unreality 非真实感;非现实感
feeling pin 探针
feeling the bottom 触浅【船】
feeling the current 受洋流影响
feeling the pulse with the finger-tip 指目
feeling the soil 手感土壤
feeling walk 探路航行
feel nervous 发慌
feel the helm 舵生效;舵来了
feel the way 探路航行
fee method 收费法
fee of permit 执照费;牌照费
fee of sample 样品费用
fee paid for land 征地费
fee-parking 收费停车
fee plus expense agreement 开支加酬金合约;实报实销加管理费合同
fee quota 费用定额
fee rate 费率
fee revenue 规费收入
fee scale 收费标准
fee simple defeasible 可取消的继承权
fee simple determinable 有限制条件的继承权
fee simple estate 不限制继承的不动产
fee simple subject to a condition subsequent 可终止的继承权
fee simple subject to an executory limitation 有期转移限制的继承权
fee system 建筑人员工资等级
feet 海湾;墙脚
feet board measure 木材量度;板尺
feet cube 立方英尺
feet head 英尺压差
fee to be apportioned 待摊费用
feet per bit 钻头进尺
feet per day 英尺/天;每天进尺
feet per minute 英尺/分;每分钟英尺
feet per second 英尺/秒
feet-run 长度估算
feet super 立方英尺
feet switch 地脚开关
feflex camera 反光照相机
fefluorination 去氟
Fehling's solution 费林溶液
feigned column 假柱
Feinc filter 芬克过滤机
feint 帽盖的卷边;细缝;卷折边;淡格线
feinted edge 卷折边
feitknechtite 六方水锰矿
felder 镶嵌地块
Feldman-Sereda model 费尔德曼—塞雷达模型
feldspar 钾长石;长石
feldspar-basalt 长石玄武岩
feldspar-bearing hornblendite 含长石角闪石岩
feldspar-bearing peridotite 含长橄榄岩
feldspar-bearing pyroxenite 含长石辉石岩
feldspar-bearing teschenite 含长石沸绿岩
feldspar bed 长石床层
feldspar ceramics 长石陶瓷
feldspar conversion 长石含量的换算
feldspar deposit 长石矿床
feldspar earthenware 长石质精陶;长石精陶
feldspar-free pseudoleucite microsyenite 无长石假白榴石微晶正长岩
feldspar-free pseudoleucite syenite 无长石假白榴正长岩
feldspar glaze 长石釉
feldspar glazed finish 长石釉面
feldspar-greisen 长石云英岩
feldspar grit 长石粗砂岩
feldspar in lumps 长石块;钾长石块
feldspar in powder 钾长石粉
feldspar jig 长石床层跳汰机
feldspar meal 长石粉
feldspar-microphyric basalt 长石微斑玄武岩
feldspar olivine augite-phyric basalt 长橄辉斑玄武岩
feldspar phyllite 长石千枚岩
feldspar phyric basalt 长石玄武岩
feldspar phyric rock 长石斑岩

feldspar porcelain 长石瓷
feldspar powder 长石粉
feldspar quartz schist 长石石英片岩
feldspar structure 长石结构
feldspath 长石
feldspathic 含长石的;长石质(的)
feldspathic-actinolite schist 长石阳起石片岩
feldspathic arenite 长石砂屑岩
feldspathic electric(al) porcelain 长石质电瓷
feldspathic glass 长石质玻璃
feldspathic glaze 长石质釉;长石釉
feldspathic glaze tile 长石釉砖
feldspathic graywacke 长石杂砂岩;长石玄武
feldspathic grit 长石粗砂岩
feldspathic litharenite 长石质岩屑砂岩
feldspathic lithic graywacke 长石岩屑杂砂岩
feldspathic lithic quartz sandstone 长石岩屑石英砂岩
feldspathic lithic sandstone 长石岩屑砂岩
feldspathic lithwacke 长石质岩屑瓦克岩
feldspathic polylitharenite 长石复岩屑砂岩
feldspathic porcelain 长石瓷
feldspathic porphyroblast 长石质变斑岩
feldspathic quartzite 长石石英岩
feldspathic quartz graywacke 长石石英杂砂岩
feldspathic quartz sandstone 长石石英砂岩
feldspathic sand 长石砂
feldspathic sandstone 长石(质)砂岩;花岗岩砂岩
feldspathic shale 长石质页岩;长石页岩
feldspathic sublitharenite 长石亚岩屑砂屑岩
feldspathic wacke 长石质瓦克岩
feldspathization 长石化(作用)
feldspathoid 似长石;长石类矿物
feldspathoidal basalt 似长石的玄武岩
feldspathoid-bearing kersantite 含似长石云斜煌岩
feldspathoid-bearing minette 含似长石云煌岩
feldspathoid mineral 似长石矿物
feldspatite 长石岩类
Feld washer 菲尔德煤气洗涤机
Felici balance 电感测量电桥
Felici generator 费利西静电高压发生器
feling silk 反手双股丝线
felite 斜方硅钙石;水泥熟料中的矿物成分
fell 咬口折缝;砍伐(丘陵);荒山;丘陵沼泽地;兽皮;采伐;伐倒木
felled sale 伐倒出售
felled timber 筏倒的木材;伐倒的树木
Fellenius method of slices 费伦纽斯条分法
Fellenius solution 费伦纽斯解法
feller 采伐者;采伐机;筏木机;伐木工
fell-field 荒高地;稀矮植物区;寒漠
Fellgett advantage 费尔盖特增益
felling 制毡法;伐木;二重接缝
felling area 伐区
felling area statistics 采伐面积统计
felling axe 筏木斧;伐木斧
felling cycle 轮伐期;采伐(周)期
felling cyclel logging cycle 采伐间隔期
felling dog 支锯片
felling industry 采伐工业
felling machine 筏木机;伐木机
felling marks 分匹色纬
felling operation 筏木作业;筏木工作
felling plan 采伐一览表
felling point 伐点;伐倒点
felling road 伐木道
felling saw 筏木锯;伐木锯;伐树锯
felling shake 伐木裂纹
felling subsere 采伐亚演替系列
felling work 筏木作业;伐木工作
felloe 轮缘;轮辋;车轮外缘;车轮外围;车轴辘
felloe band 载重带
felloe plank 桥梁护木;护木
felloe plate 轮缘板;辋板
fellow 正会员
Fellow of American Society of Appraisers 美国财产评估员协会会员
Fellow of Chartered Accountants 特许会计师学会会员(英国)
Fellow of Geology Society 地质学会会员(英国)
fellow passenger 同船人;同车人
Fellow's cutter 费洛插齿刀
Fellow's gear shaper 费洛刨齿机

fellowship grant 研究人员补助金
fellow trader 同行
fellow-trader commission 同业佣金
fellow-trader discount 同业折扣
fellow-trader price 同业价格
fellow worker 同事
Fell system 菲尔系统
felly 轮缘;车轮外缘
felsic 长英质的;长英矿物
felsic hornfels 长英质角岩
felsic mineral 长英质矿物
felsite 致密长石;长英岩;霏细岩
felsite porphyry 霏细斑岩
felsitic 霏细状的
felsitic porphyry 霏细状斑岩
felsitic rhyolite 霏细流纹岩
felsitic texture 霏细结构
felsoandesite 霏细安山岩
felsobanyaite 斜方矾石
felsoblastic texture 霏细变晶结构
felsoclasic texture 霏细碎裂结构
felsophaerite 霏细球粒
felsophyric texture 霏细结构
felspar 钠长石
felstone 霏细岩;致密长石
felsyte 霏细岩
felt 油毛毡;毡;棉毡;毛毡
felt adhesive 油毛毡胶黏剂
felt and gravel roof(ing) 油毡砂砾屋面;油毡豆石屋面;油毡撒缘豆砂屋顶;油毡石子屋顶;撒豆石油毡屋面
felt area 有感区
felt area of earthquake 地震波及区
felt-backed lino(leum) 毡底油地毡;毛毡衬里的亚麻油毡
felt back(ing) 毡垫底;毡衬里;油毡衬背
felt base 毛毡垫;毛毡底
felt-base floor cover(ing) 毛毡衬里的楼面覆盖层
felt-base rug 带毡底的地毯;毡底地毯
felt blistering 油毡起鼓
felt buff 重缩毡
felt buffer 毛毡缓冲垫
felt calender 毛毡式砑光机;毛毡滚筒
felt camp 毡包
felt cardboard 毛毡提花纹板
felt carpet 油毡覆盖层;油毡覆(盖)面层;毛毡地毯
felt cement 油毛毡胶黏剂
felt chamber 毛腔
felt cleaner 毛布洗涤器
felt cloth 薄毡料;毡布
felt collar 毡圈;毡环;管座勒脚
felt collar of pipe 管座勒脚
felt conditioner 毛布洗涤器
felt covering 毛毡衬面
felt cylinder 毡圈;毡环;毛毡圆筒
felt damp-proof course 油毡防潮层
felt deadener 吸声板;毛毡隔声板
felt draught excluder 防风毡条
felt drying machine 包毡滚筒烘干机
felt dust packing 毛毡填充物
felt earthquake 有感地震
felted fabric 缩绒织物
felted fabric backing 毛毡织品;衬垫
felted fabric base 毡基;毡垫;毡衬
felted fabric board 毡片;贴毡木板
felted fabric cylinder 毡圆;贴毛毡的圆筒
felted fabric foil 铁毛毡的金属箔
felted fabric for roofing 屋面毡
felted fabric insert 毡垫片
felted fabric insulating strip 毡绝缘条
felted fabric jute 毡衬黄麻
felted fabric mat 毡垫层
felted fabric pad 毡垫;绝缘垫
felted fabric panel 有毡垫的镶板
felted fabric roller 毡辊;绒辊
felted fabric seal 毡密封
felted fabric sheet 毡片
felted fabric strip 毡条
felted fabric underlay 毡垫底
felted fabric washer 毛毡垫圈
felted fabric web 毡幅;毛毡织物
felted membrane 毛毡滤清器
felted mineral core 粘结矿物纤维芯板
felt element 毛毡滤芯;毛毡零件;毛毡过滤元件

felt filter 毡滤器；毛毡滤芯滤清器
felt finger 毛刷
felt floorcovering 毛毡地毯
felt forming machine 制毡机
felt gasket 毡垫圈；毡衬垫；毛毡垫圈
felting 油毡铺贴；油毡铺砌；毡化
felting down 黏脱下来；消光法；去光泽法；黏织在一起
felting machine 制毡机
felting material 制毡材料
felting paper 毡纸；绝缘纸
felting product 毡制品
felting property 毡绒性质
felt insert 油毛毡嵌入物
felt lap 毛毡抛光机
felt layer 毡垫；毡层；铺毡机
felt laying 铺贴油毡
feltless polishing bob 抛光用毛毡
feltless retainer 毡护圈
feltless ring 毡垫圈
felt-lined 毛垫毡
felt mark 抄网痕
felt nail 大帽钉；油毡钉；毡钉
felt-neoprene seal 氯丁橡胶毡密封
felt oil ring 毛毡油环
felt packing 毛毡填料；毛毡衬垫；填密毡
felt pad 绝缘垫；毛毡坐垫
felt pad bearing 油毛毡垫层支座
felt paper 油毡纸；毡纸；绝缘纸
felt pen 毡笔
felt polisher 毛毡抛光机
felt polishing bob 擦光毡；布轮
felt polishing disc 抛光毡轮；毛毡抛光圆盘
felt press machine 毛毡压烫机
felt retainer 毡圈；毛毡封座圈
felt ring 毡阻
felt-ring seal 毡圈密封；毡环密封
felt roll 毡辊
felt seal 毡密封；密闭毡条；毛毡封
felt side 纸正面；（纸的）正面；下面
felt stretcher 毛毡伸张器
felt strip 毡条；毡带；油毡条
felt tarpaulin 防雨毡
felt the lead 感觉水砣绳下坠情况（机械测深）
felt thermal insulation 保温毡
felt tightener 毛毯张紧器
felt underlayment 沥青浸渍油毡防水层
felt underlining 油毡衬底
felt undersarking 油毡衬底
felt washer 油毡钉；毡垫圈
felt washer pad 油毡衬垫
felt washing device 毛毯洗涤装置
felt wheel 毡轮
felt widening roll 麻花辊
felt wiper 毡刷；毛毡擦试器
felt wood 银光纹木材；经面纹木材；辐射纹木料
felt wrapped roll 压光辊
felty 毡状
felty body 毡状体
felty texture 毡状结构
female 凹形物
female adapter 压环；内螺纹接头；内螺纹过渡管接头；凹面接合器
female bend 包管弯头
female branch tee 内螺纹三通
female cap 凹形盖
female cavity 阴模模穴；阴模模槽
female center 反顶尖
female center plate 有孔中心板
female changing room 女更衣室
female cone 内锥；锥孔；空心圆锥体；内圆锥；锚杯（预应力）；凹锥
female connection 内螺纹连接；母扣连接
female connector 内孔连接器；套筒接合器
female contact 塞孔接点
female coupling 内丝扣套管；内丝扣接头；内螺纹连接管
female coupling tap 母锥；接头打捞母锥
female die 阴模；凹模
female drawing room 女休息室
female drill rod tap 钻杆打捞母锥
female elbow 内螺纹弯管接头
female ell 内螺纹弯头；母扣弯头
female end 承口端

female end of pipe 管（子）承口；管（子）承头；管子承端；内螺纹管口；陶管小端头
female fishing tap 打捞母锥
female fitting 阴螺纹管接头；内螺纹配件；内螺纹接头；内螺纹管接头配件；凹形配件
female flange 阴法兰；阴凸缘；带槽法兰；带槽凸缘；凹面法兰；凹面凸缘
female ga(u)ge 缺口样板；外测量规
female hairdressing shop 女理发店
female joint 钟口接头；嵌合接头；套筒接合；承口；承插式接头；插承接合
female lug 凹耳
female member 包容零件
female mo(u)ld 凹模；阴模
female parent 母本
female parent line 母本品系
female plant 雌株
female plug 插座
female receptacle 插座；插孔板
female rest room 女休息室
female rod coupling tap 钻杆接头打捞母锥
female room 女室
female rotor 凹形转子
female sapphire 淡色蓝宝石
female screw 阴螺纹；阴螺旋；阴螺钉；螺母；螺帽；内螺旋；内螺纹（螺栓）
female spanner 套筒扳手
female spline 内花键
female square 内直角尺；内正方形
female support ring 内螺纹过渡接头密封环
female surface 包容面
female swivel 阴转环
female templet 阴模板
female thread 阴螺纹；内螺纹
female thread ga(u)ge 螺纹量规；内螺纹量规
female union 管（子）内接头；内螺旋联管节（双向）；内螺纹联管节
female worker 女工
femerell 烟窗；通天用天窗；屋顶天窗；屋顶排气口；屋顶排气笼
femic 铁镁质
femic constituents 铁镁质组分
feminine ruby 淡红色红宝石
femitron 场发射电子微波管
Fe-Mn hydroxide coating 铁锰氢氧化物复膜
Fe-Mn hydroxide concretion 铁锰氢氧化物结核
femora 凹槽间距（陶立克柱顶中楣三陇板上）
femto 毫微微
femtoampere 毫微微安
femtogram 毫微微克
femtometer 非姆托米
femtovolt 毫微微伏
femtowatt 毫微微瓦特
femur 凹槽间平面（陶立克柱顶中楣三陇板上）；凹槽间分隔片（陶立克柱顶中楣三陇板上）；三槽板间分隔片
fen 淹水沼泽；沼泽群落；沼泽（地）；碱沼
fenadiazole 酚二唑
fenaksite 硅铁钠钾石
fenalamide 苯丙酰胺酯
fenamic acid 灭酸
fencder rod 防护杆
fence 翼刀；栅栏；篱笆；警戒线；警戒网；围墙；篱；围栏；电子篱笆；导流栅
fence bar 铁栅
fence diagram 三维地震剖面网络图；栅状图；立体投影图；透视断块图；栅栏图
fenced-off 无栅栏
fenced-off area 禁区
fence effect 栅栏效应
fence gate 栅门
fence height 篱笆高度；围墙高度
fence hurdle 防卫栅栏
fence in 用栅围起；用栅围进
fence jack 水泵拉杆拧紧工具
fence line 栅栏线
fence material 栅栏材料
fence nail 修篱笆用的钉子
fence net 墙网
fence of boards 木板栅栏
fence off 用栅隔开
fence of fascine 沉排篱护
fence of wheel 装轮子的栅栏
fence picket 栅栏柱

fence planting 篱垣种植
fence post 栅栏柱；围（栏）柱；篱笆桩
fence rail 防护梁
fence row 篱笆行
fence signs 篱笆标志
fence silo 围栏青储［贮］塔
fence stake 栅栏桩；栅栏柱；护栏柱
fence wall 围墙
fence wire 围栏铁丝
fence wire netting 围栏铁网
Fenchel wet expasion test 芬切尔湿伸长试验
fencing 围垣；围栏；筑围墙；筑墙材料；围墙
fencing accessories 栅栏附属物
fencing piste 围栏小路；击剑场地
fencing post 栅栏柱
fencing stake 栅栏立柱
fencing wall 护墙；围墙
fenclofenac 芬氯酸
fen colter 沼泽地用开沟器
fend 闪避
fender 炉围；靠帮；缓冲材；汽车挡泥板；排障器；围护物；舷板旁小碰垫；挡泥板；防御物；防冲物；防冲设备；护舷木；护舷材
fender anti-squeak 保护板减声片
fender apron 保护板；防护板；挡泥板
fender bar 固定护木护栏；固定护舷材；护舷材
fender beam 护舷梁；护舷（垫）木；护木；挡木筏
fender block 防撞缓冲块
fender board 挡泥板；挡板；护板；保护板
fender bolt 碰装栓
fender brace 挡（泥）板拉条；保护板拉条
fender bracket 保护架；保护板架
fender buffer 防撞缓冲装置
fender column 防撞柱；防护柱
fender course 防撞层
fender curb 路缘石
fender dolphin 防撞墩；防冲簇桩
fender filler 护板嵌料
fender frame 护舷构架
fender guard 固定护舷木护栏；固定护舷材；保险杠
fender guide 保护板导杆
fender hardware 护木固紧铁件；防撞装置的固紧件
fendering 碰垫
fendering device 防撞装置
fendering installation 防撞装置
fendering rail 防撞栏杆
fendering system 码头防撞设施；码头防撞设备；碰垫系统；防撞系统；防冲装置【港】
fendering unit 防撞装置
fender lamp 挡泥板灯；保护板灯
fender layout of berth 泊位护舷布置
fender log 护木
fender pier 码头防冲桩；护墩；护堤；防冲突堤
fender pile 靠船桩；码头碰装桩；码头护木；缓冲棒；护（舷）桩；护墩桩；防撞桩；防御桩；防冲桩；（公路安全带上的）围护桩；码头堤岸护舷桩
fender plank 护木（桥梁）
fender post 码头堤岸防护桩；围护桩；防撞柱；防护柱
fender rail 栏杆；护舷材
fender rattan 藤条碰垫
fender reaction 护舷反力
fender rope 软碰垫
fender Seibu 西武式护舷
fender set 挡泥具
fender spar 浮式护舷（木）；浮护木
fender splash shield 保护板防溅罩
fender strake 外壳加强列板
fender structure 围护结构
fender support arrangement 护舷支座布置
fender system 防冲设备；保护装置；防冲装置【港】
fender wall 支承壁炉矮墙；护墙；壁炉矮砖墙
fender walling 横护（舷）木
fender washer 护舷垫圈
fender welt 保护板的衬板
fender wheel 轮式碰垫
fending groin 防护堤；防冲堤
fending groyne 防护堤
fend off 垫开；挡开
fenestella 小窗；小型窗；小窗洞；窗状壁龛；［复］fenestallae
fenestra 小洞；窗孔；［复］fenestrae

fenestra joint 金属窗条接头(一种专利品)
fenestral 小型窗;小窗洞
fenestral fabric 格状组织
fenestral structure 窗孔构造
fenestra ovalis 卵圆窗
fenestra rotunda 圆窗;正圆窗
fenestrate 安窗的
fenestration 门窗布局;窗户配列;外墙窗洞组合
fenestration design 门窗布局设计
fenestration knife 开窗刀
fenestration scoop 开窗匙
fenestrato 组合窗
fenestra vestibuli 卵圆窗
fenestrule 小窗
fenetration 窗配合
fenghuanglite 凤凰石
Fe-Ni crosslinked modified bentonite 铁镍交联改性膨润土
Fenilon fibre 费尼龙纤维
fenipentol 苯戊醇
fenite 长霓岩
fenland 沼(泽)地;干沼泽(地);排水沼泽
fenmetramide 苯甲吗啉酮
fenoprofen 苯氧苯丙酸
fenoxypropazine 苯氧丙肼
fen peat 沼泽泥炭土;(沼)泽地泥炭;低位泥炭
fense 警戒线
Fenske equation 芬斯克方程
Fenske helix packing 芬斯克螺旋型填料
Fenske-Underwood equation 芬斯克—安特乌得方程式
fen soil 沼泽地土壤;沼地土;沼泽土
fenson 除螨酯
fenster 构造窗
fensulfothion 丰索磷
fenthion 倍硫磷
Fenton and photo assisted Fenton oxidation 芬顿和光辅助芬顿氧化
Fenton bearing metal 锌锡轴承合金
Fenton chemistry 芬顿化学
Fenton electrochemical treatment 芬顿电化学处理
Fenton-like reaction 类芬顿氧化
Fenton oxidation-coagulation process 芬顿氧化混凝工艺
Fenton oxidation technology 芬顿氧化技术
Fenton pretreatment 芬顿预处理
Fenton process 芬顿工艺
Fenton reaction 芬顿反应
Fenton's peroxidation 芬顿过氧化反应
Fenton's reagent 芬顿试剂
Fenton's reagent method 芬顿试剂法
Fenton system 芬顿体系
fenxizisu 风信子素
Fenzel glasses 芬泽尔眼镜
feoff 封地
feoffee 不动产承受人
feoffer 不动产让与人
feoffment 不动产交付证
feoffor 不动产让与人
feral(l)ite 铁铝岩;铁铝土
ferbam 福美铁
ferberite 钨铁矿
Ferderal Water Quality Administration[USEPA] 美国环境保护局联邦水质管理局
ferdisilicite 二硅铁矿
feretory 神龛;棺架
Feret's diameter 费雷特直径
Ferett triangle 三角坐标图
Fergana old land 费尔干纳古陆
ferghanite 水钒铀矿
fergusite 假白榴等色岩
fergusonite 褐钇铌矿;褐钇铌(钶)矿含量
fergusonite ore 褐钇铌矿矿石
fermate 福美铁
Fermat's conjecture 费马推测;费马猜想
Fermat's last theorem 费马最后定理
Fermat's principle 费马原理
Fermat's spiral 抛物螺(旋)线
Fermat's theorem 费马定理;费尔马定理
ferment 酝酿;发酵;酵素
fermentable 可发酵的
fermentating food 发酵食品
fermentating organism 发酵生物
fermentation 发酵

fermentation alcohol 发酵酒精
fermentation and acidogenesis of sludge 污泥发酵产酸
fermentation assimilation 发酵同化作用
fermentation brewing industry 发酵酿造业
fermentation chamber 发酵室
fermentation energy 发酵能
fermentation engineering 发酵工程
fermentation gas 生物气
fermentation heat 发酵热
fermentation industry 发酵工业
fermentation industry by-product 发酵工业副产物
fermentation insurance 发酵保险
fermentation liquid 发酵溶液
fermentation media 发酵介质
fermentation of waste liquor 废液发酵
fermentation plant 发酵装置;发酵(工)厂
fermentation powder 发酵粉
fermentation process 发酵过程
fermentation septization process 腐化发酵法
fermentation tank 化粪池;发酵池;发酵罐;发酵槽
fermentation treatment 发酵处理
fermentation tube 发酵管
fermentation tun 发酵槽
fermentation waste 发酵废物
fermentation wastewater 发酵废水
fermentative bacterium 发酵性细菌
fermented soil 熟土
fermenter 发酵罐
fermenting bacterium 发酵细菌
fermenting plant 发酵工厂;发酵厂
fermenting pond 化粪池
fermenting tank 化粪池;发酵池
Fermi 费米;毫微微米
Fermi-Dirac distribution function 费米—狄拉克分布函数
Fermi-Dirac distribution law 费米—狄拉克分布律
Fermi-Dirac gas 费米—狄拉克气体
Fermi-Dirac nucleus 费米—狄拉克核
Fermi-Dirac statistics 费米—狄拉克统计;费米—狄拉克统计法
Fermi's age 费米年龄
Fermi's age model 费米年龄模型
Fermi's breeder reactor 费米增殖反应堆
Fermi's characteristic energy level 费米特征能级
Fermi's constant 费米常数
Fermi's contact field 费米接触场
Fermi's contact interaction 费米接触相互作用
Fermi's decay 费米衰变
Fermi's distribution 费米分布
Fermi's distribution function 费米分布函数
Fermi's energy 费米能级;费米能
Fermi's gas 费米气体
Fermi's golden rules 费米黄金法则
Fermi's hole 费米空穴
Fermi's interaction 费米相互作用
Fermi's level 费米能级;费米数量级
Fermi's level diagram 费米能级图
Fermi's liquid 费米液体
Fermi's momentum 费米动量
Fermi's potential 费米势
Fermi's reactor 费米反应堆
Fermi's resonance 费米共振
Fermi's selection rules 费米选择定则
Fermi's sphere 费米球
Fermi's surface 费米面
Fermi's temperature 费米温度
Fermi's threshold 费米阈
Fermi's transition 费米跃迁
fermitron 场射管
fermium 镄
Fermi-Yang model 费米—杨(振宁)模型
fermorite 锶砷磷灰石;锶磷灰石
fern 凤尾草
fern basket 芒草篮
fern cushion 芒垫
Fernichrome 铁镍钴铬合金
fernico 铁镍钴合金
Fernico alloy 非尔尼可铁镍钴合金
Fernie shale 费尔尼页岩
Fernite 费尔奈特耐热耐蚀合金;非尔奈特铁镍铬合金
fern-leaf crystal 枝晶;锯齿状晶体
fernleaf hedge bamboo 凤尾竹
fern-like pattern 羊齿形花纹

fernophrastite 纤钒钙石
fern plaited rope 芒制草绳
fern-vitrite 蕨类微镜煤
ferocyhyte 六方纤铁矿
ferodo 摩擦片
ferractor 铁氧体磁放大器;铁电振荡器
Ferrair cement 铁矿水泥;矿渣水泥;抗硫酸盐水泥
ferralarch 自然演替系列
ferrallitic 铁铝的
ferrallitic soil 铁铝土
ferrallitization 铁铝富化作用
ferralsol 铁铝土
Ferranti effect 费兰蒂效应
Ferranti-Hawkins protective system 费兰蒂—霍金斯保护装置
Ferranti meter 费兰蒂电度表
Ferranti portable visco(si)meter 费兰蒂轻便式黏度计
Ferranti rectifier 费兰蒂整流器
Ferrari's ga(u)ge 费拉里斯计
Ferrari's instrument 费拉里感应测试仪;感应式仪表
ferrarisite 费水砷钙石
Ferrari's motor 费拉里电动机
ferrate 高铁酸盐;铁酸盐
ferrate oxidation 高铁酸盐氧化
ferrate pre-oxidation 高铁酸盐预氧化
ferratogen 核铁质
ferrazite 磷钡铅石;钡铅磷矿
ferreed(relay) 铁簧继电器
ferreed switch 铁簧开关
ferreous charry micro-electroanalysis 铁炭化微电解分析
Ferrero's formula 菲列罗公式
ferriage 轮渡费;渡船费;摆渡
ferriallophanoid 红色黏土
ferri-alluaudite 铁钠磷锰石
ferriammonium chromate 铬酸铵铁
ferriamphibole 高铁闪石
ferrian-cassiterite 铁锡石
ferri-annite 高铁铁云母
ferribacteria 铁细菌
ferribeidellite 铁拜来石
ferric 含铁的;铁的
ferric acetate 醋酸铁
ferric acid 高铁酸
ferric alum 铁明矾
ferric ammonium citrate 柠檬铁铵
ferric ammonium oxalate 草酸铁铵
ferric and alumin(i)um oxide content 氧化铁氧化铝总含量
ferric arsenate 砷酸铁
ferric bacteria 铁细菌
ferric blast furnace 焦炉型高炉
ferric carbide 碳化铁
ferric-carbon microelectrolysis-coagulation sedimentation process 铁炭微电解混凝沉淀法
ferric cement 铁质胶结物
ferric charge 含铁料
ferric chloride 六水三氯化铁
ferric chromate 铬酸铁
ferric citrate 柠檬酸铁铵
ferric compound 铁化合物
ferric cyanide 氰化铁
ferric cyanide blueprint 铁盐晒图
ferric ferrocyanide 亚铁氰化铁
ferric flouride 氟化铁
ferrichromspinel 铁硬铬尖晶石
ferric humate 腐殖酸铁
ferric hypophosphite 次磷酸铁
ferric induction 铁磁感应
ferric iodide 碘化铁
ferric ion 高铁离子
ferric iron 三价铁;二价铁
ferric iron oxide 氧化铁
ferric metasilicate 偏硅酸铁
ferric mineral spring water 铁质矿泉水
ferric nitrate 硝酸铁
ferric nuclein 核铁质
ferricopiapite 高铁叶绿矾
ferric oxalate 草酸铁
ferric oxide 西红粉;氧化铁;三氧化二铁
ferric oxide hydroxide 水合氧化铁
ferric oxide powder 氧化铁粉
ferric oxide rouge 氧化铁红;红土子

ferric protoporphyrin 高铁原卟啉
ferric pyrophosphate 焦磷酸铁
ferric red 过氯乙烯防腐瓷漆
ferric red oxide 赤色氧化铁
ferric resinate 树脂酸铁
ferricrete 铁质砾岩;铁质壳;铁结砾岩
ferricrust 硬铁核壳;铁质结壳
ferric salt 高铁盐;铁盐
ferric salt reducer 铁盐减薄液
ferric siallite 铁质硅铝土
ferric-siallitic soil 铁质硅铝土
ferric stearate 硬脂酸铁
ferric subsulfate solution 次硫酸铁溶液
ferric sulfate 硫酸铁
ferric sulfide 硫化铁
ferric sulfocyanate 硫氰酸铁
ferric sulphate 硫酸铁
ferric sulphate coagulant 硫化铁混凝剂
ferric tape 铁带
ferric type expansive cement 氧化铁型膨胀水泥
ferric vanadate 钒酸铁
ferricyanide 高铁氰化物;氰酸盐;铁氰化物
Ferri diffuser 费里扩散器
ferridravite 高铁镁电气石
ferrierite 镁碱沸石
ferriferous 含铁的
ferriferous fluxing hole 铁质熔洞
ferriferous gneiss 铁质片麻岩
ferriferous sediment 含铁沉积物
ferriferrous cyanide 氰化正亚铁;铁蓝
ferrihalloysite 铁镁埃洛石
ferrihydrite 水铁矿
ferrikatophorite 高铁红闪石
ferrikinetics 铁循环;铁动态
ferrimagnet 铁氧磁材料
ferrimagnetic limiter 亚铁磁性限幅器
ferrimagnetic material 铁氧磁材料
ferrimagnetic resonance 亚铁磁共振
ferrimagnetics 亚铁磁体
ferrimagnetism 亚铁磁性;亚铁磁体;铁氧体(磁性)
ferrimagnetism mineral 亚铁磁性矿物
ferrimanganese 铁锰齐
ferrimanganic 铁锰
ferrimolybdite 高铁钼华;钼华;水钼铁矿
ferrimontmorillonite 铁蒙脱石
ferrimuscovite 铁白云母
ferrinatrite 针钠铁矾
ferriparaluminite 铁副矾石
ferriphengite 铁硅白云母
ferriphlogopite 铁金云母
ferriporphyrin 高铁卟啉
ferripyrophyllite 铁叶腊石
ferri-sericite 铁绢云母
Ferri's induction 起始感应
ferrisodium 铁钠
ferristor 铁磁电抗器
ferris wheel 垂直转动的转轮
ferris wheel feeder 轮斗式喂料机
ferris wheel slurry feeder 转轮式水泥浆供料机
ferrisymplesite 非晶砷铁石
ferrite 铁素体;铁氧体;铁酸盐;纯铁(体);纯粒铁
ferrite absorbent material 铁氧体吸收材料
ferrite band 条状铁素体
ferrite banding 脱碳层;条状铁素体
ferrite bar 铁氧体磁棒;铁氧体棒
ferrite bar antenna 铁氧体棒(形)天线
ferrite bead 铁氧体磁环
ferrite-bead memory 铁氧体珠存储器
ferrite cement 铁酸盐水泥
ferrite ceramics 铁氧体陶瓷
ferrite core 铁氧体磁芯
ferrite-core memory 铁氧体磁芯存储器
ferrite device 铁氧体器件
ferrite electric(al) effect 铁电效应
ferrite filled waveguide 铁氧体波导管
ferrite film 铁氧体薄膜
ferrite garnet 钇铁石榴石
ferrite grain 铁素体晶粒
ferriteharmonic generator 铁氧体参量倍频器
ferrite keeper 铁氧体保通片
ferrite lamina 铁氧体薄片
ferrite-magnet field 铁磁磁场
ferrite memory 铁氧体存储器

ferrite memory core 铁氧体存储[贮]铁芯
ferrite microwave device 铁氧体微波器件
ferrite mineral 铁酸盐矿物
ferrite parameter amplifier 铁氧体参量放大器
ferrite phase 铁酸盐相
ferrite phase shifter 铁氧体移相器
ferrite-plate memory 铁氧体板存储器
ferrite-plate storage 铁氧体板存储器
ferrite process 铁氧体法
ferrite region 铁素体界
ferrite rod 铁氧体磁棒
ferrite-rod antenna 铁氧体棒(形)天线
ferrite-rod memory 铁氧体磁杆存储器
ferrite stainless steel 铁氧体不锈钢
ferrite steel 铁素体钢
ferrite storage 铁氧体存储器
ferrite switch 铁氧体开关
ferrite technique 铁氧体法
ferrite treatment 铁氧体处理
ferrite yoke 铁氧体磁轭
ferrithorite 铁针石;铁钍石
ferritic cement 铁质水泥
ferritic stainless steel 铁素体不锈钢
ferritic steel 铁素体钢
ferritising 铁素体化
ferritization 褐铁矿化
ferritizing 铁素体化
ferritremolite 高铁透闪石
ferritungstite 高铁钨华;铁钨华
ferroactinolite 铁阳起石
ferroakermannite 铁镁黄长石
ferroalloy 铁合金;铁基合金
ferroalloy furnace 铁合金炉
ferroalloy metal 铁合金金属
ferroalloy powder 铁合金粉末
ferroalloy process 铁合金法
ferroalloy work 铁合金厂
ferro-alumin(i)um 铝铁(合金);铁铝合金
ferro-alumin(i)um-manganese steel 铁铝锰钢
ferroamphibole 低铁闪石
ferroan dolomite 含铁白云石;富铁白云石
ferroan-magnesian retgersite 铁镁镍矾
ferro-anthophyllite 铁直闪石
ferroantigorite 铁叶蛇纹石;陕叶蛇纹石
ferroaugite 富铁辉石
ferrobarroisite 铁冻蓝闪石
ferroboron 铁硼合金
ferrobustamite 铁硅灰石
ferrocal 非劳克铝合金
ferrocalcite 铁方解石
ferrocarpholite 铁柱石
ferrocart core 铁粉芯
ferro-cement 钢丝网水泥;铁矿渣水泥
ferro-cement agricultural boat 农用钢丝网水泥船
ferro-cement corrugated sheet 钢丝网水泥波瓦
ferro-cement fishing boat 钢丝网水泥渔船
ferro-cement flume 钢丝网水泥渡槽
ferro-cement impregnated with polymer 聚合物浸渍钢丝网水泥
ferro-cement knock down pusher barge 钢丝网水泥分布顶推驳船
ferro-cement mobile home 钢丝网水泥活动房屋
ferro-cement modulus building 钢丝网水泥定型房屋
ferro-cement panel 钢丝网水泥板
ferro-cement pipe 钢丝网水泥管
ferro-cement ship 钢筋混凝土船
ferro-cement sluice 钢丝网水泥闸门
ferro-cement transport ship 钢丝网水泥运输船
ferrocene 二茂络铁
ferrocenyltriphenylsilane 茂铁三苯硅烷
ferrocerium 铈铁
ferro-chrome(iron) 铬铁;铁铬合金
ferro-chrome salts 木质磺酸铁铬
ferro-chrome silicon 硅铬铁
ferro-chromium 铬铁合金;铬铁
ferro-clad brick 包铁的砖
ferro-clad refractory product 含铁耐火制品
ferroclinoholmquisite 斜铁锂闪石
ferrocobalt 钴铁
ferrocobaltite 铁辉钴矿
ferrocolumbite 铌铁矿;铌铁矿含量
ferro-compound 二价铁化合物
ferro-concrete 钢丝网混凝土;钢筋混凝土

ferro-concrete pipe 钢筋混凝土管
ferro-concrete vessel 钢筋混凝土船
ferro-crete 含铁硅酸盐水泥
ferrocyanic acid 亚铁氰酸
ferrocyanide 亚铁氰化物
ferrocyanide blue 氰亚铁酸蓝
ferrocyanide chloride 氰亚铁氯化物
ferrocyanide ferrous article 氰亚铁的铁制件
ferrocyanide floor hardener 铁素体楼面硬化剂
ferrocyanide metal 氰化亚铁金属
ferrocyanide orthosilicate 铁橄榄石
ferrocyanide pipe 铁管
ferrocyanide process 亚铁氰化物法
ferrocyanide product 铁制品
ferrocyanide sulphate 硫酸铁
ferrocyanide timber connector 连接木材的铁件
ferrocyanide titanate 钛榴石
ferrocyanide trap 铁制垃圾臭气阻闭器
ferrocyanide wire 铁丝
ferrod 铁氧体棒(形)天线;铁磁杆
ferrodistortive antiferro-electrics 铁畸变性反铁电体
ferrodistortive ferro-electric(al) 铁畸变性铁电体
ferrodo 摩擦插片;摩擦材料
ferrodolomite 铁白云石
Ferrodur 混凝土防水剂(一种专卖品)
ferro-dynamic(al) ceramics 铁磁电动陶瓷
ferrodynamic(al) instrument 铁磁电动系仪表;铁磁电动式仪表;动铁式仪表
ferrodynamic(al) relay 动铁式继电器
ferrodynamometer 铁磁式功率计
ferro-eckermannite 铁铝钠闪石
ferroedenite 低铁淡闪石
ferro-elastic 铁弹性的
ferro-elastic crystal 铁弹性晶体
ferro-elastic effect 铁弹性效应;铁弹效应
ferro-elastic material 铁弹性材料
ferro-elastics 铁弹体
ferro-electric(al) 铁电的;铁电性的
ferro-electric(al) amplifier 铁电(式)放大器
ferro-electric(al) axis 铁电轴
ferro-electric(al) carrier 铁电载流子
ferro-electric(al) cell 铁电元件
ferro-electric(al) ceramic display 铁电陶瓷显示
ferro-electric(al) ceramics 铁电陶瓷
ferro-electric(al) compound 铁电化合物
ferro-electric(al) condenser 铁电式电容器
ferro-electric(al) converter 铁电变换器
ferro-electric(al) crystal 铁电晶体
ferro-electric(al) Curie point 铁电居里点
ferro-electric(al) domain 铁电畴
ferro-electric(al) ferrite 铁电铁氧体
ferro-electric(al) film 铁电膜
ferro-electric(al) glass-ceramic composition 铁电玻璃陶瓷组合体
ferro-electric(al) hysteresis 电滞回线
ferro-electric(al) induced phase transition 铁电感应相变
ferro-electric(al) material 铁电材料
ferro-electric(al) memory 铁电体存储器;铁电存储器
ferro-electric(al) memory element 铁电存储矩元
ferro-electric(al) memory matrix 铁电存储矩阵
ferro-electric(al) modulator 铁电调制器
ferro-electric(al) ordering 铁电有序化
ferro-electric(al) parametron 铁电参量器
ferro-electric(al)-photo-conductor device 铁电光导器件
ferro-electric(al) property 铁电特性
ferro-electric(al) shift register 铁电体移位寄存器
ferro-electric(al) shutter 铁电光闸
ferro-electric(al) state 铁电态
ferro-electric(al) storage 铁电存储器
ferro-electric(al) transformer 铁电变压器
ferro-electricity 铁电性;铁电
ferro-electrics 铁电质;铁电体;亚铁电晶体
ferro-electronic 铁电电子性
ferrofining 铁精制
ferrofluid 铁磁流体
ferro-friedelite 铁红锰矿
ferrogabbro 铁辉长岩
ferrogedrite 铁铝直闪石
ferro-glass 钢丝玻璃;络网玻璃
ferroglaucophane 铁蓝闪石
ferro-graph 铁粉记录图;铁磁示波器

ferro-graphy 铁粉记录术
ferrogum 橡胶磁铁
ferrogunite 喷射钢丝网水泥
ferrohastingsite 低铁钠闪石
ferrohexahydrite 六水绿矾
ferroholmquistite 铁锂闪石
ferrohornblende 铁角闪石
ferrohumite 铁斜硅镁石
ferrohydrite 褐铁矿
ferrohypersthene 铁紫苏辉石
ferro-in-clave 波纹片
ferrojacobsite 铁锰尖晶石
ferro-johannsenite 铁锰钙辉石
ferrokaersutite 铁钛闪石
ferrokinetics 铁循环
ferrolite 铁矿岩;铁矿石;铁屑集料
ferrolithic concrete 费洛里特混凝土
ferrolizardite 铁钠蛇纹石
ferrolum 包铅钢板
ferromagnesian 含铁镁的
ferro-magnesian retgersite 铁镁镍矾
ferromagnesite 低铁菱镁矿
ferro-magnesium 铁镁齐
ferro-magnesium catalyst 铁镁催化剂
ferro-magnesium material 铁镁质
ferro-magnesium mineral 铁镁矿物
ferro-magnet 磁铁;铁磁体
ferro-magnetic 强磁性;铁磁性;铁磁物;铁磁的
ferro-magnetic alloy 铁磁性合金
ferro-magnetic amplifier 铁磁放大器
ferro-magnetic anisotropy 铁磁各向异性
ferro-magnetic armature 铁磁电枢
ferro-magnetic bar 铁磁性棒
ferro-magnetic body 铁磁体
ferro-magnetic ceramic 铁磁陶瓷;陶瓷磁体
ferro-magnetic colloid 铁磁性胶质
ferro-magnetic core 铁磁芯
ferro-magnetic crack detector 铁磁探伤器
ferro-magnetic crystal 铁磁晶体
ferro-magnetic Curie point 铁磁性居里点
ferro-magnetic excitation 铁磁励磁
ferro-magnetic film 磁性薄膜
ferro-magnetic glass-ceramics 铁磁性微晶玻璃
ferro-magnetic material 铁磁性物质;铁磁(性)材料;亚磁性材料;强磁性材料
ferro-magnetic modulator 铁磁调制器
ferro-magnetic moment 铁磁矩
ferro-magnetic nuclear resonance 铁磁核共振
ferro-magnetic oxide mineral 铁磁氧化矿物
ferro-magnetic oxide powder 铁氧体磁粉
ferro-magnetic parametric amplifier 铁磁参量放大器
ferro-magnetic powder 铁磁粉
ferro-magnetic relay 铁磁继电器
ferro-magnetic resonance 铁磁共振
ferro-magnetic resonance line width 铁磁共振线宽
ferro-magnetics 铁磁质;铁磁学;铁磁体
ferro-magnetic shield 铁磁屏蔽
ferro-magnetic shift register 铁磁体移位寄存器
ferro-magnetic sorbent 铁磁吸着剂
ferro-magnetic store 铁磁存储器
ferro-magnetic substance 强磁性物质;铁磁(物)质
ferro-magnetic suspension 铁磁性悬浮液
ferro-magnetic thin film 铁磁性薄膜
ferro-magnetic transition 铁磁转变
ferro-magnetism 铁磁学;铁磁性
ferro-magnetism mineral 铁磁性矿物
ferro-magnetography 铁磁性记录法
ferro-magnon 铁磁振子
ferro-manganese 锰铁;铁锰合金
ferro-manganese alloy 锰铁合金
ferro-manganese iron 锰铁(合金)
ferro-manganese-silicon 锰铁硅
ferro-manganese steel 锰钢
ferro-manganin 铁锰铜合金
ferro-manganoan precipitate 铁锰沉淀(物)
ferrometer 磁性测绘器;磁导计;铁磁计
ferromolybdenum 钼铁(合金)
ferron 非朗铁镍铬合金
ferro-nickel 铁镍合金;镍铁
ferro-nickel ingot 镍铁锭
ferro-nickel iron 铁镍合金;镍铁
ferronickelpaltinum 铁镍铂矿
ferrooxidant 铁氧化剂

ferropargasite 铁韭闪石
ferro-pennantite 铁锰鳞绿泥石
ferropericlase 铁方镁石
ferropexy 铁固定
ferrophalt 加筋沥青
ferrophosphor 高磷铸铁
ferrophosphorous aggregate 磷酸亚铁集料;磷酸亚铁骨料
ferro-phosphorus 磷铁(合金)
ferropicotite 铁铬尖晶石
ferro-porit bearing 渗硫铁系含油轴承
ferro-Portland cement 含铁波特兰水泥
ferroprobe 铁探头;铁磁探测器
ferroprussiate paper 蓝晒图纸
ferro-pseudobrookite 低铁假板钛矿
ferropyr 铁铬铝电阻丝合金
ferroresonance 铁磁谐振;铁磁共振
ferro-resonant computing 铁磁共振计算
ferro-resonant flip-flop 铁磁共振触发器
ferrorichterite 铁钠透闪石
ferro-rich ultramafic rock 富铁质超镁铁岩
ferrosalite 铁次透辉石
ferro-sand 铁砂
ferroschallerite 铁砷硅锰矿
ferro-selenium 铁硒合金;硒铁
ferroselite 白硒铁矿
ferro-shotcrete 喷射钢筋混凝土
ferrosifumaras 富马酸铁
ferrosil 热轧硅钢板
ferro-silicate 铁硅酸盐
ferro-silicium 硅铁;铁硅合金
ferro-silico alumin(i)um 硅铝铁
ferro-silico-manganese 硅铁锰
ferro-silicon 硅铁(合金);硅钢高硅铸铁
ferro-silicon alloy 硅铁合金
ferro-silicon-alumin(i)um 硅铝铁
ferro-silicon calcium alloy 硅钙合金
ferro-silicon-nickel 硅镍铁
ferro-silicon slag 硅铁渣
ferro-silicon-titanium 硅钛铁;高硅钛铁
ferrosilite 硅酸铁;铁辉石
ferrospinel 铁磁尖晶石
ferrostan 电镀锡钢板
ferrostatic pressure 钢水静压力;铁水静压力
ferrosteel 钢性铸铁;灰口铸铁
ferrostilpnomelane 低铁黑硬绿泥石
ferrostrunzite 铁施特伦茨石
ferrotantalite 钽铁矿;钽铁矿含量
ferro-tantalum 钽铁
ferrotellurite 铁黄碲矿
ferrotemp pyrometer 测铁水高温计
ferrotephroite 铁锰橄榄石
ferrothorite 铁针石
ferrotitanium 钛铁合金;钛铁
ferrotremolite 铁透闪石
ferrotschermakite 铁钙闪石
ferrotungsten 铁钨合金;钨铁
ferrotychite 碳铁钠矾
ferrotype 铁版照相
ferro-ultramafic rock 铁质超镁铁岩
ferrouranium 铁铀合金;铀铁
ferrous 亚铁的;铁的;二价铁(的)
ferrous acetate 乙酸亚铁
ferrous acid 铁酸
ferrous ammonium sulfate 硫酸亚铁铵
ferrous analysis 钢铁分析学
ferrous and non-ferrous metals 黑色及有色金属
ferrous and non-ferrous rivet 黑色及有色金属铆钉
ferrous and non-ferrous screw and nut 黑色及有色金属螺钉和螺母
ferrous arsenate 砷酸亚铁
ferrous bicarbonate 碳酸氢亚铁
ferrous carbonate 碳酸亚铁
ferrous chloride 氯化亚铁
ferrous component 钢铁件
ferrous cyanide 氰化亚铁
ferrous dense material 铁质密致材料
ferrous die casting 铁类金属压模铸造
ferrous disulfide 二硫化铁
ferrous floor hardener 地面硬化用含铁金属粉末
ferrous fluoride 氟化亚铁
ferrous fumarate 富马酸亚铁;富马酸铁
ferrous hydroxide 氢氧化亚铁
ferrous hypophosphite 次磷酸亚铁

ferrous industry 黑色金属工业
ferrous ion 亚铁离子
ferrous iron 亚铁;二价铁
ferrous-manganese nodule 铁锰质结核
ferrous-manganese stripe 铁锰质条纹
ferrous material 黑色金属材料;钢铁材料
ferrous metal 黑色金属;铁类金属;铁金属
ferrous metal analysis 黑色金属分析
ferrous metal commodities 黑色金属矿产
ferrous metal extrusion press 黑色金属挤压机
ferrous metallurgy 钢铁冶金(学);黑色冶金学
ferrous metal pipe 黑色金属管
ferrous metal protection characteristics 铁金属保护特性
ferrous metasilicate 硅酸亚铁
ferrous nitrate 硝酸亚铁
ferrous oxalate 草酸亚铁
ferrous oxide 氧化亚铁
ferrous phosphate 磷酸亚铁
ferrous powder 铁粉
ferrous rhodanate 硫氰酸亚铁
ferrous rolling mill 钢材轧机
ferrous silicon 硅铁
ferrous sulfamate 氨基磺酸亚铁
ferrous sulfate 硫酸亚铁
ferrous sulfate poisoning 硫酸亚铁中毒
ferrous sulfide 硫化亚铁
ferrous sulphate 硫酸亚铁
ferrous titanate 钛酸亚铁
ferrous trap 铁存水弯
ferrous waste 废钢铁
ferrovanadium 钒铁合金;钒铁合金
ferrovanadium steel 钒钢
ferrowinchite 铁蓝透闪石
ferrowollastonite 铁硅灰石
ferrox 草酸亚铁
ferroxcube 立方结构铁氧体
ferroxplana 高频铁氧磁性材料;超高频软磁铁氧体
ferroxyl indicator 铁锈指示剂
ferroxyl test 铁锈试验
ferrozincite 铁红锌矿
ferrozirconium 锆铁
Ferrozoid 非劳左特铁镍合金
ferruccite 氟硼钠石
ferruginosity 含铁性
ferruginous 含铁的;铁质;铁锈色的
ferruginous-bearing mudstone 含铁质泥岩
ferruginous bonding layer 铁质胶结层
ferruginous bonding sand 铁质胶结砂
ferruginous breccia 铁质角砾岩
ferruginous cement 含铁水泥;铁质水泥;铁质胶料;铁质胶结物;铁质胶结
ferruginous chert 铁质燧石
ferruginous clay 含铁黏土;铁质黏土
ferruginous coefficient 铁质系数
ferruginous concretion 铁质结核
ferruginous conglomerate 铁质砾岩
ferruginous deposit 含铁沉积物
ferruginous discharge 含铁水排放;含铁废水
ferruginous dross 含铁浮渣
ferruginous environment 铁质环境
ferruginous incrustation 铁质结皮作用
ferruginous laterite 铁质砖红壤;铁质红土
ferruginous lateritic soil 铁质砖红壤性土
ferruginous limestone 含铁石灰石;铁质石灰石
ferruginous manganese ore 高铁锰砂
ferruginous mudstone 铁质泥岩
ferruginous quartz 铁石英
ferruginous quartzite 含铁石英岩
ferruginous quartz sandstone 铁质石英砂岩
ferruginous rock 铁质岩
ferruginous rock facies 铁质岩相
ferruginous sand 赭砂;含铁砂;铁锈色砂
ferruginous sandstone 含铁砂岩;铁质砂岩
ferruginous sediment 铁质沉积物
ferruginous shale 铁质页岩
ferruginous siltstone 铁质粉砂岩
ferruginous soil 铁质土;富铁土
ferruginous spring 含铁矿泉;铁质泉
ferruginous thermal water 含铁热水
ferruginous tuffite 铁质沉凝灰岩
ferrule 线圈管;装金属箍;箍;金属箍(圈);金属包头;接头阀,嵌饰,稳定器;套箍;套圈
Ferrule alloy 费鲁尔合金

Ferrule brass 费鲁尔铅黄铜
ferrule of pile 桩头铁圈;桩箍
ferrum pulveratum 铁粉
Ferry 费瑞铜镍合金
ferry 轮渡;渡轮;渡口;渡船;摆渡
ferry berth 渡船码头
ferry boat 轮渡;渡轮;渡船;摆渡船
ferry-boat fleet 渡船队
ferry-boat service 渡船营运;渡船交通
ferry-boat traffic 轮渡运输;轮渡交通
ferry bridge 轮渡引桥;吊桥;浮桥
ferry bridge of automobile 汽车轮渡栈桥
ferry cable 浮桥缆索;渡船缆索
Ferry cell 费瑞电池
ferry crossing 渡口
ferry dock 轮渡码头
ferry fee 渡船费;摆渡费
ferry flat 平底渡船;方驳式渡船
ferry harbo(u)r 轮渡港(口)
ferry house 轮渡码头候船室;轮渡码头管理处;轮渡候车室
ferry(ing) equipment 轮渡设备
ferry landing 轮渡码头
ferry landing pier 轮渡码头;渡口
ferry landing stage 轮渡码头;渡口
ferry launch 交通艇
ferry line 浮桥缆索;渡船缆索;渡船航线
ferry man 渡船工人;轮渡工人
Ferry metal 费瑞电阻丝合金
ferry passenger 渡船乘客
Ferry pathway 法理途径
ferry place 渡口
ferry rack 轮渡码头靠船架
ferry ramp 轮渡引桥;轮渡码头引桥;渡轮引桥
ferry rope 过河缆索;渡船缆索
ferry slip 轮渡斜引道;轮渡码头港池
ferry station 轮渡站
ferry steamer 轮渡式船舶;轮渡;渡轮
ferry terminal 水陆码头;轮渡码头;水陆联运站;车船联运港
ferry traffic 摆渡交通;滚装运输;轮渡运输;轮渡交通;开上开下运输;水平装卸法
ferry transfer bridge 轮渡码头引桥
ferry-type boat 轮渡式船舶
fersilicite 硅铁矿
fersmanite 硅钛钙石;硅钠钛钙石
fersmite 铌钙石;铌钙矿
fertile absorber 有效吸收剂
fertile branch 结果枝
fertile element 可转换元素
fertile farmland 良田
fertile irrigation 施肥灌溉
fertile isotope 再生同位素
fertile land 沃地;高产地;沃沃土地
fertile material 母体材料;燃料
fertile peat pot 含肥泥炭盆
fertile plain 丰产地
fertile soil 沃土;沃地;肥(沃)土
fertile telome 实顶枝
fertile uranium 能增生的铀
fertiliser 肥料
fertility 土地生产力;繁殖力
fertility erosion 肥力侵蚀
fertility evaluation 肥力评价
fertility natality 出生率
fertility of seawater 海水肥力
fertility rate 生育率
fertilize land 肥沃土地;肥地
fertilizer 化肥;肥料
fertilizer apparatus 排肥器;排肥盘
fertilizer application 施肥
fertilizer application by sprinkling 用喷洒法施肥
fertilizer applicator 施肥机
fertilizer attachment 施肥附加装置
fertilizer can 排肥箱
fertilizer clutch 排肥器离合器
fertilizer contamination 化肥污染
fertilizer disk 排肥盘
fertilizer distributor 排肥装置;撒肥机
fertilizer divided into separate applications 分几次施肥料
fertilizer drill 施肥机
fertilizer drill unit 施肥装置
fertilizer element 肥料元素;肥料要素

fertilizer feed 排肥盘
fertilizer feed drive 排肥器传动装置
fertilizer formula 肥料配方
fertilizer gate 排肥闸门
fertilizer gearing 排肥器传动装置
fertilizer irrigation 施肥灌溉
fertilizer machinery 施肥机械
fertilizer mechanism 排肥机构
fertilizer mineral 肥料工业原料矿物
fertilizer placement drill 追肥器
fertilizer plant 化肥厂;肥料工厂
fertilizer pollution 肥料污染
fertilizer quantity lever 排肥量调节杆
fertilizer quantity regulator 排肥量调节器;施肥量调节器
fertilizer test 肥料试验
fertilizer treatment 肥料处理
fertilizer type explosive 肥料型炸药
fertilizer unit 排肥装置
fertilizer works 肥料工厂
fertilizing action 肥料作用
fertilizing equipment 施肥器具;施肥机具
fertilizing ingredient 肥料成分
fertilizing value of activated sludge 活性污泥肥效值
fertilizing value of sewage 污水肥料价值
fertilizing value of sludge 污泥肥料价值
Ferula 阿魏属
ferulaldehyde 阿魏醛
ferulene 阿魏烯
ferulic acid 阿魏酸
fervanite 水钒铁矿
Fery cell 费里电池
Fery prism 费里棱镜
Fery pyrometer 费里高温计
Fery radiation pyrometer 费里辐射高温计
Fery spectrograph 费里光谱仪
Fessenden oscillator 费森登振荡器
Fessler compound 费斯勒化合物
festival hall 大会堂
festival light 节日灯
festival room 宴会厅
festivals 节日;节假日
festival theater[theatre] 大会场
festoon 缘垛;花彩褶(冰缘);花彩(弧);垂花饰【建】
festoon chains 花环链幕
festoon cloud 花彩云
festoon cross-bedding 花彩交错层理
festoon cross-stratification structure 花彩弧状交错层理构造
festoon curtain 垂花幕
festoon curve 尖浪形曲线
festoon drape 垂花幕
festoon dryer 悬挂式干燥室;长环悬挂烘燥机
festoonery 花彩装饰;彩饰
festooning 悬挂式干燥
festoon light 垂花饰【建】;彩灯
festoon lighting 灯饰;带式装饰照明;垂带装饰照明;霓虹灯;灯彩链
festoon of island 弧内列岛
festoon-shape curve 花环曲线;垂花曲线
festoon staining 外墙雨水污纹;垂花污纹;花纹形玷污
fetch 转航【船】;浪程;海湾对岸两点的距离;取数;取出;对岸距离;读取;风区;风浪区(长度);风程
fetch a high price 卖得高价
Fetcham injector 菲特切尔曼注射器(一种专利的木材防腐剂注射器)
fetch area 风区
fetch(a)way 摇落;开始航行;松脱
fetch bit 取位;取出位;按位取数
fetch cycle 取(出)周期;取得循环
fetch front 风区前沿
fetch graph 风区图表;风区图
fetch headway 开始航行;前进
fetch home 恢复原状;收紧锚(索)
fetch home the anchor 绞锚
fetching 捞
fetch length 吹送距离;吹程;风区长度
fetch-load trace 取装跟踪;取装装入跟踪
fetch off 由岸上搬至船
fetch of snow 吹程
fetch of wave 波浪行程;波距

fetch operand 取操作数
fetch policy 读取规则
fetch protection 取保护
fetch rear 风区后沿
fetch rule 读取规则
fetch sequence 取出序列
fetch stage 读数阶段
fetch sternway 开始后退
fetch-store execution ordering 读写次序
fetch time 指令读出时间
fetch to 向风;上风
fetch unit 读取部件
fetch up 紧急停止
fetch violation 取数违法
fetch way 开始航行;松脱
fetch width 风区宽度
fetid bituminous limestone 臭沥青石灰岩
fetid odo(u)r 腐臭味
fetishism 拜物教
fetishists 拜物教徒
fetron 高压结型场效应管
fettered gyroscope 指北陀螺仪
fettle 修坯;修炉;修补;除屑
fettler 清整工;清理工;调整工;梳理工废料清除工
fettler material 补炉材料
fettlin 修坯
fettling 修整;修补炉衬;修补炉壁;铸件清理;清砂;涂炉材料;补炉
fettling bench 清理台
fettling chisel 凿断凿;修理凿;切边凿
fettling comb 抄针耙
fettling disk 修整圆盘;刮油涨环
fettling hammer 修缮(用)锤
fettling hole 炉底孔
fettling knife 修坯刀具;修补刀
fettling machine 修坯机;铸件清理机;补炉机
fettling magnesite grain 冶金镁砂
fettling material 补炉材料
fettling table 清理台
feu 黏土层;下覆岩石
feud 封地
feudal aristocracy 封建贵族
feudal economy 封建经济
feudal land rent 封建地租
feudal lord 封建领主
feudal system 封地制
feudal town 封建时代建立的城市;城堡;封建(时代的)城市
Feussner prism 费斯那棱镜
Feussner spark generator 费斯那火花发生器
Fe-vaesite 铁方硫镍矿
feverfew 白菊花
feverish 发烧的
feverish market 变动不定的证券市场
few-group analysis 少群分析
few-group diffusion theory 少群扩散理论
few-region approximation 少区近似
few stay system 少索体系(斜拉桥)
fexible unit 通用装置
Feynman diagram 费因曼图
Feynman-Gell Mann hypothesis 费因曼盖尔曼假设
Feynman-Gell Mann universal formula 费因曼—盖尔曼普适公式
Feynman integral 费因曼积分
Feynman propagator 费因曼传播子
Feynman's superfluidity theory 费因曼超流性理论
fezatione 苯噻硫酮
f fire-resistant clothing 防火衣;防火服
fiar curve 光滑曲线
fiard 低狭湾
fiat money 法令货币
fiat standard 法定标准;法定本位
fiber 纤维
fiber assembly 纤维束
fiber backing 纤维背衬
fiber ball filter material 纤维球滤料
fiber balling 纤维结块
fiber base 纤维衬底
fiber blanket pipe 纤维毡型管套
fiber board 纤维板;硬化纸板;木丝(板)
fiberboard box 纤维箱;纤维板箱;纤维硬纸匣
fiber board can 纤维板箱罐
fiber board ceiling 纤维板平顶;纤维板吊顶

fiberboard mill wastewater 纤维板厂废水
fiber board nail-based sheathing 可钉纤维衬板
fiber breakage 断头率(纤维)
fiber buckling 纤维层曲
fiber buffer 光缆填料；光纤缓冲层
fiber building board (压制)纤维板；木丝板；纤维壁板
fiber building mat 纤维垫板
fiber bundle 纤维束；纤维丛；光(学)纤(维)束
fiber bush 夹布胶木衬套；纤维衬套
fiber cable assembly 光缆组件
fiber can 纤维罐；纤维板桶
fiber cement 纤维(增强)水泥；石棉水泥
fiber cement board 纤维水泥板；水泥纤维板
fiber cement composite 纤维增强水泥
fiber cement slab 纤维水泥板
fiber ceramics 纤维陶瓷
fiber composite 纤维复合材料
fiber concentration by weight 纤维重量率
fiber concrete 纤维性混凝土；纤维混凝土
fiber conduit 硬纸(板)导管；纤维管(道)；安装纤维套(管)
fibercone press 双锥辊挤浆机
fiber connector 光纤(活动)连接器
fiber container 纤维容器
fiber content of asbestos ore 含棉率
fiber content of ore 含棉率
fibercord 纤维绳
fiber core 纸板筒；纤维芯
fiber count 码数；光纤表
fiber covered plywood 纤维合成层板；纤维面层板
fiber crop 纤维作物
fiber cutter 纤维切割器
fiber diameter 纤维直径；光纤直径
fiber diameter index 纤维直径指数
fiber disc sander 布轮磨光机
fiber doormat 擦鞋垫
fiber drum 纸桶
fiber duct 纤维导管模具；纤维导管；硬纸导管；硬纸管状模具
fiber dust 纤维性粉尘；纤维性尘埃
fibered 筋灰
fibered asphalt pavement 纤维沥青路面
fibered glass 玻璃纤维
fibered plaster 麻刀灰浆；石膏纤维灰浆；纤维灰泥
fiber elastic 纤维弹性的
fiber electrometer 悬丝静电计
fiber elongation method 拉丝法(测定玻璃黏度方法)
fiber elongation visco(si)meter 纤维伸长法黏度计
fiber entanglement 纤维缠结
fiber feedstock area 供丝区
fiber fill 填充用纤维；纤维填塞物
fiber filler 纤维填料
fiber filter 纤维过滤器
fiberflock 纤维绒
fiber flying 飞丝
fiber forming efficiency 成棉率
fiberglass bag filter 玻璃纤维过滤袋
fiberglass base-glass mat 玻璃纤维基薄屋面板
fiberglass batt 玻璃棉絮
fiberglass batt insulation 玻璃纤维保温材料
fiberglass blown wool 喷吹施工玻璃棉
fiberglass bolt 玻璃纤维锚杆
fiberglass braided cable jacket 玻璃纤维编织电缆套
fiberglass braided wire 玻璃丝编织线；玻璃丝编织电线
fiberglass carpet backing 玻璃纤维地毯背衬
fiberglass coated fabric 涂层玻纤织物
fiberglass coating 玻璃纤维涂层；玻璃纤维套
fiberglass cored bituminous felt 玻璃纤维夹心沥青油毛毡
fiberglass covered wire 玻璃丝包线
fiberglass curtain 玻璃纤维窗帘
fiberglass drum 玻璃钢转鼓
fiberglass fabric 玻璃纤维织物
fiberglass filter 玻璃纤维滤布
fiberglass filtering medium 玻璃纤维滤料
fiberglass fishing boat 玻璃钢鱼船
fiberglass forms 玻璃纤维模板
fiberglass furniture 玻璃钢家具
fiberglass grid 玻璃纤维格栅
fiberglass home insulation 住宅用玻璃棉绝热材料

fiberglass insulated wire 玻璃纤维绝缘电线
fiberglass insulator 玻璃纤维绝缘体
fiberglass magnet wire 玻璃丝包电磁线
fiberglass mat(te) 玻璃纤维毡
fiberglass mo(u)lded grating 玻璃纤维模塑格栅
fiberglass packing 玻璃纤维填料；玻璃纤维垫
fiberglass pipe 玻璃纤维管
fiberglass pipe section 玻璃纤维棉管壳
fiberglass plant 玻璃纤维工厂
fiberglass plastic 玻璃纤维塑料
fiberglass reinforced bituminous felt 玻璃纤维油毡
fiberglass reinforced elastomer 玻璃纤维增强弹性体
fiberglass reinforced plastic container 聚酯玻璃钢集装箱
fiberglass reinforced plastic form 玻璃纤维增强塑料模；玻璃钢模
fiberglass reinforced plastic hull 玻璃纤维增强塑料船体
fiberglass reinforced polyester resin 玻璃丝增强聚酯树脂
fiberglass roofing felt 玻璃纤维屋面毡
fiberglass roofing mat 玻璃纤维屋面毡
fiberglass rotor blade 玻璃纤维旋翼桨叶
fiberglass screening 玻璃纤维窗纱
fiberglass sheet 玻璃纤维板
fiberglass shingle 玻璃纤维(沥青)瓦
fiberglass sleeving 玻璃纤维套管
fiberglass staple fiber 定长玻璃纤维
fiberglass tarpaulin 玻璃纤维帆布
fiberglass thermoplastics 玻璃纤维增强热塑性塑料
fiberglass tile 纤维玻璃钢瓦
fiberglass tunnel 玻璃纤维隧道；玻璃纤维管道
fiberglass ventilation pipe 玻璃纤维风筒
fiberglass wall covering 玻璃纤维贴墙布
fiberglass wastewater 玻璃纤维废水
fiberglass wick 玻璃纤维灯芯
fiber-grade monomer 化纤单体
fiber hybrid 混杂纤维
fiber in combination 连生纤维
fiber-increasing coefficient 增棉系数
fibering 成纤
fibering machine 成纤机
fibering viscosity 成纤黏度
fiber intensity 纤维强度
fiberized asbestos 纤维化石棉
fiberizer 开棉机
fiber lamp 纤维导光灯
fiber length 纤维长度
fiber metallurgy 纤维冶金
fiber netting 纤维网
fiber optic(al) cable 光缆；光导纤维电缆
fiber optic(al) cable connector 光缆连接器
fiber optic(al) communication 光纤通信
fiber optic(al) communication system 光纤通信系统
fiber optic(al) cranioscope 光学束脑内窥镜
fiber optic(al) dot 纤维光点
fiber optic(al) gyro 光纤陀螺
fiber optic(al) image transfer device 光纤维传象装置；纤维光学图象传递装置
fiber optic(al) ribbon 光学维带
fiber optic(al) ring 光纤环路
fiber optic(al) scanner 纤维光学扫描器
fiber optic(al) waveguide 光纤波导
fiber optics 光导纤维
fiber optics cable 光导纤维电缆
fiber optics illuminator 纤维照明系统
fiber optics power meter 光导纤维功率计
fiber optics recorder 光学纤维记录器
fiberoptronics 纤维光电子学
fiber orientation 纤维取向
fiber packing 纤维填料；纤维衬垫；纤维密封垫；纸板衬垫
fiber packing fraction 纤维填充系数
fiber pan 硬纤维模板；纤维模板
fiber pen 纤维光笔
fiber pipe 纤维管(道)
fiber placement 纤维配置
fiber plaster 纤维石膏
fiber plug 纤维栓塞
fiber prepreg 纤维预浸料
fiber profile 纤维纵切面；光纤剖面
fiber protrusion 玻璃纤维毛刺

fiber-reinforced 纤维补强的
fiber-reinforced cement 纤维增强水泥
fiber-reinforced ceramic material composite 纤维增强陶瓷基复合材料
fiber-reinforced ceramics 纤维增强陶瓷
fiber-reinforced composite material 纤维增强复合材料；纤维强化复合材料
fiber-reinforced concrete 纤维增强混凝土；纤维加强混凝土；纤维混凝土
fiber-reinforced gypsum board 纤维增强石膏板
fiber-reinforced low-pH cement construction board 纤维增强低碱度水泥建筑平板
fiber-reinforced materials 纤维增强材料；纤维加强材料；纤维补强材料
fiber-reinforced metal 纤维增强金属材料
fiber-reinforced metal matrix composite 纤维增强金属基复合材料
fiber-reinforced pipe 纤维管
fiber-reinforced plastic blower 纤维增强塑料鼓风机
fiber-reinforced plastic fan 纤维增强塑料风扇
fiber-reinforced plastic product 纤维增强塑料产品
fiber-reinforced plastic rod 纤维增强塑料筋
fiber-reinforced plastics 纤维加强塑料；纤维增强塑料
fiber-reinforced polester 纤维增强聚酯
fiber-reinforced polyester sheets 纤维增强聚酯片材
fiber-reinforced polymer concrete 纤维增强聚合物混凝土
fiber-reinforced polymer sheet 纤维增强聚酯片材
fiber-reinforced reinforcement 纤维增强
fiber-reinforced stress 纤维应力；纤维强度
fiber-reinforced thermoplastics 纤维增强热塑性塑料
fiber-reinforced tube 纤维管
fiber-reinforced underlay 纤维底层
fiber-reinforced wallboard 纤维墙板
fiber-reinforced washer 纤维垫圈；纤维衬垫
fiber-reinforced (wiring) conduit 纤维管道
fiber reinforcement 纤维增强；纤维配筋
fiber reinforcement refractory 纤维增强耐火材料
fiber-resin depositor 纤维树脂施敷器
fiber root 纤维根
fiber rope 纤维缆(索)；纤维绳；麻绳
fiber rug 纤维地毯；棉经纸纬地毯
fiber saturation point (木材的)纤维饱和点
fiber show 露丝
fiber silk 人造丝
fibers in the fibrous layers 纤维层内纤维
fiber size for reinforcement 增强型浸润剂
fiber steel 钢纤维
fiber strengthened 纤维强化的
fiber stress at proportional limit 比例极限纤维应力
fiber tear 纤维撕裂
fiber-to-the-curb 光纤到路边
fiber type premo(u)lded joint 纤维式预塑缝
fiber unit 光纤单元
fiber wall paper 纤维墙布
fiber wash 纤维散失
fiber washer 纤维垫圈；纸板垫圈
fiber waveguide 光波导
fiber with bent ends 带弯曲端头纤维
fiber with hooked ends 带弯钩端头纤维
fiberwood 纤维板
fiber wool 玻璃纤
fibestos 塑胶(一种醋酸纤维素)
Fibonacci coefficient 斐波纳契系数
Fibonacci function 斐波纳契函数
Fibonacci generating function 斐波纳契数生成函数
Fibonacci merge 斐波纳契归并法
Fibonacci method 斐波纳契法；黄金分割法
Fibonacci number 斐波纳契数
Fibonacci number system 斐波纳契数系
Fibonacci polynomial 斐波纳契多项式
Fibonacci search 斐波纳契寻优法；斐波纳契搜索；斐波纳契检索法
Fibonacci sequence 斐波纳契序列；斐波纳契数列
Fibonacci series 斐波纳契级数；菲波那奇数列
Fibonacci series sorting 斐波纳契数分类
fibore 基础轴承
fibrae radials 辐状纤维
fibrae zonulares 小带纤维
fibrage 纤维层；纤维编织

fibrated 纤维化的;含纤维的
fibrated asphalt emulsion 纤维化沥青乳液
fibrated bitumen emulsion 加纤维的沥青乳化液
fibrated composition 纤维成分
fibrated compound 纤维合成物
fibrated concrete 纤维性混凝土
fibrated concrete product 纤维混凝土制品
fibrated concrete slab 纤维混凝土板
fibrated concrete tube 纤维混凝土管
fibrated dampproofing 纤维防潮的
fibrated emulsified asphalt 含纤维的乳化沥青
fibration 纤维形成
fibrator 纤维离解机
fibratus 毛状云
fibre = fiber
fibre absorption 光纤吸收
fibre accelerating mode 纤维加速方式
fibre acoustic(al) sensor 光纤声传感器
fibre active connector 光纤有源连接器
fibre adhesive 纤维状黏合剂
fibre air-entrained concrete 纤维加气混凝土
fibre alignment 纤维排列
fibre alignment connector 光纤对准连接器
fibre amplifier 光纤放大器
fibre analysis 纤维分析
fibre array 纤维排列
fibre array diagram 纤维排列图
fibre axial displacement sensor 光纤轴向位移传感器
fibre axis 纤维轴线
fibre ball filtration 纤维球过滤
fibre basket 纤维篮
fibre block 纤维垫块
fibre blowing 纤维吹制法
fibreboard case 纤维板箱
fibre board finishing 纤维板贴面
fibreboard panel 硬质纤维板;纤维板
fibre bonding 纤维黏合
fibre borescope 光纤管道镜
fibre box 纤维板箱
fibre box package 纤维板箱包装
Fibre Building Boad Development Organization 纤维建筑板发展组织(英国)
fibre building sheet 建筑用纤维板
fibre bundle transfer function 光纤束传递函数
fibre cable 光纤维电缆
fibre cam 纤维板箱
fibre circuit 光纤线路
fibre cladding 光纤包层
fibre clad rope 油麻绳包裹的钢索
fibre classifier 纤维筛分仪
fibre coarseness 纤维粗度
fibre coating 光纤涂层
fibre coating material 光纤涂覆材料
fibre coating technique 纤维涂层技术
fibre collecting zone 纤维沉降区
fibre collection hood 纤维沉降室
fibre composite material 纤维增强复合材料
fibre composition 纤维配比
fibre concentrator 光纤集线器
fibre contact point 纤维接触点
fibre coupled diode-pumped laser 光纤耦合二极管抽运激光器
fibre coupled laser diode 光纤耦合激光二极管
fibre coupled power 光纤耦合功率
fibre coupler 光纤耦合器
fibre coupling 光纤耦合
fibre cross-talk 光纤串扰
fibre cut 纤维切断
fibre cutting 纤维切断
fibre cutting machine 纤维切断机
fibre cutting tool 光纤切割工具
fibre damage 纤维损伤
fibred asphalt pavement 纤维沥青路面;加纤维沥青路面
fibre debris 纤维碎片
fibre delay line 光纤延迟线
fibre density 纤维密度
fibre deposition 纤维凝聚
fibre detector coupling 光纤检波器耦合
fibre diagram 纤维长度分布图
fibre diagram machine 纤维长度分析仪
fibre diameter monitor 光纤直径监控器
fibre dimensional stability 光纤尺寸稳定性

fibre dispersion 光纤色散
fibre displacement modulated sensor 光纤位移调制传感器
fibre distribution 纤维分布
fibre distribution box 光纤分线盒
fibre distribution frame 光纤配线架
fibred manifold 纤维流形
fibre drag 纤维间分离阻力
fibre drawing 光纤拉制
fibre drawing furnace 光纤拉丝炉
fibre drum 纸桶
fibre dust 纤维性粉尘
fibre eigenvalue equation 光纤本征值方程
fibre elasticity test 纤维弹性试验
fibre end 纤维端
fibre end density 纤维端密度
fibre entanglement 纤维缠结作用
fibre entry point 纤维进入点
fibre extent 纤维延伸度
fibre extracting machine 剥纤维机
fibre faceplate 光纤面板
fibre feed arm 丝嘴
fibre fibre friction 纤维间摩擦
fibre filler 纤维填料
fibre film bundle 光纤维膜束
fibre filter paper 纤维滤纸
fibre fineness 纤维细度
fibre fineness indicator 纤维细度指示器
fibre finish 纤维处理剂(润滑抗静电)
fibre flow uniformity 纤维流匀度
fibre forming 纤维成型
fibre forming polymer 纤维聚合物
fibre fracture 纤维裂缝
fibre furnish 纤维配比
fibre fusion 纤维熔合
fibre gasket 纤维衬垫;硬纸垫密片
fibre gear 纤维材料齿轮;树脂纤维齿轮
fibreglass 玻璃丝
fibreglass aerocore 埃罗科尔玻璃纤维
fibreglass epoxy 环氧玻璃钢;玻璃纤维增强环氧树脂;玻璃纤维钢板;环氧纤维玻璃钢
fibreglass epoxy antenna 玻璃钢天线
fibreglass insulation 绝热玻璃纤维
fibre glass insulation material 玻璃纤维绝缘材料;玻璃纤维绝热材料;玻璃纤维保温材料
fibreglass optics 玻璃纤维光学
fibreglass panel 玻璃纤维板
fibreglass printed circuit board 玻璃纤维印刷电路板
fibreglass reinforced aluminium foil 玻璃纤维加固铝箔
fibreglass reinforced metal 纤维增强金属
fibreglass reinforced nylatron cage 玻璃纤维增强尼龙保持架
fibreglass reinforced plastics 纤维玻璃增强塑料;玻璃丝增强塑料;玻璃钢
fibreglass reinforced polyester 玻璃纤维增强聚酯
fibreglass reinforcement 玻璃纤维增强;玻璃加强纤维
fibreglass staple fiber 定长玻璃纤维
fibreglass steel cooling tower 玻璃钢冷却塔
fibreglass tendon 玻璃丝刀筋
fibreglass tile 玻璃钢瓦
fibre glide bearing 纤维滑动轴承
fibre grade 纤维级
fibre grease 纤维状润滑脂;纤维润滑脂
fibre grinding wheel 纤维砂轮
fibre guide 纤维波导管;光纤波导
fibre gypsum board 纤维石膏板
fibre hammer 硬化纸板锤
fibre harness 光纤捆束
fibre hazard 光纤危害
fibre holder 光纤夹持器
fibre hybrid composite 混杂纤维复合材料
fibre identification 纤维鉴别
fibre identity period 纤维等同周期
fibre impact tester 纤维冲击强力试验仪
fibre in compression 压力纤维
fibre incrustation 纤维外皮
fibre insulated wire 纤维绝缘线
fibre insulating(grade) material 纤维绝缘材料;纤维绝热材料;纤维保温材料
fibre insulation board 软质纤维板
fibre insulation layer 纤维绝缘;纤维绝热层;纤维保温层

fibre insulator 纤维绝缘体
fibre inteference figure 纤维干涉图
fibre integrity 纤维完整性
fibre in tension 纤维拉伸
fibre in the loop 光纤环路系统
fibre jacket 光纤套层
fibre junction 光纤接续
fibre laser 光纤激光器
fibre laydown hood 纤维沉降室
fibre length 纤维长度
fibre length array 纤维长度排列
fibre length control 纤维长度的控制
fibre length distribution 纤维长度分配
fibre length tester 纤维长度试验仪
fibre length variation 纤维长度变化
fibre light guide 纤维光导;光纤波导
fibre link 光纤线路;纤维通信线路
fibrelock fabric 纤维锁结织物
fibre longitudinal view 纤维纵视图
fibre loss 纤维损耗;光纤损耗
fibre loss modulated fibre optic(al) sensor 光纤损耗调制光纤传感器
fibre lubrication 纤维润滑
fibre mat 纤维层
fibre material 纤维材料
fibre matting 纤维缠结
fibre media 纤维介质
fibre membrane 纤维薄壁
fibre merit figure 光纤品质因数
fibre metal 纤维金属
fibre metallurgy 纤维冶金
fibremeter 纤维定量喂给控制装置
fibre migration 纤维转移
fibre mist eliminator 纤维雾消除剂
fibre modification 纤维改性
fibre moisture regain tester 纤维回潮测试仪
fibre morphology 纤维形态学
fibre morphometer 纤维形态测视仪
fibre multiplexer 光纤多路复用器
fibre network yarn 纤维网络纱
fibre node 光纤节点
fibre of common mugwort 艾绒
fibre optic(al) 光导(纤维)的
fibre optic(al) accelerometer 光纤加速度计
fibre optic(al) AM laser transmitter 光纤调幅激光发射器
fibre optic(al) analog(ue) receiver 光纤模拟接收机
fibre optic(al) analog(ue) transmitter 光纤模拟发射机
fibre optic(al) array 纤维光学集束
fibre optic(al) attenuation tester 光纤衰减测试仪
fibre optic(al) attenuator 光纤衰减器
fibre optic(al) block 纤维光块
fibre optic(al) borescope 光纤管道镜
fibre optic(al) bronchoscope 光纤支气管镜
fibre optic(al) bundle 光纤束;光导纤维束
fibre optic(al) butting connector 光纤对接连接器
fibre optic(al) cable 光纤电缆;光缆;光导纤维电缆;纤维光缆
fibre-optic(al) camera 光导纤维摄像机
fibre-optic(al) catheter 光纤导管;纤维光学光导管
fibre optic(al) cathode ray tube 光纤阴极射线管
fibre optic(al) chemical sensor 光纤化学传感器
fibre optic(al) colo(u)r centre sensor 光纤色心传感器
fibre optic(al) communication 光纤通信
Fibre optic(al) Communication Information Society 光纤通信信息协会
fibre optic(al) concentration sensor 光纤浓度传感器
fibre optic(al) coupler 光纤耦合器
fibre optic(al) coupling 光纤耦合;纤维光学耦合
fibre optic(al) current sensor 光纤电流传感器
fibre optic(al) data coupler 光纤数据耦合器
fibre optic(al) data link 光纤数据联路;光纤数据线路
fibre optic(al) device 纤维光学装置
fibre optic(al) digital receiver 光纤数字接收机
fibre optic(al) distribution frame 光纤配线架
fibre optic(al) dosimeter 光纤剂量计
fibre optic(al) endoscope 光纤内窥镜
fibre optic(al) Ethernet transceiver 光纤以太网

收发信机
fibre optic(al) faced tube 纤维光学屏面管
fibre optic(al) faceplate 纤维光学面板
fibre optic(al) fluorescence thermometer 光纤荧光温度计
fibre optic(al) flyingspot scanner 光纤飞点扫描器
fibre optic(al) gas analyzer 光纤气体分析仪
fibre optic(al) guidance 光纤制导
fibre optic(al) gyroscope 光纤陀螺仪
fibre optic(al) helical microbend sensor 光纤螺旋微弯传感器
fibre optic(al) hydrophone 光纤水听器
fibre optic(al) identifier 光纤识别器
fibre optic(al) illuminator 纤维光学照明器
fibre optic(al) image dissection camera 光纤析像照相机
fibre optic(al) interface device 光纤接口设备
fibre optic(al) interferometer 光纤干涉仪
fibre optic(al) interferometric sensor 光纤干涉测量传感器
fibre optic(al) isolater 光纤隔离器
fibre optic(al) light guide 光纤光导
fibre optic(al) link 光纤连接
fibre optic(al) liquid level alarm 光纤液位报警器
fibre optic(al) liquid level sensor 光纤液位传感器
fibre optically coupled cascaded image intensifier 光纤耦合级联式图像增强器
fibre optic(al) magnetic field sensor 光纤磁场传感器
fibre optic(al) magnetostrictive sensor 光纤磁致伸缩传感器
fibre optic(al) member 纤维光学元件
fibre optic(al) memory 光纤存储器
fibre optic(al) mixer 光纤混合器
fibre optic(al) mode converter 光纤模式变换器
fibre optic(al) modem 光纤调制解调器
fibre optic(al) multimeter 光纤万用表
fibre optic(al) multiplexer 光纤多路复用器
fibre optic(al) multiport coupler 光纤多端口耦合器
fibre optic(al) nuclear hardening 光纤核子硬化
fibre optic(al) oscilloscope recording tube 纤维光学记录示波管
fibre optic(al) patch panel 光纤接线板
fibre optic(al) penetrator 光纤贯穿器
fibre optic(al) phase modulator 光纤相位调制器
fibre optic(al) photodetector 光纤光电检测器
fibre optic(al) polisher 光纤抛光机
fibre optic(al) Raman laser 光纤拉曼激光器
fibre optic(al) receiver 光纤接收机
fibre optic(al) rediation sensor 光纤辐照传感器
fibre optic(al) reflective sensor 光纤反射传感器
fibre optic(al) repeater 光纤中继器
fibre optic(al) ring media access controller 光纤环媒体存取控制器
fibre optic(al) ring resonator 光纤环形谐振器
fibre optic(al) rod coupler 光纤棒状耦合器
fibre optic(al) rod multiplexer-filter 光纤棒状复用滤光器
fibre optic(al) scan converter 光纤扫描变换器;纤维光学扫描变换器
fibre optic(al) scrambler 光纤置乱器;光纤搅频器
fibre optic(al) screen 纤维光学屏幕
fibre optic(al) sensor 光纤传感器
fibre optic(al) sheath 光纤包皮
fibre optic(al) spectral analyzer 光纤光谱分析仪
fibre optic(al) spectrometer 光纤分光仪
fibre optic(al) spectrophone 光纤光声仪
fibre optic(al) splice 光纤接头
fibre optic(al) strain sensor 光纤应变传感器
fibre optic(al) telecommunication cable 光纤通信光缆
fibre optic(al) terminus 光纤终端
fibre optic(al) test instrument 光纤测试仪
fibre optic(al) thermometer 光纤温度计
fibre optic(al) transceiver 光纤收发两用机
fibre optic(al) transmission equipment 光纤传输设备
fibre optic(al) transmission system 光纤传输系统
fibre optic(al) transmitter 光纤发射机
fibre optic(al) unit 光纤单元
fibre optic(al) video modem 光纤视频调制解调器
fibre optic(al) visible-infrared spinscan radiometer 光纤可见红外自旋扫辐射仪
fibre optic(al) winder 光纤卷绕机

fibre optics 纤维光学技术;纤维光学;光学纤维;光导纤维
fibre optics amplifier 纤维光学放大器
fibre optics camera 纤维光学照样机
fibre optics connector adaptor 光纤连接器转接器
fibre optics connector interface 光纤连接器界面
fibre optics coupled image amplifier 光纤耦合图像放大器
fibre optics coupling 纤维光学耦合
fibre optics faceplate 纤维光学面板
fibreoptics image dissection camera 光纤析像摄像机
fibre optics intensifier 纤维光学增强器
fibre optics scan system 光纤扫描系统
fibre optics scrambler 光纤保密器
fibre orientation 纤维取向
fibre packing density 纤维填充密度
fibre paper 水合纤维纸板
fibre pattern 纤维排列;光纤图形
fibre peat 纤维泥炭
fibre periphery 纤维周边
fibre plate 纤维片
fibre porosity 纤维的多孔性
fibre preform 光纤预制件
fibre press 纤维压榨机
fibre pulse compression 光纤脉冲压缩
fibre radiation damage 光纤辐射损害
fibre reactive dye 纤维活性染料
fibre recovery 纤维回收
fibre recovery machine 纤维回收机
fibre-reinforced composite material 纤维增强复合材料
fibre-reinforced plastics 纤维增强塑料
fibre-reinforced plastic speed boat 玻璃钢快艇
fibre-resin depositor 纤维树脂施敷器
fibre ribbon 光纤带
fibre ribbon connector 光纤带缆连接器
fibre ring 纤维环
fibre ringer 光纤信号器
fibre ring interferometer 光纤环形干涉仪
fibre rope eye splicing 纤维绳眼环插接
fibre rope long splicing 纤维绳长插接
fibre rope short splicing 纤维绳短插接
fibre rug 棉经纸纬地毯
fibre sanding disk 纤维砂轮
fibre saturated level 纤维饱和度
fibre saturated point 纤维饱和点(木材)
fibre saturation point (木材的)纤维干湿饱和点;纤维饱和点(木材)
fibre scattering 光纤散射
fibre scintillator 纤维闪烁体
fibre scope 纤维(光学)镜;纤维彩色图像器
fibre selector 光纤选择器
fibre sensor 纤维传感器
fibre separation index 纤维分离度
fibre separator 纤维分选器
fibre setter 纤维定形机
fibre sheath 光纤包皮
fibre sheet 纤维纸板;纤维网
fibre shuffling 纤维推移
fibre slippage 纤维滑溜性
fibre sorter 纤维长度测试仪
fibre source 光纤光源
fibre space 纤维空间
fibre splicing technique 光纤联结技术
fibre staining 纤维着色
fibre strain 纤维应变;光纤应变;光纤变形
fibre strain-induced sensor 光纤应变传感器
fibre strand 纤维辫
fibre strength 纤维强度
fibre strengthened metal 纤维强化金属
fibre stress 纤维强度;纤维应力;纵向拉应力;顺纤维方向应力
fibre stress formula 纤维应力公式
fibre stripper 光纤剥皮器
fibre structure 纤维构造
fibre stuff 纤维浆料
fibre suspension 纤维悬浮液
fibre system 光纤系统
fibre taper 纤维轴向不均匀性
fibre tendering 纤维损伤
fibre tensile strength 纤维抗拉强度;光纤抗拉强度
fibre tensile strength tester 纤维拉伸强力试验机
fibre term 纤维项

fibre texture 纤维结构
fibre to-fiber bond 纤维间结合键
fibre to-fibre contact 纤维间接触
fibre tow 纤维屑
fibre transfer 纤维转移
fibre transfer function 光纤传递函数
fibre trap 光纤阱
fibre tube 硬纸板管;丝管;纤维管(道)
fibre turf 纤维泥炭
fibre twisted pair 光纤双绞线对
fibre underlay 纤维底层
fibre video trunk 光纤视频干线
fibre wallboard 纤维墙板
fibre wall thickness 纤维壁厚
fibre washer 纤维垫圈;纤维垫片;纤维垫板
fibre wax 纤维蜡
fibre web 纤维网
fibre web compensation device 纤网补偿装置
fibre weight 纤维量
fibre withdrawal force 纤维抽出力
fibre yarn 纤维丝
fibril 纤丝;细纤维;小纤维;微丝
fibril angle 纤丝角(度)
fibrillar 纤丝的
fibrillary 纤丝的
fibrillary contractions 纤维性收缩
fibrillation 形成原纤维;纤维性颤动;原纤化作用
fibrillolytic 溶解原纤维的
fibring 纤维表示
fibrinoid degeneration 纤维素样变性
fibrinous 纤维素性的
fibrinous exudate 纤维素性渗出物
fibro 澳大利亚石棉水泥板
fibroblastic 纤维变晶状;纤维变晶状
fibrocalcific 纤维钙化的
fibrocalcification 纤维钙化
fibrocrystalline 纤维结晶状
fibrofelt 纤维毡
fibroferrite 纤铁矾
fibroglia 纤维胶质
fibroglia fibrils 纤维胶质原纤维
fibrogram 纤维图
fibroid 类纤维的
fibroid induration 纤维硬结
fibrolaminar 纤维层的
fibrolite 纤维板;夕线石;硅线石
fibrolit rock 细砂线石岩
Fibronia 草制纤维
fibrooptic(al) recorder 纤维光束记录器
fibroplasia 纤维增生
fibroplastic 纤维形成的
fibroplastic tile 纤维性塑料砖;韧塑性砖
fibroreticulate 纤维网的
fibrosampler 纤维取样器
fibrosclerosis 纤维硬化
fibrosis 纤维化;纤维变性
fibrotile 石棉水泥波形瓦
fibrous 纤维状的;纤维的
fibrous aggregate 纤维状集合体;纤维质骨料
fibrous alumina 纤维状氧化铝
fibrous ankylosis 纤维性强直
fibrous asphalt compound 纤维沥青合成物
fibrous assembly 纤维状集合体
fibrous bakelite 纤维酚醛树脂
fibrous bituminous compound 纤维沥青合成物
fibrous blastic texture 纤维变晶结构
fibrous body 毡状体
fibrous braiding 纤维编包
fibrous calcite 纤维状方解石
fibrous capsule 纤维包膜
fibrous carbon 纤维状碳
fibrous cellulose 纤维状纤维素
fibrous coat 纤维层
fibrous composite (material) 纤维复合材料
fibrous composition 纤维成分;纤维配比
fibrous composition seal(ing) ring 纤维密封环
fibrous composition washer 纤维垫圈
fibrous compound 纤维合成物
fibrous concrete 纤维增强混凝土;纤维(性)混凝土
fibrous covering 纤维绝缘层
fibrous crystal 纤维状结晶
fibrous degeneration 纤维变性
fibrous dust 纤维性粉尘;纤维性尘埃
fibrous encapsulation 纤维包裹

fibrous eutectic alloy 纤维状共晶合金
fibrous failure 纤维状断口
fibrous-felted board 纤维毡板;纤维铺装板
fibrous-felted panel 纤维毡板
fibrous filler 纤维填料
fibrous filter 纤维质过滤嘴;纤维材料过滤器
fibrous fleece 纤维网
fibrous fracture 纤维状裂纹;纤维(状)裂缝;纤维状断裂;纤维状断口;纤维裂面;纤维裂痕
fibrous glass 玻璃丝板;纤维玻璃;玻璃纤维
fibrous glass form board 玻璃纤维型板;玻璃纤维模板
fibrous glass insulation 玻璃纤维绝缘;玻璃棉保温材料
fibrous glass lagging 玻璃纤维绝缘套
fibrous glass reinforced plastics 玻璃纤维增强塑料
fibrous glass-reinforced plastics laminate 玻璃纤维增强塑料薄板
fibrous glass reinforcement 玻璃纤维增强料
fibrous glass strip 玻璃钢薄板
fibrous granoblarblastic texture 纤维花岗变晶结构
fibrous granularblastic texture 纤维粒状变晶结构
fibrous grease 纤维状润滑脂
fibrous gypsum 纤维石膏;纤石膏
fibrous gypsum board 纤维石膏板
fibrous gypsum ore 纤维石膏矿石
fibrous heat insulator 纤维状保温材料
fibrous insulant 纤维绝热材料
fibrous insulating material 纤维状绝缘材料
fibrous insulation 纤维隔(离)层;纤维绝缘;纤维隔热(层)
fibrous iron 纤维状铁;纤维断口铁
fibrous jointing material 纤维连接材料
fibrous layer 纤维层
fibrous lime mortar 掺纤维石灰砂浆
fibrous limestone 纤维石灰岩;纤维头石灰岩
fibrous loam soil 松软壤土;成层黏土质土
fibrous long spacing 纤维长间距
fibrous magnesium silicate 硅酸镁纤维
fibrous mat dust filter 纤维层过滤除尘器
fibrous material 纤维材料;纤维质材料
fibrous mat filter 纤维垫过滤器
fibrous matter 纤维物质
fibrous membrane 纤维膜
fibrous metal 纤维状金属
fibrous mor 纤维状的森林粗腐殖质
fibrous network structure 纤维状网络结构
fibrous packing 纤维填料
fibrous paper 纤维纸
fibrous particle 纤维状颗粒
fibrous peat 纤维(状)泥炭;纤维性泥岩;纤维泥浆
fibrous plaster 纤维石膏板;纤维泥灰;纤维灰泥;纤维灰浆抹灰工;纤维灰浆;麻刀灰
fibrous plasterer 纤维石膏工;纤维灰浆抹灰工
fibrous plastering 纤维灰浆抹面;纤维灰浆磨面
fibrous pore 纤维孔
fibrous powder 纤维状粉末
fibrous raw material 纤维原料
fibrous recovery 纤维回收
fibrous red iron ore 纤维赤铁矿
fibrous refractory 纤维状的耐火材料
fibrous reinforcement 纤维增强;纤维加强
fibrous reinforcement material 纤维增强材料
fibrous reticulum 纤维网
fibrous ring 纤维环
fibrous rock 纤维状岩石
fibrous rupture 纤维状断口
fibrous septum 纤维间隔
fibrous shotcrete 纤维喷浇混凝土
fibrous silicate 纤维状硅酸盐
fibrous sinter 纤维状泉华
fibrous slab 纤维板
fibrous slurry 纤维泥浆
fibrous soil 纤维性土(壤);成层土(壤)
fibrous stratum 纤维层
fibrous strength 纤维强度
fibrous-stroked dressing 细纹修琢(石面)
fibrous structure 纤维构造;纤维构造
fibrous surface 纤维状表面
fibrous tactoid 纤维质胶体
fibrous talc 纤维型滑石粉
fibrous tank 纤维玻璃池
fibrous texture 纤维状结构;纤维结构
fibrous thermal insulation 纤维热绝缘层

fibrous turf 纤维泥炭;纤维泥浆
fibrous type 纤维型
fibrous union 纤维黏合
fibrous waste 纤维废水
fibrous weld 纤维状焊缝
fibrous wool 纤维状绒毛
fibrous zeolite 纤维沸石;杆沸石
fibrous zone 纤维带
fibula 饰针;采占曲铁;饰针;采石曲铁
ficelle 绳子式的;灰褐色的
fiche duplicater[fiche duplicator]胶片复印机
fichtelite 朽松木烷;枞脂石;斐希德尔石;白脂晶石;澳松石
Fick's diffusion law 菲克扩散定律
Fick's equation 菲克方程
Fick's first law 菲克第一定律
Fick's formula 菲克公式
Fick's law 菲克定律
Fick's law of diffusion 菲克扩散定律
Fick's principle 菲克原理
ficosapentenoic acid 二十碳五烯酸
fictile 型造品;陶器的;陶制品;黏土塑造的
fiction 虚构
fictious dielectric(al) constant 虚拟介电常数
fictious transaction 买空卖空
fictitious 假想的
fictitious account 挂名账户;虚拟账目
fictitious assets 挂名资产;虚构资产
fictitious bargain 买空卖空
fictitious bill 空头支票;空头票据
fictitious bill of lading 空头提单
fictitious binding energy 虚设结合能
fictitious boundary 虚想边界
fictitious capital 虚构资本
fictitious correction 虚拟校正
fictitious cost 虚构成本
fictitious craft 虚构船舶
fictitious data 假设数据
fictitious date 假年龄
fictitious deed of trust 通用信托契据;可续编性信托契据
fictitious design load 虚拟设计荷载
fictitious design load factor 虚拟设计荷载系数
fictitious dividend 虚构股息;空头股利
fictitious document 通用文件;可续编性文件
fictitious drag coefficient 假阻力系数
fictitious eccentricity 假想偏心距
fictitious equation 虚拟方程
fictitious equator 虚拟赤道;虚拟赤道;假赤道
fictitious existence 虚构存在
fictitious factor 虚假因素;虚构因素
fictitious force 虚拟力
fictitious front 假锋【气】;伪锋【气】
fictitious graticule 虚拟线栅;虚构图网;经纬网
fictitious instrument 通用契据;可编性契约
fictitious latitude 虚构纬度;假(想)纬度;假设纬度;假定纬度
fictitious lift coefficient 假升系数
fictitious load 虚载(荷);虚(拟)荷载;假设荷载;假(设)负载;模似负载
fictitious longitude 虚构经度;虚拟经度;假设经度;假定经度
fictitious loxodrom 假等角航线
fictitious loxodrome 虚构恒向线;虚等斜线
fictitious loxodromic curve 虚等斜线
fictitious magnetic pole 假想磁极
fictitious mathematical surface 假设数学面
fictitious mean sun 假平太阳
fictitious meridian 虚子午线
fictitious model date 假模式年龄
fictitious molecular weight 虚拟分子量
fictitious mortgage 通用抵押契约;可续编性抵押契约
fictitious name 营业名称;企业名称;店号
fictitious observable 虚拟观测量
fictitious observation 虚拟观测
fictitious observation equation 虚拟观测方程
fictitious paper 空头支票;空头票据
fictitious parallel 虚构纬线;假定纬圈
fictitious pendulum 虚摆
fictitious person 法人
fictitious pole 虚构极;假极
fictitious potential 虚位
fictitious power 虚功(功)率

fictitious prestressing 虚构预应力
fictitious price 虚价
fictitious primary 假想基色
fictitious profit 虚假利润
fictitious quotation 虚构的价格
fictitious rhumb line 虚构恒向线;虚等斜线;假恒向线
fictitious source 虚拟源
fictitious spectral density 虚拟谱密度
fictitious stretching 模似张拉
fictitious tensioning 模拟拉紧
fictitious time 假时
fictitious trading 诈欺交易
fictitious transaction 虚假交易
fictitious use 虚假需求;虚构需求
fictitious year 虚构年;假年
fictive signal(l)er 模似信号机
fictive temperature 假想温度
fid 支撑材;桅栓;钉子
fiddle 台座;餐具框
fiddleback 细波花纹
fiddleback chair 提琴式背靠椅
fiddle back figure 虎皮纹;提琴背(板)纹
fiddleback grain 提琴背板木纹
fiddle block 提琴形滑车;提琴式滑车
fiddle bow 曲线型船首
fiddle drill 弓钻;弓转钻
fiddler's gear 起重机滑轮
fid(d)ley 锅炉舱棚;炉舱棚顶
fid(d)ley deck 炉舱棚顶甲板
fid(d)ley grating 炉舱棚顶格子板
fid(d)ley hatch 炉舱棚顶口
fidelity 确限;重现精度;逼真度;保真度
fidelity bond 信用债券
fidelity criteria 保真度标准
fidelity criterion 保真度判据
fidelity defect 保真度不足
fidelity factor 保真度;重现率;逼真率;保真度系数
fidelity guarantee insurance 职工保证保险
fidelity rebate system 运费回扣制
fidibus 点火纸捻
Fidler-Maxwell kiln 菲德勒—麦克斯韦隧道窑
Fidler's gear 辘轳(俚语);起重滑车;菲德勒式起落架
fidley 舱梯构架
fiducial axis 框标轴;基准轴(线)
fiducial belt 置信带
fiducial board 框标板
fiducial business 信托业务
fiducial center 框标中心
fiducial confidence ellipse 置信椭圆
fiducial distribution 置信分布
fiducial index 置信指数
fiducial indicator 零位指示器;基点指示器
fiducial interval 置信区间
fiducial interval range 置信区间范围
fiducial limit 置信(界)限;置信极限;置信范围;可信极限;可靠界限;可靠极限
fiducial line 框标线;基准线
fiducial mark 准标;框标;基准符号;基准标(志);坐标点
fiducial marks on photograph 像片框标
fiducial measurement 框标测量
fiducial plate 框标板
fiducial point 准点;置信点(统计学);零点;框标点;基准点;参考点
fiducial point location 框标点位置
fiducial point separation 框标点间距
fiducial point spacing 框标点间距
fiducial probability 置信概率
fiducial range 置信范围
fiducial shaft 框标轴
fiducial temperature 基准温度
fiduciary 受托人
fiduciary accounting 财产信托会计;信托会计
fiduciary business 信托业务
fiduciary capacity 受托人资格
fiduciary circulation 纸币流通
fiduciary contract 信托契约
fiduciary contribution 信用投资;信托投资
fiduciary currency 信用纸币
fiduciary debt 信用贷款
fiduciary estate 信托资产;信托财产
fiduciary guardian 受托监护人
fiduciary institution 信用机构;信托机构

fiduciary issue 信用发行
fiduciary level 可靠度;标准电平
fiduciary loan 信用借款;信用贷款
fiduciary media 信用媒介
fiduciary money 信用货币
fiduciary overdraft 信用透支
fiduciary property 受托保管的财产
fiduciary reserve 保证发行准备
fiduciary service 信托服务
fiduciary work 信托业务
fiedlerite 水氯铅矿
fief 采邑;封地
field 信息组;现场的;原位;园地;域;领域;矿田;记录区;田间;田地;场;半帧
field accelerating relay 磁场加速继电器
field acceptance test 现场验收试验
field-acquired geologic(al) profile 实测地质剖面图
field activity 野外工作
field adjusting device 现场调整装置;野外校正装置
field adjustment 磁场调节;磁流调节
field aerated concrete 现浇加气混凝土
field aerodrome 野战机场;野外机场
field-aided diffusion 场助扩散
field alignment 场对中;场调整
field alignment error 视界校准误差;电场调整误差
field Alpha card survey 野外阿尔发卡法
field alterable control element 场可变控制元件
field ampere-turns 磁场安匝
field amplifier 激磁电流放大器
field amplitude 场幅度
field amplitude adjustment 场幅度调整
field amplitude control 场幅度调节
field analysis 野外分析
field angle 张角;视场角;波束角
field annealing 磁致热处理;磁场致退火
field annotation 野外注记;野外调绘
field apparatus 现场设备;外业装备
field apparatus for ground photogrammetry 地面摄影测量仪
field apparatus for ground photogrammmetry 地面摄影测量现场设备
field application 现场涂装;野外应用;野外使用
field application relay 励磁继电器;激励继电器;供磁继电器
field applied 工地制作;工地使用
field-applied plywood panel 现场装配胶合板
field architectural concrete 现浇混凝土
field arrangement 野外排列方式
field array of seismic sensor 地震仪野外台阵
field assay 野外测试
field assembly 现场装配;现场组装;现场拼装
field assembly of vessels 现场安装容器
field assembly yard 工地装配场
field assisted photocathode 场辅助光阴极
field astronomical observation 野外天文观测
field astronomical point 野外天文观测点
field astronomical station 野外天文观测点
field astronomy 野外天文学
field atmospheric pollution 现场大气污染
field auditor 现场核数员
field automation 现场作业自动化;野外作业自动化;油田作业自动化
field axis 磁极轴线
field bail 野外挤奶装置
field balance 野外磁秤
field balancing technique 现场平衡法
field beam test 现场梁抗弯试验
field bearing test 现场荷载试验;现场承重试验;现场承载试验;工地承载试验
field behavio(u)r 热田性状
field bend 现场制弯头;场弯曲
field bend correction 场弯曲校正
field bending 工地弯制钢筋;现场弯筋作业
field biased betatron 附加励磁电子感应加速器
field bindweed 田旋花
field blackout period 场消隐周期
field blanking 场扫描回程熄灭
field blanking impulse 场熄灭脉冲
field blasting test 现场爆炸试验
field board 外业测图板
field bobbin 磁极线圈架;磁场线圈座
field bolted 现场栓接的;现场栓接
field bolt(ing) 现场安装栓;工地栓接
field book 现场记录簿;野外记录簿;野外记录本

工地记录簿;外业手簿;测量记录本;测绘本
field border plantings 田边植物
field boundary 农田边界
field branch 野外分局
field breaker non-linear resistor 磁场断路器非线性电阻
field-breaking resistance 消磁电阻
field-breaking switch 削磁开关;消磁开关
field break-up switch 励磁分段开关
field brightness 场亮度
field broadcasting unit 流动广播车
field bulk plant 工地储罐区
field bulk station 工地储罐区
field bund 地梗
field burning 田野燃烧;烧荒
field bus 现场总线
field cable 军用电缆
field calculation 现场计算;野外计算;外业计算
field calibration 野外检定;声场校正
field camera 现场摄像机;野外摄影机;地面测量摄影机;便携式外景摄像机;便携式(电视)摄像机
field camp 野外帐篷
field canteen 工地食堂
field capacity 最高有效水量;自然层土的含水量;土壤毛细含水量;田间(毛细管)含水量;间间持水量;田间保水量;天然含水量
field capillary moisture capacity 田间毛细管含水量
field care 田间管理
field carrying capacity 土壤持水量;田间含水能力
field changes 场强跃变
field chapel 农村小教堂
field check 现场检验;现场复核;野外检验;野外检核;野外检查;野外复核;外业检核
field chipping 场地修整
field church 农村教堂
field circuit 磁场电路;场电路
field circuit breaker 励磁断路器;灭磁开关
field circuit earth fault protection 励磁回路接地故障保护(装置)
field classification 野外分类;野外调绘
field coat 工地涂料;工地涂层
field-coated
field coating 现场涂装;现场施工(指涂料方面的)施工
field coefficient of permeability 原位渗透系数;野外渗透系数;田间土壤渗透系数
field coenergy 磁场同能
field coil 励磁线圈;激励线圈;磁化线圈;磁场线圈;磁场激发线圈;场线圈
field coil bracer 磁场线圈撑块
field coil flange 磁极线圈托板
field collection 外业收集
field compaction 野外压实
field compaction curve 现场压实曲线
field compaction of soil 土的野外压实
field compaction test 现场压实试验;现场碾压试验
field compaction trial 野外压实试验
field compactive effort 原位压实力
field comparator 基线检定场
field compilation of geographic(al) names 地名实地调查
field compilation survey 野外编制测量
field completion 野外补点;外业检测
field component 场分量
field compression curve 现场压缩曲线
field compression test 野外压实试验
field compressometer 螺旋荷载板试验仪
field computation 现场计算;野外计算;外业计算
field computer system 野外计算机系统
field concrete 现拌混凝土;现浇混凝土;现场拌制混凝土
field concrete balcony 现浇混凝土阳台
field concrete cable duct 现浇混凝土电缆管道
field concrete eaves unit 现浇混凝土屋檐
field concrete filling 现浇混凝土填充
field concrete floor 现浇混凝土楼面
field concrete frame 现浇混凝土构架
field concrete rib(bed) floor 现浇混凝土带肋楼面
field concrete shell 现浇混凝土薄壳
field concrete stair(case) 现浇混凝土楼梯
field concrete structure 现浇混凝土结构
field condition 现场条件;工地条件
field connection 现场装配;现场拼接;现场联结;现场连接;现场接线;现场安装;工地装配;工地安装

field consolidation time 原位固结时间
field construction 现场施工
field construction meeting 现场施工会议
field contouring 野外测绘等高线
field control 现场控制;现场管理;野外控制;激励控制;外业控制;地面控制;磁场调整;信息组控制
field control bar 场控制棒
field control converter 磁场调整换流机
field controlling point of aerophotogrammetry 航测外控制点
field control motor 可调磁场型电动机;磁场可控式电动机
field control point 野外控制点
field control station 现场控制台
field control testing 现场控制试验;工地控制试验
field conveyer system 堆场输送机系统
field conveyor system 堆场输送机系统
field core 磁极铁芯
field correction 野外修测透写图;野外校正;野外改正
field correlation 野外对比
field cost engineer 现场费用工程师
field coupling 场耦合
field coverage 视野;视界;视场
field crop 农作物;大田作物
field cube test 现场混凝土立方体强度试验
field-cured cylinder 现场养护的(混凝土)圆柱试体;现场养护的(混凝土)圆柱体试块
field curing 工地养护;室外养护;现场养护
field current 激励电流;场电流
field curvature 像场弯曲;像场曲面;场曲
field curvature corrector 场曲校正器
Field cycle 菲尔德循环
field data 现场资料;现场数据;应用数据;野外资料;野外数据;工作数据;工程数据;外业资料;外业数据;实测资料
field data card 野外数据卡
field data code 军用数据码
field data sheet 野外数据单
field day 展览日;野外演习日
field deciphering 野外辨认
field declarator 字段说明符
field definition 信息组定义
field deflection 帧偏转;场偏转
field deformation 场形变;场变形
field delimiter 字段分隔符
field density 现场密(实)度;原状密(实)度;原位密度;原始密度;感应密度;电场强度;磁通密度;磁感应密度;场强;场密度
field density determination of soil 土的原位密度测定
field density strength of field 场强度
field density test 现场密(实)度试验
field depot 工地仓库
field description 现场说明;田间描述
field descriptor 字段描述符
field design 现场设计
field desorption 场解吸法;场解吸
field desorption mass spectrometry 场解吸质谱法
field desorption microscope 场致解吸显微镜
field determination of hydraulic conductivity 测定田间水力传导性
field diaphragm 视域光阑;视场阑
field digital system 野外数据采集系统
field director 现场指挥员
field discharge 励磁放电;灭磁
field-discharge resistance 消磁电阻
field-discharge switch 消磁开关
field discrimination 现场辨别
field displacement 场位移
field disposal 野外处置
field distortion 磁通分布畸变;磁场失真;场失真;场畸变
field distribution 场分布
field distribution curve 场分布曲线
field ditch 支沟;垅沟;毛渠;毛沟;田间排水梁;田间灌水渠;田间灌溉沟;田沟
field ditch system 田间渠系;田间沟网
field-diverter rheostat 磁场分流变阻器
field divider 场分频器
field diving system 现场潜水系统
field document 野外资料;外业资料;外业手簿
field domain 场畴
field drag 重型耙
field drain(age) 农田排水;场地排水;畦沟;田间

排水(支)沟
field drainage system 热田排放系统
field drain pipe 农田排水管(道)
field drilling operation 野外钻探工作
field-drying transformer 场干燥变压器
field duty(of water) 田间用水量;田间灌溉率
field economizing relay 弱励磁继电器;弱激磁继电器
fielded panel 装配镶板;中凸形镶板;鼓面镶板;突面镶板;隆起镶板
fielded system 工地系统
field effect 电场效应;场效应
field effect amplifier 场效应放大器
field effect capacitor 场效应电容器
field effect controlled switch 场效应控制开关
field effect device 场效应器件
field effect diode 场效应二极管
field effect display 场效应显示
field effect display device 场效应显示放电器件
field effect high frquency power transistor 场效应高频功率晶体管
field effective pressure 原位有效压力
field effect photo electric(al)transistor 场效应光电晶体管
field effect phototransistor 场效应光电晶体管
field effect quadrode 场效应四极管
field effect semiconductor laser 场效应半导体激光器
field effect tetrode 场效应四极管
field effect transistor resistor 场效应晶体变阻器
field effect transistor 电场效(应)晶体管
field effect transistor high frequency amplifier circuit 场效应晶体管高放电路
field effect transistor mixer circuit 场效应晶体管混频电路
field effect transistor preamplifier 场效应晶体管前置放大器
field effect transistor probe amplifier 场效应晶体管探头放大器
field effect transistor resistor 场效应晶体管电阻器
field effect varistor 场效应变阻器
field elbow 现场制弯头;现场制 L 型短管
field electron microscope 场致发射电子显微镜
field elevation 外业高程
field emission 静电发射;场致放射;场致发射;场地发射
field emission energy analyzer 场致发射能量分析仪
field emission microscope 场致发射显微镜;场发射显微镜
field emission microwave device 场致发射微波器件
field emission source 场发射源
field emission spectroscopy 场致发射光谱学
field emission tube 场致发射管
field end 管子螺纹端
field energy 电场能量;场能
field engineer 维护工程师;安装工程师;现场工程师;驻场工程师;(施工单位的)工地工程师
field engineering 现场;安装技术;安装工程
field engineering design change 现场工程设计更改
field engineering design change schedule 现场工程设计更改计划表
field enumeration 现场计数
field equipment 室外设备;野外装备;外业装备;外业设备;田间设备
field equipment of conditional transfer of control 遥控分机
field-erected 现场安装的
field erection 现场装配;现场安装;工地拼装;工地架设
field-established 野外测设的
field estimator 现场估算师
field evaluation 现场评价;现场鉴定;现场估价;野外评价;野外估算
field evidence 现场数据;野外证据;野外观测结果
field examination 现场检查;野外检查;外业检查;田间检查
field excavation 田野发掘
field excitation 场激发
field exciter 场激励器
field exercise 野外演习
field experience 野外试验;野外经验

field experiment 现场试验;现场实验;现场测试;野外试验;野外实验;工地试验;田间试验;大田试验
field exploration 现场踏勘;野外勘察
field exploration program(me)现场勘察程序
field exploration system 野外勘探系统
field extension well 钻探延伸油井
field fabricated 现场装配的;现场加工制造的
field fabrication 现场装配;现场制作
field factor 场形系数
field factory 工地工厂
field failure 磁场失效
field finish 工地修整
field firing range 战斗射击场
field flatness 场地平整度
field flattener 像面平整镜;平像物镜;视场致平器
field flattening lens 平场透镜
field flow fractionation 场流分级法
field flutter 场振动变度;场颤
field flux distribution 场通量分布
field flyback 场逆程
field forces 区力
field-forcing 强迫励磁
field-forcing relay 强化磁通变动率继电器
field forge 轻便锻炉;锻工场
field form 磁场形式
field form entry loss 场畸变附加损失
field fracture 粗粒断口
field frame 磁极框架
field frequency 场频
field frequency control 帧频控制;激励调整
field frequency programmable tone transmitter 现场频率可编程序音调传感器
field funnel 野外漏斗式黏度计
field gating circuit 场选通电路
field generator 场频信号发生器
field geologic(al)map 野外地质图
field geologic(al)record 野外地质记录
field geologic(al)work 野外地质工作
field geologist 野外地质人员
field geology 野外地质学
field geology feature 现场地质特征
field geometry 矿场机械形状
field geophysicist 野外物探人员
field glass 野外镜;望远镜
field glasses 双眼望远镜;双筒望远镜
field gluing 现场胶粘
field going to water 开始涌水矿床
field gradient 场梯度
field graduation E 形标尺分划
field groundwater velocity 实际地下水流速;地下水实在流速;地下水实际流速;地下水实地流速
field growth period 本田生长期
field guide 现场指导;野外指南;外业手册
field guidebook 野外指南
field hand 农业工人;建筑工地工人
field heat 田间热
field hospital 野战医院
field house(运动场的)更衣室;储藏室
field hydrological observation 现场水文观测
field hydrological survey 现场水文观测
field ice 冰原冰;大浮冰
field identification 信息组标识;现场鉴定;野外鉴定;外业判读
field identification procedure 野外鉴定法
field identification procedure of soil 土的现场鉴别法
field identification test 野外鉴定试验
field image 视场像
field impregnation 现场浸渍
field incidence 场入射
field-incidence transmission loss 场入射隔声量
field index 场指数
field-induced photoemission 场发光电发射;场致光发射;场感应发光
field infiltration 原位渗入;田间下渗;田间渗入;田间入渗
field information 场信息
field infrared source 野外用红外辐射源
field inquiry 实地调查
field inspection 现场踏勘;工地检查;现场视察;现场检验;现场检查;野外检查;野外观测;工地视察;外业检测;田间检查
field installation 野外设备
field instrument 野外作业用具;外业仪器

field integration mode 场积累方式
field-intensifying metal 强化电场金属层
field intensity 场强度;场强;场的强度
field intensity between turns 匝间电场强度
field intensity curves 电波场强曲线
field intensity distribution 场强分布
field intensity indicator 场强仪
field intensity measurement 场强测量
field intensity measuring instrument 场强测量仪
field intensity of electricity 电场强度
field intensity of high frequency 高频场强
field interference comparator 野外光干涉比长器
field interval 场扫描消隐时间;场期间
field-interval noise lines 场期间噪声行
field investigation 现场勘察;现场调查;运转试验;野外研究;野外勘察;野外调查;工地调查;田间调查;实地调查
field investment 矿田投资额;矿产地投资额
field ion emission engine 场电离放射发动机
field ion emission microscope 场离子发射显微镜
field ionizaiton ga(u)ge 场致电离真空规
field ionization 场致电离
field ionization ion source 场致电离离子源
field ionization mass spectrometry 场离子化质谱法;场电离质谱法
field ionization source 场电离源
field ion mass spectrometer 场离子质谱计
field ion mass spectroscopy 场离子质谱学
field ion microscope 场离子显微镜;场致离子显微镜
field ion microscopy 场显微术;场离子显微术
field ion transmission microscope 场致离子透射显微镜
fieldistor 场控晶体三极管;场控晶体管
fieldistor channel 场效应晶体管沟道
field joint 现场结合;现场接合;安装接头;安装焊缝
field keystone 场梯形失真
field kiln 露天窑炉
field laboratory 现场试验室;野外试验室;野外实验室;工地试验室;工地实验室;田间试验室
field lateral 农沟;田间支渠;支渠;农渠
field layer 草本层
field layout 实地放样;土地规划图
field leakage coefficient 磁漏系数
field leakage flux 磁场漏磁通
field length 信息组长度;域长;字段长度
field lens 向场透镜;野外用镜;物镜;场透镜
field lens iris 场镜可变光栏
Fieldler's theory of leadership 菲德勒领导理论
field level(l)ing 野外水准测量
field light 飞行场地标灯
field lightening truck 野外照明车
field lighting 工地照明
field lightweight concrete 现浇轻质混凝土
field line 场力线;场电力线
field line annihilation 磁力线重接
field line reconnection 磁力线重接
field loading test 现场荷载试验;原位承载试验;野外荷载试验
field localizer 导航台
field location 线路点;野外定线;放样
field location work 野外定线工作;野外定位工作;实地定线;实地定位工作
field locking 场同步;场锁定
field loco(motive)轻便机车
field longevity 热田寿命
field loss relay 失磁继电器
field luminance 适应亮度;视场亮度
field luminescence 场致发光
field machinery 田间作业机械
field magnet 激磁磁铁;场磁铁;场磁体
field magnet distortion 磁极畸变
field magnetomotive force 磁场磁动势
field mains 集输管线
field maintenance 现场维修
field maintenance unit 野外维修队
field man 野外工作人员;工地代表
field management 现场管理;工地管理
field manual 现场手册;野外指南;外业手册
field manufacturing 现场制造
field map 油田图;野外原图;野外草图;外业原图;实测原图
field mapping 野外制图;野外填图;野外绘图;野草图绘制;田间测图

field mark 信息组标记
field matrix 场矩阵
field measurement 现场测试;现场测量;现场测定;野外测量;野外测定;实测
field measuring technique 现场测量法
field mechanism 地轮起落机构
field mesh 场网
field mesh electrode 场栅电极
field meter 场强计
field method 现场调查;野外方法;区域法;实地调查法
field microscope 野外显微镜
field mix 现场配料;现场配合;现场搅拌;工地搅拌;工地拌和混和料
field mixing test 现场拌和试验
field modulus 现场测量模量
field moisture (capacity) 田间持水量;田间保水量;原状含水量;田间土壤持水量;天然含水量;未夯实土壤水能力;田间含水量;天然湿度
field moisture content 原状含水量
field moisture deficiency 田间含水量差值(实际含水量与饱和含水量之差以水深计)
field moisture-density test 现场含水量密度试验
field moisture equivalent 现场湿度当量;原位水当量;野外含水量;工地含水量;田间水分当量;田间含水量;天然含水量;原地土壤含水当量;土夯实土吸水能力
field-molded sealant 现场模塑填缝料
field monitoring 现场监测;野外监测
field monitoring tube 现场监视管
field-mounted 现场安装的
field mounting 工地装配
field mouse 田鼠
field mower 割草机
field name 信息组名;区段名
field neutralizing magnet 磁场中和磁铁
field note 现场记录;野外记录;工地通知书;工地记录本;工地记录;外业记录
field number 视场直径
field nursery 林间苗圃
field observation 地形观测;现场观察;现场观测;原型观测;野外观察;野外观测;田间观察;实地观察;地形摄影
field observer 现场观察员
field of application 应用范围
field of circulation 流通领域
field of coefficient 系数域
field of conjugate action 啮合面
field of constants 常数域
field of curvature 曲率场
field of definitions 定义域
field of density 密度场
field of divergence 散度场;辐散场
field office 建筑工地办事处;工地办公室
field office trailer 工地办公室拖(挂)车
field of fire 射击区
field of flow net 流网场
field of force 力场
field of gravity 重力场
field of ice 冰原;冰场
field of integration 积分域
field of load 荷载场;受力范围
field of microscope 显微镜视野
field of plane on a manifold 流形上的平面场
field of points 点场
field of pressure 压力场;气压场
field of search 搜索区
field of seaweed 一片海藻;海藻区
field of standardization 标准化领域
field of temperature 温度场
field of the external source 外源场
field of the internal source 内源场
field of turbulent flow 紊流场
field of use 使用范围
field of vector on a manifold 流形上的向量场
field of view 视域(大小);视野;视界;视场
field of view meter 视场仪
field of view number 视场数
field of vision 视野;视场
field of vorticity 涡旋场;涡度场
field of water-level falling velocity 水位降速场
field of welding temperature 焊接温度场
field operation 现场作业;现场运行;现场灭火活动;野外作业;野外操作;外业;外勤工作;室外操作

field operation division 野外作业处
field operator 场算符;现场运行人员
field optic(al) collimator system 场致光学准直仪系统
field order 现场工程修改通知;工地通知书;变更通知;现场更动通知
field outfit 外业装备
field overhead 工地管理费;井场管理费
field oxygen transfer efficiency 实际转氧效率
field-painted 现场油漆的
field painting 现场涂装;现场刷漆;现场油漆
field particle 场粒子
field party 勘测队;野外勘测队
field path 工地人行通道
field path bridge 工地人行(便)桥
field pattern 场型;场图;辐射方向图
field pattern-propagation factor 场方向性相对因数
field pendulum 野外振摆仪
field performance 现场使用性能;现场功能;油田特性;野外使用性能
field permeability coefficient 野外渗透系数
field permeability test 现场渗透试验
field permeability testing 原位渗透(性)试验
field permeability test method 现场渗透试验法
field personnel training 现场人员训练
field petrol depot 野外油料库
field petroleum officer 野外油料管理员
field phase control 场相位调节
field phasing 帧定相;场定相
field photogrammetric apparatus 地面摄影测量仪
field photogrammetry 地面摄影测量学
field photo interpretation 像片调绘
field pick-up 室外摄影;实况转播
field piece 场镜
field pipe bend 现场制弯头
field pipeline 现场中管道
field placed roller-compacted concrete 现场浇筑的碾压混凝土
field plan 田间试验计划
field plant 现场设备
field plate loading test 现场荷载试验
field playback 现场回放装置
field plot 径流实验小区;田间地块;田间实验小区;实验小区;场标绘图
field plot experiment 田间小区试验
field plotter 野外图板
field plotting 野外测图
field plot trial 田间小区试验
field polarograph 现场极谱仪
field pole 励磁极;磁场线圈架;磁场(磁)极;场磁极
field pole piece 工地用杆件
field portable railway 工地轻便铁道
field position 外业计算的点位
field potential 油田每日产量;场势
field power supply 励磁电源;场激励电源
field practice 田间管理
field pressure 油田压力
field printing plant 野战制印设备
field processing 现场加工
field production 大田栽植
field production meeting 现场施工会议
field-programmable logic array 场致程序逻辑阵列
field program(me) 野外作业计划;实地方案
field programming 现场程序编制
field propagation of bubble 磁泡磁场传输;磁泡场传输
field proportioning 工地配合(比)
field protective relay 磁场保护继电器;场保护继电器
field proven 现场验证过的
field proven hardware 现场验证过硬件
field pulse 场脉冲
field pump 油田用泵;野外用泵
field pumping test 现场抽水试验;原位抽水渗透性试验;野外抽水试验;田间抽水试验
field quality control 现场质量控制
field quantization 场量子化
field quenching 现场淬火
field radiation device 野外放射仪
field railway 现场轻便铁路;工地铁路;工地(轻便)铁道;工地临时铁路;施工轻便铁路
field railway system 轻便铁路系统
field railway track 轻便铁路线

field range 视野范围;视场范围
field rate 场速率;场频;钻孔预算价格
field ray 场射线
field recognition 野外鉴定
field recognizer 场测定器
field recompression curve 现场再压缩曲线
field reconnaissance 现场踏勘;野外踏勘;野外普查;野外查勘;实地查勘
field record book 外业记录手簿
field record(ing) 野外记录;现场记录;外业记录
field recording book 野外记录薄;野外记录本
field recording card 野外记录卡
field recording data 野外记录数据
field rectifier 励磁整流器;外业纠正仪;磁场整流器
field reduction 场衰减
field reflectance spectrometer 野外反射光谱仪
field region 场效应区
field regulator 励磁调节器;磁场调节器;场强调节器
field reinforced concrete 现浇钢筋混凝土
field reinforced concrete floor 现浇钢筋混凝土楼面
field reliability 野外可靠性
field removal gate 场消除门
field repetition rate 场重复频率;场频
field-replaceable unit 现场可更换部件
field report 实地考察报告
field reporting 工地报告
field representative 现场代表;驻场代表;驻工地代表;工程现场代表
field research 野外研究;实地调查研究
field resistivity measurement 现场电阻率测定
field result 野外(试验)结果
field retrace 帧回描;场扫描回程;场回描
field reversal 磁场倒转
field review 现场复查;野外复查
field rheostat 励磁变阻器;激磁变阻器;磁场变阻器;磁变阻器;场用变阻器
field ridge 田垄
field ring 机座环形部分;机座环
field rivet(ing) 现场铆接;工地铆接;工地铆钉;安装铆接
field road 野外道路
field rod 标杆;标尺
field running test 实地运行实验
field sample 野外样品
field sample number 样品野外编号
field sampling 工地取样
fields and gardens 田园
field scan 帧扫描;场扫描
field scanner 场扫描器
field scanning rate 场扫描速率;场扫描率
field scanning sensor 场扫描传感器
field's discontinuity 场不连续性
field section 野外科
field seepage test 现场渗透试验
field seminar 野外实习
field separation 野外分层
field separator 字段分隔符
field sequence 场顺序
field-sequential 帧序制;场序制
field-sequential camera 场序制电视摄像机;场序制彩色电视摄像机
field-sequential colo(u)r transmitter 场序制发射机
field-sequential image 场序制成像;场序发送图像
field-sequential system 场顺序制;半帧序制;场序制
field-sequential transmitter 场序制发射机
field service 现场使用;现场服务;野外勤务
field service compressor 现场用压缩机;野外用压缩机
field service manager 现场经理;工地服务主任
field service unit 野外作业设施;现场服务设施
field setter 金刚石镶嵌工
field setting 定植
field setup 野外观测装置
field shaper 磁场造型器
field shear box 野外剪力仪
field shear test 野外剪切试验
field sheet 观测记录表;外业原图;测图板
field shelling 田间脱粒
field shetching 实地描绘;现场绘制草图;草测
field shop repair semitrailer 野外修理用的半拖车
field shunt 励磁分路
field signal level 场强信号电平

field simulation 现场仿真
field site precast concrete factory 现场预浇混凝土厂
field site reconnaissance 现场场地勘察
field situation 现场情况
field sketch 地志资料
field sketching 野外素描；现场手绘；现场（绘制）草图；野外描绘；野外草图；草测
field skip system 跳场制式
field sod 田野草皮；牧场草皮
field soil 原位土；田间土壤
field soil-shearing test 原状土剪切试验
field soil test 原位土工试验
field sounding test 现场触探试验
field sound transmission class 场声传输类；场声传输级
field spectral measurement instrument 野外波谱测量仪器
field spectrophotometer 野外分光光度计
field spectroradiometer 野外光谱幅射计
field spectroscopy 野外光谱学
field spider （水轮发电机的）转子支架；凸极转子；磁极（星轮）支架；磁极星轮
field splice 现场拼接
field-splitting microscope 分离视场显微镜
field-splitting switch 分场开关
field spool 励磁线圈架
field spool insulation 极身绝缘
field sprinkler infiltrometer 田间洒水装置渗透仪
field sprinkler system 田间喷灌系统
field stack 谷堆
field staff 建筑工地工作人员；现场人员
field staking-out 现场定线
field standard 外业标准
field standardization of tape 卷尺野外检定
field stars 场星
field station 现场操作站；野外工作站；工地储罐区；试验站
field statistics 现场统计
field stereoscope 野外立体镜；外业立体镜
field stone 圆石；散石；大卵石；大块石
fieldstone fireplace 砾石砌的火炉
field stop 野外光圈；视场光阑；场阑
field storage 野外储[贮]存；野外仓库；工地存料；田间储藏
field stratum 地面植被层
field strength 原位强度；场强；(的)强度
field strength contour map 等场强线图
field strength meter 场强仪；场强计
field strength pattern 场强图
field strength receiver 场强接收机
field strength standard 场强标准
field stress tensor 场应力张量
field strip 拆卸作业
field structure 磁极结构；场结构
field study 现场研究；野外研究；实地研究
field supervision 现场监督
field supervision of construction 现场施工监督
field suppressor 磁场抑制器
field survey 现场调查；野外调查；野外测量；外业
field survey data 野外测量数据
field survey depot 野外测量基地；外业基地
field survey design 现场调查设计
field-surveyed data 实测资料
field surveying 现场测量；野外测量法
field surveyor's plot 外业测量图幅
field survey pamphlet 野外测量手册；野外测量技术规定
field survey station 野外测量站；外业基地
field-swept 场扫描
field switch 励磁开关；场开关
field synchronizing impulse 场同步冲；场频率同步脉冲
field tank 工地水罐；石油储存罐
field tank farm 工地储罐区
field technic 工地操作技术
field technique of radioactive 放射性野外技术
field telephone 野外用电话；军用电话机
field television video recording 场电视录像；半帧式电视录像
field temperature 室外温度
field temperature sensor 室外温度探测器
field terminal 磁极线圈出线端
field terrace 梯田

field tested 现场试验的
field test(ing) 现场试验；田间试验；原位试验；野外试验；工地试验；田间测定
field test kit 野外分析箱
field test laboratory 工地试验室
field test of steel tape 钢尺野外比长
field test procedures 现场试验程序
field test site 野外试验场
field tests of stability 稳定性的野外检验
field theodolite 野外经纬仪
field theory 域论；场论；场理论
field theory of earthquake 地震场论
field thermometer 野外温度计
field thresher 田间脱粒机
field tile 农田排水瓦管；砌底砖
field tilt 场频锯齿形补偿信号
field-time distortion 场时失真
field timing 野外计时
field-to-field survey 分界测量
field tolerance 野外容许偏差
field topographic(al) screw 地形测量队
field topographic(al)(survey)team 地形测量队
field topographic(al) unit 野外地形测量队；地形测量队
field transfer matrix 场转换矩阵
field transistor 场控晶体管
field transmission loss 场声传输损失
field transport 野外运输
field trial 现场试验；野外试验
field triangulation 野外三角测量
field trip 野外踏勘；野外实习
field tube 力线管；双筒式锅炉管；场示管
field unit weight 原位重度
field usage 现场使用；野外使用
field use 现场使用；野外应用；野外使用
field validation 现场验证
field value 外业值
field vane shear test 现场十字板剪切试验
field vane test 现场十字板试验
field variable 场变量
field variable pattern 场变量的模式
field velocity 有效流速；实地流速
field velocity of groundwater 地下水的实际流速
field verification and image-geologic(al) mapping 野外验正及象片地质图编绘
field void ratio 原位孔隙比
field voltage 励磁电压；场电压
field warehouse loans 存货抵押贷款；就地仓库贷款
field warehousing 现场存货
field waste 田间流失；田间废水；田间灌后的废水
field water condition 农田水情
field water permeability test 路面现场透水性试验
field water retaining capacity 持水度；野外持水率；野外持水量
field wave 励磁波；场波
field waveguide 场波导
field weakening 磁场削弱
field weakening control 减弱磁场控制
field weakening ratio 磁场削弱率
field welded 现场焊接的
field weld(ing) 现场焊接；工地焊接
field welding technique 工地焊接技术
field well 生产井
field winding 励磁绕组；激励绕组；磁极绕组；磁场绕组
field wire 被覆线
field within a node 一节点内的场
field woodrush 地杨梅
field work 野外测量；现场作业；现场制作；野外制作；野外工作；外业；室外工作；实地调查；野外作业
field work allowance 野外工作津贴
field worker 现场调查员；实地调查员
field working 野外作业
field work of electric(al) prospecting 电法勘探野外工作
field work of magnetic survey 野外磁测工作
field work order 现场布置任务通知单；现场布置任务通知书
field work shop 工地车间
field work standards 现场操作规范
field worthiness 野外适用性
field yoke 磁轭
fiendish weather 险恶天气；极坏天气
fierce and frightening storm 惊涛骇浪

fierce clutch 快速离合器
fieri facias 扣押债务人动产令
fiery finish 火焰加工；燃烧修饰法
fiery mine 瓦斯气体
fiery steel 过烧钢
Fierz interference 菲尔兹干涉
Fiesen's light 芬森弧光灯
fiexible factory overhead budget 弹性间接生产费预算
fife rail 船尾楼围栏
fifteen percentile speed 第15%位地点车速
fifteen points moving average 15 点移动平均
fifteen puzzle 十五迷宫
fifth generation computer 第五代计算机
fifth generation container 第五代集装箱
fifth generation container berth 第五代集装箱泊位
fifth wheel 现场无需的设备或人员（俚语）；第五轮；连接轮（汽车尾部）；半拖车接轮
fifth wheel attachment 转向轮；多余的东西；承重旋转接头；备用轮；半挂车连接轮
fifth wheel king pin （车的）中心销
fifth wheel load （汽车的）第五轮偏距
fifth wheel tester 第五轮仪
fifty-fifty 各半
fifty percent kill 半致死浓度；半致死剂量
fifty percent particle diameter 等重粒径；等量粒径
fifty percent rule 50%的规则
fifty percent tolerance dose 半数耐受量
fifty-percent zone 半数必中界
fifty seconds per sample 每个样品用 50 秒
fight aerotrace map 飞行舰迹图
fight against the flood emergency 抗洪抢险
fighter 战斗机；防护剂
fighter station 战斗机基地
fighting equipment 战斗装备
fighting gallery 战斗陈列馆
fighting ship 战斗船
fighting strength 战斗力
fighting top 战斗桅楼
fig-tree 榕树
fig-tree wax 无花果树蜡
Figuier's gold purple 菲格尔金紫色料
figuline 陶土；瓷土；浮雕装饰
figurability 能成型性
figurate(d) 定形的
figurate number 形数
figurate stone 异形砖；琢纹石
figuration 形状；图案装饰法；图案表现法；数字形式；成型
figuration drawing 外形图
figurative 比喻的
figurative art 形象艺术；造型艺术
figurative character 象征特性
figurative constant 象征常数
figurative design 象征性设计；象征的设计；造型设计
figurative element 象征元；象征特性
figurative graph 象形图
figurative prints 绘画格调印花布
figuratrix 特征表面
figure 影像；价格；计价；花纹；图形；图案；图；数字；数码
figure-8 knot 八字形结
figure adjustment 图形校正；图形平差；数值调整
figure condition 图形条件
figure coordinate conversion 图形坐标变换
figured 有花纹的；有图案的；塑造
figured bar 异形钢筋
figured bar iron 异形钢；型钢
figured cloth 提花织物
figured contour 注记等高线
figured effect 花纹效应
figured fabric 提花织物
figured glass 图案玻璃；压花玻璃
figure dimension 外形量规
figured iron 型铁
figure disc 数字盘
figured pattern 提花图案
figured pique 提花凸纹织物
figured plate glass 压花(平板)玻璃；图案(平板)玻璃
Figure-drawing brick of Han dynasty 汉书像砖
figured rep 变化楞纹组织
figured rolled glass 压花玻璃；滚花玻璃；图案玻璃
figured roller glass 压花玻璃

figured roll glass 压花玻璃
figured sheet glass 压花玻璃;图案玻璃
figured steel 异形钢
figured twills 花纹斜纹
figured veneer 图案镶板;装饰木纹镶板;花纹单板;单板;带图案镶板
figured wood 富纹木
figure glass 花纹玻璃
figure ground articulation 图形背景清晰度
figure-ground discrimination 图形背景辨别
figure identification 图形识别
figure index 性能指数;性能指标
figure information 图形信息
figure keyboard 字符键盘
figure lamp 显字灯
figure notation 数字标记
figure of airborne electromagnetic anomaly and geology 航电磁异常地质综合图
figure of apparent resistivity 视电阻率图
figure of Buddha 佛像;佛堂
figure of conductivity 电导图
figure of confusion 弥散圈
figure-of-eight 八字形
figure of form 形心
figure of local anomaly 局部异常图
figure of loss 能量损耗系数
figure of mark pronunciation 数字拼读法
figure of merit 性能指标;性能因数;优质指数;优质因数;优值;优良指数;质量参数;灵敏值;灵敏度;品质优值;品质因素;品质因数
figure of merit curve 质量曲线
figure of merit frequency 质量因数频率
figure of merit of acousto-optic(al) material 声光材料优值
figure of merit of satellite earth station 卫星地球站质量因素
figure(of)noise 噪声系数;噪声指数
figure of speech 比喻
figure of test of air pumping number 抽气次数试验图
figure of the earth 地球形体;地球(的)形状
figure of variation of alpha-intensity 阿尔法强度变化图
figure of variation with depth of oxygen 氧气随深度变化图
figure of variation with time of oxygen 氧气随时间变化图
figure out 算出
figure out at 总共
figure pattern 数字模式
figure portal 绘花大门;花纹大门
figure post 人像柱
figure punch 号码机;数字冲压机
figurer 陶器的图案描绘者
figure reading electronic device 电子读数器
figure-sculpture 花纹雕塑
figure sheet 纹板
figure shift 跳格符号;数字换档;数码键位;变数字位;变换符号
figure shown by number 数码表示的图幅
figure signal 变换信号
figures shift 符号变换
figure S structure 扭麻花构造
figurine 仙人砖;压花;人物造型;小塑像;小雕像
figuring 用图表示;用数字表示;施工详图尺寸;根据图纸;施工详图尺寸;打号;编制工料预算;施工详图(增补)尺寸
figuring of mirror 镜面修琢
figuring of surface 表面修琢
Fiji plateau 斐济海台
Fiji Sea 斐济海
Fikentscher's viscosity formula 菲肯歇尔黏度式
filament 细丝;丝状体;丝状结构;丝极;长纤维;长丝;玻璃单丝;白热丝;暗条
filament acetate yarn 醋酯纤维纱线
filament activation 灯丝激活
filament activity 丝极放射性
filament-activity test 灯丝效率试验
filamental activated sludge 丝状活性污泥
filamental flow 线流,层流
filamentary 丝状的
filamentary cathode 丝状阴极;灯丝(式)阴级
filamentary coupling 丝状耦合
filamentary fibril 丝状原纤维

filamentary region 线状通道;丝状区
filamentary silver 银丝
filamentary spark 丝状火花
filamentary structure 纤维状构造
filamentary subminiature tube 超小型直热式电子管
filamentation 丝状形成
filament band 流束;丝带;射流
filament battery 甲电池(组)
filament blended yarn 长丝混合纱
filament bombardment 冲击灯丝
filament burnout 灯丝断丝
filament cathode 直热式阴极;丝状阴极
filament characteristic 灯丝特性
filament control 灯丝调节
filament cooling spray 单丝冷喷
filament crystal 丝晶
filament current 灯丝电流
filament data 灯丝电路参数
filament electrometer 细丝静电计
filament emission 灯丝发射
filament fiber 长丝纤维
filament for electric(al) lamp 电灯丝
filament form 灯丝形状
filament furnace 灯丝炉
filament fuse 灯丝电路熔断丝
filament generator 灯丝电源发生器;灯丝电源发电机
filament lamp 灯泡;白热丝灯;白炽灯
filament lasing 丝状激光发射
filament length 丝极长度
filament line 流线;水条线
filament mat 单丝毡
filament material 灯丝材料
filament nylon rope 长纤维尼龙绳
filament of chromsphere 色球暗条
filament of oil 石油束
filament of water 线流;流线
filament oseillation 暗条振动
filamentous 丝状的
filamentous activated sludge 丝状活性污泥
filamentous algae 丝状藻
filamentous bacterial sludge 丝状菌污泥
filamentous bacterial sludge bulking 丝状菌污泥膨胀
filamentous capsid 丝状衣壳
filamentous flow 线流
filamentous form 丝状体
filamentous fungi 丝状真菌
filamentous microorganism 丝状微生物
filament plate 丝板
filament potentiometer 灯丝分压器;灯丝电位计
filament pyrolyzer 热丝热解器
filament regulator 灯丝调节器
filament-reinforced metal 纤维加强金属
filament resistance 灯丝电阻
filament resistor 丝极电阻器;灯丝电阻器
filament rheostat 丝极变阻器;灯丝变阻器
filament saturation 温度饱和
filament supply 灯丝电源
filament support 灯丝支架
filament switch 灯丝开关
filament temperature 灯丝温度
filament transformer 灯丝变压器
filament type cathode 直热式阴极
filament voltage 灯丝电压
filament voltmeter 灯丝伏特计
filament waste 碎片废料;长纤维废料
filament wattage 灯丝功率
filament width of sidelight 舷灯光源宽度
filament winder 绕丝机
filament winding 灯丝(电源)绕组;单丝缠绕法;长纤维卷绕法
filament winding process 单丝缠绕法
filament winding tank 绕丝油箱
filament-wound case 缠丝壳体
filament-wound glass fiber reinforced thermosetting resin pipe 纤维缠绕法玻纤增强热固体树脂管
filament-wound glass-reinforced plastics 长纤维缠绕玻璃钢
filament-wound pipe 丝缠管
filament yarn 长丝纱
filament yarn fabric 长丝织物

filar 丝状的;丝状
filar cross 十字丝
filar evolute 线渐屈线
filar eyepiece 十字丝目镜
filariasis 丝虫病
filar involute 线渐伸线
filar micrometer 游丝测微器;动丝测微计
filature 制丝厂;缫丝机;缫丝厂
filature gum waste 厂丝吐
file 卷宗(文件);讲义夹;文卷;文件存储[贮]器;文件;档案;锉(刀)
file access 文件访问
file access attribute 文件存取属性
file access control 文件访问控制
file access method 文件访问方法;文件访问法
file access time 文件存取时间
file address 文件地址
file addressing 文件编址
file allocation 文件分配
file allocation table 文件分类表
file alloy 锉刀合金
file analysis 文件分析
file an application for dissolution 提出解散申请书
file attribute 文件属性
file backup 后援文件;文件备份
file bank 锉坯
file brush 锉刷
file buffer 文件缓冲区
file cabinet 文卷柜;档案柜
file call back 文件调出
file card 锉刀刷;锉刀钢丝刷;档案卡
file carrier 锉柄
file catalog 文件目录
file chisel 锉凿
file cleaner 锉用钢丝刷;锉刀钢丝刷
file cleaning card 锉刀刷;锉刀钢丝刷
file computer 信息统计机;情报统计机;文件计算机;编目计算机
file condition 文件访问条件
file consolidation 文件整理
file control 文件控制
file control block 文件控制程序块
file control system 文件控制系统
file conversion 文件转换
file copy 文件复制
file cover 文件夹;档案袋
file creation 文件建立;建立资料档案
file cut 刻锉纹;锉刀锉纹
file cutting anvil 錾座;錾锉砧
file cutting machine 锉纹加工机;锉刀切削机床
filed 场地
filed density 原位土密度
file definition entry 文件定义项
file definitionmacroinstruction 文件定义宏指令
file definition name 文件定义名
file deletion 文件删除
file description attribute 文件描述属性
file description entry 文件描述项
file descriptor 文件描述符
file design 文件设计
file designate 文件指配
file destination 目标文件
file directory 文件说明书;文件目录
file directory access 文件目录存取
filed lens 场镜
file drum 文件存储器;存储磁鼓;磁鼓文件
file dump 文件转储
file dust 锉屑
filed work 现场工作
file edit 文件编辑
file editor 文件编辑程序
file enable input 允许文件输入
file end flag 文件终止标志;文件终结标志
file finishing 锉光;锉削
file for information interchange 信息互换文件
file format 文件格式
file format conversion 文件格式转换
file function 文件功能
file generation 文件生成
file hammer 锉锤
file handle 锉刀柄
file handling 文件处理
file hardness 锉试硬度;锉刀硬度
file-hard shell 硬皮

file header 文件头;文件标题
file header label group 文件标题标记组
file holder 锉刀夹具;锉刀柄
file identification 文件识别;文件标识;文件标号
file identifier 文件标识符
file income tax return 提送所得税申报表
file insertion 文件插入
file label 文件标记;文件标号
file label control 文件标号管理
file layout 文件格式(设计);文件存储形式;存储形式
file limit 文件存储容量范围
file link 文件链接
file list 文件表
file load factor 文件负载率
file maintenance 卷宗保持;文件更新
file maintenance machine 文件处理机
file management 文件管理;档案管理
file management facility 文件管理设备
file management routine 文件管理程序
file management system 文件管理系统
file managemnent subsystem 文件管理子系统
file manager 文件管理器;文件管理程序
file manipulation 文件处理
file mark 文件标志;锉刀标志
file memory 文件存储
file merge 文件合并
filemot 枯叶色;黄褐色
file name 文件名称
file name change 文件名更改
file number 文件号
file number of report ratification 报告批准文件号
file of a map 图历薄
file of cartographic(al) digitized data 地图数字化数据文件
file of information 文件
file of precision 精密锉
file opening 开启文件;打开文件
file organization 文件组织;文件存储形式
file-oriented programming 面向文件的程序设计
file-oriented system 面向文件的系统
file over 滞后
file parameter block 文件参数块
file pockets addressing 文件袋选址
file polling 文件查询
file printout 文件打印
file processer[processor] 文件处理机
file processing 文件处理
file protection 文件保护
file protection device 文件保护装置
file protection mode 文件保护方式
file protection ring 文件保护环
file purging 文件清除;文件净化
file recovery 文件恢复
file reel 一盘文件带
file reference 文件参考
file reorganization 文件改编
file resident 文件驻留
file retention period 文件保留期
file retrieval 文件检索
file room 档案室
file routine 文件程序
files 案卷
file scan 文件扫描
file scratch test 锉刀划痕试验
file searching 文件检索
file section 文件段
file security 文件保密
file security control 文件安全控制
file separator 文件分隔符
file separator character 文件分隔字符
file shared 共享文件
file sharing 共用文件;共享文件;文件共用;文件共享
file signal 档案标记
files of forest resources 森林资源档案
files of luggage and parcel accident 行包事故案卷
file specification 文件说明表
file steel 锉刀钢
file steel brush 钢丝刷
file storage 文件存储器
file storage station 文件存储站
file store 文件存储器
file stroke 锉程
file-structure device 带文件结构的设备
file symbol 文件符号

file system 文件系统;文件档案系统
file test 锉刀试验
file tester 锉刀试验机
file title 文件标题
file to file program(me) 文件到文件连接程序
file tooth forming tool 锉齿成型刀具
file transfer 文件转移;文件传输
file transfer protocol 文件传送协议;文件传送规约;文件传递协议
file translation language 文件翻译语言
file transmission 文件传送
file type 文件类型
file unit 文件单元
file unsafe logic 文件不可靠逻辑
file update master program(me) 文件更新主程序
file updating 文件更新
file utility 文件管理程序;文件辅助程序
file utilization 文件利用率
file vault 档案保险库
file window 文件窗口
filial 分局的
filial set 分枝组
filifor corrosion resistance 耐红丝腐蚀性
filiform 线状的;纤维状的;丝状
filiform apparatus 丝状器
filiform corrosion 线状腐蚀;纤维状腐蚀;起红丝;丝状腐蚀
filiform corrosion resistance 耐丝状腐蚀(性)
filiform corrosion under film 漆膜下的丝状锈蚀
filiform lapilli 火山毛
filiform pulse 丝状脉
filigree decorative fixture 银丝嵌饰件;金丝嵌饰件
filigree effect 遮嵌效果(混凝土表面用金属栅)
filigree enamel with copper body 铜胎掐丝珐琅
filigree enrichment 银丝嵌饰;金丝嵌饰
filigree floor 嵌金属条的楼面
filigree girder 嵌金属条的梁
filigree glass 精细的玻璃;银丝玻璃;嵌丝玻璃
filigree rib(bed) floor 嵌丝的带肋楼面
filing 存档;编订;编档
filing a claim 提送赔偿要求
filing a claim for recovery 提出追偿
filing block 錾锉座;锉座
filing board 锉板
filing box 档案盒
filing cabinet 卡片(目录)箱;文件柜;档案柜
filing case 档案柜
filing chock 填补
filing clerk 档案管理员
filing date of geologic(al) report 地质报告日期
filing equipment 档案设备
filing hose 注入软管
filing lathe 锉刀车床
filing machine 锉锯(齿)机;锉床
filing of a seal-impression 提送印鉴
filing piece 填补材
filing room 档案室
filings 金属屑;锉屑
filings coherer 铁粉检波器;锉屑检波器
filing status 申报纳税身份
filings/wastewater ratio 金属粉与污水比
filing system 文件形成系统;文件编排系统;档案制度;档案系统
filing table 锉工工作台
filing time 归档时间
fill 填土;路堤;注满;人工填土;填料;填块;供料;填方;充填
fillability 满斗率
fill and drain valve 注入与泄放阀
fill and draw contact bed 间歇式接触滤池
fill and draw elutriation 注排淘洗;间歇式陶洗
fill and draw intermittent method 间断注水法
fill and draw reactor 充排式反应器
fill and draw system 间歇式;间歇式处理系统
fill and draw tank 间歇式沉淀池;间歇处理池
fill and grade 回填并整平(地面)
fill and pressure relief 加料和卸压
fill an order 供应定货;供应订货;交付定货
fill area 填充区域
fill area colo(u)r index 填充区域颜色索引
fill area facility 填充区设备
fill area index 填充区索引
fill area interior style 填充区内部类型
fill area representation 填充区表示法

fill a vacancy 补缺
fill backward 返填;返回填补
fill base 填土基层
fill block 填块;填塞块;垫块
fill boarding 垫板
fill by digging other part of rock 挖填
fill by gravity 自流充灌
fill by siphon 虹吸充灌
fill character 填充字符【计】
fill compaction 填土压实;路堤压实
fill concrete 回填混凝土;填筑混凝土;充填混凝土
fill construction 填土施工;填土工程
fill construction area 填土工程面积
fill crest 填土顶部;路堤顶部
fill dam 土(石)坝;填筑(土)坝;填土坝;堆积坝;堤
fill dam body 土坝坝身
fill dam construction 土坝施工
fill dam site 土坝位置
fill dam subsoil 土坝天然地基;土坝底土
fill dam width 土坝宽度
fill-dike reactor 充排式反应器
fill earth 填方(用)土;路堤土
filled and grooved joint 插榫平接
filled array 连续天线阵
filled bag closing machine 满包封口机
filled band 满带;满充能带;填充区域
filled-band level 满带能级
filled basin 淤积盆地
filled bitumen 加填料沥青;充填沥青
filled board 夹心板
filled cavity 填充穴
filled circle 填充圆
filled column 填实式管柱;填料柱;填料塔;填充管柱
filled composite 填充材料;充填材料
filled corner 修圆角
filled crib 填石木垛
filled cylinder 填充式管柱
filled degree of discontinuity 结构面充填程度
filled deposit 充填矿床
filled discontinuity 具填充物的软弱结构面;具填充物的不连续面
filled dry standpipe system 充填的干管系统
filled dyke 充填岩墙
filled earth thoroughly compacted in layers 分层夯实填土
filled entry 填有内容的登记项
filled fissure 充填裂隙
filled foundation 回填地基
filled gold 填金
filled grade 填充量
filled ground 回填地基;填土地基
filled-in ground 填土地基
filled-in pier 实体突堤
filled insulation 绝热填料
filled jetty 填石突堤
filled joint 填满板缝;填缝接合;填充缝
filled-lake plain 淤塞湖平原
filled land 填筑地
filled level 满带能级;满充能级
filled lower defect level 满充低缺陷能级
filled opening 填充裂隙
filled particleboard 填充木屑板
filled pipe column 混凝土管柱;填实管柱;填实管柱;填充式管柱;填充混凝土式管桩
filled pipe pile 填实管桩
filled plastics 填充塑料
filled plate 填板
filled salt lubrication 充盐润滑
filled shell 满充壳层
filled soil 填土
filled spandrel arch 实腹拱
filled spandrel arch bridge 实腹拱桥
filled spandrel stone arch bridge 实腹石拱桥
filled state 满态
filled stiffener 装有填料的加劲杆
filled stone 填充石(料)
filled-system thermometer 充填式温度计
filled thermoplastic 填充的热塑性塑料
filled thermoset 填充的热固性塑料
filled timber crib 填石木垛
filled tope 已回填砂石场
filled tower 填料塔
filled-type arrester 充填型阻火器
filled-up ground 填方路基;填方地面;填土地基;

填高地
filled valley 淤塞谷;淤积谷
filled vulcanizate 填料硫化胶
filled wall 填充墙
filled weight 加注后重量
filled with fluid 装满液体
filler 注入器;留充木;进位填充数;金属芯子;浇铸机;填隙料;填泥;填料;填充用原料;填充物;填充位;填充数;填充符;充填器;充填材料
filler absorption 填料混入
filler acceptance 填料可混入量
filler adding 填料添加
filler aggregate 填充骨料
Filler arc welding 菲拉电弧焊距
filler bead 焊珠
filler bending tester 垫片弯曲试验机
filler bin 粉料仓
filler block 止水塞;楼盖填充块体;间隔铁(护轮机);间隔垫块;混凝土填块;衬块;块
filler bowl 滤杯
filler box 注入口
filler brick 填充砖;填充块;挂钩砖;间隙砖
filler bucket elevator 填料斗式升运机
filler cap 漏斗盖;加油口盖
filler cargo 填隙货物
filler clay 填料黏土
filler coat 嵌填层(涂料);填孔涂层;第一道涂层
filler concrete 回填混凝土;填筑混凝土;填缝混凝土;填底混凝土;衬填混凝土
filler concrete panel 隔热轻集料混凝土板
filler content 填料含量
filler course 填充层
filler elevator 石粉提升机
filler fibre 填料纤维
filler floor 填充楼板
filler for brush application 用刷涂的填料
filler for cement 水泥填料;水泥惰性掺和料;水泥掺和料
filler frame 充填框
filler gate 充水闸门
filler-grade talc 填料级滑石
filler gypsum 填充石膏
filler hoist 升料器
filler hole 注水孔;加油孔;加水孔
filler hose 加油软管;加水软管
filler joist 填块格栅(防火地板中用的);加固格栅
filler joist floor 密肋空心砖楼板;格栅填料楼板
filler line 充注管
filler loading 填料含量;填充(料用)量
filler locking cap 加油口锁帽
filler masonry work 圬工嵌缝工作
filler material 填料;填充料
filler metal 焊条;填隙金属;填料金属;充填金属
filler mineral commodities 矿物填料矿石
filler neck 漏斗颈;接管嘴
filler nozzle 加油喷嘴
filler-on 加料工
filler opening 注入口;填充孔;注入孔
filler panel 填板
filler piece 填隙片;填充片;填充管片;垫片
filler pigment 填充颜料
filler pipe 注油管;注入管
filler plate 密封垫;填隙板;填(料)板
filler plug 注液孔塞;注入塞
filler plugs for trepanned holes 锥孔的管塞
filler point 填孔导纱器
filler powder 填料粉
filler reclamation 填料回收
filler reinforcement 填充剂增强;填料补强;填充补强
filler reinforcement effect 填料增强效应;填料加强效应
filler replacement 填料换装
filler reservoir 填料储器
filler retention 填料保持
filler ring 填充垫环;垫圈;倒角环
filler rod 焊条;嵌头;填丝;填料棒;充填棒
filler scale 石粉计量器
filler seam 填密缝
filler sheet 填板;垫板
filler silo 填料仓;石粉筒仓
filler slab 填板;承托板(陶土块、栏栅顶条)
filler specks 填料斑(瑕)
filler spreader 粉料撒布机
Filler's theorem 菲勒定理

filler stone 填缝石;嵌缝石;拱顶石
filler storage unit 填充料储[贮]存装置;石粉储[贮]藏装置
filler strip 嵌条
filler thread 衬垫粗线
filler tool 填塞工具
filler trough 填料槽
filler tube 注水管;漏斗管
filler type wire rope 充填式钢丝绳
filler wad 填塞垫
filler wall 柱间墙;柱间(填充)墙;填充墙
filler weigh hopper 石粉秤量斗
filler weld 填角焊缝
filler well 加油孔
filler wire 焊丝;填丝;填充焊丝;充填金属丝;充填钢丝
filler wire construction 嵌填金属丝工程
filler wire rope 嵌填金属丝缆索
fillet 芯皮黏合;圆阴角;整流片;连接边;角焊缝;肩角;内圆角;嵌头;钎角;平缘;平线脚;突出横饰线;填角;带状突起;齿根过渡曲面;承托;八字抹角;圆带
fillet and groove joint 企口接合
fillet and round 内圆角与外圆角
fillet chisel 圆角凿;平边凿
fillet chronograph 纸带记录时器
filleted corner 内圆角
filleted joint 嵌条接合
fillet ga(u)ge 圆角规
fillet gutter 狭条水槽;狭条排水槽
filleting 圆角;钢丝针布;角隅填密法;抹八字角;嵌缝(法);倒角
fillet in normal shear 正面焊缝
fillet in parallel shear 侧面焊缝
fillet lightning 带状闪电
fillet mo(u)lding 平条线脚;贴角条
fillet of concrete 混凝土嵌条
fillet of plaster of Paris 熟石膏平缘
fillet of screw 螺纹圈
fillet plane 圆角刨;圆缘刨
fillet radius 内圆角半径;齿根圆弧半径
fillet(s) and rounds 内圆角与外圆角
fillet sealant joint 填角密封接缝
fillet section 嵌条断面
fillet shape 嵌条外形
fillet shears 钢坯剪床
filletster 有槽圆头螺钉;槽刨
fillet surface 倒角曲面
fillet trim 嵌条镶边
fillet unit 嵌条单件
fillet weld 角焊(缝);填角焊;贴角焊缝;贴角焊(接);楞边焊缝;角隅焊;角肩焊
fillet welded joint 填角焊缝
fillet welding 贴角焊(接);条焊;角焊;角焊接
fillet welding in the downhand position 船形角焊
fillet welding in the flat position 船形焊
fillet welding in the horizontal position 横角焊
fillet welding in the vertical position 立角焊
fillet welding seam 贴角焊缝
fillet weld in normal shear 正面角焊缝
fillet weld in parallel shear 侧面填角焊(缝);侧面角焊缝
fillet weld in the flat position 角接平焊;水平角焊缝
fillet weld in the horizontal position 横向角焊缝
fillet weld size 焊角尺寸;填角焊缝尺寸
fillet winding machine 钢丝针布包卷机
fill excavation 填方开挖
fill factor 装填因数;填充系数;装料因子;充填因数
fill-flank trap 丘翼圈闭
fill foot 路堤底脚
fill fracture 填充的破裂口
fill gap 填补空白
fill height 填土高度;路堤高度;填高
fillieite 粒磷锰矿
fill in 填入;填充;替工;塞进;装满;临时填补位;插进;填写;填补物
fill-in brickworks 填充砖墙
fill-in fill terrace 填塞堆积阶地
filling 供料;填方;填塞;充填;装填;装满的;灌注法;浇入;嵌木塞泥脂;填腻子;打填孔料
filling abrasive material 充填材料用磨头
filling agent 填充剂
filling and drainage valve 装卸阀(门)
filling and emptying (船闸的)充水和放水

filling and emptying characteristics 灌水泄水特征
filling and emptying culvert 输水廊道
filling and emptying device of shiplift 升船机充水泄水装置
filling and emptying system 灌水泄水系统;输水系统;充水泄水系统;(船闸的)充水和放水系统
filling and emptying system lock 船闸输水系统
filling and emptying system short culvert 短廊道输水系统
filling and emptying valve 输水阀(门)
filling aperture 注入孔;加油孔
filling area 充填区;编档项
filling arm 装料臂
filling auger 装载螺旋
filling-backed fabric 纬背组织织物
filling balance 灌装秤
filling band 纬向条痕
filling band in shade 纬色档
filling bar 纬向条痕
filling bench 装模台
filling board 嵌缝板
filling cabinet 卡片箱
filling capacity 充填容积
filling cells 充实体
filling chock 衬垫
filling coefficient 充盈系数
filling colo(u)r 调和颜料;填充色料;底层涂料
filling composition 嵌缝料成分
filling compound 填物物质;油膏;抹平料;填(嵌)料
filling concrete 充填混凝土;填缝混凝土
filling condition of fault 断层充填情况
filling condition of karst 洞穴充填情况
filling connection 填密接头
filling conveyer[conveyor] 装料输送器
filling course 填层
filling criteria 编档准则
filling culvert 灌水廊道;进水涵管;进水涵洞;送水暗渠;输水涵洞;船闸文涵;充水廊道;文水涵洞
filling current 进入水流
filling curve 充水过程线
filling cycle 装载循环;充填循环
filling cyclone 填塞气旋
filling defect 充盈缺损
filling degree 装料程度
filling density 装灌密度;填充密度;充填密度
filling device 装填装置;(泵的)充水装置
filling drained sand cushion 填排水砂垫层
filling element 填料;填塞料
filling-emptying system 灌水泄水系统
filling end wall 投料端墙
filling engine 装料机
filling-faced fabric 纬面织物
filling facility 装料设施
filling flexible conduit 装料软管
filling floating dry dock 灌水下沉的浮船坞
filling for groin 丁坝填筑
filling for groyne 丁坝填筑
filling frame 直接纬纱并砂机;充填框
filling funnel 注液漏斗;漏斗
filling gallery 充水廊道
filling gas 充填气体
filling ground 充填地面;填充地基
filling ground by pumping dredged material 吹填
filling gun 注射枪
filling hatch 装货舱口
filling height 装料高度
filling hierarchy 编档等级
filling hole 注入孔
filling hose 装料软管
filling house 装油房
filling hydrograph 充水过程线
filling impeller 包装机灌装叶轮
filling-in 填充;糊版;堵版;填入;填满;放入
filling-in measurement 补测
filling-in panel 装饰板;墙板护板
filling-in paste 填充糊
filling-in piece 支顶木撑;填塞块;填孔木楔
filling installation 装料设施
filling instrument 充填器械
filling-in work 填土工作;填土工程
filling irregularity 纬档
filling knife 填塞刀;填腻子用刮刀;填缝刀
filling knitting 纬编
filling layer 填充层

filling level 充装液位
filling level measuring equipment 料位测量设备
filling lime 填充石灰
filling limit 储[贮]藏限度;蓄水量限度;填充限度
filling limit device 限量灌瓶器
filling line 装油管线
filling machine 装漆机;装料机;灌注机(器);加料机;充填机械
filling machine for collapsible tube 软膏锡管填充器
filling main 装油干线
filling mark 纬线印痕
filling material 镶补料;填充物;填(充材)料;充填物;掺加物;充填材料
filling material for optic(al) fiber cable 光缆缆芯填充料
filling method 充填采矿法
filling nature of joint 充填特征
filling net 填充网
filling notch 装填槽(轴承滚珠)
filling of brake pipe 制动管充气
filling of brake reservoirs 制动用的风缸充气
filling of coated abrasives 研磨层的填塞;研磨(涂)层堵塞
filling of crack 填隙
filling of dredged material 用挖泥材料填筑的路堤
filling of dry dock 干坞灌水
filling of reservoir 水库淤填;水库淤积;水库蓄水
filling of tank cars 槽车装料
filling of the award 申请仲裁
filling of voids 孔隙填充
filling oil cover 注油盖
filling opening 注入孔;注入口;灌水口;填充孔;充水口
filling operation 装料作业;充填工作
filling operator 装料工
filling paste 糊状填料
filling phase 充盈期
filling pick-up 拾鱼
filling piece 填隙片;垫片;填塞片;嵌片
filling pile 灌注桩;填补桩
filling pipe 装料管;装灌管;注水管;注入管;充管
filling pipeline 装料管线
filling place 装料平台
filling plant 装料设施;灌瓶站;灌瓶车间;灌瓶厂;填充植物
filling plug 注油孔塞;注入塞
filling pocket 投料口
filling point 填充点;加油站;装料点
filling port 注入口;冲水口;充水口;充水孔
filling position 装料位置;填土位置;充填位
filling powder 嵌缝粉
filling power 填充能力
filling pressure 填土压力;充气压力;填料压力;充装压力;充盈压
filling process 填土过程;充填过程
filling property of basin 盆地充填物性质
filling provision 加注设备
filling pump 装灌泵;注(液泵装)油泵;注入泵;冲水泵;充填泵;充水泵
filling putty 饰面工程用油灰;填孔腻子
filling ramp 灌装台(液化气)
filling rate 装舱率
filling ratio 填充率;充装系数;充填率
filling ratio of grinding compartment 粉磨仓填充率
filling ratio of mill 磨机填充率
filling reactor 填料反应器
filling regime 灌水状况
filling release wave 蓄汇波
filling resistance 满斗阻力
filling ring 喂料圈;垫圈
filling riser 灌装提升管
filling rivers with sand and mud 使泥沙充塞了河流
filling rock 填充用块石
filling rule 编档规则
filling rut 补修车辙
fillings 填充物
filling sand amount 灌砂量
filling screw 螺纹规
filling section 充填区;编档段
filling sequence 充填序列
filling sieve 注入滤网

filling sleeve 灌装软管
filling slippage 纬滑
filling slot 装填槽(轴承滚珠);装球缺口
filling smash eliminator 崩纬防止装置
filling snarl 纬圈
filling space 充填空间
filling spout 填料喷注;装料嘴;装管
filling spout cap 加油盖
filling stage 灌浆阶段
filling station 装料站;加油站
filling stone 填充石(料)
filling streak 纬向条花
filling symbol 填充符号
filling system 灌水系统
filling table 装油图表
filling-tear resistance 纬向撕裂强度
filling the tail void 盾尾空隙充填
filling time 装填时间;灌装时间;灌水时间;填充时间;充水时间;补炉时间
filling tissue 补充组织
filling to capacity 装满
filling trestle 填土栈桥
filling tube 加料管;灌注管;装料管;灌装管
filling twill weave 纬向斜纹
filling unit 编档单位
filling up 淤塞;淤高;泥沙淤积;堵版;充满;填塞;归档;加注
filling-up area 加油点
filling-up device 填充设施
filling-up dock 灌水船坞
filling valve 灌水阀门;供给阀充气阀;给油阀;进油阀;加注阀;充水阀
filling volume 填充数量
filling water level 充水高度
filling wind 纬管式成型
filling winder 络纬机
filling-wise 纬向
filling with concrete 用混凝土填充;用混凝土嵌缝
filling with putty 嵌油灰
filling with water 用水装满;用水填充
filling work 填土工作;填土工程
filling yarn 纬线;纬纱
fill in layers 分层填筑
fill-in pier 实体突堤
fill insulation 填充绝缘;填充隔热
fill insulation material 填塞绝缘材料
fill intensity 填筑强度
fill-in-the-blank system 填充脉冲间歇系统
fill-in well 注入孔;补充井
fillister 开槽;槽(刨);槽口;凹刨;凹槽
fillistered head 凹槽螺丝头
fillistered joint 企口接合;凹槽接合;凹槽缝;槽舌接(缝)
fillister head 凹槽螺丝头;圆头(螺栓或螺钉);凹槽螺帽
fillister head bolt 槽头螺栓;凹槽帽螺栓
fillister head of rivet 铆钉头半圆头
fillister head screw 圆顶柱头螺钉;有槽圆头螺钉;凹槽帽螺钉;有槽凸(圆)头螺钉
fillister plane 槽刨;凹刨;凹槽刨
fillister serewhead 凹槽螺帽
fill layer 填土层
fill length 填土长度;填土长度;路堤长度
fill lift 填土升运机
fill light 补充照明灯光
fill lighting 附加照明
fill line 充填输送管
fill material 填缝材料;充填材料;回填材料
fill material for reclamation 填筑的料;填海造陆用的填筑料
fill mechanism 装料机构
fill metal 焊积金属
fill mode 填充方式
fill model 填充模式
fill-net 填充网
fill network 填充网
fill nitrogen preservation 充氮保藏法
fill opening 注入孔;装料孔;加料口;填料孔
fill orifice 灌注孔;填孔
fill pile 填塞桩
fill plate 填板
fill plug 注水塞;加料口塞
fill port 装料口
fill quantity 填方数量

fill section 路堤截面;路堤断面;填土截面;填土部分;填上部分;填方(断面);堤的断面
fill settlement 填沉;路堤沉陷;路堤沉降;填土沉陷;填土沉降
fill settling 充填下沉
fill slide 填方滑坡;路堤滑坡
fill slope 路基斜坡;路堤边坡;填土边坡;填坡
fill sloping with grader 用平地机修筑路堤边坡
fill soil 路堤土
fill stake 填土标桩;路堤标桩
fill subsidence 填土沉陷
fill toe 填土坡脚;路堤坡脚;填方坡脚
fill type 人工堆积型
fill-type dam 填筑式堤坝;堆积坝
fill-type insulation 松填材料法隔热;填塞式绝缘;填塞绝缘;填充型绝缘;填层绝缘
fill unloading siding 路堤卸车线
fill-up 填满;混凝土构件平浇法;填补
fill-up carbon 增碳
fill-up concrete block 填充式混凝土砌块
fill-up ground 填筑地;填土地基
fill-up hole 填塞孔
fill-up line 注水管线;注水管路
fill-up to grade 填满到坡度线;填到设计标高
fill-up to the design level 填到设计标高
fill valve 装料活门;装料阀门;注入阀;进料阀(门)
fill width 填土宽度;路堤宽度
fill with cement 用水泥填充
fill work 填土工程;路堤工程
fill yards 压实后立方码(土料的);填筑立方码;填方数量
film 胶片;膜;软片;涂层;镀层;底片;薄膜;薄层
film absorber 薄膜吸附剂
film adhesion 膜层附着力
film adhesive 膜状胶黏剂;薄膜黏合剂;薄膜胶(黏剂)
film advance 胶片进片
film advance knob and crank 卷片钮和扳手
film advance lever 胶片推杆;输片杆
film advancing wheel 输片轮
film-air interface 膜层—空气界面
film analyzer 胶片分析器
film analysis 薄膜分析
film applicator(blade) 制膜器;刮膜机;漆膜涂布器
film applicator coating 贴膜法
film-a-search 胶片检索
film astrip 卷片式幻灯片
film attachment 胶片卷轴
film badge 感光测量器;胶片式射线计量器;胶片剂量计;测辐射(的)软片
film balance 膜天平
film band 摄影胶带
film base 片基
film-based fibre 薄膜纤维
film blanket 薄膜铺盖
film boiling 膜(状)沸腾;膜态沸腾
film-boiling heat transfer 薄膜沸腾传热
film boiling range 膜态沸腾区
film bond(ing) failure 涂层黏着力破坏;胶膜剥落
filmbook 显微图书;缩微图书
film build 成(厚)膜
film-builder 成膜物质
film building 构膜的;膜的构成;成膜
film cabinet 胶片储[贮]柜
film camera 胶片摄影机
film capacitor 薄膜电容器
film capacity 胶片盒容量
film card reader 缩微阅读器
film carrier 载膜
film cartridge 软片暗盒;胶片暗盒
film cassette 胶卷暗盒
film caster 漆膜涂布器
film casting 铸刮膜
film cement 接片胶水
film chamber 胶片储存室
film changer 胶片暗盒
film changing magazine 换暗盒
film channel 胶卷槽
film chip 胶卷小片
film circuit 薄膜电路
film clip 胶片夹
film-coated paper 薄膜涂布纸
film coater 涂膜器
film coating 胶片涂膜;涂膜;薄膜敷层;薄膜包衣

film coating machine 镀膜机
film coefficient 膜系数;膜层散热系数
film coefficient of heat transfer 薄膜导热系数
film color(u)膜状色
film composite insulation 绝缘薄膜复合制品
film composition 薄膜组成
film condensation 膜状凝结;膜状冷凝;膜态冷凝;膜(式)冷凝
film conductance 面层导热率;薄膜导热率
film contact face 胶片接触面
film contact screen 接触网(目)板
film container 胶片储[贮]柜;胶片盒
film continuity 薄膜连续性
film contraction 胶片收缩
film contrast 底片对比率
film coolant injection hole 薄膜冷却液喷孔
film coolant inlet 薄膜冷却液进口
film-cooled combustion chamber 薄膜冷却燃烧室
film-cooled engine 膜冷式发动机
film cooling 气膜冷却;薄膜(式)冷却
film corner cutter 胶片切角器
film counter 胶片计数器
film cracking 漆膜裂纹
film crystal oscillator 薄膜晶体振荡器
film curing (混凝土的)薄膜养护
film cut-off frequency 胶片的截止频率
film cut-out 薄片击穿熔断丝
film damage 胶片损坏
film decreaser 薄膜衰减器
film defect 漆膜缺陷;漆膜病态
film deformation 薄膜变形
film density 氧化膜密度;胶片密度;胶片黑度
film detachment 涂层剥落
film-detector 漆膜检测仪
film detonation 油料薄膜层爆轰
film development outfit 胶片显影设备
film dielectric(al)four-gang variable capacitor 薄膜介质四连可变电容器
film dielectric(al)variable capacitor 薄膜介质可变电容器
film diffusion 薄膜扩散
film discontinuity 漆膜的不连续性
film dislodgement 涂层脱落
film displacement 涂层脱落
film distillation apparatus 液膜蒸馏器
film distortion 胶片变形
film distribution 薄层分布
film distribution corporation 电影发行公司
film dosimeter 胶片剂量计
film dosimetry 胶片剂量测定法
film drainage rate 泡沫液体排出(岩粉)速率
film drawer 薄膜拉伸机
film drum 输片鼓轮
film dryer 薄膜干燥机;干片器;胶片干燥箱;胶片干燥剂
film drying 胶片晾干;胶片干燥
film drying equipment 胶片干燥设备
film drying outfit 胶片干燥设备
film dust 胶片屑
film editing 胶片剪辑;剪辑
film editor 编片机
film effect 薄膜效应
film elasticity 膜弹性
film elevating mechanism 胶片升降装置
film emission 薄膜型发射
film emulsion 胶冲乳剂
film end 片头
film evapo(u)ration 薄膜蒸涂;薄膜蒸发
film evapo(u)rator 薄膜蒸发器;薄膜式蒸发器
film expansion 胶片伸长
film exposure 胶片曝光量
film exposure counter 胶片曝光计算表
film factor 胶片因数
film-fiber 薄膜纤维
film fiche 胶片
film flatness 胶片平整度
film flattening 胶片展开;胶片压平
film flattening mechanism 胶片压平装置
film flo(a)tation 表膜浮选
film flow 膜态流动;(土内水分的)毛细管流动;水膜流;薄膜水流;薄层流动
film fogging 胶片灰雾;底片模糊
film force 水膜力
film for industrial radiography 工业 X 射线胶片

film format area 胶片幅面
film formation 成膜;漆膜形成;薄膜形成
film-formative wire 成膜金属线
film former 主要成膜物;成膜物(质);薄膜形成器;成膜剂
film-former forming 成膜的
film-forming 成膜性;形成薄膜的;成膜;薄膜形成的
film-forming ability 成膜能力
film-forming agent 成膜剂
film-forming characteristics 成膜特性
film-forming component 成膜剂
film-forming compound 成膜化合物
film-forming emulsifier 成膜乳化剂
film-forming fluoroprotein agent 成膜氟蛋白泡沫液
film-forming fluoroprotein foam fire extinguisher 成膜氟蛋白泡沫灭火器
film-forming material 成膜物;成膜物质
film-forming matter 成膜物质
film-forming medium 成膜介质
film-forming polymer 成膜聚合物
film-forming seal(er) 成膜封闭料
film-forming surfactant 成膜表面活性剂
film-forming temperature 成膜温度
film frame 胶片帧面;胶卷画面
film gate 镜头窗框;胶片孔
film ga(u)ge 薄膜式测量器
film glass 薄膜玻璃
film glue 薄膜胶黏剂;薄膜胶
film gluing lacquer 胶片黏漆
film grain noise 胶片颗粒噪声
film grain size 胶片颗粒大小
film granularity 胶片颗粒度
film growth 薄膜生长
film growth rate 薄膜生长速率
film guide frame 胶片保护夹架
film handling equipment 胶片处理设备
film hanger rack 胶片架
film-hardening agent 漆膜固化剂
film hardness 漆膜硬度;涂膜硬度
film hazing 漆膜起雾
film head 片头
film healing 膜修复
film heat conductance coefficient 膜导热系数
film heat transfer 膜放热
film heat transfer coefficient 膜传热系数
film holder 胶片架;胶片盒;底片夹
film hybrid integrated circuit 膜混合集成电路
film identification 胶片鉴定
filmily filling 薄膜式充填的
filminess 膜的状态
filming 膜(的)形成;摄影;薄膜形成;镀膜
filming-inspector 漆膜检测仪
filming integrity 漆膜连续性
film inspection 胶片检查
film integrated circuit 薄膜集成电路
film integrity 漆膜整体性
film intensity measuring device 膜层强度测定仪
filmistor 薄膜电阻
film jacket 片夹
film laboratory 洗印车间
film library 电影资料馆
film lightning arrester 膜片避雷器
film-like domain 薄膜域
film liner 薄膜护面;薄膜衬里
film loader 胶卷盒
film-loading 装片
film lubrication 油膜润滑
film marker 胶片符号
film memory 照相胶片存储器;薄膜存储器
film metallized bag 镀金属薄膜袋
film-metering device 薄膜测定装置
film moisture 薄膜水(分)
film mottle 片斑
film moving mechanism 卷片装置
film negative 软片;底片;负片
film notcher 片边缺口机
film notches 边缘切口
film number 胶片号码
film-numbering table 胶片编号台
film of oxide 氧化膜
film of paint 漆皮;涂料薄膜
film of survey (测斜仪的)摄影底片
filmogen 成膜涂剂;成膜剂

filmometer 漆膜抗张强度测定器
film opaque 修版液
film optic(al)modulator 薄膜光调制器
film optic(al)multiplexer 薄膜光复用器
film optic(al)sensing device 胶片阅读机;胶片光学传感器
film optic(al)switch 薄膜光开关
film optic(al)waveguide 薄膜光波导
filmorex system 缩微胶片电子检索法
film orientation 薄膜分子的定向
film overlay 薄膜重叠
film pack 胶片包装;摄影暗盒
film pack adapter 胶片盒转接器
film package 胶片包装
film packing 薄膜式淋水填料
film pack magazine 散页胶片包装盒
film-penetration theory 薄膜浸透理论
film penetration tube 电子透过记录器
film perforation 胶片穿孔
film photoconductor 薄膜光电导体
film piece 胶膜片
film plane 胶片面
film plane indicator 胶片平面指示器
film plasticizer 薄膜增塑剂
film plating machine 镀膜机
film positive 胶片正片
film preparation 膜制备;薄膜制备
film preservatives 漆膜防霉剂
film pressure 膜压(力);水膜压力
film-pressure plate 胶片压平板
film process 膜状过程;膜的形成过程
film processing 胶片处理;胶片冲洗;底片处理
film processor 胶片显影冲洗机;胶片冲洗机
film projector 放映机
film-proof 胶膜稳定的
film-proof rust 胶膜污斑
film-proof stripping 薄膜脱落
film-proof thickness 薄膜厚度
film Q-switching 薄膜 Q 开关
film reactor 膜式反应器
film reader 显微胶卷阅读机;胶片阅读器;缩微胶片阅读机
film record 缩微胶片记录
film recorder 胶片记录器
film recorder scanner 缩微胶片记录阅读器
film recording accelerograph 胶卷记录加速度仪
film recording oscilloscope 胶片记录示波器
film repair machine 补片机
film reproducer 胶片复制装置
film resilience 膜回弹性
film resistance 膜热阻;膜层阻力
film resistance meter 薄膜电阻测试仪
film resistor 膜式电阻器;薄膜电阻(器)
film resolution 胶片析象能力;胶片分辨率
film response 胶片响应性
film-return magazine 倒片盒
film reversal 胶片反转
film reversing controller 倒片控制器
film rewinder 倒卷机
film rewinding knob 倒片钮
film ribbon 薄膜打字机带
film ring 感光测量环;胶片环
film roll 胶卷
film rupture 膜破裂
film rust 锈膜
film safety 胶片安全
film scanner 胶片扫描器;缩微胶片扫描装置
film scraper 刮刀片
film scratching test 漆膜划伤试验
film screen 银幕
film section 胶片部分
film sensitivity 胶片感光度
filmsetting 薄膜排印
film shrinkage 胶片变形
film shrink packager 薄膜热收缩包装机
film size 胶片尺寸
film sizing 流膜分级
film sizing table 薄膜分级式冲洗淘汰盘
film solid media 固定培养基
film speed 胶片转动速率;胶片速度;胶片感光度
film speed dial 胶片感光指数表盘
film speed system 胶片感光速度系统
film splitter 薄膜分切机
film spool 卷片盘;胶片卷轴

film spot 膜点
film stability 胶片稳定性
film stack 薄膜叠式存储器
film stock 胶片材料
film storage 影片库;照相胶片存储器;胶片储藏
film storage album 底片夹
film storage unit 胶片存储单元
film strength 油膜强度;膜(的)强度;薄膜强度
film strength agent 油膜强度添加剂
film-stress interferometer 薄膜应力干涉仪
film strip 胶卷;片条;片带;摄影软片;长条软片;薄膜条;幻灯(卷)片;缩微影片条
film stripping test 膜剥裂试验;漆膜剥离强度试验;薄膜剥落试验
film structure 胶片结构
film studio 摄影棚;电影制片厂
film studio waste 电影制片废水
film sulfonator 膜式磺化器
film super-small variable capacitor 薄膜介质超小型可变电容器
film supply 供片盒
film supply indicator 卷片指示器
film supporting unit 承片框
film surface 胶片面
film surfaced hardboard siding 薄膜饰面硬纸壁板
film take-up spool 胶卷收片轴(测斜仪)
film tape 胶卷
film technique 成膜技术
film temperature 膜(层)温度;薄膜温度
film tensiometer 薄膜传感式张力计
film tension 胶片张力;胶片压平
film test 膜试验;漆膜试验
film tester 漆膜试验计
film theory 薄膜理论(一种薄壳设计的理论)
film thermocouple 薄膜热电偶
film thermopile 薄膜热电堆
film thickness 膜厚度;漆膜厚度;涂膜厚度;薄膜厚度
film thickness ga(u)ge 膜厚计;漆膜厚度测定仪;漆膜测厚仪;湿膜厚度计
film thickness indicator 膜厚指示器
film thickness measuring 膜厚测量;膜厚测定法
film thickness measuring device 膜厚测定仪
film thickness monitor 膜厚监测仪
film thickness test 漆膜厚度检验
film thickness tester 膜厚测定仪
film thickness uniformity 膜厚均匀性
film throughput 胶片通过量
film titler 胶片编号器
film toughness 膜强度;漆膜韧性
film traction mechanism 胶片牵引机构
film transfer function 胶片的传递函数
film transport 胶片传送装置;输片机构
film transport indicator 卷片指示器
film transporting mechanism 卷片传输装置
film transporting system 输片系统
film transport knob 卷钮
film type and size 胶片型号尺寸
film type boiler 膜式气化炉
film-type condensation 膜式冷凝
film type cooler 膜式冷却器
film type evapo(u)rator 薄膜式蒸发器;液膜蒸发器
film uniformity 膜均匀性
film varistor 膜式压敏电阻器
film vault 影片库
film velocity 输片速度
film viewer 胶片阅读器;胶片观察器;底片观察用光源
film viscosity 油膜黏度;膜黏度
film washer 洗片罐
film water 薄膜水
film width 胶片宽度
film winder 卷片旋钮
film window 胶片曝光器
film-wind tap 卷片扳手
film-wiper 胶片刮水器
film wise condensation 膜状冷凝
film-wise operation 膜式操作
film with ultra-thin emulsion 超薄乳剂浸胶片
film wrapping machine 薄膜裹包机
film writing station 胶片绘图台
filmy 薄膜的
filmy replica 薄膜复制品

filoscelle 绣花丝线
filosus 毛状云
filter 过滤机;过滤程序;滤子【地】;滤清器;滤器;滤光镜;滤光材料;滤池滤纸;滤波器;滤波程序;填料;渗滤器;射线过滤板;补偿器;滤波;滤色镜
filterability 过滤性;过滤能力;过滤率;过滤本领;滤过性;滤过率;可滤(过)性;可滤性
filterability index 过滤性指数;滤过性指数
filterability of viscose 黏胶过滤性
filterable 过滤的;可过滤的
filterable bacteria 滤过性细菌
filterableness 过滤性;滤过率;可滤性
filterable organism 滤过性有机体
filterable reactive phosphorus 滤过性活性磷
filterable stage 滤过阶段
filterable unreactive phosphorus 滤过性惰性磷
filterable virus 滤过性病毒
filter absorber 滤光镜
filter action 滤光作用;滤波作用
filter adsorber 滤池吸附器
filter aid 助滤器;助滤剂
filter algorithm 筛选算法
filter amplifier 滤波放大器
filter analyser 滤色分析器
filter analysis 滤色分析
filter apparatus 滤器
filter area rating 过滤面积额定能力
filter atlas 滤光片集
filter attenuation 滤波衰减
filter attenuation band 滤光器减光带;滤波衰减带;滤波器衰减带
filter backwashing intensity 滤池反冲强度
filter backwashing system 滤池反洗系统
filter backwashing wastewater 滤池反冲洗废水
filter bag 过滤袋;滤袋
filter bag cap 过滤袋插帽
filter bag holder 过滤袋框架
filter bag retainer 过滤袋支承架
filter bank 过滤机组;滤波器组
filter-bank spectrometer 滤波频谱仪
filter barge 隔油驳船;滤油驳船
filter base 滤床;过滤层
filter basin 过滤池
filter bauxite 铝矾土滤质
filter bed 滤床;过滤池;滤水池;滤砂池;滤垫层;渗透层;沉沙池;反滤层
filter bed depth 滤层高度
filter belt press 带式压滤机
filter blanket 过滤层;滤水垫层;渗透垫层;反滤铺盖;滤毡
filter blanket of gravel 滤池砾石过滤层
filter blinding 滤布堵塞;滤器堵塞
filter block 空心陶土过滤器;过滤块;滤块;空心陶土过滤块
filter board 过滤板;滤板
filter body 过滤器体
filter bottom 过滤器底板;滤底;滤池底
filter bowl 滤罩;滤(油)杯;滤清器壳
filter bowl gasket 滤杯垫圈
filter bowl retainer 滤杯护圈
filter box 过滤(器)箱;过滤盒;滤池体
filter bulb 滤球
filter bypass valve 过滤器旁通阀
filter cake 滤后沉淀;滤(层泥)饼;孔壁泥皮;井壁泥皮
filter cake breaker 滤饼破碎机
filter cake shredder 滤饼破碎机
filter cake texture 泥饼结构
filter cake thickness 滤饼厚度
filter cake washing 滤饼洗涤
filter cake washing circuit 滤饼洗涤流程
filter candle 滤烛;过滤芯;过滤管;滤芯;滤棒(柱);多孔陶瓷过滤芯
filter capacitor 滤波器电容器;滤波电容器
filter capacity 过滤能力;过滤量;滤波能力
filter car 滤水车
filter cartridge 滤芯;滤筒;过滤器芯子;过滤盒
filter casing 滤壳
filter cell 过滤器元件;滤匣;滤室;滤格;滤池
filter center 防空情报鉴定中心
filter chamber 过滤室
filter change 更换滤色镜
filter characteristic 滤波器特性曲线;滤波器特性
filter-charging current 过滤器充电电流

filter choke 滤波器扼流圈;滤波扼流圈
filter choking 滤池堵塞
filter chute 滤槽
filter circuit 滤波电路
filter cleaning water 滤池洗涤水
filter clogged switch 过滤器堵塞开关
filter clogging 过滤器堵塞;滤层堵塞
filter cloth 滤布;滤网;滤布
filter cloth type dust meter 滤网式尘埃计
filter coating 滤光膜
filter coefficient 滤镜系数;滤波系数
filter coefficient of direct development 正演滤波系数
filter coefficient of electrode configuration transform 电极装置转换滤波系数
filter coefficient of inverse development 反演滤波系数
filter coke 过滤用焦炭
filter collective system 滤池集水系统
filter collector 滤尘室;滤尘器;过滤除尘器
filter colo(u)r 滤色镜色彩
filter condenser 滤波电容器
filter conditioning 滤池调节
filter cone 过滤斗
filter configuration 滤池排列
filter controller 滤池控制器
filter cooler 滤器冷却剂
filter correction 滤波器校正
filter-coupler-switch-modulator 滤波耦合开关调制器
filter course 过滤进程;过滤层;滤层
filter cover 滤波器罩
filter crib 过滤木笼
filter crop 过滤作物
filter crucible 过滤坩埚
filter crystal 滤波晶体
filter cycle 过滤周期;滤池周期
filter cylinder for sampling 滤筒采样管
filter dam 滤水隔墙;透水坝
filter dehydration 过滤脱水(法)
filter design 滤波器设计
filter diaphragm 滤膜;滤机隔膜
filter diaphragm cell 滤膜电池
filter difference technique 滤光器差值法
filter disc 滤光盘
filter discrimination 滤波器滤波能力;滤波器分辨力;滤波能力
filter discrimination ratio 滤波器分辨率
filter disk assembly 滤片组
filter-down theory of economic progress 经济进步的最终渗漏论
filter drain 滤水沟;滤水暗管;滤池排水;滤层排水;盲沟
filter drainage system 滤池排水系统
filter drum 滤鼓
filter dust separator 过滤除尘器
filtered accumulation 筛选累加
filtered air 过滤的空气;滤过的空气
filtered air helmet 过滤空气面具
filtered anaerobic baffled reactor 过滤式厌氧折流板反应器
filtered beam 过滤光束
filtered cartridge 滤油芯子
filtered differential group 过滤微分群
filtered estuarine waters 滤清的河口水域
filtered hologram 已滤波全息图;滤波全息图
filtered image 已滤像;滤波图像;滤波成像
filtered open-flame carbon-arc exposure apparatus 过滤明碳弧灯照射老化仪
filtered particle testing 粒子过滤检验法
filtered photograph 滤色相片
filtered Poisson process 滤波泊松过程;过滤泊松过程
filtered radiation 滤过辐射
filtered response 滤波器响应
filtered sewage 滤清的污水;过滤的污水
filtered signal 滤过的信号
filtered solid collected in frame 滤饼
filtered solid outlet 滤饼出口
filtered solution 滤过液体
filtered state 过滤状况;滤过状况
filtered stock 滤清的油料;过滤的油料
filtered total organic carbon 滤过性总有机碳
filtered wastewater 过滤废水;滤过的废水;已过

废水
filtered water 滤清的水;过滤的水;滤过的水;滤灌合滤芯
filtered-water reservoir 过滤水池;清水池;过滤水储[贮]水池
filtered water valve 清水阀
filtered white noise 过滤白噪声;滤波白噪声
filter effect 滤色效应;滤镜效应
filter efficiency 过滤效率;过滤速度;过滤能力;滤器效率;滤池效率;滤波能力
filter effluent 滤器出流液;滤出液;滤池出水;滤器流出物
filter element 过滤元件;滤芯;滤光元件;滤波(器)元件
filter fabric 滤布;过滤织物
filter fabric mat 土工布反滤层;反滤沉积层
filter fabric soil retention test 滤层织物阻土试验;土工布反滤层试验
filter factor 渗透系数;滤色镜系数;过滤系数;滤镜系数;滤光因数;滤光系数;滤光片倍数
filter feeder 滤食性生物;滤器
filter feeding organism 滤食性生物
filter felt 过滤毡
filter film 过滤薄膜;滤膜
film film cultivation 薄滤膜培养法
filter flask 滤瓶;滤瓶;吸滤瓶
filter flies 滤池蝇类
filter floor 透水地面;滤池底
filter flow element 滤池流量元件
filter fluorimeter 滤色荧光计
filter fluorometer 滤色荧光计;滤光荧光计
filter fly 滤器蝇;滤程
filter frame 滤材框架
filter freezing 过滤冷冻
filter funnel 吸滤漏斗;过滤漏斗
filter gallery 滤水渠;滤池廊道;滤槽
filter gauze 滤网
filter glass 滤色玻璃;滤光玻璃;黑玻璃
filtergram 太阳单色像;单色像
filter grating 滤色光栅
filter gravel 过滤砂砾;滤水砾石
filter holder 滤光片架
filter hood 过滤器罩
filter hose 滤池软管
filter house 过滤站
filter housing 油滤壳体;滤池遮盖
filter humus 滤池腐殖质
filter hut 滤波器盒
fiter influent 滤进水
filtering 选出;过滤(的);滤清;滤光;滤除;渗透;滤波
filtering adjustment 滤色调整
filtering agent 过滤剂
filtering aid 助滤剂
filtering algorithm 滤除算法;滤波算法
filtering apparatus 过滤(仪)器
filtering area 过滤面积
filtering basin 滤水池;过滤层
filtering basket 滤框
filtering bed 过滤层;滤垫;滤床;滤层
filtering cartridge 滤筒
filtering cell 过滤室
filtering centrifuge 过滤离心机
filtering chamber 过滤室;过滤间
filtering clay 过滤用白土
filtering cloth 滤布
filtering cone 过滤漏斗
filtering course 滤层
filtering disc 滤片
filtering disk 滤片
filtering drum 滤筒
filtering earth 过滤用白土
filtering effect 滤色效应
filtering equipment 过滤设备
filtering factor 滤波因子
filtering fineness 过滤细度
filtering flask 吸滤瓶;过滤瓶
filtering flask with side tubulature hard glass 带支管硬质玻璃过滤瓶
filtering flask with side tubulature soft glass 带支管软质玻璃过滤瓶
filtering flow 滤流;渗流
filtering frame 滤框
filtering funnel 过滤漏斗;滤液漏斗
filtering glass 滤光玻璃

filtering head 过滤水头
filtering image 滤像
filtering jar 滤缸
filtering layer 反滤层;滤层
filtering leaf 滤片
filtering machine 过滤机
filtering mask 滤光片
filtering mass 滤料
filtering mat 滤层;滤垫
filtering material 过滤材料;滤材;渗漏材料
filtering material layer 滤料层
filtering material sampling 滤料采样
filtering medium 过滤介质;滤料;滤剂;滤质
filtering medium resistance 过滤介质阻力;滤材阻力
filtering method 过滤法
filtering method in seismic recording 地震记录滤波法
filtering mouth parts 滤口器
filtering of graded density 分级比重过滤
filtering operation 筛选操作
filtering operator 滤波算子
filtering operator length 滤波算子长度
filtering pad 滤垫
filtering parameter 筛选参数
filtering plant 过滤设备
filtering process 过滤过程
filtering process after stack 迭后滤波处理
filtering rate 过滤速率;滤(过)率;滤波速率
filtering ratio 过滤率
filtering resistance 过滤阻力
filtering screen 滤网
filtering strip 滤片
filtering surface 过滤(表)面;滤面
filtering thickener 过滤增稠器;过滤增稠剂
filtering treatment 过滤处理
filtering velocity 过滤速度;滤速
filtering velocity regulator 滤速调节器
filtering washer 洗滤器
filtering wash(ing) 洗滤
filtering well 过滤井
filter laid in the middle of aquifer 过滤器在含水层中部
filter lane 分流车道
filter layer 滤层;过滤层;滤光层;滤垫层;倒滤层;反滤层
filter leaching 过滤浸出
filter leaf 过滤叶片;滤叶
filter leaf for residual 残液过滤叶片
filter leaf group 过滤叶片组
filter leaf nozzle 过滤叶片接口
filter leaf test 滤叶试验
filter lens 滤色镜;滤光(透)镜
filter line 渗透线
filter liquor 滤液;滤光液
filter loading 过滤载荷;滤器负荷;滤池负荷
filter loss 渗漏量;滤失量;失水量(泥浆)
filter loss agent 失水量调节剂(泥浆)
filter mat 滤水毡垫
filter material 滤料;渗滤材料;反滤料
filter media conditioning stage 滤池介质调理阶段
filter medium 滤材;滤器填料
filter-medium characteristic 滤料特性
filter membrane 滤膜
filter membrane method 滤膜法
filter membrane sampler 滤膜采样器
filter membrane sampling 滤膜采样
filter milk claw 滤净集乳器
filter mount 滤光套
filter net 滤网
filter network 滤波网络
filter nozzle 滤池喷嘴;滤池喷头
filter of twined wire and wrapped net 缠丝包网过滤器
filter of wrapped net 包网过滤器
filter opening 滤水孔;滤孔
filter operating table 滤池操作台
filter operation 过滤操作
filter ophthalmic lamp 滤色检眼灯
filter oscillator 滤波振荡器
filter pack 组合滤色片;过滤组合件;滤光片组;砾石填料层;填料层;反滤层
filter pad 过滤垫;滤板
filter paper 过滤纸;滤纸

filter paper analysis 滤纸分析
filter paper box 滤纸箱
filter paper counter 滤纸放射性计数管
filter paper disk 纸质滤盘
filter paper drain 滤纸排水
filter paper holder 滤纸采样夹
filter paper method 滤纸法
filter paper microscopic test 滤纸显微镜试验
filter paper sampler 滤纸采样器
filter pass band 滤波器通带;滤波通带
filter passer 滤过性病毒;滤器穿透菌
filter period 过滤周期;滤池周期
filter photocell 滤光光电管
filter photometer 滤色光度计;滤光光度计
filter photometry 滤色光度法
filter pipe 滤管;井管滤网
filter pipe of single-size material concrete 均粒骨料混凝土滤管
filter pipet 滤波吸管
filter plant 过滤植物;过滤设备;过滤车间;滤水设备;滤水池
filter plate 滤光板;滤板
filter plexer 滤声器;滤波式双工器;滤波天线共用器;吸声器
filter pocket 滤袋;滤槽
filter polarizer 滤色偏振片
filter ponding 滤池积水;滤池成塘
filter ponding control 滤池积水控制
filter pooling 滤池积水
filter pore size 过滤器孔径
filter pore size distribution 过滤器孔径分布
filter preconditioning 滤池预调理
filter press 压滤机压滤器;压(力)滤器;滤油机;滤压机;压力过滤机
filter press action 滤压分异作用【地】
filter press cake 压滤(泥)饼
filter press cell 压滤式电池
filter press cloth 压滤布
filter pressed sludge 压滤污泥
filter pressing 压滤作用;压滤制;压滤脱水;压滤
filter press(machine) 压滤机
filter press plate 压滤机板
filter press type electrochemical reactor 压滤型电化学反应器
filter process 过滤法
filter pulp 滤浆
filter pump 过滤泵;滤液用泵;滤(池)泵;金属滤泵
filter rack 过滤机架
filter range 滤波范围
filter rate 滤速;滤率
filter rate controller 滤速控制器
filter rating 过滤器额定能力
filter ratio 滤水比;滤池系数比
filter reactor 滤波电抗器
filter record position 过滤记录台
filter regulator 滤池调节器
filter rejection band 滤波器阻带
filter residue 滤渣
filter resistance 滤波电阻
filter run 过滤周期
filter sand 滤砂
filter sandstone 滤水砂岩;滤水砂石;透水砂岩;渗滤砂岩
filter scan tube 滤光扫描管
filter screen 过滤筛网;滤网;滤筛;滤光网孔
filter screenoscope 滤色镜
filter section 滤波段;滤波器节;滤波段
filter separator 过滤分离器;过滤除尘器
filter septum 过滤膜
filter sheet 滤布
filter sieve 过滤筛
filter-silencer 过滤式消声器
filter skin 滤水表层;滤光器
filter slag 滤渣;滤水熔渣
filter slime 滤渣;滤池黏泥;滤池黏膜
filter slot 滤波隙
filter sludge water 清水污泥水;过滤污泥水
filter spectacles 滤光片眼镜
filter spectro-photometer 滤光器分光光度计
filter station 过滤厂
filter sterility test 滤器无菌检查法
filter sterilized 过滤法消毒;采用过滤方法消毒的
filter-sterilizer 过滤消毒器
filter stick 滤棒

filter stock 滤过机油
filter stop band 滤光器不透明带;滤波器阻带
filter strainer 过滤器;滤网;滤管;滤池滤器
filter stratification 滤床分层
filter strip 滤(色)条;滤色片;沉沙畦条;放淤畦条
filter surface 过滤面
filter table 平面过滤器;平面过滤机
filter tank 过滤池;滤箱;滤池;滤水器;过滤槽
filter technique 滤波技术
filter theory 滤波理论
filter thickener 过滤浓缩槽;过滤浓缩机
filter tip 过滤嘴
filter toe 滤水坡脚;坝趾反滤层;坝趾倒滤层
filter transfer function 过滤转换函数
filter transmission 滤光镜透射率
filter transmission band 滤光器透射带;滤波(器)通带
filter tube 过滤管;滤管;渗滤管
filter turn 穿空当转弯
filter type 过滤形式;过滤方式
filter type breathing mask 过滤式防毒面具
filter type respirator 过滤式呼吸器
filter type thickener 过滤式浓缩机;滤过式浓缩机
filter type ventilation 过滤式通风
filter underdrains 滤池地下排水系统;滤池底排水道
filter underdrain system 滤池配水系统;滤池地下排水系统
filter unloading 滤池卸模
filter vacuum receiver 滤液真空接收器
filter valve 过滤阀
filter vat 过滤桶
filter velocity controller 滤速控制器;滤速调节器
filter vent 滤口
filter wash 滤池冲洗
filter washer 滤池冲洗器
filter washing 洗池;滤池冲洗
filter washing by reversing of water 滤池反冲洗
filter washing liquor inlet 滤布洗涤液入口
filter washing water 洗池水
filter washing water inlet 清洗液入口
filter wash-water consumption 滤池冲洗水量
filter wastewater 洗池水;滤水废水
filter well 透水井;滤水井;排水砂井;过滤井;渗水井
filter wheel 滤色轮;滤光盘;滤光轮
filter without agitation 不搅动滤池
filter with removable filtering element 带活动滤心的过滤器
filter yield 滤池产量
filter zone 过滤地带;反滤带
filth 污秽物;污垢
filthy 污浊
filthy water 脏水
filtrability 滤过性;滤过率;可滤性
filtrable 滤得过的;可滤过的
filtrable agent 可滤过剂
filtrable solids 可过滤固体
filtrate 滤出水;滤液;滤清;滤出液
filtrate and washing water receiver 滤液与洗液受槽
filtrated brine tank 过滤盐水槽
filtrated factor 滤液因素
filtrated film type 滤片类别
filtrated mother liquor 过滤母液
filtrated stock solution 过滤母液
filtrate flow element 滤出流量元件
filtrate hardness 滤液硬度
filtrate manifold 滤液总管
filtrate outlet 滤液出口
filtrate receiver 滤液接收器;滤液槽
filtrate tank 滤液储[贮]罐
filtrate to waste 初滤排水
filtrate vacuum receiver 滤液真空接收器
filtrating bed 滤床
filtrating device 过滤装置
filtrating plant 过滤装置
filtrating unit 过滤装置
filtration 过滤作用;过滤;滤清;滤料;滤过作用;滤过;滤光;滤除;滤波
filtration accuracy 过滤精度
filtration aid 助滤剂;滤剂
filtration and separation 过滤与分离
filtration area 过滤面积
filtration barrier 滤过屏障
filtration bed 滤水层;渗水河床

filtration bio-reactor 过滤生物反应器
filtration block 过滤块
filtration capacity 过滤能力
filtration centrifuge 过滤式离心机
filtration characteristic 过滤性能;过滤特性
filtration-chlorination 过滤氯化法
filtration coefficient 渗透系数;过滤系数;滤过系数
filtration coke 过滤用焦炭
filtration column 过滤柱
filtration combustion 过滤燃烧
filtration constant 滤过常数
filtration control agent 滤失控制剂
filtration crib 过滤木笼
filtration dehydration 过滤脱水(法)
filtration device 过滤装置
filtration differentiation 过滤分异;滤波分异
filtration effect 过滤效果
filtration efficiency 过滤效率;滤过效率
filtration end point 过滤终点
filtration enrichment 过滤浓缩法
filtration equipment 过滤装置;过滤设备
filtration equipment for particle removal 过滤除尘器
filtration erosion test 渗透变形试验
filtration fabric 过滤布;滤布
filtration field 过滤场
filtration formula 过滤公式
filtration fraction 滤过分数
filtration glass fabric 过滤用玻璃纤维织物
filtration loss 滤失量
filtration loss quality of mud 泥浆失水量特性
filtration mat 过滤毡
filtration mechanism 过滤机理
filtration medium 过滤介质;滤料
filtration membrane 滤过膜
filtration method 过滤法;滤法
filtration of sieve cylinder 网筒过滤
filtration of sound 滤声;声滤
filtration of water 水的过滤
filtration on paper 纸上过滤
filtration packing 过滤填层
filtration paper 滤纸
filtration parameter 渗透参数
filtration path 渗透途径;渗漏线
filtration performance 过滤性能
filtration plant 过滤设备;过滤车间;过滤厂;滤水站;滤水厂;净水厂
filtration pore 滤孔
filtration precision 过滤精度
filtration pressure 过滤压力;滤压(力);滤过压
filtration process 过滤过程
filtration pump 过滤泵
filtration quality 失水性
filtration rate 过滤(速)率;滤(过)速度;滤过率
filtration ratio 滤过比
filtration reabsorption theory 滤过重吸收学说
filtration residue 滤渣
filtration resistance 过滤阻力
filtration slit 过滤隙
filtration slit membrane 裂隙滤过膜
filtration speed 过滤速度
filtration spring 过滤泉;滤清泉;渗流泉;渗出泉
filtration station 过滤站
filtration sterilization 过滤除菌
filtration surface 过滤表面
filtration system 过滤系统
filtration tank 污水滤池
filtration velocity 过滤速度;渗透速度
filtration virus 过滤性病毒
filtration washing 过滤洗涤
filtration water 过滤水
filtration water separator 过滤水分离器
filtration with fixed rate 定速过滤
filtration with fixed velocity 定速过滤
filtration with varying rate 变速过滤
filtrator 过滤仪;过滤器;滤清器
Filtrol 菲特罗牌膨润土
Filtrol fractionation process 费尔特洛尔白土接触精制蒸馏过程
filtro plate 滤板
filtros 滤石
filtrum 滤器
filty 沼气;甲烷;潮湿气候
filum 线状组织

filum terminale 终丝
fimbria 菌毛
fimetarius 粪生的
fiml potential 膜电位
fimmenite 赤杨花粉泥炭
fin 翼片;模缝脊;毛刺;合缝线印;鳍状物;凸片;水平舵;翅片;飞刺
finace corps 财团
final 最终的;最终边;最后进场;最后的
final acceptance 最终验收;最后验收;竣工验收;竣工验讫
final acceptance certificate 最终验收证
final acceptance line 最终验收线
final acceptance test 最终验收试验
final acceptance trial 最后交接试航
final account 施工决算;决算账户;竣工结算;决算账;决算(表)
final account for completed project 竣工结算
final accounting of revenue and expenditures 收支决算
final accounting statements 会计决算报表
final account of state revenue and expenditures 国家决算
final accounts of maritime engineering 水运工程决算
final accounts of the project 工程决算
final act 最终条例;最后决议
final actuation time 最后激励时间
final address 最终地址
final adjustment 最后调整;最后调节
final aftershock 最终余震
final age 伐期龄(森林)
final agreement 最后协议;最后协定
final air filter 终端空气过滤器
final alignment 最终对准
final amplifier 终端放大器;末级放大器
final anode 最后阳极
final-anode voltage 末级阳极电压
final approach 最终进场;最后进场
final approval 最后核准
final arrival notice of ship 船舶到港确报
final articles 最后条款
final ascent 最后上升
final assembly 最末组件;最后组装;最后装配;总装;完成装配;输出装置
final audit 全面审计;期末审计
final award 最后裁决
final backfill 最终回填料;后回填
final balance 期末余额
final bearing wear 轴承最终磨损量
final belting 最后拖光(混凝土路面)
final beneficiary 最终受益人
final blading (平地机的)最后刮平
final blanking 最后消隐
final blast 最终鼓风
final boiling point 终沸点;干点;完全蒸发时的温度
final boost 最后增压
final budget 核定预算
final building cost 工程决策
final bulking 最终膨胀
final capstan 成品卷筒
final carry digit 最后移位数;终端进位;尾端进位数
final catastrophic failure 最终毁坏性破裂
final certificate 最终证书;最终品质证明书;决算书;决算单
final certificate of payment 最终支付证书
final certificate of quality 最终品质证明书
final chart scale 成图比例尺;成比例尺
final check(ing) 最后检查;最后检验
final check picture 最后核对图像
final circuit 末级电路
final clarification 最后分类;最后澄清
final clarification tank 二次澄清池;最后澄清池
final clarifier 最终澄清池;最终沉淀池;终段澄清器;二次沉淀池
final cleaner 第二滤清器
final cleaning 最后清洗
final cleanup 最后清理工作;竣工清理
final clearing 最终精选;主伐;清理伐
final clearing age 主伐龄
final clearing of ice 终冰
final closing 最后涂层;最后结账
final closing bracket 最终闭括号
final closure 最后合拢;合न口;合拢(指堤坝等);

堵口
final coat 最后涂层;末道涂层
final coat exterior plaster 最后的抹灰;末道表面粉刷
final coat external plaster 最后外层涂浆
final coat external plaster stuff 末道涂层混合料
final coating 最后涂层;罩光漆;末道漆
final coating cement 最后涂的水泥层;表面水泥
final coat mixed plaster 末道混合涂料
final coat paint 末道油漆
final coat plaster 最后表层涂料
final collector 末级集电极
final colo(u)r picture 最终彩色图像
final common path 最后通路;最后公路
final compilation 出版原图
final completion 最终完工;最后竣工;竣工;圆满竣工
final completion certificate 最终竣工证书
final completion stage 最后竣工阶段
final composing 最后排样
final concentrate 最终精矿
final condenser 最终冷凝器
final condition 最终状况;边界条件
final connection 最后连接
final construction cost 最终施工费用;总施工费用
final construction report 竣工报告
final consumer 最终消费者
final consumer price 最后消费价格
final consumption 最终消费
final consumption demand 最终消费需求
final contact switch 最后接触开关;精磨用接触开关
final continuation line 最终继续行
final contraction value 最终收缩度值
final contract report 最终合同报告
final control element 末级控制元件
final controlled condition 最终受控条件
final controlled variable 最终受控量
final controlling element 末级施控元件
final cooling 最终冷却
final copy 出版原图
final cost 终值;决算成本;工程决算;最后成本;总造价;总成本;决算;成本
final cost estimate 最终费用概算
final course 终航向
final crop 主伐林
final crushed product 终碎产品
final crushing 最后破碎
final cure 最后固化
final curing 终塑化;终固化;后期养生;后期养护
final cut 最后开挖线;最后开挖深度
final cut-off 最后切断;停车
final cutting 最后切断;终伐;后伐;森林主伐
final data reduction 最后结果处理
final date 最后日期
final decay 后期衰变
final decision 最终决算
final declaration 确定申告
final decline 终降
final decree 最后判决;终审判决
final deflection 最后方向修正量
final delivery elevator 最终卸载升运器
final delivery schedule 最终发运计划
final demand 最终需求
final design 最终设计;最后设计;施工设计
final design review meeting 最终设计审查会
final destination 最终到达港;目的港
final diameter 最终直径;最后旋回圈半径;满舵回转直径;回旋直径
final diameter of hole 终孔直径【岩】
final digit code 最终数位码;最后数字码
final disposal 最终弃置;最终处置
final distribution 最终分配;最终分布
final dividend 最终股利;最末股息;最后红利;最后股息;终结股利
final documentation 综合编录
final draw 最终放矿
final drawing 最终图样;清绘原图
final dredged depth 最后开挖深度
final dredged level 最终挖成标高;最终疏浚标高;最终开挖标高;挖泥竣工高程;挖泥竣工标高
final drinking water quality 最终饮用水水质
final drive 终传动;末级传动;最终传动
final drive casing 最终减速器壳

final drive gear 末端传动齿轮(箱);最终传动机构;传动链末端传动齿轮
final drive guard 末级传动防护器
final drive hub 主传动轮毂
final driver 末级激励器
final drive reduction 最终传动减速
final drive shaft 主传动轴;终传动轴
final drive sprocket 终传动链轮
final drive spur gear 最终传动圆柱直齿轮
final drive transmission 末级传动牙齿箱
final drive wear guard 末级传动磨损防护器
final drying 最终干燥
final effluent 最终流出物;最终出水;最终出流物;最后流出物
final elevator 最终升运器
final encapsulation 封口
final engineering 进行定案设计;定案设计工作
final engineering cost 工程决算
final engineering report 最终工程报告
final environmental impact statement 最终环境影响鉴定;最终环境影响报告书
final environmental impact statement supplement 最终环境影响鉴定附录;最终环境影响报告书附录
final environmental statement 最终环境鉴定;最终环境报告书
final equilibrium cocnentration 最终平衡浓度
final equilibrium state 最终平衡态
final estimate 决算;结算
final estimate survey 最终估价测量
final etching 最后腐蚀
final evapo(u)ration 最后蒸发
final evapo(u)ration rate 最终挥发速率
final examination 最后检验
final examiner 成品检验员
final excavation limit 最终开挖线
final exploratory report 最终勘探报告
final felling 终伐;后伐;森林主伐
final figure 确报数
final film positive 出版原图正片
final filter 终滤器;终端过滤器;二次过滤器
final financial result 最终财务成果
final finish 精整加工
final finishing 最终精加工;终饰;最终修整;最终修饰;最终的精整加工;精整
final finsh 最后精加工
final firing 最终烧成
final flowsheet 最终流程
final forging temperature 终锻温度
final form 最后形状
final form of equation system 方程组最终形式
final frequency 最终频率
final ga(u)ge length 最终计算长度
final gear 末端传动齿轮
final gear reduction 最终齿轮减速
final geologic(al) report 最终地质报告
final gettering 最后吸气
final goods 最终产品
final grade 最终坡度
final grade line 最后采用的坡度线
final grading 最后整型;最后平整
final grain size 最终粒度;最终颗粒大小
final great-circle course 最终大圆航向
final grind 磨光
final grinder 末道研磨机;微细粉碎机
final grinding 精研磨;精细碾碎;细磨;终磨
final guidance 末段制导
final-hardened and tempered steel wire 调质钢丝
final hardness 最终硬度
final heading 最终航向【航空】
final heating 最终加热
final heave 最终隆起量
final height of water head 最终水头高度
final hole diameter 终孔直径【岩】
final host 终宿主;终寄主
final humidity 最后湿度
final ignorance 最终不知
final impulse operating relay 终端继电器;末端脉动式继电器
final incidence 最后负担
final individual coat 最后单件上涂料
final infiltration capacity 最终渗水量;最终渗能力
final injection pressure 注浆终压
final inspection 工程验收;最终检验;最终检查;最后

检验;最后检查;竣工检验;竣工检查;交货检验
final inventory 期末存货
final invoice 最终发票
final isochrone 最终等时线
finalize 最后确定
finalized design 定型设计
finalized product 定型产品
finalized run 出版原图
final lacquer 末道漆
final landing date 卸空日期
final laying 最终瞄准
final layout 最后布置
final level(l)ing 最后平整
final limited date 最后截止日期
final lining 二次支护
final link 最终链路;终端链
final literal material 最终文字材料
final load 最后荷载;有效荷载;决定性荷载;最后负荷
final location 最后定线
final location survey 最后定线测量
final lock mechanism 尾部闭锁机构
finally formed hole 终孔
final magnification 最终放大率;末级放大
final manuscript 清绘原图
final map 最终图件
final mass 最终质量;终点起爆
final maturities 最后偿还期
final maturity 贷款期限
final mean annual increment 最终平均年生长量
final merge phase 最终归并阶段
final mining depth 开采最终深度
final moisture 最终水分
final moisture content 最终湿度
final molasses 废糖蜜
final molasses storage tank 废蜜塔
final moment 最终力矩;有效力矩;决定性力矩
final mo(u)lder 最后滚边机
final movement 最终位移
final negative carry 最后负进位;终点反向进位
final net profit and loss section 期末净损益部分
final nozzle 排气喷嘴
final open-pit boundary 最终露天矿边界
final open pit edge 露天矿最终边帮
final orbit 终轨
final original 制印原图
final outcome 最终结果;最后结局
final output 最终输出;终端输出;末级输出
final particle 最后粒子
final particle size 最终粒度;最终颗粒大小
final pass 最后的焊道;最轧孔型;成品道次
final payload 净有效负载
final payment 最后付清;最终付款;末次付款;尾款清讫;付清尾款
final payment certificate 最终付款证书
final penetration 最终贯入度;最后贯入度
final performance 最终性能;临界性能
final pigging 最后一道清管【给】
final pit boundary 最终露天矿边界
final pit of open mining limit 露天开采最终境界
final pit slope 露天矿最终边坡
final plan 最终方案
final planetary gearing 二次周转轮系
final planning 最后规划
final plat 批准图;最终地段墙
final plot 出版原图
final polishing pass 精整孔型
final port 最后目的港;终点港;目的港
final port of designation 最后目的港
final port of discharge 最后卸货港
final position 终点位置;最终位置
final posterior distribution 最终后验分布
final power amplifier 末级功率放大器
final pressure 最终压力;最后压力;终压(力)
final pressure distribution 最终压力分布
final prestress 有效预应力;最终预应力(值)
final prestressing 有效预应力
final prestressing force 最终预压力
final price 最终价格;最后价格
final process 最后程序
final processing 最终加工
final product 最终制品;最终成品;最后结果;终端产品;最终乘积;最后产品
final product storage 成品库

final profile 最终剖面;最后轮廓线;最后断面
final program(me) 最后程序
final project 最终方案;最后设计
final proof 最后校样
final proofing 最后检验
final prospecting report 最终普查报告
final protocol 最后议定书
final provisions 最后条款
final pulse 完成脉冲
final purification 最终净化(作用);最后净化;最后精炼
final purification plant 最终净化装置
final purifier 最终净化装置
final quantity 终值;答案
final reaction 最终反应
final reading 最终读数;最后读数;终读数;末(次)读数
final recleaner flo(a)tation 最终再精选;最后(再)浮选
final recovery 最后恢复
final recovery parachute 回收系统
final reduction gear 最终减速齿轮
final refining 最终精制;最后精制;最后精炼
final rejection 确定报废
final rendering 末道抹灰
final rendering mix(ture) 末道抹灰混合料
final rendering stuff 末道抹灰材料
final repair 最后维修
final report 最终报告;最后报告;总结报告;总结;决算书;决算报告;结算报告
final residuum 最后残渣
final resistance of filter 过滤器终阻力
final response 最终特性
final result 最终结果;最终答案;最终成果;最后结果
final retarder 目的制动缓行器
final retention 最后持水量
final retention time 后期保留时间
final retention volume 最后保留体积;后期保留体积
final return 最终利润;最后回报;确定申报
final rinse 最后冲洗
final rinse tank concentration 末级清洗槽浓度
final rolling 终碾;末次碾压
final route 最终路由;最末路由
final ruling 最后裁决
final run 最后的焊道;抛撒飞行
final safety trip 终端安全释放机构
final sag 最终弧垂
final sailing 最终启航;最终开航
final sale 最终销售
final sample 最终油样
final saturation 最后饱和
final save-all cell 最后捕收槽
final scheme 最终设计图;最后草图
final scraping 最后一道清管【给】
final screening plant 精筛机
final sedimentation 最终沉淀(作用)
final sedimentation tank 最终沉淀池;最终沉淀槽;最后沉积池;最后沉淀池
final selector 终接选择器;终接器
final semicolon 最终分号
final separating tank 最后分离池
final separator 最终分离器
final serviceability index 最终使用能力指数
final set 最后贯入度;终集(合);终变制定
final set of pile 桩的最终下沉
final setting 终凝
final setting basin 最终沉淀池
final setting curve 终凝曲线(混凝土)
final setting time 终凝时间(水泥)
final settled account statement 总结账单
final settlement 施工决算;竣工决算;决算;最终沉陷(量);最终沉降(量);最终沉降
final settlement of accounts 决算
final settling 最终沉陷(量);最终沉降(量)
final settling basin 最终沉降池
final settling tank 最终沉淀池;最终沉淀槽;最后澄清槽;末级沉淀池
final shape 最后形状
final shaping 最后整型;最后定型
final shear stress 终切
final shipping instruction 最后船运通知;最低船运通知
final shut-in pressure 最终封井压力
final size 最终尺寸

final sizing 最终筛分;最后筛分;最后分级
final slag 终渣
final slip 最终转差率
final soil moisture 最终土壤水分
final soluble cocnentration 最终可溶解浓度
final solution 最终溶液
final speed 最终速度;终速度
final spinning block 旋压成型模
final stage 最后阶段;末级
final stage deaerator 末级除氧器
final stage engine 末级发动机
final stage of acceleration 最后加速阶段
final stage of design construction 设计施工最终阶段
final stage rocket engine 末级发动机
final state 最终状态;终止状态;终态
final state interaction 终态相互作用
final statement 最终报表;最终报告;最后结算;决算(报)表;决算;结算单;检验付款申请
final static level 最终静水位
final stock 最终库存
final store 最终库存
final strength 极限强度;最终强度
final stress 最后应力;最后应力值;有效预应力
final stucco stuff 末道粉墙泥灰料
final sub-circuit 终端支电路
final subsidence 最终下沉;最终沉陷(量);最终沉降(量)
final sum 结算额
final supply 最终供应
final survey 最终查勘;最终测量;终测;竣工测量;定线测量;定测
final tailings 最终尾矿;废弃尾砂
final tank 终端沉淀池;终点油库
final temperature 最终温度;最后温度;终止温度;终点温度;出口温度
final temperature difference 终温差
final temperature of freezing 最后冻结温度
final test 最终试验;最后试验;最后检验;最后检查
final text 定稿
final text submitted for approval 报批稿
final thermomechanical treatment 最终形变热处理
final thickness 最后厚度
final thin 最薄的
final time 终凝时间(水泥)
final tooth wearing height 牙齿最终磨损量
final total 总计
final total settlement 最终总沉降量
final transmission 主传动
final treatment 最终(净化)处理;最后处理;后处理
final treatment of pollutant 污染物最终处理
final treatment process 最终处理过程
final trunk 末级中继线
final trunk group 最末中继线群
final turning diameter 终回转直径
final unloaded sag 最终无载垂度
final unloading sag 卸载导线下垂(量)
final up-run purge 二次上吹扫
final utility 最终效用
final utility theory of value 最终效用价格理论
final vacuum 极限真空度;后排气
final valuation 最后估价
final value 终值;终价
final value estimate 最后估价
final value of annuity 年金终值
final value theorem 终值定理
final vapo(u)rization temperature 最终气化温度
final varnishing 最后上蜡;上末道清漆
final velocity 最终速度;终速度;末速(度)
final version 定稿
final video-amplifier 末级视频放大器
final view 终审结果
final void ratio 最终孔隙比
final voltage 终止电压
final volume 最终容积
final wall bench 最终边帮台阶
final water 最终水分
final water-cement ratio 最终水灰比
final water level 末刻水位
final water saturation 最终含水饱和度
final weight 最后重量
final weight of sample 样品最终重量
final welding 最后焊接;终焊
final width of bank ruin of reservoir 最终坍岸带宽度
final word 终端信息;末端信息
final work 最后工作
final working 最后加工
final wrap 末层包布
final yield 最后产率;主伐收获
finance 资金;通融资金;出资;筹资;财政;财务
finance act 财政法
finance and economics 财经
finance and trade 财贸
finance bill 金融票据;通融汇票;财政法案
finance capital 金融资本
finance charges 贷款费用;信贷费用;财务费用
finance company 信贷公司;财务公司
finance contract 信贷合同
finance corporation 金融公司
finance cost 财务成本;理财成本
finance fee 贷款附加费
finance function 财务机能;财务功能
finance house 金融公司;财务公司
Finance House Association 贸易协会
finance house deposit market 财务公司存款市场
finance image processor 财务图像处理程序
finance law 财政法
finance lease 融资租赁;财务租赁
finance manager 财务经理
finance model 财务模型
finance sharing 财政分成
financial ability 财力
financial account 财务账户;财务账单
financial accountant 财会员
financial accounting 财务会计
Financial Accounting Standards Board 财务会计标准委员会(美国)
financial adjustment 财务调整
financial administration 财务行政
financial adviser 财政顾问
financial affairs 金融事务;财务
financial agreement 金融协定
financial aid 补贴建设资金;财政援助
financial aim 财务目标
financial allocation 财政分配;财政拨款
financial analysis 财政分析;财务分析
financial analysis program(me) 财政分析程序
financial analysis ratio 财务分析比率
financial analyst 财政分析家
Financial and Economic Board 财政经济委员会
financial and material assistance 财政和物质援助
financial and monetary restructuring 财政金融体制改革
financial and monetary sector 财政金融部门
financial application software 财务应用软件
financial appraisal 财务评估;财务评价
financial appropriations 财政拨款
financial arrangement 财务安排
financial assets 财务资产
financial assets management company 金融资产管理公司
financial assitance 财政资助
financial audit 财政审计;财务审计
financial audit work 财务审计工作
financial autonomy 财务自主;财务独立
financial backing 财政支援
financial balance sheet 财务平衡表
financial balance statement 财政收支平衡表
financial benefit 财政收益
financial bond 金融债券
financial bonus 财政红利
financial book 账簿
financial budget 财政预算;财务预算
financial budget expenditures 财政支出
financial capability 财力
financial capacity 财政的负担限度;财(务能)力
financial center 金融中心
financial charges 融资费用
financial claim 债权
financial committee 财务委员会
financial community 金融界
financial company 金融公司
financial condition 财务地位
financial contribution 出资;财政捐款
financial control 财务控制;财会监督
financial controlling system 财务控制制度;财务控制系统

financial conversion cost of foreign exchange 财务换汇成本
financial corporation 金融界;信贷公司;财务公司
financial cost 融资费用;财务支出优价;财务成本
financial covenant 财务限制手段
financial credit 财政信贷;财务信用
financial credit guarantee 财务信用保证
financial crisis 金融危机;财政危机
financial decision 财务决策
financial deficit 财政赤字;财务亏空;财务不足
financial difficulty 财政困难
financial discipline 财务纪律
financial disclosure 财务状况的披露;财务公开
financial duty 财政关税
financial economics 财务经济学
financial-economic system 财经体制
financial embarrassment 财务拮据
financial evaluation 财务评价
financial evaluation of projects 项目财务评价;项目财务估价
financial examination 财务检查
financial expansion and contraction 财政的伸缩
financial expenses 财政费;财务费用
financial expense audit 财务费用审计
financial expense statement 财务费用表
financial failure 财务失败
financial feasibility 财政的可行性;财务可行性
financial feasible 财务上可行的
financial flow 资金流动
financial forecast 财务预测
financial forecasting 财务预测
financial fund 财政资金
financial gain 经济获益
financial gearing 财务杠杆
financial group 财团
financial guarantee 财政担保;财务保证;财务托付;财务担保
financial guaranty 财务担保书
financial incentive 财务激励
financial incentive system 财务奖励制度
financial income 财政收益
financial indemnity 财政补偿
financial index 财务指标;财政指标
financial information 金融信息
financial information intermediaries 财务信息中介
financial information system 财务信息系统;财务情报系统
financial in solvency 无力支付
financial institution 金融组织;金融机构
financial institutions for small businesses 小企业金融机构
financial integrity 财务完整性
financial intermediary 金融中介;金融居间人
financial internal rate of return 财务内部收益率
financial investigation 财务调查
financial investment 金融投资;财政投资;财务(性)投资
financial lease 资本租赁;融资租赁;财务租约;财务租赁
financial lease accounting 财务租赁
financial leasing 融资租赁
financial leverage 用贷款投资;财务杠杆(作用)
financially sound 财力雄厚
financial magnate 财政巨头;财阀
financial management 财务管理
financial management at different levels 财政分级管理
financial management information system 财务管理信息系统
financial management review 财政管理检查
financial management system 财务管理系统
financial mangement system of railway transport enterprise 铁路运输企业财务管理体制
financial market 金融市场
financial monopoly 财政金融垄断;财力垄断
financial net present value 财务净现值
financial net present value of foreign exchange 财务外汇净现值
financial objective 财政指标
financial officer 财务官员
financial oligarch 财阀
financial organization 金融组织;财政机关
financial outlays exceeding state revenues 财政上分配过头

financial outturn 财务成果
financial period 财务期间
financial plan for construction technology 施工技术财务计划
financial plan(ning) 财务计划;制订财务计划;财务策划
financial policy 金融政策;财政政策;财务政策
financial policy for environmental protection 环境保护金融政策
financial position 财务状况;财务地位
financial position statement 资产负债表
financial preference 财政优惠
financial price 现行价格;财务价格
financial program(me) 财务计划
financial projection 财务预测
financial proposal 带报价的建议书
financial provision for renewal 更新的财政准备
financial prudence 财政稳健政策
financial rate 金融汇率
financial rate analysis 财务比率分析
financial rate of return 财务收益率;财务回收率
financial ratio 财务指数;财务比率
financial ratio analysis 金融比率分析;财务比率分析
financial reconstruction 财界重建
financial regulations 财务条例
financial rehabilitation 财界恢复
financial reimbursement 经济赔偿
financial relation 财务关系
financial report 财务报告
financial reporting 财务报表
financial reserve 财政后备
financial resource 资金来源;财(政资)源;财力
financial resources of the people 民力
financial resource transfers 财政资源的转移
financial responsibility 财务责任
financial restructuring 财务体制改革
financial result 财务成果
financial retrenchment 财政紧缩
financial return 利润;财务收入
financial revenue 财政收入
financial review 财务评论
financial risk 财务风险
financial rotation 理财轮伐期
financial sanction 财政制裁
financial secrecy 财务秘密
financial settlement 财务结算;财务清算;财务结算
financial sheet analysis 财务报表分析
financial situation 财务状态
financial situation statement 财务情况说明书
financial solvency 财政实力;财务实力;财务偿付能力
financial squeeze 财政困难
financial stability 财政稳定
financial standards 财务标准
financial state 财务状况
financial statement 借贷对照表;决算表;财务报告;财政决算;财务报表
financial statement analysis 财务报表分析
financial statistics 金融统计
financial status 财务状态;财务状况
financial straits 财政困难
financial strategy 财务战略;财务策略
financial strength 财政实力
financial stringency 金融紧缩;财政紧迫
financial structure 财务结构
financial structure ratio 财务结构比率
financial subsidy 财政补贴
financial sufficiency 财政充裕
financial supervision 财政监督;财务监督
financial supervision and control 财政监督管理
financial support 财政支援;财政支持;财务援助;财务支持
financial support fund 财政援助基金
financial surplus 财政结余
financial system 财务体系
financial target 财务指标
financial tariff 财政关税
financial taxation 财政税收
financial transaction 财务事项;财务交易
financial value 财务价值
financial viability 财务活力;财务生存能力;财务可行性
financial work 财政工作
financial year 会计年度;财政年度;财务年度

financial yield 理财收获
financier 通融资金
financing 资金供应;资金筹措;集资;融资;通融;提供资金;出资;筹资;筹集资金
financing accounts 融资账户
financing agreement 融资协议;融资合同
financing alternative 资金筹措方案;集资方案
financing a project 工程项目筹资
financing arrangement 资金供应办法;融通资金安排;提供资金的办法;筹资安排
financing bank 提供资金的银行
financing body 资助机构;筹资机构
financing charges 借款费用
financing contract 融资合同
financing cost 资金筹措费用;融资成本;筹资成本;财资成本
financing decision 筹资决策
financing fee 押金回扣;集资费用
financing gap 财政缺口
financing lease 融资租赁;财务租约;财务租赁
financing leasing 融资租赁
financing management of railway material supply and marketing enterprise 铁路物资供销企业财务管理
financing management rule of railway material supply and marketing 铁路物资供销财务管理的原则
financing means 融资手段
financing mechanism 集资途径;集资方式
financing of foreign trade 外贸融资
financing of housing 住房筹资
financing of instalment sales 分期销售的资金融通
financing of overseas investment and enterprise 海外投资和海外资金筹措
financing of plant equipment export 工厂设备输出资金筹措
financing of projects 工程集资;工程贷款;工程筹款
financing orientation 财务导向
financing package 项目中的经济效益;一整套筹资措施;一揽子集资方案
financing process 资金筹集方法
financing statement 资金筹集报表
financing technique 集资业务;筹资方法
fin and tube type radiator 圆翼形散热器;片管式散热器
fin-cooled cylinder 散热片冷却汽缸
fin cooler 插片式冷却器
fin crack 飞翅裂缝
find 发现
find a market 开辟市场
find an anticipation 发现占先
finder 寻星镜;寻像器;寻迹器;选择器;中间人;观察装置;瞄准装置;瞄准器;取景器;探测器;定位程序;发现者
finder adapter 瞄准镜
finder aperture 瞄准装置孔径;瞄准器孔径
finder charge 试验荷载
finder hood 瞄准装置遮光罩
finder key 格网编号
finder lens 瞄准装置透镜;瞄准透镜;取景镜片;探测器透镜
finder light 取景器光线
finder mask 寻像护罩
finder screen 寻像屏
finder's fee 中人佣金
finder telescope 瞄准望远镜
finding 寻线;探测;查找;测定
finding action 寻线动作
finding chart 证认图
finding circuit 闭锁电路
finding report 调查报告
findings 研究结果;观测数据
findings of audit 审计结果
findings of test 试验结果
finding speed 测定速度
finding the course made good by three bearing of single object 单物标三方位求航迹向
Findlay cell for caustic soda 芬德来制苛性钠电池
Findlings quartzite 菲德林胶结石英岩
find meteorite 寻获陨石
find no market of ore products 矿产品无销路
find out 查明
fin drum dryer 翅片转鼓干燥器
finds herself 船磁消失

fine 细纹的;细粒物料;细粒的;细砾;精致的;精细;罚款;罚金;细微
fine adjustment 细调(节);精密校正;精调(整);微调
fine adjustment knob 细调节器;载物台细调螺旋
fine adjustment screw 精密校正螺钉;微调螺旋;微调螺钉
fine admixture 细混和物
fine agglutinant 细聚集体
fine aggregate 细粒集合体;细集料;细骨料
fine aggregate concrete 细石混凝土;细骨料混凝土
fine aggregate mixture 细集料混合物
fine aggregate type of road 细集料式道路
fine air bubble 小气泡
fine air bubble aeration 小气泡曝气
fine air bubble aeration system 小气泡曝气系统
fine alignment 精确定线
fine alumin(i)um 纯铝
fine angular 细颗粒的
fine annealing 精密退火
fine antenna 精测天线
fine arts 美术造型艺术(绘画);美术
fine arts shop 美术品商店
fine arts studio 美术室;美术馆
fine asbestos 优质石棉
fine asphalt 优质沥青
fine asphalt carpet 优质配(地)沥青磨耗层
fine asphalt concrete 优质沥青混凝土;细粒(石油)沥青混凝土;细级配(地)沥青混凝土
fine asphalt mix(ture) 细级配(地)沥青混合料
fine asphalt surfacing 优质(地)沥青混凝土路面;细级配(地)沥青混凝土路面
fine asphalt tile 优质沥青砖;细级配(地)沥青砖
fine away 削尖
fine axed stone 细剁石面
fine azimuth transmitting selsyn 方位角自整角发送机;方位角自动同步传感器
fine balance 细平衡;精密天平;精密平衡;精密调节;精调(整)
fine ballast 细石渣
fine bed 薄岩层;薄地层
fine benthic organic matter 水底细粒有机物
fine blanking 精密冲裁
fine blanking press 精密落料冲床
fine blast 精喷砂
fine-bore 精密镗孔
fine boring 精镗
fine boring machine 精密镗床
fine breaker 细料轧制机;细料轧碎机
fine breaking 细料轧制;细料压碎
fine breaking machien 细料轧制机
fine break-up 细粒分散
fine breeze 细粉;煤尘
fine brick 平薄砖;精制砖
fine broken rock 细碎石块
fine broken stone 细碎石料
fine bronze wire netting 细青铜丝网
fine bubble aeration 微泡曝气
fine bubble diffuser 微泡扩散器
fine bulk material 细粒松散材料
fine bur 细纹钻
fine calibration 精密校准
fine cargo 易碎货物;净货;精致货;好货
fine casting 精细浇铸
fine ceramics 细陶瓷
fine ceramisite 陶砂
fine chamber (轧石机的)细轧室;细碎室;细粒度破碎腔室
fine chemicals 精细化学品
fine chilled iron sand (锯解石料时掺用的)铁砂
fine chipboard 高级刨花板
fine chip(ping)s 细石屑
fine clarification 细净化区
fine clastic soil 细碎屑土
fine clastic texture 细碎屑结构
fine clay 细泥;细(粉状)黏土;细白土;活性白土细粉
fine clay slurry 细致的黏土悬浮物;细黏土悬浮物;细泥浆
fine cleaner 细滤清器;细滤器
fine cleaning 细净化
fine clinker 精制硬砖;精制缸砖
fine coal 粉煤
fine-coal cake 粉煤滤块
fine-coal dewatering 粉煤疏干

fine-coal filtering 粉煤过滤
fine coarse aggregate ratio 细粗集料比
fine coke 细炭粉末
fine coke breeze 细焦粉;焦尘
fine cold asphalt 冷沥青路面
fine cold asphalt mixture 冷铺细粒地沥青混合料
fine collimation 精确准直
fine collimation apparatus 精细准直仪器
fine comminution 细磨;细粉磨
fine component 细粒部分;细粒组分
fine concrete 细集料混凝土;细骨料混凝土
fine concrete aggregate (混凝土的)细集料
fine concrete floor screed 细集料混凝土楼面找平层
fine concrete mix 细骨料混凝土混合料
fine concrete screed material 细集料混凝土楼面找平层材料
fine concreting 浇筑细集料混凝土
fine constituent 细微成分
fine constitution 细致体质
fine control 细调;高精确度调整;均匀调节;精密控制;精调(整)
fine control sensitivity 精细控制灵敏性
fine copper 纯铜
fine cord 细条纹
fine cordage 细麻绳
fine coupling 精细耦合
fine crack 细裂纹;细裂缝;发裂
fine crumb 细屑粒;细团块
fine-crushed 细碎的
fine-crushed graded gravel 细碎级配砾石
fine-crushed road 细碎石路
fine-crushed rock 细碎石块;细碎石
fine-crushed rock road 细碎石路
fine-crushed sand 轧制细砂
fine-crushed stone 细碎石料
fine crusher 细料轧碎机;细碎(破碎)机;精压碎机
fine crushing 细轧;细(破)碎;精碎;第三级破碎
fine crushing chamber 细碎室
fine crushing department 细碎工段
fine crushing machine 细碎机;细料轧碎机;细轧碎机
fine crushing rolls 细碎辊碎机
fine crystalline 细晶的
fine crystalline dolomite 细晶白云岩
fine crystalline limestone 细晶灰岩
fine crystalline texture 细晶结构
fine crystal superplastic forming 细晶超塑成型
fine cut 细切割;精削
fine cut burr 细纹刻石刀
fine cut file 细锉;精加工锉
fine cut stern 瘦削型船
fine cutting 细刻磨
fine data channel 精确数据通道
fine day 晴天
fine definition 高清晰度
fine definition image(ry) 高清晰度图像
fine delay 微小延迟
fine delay dial 延时微调刻度
fine deposit 细粘土;细泥沙
fine detail resolution 细节清晰度
fine digital processing 精细数字处理
fine discharge 细粉卸出
fine dispersed system 细分散系统
fine dispersion 精细分散
fine distinction 细微的差别
fine divide suspended matter 细悬浮物
fine dot raster 细网目片
fine dot shadow mask 细点荫罩
fine dotted line 细虚线
finedraw 细缝;拉细丝
fine drawing 清绘
fined steel 优质钢
fined tube exchanger 翅管换热器
fine dust 细粉尘;细粉;岩粉;微尘;石粉
fine earth 细土;细漂白土粉;细粒土(壤)
fine earthware 精陶
fine-edge blanking 精密落料;精密冲裁
fine emery powder 细金刚砂
fine emulsion 细乳状液;细滴乳液
fine end 细节
fine-entrance ship 尖首船
fine etching 精密蚀刻;精细蚀刻法
fine faience 精陶
fine faience tile 精制釉陶面砖;浇筑釉陶锦砖

fine feed 细料进给;细进刀;精细进给
fine feed adjustment tank 细调料槽
fine feed curve 细料进给曲线
fine feed range 细料进给范围
fin efficiency 散热片效率;翅片效率
fine fiber 细纤维
fine-fibered 拉成纤维的
fine file 细锉刀;细锉
fine fill 细充填料;细充填料
fine filler 细填充料
fine filler cement 微骨料水泥
fine filler fly ash Portland cement 微骨料煤粉硅酸盐水泥;微骨料煤粉波特兰水泥
fine filler pozzolana Portland cement 微骨料火山灰质硅酸盐水泥;微骨料火山灰质波特兰水泥
fine filling 细纬
fine filling bar 细纬档
fine filter 细滤器;细滤池;细过滤器
fine filtering 细滤
fine finish 精加工;高级精加工;高级光洁度;精细抛光;超精加工
fine finishing 精加工
fine finishing cut 精密完工切削
fine fissure 细裂纹;微裂(缝);发裂
fine fit 精确配合;二级精度配合
fine flake 细刨花
fine flaw 细裂缝
fine focus 细焦点
fine focused 准确聚焦的
fine focusing 细聚焦;精确对光;精确调焦
fine focusing adjustment knob 细调焦距螺旋
fine focusing collar 精密聚焦环
fine focusing unit 精确聚焦单元
fine-focus X-ray tube 细聚焦X射线管
fine forceps 精细镊
fine for delaying payment 滞纳金
fine for paying late 滞纳金
fine for store up 等存费
fine fraction 细粒级;细(颗)粒部分
fine fraction curve 细粒级曲线
fine fraction range 细粒级范围
fine furnace black 细炉黑;细粒炉黑
fine gas bubble 小气泡
fine gas cleaning 气体高度净化
fine gathering 细褶裥
fine ga(u)ge screen 细孔筛
fine glass 玻璃细片
fine glass rod 细玻棒
fine gold 纯金
fine goniometer 精密测角仪
fine grade 二级精度
fine-graded 细粒级的;细粒的;细级配的
fine-graded aggregate 细级配集料;细级配骨料
fine-graded asphalt concrete 细粒沥青混凝土
fine-graded asphaltic concrete 细粒(石油)沥青混凝土;细级配(地)沥青混凝土;细粒级沥青混凝土
fine-graded asphaltic concrete carpet 细级配(地)沥青混凝土磨耗层
fine-graded asphaltic concrete pavement 细级配(地)沥青混凝土路面
fine-graded asphaltic concrete tile 细级配(石油)沥青混凝土瓦;细级配(地)沥青混凝土瓦
fine-graded bituminous concrete 细级配沥青混凝土
fine-graded bituminous concrete pavement 细粒式沥青混凝土路面
fine-graded material 细级配材料
finegrader 精密平地机
fine grading 仔细平整土地;细致地摊平路面;细粒分级;细级配;细致路面拉坡(纵坡)
fine grain 细(致)纹理;细纹;细(颗)粒;细晶粒;微粒;粉质砂土
fine grain analysis 细颗粒分析
fine grain ceramics 细粒陶瓷
fine grain developer 微粒显影液
fine grain development 微粒显影
fine-grained 细粒的;细纹的;细料的;细颗粒的
fine-grained aggregate 细粒集料
fine-grained alluvial deposit 细粒冲积物
fine-grained artificial stone 细粒人造石
fine-grained asphaltic binder course 细粒(地)沥青混凝土联结层
fine-grained basalt 细粒玄武岩

fine-grained building stone 细纹建筑石料
fine-grained carbon 细粒碳
fine-grained cast stone 细粒铸石
fine-grained chip(ping)s 细粒石屑
fine-grained clastics 细粒碎屑
fine-grained concrete 细粒混凝土
fine-grained fracture 细粒裂纹;细粒状断口;细晶断口;细纹裂面;细纹继裂面;细裂纹面
fine-grained granite 细粒花岗岩
fine-grained gravel aggregate 细粒砾石集料;细粒砾石骨料
fine-grained ingot 细晶锭
fine-grained lightweight concrete aggregate 细粒轻质混凝土集料;细粒轻质混凝土骨料
fine-grained limestone 细粒灰岩
fine-grained material 细粒料
fine-grained meander belt deposit 细粒曲流带沉积
fine-grained micrinite 细微粒体
fine-grained monolith 细粒岩块
fine-grained mortar 细粒砂浆;细粒灰浆
fine-grained mortar with synthetic resin dispersion 含有合成树脂悬浮物的细粒砂浆
fine-grained patent stone 细粒铸石;细粒人造石
fine-grained perlite 细粒珍珠岩
fine-grained picture 细粒图像
fine-grained powder 细粒炸药;细粒粉末
fine-grained reconstituted stone 细粒再加工石料;细粒人造石
fine-grained rock 幼粒岩石
fine-grained salt 细粒盐
fine-grained sand 细粒砂;细砂
fine-grained sandstone 细粒砂岩;细粒砂(石)
fine-grained soil 细粒土(壤);细颗粒土;细碎屑土;细砾土
fine-grained steel 细晶粒钢
fine-grained stone 细纹理石料
fine-grained structure 细粒状构造
fine-grained texture 细密结构;细粒结构
fine-grained wood 细纹木(材)
fine-grained wood wool board 细纹木丝板
fine-grain emulsion 微粒乳液;微粒乳剂
fine grain film 微粒胶片
fine graininess 细粒度
fine grain mixture 细颗粒混合物
fine grain screen 细孔筛
fine grain size 细粒(径)
fine grain soil 粉质土
fine granular 细团粒;细颗粒状的;细颗粒;细晶粒的
fine granular cast 细颗粒管型
fine granular crystalloblastic texture 细粒变晶结构
fine granularity 良好颗粒性
fine granular psamitic texture 细粒砂状结构
fine granular texture 细密结构;细粒结构
fine granule 微粒剂
fine graphite 细石墨
fine gravel 细卵石;细砾(石)
fine gravel filter 细砾石滤池
fine gravel for concrete 混凝土用细砾石
fine gravel ore deposit 细砂矿床
fine grinder 细研磨机;精细研磨机;微细碎机
fine grinding 细(粉)磨;精细研磨;精磨;磨细
fine grinding belt 精磨带;抛光带
fine grinding chamber 细磨仓
fine grinding compartment 细磨分隔仓;细磨仓
fine grinding mill 细磨机;精细研碎机
fine grinding wheel 细砂轮
fine grit wheel 细砂轮
fine groove 密纹
fine grooved grating 细槽光栅
fine ground 细磨过的
fine-ground cement 细磨水泥
fine gyrasphere crusher 旋转式细料轧碎机
fine gyratory crusher 细碎用旋回圆锥式破碎机
fine hackle 细锯纹区
fine indented cut plating iron 细牙槽刨铁
fine index 细索引
fine index for Y Y 坐标细分划盘
fine ingredient 细成分
fine iron ore 细铁矿
fine jaw crusher 颚式细料轧碎机;细碎(用颚)式破碎机
fine jiging 细粒跳汰
fine lapping 细研磨
fine level(l)ing 精密水准测量

fine lightweight concrete aggregate 细轻质混凝土集料;细轻质混凝土骨料
fine limestone chip(ping)s 细粒石灰石石屑
fine line 细线(道);细条纹;细实线
fine-line compound veneer 细纹合成饰面板
fine-line graver 细线刻针
finely banded coal 细条带状煤
finely banded structure 细条带状结构
finely broken stone 细碎石
finely crushed glass 轧碎玻璃;粒状玻璃;细碎玻璃
finely crushed rock 小碎石
finely crystalline 细晶质的
finely detritus groundmass 细碎屑基质
finely dissected topography 细切地形
finely dissected topography topographic(al) details 地形碎部
finely divided image forming particles 高分辨率成像微粒
finely divided iron powder 极细铁粉
finely divided materials 细填充料
finely divided mineral admixture 细磨矿物掺和料
finely divided particle 微粒子
finely divided resin 细粒树脂;微粒树脂
finely divided scale 精密刻度尺
finely divided solid 细散固体
finely divided suspension 微细悬浮液
finely finished work 精细完工件
finely graded aggregate 细级配集料;细级配骨料;高细粒含量集料;高细粒含量骨料
finely granular 细颗粒的;细粒状
finely granular structure 细团粒状结构
finely ground 精碾的;细磨的;细粉磨的
finely ground barite 细磨重晶石
finely ground baryte 细磨重晶石
finely ground cement 细磨水泥;细碾水泥
finely ground charcoal 木炭粉
finely ground coke 细焦粉
finely ground colloidal suspension 微细胶状悬浮剂
finely ground fire clay 细耐火泥
finely ground grain 细磨颗粒
finely ground lime 细磨石灰
finely ground ore 细磨矿石
finely ground quartz 细磨石英
finely ground rock 细磨碎石
finely laminated 细层纹的
finely laminated clay 薄成层黏土;薄层状黏土
finely laminated rock 细层纹岩石
finely porous 细孔隙的;微孔的
finely porous mat 细孔垫布
finely powdered 细粉状的
finely powdered clinker 细磨熟料
finely pulverized blast furnace slag 磨细粒化高炉渣;细磨高炉矿渣
finely pulverized fuel 细粒雾化的燃料
finely pulverized powder 微细粉末
finely squamose 细鳞片状
finely stratified 薄层理的;薄层的
finely stratified clay 薄成层黏土
finely striated coal 细线理状煤
finely toothed 带细齿的
fine machining 精加工
fine manufactured sand 细人造砂
fine marble chip 细大理石渣
fine material 细粒材料
fine measurement 精密测量;精测
fine measuring instrument 精密量具;精密测量仪器
fine measuring scale 精测尺
fine measuring screw 测微螺旋
fine mechanics 精密机械
fine melt 全熔(炼)
fine melt resin 全熔树脂
fine mesh 小网眼;细(眼)网目;细孔;细筛孔
fine-mesh barrier grid 细网抑制栅
fine-meshed 细筛孔的;带细孔的
fine-meshed carriageway grid 细网格式车行道铺面
fine-meshed screen 细孔网
fine-mesh filter 细网(眼)滤器;细孔滤网滤清器
fine-mesh grid 细网栅;密网栅极
fine-mesh screen cloth 细孔筛布
fine-mesh sieve 细孔筛
fine-mesh silk 编细孔网用的丝
fine-mesh solid support 细筛目载体
fine-mesh wire 细目网

fine-mesh wire netting 编细孔金属丝网
fine metal 细石料;精炼纯金属;纯(净)金属
fine metal mesh 细金属网眼
fine mica cloth 细云母布
fine mill 光铣
fine mineral surfacing 细矿物面材;细矿物撒布料
fine mix 细粒混合料
fine moist rale 细湿罗音
fine mortar 细砂浆
fine motion 微动
fine motion screw 微动螺旋;微动螺钉
fine movement 精细动作;微动
fine mull 细腐熟腐殖质
fine nail 小钉
finend duplex tube 细头二联管
finend pipe 细头管
finend strip(electric)heater 细头电热丝加热器
finend tube 细头管
fine needle cordoroy 细条灯芯绒
fineness 细度;粒子细度;金银成色;纯度;粉料细度;细微
fineness coefficient 肥瘦系数;肥瘠系数
fineness degree 细度;光洁度;纯度
fineness determination 细度测定;光洁度测定;纯度测定
fineness factor 细度因子;细度系数
fineness ga(u)ge 细度计;刮板细度计
fineness grind ga(u)ge 刮板细度计
fineness limit 细度极限;纯度极限
fineness-maturity tester 细度成熟度试验仪
fineness meter 细度测定仪
fineness modulus 细度系数;细度模数;细度模量;粒度模数
fineness modulus method of proportioning 配合比细度模量法
fineness modulus of sand 砂的细度模数
fineness number 细度指数(重塑土某一强度时的含水率)
fineness of aggregate 骨料细度;集料细度
fineness of cement 水泥的细度
fineness of gold 金成色
fineness of grain 细晶粒度;颗粒细度;晶粒细(化程)度
fineness of grind 研磨细度;轧浆细度;磨版细度
fineness-of-grind ga(u)ge 研磨细度规
fineness of grinding 磨(碎)细度;磨粒细度;磨粉细度;粉磨细度
fineness of grinding cement 水泥粉磨细度
fineness of lime 石灰细度
fineness of powder 细粉细度
fineness of scan(ning) 扫描密度
fineness parameter 细度参数
fineness ratio 细度比;粒度比;长细比;长宽比(率);长径比
fineness regulator 磨碎细度调节器
fineness specification 细度规格
fineness test 细度试验
fine nozzle 细孔喷油嘴
fine on the bow 很靠近船首方向
fine ore 细矿石;粉矿
fine ore bin 细矿仓
fine ore storage 细矿仓
fine outlet 微粉出口
fine pan 细粒物料淘洗盘
fine paper 高级纸(张);可靠证券
fine particle 细粒子;小质点;细(颗)粒;微细细粉颗粒;微粒
fine particle board 细粒板
fine-particle filtration 细粒杂质的过滤
fine particle silica 微粉二氧化硅
fine particulate mass 细颗粒物
fine particulate organic matter 细颗粒有机物
fine paste 细研磨膏
fine pattern 精致图;精细图案
fine pearlite 细片状珠光体;精细珠光体
fine pencil 画细线用的铅笔
fine perlite 细珍珠岩
fine phase shifter 精密调相器
fine pick 细纬
fine picked stone 细凿石
fine picked stone finish 细凿石面
fine pile 细桩
fine pitch 小螺距;细距
fine-pitch cutter 小节距齿铣刀;小齿铣刀

fine-pitch(ed)thread 细牙螺纹
fine-pitch gear 小模数齿轮
fine-pitch mask 细节距影孔板
fine-pitch screw 细牙螺钉
fine-pitch thread 细牙螺纹
fine plain 细平布
fine plain emery wheel 细砂轮
fine planer 精加工刨床
fine plaster throwing machine 稀灰泥喷射机
fine-pointed dressing 细琢修整;细琢石;细琢(石面)
fine-pointed finish 细琢;细凿修整
fine-pointed stone 细凿石
fine-pointed stone finish 细凿石面
fine polishing 细抛光;精抛光
fine porcelain 细瓷
fine porcelainized stoneware 细瓷化炻器
fine pore 小孔隙;细孔
fine-pored 细孔的;微细孔隙的
fine porosity 微气孔群
fine pottery 精陶
fine pottery fracture 细瓷状断口
fine pouring 稀薄灌注
fine powder 细粉;微细粉末
fine powder product 粉料成品
fine pressure 净压(力)
fine processing 精密加工
fine product 细粒产物
fine product collector 粉料成品收集器
fine proximity 细邻近
fine pruning 精剪
fine pulverizer 精细磨粉机
fine purification 精制
fine quality 品质优良
finer 精炼炉
finer abrasive 细磨料
fine rack 细格栅
fine-range scope 距离精测器;精确距离显示器
fine ranging 精密测距
fine rate 优惠贴现率;优惠利率;罚款率
fine reading 细读数
fine ream 细条纹
fine reduction 细碎(作用);细粉碎
fine reduction gyratory 细碎旋回破碎机
fine reduction gyratory crusher 细碎用旋回圆锥式破碎机
fine reed 细筘
fine regulation 细调
fine resolution 细分解;高鉴别力
fine retic(u)le 精密十字标线;精密调制盘
fine-ribbed fabric 细罗纹织物
fine rib machine 细罗纹机
fine ridge 密间距薄水垄沟
fine road-metal 细筑路碎石
fine roller mill 辊式细磨机
fine rolling 精轧
fine-rubbed finish 细磨加工面;磨光面
finery 装饰品;装饰;盛装;精炼炉;木炭精炼炉
fines 细屑;细粒子;细粒土(壤);细骨料;细粉(料);微粒
fine sand 细砂粒;细砂
fine sand beach 细砂海滩
fine sand blast 精细砂磨;细喷砂
fine sand cement mortar 水泥细砂浆
fine sand concrete 细砂混凝土
fine sand concrete aggregate 细砂混凝土集料;细砂混凝土骨料
fine sand feeder 细砂注入机
fine sand foundation 细砂地基
fine sand loam 细亚砂土;细砂壤土;细砂炉姆
fine sand reservoir 细砂岩储集层
fine sandstone 细砂岩
fine sandy clast 细砂屑
fine sandy clay wares 细砂陶器
fine sandy loam 细亚砂土;细砂壤土;细砂炉姆;黏质粉砂
fine saw 细锯
fine scale 精密标度
fine-scaled distribution 细分度的分布
fine scanning 高质扫描
fines content 精细颗粒含量;细粒含量
fine scratch 轻划伤
fine screen 细网栅;细筛(网);细滤网;细孔筛;细格栅;精筛;网目板

fine-screen halftone 细网点印版
fine screening 精筛选;精筛
fine-screen shaker 细网振动筛
fine screw 细纹螺旋
fine screw tap 细牙螺丝攻
fine scrubbed concrete surface 仔细刷过的混凝土表面
fine sediment 细黏土;细沙;细沉积物
fine-sediment load 冲泻质(泥沙);细悬移质
fine seed 小气泡(小于0.2毫米);灰泡
fine selsyn 精调自动同步机
fine separation 细分选;细选粉
fine set(ting) 精确安装;细调;精确装配;精密整定;精密调整;紧密调整
fine setting device 微调装置
fine settling 精确安置
fine sewage screen 细孔污水滤网
fines forfeits and penalty receipt 罚款收入
fine shaped ship 瘦型船
fine sheet asphalt 细片地沥青
fine shellac 上等紫胶
fine shelly sand 含贝壳的细砂
fine ship 瘦型船
fine sieve 细(眼)筛;细屑粒
fine-sight district 风景区;景色美丽的地区
fine silk net 细筛绢网
fine silk sieve 高级桑蚕丝筛网
fine silt (原砂的)细泥;细淤泥;细粉土;细粉砂
fine silt analysis 细粉粒分析
fine silt bond 细粉粒黏结
fine siltstone 细粉砂岩
fine silty clast 细粉砂屑
fine silty sand 细粉砂
fine silver 纯银
fine size grading 细粒分级;细级配
fine sizing 细粒筛分;精筛
fine slip band 细滑移带
fine slit texture 细粉砂结构
fine slurry 细料浆
fine soil 小团粒结构土壤;细粒土(壤)
fine soil grain 细土粒
fine solder 细焊料;锡铅合金焊料
fine solid 微细固体
fine sort 精细排列
fine-sorted material 细粒材料
fines output 细粉产量
fines overflow 细料浆溢流
fine spectrum analysis 精细频谱分析
fine spray irrigation 雾灌
fine-spray nozzle 细雾滴喷嘴
fines return 粉末放出
finesse 等强干涉束有效数
fine start 微启动
finest concrete 特细集料混凝土;特细骨料混凝土
fine steel 优质钢;合金钢;特殊钢
fine steel wire 细钢丝
fine stern 尖尾(船型)
fine stern ship 尖尾型船
finest grade of graphite 最纯的石墨
finestiller 精馏器
fine-stippled sandblasted finish 细喷砂面;细喷砂面
fine stipple finish 细拉毛饰面
finest material 精细颗粒材料
fine stone 细磨石
fine stone crusher 石材二次破碎机
fine stoneware 细炻器
fine stopper 细填料
finest particle 精细颗粒
fine strainer 细滤器
fine streak 细条纹
fine streamline body 良流线形体;良流线体
fine stream sand 细河砂
fine strip memory 微带存储器
fine structure 细微构造;精细结构;微细结构
fines tructure band 精细结构带
fine structure constant 精细结构常数
fines tructure mesh 小孔网;细网;密网筛
fine structure parameter 精细结构参数
finest size 最小尺寸
fine stuff 细料;细灰浆;细灰缝;抹面细灰浆
fine stuffing 细填料
fine surface mulch 薄层覆盖
fine surface mulch of gravel 细砾薄表层

fine suspended load 细悬移质
fine suspended material 细悬浮泥沙
fine suspended sediment 细悬浮泥沙
fine suspended solid 细悬浮固体
fines work-up 粉末再加工
fine talc 细云母石;细滑石(粉)
fine tar concrete 细柏油混凝土
fine texture 密致结构;显微结构;细致纹理;细微结构;细密结构;细(粒)结构;组织(构造)致密;致密结构;金相组织;结构微密
fine textured 细密组织的
fine-textured soil 土壤细粒结构;细质(地)土壤;细结构土;细粒土(壤)
fine-textured wood 细质木材;细纹木
fine texture topography 细密地形;地形碎部
fine thermal black 细热裂黑;细粒子热裂炉黑
fine thread 细牙螺纹;细纹
fine thread die 细纹螺丝板牙
fine thread screw 细纹螺钉
fine thread tap 细牙丝锥
fine tilting screw 微倾螺旋
fine tin 精炼锡;纯锡
fine-to-coarse-grained 细至粗粒的
fine-to-medium-grained 细至中粒的
fine-toothed cutter 细齿铣刀
fine-toothed reversible ratchet 细齿可逆棘轮;可换向的细齿棘轮
fine-toothed saw 细齿锯
fine-toothed saw blade 细齿锯条
fine topology 细拓扑
fine traces analysis 精细叠加道分析
fine transition 细粒反滤料;细粒的过渡层
fine trash rack 密拦污栅
fine tremor 细震颤;频视震颤
fine triangular waveform generator 精密三角形波发生器
fine tuff 细粒凝灰岩
fine tuning 细调谐;精细调谐;精密调谐;微调
fine tuning control 细调
fine turning 精车
fine turning lathe 精密车床
fine type 细致型
fine type letter 白体字
fine vacuum 高真空
fine water spray nozzle 细喷雾枪
fine weather 晴天
fine weather effect 晴天效应
fine weighing 精称
fine welding 精密焊接
fine wheel 细砂轮
fine wire 细金属丝
fine wire drawing machine 精拉丝机
fine wire netting 细金属丝网
fine wire rope 细丝钢丝绳
fine wire screen 细丝网筛;细孔金属丝筛网
fine wire welding 细丝焊
fine wood 细纹木(材);细木
fine wood-work 细木作
fine workmanship 细工;精巧制作;精工
fine zero adjustment 零位精确调整
fine zero control 零点微调作用
fine zone 细骨料区
fin fan exchanger 翅扇式热交换器
fin filter 薄片过滤器
fin-fitting machinery 减摇鳍传动机构
finger 一指阔;一指长;指状元件;指头;指示针;指示服务;指梁;测厚规;分指移动
finger-action tool 指状抓手
finger baffle 导向隔板
finger-bag 指套
finger bar 挂叉板;刀架
finger-bar cutter 带护刃器梁的切割器
finger-bar trap 指状沙坝圈闭
finger basin (突堤码头区的)指形港池
finger basket 捞爪;捞环;多通管接头
finger belt 钉齿皮带
finger bit 指状钻头
finger blade agitator 指形叶片搅拌机;栅桨式搅拌机
finger-board 道路指向牌;指架台;挂叉板;指板;键盘
finger board of road 道路指向牌
fingerbreadth 指幅
finger brush 指形刷
finger buff 指形折布抛光轮

finger cam 齿凸轮
finger-car 子母车
finger chute 指状轨条溜台;指形滑槽
finger chute gate 指状溜口闸门
finger citron flower 佛手花
finger citron sliced 佛手片
finger clamp 指形压板;带爪(的)压板
finger combing 手指梳花
finger contact 指形触点
finger control 手动控制
finger-cot 指套
finger cracking test 指形抗裂试验
finger cutter 指状铣刀;指形铣刀
finger distance 指距
finger dock (突地码头区的)指形港池;指形船坞
finger door 指状闸门
finger drain 指状排水系统
fingered 指状
finger electrode 指状电极
finger feed 指式耙肥器;机械手送料
finger gate 指状浇口
finger ga(u)ge 厚度规;测厚规
finger grass 指形禾草
finger grate 指杆筛
finger grid 指杆筛
finger grip (塔上的)钻杆夹;指状抓取器;夹具;打捞器(钻杆);打捞工具
finger guard 指套
finger gull(e)y 指形排水沟
finger gullying 指形排水沟;初期沟蚀
finger head 飞机场起飞点
finger hole 指孔
fingering 指法符号
finger-input system 手指输入装置
finger interface 指体界面
finger jetty 指形突(堤)码头
finger joint 指形接合;竖接缝;梳形接合;插榫对接;指接;榫接头
finger-jointed 榫接木工作业
finger lake 指状湖;指形湖;长湖
finger lamp 指形灯
finger layout 飞机场主楼到起飞点有盖顶的过道
finger-length measurement 指形测量指示法
finger lever 指状手柄;指形手柄
finger lifter 指状岩芯提取器
fingerling 小鱼;鱼种;幼鱼;微小物
fingerling tag 鱼种标志
finger mark 指印;指纹;手指印;指印
finger microcondenser 微量指形冷凝管
finger-nail indentation test 指甲刻痕试验
finger-nail plate (门上的)手污防护板
finger-nail post 指路牌
finger-nail system 起飞指示系统
finger-nail test 刻划硬度试验
finger notch 指状凹口
finger nut 指形螺母
finger of bit 钻头销
finger of drag bits 刮刀钻头销
finger out 伸出(突堤向水域)
finger out into water 伸入水域
finger pad 指垫
finger-paint 指涂;指画
finger painting 指涂法;指画法;手指画
finger phenomenon 伸指现象
finger piece 指状物
finger pier 指形突(堤)码头;窄突堤;突码头;突堤(式)码头
finger pin 探钩
finger plan 指形平面
finger plate 指孔盘;门推板;门上把手护板;回转板;推手板;防污板;指板;(门上的)手污防护板;门锁孔盖;门上把手锁眼处防指污的板;路标
finger post (解决问题的)线索;指向柱;指路牌
finger press fit 轻压配合
finger print 指印;指纹(图案);手印
fingerprint chromatogram 指纹色谱图
fingerprinted hydrocarbon 指纹状烃
finger printing 指纹打印
fingerprint map 指纹图
fingerprint removal test 消除指印纹(防锈)试验
fingerprint remover 消除指印防锈油;指印消除型防锈油
fingerprint technique 指纹技术
finger pull (玻璃扯门的)指槽

finger rack 指杆架
finger rafted ice 堆积指状冰
finger rafting 堆积指状冰
finger raise 指状格条天井;指形格条天井
finger reel 搂齿式拨禾轮
finger retarder 指状缓速器
finger-ring badge 指环剂量计
finger rotary detachable bit 螺旋钻的可卸钎头;麻花(钻)活钻头
fingers 指粒
finger setting 销定位
finger sgraffito 手指刮花
finger-shape condenser 指形冷凝管
finger slip 弧形磨石;弧口凿磨石
fingers of fire 火舌
finger spacing 指状闸门间距
finger spring seal 指状弹簧油封
finger-stall 指套
finger stop 指状限位器;指形制动销;指档;手动限位器
finger stoped plate 指形板
finger system 指掌形(平面设计)系统
finger tab-out test 指黏试验
finger-tight 手拉紧
finger tilt mechanism 钩式翻钢机
finger-tip control 手指操纵;指拨控制;单指调整器;单锁调整器;按钮控制;按钮操纵
finger tip pressing technique 押手
finger-type contact 指型触点;指形触点
finger-type control rod 指形控制棒
finger-type core lifter 框式岩芯卡取器;指状岩芯提取器
finger-type expansion joint 指形膨胀缝;梳形膨胀缝
finger-type gear-milling cutter 指形齿轮铣刀
finger-type shoe 插销式管靴
finger valve 指阀
finger weeder 指状簧齿除草机
finger wharf 突堤(式)码头
finger-wheel rake 指轮式搂草机
finger-wiped joint 黄铜焊接头;插入焊接头
fin heating tube 翅片加热管
fin heat resistance 翅片热阻
fin height 翅片高度
finial 叶尖饰;尖顶饰;端饰;攒尖饰
finial pinnacle 攒尖饰
finimeter 储量计
fining 精炼;澄清
fining agent 净化剂;澄清剂
fining-away 偏斜
fining cell 澄清室池
fining furnace 精炼炉
fining-off 涂刷灰膏涂层;末道粉刷;渐渐地稀薄;细加工
fining of riverbed 河床细化
fining period 精炼期
fining pointing 精密照准
fining test 指形抗裂试验
fining-upward sequence 向上变细层序
finish 细剁石面;最后修饰;装修;终饰材;整饰;罩面漆;光洁度;光柒;精饰;结束;内部装修材料;末道漆;末道工序;面漆;完工;完成;涂炉;光制;饰面
finishability 易修整性;终饰性;可修整性;精加工性;表面易抹性
finishability of concrete 混凝土的可修整性
finish a borehole 钻孔完成
finish all over 全部完工;全部完成;全部竣工;完全结束
finish allowance 加工裕量
finish and service 竣工
finish a well 钻孔完成
finish barefooted 裸眼完井;裸眼成井
Finish birch panel forms 芬兰桦木板模
Finish birch plywood 芬兰桦木胶合板
finish blanking 刃口冲裁模
finish bottoming 整形加工
finish builder's fittings 建筑装饰附件
finish builder's hardware 建筑装修小五金
finish burned clinker 烧成熟料
finish by screeding board 刮板抹平(混凝土)
finish carpentry 细木工作业;木工装修作业
finish casing 框饰;贴脸;台口线
finish cement 修边胶;罩面胶(泥);饰面水泥

finish coat 修饰涂层;罩面;罩光漆;面层抹灰;面漆
finish coat floating 面涂抹面
finish coating 油漆罩面;装饰涂料;涂末道漆;表面涂漆;面漆
finish coat paint 罩面漆
finish coat plaster 末道粉刷层
finish coat resistance test 罩面抗性试验
finish composition 装饰材料成分
finish compound 修边胶;罩面胶
finish construction survey 竣工测量
finish cooling zonal final cooling zone 最终冷却带
finish cover pass 最后的焊珠;最后的焊道
finish-curl die 卷曲成型模
finish department 修整仓;尾仓
finish depth 加工深度
finish draw 精拉
finish drawing 竣工图
finished 光制的;完成的
finished aggregate 成品集料;成品骨料
finished attic 修整的屋顶层房间
finished bag 成品袋
finished black plate 黑钢皮
finished blend 混成油
finished bolt 光制螺栓;精制螺栓
finished building 建成建筑物
finished building fabric 预制构件
finished cement 成品水泥;水泥成品;袋装水泥
finished concentrate 最终精矿
finished concrete 饰面混凝土
finished construction 已完成工程
finished cost 完工成本
finished crushing 最终破碎
finished cure 后硫化
finished depth 竣工水深
finished dimension 完成尺寸
finished edge 光边;加工(的)坡口
finished enamel 表面瓷漆
finished fabric 经处理织物;处理布;成品布
finished face 装修层;完整面
finished factory 加工厂
finished floor 磨光地面;磨光地板;修整过的楼面;完工的地面板
finished floor level 楼面竣工标高
finished glass 成品玻璃
finished goods 制成品;加工品;成品
finished goods on consignment 托销成品
finished goods on hand 现存制成品
finished goods warehouse 成品仓库
finished grade 竣工坡度;终饰坡度;修整好的面层;修整过的坡度;已修整(的)坡度;终饰等级;建成的地面
finished grade conveyer 成品输送机(粉碎过筛线设备)
finished ground level 修整过的地面高度;竣工地面高程
finished hexagon head bolt 光六角螺栓
finished hole size 终孔尺寸
finished ingot 成品锭
finished interior 内装修作业
finished layer 装修层
finished length 竣工长度
finished lens 成品透镜
finished line 竣工线
finished low-level 低标准修整
finished machine drawing 铸件工艺图
finished market product 成品商品
finished meal 入窑生料
finished nut 精制螺母
finished O. D. ga(u)ge 最终外径尺寸
finished ore 精矿;精选矿石
finished outside diameter ga(u)ge 最终外径尺寸
finished-parts storage 成品库
finished piece(s) 完工件
finished pigment 经表面处理的颜料
finished plate 精整钢板
finished powder metallurgical product 粉末冶金成品
finished primer 末道底漆
finished product 成品;制成品
finished product conveyer[conveyor] 成品输送机(粉碎过筛线设备)
finished production in current period 本期完工产量
finished product looper 油毡停留机;成品停留机

finished product section 最终成品断面
finished products inventory 制成品盘存
finished product storage 成品库
finished roll 光制轧辊
finished roller 磨光滚筒
finished sand 细粒砂
finished sheet 精整薄板
finished side leather 成品侧边皮革
finished size 装修后尺寸;完成尺寸;成品尺寸
finished slurry 入窑料浆;配好料浆
finished small hexagon flat nut 光小六角扁螺母
finished small hexagon nut 光小六角螺母
finished small stone 细加工条石
finished square nut 方光螺母
finished stack 竣工标桩
finished standard 竣工标准
finished steel 精制钢;成品钢
finished stope 采全区;采空场
finished storage silo 水泥库;成品储库
finished string 已装修的楼梯斜梁
finished strip 成品带钢
finished surface 已修整表面;已加工面;竣工路面;竣工的路面;加工表面;完工面;完成面;完成的路面
finished thickness 竣工厚度
finished voyage 航行结束
finished washer 光制垫圈
finished water 成品水
finished-water reservoir 清水池
finished weight 成品重量
finished width 竣工宽度
finished work 已加工工件
finish enamel 修补用瓷漆
finisher 修整器;修整机;修整工;终锻模;磨面机;炉前工;精整工;精加工;精加工工具;抹面机;调整工;玻璃吹工
finisher belt grinder 砂布带打光机
finisher feeding truck 路面修整机的供料车
finisher frame 修整机框架(路面)
finish facing 精加工端面
finish floor 楼板面层;楼房装修层地板
finish flooring 楼面装修材料
finish foil mill 铝箔精轧机;箔精轧机
finish forge 精锻
finish-forging 终锻
finish ga(u)ge 终检
finish grade 修整好的面层;竣工坡度;最终坡度;最后规定级别
finish grading 最后整型;最后整平
finish grinding 磨光;细磨;终磨;终粉磨;精磨
finish grinding circuit 细磨回路
finish grinding mill 水泥磨机;成品磨机
finish hardware 光制小五金;装饰小五金;精制小五金
finish hardware for gate 门上装饰小五金
finish impression 终锻模膛
finishing 光制;饰面;修整;修饰;装修工程;终饰【建】;竣工;精加工;抹面;涂装;带式磨光;擦光;表面修饰
finishing acceptance 竣工验收
finishing agent 修饰剂;颜料表面处理剂;整理剂;处理剂;表面处理剂
finishing allowance 加工留量
finishing area 加工面积
finishing barrel 清理滚筒
finishing beater 成浆机
finishing belt 最后加工地带(路面);精磨带;抛光带
finishing bid 备妥投标文件
finishing bit 精铣铣刀;终孔钎头;光制刀尖;光削刀具;精加工钻头
finishing block 拉细丝机
finishing board 装饰板规板;修整板
finishing bolt 光螺栓
finishing broach 精削拉刀
finishing building 竣工建筑物
finishing by belt 用带修饰表面;用带抛光表面
finishing carpentry 细木工;细木工
finishing chemicals 表面处理剂
finishing chip 精抛光;最后修琢;精加工
finishing cloth 擦布
finishing coat 装饰涂层;终饰层;罩面(层);饰面层;修饰层;路面罩面;末道涂层;末道粉刷层;(涂料的)外层;饰面涂层;表面修整层;罩面抹灰层
finishing compound 整饰化合物;罩面抹灰料

finishing cut 精加工;完工切削
finishing cutter 精加工铣刀
finishing department 精整工段
finishing die 精拔拉模;成型(压)模
finishing drawing frame 末道并条机
finishing drill 修准钻
finishing effluent 精炼废水
finishing effluent treatment 精炼废水处理
finishing equipment 精整设备
finishing face 装修面
finishing feed 光制精加工进刀;精加工进给
finishing file 精加工锉
finishing fire retardant paint 饰面型防火涂料
finishing fixture 涂饰夹具
finishing float 修整镘板
finishing grind(ing) 最后研磨
finishing groove 终止纹槽;精轧孔型;精加工槽
finishing groover 精轧槽
finishing hardware 精制小五金
finishing hydrate 墙面粉刷用的熟石灰物
finishing hydrated lime 饰面熟石灰
finishing impression 终锻模膛
finishing jig 精选跳汰机
finishing knife 整修刀;(泥工的)涂灰浆刀;磨平刀;修整刀具
finishing lacquer 硝基清漆;挥发性漆
finishing lathe 精密车床
finishing layer 盖层;终饰层;罩面(路面)
finishing lime 细石灰;装修用石灰
finishing line 终点线;精整作业线
finishing machine 修整机;修坯机;整机;路面修整机;整整机;精整机;精加工机床;抹面机;磨光机
finishing material 装修材料;装饰材料;完工物料;饰面材料;最后合金料
finishing metal 炼金属
finishing method 抛光方法
finishing mill 细(粉)磨机;精制机;精轧机
finishing mill group 精轧机组
finishing mill line 精轧机列
finishing mortar 抹面灰浆
finishing nail 饰面钉;装修(用)钉
finishing nut 光制螺母
finishing of coat of plaster 抹灰面层
finishing off 细木饰面工作;装修磨光作业
finishing of wall 墙体饰面;墙面装修
finishing operation 整饰操作;精整(修饰);精整修理
finishing oval pass 椭圆精轧孔型
finishing paint 罩面漆;饰面漆
finishing pass 过筛产品;精轧孔型;精轧道次
finishing peg 回填(土)标桩
finishing planer 精加工刨床
finishing plant 精制工厂;后处理工厂
finishing point 终点
finishing polish 精修抛光;精抛光
finishing procedure 修整手续;修整程序
finishing process 修整工作
finishing rate 最终充电电流
finishing reamer 精绞刀
finishing roll 精轧轧辊
finishing rolling mill 终轧机;精密轧机
finishing room 修整(车)间;油漆(车)间;成品(车)间
finishing screed 整平板;修整刮板;磨光梁
finishing screen 终筛;细筛
finishing size 加工尺寸
finishing stake 修整标桩;竣工桩
finishing steel 最后的钻杆;终孔钎子
finishing stock 精加工余量
finishing stove 加工炉
finishing strip 磨光带
finishing superheater 末级过热器
finishing system 饰面涂漆
finishing table 最后精选摇床;精加工台
finishing temperature 终轧温度;(焊接的)终了温度;完工温度;最终温度
finishing test 底面漆配套试验
finishing thresher 清选脱粒机
finishing time of test 试验结束时间
finishing tool 终饰插刀;精车刀(具);抹子;装饰工具;终饰工具
finishing tooth 精削齿;精密切齿
finishing tooth of broacher 拉床精削齿
finishing touch 最后精整;最后加工;润饰

finishing tough 最后精整
finishing trade 终饰行业;扫尾工序
finishing trowel 整饰抹子;光面镘刀;铁泥刀
finishing varnish 饰面清漆;罩面清漆;末道清漆
finishing washer 光制垫圈
finishing waste 精炼废水
finishing wastewater treatment 精炼废水处理
finishing wheel 抛光砂轮
finishing work 修饰工作;最后加工;最后工序;精整工作;结尾工作;装修工程
finish lamp 操作结束检验灯
finish lapping 精研
finish layer 表面涂层
finish lead 线圈出线端
finish machining 精加工
finish mark 光洁度符号;加工符号
finish mass 饰面材料
finish mill 细磨木浆;终粉磨机
finishmo(u)ld 玻璃瓶颈部成型模
finish nail 暗钉
finishness 光洁度
finish of butt joint 对接头加工
finish on a wall 砌面
finish one side 单面光制
finish plane 光刨
finish plaster 最后刷白;最后抹灰罩面;抹灰罩面;抹灰面层;面层粉刷石膏;饰面粉刷;石膏腻子
finish product 光制品
finish raw grinding 原料细磨
finish raw mill 原料细磨机
finish ream 光铰
finish rescreening 最终再筛分
finish roll 给油辊
finish rolling 终压;打光压实;最后碾压;精轧
finish sag 装饰面下陷
finish sarface 加工面
finish size 加工尺寸;完工后尺寸
finish skip 板面装饰缺空;板面装饰空白
finish spreader 路面材料撒布器;路面材料撒布机
finish string 装饰楼梯斜梁;装饰斜梯梁
finish surface 光制表面
finish the mo(u)ld 修型
finish tile 墙面砖
finish time 终止时间;结束时刻;完成时间;结束时间
finish to ga(u)ge 终轧;按样板加工
finish tools 瓶颈加工工具
finish to size 按尺寸加工
finish trowel 压光抹子;抹光镘刀
finish turning 精车
finish turning lathe 精加工车床
finish turn inspection 完工检查
finish up a job 工作结束
finish vanish 上光漆
finish varnishing 最后上的漆;末层涂漆
finish with engine 主机用毕【船】
finish with terrazzo 水磨石饰面
finish work 终饰;修整;竣工
finite 有限性
finite Abelian group 有限阿贝耳群
finite additivity 有限可加性
finite aggregate 有限集合;有穷集
finite algebra 有限代数
finite algebraic extension 有限代数扩张
finite algebraic number field 有限代数数域
finite amplitude 有限振幅
finite amplitude depth sonar 有限振幅测深声呐
finite amplitude wave 有限振幅波
finite angle 有限角
finite aquifer 有限含水层
finite automat 有限自动机
finite automat inference 有限自动机推断
finite automaton 有限自动机
finite baffle 有限屏幕
finite basis 有限基
finite basis theorem 有限基底定理
finite beam 有限梁;有限长梁
finite beam on elastic foundation 弹性基础上定长梁
finite beam source 有限射束源
finite bending 有限弯曲
finite branch 有限分枝
finite cardinal number 有限基数
finite chain 有限链
finite character 有限特征

finite circuit 有限回路;有限环道
finite clipping 有限限幅
finite closed aquifer 有限闭合含水层
finite closed interval 有限闭区间
finite cochain 有限上链
finite collection 有限集合;有限集【数】
finite commutative field 有限交换域
finite compressible layer 有限压缩层
finite concentration 一定浓度
finite condition 有限条件
finite configuration 有穷格局
finite continued fraction 有限连分数
finite correction 有限修正
finite cosine transform 有限余弦变换
finite counting automat 有限计算自动机
finite covering 有限覆盖
finite cyclic(al) group 有限循环群
finite cylinder 有限圆柱体
finite cylindric(al) reactor 有限圆柱形堆
finite decimal 有尽小数
finite decomposition 有限分解
finite deflection 有限挠度
finite deformation 有限形变;非无穷小变形
finite degrees of freedom 有限自由度
finite delay time 有限延迟时间
finite depth 有限深度
finite description 有限描述
finite difference 有限差数;有限差(分);差分
finite difference approximation 有限差(分)近似(法);有限差分逼近;差分逼近
finite difference calculus 有限差分法
finite difference depth migration 有限差分深度偏移
finite difference energy method 有限差分能量法
finite difference equation 有限差分方程
finite difference equation model 有限差分公式模型
finite difference expression 有限差分表示式;差分表达式
finite difference formula 有限差分公式
finite difference method 有限差分法
finite difference migration 有限差分偏叠加
finite difference modelling 有限差分模拟(法)
finite difference operator 有限差分算子;差分算子
finite difference scheme 有限差分格式;差分格式
finite difference solution 有限差分解;差分解
finite difference synthesized record 有限差分合成记录
finite difference theory 有限差分理论
finite difference water quality model 有限差分水质模型
finite differencing scheme 有限差分格式
finite dimensional algebra 有限维代数
finite dimensional complex manifold 有限维复流形
finite dimensional control 有限维控制
finite dimensional distribution 有限维分布
finite dimensional Euclidean space 有限维欧几里得空间
finite dimensional linear space 有限维线性空间
finite dimensional manifold 有限维流形
finite dimensional mapping 有限维数映射
finite dimensional module 有限维数模
finite dimensional projective geometry 有限维射影几何
finite dimensional space 有限维空间
finite dimensional subspace 有限维子空间
finite dimensional vector space 有限维向量空间
finite discontinuity 有限间断;有限不连续性
finite discrete system 有限离散系统
finite dislocation model 有限错动模型
finite displacement 有限位移
finite displacement theory 有限位移理论
finite distance 有限远距离
finite disturbance 有限扰动
finite duration 有限持续时间
finite elastic layer 有限弹性层
finite elastic-plastic theory 有限弹塑性理论
finite elastoviscoplasticity 有限弹黏塑性
finite element 有限元(件);有限元
finite element analysis 有限元分析;有限单元分析
finite element approach 有限单元逼近
finite element approximation 有限单元逼近;有限单元近似解
finite element domain 有限单元域

finite element force method 有限单元法力法
finite element formula 有限单元公式
finite element formulation 有限单元公式
finite element framework 有限单元结构
finite element grid 有限元网络
finite element idealization 有限单元理想化
finite element mesh 有限元网格;有限单元网格
finite element method 有限元法;有限(单)元法
finite element mode 有限单元体模式
finite element model 有限单元体模型
finite element modelling 有限单元模拟
finite element network 有限元网络;有限单元网络
finite element numerical modeling 有限单元数值模拟
finite element programming 有限元程序设计
finite element simulation 有限元模拟
finite element solution 有限单元解法
finite element subspace 有限单元子空间
finite element system 有限单元系统
finite element technique 有限单元技术;有限(单)元法
finite element unit 有限元的单元体
finite energy correction 有限能量值;有限能量修正
finite energy resolution 有限能量分辨率
finite extension 有限扩张
finite extent 有限域
finite field 有限域
finite filtration 有限滤子
finite Fourier series 有限傅立叶级数
finite Fourier transform 有限傅立叶变换
finite function 有限函数;有穷函数
finite gain 有限增益
finite Galois extension 有限伽罗瓦扩张
finite game 有限对策;有限博奕
finite gap 有限宽度隙
finite geometry 有穷几何
finite graph 有穷图;有限图
finite group 有限群
finite group of automorphism 有限自同构群
finite group of outer automorphism 有限外自构群
finite Hankel transform 有限亨克尔变换
finite heat conductivity instability 有限热导率不稳定性
finite increment 有限增量
finite index 有限指数
finite induction 有限数学归纳法;有限归纳法
finite ingot 有限长锭料
finite input source 有限输入源
finite iteration 有限迭代
finite integral transform 有限积分变换
finite intersection property 有限交性
finite interval 有限区间
finite journal bearing 有限径向滑动轴承
finite jump 有限跳跃
finite Larmor radius stabilization 有限拉莫尔半径稳定法
finite lattice 有限点阵
finite layer of soil 有限土层
finite length 有限长度
finite length effect 有限长效应
finite length fault 有限长断层
finite length function 有限长函数
finite length jet 有限射流
finite length launcher 有限长度的发射装置
finite length string 有限长度串
finite lens length 有限透镜长度
finite liability 有限责任
finite life 有限耐久性
finite life design 有限寿命设计
finite life region 有限寿命区
finite lifetime 有限寿命
finite limit 有限限度
finite linear viscoelasiticity 有限线性黏弹性
finite linewidth 有限线宽(度)
finite lot 有限量物料
finitely Abelian group 有限交换群
finitely additive class 有限加性类
finitely additive measure 有限加性测度
finitely additive set function 有限加性集函数
finitely closed subcategory 有限闭子范畴
finitely cocomplete category 有限共完全范畴
finitely defined 有限定义
finitely equivalent sets 有限等价集

finitely generated extension 有限生成扩张
finitely generated field 有限生成域
finitely generated free modules 有限生成自由模
finitely generated group 有限生成群
finitely generated module 有限生成模
finitely generated object 有限生成对象
finitely generated projective modules 有限生成射影模
finitely generated ring 有限生成环
finitely presented functor 有限出现函子
finitely presented group 有限出现群
finitely valued function 有限值函数
finite mapping 有限映射
finite Markov chains 有限马尔可夫链
finite mathematics 有限数学;有穷数学
finite matrix 有限矩阵
finite measurable function 有限可测函数
finite measure 有限测度
finite measure space 有限测度空间
finite medium 有限介质
finite memory 有限存储器
finite memory automaton 有限存储自动机
finite memory filter 有限存储滤波器
finite module 有限模
finite moment theorem 有限矩定理
finite morphism 有限射
finite motion 有限运动
finite movement source model 有限移动源模型
finite moving source 有限运动源
finite multiplicative group 有限乘法群
finite multiplier 有限乘数
finiteness 有限性
finiteness condition 有限性条件
finiteness problem 有限性问题;有限问题
finite nilpotent group 有限幂零群
finite nonempty set 有限非空集
finite nonempty subset 有限非空子集
finite nuclear size effect 有限核大小效应
finite nucleus 有限核
finite number of nodal point 有限节点数
finite object distance 有限目标距离
finite object point 有限远物点
finite observation time 有限观测时间
finite open interval 有限开区间
finite open refinement 有限开加细
finite-orbit instability 有限轨道不稳定性
finite order 有限价
finite ordinal number 有限序数
finite part 有限部分
finite partition 有限划分
finite past 有限的过去
finite permutation group 有限置换群
finite perturbation theory 有限微扰理论
finite pile 有限桩
finite plane 有限平面
finite plastic domain 有限塑性范围
finite point 有限点
finite-pole-width magnet 有限极宽磁铁
finite polyhedron 有限多面体
finite population 有限总体
finite population of customer 顾客的有限总体
finite positive entropy 有限正熵
finite precision 有限精确度
finite precision number 有限精度数
finite presentation 有限表示
finite prime divisor 有限素因子
finite probability 有限概率
finite process 有限程序
finite progression 有限级数
finite projection 有限射影
finite projective geometry 有限射影几何
finite projective geometry codes 有限射影几何码
finite projective plane 有限射影平面
finite purely inseparable extension 有限纯不可分扩张
finite quantity 有限量
finite rank 有限秩
finite rate 有限率
finite rate of increase 有限增长率;周限增长率
finite reactor 有限堆
finite reflux 有限回流;实际回流
finite region 有限区域
finite representation 有限表示
finite resolving power 有限分辨率

finite restriction 有限限制
finite ring 有限环
finite rod bundle 有限棒束
finite rotation group 有限旋转群
finite sample behavio(u)r 有限抽样行为
finite scheme 有限概型
finite segmenting 有限分段
finite separable extension 有限可分扩张
finite sequence 有限序列
finite series 有限级数
finite series solution 可限级数解
finite set 有限集合;有限集【数】;有穷集
finite shield build-up factor 有限屏蔽层积累因数
finite simple group 有限单群
finite simplicial complex 有限单纯复形
finite sine transform 有限正弦变换
finite-size panel 有限尺寸板
finite slice method 条分法
finite solution 有限解【数】;定解
finite solvable group 有限可解群
finite space 有限空间
finite specification 有限说明
finite spin 有限自旋
finite standpoint 有穷观点
finite state algorithm 有限态算法
finite state automaton 有限自动机
finite state channel 有限状态信道
finite state device 有限态装置
finite state diagram 有限状态图
finite state machine 有限自动机
finite state recognizer 有限态识别程序
finite state stochastic games 有限态随机对策
finite state table 有限态表
finite state technique 有限态技术
finite straight line 有限直线
finite strain 有限应变
finite strain theory of elasticity 有限形变弹性理论
finite stress 有限应力
finite strip(e) method 有限条(分)法
finite subadditivity 有限子可加性
finite subcomplex 有限子复形
finite subfamily 有限子族
finite subgroup 有限子群
finite subset 有限子集
finite substitution 有限置换
finite sum(mation) 有限和
finite switching time 限定开关时间
finite system of generators 有限生成系
finite table 有限表
finite thin sheet 有限薄片
finite time 有限时间
finite time average 有限均值
finite time filtering 有限时间滤波
finite trace 有限迹
finite transcendence degree 有限超越次数
finite transducer 有限转换器
finite triangulation 有限三角剖分
finite tube bundle 有限管束
finite twisting 有限扭转
finite type 有限型
finite universe 有限全域
finite value 有限值;有限数值
finite variance 有限方差
finite volume coast and ocean model 有限体积近海模型
finite volume method 有限容积法
finite volume water quality model 有限元水质模型
finite water depth 有限水深
finite water depth terminal 有限水深的码头
finite wave train 有限波列
finite wedge 有限尖楔
finite width 有限宽度
finite-width effect 有限宽效应
finitism 有限论
finitist 有穷论者
finitistic space 有穷论的空间
finitude 有限
fin keel 鳍状龙骨
Fink process 氢氧保护热镀铝法
Fink-ring 调整环
finks 破坏罢工者
Fink truss 芬克式桁架;芬克桁架;法国式桁架
Finland architecture 芬兰建筑
finless(black)porpoise 江豚

finned 突肋
finned air cooler 带散热片的空气冷却器
finned coil 带散热片的螺旋管;翅片盘管;有翅盘管
finned cooler 散热片冷却器;带叶片冷凝器;翅形冷却器;翅形冷凝器
finned drum 翅片转鼓
finned duplex tube 双翅管
finned evapo(u)rator 翅片式蒸发器
finned heater 翅管加热器;翅片加热器
finned heat exchanger 翅片式换热器;翅片换热器
finned heating tube 片式供暖管
finned length 带肋长度
finned pile 带翅桩
finned pipe 翅片管
finned plate radiator 肋片平板式散热器;片板式散热器
finned radiator 暖气片
finned roller 翅辊
finned strip electric heater 翅片式电热器
finned strip heater 翅片式加热器
finned strip heating tube 翅片式供暖管
finned strip tube 翅片管
finned surface 翼片热面;翅面
finned tube 翼(型)管;翼形管;翅(片)管
finned tube exchanger 翅片管换热器;翅管交换器
finned tube heater 翅片管加热器
finned tubes exchanger 翼换热器
finned tubular radiator 翅片管散热器
finned type heat exchanger 翅片式热交换器
finned type heating coil 片式供暖盘管
finnemanite 砷氯铅矿;菲氯砷铅矿
finning 肋材装配;筋条加强;用肋加固
Finn oil 芬兰油
finny creature 有鳍动物
fin of radiator 散热片翅板
fin oil 松浆油
fin panel casing 膜式水冷壁
fin pitch 翅片距
fin radiator 翅式散热器
Finsen lamp 费生弧光灯;水银弧光灯
finsen light 紫外线灯;水银灯
fin shield 片状冷却器;插片冷却器
fin sign 鳍状标牌
Finsler geometry 芬斯拉几何
Finsler structure on a manifold 流形上的芬斯拉结构
fin stabilizer 减摇鳍;稳定鳍;防摇鳍
Finsterwalder prestressing method 芬斯脱华特预(加)应力(方)法
fintie number of steps 有限步数
fin-tilting machinery 水平翼转舵机械
fin-tip 直尾翅梢
fin tube 鳍状管;翅片管
fin tube coil 翅片盘管
fin tube exchanger 翅片管换热器
fin tube heater 翅管供热器;翅片管加热器
fin tube radiation 翅管散热
fin tube type heat exchanger 翅管式热交换器
fin tube type metallic recuperator 翅管式金属换热器
fin wall 翼缘墙
fin waveguide 带叶片波导;带翅波导
fin whistle 波纹金属涵管
fioraceta 变形醋酯纤维
fiord 峡湾海岸线;峡湾
fiord ice 在峡湾形成的冬冰
fiorite 硅华
fir 云杉;冷杉
fir cone gasket 冷杉球果形密封垫;拱式弹性密封垫
fire 火灾;炽热体;放炮警告
fire academy 消防学院
fire access 灭火口
fire accident 火灾事故;火警事故
fire achieves 火灾档案
fire action 耐火效果
fire adjustment 发火调整
fire administration 消防局
fire agent 消防部门
fire agriculture 烧荒农业
fire a hole 炮眼点火放炮;爆破炮眼
fire airplane 消防飞机
fire aisle 阻火通道
fire alarm 消防警报装置;火灾警报;火灾报警(器);火警预报器;火警(警报);火警报警器;瓦斯警报器

fire alarm bell 火警铃
fire alarm box 火灾报警箱;火警箱
fire alarm call 报火警
fire alarm casing 火灾报警盒
fire alarm communication and signal system 火警通信和信号系统
fire alarm control and indicating equipment 火灾报警控制和显示设备
fire alarm control panel 火灾报警控制板
fire alarm control system 火灾报警控制系统
fire alarm detector of ion type 离子式火灾报警探测器
fire alarm detector of thermal type 消防感温式报警探测器
fire alarm device 火灾警报装置
fire alarm dispatcher 火灾警报调度员
fire alarm equipment 火灾警报装置
fire alarm evacuation signal 火警疏散信号
fire alarm headquarter 消防通信调动中心
fire alarm inidication device 火警显示装置
fire alarm network 火灾报警网
fire alarm receiving station 火警接收站
fire alarm remote indicating equipment 火警远距离显示设备
fire alarm signal 火灾报警信号;火警信号
fire alarm signaling device 火灾信号发生装置
fire alarm signalling apparatus 火警信号机
fire alarm signal station 火警信号站
fire alarm sounder 火灾警报器
fire alarm sounding system 火警报警系统;声响火(灾报)警系统
fire alarm system 火灾警报系统;火灾报警系统;火警(警报)系统;防灾报警系统
fire alarm thermostat 火灾热动开关
fire alarm zone 火灾警报区
fire algae 火藻
fire analysis 火情分析
fire analysis of steel building system 钢结构建筑物系统的火灾分析
fire and ambulance station 消防及救护站
fire and bilge pump 消防污水两用泵
fire and causality insurance 火灾和意外保险
fire and explosion safety 防护和防爆安全
fire and explosition prevention 防火防爆管理
fire and explosive hazard 火灾和爆炸危险
fire and marine insurance 火灾及海上保险;物业水火保险;物产(水火)保险
fire and rescue bill 消防救生部署表
fire and smoke transport 火和烟迁移
fire and snow melting system 消防及融雪系统
fire an engine 开动发动机
fire annihilator 灭火器
fire annihilator pipe 消防管
fire apparatus 消防设备;消防器材;灭火器材;灭火器
fire approach suit 隔热服
fire area 防火区间;各层消防面积;防火面积;防火带中附防火墙隔开的区段;避难区(域)
firearmor 镍铬铁锰合金
fire arrestor 阻火器
fire arrow 火箭
fire assay 火法试金;火法化验;试金分析法
fire assay through sulfide button 锍试金
fire assembly 防火组件;防火装置;消防设施组件;防火门窗配件
fire assembly automatic 自动防火(门)组合装置
fire atlas 防火图(表)集
fire attack 火灾扑救
fire attack line 灭火水带线
fire attack strategy 灭火战略
fire avalanche 火山崩流
fire axe 消防用斧;消防斧
fire back 炉壁的背墙砖;壁炉背墙
fire-back boiler 炉征热水器;回火锅炉;厨灶锅炉
fireball 火团;火流星
fireball formation 火球的形成
fireball radius 火球半径
fire bank 防火堤
fire bar 炉栅;炉条;炉排;炉算;加热条
fire barrier 防火设施;挡火墙;防火间隔;隔火设施
fire-barrier ceiling 防火天花板
fire-barrier material 挡火材料
fire-barrier sealant 防火墙密封胶

fire basket 烘篮
fire bat(t) 耐火砖坯;火拍
fire beater 拍火器;打火扫把
fire bed 火床;火层;炉中火床
fire behavio(u)r 火灾特性
fire behavio(u)r analyst 火灾行为分析员
fire behavio(u)r forecast 火灾行为预测
fire bell 火(报)警钟;火警铃
fire belt 防火带
fire bill 消防部署
fire black-out 火基本扑灭
fire blanket 消防毯;灭火毯;防火毯
fire block 阻火隔板;防火块;耐火砌块
fire block gypsum board 隔火石膏板
fire-board 壁炉板;挡炉板
fire-board board 挡炉板的木板
fire-board ceiling 挡炉板顶盖板
fire-board finish 挡炉板装饰
fire boat 消防艇
fire bond 耐火材料黏结剂
fire boss 灭火总指挥;扑火总指挥;防火监护员
fire bossed 定期接受防火检查的
fire box 炉膛;火箱;燃烧室
firebox boiler 火室式锅炉
firebox door 火箱门
firebox sheet 火箱板
firebox shell 火箱壳
firebox steel 火箱钢
fire branch 消防枪;消防水枪
fire-brand 燃烧的火头;燃木
firebreak 隔火设施;防火空隙地带;防火隔墙;防火隔断;防火带;隔火路;挡火墙;挡洪墙;挡火线
firebreak forest 防火林
fire breakout 起火
firebreak partition(wall) 防火隔墙
fire-breeding 有火灾征兆
fire brick 黏土(耐火)砖;耐火砖;火砖
firebrick aggregate 耐火砖集料;耐火砖骨料
firebrick arch 耐火砖拱
firebrick lid 耐火砖盖
firebrick lined 耐火砖衬砌;火砖砖砌
firebrick lined chimney 耐火砖内衬烟囱
firebrick lining 耐火砖衬里;耐火砖衬
firebrick structure 耐火砖结构
firebrick support 耐火砖载体
fire bridge 火箭拱;火墙;桥墙
fire brigade 消防队;消防处
fire brigade access window 消防用窗
fire brigade control room 消防队控制室
fire brigade hose reel 消防车胶管卷
fire brigade vehicle 消防车;救火车
fire broom 灭火帚
fire bucket 消防桶;消防水桶;防火水桶
fire buff 消防队之友
firebug 纵火犯
fire building 首先起火的建筑物
fire bulkhead 隔火墙;挡洪墙;防火隔板;挡火(隔断)墙
fire burning index 燃烧指数
fire button 灭火按钮
fire cabinet 消火栓箱
fire cache 消防器材储藏处
fire call 火灾报警
fire call receiving 消防通信调度室
fire camp 消防营地;打火营部
fire cannon 灭火炮
fire canopy 防火挑板
fire cap 消火栓管嘴帽
fire casualty 火灾伤亡者
fire cause 起火原因
fire cause class 火灾类别
fire cell 燃烧室
fire cement 耐火水泥;耐火胶结材料
fire chamber 火箱;燃烧室
fire check 阻火器;热裂纹;坯裂;防回火装置
fire check door 挡火门;防火门;半防火门;阻燃门
fire check key 高温锥
fire chief 消防队长;消防部门主管人
fire chute 避难滑道
fire cistern 消防储[贮]水池;消防水池
fire classification 火灾分类
fire clay 烧焦(黏)土;耐火土;火泥
fireclay-base refractory mortar 以耐火黏土为主成分的耐火砂浆
fireclay bath tub 耐火泥浴盆
fireclay body 火泥制品;火泥本体
fireclay brick 火砖;耐火(黏土)砖;火泥砖
fireclay bushing furnace 陶土坩埚拉丝炉
fireclay chamotte 耐火黏土熟料
fireclay concrete 耐火混凝土
fireclay container 烧钵;火泥容器
fireclay crucible 耐火坩埚
fireclay deposit 耐火黏土矿床
fireclay disk 耐火土盘
fireclay goods 耐火泥商品
fireclay grog refractory 耐火泥熟料
fireclay insulating refractory 隔热耐火材料
fireclay lining 火泥炉衬;耐火泥衬;耐火黏土衬里
fireclay mineral 耐火泥矿材;耐火泥矿物
fireclay mortar 耐火砂浆;耐火(泥)灰浆
fireclay mo(u)ld 耐火黏土模
fireclay nozzle 耐火喷嘴
fireclay plastic refractory 塑性耐火材料
fireclay refractory 黏土质耐火材料
fireclay refractory material 耐火砖
fireclay sanitary ware 耐火黏土卫生器
fireclay sink unit 耐火泥排水落水器
fireclay sleeve 耐火泥釉砖
fireclay smog 耐火黏土熟料
fireclay ware 耐火泥制件;耐火泥桶
fire climate 火险气候
fire clock 火钟
fire-cloud furnace 碳粉电气淬火炉
fire coal 取暖用煤;取暖煤
fire-coat 鳞皮;氧化膜;耐火涂层;氧化皮
fire-coat cement 耐火水泥
fire-coat mortar 耐火砂浆
fire-coat paint 耐火漆
fire cock 消火栓;消防栓;消防龙头;防火开关
Fire code 法耳码
fire code 消防规范
fire coinsurance 火灾共同保险
fire combat station 消防岗位
fire command center 消防指挥中心
fire commander 火场指挥员
fire command station 消防指挥站
fire command system 消防指挥系统
fire committal button 灭火按钮
fire community 消防界
fire company 消防队;火灾保险公司
fire compartment 防火分隔间;防火分区
fire compartmentation 防火分区法;防火分隔
fire compartment of building 建筑防火分区
fire concentration 火灾集中
fire connector 消防栓连接器
fire conservancy 防火区
fire consumption 消防用水量
fire containment 遏制火灾
fire control 消防;灭火;火灾控制;防火;燃烧控制
fire control access door 消防设施门
fire control car 消防车
fire control computer 火力控制计算机
fire control damper 防火控制挡板
Fire Control Digest 消防文摘(期刊)
fire control improvement 防火设施
fire control instrument 射击控制仪器
fire control line 火灾控制线;防火线
fire control measure 防火措施
fire control operator 消防人员
fire control order 消防秩序
fire control panel 火力控制面板;防火控制盘
fire control plan(ning) 灭火作战计划;防火规划
fire control portable equipment 便携式消防设备
fire control quadrant 射击控制象限仪
fire control regulation 消防条例
fire control sonar 攻击声呐;射击控制声呐
fire control system 消防系统
fire control unit 防火装置
fire co-operator 义务护林员
fire-cord decoration 流纹装饰;熔楞装饰
fire core 耐火心板
fire cover 消防力量
fire crack 碎裂;干裂;炽裂;加热裂纹;热裂;燃烧开裂
fire crackers pollution 烟花爆竹污染
fire cracker welding 鲔焊
fire cracking 热裂;退火裂(纹)
fire crack mark 燃烧开裂痕迹
fire-crash water tender 救援水罐消防车
fire-curing 烟熏烤干
fire curtain 火灾幕(帘);防火挡板;防火(卷)帘
fire curve 火灾曲线;时间燃烧温度曲线
fire cut 梁端斜面;断火线条;避难间;端头小梁
fire cutoff 挡火物
fire-cutting partition 耐火墙;隔火墙;防火墙
fire cycle detector 循环式火灾探测器
fire cycle system 自动开关式喷水灭火系统
fire dam 防火墙
fire-damaged 烧损的
firedamp 沼气;甲烷;爆炸气体;密封防火墙;瓦斯
firedamp alarm 沼气警报
firedamp cap 焰晕;蓝色焰晕
firedamp content 沼气含量
firedamp detector 沼气探测器;可爆空气探测器;沼气检定器;沼气检测仪
firedamp drainage 沼气排放;排放瓦斯
firedamp drainage drill 排放瓦斯钻机
fire damper 耐火风门;火灾阻止器;火灾遮断器;挡火板;防火阀;防火挡板;防火门
fire damper in vent duct 通风管道中的防火阀
firedamp explosion 沼气爆炸
firedamp feritilizer 沼气肥
firedamp fringe 瓦斯边界区;瓦斯边界层
firedamp ignition 沼气点燃
firedamp indicating detector 沼气示意器
firedamp indicator 沼气检定器;瓦斯指示器
firedamp layer 瓦斯积聚层
firedamp limit 瓦斯极限含量
firedamp machine 防火式电机
firedamp migration 瓦斯迁移
firedamp pressure-chamber method 瓦斯压力室排放法
firedamp probe 瓦斯探针
firedamp proof machine 沼气式电机
fire damp proof motor 防火式电机
firedamp protection 沼气保护
firedamp reforming process 沼气重整法
firedamp testing 瓦斯测定
fire danger 火灾危险
fire danger board 火险告示牌
fire danger class 火险等级
fire danger index 火险指数;火险指标
fire danger meter 火险预警仪;火险计算尺
fire danger rating 火险评级;火险评分
fire danger rating area 火险等级区
fire danger scale 火险计算表
fire danger station 火险观测站
fire dange scale 火险等级表
fire data 火灾数据
fire database 火灾数据库
fired block 烧结砌块
fire boiler 煤气锅炉
fired brick 窑制砖;煅烧黏土砖;烧结砖
fired brick of colliery waste 烧结煤矸石砖
fired ceramic coating 烧结陶瓷涂料;烧成陶瓷涂料
fired clay 烧黏土
fired clay brick 普通黏土砖
fired clay curved roof(ing) tile 耐火泥屋面曲瓦
fired clay curved tile roof 耐火泥曲瓦屋顶
fired clay hip tile 耐火泥屋脊瓦
fired clay light(weight) aggregate 耐火泥轻质料;耐火泥轻质骨料
fired clay product 耐火泥制品
fired clay ridge tile 耐火泥脊瓦
fired clay tile 烧结黏土瓦
fired damp 矿井气
fire death 火灾死亡
fire deaths 死亡人数
fire demand 消防用水(量)
fire demand of water 消防需水量
fire demand rate 消防需水率;消防需水(流)量
fire department 消防机构;消防处;消防部门(的设备)
fire department access 消防通道;消防出入口
fire department access point 消防队进入点
fire department connection 消防连接口;消防处接头
fire department key box 消防队锁匙箱
fire department operation 消防部门业务
fire department personnel 消防部门工作人员
fire department(pumper)connection 水泵接合器

fire department system 消防部门数据
fire department vehicle driver 消防队车辆驾驶员
fire despatch 消防调度
fire detect 激发检测
fire detecting and extinguishing apparatus 火灾探测和灭火装置;火警探测及灭火装置
fire detecting and extinguishing device 火警探测及灭火装置
fire detecting area 火灾探测区
fire detecting arrangement 火警探测装置;探水装置
fire detecting cabinet 火灾报警箱;火警探测箱
fire detecting device 火警探测装置
fire detecting system 自动火灾警报装置;火灾探测系统;火灾检测系统
fire detection 火灾侦察;火警探测;火灾探测
fire detection alarm system 火灾探测报警系统
fire detection coverage 火灾探测覆盖范围
fire detection device 火警装置
fire detection relay 火灾探测继电器
fire detection system 火灾探测系统;火感系统
fire detection system test switch 火警探测系统试验开关
fire detector 火灾探测器;火灾检测器;火警探测装置;火警探测器;火警监测器;火感器;探火器
fire detector apparatus 火灾探测装置
fire detector element 火灾探测元件
fire detector switch 火警探测器开关
fire detector system 火灾探测系统
fire detector unit 测火器
fire determination 火灾判断
fire development 火灾发展
fire devil 路面加热机;火盆;火焰漩涡
fired fly ash brick 烧结粉煤灰砖
fired guard 隔火道
fire dike 防火堤
fire direction net 火力方向网
fire disaster 重大火灾;火灾
fire discharge 消防流量
fire dispatch 消防调度
fire display panel 火灾显示盘
fire distance 防火间距
fire distribution 火力分配
fire district 防火区
fire division 防火隔板;防火间隔层
fire division floor 隔火楼板
fire division masonry wall 防火砖石分隔墙
fire division wall 防火墙;隔火墙;防火隔墙
fired material 烧过的材料
fire dog 壁炉柴架;炉壁柴架
fire door 保险门;消防门;户门;耐火门;防火门
fire door assembly 防火门组(装)件;防火门组合件
fire door hardware 防火门五金配件
fire door latch 防火门搭扣;防火门插销
fire door protection ring 炉门门圈;炉门保护环
fire door rating 门窗耐火等级;防火门等级
fire door shield 炉门护板
fired pillar 垂直火道
fired pin 射钉
fired pipe 消防喷射(水)管
fired pressure vessel 直接火焰加热压力容器
fired process tubular heater 直接火操作的管式加热炉
fired product 消防制品
fire drainage 排放瓦斯
fire drain valve 消防栓泄水阀
fire drencher 消防筒
fire drill 消防演习
fire drill tower 消防练习塔
fire driveway 消防车道
fired shale 煅烧过的页岩土
fired shale product 煅烧过的页岩土制品
fired state 发射状态;煅烧过的状态;点火状态
fired strength 耐火强度(混凝土燃烧后的抗压、挠曲强度);烘干强度
fired strength of concrete 混凝土耐火强度
fired talc 焙烧滑石
fired-to 有自由面的爆破
fired unit weight 燃烧后的单位重(耐火混凝土);烘干重度
fire durability 耐燃性;耐火性
fire duration 耐火时间;火灾延续时间;燃烧持续时间
fire-duration test 燃烧持续试验
fire duty 直接的灭火行动
fire dyke 防火堤

fire dynamics 火灾动力学
fire ecology 火灾生态学
fire education 消防教育
fire effect 火烧结果
fire effect model(l)ing 火灾影响模拟
fire effluent 火灾气流
fire elevator 消防电梯
fire embankment 防火墙;防火塔;防火堤
fire emergency elevator 消防紧急电梯
fire emergency management 消防紧急管理
fire emergency services 消防应急服务
fire enclosure 火区密封
fire end 火端;热接点
fire-end bright 火焰明亮
fire-end by black smoke 火焰带黑烟
fire endurance 耐火时限;火灾持续性;耐火时间;耐火极限
fire endurance calculation 耐火极限计算
fire endurance rating 耐火等级;耐火率
fire engine 消防器;消防机;消防车;救火机;救火车;灭火泵
fire engine hose 消防胶管
fire engine house 消防车库;救火车房
fire engine pump 车载消防泵
fire engine room 消防车库
fire entry suit 避火服;救火衣
fire environment 火灾环境
fire equipment cabinet 消防装备箱
Fire Equipment Manufacturers Association 消防设备制造商协会
fire escape 消防安全出口;疏散楼梯;防火应急出口;安全梯;安全出口;太平门
fire escape chute 救生滑槽;安全滑槽;安全倾卸槽
fire escape corridor 太平通道
fire escape ladder 应急梯;安全梯;太平梯;救火梯
fire escape roof(walk)way 屋顶太平走道
fire escape route 消防应急通道
fire escape stair(case) 避难梯;安全出口楼梯;太平(楼)梯;防火安全楼梯;安全梯;疏散梯;防火(楼)梯
fire escape staircase well 太平楼梯井
fire escape trailer 带消防梯的挂车
fire escape tube 安全出口通道
fire escape window 太平窗;紧急出口窗
fire evaluation test 燃烧评定试验
fire evidence 火灾证据
fire-evil 火邪袭
fire excape 安全滑梯
fire exception 火灾不在内
fire exercise 消防演习
fire exit 消防安全门;紧急出口;太平门;安全门;安全出口
fire exit bolt 太平门栓;安全门闩;安全出口栓
fire exit drill 消防疏散训练
fire exit hardware 太平门五金配件;安全门五金配件
fire experiment 耐火试验
fire explosion 瓦斯爆炸
fire exposure 暴露于火
fire exposure severity 暴露于火灾的严重程度
fire exposure severity test 受火严重性试验
fire exposure time 暴露于火的时间
fire extend 着火范围
fire extinciton by foam 泡沫灭火
fire extinction 灭火
fire extinction engineering 灭火工程
fire extinction equipment 灭火设备
fire-extinguisher 消火器;灭火器;灭火机
fire-extinguisher agent 灭火器用灭火剂
fire-extinguisher box 灭火器箱
fire-extinguisher bracket 消防用具支架;灭火器架
fire-extinguisher cabinet 灭火器橱;灭火器箱;灭火器柜
fire-extinguisher cock 灭火器开关
fire-extinguisher dry chemical 灭火器干粉
fire-extinguisher equipment 灭火器设备
fire-extinguisher fluid 灭火液;灭火器液
fire-extinguisher handle 灭火器手柄
fire-extinguisher of self-sensitive chemical 化学自动灭火器
fire-extinguisher switch 灭火器喷射开关
fire-extinguisher symbol 灭火器符号
fire-extinguisher system 消防系统;灭火系统
fire-extinguisher transfer switch 灭火器转换开关
fire-extinguisher trolley 灭火器推车

fire-extinguishing 灭火
fire-extinguishing agent 灭火剂
fire-extinguishing apparatus 灭火装置;灭火设备;灭火器
fire-extinguishing appliance 消防设备;灭火设备
fire-extinguishing bomb 灭火弹
fire-extinguishing bottle 灭火剂钢瓶
fire-extinguishing bullet 灭火弹
fire-extinguishing cart 灭火手推车
fire-extinguishing chemical 灭火剂
fire-extinguishing composition 灭火混合物
fire-extinguishing device 灭火设备
fire-extinguishing drill 灭火演习
fire-extinguishing effectiveness 灭火效能
fire-extinguishing efficiency 灭火效能
fire-extinguishing engineering 灭火技术
fire-extinguishing equipment 灭火设备;消防设备
fire-extinguishing fan 灭火风机
fire-extinguishing fluid 灭火液
fire-extinguishing foam 灭火泡沫(材料);泡沫灭火剂
fire-extinguishing gas 灭火气体
fire-extinguishing in cargo ship 货轮灭火设备
fire-extinguishing installation 消防设备;灭火装置;灭火设施
fire-extinguishing line 消防路线
fire-extinguishing method 灭火方法
fire-extinguishing performance 灭火性能
fire-extinguishing plant 消防设备;灭火装置
fire-extinguishing property 灭火性能
fire-extinguishing pump 灭火水泵;消防(用)泵
fire-extinguishing run 消防通道
fire-extinguishing sand 消防用砂;灭火砂
fire-extinguishing system 灭火系统;消防系统
fire-extinguishing tanker 消防车;水罐消防车
fire-extinguishing test 灭火试验
Fire Extinguishing Trade Association 灭火器行业协会
fire-extinguishing tube 灭火管
fire-extinguishing vapo(u)r 灭火蒸气
fire facade 防火栅;防火屏
fire face 燃烧面
fire factor 燃烧因素
fire fall 火烧倒塌
fire fan 消防爱好者;轻便鼓风机
fire fatality 火灾死亡
fire fighter 消防战斗员;消防队员;扑火队员;消防人员
fire fighter gypsum board 消防石膏板;防火石膏板;耐火石膏板
fire fighter helmet 消防头盔
fire fighter professional qualification 消防队专业资格
fire fighter's boots 消防靴
fire fighter's elevator 消防电梯
fire fighter's glove 消防手套
fire fighter's protective clothing 消防战斗服
fire fighter truck 消防车
fire fighter uniform 消防员制服
fire fighting 灭火战斗;火灾扑救;打火;消防
fire-fighting access 消防通道
fire-fighting access lobby 灭火通道大厅
fire-fighting amphibian 水陆两栖消防车
fire-fighting and heat isolating paint 防火隔热涂料
fire-fighting apparatus 灭火设备
fire-fighting appliance 消防设备;消防器材
fire-fighting behavio(u)r 灭火行为
fire-fighting boat 消防船
fire-fighting bomb 灭火弹
fire-fighting by equal pressure 均压灭火
fire-fighting clothing 消防战斗服
fire-fighting command car 消防指挥车
fire-fighting craft 消防艇
fire-fighting crew 消防队
fire-fighting device 消防设施
fire-fighting efficiency 灭火效率
fire-fighting elevator 消防电梯
fire-fighting equipment 灭火设备;消防(器材)设备;消防灭火设备;消防灭火工具设施
fire-fighting equipment of port 港区消防设施
fire-fighting exit staircase 消防出口楼梯
fire-fighting fabric 消防织物
fire-fighting facility 灭火设备
fire-fighting finish 火抛光

fire-fighting finished 火焰抛光过的
fire-fighting foam 灭火泡沫
fire-fighting foam agent 灭火泡沫液
fire-fighting footwear 消防靴
fire-fighting force 灭火力量
fire-fighting gable 防火山墙
fire-fighting gang 消防队
fire-fighting grading 消防等级
fire-fighting grenade 灭火手雷
fire-fighting halon 灭火用哈龙
fire-fighting hose 消防水(龙)带;消防胶管
fire-fighting ladder 消防梯
fire-fighting lance 消防水枪
fire-fighting line 灭火水带线路
fire-fighting method 灭火(方)法
fire-fighting monitor 消防网;消防喷射器;消防炮;架式水喷器
fire-fighting operation 消防作业
fire-fighting order 灭火规定
fire-fighting party 消防队
fire-fighting plan 消防设备布置图
fire-fighting pool 消防水池
fire-fighting powder 灭火干粉
fire-fighting procedure 消防措施;灭火方法
fire-fighting process 消防作业
fire-fighting pump 消防泵
fire-fighting pumping room 消防泵房
fire-fighting pump set 消防泵组
fire-fighting resources 灭火资源
fire-fighting ring main 灭火环形主管
fire-fighting safety 灭火安全
fire-fighting safety service 消防安全部门
fire-fighting school 消防学校
fire-fighting service 消防勤务
fire-fighting ship 消防船
fire-fighting staircase 消防楼梯
fire-fighting stairway 消防楼梯
fire-fighting station 消防部署
fire-fighting strategy 灭火战略
fire-fighting supply 消防供水
fire-fighting system 消防制度;消防系统
fire-fighting tactics 灭火战术
fire-fighting team 消防队;救火队
fire-fighting technique 灭火技术
fire-fighting trailer 消防拖车
fire-fighting train 消防列车
fire fighting truck 救火车
fire-fighting tug 消防拖轮
fire-fighting tunic 消防战斗服
fire-fighting turnout 消防战斗服
fire-fighting unit 消防装置
fire-fighting vehicle 消防车
fire-fighting vehicle access 消防车通道
fire-fighting water supply 救火供水
fire-fighting window 消防孔;消防窗
fire-fighting with fire-extinguisher 灭火器灭火
fire-fighting with inert gas 惰性气体灭火法
fire-fighting work 消防作业
fire finder 火灾寻找器;火灾探测器
fire-finder map 火灾位图
fire finisher 火抛光机
fire finishing 火抛光(玻璃制品)
fire flap 防火挡板
fire float 消防船
fire float pump 消防艇消防用泵
fire-flood cycle 火灾洪水周期
fire floor 火烧层
fire flow 消防流量;火舌
fire-fly glass 荧光玻璃
fire foam 灭火沫;灭火泡沫(剂);泡沫灭火剂
fire foam-producing machine 灭火泡沫发生器
fire foe 阻燃织物(商品名)
fire forest belt 防火林带
fire fountain 火泉(火山暂时喷发);熔岩喷涌
fire frame 炉门框
fire front 火灾前沿
fire gable 防火山墙;封火山墙;风火山墙
fire gas 易燃气体;可燃气体
fire-gas detector 可燃气体探测器
fire-gas-explosion 易燃气体爆炸
fire gas warning system 火气警系统
fire goggles 消防护目镜
fire grade 防火等级
fire grading 建筑物耐火性;耐火等级;火灾等级;防火等级
fire grading of building 建筑物防火等级
fire grading period 防火分级时间
fire grate 炉算;炉栅;炉排;炉算子;炉算
fire grate bar 炉条
fire grate segment 炉排片
fire groun command 火场指挥员
fire ground announcement 火场广播
fire ground communication 火场通信
fire ground factor 火场因素
fire ground network 火场指挥网
fire guard 防火栅;炉挡板;火炉栏;人工防火线;防火员;防火监视哨
fire gun 消防水枪;灭火枪
fire-gutted 煅烧
fire-gutted structure 煅烧的结构
fire hat 消防帽
fire hazard 火灾隐患;火灾危险性;火灾危害性;灾;火险
fire hazard classification 火险分类;火灾危险(性)分类;火灾(危害性)分类
fire hazard index 火灾指数
fire hazardous 易着火
fire hazardous area 火险地区
fire hazard reduction 降低火灾危险度
fire hazard test 火险试验
fire headquarter 火场指挥所
fire-hearth 火源;火床
fire heating rust removing 火焰加热法除锈
fire helmet 救火头盔
fire hole 炉口;火孔;点火孔
fire hole ring 炉口衬圈
fire hook 消防钩;火钩
fire hose 消防水龙带;消防软管;消防龙头;救火水龙头;救火水龙带;救火软管;灭火水龙带;水龙带
fire hose and extinguishing cabinet 灭火器柜;灭火器橱
fire hose and fire plug cabinet 消火栓柜
fire hose box 消防水龙带箱
fire hose cabinet 灭火水龙带柜;消防水龙带箱;消防柜
fire hose connection 灭火水龙带接头
fire hose connection in the tunnel 隧道内灭火皮管连接
fire hose coupling 消防水龙(带)接头;消防水带接口
fire hose fitting 消防水带配件
firehose instability 水龙带不稳定性
fire hose nozzle 消防水龙带喷嘴;消防枪;消防带喷口;灭火水枪
fire hose rack 灭火水龙带支架;水带架;消防软管架
fire hose reel 消防水带卷盘
fire hose station 消防站;灭火皮带管站
fire hose with inside lined rubber 内衬胶消防水龙带
fire house 消防站;消防车库;救火房
fire hydrant 灭火龙头;火栓;消防(水)龙头;消防栓;太平龙头;防火水龙头
fire hydrant box 消防栓箱
fire-hydrant cabinet 消防栓箱
fire hydrant chamber 消火栓井;消防栓箱;消防栓井;灭火栓箱
fire ignition due to spontaneous combustion 自燃造成的火灾
fire ignition energy 引燃能量
fire ignition sequence 着火顺序
fire incident data organization system 火灾事件数据组织系统
fire indemnity 火灾损失赔偿
fire indicating lamp 火灾确认灯
fire indicating unit 火灾显示装置
fire indicator 火警指示器;火警信号器
fire information field investigation 消防情报实地调查
fire information retrieval system technique 火灾情报检索系统技术
fire inhibitor 火焰抑制剂
fire injury 烧伤
fire inspection 防火检查
fire inspection reporting system 防火检查报告系统
fire inspector 防火检查员
fire installation 消防装置;消防设备
fire insurance 火灾(保)险;火险;保火险
fire insurance contract 火险契约

fire insurance on stock and supply 库存物资火险
fire insurance policy 火灾保险单
fire insurance premium 火灾保险费
fire insurance rate 火灾保险费率
fire insurance rating 火险等级评定
fire insurance surveying 火灾保险测定
fire integrity 耐火完整性
fire in the open 露天火灾
fire in tunnel 洞内火灾
fire investigation 火灾调查
fire investigation program(me) 火灾调查计划
fire investigator 火因调查员
fire irons 炉用具;生火工具;壁炉用具
fire isolation 火区隔离
fire killed timber 防火木料;防火木材
fire knee 防火漆墙(一排房屋中相邻房屋隔墙在屋顶上的凸出部分)
fire knockdown 控火
fire ladder 防火梯;消防梯
fire lagging 防火绝缘层
fire lamp 火警指示灯
firelamp accumulation 瓦斯积聚
fire lane 消防车通道;防火巷道;防火隔离线;防火带
fire launch 消防艇
fire layer 沼气聚积层;燃烧层
fire-leading 引火
fire legislation 消防法规
fireless cooker 无火炊具
fireless locomotive 灭火机车;无火机车【铁】
fireless run 无火始动
fireless steam loco(motive) 无火蒸汽机
fire life 火灾延续时间
fire lift 消防电梯
fire lift priority switch 消防电梯优先开关
fire lift switch 消防电梯开关
firelight 炉火;火光
fire lighter 引火物;点火剂
fire lighting 点火
fire limit 消防线;火灾危险界限;火警线;消防界线(防火线);消防分区
fire line 消防专用管道;消防管线;火灾最前线;火灾现场警戒线;防火线;防火沟
fire lintel 货物起卸口防火过滤
fire load 消防负荷;火荷载;火负荷;燃料荷载
fire load density 火灾负荷密度
fire loading 消防负荷;防火负荷;火灾荷载;可燃料重量(每平方英尺建筑面积中含有可燃材料的重量)
fire lobby 消防走廊
fire loss 火灾损失
fire loss adjustment 火灾损失调整
fire loss information 火灾损失资料
fire loss insurance 火灾损失保险
fire loss investigation 火灾损失调查
fire loss prevention 火灾损失预防
fire loss statistics 火灾损失统计
fire loss volume 烧失量
fire main(pipe) 消防(水)总管;消防干管;灭火主管道
fire main system 消防总管;消防水管系统;消防管路系统
fire main with fire valve 带消火阀的消防总管
fireman 消防员;消防人员;加煤工;煤矿井下检查员;爆破工
fire management 消防管理
fire management note 消防管理备忘录
fire management plan 消防管理计划
fire management system 消防管理系统
fireman hose (消防员的)水龙带
fireman's axe 消防斧
fireman's escape 轴隧应急出口;安全梯出口
firemanship 消防实践
fireman's outfit 消防员装备
fireman's platform 消防员的台架;司炉工作站位;司炉工作台
firemans's cabin 井下消防站
fireman's switch 消防开关
fireman's uniform 消防员制服
fire mantle 防火林带;防火隔离带
fire mark 火烛
fire marshal 消防队长;防火部门主管人
fire mask 消防面具
fire metallurgy 火法冶金
fire model 火灾型式;火灾模型;点火模型

fire modeling 火灾模拟
fire modeling computational system 火灾模拟计算系统
fire monitor 消防炮;消防监视站;架式水喷器
fire monitoring 火灾监测
fire mortar 耐火砂浆
fire museum network system 消防博物馆网络系统
fire-new 崭新的;新制的;全新的
fire nose 救火皮带
fire nozzle 灭火水枪;消火水枪;消防水枪;消防喷嘴
fire occurrence map 火灾发生图
fire office 火灾保险公司
fire-on 彩烧
fire opal 火蛋白石
fire orders 五类典型柱石
fire ordinance 消防条例
fire origin 起火源
fire originating unit 起火单位
fire originator 起火单位
fire out 放火烧除
fire outbreak 火灾爆发
fire out time 熄灭时间
fire over 空运转;空热耗
fire pack 防火背包
fire pan 火盆;灰槽
fire parameter 火灾参数
fire partition 防火(间隔)区段;隔火墙;耐火隔墙;防火隔墙;防火隔断
fire partition opening 防火隔墙通道
fire partition wall 消防隔墙;防火间壁墙
fire passage 消防通道
fire path 救火路线;火灾蔓延通道;防火通道
fire patrol 消防巡逻;火灾巡查;防火安全检查
fire patrolman 消防巡逻队员;消防队员
fire patrol system 消防巡查制度
fire penetration 火灾蔓延
fire penetration test 燃烧穿透测试
fire performance 耐火性能;防火功能
fire performance-based code 基于耐火性能的规范
fire peril 火灾危险
fire permit 动火许可证
fire pipe 消防管
fire Pitometer 消防栓毕托计
fireplace 火室;壁炉;壁垒
fireplace brick 火炉炉砖
fireplace chimney 火炉烟囱
fireplace construction 火炉构造
fireplace flue 火炉风道
fireplace grate 火炉炉算
fireplace hearth 壁炉膛
fireplace insert 嵌入壁炉采暖炉;护膛衬板
fireplace leaning to wall 靠墙壁炉
fireplace lounge 炉边躺椅
fireplace masonry work 砌炉圬工工作
fireplace material 砌炉材料
fireplace opening 火炉炉孔
fireplace recess 火炉炉膛
fireplace stove 壁炉式火炉
fireplace tiles 壁炉炉衬
fireplace unit 壁炉整套装置
fireplace wall 壁炉墙
fire plant 消防装置
fire plate 防火板
fire plug 消防栓;消防龙头;太平龙头;消火栓
fire point 着火点;火点;燃(烧)点;起爆点
fire point of asphalt 沥青燃点
fire poker 火钳
fire policy 火灾保险单
fire-polish 火抛光(玻璃)
fire polished 火抛光的
fire polisher 火抛光机
fire polishing (玻璃的)烧边;火抛光
fire pot 火炉;燃烧室;火箱
fire potential 火险隐患
fire power 火力
fire practice 消防演习
fire precaution 消防措施;防火装置;防火办法;防火措施
fire precaution measure 防火措施
fire pressure 消防水压
fire presuppression 防火准备措施
fire protection zone 防火圈
fire preventing 防火措施
fire prevention 火灾预防法;防火

fire prevention and smoke exhaust damper 防火排烟阀
fire prevention greening 防火绿地
fire prevention measure 防火措施
fire preventive appliance 防火器械
fire preventive device 防火设备
fire pricker 火镰钩
fire progress map 扑火进度图
fire-proof 防火的;耐火的
fire-proof aggregate 耐火集料;耐火骨料
fire-proof air-tight door 防火密闭门
fire-proof architectural board 防火建筑平板
fire-proof asbestos cement board 防火石棉水泥板
fire-proof asbestos clothing 防火石棉衣
fire-proof asbestos fiber 防火石棉纤维
fire-proof brick 耐火砖
fire-proof building 防火建筑;耐火建筑(物);耐火房屋
fire-proof bulkhead 防火隔堵;防火舱壁;防火(隔)墙;耐火隔壁;耐火舱壁
fire-proof cable 防火电缆
fire-proof cable and wire 耐火电缆电线
fire-proof casing 耐火衬套
fire-proof cast stone 耐火铸石
fire-proof cement 耐火水泥;耐火胶结材料
fire-proof coating 耐火涂料
fire-proof cock 防火开关
fire-proof concrete 耐火混凝土
fire-proof construction 防火建筑;防火构造;耐火结构;耐火建筑(物);耐火构造
fire-proof construction(al) material 耐火结构材料
fire-proof covering 防火套
fire-proof curtain 防火屏障;防火幕(帘)
fire-proof damper 防火阀
fire-proof district 防火地区
fire-proof door 耐火门
fire-proof dope 耐火涂料
fire-proofed wood 耐火材
fire-proof engine 防火式电机
fire-proof fabric 防火织物
fire-proof fibre 防火纤维
fire-proof floor 防火楼板;耐火楼板
fire-proof floor spring door 防火地弹门
fire-proof floor with suspended ceiling 有吊顶的耐火楼板
fire-proof grease 耐火润滑脂;耐火润滑油;不易燃滑脂
fire-proof gypsum sheathing 防火石膏衬板
fire-proofing 防火(措施);防火层厚度;防火法;防火处理;耐火装置
fire-proofing adhesive 耐火黏合剂
fire-proofing agent 阻燃剂;防火剂
fire-proofing cement 耐火水泥
fire-proofing chemical 耐火剂
fire-proofing coat 耐火面层
fire-proofing compound 防火化合物
fire-proofing dope 防火添加剂
fire-proofing finish 防火整理
fire-proofing hanging 耐火悬件(指帘帷等)
fire-proofing impregnation 防火浸渍
fire-proofing installation 防火设施
fire-proofing material 耐火材料;阻燃材料
fire-proofing measure 消防措施
fire-proofing mortar 耐火砂浆
fire-proofing paint 耐火油漆
fire-proofing plaster 耐火涂层;耐火灰浆
fire-proofing plywood 防火胶合板;耐火胶合板
fire-proofing preparation 防火制剂
fire-proofing protection 消防措施
fire-proofing pump 消防泵
fire-proofing reagent 防燃剂
fire-proofing tile 防火砖;耐火砖(瓦)
fire-proofing wood 防火木材;防火木材;耐火材
fire-proof installation 消防装置;防火装置
fire-proof insulation 防火绝缘
fire-proof intermediate floor 耐火中间楼面
fire-proof life-belt 耐火救生带
fire-proof lifeboat 耐火救生艇
fire-proof life-line 耐火救生绳
fire-proof line 防火绳
fire-proof lining 耐火盖层
fire-proof machine 防火机器;防火电机;防爆式电机
fire-proof machinery 防火机械
fire-proof mat 耐火席子

fire-proof material 防火材料;耐火材料
fire-proof membrane 防火薄膜
fire-proof nature 防火特性
fire-proofness 耐火性
fire-proof paint 防火涂料;防火漆;耐火涂料;耐火漆
fire-proof paint coat(ing) 耐火涂料涂层
fire-proof panel 防火板
fire-proof paper 防火纸
fire-proof partition 防火分隔;防火隔墙
fire-proof petrol tank 耐火汽油桶
fire-proof prefabricated element 工厂预制耐火构件
fire-proof resistance test 耐火性试验
fire-proof roll screen 防火卷帘
fire-proof roof floor 耐火顶层楼面
fire-proof safe 防火保险柜
fire-proof sealing material 防火封堵材料
fire-proof sheathing 防火石膏衬板;防火衬板
fire-proof slab floor 耐火楼板
fire-proof stair(case) 耐火楼梯
fire-proof structure 防火结构;耐火结构
fire-proof suit 防火服
fire-proof system 防火系统
fire-proof tank 防火油箱
fire-proof timber 耐火木材
fire-proof treatment 耐火处理
fire-proof trim 防火面饰料
fire-proof uncovered floor 耐火露天楼面
fire-proof wall 隔火墙;防火墙;防火壁;耐火墙
fire-proof wood 耐火木材;耐燃木材
fire-proof wood(en) door 木质防火门
fire-proof zone 防火地区
fire propagation 火灾波及
fire propagation index 火焰传播指数
fire property 火灾性质
fire-protecting arrangement 消防设备
fire-protecting door 防火门
fire-protecting paint 防火涂料
fire-protecting paint for building 建筑防火涂料
fire-protecting performance 耐火性能;防火性能
fire-protection 消防;护林防火;防火
fire-protection appliance 消防用具
fire-protection clothing 消防服
fire-protection coating 防火涂料
fire-protection code 消防规范
fire-protection criteria 消防标准
fire-protection design 防火设计
fire-protection device 防火装置
fire-protection distance 防火距离
fire-protection district 消防区
fire-protection drawing 消防图
fire-protection eaves 防火挑檐
fire-protection encasement 消防包装
fire protection engineer 消防工程师
fire protection engineering 消防工程
fire-protection equipment 消防设备
fire-protection expenses 防火费
fire-protection flap 防火挡板
fire-protection haunching 消防护拱
fire-protection industry 消防产业
fire-protection installation 防火设备
fire-protection layer 防火层
fire-protection map 消防车行程图
fire-protection of building 建筑防火
fire-protection of structural steelwork 钢结构的防火处理
fire-protection organization 消防组织
fire-protection paint 防火涂料;防火漆
fire-protection planning 防火设计
fire-protection professional 消防专业人员
fire-protection propaganda 消防宣传
fire-protection rating 防火分级;耐火等级;防火等级
fire-protection requirement 防火要求
fire-protection ring 防火圈
fire-protection safety sign 消防安全标志
fire-protection service 防火设施;消防设施
fire-protection sheath coat 防火保护面层
fire-protection shutter 卷帘式防火门
fire-protection signal(l)ing system 消防信号系统
fire-protection sprinkler system 喷射消防系统;喷水灭火系统;消防喷淋系统
fire-protection sprinkler valve 灭火喷洒阀
fire-protection standard 防火标准
fire-protection strip 防火巷道;防火隔离带
fire-protection suit 救火衣;防火衣;防火服

fire-protection switch 防火开关
fire-protection system 消防系统;防火系统
fire-protection tree 防火树
fire-protection valve 灭火喷水阀;防火阀
fire-protection water supply 消防水源
fire-protection water tank 消防水罐
fire-protection zone 防火区
fire-protective lining 防火衬砌;防火衬层
fire-protective paint 防火漆
fire pump 消防水泵;消防泵;灭火泵
fire pump and sprayer 灭火喷射器
fire pump drive 灭火水泵驱动
fire pumper 消防车
fire pump room 消防泵房
fire pump station 消防泵站
fire pump with drive 带发动机的消防泵
fire quarters 消防部署
fire radio communication network 消防无线电通信网
fire rake 救火钩
fire rate 耐火分级
fire-rated 耐火分级的;防火分级的
fire-rated board 防火板
fire-rated door 测定耐火等级的门;标准防火门
fire-rated form board 耐火模板
fire-rated form(work) 耐火模板
fire-rated partition 耐火隔板
fire-rated penetration 防火洞孔
fire-rated systems using plywood 胶合板防火等级系列
fire-rated window 防火规范窗
fire rating 耐火等级;防火等级;耐火率
fire rating test 耐火等级试验
fire red 颜料火红;炉红;氮化对位红
fire reel 救火皮带卷轴
fire refined metal 火法精炼金属
fire refining 火法精炼
fire refining process 火法精炼法
fire region 防火区
fire regulation 消防规则;火力调节
fire regulator 火力调节器
fire research 消防研究
fire research station 消防研究机构
fire-resistance 抗燃烧性;抗火性;防火性能;防火(能力);耐火性
fire-resistance belt(ing) 耐火运输带
fire-resistance class 耐火等级
fire-resistance classification 耐火性分级;防火分级
fire-resistance duration 耐火极限
fire-resistance grading (结构件的)耐火等级;(房屋的)防火等级;耐火性分级
fire-resistance grading period 耐火等级期限
fire-resistance gypsum plaster board 耐火纸面石膏板
fire-resistance hour 耐火时间
fire-resistance mortar 耐火砂浆
fire-resistance period 耐火期;耐火时间
fire-resistance rating 耐火值;耐火额定值;耐火等级
fire-resistance requirement 防火要求
fire-resistance rolling shutter door 防火卷帘门
fire-resistance steel door 钢质防火门
fire-resistance test 耐火性试验;耐火试验
fire-resistance wall 防火墙
fire-resistant 抗火的;耐火的;耐高温的;阻燃的;抗燃的;防火的
fire-resistant belt(ing) 耐火运输带
fire-resistant board 防火板;耐火板
fire-resistant building member 建筑物耐火构件
fire-resistant cable 耐火电缆
fire-resistant chemical 耐火剂;耐火化学材料
fire-resistant clamp 耐火卡箍
fire-resistant cloth 防火布
fire-resistant coating for steel structure 钢结构防火涂料
fire-resistant concrete 耐火混凝土
fire-resistant construction 耐火结构
fire-resistant conveyer belt 防火运输带
fire-resistant door 防火门;耐火门
fire-resistant finish 防火罩面漆
fire-resistant floor 耐火楼面
fire-resistant fluid 耐燃液体;不燃液
fire-resistant foundation (由砖石或混凝土砌筑的)建筑物基础;耐火基础
fire-resistant fuel 防火燃料

fire-resistant grease 耐火润滑脂;耐火润滑油
fire-resistant housing 耐火机壳
fire-resistant hydraulic fluid 耐燃液压油
fire-resistant hydraulic liquid 耐火液压油
fire-resistant(in)filling 耐火填料
fire-resistant material 耐火材料
fire-resistant material engineering 防火材料工程
fire-resistant member 耐火构件
fire-resistant oil 抗燃油
fire-resistant packing 耐火填料
fire-resistant paint 耐火漆;耐火油漆
fire-resistant partition(wall) 防火隔墙
fire-resistant property 耐火特性
fire-resistant rating 耐火等级
fire-resistant roof 耐火屋顶
fire-resistant shield 耐火屏蔽;耐火隔板
fire-resistant shutter 耐火百叶窗;耐火卷帘门
fire-resistant stair(case) 耐火楼梯
fire-resistant structure 耐火结构
fire-resistant tarpaulin 防火帆布
fire-resistant wall 防火墙;防火隔墙
fire-resistant wall board 耐火墙板
fire-resistant wire 耐火绝缘导线;耐火导线
fire-resisting 耐火(的);防火的;阻燃的;抗火的
fire-resisting beam 耐火梁
fire-resisting bulkhead 阻燃舱壁;耐火舱壁
fire-resisting cable 耐火电缆
fire-resisting casing 耐火罩;耐火箱;耐火套;耐火壳;耐火盒;耐火包装
fire-resisting ceiling 耐火天花板;耐火顶棚;耐火顶板
fire-resisting cement 耐火水泥;耐火胶结材料
fire-resisting concrete 耐火混凝土
fire-resisting construction 耐火构造;防火结构
fire-resisting covering 耐热包皮
fire-resisting damper 耐火风门;防火阀;防火挡板
fire-resisting division 防火分隔
fire-resisting door 耐火门;防火门;防火门扇
fire-resisting dope 耐火涂料
fire-resisting duct 耐火管道
fire-resisting finish 耐火饰面;防火罩面漆;耐火油漆涂料
fire-resisting floor 耐火地面;耐火地板;耐火楼板
fire-resisting floor wall 耐火地板墙
fire-resisting glass 耐火玻璃
fire-resisting glazing 耐火玻璃(窗);耐火釉
fire-resisting layer 防火层
fire-resisting magazine 防火暗舱
fire-resisting material 耐火材料;防火材料
fire-resisting paint 耐火油漆;耐火漆;防火涂料;防火漆
fire-resisting partition 耐火隔墙
fire-resisting plastics 耐火塑料
fire-resisting property 耐火性能;防火性能
fire-resisting rolling shutter 防火卷帘门
fire-resisting roof 耐火屋顶;耐火天花板
fire-resisting sand 防火砂
fire-resisting shaft 耐火竖井
fire-resisting shutter 耐火卷帘门;耐火百叶窗
fire-resisting stair(case) 耐火楼梯
fire-resisting stanchion 防火窗间小柱
fire-resisting steel sliding door 防火钢推拉门
fire-resisting structure 耐火结构
fire-resisting timber 耐火木材
fire-resisting timber door 防火木门
fire-resisting trunking 耐火管道
fire-resisting wall 耐火墙
fire-resisting window 防火窗
fire-resisting wood 耐火木材
fire-resistive 耐火的
fire-resistive ceiling 耐火顶棚;耐火顶板
fire-resistive coating 耐火罩面
fire-resistive construction 结构防火构造;耐火构造;防火构造
fire-resistive file room 耐火的档案室
fire-resistive floor 耐火楼板
fire-resistive flooring 耐火地面;抗火楼板
fire-resistive material 耐火材料;防火材料
fire-resistive partition 防火隔断
fire-resistive protection 隔热保护;防火保护
fire-resistive rating 耐火等级(测定)
fire-resistive separation 防火隔墙
fire-resistive suspended ceiling 耐火吊顶
fire-resistive type 耐火类型;防火类型

fire-resistive wall 耐火墙
fire responsibility system 防火责任制度
fire responsibility system for assigned posts 岗位防火责任制
fire retardance 阻燃性
fire retardancy 防火性
fire-retardancy test 阻燃试验
fire-retardant 防火剂;防火的;阻火的
fire-retardant additive 阻燃添加剂
fire-retardant adhesive 抗热黏合剂;阻燃黏合剂;滞燃黏合剂
fire-retardant agent 阻燃剂;延缓燃烧剂
fire-retardant barrier 阻燃隔离层
fire-retardant building member 建筑物耐火构件
fire-retardant building unit 建筑物耐火构件
fire-retardant ceiling 阻燃天花板;阻燃顶棚;阻燃顶板;延迟燃烧天花板;延迟燃烧顶棚
fire-retardant chemicals 阻燃剂;阻火剂;滞燃化学剂;阻燃化学物质;阻燃化学品;防火化学制剂
fire-retardant chemistry 阻燃化学
fire-retardant coating 耐燃涂层;阻火涂层;阻燃涂料;阻燃层;滞燃罩面;耐燃罩面;防火涂料
fire-retardant construction 阻燃结构;耐火结构
fire-retardant core 阻燃芯(体)
fire-retardant door 耐火门;防火门
fire-retardant duct insulation 阻燃管道的绝热;(防火材料保护的)风道隔热
fire retardant equipment 防火设备
fire-retardant fastener 防火搭扣
fire-retardant finish 阻燃涂层抛光;滞燃饰面;耐火饰面;阻燃装修
fire retardant flight suit 防火飞行服
fire-retardant foam 阻燃泡沫材料
fire-retardant for cellulosic material 纤维素材料的阻燃
fire-retardant glass 防火玻璃
fire-retardant lath 阻燃型条板
fire-retardant lumber 阻燃木材;耐火木料
fire-retardant material 阻燃材料
fire-retardant mechanism 阻燃机理
fire-retardant member 耐火构件
fire-retardant(mixed)plaster 耐火灰浆
fire-retardant paint 耐火(油)漆;耐火涂料;阻燃漆;防火涂料;防火漆
fire-retardant paper 阻燃浸渍纸
fire-retardant plasticizer 阻燃性增塑剂
fire-retardant polyester resin 阻燃聚酯树脂
fire-retardant preservative 防火防腐剂
fire-retardant pressure impregnation 阻燃剂压力浸渍
fire-retardant rating 耐火等级(测定)
fire-retardant resin 阻燃性树脂;滞燃树脂
fire-retardant roof cover 阻燃屋面覆盖层
fire-retardant roof coverings 阻火屋盖
fire-retardant sand 阻燃砂
fire-retardant structural system 滞燃结构体系
fire-retardant suit flight 阻燃飞行服
fire-retardant test 阻燃试验
fire-retardant treated 滞燃处理
fire-retardant treated lumber 阻燃处理的木材
fire-retardant-treated wood 阻燃处理木材
fire-retardant treatment 滞燃处理;难燃处理;延迟燃烧处理;耐火处理
fire-retardant type flame arrester 耐燃型阻火器
fire-retardant unit 耐火构件
fire-retardant wall covering 阻燃性贴墙布
fire-retardant wood 阻燃木料;滞燃木材
fire-retardation 阻燃作用
fire-retarded 用耐火材料保护的
fire-retarding agent 防火剂
fire-retarding bulkhead 阻燃舱壁
fire-retarding chemical 阻燃剂
fire-retarding component 耐火构件
fire-retarding curtain 挡火幕
fire-retarding division 阻燃分隔
fire-retarding glazing 滞燃玻璃;耐火玻璃
fire-retarding liquid coating material 耐火液体涂层材料
fire-retarding material 耐火材料
fire-retarding(mixed)plaster 耐火灰浆
fire-retarding paint 耐火(油)漆;阻燃漆;防火漆;防火涂料
fire-retarding system 滞燃系统
fire-retarding treatment 阻燃处理;滞火处理

fire-retarding treatment of timber 木材阻燃处理
fire-retarding unit 耐火建筑构件;耐火构件
fire ring 燃烧环
fire risk 火源危险;火灾危险(性);火灾危害性;火险
fire risk analysis 火灾风险分析
fire risk extension clause 火险自然扩展条款;火险附加条款
fire risk management 火险管理
fire risk measure 火灾危险性量度
fire risk on freight 货运火险
fire risk only 只承担火险
fire road 救火通路;消防通道
fire rod 点火棒
fire rolling shutter 防火可卷百叶窗
fire room 火室;锅炉房;锅炉间;锅炉舱
fire rope 点火绳
fire route 救火路线
fire run 火头迅速推进
fire runner 消防安全检查员
fire safe 无火险的;安全防火的
fire safe material 耐火材料;防火材料
fire safety 防火安全
fire safety building code 防火安全建筑物规范
fire safety certificate 消防安全合格证书
fire safety clothing 消防安全服
fire safety committee 防火安全委员会
fire safety design 防火安全设计
fire safety education 消防安全教育
fire safety engineering 防火安全工程
fire safety equipment 防火设备
fire safety evaluation system 防火安全评估体系
fire safety furniture 防火安全家具
fire safety problem 防火安全问题
fire safety requirement 用火安全要求
fire safety science 防火安全科学
fire safety shutter 防火安全板
fire safety system 防火安全系统
fire safety system analysis 防火安全系统分析
fire safety system approach 消防安全系统法
fire safety week 消防安全周
fire sale 火灾受损物资拍卖
fire sand 耐火砂;氮碳化硅(制造碳化硅的中间产物);防火砂;消防砂
fire sand box 灭火砂箱
fire scale 耐火氧化皮
fire scanning 火情扫描
fire scheme 防火设计
fire science 消防科学
fire screen 消防;防火栅;防火屏(障);隔热屏(障);火炉栏;火隔;挡火屏;挡火网;壁炉栏
fire seal 防火密闭墙
fire sealing 封闭火区
fire-sealing gable 封火山墙;风火山墙
fire season 火灾季节
fire seat 火源位置
fire security measures 防火安全措施
fire security routine inspection 消防安全例行检查
fire separation 火灾隔离带宽度;防火间距;防火分隔
fire separation wall 挡火墙
fire sequence 火灾环境信号显示顺序
fire service 消防业;消防设施;消防工作;消防岗位;消防队
fire service act 消防条例
fire service connection 消防连接管;消防泵接水口
fire service control panel 消防控制盘
fire service control room 消防控制室
fire service detector check meter 消防(用)检测水表
fire service equipment 消防装置
fire service expert 消防工作专家
fire service implement 消防器材
fire service information 消防业务信息
fire service inspection system 消防检查体系
fire service instructor 消防教官
fire service inventory 消防用品
fire service ladder 消防梯
fire service location package 消防机构位置软件包
fire service main 消防水管
fire service meter 消防水表
fire service pipe 消防管
fire service pump 消防水泵
fire service rcognition day 消防服务表彰日
fire service system 消防部门系统
fire service training 消防训练

fire service water supply 消防水源
fire severity 火灾损害潜势;火灾烈度
fire severity concept 火灾烈度概念
fire shaft 垂直火道
fire shelter 防火屏(障)
fire shield 看火罩
fire shovel 火铲;煤铲
fire shrinkage 烧缩;焙烧收缩
fire shutter 耐火百叶窗;防火卷帘门;防火活门;防火百叶窗
fireside lounge 炉边休息处;炉边躺椅
fire sighting 看火
fire signal 火警信号
fire siren 火警笛
fire site 火场
fire size 消防栓尺寸
fire size class 火灾规模类别
fire slash 火灾迹地
fire slice 长柄火铲
fire-smoke detection 烟火探测
fire smothering 窒息灭火
fire-smothering blanket 灭火毯
fire-smothering gas 灭火气;惰性灭火气体;窒息灭火气体
fire-smothering gear 窒息性气体灭火装置;气体灭火装置
fire-smothering steam 窒息灭火用蒸汽
fire-smothering system 窒息灭火系统
fire source 火源
fire space plate 火警地区指示牌
fire speckling 火刺
fire spread index 火势蔓延指数
fire spread(ing) 火势蔓延
fire sprinkler system 喷洒灭火系统;消防喷水系统
fire-sprinkling system 喷洒灭火系统;喷水灭火系统;灭火洒水系统
fire stability 耐火稳定性
fire stage 火灾阶段
fire stain 焦斑;火灾焦斑
fire staircase 消防楼梯间
fire standpipe 消防竖管
firestat 恒温器
fire station 消防站;井下消防站
fire statistics 火灾统计
fire statute 消防法规;火灾法则
fire stone 耐火粘土;耐火(岩)石;耐火硅质岩;火石;黄铁矿结核;燧石
fire stop 止火板;烟囱封堵;夹板墙挡板条;挡火织物;挡火撑;穿堂火遏止物(防止火蔓延);阻火;拦火物;挡火物;挡火层;防火墙;挡火物
fire stopping 阻火法;防火阻隔措施;安装挡火物
fire stop product 阻火产品
fire storage 消防储水
fire storm 火力强攻;火暴;风暴性火灾
fire stream 消防水流;消防喷水流
fire strength 火灾强度
fire suction hose 消防吸水管
fire suit 消防服
fire superintendent 消防监督
fire superiority 火力优势
fire supervisor 防火负责人
fire supply 消防给水
fire suppressant gas 灭火气体
fire suppression 消防;灭火;扑火
fire suppression bottle 灭火瓶;灭火器
fire suppression control valve 灭火控制阀
fire suppression organization 灭火组织
fire suppression ring 灭火圈
fire suppression system 灭火系统
fire suppressor 防燃剂
fire survey 火情勘查
fire-susceptible 易燃的
fire swatter 打火扫把
fire switch (发动机的)起动开关;点火电门
fire symbol 消防标志
fire system 消防系统
fire tactics 灭火战术
fire-tank wagon 水罐消防车
fire telephone line 消防电话线
fire temperature 着火温度
Fire Temple 火焰庙(伊朗)
fire tender 消防车;消防船;救火机
fire terrace 防火平台;防火平顶
fire test 着火温度试验;着火点测定;耐火试验;焚烧试验
fire test exposure-severity 火焰照射强度试验;火焰照射强度测试;燃烧试验暴露于火的强度
fire test procedure 燃烧试验程序
fire-test-response characteristic 火灾试验响应特性
fire tetrahedron 燃烧四面体
fire the charge 引燃炸药
fire the hole 爆破炮眼
firethorn 火棘属植物
fire tile 防火瓦
fire tongs 火钳
fire tool 司炉用具
fire-tool cache 配套储[贮]藏的消防工具;灭火器材储[贮]藏所
fire tower 消防塔;防火防烟分隔楼梯间;消防瞭望台;瞭望塔;火警瞭望塔;火警监视台;火警观察塔;望火塔;防火塔
fire trail 防火道
fire training facility 消防训练设施
fire trap 引起火灾的垃圾堆;引起火灾的废料堆;易失火的建筑物;高火热;无太平门建筑物;易燃废物堆
fire trap area 高火险区
fire tray 燃烧盘
fire trench 防火壕;防火沟
fire triangle 燃烧三要素
fire trier 沼气检查员
fire trough 燃烧道
fire truck 救火车;消防车
fire truck reel 消防车胶管卷
fire tube 烟管;(锅炉内的)火管
fire tube apparatus 火管装置
fire tube boiler 火管式锅炉;火管锅炉
fire tube boiler survey 火管锅炉检验
fire tube heater 火管加热炉
fire tube steel boiler 火管锅炉
fire tube test 火管试验
fire tube-water tube boiler 水管火管合并式锅炉
fire under control 得到控制的火灾
fire underwriter 火灾保险业者
fire unit 灭火装置
fire valve 防火阀;消防阀;灭火阀;熔丝阀
fire vent 排烟口;排烟孔道
fire ventilation 火灾排烟通风
fire ventilator 火灾排烟通风机
fire venting 消防通气孔
fire vessel 消防船;救水船
fire victim 火灾受害者
fire viewer 沼气检查员
fire wagon 救火机;沥青熔锅车
fire walker 消防巡逻员;火灾警戒员
fire wall 防火隔板;隔火墙;炉壁;绝热隔板;挡火墙;挡当墙;风火墙;防火(隔)墙
firewall protection 防火保护墙
fire warden 消防管理员;地区防火指挥员
fire warden line 防火线
fire warning 火警(信号)
fire warning device 火灾报警器
fire warning light 火警灯
fire warning light test button 火警信号灯检查按钮
fire warning sensor 火警信号传感器
fire warp 火警曳缆
fire waste 烧损
fire watch 消防值班;消防巡查;防火检查
fire watcher 消防值班员;火灾警戒员
fire water 消防(用)水
fire water box 消防水箱
fire water line 消防供水线
fire water pond 消防水槽;消防水池
fire water pressure 消防水压(力);灭火水压力
fire water storage tank 消防用水储罐
fire water supply 消防供水;消防给水
fire water(supply)duration 消防供水持续时间
fire water system 消防用水系统
fire water tank 消防水箱
fire weather 火险天气
fire weather forecast 火险天气预报
fire weather station 防火气象站
fire welding 锻接;锻焊
fire well 消防水井
fire whirl 火焰漩涡
fire wind 大火风暴引起的风
fire window 消防窗;防火窗
fire window assembly 装配成的防火窗

fire wire 耐火线
fire-wire control unit 火灾报警缆线控制装置
fire-wire electric(al) cable 缆式火灾探测器
fire-wire sensing element 缆式火灾探测敏感元件
fire wood 薪炭火;木柴;燃材木柴;柴(草)
firewood chopper 柴刀
firewood forest 薪炭林
firewood stove 柴炉
fire work 消防工作;焰火具
firework display 放焰火
fire work period 林火第一工作日
fire works 烟火;花爆;焰火
fire wound 火伤
fire zone 防火区;防火分区
fire zoning 防火分区
fir fixed (成框架或桁架)暂时拼装;暂时固定;未刨屋架木材;(只用钉子钉住的)未刨木材;定位木条
firing 引燃;灼烧;开火;解雇;起火;起爆;添煤;烧制;烧成
firing angle 引燃角;点火角;柴油机喷油提前角
firing at one 一次点火起爆
firing azimuth 射击方位角
firing base 射击底座;发射台
firing bay 热试验间
firing behavio(u)r 燃烧行为
firing box 燃烧室;发射箱
firing brick 耐火砖
firing button 引燃按钮
firing cable 引爆电缆;引火线;起动电缆;导火索;爆破用电系缆
firing chamber 火室;燃烧箱;燃烧室;焙烧室
firing chart 火力配制图
firing circuit 引燃电路;起动电路;点火电路
firing constriction coefficient 烧成收缩系数
firing cost 爆破费用
firing crack 烧裂;烧成裂隙;烧裂缝
firing crew 消防队(员);发射班
firing current 开启电流
firing defect 点火不足;焙烧缺陷;烧成缺陷
firing delay 点火延迟
firing device 燃烧装置;燃烧设备;起爆器
firing door 炉门;火门
firing door burner 炉门燃烧器
firing drawing-chamber 烧炉
firing effect 燃烧效果
firing equipment 加热设备;明火加热设备
firing expansion 烧成膨胀
firing failure 引燃失败
firing fan 一次风鼓风机
firing gang 消防队
firing gear 机械添煤传动机构;击发装置
firing group by group 成组点火起爆
firing gun 射孔枪
firing hammer 发火撞锤
firing hole 点火孔
firing hole layout 爆破孔布置
firing hood 灼烧帽盖;窑炉炉头;射击挡板
firing-in butt 军事靶场
firing initial energy 点火起始能
firing installation 点火装置;焙烧装置
firing interval 发火间隔
firing jack 射击调平装置
firing key 点火键;引爆电键;起动开关
firing level 燃烧面
firing lighting 点火
firing-line (焊接管道接缝的)主要焊接工班;修理管线区域;石油管线工地;火线
firing line location 火线位置
firing lock 击发机构
firing machine 加煤机
firing mechanism 击发机构
firing method of explosion crater 药包起爆方法
firing nozzle 燃烧器喷嘴
firing of mine 矿井着火
firing of refractories 耐火材料焙烧
firing one by one 逐个点火起爆
firing order 引燃顺序;点火次序;发射次序;发火次序
firing out 放火烧除
firing party 射击队
firing pattern 引爆方式;爆破孔布置形式
firing pin 击针
firing-pin group 击发装置
firing plate 装料板
firing platform 起动台;发射台架

firing point 着火点;燃烧点;起燃点;起动点;导通点;发火点
firing potential 引燃电位;点火电位
firing practice, exercise area 打靶和演习区
firing pressure 燃烧压力
firing process 点火步骤;焙烧过程;一次烧成
firing pulse 引燃脉冲;点火脉冲
firing range 射击距离;引火点范围;着火范围;火炮射程
firing rate 加热速度;加热料速率;热输入;燃烧(速)率;燃料用量;热负荷
firing residue 烧渣
firing ring 测温环;测热圈
firing rotary kiln 回转窑点火
firing rule 激发规则
firing run 点火试验
firing sector 烧火间
firing sequence 炮眼爆破次序
firing shrinkage 烧缩;热缩;退火收缩;烧成收缩;焙烧收缩
firing space 炉膛
firing speed 发火速度
firing spring retainer 击针簧挡板
firing stroke 膨胀冲程;爆炸冲程;工作冲程;燃烧冲程;点火冲程
firing switch 起动开关
firing system 燃烧系统;燃烧加热系统
firing table 射表
firing table elevation 射表仰角
firing team 射击队
firing technique 点火技术
firing temperature 着火温度;点火温度
firing test 点火试验阀;爆炸试验
firing test data 试车数据(发动机)
firing time 燃烧时间;点火时间
firing time test 燃烧速度试验
firing to cement clinker 烧成水泥熟料
firing tools 点火工具
firing torque 发火转矩
firing tower 发射台
firing under fluid 液下爆破
firing unit 燃烧装置
firing up 点火升温
firing vehicle 火箭发射装置
firing voltage 开始放电电压;点火电压
firing zone 燃烧区;燃烧带
firing-zoning 防火分区制
firkin 小桶;加仑小桶(1加仑小桶=9加仑)
fir-like spatter dash 皮毛石
firm 牢靠的;牢固;可靠的;坚固(的);坚固;商号;厂商
firm acceleration 稳定加速
firm agreement 不能撤消的协议
firmament 晴夜空;太空
firm and impervious joint 牢固接合
firmary 医务所
firm bargain 确定(的)买卖;实盘交易
firm base 坚硬地基
firm belief 确信
firm bid 确定出价;递实盘
firm bottom 坚硬地层;坚实地基;稳固底板;硬土
firm capacity 正常输出(功率);可靠容量;保证出力
firm clay 硬黏土
firm commitment 坚定承诺;包销承诺
firm-commitment offering 确定承诺出价
firm contract 约束性合同;经过确认具有束缚性的契约;确定的契约
firm demand 固定负载;固定负荷
firm discharge 安全流量;保证(使用)流量;固定流量;牢靠出流量;可靠流量
firm energy 可靠电能;保证电能
firmer chisel 木(工)凿;木柄套箍木凿;榫(孔)凿
firmer chisel set German type 德式套装(木)工凿
firmer gouge 木工用弧口凿;半圆(式)木凿
firmer gouge chisel 半圆木凿
firmer tools 木工作业工具
firm fixed price contract 固定承包价合同
firm ground 硬(土)地基;硬地面;坚硬基岩;坚实地基;坚实地基;坚固地基;稳固岩层;稳固地基;稳定地基
firm ice 固结冰;坚冰
firming agent 固化剂
firm joint 两脚规
firm-joint cal(l)iper 拧紧铰接卡钳;固接卡钳;卡钳

firm knot 连贯节;坚朽节;硬节
firm labor turnover 厂商劳动力周转率
firm lump-sum price 固定总价
firmly 坚固地
firmly believe 认定
firmly cemented 胶结坚固的;胶结牢固的
firmly related 坚相关
firmness 坚牢;坚固性;稳固
firmness of soil 土壤压实性;土壤压密性;土的压密性
firm offer 固定报价;坚定邀约;确定的要约;确定报价;实盘;实价;不能撤销的发价
firm of specialists 专业公司
firm order 确定订货
firm output 正常输出(功率);固定出力;可靠出力;恒定输出;稳定输出;保证输出功率;保证出力
firmoviscosity 固黏性
firm peak capacity 正常最大输出;可靠峰值容量;恒定最高容量;恒定峰值功率;稳定最大输出
firm peak output 稳定最大输出;稳定峰值功率输出
firm position 坚位置
firm power 可靠电力;可靠出力;基本电力;稳定功率;保证出力;安全功率
firm power energy 稳定功能
firm price 合同约定价格;确定的价格
firm-price bid 固定报价
firm-price offer 固定价格报价
firm-price porposal 固定价格投标
firm-price tender 固定价格投标
firm production function 厂商生产函数
firm quotation 确定报价;实盘
firm quote 确定报价
firm red heart 红心腐
firm rock 坚岩;稳固岩石
firm sale 确定销售
firm sale contract 确定销售契约;确定(的)销售合同
firm's car 公司用车
firm's dwelling unit 企业(职工)住宅
firm site 硬土场地
firm soil 硬土
firm soil stratum 硬土层
firm stratum 坚实土层
firm time 变硬时间
firm-to-firm variation 厂商间差异
firm top 稳固顶板
firm wall 稳固围岩
firm ware 固件;稳固设备;微程序语言;稳固件;程序包
firmware development 固件开发
firmware device 固件设备
firmware expansion module 固件扩充模件
firmware microcode 固件微代码
firmware microinstruction 固件微指令
firmware monitoring 固件监视
firmware option 固件选择
firmware structure 固件结构
firmware support 固件支撑
firm wood 硬木料
firm yield 可靠供水量;可靠产量;稳定产量
firn 永久冰雪;粒雪;万年雪
firn area 永久冰雪区;粒雪区;万年雪区;冰雪区
firn basin 粒雪盆(地);雪窖;冰窖
firn cover 永久积雪;粒雪覆盖;万年积雪
firncrash 雪冰崩落
firn edge 粒雪界;冰雪界线
fir needle oil 冷杉叶油
firn field 永久雪原;粒雪原
firn glacier 粒雪冰川
firng-pin 撞针
firn ice 万年雪冰;粒雪冰;冻结雪冰
firn ice avalanche 粒雪冰崩
firnification 粒雪形成(过程)
Firnkas system 弗恩加斯构造体系(采用预制承重墙,预应力楼板)
firn limit 雪线;永久积雪线;永久积雪(界)限;粒雪线
firn line 雪线;永久积雪线;永久积雪(界)限;粒雪线;粒雪界
firn-line zone 永久积雪区
firn snow 粒雪;陈雪
firn wind 冰川风
fir oil 冷杉叶油
fir pine 胶冷杉
fir pole 杉篙
firring 楔条;长条楔形木板;外墙内的空间;钉楔条;

垫平(块)
firry 冷杉木制的;多冷杉的;枞木制的
fir-shaped crown-bit 多级钻头
first 首要
first access point 第一入口处
first address 直接地址
first advance 超前工作面
first aid 紧急救护;急救;事故抢救
first-aid apparatus 急救设备
first-aid appliance 急救用品
first-aid attendant 基本救援人员
first-aid box 急救箱
first aid dressing 急救包扎
first-aid equipment 急救设备;基本救援设备
first aider 急救员
first-aid extinguishing 自救灭火
first-aid hose 小口径胶管水带;应急水龙带
first-aid injury 急救创伤
first-aid kit 急救(药)箱;急救包
first-aid measures 抢救措施
first-aid medicine 急救药品
first-aid organization 急救组织
first-aid outfit 小型水罐消防车;急救医疗器具
first-aid post 急救站
first-aid repair 紧急修理;紧急抢修;抢修;初步修缮;初步修理
first-aid room 急救室
first-aid standpipe 应急(消防)立管
first-aid station 救护站;急救站
first-aid supply 急救材料
first-aid team 救护队
first-aid to the injured 伤员急救
first-aid treatment 急救;急救措施
first alarm 一级火警;首次火警
first alarm assignment 第一出动
first-angle projection 第一角投影(法);第一象限投影法;首角投影法
first annealing 初次退火
first anode 第一阳极
first answer print 第一部完成拷贝
first application 第一次浇(沥青);初浇
first approach section 第一接近区段
first approximation 一级近似;一次近似值;首次逼近;第一近似值;第一次近似;初始近似;初次逼近;初步近似(值)
first approximation chart 古海图
first arrival 初至
first arrival correlation 初至对比
first arrival first attack 初期灭火
first arrival muting 初至切除
first arrival time 初至时间
first article configuration inspection 初样概略检验
first attack 初次发病
first attendance 头车
first audit 初次审计
first axiom of theology 流变学第一公理
first azimuth method 第一方位角法
first backcross generation 第一回交子代
first base 底基层;次基层
first batch 首批
first batch of newly designed products 初次产品
first beater 第一逐稿轮
first bid 头标;首次报价
first bit 钻子;钻头;小钻
first bite 初次腐蚀
first border 舞台台口排灯
first bottom 一级河床;漫水阶地;洪泛平原;河漫滩;第一河底;泛滥平原
first boundary-value problem 第一类边值问题
first boundary volume problem of potential theory 狄利克雷问题
first bounded interval measure 首次有界区间度量
first bower anchor 船首右大锚
first break 初至;波的初至
first break refraction method 初至折射法
first breaks aligned 初至对齐
first break time 初至时间
first category of commodities 一类商品
first category of goods 一类物资
first category vessel 一类容器
first charge 第一次投料
first choice 首选
first-class 第一流的;头等;上等的
first-class brick 一级砖
first-class cabin 一等舱;头等舱
first-class car 头等车【铁】
first-class certificate 一等证书
first-class charter 一级租船人
first-class commercial paper 一等商业票据
first-class current 第一类流
first-class diver 一级潜水员
first-class highway 一级公路
first-class mail 密封邮件
first-class maintenance 一级保养;一级维护
first-class ore 一级矿石
first-class paper 头等证券;头等票据
first-class passenger 一等舱旅客
first-class quality 头等质量;头等货
first-class road 一级道路
first-class wood pile 一级木桩
first coat 第一道涂工(油漆等);头道漆;头道灰;一度油漆;底涂(层);底漆;头道抹灰;刮糙
first coat in plastering work 底涂磨面底层;抹灰工作头道打底
first coat of plaster 头道灰
first come first-served 先到者先接受服务原则
first come first-served system 先到先办制
first come first-service 先来先服务
first coming pressure 首次来压
first consolidation 最初固结;初始固结
first contact 初亏
first contact of umbra 初亏
first copy-out time 首次拷贝时间;第一份复制品输出时间
first cost 原价;最初费用;最初成本;造价;开办费;建造费(用);基建费(用);初投资费用;初期投资;初期费用;初次费用
first cost method 最初成本法
first cost of locomotive and vehicle 机车车辆购置费
first-cost technology 一次性投资工艺
first countable topological space 第一可数拓扑空间
first course 头道锉纹
first crack 初始裂缝;初裂
first cracking strength 初裂强度
first crack stress 初裂应力
first crush 初始压碎
first crushing 初碎
first curing 第一次养护
first curvature 第一曲率
first curvature radius 第一曲率半径
first curvature vector 第一曲率向量
first data multiplexer 第一数据多路转接器
first degree 最低级的
first-degree burn 一度烧伤
first-degree discrimination 一级差别待遇
first demand 首次要求
first departure section 第一离去区段
first derivative 一阶导数
first derivative method 一次微分法
first derivative spectrum 一级导数光谱
first derived curve 导曲线
first detail group 第一细目栏
first detector 混频器;第一检波器
first developing 首次显影
first difference 一阶有限差;一阶差分;一次差
first difference method 一阶差分法
first difference of variables 变量的一阶差分法
first difference series 一阶差分数列
first difference transformation 一阶差分变换
first dilution ratio 一次贫化率
first discharge rate 初排出速率
first discolo(u)ration 初期变色
first dollar coverage 全额赔偿
first drafting 底图
first drawing frame 头道并条机
first dressing 第一次清理
first driver 初级激励器
first driver unit 初级主振部分
first eccentricity 第一偏心率
first electron lens 阴极透镜;第一电子透镜
first element of chain 链的首元素
first engineer 大管轮【船】
first entry table 初始输入表
first environment 第一环境
first equation of Maxwell 麦克斯韦第一方程
first estate 第一产业
first estimate-second estimate method 中天时间逐次逼近算法
first etching 初次腐蚀
first excursion probability 首次偏移概率
first felling-line 初次疏伐线
first filial generation 子一代
first filling 第一次投料;初期蓄水;初次蓄水
first filter 粗滤器
first fire 起爆器
first fire composition 点火药
first fit 首次适合;首次满足
first fit method 首次满足法
first fixer 粗木工(匠)
first fixings 初级装配;预装配;细木工程准备工作;嵌固件;木构件
first floor 底层;首层(美国);一层(指楼房);第一层楼(美国)
first floor plan 首层平面(美国);一层平面图(美国);第一层平面图(美国);二层平面(英国);第二层平面(英国)
first-flush-effect 初期雨水冲洗影响
first flush load 初次(暴雨)冲洗负荷
first frequency doubler 第一频率倍增器
first Fresnel zone 第一菲涅耳区
first frost 初霜
first fruits 初年度收益;初步成果
first Gaussian fundamental quantity 第一高斯基本量
first gear 一挡
first-gear engagement 一挡齿轮啮合
first generation 第一代;原件第一代
first-generation computer 第一代计算机
first generation container berth 第一代集装箱泊位
first glazing 新装玻璃
first-grade 甲级;头等的
first grade fuel 一级燃料
first-grade gravity base station for exploration geophysics 物探重力一级基点
first grade of integrated wastewater discharge standard 污水综合排放一级标准
first grade of national discharge standard 国家一级排放标准
first grade structure 一级构造
first grade timber 一级木材
first grading (路基的)初次整形;初次整平;初次平整;初步土方修整;初步土方平整
first grid 第一栅极
first growth 原始森林;处女林
first guess field 初估值场
first gust 首次阵风
first hand 直接调查;实地考察;第一手的
first-hand data 第一手资料;原始资料
first-hand information 原始资料;直接信息;第一手材料;第一手资料
first-hand material 第一手材料
first hand tap 初攻丝锥
first harmonic 一次谐波;基(谐)波
first harmonic oscillation 基本振荡
first harmonic resonance 基波谐振;基波共振
first horizontal derivative 水平一阶导数
first ice 初冰
first ice date 初冰出现期
first impression 初步印象
first-in-chain 链中第一单元;链首(部);链的首元素
first-in first-out 上推表;先到先服务;先进先出
first-in first-out method 先进先出法
firstings 灰浆试配
first-in last-out 先进后出
first inspection 首次检验
first instant 一开始
first in still here 先进留库盘存法
first in system-first served rule 先到先做规则
first integral 初积分
first intermediate host 第一中间宿主
first intermediate roll 第一层中间辊
first internode length 第一节间长度
first investigation 直接调查;实地考察
first ionization constant 一级电离常数
first iron 初铁
first item list 首项条目;首项表
first kind collision 第一类碰撞
first Lagrangian point 第一拉格朗日点
first latex rubber 头等橡胶
first law of motion 牛顿第一定律;第一运动定律

first law of the mean 第一中值定理
first law of thermodynamics 热力学第一定律
first leaky system 第一越流系统
first-level address 一级地址；直接地址；第一级地址
first-level calibration 一级刻度
first-level definition 第一级定义
first-level interrupt handler 一级中断处理程序
first-level lift 第一水平提升高度
first-level message 一级消息
first-level message member 一级消息成员
first-level outcome 一级结果
first-level purchasing and supply station 一级采购供应站
first-level statement 一级程序设计语句
first-level storage 一级存储器
first lien 第一留置权；首先留置权
first light 航用晨光时；熹微光
first line 最优良的；一级品；高档商品；第一线(的)
first line form advance 首行进页
first line management 初级管理人员
first line manager 第一线管理人员
first line selector 第一寻线器
first line servicing 一线检修
first line switch 第一预选器
first longitudinal vein 第一纵脉
first loop feature 首环功能部件；第一环路适配器
first lower punch 第一下冲模
first lunar meridian 月球本初子午线
first management step 第一个管理步骤
first market 初次市场
first mate 大副【船】
first meiotic division 第一减数
first member 左端
first member of equation 方程的左边
first meridian 零度经线；起始子午线；本初子午(圈)线
first minor 初余子式
first mixer 粗木工(匠)
first mode of vibration 第一振型
first mode resonance 第一型共振
first module program(me) 第一组件程序
first moment 一阶矩；一次矩
first moment area method 第一力矩面积法
first moment matrix 一阶矩矩阵
first moment of area 截面静矩
first momnet of mass 惯性静矩
first mortgage 优先抵押权；第一抵押
first mortgage bond 一类抵押债券
first motion 直接运动；初动
first-motion direction 初动方向
first-motion drive 直接传动(的)
first-motion hoist 直接传动提升机
first-motion studies 初动研究
first-motion wave 初至波
first motor man 机匠长
First National City Bank[Citibank] 花旗银行
first newel(post) 端柱(螺旋楼梯)
first night watch 头班；第一夜班(下午8~12点)
first nodal point 第一中心点
first non-equilibrium pumping-test formula 第一非平衡状态下抽水试验公式
first normal form 关系第一范式；第一规式；第一范式
first normal move-out stack 初次动校叠加
first notice day 通知期限的第一天
first of a kind plant 一等工厂
first-of-chain 链首(部)
first officer 大副【船】
first open water 开封初期；开冻日；首航；第一解冻日；初次解冻日；不负担港口解冻之风险
first operand 第一操作数
first-order 一级；一等；一阶；第一级；初指令；第一阶
first-order aberration coefficient 初级像差系数
first-order accuracy 一等精度
first-order adsorption rate constant 第一级吸附速率常数
first-order approximation 第一级近似
first-order autocorrelation coefficient 一阶自相关系数
first-order autoregressive equation 一阶自回归方程
first-order autoregressive scheme 一阶自回归型式
first-order benchmark 一等水准点
first-order benchmark level 一等水准仪

first-order condition 一阶条件
first-order constraint qualification 一阶约束品性
first-order control frame 一等控制网
first-order control network 一等控制网
first-order correction 一阶修正量；一级修正
first-order derivative 一阶导数；一次导数；第一次导数
first-order derivative spectrophotometry 一阶导数分光光度法
first-order design 一阶设计
first-order difference 一阶差分
first-order difference equation 一阶差分方程
first-order differential equation 一阶微分方程
first-order differential rectification 一阶微分修正
first-order diffraction spot 第一阶衍射斑点
first-order dispersive filter 第一阶波散滤波器
first-order effects 一阶效应
first-order elimination model 一级消除模型
first-order equation 一阶方程；一次方程(式)
first-order error 一阶误差；一次误差
first-order expression 一阶表达式
first-order factor 一次因子
first-order factor analysis 一阶因素分析
first-order feedback system 一阶反馈系列
first-order fluorescence 一次荧光
first-order force 一力力
first-order functional value 一阶函数值
first-order geodetic network 一等大地网
first-order homogeneous nonstationary process 一阶齐次非平稳过程
first-order instrument 一等测量仪器【测】；精密仪器
first-order kinetic reaction 一级动力学反应
first-order kinetics 一级消除动力学
first-order level(l)ing 一等水准(测量)
first-order level network 一等水准网
first-order linear function 一阶线性函数
first-order logic 一阶逻辑
first-order loop 一阶环路
first-order Markov process 一阶马尔可夫过程
first-order Markov scheme 一阶马尔可夫型
first-order Markov sequence 一阶马尔可夫序列
first-order metric(al)ga(u)ge 一级线纹米尺
first-order necessary condition 一阶必要条件
first-order of solution 第一次近似值
first-order perturbation 一阶摄动
first-order phase transition 一级相变；第一级相变
first-order point 一等(测)点
first-order precision 一等精度
first-order predicate calculus 一阶谓词演算
first-order predicate logic 一阶谓词逻辑
first-order rate process 一级速度过程
first-order reaction 一级反应
first-order reflection 一级反射
first-order river 一级河流
first-order second moment method 一级二阶矩法
first-order segregate 初级分离
first-order serial correlation 一阶序列相关
first-order spectrum 一级光谱
first-order staff 一等水准尺；精密水准(标)尺
first-order station 一等(测)站；一等(测)点
first-order stationary 一阶平稳
first-order stream 一级河流
first-order subroutine 一级子程序；第一级子程序
first-order system 一阶系统
first-order term 一次项
first-order test 一级检验
first-order theory 一阶理论；线性化理论；一级近似理论；高斯光学；第一级近似理论
first-order transition point 第一级转变点；第一级跃迁点
first-order traverse 一等导线
first-order traverse point 一等导线站；一等导线点
first-order traverse station 一等导线站；一等导线点
first-order triangulation 一等三角测量法；一等三角测量
first-order triangulation chain 一等三角锁；一等三角链
first-order triangulation net 一等三角网；一等三角测量网
first-order triangulation point 一等三角测量点
first-order wave 一级波
first-order weather station 一级气象站；一等测候站
first-overkettle 除铜精炼锅
first overtone 一次倍频

first page indicator 首页指示器
first-party audit 第一方审核
first pass 一遍通过；第一孔型；第一遍(扫描)；初次通过
first passage time 首次经过时间；初过时间；初通时间
first-pass own code 第一遍扩充工作码
first-pass sorting 第一遍分类
first phase 初相；第一期(的)
first phase design 第一次设计
first phase of construction 一期工程
first phase preparations 前期准备工作
first pickling 初次酸洗
first piece 粗加工工件
first point of Aries 春分点
first point of Cancer 夏至点
first point of Capricornus 冬至点
first point of Libra 秋分点
first polish 底层抛光；底层磨光
first pollution 第一污染
first pour 第一期浇灌
first power 一次幂；一次方；开始发电；第一批机组发电
first preliminary tremor 第一初期微震
first pressing 第一次压制
first pressure 初始压力
first principal strain 第一主应变
first principal stress 第一主应力
first principle 基本原则；首要原则
first print 试印样
first priority 绝对优先项目
first probability distribution 第一概率分布
first proof 初样
first prospecting stage 初勘阶段
first put on trial sale 先行试销
first quality 优质；一级质量；一等品；上等品质
first quality brick 一等砖
first quality ware 一级品
first quarter 上弦月；上弦
first quarter neap tide 上弦小潮
first quartile 第一四分位数
first radiation constant 第一辐射常数
first-rank galaxy 一级星系
first rate 最上等的；头等的；第一流的；上等的
first-rate quality 头等质量
first-rater 头等货
first reaction of cyanide decomposition 氰化物分解一次反应
first record of a set 集的第一记录
first reduction gear pinion 第一级减速小齿轮
first reduction gear wheel 第一级减速大齿轮
first refusal(right) 优先取舍权；优先决定权
first-remove subroutine 第一级子程序
first responder 第一出动人员
first response district 责任区
first revisal 初次改样
first riverbed 一级河床
first roller 第一机架辊
first run 正面印刷
first runnings 初馏物
first-run slag 初流渣
firsts 一级品；一等品
firsts and seconds 一级和二级
first-scattering angle 第一散射角
first/second-order leveling 一二等水准测量
first separator 第一次分裂装置
first septum 第一隔
first set graft reaction 初次移植物反应
first set rejection 初次排斥反应
first several modes 最低几个振型
first side 先加工面
first signal 第一信号
first signal system 第一信号系统
first sintering 第一次烧结；初次烧结
first slab 第一块铺石板
first space laboratory project 第一次空间实验项目
first speaker 发话端
first speed 一挡；初速
first speed gear 一挡齿轮
first squeeze 初始压碎；初始挤压
first-stage 第一期(的)；第一阶段；初期
first-stage anaerobic biofilter 第一级厌氧生物滤池
first-stage anaerobic treatment 第一级厌氧处理
first-stage annealing 第一阶段退火

first-stage biochemical oxygen demand 一级生化需氧量;第一阶段生化需氧量;第一级生化需氧量
first-stage biological aerated filter 第一级曝气膜生物反应器
first-stage biological contact oxidation 第一级生物接触氧化
first-stage biological treatment process 第一级生物处理工艺
first-stage cofferdam 第一期围堰
first-stage concrete 第一期混凝土
first-stage construction 一期工程
first-stage development 第一阶段施工;第一阶段发展
first-stage graphitization 第一阶段石墨化
first-stage installation (电站的)第一期装机
first-stage intermittent aeration membrane bioreactor 第一级间歇曝气膜生物反应器
first-stage ionization 一级电离
first-stage lining 第一次衬砌
first-stage of creep 一段蠕变
first-stage of primary design 初设一期
first-stage oxidation treatment 一级氧化处理
first-stage oxidizing process 第一级氧化工艺
first-stage polymerization 初级聚合
first-stage project 一期工程
first-stage reduction 一次缩小
first-stage rotating biological contactor 第一级生物转盘
first-stage terrace 一级阶地
first-stage/three-yard marshalling station 一级三场编组站【铁】
first step 首阶梯楼梯踏步;第一级
first step of refusal 优先取舍权
first stone 基石
first stop marker 第一停车信标
first stor(e)y 二楼(英国);第一层楼(美国)
first stress level 初应力大小
first stud gear 第一变速齿轮
first summation 一次和分
first-surface decorating 正面装饰
first-surface mirror 外表面镀膜反射镜;表面镀膜镜;前表面反光镜
first-surface painting 正面涂漆
first surplus reinsurance treaty 第一益额分保合同
first swing 初动
first symbol 开头符号;首符
first tap 头攻丝锥;粗加工丝锥;初加工丝锥
first taxonomic classification 第一个分类学上的分类
first term 开头项;首项
first terminal 开头终结符
first time step 起步
first-to-market 首先进入市场
first treatment 一级处理;初步处理
first trial cut 第一次试切削
first trust deed 优先贷款信托文契
first twist 初捻
first type surface 主表面
first undercoat 第一层底漆
first user (住房落成后的)第一住户;(产品的)第一使用者
first variation 初级变分
first vertical derivative 垂向一阶导数
first vibration mode 第一振型
first virial coefficient 第一维里系数
first visible crack 可见初始裂缝
first watch 头班
first-water 最高质量或光泽最纯(钻石或珍珠等宝石);(钻石等的)第一水;最优秀;第一流
first watering 初灌
first weight 初始压碎;初始加荷;顶板初次来压
first winding 初级绕组
first working 采准工作
first work piece 第一个工件
First World Climate Conference 第一次世界气候会议
first-year commission 初年度代理权
first year ice 当年冰
first yield 首次屈服
fir tamarack 落叶树
firth 狭窄海湾;港湾;河口(湾);三角港
fir-tree bit 多刃式扩孔旋转钻头(刃排列在塔形钻头后面)
fir-tree connection 枞树形叶根固定

fir-tree crystal 枝晶;树枝晶
fir wood 冷杉木;枞松木
fiscal action 财政措施
fiscal agency agreement 财务代理协议;财务代理契约
fiscal agent 财政代理人;财务代理人(一般指银行或信托公司)
fiscal and monetary policy 财政与货币政策;财政金融政策
fiscal audit 财政审计
fiscal barrier to trade 财政贸易壁垒
fiscal budget 财政预算
fiscal burden 财政负担
fiscal capacity 财政能力
fiscal capital 金融资本
fiscal charges 财务支出
fiscal closing date 会计期终结账日
fiscal committee 课税委员会
fiscal concept 财政观念
fiscal control 财政监督;财务控制;财务监督
fiscal decentralization 财政的地方分权
fiscal discretionary power 财政决定权
fiscal dividend 财政支用余额(美国);财政盈余
fiscal drag 财政拖累
fiscal expenditures 财政支出
fiscal federalism 财政联邦主义
fiscal function 财政功能
fiscal illusion 财政幻觉
fiscal imbalance 财政不平衡
fiscal incentive 财政鼓励
fiscal institution 财税机构
fiscal insurance 赋税保险
fiscal investment and loans 财政投资及贷款
fiscal issuance 财政发行
fiscalist 财政主义者
fiscal jurisdiction 财政管辖权
fiscal leverage 财税杠杆
fiscal levy 财政税收
fiscal-monetary mix 财政金融政策相结合;财政货币混合政策
fiscal monopoly 财政独占
fiscal multiplier 财政乘数;财金乘数
fiscal or monetary action 财政或货币措施
fiscal period 财务期间
fiscal plan 财务计划
fiscal policy 财政政策
fiscal privilege 财政特权
fiscal process 财政手续程序
fiscal relation 财政关系
fiscal resources 财源
fiscal responsibility 财政责任;财务责任
fiscal restraint 财政约束;财政紧缩
fiscal revenue 财政收入
fiscal solvency 财务实力
fiscal subsidy 财政补贴
fiscal system 财政制度
fiscal taxation 财政税收
fiscal welfare benefits 财政福利津贴
fiscal year 会计年度;财政年度;财务年度
fischerite 柱磷铝石
Fischer's yellow pigment 钴黄颜料
fisetin 漆树黄酮
fish 加副木夹牢;钓锚器;吊锚工作
fishable population 可捕种群
fishable size 可捕鱼类大小
fish adhesive 鱼胶
fish-aggregating deivce 鱼类聚集装置
fish attracting device 诱鱼设备
fish auction 鱼市场
fish avoidance reaction 鱼类逃避反应;鱼回避反应
fish back 锯齿板;梳齿板
fish back spring leaf 鱼背弹簧片
fish backs spreader 鱼背展幅机
fish-bait hold 鱼耳舱
fish bar 鱼尾板;夹杆
fish barrier 拦鱼栅
fish basket 鱼篓
fish beam 夹板接合梁;鱼尾板连接梁;鱼腹式梁
fish behavior 鱼的习性
fish-bellied bar 鱼腹杆;等强度梁
fish-bellied beam 鱼腹式梁;鱼腹梁
fish-bellied girder 鱼腹式大梁;鱼肚梁
fish-bellied purlin(e) 鱼腹式檩条
fish-belly 鱼腹式的

fish-belly bottom flange 鱼腹式量下凸缘
fish-belly flat car 鱼腹平车
fish-belly gate 鱼腹式大门;上开式大门
fish-belly girder 鱼腹式大梁
fish-belly hinged-leaf gate 鱼腹式大门;上开式大门
fish-belly rail 鱼腹式横梁
fish block 吊锚滑车
fish bolt 接头(夹板)螺栓;鱼尾(板)螺栓;对接螺栓;燕尾螺栓;开脚螺栓
fishbolt hole 鱼尾板螺栓孔
fishbolt nut 鱼尾板螺母
fishbone 鱼骨形断裂;鱼骨形
fishbone antenna 鱼骨天线;鱼骨形天线
fishbone cracking test 鱼骨形抗裂试验
fishbone earth 鱼骨形接地
fishbone receiving antenna 鱼骨形接收天线
fishbone-stitch 鱼骨形针迹
fishbone type antenna 鱼骨形天线
fish boom 收锚杆
fish borne food poisoning 鱼类食物中毒
fish bowl planning 玻璃鱼缸式规划(一种接受公众监督具有相当透明度的规划方法)
fish box area.
fish box mend room 鱼箱修理间
fish box yard 鱼箱堆场
fish bred at the bottom of sea 海底繁殖鱼类
fish breeding and poultry raising 养鱼业
fish breeding ground 鱼类繁殖场
fish breeding installation 养鱼设备
fish canal 鱼道
fish canning factory ship 鱼罐头加工船
fish carrier 运鱼船
fish catch 渔获量;捕鱼量
fish chair 轨座
fish channel 鱼道
fish chute 鱼滑槽;过鱼道
fish collection facility 集鱼设施
fish collection ship 集鱼船
fish collection system 集鱼系统
fish concentration 鱼类群聚;鱼类集群
fish corral 鱼栅
fish crop 鱼产量
fish cultivating 鱼类养殖
fish cultivation in rice field 稻田养鱼
fish culture 鱼类养殖;养鱼
fish culture in paddy field 稻田养鱼
fish culture in reservoir 水库养鱼
fish culture in river 河道养鱼
fish culture in stream 河道养鱼
fish davit 收锚杆;半吊杆
fish detection 鱼群探测
fish detector 鱼群探测装置;鱼群探测仪
fish diagram 鱼图
fish diversion 鱼群导流
fish dock 渔船码头;鱼船码头
fished beam 接合梁
fished bolt 夹板螺栓
fished joint 夹板结合;夹板接头;夹板接合;鱼尾板连接;鱼尾板接口;鱼尾板接合
fished resources 捕捞资源
fished stock 捕捞资源
fish elevator 鱼升道
fish entrance 鱼入口
fisher 渔船;捕鱼器具
fish eradicant 鱼类铲除剂
Fisher carbonhydrogen-nitrogen analyzer 费歇尔碳氢氮分析仪
fisher carrier 鱼类运输船
Fisher diameter 费歇尔直径
Fisher Equation 费歇尔方程式
Fisher exact probability test 费歇尔恰当概率试验;费歇尔精确概率分布检验
fisherfolk 渔工;渔夫
Fisher-Hepp rearrangement 费歇尔—海勃重排
Fisher-Hinnen method 费歇尔—欣纳法
fisheries control 捕捞控制
fisheries expert 渔业专家;水产专家
fisheries habitat 鱼类生境
fisheries institute 水产学院
Fisheries Law of the People's Republic of China 中华人民共和国渔业法
fisheries legislation 渔业法规
fisheries oceanography 水产海洋学
fisheries water 水产用水

Fisher-Irwin test 费歇尔—欧文检验
Fisher level 费歇尔水准仪
Fisher-MacMichal visco(si)meter 费歇尔—麦克米查尔黏度计
fisher(man) 渔民;渔人;渔夫
fisherman's anchor 渔具锚
fisherman's bend 渔人结;锚结
fisherman's fender 软碰垫;索环碰垫
fisherman's knot 渔网结
fisherman's soap 渔民皂
fisherman staysail 渔人支索帆
fisherman's walk 甲板上狭小场所
Fisher model 费歇尔模型
Fisher particle size 费歇尔粒度
Fisher quantity index 费歇尔物量指数
Fisher's acid 费歇尔酸
Fisher's correlation-coefficient transformation 费歇尔相关系数变换
Fisher's distribution 费歇尔分布
Fisher's ellipsoid 费歇尔椭球体
Fisher's equation 费歇尔交换方程式
Fisher's esterification 费歇尔酯化作用
Fisher's F distribution 费歇尔F分布
Fisher's index 费歇尔指数
Fisher's indole synthesis 费歇尔吲哚合成(法)
Fisher's osazone reaction 费歇尔成脎反应
Fisher-Speier esterification 费歇尔酯化作用;费歇尔—斯皮尔酯化作用
Fisher's phenylhydrazone reaction 费歇尔苯腙反应
Fisher's projection formula 费歇尔投影式
Fisher's reagent 费歇尔试剂
Fisher's reagent for water 费歇尔测水试剂
Fisher's spheroid 费歇尔椭球体
Fisher's synthesis 费歇尔合成
Fisher's test 费歇尔检验法
Fisher's transformation 费歇尔变换
Fisher's Tropsch process 费歇尔—特罗普歇法
Fisher sub-sieve sizer 费歇尔微筛分粒器;费歇尔微粒测定仪;费歇尔微粒测量仪
Fisher-Tropsch hydrocarbons 费托法合成的烃类
Fisher-Tropsch method 费托法
Fisher-Tropsch process 费托法
Fisher-Tropsch reaction 费托反应;费托法反应
Fisher-Tropsch synthesis 费托法合成
Fisher-Tropsch wax 费托合成过程中得到的蜡
fishery 渔业(学);渔权;渔场
fishery administration 渔政;渔业管理(机构)
fishery administration ship 渔政船
fishery agreement 渔业协定
Fisher-Yates test 费歇尔—耶茨检验
fishery base 渔业基地
fishery biology 渔业生物学
fishery chart 渔场图
fishery conservation 渔业保护;水产资源保护
fishery conservation zone 渔业资源保护区;鱼类保护区
fishery examination boat 渔业实验船
fishery fleet 渔船队
fishery forecast 渔情预报
fishery guidance ship 渔业指挥船;渔业指导船
fishery harbo(u)r 渔业港
fishery harvesting 渔获
fishery hazard 渔业公害
fishery hydrography 渔业水文学
fishery industry 渔业;水产业
fishery inspecting vessel 渔政船
fishery inspection ship 渔业监督船
fishery law 海捞法规;捕捞法规
fishery loss 渔业损失
fishery management 渔业经营;渔业管理
fishery mother-ship 渔业母船
fishery organization 渔业组织机构;渔业组织单位;渔业组织部门
fishery plan 渔业计划
fishery policy 渔业政策
fishery port 渔港
fishery processing ship 鱼类加工船
fishery production 渔业生产;渔业产量
fishery program(me) 渔业计划;渔业规划
fishery protecting cruiser 护渔巡洋舰
fishery protection 渔业保护;水产资源保护
fishery protection ship 渔业保护船
fishery quay 渔码头;渔船码头

fishery research 渔业研究;渔业分析;渔业调查;渔场考查
fishery research ship 渔业研究船
fishery research trawler 渔业研究用拖网渔船
fishery research vessel 渔业调查船
fishery resources 渔业资源;水产资源
fishery right 渔业权
fishery supervision and management 渔业监督管理
fishery supply ship 渔业补给船
fishery survey ship 渔业调查船
fishery training ship 渔业训练船
fishery vessel 渔业船
fishery water 渔业用水
fishery water body 渔业水体
fishery water quality assessment 渔业用水质量评价
fishery waters 渔业水域
fishery wharf 渔港
fishery zone 捕鱼区
fish eye 鱼眼;白点;生灰斑
fish-eye camera 水下照相机;鱼眼照相机;水中照相机;水下摄影机;超广角摄影机
fish-eye lens 鱼眼透镜
fish eye stone 鱼眼石
fish-eye type objective lens 鱼眼型物镜
fish eye view 底视图
fish facility 鱼道
fish factory ship 鱼品加工船;鱼获物加工船
fall fall 收锚滑车索
fish farm 鱼类孵化场;鱼场;养鱼场;鱼苗育苗场
fish farming 鱼类养殖
fish-farming with sewage 污水养鱼
fish fauna 鱼类区系
fish finder 鱼群探测仪
fish flour 鱼粉
fishfood organism 鱼食生物
fish for 捞回
fish fry 鱼苗
fishgarth heck 鱼栅
fish gelatin 鱼胶;鱼固胶
fishgig 鱼叉
fish glue 鳔胶;鱼胶
fishgraph 鱼探仪
fish guiding device 导鱼设备
fish habitat protection 鱼类生境保护
fish handling and processing shed 鱼类加工车间;水产加工车间
fish handling facility 运鱼设备
fish harvesting 渔业收获
fish hatchery 鱼类养殖场;鱼类孵化场
fish haven 渔堰;鱼苗育苗场
fish hawk 鱼鹰
fish hoist(er) 升鱼机
fishhold 鱼舱
fish hold bilge 鱼舱污水沟
fishhook needle 钩形缝合针
fishing 渔业;打捞;捕鱼
fishing and catching gear 渔捞设备
fishing and factory ship 捕鱼及加工船
fishing and freezing vessel 鱼捞冷冻船;捕鱼兼冷藏船
fishing area 渔区
fishing a spar 裂纹木杆加副木夹牢
fishing a yard 桁上缠小夹条
fishing bank 鱼群集中地;鱼礁;鱼场
fishing ban period 禁渔期
fishing base-plate 鱼尾板
fishing basket 捞爪;打捞篮
fishing boat 鱼船;捕鱼作业标
fishing boom 吊鱼吊杆
fishing box 接线盒;连接箱
fishing catch 一把抓
fishing chart 渔业用图;鱼场图
fishing chief 渔船船长
fishing collar 打捞钻铤
fishing depot ship 渔业母船
fishing die(s) 打捞卡瓦;铅模打印
fishing dory 小型平底渔船
fishing dust 捞砂【岩】
fishing effort 捕捞效果
fishing fleet 渔船队
fishing for casing 套管打捞;打捞套管
fishing for tubing 打捞油管

fishing garth 渔堰
fishing gear 打捞装置;捕捞设备
fishing ground 渔区;渔场;鱼场
fishing harbo(u)r 渔港
fishing head 打捞头
fishing hook 打捞钩
fishing jar 渔罩;渔容器;渔篓;打捞振击器;打捞振动杆
fishing job 渔业;捕捞;打捞作业【岩】
fishing junk 渔帆船
fishing lamp 渔船工作灯;捕鱼灯
fishing light 渔灯;渔船工作灯;捕鱼灯
fishing limit 渔区范围
fishing line 渔网绳;钓鱼线;钓绳
fishing magnet 磁力打捞器
fishing(male)templet 鱼尾样板
fishing monitoring 渔业监测
fishing motor boat 机动渔船
fishing neck 打捞颈部
fishing net 渔网
fishing net machine 渔网编织机
fishing nipple 打捞公锥
fishing operation 打捞作业【岩】;打捞落鱼;捕鱼工作
fishing pan 蒸盐锅
fishing patrol boat 渔业监督船
fishing permit 捕捞许可证
fishing place 渔场;捕鱼场所
fishing port 渔港
fishing port district 渔港区
fishing port law 渔港法
fishing quay 渔(用)码头;渔船码头
fishing rack 打捞工具船
fishing reef 鱼礁
fishing region 渔区
fishing right 捕鱼权
fishing robot 打捞机器人
fishing rod 钓鱼杆;打捞杆
fishing rod cloth 钓鱼杆基布
fishing season 渔汛;捕鱼期;捕鱼季节
fishing service 打捞作业队
fishing service zone 渔港后勤区
fishing ship 渔船
fishing site 捕鱼位置
fishing smack 小渔船;渔船;鱼船
fishing socket 打捞母锥
fishing space 鱼尾板空间
fishing spear 渔枪;渔矛;打捞矛
fishing stores 捕鱼物料
fishing strake 渔栅
fishing string 反丝扣钻杆柱
fishing surface 鱼尾板安装面
fishing tackle 渔具;钓鱼具;打捞索具
fishing tap 打捞丝锥;打捞公锥
fishing taper 打捞丝锥
fishing technology 捕鱼工艺
fishing time 打捞作业时间
fishing tongs 捞取钳
fishing tonnage 渔船吨位
fishing tool 钻杆打捞器;事故处理工具【岩】;打捞工具
fishing vessel 渔轮;渔船
fishing vessel base-ship 渔船移动基地
fishing vessel clause 渔船条款;捕鱼船条款
fishing waters 渔场;鱼场;捕鱼水域
fishing wharf 渔业船;渔船头;鱼船码头
fishing wire 鱼丝;螺旋形钢丝
fishing work 打捞作业(钻探技术);捕捞工作
fishing zone 渔(业)区;鱼区;捕鱼区
fish job 打捞工作
fish joint 镶接头;甲板栏杆固定板;夹板接合;鱼尾板接合
fish kill 屠杀鱼群;鱼灾
fish ladder 鱼梯;鱼道;通鱼路;鱼鳞梯
fish landing 卸鱼码头;鱼(船)码头;装鱼码头
fish landing and products processing area 卸鱼及鱼加工区
fish landing capacity per day 泊位日卸鱼能力
fish lead 铅鱼
fish-lens 鱼眼透镜
fish lift 升鱼机
fish-line conductor 螺线形导线
fish locating 探鱼顶
fish lock 鱼闸;过鱼闸

fish lore 鱼类知识
fish-luring light 诱鱼灯
fishman 打捞工
fish manure 鱼肥
fish meal 鱼粉
fish meal carrier 鱼粉船
fish-meal hold 鱼粉舱
fish migration 鱼类洄游
fishmonger 鱼市
fish mo(u)lds 鱼霉
fish mouth 鱼口状开口;鱼口形开口
fish mouthing 裂痕
fish-mouth-shaped dike[dyke] (治鱼工程中的)鱼嘴
fish mouth toe close 鱼嘴式缝头
fish net buoy 渔网区浮标;网具浮标
fish net mesh size 鱼网孔径大小;鱼网规格
fish noise 鱼群噪声
fish offal 鱼类废弃物
fish oil 鱼油
fish oil fatty acid 鱼油脂肪酸
fish oil soap 鱼油皂
fish-oil stuffing 鱼油加脂
fish-oil tannage 鱼油鞣
fish orientation 鱼的定向
fish paper 鱼皮纸;鱼膏纸;青壳纸
fish parasite 鱼的寄生物
fish pass 鱼道;过鱼道;放鱼道
fish pass facility 过鱼设施
fish-passing facility 鱼通过设施;鱼道设施
fish pass structure 过鱼建筑物
fish phase diagram 鱼图
fish piece 鱼尾板;接合板
fish pier 渔船码头;鱼(船)码头
fish-pit 鱼溜
fishplate 镶接板;鱼尾板;接合板;夹(接)板;接头夹板(鱼尾板)【铁】;接(合)夹板;铺ရ托梁
fishplate and bolts splice for timber pile 鱼尾木夹板和螺栓接长木桩
fishplate bar 鱼尾板;接板
fishplate bolt 鱼尾板螺栓;对接轨条螺栓
fishplate friction 鱼尾板摩擦力
fishplate joint 鱼尾板接合;鱼尾板连接;鱼尾板接头
fishplate mill 鱼尾板轧机
fishplate pass 接夹板孔型;鱼尾板孔型
fishplate rail joint 鱼尾板钢轨接头
fishplate splice 鱼尾板镶接;鱼尾板拼接;鱼尾板接合
fish poison 鱼毒
fish poisoning 鱼中毒
fish-pole antenna 钓竿状天线
fish pollutant 鱼污染物;鱼类污染物
fish pomace 鱼渣
fish pond 鱼塘;鱼池;养鱼塘;养鱼池
fish population 鱼类种群
fish population dynamics 鱼类种群动态
fish pound 拦鱼堰
fish processing 鱼类加工;鱼加工
fish processing plant wastewater 鱼类加工厂废水
fish processing ship 鱼品加工船;鱼加工船
fish protein concentrate 鱼粉
fish pump 吸鱼泵
fish quality standard 鱼品质量标准
fish rack 鱼栅;拦鱼栅
fish reconnaissance boat 鱼群侦察船
fish remediation 鱼类恢复
fish reproduction 鱼类繁殖
fish resources 鱼类资源
fish rib 鱼肋
fish-roe crackle 鱼子纹
fish-roe glaze 鱼子釉
fish-roe yellow 鱼子黄
fishroom 鱼舱
fish salad 鱼片色拉
fish sample 鱼类样本
fish scale 鱼鳞状脱皮;鱼鳞物;鱼鳞饰;鱼鳞;鱼鳞形垢;鳞斑
fish scale fracture 鱼鳞状断口;鳞状断口;鳞形裂缝
fish scale-like mark 洞壁鱼鳞痕
fish scale pit 鱼鳞坑
fish scale powder 鱼鳞粉;珠光粉
fish-scale tiling 鱼鳞形盖瓦
fish scaling 起鳞;剥釉;爆鳞
fish-scaling effect 起鳞现象

fish school 鱼群
fishscope 鱼群示波器
fish scrap 鱼渣
fish screen 鱼栅;鱼筛;拦鱼栅网;护鱼隔网(堤坝放水口道用)
fish searching 鱼群侦察
fish searching ship 鱼群搜索船
fish seed 鱼种
fish shackle 锚杆环;平衡环;吊锚圈
fish shelter 鱼礁
fish-skin 鱼皮
fish soluble 鱼汁
fish sound 鱼声
fish stair(case) 鱼道
fish stakes 鱼栅;拦鱼网栅
fish stearine 鱼油硬脂
fish stock 鱼群;鱼类资源;鱼类(群体)
fish tackle 收锚复绞辘
fish tag 鱼类标志
fishtail 划伤;鱼尾状楔块;(锚固件的)鱼尾端;(拔钉子的)锤头;鱼尾状的;鱼尾槽;鱼尾状物;鱼尾飞边
fishtail bit 鱼尾钻;鱼尾形钻头
fishtail bolt 鱼尾螺栓;开脚螺栓;鱼尾紧固螺栓;燕尾螺栓
fishtail burner 蝙蝠翼式灯
fishtail cutter 鱼尾铣刀
fish-tailed downcomer 鱼尾喷管
fish-tailed end 燕尾开脚
fish-tailed fixing 鱼尾形紧固件
fishtail end 轧件的鱼尾端
fishtail head 鱼尾灯头
fish-tailing 鱼尾状左右摆动
fishtail jet 鱼尾式喷嘴
fishtail plate 鱼尾板
fish tail regulator 鱼尾杆调节器
fishtail spray 鱼尾喷嘴
fishtail train 鱼尾形拖裾
fishtail twin 鱼尾形双向阀
fishtail-type kneader 鱼尾形捏合机
fishtail wind 不定向风
fish tallow 鱼脂;鱼蜡
fish tankage 鱼渣粉
fish tape 波纹状板带;敷线用牵引线
fish top 鱼尾式灯头
fish toxicity test 鱼类毒性试验
fish track 鱼形径迹
fish transport ship 运鱼船
fish trap 渔栅
fish-type body 流线体
fish unloading machinery 卸鱼机械
fish viewing window 观鱼窗
fishware 鱼舱
fish waste 鱼废物
fish-water separator 鱼水分离器
fish way 鱼道;过鱼道;放鱼道
fishway exit 鱼道出口
fishway pump 鱼道充水泵
fish weir 渔堰;鱼栅
fish well 鱼舱
fish wheel 捕鱼水车
fish winding 螺旋绕组
fish wire 电缆牵引线
fish wool 海丝
fish works 鱼类制品厂
Fisk window 费希尔科窗(一种专利隔声商品)
fisser 裂变物质
fissible fuel 裂变燃料
fissible material 可裂变材料;分裂性物质
fissile 裂变性;可裂的
fissile component 可裂变成分
fissile isotope 可裂变同位素;可分裂同位素
fissile material 裂变材料;可裂变物质;核燃料
fissile material production reactor 裂变材料生产堆
fissile phase 裂变相
fissile region 裂变区
fissile shale 易劈页岩;易剥裂页岩;泥片岩
fissile structure 剥离构造
fissility 易裂性;可劈性;可裂性;可裂性;分裂性
fission 裂开;裂变;剥离;剥开
fissionability 裂变能力;裂变度
fissionable 裂变的;可裂变的;可分裂的
fissionable atom 裂变原子
fissionable material 裂变物质;裂变材料;核分裂

性物质
fissionable substance 裂变物质
fission chain 裂变链
fission chamber 裂变室;裂变电离室
fission chemistry 裂变化学
fission contribution 裂变作用
fission corrosion 裂变腐蚀
fission counter 裂变电离室
fission cross-section 裂变截面
fission dadiochemistry 裂变放射化学
fission detector 裂变径迹探测器
fission energy 裂变能(量)
fissioner 裂变材料;可裂变材料
fission fraction 裂变份额
fission fragment 裂变碎片;分裂碎块
fission fragment track discriminator 裂变碎片径迹甄别器
fission fuel 裂变燃料;核燃料
fission gas 裂变气体
fission heat 裂变热
fissioning uranium 裂变铀
fission level 裂变能级
fission material 裂变物质
fission neutron 裂变中子
fission parameter 裂变参量
fission poison 裂变毒物
fission process 裂变过程
fission product 裂变产物
fission product contaminant 裂变产物污染物
fission product damage 裂变产物损害
fission product fallout 裂变产物落下的灰;裂变产物沉降物
fission product poisoning 裂变产物中毒
fission product waste 裂变产物废物
fission reaction 裂变反应;分裂反应
fission reactor 核反应堆
fission segment 裂变碎片
fission spectrum 裂变谱
fission theory 分裂字说
fission threshold 裂变阈
fission track ages 裂变径迹年龄
fission track count 裂变径迹计数
fission track dating 裂变径迹测年;裂变轨迹年代测定法
fission track dating method 裂变径迹测年法
fission track method 裂变径迹法
fission-type reactor 核裂变反应堆
fission weighing 裂变权重
fission yield 裂变产额
fissium 裂变产物合金
fissura petro-occipitalis 岩枕裂
fissura secunda 次裂
fissura sphenopetrosa 蝶岩裂
fissuration 龟裂;裂隙形成;形成裂缝
fissura tympanosquamosa 鼓鳞裂
fissure 隙缝;裂隙;裂纹;裂痕;裂缝;微缝;分裂
fissure bark 片裂树皮
fissure basalt 裂隙玄武岩
fissure cave 裂隙洞
fissure cavity 裂洞
fissure-clay aquifer 裂隙黏土含水层
fissure conduit 裂隙通路;裂隙涵管
fissured 裂缝的
fissured acoustic(al)tile 裂纹状吸声板;毛毛虫状吸声板
fissured aquifer 裂隙含水层
fissured clay 裂隙黏土;裂缝化黏土;裂缝化黏土
fissured degree of rock 岩石裂隙化程度
fissure density 裂隙密度
fissure development locus 裂隙发育部位
fissure diagram 裂隙统计图
fissure drainage 裂隙排水
fissured rock 有裂缝的岩石;裂隙岩石;破碎岩
fissured structure 裂缝结构;裂隙构造
fissured tongue 沟纹舌
fissured volcano 裂隙式火山
fissured zone 裂隙地带
fissure eruption 裂隙(式)喷发;裂缝喷发
fissure factor of core 岩芯裂隙率
fissure fault 裂隙断层;裂缝断层
fissure filling 裂隙填充;裂缝填充;裂缝充填(物)
fissure flow 裂隙(式熔岩)流;裂缝喷溢熔岩流
fissure grouting 裂隙灌缝;裂隙灌浆
fissure in bed rock 基岩裂隙

fissure in ground 土裂隙;土裂缝
fissure intersection 裂隙交切
fissure-karst aquifer 裂隙岩溶含水层
fissure-karst water 裂隙岩溶水
fissure lime stone land 石灰岩裂隙地
fissure medium 裂隙介质
fissure microphotometer 裂隙显微光度计
fissure observation 裂缝观测
fissure occupation 裂隙充填
fissure of coal 煤的裂隙
fissure of displacement 错缝
fissure permeability 裂隙渗透率
fissure plane 裂隙面
fissure porosity 裂隙孔隙度
fissure ratio 裂隙率
fissure ratio along a line 线裂隙率
fissure ratio on plane 面裂隙率
fissure round bur 裂圆钻
fissure sealant 窝沟封闭剂
fissure spacing-span ratio 隙跨比
fissure spring 裂隙泉;裂缝泉
fissure system 裂隙组;裂隙系
fissure tectono-geochemistry 裂隙构造地球化学
fissure type 裂隙状;裂隙式
fissure vein 裂隙脉;裂缝(矿)脉
fissure volcano 裂隙式火山
fissure water 裂隙水;裂缝水;缝隙水
fissuring 节理;裂隙
fist products 拳头产品
fist-shape beam 霸王拳【建】
fistuca 打桩锤;打桩器(古代)
fistula of perineum 海底漏
Fit 非特
fit 修齿;密接;配套;吻合
fit bolt 配合螺栓;接触螺栓;固定螺栓
fitch 组合量板;镶板;扁漆刷;小毛刷;长柄小刷;长柄漆刷
fitch liner 小笔刷
fit clearance 配合间隙
fit currency 可使用的通货
fit diver 适潜潜水员
fit for demolition 宜于拆毁的
fit for depasturing 适于放牧
fit for human habitation 适于居住的
fit for testing 适于试验的
fit for wrecking 宜于拆毁的
fit grafting 合接
fit hypothesis 拟合假设
fit into 装入
fit joint 紧套接口;套筒连接
fit key 配合键
fitment 设备;配件;附件;家具
fitness 适合度;适当性
fitness figure 适合度因数
fitness of environment 环境适度;环境适应(性)
fit of the plunger 柱塞配合
fit on 装配上;外配合
fitoncidin 葱素
fit on pad (便桶用的)硬木座圈
fit out 装备;配备;装配
fit-over 装上;安上
fit quality 配合质量
fit rod 量孔深度的杆
fit strip 镶条;配合垫板
fit system 配合制度
fittage 装配任务;杂费;间接费(用);填舱货
fitted assembly 装配;拼装
fitted bearing 配合轴承
fitted block 砌合方块
fitted bolt 配合螺栓
fitted capacity 装配容量
fitted carpet 全室地毯
fitted curve 拟合曲线
fitted freely 自由配合的
fitted furniture 全套家具
fitted linear regression line 拟合的线性回归直线
fitted model 拟合模型
fitted multiple regression equation 拟合多重回归方程
fitted outside 外装式
fitted pile shoe 贴合桩靴
fitted pintle 配合舵杆;舵向指针
fitted position 装配尺寸
fitted regression line 拟合的回归直线

fitted value 拟合值
fitted vehicle 装有自动制动机的车辆
fitter 修配工;修理工;装配工;钳工;设备安装工
fitter file 装配工锉
fitter for heating installations 水暖工;供暖装置装配工
fitter's bench 装配工作台;钳工工作台
fitter's hammer 钳工锤;台用锤
fitter's scaffold(ing) 装配工的脚手架
fitter's tools 装配工具
fitter's work 装配工作;设备安装工作;设备安装工程
fit the bill 适应要求
fit tightly 紧密配合
fitting 选配;组装;装配;管件;零配件;卫生管道配件;附件
fitting a curve 拟合曲线
fitting allowance 装配余量;调整余量
fitting arrangement 安装系统图
fitting arrangement drawing 附件装配图
fitting assembling 装配;零件装配
fitting a straight line 拟合直线
fitting attachment 连接;接线;装配附件
fitting by eye 目视拟合
fitting check 拟合检验
fitting clearance 安装间隙
fitting coefficient 拟合系数
fitting condition 镶合条件;拟合条件
fitting constant 拟合常数
fitting criterion 拟合准则
fitting curve 选配曲线;拟合曲线;曲线拟合
fitting data 拟合数据
fitting drawing 装配图(样)
fitting for grain 铺舱作业
fitting freely 自由配合的
fitting-in 配合
fitting instruction 装配说明书;安装说明书
fitting joint 装配连接;装配接头
fitting key 安装键
fitting limit 装配范围;配合极限;安装范围
fitting lubricant 安装润滑剂
fitting lubrication 合适的润滑作用
fitting mark 装配标记;对准标;安装标记
fitting metal 配件
fitting method 拟合方法
fitting of a secondary degree parabola 二阶抛物线拟合
fitting of a straight-line 直线拟合
fitting of distribution 分布拟合
fitting of forgeable cast iron 可锻铸铁接头配件
fitting of function 函数拟合
fitting of malleable cast iron 可锻铸铁接头配件
fitting of screen bottom 栅底安装(指水厂、污水厂);筛底安装(指水厂、污水厂)
fitting of stuffing 密封填料
fitting of tire 轮胎装配
fitting of variogram 变差函数拟合
fitting on 安装装配
fitting out 装备;舾装;装具
fitting out a ship 装配船只
fitting-out basin 船只装配码头;船只装配船坞;舾装泊位(在船坞附近的水区);装配港池
fitting-out berth 舾装泊位(在船坞附近的水区);舾装码头;装配泊位
fitting-out bolt 装配螺栓
fitting-out crane 舾装起重机;装配用起重机
fitting-out dock 舾装船坞;舾装船坞
fitting-out number 舾装数
fitting-out pier 舾装码头;舾装码头
fitting-out port 舾装港口;船只停靠港口;补给港口
fitting-out quay 补给码头;舾装码头;装配码头;安装码头
fitting-out shop 舾装车间
fitting-out vessel 舾装船舶
fitting-out wharf 舾装码头;舾装码头;安装码头
fitting overall chain 安装防滑链
fitting parts 配件;装配件;零件
fitting piece 配件;配合件
fitting pin 锁紧销
fitting pipe 管接头配件;连接管
fitting polynomial 选配多项式
fitting procedure 装配程序;安装程序
fitting process 装配过程
fittings 连接角钢;器材;配件;附属装备角铁;附加装置

fittings and fitments 房屋设备安装及装修;安装及装修(房屋设备)
fitting screw 装配螺钉;紧配连接螺栓
fitting sequence 安装顺序
fittings facotry 配件工厂;建筑配件工厂
fittings for pipe 管子配件
fittings for single trap siphon 提水虹吸式配件
fitting shop 装配厂;装配(车)间
fittings locker 船舶属具库
fittings microscope 吻合显微镜
fittings of tube 管子配件
fitting strip 轴瓦调整垫片;夹板
fitting surface 装配面;配合面;安装面
fitting temperature 装配温度
fitting tight 装配密封
fitting tolerance 配合公差
fitting union 管套节;联管节
fitting-up 结构部件成形;结构部件安装;安装
fitting-up bolt 装配螺栓
fitting-up gang 安装队;装配工作队
fitting-up inspection 组对检查
fitting-up joint 连接处的装配
fitting work 装配工作;装配工种;设备安装工作;装修工程
fitting workshop 装配工间
fit tolerance 配合公差;装配公差
fit-up 活动模壳;(混凝土模壳的)大型材料;活动模板;临时戏台;活动舞台
fit-up gap 装配间隙
fit-up inspection 装配检查
fit value 适合值
Fitzgerald contraction 菲茨杰拉德收缩
Fitzgerald-Lorentz contraction 菲茨杰拉德-洛仑兹收缩
Fitz mill 菲兹微粉碎机
Fitz-Simons visco(si)meter 菲茨黏—西门司黏度计
five-address instruction 五地址指令
five-axis method 五轴法
five-bay 五开间的
five bearing crankshaft 五主轴颈曲轴
five-bed 五床位的
five bit code 五单位码
five-blade propeller 五叶推进器
five bowl universal calender 五滚筒式研光机
five-centered arch 五心拱
five-channel scanning radio meter 五通道扫描辐射计
five coat system 五涂层结构
five-colo(u)red ware 五彩器皿
five-corbiestep gable with bargeboard 五花山墙
five-core cable 五芯电缆
five cycle of a recurrent selection program (me) 五次轮回选择方案
five-day biochemical oxygen demand 五日生化需氧量
five-day BOD[Biological Oxygen Demancl] 五日生化需氧量
five-day forecast 五日预报
five-digit numbering 五位制
five digit system 五位制
five-drum winch 五鼓卷扬机;五鼓绞车
five-element lens 五合透镜
five figure system 五位制
five-fixed freight train 货运五定班列
five-fixture bathroom 五件套浴室
five-foiled arch 五叶形饰的拱
fivefold 五倍
fivefold symmetry 五重对称
five-fole coincidence 五重符合
five heddle satin 五枚缎纹
five-island berth 五岛式泊位
five-leaf sliding folding shutter door 折叠式滑移五页门
five leaf twill 五枚斜纹
five-legged transformer 五心柱形变压器
five-lens aerial camera 五镜头空中照相机
five lens camera 五镜头摄影机
five level 五单位
five-level code 五电平码;五单位码
five-level section 五点段面
five-level startstop operation 五电平起止操作
five limbs type core (变压器的)五柱式铁芯
fiveling 五单体孪晶

five-lobe tracery 五叶窗花格
five-master 五桅船
five-membered ring 五节环
five-method of rock piling 掇山五法
five-minute oscillation 五分钟振荡
five orders 五类典型柱;五柱式
Five Pagoda temple 五塔寺
five-panel(led)door 五块镶门板;五合板门
five parts formula 球面五联公式;球面正余弦公式
five pavilion bridge 五亭桥
five percent discount for cash 现金付款95折优待
five percent method of depreciation 半面折旧法
five percent off 95折
five percent of the sum 总和的5%
five-petaled flowers 五瓣花
five ply 五层作(法)
five-ply door 5层板门
five plywood 五夹板;五合板
five pointed star 五角星
five point method 五点法
five-point plug 五脚插头
five points moving average 5点移动平均
five-purlin(e)beam 五架梁
five-range 5挡
five-range transmission 5速齿轮箱
five-roll Abramsen machine 艾布拉姆森式5辊矫直机
five-roller 5辊滚压机
five-roller mill 5辊轧机
five-roller stretcher 5辊拉伸机
five roll idler 5节托辊
five-roll machine 5辊矫直机
five-roll mill 5辊滚压机
five-room(ed)dwelling unit 5居室单元
five-sheave block 5轮滑车
five-sided prism 五角棱镜
five-sided square 五角棱镜直角器
five-sided stone bridge 五边形石桥
five small industry 五小工业
five speed 5挡
five-speed gear box 5速齿轮箱
five-speed transmission 5挡变速器
five-spot flood system 5点注水系
five spot flow formula 5点井网流动公式
five-spot method 五点法
five-spot pattern 5点井网;五点布井法
five-spot water flooding 5点井网注水
five-stand tandem mill 5机座串列式轧机
five strand crabber's eye sennit 5股编绳索
five stretcher courses to one header course alternately 五顺一丁砌法
five-throw crankshaft 5曲柄曲轴
five-throw turnout 5开道岔
five-tier drier 5层干燥机
five-tine rock grapple 5齿岩石抓斗
five unit alphabet 5单位码
five unit code 5单元码
five unit start-stop apparatus 5单位起止式设备
five-wire system 五线制
five-year plan 五年计划
five-zero-one-one urea formaldehyde adhesive 5011脲醛树脂胶
five-zero-two ethyl quick setting adhesive 502速凝胶
fix 确定;定影
fixable 可装设的;可固定的
fixanal 配定标准液
fix a price 决定价格
fixation 固土作用;固定(作用);固定装配;定影;安置
fixation accelerator 促染剂
fixation disc 固着盘
fixation forceps 固定镊
fixation hook 固定钩
fixation liquid 固定液
fixation mark 固定标记
fixation measuring microscope 定位量测放大镜
fixation method 固定方法;固定法
fixation of atmospheric nitrogen 空气氮的固定
fixation of hazardous waste 危险废物固定
fixation of sludge 污泥固定
fixation of solar energy 太阳能固定
fixation of wastes 废物固定
fixation phenomenon 补体固定现象
fixation point 注视点

fixation reaction 固定反应;结合反应
fixation screw needle 固定螺丝针
fixation time 定影时间
fixation with bamboo splints 竹片固定
fixation with paper pad 纸垫固定
fixation with steel pin 钢丝针固定
fixation with steel wire 钢丝固定
fixative 固着剂;固定液;固定剂;定色液
fixator 固定器
fix by position lines 坐标定位
fix by radar bearing and distance 雷达方位距离定位;雷达方位距离船位
fix by sounding 测深辨位
fixed 固定的;不挥发
fixed acid 固定酸;非挥发性酸;不挥发(性)酸
fixed acoustic(al)buoy 固定音响浮标;固定声学浮标
fixed action 固定作用
fixed-action pattern 固定动作模式
fixed activated carbon bed 固定活性炭床
fixed activity 固定活度
fixed address 固定地址
fixed adsorbent 固定吸附剂
fixed adsorbent bed 固定(式)吸附剂床
fixed aerial 固定天线
fixed air 固定空气;不流动空气
fixed air gap 固定空隙
fixed ammonium 铵固定
fixed amplitude 稳幅
fixed analog(ue) 固定模拟
fixed analysis of variance 固定方差分析
fixed anchor 固定锚
fixed anchorage 固定抛锚地;固定锚地;固定泊位
fixed and flashing light 固定式闪光灯;固定光和闪光;定闪光;定光夹闪光
fixed and group flashing light 定联闪光;定光夹群闪光
fixed and group occulting light 定联明暗光
fixed and group of flashing light 固定和组合闪光灯
fixed angle sounding 固定角定位测深法
fixed annuity 确定年金
fixed anode 固定阳极
fixed antenna 固定天线
fixed antenna direction finder 固定天线测向器
fixed aperture 固定光圈;固定光阑
fixed appliance 固定设备
fixed arch(at both ends) 固定拱;无铰拱;固端拱
fixed arch bridge 无铰拱桥
fixed arcing contact 静触头
fixed arc metallizing gun 固定式电弧喷涂枪
fixed area 固定区(域);固定磁鼓存储面积
fixed-area nozzle 固定截面喷嘴
fixed array method 固定阵列法
fixed array multielement lidar 固定阵列多元激光雷达
fixed array multilaser radar 固定阵列多元激光雷达
fixed ash 内在灰分
fixed assembly line 不变流水线
fixed assets 固定资产
fixed assets accounting 固定资产核算
fixed assets appraisal 固定资产的估计
fixed assets deppreciation audit 固定资产折旧审计
fixed assets formation 固定资产组成
fixed assets for non-production purposes 非生产用固定资产
fixed assets for rent 租出固定资产
fixed assets in liquidation 清理中固定资产
fixed assets internal control system audit 固定资产内部控制制度的审计
fixed assets internal transfer audit 固定资产内部转移的审计
fixed assets in use 使用中的固定资产
fixed assets investment 固定资产投资
fixed assets leasing audit 固定资产租赁的审计
fixed assets ledger 固定资产分类账
fixed assets management of rights of railway bureau 铁路局对固定资产管理的权限
fixed assets management responsibility of railway bureau 铁路局对固定资产管理的责任
fixed assets not needs 不需用的固定资产
fixed assets not used in production 非生产用固定资产
fixed assets of railway transport enterprise 铁运输企业固定资产
fixed assets of transportation 运输业固定资产
fixed assets physical form audit 固定资产实物形态的审计
fixed-assets ratio 固定资产比率
fixed assets register 固定资产登记表
fixed assets repairs cost audit 固定资产修理费的审计
fixed assets schedule 固定资产明细表
fixed assets scraping and liquidation audit 固定资产报废清理的审计
fixed assets sealed and stored 封存固定资产
fixed assets statement 固定资产表
fixed assets tax 固定资金占用税
fixed assets tax rate 固定资金占用税率
fixed assets to capital ratio 固定资产对资本比率
fixed assets used in production 生产用固定资产
fixed assignment 固定赋值
fixed attenuator 固定衰减器
fixed attribute 固定属性
fixed automatic block 固定式自动闭塞
fixed automaton 固定性自动机
fixed awning 固定遮篷;固定天篷
fixed axis 固定轴
fixed axle 支重桥;固定轴
fixed axle gate 定轴平板闸门
fixed ballast 固定压载(物)
fixed-bar grille 固定铁格栅
fixed base 固定碱;固定基面;固定底座;定基;柱固定底座
fixed base crane 固定式起重机
fixed base current bias 固定基极偏流
fixed baseline 固定基线
fixed-basket type 网篮式
fixed-bat reel 固定拨禾轮
fixed beacon 固定信标;固定立标;固定灯标;灯塔(航海用)
fixed beam 固定梁;定端梁
fixed beam at one end 一端固定梁
fixed-beam ceilometer 固定光束云幕仪
fixed bearing 固定支座;固定支承
fixed bearing end 固定轴承端
fixed bearing for bridge 桥梁固定支座
fixed bearing with curved steel plates 弧形支承钢板固定支座
fixed bed 固定床;固定层;定床
fixed bed absorber 固定床吸附器
fixed bed absorption 固定床吸附
fixed bed activated carbon contactor 固定床活性炭接触器
fixed bed adsorption column 固定床吸附柱
fixed bed bioreactor 固定床生物反应器
fixed bed catalyst 固定床催化剂
fixed bed catalyst chamber 固定床催化反应器
fixed bed catalytic cracking 固定床催化裂化
fixed bed catalytic ozonation process 固定床催化臭氧氧化工艺
fixed bed catalytic process 固定床催化过程
fixed bed catalytic reactor 固定床催化反应器
fixed bed drying system 固定床式干燥系统
fixed bed flow test 定床水流模型试验
fixed bed gas absorber 固定床气体吸附剂
fixed bed gasification 固定床气化
fixed bed groundwater 固定
fixed bed hydroforming 固定床氢重整;固定床加氢重整
fixed bed ion exchange 固定床离子交换
fixed bed ion exchanger 固定床离子交换器
fixed bed model 固定(河)床模型;定床模型
fixed bed model test 定床模型试验
fixed bed operation 固定床操作
fixed bed process 固定床过程
fixed bed reaction equipment 固定床反应设备
fixed bed reactor 固定床反应器
fixed bed river engineering model 定床河工模型
fixed bed river model 定床河道模型
fixed bed scrubber 固定洗涤器
fixed bed sedimentation model 定床淤积模型
fixed bed-type milling machine 固定床身式铣床
fixed bed unit 固定床设备
fixed belt conveyer[conveyor] 固定带式运输机
fixed belt conveyer for feeding salt 上盐固定式皮带运输机

fixed benchmark 固定水准基点
fixed bias 固定偏压
fixed binary 定点二进制
fixed binocular loupe 固定式双筒放大镜
fixed biofilm 固着生物膜
fixed biofilm system 固着生物膜系统
fixed biomass carrier 固着生物质载体
fixed blade 固定叶片;固定轮叶
fixed blade propeller 固定桨叶螺旋桨
fixed blade propeller turbine 轴流定桨式水轮机;定轮叶桨式水轮机
fixed blade propeller type turbine 定轮叶旋桨式水轮机;定桨式水轮机
fixed blade propeller type wheel 定桨式水轮
fixed blade propeller wheel 定轮叶旋桨式水轮
fixed blade turbine 定轮叶式水轮机
fixed block 固定块;固定滑车;定滑轮
fixed block architecture 固定块结构
fixed block architecture device 固定块结构设备;定长块结构设备
fixed block format 固定数字组格式
fixed block length 固定块长度
fixed block slider crank mechanism 固定滑块曲柄机构
fixed block transmission 固定块传输;固定长块传送
fixed bollard 固定系船柱
fixed bolt 固定锚杆;固定螺栓
fixed bond 定期债券
fixed boom 固定式悬臂
fixed boom crane 定臂式起重机;固定臂(式)起重机
fixed boundary 固定边界
fixed bow fins 固定船首动水翼
fixed breakage point 固定断裂点
fixed breakwater 固定式防波堤
fixed bridge 固定桥
fixed bridge scraper 固定桥式刮泥器;固定桥式刮泥机
fixed brush 固定电刷
fixed brush type polyphase series motor 固定电刷式多相串激电动机
fixed budget 固定预算
fixed buffer 固定减振器
fixed bunker 固定燃料仓
fixed buoy 固定浮标
fixed bushing 固定轴衬
fixed cableway 固定缆索道
fixed-cal(l)iper disk brake 固定夹钳式制动器;固定卡钳式钢盘制动器
fixed camera mounting 固定摄影机座架
fixed camouflage 永久性的伪装
fixed cantilever 固定托架
fixed capital 固定资金;固定资本;固定基金
fixed capital cost 固定资本值;固定投资费
fixed capital efficiency 固定资产效率
fixed capital investment 固定资本投资
fixed capital management 固定资金管理
fixed carbon 焦渣;固定碳
fixed carbon content 固定碳含量
fixed casement window 固定窗
fixed catalyst 固定催化剂
fixed cavitation 固定空穴;固定空蚀
fixed cavity 固定空腔谐振器
fixed-cement factor method 固定水泥系法;水泥系数法
fixed center 固定中心;固定顶尖
fixed-center change gear 固定中心距变速齿轮
fixed-center change gear pump 固定中心距变速齿轮泵
fixed center distance 固定顶尖距
fixed channel 固定河槽;稳定河槽
fixed charge 固定电荷;固定支出;固定开支;固定费(用);固定的消耗量;固定成本
fixed-charge model 固定电荷模型
fixed-charge problem 固定费用问题
fixed-charge ratio 固定费用比率
fixed chute 固定溜槽
fixed claw 固定爪
fixed clock time control 定时控制;定间隔时钟控制方式
fixed coaxial attenuator 固定同轴衰减器
fixed coil 固定线圈
fixed-coil antenna 固定环形天线
fixed coil indicator 定圈式指示器

fixed column 固定柱
fixed combustion source 固定床燃烧室
fixed compression chamber 固定式加压舱
fixed condenser 固定式电容器;定容式冷凝器
fixed consumption 固定的消耗量
fixed contact 固定接点;固定触点
fixed control 定位操纵
fixed control area 固定控制区
fixed converter 固定变换器
fixed correction 固定修正量
fixed cost 固定成本;不变费用;固定费(用)
fixed cost contract 固定费用合同
fixed cost of location 建厂区的不变资本
fixed cost when business shut down 经营关闭时的不变成本
fixed cost when operating 经营中的不变成本
fixed counter weight 固定平衡重
fixed coupling 固定联轴节
fixed-course detector 固定航向指示器
fixed cover 固定盖
fixed cover sludge digestion tank 固定盖式消耗池
fixed crane 固定起重机
fixed crest dam 固定坝(坝顶上无滑动闸门)
fixed crest spillway 固定堰顶式溢洪道
fixed crest weir 固定堰顶式堰
fixed crossing clamp 固定型交叉线叉
fixed crystal 固定晶体
fixed crystal detector 固定晶体检波器
fixed cutter 固定刀片
fixed cycle 固定周期
fixed cycle operation 定周期操作;定时操作;固定周期运算;固定周期操作
fixed cycle signal 固定周期交通信号;定周期(交通)信号
fixed dam 固定挡板;固定坝
fixed data 固定数据
fixed data-name 固定数据名
fixed-date delivery 定期交货
fixed datum 固定起始值;固定基面;固定基准(线)
fixed debt 固定债务
fixed decimal 固定小数;固定十进制;定点
fixed decimal point 固定十进制小数点
fixed delay 固定延误;固定时延
fixed-delay output 固定时延输出
fixed delivery pump 定容量泵
fixed denture 固定义齿
fixed deposit 定期存款
fixed derrick 固定吊杆;固定式悬臂起重机
fixed derrick crane 固定式动(力)臂起重机
fixed deviation (无线电测向仪的)固定自差
fixed dewatering screen 固定脱水筛
fixed die half 定模(压铸用)
fixed discharger 固定放电器
fixed disk 固定盘
fixed displacement 定量泵
fixed-displacement motor 定量马达;定容量马达
fixed-displacement pump 固定排量泵
fixed-displacement-variable pressure pump 定排量变压力泵
fixed distance 固定距离
fixed distributor 固定配水器;固定配电器;固定布水器
fixed distributor of biologic(al) filter 生物滤池固定布水器
fixed drawbar 固定的牵引装置
fixed drilling platform 固定式钻进平台
fixed drip tray 固定式聚漆盘
fixed dry chemical system 固定式干粉灭火系统
fixed dune 固定沙丘
fixed earthing 固定接地
fixed earth station 固定地球站
fixed earth support 下端固定支承
fixed eccentric 固定偏心轮
fixed echo 固定目标回波;固定回波
fixed echo suppressor 固定回波消除器
fixed edge 固定边(缘)
fixed-edge-slab 嵌固板;固定边缘的板
fixed effectiveness approach 效用固定分析法
fixed-electrode method 固定电极法
fixed electrolevel inclinometer 固定电位测斜仪
fixed element 定位元素
fixed elevation 固定高度;固定标高
fixed eluant 固定洗提液
fixed encoded theodolite 固定编码经纬仪

fixed end 固端;定端;定端
fixed-end arch 固端拱;无铰拱;定端拱;无角拱
fixed-end arch bridge 固端拱桥
fixed-end beam 定端梁;固(定)端梁
fixed-end bending moment 固端弯矩
fixed-end column 固端柱;固定端柱
fixed-end condition 固端条件
fixed-end drum 端部固定的桶
fixed-ended beam 固端梁
fixed-ended girder 固端大梁
fixed-end fixity 定端稳定
fixed-end moment 固定力矩;固(定)端弯矩;固(定)端力矩
fixed-end pile 底端固定的桩
fixed-end restraint 固端约束
fixed-end stiffness 定端刚度
fixed-end support 固定端支承
fixed engine 固定式发动机
fixed equalizer 固定均衡器
fixed equipment 终端硬锚导线;固定(式)设备
fixed equipment circuit 家用固定设备线路
fixed error 固定误差
fixed exchange rate 固定汇率
fixed exit nozzle 定出口截面喷嘴
fixed expansion 固定膨胀
fixed expenditures 固定开支
fixed expenditures not related to traffic volume 与运输量无关的固定支出
fixed expenses 固定费用;固定开支
fixed eyepiece 固定目镜
fixed fee 固定费(用)
fixed-feed grinding 固定进给磨削
fixed fee for the use of a site 场地使用费
fixed fence 固定栅栏
fixed fiber optic(al) connector 固定光纤连接器
fixed fiber optic(al) connectors receptacle 固定光纤连接器插座
fixed field 固定信息组;固定区(域)
fixed field accelerator 固定磁场加速器;固定场加速器
fixed field format 固定域格式
fixed film biological reactor 固定膜生物反应器
fixed film denitrification 固定膜反消化
fixed film nitrification 固定膜消化
fixed film process 固定膜法
fixed film reactor 固着膜反应器
fixed film resistor 固定薄膜电阻器
fixed film system 固定膜系统
fixed fin 垂直安定面
fixed fire escape 固定太平梯
fixed fire extinguisher 固定灭火器
fixed fire extinguishing system 固定(式)灭火系统
fixed fire pump 固定消防泵
fixed fire suppression device 固定灭火装置
fixed fire suppression system 固定式灭火系统
fixed flange 固定法兰(盘);固定凸缘
fixed flashing light 固定航标灯;固定闪光灯
fixed flat bed platen 固定式平台压板
fixed flow rate 固定流量
fixed foam generator 固定泡沫混合发生器
fixed focal length 固定焦距
fixed focus 固定焦点
fixed-focus camera 固定焦距摄影机;定焦(距)照相机;定焦(距)摄影机
fixed-focus lens 固定焦距物镜;定焦点透镜
fixed-focus objective 固定焦距物镜
fixed-focus pyrometer 恒焦距高温计
fixed form 固定模板;固定方式;定形
fixed format 固定(长)格式
fixed format file 固定格式文件
fixed format input 固定格式输入
fixed format message 固定格式消息
fixed form coding 固定形式编码;固定式编码
fixed form operation 固定方式操作
fixed formwork 固定模板(工程)
fixed frame 固定框子;固定框
fixed framed arch 构架式固端拱;构架式固定拱
fixed frequency 固定频率
fixed frequency agile racon 固定频率雷康;固定频率雷达应答器
fixed frequency cyclotron 固定频率回旋加速器
fixed frequency filter 固定频率滤波器
fixed frequency generator 固定频率发生器;固定频率发电机

fixed frequency multivibrator 固定频率多谐振荡器
fixed frequency oscillator 固定频率振荡器
fixed frequency radar responder beacon 固定频率雷达应答标
fixed frequency source 固定频率源
fixed frequency transmitter 固定频率发射机
fixed frog 固定辙叉
fixed froth installation 泡沫灭火装置
fixed fuel tank 固定燃料箱
fixed function 固定功能
fixed function generator 固定函数发生器;固定功能生成程序
fixed funds 固定基金
fixed-gain damping 固定增益阻尼
fixed gantry 固定高架起重机
fixed gantry crane 固定(式)门(式)起重机;固定(式)龙门起重机;固定高架起重机
fixed gas 固定气体;难于凝聚的气体;不凝气
fixed gate 固定选通脉冲
fixed-gate generator 固定选通脉冲发生器
fixed ga(u)ge 固定水尺;固定规
fixed generating equipment 固定发电设备
fixed-geometry channel 断面形状稳定的河槽
fixed geothermal power station 固定式地热电站
fixed girder 定端大梁;定端梁
fixed glass 地平镜
fixed glass louver 固定玻璃百叶窗
fixed glass wall 固定玻璃壁
fixed glazing 固定安装的玻璃
fixed grain 固结磨粒
fixed grate 固定炉栅;固定炉箅;固定格栅
fixed grate incinerator 固定炉箅焚烧炉
fixed grid baffles 固定叶片
fixed grizzly 固定结筛;固定格栅;固定格筛;固定(箅)条筛
fixed groove mercury barometer 定槽水银气压表
fixed ground connection 固定接地
fixed grounding 固定接地
fixed groundwater 约束地下水;固定地下水;结合水;束缚水;非重力式地下水
fixed guide vane 固定导叶
fixed guide way transit system 有轨快速客运系统
fixed guide wheel 座环
fixed gun 固定式机关枪
fixed gun sight 固定炮瞄准具
fixed hammer 固定锤头
fixed handle 固定手柄;固定柄
fixed-handle circuit breaker 固定跳闸;固定断路器
fixed head 固定头盖板;固定头;固定磁头
fixed header prefix 固定标题前缀
fixed head magnetic drum 固定头磁鼓
fixed head sprinkler 固定喷头喷灌机
fixed head storage 固定头存储器
fixed heater 固定加热器
fixed hinge 定铰;固定铰(链)
fixed hinged support 固定铰支座
fixed hopper 固定式料斗
fixed horizontal polarization 固定水平偏振
fixed horizontal screen 固定水平筛
fixed hydrocarbon 固定碳氢化合物;非挥发性
fixed hydrophone 固定水听器
fixed ignition 固定点火
fixed image graphics 固定影像图形技术;固定图像图形学
fixed impeller straight through cyclone 固定叶轮直通式旋风分离器
fixed income 固定收入
fixed increment rule 固定增量规则
fixed index 固定式照准器;固定式瞄准器;固定瞄准器
fixed-in door frame 后装门框
fixed inductance 固定电感
fixed inductor(coil) 固定感应线圈
fixed information 固定情报
fixed information file 固定信息存储器
fixed infrared seeker head 固定红外寻的头
fixed insertion 固定插入
fixed insertion editing 固定插入编辑
fixed installment depreciation method 分期定额折旧法
fixed installment method 分期定额法
fixed interconnection matrix 固定关联矩阵
fixed interconnect wiring 固定互连布线
fixed interest security 定息债券

fixed interval 固定间隙;固定间隔
fixed investment 固定投资额
fixed investment in residential structures 居住用建筑物的固定投资额
fixed investment trust 固定性投资信托
fixed-in window frame 后装窗框
fixed jaw 固定爪;固定颚板
fixed jaw crushing plate 固定的碎矿颚板
fixed jet distributor 固定喷射配水器;固定喷射布水器;固定喷洒配水器;固定喷洒布水器
fixed jib 固定臂
fixed jib crane 固定式悬臂起重机
fixed joint 固定连接;固定接头;刚性连接;刚结点
fixed key word mode 固定关键字方式
fixed-knife planer 固定刀式刨床
fixed knot 固定节;连贯节
fixed ladder 固定梯
fixed landing bridge 固定式登陆桥;固定跳板
fixed launcher 固定发射装置
fixed layer 固定层
fixed lead 固定超前
fixed lease 固定租赁
fixed leg 固定柱
fixed legal ratio 法定比率
fixed leg tripod 定腿三角架
fixed-length 固定长度;定长
fixed-length block 固定长块;定长块
fixed-length code 固定长度编码
fixed-length data 定长数据;固定长数据
fixed-length data area 固定长度数据区;定长数据区
fixed-length field 固定长度字段;固定长度信息组;定长字段;定长域;定长信息组
fixed-length file 定长文件
fixed-length file record 固定长度文件记录;定长文件记录
fixed-length format 固定长度格式
fixed-length record 定长记录;等长记录;固定长度记录
fixed-length record file 固定(长)记录文件;固定(长)度记录文卷
fixed-length record format 固定长度记录格式
fixed-length record system 固定长记录系统;定长记录系统
fixed-length spreader 定长吊架
fixed-length string 定长串
fixed-length symbol 定长符号
fixed-length variable 定长变量
fixed-length word 固定长字;定长字
fixed lens 固定透镜
fixed letter of revenue 确定收入单证
fixed level 稳定水位
fixed liability 固定债务;固定负债
fixed light 固定舷窗;固定光;固定灯(光);固定窗;稳定光;定光(灯)
fixed light intensity 定光光强
fixed-light position 固定光位
fixed limit of construction cost 最高允许(工程)造价;工程费用的固定限额;控制造价
fixed line number 固定行数
fixed line observation 断面观测;定线观测
fixed link 固定连接;固定的链节;固定杆
fixed linkage 固定式悬挂装置
fixed link pack area 固定连接装配区
fixed load 固定载荷;固定负载;固定负荷;不变荷载
fixed load demand 固定负载;固定负荷
fixed load moment 死荷载力矩;固定荷载力矩
fixed load stress 固定荷载的应力
fixed loan secured 有担保定期贷款
fixed loan unsecured 定期信用贷款
fixed logic 固定逻辑
fixed loop aerial 固定环状天线;固定环形天线
fixed loop antenna 固定环形天线
fixed loss (与负载无关的)固定损失
fixed low point 固定低点
fixed mandrel 定径心轴
fixed mark 固定标志
fixed marker 固定标志
fixed market 固定市场
fixed mark on water 水中固定标志
fixed mask 固定屏幕
fixed mast (桅杆起重机的)固定桅杆;固定式桅杆
fixed matter 固定物质
fixed mean pole 固定平极
fixed member 固定成员

fixed membership class 固定属籍类别
fixed memory 固定存储器
fixed memory apparatus 固定存储装置
fixed metal louver 固定金属百叶窗
fixed milk line 固定式送奶管路
fixed mine wagon 固定式矿车
fixed mining car 固定式矿车
fixed mirror 固定镜;定镜
fixed mixer 固定式搅拌机;固定式拌和机;不倾式搅拌机
fixed mode 固定模式;固定方式
fixed-mode record 固定方式记录
fixed moisture 固定水分;固定湿度;结合水;束缚水
fixed moisture content 固定含水量
fixed moment 固端弯矩
fixed monitor 固定式监视器
fixed monument 固定标志桩
fixed mooring 固定系缆;固定系泊;锚定块体;码头固定系缆
fixed mooring berth 固定系船码头;固定系泊泊位;固定式系船处;靠船墩式泊位
fixed mooring quay 固定系船码头
fixed mooring wharf 固定系船码头
fixed mo(u)ld 固定模
fixed mount 支架;固定架
fixed mounting base 固定支承座
fixed name 固定名;固定的名字
fixed-needle surveying 固定磁针测量
fixed-needle traverse 固定指针导线
fixedness 不挥发性
fixed net 固定鱼网
fixed neutral density filter 固定中密度滤光器
fixed nitrogen 固定氮
fixed nozzle 固定喷嘴;固定喷头
fixed number of labo(u)r 劳动定员
fixed number of persons 定员载客
fixed number testing 定式试验;定数检验
fixed number testing plan 定数计划试验
fixed object 固定目标
fixed observation point 固定性观测点
fixed observatory 固定观测台
fixed oceanographic(al) stations of the world 世界定点海洋观测站
fixed oceanographic(al) station 固定海洋观测站;定点海上观测站
fixed oceanographic(al) weather station 固定海洋气象站
fixed off-set frequency agile racon 固定偏频雷达应答器
fixed offshore platform 固定式海上平台
fixed oil 硬化油;固定油;不挥发(性)油
fixed operating sleeve 固定连接套
fixed optics 固定光学
fixed-orbit accelerator 固定轨道加速器
fixed ordering procedure 固定排序过程
fixed order-quantity system 固定订货量系统
fixed orifice 固定节流孔
fixed overflow 固定溢出;定点溢出
fixed overflow dam 固定式溢流坝
fixed overhead 固定间接费;固定管理费
fixed overhead price variance 固定制造费用价格差异
fixed overhead quantity variance 固定制造费用数量差异
fixed overhead rate 固定间接费率
fixed overhead variance 固定间接差异
fixed overlayable segment 固定可覆盖段
fixed oxygen 固定氧
fixed packing 固定式密垫
fixed page 固定页面
fixed parabolic reflector 固定抛物反射器
fixed par of exchange 汇兑的法定平价
fixed part 固定(零)件;固定部件;固定部分
fixed partition 固定分区
fixed partition cell 固定隔板沉淀池
fixed partition wall 固定隔墙;固定隔板
fixed pattern method 固定图案法
fixed pattern noise 固定图形噪声
fixed payment tariff 定额收费表
fixed percentage depreciation method 定率折旧法
fixed percentage of a declining balance 定率递减余额
fixed percentage of a declining method 定率递减法

fixed-percentage-on-declining balance depreciation method 定率递减折旧法
fixed period 固定周期
fixed periodic(al) block 固定的轮伐更新区
fixed permanent segment 固定永久段
fixed permeate flux 固定渗透通量
fixed phase 固定相;参考相位
fixed phase reference 固定相位基准
fixed phase relationship 固定相位关系
fixed phase shift 固定相移
fixed phase shifter 固定移相器
fixed phase wave 固定相波
fixed phasing sequence 固定相序
fixed pile 嵌固桩
fixed pillar slewing crane 定柱旋臂起重机
fixed pin 销固定;固定销
fixed piston 固定活塞
fixed piston sampler 固定活塞式取样器
fixed pitch 固定螺距
fixed pitch propeller 固定螺距推进器;定(螺)距螺旋桨
fixed pivot 支枢
fixed plant 固着植物;固定设备
fixed plate 固定板块;固定板
fixed platform 固定式平台;固定平台
fixed pneumatic fender 固定式充气护舷
fixed point 观测点;固定点;常点;测标位置;不动点;标定点;定(位)点
fixed-point arithmetic 定点运算
fixed-point binary 定点二进制
fixed-point binary constant 定点二进制常数
fixed-point buffer 定点缓冲剂
fixed-point calculation 定点运算;定点计算
fixed-point computer 定点计算机
fixed-point constant 定点常数
fixed-point data 定点数据
fixed-point decimal constant 固定小数点十进制常数;十进定点常数
fixed-point detection 固定点(火灾)探测
fixed-point double word-length arithmetic 定点双字长运算
fixed-pointer format 定点制格式
fixed-point ga(u)ge 定点测针
fixed-point machine 固定辙器;固定辙机
fixed-point mathematics 定点运算;定点数学
fixed-point method 定点(方)法;不动点方法
fixed-point notation 定点记数法;定点表示法
fixed-point observation 定点观测
fixed-point operation 定点运算;定点操作
fixed-point part 小数部分;尾数;定点部分
fixed-point representation 定点表示法;定点表达形式
fixed-point representation of a number 定点制数表示法
fixed-point representation system 定点表示制
fixed-point sampling 定点取样
fixed-point simulation 定点仿真
fixed-point system 定点系统
fixed-point theorem 不动点定理
fixed-point tidal station 定点验潮站
fixed-point value 定点值
fixed polynomial 固定多项式
fixed portion 固定部分
fixed position 固定点位
fixed-position addressing 固定位置访问;定位选址
fixed-position welding 定位焊接
fixed post 固定标志桩
fixed potassium 固定态钾
fixed potential 固定电势
fixed power 固定功率
fixed-power microscope 固定倍率显微镜
fixed preassigned multiple access 固定预分配多址
fixed preparation 固定标本
fixed pressure detector 定压式爆炸探测器
fixed pressure ga(u)ge 固定气压计
fixed pressure operation 定压运行
fixed pressure plate 固定压紧板
fixed price 官价;固定价格;不变价格
fixed-price competition 固定价格竞争
fixed-price contract 总价不变合同;固定总价合同;固定价格合同;定价合同
fixed-price movement 标价变动
fixed-price property 固定资产
fixed-price proportion 固定价格比率

fixed-price tender 固定价格投标
fixed primary annulus 行星传动第一级固定齿轮
fixed priority 固定优先级
fixed priority scheme 固定优先方式
fixed prism 固定棱镜
fixed-product area 固定乘法区;固定乘积存储区;确定乘积区
fixed production platform 固定式生产平台
fixed-program(me) 固定程序
fixed-program(me) machine 固定程序机
fixed propeller runner 固定的螺栓桨式转子
fixed property 不动产
fixed proportion 固定比例;定比
fixed proportion mixing method 定比混合法
fixed prosthesis 固定修复
fixed pulley 定轮;固定滑轮;紧轮;定滑轮;定滑车
fixed pump 固定的泵;固定泵
fixed quantity 定量
fixed quantity feeder 定量下料器
fixed quantity ordering system 固定量订货方式
fixed quantity price contract 定量工程量总价合同
fixed quantity water meter 定量水表
fixed quay crane 固定的码头起重机;固定式码头起重机
fixed quick shear test [r-test] 固定快剪试验
fixed quotas for revenues and expenditures 财务包干
fixed radio access 固定无线存取
fixed radio station 固定无线电台;固定电台
fixed radix notation 固定基数记数法;定根值记数法
fixed radix numeration system 固定基数记数体系;固定基记数制
fixed radix scale 固定基数记数法
fixed range marker 固定距离圈;固定距标圈;固定距标框
fixed-range mark generator 固定距离标记发生器
fixed range rings 固定距离圈;固定距标圈;固定距标
fixed rate flow 定速流动;定量流动;等流量
fixed rate mortgage 固定利率抵押;固定利率抵押贷款
fixed rate of interest 定息
fixed ratio 定比
fixed ratio meter 固定比例计
fixed ratio pantograph 固定比例缩放仪
fixed ratio printer 固定比率打印机
fixed ratio transmission code 定比传输码
fixed reactor 固定式反应器
fixed receiver 定收
fixed reference line 固定参考线
fixed reference point 固定参考点
fixed repair shed 定修库
fixed residue 固定残渣;不易挥发的残渣
fixed residue of water 水的固定残渣
fixed resistance 不可变电阻
fixed resistor 固定电阻器
fixed retainer 固定保持器
fixed retaining wall 固定(式)挡土墙;顶端锚固的挡土墙
fixed reticle 固定十字标线;固定分划板;固定调制盘
fixed reticule 固定十字丝
fixed ring 固定圈;固定环
fixed-ring consolidometer 固定环式固结仪
fixed river bed 固定河床
fixed roll(er) 支承轧辊;固定滚轴;固定滚筒;固定辊
fixed roller gate 定轮闸门;定轮平板闸门
fixed roof tank 带固定盖的桶
fixed roof type electric(al) arc furnace 固定炉顶式电弧炉
fixed rope 固定索
fixed routine 固定例行程序;固定程序
fixed routing per call 每次呼叫固定路径选择
fixed routing per call method 每次呼叫固定路径选择法
fixed royalty 固定提成
fixed sample-size 固定样本量
fixed sand by bedding clay 黏土固沙
fixed sand by bedding stone 铺石固沙
fixed sand by covering asphalt 沥青固沙
fixed sand by mat shape grass land 铺草席固沙
fixed sand by vegetation 植被固沙
fixed sash 固定玻璃(墙上);固定框扇;固定窗扇;固定窗框
fixed satellite 地球同步卫星;固定卫星

fixed scale royalty 固定提成率
fixed scarifying attachment 固定的翻松路面附件;定位松土机
fixed-scattering angle spectrometer 固定散射角能谱仪
fixed schedule 固定的目录
fixed schedule of charges 固定的收费时间表
fixed screen 固定纱窗;固定的筛子;固定的屏幕;固定的滤网;固定的滤光器;固定的滤光片;固定的晶格;固定的格栅;固定筛
fixed screen window 固定纱窗
fixed scuttle 固定舷窗
fixed seal arrangement 固定密封装置
fixed sea-mark 固定海上标志
fixed seat 固定座椅;固定座
fixed seating 固定座位
fixed section 固定断面
fixed seismometer 固定地震计
fixed sequence of code 固定代码顺序
fixed sequence phasing 定序信号相
fixed sequence robot 固定程序机器人;工业用简易机器人
fixed sequence wiring analyzer 固定顺序接线分析仪
fixed service 固定业务
fixed service satellite communication 固定业务卫星通信
fixed set order 固定系次序;固定谱系顺序
fixed set-point control 定值调节
fixed ship way 固定滑道【船】
fixed shoe of bridge 固定桥座
fixed shore 固定竖撑
fixed shore ramp 固定式岸上坡道
fixed shutter 固定百页窗
fixed shuttering 固定模板
fixed side light 固定舷窗
fixed sidereal day 固定恒星日
fixed-sieve jig 定筛跳汰机;定筛箕淘机
fixed sight 固定瞄准具
fixed sign 固定标志
fixed signal 固定信号
fixed simple frame 固定简单框架
fixed site monitor 定位监测器
fixed-size command 固定长度指令;固定长度命令
fixed-size location 定长单元
fixed-size node 固定尺寸节点;定长节点
fixed size record 固定长记录;定长记录
fixed skylight 固定天窗
fixed slipway 固定滑道
fixed slit 固定狭缝
fixed solid resisior 固体电阻器
fixed solids 不易挥发的固体物;固定固体;不挥发固体
fixed solution 固定溶液;固定解
fixed sonar 定位声呐
fixed source 固定源
fixed source entropy 固定源熵
fixed source field 定源场
fixed source-loop method 大定源回线法
fixed spacing 固定间隔
fixed span 固端跨;固定跨(度)
fixed spark-gap modulator 固定火花隙调制器
fixed speed 固定转速
fixed spheric(al) radio telescope 固定球面射电望远镜
fixed-spindle gyratory crusher 固定轴旋转圆锥破碎机;定轴式回旋破碎机
fixed spot welder 固定式点焊机
fixed spray 固定喷雾(器);固定喷洒器
fixed spreader 固定式吊具
fixed sprinkler 固定式喷灌机
fixed sprinkler system 固定喷水灭火系统
fixed spud 固定钢桩【疏】
fixed square search 定点展开方形搜索;地理正方形搜索
fixed stacker 固定式堆833机
fixed staff 固定水尺;固定标尺
fixed star 恒星
fixed star photograph 恒星像片
fixed station 固定台;固定电台
fixed station data collector 固定站数据搜集器
fixed station monitoring 固定监测站
fixed stay 固定支撑
fixed steel offshore platform 近海固定式钢质平台

fixed steel window shutter 固定钢百叶窗
fixed step point control 定值调节
fixed storage 固定存储器
fixed structure 固定结构;固定建筑物
fixed-structure channel marker 固定航道标志
fixed-structure light beacon 固定灯标
fixed suction installation 固定吸水设施
fixed sulfur 固定硫
fixed sum data 定和数据
fixed supply 定量供应
fixed support 固定支座;固定支架;固定支点;嵌固支点
fixed-surface jet washer 固定式表面喷射冲洗装置
fixed survival craft radio station 机动艇固定无线电台;机动艇固定无线电设备
fixed suspended solids 不挥发悬浮固体;固定悬浮物
fixed suspender 固定悬挂件
fixed symmetrical arch 固定对称拱
fixed system 固定系统
fixed table 固定工作台
fixed tangential rocker bearing 固定切向的摇杆轴承
fixed-tank vehicle 固定式槽车
fixed target 固定目标;固定舰标
fixed tariff duties 协定关税
fixed telemetering land station 固定遥测地面台
fixed temperature detector 定温探测器
fixed temperature detector using fusible alloy 易熔合金定温火灾探测器
fixed temperature detector using thermocouple 热电偶定温火灾探测器
fixed temperature detector with heat sensitive resistance 热敏电阻定温火灾探测器
fixed temperature heat detector 固定温感式火灾探测器
fixed term 定期汇价的不变部分
fixed-term appointment system 任期制
fixed term bill of exchange 定期合同
fixed term contract 定期合同
fixed term deposit 定期存款
fixed terminal time 固定终点时间
fixed termination 硬锚
fixed termination conductor 终端硬锚导线
fixed termination contact wire 端部固定接触线
fixed term insurance 定期保险
fixed testing plant 固定监测设备
fixed the price 规定价格
fixed thread 固定丝;定丝
fixed thrust 恒定推力
fixed thrust roller 固定回转堆料机
fixed-tiltable reflector 半自动反射面
fixed-time broadcast 定时广播
fixed-time call 定时通话;定时呼叫
fixed-time delay 定时延迟
fixed-time delay of slave station 副台固定时延
fixed-time humidity 定时湿度
fixed-time increment 定时时间增量
fixed-time incrementing 定时时间增长
fixed-time-lag 定时限;定时滞后
fixed-time linear system 线性时不变系统
fixed-time mode 定向放行式
fixed-time signal 定时信号
fixed-time temperature 定时温度
fixed-time testing 定时试验;定时检验
fixed-time testing plan 定时试验计划;定时试验方案;定时测定方案
fixed-time traffic signal 定时交通信号
fixed timing mark 固定记时标记;定时标记
fixed to floating point conversion 定点浮点转换
fixed tooth 恒齿
fixed tooth plate 固定齿板
fixed tower 固定塔
fixed tower system 固定塔桩系统;固定水准点塔系统
fixed trace 固定迹线
fixed tracking station 固定跟踪站
fixed transmembrane pressure 固定流通膜压
fixed transmitter 固定发射机
fixed transom 固定亮子;固定气窗
fixed tray 固定盘
fixed tremie concrete 固定导管法水下灌注混凝土
fixed trip 固定跳闸;固定断路器
fixed trip circuit breaker 手动跳闸断路器
fixed tripper 固定的卸料装置;固定的铁路信号装置
fixed trip switch 手动跳闸开关
fixed tube sheet 固定管板
fixed tube sheet exchanger 固定管板换热器;固定管板式换热器
fixed tube sheet heat exchanger 列管式固定管板换热器
fixed tube sheet type 固定管板式
fixed tube sheet type heat exchanger 固定管板式热交换器
fixed-tuned amplifier 固定调谐放大器
fixed-tuned crystal detector 固定调谐晶体检波器
fixed-turned loop detector 定调式线圈检测器
fixed twopoint suture 二定点缝合法
fixed type crane 固定式起重机
fixed type double-pipe heat exchanger 固定式套管换热器
fixed type entry guide 固定型导架【船】;固定式导口
fixed type heat exchanger 固定管板式热交换器
fixed type metal-clad switchgear 固定型铠装开关装置
fixed type rigid coupling 固定式刚性联轴节
fixed unit price contract 固定单价合同
fixed value 固定值;定值;不变值
fixed vane 固定轮叶;定叶片
fixed vane turbine 定叶式水轮机;定桨式水轮机
fixed variable 固定变量
fixed-variable approach to know cost behavior 了解成本动态固定变异法
fixed vertical polarization 固定垂直偏振
fixed voltage 固定电压
fixed wages 固定工资
fixed water 死水;不流动的水;固着水;固定水分
fixed water cooled burner 固定式水冷燃烧器
fixed water spray system 固定喷雾灭火系统
fixed wave form generator 固定波形发生器
fixed wave number 固定波数
fixed way 固定台架;滑道
fixed weighting 固定加权
fixed weight moment 死荷载力矩
fixed weight stress 死荷载应力
fixed weight type hand roller 固定重量型手压辊
fixed weir 固定溢流堰;固定堰;固定式堰
fixed welding machine 固定式焊机
fixed wheel base 固定轴距
fixed wheel gate 定轮闸门
fixed width finisher 定宽修整器
fixed window 稳固窗;固定窗
fixed wing aircraft 固定翼飞机
fixed wing base manager 固定翼飞机基地经理
fixed wire 固定丝
fixed-wire logging 固定钢丝集material
fixed wiring method 固定布线法
fixed with bolt 用螺栓扣紧;用螺栓固定
fixed word length 固定字长
fixed wrench 固定扳手
fixed zero system 固定零点系统
fixed zoom lens 固定的变焦距物镜
fixer 砌墙工人;(现场的)抹灰工;胶结剂;定色剂;修车工;固定器;固定剂;胶粘剂;定影液;定影剂
fixer network 定位网络
fixer's bedding 圬工石灰浆;圬工石灰膏;打底石灰膏
fixer system 定位网络
fixer upper (需要大修后按市价出售的)房地产
fix-fix coupler type 嵌插型连接机构
fix-glass storefront 玻璃店面
fixing 固定操作;固定;定影;成交
fixing accessories 固定辅助设备
fixing aeration 固定曝气
fixing agent 固色剂;固定剂;定影剂
fixing aid 定影助剂;定位(辅助)设备
fixing angle 定位角
fixing base 固定基础;固定地基
fixing bath 定影液;定影槽
fixing batsh production on a periodic basis 以期定量
fixing block 木落砖;受钉砌块;受钉块
fixing bolt 固定锚杆;螺栓标志
fixing bracket 固定托架;托座;牛腿;壁灯架
fixing brake 止动闸;上动固定闸;上动防松闸
fixing brick 受钉砖块;木落砖;受钉砖;受钉块
fixing by cross bearings 方位定位
fixing by distance 距离定位
fixing by horizontal angles 水平夹角定位
fixing by landmarks 陆标定位
fixing by nails 用钉固定
fixing by very high altitude sun sights 太阳特大高度求船位
fixing by wedges 用楔固定
fixing cabin 固定型小室
fixing clearance 安装空隙;安装净空
fixing clip 固定卡;固定夹
fixing collar 加固圈
fixing compound 镶玻璃油灰
fixing concrete 装修加固混凝土
fixing cord 固定线
fixing degree 固定程度
fixing development 定显影
fixing device 固定装置;紧固件
fixing dimension 固定尺寸;装配尺寸
fixing fillet 嵌缝木条;固定嵌条;木针;受钉嵌条
fixing flange 固定翼缘
fixing foot 固定撑脚
fixing frame 固定框架
fixing-hardening bath 坚膜定影液
fixing hole 固定孔;安装孔
fixing instruction 安装说明
fixing length 固定长度
fixing letter 成交函
fixing level 定影水平;定影级别
fixing lug 固定扣钉;固定耳
fixing material 固定材料
fixing means 固定手段
fixing method 固定方法
fixing model scale 模型比例尺确定
fixing moment 固定力矩
fixing mortar 固定砂浆
fixing nut 固定螺母
fixing of budget 编制预算
fixing of position 定船位
fixing pad 受钉压缝条;受钉嵌条
fixing piece 固定块;固定件
fixing plate 固定板
fixing point 固定点;定点
fixing powder 固色粉;定影粉
fixing price according quality 按质论价
fixing process 定影法
fixing profile 固定外形
fixing rail (钢结构的)固定用横档
fixing ring 固定环
fixing route bus 定线公共汽车
fixings 嵌固件
fixing salt 定影剂
fixing screw 固定螺钉;定位螺钉
fixing section 固定断面
fixing shape 固定形状
fixing slip 木针;固定嵌条;固定嵌带
fixing solution 固定液;定影液
fixing stirrup 固定卡子;固定箍
fixing strip 固定条带;钉板条
fixing system 固定系统
fixing technique 固定技术
fixing the tunnel alignment 隧道定线
fixing the tunnel levels 定隧道高程;定隧道标高
fixing tray 定影盘
fixing trim 固定镶边
fixing unit 固定单元
fixing-up expenses 出售房前的修理费
fixing wall 固定墙
fixing web 固定腹板
fixing wedge 固定楔块
fixing wire 固定铁丝
fixing work 固定工作;定位工作
fixing work according to quotas 定额包工
fix-it shop 修理店
fixity 稳定性;凝固性;固定性;固定度;不挥发性
fixity at the connection 连接稳固性
fixity depth 嵌固深度
fixity involatile 不挥发性
fix-pakced bed photocatalystic reactor 固定式填充床光催化反应器
fix-pin butt 固定轴铰链
fix-pitch airscrew 定距螺桨
fix-pitch propeller 定距螺桨
fix planting 定植
fix point 定位点

fixpoint induction method 不动点归纳法
fixpoint theory 不动点理论
fixpoint theory of program(me) 程序不动点理论
fix rates 确定费率
fix screw 固定螺钉
fix stopper 固定挡销;定位销
fix the price 定价
fixton pad 座盖
fixture 固定(装置)物;工件夹具;紧固;夹具;卡具;附属装置
fixture assembly 夹具组合件
fixture block 卡块
fixture branch 卫生设备分支管;固定分支;卫生器(具)支管;(供水管的)固定装置分支;分支管
fixture cut-out 线盒熔丝
fixture date 成交期
fixture drain 卫生设备给水管;排水连接管;固定装置的下水道;卫生器(具)排水管
fixture joint 固定连接
fixture note 租船成交备忘录;定期票据
fixture of ro/ro ship 滚装船固定装置
fixtures of handrail 扶手卡
fixture-supply pipe 卫生器(具)给水管;卫生设备给水管;固定装置给水管
fixture trap 卫生设备疏水器
fixture type 成交类型
fixture unit 卫生器具当量;卫生器具单位
fixture-unit flow rate 卫生设备单位流率
fixture vent 器具通气管;卫生通气管
fixture wire 设备引线;灯头线
fixture with integral mounting rail 有整体拼装轨道的夹紧装置
fix-type suspension 刚性悬架
fix with plugs 栓连接
fix with rivet 铆钉固定
fix with screw 螺钉固定
Fizeau fringe 斐索(干涉)条纹
Fizeau interferometer 斐索干涉计
Fizeau toothed wheel 斐索齿轮法
fizelyite 菲辉锑银铅矿
fizz 漏气;充气饮料
fjard 低狭湾
fjeld 冰蚀高原(北欧等地);冰蚀高地(北欧等地)
fjord 峡湾型海湾;峡湾;海礁
fjord coast 峡湾型海岸线;峡湾型海岸
fjord coastline 峡湾型海岸线
fjord like 峡湾型的;港湾式的
fjord shore 峡湾型海岸
fjord shoreline 峡湾型海岸线
fjord-type coast 峡湾型(型)海岸
fjord type coastline 峡湾型海岸线
fjord valley 峡湾谷
FK3 Catalogue 第三基本星表
F-K migration with variable velocity 变速的 F-K 偏移
flabbiness 松弛;软弱;坯体塌软
flabby 松软的
flabby cast 软注坯
flabellate 扇形
flabelliform 扇形
flab slab (带柱帽的)无梁楼盖
flaccid 弛缓的
flag 镜头遮光罩;旗(帜);铺路石板;特征位标记;多摄毛刷;船旗;薄板;标帜;标志;标识;标记
flag alarm 报警信号器
flag arrangement 旗形装置
flag bit 特征位;标志位
flagboat 旗艇
flag bridge 司令台;司令桥楼
flag build 记号的形成
flag buoy 带旗浮标
flag button 特征符按钮
flag byte 标志字节
flag chest 旗箱
flag clip 旗缆铜扣
flag cloth 旗布
flag code 标记码
flag condition 标志条件;标记条件
flag control 标志控制
flag crown form 旗冠型
flag data 标志数据;标记数据
flag dressed vessel 挂彩旗的船
flagelliform 鞭毛状的
flagellum 鞭毛

flagellum staining 鞭毛染色法
flag event 标记事件
flag flipflop 标志触发器;特征位触发器
flag float 信号浮标
flag form 旗型
flag gate 卧倒闸门
flagger 信号旗
flagger vest (交通警察或工程人员的)防护衣
flagging 下垂的;松弛的;铺砌石板;铺路;石板路(面)
flagging preposition 标志介词
flagging stone 石板(铺路用);扁石
flaggy 充满石片的(土壤);薄砂岩的;薄层状;板层(状)
flaggy rock 薄层状岩体
flag hooks 旗钩
flag indicator 旗语信号指示器;旗号;动作指示器;标志显示
flag input 标志输入
flag kit 旗帜箱
flag line 旗绳;标志线路
flag locker 旗箱;旗柜
flag logic function 标志逻辑函数
flag lot 旗形地块
flagman 旗工;信号旗手;司旗员;测旗工;扬旗工【铁路】;信号工;交通控制员
flag manifold 旗流形
flag of convenience 方便旗【船】
flag of distress 求救信号
flag operand 特征位操作数;标志操作数
flag operation 标志操作
flag paving 石板路面
flagpole 标杆;旗杆;花杆(测);视距尺;测视信号;条状信号
flagpole antenna 金属杆天线
flagpole shadow path 棒形图
flag post 旗杆
flag register 标志寄存器;标记寄存器
flag rod 花杆【测】
flags 铺石路(面);铺石板路
flag sense 标志读出
flag sequence 标志序列;标志顺序
flag shift operation 标志移位操作
flag-ship 旗舰
flag signal 旗语;手势信号
flag signal(l)ing 旗号通信
flag slab 毛板
flagstaff 旗杆
flagstaffite 柱晶松脂石
flag state 船旗国
flagstone 扁石;面砖;铺路石板;铺砌石板;石板(道);薄层(砂)岩;板状(砂)岩;板石;板层(砂)岩;铺路石
Flagstone 赫德森河蓝灰砂岩
flagstone flooring 石板地面(室内)
flagstone path paved at random 随意组合方石板路
flagstone pavement 石板铺砌;石板路面
flagstone paving 用石板路砌;石板铺砌;石板路面
flagstone pitching 石板铺砌
flagstone walk 石板路
flag stop 标志停车点
flag surtax 不同国籍进口货附加税
flag switch 旗形开关
flag toggle 旗绳木扣
flag tower 旗塔
flag wagging 手旗信号
flag word 标志字
flail 扑水器;甩刀式切碎装置;扫雷装置;碎石锤
flail arm 连枷臂
flail cutter 甩刀式切碎机
flail hanger retainer bolt 甩刀固定螺栓
flail knife 甩刀;甩板
flail mechanism 甩刀装置
flail mower 甩刀式割草机
flail rotor 甩刀式旋转切碎机;甩刀式滚筒
flail row cleaner 甩刀式清垄器
flail-type beater 连枷式击碎轮
flail-type blade 甩刀式刀片
flair 鉴别力
flaired head 倒锥形柱头
flakboard 压缩板
flake 小片;舷侧踏板;一层绳;鳞片;片状;片材;水带的带束叠装;石片;吊板;成片剥落;薄片

flake alumin(i)um 片状铝粉;薄铝片
flake alumin(i)um pigment 薄片铝制的颜料
flake alumin(i)um powder 薄片状铝粉
flake asbestos 石棉粉;石棉薄片
flake-bark 片状树皮
flake board 碎木板;碎料板;木屑刨花板
flake calcium chloride 片状氯化钙
flake caustic 片状烧碱
flake composite 薄片组合件
flake copper 片状铜粉
flake crack 白点
flaked alumin(i)um powder 薄片状铝粉
flaked asbestos 片状粉末石棉;薄片石棉
flaked graphite 薄片石墨
flaked hose 带束叠装的水带
flaked ice plant 片冰设备;片冰厂
flaked iron oxide 薄片氧化铁
flaked mica 鳞片状云母
flaked powder 片状炸药;片状粉末
flaked rail 剥离的钢轨
flake film 薄片
flake glass 鳞片玻璃;(装饰用的)薄片碎玻璃
flake glass coating 薄玻璃涂层
flake glass reinforcement 片状玻璃加强
flake gold 片金
flake gold powder 片状金粉
flake graphite 鳞片石墨;片状石墨;石墨粉
flake ice 片冰;薄片冰;刨花冰
flake ice maker 片冰机;刨花冰制冰机
flake iron powder 片状铁粉
flake lead 铅粉;铅白油涂料
flakelet 小片
flake-like powder 片状粉末
flake litharge 鳞状氧化铅;鳞状密陀僧
flake metal powder 片状金属粉
flake off 片落;脱皮;呈鳞片状剥落;剥掉;剥落(掉)
flake of rust falling from old iron 从旧铁上落下的铁锈层
flake of sluge 泥饼
flake pigment 片状颜料
flake powder 片状炸药
flaker 结片机;刨片机;包片机
flake runoff corrosion 片状径流溶蚀
flake-shaped particle 片状颗粒
flake-shaped structure 鳞片状构造
flake shellac 片紫胶;片胶
flake structure 薄片结构
flake surfaced asphalt felt 片料覆盖油毡
flake texture 片状结构
flake tool 刮削器
flake white 铅白;碳酸铅白;白绘画颜料
flake yarn 鳞片线
flakiness 片状;片层分级;成片性;扁平度
flakiness index 剥落指数
flakiness ratio 宽厚比
flaking 制片;鳞片;片状脱落;片状剥落;剥落;脱片;掉皮;成片剥落;超薄片切削;草屋顶的底层
flaking gel coat 胶化剥落
flaking machine 轧片机;刨片机
flaking mill 制片机
flaking of brickwork 砖面剥落
flaking resistance 抗片状剥落性;耐片落性;耐剥落性
flaking window 边窗
flak jacket 避弹衣
flaky 有白点的;易剥落的;鳞状的;片(的);成片的;薄片状
flaky aggregate 片状集料;片状骨料
flaky and elongated aggregate 细长片状骨料
flaky black carbon 片状炭黑
flaky constituent 片状成分
flaky fir 鳞皮冷杉
flaky grain 片状颗粒
flaky graphite 片状石墨;薄片状石墨
flaky material 片状材料
flaky particle 片状颗粒
flaky resin 片状树脂;粉状树脂
flaky snow 片状雪
flaky structure 薄片状构造
flaky texture 鳞状结构;薄片状结构
flaky web 厚度不匀纤维网
flam 外展(船首)
flambeau 装饰烛台;火炬;燃烧废气的烟囱

flamboy 警告火焰
flamboyance 火焰式建筑风格
flamboyancy 火焰式建筑风格
flamboyant arch 焰式拱；火焰式拱
flamboyant architecture 焰式建筑；火焰式建筑
flamboyant finish coating 罩在光亮的底层漆或金属表面上的光亮清漆或有色透明漆
flamboyant period 火焰式曲线时期
flamboyant red glaze 火焰红釉
flamboyant rose window 焰式圆花窗；火焰式圆花窗
flamboyant structure 辉耀构造
flamboyant style window 火焰式(窗)
flamboyant tracery 焰式窗(花)格；火焰式窗(花)格
flamboyant tracery window 火焰式花格窗
flamboyant tree 凤凰树；凤凰木
flame 火焰
flame ablation 熔化烧蚀；溶化烧蚀
flame absorption 火焰的吸收
flame accelerator 促燃剂
flame adjustment 火焰调节
flame amphogneiss 火焰状混合片麻岩
flame analysis 焰色分析；火焰分析
flame anchor 火苗
flame angle 火焰角
flame arc 焰弧
flame arc lamp 焰弧灯；弧光灯
flame area 火焰面积
flame arrester 阻焰器；阻火器；灭火器；火焰消除器；防焰罩；防火星网
flame arrester for oil-tanker 油船用阻火器
flame arrester for petroleum tank 石油储罐阻火器
flame arrester for pipe 管道阻火器
flame atomic absorption spectrophotometry 火焰原子吸收分光光度法
flame atomization 火焰原子化法
flame atomizer 火焰雾化器
flame attachment 火焰附加器
flame attenuating process 火焰牵伸工艺
flame attenuation 火焰拉长；火焰吹拉
flame attenuation burner 长火焰喷嘴
flame augmentation 火焰加强
flame axis 焰轴
flame azalea 火焰杜鹃花
flame background 火焰背景
flame background noise 火焰本底噪声；火焰背景噪声
flame band 火焰辐射带；火焰带
flame base 焰底
flame black 焰黑
flame blowing 火焰喷吹法
flame blowing process 火焰喷吹法
flame blow-off 火焰吹灭
flame blow-off factor 火焰吹灭因数
flame body 焰体
flame brazing of alumin(i)um bar section 铝型材火焰钎焊
flame breakdown 火焰中断
flame break point 火焰断裂点
flame breakthrough 烧穿
flame bridge 焰桥
flame bucket 火焰反散(射)器
flame bulb 焰球
flame bush 火焰舌
flame cap 焰晕；蓝色焰晕
flame carbon 发弧光炭棒；发光炭精棒
flame carbon arc lamp 发光炭精棒弧光灯
flame catcher 消焰器
flame chamber 火焰室
flame characteristic 火焰特性
flame checking 阻燃整理
flame chemiluminescence detection 火焰化学发光检测
flame chilling 火焰熄灭
flame chipping 火焰熄灭；烧剥
flame circumstance 火焰情况
flame cleaner 火焰除锈器
flame cleaning 火焰净化法；火焰清洗；火焰清理；火焰清洁(法)；火焰清除法；火焰除锈(法)；火焰除漆(法)
flame cleaning blowpipe 火焰表面清理焊炬
flame coal 焰煤；长焰煤
flame coating 火焰喷镀
flame colo(u)ration 焰色

flame colo(u)ration test 焰色试验
flame colo(u)red 橘黄色火焰；橘黄(色)
flame colo(u)r test 焰色试验
flame combustion 热力燃烧
flame cone 焰芯
flame configuration 火焰场分布
flame contact ignition 火焰接触点火
flame-cored carbon 焰炭碳棒
flame couple 热电偶
flame coverage area 火焰覆盖面积
flame covered 火焰覆盖的
flame covered area 火焰覆盖面积
flame cross-section 火焰横截面
flame cultivator 喷焰除草机
flame-curing method 火焰固化法
flame cut 气炬切割；火焰切割
flame cut edge 焰边切
flame cutter 切割吹管；割矩；火焰切机；火焰切割器；气割机
flame cutting 火焰切割；火焰切割法；气割
flame cutting method 火焰切割法
flame damper 消焰器；灭焰器；灭火器；防火器
flame depth 火焰深度
flame descaling 火焰除鳞；喷焰除鳞
flame desiccating 喷焰除鳞
flame detection apparatus 火源监测仪
flame detection device 火焰探测装置
flame detector 火焰检测器；火焰探测器
flame dike 挡火墙
flame distribution 火焰分布
flame diverter 折焰器
flame dolomite 焰白云石
flame drifting 火焰发飘
flame drilling 火钻钻进
flame drip 燃烧溶滴
flame drop-back 火焰后缩
flamed slab 烧毛板
flame duration 燃烧时间
flame ejaculation 火焰喷射
flame emission 火焰发射
flame emission chopper 火焰发射断续器
flame emission detector 火焰发射探测器；火焰发射检测器
flame emission photometry 火焰发射光度法
flame emission spectrometer 火焰发射光谱仪
flame emission spectrometry 火焰发射光谱法
flame emission spectrophotometer 火焰发射分光光度计
flame emission spectrum 火焰发射光谱
flame emissivity 火焰发射率；火焰发射度
flame enrichment 加火
flame envelope 火焰的包围物；火焰包络线
flame equation 火焰方程
flame erosion resistance 抗火焰冲刷性
flame exposure test 暴露于火焰试验
flame extension 火焰扩展
flame extinction 火焰熄灭
flame extinction time 灭火时间
flame extinguishing ability 灭火能力
flame extinguishing concentration 灭火浓度
flame eye 火焰监视器
flame failure 熄火
flame failure control 火焰防灭控制；火焰调节器
flame failure control device 火焰中断控制器
flame failure controller 火焰消失控制器
flame failure control valve 灭火剂控制阀
flame failure detection 火焰失效探测；火焰观测
flame failure device 熄火安全装置；熄火保护装置
flame failure protection 灭火保护
flame fire detector 火焰消防探测器；火焰探测器
flame flash arrestor 火焰防止器；火花消除装置
flame flash-back 逆燃；火焰闪灯
flame flicker detector 火焰闪烁探测器
flame formation 火焰的形成
flame fringe 火焰边缘
flame front 火焰前缘；火焰锋
flame front area 火焰前锋面积
flame front velocity 火焰前锋速度
flame fuel 火焰燃料
flame furnace 火焰窑；反射炉
flame fusion method 熔法
flame-generated 火焰引起的
flame gouging 火焰挖槽法；火焰清铲；火焰表面切割
flame grooving 气割开坡口

flame gun 火焰喷射机；火焰喷枪；喷灯
flame hardening 火焰表面硬化；火焰(表面)淬火
flame hardening blowpipe 火焰表面淬火烧炬
flame hardening machine 火焰硬化机；淬火硬化机
flame hardnessing 火焰淬火法
flame heater 火焰加热器
flame holder 火焰稳定器
flame hole 火焰孔
flame igniter 点火装置
flame ignition 火焰点火
flame ignitor 点火器
flame impingement 火焰冲刷
flame incineration 火焰焚烧
flame infrated emission detector 火焰红外发射检测器
flame inhibition 火焰抑制
flame inhibitor 火焰抑制剂
flame ionization 火焰电离
flame ionization analyzer 火焰电离分析仪
flame ionization analyzer and detector 火焰电离分析检测器
flame ionization chromatograph 火焰电离色谱仪
flame ionization detector 火焰离子化检测器；火焰电离检测器
flame ionization detector capillary gas chromatography 火焰电离检测器毛细管气相色谱法
flame ionization gas chromatography 火焰电离气相色谱法
flame ionization ga(u)ge 火焰电离计
flame ionization method 火焰电离法
flame ionization method survey 火焰电离法检漏
flame ion mass spectrometry 火焰离子质谱法
flame jet 火焰喷口
flame-jet drilling machine 火焰喷射钻进机；火焰喷射凿岩机
flame kernel 火焰中心
flame lamp 火焰灯
flame leaf birch 亮叶桦
flame length 火焰长度
flameless atomic absorption spectrophotometer 无焰原子吸收分光光度计
flameless atomic absorption spectrophotometry 无焰原子吸收法
flameless atomic absorption spectroscopy 无焰原子吸收光谱法；非火焰原子吸收光谱法
flameless atomization 无火焰原子化法
flameless atomizer 无烟雾化器
flameless burner 无焰烧嘴；无焰燃烧器
flameless burning 无焰式燃烧
flameless catalytic combustion 无焰催化燃烧
flameless catalytic gas heater 无火焰催化气体加热器
flameless combustion 无焰燃烧
flameless explosive 无烟火药
flameless ionization detector 无(火)焰电离检测器
flameless powder 无焰火药
flameless procedure 无焰法；非火焰法
flame lighter 点火器
flame-lift 离焰
flame-like 火焰状
flame-like tracery 火焰式窗格
flame-like tracery window 火焰式花格窗
flame luminosity 火焰亮度
flame machining 火焰表面加工
flame melting method 气炼熔融法
flame microphone 火焰传声器
flame monitor 火焰监视器
flame movement 火焰的移动
flamenco pigment 斑彩(珠光)颜料
flame noise 燃烧噪声
flame nozzle 火焰喷嘴
flame of shot 炮焰
flameout 熄火
flame pass 火焰通路；火焰通道
flame path 火焰通路；火焰路线
flame photometer 火焰光度计
flame photometric(al) analyzer 火焰光度分析仪
flame photometric(al) detection 火焰光度检测；火焰电离检测
flame photometric(al) detector 火焰光度检测器
flame photometry 光焰光度分析；火焰光谱法；火焰光度学；火焰光度术；火焰光度分析；火焰光度法；火焰测光法
flame photometry detection 火焰电离检测

flame photometry detector 火焰光度检测器
flame pipe 火焰管
flame planer 龙门式自动气割机
flame plating 火焰喷镀;喷涂;爆震喷涂
flame point Tag 泰格法闪点
flame polishing 火焰抛光
flame port 火孔
flame-pretreatment 塑料火焰预处理
flame projector 火焰喷射器;喷火器
flameproof 隔爆的;防火焰;防火的
flameproof construction 防爆结构
flameproof diesel loco(motive) 防火柴油机车
flameproof electric(al) motor 防爆电(动)机
flameproof enclosure 机械防火罩;机械防火外壳;防爆罩(采矿);防火外壳
flameproof engine 防火式电机
flameproof fabric 防火织物
flameproof glass 防火玻璃;耐热玻璃
flameproofing 耐热的
flameproofing agent 阻燃剂;耐火剂
flameproofing wall 耐火墙;防火墙
flameproof lamp 防爆灯;火焰安全灯
flameproof luminaire 防火照明设备;防爆灯;防火照明灯
flameproof machine 防爆式电机
flameproof material 耐火材料
flameproof mine locomotive 防火矿用机车
flameproof motor 防爆耐火电动机;防爆电动机
flame proofness 耐火性
flameproof organic fiber 防燃有机纤维
flameproof paint 耐火漆
flameproof switch 防爆开关;防爆式开关
flameproof textile 防燃织物
flameproof three-phase induction motor 防爆型三相感应电动机
flameproof transformer 防爆式变压器
flameproof ware 防火器皿;耐热器皿
flame propagation 火焰分布;火焰传布;火焰传播
flame protection 防爆保护装置
flame pyrometer 火焰高温计
flame quenching 火焰淬火
flamer 火焰喷射器
flame radiation 火焰辐射;火焰的辐射
flame reaction 焰色反应;火焰反应
flame reactor 火焰反应器
flame regulation 火焰调节
flame resistance 抗燃性;耐燃性;耐火性;耐火度;阻燃性
flame resistant 阻燃性;阻燃剂
flame-resistant cable 耐火电缆
flame-resistant coated kraft paper 防火涂料覆牛皮纸
flame-resistant fabric 耐火织物
flame-resistant glass 耐火玻璃
flame-resistant material 阻燃材料
flame-resistant plastic 耐火塑料
flame-resistant polyester resin(a) 耐火聚酯树脂
flame-resistant tarpaulin 阻燃盖布;耐火盖布
flame-resisting 耐火的
flame resisting wall 阻火墙
flame resistivity 阻燃性
flame resonance spectrometer 火焰共振分光计
flame-retardant 耐燃剂;阻燃剂;阻燃物;耐火处理;火焰抑制剂;阻燃纤维板;阻燃(烧)剂;阻燃的
flame-retardant asphalt felt 阻燃性油毡
flame-retardant carpet cushion 阻燃地毯垫
flame-retardant chemical 阻燃剂
flame-retardant coating 阻燃涂料;阻燃涂层;耐火层
flame-retardant construction 耐燃构造
flame-retardant contact adhesive 防火性万能胶
flame-retardant fabric 阻燃织物
flame-retardant fiber 防燃土工布;阻燃纤维
flame-retardant for textile 纺织品阻燃剂
flame-retardant grade 阻燃性等级
flame-retardant of plastics 塑料阻燃剂;塑料火焰抑制剂
flame-retardant paint 耐燃漆;阻燃涂料;阻燃漆
flame-retardant panel 阻燃火板
flame-retardant partition 防火隔墙;阻燃隔墙
flame-retardant plastics 阻燃性塑料
flame-retardant resin 阻燃树脂
flame-retardant-treated wood 阻燃处理木材
flame-retardant treatment 阻燃处理
flame-retarded ABS polymer 阻燃丙烯腈=丁烯=苯乙烯聚合物
flame-retarded resin 阻燃树脂
flame retardent 灭火剂
flame retarder 阻燃烧剂
flame-retarding cable 阻燃电缆
flame-retarding glass 耐热玻璃;防火玻璃
flame-retarding polymer 阻燃聚合物
flame retention 火焰稳定;火焰保持力
flame reversal 火焰换向;换火
flame safeguard 熄火安全装置;熄火保护装置;燃烧安全装置
flame safety lamp 保险灯
flame scaling 锌镀层火焰加固处理;热浸镀锌(钢丝)
flame scarfing 火焰清理
flame screen 火星防护网;防火网
flame seal 火焰封接
flame seal galvanizing 钢丝火封软熔热镀锌法
flame sensor 火焰检测器;火焰传感器
flame shape 火焰形状
flame shield 火焰反射器;遮火罩;耐火墙;火焰防护;挡火墙
flame soldering 低温火焰罩
flame space 火焰空间
flame spectrometer 火焰光谱仪;火焰分光计
flame spectrometric analysis 火焰光谱分析
flame spectrometry 光焰分光;火焰光谱法;火焰谱术
flame spectrophotometer 火焰分光光度计
flame spectrophotometry 火焰光度分析;火焰分光光度术;火焰分光光度法
flame spectrum 火焰光谱
flame speed 火焰蔓延速度;火焰传播速度
flame speed classification 火焰扩散速度分级
flame spray coating 火焰喷涂法;火焰喷涂
flame-sprayed 火焰喷敷的
flame-sprayed ceramics 火焰喷溅陶瓷;火喷涂层陶瓷
flame-sprayed spread 火焰喷出扩张
flame-sprayed travel 火焰喷出传播距离
flame sprayer 火焰喷射机
flame spray gun 火焰喷涂枪;火焰喷枪
flame spraying 火焰喷涂(法);火焰喷射(法);火焰喷镀;热熔喷镀法
flame spraying ceramics 火喷涂层陶瓷
flame spraying equipment 火焰喷涂装置;火焰喷涂设备
flame spraying pyrometer 火焰喷射高温计
flame spray powder coating 粉末火焰喷涂法
flame spread 火焰蔓延;播焰
flame spread classification 火焰扩散分级
flame spread classification test 耐火等级试验(建筑材料)
flame spreader 火焰扩张器
flame spread index 火焰扩散指数;火焰蔓延指数
flame spreading factor 火焰蔓延率
flame spreading rate 火焰蔓延率
flame spreading rating 火焰蔓延指数;火焰蔓延等级
flame-spread rating 火焰扩散等级
flame spread retardancy 阻燃性
flame stability 火焰稳定性
flame stabilization 火焰稳定
flame-stabilized burner 稳(定)百燃烧器
flame stabilizer 火焰稳定器
flame stack 炼厂燃管;火舌管
flame sterilization 火焰灭菌
flame strand annealing 喷焰连续退火
flame structure 火焰状构造;火焰构造
flame temperature 火焰温度
flame temperature detector 火焰温度检测器
flame test 焰色试验
flame thermionic detector 火焰热离子检测器
flame thermocouple detector 火焰热电偶检测器
flame thrower 火焰喷射器;喷火器
flame-thrower nozzle 火焰喷嘴
flame-throwing drill 火焰钻机;火力凿岩机
flame trap 吸焰器;阻火器;隔焰器;火焰阻止罩;防焰器;回火装置
flame travel 火焰行程;火焰的移动
flame treating 火焰处理
flame treatment 火焰处理
flame tree 凤凰树;凤凰木
flame tube 炉胆;火管
flame tube interconnector 联焰管
flame type heater 火焰式加热器
flame-type recombiner 火焰式复合器
flame velocity 火焰传播速度
flame ware 烹饪器皿;耐火器皿;厨用玻璃器皿;耐热玻璃器皿
flame weeder 火焰除草器;喷焰除草机
flame weeding 火焰除草
flame weld 火焰焊;熔焊
flame welding 气焊;火焰焊接;熔气焊;烧焊;熔焊
flame wool 火焰法玻璃棉
flame zone 火焰带
flamge gas 火焰气
flaming 喷火
flaming durability 发焰延长时间
flaming test 冲头扩孔法钢材延性试验
flaming weeder 火焰除草机
flammability 易燃性;可燃性
flammability index 可燃性指数
flammability limit 可燃性极限;自燃的极限
flammability point 燃烧点
flammability resistance 防火性
flammability test 可燃性试验
flammable 可燃的
flammable air-vapo(u)r mixture 易燃空气油气混合物
flammable inhibitor 发火抑制剂
flammable limit 着火浓度极限;爆炸极限
flammable liquid 可燃液体
flammable material 易燃品;易燃材料
flammable mixture 可燃混合物
flammable range 可燃区间
flammable refrigerant 易燃制冷剂
flammable solid 易燃固体
flammable vapo(u)r 可燃蒸气
flammable vapo(u)r concentration 可燃蒸气浓度
flammable waste 可燃烧废物;可燃废物
Flamsteed's number 弗兰姆斯蒂数
flan 弗兰风
flanch (烟囱防水的)凸缘
Flanders storm 弗朗德风暴
Flandrian transgression 弗朗德里安海侵
flane stabilizer 钝体
flange 接盘;合缝凸出;弯边;突边;凸缘(盘);凸边;法兰;安装边缘
flange adapter 法兰接头
flange and elbow bend pipe 法兰盘弯管
flange and flare elbow 法兰扩缘弯头
flange and flare piece 法兰扩缘弯头;法兰扩缘短管
flange and plain connecting piece 法兰与平口连接件
flange and spigot bend 单盘插口弯管;法兰套管弯头
flange angle 翼角钢;翼缘角钢;凸缘角铁
flange angle iron 翼角铁
flange angle steel 凸缘角钢
flange area 凸缘面积;法兰面积
flange area method 翼缘面积法
flange back 凸缘衬圈
flange back case 凸缘后盖表壳
flange back face 止推挡边后端面
flange bar 带翼缘的杆件
flange base line 轮缘高度测定线
flange beam 工字钢;工字钢梁;工字梁;加翼梁
flange bearing 带法兰盘轴承
flange bend 法兰弯头
flange body 凸缘体
flange bolt 凸缘螺栓;法兰螺栓;法兰连接螺栓
flange brace 翼缘连系构件;翼缘紧固件
flange bracing 弦杆加劲;纵向连杆
flange breech(es) fitting 法兰叉形管
flange buckling 翼缘屈曲
flange butt joint 卷边对接接头
flange cast iron stop valve 法兰铸铁截止阀
flange cast steel sluice valve 法兰铸钢闸阀
flange chuck 凸缘卡盘;凸缘花盘
flange clearance 轮缘槽
flange combination 法兰组合
flange connected 凸缘连接的
flange connection 凸缘接合;凸缘连接;法兰连接
flange connector 凸缘连接器
flange cooled cylinder 散热片汽缸
flange cooler 盘式冷却器
flange cork closure 凸缘软木塞封口

flange coupling 凸缘结合;凸缘联轴节;法兰联结;法兰接合;翼缘联结;法兰联轴节
flange coupling adapter 法兰连接接头
flange cover plate 梁翼盖板
flange crimping 凸缘卷曲法
flange cross-section (梁的)翼缘横截面;翼缘横断面
flange curling 翼缘卷曲
flange cut 翼缘切口;翼缘切割
flange cut-off valve 法兰截止阀
flanged 有翼缘的;装有法兰的;带凸缘的;带法兰的
flanged ball valve 法兰球阀
flanged bar 带翼缘的夹板
flanged beam 折边梁;凸缘梁;工字梁;加翼梁
flanged bend 法兰弯头
flanged bolt 凸缘螺栓
flanged bottom 偏底;凸缘底
flanged bracket 折边肘板
flanged branch 法兰接头支管
flanged brass valve 凸缘铜阀
flanged breather valve 法兰呼吸阀
flanged bushing 凸缘衬套
flanged butt weld 弯边对接焊
flanged cast-iron pipe 法兰铸铁管
flanged clamping plate 凸缘夹紧板
flanged conic(al) head 折边的锥形封头
flanged connected 法兰连接的;凸缘连接的
flanged connection 法兰盘连接;法兰连接;凸缘连接
flanged coupling 法兰连接;凸缘连接
flanged cross 法兰四通
flanged cup 带外止动挡边的外圈
flanged cylinder 凸缘汽缸
flanged diaphragm ga(u)ge 法兰式膜片压力表
flanged differential pressure transmitter 法兰式差压发送器
flanged edge 弯边;拆缘
flanged edge joint 凸缘边接合
flanged edge weld 边缘焊;卷边焊
flanged elbow 法兰弯头
flange design bolt load 法兰设计螺栓荷载
flanged expansion joint 伸缩接头;伸长接头;法兰盘伸缩接头
flanged finish 飞边瓶口
flanged fitting 法兰管件
flanged gasket 凸缘衬垫;法兰衬垫
flanged gate valve 法兰闸阀
flanged girder 工字大梁
flanged joint 凸缘联轴节;法兰接头;法兰接合;卷边接头;凸缘式联轴器
flanged knee 折边肘板
flanged lip 凸缘边
flanged motor 凸缘电动机
flanged nozzle 法兰喷嘴;法兰喷嘴
flanged nut 法兰螺帽
flanged packing 凸缘密封垫;法兰衬垫
flanged pipe 带法兰管;法兰管
flanged pipe socket 凸缘承窝
flanged pipe spigot 凸缘套管
flanged pipe with threaded flanges 螺纹法兰的法兰管
flanged piston 凸缘活塞
flanged plate 翼缘板;翻边板
flanged plug 法兰塞
flanged press finish 飞边瓶口
flanged pressure pipe 凸缘压力管;法兰压力管
flanged pulley 凸缘皮带轮
flanged radiator 凸缘片式散热器;凸缘散热器
flanged rail 宽底钢轨
flanged rail wheel (火车的)凸缘轮
flanged reducer 法兰大小头;法兰减缩管
flanged return bend 法兰回转弯头
flanged ridge tile 凸缘脊瓦;凸缘顶砖
flange drilling 法兰钻孔
flanged ring 凸缘环
flanged rod 法兰杆
flange drum 无箍桶
flanged seam 凸缘接合;凸缘接缝
flanged seam riveting 凸缘接缝铆钉
flanged section 凸缘型钢
flanged shaft coupling 凸缘联轴节
flanged shear wall 带翼缘抗剪墙
flanged sheet 折缘板
flanged spool piece 凸缘弹簧弯头;双法兰短管

flanged T 法兰三通(管)
flanged tee 法兰三通(管)
flanged test 凸缘试验
flanged tube 法兰管
flanged-tube radiator 凸缘管散热器
flanged-type liner 凸缘式衬筒
flanged union 法兰联管;凸缘管接
flanged valve 法兰阀
flanged wall piece 法兰墙穿管;法兰穿墙管
flanged wheel 有缘轮;带凸缘轮
flange(d)wheel flange pulley 凸缘轮
flanged yoke 凸缘叉臂;法兰叉
flange eccentric reducer 法兰偏心渐缩管
flange element 凸缘元件
flange end 法兰端(头)
flange engine bearing 凸缘的主轴瓦
flange face 法兰结合面;法兰面
flange facing 法立接触面;法兰压紧密封面
flange fittings 法兰式管接头;法兰配件
flange focal distance 肩焦距;基面焦距;法兰焦距
flange for air pump 气泵法兰
flange for direct drive 法兰直接驱动
flange form 法兰式
flange for upper shell 上筒体法兰
flange gasket 凸缘垫圈;法兰衬垫;法兰衬垫
flange gasket prestress 法兰垫初期应力
flange gasket residual stress 法兰垫残余应力
flange ga(u)ge 凸缘规
flange girth 翼缘
flange grommet 凸缘垫环
flange height 轮缘高度
flange ice 岸冰
flange joint 凸缘联轴节;法兰接合;凸缘连接;凸缘接合;法兰盘连接;法兰盘接合;法兰连接;法定连接
flange joint of pipe 管道法兰螺栓连接
flange key 法兰扳手
flange length of head 封头法兰长度
flangeless 无突缘的;无凸缘的
flangeless liner 无凸缘衬筒
flange liner 凸缘衬套
flange lubrication device 轮缘润滑装置
flange lubricator 轮缘涂油器
flange machinery 起重机械
flange member 凸缘构件
flange moment 法兰力矩
flange moment under operating condition 法兰操作弯矩
flange moment under pretightening condition 法兰预紧弯矩
flange motor 凸缘底座电动机
flange mo(u)ded packing 帽形填密件
flange mount 凸缘安装;凸缘架
flange mounted 支于凸缘架的;凸缘紧固的;凸缘架的
flange-mounted magneto 凸缘磁电机
flange mounted type 法兰安装方式
flange mounting 凸缘式安装件
flange nozzle 凸缘喷嘴
flange nut 圆缘螺母;圆缘螺母;凸缘螺母;带缘螺母;法兰螺帽
flange of beam 梁翼缘;梁翼
flange of brasses 黄铜凸缘
flange of bush(ing) 衬套凸缘
flange of coupling 联轴器凸边;联轴节凸缘
flange of lower shell 下筒体法兰
flange of pipe 管子凸缘;管子法兰(盘)
flange of steel beam 钢梁翼缘
flange of valve 管阀凸缘
flange of wheel 轮缘
flange outer diameter 法兰外径
flange pass 凸缘孔型
flange pipe 法兰管;凸缘管
flange plate 翼缘加劲板;翼缘板;凸缘板;法兰盘;法兰板
flange protection 法兰保护
flange pulley 凸缘皮带轮
flange quality plate 压力容器级钢板
flanger 起梭机;弯边机;凸缘机
flange rail 平底钢轨
flange rating 法兰压力等级;法兰等级
flange resistance 轮缘阻力
flange rib 凸缘肋
flange ring 凸缘环
flange rivet 翼缘铆钉;凸缘铆钉

flange rod 翼缘杆
flanger turner 翻边机
flanges bolted design 法兰螺栓设计
flange seal 凸缘密封
flange section 凸缘断面
flange shaft coupling 法兰联轴节
flange slab 翼缘板;法兰板
flange sleeker 法兰杆
flange sleeve 凸缘套
flange slope 凸缘斜坡
flange spigot bend 单盘插口弯管
flange splice 翼缘镶板;翼缘拼接板
flange splice plate 翼缘镶板
flange spout 凸缘喷嘴
flange spreader 法兰扩张器
flange spring washer 法兰弹簧垫圈
flange steel 翼缘钢;凸缘钢
flange stiffening 凸缘加强;凸缘加劲
flange stop valve 法兰截止阀
flange strengthening 翼缘增强;翼缘加固
flange stress 法兰应力
flange stripper guide 出口护板
flange tap 凸缘分接头
flange tee 丁字凸缘;T形凸缘;二重交错凸缘
flange test 卷边试验
flange thickness 法兰厚度
flange tool 法兰杆
flange-to-web-joint 翼缘与腹板联结
flange-to-web weld 卷边焊缝
flange tube 凸缘管
flange type everlasting blow-off valve 凸缘式连续放泄阀
flange type motor 法兰式电动机
flange type shaft coupling 凸缘联轴节
flange union 法兰盘连接;法兰接头;法兰联管
flange up 作凸缘;镶边;卷边
flange washer 凸缘垫圈
flange way 轮缘槽;凸缘沟
flangeway clearance 轮轨游间
flangeway depth 轮缘槽深度
flangeway filler 轮缘槽间隔铁
flangeway groove 轮缘槽
flangeway width 轮缘槽宽度
flange weld 卷边焊缝
flange with welded neck 焊上颈圈的法兰
flange wrench 凸缘扳手
flange yoke 凸缘叉
flanging 摺翼缘;折缘;折边;卷(折)边
flanging angle 卷边角
flanging die 凸缘冲模
flanging machine 翻边机;折边压床;折边机
flanging press 压边机;折边(压)机;弯边(压力)机;翻边压力机
flank 肋部;船首顶流摆开;齿腹;齿侧;侧翼;侧(面)
flank ahead 加速前进【船】
flank angle 螺纹侧面角;侧面角(螺纹)
flank correction 齿面修正
flank development program(me) 侧翼式开拓系统
flanked 置于侧面的
flanker 侧堡
flank escape 侧翼漏头
flank face 刀后面
flank fire 侧面火
flank fire suppression 侧面扑火
flank hole 探水角孔;超前探眼;侧向探眼;侧面(钻)孔
flanking column 翼柱
flanking moraine 侧岸冰碛
flanking movement 侧向运行
flanking of sound 声的侧传
flanking on wheel tread 踏面剥离
flanking path 侧传途径
flanking rudder 倒车舵
flanking sound 侧传声
flanking sound transmission 侧向传声
flanking tower 翼塔;侧塔
flanking transmission 侧面传播;迂回传播
flanking transmission of sound 侧向透声;侧向传声;声音迂回传播;声音侧面传播
flanking window 边窗(大门);通道窗;侧向窗;走廊窗;侧翼窗
flank interference 齿面干涉
flank masonry wall 圬工墙;侧墙

flank meristem 侧面分生组织
flank moraine 侧碛
flank observation 侧方观测
flank of a fire 火侧翼
flank of play 老矿区浅孔钻进
flank of thread 螺纹面
flank of tool 刀具侧面
flank of tooth 齿面
flank profile 齿廓
flank protection 侧面保护
flank radius 侧面曲率半径
flanks 拱侧翼;石侧面
flank separation zone 侧分带
flank-siding 沿路边车道交通
flank traffic 外侧交通(沿道边车道的交通)
flank wall 侧面墙;山墙;侧(边)墙;边墙
flank wall balcony 侧阳台;侧跳台
flank wear 齿侧面磨损,侧面磨损
flank well 矿区边缘井
flannel 法兰绒
flannel disc 法兰绒擦光盘
flannelet(te) 棉法兰绒;绒布
flannelling 绒布擦坯
flanning 窗框两侧斜边;外张的;壁炉框两侧斜边;八字形的
flap 小舱室;座盖;转板;折翼;折板;铰链板;绞链板;活叶;活动板;片状物;拍动;拍打;短甲板;挡水板;副翼板;翻褶;翻板;活瓣
flap actuating gear 挡板导向拉杆
flap attenuator 片型衰减器;刀型衰减器;刀形衰减器
flap chair 折叠椅
flap check valve 阀瓣式止回阀
flap dock gate 卧倒坞门
flap-door 吊门;活板门;翻动闸门
flap gate 舌板闸门;铰链式闸门;逆止闸门;倾倒式闸门;卧式闸门;翻板(节制)闸门;舌曲(倾倒式)闸门
flap hinge 明合叶;翻板合叶;明(板)铰链
flap installation 车板安装
flap lubricator 垂口润滑器
flap panel 旋转板;襟翼板
flapped hydrofoil 带襟翼水翼船
flapper 活动挡板;号牌;片阀;有铰链的门;活瓣;挡板;插板
flapper nozzle 挡板喷嘴
flapper-type rain cap 挡板式压顶;挡板式墙帽;垂边式压顶;垂动式墙帽
flapper valve 挡止回阀;挡板阀;挡瓣阀;挡板阀;瓣阀
flapper-valve controller 挡板阀控制器
flapping angle 翼动角
flapping of a sail 顶风帆
flapping screw 带舌环螺钉
flapping sound 拍击音
flap pipe section 开口管壳
flappong 摇摆运动
flap-seat 折椅
flap-shaped incision 瓣状切开
flap shutter 瓣装挡板
flap structure 卷滑构造
flap table 折页桌
flap tile 折瓦;屋顶脊谷用瓦;凹瓦
flap trap (排水口的)逆止吊门;止回阀;逆止阀;吊门逆止阀;瓣阀
flap-type attenuator 刀形衰减器
flap type rudder 可变翼形舵;襟翼舵
flap(up)seat 折叠式座(位)
flap valve 舌形阀;止回阀;平板阀(门);片(状)阀;闪动阀;翻板阀;瓣阀;活形阀
flap valve assembly 活门栅
flap valve grid 进气活门栅
flap weir 活动水闸
flap wheel 翼片砂轮;扬水机;扬水轮
flare 向外扩伸;耀斑;喇叭天线;火舌;火炬;火号;喷焰;物镜的光斑;外展(船首);外倾,外飘,外扩;光斑;闪光;色球爆发;端部展张,船舷外倾;潮红;放空燃烧装置
flare adapter 照明弹架
flare-aircraft 照明机
flare angle 斜展角;张角;展开角;扩张角;外倾角;外飘角
flare bed 耐火材料的内衬导管
flare bevel 斜展角(护轮轨)
flare block 斜垫块;护轨斜垫块

flare bomb 照明弹
flare burner 火炬燃烧嘴
flare class 耀斑级
flare compensation 寄生光斑补偿
flared 钟形
flared access 喇叭形进口;喇叭口形入口
flared bow 外倾船首型;外飘船首型
flared column head 漏斗式柱头;喇叭形柱头;喇叭柱柱顶
flared crossing 展宽式交叉(路口);漏斗式交叉(口)
flared deflection yoke 喇叭式偏转系统;扩展式偏转系统
flared flange method 蝶形孔型系统设计
flared haunch 扩口拱脚石
flared header 半头黑砖;半黑头砖
flared-head gyratory 头部为喇叭形圆锥破碎机
flared inlet 喇叭形进水口
flared intersection 展宽式交叉(路口);漏斗式交叉(口);拓宽式交叉口
flared joint 扩口接头;扩口接合;胀接
flared outlet 喇叭形出(水)口
flared radiating guide 喇叭形辐射波导
flared ship sides 外倾舷侧
flared support 喇叭状支承;扩大支承面
flared tube 扩口管
flared union 扩口联管节
flared up tube 扩口管
flare factor 扩张系数;扩展因数;蜿展因数
flare fitting 扩口连接;锥度管子配件;喇叭口套管连接;喇叭口配件;扩口接头
flare-flash 耀斑闪光
flare gas 闪光气体;气焊气体;火焰气体;火炬气
flare gun 信号手枪
flare header 深色钉砖;半头黑砖
flare indicator 耀斑指示器
flare kernel 耀斑核
flare kiln 瓶颈式窑
flare light 闪烁光
flare-like brightening 似耀斑增亮
flare-like phenomenon 似耀斑现象
flare line 火炬管线
flare loop 耀斑环
flare nimbus 耀斑暗晕
flare onset 耀斑激发
flare opening 喇叭口;护轨端开度
flare-out 拉平;拉直
flare out bow 外倾船首;外飘船首
flare package 闪光装置
flare path 照明跑道
flare pit 燃坑
flare point 着火点;燃烧点;燃点
flareproof glass 防眩光玻璃
flare puff 耀斑喷焰
flare ribbon 耀斑带
flare signal 闪光信号
flare spectrum 耀斑光谱
flare spot 晕圈状光;耀斑;寄生光斑
flare stack 放空烟囱(炼油厂、石油化工厂);火炬烟囱;火把烟囱
flare stack ignition device 火炬烟囱点火装置
flare stop 光阑
flare surge 耀斑冲浪
flare system 废气处理系统
flare to waste 焚毁
flare track 八字线【铁】
flare triangulation 闪光三角网;闪光三角测量
flare-tube fitting 喇叭管接头
flare-type fitting 端部张开型接头
flare up fire 火焰信号
flare up light 火焰信号;突然发出火焰
flare veiling glare 杂光
flare voltage 闪光电压
flare wall 翼墙;喇叭形墙;斜翼墙
flare wave 耀斑波
flare welding 喇叭形坡口焊接
flare wing-wall 斜翼墙;喇叭形翼墙
flare wing-walled abutment 八字形桥台
flaring 喇叭形的;喇叭口状的;扩口的;蜿蜒;凸缘;天然气火炬
flaring angle 喇叭锥顶角
flaring chromosphere 耀现色球
flaring cup grinding wheel 碗形砂轮
flaring draft pipe 喇叭口尾水管

flaring funnel 喇叭形漏斗
flaring gate pier 宽尾墩
flaring head 喇叭形端部;扩口端部
flaring inlet 喇叭形入口;喇叭式进水管
flaring machine 旋转扩口机
flaring pier energy dissipation 宽尾墩消能
flaring pipe 喇叭形管
flaring side 外张舷;外倾舷;外飘舷
flaring test 扩口试验
flars-and-officers block 居住办公两用建筑物
flaser 压扁
flaser bedding 压扁层理;脉状层理
flaser bedding structure 脉状层理构造;伪流动构造
flaser gabbro 压扁辉长岩
flaser structure 压扁构造;伪流动构造
flase symbol 假符号
flase threading(for fishing) 假螺纹(用于打捞)
flash 溢料;模锻飞边;毛口;毛边;焊瘤;去毛刺;强脉冲;闪现;闪光焊毛刺;飞刺;飞翅
flash agitator 急骤搅拌器
flash alarm 闪光报警器
flash alloy 闪光合金
flash and show a light 闪示一下信号灯;闪示和出示灯光
flash and strain 飞边
flash annealing 表面退火
flash antenna 平面天线
flash apparatus 闪光器;闪光读数器
flash arc 火花弧;闪光电弧
flash back 逆风;反闪;反点火;逆火;(气焊的)回火;闪回
flashback arrester[arrestor] 回火制止器;回火熄灭器;回火保险器;回火防止器
flashback chamber 洗气室;回火水封室;水封隔间
flashback fire 反闪火焰
flashback phenomenon 反闪现象
flashback tank 水封箱
flashback voltage 反闪电压
flash baking 快速烘烤;快速烘干;快速烘焙
flashbang 声光时差
flash barrier 隔弧板;瞬时遮光板;瞬时屏蔽;瞬时隔离罩;瞬时隔板;闪光屏蔽;闪光(保护)挡板
flash blindness 闪光盲
flash board 决泄板;泄水闸板;挡水闸板;临时挡水闸门;(调节水位的)坝顶闸板
flashboard check gate 自动翻板闸门;闸板式节制闸门;决泄板节制闸;翻板(节制)闸门
flashboard check gate of crest 坝顶翻板带节制闸门
flash boiler 快热锅炉;速发气锅炉;闪蒸锅炉
flash bomb 闪光弹
flash bottoms 闪蒸残渣
flash box 扩容器;膨胀箱
flash bulb 闪光泡;闪光灯;镁光灯
flash burn 火焰烧伤;闪燃;闪光烧伤;电弧灼伤
flash burr 毛刺
flash butt weld 闪光焊;闪光对焊
flash butt welder 闪光对焊机
flash butt welding 电阻闪光焊接;闪光对(接)焊;电弧对接焊;闪光焊;火花对头接焊接法
flash calcining 快速煅烧;气流煅烧
flash camera 闪光照相机
flash card 亮卡;闪光(记录)卡
flash-card indexing 闪光卡片索引
flash chamber 内蒸馏;内蒸阶;薄罩面;(覆盖混凝土表面微疵的)薄层喷射混凝土;蒸发器;扩容箱;闪(发)蒸(发)室
flash clearance 溢料间隙
flash coat 喷浆盖层;速凝喷射层;闪光覆层;表面处理的涂层
flash coating 金属喷镀层;极薄镀层;闪光涂层;薄镀层
flash colo(u)r 瞬变色
flash combustion 急骤燃烧法
flash compressor 火焰闪光抑制剂
flash condenser 闪蒸冷凝器
flash connections 闪光插头及插座
flash cooler 闪蒸冷却器
flash cooling 快速冷却
flashcube 立体闪光灯
flash cup 闪点杯;测闪点杯
flash curve 闪蒸曲线;闪光特性曲线
flash defilade 闪光遮蔽
flash depressor 火焰闪光抑制剂

flash desorption 闪脱
flash desorption spectroscopy 闪光解吸光谱术
flash detonator 火雷管
flash dewax 闪烧脱落
flash distillation 急骤蒸馏(法);突然蒸发;闪蒸(馏)
flash distillation column 闪蒸塔
flash distillation plant 闪蒸装置
flash down 压力迅速下降;闪降
flash drum 闪蒸罐;闪蒸槽
flash dry 快闪;闪干;急骤干燥
flash dryer 快速干燥器;急骤干燥器;气流干燥器;快速烘干机(立筒式);快干机;烘干破碎机;气力输送烘干机;飘悬干燥机;闪(蒸)干(燥)器
flash drying 快速烘干;快速干燥;气流干燥;瞬间干燥;闪速干燥;油漆迅速干燥法;急骤干燥
flash drying machine 飘悬干燥机
flash drying system 瞬间干燥系统;快速干燥系统
flash drying with hot air 热气速干
flash duration 闪光时间;闪光持续时间
flashed brick 青砖;多色砖;边缘烧黑砖
flashed glass 贴色玻璃;套色玻璃;闪光玻璃
flashed opal 套料乳白玻璃
flashed opal glass 镶嵌用乳白玻璃;套料乳白玻璃
flashed opal plat glass 套料乳白平板玻璃
flashed tile 闪光砖
flashed vapo(u)r 扩容蒸气
flash effect 闪光效应
flasher 自动断续开关;扩容器;闪烁器;闪烁开关;闪烁光源;闪光仪;闪光器;闪光灯
flasher bulb 闪光泡
flasher lamp 闪光转向信号灯
flasher mechanism 闪光机构
flasher relay 闪弧继电器;闪光继电器
flashes per second 每秒闪光次数
flash evapo(u)ration 骤蒸;快速蒸发;急骤蒸发;闪蒸
flash evapo(u)ration technique 快速蒸发技术
flash evapo(u)rator 闪蒸蒸发器;闪蒸工发器
flash exposure 闪烁曝光;闪光曝光
flash factor 闪光因数
flash figure 闪像;闪图;迅变干涉图
flash filament method 闪烁灯丝法
flash filament technique 闪烁灯丝法
flash film concentrator 急骤薄膜式浓缩器;闪蒸薄膜浓缩器
flash film evapo(u)rator 急骤薄膜式蒸发器;闪蒸薄膜蒸发器
flash fire 急剧燃烧;闪燃;闪火
flash fire propensity 闪火倾向
flash flo(a)tation 闪速浮选
flash flood 骤发洪水;河水暴涨;山洪暴发;暴(发)洪(水)
flash-flood warning 山洪警报;暴洪警报
flash flow 暴洪水流
flash fuel 速燃物
flash gas 扩容气;闪蒸气(体);闪发气体
flash gasoline 闪蒸汽油
flash gas refrigeration 闪蒸气体冷冻
flash generator 闪击式高压发生器
flash getter 蒸散式消气剂
flash glass 有色玻璃
flash groove 溢料缝
flash guard 防弧装置;防弧器
flash gun 闪光枪;闪光器;闪光粉点燃器;闪光操纵装置
flash heat 闪烁热
flash heater 急速加热器;瞬时蒸发加热器;闪蒸加热器
flash heating 快速加热;急骤加热
flash heating sintering 快速加热烧结
flash hider 消焰器
flash hole 传火孔
flash information service 特快信息服务
flashing 局部发亮;河水暴涨;喷溅;水冲;烧化;闪蒸;闪烁;闪弧;闪光;电弧放电;挡水板;防水片;泛水;发生火花;不均匀光泽
flashing allowance 烧化余量;闪光余量
flashing amber 黄色闪光灯
flashing amber lantern 黄琥珀色闪光灯
flashing apparatus 闪蒸设备;闪光装置
flashing arrow 闪光指示箭头
flashing beacon 闪光信号灯;闪光(立)标灯
flashing block 陶瓦块;泛水砌块

flashing board 防雨板;泛水板
flashing cement 泛水胶泥;披水黏结剂;防水胶泥;防漏黏结剂;泛水黏结剂
flashing chamber 闪蒸室
flashing column 闪蒸塔
flashing composition 引爆剂
flashing compound 防水填缝剂;引爆剂(爆破工程)
flashing current 闪光电流
flashing direction signal 闪光指向信号;闪光指路信号
flashing discharging tube 闪光放电管
flashing expansion 闪发膨胀
flashing feature 闪光装置
flashing flow 闪流
flashing geodetic satellite 大地测量闪光卫星
flashing head 闪光头
flashing hook 设在墙上的吊钩
flashing index 闪光指示
flashing indication 闪光信号
flashing indicator 闪光指示器
flashing jack 闪光插座
flashing joint 泛水接头
flashing key 闪烁电键;闪灯电键
flashing light 脉冲光源;闪光灯(标);闪光(体);电筒
flashing light alarm 闪光报时
flashing light for collision avoidance 避让闪光灯
flashing light indicator 闪光指示器
flashing lightning 闪电
flashing-light satellite 闪光卫星
flashing lightsignal 闪光信号
flashing light signal(l)ing 闪光通信
flashing light system 闪光灯系统;闪光装置
flashing liquid vessel 闪蒸液器
flashing machine 闪光机
flashing material 防雨材料;挡水材料
flashing mechanism 闪光机构
flashing membrane 泛水用片材;泛水膜
flashing method 防雨方法;挡水方法
flashing of a vent(pipe) 排气管防雨盖板;排气管挡水盖板
flashing off 急骤溜掉;(耐火材料的)软溶
flashing oscillator 闪光振荡器
flashing piece 防水片;挡水片
flashing plant 闪蒸装置;快速蒸发;减压裂化
flashing point 闪燃点;引火点;闪(光)点
flashing point tester 闪点试验仪;闪点试验器;闪点测定仪;闪点测定器
flashing potential 着火电位
flashing power source 闪光电源
flashing pressure 闪蒸压力
flashing rate 闪光频率
flashing reactor 闪蒸反应器
flashing red 红色闪光灯
flashing reglet 泛水槽嵌入物;泛水槽嵌入件
flashing relay 闪光继电器
flashing rhythm of light 灯光节奏
flashing ring 阻水环;管道套圈
flashing sequence 闪光序列
flashing-setting agent 促凝剂
flashing sign 闪烁信号灯
flashing signal 闪光信号
flashing strip 挡水条板
flashing temperature 闪蒸温度;发火温度
flashing test 击穿试验;高压绝缘试验;闪点(的)测定
flashing tile 泄水瓦(管);瓦管
flashing time 闪光时间
flashing to atmosphere 闪蒸至常压
flashing trafficator 闪光指示灯
flashing unit 闪光器
flashing valley 泛水沟;披水沟
flashing valve 冲水阀
flashing warning lamp 闪光警报灯
flashing yellow 黄色闪光灯
flash intensity 闪光强度
flash irradiation 闪照射
flash item 快速产品项目;简短项目
flash joint 闪光焊接接头
flash joint of rods 钻杆的平头接合
Flashkut 落锤锻造钢
flash lamp 信号灯;手电筒;闪光泡;闪光灯(泡);小电珠

flash lamp bulb 闪光灯泡
flash layer 闪光层
flash length 闪光长度
flashless die forging 无飞边模锻
flashlight 闪光(灯);手电筒;闪光信号灯
flashlight battery 手电筒电池
flashlight bomb 闪光弹
flashlight bracker 闪光灯架
flash-light illumination 闪光照明
flashlight method 闪光灯法
flashlight powder 闪光粉
flashlight source measurement 闪光源的测量
flashlight torsionmeter 闪光扭力仪
flash line 突起线
flash load 瞬时载载;瞬时负荷
flash loss 闪光留量
flash magnetization 脉冲电流磁化;瞬时磁化;闪磁化
flashmatic 闪光曝光器
flash melting 软熔发亮处理
flash-melting process 闪光融熔法
flash memory 快闪存储器
flash meter 闪光仪;闪光曝光计
flash mixer 快速搅拌器;快速混合器;急骤搅拌器;急骤混合器;急速搅拌器
flash mixing 快速混合
flash mix tank 快速混合池
flash mo(u)ld 溢式压塑模;溢式模具;溢流模;溢出式(铸)塑模;平线脚;敞开式模具
flash naphtha 闪蒸石脑油
flash-off 闪蒸出;急骤溜掉
flash-off area 油漆快干段(装配线)
flash-off steam 扩容蒸汽;闪蒸蒸汽
flash-off time 气散时间;闪蒸时间
flash oil 闪蒸油
flash-o-lens 闪光透镜
flashometer 闪光仪
flashout 闪蒸排出
flashover 击穿;跳火;闪络;电弧闪击;飞弧;殉爆
flashover characteristic 放电特性
flashover contact 闪络接触
flashover discharge 飞弧距离
flashover distance 闪落距离
flashover ground current 闪络接地电流
flashover ground relay 闪络接地继电器
flashover potential 闪络电位
flashover protection 防止飞弧
flashover relay 闪络继电器
flashover sign 跳火信号灯;闪络信号
flashover signal 闪络信号
flashover strength 闪络强度;飞弧强度
flashover test 高压闪络试验;闪络试验;下弧试验
flashover voltage 击穿电压;闪络电压;飞弧电压
flashover welding 闪光焊
flash paper 闪光纸
flash pasteurization 高温瞬间灭菌法
flash pasteurizer 瞬间杀菌机
flash period 闪光周期
flash phase 闪光相
flash photography 频闪摄影术;闪光照像;闪光照相术;闪光摄影术
flash photolysis 闪光光解
flash-photometry 闪光光度学
flash photo printer 闪光摄影印刷机
flash pin ga(u)ge 探销式塞规
flash pipe 闪蒸管;闪燃管
flash pistol 闪光器
flash plate 薄镀层
flash plating 薄镀层;薄镀板;薄镀
flash point 着火点;闪(火)点;引火点;暴发点
flash point apparatus 闪点仪(开口杯式);闪点测定仪
flash point determination 闪点确定;燃点确定
flash point of asphalt 沥青闪点
flash point of Cleveland open cup 克利弗兰开杯法闪(火)点
flash point Tag open-cup 泰革敞杯法闪(火)点
flash point test 闪火点试验;闪点试验
flash point tester 闪点仪(开口杯式)
flash point testing apparatus 闪点试验仪器
flash point yield curve 闪点产率曲线
flash polymerization 猝发聚合反应;暴聚
flash pot 闪蒸罐
flash pressure 扩容压力

flash process 闪蒸过程
flash pyrolysis 闪热解
flash quenching 喷风淬火;喷射淬火
flash radiography 闪光射线分析;闪光高速摄影
flash range 闪蒸温差
flash ranging 光测;频闪测距;闪光测距
flash ranging adjustment 光测修正
flash-recall 闪灯式二次呼叫
flash relay 闪光继电器
flash removal 去除毛刺
flash removed 除毛刺
flash report 要况报告
flash ridge 溢料面;溢出埂;绉纹边缘
flash roast 漂悬焙烧法
flash roaster 悬浮焙烧炉;快速煅烧炉;飘悬焙烧炉;漂悬焙烧炉;闪速焙烧炉
flash roasting 悬浮焙烧;闪火焙烧
flash ruby 闪光红宝石玻璃;套红料
flash runoff 暴发径流
flash rusting 瞬时锈;瞬蚀
flash section 闪蒸段;闪蒸部分
flash separation 闪蒸分离
flash set 瞬时凝结;骤凝;快速凝固;(混凝土的)急凝;速凝;闪凝
flash set (水泥浆的)瞬凝
flash setting 急骤凝固;瞬时凝结;闪凝
flash-setting agent 急凝剂;速凝剂
flash-setting alumina cement 瞬凝矾土高铝水泥
flash-setting cement 瞬凝水泥;速凝水泥
flash-setting gypsum 瞬凝石膏
flash shot 闪光摄影
flash-signal lamp 闪光信号灯
flash sintering 快速烧结
flash smelting 闪速熔炼
flash smelting process 闪速熔炼法
flash socket 闪光灯座
flash spectroscopy 闪光光谱学
flash spectrum 闪光谱;闪光谱
flash-spinning technique 闪纺技术
flash spotting 光测法;闪光观测
flash star 闪星
flash steam 扩容蒸汽;闪发蒸汽;二次蒸汽
flash-steam generator 闪蒸水蒸汽发生器
flash stream 暴洪
flash sump 闪发蒸发室集水底壳
flash synchronization 闪光同步
flash-synchronized shutter 闪光同步快门
flash synchronizer 闪光同步机构
flash system 闪系列
flash tank 二次蒸发箱;扩容器;膨胀箱;疏水膨胀器;闪蒸箱;闪蒸罐
flash technique 光泽技术;闪光技术
flash temperature 瞬现温度;闪蒸温度;闪点温度
flash test 高压绝缘试验;瞬间高压试验
flash testing equipment 瞬间试验机
flash time 闪光时间
flash to 迅速变成;急骤蒸发成
flash tower 闪蒸塔
flash trap 闪蒸室
flash treatment 被膜处理
flash triangulation 闪光三角测量
flash trimmer 去毛刺机
flashtron 气体放电继电器
flash tube 引火管;闪蒸管;闪光灯
flash tube excitation source 闪光管激发源
flash tube ignition 闪光管点火
flash tube pumping 闪光管抽运
flash tube stroboscope 闪光管频闪仪
flash type 闪跃型
flash type distiller 闪发式蒸馏器
flash type evapo(u)rator 闪发式蒸发器
flash type stroboscopy 闪光管频闪(管测)仪
flash type stroboscopy 闪光管型频闪观测法
flash unit 闪光设备;闪光器;闪光(灯)装置
flash up 闪发
flash valve 闪阀
flash vapo(u)rization 真空闪蒸
flash vapo(u)rization curve 闪蒸曲线
flash vapo(u)rization inlet 闪蒸入口
flash vapo(u)rization point 闪蒸温度
flash vapo(u)rizer 闪蒸器
flash vessel 扩容器;闪蒸室;闪蒸器
flash wall 隔墙;挡火墙;(倒焰窑火箱部位)
flashway 泄水道;灌水泄水通道;灌水通道(船闸);冲沙道
flash weld 闪光焊;电弧焊;闪速对焊
flash welder 电弧焊机;闪光焊机
flash welding 闪光对焊;火花对焊;火花电弧焊;焊;闪光电弧焊;电弧对焊
flash-weld tool joint 弧焊工具接头
flash X-ray tube 闪光摄影 X 光管
flashy flood 暴雨;暴发洪水
flashy flow 汹涌急流;湍急水流;暴洪水流
flash yield curve 闪蒸曲线
flashy load 瞬间荷载;瞬时荷载;瞬时负荷
flashy regime of river 河流的暴涨性
flashy river 山溪性河流;暴洪河流
flashy stream 有暴雨河流;山溪性河流;暴洪河流;瞬发洪河流;暴洪河川;暴洪
flash zone 闪蒸段
flasing over 飞弧
flask 型盒;细颈钢瓶;造型箱;造型砂箱;曲颈瓶;瓶(形物);烧瓶;长颈瓶
flask bar 箱带;砂箱筋
flask board 底板;托模板;底模
flask clamp 砂箱夹
flask conveyor[conveyer] 砂箱传送带
flask drop down machine 落箱机
flasket 小瓶;长形浅篮
flask filler 砂箱填料
flask heater 烧瓶加热器
flask holder 烧瓶(支)架
flasking 装型盒
flask joint 分箱面
flaskless mo(u)ld 无箱铸型
flask line 分模线
flask liner 箱衬板
flask mo(u)lding 砂箱造型
flask pin 砂箱定位销
flask rammer 平头捣锤
flask separator 分箱机
flask shaker 落砂机
flask-shaped 瓶形的
flask-shaped heart 瓶状心
flask trunnion 砂箱轴;砂箱耳
flask with round bottom and long neck 长颈细口圆底烧瓶
flask with side arm 蒸馏烧瓶;侧臂式烧瓶
flat 哑色的;磨刻面;浅箱;平滩;平的;平板状(的);无明暗差别的;无光泽的;无光的;套房;水位标线间面积;晒板台纸;断坪;次级金刚石;舱内甲板;玻璃压板;扁平的;板片
flat abrasion 平面磨蚀
flat abrasion test 平磨试验
flat air bearing 平式空气轴承
flat alkyd enamel 平光醇酸瓷漆
flat alumin(i)um alloy plate 铝合金平片
flat amplifier 平直放大器
flat and edge method 平竖轧制法
flat-and-edging method of rolling 平竖轧制法
flat and meter rate schedule 按表收费制
flat and round steel-chain 平圆式钢链
flat angle 平角
flat anvil 平砧
flat approaching grade 平坡道口
flat arch 单砖拱;平砼;平式炉顶;平拱;坦拱;扁拱
flat arch bridge 坦拱桥
flat arched girder 平拱梁
flat arched girder roof 平拱梁屋顶
flat arch lintel 平拱过梁
flat area 平面面积
flat armature 扁平衔铁;边衔铁
flat asbestos sheet 石棉平板
flat-back car 平背车
flat back pattern 平接模
flat back stope 逆台阶工作面
flat bag filter 扁袋集尘器
flat band 平缓弯段;大半径弯段;索石;素石
flat band method 平板法
flat band point 平带点
flat band voltage 平带电压
flat bank revetment 缓坡护岸
flat bar 带钢;扁铁;扁条;扁(条);扁杆;扁材;板片;条钢
flat bar iron 窄扁钢
flat bar joint 长方形鱼尾板
flat bar keel 平板龙骨
flat bar screen 平格栅;平格筛;扁钢格栅

flat bars lacing 扁钢缀条
flat base 平底;平板底座
flat baseline 平坦基线
flat base rectifier 扁平接触冷却整流器
flat base rim 平轮缘;平底轮缘;平轮轮辋
flat bastard file 粗齿扁锉
flat battery 扁电池
flat beam 平直束;扁平梁
flat bearing 扁柱;平支座;平面支承;平导板;双脚支柱
flat bearing structure 平面支座结构
flat beater 平板式打夯机;平板打夯机
flat bed 平层;平坦河床;平底座;水平层;平板挂车;平河床;平板(车);平台(式);绘图平台
flat-bed chromatogram 平板色谱
flat-bed chromatography 平床色谱法
flat-bed company 平板挂车汽车运输连
flat-bed container 平床式集装箱;平床货柜
flat-bed crawler truck 履带式平板货车
flat-bed digital plotter 平台数控绘图机
flat-bedding 平坦层理
flat-bed glazer 平板式上光机
flat-bed offset(printing)machine 平台胶印机
flat-bed plotter 平台式绘图机;平板绘图仪;平台式自动绘图机
flat-bed press 平板热压机
flat-bed roof 平台式屋顶
flat-bed scanner 平台扫描仪
flat-bed screen printing 平板网印
flat-bed tandem trailer 纵列平板拖车
flat-bed trailer 平板拖车;平板式拖车;平板(式)拖车;平板半挂车
flat-bed truck 平板式运货汽车;平板车;平台式载货汽车;平板货车
flat beetling 平幅捶布
flat belt 扁皮带;平带;平皮带
flat-belt conveyer 平型带式输送机;平带输送机;平(皮)带传送机
flat-belt drive 平皮带式传动;平皮带传动
flat-belt grain conveyer 平带式谷物输送器
flat-belt pulley 平皮带轮
flat bend 平缓弯段;大半径弯头
flat-bend test 板material弯曲试验
flat billet 平错齿坯
flat bimodule 平坦双模
flat bit 平钻;扁钻
flat bit tongs 平口钳;扁嘴钳
flat blade 平叶片;扁叶片
flat-blade fan 平扇叶风扇
flat-blade turbine 平叶涡轮
flat blank 扁坯;板坯
flat block 底层房屋;装配平台
flat bloom 扁锭
flat board 平板
flat board to the tool grade 工具等级型平板
flat boat 平底船;方驳
flat-bodied 体侧平直的
flat bog 平地沼(泽);低沼泽(地);低位沼泽;低洼地
flat bogie wagon 平车;低边敞车
flat bone 扁平骨;扁骨
flat bottom 平底
flat-bottom barge 平底驳船
flat-bottom beaker 平底烧杯
flat-bottom bin 平底仓
flat-bottom car 平底车
flat-bottom cart 平板车
flat-bottom crown 平面金刚石钻头;平端面钻头
flat-bottom diamond crown 平端面金刚石钻头
flat-bottom ditch 梯形沟;平底沟
flat-bottom draw value 平底拉延值
flat-bottomed 平底的
flat-bottomed bin 平底箱
flat-bottomed boat 平底船
flat-bottomed clarifier 平底沉淀池;平底澄清池
flat-bottomed flume 平底测流槽
flat-bottomed land 平原地带
flat-bottomed pontoon 平底浮筒平台
flat-bottomed rail 平底钢轨;宽底钢轨
flat-bottomed ship 平底船
flat-bottomed silo 平底仓
flat-bottomed skip 平底混凝土料斗;平底混凝土料车
flat-bottomed valley 平底谷
flat-bottomed vessel 平底船

flat-bottom furnace 平炉底炉膛
flat-bottom gondola 平底高边敞车
flat-bottom hole 平底孔
flat-bottom punch 平底冲头
flat-bottom rail 平底钢轨;平底轨
flat-bottom slot 平底槽;平底池
flat-bottom tank 平底罐
flat-bottom tappet 平底挺杆
flat-bottom washing bottle 平底洗瓶
flat-bottom zone 平坦地区
flat bow 平弓;平板艇
flat braid 平编带
flat-braided cord 双芯扁软线;平打绳;扁形编织绳
flat-braided elastic 平编宽紧带
flat brass 扁黄铜
flatbreaking 平耕
flat brick 扁砖;边砖
flat brick arch 平砖拱
flat brick lintel 平砌砖过梁
flat brick pavement 平铺砖路面
flat brick paving 块砖平铺;平砖路面
flat broach grinder 平面拉刀磨床
flat brush 油漆刷;涂料刷;扁形刷子;扁形毛笔;扁刷
flat building 底层建筑物;小公寓房屋;平房
flat-built tire 平面轮胎
flat bulb iron 球扁铁;扁球铁;球(头)扁钢
flat bulb steel 球扁钢;扁球钢
flat bundle test 平列纤维束强力试验
flat burner 扁平喷嘴
flat bus-bar 扁母线
flat bush 平钻;平面轴套
flatbusting 平耕
flat butted seam 对头接缝
flat cable 带状电缆;并排线;扁形电缆;扁(平)电缆;扁平导线电缆
flat cable head 带状电缆接头
flat calm 完全无风
flat cancellation 免费注销
flat cap 小四开图纸;软水手帽;顶圆帽
flat car 卡车;货车;平(板)车;低边敞车;敞车;平敞车
flat-card resolver 平片式解算器
flat-card resolving potentiometer 平片式解算电位器
flat car for container 装运集装箱平车
flat car for trailer 装拖车用平板车
flat carriage 平板车
flat carving 平雕;平雕刻工艺;底座切雕工艺
flat ceiling 平顶;海墁天花
flat cement(roofing)tile 水泥平瓦
flat chain system 花环平挂;平挂链
flat champignon rail 平头钢轨
flat-channel amplifier 平直通路放大器;平直特性放大器
flat-channel noise 平直幅频起伏噪声;平路噪声
flat characteristic 平态特性;平顶特性
flat characteristic curve 平直特性曲线
flat chassis 平底盘
flat chisel 钢凿;平凿;扁凿;扁铲
flat-chord truss 平弦桁架
flat circular stone 平的圆石碾
flat clamp 平压铁;平压板
flat clamping plate 平压板
flat classification yard 平面调车场【铁】
flat clay roof(ing)tile 黏土平瓦
flat clay tile roof 黏土平瓦屋顶
flat clearer 盖板清洁器
flat clipping machine 包盖板机;包盖板裁剪机
flat coarse file 粗齿扁锉
flat coast 平坦海岸;低平海滨;低平海岸
flat coat 无光漆层;中层漆;旧漆层;中间涂层;平光涂层;无光涂层
flat coaxial transmission line 带状传输线
flat coil 平线圈;扁绕线圈
flat coining 平面精压
flat cold rolled bar 冷轧扁钢
flat cold rolled sheet 冷轧薄板
flat colo(u)r 普染色
flat-comb binder 扁平梳状装订器
flat commission 统一手续费;统一佣金
flat composition crown 平路拱
flat compositron (电机的)平复励绕
flat-compound characteristic 平复励特性
flat-compound dynamo 平复励发电机

flat-compound excitation 平复励磁
flat-compound generator 平复励发电机
flat-compounding 平复励
flat concrete column 混凝土扁柱
flat concrete roof 混凝土平屋顶
flat-conductor cable 扁平导线电缆
flat connection 平坦连通
flat connector 平板接头
flat construction 平屋顶构造
flat conveyor belt 平型运输带
flat copper 条铜;扁铜
flat corbel-table(frieze)平挑檐
flat cost 直接费(用);工料费;净成本;纯(粹)成本
flat cotter 扁销
flat countersink 平埋头钻
flat countersunk head rivet 平埋头铆钉
flat countersunk head screw 平顶埋头螺钉
flat countersunk rivet 埋头平顶铆钉
flat counter tube 平板型正比计数管
flat country 平原(地区);平坦地区;平地;低地
flat-country test 平原地带试验
flat course 顺砖皮
flat cover-degree 平面盖度
flat covering roller 光面压土轮
flat cradle vault 扁筒形拱顶
flat crank 平曲柄
flat crest 平顶;扁牙顶
flat-crested measuring weir 平顶量水堰
flat-crested spillway 平顶式溢洪道
flat-crested weir 宽顶堰;平顶堰
flat crown 平坦路拱;平路拱;平顶树
flat-crowned sweep 高速耕耘铲
flat crystal 平面晶体;片状晶体
flat-crystal spectrometer 平面晶体分光计
flat cumulus cloud 平积云
flat cup nut 扁阿盖螺母;扁帽盖螺母
flat cupola 扁圆屋顶
flat cure 平铺硫化
flat curing 平幅焙烘
flat curve 缓曲线;平直曲线;平顺曲线;平缓曲线;平滑曲线;扁曲线
flat cut 向下打眼钻爆法;弦向切削;弦切
flat cutting 平剖;顺纹锯木;弦向锯开;平切;平锯
flat cylindrical vault 扁圆筒形拱顶
flat deck buttress dam 平板支墩坝
flat deck car 平板车
flat deck poultry cage 平列鸡笼
flat deck roof 平顶盖
flat demand rate 固定需量收费制
flat design 平板式
flat device 平坦型器件;扁平器
flat diaphragm 平膜片
flat die 搓丝机;搓丝板;平砧;平(钢)模
flat-die forging 自由锻造
flat-die hammer 自由锻锤
flat die thread rolling 平模滚轧螺纹法;搓丝
flat dilatometer 平板膨胀仪;扁式松胀仪;扁式膨胀仪;扁式旁压仪
flat dip 微倾斜
flat discount rate 平贴现率
flat disc 平圆盘
flat disc turbine agitator 圆盘平直涡轮式搅拌器
flat dolly 扁头铆顶
flat dome 扁穹隆
flat dormer(window)平屋顶窗;平老虎窗
flat double guides 双轨导向器
flat down 平面朝下
flat-down method 六角钢平轧法
flat draghead 平耙头
flat drainage 平面排水
flat drawn 垂直引上的平板玻璃
flat-drawn glass 拉制玻璃
flat-drawn sheet glass 拉制玻璃;普通平板玻璃
flat drill 平钻;扁钻;直三角钻
flat drill bit 扁钻头
flat drill steel 钻探用扁钢
flat drop plate 平落式槽口排种盘
flat drop round-hole plate 平落式圆孔排种盘
flat duct 平导管
flat ear 扁平耳
flat earth model 平地模型
flat edge and bevel 修边和角
flat-edge die 平底冲头
flat edge file 平边锉

flat edge trimmer 修边机
flated pointed tool 平口刀具
flat element 平面元
flatels (旅馆或公寓的)设备齐全的楼层
flat enamel 无光泽的搪瓷
flat enamel brush 瓷漆板刷
flat-end 平端
flat-ended horizontal cylindrical drum 平底卧式圆筒形罐
flat-ended tube 平底管
flat end needle rollers 平头滚针
flat end set screw 平端固定螺钉
flat engine 平置式发动机;卧式发动机
flat entrance 住宅入口;公寓入口
flat entrance door 住宅大门;公寓大门
flat etching 初次腐蚀
flat facade 平正面
flat face 平端面;平坦面;平面
flat-faced bit 平端面金刚石钻头;平面金刚石钻头;平端面钻头
flat-faced crown 平端面钻头
flat-faced fillet weld 平顶角焊缝焊接;角焊
flat-faced flange 平面法兰;平面凸缘
flat-faced opposite anvil 平面对等砧装置;布里奇曼钻压
flat-faced pulley 平面皮带轮
flat-faced sheave 平面轮
flat-faced tube 平面板式管
flat-faced wheel 平面砂轮
flat factor of safety 静力安全系数(提升钢丝绳)
flat fading 平匀衰落;平衰落;按比例衰减
flat fading channel 平坦衰落信道
flat-fan nozzle 扇形雾锥喷嘴
flat fee contract 预先规定合同价格
flat-fell seam 平式接缝
flat field 平面场;平场
flat field camera 平场照相机
flat field correction 平场改正
flat field generator 平面场发生器
flat field lens 平扫描场透镜
flat field magnet 平面场磁铁
flat field objective 平场物镜
flat field photometry 均场测光
flat field uniformity 平场均一性
flat file 平面文件;平锉;扁锉;单调资料
flat fillet weld 平填角焊缝;平面填角焊
flat fillet weld in front 正面贴角焊缝
flat film 平面薄膜;散张胶片
flat filter 宽滤波器;平面过滤器;平面过滤机
flat finish 无光泽面层;平光面漆;无光饰面;无光面漆
flat finisher 平面修整机
flat finishing 平面涂装法
flat finish paint 无光泽面层油漆
flat finish varnish 本色清漆
flat fishplate 平鱼尾板
flat fixed film 平面固定膜
flat flame 平焰
flat-flame burner 蝙蝠翼式灯;平焰式燃烧器;平面火焰喷燃器;平焰烧嘴
flat flange joint 平法兰连接;平凸缘连接
flat flexible cable 扁形软性电缆
flat flood plain 平坦河漫滩
flat floor 平炉底;平肋板;平底;平板楼板
flat floor space 公寓楼层空间
flat flux 平坦通量
flat-flux reactor 平坦通量反应堆
flat folded seam 平折缝
flat follower 随动圆片
flat foolscap 大八开(图)纸
flat foot 平底脚
flat for attendant on a train 乘员公寓
flat forge tongs 平锻钳
flatform and stake racks truck 平台栏车
flat forming tool 棱形成型车刀
flat-form lorry 平板车
flat-form process 平模流水法
flat-form tool 平口成型刀具
flat-form training 扁平整枝
flat-form truck 平板车
flat foundation 浅基(础)
flat-four engine 卧式四缸发动机
flat frame 平面框架
flat frame element 平面框架杆件

flat frame structure 平面框架结构
flat frequency content 平直频率成分
flat frequency response 平坦频率响应曲线
flat fronted vehicle 平头汽车
flat function 平坦函数
flat fuse wire 扁熔断丝
flat gain 平坦增益
flat gain capacitor 线性增益调节电容器;平调电容器
flat gain control 平坦增益控制
flat gain control capacitor 平坦增益调节电容器
flat gain master controller 平调主控制器
flat gain regulation 平增益调整;平调
flat-gap process 狭缝法
flat garden 平地花园
flat gate 扁平内浇口;扁浇口
flat ga(u)ge 样板;扁形规;板规
flat gauze oxidizer 平网氧化器
flat-geometry light emitter 平坦型光发射器
flat-geometry light source 平坦型光源
flat glass 平面玻璃;平板玻璃
flat glass furnace 平板玻璃窑炉
flat gloss 无光泽
flat gloss oil paint 无光泽油漆;平油漆
flat gouge 弧口薄凿
flat grade 缓坡;平缓坡度;平缓坡道;平缓比降
flat gradient 缓(斜)坡;平缓坡度;平缓坡道;平缓比降
flat gradient for starting 起动缓坡【铁】
flat gradient section 缓坡地段
flat grain 木材横纹切割;平(直)纹理;平展纹理;平型木纹
flat-grained lumber 直锯平纹木材
flat-grained timber 平纹木材
flat-grained lumber 弦切板
flat-grained shingle 平纹屋顶板
flat grate 平栅炉;平面光栅;平(炉)栅
flat grate producer 平面光栅制造器
flat-grid strain ga(u)ge 平面栅状应变仪
flat grinder 平面磨床
flat-grinding machine 平磨机
flat grooved rail 平槽轨
flat ground 平坦地区;平地
flat-growing vegetables 平地栽培植物
flat guard rail section 平护栅断面
flat guide 平面导轨
flat guide plate 平面导板
flat gutter 平檐(雨水)沟
flat hammer 平锤;扁锤
flat hand 扁平手
flat hand file 平手锉
flat handle 扁平手柄
flat handle drum 扁提桶
flat handrail section 平扶手断面
flat hanger 平面吊架
flat harder 平板式振动压毡机
flat hat 软水手帽
flat head 平头;平封头;平头埋头螺钉
flat-head bit 平头钻(头)
flat-headed 平头的
flat-headed bolt 平头螺栓
flat-headed counter-sunk bolt 平顶埋头螺栓
flat-headed hook 扁头鱼钩
flat-headed nail 扁头钉
flat-headed piston 平顶活塞
flat-headed rail 平头(钢)轨
flat-head grooved bolt 平头槽螺栓
flat-head piston 平顶活塞
flat-head pliers 平头钳
flat-head rivet 扁头铆钉;平头铆钉
flat-head screw 扁头螺钉;平头螺钉
flat-head straight neck rivet 平头直颈铆钉
flat-head type 平头式
flat-head valve 平头气门
flat-head wire nail 平头圆铁钉
flat hearth generator 平炉床发生器;平炉床
flat hearth type mixer 平床混合炉
flat heavy plate 重平楼盖板;重平板
flat hinge 平铰;平
flat hole 平炮眼;水平炮眼
flat hook 扁平钩
flat hoop iron 平箍钢;带钢
flat horizontal surface 扁平面
flat hot-rolled bar 热轧扁钢
flat ice 平坦海冰;平坦冰

flat idler 水平托辊
flat-image amplifier 平面图像放大器
flat-image field 平面像场
flat in 横插进去
flat inclined coast 曲折平地海岸
flat inclined mirror 斜面镜
flat interest charges 粗利率
flat interest rate 统一利率
flat interlocking(clay)tile 咬接平瓦
flat interlocking tile 连锁平瓦;平联锁瓦;联锁平瓦
flat invert 平缓仰拱
flation 平稳时期
flation price 平稳价格
flatiron 扁铁;扁条铁;平顶山(脊);熨斗(形山)
flat iron and angle iron cutting machine 扁铁及三角铁切割机
flat iron bar 扁铁条
flat iron(bars)lacing 扁铁缀条
flat-iron butt joint 扁铁铰链连接
flat-iron collier 熨斗式运煤船
flat-iron shape 熨斗状外形
flat iron wheel 扁铁轮
flat jack 平板千斤顶;测量岩压水力囊;扁千斤顶;液压千斤顶
flat jack method 扁千斤顶法
flat jack process 扁千斤顶法
flat jack slot test 狭缝法试验
flat jack technique 液压钢枕法;扁千斤顶法
flat jack test 扁千斤顶试验
flat jewel 平钻
flat joint 平节理;平灰缝;平缝
flat joint bar 长方形鱼尾板
flat jointed 平连接的
flat-joint jointed 平缝砌合;凹槽平缝;窄槽平缝接合的
flat-joint jointed pointing 窄槽平缝接合勾缝
flat-joint point(ing)勾平缝;平嵌砖缝
flat jumper 扁凿
flat keel 平龙骨;平底;平板龙骨
flat-kernel seed 不实粒
flat key 平键;扁锁匙
flat kitchen 住宅厨房;公寓厨房
flat knot 平结
flat lacquer 无光(泽)喷漆;无光泽蜡克;无光硝基漆;无光挥发性漆
flat lake coast 低平湖岸
flat laminated glass 平夹层玻璃
flat lamp core 扁灯芯
flat lampshade 散射型灯罩
flatland 平原(地);假设的二维世界;平地
flatlander 平槽
flat-land railway 平原地区铁路
flat-land section 平原河段
flat lapping block 精研平台
flat layer 平层的
flat layered pile (集料的)平层料堆
flat-layer technique 平层技术;平层法
flat lead 铅皮;铅薄板
flat leakage 平漏
flat leakage power 平坦漏过功率
flat lease 定期定额付租租契
flatless card 无盖梳棉机
flatlet 小型套房;单间紧凑的小公寓
flat letter 八开图纸
flat light 平淡照明;平淡光;单调光
flat-lighting 平淡照明;单调光
flat light truss 轻型平屋架
flat line 平坦线路
flat linear electric(al)motor 平板型的直线电机
flat linear induction sodium pump 平面式直线感应钠泵
flat liner 平衬板
flat link chain 扁环节链;板链
flat link of chain 链条炉排的链节
flat links and links knitting machine 平板机
flat list prices 统一价格表
flat locking rod 扁锁闭杆
flat-lock seam 企口缝
flat-lock seam roofing 咬合缝屋面
flat-lock soldered seam 锡焊咬合缝
flat long slotted crosshead 长槽平十字头
flat lorry 平板车
flat low-lying coast 低平海滨;低平海岸
flat lump hammer 手锤(非金属大锤)

flatly 平伏的
flatly cambered rammer 平凸面夯击器
flat-lying 平缓的;平伏的;平伏
flat-lying bed 平伏层
flat-lying deposit 平伏矿床
flat-lying formation 平卧地层
flat-lying joint 平节理
flat-lying seam 平伏矿层
flat-lying sedimentary rock region 平缓沉积岩层区
flat magnetic axis 平面磁轴
flat magnetic field 平面磁场
flat market 萧条市场
flat marshalling yard 平面调车场【铁】
flat mass 平伏矿床
flat mattock 扁头鹤嘴锄;扁斧
flat maximum 平顶最高值
flat measures 平伏层
flat metal trim 金属镶边条;金属贴脸
flat mild steel bar 扁型软钢
flat mill 平磨
flat minimum 平底最低值
flat mode 平坦模型
flat model 可置平模型
flat module 平坦模
flat modulus 平坦定伸
flat money 法定货币
flat monofilament 扁平单丝
flat morphism 平坦射
flat mo(u)lding 平线脚装饰;平线脚
flat mouth tongs 平口钳
flat mushroom 扁平蘑菇
flat nail 平头钉子;扁头钉
flat negative 软调底片
flatness 平直(程)度;平坦(度);平面性;平面度;平滑性;无光;扁平
flatness error 平面度误差
flatness inspection devices 平面度测量装置
flat ness measuring instrument 平直度测量仪
flatness of aggregate 集料的扁平度
flatness of field 视界平淡
flatness of pebble 卵石扁平度
flatness of production frontier 生产边界的平直性
flatness of the response 平响应曲线;平面特性
flatness of wave 脉冲平顶
flatness ratio 平滑比;扁平度;平比
flat netted board 平面网板
flat niche 平壁龛
flat nippers 平钳
flat noise 白噪声
flat noise generator 平滑噪声发生器;白噪声发生器
flat-nose bit 平端面钻头;平面金刚石钻头;平端面金刚石钻头
flat-nosed stitch cam 平头弯纱三角
flat-nose pliers 鸭嘴钳;平头钳;平口钳;扁嘴钳
flat-nose pliers with chrome plated head 电镀头扁嘴钳
flat nozzle 平齐式接管
flat nut 平头螺母
flat nutty structure 核状结构
flat of bottom 平底部分;船底平板
flat of thread 螺纹面
flat oil paint 平光油性漆;晦漆;油质无光油漆;无光油性漆
flat oil stone 扁油石
flat-operated relief valve 浮筒操纵安全阀
flat optic(al)cable 扁平光缆
flat optic(al)tool 光学玻璃平面研磨盘
flat ornament 平面纹饰
flat-out 全速
flat pack 扁平组件;扁平封装;扁平包装
flat package 扁平组件;扁平外壳
flat-pack integrated circuit 扁平封装集成电路
flat paint 无光涂料;平光漆;平光涂料;无光泽油漆;无光(油)漆
flat paint brush 油漆刷;扁漆刷;扁(平)油漆刷
flat pallet 平台板;平(板)托盘
flat pan 平盘
flat pane hammer 琢石锤;尖锤;斧锤
flat panel display 平板显示
flat pannel display 平面显示
flat parabola 扁抛物线
flat parabola arch 扁抛物线拱
flat parabola arched girder 扁抛物线拱梁
flat parallel system 平面平行系统

flat particle 片状颗粒;扁集料;扁骨料
flat partition board 平隔板
flat pass 平轧道次;平底孔型;扁平孔型
flat passage 平通管路
flat pass-band 平通带
flat pattern 展开图;平型
flat pattern generation 展开图生成
flat peak 平顶峰
flat peen hammer 琢石锤;尖锤;斧锤
flat pein hammer 琢石锤;尖锤;斧锤
flat per cent 固定百分率
flat pick 平头镐
flat picker 平扁块拣选筛
flat piece 扁集料;扁骨料;大于规定值的骨料颗粒;扁平颗粒
flat pieces of aggregate 集料扁平颗粒;集料片状颗粒
flat-pin plug 扁(脚)插头
flat pin receptacle 扁孔插座
flat pivot 平头枢轴
flat plane antenna 平面天线
flat plane scanning method 平面扫描法
flat planting 平畦植法;平播
flat plaster ceiling 抹灰平顶
flat plate 平板
flat plate cascade 平板叶栅
flat plate closure 平板封头
flat plate collector 太阳能吸收器;太阳能转换器;平板型收集器
flat plate diffuser 平板扩压器;平板扩散器
flat plate display 平板显示器
flat plate drag 平板阻力
flat plate floor 平板楼盖
flat plate flow 平板绕流;薄层水流
flat plate foundation 平板基础
flat plate heat exchanger 平板式换热器
flat plate keel(son) 平板内龙骨;平板龙骨
flat plate lamp 平电极管;平板电极管
flat platen 平砧
flat platen pressed 平板压制
flat platen-pressed particle board 平压刨花板
flat plate pressing machine 平板压烫机
flat plate printing 平板印花
flat plate radiometer 平板辐射计
flat plate rudder 平板舵;单板舵
flat plate shoe 平板柱靴
flat plate solar collector 平板日光搜索器
flat plate structure 平板结构
flat plate vibrator 平板式振动器;平板式振捣器
flat pliers 扁嘴钳;平嘴镊;平口钳
flat plug 扁插头
flat plug ga(u)ge 扁形测孔规
flat pointed tool 平口刀具
flat pointing 勾填平缝;平勾缝
flat point screw 平推端螺钉
flat point section 平点剖面
flat point set screw 平端定位螺钉
flat policy 不含共保条款的保险单
flat pontoon 平的浮动头;平的浮船;平底浮筒
flat porcelain enamel 无光泽搪瓷
flat position 平焊位置
flat position welding 平位焊;俯焊;平焊(接);顶面平卧焊
flat position welding flange 平焊钢法兰;平焊钢凸缘
flat position welding of fillet weld 角焊缝平焊;船形角焊
flat positive peak 平顶正峰
flat power 平坦功率
flat-press board 平压板;平压碎料板;平压刨花板;平压颗粒板
flat pressing 平铺熨烫
flat press seal 平压密封
flat-pressure-response microphone 平直压强响应传声器
flat price 一揽子价格;平价;统一价(格);统售价
flat printed coil 平板印刷线圈
flat probe 平探头
flat product 扁平带材;扁平箔材;扁平板材;扁产品;扁平轧材
flat profile 平面曲线
flat profile tyre 扁平轮胎
flat pulse 平脉冲;平顶脉冲

flat punch 平冲头
flat purl machine 平型回复机
flat pyramid structure 扁平式金字塔结构
flat rack 平板集装箱;板架
flat rack container 搁架式集装箱(只有柱架及底板);平板货柜;板架集装箱
flat rack folding container 平架集装箱
flat raft 平排;平筏
flat rail 平轨;扁轨
flat rainwater gutter 平檐雨水沟
flat raised band 带形花边
flat rasp 平粗齿木锉
flat rasph 扁粗锉
flat rate 统一税率;普通资费;固定费率;均一费率;毛估流量收费;平价;统一收费率;统一价(格);统一汇率;统一费率;包价收费;包灯收费制;按时计价;按时计费制;按单位时间计价;包价
flat rate allowance 定额津贴
flat rate of negotiation 押汇平汇率
flat rate reduction 统一费率减少
flat Rayleigh fading 平坦瑞利衰落
flat reaming head 平夹式扩孔钻头
flat recess (装饰墙面的)假门;(装饰墙面的)假窗
flat reducer 还原剂;变径管(大小头);减压阀
flat region 曲线平坦段;平直区;平原区;平坦区
flat relay 扁型继电器
flat relief 平浮雕
flat repeater 扁钢围盘
flat response 平坦响应
flat-response counter 平稳反应计数器;平衡反应计数器;水平响应计数器
flat response curve 平坦响应曲线
flat retractor 扁平牵开器
flat ribbon 平带
flat ribbon cable 扁平带状电缆;扁平编织线
flat ribbon connector 扁平带状插头座
flat rib expanded metal 带扁平肋的网眼钢板
flat rib lath(ing) 平肋板条
flat rib metal lath 带肋金属拉网;肋板型金属条板
flat ridge 垣(一种黄土高原的地形)
flat riffle 平格槽缩样器
flat rim 平轮辋
flat ring dynamo 平面环形电枢发电机
flat riser 直升飞机;垂直起落飞机
flat riverbed 平坦河床;平河床
flat rock 平岩
flat rocker column 平头摆杆摇座;平头摆杆摇柱
flat rod 超公差线材
flat roll 光面压土器;平面轧辊
flat rolled 压延板材;平轧的;压延的
flat rolled iron 扁轧铁
flat rolled iron lacing 扁轧铁缀条
flat rolled steel 扁轧钢;扁钢
flat rolled steel bar 压延钢条
flat roller 平滚筒;水平托辊
flat-rolling 扁平孔型轧制
flat-rolling mill 扁钢轧机;扁材轧机
flat roof 平屋顶
flat roof deck 平屋面
flat roof dormer 平屋顶气窗
flat roof drainage 平屋顶排水(设备);平屋面排水
flat-roofed 平顶的
flat-roofed annexe 平屋顶配房;平屋顶附属房屋
flat-roofed block 平屋顶房屋
flat-roofed building 平屋顶建筑物;平屋顶建筑
flat-roofed dormer-window 平屋顶气窗;老虎天窗
flat-roofed house 平屋顶房屋
flat-roofed unit 平屋顶构件
flat roof extract ventilation duct 平屋顶抽气通风管道
flat roof fan 平屋顶风机
flat roof fascia board 檐口平顶
flat roof form 平顶模板
flat roof glazing 平屋顶镶玻璃
flat roof gutter 平屋顶排水槽
flat roof hatch 平屋顶天窗
flat roofing insulating comound 平屋顶绝热组合物
flat roofing sheet 屋面平板
flat roofing tile 平瓦
flat roof insulating compound 平屋顶绝缘物;平屋顶保暖物
flat roof insulating material 平屋顶绝缘材料;平屋顶保暖材料
flat roof insulating slab 平屋顶绝缘板;平屋顶保暖板
flat roof insulation 平屋顶绝缘;平屋顶保暖;平屋顶保温
flat roof mirror 平屋脊镜
flat roof outlet 平屋顶排水口
flat roof resonator 平顶谐振器
flat roof skylight 平顶天窗
flat roof slab 平屋顶面板
flat roof system 平屋顶体系
flat roof ventilator 平屋顶通风机
flat roof waterproofing 平屋顶防水
flat roof with air circulation 通风的平屋顶
flat rope 多股绳索;多股平索;扁绳;扁平钢丝绳
flat-rope reel 扁钢丝绳绞轮
flat rotation system 平面旋转系统
flat rubber belting 橡胶带;平带
flat rubber product 平型橡胶制品
flat rubber strip 扁橡胶条
flat run board 平滑板
flat runner 平滑板
flat-running belt conveyer 水平胶带输送机
flat-running reclaiming belt 水平胶带取料机
flat run panel 平滑镶板
flat run sheet 平滑模板
flats 公寓;浅滩
flat saddle key 平鞍形键
flats-and-officers building 公寓和办公室建筑(物);公寓和办公大楼
flat-sandwich multiconductor cable 带状多芯电缆
flat-sawed 弦向锯切的
flat-sawed lumber 顺锯板木材
flat sawing 顺纹锯法;顺锯法;平锯
flat-sawn 弦向锯切的;弦截
flat-sawn grain 弦切纹理
flat-sawn lumber 平锯木材
flat-sawn timber 顺锯材;直锯材;平锯材
flat scanning 平面扫描
flat scarf 平嵌接;平合嵌接
flatscope 平坦度仪;平滑仪
flat scraper 平面刮刀
flat screen 平幕;平筛;平板筛
flat screen of bars 扁钢格栅
flat screen printing machine 平板筛网印花机
flat screw 平头螺钉
flat seabed 平坦海底;平坦海床
flat seal (阀的)平滑度
flat sealing ring 平压密封环
flat seam 平接缝;平接板;平合缝;平缝
flat seaming 平缝法
flat seam needle 平缝帆针
flat seam roofing 平咬合缝屋面;平缝屋面
flat seat 平座
flat-seated swell 平位膨胀
flat-seated valve 平座阀(门)
flat seat washer 平座圈
flat section 平底框架
flat section ingot 扁锭
flat sector magnet 平扇块磁铁
flat seizing 平扎绳
flat semi-finished product 扁钢半成品
flat sennit 平编
flat set 平整定形
flat sewing 平缝
flat shape 平面形状
flat-shaped aggregate 扁形集料;扁形骨料
flat-shaped ingot 扁钢锭
flat sheet 平面图;平板
flat-sheet circle 圆形平板
flat sheet horizontal layout 平面布置
flat-sheet material 平板材
flat-sheet membrane bioreactor 平薄膜生物反应器
flat sheet metal 平铁皮
flat sheet metal lath(ing) 扁金属板条
flat sheet transfer unit 平板转移印花机
flat shell 扁壳体
flat shell roof 扁壳屋顶
flat shim 平垫片
flat shoal 平坦浅滩;平滩;低平浅滩
flat shoe 平履带板
flat-shoe-shaped draghead 平鞋形耙头
flat shore 平坦海岸;低岸;岸边浅滩
flat shunting 平面调车;水平调车
flat shunting neck 平面牵出线
flat shunting yard 水平调车场

flat side of a hammer 锤面
flat sieve 平筛
flat silver 成套银餐具
flat site 平坦地段
flat skin plate 平薄叶板
flat skylight 平天窗;水平式天窗
flat slab 平板;无梁板;双向板
flat slab beamless construction 无梁平板构造
flat slab bridge 平板桥
flat slab buttress dam 平板式支墩坝;平板式扶垛坝
flat slab capital construction 无梁板柱结构
flat slab ceiling 无梁天花板
flat slab column construction 无梁板柱结构
flat slab construction 无梁楼板建筑;无梁楼板构造;无梁平板构造
flat slab dam 平板坝
flat slab deck 平板式挡水面板
flat slab deck dam 平板支墩坝;平板式挡水板坝;平板坝
flat slab floor 平板式楼板;无梁楼盖;无梁楼板;无梁板
flat slab floor construction 无梁楼板结构
flat slab lavatory(basin) 平板铺的盥洗器
flat slab structure 无梁板结构
flat slat 平板条
flat slide valve 平座阀(门);平滑阀
flat slope 缓坡;平坡;平缓坡度;平缓坡道;平缓比降
flat slotted head 平槽头
flat slotted head bolt 平槽头螺栓
flat socket 扁插座
flat soldered seam 锡焊平接缝
flat source 平坦分布源
flat space 平坦空间
flat space-time 平直时空
flat spade 平铲
flat spectrum 平谱
flat-spectrum source 平谱源
flat speed profile 平坦的速度分布线
flat sphere pair 平坦球对
flat spherical shell 扁球壳体
flat spike grid 扁头钉钉成的栅格
flat spin 扁平自旋
flat spiral 平螺旋线;光滑旋管;平面螺管
flat-spiral auger shoe 扁平螺旋钎头
flat spot 平坦点;无偏差灵敏点;无光斑点;擦伤(路面);暗斑;无光泽油漆作业
flat-spotting 平点
flat-spotting of tyres 轮胎接地点扁平化
flat spring 叶片弹簧;平面簧;片(弹)簧;扁(弹)簧;板弹簧
flat spring coupling 簧片联轴器
flat spring steel 扁簧钢;扁弹簧钢
flat square 扁钢角尺
flat square nut 扁方形螺母
flat stacking 平推法;平垛法;平堆法
flat-staggered amplifier wideband amplifier 宽频带放大器
flat state of stress 平面应力状态;两维应力状态
flat steel 扁钢
flat steel bar 扁钢条
flat steel fibre 扁平钢纤维
flat steel lawn rake 草地扁钢耙
flat steel plate 平钢板
flat steel profile 扁钢外形
flat steel roller 平轮压路机;光轮压路机
flat steel section 扁钢断面
flat steel segment 平板钢管片
flat steel wire 扁钢丝
flat stern 方形船尾
flat stern ship 方形尾船
flat stitching 平订
flat stock 平形工作物
flat-stock anchor 平销锚栓;平销锚栓
flat stone chisel 扁凿;扁石凿
flat stone mill 平面石磨;平板石磨
flat straight coast 直线平坦海岸
flat strainer 平板筛浆机
flat-strand rope 异形股钢丝绳
flat-strand wire rope 扁股钢丝绳
flat stress problem 平面应力问题
flat strip net 扁条状网
flat strip slate pencil 方条滑石笔
flat strip steel 平钢带
flat structure and tall structure 平结构和高结构

flat stud tube 翅片管
flat submanifold 平坦子流形
flat suction box 平板吸水箱
flat supporting structure 平面支承结构
flat surface 平直表面;平整表面;平面
flat-surfaced probe 平面探头;平面传感器
flat-surfaced screen 平面筛网
flat surface finishing machine 平面精加工机床
flat surface glazed coat(ing) 无光泽表面涂层
flat surface glazing 无光泽表面涂油
flat surface grinding 磨平面;平面磨削
flat surface sensor 平面传感器
flat suspended main crown 平吊顶
flat suspension rod 平面吊杆
flat switchboard 平面交换机
flats with shops 公寓带商店
flat table guide 平台导承
flat table pantograph engraving machine 平台缩小雕刻机
flat tank 平面双层底舱
flat tapered 扁锥形的
flat taper file 平面锥锉;尖细锉
flat target 平面靶
flatted arch bridge 平拱桥
flatted factory 多层出租厂房;平房工厂;分层工厂
flatted tracks 暗轨铁路;暗轨铁道
flatted truck 平板货车
flat-temperature profile 均匀温度分布;平坦温度分布曲线
flat-temperature zone 恒温区
flat tempered glass 平钢化玻璃
flatten 压平;展平;变平
flatten close test 钢管压扁试验;密贴压扁试验
flattened and curled shape 变平或变弯曲的形状
flattened arch 平拱;扁拱
flattened both on the outer surface 内外表面都略呈扁平形
flattened door knob 扁球形门把
flattened expanded(metal)mesh 扁孔金属网
flattened half-round steel 扁半圆钢
flattened knob 扁球形把手
flattened model 已置平模型
flattened rayon 扁平人造丝
flattened region 展平区
flattened rivet 扁铆钉;平头铆钉
flattened rivet head 扁铆钉头
flattened roll 被压扁的轧辊
flattened round bar 扁圆钢
flattened spheroid 扁球形;扁球体
flattened square head bolt 平顶方头螺栓
flattened stipple finish 树皮拉毛饰面
flattened strand 扁平的绞合线
flattened strand cable 扁平钢丝缆
flattened strand rope 扁股钢丝绳;外面整平的绞合绳;扁平股索
flattened strand wire rope 平面股钢丝绳
flattened thread 钝螺纹
flattened wire 压扁丝
flattener 压延工;压平器;平锤;矫直机;扁条拉模;伸展机;锤平器;压延机
flattening 压延;压扁作用;矫平;平均化;平化;摊平;扁平;扁度
flattening agent 消光剂;平光剂;(涂料的)退光剂
flattening close test 密贴压扁试验
flattening coefficient 扁平系数
flattening effect 扁平化效应
flattening factor for the earth 地心扁率
flattening fold 压扁褶皱
flattening hammer 整平器;修平器
flattening index 平整指数
flattening iron 摊平铁器;摊平玻璃的铁器
flattening kiln 摊平炉(平板玻璃);摊平窑
flattening machine 整平机
flattening material 展平材料
flattening mill 轧平机
flattening of ellipsoid 椭球扁率
flattening-off 拉平
flattening of the earth 地球椭率;地球扁率
flattening of the earth-ellipsoid 地球椭圆扁率
flattening out 逐渐展平(洪水波在扩散过程中);推平;拉平;碾平
flattening-out of flood wave 洪水波展平
flattening oven 平板炉;摊平炉(平板玻璃)
flattening pressure 压平力

flattening radius 展平区半径
flattening stone 展平石块
flattening table 摊平台
flattening test 压扁试验
flattening tool 摊平锹;摊平工具
flatter 压平器;压平机;扁条拉模;摘钩工;平面锤;扁平槽
flatter distribution 扁平形分布
flattered rivet 扁头铆钉
flat terrain 平坦地形;平坦地(带);平地
flatter ribbon-like fibre 扁带状纤维
flat thin tubesheet 平板形薄管板
flat thread 平螺纹;方螺纹
flat threaded steel 扁螺纹钢
flat tie plate 平垫板
flat tile 平瓦片;平瓦
flat tile roof 平瓦屋顶;平瓦房顶
flat timber roof 平木板屋顶;木板屋顶
flatting 平化;变平
flatting agent 减光剂;消光剂;平光剂;(涂料的)退光剂
flatting down 磨光;磨平(油漆作业);消去光泽;磨退;磨掉光泽
flatting effect 消光效应
flatting efficiency 消光效率
flatting hammer 矫平锤
flatting mill 压平机;压扁机
flatting oil 褪光油;消光油
flatting operation 平面磨削
flatting pigment 消光颜料
flatting process 平面磨削过程
flatting putty 平光油灰
flatting varnish 没光漆;减光漆;清漆磨光作业;罩光清漆;抛光清漆
flat tip 平头电极
flat tip electrode 平头电极
flat tire 瘪气轮胎
flat tongs 平口钳
flat tool 平面研磨盘;平口刀具
flat-toothed belt 平型传动带;齿形传动带
flat top 航空母舰;平顶;平顶建筑物
flat-top antenna 平顶天线
flat-top apron conveyer 平板运输机
flat-top barge 平顶驳船
flat-top beam 平顶光束
flat-top chain conveyor 平台承重链式输送机
flat-top coiler 平顶圈条器
flat-top conveyer 平顶输送机
flat-top die 平顶模
flat-top distortion 平顶畸变
flat-top hill 平顶山
flat-top mountain 平顶山
flat topography 平坦地区;平原;平坦地(带)
flat-top operation 平顶波工作
flat-top peak 平顶峰
flat-topped 平顶的
flat-topped annexe 平顶附属房屋
flat-topped basilica 平顶长方形建筑物
flat-topped block 平顶房屋
flat-topped building 平顶建筑物
flat-topped cam 平顶凸轮
flat-topped crest 平顶山脊
flat-topped culvert 平顶涵洞;平顶箱涵
flat-topped curve 平顶曲线
flat-topped house 平顶房屋
flat-topped pulse 平顶脉冲;方脉冲
flat-topped ridge 平梁
flat-topped rock 平顶岩
flat-topped structure 平顶结构
flat-topped unit 平顶建筑构件;平顶构件
flat-topped voltage pulse 平顶电压脉冲
flat-topped wave 平顶波
flat-topped waveform 平顶波形
flat-topped weir 平顶堰
flat top(ping) 平顶
flat-top piston 平顶活塞
flat-top response 带通响应;平顶响应
flat-top ridge 山岭
flat-top roller chain 板条式滚子链
flat-top structure 平顶建筑物
flat-top wave 平顶波
flat-top wire cloth 平面金属丝筛布
flat tower bolt 扁插销
flat track 平轨道

flat track shoe 扁平履带片
flat trades 平价交易
flat trailer 平板挂车
flat trailer truck 平台挂车
flat trajectory fire 低伸火力
flat transmission 平直传送；平发送
flat transmission belt 平型传动带
flat transparency 平面透明度
flat tree 平顶树
flat trough 坦坡风
flat trowel 平镘
flat truck 平板（货）车
flat tube 扁平管
flat tube radiator 扁管式散热器；扁管散热器
flat-tuning 粗调谐
flat turn 平转
flat-turret lathe 平转塔六角车床；平转塔车床
flat twin cable 双芯扁电缆；扁形双芯电缆；扁平双芯电缆
flat twin engine 平列双排发动机
flat twin type engine 平列双排发动机
flat twist drill 平底麻花钻；扁麻花钻
flat type alumin(i)um wire 扁铝线；扁平铝线
flat type copper wire 扁铜线
flat type lamp shade 扁灯罩
flat type magnetic thin-film memory 扁平型磁膜存储器
flat type mixing plant 平台式搅拌设备；平台式拌和设备
flat type pile 平板型板桩
flat type pile shoe 平板桩靴
flat type piling bar 平板桩
flat type relay 扁型继电器；扁平型继电器
flat type segment 平板形管片
flat type stranded wire 扁形多股绞合线
flat type travel(l)ing mixer 平台式移动搅拌机
flat type zincmanganese dioxide dry cell 扁平式锌锰干电池
flatulence 胀气；气体浮聚
flatulent 气胀的
flat underside 平底板（车身）
flatus discharged 浊气
flat valley 浅谷；平谷
flat valve 平面座阀；平阀；配流阀
flat varnish 平光清漆；无光（清）漆；消光漆；清漆磨光打底
flat vault 扁平穹隆
flat vaulted ceiling 平弧屋顶；平弧形天花
flat-vee weir 薄壁V形切口量水堰
flat velocity distribution 平均速度分布
flat velocity profile 平均速度分布图
flat vertical gate 直升平板闸门；提升平板闸门
flat vibrating screen 平板振动筛
flat vitreous enamel 无光泽珐琅
flat vitriflable colo(u)r 无光泽的裂化颜料
flat voltage 稳定电压
flat V weir 薄壁V形切口量水堰
flat wagon with fixed sides 侧墙板固定的平车
flat wagon with hinged sides 侧墙板铰接的平车
flat wagon with mechanical sheeting 带机械护墙板的平车
flat wall 光面墙；平直模壁；平直钢锭模壁
flat wall brush 扁墙刷；扁平墙壁刷
flat wall paint 无光泽的墙漆；无光墙（壁）漆
flatware 浅平皿；爆破餐具；浅皿；盘碟
flat warehouse 平房谷仓；平房仓库
flat washer 平垫圈
flat wave 平顶波
flat way 平面朝下；扁平导轨
flat web piling 平腹板桩
flat web sheet pile 扁腹钢板桩
flat web sheet piling 扁腹钢板桩
flat web sheet piling wall 平腹板桩墙
flat web steel sheet pile 平板型钢板桩
flat weighting 平坦加权
flat weld(ing) 俯焊；搭接焊；平卧焊；平焊（接）
flat wheel 光面压路机
flat-wheel roller 光面路机；光轮压路机；光轮压碾；平轮压路机
flat-white nondirectional screen 均匀白色漫射屏
flat window on pitched roof 斜屋顶坡窗
flat wire 扁平线；扁钢丝
flat wire bar system 扁平线条方式
flat wire cloth 扁钢丝布

flat wired glass 装甲玻璃板；嵌丝玻璃板
flat wire fabric 扁钢丝织物
flat wire mill 扁线材轧机；扁钢丝轧机
flat wire rope 扁钢丝绳
flat wire woven fabric 扁钢丝编织物
flatwise 平放
flat-wise bend 平直弯曲
flatwise coil 扁平线圈
flat wood(en) roof 木板屋顶
flatworm 涡虫；扁虫
flat wrench 扁平状扳手
flat yard 平面调车场【铁】
flat yarn 变形加工用原丝
flat yield 统一价（格）；统扯收益
flaunch 凸缘（烟囱防水）
fla(u)nching（烟囱防水的）凸缘；烟囱顶泛水
flava 黄色的
flavanthrene yellow 黄烷士林黄
flavanthrone 黄蒽酮
flavanthrone yellow 黄烷士酮黄；黄光还原黄
flavescent 带黄色的；变成黄色的
Flavian Amphitheatre 弗拉维式建筑（古罗马）
Flavian architecture 圆形大露天剧场（古罗马）
flavin 黄素
flavochrome 黄色素
flavonoid 黄酮类
flavor 气味
flavorful compound 香料化合物
flavoring agent 矫臭剂
flavor profile analysis 气味廓线分析
flaw 瑕疵；一阵烈风；一阵狂风；裂纹；裂痕；挨断层；狂阵风；缺陷；疵点；发裂纹（钢材）；冰裂
flaw activity 裂缝活性
flaw dotootability 探伤能力
flaw detection 裂缝检验；裂缝检查；探伤（法）；疵伤检验
flaw detection data 缺陷监测数据
flaw detection sensitivity 探伤灵敏度
flaw detector 故障检测器；（钻眼内的）裂缝探测器；（钻眼内的）裂缝检查器；探伤仪；探伤器
flaw distribution 缺陷分布
flaw echo 缺陷回波
flawed article 副号
flaw fault 横推断层
flaw ice 岸冰
flaw in castings 铸件裂痕
flaw indication 缺陷指示；缺陷信号
flaw lead 冰间航道
flawless 无缺陷的；无缺点的；无裂纹（缝）；无裂缝的
flawless solid 无裂隙固体
flaw location 缺陷定位；探伤定位
flawmeter 探伤仪
flawmetering 探伤
flaw-piece 边皮
flaw sensitivity 缺陷灵敏度
flaw size 裂纹的尺寸；裂缝尺寸
flawy 有裂缝的
flax 亚麻线；亚麻纤维；亚麻
flaxboard 亚麻板；亚麻刨花板
flax brake 剥麻机
flax-breaking roller 亚麻碎茎辊
flax buncher 亚麻割捆机
flax burlap 亚麻织物；亚麻布
flax burlap mat 亚麻布坐垫
flax canvas 亚麻帆布；麻帆布
flax carpet 亚麻地毯
flax common 长茎亚麻
flax degumming 亚麻脱胶
flax dressing 打麻
flax dust 亚麻尘
flaxen 淡黄色的；亚麻色的
flax felt 亚麻毡
flax fiber 亚麻纤维
flax fiber board 亚麻纤维板
Flax hair-cords 麻纱
flax hards 亚麻短纤维；亚麻粗纤维
flax hessian 亚麻粗布
flax hessian drag 亚麻粗布刮路机
flax hessian drag finish 亚麻粗布刮路机整修混凝土路面
flax mill wastewater 亚麻厂废水；麻纺厂废水
flax oil 亚麻油
flax ornament 麻叶形纹饰

flax puller 拔麻机
flax pulling 拔麻
flax retting 沤麻
flax rope 亚麻绳；麻绳
flax sacking 亚麻袋布；粗麻袋
flaxseed 亚麻仁；亚麻籽
flaxseed coal 碎无烟煤
flaxseed oil 亚麻油；亚麻仁油；亚麻籽油
flax shives 亚麻屑；亚麻皮
flax silk 麻丝
flax soaked with bitumen 沥青浸渍麻布
flax stock 亚麻干茎
flax straw 亚麻稿秆
flax tow 麻麁纤维；亚麻短纤维屑；梳出的麻屑
flax-waste 亚麻碎屑
flax wax 亚麻腊
F-layer F层
flaying knife 剥皮刀
flea 低级旅店
flea bag 盖茅草屋顶
flea beetle 跳甲
flea fair 跳蚤市场
fleaking 睡袋；铺草屋顶芦苇垫底；低级旅店；单间公寓
fleam 锯齿夹角
flea market 露天旧货市场；旧货市场；跳蚤市场
fleam-tooth 尖角锯齿；等腰三角形锯齿
flea-size motor 超小型电动机
flecainide acetate 醋酸氟卡胺
fleche 顶尖装饰；尖顶塔
flechette 剪弹
fleck 雀斑；小斑点；斑影；斑点
flecked 色污斑点
flecked surface finish 斑纹饰面
fleck scale 鳞斑
fleckschiefer 微斑板岩
flecnode 拐结点
Fle-con 弗雷康（一种包装袋）
flection curve 弯曲曲线
flection formula 曲率公式
flection limit 弯曲极限；挠曲极限
flection load 弯曲荷载；挠曲荷载
flection stress 弯曲应力
fleding 折翠
fleece-wool 羊毛织品
fleecy sky 卷毛云天
fleet 运输船队；舰队；滑过；船队；车队；变换位置
fleet air arm 海军航空兵
fleet angle 绳索的最大偏角（绞车）
fleet base 船队基地
fleet by tow 拖带船队
fleet capacity 车队容量
fleet data management system 船队数据管理系统
fleeter ring 拔丝圈
fleet floor flo(a)tation 快速浅层气浮法
fleeting 编队【船】
fleeting anchorage 编队锚地
fleeting angle 钢丝绳与绞车卷筒轴线的倾角；传动角度
fleeting area 驳船集结区；编队区
fleeting move 追踪运行
fleeting of dredge(r)s 挖泥船编队
fleeting operation 追踪运行
fleeting tackle 水平滑车；辅助滑车
fleeting target 瞬间目标
fleetline body 流线型车身
fleet lockage 原（船）队过闸；整队过闸
fleet management 快速管理
fleet mix 船队编队；船队
fleet movement schedule 船队运行时间表；运行计划
fleet net 船队（信息）网
fleet of barge 驳船队
fleet operation 汽车运输公司
fleet replenishment tanker 舰队供油船
fleetsatcom 舰队卫星通信系统
fleet satellite communication 船队卫星通信
fleet testing 在使用中试验；钢绳试验；使用性试验
fleet the block 扯开滑轮组
fleet through hoisting 滑过式升降机
fleet train 舰队后勤船只
fleet tug 舰队拖轮；船队拖轮
fleet water 枯水；浅水

Fleetweld 电焊电极(一种专利产品)
fleet wheel 钢绳滑轮;钢架滑轮;钢丝绳滑轮
fleischerite 费水锗铅矾;纤锌矿
Fleming's cement 弗莱明水泥
Fleming's rule 右手定则
Fleming's valve 弗莱明管
Flemish bond 一顺一丁砌砖法;梅花砌砖法;荷兰式砌合;荷兰式砌法
Flemish brick 黄色硬砖(铺面用);铺面黄色硬砖
Flemish chimney 荷兰式烟囱
Flemish coil 平盘绳
Flemish cross bond 荷兰式交错砌合(法)
Flemish diagonal bond 荷兰式对角砌合法
Flemish dormer-window 荷兰式气窗;荷兰式天窗
Flemish double-cross bond 荷兰式双错位砌合(法)
Flemish double-stretcher bond 两顺一丁砌法;一丁两顺砌筑法
Flemish down 平盘绳
Flemish eye 绕匝索环
Flemish fake 平盘绳
Flemish foot C 形或 S 形旋纹酒桶状家具脚
Flemish garden bond 每皮三顺一丁砌合;三顺一丁砌体
Flemish garden wall bond 荷兰园墙式砌合法;三顺一丁砌法
Flemish header 荷兰式丁砖(层)
Flemish knot 荷兰式绳结
Flemish tile 荷兰式空心砖
Flemish window 荷兰式窗子
flenu coal 长焰烟煤
flesh 肉色
flesh blond 肉棕色
flesh colo(u)r 肉色
Flesh-Demag process 富雷许—德马格制气法
flesh tone 肤色
fletcherite 硫铜镍矿
Flettner rudder 襟翼舵
Flettner windmill 弗雷特纳风车
fletton 斑红色;费莱顿砖
fletton brick 橙黄色页岩砖
fleur (脊瓦上的)饰件;花的图样;粉状填料;粉状填充物
fleur-de-lis 鸢尾花,百合花形纹章;鸢尾花,百合花饰
fleuret 小花形装饰
fleuron 叶饰;花形图案装饰;百合花饰
flewing 向里斜展(窗框侧壁)
flex 麻类;花线;护套线;皮线
flex abrasion 挠曲磨损
flex action 弯曲作用
flex arc welder 高频电流稳弧焊机
flex cock 软性开关
flex-combination wrench 回旋式两用扳手
flex connector 软管连接器
flex cord 软线;皮线
flex crack(ing) 挠裂;疲劳裂纹;反复挠曲开裂;挠曲裂缝
flex cracking resistance 抗反复挠曲龟裂性
flex cycle 挠曲循环
flex-duct 空调软管;软风管;柔性管道;软通风管
flexer 挠曲试验机
flex fatigue 挠曲疲劳
flex-head wrench 回旋头扳手
flex hook 电源线挂钩
flexibilitas cerea 蜡样屈曲
flexibility 易弯(屈)性;灵活性;可弯曲性;挠(曲)性;揉曲性;柔(顺)性;柔曲性;挠度;屈曲性;通融性;适应性;变现速度
flexibility agent 助柔剂;增柔剂
flexibility coefficient 挠曲系数
flexibility conic(al) mandrel 锥形心轴弯曲柔韧性
flexibility coupling 柔性联轴节
flexibility cylindric(al) mandrel 圆柱形心轴弯曲柔韧性
flexibility dolphin 柔性靠船架
flexibility drive system 柔性传动系统
flexibility factor 柔度系数;挠曲系数;挠度系数;(管道的)相对弹性挠度
flexibility-fender system 柔性护舷系统;柔性护木系统
flexibility influence coefficient 柔度影响系数
flexibility in planning 计划灵活性;规划灵活性
flexibility matrix 挠度矩阵;柔性矩阵;柔度矩阵

flexibility method 柔度法;挠度法
flexibility modifier 柔性调节剂
flexibility number 挠度数
flexibility of execution 弹性执行
flexibility of malthoid 油毡柔度
flexibility of operation 加工适应性;操作灵活性
flexibility of piping system 管路的挠性
flexibility of production planning 生产计划的灵活性
flexibility of spring 弹簧挠性;弹簧挠度
flexibility principle 弹性原则
flexibility ratio 灵活度;调速伸缩率
flexibility spline 柔性塞缝片
flexibility test 挠度试验;柔性试验;柔韧性试验
flexibilizer 增塑剂;增韧剂;软化剂
flexible 易弯的;灵活预算;可弯曲(的);可塑造;能伸缩;柔性的
flexible access system 灵活接入系统
flexible access terminal 灵活接入终端
flexible adhesive 柔性黏合剂;柔性胶黏剂
flexible agent 助柔剂
flexible air lift mud pipe 挠性气升吸泥管
flexible air line 压缩空气软管
flexible appliance connector 挠性设备连接器
flexible apron 柔性护面;柔性海漫
flexible arch 柔性拱
flexible arch(ed) bridge 柔性拱桥
flexible armo(u)red cable 柔性铠装电缆
flexible armoured conveyer 可弯曲铠装输送机
flexible auger 挠性螺旋
flexible auger conveyer 挠性螺旋输送器
flexible automation 灵活(可调)自动化;柔性自动化
flexible axle 挠性轴;软轴;柔性轴
flexible bale collector 挠性草捆收集器
flexible base 柔性基础;柔性基层
flexible base course 柔性基层;柔性底层
flexible base materials 柔性基层材料
flexible base pavement 柔性基层路面
flexible bearing 挠性轴承
flexible bed-sweeping 软式扫床
flexible bellows 伸缩膜盒
flexible bend 软弯头;挠性弯头;柔性弯头
flexible blade coating 浮动刮刀涂布
flexible block 柔性垫块
flexible board 柔性板
flexible bodywork 铰接式车箱;挠性车箱
flexible breakeven pricing 变动损益平衡点定价
flexible breakwater 挠性防波堤
flexible budget 可变预算;弹性预算
flexible budgeting 伸缩预算
flexible budget method 可变动预算表法
flexible bulkhead 柔性岸壁
flexible bus 软总线
flexible busbar 软铜排
flexible bush 挠性衬套
flexible cable 挠性(电)缆;软性电缆;柔性索缆;柔性缆索;柔性钢索;柔性电缆
flexible cage guide 钢丝绳罐道;挠性罐道
flexible cargo hose 输油软管
flexible carpet 软性磨耗层;柔性面层
flexible carriageway 软性车行道
flexible casing 挠性轴套管(机械传动用)
flexible cell 橡胶容器;软油箱
flexible chain 挠性链
flexible channel multiplier 适应性通道倍增器
flexible choker 柔性挡板
flexible chord 柔性弦;柔性弦;电线
flexible circuit conductor 软导线
flexible cladding 柔性面层
flexible closed screen 挠性封闭屏
flexible coatin 挠性涂层
flexible coefficient 柔性系数
flexible collodion 柔性火棉胶
flexible column 柔性柱
flexible compressed-air line 软性压缩空气管线;软性压缩空气管道
flexible concrete 柔性混凝土
flexible conductor 柔性导体
flexible conductor bus 软导线母线
flexible conduit 蛇皮管;挠性(导)管;软性导线管;软管;软性涵管;柔性管道;蛇(纹)管
flexible connection 挠性联结;软式系统;软连接;柔性联轴节;柔性接头;柔性接头
flexible connection flexible joint 挠性连接

flexible connector 软接头;可伸缩接头;活动接头;挠性连接器;活络接头;柔性连接器;弹性接头;弹簧接头;伸缩节;软管连接器
flexible construction 柔性结构;柔性构造
flexible container 集装袋;挠性容器;柔性集装器
flexible control 挠性控制;挠性操纵
flexible control cable 挠性控制索
flexible control system 可调节控制系统
flexible conveyer[conveyor]挠性运输机
flexible copper cord 挠性铜线;铜软电缆
flexible cord 塞绳;绝缘软线;花线;护套线;软线;皮线
flexible core wall 柔性芯墙
flexible coupler 活头车钩
flexible coupling 柔性接头;挠性联轴节;软连接;柔性耦联;柔性联轴节;弹性联轴器;弹性联轴节;弹性连接
flexible cover 柔性盖层
flexible crawler tractor 挠性履带式拖拉机
flexible culvert 柔性涵洞
flexible curtain 挠性幕罩(气垫车)
flexible curve 挠性曲线样板
flexible curve rule 柔性曲线规
flexible dam 软料坝;柔性坝
flexible damp course 柔性防潮层
flexible derricking mechanism 柔性变幅机构
flexible diaphragm 柔软膜片
flexible die 柔性模(具)
flexible die forming 软模成形
flexible dimension 挠性尺寸
flexible disc coupling 柔性盘联轴节
flexible dolphin 挠性护柱;挠性护墩桩;柔性系船墩;柔性墩;柔性靠船架
flexible door 双开弹性门;挠性门;弹性门(采用橡胶等材料做门扇)
flexible drill steel 挠性钢钎;柔性钢钎
flexible drill stem 挠性钻杆
flexible drill stem storage basket 挠性钻杆储[贮]藏架
flexible drill stem storage reel 挠性钻杆卷筒
flexible drive 软性驱动;挠性传动;柔性传动(装置)
flexible driven gear 挠性从动齿轮
flexible-drive screen 柔性驱动筛;柔性驱动屏幕
flexible drop chute 帆布橡胶排水管;柔性下料管;软溜槽
flexible duct 软管;柔性管道
flexible elastic condition 可挠弹性状态
flexible elastomeric cellular thermal insulation 柔性弹性体泡沫绝热材料
flexible electrode 软焊条
flexible exchange rate 弹性汇率;伸缩(性)汇率;浮动汇率;可变动的汇率
flexible exhaust pipe 软性排水管;软性废气管;排水软管
flexible expanded plastics 软性泡沫塑料
flexible expansion piece 挠性膨胀节
flexible extension 挠性接头部件
flexible facing joint 柔性护面接缝;坡底柔性接缝
flexible factor 柔性系数
flexible fastening 弹性固定;弹性扣件
flexible fender 柔性垫
flexible fender cushion 柔性防撞垫;柔性防冲衬垫
flexible fender system 柔性护舷系统;柔性护木系统
flexible fiber 柔性纤维;弯曲纤维;挠性纤维
flexible fiber scope 软纤维内窥镜
flexible file 挠性锉刀
flexible filling blade 嵌填用韧性刮铲
flexible finger 挠性指杆
flexible first storey 柔性底层
flexible first storey construction 柔性底层结构
flexible fixing 弹性连接
flexible flat drill steel 挠性扁钢钎
flexible floating breakwater 挠性浮式防波堤
flexible floor 柔性楼盖
flexible flooring 柔性地板材料;塑料地板材(料)
flexible foam 柔性泡沫材料;柔性泡沫
flexible foamed plastics 软性泡沫塑料
flexible foamed polyurethane 软性泡沫聚氨酯甲酸乙酯
flexible foam lagging 弹性泡沫保温套
flexible folding wing 挠性折叠翼
flexible footing 柔性基础
flexible foundation 柔性基础;弹性基础

flexible foundation beam 软性基础梁
flexible foundation slab 软性基础板
flexible frame 挠性框架;柔性框架
flexible freeze 灵活冻结
flexible freight container 集装袋
flexible fuel tank 挠性燃料桶
flexible full track 柔性全履带车
flexible gear 机动性传动装置;挠性连接传动;挠性传动机构
flexible gearing 挠性连接传动;挠性传动装置
flexible gear-shift lever 挠性变速杆
flexible glue 软胶;揉曲性黏合剂
flexible grain conveyer 谷物挠性输送器
flexible graphite 柔性石墨
flexible graphite gasket 柔性石墨垫片
flexible grinder 软轴磨砂轮
flexible grinding circuit 可变粉磨流程
flexible guide 挠性罐道
flexible gyroscope 挠性陀螺(仪)
flexible handle 活动把手
flexible handling of space 空间的灵活布置
flexible hanger 软吊线
flexible harrow 挠性链齿耙;活动耙
flexible head coupler 活头车钩
flexible heater 蛇形管加热器
flexible highway 软性路面公路
flexible hinge 挠性节
flexible hitch 挠性挂接装置;浮动式悬挂装置
flexible hollow waveguide 挠性空腔波导管
flexible hose 挠性软管;软管
flexible hose connection 柔性管连接
flexible hose covered with plastic film 包塑软管
flexible hose pump 挠性软管泵;无脉动泵
flexible hours 弹性工时
flexible idler 挠性托轮
flexible image guide 柔性传像束
flexible imaging boundle 挠性导像光缆
flexible-impeller (rotary) pump 挠性叶轮泵
flexible industrial door 工业用双开弹簧门
flexible influence factor 柔性影响系数
flexible information 动态信息
flexible information structure 动态信息结构
flexible insulated hose 绝缘软管
flexible insulation 软性绝缘;柔性线绝缘;软绝缘
flexible interest rate 弹性利率
flexible intermediate bulk container 可调式联运散货集装箱;挠性中间散料容器
flexible job system 弹性工作制度
flexible joint 挠性接头;活络接头;活接(头);活动接头;活动接合;软接头;柔性联轴节;柔性节点;柔性接头;柔性缝;弹性联轴器;弹性联轴节;伸缩接头
flexible joint between elements 管段柔性接头
flexible joint of discharge pipeline 排泥管挠性接头
flexible jumper 挠性接合器;软跳接
flexible knife 软性刮平整修刀;挠性刀具
flexible lamp 活动电灯
flexible lamp cord 电灯花线
flexible layer 柔性层
flexible lead 软导线
flexible leg 柔性支腿
flexible life 挠曲寿命;弯曲疲劳寿命
flexible lightguide 挠性光导
flexible limit 挠曲限度
flexible lining 柔性衬砌
flexible-lip seal 弹簧唇形密封
flexible load 柔性荷载
flexible loading plane 柔性加荷面
flexible loan insurance plan 软性贷款保险计划
flexible logic (al) cell 可变逻辑元件
flexible loss and profit 变动损益
flexible lubricator 挠性润滑器;柔性(连接)润滑器
flexible manufacture system 灵活加工系统;可调加工系统
flexible manufacturing system 柔性加工系统;弹性制造系统
flexible markup practice 伸缩加成定价法
flexible markup pricing 弹性定价法;可变成本加成定价法
flexible mat 挠性垫;柔性毡
flexible material 柔性材料
flexible mat floating breakwater 柔性垫块浮体防波堤

flexible matrix 柔性矩阵
flexible mattress 柔性梢排;柔性垫子;柔性沉排
flexible member 柔性构件
flexible membrane 柔性膜
flexible membrane liner 柔性薄膜衬垫
flexible membrane lining 柔性薄膜衬垫
flexible membrane mo(u)lding 挠性袋模塑
flexible metal 柔性金属
flexible metal bellows 挠性金属膜盒
flexible metal conduit 蛇皮管;柔性金属管道
flexible metal flashing (piece) 铁皮泛水;金属皮泛水;薄钢护板
flexible metal for roofing 屋面用软金属皮;盖屋顶薄钢板
flexible metal hose 金属软管;挠性金属管;软金属管;柔金属软管;柔性金属管
flexible metal lic conduit 金属软管;金属(软管)蛇管
flexible metallic hose 金属软管;钢丝橡胶管;挠性金属管;软钢管
flexible metallic tube 金属软管;金属软管蛇管
flexible metallic tubing 金属软管蛇管;金属软管
flexible metal masonry wall-flashing (piece) 砖石墙金属泛水;护墙薄钢板
flexible metal pipe 金属软管
flexible metal roofing 柔性金属板屋面;柔性金属屋面
flexible metal sheet 柔性金属板
flexible metal sheet roof cladding 铁皮屋顶;金属皮屋顶;薄钢板屋顶覆盖层
flexible metal sheet roofing 铁皮屋顶;金属皮屋顶;薄钢板屋顶覆盖层
flexible metal sheet roof sheathing 铁皮屋顶;薄钢板屋顶覆盖层;金属皮屋顶;薄钢板屋顶覆盖层
flexible metal shoe 挠性金属密封套
flexible metaltape antenna 软金属带天线
flexible metal tube 金属软管
flexible mica 挠性云母
flexible milk line 输奶软管管路
flexible minerals 挠性矿物
flexible mooring dolphin 柔性系船墩
flexible mortar 柔性砂浆
flexible mo(u)ld 铸塑软模具;软模;柔性模(具)
flexible-mounted engine 安装在弹性座垫上的发动机
flexible mud hose 泥浆软管
flexible multiplexer 灵活复用器
flexible name 可变名(称)
flexible neoprene seal 弹性氯丁橡胶条
flexible neoprene (swing) door 柔性氯丁橡胶弹簧门;柔性氯丁橡胶摆(动式)门
flexibleness 易弯性;挠性;柔软性
flexible node 柔性节点
flexible non-metallic tubing 非金属软管
flexible nozzle 可调喷嘴;挠性喷嘴
flexible oil barge 挠性水运油囊;塑性油驳
flexible oil piston ring 挠性油环
flexible operation 挠性操纵
flexible O-ring injection fittings 挠曲O形圈注射配件
flexible overlay 柔性盖层
flexible packaging 挠软封袋法
flexible partition 软性隔壁
flexible pavement 柔性铺面;柔性面层;柔性路面
flexible pavement design 柔性路面设计
flexible pavement design method 柔性路面设计方法
flexible pavement subgrade 柔性路面地基
flexible pier 柔性墩
flexible pile 挠性桩
flexible pipe 挠性(导)管;软管;柔性管;弹性管;蛇管
flexible pipe coupling 软管接头
flexible pipe joint 挠性钻杆接头
flexible pipeline pig 挠性管道清洁器
flexible pipe section 柔性管套
flexible pipe union 柔性联管节
flexible plan 弹性计划
flexible plastic disc 柔性塑料碰盘
flexible plastic door 柔性塑料摆动式门
flexible plastic foam 柔性泡沫塑料
flexible plastics 软质塑料;柔性塑料
flexible plate 柔性板;柔性止水片
flexible plate coupling 挠性板联轴节

flexible plate loading method 柔性板荷载法
flexible plate supersonic nozzle 挠性板超音速喷管
flexible platform 柔性承台
flexible platform on piles 柔性桩台
flexible plug and socket connector 柔性插销及套管连接器
flexible plunger method 软冲头法(玻璃钢)
flexible plywood 柔性胶合板
flexible point 灵活点
flexible polyethene 挠性聚乙烯
flexible polyethene pipe 柔性聚乙烯管;聚乙烯软管
flexible polyhedron search 可伸缩多面体搜索
flexible polythene 柔性聚乙烯
flexible polythene pipe 柔性聚乙烯管;聚乙烯软管
flexible polyurethane 柔性聚亚氨酯;柔性聚氨基甲酸乙酯
flexible polyurethane bonding 挠性聚氨基甲酸接合剂
flexible polyurethane foam 软泡沫聚氨酯塑料
flexible polyvinyl chloride film 柔性聚氯乙烯薄膜
flexible polyvinyl chloride resin 柔性聚氯乙烯树脂
flexible polyvinyl chloride (swing) door 柔性聚氯乙烯弹簧门;柔性聚氯乙烯摆(动式)门
flexible precast concrete pile dolphin 柔性预制混凝土桩靠船架
flexible pressure 挠性压力
flexible price 可变动的价格;伸缩性价格
flexible-price policy 弹性价格政策
flexible prime rate 可变基础利率
flexible probe 软质探针
flexible programming system 可变程序设计系统
flexible progressive system 可变连续通行系统;灵活性推进式信号系统
flexible protective casing for pipe 挠性保护管套
flexible PVC 软聚氯乙烯
flexible rapier 挠性剑杆
flexible rate 可变利率;可变汇率;伸缩汇率;浮动利率
flexible rate mortgage 弹性利率抵押贷款
flexible reflector curtain 挠性天线反射器;软性反射幕;柔性天线反射器
flexible reinforced (concrete) foundation beam 柔性钢筋混凝土基础
flexible reinforcement 柔性钢筋
flexible relieving platform 柔性承台
flexible repeater housing 可挠增音机箱
flexible reserve 可变准备
flexible resilient 挠弹性
flexible resin (a) 挠性树脂
flexible resistor 柔性电阻器;柔软电阻
flexible retaining wall 柔性挡土墙
flexible revetment 柔性护坡
flexible rib 可伸缩肋
flexible ribbon steel 挠性钢带;挠性带状钢钎
flexible rigidity 挠曲刚度
flexible road 柔性道路
flexible road construction 柔性道路施工
flexible road industry 柔性道路工程公司
flexible rod 柔性杆
flexible rolled up tank 折叠式油罐
flexible roller 挠性滚子;(轴承的)弹簧滚柱
flexible roller bearing 螺旋滚柱轴承;挠性滚柱轴承;柔性滚动轴承
flexible roller cage 挠性滚子保持架
flexible roller-mulcher-tiller 挠性松土覆盖镇压机
flexible rope 挠性索
flexible rotor 挠性转子
flexible rubber 可伸缩橡胶管
flexible rubber gasket 柔性橡胶垫圈
flexible rubber-insulated wire 橡胶软线
flexible rubber pig 挠性橡胶清管器
flexible rubber pipe 橡胶软管
flexible rubber (swing) door 柔性橡胶弹簧门;柔性橡胶摆(动式)门
flexible rule (r) 卷尺;软尺
flexible safety barrier 挠性安全护栅
flexible safety fence 柔性护栏
flexible sandstone 挠性砂岩;柔性砂岩
flexible schedule 弹性工作时间
flexible screw conveyer 挠性螺旋输送器
flexible seal 挠性密封;柔性止水;弹性密封
flexible sealant 软密封剂
flexible seamless tubing 挠性无缝管;柔性无缝管(道);无缝软管

flexible shaft 可弯轴;可变轴;挠性轴;软轴;柔性轴
flexible shaft concrete grinding machine 软轴混凝土研磨机
flexible shaft coupling 软轴连接
flexible shaft drive (钢索的)挠性轴传动;软轴传动
flexible shaft drive internal vibrator 软轴驱动插入式振捣器(混凝土捣实用)
flexible shaft grinder 软轴砂轮机
flexible shaft grinding machine 软轴研磨机
flexible shafting 挠性轴系
flexible shaft internal vibrator 软轴内插式振动器
flexible shaft level controller 挠性轴的水准控制器
flexible shaft level indicator 挠性轴的水准控制器
flexible shaft machine 挠性轴机床
flexible shaft mechanical transmission 挠性轴机械传动
flexible shaft protecting 软轴套
flexible shaft transmission 挠性轴传动;软轴传动
flexible shaft type 挠性轴式
flexible sheet 柔性板;弹性模片
flexible shock absorber 柔性减振器
flexible shroud 柔性护罩
flexible silicon strain ga(u)ge 柔性硅应变仪
flexible single crossover turnout 单渡线可挠型道岔
flexible single turnout 单开可挠型道岔
flexible slab 软板;柔性板
flexible solid 柔性固体
flexible spindle 分离锭胆式自调中心锭子
flexible spline 柔性夹板;柔性塞缝片
flexible spout 柔性卸料槽;柔性料槽
flexible sprayer 软性喷射机
flexible sprocket-wheel pulverizer 挠性星轮碎土器
flexible stage 活动舞台
flexible stainless-steel bellows 挠性不锈钢密封盒
flexible standard 弹性标准
flexible standard budget 弹性标准预算
flexible stay 挠性撑条
flexible stay bolt 伸缩撑螺
flexible steel 挠性钢钎
flexible steel arch 柔性钢结构拱;可挠曲的钢结构拱
flexible steel cylinder dolphin 柔性钢筒靠船墩
flexible steel in flat section 挠性扁钢钎
flexible steel pile dolphin 柔性钢桩靠船墩
flexible steel wire rope 软钢丝绳
flexible stem electrodrilling 柔杆电钻钻进
flexible stor(e)y 柔性楼房;柔性楼层
flexible structure 柔性结构;柔性构造
flexible strut 挠性支撑
flexible substratum 柔性底层
flexible suction tube 挠性吸泥管
flexible support 挠性支持;柔性支架;柔性支承
flexible supported end 柔性支承端
flexible supported system 柔性支承系统
flexible supported wall 柔性支承墙
flexible surface 柔性面层;柔性路面
flexible surfacing 柔性面层;柔性表面
flexible surfacing design 柔性表面设计
flexible surfacing design method 柔性表面设计法
flexible suspension 柔性悬索;柔性悬挂;柔性悬浮;弹性吊架
flexible swing door 柔性弹簧门;柔性摆(动式)门
flexible switch 可变式尖轨
flexible symbol 可变符号
flexible system program(me) list 可变系统程序表
flexible tacties 机动灵活的政策
flexible tank pressurization 柔性容器增压
flexible tariff 灵活关税;可变税率;可变关税;伸缩关税
flexible tie-back bulkhead 柔性岸壁
flexible timber pile dolphin 柔性木桩靠船架
flexible time 弹性工作时间
flexible tolerance algorithm 可伸缩容限法
flexible tower 挠性铁塔
flexible transition coupling 柔性过渡接头
flexible transmission 挠性传动
flexible transport 缆车运输;皮带运输;无轨运输;无轨电车交通
flexible trust 弹性信托
flexible tube 挠性管;软管;柔性管
flexible tube auger conveyer 挠性管式螺旋输送器
flexible tube valve 挠性管阀
flexible tubing 软绝缘管;柔性非金属管;挠性管路;柔性(连接)管

flexible tunnel lining 柔性隧道支护
flexible type 挠性型
flexible type of roller bearing 挠性滚柱轴承
flexible type shield 柔性掩护支架
flexible type vibrator 挠式振动器
flexible understructure 柔性下部结构
flexible unit 柔性组件;通用设备
flexible universal joint 挠性万向节;软性万向节
flexible urethane pig 挠性聚氨酯清管器
flexible use of tariff protection 灵活运用关税保护
flexible-van container car 活动货箱型集装箱车
flexible-van system 活动货箱型方式;活动货箱方式
flexible varnished tubing 漆包软管
flexible ventilation ducting 通风软管
flexible vibrator 挠性振捣器;插入式振捣器
flexible vinyl flooring 软质乙烯地板材料
flexible volume method 柔性体积法
flexible wall 挠性腹板;柔性墙
flexible wall covering 挠性墙纸
flexible water-proof 柔性防水
flexible waveguide 可弯曲波导管;可弯曲波导;可挠波导;挠性波导管;软波导(管);柔性波导(管)
flexible welding rod 软焊条
flexible wheel base 变轮距;挠性轮距
flexible wing 可折叠翼
flexible wire 挠性线;花线;护套线;软线;皮线
flexible wire cable 软性钢索
flexible wire for electric(al) welding 电焊软线
flexible wire release 线条快门开关
flexible wiring 软性布线
flexible wood pile dolphin 柔性木桩靠船架
flexible work hours system 弹性工时制
flexible working hours 机动工时;弹性工时
flexibly mounted equipment 柔性装配设备
flexiboard 柔性板
flexichoc 真空聚爆式震源
Flexicore slab 弗莱克芯西科板(混凝土空心浇筑整体式构件)
flexile fatigue test 弯曲疲劳试验
flexility 柔度
fleximer 无缝柔性地面;柔质乳胶;无缝铺地作业;水泥橡胶乳胶混合料
fleximeter 挠度计;柔曲计;弯曲应力测定仪
flexing action 挠曲作用;弯折作用
flexing brick 软曲砖
flexing concrete slab 挠曲的混凝土板
flexing curve 弯曲线;挠曲线
flexing cycle 弯曲周期
flexing elasticity 弯曲弹性
flexing formula 挠曲公式
flexing life 弯曲寿命;弯曲寿命;耐弯曲性;耐挠曲性
flexing limit 弯曲极限;挠曲极限
flexing load 弯曲荷载
flexing machine 挠曲试验机;挠曲机
flexing property 弯曲性
flexing resistance 抗挠阻力;抗挠(曲)阻力;抗挠曲性;抗挠(强度);耐弯曲性;弯曲阻力
flexing roll 弯折辊
flexing strength 挠曲强度
flexing stress 挠曲应力
flexing test 挠曲试验
flexion 屈曲
flexional elasticity 弯曲弹性;弯曲变形
flexional symbols 变形符号
flexion and extension 屈伸运动
flexion curve 弯曲线;挠曲线
flexion deformity 屈曲畸形
flexion formula 挠曲公式
flexion load 弯曲荷载
flexion of surface 曲面(的)拐度
flexion reflex 屈曲反射
flexion spring 挠曲计
flexion stress 挠曲应力
flexiplast 挠性塑料;柔性塑料
flexiplastics 挠性塑料
flexiport 移动式微型港口
flexitime 弹性上班时间;可伸缩的劳动时间
flexitime system 弹性工作时间制度
flexivity 挠度
flex-job system 弹性工作制度
flexline oil piston ring 挠曲线活塞刮油环
flexlock 柔性止水缝
flexode 变特性二极管;变结二极管
flexographic(al) ink 橡胶凸版油墨

flexographic(al) press 曲面印刷机
flexographic(al) ink 苯胺印刷油墨
flexography 曲率图;苯胺印刷术;苯胺印刷
flexometer 挠度器;挠计;挠度仪;挠度计;曲率仪;曲率计
flexonics 挠性结合
flexotir 爆炸笼
flexowriter 快速印刷装置;多功能打字机
flexplast 柔性塑料
flex plate 柔性板
flex plug and socket 活动插销座
flex plug and socket connector 柔性插销与管套连接器
flex point 拐点
flex rotor 柔性转子
flex steel tape 钢卷尺
flex stiffness 挠曲劲度
flextensional mode 弯曲振动方式
flextensional transducer 弯张换能器
flex tester 板材弯曲试验器;板材弯曲试验机
flextime 弹性工作时间
flex-toe claw bar 活动趾撬棍
Flextol 弗莱斯克托(为电动工具系列品牌)
flexuose 弯曲的;锯齿状的;动摇不定的;波状的;曲的;多曲折的
flexuosity 屈曲;弯曲率;波状
flexuous 弯曲的
flexural 弯曲的
flexural-aileron flutter 弯扭副翼颤振
flexural and torsional combined stresses 弯曲和扭转组合的应力
flexural beam 抗弯梁;弯曲梁
flexural bending 受挠曲弯曲;柔性弯曲
flexural bond 弯曲结合力;受弯引起的黏附力破坏;挠曲握裹力;弯曲徐变
flexural bond test 弯曲黏附性试验;受弯的黏附力试验
flexural buckling 弯曲屈曲;受弯引起的失稳;受弯引起的屈曲;挠性屈曲
flexural building 柔性受弯建筑物;受弯建筑物
flexural center 挠曲中心;弯曲中心
flexural compressive failure 弯压破坏;挤压裂纹;挤压断裂
flexural compressive strength 弯曲抗压强度
flexural constant 弯曲常数
flexural crack 弯曲裂纹;弯曲裂缝
flexural cracking 弯曲裂缝;弯曲裂缝
flexural damping 弯曲阻尼
flexural deflection 弯曲变形
flexural deformation 挠曲变形;弯曲变形
flexural displacement 弯曲位移
flexural elasticity 弯曲弹性
flexural endurance 弯曲疲劳;弯曲耐久力;弯曲持久性;挠曲疲劳
flexural endurance limit 弯曲持久性极限
flexural equation 弯曲方程
flexural failure 弯曲破坏
flexural fatigue 挠曲疲劳;弯曲疲劳
flexural fatigue endurance 弯曲疲劳耐久性
flexural fatigue endurance limit 弯曲疲劳耐久性极限
flexural fatigue limit 弯曲疲劳极限
flexural fatigue resistance 弯曲疲劳抵抗力
flexural fatigue strength of concrete 混凝土的弯曲疲劳强度
flexural fiber stress 挠曲纤维应力
flexural fold 弯曲褶皱【地】
flexural force 挠曲力
flexural glide 弯曲滑动
flexural index 弯曲指数
flexural limit 弯曲极限
flexural load 弯曲荷载;挠曲荷载
flexural loading test 振荡受力疲劳试验
flexural loop joint 受弯环接头
flexurally rigid 抗弯的;刚坚不可弯曲的
flexural measurement 弯曲测量
flexural member 挠性构件;受弯构件
flexural meter 挠度仪
flexural mode 弯曲振动方式
flexural mode of vibration 弯曲振型
flexural mode vibration 弯曲型振动
flexural modulus 抗弯模数;抗弯模量;弯曲模量;弯折模量;弯曲模数;弯曲模量
flexural moment 挠曲力矩

flexural number 弯曲指数
flexural oscillation 弯曲振荡；弯曲摆动
flexural oscillation failure 弯曲振动破坏
flexural oscillation strength 抗弯曲振动强度
flexural oscillation test 弯曲振动试验
flexural oscillator 弯曲形振荡器
flexural resistance 抗弯曲强度；抗弯强度
flexural resonance 弯曲共振
flexural resonance method 弯折共振试验法
flexural rigidity 线刚度；弯曲硬度；抗弯刚性；抗挠刚度；抗挠刚度；挠性刚度；弯曲强度；绕曲刚度；弯曲刚度
flexural shear crack 弯曲剪切裂纹；弯曲剪切裂缝；受弯的剪切裂缝
flexural shear failure 弯曲剪切破坏
flexural slip 挠褶滑动；挠滑动；曲滑；弯曲滑动；层面滑动
flexural spring 弯曲弹簧
flexural stiffness 挠曲劲度；抗弯劲度
flexural strain 挠曲应变；弯曲应变
flexural strain energy 弯曲应变能
flexural strength 挠曲强度；抗折强度；抗弯强度；抗挠强度；弯曲强度
flexural stress 抗弯应力；挠(曲)应力；挠度应力；弯(曲)应力
flexural stress distribution 弯曲应力分布
flexural stress formula 弯曲应力公式
flexural tensile failure 弯曲拉伸破坏
flexural tensile strain 弯拉应变
flexural tensile strength 弯拉强度
flexural tensile stress 弯拉应力
flexural tensile test 弯曲拉伸试验
flexural test 挠性试验；挠曲试验；弯曲试验
flexural torsion 弯曲扭动；扭曲
flexural-torsional buckling 弯曲扭折；弯曲扭断；弯扭失稳；弯扭屈曲
flexural-torsion flutter 弯曲扭颤振
flexural vibration 挠性振动；挠振动；曲线式振动；弯曲振动；弯挠振动
flexural vibration failure 弯曲振动破坏
flexural vibration strength 弯曲振动强度
flexural vibration test 弯曲振动试验；弯曲振荡试验；弯曲摆动试验
flexural wave 弯曲波
flexural yielding 弯曲屈服
flexural zone 挠折带
flexure 接触导线弛度；挠曲；柔曲；屈曲节；曲率；曲；弯沉；单斜挠褶
flexure bar 弯曲振动棒
flexure coast 单褶海岸
flexure crystal 弯曲晶体
flexure curve 弯曲线；挠曲线
flexure(d)fold 弯曲褶皱【地】
flexure error 弯曲误差
flexure error of division face of staff 标尺分划面弯曲差
flexure fault of the crust 地壳挠折
flexure-flow fold 弯流褶皱
flexure fold(ing) 挠曲褶皱(作用)；弯曲褶皱作用
flexure formula 挠曲公式；弯曲公式
flexure graben 弯曲地堑
flexure hinge 柔性铰
flexure limb 翼部挠曲
flexure limit 弯曲极限；挠曲极限
flexure member 挠性构件；柔性构件
flexure meter 弯曲试验机；挠曲计；挠度计；曲率仪；曲率计
flexure modulus 抗弯模数；抗弯模量
flexure moment 偏转力矩；弯曲力矩
flexure of sedimentary cover 盖层挠曲
flexure of sound wave 声波曲射
flexure of the crust 地壳挠曲
flexure of the tube 镜筒弯沉
flexure operation 弯曲动作
flexure plane 弯曲面
flexure produced by axial compression (轴向压力引起的)纵向挠曲
flexure reinforcement 受弯钢筋
flexure resistance 抗弯刚性
flexure-shear interaction 弯剪相互作用
flexure ship flooding 弯滑褶皱
flexure slip fold 层面滑动褶轴
flexure strength 抗弯强度；抗挠强度
flexure stress 抗弯应力；抗挠应力；挠曲应力；弯曲应力

flexure strip 弹性片
flexure test 挠性试验；挠曲试验；弯曲试验
flexure test bar 抗弯试棒
flexure test beam 弯曲试验梁
flexure test machine 挠性试验机；挠曲试验机；弯曲试验机
flexure theory 弯曲理论
flexure torsion 挠曲扭转
flexure type 弯曲型
flexure under lateral 侧向挠曲
flex-vent oil ring 挠性油环
flex-wing 可折三角形机翼；柔性翼
Flexwood 弗莱斯伍德(一种具有柔性和防水性能的极薄木片)
flich 料方
flick diaphragm 跳跃膜片
flicker 闪烁现象；闪烁
flicker-beam 闪变光束
flicker blade 防闪烁叶片
flicker-brightness performance 闪烁与亮度性能
flicker control 振动调整；闪烁控制；闪动控制
flicker counter 闪烁计数器
flicker disturbance 闪烁干扰
flicker effect 闪烁效应；闪变效应
flicker factor 闪烁系数
flicker frequency 闪烁频率；闪变频率
flicker fusion 闪光融合
flicker-fusion frequency 闪烁融合临界频率；光闪频率；停闪频率
flicker ground 间歇接地
flickering 闪光；闪变
flickering device 频闪装置；闪视装置
flickering flame 闪烁光焰
flickering lamp 闪光灯
flickering light 增减光度的闪光；闪烁光
flickering searchlight 闪动探照灯
flickering signal 闪光信号
flicker method 光闸法立体观察；交替闪视法；闪变法
flicker noise 闪烁噪声；闪变噪声；闪变效应噪声
flicker phase 闪变相位
flicker photometer 闪烁光度计；闪光光度计；闪变光度计
flicker photometry 闪变光度术
flicker principle 闪视法
flicker rate 闪烁率
flicker relay 闪光继电器
flicker sensation 闪光感觉
flicker spectrophotometer 闪变分光光度计
flicker system 闪变系统
flicker test 闪变试验
flicker threshold 闪烁阈；闪烁限度；闪烁门限
flicking light 闪烁光
flick noise 闪光噪声
flier 悬垂顶撑；整速轮；快车；梯级；踏步板；手轮；飞行器；飞轮
flier arch 拱扶垛；扶壁；飞拱
flies 舞台侧翼；舞台上部布景控制处；舞台上空
flight 行程；楼梯级；阶梯步级；梯级；梯段；射程；船舷弧线的突升；飞行
flight acceptance test 飞行验收试验
flight altitude 航高；飞行姿态；飞行高度
flight announcement center 飞行广播中心
flight auger 螺杆式螺旋钻；链动螺钻
flight balance 飞行平衡
flight belt 抛掷式充填机；抛掷机皮带；抛掷式填充机
flight block 摄影分区
flight calculator 风速计算器
flight characteristic 飞行特征
flight chart 航空地图
flight condition 飞行条件
flight control 飞行管制
flight controls 操纵系统(飞机)；飞机操纵系统
flight control system 飞机操纵系统
flight conveyer 链动输送机；刮板运输机；刮板式输送机；链板运输机；链板式输送机；刮板运输器；刮板(式)输送机；链式刮板输送机；链板式输送机
flight crew 飞行人员
flight deck 飞行甲板
flight direction 飞行方向
flight distance 飞程
flight elevator 刮板升运器
flight engineer 随机工程师

flight feeder 刮板供料机
flight from cash 抛售现金
flight from the dollar 抛售美元
flight gyro 地平陀螺仪
flight header 梯段横梁
flight height 绝对高度；航高；飞行高度
flight hole (木材的)虫眼；飞出孔
flight hour 飞行小时
flight indicator 陀螺地平仪
flight information and air facility data 飞行资料与航空站设备数据
flight lead 螺距；丝距
flight length 梯段长度
flight line 飞行线路；机场维护工作区；航线【航空】；航空线
flight line aerial photography 航线航空摄影
flight line kilometers 测线公里
flight line of aerial photography 摄影航线
flight line space 航线间距
flight link 刮板链节
flight locks 多级船闸
flight log 航空志；飞行记录装置；航行记录簿
flight map 飞行图
flight monitor 飞行监测器
flight navigator 飞行导航仪
flight number 航次(指飞机)
flight of capital 资本外流
flight of locks 连箱船闸；多级闸门；梯级(式)船闸；多级船闸；船闸梯级
flight of stair(case) 梯段
flight of terraces 梯级台地；多级台地
flight operation system for noise abatement 低噪声飞行方式
flight panel 航行驾驶仪表板
flight passenger 班机旅客
flight path 飞行路线；航向线【航空】；航迹【航空】
flight path accelection 航道加速度
flight path analyzer 航迹分析器
flight path angle 航迹角
flight path axis 航迹轴
flight path computer 航线计算器
flight path deviation 飞行路线偏离
flight path deviation indicator 航迹偏差指示器；飞行路线偏差指示器
flight path planning equipment 飞行路线计划设备
flight path recorder 航迹记录器【航空】
flight path recorder chart 航迹记录图【航空】
flight path slope 航迹角
flight personal 飞行人员
flight photo(graph) 航空照片；飞行照片
flight plan 梯级规划；飞行计划图
flight planning 飞行计划
flight profile 空中断面记录图；飞行剖面图
flight quality 飞行质量
flight quality of aerial photography 航摄飞行质量
flight record 飞行记录
flight recorder 飞行自动记录仪
flight-refuel 空中加油
flight-remote control device 飞行遥控装置
flight report 飞行报告
flight research center[centre] 飞行研究中心
flight rise 梯段步高
flight route 航空路线
flight run 梯段步距
flight scale 航摄比例尺
flight scraper 刮板式刮泥器；链板式刮泥器
flight sewer 阶梯式污水管道；阶梯式管道；跌落式污水(管)道；跌落式排水管(道)
flight simulation test 模拟飞行测试
flight slab 梯段板
flights lock 船闸梯级
flight stairway 过街楼
flight strip 飞机着陆场；空降场；简易跑道；简易机场；机场起飞跑道；航线【航空】；航摄带；航带；起飞跑道；条幅式侦察照片
flight strip design 航带设计
flight strip region aerial photography 带区航空摄影
flight strip survey 航线测量【航空】
flight tariff 可变关税
flight test 飞行试验
flight test engine 试飞发动机
flight testing 试飞
flight tip 螺旋尖梢

flight trace of photography 摄影航迹
flight travel speed 刮板运行速度
flight unit 飞行组
flight velocity 航行速度(指飞行物)
flight visibility 空中能见度;飞行能见度
flight weather 飞行天气
flight width 梯段宽度
flimsy evidence 不可靠证据
flimsy ground 软弱地基
flinch 法兰
flinching 削成斜面(木端)
flinders 破碎片;破片
Flinder's bar 垂直软铁校正棒;弗林特棒
flinger 抛油圈;抛油环;抛射器;抛射机
flinger ring 挡液环
fling-off cooling 冲离冷却
flinkite 褐砷锰石;褐砷锰矿
Flinn diagram 费林图解
flint 细粒砂岩;火树石;火石坚硬;火石
flint abrasive 木砂纸用磨料;粗磨料
flint aggregate 坚硬集料;坚硬的骨料
flint ball 燧石球
flint brick 坚硬砖;燧石砖
flint clay 硬黏土;硬质黏土;焦宝石;燧土
flint colo(u)red paper 彩色蜡光纸
flint container glass 无色瓶罐玻璃
flint-containing limestone 燧石灰岩
flint crush rock 超糜棱岩化
flint dry 高温烘燥
flint faience 高硅陶器;燧石陶器
flint fireclay 火石质耐火黏土;硬质耐火黏土
flint glass 火石玻璃;含铅玻璃;铅玻璃;无色玻璃;燧石玻璃
flint glass paper 粗砂纸
flint glass prism 火石玻璃透镜
flint-glazed paper 蜡光纸
flint glazing machine 磨光机
flint gray 燧石灰色
flinting ground 坚硬地面
flint lens 火石玻璃透镜
flint mill 砾磨机;燧石球磨
flint optic(al) glass 火石光学玻璃
flint paper 研光纸;粗砂纸
flint pebble 燧石子;粒状燧石;石球
flint rubble 燧石渣;燧石块
flint sand 火石砂
flints clay 燧石黏土
Flintshire process 富林脱歇尔炼铅法
flint stone 燧石;火石;打火石
flint stones tube mill 燧石管磨机
flint wall 坚硬墙;火石墙
flint walling 燧石圬工墙(一种组合圬工墙)
flinty 燧石的
flinty ground 坚硬地面;石质硬底子;粗砂
flinty hardness 淬火玻璃硬度
flinty limestone 燧石质石灰岩
flinty slate 坚硬板;燧石板岩
flinty soil 含硅土
flip-and-flop 双稳态多谐振荡器
flip-and-flop generator 双稳态触发器
flip bucket 排流鼻坎;挑流消能坎;挑流鼻坎
flip chart 卡片簿;翻转图;活动挂图
flip chip 叩焊晶片;倒装晶片;倒装(芯)片;倒装法
flip-chip bonder 倒装式焊接器;倒装焊接器
flip-chip bonding 叩焊;倒装式接合;倒装焊接
flip-chip method 倒装焊接法
flip-chip ultrasonic bonder 倒装超声焊接器
flip coil 弹回线圈
flip-flop 翻转电路;角异色性;双稳态多谐振荡器;双稳触发器;触发器;触发电路;翻转器
flip-flop amplifier 附壁放大器
flip-flop circuit 双稳态多谐振荡电路;触发(器)电路
flip-flop column system 回转柱系统
flip-flop control 双稳态控制
flip-flop counter 触发计数器;触发计数管
flip-flop element 触发器单元
flip-flop frequency divider 触发电路分频器
flip-flop generator 双稳态多谐振荡器
flip-flop intergrated circuit 触发器集成电路
flip-flop mirror 翻动转镜
flip-flop number 触发计数器;触发计数
flip-flop register 触发(式)寄存器;触发(器)式寄存器;触发器存储器
flip-flop ring 触发计数环

flip-flop shift register 触发移位寄存器
flip-flop sign 符号触发器
flip-flop stage 触发板;触发级
flip-flop storage 触发器存储器
flip-flop transition 翻转时间
flip-flop transition time 双稳态触发器翻转时间
flip-flop type entry guide 翻转型导口
flip-flop type system 翻转式系统
flip open cutout fuse 跳开式熔断器
flip-over bucket 翻斗
flip-over bucket loader 高架铲运机;翻斗式装载机
flip-over process 倒逆过程
flipper 橡胶蹼;圆木装卸机;对位舌板
flipper guide 鳍形导杆;对位舌板;侧导板;翻转导手【机】
flipper-turn 快速转弯
flipping screens 切换屏幕
flip-plug adaptor 可拆式电缆盒接头
flip pump 倒装泵
flip trajectory bucket dissipater 挑流式消能设备;挑流式消能工
flip-up target 提升靶
flist 弗利斯特雨
flitch 组合贴板;毛方;厚条板;桁板;贴板;单板料片;单板堆;背板(打桩)
flitch chunk 厚板;组合厚板
flitch dam 木板坝;单板坝
flitched beam 钢木组合板梁(两根木梁中夹有钢板,作成一个构件);组合板梁;合板梁
flitched (plate) girder 钢木组合大梁(由方木与金属构成);组合板大梁;钢木夹合梁
flitch matching 单板匹配(胶合板)
flitch plate 组合(垫)板;组合梁夹铁(夹在组合梁中起加强作用的钢板);夹合钢板
flitch-plate girder 钢木组合大梁(由方木与金属板构成);贴板梁
flitch-trussed beam 组合桁架梁;桁木桁架梁
flitered flue gas 滤过的烟气道
fliter effluent 滤器流出液
flit-gun 家用喷雾器
flit-plug 可拆卸插头
flitter 金属碎片;金属箔;片状金粉;闪流剂;闪光颜料
flitter gold 黄铜箔
flivver 海军小艇
floaging carrier 浮动载波体制
float 小排;转石;路面整平器;送送;开设;木抹子;镘刀;画面抖动;飘浮(状);低架平板车;浮子;浮体;浮囊;浮动
floatability 可浮(选)性;抗沉性;漂浮性;浮动性
floatable material 可漂浮物质
floatable solid content 悬浮性固体量
floatable system 漂浮体系
float accumulator 浮充蓄电池(组)
float-actuated liquid level detector 浮子式液位检测器
float actuated recorder 浮动式自记水位计
floatage 漂浮(物);木材流送;船体吃水线以上部分;浮力
floatainer 流动集装箱;漂浮货柜
float a loan 发行债券
float and set 三道抹灰;三层抹灰
float-and-sink analysis 浮沉分析
float-and-sink analysis number 浮沉试验编号
float-and-sink analysis of fines 煤泥浮沉试验
float-and-sink analysis of fines result table 煤粉浮沉试验结果表
float-and-sink analysis report table 浮沉试验报告表
float and sink method 浮沉法
float-and-sink process separation 重介质选矿(法)
float-and-sink sample 浮沉样
float-and-sink testing 浮沉试验
float and sink treatment 上浮下沉处理
float-arm 浮子臂
flo(a)tation 悬浮;漂浮(法);浮(游)选;浮力作用;浮集法;浮动
flo(a)tation agent 助浮剂;浮选剂
flo(a)tation analysis 漂浮分析
flo(a)tation apparatus 浮选装置
flo(a)tation area 漂浮面积
flo(a)tation balance 浮力秤
flo(a)tation basin 气浮池
flo(a)tation cell 浮选机;浮选槽(室)

flo(a)tation cell feed box 浮选机入料箱
flo(a)tation center 浮选水面形心
flo(a)tation characteristic 漂浮特性
flo(a)tation circuit 浮选流程;浮选回路
flo(a)tation coagulant 浮选凝结剂;浮选混凝剂
flo(a)tation coefficient 漂浮系数
flo(a)tation collar 安全环状浮袋
flo(a)tation collection agent 浮选促集剂
flo(a)tation collector 浮沫收集器;浮选促集剂;浮选捕收剂
flo(a)tation column 浮选柱
flo(a)tation concentraction method 漂浮浓聚法
flo(a)tation concentrate 浮选精矿
flo(a)tation concentration 浮选浓缩
flo(a)tation concentration method 浮选浓缩法
flo(a)tation constant 漂浮常数
flo(a)tation cost of bonds 发行债券费用
flo(a)tation crust 气浮池浮渣(水处理);浮选器浮层(选矿)
flo(a)tation depth 浮体吃水深度
flo(a)tation device 浮选装置
flo(a)tation equipment 浮选装置
flo(a)tation frother 浮选起泡剂
flo(a)tation gear 浮筒式起落架;浮水装置
flo(a)tation jacket 救生背心
flo(a)tation leaching method 浮游选矿法
flo(a)tation line 浮吃水线;浮起吃水线
flo(a)tation machine 浮选机
flo(a)tation matter 漂浮物
flo(a)tation method 漂浮浓集法;浮选(浸出)法;菲勒本法
flo(a)tation mill blower 浮选用鼓风机
flo(a)tation modifying agent 浮选改良剂;浮选调整剂
flo(a)tation of loan 融资
flo(a)tation oil 浮选油
flo(a)tation-oil feeder 浮选加油器
flo(a)tation plant 浮选装置;浮选车间;浮选厂
flo(a)tation pond 气浮池
flo(a)tation process 气浮法;浮选法
flo(a)tation promotor 浮选促进剂
flo(a)tation reagent 浮选剂
flo(a)tation reflux 气浮回流
flo(a)tation saveall 飘浮式回收机
flo(a)tation section 浮选车间
flo(a)tation separation 浮选分离;浮上分离
flo(a)tation sponge 漂浮性海绵胶
flo(a)tation stability 浮选稳定性
flo(a)tation suit 救生衣
flo(a)tation table 浮选摇床;浮选淘汰盘
flo(a)tation-tabling 摇床浮选;台浮
flo(a)tation tailings 浮选尾矿
flo(a)tation tank 气浮池;浮选池;浮箱;浮力舱;浮动储[贮]油罐
flo(a)tation thickener 浮选浓缩池
flo(a)tation thickening 气浮浓缩
flo(a)tation tire 高通过性宽断面轮胎
flo(a)tation treatment 浮选处理
flo(a)tation treatment of wastewater 废水上浮处理
flo(a)tation treatment technique 气浮处理技术
flo(a)tation unit 浮选装置
flo(a)tation wastewater 选矿浮选废水
flo(a)tation wheel 越野行走轮
float attribute 浮动属性
float-balanced shiplift 浮筒式升船机
float ball 浮球
float barograph 浮子气压记录仪;浮子气压计
float batch 料山;料堆
float bath 金属液槽;浮抛窑
float block 浮砖
float board 轮叶;木排;(水车的)蹼板;渡船;承水板;筏
float boat 漂浮艇
float bond 发行债券
float bowl 浮筒
float bracing 浮筒支撑;浮力拉条
float brick 浮砖
float bridge 浮桥;固定浮坞;轮渡码头引桥
float buoyancy 浮筒浮力
float-cage type ga(u)ge 浮动箱式液位计
float caisson 浮式沉箱
float case 沉箱

float cell 浮槽分级机
float chamber 金属液槽;浮子室;浮厢;浮筒室;浮抛窑;浮标
float chamber cap 浮室盖
float chamber spring ring 浮筒室弹簧圈
float charger 浮动充电器
float coal 精煤
float coat 水泥抹平;抹灰层;抹面层;抹面灰浆;第二道抹层
float coefficient 浮标系数
float collar 带回压阀的套管接箍;浮籀
float control 浮标控制
float-controlled drainage pump 浮子控制排水泵;自动排水泵
float-controlled switch 浮子控制开关
float-controlled valve 浮子控制阀;浮子阀
float controller 浮子控制器;浮子调节器
float control method 浮标控制法
float counter balance 浮筒杠杆
float coupling 浮籀
float course 浮标测流路线
float crosshead pin 浮动十字头销
float current meter 漂浮式流速计;浮标流速计
float-cut file 斜纹锉;单纹锉(刀)
float displacement 浮筒排水量;浮标排水
float diving 浮标潜水
floated asphalt 镘整地沥青;摊铺地沥青
floated bright 净水流送的
floated coat 抹平面;抹灰层
floated concrete 流态混凝土;抹(平)的混凝土;抹面混凝土;镘(整)混凝土;镘光混凝土;轻质混凝土
floated filler 轻质填料;抹平填料
floated finish 抹灰饰面
floated glass 厚玻璃
floated gyro(scope) 浮子式陀螺仪;悬浮式陀螺仪
floated ice 覆水冰
floated loan 分配贷款
floated screening 木蟹找平
floated surface 墁平面;抹平面;抹平的表面;镘(平)表面
floated work 抹灰作业;抹灰工程;抹灰工作
float electrode 活动电极
floater 临时工;流动工;抹灰工具;镘工;漂流瓶;漂浮物;浮子;浮砖;浮体;浮水树种;浮水材;浮动钻井装置;浮式设备;浮杆;浮动利率票据;粉煤灰漂珠;债券发行人;抹平子
floater chamber 浮子井
floater crane 水上起重机;浮动起重机
floater door 挡砖进出口
floater extending deep down into the glass 浸没式挡砖
floater guide block 挡砖
floater hole 挡砖放入口
floater lug 挡砖凸口
floater net 浮筒救生网
floater notcher 挡砖凹口
floater roll arm 从动轧辊支架
float expansion valve 浮球调节阀
float factor 浮动因子
float feed 浮子调节进料;浮筒式进给
float feed apparatus 浮筒式给水器
float fill 水力冲填
float finish 镘具整修;浮镘饰面;用镘修整;压实赶光;木抹修整;抹光;镘修整;镘刀饰面
float finish floor 抹平地面
float gate 浮门
float ga(u)ge 油舱量尺;浮子液面尺;浮子水位(计);浮子式液位指示器;浮子式液面计;浮子水位计;浮子式水尺;浮子式浮尺;浮筒式液位计;浮式水标尺;浮标水尺
float ga(u)ging 浮子测流法;浮标测流
float gear 浮筒起落装置
float gimbal 浮动式万向支架
float gimbal structure 浮式常平架结构
float glass 浮法玻璃;浮法平板玻璃
float glass bath 浮法玻璃的浮抛窑
float glass process 浮法玻璃生产法
float glass processing 浮法玻璃深加工
float glass thickness control 浮法玻璃厚度控制
float governor 浮子调节器
float guide 浮筒导杆
float head 浮动头
float hinge pin 浮子绞链销
floating 抹平;漂游;图像抖动;图案走样(陶瓷表面

彩饰缺陷);提浆抹光;浮游的;浮动(的);浮充
floating absorbent 漂浮吸收剂;浮动吸收剂
floating action 无静差作用;无定向作用;浮动作用;不稳定作用
floating action type servomotor 浮动式伺服电动机
floating address 可变地址;浮动地址
floating address register 浮动地址寄存器
floating aerator 漂浮充气器;浮动式曝气器
floating aerodrome 航空母舰;浮机场
floating aid 浮工导航机具;浮动标志
floating algae 浮游藻类
floating amplifier 隔离放大器
floating anchor 流锚;海锚;浮锚
floating and self-erecting platform 海洋平台;浮动并自行装配的站台;浮动并自行装配的平台;海上钻探平台;海上钻井平台
floating and tense pulse 脉浮紧
floating angle method 浮角装板法
floating animal 漂浮动物
floating aquatic 漂浮水生植物;浮水生物
floating area 抹灰面积
floating arm drawoff 浮臂泄水
floating arm engraver 悬臂式刻图仪
floating assets 流动资产
floating attachment 浮动悬挂装置
floating auger 浮动螺旋
floating axle 浮轴;浮(式)(半)轴;浮动轴
floating balance 悬浮摆轮;浮动摆轮
floating balancing ship lift 浮筒平衡梁(式)升船机
floating ball 球形浮标;浮球
floating bamboo fender 竹捆浮碰垫
floating bank 浮煤式锅炉
floating barge 浮驳
floating barge elevator 钻井浮船升降机
floating barrier 浮栅;浮动挡板
floating base 悬空基极;弹簧模座;浮底座
floating battery 浮置蓄电池;浮置电池组;浮动蓄电池;浮充电池组
floating bay 一次抹灰面(两块整平板之间的墙面);灰饰面积;抹灰格间
floating beacon 测量浮立标;浮动信标;浮灯标;浮标
floating bearing 浮动轴承;浮动支座
floating bearing end 浮动轴承端
floating bed 浮床
floating-bed scrubber 浮动涤气器
floating bell pressure ga(u)ge 浮钟式压力计
floating belt 移动皮带
floating belt conveyer 浮式皮带输送机
floating blanket 悬浮层
floating block 可移动滑车;横架起重滑车;浮动滑车
floating boat dock 小浮码头
floating body 漂浮物体;浮体
floating bog 浮沉沼泽;浮草沼泽
floating bollard 浮式系船柱;浮式系船栓
floating boom 导污浮架;导冰浮架;浮栅;浮式碰垫;浮护木;浮杆起重架;浮动围埂;水拦油栅;浮拦木
floating booster station 接力泵船;水上接力泵站
floating bottom 浮底
floating box 浮箱
floating brake lever 浮装闸杆
floating brake shaft 浮动闸轴
floating breakwater 挡浪浮堤;浮式防波堤;防浪木排
floating brick 浮砖;轻质砖
float(ing) bridge 浮桥
floating bucket elevator 浮式链斗提升机
floating buffer 浮动缓冲区
floating building 浮飘建筑;浮动建筑物
floating bulkhead 浮闸木;浮箱式闸门
floating bulk transfer station 浮式过驳平台
floating buoy 浮筒;浮标灯;浮标
floating bush 浮动衬套
floating bush bearing 浮动衬套轴承
floating caisson 浮式沉箱;浮式沉井;浮运沉箱;浮箱;坞门;浮动沉箱
floating caisson guidewall 浮式沉箱导墙;浮式导航墙
floating-caliper disc brake 浮动卡钳式钢盘制动器;浮动夹钳盘式制动器
floating camel 浮柜式护木;浮箱;浮筒;浮柜
floating camel fender 浮箱防冲器;浮柜式护舷;浮(柜式)护木;浮柜式防冲器;浮碰垫
floating capital 游资;流动资金;流动资本
floating card 漂浮卡片

floating cargo 业经装船之货;路货;海运货物;漂散货物;未到货;途中货物
floating cargo landing stage 运货浮动平台
floating car method 行驶车辆观测法;浮动车观测法
floating carrier amplifier 浮动载波放大器
floating carrier modulation 浮动载波调制
floating carrier system 浮动载波制
floating carrier transmitter 浮动载波发送机
floating carrier wave 载波不定波
floating ceiling 悬浮式顶棚
floating center 中间游动盘
floating chain 浮体锚链
floating chamber 浮箱
floating channel 浮动通道
floating charges 浮动费用;流动费用;流动抵押品(专指农场贷款);流动质权;浮(动)充电
floating charge power supply 浮充供电
floating charger 浮充电机
floating chase 浮动版架
floating city 水上城市
floating clamp 浮动夹头
floating clause 浮泊条款
floating coat 墁涂;中涂层;底或中道灰泥;二度漆;二道油漆;二道抹层次;二道抹灰
floating column 浮柱
floating commutator 浮置整流器
floating compass 浮动罗盘
floating concrete dry dock 浮式混凝土干船坞
floating concrete factory 水上混凝土拌和厂
floating concrete-mixer 混凝土搅拌船
floating concrete mixing plant 混凝土搅拌船;水混凝土搅拌厂
floating concrete plant 浮式混凝土工厂
floating concrete screed 浮动水泥地面;不固定混凝土找平层
floating condition 浮运条件;浮动状态
floating conductor 漂浮导体
floating construction 浮动构造;浮隔构结构;浮式构造;浮动结构;浮动建筑物
floating construction method 浮运施工法
floating container 流动集装箱;漂浮货柜;浮运集装箱;浮式集装箱;浮动集装箱
floating container ship 浮装式集装箱船
floating control 浮点控制;无定位调节;无差调节
floating controller 无静差控制器;浮动控制器
floating conveyer 浮动式输送机
floating core rod 浮动芯棒
floating coupling 浮动式联轴节;浮动式接头
floating cover 浮тру摊铺式面层;浮式顶盖;浮起盖;浮动盖(板)
floating cover sludge digestion tank 浮盖式消化池
floating craft 小型船舶;船舶(尤指小型船舶);浮式施工机械;浮船
floating crane 起重(机)船;水上起重机;浮筒起重机;浮体起重机;浮式起重机;浮式起重船浮吊;浮动起重机;浮吊
floating crane for cargo handling 装卸用浮式起重机
floating crane for erection work 建筑安装用浮式起重机
floating crane for salvage 救捞起重机
floating crane for salvage work 救援用浮式起重机
floating crane ship 浮运式起重机
floating crucible 浮置坩埚
floating currency 流动货币;浮动通货
floating cutterbar 浮动式切割器
floating dam 浮闸;浮式挡水建筑物;浮基坝;浮坝;筏基坝
floating datum 浮动基准面
floating debenture 流动债券;浮动债券
floating debris 漂流物;漂浮物;漂浮碎屑
floating debris pass 漂浮物过道;排漂通道
floating debt 短期债务;流动债务;流动债券;浮动债务
floating decimal 浮动十进制小数
floating delivery pipe 浮排泥管
floating deposit 浮动性存款
floating derrick 起重船;水上吊机;动臂起重船;浮式(桅杆)起重机;浮式吊机;浮动起重机;浮吊(起重机);悬臂起重(机)船
floating device 浮式设备
floating device for measuring evapo(u)ration 漂浮式蒸发量测装置

floating dial compass 动圈罗盘仪
floating die 可动压模;弹簧压模;浮动模型
floating die assembly 弹簧模具;浮动模具
floating die press 弹簧模压机
floating digester cover 消化池浮盖
floating disc 浮动盘
floating disc compass 浮动罗盘
floating discharge line 浮式排水管道
floating discharge pipe 水上排泥管;浮排泥管
floating discharge pipeline 水上排泥管线;浮排泥管线
floating display 浮动显示
floating distillation tower 浮阀塔
floating diving bell 浮式潜水钟;浮式潜水罩
floating dock 浮坞;浮(式)码头;浮船坞
floating dock anchorage 浮船坞锚定
floating dock berth 浮式坞泊位;靠浮船坞的栈桥码头
floating dock construction 浮坞的建造
floating dock fixed by rigid boom 浮船坞刚杆固定
floating dock mooring 浮坞锚泊装置
floating dock side wall 浮船坞坞墙
floating dock siting 浮坞选址
floating dollar sign 浮动美元符号
floating-dot 浮点
floating downward 下浮
floating drawbar 浮动拉杆
floating dredge(r) 挖泥船;浮动式挖泥船
floating drift 漂浮物
floating drift tube 浮动漂移管
floating drift tube klystron 漂移线调管
floating drill barge 钻探船
floating drilling platform 浮式钻探平台;浮式钻井平台
floating drilling rig 浮泊钻探设备
floating drilling ship 浮式钻探船
floating drilling vessel 钻探船;钻井浮船
floating driver 水上打桩机
floating driving plant 浮式打桩装置;浮式打桩设备
floating drum 浮鼓
floating drum gate 浮动式鼓形闸门
floating dry-dock 干船坞;浮船坞;浮式干船坞
floating dust 飘尘;浮尘;飞尘
floating dust pollution 飘尘污染;漂尘污染
floating earth 流动(也称流砂);浮土;浮动(接地)
floating electrode 浮动电极
floating elevator 浮动式升降机
floating embankment 浮堤
floating end joint 浮置端缝
floating engine 浮动式发动机;浮动式电动机
floating equipment 浮式设备;水面设备;浮水设备;浮动性设备
floating erection 浮运架设(法)
floating evapo(u)ration pan 漂浮式蒸发皿;浮式蒸发仪
floating evapo(u)ration station 漂浮蒸发站
floating evapo(u)rimeter 漂浮式蒸发计;浮式蒸发仪
floating excavator 浮式挖掘机
floating exchange 浮动外汇
floating exchange rate 浮动汇率
floating exchange rate system 可变汇率制;浮动汇率制度
floating executive program(me) 浮动执行程序
floating fault block 浮动断块【地】
floating fender 浮沉(式)防冲材;浮式碰垫;浮式护舷(木);浮式防撞装置;浮式防冲器;浮护木;浮碰垫
floating fine particle 漂浮细粒;浮游微粒子;浮选微粒
floating fish factory 鱼获物加工船
floating fixture systems 浮动夹具系统
floating floor 橡胶夹层地板;隔声地板;架空地板;夹层地板;积木地板;浮筑(式)地板;浮式楼地面;浮式地板;浮动地板;浮床
floating flooring finish 夹层地板终饰;夹层地板修整
floating floor screed 浮动地板准条;浮动地板刮条
floating foam 浮动泡沫;浮动的泡沫
floating forest 漂浮森林
floating foundation 浮力基础;浮基(础);浮筏基础;筏片基础;筏基
floating foundation costae girder 有肋梁筏片基础
floating foundation dam 浮基坝
floating foundation non-costae girder 无肋梁筏片基础

floating fund 浮游资金
floating gang 杂工队;流动工队;飞班
floating gantry 龙门浮吊;浮式龙门吊
floating garden 漂浮庭园
floating gate 浮动闸门;浮门;浮坞门;浮式闸门;浮动栅
floating-gate amplifier 浮栅放大器
floating ga(u)ge 浮表
floating ga(u)ging 浮标测流(速)
floating globe valve 浮球阀
floating grab(bing) installation 移动抓取设备
floating grade 溜行坡度;滑行坡度
floating grain elevator 浮式卸粮机
floating grid 漂浮栅极;浮置栅极;浮动栅极
floating gudgeon pin 浮动活塞销
floating gyro 悬浮陀螺仪;浮动陀螺仪
floating habit 深水习性;浮水习性
floating harbo(u)r 漂浮式防波港;浮式港口;浮式防波堤港(湾);浮港(用浮筏围成的避浪港);浮动港;浮动避风港
floating head 浮头盖;浮动头
floating head backing device 浮头勾圈;浮动背衬环
floating head condenser 浮头式冷凝器
floating head flange 浮动头法兰;浮动头凸缘
floating head kettle type reboiler 釜式浮头再沸器
floating head magnetic drum 浮动头磁鼓
floating head type heater 浮式管板式加热器
floating-head type heat exchanger 浮头式换热器
floating holder 活动夹具;活动夹持器;浮动刀夹
floating hose 浮管;浮动软管
floating hospital 医疗船
floating hostel (水上的)浮动旅店
floating house 水上住宅
floating household 流动户口
floating hydrocarbon 浮烃
floating ice 大浮冰;浮冰
floating ice-boom 挡冰栅
floating ice group 浮冰群
floating ice jam 浮冰拥塞
floating ice sluice 排浮冰闸;排冰道
floating in 浮运进入;浮运到位
floating in casing 带回压阀的套管柱下到钻孔中
floating indebtedness 流动债务
floating input 浮置输入
floating insertion 浮动插入
floating insertion character 浮动插入字符
floating insertion editing 浮动插入编辑
floating inspector 流动检查员
floating installation 浮动装置
floating instrument platform 竖立船;浮动(水文)观测平台
floating intake 活动式进水口
floating interest rate 浮动利率
floating into position 浮送就位
floating island 浮岛(整块带浮丛林漂浮海面);水上浮动花园
floating item 浮动价格商品
floating jet stenter 气垫式喷嘴热风拉幅机
floating jetty 浮式栈桥;浮式突码头
floating jib crane 水上悬臂起重机
floating key 浮键
floating knife coater 活动刮涂机;浮刀刮涂器
floating knife roll coater 浮动刮刀辊涂机
floating labor 流动劳动力
floating lamp 浮标灯
floating landing pier 浮式栈桥
floating landing stage 登岸平台;简易浮码头;浮栈桥;浮码头;浮动平台
floating-leaf plant 浮叶植物
floating lending rate 浮动贷款利率
floating level meter 浮筒液位计
floating level sensor 浮球式液位传感器
floating level transducer 浮式液位传感器
floating lever 浮动杠杆;流动杠杆;浮动杆
floating lever rod 浮动杠杆
floating liablility 流动负债
floating lien 浮动留置权
floating light 救生圈火号;桅顶灯;灯船;浮灯标;浮标灯
floating line 水面管线
floating line pumping 泵到罐输送
floating loan 浮动贷款;短期贷款
floating lock 浮闸

floating log 漂浮圆木;漂浮木材;浮木
floating lysimeter 漂浮式土壤蒸发器
floating machine 抹平机;浮充机
floating machine shop 机器修理船;浮动修理所
floating marine navigation aid 浮游航标;海上助航浮标
floating mark 浮标;(测距仪或测高仪的)立标;水上标志;浮游测标;浮动测标;浮(动)标志
floating mark buoy 浮动标识浮标
floating marker 浮标
floating markup 浮动加码
floating mat 浮排
floating material 漂浮物料
floating matter 漂浮物
floating media filter 漂浮介质滤池
floating method 漂浮法;浮运法
floating mine 浮动水雷
floating mining plant 漂浮式采矿装置
floating mixer 搅拌船(混凝土);混凝土搅拌船
floating money 游资;浮动资金
floating monobuoy 单点系泊浮筒
floating moor 浮动沼泽;浮动泊位
floating mooring 浮式系船
floating mooring bitt 浮式系船柱
floating mooring equipment 浮式系船设备
floating mooring ring 浮式系船环
floating mortgage 总额抵押;浮动抵押
floating mud 浮泥
floating (net) cage 浮式网箱
floating neutral 浮置中线;浮接中心线;不接地中性线
floating number 浮点计位数;浮点数
floating oceanographic(al) and meteorologic(al) automatic station 漂浮式海洋气象自动观测站
floating ocean research and development station 大洋研究和开发浮动站
floating of concrete pavement 混凝土路面的镘平
floating oil 漂油;浮油;浮选油
floating oil barrier 漂油栅;浮油栅
floating oil filter 浮动滤油器
floating oil(pipe)line 水上输油管线
floating oil-production-equipment barge 浮式石油钻井平台
floating oil production platform 浮式采油平台
floating oil production system 浮式采油系统
floating opening bridge 开合浮桥
floating-operated valve 浮球控制阀
floating optimal sample size 浮动最优样本容量
floating overflow weir 浮动溢流堰
floating pan 浮式蒸发仪;浮式蒸发皿;浮(式)皿
floating panel 浮动导流板
floating parquet(ry) 浮动镶花地板
floating partition 活动隔墙;活动隔断
floating passenger landing 浮栈桥
floating passenger landing stage 人行浮式码头
floating peat 漂浮泥炭
floating pier 趸船码头;浮栈桥;浮式栈桥;浮桥码头;浮码头;浮动码头
floating pile 悬浮桩;摩擦桩;浮桩;浮式桩;浮承桩
floating pile driver 打桩船;浮式打桩机;水上打桩机
floating pile driving plant 浮动打桩装置;水上打桩设备
floating pile foundation 悬浮桩基;摩擦桩基(础);浮桩基础
floating pileline (装在驳船上的)浮动管线;(装在驳船上的)浮动管路
floating piling 水上沉桩
floating piling plant 水上打桩设备
floating pilot lamp 浮动信号灯
floating pipe 浮动管(道)
floating(pipe)line 水上排泥管线;海上输油管道;漂浮管道;排泥浮管;水上管道;水上管线;水面管线;浮式管线;浮式管路;浮式管线
floating pipeline system 浮式管线系统
floating piston 浮动活塞
floating piston pin 浮式活塞销;浮动(式)活塞销
floating plane 浮筒式水上飞机;浮面
floating plant 漂浮水植物;漂浮植物;浮式水上施工机械;水上机械设备;浮式植物;浮式机械施工机械
floating plant system 水上机械施工法
floating plate 浮板
floating platen 浮动压板
floating platform 浮式平台;浮动平台

floating plug 浮塞;浮动塞棒
floating plunger positioner 浮子定位器
floating pneumatic conveyer 悬浮气力输送机
floating pneumatic fender 浮式充气护舷
floating pneumatic wheel fender 充气气轮胎浮碰垫
floating-point 浮点小数点;浮点(法)
floating-point accumulator 浮点累积器
floating-point adder 浮点加法器
floating-point addition 浮点加
floating-point arithmetic 浮点运算(法);浮点算法
floating-point arithmetic library 浮点运算程序库
floating-point arithmetic package 浮点运算程序包
floating-point arithmetic routine 浮点运算程序
floating-point arithmetic sign 浮点算术符号
floating-point base 浮点基数
floating-point basis 浮点基
floating-point binary 浮点二进制
floating-point binary constant 浮点二进制常数
floating-point buffer 浮点操作数缓冲器
floating-point calculation 浮点计算
floating-point code 浮点编码
floating-point coding compaction 用浮点编码的数据精简法;浮点编码压缩;浮点编码精简法
floating-point coefficient 尾数;浮点系数;浮点尾数
floating-point compaction 浮点简缩
floating-point computation 浮点运算;浮点计算
floating-point computer 浮点计算机
floating-point constant 浮点常数
floating-point control 浮点控制
floating-point control field 浮点控制字段
floating-point coprocessor 浮点协处理器
floating-point data 浮点数据
floating-point data item 浮点数据项
floating-point decimal 浮点十进制小数
floating-point decimal arithmetic 浮点十进制算术运算
floating-point decimal constant 浮点十进制常数
floating-point decimal subroutine 浮点十进制运算子程序
floating-point decimal system 浮点十进制
floating-point divide by zero trap 浮点被零除自陷
floating-point divide exception 浮点除法异常;浮点除法例外
floating-point division 浮点除(法)
floating-point exponent 浮点阶
floating-point feature 浮点特性;浮点功能部件
floating-point format 浮点格式
floating-point gain amplification 浮点增益放大器
floating-point hardware 浮点硬件
floating-point input format 浮点输入格式
floating-point instruction group 浮点指令组
floating-point instruction set 浮点指令系统
floating-point integer 浮点整数
floating-point literal 浮点文字;浮点表示
floating-point logic machine 浮点逻辑机
floating-point mathematics 浮点数学
floating-point mo(u)ld 浮点模型
floating-point multiplication 浮点乘法
floating-point normalize control 浮点规格化控制
floating-point notation 浮点记数法;浮点记法
floating-point number 浮点数;浮点计数器
floating-point operation 浮点运算;浮点操作
floating-point operation arithmetic 浮点算术运算;浮点算术操作
floating-point overflow 浮点溢出
floating-point overflow bit 浮点溢出位;浮点上溢位
floating-point overflow trap 浮点溢出自陷
floating-point package 浮点程序包;浮点包
floating-point precision 浮点精度
floating-point processor 浮点处理机
floating-point radix 浮点数系基数;浮点基数
floating-point register 浮点寄存器
floating-point representation 浮动(小数点)表示法;浮点表示
floating-point representation of a number 浮点制表示法
floating-point routine 浮点(例行)程序;浮点程序包
floating-point shift 浮点移位
floating-point sign 浮点符号
floating-point status vector 浮点状态向量
floating-point subroutine 浮点子程序
floating-point symbolic address 浮点符号地址
floating-point system 浮点制;浮点系统

floating-point transformation 浮点转换
floating-point underflow 浮点下溢
floating-point underflow trap 浮点下溢自陷【计】
floating policy 流动保险单;长期有效保险单
floating pollutant 漂浮污物;浮动污染物
floating pontoon 趸船;浮坞门;浮船
floating pontoon for small craft 小船浮码头
floating pontoon pumping station 浮船式泵站
floating pontoon wharf 趸船码头;浮(船)码头
floating population 住所不定者;住所不定人口;流动人口;水上人口;常住无户籍人口;浮动人口
floating port 浮式港口;浮动避风港
floating position 漂浮位置;浮动位置
floating post 移动支柱;浮柱;浮式系船柱
floating post type 浮动柱型
floating potential 漂游电位;浮漂电势;漂移电位
floating potential instrumentation 浮动电位仪
floating pound 浮动英镑;浮动的英镑
floating power 悬浮力
floating power barge 水上发电站;水上动力站
floating power unit 浮动力装置
floating price 浮动价格
floating prime rate 浮动基本利率
floating prime rate of interest 浮动优惠利率
floating primrose willow 漂浮水龙
floating process agent 浮选药剂
floating production 浮动分配生产
floating production storage and off loading system 浮式原油生产储存装卸装置
floating production system 浮式生产系统
floating pulse 脉浮
floating pump 泥浆泵;吹洗泵;浮(动)泵
floating pump assembling unit 浮动式水泵机组
floating pump assembly unit 浮船式水泵机组
floating pumping station 泵船;浮动式泵站
floating pump type aerator 浮泵式曝气池
floating purchasing power 浮动购买力
floating pyrometer 浮式高温计
floating raft construction 浮筏构造
floating raft foundation 浮筏基础
floating rail 活动护栏
floating raise of wages 浮动升级
floating ramp 浮动坡道
floating rate 蓄电池浮充率;可变利率;可变汇率;(蓄电池的)浮充率;浮动利率;浮动汇率
floating rate bond 浮动利率债券
floating rate certificate of deposit 浮动利率存单
floating rate drop lock bonds 闭锁最低利率浮动债券
floating rate market 浮动利率市场
floating rate note 浮动利率债券;浮动利率本票;浮动利率票据
floating rate system 浮动率制度
floating reamer 浮动铰刀
floating reamer holder 浮动铰刀刀夹
floating refinery 浮动炼油厂
floating region 浮区
floating regulator 可调电压稳压器
floating rental rate 浮动租金率
floating reservoir 浮式油罐
floating response 漂浮响应;无静差作用
floating reticle 浮动标线
floating rib 浮肋;背浮肋
floating rig 钻探船;水上机具;浮式机具;钻探机船
floating ring 漂浮环;浮环;浮动环
floating-ring clutch 浮环离合器
floating-ring consolidation instrument 浮环式固结仪
floating-ring seal 浮环密封
floating-ring shaft 浮环套管
floating-ring test instrument 浮环式试验仪
floating rock crusher 凿岩船
floating rod 锚杆;刮尺
floating roll 浮动滚筒;浮动辊
floating-rolled steel 辊轧扁钢;滚轧扁钢
floating roller 活动辊
floating roof 浮置上盖
floating roof(ing) screed 浮顶找平
floating roof reservoir 浮顶储[贮]罐
floating roof tank 浮顶式罐;浮顶水池;浮顶(油)罐;浮盖池
floating root 浮根
floating rope 动轮钢丝绳
floating rule 抹灰面刮尺;米厘条;镘板;抹灰面用

嵌线条;刮杆;刮尺
floating sand 流沙(也称流砂)
floating sand plant 水上制砂工厂
floating scale lever 浮秤杠杆
floating scraper 浮式刮刀
floating screed 冲筋(标志抹灰厚度的窄条);浮式找平层;抹灰准条
floating scum 浮渣
floating scum-board 托灰板;撇木板(用于污水池中撇开泡沫悬渣,能升降的隔板)
floating seal 浮动油封
floating seal arrangement 浮动密封装置
floating sealed piston 滑动密封活塞;浮动密封活塞
floating sea-mark 浮动航标
floating seaweed 漂流海藻
floating section 浮选在水面的管段
floating seedling 浮苗
floating set 浮充机组
floating shaft 浮动轴
floating shear legs 人字吊杆起重船
floating shears 人字吊杆起重船;浮吊
floating sheave 浮滑轮
floating shed 水上仓库
floating ship lift 浮式升船机;浮升式升船机
floating shoe 沉箱刃脚
floating side guards 活动侧导板
floating side up 侧浮
floating sign 浮动符号
floating signal 浮动信号
floating skimmer 浮式撇油器;浮式撇取器
floating slab 悬空板;浮置板;浮板
floating slag 漂浮矿渣;浮渣
floating-sleeve bearing 浮动衬套轴承
floating sludge 漂浮污泥;浮(污)泥;悬浮污泥
floating soap 浮水皂
floating solid 飘浮固体;漂浮固体;浮选固体
floating space simulation 漂浮空间模拟
floating speed 上浮速度
floating sphere 浮球
floating-spot 浮点
floating spray column 浮动喷射塔
floating spread 浮动加码
floating spring 浮动弹簧
floating stability 漂浮稳定性
floating stage 舷侧工作木排;漂游植物阶段
floating steam trap 浮式隔汽具
floating steel crane 水上起重机;浮式钢吊车
floating stock 发行的股票;流通股;流动股票;浮动股票
floating strainer (水泵吸水的)浮式滤头;浮式滤网
floating structure 浮式建筑物;浮飘建筑
floating subroutine 浮点子程序
floating suction 浮式吸水管;浮式虹吸
floating suction pipe 浮动吸入管
floating supply 流动库存;浮动供给
floating switch 浮球开关
floating symbolic address 浮动符号地址
floating system 浮动系统
floating systematic error 浮动系统误差
floating tail 无升力尾翼
floating tap chuck holder 浮动丝锥卡盘架
floating temperature control 浮动温控
floating test 膨胀试验;浮充试验;浮杯试验
floating-test-car 浮动试验车
floating the casing 浮力下(套)管
floating tide ga(u)ge 浮标式自记验潮仪
floating timber fender 浮式护岸木
floating tissue 漂浮组织
floating tool holder 浮动刀架
floating tooling 浮动刀具
floating town 海上城市
floating trade 海上贸易
floating trash 漂浮废物;浮游渣滓;浮游垃圾
floating triangulation 浮标法三角测量
floating tube 浮动管(道);浮式管道
floating-tube core barrel 单动双层岩芯管(内管不动);单动双层取芯器(内管不动)
floating tubesheet 浮头管板
floating tunnel 悬浮隧道
floating turntable 浮动回转台
floating type bearing 浮动式轴承
floating type controlled system 浮动控制系统
floating type gyroscope 悬浮陀螺仪
floating type thermometer 浮式温度计

floating type wharf 浮式码头
floating unit price semi-piecework wage 浮动单价半计件工资
floating-up 起浮
floating value 浮动价值
floating valve 浮阀;浮动阀;浮球阀
floating valve gear 浮阀装置
floating vegetation 漂浮植物;浮水植物
floating velocity 漂流速度;漂浮速度
floating vessel 浮船
floating voltage 空载电压
floating vote 浮动投票
floating wages 浮动工资
floating wages linked with performance 效益浮动工资
floating warehouse 浮动货栈;浮动仓库
floating washer 漂浮式洗矿机
floating weight 移动式重锤
floating wharf 浮码头
floating wheel 浮轮;摆轮
floating wire 引张线
floating wood floor 夹层木地板;浮式木地板
floating workshop 修理船;水上修理车间
floating wreckage 失事船的漂浮物
floating zenith telescope 浮动天顶仪
floating zero 浮动原点;浮动零点
floating zone 浮区
floating zone melting 浮区熔融;浮区熔化
floating zone melting growth 悬浮区熔生长法
floating zone melting method 浮动区域熔炼法
floating zone process 浮动区域精炼法
floating zone refining 浮区提纯
floating zone refining method 区熔提纯法
floating zone technique 浮区法
float into position 浮移到位
float jet tray 浮动喷射塔盘
float landing gear 浮式起落架
floatless level controller 无浮子液面控制器
floatless liquid-level controller 无浮子液面控制器
float level 浮子水准;浮筒水准线
float level controller 浮子液面计;浮面控制仪;浮动式液面调节器
float level ga(u)ge 浮子液位测量仪
float level indicator 浮子水位指示器
float level meter 浮子液位计
float lever 浮筒杆
float line roller 浮子绳滚筒
float magnetism type liquid level ga(u)ge 浮筒磁力式液位计
floatman 火车渡船管理员
float manometer-type flow indicator 浮子压力计式流量指示表
float mareograph 浮式自记潮汐仪
float marker device 浮标装置
float material (浮选中的)轻料;浮料
float matter 浮物
float matter curve 浮物曲线
float measurement 浮子测流法;浮标测流(速);浮标测定
float measuring cross-section 浮标测流断面
float mechanism 浮阀
float meter 浮子(式)流量计;浮式液面计;浮标测量计
float method 浮子测流法
float method of measuring discharge 浮标测流量法
float mineral 浮散矿物
float needle 浮(子)阀针;浮针
float needle valve 浮子针阀
float oil washing sample 浮油选样
float on/float off 滑上滑下;浮上浮下;浮进浮出
float on/float off ship 浮式式集装箱船
float on/float off system 载驳运输方式;浮进浮出方式
float-operated 浮子控制的;浮子操纵的
float-operated explosion-proof mercury switch 浮子操动防爆水银开关
float-operated ga(u)ge 浮子操纵水表;浮子操纵标;浮标水位计;浮标水尺
float-operated regulating valve 浮标操纵调节阀
float-operated switch 浮子控制开关
float-operated trap 浮筒隔汽具
float-operated valve 浮球阀;浮球控制阀
float ore 漂移矿石;漂流矿石

float-out 出坞
float out to location 浮移到位
float out to site 浮移到现场
float pan 准条;刮板;浮式蒸发皿;冲筋
float period 浮动期
float plane 浮筒式飞机
float plate 砑光机;浮动压板
float plate assembly 浮动板部件
float position 上举位置(挖土机前端铲斗);上浮位置(挖土机前端铲斗)
float-process 浮法生产
float product 轻产品;浮选产物
float rain-ga(u)ge 浮子式雨量计
float regulator valve 浮球调节阀
float relay for transformer 变压器浮子继电器
float resistance 浮筒阻力
float response 漂浮响应
float ring consolidometer 浮环式固结仪
float road 浮运路
float rod 测速杆;浮(标)杆;标式浮子
float rod correction 浮杆修正
float-rolled steel 辊轧扁钢
float rudder 水中舵
float run 浮标行距;浮标行程(指上下测流断面的间距);浮标测距
floats 脚光
float sample 浮煤样
float scaffold 船式脚手架;吊板式脚手架
float seaplane 浮筒式水上飞机
float selecting engine 浮选机
float shaft 浮动轴
float shoe 具有止回阀的(井)管脚;浮动闸瓦承座;浮鞋
float-sink analysis 浮沉分析
float skimming device 刮泡装置
float sludge 悬浮污泥
floatsman 磨石工
float spindle 浮针
float spindle guide 浮筒针导承
float steam trap 浮式隔汽阀
float stone 磨石铁(用于磨砖上粗面痕迹);镪光石;悬粒灰岩;磨(砖)石;轻石;漂石;铁磨砖石;多孔蛋白石;浮(移)石;多孔石英;多孔岩石;漂浮砾石灰岩
float stone grinding 用浮石粉抛光
float stop(valve) 浮球阀;浮止阀
float strut 浮筒支柱
float study 浮标研究
float switch 浮子开关;浮筒开关;浮控开关;浮动开关
float tank 浮子水箱;浮选箱;浮式箱;浮抛窖
float tape 浮尺(自记水位计)
float technic 浮选技术
float technique 浮标法
float test 流质试验;流动性试验;漂浮试验;浮选试验;浮标试验
float thrower 浮标投放器
float timber 浮运木材;浮木
float tracking 浮标跟踪
float trap 浮子凝汽阀;浮式凝汽器;浮动阀
float tube 浮运进水管;浮筒管;浮标管
float-type carburettor 浮动式汽化器
float-type compass 浮子式罗盘
float-type controller (检查液面的)浮式仪表;浮标式仪表
float-type flow meter[metre] 浮筒式流量计;浮子(式)流量计;浮标式流速计
float-type ga(u)ge 浮子式流量计;浮标式测流量仪
float-type gyroscope 液浮陀螺仪
float-type interceptor 浮式截水器;浮式截流器;浮子截流器
float-type level controller 浮标控制仪;浮子式液面控制器
float-type level ga(u)ge 浮子式液位计;浮标液面计
float-type level regulator 浮标液面控制器
float-type limnimeter 浮式自动湖泊水位计
float-type liquidometer 浮式式液位计
float-type manometer 浮子式压力计
float-type meter 浮子式水表
float-type pneumatic measuring instrument 浮子式气动测量仪
float-type pneumatic water feeder 浮子式气压加水器
float-type pressure ga(u)ge 浮子式压力计

float-type rain ga(u)ge 浮子式雨量计;浮筒式雨量计
float-type specific gravity meter 浮子式比重计
float-type steam trap 浮球疏水器
float-type tank ga(u)ge 浮子式油罐液面计;浮子式验潮仪
float-type transmitter 浮子式传感器
float-type visco(si)meter 浮标式黏度计
float-type water-level recorder 浮式自记水位仪
float-type water stage 浮子式水位
float-type water stage recorder 浮筒式自记水位计;浮子自记水位仪;浮标式自记水位仪
float under carriage 浮筒式起落架
float upward 上浮
float valve 浮子水位控制阀;浮子阀;浮球阀
float valve needle 浮动阀的针
float valve tray 浮阀塔板
float velocity 浮标流速
float vibrator 振捣板;平板振动器;平板振动机;浮式振动器;表面振动器
float visco(si)meter 浮子黏度计;浮标黏度计
float viscosity 在浮标黏度计内测定的黏度;浮测黏度
float water level indicator 浮子水位指示器
float well 静水井;浮尺井
float wheel 浮轮
float with cement 用水泥抹平;用水泥抹光;水泥抹光
float wood 浮云木材
float work 抹面;抹灰工作
float-zone crystal 浮区晶体
float-zone method 浮区法
floc 絮体;絮状物;絮片;絮凝物;矾花
floc barrier 絮体拦栅
floc basin 絮凝池
flocbed 絮凝沉降层;絮凝层
floc break-up 絮体破碎
floc carryover 絮体带出
floc characteristic 絮凝物特性
floccing 絮凝物;絮凝处理
floccose 絮状的
flocculability 絮凝性
flocculant 絮凝剂;絮聚剂
flocculant aid 助絮凝剂
flocculant dosage 絮凝剂投加量
flocculant of clay 黏土絮凝剂
flocculant-producing microbe 絮凝剂产生菌
flocculate 结成小团块;絮凝作用;絮凝物
flocculated activated biosolid 絮凝活性生物固体
flocculated clay buttress bond 絮状支托粘结
flocculated colloid 絮凝胶体;凝絮胶体
flocculated flo(a)tation 絮凝浮选法
flocculated latex crumb 絮凝胶乳浆
flocculated mixing 絮凝搅拌
flocculated sediment 絮凝状泥沙;絮凝状沉积物;絮凝沉积物
flocculated sludge 絮状污泥;絮凝渣;絮凝(性)污泥
flocculated structure 絮凝结构
flocculater clarifier 絮凝澄清器
flocculating 絮凝化
flocculating activity 絮凝活性
flocculating admixture 絮凝添加剂
flocculating agent 絮凝(助)剂;絮化剂;凝聚剂
flocculating chamber 絮凝室;絮凝区;絮凝池
flocculating colloid 絮凝胶体
flocculating condition 絮凝状况
flocculating constituent 絮凝组分;絮凝体;絮凝成分
flocculating-decolo(u)rizing agent 絮凝脱色剂
flocculating effect 絮凝效应
flocculating efficiency 絮凝效率
flocculating mechanism 絮凝机理
flocculating mixing test 混凝搅拌试验
flocculating phenomenon 絮状集现象
flocculating property 絮凝性能
flocculating sedimentation 絮凝沉淀
flocculating settling 絮凝沉降;絮凝沉淀
flocculating tank 絮凝池
flocculating unit 絮状沉淀单位;絮凝装置;絮凝沉降单位
flocculation 絮状作用;絮凝(作用);絮化作用;结絮(作用);胶凝作用;凝聚作用
flocculation accelerator 絮凝促进剂;絮凝加速剂;絮凝促凝剂

flocculation agent 絮凝剂
flocculation and turbidity reaction 絮浊反应
flocculation basin 絮凝池;絮凝澄清池
flocculation-biological contact oxidation process 絮凝生物接触氧化法
flocculation body 絮凝体
flocculation deposition 絮凝沉淀
flocculation dewatering technology 絮凝脱水技术
flocculation efficiency 絮凝效率
flocculation factor 絮凝因素;絮凝系数
flocculation-flo(a)tation 絮凝浮选法
flocculation-flo(a)tation adsorption 絮凝—气浮吸附
flocculation-flo(a)tation method 絮凝—气浮法
flocculation inhibition test 絮状抑制试验
flocculation kinetics 絮凝动力学
flocculation limit 絮凝限度;絮凝(极)限;上液限
flocculation mechanism 絮凝机理
flocculation morphologic property 絮凝形态特征
flocculation of the vehicle 油墨的肝化
flocculation oil removing method 絮凝脱油法;石油絮凝脱水法
flocculation plant 絮凝装置
flocculation point 絮凝点
flocculation precipitation 絮状沉淀反应
flocculation process 絮凝法
flocculation property 絮凝性能
flocculation ratio 絮凝率;絮凝比;絮凝系数
flocculation reaction 絮凝反应
flocculation reaction tank 絮凝反应池
flocculation sedimentation 絮凝沉淀
flocculation-sedimentation aids 絮凝助沉
flocculation-sequencing batch reactor activated sludge process 絮凝序批间歇式反应器活性污泥法
flocculation settler 絮凝沉降池
flocculation settling 絮凝沉淀
flocculation tank 絮凝池
flocculation test 絮状试验;絮状沉淀试验;絮凝试验
flocculation time 絮凝时间
flocculation-turbidity test 絮浊试验
flocculation unit 絮凝装置;絮凝单位
flocculation value 絮凝值
flocculation zone 絮凝区
flocculator 絮凝器;絮凝搅拌器;絮凝搅拌机
flocculator settler 絮凝澄清器
floccule 絮状物;絮凝粒
floccule body 絮凝体
flocculence 絮状沉淀法,棉絮状
flocculent 絮状的;絮状沉淀;絮凝的
flocculent deposit 絮凝状沉积物;絮凝状沉淀
flocculent gel 絮凝胶
flocculent gypsum 絮凝石膏;毛絮石膏
flocculent layer 絮凝沉淀层
flocculent precipitate 絮状沉淀物;絮状沉淀
flocculent settling 絮状沉降;絮凝沉降;絮凝沉淀
flocculent sludge 絮状污泥;絮凝性污泥
flocculent soil 絮凝土
flocculent structure 絮状结构;絮凝结构;毛絮构造;海绵状构造
flocculent texture 海绵状结构;团聚结构
flocculent turbidity 絮状混浊
floccules 絮凝物
flocculoreaction 絮状反应;絮凝反应
flocculus 絮状体;絮片
floccus 絮状云;毛丛
flocflo(a)tation 絮凝浮选
floc formation 絮体生成;絮凝物形成;絮凝体形成
floc-forming bacteria 絮凝物生成菌;成絮细菌
floc-forming chemical agent 化学成絮剂
floc-forming(chemical)reagent 绒聚剂;絮凝剂
flock 絮体;絮状沉淀;絮团;絮凝体;植绒墙纸;毛絮;绒屑
flock bed 毛屑垫床
flock binder 植绒黏合剂
flock bond strength 植绒黏合强力
flock coating 植绒涂料;植绒涂层
flock dot 植绒点子
flocked carpet 植绒地毯
flocked fabric 植绒织物
flocked goods 植绒织物
flocked yarn 植绒纱
flock finishing 植绒整理;植绒涂装
flock finish method 植绒涂装法

flock gun 植绒喷枪
flocking 植绒(涂装)法;植绒
flocking machine 植绒机
flocking process 植绒法
flock mattress 毛屑床垫
flockmeter 纤维流量计
flock of sheep 白浪(花)
flock paper 植绒纸;植绒壁纸;糊墙花纸
flock point 絮化点
flock printed sheer 植绒薄绸
flock printing 植绒印花
flocks 絮化
flocks and hards 纤维屑(塞缝隙用)
flock settler 絮凝沉降池
flock spraying 喷絮植绒;喷绒
flocky 絮状的;絮凝的
flocky precipitate 絮状沉淀物
floc nucleus 絮体核心
flo-con carrier system 浮箱式载驳方式
floc point 絮凝点
flocrete 流态混凝土
flocs 棉絮
floc strength 絮体强度
floc structure 絮凝物结构
floc test 絮凝试验;凝絮试验
floc water 絮体水
floc weighing agent 絮体加重剂
Flodin process 弗洛丁电炉炼钢法
floe 大片浮冰;大浮冰;浮冰块
floe belt 浮冰带
floeberg 浮冰堆;小冰山;堆积浮冰;浮冰山;冰丘群
floe ice 大浮冰;浮冰
floe ice concentration 浮冰密集度
floe-sheet of metallurgy 冶金流程
floe till 浮冰冰碛
flogging 手工磨平地板;清理地面;磨光木地板
flogging chisel 粗工重型冷錾;大凿
flokite 发光沸石;发沸石
flomerule 团伞方序
flong 作纸型用的纸;纸型用纸
flong paper 字板纸
floocan 脉节理黏土
flood 满溢;水灾;大(洪)水
flood abatement 灭洪
floodability 可浸性;浸没性
floodable length 可浸长度
flood absorption capacity 蓄洪能力;洪水收容能力
flood alleviation 缓洪
flood amplitude 洪水幅度;洪水波幅
flood and ebb 涨潮与落潮;潮汐
flood arch 防洪拱洞
flood area 洪水掩盖面积;洪泛区;淹水区;淹没面积;洪水面积;泛区;淹没区
flood axis 涨潮流轴线;洪水轴线;洪水流向
flood back 溢出回流
flood bank 洪水河岸;防洪堤
flood barrier 拦坝;拦河坝;防洪闸(门);防洪设施;防洪堤
flood basalt 高原玄武岩
flood basin 蓄洪水库;蓄洪区;蓄洪库;洪水盆地;洪水面积;洪水河滩;洪泛区;泛滥盆地;泛滥盆地;泛洪区
flood bed 洪水河槽;洪水河床;洪水泛滥地;河滩地;河漫滩
flood benefit 防洪受益
floodboard 泄水闸板
flood borne sediment 洪水携带的沉积物
flood breadth of river 洪水河面宽度
flood bridge 洪水桥
flood by-pass 分洪渠(道);分洪道
flood calamity 洪涡;洪水灾害;洪灾
flood canal 溢洪河道;洪水渠道
flood carrying capacity 泄洪能力;排洪能力
flood casting 放浇注;放浆浇注
flood catastrophe 大水灾
flood channel 涨潮水道;行洪河槽;涨潮航道;涨潮海峡;洪水河槽;洪泛河槽
flood characteristic 洪水特征值;洪水特性
flood characteristic curve 洪水特性曲线
flood coat 沥青漫涂层;漫水层;流水层;沥青防水涂层
flood coat for chip(ping)s 碎石填料;碎石填料
flood cock 船底塞;溢流开关
flood coefficient 洪水系数

flood control 洪水控制;防汛;防洪
flood-control basin 调洪洼地
flood-control benefit 防洪效益
flood-control capacity 防洪库容
flood-control channel 蓄洪河槽;防洪渠;调洪河槽
flood-control dam 蓄水闸;蓄洪坝;拦洪坝;防洪坝
flood-control engineering 防洪工程(学)
flood-control facility 防洪设施
flood-control headquarters 防汛指挥部
flood-control level 防洪水位
flood-control limiting level 防洪限制水位
flood-control map 防洪地图
flood-control operation 防洪运用;防洪调度
flood-control planning 防洪规划
flood-control project 防洪工程
flood-control regulation 防洪法规
flood-control reservoir 蓄洪水库;调节水库;调洪水库;防洪水库
flood-control reservoir capacity 防洪库容
flood-control storage 防洪调节;防洪库容;防洪蓄水(量)
flood-control system 洪水控制系统;防洪系统
flood-control systems for major cities 重要城市防洪体系
flood-control works 防洪工程
flood coverage 水驱面积
flood crest 峰顶水位;洪水峰顶;洪峰
flood crest attenuation 洪峰降低
flood crest forecast 洪峰预报
flood crest profile 洪峰纵剖面
flood crest reduction 洪峰削减(量)
flood crest stage 洪峰水位
flood crest stage ga(u)ge 洪峰水位计
flood crest travel 洪峰行径;洪峰行进;洪峰前进;峰顶行进
flood current 涨潮流;洪流
flood curve 洪水曲线
flood cycle 洪水周期
flood dam 蓄水坝;壅水坝;防洪堤坝;防洪坝
flood damage 淹没损失;洪灾损失;洪灾;洪水造成的损失;洪水灾害;洪水损失;水害;水毁
flood damage rate 洪灾损失率
flood danger 洪水险情
flood debris 洪水漂浮物
flood decline 洪水消落;洪水退落;退洪
flood defense(works) 防洪设施;防洪工作;防洪工程;防洪堤
flood deposit 洪积物;洪水堆积物
flood deposition and low water erosion 洪淤枯冲
flood-depth of river 河流洪水深度;河流的洪水深度
flood-depth of stream 河流洪水深度
flood detention 滞洪
flood detention area 滞洪区
flood detention basin 滞洪区
flood detention dam 滞洪坝;拦洪坝
flood detention district 滞洪区
flood detention project 滞洪工程
flood detention reservoir 蓄洪水库;滞洪水库
flood detention storage 拦洪库容
flood detention work 滞洪工程
flood dike 防洪堤
flood disaster 洪灾;洪水灾害
flood disaster investigation 洪灾调查
flood discharge 泄洪;洪水流量;洪水量
flood discharge forecast 洪水流量预报
flood discharge of river 河川洪流量
flood discharge outlet 溢洪道
flood discharge structure 泄洪建筑物
flood discharge tunnel 泄洪隧洞;泄洪隧道
flood discharging channel 排洪渠(道)
flood distribution 洪水分布
flood distribution gate 分洪闸
flood diversion 分洪
flood diversion and storage project 分蓄洪工程
flood diversion area 分洪区
flood diversion area management 分洪区管理
flood diversion area operation 分洪区运用
flood diversion channel 分洪道
flood diversion construction 分洪建筑物
flood diversion facility 分洪设施
flood diversion gate 分洪闸
flood diversion operation 分洪运用
flood diversion project 分洪工程
flood diversion sluice 泄洪闸;分洪闸

flood diversion works 分洪工程
flood drainage way 排洪道
flood duration 洪水历时
flooded 洪水淹没的
flooded area 淹没区；淹没面积；洪水淹没区；洪（水）泛（滥）区；洪泛面积；受淹面积；泛滥区（域）
flooded bed scrubber 溢流床洗涤器
flooded carburet(t)or 溢流汽化器
flooded chiller 泛溢式冷凝器
flooded cooler 满溢式冷却器
flooded dissolver 溢流式溶解器
flooded evapo(u)rator 满液式蒸发器；全浸式蒸发器；溢流式蒸发器
flooded fixture 溢流器具
flooded floating dry dock 灌水下沉的浮船坞
flooded geyser 漫流式间歇泉
flooded ice 淹没冰；流冰；覆水冰
flooded land 受淹地区
flooded lubrication 溢流润滑
flooded periphery 淹没区；淹没面积
flooded roof 漫水屋顶
flooded scrubber 淹没式洗涤器
flooded soil 渍水土(壤)；淹水土壤
flooded submersible 注水式潜水器；可充水式潜水器
flooded system 满溢式系统；充溢系统
flooded tank 注水舱
flooded waterplane 浸水状态水线面
flood embankment 防洪堤
flood erosion 洪水侵蚀；洪水冲刷
flood estimation 洪水(量)估算；洪水预估；洪水预测；洪水预报
flood evaporator 淹没式蒸发皿
flood event 洪水险情；洪水事件；洪水事故
flood fall 洪水消落；洪水退落；退洪
flood fencing 挡洪栅
flood fever 洪水热
flood fighting 抗洪抢险；防汛抢险
flood flanking 堤防
flood flow 洪水流量；洪水径流；洪流
flood flow formula 洪流计算公式
flood flukes 肝蛭
flood forecasting 洪水预估；洪水预测；洪水预报
flood forecasting scheme 洪水预报方案
flood forecast system 洪水预报系统
flood formula 洪水推算公式；洪水计算公式
flood-free bank 不淹河岸；不受淹河岸
flood frequency 洪水频率
flood frequency computation 洪水频率计算
flood frequency curve 洪水频率曲线
flood frequency relationship 洪水频率关系
flood gate 泄洪闸门；泄洪道；挡潮闸(门)；防洪闸(门)；防洪门；防潮门；溢洪道；船坞浮闸门；防潮水闸
flood ga(u)ge 洪水计
flood gun 浸没电子枪
flood havoc 洪水灾害；水灾
flood hazard 洪水险情；洪患；水灾风险
flood height 洪水位
flood hydrograph 洪水过程(曲)线
floodibility test 不沉性试验
flood ice flood 凌汛
flood icing 冰泉
flood information 洪水资料；洪水信息；洪水数据；洪水情报；洪水报导
flooding 溢流；水灌；注水；浸水；漫游；满屏幕；洪水泛滥；水驱；泛墨；发洪水
flooding agent 溢流剂
flooding angle 进水角
flooding area 淹没区；淹没面积；洪水淹没面积；泛滥区(域)；洪水淹没区
flooding a stranded vessel to prevent pounding or shifting 灌水固位【救】
flooding by system of squares 方畦淹灌
flooding calculation 可浸长度计算
flooding cock 溢流旋塞
flooding curve 可浸长度曲线
flooding dam 壅水坝
flooding damage 水涝害
flooding detector 漏洞探测器
flooding dock 满溢船坞；灌水船坞
flooding duct 充水导管
flooding factor 淹没系数

flooding finder 漏洞探测器
flooding frequency 洪水泛滥频率
flooding from contour ditch 等高沟淹灌
flooding ice 冰泉
flooding injury 水涝害
flooding irrigation 溢流灌溉
flooding irrigation method 淹灌法；漫灌法
flooding method 浮动方法
flooding nozzle 冲洗喷嘴
flooding of pumphouse 泵房进水
flooding period 汛期时间
flooding phenomenon 浮色现象
flooding pipe 溢流管
flooding plain 洪水泛滥平原
flooding plant 灌水设备
flooding plug 溢流塞
flooding point 溢流点
flooding rate 溢流速度
flooding river 泛滥河流
flooding river channel 泛滥河道
flooding routing 扩散式路径选择(网)
flooding sluice 防洪闸门
flooding system lubrication 溢流润滑；浸油润滑
flooding test 水密门试验；充水试验
flooding type infiltrometer 漫灌式下渗仪；漫灌式入渗仪
flooding valve 溢流阀
flooding velocity 溢流速度
flooding wave 洪水波
flooding well 注水井
flooding zone 洪水淹没区；洪水影响范围；泛滥区(域)
flood insurance 洪水保险；水灾保险
flood intensity 洪水强度；洪峰流量
flood interval 最大涨潮流间隙；涨潮流月潮间隙；涨潮间隔
flood inundation period 洪水淹没期
flood investigation 洪水调查
flood irrigation 淹灌；漫灌
flood irrigation of sewage 污水漫淹灌溉
flood irrigation system 漫灌系统
flood lamp 泛光灯
flood land 漫滩(地)；漫滩池；洪灾区；洪水滩；洪泛区；洪滩；河漫滩(地)；水淹地(区)；潮漫滩
flood level 溢流水位；洪水位
flood level duration curve 洪水位过程线
flood level mark(ing) 满潮标记；洪水痕迹；洪水标记
flood-level rim 溢水道；溢流边缘
flood light 投光灯；聚光灯；探照灯；散光；泛光灯；汛光灯；强光灯
floodlighting 投光照明；泛光照明；汛光照明；强力照明；散光照明
floodlighting system 聚光照明装置；泛光照明系统
floodlight projector 泛光灯；强力探照灯
floodlight scanning 空间扫掠
floodlight stand 泛光灯架
flood line 高潮线；高水位边线；洪水线
floodlit court 灯光球场
floodlit field 灯光球场
flood losses 洪水淹没损失；洪水损失；洪泛损失
flood loss management 洪水损害管理
flood loss rate 洪水损失率
flood-lubricated bearing 液体滑润轴承；浸油润滑轴承
flood lubrication 淹没润滑；浸油润滑；浸入润滑
flood mark 高潮线；涨潮标志；高水位标志；洪水线；洪水标记；泛滥标
flood measuring post 测洪柱
flood meter 涨潮水位计
flood mitigation 减洪
flood of the suction of the pump 灌满泵的吸水部分(开泵前)
floodometer (涨潮时的)水量记录计；潮洪水位测量仪；洪水(水位)计
flood opening 泄洪孔
flood-out pattern 清扫图形
flood pattern 注水系统；注水井网；洪水模式
flood peak 洪峰
flood peak discharge 洪峰流量
flood peak flow modulus 洪峰流量模数
flood peak flow velocity 洪峰流速
flood peak forecast 洪峰滞时

flood peak lag 洪峰滞后
flood peak level 洪峰水位
flood peak rate 洪峰流量
flood peak-rate method 洪峰法
flood peak-reduction factor 洪峰折减因数
flood peak runoff 洪峰水量
flood peak stage 洪峰水位
flood peak time 洪峰时段
flood performance 注水性能
flood period 汛期；洪水期
flood periphery 洪水边缘
flood-plagued area 洪灾区
flood plain 漫滩(池)；洪水漫滩河槽；洪积平原；洪泛区；洪泛平原；淹泛地；潮漫滩；泛滥平原
flood plain accretion 河漫滩淤积
flood plain area 泛滥平原地区
flood plain bench 漫滩阶地；洪泛平原阶地
flood plain clay 洪泛平原淤积土
flood plain deposit 漫滩堆积(物)；漫滩沉积(物)；洪泛区沉积土；洪泛平原淤积土；泛滥平原沉积
flood plain discharge 洪泛平原流量
flood plain facies 河漫滩相
flood plain insurance 洪汛；洪泛区保险；防洪保险
flood plain lake 洪泛平原湖；河漫滩湖；河漫湖泊
flood plain lake deposit 洪泛湖泊沉积
flood plain lake facies 河漫湖泊相
flood plain landform 河漫滩地形
flood plain lobe 洪泛蛇曲带；河曲带；舌瓣状河漫滩
flood plain management 洪泛区管理；泛滥平原管理
flood plain marsh 洪泛平原沼地；泛滥平原沼泽；河漫沼泽
flood plain marsh deposit 河漫沼泽沉积
flood plain marsh facies 河漫沼泽相
flood plain microrelief 洪泛平原微地形
flood plain regulation 洪泛平原整治；河漫滩整治；河漫滩整理
flood plain scour 洪泛平原冲刷；滩地冲刷
flood plain scroll 洪泛平原河弯；曲流迂回扇；迂回扇
flood plain sediment 漫滩堆积(物)；河漫滩沉积物
flood plain shape 河漫滩形态
flood plain splay 漫滩冲积扇
flood plain storage 洪水河槽调蓄；滩地调蓄
flood plain terrace 漫滩阶地；洪泛平原阶地
flood plain texture 河漫滩结构
flood plain thickness 河漫滩厚度
flood plain type 湖沼式；汇区式；漫滩式(指水库形式)
flood plain type reservoir 湖沼式水库
flood plain valley 河漫滩河谷
flood plain zoning 洪泛区分区
flood plane 洪水水面；洪水面
flood play sediment 漫滩冲积物
flood pool 泛滥盆地；防洪水库
flood pot experiment 注水试验
flood pot test 水驱油试验
flood prediction 洪水预估；洪水预测；洪水预报
flood prediction service 洪水预报机构；洪水预报工作；洪水预估工作
flood-predominating channel 涨潮主槽
flood-prevention 防汛；防洪工作；防洪
flood-prevention equipment 防浸水设备
flood prevention reservoir 防洪水库
flood probability 洪水几率；洪水概率
flood producing storm 造洪暴雨
flood profile 洪水纵断面；洪水剖面；洪水过程(曲)线
flood projection 泛光投影；泛光投射
flood prone 洪涝
flood-prone area 易(受洪水)泛滥区；洪水多发区
flood-proof electric(al) equipment 防浸式电气设备；防溅式电气设备
flood-proofing 防汛抢险；防汛；洪水防护(措施)
flood-proofing law 防洪法规
flood-proofing organization 防汛组织
flood protected area 防汛保护区
flood protection 防汛；防洪(工作)；防洪保护
flood protection embankment 防洪堤
flood protection material 堵漏器材；防洪器材
flood protection measure 防洪措施；防汛措施
flood protection of tunnel 隧道洪水防治
flood protection works 防洪设施；防洪建筑物；防洪工程；防护工程

flood pump 溢流泵;疏水泵
flood quay 高潮码头
flood reccurrence interval 洪水重现期
flood recede 洪水消退;洪水退落;洪水减退
flood recession 洪水消退;洪水退落;洪水减退
flood record 洪水实测资料;洪水记录;洪水记录资料
flood reduction 减洪;减低洪水
flood regime(n) 洪水情况
flood region 洪水地区;洪泛区;洪水区
flood regulating level 调洪水位
flood regulation 洪水调节;调洪
flood-releasing capacity of spillway 溢洪道泄洪能力
flood-releasing front 泄洪前缘
flood-releasing work 泄洪建筑物
flood relief 洪水解救;泄洪
flood relief bridge 排涝桥
flood relief channel 泄洪(渠)道;分洪渠(道);排洪渠(道)
flood relief loan 水灾贷款
flood relief sluice 泄洪闸
flood relief works 泄洪工程;洪水解救
flood-reporting station 水情站;报汛站
flood reservoir 蓄洪水库;分洪水库
flood resistant 耐涝的
flood retarding basin 滞洪区
flood retarding effect 滞洪效应
flood retarding project 滞洪工程;拦洪工程;拦河工程;防洪工程
flood retention 蓄洪;滞洪;拦洪
flood retention basin 蓄洪水库;滞洪洼地;拦洪水库;滞洪区
flood retention capacity 水库防洪库容
flood retention work 蓄洪工程
flood rise 洪水上涨
flood risk mapping 洪患地区图
flood routing 洪水追踪法;洪水演算;洪水行程
flood routing flooding 洪水泛滥
flood routing process 洪水演算法;洪水演进法
flood routing through reservoir 水库调洪演算
flood runoff 洪水径流;洪峰径流
flood season 汛期;洪汛期;洪水期;洪水季(节)
flood seasonal distribution 洪水季节分布
flood section 洪水断面
Flood's equation 弗勒德方程
flood series 历年洪水资料;洪水系列
flood severity 洪水严重程度
flood shape 洪水形态
floodside of dam 迎水坡面
flood slack 高平潮
flood-source area 洪源区;成洪地区
flood-source region 洪源区
flood span 洪水跨度;高水位跨度(桥梁)
flood speed 涨潮流速;洪水流速
flood spillway 溢洪道
flood spreading 洪水散布;泛滥散布
flood stage 洪水位(线);危险水位
flood stage forecast 洪水水位预报
flood stage of lake 湖水位
flood storage 蓄洪;洪水调节;调洪库容;防洪蓄水;防洪库容
flood storage area 蓄洪区
flood storage basin 蓄洪水库
flood storage capacity 洪水期储[贮]水量;调洪能力;调洪库容;防洪库容
flood storage project 蓄洪工程
flood storage reservoir 蓄洪水库
flood storage with land reclamation 蓄洪垦殖
flood storage work 蓄洪工程
flood stream 涨潮流;洪水河流
flood strength 最大涨潮流(速);涨潮流速;潮流强度
flood stricken area 洪泛区
flood subsidence 洪水退落
flood subsidence rate 落洪率;洪水消退速率
flood surcharge 洪水超高
flood surge 涨水波浪;洪水涌浪;洪峰
flood survey 洪水调查
flood synchronization 洪水同步(化)
flood synthesis 洪水组合;洪水综合法
flood terraces 谷底阶地
flood tidal delta 涨潮三角洲
flood tidal deltaic deposit 涨潮三角洲沉积
flood tide 涨潮(流);高峰(指洪峰);高潮(位);巨量;升潮;大潮

flood tide channel 洪水河道;高潮河槽;涨潮水道;涨潮航道
flood tide current 涨潮流
flood tide depression 涨潮低降
flood tide gate 涨潮闸门;涨潮坝门;外闸门;挡潮闸(门);防潮闸(门)
flood tide range 涨潮潮幅;涨潮潮差
flood tide stream 涨潮流
flood tide volume 涨潮量
flood times 洪水期
flood to peak interval 涨洪时段;起峰时段;涨潮时段
flood trash 流失木
flood tuff 熔结凝灰岩
flood type 洪水类型
flood valve 溢流阀;舷侧通海阀
flood velocity 洪水流速;洪水速度
flood volume 洪水总量;洪水体积;洪水(水)量;洪量
flood wall 挡水墙;防汛墙;防洪石堤;防洪墙;防洪堤;防洪岸壁
flood warning 洪水警报
flood warning service 洪水报警机关;防汛机构
flood warning system 洪水报警系统
flood water 洪水;注水;升潮
floodwater dam 防洪坝
floodwater division 分洪
floodwater level 洪水位
floodwater retarding dam 拦洪坝
floodwater retarding reservoir 滞洪蓄水库
floodwater retarding structure 蓄洪建筑物;蓄洪构筑物;防洪建筑物;滞洪建筑物
floodwater zone 洪水(淹没)区;洪泛区
flood wave 洪波
flood wave recession 洪水波消退
flood wave routing 洪水波演算
flood wave subsidence 洪水波消退
flood wave transformation 洪水波变形
flood wave velocity 洪水波速度;洪峰速度
flood way 分水道;泄洪道;分洪(河)道
floodway channel 分洪河槽
floodway district 行洪区
flood with a recurrence interval of more than one hundred years 百年不遇洪水
floodwood 漂流木材;浮木
flood years 洪涝时期;洪涝年份
flood zone 洪水泛滥地带;洪水区
floolower ring 附环
floor 箱底(集装箱);沟低;楼面;楼层;楼板;肋板;交货场地;平甲板;台面;地面【建】;底(盘);船底肋板;层(楼)
floor adhesive 楼层地面黏合剂;楼板黏合剂
floor-affixed chair 固定椅
floorage 建筑面积;使用面积;地板面积
floor angle 肋板角钢
floor arch 平拱;平背拱;地板拱
floor arch lintel 平拱过梁
floor area 楼面面积;楼层面积;楼板面积;建筑各层面积;仓库面积
floor area for residential dwellings 住房面积
floor area of a dwelling 住房地板面积
floor area of a room 房间面积
floor area of building 房屋面积;建筑楼层面积
floor area ratio 建筑系数;容积率;建筑面积比
floor arming system 楼板框架体系
floor assembly 地板组合;地板总装
floor audit 基层审计
floor baffle 护捆上的消能设备
floor bar 枕木;底梁
floor batten 楼面板条;木楼板龙骨
floor bay 地板开间;地板架间
floor beam 楼面梁;楼板梁;横梁;地板梁
floor beam of bridge 桥面横梁
floor beam of bridge deck 桥面横梁
floor bearer 楼板垫块;(集装箱的)底梁
floor bench scale 地面台秤
floor binder 楼板系梁
floor binding joist 楼板格栅
floor block 地板块件;地板砌块;地板装的滑车;地板预制块;地板木(块);消力墩
floor board 楼(面)板;肋板材;桥面板;台面厚木板;地板用木材;舱底板;铺(地)板;铺地块体
floor boarding 铺地板;楼面块体
floor boarding joist 上梁;楼面板托梁;顶棚梁

floor bonding adhesive 楼板黏合剂
floor bonding agent 楼板黏合剂
floor bonding medium 楼板黏合剂
floor box 地板万能插口;(地板中的)出线盒;地面出线盒;地板插座
floor boy 勤杂工
floor brad 地板钉
floor branch 楼层分支
floor breaks (废矿的)底裂;底板裂开;底板断裂;底板裂隙
floor breakthrough 楼板贯穿断裂
floor brick 地面砖;铺地砖;铺地建筑砖
floor broker 交易所场内经纪人;现场经纪
floor brooding 地面育雏
floor brush 地板刷
floor buffer 楼板减振垫
floor buffer system 楼板减振垫做法
floor building component 楼板构件
floor building member 楼板构件
floor building unit 楼板构件
floor burning appliance 设在楼板上的燃烧装置;设在地板上的燃烧装置
floor burst 底板岩爆;底板突起;底板突出;底板破裂;底板隆起
floor care 楼板保养;地板保养
floor carpet 地毯
floor cast(ing) 楼板浇筑
floor ceiling 舱底铺板
floor ceiling joist 楼层平顶格栅;楼层顶棚龙骨
floor cement(ing agent) 楼面地板胶结剂;楼板胶结剂
floor center 地板支架;楼板对中
floor charging machine 地行行式装料机
floor chart 平面布置图;布置图
floor chisel 地板垫木;地板填隙铁;铺地板錾紧凿
floor clamp 安板夹子;楼板夹钳;钻杆夹持器;紧板铁马
floor clay block 楼板砖;焙烧的铺地砖
floor clay brick 楼板砖;焙烧的铺地砖
floor cleaner 楼板吸尘器;楼板清洗剂;楼板清洁器;地板吸尘器;地板清洗剂;地板清洁器;地板洗净剂;地板清洁剂
floor clearance 门下间隙
floor clear varnish 楼板清漆
floor clip 木地板固定条;混凝土楼面嵌固条;地板夹子;木楼板龙骨夹
floor clock 落地钟
floor closer 地面关门器;地龙;地面门档;地弹簧;地面闭锁器
floor cloth 拖布;铺地板厚漆布;墩布;铺地织物;铺(地)面织物;地板布;铺地板的漆布
floor clown-to-ground openwork screen 落地钟
floor coat 楼板涂层;楼板面层
floor coating 地板涂层;地板涂层;地面涂料
floor column 楼板柱;楼板支承柱
floor combustion method 炉底燃烧法
floor component 楼板构件
floor composed of large units 大型构件组成的楼板
floor compound unit 楼板复合构件
floor concrete 底板混凝土
floor construction 楼面构造;楼盖构造;桥面构造;地板构造
floor contact 平台触点;地板触点;分层触点
floor contamination indicator 地面污染指示器
floor contamination meter 地面污染计
floor controlled monorail system 地面操纵单轨系统
floor controlled overhead crane 地面操纵桥式起重机
floor conveyer 地面输送带
floor cover 楼面护层;楼面覆盖(面)层
floor covering 楼面修整;楼面饰面石板;楼面面层;地毯;地面覆盖层;地面覆盖(面)层;地板面层;地板革;楼面覆盖(面)层;地板蒙布
floor covering coat 楼板涂层;地板涂层
floor covering emulsion 楼板修饰用乳胶;地板修饰用乳胶
floor covering felt 楼板饰面毡;地板覆盖层油毛毡
floor covering hardboard 楼板饰面硬质纤维板;地板覆盖层的硬纸板
floor covering lino(leum) 楼板饰面油地毡;地板修饰用的油毡
floor covering material 楼板饰面材料;地板面层

修饰用的材料;地板覆盖材料;敷地材料
floor covering plastics 楼板饰面塑料;地板面层修饰用的塑料
floor covering rolls 楼板覆盖面层碾辊
floor covering screed material 楼板面层找平材料
floor covering sheet 楼板面盖板
floor covering slate 地板面层铺盖的石板
floor covering tile 地板铺盖的瓷砖
floor covering work 楼板覆盖面工作
floor crack 底板裂隙
floor cramp 紧板铁马;地板夹;楼板扣钉
floor cross-section 楼板剖面;地板横截面
floor culvert 闸底涵洞
floor current 沿地板面气流
floor cut 掏底
floor damp-proof course 地面防潮层
floor damp-proofing 地面防潮
floor depth 楼板厚度;楼面高度;桥面(系)高度
floor design 楼板设计
floor diaphragm 楼面隔板;楼层横隔板
floor disc 楼面圆板
floor dog 紧板铁马
floor door catch 楼板门后夹;地板门后夹
floor door closer 楼层关门装置
floor door closing device 楼面制门器;地面停门器;地面制门器;地面停门器
floor drain 楼面排水;地面排水(管);地漏
floor drainage 桥面排水
floor dresser 地板装修器
floor dryer 底部通风式干燥器
floor duct 楼板层管道
floor duct work 地下管道工程
floor dunnage 垫舱板
floor duty 在办公室值班(房地产经纪人事务所)
floored warehouse 铺地板仓库;铺设地板的仓库;地板仓库
floor elevation 楼面标高;底板高程
floor emulsion 楼板乳化涂料;地板乳胶
floor enamel 地板瓷漆
floor engine 扁平型发动机
floorer 铺地板工人
floor expansion joint 楼板伸缩缝
floor fan 落地扇;落地风扇
floor fan without timer 不定时落地扇
floor felt 楼板饰面毡;地板油毛毡
floor fill 楼面填充
floor filler(block) 楼板层填块;地面填料
floor filler slab 楼板填充板;地板填充板
floor filler tile 楼板填充瓷砖;地面填充砖块
floor filling 楼板填料
floor filling material 楼板层填充材料;地面隔层填料
floor finish 室内地面;楼面装修;楼板材料;楼面修整;楼地面装修;地板涂漆;地板涂料;面层
floor finish coat 楼板饰面层;地面终饰层
floor finish emulsion 楼板饰面乳化涂料;地面修整用乳胶
floor finish hardboard 楼板饰面硬质纤维板;地面修整用硬板皮
floor finishing 面层材料;楼面终饰;地面终饰
floor finish lino(leum) 楼板饰面油毡;地面修整用油毡
floor finish(material) 楼板饰面找平材料;地面平整材料;楼面终饰材料;地面最后加工材料
floor finish screed(topping) 楼板饰面找平;地面平整
floor finish work 楼面修整工作;楼板饰面工作;地面修整工作
floor fire 楼层供暖炉;房屋楼层的暖气炉
floor firing furnace 炉底燃烧炉
floor flocks 地板上的絮状沉淀
floor-flush tow conveyer 沉轨牵引式输送机
floor for animal shelter 畜舍地板
floor form(work) 地板模架;楼板模板;楼板模架;地面模板
floor formwork handling rig 地板模板输送装置;地板模板输送设备
floor frame 底部肋骨
floor frame work 楼面构架(工程)
floor framing 楼盖构造;楼板骨架;地板构造;楼面构架;地板构架
floor framing plan 楼板构架平面图;楼板结构平面图
floor framing system 楼盖构造体系
floor function 地板函数

floor furnace 地板炉;地下热风炉;地板取暖器
floor furring 楼板垫木
floor girder 桥面板主梁;楼盖主梁;桥(面)板大梁
floor glue 地板胶
floor grating 楼地板算条;地面格筛
floor grid 桥面板系;楼地板算条;地面网栅
floor grid plane 楼板层算条平面;地面网栅平面
floor grill(e) 地板格栅板;地面回风口;楼板算子
floor grinder 固定式砂轮机;地板磨光机
floor groove 地面沟槽;底板上的沟槽;底槽
floor ground plan 楼层平面图;房屋楼层平面布置图
floor guide 地面活动门导槽;地面导槽;楼板滑面槽;地板上隔舱壁槽;楼面导轨;楼板门槽;地板中的拉门导槽;地板导槽
floor gull(e)y 楼板排水沟;地面排水沟
floor hanger 楼盖结构的吊筋;支承地板托梁的镫铁
floor hardener 地面硬化剂;铁屑;水泥地面坚硬材料
floor hardness tester 地板硬度测试仪
floor hardtimber 楼板铺用硬木;地板用硬木
floor hardwood 楼板铺用硬木;地板用硬木
floor hatch 楼板上下口
floor head 肋板头;肋板端
floor header 楼梯平台格栅
floor-heated 地板加热的
floor heater 地板取暖器
floor heating 楼板暖气;楼板供暖;地板供暖;地板采暖
floor heating by air 楼板层热风采暖;地板层热风采暖
floor heating by electricity 用电的地板采暖(系统)
floor heating cable 地板采暖电缆
floor heating panel 楼板加热板;地板加热板
floor heating plate 地板采暖板
floor heating system 楼板供暖系统;地板供暖系统;地板采暖系统
floor heave 地板隆起;底鼓;底板隆起
floor height 楼层高度;地板高度;底板高度(汽车);层高【建】
floor hinge 地板弹簧铰链;楼板铰链;装在门底与地板之间的铰链
floor holder 地板门挡
floor hole 楼板(管道)孔;楼板孔洞(管道孔)
floor hollow brick 空心楼板砖
floor hollow tile 楼板空心砖
floor hopper truck 底部卸料式货车;底卸式货车;底卸式自卸汽车
floor indicator 电梯楼层按钮板;楼层指示器
flooring 地面材料;桥面板;桥面安装;铺地面;铺地(板);室内地面;地面材料;地板材料
flooring adhesive 楼板黏结剂;地板黏合剂;地板胶黏剂
flooring batten 铺楼面用板条;(在地板下面的)地板条
flooring block 铺地(板)块;铺设地板块;铺地板(木)块
flooring board 楼面板;铺面板;铺板
flooring bonding adhesive 楼板黏结剂;地板黏结剂
flooring bonding agent 楼板黏结剂;地板黏结剂
flooring brad 地板角钉;(扁头的)地板钉
flooring brick 地面砖;铺地砖
flooring care 楼板保养;地板保养
flooring cement 地板胶
flooring clay brick 铺地黏土砖
flooring(clear) varnish 地板清漆;地板凡立水
flooring clip (固定在楼面上的)地板夹子;木楼板龙骨夹
flooring coat 楼板涂层;地板涂层
flooring dish 碟形地面
flooring door catch 室内地面门后夹;楼面制门器
flooring emulsion 楼面用乳化液;地面乳剂
flooring emuslion 楼板乳胶
flooring felt 铺楼面油毛毡;楼面油毛毡;地面(油毛)毡;铺地用毯
flooring finish 楼面饰面;地板木材加工修饰
flooring finish lino(leum) 楼面油毛毡;地面油毡
flooring finish material 楼面材料;地板面层材料
flooring finish plastics 楼面用塑料;地板面层塑料
flooring finish screed(topping) 楼地面找平层;地面面层
flooring finish tile 铺地瓷砖;楼面用瓷砖
flooring finish work 地面修整工作;地面饰面工作
flooring grating 楼面算条;地面格筛

flooring grinder 地板研磨机
flooring guide (楼板、地板上的)拉门导轨
flooring gulley 楼板排水沟;地面排水沟;地面集水沟
flooring hardboard 楼面用硬质纤维板;地面硬质纤维板
flooring hardener 地板硬化剂
flooring hardness tester 地板硬度测定仪
flooring layer 楼板铺砌工;楼面面层
flooring lino(leum) 楼面油毡;地面油毡
flooring marble 铺地大理石;楼板用大理石
flooring mastic 楼面油灰;楼板腻子;地板油灰;地板腻子
flooring material 楼面材料;地面铺盖料;铺地材料
flooring mosaic 铺地镶嵌锦砖;铺地马赛克
flooring nail 地板钉;楼板钉;铺面螺钉
flooring off 以垫舱货垫舱;底层舱装货
flooring outlet 楼面出水口
flooring paint 楼板油漆;楼板涂料;地板油漆;地板涂料
flooring paper 地毯用纸
flooring pattern 楼地面图案
flooring plaster 楼面石膏粉饰
flooring plastics 铺地塑料;楼板用塑料
flooring plastic sealing 楼面塑料密封;地板塑料密封
flooring plug socket 楼面插座;地板插座
flooring polish 楼面磨光;地板抛光剂
flooring power point 楼板电源插座
flooring protection agent 楼板防护剂
flooring quarry(tile) for animal shelters 用于畜舍的铺地缸砖
flooring radiant panel 铺地辐射板
flooring roller 地面滚压机;地面辊压器;地板碾辊
flooring saw 楼面锯;地板锯
flooring screed material 楼面找平材料;地板面层
flooring seal(er) 楼面密封剂;地板密封剂
flooring slab 地板板皮科
flooring slate 铺楼面用石板;铺地石板
flooring socket 楼面插座
flooring softtimber 铺楼面软木;铺地面软木
flooring softwood 铺楼面软木;铺地面软木
flooring tile 铺地瓷砖;地面砖
flooring track 楼面导轨
flooring trowel 地板磨光镘刀
flooring varnish 楼面清漆
flooring wax 地板蜡
flooring wearing course 楼面磨耗层
flooring wear(ing) surface 楼面磨损面;地面磨损面
floor inlet 楼面进水口;地面进水口
floor in prestressed clay 预应力黏土地面
floor inspection 现场检验;就地检验
floor insulation 楼面绝缘;楼板隔声层;楼板隔热层;地面绝缘
floor iron 底铁
floor joint 楼板接头;楼板接头;地面接头
floor joist 桥面横梁;楼板横梁;地板托梁;地板格栅;楼盖格栅;地板格栅;楼板格栅;基台梁;木楼板龙骨;地板龙骨(美国)
floor joist bearing 楼板格支承;地板托梁支撑
floor knob (装在地板上的)门挡块;门碰头
floor lacquer 楼面硝基漆;楼板喷漆;楼板腊克;地面硝基漆;地板喷漆;地板腊克
floor lamp 落地台灯;落地灯;立灯
floor landing 楼梯平台;楼面平台
floor lay 垫层;底层
floor layer 楼板面层;地板面层
floor-layer's labo(u)rer 铺地面和墙面砖工的助手
floor laying work 地面铺设工程
floor length 平台长度;地板长度
floor level 平台标高;楼面高度;楼板标高;楼盖水平面;楼面平面;室内地面平面;地面高度;地平平面;地板;底层;底板
floor lift 底鼓;底板上升(采矿)
floor lift truck 地面起重车
floor light 落地灯;楼面采光口;地板采光窗;辅助照明
floor line 楼板线;楼面线;楼地板线;地面标线
floor lining 楼板衬垫料
floor lino(leum) 楼面油地毡;地面油地毡
floor live load(ing) 楼面活荷载;地板活荷载;楼面活荷载
floor load(ing) 楼板荷载;楼板荷载;地板荷载;底板荷载;底部负荷

floor loading capability 箱底承载力
floor loan 部分抵押贷款
floor lowering 落底
floor lug 肋形角钢
floor machine 木地板刨平机；铺地板机
floorman 钻工（井口）
floor marble 地面大理石
floor mastic 地板油灰
floor mat 地板毡；地板铺垫；草地席；草地毯
floor material 楼板材料；地板材料
floor member 楼板构件；地板构件
floormen 机台工人【岩】
floor monitor 地面监测器；地板放射性监测器
floor mosaic 铺地马赛克
floor mo(u)ld 地铸模；地坑铸型
floor mo(u)lder 大件造型工
floor mo(u)lding 踢脚板底缝压条；地面周边饰条；踢脚压条；地面造型
floor-mounted drinking fountain 安装于楼板上的饮用喷水器
floor-mounted fire warning device 安装于楼板上的火警装置
floor-mounted(hot)water heater 安装于楼板上的热水器
floor-mounted unit heater 安装于楼层上的采暖机组；安装于楼板上的单元采暖器
floor nail 地板钉
floor of deep ocean 深海底
floor of dowelled wood(en)beams 板销木梁楼板层；暗销钉合木梁
floor of gallery 矿坑底板
floor of lock chamber 闸室底
floor of nose 鼻底
floor of trench 地沟底
floor of wound 创底
floor oil 地板油
floor-on-grade 地坪
floor opening 楼板出入孔；楼板洞
floor-operated crane 楼面上操纵的起重机
floor outlet 路面出水口；地面出(水)口；地面出线口
floor pack 底板分块
floor pad 楼板衬垫；地板衬垫
floor paint 地板(油)漆；地板涂料
floor pan 底部垫
floor panel 地板节间；楼板单元；地板大方格块；（由墙，柱或梁支承的）预制装配式楼板；大型楼板
floor panel heating 楼板辐射采暖；地板辐射采暖；底部辐射采暖
floor pattern magnetotelephone switchboard 磁石落地式电话交换机
floor pedestal 落地式轴承架
floor pit 地板坑
floor plan 楼面布置图；楼层平面图；平面布置图；地面标高图
floor plank 桥面板
floor plank for timber bridge 木桥面板
floor plan layout 平面布置
floor plastics 地板塑料
floor plate 肋板；基础板；铺板；平甲板的板材；地板，船底铺板；波纹板；网纹钢板；楼板上平垫木；垫板
floor platform 桥面
floor plug 地板电(源)插座；地板(电门)插座；落地插座；地面插座
floor polish 地板蜡
floor polisher 地板磨光机；楼板擦光器；地板打蜡器；地板擦光器
floor polishing 楼地面上光
floor post 楼梯平台支柱
floor pressure arch 底板压力拱；底板拱
floor price 最低价；最低价（格）；最低（出口）价格；底价
floor-proofing of ground 地面防潮
floor protection agent 楼板防护剂；地板防护剂
floor push 闸刀开关；脚踏开关
floor quarry 铺地缸砖；铺地砖
floor radiant heating 底板辐射采暖
floor rammer 长形夯砂锤；长型夯砂锤；地面夯实机
floor receptacle 地板（电源）插座；地面插座
floor receptacle outlet 地面插座出口线
floor recess 楼层壁龛；底槽
floor reel 地上卷取机
floor reflection factor 楼板反射因数；楼板反射系数；地板反射因数；地板反射系数

floor register 楼面出风板；脚面通风器；地面出风口；楼面通风板
floor reinforcement 楼板钢筋
floor relay 底层继电器；分层继电器
floor response spectrum 地面响应谱
floor rib 肋构楼板；楼板肋
floor ribband 地板防滑材【船】
floor rider 肋龙加强材
floor rug 地毯
floor runner track 附地隔墙龙骨；隔墙地龙骨
floor sand 填砂；底砂；背砂
floor sander 地板打光机；地面喷砂打光机
floor sander-polisher 地板打磨器
floor sanding machine 地板磨光机
floor sanding paper 粘砂刚纸
floor scheme 楼板结构；地板结构
floor scraper blade 地面刮平机刮片；地面刮平机刮板；底部刮片（混凝土拌和机）
floor screen 炉底水帘管；水筛炉底
floor sealer 地板罩面料；地板用封固底漆
floor seal(ing) 地板密封
floor section 车厢地板
floor sheen 地板表面光泽
floor sheet 踏板
floor shift 落地式变速操纵杆
floor shuttering 地板模壳；地板模板
floor sill 基台；木底座；底基；栅子地板底梁；楼木底座；地面门坎
floor skin 楼板外层；地板外层
floor slab 铺地石板；楼面石板；地面石板；楼板，地板；桥面板；路面板；钢筋混凝土结构地面板
floor sleeper 楼面地板小格栅；木楼板龙骨
floor sleeve 楼板套管；楼面插管
floor slope 地板坡度；底帮
floor socket outlet 地板插座；地面插座
floor soffit 平顶下部；顶棚；地板下部
floor softtimber 地板软木材
floor softwood 地板软木材
floor space 占地面积；楼面面积；楼层面积；肋板间距；居住面积；居室面积；建筑面积；底面积；车间面积；仓库面积
floor space completed 竣工面积
floor space index 建筑面积指标；容积率
floor space per person 每人有效建筑面积；人均使用面积
floor space under construction 施工面积
floor span 楼板跨度；格栅跨度；桥面跨径；桥面跨度
floor special purpose receptacle 地面专用插座
floor spring 地板弹簧铰链；门簧铰链；地弹簧
floor springs and checks 自动开关地弹簧
floor squeeze 底板隆起
floor stand 地轴(承)架；落地支架；地板支架（钻塔）
floor standard 地板标准
floor stand grinder 台式砂轮机
floor stilt 门框垫块
floor stirrup 楼板钢筋；地板钢筋箍
floor stop 门碰头；(地板上的)门挡；地面门碰头；地板门挡
floor stress during operation 楼板在使用中的应力；地板的作用应力
floor stress when climbing 楼板在吊装时的应力
floor-stringer of bridge deck 桥面纵梁
floor structural system 楼板结构系统
floor strutting 地板格栅支撑；楼盖梁支撑；楼面加劲条；楼盖梁加劲条；木楼板格栅撑
floor support 楼板柱
floor supported S/R machine 地面支承型有轨巷道堆垛起重机
floor surfacing 地板整饰材料
floor suspender 楼板吊杆；桥面吊杆
floor swab 起模用毛笔
floor sweeper 地面扫除器
floor switch 楼层开关；平台开关；地板开关；层位开关
floor system（桥的）楼面系统
floor system of bridge 桥面系
floor tank 灌水底舱
floor telephone outlet 地面电话出口线
floor temperature 楼面温度
floor thickness 楼板厚度；地板厚度
floor-through 公寓房间；全层公寓
floor-through wax 全层公寓；打蜡地板
floor thrust 底板逆冲断层；底板冲断层

floor tile 楼盖砌块；地面砖；地板砖；花砖；铺地(面)砖；底879瓦砖
floor tile cutter 地砖切割机
floor tile press 地砖压机
floor tile production line 地板砖生产线
floor tiler's trowel 铺瓷砖工用的镘刀
floor timber 地梁；地板木材；底梁；肋根材；底肋材
floor time 停机时间
floor-to-ceiling dimension 楼面至顶棚尺寸
floor-to-ceiling height 室内净高；楼面至顶棚高度
floor-to-ceiling loan 分阶段贷款
floor-to-ceiling window 整层高的窗；空间高度的玻璃窗
floor-to-floor dimension 楼面至楼面尺寸
floor-to-floor height 楼面至楼面高度；楼层高（度）
floor-to-floor time 时间间隔
floor to head height dimension 地板至开口顶端尺寸；地板至窗口顶端尺寸
floor topping 楼板面层
floor-to-roof dimension 楼面至屋顶尺寸
floor-to-roof height 楼面至屋顶高度
floor-to-wall joint 楼板与墙的接缝
floor track 楼地板上的导轨
floor trader 交易场所内的商人；现场客
floor trap 地板孔
floor trowelling machine 地面抹光机
floor truck 轻便手推车
floor tube 炉底管
floor type 落地(固定)式；立式；地面种类
floor type borer 落地镗床；落地式镗床
floor type boring machine 落地镗床；落地式镗床
floor type carryall 地行式运输机；地行式铲运机；落地式运输机
floor type furnace charger 地面炉料加料机
floor type grinding wheel 落地砂轮机
floor type heater 地板炉；地下热风炉
floor type pressure relief valve 地面式减压阀
floor type sensitive drill press 落地式灵敏钻床
floor type switchboard 固定式交换机
floor underlayment 楼面底板；地板垫层
floor unit 楼地板构件；楼板构件；地板构件
floor value 最低限额价值
floor varnish 地板清漆
floor vault 平顶拱
floor ventilation 地板通风
floor warning 地面供暖
floor warning by electricity 地板电热
floor warning cable 楼板层加热电缆
floor warning installation 楼地板加热装置；楼层供暖装置
floor warning system 楼板层采暖系统
floor wax 地板蜡
floor way 桥梁的桥面系统（包括桥面及其支承部件）；桥面系统
floor weight 楼板重量
floor width 楼梯平台宽度
floor window 落地窗
floor with precast beams placed close together 紧急排放预制梁的楼板
floor wood 地板木材
floor yellow 地板黄
floor zone 楼层范围
flop 全部功能失效；突然转变；随角异色
flop colo(u)r 掠射角观察的颜色
flop damper 烟道调节闸门；重力风门
flop effect 随角异色效应
flop gate 自动放水闸门；翻板门；导向闸门；扑拍门扇
flop house 廉价住所；廉价住房；低级旅馆
flop-in method 增加法
flopover 电视图像上下跳动；触发器
flopper 波浪边（板材缺陷）；薄板上皱纹
flopping head 渐落水头
floppy disc 软盘；柔性塑料碰盘
floppy disk 软盘
flop valve 瓣阀
Floquet theorem 弗洛盖定理
flora 植物群；植物区系；菌群；区系；地区性植物；窗外花台
floral 花卉的
floral apex 花原端
floral axis 花轴
floral border blanket 花边毡

floral decoration 花纹装饰;花卉装饰
floral decorative fixture 花纹装饰用具
floral design 花卉装饰设计;花卉图案;花纹图案
floral diagram 花卉图式;花卉图案
floral element 植物群系成分
floral enrichment 花纹美化装饰
floral envelope 花被
floral form 花式
floral formula 花程式
floral-hoop-like conduit 花环状通道
floral landscape 植物景观
floral leaf 花叶
floral mechanism 花的机制
floral motifs 花卉花纹
floral organ 花瓣
floral ornament 花纹装饰;花卉装饰
floral part 花瓣
floral pattern 花饰;花卉图案
floral pattern glass 压花玻璃;花纹玻璃;冰花玻璃
floral-pendant gate 垂花门
floral province 植物群区
floral scroll 漩涡形彩纹;涡卷线脚;卷草纹
floral shoot 花柄
floral stage 植物群阶【地】
floral zone 植物群带;植物带
Florence Cathedral 佛罗伦斯大教堂
Florence leaf 装饰用的黄色合金箔
Florence oil 橄榄油
florencite 磷铝铈石
Florentine arch 两心拱;佛罗伦萨拱
Florentine blind 佛罗伦萨式窗遮帘
Florentine glass 佛罗伦萨玻璃
Florentine lake 紫红色淀
Florentine lapin 弗罗伦廷宝石
Florentine mosaic 佛罗伦萨马赛克
Florentine mosaic work 佛罗伦萨马赛克饰面
Florentine Renaissance 佛罗伦萨文艺复兴式
Florentine style 弗罗伦萨风格
florescence 开花期
Flores Sea 弗洛勒斯海
floret 小花
Florey unit 弗洛里单位
floriated 花的;华丽的;花形的
floriated decoration 花纹装饰
Floricin oil 弗洛里辛油
floriculture 种花;花卉园艺学;花草栽培
floriculturist 种花工;花卉栽培工
florid 华丽的
Florida phosphate 佛罗里达磷酸盐
florid architecture 装饰性建筑;彩饰建筑
florid complexion 红红的面色
floridean starch 红藻淀粉
florid Gothic 华丽哥特式
floridin 活性白土;漂白土
floriferous starch 红藻淀粉
florilegium 选集
florisil 硅酸镁载体
florist 花匠;花店
florizone 植物带;植被地带
florology 植物区系学
florovite 符硼镁石
flory-boat 花纹舢板
Flory constant 弗罗利常数
Flory cross 花饰十字架;弗罗利十字架
Flory temperature 弗罗利温度
flos ferri 霰石华;文石华;铁华
Flos Mume 梅花
flospinning 离心成型(法)
floss 木棉;绵矿
flossa 佛罗萨地毯
floss hole 烟道孔
floss-silk 丝绵
flotability 浮游度
flotator 浮选机
flotiated 华丽的
flotilla 小舰队;小船队;驱逐舰队;艇队;船队
flotilla leader 先头舰;领航驱逐舰
Flotrol 恒电流充电机
flots 重叠牙边
flotsam 零碎物(件);漂浮物;漂浮残骸;失事船浮出的货物;浮货;废料
flounce 荷叶边
flounder plate 眼板
flour 研磨成粉;岩粉;磨成粉;粉末

flour adhesive 面粉黏结剂
flour all risks clause 面粉全险条款
flourbaryt 萤石
flour beetle 粉甲虫
flour bolt 面粉筛
flour-chrome-arsenate-phenol 铬砷酚合剂(木材防腐剂)
flour content 粉状颗粒含量
flour dust 面粉尘
flourescent dye 荧光染料
flour filler 细(粉状)填料;粉状填料
flour gold 粉金
flourinated ethylene-propylene 全氟乙丙烯
flourish 蔓叶花样;花饰
flourishing 繁华
flour lime(stone) 细石灰;石灰粉;粉石灰
flour mill 磨面机;磨粉机;面粉磨;面粉厂
flour milling machine 磨面机;磨粉机;和面机
flour of emery 细金刚砂粉
flouromer 量粉计;澄清器;澄粉器
flouroscopy 荧光检查
flour packer 面粉装袋机
flour paste 浆糊
flour sacker 面粉装袋机
flour sieve 面粉筛
flour sling 网带吊货兜
flour substance 粉状物质
floury alumina 面粉状氧化铝
flour yield 出粉率
floury soil 细砂土壤;粉砂
floury structure 粉末状结构
floury texture 粉末质地
flous 氟硅酸盐
flow 流动;塑变;事务系列
flowability 流动性;流动能力
flowability of concrete 混凝土流动性
flowability of mo(u)lding compound 模压塑料流动性
flowability of plastic 塑料流动性
flowable formulation 流动剂
flow acoustics 流动声学
flow-adjusted concentration 调节液流浓度
flow adjustment factor 流量调节因子
flowage 流动状态;流动特性;柔流
flowage damage 淹没损失
flowage differentiation 流动分异(作用);流动分析(作用)
flowage easement 溢流地役权(水在他人土地流过之权)
flowage fold 流状褶皱
flowage folding 柔流褶皱
flowage line 流边线;水面线;泛滥边线
flow agent 流动剂
flowage of river 河流流动特性
flowage of stream 河流流动特性
flowage prevention 防止淹没;防止泛滥
flowage right 淹地权
flow aid 助流剂;流动性助剂
flow analogy 水流模拟
flow analysis 流分析;流动分析法;水流分析
flow-and-jam detector 流阻检测器
flow-and-plunge structure 流状柱状构造
flow and sediment data 水沙资料
flow and slump test 流动和坍落度试验
flow angle 流倾角;气流角
flow area 水流断面;流域;过水面积;流通面积
flow around a body 绕流;物体绕流
flow axis 流动轴线;水流轴线
flow-back 回流
flow-balancing method 流量平衡法
flow-balancing tank 流量平衡池
flow band 流带
flow banding 流纹带;流动条带
flow beam 流束
flow bean(choke) 节流嘴;自喷油嘴
flow behavio(u)r 流动特性
flow behavio(u)r index 流动特性指数
flow block 溢料口砖;通路流槽砖
flow blue 流青
flow bog 浮沉沼泽;浮沉泥炭沼泽
flow brazing 浇注钎焊;波峰钎焊
flow breccia 流状角砾岩
flow brightening 软熔发亮处理;软钎焊涂层
flow-brightening process 流动光亮法

flow button test 熔球试验
flow by gravity 自然流动;自流;重力流动
flow by heads 自喷
flow by-pass 旁流
flow calorimeter 流量热计
flow calorimetry 流量测热法
flow camera 连续摄影机
flow capacity 泄水量;泄水能力;泄流能力;过水能力;流通能力;排水能力;阀的容许流量
flow capacity of control valve 调节阀流通能力
flow capacity of water meter 水表的流通能力
flow cascade 梯级跌水
flow cast 流铸型;流型
flow catch(er) 喷流挡板
flow cell 流通池;流路池;流动元件;流动池
flow channel 液流通路;流槽;河槽;水流(通)道
flow characteristic 流动特性;水流特性
flow characteristic of control valve 调节阀流量特性
flow characteristics coefficient 水流特性系数
flow characteristics of mud 泥浆的流动特性
flow characteristics of powder 粉末流动性
flow characteristic of regulating valve 调节阀流量特性
flow chart 作业图;流动图表;流程图;框图;生产过程图;程序(方)框图;操作程序图表;流量图;程序图表
flowchart connector 流程图连接符
flowchart convention 流程图约定;流程图规则
flowchart for standard development 标准编制流程图
flowchart graph 框图形成的图
flowcharting 拟制流程图;流(程)图编制;流程图的绘制;流程图表示
flowchart microprogramming language 流程图微程序设计语言
flowchart of data by simulation model 模拟模型数据流程图
flowchart of national economic benefit and cost 国民经济效益费用流量表
flowchart of ore processing 选矿流程图
flowchart of sample treatment 样品加工流程图
flowchart package 流程图部件
flowchart program(me) 流程图程序
flowchart program(me) scheme 框图型程序模式
flowchart schema 流程图模式;框图模式
flowchart symbol 流程图符;流程图符号;程序框图符号
flowchart technique 流程图技术
flowchart template 流程图模板
flowchart text 流程图正文;流程图(文字)说明;流程图文本
flow chip 带状切屑
flow choking 气流壅塞;气流堵塞
flow circuit 水流回路;输油管网
flow clarification 流倾沉淀法;流动澄清法
flow classification 水流分类
flow cleavage 流状劈理;流劈理
flow cleavage belts 流劈理带
flow coater 流动涂布器
flow coat(ing) 流涂;浇涂;流动涂层;淋釉法
flow coat method 流涂法;浇涂法
flow coefficient 流量系数;流动系数;径流系数
flow colo(u)rimeter 流动式比色计;流量比色计
flow column 油管柱
flow combine 集流
flow comment 流注释
flow compensator 流量平衡器;流量补充器
flow component 径流成分;水流成分
flow concentration 集流;汇流
flow-concentration curve 流量密度曲线
flow-concentration-time diagram 流量-浓度-时间图
flow condition 流态;流动条件;水流条件
flow condition for navigation 通航水流条件
flow conditioning agent 流动调节剂
flow cone 浆料稠度测定器(压力灌浆);流动锥
flow continuity 连续性流动
flow control 液流调节;信息流控制;流量控制;流量调节;流(程)控制;进料控制;水流控制
flow control additive 流动调节剂
flow control agent 流控剂;流动控制剂
flow control device 流量调节器

flow control gate 流量控制闸
flow controller 流量控制器;流量调节器
flow control pump 可控流量泵;计量泵
flow control switch 流量调节开关
flow control system 流量控制系统;流动控制系统
flow control valve 流量控制阀;流量调节开关;流量调节阀
flow converter 液力变扭器
flow conveyer 连续流(刮板)输送机;流化输送机
flow cooling 流行冷却;流动冷却
flow corrective cone 流动校正锥体
flow counter 流通式计数器;流量型计数器
flow cup 溢出杯;流出杯;溜杯
flow curtain electrophoresis 流幕电泳
flow curve 流变曲线;流量曲线;流动曲线(液限试验);气流曲线;速流曲线
flow data 流量资料;径流资料;水流资料;水流数据
flow deficiency 流量短缺;流量不足
flow deflection 气流偏转;水流偏转
flow deflector 导流片
flow demand 需要流量;要求流量
flow-density curve 流量密度曲线
flow depth 流深;流动深度;水深;水流深度
flow detector 流动式检测器
flow diagram 工艺流程图;流线图(谱);流量图;流动曲线图;工程图;流程示意图;框图;程序(方)框图
flow diagram of bridge construction 桥梁施工流程图
flow diffusion 扩散流;水流扩散
flow direction 流向;流体流向;流程方向
flow direction(al) vane 风向标
flow-direction indicator 流向仪
flow direction measurement 流向测量
flow direction of underground river 暗河流向
flow direction probe 流向测量器
flow-direction vane 流向仪
flow discharge 流量;泄水能力
flow distortion 气流畸变;气流变形;水流畸变;水流复形;水流变形
flow distributing and collecting valve 分流集流阀
flow distributing device 配水装置
flow distributing synchronous valve 分流同步阀
flow distribution 流量分配;流量分布;水流分布
flow distribution curve 流量分布曲线
flow distribution plate 分流板
flow distributor 流动分布器;导流筒
flow divergence 气流扩散
flow-diversing system 导流系统
flow diverter 偏流器
flow divider 分流阀;流量分配器;分流器
flow divider combiner 分流集流阀
flow divider valve 分流阀
flow dividing and collecting valve 分流集流阀
flow dividing valve 分流阀
flow down 流下
flow down vigorously 一泻千里
flow duration 径流历时;出流历时
flow duration curve 流量过程线;流量历时曲线
flow duration diagram 流量历时图
flow duration relationship 流量历时关系
flow dynamic(al) axis 水流动力轴线
flow dynamics 流体动力学;水流动力学
flowed energy 气流能量
flowed in mud 流入淤泥;流入泥沙
flowed in soil 流入淤泥;流入泥沙
floweing period 开花期
flow energy 流动能(量)
flowe plant 显花植物
flow equalization 均流
flow equalize 均流
flow equalizer 流量均衡器
flow equation 流量方程;流动方程
flow equilibrium 流量平衡;流动平衡
flow equivalent diameter 等流量当量直径
flower 细粉末;铸件的氧化物色斑;花饰;排版花饰
flower abscission 落花
flowerage 花饰
flower and plant shop 花木商店
flower arm 翘;华拱
flower arrangement receptacle 花插
flower arranging 花卉装饰;花卉布景
flower basin 花盆
flower bed 花圃;花坛;花台

flower bed surround 花坛墙
flower border 境边花坛;华境
flower bowl 花圃;花盘
flower box 花匣;花盒;花箱
flower-brick 插花方砖
flower bud 花芽
flower clock 时钟花坛
flower cluster 团伞花
flower container 花箱;花罐
flower corsage 花卉装饰
flower culture 花卉栽培
flower decoration 花卉装饰
flowered 花纹装饰的
flower envelope 滑被
flowerer 描花工
floweret 小花;小花饰【建】
flower festoon 花饰;垂花饰
flower forcing 催花
flower formula 花程式
flower garden 花园;庭园
flower growing 花草栽培
flower hedge 花篱
flower ice 雪花覆盖冰
flowering 水华;成熟
flowering cherry 樱花树
flowering glume 花颖片
flowering peach 碧桃
flowering plant 开花植物
flowering shrub 观花灌水
flowering straw 花草
flowering tree 观花树木
flower leaf 花叶
flower-like design 花型图案;花纹
flower model 花状模型
flower motif 花纹图案;花的主题
flower nursery 花圃
flower of antimony 氧化锑
flower of iron 霰石华;文石华;铁华
flower of salt 盐析作用
flower of sulphur 硫华
flower of trough 花盆;花缸;花槽
flower of winds 海图上风向花
flower of zinc 锌华
flow erosion 流动浸蚀
flower period 花期
flower piece 花朵装饰;花的装饰布景
flower pot 喇叭皮带线盆;花盆
flower pot bell shape 复线金钟盆
flower pot for narcissus 长方水仙盆
flower production 花卉生产
flower receptacle with several mouths 多嘴花囊
flowers and plants 花卉
flower shaped ornament 花型纹饰
flower shelf 花架
flower shoot 花枝条
flowers of salt 盐花
flowers of sulphur 硫华
flowers of zinc 锌华;氧化锌
flower stalk 花序梗;花柄
flower terrace 花坛
flower tub 花盆;花罐
flower vase 花瓶
flower window 花饰窗;花窗;用花装饰的窗子
flower work 花饰;花环
flow expansion 气流膨胀;水流扩张;水流扩散
flow expansion area 水流开展区
flow factor 流量因数
flow failure 流动破坏
flow fan 排风扇;风扇
flow feeder 供液机;流量喂料机
flow field 流线谱;流场
flow field flow fractionation 流动场流动分级
flow field simulation 流场模拟
flow figure 应变图
flow filament 流线;流束
flow filter 水流过滤装置
flow fluctuation 流量波动;水流起伏;水流脉动
flow fold 流状褶皱
flow force 水流力
flow forecast 流量预报;径流预报
flow form 流型;流态;流动形态
flow forming 旋压成型法;流动旋压
flow formula 流量公式;流量计算公式
flow frequency 流量频率

flow frequency curve 流量频率曲线
flow friction 流动摩阻;流动摩擦;水流(摩)阻力;水流摩擦
flow friction characteristic 流阻系数;流动摩擦系数;流动摩擦特性
flow from a pump 泵的排量
flow function 流函数
flow gas velocity 气流速度
flow gate 自喷闸门;内浇口
flow ga(u)ge 水表;水速仪;流量指示计;流量计;测流规
flow ga(u)ging 流量测定;测流
flow ga(u)ging station 测流站
flow ga(u)ging weir 测流堰
flow geometry 水流几何学
flow glass 流动观察玻璃
flow governor 流量调节器
flow gradient 流量梯度;泄水道坡度;水流坡度
flow graph 信号流图;流向图;流量曲线;流动图
flow graph algorithm 流图算法
flow graph analysis 流向图分析
flow graph of underground river in karst area 岩溶暗河流量过程曲线
flow graph reducibility 流向图可约性
flow graph schematic model 流程图图解模型
flow-guiding screen 导流屏
flow-guiding structure 导流建筑物
flow gun 液流喷射器
flow harden 冷扎硬化;冷变形硬化
flow head 自喷井口装置;井口自喷装置
flow hole 流液洞
flow homogenizer 流量均化器
flow hydrograph 流量水文过程线;流量过程(曲)线
flow ice 流动冰块;流冰
flow improver 流动改进剂
flow in 流入
flow inclination 流倾角;水流坡降
flow in continuum 连续流(动)
flow index 流性指数;流度指数;流动指数
flow indication 流量指示
flow indicator 流速计;流量指示器;流量指标;流量计;进料量指示器
flow-induced oscillation 水流引起的振荡
flow-induced vibration 流激振动;流动引起的振动
flowing 流动的
flowing artesian 自流承压水
flowing artesian well 自流井
flowing avalanche 流动崩坝;滑动崩塌
flowing a well 开井
flowing bore 自流井
flowing borehole 自流钻孔;流水钻成井孔
flowing bottom hole pressure 井底动压力
flowing by head 间歇自喷
flowing chromatogram 流动色谱图
flowing clay 流土
flowing colo(u)r 流色;流动比色
flowing colo(u)rimeter 流动比色计
flowing concrete 流动混凝土;流态混凝土
flowing contours 流线型
flowing deformation 流变
flowing direction of subsurface water 地下水的流向
flowing end 取料端
flowing-film concentration 层流薄膜选矿法
flowing full 满流
flowing furnace 熔化炉
flowing gas casting 气流浇铸
flowing gas factor 自喷油气比
flowing gas-oil ratio 自喷油气比
flowing geyser 无定期喷泉;间歇泉;(非周期性的)间歇喷泉
flowing glaze 流釉
flowing ground 流动性地基
flowing irrigation method 淹灌法
flowing juice state 流动果汁状
flowing life 井的自喷期
flowing limit 流限
flowing line 流线型;排水管线;出水管线;出水管道
flowing liquid colo(u)rimetric detector 流液比色检测器
flowing method 流涂法
flowing mouth 出水口
flowing of well 井喷
flowing plan form 流动平面图形

flowing plant 空压机站;输送站
flowing power 流动性
flowing pressure 流水压力;动水压(力);自喷压力;流动压力;喷出压力
flowing-pressure gradient 液压梯度
flowing-pressure survey 测压
flowing production 自喷生产
flowing production rate 自喷产量
flowing programming 流动程序
flowing property 流动性
flowing resources 可再生资源;可恢复资源;可更新资源;可复用资源
flowing rock formation 塑性岩层
flowing sand 流沙(也称流砂)
flowing sand microflocculation filtration 流沙微絮凝过滤
flowing sheet water 地表径流(量)
flowing soil 泥流
flowing space 流动空间
flowing spring 自流泉
flowing storage 产流蓄水量
flowing style 流线型
flowing surface water 地表径流(量)
flowing suspension 流动悬浮物
flowing-temperature factor 流动温度因数
flowing test 放喷
flowing-through chamber 穿流室;通流室
flowing-through channel 流通渠道;连通槽(沟道窘井底部)
flowing-through period 过流时间;流过时间
flowing-through pipe 通气管
flowing-through time 过流时间;流过时间
flowing tide 涨潮
flowing time in grit chamber 沉沙池流行时间
flowing tracery 流线型窗花格;曲线窗花格;气流窗花格
flowing tubing head pressure 自喷井管头压力
flowing varnish 流涂清漆
flowing water 流水;流动水;活水
flowing water environment 流水环境
flowing water well 自流井
flowing wave 随波
flowing well 自流井;自喷(油)井
flowing zone test 自喷层测试
flow injection 流动注射
flow injection analysis 流动注射分析法
flow injection analyzer 流动注射分析仪
flow injection atomic absorption spectrometry 流动注射原子吸收光谱法
flow injection catalytic spectrophotometry 流动注射催化分光光度法
flow inlet 供液口;供浆管
flow in open air 明流
flow instability 流动不稳定性
flow instrument 流量仪;流量表
flow integrating system 流量积分系统
flow intensity 水流强度
flow intensity index 水流强度指数
flow in three-dimension 三维流(动)
flow into the market places 进入消费领域
flow irregularity 流动不均匀性;流动不规则性;水流不规则性
flow kentic energy 水流动能
flow landslide 流动塌方
flow layer 流动层;流层
flow level(l)ing 匀饰性;流展性;流平性;均涂性
flow limit 液限;流动屈服值;流动极限;屈服点;塑性流动值
flow limiting 流量限制
flow-limiting nozzle 限流嘴
flow-limiting passage 限流通路
flow line 流(向)线;流纹;流送管;流水作业;流动线;流程(图连);变形流线;晶粒滑移线;交通流线;熔缝;水流线;底流线;出油管(线);供水路;流水(作业)线;蠕变曲线
flowline aqueduct 渡槽;非满流高架水管;流线渡槽
flowline assembly 流水作业线装配
flowline conveyer 流线输送机
flowline conveyer method 流水作业输送法;流线型输送法;流线输送法
flowline map 流向线图
flowline method 流水作业法;动线法
flowline plan 流线图(谱)
flowline pressure 送水管压力;输油管压力;流管压力
flowline procedure 流水作业法
flowline production 流水作业(法);流水线生产
flow load 流动荷载
flow loss 流动损失
flow machine 液流供料机
flow map 动线地图
flow mark 皱皮;流线印痕;流纹;流痕;荒痕;波纹;波流痕
flow mark karst shape 流痕类
flow mass 流量累积;累计排水量;累积流量;质量流量
flow mass conditioning 量调节
flow mass curve 累计流量曲线;流量累积曲线;流量—质量曲线
flow matrix 流动矩阵;流程矩阵
flow meadow 水草地
flow measurement 流量量测;流量测量;流量测定;流动测量;测流
flow measurement by current meter 流速仪测流(法)
flow measurement during freezing 冰期测流
flow measurement weir 量水堰
flow measuring device 流量计;测流设备;测流装置
flow measuring float 测流浮子;测流浮标
flow measuring flume 流量测验槽;流量测量槽;测流(水)槽
flow measuring instrument 流量测量仪表
flow measuring probe 流量测量器
flow measuring site 测流现场
flow measuring structure 测流建筑物;测流构筑物
flow measuring unit 流量测量器
flow mechanics 流体(动)力学;流动力学;流动机构;水流力学
flow mechanism 流动机理
flow medium 流体介质
flow meter 燃气表;流速仪;流量系数;流量计;流量表;计量表(指仪表);流速计
flow meter and transmitter 水流计及变送器
flow metering valve 流量控制阀
flow meter pacing system 流量计调整系统
flow meter readout 流量计读出器
flow method 流水作业法;流动法
flow mixer 流体混合器;流动混合器;连续混砂机
flow mixing 流动混合
flow model 流线型模式;流动模型;气流模型;水流模型
flow modeling 水流模拟
flow model test 水流模型试验
flow modifier 流动改性剂
flow momentum 水流动量
flow momentum equation 水流动量方程
flow monitor 流量监测器
flow motion 水流运动
flow mo(u)lding 传道模压法
flow movement 水流运动
flow naturally 自然喷出
flow net 流(水)网;流动网;水网;渗流网
flow net analysis 流网分析
flow net concept 流网概念
flow net method 流网法
flow net plan 平面流网图
flow net profile 剖面流网图
flow net solution 流网解
flow network 流程网络;流路网;流动网络
flow noise 流动噪声;水流噪声
flow noise in duct 管道噪声
flow noise spectrum 流动噪声谱
flow non-uniformity 气流不均匀性;水流不均匀性
flow nozzle 管嘴;流量(计)喷嘴;流量测量喷嘴;测流嘴
flow nozzle meter 管嘴流量计;管嘴水计;流量管嘴水表;流量喷管水表
flow number 流速
flow observation 流谱观察(气流);流量观察;水流观测
flow observation of surface water 地表水流量观测
flow obstacle 流动障碍(物)
flow obstruction cover 水流干扰覆盖层
flow of benefits 效益流程
flow of capital 资本流动
flow of cash 现金流动
flow of catchment 流域径流;排水
flow of concessional fund 优惠性资金流入
flow of control 控制权流向;控制流向
flow of cost 成本流转;费用流程
flow of curves 曲线流
flow of dissolved matter 溶质径流
flow of dissolved substance 溶质径流
flow of electron 电子移通
flow of energy 能流
flow of event 事件流
flow off 流(出);径流;溢流(口);流走
flow of fluid 流体流动
flow of forces 力的流线
flow of fund analysis 资金流量分析
flow of funds 资金流转;资金流通;基金流量
flow of funds accounting 货币流量会计
flow-off volume 流失量
flow of gas 气流
flow of goods traffic 货流
flow of ground 土地流动;土的塑性变形;土的隆起;表径流
flow of groundwater 地下水流量
flow of heat 热流
flow of high Froude number 高傅汝德数水流
flow of income 收入流量
flow of investment 投资量
flow of investment fund 投资资金的流量
flow of liquid 液体流动
flow of metal 金属流变;金属流纹;金属变形
flow of momentum 动量变化
flow of operations 营业流程
flow of pedestrians 行人流;人流
flow of personnel 人才流动
flow of rail head 轨头肥边
flow of resources 资源流动
flow of sap 液流
flow of seepage 渗流
flow of services 服务流量
flow of solid matter 固体流动;固体径流;泥沙径流;输流
flow of steams 流动
flow of stresses 应力流线
flow of structure 流状构造
flow of the slag 熔渣流动性
flow of traffic 交通量;船流(量);车流(量)
flow of veneer 胶合板带饰
flow of water 水流
flow of work 工作流程
flow operation 操作过程
flow orifice 孔板流量计
flow out 流出;外流
flow out diagram 流出量图;出流曲线图
flow out in shape of stream 呈股流出
flow outlet 流出
flow outside boundary 边界外流动
flow out state of spring 泉水流出状态
flow over 横流
flow over a body 绕流
flow over weir 堰流
flow parameter 水流参数
flow partly full 半满流
flow passage 流管;流道;渠道
flow passage of spring 泉水流出通道
flow pass reservoir 水库径流
flow path 流径;流线;流迹;流动(通)道;流程;渗流路线;渗流路径
flow path of flood and ebb 涨落潮流路
flow pattern 流型;流线谱;流态;径流运动方式;气流模式;水流形式
flow pattern of the atmosphere 大气流型
flow per minute 每分钟流量
flow phase 流动相
flow phenomenon 塑流现象
flow pipe 管嘴;供水管;循溢水管;流管;送水管;出水管;分配管(道)
flow pipe of heat supply 供热水管
flow plan 输送线路图
flow plane 流动平面;气流平面;水流(平)面
flow plasticity 流动塑性
flow plug flumes 自喷油嘴
flow point 液化点;流点;开始软化温度
flow-point test 流点试验
flow potential 水流势能
flow powder 流粉
flow pressure diagram 流量压力曲线
flow pressure response 流体压力反应

flow process 流动过程;滴料供料法
flow process chart 工艺流程图;工艺过程卡(片);流转过程图;流动程序图;流程(程序)图;加工流程图
flow process diagram 过程流程图;流程图
flow process line 流水作业线;流水生产线
flow process of authority and responsibility 权责流动过程
flow process of construction 流水作业施工
flow production 流水作业的大量生产;流水作业
flow production line 流水生产线
flow profile 水流纵剖面(图)
flow profile in open channel 明槽水面线
flow programmer panel 程序变流盘
flow programming 程序变流
flow promoter 涂层和易剂(粉光抹平掺和剂);增流剂;流动性促进剂
flow property 流动性质
flow proportional counter 流量比例曲线
flow pulsation 流量脉动;水流脉动
flow quality 流动特性;气流品质
flow quantity 流量
flow range 流动范围
flow rate 流量;移动率;流速;流(动速)率;排量;出料率;比移值
flow rate classify of spring 温泉的流量分类
flow rate display 流率显示
flow rate equation 流率方程
flow rate fluctuation 流量起伏
flow rate formula in unit length 单位长度流量公式
flow rate formula per unit width 单宽流量公式
flow rate impulse 流速脉冲;调整流量技术
flow rate indicator 流量指示器;流速指示器;流率指示器
flow rate measurement 流率测量
flow rate of carrier gas 载气流速
flow rate per second 秒流速;每秒流量
flow rate transmitter 流量传送器
flow rating curve 流量率定曲线;流量关系曲线;水位—流量关系曲线
flow rating pressure 流量额定压力
flow ratio 流量混合比;流量比
flow ratio control 流量比控制
flow ratio controller 流量比率控制器
flowrator 转子流量计;流量计;浮标式流量计;变截面流量计
flow reactor 连续(式)反应器;水流反应器
flow recession curve 亏水曲线;退水曲线
flow record 流量记录
flow recorder 流量记录器;自记流量计;流量记录仪;流量表
flow recording controller 流量记录控制器
flow reduction device 节水器;节水装置
flow reduction system 节水系统
flow regime 流(动状)态;水流状态;水流状况;水流体系;水流情势;水(流)情(况);水流动态
flow regime classification 熔体动态分类
flow regulating 流量调节
flow regulating device 流量调节装置
flow regulating valve 流量调节阀;节流阀;调节阀
flow regulation 流量调节;水流调节;水流调节
flow regulation cock 流量调节龙头
flow regulator 流量调节装置;流量调节器;水流调节器
flow relation 相应流量关系
flow relay 流通继电器;流动继电器
flow resistance 流(动)阻(力);抗流阻;水流阻力;水力摩阻;电流阻力
flow resistance bed 稳定河床
flow resistance factor 水流阻力系数
flow resistant bed 坚固河床
flow resource 流转资源
flow reversibility 流向可逆性
flow rock 流岩
flow roll 流滚
flow route 流通路线
flow route of passengers and baggages 旅客及行旅流程
flow routing 流量演算;河槽洪水演进
flow ruler 流动法则
flow scheme 工艺流程图;流程
flow section 过水断面;流动截面(积)
flow-sensing unit 流动传感器
flow sensor 流量传感器

flow separation 流动分离;径流分割;气流离体;气流分离;水流分离;边界层分离
flow separation phenomenon 水流分离现象
flow separator 气流分离器
flow sheet 作业图;工艺流程图;流向图;流程图;流程表;生产过程图;生产程序图;程序(方)框图;操作程序图
flow sheet of activated sludge process 活性污泥(法)工艺流程
flow-sheet of mineral dressing 选矿流程
flow side 流动侧面
flow similarity criterion 水流相似准则
flow simulation 流量系列模拟;径流模拟;气流模拟;水流模拟
flow slide 流动滑坡;流动滑块;泥流型滑坡;泥流;塑流型滑坡;崩泻
flow snubber 限流器
flow soldering 流体焊接;射流焊接;波峰钎焊
flow solder method 滚动焊剂法
flow specification 流说明;处理过程明细
flow speed controller 流速度控制器
flow splitting device 分流器
flow splitting system 分流装置
flow stability 流态稳定性;流动稳定性;水流稳定性
flow stage relation 水位流量关系
flow state 流动状态
flow statement 流程表
flow state of groundwater 地下水流态
flowstone 流石类;流石
flow stowing 自流充填(法);水沙充填;水力充填
flow stream 水流
flow streamline 流线;水流线
flow strength 流速水头;流动强度;水流强度;水流能量
flow stress 塑性变形应力;塑流应力;流动应力;流变应力;屈服应力
flow string 自喷管柱;采油管柱;采油管
flow structure 流状构造;流纹构造;流动构造;水流结构
flow structure in river 河道水流结构
flow summarizer 流量累积器
flow summation curve 流量累积曲线
flow superposition method 流量叠加法
flow surface 过水面;流动面
flow survey 流动测量
flow survey of river 河水流量测量
flow survey plane 测流(平)面
flow switch 流量开关;气流换向器;水流开关
flow system 流水作业方式;流动系统;水流系统
flow table 流盘;振动台;流水槽;流动台;流程表;跳桌;稠度试验台
flow table apparatus 流动稠度试验台
flow table test 流动台试验;跳桌流动试验;稠度试验台;稠度台试验
flow table with cone 带锥体的流动稠度试验台
flow tank 储油罐;沉降箱
flow technology 流式技术
flow temperature 流动温度;流点温度;倾点温度
flow temperature of ash T three 灰流动温度
flow temperature point 流动温度点
flow test 流通试验;流量试验;流度试验;(混凝土的)流动性试验;流动(度)试验;流动测试;碾压试验;倾动试验
flow test connection 水流试验接头
flow tester 流动性试验机
flow test of concrete 混凝土流动性试验
flow texture 流状结构
flow-through 溜槽(混凝土);滑槽(混凝土);流过
flow-through accounting 实缴税收会计核算
flow-through bioassay 流水式生物测定;连续流动生物测定;连续流动活体鉴定
flow-through branch 支管水流
flow-through chamber 流过室
flow-through colo(u)rimeter 流通式比色计
flow-through curve 水流动曲线
flow-through detector 直通型检测器
flow-through electrophoresis 流通电泳(法)
flow-through ionophoresis 流通离子电泳
flow-through method 跨期所得税当期计入纳税法
flow-through period 水流流动时间;流过时间
flow-through rate 流过率
flow-through run 主管水流
flow-through system 径流系统
flow-through test 流水式试验

flow-through time 流过时间
flow-through toxicity test 流水式毒性试验
flow time 排放时间;流动时间
flow time of viscosity 黏度流出时间
flow tracer 信息流跟踪程序
flow tracer routine 追踪程序
flow tracing 流量示踪
flow tractive force 水流推移力
flow transition 明满过渡流
flow transmitter 流量传送仪;流量传感器
flow trough 溜槽(混凝土);滑槽(混凝土)
flow tube 流束;流管;测流管
flow turbulence 流体扰动;气流紊流度;水流紊动
flow turbulence condition 水流紊动条件
flow turbulence phenomenon 水流紊动现象
flow turbulent phenomenon 水流紊动现象
flow turning 变薄旋压
flow-type calorifier 流过式加热器;直流式加热器;连续流式热风炉;连续流式加热器
flow-type feeder 恒压流量给料器
flow-type gas water heater 直流式煤气热水器
flow-type water heater 直流式热水器
flow ultramicroscope 流动超显微镜
flow under pressure 压力流
flow under sluice gate 闸孔出流
flow uniformity 流动均匀性;气流均匀性;水流均匀性
flow uniqueness 水流均一性
flow value 流值;流动值
flow valve 流量开关
flow variability 流量变率
flow variable 流量变量
flow variation 排放改变;流量变化
flow variator 气流调节器
flow velocity 流速;流动速度;水流速度
flow-velocity component 流速分量
flow-velocity curve 流速曲线
flow-velocity measurement 流速测量
flow-velocity of drill rod 钻杆内流速
flow-velocity of subsurface water 地下水的流速
flow-velocity of underground river 地下暗河流速
flow visco(si)meter 流量式黏度计
flow visualization 流动显形;流动显示;流场显示
flow volume and distribution figure 流量分布图
flow washer 洗衣机;洗涤机
flow water 排出水;流动水
flow-water quality coupled model 流量水质耦合模型
flow-weighted average loading 流量加权平均负荷
flow welding 铸焊;流注焊接;流体焊接;浇焊
flow well 自流井
flow width 水流宽度
flow with hyper-concentration of sediment 高含沙水流
flow with water table 潜水流;无压地下水流;非承压(地下)水流
flow wrinkle 流动皱纹
flow zone 流动区
Flrida clay 漂白土
flschesserite 硒金银矿
fluate 氟化物(防止建筑材料表面氧化)
fluate coat 氟硅酸盐涂层
fluate hardener for gypsum 用作石膏硬化剂的氟化物
fluate treatment 氟硅酸盐处理
fluavil 固塔树脂
flucan 脉壁黏土
fluckite 砷氢锰钙石
flucloxacillin 氟氯恶西林
flucrylate 氟氰丙烯酯
fluctrating electric(al) field 起伏电场
fluctuate 波动
fluctuating acceleration 纵向加速度;变加速度
fluctuating backwater area 回水变动区;变动回水区
fluctuating backwater area of reservoir 水库回水变动区
fluctuating coefficient of goods flow 货流波动系数
fluctuating coefficient of passenger flow 客流波动系数
fluctuating component 起伏成分
fluctuating current 脉动电流;波动电流
fluctuating data 起伏数据;不规则输入数据
fluctuating demand 波动需求

fluctuating echo 起伏回波
fluctuating effective rate of return 波动实际收益率;变动的实际盈利率;变动的实际收益率
fluctuating electric(al)field 变动电场
fluctuating(exchange)rate 波动汇率
fluctuating flow 脉动水流;起伏流动;波动水流;波动流动
fluctuating flow rate 波动流量
fluctuating force 脉动力;起伏力
fluctuating frequency 起伏频率;波动频率
fluctuating income 波动收入
fluctuating interest rate 波动利率
fluctuating light beam 脉动光束
fluctuating load 不稳定荷载;不稳定负载;脉动荷载;脉动负载;起伏负荷;波动荷载;波动负载;变化载荷;变动载荷;变动荷载;变动负载
fluctuating loudness 不稳定响度;波动响度
fluctuating market 波动行市;变动的行市
fluctuating nappe 波动水舌
fluctuating par 变动平价
fluctuating power 振荡功率;波动功率
fluctuating pressure 脉动压力;波动压力
fluctuating pressure distribution 脉动压力分布
fluctuating price 浮动价格
fluctuating price contract 波动价格合同;变动价格合同
fluctuating rate 波动利率;波动汇价
fluctuating ratio 波动率;变动率
fluctuating reach 游荡性河段
fluctuating signal 脉动信号;起伏信号
fluctuating stress 交变应力;脉动应力;波动应力;变化应力
fluctuating synchronism 同步性起伏
fluctuating target detection 起伏目标探测
fluctuating temperature 脉动温度;波动温度
fluctuating tensile load 脉动拉伸荷载
fluctuating value 起伏值;波动值
fluctuating variation 参差变异
fluctuating velocity 脉动流速
fluctuation 涨落;增减;起伏现象;起伏;市价变动;升降;波动;变动
fluctuational belt of water table 地下水位变动带
fluctuation belt 波动带;变动区
fluctuation belt of water table 地下水变动带;水位变动带
fluctuation clause 波动条款
fluctuation correction factor 涨落校正因素
fluctuation damping parameter 起伏阻尼参数
fluctuation data 起伏数据
fluctuation-dissipation theorem 起伏散逸定理
fluctuation dynamics 波涛动力学;波动动力学
fluctuation echo 起伏回波
fluctuation effect 波动效应
fluctuation equivalent power 起伏等值功率
fluctuation in atmospheric(al)pressure 气压升降
fluctuation income 波动收入
fluctuation in discharge 流量变化
fluctuation in exchange 汇兑涨落;汇兑变动
fluctuation in price 价格涨落;价格变动;物价波动
fluctuation in stage 水位涨落;水位变动
fluctuation in wage rates 工资浮动
fluctuation margin 波动界限;波动幅度
fluctuation noise 浮动噪声;起伏噪声;无规(则)噪声
fluctuation noise level 起伏噪声水平
fluctuation of climate 天气变动
fluctuation of current 电流波动
fluctuation of density 密度起伏
fluctuation of discount rate 贴现率的波动
fluctuation of energy 能量变化
fluctuation of glacier 冰川进退;冰川变动
fluctuation of groundwater level 地下水位变幅
fluctuation of level 液面升降
fluctuation of load 更迭荷载;负载波动;负荷波动
fluctuation of metal line 玻璃液面波动
fluctuation of piezometric surface 测压水位
fluctuation of pressure 压力波动
fluctuation of pressure in boundary layer 附面层压力起伏
fluctuation of river level 河流水位变化
fluctuation of sea level 海平面变动
fluctuation of service 运转不稳定;运行不稳定;操作不稳定
fluctuation of speed 转速波动;速度增减
fluctuation of stock 资源波动;储量变化
fluctuation of temperature 温度(的)波动;温度不稳定
fluctuation of water-level 水位变化
fluctuation of water quality 水质波动;水质变化;水质变动
fluctuation of water table 潜水位波动;潜水面升降;地下水平面升降
fluctuation par 波动平价
fluctuation period 脉动周期
fluctuation quantity 波动量
fluctuation range 波动幅度
fluctuation rate 起伏率;波动率
fluctuation ratio 波动比;波动率;波动比;变动率
fluctuation reserve 物价变动
fluctuation spectrum 涨落谱
fluctuation spread 波动幅度
fluctuations trade 贸易波动
fluctuation test 彷徨变异测试;变异反应试验
fluctuation theory 涨落理论;起伏理论
fluctuation value of pumping discharge 抽水孔流量波动值
fluctuation value of water table 抽水孔水位波动值
fluctuation velocity 脉动速度;脉动流速;涡动速度
fluctuation voltage 起伏电压;波动电压
fluctuation within a narrow range 小幅度波动
fluctuation zone 变动区
flue 炉胆;暖气管;气体管
flue analyzer 废气分析器
flue arch 烟道碹拱;烟道碹;烟道拱
flue area 烟道面积
flue auger 烟道通条
flue baffler 烟气折流器;烟道地板;烟道挡板;折流板(烟道)
flue bend 烟道弯管
flue block 烟道座;烟道砖;烟道预制块;烟道砌块
flue blower 吹灰机
flue boiler 燃烧管锅炉
flue brick 烟道砖
flue bridge 烟通耐火拱顶;烟道耐火拱顶;耐火砖拱;通道拱顶
flue brush 烟道刷
flue cement 转窑水泥飞灰
flue cinder 均热炉渣
flue collar 烟道接口;烟道套筒;(锅炉上的)烟道接头;烟道箍
flue collector 烟气捕收管;主烟道;主烟囱
flue connection 烟道连接
flue-cured tobacco 烟熏烟
flue curing 烟熏干燥法;热气烤干
flue damper 烟道闸;烟道(调节)挡板;烟囱闸板;调风闸;风挡
flue door 烟道门
flued gas appliance 半密闭式燃(气用)具
flued opening 翻边开孔
flue dryer 烟道干燥器
flue duct 烟尘
flue dust 烟道尘;烟(道)灰;烟囱灰;转窑水泥飞灰;管内烟灰
flue dust chamber 烟气降尘室
flue dust collector 烟气除尘器;烟道除尘器;烟灰收集器
flue dust pocket 烟道沉尘室
flue dust retainer 烟道吸尘装置;集灰斗
flue-dust separator 烟灰分离器
flue effect 抽吸作用;烟囱效应
flue exhauster 烟道排风机
flue-fed apartment-type incinerator 烟道进料型公寓焚化炉
flue-fed incinerator 烟道进料焚化炉
flue for gas appliance 煤气设备的烟道
flue furnace 火管燃烧室
flue gas 烟气;烟道气;烟道排气
flue gas analysing apparatus 烟道气分析装置
flue gas analysis 烟气分析;烟道气分析
flue gas analysis meter 烟道气分析器
flue gas analyzer 烟气分析仪;烟气分析器;炉烟分析器
flue gas baffle 烟气挡板
flue gas blower 烟道气鼓风机
flue gas boiler 烟道气锅炉
flue gas chamber 烟气室
flue gas cleaner 烟气净化器
flue gas cleaning 炉气净化
flue gas cleaning system 烟道气净化系统
flue gas condensate 烟道气的凝结物
flue gas conditioning 烟气调质
flue gas cooling 烟道气的冷却;废气冷却
flue gas desulfurization 烟气脱硫;烟道脱硫
flue gas desulfurization by ammonia absorption process 氨吸收法烟气脱硫
flue gas desulfurization by ammonium phosphate process 磷铵肥法烟气脱硫
flue gas desulfurization facility 烟道气脱硫装置
flue gas desulfurization process 烟道气脱硫过程
flue gas desulfurization system 烟气脱硫系统;烟道气脱硫系统
flue gas desulfurization unit 烟气脱硫装置
flue gas desulfurization with aluminium subsulfate 碱式硫酸铝法烟气脱硫
flue gas desulfurization with lime and/or limestone 石灰/石灰石法烟气脱硫
flue gas desulfurization with sodium citrate 柠檬酸钠法烟气脱硫
flue gas desulurization with magnesium oxide 氧化镁法烟气脱硫
flue gas dust collector 废气集尘器
flue gas dust removal 烟道煤灰的清除;废气除尘
flue gas exhauster 抽烟气机
flue gas explosion 烟气爆炸
flue gas extinguisher system 烟气灭火系统
flue gas fan 烟气引风机
flue gas fire extinguisher 烟气防火装置
flue gas holder 烟道气气柜
flue gas leading 废气管
flue gas loss 烟道气热损失;排烟损失
flue gas purifier 烟气净化器
flue gas purifying installation 烟气净化装置
flue gas recirculation 烟气再循环
flue gas recirculation furnace 烟道气循环炉
flue gas re-heating 烟道气再热
flue gas return 烟道气循环
flue gas sampler 烟道气采样器
flue gas scrubber 烟道气洗涤器
flue gas system 烟气填充防火系统;烟气防火装置
flue-gas temperature 烟气温度
flue gas tester 烟道气检验仪
flue gas thermometer 烟道气温度计
flue gas treatment 烟道气处理
flue gas valve 烟道气开关;烟道气阀门
flue gas velocity 排烟速度
flue gas vent 烟道气的放空口
flue gas washer 烟道气清洗器
flue gas washing 烟气清洗
flue gas waste boiler 烟气废热锅炉
flue gas weight 烟道气重量
flue gathering 烟道集流
flue grouping 烟道组(合);烟道集合
flue header 烟管
flue heated hot bed 管道加热温床
flue heating 管道加热
flue heating surface 焰管受热面;炉胆受热面;火管受热面;烟道受热面
flueing 向里斜展(窗框侧壁,平面成八字形)
flueing soffit 弯曲的平底面(弯曲楼梯)
flueless 无烟道的
flueless appliance 无烟道燃具;无烟道燃具
flueless gas appliance 开放式燃(气用)具
flueless heater 无烟道采暖炉
flue liner 烟道衬砖;烟道衬块
flue lining 耐火管材;锅炉烟道衬;烟道内衬;烟道衬砖;烟道衬里;烟囱内衬
fluellite 氟铝石;氟磷石
flue loss 烟道损失
flue man 烟道工
fluence 注量;积分通量
fluency 纯熟度
fluent metal 液态金属
flue opening 烟道口;烟道孔
flue outlet 烟道出口
flue outlet draft 烟道出口抽力
flue pipe 烟筒(管);烟道气管;烟道
flue pipe brick 套接瓦筒;套接短管
flue resistance 烟道阻力
fluerics 射流学
flueries 纯射流技术
flue shutter 烟道闸板;烟囱闸板
flue side 烟道管壁

flue surface 烟道受热面
flue tile 烟道瓦筒
flue tube boiler 烟管锅炉
flue tube sheet 烟管板
flue utilization 烟气热量的利用
flue wall 烟道墙
flueway 烟道净空
fluffer 纤维分离机;搅打疏松器;疏解机
fluffing 起毛
fluffing machine 磨面机
fluffless rag 无绒毛布片
fluff point 疏松点
fluff up 翻松
fluffy 易碎的;木面起毛;绒毛状的;松散的;松软的
fluffy black 飞扬性炭黑
fluffy dust 毛状灰尘
fluffy mud 充气泥浆
fluffy point 疏松点
fluid 流质;流体
fluid acceptance rate 流体回灌率
fluid actuated 流体驱动的
fluid air preheater 流态化空气预热器
fluidal 流体的
fluidal arrangement 流状构造
fluidal cast 流动模
fluidal disposition 流动状态
fluidal structure 流动构造;流状构造;流纹构造
fluidal texture 流动结构;流状结构
fluid amplification 射流放大
fluid amplifier 流体放大器;射流放大器
fluid analyzing apparatus 流体分析仪
fluid-applied plastic roof coating 液浇塑胶屋面涂层
fluid area 水热区
fluid aspirator 吸液罩
fluid at rest 静止液体
fluid balance 液体平衡
fluid bath 液槽
fluid bearing 液压轴承;流体轴承
fluid bed dryer 液床烘干器;沸腾式干燥器
fluid bed drying 沸腾干燥法
fluid behavior 流体行为
fluid bituminous material 液体(地)沥青材料;液态沥青材料
fluid body 液体;流体
fluid boundary layer 流体边界层
fluid bowl 溶槽
fluid brake 流体闸
fluid bulk temperature 流体平均温度
fluid capacity 冲洗流量
fluid capital 流动资金;流动资本
fluid capture cross section 流体的俘获截面
fluid carbon 挥发性炭
fluid cargo 液体货(物)
fluid carrier 釉浆输送机
fluid catalyst 流化床催化剂
fluid catalytic cracker 流体床催化裂化设备
fluid catalytic cracking 流化床催化裂化
fluid cement grout 流态水泥砂浆;水泥浆
fluid chamber 液压油腔
fluid char adsorption process 活性炭流化吸附法
fluid chemistry 流体化学
fluid classification 流体分级
fluid clutch 液压离合器;液力耦合器;液力离合器;液力传动装置;流体离合器
fluid coal 流态煤粉;流化煤
fluid coefficient 流体系数
fluid coker 流化焦化器
fluid coking 流态化炼焦
fluid coking unit 液化焦化装置
fluid column 液柱;流体柱
fluid column roaster 流化柱焙烧炉
fluid combustion 液体燃料的燃烧
fluid compartment 液体空间;隔室;隔舱
fluid compass 液体罗经;湿罗盘
fluid-compressed 液态压缩
fluid-compressed steel 液态压缩钢
fluid compression process 液态压制法
fluid computer 流体计算机;射流计算机
fluid concrete 易流混凝土;流态混凝土
fluid conditioner 流体调节器
fluid conductivity 水力传导系数
fluid conductivity of well 井底流体渗流强度
fluid contact 流体接触面

fluid contacting apparatus 流体接触装置
fluid contacting unit 流体接触装置
fluid container 液体容器;流体容器;减震箱
fluid control 液压空气液压控制
fluid-controlled valve 液力控制阀;流体控制阀
fluid converter 液力变换器
fluid conveyer 流体输送机
fluid-cooled electrical machine 液冷电机
fluid-cooled winding 液冷绕组
fluid couplant 液体耦合剂
fluid coupling 液压联轴器;液力耦合器;液力联轴节;流体联轴节;流体连接
fluid-coupling noise 液压联轴节噪声;伴流噪声
fluid crystal display 液晶显示(器)
fluid culture 液体培养
fluid current 液体流;流体流
fluid current meter 液体流速计
fluid curtain 液体帘
fluid curve 液体曲线
fluid cut 液体冲蚀
fluid cycle 液体循环;液力循环
fluid cylinder 液压缸;泵缸
fluid cylinder liner 泵缸套
fluid damper 液压减振器;流体减振器
fluid damping 液体阻尼;流体节制
fluid deflection 流体折转角
fluid deicer 流体除冰器
fluid delivery 液体输送
fluid density 流体密度
fluid density meter 液体密度计;液体密度表;流体密度计
fluid die 液压模;液体模具
fluid differential pressure ga(u)ge 液体压差计;液体差压计
fluid director 导流器
fluid dirt level 液体脏污程度
fluid discharge 液体排出;液体流量
fluid discharge point 流体排放点
fluid displacement 流体驱替
fluid displacement tachometer 流体转速计
fluid distribution in an oil reservoir 油层内流体分布
fluid distributor 流体分配器
fluid ditch 液槽
fluid dram 液量打兰;液打兰(药量单位)
fluid drive 液压驱动;液压力;液压传动(器);液力传动(装置);流体驱动;流体(动力)传动
fluid(drive) coupling 液压联轴节
fluid drive fan 液动风扇
fluid-driven pump 液压传动泵
fluid drop 液滴
fluid dune 流动沙丘
fluid dynamic(al) behavior 流体动力特性
fluid dynamic(al) process 流体动力过程
fluid dynamic(al) research 流体动力研究
fluid dynamics 流体动力学
fluid dynamics of multiphase 多相流体力学
fluid echoless segmen 液性平段
fluid elastic instability 流体弹性不稳定
fluid electrode 液体电极
fluid element 流体元;流体微分体;流体单元
fluid emission rate 流体发射速度
fluid end of pump 泵的流体端
fluid energy mill 液力磨矿机;流能磨;气流粉磨机;喷射式磨机;液压磨矿机
fluid erosion 流体侵蚀
fluid erosion of matrix 液体对基底冲蚀
fluid extract 液体栲胶;流浸膏(剂)
fluid exudation 液体渗出
fluid-filled column 充水柱子
fluid-filled working chamber 液体工作室
fluid film 液体薄膜;液膜;润滑油膜
fluid film bearing 液膜轴承;流体膜轴承
fluid film lubrication 液膜润滑
fluid fire extinguishing 液体灭火器;液体灭火机
fluid floated gyroscope 液浮陀螺仪
fluid flow 液体流(量);液流;流体流(动)
fluid flow analogy 液流模拟;流体流动模拟;流动动力模拟;水动力模拟
fluid flow equation 液体流动方程
fluid flow indicator 液体流量指示器
fluid flow mechanics 流体力学
fluid flowmeter 液体流量计

fluid flow passage 流体通道
fluid flow pattern 液流图
fluid flow pump 液流泵
fluid flow regulator 流体流量调节器
fluid flux 流体通量
fluid flywheel 液压飞轮;液力飞轮;流体飞轮
fluid-flywheel clutch 液力耦合器
fluid force chemistry 流体力化学
fluid friction 液体摩擦;流体摩擦;黏性摩擦
fluid friction bearing 液体摩擦轴承
fluid friction brake 液体摩擦制动
fluid fuel 液体燃料;液态燃料
fluid fuelled reactor 液态燃料反应堆
fluid fuel reactor 流体燃料反应堆
fluid ga(u)ge 液体指示器
fluidgenic escape structure 流体逸出结构
fluidgenic inclusion 流体包(裹)体
fluidgenic percolator 流体渗透仪
fluidgenic permeability 流体渗透率
fluidgenic pollutant 流体污染物
fluid geometry 流体几何条件
fluid governor 液体调节器
fluid gravity flow 流体重力流
fluid handling 液体的送运
fluid head 液压头;液体压头;流体高差
fluid heating furnace 流体加热炉
fluidic 流控的;射流的
fluidic accelerometer 射流加速计;射流加速度计
fluidic amplifier 射流放大器
fluidic batching counter 射流间歇计数器
fluidic circuit 射流电路
fluidic control 射流控制
fluidic device 射流装置
fluidic element 射流元件
fluidic feedback 射流反馈
fluidic generator 射流发生器
fluidic hardware 射流器件
fluidic module 射流组件
fluidic phase detection circuit 射流相位检测回路
fluidic power 流体动力
fluidic pulse shaper 射流脉冲整形器
fluidic relay 射流继电器
fluidics 流体学;流控学;射流学;射流技术
fluidic sensor 射流传感器
fluidic stabilizer 射流稳定器
fluidic stepping motor 射流式步进电动机
fluidic switch 射流开关
fluidic system 射流系统
fluidic technology 流控技术
fluidic type hydro-percussive tool 射流式冲击器
fluidic volume flowmeter 射流体积流量计
fluidifier 流化剂;浆体流化剂;压力灌浆外加剂
fluidifying 流化
fluidimeter 黏度计;流度计;夹套式黏度杯
fluid immersion effect 流体浸入影响
fluid impendance 流体阻抗
fluid inclusion 液体包裹体;液包体;流体包(裹)体;气液包裹体
fluid inclusion determining pressure method 包体测压法
fluid index 流动指数
fluid influx 流体流入
fluiding flow 流体化流
fluid injection 流体注入
fluid insulation 液体绝缘
fluid interchange 流动换热
fluidisaton 流态化
fluidised bed 流沙地层;流化床;流动层
fluidised bed coating technique 流化床涂层工艺
fluidised bed cooler 流化床冷却器
fluidised bed furnace 流化床炉
fluidised bed jet mill 流化床喷射管
fluidised bed process 流化床法
fluidised bed reactor 流化床反应器
fluidised combustion 流化燃烧
fluidised incineration 流化焚烧
fluidised incinerator 流化焚烧炉;流动焚化炉
fluidity 流度;流动性;流动度
fluidity improver 流动性改进剂
fluidity index 流动性指数
fluidity meter 流度计
fluidity of decision situation 流动性决策形式
fluidity of liquid 液体流动性
fluidity of mortar 砂浆流动度

fluidity of the slag 熔渣流动性
fluidity-temperature relationship 流度温度关系
fluidity test 流动性试验
fluidizable particle size 可流态化的颗粒大小
fluidizaing reactor 沸腾焙烧炉
fluidization 流体化;流态化(作用);流化(作用)
fluidization cooler 沸腾冷却器
fluidization dryer 流态床烘干机;流态化烘干机;流态化干燥器
fluidization installation 流态化装置
fluidization number 流态化数
fluidization of coal 煤流态化
fluidization of solid 固体流态化
fluidization point 沸腾点
fluidization quality 流态化质量
fluidization roasting(process) 沸腾焙烧法
fluidization state 流态
fluidization technique 流化技术
fluidization tower 流化塔
fluidized absorption 流化吸附
fluidized activated carbon bed 流化活性炭床
fluidized area 流化区
fluidized bed 流态化床;流动床;流动层;浮动床【给】;沸腾床;沸腾层
fluidized bed adsorber 流化床吸附器
fluidized bed apparatus 流化床设备
fluidized bed biofilm 流化床生物膜
fluidized bed biofilm reactor 流化床生物膜反应器
fluidized bed bioreactor 流化床生物反应器
fluidized bed calciner 流态床分解炉
fluidized bed cement kiln 沸腾层水泥煅烧窑
fluidized bed coating 流化床涂装法;流化床涂层法
fluidized bed coking 流化床焦化
fluidized-bed combustion 流态化燃烧;沸腾燃烧;流化床燃烧
fluidized bed combustion process 流化床燃烧法
fluidized bed combustor 流化床燃烧装置
fluidized bed cooler 沸腾层冷却机
fluidized bed copper oxide process 氧化铜流化床法
fluidized bed dip coating 流化床浸涂;沸腾床浸涂
fluidized bed dipping 流化床浸涂法;沸腾床浸涂
fluidized bed dryer 流化床式干燥器;沸腾床干燥器;流态化烘干机;流态化干燥器
fluidized bed drying 流化床干燥法
fluidized bed electrolysis 流化床电解
fluidized bed firing furnace 流化床燃烧炉
fluidized bed furnace 流态砂浴炉;流(态)化炉;流化床炉;流动粒子炉
fluidized bed granulation 流态化造粒;流化床造粒
fluidized bed incinerator 流化床燃烧炉;流化床焚烧炉
fluidized bed iron exchange 流化床离子交换
fluidized-bed kiln 沸腾煅烧炉
fluidized bed mixer 流化床混合器;流化床混合机;气流混合器
fluidized bed painting 流化床涂装
fluidized bed pelletizer 流动层造粒机
fluidized bed photocatalytic reactor 流化床光催化反应器
fluidized bed plant 流化床设备
fluidized bed process 流化床过程;流化床法
fluidized bed reactor 流化燃料反应堆;流化床反应器;流化床反应堆
fluidized bed recovery process 流化床回收过程
fluidized bed roaster 流化床焙烧炉
fluidized bed roasting 流化床焙烧;沸腾层焙烧
fluidized bed roasting reaction 流化床焙烧反应
fluidized bed scrubber 流化床洗涤器
fluidized biofilm process 流化生物膜法
fluidized calcination 硫化煅烧
fluidized carbon bed 流化炭床
fluidized catalyst 流化催化剂
fluidized coal 流态煤粉
fluidized coating 流化涂装
fluidized column 流化塔
fluidized conveying 悬浮输送;流态化输送
fluidized conveyer 悬浮输送机;流态化输送机
fluidized discharge 流态化卸料
fluidized drier 流化干燥器
fluidized drying 流化干燥
fluidized electrochemical reactor 流化电化学反应器
fluidized flow 液化流;流体化流

fluidized flow deposit 流化流沉积
fluidized fluosolid roaster 沸腾床焙烧炉
fluidized furnace 流动粒子炉;流床式炉
fluidized gasification 流化床气化
fluidized gasification system 流化床气化系统
fluidized incineration 流化焚烧
fluidized incinerator 流化焚烧炉;流动焚化炉
fluidized material 流态化材料
fluidized material cooler 流态化物料冷却器
fluidized particle quenchine 流动粒子淬火
fluidized pneumatic conveyor[conveyer] 流态化气动输送机
fluidized purification 流化净化
fluidized reactor 流化燃料反应堆
fluidized sand 流态砂
fluidized sand bed furnace 流化砂床炉
fluidized solid 流化固体
fluidized solid kiln 沸腾层窑;流化层窑
fluidized solid reactor 流化固体反应堆
fluidized spray dryer 流化喷雾干燥器
fluidized state 流化状态
fluidized system 流化床系统
fluidized trough conveyer 流态化输送料槽
fluidize flour 气流输送粉末
fluidizer 流化装置
fluidizing agent 液化剂;增塑剂;流化介质;流化剂
fluidizing air 流态化空气
fluidizing air system 流化床送风系统
fluidizing box 液化箱;曝气箱
fluidizing calciner 流动煅烧炉(石灰)
fluidizing chamber 流体化室
fluidizing conveyer 空气输送斜槽
fluidizing cooler 流化式冷却器;沸腾冷却器
fluidizing dryer 流态床烘干机;流态化烘干机;流态化干燥器;流化式干燥器
fluidizing gas 流态化气体
fluidizing panel 多孔充气板
fluidizing reactor 流态化反应器
fluidizing reagent 流化剂
fluidizing system 流态化系统;射流装置
fluid jet 流体喷射
fluid jet stream 流体射流
fluid kinematics 流体动力学
fluid kinetics 流体动力学
fluid level 液平面;液面
fluid level control 液位控制
fluid level controller 液位控制器
fluid level ga(u)ge 液位指示器;液位计
fluid level ga(u)ge rod 液面测杆
fluid level indicator 液位指示器;液面指示器
fluid line 流体线
fluid liner 泵缸套
fluid link 液压连接压流网绞盘
fluid-link steering 液压转向
fluid load 流体负荷
fluid logic 流体逻辑;流控逻辑
fluid loss 流体流失
fluid loss additive 降失水剂;失水量添加剂(泥浆)
fluid loss agent 脱液剂
fluid loss reducing agent 失水量降低剂(泥浆)
fluid loss test 漏失量试验
fluid loss volume 失水量
fluid low 液体不足
fluid lubricant 液体润滑剂
fluid lubrication 液体润滑作用;液体润滑
fluid lubrication bearing 液体润滑轴承
fluid mass 液体质量;流体质量
fluid measure 液量
fluid measurement 流量测定
fluid mechanics 液体力学;流体力学
fluid mechanics principle 流体力学原理
fluid mechanism 流体机理
fluid medium 液体培养基;液体介质;质点;流体介质;流态介质
fluid meter 流体计量器;流度计
fluid metering 流量测定
fluid mosaic membrane 流体镶嵌膜
fluid mosaic model 液态镶嵌模型;流体镶嵌模型;流动镶嵌模型
fluid mosaic theory 液态镶嵌学说
fluid motion 流体运动
fluid motion equation 流体运动方程
fluid motor 液压马达;液力发动机;液动马达

fluid mud 浮泥
fluid mud layer 浮泥层
fluid name 液体名称
fluid network 流体网络
fluid network analyser 流体网络分析器
fluid nozzle 流体喷头
fluid nutrient medium 液体培养基
fluid oil 液态油;润滑油
fluidometer 流度计
fluid operated devices 流体操纵部件
fluid operated pump 液压泵;液力泵
fluid origin 铸造成的
fluid oscillation 流体振荡
fluid ounce 液(量)盎司
fluid outlet angle 液体流出角
fluid overpressure 流体超压
fluid packed pump 液体封闭器;液泵
fluid paraffin 液体石蜡
fluid particle 液体元粒子;流体质点
fluid passage 液道;水槽
fluid pellet system 液力散弹钻进方式
fluid penetration 液体渗透
fluid permeability 液体渗透率
fluid phase 液相;流体相
fluid pipeline 液体管道
fluid poison 液态毒物
fluid pollutant 液态污染物
fluid population 流动人口
fluid pore pressure 流体孔(隙)压力
fluid potential 流体势
fluid potential and field of force 流体的势和力场
fluid power 液压功率;液压;液体动力
fluid power drill 液压凿岩机
fluid power drive 液压驱动
fluid power motor 液压发动机;液力马达;液力电动机
fluid power system 液压驱动系统
fluid power transmission 液体变速装置;液体变速机;液力传动
fluid pressure 静水压力;液压系统压力;液压;液体压力;流体压力;静液压力
fluid pressure cell 液压元件
fluid pressure governor 液压调节器
fluid pressure-induced faulting 流体压力诱发的断层作用
fluid-pressure laminating 液压层合模塑
fluid pressure line 液压系统压力管路;液压管路
fluid pressure meter 流体压强计;流体压力计
fluid pressure motor 液压电(动)机
fluid pressure mo(u)lding 液压造型
fluid pressure operated jack 液压千斤顶
fluid pressure reducing valve 液压降低阀
fluid pressure warning device 液压安全装置
fluid pump 液压泵;液泵;流体泵
fluid pumping conveyer 液体泵吸输送器
fluid punch 液体凸模
fluid purification system 液体净化系统
fluid radiation 液体辐射
fluid rate 流速;冲洗液流速
fluid region 流体区域
fluid resin 松浆油
fluid resistance 流阻;流体阻力
fluid resistivity 流体电阻率
fluid-retaining structures 挡液体结构物
fluid rheology 流体流变学
fluid rubber 液体橡胶
fluid sampling 流体取样
fluid sampling apparatus 流体采样设备
fluid sampling bomb 流体取样弹
fluid saturation 液体饱和度;流体饱和度
fluid savings 流动储蓄
fluid seal 液封;流体封闭
fluid sealant 液体密封剂;密封胶
fluid seal gas holder 湿式储气柜
fluid sensor 液控传感器
fluid separation device 流体分离装置
fluid servo-motor 液压伺服马达
fluid site 水热区
fluid slag 流动渣
fluid slippage 液体漏失
fluid slurry 稀料浆;水分大的料浆
fluid soil 液态土
fluid-solid interface 液固界面
fluid solid kiln 流态化窑

fluid solid process 流态化法;沸腾法
fluid solid system 液固系统;流体固体系统
fluid sonolucent area 液性暗区
fluid sonolucent point 液性暗点
fluid space 液体空间
fluid speed meter 液体流速计
fluid start 液力起步
fluid state 流体状态
fluid state laser 液体激光器;流体激光器
fluid static pressure 静水压(力)
fluid static pressure meter 流体静压力计
fluid statics 流体静力学
fluid strata 流体层
fluid stream 液流
fluid stress 流体应力
fluid structure 流体结构
fluid structure interaction 流体结构相互作用
fluid supply suspension relay 断流继电器
fluid supply tank 液体供应箱
fluid surface 流体面
fluid suspension culture 液体悬浮培养
fluid swivel 液压旋转头
fluid-tamping 液压捣实(孔内爆破时)
fluid tank container 流体集装箱
fluid technique 流态化技术
fluid temperature 流体温度
fluid temperature range 流态时的温度范围
fluid-tight 不透水的;不漏流体的
fluid-tight contact 液密接触
fluid-tight joint 液密接头
fluid tightness 液体密封性
fluid ton 流体吨
fluid toxoid 液体类毒素
fluid transmission 液压传递;液力传动系;流体变速机
fluid transmission gear 液压传动齿轮;流体传动齿轮
fluid trap 流体圈闭
fluid unit 流化床设备
fluid valve 液流阀;水力阀
fluid valving 射流开关
fluid velocity 流动速度
fluid velocity of an underground river 暗河水流速
fluid velocity profile 液体流速分布图
fluid viscosity 流体的黏滞性
fluid viscosity ratio 流体的黏度比
fluid volume 液量;排量;冲洗液量;泵量
fluid volume meter 液量计
fluid wash 流体冲蚀
fluid water phase 水液相
fluid wave 液波
fluid wax 液体石蜡;液蜡
fluidway 冲洗槽
fluid wedge 液楔;流楔
fluid welding 流焊
fluid whirl 流体漩涡
fluid zone 水热区
fluigram(me) 立方厘米液体
fluing 向里斜展(窗框侧壁);厚墙窗洞斜边;窗洞八字做法
fluing arch 八字拱(前拱半径大于后拱口);斜面拱
fluing soffit 梯底背面
fluke 倒钩;锚爪
fluke anchor 开爪锚
fluke angle 锚爪转角
fluke to shank angle 锚爪与锚柄夹角
fluke turned anchor 折嘴锚(用于固定的锚碇)
fluking 乘风航逡
flume 斜槽;峡沟;流斜槽;流水槽;水滑道(运木材用);水槽
flume chute 斜槽
flume crossing 跨河渡槽;渡槽交叉口
flumedroxone 氟美烯酮
flume experiment 水槽试验
flume flow meter 槽式流量计
flume model 夹具模型;沟槽模型
flume test 水槽试验
flume wastewater 斜槽废水
flume with both side contraction 两侧束狭测流槽
flume with side contraction 侧向测流槽
flumina pilorum 毛流
flunkey 非熟练工人
fluoantigorite 氟叶蛇纹石
fluoborate 氟硼酸盐

fluoborate glass 氟硼酸玻璃
fluoboric acid 氟硼酸
fluoborite 氟硼镁石
fluocerite 氟铈矿
fluochrystile 氟钎蛇纹石
fluoform 氟仿
fluohydrocarbon 氟代烃
fluolite 松脂石
fluomethane 氟代甲烷
fluon 氟隆;聚四氟乙烯(树脂)
fluon bottle 聚四氟乙稀瓶
fluon seal 聚四氟乙烯密封
fluon-to-stainless steel 聚四氟乙稀不锈钢
fluophosphate glass 氟磷酸玻璃
fluoplatinic acid 氟铂酸
fluoracyl chloride 氟乙酰氯
fluorane 荧烷
fluorapatite 氟磷酸钙;氟磷灰石
fluorapatite crystal 氟磷酸钙晶体
fluorapatite laser 氟磷酸钙激光器
fluorapophyllite 氟鱼眼石
fluorate 氟化
fluorating 氟化
fluorbaryt 萤石
fluor borate glass 氟硼酸盐玻璃
fluor chrome arsenate phenol 防腐盐;氟铬砷酚合剂(木材防腐剂)
fluor crown 冕牌萤石玻璃;氟冕玻璃
fluor crown glass 氟冕光学玻璃
fluorecence spectrum 荧光光谱
fluorellestaadite 氟硅磷灰石
fluoremetry 荧光测定法
fluorene 芴
fluorescamine 胺荧
fluorescein(e) 荧光素;荧光黄
fluorescein(e) dye 荧光素染料
fluorescein(e) paper 荧光素试纸
fluorescein(e) test 荧光素试验
fluorescence 荧光(性)
fluorescence ability 发荧光能力
fluorescence alteration 荧光衰变
fluorescence analysis 荧光分析法;荧光分析
fluorescence band 荧光带
fluorescence bleach 荧光漂白
fluorescence branching ratio 荧光分支比
fluorescence causing substance 荧光产物
fluorescence centre 荧光中心
fluorescence chromatogram 荧光色谱(图)
fluorescence chromatography 荧光色谱法
fluorescence coating 荧光层
fluorescence colo(u)r 荧光颜色;荧光色
fluorescence colo(u)r of petroleum 石油的萤光色
fluorescence conversion efficiency 荧光转换效率
fluorescence cycle 荧光循环
fluorescence decay time 荧光余辉时间
fluorescence detector 荧光检测器
fluorescence determination 荧光测定
fluorescence discharge 荧光放电
fluorescence discharge lamp 荧光灯
fluorescence effect 荧光效应
fluorescence efficiency 荧光效率
fluorescence emission 荧光发射
fluorescence emission spectrum 荧光发射光谱
fluorescence excitation spectrum 荧光激发光谱
fluorescence improver 荧光增进剂;荧光促进剂
fluorescence indicator 荧光指示剂
fluorescence indicator analysis 荧光指示分析
fluorescence intensity 荧光强度
fluorescence-labeled water treatment agent 荧光标识处理剂
fluorescence light 荧光
fluorescence line width 荧光线宽
fluorescence method 荧光法
fluorescence microscope 荧光显微镜
fluorescence microscopy 荧光显微术;荧光显微法;荧光镜检法
fluorescence microwave double resonance 荧光微波双共振法
fluorescence peak 荧光峰
fluorescence photography 荧光照相术
fluorescence polarization 荧光偏振
fluorescence polarization microscope 荧光偏振显微镜
fluorescence quantum counter 荧光量子计数器

fluorescence quantum efficiency 荧光量子效率
fluorescence quenching method 荧光淬灭法
fluorescence radiation 荧光辐射
fluorescence reaction 荧光反应
fluorescence screen 荧光屏
fluorescence spectrography 荧光光谱法
fluorescence spectrometer 荧光分光计
fluorescence spectro photometer 荧光分光光度计
fluorescence spectrophotometry 荧光分光光度法
fluorescence spot-out method 荧光点滴法
fluorescence standard substance 荧光标准物
fluorescence state 荧光态
fluorescence stimulation 荧光放射增强
fluorescence strength 荧光强度
fluorescence survey 荧光测量方法
fluorescence thin layer 荧光薄层
fluorescence titration 荧光滴定
fluorescence titrimetric method 荧光点滴法;荧光滴定法
fluorescence X-ray counting 荧光 X 射线计数
fluorescence yield 荧光效应;荧光产额
fluorescent 荧光的
fluorescent additive 荧光添加剂
fluorescent analysis 荧光分析法;荧光分析
fluorescent analysis spectrum 荧光分析光谱
fluorescent anode 荧光阳极
fluorescent bacteria 荧光细菌
fluorescent bed-material tracer 荧光床料示踪剂
fluorescent bleach 荧光漂白
fluorescent body 荧光小体
fluorescent brightening agent 荧光增白剂
fluorescent center 荧光中心
fluorescent characteristic 荧光特征(曲线);荧光特性(曲线)
fluorescent characteristic X-ray 荧光标识 X 射线
fluorescent coating 发光涂层;荧光涂层
fluorescent colony 荧光菌落
fluorescent colo(u)r 荧光色材
fluorescent composition 荧光混合物
fluorescent compound 荧光化合物
fluorescent crystal 荧光晶体
fluorescent detection 荧光探伤
fluorescent digital display tube 荧光数码管
fluorescent display 荧光显示
fluorescent display device 荧光显示器件
fluorescent dyed sand 荧光染色砂
fluorescent effect 荧光效应
fluorescent effect correction 荧光效应修正
fluorescent electrolyte 荧光电解液
fluorescent enamel 荧光搪瓷
fluorescent energy 荧光能量
fluorescent face 荧光面
fluorescent fault detector 荧光探伤仪
fluorescent film 荧光膜
fluorescent film technique 荧光薄膜技术
fluorescent fittings 荧光灯装置
fluorescent fixture 荧光照明装置;荧光灯具
fluorescent flag 荧光指示器
fluorescent glass 荧光玻璃;发光玻璃
fluorescent glass detector 荧光玻璃探测器
fluorescent high-pressure mercury lamp 高压荧光汞灯
fluorescent illumination 荧光照明
fluorescent indicator 荧光指示器;荧光指示剂;荧光指示计
fluorescent indicator adsorption analysis 荧光指示剂吸附分析
fluorescent indicator adsorption method 荧光指示剂吸附法
fluorescent indicator adsorption technique 荧光指示剂吸附术
fluorescent ink 荧光油墨
fluorescent inspection 荧光探伤;荧光检验;荧光检查
fluorescent intensifying screen 荧光增感屏
fluorescent lamp 荧光灯;日光灯
fluorescent lamp ballast 荧光灯镇流器
fluorescent lamp ballast with emergency function 带应急功能的荧光灯镇流器
fluorescent lamp sealing machine 荧光灯封接机
fluorescent lamp stabilizer 荧光灯稳压器
fluorescent lamp starter 荧光灯启辉器
fluorescent layer 荧光层
fluorescent lifetime 荧光寿命

fluorescent light 日光灯;荧光灯
fluorescent light fixture 荧光照明设备
fluorescent light illuminator 荧光灯具
fluorescent lighting 荧光(灯)照明;日光灯照明
fluorescent lighting fixture 荧光照明设备;荧光灯照明设备
fluorescent lighting source 荧光照明光源
fluorescent light source 显微镜荧光灯源
fluorescent light trap 荧光诱蛾灯
fluorescent line 荧光谱线
fluorescent line width 荧光线宽
fluorescently labeled bacteria 荧光标记细菌
fluorescent magnetic inspection 荧光磁性检测
fluorescent magnetic particle inspection 磁性荧光法检查
fluorescent magnetic particle (powder) 荧光磁粉
fluorescent magnetic powder detector 荧光磁粉探伤机
fluorescent map 荧光地图
fluorescent material 荧光物质;荧光材料
fluorescent mercury discharge lamp 荧光汞放电灯;水银荧光灯
fluorescent mercury lamp 荧光汞灯;荧光水银灯;水银荧光灯
fluorescent metabolite 荧光代谢物
fluorescent method 荧光法
fluorescent microscope 荧光显微镜
fluorescent monomer 荧光单体
fluorescent noise generator 荧光噪声发生器
fluorescent oil 荧光油
fluorescent O-lamp 圆形荧光灯
fluorescent optics 荧光光学
fluorescent paint(ing) 荧光涂料;发光漆;荧光漆
fluorescent particle atmospheric tracer 荧光粒子大气示踪剂
fluorescent particle technique 荧光粒子技术
fluorescent particle tracer study 荧光质点示踪研究
fluorescent penetrant 荧光渗透剂
fluorescent penetrant method 荧光探伤法;荧光探测法
fluorescent-penetrate test 荧光穿透试验
fluorescent photography 荧光摄影
fluorescent photometer 荧光光度计
fluorescent pigment 荧光颜料
fluorescent plastics 荧光塑料
fluorescent plate 荧光板
fluorescent plate technique 荧光板技术
fluorescent polarization 荧光偏振
fluorescent polymer 荧光聚合物
fluorescent powder 荧光粉
fluorescent probe 荧光探头;荧光探测器;闪烁探测器
fluorescent probe technique 荧光探针技术
fluorescent pumping 荧光抽运
fluorescent quantum efficiency 荧光量子效率
fluorescent quenching 荧光淬火
fluorescent radiation 荧光性辐射;荧光辐射
fluorescent reagent 荧光试剂
fluorescent reflector lamp 反光荧光灯;荧光反光灯
fluorescent sand 荧光沙
fluorescent scale 荧光表盘
fluorescent scanning technique 荧光扫描技术
fluorescent scattering 荧光散射
fluorescent screen 荧光屏;荧光板
fluorescent show 荧光显示
fluorescent spectrometry 荧光光谱法;荧光分光光度法
fluorescent spectroscopy 荧光光谱法
fluorescent spectrum 荧光光谱
fluorescent staining 荧光染色法;荧光染色
fluorescent staining technique 荧光着色技术
fluorescent strip 荧光带
fluorescent substance 荧光物质;荧光体
fluorescent technique 荧光技术
fluorescent test 荧光探伤
fluorescent tracer 荧光示踪剂
fluorescent tracer technique 荧光示踪技术
fluorescent tracing 荧光示踪
fluorescent tracing unit 荧光示踪装置
fluorescent transition 荧光跃迁
fluorescent tube 荧光管;荧光灯(管);日光灯管
fluorescent ultraviolet condensation light and water-exposure apparatus 荧光紫外线冷凝光照水淋耐候仪

fluorescent water soluble polymer 荧光水溶性聚合物
fluorescent water treatment polymer 荧光水处理剂
fluorescent whitener 荧光增白剂
fluorescent whitening agent 荧光增白剂
fluorescent whitening dye 荧光增白染料
fluorescent X-ray 荧光X射线
fluorescent X-ray spectrographic analysis X射线荧光光谱分析
fluorescent X-ray spectrometry 荧光X射线光谱分析
fluorescent X-ray spectroscopy 荧光X射线光谱学
fluorescent X ray spectrum X射线荧光光谱
fluorescent yellow 荧光黄
fluorescent yield 荧光产额
fluorescer 荧光剂;荧光增白剂
fluorescin 荧光生
fluorescopy 荧光学
fluorexone 荧光素配位剂
fluorgypsum 氟石膏
fluorhydrazine 氟化肼
fluorhydric acid 氢氟酸
fluoric 含氟的;氟代的
fluoric acid 氟酸
fluoric crown 含氟冕牌玻璃
fluoric ether 氟代烷烃;氟代酸的酯
fluoridation 加氟作用;氟化反应
fluoridation of drinking water 饮用水加氟;饮水加氟
fluoride 氟化物
fluoride concentration status of groundwater 地下水含氟浓度状况
fluoride-containing acidic wastewater 含氟酸性废水
fluoride-containing concentration 含氟浓度
fluoride-containing groundwater 含氟地下水
fluoride-containing nitro-imidazole compound 含氟硝基咪唑化合物
fluoride-containing waste(water) 含氟废水
fluoride distribution 氟化物分布
fluoride film 氟化物薄膜
fluoride-free glass 无氟玻璃
fluoride glass 氟化物玻璃;氟化玻璃
fluoride incidence 氟化物影响范围
fluoride in drinking water 饮水中的氟化物
fluoride ion 氟根离子
fluoride ion removal 除氟离子
fluoride opal glass 氟化物乳白玻璃
fluoride ore 氟化物矿石
fluoride poisoning 氟化物中毒
fluoride pollutant 氟化物污染物
fluoride pollution 氟化物污染
fluoride removal 除氟化物
fluoride removal technique 除氟化物技术
fluoride selective electrode 氟化物离子选择性电极
fluoride single crystal 氟化物单晶
fluoride waste 含氟废料;除氟
fluoridize 加氟
fluorigenic labeling technique 荧光生成标记技术
fluorigenic reaction 荧光发生反应
fluorime 荧光胺
fluorimeter 荧光计;荧光光度计
fluorimetric analysis 荧光分析
fluorimetric method 荧光法
fluorimetry 荧光分析法;荧光测定(法)
fluorinate 荧光体
fluorinated crown 氟铬黄
fluorinated ether 氟化醚
fluorinated ethylene propylene 聚氟乙烯丙烯;氟化乙丙烯
fluorinated ethylene propylene resin 氟化乙丙烯树脂
fluorinated hydrocarbon 氟化烃;氟化碳氢化合物
fluorinated hydrocarbon propellant 氟化烃抛射剂
fluorinated paraffins 氟化链烷烃
fluorinated plastic 氟化塑料
fluorinated polymer 氟化高聚物
fluorinated rubber 氟化橡胶
fluorinated silicone rubber 氟硅橡胶
fluorinated surfactant 氟化表面活性剂
fluorinated thermoplastics 热塑性氟塑料;氟化热塑料
fluorinated vinylidene chloride 氟化偏氯乙烯

fluorinated water 含氟水
fluorination 氟化作用
fluorination process 氟代过程
fluorination reactor 氟化器
fluorine buffer 氟缓冲
fluorine cleanup reactor 氟吸收反应器
fluorine compound 氟化物;氟化合物
fluorine concentration 氟化合物浓度
fluorine-contained polymerisate fiber 含氟聚合物纤维
fluorine-containing waste 含氟废物
fluorine crown 氟铬黄
fluorine dating 氟年代测定法
fluorine-doped silica glass fiber 掺氟石英玻璃光纤
fluorine drilling 氟素钻进
fluorine emission 含氟排放物;氟排放物
fluorine fluxing agent 氟熔剂
fluorine iodine 氟化碘
fluorine ion conductor 氟离子导体
fluorine method 氟法
fluorine poisoning 氟中毒
fluorine reagent colo(u)rimetry 氟试剂比色法
fluorine refrigerant 氟致冷剂
fluorine rubber 氟橡胶
fluorine-silica fiber 掺氟石英光纤
fluor iodine 氟化碘
fluorion 氟离子
fluorite 萤石含量;萤石;氟石
fluorite deposit 萤石矿床
fluorite granite 萤石花岗岩
fluorite greisen 萤石云英岩
fluorite gypsum 萤石膏
fluorite-lens 萤石棱镜
fluorite muscovite greisen 萤石云母云英岩
fluorite objective 萤石物镜
fluorite optics 萤石光学
fluorite ore 萤石矿石
fluorite ore associated by sulfide mineral 硫化物型萤石矿石
fluorite-prism 萤石棱镜
fluorite quartz greisen 萤石石英云英岩
fluorite quartz ore 萤石石英矿石
fluorite structure 萤石型结构
fluoritization 萤石化
fluorizated 氟化的
fluorizating agent 氟化剂
fluormica 氟云母
fluoroacetate 氟乙酸盐
fluoroacetic acid 氟乙酸
fluoro-acetic fluoride 氟代乙酰氟
fluoro-acid 氟代酸
fluoroacrylate cladding fiber 氟丙烯酰酯包层光纤
fluoroalkane 氟代烷烃
fluorobenzene 单氟代苯;氟苯
fluorocarbon 碳氟化合物;氟碳化合物
fluorocarbon coating 氟碳树脂涂料
fluorocarbon elastomer 氟烃弹性体
fluorocarbon fiber 氟烃纤维
fluorocarbon membranefilter 氟碳膜式过滤器
fluorocarbon oil 氟代烃油
fluorocarbon plastics 氟碳塑料
fluorocarbon resin 碳氟树脂;氟烃树脂;氟碳树脂
fluorocarbon surfactant 氟碳表面活性剂
fluorocarbonzl 二氯氟甲烷
fluorocarboxylic acid 氟代羧酸
fluorochemicals 含氟化合物
fluorochlorocarbon 氟氯碳化物
fluoro chlorohydrocarbons 氟氯烃类
fluorochloromethane 氟氯甲烷
fluorochrome 荧色物;荧光染料
fluorod 荧光测量棒
fluorodensitometry 荧光密度测定;荧光光密度分析法
fluorodi chloromethane 二氯氟甲烷
fluoroelastomer 氟基弹性材料
fluoroelastomer coating 氟弹性体涂料
fluoro epoxy ether 环氧氟醚
fluoroethane 氟代乙烷
fluoroethanoic acid 氟乙酸
fluoroethyl 氟代乙酯
fluoroethylene resin 氟乙烯树脂
fluorofibre 含氟纤维
fluorogen 荧光团
fluorogenic substrate 荧光底物

fluorogram 荧光谱图;荧光屏照片
fluorograph 荧光图像摄影;荧光图
fluorographic study 荧光仪检查
fluorography 荧光(图)照相术;荧光屏摄影术
fluoro gum 氟橡胶
fluorohydrocarbon 氢氟碳化合物;氟代烃
fluorohydrocarbon plastics 氟烃塑料;氟代烃塑料
fluorol 氟化钠
fluorometer 荧光流量表;荧光计;氟量计
fluoromethane 氟代甲烷
fluoromethylation 氟甲基化
fluorometric assay 荧光测定
fluorometric bituminological analysis 荧光沥青分析
fluorometric titration 荧光滴定
fluoronitrobenzoic acid 氟硝基苯甲酸
fluoro(o)lefins 氟代烯烃
fluorophenol 氟代苯酚
fluorophlogopite 氟金云母
fluorophor 荧光体
fluorophore 荧光团
fluorophoshoric acid 氟磷酸
fluorophotometer 荧光光度计
fluorophotometric 荧光光度
fluorophotometric analysis 荧光光度分析
fluorophotometric titration 荧光光度滴定
fluorophotometry 荧光光度测定法
fluoroplastic film 含氟塑料薄膜
fluoroplastics 含氟塑料;氟塑料
fluoropolyene 氟多烯
fluoropolymer 含氟聚合物;氟聚合物;氟化高聚物
fluoropolymer insulated cable 氟聚合物绝缘光缆
fluoropolymer resin 氟聚合物树脂
fluoro-propane 氟代丙烷
fluoro protein foam 氟蛋白泡沫
fluoro-protein foamite 氟蛋白泡沫灭火液
fluororesin 氟树脂
fluororoentgenography 荧光X线照相术
fluororubber 氟橡胶胶粘剂;氟橡胶
fluorosclerosis 氟性硬化
fluoroscope 荧光镜(屏);荧光检定仪;荧光检查器;X光透视
fluoroscopic device 伦琴射线荧光镜
fluoroscopic examination 荧光透视检查
fluoroscopic image intensifier 荧光图像增强器
fluoroscopic inspection 荧光检验
fluoroscopic viewing 荧光透视
fluoroscopy 荧光学;透视;放射检查;X透视检查
fluorosilane 氟代硅烷
fluorosilicate crown glass 氟硅冕玻璃
fluorosilicone rubber 氟硅酮橡胶
fluorosis 氟中毒
fluorosis of bone 氟骨症
fluorosis of the teeth 牙氟中毒
fluorotoluene 氟甲苯;氟代甲苯
fluorous 氟的
fluorous salt of copper 氢氟酸铜;含氢氟酸的铜盐
fluorphosphate glass 氟磷酸盐玻璃
fluor reagent 氟化剂
fluor reagent spectrophotometric method 氟试剂分光光度法
fluor reagent spectrophotometry 氟化剂分光光度法
fluor resin 氟树脂
fluorspar 萤石;氟石
fluorspar of baryte 氟重晶石
fluorspar powder 氟石粉
fluorubin 荧红环
fluorubine pigment 荧光玉红类颜料
fluosilicate 氟硅酸盐
fluosilicate coat 氟硅酸盐涂料;氟硅酸盐涂层
fluosilicate hardener for gypsum 用作石膏硬化剂的氟硅酸盐
fluosilicate of lead 铅氟硅酸盐
fluosilicate sealing 氟硅酸盐密封处理
fluosilicate titanium 氟硅酸钛玻璃
fluosilicate treatment 氟硅酸盐处理
fluosilicic acid 氟硅酸
fluosilicone coating 氟硅酮涂料
fluosolid 流化层
fluosolid process 流态化焙烧法;沸腾焙烧法
fluosolid roaster 流态化焙烧炉
fluosolid roasting 沸腾焙烧
fluosolids system 流态化焙烧法

fluospectrophotometer 荧光分光光度计
fluostannate 氟锡酸盐
fluosulfonic acid 氟磺酸
fluothane 氟烷
fluoxyth 氟酚合剂(一种木材防腐剂)
fluralsil 氟硅化锌(一种木材防腐剂)
fluram 胺荧
flurent 冲积新成土
flurosion 河流浸蚀作用
fluroxene 氟乙烯醚
flurry 雪崩风;阵风;风雪机;风雪(风)
flusch 松碎沉积;松散沉积
flush 溅湿地;埋入的;齐头排放;齐平(指勾缝);平铺;冲水;潮湿地;潮红;奔流;暴涨
flush air intake 戽斗形进气口;平直进气口
flush away 洗去;洗涤;冲去
flush base 平踢脚板
flush bead 平的半圆线脚;平的半圆木条;平圆线脚;平线条;平焊缝;平的串珠线脚【建】;凹圆线条
flush bead mo(u)lding 平圆线脚
flush boarding 横钉平装护墙板
flush bolt 皿头螺栓;埋头螺栓;埋入插销;平头螺栓;平(头)插销;暗插销
flush bolt backset 平插销的退进
flush-bonding 嵌入式
flush box 冲洗水箱
flush bushing 无凸缘衬套
flush button 冲水按钮
flush channel 冲洗槽
flush cistern 厕所水箱
flush closedown 清仓关闭
flush coaming 平缘材;平围板
flush coat 沥青面层;磨光层;平齐层
flush coater 喷洒机
flush colo(u)r 底色
flush concrete curb 平埋混凝土路缘
flush corner joint 端面角接头
flush coupled 外平连接的(钻杆)
flush coupled casing 外平连接套管;平接套管
flush coupled-type drill string 平接钻具
flush coupling 锁接头
flush coupling joint 平接接头
flush cover 平舱口盖
flush cupboard catch 齐平橱柜门勾;埋平橱柜门闩
flush-cup pull 齐平凹拉手;杯形拉手
flush curb 平(埋)缘石;平埋路缘;隐蔽式道牙
flush cut 平面切削
flush cut joint 平头接头;平头接合
flush-cutting saw with wood(en) handle 木柄平头锯;木柄钢锯架
flush dam 泄水坝
flush deck 平甲板
flush deck barge 平甲板驳船
flush decker 平甲板船
flush deck socket 水平连接器(集装箱)
flush deck vessel 全通甲板船;平甲板船
flush distillation 一次蒸发
flush door 光面门;全板门;平面门
flush drill pipe 埋接钻杆
flush eaves 封护檐;平头檐
flushed colo(u)r 底色;挤水颜料浆
flushed joint 齐平接缝;平灰缝;平缝
flushed paste 挤水色浆
flushed pigment 挤水颜料
flushed zone 冲洗带
flush-encased block 嵌装锁
flush-encased dead bolt block 嵌装式无弹簧门锁;嵌装螺栓暗保险锁
flush end 平端
flush equipment 冲洗设备
flusher 加机油装置;喷洒装置;冲洗装置;冲洗器;冲水器
flusher cleaning attachment 冲洗装置
flush-filled 平嵌的
flush-filled joint 平嵌接缝;平嵌灰缝
flush filling 平齐装科
flush filter plate 平槽压滤板;平槽滤板
flush fixing 齐平接头;齐平接榫
flush floor base 平踢脚板
flush fluid loss 冲洗液消耗量
flush gable roof 硬山
flush gallery 冲砂廊道;冲沙廊道
flush gate 冲泄闸;冲刷闸;冲砂闸门;冲淤闸门;冲淤水闸;冲洗闸(门)

flush girt 活动地板边撑
flush glazing 平装玻璃
flush handle 嵌入式门环;暗门柄;平顺门把
flush head 埋头;通水接头
flush headed bolt 埋头螺栓
flush head rivet 埋头铆钉;皿头铆钉;平头铆钉
flush head screw shackle 平头螺栓卸扣
flush human sewage 人类洗涤污水
flush hydrant 路面消防栓;平头消防栓
flushing 洗井(石油钻探时);油井注水;挤水作用;洒水;吹洗;冲刷;泛油(路面泛出多余沥青);发红
flushing ability 冲洗功能
flushing action 冲洗作用
flushing agent 冲洗剂
flushing air 冲洗空气;冲洗的空气
flushing and ramming 水冲法打桩
flushing asphalt 沥青泛油(路面)
flushing basin 冲洗池
flushing box 冲洗水箱;冲洗槽
flushing canal 排水沟;冲洗沟;冲砂渠;冲洗槽;冲沙渠
flushing chamber 冲洗室
flushing channel 冲沙渠
flushing cinder 冲渣
flushing circuit 清扫管线;冲洗管线
flushing cistern 冲洗水箱
flushing cistern 冲水水箱
flushing conduit 清扫管线
flushing core 用水冲击岩芯
flushing coupling 齐平式连接
flushing culvert 冲沙涵洞
flushing device 平水设备(自记水位计观测井筒);冲洗装置
flushing duct 冲洗管道
flushing equipment 冲洗设备
flushing fluid 冲洗液;洗井液
flushing function 冲洗功能
flushing gallery 冲洗坑道
flushing gas 冲洗气体
flushing gate 冲洗闸(门);冲淤闸门;冲洗水闸;冲泄闸门;冲沙闸(门)
flushing gutter 冲砂沟;冲沙沟
flushing hole 渣口;渣孔;冲洗孔
flushing hose 清洗用软管
flushing interval 冲洗周期
flushing lever 冲洗扳手
flushing line 冲洗管道;冲洗管线;冲洗管路
flushing liquor 冲洗液
flushing machine 冲洗机;冲水机
flushing main 冲洗总水管
flushing manhole 冲洗井;冲洗检查井;冲洗孔
flushing material 冲洗材料
flushing medium 冲洗介质
flushing nozzle(of the bit) (钻头上的)冲洗液喷嘴
flushing of asphalt 沥青泛油【道】
flushing of core 冲刷岩芯
flushing oil 冲洗用油;洗液;洗涤油
flushing out 冲洗
flushing-out of core (反循环取芯法的)冲击岩芯
flushing out valve 冲泄阀
flushing pan 抽水马桶
flushing pipe 冲洗水管;水力充填管;冲洗管(道)
flushing piping 冲洗管
flushing port 冲洗孔
flushing process 脱水油墨制造法
flushing process pigment 用挤水法生产的颜料;挤水法颜料
flushing pump 冲洗泵
flushing regime 冲洗方法
flushing regulator 冲洗调节器
flushing siphon 冲洗虹吸管
flushing sluice 污水闸;排沙闸(门);冲洗闸(门)
flushing sprinkler for wharf and roadway 码头及道路冲洗喷头
flushing surface 冲洗面
flushing system 冲洗管系
flushing tank 冲洗水箱;冲洗池;冲洗水柜;冲洗井;冲水(水)箱;冲水池;冲水槽
flushing tap 洒水龙头
flushing test 冲水试验;冲沙试验
flushing time 冲洗时间
flushing trough 冲洗水槽;自动冲水槽(公共厕所)
flushing tube 冲洗管

flushing tubing 冲洗管
flushing tunnel 冲砂隧洞;冲沙隧洞;冲沙隧道
flushing up 齐平接缝;(砌体)勾平缝
flushing valve 冲泄阀;冲洗阀
flushing washing 强冲洗
flushing water 冲洗液;冲洗用水;冲沙水
flushing water loss 冲洗液漏失
flush inlet 齐平式进水口;不前伸(的)进口
flush joint (砌体的)勾平缝;齐平接缝;平直接合对结;平头接头;平头接合;平贴接合;平(灰)缝;端接
flush-joint casing 平接式套管;外平套管
flush-joint drill pipe 外平连接钻杆
flush-joint drive pipe 厚壁内平打入套管
flush-jointed casing 平接套管
flush-joint liner 内平接头连接套管
flush-joint of rods 钻杆的内平连接
flush-joint pipe 内平管(钻杆)
flush-joint pointing 平勾缝
flush kerb 齐平路缘;平埋侧石;路边侧石;平埋路缘
flush lawn hydrant 齐草地的配水龙头
flush-level powder fill 平齐装粉
flush liberation 猛释气体;减压释气
flush light 吸顶暗灯
flush light panel 筒灯
flush masonry jointing 砖石平接灰缝
flush median strip 与路面铺砌平齐的中心分车带;平齐中心带
flush mixer 冲洗混合器
flush mounted 暗装的;嵌入式的;齐平安装的
flush-mounted array 平镶阵
flush-mounted drawout relay 平装抽出式继电器
flush mounting 埋入装置;嵌入装置;平齐安装;平(口)装;水平校验装置
flush-mounting frame 与墙齐平安装的框架
flush nut 平顶螺母
flushometer 冲洗阀;冲水阀
flushometer valve 冲洗阀门;冲水阀;冲洗定量阀
flush out 析出(量);浸出(量);清洗;冲走;冲掉
flush-out valve 冲洗阀
flush panel 齐平镶板(与框架齐平);平镶板;平光镶板
flush paneled door 光面镶板门;平镶板门
flush paragraph 不缩排段
flush perforation (管壁未穿透的)失败射孔
flush period 降雨期;泛滥期
flush pipe 溢水管;冲水管
flush plate 平板盖;满版图;平装开关面板;平贴盖板;平(槽)板
flush plate filter press 平槽压滤机
flush plating 平接板
flush plug 嵌入式插座
flush plug consent 嵌入式插座
flush plug receptacle 埋装式插座;埋入式插座
flush plug socket 嵌入式插座
flush point 平缝;平灰缝
flush practice 出渣;冲渣操作
flush producer 高产油井
flush production 高峰产量;初期产量
flush pump 水泵
flush pump ga(u)ge 水泵表
flush quenching 溢流淬火;喷水淬火;冲水淬火
flush receptacle 墙插座
flush recessed 嵌入式的
flush regulator 冲洗调节器
flush ring 活动拉环;平拉环
flush rivet 埋头螺钉;平头铆钉
flush-riveted construction 埋头铆接结构
flush-riveted covering 埋头铆壳
flush rivet(t)ing 埋头铆接;平铆(接);平光铆接
flush scuttle 甲板煤舱口;甲板煤舱孔
flush seal 薄层沥青封面(道路工程中)
flush-set diamonds 孕镶金刚石
flush-sided 平边的
flush-sided body 平边车身
flush siding 平接企口墙板
flush sintering 快速粉末冶金
flush snap-switch 埋装式活动开关
flush socket 暗装插座;埋装插座;平装插座
flush soffit 平齐底面(三角形的截面梯级);接连底面;平天花板底面
flush stripe 平式分隔带
flush surface ga(u)ge 铣度表面丝锥
flush switch 嵌入(式)开关;埋装开关;埋入式开关;平装开关
flush system 平镶式;平铺式
flush tank (沟渠的)冲洗箱;便器水箱;抽水箱;冲洗槽;冲水箱;冲水池;冲水槽
flush toilet 冲水式厕所;有水厕;抽水马桶;抽水大便器
flush track 暗轨
flush trim die 修边模
flush trimmer 整音器;剔除器
flush type 埋入式;嵌装式;齐平式;平装型;埋装式;平整式
flush type circuit 嵌入式电路;齐平式电路
flush type construction 埋接式结构;嵌入式结构
flush type hydrant 平头消防栓
flush type instrument 平装型仪表;埋装式仪表;嵌装式仪表;嵌入式仪表
flush type meter 嵌入式仪表
flush type receptacle 埋入式插座
flush type releasing spear 冲洗型可退式打捞矛
flush type slab handle 埋入式板把手;齐平式板把手
flush valve (厕所水箱内的)冲泄阀
flush visible under-face 平齐底面;平接底面
flush wall box 埋入式电气盒
flush water 洗涤水;冲洗水;冲水
flush water pump 冲洗水泵
flush weir 泄砂堰;泄水堰;泄水坝;泄沙堰;溢水堰;溢流堰
flush weld 光焊;削平补强的焊缝;齐平焊;平焊接;平焊缝;无加强高的焊缝
flush with the ceiling 与天花板平齐;与顶棚平齐
flush with the wall 与墙齐平
flute 柱槽;沟槽;刃沟;排屑槽;竖沟;出屑槽;波纹;凹槽(饰)
flute by bar 脊旁凹槽;坝旁凹槽
flute cast 流痕;槽铸型;槽形印模;槽模
fluted 带槽纹的
fluted bar 凹面方钢
fluted bar iron 凹面方钢
fluted bead 外圆角秋叶
fluted board 空心板
fluted brass tube 槽纹黄铜管(主要用于电气和楼梯扶手)
fluted bulkhead 波形舱壁;凹槽舱壁
fluted chucking reamer 机用带槽铰刀
fluted column 凹槽柱(身);有凹槽柱身
fluted copper tube 槽纹铜管(主要用于电气和楼梯扶手)
fluted core fabric 槽芯织物
fluted coupling 稳定器
fluted cutter 槽形铣刀;槽式铣刀
fluted disc 齿盘
fluted drainer 凹槽滴阴沟;凹槽滴干板
fluted finish 波纹终饰
fluted formwork 有凹槽的模板
fluted funnel 槽沟漏斗
fluted glass 瓦楞玻璃;瓦槽玻璃;槽纹玻璃;凹槽玻璃
fluted hollow shape 带槽空心型材
fluted micrometer 沟槽千分尺
fluted mo(u)ld 波形模
fluted mo(u)lding 凹槽线脚【建】;凹槽饰
fluted nut 槽顶螺母;槽顶螺帽
fluted reamer 带槽铰刀;槽式铰刀
flute drill 笛形钻
fluted roll 瓦楞辊;槽辊
fluted roll drill 槽轮式条播机
fluted roller 起凹槽;槽纹(压延)辊
fluted roller glass 槽纹玻璃
fluted roller shaft 槽轮轴
flutedroll mill 槽纹辊
fluted shaft 开槽轴;槽轴
fluted sheet 压槽板;压槽窗玻璃;波纹板;槽形铁皮;槽纹板;凹槽玻璃
fluted spectrum 条段光谱
fluted square pass 凹边方孔形
fluted tapered shell pile 凹槽锥形壳桩
fluted tile 有凹槽的瓷砖
fluted tube 带槽管材;槽纹管
fluted twist drill 麻花钻
fluted-wheel drill 槽轮式条播机
flute instability 槽形不稳定性;槽纹不稳定性
flute length 槽长
fluteless tap 无槽丝锥
flute mark 槽痕;凹槽痕迹
flute of drill 钻头排屑槽
flute pitch 槽距
flute profile 槽形
flute rill mark 槽形细流痕
fluting 折断;弯折;柱槽;折纹;开槽;磨刻槽纹;切槽;带槽的材料;槽痕风化;凹槽形成作用;凹槽(饰)
fluting board 瓦楞原纸
fluting cutter 槽铣刀
fluting cutter for tap 螺丝攻槽铣刀
fluting of column 柱槽
fluting plane 凹刨
flutter 脉动干扰;频率抖动;颤振;颤动
flutter amplitude 颤幅
flutter calculation 颤动计算
flutter checker 颤动检验器
flutter coil 颤动校准线圈
flutter component 脉动分量;颤动分量
flutter computer 颤动模拟计算机;颤动计算机
flutter control 抖振控制
flutter depth 颤动深度
flutter echo 多次回声;颤动回声
flutter fading 振动衰减;散乱反射衰减;颤动衰减
flutter failure 振动破坏;颤振破坏
flutter feeder 抖动给料器
flutter generator 脉动发生器;颤振发生器
fluttering characteristic 颤振特性
fluttering critical wind velocity 颤振临界风速
fluttering rate of magnetic recorder 磁记录器抖晃率
flutter meter 频率颤动测试器
flutter rate 振动率;颤动率
flutter relay 振动式继电器
flutter simulator 颤动模拟器
flutter speed 颤振速度
flutter test data 振动试验数据
flutter valve 翼形阀;波动阀
flutter wheel 翼形水轮
fluvial abrasion 河流磨蚀;河流冲蚀;水流冲刷
fluvial action 河流作用
fluvial anadromous fish 潮河产卵鱼类
fluvial barge 内河驳船
fluvial bog 河滩沼泽;河边低地
fluvial channel 河道;河槽
fluvial conglomerate 河流砾岩
fluvial current 河水流;河水泛滥
fluvial cycle of denudation 流水剥蚀循环;河流剥蚀循环
fluvial cycle of erosion 河蚀旋回;河流侵蚀循环;河流冲刷循环;常态旋回
fluvial denudation 流水剥蚀;河流冲刷作用;河流剥蚀(作用)
fluvial deposit 河流沉积(物);河川沉积物;河成积(物)
fluvial dynamics 河流动力学
fluvial ecosystem 河川生态系统
fluvial erosion 流水侵蚀;河流侵蚀;河流冲刷作用;河流冲刷;河流冲蚀
fluvial erosion basis 河流侵蚀基准面
fluvial facies 河流相
fluvial flat 河积平原
fluvial geological process 河流地质作用
fluvial geomorphic cycle 河流地貌循环
fluvial hydraulics 河流水力学;河道水力学;河川水力学
fluvial index 流动指数
fluvial marine sediment 河海沉积(物)
fluvial material 河成泥沙
fluvial morphology 河貌学;河流形态(学);河流地貌(学);河川形态学;河川地貌学
fluvial mud 河泥
fluvial outwash 河流冲积
fluvial plain 河积平原;河成平原;冲积平原
fluvial process 造床过程;流水动力;河流动力;河道演变;河道变迁;河床演变;河成过程;冲积过程;成河过程
fluvial process observation 河床演变观测
fluvial process survey 河床演变观测
fluvial regime(n) 河性;河势;河(流状)况;河流情态;河流情势;河流情况
fluvial river 内陆河(流);河流的
fluvial sand 河沙(也称河砂);冲积沙
fluvial sediment 河水沉积物;河流泥沙;河流沉积物;河川沉积物

fluvial sedimentation model 河流沉积模式
fluvial sediment kinematics 河流泥沙运动学
fluvial sediment transportation 河流泥沙输送
fluvial sequence 河流层序
fluvial soil 河流冲积土;河口冲积土
fluvial stream 无潮汐影响的河流
fluvial system 河流体系
fluvial terrace 河流阶地;河成阶地
fluvial tide 河潮
fluviatic 河成的
fluviatile 河川活动;河成的
fluviatile cycle of erosion 河蚀旋回;河流侵蚀循环
fluviatile dam lake 河流堰塞湖
fluviatile deposit 河流沉积
fluviatile erosion 河流(沉积)的冲蚀;河流侵蚀
fluviatile facies 河相
fluviatile lake 河迹湖;河成湖
fluviatile plain 河成平原
fluviation 流水作用;河流作用;河流活动;河川作用;河川活动;河成作用;成湖作用
fluvioclastic rock 河成碎屑岩
fluvioeolian 河风(成因)的
fluvioeolian deposit 河风沉积
fluviogenic soil 冲积土(壤)
fluvioglacial 河流冰川的;河冰的;冰水作用的;冰水(生成)的;冰川水形成的;冰川和河流的
fluvioglacial accumulation 河流冰川堆积;冰水堆积
fluvioglacial deposit 冰水沉积(物)
fluvioglacial drift 冰水漂积物
fluvioglacial landform 冰水河流作用形成的地形
fluvioglacial river 冰水河
fluvioglacial stream 冰水河流;冰川水河流
fluvioglacial terrace 冰水阶地
fluviograph 自记水位仪;自记水位计;河流水位自动测录仪;水位仪;水位计
fluviokarst 流水岩溶;河成喀斯特
fluviolacustrine 河湖生成的;河流湖泊的;河湖的
fluviology 河流学;河川学
fluviomarine 河海栖的;河流海岸的;河海的;河海沉积的
fluviomarine deposit 河海沉积(物)
fluviomaritime section of stream 感潮河段;河流的近海河段
fluviometer 河流水位仪
fluviomorphological process in river 河床演变过程;河床形成过程
fluviomorphology 河相学;河貌学;河流形态(学)
fluviosol 冲积土(壤)
fluvioterrestrial 陆上河水的;陆地河流学;河陆栖;淡水的
fluvioterrestrial deposit 陆上河水沉积(物)
flux 渣化;造渣;焊剂;熔体;熔剂;溶剂气焊;溶剂气割;焊剂;通量
flux albedo 通量反照率
flux analysis 流束分析
flux and reflux 潮水涨落
flux bar 磁通量条
flux bath 溶剂槽
flux block 液面砖;池壁砖
flux boundary 通量界限
flux-coated electrode 涂熔剂焊条;造渣型焊条;涂有熔剂的焊条;涂药焊条
flux concentration 通量集中
flux concentrator 通量集中器
flux constituent 焊剂成分
flux-cored electrode 管状焊条;包芯焊条
flux-cored solder wire 中间有焊剂的空芯焊丝
flux-cored welder 包芯焊条焊机
flux-cored welding 包芯焊条焊接
flux-cored wire 管状焊丝
flux core-type electrode 熔剂芯焊条;溶剂芯焊条
flux counter 通量计数器;磁通计数器
flux cover 渣壳;熔剂覆盖层
flux curve 光通量曲线;磁通曲线
flux-cutting law 磁通切割律
flux density 流量密度;流线通密度;气流密度;通量密度;磁通密度
flux density scale 流量密度标
flux density spectrum 流量密度谱
flux depression 流量衰减
flux-depression correction 通量压低校正
flux-depression modification 通量减弱修正
flux detector 通量探测器
flux differential relay 磁通差动继电器

flux dip 通量坑
flux distortion 通量畸变
flux distribution 通量分析;通量分布;磁通分布
flux divergence 通量散度;通量辐散
fluxed asphalt 加过稀释剂的沥青;软制(地)沥青
fluxed bitumen 软制沥青
fluxed bituminous material 软制沥青材料
fluxed electrode 焊剂焊条;熔剂焊条;涂药焊条
fluxed lake asphalt 软制湖沥青;加过稀释剂的湖沥青
fluxed native asphalt 软制天然(地)沥青
flux encased electrode 包药加强焊条
flux envelope 渣壳
fluxes 焊剂
flux factor 熔解系数;熔剂因子
flux flattening 通量补偿
flux flint 熔剂燧石岩
flux force-condensation scrubber 通量力冷凝洗涤器
flux function 通量函数
flux gate 通量门;磁通量闸门
flux-gate compass 磁通量闸门罗盘;磁通量阀门罗盘
flux-gate detector 磁通量闸门检测器
flux-gate magnetometer 磁通门磁力仪;磁(通)门磁强计
flux-gate sensor 磁选通器传感器
flux-gate spinning magnetometer 磁通门式旋转磁力
fluxgraph 通量记录仪;磁通仪
flux-gravity diagram 流量重力图
flux-grown 熔盐法;熔融生长的
flux-grown ruby 熔融生长红宝石
flux growth 熔化生长
flux growth method 熔剂法
flux guide 磁通控制器;磁导
flux gun 熔剂枪
flexibility 助熔性
flux inclusion 焊剂夹杂
fluxing 助熔;造渣;熔解;熔剂处理
fluxing action 助熔作用
fluxing agent 稀释剂;助熔剂;焊药;软制剂;熔剂
fluxing asphalt 软制(地)沥青
fluxing effect 造渣作用
fluxing hole 熔洞
fluxing lime 稀释用石灰
fluxing medium 助熔剂;造渣剂
fluxing mineral 熔剂矿物
fluxing oil 稀释油;软制油
fluxing ore 助熔矿石
fluxing power 助熔能力;渣化能力;结渣率
flux ingredient 焊剂成分
fluxing temperature 熔解温度
flux injection cutting 喷燃剂切割
flux invariant 通量不变量
fluxion 流数
fluxion calculus 微积分
fluxion gneiss 流状片麻岩
fluxion structure 流状构造;流纹构造
fluxion texture 流状结构;流纹结构
flux jumping 磁通跳跃
flux leakage 磁通量漏泄;磁漏
flux level 液面(熔融玻璃);通量级;玻璃液面
flux lime 石灰石
flux line 液面线;力线;通量线;玻璃液面
flux-line attack 液面线侵蚀
flux-line block 池墙上部砖;玻璃熔液面的池壁砖;液面池壁砖
flux-line corrosion 液面线侵蚀;料液面侵蚀
flux-line waterbox 玻璃液面冷却水包
flux linkage 磁链匝连数;磁链
flux-linkage equation 磁链链方程式
flux map 气流图;通量图
flux mapping 通量测绘
flux mapping system 通量测绘系统
flux material 熔剂;焊剂;助熔剂
flux measurement 通量测量
flux-measuring channel 通量测量道
flux metal 助熔金属
fluxmeter 麦克斯韦计;通量计;磁(通)量计;辐射通量计
flux method 流明法;通量法
flux monitor 通量监测器;磁通监控仪
flux monitoring 通量监测

flux-monitoring foil 通量监测箔
fluxo-accumulation association 滑塌堆积组合
flux of energy 能量通量
flux of force 力通量;力束
flux of force lines 力线束
flux of light method 光通量法;光束法
fluxograph 流量记录仪;流量记录器;河流水位计
fluxoid 全磁通
flux oil 助熔油;软制油;焊剂;沥青稀释油;半柏油;稀释剂;稀释油
fluxon 磁通量子
fluxoturbidite 滑塌浊积岩
flux oxygen cutting 氧熔剂切割
flux oxygen cutting method 氧溶剂切割法
flux paste 稀释剂;软化浆;软化膏
flux path 磁通路线;磁通轨迹
flux pattern 通量图;通量分布图
flux peak 最大通量;通量峰
flux peaking 局部通量剧增
flux penetration 通量穿透
flux period 通量变化周期;磁通量变化周期;反应堆周期
flux pining 钉扎
flux plot 通量图
flux plot hole 通量测绘孔
flux plotting 通量作图
flux pump 磁通泵
flux quantum 磁通量子
flux quartzite 熔剂石英岩
flux rate 通量率
flux rating 通量额定值
flux ratio 通量比
flux raw material commodity 熔剂原料矿产
flux recovery 通量恢复
flux residue 焊渣;残留物;滤渣;残渣
flux roll 熔剂辊
flux sampler 通量采样器
flux scale 流量标
flux scanning channel 通量分布测量管道
flux screw 磁通调整螺钉
flux-second 通量每秒
flux sensor 磁通传感器
flux shielding 渣保护
flux silicarenite 熔剂石英砂岩
flux silica sand 熔剂石英砂
flux sleeve 导磁套
flux solder 熔焊料
flux step method 通量分步法
flux trap 通量阱;磁通阱
flux trapping 通量阱
flux trap reactor 通量阱(反应)堆
flux-trap region 通量阱区
flux traverse 切割射线流
flux-traverse measurement 通量分布测量;通量穿透测量
flux tube 通量管;磁流管
flux unit 流量单位
flux valve 流量阀
flux vein quartz 熔剂脉石英
flux voltmeter 磁通伏特计
flux-wall guided transfer 渣壁过渡
flux wave 通量波
fly 旗尾;配合手轮
fly about 风向转变
fly a kite 发通融票据;开空头支票
fly-ash 飘尘;飞灰;飞尘;浮尘;粉煤灰
fly-ash adsorption 粉煤灰吸附
fly-ash aggregate 粉煤灰集料;粉煤灰骨料
fly-ash brick 粉煤灰砖
fly-ash cast stone 粉煤灰铸石
fly-ash cement 飞灰水泥;粉煤灰水泥;烟灰水泥
fly-ash cement concrete 粉煤灰水泥混凝土;烟灰水泥混凝土
fly-ash collector 粉煤灰收集器;飞灰收集器
fly-ash concrete 粉煤灰混凝土
fly-ash concrete block 粉煤灰混凝土砌块
fly-ash control 飘尘控制
fly-ash crushed-stone pile 粉煤灰碎石柱
fly-ash disposal 飞灰处理
fly-ash fill 填粉煤灰;粉煤灰填筑(路堤)
fly-ash filter 飞灰过滤器
fly-ash-lime block 粉煤灰石灰砌块;飞灰石灰砖
fly-ash mark 飞灰变色
fly-ash monitor 飞灰监测器

fly-ash pond 飞灰池
fly-ash pond water 飞灰池水
fly-ash porous ceramsite 粉煤灰多孔陶粒
fly-ash Portland cement 波特兰粉煤灰水泥；粉煤灰硅酸盐水泥
fly-ash removal apparatus 除灰装置
fly-ash separator 飞灰分离器
fly-ash settling chamber 飞灰沉降室
fly-ash sewage sludge mixture 粉煤灰—污水污泥混合物
fly-ash stabilized soil 粉煤灰加固土
fly-ash structure 粉煤灰结构
fly-ash-type pozzolan 火山灰
Flyaside 弗莱阿赛德（一种专利的滑动门装置产品）
fly at zero 超低空飞行
fly auger drilling 连续螺旋钻进
flyaway kit 随机器材包
flyback 快速回零；光的回程；逆行；回扫（描）；回描；回零；倒转
flyback action 归零动作，快速回零机能
flyback blanking 逆程消隐
flyback circuit 回扫电路
flyback converter 逆向变换器；回扫电压变换器
flyback flux 回扫通量
flyback line 回描线；回程线
flyback period 回描周期；回描时间；返程周期
flyback power supply 回描电源
flyback pulse 回扫脉冲；回描脉冲；扫描逆程脉冲
flyback ratio 回扫率；回描率
flyback time 回扫（描）时间；回描时间；回程时间；返程周期
flyback timing 退回计时法
flyback transformer 回授变压器；回扫变压器；反馈变压器
flyback voltage 回扫电压；冲击激励电压
flyback waveform 回扫波形
flyball 带球形重锤的离心式调速器；飞球；飞锤
flyball arm 承重球杆
flyball governor 飞球式调速器；飞锤式调速器
flyball regulator 离心调节器
flyball tachometer 离心式转数计；飞球式转速计
fly-bar 刀片；舞台吊杆
fly block 滑动车
fly-boat 快艇
fly-boat fleet 快艇舰队
fly-by 低空飞越
fly-by-wire 遥控自动驾驶仪
fly-coasting 溜放调车
fly contact 轻动接点
fly-cruise 空海联航
fly-cutter 高速切削刀
fly-cut(ting) 快速切削；飞刀切削
fly destroying preparation 灭蝇制剂
fly door（舞台的）天桥
fly drill 手拉钻；飞轮手钻
flyer 整速轮；快车；平行梯级；梯级；手轮；锭壳；飞行器；飞轮
flyer doffing spinning frame 半自动落纱翼锭细纱机
flyer lead 翼导
flyer spindle 翼锭
flyer winding frame 翼锭络筒机
fly floor 舞台天桥
fly gallery 舞台天桥；（舞台的）天桥；布景长廊
fly gate 溜槽口活门；两开门；铰链阀
Flygt pump 弗利特泵
fly hand 收纸叼纸牙
fly-headed screw 蝶形头螺钉
flying 飞散
flying accident 飞行事故
flying altitude 飞行海拔高度
flying altitude above ground 相对航高
flying altitude mean sea level 绝对航高
flying altitude of barometer 气压计飞行高度
flying anchor 海锚
flying angle 飞行迎角
flying arch 拱式扶墙；吊篙；矮篙
flying arch method 先拱后墙法；比利时（开挖）法
flying average altitude 飞行平均高度
flying boat 船身式水上飞机；飞艇；飞船
flying bond 美式砌合；每皮二顺一丁砌合；跳丁砖砌合；跳丁砌法
flying bridge 天桥；跨线桥；悬索桥；最上层船桥；架空小桥；渡船；浮桥；浮船坞飞桥
flying buttress 拱形扶壁；拱式支墩；拱式扶垛；飞拱；飞扶垛；飞扶壁
flying buttress arch 拱扶垛
flying carrier 航空母舰
flying chip 飞屑
flying clock 航空钟；飞钟；飞行钟
flying connection（利用方钻杆鼠洞接长钻杆的）快速连接法
flying control system 飞机操纵系统
flying crossing 不停车会车【铁】
flying cut-off 剪铁条机
flying cutter 飞刀
flying day 飞行日
flying deck 顶甲板
flying design 飞行计划
flying dial ga(u)ge 连续式测厚千分表
flying dish 翼圆饰；有翼的圆形装置【建】
flying dust 飞尘
flying dutchman 侧引滑车
flying facade 假门面；屋顶面墙
flying falsework 悬空工作架；悬空脚手架；悬挂（式）脚手架
flying fence 飞越栅栏
flying ferry 钢索拖轮；缆车渡；渡运飞机
flying fiber 飞丝
flying field 简易机场；简易飞机场
flying fish 飞鱼；飞鱼
flying-fish sailor 无航海经验的水手
flying Flemish bond 每皮两顺一丁砌砖法；两顺一丁砌法；每层两顺一丁砌法
flying form 飞横；悬空模板
flying fox 架空索道运载器；缆索渡；架空索道载送器
flying gangway 悬吊式跳板
flying height 航高；浮动高度
flying height above sea level 绝对航高
flying height above terrain 相对航高
flying height above the ground 相对航高
flying height of photography 摄影航高
flying-in weld 死口焊口
flying item 飞行项目
flying lead 跨线
flying level（无水准点的）快速水准测量；手提式水准仪；快速测高
flying level(l)ing 快速水准测量；简易水准测量
flying-lift of gas holder 飞塔节
flying light 高浮（空载船舶）；空载停泊
flying line of photography 摄影航线
flying load 飞行荷载
flying lorry 运货飞机
Flying Mercury 麦丘利神塑像
flying micrometer 连续式测厚仪；快速测微仪
flying mooring 八字锚系泊（船头抛的）
flying nurserier 流动苗圃
flying-off 断裂
flying panel 座舱仪表板
flying paper 快速移动纸
flying passage 船员步桥
flying-point scanning tube 飞点扫描管
flying press 螺纹摩擦压力机
flying print 高速打印
flying printer 高速打印机
flying rafter 飞椽；飞子；飞檐椽
flying range 航程（航空器）
flying relative altitude 飞行相对高度
flying rope 牵引索；牵引绳
flying route 飞行路线
flying sand 飞砂
flying sandbank model 飞沙堆模型
flying sap 活动坑道
flying scaffold 挑出的脚手架；悬空脚手架；悬挂（式）脚手架；舷边作业吊架；挑出式脚手架；吊脚手架
flying school 航空学校
flying screed 挑出的整平板（混凝土）
flying season 飞行季节
flying shear 飞剪
flying shear cutter 飞剪式碎边剪
flying shear line 飞剪作业线
flying shears 飞剪机
flying shelf 悬空的工作架；悬空架；悬挂（式）脚手架
flying shore 墙撑（墙间的，用于房屋修理等）；悬空顶撑；横支撑木；横撑
flying shoring 横支撑工程
flying speed 航速（指航空器）
flying speed triangle 航行速度三角形
flying spot 扫描（射）点；浮动光点；飞（光）点
flying-spot camera 飞点式照相机；飞点扫描摄像机
flying-spot film reader 飞点式胶卷读出器
flying-spot microscope 扫描显微镜；飞点（扫描）显微镜
flying-spot pattern generator 飞点式测试图信号发生器
flying-spot recorder 飞点记录器
flying-spot scanner 飞点自动检测器；高速点扫描器；飞点扫描设备；飞点扫描器
flying-spot scanning 飞点扫描（法）
flying-spot scanning digitizer 飞点扫描数字化器
flying-spot scanning system 飞点扫描系统
flying-spot scanning tube 飞点扫描管
flying-spot system 飞点系统
flying-spot tube 扫描管；飞点管
flying-spot video generator 飞点式视频信号发生器
flying stair(case) 无柱的楼梯
flying switch 牵引溜放调车
flying system 吊景系统
flying test bed 飞行试验台
flying theodolite 飞行经纬仪
flying time 飞行时间
flying track of photography 摄影行迹
flying tuck 旋转式折页刀
flying vein 分支矿脉
flying veins 分枝矿脉
flying weather 飞行天气
flying windmill 直升机（俚语）
flying wire 承力缆索；承力索；悬索
fly-jerking 溜放调车
fly jib 副臂；飞臂；起重机臂；吊臂
flyjib nut 吊车臂螺帽
fly ladder 天桥爬梯；云梯顶部
fly-leaf（铰链接合中的）活动页；（铰链接合中的）活动板；扉页
fly lens 蝇眼透镜
fly level（无水准点的）快速水准测量
fly lighting gallery 灯光天桥
fly line 舞台吊索
fly loft 舞台上空
fly net 漂网；无底网；防虫网
Flynn's classification schema 弗林分类法
fly nut 翼形螺母；翼形螺帽；蝶形螺母
fly-off 利用蒸发池排水
fly-off water 植物截留的水；蒸发用的水
flyover 过街楼；立体交叉；跨线桥
flyover bridge 跨线桥；铁路在下交叉
flyover crossing 高架桥；立体交叉路（口）；立交桥；立交路口；跨线桥；上跨交叉；跨线交叉；立体交叉；跨线桥简易立体交叉
flyover culvert 立交涵洞
flyover junction 立体交叉路（口）；立交路口
flyover ladder 跨越梯
flyover road 架空通道；架空道路
flyover roundabout 跨线（桥）环形立体交叉；跨线桥；环形立体交叉（系统）；有фор线桥的环形立交
flyover structure 跨线桥结构；架空结构
flyover structure above channel(river) 跨河建筑物
flyover to 飞渡
fly-past 跨度
fly pinion 制动器轴小齿轮
fly press 飞轮式（螺旋）压力机；螺旋压机；螺旋（式）压力机；螺杆压机
fly-proofing 飞虫防护（物）；飞虫防护（剂）；防蒸发的；防蝇的
fly rafter 横橡；装饰的山墙封檐板
fly rail（舞台天桥上的）栏杆；（折叶板的）活络撑脚；折叶板
fly rock（爆破后的）岩石碎块；（爆破引起的）散飞石块
flysch 浊积岩；厚砂页岩水层；复理石；复理层【地】
flysch association 复理石组合
flysch facies 复理石相
flysch formation 复理层建造
flysch geosyncline 复理石地槽
flysch nappe 复理式推复体
flyschoid formation 类复理石建造
flysch sediment 复理式沉积物
fly screen 防蝇纱（窗）；防蚊纱纱；门窗铁纱装置
fly screw nut 蝶式螺帽

fly's eye electron optics 蝇眼电子光学
fly's-eye lens 蝇眼透镜;复眼透镜
fly's eye lens plate 复眼微透镜板
flysh 浊积岩;复理层【地】
fly sheared length 飞剪剪切后的定尺寸长度
fly-shunting 溜放调车(法)
fly-shunting track 溜放线
fly spray 灭蝇剂
fly spring 跳簧
fly stair(case)(舞台的)天桥楼梯
fly switching 飞钩
fly tipping 乱倒;无控制倾弃;无节制地倾弃
fly tower 屋顶塔楼;舞台塔
fly trap 捕蝇草;诱蝇笼
fly up in the wind 船首急速转向风
flyway 候鸟迁徙所经路径;舞台花道(自舞台向观众席伸出);舞台道
fly wheel 飞轮;惯性轮
flywheel action 飞轮作用
flywheel arm 飞轮轮辐
flywheel-assisted reduction gear unit 带飞轮减速机
flywheel balancing 飞轮平衡
flywheel belt pulley 飞轮滑轮
flywheel casing 飞轮壳
flywheel chopper 轮刀式切碎机
flywheel circuit 惯性同步电路
flywheel clutch 飞轮离合器
flywheel controlled radiator bleeder 惯性控制的除气装置
flywheel crowned for flat belt 有冠飞轮
flywheel cutter 轮刀式切碎机;飞轮切割机
flywheel damping effect 飞轮缓冲效应
flywheel dowel retainer 飞轮定缝销钉
flywheel drill 飞轮手钻
flywheel drive 飞轮传动
flywheel effect 飞轮效应
flywheel frame 飞轮架
flywheel friction welding 惯性摩擦焊
flywheel gear 飞轮装置
flywheel generator 飞轮发电机
flywheel governor 飞轮调节器
flywheel guard 飞轮罩
flywheel horsepower 飞轮的马力;飞轮的功率
flywheel housing 飞轮罩
flywheel housing gasket 飞轮壳衬垫
flywheel jet control 飞轮喷射控制
flywheel key 飞轮键
flywheel knife 轮刀
flywheel magneto 飞轮式永磁发动机
flywheel magnetogenerator 飞轮永磁式发电机
flywheel mark 飞轮记号
flywheel mass 飞轮质量
flywheel mechanical clutch press 飞轮机械离合器压机
flywheel moment 回转力矩;飞轮力矩
flywheel pilot flange 飞轮定心凸缘
flywheel pit 飞轮坑
flywheel plate 飞轮摩擦板
flywheel press 飞轮式压力机
flywheel pulley 滑轮飞轮
flywheel pump 带飞轮的泵;飞轮泵
flywheel reduction gear 带飞轮的减速器
flywheel rib 飞轮辐
flywheel rim 飞轮缘
flywheel ring gear 飞轮齿圈;飞轮环齿轮
flywheel rotor 飞轮转子
flywheel shaft 飞轮轴
flywheel shear bolt 飞轮剪断安全螺栓
flywheel sheave 滑车飞轮
flywheel starter 飞轮式起动器
flywheel starter gear 飞轮起动器齿轮
flywheel synchronization 飞轮同步
flywheel test 飞轮试验
flywheel thrust 飞轮止推环
flywheel time base 飞轮时基;非同步扫描
flywheel type alternator 飞轮式交流发电机
flywheel type friction welding 储[贮]能摩擦;储能摩擦焊
fly wire 悬索;(覆盖墙板接缝用的)细金属网;板缝盖网
FM receiver 调频接收机
FM transmitter 调频发射机
fmue hood 排风柜

fnsuma 日本式隔窗;(日本房屋中的)拉门
f-number 光圈数;光圈号数
foalsfoot 叶眼蕺
foam 海绵状的;起泡沫;泡沫
foamability 发泡性
foamable polystyrene 可发性聚苯乙烯
foam adhesive 泡沫胶黏剂
foam analysis 泡沫分析
foam and surface activating agent 泡沫和表面活化剂
foam annihilator 泡沫灭火器
foam application 泡沫的应用
foam arrested dust 泡沫捕尘
foam article 泡沫制品
foam-ash-silicate concrete 泡沫的粉煤灰硅酸盐混凝土
foam asphalt 泡沫地沥青
foam-back 泡沫背衬
foam-backed textile 泡沫塑料衬里织物
foam-back fabric 泡沫塑料面织物
foam backing 泡沫塑料衬里
foam board 泡沫板;泡沫塑料板材
foam bonding 泡沫材料黏合
foam booster 泡沫促进剂
foam breaker 破泡剂;泡沫抑制剂
foam breaking coil 泡沫破碎盘管
foam bubble 泡沫气泡
foam carbon 泡沫炭
foam carpet 泡沫垫层
foam casting 泡沫铸塑
foam catcher 泡沫捕集器
foam cell 泡沫气孔
foam cement concrete 泡沫水泥混凝土
foam cement screed 泡沫水泥地面
foam ceramics 泡沫陶器;多孔陶瓷
foam chamber 泡沫室
foam characteristic 泡沫特性
foam chute 泡沫溜槽
foam circulation fluid 泡沫循环液
foam clay 泡沫黏土;多孔性黏土
foam coated fabric 泡沫涂层织物;泡沫胶布
foam cock 泡沫旋塞;排污旋塞;吹洗锅炉开关
foam collapse 泡沫结构破坏
foam column 泡沫塔;泡沫发生塔
foam compound 泡沫材料
foam concentrate 泡沫液
foam concrete 泡沫混凝土
foam concrete(building)block 泡沫混凝土(砌)块
foam concrete filler 泡沫混凝土填料
foam concrete screed 泡沫混凝土地面
foam concrete tile 泡沫混凝土砖;泡沫混凝土瓦
foam connection 泡沫管接头
foam control 泡沫控制
foam control agent 控泡剂;泡沫控制剂
foam control agent powder 消泡剂;泡沫控制粉剂
foam core 泡沫塑料芯板;发泡塑料芯layer;泡沫芯层
foam-cored laminate 芯层发泡层压制品
foam core sandwich panel 泡沫芯夹层板
foam couple 泡沫管接头
foam crust 泡沫状结壳
foam-cushioned hardwood flooring 泡沫衬垫硬木地板
foam cushion method 泡沫垫层法
foam degumming 泡沫精练
foam degumming apparatus 泡沫精练机
foam density 泡沫密度
foam depressant 消泡剂;抑泡剂
foam dome 泡沫塑料穹隆
foam drilling 泡沫钻进;泡沫集尘钻眼法;泡沫冲洗钻进
foam drying 泡沫干燥
foam dust separator 泡沫除尘器;沉浸除尘器
foam dust suppression 泡沫抑尘;泡沫防尘
foam dy(e)ing 泡沫染色
foamed adhesive 发泡胶黏剂;泡沫黏合剂;胶黏剂
foamed adhesive for impregnating carpet 地毯泡沫浸渍黏合剂
foamed aluminium 泡沫铝
foamed asphalt 泡沫(石油)沥青
foamed blast-furnace slag 泡沫状鼓风炉渣;泡沫炉渣;水淬炉渣
foamed blast-furnace slag concrete 发泡高炉熔渣混凝土

foamed blast-furnace slag concrete building block 高炉泡沫熔渣混凝土砌块
foamed blast-furnace slag powder 高炉泡沫熔渣粉
foamed board 泡沫(塑料)板
foamed cement concrete 水泥泡沫混凝土
foamed cement screed 发泡水泥地面
foamed ceramics 泡沫陶瓷
foamed clay 轻质泡沫黏土制品;发泡黏土
foamed coating 泡沫涂料;泡沫涂层
foamed concrete 泡沫混凝土;多孔混凝土
foamed concrete block 泡沫混凝土砌块
foamed concrete(building)tile 泡沫混凝土建筑瓦
foamed concrete filler 泡沫混凝土填料
foamed concrete mixer 泡沫混凝土搅拌机
foamed concrete screed 泡沫混凝土地面
foamed concrete slab 泡沫混凝土板
foamed concrete wall block 泡沫混凝土墙块
foamed concrete wall slab 泡沫混凝土墙板
foamed corundum brick 泡沫刚玉砖
foamed cupola 泡沫塑料圆屋顶
foamed glass 多孔玻璃;泡沫玻璃
foamed glass aggregate 泡沫玻璃骨料
foamed glass block 泡沫玻璃块
foamed glue 泡沫胶
foamed-in-place filler 泡沫塑料填料
foamed-in-place insulation 现场发泡绝热塑料
foamed-in-place insulation material 现制泡沫保温材料
foamed-in-situ plastics 现场发泡塑料
foamed insulating material 泡沫保温材料
foamed insulation 泡沫绝缘;泡沫隔声层;泡沫隔热层
foamed latex 泡沫胶乳;泡沫乳胶
foamed lava 泡沫熔岩
foamed lava concrete 泡沫熔岩混凝土
foamed lava concrete wall slab 泡沫熔岩混凝土墙板
foamed lava gravel 多孔火山岩砾石
foamed lightweight concrete 泡沫轻质混凝土
foamed metal 泡沫金属
foamed mortar 泡沫砂浆;泡沫灰浆
foamed mud 泡沫泥浆
foamed natural rubber 泡沫天然橡胶
foamed neoprene 泡沫氯丁橡胶
foamed pearlite 泡沫珍珠岩
foamed plaster mo(u)lding 泡沫石膏造型
foamed plastic ball 泡沫塑料球
foamed plastic board 泡沫塑料板
foamed plastic concrete 泡沫塑料混凝土
foamed plastic cupola 泡沫塑料圆屋顶
foamed plastic cushioning 泡沫塑料衬垫
foamed plastic cylinder 泡沫塑料圆筒
foamed plastic dome 泡沫塑料穹隆
foamed plastic insulated hold 泡沫塑料隔热舱
foamed plastic insulation 泡沫塑料绝缘
foamed plastic pattern 泡沫塑料模
foamed plastic plaster baseboard 泡沫塑料石膏底板
foamed plastic roller 泡沫塑料滚压装置;泡沫塑料滚筒
foamed plastics 泡沫塑料;发泡塑料;塑料泡沫
foamed plastic sandwich construction 泡沫塑料夹层结构
foamed plastic seal 泡沫塑料密封
foamed plastic sealing strip 泡沫塑料密封条
foamed plastic sheet 泡沫塑料薄板
foamed plastic waterbar 泡沫塑料止水条
foamed polyethylene 泡沫聚乙烯
foamed polystyrene 发泡聚苯乙烯;泡沫聚苯乙烯
foamed polystyrene board for thermal insulation purposes 热绝缘泡沫聚苯乙烯板;泡沫聚苯乙烯隔热板
foamed polystyrene feeder 泡沫塑料冒口
foamed polystyrene pattern 泡沫塑料模型
foamed polystyrene tile 聚苯乙烯泡沫塑料片砖;聚苯乙烯泡沫塑料片瓦
foamed polyurethane 泡沫聚亚氨酯;聚氨基甲酸酯泡沫塑料
foamed polyurethane strip 泡沫聚亚氨酯条;聚氨基甲酸酯泡沫塑料条
foamed polyvinyl chloride 聚氯乙烯泡沫塑料;泡沫聚氯乙烯
foamed resin 泡沫树脂

foamed rigid polyvinyl chloride 硬质泡沫聚氯乙烯;硬质聚氯乙烯泡沫塑料
foamed rubber 泡沫橡胶
foamed rubber latex 泡沫橡胶乳汁;泡沫橡胶浆
foamed rubble 泡沫循环
foamed runway 敷泡沫层的跑道
foamed seal 泡沫塑料密封
foamed sheet 泡沫塑料薄板
foamed slag 高炉泡沫矿渣;泡沫熔渣;泡沫炉渣;多孔矿渣
foamed slag aggregate 发泡矿渣骨料
foamed slag brick 泡渣砖
foamed slag concrete 泡沫熔渣混凝土;泡沫混凝土;泡沫炉渣混凝土
foamed slag concrete block 多孔矿渣骨料混凝土砌块
foamed slag concrete hollow block 泡沫矿渣混凝土空心砌块
foamed slag concrete plank 泡沫熔渣混凝土厚板;泡沫矿渣混凝土厚板
foamed slag concrete pot 泡沫熔渣混凝土套管;泡沫矿渣混凝土套管
foamed slag concrete slab 泡沫熔渣混凝土板;泡沫矿渣混凝土板
foamed urea-formadehyde 泡沫脲甲醛
foamed vinyl resin 泡沫乙烯树脂
foamed water bar 泡沫塑料止水条
foamed waterstop 泡沫塑料止水器;泡沫塑料止水剂;泡沫塑料水密封
foamed wool 发泡棉
foam end point 泡沫终点
foam-entraining admixture 泡沫掺加剂
foamer 泡沫剂;发泡剂
foamer meter 泡沫测定仪
foamet 泡沫分离法
foamex 发泡树脂
foam extinguisher 泡沫灭火器
foam extinguishing agent 泡沫灭火剂
foam extinguishing system 泡沫灭火系统
foam fabric 泡沫胶布
foam factor 发泡倍数
foam feed 泡沫进给;泡沫给进(钻眼防尘用)
foam-filled block 充泡沫料的砌块
foam-filled fender 充填泡沫护舷;充泡沫护舷
foam-filled floating fender 浮式泡沫护舷
foam-filled gypsum block 充气石膏砌块
foam-filled gypsum board 充气石膏板
foam fillings 泡沫充填料
foam filter 泡沫过滤器
foam fire extinguisher 泡沫灭火器;泡沫灭火机
foam flame bonding 泡沫塑料热熔粘贴
foam float 泡沫浮子
foam flo(a)tation 泡沫浮选(法)
foam floating roof 泡沫浮层
foam flow 泡沫流
foam formation 泡沫生成
foam-forming admixture 泡沫剂;发泡剂
foam-forming fire retardant agent 泡沫灭火剂;发泡防火剂
foam-forming liquid 起泡剂
foam fractionation 泡沫分级(器);泡沫分级分离
foam-gas concrete 泡沫加气混凝土
foam gasket 泡沫密封垫
foam-generating unit 泡沫发生器
foam generator 泡沫发生器
foam glass 泡沫玻璃
foam glass block 泡沫玻璃砖
foam glass brick 泡沫玻璃砖
foam glue 泡沫黏合剂;泡沫黏合剂;发泡黏合剂;发泡胶黏剂
foam gluing 泡沫胶合
foam go-devil 泡沫清管器
foam height 泡沫高度
foam hydrant 泡沫消防龙头;灭火器泡沫的给水栓;泡沫(灭火器)消防栓
foamicide 破泡剂
foam impression 泡沫痕
foam in aeration tank 曝气池泡沫
foam index 泡沫指数
foaminess 起泡度
foaming 起泡;发泡
foaming ability 起泡沫(能)力;发泡能力
foaming accelerator 发泡促进剂
foaming action 发泡作用;起泡作用

foaming activated sludge 发泡活性污泥
foaming activated sludge plant 发泡活性污泥处理设备
foaming adjutant 起泡剂
foaming agent 起泡剂;泡沫剂;发气剂;发泡剂
foaming and separation method 起泡分离法
foaming capacity 起泡沫(能)力
foaming chemical 发泡剂
foaming coefficient 起泡系数
foaming concrete 泡沫混凝土
foaming equipment 泡沫涂层装置
foaming flo(a)tation method 起泡分离法
foaming in place 现场发泡
foaming in situ 现场发泡
foaming in sludge digestion 污泥消化发泡
foaming line 泡沫线
foaming method 泡沫法
foaming polyurethane 发泡聚氨脂
foaming potential of sludge 污泥发泡势
foaming power 发泡能力
foaming process 起泡作用;泡沫法
foaming slag 轻质渣;泡沫渣
foaming space 泡沫空间
foaming stability test 泡沫稳定性试验
foaming station 发泡站
foaming structure 泡沫结构
foaming substance 起泡物;发泡(性)物质
foaming test 起泡试验;发泡试验
foam inhibiting agent 抑泡剂;阻泡剂
foam inhibitor 抑泡剂;泡沫抑制剂;防泡沫剂
foam injection 泡沫注射
foam-in-place 现场发泡
foam installation 泡沫灭火装置
foam insulation 泡沫绝缘;泡沫保温
foamite 灭火药沫;泡沫灭火剂
foamite extinguishing method 泡沫灭火法
foamite system 泡沫灭火装置;泡沫灭火系统
foam jet 泡沫喷嘴
foam killing agent 消泡剂;破泡剂
foam-laminated fabric 泡沫层织物;泡沫塑料面织物
foam laminate fabric 泡沫塑料层压织物
foam laminating 泡沫塑料层压黏合
foam laminating machine 泡沫塑料层压机
foam lance 泡沫喷枪
foam latex 泡沫乳胶
foam layer 泡沫层
foam life 泡沫寿命
foam line 浪花线;泡沫(界)线
foam main 泡沫总管
foam maker 泡沫发生器;发泡器
foam-making duct 泡沫生成管
foam mark 泡沫痕
foam mat 泡沫材料
foam material 泡沫材料
foam metal 泡沫金属
foam meter 泡沫仪
foam mixing chamber 泡沫搅拌室;泡沫发生室
foam modifier 泡沫调节剂
foam mortar 泡沫砂浆;泡沫灰泥
foam mo(u)ld 泡沫塑料模
foam neoprene 泡沫氯丁橡胶
foam number 泡沫值
foam over 泡沫携带
foam persistence 泡沫持久性
foam phase separation 泡沫相分离法
foam pig 泡沫清管器
foam pipe 泡沫管
foam pipe lagging 泡沫保温管套
foam plaster base 泡沫石膏基底
foam plastics 泡沫塑料
foam polyurethan(e) coating 聚氨酯泡沫涂料
foam powder 发泡粉
foam-producing ability 造泡能力
foam-producing matter 发泡物质;产泡沫物质
foam profile 泡沫分布型
foam-promoting builder 泡沫促进剂
foam-proof agent 防泡剂
foam proportioner 泡沫比例器
foam pump 泡沫泵
foam recycle 泡沫循环
foam reducing composition 抑泡剂
foam removal 泡沫去除
foam reservoir mo(u)lding 泡沫储[贮]液模塑工艺

foam-resistant agent 防泡剂
foam retention 泡沫保持能力
foam ribbon 泡沫束带氏;泡沫带饰
foam rubber 多孔橡胶;橡胶海绵;海绵橡胶;泡沫橡胶
foam-rubber backing 泡沫橡胶底衬
foam-rubber cushion 海绵垫
foam rubber cutter 泡沫橡胶切割器
foam rubbered seat 泡沫橡胶座
foam rubber fender 漂浮橡胶护舷
foam rubber product 泡沫橡胶制品
foam rubber seat 泡沫橡胶座
foam rubber tyre 泡沫橡胶轮胎
foam scraper 泡沫清管器
foam scrubber 泡沫洗涤器;泡沫吸收器
Foamseal 富姆泡沫止水剂
foam separation 泡沫分离
foam separation process 泡沫分离法
foam separator 泡沫分离器
foamsil 泡沫石英玻璃
foam-silicate 泡沫硅酸盐
foam-silicate concte 泡沫硅酸盐混凝土
foam-silicate concte slab 泡沫硅酸盐混凝土板
foam slab machine 泡沫塑料成型机
foam slag 泡沫矿渣
foam-slag silicate concrete 泡沫控制硅酸盐混凝土
foam slump 泡沫沉陷距
foam smothering system in machinery space 机舱泡沫灭火系统
foam solution 泡沫溶液
foam sponge indentation tester 泡沫胶耐压试验机
foam sponge mo(u)ld 泡沫胶硫化模
foam sprinkler 泡沫喷洒器
foam stability 泡沫稳定性
foam stabilizer 泡沫稳定剂
foam suppressant 消泡剂
foam suppressing agent 抑泡剂
foam suppressor 消泡剂;抑泡剂;泡沫抑制剂
foam synthetic(al) resin 泡沫合成树脂
foam test 泡沫试验
foam-tester 泡沫试验器
foam-thermal insulation 泡沫绝热材料
foam-to-fabric process 泡沫成布法
foam tower 泡沫塔
foam tower scrubber 泡沫塔除尘器
foam trap 泡沫收集器
foam type 泡沫式灭火器
foam type fire extinguisher 泡沫型灭火器
foam type fire extinguishing system 泡沫型灭火系统
foam value 泡沫值;泡沫数值
foam volume 泡沫体积
foam water cannon 泡沫消防炮
foam with open cells 开孔泡沫
foamy 多泡沫的
foamy adhesive 多泡沫的胶黏剂
foamy lava 泡沫熔岩
foamy meltdown 泡沫的消散
foamy sealant 泡沫密封层
foamy structure 泡沫结构
foatrope knot 握索结
FOB C_5 包括5%佣金的离岸价
FOB destination 目的地交货
foble board 由四根钻杆组成的立根高度塔板
FOB plane 飞机上交货价
FOB port of shipment 装运港离岸价
focal 焦点的
focal adjustment 焦距调整
focal aperture 聚焦孔径;焦斑口径
focal area 震源区;聚焦区;聚焦面积;焦点区;焦点范围;焦斑面积
focal area for sound 声音焦点
focal axis 焦轴
focal center length 光心高度
focal chord 焦弦
focal circle 聚焦圆;焦圆
focal coordinate 震源坐标
focal curve 焦点曲线
focal cusp 焦会线
focal date 假定起息日
focal depth 震深;震源深度;焦深;焦点深度
focal dimension 震源大小
focal distance 震源距(离);焦距
focal distance of photographic(al) lens 摄影物

镜焦距
focal distance ratio 焦距比
focal eikonal 焦点程函
focal image 焦面像
focal interval 焦间节
focal involution 焦点对合
focal isolation 焦聚分光;焦点隔离法
focalization 焦距调整;对焦
focalizer 聚焦系统;聚焦设备;聚焦器;对焦仪
focal length 焦距;焦点距离
focal length calibration 焦距校准
focal length determination 焦距测定
focal length equivalent 等效焦距
focal length of aerial camera 航摄仪焦距
focal length of lens 透镜焦距
focal length range 调焦范围
focal length ratio 焦距比
focal length scale 焦距刻度
focal length setting 焦距调节
focal length spindle 焦平面主轴
focal length tolerance 焦距容限
focal line 焦线;焦距线
focal mechanism 震源机制
focal mechanism solution 震源机制解
focal monochromator 聚焦单色仪
focal plane 焦平面标志;焦(平)面;焦点(平)面
focal plane aberration 焦平面像差
focal plane arrays 焦平面阵列
focal plane camera 焦平面快门照相机;焦面摄影机
focal plane filter 焦面滤光片
focal plane frame 焦面框
focal plane glass 焦平面玻璃
focal plane modulator 焦平面调制器
focal plane plate 焦面玻璃;焦面板;焦距玻璃
focal plane scanning system 焦平面扫描系统
focal plane shift 焦面位移
focal plane shutter 焦平面快门;焦面快门
focal plane slit 焦面狭缝
focal point 重点;震源点;联络点;焦点;燃点
focal point investigation 重点调查
focal point of stress 应力集中点
focal point of subsidence 作用集中点(采矿);影响集中点(采矿);沉陷集中点
focal point of working 工作重点;采掘中心(采矿);作业中心;开挖中心;回采中心
focal point triangulation 焦点三角测量
focal position 焦点位置
focal power 光焦度;度;倒焦距
focal process 震源过程
focal radii 焦点半径
focal ratio 焦比
focal reducer 缩焦器
focal region 震源区;汇点区
focal setting 焦距装定;焦距调整
focal shift 焦点位移
focal shutter 焦面快门;幕焦快门
focal sphere 震源范围;焦球
focal spot 焦点;焦斑
focal surface 焦面
focal-to-site distance 震源至场地距离
focal-to-site station distance 震源至台站距离
focal variation ratio 变焦距倍率;变焦倍比
focal volume 震源体积
focal zone 震源区;聚焦带
foci-finder 焦点检定器
focimeter 焦距测量仪;焦距计;焦点(测定)计
foco-collimator 测焦距准直仪;测焦距准直光管
focoid 虚圆点
focometer 焦距仪;焦距计
focometry 测焦距术
focus 中心点;焦点;集中;聚焦
focus at infinity 无限远调焦
focus-coil assembly 聚焦线圈系统;电磁聚焦组件
focus-coil axis 聚焦线圈轴
focus coil housing 聚焦线圈壳
focus control 焦点调整;调焦装置
focus control system 聚焦控制系统
focus correction 焦点校正
focus current regulator 聚焦电流稳定器
focus-deflection coil 聚焦偏转线圈
focus drum 聚焦筒
focused beam 聚焦光束
focused condition 聚焦条件
focused Gaussian laser beam 高斯聚焦激光束
focused image holography 聚焦像全息摄影
focused log 聚焦测井
focused pollution index 主要污染指数;主要污染指标
focused reflector 焦正反射面
focused transducer 聚焦型探头
focuser 聚焦器;焦距放大镜
focus-film distance 焦片距
focus for infinity 无限远焦点
focus glass 调焦玻璃
focus group interview 重点集体面谈话法
focus-image distance 焦像距
focusing 聚焦(法);调焦;对焦(点);对光
focusing accuracy 聚焦精度
focusing action 聚焦作用;调焦作用
focusing adjustment 调焦
focusing aid 聚焦装置;对光辅助装置
focusing arrangement 聚焦装置
focusing barrel 调焦镜筒
focusing camera mirror 照相机聚焦镜
focusing coil 聚焦线圈
focusing condition 聚焦条件
focusing criterion 聚焦准则
focusing cup 聚焦杯
focusing current 聚焦电流
focusing defect 聚焦缺陷
focusing device 聚焦装置;聚焦器;调焦装置;对焦仪
focusing diffratometer 聚焦衍射计
focusing distance 聚焦距离;调焦距离
focusing electrode 聚焦电极;聚焦电极
focusing error 对光误差
focusing eyeglass 图像聚焦检验器
focusing eyepiece 调焦目镜
focusing factor 聚焦系数
focusing foil 聚焦箔
focusing gear 调焦传动齿轮
focusing geometry 聚焦几何形状
focusing glass 对光镜;图像聚焦检查镜;对光放大镜
focusing knob 对光钮
focusing knurl 调焦滚花螺旋
focusing laser Doppler velocimeter 聚焦激光多普勒测速计
focusing lens 聚焦透镜
focusing lever 调焦拨杆
focusing magnifier 对光放大镜
focusing mark 聚焦标记;聚焦标;对光标
focusing mechanism 调焦装置;调焦机理
focusing microscope 调焦显微镜
focusing mirror 聚焦反射镜
focusing mount 聚焦架
focusing movement 调焦
focusing negative 聚焦底片
focusing nut 调焦螺母
focusing of energy 能量聚集
focusing of ion beam 离子束聚焦
focusing optic(al) fiber 聚焦光纤
focusing plane 调焦平面
focusing projector 调焦投影器
focusing ring 对光环;聚焦环;调焦圈;调焦环
focusing scale 焦距刻度;焦距标度
focusing screen 聚焦屏;对焦屏
focusing screw 聚焦螺旋;聚焦调节螺丝;调焦螺旋;对光螺旋;对光螺丝;定焦螺丝
focusing slide 调焦滑筒
focusing solar power plant 聚焦式太阳能发电厂
focusing Soller-slit system 苏莱尔调焦狭缝系统
focusing spectrometer 聚焦分光计
focusing surface 聚焦面
focusing system 聚焦系统
focusing telescope 聚焦节距镜
focusing tube 聚焦筒
focusing unit 聚焦装置
focus knob 聚焦旋钮
focus lamp 聚焦灯
focus level 聚焦程度;焦点水平
focus mask 聚焦栅极;聚焦网
focus modulation 聚焦调制;聚焦调整
focus mount 聚焦框架
focus-object distance 焦物距
focus of infection 传染中心
focus of lens 透镜的焦点
focus-out 散焦
focus pack 聚焦组件
focus perspective 焦点透视
focus point 聚焦点
focus projection and scanning 聚焦投影和扫描法
focus range 聚焦范围
focus regulation 聚焦稳压
focus regulator 聚焦稳压器
focussed spot diameter 焦斑直径
focus set 聚焦调整
focus setting 调焦;焦点设置
focus shift 焦点移动;焦点位移
focus(s)ing glass 调焦屏
focus(s)ing lens 调焦透镜
focus test negative 聚焦检验底片
focus-to-film distance 焦点到底片距离
focus vesiculosus 墨角藻
fodder 饲料;铅锭标准重量;素材;饲料
fodder chopper 铡刀
fodder cutter 饲料切碎机
fodder distributor 饲料分配器
fodder grass 牧草
fodder grinder 饲料粉碎机
fodder mixing machine 饲料搅拌机
fodder rack 饲料架
fodder tower 饲料塔
foehn 焚风
foehn air 焚风空气
foehn cloud 焚风云
foehn cyclone 焚风气旋
foehn island 焚风岛
foehn nose 焚风鼻
foehn pause 焚风停顿
foehn period 焚风时期
foehn phase 焚风阶段
foehn phenomenon 焚风现象
foehn storm 焚风风暴
foehn trough 焚风槽
foehn wall 焚风云壁
foetial 臭的
fog 胶片走光;灰雾(摄影);发雾;二级能见度(能见距离2.5链)
fog alarm 雾航音响信号
fog alert system 雾警报装置
fog applicator 烟雾发生器
fog bank 雾阵;雾堤
fogbeam 雾光(汽车前灯防雾光束)
fog bell 警钟雾号;雾钟
fog belt 雾区;雾带;多雾地区
fog bound 因雾受阻
fog bow 雾虹
fog buoy 雾中拖标;雾号浮标;雾标
fog cabinet 雾室
fog chamber 云室
fog channel 雾道
fog climax 雾障演替顶极
fog coat 喷雾涂层;雾化沥青封面
fog-cooled reactor 雾冷堆
fog crystal 雾晶
fog-cured 喷雾养护的;喷雾管嘴;雾室养护的
fog curing 喷雾养护;雾室养护;雾气养护
fog day 雾日
fogday traffic 雾中行车
fog density 灰雾密度;雾密度
fog-density indicator 雾浓度指示器;雾密度指示器
fog deposit 冻雾覆盖层
fog detecting light 雾情探测灯
fog detection 雾探测
fog detector light 雾情探测灯
fog diaphone 低音雾号
fog dispeller 消雾剂
fog dispersal 消雾;雾消散
fog disperser 消雾剂;散雾剂
fog dissipation 雾消散
fog-dog 雾峰中明亮部分
fog drip 雾雨;雾滴;树上雾滴
fog drop 雾滴
fog duration 雾持续时间
fog dust 雾烟
fog easter 雾虹
fogeater 雾中升起的满月
fog filter 雾镜
fog forest 雾林
fog formation 雾生
fogged coat 虚枪涂层;湿罩光
fogged metal 晦面金属;发雾金属
fog generator 烟雾发生器

fogger 烟雾发生器;润湿器
fogging 失去光泽;凝化;模糊;迷惑试验;起雾;成雾
fogging and spraying machine 喷烟喷雾两用机
fogging machine 喷烟器
fogging oil 雾化油
foggite 羟磷铝钙石
fog gong 警钟雾号;雾锣
fog gun 烟雾喷枪;雾炮
foggy 有雾的
foggy day 有雾日雾天;雾日
fog horizon 雾层顶
fog horn 雾天音响信号喇叭;雾角;雾号;雾笛;雾喇叭
fog investigation dispersal operations 研究性的消雾系统
fog laden air 带雾空气
fog-lamp 雾天行车灯
foglie 木叶装饰的
fog light 雾天灯
fog machine 烟雾机;雾烟机
fog mark 雾标
fogmeter 雾量计
fog nautophone 电动高音雾号
fog nozzle 喷雾管嘴;喷雾嘴
fog patches 碎雾
fog pollution 雾害
fog precipitation 雾雨
fog preventing agent for cleaning mirror face 镜面清洁防雾剂
fog prevention forest 防雾林
fog quenching 喷雾淬火
fog rain 雾雨
fog reactor 雾冷反应堆
fog reed 雾笛
fog region 雾区;多雾地区
fog resistant glass 抗雾玻璃
fog room 湿养护室;养护室;喷雾室;雾室
fog scale 雾级标度
fog seal 雾状封面(路面);薄层封面(路面)
fog seal coat 雾状封面(路面)
fog shower 雾雨
fog signal 雾信号;雾(情)信号;雾号;雾笛;雾标
fog signal emitter 雾号发声器
fog signalling plant 雾号站
fog signal station 雾信号台;雾号台
fog siren 雾汽笛;雾笛
fog solution 雾气溶液
fog spar 雾中拖锚
fog spray 喷丸器;喷雾
fog spraying 喷雾
fog spraying curing 喷雾养护
fog track 雾径迹
fog trumpet 雾喇叭
fog two-phase flow 雾状两相流动
fog-type insulator 耐雾绝缘子;防雾绝缘子;防雾绝缘器
fog-type nozzle 雾式喷嘴
fog value 灰雾值
fog warning 雾警
fog warning apparatus 雾警设备
fog whistle 雾哨;雾笛
fog wind 雾风
fog zone 多雾地区
foidite 似长石岩
foil 叶形装饰;金属薄片;花瓣形;配景;探测片;衬线;衬托;薄金属(片);箔材;箔
foil activation measurement 箔活化测量
foil-activation technique 箔激活技术
foil aperture 全口径
foil array 箔片束
foil backed gypsum board 滤箔石膏板
foil backed plasterboard 铝箔衬背石膏板
foil bag 箔衬袋
foil bearing 箔带轴承
foil-borne 翼航状态
foil changer 箔交换器
foil coating 金箔涂料
foil coil 箔线圈
foil conductor 箔导体
foil count 探测片计数;薄片放射性计数
foil cover(ing) 金属箔蒙皮;金属箔保护层
foil craft 水翼艇
foil decorating 叶形装饰(的);薄片装饰(的)
foil detector 金箔探测器;探测片;箔探测器

foil dosimeter 箔剂量计
foiled arch 叶形饰拱
foil encapsulated strain ga(u)ge 金属箔屏罩应变计
foil faced fiber glass 铝箔面纤维玻璃
foil-free dry 不粘箔干
foil fuse 箔熔断丝
foil gasket 箔垫圈
foil ga(u)ge 应变片
foil holder 探测片支架
foil holding rod 探测片支架杆
foiling 叶形饰
foil insert(ion) 金属箔嵌衬(物)
foil insulant 绝缘膜;薄绝缘材料;薄绝热;箔保温材料
foil insulation 金属箔绝缘;箔绝热
foil jacket 金属箔外套;金属箔蒙皮
foil laminating machine 压金属箔机
foil-like finish 金属箔状装饰
foil maker 制箔机
foil mica capacitor 云母电容器
foil mill 箔材扎机
foil mionitor 箔监测仪
foil nozzle 导叶通道
foil paper 金属箔墙纸;衬底纸;箔纸
foil pliers 箔镊
foil-polystyrene board 聚苯乙烯金箔胶合板
foil polystyrene laminate 聚苯乙烯金箔层合板
foil printing machine 箔片印刷机
foil resistance strain ga(u)ge 箔阻应变规
foil rolling mill 制箔碾压工厂
foil sample 箔样品
foil sampler 金属薄片取土器;环刀取样器;环刀取土器(活塞式);衬片取样器;薄钢带取土器;薄壁取样器;箔钢取土器
foil sheet(ing) 金属箔薄片;箔叶;箔片
foil-shrink film bag processing machine 箔收缩型薄膜袋加工机
foil soil sampler 薄壁取土(样)器
foil support 探测片支架
foil surface insulation 金属箔表面绝缘
foil target 箔靶
foil thermophone 箔式热致发声器
foil-type safety valve 箔式安全阀
foil-type strain ga(u)ge 箔式应变片;箔式应变计
foil wallpaper 薄金属片墙纸
foil winding 箔式线圈;箔绕组
foil wrapping machine 箔片包装机
foist 蒙混;轻型驳船
Foitzik photoelectric(al) photometer 福伊泽克光电光度计
Foke block 福克块规
Fokker-Planck equation 福克—普朗克方程
fold 皱褶;皱折;褶曲;倍;褶皱;折叠;折边
foldable grandstand 可装折的看台
foldable screen 可折屏风
fold amplitude 褶皱波幅
fold angle 折角;折合角
fold apex 褶皱顶
fold arc 褶皱弧
fold axis 褶皱轴
fold back 折起来;扭曲;向后折叠
fold back circuit 监听线路
fold basin 褶皱盆地
fold belt 褶皱带;造山带
fold boat 橡胶帆布艇
fold breccia 褶皱角砾岩
fold coast 褶皱海岸
fold core 褶皱核
fold crack 折裂;折合处裂缝;折叠裂缝
fold domain 折叠区域
fold-down bed 折叠床
fold down(wards) 朝下折叠
folded 褶皱的;折叠的
folded and grooved seam 折叠和槽式接缝
folded antenna 折叠天线
folded area 褶皱面;褶板;折板(屋顶)
folded basement 褶皱基底
folded base ring on end supports 终端支柱上的弯折圈;端柱上的折叠底座圈
folded buff 折叠布轮抛光
folded cantilever contact 折叠悬臂式触点簧片
folded casement 折叠玻璃窗

folded cavity 折叠共振腔;折叠空腔
folded chain 褶皱山脉
folded concrete 折叠式混凝土
folded continuous loop 折叠式连续环
folded cylinder TR tube 折叠圆柱天线开关管
folded cylindrical surface 褶皱的圆柱面;折叠式圆柱面
folded dipole 折合偶极子;折叠偶极子;折叠偶极(天线)
folded dipole antenna 折合振子天线;褶叠偶极(天线)
folded dipole reflector 折叠偶极子反射器
folded doublet 折叠对称偶极天线
folded fault 褶皱断层
folded filament yarn 双股丝
folded filter 折叠式过滤器
folded flooring 地板折接铺设;折叠式楼板
folded fold 重复褶皱
folded form 折叠型
folded leaf 折叠式插页
folded light beam 折叠光束
folded media type air filter 折叠型空气过滤器
folded migmatite 褶皱状混合岩
folded mountain 褶皱山
folded optic(al) path 折曲光程
folded outer edge bar 褶皱外缘坝
folded over joint 折接
folded plate 折合板;多棱薄壳;褶板;折板
folded plate action 折板作用
folded plate concrete roof 折板混凝土屋顶
folded plate construction 折板建筑;折板结构
folded plate cupola 折板式圆屋顶
folded plate dome 折板穹隆
folded plate roof 折板屋顶
folded plate segment 折叠扇形板;折叠弓形板
folded plate shell roof 折叠板壳屋顶
folded plate structure 折板(式)结构;折板结构;折板建筑
folded plate theory 折板理论
folded region 褶皱区
folded region karst 褶皱区岩溶
folded septum envelope 折叠式滤芯袋
folded sheet metal 挤压的波钢板
folded slab 折板
folded slab cupola 折板(式)圆屋顶
folded slab dome 折板穹隆
folded slab roof 折板屋顶
folded slab segment 折叠扇形板;折叠弓形板
folded slab structure 折板结构
folded staple fiber yarn 定长纤维合股纱
folded strata 褶皱的地层
folded structure 褶皱构造;折(板)结构
folded table 符号监测表
folded thermocouple 折叠式热电偶
folded tissue 折叠组织
folded top antenna 折顶天线
folded vernier 双读游标;重叠游标
folded yarn 并捻纱
fold elements 褶皱要素
folder 纸夹;织物折断强度试验器;折页器;折页机;折叠器;折纸机;折布机;落布架;夹;文件夹;速缝机;断裂试验仪
folder gluer 折叠胶黏机
fold facing 褶皱面向
fold fault 褶皱断层
fold-fault mountain 褶皱断层山
fold frequency 褶皱频率
fold generation 褶皱世代
fold increase 成倍增加
folding 褶皱作用;折叠式的;叠合;折叠;折边
folding altar 折翼祭坛
folding amplifier 折叠放大器
folding anchor 折爪锚
folding back 折叠式靠背;折叠式椅背
folding basin 折叠式洗脸台
folding bed 折叠床
folding belt 褶皱带
folding blade 折页刀
folding board 罐托
folding boat 折叠式小艇;橡胶帆布艇
folding boom 折叠式吊杆
folding bridge 开合桥
folding bridge bay 折叠式桥节
folding camera 折叠式照相机

folding casement 折(叠门)窗;双扇窗;窗扇;折叠玻璃窗;折叠窗扇
folding chair 折椅;折叠椅
folding concrete form 折叠式混凝土模板
folding container 折叠式集装箱;可折叠式集装箱
folding conveyer 可折式输送机
folding conveyer belt 折叠式传送带
folding cylinder 折页滚筒
folding door 双扇门;折门;折叠式车门;折叠门
folding door box 折叠门箱
folding door fitting 折叠门配件
folding door furniture 折叠门装置
folding door hardware 折叠门五金;折叠门配件
folding drilling mast 折叠式钻杆
folding earthquake 褶皱地震
folding easel 折叠画架
folding elevator 折叠式升运器
folding engine bonnet 折叠式发动机罩
folding fan with bone rib 白骨折扇
folding fin 折叠式尾翅
folding fissure 褶皱裂缝
folding foot rest 折叠式搁脚板
folding frequency 折叠频率;卷叠频率
folding furniture 折叠家具
folding gate 折叠门;折叠大门;活栅门;铁栅栏门
folding handle 折叠式手柄
folding-hanged 折叠铰链
folding hatch cover 折叠式舱(口)盖
folding head 折叠式车顶
folding-hinged door 铰链门
folding hoop 折合箍
folding horst 褶皱地垒
folding jaws 折页滚筒叼牙
folding-jib gantry 折臂式高架起重机
folding key 折叠钥匙
folding ladder 人字梯;折梯;折叠梯
folding lash plate 折叠式连接件(集装箱)
folding lattice gate 折叠式花格大门;收缩式栅栏门
folding lavatory 折式盥洗台
folding layers 褶皱层
folding leaf 折合表尺板
folding leg 折合架;折叠架
folding length 折叠链长度
folding(life)boat 折叠式救生艇
folding litter 折叠式担架
folding machine 折边机;折弯机;折页机;折叠机
folding magnifier 折叠式放大镜
folding map 折叠式地图
folding mast 折叠桅杆;折叠式桅杆
folding mirror 折叠式反射镜
folding mirror stereoscope 折叠式反光立体镜
folding mo(u)ld 折叠式模板
folding movement 褶皱运动
folding name 褶皱名称
folding nappe 褶皱推覆体;褶皱推复
folding obstetric table 折叠式产床
folding of drawings 折叠图纸
folding panel 折叠式镶板门
folding partition(wall) 折叠式隔墙;折叠式隔断;折叠(式)间壁;折隔屏;折叠式板壁;折隔断
folding period 褶皱期
folding phase 褶皱幕
folding plate camera 折叠式硬片照相机
folding plates 折页插图;折页板
folding platform 折叠式平台
folding pocket magnifier 折叠式袖珍放大镜
folding pocket measure 折尺
folding pocket rule 可折(袖珍)尺
folding press 折边机;折叠机
folding ramp 折叠式坡道
folding ratio 叠合率
folding rod 折叠杆
folding rollers 折页辊
folding roll-film camera 折叠式胶片照相机
folding roof 折顶
folding roof body 折顶车身
folding ruler 折尺;花杆
folding sash 折窗扇;折窗框
folding scale 折尺
folding scheme 折叠方案
folding scraper 折合式耙斗
folding screen 折叠屏风;折叠屏风;册屏
folding seat 活动椅;折叠椅

folding seat bed 两用床
folding shield 折叠式护板
folding shutter 折叠式百叶窗;盒式百叶窗;箱型百叶窗;箱形百叶窗;折叠式百叶
folding shutter door 折叠式门;卷帘式门
folding sight 可折叠的表尺
folding sliding door 折叠式拉门;折叠式滑门;平拉门
folding sliding grille 折叠式活动格栅
folding sliding shutter door 折叠式活动百叶门;折叠式活动百叶窗
folding square 可折直角尺;可折斜角规
folding stabilizer 可折撑件;铰接撑件
folding staff 折叠式标尺;折叠标尺;折标尺
folding stair(case) 阁楼活动楼梯;折梯;折叠梯
folding steel chair 钢折椅
folding step 折叠式踏步;折叠式踏板;折叠式登车阶梯
folding stirrup 弯折钢筋箍
folding-stock anchor 折杆锚
folding stool 折凳
folding stratum 褶皱层
folding strength 耐折强度;耐折度;耐揉度
folding stress 折弯应力
folding table 折桌;折叠桌
folding tank 可折叠罐
folding tent trailer 折叠帐篷拖车
folding test 弯折试验;折叠试验;折弯试验
folding till(top) 折叠式车篷
folding tongs 折叠钳
folding top 折叠式车顶
folding top bow 折叠式篷顶的弧拱
folding top structure 折叠式篷顶架
folding tractor strakes 折叠式附加轮籁抓地齿
folding trestle 折叠式里脚手
folding tripod 折叠式三脚架
folding triptych 可折叠的三张相连的图画
folding type recording chart 折叠型记录纸
folding wall 折叠隔墙
folding wedges 松紧楔;双楔;对楔;对顶楔
folding window 折叠窗
folding windshield 折叠式风挡
folding wood(en)rule 木折尺
fold intensity 褶皱强度
fold joint 褶皱接缝
fold kern 褶皱核
fold limb 褶皱翼
fold line 折线;折叠线
fold mountain 褶皱山脉;褶曲山
fold-mullion 褶皱式窗棂
fold of traverse stacking 横向叠加次数
fold-out 折叠式插页
foldover 折叠;叠影;重影
fold scale 褶皱尺度
folds in rock beds 岩层褶皱
fold system 褶皱系;褶曲系
fold tectono-geochemistry 褶皱构造地球化学
fold test 褶皱检验
fold-top car 篷车
fold train 褶皱链
fold trap 褶皱圈闭
fold-type boudin 褶皱型石香肠
fold up(wards) 倒闭;朝上折叠;折叠起来;停业;关闭
fold vergence 褶皱倒向
fold wavelength 褶皱波长
fold zone 褶皱带
foliaceous 叶片状
foliage 叶饰
foliage and strapwork 叶片花饰与带籁线条
foliage capital 叶饰柱头;叶片花饰的柱顶
foliage cusp 叶片花饰的尖顶
foliage cutter 切叶机
foliage frieze 叶片花饰壁缘
foliage plant 观叶植物
foliage scroll 叶片漩涡形装饰
foliage tree 观叶树
foliar analysis 叶片分析
foliar transpiration 叶面蒸腾;叶面散发
foliate 叶片状
foliated 叶片状;叶片的;片状
foliated arches 叶状拱
foliated capital 叶饰柱头
foliated clay 片状黏土

foliated coal 片状煤;叶片状煤
foliated copper 薄铜片;薄铜皮
foliated cusp 叶片花饰的尖顶
foliated fracture 叶片状断口
foliated frieze 叶片花饰壁缘
foliated glass 片状玻璃
foliated granite 叶片状花岗岩
foliated grit(stone) 片状砂砾岩;片状粗砂岩
foliated gypsum 片状石膏
foliated ice 片状冰
foliated ice wedge 叶状冰楔
foliated joint 嵌接
foliated rock 叶状岩石;叶片状岩石;层状岩体;层状岩石
foliated sandstone 片状砂岩;片状砂砾岩
foliated scroll-handle 薄片漩涡状手柄
foliated structure 叶片状构造;叶片构造;菊状结构;薄片状结构;薄片状构造
foliated talc 叶片滑石;片状滑石
foliated tellurium 叶碲矿
foliated texture 叶片状结构;叶片结构
foliation 叶状饰;叶理;叶瓣饰;制箔;褶纹;片理;成片;成层;放叶;剥理;板理
foliation fissure 叶理裂缝
foliation fold 面理褶皱
foliation plain 花叶形装饰;层状沉积平原
foliation plane 片理面【地】;叶理面
foliation structure 叶状构造;叶理构造
folin phenol active substance 福林酚活性剂
folio 海图夹;文件夹;对折纸;对折(本)
folio folding machine 对开折页机
folio label 海图夹标签
folio list 海图夹目录
folio number 海图夹编号
folio paper cutter 对开切纸机
folio reference 账页参考
foliose lichen 叶状地衣
folio verso 见本页背面
folium 叶片;叶理;叶状层;叶形线;[复]folia
folium loti 荷莲叶
folium of Descartes 笛卡儿叶形线
folk art 民间艺术
folk arts and crafts 民间工艺品
folk blue and white 民间青花
folk custom 民间习俗;民俗
Folkestone beds 福克斯通层
folk house 民居
folk kiln 民窑
folklore 民族款式;民俗学
folk prescription 偏方
folkstyle 民族款式
folkways 民俗
follicle population 毛孔密度
follow 遵循;跟随;推杆;伴随
follow block 送桩木;抵板;桩帽;桩籁;送桩木
follow board 模板;造型底板;模型托板
follow course 顺直航向
follow current 持续电流
follow die 系列压模;顺序模
follow-down drilling 跟管钻进
followed biological treatment 后续生物处理
followed either by plowing or disking 耕地或耙地之后
followed link 被选择过的链
follower 转发器;跟踪系统转发器;跟踪机构;跟随器;挤水棒;压杆;替灯;替打装桩;随进钻夹;随动件;随动机构;随动棒;送柱;输出器;从动轮;从动件;从动机械;从动机构;重发器;(合同的)附页;被动轮
follower amplifier 跟踪放大器
follower arm 进料杆
follower cam 从动齿轮
follower chart 钻孔结构图
follower drive 随动件拖动
follower eccentric 进程偏心轮;从动偏心轮
follower fixture 随行夹具
follower flange 填料压盖法兰;填料压盖凸缘
follower gear 随动齿轮;从动齿轮;从动齿轮机;被动齿轮
follower gland 外侧填料盖
follower head 跟踪头
follower lever 从动杆
follower motor 随动电(动)机
follower pile 送桩;替打垫桩

follower plate 填料函压盖;随动板;仿形圆盘
follower plug 随动木塞
follower pulley 从动皮带轮
follower rail 外侧导轨
follower rest 随行中心架;随行刀架;随动刀架;跟刀架
follower ring 随动圈;从动环;随动环
follower ring and washer 从动环及垫圈
follower-ring valve 附圆环阀
follower spring 随动弹簧
follower stud and unt 从动板柱螺栓及螺母
follower unit 跟踪装置
follower wheel 随动轮;从动轮
follow focus attachment 跟踪调焦装置
follow focus lens 跟焦镜头
following 遵循;随动;顺次;按照
following at a distance 尾随行驶
following black 后沿变黑
following chart 流程图
following control system 随动控制系统
following current 顺流
following device 仿形装置
following dirt 伪顶
following edge 下降边;后缘;后沿;推进器后叶缘
following error 跟踪误差;随动误差
following gamma examination 顺便伽马检查
following grouting 采后灌浆
following in range 远距离跟踪
following limb 跟踪边缘
following mechanism 跟踪装置
following mining system 前进式开采系统
following motion 跟踪运动
following movements 追踪列车
following railway 经由铁路
following range 跟踪范围
following range of synchronization 同步保持范围
following rate 跟踪速率
following response 后随反应
following rest 随行扶架
following sea 尾随流;尾随浪;送尾流;顺流;顺浪
following settlement 后续下沉
following shot 追踪摄影
following stone 伪顶;随矿物落下的顶板岩石
following swell 顺浪
following system 跟踪系统
following system of land classification 下列土地分级法
following the track 随铺轨道(坑道掘进后)
following tide 顺潮
following train 续行列车;跟随列车
following-up mechanism 伺服系统
following white 后沿变白;拖白边
following wind 送尾风;顺(水)风【气】
follow-on 改进型的;改进型
follow-on current 持续电流
follow-on die 连续冲模
follow-on irrigation 轮灌
follow-on subassembly 随动组件
follow scanner 跟踪扫描器
follow-scene 移动摄影
follow shot 跟踪摄影;全景镜头
follow spot light 追光灯;跟踪聚光灯
follow suit 仿效
follow the example of 仿效
follow the instructions 按照技术指导
follow the international practice 遵守国际惯例
follow-the-pointer dial 指针重合式刻度盘
follow-the-pointer indicator 指针重合式指示器
follow-through period 经流时间
follow town (大城市的)卫星城镇
follow treaty 守约
follow-up 硬反馈;增援;进度控制;进度检查;随动;伺服
follow-up amplifier 跟踪放大器;随动放大器
follow-up coil 螺线管;随动线圈
follow-up control 随动调节;跟踪控制;跟踪调节;跟随控制;随动控制;从动操纵
follow-up control system 随动控制系
follow-up device 跟踪装置;随动设备
follow-up evaluation 跟踪评估
follow-up gear 随动装置
follow-up grouting 随后注浆(法);采后灌浆
follow-up hand feed 手摇走刀;手动进给
follow-up hole 续钻小一号尺寸钻孔

follow-up instruction 补充指示
follow-up investment 后续投资
follow-up log 随钻测井
follow-up mechanism 随动系统;随动机构;从动机构
follow-up motor 随动电(动)机
follow-up point 随动指针
follow-up pointer 从动指针;随动指针
follow-up potentiometer 随动电位计;随动系统电位计;伺服系统电位计;反馈电势计
follow-up pressure 自动加压;随动增压;随动加压
follow-up pulley 随动轮;从动(滑)轮
follow-up reading 跟踪读数
follow-up readout 跟踪读数
follow-up seal coat 第二次封层
follow-up servo mechanism 跟踪伺服机构
follow-up sleeve 随动套筒
follow-up speed 随动速度
follow-up study 追踪调研;随访研究;随动研究
follow-up survey 追访测量
follow-up switch 随动开关
follow-up system 随动系统
follow-up testing 后续试验
follow-up theory 随动系统理论
follow-up transformer 随动变压器
follow-up unit 随动部件
follow-up valve 随动阀
follow-up well 后续井;仿效井
fomal description 格式说明
fomentation 热敷
fomerell 灯塔上灯室的顶
fomes 污染物质;污染物
fomite 污染物质;污染物
fonctionelle 泛函数
fondant 熔体
fondoform 洋底地形
fondothem 洋底沉积
fondu 高炉水泥;高铝水泥
font 洗礼盘;铅字
Fontainebleau Palace Garden 枫丹白露宫园(法国)
Fontainebleau sands 枫丹白露砂层
Fontainebleau sandstone 枫丹白露砂岩
fontanel 囟门
font cavity 铸口;缩孔
font library 词库
font type 铅字
Foochow pole 福州筒木
food 食品
food additive 食品添加剂
Food and Agricultural Organization of United Nations 联合国粮农组织
Food and Agriculture Organization of the United Nations 联合国与农业组织
food and beverage industry 饮食工业
food and drink 饮食
Food and Drug Administration 食品与药品管理局
food bank 防洪堤岸
food base 粮食基地
food calling 诱食
food can varnish 食品罐头清漆
food chain 食物链
food chemistry 食品化学
food company 食品公司
food composition 食品组成
food consumption 采食量
food contamination monitor 食物污染监测器
food conveying elevator 食梯
food crisis 粮食危机
food crop 粮食作物
food cycle 食源体系;食物循环;食物链
food detection 食品检测
food essence 精微
food expenses 伙食费
food for long voyage 远航食品
food habit 饮食习惯;食性
food-handler 食品处理者
food handling 食品加工
food-handling truck 食品运输车
Food Hygiene Law of the People's Republic of China 中华人民共和国食品卫生法(试行)
food industry 食品工业
food industry wastewater 食品工业废水
food inspection 食品检查

food lift 食梯
food locker 食品柜
food market 菜市场
food-microorganism ratio 食物—微生物(量)比
food-mill wastewater 食品厂废水
food monitoring house 食品监测室
food organism 饵料生物
food patch 饲料地
food plant 食用植物
food pollution analysis 食物污染分析
food preservation 食品保藏
food preservative 食品防腐剂
food processing 食品加工
food processing and manufacturing 食品加工和制造
food processing factory 粮食加工厂;食品加工场
food processing industry 食品加工工业
food processing plant 食品加工厂
food processing wastewater 食品(工业)废水
food procuring contrivance 摄食方法
food products factory 食品厂
food prohibited for production and marketing 禁止生产经营的食品
food refrigerating room 食品冷冻间
food room 食品储[贮]藏室
food science 食品科学
food security 粮食保障
food service industry 食品行业
food soil 食物类污垢
food stability 食品储[贮]藏性能
food storage 食品储[贮]藏
food stores 食品店
foodstuff 食品
food stuff industry 食品工业
foodstuff rubber 食品工业用橡胶
foodstuff subsidization 营养补助费;食品津贴;食品补助费
food substance 食品
food supplement 食品增补剂
food supply 食品供应
food technology 食品工艺学;食品工艺杂志
food warmer 食品保温箱
food waste 食余残渣;食物废物;厨房垃圾
food waste grinder 食物废渣粉碎机
food web 食物网
food wrapper 食品包装纸
fool proof 极简单的;有安全装置的;防止错误操作的;安全装置;绝对安全;极安全的;确保安全的
foolproof apparatus 防止误操作设备;安全自锁装置
foolproof elevator 保险电梯
foolproofing 绝对安全性
foolproofness 安全装置
foolscap 大裁
foolscap-and-half 大四开图纸
foolscap paper 大页书写纸
foolscap-third 四开图纸
fool's gold 愚人金;黄铜矿;黄铁矿
foorage 板英尺含量
foot 最下部;支承基面;底足(器皿);帆的下缘;英尺
foot accelerator 脚踏型加速器
foot adjusting screw 底脚调整螺钉
footage 英尺数;验收丈量;总尺码;总长(度);进尺量;进尺【岩】;山前带;尺码
footage circulation 循环进尺
footage cost of tunneling 巷道进尺成本
footage counter 尺码计数器;长度计数器
footage(drilled)per bit 钻头进尺
footage indicator 尺码指示
footage measurement of tunnel 隧道进尺丈量
footage number 尺标
footage per round trip 回次进尺
football field 足球场
football helmet 橄榄球头盔
foot bar 踏杆
foot base 勒脚线条
foot bath 洗脚池;脚盆
foot bearing 底轴承;底脚支承
foot bench 搁手凳;搁脚凳
foot block 踢脚板固定块;墙裙座;线脚托座;柱脚石;柱脚块;木柱座;木柱承板;顶尖座;垫木
foot blower 脚踏鼓风机
foot board 脚磴(舢板上);(车辆的)踏板;上杆脚板;上杆钉;地角板;踏脚板;踏板
foot board measure 板英尺(度量木材的)

foot board rubber pad 脚踏板橡胶垫
foot boat 载人渡船
foot bolt 脚踏门闩;基础栓;脚踏门锁;门底脚螺栓;踏钮;地脚螺栓;竖直螺栓;基础螺栓
foot brake 脚(踏)闸;脚踏(式)制动器
foot brake rod 脚踏闸杆
foot bridge 小桥;人行桥;天桥;步行桥
foot button 脚踏按钮
foot-candle 英尺·烛光;尺烛光;光距
foot-candle meter 英尺烛光计;光度计
foot cave 脚洞
foot clamp 钻床夹持器
foot clutch 脚踏式离合器
foot control 脚控制
foot controlled valve 脚控阀门
foot crossing 过街通道;人行横道
foot crossing between platform 站台间人行道
foot cube 立方英尺
foot cut 椽脚切口(搁接墙顶横木);下脚切口
foot delta 足状三角洲
foot disc 脚盘
foot drill 踏钻
foot drive 脚推传动;脚踏传动
footed bowl with cover 带盖有脚碗
footed pile 盘脚桩
footed tumbler 带脚杯
footeite 铜氯矾
foot end 植物地下部分
foot engine 脚踏牙钻车
footer 柱脚垫木
footer mo(u)ld 压印底圈的模型
foot face 钝边
footfeed 加速踏板
foot-fishing 接轨夹板
foot flower 脚踏鼓风机
foot ga(u)ge 检潮标;水(位)尺;深度计
foot gear control 脚控传动
foot glacier 足状冰川
foot grating 格子踏板;踏脚格子
footgrip 防滑条
foot guard 护脚档
foot hammer 锻工锤
foot hanger 踏板吊架
foot hill 丘陵地带;山麓小(山)丘;山麓(丘陵);山边丘林地带
foothill belt 山麓带
foothill freeway 山麓公路
foothill region 山麓区
foothill settlement 山麓聚落
foot hinge of arch 拱铰
foot hold 立足处;立足点;踏板;垫轴架;支柱
foothold of driller 钻工工作平台
foot hole 土壤中的炮眼;踏脚孔
foot hump retarder 驼峰尾部缓行器
footing 总额;立足点;基脚;基础;合计;壳脚圈梁;舢板底板;垫层;底座;大方脚;编制;底脚
footing adjustment 底脚调整
footing anchorage 基础锚固
footing beam 主樑椽脚系梁;屋架拉梁;(屋盖的)连系梁;基础梁;地基梁;底脚梁
footing block 基底块体
footing brick masonry work 基础砖砌筑工作
footing brickwork 基础砖砌体
footing cap 地脚帽
footing clinker 基础烧结渣
footing concrete 基础混凝土
footing condition 地基条件
footing course 基层;垫层;底层
footing depth 基础深度;大方脚埋深
footing ditch 基础沟槽
footing drain 基础排水(管)
footing drawing 基础图
footing dressing 基础整修;墙脚处理
footing elevation 地基标高
footing excavation 基脚开挖;底脚开挖;基础开挖
footing form 底座模板;底脚模板;底座模壳;基础模板
footing foundation 底脚基础;柱基(础);扩大底座基础
footing girder 地基大梁
footing grid plane 基础网络平面
footing load 基脚荷载;基础荷载;底脚荷载
footing masonry work 基础砖石建筑工作
footing material 底脚材料;基础材料
footing of bridge abutment 桥台底板
footing of embankment 堤脚
footing of foundation 基础底座
footing of sun creen wall 遮光墙基础
footing of wall 墙基脚;墙基;墙底脚
footing on piles 桩的基脚;桩的底座
footing piece 墙撑柱撑垫;垫件;底板;底脚块
footing pier 基础结构底座
footing pressure 基础压力
footing reaction 地基反力
footing reinforcement mat 基础钢筋网
footing resistance 杆塔基础电阻
footing rotation 基础转动
footings 结算总额
footing screw 地脚螺栓
footing settlement 基础沉降
footing shell 基础外壳
footing step 基础台阶
footing stone 底座石;基石
footing trench 基础沟槽
footing type 地基类型
footing vault 基础拱顶
footing waling 垫板
footing wall 地基墙
footing wall brick 基础墙砖
footing with holes or notches 有孔洞或缺口的基础
foot iron 铁爬梯;踏脚铁条
foot-Lambert 英尺·朗伯
foot lever 脚踏操纵杆
foot lever press 脚控压力机;脚踏压力机
footlid 柱脚垫木
foot lift 踏板式起落机构
foot lifter 脚踏提升器
footlight 脚光;舞台前灯;舞台灯
footlight spot 脚光聚光灯
footline 末行;底线;栏杆下横档;下栏索
foot liner 金属垫片;脚衬片
footling 纵底板条(舢板)
footlog 独木桥
foot-loose industry 无导向工业;随意区位工业;选址不受限制的工业
foot lug 爪;支管;销钉;舌片
foot mark 足迹;脚印
foot masher 脚踏式搅碎机
foot mat 基础钢筋网;底垫层
foot mattress 护底沉排
foot measure 英尺计量
foot-meter rod 英尺米双面标尺
foot moment 基础力矩
foot-mounted motor 落地安装型电动机;底座安装型电动机
foot-mounting cubicle 落地型开关柜
foot mount type 底座安装式
footnote 脚注;附注;下标;注脚
foot of a hill 山脚
foot-of-hole 炮眼进英尺
foot of mountain 山麓
foot of pile 桩角;桩尖
foot of staff 尺垫【测】
foot-operated brake 脚踏制动
foot-operated door holder 脚踏定门器
foot operated faucet 脚踏式水龙头
foot-operated soap dispenser 脚踏给皂水器
foot-operating lathe 脚踏车床
footpace (楼梯的)休息平台;梯台;步行的跨度;步测
foot pad 支架脚垫
footpad suction unit 脚踏吸引器
foot pass between platform 平过道【铁】
foot path 小道;梯子;步道;小路;小径;梯子;踏板;人行(小)道
footpath bridge 步道桥
footpath concrete flag(stone) 人行道混凝土铺面板
footpath concrete paving flag 热镀锌混凝土铺路板
footpath edging 人行道边缘
footpath flag 人行道铺路板
footpath flagstone 人行道石板;人行道铺路石板
footpath paving 人行道铺装;人行道铺面;人行道路面
footpath platform (桥式吊车的)走道平台
footpath roller 人行道滚压机
foot pedal 脚踏开关;脚踏板
foot pedal hanger 踏板吊架
foot per hour 英尺/小时
footpiece 地梁;底木;支架底梁;接梁;垫木;导流片;套管靴;底脚块
foot pin 脚钉;斗臂主销;地脚钉;地钉;尺垫【测】
foot pin for level(l)ing rod 水准标桩
foot plank 脚手板;人行道步板;桥(梁)面步行板;跳板;桥面人行板
foot plate 水准尺座;脚系(屋架支撑处的系板);基础板;底板;标尺座;支柱垫板;钢柱脚板;脚踏板;脚板;踏腰线板板;尺垫【测】
foot platform 工作平台
foot point 垂足
foot-pound 英尺·磅;尺·磅
foot-poundal 英尺·磅达
foot-pound-second 英尺·磅·秒
foot-pound-second system 英尺一磅一秒单位制
foot press 脚踏压力机;脚踏挤压机
foot press switch 脚踏开关
footprint 足迹;轨迹;基底面
foot-protecting concrete 固基混凝土
foot protection 基础加固工程;护坡;坡脚保护
foot pump 脚踏打气泵;脚(踏)泵;踏板式泵;脚定位手泵
foot push 脚踏按钮
foot rail 横挡(家具腿支撑);船尾线脚;船尾橡材;直达采区巷道;脚蹬棍
foot rest 脚踏板;搁脚板;脚架;脚凳
foot rest crank arm 脚踏曲柄臂
foot rest lever 脚踏杆
foot rest rubber 脚踏橡胶
foot rill 直达采区巷道;水平坑道
footring 底足圈;脚卷梁;穹顶脚圈梁;脚圈梁
foot rod 地角杆
foot roll 底辊
foot rope 踏脚索;底帆边索;脚索;脚缆
foot rot 基腐病
foot rot of flax 亚麻根腐病
foot rule 英制尺;脚线;底线
foot ruler 英制直尺
foot run 沿尺;英尺程
foots 下脚;油脚;渣滓;沉积物
foot scraper 脚踏垫;刮鞋板;脚刮板
foot screw 校平螺丝;地脚螺杆;脚螺旋;地脚螺钉;地脚螺栓;地脚螺钉;底脚调整螺钉;底脚螺旋;底脚螺钉
foot-second 英尺·秒
foot section 尾部
foot shears 脚踏剪床
foot shoe 带反向阀套管靴
foot side 下盘;矿床底板;基础墙;底垫;底帮
foot slope 麓坡
foots oil 油脚;蜡下(油)
foot spar 脚蹬(舢板上)
foots scraper 踢脚垫
foot stalk 梗柄
foot stall (柱墩的)勒脚;脚踏板;座底;柱基(础);柱墩;基脚;台座
foot stall of column 柱基(础)
foot steering brake 脚踏转向制动器;脚刹车
foot step 脚踏子;脚蹬;桄座;台阶板;踏板;垫轴台;踏板(楼梯)
footstep bearing 止推轴承;立轴下轴承;立轴承;臼形轴承
footstep pillow 立轴承垫座;承力瓦块球面座
footstep sound attenuation 踏板声防止;脚步声衰减
footstep sound insulation 脚步声绝缘;隔绝脚步声
footstep sound insulation board 脚步声绝缘板;脚步声隔声板
footstep sound intensity 踏板声强度;脚步声强度
footstep sound measurement 脚步声测量
footstep sound reduction index 脚步声衰减指数
footstep sound transmission 脚步声传递
footstep sound transmission loss 脚步声传递损失
footstep switch 脚踏开关
footstep traffic 步行交通
footstep valve 脚踏阀门
footstock 后顶针架(车床);尾座;定心座;顶座
footstock lever 踏杆
footstone 山墙下端支承压顶的石块;起拱石;基石;山墙斜坡底部斜高石;土墓基石
footstool 脚凳;踏凳;踏脚凳;轻便梯凳
foot street 步行街(道)
foot subway 人行地下道
foot super 平方英尺
foot switch 脚盘开关

foot the bill 负担费用
foot thresher 脚踏脱粒机
foot throttle 脚踏风门
foot-tight industry 选址受限制的工业
foot traffic 行人交通;步行交通
foot treadle 踏板
foot tunnel 人行隧道;人行地下道
foot up 总计
foot valve 吸水阀;止回阀;管底阀;脚(踏)阀;底阀;泵吸管端逆止阀;背压阀
foot valve cage 底阀罩
foot valve seat 底阀座
foot valve with strainer 有筛网的底阀
foot walk 人行道
foot wall 下盘;矿脉基座;矿床底板;基础墙;底垫;底帮;坝趾齿墙;底壁
foot-wall active downthrow 下盘主动下投
footwall blast 基础墙爆破
footwall shaft 下盘斜井
footway 小路;小径;人行小道;人行(通)道;梯子间;梯子道
footway cantilever 人行道悬臂
footway cantilever bracket 人行道悬臂托架
footway cantilever bracket of bridge 桥梁人行道悬臂
footway-crossing 人行横道
footway framing 人行道构架
footway framing of bridge 桥梁人行道构架
footway index 人行道指数
footway in tunnel 隧道的人行道
footwear 防护鞋
foot wheel 脚踏轮
foozle 制造误差;废品
Foppl network dome 弗泼尔网状穹隆
forage 饲料;草料;采食牧草
forage blower 饲料抛送机
forage box 运饲料车厢
forage chopper 饲料切碎机
forage-chopper cutterhead 饲料切碎装置
forage chopping equipment 饲料切碎机具
forage crop 饲料作物
forage cutter 铡草机
forage drier 饲料干燥机
forage equipment 饲料调制设备
forage fodder 草料
forage grass 牧草
forage harvester 饲料收获机;饲料收割机
forage harvesting and silage equipment 饲料收获和青储[贮]设备
forage kitchen 饲料调制室
forage poisoning 牧草中毒
forage ratio 饲料比率
forage seed breeding farm 牧草良种繁殖场
forage unloader box 自卸式饲料分送器
Foraky boring method 佛拉奇钻孔法
Foraky freezing process 佛拉奇冻结法
for-all structure 全操作结构
foramen caecum 盲井
foramen ovale 卵圆孔;第二中隔孔
foramen petrosum 岩孔
foraminate perforation 麻黄式穿孔
foraminate perforation plate 麻点式穿孔板
foraminifer 有孔虫
Foraminifera 有孔虫目
foraminiferal compensation depth 有孔虫补偿深度
foraminiferal faunal province 有孔虫动物地理区
foraminiferal limestone 有孔虫石灰岩
foraminiferal ooze 有孔虫软泥
foraminiferan 有孔虫
foraminite 有孔虫岩
foraminiteral limestone 有孔虫(石)灰岩
foram lysocline 有孔虫溶跃层
for a short time 临时的
forb 杂草;非禾本草植物
forbearance 债务偿还期的延缓;延展期限
forbearance agreement 延期偿款协议
forbesite 纤砷钴镍矿
forbidden 禁用
forbidden area 禁地
forbidden band 禁带
forbidden cell 禁用单元
forbidden character 禁用字符
forbidden code 禁用代码
forbidden combination 禁止组合;禁用组合;非法组合
forbidden combination check 非法组合校验;禁用组合校验
forbidden decay 禁带衰减
forbidden digit 禁止组合数码;禁止数字;禁用数字;非法数字;不合法数码
forbidden digit check 禁止数字检查;禁用数字校验;禁用数位检验;非法数字校验
forbidden emission 禁带发射
forbidden fishing zone 禁渔区
forbidden gap 禁隙
forbidden gap energy 禁带隙能
forbidden harmonics 禁用调和函数
forbidden line 禁(戒)线;禁带谱线
forbidden region 禁区;禁带
forbidden transition 禁区跃迁;禁戒跃迁;禁带跃迁
forbidden zone 禁区界线
forbidden zone for fishing 禁渔区
forbidding subgraph 禁用子图
forble board 二层台
for buman environment 适合人类的环境
for buyers account 由买方负担
for cash 付现钱;付现金;付现交易;付现
force 力(量);强行置码;势力
force acceleration 生产加速度的力
force account 自营项目开支;计工制;成本加费用账款;包工制;自办工程账(政府自办工程的估价计算);成本加利润账;计工的
force-account basis 计工制
force-account construction 计工制工程;计工建筑
force-account system 计工制
force-account works 计工制工作;自营工程;包工工程
force administration data system 强制管理数据系统
force against 压紧
force a kiln 强化窑煅烧
force amplitude 力幅
force and splash lubricating 压力和飞溅润滑
force application 力的作用;力的运用;施力
force area 受力面积
force-balanced accelerometer 力平衡式加速度计
force bearing ring 承力环
force boundary 力边界
force cell 测力传感器;测力计
force-circulation 压力循环
force circulation air cooler 强制循环空气冷却器
force closure 力锁合
force coefficient 力系数
force component 分力
force condition 应力分布;力的作用条件
force constant 力常数
force couple 力偶
force cup 橡胶吸水器;橡皮吸水泵;撅子(排堵橡胶碗);排污杯;揣子【建】
forced 增压;强制的
forced absorption 受迫吸收
forced action mixer 强制式搅拌机;强制式拌和机
forced aeration 强迫掺气
forced air 加压气流;强制空气;强制风;强迫通风
forced air blast 鼓风;鼓风
forced air circulation 强制(空气)循环;强迫通风
forced-air-cooled oil immersed transformer 强制风冷油浸变压器
forced air-cooled tube 强迫风冷管
forced air cooling 压风冷却;强制(式)风冷;强制空气冷却;强迫风冷;强力风冷却;风冷却;冷风冷却
forced air cooling tube 强迫风冷管
forced-air dehydrator 强制通风式干燥机
forced air drying (木材的)强制气干
forced air furnace 鼓风加热炉;鼓风炉
forced air heating 热风采暖设备;压力热风供暖;压风加热;强制送风采暖;强制热风供暖;强制空气加热
forced air pressure discharge 风压释放
forced air supply 强制(式)供气;压力供气;人工通风;强制送风;强制式通风
forced-air ventilating system 压力通风系统
forced air ventilation 强制空气通风
forced alternating current 强迫交变电流;受迫振荡电流
forced approach and run alongside 强行靠泊
forced auction 强制拍买
forced board 压制板
forced bond 强迫债券
forced caving 强制崩落
forced centering 强制对中
forced centering device 强制对中装置
forced centring 强制对中
forced choice approach 被动选择法
forced circulating reboiler 强制循环再沸器
forced circulation 强迫循环;压力环流;压力环境;强制循环;强制流动
forced circulation air cooler 鼓风冷却循环系统
forced circulation boiler 强制循环锅炉
forced circulation by pump 用泵生产强制循环
forced circulation central heating 受迫循环集中供热
forced circulation cooling 强制循环冷却
forced circulation evapo(u)ration 强制循环蒸发
forced circulation evapo(u)rator 强制循环蒸发器;压力循环蒸发器
forced circulation furnace 强制循环的热处理炉
forced circulation kiln 强制循环干燥窑
forced circulation register 强制循环调风器
forced circulation steam generator 强制循环蒸汽发生器
forced circulation system 机械循环系统;压力循环系统;强制循环系统;强迫循环系统
forced circulation type evapo(u)rator 强制循环式蒸发器
forced coding 强制编码
forced coding program(me) 强制编码程序
forced commutation 强制换向;强迫整流
forced concrete mixer 强制式混凝土搅拌机
forced concrete mixing plant 强制式混凝土搅拌设备
forced congruence 强迫一致
forced convection 强制对流;强迫对流;外力对流;受迫对流
forced convection air heater 暖风机
forced convection boiling 强迫对流沸腾;强制对流沸腾
forced convection cooling 强制对流冷却;强迫对流冷却
forced convection heater 压力对流加热器;强制对流加热器
forced convection heat transfer 强制对流换热;强迫对流换热
forced convection lehr 强制对流退火窑
forced convection oven 强制对流式烤箱
forced convection vapo(u)rization 强制对流汽化
forced convergence method 强迫收敛法
forced cooling 强制冷却;强迫冷却
forced cooling of meniscus 弯月板强制冷却
forced crystallization 强迫结晶
forced current drainage 强制排流
forced damped vibration 强制减振;强迫阻尼振动;强迫衰减振动
forced decision 强制裁决法
forced deposit 强制存款
forced diagram 力图
forced-dipole transition 受迫偶极子跃迁
forced discharge 强迫卸货
forced display 强制显示;强行显示;强加显示
forced draft 压力送风;压力气流;正压送风;鼓风(式);人工通风;强制通风;强迫气流
forced draft blower 压力通风机
forced-draft boiler 强制通风锅炉
forced draft burner 强制通风燃烧器
forced draft cooling 强制通风冷却
forced draft cooling tower 强制通风凉水塔;强制通风冷却塔;送风式冷却塔
forced draft degasifier 强制通风脱气塔
forced draft duct 强制通风管道
forced draft fan 压力送风机;强制引风机;强制通风机;增压风扇
forced draft front 强制通风炉口
forced draft kiln 强制循环干燥窑
forced draft mechanical cooling tower 鼓风机械通风冷却塔
forced draft shaft 压力通风井;压力抽风井
forced draft water-cooling tower 强制通风水冷却塔
forced drainage 泵站排水;强制排水;强制排流
forced drainage system 强制排水系统;泵站排水系统

forced draught 压力通风;强制排烟;强迫通风;强力通风
forced draught aerator 压力通气式曝气器
forced draught air cooler 强制通风空气冷却器
forced draught blower 加压通风机;送风机
forced draught cooling 强制通风制冷;强迫通风冷却
forced draught cooling arrangement 强制通风冷却装置
forced draught cooling tower 送风式凉水塔;强制通风冷却塔
forced draught fan 压力送风机;压力抽风机;鼓风机;强压送风机
forced draught filter 压力通风集尘器
forced draught furnace 压力通风炉
forced draught kiln 强制循环干燥窑
forced draught type 强制通风式
forced drausht cooling tower 压力通风冷却塔
forced drop shaft 水压沉井凿井
forced dry 强制干燥
forced drying 快速烘干;强力干燥;加温干燥;干烘油漆法;强制干燥
forced drying lacquer 强制干燥挥发性漆
forced drying temperature 强制干燥温度
force decomposition 力的分解
force-deformation diagram 力变形图
force-deformation relationship 力变形关系
forced electric(al) drainage 强制排流器
force density 力密度
forced entrance 挤进
forced fan 压力通风风扇;压力送风机;增压风扇
forced feed 压力加料;压力送料;压力进料;压力进给;加压进料;强制进给
forced feed drilling 加压钻进
forced feed lubrication 强制润滑;压入润滑(法)
forced feed lubricator 强制供油润滑器
forced feed oil system 强制供油系统
forced feed system 强制进料系统
forced filtration 加压过滤
forced filtration method 加压过滤法
forced filtration velocity 强制过滤速度
forced firing 逼火
forced fit 妥帖装配;压力组装
forced-flame drilling 强制火焰钻进
forced flexural wave 强迫弯曲波;受迫弯曲波
forced flow 有压流;加压流;强制流(动)
forced-flow boiler 压力循环锅炉
forced-flow liquid chromatography 强制流动液体色谱法
forced-flow once-through steam generator 增压直流锅炉
forced fluidized bed 搅拌流化床
forced frequency 强迫频率;受激振荡频率
forced frugality 强制节约
forced fuel feed 压力燃料供给;压力给油;强制式加燃料
forced gear engagement 强制啮合齿轮
forced harmonic motion 强迫谐振
forced harmonic vibration 强迫和谐振动;受迫振动
forced heat convention 强制对流传热
forced heating 受迫供热;受迫采暖
forced heating installation 受迫供热装置;受迫采暖装置
forced heating system 受迫供热系统;受迫采暖系统
forced heir 法定继承人
forced heirship 法定继承权
forced hot water heating 压力(式)热水供暖;强制热水供应;强制热水供暖
force diagram 作用力示意图;力(线)图;力的图解;受力图
forced ignition 强制点火
forced-in air 强制鼓风
forced induction engine 增压发动机
forced inspiratory volume 强力吸气容积
forced insurance 强制保险
forced interruption 强行断电
force direction 力的方向
force-displacement diagram 功图;力位移关系曲线图
force-displacement relationship 力位移关系(曲线)
force distribution 力的分布
forced jump 强迫跳伞
forced kiln operation 强化窑操作;强化煅烧

forced labo(u)r 强制劳动;强迫劳动
forced landing 强迫降落
forced liquidation 强迫清算
forced loan 强制贷款;强迫借贷
forced locking device 自动锁紧装置;强制闭锁装置
forced lubrication 强制润滑(法);压油润滑;压力润滑;加压润滑
forced lubrication pump 压力润滑泵
forced lubricator 压油润滑器
forcedly air-cooled engine 强制风冷式发动机
forced magnetization 强制磁化
forced magnetostriction 强制磁致伸缩;强迫磁致伸缩
forced main 压力干管
forced migration 强迫迁移
forced(mixing type)mixer 强制式搅拌机;强制式拌和机
forced modelocking 强迫锁模
forced motion 强迫运动
forced movement 强制运动;强迫运动
forced noise 强制噪声;强力噪声
forced non-liner oscillation 非线性强迫振荡
forced nutation 受迫章动
forced-off command 强迫让道指令
forced-oil-air circulation 强制油循环
forced-oil-air cooling 强制油循环吹风冷却
forced-oil circulation 强制油循环
forced-oil-cooled 强制油冷却
forced-oil-cooled bushing 强迫油冷式套管
forced-oil-cooling 强迫油冷
forced-oil-cooling transformer 强迫油冷变压器
forced-oil forced-air cool 强迫油冷强迫风冷式冷却
forced-oil transformer 强制油冷式变压器
forced one-pipe heating 单管受迫供热;单管受迫采暖
forced O-ring seal 强制O形环密封
forced oscillation 强制振荡;强迫摆动;强迫振动;强迫震荡;强迫振荡;受迫振荡
forced outage 事故停电
forced outage rate 强迫停机率;事故停机率
force down 压下
force down price 压价;杀价
forced payment 强迫付款
forced pipe 压力小管
forced position 强迫体位
forced posture 强迫体位
forced power transmission 强行送电
forced price 受约制的价格
forced process 强制过程
forced production 强化开采
forced pump 压力水泵
forced quotation 限价
force-draft cooling system 强制通风冷却系统
forced recirculating dry kiln 强制循环式干燥窑
forced release 强制脱扣;强制脱钩;强迫释放
forced response 强制响应;强迫反应
forced reversing 强迫换向
forced rolling 强制横摇
forced rotation 受迫转动
forced sale 强制拍卖
forced sale value 迫售价格价值;追卖价值;清算拍卖价值
forced saving 强制储蓄
forced screening 用力刮平;加压过筛
forced service 强制运行
forced spatial excitation 强迫空间激振
forced standing position 强迫立位
forced stroke 工作冲程
forced surface wave 强制表面波
forced surge 强迫涌浪
forced synchronization 强制同步
forced synchronizing 强制同步
forced system 压力供气设备
forced thermal scattering 强制热散射
forced tidal wave 强制潮汐波
forced torsional vibrations 强迫扭转振动
forced transducer 强制换能器
forced transition 受迫跃迁
force due to viscosity 黏性力
force due to braking 制动产生的力
force due to ship 船舶荷载
force due to sway effect of the engine and train 车辆横向摇摆力
force due to viscosity 黏性力

forced unwinding 强制退解
forced value 清算拍卖价值
forced-ventilated motor 强制通风式电动机
forced ventilating 强制通风
forced ventilation 压力通风;鼓风;机械通风;强制(式)通风;强迫通风
forced ventilation motor 强制通风电动机
forced vibration 强制振动;强迫振动;强迫摆动;受迫振动;受拉振动
forced vibration cross-spectrum analysis 受迫振动互谱分析
forced vibration frequency 强迫振动频率
forced vibration method 强制振动法
forced vibration torsion pendulum 强制振动扭摆
forced vibrator 强迫振动器
forced vibratory compaction 强制振动捣实(法)
forced vortex 有压涡流;强迫漩涡;强制进料
forced vortex motion 强迫涡流运动
forced warm air furnace 受迫鼓热风炉;强制鼓热风炉
forced warm air heating 受迫热风采暖;强制热风供暖
forced warm air heating plant 热风供暖站;热风采暖设备
forced warm air heating system 热风采暖系统
forced water circulation 强制水循环
forced water cooling 压水冷却;强迫水冷
forced wave 压力波;强制波;束缚波;受迫波
forced working 强制工作
force eight wave 八级浪
force eight wind 八级风
force equilibrium 力平衡
force factor 力因数
force fan 增压风扇;鼓风机;增压通风机
force feed 压油润滑;强制给进;压力加料;压力送料;压力进料;压力进给;加压进料;强制进给
force feedback 力反馈
force feedback accelerometer 力反馈加速度计
force-feed gun 压注式润滑枪
force-feed loader 压入装料机;挤入装料机
force-feed lubrication 压力润滑(法)
force-feed lubrication system 压力润滑系统
force-feed lubricator 压力润滑器
force-feed main 增压输送管路
force-feed non-splash lubrication 加压不溅润滑
force-feed oiler 压力进给加油器
force-feed oiling 压力供油法
force field 力场
force-field theory 力场理论
force fit 压入配合;压配合
force-five wave 五级浪
force-four wave 四级浪
force-free edge 无外力作用边;无外力边界
force-free field 无力磁场
force-free magnetic field 无力磁场
forceful arc 强电弧
force function 力函数
force ga(u)ge 测力计
force-in air 强制鼓风
force-insensitive mounting 加力不敏感装置
force intensity 力场强度
Forcel 福塞尔聚酯纤维
force-landing 强迫降落
force lift 压送泵;压力泵
force lift pump 增压泵
force-limiting device 力限器
force line 力线
force liquidation 强制清算
force-lubricated 压油润滑的
force main 压力主管道;压力干管;压力总管;承压干管
force majeure 不可抗力;不可抗拒力量;不可抗拒的因素
force majeure clause 不可抗力条款
force majeure event 人力不可抗拒事故;不可抗力事件
force majeure exception 不可抗力免责;不可抗力免除
force-measuring device 测力器
force method 力法
force moment 力矩
force-moment diagram 力矩图
force motion 力操纵机构
force motor 执行电动机

force-nine wave 九级浪
force of attraction 吸力;引力
force of breaking wave 破浪作用力
force of broken wave 已破浪作用力
force of capillary attraction 毛细管吸引力
force of cohesion 黏性力;黏聚力
force of compression 压缩力
force of constraint 约束力矩
force of crystallization 结晶力
force of deviation 偏转力
force of explosion 爆炸力
force off 压出
force-off command 强制让道指令
force of flexion 挠曲力
force of friction 摩擦力
force of gravitation 重力;引力
force of gravity 重力
force of impact 冲击力
force of impression 压印力
force of inertia 惯性力
force of labor 产力
force of non-breaking wave 未破浪作用力
force of penetration 贯入力;穿透力
force of percussion 撞击力
force of periphery 圆周力
force of support 支承力
force of texture bond 结构联结力
force of the current 流体流动的力;(电、水、气的)流的力
force of wind 风压力;风力
force on 压入
force on charge 磁力
force-one wave 一级浪
force-one wind 一级风
force out 挤出
force parallelogram 力平行四边形
force piece 工作面支架;临时超前支架
force pipe 压力水管;压送管(道);压管
force piston 阳模;模塞
force placement method 力布局法
force plane 粗刨
force plate 阳模托板;测力板
force plug 阳模;模塞
force point 施力点
force polygon 力多边形
force potential 力势
force price down 压价
forceps 镊子;焊钳;钳子
forceps holder 钳夹
forceps jar 钳镊缸
force pump 活塞泵;压力泵;加压泵;压力水泵
forcer 压花冲头;起伏机;蜗杆压榨机
force rate 力比
force ratio 力比
force resolution 力的分解
force resultant 力的合成
forces balanced in pairs 成对平衡的力
force scale 力比例尺
force screw 螺旋量
force sensing element 力敏元件
force-seven wave 七级浪
forces intersecting at an acute angle 相交成锐角的力
force-six wave 六级浪
forces network 军用通信网
forces station 军用电台
force structure analysis 部队结构分析
force structure costing 部队结构费用估计
force-three-wave 三级浪
force-three-wind 三级风
force through ram pressure 冲压
force tide 强潮
force transducer 力传感器;测力传感器
force transfer system 力传递系统
force triangle 力三角形
force twelve wind 十二级风
force-two wave 二级浪
force-two wind 二级风
force unbalance 力的不平衡
force up 上压;上挤
force up commodity price 抬价
force variation 力的变化
force vector 力矢量
force wave 压力波

forcible 强制的
forcible entry 非法侵入
forcible execution 强制性执行
forcible form of union 强制联合形式
forcible separation method 强制剥离试验
forcing 加强显影;强迫;施加压力;施工压力;塞入
forcing bed 温床
forcing coil 强迫线圈
forcing control 强行控制;强迫控制
forcing down the targets 压指标
forcing fan 压风机
forcing filter rate 强制滤速
forcing fit 强迫配合
forcing frame 促成温床
forcing frequency 扰动频率;强迫频率;受迫振动频率
forcing function 强制函数;强加函数;外力函数
forcing function(al) generator 功能发生器;正弦发生器
forcing hose 压力软管;增压器软管
forcing house 花房;强化温室;温室;玻璃房
forcing-in ventilation 压入式通风
forcing jack 装卸机
forcing lift pump 加压提升泵
forcing machine 压力机;装卸机
forcing main 加压干管
forcing method 力迫法
forcing of kiln operation 强化煅烧
forcing of oil 润滑油的加压
forcing on 压上
forcing open of the point 挤岔
forcing pipe 压力(水)管;加压(干)管
forcing pump 压力(水)泵;加压泵
forcing screw 压紧螺钉;加压螺钉;起重器
forcing tube 压力水管
forcing up the target 压指标
forcipate(d) 钳形的
forcipression 钳压法
forcipressure 钳压法
for civil use 民用
for clause 循环子句
for configurations qualified through testing 结构已经试验证明合格
for consideration of safety 为安全计
for copper pipe ring 铜管接环
ford 徒涉处;浸水路面;过水路面;可涉水而过的地方;浅滩;渡口;徒步处;津流
ford a stream 涉水过河
Ford cup 福特黏度计
for declaration purpose only 只用于申报;仅限于申报
for deposit only 仅限转账
for details see attached table 详见附表
Ford Foundation 福特基金会
fording depth 涉水深度
Fordism 福特制
Ford louse 福特汽车
ford road 涉水路
Ford's curve 福特曲线
Ford's system 福特制;福特管理制
Ford's viscosity 福特黏度
Ford's viscosity cup 福特(黏度)杯
fore 前部;船首
fore anchor 头锚;船首锚
fore and aft 纵向(的);纵长,在船头船尾的;首尾向的;从船头到船尾的;船首尾向
fore-and-aft acceleration 航向加速度
fore-and-aft axis 纵轴
fore-and-aft balancing 纵向平衡
fore-and-aft beam 纵梁;纵桁
fore-and-aft bridge 步桥;纵向天桥
fore-and-aft bulkhead 纵舱壁【船】
fore-and-aft control 纵向操纵
fore-and-aft correctors 纵向校正磁棒
fore-and-aft diaphragm 纵向隔板
fore-and-aft distance 纵向距离
fore-and-after 舱口盖纵梁;纵后;舱口纵桁
fore-and-after clearance 前后间距(工作部件)
fore-and-aft force 纵向力
fore-and-aft gangway 纵向步桥
fore-and-aft inclinometer 前后倾斜计
fore-and-aft level 纵水准仪
fore-and-aft level(l)ing 纵向调平
fore-and-aft line 首尾线

fore-and-aft magnet 纵向自差校正磁铁
fore-and-aft mooring 首尾系泊
fore-and-aft motion 纵向操纵装置
fore-and-aft oscillation 纵向振荡
fore-and-aft overlap 航向重叠
fore-and-aft plane 纵向平面
fore-and-aft range mark 首尾导标
fore-and-aft rigged vessel 纵帆船
fore-and-aft road 纵板道
fore-and-aft runner 纵骨;纵材
fore-and-aft sail 斜帆;纵帆
fore-and-aft shift 纵向水平移动;前后移位
fore-and-aft stay 前后支柱
fore-and-aft tilt 纵向倾斜;航向倾斜
fore-and-aft tip 航向倾斜角
fore-and-aft traction 前后方向牵引力
fore-and-aft tree 界标树
fore-and-aft trim 前后平衡调整
fore and back level(l)ing 前后视水准测量
fore apron 闸前护栏;前台;房前院地
fore-arc area 弧前区
fore-arc basin 弧前盆地
fore-arc downwarped extracontinental basin 弧前型陆外下陷盆地
fore-arch 前拱
forearm 前臂
forearm connection 前支架连接
forearm length 前臂长
fore axle 前轴
fore back spring 前倒缆
forebay 储[贮]水箱;风雨檐;前池;前舱;上游河段;船首前池;池窑前炉;池窑供料道;前室
forebay apron 前池护坦
forebay area 前池区
forebay channel 前池渠道
forebay elevation (船闸的)上游水位
forebay of pumping station 泵站前池
forebay reservoir 前池水库
forebeach 前滩
fore beak 前尖嘴
fore bearing 前轴承
fore-blow 预吹
fore-board 前甲板
forebode 预示
forebody 前体;船首;机身前部;前部船体
foreboom 前桅张帆杆
fore breast(line) 前横缆;掘进工作面
fore bridge 前桥楼;桥楼
forebuilding 仲山的建筑
fore cabin 前客舱;前舱
fore carriage 前轮架;前车架;前导轮架
forecast 预测;预报
forecast accuracy 预报准确率;预报准确性;预报准确率;预报准确度;预报精度
forecast amendment 预报改正;预报订正
forecast area 预报区域;预报区
forecast balance sheet 预测资产负债表
forecast bulletin 预报公告;天气公报
forecast cargo traffic 预测货运量
forecast chart 预图;预报图(表)
forecast consumption 计划消耗
forecast curve 预报曲线
forecast cycle 预报周期
forecast demand 预估的要求
forecast district 预报区
forecast earthquake 预测地震
forecast economic development 经济发展预测
forecasted flow 预报流量;预报径流
forecasted hydrograph 预报(水文)过程线
forecasted overhead 预测间接费用;预测间接成本
forecast equation 预报方程
forecaster 预报员;预报因子
forecast error 预报误差
forecast estimation 预报估计;预报估测
forecast for construction 施工预报
forecasting 预报方法;预测;预报
forecasting by mathematic(al) statistics 数理统计预报方法
forecasting center 预报中心;气象中心;天气预报
forecasting criterion 预报判据
forecasting demand 预估需水量;预测需水量
forecasting error 预测误差
forecasting flow 预测流量
forecasting function 预报函数

forecasting grade 预行分级
forecasting mineral resources 预测矿产资源量
forecasting of basin flow concentration 流域汇流预报
forecasting of cost 成本预测
forecasting of demand 需求预测
forecasting of price 价格预测
forecasting of production and marketing 产销预测
forecasting of profit 利润预测
forecasting of reservoir inflow 进库流量预报；水库来水预报
forecasting of streamflow 河川流量预报
forecasting of waterborne traffic 水运量预测
forecasting oil slick 油膜预报
forecasting period 预见期；预报期间
forecasting process 预测过程
forecasting risk 预测风险
forecasting runoff 预测径流(量)
forecasting sequence 预测序列；预测顺序
forecasting service 预察工作；预报机构；预报工作
forecasting settlement 沉陷的预测；沉陷的预报；预测沉降
forecasting station 预报站
forecasting technique 预测技术；预报技术；预报方法
forecastle 艏楼；首楼【船】；船首楼栏杆；船首楼
forecastle break bulkhead 船首楼后端舱壁
forecastle deck 船首楼甲板
forecastle erection 艏楼安装
forecastle port 船首楼左边
forecastle port locker 船首楼左边房间
forecastle starboard 船首楼右边
forecastle starboard locker 船首楼右边房间
forecast map 预图
forecast method 预报方法
forecast model 预报模型；预报模式
forecast of business condition 商情预测
forecast of cold wave 寒潮预报；寒潮警报
forecast of debacle 解冻预报
forecast of earthquake 地震预测；地震预报
forecast of epiphytotic disease 植物病害流行预测
forecast of epiphytotics 流行病预报
forecast of ice 冰情预报
forecast of ice condition 冰情预报
forecast of ice-jam flood 凌汛预报
forecast of ice-jam stage 壅冰水位预报
forecast of melting snow runoff 融雪径流预报
forecast of minimum water supply 枯水径流预报
forecast of pest and disease 病虫害预报
forecast of reservoir inflow 进库流量预报；入库流量预报
forecast of runoff 径流预报
forecast of sea fog 海雾预报
forecast of settlement 沉降预测；沉降预报
forecast of snowmelt runoff 融雪径流预报
forecast of soil moisture 土壤水分预报；墒情预报
forecast of spring freshet 春汛预报
forecast of streamflow 河流流量预报；河川径流预报
forecast of water supply 供水预报
forecast of wave decay 波浪衰减预报
forecast performance of model 模型预测效能
forecast period 预测期；预报周期；预报期
forecast price 预测价值；预测价格
forecast probability 预报概率
forecast-reversal test 预报正反检验
forecast terminology 预报术语
forecast-type budget 预测式预算
forecast value 预报值
forecast variable 预测变量
forecast variance 预测差异
forecast verification 预报检验
forecast zone 预报区
fore-chamber 前室
forechurch 教堂前厅
foreclose 预先解决；没收担保品；取消赎取权
foreclosure 取消抵押品赎回权
foreclosure sale 已没收担保品的销售；没收担保品的销售
foreclosure value 已没收担保品价值；没收抵押品价值；没收担保品价值
fore cooler 前冷却器；预冷却器；预冷器
forecooling 预冷
fore-cooling room 预冷室

fore-course 前桅主帆
forecourt 前庭【建】；屋前小院
foredate 填早日期
foredated 预先填上日期的；预填日期
foredated bill 填早日期票据
foredated check 预先填上日期的支票(票上日期早于实际出票日)；填早日期支票
fore-deck 前舱面；前甲板
foredeep 狭长深海槽；陆外渊；前渊；外地槽；深长海沟
foredeep type formation 山前坳陷型沉积建造
fore door 住房前门
fore draft 前吃水；前部船体吃水；艏吃水；船首吃水
foredrag 前部阻力
fore draught figures 船首水尺
fore drift 超前平巷
foredune 沿岸沙丘；沿岸沙垄；水边低沙丘；岸前沙丘
fore-end 前端；前端部分
fore-engined ship the middle of ship 艟前机型船
forefan depression of alluvial fan 冲积扇扇前洼地
forefather 前人
forefeel 预感
forefield 前方场地
forefoot 艏踵；龙骨前端部；前脚(船首柱底部)
forefoot knee 前脚肘
forefoot plate 前脚板
fore frame 前肋骨
forefront 最前部(前线)；最前部前面；最前部前列；最前线
foregallery 前廊
fore gate 总入口；前进大门；船闸上游闸门
foregeability 可锻性
fore-gear 前桅卷帆索
foregift 押租；承租人预付的租金
foregoing 前述的；前面的
foregone benefits 利益损失
foregone conclusion 意料结果；定局；必然结果
foregone earning 放弃的收入
foregone income 放弃的收入
foregrinding 预先磨碎
foreground 最引人注意的位置；最显著的地位；前台；前景
foreground-background processing 前后台处理
foreground-background program(me) 前台后台程序
foreground camera 前景摄像机
foreground collimating mark 前景标定点
foreground collimating point 前景框标
foregrounding 前台设置
foreground-initiated background job 前台启动的后台作业
foreground initiation 前台启动
foreground initiator 前台启动程序
foreground job 前台作业
foreground message processing program(me) 前台信息处理程序
foreground mode 前台方式；前台操作方式
foreground monitor 前台监控程序
foreground partition 前台区；前台划分；前台部分
foreground picture 前景图
foreground point 前景点
foreground processing 优先处理；前台处理
foreground program(me) 前台程序
foreground region 前台区
foreground rotation 前景旋转
foreground routine 前台例行程序
foreground scheduler 前台调度程序
foreground signal 前景信号
foreground star 前景星
foreground task 前台任务
fore-guy 前支索
forehammer 大手锤；手用大锤；手用大槌
forehand 正手的；手挡
forehand solids 外来的杂物
forehand welding 前进焊；向左焊；向前焊；左向焊；左焊法；正手焊法；正手焊
fore harbo(u)r 外港
forehatch 前舱(口)
forehead 前额；前部；额
forehead lamp 额灯
forehead pad distance 前额支撑垫
forehead-plasty 额成形术
fore head spring 前倒缆
forehead support 额靠

forehearth 供料通路；料道；玻璃窑工作带
forehearth arch 通路碹
forehearth branch 通路分支
forehearth casing 供料道外壳；池窑料道罩；池窑工作部罩
forehearth colo(u)ration 料道着色
forehearth connection block 前炉连接砖；通路连接砖
forehearth entrance block 通路入口砖
forehearth limb 通路分支
forehearth section 通路段
forehearth side rail 通路段
fore hold 前货舱；前舱；前部船舱
fore hook 尖蹼板
foreign 外国的
foreign affairs 外交
foreign affiliates 外国子公司
foreign agency 国际代理机构；商行在国外的代理处
foreign aid program(me) 外援计划
foreign and international bond markets 外国和国际债券市场
foreign applicable standards 国外公认的有关标准
foreign atom 掺杂原子
foreign balance 国际结算平衡表；外汇结存
foreign bank bill 外国银行票据
foreign bidder 国外投标人；国外报价人
foreign bill 国外汇票
foreign bill of exchange 外汇单；外国汇票
foreign bills payable in gold 以黄金支付的外国汇票
foreign body 掺杂物；异体；异物；杂物；外来物；掺和物
foreign body finder 异物探寻器
foreign-body locator 异物探测器
foreign bond 国外债券；外国债券
foreign bond markets 外国债券市场
foreign borrower 外国借款人
foreign borrowing 外国借款
Foreign Broadcasting Information Service 对外广播新闻局
foreign bureau incurred charge payable 应付代收外局款
foreign bureau incurred charge receivable 应收外局代收款
foreign capital 外资；外国资本
foreign capital flows 外国资本流量
foreign capital inducement 吸收外资
foreign capital inflow 外资流入
foreign capital intake 吸收外资
foreign capital investment 外资投资
foreign capital investment target 外资投向
foreign car 外国汽车
foreign cash investment 外币现金投资
foreign check 外国支票
foreign coal 外来煤；外矿送选的煤
foreign collections 向国外托收出口票据
foreign colo(u)r 洋彩
foreign commerce 外贸；对外贸易
foreign commercial bank loans 国外商业银行贷款
foreign company 外国公司
foreign component 外来物质；外来成分
foreign contracted project 对外承包工程
foreign corporation 外国公司
foreign crystal 异种晶体
foreign cullet 厂外碎玻璃
foreign currencies earned through trade 贸易外汇
foreign currency 外币
foreign currency allotment 外币配额
foreign currency bill 外币汇票
foreign currency bills payable 可付外币；外币付款票据
foreign currency conversion 外币的折算
foreign currency conversion rate 外币折合率
foreign currency credit 外币信贷
foreign currency dealing 炒买炒卖外汇
foreign currency debentures 外币债券
foreign currency deposit 外币存款
foreign currency deposit account 外币存款户
foreign currency earnings 外汇收入
foreign currency exchange 外币兑换
foreign currency futures 外币期货
foreign currency holdings 外币持有额
foreign currency option 外币期权

foreign currency payment instruments 外币支付凭证
foreign currency per U.S. dollar 每美元外币价
foreign currency position 外币头寸
foreign current 外部电流
foreign debt 外债
foreign deposit 国外存款
foreign drain 现金外流
foreign duty pay 国外勤务津贴
foreign ecologic(al) and technical exchange 对外经济技术交流
foreign economic contract law 涉外经济合同法
foreign element 外业因素;外来因素
foreign enterprise 国外企业
foreign equity holding 外国股份
foreign exchange 外汇;外币
foreign exchange and cooperation of education 教育对外交流与合作
foreign exchange arbitrage 外汇套利
foreign exchange assets 外汇资产
foreign exchange balance 外汇结存
foreign exchange bank 外汇银行
foreign exchange broker 外汇经纪人
foreign exchange concentration system 外汇集中制度
foreign exchange control 外汇管制;外汇管理
foreign exchange control regulation 外汇管制法规;外汇管理条例
foreign exchange cost 外汇成本
foreign exchange cushion 外汇缓冲
foreign exchange exposure management 外汇风险暴露管理
foreign exchange fluctuation 外汇浮动
foreign exchange fluctuation insurance 外汇波动保险
foreign exchange future 外汇期货
foreign exchange gains or loss 外汇损益
foreign exchange illusion 外汇幻觉
foreign exchange instrument 外汇票证
foreign exchange license 外汇许可证
foreign exchange line 国际信息交换线路
foreign exchange loan 外汇贷款
foreign exchange management 外汇管理
foreign exchange market 外汇市场
foreign exchange market closed 已关闭的外汇市场
foreign exchange option 外币期权
foreign exchange parity 外汇平价
foreign exchange permit 外汇许可证
foreign exchange policy 外汇政策
foreign exchange position 外汇头寸
foreign exchange quotation 外汇牌价;外汇报价
foreign exchange rate 外汇牌价;外汇汇率
foreign exchange receipt 外汇收入
foreign exchange receipt and disbursement 外汇收支
foreign exchange regulation regulations 外汇管理条例
foreign exchange reserve 外汇储备
foreign exchange retail market 外汇零售市场
foreign exchange retaining system 保留外汇制
foreign exchange service 局外交换业务;区外交换服务;外部交换业务;国际电报电话局
foreign exchange settlement 外汇结算
foreign exchange speculation 外汇投机
foreign exchange trader 外汇交易员
foreign exchange transaction 外汇业务;外汇交易
foreign exchange wholesale market 外汇批发市场
foreign field 外磁场
foreign firm 外国商行;外国企业
foreign flavor 不正常气味
foreign freight agent 外国货运代理人
foreign frequency 外部频率
foreign-funded enterprise 外资企业
foreign-funded firm 外资商行
foreign general agent 国外总代理人;外国一般代理商
foreign going vessel 国外航行船;国际航行船
foreign goods 外来商品;舶来品
foreign government bond 外国政府公债
foreign grog 外来熟料
foreign host 外来主机
foreign imperfection 外来的不完整性
foreign impurity 外来夹杂物
foreign inclusion 外来杂质;外来夹杂物;外来包体【地】;捕虏体【地】
foreign interstitial 外来填隙
foreign-invested enterprise 外资企业
foreign investment 国外投资;外国投资
foreign investment law 外资法
foreign investment policy 外商投资政策
foreign ion 杂质离子
foreignism 外来语
foreignize 外国化
foreign joint venture 外国合营(者)
Foreign Judgment Reciprocal Enforcement Act 外国判决互惠执行法案
foreign juridical person 外国法人
foreign labo(u)rer 外来劳动力;外国工人
foreign law 外国法律
foreign literature 国外文献
foreign load 国外贷款
foreign loan 外国贷款;外债
foreign map 国外地图
foreign market 国外市场
foreign market value 国外市场价值
foreign material 异物;夹杂物;外来材料;掺混物
foreign matter 异物;杂质;外来杂质;外来物(质)
foreign matter equipment and material 进口器材
foreign median massifs 国外的中间地块
foreign medium 外部媒体
foreign merchant 外商
foreign money checktake 外币核收
foreign nationals 外国侨民
foreign navigation 国外航行;国际航行
foreignness 外来性
foreign nitrogenous compound 外来含氮化合物
foreign object 外物
foreign odo(u)r 异臭;恶臭;不适的气味
foreign oil 外国石油
foreign order 国外定货单
foreign ore 外国矿石;舶来矿石
foreign-oriented economy 外向型经济
foreign particle 杂质粒子;杂粒;外来物
foreign patent 外国专利;外国注册
foreign payment 国外支付
foreign payment instrument 外币支付票据
foreign plant 外来植物
foreign policy 对外政策
foreign regional tectonics 国外区域构造
foreign remittance 国外兑
foreign revenue 外国税收
foreign securities 外国证券
foreign selling 外销
foreign solids 外来的固体杂质
foreign source data 外来源数据
foreign standards and codes 国外标准规范
foreign state loans 外国政府贷款
foreign substance 杂质;外来物(质)
foreign tax 外国税赋
foreign tax credit 可减免国外纳税额;外国税收抵免
foreign technology devision 国外技术部
foreign timber trade 外材贸易
foreign trade 外贸;对外贸易
foreign trade agency 外贸机构
foreign trade centres 外贸中心
foreign trade control 对外贸易管理
foreign trade custom 外贸惯例
foreign trade dependence 贸易依存度
foreign trade finance 贸易融资
foreign trade harbo(u)r 外贸港(口)
foreign trade multiplier 外贸乘数
foreign trade policy 对外贸易政策
foreign trade port 外贸港(口);外贸
foreign trade right 外贸权
foreign trade surplus 对外贸易顺差
foreign trade transaction 对外贸易交易
foreign trade zone 对外贸易区;外贸区
foreign traffic 外路运输
Foreign Tribunals Evidence Act 外地裁判庭证据法案
foreign voyage 国际航程
foreign water 客水;外来水
foreign white phosphorus 外来白磷
forelady 女工长
foreland 陆岬;岬;海角;前沿地;前缘地;前沿地;(码头的)前沿;前麓地;前陆;前地;滩地
foreland basin 前陆盆地
foreland-dipping duplex 前倾双冲构造
foreland facies 陆架相;前沿地相
foreland folding 前陆褶皱作用
foreland grit 前陆粗粒砂岩
foreland sequence 前沿地层序
forelift 前桅帆桁吊索
foreline 前级管道;前级真空管线
foreline flange 前级管道法兰
foreline hose 前级管道软管
foreline tank 前级管道罐
foreline trap 前级管道阱
foreline valve 前级管道阀
forellenstein 橄长岩
forelock 楔栓;楔;开尾销;开口销;锚杆销;栓;扁销
forelock key 销;键;开口键
Forel scale 福莱尔水色等级
Forel's colo(u)r standard 福莱尔标准比色计
fore-main shock earthquake 前震主震型地震
foreman 装卸指导员;工头;工(段)长;工班长;领工员;领班
foreman bricklayer 瓦工工长
foreman carpenter 细木工领班;木工监工
foreman carpenter and joiner 木工工长
foreman driller 钻探队长;钻工领班;钻工工头
foreman glazier 玻璃工工长
foreman plasterer 抹灰工工长
Foreman Series 福曼统【地】
foremanship 基层管理能力
foreman steel erector 钢结构安装工领班;钢结构安装工监工
foreman stevedore 码头搬运领班;码头搬运工头
foremast 前桅
foremast deck 锚甲板;船首上甲板
fore masthead light 前桅航行灯
foremastman 普通水手
fore measurement 前测
foremelter 预熔器
foremost frame 最前肋骨
foremountain coall-bearing formation 山前含煤建造
fore mountain depression 山前坳陷
forenoon watch 午前班
forenotice 预先警告;预告
Foren process 福伦轧管法
forensic chemistry 法医化学
fore observation 前视(测量前进方向中的下一站)
fore optics system 输入光学系统
forepale 板桩支架
forepart 前部
forepart construction 船首结构
forepeak 艏尖舱;首尖舱
forepeak bulkhead 防撞舱壁;尖舱舱壁;艏尖舱舱壁
forepeak tank 前尖舱;头尖舱;船首尖舱
fore perpendicular 首垂线
forepiece 前部件
forepiling plate 密排插板
foreplane 中刨;粗刨
foreplate 前板
forepole 超前支护;超前支架;超前梁;超前伸梁;(隧道中的)矢板;前支护桩;坑道顶板;前探梁;挡土桩;挡土板桩;板桩支架
forepoling 插板支撑;预支掘进;矢板;挡土木桩;挡土板桩;搭接插板挡土法;超前伸梁掘进法;插板法
forepoling arm 超前支架前伸梁
forepoling board (隧道中的)矢板;前支护板
forepoling for tunnel roof 护顶背板
forepoling girder (隧道中的)矢板梁;板桩用大梁
forepoling support 及时支撑
fore poppet 前垫架;首支架【船】
fore poppet pressure 首支架压力【船】
fore pressure 前压力;前级压强;前级压力
forepressure breakdown 前级压强破坏
forepressure characteristic 前级压强特性
forepressure ga(u)ge 前级真空规
forepressure measurement 前级压强测量
forepressure side 前级压强端
forepressure tolerance 前级耐压
forepump 预抽(真空)泵;前级泵;前级真空泵
forepumping 前置抽气;前级抽气
forequarter 前部
fore rake 前倾
forereach 赶上;继续前进;超过;超出
fore-reef facies 礁前相
fore-reef zone 礁前带

fore rigging 前桅索具
fore-roof 护棚
foreroom 前堂
fore-royal 前桅顶帆
forerunner 先行者;先驱;预兆;祖先;接近某物的预兆;前震;前兆;前驱涌
forerunner earthquake 前震
forerunner industry 先导产业
forerunner of strong earthquake 强震前兆
fore running 前馏分;初馏物
foresail 前桅帆
forescatter 前向散射器;前向散射
forescattering angle 前向散射角
foresee 预见
foreseeability 预见性
foreseeable damages 可预见的损失
foreseeable loss 可预见的损失
foreseen damages 可预见的损害赔偿
foreset 工作面支架;临时超前支架
foreset bedding 交错层理;前积层(理)
foreset deposit 前积层沉积
foreset laminae 前积纹层
foreset slope 前积坡
foreshadow 预示
foreshadowing 预测;铺垫
fore shaft 锁口
foreshaft 竖井口
foreshaft sinking 上部井筒凿井
foresheet 前桅帆脚索
fore shift 早班
foreshock 前震
foreshock activity 前震活动性
foreshock-after-shock pattern 前震余震型
foreshock-main-shock type 前震主震型
foreshock wave 前震波
foreshore 前撑柱;涨滩;前演;前滩;前滨;前岸;低海滩;潮间岸滩
foreshore deposit 前滨沉积
foreshore feature 前演要素
foreshore flat 干出滩
foreshore step 海滨阶地
foreshorten 按照透视法缩小
foreshortened length 折合长度;缩减长度
foreshortening 透视缩图;用远近法缩小绘画;透视收缩;投影缩减;缩短投影;按透视法缩小绘制
foreshortening effect 投影缩减效应
foreshot 初馏物
foreside 前沿
foreside pontoon 前边浮箱
foresight 前视;前瞄准器;遇见;描准器;前视
foresight reading 前视读数【测】;前方交会
fore-skysail 前桅天帆
foreslope 前缘斜坡;前缘边坡;前坡
fore spring 前倒缆;艏倒缆
fore spring line 前倒缆;艏倒缆
forest 林区;林木;树林;森林(地带)
forest acreage 森林覆盖率
forest administration 森林管理学
forest aerial photogrammetry 森林航测
forest aerial photography 森林航空测量
forest aerosurveying 森林航空测量
fore staff 十字测天仪
forest age 林龄
forestage 前级;舞台台唇;台唇
forest airfield 林间机场
forestall 抢先
forestall trouble 防止事故
forestal resources 森林资源
fore stand and end plate nozzle 前后端板接管
fore stand-hanging hook 前端板吊钩
fore stand leg 前端板支腿
fore stand plate 前端板
fore stand side 前端板侧
forest animal 森林动物
forest area 森林区;森林面积
forest area per capita 人均林地面积
forest aspect 林相
forest assessment 森林资源清查
forestation 植林;造林法
forestation coefficient 森林覆盖度
forestation protection 植树保护
fore stay 前斜缆;前稳索;前拉索;前拉杆;前(桅)支索
forest belt 林带;森林带

forest biology 森林生物学
forest biomass 森林生物群
forest bog 林沼;森林沼泽
forest botany 森林植物学
forest boundary 林区界;森林边界
forest bridge 林区桥梁
forest canopy 林冠覆盖;林冠;森林遮盖
forest capital 森林资本
forest cemetery 森林墓地
forest clearing 林中空旷地
forest climate 森林气候
forest climax 森林顶极
forest concession 林业产销证;森林采伐权
forest conservancy 森林养护;森林覆盖;森林保护(区)
forest conservation 森林护养;森林保护;封山育林
Forest Conservation Society of America 美国森林保护协会
forest consolidation 林区整理
forest-cost-capital 森林生产资本
forest court 森林法院
forest cover 树林植被;森林植被;森林覆盖
forest coverage 森林覆盖率
forest coverage rate 森林覆盖率
forest covered area 森林覆盖面积
forest cover statistics 森林覆盖面积统计
forest cover type 森林覆被型
forest crematorium 森林火葬场
forest crop 森林作物
forest culture 植林;造林
forest cutting 森林砍伐;伐林
forest damage 森林损害
forest decay 森林逐渐失去活力;森林退化
forest denudation 森林滥伐
forest depot 储[贮]木场;集材场
forest description 森林调查
forest deterioration 森林逐渐衰败;森林破坏
forest devastation 森林破坏;森林荒废
forest dieback 森林(顶)梢枯死
forest district 林区;森林区(划)
forest-division 森林区划
fore steaming light 前桅航行灯
forest ecologic(al) system 森林生态系统
forest ecology 森林生态学
forest economics 森林经济学
forest economy 森林经济
forest ecosystem 森林生态系统
forested area 植树面积;绿荫面积;林荫面积
forest edge 林缘
forested gley soil 森林潜育土
forested region 绿荫地区;森林区
forest environment 森林环境
forester 林业工作者;林务员
foresterite 橄榄石砂
forest estate 森林产业
forest esthetics 森林美学
forest examiner 森林技师
forest expectation value 森林期望价值
forest experimental station 林业试验站
forest experiments officer 森林技师
forest experiment station 森林试验站
forest fatal shade 森林致死遮荫度
forest fertilization 森林施肥
forest finance 森林财政
forest fire 森林火灾;林火
forest-fire cloud 森林火灾云烟
forest-fire control 森林火灾防救
forest-fire insurance 森林火灾保险
forest-fire meteorology 森林火灾气象学
forest-fire prevention 森林防火
forest floor 林地覆被物;森林地表;林床;森林覆被(物);森林地面;森林地被物
forest floor detention storage 林床滞留水量;林床拦蓄(量)
forest for conservation of water supply 水源涵养林;水源保护林
forest for erosion control 土砂流失防护林;防止土壤侵蚀的森林;防护林
forest form 林相
forest form map 林相图
forest for protecting slopes from soil erosion 护坡林
forest for protection against soil denudation 防沙林;防风固沙林

forest for public health 保健林区
forest for special purpose 特种用途林
forest for special use 特种用途林
forest for water and soil conservation 水土保持林
forest garden 林木育种园;示园;森林树木园
forest geobotany 森林地植物学
forest grassland 林间草地
forest grazing 林间放牧
forest green 森林绿
forest grower 造林者
forest growing stock 森林蓄积量
forest-guard system 森林巡护制度
forest habitat 森林生境
forest highway 林区公路;森林公路
forest humus 森林腐殖质
forest humus soil 森林腐殖土
forest husbandry 森林管理
forest hydrology 森林水文学
forest hygiene 森林卫生
forest hygienics 森林卫生学
forestick 挡木
forest improvement 森林改进伐
forest-income tax 森林所得税
forest industry 森林工业
forest influence 森林影响
foresting 植林;造林
forest in grasslands and deserts 草原和沙漠地带的森林
forest inspection bureau 森林监督局
forest inventory 森林资源清查;森林资源调查;森林存量
forest inventory control 森林资源管理
forest land 林地;森林带
forest landscape 森林景观
forest law 森林法
Forest Law of the People's Republic of China 中华人民共和国森林法
forest limit 森林界限
forest limit temperature 森林极限温度
forest line 森林界线;森林(分布)线
forest litter 落叶层;林地植被;枯枝落叶层;森林地被物
forest logging railway 森林铁路
forest management 森林经营学;森林经营;森林管理
forest management erosion 森林管理与侵蚀
forest management system 森林管理系统
forest management unit 森林经营单位
forest manager 森林管理人
forest map 森林图;森林地图
forest mapping 森林制图
forest marble 树景大理岩;自然图案大理石
forest-master 森林技师
forest measurement 测树学
forest meteorology 森林气象学
forest microclimate 森林小气候
forest moss peat 森林苔藓泥炭
forest-nursery 森林苗圃
fore-stock 前支架
forest of columns 柱群;柱林;柱建筑群
forest of conifer species 针叶林
forest of dean 灰色砂岩(产于格拉撒斯特郡)
forest offence 违犯林业规章行为
forest on land liable to inundation 泛区森林
forestope 前探梁;超前伸梁;梯段开挖掌子面(隧洞工作面);前超梁
forest opening 林中空旷地
forest outlier 孤存林;残遗孤林
forest park 森林公园
forest peat 森林泥炭
forest percent 森林覆盖率;森林造林率
forest pest 森林害虫
forest photogrammetry 森林摄影测量
forest planning 森林规划
forest plant 森林植物
forest plantation 植林区;植树造林
forest planting 植树造林;植苗林;栽植造林;林业造林法;林业林
forest plot 森林小区
forest policeman 森林警察
forest policy 林业政策;森林政策
forest product 林产物;林产品;木料;木材;森林产品
forest productivity 森林生产率
forest property 森林财产;森林所有权

forest-protection 森林保护(学)
forest purification 森林净化
forest-railroad 森林铁道
forest-railway 森林铁道
forest range 护林区
forest ranger 营林员;林警;护林员
forest reclamation 森林改良
forest regeneration 森林更新
forest region 林区;森林植物带
forest regulation 私有林控制法规;森林条例
forest remote sensing 森林遥感
forest renewal 森林更新
forest rent 林租;林利
forest replanation 森林补植
forest reservation 封山育林
forest reserve 禁伐森林;护林区;森林资源;森林蓄积量;森林保护区;封山育林区
forest resource assessment 森林资源评估
forest resources 森林资源
forest resources conservation 森林资源保护
forest ride 林间车道
forest-right 森林使用权
forest road 林区道路;林间路;森林(道)路
forestry 林(业)学;林业;林地;森林学
forestry adviser 林业顾问
forestry area statistics 森林面积统计
forestry ax(e) 伐林斧
forestry base 林业基地
forestry centre 林场
forestry community 森林群落
forestry fund 林业基金
forestry legislation 林业立法
forestry machinery 林业机械
forestry station 林场
forestry torch 林业点火枪
forestry waste 林业废物
forestry worker 林业工作者
forest sanitation 森林环境卫生
forest scenic spot 森林风景区
forest-school 森林学校
forest-servitude 森林使用权
forest settlement 森林产权判定
forest shelter belt 防护林带;防风林带
forest site indicator 森林立地指示植物
forest soil 森林土壤;森林土
forest soil reclamation 森林土壤改良
forest species 森林树种
forest squatter 林内潜居民
forest staff 森林职员
forest stage 森林阶段
forest stand 林分
forest statistics 森林统计学
forest steppe 森林草原
forest steppe belt 森林草原带
forest succession 森林演替
forest supervision 森林监督管理
forest survey 林业调查;林业测量;森林调查
forest surveying 森林测量
forest swamp 森林沼泽
forest swamp deposit 森林沼泽沉积
forest swamp facies 森林沼泽相
forest technology 森林工艺学
forest thinning 森林抚育间伐
forest-to-coal process 森林成煤过程
forest track 林区分界线;森林(道)路
forest trail 林荫小径
forest-tramway 森林轨道
forest tree 林木;森林树木
forest-tree nursery 森林苗圃
forest tundra 森林冻原
forest type 林型
forest type science 林型学
forest typology 林型学
forest utilization 森林利用学
forest valuation 林价算法;森林估价
forest vegetation 森林植物群
forest village 森林村
forest-warden 森林巡视
forest watersned 森林流域
forest weather station 森林气候观测所
forest wind 森林风
forest worker 林业工作者
forest year 森林事业年度
forest zone 林区;森林带

forest zoology 森林动物学
fore support 超前支架
foretop 前桅楼;前探梁
foretopman 前桅哨
foretopmast 前桅的中段
foretopsail 前桅中桅帆
fore treatment 前处理
fore trigger 前扳机
foretruck 前桅冠
foreturn 前面绕圈
fore-vacuum 前级真空;预真空
fore-vacuum condenser 前级真空冷凝器
fore-vacuum connection 前级真空连接
fore-vacuum cooler 前级真空冷却器
fore-vacuum cylinder 前级真空腔
fore-vacuum ga(u)ge 前级真空规;前级真空表
fore-vacuum line 前级真空管道
fore-vacuum pipe 前级真空管道
fore-vacuum port 前级真空连接口
fore-vacuum pressure 前级真空压强
fore-vacuum pump 前级真空泵
fore-vacuum pump system 前级真空抽气系统
fore-vacuum rotor 前级真空转子
fore-vacuum space 前级真空空间
fore-vacuum subassembly 前级真空组件;预真空组件
fore-vacuum system 前级真空系统
fore-vacuum trap 前级真空冷阱
fore-vacuum tubing 前级真空软管
fore-vacuum valve 前级真空阀
fore-vacuum vessel 前级真空容器;预真空容器
fore(ward) reading 前视读数
forewarmer 预热器
forewarn 预先警告
forewarning 前兆
fore wheel 前轮
forewing 前置机翼
forewoman 女领班;女监工员;女工头;女工长
foreyard 前院;前桅下帆桁
foreyn 污水池;排水管;粪坑
forfeit 没收物;没收;罚款;罚金;被没收
forfeited securities 没收抵押品
forfeited securities arising from loans 没收押品;没收抵押品
forfeited share 被没收的股份
forfeited stock subscription 没收股款
forfeit for breach of contract 违约金
forfeit stock 被没收的股份
forfeiture 没收物;没收财产;罚款;罚金
Forfeiture Act 没收法例
forfeiture of property 财务没收
forge 假造;锻造;锻铁炉
forgeable 可锻的
forgeable cast iron 可锻铸铁
forgeable iron 延性铁;可锻铁
forge ahead 渐渐追过前船;前进;前冲
forge ahead despite difficulties 知难而进
forge bellows 锻炉风箱;锻风箱
forge blast 鼓风锻造炉
forge coal 锻造用煤;锻冶用煤
forge cold 冷锻
forge crack failure 锻裂破断
forge crane 锻造起重机
forged 锻造的
forged alloy 锻造合金
forged alumin(i)um piece 铝锻件
forged and welded vessel 锻焊式容器
forged bar 锻造圆钢
forged bit 淬火钎头
forged blank 锻制毛坯
forged brass 锻黄铜
forged carbon steel 锻造碳钢
forged chain cable 锻造锚链
forged documents 假造的文书单据;伪造单据
forge delay time 锻焊延迟时间;加压滞后时间;锻压时间
forged endorsement 伪造背书
forged flange 锻制法兰
forged head 锻造封头
forged iron 熟铁
forged metal 锻制金属
forged nail 锻制钉
forged piece 锻件
forged pipe 熟铁管

forged product 锻件
forge drill 锻造钻头
forged signature 伪造签字
forged split T 锻钢对开三通
forged split tee 锻钢对开三通
forged steel 锻钢
forged steel magnet 锻造钢磁铁
forged steel rolls for hot rolling 锻钢热轧辊
forged steelwork 钢锻件
forged two wing bit 锻造鱼尾钻头;锻造双翼钻头
forged vessel 锻造容器
forged weld 锻压焊缝
forge furnace 锻造炉;锻炉
forge hammer 锻锤
forge-hammer foundation 锻锤基础
forge hearth 锻铁炉;锻炉
forge hot 热锻
forge iron 锻铁
forge lathe 锻件粗车床
forge length (摩擦焊时产生的)顶锻变形量
forge milling machine 重型铣床
forge over 冲过(快速冲过浅滩等)
forge pig iron 可锻生铁
forge pigs 锻铁
forge plant 锻造厂
forge press 锻压机
forge press machinery 锻压机械
forger 锻工
forge rate 顶锻变形速度
forge roll 辊锻机
forge rolling machine 辊锻机
forgery 伪造物;伪造品;伪造
forgery bond 假造债券
forge scale 氧化皮;锻铁鳞;锻鳞
forge shop 铁匠铺;锻工车间;锻工场
forge steel 锻钢
forge test 锤击试验;可锻性试验
forge time 锻压时间
forget-me-not 勿忘草【植】
forge tongs 锻工钳
forge weld 锻压焊缝
forge-welded(monolayered) cylinder 锻焊式单层圆筒
forge welding 锻接焊;锻接;锻焊
forge work 锻工(工作);锻造(厂)
forging 锻造;锻
forging alloy 可锻合金
forging alumin(i)um piece 锻制废水
forging and pressing equipment 锻压设备
forging and stamping 锻压
forging brass 可锻黄铜;锻造黄铜
forging bursting 锻裂
forging characteristic 锻造特性
forging crane 锻造起重机
forging department 锻造车间
forging die 锻模
forging dimension 锻造尺寸
forging drawing 锻造图;锻件图
forging equipment 锻制设备;锻造机;锻压设备;锻压机
forging factory 锻造厂
forging flow lines 锻造流线
forging furnace 锻造炉;锻炉
forging-grade ingot 锻造用钢锭;锻用钢锭
forging hammer 锻工锤;锻锤
forging hammer foundation 锻锤基础
forging iron 锻铁
forging line 锻造生产线
forging machine 锻压机;锻机
forging manipulator 锻造操纵器;锻造操纵工;锻造翻钢机;锻造操作机;锻造操纵机
forging method 锻造法
forging mo(u)ld 锻模
forging operation 锻造操作
forging plane 锻造面
forging plant 锻造车间
forging press 锻造压力机;锻压机
forging pressure 锻造压力
forging property 锻造性质
forging pump press 锻造水压机
forging quality 可锻性
forging quality plate 可锻钢板
forging quality steel 可锻优质钢板;锻造用钢胚

forging range 锻造温度范围
forging ratio 锻造比
forging reduction 镦粗
forging rolls 轧锻机;辊轧机;锻制轧辊
forging shop 锻工车间
forging steel 锻钢
forging steel valve 锻钢阀
forging stock 锻坯
forging strain 锻应变
forging stress 锻造应力
forging swaging machine 锻造机
forging temperature 锻造温度
forging temperature interval 锻造温度范围
forging test 锻造试验;锻压试验;锤压试验;锤击试验
forging thermit 锻接铝热剂
forging tubesheet 锻造管板
forgive a debt 免偿债务
forgive debt 免除债务
forgiveness of debts 免除债务
forgiveness of liabilities 免除债务
forgy handle 杠杆齿条式千斤顶手把
for-hire 出租的
fork 货叉;河流的主要支流分叉;弹簧叉;双叉挂车夹;导向器;岔流;岔;插销头;叉形夹;叉;分岔;边拉伸器
fork and blade connecting rod 叉形中间连杆
fork arm 叉杆
fork arrangement 叉形装置
fork bearing 叉轴承
fork budding 钩形芽接
fork catch 叉形门扣;(门的)叉形拉手;叉形档
fork center shaft 转向节中心轴;轮叉中心轴
fork chain 叉链
fork channel blockage 堵塞汊道
fork channel closure 堵塞汊道
fork chuck 叉形卡盘
fork clamp 叉形挟钳
fork connection 叉状连接;叉式连接法;叉式接法;分岔连接
fork connector 叉形接头
fork coupler 叉形联结装置
fork dispatcher 分叉分派程序
forked 叉状的;叉根的;分叉的
forked abutment 叉形桥台
forked axle 叉轴;叉形轴
forked bar 叉形杆件
forked bed 叉形座
forked branch 分叉树枝
forked chain 支链
forked circuit 分支电路;分岔电路;分叉电路
forked clamp 叉形压板;叉形夹板
forked connecting rod 叉式连杆
forked crosshead 叉形十字头
forked driving rod 叉形传动杆
forked echo 分岔回波
forked element 叉形接头
forked end 叉端
forked end lever 叉头杆
forked frame 叉形架
forked growth 叉状生长;分叉生长
forked head 叉形箭头
forked hose 叉式软管;叉式皮龙管;分叉软管
forked joint 锯齿形接合;叉形接头
forked journal 叉形轴颈
forked lever 叉形杆;叉杆
forked lightning 叉状闪电;分叉闪电
forked line 端部分叉的绳索
forked link 叉形连杆
forked long slotted cross head 叉形长槽十字头
forked loop 叉形环;双环
forked loop bar 叉形眼杆
forked mortise and tenon joint 叉形镶榫接;叉形公母接;叉形槽舌接
forked mounting 叉式装置
forked ordinary crosshead 叉形十字头
forked piles anchorage 锚定叉桩
forked pipe 叉形叉管;分叉管;叉管
forked repeater 双路转发器
forked road 岐路;岔路;分叉道路
forked rod 叉头杆
forked screw driver 叉形螺旋起子;叉形螺丝刀
forked ship incline 人道式斜面升船机
forked siamesed 叉形的

forked spanner 叉形扳手
forked spring carrier arm 叉形簧架
forked standard 叉座
forked strap 叉形系铁;叉形金属带
forked tenon 交叉榫接;丁字榫接;叉形榫
forked tie 叉形枕木;叉形锚定板;叉形拉杆
forked tube 叉形管;分叉管
forked vein 二叉叶脉
forked wood 叉形木材
fork end 叉端
fork expander bolt 车前叉调整螺丝
fork extension 铲车叉套管;叉齿套
fork extension sleeve 叉齿套
fork eye 叉形眼
fork factor(of tributary) 分汊系数;分流系数
fork file 平圆角锉
fork frequency 音叉频率
fork gap 叉隔
fork ga(u)ge 叉规;分岔标准尺;分叉标准尺
fork grafting 叉状嫁接
fork grip 叉形夹钳
fork groove 叉形槽
forkgrooving machine 开槽机;铲沟机
fork guard 保险叉
fork handle 叉柄
fork hay onto a stack 叉草上垛
fork head 护轮架;翻砂架;叉形头
fork hoe 叉锄
fork hook 叉钩
fork hose 叉式软管;叉式皮龙管
forking 分叉
forking bed 分叉岩层
forking lath 交叉板条
fork intersection 叉形交叉(口)
fork in the road 三岔路口;三叉路口
fork joint 叉形连接;叉形接头
fork junction 叉形交叉(口)
fork lift 铲车;叉形起重机;叉架起货机;升降货车;叉式万能装置;叉式万能升降;叉式升降机;集装箱叉车;叉式装卸机;叉式装卸车;叉举;叉车
forklift attachment 叉形抓爪
forklifter 万能升降叉;叉车;叉式升降机
forklift for pallets 集装箱式叉式运货
forklift hoist 叉式起重机;叉式铲车
forklift loader 叉式装载机
forklift mast 叉式起重杆
forklift pocket 叉槽
forklift truck 提升叉车;叉式装卸车;叉式提升搬运车;叉架式运货车;叉(铲)车;铲车;叉式扬汽车;叉式起重车
forklift with diesel engine 柴油驱动叉车
fork link 叉形连接杆;叉形杆
fork load unload car 叉式装卸车
fork mounting 叉形座架
fork pin 叉形销;叉销
fork pocket 叉槽
fork process 分叉进程
fork rod 叉形棒;叉棍
forks arm 叉形臂
fork section 交叉段
fork shaft 叉轴
fork shaped 叉形的
fork-shaped tee 叉形三通
fork-shaped wharf 突堤式码头
fork spanner 叉形扳手
fork stacker 叉式装载机;叉式铲车;叉车
forkstaff plane 凸圆线刨;突缘线刨;突圆线刨
fork standards 叉架头部
fork stripper 叉形卸料板
fork structure of bubble device 磁泡器件叉形结构
fork-tail 叉尾
fork terminal 叉形引出线
fork the end of bar 钢条端分叉
fork the hole (在完成井中的)钻旁孔
fork tone 音叉音
fork-tone modulation 叉音调制
fork tongs 叉式钳
fork track 叉式起重机
fork truck 叉式起重车;叉式自动装卸车;叉式货车;叉车
fork truck attachment 叉式自动装卸车附件
fork truck scoop 叉形卡车铲;叉车铲(斗)
fork-type digger 叉形挖掘机;叉式挖掘机
fork-type hay loader 叉式装干草机

fork-type washing machine 叉式洗涤机
fork wrench 义形扳手
for light and air 供采光通风用
for list 循环元素表;循环表
for livestock 兽用
for love or for money 现款交易
form 形状;形式;用纸;样式;格式;类型;模塑;盒子板;体型;打印纸;表格;表单;模壳
formability 可塑性;易成型性;可成型性;成型性
formable sheet 可成型板
form a complete production network 配套生产
form a complete set 配套
form a complete set of 配套组合
form address 格式地址
form advance 进页;换页
form a flute 凿槽
form agent 成型剂
form a groove 设计孔型
form aid 制模剂;制模工具
Formal 福默尔
formal 形式的;正式的;克式量;缩甲醛
formal acceptance 正式验收
formal acceptance test 正式验收试验
formal-actual parameter correspondence 形式实参对应
formal address 形式地址
formal advertising 正式广告;正式公告;公告
formal agreement 正式协议;正式协定
formal algebra 形式代数
formal and spatial concept 形式与空间概念
formal approach 形式方法
formal approval 正式批准
formal approximation 形式近似;形式逼近
formal argument 形式变元
formal axiomatics 形式公理学
formal bond 形式键
formal cause 形式原因
formal claim 正常索赔
formal class 形式类
formal concentration 克式量浓度
formal construction 形式构造
formal contract 正式契约;正式合同
formal control 形式控制
formal copper wire 聚乙烯醇缩甲醛铜线
formal declarer 形式说明词
formal deduction 形式推导
formal definition 形式定义;格式定义
formal degree 形式次数
formaldehyde 蚁醛;甲醛
formaldehyde azo test 甲醛偶氮试验
formaldehyde resin 甲醛树脂
formaldehyde test 蚁醛试验
formaldehyde-treated wood 甲醛处理过的木材
formaldehyde wastewater 含甲醛废水
formal denotation 形式标志
formal description 形式描述
formal design specification 形式设计说明
formal discipline 形式训练
formal disoretization error 形式离散误差
formaldoxime 甲醛肟
formale bottle 聚乙烯醇瓶
formale copper wire 聚乙烯(绝缘)铜线
formal effective 正式生效
formal entry 正式进口
formal environmental education 正式环境教育
formal flower bed 整形花坛
formal garden 规则式庭园
formal garden style 规整式园林
formal group 形式群
formal implication 形式蕴涵
formalin 甲醛(水)溶液;特性周波带;福尔马林
formal inference 形式推理
formal inspection 正式检查
formalin sterilizing unit 福尔马林消毒装置
formal integration 形式积分法
formal investigation 正式调查
formalism 形式主义;形式论;公式化
formalities 正式手续;例行手续
formality 形式性;克式浓度;手续
formalization 形式体系化;形式化;定型;定形
formalize 形式化
formalized arithmetic 形式化算法
formalized computer 形式化计算机
formalized computer program(me) 形式化计算

机程序
formalized model 形式化模型
formal language 形式语言
formal law 成文法
formal lawn 整形草坪
formal leading coefficient 形式的首项系数
formal logic 形式逻辑
formal lower-bound 形式下界
Formall process 落锤深冲法;福马尔落锤深冲法
formally adjoint operator 形式伴随算子
formally real field 形式实域
formal method 形式方法
formal model 形式模型
formal notation 形式标志
formal notice 正式通知
formal notion 形式概念
formal operations 形式操作
formal parameter 形式参数;形参
formal parameter call 形式参数调用
formal parameter list 形式参数表
formal parameter part 形式参数部分
formal parsing 形式分析
formal parsing algorithm 形式分析算法
formal plan 正式计划
formal potential 表观电位
formal power series 形式幂级数
formal power series field 形式幂级数域
formal power series ring 形式幂级数环
formal product 形式产品
formal proof 形式证法
formal pushdown automaton 形式下推自动机
formal receipt 正式收据
formal record 形式记录;正式记录
formal representation 形式表达法
formal resin 聚乙烯醇缩甲醛树脂
formal ring 形式环
formal road system 规整式道路系统
formal row 形式行
formal rule 形式规则
formal scheme 形式概型
formal solution 形式解
formal specification 形式说明
formal spectrum 形式谱
formal stratigraphic(al) unit 正式地层单位
formal symmetrical elevation 形式对称的立面图
formal system 形式系统
formal talk 正式会谈
formal theory 形式理论
formal transformation 形式变换
formal treatment 形式处理
formal undecidable proposition 形式不可判定命题
formal unit 正式地层单位;正式单位
formal upper bound 形式上限
formal variable 形式变量
formal verification 形式证明
formal visit 正式访问
formal warning 正式警告
formamidoxime 氨基甲肟
formamyl 氨基甲酰基
form anchor 模板锚定物;模板锚定器;模板锚定板;模板锚固装置
form and structure zoning 形态分区
formanilide 甲酰替苯胺
formaniline 甲醛缩苯胺
formanite 黄钇钽矿
formant frequency 共振峰频率
form assembly 模板装配
format 形式;信息编排;共振峰;格式;排式
format analyzer 格式分析器
format character 格式符
format chart 格式图
format check 格式检验;数据控制程序(的)检验
format code 格式代码;形式代码
format control 格式控制;数据安排形式控制
format control card 格式控制卡片
format description 格式描述
format description type 格式说明类型;格式描述类型
format designator 格式标志符
format dimensions 规格尺寸
format display 格式显示
formate 蚁酸盐;甲酸盐
format effector 格式明细符;格式控制(字)符;布局控制字符

format effector character 格式有效符;格式生效符
format error 格式误差;格式错误
format field descriptor 格式域描述符
format for construction specification 建筑施工手册
format identification 格式识别;格式标识
format implicit address instruction 格式隐地址指令
format instruction 格式指令
formatio alba 白网状结构
formation 基床;形(成)层;组成;组【地】;构造水;构造;路基表面;建造;群系;生成;队形;地层;成型;成林措施
formation accuracy 路基标高的正确性
formational geology 构造地质学;建造地质学
formation analysis log 地层分析测井图
formation anchorage 编队锚地
formation and development 形成和发展
formation capture cross section 地层的俘获截面
formation characteristics 地层特性
formation compressibility 地层压缩系数
formation condition 生成条件;路基情况;地基情况;岩层状态
formation constant 形成常数
formation contaminants 地层污染
formation correlation diagram 地层对比图
formation deflecting rate 地层造斜率
formation densilog 地层密度测井
formation densilog curve 地层密度测井曲线;层分辨率密度测井曲线
formation densilog tool 地层密度测井仪
formation density compensated tool 补偿地层密度测井仪
formation density log tool 地层密度测井仪
formation depth of intrusive body 侵入体形成深度
formation dip angle 地层倾斜角
formation drilling 地层试探钻孔
formation elevation 路基高程;地基高程
formation entry 砂侵
formation evaluation 地层评价
formation expenses 开办费
formation factor 形成因素;岩层产状要素;结构系数;建造因素;地层因素
formation factor of corresponding matrix porosity 对应骨架孔隙度的地层因数
formation factor of sand laminate in shaly sand 泥质砂岩中纯砂岩夹层的地层因数
formation fluid sampler 地层流体取样器
formation fracture 地层裂缝
formation fracturing 地基钻孔;地层断裂
formation gas-oil ratio 岩层油气比
formation geometric factor 地层的几何因子
formation hardness 岩层硬度
formation interval tester 井段地层测试器;地层间隔测试器
formation irregularity 地基不规则性
formation level 路肩高程;路肩标高;路基面;路基标高;施工基面(标高);表面的竣工高程
formation line 施工线
formation lithology 地层岩性
formation map 地层图
formation microscanner 地层微扫描器
formation multitester 多次地层测试器
formation name 地层名称
formation occurrence 地层产状
formation of alkali soils 碱土形成
formation of bed 河床形成
formation of calcium silicate 硅酸钙的形成
formation of capital 资本形式
formation of cement compounds 水泥矿物的生成;水泥化合物的形成
formation of country 区域构造
formation of cracks 龟裂的形成;断裂的形成;裂纹形成
formation of fleet 编队【船】
formation of freight trains 货物列车编组
formation of gas 气体的形成
formation of gas bubble 气泡的形成
formation of groundwater chemical composition 地下水化学成分的形成
formation of hoar frost 结白霜
formation of ice 结冰;冰的形成
formation of image 成像
formation of lump 形成大块;结块

formation of precipitation 降水(的)形成
formation of pusher train 顶推船队队形
formation of rings 圈形成
formation of rust 生锈
formation of scale 水垢生成
formation of slag 造渣
formation of supply in general 一般储备的形式
formation of towing train 吊拖船队队形
formation of truck-haul freight trains 货物列车编组最佳方案
formation of washboard waves 洗衣板式波浪的形成(路面)
formation of wave 波浪形成
formation of weld 焊缝成型
formation packer 地层封隔器
formation plugging 地层封堵
formation predict 地层预告
formation pressure 地层压力
formation pressure at the n production stage n生产阶段的地层压力
formation pressure control 地层压力控制
formation protection layer 地基保护层
formation rate 形成率
formation resistance factor 地层电阻系数
formation resistivity 地层电阻率
formation resistivity factor 地层电阻率系数
formation rule 形成规则
formation sample 岩样;地层岩样
formation sample analysis 岩样分析
formation shut off 地层封堵
formation siding 编组站【铁】
formation solubility 地层溶解度
formation strength 地基强度;地层强度
formation stress field 构造应力场
formation temperature 形成温度;地层温度
formation temperature of sampling point 采样点地层温度
formation tester 地层试验器;地层测试器
formation test(ing) 地层试验
formation testing operations 地层探测操作;地层试验操作
formation thickness estimates 地层厚度估算
formation time 形成时间
formation time lag 形成时滞
formation transformer 电成型用变压器
formation type 生物群系型
formation voltage 形成电压;化成电压
formation volume factor 地层体积系数
formation water 地层水
formation water production rate 地层水产量
formation water resistivity 地层水电阻率
formation width 地基宽度
formation work 地基整平;编队作业
formation yard 装配场
format item 格式元素;格式项
format item specification 格式项说明
formative 形成的
formative arts 造型艺术
formative discharge 造床流量
formative factor 形成因素
formative gear 成型齿轮
formative radius 形成半径;成型半径
formative region 形成区
formative study of development strategies 发展战略学
formative substance 形成物质
formative technology 造型工艺
formative tissue 形成组织
formative voltage 形成电压
format label 格式标号
formatless input 无格式输入
format line 格式行
format list 格式表
format memory data 格式存储数据
format of data directed output 定数数据输出格式
format of edit-directed transmission 定向编辑传输格式
format order 格式指令
format-pattern 格式图像
format primary 格式初等量
format procedure 格式过程
format program(me) 格式程序
format recognition 格式识别
format secondary 格式二次项

format select 格式选择
format service program(me) 格式服务程序
format set 格式组;格式集
format sheet 格式纸;格式页;格式图表
format specification 格式说明;格式规定
format specifier 格式符
format standard 格式标准
format storage 格式存储
format string 格式行;格式串
format structure 格式结构
format symbol 格式符号
formatted data 格式数据
formatted data set control block 格式化数据集控制块
formatted display 格式化显示
formatted dump 格式化转储
formatted file 格式文卷;格式(化)文件
formatted image 格式化图像
formatted input-output 格式化输入输出
formatted logic record 格式逻辑记录
formatted program(me) 格式化程序
formatted read statement 有格式读语句
format(ted) record 格式(化)记录
formatted request 格式化请求
formatted screen image 格式化屏幕映像
formatted software capacity 格式化软件容量
formatted system services 格式化系统服务(程序)
formatter 格式化程序
format text 格式正文;格式原文
formatting 格式化;格式编排;编排格式
format translate 格式翻译
format verification device 格式检查装置
format vocoder 共振峰声码器
Formazin turbidity unit 福尔马肼(浊)度
form board(ing) 模壳板;木模板
form board test 形板测验
form brace 装配支柱;模板支撑;模板拉条
form bracing 模板支撑
form broach 成型拉刀
formbuilding 构型法
form burn 表面走形
form cage 模板骨架
form clamp 模壳边撑;模板夹具;模板边撑
form class 形级
form coating 模壳涂料;模板涂油;模板涂料;模板隔离剂
form code 形式码
form coefficient 船型系数
form concept 形式概念;立体感
form concrete vibrator 附着式混凝土振动器
form contract 标准合同
form control buffer 格式控制缓冲
form control image 格式控制图像
form control template 靠模控制样板;仿形控制样板;仿形靠模板
form copying 靠模法;仿形加工法
form-copying type gear grinder 仿形齿轮磨床
form correction repair 整形修理
form cost 模板费
form crushing roll 成型辊
form culture 整形栽培
form cutting 成型刀削;成型切削
form cycling rate 模板周转速度;模板周转率
form diameter 外形直径;成型直径
form dimension 定形尺寸
form draft 型吃水
form drag 形阻;形状阻力;型阻力
form drag of sand wave 沙波形阻力
form drawing 模板图;木模图
form dresser 整形器
form duct 预留槽
forme 印板
formed 有模板的;用模板的;成型(加工)的
formed bellows 沉积波纹管
formed body 压型体
formed coil 模绕线圈;嵌绕线圈;成型线圈
formed conductor 成型导体
form(ed) cutter 成型铣刀;成型刀具
formed cutter milling 成型铣刀铣削
formed ditch digging bucket 沟渠成型挖掘铲斗
formed drain 预制排水管
formed fabric 成型法非织造织物
formed in dry state 在干燥状态下形成的;干燥成型的
formed in place 就地成型的
formed mild steel pan (板式给料机中的)压模钢链板
formed plastics 成型塑料
formed plate 成型钢板;铸制极板
formed plywood 模压胶合板;成型胶合板
formed punch 冲头
formed rope 六股钢丝缆
formed rubber tank 可任意形状的软油罐
formed section 冷弯型钢
formed steel 型钢
formed steel base 型钢底座
formed steel construction 型钢结构
formed stool 成型粪;成型便
formed threading tool 成型螺纹铣刀
formed tooth 成型齿;成型铲齿
formed turning tool 成型车刀
formed winding 成型绕组
form effect 形状效应
form element 形状要素
former 线圈架;框架;模子;模型;模板工程;手动模刀;打桩套管;成型设备;成型器;成型机;成型刀;幅板
former bar 放大尺
former block 定形块;修形砌块;冲模
form erecting 支模
form erection 模板安装
formerets 穿半肋;穿半筋(土木建筑);附墙拱肋
former glaciation 古冰川作用
former glacier 古冰川
forme rollers 印版墨辊
former plate 仿形样板;仿形样板
former rail 前导轨
former rib 保形肋
former roll 成型轧辊
former stress field 前期应力场
former winding 模压成型绕组;模绕组;模绕法
former-wound coil 型卷线圈;模绕线圈;模拟线圈
form face 模板面
form factor 形状因子;形状因素;形状因数;形状系数;曲线形式系数;成形系数;波形因素;波形因数
form factor of the weld 焊缝成形系数
form feed character 格式馈送符号;格式馈给符号
form feed(ing) 格式馈送;格式馈给;格式进纸;打印式输送
form fixer 模板工;模板工
form flash 格式投影;格式装饰;格式显示;格式闪烁
form freeboard 型干舷
form function 形状函数
form genus 形态属
formgrader 模槽机
form grinder 成型磨床
form grinding 样板研磨;成型磨削
form grinding process 成型磨削过程
form hachure 地貌晕翁
form handling 模板拆装工作
form hanger 模板支撑;模板挂钩;模板吊架
form hardware 模壳小五金
form height 形高
formica 胶木
formic acid 蚁酸;甲酸
Formica knife 弗米加爪形刀具
formicary 蚁穴;蚁山;蚁巢
formiciasis 蚁咬病
formidable natural condition 恶劣的自然条件
form index 形态指标
forming 冶成;模压件;成型;非切削成型;仿形;编成
forming after peak temperature 峰后定形
forming angle 成型角
forming attachment 靠模附件;仿形附件
forming between profile 模夹变形丝
forming block 滚料板
forming by turning 车坯成型
forming chamber 成型室
forming conveyer 成型输送带
forming crew 制模班组
forming current 成型流
forming cutter 成型刀
forming cycle 混凝土模板周转
forming defect 成型缺陷
forming die 成型模具;成型模;成型钢模
forming dresser 成型修整器(砂轮)
forming electrode 聚焦极
forming element 模壳元件
forming fare 形成面
forming gang 制模班组
forming gas 合成气体
forming head 成型封头
forming hood 成型罩
forming job 模壳作业
forming lapping 成型研磨
forming lathe 仿型车床;仿形车床
forming limit 成型极限
forming limit diagram 成型极限图
forming machine 铺装机;定型机;成型机;成板机
forming method 立模方法;成型方法
forming mill 冷弯机
forming of blocks 块材制作
forming of glass 玻璃成型(法)
forming of soil 土壤的形成
forming of tiles 制瓦
forming operation 成形操作;形成操作
forming party 制模班组
forming plate 整平板(路面)
forming pliers 轧印钳
forming press 压力机;成形压力机
forming pressure 制模压力;成型压力
forming process 形成过程;成型过程;成型工序;成形法
forming property 成型性能
forming punch 成型冲头
forming rest 成型刀架
forming roll(er) 压延辊;成型辊
forming rolls 成型轧辊
forming section 成型区;成型部
forming station 成型站
forming system 模壳系统
forming team 制模班组
forming technique 成型技术
forming time classify of sinter 泉华形成时间分类
forming tip 漏嘴
forming tool 成型工具;样板刀;成型刀具;成型车刀
forming tube 绕丝筒
forming tunnel 成型隧道;成丝通道
forming turret lathe 仿形六角车床
forming twister 模压捻线机
forming wire 成型网
forming work 成型加工
form insulation 模板保温;模板绝热材料
form insulation material 模板保温材料
form into a paste 调制成浆状
Formite 佛麦特钨铬钢
form joint 模板接缝
form jumbos 支架钻车
form lacquer 模板防粘涂料
form lagging 模板用木板
form lateral pressure 模板侧压力
form laying 立模;模板安放
form layout 模板布置
formless contract 非正式合同;不定型合同
formless paving machine 滑模铺路机
formless stage 树木不整形期
form letter 打印文件
form line 拟构等高线;地形线;地表形态线
form liner (模板上的)塑面线条;模板贴料;模(板)垫(条);模(板垫)衬
form lining 模板涂层;模板内衬;模板衬料;模板衬垫
form-loss coefficient 形状损失系数
form loss of head 形状水头损失
form lube 脱模润滑剂;模板润滑油
form lumber 木模(板);模板材
form matter 模型材料
form milling 成型铣削
form milling cutter 成型铣刀
form nail 模板钉;双帽钉
form of acknowledgment 收函通知书
form of a dualistic property right 二元产权形式
form of a monistic property right 一元产权形式
form of agreement 协议书表格
form of application 申请书格式;申请表格
form of budget 预算格式
form of business organization 业务组织形式
form of car ferry 汽车渡船型
form of cartels 卡特尔类型
form of construction work 施工方式
form of contract 合同形式;合同类型;合同格式
form of detailed schemes 细目式计划

form of distribution value 配气阀形式
form of emanation source 射气源形态
form of exclusive private property 排他性的私有财产形式
form of gas occurrence 瓦斯存在形式
form of gas outflow 瓦斯涌出形式
form of gear tooth 齿轮齿形
form of grain arrangement 颗粒排列方式
form of groundwater existence 地下水存在形式
form of manufacturer's data report 制造厂的数据报告格式
form of Moho-discontinuity 莫霍面形态【地】
form of note 手簿格式
form of piece-wage 计件工资形式
form of pollutant 污染物的形态
form of products 产品方案
form of proxy 委托书
form of radiator 辐射体形态
form of reinforcement 配筋形式
form of report 报告书的格式
form of shell structure 壳体结构的模板
form of single vein 单脉形态
form of taxation 课税形式
form of tender 投标表格;投标承包方式;投标书格式;投标书;投标格式
form of the physical store 存储的物理形式
form of thread 螺纹牙形
form of time wage 计时工资形式
form of traction 牵引种类【铁】;牵引方式
form of transport 运输方式
form of treaty 合同方式
form of value 价值形态
form of ventilation 通风方式
form oil 模板涂料;模板(用)油;脱模油
formol 甲醛
for money 现款交易(伦敦证券所用语)
form on traveler 移动模架
formothion 安果
form overlay 格式重叠;多格式重叠打印功能
form panel 模板;钢镶合板;模板
form parameter 形状参数
form paste 脱模剂;脱模膏;模壳膏;模板膏
formpiston 模塞
form placing 支模板;立模;架设模板;模壳设置;设置模壳;设置模板;安装模板
form planing 仿形刨法
form plate 模板
form point 形点
form-point height 形点高
form pressure 模板压力
form printing machine 表格印刷机
form project 模板工程
form quotient 形率
form ratio (河流的)宽深比;深宽比
form receiving tray 落纸架;收纸架
form release agent 隔模剂;脱模剂
form release compound 脱模剂
form relieved cutter 铲齿铣刀
form-relieved hob 铲齿滚刀
form-relieved tooth 铲齿
form removable 模板拆除
form removal 模壳拆除;(混凝土)脱模;拆除模壳;拆除模板
form resistance 形状阻力;船型阻力
form resistance of bed wave 沙波形状阻力
form-retentive 形状保持不变的;模板不变的
form-retentiveness 形状保持不性的;模板不变性
form rib 保形肋
form roller 送墨辊;上墨辊
form rolling 成型辊轧
form roughness 形状糙率;形态糙度
forms agent 制模剂
forms at glance 地形一览
form scabbing 拆模工艺;拆模伤斑;模板疤
forms design 报表设计
form sealer 模板密封材料
form sense 形态感
form separation method 形态分离法
form separator 形态分离仪
form-set 整套模板;形态层组
form setter 架模工;模壳工;模板工
form setting 装置模壳;支模板;架设模板;模板安装;安装模壳
forms facing 模板外壳;模板表面

forms handler (隧道、坑道施工用)模壳工程安装车;(隧道、坑道施工用)模板工程安装车
form shaping 仿形刨法
form sheathing 模壳衬板;模板衬板
form sheet 格式表
form side pressure 模板侧压力
form skip 记录纸进给
forms lining 模板内衬;模板衬里
forms of art 艺术形式
forms of investment 出资方式
forms of train working diagram with its vertical lines divided into columns of two or ten minutes, half hour and hour 列车运行图格
forms oil 脱模润滑油
forms overlay 多格式重叠
form spacer 模板间隔物;模板定距标尺
forms paint 模具油漆
form spreader 混凝土摊铺机;模板撑杆;模板撑挡
forms sealer 制模密封剂;模板封闭剂
forms slice 表格复片
form stability 型状稳定性
form steel 铸钢
forms tie 模板系杆
form stop 模板堵头;(浇混凝土的)堵头板;模板端头;模板挡
form strength 模壳强度
form stress factor 形状应力因数
form stripping 模板拆除;(混凝土)脱模;地形剥裂;拆模(板)
form stripping agent 脱模剂
form sulphur content 形态硫量
form surface 形态面
forms wax 制模蜡;脱模蜡
form tamper (道路)模板夯实机
form tie 模板支撑;模板拉杆;磨模拉铁
form tie assembly 模板支撑装配;模板支撑构件
form tolerances 形状公差
form tool 样板刀;成型刀
form tracers 形状寻迹器
form traveler 模板移动装置
form truing 砂轮成形修整
form turn 成型车削
form-tying 模板支撑装配;模板钉结
form type code 格式码
formula 公式;配方;反应式
formula adjustment 配方调整
formula calculation 配方计算
formula calculator 公式计算器
formula change 配方变更
formula error 公式误差
formula for interpolation by divided differences 均差内插公式
formula for rainfall demand 雨量公式
formula for theoretic(al) gravity 理论重力公式
formula funding 按公式拨款
formula ingredient 配方用料
formula language 公式语言
formula manipulation 公式控制;公式操作
formula model 公式模型
formula of flow 流量(计算)公式
formula of mean latitude 中纬度公式
formula of measurement 丈量公式
formula of partial penetrating well with flat bottom 非完整平底井公式
formula of pile driving 打桩公式
formula of pumping well with fixed drawdown 定降深抽水井公式
formula of sectional flow rate 断面流量公式
formula of well with semi-spheric(al) bottom 半球形井底公式
formula price 理论价格
formula ratio 配方比(率)
formula recognition 公式识别
formular investing 方程式投资
formulary 公式汇编;配方手册
formula sign 配方设计
formulate 制定;阐述
formulate alternative 制定可行(性)方案
formulated policy 确定政策
formulated products 按配方制造产品;配制产品
formulating 配方设计
formulating of recipe 配方设计
formulation 制剂;公式化;列方程式;列совмест公式;配方
formulation of decision situation 决策形势的构成

formulation of equation 列出方程
formulation of stocks 库存组成
formula to compute permeability 计算渗透系数公式
formulator 配方设计师
formula transformation 公式变换
formula translator 公式译码机;公式翻译程序
formula variation 配方差异
form utility 形态效用;形式效用
Formvar 弗姆(一种绝缘材料)瓦;聚醋酸甲基乙烯酯(一种绝缘材料);热塑树脂
form variation 型变异
form-vibrated concrete 模板振捣混凝土(与插入振捣混凝土不同)
form vibrator 模壳振动器;模板振动器;外部振动器;外部振动器;附着式振动器
form vision 形态视觉
form winding 模绕法
form window 地形开口
formwork 模壳(工作);灌注水泥的模架;制模工作;盖板;模板(工程)
formwork agent 制模剂
formwork aid 制模剂;制模工具
formwork board 模板
formwork board cleaning machine 模板洗涤机;模板清洁器
formwork bottom 模板工程的底板
formwork cleaner 模板清除器;模板清除机
formwork compound 脱模剂
formwork contractor 模壳承包商;模板承包商
formwork deflection 模板弯曲
formwork drawing 模壳图(样);模板图(样);支架图(样)
formwork element 模壳部件
form work engineering 模板工程
formwork erection 模板装配;立模
formwork erector 模板装配工
formwork fixture 模板固定件
formwork for precast work 预制构件模板
formwork grease 脱模膏;模板润滑膏;模板牛油
formwork handler 模板工程安装车
formwork joint 模板接缝
formwork joist 模板格栅
formwork jumbo 模板台车
formworkless 无模板的;不用模板的
formwork liner 模板衬里
formwork lining 模板内衬
formwork member 模板构件
formwork movement 模板移动;模板升高
formwork oil 隔离剂;脱模润滑油;模板润滑油
formwork oil atomizing device 模板油喷雾器
formwork paint 模板油漆
formwork paste 制模剂
formwork plan 模板设计;模板规划
formwork planning 模板布置
formwork plate 模板
form(work) plywood 模板用胶合板
formwork reaction modulus 模板反力模量
formwork removal 脱模;拆除模板
formwork safety 安全模板;安全支架;安全模壳
formwork sealer 制模密封剂
formwork setter 模板安装工
formwork setter's hammer 模板工锤子
formwork sheet(ing) 制模薄板
formwork sideboard 模板的侧板
formwork specifications 模壳工程规范
formwork stripping 拆除模板
formwork supports 模板支撑
formwork tie 模盒紧固拉杆;模板拉铁;模板拉杆
formwork tool 立模工具
formwork transport wagon 模板运输车
formwork uprighting survey 立模测量
formwork vibration 模板振捣;附着式振捣
formwork vibrator 附着式振捣器;模板振捣器
formwork wax 制模蜡
formwork work 模板工程工作
formwork yoke 模板卡箍;模板箍圈
form-wound motorette 模绕线圈试验装置
formy 间型(中间极小两端开阔)
fornacite 砷铬铜铅石
fornicate 弯曲的;穹隆状的;弓状的;拱形的;穹隆形的
fornication 拱顶或拱的构造
fornicolumn 穹隆柱

fornicommissure 穹隆连合(体)
fornix 穹形鳞片；穹束；穹隆
for order 准备出售
for our information 供我参考
for our reference 供我参考
Forrel cell 福雷尔环流
Forrester machine 福雷斯特浮选机
for shares 认购权
forst area 林业区域
forsterite 镁橄榄石
forsterite-based brick 镁橄榄石耐火砖
forsterite brick 镁橄榄石砖
forsterite ceramics 镁橄榄石(陶)瓷
forsterite diopside marble 镁橄榄石透辉石大理岩
forsterite marble 镁橄榄石大理岩
forsterite refractory 镁橄榄石耐火材料
forsterite refractory brick 镁橄榄石耐火砖
forsterite refractory product 镁橄榄石耐火材料制品
forsterite whiteware 镁橄榄石白坯陶瓷器皿
Forstner bit 弗斯脱纳钻
Forsythia 连翘属
fort 要塞；炮台；城堡；堡垒
fort agate 城塞玛瑙
Fortal 佛达尔铝合金
fortalice 外堡；小堡垒
forthability index 起泡指数
for the carrier 交承运人
for the consignee 交收货人
for the record 备案
forth fire extinguisher 泡沫灭火器；泡沫灭火机
fortifiable 宜于设防的；宜于加固的
fortification 要素；设堡；碉堡；要塞建筑；筑城；防御工事
fortification intensity 设防烈度
fortification(masonry)wall 加固墙；防御墙
fortified 设防的；加固的
fortified acid 强化酸
fortified bread 强化面包
fortified church 设防教堂
fortified feed 富化饲料
fortified glue 强化胶；增强胶
fortified monastery 设防寺院
fortified paint 加固涂料
fortified palace 设防宫殿
fortified place 设防地区
fortified port 军港
fortified residence 设防的居民区
fortified resin 强化树脂；加固树脂
fortified rosin 强化松香
fortified rosin size 强化松香胶料
fortified SB latex 强化丁苯胶乳
fortified stronghold 设防要塞
fortified structure 防御建筑物
fortified town 设防城市
fortified town gate 加固的城门；防御城门
fortified tyre 加强轮胎
fortified zone 设防地带；要塞地带
fortifier 增强剂；胶粘加强剂；强化物；补强剂
fortify 使坚固；加强；构筑工事；设防
fortifying 增强的；强化的
fortilage 外堡；碉堡
Fortin barometer 福丁气压计；福丁气压表
fortlet 小堡
fortnight 两星期；双周
fortnightly 双周刊
fortnightly component 双周分潮；半月分潮
fortnightly constituent 半月分潮
fortnightly inequality 半月差
fortnightly periods 节气
fortnightly tide 双周分潮；半月分潮；半月潮
fortran 公式变换语言
Fortrat parabola 福特雷脱抛物线
Fortrel 福特勒尔(一种聚酯纤维)
for trepanning plug sections 用于穿孔螺塞部分
fortress 堡垒；炮台
fortress castle 城堡
fortress castle architecture 城堡建筑
fortress castle-chapel 城堡式小教堂；城堡式礼拜堂
fortress castle-church 城堡式教堂
fortress castle-ditch 护城河；护城壕
fortress castle-gate 城堡之门
fortress castle-tower 城楼；城堡塔楼
fortress castle wall 城堡墙
fortress-chapel 要塞小教堂；要塞礼拜堂

fortress-church 要塞教堂
fortress-house tower 瞭望塔
fortress-like 要塞式的
fortress masonry 要塞砖石建筑
fortress-monastery 设防寺院；加固的寺院
fortress-town 要塞城；设防城市
fortress wall 堡垒墙
fortress well 要塞井；设防的井
fortuitous 偶然的；不规则的
fortuitous distortion 偶然畸变；偶发畸变；不规则失真；不规则畸变
fortuitous event 意外事故
fortuitous phenomenon 偶然现象
fortuity 意外事故；偶然性
Fortuna minimeter 可调倍率式杠杆比较仪
Fortunes Poulownia 泡桐
forty feet pull tubular steel derrick 四十英尺提升管子钻塔(1英尺=0.3048米)
forty-five degree lateral reducing on branch 分叉缩径四十度分叉管
forty-five degree lateral reducing on one run and branch 一头和分叉的四十五度Y形管
forty-five degree swan-neck 弯管(如楼梯扶手、管子等)；四十五度弯头
forty-four degree angle branch 四十五度支管
forty-four degree bend 四十五度弯头
forty-four degree lateral 四十五度Y形管
forty-four degree miter joint 四十五度削接头
forty hours week 每周四十小时工作制
forty-nine cm brick wall 两砖墙
forty-nine points moving average 四十九点移动平均
forulic acid 阿魏酸
forum 座谈会；论坛；讲座
Forum of Trajan (古罗马帝王的)特拉扬尼广场
forum Romanum 古罗马广场
forvacuum pump 预抽真空泵
for vessels under external pressure 用于外压容器
forward 向前的；转寄；正向；正方向；进带；前方；期货；船首部分
forward abstract 发送摘要
forward acceleration 纵向加速度；前进加速度
forward accounting 预期会计
forward aid program(me) 运期援助计划
forward airfield 前进机场
forward and backward movement knob 载物台前后运动螺旋
forward and reverse 正反向
forward and reverse lever 前进回动杆
forward angle 前方位角
forward-angle counter 向前散射粒子计数管
forward applied voltage 正向外加电压
forward arrivals 先到货物
forward azimuth 前方位角
forward backward counter 正反向计数；可逆计数器；加减计数器
forward bearing 向前方位；前象限角
forward bias 压向正偏；正向偏置；正向偏压
forward-biased rectifier 正向偏压整流器
forward bitt 船首系缆桩
forward blocking interval 正向关断期间；正向闭锁期间
forward blocking state 正向闭锁状态
forward booking 订舱
forward bow spring 舷斜缆；首斜缆
forward breakdown 正向击穿
forward breakover 正向转折
forward breast line 首横缆
forward bridge 舰桥楼
forward brow 船首跳板
forward burning 向前燃烧
forward business 期货交易
forward cancelled 期货取消
forward cargo hold 船头货舱
forward chaining 正向链接；前向链
forward channel 正向信道；前向信道；前方信道；单向通道
forward characteristic 正向特性；正向传输特性
forward charges 发运费用
forward circuit 正向电路
forward commitment 预发承付款项
forward commitment authority 预约付款权力
forward component 正向分量
forward conductance 正向电导

forward conducting region 正向导电区
forward conduction 正向传导
forward contract 运期契约；运期合同；预约；期货契约；期货合同；期汇契约
forward control 正向控制
forward controlling element 正向施控元件
forward converter 正向变换器
forward conveyer pawl 输送器前进爪
forward counter 正向计数器
forward counting 顺向计数
forward crosstalk 前向串扰
forward current 正向电流；前向电流
forward data 传来数据
forward decay 正向衰变
forward definition chain 向定义链
forward delivery 远期交货；定期交货
forward derivative 前向导数
forward diagonal 正向对角线
forward difference 前向分差；前向差分
forward difference method 向前微分法；向差分法；前向差分法
forward difference operator 前向差分算子
forward differential equation 前向微分方程
forward differential resistance 正向微分电阻
forward direct current resistance 正向直流电阻
forward direction 正向；前向
forward dome 前端圆头
forward draft 前吃水；前部船体吃水；艏吃水；船首吃水
forward drive 正车
forward drop 正向电压降
forward dump 前卸料车
forward dynamic(al) programming 前向动态规划
forward electric(al) fence 前斜电篱笆
forward engine room 前机舱
forward entropy 前熵
forward equation 前向方程
forwarder 运输业者；运输行；货运代理行；货运代理人；输送器；代运人；传送装置；承运人；发运代理人
forward error analysis 前向误差分析
forward error correcting 前向纠错
forward error correcting system 前向纠错制
forward error correction 前向纠错
forward error correction mode 前向纠错方式
forwarders 短材集运机
forwarder's bill of lading 发运代理提单
forwarder's certificate of receipt 发运代理收据
forward excavator 正铲挖土机
forward exchange 远期外汇；远期汇率；期货外汇
forward exchange contract 期货汇兑合同
forward exchange market 外汇期货市场
forward exchange rate 期汇汇率
forward exchange transaction 外汇期货交易；期汇业务
forward expenses 运输费(用)
forward-extrude 正向压挤
forward extrusion 正向压挤；正(向)挤压；顺向挤压
forward feed of material 材料的预约供应
forward-field impedance 正向磁场阻抗
forward-field torque 正向转矩
forward financial statement 预计财务报告；预计财务报表
forward flow 向前流动；进口气流；前进流；平直流；顺流
forward flow zone 射流区
forward funding 提供远期资金
forward gain 正向增益
forward gear 前进装置；前进档；前进齿轮
forward gear train 前进齿轮(传动)系
forward ground 前台
forward guy 前张索
forward hauling winch 前移绞车
forward heat shield 前置防热板
forward ice belt region 船首部冰带区
forwarding 运送
forwarding accounting system 发送核算制
forwarding agent 运输代理人；运输代办行；运输代办人；转运公司；货运代理行；货运代理人；承运人
forwarding agent's certificate of receipt 运输行收货凭证
forwarding agent's notice 代运通知书
forwarding agent's receipt 代运人收据
forwarding book 发运簿；发送簿

forwarding broker 运输代理人;运输代办人
forwarding declaration 发送声明书;发送声明书
forwarding line 发车线【铁】
forwarding operation 代运业务
forwarding order 运输委托书
forwarding railway 发送铁路
forwarding receipt 中转收据
forwarding siding 发车线【铁】
forwarding station 发送站
forwarding track 运送轨道;发车线【铁】
forwarding transport service 运输业务
forwarding unit 发货单位
forwarding yard 出发场
forward insertion gain 正向插入增益
forward integration 向前整合;向前合并;后延联合
forward interpolation 向前内插;向前插值;前向内插
forward intersection 前方交会
forward intersection by plane-table 平板仪前方交会
forward journey 送进孔型
forward lap 前重叠
forward lead 前移;顺向位移
forward lead distance 前移距
forward leakage current 正向漏电流
forward linkage 向前边锁决
forward linkage effect 向前连系效果
forward linking 前向联结;前面信号联结
forward looking 前视
forward looking infrared 前视红外线
forward looking infrared system 前视红外系统;红外(线)前视系统
forward looking laser sensor 前视激光传感器
forward looking radar 前视雷达
forward loss 飞向损耗
forward masthead light 前桅灯
forward maturities 期货交易到期日
forward mean current rating 正向平均电流定额
forward mean voltage drop 正向平均压降
forward merge 正向归并
forward mode 正向模式
forward motion compensation 航向移动补偿;前移补偿
forward-mounted 前部安装的
forward movement of film 胶片前移
forward moving distance 前移距
forward mutation 正向突变
forward mutual admittance 正向互导纳
forward nodal point 前节点
forward oblique 机首倾斜
forward observer laser rangefinder 前方观察员激光测距机;前向观察激光测距仪
forward offset voltage 正向低偿电压
forward of the beam 正横前面
forward option 期货期权
forward or backward search buttons 进或退搜索线
forward order 托运单
forward-order current 正序电流
forward overlap 航向重叠;前后重叠
forward pass 往来推算;送进孔型
forward path 正向通路;正向通道;前进路
forward perpendicular 艏垂线;首垂线;船首垂线
forward planning 预先计划
forward play 顺向张力
forward polyphase sort 正多相分类
forward power 正向传输功率
forward power meter 正向功率计
forward pressure 前进压力
forward price 期货价(格)
forward probe 超前探测
forward problem of gravity anomaly 重力异常正演问题
forward progression lap winding 右行叠绕组
forward purchase 预购
forward purchasing contract 预购合同
forward quarter 船前部
forward quotation 期货报价(单)
forward rake 前倾
forward rate 远期汇率;期货价(格);期货汇率
forward ray cone 前向射线锥
forward read 正读
forward reading 正向读出
forward recovery time 正向恢复时间
forward recovery voltage 正向恢复电压

forward reference 向前引用
forward reflection 前向反射
forward reflection radiography 向前反射照相术
forward refuge compartment 前部脱险舱
forward region 正向区
forward repositioning 正向再定位
forward resistance 正向电阻
forward-reverse primary transmission 前进-倒退式主要传动
forward-reverse transmission 前进-倒退式传动
forward roll entering water 前滚式潜水法
forward roller-coating 顺向辊涂
forward-rotating wave 正向旋转波
forward rule 正向规则
forward run 向前跑;(火车)上行
forward scan 正向扫描
forward scatter 向前散射;正向散射;前向散射
forward-scattering angle 向前散射角;前向散射角
forward-scattering collision 向前散射碰撞
forward-scattering feedback 向前散射反馈
forward scheduling 预先调度法;提前调度法
forward selection 预选择;预选;前向选择
forward sequence 正向顺序
forward shift 前移;顺向位移
forward shift distance 前移距
forward shifting distance indicator 前移距指示仪
forward shipment 远期装运
forward shock 头部激波
forward shovel 正向铲;正铲挖土机;正铲挖掘机
forward shovel attachment 正铲挖土机附件;前端铲运配件
forward shovel bucket 正铲挖土机铲斗;前端铲斗
forward shovel fitting 正铲挖土机配件;前端铲运器件
forward side anchor 前边锚
forward sight 前视
forward signal 正向信号
forward signal(l)ing path 正向传信通路
forward skidding 向前滑行;刹车滑行
forward slip 前滑
forward solution procedure 前向解方法
forward space 正向空间
forward speed 正车航速;前进速度;推进速度
forward spinning 正旋
forward spring 前倒缆;前池;艏倒缆;首倒缆
forward spud 前桩【疏】;前钢桩【疏】
forward-step signal 正走一步信号
forward stepwire regression 前向逐步回归
forward stepwise method 前向逐步法
forward stroke 前向冲程;正程;前进行程;前进冲程
forward-stroke interval 正程间
forward substitution 前向替换
forward summary 发送总表
forward superstructure 前部上层建筑
forward switching loss 正向开关损耗
forward thrust 给进压力
forward tilting of the bucket 料斗前倾
forward tipping 向前倾斜的
forward-tip(ping) bucket 前倾的料斗
forward torque 正向转矩
forward trading 期货交易
forward transadmittance 正向跨导纳
forward transfer characteristic 正向转移特性
forward transfer element 正向传递单元
forward transfer function 正向传递函数
forward transfer impedance 正向转移阻抗
forward transfer signal 正向转接信号
forward transformation kernel 正向变换核
forward transmission resistance capacitor network 正向传输阻容网络
forward travel 前进运动
forward type 平头型
forward type cycle 前向式循环
forward velocity 前进速度
forward velocity of crest particles 波顶质点前进速度
forward velocity tracing 前向流速曲线
forward vision 前视
forward voltage 正向电压
forward voltage drop 正向电压降
forward wake 正伴流
forward wave 直达波;正向波;前向波;前冲波
forward wave amplifier 前向波放大器
forward welding 向前焊;左向焊;左焊法;正手焊;

前倾焊;前进焊
forward wind key 速进键
forward wound winding 正绕绕组
for your attention 请你注意
for your information 供你方参考;仅供参考
for your information and guidance 望参照执行
for your reference 供你方参考
Fosalsil 佛撒尔西尔式绝热材料
foshagite 傅硅钙石;变针硅钙石
foshallasite 鳞硅钙石;水硅灰石
Fossa Magna 大海沟带(日本)
fossa trochanterica 转子窝
foss(e) 城壕;护城河;壕;海渊
fosse moat 壕堰
fossil 化石【地】;含化石的
fossil anomaly 化石异常
fossil assemblage 化石组合
fossil bacteria 化石细菌
fossil charcoal 丝炭
fossil community 化石群落
fossil compound 化石化合物
fossil copal 黄脂石
fossil crater 化石坑;化石环形山
fossil delta 古三角洲
fossil diagenesis 化石成岩作用
fossil diversity 化石分异性
fossil energy 矿物能
fossil erosion surface 古侵蚀面
fossil facies 化石岩相
fossil fauna 化石动物区系
fossil field model 化石场模型
fossil flax 石棉
fossil flour 硅藻土;化石粉
fossil forest 石化的树林;化石的树林
fossil fuel 矿物燃料;化石燃料
fossil fuel plant 火电厂
fossil geochronometry 化石地质年代测定法
fossil geothermal system 古地热系统
fossil glacier 古冰川
fossil grain 化石颗粒
fossil ground water 原生水;古地下水
fossil gum 化石树胶
fossil gum resin 化石树胶脂
fossil ice 化石冰
fossil identification 化石鉴定
fossiliferous 含化石的
fossilification 化石作用;化石化;陈腐化
fossilization 化石(化)作用;化石化;陈腐化
fossilized material 化石化材料;腐化材料
fossil karst 古岩溶;古喀斯特;化石岩溶
fossil lake 古湖
fossil landscape 古地形
fossil landslide 古滑坡
fossil magnetism 化石磁性
fossil man 化石人
fossil marble products 化石大理石产品
fossil meal 硅藻土;化石粉
fossil oil 矿油;石油
fossil oil pool 古油藏
fossil peneplain 古准平原
fossil plain 古侵蚀平原;古平原
fossil reef 古礁;化石礁
fossil remains of seismicity 古地震遗迹
fossil resin 琥珀;化石树脂
fossil river course 古河道
fossil rock 石灰石
fossil soil 古土壤;化石土
fossil stream course 古河道
fossil submarine canyon 古海底峡谷
fossil time 化石地质年代
fossil trace 化石足迹
fossil water 原生水;矿物油;矿物水;埋藏水;化石水;封存水
fossil wax 地蜡;矿物蜡
fossil wood 木化石
fossorial 穴居型
fosted 部分生烧的
foster new growth areas in economy 培育新的经济增长点
Foster pyrometer 福斯特高温计
Foster's criteria 福斯特准则
Foster-Seeley discriminator 相移鉴频器
Foster's formula 福斯特公式
Foster-Wheeler process 福斯特-威洛蒸气转化法

fota 厚实棉布
foteign body gouge 异物凿
fother 海上堵漏
fotoceram 感光微晶玻璃;光敏玻璃陶瓷
Fotoform process 光学蚀刻法
fotomicrograph 显微照片
Fottinger coupling 弗廷格联结器
Fottinger speed transformer 弗廷格变速器
Fottinger transmitter 弗廷格联结器
Foucault's chart 傅科图案
Foucault's current 涡电流
Foucault's current brake 涡流制动器
Foucault's current effect 涡流效应
Foucault's grating 傅科光栅
Foucault's gyroscope 傅科回转仪
Foucault's knifeedge test 傅科刀口检验
Foucault's measurement 傅科测量
Foucault's method 傅科法
Foucault's mirror 傅科镜
Foucault's pendulum 傅科摆
Foucault's prism 傅科棱镜
Foucanlt resolution target 傅科分辨率检验板
Foucault's rotating-mirror method 傅科转镜法
fougasse 定向地雷
foul 污浊;污秽物;污垢
foul air 浊气;污浊空气;污秽空气
foul air chimney 排污浊空气的烟囱;排废气烟囱
foul air duct 烟道;(隧道通风系统中的)吸风管线;污浊空气导管
foul air floor duct 楼层排污浊空气管道;楼层排废气管道
foul air floor flue 楼层排污浊空气管道;楼层排废气管道
foul air flue 浊气道;污浊空气管道;通气道;回风道
foul air grate 排污浊空气格栅;排废气格栅
foul air hole 排污浊空气孔;排放废气孔
foul air opening 排污浊空气孔;排放废气孔
foul air pipe 排污浊空气管;排废气管
foul air shaft 排放废气竖井
foul anchor 链缠的锚;绞缠锚;被锚链搅缠的锚
foulant 污浊物
foulant solubility 污浊物增溶剂
foul area 多礁区;险区;险恶地;多暗礁区域;不良锚地
foul beach 危险海滩
foul berth 回转不便锚泊地;危险泊位;不安全泊位
foul bill of health 不洁检疫证书
foul bill of lading 不洁提单
foul bottom 污底;不宜抛锚的河底;不宜抛锚的海底;不良锚地
foul cable 锚链绞缠
foul clay 不合格砖泥;缺乏助熔质砖泥
foul coast 危险海滩;危险海岸;多礁海岸;多暗礁的海岸;不适于航行的海岸
foulding community 污着群落
foul drain 排水管;排水沟;污浊下水道;阴沟
fouled ballast bed 脏污道床
fouled membrane 污染膜
foul gas 恶臭气体;不凝性气体
foul ground 多暗礁的海底;不宜抛锚的河底;不宜抛锚的海底;不良锚地
foul hawse 锚链绞缠
foul holding ground 不能抛锚的乱石底;不能锚的海底
fouling 结污;绞缠;粘垢;污着;污损;污底;污垢
fouling agent 污秽物
fouling allowance 污底裕度
fouling amount 污垢量
fouling by marine organism 海洋生物污损
fouling coefficient 污垢系数
fouling community 污损群落
fouling complex 污损群落;污损复合体
fouling control 污垢控制
fouling control agent 污垢控制剂
fouling deposit 污垢沉积物
fouling factor 污垢系数;沾污系数;生垢因数
fouling film 污损膜;污塞层
fouling index 沾污指数
fouling inhibition 污垢抑制
fouling inhibitor 污垢抑制剂
fouling mitigation 减少污着
fouling monitoring 污垢监控
fouling of cutterhead 铰刀头堵塞
fouling of heat exchangers 换热器结垢

fouling of heating 受热面的积灰
fouling of heating surface 受热面积灰
fouling of ship 船底污损
fouling organism 污着生物;污损生物;污染生物;附植生物
fouling organism community 污着生物群落
fouling point 警冲标;计算停车点【铁】;安全界限点
fouling post 警冲标
fouling product 污着物;污垢物;恶臭物;污垢物
fouling rate 沾污速率
fouling resistance 污垢热阻
fouling resistance reverse osmosis 抗污着反渗透
fouling resistance reverse osmosis membrane 抗污着反渗透膜
fouling resistant 耐污染
fouling thermal resistance 污热阻
fouling-up 弄脏
fouling velocity 污着速度
foul land 污秽土地
foul mate receipt 不洁收据
foul mate's receipt 附条件的大副收据
foul patch 险恶地;不良锚地
foul pollution 极度污染;秽臭污染;恶臭污染
foul proof 毛校样
foul rope 纠缠的绳索
fouls 煤层尖灭
foul sewer 阴沟;排污管;污水管;污水道
foul ship 污底船
foul shipping order 附条件的装货单;不洁装货单
foul smell 恶臭味
foul-smelling gas 恶臭气体
foul solution 污染溶液
foul the core (被泥浆污染的)污染岩芯
foul tide 逆潮(流)
foul-up 弄乱;搞糟
foul vessel 重油船;不清洁船
foul water 秽臭水;混水;污水;臭水
foul water disposal facility 污水处理设备
foul water gallery 污水通道
foul water pipe 污水管
foul water(pipe)trench 污水管道沟
foul water purification 污水净化
foul waters 危险水域;危险海域;危险海区;多暗礁水域
foul water sewer 阴沟
foul water trench 污水坑;污水沟
foul water tunnel 污水通道
foul weather 险恶天气;坏天气;不良气候
foul wind 逆风;顶头风;恶风
founctional parameter 机能参数
found 熔制;打基础
foundaer polyp 原生珊瑚体
foundation 基金(会);基础;巢础;房基
foundational environmental geologic(al) survey based on remote sensing 遥感环境基础地质调查
foundational wave zero sequence voltage 基波零序电压
foundation analysis 基础分析
foundation anchor 基础锚栓;基础螺栓;地脚螺栓
foundation anchorage 基础锚固
foundation anchoring 基础锚固
foundation area improved by plastic drain 塑料排水板处理
foundation arrangement drawing 基础布置图
foundation base 基؛؛基础底面
foundation beam 基础梁;地梁;地基梁
foundation bearerl ground beam 地基梁
foundation bearing capacity 地基承载力
foundation bed 基础底面;基座;基底(层);基床;基础垫层
foundation behaviour 基础性状
foundation binding 基础垫层
foundation block 地基砌块;基础砌块;护脚块体
foundation bolt 基础螺栓;地脚丝;地脚螺栓;底脚螺栓
foundation bolt for motor 电动机地脚螺栓
foundation bolt with nose 凸缘基础螺栓;前缘地脚螺栓;地角基础螺栓
foundation breath 基础阔度
foundation brick 基础砖砌体;基础砖工;基础砌砖
foundation brick masonry work 基础砖砌筑工作
foundation brickwork 基础砖砌体;基础砖工
foundation brickworker 基础砖工

foundation building pit 基坑
foundation built-in 基础嵌入
foundation by means of grouting 灌浆加固的基础
foundation by means of injecting cement 基础灌浆加固
foundation by pit sinking 挖坑沉基;沉井基础
foundation by timber casing with stone filling 木框石心基础
foundation calculation 地基验算
foundation ceremony 奠基典礼
foundation clinker 基础熔渣;基础煤渣
foundation coating 基础涂料
foundation compliance 地基可塑性
foundation comprehensive resilient modulus 地基综合回弹模量
foundation concrete 基础混凝土
foundation condition 地基条件
foundation consolidation 地基固结;地基(用固结灌浆)加固
foundation consolidation strength 地基固结强度
foundation construction 基础建筑;基础施工
foundation course 基层;勒脚层;基础层
foundation cross 格床基础;基础交叉线
foundation cut-off 地基截水(墙)
foundation cut(ting) 基础开挖
foundation cylinder 基础井筒;基础圆柱
foundation damp proofing course 基础防潮层;地基防潮层
foundation deformation 基础变形;地基变形
foundation deformation modulus 地基变形模数;地基变形模量
foundation depth 基础深度
foundation design 基础设计;地基设计
foundation designed for seismic load 抗震基础
foundation detail 基础图样
foundation detailing 基础细部构造
foundation displacement 基础位移
foundation ditch 底坑;基坑;基础沟;基槽
foundation drain 基础排水;基础排水系统;基础排水管
foundation drainage 基础排水;地基排水
foundation drainage hole 基础排水孔;地基排水孔
foundation drainage system 地基排水系统
foundation drainage tile 基础排水(瓦)管
foundation drain hole 基础排水孔
foundation drawing 基础图;地基图
foundation drilling 地基钻探
foundation elevation 基础标高
foundation embankment 基础埋置深度;基础埋置
foundation embedment 基础埋置深度
foundation embedment pattern 基础埋置方式
foundation engineering 基础工程(学)地基工程
foundation engineering practice 基础工程实践
foundationer 提供基金者
foundation excavation 地基开挖
foundation exploration 地基勘探;基础勘探;基础勘查
foundation exploration by test pits 用试坑作地基勘探
foundation exploration drilling 地基勘测钻探
foundation exploration of general construction 一般建筑物地基勘察
foundation failure 基础破坏;地基失效;地基失稳;地基破坏
foundation faulting 基底断裂作用
foundation filling 木框石心基(础)
foundation folding 基底褶皱作用;基底褶皱
foundation for beehive 蜂房座
foundation forms 基础模板
foundation for road 路基
foundation frame 机座骨架;机座底架;基础构架
foundation framework 基础框架工程;基框(工作);基础模板(工作)
foundation gallery 基础廊道
foundation geology step 基建地质阶段
foundation girder 基础梁;地基大梁;基础主梁
foundation grid plan 基础栅格平面;基础网格平面
foundation grill 基础格床
foundation grillage 基础格床
foundation ground 地基
foundation grouting 基础灌浆
foundation hoist 基础起重机;基础绞车
foundation hole drilling 基础施工钻孔
foundation improvement 地基加固

foundation-induced vibration 基座引起的振动
foundation injection 基础喷射;基础灌浆
foundation in rock 岩中基础;岩基
foundation investigation 基底勘察;地基及基础研究;地基调查;地基察勘;地基查勘
foundation investigation by test pits 用试坑研究地基
foundation investigation drilling 地基钻探研究
foundation isolation 基础隔振
foundation isolation plate 基础隔板
foundation joint 基础接头;基础接缝;基础缝
foundation layout plan 基础布置方案;基础布置平面图
foundation leakage 地基渗漏
foundation level 基底标高;基础水平;基础标高;地基高程;地基标高
foundation light 基本光;衬底光
foundation line 开挖线;基础线
foundation load 基础荷载
foundation loading 基础加荷;基础荷载
foundation map 基础图
foundation masonry 基础圬工
foundation masonry wall 基础圬工墙;基础砌筑墙
foundation masonry work 基础砌筑工作
foundation mat 基础底板;基褥;基垫层;基础筏板;基础底板;筏基
foundation material 基础材料
foundation mat tress 基础底板;沉褥基础;沉排基础
foundation medium 基础介层
foundation medium of soil 基土介层
foundation modulus 地基系数;基础模数;基础模量;地基模数;地基模量
foundation nursery 基础种圃
foundation of dam 坝基
foundation of masonry 砌体基础;圬工基础
foundation of wall 墙基础
foundation on caisson 沉箱基础;沉井基础
foundation on raft 浮筏基础;筏形基础;筏基(础)
foundation on wells 井筒基础;沉井基础
foundation pad 基础垫层
foundation pier 桥墩;基座;基墩;桥台
foundation pile 基桩;基础桩
foundation pile hole 基桩孔
foundation piping 地基管涌现象
foundation pit 基坑验槽;基坑
foundation pit well point drainage method 基坑井点排水法
foundation plan 基础平面图
foundation planting 基础设施;房屋旁边种植;基础种植
foundation plate 底板;基础板;滑道牵板
foundation practice 基础施工技术;基础处理方法
foundation pressure 基底压力;基础底面(土)压力
foundation problem 基础问题
foundation raft 基础底板;基础筏
foundation reaction 地基反力
foundation reconnaissance 基础勘测
foundation reconnaissance by test pits 用试坑勘测地基
foundation reconnaissance drilling 地基勘测钻探
foundation reinforcement 基础中的钢筋;基础钢筋
foundation restraint 基础约束
foundation rigidity 基础刚度
foundation ring 基础环;底圈;底环
foundation rock 基岩;基底岩
foundation rocking 基础摇摆
foundation rock sample 基岩样品
foundation sampling 基岩取样
foundation scouring 基础冲刷
foundation screw 基础螺栓;地脚螺丝
foundation seed farm 基础种场
foundation set 基组
foundation settlement 基础下降;基础沉陷;基础沉降;地基下沉;地基沉陷;地基沉降量;地基沉降
foundation shell 基础壳体
foundation sill 基础梁;基础槛;底槛
foundation slab 基础模板;基础(底)板;护底;底板;承台
foundations of mineralization prediction 成矿预测基础
foundation softening 地基软化
foundation soil 地基土(壤);基土;基础土(壤);地基;持力层
foundation soil profile 基础土层剖面;地基土剖面图

foundation soil replacement by sand 地基的挖土换砂
foundation soil sample 地基土壤样品
foundation soil stratum 地基土壤层
foundation spring stiffness 基础弹性刚度
foundation stab 基础齿墙
foundation stability 地基稳定性
foundation stabilization 地基稳固化;地基加固
foundation stable fluid 地基稳定液
foundation standard 基础标准
foundation stiffness 基础刚度
foundation stone 根底;基石;奠基石;屋基石
foundation stone laying ceremony 奠基仪式
foundation stratum 地基岩土层;地基土层;基础地层
foundation strength 地基强度
foundation stress 基础应力
foundation structure 基础结构
foundation structure of rail track 轨道基础结构
foundation support 基座
foundation system 基础体系;基础方法
foundation testing 基岩试验
foundation tie-beam 基础系梁
foundation tie rod 基础拉杆
foundation treatment 基础处理;地基处理
foundation trench 基坑;基槽
foundation type 地基类型
foundation underpinning method 基础托换(法)
foundation under rail 轨下基础
foundation under water 水下基础(工程)
foundation vault 基础拱顶
foundation vibration 基础振动
foundation wall 基墙;基础墙;地基墙
foundation wall brick 基础墙砖
foundation-wall keyway 基础墙接缝(槽)
foundation wash 基础冲溃
foundation water pressure 基础(上)浮力;基础下水压力
foundation well 基础井;基础井筒
foundation width 基础宽度
foundation with holes or notches 有孔洞或缺口的基础
foundation with interbedded argillous soil 粘土夹层地基;软弱夹层地基
foundation work 基础工作;基础工程
foundation work under compressed air 压缩空气下的基础工程;气压沉箱施工
foundation yielding 地基屈服
founded 沉箱已就位
founded caisson 沉箱就位
founder 陷落;沉没;铸工;熔制工;倒塌;创始人;翻砂工(人);发起人
founder by the head 艏沉没
founder head down 俯首沉没
foundering 陷落;岩浆蚀顶作用
foundering ship 正在沉没的船
founder-member (团体等的)发起人;(团体等的)创办人
founderous 泥泞的;沼泽地的
founder's black 铸工用黑色碳粉
founder's bonds 发起人公司债券
founder's shares 发起人股票
founder's stock 发起人股份
founder's type 标题活字
found in digestive juice 存在于某种消化液
founding 铸造;铸熔术;铸件;熔铸;奠基;翻砂
founding depth 基础深度
founding furnace 铸造炉;熔炉
founding hospital 育婴堂
founding level 基础高程;基础标高;(沉箱)沉降基准面
founding process wastewater 铸造生产的废水;铸造废水
founding property 铸造性能
foundling hospital 育婴堂
found on a layer of tipped sand 建于垫砂层上
foundry 铸造(车间);翻砂间;翻砂车间;铸工场
foundry alloy 铸造合金;母合金
foundry car 铸罐车
foundry cast-iron 铸造用铁
foundry chisel 清理浇铸口錾
foundry clay 铸造用黏土;铸型黏土
foundry clay ore 铸型黏土矿石
foundry coke 铸造用焦(炭);铸造焦炭;冲天炉焦

foundry core 翻砂型心
foundry core sand 型芯砂
foundry cupola 铸造用化铁炉;化铁炉
foundry defect 铸造缺陷;铸疵
foundry equipment 铸造设备;铸造车间设备
foundry facing 石墨涂料;铸模面料
foundry fan 铸造用鼓风机;铸工鼓风机
foundry flask 砂箱
foundry floor 造型工地
foundry furnace 铸造熔化炉;铸造炉
foundry goods 铸件
foundry hand 铸工
foundry industry 铸造业;铸造工业
foundry ingot 铸铁锭;生铁
foundry iron 生铁;铸造用生铁;铸铁
foundry jolter 造型震实机
foundry ladle 浇桶;浇包;铁水包
foundry layout 铸造车间设计
foundry loam 造型黏土
foundry loss 铸造废品
foundry machinery 铸造机械
foundryman 铸造工;铸工
foundryman's fever 铸造热;铸工热
foundry mo(u)ld 铸型;铸模
foundry pig 生铁;铸造生铁
foundry pig iron 铸件;铸造用生铁
foundry pin iron 铸铁
foundry pit 铸坑
foundry practice 铸造生产;铸造学;铸工实习;翻砂业务
foundry process wastewater 铸造作业废水
foundry production line 铸造生产线
foundry proof 最后校样
foundry raw material commodities 铸型原料矿产
foundry refractory 铸造用耐火材料
foundry resin 铸造用树脂
foundry return 回炉料
foundry sand 型砂;铸造用砂;铸型砂;铸模砂
foundry scale 铸件鳞片
foundry scrap 铸造废铁;铸造废料;废铁
foundry shop 铸工场;铸造车间;铸造厂;铸工车间;翻砂车间
foundry slag 炉渣
foundry sleeker 光砂刀
foundry slewing crane 翻砂间旋臂起重机
foundry stove 铸造烘炉
foundry technique 铸造工艺
foundry term 铸造术语
foundry wastewater 铸造厂废水
foundry weight 压铁
foundry work 铸造工作;铸工工作;铸工
foundry worker 铸造工人;翻砂工(人)
foundry worker's pneumoconiosis 铸工尘肺
found stone 毛石;基石;奠基石;粗块石
fountain 液体储藏器;喷(水)泉;喷水器;水斗
fountain aerator 喷泉式曝气池;喷浆式曝气池
fountain basin 喷水池;喷泉池
fountain brush 在把柄上装有墨槽或油漆槽的划线刷或漆刷;喷水器
fountain effect 喷注效应
fountain failure 涌溃
fountain flow 喷泉流
fountain for ceremonial ablutions (清真寺)宗教洗礼用泉水
fountain geyser 喷溅式间歇泉
fountain head 泉源;喷水头;水源;源头
fountain jet 喷泉射流
fountain model 喷泉模型
fountain pan 给漆盘
fountain pen 自来水笔;钢笔
fountain pen barrel 钢笔杆
fountain pen ink 钢笔墨水
fountain pen nib 钢笔尖
fountain pen with stainless steel cap 不锈钢套钢笔
fountain pen with stainless steel nib 不锈钢笔
fountain pipe 喷水管;喷泉管
fountain plaza 喷泉广场;喷泉场地
fountain pool 喷水池
fountain roller 给漆辊;上墨辊
fountain site 喷泉场地
fountain solution 调湿溶液
fountain-solution density 湿润液浓度
fountain syringe 自流注射器

fountain water 泉水
four-abreast hitch 四马一套
four acceleration 四元加速度
four accelerator 四元加速器
four-address 四地址
four-address code 四地址码
four-address instruction code 四地址指令码
fouramine 四胺
four arithmetic(al) operation 四则运算
four armed spider 四爪卡盘
four-arm junction 四接合
four arm mooring 四向系泊；四向锚泊法
four-arm pickling machine 四摇臂式酸洗机
four-arms intersection 四路交叉口
four-arm spider 星形轮；十字叉
four-aspect automatic block 四显示自动闭塞【铁】
four-axis method 四轴法
four-axis mounting 四轴装置
four-axis spot wobble 四轴光点颤动
four-axis system 四轴系统
four-axle 四轴
four-axle car 四轴车
four-axled coach 四轴客车
four-axled wagon 四轴货车
four axle tractor 四轴牵引车
four-axle truck 四轴载重汽车
four-axle vehicle 四轮车辆
four-ball apparatus 四球机
four-ball extreme pressure lubricant tester 四球式极压润滑剂试验机
four-ball extreme pressure tester 四球极压试验机
four-ball load carrying capacity test 四球机负荷能力试验
four-ball machine 四球摩擦机
four-ball tester 四球试验机；四球润滑剂性能测定仪
four-ball test rig 四球试验机
four-ball top 四球角锥
four-band multispectral data 四波段光谱数据
four-banger 四汽缸发动机
four-bar chain 四杆链
four-bar guide 四柱导承
four-bar linkage 四杆联动机构；四联杆机构
four bar linkage mechanism 四联杆机构
four-bar motion 四杆运动
four barrel 四腔式；四镜头
four-bar steering linkage 四杆转向联动装置
four basic colo(u)rs 四原色
four basic fields 四个基本学科
four-bay 四跨度的；四跨的；四间的
four-bearing crankshaft 四轴承曲轴
four-bed 四砖层的；四床的
four-bit slice circuit 四位片电路
four-bit slice microprocessor 四位片微处理机
four-bladed bit 四叶钻头；四叶片刀头；十字钻头
four-bladed capacitor 四叶电容器
four-bladed circular grab 四瓣抓土机
four-bladed fan 四叶风扇
four-blade-drag bit 四刮刀钻头
four-bladed vane 十字板(仪)
four-bladed ventilator 四叶风扇
four-blade propeller 四叶螺旋桨
fourble 四联(钻)管
fourble board (在四根钻杆组成立根的高处的)塔板
four body problem 四体问题
four-bolt splice bar 四栓鱼尾板；四孔鱼尾板
four-boss breaker cam 四楞凸轮
four-bowl calender 四辊压延机
four busbar regulation 四汇流条调整
four-by-four vehicle 双轴四轮驱动车辆
four-by-two vehicle 双轴两轮驱动车辆
four cable crane 四缆索起重机
four-cant 四股绳；四股
Fourcault 弗克法拉制平板玻璃法
Fourcault glass 弗克法拉制的平板玻璃
Fourcault process 弗克法(垂直有槽引上拉制平板玻璃法)
four cavity klystron 四腔速调管
four cell placer jig 四箱砂矿跳汰机
four-center approximate method 四圆心近似法(画椭圆用)
four-centered arch 四心(桃尖)拱；扁平拱
four-centered pointer arch 四心尖顶拱
four-centered Tudor arch 四心直线尖顶拱；四心都德式拱
four-centre reaction 四中心反应
four-center spiral 四心螺线
four-centre transition state 四中心过渡状态
four-centre type reaction 四中心反应
four channel burner 四通道燃烧器
four channel multiband camera 四通道多光谱带照相机
four-channel multiplier 四通道乘法器
four-channel radio terminal set 四路无线电通信终端设备
four-channel switch 四路转换开关
four-channel time division multiplex 四路时分多路复用
four-channel union 四通接头
four-channel valve 四通阀
fourchite 无橄榄沸碱煌岩；钛辉沸煌岩
four-choke carburettor 四腔化油器
four-coat system 四层涂刷法
four-colo(u)r gold 四色金
four-colo(u)r offset press 四色胶片机
four-colo(u)r photometry 四色测光
four-colo(u)r photometry system 四色测光系统
four-colo(u)r printing 四色印刷
four-colo(u)r problem 四色问题
four-colo(u)r reproduction 四色复制
four-colo(u)r separation process 四色分色制版法
four-column hall 四柱厅
four-column hydraulic press 四柱水压机；四柱式液压机
four-compartment aggregate feeder 四室骨料喂送器
four-compartment bin 四室料仓
four-compartment mill 四屋式磨碎机；四室式磨碎机
four-compartment revolving door 四翼式转门
four-component alloy 四元合金
four-component objective 四组元物镜
four-component quartz tube strainmeter 四分量石英管应变计
four-component system 四组分系统；四元(体)系
four-core cable 四芯电缆
four-core wire 四芯线
four-cornered bar 四角型钢；四角拉杆
four-cornered column 四角柱；四角支柱
four-cornered tower 四角塔
four corners 四角；四岔路口；十字路口；方形
four-course MF radio range 四航向中频无线电信标
four-course radio beacon 四航向无线电信标
four-course radio range 四航向无线电信标
four-course radio range station 四航道无线电导航台
four-crank press 四曲柄压力机
four crossway 十字路
four current 四元电流
four-current density 四元电流密度
four-cut finish 粗凿石面
four cutter bit 四牙轮钻头；四齿轮钻头
four-cycle 四冲程循环的；四冲程循环
four-cycle diesel 四冲程柴油机
four-cycle engine 四冲程发动机
four-cylinder engine 四汽缸发动机
four-cylinder locomotive 四汽缸机车；四缸机车
four-cylinder steering gear 四缸操舵装置
four-cylinder triple-expansion engine 四缸三胀式蒸汽机
four-cylinder turbine 四缸汽轮机
four-decker 四层甲板船
four-decker cage 四层罐笼
four-deck grader 四筛分级器；四层筛分器
four-deck jumbo 四层盾构(隧道)
four-deck screen 四层筛；四筛分选；四层筛子；四层筛网
four-density 四维密度
four-digit display 四位数字显示
four-digit system 四进位制；四单元制
four-dimensional 四维的
four-dimensional data assimilation 四维资料同化
four-dimensional fundamental form 四维基本形
four-dimensional geometry 四维几何
four-dimensional holographic recording 四维全息记录
four-dimensional space 四维空间
four-dimensional vector 四维矢量
four direction side loader 四向装卸叉车
four-disc reaming bit 四翼圆盘式扩孔钻头
four-drag-blade bit 四刮刀钻头
four-drill rig 四凿岩机钻车；四机凿岩台车
Fourdrinier board machine 长网纸板机
Fourdrinier machine 长网成型机
Fourdrinier type drying machine 长网干燥机
Fourdrinier yankee machine 长网单缸造纸机
four-drum hoist 四筒绞车
four drum revolving reel 四轴卷纸机
four-drum steam hoist 四滚筒汽汽提升机
four-edged bit 四翼钻头；十字钻头
four-effect evapo(u)rator 四效蒸发器
four-electrode method 四极法
four-electrode system 四电极系统；四点法
four-element 四透镜
four-element detector 四元件检测器
four-element grid 四元网络
four element infrared objective 四组元红外物镜
four-element theory 四元素说
four-engine single-shaft system 四机单轴式
four-entry pallet 四通制模板；十字形制模板
four essential crustall-wave systems of the earth 四大地壳波系
four eyed joining piece 四眼联结板
four family dwelling 四合院
four feeder wire clamp 四线馈线线夹
four feed lubricator 四路给油润滑器
four-figure code 四位数代码；四位码
four-figure display 四位数字显示
four figures 四位数
four figure system 四(进)位制；四单位制
four-fixture bathroom 四件套浴室
four-flanged disc shutter 四叶片快门
four-flanged shutter 四叶片快门
four-fluked anchor 四爪锚
four-flute taper-shank core drill 四槽锥柄心孔钻床
fourfold 四倍
fourfold axis 四重轴
fourfold axis of symmetry 四次对称轴；四重对称轴
fourfold block 四折叠板；四轮滑车；四车滑车
fourfold coordination 四次配位
fourfold correlation 四项相关
fourfold coupling 四重联轴节
fourfold interlacing 隔三行扫描
fourfold purchase 四轮滑车
fourfold rule 四折木尺
fourfold symmetry 四重对称
fourfold table 四格表
fourfold wood rule 四折木尺
four-fole node 四重节
four-force 四元力
four frequency diplex 双路移频制；四频双工制
four fundamental rules 四则运算
four gable form 四山墙形式；双曲线抛物面
four gimbaled 四框架的
four-groove drill 四槽钻头；四槽扩孔钻
four-harness twill 四页斜纹
four-head pressure 四头翻边机
four hearth furnace 四室炉
four-high mill 四辊轧机
four-high non-reversing cold rolling mill 不可逆式四辊冷轧机
four-high reversing cold mill 可逆式四辊冷轧机
four-high reversing cold strip mill 四辊可逆式带材冷轧机
four-high roller level(l)er 四重式多辊矫直机
four-high rolling mill 四辊式轧机
four-high rougher 四辊粗轧机
four-high setup 四辊式装置
four-high tandem cold mill 连续式四辊冷轧机
four-horse(d) chariot 四轮马车雕饰
four-horse team 四套马
four-hour shift 四小时轮班(制)
four hour varnish 四小时快干漆
four-hundred-day clock 周年钟；四百天钟
Fourier analysis 谐波分析；展成傅立叶级数；调和分析；傅立叶分析
Fourier analysis in two-dimensions 二维傅立叶分析
Fourier analysis method 傅立叶分析法
Fourier analysis-synthesis 傅立叶分析合成
Fourier analyzer 傅立叶分析仪；傅立叶分析器

Fourier-Bessel integral 傅立叶—贝塞耳积分
Fourier-Bessel series 傅立叶—贝塞耳级数
Fourier-Bessel transform 傅立叶—贝塞尔变换式
Fourier coefficient 傅立叶系数
Fourier decomposition 傅立叶分解
Fourier expand 傅立叶展开
Fourier expansion 傅立叶级数展开
Fourier filtered image 傅立叶滤波图像
Fourier filtering 傅立叶滤波
Fourier frequency 傅立叶频率
Fourier half-rang series 半幅傅立叶级数
Fourier heat equation 傅立叶热传导定律
Fourier hologram 傅立叶全息图
Fourier image technique 傅立叶成像技术
Fourier integral 傅立叶积分
Fourier integral analysis 傅立叶积分分析
Fourier interferometric spectrometer 傅立叶干涉光谱仪
Fourier inversion 傅立叶逆变换
Fourierism 傅立叶思想
Fourier-Legendre series 傅立叶—勒让德级数
Fourier-Mellin transform 傅立叶—梅林变换式
Fourier optic(al) imaging system 傅立叶光学成像系统
Fourier optics 傅立叶光学
Fourier phase spectrum 傅立叶相位谱
Fourier principle 傅立叶原理
Fourier projection 傅立叶投影
Fourier representation 傅立叶表示法
Fourier series 傅立叶数列;傅立叶级数
Fourier's law 傅立叶定律
Fourier space 傅立叶空间
Fourier spectrometer 傅立叶频谱仪;傅立叶光谱仪
Fourier spectroscopy 傅立叶光谱学
Fourier spectrum 傅立叶谱;傅立叶光谱
Fourier spectrum diagram 傅立叶频谱图
Fourier's theorem 傅立叶定律
Fourier Stieltjes series 傅立叶—司蒂吉斯级数
Fourier Stieltjes transform 傅立叶—司蒂吉斯变换
Fourier synthesis 傅立叶综合法
Fourier transform(ation) 傅立叶变换(式)
Fourier transformation filtering 傅立叶变换滤波
Fourier transformer 傅立叶变换器
Fourier transform hologram 傅立叶变换全息图
Fourier transform holography 傅立叶变换全息术
Fourier transform infrared spectrograph 傅立叶变换红外光谱仪
Fourier transform infrared spectrometer 傅立叶变换红外分光光度计
Fourier transform infrared spectroscopy 傅立叶变换红外分光镜
Fourier transform interferometer spectrometer 傅立叶变换干涉分光计
Fourier transform lens 傅立叶变换透镜
Fourier-transform nuclear magnetic resonance 傅立叶变换核磁共振
Fourier transform operator 傅立叶变换算符
Fourier transform pairs 傅立叶变换对
Fourier transform plane 傅立叶变换平面
Fourier transform spectra 傅立叶变换光谱
Fourier transform spectrometer 傅立叶变换分光仪
Fourier transform spectroscopy 傅立叶变换光谱学
Fourier transform ultraviolet and visible spectroscopy 傅立叶变换紫外—可见光谱法
Fourier variable 傅立叶变量
Fourier wave 傅立叶波
four-in-line 四缸直列
four-in-line engine 四缸直列发动机
four in one bucket 四合一铲斗
four-jaw 四爪
four-jaw chuck 四爪(卡)盘
four-jaw concentric chuck 四爪同心卡盘
four-jaw independent chuck 四爪卡盘;四爪分动卡盘
four-jaw independent lathe chuck 四爪单动车床卡盘
four jaw large hole chuck 大通孔四爪单动卡盘
four jaw plate 四爪卡盘
four-knife edge weigh(ing) gear 四刀边称量机构(混凝土搅拌厂设备)
four-lane bridge 四车道桥梁
four-lane divided roadway 四车道分隔道路
four-lane highway 四车道公路
four-lane road 四车道公路;四车道路

four large stabilizing 四大稳
four-layer device 四层器件
four-layer switch 四层开关
four-layer winding 四层绕组
four-leaf around corner sliding shutter door 四扇式转角推拉(遮蔽)门
four-leaf sliding folding shutter door 四扇式折叠推拉(遮蔽)门
four-leaf twill 四页斜纹
four-leaved flower 四叶花饰;四叶花雕饰(盛饰时期的建筑特征之一)
four-leaved rose curve 四瓣玫瑰线
four-leaved tracery 四叶形窗花格
four legged gravity escapement 四脚重力擒纵机构
four-leg intersection 四路交叉;四岔交叉;四路相交的交叉
four-leg platform 四柱腿平台
four-leg sling 四脚吊绳;四脚吊链
four-lens 四透镜
four-lens objective 四透镜物镜;四合物镜;四合镜物镜
four-lens zoom system 四透镜变焦系统
four-level 四能级
four-level emitter 四能级发射体
four-level fluoresent crystal 四能级荧光晶体
four-level fluoresent solid 四能级荧光固体
four-level generator 四能级振荡器;四级振荡器
four-level interchange (道路交叉处)四层立交构筑物
four-level material 四能级材料
four-level super crossroad 四层立体交叉口
four-level system 四能级系统
four-light 四窗
four-light body 四侧有窗车厢
four-limbed 四芯柱;四铁芯;四插角的
four line high draft system 四罗拉式大牵伸装置
fourling 四晶
four-link mechanism 四联杆机构
four-lipped cam 四凸耳凸轮
four lobe cam 四角形凸轮
four-lobe tracery 四角形窗花格
four-log increase 四对数增加
fourmarierite 红铀矿
four-masted bark 四桅帆船
four-master 四桅船
four-membered ring 四元环
four-months frost-free period 四个月无霜期
four-motor travel(l)ing crane 四马达移动起重机
Fourneyron wheel 福内朗式水轮机
four-node flat rectangular shear panel 四节点平矩形剪切面
four-node oscillation 四波节振动
four of the most important environmental factors 最重要的四种环境因素
four over five draft 四上五下牵伸
four-over-four array 八振子天线阵
four-panel door 四镶板门;四板门
four panel(l)ed door 四镶板门;四镶板门
four parameter model 线性粘弹性流变模型;四参量模型
four-part alloy 四元合金
four-part cross-rib(bed) vault 四段横肋拱顶
four-part line 经滑轮分为四股的缆索;四滑轮系统
four parts formula 四联公式
four-part vault 四分穹隆
four-pass design (锅炉的)四通道设计
four-passenger body 四座车身
four paws 四爪链钩
four-pendulum apparatus 四摆仪
four-phase cycle 四相循环
four-phase dynamic logic circuit 四相动态逻辑电路
four-phase modulation 四相调制
four-phase single stage torque converter 四相单级液力变矩器
four-phase square-wave generator 四相方波发生器
four-phase stepper motor 四相步进电动机
four-phase system 四相制
four-phase two stage torque converter 四相双级液力变矩器
four-piece butt matching 菱形拼合;四片对拼法;四片对接镶嵌
four-piece rim 四构件式轮辋

four-piece set 四构件支架(包括帽梁、支柱及槛梁,用于软弱地层中井巷的超前开挖);四构件棚子
four-piece wheel 四构件轮辐式车轮
four-pin base 四脚管底
four-pin clevis link 四孔联板
four-pin driven collar nut 四锥孔圆缘螺母
four-pin driven nut 四锥孔螺母
four-pin plug 四柱插销;四脚插头
four pin strap 四孔联板
four-ply built-up roof cover(ing) 四层组合覆盖屋面
four ply cloth 四层织物
four-ply seam 四层接合缝
four-ply tyre 四层帘布层轮胎
four-point 四探针
four point bearing 四点方位;四点定位法
four-point bit 四翼钻头;四刃钎头
four-point contact ball 四点接触球轴承
four-point control method 四点控制法
four-point(ed) bit 十字钻探钻头;十字扁铲
four-point linkage 四点悬挂装置
four-point load(ing) 四点荷载
four-point method 四电极系统;四点法;四标三角法
four-point mounting 四点悬挂法;四点安装法
four-point nose 四点测角头
four-point pigsty(e) 四点木垛
four-point press 四曲拐曲轴压力机;四点曲轴压力机
four-point probe 四探针
four-point probe measurement 四探针测量
four-point probe method 四点探针法
four-point resistivity test system 四探针电阻率测试系统
four-point small angle diagram 四点小角衍射图
four-point starting box 四点起动箱
four-point suspension 四点悬式冲床
four-point suspension mounting 四点悬吊装置
four-point suspension spreader 四点吊吊具
four-point switch 四接点开关
four-pole 四端网格
four-pole admittance 四端网络的导纳
four-pole circuit 四端电路
four-pole motor 四极电动机
four-pole network 四极网络;四端网格
four-pole relay 四极继电器
four-pole turbo-generator 四极汽轮发电机
four-port pump 四通阀
four-position lens turret 四位置透镜转头
four-post car lift 四柱汽车举升器
four-poster 四柱大木;四柱大床
four-poster bedstead 四柱床架
four-post hydraulic press 四柱式水压机
four potential 四元电位
four principal foum 四种主要形式
four principal vertical position 四个主要垂直装置
four-probe arrangement 四探针装置
four-probe method 四探针法;四探极法
four-quadrant construction 四象限结构
four-quadrant diagram 四象限图
four-quadrant multiplier 四象限乘法器
four-quadrant operation 四象限运算
four-quadrant rectifier 四象限整流器
four-quad solid-state sensor 四象限固态传感器
four quarters 整砖;整块砖
four-race roller bearing 四排滚棒式轴承
four-rake classifier 四耙式分级机
four-ram steering gear 四柱塞操舵装置
fourre 热带密灌丛
four ring 四节环
four ring snaffle bit 双环衔铁
four-roll calender 四辊压延机
four-roll crusher 四辊(式)破碎机;四辊碾碎机
four-roller bearing 四辊轴承
four-roller bit 四牙轮钻头;四辊钻头
four-roller mill 四辊式轧机
four-roll mill 四辊磨
four-room(ed) flat 四室套房;四室公寓
four-rope friction winder 四绳索摩擦提升机
four-rope grab 四索抓斗
four-rope log grapple 四索木材抓斗
four-rope rod type grab 四索杆式抓斗
four-rope suspension grab(bing) 四索悬挂抓斗
four-rope winding 四绳提升
four-rotor engine 四缸转子发动机

four sampling intervals per decade 四分度取样
four-screw driver 四用螺钉起子
four seat crew capsule 四座乘员分离舱
four seater 四座式;四座小客车
four-section support 四构件支架
four-segment circular grab 四叶片圆抓斗
four sheave travelling block 四滑轮游动滑车组
four-sheet mineral 四片层矿物
four-sided bar 四角型材;四角嵌条
four-sided column 四棱支柱;四角柱
four-sided figure 四边形
four-sided gateway 四面通道
four-sided pier 矩形支柱;矩形墩
four-sided planing 四面刨床
four-sided rotor 四角转子
four-sided structural sealant glazing 四边结构密封镶嵌玻璃
four-sided tower 四角塔;方塔
four silent-speed gearbox 四速无声变速箱
four socket adapter 四插座转接器
foursome 四人一组
foursome cabriolet 四座篷顶小客车
four-span 四跨度的;四跨的
four species 四则
four speed gear 四速齿轮
four speed gear box 四档变速箱;四速变速箱
four speed gear shift 四档变速
four speed shutter 四速快门
four speed transmission 四档变速器
four-spigot classifier 四室分级机
four-spindle automatic 四轴自动
four-spindle lathe 四轴车床
four-spindle numerically controlled lathe 四轴数字控制车床
four-spindle semiautomatic 四轴半自动
four spout packing machine 四嘴包装机
four-square lock 正方形锁
four square tool 四棱刀
four stack diffusion furnace 四管扩散炉
four-stage 四级
four-stage blower 四级增压器
four-stage compressor 四级压缩机
four-stage mass spectrometer 四级质谱计
four-stage mass spectrometry 四级质谱法
four-stage method 四阶段法
four-stage preheater 四级预热器
four-stage pump 四级泵
four-stage supercharger 四级增压器
four-stage tandem accelerator 四级串列加速器
four-stand tandem cold strip mill 四机座连续式带材轧机
four-stand tandem mill 四机座连续式轧机
four-start 四通
four-start spiral 四头螺旋
four-start worm 四线蜗杆
four-step heat treatment 四步热处理
four-step rule 四步法
four-strand double crossing over 四线双交换
four-strand mill 四线式轧机
four-strand rope 四股绳;四股钢丝
four-strand round sennit 四股编绳法
four-strand stage 四线期
four-stroke 四冲程的
four-stroke cycle 四行程循环;四冲程循环
four-stroke diesel engine 四冲程柴油发动机
four-stroke engine 四行程发动机;四冲程发动机
four-stroke marine diesel 四冲程船用柴油机
four-stroke oil engine 四冲程柴油机
four-strokes engine 四冲程引擎
four-tailed bandage 四头带
four-tape sort 四带分类
four-terminal attenuation 四端网络衰减量;四端网络衰耗
four-terminal circulator 四通路循环器
four-terminal constant 四端网络常数
four-terminal device 四端器件
four-terminal equation 四端网络方程
four-terminal interstage network 四端级际网络
four terminal network 四端网格
four-terminal oscillator 四端网络振荡器
fourth class highway 四级公路
fourth contact 复圆月蚀;复圆日
fourth dimension 第四维
fourth engineer 四轨

fourth environmental disease 第四公害病
fourth generation computer 第四代计算机
fourth generation container 第四代集装箱
fourth generation container berth 第四代集装箱泊位
fourth generation container terminal 第四代集装箱码头
fourth harmonic generator 四次谐波发生器
four-thirds law 四分之三幂定律
fourth normal form 第四范式
fourth officer 驾助;四副;四幅
fourth order 四等
fourth-order benchmark 四等水准点
fourth-order leveling field book 四等水准观测手薄
fourth-order reaction 四级反应
fourth-order triangulation point 四等三角点
fourth pinion 第四小齿轮
fourth proportion(al) 比例第四项
fourth rail 第四接触轨
fourth-rail in sulator 第四接触轨绝缘子
four-throw 四曲柄
four-throw crank shaft 四driveshaft曲柄轴
fourth speed 四档速度
fourth speed gear 第四速齿轮
fourth spudding date 第四开日期
fourth wire 中性线;第四线
four-tine grapple 四叉抓斗
four-track line 四线铁路
four-track railway 四线铁路
four-track section 四线区段
four-tube packing machine 四管包装机
four-turn road 四个来回路
four types of soil 四类土壤
four-tyre car 四轮胎小客车
four-unit code 四元码
four-unit sliding door 四扇一组(的)拉门;四扇整体门
four vector 四元向量;四元矢量
four-vector potential 四元矢量势
four-vertex theorem 四顶点定理
four-wave mixing 四波混频;四波混合
four-wave sum-mixing 四波求和混频
four way 四向的;四通(的);十字路;四路交叉
four-way arch 四向拱
four-way bit 四翼刮刀钻头
four-way branch 四通管
four-way change-over valve 四路转换阀
four-way cock 四通栓塞;四向小龙头;四通旋塞;四通龙头;四通开关
four-way connection 四通连接;四通接头
four-way control 四向调节
four-way coupling 四通接头
four-way iron plug 四脚铁插头
four-way jack 四线塞孔
four-way joint 四通接头
four-way joint box 四路联箱
four-way junction 四通(管接头)
four-way junction box 四路接线箱
four-way launching 四滑道下水
four-way level(l)ing 纵横向调平
four-way pallet 四通制模板;十字形制模板
four-way piece 四通管头
four-way pipe 十字形管;四通管
four-way plug valve 四通塞阀
four-way port 四通阀
four-way radar surveillance 四面雷达监视
four-way reinforcement 四向配筋
four-way reinforcing 四向配筋的
four-way rim wrench 四通轮缘扳手
four ways 四岔路口
four-way solenoid valve 四通电磁阀
four-way stop 四向停车
four-way stop sign 四向停车标志
four-way switch 四向开关;四通开关;四路开关
four-way system of reinforcement 四向配筋
four-way tap 四通开关;四通阀
four-way tee 四通接头;十字头
four-way tool block 四通刀架
four-way tool post 四向刀架
four-way tube 四通管
four-way union 四通管接;四向管接;四通联管节;四通接头;(连接管子的)十字接头
four-way valve 四通阀
four-way wrench 四用扳手

four wedge joint 四楔结合
four-wheel 四轮
four-wheel anti-skid system 四轮防滑装置
four-wheel automobile 四轮汽车
four-wheel brake 四轮制(动)器;四轮刹车
four-wheel braking 四轮制动设备;四轮制动
four-wheel car 四轮车辆
four-wheel coupled 四连轴式;二联轴式
four-wheel drive 四轮驱动
four-wheel drive and steer grader 四通驱动与操纵的分级器
four-wheel drive bogie 四轮驱动台车
four-wheel drive lorry 四轮驱动货车
four-wheel drive self-propelled vehicle 四轮驱动自行车辆
four-wheel drive tractor 四轮驱动拖拉机
four-wheel-drive tractor shovel 四轮驱动拖拉铲运机
four-wheel drive truck 四轮驱动载货汽车
four-wheel drive vehicle 四轮驱动车辆
four-wheeled chassis 四轮的机架;四轮的底盘
four-wheeled crab 四轮的蟹爪式起重机
four-wheeled drive loader 四轮驱动的装载机
four-wheeled drive tractor 四轮驱动的拖拉机
four-wheeled fork lift truck 四轮叉车起吊卡车
four-wheeled hand truck 四轮手推车
four-wheeled prime mover 四轮原动机
four-wheeled rubber-tyred roller 四轮橡胶拖拉机
four-wheeled scraper 四轮括土机;四轮铲运机
four-wheeled side tipping trailer 侧倾卸式四轮挂车
four-wheeled steer(ing) 四轮导向的;四轮操纵的
four-wheeled tractor 四轮拖拉机
four-wheeled vehicle 四轮汽车
four-wheeler 四轮车
four-wheel garage jack 四轮修车起重器
four-wheel grader 四轮平地机
four-wheel landing gear 四轮起落架
four-wheel motor vehicle 四轮机动车辆
four-wheel screw jack 四轮修车起重器
four-wheel steering 四轮转向
four-wheel steering tractor 四轮转向拖拉机
four-wheel timber trailer 四轮运木挂车
four-wheel tractor 四轮拖拉机
four-wheel trailer 四轮挂车;四轮拖车
four-wheel truck 两轴转向架
four-wheel vehicle 四轮汽车
four-wheel wagon 四轮(拖)车
four-wing bit 四翼缘钻头;四翼缘刀具
four-wing drilling bit 四翼钻头
four-winged steel bit 十字形钢钻头
four-wing reaming bit 四翼扩孔钻头
four-wing revolving door 四扇式转门
four-wing rotary bit 四翼钻头
four-wire 四线的
four-wire arm 四线横担
four-wire carrier system 四线载波制
four-wire channel 四线通道
four-wire circuit 四线制电路;四线电路
four-wire circuit system 四线电路制
four-wire connector 四线连接
four-wire cross bar switching system 四线纵横交换制
four-wired cable 四芯电缆
four-wire line 四线制传输线;四线线路
four-wire multiplex facility 四线制多路复用设备
four-wire prestressed jack 四线预加应力千斤顶
four-wire repeater 四线中继器;四线制增音器
four-wire side circuit 四线实线电路
four-wire switching 四线制交换;四线交换
four-wire switching centre 四线制交换中心
four-wire system 四线制;四线系统
four-wire terminating set 四线制终端设备
four-wire three-phase system 四线三相制
four-wire type circuit 四线式电路
four wont way 十字路
Fout's staining method 富特染色法
fowan 福万风
foward feed of material 进料的预先供应
Fowler function 福勒函数
Fowler position 福勒位置
fowl house 家禽舍;家禽饲养房
fowl manure 家禽粪
fowlrun [英]养鸡场

fox 杂用小绳
fox bolt 开尾地脚螺栓;端缝螺栓;叉端螺栓;开尾螺栓
foxes 植绒帆布;绳屑;绳毛绒
fox(e)y 木材腐朽
fox grape 美洲葡萄
Fox Hills Sandstone 福克斯山砂岩
foxhole 散兵坑
foxic waste 毒性废料
foxiness 变色;木材腐朽;褐斑;腐化变色
fox lathe 狐狸车床
foxtail 销栓;狐尾梢
fox-tailed wedge 紧榫接;扩裂楔
foxtail joint 楔形接合
foxtail millet 谷子
foxtail saw 狐尾锯
foxtail wedge 扩裂楔;紧榫接;暗楔;狐尾楔
foxtail(wedge)joint 楔连接
foxtail wedging 狐尾楔紧固;狐尾榫(接);暗装紧固;狐尾楔栓(牢);扩张楔;暗楔法
fox trip spear 螺旋式打捞器
fox wedge 扩张楔;扩裂楔;紧榫楔
fox wedging 狐尾榫
foxy 红褐色;赤褐色的
foyaite 流霞正长岩
foyamelilitite 流霞黄长岩
foyer 门厅;门廊;剧场休息室;(客船的)外休息厅
foyer of theatre 剧场休息室
foyer wall 门厅墙壁
FPA clause 平安险条款
Fraas breaking brittle point 弗拉斯脆化断裂点
Fraas breaking point 弗拉斯断裂点
Fraas brittle point 弗拉斯脆点
Fraas brittle temperature 弗拉斯发脆温度
Fraas tester 弗拉斯试验器
fracitonal exponent method 分式指数法
frac-shot 堵裂粉
fractable 盖顶;山墙端盖顶石
fractal analysis 分形分析
fractal and chaos image 分形和混沌图像
fractal dimension 分形维数
fractal flow characteristic 分形水流特征
fractal geometric morphology with periodicity 具有同期性的分形几何图形
fractal geometry 分形几何
fractal graphics 分数维图形学
fractal image 分形图像
fractal image system 分形图像系统
fractal Levy motion 分形列维运动
fractal net work 分形网络
fractals 分形分析
fractal structure 分形结构
fractal theory 分形理论
fractile 分位数值;分位数;分位点
fraction 一部分;馏分;零组;粒组;粒度级分;级分;生存率;分效;分数(部分);分式【数】;分馏物;部分;百分率
fractional 馏出蜡;分数的;分级的;分段的;分成几份的
fractional amount 零数
fractional anaerobic 分步厌氧
fractional analysis 粒级分析;筛析;分组分析;分馏分析;分化分析;分次分析法
fractional appraisal 房地产分部分估价
fractional area of contact 接触面积率
fractional arithmetic 小数运算;分数运算
fractional bank note 小额银行券
fractional Brownian motion 分形布朗运动
fractional Brownian motion process 分形布朗运动过程
fractional card 部分倒用卡
fractional catabolic metabolism rate 部分分解代谢率
fractional centrifugation 分部离心分离法
fractional charge 分数电荷
fractional classification 尺寸分级
fractional coefficient 分数系数
fractional collection efficiency 分级除尘效率
fractional column 分馏塔
fractional combustion 分级燃烧
fractional computer 分数计算机
fractional condensation 分凝作用;分级冷凝
fractional condenser 分级冷凝器
fractional condensing tube 分凝管

fractional condensing unit 小型压缩机冷凝机组
fractional coordinates 分数坐标
fractional crystallization 结晶分异作用;分馏结晶作用;分离结晶(作用);分级析晶;分级结晶(法);分步结晶;分别析晶
fractional culture 分级培养法
fractional device 分馏装置
fractional diameter 中间直径
fractional differentiation 分数微分
fractional dimension space 分数维空间
fractional distance point 分距点
fractional distillation 精馏作用;精馏(法);分馏作用;分馏精馏;分馏(法);分级蒸馏(法);分部蒸发
fractional distillation process 分馏过程
fractional distillation product of membrane 膜分馏物
fractional distillation test 分馏试验
fractional distilling flask 分馏烧瓶;分馏瓶
fractional distilling tube 部分冷凝器
fractional distortion 部分畸变
fractional dose 分剂量;分数剂量
fractional driving 分级传动
fractional effective dose 部分有效剂量
fractional efficiency 局部过滤效率;分馏效率
fractional electric(al)charge 分数电荷
fractional electric(al)motor 小功率电动机;分马力电动机
fractional energy-saving 节能率
fractional equation 分数方程;分式方程
fractional error 相对比例误差;部分误差;相对误差
fractional exponent 分指数;分式指数;分式指数
fractional expression 分数式
fractional extraction 分馏萃取;分级萃取;分步抽提;分布抽提
fractional fixed point 小数定点制
fractional fix-point 分式定点
fractional fix-point method 分式定点法
fractional flow 部分液流
fractional free distribution 小额免费分配
fractional free issue 小额免费发行证券
fractional frequency change 频率微变
fractional frequency offset 分数频率补偿
fractional function 分数函数
fractional harmonic 分数谐波
fractional harmonic wave 次谐波
fractional horsepower air compressor 小功率空气压缩机
fractional horsepower asynchronous motor 分马力异步电动机
fractional horsepower light metal induction motors 铝合金壳分马力感应电动机
fractional horsepower motor (小于一马力的)小功率发电机;分数功率电动机;小马力电动机;分(数)马力电动机;低功率电动机
fractional index 分式指数
fractional instantaneous phase measurement 部分瞬时相位测量
fractional integral 非整数次的积分;分数次积分
fractional integration 分数次积分
fractional interest 部分利益
fractional ionization 分级电离
fractional iteration 分数迭代;分式迭代
fractional Levy motion 分形列维运动
fractional linear combination 分式线性组合
fractional linear displacement 分式线性代换
fractional linear substitution 分式线性代换
fractional load 轻载;部分货载;部分荷载;部分负载
fractional loading coil 分数加感线圈
fractionally charged particle 分数电荷粒子
fractional magnetic charge 分数磁荷
fractional melting 分熔;分熔溶融;分步熔化
fractional Miller indices 分数密勒指数
fractional modulation 分数调制
fractional module 分数模
fractional moist heat sterilization 间歇热灭菌法
fractional money 辅币
fractional mutant 部分突变型;部分突变体
fractional-mu tube 分数放大系数管
fractional neutralization 分步中和
fractional note 小额银行券
fractional number 分数;分式【数】
fractional-octave 分倍频程
fractional of technical podwer 工业粉末分级
fractional order 分数阶

fractional part 小数部分;分数部分
fractional pitch 分数螺距;分数极距
fractional-pitch winding 分数节距绕组;分时绕组;分距绕组;分节绕组
fractional pointer 分巷指针
fractional point of distance 分距离点
fractional polarization 部分偏振
fractional porosity 相对孔隙率;相对孔隙度;孔隙率分数;部分孔隙率
fractional power 分数幂
fractional precipitation 分级沉淀;分段沉淀;分(步)沉淀
fractional program(me) 分数规划;分式规划
fractional programmig 分数规划
fractional progression 分数级数
fractional purging 分级清洗
fractional rate 部分利率
fractional rating 分数功率
fractional replication 分数配置
fractional representation 小数表示
fractional reserve 部分准备
fractional reserve banking 部分准备金银行制度
fractional reserve banking system 部分准备金银行制度
fractional root 分数根
fractional sampling 土的粒度分级取样;分数采样法;土壤粒度分级取样
fractional saturation 部分饱和
fractional scale 分数比例尺
fractional scanning 局部扫描;分段扫描
fractional sedimentation method 分离沉淀法;部分沉淀法
fractional sine wave 分数正弦波
fractional slope 用分数表示的坡度
fractional-slot 分数槽
fractional slot winding 分数槽绕组
fractional solids content 粉体颗粒分数
fractional solution 分熔;分级溶解
fractional step method 分裂法
fractional sterilization 间歇灭菌(法);分段灭菌;分次灭菌
fractional still head 分馏头
fractional township normal town 都市规划小区
fractional turn 分级转动
fractional ultrafiltration 分步超滤法;分布超滤法
fractional unemployment 短期失业
fractional variation 百分比变化
fractional void 孔隙率;空隙分数
fractional volume of clean sandstone 纯砂岩的相对体积
fractional volume of shale 泥质的相对体积
fractional volume of silt 细岩砂的相对体积
fractional weights 小数砝码
fractional yield 相对产量
fractionate 分级
fractionate contraction 不全收缩
fractionated dose 分级剂量;分次剂量
fractionated gain 分部增益
fractionated irradiation 分次辐照
fractionated oil 分级油
fractionated precipitation 分级沉淀
fractionated sedimentation method 分步沉降法
fractionating 摩擦
fractionating column 分馏柱
fractionating device 分馏装置
fractionating diffusion pump 分馏扩散泵
fractionating efficiency 分馏效率
fractionating plate 分馏塔的塔盘
fractionating process 分馏(法)
fractionating pump 分馏泵
fractionating system 分馏系统
fractionating tower 精馏塔;分馏塔
fractionating tray 分馏塔的塔盘
fractionating tube 分馏烧管;分馏管
fractionation 精馏;化学分离法;分数化;分馏(法);分级分离;分部分离
fractionation by adsorption 吸附分离
fractionation dose 剂量分次给予
fractionation efficiency of tower 塔的分馏效率
fractionation factor 分级因数
fractionation method 分段法
fractionator 气体分离装置;分馏器
fraction by volume 容积分数
fraction collector 馏分收集器;分馏装置;分部收集

器;部分收紧器;部分收集器
fraction decomposition 分式分解
fraction defective 次品率;废品率;不良率;不合格率
fraction-failed test 局部破坏试验
fraction gear 组合齿轮
fractioning tower 分馏塔
fraction in lowest terms 最简分式
fraction of critical damping 临界阻尼百分比
fraction of mineral aggregate 矿物骨料的破碎;矿物骨料的粒化
fraction of particle size 粒径组
fraction of petroleum 石油馏分
fraction of the year 年分
fraction order 分数级数
fraction organic carbon 粉末有机碳
fraction radiation 分剂放射
fraction surviving 存活率
fraction void 疏松度
fraction weight 分级重量
fractoconformity 破碎整合【地】
fracto-cumulus 碎积云
fractograph 断口组织的照片;断口组织
fractographic(al) pattern 破断模型
fractography 断口组织检验;断口形貌学;断口显微观察术;断口显微分析
fracton 分形振子
fracto-nimbus 碎云雨
fracto-stratus 碎层云
fractural diamond 金刚石碎屑
fractural stage design method 破损阶段设计法
fractural structure 断裂构造
fractural zone 断裂带
fracturation of rock 岩层断裂
fracture 折断;裂痕;裂缝;破裂;破坏;断裂;断片;断面;断裂;断口
fracture analysis 断裂分析;断口分析
fracture analysis diagram 断裂分析图;断口分析图
fracture and fault 裂隙和断层
fracture angle 破裂角
fracture appearance 破裂外观;断口外观
fracture area 断口面积
fracture around underground opening 围岩断裂
fracture arrest 断裂阻延;止裂
fracture arrest temperature 裂纹终止温度
fracture behavio(u)r 裂隙性态;破裂性状
fracture bore 破裂孔
fracture boundary 断裂边界
fracture by fatigue (材料)疲劳破裂
fracture characteristic 断裂特征
fracture cleavage 破劈理;破解理
fracture cleavage belts 破劈理带
fracture condition 破坏状况
fracture conduit 裂缝形通道
fracture cone 破裂锥形面
fracture criteria 破裂准则
fracture criterion 破裂准则;破裂标准;断裂判据
fracture crystal 断裂晶体
fractured chalk reservoir 裂缝性白垩储集层
fractured chert reservoir 裂缝性燧石储集层
fractured earth movement 破裂性地壳运动
fracture density 裂隙密度;裂缝密度
fracture density log 裂缝测井
fracture density of a fracture network 裂缝网络密度
fracture density of a fracture system 裂缝系密度
fractured face 破碎裂面;破碎面
fractured fault zone 断面破碎地段
fractured formation 裂隙岩层
fractured ground 裂隙地层
fracture dimension 破裂尺寸
fracture direction 裂隙方向
fractured load 破坏荷载
fractured member 破裂构件
fracture dome 松裂穹
fractured part 断裂零件
fractured reservoir 裂缝储集层
fractured shale reservoir 裂缝性页岩储集层
fractured surface 断面;断裂表面;裂面;破裂面
fractured trap 裂缝圈闭
fracture dynamics 断裂动力学
fracture dzone 裂隙区;断口带;破碎带;破裂带
fracture edge 破裂缘
fracture face 断口面
fracture failure criterion 断裂失效准则

fracture filling [fissure and coating] 裂隙充填物
fracture flow capacity (水力压裂时)裂隙的流过能力
fracture fluid 压裂液
fracture formation 裂缝层
fracture frequency 破裂频率;断裂频率
fracture gap opening 裂缝宽度
fracture genesis earthquake 断层成因
fracture grade 断口等级
fracture grouting 劈裂注浆;劈裂灌浆
fracture initiation resistance 断裂初始抗力
fracture intensity 破裂强度
fracture interval 裂缝间距
fracture lance 裂隙矛
fracture limit 断裂极限
fracture line 破裂线;断裂线
fracture linear density 裂缝线密度
fracture line theory 断裂线原理
fracture load 断裂荷载;断裂负载;断裂负荷;破坏荷载
fracture log curve 裂缝测井曲线
fracture mechanic property of rock 岩石断裂力学性质
fracture mechanics 断裂力学
fracture mechanism 破裂机制;破裂机构;断裂机理
fracture method 破碎方法
fracture mirror 断裂镜面
fracture model 断裂模型
fracture nature 断裂性质
fracture number 断口度数
fracture observation 断裂观测
fracture of coal 碎煤
fracture of facet 棱面断口
fracture opening 断裂空隙
fracture orientation 裂隙方向
fracture origin 断裂源
fracture origin analysis 裂缝原因分析
fracture parameter 裂缝参数
fracture path 断裂路径
fracture pattern 断裂花纹;破裂形状;破裂模式;碎块形状;断裂形式;断裂方式
fracture permeability 裂缝渗透率
fracture phenomenon 地裂现象;破裂现象
fracture plane 裂隙面;破裂面;断面;断裂面
fracture plane inclination 断裂面倾斜
fracture plate 破裂板
fracture point 破裂点
fracture porosity 裂隙率;裂缝孔隙度
fracture potential 断裂位能
fracture power 断裂功
fracture probability 断裂概率
fracture process 断裂过程
fracture profile 断裂剖面
fracture-propagating flaw 破裂传播裂隙
fracture propagation 裂缝传播;裂口扩展;断裂蔓延
fracture reservoir 断裂型热储
fracture resistance 断裂阻力
fracture rock 破裂岩石
fracture safe design 断裂安全设计
fracture section 破裂断面;断裂剖面;断裂截面
fracture-shaped sink hole 裂隙状落水洞
fracture spacing 裂隙分布
fracture speed 裂纹扩展速度;断裂纹扩展速度
fracture spring 隙泉;裂隙泉
fracture state 断裂状态
fracture strength 抗裂强度;抗断强度;抗动载强度;断裂应力;断裂强度
fracture stress 破碎应力;断裂应力
fracture structure 断裂结构
fracture surface 破裂面;断面;断裂(表)面;断口
fracture surface energy 断裂表面能
fracture surface examination 断口检查
fracture surface marking 裂纹表面标记
fracture susceptibility 开裂敏感性
fracture system 裂纹体系;破裂系;断裂系(统)
fracture system map 断裂系统图
fracture tectono-geochemistry 断裂构造地球化学
fracture test (混凝土)破坏性试验;破坏试验;断口试验
fracture texture 块态结构;断口结构;断口构造
fracture theory 断裂理论
fracture toughness 临界应力强度系数;破坏韧性;断裂韧性;断裂韧度
fracture toughness property 断裂韧性

fracture toughness specimen 断裂韧性试样
fracture toughness test 断裂韧性试验
fracture transition of elastic 弹性断裂转变温度
fracture transition temperature 断裂转变温度;断口转变温度
fracture treatment (地层水力)压裂处理
fracture valley 断裂谷
fracture water 裂隙水
fracture wear 断裂磨损;断裂磨耗
fracture width 裂缝宽(度)
fracture work 断裂功
fracture zone 裂隙区;裂缝带;裂裂区;断裂带;层破碎带;不规则的海底断层区
fracture zone lineament 破碎带线性体
fracturing agent 破碎剂
fracturing-anticalinal trap 裂缝—背斜圈闭
fracturing anticlinal pool 裂缝性背斜油藏
fracturing by vibration (石油开采)振动压裂
fracturing fluid 压裂液添加挤;压裂液
fracturing fluid tank truck 压裂液腱车
fracturing grouting 劈裂灌浆
fracturing load 破坏荷载
fracturing of separate layers 分层压裂
fracturing semi-diameter of crater 漏斗破坏半径
fracturing technology 水力压裂工艺
fracturing trap 裂缝圈闭
fracturing truck 压裂泵车
fracturing unit 压裂设备
fractus nimbus 碎雨云
fragile 易碎品;易碎的;脆(性)的;脆弱的
fragile article 易碎品
fragile cargo 易损货物;易碎货物
fragile fibre 脆性纤维
fragile goods 易碎货物
fragile material 脆性材料
fragileness 易碎性
fragile powder 脆性粉末
fragile support 脆性载体
fragile target 脆靶
fragility 易碎性;裂性;碎性;脆性;脆弱;脆度
fragility index 脆性指数
fragility test 脆性试验
fragipan 脆磐
fragipan soil 脆磐土
fragment 裂片;裂变碎片;破片;片断;片段;碎屑(状的);碎片;碎块;断片;段落;存储区未用满部分;分段存储
fragmental 碎屑状;碎屑质的【地】;碎片的;分裂成碎片的
fragmental data 不完全资料
fragment(al) debris 碎岩屑
fragmental deposit 碎屑沉积(物)
fragmental ejecta 喷屑
fragmental ion 零碎离子
fragmental reservoir 碎屑岩储集层
fragmental rock 碎屑岩
fragmental structure 碎屑构造;碎片状结构
fragmental texture 碎屑结构
fragmental volcanic rock 碎屑火山岩
fragmentary 脆性的;残缺
fragmentary data 不完全资料;不完全数据
fragmentary debris 碎岩屑
fragmentary equivalent form 局限性的等价形式
fragmentary hydrologic data 间断水文资料;不连续水文资料
fragmentary material 碎屑物质
fragmentary product 碎屑产物
fragmentary rock 碎屑岩
fragmentary sample 残缺样本;不完全样本
fragmentary system 碎裂体系
fragmentation 破碎(作用);碎裂;断裂;存储碎片;存储区未用满部分;存储器中未用满部分;存储残片;分区输入程序;分割;分段存储
fragmentation bomb 杀伤炸弹
fragmentation emission test 破片飞散试验
fragmentation hypothesis 分离说
fragmentation mechanism of rock 岩石破碎机理
fragmentation method 碎裂法
fragmentation model 碎裂模型
fragmentation nucleus 碎核;破碎核
fragmentation of blasted rock 破碎度
fragmentation pathway 裂解途经
fragmentation pattern 裂解方式
fragmentation problem 断片问题

fragmentation protective body armo(u)r 防弹片护身装甲
fragmentation reaction 碎裂反应
fragmentation test 碎片状态试验
fragment code 分段码
fragmented bortz 不纯金刚砂
fragmented rock 危石
fragmented transport 分段运输
fragment emission 破片飞散
fragmenting 破碎
fragmenting of continents 大陆分裂作用
fragment ion 碎片离子
fragmentiser 碎裂机
fragment notation 分段表示法
fragment of brick 砖块
fragment offset 分段差距
fragment of mother rock 母岩碎屑
fragment of pottery 瓷器残片
fragmentography 碎片谱法
fragment peak 碎片峰值
fragments of rock 碎石渣
fragments of the crust 地壳块体
fragrance 香味;香料;芳香
fragrant 芳香
fragrant birch 香桦
fragrant epaulette tree 白辛树
fragrant garden 芳香花园
fragrant plant 芳香植物
fragrant wood 沉香木
Frahm frequency meter 弗拉姆频率计
frail 单薄的;脆弱的;薄弱
frail construction 单薄结构
frailty 弱点
fraipontite 锌铝蛇纹石
fraise 扩孔钻(岩石钻孔扩大直径);铣刀
fraise adapter 铣刀附件;铰刀附件
fraise arbor 铣刀杆
fraise unit 铣削动力头
fraising 铰孔;切环槽
frake 西非橄榄树
Fralick's fluid 弗雷利克液
Fram 法拉姆式楼板
Fra Mauro formation 弗拉磨拉建造【地】
framboid 微球团;微球粒
frame 信息图像;帧;构梁;构架;肋骨;架(子);机壳;农场;画面;画幅;汽车大梁;车架;巢框;幅;边框;框架
frame access function 框架的存取函数
frame action 框架作用
frame aerial 框形天线
frame aerial camera 画幅航空摄影仪;分幅航空摄影
frame alignment 帧同步;帧定位
frame alignment recovery time 帧定位恢复时间
frame alignment signal 帧定位信号
frame alignment time slot 帧定位时(间)隙
frame amplifier 帧信号放大器;帧放大器
frame analysis 框架分析
frame and brick veneer construction 框架嵌砖结构;框架嵌砖建筑
frame and panel construction 框架墙板(装配)结构;框架墙板(装配)建筑
frame and shear wall structure 框架剪力墙结构
frame and skid mounted drill 装钻机的机架和滑撬
frame angle bar 肋骨角材
frame antenna 框形天线
frame a plan 制订计划
frame area 像幅;帧面积;保护地
frame assembly 框架总成
frame axis 框架轴
frame bar 分帧线;分格线
frame base 框架底板;机架底板
frame basic function 框架的基本函数
frame beam 机架梁;机座肋条;机架横梁
frame bender 肋骨冷弯机
frame bent 排架
frame bevel (船首尾)斜面肋骨
frame blanking 帧回描熄灭
frame-blanking amplifier 帧熄灭放大器
frame board 腰线板;踏腰线板板
frame body plan 肋骨横剖面图
frame border 帧边缘
frame brace 底板拉条
frame bracket 肋骨肘板
frame bridge 框架(式)桥

frame buffer 图像缓冲器;帧缓冲器;画面缓冲区
frame buffer memory 帧缓冲存储器
frame building 框架建筑;构架建筑
frame-by-frame exposure 逐帧照射
frame-by-frame picture recording 分解法图像录制
frame camera 分摄像机;分格摄影机;分幅摄影机
frame center crank 叉形架
frame clearance 门框间隙
frame coil 定心线圈
frame column 框架柱
frame component 框架构件
frame connection 框架连接
frame construction 骨架结构;构架结构;框架(式)结构;框架建筑;框架构造
frame construction type 框架结构型式;框架建筑型式
frame contract 标准合同
frame corbel 箍式悬挑;架式悬挑
frame corner 桁架角隅
frame counter 像片计数器
frame cover 护板
frame-covered structure 全由构架组合的结构
frame crane 龙门(式)起重机;固定式门式起重机;龙门吊
frame crank press 单柱曲柄压力机
frame cripping 框架断裂;框架变形
frame cross beam 车架横梁
frame cross member 车架横梁
frame culture 温床栽培
frame cushion 车架缓冲装置
frame cutting 平行式剪切
framed 构架式的;构成的;框架的;榫构合
framed abutment 框式桥台
framed and braced door 斜撑构成门;框架门
framed-and-ledged-and-braced door (门扇带框的)拼板斜撑门
framed and ledged door 框架直拼板门;横架构成门;(带框的)拼板门
framed and panel(l)ed door 门档镶板门
framed and two-panel(l)ed door 框档双壮镶板门
framed arch 构架拱;构成拱;桁拱
framed beam 构架梁
framed bearing structure 框架承重结构
framed bent 框式桥台;框架桥台;排架
framed building 骨架式房屋;构架式建筑物;构架建筑;构架房屋;框架(式)建筑
framed cellular floor 框格型楼板
framed connection 框架结合;梁柱框架式联结
framed construction 框架结构;构架建筑;构架构造
framed construction hollow door 框架结构空心门
framed construction solid door 框架结构实心门
framed dado 框形墙裙
framed dam 框架式坝;框架坝
framed dome 构成圆顶
framed door 框档门;直拼撑框架门
frame deflection 帧偏转
frame deflector coil 帧偏转线圈
frame description 框架的描述
frame detector 帧检测器
framed floor 格栅地板;构架桥面;构架地板;构架底板;构架桥面;构架楼板;双层龙骨地板
framed floor cover(ing) 构架楼板覆盖层;构架地板覆盖层
framed flooring 构架楼板;构架地板
framed gable wall 框架山墙
framed girder 桁架梁;析梁;构架(大)梁
framed ground 构架楼板;门窗框木砖;门窗柜木砖;门侧柱;构架地面
framed house 骨架式房屋;构架房屋;框架房屋
frame diagram 框架图
frame diaphragm 内门框隔板
frame intake 帧式进水口
frame direction-finder 框形测向器
frame distortion 帧畸变;图像畸变;框架扭曲;框架变形
frame divider 帧分频器
framed joist 构架式格栅
framed lining 门框装修
framed load-bearing structure 框架承重结构
framed mattress 构架式底板;框架沉排;框格(式沉)排;木框式沉排;编篱
framed mirror 镶框镜
framed door 框架门
framed opening 加边框洞口;门窗口

framed partition(wall) 框架式隔墙;桁架式间壁;构架(式)间壁;木隔墙;框架间壁;有边框隔墙;构架(面板)隔墙;构架隔断
framed revetment 框架式护岸;框架式挡土墙;框格排护岸;编篱护岸
frame drive pulse 帧频起动脉冲
framed roof 檩条椽子屋顶;构架屋顶
framed screen cloth 装在框架上的筛网布
framed square 未装修的镶板门
framed structure 构架结构;桁架结构
framed structure analysis program(me) 框架结构分析程序
framed supporting structure 框架支承结构
framed timber door 框架大门
framed trestle 构架栈道;构架式栈桥;构架式栈道
framed tube 框筒
framed tube-core structure 筒中筒结构
framed tube structural system 框筒结构体系
framed tube structure 框筒结构;单框筒结构
frame duration 帧周期
framed view 框景
framed wall(ing) 构架墙;立筋墙
framed weight-carrying structure 框架承重结构
framed weir 构架式堰;框架堰
framed work wall 构架墙
frame erecting gear 构架架设的机具;构架架设的齿轮
frame erection 肋骨装配;架肋骨
frame error 帧差错
frame facet 框架的侧面
frame factor 结构系数
frame field 帧场
frame filter press 框式压滤机
frame finder 框架取景器
frame flyback 帧回描;帧回程
frame flyback suppression 帧回描熄灭
frame flyover bridge 框架式地道桥
frame for fittings 门窗框
frame fork 车架叉
frame formula 框架公式
frame foundation 框架地脚;架座;机架地脚
frame fracture 框状断口
frame frequency 帧频
frame gasket 门框衬垫
frame girder 框架梁;机架横挡;框架横梁;构架梁
frame gland 托架压盖
frame grab 帧取样;图形捕获
frame grabber 帧接收器
frame ground 机架地线
frame grounding 机架接地
frame-grounding circuit 机架接地电路
frame grounds 门窗柜木砖
frame handler 帧控制程序
frame head 上段肋板
frame height 帧面高度
frame high 框口净高
frame hinge 框架铰接
frame hold 取景调节器;摄像调节器
frame holder 承片框
frame house 框架房屋;木构架房屋;木板房(屋)
frame hysteresis 帧磁滞
frame inheritance 框架的继承
frame inner bearing 构架内部支承
frame instability 框架不稳定性
frame jet 框架千斤顶
joint 车架接合
frame-leakage current protection 框架漏电保护(装置)
frame-leakage protection 构架漏电保护装置
frame leg 构架支柱;构架腿
frameless 无框架的;无构架的
frameless automobile 无车架小客车
frameless body 无车架车身
frameless chassis 无框架底盘
frameless construction 无车架结构
frameless door 无框门;无框玻璃门
frameless glass 无框架的窗玻璃
frameless glass door 无框玻璃门;无边框玻璃门
frameless sliding window 无框推拉窗;无框扯窗
frameless sulky plough 无架乘式犁
frameless sulky plow 无架乘式犁
frameless tanker 无架式油灌车
frameless trailer 无架挂车
frameless vehicle 无车架车辆

frameless window 无框玻璃窗
frame level 框式水平仪
frame level data 帧级数据
frame level interface 帧级接口
frame level procedure 帧级过程
frame lifting 顶框脱模
framelift mo(u)lding machine 框式造型机
frame light beacon 框架形灯桩
frame-like(load)bearing structure 框架式承载结构
frame-like supporting structure 框架式支承结构
frame limiting 帧幅限制;画面限制
frame line 分帧线;分格线
frame linearity 帧直线性
frame liner 肋骨衬条
frame lines 型线;肋骨线;船体肋骨线
frame load-bearing system 承重框架系统
frame loading 框架加载;构架荷载
frame lock 车锁
frame member 构件
frame memory 帧存储器
frame modulus 肋骨剖面模数
frame monitoring tube 帧监视管
frame mo(u)ld 构架模型;肋骨样板
frame number 帧号码;帧编号
frame of axes 坐标系(统);坐标架
frame of boring rig 钻井塔
frame of bridge 桥梁框架
frame of expendable pallet 简易托盘架
frame of one bay 简单构架;单跨框架
frame of one span 简单构架;单跨框架
frame of reference 空间坐标系;基准坐(标)系;读数系统;参照坐标系;参照系;参照构架;参考(坐标)系;参考架;参考格网
frame of roof 屋架
frame of the bit 钻头体
frame out 离开屏幕
frame output valve 帧信号输出管
frame page 帧页面
frame-panel building 骨架板柱式房屋;框架墙板(装配)建筑;间架式建筑
frame panel interaction 框架墙板共同作用
frame parts 车架零件
frame pattern 帧模式
frame period 帧周期
frame photography 框幅摄影;分格摄影术;分幅摄影术;分层摄影术
frame pier 框架墩
frame pitch 像幅距;帧距
frame pivot bearing 立式推力轴承;框架中支柱枢轴
frame plane 框形平面;框架平面
frame planer (移动式的)龙门刨床
frame plate 框架板
frame plough 架式犁
frame plow 架式多铧犁
frame post 框架柱;框架支柱;构架支柱
frame problem 框架问题
frame process 图像处理功能
frame pulley 窗框滑轮
frame pulse 帧脉冲
frame pulse pattern 帧脉冲同步码
framer 组织者;制造者;帧调节器
frame raising gear 构架提升机具;构架安装机具
frame rate 帧频;画面更新率
frame relay 帧中继
frame representation 框架表示
frame resistance 线绕可变电阻
frame retrace 帧回描
frame retrace time 帧逆程时间;帧回描时间
frame rider 框架横架
frame rigidity 框架刚度
frame rod 框架构件;框架杆条
frame roll 帧滚动
frame sample 帧取样
frame saw 弓锯;框(架)锯;排锯;多锯条框锯机
frame saw blade 框锯条
frame-saw file 框锯锉;锯子锯锉
frame scan 帧扫描
frame scanning 帧扫描
frame-scanning speed 帧扫描速度
frame-scanning period 帧扫描周期
frame section 框架截面
frame seeding 温床播种
frame set 框式支架
frame shape 框架形状;框架外形

frame shears 平行式剪切机
frame-shear wall 框架剪力墙
frame shift 移码
frame side member 车架道梁;车架边梁
frame sign 帧大小
frame size 像幅大小;像幅;祯幅
frame slip 帧滑动
frame slip packing piece 肋骨衬条
frame space 肋距;肋骨间距;构架间距
frame spacing 肋距
frame span 框架跨度
frame splice 车架镶接
frame split 分片机架接缝
frame stand 框架座
frame start 帧起始
frame stations 型线;肋骨线;船体肋骨线
framesticks 夹立板
framestone 骨架岩;生物构架石灰岩
frame stop 框子挡块;框架挡销
frame storage 帧存储
frame store 帧存储
frame structure 构架结构;帧结构;框架结构
frame superstructure 上部框架结构
frame support 构架支座;框式支架
frame supported shear wall 框支剪力墙
frame suspended motor 构架悬挂电动机;底座悬挂式电动机;底架悬挂电动机
frame sway mechanism 框架倾斜机理
frame sweep unit 帧扫描装置
frame synchronization 帧同步
frame synchronization code 帧同步码
frame synchronization control 帧同步控制
frame synchronization logic 帧同步逻辑
frame synchronizer 帧同步器
frame synchronizing pulse separator 帧同步脉冲分离器
frame synchronizing signal 帧同步信号
frame system 框架系统;构架结构
frame table 框表
frame theory 框架理论
frame thrust 框架推力
frame tilt 帧倾斜;帧频锯齿波补偿信号
frame timber 肋骨角材;肋材;模板框架横木
frame time 帧周期;成帧时间
frame time base 帧扫描;图像扫描电路
frame timing 帧计时
frame-to-frame correlation 帧间相关性
frame-to-frame differences 帧间差
frame-to-frame jitter 镜头间颤动
frame-to-frame jump 帧间跳动
frame-to-frame linear prediction 帧间线性预测
frame-to-frame playback 逐帧直接再现
frame-to-frame response 帧间响应
frame-to-frame step response 帧间跳动响应;帧间阶距
frame-to-frame variation 帧间变化
frame transfer 帧转移;帧传递
frame-tube 框筒
frame-tube structure 框架-筒体结构
frame-tube system 框筒结构体系
frame type bridge 框架桥;刚架(式)桥
frame-type building 框架建筑
frame-type core box 框架式芯盒
frame type filter 板框式压滤机
frame type finder 框式取景器
frame type foundation 框架式基础
frame type house 框架式房屋
frame type of construction 框架建筑形式
frame-type switchboard 骨架式配电盘
frame unit 框架单元
frame wall 框架墙
frame-wall structure 框架墙结构
frame waveform 帧信号波形
frame weir 框架式堰
frame wheel carrier 备用轮架
frame with three hinges 三铰框架
frame(wood)construction 木构架建筑(物)
frame work 体制;机壳;骨架(构);构筑物;基础工作;构架(工程);框架(工程);桁架(结构);外墙
framework analogy 构架模拟;构架类比
framework bay 框架跨度;框架间距
framework bogie 构架式转
framework box 框架木箱
framework calculation 框架计算

framework construction 构架建筑
framework construction method 构架施工法
Framework Convention on the Conservation of Climate 保护气候框架公约
framework filter 骨架过滤器
framework for environment action 环境行动纲领
framework method 构架法;拱架法
framework motif(pattern) 架状基型
framework of body 车身构架
framework of competition 竞争结构
framework of control 控制网[测]
framework of faults 断裂体系;断裂格架;断裂系统
framework of filter 过滤器骨架
framework of fixed points 控制网【测】
framework of flooring 楼层骨架
framework of geodetic control 大地控制网
framework of steel reinforcement 钢筋骨架
framework plan 规划大纲;框架计划
framework silicate 网硅酸盐
framework silicate structure 架状硅酸盐结构
framework space 骨架空间
framework timber rail 构架木栏杆
framework tube 骨架管
framework wall 隔墙
Fram floor 防火楼板(一种专利产品)
framing 帧型调整;骨架;构造框架;构架;框架;桁架结构;图像定位;成帧
framing anchor 骨架锚固件;连接锚板
framing bit 帧指示器;分帧位
framing camera 分帧照相机;分幅照相机;分幅摄影机
framing chisel 粗木工凿
framing clip 框架夹片
framing column 框架支柱;构架支柱
framing component 框架构件
framing control 帧调整;居中调节;成帧调节
framing disable 成帧阻塞
framing drawings 安装图说明书
framing error 成帧错误
framing indication 帧指示
framing lens 分幅透镜
framing lumber 构架材
framing magnet 成帧磁铁
framing mask 限帧框;图像边框
framing member 框架构架;框架构件;构架件
framing of images 影像成帧
framing of port mouth 小炉口框架
framing photography 分幅照相术;分幅照相法
framing plan 构架平面图;构架平面布置;结构平面图;船体结构图;船体构架图
framing post 构架柱
framing scaffold 脚手架
framing set 框式支架
framing sheet pile 有撑架的板桩;框架板桩
framing signal 定位信号
framing square 构架(直)角尺;木工角尺
framing stair(case)well beam 楼梯井框架梁
framing steel 结构钢
framing structure 空间结构;构架结构
framing system 框架系统;构架系统;肋骨结构
framing table 构架台;构架平台;矿泥分选盘
framing thermal imager 分幅热像仪
framing timber 建筑木材;构架木材
framing window pulse 成帧窗孔脉冲
France flexible pavement design method 法国柔性路面设计法
francevillite 钒钡铀矿
Franchet lustre 弗朗舍光泽彩料
franchise 相对免税率;允差;专营权;经营特许权;经销权;免赔率;免赔额;特许(权);特性经营权
franchise bond 特许权保证书
franchise chain 特许连锁
franchise clause 免赔(率)条款;免赔额条款;保险免赔率条款
franchise collection 有证垃圾收集箱;特许收集
franchised dealer 特约经销商;特许零售商
franchise deal 特许交易
franchise distribution network 特许销售网
franchised shop 特许商店
franchised store 特许商店
franchisee 特许被授予人
franchiser 特许授予人
franchise stamp 特许邮票
franchise system 特许制;特许体系

franchise tax 特许税;特许权税
franchising 特许;出卖产销权
franchising arrangement 特许安排
franchisor 特许授予人
Franciscan church 弗朗西斯教堂
Francis flow formula 弗兰西斯堰流公式
Francis formula 弗朗西斯公式
Francis reversible pump turbine 轴向辐流可反转的水泵水轮机
Francis turbine 轴向辐流式涡轮机;轴向辐流式透平;轴向辐流式水轮机;辐向轴流式水轮机;弗兰西斯水轮机
Francis type runner 弗兰西斯式水轮机转轮
Francis water turbine 弗兰西斯水轮机;轴向辐流式水轮机
Francis weir formula 弗兰西斯溢流堰公式
Francis wheel 弗兰西斯水轮
francium 钫
Franck-Condon factor 弗兰克—康登因子
Franck-Condon overlap factor 弗兰克—康登重叠因数
Franck-Condon potential curve 弗兰克—康登势能曲线
Franck-Condon principle 弗兰克—康登原理
franckeite 辉锑锡铅矿
Franck-Hertz experiment 弗兰克-赫兹实验
Franco 弗兰可高速钢
franco 运费准免;免费
francoanellite 磷铝钾石
franco boarder 边境交货条件
franco border 边境交货价格条件
franco frontier 边境交货价格条件
franco invoice 全部费用在内发票;法国式发票
Franconian stage 弗朗康阶【地】
Francon interference ocular 弗朗康干涉目镜
franconite 水泥钠矿
franco quay 目的港码头交货价
franco rendu 目的地买方指定地点交货价
Franc zone 法郎区
frangibility 易碎性;脆性;脆度
frangible 脆的
frangible coupling 易卸接头;易分离耦合;截断连接
frank 免税邮寄;免税递送;免费通行
frankdicksonite 氟钡石
franked income 免税收益;免税所得
Frankfort black 植物炭黑;木炭黑;德国黑颜料;法兰克福黑(颜料)
Franki displacement caisson 富兰基(式)位移沉箱
frankincense 结晶松香;乳香
frankincense oil 蓝刃油;乳香油
franking 加强(榫跟件);斜角接头;印标志;盖章;盖邮戳的;盖印
Franki pile 富兰基式(现场)灌注桩;现场灌注桩基桩
Franki pile driver 富兰基(混凝土)打桩机
Franki tube 富兰基管
Frankland's method 富兰克法
Franklin antenna 富兰克林天线
Franklin centimeter 富兰克林厘米
Franklin equation 富兰克林方程
Franklinian 富兰克林阶【地】
Franklinian geosyncline 富兰克林地槽【地】
franklinic electricity 摩擦电
franklinite 锌铁尖晶石
Franklin lifebuoy 铜壳救生圈
fransmission-type radio isotope ga(u)ge 透明型放射同位素测量计
fransoletite 水磷铍钙石
franzinite 弗钙霞石
Franz Keldysh effect 弗朗兹—凯尔迪什效应
frap 用索缚紧
frapping line 止荡绳
frapping turns 扎骨
Fraras breaking point 弗雷拉斯脆点
Fraras breaking point test 弗雷拉斯断裂点试验
Frary metal 弗雷里(铅—碱土金属轴承)合金
Frasch process 弗赖什采磺法
Fraser's air sand process 弗雷泽风砂分选法
Fraser Sweatman pin safety system 弗雷泽斯韦特曼氏针形安全系统
Frasnian stage 弗拉斯阶【地】
frater for lay brethren 修道院世俗斋堂
frater house 寺院食堂;大食堂
frater lay brethren 杂役僧侣斋堂
fraternity house 临时寄宿

Fratol 氟乙酸钠
fratry 寺院食堂
Fraude's reagent 高氯酸
fraudulent gain 非法收入
fraudulent income tax returns 虚假的所得税申报书
fraudulent misrepresentation 欺诈性误述(船租)
fraudulent sale 诈骗性买卖
fraudulent sale practice 诈骗性销售营业
fraueneis 透明石膏
Fraunhofer approximation 夫琅和费近似
Fraunhofer binary hologram 夫琅和费二进制全息图
Fraunhofer condition 夫琅和费条件
Fraunhofer corona 夫琅和费日冕
Fraunhofer diffraction 夫琅和费衍射
Fraunhofer diffraction and spectral analysis 夫琅和费衍射和频谱分析
Fraunhofer diffraction formular 夫琅和费衍射公式
Fraunhofer diffraction fringe 夫琅和费衍射条纹
Fraunhofer diffraction hologram 夫琅和费衍射全息图
Fraunhofer diffraction pattern 夫琅和费衍射图
Fraunhofer diffraction region 夫琅和费衍射区
Fraunhofer doublet 夫琅和费双线
Fraunhofer grating 夫琅和费光栅
Fraunhofer hologram 夫琅和费全息图
Fraunhofer holography 夫琅和费全息照相
Fraunhofer intensity distribution 夫琅和费光强度分布
Fraunhofer lines 太阳光谱线;夫琅和费谱线
Fraunhofer line spectrum 夫琅和费线光谱
Fraunhofer region 夫琅和费区
fraxetin 白蜡树内酯
fraxinellone 白蜡树酮
Fraxinus excelsior 欧洲白蜡树
fraxitannic acid 白蜡树皮单宁酸
fray 磨损处;磨损(帆布);擦散(磨损);(绳子等)擦断
frazil 潜冰;屑冰;冰屑;冰晶
frazil ice 脆冰;松冰团;水内小冰块;水内冰;底冰;初冰;冰片
frazil ice forecast 初冰预报
frazil ice frazil 针冰;(河流的)底冰
frazil river 挟屑冰河流
frazil slush 冰屑(泥)浆;冰花
frazil stream 挟屑冰河流
freak drought 畸形干旱
freak ovbservation 反常观察值
freak range 不稳定接收区
freak result 反常结果
freak storm 反常暴风
freak wave 异常波;畸形波
freboldite 六方硒钴矿
freckle defect 误置缺陷
Frederick sburg Series 弗雷德里克斯堡统【地】
Fredholm determinant 弗雷德霍姆行列式
Fredholm integral equations 弗雷德霍姆积分方程
Fredholm operator 弗雷德霍姆算子
Fredholm theorem 弗雷德霍姆定理
Fredholm theory 弗雷德霍姆理论
free 游离;自由(的);正赞后受지;免费;单体的
free Abelian group 自由交换群
free acceleration test 自由加速试验;轻便快速测试装置
free access 自由进入
free access floor 活地板;活动地板
free acid 游离酸
free acidity 游离酸度
free action 自由作用
free admission 免费入场券
free admittance 自由导纳
free aerostat 自由高空气球
free agent 免费代理程序
free aid 不规定用途的援助
free air 自由空气;自由空间;自由大气;大气
free air anomaly 自由空气异常;重力自由空间异常;空间异常
free air capacity 自由排气能力;大气排气量
free air cogeoid 空间归算的调整大地水准面
free air condition 自由空气条件
free air correction 自由空气校正;自由空气改正;自由空间校正
free air correction value 自由空间改正值
free air delivery capacity 自由气生产能力

free air diffuser 进气道;进气扩散道
free air displacement 自由排气能力
free air dose 自由空气剂量
free air dryer 自然通风干燥器
free air facility 空气动力试验设备
free air flow 自由气流
free air gravity anomaly 自由空气重力异常;海平重力异常
free air ionization chamber 自由空气电离室
free air overpressure 自由空气超压力
free air pressure 自由空气压力
free air reduction 正常大气归算;空间改正
free air space 自由气隙
free air temperature 自由大气温度
free air temperature ga(u)ge 气温计
free air test 大气中试验
free air thermometer 大气温度计
free air turbulence 大气湍流
free-air value 大气值
free alkali 游离碱
free allocated-in 无偿调入
free allocated-out 无偿调出
free allowance 免费额
free alongside(at)quay 码头交货
free alongside ship 启运地船边交货价;船边交货
free alternating current 自由振荡电流;自由交变电流
free alternation 自由振荡
free alumina 游离氧化铝
free ammonia 氨气
free ammonia and saline ammonia 游离氨和盐氨
free aperture 自由孔径
free aquifer 自由含水量;自由含水层
free arc 自由弧
free area 自由(流通)面积;有效面积;管道过水面积;空闲方
free area list 自由区表
free area routine 自由区程序
free articles 免税货物
free ascent 自由上浮
free-ash coal 无灰煤
free-ash coke 无灰焦
free asphalt 游离(地)沥青
free assets 自由资产
free association 自由联想
free associative algebra 自由结合代数
free associative ring 自由结合环
free-astray 误运免费送至到达站
free atmosphere 自由空气;自由大气(层);大气
free atom 单体原子
free at quay 码头交货价(格);码头交换价(格)
free attenuation vibration 自由衰减振动
free at wharf 码头交货价格条件;码头交换价格条件
free available chlorine 游离性有效氯
free available residual chlorine 游离性有效余氯
free axis 虚轴;自由轴
free axle 自由轴;活动(支承)轴
free baggage allowance 免费行李;免费随带行李
free balloon 自由气球
free bar 自由杆
free base 游离碱
free beam 自由梁;简支梁
free bearing 自由座;铰座;球形支座
free bend ductility 自由弯曲延展性
free bending moment diagram 自由弯矩图
free bend test 自由弯曲试验
free berth 空泊位
free black 粉末炭黑;飞扬性炭黑
free block 自由块;空闲块
free block queue element 自由块队列元素
free-blow air conditioner 自由通风式空调机组;自由吹送式空调器
free blowing 自由吹制成型;无模人工吹制
free blown 自由吹制
free-blown glass 自由吹制玻璃;无模人工吹制玻璃
freeboard 超高(指超出水面);舱面出水高度;干舷高(度);干舷;地面至车架底面的净空;出水高度;保护高;保护板
freeboard and room 免费伙食与住宿
freeboard assignment 干舷勘划
freeboard certificate 载重线证书;干舷证书
freeboard coefficient 干舷系数
freeboard deck 干舷甲板
freeboard depth 干舷深度

freeboard depth ratio 干舷型深比
free boarder 边境交货条件
freeboard length 干舷长度
freeboard level 超高水位
freeboard mark 载重线标志;干舷标志
freeboard of caisson 沉箱干舷高度
freeboard of channel 明渠干舷
freeboard of lock wall 闸墙超高
freeboard paint 干舷漆
freeboard ratio 干舷比
freeboard regulation 干舷规则
freeboard storage (水库)超高库容;超高部分蓄水
freeboard zones 干舷区;干舷界限
free body 自由体;孤立体;隔离体;分离体
free body balance equation 区域平衡方程
free body diagram 自由体图解;自由体受力图;孤立体图;分离体图;隔离体图
free bollard 活动系船柱
free Boolean algebra 自由布尔代数
free border 边境交货价格条件
free boundary 无荷载边界;自由边界
free boundary electrophoresis 自由界面电泳
free boundary problem 自由边界问题
free bound factor 自由束缚因子
free box wrench 活套筒板手
free breakage 自然崩落
free bulkhead 活动挡土墙;活动舱壁;活动岸壁
free burden 自由面崩落岩(石)层
free-burning arc 明弧
free-burning coal 易燃煤;非粘结煤;不结焦煤
free burning fire 自由蔓延火
free burning mixture 自燃混合气
free burning rate 自由燃烧率
free calcium hydroxide 游离氢氧化钙
free calcium oxide (水泥中)游离氧化钙;游离石灰
free calcium oxide content 游离氧化钙含量
free cannon pinion 自由分轮
free cantilever casting 悬臂浇注;悬浇
free cantilever erection 悬臂拼装;悬拼
free capital 游资
free carbide 游离碳化物
free carbon 游离碳;自由碳;单体碳
free carbon dioxide 游离二氧化碳
free cargo 免税货物
free car park 免费停车场
free carrier 向运送人交货条件;自由载流子;货交承运人
free carrier absorption 自由载流子吸收
free carrier concentration 自由载流子浓度
free caving 易坍塌的
free cementite 游离渗碳体;自由渗碳体
free-center-type clamp 线路释放线夹
free-central placentation 特立中央胎座式
free chain complex 自由链复形
free channel 明渠;无闸坝航道;畅通无阻的航道
free charge 自由电荷;免费
free charge transfer 自由电荷传递
free chloride 游离氯
free choice of goods 选购商品;选购品
free circulation 自由循环
free city 自由市;自由城市
free-clamped end condition 自由嵌固端条件;一端固定一端自由条件
free-coasting payload 惯性飞行时的有效荷载
free column volume 自由柱容
free commodities 免税商品;免税品;免税货物
free commutative group 自由交换群
free competition 自由竞争
free complex 自由复形
free component 自由分量
free consumption entry 免消费税输入申报单
free contract 目的地交货合同
free contraction 自由收缩
free contraction joint 自由收缩缝
free control interval 空闲控制区间
free convection 自由对流;自然对流;热力对流;热对流
free convection boiling 自然对流沸腾
free convection boundary layer 自由对流边界层
free convection cooling 自然对流冷却
free convection heat transfer 自然对流传热
free convection number 格拉肖夫数
free convertibility 自由兑换
free cooling shrinkage 自由冷缩

free core pool 自由存储区
free corner 不连角隅
free crest of spillway 自由溢流堰顶;无闸堰顶;不没式溢流堰顶
free cross-section 自由截面
free cross-section(al) area 自由截面积
free cross slide 自由横滑板
free crushing 自由压碎;自由破碎
free currency 自由货币
free current operation 自然励磁
free cursor 软线跟踪头
free curve 自由曲线
free cutting 自由切削;高速切削;无支承切割
free cutting bit 自磨钻头
free cutting brass 易切削黄铜;自由切削黄铜
free cutting bronze 快削青铜
free cutting machinability 易切削性
free cutting steel 易切削钢;自由切削钢;高速切削钢
free cutting steel wire 易切削钢丝
free cyanide 游离氰化物
free cyclic(al) group 自由循环群
free cylindric(al) vortex 自由柱状漩涡
free damped 自由衰减的
free damping 自由衰减
free damping oscillation vibrometer 自由衰减振荡示振仪
free decomposition 自由分解
free deformation 自由形变
free degree 自由度
free delivered 已准出舱;准许交货的;目的地交货价(格);目的地交货
free delivery 免费邮送;免费送货;免费交货;免费递送;无条件交货
free delivery-type air conditioner 自由吹送式通风空调器
free delivery-type air conditioning unit 自由吹送式空调机组
free delivery-type air cooler 自由吹送型空气冷却器
free demonstration videotape 免费的示范录像带
free depot 空闲的仓库
free depreciation 自由跌价;任意折旧计算
free derivative 自由导数
free despatch 免付速谴费
free diffusion 自由扩散
free digging rate 自然挖取效率;正常挖取效率
free discharge 自由泄流;自由流量;自由流出;自由放电;自由出流;免费卸载
free discharge valve 自由泄流阀
free discharging jet 自由喷射流
free dispach 免费发送
free dispatch 免付速谴费;免费递送
free dissolved electrons concentration 溶解自由电子浓度
free distance 空隙;间隙
free distributive lattice 自由分配格
free diving 自由潜水
free dock 码头交货
freedom 自由性;自由度
freedom degree 自由度
freedom drom jamming 抗干扰性
freedom form vibration 抗振性
freedom from bias of the instrument 仪器的抗偏差性
freedom from concrete (混凝土)无气泡
freedom from corrosion 无腐蚀;不腐蚀
freedom from cracking(and crazing) 无龟裂;不裂缝
freedom from defects 无缺点;无疵病
freedom from distortion 无畸变
freedom from ground connection 无地线;不接地
freedom from odo(u)r 无气味
freedom from slip(page) 无滑动
freedom from taxation 免税数
freedom from warpage 无翘曲;无扭曲
freedom memorial 自由纪念碑
freedom of contract 签约自由;契约自由
freedom of creep 无蠕变
freedom of entry 进入市场的自由;参加市场自由
freedom of motion 运行自由度;运动自由度;自由度
freedom of navigation (国际河流或公海上的)自由通航权;航行自由
freedom of open sea 公海自由;(航行权)
freedom of rectifier 纠正仪自由度
freedom of seas 公海自由;(航行权)

freedom of the high sea 公海自由;(航行权)
freedom of the seas 公海航行权
freedom of trade 贸易自由
freedom to trade 贸易自由
free drainage 自通排水;自流排水;天然排水
free draining 自流排水;天然排水;排水性能良好的
free draining model 自由穿流模型
free draining molecule 自由穿流分子
free drawbar 铰接牵引杆
free drift method 自由转换法
free drop 自由空投物;跌水
free drop catch 自由下降挡
free drop concrete mixing plant 自落式砼搅拌机;自落式混凝土搅拌机
free drop mixer 自落式搅拌机
free earth support 下端自由支承
free eccentric wheel 顺车偏心轮
free echoes 自由回声
free economic zone 自由经济区
free edge 自由边(缘);毛边;无支承边
free edge diaphragm 自由边缘膜片
free education 公费教育
free efflux 自由射流;自由出流
free electric(al) charge 自由电荷
free electricity 自由电荷
free electrolyte 游离电解质;自由电解质;分离电解液
free electromagnetic field 自由电磁场
free electronc concentration 自由电子浓度
free electrophoresis 自由电泳(法)
free element 单体元素
free elevator 可移动升降梯
free end 活动支座;悬空端;自由端;暴露面;游离端
free end bearing 松端支承;自由支承;自由端支承;简单支承;伸出支座轴承
free-end bridge 单端固定桥
free end column 自由端柱
free end pile 桩顶自由桩
free end-play 自由轴向窜动
free end spring 自由端游丝
free end travel 自由端位移
free end triaxial test 自由端三轴试验
free energy 自由能(量);热力势
free energy change 自由能变化
free energy criterion 自由能判据
free energy curve 自由能曲线
free energy functional 自由能泛函
free energy of formation 生成自由能
free energy of melting 熔化自由能
free energy scale 自由能标度
free enterprise system 自由企业制度
free entries 免税品
free entry 免税进口申报单;免税报单
free environment 自由环境
free-escape 太平门
free escapement 自由式擒纵机构
free exchange rate 自由浮动汇率
free expansion 自由膨胀
free expansion honing 定压珩磨
free expansion joint 自由伸缩接头
free face 自由面;直坡;临空面;暴露面
free fall 自由下落;自由投放;自由落料;自由降落
free fall acceleration 自由下降加速度;自由落体加速度
free fall aerator 自由跌落曝气器
free-fall boring 钢绳冲击钻进
free fall corer 自由下落取样管;自由下落取心管
free-fall drill 钢绳冲击(式)钻机
free faller 自由落体
free fall hammer 自由落锤
free falling 自由坠落
free falling apparatus 冲击设备
free falling body experiment instruments 自由落体实验装置
free falling catch 自由下落挡
free falling classifier 自由下沉式分级机;自由沉降分级机
free falling velocity 自由沉降速度
free fall law 自由落体定律
free fall lifeboat 自由落水救生艇
free fall method 自由降落法
free fall mixer 阶梯式拌和机;自由下落式拌和机;自由下落搅拌机;自由式搅拌机
free fall model 自由落体模型

free fall of ram 自由落锤
free fall rocket core sampler 自动下沉采泥器;自动上浮采泥器;推进上浮取样管
free fall time 自由下落时间
free fall type mixing drum 自由下落式拌和鼓
free fall velocity 自由落体速度;自由降落速度
free fall weir 自由溢流堰;自由式溢流堰;自流式溢流堰
free fatty acid 游离脂肪酸
free fatty acid crystal 游离脂酸结晶
free fed 自由进料
free feed 自由进料
free feeding 自由给料
free ferrite 游离铁素体
free fiber 自由棉
free field 自由(音)场;自字段;备注栏
free field calibration 自由场校准
free field chamber 自由场室;消声室
free field characteristic 自由场特性曲线
free field condition 自由场条件
free field correction 自由场校准
free field correction curves 自由场修正曲线
free field current response 自由场电流响应
free field current sensitivity 自由场电流灵敏度
free field design spectrum 自由场设计谱
free field format 自由段格式
free field frequency response 自由场频率响应
free field ground motion 自由场地面运动
free field motion 自由场运动
free field of sound 自由声场
free field particle velocity 自由场质点速度
free field power response 自由场功率响应
free field power spectrum 自由场功率谱
free field reciprocity calibration 自由场互易校准
free field response 自由场响应
free field room 自由场室;消声室
free field sensitivity 自由场灵敏度
free field storage 自由字段存储
free field stress 自由场应力
free field voltage response 自由场电压响应
free field voltage sensitivity 自由场电压灵敏度
free film 游离漆膜;非支撑膜
free filter 自由滤子
free fit 自由配合;松配(合)
free flame 自由火焰;活火头
free flexural wave 自由弯曲波
free float (工程进度上的)自由浮动;协调工作(建筑施工充分配合各项工种,抢时间开工)
free floating gate 自由式浮箱门
free floating growth 自由浮动生长
free floating platform 自由浮动平台
free floating suction dredge(r) 自航耙吸式挖泥船
free flooded transducer 自由浸沉式换能器
free flooding ballast 自由灌水舱
free flooding irrigation 淹灌;自由漫灌
free flooding opening 开式舷孔;开式排水孔
free flooding tank 与海水相通的水舱
free floor 活动铺板
free floor of bridge 桥面活动铺板
free flow 自由(水)流;明流;无压流
free flow cargo 流散性货物
free flow condition 自由流态
free flow conduit 无压管道
free flowing 自由流动;自重流动;高流动性
free flowing black 无尘炭黑
free flowing bore 涌泉井;自流井
free flowing bulk material container 粉状货集装箱
free flowing castable 自流浇注料
free flowing channel 明流渠(道);无压水槽;明渠渠道;明流槽
free flowing granules 自由流动颗粒;无尘颗粒
free flowing lock 开通闸
free flowing material 流动性材料;松散材料
free flowing powder 自由流动粉末;无尘粉末
free flowing product 流动性物料
free flowing river 原态河流;无坝河流
free flowing steam sterilization 常压蒸汽消毒;常压蒸汽灭菌
free flowing stream 原态河流
free flowing stretch 无闸坝河段
free flowing traffic 畅行车流
free flowing well 自流井
free flow layout 畅流型方案
free flow of motor 自由车流

free flow operating speed 自由行驶车速;自由行车速率
free flow speed 自由车速
free flow tunnel 明流隧洞;无压隧洞;无压隧道
free flow tunnel development 明流隧洞(引水)式电站
free flow valve 易流阀;自由流动阀
free flow viscosimeter 自由流动黏度计
free flow weir 自由溢流堰
free fluid index 自由流体指数
free fluidity concrete 高流动度混凝土
free foehn 自由焚风
free-for-all 对公众开放的
free foreign agency 国外代理免费
free foreign exchange 自由外汇
free forging 自由锻造
free form 自由形式;不定格式
free format 自由格式
free format input 自由格式输入
free format source code 自由格式源代码
free form coding 自由型编码
free form data 自由形式数据
free form design 自由式造型设计
free forming 自由成型
free-form operation 不定格式操作
free-form roof structure 自由形式的屋顶结构
free-form surface 自由曲面
free forwarding station 发送站交货;发送站费用已付
free-free beam 悬臂简支梁
free frequency 自由频率;固有频率
free from acid 无酸的;不含酸的
free from all average 一切海损均不赔偿;全损险;全损赔偿
free from average 海损不保;不包括海损险
free from chlorides 无氯化物
free from commitment 不承担义务的
free from corrosive substances 无腐蚀物的
free from damage 损价不保;不包括损坏
free from damage on the machined surface 机加工面无损伤
free from dust 无尘的;不含粉尘的
free from duty 税费不保;不包括税费
free from encumbrance 无债务纠纷
free from error 无误差的
free from extreme deformation 无极端变形
free from flaw 无裂缝
free from general average 共同海损不保
free from glare 无眩光
free from incumbrance 无债务纠纷
free from particular average 平安险;单独海损;不担保单独海损
free from particular average absolutely 单独海损绝对不保
free from rats 无鼠的
free from scale 无氧化皮的;无铁鳞的
free from volume increase 无体积膨胀
free function 自由函数
free game 自由博弈;无规博弈
free gas 游离瓦斯;游离气;自由气(体)
free gas cap 游氣气顶
free gas saturation 自由气饱和率
free gear 空套齿轮
free-glass clinker 无玻璃体熟料
free gliding 自由滑动
free gold 游离金
free goods 免征进口税的货物;免税品;免税(进口)货物;免费商品
free graphite 游离石墨
free grid 自由栅(极);浮置栅极
free ground water 自由潜水;无压地下水;非承压地下水;自由地下水
free ground water table 无压地下水位
free groupoid 自由广群
free guide vane 自由导叶;失控导叶
free gyroscope 自由(度)陀螺仪;三自由度陀螺仪
free hand 自由行动;脱扣手柄
free hand curve 徒手绘曲线
freehand drawing 手绘图;单图;徒手画(的草图);示意图;草绘(稿图)
freehand field map 野外手图
freehand geologic(al) profile 随手地质剖面
freehand grinding 手持磨削;手工研磨;手工磨刻
free handle 自由跳闸;活动把手

free handle breaker 自动断路器
freehand line 徒手画的线
freehand motion 随手移动
freehand profile 手勾剖面;信手剖面图
freehand shaping 手工成型
freehand sketch 徒手素描;手(工)绘草图;徒手(画的)草图;手勾草图
freehand sketching 随手画草图
freehand tool holder 悬臂架
free hanging loop 自由悬挂活套
free harbo(u)r 自由港
free harmonic vibration 自由谐振;自由和谐振动
free haul 免费运输;免费运价;免费搬运;免税运距;免费运送;无偿运距;不加偿运程
free-haul distance 免税运距;免费运距
free-haul traffic 免费运输
free-haul yardage 免费土方运输;不加运费的土方量
free head 自由水头;无压水头
free headroom 净空高(度)
free health service 免费医疗
free heat 自由热
free height 自由高度;净空高(度)
free height of blade 叶片有效高度
free height under the bridge 桥下净空高度
freehold 永久产权;世袭不动产;自由土地保有权;不动产所有权
freeholder 世袭地的保有权;不动产的所有权;自由土地保有人
freehold flat 自产共管公寓
freehold property 自由保有地产
free hole 自由空穴
free hole absorption 自由空穴吸收
free honing 无支撑珩磨
free house 免费运送到家;自由售酒商店
free hydraulic jump 自由水跃
free hydrochloric acid 游离盐酸
free hydrochloric acid determination 游离盐酸测定
free hydrogen 游离氢
free hysteresis 灭磁滞
free impedance 自由阻抗;短路输入阻抗
free import 免税进口
free importation 免税进口
free imports 免税进口货
free in 船方不承担装货费;舱内交货价格
free in and out 自埋装卸;船方不承担装卸费;不管装和卸
free income tax 免付所得税
free incribing 自由内接
free index 自由指标
free inertial flow 自由惯性流
free inflow 自由进流
free influx 自由进流
freeing 解脱
freeing of the nappe 水舌脱离;射流分离(现象)
freeing pipe 排水管;排气管;逸气管
freeing port 舷墙排水口;排水口【船】
freeing preparatory 初步除气
freeing scuttle 排水门;放水口
freeing tanks of gas 清除气油
freeing wheel 修整轮
free in harbo(u)r 港内交货
free input-output mode 自由输入输出方式
free instrument 自浮海底取样机
free insurance 免费保险
free into barge 交到驳船价格
free into bunker 交到船上燃料舱价格
free investment 自由投资
free ion 游离离子
free iron 游离铁
free iron in soil 土壤中游离铁
free isotope 游离同位素
free issue 自由发行
free jet 自由射流;自由喷流
free jet blast plant 无遮喷丸清洗装置
free jet chute 自由射流斜槽
free jet nozzle 自由射流喷嘴
free jet stream 自由射流
free jet turbine 自由射流式涡轮机
free-jet-type turbine 射流式水轮机
free-joint 万向节;万向接头
free jump 自由水跃
free knot 自由结点
free lamp circuit 示闲灯电路

free-lance 个体职业者
free-lance architect 独立营业的建筑师
free lane change 自由变换车道
free lateral acceleration 自由横向加速度
free lattice 自由格
free length 自由长度
free level 自由水位；无压水平面
free level tunnel 明流隧洞；无压隧洞；无压隧道
free lever swinging socket wrench 活柄活动套筒扳手
free levitation method 自由悬浮法
free lift 自由升力；自由起升高度
free lifting force 自由上升力
free-lift mast 自由上升桅杆
free-lift mast forklift truck 自由上升杆铲车；推杆铲车
free lime 游离石灰
free line 空线
free line net 自由式路网
free line signal 空线信号
free linkage 浮动悬挂装置
free lipid 游离类脂物
free liquid 自由液体
free list 自由表；免税商品目录；免税商品表；免税进口货单；免税货物(明细)表；免税货名表；免税货单；免收进(出)口税的货物单；免费入场名单；海关免税品(目录)
free living form 自由生活类型
free living generation 自生世代
free loan 无息贷款
freely adapted Gothic manner 自由式哥特建筑风格
freely adapted style 自由风格
freely articulated (ground) plan 明晰的平面图
freely burning fire 自燃火灾
freely convertible currency 自由兑换货币
freely falling body 自由落体
freely floating exchange rate 自由浮动汇率
freely floating pontoon 自由浮动沉箱；自由浮式驳船；自由浮动沉箱式闸门；滑动型沉箱(式闸门)
freely floating system 自由浮动汇率制度
freely flowing 自由流动的
freely-hanging plummet 悬挂垂球
freely humic acid 游离腐植酸
freely market system auction 自由市场制拍卖
freely movable bearing 自由移动支承；自由滑动支承；活动支座
freely moving carriage 自由活动托架
freely moving traffic 畅通的交通
freely supported 自由支承的；简支的
freely supported beam 自由支承式梁；简支梁
freely supported beam bridge 简支梁桥
freely supported end 自由端
freely supported structure 简单支承结构
freely suspended charge 自由悬挂装药
free machining 高速切削
free-machining steel 高速切削钢；易切削钢
free magnesia 游离氧化镁
free magnetic charge 自由磁荷
free magnetic pole 自由磁极
free magnetism 自由磁性
free magnetization condition 自由磁化状态
free maintenance 免费保修期
free maintenance period 保修期(缺陷责任期)
Freeman-Nicbols roaster 弗里曼-尼科耳型焙烧炉
free market 自由市场；农贸市场
free market economy 自由市场经济
free market price 自由市场价格
freemason 毛石圬工
free massing 群体的自由组合
free meander 自由曲流
free meandering channel 自由蜿蜒河道
free medical care 公费医疗；免费医疗
free medical service 公费医疗；免费医疗
free medical treatment 免费医疗
free memory selector 空存储选择器
free metal 游离金属
free metal ion 游离金属离子
free mill 自由粉碎机
free-milling 易选的
free-milling gold 易汞齐化金
free-milling ore 易选矿石
free mobility 自由可动性
free module 自由模

free moisture 骨料表面湿度(未经吸收的湿度，自由水)；游离水分；自由水分；自由湿气；非结合水
free moisture of aggregate 骨料表面湿度
free molecular flow 自由分子流
free molecular flow force 自由分子流作用在物体上的力
free molecule diffusion 自由分子扩散
free molecule flow 自由分子流
free moment 自由力矩
free monoid 自由独异点
free monomer 单体
free mooring arrangement 游离系泊装置；活动系泊装置
free motion 自由运动
free motional impedance 自由动态阻抗
free motion of a liquid 自由流
free motion of a stream 自由流
free mouth 开敞河口；(无拦门沙的)敞开河口
free movement 自由运动
free movement of vessel on supports 支座上的容器活动不受压缩
free-moving traffic 无阻碍交通；畅行交通；畅行车流；畅通的交通
free-moving traffic capacity 畅行交通量
free multilateral foreign trade 自由的多边贸易
free nappe 自由水舌；通气水舌
free nappe weir 自由水舌溢流堰
free needle survey 罗盘测量
freeness 自由度
freeness number 排水度数
freeness tester 打浆度测试器
free net (work) 自由水网
free network with rank deficiency 秩亏自由网；秩亏网
free objective 可伸缩物镜
free occurrence 自由出现
free of address 免运费回扣(指租船)；免付委托佣金；免付洽租佣金
free of all average 一切海损不保；只赔全损；免去海损估价；全损赔偿
free of all average and salvage charges 一切海损及救助费均不负责
free of blowholes 无气泡
free of bubbles 无气泡
free of capture, seizure, riots and civil commotion 掳获、捕获、暴动和内乱不保
free of cavitation 无空蚀；不产生空蚀的
free of charges 盈利；免税；免票；免费
free of cost 免费奉送
free of cracks 无裂缝
free of customs 免付关税
free of damage 损害不赔
free of damage absolutely (船体)损坏不赔
free of duty 免税进口；免税；免付关税
free of duty entry 免税通过
free of embrittlement 不脆化、不脆变
free of foreign capture 国外虏获不保
free of franchise clause 免贴额条款
free of freightage 不包括运费
free of frost 无霜
free of general average 共同海损不赔
free of heart center 无心材
free of income tax 免所得税；免纳所得税
free of interest 免息；免收利息；不计利息
free of knots 无节疤
free of losses 无损失；无损耗
free of particular average 平安险
free of particular average English condition 英国条件平安险
free of pollutant mass 污染物团释放；污染物释放
free of reported casualty 据称伤亡不保；不负申报海滩费
free of riots 暴动不保
free of riots and civil commotions 暴动和内乱险不保
free of seizure 捕捉险不保
free of shrinkage cracks 无收缩裂纹
free of stamp 免印花税；免付印花税
free of stress 无应力
free of tax 免税
free of turn 不按到港顺序
free on board 积载费及平舱费在内的离岸价格；船上交货价格；船上交货；离岸价(格)；就船交货价格

free on board airport 出口地机场交货价
free on board and commission 代理费在内的力岸价格
free on board cars LCL [less than carload lot] 开往指定出口港货车上交货价
free on board destination 目的地交货价(格)；目的地交货
free on board harbo(u)r 起运港船上交货价格
free on board in harbo(u)r 港内船上交货价；起运港船上交货价格；发货地价格
free on board invoice 交货至船上发单
free on board liner terms 按班轮条件船上交货价
free on board plane 飞机上交货条件；飞机上交货价
free on board plant 工厂出厂价
free on board point-of-production pricing 离岸价格生产地定价法
free on board pricing 按船上交货定价
free on board quay 码头上交货价；起运码头船上交货价；启运港码头交货
free on board shipping point 寄发地交货；起运地点交货价格；启运点交货价
free on board stowed 离岸价包括理舱费；船上货包括堆装价格；舱底交货
free on board trimmed 平舱费用在内的船上交货价
free on board unstowed 离岸价不包括理舱费
free on car 卡车交货价(格)
free on carriage paid to 运费付至……
free oneself 解脱
free on lighter 驳船交货价格
free on place 飞机上交货价
free on quay 码头上交货；码头交换价(格)
free on rail 火车上交货(价)；铁路上交货价；铁路交货(价)；铁路付款交货；铁路费用已付
free on ship 船上交货价格
free on side 目的港船上交货
free on steamer 汽船交货价格条件；启运港船边交货(价格)；船上交货价格
free on train 火车上交货(价)；铁路敞车交货
free on truck 卡车交货价(格)；货车上交货；敞车交货(价格)；车上交货(价格)
free on vessel 船上交货
free on wagon 货车上交货价
free-on-wheels capacity (汽车起重机的)抬起轮胎的起重量
free-open-stack system 自由开架式
free-open-textured 结构松散的；松散结构的
free-open-textured sand 松散砂
free orifice 自由出流孔
free oscillating system 自由荡动系统
free oscillation 自由振荡
free oscillation of the earth 地球的自由振荡
free oscillation period 自由振荡周期
free out 船方不负担卸货费用；船方不承担卸货费
free outfall 自由溢流(口)；自由出入的河口；开敞河口
free outflow 自由流动；自由出流
free outlet 自由出流(口)
free outline 自由轮廓
free overboard 输入港船边交货价格；到港价格
free overfall 自由溢水门；非淹没水门；跌水；自由溢流；自由降落
free overfall crest 自由溢流堰顶
free overfall gate 自由溢水门
free overfall jet stilling basin 自由溢流式消力池；自由溢流式静水池
free overfall weir 自由溢流堰
free overflow 自由溢流
free overflow jet 自由溢水射流；自由降落水舌
free overflow weir 滚水坝
free overside 船上交换价格；目的港船上交货价；输入港船边交货价格；到港价格
free oxides test 游离氧化物试验
free oxygen 单体氧
free pass 免票
free passing of cars 允许超车；自由超车
free path 自由行程
free pattern 自由图案
free pay 免税数
free pendulum 自由摆
free-pendulum clock 肖特钟
free penstock 自由(支承)式压力钢管
free perimeter 免税区；免税地带
free period 自由周期；自然周期；免费期间
free permittivity 固有电容率

free phenol 单体酚
free piercing 自由冲孔
free pile 自由桩
free pilotage 任意引航制
free piston 自由活塞;气举活塞
free-piston air compressor 自由活塞式空气压缩机
free piston compressor 自由活塞压缩机;自由活塞式压缩机
free piston drive sampler 自由活塞打入式取土器
free piston engine 自由活塞式发动机;自由活塞发动机
free piston gas generator 自由活塞气体发生器
free piston gas generator engine 自由活塞煤气发生炉发动机
free piston gas generator turbine 自由活塞煤气发生炉透平
free piston gasifier 自由活塞式气体发生器;自由活塞燃气发生器
free piston gas turbine 自由活塞式燃气轮机;自由活塞式燃气轮机装置
free piston generator 自由活塞式发电机;自由活塞式煤气发生器
free piston installation 自由活塞装置
free piston machinery 自由活塞机械
free piston plant 自由活塞装置
free piston pump 气举活塞泵
free piston pumping 气举活塞泵送
free pole 自由磁极
free pool 空闲库;空闲池
free pore space 自由孔隙
free porphyrin 游离卟啉
free port 自由贸易港;自由港;免税口岸;排水口;无税港口
free position 空档(位置);任意状态
free potential grid 闲栅;自由电位栅极
free pouring 自由流动的
free pouring density 自由堆积密度
free power turbine 自由涡轮
free pratique 检疫证(书)
free precession magnetometer 自由旋进磁力仪
free pressing 直接加压
free price 自由价格
free product 自由积
free progressive wave 自由行波;自由前进波
free publicity 免费宣传
free public transportation 免费公共交通
free pulley 空转轮
free punch 无导向凸模
free quasi-particle approximation 自由准粒子近似
free radiation 自由辐射器;自由辐射
free radiator 自由辐射器
free radical 游离基;自由基
free radical addition 游离基加成
free radical mechanism 游离基机制
free radical polymerization 游离基引发聚合
free radical reaction 自由基反应
free radical substitution 游离基取代
free range 自由放牧场
free recoil 自由后座
free recoil mount 自由后座炮架
free recombination 自由重组
free redemption 自由偿还制度
free reeds 自由簧片
free reserve 自由准备金
free residual chloride 游离性余氯
free residual chlorination 游离性余氯化
free residual chlorine 游离余氯;自由剩余氯
free resolution 自由分解
free resource 免费资源
free retaining wall 基底稍斜倾的挡土墙;自由式挡土墙;安全挡土墙
free revolving action 自由旋转作用
free rheological law 自由流变定理
free ride 自由通过
free rider 免费享用公共货物者
free riding 免费乘车
free right turn 随时右转
free ring 自由环
free river 无闸坝河流;畅流河道
free river reach 无闸坝河段
free road 不收费道路
free road system 自由式道路系统
free rocking 自由摇摆
free roller gate 滚轮闸门;辊轮闸门;活动大门
free rolling 自由横摇
free rolling action 自由滚动作用
free rolling test 滑行试验
free rolling tyre 自由轮轮胎
free rolling wheel 空转轮
free rotation 自由旋转
free rotation of bogie 转向架自由转动
free rotor gyroscope 自由转子陀螺仪
free rotor gyro-stabilized inertial reference platform 自由转子陀螺稳定的惯性参考平台
free routing 自由路径选择
free run 自由滑动距离
free running 自由振荡;自由行驶;自由航行;不加荷运转
free running bin capacity 料仓能容量
free running blasting 粉状炸药
free running blocking generator 自激间歇振荡器
free running blocking oscillator 自激间歇振荡器
free running circuit 自运转电路;不同步电路
free running differential 自激差速器
free running fit 自由配合;轻配合;轻转配合
free running frequency 自然频率;固有频率
free running juice 自流液计
free running laser 自激激光器
free running laser mode 自由振荡激光模式
free running local synchronizer oscillator 自激本机同步振荡器
free running multivibrator 自激多谐振荡器
free running nozzle 自流式水口
free running of sand 泥沙的自由流动
free running piston 自由运转的活塞;无推杆活塞
free running revolutions 空车转数
free running sawtooth generator 自激锯齿波发生器
free running scan 自激扫描
free running silo capacity 自由控制的筒仓容量
free running slag 易流渣
free running speed 空转转速;空转速度;平衡速度;无载转速;定常速度
free running sweep 自激扫描
free running velocity 空转速度
free sailing ship model 自由航行船模
free sailing test 自由航行试验
free sample 免费样品
free sampling 任意抽样
free sand blasting 无遮喷砂法
free scale photoplan 自由比例尺像平面图
free schema 自由模式
free scheme 自由图式
free school of architecture 义务建筑学校
free screening 自由筛选;自由筛分;自由屏蔽
free section 可拆部分
free sedimentation 自由沉降
free semi-group 自由半群
free service 免税服务;免费服务
free service and repairs 免费服务和修理
free settling 自由沉降;自然沉降;易沉降的
free settling classifier 自由沉降分级机
free settling hydraulic classifier 自由下沉水力分级机;自由沉降水力分级机
free settling particle 自由沉降颗粒
free settling ratio 自由降落系数
free settling tank classifier 自降分级器
free settling tube 自由沉降管
free sheet of water 水舌(堰口);水的自由流动层
free shellac 散片紫胶
free silica 游离二氧化硅
free silicic acid 游离硅酸
free silicon dioxide 游离二氧化硅
free site 简单现场;独立工地
free size 自由尺寸;补偿尺寸
free slab length 自由板长度
free software 自由软件
free soil 松散土(壤);疏松土(壤)
free solution electrophoresis 自由溶液电泳(法)
free sound field 自由声场
free space 自由空隙;自由空间;空闲空间
free space attenuation 自由空间衰减
free space blasting 自由空间爆破
free space correction 自由空间改正
free space diagram 空间图
free space field 自由空间电场
free space field intensity 自由空间(电)场强度
free space impedance 自由空间阻抗
free space loss 自由空间损耗
free space path loss 自由空间路径损耗
free space pattern 自由空间辐射图
free space power 自由空间功率
free space propagation 自由空间传播
free space propagation condition 自由空间传播条件
free space radar equation 自由空间雷达方程
free space radiation pattern 自由空间辐射方向图
free space range 自由空间搜索距离;自由空间中的作用距离
free space wave 自由空间波
free space wavelength 自由空间波长
free span dome 自由跨距圆顶
free spectral range 自由光谱区
free speed 自由速率
free spillway 自由(式)溢洪道;开敞式溢洪道
free spill weir 自由式溢流堰
free spiral vortex 自由螺旋形漩涡
free spooling (卷筒)自由缠绕
free spring 自由簧;无约束游丝
free staff 精木料
free stail barm 不拴系牛舍
free-standing 悬臂的;独立结构;不需拉杆的;独立式
free-standing access 独立式出入口
free-standing assembly hall 独立式会堂
free-standing awning 独立遮篷
free-standing (bath) tub 四周不靠墙的浴盆;独立式浴盆
free-standing bell tower 独立钟楼
free-standing block 一组独立式房子;独立式住宅;独立式大楼;独立家屋
free-standing building (经营单一业务的)商业建筑;独立式建筑;独立式房屋
free-standing cantilever 独立式悬臂桥墩
free-standing chaitya (hall) 独立式岩窟寺院会堂
free-standing chapel 独立式小教堂
free-standing chimney 自立式烟囱;独立(式)烟囱;独立(大)烟囱
free-standing column 独立柱
free-standing cooker 独立式炉灶
free-standing cubicle 自立式柜
free-standing fuel element 自立型燃料元件
free-standing fuel tank 日用油箱;日用油柜
free-standing gravity davit 轻型重力式艇架
free-standing heater 独立式供暖器
free-standing masonry wall 独立式砌筑墙;独立的圬工墙
free-standing mast 无绷绳固定的桅杆
free-standing new town 独立新城
free-standing partition (wall) 自立式瓷砖隔墙
free-standing pile 独立桩
free-standing shaft (具有大烟道的)独立式烟囱
free-standing steel scroll 自由支承式钢蜗壳
free-standing support 独立支柱
free-standing surface silo 独立式地面圆仓;独立式地面筒仓
free-standing surge tank 独立式调压塔;自由支承调压塔
free-standing tower structure 独立式塔形结构
free-standing wall 独立式屏蔽墙
free state 游离状态;自由状态;自由态
free statement 释放语句
free steering method 自由执行法
free steering sequence 自由执行序列
free stem 优先靠码头
free stock 共砧
freestone 易切岩;易切石料;易劈岩;乱石;毛石;软性石;粗石
freestone masonry 毛石圬工;毛石砌体
free storage 空闲存储器
free storage area 自由存储区
free storage block 自由存储块
free storage list 自由存储表;空闲存储表
free storage of articles 免费存放物品
free storage period 免费堆存期;免费保管期
free store stack 自由存储栈
free stream 自由气流;无闸坝河流
free stream direction 自由气流方向
free stream dynamic(al) pressure 自由流的动压力
free stream flow 自由流;无扰动水流;迎面流
free streamline 自由流线
free streamline theory 自由流线流动理论;自由边界流动理论

free stream Mach number 自由流马赫数
free stream pressure 自由流压力
free stream static pressure 自由流静压力
free stream total head 自由气流总压头
free stream turbulence 自由流紊动;自由流湍流
free stream velocity 自由气流速度;自由流动速度
free strip 自由带
free stroke 自由行程
free stuff 软性木材;无疵材;净料;精木料
free style 自由式
free style road system 自由式道路网
free subsidence 自由下沉
free sulfur 单体硫
free support 自由支座;活动支座
free surface 临空面;游离面;自由液面;自由面;自由表面;潜水面(指地下水)
free surface amplification 自由表面放大(作用)
free surface correction 自由液面修正
free surface culvert 无压涵洞
free surface effect 自由液面效应;浅浸效应
free surface energy 液面自由能量;自由表面能;表面张力
free surface flow 自由面流;无压流;自由面水流;重力流;明渠流;明渠道流
free surface flow channel 自由面水流渠;明渠流渠道;明渠
free surface flow line 自由面水流线;重力流管道
free surface flow tunnel 自由面水流隧道;重力流隧洞;渠流隧洞
free surface instability 自由面不稳定性
free surface kinematic condition 自由面运动状态
free surface moisture 自由表面水分
free surface of flow 水流自由表面
free surface of ground 地表
free surface of liquid 自由液面
free surface of scum chamber 浮渣室自由表面
free surface pump 自由液面泵
free surface resistance 自由液面阻力
free surface seepage 无压渗流
free surface slope 自由水面比降
free surface tunnel 自流式隧道
free surface turbulent flow 自由面紊流
free surface velocity 自由面速度
free surface vortex 自由面涡漩
free surface wave 自由表面波
free surge 自由涌浪
free surplus 任意拨盈余;未指拨盈余
free swell 自由膨胀
free swelling index 自由膨胀指数;自由膨胀序数
free swelling rate 自由膨胀率
free swell test 自由膨胀试验
free swing 自由摆动
free swinging flail 自由摆动式甩刀
free swinging knife 铰接刀;甩刀
free swing period 自由摆动周期
free swiveling hydrodynamic(al) fairing 自由旋转导流罩
free symbol 自由符号
free symbol sequence 自由符号序列
free tar 游离柏油
free task control block 自由任务控制块
free term 自由项
free term of baseline condition 基线条件自由项
free term of coordination condition 坐标条件自由项
free term of side condition 极条件自由项
Free test 弗里试验
free-thaw 冻结熔化
free ticket 免票
free tidal wave 自由潮汐波
free time 预备时间;空闲时间;免税期;免费时间;免费(堆存)期;非工作时间
free time of cargo 货物免费保管时间
free to contract and expand 自由胀缩
free top 简支顶盖
free topological algebra 自由拓扑代数
free topological group 自由拓扑群
free torsion(al) vibration test 扭转自由振动试验
free tower 跳伞塔
free track 闲置车道
free trade 自由贸易;贸易自由
free trade agreement 自由贸易协定
free trade area 自由贸易区
free trade wharf 自由贸易港口码头

free trade zone 自由贸易区;进出口加工区
free traffic 自由行车;畅通无阻的交通;畅通的交通
free transformation group 自由变换群
free translation 意译
free transmission 空行程传动
free transmission range 自由传输范围
free travel (履带式车辆)间隙;自由运行;免费旅行
free travel(l)ing wave 自由行波
free traverse 自由导线【测】;不闭合导线【测】
free trial 免费试用
free trimming 不负责平舱费
free trip 自由跳闸;自动脱扣
free turbine 自由涡轮
free turbine design 自由涡轮设计
free turbine engine 自由涡轮燃气轮机
free turbulent shear flow 自由剪切紊流
free turning (船舶)自由掉调
free-type subsurface hydraulic pump 液压活塞潜水泵
free ultrafilter 自由超滤子
free union 自由并集
free unwinding 自由退解
free valence 自由价【化】
free valve mechanism 活动阀机构
free variability 自由变异性
free variable 自由变元;自由变数;自由变量
free variation 自由变异;自由变分
free vector 自由向量;自由矢量
free vector space 自由向量空间
free vehicle bottom water 自动式底层水取样器
free vehicle bottom-water sampler 自由式底层水取样器;自动底层水取样器
free ventilation 自由通风
free venting 自由通风
free vertical strain 自由竖向应变
free vibration 自由振动;自由振荡
free vibration characteristic 自振特性
free vibration characteristic parameter 自振特性参数
free vibration column test 自振柱试验
free vibration frequency 自振频率
free vibration method 自由振动法
free vibration mode 自由振动形式
free volume 自由体积
free volume theory 自由体积理论
free vortex 自由漩涡;自由涡流
free vortex blade 自由涡流叶片
free vortex blading 自由漩涡叶片
free vortex flow 自由涡流
free vortex sheet 自由涡流面
free wall 自由壁
freeware 免费软件
free warehouse 普通仓库
free waste weir 自由水溢流堰;开敞式溢流堰
free water 游离水;自由液面水;自由(流动)水;重力水;开敞水面;不受约束的水
free water/cement ratio 自由水灰比
free water clearance 自由水清除率
free water content 过剩水(水泥砂浆水量过多);自由含水量
free water damage 自由液面水损害
free water effect 自由液面水影响
free water elevation 自由水位;自由水高程;地下水位
free water in wood 木材的自由水
free water level 自由水面;自由水位
free waters 开敞水域;天然水域;敞开水域
free water surface 自由水面
free water surface correction coefficient 自由液面水修正系数
free water surface evapo(u)ration 自由水面蒸发
free water surface graph 自由水面坡线图
free water surface slope 自由水面比降
free water surface system 自由水面系统
free water table 自由潜水面;自由(地下)水位;无压地下水位
free waterway 畅通无阻的航道
free wave 余波;余摆波;自由前进波;自由波
free wave absorption coefficient 自由波吸收系数
free wave pattern 自由波型
freeway 快车道;高速公路;高速干道;高速道路;快速(干)道;超速干道
freeway bridge 应急桥
freeway frontage road(way) 高速干道辅助道路

freeway green belt 高速公路绿化带
freeway interchange 高速公路互通式立交
freeway operation 高速公路管理
freeway overpass 立体交叉桥
freeway ramp 高速公路坡道;高速公路匝道
freeway surveillance 高速公路监控;高速干道监视
freeway system 高速公路系统
freeway-to-freeway interchange 高速干道互通式立体交叉
freeway traffic 高速公路运输
freeway transportation 高速公路运输
free weir 不受淹没堰;自由堰;自流式溢流堰
free wheel 惯性滑行;自由(转)轮;飞轮
freewheel assembly 自由组合件;自由轮装配
free wheel device 自由轮离合器;超越离合器
free wheeled vehicle 无轨车辆
freewheeling 空程;自由轮传动;单向转动
freewheeling brake 惯性滑行制动器
freewheeling clutch 活转离合器;单向离合器;空程离合器;超越离合器;自由回转离合器
freewheeling drum 自由绕线盘;自由轮绕筒
freewheeling mechanism 空转机构
freewheeling roller clutch 自由回转滚柱式离合器
freewheeling system 自由轮体系
freewheel mechanism 自由回转机构
free wind 顺风
free wool 纯净羊毛
free working 自由开挖面;自由工作面;可采的;开采;回采空间
free world trade 世界自由贸易
free wrench handle 棘轮扳手
free yardage owing defects 标疵扣尺
freezable water 可冻结水
freeze 滞塞;烧焊;冻结
freeze box 防冻箱
freeze casting 冷冻注浆;冷冻浇注法
freeze casting method 冰冻铸法
freeze chemical method 冷冻化学法
freeze coagulation 冻结
freeze concentration 冰冻凝结浓缩
freeze condensation 冷冻浓缩(法)
freeze desalination 冷冻脱盐
freeze desalting process 冷冻脱盐法
freeze dried 冻干的
freeze drier 冷冻干燥器;冷冻干燥机;冻干装置
freeze drying 干冻(结);冷冻干燥(法);冻干
freeze drying food 冻干食物
freeze drying lyophilization 冷藏干燥
freeze drying process 冷冻干燥加工;冰冻干化法
freezed storage room 冷冻间
freeze etching 冷冻刻蚀(术);冻结腐蚀;冻蚀法;冰冻蚀刻
freeze-etching technique 冰冻蚀刻法
freeze formation 冰冻地层
freeze free period 解冻期;不冻期
freeze hole 冻结孔
freeze in 卡住;卡钻;冻住
freeze-in impurity 凝入杂质
freeze injury 冻害
freeze mode 冻结状态;保持式
freezen damage 冻害
freeze-out gas collector 气体冻干捕集器;冻干气体捕集器
freeze-out sampling 冻干采样
freeze over 全面冻结;封冻
freeze pipe 冰冻管
freeze plug 防冻塞
freeze point 卡钻部位;冰点
freeze point indicator 钻具卡住部位指示器
freeze-proof 抗冻性;抗冻的;(防冻(性)
freezer 致冷器;制冷机;冷却器;冷凝器;冻冷器;冷冻间;冷冻机;冻结设备;冻结器;冻冻间;冻结柜;电冰箱;低温冷库;冰箱
freezer burn 冻结器冻伤
freezer drum 制冷器鼓筒
freeze resistance 耐寒性;耐冻性;防冰冻性
freezer-free period 解冻期
freezer locker 冷藏库小室;冷藏柜
freeze road 冻板路
freezer room 冷冻间
freezer ship 冷藏船
freeze sinking 冻结凿井法
freeze-thaw 冻融
freeze-thaw action 冻融作用

freeze-thaw condition 冻融条件
freeze-thaw consolidation 冻融固结
freeze-thaw cycling 冻融循环
freeze-thaw damage 冻融破坏
freeze-thaw durability 冻融耐久性
freeze-thaw durability test 冻融耐久性试验
freeze thaw protection 冻融保护
freeze-thaw resistance 抗冻融性(能)
freeze-thaw stability 冻融稳定性
freeze-thaw stabilizer 冻融稳定剂
freeze-thaw stable 耐冻耐熔;冻融稳定的
freeze-thaw technique 冰冻解冻法
freeze-thaw test 冻融试验
freeze-up 封冻;冰塞
freeze up date 冻结日期;封冻日期
freeze-up days 封冻日数
freeze-up forecast 冻结预报;封冻预报
freeze-up period 封冻期;冰封期
freezing 冷凝;冷冻;结冰(法);凝固;冻结;封冻;冰冻
freezing action 凝结作用;冰冻作用
freezing agent 制冷剂;冷冻剂
freezing and drying 冰冻干燥法
freezing and thawing 冻融
freezing and thawing count 冻融循环次数
freezing and thawing cycle 冻融循环
freezing and thawing test 冻融(循环)试验;冻解试验
freezing apparatus 冷藏器;冻结装置
freezing board 冻结板
freezing borehole 冻结钻孔
freezing borehole pattern 冻结钻孔布置
freezing breakage point of insulating gel 绝缘胶冻裂点
freezing by expansion 膨胀凝固
freezing cargo 冷冻货(物)(冷藏船装运30华氏度以下货物)(30华氏度≈-1摄氏度)
freezing carrier 冷货运输船
freezing chamber 冷冻室;冷藏室;冰冻室
freezing coefficient 冷冻系数
freezing cofferdam 冻结式围堰
freezing constant 凝固点降低常数
freezing(construction)method 冻结施工法
freezing curve 冷却曲线;凝固曲线;冰冻曲线
freezing damage 冻害
freezing days of surface water 地表水冻结天数
freezing depth 冰冻深度;冻层深度
freezing drizzle 雨凇(下降时呈液态,着地后就冰冻);过冷毛毛雨;冻雾雨;冻毛毛雨
freezing drying microtomy 冷冻干燥切片法
freezing duration 冻结持续时间
freezing effect 冻结效应
freezing equipment 冻结装置;冻结设备
freezing factory ship 冷冻加工船
freezing fog 雾凇;冻雾
freezing force 冻结力
freezing goods 冷冻货(物)(冷藏船装运30华氏度以下货物)(30华氏度≈-1摄氏度)
freezing heave 冰冻隆起
freezing height 冻结高度
freezing hole 冻结孔
freezing in 凝入
freezing index 冻结指数;冰冻指数
freezing injury 冻伤;冻害;冬害
freezing injury of plant 植物冻害
freezing in reaction 凝入反应
freezing interface 凝固界面;凝固分界面
freezing isoline 等冻结线
freezing level 结冰高度;凝结高度;凝固高度;凝固程度;冻结高度;冻结程度;冰冻线(标高)
freezing-level chart 冻结高度图
freezing level in soil 土壤中的冰冻标高
freezing line 冰冻线
freezing locker 冷藏库
freezing machine 冷凝机;冷冻机
freezing mechanism 凝固机理
freezing-melting compression strength 冻融抗压强度
freezing-melting resistance 耐冻融性
freezing method 凝固法;冷冻凿井法;冻结法;冻固法
freezing method of sinking foundation 土壤下沉基础冻结法(用于流砂地基等);土壤冻结下沉基础

freezing method of tunnel(l)ing 隧道冻结施工法;隧道冻结法施工
freezing microtome 冷冻切片机;冻结切片机;二氧化碳冷冻切片机
freezing mixture 冷却剂;冷凝剂;冷冻剂;冷冻混合物;冻结混合物;冰冻(合)剂
freezing nucleus 冻结核;凝固核
freezing of a furnace 结炉
freezing of assets 资产冻结
freezing of drill exhaust 凿岩机排气管的冻结
freezing of lines 管路冻结
freezing overburden 覆盖层冻结法(钻进)
freezing period 结冰期;冻结期;冻季;冰冻期
freezing pipe(line) 冻结管(路)
freezing plant 冷冻厂
freezing point 析出点;冷凝点;冻冷点;凝固点;(结)点;冰点
freezing-point curve 凝固点曲线
freezing-point depressant 冰点降低剂;降冰点外加剂
freezing-point depression 凝固点降低;冰点下降度;冰点降低
freezing-point lowering 冰点降低
freezing-point of liquid 液体冰点
freezing-point test 冰点试验
freezing precipitation 降下冻雨或冰雾雨;冻降水
freezing process 冷冻法;冻结过程
freezing process for shaft sinking 冰冻法凿井
freezing processing 冻结工艺;冻结法
freezing prove technique 防冻技术
freezing rain 雨凇(下降时呈液态,着地后就冰冻);过冷雨;冻雨
freezing range 凝固区间
freezing rate 凝固速度
freezing reaction 凝固反应
freezing resistance 耐寒性;耐冻性;耐冬性
freezing room 冷冻间;冷冻舱;结冻间
freezing salt 冷冻盐;粗盐
freezing sampler 冻结取土器;冻结取样器
freezing season 结冰期;上冻期;冻结期;冻季;害;冰冻季节
freezing shaft 冻结法开凿的竖井
freezing solidity 冻结法
freezing speed 冻结速度
freezing spray 冻结的喷雾
freezing stage 冷冻台
freezing state of spring 泉水冻结状态
freezing stratum 冻土层
freezing stress 冻结应力
freezing stress method 冻结应力法
freezing temperature 冷凝温度;凝固温度;凝固点;冻结温度;冰点
freezing test 冷冻试验;抗冻性试验;耐寒性试验;人工低温试验;冻结试验;冰点测定;冰冻试验;冰点(的)测定
freezing thaw durability 冻解久性
freezing-thawing cycle 冻融循环
freezing-thawing test 冻融试验
freezing time 凝固时间;冻结时间
freezing time of spring 泉水冻结时间
freezing tunnel 隧道式冻结间
freezing unit 冷冻装置
freezing(water)level 封冻水位
freezing weather 冰冻天气
freezing with liquid nitrogen 低温液化氮方式
freezing zone 冻土层;冻结带;冰冻带
free zone 自由区;自由领域;进港自由区;免税区;免收关税地区
freibergite 银黝铜矿
Freiburg Cathedral 弗莱堡大教堂
Freidel-Crafts catalyst 弗瑞迪-克莱福特催化剂
freieslebenite 柱硫锑铅银矿
freight 运输;运费;装载吨(1装载吨=40立方英尺);海运运费;水脚;舱底交货的到岸价格(1立方英尺≈0.028立方米)
freight, insurance paid to 运费和保险费付到
freight absorption pricing 津贴运费订价法
freight account 运费账单;运费置清单;装货清单
freightage 租运;货运价;货运费;货运;船货
freight agent 运行行;运货代理商;转运公司;货运代理行;货运代理人
freight agreement 运费协定
freight all kind rate 综合运费率
freight all kinds 不分品种运价;不分货种运输

freight all kindsrate 同一运费率;同一费率
freight amount 运费额;货运量
freight and cartage 水陆运费
freight and commission 货价加运价和佣金;货价加运费和佣金
freight and demurrage 运费和延滞费
freight and miscellaneous charges 运杂费
freight area 货运区
freight at destination 到付运费;到达付运费
freight at risk 有风险的运费
freight base 运费标准
freight base station 货运基地车站
freight basis 运费制
freight berth 货船泊位
freight bill 运货单;运费清单;运费单;装货清单
freight boat 运货船
freight broker 运费经纪人;运输纪人;运货经纪人
freight by measurement 容积运费
freight by plane 保险加空运费价格
freight by weight 计重量运费
freight canvasser 揽货员
freight car 卡车;行李车厢;运货卡车;货车
freight cargo 货物
freight car inspection depot 货物列车检验所【铁】
freight car kilometers 货车公里
freight car repairing depot 站修所【铁】
freight carriage 货运
freight carrier 货船
freight car scale 货车轨道衡
freight car technical handing-overpost 车辆技术交接所
freight center 货运中心
freight charges 运费;货运价;货运费
freight charges for water transport 水运运费
freight charges on tapering basis 按递远递减计算的运费
freight circulation center 货物流通中心
freight circulation trip 货物流通出行
freight claim 运货索赔
freight classification 货种
freight classification yard 货运分类堆场
freight clause 运费条款;运费支付条款;租船支付条款
freight clerk 押运员;货载管理员
freight collect 运费由提货人支付;运费由提货人负担;到付运费;报关行
freight collision clause 碰船货损条款
freight conference 运价公会
freight consolidation building (集装箱码头的)装箱拆箱库
freight container 货运集装箱;货箱
freight container for international trade 国际集装箱
freight container traffic 集装箱运输
freight control computer 货运控制计算机
freight conveyer 货运输送机
freight cost 运费
freight department 运输处
freight depot 货(运)站;货栈;货物车站
freight elevator 运货电梯;货运电梯;运货升降机;货物升降机;货梯;载货电梯
freight equalization 平衡运费到价
freighter 租船人;货主;运船;散货船;承运人
freighter aircraft 货运飞机
freight ex ship's hold 船舱底交货的到岸价格
freight flow 货流
freight flow diagram 货流图
freight flow draft 货流图
freight flow drawing 货流图
freight flow statistics 货流统计
freight flow survey 货流调查
freight flow volume 货流量
freight forward 运费由提货人支付;运费由提货人负担;到付运费;报关行
freight forwarder 运输经纪人;运输代理行;运输代理人;运输代办人;货运代理行;货运代理人;承运人
freight free 免运费;免收运费
freight goods 货物
freight handler 装卸公司;装卸工(人)
freight handling 货物搬运
freight handling area 装卸操作区
freight handling facility 装卸设备;货物装卸设施

freight handling plan 货物装卸计划
freight haulage 货运;货物运输
freight haulage plan 货物运输计划
freight house 货仓;货栈;货房
freight house canopy 货仓棚
freight index 运价指数
freight in full 全包运费
freighting 海上运输契约
freighting voyage 运货航次
freight insurance 运费保险;成本加运费保险费价格
freight invoice 货票
freight inward 进货运费
freight label 货物标记
freight land terms 卸货费在内的到岸价格条款
freight lift 运货电梯;货运电梯;货物起卸机;货物电梯
freight liner 货运列车;货(运)班轮;货运班车;定期直达列车;集装箱火车;货柜火车
freight liner service 定期直达列车运输
freight liner train (火车)货运班车
freight list 运价表;运费(明细)表;货物清单
freight locomotive 货运机车
freight manifest 载货表清单
freight market 运输交易市场;货运市场
freightment 货运交付业务
freight note 运费账单;运费(清)单;装货清单
freight of all kinds 铁路综合运价
freight office 货运办公室
freight of goods 货物运输
freight of shipping 水运费
freight on board destination 目的地交货价(格)
freight on inter branch transfers 分店间送货运费
freight operation at originated station 货物发送作业
freight-oriented railway network 货运铁路网
freight-out 销货运费
freight out and home 往返运费;出口和回程货物
freight outward 销货运费
freight paid 运费已付;运费付讫
freight paid in advance 预付运费
freight payable at 交付运费地点
freight payable at destination 货到付运费;到付运费
freight penalty 运费罚金
freight platform 装卸台;货运站台;货物站台
freight platform hoist 平台式货物提升机
freight point to point cost 货运点到点成本
freight policy 运费保险单;海运保险单
freight prepaid 运费预付
freight railroad 货运铁路
freight railway 货运铁路
freight rate 运价(率);货物运价(率);运(输)费率
freight rate of water transport 水运运价
freight rate structure 货运运价结构
freight rebate 运费回扣
freight receipt 运费收据;海上货运单
freight release 交货许可证;提单限额超装许可证
freight rocket 货载火箭
freight section 货位
freight securing value charges 货物保价费
freight service 货运业务
freight shed 货棚;仓库
freight ship 运货船;货轮;货船
freight special line 货运专线
freight station 货运车站;货站;运送站;货物站
freight steamer 货船
freight stock 货车
freight tariff 运价表;运货价目表;运费费率表;货物运价表
freight terminal 货运(总)站;货运终点站;货运中转基地;货运枢纽;货运码头
freight to be collected 运费待收
freight to be paid after discharge at destination 货到埠卸载时付给运费
freight to be paid at destination 运费到付
freight to collect 运费到付;运费待收;货到收运费;到货运费
freight ton-kilometer 货物周转量;货物吨公里
freight ton(nage) 载货吨(位);载货量;计费吨;货运吨;体积吨;运费吨
freight track [美]货运线;货运线
freight traffic 货运量;货运量;货运(交通);货物运输
freight traffic accident 货运事故
freight traffic facility 货运设施

freight traffic flow 货流
freight traffic line 货运线
freight traffic manager 货运经纪人
freight traffic only line 货运专线
freight traffic plan 货物运输计划
freight traffic revenue 货运收入
freight traffic statistical index 货物运输统计指标
freight traffic statistics 货运统计
freight traffic unit cost 货运单位成本
freight traffic volume 货运交通量
freight train 运货列车(美国);货运列车;货物列车;货车【铁】
freight train formation plan 货物列车编组计划
freight train locomotive 运货机车
freight train path 货物列车运行线
freight trains grouping plan 货物列车编组计划
freight train transfer increased revenue 货车中转加给收入
freight transfer platform 货物转运站台
freight transfer station 货物中转站
freight transit speed 货物送达速度
freight transport 货物运输
freight transport contract 货物运输合同
freight transport expenses 货运支出
freight transport expenses on every kilometre 单元货运支出
freight transport loss and indemnity rate 货物运输损失赔偿率
freight transport plan 货物运输计划
freight transport settlement 货运清算
freight transport turnover 货物运输周转量
freight truck 运货卡车;货车转向架;货车
freight turnover 货运周转量;货运装卸;货物周转量
freight unit 货运计算单位
freight vehicle 运货(汽)车;货运汽车;货运车(辆);货车
freight volume 运量;货物运输量
freight volume of water transport 水运运量
freight volume of waterway transportation 河运运量
freight wagon 货车
freight yard 港区货场;货运调车场;货运场;货场;堆货场(地)
freirinite 砷钙钠铜矿
freize of the Greek Doric order 希腊陶立克式壁缘
fremitus 震颤
Fremont impact test piece 弗雷蒙冲击试验片
fremontite 钠磷锂铝石
Fremont test 富利蒙特冲击试验
Fremy's salt 弗里米盐
french 法兰西
French-American mid-ocean undersea study 法—美大西洋中部海底研究(计划)
French arch 法式拱;法兰西式平拱;法国式拱
French architecture 法国式建筑;法国建筑;法式建筑
French blue 佛青;法国蓝
French bolt 法兰西螺栓
French building 法国式房屋
French casement 法式玻窗;对开落地长窗;玻璃门;落地窗;玻璃落地窗
French chalk 滑石粉;滑石
French chalk surfacing 滑石粉镀面;滑石粉表面加工
French Chippendale 法国齐本德尔式家具
French classicism 法兰西古典主义
French coil 并列排绳法
French Community 法兰西共同市场
French curve 曲线规;曲线板;铁道弯尺
French degree 法国度
French door 法国式双扇玻璃门;玻璃(落地)门;法式(两用)门
French door lock 落地窗锁
French drain 干砌石排水沟;乱石盲沟;毛石排水沟;盲沟;填石盲沟;石砌排水沟
French drainage 盲沟排水
French embossing 多种深度结合的蚀刻
French escutcheon 法式锁孔盖板
French fake 并列排绳法
French flier 法(国)式梯级;转折梯段的踏步
French folio 稿纸
French formal garden 法国巴洛克式庭园
French gardens 法国园林
French gold 铜锌锡合金

French Gothic(style) 法国哥特式建筑风格
French green energy plan 法国绿色能源计划
French grey 浅灰色
French groove 法国式书脊槽
French hardware 法式装修五金
French interlocking roofing tile 法国式扣搭屋面瓦
French joint 法国式书脊槽
frenchman 勾缝溜子;勾缝刀;接头修整工具
French marigold 金盏花;万寿菊
French method roofing 斜铺板材屋面;菱形铺瓦屋面;(板材屋面的)菱形(斜向)铺法
French nail 法国半圆头钉
French nail with round head 圆头销钉;圆头钉
French National Institute of Industrial Property 法国工业产权局
French Normandy (铺木瓦斜屋顶与塔楼的)法国建筑型式
French ochre 法国赭色;法国赭石
French(oil of)turpentine 法兰西松节油
French order 法国柱式
French polish 紫胶清漆;罩面光漆;泡力水;法国抛光漆;虫胶醇溶液;法国罩光漆;法国擦亮剂;清漆打磨光面工艺
French polishing 虫胶清漆;法式抛光
French process oxide 法国法氧化锌
French pusley 半边莲
French red 红彩
French Renaissance 法国文艺复兴式
French Romanesque 法兰西仿罗马风格
French roof 折线式屋顶;复折(式)屋顶;法(国)式屋顶
French roofing tile 法式屋瓦
French sash 落地长窗;落地(铰链)窗;铰链窗
French scaffold 移动式脚手架;移动脚手架
French scarf(joint) 法国式斜嵌接;法国式斜接接头
French scarf(joint)with wedge 法国式斜口接合;法国式楔形斜面接合
French scroll 法国涡饰
French shank 法式球形柄
French shroud knot 三股绞结
French slating 菱形铺瓦屋面
French spindle 法式木工造型机具
French spirit of turpentine 法兰西松节油
French spring 法式弹簧
French Standard 法国标准
French standard thread 法国标准螺纹
French stuc 人造假石;粉浆拉毛饰面
French stucco 粉浆拉毛饰面;人造假石面;法国粉浆饰面
French tile 马赛克;法国槽瓦
French truss 法(国)式桁架;芬克式桁架
French turpentine 法国松节油
French type claw hammer 法式羊角锤
French type machinist's hammer 法式钳工锤
French varnish 虫胶漆;揩涂清漆;擦涂清漆;法国清漆
French verones green 维罗纳绿
French whipping 反手结编扎法
French white 净白铅;滑石粉;法国白
French window 法式窗;玻璃门;落地(长)窗
French window lock 落地窗锁
Frenet-Serret formulas 弗雷涅-塞雷公式
Frenkel defect 弗兰克尔缺陷
Frenkel disorder 弗兰克尔位错
Frenkel exciton 弗兰克尔激子
Frenkel mixer 弗兰克尔螺旋搅拌机
Frenkel pair 弗兰克尔对
Frenkel type 弗兰克尔型
frenulum 下系带;系带;连接柱
Freon 氟利昂;氟氯烷致冷剂;氟冷剂
freon ammonia refrigeration equipment 氟氨冷冻机械
freon brine refrigerating installation 氟利昂盐水制冷装置
freon centrifugal compressor 氟氯烷(冷却剂)离心式压缩机
freon compressing and condensing unit 氟利昂压缩冷凝机
freon compressing unit 氟压缩机
freon compressor 氟立昂压缩机
Freon gas 氟利昂气
Freon leak detector 氟利昂探漏器
Freon pipe 氟利昂管
freon refrigerator 氟利昂制冷机

frequence 恒有度
frequency 频率;频度;频带;次数
frequency acceptance band 接收频带
frequency accuracy 频率准确度
frequency adjustment 频率调整
frequency agile racon 鉴频雷康;鉴频雷达应答器
frequency agility 频率捷变
frequency aging 频率陈化
frequency alias 频率混淆
frequency allocation 频率分配;频率分布;频段分配
frequency amplitude curve 频率振幅曲线
frequency analysis 频谱分析;频率分析
frequency analysis compaction 频率分析精简法
frequency analysis technique 频率分析技术
frequency analyzer 频率分析仪;频率分析器
frequency and time standard 频率及计时间标准
frequency astigmatism conductivity 频散电导率
frequency astigmatism resistivity 频散电阻率
frequency attenuation 频率衰减
frequency band 频带;波段
frequency band compression 频带压缩
frequency band level 频带级
frequency band spreading 频段扩展
frequency bandwidth 频带宽度
frequency-basis flood 根据频率计算洪水
frequency booster 倍频器
frequency breakdown 频率急降
frequency calibration 频率检校
frequency carrier deviation meter 载频偏移计
frequency changer 换频带;频率变换器;变频器;变频机
frequency-changer crystal 变频晶体
frequency-changer set 变频机组
frequency channel 频率通道;频段;频道
frequency character(istic) 频率特征;频率特性
frequency characteristic gradient 频率特性的陡度
frequency characteristic run 频率特性试验
frequency character of secondary magnetic field 二次磁场的频率特性
frequency chart 频数图;频率图
frequency class 频率级
frequency code 频率码;频率电码
frequency coding 频率密码
frequency coding system 频率编码制
frequency comparator 频率比较器
frequency compensation 频率补偿
frequency component 频率分量;频率成分
frequency constant 频率常数;频率常量
frequency content 频率成分
frequency control 频率控制;变频调速
frequency control(led) actuator 频率控制传动装置
frequency control system 频率控制系统
frequency conversion 频率转换;频率变换;变频
frequency conversion circuit 频率变换电路
frequency-conversion crystal 变频晶体
frequency conversion effect 变频效应
frequency conversion loss 变频损耗
frequency converter 频率转换器;变频器;变频机
frequency converter tube 变频管
frequency convolution theorem 频率卷积定理
frequency correction circuit 频率校正电路
frequency correlation 频率对比
frequency counter 计数式频率计
frequency coverage 频谱范围;频率范围
frequency crossing 频率交叉
frequency curve 频数曲线;频率曲线;次数曲线
frequency curve of water supply 供水频率曲线
frequency cutoff 频率截止
frequency deceleration 频率减慢
frequency decomposition 频率分解
frequency demodulation 鉴频;频率解调
frequency demultiplication 频率分频
frequency demultiplier 降频器;分频器
frequency density 频率密度
frequency dependence 频率相关;频率关系式
frequency-dependent duration characteristic 与频率有关的持续时间特征
frequency dependent weighting function 频率加权函数
frequency derived channel 频分信道
frequency detector 鉴频器;检频器
frequency deviation 频偏;频差
frequency device 频率敏感器件
frequency diagram 频率图解;频率图

frequency difference 频差
frequency difference detector 频差检波器
frequency difference resonance 频差共振
frequency differential relay 频率差动继电器
frequency discriminating optic(al) chopper 鉴频光学限制器
frequency discrimination 鉴频;频率鉴别;频率检波
frequency discriminator 鉴频器;频率检波器
frequency dispersion 频散
frequency distortion 频率失真;频率畸变
frequency distribution 频数分布;频率分布;次数分布
frequency distribution curve 频率分布曲线
frequency distribution diagram 频率分布图
frequency distribution histogram 频率分布直方图
frequency distribution of earthquake 地震频度分布
frequency distribution of isotope composition 同位素组分频率分布图
frequency distribution table 频数分布表
frequency diversity 频率分集
frequency divider 分频装置;分频器
frequency divider stage 分频级
frequency dividing circuit 分频电路
frequency dividing ratio 分频比
frequency division 频率分隔;频率分割;分频
frequency division audiometer 分频听力计
frequency division modulation 分频调制
frequency division multiple access 频分多址接入
frequency division multiplex(ing) 分频多路传输;频分多路复用;频分复用
frequency division multiplex system 分频多路传输方式
frequency division multiplier 分频乘法器
frequency division of noise 噪声频程
frequency division system 分频制
frequency domain 频域;频率域;频畴
frequency domain analysis 频域分析;频率域分析法
frequency domain analyzer 频域分析器
frequency domain criterion 频域判据
frequency domain deconvolution 频率域反褶积
frequency domain equalizer 频率域均衡器
frequency domain filter 频率域滤波
frequency domain method 频(率)域法;频率范围
frequency domain model 频域模型
frequency domain time series 频域时间数列
frequency-doubled 倍频的
frequency-doubled effect 倍频效应
frequency doubler 二倍增频器;倍频器
frequency doubling 倍频
frequency drift 频漂;频率漂移
frequency drift error 频漂误差
frequency effect 频率效应;倍频带效应
frequency eliminator 除频器
frequency emphasis filtering 频率加强滤波
frequency energy distribution 频率能量分布
frequency equalizer 频率均衡器
frequency equation 频率方程(式)
frequency error 频率误差
frequency exchange signaling 频率交换信号传输
frequency factor 频率因素;频率因数
frequency field 频率域
frequency field function 频率场函数
frequency filter 滤音器;频率滤波器
frequency filtering 频率滤波
frequency fine tuning 频率微调
frequency flood 普通洪水;常遇洪水
frequency flutter checker 频颤检验器
frequency frogging 频率交换
frequency function 频数函数;频率函数
frequency generator 频率产生器
frequency government 频率管理
frequency graph 频率图
frequency group 频群
frequency halving 分频
frequency-halving circuit 分频电路
frequency harmonic 谐和频率
frequency histogram 频率直方图;次数直方图
frequency histogram showing the distribution of elements 元素分布频率直方图
frequency-hopping 跳频
frequency-hopping spread spectrum 跳频扩展频谱

frequency hysteresis 频率滞后
frequency-identification unit 波长计
frequency-independent antenna 非频变天线
frequency index 频率指数
frequency indicator 频率计;频率表;示频率;单频指示器
frequency influence 频率影响(对仪表读数准确度)
frequency in pitch 纵向振动频率
frequency in roll 横向振动频率
frequency instability 频率不稳定度
frequency-intensity wind diagram 风频率强度图;风力频率图
frequency interlace 频率交错
frequency interlace technique 频率交错技术
frequency interlacing 频率交错
frequency interleave 频率交错
frequency-interleaved pattern 交错频率信号图
frequency interleaving 频率交错法
frequency interleaving system 频率交错制
frequency interpretation of probability 概率的频率解释
frequency interval 频程
frequency inversion 频率反演;频带倒置
frequency jitter 频率抖动
frequency keying 频率键控
frequency keying method 频率键控法
frequency level 频率级
frequency limit 频率界限
frequency limitation 频率极限
frequency lock indicator 频率同步指示器
frequency magnitude 频度震级
frequency magnitude coefficient 频度震级系数
frequency magnitude parameter 频度震级参数
frequency marker oscillator 频标振荡器
frequency matching 频数配合
frequency measuring bridge 频率测量电桥;测频电桥
frequency measuring counter 测频计数器
frequency measuring equipment 频率测量设备
frequency measuring station 频率测试台
frequency meter 计数频率计;频率计;频率表;测频仪
frequency-meter of network 电力网周率表
frequency mixer 混频器
frequency mixing 混频
frequency-modulated carrier 调频载波
frequency-modulated fuze 调频引信
frequency-modulated generator 调频振荡器
frequency-modulated oscillator 调频振荡器
frequency-modulated pumping 频调抽运
frequency-modulated signal 调频信号
frequency-modulated sonar 调频声呐
frequency-modulated sound 频调声;调频声
frequency-modulated synchrotron 调频同步加速器
frequency-modulated system 频率调制系统;调频系统
frequency-modulated telecontrol system generator 调频遥控系统发生器
frequency modulated torque regulator 调频扭力仪
frequency-modulating blade 调频叶片
frequency modulation 频(率)调制;频率调节;调频
frequency modulation air-conditioner with fuzzy control 变频式模糊控制空调器
frequency modulation characteristic 调频特性
frequency modulation chart 调频图
frequency modulation deviation 调频频偏
frequency modulation distortion 调频失真
frequency modulation generator 调频信号发生器
frequency modulation index 调频指数
frequency modulation jamming 调频干扰
frequency modulation noise 调频噪声
frequency modulation noise level 调频杂波电平
frequency modulation pulse compression 调频脉冲压缩
frequency modulation radio altimeter 调频无线电测高计
frequency modulation receiver 调频接收机
frequency modulation rejection 调频抑制
frequency modulation system 调频制;调频系统
frequency modulation telemetry transmitter 调频遥测发射机
frequency modulation terminal equipment 调频终端设备

frequency modulation threshold 调频门限
frequency modulation transmitter 调频发射机
frequency modulation tuner 调频调谐器
frequency modulation type altimeter 调频式测高计
frequency modulator 调频器
frequency moment 频数距
frequency monitor 频率监视器;频率监测器
frequency multiplexing 频率多路传输
frequency multiplexing technique 多频(率)多路技术
frequency multiplication 频率倍增;多倍频率效应;倍频
frequency multiplier 倍频器
frequency multiplier chain 倍频链
frequency multiplier circuit 倍频电路
frequency multiplier klystron 倍频速调管
frequency number 频数;频率数
frequency odulation telecommunication system 调频通信制
frequency of access 存取频率
frequency of bucket emptying per minute 每分钟倒斗次数
frequency of collision 碰撞频率
frequency of converter 变频器
frequency of cyclic(al) loading 周期性荷载频率
frequency of determined flow over cross-sections 断面测流次数
frequency of discharge observation 流量观测频率
frequency of disease development 发病频率
frequency of dosing 投配频率
frequency of drawdown 水位降深次数
frequency of drought 干旱频率
frequency of earthquake 地震频度
frequency of earthquake occurrence 地震发生频率
frequency of echo sounder 回声测深仪频率
frequency of eddies 涡旋次数
frequency of exceedance 超过频率;保证率
frequency of fault 断层频度
frequency of fissures 裂缝频度
frequency of flood 洪水频率
frequency of flooding period 汛期次数
frequency of flows 流量频率
frequency offset 频偏;频率偏移
frequency offset error 频偏误差
frequency of ground shaking 地震动(态)频率
frequency of infinite attenuation 无限衰减频率
frequency of injuries 负伤频率
frequency of irrigation 灌水频率;灌水次数;灌溉次数
frequency of joints 节理频数;节理频率;节理频度;节理密度
frequency of lineament 线性体频数
frequency of maintenance 维修频率;维修次数
frequency of measurement 观测频率;观测频度
frequency of movement 活动次数
frequency of occurrence 发生频率;出现频率
frequency of occurrence of wave height 波高出现频率
frequency of operation 运用频率;工作频率
frequency of optimum traffic 最佳传输频率
frequency of oscillation 振动频率;振荡频率
frequency of overfall 溢流频率
frequency of overflow 溢流频率
frequency of penetration 穿透频率
frequency of periodic(al) load 周期性荷载频率
frequency of precipitation 降水频率;降水次数
frequency of purchase 采购频率
frequency of radon counting rate 氡计数频率
frequency of rare floods 稀遇洪水频率
frequency of reservoir earthquake 水库地震频度
frequency of resonance 谐振频率
frequency of sampling 取样频率;抽样频率;抽样频度;采样时距;采样频率
frequency of service 行车频率;公共交通班次频率;发车频率;班次密度
frequency of ship's arrival and departure 到发船密度
frequency of shock 震动频率;振动频率
frequency of soundings 测深点密度
frequency of stage 水位频率
frequency of stroke 冲次
frequency of the carrier wave 载波频率
frequency of transfer state clock 传送状态钟频率

frequency of usage 利用率
frequency of valve operation 阀的操作次数
frequency of vibration 振动频率
frequency of washing 冲洗频率
frequency of watering 灌水频率
frequency of water-level observation 水位观测频率
frequency of wind direction 风向频率
frequency of working 工作频率
frequency optimum traffic 最佳工作频率
frequency overvoltage 工频过电压
frequency parameter 频率参数
frequency period 频率周期
frequency plan 频率计划
frequency plot 频率图
frequency polygon 频率(分布)多边形;频率多边形
frequency pool 频率池
frequency prediction 频率预测
frequency programming 频率程序设计
frequency programming system 频率程序系统
frequency pushing 频率推移;推频
frequency pushing factor 频率推移系数
frequency range 波段;频谱范围;频率范围;频带;频段
frequency ratio 频数比;频率比;频比
frequency reference 基准频率;参考频率
frequency regulating plant 调频电厂
frequency regulation 频率调节
frequency regulator 频率调节器;调频装置;调频器
frequency relay 频率继电器
frequency resolution 频率分解;频率分辨率
frequency response 谐波响应;频率响应;频率反应;瞬变过程特性
frequency response analysis 频度特性分析
frequency response approach 频率响应计算法
frequency response characteristic 谐波响应特性(曲线);频率响应曲线;频率反应特性
frequency response curve 频率响应曲线;频率反应曲线
frequency response function 谐波响应函数;频率响应函数;频率反应函数
frequency response method 频率响应法
frequency response of tape 带频响应
frequency sampling theorem 频率抽样定理
frequency scale 频率刻度;频标
frequency scan 频率扫描
frequency scrambler 扰频器;倒频器
frequency selecting amplifier 选频放大器
frequency selecting connector 选频连接器
frequency selection 选频
frequency selective amplifier 选频放大器
frequency selective damping material 选频阻尼材料
frequency selective level indicator 选频电平指示器;选频式电平表
frequency selective network 选频网络
frequency selective system 选频系统
frequency selective voltmeter 选频伏特计
frequency selector 频率选择器
frequency separating filter 分频滤波器
frequency series 频数数列;频率级数
frequency-shared channel amplifier 分频通道放大器
frequency shift 移频;频移;频偏
frequency shift automatic block 移频自动闭塞
frequency shift converter 移频变换器
frequency shifting 频率移动;频率位移
frequency shift keyer 频率键控器
frequency shift keying 移频键控;频移键控;频移调制;调频器;数字调频
frequency shift system 移频系统;频移系统
frequency shift transmitter monitor 移频发射机监控装置
frequency shift value 频偏值
frequency smoothing 频率平滑
frequency sounding curve of axis dipole array 轴向偶极频率测深曲线
frequency sounding curve of equator dipole array 赤道偶极频率测深曲线
frequency sounding instrument 频率测深仪
frequency sounding method 频率测深法
frequency space 频率空间
frequency spectrograph 频谱仪
frequency spectrometer 频谱仪

frequency spectrum 频谱;频率谱
frequency spectrum analysis 玻璃频谱分析
frequency spectrum analyzer 频谱分析仪
frequency spectrum compression 频谱压缩
frequency spectrum designation 频段名称
frequency spectrum direct vision method 频谱直视法
frequency spectrum intensity 频谱密度
frequency splitting 频率分隔;分频
frequency stability 频稳度;频率稳定性;频率稳定度
frequency stabilization by quartz resonator 石英稳频法
frequency stabilization of laser 激光频率稳定
frequency stabilized carbon dioxide laser 稳频二氧化碳激光器
frequency stabilized He-Ne laser 稳频氦氖激光器
frequency staggering 频率参差;带带参差
frequency standard 频率标准
frequency step 频率阶跃
frequency step down 频率降低
frequency study 工作的抽样检验
frequency surface 频率面;次数曲面
frequency sweep 频率扫描
frequency sweep generator 扫频发生器;摇频振荡器
frequency swing 频率摆动
frequency switching 频率开关
frequency synthesis technique 频率综合技术
frequency synthesizer 频率合成器
frequency table 频数表
frequency teller 频率计
frequency temperature characteristic 频率温度特性
frequency theory of distribution 分布的频率理论
frequency theory of nerve impulse 冲动频率论
frequency times 频率次数
frequency time spectral density function 频率时间谱密度函数
frequency to free vibration 自振频率
frequency tolerance 频率容限
frequency to voltage converter 频率电压变换器
frequency track circuit 频率式轨道电路
frequency tracker 频率跟踪器
frequency transfer function 频率传递函数
frequency transformation 频率变换
frequency transformer 频率转换器;变频器
frequency translating transponder 变频转发器
frequency transmission 频率传输
frequency transmission curve 频率传输曲线
frequency tripled effect 三倍频效应
frequency tripler 三倍倍频器
frequency variator 调频器
frequency vector 频率向量
frequency view of probability 概率中的频数观点
frequency wave number analysis 频率波数分析
frequency wow 频率颤动
frequeney division multiple 分频多路
frequent 频繁的
frequent core blocking 经常产生岩芯堵塞的
frequent curve line 多曲线线路
frequented coast 不定岸线
frequent flood 常遇洪水
frequent hoeing 频繁锄地
frequent industrial inquiry 频繁产业调查
frequently occurred earthquake 多遇地震
frequently encountered disease 多发病
frequent strikes 频繁罢工
frequent term 常用词
frequent value of action 作用频遇值
fresco 石膏装饰板;壁画施工法;壁画
fresco colo(u)r 壁画用水性颜料;壁画色
frescoed 作壁画的
frescoed decoration 壁画装饰
frescoing 壁画法
fresco of griffins 鹰头狮身像壁画;翼狮像壁画
fresco painter 壁画油漆工;壁画家
fresco painting 壁画油漆
fresco plaster 壁画涂层;壁画灰泥
Frese's flow formula 弗雷塞堰流公式
fresh 新鲜气团;淡水水流
fresh adsorbent 强吸附剂
fresh air 新风;一次空气;全新风;新鲜空气
fresh-air breathing apparatus 氧气供给器
fresh-air circulation 新鲜空气循环;新鲜空气环

流;新风循环
fresh-air duct 净气管;新鲜空气管道
fresh-air duct pump 新风管道系统
fresh-air fan 送风机;新风机;净气风扇
fresh-air flue 新鲜空气管道;净气管;净气道
fresh-air handling unit 新风机组
fresh-air heater 新风加热器
fresh-air heating appliance 新风加热设备;新风加热器
fresh-air heating supply 净气供暖
fresh-air inlet 新鲜空气进口;新鲜空气入口;吸气口;净气进口;进气口;进气孔;送风机进口
fresh-air inlet tower 新风进风塔
fresh-air intake 净气进口;新鲜空气入口;室外空气进口
fresh-air intake stack 新风进口竖管
fresh-air intake tower 新风进风塔
fresh-air louvers 净气百叶窗
fresh-air make-up 新鲜空气补充;新风补偿量;补给新风
fresh-air mask 新鲜空气面具;氧气面罩
fresh-air operation 新风处理
fresh-air preheater 新风预热器
fresh-air proportion 新风比例
fresh-air raise 入风天井
fresh-air ratio 新风比
fresh-air requirement 新风量
fresh-air sports 户外运动
fresh-air supply 净气供应
fresh-air supply to roof 屋顶自然通风
fresh-air system 新风系统
fresh-air unit 新鲜空气导入装置
fresh-air volume 新风量;进风量
fresh-air warmer 新鲜空气加热器;新风加热器
fresh-air warming 新鲜空气加热;新风加热
fresh-air warming appliance 新风采暖设备;新风采暖器
fresh and live freight 鲜活货物
fresh and live goods 鲜活货物
fresh and/or rain water damage 淡水和/或雨淋损失
fresh arrival 新到货(物)
fresh bedrock 新鲜基岩;未风化基岩
fresh bit 翻新钻头
fresh breeze 劲风;清(新微)风;清劲风;五级风
fresh catalyst 新鲜催化剂
fresh catalyst hopper 新鲜催化剂料斗
fresh cement mortar 新拌水泥砂浆;未结硬的水泥砂浆
fresh charge 新鲜充量
fresh concentration 起始浓度
fresh concrete 新浇混凝土;新拌混凝土;未硬结混凝土
fresh-core technique 新鲜岩芯技术
fresh cracked gas 新鲜裂化气
fresh crude oil 新鲜原油
fresh cut 新伐
fresh deposit 淡水沉积(物)
fresh driver 新驾驶员
freshen 搬位(指绳索、锚链的掉头或改变受磨部位)
freshen a hawse 改变锚链在锚链筒部位
freshen a rope 改变绳索摩擦部位
freshened sea water 淡化海水
freshener 新手
freshening 淡化
freshening wind 增强中的风
freshen the ballast 压载水舱换水
freshen the way 船速增加
freshet 汛;洪涨;洪水;山洪;淡水河流;春汛;发洪水;暴涨(河流)
freshet canal 泄洪渠;泄洪道
freshet discharge 汛期流量
freshet flow 汛期流量
freshet hydrograph 洪水过程(曲)线;春汛过程线
freshet period 汛期;洪水期;春汛期
fresh fan 净气门扇
fresh feed 新鲜原料
fresh feed conversion 新鲜原料转化率
fresh feed pump 进料泵
fresh feed surge drum 新鲜原料收集器
fresh film load 换片
fresh fish carrier 鲜鱼渔船
fresh fish hold 鲜鱼舱

fresh flue 新风道
fresh fruit 新鲜水果
fresh fuel 新燃料
fresh gale 强风(六级风);大风
fresh gas supply 净气供应
fresh goods 鲜活商品
fresh granite 新鲜花岗岩
fresh green 翠绿色
fresh grid 新坐标格网
fresh groundwater 地下淡水;新地下水
fresh ground water in littoral deposit 滨海地下淡水
fresh ice 新(形成的)冰;新结冰;新成冰;淡水冰;不含盐分的冰
fresh information 新信息
fresh intake 新鲜入风
fresh iron 初熔铁
fresh keeping method 鲜储[贮]法
fresh lagoon facies 淡化泻湖相
fresh lake 淡水湖
fresh latex 新鲜胶乳
fresh lime 新灰
fresh louvers 净气百叶窗
freshly eluted resin 新淋洗过的树脂
freshly exposed surface 新鲜露头面【地】
freshly fallen snow 初雪;初落雪;新雪
freshly frozen ice 初期冰;新结冰;新成冰
freshly made cut 新挖路堑
freshly made fill 新填路堤
freshly made product 新制成品
freshly made slab 新制板材
freshly maintained rail 新维修轨道
freshly mixed 新拌和的
freshly mixed concrete 新拌混凝土
freshly mixed concrete density 新拌混凝土的密度
freshly mixed mortar 新拌砂浆;新拌灰浆
freshly placed 新浇的
freshly placed concrete 新浇混凝土;新筑混凝土
freshly plowed soil 新耕土
freshly precipitated hydrous iron oxide 强沉淀水合氧化铁
freshly precipitated hydrous oxide 强沉淀水合氧化物
freshly set mortar 新凝泥浆;新凝(结)砂浆;初凝(水泥)砂浆
fresh material 新材料;新料
fresh mineral 新采矿物
fresh money 新钱
fresh mulch 新草覆盖;初成残草覆盖
freshness 新鲜度;新鲜
freshness of water 水的淡性
fresh oil 新油
fresh oil spills 新鲜漏油
fresh-oil system 新鲜润滑油系统
fresh oil tank 新鲜润滑油箱
fresh outcrop 新露头
fresh paint 新油漆;新涂未干的油漆
fresh paste 水泥净浆
fresh product 新制产品
fresh red 鲜红(明代早期铜红釉)
fresh rock 未风化岩石;润湿的岩石;新鲜岩石
fresh roof 新暴露顶板
fresh rubber 新胶
fresh run 新进入淡水河的海鱼
fresh-salt water interface 咸淡水界面;淡咸水分界面
fresh sample 新样品;新鲜样品
fresh sand 新砂
fresh sea water 淡咸水
fresh sewage 新鲜污水
fresh shot 海洋中的淡水流;淡水水流;淡水冲出的海口区
fresh sludge 新鲜污泥;原污泥;绿污泥;未结硬污泥;生污泥
fresh snow 刚下的雪;新下的雪;初雪
fresh snow layer 新雪层
fresh soil 荒地
fresh state 新鲜状态
fresh steam 新鲜蒸汽
fresh surface 新生表面
fresh surface waters 淡地表水体
fresh system 新风系统
fresh troops 生力军
fresh vegetables 新鲜蔬菜

fresh waste 新鲜废物
fresh wastewater 新鲜废水;原废水;生废水
fresh water 新鲜水;工艺水;淡水(的)
freshwater alga 淡水藻
freshwater algae 淡水藻类
freshwater allowance 淡水吃水余量
freshwater alluvium 淡水冲积层
freshwater and marine organisms 淡水生物和海洋生物
freshwater animal 淡水动物
freshwater aquiculture 淡水养殖
freshwater aquifers 淡水含水层
freshwater area 淡水区域
freshwater barrier 淡水屏障;淡水夹层;淡水隔水层
freshwater biological resources 淡水生物资源
freshwater biology 淡水生物学
freshwater bivalve 淡水双壳贝类;淡水牡蛎
freshwater body 淡水水体
freshwater circulating pump 淡水循环泵
freshwater community 淡水生物群系;淡水群落
freshwater contaminant 淡水污染物
freshwater contamination 淡水污染
freshwater cooler 淡水冷却器
freshwater cooling pump 淡水冷却泵
freshwater damage 淡水损害;淡水濡损
freshwater damage clause 淡水险条款
freshwater degradation 淡水恶化
freshwater deposit 淡水沉积(物)
freshwater discharge pump 淡水抽出泵
freshwater distiller 淡水蒸馏器
freshwater distilling plant 制淡水设备
freshwater dolphin 淡水豚(类)
freshwater draft 淡水吃水(深度)
freshwater drain collecting tank 淡水排泄污水舱
freshwater ecology 淡水生态学
freshwater ecosystem 淡水生态系统
freshwater ecosystem condition 淡水生态系统条件
freshwater environment 淡水环境
freshwater estuary 淡水河口
freshwater extraction pump 淡水排出泵
freshwater facies 淡水相
freshwater fauna 淡水动物群;淡水动物区系
freshwater filter 淡水过滤器
freshwater fish 淡水鱼
freshwater fish culture 淡水养鱼
freshwater fishery 淡水渔业;淡水鱼场;淡水养鱼业
freshwater fishes 淡水鱼类
freshwater flora 淡水植物群;淡水植物区系
freshwater flow 淡水流量
freshwater formation 淡水建造
freshwater freeboard 淡水干舷
freshwater generator 海水淡化器;淡水处理装置
freshwater geomagnetic electrokinetograph 淡水地磁动电测流器
freshwater habitat 淡水生境
freshwater ice 淡水冰
freshwater in lake 湖泊淡水
freshwater lagoon 淡水泻湖
freshwater lake 淡水湖
freshwater lake deposit 淡水湖泊沉积
freshwater lake facies 淡水湖泊相
freshwater lens 淡水透镜(体)
freshwater limestone 淡水石灰石;粗糙石灰石;淡水石灰岩
freshwater line 淡水管道
freshwater line pressure pipe 淡水压力管
freshwater load in summer 夏季淡水满载水线
freshwater load line 淡水载重(水)线
freshwater macro-invertebrate 淡水性大型无脊椎动物
freshwater makeup pump 淡水补充泵
freshwater mark 淡水载重(水)线
freshwater marsh 淡水沼泽;淡水草沼
freshwater marsh wetland 淡水草沼湿地
freshwater mass 淡水团
freshwater monitoring 淡水监测
freshwater mud 清水泥浆;淡水泥浆
freshwater organism 淡水生物
freshwater origin 淡水成因
freshwater overflow zone 淡水溢出带
freshwater pearl 淡水珍珠
freshwater peat 淡水泥炭
freshwater pipe 淡水管
freshwater piping 淡水管系

freshwater pisciculture 淡水养鱼业
freshwater plankton 湖沼浮游生物
freshwater plant 淡水植物;淡水工厂;淡水处理厂;海水淡化(工)厂
freshwater pollutant 淡水污染物
freshwater pollution 淡水污染
freshwater pool 淡水池
freshwater pump 淡水泵
freshwater reservoir 淡水蓄水库;淡水库
freshwater resources 淡水资源
freshwater river 淡水河流
freshwater runoff 淡水径流(量)
freshwater sand 淡水砂(层)
freshwater-seawater mixing system 淡水—海水混合体系
freshwater sediment 淡水沉积(物)
freshwater shell button 淡水贝壳钮扣
freshwater shrimp 淡水虾
freshwater species 淡水物种
freshwater stay 临时支索;横牵索;桅间索
freshwater stream 淡水河流
freshwater supply 淡水供应
freshwater supply ship 淡水供应船
freshwater swamp 淡水沼泽;淡水林沼
freshwater swamp deposit 淡水沼泽沉积
freshwater system 清水系统;淡水系统
freshwater tank 淡水箱;淡水柜;淡水池;淡水槽;淡水舱
freshwater waters 淡水水体
freshwater wetland 淡水湿地
freshwater wetland system 淡水湿地系统
freshwater zone 淡水区域
fresh weight 鲜重
fresh wood 新伐木材;未干木材
fresh zephyr 强烈西风
Fresnal lens 弗列斯诺尔阶梯光栅镜头
fresnel 菲涅尔(频率单位,百亿赫兹)
Fresnel-Arago law 菲涅尔—阿喇戈定律
Fresnelized field flattener 菲涅尔平场透镜
Fresnel-Kirchhoff diffraction formula 菲涅尔—基尔霍夫衍射公式
Fresnel-Kirchhoff integral 菲涅尔—基尔霍夫积分
Fresnel-Kirchhoff theory 菲涅尔—基尔霍夫理论
Fresnel's annular zone objective 菲涅尔环带物镜
Fresnel's approximation 菲涅尔近似
Fresnel's biprism 菲涅尔双棱镜
Fresnel's construction 菲涅尔波带设计
Fresnel's diffraction 菲涅尔衍射
Fresnel's diffraction effect 菲涅尔衍射效应
Fresnel's diffraction fringes 菲涅尔衍射条纹
Fresnel's diffraction pattern 菲涅尔衍射图
Fresnel's diffraction wave 菲涅尔衍射波
Fresnel's double mirror 菲涅尔双反射镜
Fresnel's drag coefficient 菲涅尔曳引系数
Fresnel's ellipsoid 菲涅尔椭球
Fresnel's equation 菲涅尔方程
Fresnel's equation of wave normal 菲涅尔波面法线方程
Fresnel's fringe 菲涅尔条纹
Fresnel's hemilens 菲涅尔半透镜
Fresnel's hologram 菲涅尔全息图
Fresnel's holographic system 菲涅尔全息系统
Fresnel's holography 菲涅尔全息照相;菲涅尔全息术
Fresnel's integral 菲涅尔积分
Fresnel's knife-edge test 菲涅尔刀口试验
Fresnel's lens 菲涅尔透镜;菲涅尔聚光镜;螺纹透镜
Fresnel's mirror 菲涅尔镜
Fresnel's number 菲涅尔数
Fresnel's pattern 菲涅尔方向图
Fresnel's reflectance loss 菲涅尔反射损耗
Fresnel's reflection 菲涅尔反射
Fresnel's reflection coefficient 菲涅尔反射系数
Fresnel's reflection formula 菲涅尔反射公式
Fresnel's reflection loss 菲涅尔反射损耗
Fresnel's region 菲涅尔区
Fresnel's region of diffraction 菲涅尔衍射区
Fresnel's rhomb 菲涅尔菱形(镜)
Fresnel's rhombus 菲涅尔菱体
Fresnel's spotlight 菲涅尔投光灯
Fresnel's stepped lens 菲涅尔阶梯透镜;菲涅尔分步透镜
Fresnel's theory of double refraction 菲涅尔双折射理论

Fresnel's zone 菲涅尔区(间隙);菲涅尔带
Fresnel's zone construction 菲涅尔波带法
Fresnel-transform hologram 菲涅尔变换全息图像
Fresnel-zone half plate 菲涅尔带半波片
Fresnel-zone lens 菲涅尔带透镜
Fresnel-zone plate 菲涅尔带片
Fresno 菲涅斯诺修面机(一种混凝土修面机)
fresno 天蓝符山石
fresnoite 硅钛钡石
Fresno slip 马拉斗式刮土机;马拉斗式铲土机
Fressinet concrete hinge 弗列西涅特预应力混凝土铰接结构
Frestex 弗列斯塔涂料(一种专卖的塑性涂料)
fret 格子细工装饰;基粒间膜;磨损处;磨耗;回纹饰;万字浮雕;雕花
fret lead 嵌彩色玻璃窗的铅条骨架;铅条条
fret mill 边碾
fret saw 线锯;钢丝锯;雕花锯;绊锯;细工锯;嵌锯
fret saw blade 钢丝锯条
fret saw frame 钢丝锯架
fret-sawing machine 螺纹锯床;电蚀锯床
fret saw with clamp 伐木锯
frettage 擦伤腐蚀
fretted channel 海水冲成的渠道
fretted lead 窗嵌玻璃铅条;花饰铅条;嵌窗铅条;铅棂条
fretted rope 磨损了的绳子
fretted terrain 回纹岩层
fretting 路面损失;微振磨损;剥蚀;表面侵蚀;摩擦腐蚀
fretting corrosion 磨损;磨耗;磨蚀;擦伤腐蚀;松散;路面磨损;摩擦腐蚀
fretting erosion 磨损腐蚀
fretting fatigue 磨损疲劳;磨蚀疲劳
fretting oxidation 磨损氧化
fretwork 万字细工;透雕细工;风化粒状岩石;凸花细工;粒状岩石风化;回纹(细)饰;格子细工;雕花细工;浮雕细工
fretwork saw blade wire 钢丝锯条用钢丝
fretwork weathering 格子状风化;蜂窝状风化;蜂窝式风化
Fretz-Moon method 弗雷茨穆恩连续式炉焊管法
freudenbergite 黑钛铁钠矿
Freundlich adsorption coefficient 弗兰德利希吸附系数
Freundlich adsorption isotherm 弗兰德利希吸附等温线
Freundlich equation 弗兰德利希方程
Freundlich isotherm 弗兰德利希等温线
Freundlich isotherm equation 弗兰德利希等温方程
Freundlich model 弗兰德利希模型
Freundlich's adsorption isotherm 弗兰德利希吸附等温线
Freund's adjuvant 弗罗因德佐剂
Freund's complete adjuvant 完全弗罗因德佐剂
Frey automatic cutter 弗雷自动切砖机
Frey cutter 弗雷切砖机
Freysinnet 弗雷西奈式后张法
Freysinnet anchoring system 弗雷西奈式锚定系统(预应力混凝土中用)
Freysinnet bearing 弗雷西奈支座
Freysinnet parallel-wire cable 弗雷西奈平行线缆
Freysinnet type jack 弗雷西奈式千斤顶;扁千斤顶
Freyssinet cone anchorage 弗雷西奈式圆锥锚(头);圆锥锚具;弗氏锚
Freyssinet double acting jack 弗雷西奈式双动千斤顶(预应力张拉用)
Freyssinet jack 液压钢枕;压力枕;弗雷西奈式双动千斤顶(预应力张拉用)
Freyssinet prestressing system 弗雷西奈式张拉系统
Freyssinet system 弗氏拉张系统;弗雷西奈式系统
Freyssinet system of prestressing 弗雷西奈预应力法
friability 易碎性;易剥落性;松散性;脆(碎)性
friability of coal 煤脆性
friability test 脆性试验;脆性试验
friability tester 脆碎性测定器
friable 易碎的
friable alumina 脆性氧化铝
friable asbestos material 脆性石棉材料
friable calcite 脆性方解石
friable clay 易碎粘土
friable core 易碎岩芯

friable formation 破碎地层;松散地层
friable ground 破碎地层
friable gypsum 脆性石膏
friable hard formation 硬脆地层
friable iron pan 脆性铁盘
friable lignite 脆性褐煤
friable marble 易碎大理石;脆性大理石
friable material 易碎材料
friable metal 脆性金属
friable particle 易碎颗粒;脆度系数
friable rock 易碎岩石;酥炸岩;松散岩石
friable sand 散粒砂
friable shale 脆性页岩
friable soil 酥性土;松散土(壤);脆性土
friable state 松散状态;松散状况
friar's cloth 粗重方平棉布
friary 男修道院
friatest 金刚石强度测定仪
Friberg's theory 弗里贝格理论(关于水泥混凝土路面传力杆受力的理论)
fricative 摩擦的
fricative track 摩擦痕;颤痕
frichophore 毛腔
Frick alloy 铜锌镍合金
Fricke dosimeter 弗里克剂量计
Frick's alloy 弗里克铜锌镍电阻丝合金
friction 磨擦;阻;摩擦(力);冲突
frictional 非黏结性的
frictional acceleration response spectrum 摩擦加速度反应谱
frictional action 摩擦作用
frictional adjuster 摩擦力调节器
frictional angle 摩擦角
frictional area 摩擦面积
frictional arm method 摩擦臂测定法
frictional attenuation 摩擦衰减
frictional axis 摩擦轴
frictional back gear 摩擦背轮
frictional band 摩擦带
frictional band brake 摩擦闸箍
frictional band clutch 摩擦带离合器
frictional based twisting system 摩擦式加捻体系
frictional bearing 摩擦轴承
frictional behavio(u)r 摩擦性能
frictional belt 摩擦带;(船航行时随船体移动的)边界水层;边界层
frictional block 摩擦块
frictional boundary layer 摩擦边界层
frictional brake 摩擦制动(器);摩擦闸(瓦);摩擦刹车;摩擦测力计;摩阻刹车
frictional brake control system 摩擦制动系统
frictional brake drum 摩擦制动鼓
frictional brake handle 摩擦闸摇柄
frictional breccia 摩擦角砾岩
frictional buffer 摩擦式缓冲装置
frictional calendar 摩擦轧光机
frictional calendaring 摩擦压制;摩擦轧光
frictional catch 摩擦挡;摩阻挡
frictional chuck 摩擦夹盘
frictional circle 摩擦圆盘;摩擦圆
frictional circle analysis 摩擦圆分析法
frictional circle clutch 摩擦圆盘离合器
frictional circle method 摩擦圆法
frictional clamp 摩擦制动装置;摩擦夹钳
frictional clamping plate 摩擦夹板
frictional cleaner 摩擦式清洗器
frictional clutch 摩擦联结器;摩擦合制动;摩擦离合器
frictional clutch assembly 摩擦离合器总成
frictional clutch cup 摩擦离合器外盘
frictional coat 摩擦套;摩擦保护层
frictional coefficient 摩擦系数
frictional coefficient of foundation bottom 基底摩擦系数
frictional compensation 摩擦补偿
frictional cone 摩擦轮
frictional constant 摩擦常数
frictional corrosion 摩擦腐蚀
frictional countershaft 摩擦离合器副轴
frictional coupling 摩擦联结节
frictional crack 摩擦裂隙;摩擦裂缝
frictional current 摩擦流;摩擦海流
frictional damage 摩擦损伤
frictional damper 摩擦(式)阻尼器;摩擦减震器

frictional damping 摩擦阻尼
frictional depth 摩擦深度
frictional disc seat 摩擦圆盘座
frictional disk 摩擦圆盘;摩擦片
frictional disk clutch 摩擦圆盘离合器
frictional disk saw 摩擦圆盘锯
frictional disk shock absorber 摩擦减震器
frictional dissipation 摩擦扩散
frictional dogs 摩擦爪;摩擦闸瓦
frictional drag 摩擦阻力;摩擦曳力
frictional drilling machine 摩擦变速钻床
frictional drive 摩擦传动装置;摩擦传动
frictional drive hoist 摩擦提升机
frictional driven spools 摩擦传动的筒子
frictional drop 摩擦差
frictional drum 摩擦卷筒
frictional dynamometer 摩擦功率计;摩擦测力计;摩擦测功器
frictional effect 摩擦效应
frictional electrical machine 摩擦起电器;摩擦电机
frictional electricity 摩擦电
frictional electrification 摩擦带电
frictional error 摩擦误差
frictional facing 摩擦衬片
frictional factor 摩擦因子;摩擦因素;摩擦因数;摩擦系数
frictional feeder 摩擦式自动给纸装置
frictional field 摩擦力界
frictional flow 摩擦流(动);黏性液体流
frictional force 摩擦力
frictional force of shaft 桩身阻力
frictional ga(u)ge 摩擦计
frictional gear 摩擦轮
frictional geared winch 摩擦传动绞车
frictional gearing 摩擦传动装置
frictional gearing wheel 摩擦传动轮
frictional glazed machine 摩擦抛光机
frictional glazed paper board 摩擦压光纸板
frictional glazing 摩擦轧光
frictional goods 摩擦制品
frictional governor 摩擦式调节器
frictional grip 摩擦锚夹;摩擦夹紧装置
frictional grip bolt 摩擦紧固螺栓
frictional grip bolt joint 摩擦锚栓接(头)
frictional-grooved gearing 楔形(槽)轮摩擦传动
frictional guide 摩擦导卫板
frictional hammer 摩擦锤
frictional head 摩擦(损失的)压头;摩擦水头
frictional head loss 摩擦水头损失;沿程水头损失
frictional heat 摩擦热
frictional heating 摩擦加热
frictional heat loss 摩擦热损失
frictional hinge 摩擦铰链
frictional hoist 摩擦提升机;摩擦式绞车
frictional hoisting 摩擦式提升
frictional horsepower 摩擦马力;摩擦功率
frictional igniter 摩擦点火器
frictional ignition 摩擦点火
frictional index 摩擦指数
frictional influence 摩擦影响
frictional interference 阻滞干扰
frictional jaws 摩擦夹片
frictional joining 摩擦连接
frictional joint 摩擦缝
frictional lag 摩擦滞后
frictional layer 摩擦层
frictional lever 摩擦杆
frictional lever type 摩擦杠杆式
frictional limit design 摩擦极限设计
frictional lining 摩擦片;摩擦衬片
frictional loading 摩擦荷载
frictional lock 摩擦闸
frictional loss 摩擦损失;摩擦损耗;摩耗
frictional loss factor 摩擦损失系数;摩擦损耗系数
frictional loss of(water)head 摩擦损失水头;水头摩擦损失
frictional mandrel 摩擦轴
frictional material 摩擦材料
frictional material test machine 摩擦材料试验机
frictional metal chock 摩擦式金属支架
frictional metamorphism 摩擦变质作用
frictional meter 摩擦仪(表);摩擦系数测定仪
frictional moment 摩擦力矩
frictional multiplier 摩擦倍率

frictional nigger 摩擦翻木机
frictional nip 摩擦轧点
frictional noise 摩擦噪声
frictional overheadv 摩擦性间接费用
frictional pile 摩擦桩
frictional(pile)foundation 摩擦桩基(础)
frictional plate 摩擦片;摩擦板
frictional plate clutch 摩擦片离合器
frictional point 摩擦点
frictional power 摩擦功率
frictional press 摩擦压砖机;摩擦压(力)机;摩擦压光机;摩擦压床
frictional pressure 摩擦压力
frictional pressure control 摩擦压力控制;摩擦力控制
frictional pressure drop 摩擦压降
frictional pressure loss 摩擦压力损失;摩擦预应力损失
frictional primer 摩擦点火管
frictional prop 摩擦支柱
frictional property 摩擦性质
frictional pulley 摩擦轮;摩擦滑轮
frictional pulley clutch 摩擦轮离合器
frictional pump 摩擦泵
frictional pusher 摩擦式推钢机
frictional rail 摩擦轨
frictional ratio 摩阻比
frictional release 摩擦式释放器
frictional resistance 摩擦阻力;摩擦力
frictional resistance in discharge pipeline 排泥管内摩擦阻力
frictional resistance moment 摩擦阻力矩
frictional restraint of(sub)grade 路基摩擦的约束性
frictional revolution counter 摩擦式转数表
frictional ring 摩擦环
frictional roll 摩擦卷筒;摩擦滚筒;摩擦传动辊
frictional roller 摩擦滚轴;摩擦滚筒;摩擦辊柱;摩擦(滚)轮
frictional roller drive 摩擦滚柱传动(装置)
frictional rolling hatch cover 摩擦滚动舱口盖
frictional saw 摩擦锯
frictional sawing 摩擦锯切
frictional saw machine 摩擦锯床
frictional screw press 摩擦压力机;摩擦螺旋压机
frictional secondary flow 摩阻副流;二次流
frictional sediment 无黏性泥沙
frictional separation 摩擦选矿
frictional separation method 摩擦选矿法
frictional separator 摩擦式清选机
frictional shaft 摩擦轴
frictional shear joint 摩擦剪节理
frictional shidding device 摩擦制动装置
frictional shock absorber 摩擦式减振器;摩擦减震器
frictional shoe 摩擦刹车片
frictional sleeve 摩擦套
frictional sliding 摩擦滑动(作用)
frictional slip 摩擦衬套
frictional slip coupling 摩擦滑动式离合器
frictional slope 摩擦坡角;摩擦坡度
frictional snap latch 摩擦式弹簧锁
frictional socket 摩擦打捞筒;摩擦打捞器
frictional soil 内摩(擦)阻力大的土(壤);摩阻性土壤;无黏性土
frictional speed 摩擦速度
frictional spring 摩擦弹簧
frictional start (汽车的)摩擦起步
frictional stopping device 摩擦制动装置
frictional strain ga(u)ge 摩擦式应变计
frictional strength 摩擦强度;摩阻强度
frictional stress 摩擦应力;摩阻应力
frictional support 摩擦支座
frictional surface 摩擦面;摩擦表面
frictional tachometer 摩擦式转速计
frictional tape 绝缘胶布;摩擦带
frictional test 摩擦试验
frictional testing machine 摩擦试验机
frictional texturizing 摩擦变形
frictional torque 摩擦扭矩
frictional torque testing instrument 摩擦力矩测试仪
frictional transmission 摩擦传动
frictional trip 摩擦式安全器

frictional twister 摩擦式假捻器
frictional twisting 摩擦加捻
frictional type vacuum ga(u)ge 摩擦式真空计;摩擦式真空规
frictional unemployment 摩擦性失业
frictional velocity 摩擦速度;摩擦流速
frictional vorticity 摩擦涡度
frictional wafer 摩擦片
frictional wake (船舶航行时的)摩擦拌流
frictional weir 摩擦堰
frictional welding 摩擦焊接;摩擦焊
frictional welding machine 摩擦焊接机
frictional wheel 摩擦轮
frictional wheel drive 摩擦轮驱动;摩擦轮传动
frictional wheel integrator 摩擦轮积分器
frictional wheel speed counter 摩擦轮转速计数器
frictional winch 摩擦传动绞车
frictional winder 摩擦提升机;摩擦轮提升机
frictional winding 摩擦提升;摩擦式提升
frictional windlass 摩擦式绞车;摩擦卷扬器
frictional work 摩擦功
frictional working current 摩擦电流
frictional yarn feed installation 摩擦喂纱装置
friction anchorage 摩阻锚固
friction and end-bearing pile 中间类型桩
friction and resistance coefficient of filter 过滤器摩阻系数
friction and resistance coefficient of well pipe 井管摩阻系数
frictionate 摩擦
friction ball 钢滚珠;钢珠(轴承用)
friction band 制动带
friction bearing 滑动轴承
friction-bearing surfaces 摩擦轴承接触表面
friction bevel gear 锥形摩擦轮
friction-board hammer 圆盘摩擦锤;夹板锤
friction bond 摩擦连接
friction brake 摩擦刹车;摩阻制动器
friction-braked rotation 摩擦制动旋转
friction breccia 断层角砾岩;磨擦角砾岩;擦碎角砾岩
friction burn 擦伤
friction cap 摩擦柱帽
friction catch 摩擦制动装置;摩擦搭扣;弹簧门插销;摩阻挡
friction cathead 摩擦(吊)锚架
friction-caused wear 磨损;摩擦损耗
friction cermet 金属陶瓷摩擦材料
friction change 摩阻变化
friction circle 摩阻圆
friction circle method 摩阻圆法
friction clamp 直钳
friction clutch 磨擦离合器;摩擦传动器
friction clutch of double-disk dry type 干式双片摩擦离合器
friction clutch of the multiple disk 多盘摩擦离合器
friction coat 防磨涂层;减摩涂层;耐磨涂层
friction coefficient 摩阻系数
friction coefficient of pavement 路面摩擦系数
friction colo(u)r 电光纸上色料
friction compound 擦胶剂
friction cone 锥形摩擦轮
friction-controlled steering mechanism 摩擦操纵转向机构
friction coupling 摩面连接器
friction course 摩阻面层;防滑面层
friction cracks 冰川擦口
friction curve 面层曲线
friction drive 异速传动
friction-end bearing pile 摩擦一端承桩
friction factor 摩阻系数
friction factor of pavement 路面摩擦系数
friction feeder 摩阻续纸装置
friction-free 无摩擦的
friction glazed paper 磨光纸
friction head 摩阻水头;水头摩擦损失
frictioning 刮胶;擦胶
friction in pumping 泵送的摩擦力
frictionize 摩擦
friction jewel 压力钻
friction latch 弹簧门插销;弹簧锁
friction law 牛顿摩擦定律
friction layer 地表边界层
frictionless 无摩阻的;无摩擦的

frictionless channel 无摩擦河槽
frictionless channel flow 无摩擦河槽水流
frictionless flow 理想液流；无滞性流；无粘流（动）；无磨阻流；无摩擦流（动）
frictionless fluid 理想流体；理想流动；无粘性流体；无摩擦流体
frictionless hinge 无阻铰链；无摩擦铰(链)
frictionless liquid 无阻尼流体；无摩擦液体
frictionless plane 无摩擦层面
frictionless roller 无摩擦滚轴
frictionless simulator 无摩擦模拟器
friction loss in post-tension 后张法中摩擦损失
friction loss in transition(al)(section) 过渡段摩擦损失
friction loss of discharge pipeline 排泥管线摩擦损失
friction loss of head 沿程水头损失
friction measuring weir 摩阻量水堰
friction number 摩阻数
friction of bearing 支座摩阻力
friction of distance 路程阻力
friction of motion 运动摩擦；滑动摩擦；动摩擦
friction of rest 静止摩擦；静摩擦(力)
friction of rolling 滚动摩擦(力)
friction of static 静止摩擦
frictionometer 摩擦系数测定仪；摩擦系数测定器
friction pile 摩阻桩
friction-producing layer 产生摩擦层
friction pull test 密着力试验；剥离试验
friction ratio 滚筒速比
friction reducer 减阻剂
friction reducing polymer composite 高分子减摩复合材料
friction reel 辊筒式卷纸机
friction resistance 耐磨性；摩阻
friction resistance in ventilation pipe 风筒摩擦阻力
friction resistance on rail 轨道摩擦阻力
friction resistance to sliding 抗滑摩阻(力)
friction-resistant 耐磨擦的
friction ring 挡圈
friction rock bolt 摩擦型锚杆
friction roll 压紧辊
friction roller 导缆口滚柱
friction saw 无齿圆锯
friction seal coat for steep gradient roads 陡坡道路的摩擦封面层
friction sensitiveness 炸药摩擦感度
friction sheave 刹车鼓
friction slab 摩阻板
friction-sparking 摩擦火花
friction speed 不同速度
friction spring 稳定器弹簧
friction stay 摩擦窗撑
friction test 迟钝检查
friction tester 摩擦系数测定仪
friction-tight 紧摩擦
friction top 压紧盖
friction top can 压紧罐
friction-twist draw-texturing machine 摩擦加捻拉伸变形机
friction-type 摩擦型
friction-type anchorage 摩擦锚具；摩擦式锚基；摩擦式锚定
friction-type bolt 摩擦型锚杆
friction-type connection 摩擦式连接；摩擦型连接
friction-type felt cartridge 摩擦式毡垫(翼锭细纱机)
friction-type knitting needle 摩擦式织针
friction-type mix 摩擦粉料
friction-type safety clutch 摩擦式安全离合器
friction-type shock absorber 摩擦式减振器
friction-type unit 摩擦型装置
friction variation 摩阻变化
friction velocity 剪切流速
friction wheel drive 摩擦轮驱动
friction wheel-drive concrete mixer 摩擦轮驱动的混凝土搅拌机；摩擦轮传动的混凝土搅拌机
friction wheel feeder 摩擦轮加料器
friction wheel tachometer 摩擦轮式转速计
friction yielding prop 机械山压支柱
Friday mosque 礼拜清真寺；会众清真寺
fridge 冷冻机；冷藏室；电冰箱；冰箱
Friedel and Craft's synthesis 弗瑞德—克来福特合成
friedelane 元羁烷；木栓烷
Friedel-Crafts-Karrer nitrile synthesis 弗瑞德—克来福特—卡勒成腈合成法
friedelinol 木栓醇
friedelite 红锰铁矿
Friedel's law 弗里德尔定律
friedge 冷藏库
Friedlander quinoline synthesis 弗瑞德兰德喹啉合成法
Friedmann test 弗里德曼试验
Friedmann universe 弗里德曼宇宙
friedrichite 弗硫铋铅铜矿
Friele-MacAdam-Chickering colo(u)r difference equation 费里埃尔—麦克阿达姆—切林色差方程
friendly cooperation 友好合作
friendly ice 堆积冰群
friendly numbers 亲和数
friendship hospital 友谊医院
friends of the earth 地球之友
Fries's rule 弗里斯规则
frieze 檐壁，中楣；或腰线(雕有图案、花纹等)；墙上方的水平装饰带；雕带
frieze block 檐口椽间木，阁楼椽檐板
frieze carpet 紧捻圈绒面地毯；起绒地毯
frieze-like 壁缘式的
frieze panel 最上一块门芯板；(有五块或更多镶板门的)上镶板；上冒头(门)；(有五块或更多镶板门的)顶镶板；栏扳；门眼镜框；束腰板
frieze rail (最上一块门芯板下的)中冒头；上腰板；雕刻条板的下横档
friezette 棱纹家具布
frieze with animal reliefs 有动物浮雕的壁缘
frieze with inscription 有文字雕刻的壁缘
frieze yarn 紧捻地毯毛纱
friezing 毛绒卷曲加工
frigate 海防舰；驱逐领舰；轻帆船
frig bob saw 采石用长手锯（英国矿山切割石料用）
frig(e) 冰箱；冷冻机；冷藏库
frigen 氟利根
frigerant 冷却剂
frightful billows and terrible waves 惊涛骇浪
frigidaire 电冰箱
frigidarium 低温室；冷水浴室(古罗马)
frigid climate 寒带气候；严寒气候
frigid forest-region 寒带林
frigidity 寒冷
frigid karst 寒带岩溶
frigid-temperate coniferous climate 寒温带针叶林气候
frigid-temperate zone 寒温带
frigid temperate zone mix forest illimerized soil zone 寒温带混交林灰化土带
frigid zone 冻结带；寒带
frigofuge 嫌寒植物；避寒植物
frigofuge plant 避寒植物
frigolabile 不耐寒的
frigorideserta 寒荒漠群落
frigorific mixture 冰冻混合物
frigorigraph 冷却计
frigorimeter 深冷温度计；低温计
frigory 每小时千卡(冷冻能力计量单位)
Friling draghead 弗吕林耙头
frill 张开的或打褶的装饰边
frill cuts 竖沟切口
frilled organ 饰品
frill girdling 竖沟环割
frilling 皱边；褶边
fringe 缘饰；散乱边纹；边缘；边沿
fringe amplitude 条纹幅度
fringe area 线条区；市中心外围地区；散乱边纹区；电视接收边缘区
fringe benefit 小额优惠；小额补贴；附加福利；福利金；额外福利；附加津贴；附带利益
fringe contrast 干涉条纹对比；条纹反差；条纹衬度
fringe cost 附加成本
fringe counter 条纹计数器
fringe count micrometer 条纹计数(式)干涉仪
fringe crystal 柱状晶体
fringed filter 条纹滤色片
fringed micelle 缨状胶束；穗状胶束
fringe effect 边缘效应；边际效应
fringe envelope 条纹包络
fringe field 散射场
fringe glacier 裾状冰川；边缘冰川
fringe howl(ing) 临振啸声
fringe identification 条纹证认
fringe industries 次要工业部门
fringe intensity 条纹强度
fringe issue 福利待遇条款
fringe joint 边缘节理
fringe load 高频波动负荷
fringe location 条纹定位
fringe magnetic field 边缘磁场
fringe market 边缘市场
fringe modulation 条纹调制
fringe of sea 海滨；海边；海岸
fringe order 条纹序数
fringe parking 市郊停车场
fringe pattern 干涉图样；干涉图(案)；条纹图样(等色线)；条纹图形(等色线)；等色线条纹(光弹)
fringe phase 条纹相位
fringe radiation 边缘辐射；边缘发射
fringe rate 条纹率
fringe reef 边缘暗礁
fringe region 边缘层
fringe region of the atmosphere 大气边缘层
fringe response 条纹响应
fringe rill mark 细流痕
fringe selvedge 毛边
fringe separation 条纹间距
fringe sharpness 条纹锐度
fringe spacing 条纹空间；条纹间距；条纹间隔
fringe stability 条纹稳定性
fringe stopper 条纹驻留器
fringe stopping 条纹驻留
fringe stopping center 条纹驻留中心
fringe stopping system 条纹驻留系统
fringe torque moment 边缘扭矩
fringe value (材料的)条纹值；条纹的明暗配合
fringe visibility 条纹能见度；条纹可见度
fringe-visibility spectrum 条纹可见度谱
fringe wage 小额优惠工资
fringe water 毛细管水；毛管边缘水；边缘水；(油水接触带的)边水
fringe well 边缘区油井(油田)
fringe zone 边饰带
fringing 静电场形变；加边缘；边缘通量的形成
fringing coefficient 边缘系数
fringing current 沿岸流；岸边流
fringing effect 边缘效应
fringing field 弥散场；边缘场
fringing flux 边缘通量
fringing forest 护堤林带；护堤林
fringing glacier 裾状冰川；边缘冰川
fringing reef 陆地边缘的珊瑚礁；裾礁；裙礁；贴岸礁；边(缘暗)礁；岸礁
fringing reef coast 岸礁海岸
fringing sea 边缘海
fringing straddling 离散
Frischer ring 弗里希尔(瓷)环
frise yarn 毛圈花式线
frisket 印刷器的轻质夹纸框；蒙片；防止粘釉用的纸片
frisking 搜寻
frison 长废丝
frit 釉料；制玻璃原料；熔块；烧料；玻璃原料；玻璃料
frit-glaze tile 熟釉瓷砖
frith 狭窄海湾；河口湾；三角港
frithstool 靠近祭坛的座位(古英国)
Fritillaria 贝母属
fritillarine 贝母属碱
fritillary flower 贝母花
frit-in method 熔入法
frit kiln 熔块窑
frit reaction 熔融反应
frit seal 熔接密封；低熔点玻璃封接
frit seal glass 封接玻璃
fritted china 熔块瓷
fritted glass 熔块玻璃；熔结多孔玻璃；烧结玻璃；多孔玻璃
fritted glass disc bubbler 多孔玻璃吸收管
fritted glass disk 多孔玻璃盘
fritted glass filter 熔结多孔玻璃过滤器；烧结玻璃过滤器；多孔玻璃过滤器
fritted glassware 烧结玻璃器皿

fritted glaze 熔块釉;熟釉
fritted metal 烧结金属
fritted porcelain 熔块瓷;烘炙陶瓷
fritting 烧结;预熔;熔结;烧制熔块
fritting furnace 熔化炉;烧结炉
fritting of the batch 配合料预熔
fritting way 半熔方式
fritting zone 初熔区
fritzscheite 锰钒铀云母
frizzing 缩花;爆花
Frobenius method 弗洛比尼斯法
froben tablet 风平片
frock 工装
Frodingham 弗罗丁翰钢板桩(一种专利产品)
Frodue coefficient 弗劳德数系数
froe 楔刀;劈板斧
frog 砖面凹坑;砖面凹槽;砖凹槽;辙叉;犁托;分线器;标尺台
frog angle 辙叉角;岔心角
frogback handrail 蛤蟆背形扶手
frog-belly 蛙形腹
frog board 蛙板
frog bolt 蛙叉螺栓
frog brick 凹槽砖
frog cam 形心凸轮
frog center 辙叉心
frog clamp 拉钳
frog compactor 蛙式夯实机
frog crossing 弯轨交道叉
frogeye 蛙眼
frog-eye clay 蛙眼黏土
frog-face 蛙面
frog flangeway 辙叉轮缘槽
frogging repeater 换频中继器
frog guard rail 辙叉护轨
frog heel 辙叉跟
frog heel spread 辙叉跟宽;辙叉跟端开口
frogleg 蛙腿式
frogleg windings 蛙腿式绕组;波叠绕混合绕组
frog length 辙叉长度
frogman 轻装潜水员;潜水员;蛙人
frog number 辙叉号数
frog number of points 道岔号数
frog plate 辙叉垫板
frog point 辙叉尖(端)
frog position 蛙位
frog rammer 机动夯;蛤蟆夯;跳跃式打夯机;爆炸夯;蛙式夯;蛙式打夯机
frog-shaped nose 蛙状鼻
frog storm 蛙暴
frog tamper 蛙式夯土机;蛙式捣实机
frog the spread 辙叉趾端开口
frog throat 辙叉咽喉
frog toe 辙叉趾;辙叉跟端
frog tongs 辙叉吊钳
frog-type jumping rammer 蛙式打夯机
frog wing 辙叉翼
frog wing riser 辙叉翼轨加高
Frohlich's stress concentration factor 弗罗利克应力集中系数
frolic pad [俚]夜总会
froloyite 弗硼钙石
from a rapid 退滩
fromat effector character 格式控制字符
from beginning to end 始终
from-depot 始站
from one setup 在一个站点钻一束钻孔
from recessive to dominant 从隐性到显性
from spud-in to total depth 从开孔至终孔总深度
from-station 发站【铁】
from stock 从仓库交货的
from-tank farm 始站
from taxes 免税货物
from the top downward 由上到下
from time to time 随时
frondelite 锰绿铁矿
frondescent cast 叶状铸型
frondescent furrow flute casting 叶状沟槽铸型
front 前缘;前线;前端;前面;前部;刀面;齿前;额前线
front abutment 前拱脚
front abutment pressure 前拱座压力
frontage 沿街宽度;屋向;门窗上部三角饰;门窗上壁拱;正面宽度;临街面;前沿;屋前空地;滩岸
frontage assessment 设施改进后按房地产临街部分多少加征税款
frontage carriageway 屋前临街车道
frontage line 建筑红线;临街建筑线;临岸建筑线
frontage method 街面长度计算法
frontage road 沿街辅道;沿街道路;临街道路;街面道路;集散道路
frontage roadway 沿街道路
frontage space 屋前空地
frontage street 临街道路
frontage street or road 沿街道路;辅助性道路;服务性道路;侧道
frontage traffic 临街住户屋前交通
frontal 门窗上部的三角顶饰;山头【建】;额侧的
frontal action 锋作用;锋面作用;锋面活动;锋的活动
frontal-advance performance 前缘推进动态
frontal air mass 锋区气团
frontal analysis 前缘分析法;前沿分析(法);锋面分析
frontal analysis method 前沿分析法
frontal and loader 拖拉机式装载工具
frontal angle 额角
frontal appearance 前部外观
frontal apron 冰水沉积平原;冰前平原;冰川前堆积层
frontal apron plain 冰前冲积平原
frontal arc 前弧;额弓
frontal area 最大截面;(汽车的)正面投影面积;前面
frontal base 额底
frontal cloud 锋面云
frontal column 正面柱
frontal compression 前部压缩
frontal compression belt 外缘推挤带
frontal coverage 正面对空观察
frontal cyclone 锋面气旋
frontal datum plane 正基准面
frontal drag 正面阻力
frontal drive 前缘驱动
frontal fillet weld 正面角焊缝
frontal fillet welding 正面角焊
frontal fog 锋面雾
frontal grill 前进气活门栅
frontal image 正面影像
frontal inversion 锋面逆温
frontal land 正面临街用地;临街土地
frontal lifting 锋升;锋面抬升
frontal light 正面光
frontal line 正平线;锋线
frontal mass 锋区气团;锋面气团
frontal method 波前法;迎头法
frontal moraine 前碛
frontal occlusion 锢囚锋
frontal orbital 前沿轨道
frontal part 额部
frontal passage 锋面过境
frontal plain 冰水沉积平原
frontal plane 前平面;正平面
frontal plate 额板
frontal precipitation 锋面雨;锋面降水
frontal profile 锋区剖面
frontal-profile line 铅垂线
frontal projected area 正投影面积
frontal projection 正面投影
frontal rain 锋面雨
frontal resistance 正面阻力
frontal slope 房屋正面的坡屋面
frontal strip 锋带
frontal surface 正面;锋面
frontal system (机场办理登记手续的)站前服务系统;锋系
frontal thunderstorm 锋面雷雨;锋面雷暴
frontal type precipitation 锋面型降水
frontal wave 锋面波
frontal weather 锋面(云)天气
frontal weft-insertion 正面衬纬
frontal zone 前沿地带;锋区;锋带
front-and-back shift 前后交接班
front-and-back shunting signal 双面调车信号(机)
front-and-rear axle differential 前后桥差动器
front-and-rear wheel steer(ing) 前后轮转向
front angle 前角
front anomaly 前缘异常
front apron 闸前护坦;码头前沿
front arch 盖板砖;八字砖;鞍前桥
front armo(u)r 前装甲
front axis 前轴
front axis suspension 前轴悬挂
front axle 前轴
front axle aligning tool 前轴对准工具
front axle beam 前桥梁
front axle bracket 前轴架
front axle carrier 前桥托架
front axle control lever 前轴控制杆
front axle control rod 前轴控制杆
front axle differential 前桥差速器
front axle differential case with cap 前桥差速器壳及盖
front axle drive 前轴传动;前桥传动
front axle engagement 前桥结合
front axle fork 前轴叉
front axle housing cover 前桥壳盖
front axle housing with axle shaft housing 前桥壳带半轴壳
front axle number 前轴号码
front axle oil capacity 前桥润滑油量
front axle pivot 前桥支枢
front axle pivot pin 前轴枢销
front axle propeller shaft 前桥传动轴
front axle radius beam assembly 前桥摇摆梁总成
front axle radius rod 前桥半径杆
front axle shaft 前桥半轴
front axle slip angle 前桥侧偏角
front axle stay rod 前桥撑杆
front axle steering knuckle 前桥转向节
front axle suspension 前轴悬挂;前桥悬挂
front axle tie rod 前桥横拉杆
front axle tie rod end 前桥横拉杆末端
front axle universal joint ball 前桥万向节球
front axle universal joint centre ball 前桥万向节中心球
frontback connection 正反面连接
front balcony 前阳台
front bank 开挖面
front barrel flange 料筒前凸缘;壳体前法兰盘
front bearing 前轴承
front bearing spacer 前轮轴承隔片
front bevel 前斜面
front bevel angle 前斜角
front beveled gib 前斜夹条
front bottom end rail (集装箱的)前下端梁
front brake 前刹车;前闸
front brake drum with hub 前制动鼓带轮毂
front brake limiter switch 前轮制动控制器
front brake tube T-union 前轮制动管路三通接头
front brake wheel cylinder (液压制动系)前轮制动分泵缸
front brick 正面砖
front bridle chains 前限动装置;前拉紧装置
front building 前部房屋
front bulkhead 前端舱壁;前舱壁
front bumper 前保险杠
front bumper arm 前保险杠臂
front bumper arm grommet 前保险杠套环
front bumper bracket 前保险杠支架
front bumper height 前保险杠高度
front bumper shock absorber 前保险杠减振器
front cable winch 前钢缆绞车
front cap gasket 前部帽衬
front carrier 前行星齿轮架
front case 正面壳
front-cell focusing 前透镜调焦
front-cleaned rack 正面清洁格栅
front clearance 副后角
front clearance angle 副后角
front-coated mirror 前表面镀膜镜
front coil 前圈
front column 前柱
front compression 前端压缩
front conflict 正面冲突
front connected switch 前面接线开关
front connection 正面连接;(家庭用电表等)前面接线
front connection type 前面接线式
front contact 前接点
front continuous fillet weld 正面贴角连续焊缝
front cord 正面塞线
front court 前院空地
front courtyard 前院

front cross conveyer 前部横向输送器
front cross-section 正剖面
front crushing plate 前压(碎)板
front curtain 大幕
front-cutting edge 副切削刃
front cylinder cover 前缸盖
front damper 前阻尼器
front desk 接待厅
front despatching 前方调度
front directional 前转向灯
front dispatching 前方调度
front door 正门；门；大门
front door telephone 大门口通话装置；大门口电话装置
front drive 前驱动；前传动；前端驱动；前轴驱动的
front drive axle clutch fork 前桥传动离合叉
front-drive chassis 前轮转动底盘
front-drive rake 前轮传动式搂草机
front-drive truck 前轮驱动载货汽车
front drive vehicle 前轮驱动汽车
front dump 前卸
front-dump closure 立堵截流
front dumping 前端翻卸
front-dump lorry 前卸推土车；前倾翻斗车
front-dump scraper 前卸式耙斗
front-dump truck 前卸推土车；前倾翻斗车
front dump-type ingot buggy 端卸式送锭车
front edge 前沿；石板瓦前缘
front-effect photocell 半透明光电阴极光电管
front element 前组；前缘元素
front element mount 前透镜框
front elevation 正视图；正面图；正立面
front elevation drawing 正(立)面图；立视图；立面图
front elevation view 正视图；前视图
front elevator 前置升运器
front embankment 正堤；前堤
front end 中高频端；高频端；端部；前端
front-end angle ring 前端角铁
front-end attachment 前端附加装置
front-end bucket 前装式铲车；前装式铲斗
front-end circuit 前端电路；前置电路
front-end communication processor 前端通信处理机
front-end computer 前端计算机
front-end concentrator 前端集中器
front-end conversion unit 前端转换装置；前端转变装置
front-end cost 前端费
front-end crawler shovel 前装履带式铲土机
front-end dead weight 前部死重量；前部死荷载
front-end dumper 前端自卸车
front-end equipment (起重机的)前设备；前悬挂装置；前端式设备
front-end fee (订约后先收的)一次性费用；开办费；先付费用；一次性手续费；期初费用
front-end finance 期初贷款；前端信贷
front-end frame 前端框架
front-end housing 前端盖
front-end intermodulation 前级互调
front-end investment 初期投资
front-end job 前轮保修作业
front-end lift 前端装车机
front-end loader 前卸(式)装载机；正铲装斗车；前装机；前悬装载机；前装(式)装载机；翻斗车；正向装料机；正向装料机
front-end loading 期初加重收费
front-end of spindle 主轴前端
front-end panel 前端板
front-end processing 前端处理
front-end processor 前端处理机；前端通信处理机
front-end rig 前端设备
front-end shop 前桥保修车间
front-end shovel 前装铲车
front-end sill (车身底板的)前部端梁
front-end static weight 前部静重量
front-end support 前端支持
front-end tractor loader 前装牵引装载机
front-end tuner 前端调谐器
front-end type loader 前端装载机
front-end wall 前端壁
front-end weight 前部重量
front-end work attachment 前端装备；前部装备
front engine 前发动机

front engined car 前置发动机小客车
front engine front drive vehicle 前置发动机前轮驱动式车辆
front engine location 前置发动机位置
front engine support cross member 前置发动机支承横梁
front engine vehicle 前置发动机汽车
front entrance 前门；正面入口；正门
front equipment 前端装备；前部装备
front equivalent yard 沿街后院
front exit 前方出口
front face(side) 正面；前切削面；前(端)面
front face area 迎风面积；正面面积
front face of heading 导坑前端
front factor 前沿因子
front fender 前护板；前保护板
front fillet weld 正面填角焊；正侧面填角焊缝
front fitting radius 前回转半径【机】
front fitting radius of semi-trailer 半拖车前回转半径
front fitting radius of semi-trailer tractor 半牵引车前回转半径
front flank (牙轮的)前面齿
front focal distance 前焦距
front focal length 前焦距
front focal plane 前焦平面
front focometer 前焦距计
front focus 前焦点
front fork 前轮叉
front fort 前线堡垒
front garden 前花园；宅前花园；前园
front gate 正门；前门
front ga(u)ge 前轮距；前轮轨距
front gears 正时齿轮；分配齿轮
front glass 前面玻璃；遮光玻璃；前透镜；挡风玻璃；保护玻璃
front guard 前护栅；前防护器；前保险杠
front guard frame 前保险构架
front halo 前缘晕
front hammer 捣锤
front handling mobile crane 正面吊运自行式起重机；正面吊运移动式起重机；正面吊运汽车式起重机
front harbo(u)r 外港
front head 前头部
front head air release (凿岩机钻头)前端排气孔
front-head release port 前端泄气口
front hearth 壁炉前池
front-heavy car 前桥重量分配大的汽车
front-hinged bonnet 前翻式机罩
front idler 前惰轮；前导向轮；导向轮
frontier 新领域；国境；拓宽；边境；边界；新疆
frontier defence inspection station 边防检查站
frontier depot 国境车站
frontier phenomenon 边界现象
frontier point 国境分界点(铁)；边界点
frontier science 尖端科学
frontiers of science 尖端科学
frontier station 国境站；国境车站
frontier survey 边界勘测
frontier trade 边境贸易
frontier zone traffic 边境交通
front index plate 正面分度盘
front intake door 前进口舱门
front intermontane basin 前山间盆地
frontispiece 门窗上壁拱；房屋主要立面；正面目标；弧形门楣
front jockey wheel 前导向轮；前导轮
front lamp 前灯
front lay 前沿放置；前端定位装置
front leading lamp 前导灯
front leading light 前导灯
front leading mark 前导标
front leg (三脚架的)前腿
front lens 前透镜
front lens mount 前透镜框
frontlet 额饰
front lifting 前方超吊
front light 正面光；前灯；头灯
front lighting 面光；顺光照明
front light tower 前灯桩
front line 第一线
front line airfield 前线机场
front line first aid 火线抢救

front line manager 第一线管理人员
front line of a zone lot 区划地块的界线
front line of building 房屋的前沿线
front lintel 门过梁；露明过梁；前横梁；前过梁(承托空芯墙外层的过梁)
front lip tile 前唇砖
front loader 前挖斗式装载机；前(端式)装载机；前装机
front loading 前端装载
front loading forklift 前向叉车
front loading mechanism 前装载机构
front-located ram 前置式动力油缸
front locating pad 前定位垫
front lorry 前剷推料车；前倾翻斗车
front lot line 沿街边线
front mill table 前工作辊道
front money (购置房地产的)起码资金
front moraine 前碛
front mo(u)ld 前模
front-mounted flywheel 前装式飞轮
front-mounted loader 前悬装土机
front-mounted radiator 前置散热器
front-mounted self-contained unit 前装式自备动力钻机；前部安装凿岩机和有单独压风机的自行式凿岩设备
front moving lens 前移动透镜
front mower attachment 前置割草装置
front nodal point 前节点
front of building 房屋的正面
front of column 主柱正面
front office 旅馆前厅办公室；负责人办公室
front of levee 堤坝前沿；堤前；堤防正面；堤防临河面
front of queue 队列的前端
front of saddle 溜板正面
front of station 车站正面
front of terrace 阶地前沿
front of wave 激波阵面；波头；波前
front of wave test 脉冲波试验；陡波试验；波前冲击试验
frontogenesis 锋生(作用)；锋生过程；锋面生成
frontology 锋面学
frontolysis 锋消；锋面消失
fronto(o)n 山墙顶端；三角形楣饰；三角梁；大门门头饰；回力球场；山头【建】；三角门楣
front operating aperture 前工作孔径
front outline 正视图；垂直投影；前视图；前视轮廓图
front outreach 前伸距
front overhang 前沿长度
front page 标题页
front panel 前面板；防冲钢板；防冲板
front panel command 前面板命令
front panel control 面板控制
front panel(curtain)wall 预制面板；幕墙面板
front panel function 前板功能
front panel illuminator 面板照明灯
front panel microroutine 前板微程序
front panel(prefab) 建筑物正面前覆盖板
front part actual length of turnout 道岔前部实际长度
front part theoretical length of turnout 道岔前部理论长度
front passage fog 锋际雾
front perspective 正面透视
front piled platform 前方桩基承台
front pile platform 前方承台
front pilot (拉刀的)前导部
front pinacoid 前轴面
front pitch 前节距
front porch 脉冲的前沿；前沿；前门廊
front porch interval 前基座时间
front porch switching 前沿切换
front position 前端位置
front power take-off 前部功率输出端
front power unit 前端驱动装置
front print 正面打印
front projection 直接投影
front property line 个人地产与国家公路之间的界线；地产前沿界线
front protection 前护栅；前防护器
front pull hook 前拉钩
front pump 前泵
front pusher blade 推土机前括铲；前推动器叶片
front putty 面油灰；面灰；抹面油灰；抹面腻子；底灰(指油漆)

front quarter light 近光灯
front rail 横向栏杆
front rake 前耙;前路耙;前倾角
front raking pile 前斜桩
front raking pile driving 俯打
front raking pile force 前斜桩力
front range 前山甲;前海角
front range light 前导向灯
front-rank 第一流的
front reaction 前端反应
front rearnodal point of lens 物镜前(后)节点
front-rear weight distribution 前后轴重量分配
front rig 前端设备
front rod 尖端杆
front roll 前滚轮
front roll centre 前滚动中心
front roller shaft 前导轮轴
front roll rate 前滚动率
front room 房子前部的房间
front screed (抹灰的)前样板
front screen 风挡
front-screen projection 前景放映法
front screw 前螺杆
front seat 前座
front sectional elevation 前视剖面图
front shear 正面剪切;正面冲剪
front shear line 锋面切变线
front sheet (集装箱)端壁板
front sheet-piling platform 前板桩式高桩码头
front shield 前置屏蔽
front shock absorber 波前减震器
front shoe 前瓦形支块
front shot 前视
front shovel 正铲挖掘机
front shutter 镜头快门
front side 正面;正视图;前端;前面
front side guard 轧机前的推床导板
front side spot (观众厅两侧的)耳光室
front sight 准星
front-silvered mirror 前表面镀银反射镜
front slagging 炉前出渣
front slagging spout 连续出铁槽
front slip angle 前轮滑移角
front slope 悬崖坡;迎水坡;仰坡;堑壕前崖;前坡
front slope of air temperature 气温前沿坡度
front slope of dike 坝的迎水(面)坡(度)
front snap guard 前闸轨
front span 前节跨
front spot 前光(观众厅的灯光总称)
front spout cover 供料机盖砖
front spring 前钢板
front spring eye bushing 前钢板弹簧卷耳孔衬套
front spring fourth leaf clamp 前钢板弹簧第四片卡扣
front spring pad 前钢板弹簧橡胶块
front stage 舞台前沿
front stair(case) 前楼梯
front stand 前支架;前台座
front standing pillar 前支柱
front stand wheel 前支轮
front stead 庭院;屋前基地
front steps 前门台阶
front string 露明楼梯斜梁;明露楼梯斜梁
front support rod 前支杆
front surface 前表面;端面
front surface effect 前面效应
front surface mirror 前表面反射镜;前表面反光镜
front surface of the sheet 冷面;玻璃板背面
front surface reflection 前表面反射
front suspension 前悬挂
front tarpaulin 前盖布
front tension bar 前拉杆
front terminal type 正面端子型
front tile 屋檐瓦;前面瓦
front tipper 前倾翻斗车;前翻斗矿车;前翻斗手推车;朝前卸料手推车
front-tipping bucket 前倾式铲斗
front-tipping wagon 前倾翻斗车
front-to-back effect 前后影响;前后不一致的影响
front-to-back lap 前后重叠
front-to-back ratio 正反向比
front-to-back slope 前后坡度
front tool 端面切刀
front top end rail (集装箱)前上端梁

front top rake 前顶角
front top rake angle 主前角;副前角
front-to-rear ratio 前后比
front tow hook 前拖钩
front track 前轮轮距
front truck 前转向架
front-turn signal lamp 前转向信号灯
front tyre 前轮轮胎
front vertex distance 前顶焦距
front vertex focal distance 前顶焦距
front vibrator 前振动器
front view 主视图;正视图;正面(视)图;前视图;对景图
front view of gate of ventilation duct 风道门正面
front wall (桥台的)胸墙;正面墙;前墙;前脸墙;前壁;成型端墙
front wall cell 前壁光电池
front wall of abutment 桥台胸墙;桥台前墙
front wave 波阵面
front-waveband synthesized record 前沿波带复合记录
front wearing plate 前护板
front-wheel aligner 前轮定位工具
front-wheel alignment 转向轮定位;前轮对准;前轮定位
front-wheel angle tester 前轮定位仪;前轮定位试验器
front-wheel ballast weight 前轮配重块
front-wheel bearing 前轮轴承
front-wheel bearing nut lock washer 前轮毂轴承螺帽锁紧垫圈
front-wheel brake 前轮闸
front-wheel brake control 前轮制动控制器
front-wheel brake drum grease guard 前轮闸鼓护油罩
front-wheel bump travel 前轮跳动行程
front-wheel camber 前轮外倾角
front-wheel cylinder 前轮油缸
front-wheel drive 前轮驱动
front-wheel drive car 前轮驱动小客车
front-wheel drive loading shovel 前轮驱动斗式装载机
front-wheel drive system 前轮驱动系
front-wheel drive vehicle 前轮驱动式车辆
front-wheel dust washer 前轮防尘垫圈
front-wheel felloe 前轮缘
front-wheel ga(u)ge 前轮距
front-wheel grease retainer 前轮护脂器
front-wheel house complete panel 前轮罩板总成
front-wheel hub 前轮(轮)毂
front-wheel hub cap 前轮毂盖
front-wheel inner bearing 前轮内轴承
front-wheel lean 前轮倾斜(筑路机)
front-wheel lean control housing 前轮倾斜控制箱
front-wheel lean cylinder 前轮倾斜液压缸
front-wheel lean rack 前轮倾斜支架
front-wheel lean shaft 前轮倾斜轴
front-wheel locking mechanism 前轮锁止机构
front-wheel mechanism 前轮机构
front-wheel oil seal felt washer 前轮挡油毡垫
front-wheel pitch 前轮外倾度
front-wheel pivot pin bearing 前轮支枢销轴承
front-wheel retainer 前轮限止器
front-wheel steer(ing) 前轮转向
front-wheel tread 前轮轮距
front-wheel weight 前轮荷载
front-wheel wobble 前轮摆动
front wiring 明线布置;明线布线
front yard 屋前临街庭院;前庭;前院
froodite 斜铋钯矿
frost 磨砂;冰冻;霜冻
frost action 寒冻作用;霜冻(作用);冻裂作用;冻结作用;冰冻作用
frost-action design 防冻害设计
frost-active soil 霜冻作用土壤(层);易冻土
frost area 冰冻地区
frost attack 冰冻破坏;冰冻侵蚀
frost back 倒霜
frost batter 防冻倾斜度
frostbite 冻害;霜冻;冻伤;冻疮
frostbite preventive 防冻药品
frost-bitten corner 受冻掉角
frost blanket 防霜层;防冻层
frost blanket gravel 抗冻砾石层

frost blanket sand 冰冻冲积砂(层)
frost board 防冻盖板
frost boil area 翻浆地区
frost boil(ing) 冻胀;霜凸(地);冻沸(现象);冻融(翻浆);翻浆;结冰翻浆;冻腾
frost boil soil 霜沸土
frost bottom 防冻底
frost bound 封冻(的);冰结的
frost box 防冻箱
frost bursting 融冻崩解作用
frost canker 霜瘤
frost casing 防冻壳
frost churning 融冻泥流(作用)
frost cleft 霜裂;冻裂隙;冻裂
frost climate 冰原气候
frost closure 封冻
frost condition reduced (sub) grade strength design 考虑冻害导致强度下降的地基设计
frost crack 霜裂;冻裂(缝);冰冻裂隙
frost cracking 霜冻开裂;冻裂作用
frost cracking of foundation 地基开裂
frost crazing 冷冻开裂
frost-criteria evaluation 霜冻标准的评定
frost criterion 霜冻准则
frost damage 霜冻损害;霜冻害;冻裂;冻坏;冻害;冰冻损伤
frost day 霜日
frost depth 冻结深度;冰冻深度
frost depth indicator 冻结深度指示器
frost dew 冻露
frosted 无光泽的;闷光的;磨砂显微镜载玻片;磨砂的;生烧无光的;成霜状
frosted bulb 磨砂灯泡
frosted dried film 冰花干燥薄膜
frosted face 毛化面;无光泽(表)面;霜化面
frosted finish 霜花面饰;磨砂;浸蚀饰面;雪花面饰【建】;毛面光洁度;毛化整理;霜白表面;表面带微裂纹瓶口
frosted glass 雪花玻璃;磨砂玻璃;毛玻璃;霜化玻璃;霜花玻璃;冰花玻璃
frosted glass lamp shade 毛玻璃罩
frosted glass plate 毛玻璃器皿;毛玻璃片
frosted glass window 磨砂玻璃窗
frosted glass with muslin pattern 磨砂花纹玻璃
frosted incandescent lamp 磨砂白炽灯
frosted lacquer 无光漆
frosted lamp 闷光灯泡;毛玻璃灯泡
frosted lamp bulb 磨砂灯泡
frosted lamp globe 磨砂球形灯泡;磨砂灯泡
frosted rustic work 霜花粗面装饰
frosted screen 毛屏
frosted work 仿雪花装饰
frosted yarn 霜花纱
frost effect 冻结作用;冻结效应;冰冻作用;冰冻影响
frost feather 冰羽
frost fissure 冰冻裂隙;冰冻裂缝
frost flakes 冰雾
frost flower 霜花
frost fog 冰雾
frost fracture 霜冻裂缝;冻裂缝
frost-free 无霜(冻)
frost-free depth 冰冻深度
frost-free growing period 无霜生长期
frost-free growing season 无霜期
frost-free period 无霜期
frost-free season 无霜季节;无霜期
frost front 冻结前缘
frost glass 薄片碎玻璃
frost grit spreader 防霜石屑摊铺机;防霜粒料摊铺机;防滑石屑摊铺机;防滑粒料摊铺机
frost gritter 霜冻砂子摊铺机
frost-gritting 撒砂防滑
frost hardiness 耐冻性
frost hazard 冻灾;霜冻危险
frost haze 冻霾
frost heart 冻髓(木材髓部受冻伤颜色加深);冻心材
frost heave 霜拔;冻胀丘;冻胀;冻结隆起;地坪冻胀;冰冻隆胀;冰冻隆起
frost heave board 冻害垫板
frost-heave capacity 冻胀量
frost-heave capacity test 冻胀量试验
frost-heave force 冻胀力
frost-heave mound 冻胀丘
frost-heave soil 冻胀土

frost heaving 霜拔;冻胀;冻拔;道路冻胀;冰冻隆胀
frost heaving force 冻胀力;地基冻胀力
frost heaving of foundation 地基冻胀
frost heaving soil 冻胀土
frost hole 霜冻孔
frost hollow 霜洼;霜冻穴;成霜洼地
frost index 霜冻指数
frost-induced cracking 冰冻引起的龟裂
frosting 涂成霜面;消光表面;晶纹花;结霜;磨砂面;去光泽面;去光泽;起霜花;起霜;霜冻;霜纹;霜面(化)
frosting bath 霜化槽
frosting device 除霜装置
frosting machine 磨砂机
frosting oil 晶纹油
frosting salt 霜盐
frosting surface 磨砂面
frosting varnish 无光漆;无光(泽)清漆
frosting work method 冻结法
frost in houses 室内结霜
frost injury 霜害;冻害
frost killing 冻死
frost layer 冰冻层
frost lens 冰晶
frostless 无霜的
frostless season 无霜期;无霜季节
frostless zone 无霜带
frost level 霜高
frost level indicator 结霜液面指示器
frost line 霜线;冻深线;冻结线;地冻深度;冰冻线
frost-melting period 融冻时期
frost-melt period 冻融融化时间
frost mist 霜雾
frost mound 冻丘
frost nail 马掌钉
frost patterns 霜花图案
frost penetration 冰冻深度;冻土层厚度;冻结深度;深度渗入
frost penetration depth 冰冻深度;霜冻(渗入)深度
frost-penetration zone 霜冻(渗入)地带
frost period 霜期;冰冻期
frost phenomenon 冰冻现象
frost pocket 霜袋地
frost point 霜冻点;霜点;冰点
frost-point hygrometer 露点湿度计;露点湿度表;霜点湿度计;霜点湿度表
frost-point technique 霜点方法
frost polygon 冰冻龟裂形
frost precaution 冰冻措施
frost pressure 冻胀压力
frost prevention 防霜冻;防霜;防冻
frost process 起霜工艺
frost-prone 易冻的
frost-proof 抗冻的;防冻的;防冻
frost-proof brick 耐冻砖
frost-proof calcium silicate brick 耐冻硅酸盐砖;防霜硅酸盐砖
frost-proof closet 防冻厕所
frost-proof course 防冻层
frost-proof depth 冻结层底面深度;防冻厚度;防冻深度
frost-proofer 防冻剂
frost-proof foundation 防冻基础
frost-proof horizontal coring(clay)brick 耐冻水平孔黏土砖;抗冻水平孔黏土砖;防霜水平孔黏土砖
frost-proof lime-sand brick 耐冻灰砂砖;抗霜灰砂砖
frost-proof material 耐冻材料;防冻材料
frost-proof motor 耐寒式电动机
frost-proof porous brick 耐冻多孔砖;抗霜多孔砖;抗冻多孔砖
frost-proof sand-lime brick 耐冻灰砂砖;耐冻硅酸盐砖;抗冻灰砂砖;防霜硅酸盐砖
frost-proof solid brick 耐冻实心砖;抗冻实心砖;防霜实心砖
frost-proof solid calcium-silicate brick 防霜实心硅酸盐砖
frost-proof solid sand-lime brick 抗冻实心灰砂砖
frost-proof vertical coring(clay)brick 耐冻垂直孔黏土砖;抗冻垂直孔黏土砖
frost-proof wing 防霜翅片
frost protection 霜冻防护;冻结防止;防霜(冻);防冻(措施)
frost protection blower 防冻热风吹送器
frost protection course 防冻层
frost protection design 防冻设计
frost protection layer 防冻层
frost protection liquid 防冻液;防冻剂
frost protection measure 防冻措施
frost protection powder 防冻粉
frost protection solution 防冻液
frost protective 防霜的;防冻的
frost removal 除霜器;除霜;防霜
frost removal timer 化除霜定时器
frost resistance 抗霜(冻)性;抗浆力;抗冻力;耐霜性;耐冻性;耐冬性;耐久性;防起霜性;抗冻(性)
frost-resistant 抗冻性;防霜的;防冻的
frost-resistant brick 耐冻砖;防霜砖
frost-resistant brickwork 防冻砖砌体
frost-resistant calcium silicate brick 耐冻硅酸盐砖;防霜硅酸盐砖
frost-resistant concrete 耐冻混凝土;抗冻混凝土;防冻混凝土
frost-resistant engineering brick 抗冻工程用砖;防霜工程用砖
frost-resistant lime-sand brick 抗霜灰砂砖;抗冻灰砂砖
frost-resistant masonry work 耐冻砌筑工作;防霜砌筑工作
frost-resistant material 防霜材料
frost-resistant porous brick 抗霜多孔砖;抗冻多孔砖
frost-resistant sand-lime brick 抗霜灰砂砖;抗冻灰砂砖
frost-resistant solid brick 抗冻实心砖;防霜实心砖
frost-resistant vertical coring(clay)brick 耐冻水平孔黏土砖;耐冻垂直孔黏土砖;抗冻垂直孔黏土砖;防霜水平孔黏土砖
frost-resisting 耐寒的;抗冻(的);防冻的
frost-resisting admixture 抗冻添加剂
frost-resisting brick 抗冻砖;防霜砖
frost-resisting concrete 抗冻混凝土
frost-resisting material 耐冻材料;防冻材料
frost-resisting power 耐寒力
frost-resisting property 耐冻性;抗冻性;防霜性;耐寒性
frost resistivity of rock 岩石抗冻性;岩石的抗冻性
frost resistivity test of rock 岩石抗冻试验
frost retarding layer 防冻隔离层;防冻层
frost rib 霜肿;霜冻伤疤
frost rib ridge 霜棱或霜脊
frost ring 霜冻轮;冻伤年轮
frost riving 融冻崩解作用
frost scaling 霜冻剥落;冰冻剥落
Frost's cement (用白垩和黏土制成的)弗劳斯特水泥
frost season 霜期
frost shake 冻裂
frost shattering 融冻崩解作用;冻裂作用
frost-shed 霜棚
frost shield 防霜玻璃
frost shim 冻害垫板
frost smoke 海上蒸汽雾;霜烟;霜雾;冻烟;冻雾
frost snow 冰晶
frost splitting 融冻崩解作用;冻裂
frost stirring 融冻泥流(作用)
frost strength 抗冻强度;冻结强度
frost strength test 冻土强度试验
frost susceptibility 易冻性;霜冻敏感性
frost susceptible 易冻胀的
frost-susceptible soil 易冻土
frost table 冻土面
frost thawing 冻融
frost thrust 冰冻胀力;冰冻推力
frost upheaval 冻胀;冰冻隆起
frost valve 防冻阀
frost weathering 寒冻风化;融冻崩解作用;冰冻风化(作用)
frost wedging 融冻崩解作用;冻楔作用;冰楔作用
frost work 冻裂;仿雪花装饰;(银器等的)霜花(纹)装饰;霜花
frosty weather 霜冻天气
frosty wrinkled appearance 冰纹外观
frosty-zone 霜带
frost zone 结冻区;冻结区;冻结带;冰冻区;冰冻地带;冰冻层
froth 泡沫
frothability 起泡度
frothability index 起泡性指数
froth agent 起泡剂
froth boiling apparatus 泡沫精炼机
froth breaker 消泡器;消泡器
froth buildup 泡沫堵塞
froth chromatography 泡沫色谱
froth concentrate 泡沫精矿
froth-control 泡沫控制
froth-control system 泡沫控制系统
froth cooler 泡沫接触式冷却器
froth degumming 泡沫脱胶
froth destroyer 泡沫抑制剂;泡沫消除剂
frothed glass 泡沫玻璃
frothed latex 泡沫胶乳
frother 起泡剂;起泡剂;泡沫剂;泡沫发生器
froth fire extinguishing system 泡沫灭火系统
froth flo(a)tation 泡沫浮选(法);浮选法
froth flo(a)tation hydrotator 泡沫浮选水转分选机
froth flo(a)tation method 泡沫浮选法
froth flo(a)tation process 泡沫浮选法
froth flo(a)tation separation 泡沫浮选分离
froth flo(a)tation test 浮选试验
froth flow 泡流
froth formation 起泡沫;发泡沫
froth formation action 发泡作用
froth formation in situ 现场发泡
frothily 起泡沫
frothiness 起泡沫性
frothing 起泡的;起沫;泡沫现象;翻浆;发泡
frothing agent 起泡剂;起泡剂;泡沫剂;发泡剂
frothing capacity 起泡性能
frothing collector 起泡收集器;起泡捕收剂
frothing oil 起泡沫油
frothing percentage 发泡率
frothing(re)agent 起泡剂
froth killer 泡沫消除器;泡沫消除剂
froth level detector 泡沫水平探测器
froth liquid vessel 泡沫灭火剂容器
frothmeter 泡沫计
froth nozzle 泡沫喷射器
froth-over 冒泡;泡腾
froth overflow 泡沫溢出
froth paddle 泡沫刮板
froth pit 气泡斑
froth promoter 泡沫促进剂
froth pulp 泡沫产品
froth recovery 泡沫产品回收率
froth removal zone 泡沫排除区
froth reservoir 泡沫储[贮]槽
froth rubber 泡沫橡胶
froth separation 泡沫分选
froth stiffener 泡沫强化剂
froth tower 泡沫发生塔
frothy 泡沫状的
frottage (作图时表现宽感的)擦出技法
frottage corrosion 摩擦腐蚀
Froude brake 水轮制动机
Froude number 弗劳德数;运动水流因数
Froude number 2 第二弗劳德数
Froude's criterion 弗劳德准则
Froude's dynamometer 弗劳德功率计
Froude's law 弗劳德定律
Froude's law of model test 弗劳德模型试验法则
Froude's law of model testing 弗劳德模型试验法则
Froude's law of similarity 弗劳德相似律
Froude's model law 弗劳德模型律
Froude's number 弗劳德数;弗劳德数
Froude's similarity criterion 弗劳德相似准则
Froude's similarity law 弗劳德相似律
Froude's transition curve 弗劳德缓和曲线
frow 楔刀
frowst 室内闷热;霉臭
frowy 腐木;变软发脆的陈木
frozen account 冻结账户;被冻结的存款
frozen action 冰冻作用
frozen assets 冻结资金
frozen bearing 冻结轴承
frozen bit 冻结的钻头;扎住的刀头;冻结的钻头;冻结的刀头
frozen cake 冻块
frozen capital 冻结资金
frozen cargo 冷冻货(物)(冷藏船装运华氏30度以下货物)(30华氏度≈−1摄氏度);冰冻货

frozen casing 扎住的井管;卡住的井管
frozen chamber 冷藏舱
frozen component 初凝组元
frozen composition 固定成分
frozen credit 冻结信用贷款
frozen crust 冻壳;冻结层;冰壳
frozen deposits 冻结存款
frozen depth 结冻厚度;冻深;冰冻深度;冰冻厚度
frozen desulfurization 冷冻法脱硫
frozen dew 冻露
frozen dowel bar 冻结传力杆
frozen drill pipe 扎住钻井管;冻结钻井管;被卡钻杆
frozen droplet 冻雨
frozen dynamite 冻结胶质炸药
frozen earth 冻土,冻结土
frozen earth dam 冻土坝
frozen earth storage 冻土储存
frozen equilibrium 冰冻平衡
frozen field 冻结场
frozen fish 冻鱼
frozen fish store 冷冻鱼储[贮]藏室;冻鱼库
frozen flow 冻结流
frozen fog 冻雾;冰雾;冰霜
frozen food 冰冻食物
frozen food store 冰冻食品库
frozen frazil slush 屑冰块
frozen freight 冻结货物
frozen fund 冻结款项
frozen ground 冻(土)地;冻土(层);冻结地(层);地冻
frozen-ground phenomenon 地冻现象
frozen-heave factor 冻胀率
frozen-heave force 冻胀力
frozen-heave test apparatus 冻胀检测仪;冻胀(测试)仪
frozen-in distribution 固定分布
frozen-in field lines 磁力线冻结
frozen ingot 凝固锭料
frozen injury 冻害
frozen-in magnetic field 冻结磁场
frozen iron 冷铁;凝结金属
frozen joint 冻结接头【铁】
frozen layer 冻土层
frozen loan 冻结贷款
frozen meat 冻肉
frozen meat ship 冻肉运输船
frozen mercury process 冰冻水银模熔模铸造法
frozen metal 凝固金属
frozen off 冻住
frozen overburden 永冻覆盖层
frozen pack storage 冻藏
frozen pension 一次付清养老金
frozen picture 静态图像;凝固图像
frozen plant 冻伤的植物
frozen port 封冻港
frozen precipitation 固态降水;冻雨;冻结降水
frozen price 限价;冻结价格
frozen product container 冷藏集装箱
frozen products clause 冰冻货物条款
frozen products container 冷货集装箱
frozen products insulated container 冷冻集装箱
frozen rain 冻雨
frozen runway 结冰跑道
frozen sea 冰冻海
frozen section 冰冻切片
frozen shaft sinking 冻结法井筒下沉;冰冻法凿井;冰冻法打井
frozen-shoulder 冻肩
frozen snow 冻雪
frozen soil 冻土;冻结土
frozen soil investigation 冻土调查
frozen soil mechanics 冻结土壤力学
frozen soil talik 冻土融区
frozen soil thaw compression test 冻土融化压缩试验
frozen state 冷凝态
frozen stress 冻结应力
frozen-stress method (光弹性的)冻结应力法
frozen string of casing 井管的轧住线;井管的卡住线
frozen sucker set 棒冰机
frozen-thawed soil system 冻土系线
frozen throat 冻结的流液洞
frozen transport 冷冻运输

frozen tundra 冻土带
frozen up 卡住
frozen wage 冻结工资
frozen water 固态水;冻结水
frozen wave 冻结波
frozen zone 冻结带;冻土带;冰冻带;冰冻层
froze thickness of surface water 地表水冻结厚度
Frucote 氨丁烷
fructofuranose 呋喃果糖
fructus citri sarcodactyli 佛手片
Fructus Citri Sarcodactylis 佛手
fructus industriales 一年生植物;人工产品
fructus naturales 自然产物;多年生植物;天然收成
Frue vanner 弗罗型洗选槽;弗罗型侧摇带式流槽;淘矿机
fruit 果园
fruit and vegetable drier 果蔬干燥机;水果和蔬菜干燥机
fruit and vegetable filler 果蔬装填机
fruit and vegetable peeler 果蔬去皮机
fruit and vegetable processing plant 水果和蔬菜加工厂;果品蔬菜加工厂
fruit and vegetable processing waste 水果和蔬菜加工废物
fruit bearing forest 果林
fruit bearing plant 结果植物
fruit branch 果枝
fruit carrier 鲜果船;水果专用船;水果运输船
fruit culture 果树栽培
fruit drops 蓝雀冰
fruit-effect plant 观果植物
fruit farming 果树栽培
fruit garden 果园
fruit gardening 果树栽培
fruit grower 果农
fruit growing 果树栽培
fruiticetum 灌木群落;灌木丛
fruit inception 开始奏效
fruiting branch 果枝
fruiting cane 结枝
fruit-juice tablet wrapper 果汁糖块包装机
fruit machine 吃角子老虎(赌具)
fruit of needle juniper 杜松实
fruit parlour 水果店兼冷饮店
fruit period 果期
fruit piece 果实的静物画(雕刻)
fruit retention 保果
fruits 结果
fruit setting garden 结果果园
fruit shed 水果仓库
fruit ship 鲜果船;水果运输船
fruit stand 水果摊
fruit stock 水果仓库
fruit storage 果实储[贮]藏库
fruit tree 果树
fruit vegetable drier 果蔬干燥机
fruit work 用水果装饰成弧形垂莲花彩
frunk wrapping 包干
frustrated internal reflectance 衰减全反射比
frustrated multiple internal reflectance 多次内反射装置
frustrated reflection 受抑反射;受损反射
frustrated total reflection 受抑全反射
frustrated total reflection filter 受抑全反射滤光器
frustration 航次受阻
frustration clause 航次受阻条款
frustration of contract 合同落空;契约无效;契约落空
frustum 平截头体;平截头台;平截头璇;柱身;立体角;棱锥台;井圈状;截锥(体);截头体;[复]frusta 或 frustums
frustum cone 圆锥台
frustum of a cone 圆锥截体;截锥(体);截头锥体;平截头圆锥体
frustum of parabola 抛物线截角锥体
frustum of pyramid 圆锥台;棱锥台;截头锥体;平截(头)棱锥体
frustum of sphere 球截角锥体
frustum of wedge 尖劈截角锥体
frutescent 灌木状的
frutex 灌木
fruticeta 灌木群落;灌木林
fruticose 灌木状的

fry-dry method 炒干法
fryer 油炸(煎)锅;彩色摄影照明器
frying-pan drying 炒锅干燥
frying-pan method 炒锅法
F-test of regression model 回归模型的F检验
fubrogonium iodide 呋波碘铵
Fuchs' drier 夫斯(气体)干燥器
Fuch's gold purple 福契氏紫金颜料
fuchsia 紫红色
fuchsin-aldehyde reagent 品红醛试剂
fuchsin(e) 洋红;复红;蔷薇苯胺;品红
fuchsin(e)-aldehyde reagent 品红醛试剂
fuchsin(e) contact screen 品红接触网
fuchsin(e) red 品红色
fuchsin(e)-sulphurous acid 品红—亚硫酸试液
fuchsin(e)-sulphurous acid method 品红—亚硫酸法
fuchsin(e) test 吸红试验;品红试法;吸湿试验
fuchsite 铬云母
fuchsone 品红酮
fucoid 藻状迹;管迹(遗迹化石);可疑迹【地】
fucold(al) sandstone 墨角藻状砂岩
fucoxanthin 藻褐素
fucus 墨角藻
fudge 粗制滥造
fue feeder 燃料添加器
fuel 燃油;燃料
fuel accounting 燃料核算
fuel accumulator 蓄油器
fuel actuator 燃料泵
fuel additive 燃料(油)添加剂
fuel adjustment 燃料调节
fuel admission valve 燃料进入阀
fuel-air cycle 燃料空气循环
fuel-air explosive 油气爆炸物;燃料空气爆炸物
fuel-air mixture 燃料空气混合物;燃料空气混合气
fuel-air mixture analyzer 燃料空气混合物分析器;燃油空气混合气分析仪
fuel-air ratio 油气比;燃油空气混合比;燃料空气(混合)比
fuel-air ratio control 燃料空气比控制
fuel-air ratio indicator 燃料空气比指示器
fuel-air weight ratio 燃料空气重量比
fuel alcohol 燃料酒精
fuel alloy 燃料合金
fuel analysis 燃料分析
fuel and air mixture 燃料和空气混合物
fuel and air restrictor 燃油和空气节制器;经济喷嘴
fuel and light indices 燃料与照明用电指数
fuel and lubricant 燃料和润滑剂
fuel and lubricant truck 燃料和润滑剂卡车
fuel and oil servicing truck 燃料和润滑剂供应卡车
fuel and power industries 燃料动力工业
fuel and stores 燃料与物料
fuel anti-knock quality 燃料抗爆性
fuel anti-knock value 燃油抗爆值
fuel ash 燃料灰
fuel ash deposition 燃料灰沉积
fuel assay reactor 燃料试验反应堆;燃料试验堆
fuel assembly 燃料堆
fuel atomization 燃料雾化
fuel atomizer 燃料喷雾器
fuel atomizing 燃料雾化;燃料的雾化
fuel atomizing burner 燃料雾化喷燃器
fuel availability 燃料利用(率)
fuel bag 燃油囊
fuel balance 燃料平衡表
fuel ball 核燃料芯块
fuel bank 燃料库;燃料堆
fuel barge 燃料驳船;燃料驳
fuel battery 燃料电池
fuel battery power-to-volume ratio 燃料电池功率容积比
fuel bed 燃料床;燃料层
fuel bed combustion 燃料层着火燃烧
fuel bed control 燃料层厚度调节
fuel bed depth 燃料层厚度
fuel bell 燃料钟
fuel bill 燃料费账单
fuel bin 燃料仓
fuel binder 燃料粘合剂
fuel blend 掺杂汽油
fuel blowback 燃料混合气瞬间倒流
fuel booster pump 燃油升压泵;燃料增压泵

fuel bowl 燃料杯
fuel box 燃料元件容器
fuel breeding 燃料增殖
fuel brick 煤块;燃料块
fuel briquette 燃料砖
fuel brown coal 燃料褐煤
fuel bundle 燃料棒束
fuel bunker 煤斗;燃料储[贮]槽;燃料舱;燃料仓;油船
fuel-burning area 炉排工作面积
fuel-burning equipment 燃烧设备;炉膛设备
fuel-burning lamp 燃油灯
fuel-burning power plant 火力发电厂
fuel by-pass regulator 燃料旁通调节器
fuel calorimeter 燃料热值测定器
fuel cam 燃油凸轮
fuel can 燃料包壳
fuel canal 燃料管道
fuel canister 燃油罐
fuel capacity 燃料箱容量;燃料容积;燃料能力
fuel cartridge 燃料元件
fuel cassette 燃料盒
fuel cell 燃料电池
fuel cellar 燃料地下室;燃料地窖
fuel cell bottom 燃料电池底
fuel cell car 燃料电池汽车
fuel cell catalyst 燃料电池催化剂
fuel cell ceramics 燃料电池陶瓷
fuel cell detector 燃料电池检测器
fuel cell electrolyte 燃料电池电解质
fuel cell fuel 燃料电池的燃料
fuel cell powered vehicle 燃料电池汽车
fuel cell powerplant 燃料电池动力装置
fuel cell vehicle 燃料电池车
fuel centralizer 燃料集中分配器
fuel chamber 燃烧室
fuel channel 燃料管道
fuel channel(l)ing 燃料元件装盒
fuel charge 加油量;燃料供送
fuel charger 燃油泵
fuel charging mean 燃料注入工具;燃料加料器
fuel chemistry 燃料化学
fuel circuit 燃料回路
fuel circulating pump 燃料循环泵;送油泵
fuel circulating system 燃料循环系统
fuel cladding bond 燃料包壳结合层
fuel cladding temperature meter 燃料包壳温度计
fuel clogging 燃料阻塞
fuel clump 燃料块
fuel coal 燃料煤
fuel coating 燃料涂层
fuel cock 燃油阀;燃油开关
fuel coefficient 燃料消耗系数;燃料系数
fuel coke 燃料焦炭
fuel collecting pipe 燃料收集管
fuel column 燃料柱
fuel combination 燃料混合
fuel commodities 燃料矿产
fuel compact 燃料压块
fuel compartment 燃油舱;燃料舱
fuel component 燃料组元;燃料组分
fuel composition 燃料组成
fuel conservation 节约燃料;燃料存储
fuel consumption 油耗;耗油量;燃料消耗(量);燃料耗量
fuel consumption curve 燃料消耗曲线
fuel consumption index 燃料消耗指数
fuel consumption meter 燃料消耗计
fuel consumption per ten thousand gross ton-kilometers 每万吨公里燃料消耗量
fuel consumption quota 燃料消耗定额
fuel consumption rate 燃料消耗率
fuel consumption test 燃油消耗率测定
fuel consumption trial 燃料消耗试验
fuel container 燃料箱
fuel content ga(u)ge 油量计;燃油存量表
fuel control 供油调节;燃料控制
fuel control assembly 燃料控制装置
fuel control diaphragm 燃料供给控制膜
fuel control lever 燃料控制杆
fuel control linkage 燃料供给操纵杆
fuel control package 燃料调节盒
fuel control unit 燃料控制装置
fuel control valve 燃料控制阀

fuel conversion efficiency 燃料转换系数
fuel conversion factor 燃料转换系数;燃料转换率
fuel conveyer 燃料运输器
fuel cooled nozzle 燃料冷却喷管
fuel corrosion 燃料腐蚀
fuel cost 燃料费;燃料成本
fuel cut 燃油切断装置
fuel cut-off 燃料停止输送;断油开关
fuel cutoff switch 燃料停供开关
fuel cycle 燃料循环;反应堆燃料循环
fuel cycle cost 燃料循环成本
fuel cycle economics 燃料循环经济学
fuel cycle technology 燃料循环工艺学
fuel dam 渣坎
fuel damage 燃料损伤
fuel data 燃料数据
fuel delivery 燃油供给
fuel delivery characteristics 燃油供给特性
fuel delivery line 燃料输送管路
fuel delivery pipeline 燃料输送管道
fuel delivery system 燃料输送系统
fuel densification 燃料增浓
fuel densification phenomenon 燃料密实现象
fuel depletion 燃料贫化
fuel depletion flight 燃料耗尽的发射
fuel depot 燃料库;燃料仓库
fuel depth 燃料深度
fuel desulfurization 燃料脱硫(作用)
fuel desulphurization 燃料脱硫(作用)
fuel detergency 燃料净化性
fuel detergenting 燃料净化
fuel dilution 稀释燃料;燃油稀释
fuel dilution of oil 燃油稀释机油
fuel dilution test 燃料稀释试验
fuel discharge 燃料消耗(量)
fuel discharge valve 燃料放出阀
fuel dispenser 燃料油油罐车
fuel dispensing facility 燃料分送设备
fuel distance 燃料补给距离
fuel distillation bell 燃料蒸馏罩
fuel distributing cock 燃料分配开关
fuel distribution 燃料分配;燃料分布
fuel distributor 燃料分配器
fuel door 加煤炉腔门;燃料门
fuel dope 燃油防腐剂;燃料防爆剂;燃料防爆剂
fuel doping 燃料添加剂
fuel doubling time 燃料倍增时间
fuel drain 放油
fuel drain cook bowl 燃料油泄塞杯
fuel draining plug 燃料放出塞
fuel drain plug 放油塞
fuel dribbing 燃油残滴
fuel drum 燃油桶
fueld saving 节约燃料
fuel duct 燃油导管
fuel duty 燃料供应量
fuel economizer 节油器;燃料节约器;燃料节省器
fuel economy 节约燃料;燃料烧效率;燃料经济性
fuel economy penalty 燃料经济性的缺陷
fuel efficiency 热效率;燃料效率;燃料热值
fuel efficient 节省燃料的
fuel electric(al) plant 火力发电厂
fuel electric(al) power consumption quota of locomotive 机车燃料电力消耗定额
fuel electrode 燃料电极
fuel element 燃料元件;释热元件
fuel element bond 燃料元件结合
fuel element cap 燃料元件端帽
fuel element case 燃料元件盒
fuel element cleaning 燃料元件清洗
fuel element converter 燃料元件转换器
fuel element core 燃料元件芯体
fuel element failure instrumentation 燃料元件损伤仪
fuel element flask 燃料元件容器
fuel element guide 燃料元件导向装置
fuel element hanger rod 燃料元件悬挂棒
fuel element hole 燃料元件孔道
fuel element jacket 燃料元件外壳
fuel element nozzle 燃料元件端孔架
fuel element processing 燃料元件处理
fuel element rupture detector 燃料元件破损探测器
fuel element transfer tube 燃料元件输送管
fuel encapsulating machine 燃料封装机

fuel endurance 燃料储备能力
fuel engineering 燃料工程
fuel entry 燃油引入
fuel equivalent 燃料当量
fuel evapo(u)rability 燃油挥发性
fuel evapo(u)ration rate 燃料蒸发量比;燃料产汽率
fuel evapo(u)rative emission 燃油蒸发的有害排出物
fuel evapo(u)rator 燃料蒸发器
fuel exhaustion 燃料消耗(量)
fuel expanse 燃料费
fuel fabrication plant 燃料元件生产工厂
fuel factor 热效应;燃料系数
fuel feed 加权燃料;燃料进给;燃料供给
fuel feedback control system 燃料反馈控制系统
fuel feed control 燃料供给控制
fuel feeder 燃料加料器
fuel feeding 燃料供给
fuel feeding bell 燃料进口钟
fuel feed line 燃料供给管路
fuel feed pipe 燃油输送管
fuel feed pump 供油泵;燃料供给泵
fuel feed pump plunger 油泵柱塞
fuel feed system 燃料供给系统;供油系统
fuel filling 加燃料;加油;加权燃料
fuel filling column 加油柱;燃料塔
fuel film 燃油膜
fuel filter 滤油器;燃油滤清器;燃油过滤器;燃油过滤器
fuel filter bowl 燃油滤清杯
fuel filter element 燃油滤清器滤芯
fuel filter heater 燃油滤清器加热器
fuel filter strainer 燃料滤网
fuel-fired gas-turbine plant 燃气轮机发电厂
fuel-fired heating equipment 烧油供热装置;燃油加热器
fuel-fire furnace 火焰窑
fuel firing rate 燃料燃烧率
fuel flow 燃油流
fuel flow control 燃料流量调节;油量调节
fuel flow control system 燃油流量控制阀
fuel flow divider 油量分配器
fuel flow indicator 燃料流量指示器
fuel flow meter 燃油流量计;燃料流量计
fuel flow pressure 燃油油压力
fuel flow proportioner 供油调节器
fuel flow rate 燃料流率
fuel flow totalizer 燃油累积流量计
fuel flow transducer 燃油流量传感器
fuel flow trim 燃油流量调整
fuel fog 燃料雾
fuel forest 薪炭林
fuel for household use 家用燃料
fuel gallery 燃油通道
fuel gas 可燃气体;燃(料)气;气体燃料
fuel gas system 燃气系统;燃气气系统
fuel gas treatment 燃料气的精制
fuel ga(u)ge 油量计;油位表;油量表;燃油(量)表;燃料量计;燃料表
fuel ga(u)ge adapter 油表管接头
fuel ga(u)ge breaker 油表开关
fuel ga(u)ge control switch 油表操纵开关
fuel ga(u)ge float 油表浮子
fuel ga(u)ge glass 油量计玻璃;玻璃油量计
fuel geology 燃料地质学
fuel governor 燃料流量调节器;燃料调节器
fuel grade 燃料分类;燃料等级
fuel gravity system 燃油重力自供系统
fuel gravity tank 燃料重力供油箱
fuel handling 燃料管理;燃料处理;燃料储运
fuel handling port 燃料元件装卸口
fuel hand pump 燃油手泵
fuel haul 燃料输送
fuel hauling equipment 燃料输送设备
fuel hauling truck 燃料输送卡车
fuel head 油面高度;燃油压头
fuel heater 燃油加热器
fuel hole diameter 燃料喷嘴孔径
fuel hopper 燃料斗
fuel horse power 燃料马力;燃料功率
fuel hose 燃料软管;燃油管
fuel impurity 燃料杂质;燃料的杂质
fuel indicator 燃料液位指示器;燃料(液面)指示器

fuel induction 燃料吸入
fuel industry 燃料工业
fuel inertia 燃料惯性
fueling 加燃料
fueling area 燃料加注场
fueling at sea 海上输燃油
fueling injection equipment 喷油装置
fueling injection grate 喷嘴格栅
fueling port 加燃料港
fuel injected twocycle engine 燃油喷射式二行程发动机
fuel injection 燃油喷射;燃料喷射;注油
fuel injection and exhaust system 燃油喷入与排出系统
fuel injection beginning 喷油起点
fuel injection end 喷油终点
fuel injection engine 燃料喷射式发动机
fuel injection equipment 燃油喷射装置
fuel injection needle 喷油针
fuel injection nozzle 燃油喷嘴;燃料喷嘴;喷油嘴
fuel injection pattern 燃油喷雾形状;燃料喷射图
fuel injection pipe 高压油管
fuel injection pressure 燃油喷射压力
fuel injection pump 高压喷油泵;燃油喷射泵;燃料油喷射泵;喷油泵
fuel injection pump barrel 燃料喷射泵筒
fuel injection pump camshaft 燃料喷射泵凸轮轴
fuel injection pump driving spindle 燃料喷射泵传动轴
fuel injection pump governor 燃料喷射泵调节器
fuel injection pump plunger 燃料注射泵柱塞
fuel injection pump tester 喷油泵试验台
fuel injection pump testing instrument 喷油泵试验装置
fuel injection rate 喷油率
fuel injection system 燃料喷射系统;燃料喷气系统
fuel injection timing mechanism 燃料喷射定时机构
fuel injection valve 燃料喷射阀;燃油喷射阀
fuel injection valve body 燃料喷射阀阀体;燃油喷射阀阀身
fuel injection valve nozzle 燃油喷射阀喷嘴
fuel injector 燃油喷注器;燃烧喷嘴;燃料喷射器;喷油器
fuel injector igniter 燃料喷射引路点火器
fuel injector modulation 燃油喷嘴调节
fuel inlet 燃料进口
fuel inlet fitting 燃油进口接头
fuel input 燃料加入量;燃料加入
fuel inspection 燃料检验
fuel-intake connection 供油接头
fuel inventory 燃料总投入量;燃料装载;燃料剩存量
fuel irradiation level 燃料辐照度
fuelizer 燃料加热装置
fuel jacket 燃料元件外壳
fuel jet 燃油喷嘴;燃油喷射
fuel jettison 应急放油装置;燃料放出
fuel jettisoning nozzle 燃料放泄喷嘴
fuel jettison pump 应急放油泵
fuel kernel 燃料芯核
fuel knock 燃料爆震;敲缸;气体爆击
fuel lag 燃料供给的滞后
fuel lead 燃料软管;提前供油
fuel leak 燃料泄漏;燃料漏泄
fuel-lean combustion 贫燃料燃烧
fuel level 燃油液面高度;燃料油位
fuel level adjustment 燃油油位调节
fuel level ga(u)ge 油位表;液位表
fuel level indicator 燃料油面指示器;油面指示器;燃料油位指示器
fuel level indicator lamp 燃料储量指示灯
fuel level in tank 油箱油位
fuel level line plug 油位塞
fuel level plunger 燃料油位杆;燃料油位尺
fuel level transmitter 燃料液面传感器
fuel lifetime 燃料寿命
fuel lift pump 燃油泵
fuel limiter 电动燃料控制装置
fuel line 燃油输送管路
fuel line clip 燃油输送管卡子
fuel(l)ing 加油;供燃料;燃料转注
fuel(l)ing gear 燃料接受装置
fuel(l)ing injection pattern 燃料喷油图
fuelling machine 燃料装料机

fuel(l)ing main 燃料转注干线
fuel(l)ing oil hose 燃料油管
fuel(l)ing port 燃料补给港
fuel(l)ing station 加油站;燃料供应站
fuel(l)ing vehicle 油罐车
fuel lizer 燃油加热装置
fuel loading 燃料装载
fuel lock 燃料闸门
fuel make-up 燃料补给;燃料补充
fuel management 燃料管理
fuel management chart 燃料加注表
fuel management system 燃料管理系统
fuel manifold 燃料管
fuel material 燃料
fuel matrix 燃料基体
fuel meat 燃料部分
fuel meter 燃料计
fuel metering 燃油(消耗)计量;燃料配量;燃料计量;燃料调节;调节燃料
fuel metering device 燃料流量计;燃料流量表
fuel metering needle 燃料量针
fuel metering pump 燃料计量泵
fuel metering unit 燃料用量测定仪;燃料配量装置
fuel mileage 燃料里程
fuel mixer 燃料混合器
fuel mixture 可燃混合物;可燃混合气体;燃料混合(物)
fuel mixture control 燃料混合剂调节
fuel mixture ratio 燃料混合比
fuel moderator interface 燃料和慢化剂的分界面
fuel modification 燃料改善;燃料改进
fuel need 燃料需要
fuel nozzle 燃油喷嘴;燃料喷嘴
fuel oil 燃油;燃料油;柴油
fuel oil additive 燃油添加剂;燃油添加剂
fuel oil analysis 燃料油分析
fuel oil atomization 燃油雾化
fuel oil atomizer 燃油雾化器
fuel oil atomizing steam 燃油雾化蒸汽
fuel oil barge 油驳;燃料油驳
fuel oil barrier 燃油油阻挡装置
fuel oil booster pump 燃油增压泵
fuel oil bunkering port 油港;燃料油装船港口
fuel oil burner 燃油油烧器
fuel oil burning pump 燃油燃烧泵
fuel oil change over indicator 燃油变换指示器
fuel oil compensating system 燃料油补偿系统
fuel oil consumption 燃油消耗(量);燃料油消耗量
fuel oil consumption ratio 燃料润滑油消耗比
fuel oil deep tank 燃油深舱
fuel oil distillate 燃料油馏出物
fuel oil drum 燃油油罐
fuel oil equivalent 燃料油当量
fuel oil filling system 燃料油装料系统
fuel oil filter 燃油滤器;燃油过滤器;燃料油过滤器
fuel oil-fired central heating 燃料油集中供热
fuel oil flash tower 燃料油闪蒸塔
fuel oil-gas tar 燃料油气焦油
fuel oil heater 燃油加热器
fuel oil interceptor 燃料油截流器;燃料油截流管
fuel oil meter 油规
fuel oil particle 燃油油粒;燃油油滴
fuel oil pipe 燃油油管;燃油油管线
fuel oil preheater 燃油预热器
fuel oil pressure ga(u)ge 燃料油压计
fuel oil primary filter 燃料油初(级)滤(清)器
fuel oil product 燃油成品
fuel oilproof 耐油的;抗油的
fuel oil pump 燃油油泵
fuel oil pump governor 燃油泵调节器
fuel oil pumping station 燃油泵(送)站
fuel oil purifier 燃油水分分离器;燃油净化器
fuel oil rack indicator 燃油齿条指示器
fuel oil reprocessing 燃油再处理
fuel oil residue 燃油残渣;燃油残滴;燃油残渣
fuel oil-resistant 耐油的
fuel oil return 燃料油回路
fuel oil sampling 燃料油取样
fuel oil secondary filter 燃料油二级滤清器
fuel oil service and transfer system 燃料油管系
fuel oil service pump 日用燃油泵;燃料油(供)给泵
fuel oil service supply pump 日用燃油供油泵
fuel oil service tank 燃料油储油箱
fuel oil settling tank 燃料油沉淀箱

fuel oil shift pump 燃油输送泵
fuel oil slag 燃料油渣;燃料油炉渣
fuel oil stabilizer 燃料油稳定剂
fuel oil storage 燃料油的储藏
fuel oil storage tank 燃油储[贮]罐;燃油储[贮]存柜
fuel oil store 燃料油库
fuel oil stripper 燃料油汽提塔;汽提除油器;燃料油解析塔
fuel oil supply 燃料油供给
fuel oil supply system 燃料油供应系统
fuel oil system 燃油系统
fuel oil tank 燃油箱;油舱;燃料油(储)罐
fuel oil tanker 燃料油补给船
fuel oil tank whistle indicator 燃料油罐警笛
fuel oiltight 油密的;不漏油的
fuel oil transfer pump 燃油输送泵
fuel oil volume ratio 混合油容积比
fuel oil waste 燃料废油;废燃料油
fuel oil yield 燃料油产率
fuel orifice 燃料喷射孔
fuel outlet 放油口;放油孔
fuel outlet valve 放油阀
fuel pacer system 燃油过耗警告系统
fuel particles 燃料质点;燃料颗粒
fuel passage 燃油管道
fuel pellet 燃料芯块;燃料球芯块
fuel penalty 燃料经济性的缺陷
fuel pencil 燃料元件细棒
fuel performance 燃料性能
fuel pin 燃料量孔阀针
fuel pipe 油管;燃油管道
fuel(pipe)line 燃油管线;燃油管路
fuel pipe union 燃油管联管节
fuel piping 燃料输送管系
fuel pit 燃料坑
fuel pitch operation 燃料沥青型操作
fuel planar smear density 燃料平面有效密度
fuel plate 燃料板;板状燃料元件
fuel port 燃料输入孔;燃料口
fuel port tube 燃料输料管道
fuel postinjection 补充喷油
fuel preliminary filter 燃油粗滤器
fuel preparation 燃料制备
fuel pressure 燃料压力
fuel pressure control 油压控制
fuel pressure ga(u)ge 油压计;燃料压力表
fuel pressure injection 启动注油泵
fuel pressure manifold 油泵集流腔;油泵导管
fuel pressure pump 燃油压力泵
fuel pressure switch 燃油压力开关
fuel pressure valve 油压阀
fuel priming pump 燃油泵
fuel process cell 燃料处理热室
fuel process combustion 燃料燃烧过程
fuel processing 燃料处理
fuel processing waste 燃料加工废物
fuel pulverization 燃料雾化
fuel pulverizing mill 燃料研磨机
fuel pump 燃油泵;燃料泵
fuel pump adapter 燃油泵接头
fuel pump adjusting screw 燃油泵调节螺钉
fuel pump air dome 燃油泵空气室
fuel pump analyser 燃油泵分析器
fuel pump body 燃油泵体
fuel pump bowl 燃油泵杯
fuel pump bowl clamp 燃油泵滤杯固定卡
fuel pump bracket 燃油泵支架
fuel pump case 燃油泵壳
fuel pump coupling 燃油泵联轴节
fuel pump cover 燃油泵体壳
fuel pump deliver line 燃油输送管路
fuel pump diaphragm 燃油泵膜片
fuel pump diaphragm spring 燃油泵膜片弹簧
fuel pump dip stick 燃油泵量油杆
fuel pump drive 油泵驱动
fuel pump drive shaft 燃油泵主动轴
fuel pump drive spindle 燃油泵主动心轴
fuel pump drive sprocket 燃油泵传动链轮
fuel pump eccentric 燃料泵偏心轮
fuel pump filter 燃油泵滤清装置
fuel pump gasket 油泵垫圈
fuel pump governor 油泵调节器
fuel pump housing 燃油泵壳体

fuel pump impeller 油泵叶轮
fuel pump lever 燃料泵操纵杆
fuel pump lever clamp 燃料泵操纵杆夹
fuel pump main housing 燃料泵主壳体
fuel pump metering mechanism 燃料泵调节机构
fuel pump plunger 油泵柱塞
fuel pump priming lever 油泵给油杆
fuel pump push rod 油泵推杆
fuel pump rocker arm 油泵摇臂
fuel pump rocker arm connecting arm 燃油泵摇臂联杆
fuel pump rocker arm link 燃油泵摇臂杆
fuel pump screen 燃油泵滤网
fuel pump sediment bowl 燃料泵沉淀杯
fuel pump spindle coupling 燃料泵心轴联轴节
fuel pump spindle oil thrower 燃料泵心轴甩油环
fuel pump spindle resilient member 燃料泵心轴弹性接头
fuel pump strap 燃料泵固定带
fuel pump suction valve 燃油泵进油阀
fuel pump sump 燃料泵槽
fuel pump tank 燃料泵箱
fuel pump tappet 燃料泵挺杆
fuel pump valve retainer 燃油泵阀支片
fuel pump with vacuum pump 带真空泵的燃料泵
fuel purging system 燃料净化装置
fuel purification 燃料净化
fuel purification unit 燃料净化装置;燃料净化设备
fuel quality standard 燃料质量标准
fuel quantity control 燃料油量控制
fuel quantity ga(u)ge 燃料表
fuel quantity indicator 燃油量表
fuel quantity meter 油量表
fuel rack 燃料元件架
fuel range 燃油续航力
fuel rate 燃料消耗率
fuel rate controller 燃料量调节器
fuel rating 燃料比功率
fuel ratio 燃烧比率;燃料比
fuel ratio control 燃料比(率)控制
fuel receiving and storage station 燃料接收储[贮]存站
fuel recovery 燃料回收
fuel recovery cell 燃料回收热室
fuel refreshment 燃料补充
fuel regulator 燃油调节器;燃料调节器
fuel regulator valve 燃油调节阀
fuel rejuvenation 燃料更新
fuel relocation effect 燃料再分布效应
fuel replacement energy 火电的代用电能
fuel reprocessing 燃料再处理;燃料后处理
fuel reprocessing cell 燃料后处理室
fuel reprocessing facility 燃料后处理装置
fuel reprocessing loop 燃料后处理生产线
fuel reprocessing plant 燃料后处理厂
fuel reprocessing waste 燃料后处理废物
fuel reprocessor 燃料后处理厂
fuel reproduction 燃料再生
fuel reproduction factor 燃料再生因子
fuel requirement 燃料需要量;燃料需要
fuel reserve 燃料储备
fuel reserve tank 燃料储备箱;储油箱
fuel reservoir 燃料库
fuel resistance 耐燃料油性
fuel resynthesis 燃料再合成
fuel return pipe 燃油回流管
fuel-rich combustion 富燃料燃烧
fuel ring 燃料环
fuel rod 燃料棒
fuel rod in nuclear power plant 核动力装置燃料棒
fuel rod locating plate 燃料棒定位板
fuel rod shearer 燃料棒剪切机
fuel room 燃料间
fuel runout warning device 燃料耗尽信号器
fuel salt 燃料盐
fuel salt pump 燃料熔盐泵
fuel sampler 燃料取样器
fuels and energy 燃料动力
fuel saving 节油
fuel screen 燃油滤网
fuel section 燃料段
fuel segregation 燃料分离
fuel sensitivity 燃料敏感性
fuel sensor 燃油存量传感器

fuel servicer 燃料加注车
fuel servicing truck 燃料供应车
fuel setting key 引信定秒器
fuel sheath 燃料包壳
fuel sheet 燃料片
fuel ship 燃料供应船
fuel shortage 燃料不足
fuel shut-off 燃料关闭(阀);燃料停供;切断燃料
fuel shut-off valve 防火开关
fuel shut-off valve rod 燃料关闭阀操纵杆
fuel sieve 燃料滤网
fuel slab 燃料板
fuel slag 燃料废渣
fuel slippage 燃料流动性
fuel slug 燃料块
fuel solution 燃料溶液;燃料溶体
fuels ores 燃料矿产
fuel space 燃料容积
fuel specification 燃料规格
fuel specific impulse 燃料比冲量
fuel specimen capsule 燃料样品辐照盒
fuel sphere 燃料球
fuel spillage 燃油漏失量;燃油流失量
fuel spray 燃油喷雾;燃油喷射
fuel spray characteristic 燃油喷射特性
fuel sprayer 喷油嘴
fuel spraying 燃油雾化喷射
fuel spray nozzle 喷油嘴
fuel spreading 燃油雾化
fuel stack 燃料芯块柱
fuel station 加油站;燃料站
fuel storage 燃料库;燃料储藏
fuel storage cell 燃料储[贮]存室
fuel storage hopper 燃料仓斗
fuel storage pool 燃料储[贮]存池
fuel storage stability 燃料长期储存稳定性
fuel storage tank 燃料储存箱
fuel strainer 滤油器;燃油过滤装置;燃油过滤器
fuel strainer assembly 燃料棒组件
fuel strainer foot 滤油器脚座
fuel substitution 燃料替换
fuel suction hood packing 燃料吸进帽密垫
fuel sulphur content 燃料的含硫量
fuel supplier 燃料供应者;燃料供应公司
fuel supply 供油;燃料供应;燃料供给
fuel supply ga(u)ge 给油计
fuel supply line 供油管
fuel supply pipe 燃料输送管
fuel supply point 燃料供应点
fuel supply pump 燃油供给泵;燃油输送泵
fuel supply service 燃料供应部门
fuel supply system 燃料供给装置
fuel supply tube 燃料输送管
fuel swirler 离心式喷油嘴
fuel switching away from oil 改用非油类燃料
fuel system 燃料系数;燃料供给系统
fuel system diagram 燃料供应系统图
fuel system of diesel engine 柴油机供油系统
fuel system of gasoline engine 汽油机加油系统
fuel system run 燃料系统试验
fuel system test rig 燃料系统测试装置
fuel system trouble 燃料系统的故障
fuel tank 燃油柜;燃油舱;燃料箱;油箱
fuel tankage 油箱容量
fuel tank balance tube 燃料箱平衡管
fuel tank bracket 油箱支架
fuel tank cap 油箱盖
fuel tank capacity 油箱容积
fuel tank cap packing 燃料箱盖密垫
fuel tank drain cover 燃料箱放油口盖
fuel tanker 供油船;燃油船
fuel tanker car 油罐车
fuel tanker truck 油罐车
fuel tank evacuating pump 燃油箱抽油泵
fuel tank filler 油箱加油口
fuel tank filler cap 燃油箱加油盖;燃料箱加注孔盖
fuel tank float 燃油箱浮子
fuel tank guard 油箱护板
fuel tank hanger 燃油箱吊架
fuel tank joint 燃油油箱接头;燃料油管接头
fuel tank lock 燃油箱盖
fuel tank of ship 船用油轮
fuel tank outlet 油箱出口
fuel tank preserving fluid 燃料箱油封液

fuel tank pressure relief valve 燃油箱限压阀
fuel tank pressurization 油箱增压
fuel tank rear bracket 燃油箱后支架
fuel tank selector valve 燃油箱分配阀
fuel tank strainer 燃油箱滤网
fuel tank strap 燃油箱卡箍
fuel tank strap anti-squeak 燃油箱卡箍减声器
fuel tank support 燃油箱支架
fuel tank support bar 燃油箱支杆
fuel tank truck 燃料车
fuel tank vent 油箱通风孔
fuel tap 油阀
fuel tar 燃料焦油;燃料焦炭
fuel temperature coefficient 燃料温度系数
fuel tester 燃料试验器
fuel test(ing) 燃料试验
fuel thermal efficiency 燃料热效率
fuel thickener 燃料稠化剂
fuel tight 燃料不渗透的
fuel to cavity 燃料空腔比
fuel to oil ratio 燃油润滑油比
fuel totalizer ga(u)ge 燃料总指示器
fuel trailer 燃料输送挂车
fuel transfer cask 燃料传输容器
fuel transfer pond 燃料转运池
fuel transfer pump 燃油转运泵
fuel treating equipment 燃料处理装置
fuel trimmer 燃料调节器
fuel truck 燃料输送卡车
fuel truck nozzle 加油枪
fuel tube 管状燃料;燃料管
fuel type vacuum tower 燃料型减压塔
fuel unit 燃料单元
fuel unloading machine 燃料卸料机
fuel utilization 相对燃耗;燃料利用(率)
fuel utilization factor 燃料利用系数
fuel value 燃烧值;燃料(热)值
fuel value of digester gas 消化池气(体)热值
fuel value of refuse 垃圾燃烧值
fuel value of sludge 污泥的燃烧值
fuel valve 燃油阀;燃料阀
fuel vapo(u)r 燃料蒸气
fuel vapo(u)rization 燃料蒸发
fuel vapo(u)rizer 燃料蒸发器
fuel vapo(u)r pocket 燃料蒸气泡
fuel viscosity variation 燃油黏度
fuel volatility 燃料挥发性
fuel volatility adjustment 燃料挥发度调节
fuel wood 燃材;柴;薪材
fuelwood planation 薪材造林
fuel zoning 燃料分区
fuffer limit 颤动极限
fug 室内浊气
fugacious 暂时性的
fugacity 逸性;逸度;易逸性;易逸度
fugacity coefficient 逸度系数
fugacity equilibrium model 易逸性平衡模型
fugacity-fugacity diagram 逸度—逸度图
fugacity of component 组分逸度
fugitive 早期褪色;无常的
fugitive air 矿山漏风
fugitive binder 临时结合剂;短效粘结剂
fugitive colo(u)r 易褪色的颜色;易褪色的色料;干湿变色色材;短效颜料;短效色
fugitive constituent 挥发份
fugitive dust 易散粉尘
fugitive dust emission 易散性烟尘排放
fugitive dye 易褪色的染料;褪色染料;短效染料;早期褪色
fugitive emission 易散性排放;无组织排放;短时排放
fugitive filler 短效填料
fugitive flavo(u)r 易挥发味
fugitive flux 短效溶剂
fugitive glue 短效粘合剂
fugitive lubricant 挥发性润滑剂
fugitiveness 不稳定性;不耐久性
fugitive pigment 褪色颜料
fugitive resources 短效资源;短时资源;易耗资源;不可再生资源
fugitive source 易散性污染源
fugitive species 漂泊种
fugitive tinting 易褪着色
fugitive water 渗漏水

fugitometer 染料试验计
Fujian ware 福建瓷
Fuji Electric(al) 富士电机株式会社
Fujitsu 富士通
fukaliye 福碳硅钙石
Fukien ware 福建瓷
fukinane 蜂斗菜烷
fukinone 蜂斗菜酮
fukuchilite 硫铁铜矿
fulchronograph 闪电电流特性记录器
fulcrum 支轴;支点
fulcrum arrangement 支承装置
fulcrum ball 支承球
fulcrum bar 支杆
fulcrum bearing 支(点)承座;支承
fulcrum guide 支架导承
fulcrum jack 支点千斤顶
fulcrum lever 杆杠
fulcrum of moment 力矩中心;力矩支点
fulcrum of the boom 吊杆支点
fulcrum pin 旋转轴;轴颈;(弧形闸门的)支座枢轴;支(轴)销;支承销
fulcrum pin cap 支销盖
fulcrum pin castle nut 支销槽顶螺母
fulcrum pin washer 支销垫圈
fulcrum shaft 支轴
fulcrum slide 支点滑板
Fulda faience 福尔达锡釉陶器
Fulda porcelain 福尔达瓷器
Fuld-Gross unit 富尔德－格罗斯单位
fulfil 完成
fulfil a former agreement 履行预约
fulfil ahead of schedule 提前完成
fulfill 履行
fulfill a contract 履行合同
fulfillment date 履行日期
fulfillment of schedule 完成进度
fulfillment the contract 履行合同
fulfill task on time 按时完成任务
fulfil quality requirements 符合规格
fulfil term 履行条款
fulgurate 闪烁;闪电般发光
fulguration 电干燥法
fulgurator 火焰闪烁器
fulgurite 闪电熔斑;闪电管石
fulgurometer 闪电测量仪
Fulham china 福尔汉透明瓷器
Fulham stoneware 福尔汉盐釉炻器
fuliginous 煤烟状的;煤褐色的
fuligo 煤烟
full dead annealing 完全退火
full absorption costing 全额成本法
full absorption method 全部成本法
full acoustic(al) migration 全声波偏移
full acoustic(al) wavetrains recording 声波全波列测井成果图
full actuated controller 全感应信号控制机
full adder 全加(法)器
full adder circuit 全加法电路
full additive method 全添加法
full address 全地址
full address jump 全地址转移
full adhesion 全粘着
full adjustable seat 全调节式座椅
full admission 全开吸气
full admission turbine 全部进气涡轮机;全部进气透平;全部进气汽轮机;整周进水式水轮机;全周进汽涡轮机
full advance 全截面掘进;全断面推进
full-advance method 全面前进式开采法
full-advance panel 全面前进式盘区
full-advance position 全提前点火位置
full-advance system 全面前进式开采法
full-ageing 完全老化
full ahead 前进三【船】
full allowance 满额
full analysis 全分析
full analysis of the step test 分级试验全分析
full and by 满帆顺风;船首逼风但保持满帆
full and change high water 大潮高潮间隔
full and change(of the moon) 大潮潮时
full and complete cargo 满舱满载货物;满舱满载货物
full and down 满载,满舱;十足满载(船舶载量)
full and down of vehicle (车的)合理配载

full angle 全圆角
full annealing 再结晶退火;全退火
full-antomatic buret 全自动滴定管
full aperture 全孔径
full aperture drum 活底桶;全开口式桶
full aperture kicker 大开度冲击磁铁
full appearance 出齐
full application braking 全制动(作用)
full application(of brake) 全制动
full application position 全通位置;全闭合位置
full application time of brake 制动全部作业时间
full arch 全拱
full arch gantry 双支腿门座
full arch gantry crane 双支腿龙门吊
full audit 全面审计
full auto-bonding 全自动接合
full auto-bonding system 全自动接合装置;全自动焊接装置
full automated cataloguing technique 全自动编目技术
full-automatic concrete mixer 全自动混凝土搅拌机
full automaticity 全自动
full automation 全自动化
full automation module 全自动化组件
full auto trigger 全自动触发器
full autotrophic denitrification 全程自养脱氮
full availability 全利用度
full back arbor 强力刀柄
full back cutter 强力切削工具
full balanced rudder 全平衡舵
full-bar generator 全彩条信号发生器
full bath 满槽
full bearing 全支承;满支承
full bench section 山边全挖断面
full-biased excitation 全偏激磁
full bias rail box girder 全偏轨箱形梁
full binary adder 二进制全加器
full binary tree 满二叉树
full-bin drying system 整仓干燥系统
full blast 全喷砂;全鼓风
full-blown depression 全面萧条
full-blown power plant 大型配套发电厂
full-blown sponge rubber 充分发泡海绵胶
full bobbin stop motion 满筒自停;满管自停装置
full-bodied 完满的;内容充实的;高粘度的;高增容的;高稠性的;高稠度的
full-bodied paint 高稠度漆
full-boiling point 全部馏分蒸出时的温度
full-boiling range reference fuel 全沸点标准燃油(试验用)
full bond 顶砖砌合
full bonded bolt 粘接型锚杆
full bonding 全(面)粘合;顶砖砌合
full bore 贯眼
full-bore tubing tester 贯眼油管测试器
full-bottom advance 井筒全截面掘进;全断面掘进
full bottomed ship 肥底船(水线下部分较宽大的船)
full bound 全布面装订的
full bow 肥型船首
full brake trial 全制动试验
full braking 全制动
full brick 整砖;整块砖;煤砖
full budgeting 编制全面预算
full-budgeting approach 编制全面预算方法
full calibrating 全项目标定
full call letters 完整呼号
full can stop motion 满筒自停装置
full capabillty mode 全功能方式
full capacity 满载容量;满负荷生产量;全容量;全能生产;额定生产量
full capacity tap 全容量抽头
full cap stop motion 满管自停装置
full cargo 满载
full carriageway traffic volume 全车道流量
full carrier 全载波
full carrier value 全载波值
full casing tube boring maching 全套管成孔机
full cell method 满细胞法
full-cell process (木材的)全量浸注法;满细胞法(一种木材防腐的压力处理法);(木材的)浸渍防腐法;填满细胞法;(木材防腐的)充细胞法;全吸收法
full-cell treatment 满细胞法;(木材的)浸渍防腐法

full cellularized container ship 全槽格式集装箱船
full-centered arch 半圆拱
full-centre calotte 半圆穹顶
full chained 全链式
full charge 全装料的
full charge specific gravity 全充电电池溶液比重
fullcharging 满荷
full charter 全部租佣
full chord laminar flow 全翼弦层流
full circle 全循环
full circle contact 全圆周上的接触
full circle crane 全(周)转动起重机;全旋转式起重机;全回转起重机
full circle mining of tunnel heading 全断面隧洞开挖
full circle revolving loader 全旋转式装料机
full circle shovel 旋转式铲斗车;全转式机铲
full circle slewing crane 全转式起重机;全圈旋转式起重机
full circle slewing floating crane 全回转式浮吊;全回转(浮式)起重机
full circle swinging 全回转
full circular 全周式
full clay brick 整砖;整块砖
full closed crankcase ventilating system 全封闭型曲轴箱通风系
full closure 全封闭
full clover-leaf crossing 全苜蓿叶式立体交叉
full clover-leaf interchange 苜蓿叶形立体交叉
full coat 无缺陷漆膜;厚涂清漆;厚涂层;完满涂层;涂层的最大施工厚度
full cock position 全压位置
full coinsurance 全额共同保险
full colo(u)r 全色;全彩色;彩色;五彩
full colo(u)ration 纯色
full-colo(u)r holography 全色全息术
full-colo(u)r image 全彩色影像;全色图像
full-colo(u)r laser display 全色激光显示
full-colo(u)r picture 全色图
full-colo(u)r view 彩色图像
full column resin grouted bolt 全孔树脂锚杆
full commission 全额佣金;全部编制
full compliment of accessories 全部附件保持供给
full compound key 全组合键
full-connected load 全负载;全负荷
full container 整装货柜(整装集装箱)
full container load 整箱集装箱的荷载;整箱货;整柜装;整柜载货
full container load cargo 整柜装货物
full container ship 全集装箱货轮;全集装箱船
full control method 满控制法
full control of traffic 整体交通控制
full control point 像片平高控制点
full conversion 全转换
full coordinates 全坐标
full core fibre 全芯纤维
full correlation 全相关
full correlation analysis 全相关分析
full cost 全部成本
full costing 完全成本计算;完全成本法
full cost less depreciation 全部成本减折旧
full cost pricing 全成本定价(法);按全部成本定价
full cost principle 全部成本原则
full cost system 全部成本法
full count out 满数输出
full coupon bond 全额息票债券
full-course concrete pavement 全层混凝土路面
full-course construction work 全层(混凝土)建筑工程
full covenant and warranty deed 全部契约的条款和全部授权证书
full coverage 全额担保;完全承保;定金承保
full coverage of bath 满槽
full coverage spray 充分覆盖喷施
full coverage steerable antenna 全向可控天线
full cross-section 整个截面;全断面
full cross-sectional pumping 全断面取水
full cross-sectional element 整个截面的构件;全断面的构件
full crown 全冠
full crystal 全晶质玻璃;全晶玻璃
full cubic point group 全立方点群
full cure 充分硫化
full current 全电流

full curve 连续曲线;完全曲线;实曲线
full cut-off 全停;全截止;全闭
full cut section 全挖式断面
full cycle 全循环
full damping 满阻尼;全阻尼
full data set authority 全数据集特许权;全部数据集授权
full day's operation 全天工作
full decator 加压蒸呢机
full deck vessel 全通甲板船
full decode address 全译码地址
full demand operation 全需求量运转
full density product 密实制品
full depression wing flap 全开度襟翼
full depth 大切削深度;大吃刀
full-depth asphalt 全深沥青
full-depth asphalt base 全深沥青基层
full-depth asphalt concrete pavement 全厚式沥青混凝土铺面
full-depth asphalt pavement 全深沥青路面;全厚式沥青铺面;全厚沥青路面
full-depth bituminous patch 全深沥青补坑
full-depth block 竖砌池壁块
full-depth construction 全深(度)施工;竖砌池壁结构
full-depth gear 全齿高齿轮;标准齿高齿轮
full-depth grouting 全深灌浆
full-depth internal concrete pavement vibrator 全深度插入式混凝土路面振捣器
full-depth involute system 全高齿渐开线制;全齿高渐开线制
full-depth paving 全深铺砌
full-depth tooth 全高齿;全齿高齿
full developed flow 全扩展的流
full development 全面开拓;全部展开
full deviation channel level 全频偏信道电平
full diameter 直径的全断面;全主径;大直径
full diameter core 大直径岩芯
full diameter of thread 螺纹外径
full diameter waterway 全直径水路
full diesel 纯柴油机
full digital automatic mapping system 全数字化自动测图系统
full digital mapping 全数字化摄影测图
full digital photogrammetry 全数字化摄影测量
full dimension extensible conveyor system 全面延伸式运输机系统
full dimension mining 全高开采
full dip 总弛度;(电线的)总垂度;真倾角
full dip angle 真倾角
full dip infiltration 全部浸入浸渍法
full dip process 全浸法
full discharge 整个流量;全速排放;全部流量;全部卸载
full discharge of liability 全部履行义务
full disclosure 完全公开;财务事项的充分公布;财务公开
full disk 整盘;整圆盘
full-disk buff 布抛光轮
full distance 全距离;全长
full distance test 全程试验
full dose 全量
full double-level design 全双层车站
full drainage(system) 全排水系统
full dress 挂满旗【船】
full-duplex 全双工;全双向的
full-duplex channel 全双工信道;全双工通道
full-duplex line 全双工线路
full-duplex line adapter 全双工线路转接器
full-duplex link 全双工链路
full-duplex mode 全双工工作方式
full-duplex operation 全双工操作;同时双向操作
full-duplex preformance 全双工性能
full-duplex primary station 全双工主站
full-duplex service 全双工服务
full-duplex teletype 全双工电传
full-duplex terminal 全双工终端
full-duplex transmission 全双工传输
full duration 全宽(度);全部持续时间
full earth illumination 满地照明
Fulleborn's method 浮集法
full eccentric drive shaft 全偏心传动轴
full eccentrice type 全偏心
full eccentric type crank press 偏心式曲柄压力机

full echo suppressor 全回波抑制器
fulled aperture 连续孔径
full edged 整边(木材)
full edition 详表
full electrification 全盘电气化;全面电气化
full electronic switching 全电子式交换
full electronic switching system 全电子式交换制;全电子交换机;全电子电话交换系统
full electronic telephone exchange 全电子式交换机
full employment 充分就业
full enclosed gas insulated switchgear 全封闭式开关;全封闭气体绝缘开关
full ended 首尾重线
full-ended ship 肥艏艉船
full endorsement 记名背书;完整背书;特别背书
full engagement 全齿啮合
full enriched uranium 全加浓铀
fuller 压槽模;敛缝;切分孔型;漂布工;填密;填料工;填缝凿;套(柄铁)锤;半圆形套柄铁锤;凿密
Fuller airslide 富勒气滑式
fuller block 圆底模
fuller board 压制纸板;填隙压板
Fuller faucet 富勒式水龙头;弯管龙头
Fuller grate cooler 富勒格栅式(水泥熟料)冷却器
fuller hammer 圆截面凿岩机
fullering 压槽锤开槽;捻缝(口);凿密;墙缝;嵌缝
fullering cup 密封皮碗
fullering tool 压槽锤;凿密工具;锤击工具
fullering with the core bar 芯轴拔长
Fuller Kinyon pump 水泥压力输送泵;富勒·金尼昂泵
Fuller Kinyon unloader pump 富勒·金尼昂粒料输送泵
Fuller-Lehigh mill 富勒·利弗球磨机;富氏球磨机
Fuller rule 富勒规则
Fuller's best mix curve 富勒最佳混合曲线
Fuller's clay 漂白土
Fuller's curve 富勒(级配)曲线;富氏曲线
Fuller's dome 富勒式圆屋顶
Fuller's earth 硅藻土;漂洗泥;漂白土;酸性白土
Fuller's earth deposit 漂白土矿床
fuller's faucet 弯管龙头;富勒式水龙头
Fuller's grading curve 富勒级配曲线
Fuller's ideal curve 富勒理想曲线(求材料最大密度和最小孔隙率的配比)
Fuller's maximum density curve 富勒最大密实度曲线
fuller's parabola 富勒抛物线
fuller's tool 压槽锤
full exposure 全光照
full exposure to weather 完全遭受风化;完全风化
full face 全工作面;全断面(的)
full face attack 全断面掘进
full face bit 全断钻头;不取心钻头
full face blast(ing) 全断面爆破;全工作面一次爆破
full face boring 全断面钻探;全断面钻进
full face contact 全面接触
full faced 整面法兰
full face digging 全面开挖(隧道);隧道全面开挖
full face diving helmet 全面罩潜水头盔
full face drilling 全断面钻探;全断面钻进;全断面掘进
full face driving 全断面掘进
full face excavation 全断面开挖
full face excavation method 全断面一次开挖法;全断面开挖法
full face excavation tunnel 全断面隧道开挖
full face firing 全工作面引爆法
full face machine 全断面联合掘进机;全断面掘进机
full face mask 全面罩
full face mask type 全面罩式
full face method 全断面隧道开挖法;全断面法
full facepiece mask 全面罩
full face round 全断面炮孔组
full face tunnel(l)er 全断面隧道掘进机
full face tunnel(l)ing 全断面隧道施工;全断面隧道开挖;全断面开挖;隧道全面开挖;隧道全断面掘进(法);全断面掘进
full face tunnel(l)ing method 全断面掘进法;全断面开挖法
full face work 全断面工作(隧道开挖);全断面工程(隧洞开挖)
full factor 满载率;填隙因数

full factoring 全面保付代理
full fallow 清洁休闲
full fare 全票费
full feathering 顺桨
full feathering propeller 全活叶螺旋桨
full field 全磁场
full field control point distribution 全野外布点
full field relay 满励磁继电器
full field speed 全励磁转速
full figure 全像;全图
full filled band 满带
full fillet weld 满角焊(缝);全角焊
full fillet welding 满角焊接
full financial responsibility 财务包干
full-finish 双面整理
full fire 大火阶段
full-fired colo(u)r 高温颜料
full fixed 端头完全嵌固(指框架结构节点)
full fixity 完全稳定;完全固定
full-flap 全放下襟的翼
full-fledged worker 熟练工(人)
full-flighted screw 全程螺杆
full floating 全浮动
full floating axle 全浮式轴
full-floating axle shaft 全浮轮轴
full-floating bearing 全浮轴承
full-floating coupling 全浮式联轴节
full floating feeder 全浮式喂料器;带式装卸机
full-floating gudgeon pin 全浮式活塞销
full-floating mechanical packing 全浮动机械填料
full-floating piston pin 全浮式活塞销
full floating rear axle 全浮后轴
full floating ship 肥型船
full-floating stub axle 浮动短轴
full floating wrist pin 全浮式活塞销
full floor 全楼层
full flow 有压流;总流量;满流;全流量
full flow condensate 汽轮机主凝结水
full flow core barrel 全通岩芯管
full flow endurance 全功率工作时间
full flow filter 全流(量)过滤器;满流过滤器;全流式滤清器
full flow filtering 全水流过滤;全流过滤(燃料油)
full flow filtration 全流式过滤;全流量过滤
full flow flux ceramic membrane 全水流通量陶瓷膜
full flow heat exchanger 全流式热交换器
full flow oil filter 满流(式)滤清器;全流式机油滤清器
full flow polishing demineralizer 全流量高性能离子净化器
full flow type 全流式
full flow valve 进水阀;全流阀
full fluid film lubrication 全油膜润滑
full fluid film region 全液膜润滑区
full fluid steering 全液力转向
full flush 圆形水槽
full force feed 全压力进给;压油润滑
full force feed lubrication 全压力润滑
full fore-ended ship 肥艏船
full foreign-owned enterprises 外资企业
full-forward position 全向前位置
full frame 支撑构架;整幅像片(测);联结构架;满帧;撑系框架
full frame flash exposure 全画闪光曝气
full framing 斜撑框架;满堂支架;满堂红支架
full free lift 全自动升降装卸车
full freight 全运价
full-frequency range recording 全频范围记录
full-frequency recording range 全频程记录范围
full-fresh oiling 全新鲜机油润滑
full functinal dependence 全功能相关
full funding 充分提供资金
full-fusion thermit welding 热剂铸焊
full-fusion welding 全熔合焊
full gain 总增益;满增益
full-gale 飓风力
full gantry crane 全能龙门起重机
full gas 全燃气
full gate 满开闸门;满开的;全开的;全开度
full gate opening 门孔全开度
full gate operation 闸门全开度运行
full gate turbine discharge 导叶全开时水轮机流量
full ga(u)ge 全径;全轨距

full ga(u)ge bit 全径钻头
full ga(u)ge(bore)hole 通径钻孔
full ga(u)ge branch hole 与主钻孔直径相同的支钻孔
full ga(u)ge deflecting bit 全径造斜钻头；全径偏斜钻头
full ga(u)ge drill hole 全径钻孔；满规钻孔
full ga(u)ge railroad 标准轨(距)铁路
full ga(u)ge railway 标准轨(距)铁路
full gear 全开状态
full glass curtain wall 全玻璃幕墙
full glass door 全玻璃门
full gloss 全光泽；全光；最高度光泽；顶峰光泽度
full gloss enamel 强光泽瓷漆；高光泽瓷漆；全光瓷漆
full gloss finish 全光泽精修；高光泽精修；全光饰面
full gloss latex enamel 强光泽乳胶瓷漆；高光乳胶瓷漆；全光胶乳瓷漆
full gloss oil paint 高光泽油漆
full gloss paint 全光泽油漆；高光泽油漆；全光泽涂料；高光泽油漆；高光泽涂料；全光漆
full gloss wall slab 高光泽墙板
full graphic(al)panel 全图示面板；全图示控制面板
full gravity block 全重力出矿矿块
full grey scale 全灰度等级
full grinding roller 长磨辊
full ground contact 全接地
full group sampling 整群抽样
full hank stop motion 满绞自停装置
full-hardened steel 全淬硬钢
full hardening 全硬化；全淬硬；淬透
full hardness 全干硬度
full head 全水头
full header 全丁砖砌层
full head rivet 圆头铆钉
full head room 全高度舱室
full heat protection 完全防热
full heat quantity 全热量
full-height partition 到顶隔墙
full-height tooth 标准齿
full-height window 整层高的窗；全高度的窗
full helm 满舵
full hermetic refrigerant compressor 全封闭式制冷剂压缩机
full-hill drop plate 多粒槽口式排种盘
full hole 贯眼型；贯眼；全井孔
full hole asphalt concrete 全深沥青混凝土
full hole bit 全面钻进钻头；全断面无岩心钻头
full hole casing 全孔套管；全井套管
full hole cementing 全井注水泥
full hole drill collars 全孔径钻铤
full hole drilling 全面钻进；全孔钻进
full hole flow 全孔流动
full hole joint 贯眼接头
full hole rock bit 全断面无岩心钻头；大直径全断面牙轮钻头
full hole size 全径
full hole size bit 大直径全断面钻头
full hole testing 全井径处地层测试；通径井试验
full hole thread 贯眼扣
full holetooljoint 贯眼接头；贯眼钻杆接头
full hydraulic lift 全液压举升器；全液压电梯
full hydraulic operation 全液压操作
full image rectification 全片纠正
full impregnated 全透
full impregnation 完全浸渍
fulling board 压榨机
fulling clay 漂白土；缩绒粘土
fulling machine 缩呢机
fulling mill 漂洗机
fulling roller 缩呢辊
fulling trough 缩呢槽
full install 全安装；重新安装
full installment 全额支付
full insulated joint 两边绝缘的结合
full insulated winding 全电压绝缘绕组
full insurance 全保险；全额保险；全部保险
full interest admitted 享有全部可保利益；承认全部利益；承认全部保险利益
full invested 全面投资
full-ionization 全电离
full irrigation 满灌
full-jacquard 全提花
full jetty 整片式码头
full-journal bearing 全围式滑动轴承

full key 满键
full keystone ring 全楔形环
full killed steel 全镇静钢；完全镇静钢
full labo(u)r power 全劳动力
full-laden 满载的；全载的
full-lane patch 全车道补坑
full lap 全搭接接合
full lap stop motion 满卷自停装置
full lead crystal 全铅晶质玻璃
full lead crystal glass 全铅晶质玻璃
full lean mixture 全贫混合物
full-length 全长；标准长度
full-length contact 全长上的接触；全长上的接触
full-length cylinder liner 全长式汽缸套
full-length heat-treated rail 全长淬火轨
full-length liner 全长式缸套
full-length mirror 全长镜子；通长镜子；穿衣镜
full length of station track 站场全长
full-length water jacket 全长式汽缸水套
full-length welding 满焊
full liability 全部负债；全部责任；完全责任
full liability through bill of lading 全责联运提单
full license 正式执照
full life cycle test 全生命周期测试
full life restoration 全寿命恢复
full lift valve 全升阀；全升程阀
full line 整行；全线；实线；丰满型线（船体）
full linear group 全体线性群；完全线性群
full line buffer 全行缓冲器
full line department store 大百货商店
full line mode 满行方式；全行(显示)方式
full line up 全调测
full liquid cooling 全液冷
full liquid-cooling generator 全液冷发电机
full load 整箱货；满载时；满(负)载；满负荷；满额荷载；全(负)载；全负荷
full load adjustment 满载调整
full load capacity 满载容量
full load characteristic 满载特性；全荷特性；全载特性
full load characteristic curve 满载特性曲线
full load coefficient 满载系数
full load condition 满载状态；满载情况；满负载条件
full load current 满负载电流
full load displacement tonnage 满载排水量；满载排水吨位
full load draft 满载吃水
full load(ed)current 满载电流
full load efficiency 满载效率；满负载效率
full load excitation 满载励磁；满载激励
full load field 满载磁场
full loading 全负荷；满载
full load line 满载水线；满载吃水线
full load line mark 满载吃水线吃水标志
full load loss 满载损失；满荷损耗
full load maximum 全载装的最大限度
full load meter adjustment 满载调整装置
full load need 满负荷耗量
full load operation 满负荷工作；满负荷操作
full load output 满载出力
full load period 满载阶段
full load power 满载功率
full load production standard 全负荷生产定额
full load rated speed smoke characteristic 满负荷额定转速排烟特性
full load ratio 满载率
full load running 满载运转；满载运行
full load saturating curve 满载饱和曲线
full load slip 满载转差率
full load speed 满载速度；满载航速
full load stall 满载失速
full load starter 满载起动器
full load test 满载试验；满负荷试验
full load torque 满载转矩；满载扭矩
full load traffic 满负荷交通
full load trip 满载行程
full load velocity characteristic 全负荷速度特性
full load water-line 满载吃水线
full load water plan 满载吃水线平面
full lock 全锁
full-locked coil construction rope 全封闭索股结构钢索
full logarithmic scale 全对数尺度
full longwall 全面开采法；全面长壁开采法

full look ahead carry 全先行进位
full lot 满份儿
full-louvered door 全百叶门
full magnetic controller 全磁控制器
full manned 配备齐全的(指人员)
full margin 全额保证金
full marks 满分
full matrix 满矩阵；全矩阵
full matrix method 满矩阵法
full matrix ring 全阵环
full mature stage 壮年期；盛壮年期
full mature valley 盛壮年谷
full maturity 壮年期；盛壮年期
full meander 壮年曲流；完备曲南；S形弯曲河段
full mechanization 全部机械化
full mechanization of cargo-handling 装卸工作全盘机械化
full member 正式会员
full metro 全地铁
full-mill 耐火(结构)房屋
full modulation 全调制
full moon 满月；望(月)
full moon illumination 满月照明
full moon spring tide 满月大潮
full mortise hinge 全嵌铰链
full motion video compression board 动态图压缩板
full-mo(u)ld casting process 实型铸造
full multiple 全复接
fullness 漆膜丰满度；深度；丰满度；饱满
fullness coefficient 满蓄率；丰满系数；方形系数
fullness of shade 色调丰满度；色调饱满度
full neutron flux 满中子通量
full nominal speed 全标称速度
full noon brightness 满月亮度
full numeral dial 全数字盘
full nut 全高螺母
full of liquid 液满；充满液体
full-opened corner joint 全开口角接头
full opening 全部敞开的
full open state 全开状态
full open throttle 全开节气门
full operand 全操作数
full operating condition 全运行工况
full operating range 全工作范围
full operation 全部投入使用；全部投产
full out 无缩格排版
full outflow capacity 全部泄流能力；敞泄能力
full output 最大生产率；满负荷；全部出力
full out terms 到港重量条款
full overlap method 全重叠法
full package 满管
full package counter 满筒计数器
full packed 全部包装
full packing 全填充
full-page display 全页显示
full-paid stock 全部缴清的股份
full-parabolic(al)reflector 全抛物面反射器
full pay 全薪
full payment 全付；全部付款
full payoff 完全付清
full pay out 费用全付
full pay out lease 全额租赁
full penetration 全熔透
full penetration butt weld 满对接焊
full performance 全部履行
full period service 全日服务
full phase 满相
full pickled sheet 全酸洗钢板
full pipe 满流管；满管
full pitch 高跨相等面坡；正常齿距；整节距；全距；全节距；屋架高跨相等时屋面坡度
full-pitch auger 全螺距螺旋
full-pitched coil 整节距线圈；全距线圈
full-pitch winding 全节矩绕组
full-plant discharge 水电站满荷宣泄流量；电站满载泄流量
full-plant operation 全面运转
full plate 整体夹板
full-plate watch 整圆夹板的表
full pointed rivet 满填铆接；满旗铆接
full pool level 满库水位
full pore-pressure ratio 全孔隙压力比
full portal gantry 高架起重机；龙门起重机

full potential 全电位;全电势
full power 最大推力;满功率;全功率;全部机组发电
full power ahead 全功率正车
full power astern 全功率倒车
full power condition 满功效条件
full power flight 全功率飞行
full power of management 自主经营
full power output 全功率输出
full power response 全功率响应;全功率反应
full power trial 满负荷试运转;全功率试验;全功率试车
full premium if lost 遗失后补缴全部保险费;全损时补缴全部保险费
full pressure 全压力
full pressure circulating lubrication system 全压循环润滑系统
full pressure lubricating system 全压润滑系统
full pressure lubrication 全压润滑
full pressure lubrication system 全压力润滑系统
full pressure operation 全压运转
full pressure ratio 全压比(率)
full pressure suit 密闭服;全压服
full prestressing 全预应力
full prestressing concrete 全预应力混凝土
full prestressing design 全预应力设计
full price 足价
full-price offer 原价购买
full process 全进程
full production 满负荷生产;全能力生产
full production capacity 总产量;全产量
full protection of wood 完全保护木材
full protection policy 全额负担保险单;全额
full proving 全项目标定
full pulse 洪脉
full pumping 满泵
full quenching 淬透
full race monitoring 全记录监测
full radiator 全辐射体;完全辐射体
full-radius crown 圆断面唇部钻头
full-range 满量程;满标度;全量程
full-range transmission 全变速范围的变速器
full-range tuner 全波段调谐器;全范围调谐装置
full range variable dodging 全片可变匀光
full rank 满秩
full rank adjustment 满秩平差
full raster 满光栅;全光栅
full rate 全价
full-rated capacity 全额定载货容量
full-rated speed 满载额定转速
full reach and burden 全部载货容量
full-read pulse 全选读脉冲
full recording 全自动记录
full recording mode 全记录方式
full regeneration 完全再生
full regulation 全面调节;完全调节
full release 全缓解
full-release position 全松位置;全缓解位
full release time 全缓解时间
full relief 圆雕;全痕(遗迹化石);高凸浮雕
full replacement cost 全部重置成本
full reservoir 满库
full reservoir surface 满库水位;满库水面
full-resolution picture 清晰图像;高分辨率图像
full-resolution template 全分辨能力的样板
full retard 全延迟点火
full retard position 全延迟位置
full retreat mining 全面后退式开采
full retreat mining system 全面后退式开采法
full retreat panel 全面后退式盘区
full return 全部返回(冲洗液);均可退费
full-revolution dumper 全转式翻车机
full-revolving 全回转
full-revolving back hoe excavator 全转式反铲挖土机
full-revolving crane 回转式起重机
full-revolving electric(al) shovel 全转式电铲
full-revolving loader 全旋转式装料机
full-revolving turret 全回转式转塔
full-rich position 全富位置
full rigged 装备齐全的
full rotating 全旋转式的
full-rotating crane 全周转动起重机;全旋转式起重机
full-rotating derrick 全(周)转动起重机;回转式起重机

full rotation type 全自动式
full round nose bit 圆断面唇部钻头
full-round placing system 全断面衬砌
full row rank 满行秩
full rudder 满舵
fullrun 满筒
full sail 满帆
full-scale 毛测;满刻度(的);满度;满标度;满标,等大;原尺寸的;足尺(的);全量程;全刻度;全尺寸;全标度;实物尺寸
full-scale activated sludge plant 成套活性污泥处理设备
full-scale civil management 全面行政管理
full-scale clearance 全部清除存储信息;全部清除
full-scale condition 真实条件;全尺寸条件
full-scale construction 全面施工
full-scale crash test 全尺寸碰撞试验
full-scale cycle 满标循环
full-scale deflecting force 满刻度偏转力
full-scale deflection 满刻度偏转;满度偏转;足尺挠度;全刻度偏转
full-scale detail 足尺大样
full-scale device 生产性装置
full-scale engine test 全尺寸发动机试验
full-scale enhanced biological phosphorus removal process 完全强化生物除磷工艺
full-scale error 满标误差
full-scale experiment 全尺寸试验;实物试验
full-scale flow 全尺寸流
full-scale input 全部输入信号
full-scale investigation 原型试验研究;全尺寸条件下的研究
full-scale layout 足尺放样
full-scale load(ing) test 全负荷试验;满负荷试验
full-scale lofting 实尺放样
full-scale measurement 满刻度测量;全尺寸测量
full-scale measuring range 全范围量程
full-scale membrane bioreactor 完全膜生物反应器
full-scale meter deflection 满标偏转
full-scale mock-up 全尺寸模型;实体模型
full-scale model 实尺模(型);原型模型;原尺寸模型;一比一模型;足尺模型
full-scale modeling 足尺模拟
full-scale model test 足尺模型试验
full-scale operation 全面运转;全面进行生产
full-scale output 满标输出
full-scale pattern 足尺图案
full-scale plant 工厂装置;生产性设备;生产规模厂
full-scale plant study 生产性规范研究
full-scale pretensioned 实物预拉的;实物先张的;全预张力的
full-scale production 满额生产;全面开工;全规模生产
full-scale range 满刻度量程;满刻度范围;满刻度;全量程
full-scale reactor 全尺寸堆
full-scale reading 满刻度读数;满标值;最大读数;全尺寸读数
full-scale response 满刻度响应
full-scale seakeeping trials 实船适航性试验
full-scale sensitivity 满标灵敏度
full-scale ship test 实船试验
full-scale simulation 全尺寸模拟
full-scale structure 全尺寸(建筑)结构;原型结构
full-scale styling representation mock-up 全尺寸造型模型
full-scale template 一比一样板;足尺板;足尺放样
full-scale temple 足尺样板
full-scale test 足尺试验;满载试验;全面试验;原型试验;真实条件试验;全尺寸试验;实验试验;实物试验;生产性试验
full-scale test bench 实物试验台
full-scale testing 生产性试验
full-scale travel 满刻度行程
full-scale tunnel 全尺寸风洞;实物试验风洞
full-scale turbine 原型水轮机;真机水轮机
full-scale value 满(刻)度值;满标度值;原尺寸值
full-scale water treatment 完全水处理
full scan 满扫描
full-scan television camera 全扫描电视摄像机
full scantling vessel 标准强力船;重构船;全实船
full scene 全景
full screen 全屏幕
full-screen form 全屏幕形式

full-screen panel 全屏板
full-screen processing 满屏处理
full screw bolt 全螺纹(螺)栓
full scroll(case) 整个蜗壳;全部蜗壳;完整(形)蜗壳
full sea 高涨的海浪;满潮
full seamark 高潮水位
full-seam extraction 全厚开采
full-seam mining 一次采全高;全矿层回采
full sea speed 海上全速
full seawater 纯海水
full section 全剖面;全断面(的)
full section blast(ing) 全断面爆破
full section(ed) view 全剖视图
full section excavation 全断面开挖
full section filter 整节滤波器
full section method 全断面法
full seed 饱满籽粒;饱满种子
full seeding 满播;满膘
full selected current 全选电流
full sensitivity contrast 全灵敏度对比度
full-service 经营全部业务
full-service bank 全能服务银行
full-service brake application 常用全制动
full-service lease 附加全部服务的租赁
full-service wholesaler 提供充分服务的批发商
full set 终凝;充分凝结;全组;全套
full set bill of lading 全套提单
full set of documents 全套单据
full shade 饱和色
full shaped ship 肥型船
full shaped vessel 肥型船
full sheet 全页图
full shield 全断面盾构
full shift capability 全移位能力
full shot 全景拍摄;全景镜头
full shot noise 全散粒噪声
full shroud 全齿高加强板齿轮
full-shut position 全闭状态
full-sib selection 全同选择
full-signal pulse 全尺寸脉冲
full size 足尺;原型的;实物尺寸;实际尺寸;原尺寸;最大尺寸;足尺的;真实尺寸;全尺度
full-size component 全尺寸部件
full-size construction 全尺寸建筑结构;全尺寸建筑构造
full-sized 足尺的;总容积的;原大的;实物大小的;全尺寸的;满容量的;满负荷;全轮廓的
full-sized bit 原尺寸钻头;全径钻头
full-sized brick 整砖
full-size design 实尺设计图
full-size detail 原尺寸大样;足尺大样
full-sized model 一比一模型
full-sized plate 标准尺寸的塔盘
full-sized pulse 标准脉冲
full-size draft 全尺寸图
full-size drawing 足尺图;实尺图
full-sized template 一比一样板
full-sized test 实物试验
full-sized weld 全尺寸焊缝
full-size furnace 满容量炼炉
full-size measurement 实物尺寸
full-size model 实尺模型;全尺寸模型
full-size model test 大模型试验
full-size reproduction 等尺寸复制;等大复制
full-size scanning 全尺寸扫描
full-size test 全尺寸试验
full skirted piston 全裙活塞;全侧缘活塞
full skirt piston 全筒式活塞
full slewing crane 全方位旋转式起重机
full slice system 整片式
full-slipper piston 全拖鞋式裙部活塞
full social income tax 全额社会所得税
full-span corridor panel 全跨度走廊板
full-span flap 全翼展襟翼
full-span slotted flap 全翼展开缝襟翼
full-span wooden bent centering 排架式满布木拱架
full-span wooden inclined strut centering 斜撑式满布木拱架
full spectrum seismograph 宽频地震仪
full speed 最高速度;全速
full speed again 加车【船】
full speed ahead 全速前进
full speed astern 后退三【船】;全速后退

full speed operation 全速运转;全速工作
full speed running 全速运行
full speed trial 全速试航
full spiral case 完整形蜗壳
full spiral cut 全螺旋法
full splice joint 全板接合
full spread 满帆
full stabilized conductor 全稳定导体
full stainless steel watch 全钢表
full state trading 完全国营贸易
full steerability antenna 全向可控天线
full steerable antenna 全向方向性可控的天线
full stock 全额股票
full stocking 完满立木度
full stop 完全停车
full storage 整存;满储;满蓄
full-storage system 整存系统
full stor(e)y 全楼层
full strength 最大强度
full-strength colo(u)r 全色调颜色;全色调颜料
full-strength joint 等强接头
full-strength weld 等强焊缝
full-strength welding 全强度焊接
full stroke 全行程
full stroke admission 全行程进给
full-stroke hay press 全行程的干草压捆机
full-stroke safety valve 全冲程安全阀
full subsidence 完全下沉;完全沉陷
full subtracter 全减法器
full sun 全日照
full supply level 满水水位;正常蓄水位;正常供水位;全供水位
full support 全支座;全支承
full-surfaced width 全铺路面宽度;路面全宽
full-surface hinge 大面铰链
full swing 最大摆动;(挖土机的)全回转;最大振荡;全摆幅
full swivel 全旋转
full symmetry 完全对称
full synchronizing 全同步
full term 全付条款
full terminal echo suppressor 全终端回波抑制器
full terms bill of lading 全保条件提单
full text 全文
full text retrieval 全文检索
full texture 紧密组织
full thread 全螺纹
full thread bolt 全螺纹(螺)栓
full three-dimensional migration 全三维偏移
full throttle 全开风门;全节流;节气门全开
full throttle acceleration 全油门加速
full tide 高潮;满潮
full-tide cofferdam 满潮围堰;(高潮时也不会被淹没)高潮围堰
full time 全部工作时间
full-time administrator 专职管理人员
full-time and part-time education 全日制和非全日制教育
full-time duplex connection 全部时间双工连接
full-time employment 全日制工作
full-time farming 全年耕作
full-time jobs 全日工作职位
full-time labo(u)r 全劳动力
full-time message switching 全部时间消息交换
full-time record 全期记录
full-time service 全日工作
full-time staff 专职人员
full-time storage plant 多年调节电站
full tinting strength 全着色力
full tone 全色泽;全影调
full-tone coating 原色调涂料
full-tone copy 全色复制本
full-tone original 全色原件;全色调原版
full-track 全轨;全磁迹;全履带车(辆)
full-track carrier 全履带运输工具
full tracked chassis 全履带底盘
full-track fractor 全履带式拖拉机
full-track laying vehicle 全履带车辆
full-track vehicle 全履带(式)车辆;全履带车轮
full traffic-actuated controller 全车动式信号控制器
full traffic-actuated signal 全感应信号(指交通);全车动式信号
full trailer 重型拖车;全(重)挂车;拖车
full trailer combination 重型拖车车组;全拖车车连接车

full trailer system 全拖车方式
full trailer tractor 全拖车牵引车;全牵引车
full train load 列车满载
full-transistor ignition system 全晶体管点火系
full transponder circuit 全脉冲转发
full transverse ventilation 全横向通风
full trough gliding 满槽滑动
full tub 重矿车
full tune-up 全调整
full turn-key 交钥匙统包
full type 满装式
full uniflow engine 全单流式发动机
full unit 实箱(集装箱)
full universal drill 万能钻床;通用钻床
full utilization 充分利用
full vacuum electron beam welder 高真空电子束焊机
full value insurance 全值保险
full-value letter of credit 全值信用证
full-value variable 全值变量
full-variable multiple access 全可变多址连接方式
full vector offering 全矢量插入
full vertical circle 全圆刻度垂直
full view 全视图;全貌;全景
full-view front 全视野的车身前部
full-view hologram 全视全息片;全景全息图
full-view mast A型桅杆
full-view side window 全视野侧窗
full vision cab(in) 全景驾驶室
full vision cupola 环视圆屋顶
full vision dial 全视度盘
full vision dome 全视野穹隆
full voltage 满电压;全电压
full voltage effect 全电压效应
full-voltage starting 全压起动
full voltage starting motor 全电压起动电动机
full wafer 整片
full wafer memory 整片存储器
full water 满水;充水
full water cooling 全水冷
full water cooling generator 全水冷发电机
full water test 盛水试漏
full water weight 充水时重量
full wave 全波
full-wave amplifier 全波放大器
full-wave balanced amplifier 全波平衡放大器
full-wave band receiver 全波段接收机
full-wave bridge 全波整流电桥
full-wave circuit 全波电路
full-wave compensator 全波补偿器
full-wave control 全波控制
full-wave dipole 全波振子
full-wave doublet 全波偶极子
full-wave lighting control 全波照明控制
full-wave mercury rectifier 全波汞弧整流器
full-wave oscillation 全波振荡
full-wave phase control 全波相位控制
full-wave plate 全波片
full-wave power supply 全波电源
full-wave rectification 全波整流
full-wave rectifier 全波整流器
full-wave rectifier tube 全波整流管
full-wave rectifying circuit 全波整流电路
full-wave rectifying type X-ray apparatus 全波整流式X线机
full-wave square-law detector 全波平方律检波器
full-wave thyristor power supply 全波可控硅电源
full-wave vibrator 全波振动器
full-wave voltage doubler 全波倍压器
full-wave voltage impulse (避雷器的)全波电压冲击
full-way centrifugal pump 全通路离心泵;充分式离心泵
full-way valve 直通阀;同径阀;全通道式阀;滑门阀
full weapon protection 三防(海、陆、空国防)
full weathered layer 全风化层
full-web section switch rail 特种断面尖轨
full weight 总重(量);毛重;满载重量;全重(量);饱满性
full weight load 满重荷载
full weld 满焊
full width 总宽(度);全宽(度)
full-width at half maximum 半宽(度);半极大处

全宽度;半峰全宽
full-width cab-protection plate 全宽(度)护车板
full-width compaction 全宽压实
full-width construction 全宽施工;全宽建筑
full-width hub brake 全宽度轮毂制动器
full-width paver 全宽(度)铺路机
full-width slab 全宽式混凝土路面板
full-width type 全宽型
full-width weir 满槽堰(即指全河渠宽的);全宽堰
full wind pressure 全风压
full with substance 满物料
fully acetylated cotton 高度乙酰化棉纤维
fully active homing head 主动式自动引导头
fully actuated 全部动的
fully adhered roofing 全粘屋面
fully adjustable speed drive 无级调速传动装置
fully aerated flow 完全掺力水流
fully air-conditioned 全空调的
fully amortized mortgage 全部偿清的抵押贷款
fully analytic(al) aerotriangulation program(me) 全解析空中三角测量程序
fully anchored cable-stayed bridge 外锚式斜拉桥;地锚式斜拉桥
fully anti-clockwise 逆时针满舵
fully arisen sea 充分发展的风浪;充分长成风浪
fully aroused sea 充分长成风浪
fully associative buffer storage 全相联缓冲存储器
fully automated assembly line 全自动装配线
fully automated cataloguing technique 全自动编目技术
fully automated computer program(me) 全自动计算机程序
fully automated grade crossing 全自动平交道口
fully automated refuse truck 全自动化垃圾车
fully automatic 全自动的
fully automatic arc welding 全自动电弧焊
fully automatic boiler 全自动化锅炉
fully automatic bonding 全自动接合
fully automatic bonding system 全自动接合装置
fully automatic camera 全自动照相机
fully automatic choke system 全自动阻风门控制系统
fully automatic colo(u)r densitometer 全自动浓度比色计
fully automatic compiling technique 全自动化编译技术;全自动编译技术
fully automatic compression machine 全自动压接机器
fully automatic control 全自动控制;全自动操纵
fully automatic diaphragm 全自动膜盒;全自动(可变)光阑
fully automatic door 全自动门
fully automatic drawing-in machine 全自动穿扣机
fully automatic electronic judging device 全自动(化)电子判定器
fully automatic gate 全自动闸门
fully automatic gearbox 全自动变速器
fully automatic grade crossing 全自动平交道口
fully automatic grid plotting 全自动展绘格网
fully automatic grinder 全自动磨床
fully automatic information retrieval system 全自动信息检索系统
fully automatic lathe 全自动车床
fully automatic machine 全自动机床
fully automatic pipe bender 全自动化弯管机
fully automatic plating 全自动电镀
fully automatic plough 全自动犁
fully automatic press 全自动压机
fully automatic processing 全自动化生产过程;全自动处理
fully automatic reperforator switching 全自动复凿孔机交换机
fully automatic screw machine 全自动螺栓机床
fully automatic selection 全自动拨号
fully automatic sliding door installation 全自动的滑门装置
fully automatic switching 全自动交换
fully automatic switching network 全自动交换网
fully automatic switching system 全自动交换系统
fully automatic tape relay set 全自动纸条转接设备
fully automatic technique 全自动技术
fully automatic telecontrol 全自动遥控
fully automatic toll switching system 全自动长途电话交换机

fully automatic tongs 全自动夹钳
fully automatic turntable ladder 全自动转台阶梯
fully automatic turret milling machine 全自动转塔铣床
fully automatic turret screw machine 全自动转塔式车床
fully automatic vehicle 全自动车
fully automatic watch 全自动表
fully automatic welding 全自动焊
fully automatic winding 全自动提升
fully automatic working 全自动工作;全自动操作
fully blown bitumen 深度氧化沥青
fully bonded bolt 全面粘接型锚杆
fully buffered channel 全缓冲通道
fully cavitating propeller 全空泡螺旋桨
fully cellular container ship 全蜂窝式集装箱船;全蜂窝式货柜船;导轨式全集装箱船
fully cellular container vessel 全格舱式集装箱船
fully cellularized container ship 全区划式集装箱船
fully centralized control system 全集中控制系统
fully charged 全进料的
fully charged battery 全充电蓄电池
fully clockwise 顺时针满旋
fully closed waterproofing 全封闭防水
fully compacted concrete 充分密实混凝土
fully compensated operational amplifier 全补偿运算放大器
fully compensated raft foundation 完整补偿筏基础
fully connected network 全连接网络;全互连网络
fully connected primary satellite system 全连主卫星系统
fully connected single primary satellite system 全连单用主卫星系统
fully constrained non-crystalline network 完全约束的非晶网
fully controlled bridge 全控桥
fully correlated oscillations 全相关振荡
fully cyclic peripheral point spectrum 全循环边角点谱
fully dense 全致密的
fully deoxidized steel 全脱氧钢
fully depreciated 已全部折旧的
fully depreciated assets 全部折旧完的资产;已全部折旧资产
fully depressed expressway 全堑式快速干道(净空全部位于地面以下)
fully detrinised starch 全部糊精化淀粉
fully developed 完全开拓的;全部展开的
fully developed flow 完全展开流
fully developed individual 全面发展的个人
fully developed point of bar 钢筋充分利用点
fully developed sea 充分发展的风浪;充分长成风浪
fully developed turbulence 全面紊流;完全紊流;完全紊动
fully directional interchange 全定向型互通立交;全定向互通式立交
fully directional submersible vehicle 全方位潜水器
fully directional three-leg intersection 全定向三叉互通立交
fully discharged battery 全放完电蓄电池
fully dissociated signalling 全分离信号方式
fully dissolved 完全溶解的
fully drawn texturing yarn 全拉伸变形丝
fully drawn yarn 全拉伸丝
fully eclipsed form 全重叠法
fully electrified 全盘电气化的
fully enclosed 完全封闭的
fully energized 全通电
fully ensured 有充分保证
fully equipped engine 全附件发动机
fully equipped weight 全装备重量
fully expand 完全展开
fully extended length 全伸出长度
fully factored load 最大荷载
fully filled optic(al) fibre cable 全填充式光缆
fully fixed end 全固定端
fully fixed member 全固定构件
fully flameproof motor 全防爆型电动机
fully flattened 绝对平面的
fully floating rear axle shaft 全浮式半轴
fully flooded dye machine 全浸没式染色机
fully fluorinated paraffin 完全氟化的烷烃
fully functional network 全功能网络
fully glazed 全部安装玻璃的

fully good 完好的
fully graded aggregate 全(部)级配集料;全(部)级配骨料
fully grown sea 充分长成风浪
fully guarded 全防护型的
fully guarded machine 全防护型电机
fully halogenated 全卤化的
fully halogenated alkane 全卤化烷烃
fully hardened installation 全硬化设施
fully hardening steel 全硬淬透钢
fully heat-treated rail 全面淬火钢轨
fully hydrated cement 完全水化水泥
fully hydraulic 全液压的
fully hydraulic drive 全液压驱动装置
fully impervious clothing 全密封工作服
fully impregnated insulation 全浸渍绝缘
fully independent variation 无相依变化
fully insulated winding 全绝缘绕组
fully insured 全保险;全保的
fully integrated barge 全分节驳船
fully integrated control system 全集中控制系统
fully integrated digital network 全集成数字网
fully integrated tow 全组合式驳船队;全分节顶推驳船队
fully interlocking 完全锁口的
fully invariant series 全不变列
fully invariant series of subgroups 全不变子群列
fully invariant subgroup 全不变子群
fully ionized gases 完全电离气体
fully isolated route 全封闭线路
fully jewelled 满钻
fully killed alloy 全脱氧合金
fully killed steel 镇静钢;全脱氧钢
fully laden 满装载的;满载的
fully laden ship 满载船
fully laden state 最大装载状态
fully lift position 全升举位置
fully loaded 满载的
fully loaded river 多泥沙河流
fully loaded ship 满载船
fully loaded stream 多泥沙河流
fully loaded train 满载列车
fully locked 密封的
fully locked cablewire 全锁钢丝索
fully mechanized equipment 综合机械化设备;全机械化设备
fully mixed aeration system 完全混合曝气系统
fully modulated radio frequency signal 全调制射频信号
fully mounted 全悬挂式的
fully mounted plough 全悬挂式犁
fully-normalized harmonics 完全正则调和函数
fully-normalized spheric(al) harmonics 完全规范化调和函数
fully on 全通
fully paid 全部付讫
fully paid capital stock 缴足股本
fully paid share 付讫股票;已全部缴款股票;付清股份
fully paid stock 已全部缴款股份;已缴股份
fully paid-up 已全部付清;全部付清
fully parallel associative processor 全并行相联处理机
fully parallel memory 全并行存储器
fully participating preferred stock 全参与优先股
fully penetrating well 贯穿井;全透水井;全穿透井;完整井
fully perforated tape 全穿孔纸带
fully permanent sprinkler system 全固定喷灌系统
fully portable agricultural sprinkling installation 全便携农用喷洒设备
fully portable sprinkler-system 全移动喷灌系统
fully prestressed concrete 全预应力混凝土
fully processed 全部处理的
fully put into production 全部投产
fully qualified name 全限定名;完整限定名
fully quarter-sawn 全径切的
fully recessed door handle 全凹式车门把手
fully reducible matric algebra 全可约矩阵代数
fully refined 精制的
fully reflected reactor 全反射层反应堆;全反射层堆
fully reflecting surface 全反射面
fully registered bond 不能转让的注册记名债券
fully relaxed dimension 全松弛处理后尺寸

fully restrained 全稳定的;全固定的;高度受约束的;高度克制的
fully restrained beam 固端梁;砌入梁
fully rigid framed structure 全刚性性框架结构
fully rigid framing 全刚性结构
fully rough flow 超临界水流
fully saturated green period 全饱和绿灯
fully saturated sample 完全饱和土样
fully secured 有充分保障(的);担保充分;充分担保(的)
fully secured liabilities 担保充分的负债
fully shaped collar 全成型领圈
fully shifted system 昼夜轮班工作制
fully shrouded trickle valve 全罩翼阀
fully sliding condition 全打滑状态
fully softened strength of clay 黏土完全软化程度
fully specified finite-state machine 全部说明的有限状态机
fully sprung floor 全振动地板
fully sprung seat 全减振坐位
fully stabilized satellite 全稳定卫星
fully stable 完全稳定的;全稳定的;全固定的
fully stable emulsion 慢裂乳液
fully steerable radio telescope 全动射电望远镜
fully submerged hydrofoil system 全浸式水翼装置;深浸式水翼装置
fully suspended roof 全吊挂炉顶
fully synchronized transmission 全同步变速箱
fully tax-exempt treasury securities 全部负税
fully tiled 全部铺瓷砖的
fully tracked vehicle 全履带牵引车
fully transistorized 全晶体管化
fully treated pavement structure 全深处理的路面结构
fully turnable tug 全回转型拖轮
fully variable multiple access 全可变多址
fully vented 完全通风的
fully vested 养老金计划的全部受益
fully water-cooled furnace 全水冷壁炉膛
fully water-cooled turbogenerator 全水冷汽轮发电机
fully welded seamless door 全焊无缝门
fulminate 雷酸盐
fulminating altitude anoxia 暴发性高空缺氧
fulminating anoxia 暴发性缺氧
fulminating cap 雷帽;雷管;雷汞爆管
fulminating gold 雷酸金
fulminating infections disease 烈性传染病
fulminating lead 雷酸铅
fulminating oil 雷爆油
fulminic acid 雷酸
fuloppite 柱辉锑铅矿
fulvate 富里酸盐
fulvene 富烯
fulvic acid 灰黄霉酸;黄腐(植)酸;富里酸
fulvic acid solution 富里酸溶液
fulvicin 灰黄素
fulvous 茶色的;黄褐色的
fumaramic acid 富马酰胺酸;富马酸酰胺
fumaramide 富马酰胺;富马酸二酰胺
fumarate 富马酸盐;反丁烯二酸酯
fumarhydrazide 富马酰肼
fumaric acid 富马酸;反(式)丁烯二酸
fumaric resin 富马酸(酐)树脂;反丁烯二酸(酐)树脂
fumarimide 富马酰亚胺
fumaroid 富马型;反丁二烯型
fumaroid form 富马型
fumarole 火山喷气孔;气孔;喷气孔【地】;喷气孔
fumarole activity 喷汽活动
fumarole condensate 喷汽孔凝结水
fumarole deposit 喷汽孔沉淀物
fumarole discharge 喷汽孔排汽
fumarole eruption 喷汽孔喷发
fumarole gas 喷汽孔气体
fumarolic activity 喷汽孔活动
fumarolic field 喷汽孔田
fumarolic structure 气孔状构造
fumarolic-type gas emission 喷汽孔式气体喷射
fumaroyl acetoacetate 富马酰乙酰乙酸
fumaryl chloride 富马酰氯
fumatorium 熏蒸消毒室;熏蒸器(用来消灭害虫);熏蒸场
fumatory 熏蒸室;熏蒸场

Fumazone 二溴氯丙烷(熏蒸杀虫剂)
fume 烟雾;烟气;蒸汽;微粒污染物
fume abatement 烟雾消除
fume burner 气体燃烧器
fume chamber 烟雾室;烟柜;通风柜
fume cleaning 烟雾净化;尘雾净化
fume collector 烟罩
fume combustion apparatus 烟雾燃烧设备
fume conditioning 烟气调节;气体调节
fume cupboard 烟橱;排烟柜;通风橱;抽风柜
fume damage 烟气毒害;烟害
fume detector 烟雾探测器
fumed oak 烟橡
fumed silica 烟制二氧化硅;气相法二氧化硅;气相法白炭黑
fume emission 烟气排放
fume extraction 烟雾回收
fume extractor 排烟装置;排烟设备;排尘雾设备;抽器;风抽子
fume fading 烟气褪色
fume fever 烟尘热;铸造热
fume height 烟柱高度
fume hood 烟柜;烟橱;去烟罩;通风柜;通风橱
fume incineration 烟雾焚烧;微粒污染焚烧
fume incinerator 微粒污染物焚化器
fume loss 烟雾损失
fume nuisance 烟害
fume off 排气;排出气体
fume pollution 烟害
fume-proof enamel 耐烟雾瓷漆
fume-proof paint 耐烟雾漆
fume recovery system 烟道气回收系统
fumerell 屋顶通气孔
fume-removal equipment 抽气设备
fume removal unit 除烟机构
fume resistance 抗烟性;耐烟雾性
fume-resistant 防烟的
fume-resisting machine 防烟式电机
fumerole 气孔;喷气孔【地】
fumeroot 熏烟草
fume scrubber 烟雾洗涤器;雾气洗涤器
fumes detector 烟雾信号装置;烟雾检测器
fume separator 烟气分离器
fume shape 烟雾形状
fume smoking 烟熏
fume suppressor 烟雾抑制器
fumet(te) 熏香剂
fume vent line 烟雾排出管道
fumewort 熏烟草
fumigant 熏蒸(消毒)剂;熏(杀)剂;熏船燃料;消毒烟剂;烟熏剂
fumigant poisoning 熏蒸剂中毒
fumigate 熏蒸消毒(法);烟熏
fumigating 熏蒸
fumigating insecticide 熏蒸杀虫剂
fumigating system 熏蒸法
fumigation 熏蒸消毒(法);熏蒸;熏烟消毒;熏烟;熏船;熏舱;烟下沉现象;漫烟型
fumigation action 熏蒸作用
fumigation agent 熏蒸剂
fumigation certificate 熏舱证书
fumigation chamber 熏蒸室
fumigation expenses 熏舱费(用);熏仓费用
fumigation in shoreline regions 海岸带熏烟
fumigation of ship's holds 熏舱
fumigation plant 熏舱设备
fumigator 熏蒸消毒器;烟熏器
fuming 发烟的
fuming acid 发烟酸
fuming cupboard 烟柜;通风橱;烟橱;通风柜
fuming liquids 发烟液体
fuming nitric acid 发烟硝酸
fuming-off point 爆燃点
fuming-off temperature 爆燃温度
fuming sulfuric acid 发烟硫酸
fuming sulphuric acid 发烟硫酸
fumivorous material 消烟材料;除烟材料
fumming 漆膜烘烤时起烟雾
fumulus 烟状云;缟状云
Fun City 欢乐城(美国纽约市的俚称)
function 作用;职能;功用;功能;机能;函数;操作过程
functional 功能性的;功能的;操作的;泛函的
functional absorber 空气吸声体;空间吸声体

functional accounting 职能会计
functional accumulation 机能蓄积
functional adaptation 功能适应;机能适应(性)
functional addition 水泥调质添加剂;水泥调质处理
functional address instruction 功能地址指令;操作地址指令
functional address instruction format 操作地址指令格式
functional allocation 功能分配
functional analysis 功能分析;泛函分析
functional analysis approach 功能分析方法
functional analysis system technique 功能系统分析;功能分析的系统方法
functional-analytical group 分析官能团
functional architecture 功能建筑;专用建筑;实用建筑
functional area 职权范围;职能区域;职能范围
functional arrangement 操作线路
functional array 功能阵列
functional authority 职能权限;职能权力
functional authority credentials 功能特许证;功能授权凭证;功能使用权凭证
functional averaging method 泛函平均法
functional beauty 功能美
functional block 功能器件;功能框;功能块
functional block diagram 功能块框图;原理方框图
functional body 专门机构
functional building 专用房屋;功能建筑
functional building elements 建筑功能构成要素
functional button 功能按钮
functional calculator 函数计算器;函数计算机
functional calculus 泛函演算
functional call 函数调用
functional capability (建筑构件的)功能容量
functional ceramics 功能陶瓷
functional change 功能转变;功能性变化
functional changeover 功能转换
functional character 功能字符;功能性质;功能控制符;功能符(号);控制符(号);机能特性
functional characteristics 功能特点
functional check 功能检验;功能检查
functional circuit 功能电路;解算电路
functional city 机能都市;功能城市
functional classification 功能分类
functional coating 功能性涂层;功能涂料
functional code 功能码
functional colo(u)ring 功能配色(建筑)
functional comparator 函数比较器
functional compensation 功能代偿
functional completeness 功能完整性;功能完全性;功能完备性;函数完整性
functional component 功能成分
functional composite material 功能复合材料
functional compound 官能化合物
functional computer 函数计算机
functional concept of architecture 建筑学的功能概念
functional configuration 功能结构
functional connection 功能联系
functional consolidation 职能合并
functional control 功能控制
functional control block 功能控制块
functional control device 功能控制器
functional conversion 功能转变
functional correlation 函数相关;机能相关
functional cost analysis 功能成本分析
functional counter 操作计数器
functional crystal 功能晶体
functional data 功能数据
functional data table 功能数据表
functional decentralization 职能式分权制
functional definition 功能定义
functional definition language 功能定义语言
functional definition module 功能定义(程序)模块
functional demand 功能性需求
functional department 职能部门
functional dependence 功能相关性;函数相依;函数相关;函数关联
functional dependency 函数相关性
functional depreciation 功能折旧;功能衰减
functional description table 功能描述表
functional design 性能设计;功能设计;功能(部件)设计
functional designator 函数指示符

functional design of road 道路功能设计
functional design of structures 结构的功能设计
functional determination 功能确定;合格鉴定
functional device 功能器件;功能部件
functional diagnosis 机能诊断
functional diagram 功能(设计)图;作用功能图;作用方框图;工作原理图;方块图
functional diameter 作用直径
functional differentiation 功能转化;机能分化
functional digit 功能位;功能数字(组);功能数
functional distribution 功能性分配
functional-district of water environment 水环境功能区
functional-district of water environment protection 水环境保护功能区
functional disturbance 机能障碍;机能性失调
functional division 按作用分类
functional efficiency 机能效率
functional element 功能元件;功能器件
functional element program(me) 子程序
functional enhancement package 功能增强程序包
functional equation 函数方程;泛函方程
functional error 功能误差
functional examination 功能检查法
functional expenses 业务费用
functional fiber 功能纤维
functional field 功能信息组
functional filler 功能性填料
functional fine material 功能精细复合材料
functional fitting method 函数拟合法
functional form 功能形式;功能型
functional frame for budget 预算职能结构
functional generating subroutine 功能产生子程序
functional generator 函数模拟装置;函数发生器;程序编制器
functional gradient ceramics 功能梯度陶瓷
functional gradient material 功能梯度材料
functional grille 功能格栅
functional group 功能基团;专业小组
functional grouping 功能组合
functional hall 会议—宴会厅
functional harmony 机能协调性
functional hierarchy 功能层系
functional hole 功能孔
functional identifier 功能识别符;功能标识符
functional inadequacy 功用不合
functional independent testing 功能无关检测
functional input section 功能输入部分
functional instruction address 操作指令地址
functional interchangeability 功能互换性
functional interface 功能接口
functional interleaving 功能交错;功能交插;功能交叉;操作交错(进行)
functional interpreter 功能解释程序
functionalism 功能主义
functionalism architecture 功能主义建筑
functionalist 功能主义者
functionality 官能度;功能度
functionalization 职能化
functionalized polyethylene imine water-soluble polymer 功能聚乙烯亚胺水溶性高聚物
functionalized polymer 功能高聚物
functional joint 构造缝;工作缝;伸缩接合;伸缩缝
functional key 功能键
functional keyboard 功能键盘
functional label 功能标记
functional layer 功能层;机能层
functional limit 函数极限
functional localization 机能定位
functional logic 功能逻辑
functionally designated area 功能特定区
functionally designed 配套设计的;从使用观点设计的
functionally distributed computer system 功能分布(式)计算系统
functionally distributed network 功能性分布网络;功能分布式网络
functionally distributed system 功能分布式系统
functionally dominant phenol-degrading bacteria 功能优势酚降解菌
functionally graded materials 梯度功能材料
functionally gradient ceramics 梯度机能陶瓷
functionally gradient material 梯度材料
functionally observational battery test 功能系列

试验
functional macro 操作宏指令
functional macro instruction 功能宏指令
functional management 功能管理;职能管理
functional management data 功能管理数据
functional management data services 功能管理数据服务(程序)
functional management header 功能管理报头
functional management layer 功能管理层
functional management profile 功能管理简要表;功能管理概念说明
functional material 功能材料
functional memory 功能存储器
functional microorganism 功能微生物群体
functional mobility 功能灵活性
functional mode 工作状态
functional model 函数模型
functional modular controller 功能模块控制器
functional modularity 功能模式化;操作模式化
functional modulatory 操作模式化
functional module 功能微型组件;功能模块
functional morphological analysis 功能形态分析
functional morphology 功能形态学
functional mosaic 机能镶嵌
functional multiplier 函数乘法器;函数倍增器
functional name 功能名;功能部件名
functional objective 功能目标
functional obsolescence 功能性过时;功能性陈旧;功能衰退;建筑过时
functional operation 函数运算
functional operator 功能操作符;函数运算子
functional organization 职能组织;职能机构
functional parallelism 功能并行性
functional parameter 功能参数
functional part 功能部分
functional partition 功能划分
functional partitioning microprocessor 按功能划分的微处理器
functional performance 使用性能
functional periodicity 函数周期性
functional pigment 功能性颜料
functional pins 功能引线
functional plan 功能规划;专项规划;专题规划;单项规划
functional planning 功能规划
functional planting 功能栽植
functional plot subroutine 功能绘图子程序
functional plotter 函数描图器;函数绘图仪
functional pollution criteria 机能的污染标准
functional polymer 官能聚合物;官能高聚物
functional pool 功能池
functional position 功能位
functional preselection 功能预选
functional principle 实用原则
functional processing module 功能处理模块
functional program(me) 功能程序
functional program(me) library 函数程序库
functional proof cycle 工作性能检查周期
functional protection 功能性保护
functional prototype of high definition television 高清晰度电视功能样机
functional punch 功能孔;功能穿孔
functional recorder 函数记录器
functional recovery 功能恢复
functional recovery routine 功能恢复例行程序
functional reduction 功能性复位;功能复位
functional redundancy 功能丰余性
functional region of city 城市功能分区
functional relation(ship) 函数关系
functional reliability 运行可靠性;工作可靠性;机能可靠性
functional requester 功能请求者
functional requirement 功能(性)要求;功能条件说明书;职能要求
functional residual capacity 功能性余气量;功能性残气量;有效余气量
functional response 机能反应
functional restoration 功能恢复;机能恢复
functional risk 功能风险
functional routine 功能例行程序
functional scheme 工作原理图
functional segregation 功能分化
functional series 函数序列
functional shift 功能转变

functional signal 功能信号
functional simulation 功能模拟
functional simulator 功能模拟程序
functional space 函数空间
functional space schedule 功能空间一览表
functional specialization 功能专用化;功能规约;(交通控制系统的)功能规范
functional sphere 功能区域
functional storage 功能存储器
functional stress 机能性应力
functional strobe 功能选通
functional structure 功能结构;职能式组织管理机构
functional study 功能研究
functional subprogram(me) 功能子程序;函数子程序
functional surface 功能路面
functional switch 功能(转换)开关;函数开关;逻辑开关
functional symbol 功能符号;函数符号;操作符(号)
functional table 函数表
functional test 功能试验;功能检测;功能测试;机能试验;机能检查;合格试验
functional tester 功能测试器
functional testing 功能测试
functional test procedure 机能试验程序
functional test specification 机能试验规范
functional theory 函数论
functional tolerance 功能公差
functional training 职能训练
functional transformation 函数变换;泛函变换
functional transmitter 函数传感器
functional unit 功能组件;功能装置;功能器件;功能单元;功能单位;功能部件;函数元件;函数单元;操作部件
functional unit time 功能部件耗时
functional urban district 城市功能分区
functional value 函数值
functional viability 功能健全程度;职能完整性
functional wall 功能墙
functional word 功能字
functional zone 功能区;职能区
functional zoning 功能区
functional zoning for urban land use 城市功能分区
function area of industrial engineering 工业工程的功能范围
functionary 工作人员;公职人员;公务员
functioning 行使功能
function block 功用块
function building 专用房屋
function chamber 下水道汇流井
function checkout 功能核查
function code 机器功能码;操作码
function code table 操作码表
function data table 操作数据表
function designator 操作指示符
function design of environment(al) information system 环境信息系统的功能设计
function diagram 工作图;方块图
function digit 操作位;操作(码);操作码
function driver 操作驱动器
function fitter 折线函数发生器
function generating potentiometer 非线性电位计
function generator of more variables 多变量函数发生器
function generator of two variable 双变量函数发生器
function hole 标志孔
function key 操作键
function keyboard 功能键盘;操作键盘
function letter 操作字母;操作字码
function lever 操作杆
function measured 所测的函数
function mechanism 作用机理;操作机构
function of a complex variable 单复变函数;复变函数
function of anchorage 锚定作用
function of average 平均数功用
function of civic center 市中心功能
function of control program(me) 控制程序的功能
function of dispersion 分散函数
function of exchange 交换功能;交换职能
function of function 复合函数

function of function rule 复合函数法则
function of initial time 初始时间函数
function of macroscopic regulation 宏观调节职能
function of many variations 多元函数
function of motion 运转功能
function of relative orientation 相对定向函数
function of settlement 住宅区功能;新住宅区职能
function of space coordinates 空间坐标函数
function of spatial frequency 空间频率函数
function of spectral power distribution 光谱功率分布函数
function of state 状态函数
function of structure 建筑物的功能
function of temperature 温度函数
function of the extent of reaction 反应程度函数
function of tracer migration with water 示踪剂随水运移性能
function of urban ecologic(al) system 城市生态系统功能
function of wind loading 风荷作用
function part 操作部分
function punch 标志符标志孔
function room 专用房间;小宴会厅;会议厅
function routine 操作例行程序
function subprogram(me) 操作子程序
function switch 工作转换开关;操作开关
function table 转码器;转换装置
function timer 操作计时器
function timing 操作计时
function unit 控制部件;操纵部件
functor 功能元件;算符
fund 资金;款项;基金
fund account 基金账户
fund accounting 基金会计
fund allocation 经费分配;基金分配
fundamental 基本的
fundamental absorption 基本吸收
fundamental absorption band 基本吸收带
fundamental active power 基波有功功率
fundamental affine connection 基本仿射联络
fundamental analysis 基础分析;基本分析
fundamental assumption 基本假设
fundamental astrometry 基本天体测量学
fundamental astronometry 基础天体测量学
fundamental astronomy 基本天文学
fundamental axis mode 基轴模
fundamental band region 基本能带区
fundamental base band 基本频带
fundamental benchmark 基本水准点
fundamental blade passage frequency 桨叶基频
fundamental breach 重大违约;根本违约
fundamental catalogue 基本目录
fundamental category 主范畴;基本类目
fundamental chain 主链;母链
fundamental characteristic 基本性能
fundamental circle 主大圆;基准大圆;基圆;基本圈
fundamental circuit 基本回路
fundamental circuit matrix 基本回路矩阵
fundamental code 基本规范
fundamental colo(u)r 主要颜色;基本色泽;基本色
fundamental combination for action offects 作用效应基本组合
fundamental complex 基底杂岩
fundamental component 主要成分;基波分量;基本原件;基本分量
fundamental component distortion 基波失真
fundamental construction 基本组成;基本建设;基本构造
fundamental construction clearance 基本建筑限界
fundamental coordinate system 基本坐标系
fundamental criterion 基本准则
fundamental crystal 基频晶体
fundamental current 基波电流
fundamental curve 基本曲线
fundamental cutest 基本割集
fundamental cutset matrix 基本割集矩阵
fundamental cycle 基本回路
fundamental data 基本数据
fundamental design 基本设计
fundamental deviation 基本偏差
fundamental differential equation of geodetic gravimetry 大地重力学基本微分方程
fundamental domain 基本(区)域【数】
fundamental driving force 基本推动力

fundamental duality theorem 基本对偶定理
fundamental equation 基本方程(式)
fundamental equation of hydrodynamics 流体动力学基本方程
fundamental extract circuit 基频提取电路
fundamental factor 基本因素
fundamental fault 基底断层
fundamental fiscal asymmetry 财政的基本不对称情况
fundamental force 基本力
fundamental form 基本型
fundamental forms of a surface 曲面的基本型
fundamental formula 基本公式
fundamental frequency 基音频率;基频;基波频率;基本频率
fundamental frequency band 基频谱带
fundamental frequency combining 基频合并
fundamental frequency composition 基频成分
fundamental frequency of Decca 台卡基波频率
fundamental function 基本函数;基本功能;特征函数
fundamental gas velocity of dust collector 除尘器气体基本速度
fundamental gneiss 基底片麻岩
fundamental goals 根本目的
fundamental group of a topologic(al) space 拓扑空间的基本群
fundamental harmonic 一次谐波
fundamental-harmonic pair 基波对
fundamental idea of the invention 发明的基本思想
fundamental implicate 标准外项
fundamental inputconnection matrix 基本输入联结矩阵;基本关联矩阵
fundamental interconnection vector 基本关联向量
fundamental interests 根本利益
fundamental interlocking circuit 基本联锁电路
fundamental interval 基本音程
fundamentality 基本的性质或状态
fundamental lattice 基本点阵
fundamental lattice absorption 基本点阵吸收
fundamental law 基本法;基本定律
fundamental level(l)ing 基本水准测量
fundamental logic 基本逻辑
fundamental longitudinal mode 基纵模
fundamental loop 基本回路
fundamental magnitude 基本量
fundamental matrix 基本矩阵
fundamental mesh 基本网孔
fundamental metric tensor 基本度量张量
fundamental mode 主振型;振荡主模;基(谐)型;基谐模式;基本振型;基本模式;基本(工作)方式
fundamental mode of vibration 基本振型;基谐振型
fundamental motion 基本运动;基本动作
fundamental neighbo(u)rhood 基本邻域
fundamental net(work) 基本网(络);基本控制网
fundamental niche 基础生态位
fundamental node 基本节点
fundamental norms 基本规范;基本定额
fundamental operation 基本运算
fundamental oscillation 基本振荡
fundamental parameter 基本参数
fundamental particle 基本质点;基本粒子
fundamental particle physics 基本粒子物理学
fundamental particle reaction 基本粒子反应
fundamental performance parameter 基本性能参数
fundamental period 基本周期
fundamental photometric unit 基本光度学单位
fundamental plane 基准面;基本面
fundamental plane of spheric(al) coordinates 球面坐标基面
fundamental point 控制点;基本点
fundamental power 基波功率
fundamental principle 根本原则;根本原理;基本原理
fundamental process 基本过程
fundamental product 基本积
fundamental productivity at surface 表层基本生产力
fundamental purpose 主要目的;主要目标
fundamental quantity 基本量
fundamental reaction 基本反应
fundamental reference system 基本参考系

fundamental reflection 基本反射
fundamental renewal equation 基本更新方程
fundamental research 基础(理论)研究;基本理论研究
fundamental ripple frequency 脉动基频
fundamental risk 基本危险(性)
fundamental rock 基岩
fundamental segment 基本节段
fundamental seismic degree 基本地震烈度
fundamental seismic intensity 基本地震烈度
fundamental sensation 基本感觉
fundamental sequence 柯西序列;基本序列
fundamental series 基线系
fundamental set 基本设备
fundamental sheets 基本报表
fundamentals of accounting 会计基本原理
fundamentals of electric(al) and electronic engineering 电工基础
fundamental solution 基本解
fundamental solution matrix 基本解矩阵
fundamental specification 基本规范
fundamental standard 基本标准
fundamental star 基本星
fundamental star catalogue 基本星表
fundamental star place 基本星位置
fundamental state 基本情况
fundamental strength 基本强度
fundamental strength of concrete 混凝土的基本强度
fundamental stress 最大应力
fundamental structure ga(u)ge 基本建筑限界
fundamental sum 基本和
fundamental suppression 基频抑制
fundamental surveillance 基本监测
fundamental technical data 基本技术数据
fundamental tensor 基本张量
fundamental theorem for rectangular game 矩形对策的基本定理
fundamental theorem of algebra 代数基本定理
fundamental tide 天文潮
fundamental tide level 天文潮位
fundamental tolerance 基本公差
fundamental tolerance unit 基本公差单位
fundamental tone 基本色调;基音
fundamental transformation matrix 基本变换矩阵
fundamental transition 基本跃迁
fundamental transverse mode 基横模
fundamental unit 基本单位
fundamental urban function 城市基本功能
fundamental vibration 基谐振动;基本振动
fundamental vibration-rotation region 基本振动转动区
fundamental wave 主波;基波
fundamental wavelength 基波波长
fundamental-wave resonator 基波谐振器
fundamental wind pressure 基本风压
fundamental work 分项工程
fundamental zero 基本零点
fundamentum 基本法则
fund analysis 资金分析
fund application 资金运用
fund assets 基金资产
fund availability 可用资金
fund available 可动用资金
fund balance sheet 基金资产负债表
fund borrowed 借入资金
fund collecting 集资
funded debt 基金负债;长期借款
funded deficit 基金亏损
funded liability 长期借款
funded pension plan 没有基金的退休金计划
funded period 设置基金期间
funded reserve 基金储备;投资基金;设有基金之准备
fund flow statement 基金周转表
fund for equipment investment 设备投资基金
fund for repayment of borrowings 借款偿还基金
fund group 基金分类
fund guarantee 资金担保
fund holder 基金持有人
fundible securities 可代换证券
funding 提供资本;拨款
funding bond 筹集资金债券
funding debenture 以债券偿付利息

funding decision 分配资金决策
funding depreciation 折旧基金
funding for some key projects 重点支出
funding loan 基金贷款;基本贷款
funding mortgage loan 偿债抵押债券
funding system for scientific research projects 科研基金制
fund-in-trust 信托基金
funditae 翅底片
fund liability 基金负债
fund obligation 基金负担
fund of funds 基金中的基金
fund of partly remained profit 利润留成基金
fund on hand 现有基金
fund outflow 资金外流
fund procurement 资金调度
fund raising 资金筹集;资金筹措;集资;基金筹集;筹资
fund raising activities 资金筹措活动
fund raising cost 集资成本
fund raising for railway 铁路筹资
fund raising institution 融资机构
funds 经费
funds allocated for non-productive purpose 非生产性资金
funds appropriation 资金转拨
funds at one's disposition 可自行支配的资金
funds at the disposal of enterprise 企业运用资金
funds controlled by the Central Government 中央政府性基金
funds flow 资金流转
funds for consumption 消费基金
funds for development research 发展研究基金
funds for economic growth 经济增长基金
funds for labo(u)r protection 劳动保护措施基金
funds for non-commodity items 非商品资金
funds for public undertaking 事业外
funds for trial manufacture of new products 新产品试制费
funds for use at one's own discretion 可供自己支配的资金
funds from parent 母公司拨入资金
funds from parent company 母公司拨入资金
funds handed over-to the state treasury 财政缴款
funds in hand 现有基金
funds norm for raw and processed materials 原材料及加工中物料的资金定额
funds of construction 建设资金
funds of railway construction 铁路建设基金
funds on account 账款
funds provided 提供资本
funds received for participation in loans 收到的贷放款项
funds secured from internal sources 向内筹借资金
fund statement 资金表
funds to be flowed abroad 流至国外的资金
fund surplus 基金盈余
fund turnover period 资金周转期
fundus 底
fundus camera 底部照相机;底部摄影机
funeral art 葬礼艺术;墓碑艺术
funeral basilica 葬礼用长方形建筑
funeral chapel 殡葬小礼堂;殡仪馆
funeral church 殡仪馆
funeral house 殡仪馆
funeral parlour 殡仪馆
funeral residence 殡仪馆
funerary monument 墓碑
funerary mound 坟山
funerary slab 墓穴板
funerary temple 葬礼寺庙;吊唁圣堂
fun fair 娱乐场
fungal biomass sorbent 真菌生物质吸附剂
fungal component 地衣共生菌
fungal degradation 真菌降解
fungal insecticide 真菌杀虫剂
fungal resistance 抗真菌;抗霉菌;抗霉性
fungal spore 真菌孢子
fungible 可替代的
fungible goods 可混合货品;可互换货品;可互换(的)货物
fungicidal paint 防霉漆;杀菌涂料
fungicidal property 杀菌活性

fungicidal wall paint 防霉墙壁涂漆;防霉墙壁涂料
fungicidal wash 杀菌洗涤剂
fungicide 杀真菌剂;杀霉菌剂;杀菌剂;防霉剂;防菌剂
fungicide paint 防霉漆
fungicide sprayer 杀菌剂喷洒器
fungi growth 真菌生长;生长霉菌
fungi imperfecti 不完全真菌
funginert 耐真菌的
funginite 真菌体
fungi-proofing 防菌
fungistat 抑霉菌剂;防霉菌剂;抑真菌剂;抑菌剂
fungistatic 抑制真菌的
fungistatic agent 抑霉菌剂;防霉菌剂;抑真菌剂
fungoid 真菌的
fungosclerotinite 真菌质体
fungotelinite 真菌质结构镜质体
fungus 真菌;发霉
fungus attack 真菌侵蚀;霉菌侵蚀;菌害
fungus flocculant 真菌絮凝剂
fungus infestation 霉菌侵袭;霉菌传染
fungusized 涂防霉剂的
fungus proof 防霉的
fungus resistance 抗菌性;耐霉(菌)性;防霉性;防腐性
fungus resistant outer covering 防生菌包裹层
fungus resistant plasticizer 抗霉性增塑剂
fungus resisting paint 防霉涂料;防霉漆
fungus subterraneous 隐藏的霉菌;地下霉菌
fungus test 抗霉菌试验;防霉性能试验
fun house 游艺室
funicular 索带的
funicular arch 索状拱
funicular car 缆车
funicular curve 垂曲线;悬链线;索(状)曲线
funicular curve of load 荷载索曲线
funicular diagram 索状图
funicular distribution 纤维状分布
funicular force 索力线;毛管拉力
funicular line 索线
funicular machine 静力试验机
funicular polygon 索状多边形;索多边形
funicular polygon equation 索多边形方程
funicular pressure line arch 纤维压力线拱;索压力线拱
funicular pressure line method 压力(多边)线法;纤维压力线法;索压力线法
funicular pressure line vault 纤维压力线拱顶;索压力线拱顶
funicular railroad 缆索铁路;缆索铁道;缆车道
funicular railway 缆车索道;缆索铁路;缆索铁道;缆(车)道;空中缆车;铁索吊车
funicular railway car 缆索车
funicular railway coach 缆车
funicular tension line 纤维张力线;索张力线;受拉多边线
funicular traction 缆绳牵引
funicular water 络索水;纤维水
funkhole 掩蔽部;防空壕
funkite 粒透辉石
funnel 漏斗;龙卷漏斗柱;浇孔;风蚀漏斗
funnel apron 烟囱顶罩
funnel area 烟囱面积;烟囱横断面积
funnel base 烟囱基座
funnel bonnet 烟囱顶罩
funnel bulb 漏斗形灯泡
funnel cap 烟囱帽盖;烟囱顶罩
funnel casing 烟囱外套
funnel classifier 漏斗状分级器
funnel cloud 漏斗云
funnel coupling 漏斗形连接器
funnel cover 烟囱罩布;烟囱盖
funnel damper 烟囱调节(风)门
funneled 漏斗形的
funnel effect 漏斗效应
funnel flow 漏斗流
funnel-form 漏斗状(的)
funnel gases 烟囱气体
funnel glass 显像管玻璃
funnel guide 漏斗式导丝器
funnel guy 烟囱支索;烟囱稳索;烟囱牵条
funnel hood 烟囱帽;斗形烟囱;烟囱(顶)罩
funnel light 烟囱标志灯
funnel-like 漏斗形的
funnel(l)ing 漏斗作用;狭管效应;狭管现象;漏斗缩分法
funnel(l)ing effect 漏斗效应
funnel(l)ing inside shaft kiln 立窑内塌洞
funnel(l)ing inside the kiln 塌洞
funnel mark 烟囱标志
funnel paint 烟囱漆
funnel pipe 漏斗管;铁烟囱
funnel positioner 漏斗定位器
funnel prominence 漏斗状日珥
funnel ring 烟囱牵索圈
funnel shaft 烟囱外壳
funnel-shaped 漏斗形的;喇叭形的
funnel-shaped antenna 漏斗形天线
funnel-shaped bay 漏斗(状海)湾
funnel-shaped borehole 斗形钻孔
funnel-shaped burning zone 漏斗型烧成带;喇叭口形烧成带
funnel-shaped depression 漏斗形凹陷
funnel-shaped estuary 漏斗状河口;喇叭形河口
funnel-shaped opening 漏斗形承口;漏斗形开口
funnel-shaped sintering zone 漏斗型烧成带;喇叭口形烧成带
funnel-shaped spillway 漏斗状溢洪道
funnel shroud 烟囱支索;烟囱稳索
funnel stand 漏斗架
funnel stay 烟囱支索;烟囱稳索
funnel support 漏斗架台
funnel town 漏斗形城市;漏斗式城市
funnel tube 漏斗管;长梗漏斗
funnel tube with filter 带滤器漏斗
funnel type 漏斗状;漏斗型
funnel type straight oil cup 漏斗型直线油杯
funnel umbrella 烟囱顶罩
funnel uptake 烟囱咽喉;烟囱上升道;上升烟道
funnel visco(si)meter 漏斗黏度计
funnel viscosity 漏斗黏度
funnel wire brush 烟囱刷
funnel with nozzle 喷嘴漏斗
funnel with sieve 带筛网漏斗
funny 双人双桨小艇
fuor 补强板条
fur 毛皮;皮毛;水垢
furacrolein 呋喃丙烯醛
fural 呋喃亚甲基
furaldehyde 糠醛;呋喃甲醛
furamide 糠酰胺
furan 呋喃
furanacrylic acid 呋喃丙烯酸;呋喃丙烯醛
furancarbinol 呋喃甲醇
furan carboxylic acid 呋喃甲酸;糠酸
furan chromone 呋喃色酮
furan consolidating resin 呋喃胶结树脂
furandione 呋喃二酮
furan(e) mastic 呋喃胶泥
furan(e) resin 呋喃树脂
furan(e) resin adhesive 呋喃树脂粘合剂
furan group 呋喃基
furanidine 呋喃烷
furan mortar 呋喃砂浆
furan nucleus 呋喃环
furanodienone 呋喃二烯酮
furan plastic 呋喃塑料
furan resin 呋喃树脂
furan ring 呋喃环
furan solvent 呋喃系溶剂
furapromide 呋喃丙胺
furapromidum 呋喃丙胺
furazabol 呋拉扎勃
furazolium chloride 呋唑氯铵
furbish 刷新
furbisher 磨工;抛光工
furbishing 擦亮
furbishing preparation 擦亮制剂
fur black 毛皮黑
furbuckles 扭紧螺丝
furcate 分叉的
furcation coupling 分叉耦合
furcation pipe structure 岔管结构
furcocercous cercaria 叉尾尾蚴
fur colo(u)r 毛皮颜色
fur-cutting operation 毛皮剪毛工艺
fur dyes 毛皮染料
fur farm 毛皮兽场;毛皮动物农场
furfenorex 呋芬雷司
furfuracrolein 呋喃丙烯醛
furfuracrylic acid 呋喃丙烯酸
furfural 糠醛
furfural acetone resin 糠醛丙
furfural aniline resin 糠醛苯胺树脂
furfuralcohol 呋喃基甲醇
furfuraldehyde 糠醛
furfural extraction 糠醛萃取
furfural process 糠醛萃取法
furfural resin 糠醛树脂
furfural wastewater 糠醛废水
furfuramide 糠醛胺
furfuramide resin 糠醛胺树脂
furfuran 呋喃
furfurol test 糠醛试验
furfuryl 糠基
furfuryl acetone 糠丙酮
furfuryl alcohol 氧茂甲醇;糠醇;呋喃甲醇
furfuryl resin 糠基树脂
fur goods 毛皮制品;皮毛类制品
furiani 富里阿尼风
furidarone 呋碘达隆
furildioxime 新镍试剂
furl 折叠
furlong 福浪(英制长度单位,1 福浪 = 1/8 英里);(1 英里 = 1609.3 米)
furnace 窑炉;炉子;炉栅;炉膛;炉胆;反应堆
furnace accretion 炉瘤
furnace addition 炉内加入物
furnace air 窑内空气;炉内空气
furnace annealing 炉内退火
furnace arch 炉碹;炉拱;炉顶拱;炉顶
furnace atmosphere 窑炉气氛;炉膛内空气状态;炉气;炉内气氛
furnace attendant 炉工
furnace back pressure 炉内反压
furnace bar 炉栅;炎条;炉排;炉算
furnace bed 炉床
furnace bin 炉前煤斗
furnace black 炉黑;炉法炭黑;耐高温沥青漆
furnace blast 鼓风
furnace block 炉砖
furnace body 炉体
furnace bottom 窑底;炉底
furnace bottom block 窑底砖
furnace bow down 炉弯曲向下成弓形
furnace bow up 炉弯曲向上成弓形
furnace bracing 窑体钢箍结构;炉子拉条
furnace brazing 炉中钎焊;炉内铜焊;炉内钎焊
furnace brick 高炉用耐火砖;炉砖;耐火砖;火砖
furnace bridge 高炉横烟囱;烟道隔板
furnace burdening 装炉料
furnace campaign (二次大修之间的)大修炉龄;窑令;炉龄
furnace capacity 窑炉容量;炉(子)容量
furnace carbon black 炉法炭黑
furnace casing 炉套壳
furnace casting 炼炉铸件
furnace chamber 炉腔;燃烧室
furnace charge 冷装窑;炉料
furnace charging gear 炉子加料装置
furnace charging table 装炉辊道
furnace cinder 氧化皮;炉渣
furnace clinker 高炉熔渣;炉渣结块;炉渣
furnace clinker aggregate for concrete 混凝土用炉渣骨料
furnace clinker concrete 炉渣混凝土
furnace clinker material 炉渣材料
furnace coal 炉用煤
furnace coke 冶金焦炭
furnace coke yield 高炉焦产率
furnace construction 窑炉结构;炉子结构
furnace control 窑炉控制
furnace cooling 炉冷;随炉冷却
furnace cover 炉盖
furnace cupola 冲天炉
furnace curve 窑炉操作曲线
furnace cycle 炉期
furnace damper 炉闸
furnace dead plate 固定炉板
furnace decoration 彩烧窑
furnace delivery table 出炉辊道
furnace design 窑炉设计;炉子设计

furnace discharge 窑炉卸料
furnace door 炉门
furnace draft 炉子抽(风)力;窑内抽力;炉子抽风机;炉内抽气
furnace-draft regulation 炉膛通风调节;炉膛负压调节
furnace draft regulator 炉子抽风调节器
furnace draught 窑炉抽力
furnaced serpentine 焙烧蛇纹石
furnace dust 窑内飞料;窑内飞尘;炉灰
furnace efficiency 窑炉效率;炉(子)效率
furnace engineering 筑炉工程
furnace entry table 进炉辊道
furnace evapo(u)rator 炉内蒸发段
furnace exhaust 高炉瓦斯;炉膛排烟
furnace extension 前置炉膛
furnace fill 冷装窑
furnace filling counter 装料计数器
furnace floor 炉底
furnace flue 窑炉烟道;焰管;火管
furnace for brightness 光亮炉
furnace for gaseous fuel 气体燃料炉
furnace for heat-treatment 热处理炉
furnace fuel 炉用燃料
furnace gas 炉气;炉内气体
furnace gate frame 炉门框
furnace gun 工业炉喷枪
furnace hard soldering 炉内硬焊;炉内铜焊
furnace hearth 炉缸;炉底;炉床
furnace heating 高炉暖气;炉供暖;火炉供暖;火炉采暖
furnace heating-up 窑炉升温
furnace heat liberation 炉膛容积热强度;炉膛放热
furnace heat release 炉膛释热;炉膛容积热强度
furnace hoisting machine 高炉提升机
furnace hopper 冷灰斗
furnace in fluid bed 沸腾炉
furnace inside atmosphere 炉内气氛
furnace installation 炉子装置;炉子设备
furnace insulation 窑炉隔热
furnace jacket 炉套;炉壳
furnace kiln 炉气体干燥窑
furnace length 炉子长度
furnace liberation rate 炉膛容积热强度
furnace lid 炉盖
furnace life 炉子寿命;炉龄
furnace lines 炉子断面
furnace lining 炉子衬壁;炉衬
furnace lining corrosion 抠炉
furnace load 炉膛热负荷;炉内负荷
furnace magazine 炉用垛料台
furnace maintenance 窑炉维修
furnace-man 熔制工
furnace mantle 炉壳
furnace material 炉料
furnace metal 粗金属
furnace of calcium carbide 电石炉
furnace offtake 窑炉出口烟道;炉子出口烟道
furnace oil 窑炉用油;高炉燃油;炉用油;炉用燃油;燃料油
furnace operation 窑炉操作
furnace operation curve 窑炉操作曲线
furnace output 炉子产量
furnace patenting 炉内铅浴淬火
furnace performance 窑炉熔化性能
furnace platform 炉台
furnace platform cleaner 炉台清扫器
furnace practice 炼炉操作
furnace preheating system 熔窑烤窑系统
furnace pressure 窑炉压力;炉内压力
furnace pressure controller 炉膛负压调节器;送风调节器
furnace pressure recorder 窑炉压纪录器
furnace processor 高炉产量
furnace profile 炉膛轮廓;炉膛断面
furnace protection 保窑
furnace pull 熔窑出料流量
furnace pull-out rolls 炉内拉料辊
furnace pusher 加热炉推钢机
furnace rake 炉耙
furnace rated current 炉子额定电流
furnace rating 炉膛热负荷
furnace rear 炉后
furnace reduction 炉内还原

furnace refining 炉内精炼
furnace residue 炉渣
furnace resistor 炉用电阻
furnace reversal control 窑炉换向控制
furnace rolls 炉内辊
furnace roof 炉顶
furnace room 炉子间;燃烧室;烧制车间;锅炉房
furnace run 窑炉作业周期
furnaces and kilns renovation 炉窑改造
furnace scale 炉内氧化皮
furnace seat 炉底
furnace shaft 炉体;炉身
furnace shell 炉身
furnace shut-down 停窑;停炉
furnace siege 窑底;炉基;炉底
furnace site adjustment 炉前校正
furnace site testing 炉前测定
furnace slab 蜂窝板
furnace slag 高炉渣;炉渣
furnace soldering 炉中软钎焊
furnace sorption injection 炉膛吸附剂喷入法
furnace sow 炉底结块
furnace spectrum 电炉光谱
furnace stack 炉体;炉身
furnace stage 炉台
furnace steel 熔炉钢
furnace stress relieving 炉内应力消除;炉内(加热)消除应力
furnace sump metal 炉槽金属
furnace superintendent 炉长
furnace system 炉膛设备
furnace temperature 炉温
furnace temperature gradient 炉温梯度
furnace test 燃烧炉试验
furnace thermal efficiency 炉子热效率
furnace tool 炼炉工具
furnace top bell 炉顶钟盖
furnace top distributing gear 炉顶布料器
furnace top pressure 炉顶压力
furnace top pressure power generation 高炉衬顶气压发电
furnace transformer 电炉用变压器;电炉变压器
furnace transmutation 窑变
furnace tube 炉管
furnace tube spring hanger 炉管弹簧吊架
furnace-type carbon black 炉法炭黑
furnace unit 炉子设备
furnace vessel 炉膛前床
furnace vestibule 炉子前室
furnace wall 炉壁
furnace warming 炉子加温
furnace with forced draft 有强制通风的加热炉
furnace zone 炉带
Furness 弗尼斯扣件(一种专利玻璃扣件)
Furness method 弗尼斯法【道】
furnish 装备;配备;提供
furnished flat 备有家具的出租公寓
furnished house 家庭式出租房间
furnished house for rent 全套出租
furnished room 备有家具的出租房间;带家具的房间;备有家具的出租住房
furnished rooming house 备有家具的出租住房
furnisher 给沙器;给浆筒
furnishing 器具;家具;装璜;供给;陈设
furnishing department 配料车间
furnishing material 室内装饰材料
furnishing roll 涂装供料辊;施胶辊
furnishing system 设备系统
furnish power 发电
furniture 用具;家具
furniture and fixtures 家具设施
furniture connection 家具接榫
furniture design 家具设计
furniture dimension stock 家具规格材
furniture drawer 家具抽屉
furniture factory 木器厂
furniture fitting 家具五金配件
furniture glass 家具玻璃
furniture grade 家具级(木材)
furniture guard 家具挡板
furniture hardware 家具五金
furniture industry 家具工业
furniture lacquer 家具漆
furniture layout drawing 家具布置图

furniture of rosewood 红木家具
furniture of steel 钢家具
furniture of steel and wood 钢木家具
furniture plywood 家具胶合板
furniture polish 家具擦光蜡
furniture store 家具店
furniture varnish 家具清漆
furniture warehouse 家具栈房;家具仓库
furnos treatment 炭化喷油处理
furodiazole 呋罗达唑
furofenac 呋罗芬酸
furoic acid 糠酸
furol 呋喃甲醛
furongite 芙蓉铀矿
furosemide 呋喃苯胺酸
furostan 呋甾烷
furostilbestrol 呋罗雌酚
furoylamide 呋喃甲酰胺
furoyl chloride 呋喃甲酰氯
fur polishing machine 毛皮上光机
furred 水碱;生水垢
furred ceiling 加垫平顶;加点天棚;贴条吊顶;吊顶
furred down 低吊顶
furred wall 混水墙;混水墙
furrier 毛皮加工者
furrier store 第二层楼商店
furring 钉木条;钉板条;垫高料;衬板条;长条楔形木板
furring brick 面砖;贴面砖;衬垫砖
furring channel 槽形副龙骨;小龙骨;顶棚格栅
furring clip 龙骨卡;龙骨连接件
furring compound 垫高的院子;垫高的场地;贴面用胶粘剂
furring insert 固定用龙骨的预埋件;龙骨衬垫
furring nails 抹灰柱条;板条钉
furring of bamboo 竹板条
furring piece 平屋顶梁垫木
furring strip 木条或极轻型槽钢;钉面板条
furring tile 间隔墙砖;抹灰空芯砖;护墙瓷砖;墙面瓷砖;贴面瓷砖;衬里陶砖
furrow 皱纹;沟槽水沟;(物体表面的)沟槽;垄沟;起皱纹;车辙
furrow applicator 沟底施肥机
furrow axle 沟轮轴
furrow bank 沟壁
furrow bottom 开沟犁体
furrow breaker 犁壁碎土板
furrow dam 沟堤
furrow damming 灌水垄沟堵塞
furrow drain 排水毛沟;排水垄沟
furrowed stone 槽痕琢石
furrowed surface 开槽饰面(指挖有平行槽的饰面);槽痕表面
furrowed tongue 沟裂舌
furrow field 垅田
furrow flute cast 沟槽铸型
furrowing 划槽;开沟
furrowing blade 开沟铲
furrowing body 开沟犁体
furrowing machine 开沟机
furrowing sled 挖沟刮板
furrow irrigation 沟灌
furrow lever 沟轮机构操纵杆
furrow membrane 沟膜
furrow opener drag-bar 开沟器导杆
furrow planting 沟栽
furrow press 耕沟土垄镇压轮
furrow presser 耕沟土垄压平器
furrow stream 垅沟灌水流量
furrow wheel 沟轮
furrow width 沟宽
furrow width adjuster 耕沟宽度调节器
fursalan 呋沙仑
fur scatter rug 毛皮垫
fur scouring agent 毛皮洗涤剂
fur seal 海狗
fursultiamine 呋喃硫胺
furterene 呋氨蝶啶
further 促进
further assistance 后续支援;后续协助
further development 进一步发展
further fix 后续定位
further information 详细情况
further investigations of 进一步研究

further outlook 远期天气展望;短期天气展望
further speed up environmental protection plan 进一步加快环保计划
further studies on 进一步研究
further teasing 继续租赁
furthest distance of travel 最远行走距离
furticose lichen 枝状地衣
furtrethonium iodide 呋喃碘铵
furunce maneuver 炉子操作
furutobeite 硫铅铜矿
furyl 呋喃基
furylacrolein 呋喃基丙烯醛
furylacrylamide 呋喃基丙烯酰胺
furyl alcohol 呋喃基甲醇
fusafungine 夫沙芬近
fusain 乌煤;丝炭;丝煤
fusain group 丝炭组
fusain layer 丝炭层
fusant 熔物;熔体;熔融物
fusarole 半圆粗串珠饰(古典陶立克式钟形圆饰下边圆形脚)
fuscite 中柱石
fuscous 暗褐色的;深色的
fuse 可熔片;火药líne;熔丝;熔融;熔合;熔断器;导火索;保险丝
fuse alarm 熔线报警;熔丝报警器;熔断报警器
fuse alloy 易熔合金
fuse blasting 导火线爆破
fuse block 保险丝装置;熔线盒;熔丝盒;熔断路器;保险丝盒
fuse board 熔丝盘;熔线板;熔断器盘;熔丝盘;保险丝板
fuse box 熔线盒;熔丝盒;熔断器;断流器箱;保险丝匣;保险丝盒
fuse breaking distance 熔断距离
fuse break lamp 熔丝熔断指示灯
fuse bridge 导火电桥
fuse burn-out 熔断器断丝
fuse cabinet 熔丝盒;保险丝盒
fuse cap 引爆雷管;火雷管
fuse carrier 保险丝座
fuse cartridge 熔线盒;熔丝盒;熔丝管
fuse case 熔丝盒
fuse chamber 熔丝盒;保险丝盒
fuse characteristics 熔丝特性
fuse clip 保险丝夹
fuse connection in parallel 并联结线
fuse connector plug 保险丝接头塞
fuse coordination 熔丝配合
fuse cover 熔断器盖
fuse cutout 螺旋保险;熔丝断路器;熔断开关;保险器
fused 熔融的
fused alumina 熔融氧化铝;熔凝氧化铝磨料;电熔刚玉
fused alumina aggregate 熔融矾土集料;熔融矾土骨料
fused alumina ingot 刚玉熔块
fused alumina-zirconia abrasive 锆刚玉磨料
fused basalt 熔融玄武岩
fused basic brick 熔铸碱性耐火砖
fused bath 熔池
fused bifocals 熔合双焦点透镜;熔接的双焦眼镜;熔合双焦镜片
fused block 熔块
fused bond 烧结结合剂
fused bundle 端面熔合纤维束
fused calcium magnesium phosphate 钙镁磷肥
fused carbide tipping 用硬质合金敷焊牙轮的齿
fused cast alumina brick 熔铸氧化铝(耐火)砖
fused cast basalt 熔铸玄武岩
fused cast block 熔铸块;熔铸耐火砖
fused cast brick 熔铸砖
fused cast chrome-corundum refractory 熔铸铬刚玉耐火材料
fused cast cruciform 熔铸十字型格子砖
fused cast magnesite-chrome brick 熔铸镁铬砖;熔铸碱性耐火砖
fused cast mullite brick 熔铸莫来石砖
fused cast refractory 熔铸耐火材料
fused cast refractory brick 熔铸耐火砖
fused cast zirconia-alumina-silica brick 熔铸锆刚玉砖
fused catalyst 熔融催化剂
fused caustic 熔融的苛性碱
fused cement 熔融水泥料;熔融水泥;熔凝水泥
fused charge 熔融装药
fused colophony 熔凝松香;熔凝树脂
fused colophony with lime hydrate 带熟石灰的熔凝松香;带熟石灰的熔凝树脂
fused contact phototransistor 熔合接点光电晶体管
fused copal 熔凝松脂;熔凝树脂
fused corundum 电熔刚玉
fused corundum aggregate 熔融刚玉集料;熔融刚玉骨料
fused corundum block 电熔刚玉砖
fused drier 熔融催干剂
fused dryer 熔融催干剂
fused earth substance 岩浆
fused electrolyte 熔融电解质
fused electrolytic cell 熔质电池
fused electrolytic iron 熔铸电解铁;熔融电解铁
fused ends bundle 端面熔合纤维束
fused epoxy coating 溶解溶环氧涂层
fuse detonator 火雷管
fuse detonator initiation 火雷管起爆
fused extruding spinning process 熔融挤压纺丝法
fused fibre bundles 熔接的纤维束
fused fibre optics plate 熔融光学纤维面板
fused fibre splice 熔融光纤接头
fused flux 熔炼焊剂
fused glass cell 熔化玻璃试池
fused grading shield 熔融分段屏蔽物
fused grain refractory 融熔颗粒耐火材料;熔融颗粒耐火材料
fused hearth bottom 烧结炉底
fused heterocycle 稠杂环
fuse diameter 导火线直径
fused impurity phototransistor 熔合杂质光电晶体管
fused industrial metal 熔铸工业金属
fuse disconnecting switch 熔丝隔断开关
fused junction 合金结;熔融结;熔成结
fused leafet 并合小叶
fused magnesite 电熔镁砂
fused magnesite brick 电熔镁砖
fused magnesium oxide 熔融氧化镁
fused mass 熔体;熔融物;熔融体
fused material 熔融物
fused mullite 熔铸莫来石;熔融莫来石;电熔莫来石
fused on 熔融凝住(玻璃管)
fused ore 熔化物
fused phosphate bath 熔融磷酸盐电解槽
fused phototransistor 熔合光电晶体管
fused plasmodium 并合变形体
fused plug 保险插座;安全塞;熔塞
fused polycyclic system 稠合多环系
fused potassium sulfide 硫化钾
fused product 熔解产品
fused quartz 熔融石英;熔凝石英;熔化水晶;熔化石英;溶石英
fused quartz block 熔融石英砖
fused quartz delay line 熔化石英延迟线
fused quartz glass 熔凝石英玻璃
fused refractory 熔融耐火材料
fused ribbon 熔边狭条
fused ring 稠环
fused ring compound 稠环化合物
fused salt 熔盐
fused salt bath 熔盐电解槽
fused salt chemistry 熔盐化学
fused salt electrolysis 熔盐电解
fused salt electrolysis process 熔盐电解法
fused salt electrolytic refining 熔盐电解精炼
fused salt electrorefining 熔融盐电精制
fused salt extraction 熔盐萃取
fused salt growth 熔盐生长(法)
fused salt liquid phase 熔盐液相
fused salt medium 熔盐介质
fused salt polarography 熔盐极谱法
fused salt reactor 熔盐堆;熔盐反应堆
fused sand 烧结的砂子
fused seal 熔凝
fused signal 导火信号
fused silica 硅石玻璃;熔融石英;熔融二氧化硅;熔凝氧化硅;熔凝硅石;不透明石英玻璃
fused silica blank 熔融石英毛坯
fused silica brick 熔融石英砖
fused silica cavity 熔凝氧化硅共振腔
fused silica delay line 熔化硅延迟线
fused silica fibre 熔凝硅石纤维
fused silica filament 熔凝硅石丝
fused silica filter 石英玻璃滤光片
fused silica reflector 熔融氧化硅反射器;熔凝氧化硅反射器
fused silica substrate 熔凝氧化硅基质
fused six-membered rings 稠合六元环
fused slag 熔渣
fused slurry coating technique 涂层熔烧工艺
fused spray deposit 熔融喷镀覆层
fused sprinkler head 熔断式喷水龙头
fused switch 保险开关
fused together 熔融
fused tricalcium phosphate 熔成磷肥
fused tungsten carbide tipping 碳化钨堆焊的(切削刃)
fusee 引信;均力圆锥轮;均力器;耐风火柴
fusee chain 均力圆锥滑轮链
fuse element 熔线元件;熔丝
fuse grip jaw 熔断器夹爪
fuse head 保险插销;引信头部
fuse holder 熔丝支持器;熔丝架;保险丝座
fuselage 飞机机身
fuselage axis 机身轴
fuselage cover (飞机)机身外壳
fuselage load distribution 机身荷载分布
fuselage longeron (飞机)机身纵梁
fuselage stern (飞机)机身尾部
fuselage stress diagram 机身应力图
fuselage truss wire 机身构架拉线
fuse lighter 导火线点燃器;引线点火器
fuse link 熔丝链;保险丝盒;熔融体;熔断片
fuse-link with expulsion fuse 跌开式熔断器
fusel oil 杂醇油
fusel-oil tar 杂醇油沥青
fuse mechanism 信管装置;引信;起爆机构
fuse metal 易熔金属;保险丝合金
fuse mounting 熔断器配件
fuse off 熔融后拉断;熔离
fuse of instantaneous detonating 瞬发起爆导火索
fuse panel 熔丝盘;熔断器板
fuse piece 熔片
fuse plug 熔线塞;熔丝塞(子);插塞式保险丝;保险塞;熔丝段;安全塞
fuse plug emergency spillway 自溃式非常溢洪道
fuse plug levee 易自溃堤段;保险堤段
fuse point 熔融点
fuse-primer 导火管
fuse protection 熔线保护
fuse protector 熔断器
fuser 上色辊
fuse rack 熔线架;熔丝架
fuse rating 熔线额定值;熔丝额定值;保险丝额定值
fuse refractory 熔线耐火材料
fuse-resistor 保险丝电阻器
fuseron (滞后断开电流的)特种熔线
fuse salt 熔融盐
fuse signal 熔丝信号
fuse socket 熔丝管座
fusestat 插头熔线;S形熔线
fuse strip 熔线片;熔线;片状熔丝
fuse switch 熔线开关;熔丝开关;带保险丝开关
fuse-switch unit 带保险丝开关
fuse time-current test 熔线时间电流试验
fuse tongs 熔线管钳;熔丝更换器
fuse tube 信管
fuse-type temperature meter 熔丝型温度计;热熔型温度计
fuse-type temperature relay 熔线型热动继电器
fuse unit 保险丝盒
fuse wire 熔线;熔丝;熔断线;保险丝
fuse wire insert 熔线插入件
fusibility 易熔性;可熔性;熔性;熔融性;熔度
fusibility curve 熔度图
fusibility diagram 熔度图
fusible 易熔的;可熔的
fusible alloy 易熔合金;可熔合金;低熔点合金
fusible ash 易熔灰分
fusible circuit breaker 熔丝断路器
fusible clay 易熔粘土;易熔瓷土
fusible cone 熔锥;测温锥;测温三角锥
fusible cutout 熔断器
fusible disconnecting switch 熔丝式隔离开关;熔

丝隔断开关;带保险丝的断路器;保险丝断路器
fusible foil method 熔箔法
fusible glass 易熔玻璃
fusible glaze 易熔釉
fusible link 易熔接件(安全防火用);易熔连杆;可熔连杆;熔断丝
fusible link damper 保险丝阻尼器;保险丝调节器
fusible link valve 熔丝阀;防火阀
fusible member 易熔部件
fusible metal 易熔金属
fusible metal film resistor 保险丝型金属膜电阻器
fusibleness 熔度
fusible pattern 熔模
fusible pattern mo(u)lding 熔模铸造
fusible plug 易熔塞;可熔塞;熔塞;插塞式熔断器;插入式保险丝管;保险丝插塞;保险塞;安全塞
fusible powder 可熔性粉末
fusible resistor 可熔电阻器;熔阻丝
fusible slag 易熔渣;流动性熔渣
fusible solder 易熔焊料
fusible switch 熔丝开关
fusible switch power panel 熔丝开关动力箱
fusiform 流线型的;流线形的;梭状的;纺锤形
fusiform body 纺锤体
fusiform buoy 纺锤形浮标
fusiform ray 纺锤状射线
fusiform spike 纺锤形的穗
fusing 熔融;熔接;熔合;熔断;定影
fusing agent 助熔剂;焊药;熔剂
fusing chart 熔度表
fusing coefficient 熔化系数
fusing current 熔断电流
fusing disc 熔割盘
fusing energy 熔化能
fusing factor 熔断系数
fusing heater 熔化加热器
fusing into 熔入
fusing level 定影水平;定影级别
fusing melting 熔化
fusing oven 塑化烘箱
fusing point 聚变温度;熔化(温度)点;熔点
fusing point of asphalt 沥青熔点
fusing point test 熔点测定
fusing soldering 熔焊
fusing temperature 熔化温度
fusing time 熔化时间;熔断时间
fusing together 熔凝
fusinite 丝质类;丝炭煤素质
fusinite needles 针状丝质体
fusinite-posttelinite 丝质次结构体
fusinite splitter 碎屑丝质体
fusinite-telinite 丝质结构体
fusinization 丝炭化
fusinized circleinite 丝炭化浑圆体
fusinized fungal 丝炭化真菌物质
fusinized groudmassinite 丝炭化基质体
fusinized resin 丝炭化树脂
fusinoid group 丝质组
fusion 熔体;熔融;熔解;熔化;熔合
fusionable material 热核燃料
fusion agent 融合剂
fusional 融合的;熔化的;熔合的
fusion alloying 熔配合金;熔合法配制合金
fusion analyser 熔解分析器
fusion area 熔合区
fusion-bed gasification 熔融床气化
fusion bomb 热核弹
fusion bonding 融粘
fusion-bonding process 熔融胶合法
fusion-breccia 熔融角砾岩
fusion bright coal 丝炭亮煤质煤
fusion cake 熔块
fusion cast 熔铸的;熔铸
fusion casting 熔铸(法)
fusion casting mo(u)ld 熔铸模;熔融浇注模
fusion-cast process 熔铸法
fusion-cast refractory 熔铸耐火材料;熔注耐火材料
fusion-cast refractory product 熔铸耐火材料成品
fusion-cast synthetic(al)mica 熔铸合成云母
fusion-cast tank block 电熔铸砖
fusion cone 测温锥;熔融锥
fusion crust 熔凝壳
fusion current 熔化电流
fusion curve 熔化曲线

fusion cutting 熔化切割;熔割
fusion drill 熔化钻机
fusion drilling 熔化钻眼法
fusion efficiency 熔化效率
fusion electrolysis 熔盐电解
fusion energy 聚变能;热核能
fusion energy plant 聚变核能发电厂
fusion enthalpy 熔化焓
fusion entropy 熔化熵
fusion experiment 熔化实验
fusion face 焊(缝)坡口面;熔合面;坡口面
fusion facility 聚变装置
fusion flow method 熔流法
fusion-flow test 熔球试验法
fusion formulae under high-pressure 高压熔化方程
fusion-free fiber 不熔纤维
fusion frequency 熔解频率;停闪频率
fusion fuel 聚变燃料
fusion furnace 熔化炉
fusion heat 熔热;熔解热;熔化热
fusion jointing 熔接
fusion lamination 熔融层压法
fusion length 焊缝长度
fusion-line 熔合线
fusion loss 熔化损失;配合料挥发率
fusion metallurgy 熔炼
fusion method 熔融法;熔化法
fusion mixture 熔化混合物
fusion name 耦合名称
fusion nucleus 融合核;并合核
fusion of image 影像凝合
fusion of wave 多波反应
fusion parameter 熔解参数
fusion phenomenon 融合现象
fusion piercing 熔化钻进(法);火力凿岩;熔化穿孔;热力法钻进;热力穿孔
fusion-piercing drill 熔化穿孔机;热力钻机
fusion piercing equipment 热力钻进设备;热力穿孔设备
fusion piercing method drill 热力法穿孔钻机
fusion point 核聚变温度;熔(融)点;熔化温度
fusion point of ash 灰的熔点
fusion point of coal ash 煤灰熔点
fusion pot 熔罐
fusion power plant 热核发电厂
fusion pressure welding 熔化压接
fusion process 熔炼法;熔化过程
fusion purification process 熔融提纯法
fusion pyrometer 熔融高温计;熔化高温计
fusion range 熔融范围;熔化范围
fusion reaction 聚变反应;熔合反应;热核反应
fusion reactor 聚变反应堆;核聚变反应堆;热核反应堆
fusion rebonded corundum block 电熔再结合刚玉砖
fusion reflex 合像反射
fusion splicing 熔接
fusion tank 熔融槽
fusion temperature 聚变温度;熔融温度;熔解温度;熔化温度;熔点;烧结温度
fusion test 熔融试验;熔化试验
fusion thermit welding 熔化铝热焊;热剂熔焊
fusion tube 熔管
fusion type plasma arc welding 熔透型等离子弧焊
fusion type welding 熔透型焊接法
fusion viscosimeter 熔融粘度计
fusion weapon 热核武器
fusion-welded 熔焊
fusion welding 熔焊接;熔焊;热熔焊(接)
fusion welding metal 熔焊金属
fusion welding technic 熔焊技术
fusion with borax 硼砂熔融
fusion with K_2CO_3 碳酸钾熔融
fusion with $K_2S_2O_7$ 焦硫酸钾熔融
fusion with KOH 氢氧化钾熔融
fusion with $LiBO_2$ 偏硼酸锂熔融
fusion with mixture of Na_2CO_3 and Na_2O_2 碳酸钠—过氧化钠熔融
fusion with Na_2CO_3 碳酸钠熔融
fusion with Na_2O_2 过氧化钠熔融
fusion with Na_2O_2 and KOH 过氧化钠—氢氧化钾熔融

fusion with Na_2O_2 and NaOH 过氧化钠—氢氧化钠熔融
fusion with $Na_2S_2O_7$ 焦硫酸钠熔融
fusion with NaOH 氢氧化钠熔融
fusion yield 聚变产额
fusion zone 熔融带;熔化区;熔化带;熔合部;熔焊区
fusi-resinite 丝炭化树脂体
fusite 微丝炭;微丝煤;丝炭
fusite group 丝炭组
fusoclarite 微丝炭亮煤
fusoclarodurite 微丝质亮暗煤
fusoducroclarite 微丝质暗亮煤
fusodurite 微丝质暗煤
fusoid 纺锤状的;纺锤形
fusorole 盘珠饰
fusotelite 微丝质结构镜质
fusovitrite 微丝质镜煤
fuso-xylain 微丝炭—木煤质质
fuso-xylite 微丝炭—木煤
fuso-xylite clarite 微丝炭—木煤型亮煤
fuso-xyloclarite 微丝炭—木煤型暗煤
fuso-xyloduroclarite 微丝炭—木煤型亮暗煤;微丝炭—木煤型暗亮煤
fuss type automatic voltage regulator 振动式自动电压调整器
fust 柱身
fustian 灯芯绒;粗斜纹布
fustics 黄桑木;黄木;染料桑木;染料木树;佛提树染料
fusulinid extinction 䗴绝灭
fusulini limestone 纺缍虫灰岩
fusuma 日本式隔扇
fut 砰的一声
futher work desirable 待做工作
futtock 制造肋材的弯木;复肋材
futtock band 装有挽缆插栓的桅箍
futtock chain 桅顶支索固定链
futtock hole 肋材孔
futtock plate 内龙骨翼板;桅顶支索固定板
futtock rigging 下桅盘护绳
futtock shroud 桅楼侧支索
futtock timber 复肋材
future acquired property 将获得的房地产
future activities 未来作业
future address patch 未来地址插入码
future advance 远期预付;后续贷款
future advance clause 后续贷款条款
future average daily traffic 远期平均日交通量
future city 未来城市
future contract 期货合同
future cost 预测成本;未来成本
future delivery 远期交货;期货交割
future development area 远景发展地段
future economic development 经济发展后劲
future edition 再版
future enlargement 远景扩建
future exchange 期货外汇
future expansion area 远景发展地段
future expected data 将来预期的数据
future expected population 未开发的种群
future expenses 未来费用
future extension 将来扩建
future goods 期货
future goods transaction 期货交易
future label 待标号;未来标记
future land-use pattern 土地使用远景(图)
future light cone 未来光锥
future load demand 远景负荷需要
future market 期货市场;未来市场
future operation and maintenance 远期运转维修费
future position 预测点
future potential recovery 将来潜在可采储量
future price 期货价(格)
future project for development 发展前景
future public land mobile telephone system 未来公共陆地移动电话系统
future purchase 期货购入
future quotation 期货报价(单)
future reserves 远景储量
future road 拟建道路
futures 远期交货;期货
future sale 期货销售;期货交易
futures brokerage house 期货经纪商
futures business 期货交易

future scenarios 未来情景设想
futures commission merchant 期货委托商店
futures contract 期货合同;期货交易合同
futures delta 期货变化率
futures exchange 期货交易所
futures foreign exchange contract 期货汇兑合同
future site use 未来场地使用
futures market 期货市场
futures-style option 期货式期权
futures trading 期货贸易;期货交易
future sum 期货总额
future surrounding case 将来环境角色
future traffic 远景运量
future traffic forecast 交通预报
future traffic volume 远景交通量;远景货运量;将来交通量
future traffic volume estimating 交通量预测
future unit 预留机组
future value 将来值;未来价值
future wharf line 计划码头线
future worth 将来值
future worth comparisons 将来值比较法
future worth of one per period 定期存入等量数目款项到期后的未来价值比
Futurism 未来主义;(二十世纪建筑的)未来派
futurist 未来学家
Futurist architecture 未来派建筑
futuristics 未来学
futurologist 未来学者;未来学家
futurology 未来学
fuze 引信;火药线;保险丝
fuze action 起爆
fuzee 引信;红色闪光信号
fuze setter 引信定时器
fuzz 毛丝;起毛
fuzzbox 模糊音装置
fuzzification 模糊性
fuzzification function 模糊化函数
fuzzification functor 模糊化函数因子
fuzzification process 模糊化过程
fuzzifying decision 模糊决策
fuzzifying derivation 模糊导数
fuzzifying dynamic(al) system 模糊动态系统
fuzzifying equation 模糊方程
fuzzifying extremum 模糊极值
fuzzifying integral fuzzy integral 模糊积分
fuzzifying state regulator 模糊状态整流子
fuzziness 模糊性;模糊;不清晰;不清楚
fuzzing 纸张起毛
fuzzing logic 模糊逻辑学
fuzzing mathematics 模糊数学
fuzz resistance 抗起毛性
fuzz stick 干木片
fuzzy 有绒毛的;模糊的;失真的
fuzzy algebra 模糊代数
fuzzy algorithmic approach 模糊算法
fuzzy assertion 模糊推断
fuzzy assessment method 模糊评估法
fuzzy automata 模糊自动机
fuzzy behaviour 模糊特性
fuzzy boundary 模糊界限
fuzzy cardinality 模糊基数
fuzzy Cartesian product 模糊笛卡儿积
fuzzy category 模糊范畴
fuzzy characteristics 模糊特征
fuzzy closed set 模糊闭集
fuzzy cluster 模糊聚类
fuzzy cluster analysis 模糊聚类分析
fuzzy clustering 模糊簇聚
fuzzy clustering algorithm 模糊簇聚算法
fuzzy coalition 模糊联盟
fuzzy column condensation 模糊柱形凝聚;模糊列凝聚

fuzzy column extension 模糊柱形扩张;模糊列扩张
fuzzy composition law 模糊组合法则
fuzzy comprehensive evaluation 模糊综合评估
fuzzy computer 模糊计算机
fuzzy concept 模糊概念
fuzzy connective 模糊连接词
fuzzy consensus 模糊一致
fuzzy constraint 模糊约束
fuzzy continuity 模糊连续
fuzzy continuous 模糊连续的
fuzzy control 模糊控制
fuzzy controller 模糊控制器
fuzzy convex hull 模糊凸包
fuzzy convex set 模糊凸集合
fuzzy correlation 模糊相关
fuzzy covariance 模糊协方差
fuzzy decision 模糊决策
fuzzy decision-tree 模糊决策树
fuzzy derivation 模糊导数
fuzzy diagnosis 模糊识别
fuzzy distance 模糊距离
fuzzy dynamic(al) programming 模糊动态规划
fuzzy dynamic(al) system 模糊动态系统
fuzzy entropy 模糊熵
fuzzy environment 模糊环境
fuzzy equation 模糊方程
fuzzy error 模糊错误
fuzzy evaluation 模糊评判;模糊评估
fuzzy event 模糊事件
fuzzy expression 模糊表达式
fuzzy extremum 模糊极值
fuzzy feedback control system 模糊反馈控制系统
fuzzy field 模糊域
fuzzy field of subset 模糊子集域
fuzzy filter 模糊滤波器
fuzzy finite state automaton 模糊有限状态自动机
fuzzy flow chart 模糊流程图
fuzzy formula 模糊公式
fuzzy function 模糊函数
fuzzy game 模糊对策
fuzzy grain 毛状纹理
fuzzy graph 模糊图
fuzzy group 模糊群
fuzzy identity function 模糊恒等函数
fuzzy identity morphism 模糊恒等映照
fuzzy image 模糊影像;模糊印样;模糊图像;不明显图象
fuzzy implication 模糊蕴涵;模糊隐含
fuzzy inference 模糊推理
fuzzy information 模糊信息
fuzzy interval 模糊区间
fuzzy linear ordering 模糊线性序
fuzzy linear programming 模糊线性规划
fuzzy linear regression 模糊线性回归
fuzzy literal 模糊文字
fuzzy logic 模糊逻辑
fuzzy logical control 模糊逻辑控制
fuzzy logical controller 模糊逻辑器;模糊逻辑控制器
fuzzy logical functions 模糊逻辑函数
fuzzy mapping 模糊映射;模糊变换
fuzzy mathematical programming 模糊数学规划
fuzzy matrix 模糊矩阵
fuzzy measure 模糊测度
fuzzy membership functions 模糊隶属函数
fuzzy model 模糊模型
fuzzy modus ponens 模糊假言推理
fuzzy morphism 模糊映照
fuzzy multicriteria modelling 模糊多准则建模
fuzzyness 模糊性
fuzzy noise 模糊噪声;模糊干扰
fuzzy number 模糊数
fuzzy number absolute value 模糊数绝对值

fuzzy number addition 模糊数加法
fuzzy number division 模糊数除法
fuzzy number entropy 模糊数熵
fuzzy number exponential 模糊数指数
fuzzy object 模糊目标
fuzzy objective 模糊物体
fuzzy open set 模糊开集
fuzzy optimal control 模糊最佳控制
fuzzy orbit 模糊轨道
fuzzy output map 模糊输出映射
fuzzy output transformation 模糊输出变换
fuzzy partial graph 模糊部分图
fuzzy partition 模糊划分
fuzzy partition matrix 模糊划分矩阵
fuzzy point 模糊点
fuzzy prime implication 模糊素隐含
fuzzy probability 模糊概率
fuzzy probability distribution 模糊概率分布
fuzzy probability field 模糊概率场
fuzzy probability regression 模糊概率回归
fuzzy process 模糊过程
fuzzy processor 模糊信息机
fuzzy production 模糊生成
fuzzy programming 模糊规划
fuzzy proposition 模糊命题
fuzzy prototype 模糊样品
fuzzy quantity 模糊量
fuzzy random variable 模糊随机变量
fuzzy region 模糊不清区域
fuzzy relation 模糊关系
fuzzy relation equation 模糊关系方程
fuzzy relation system 模糊关系系统
fuzzy response 模糊响应
fuzzy restriction 模糊约束;模糊限制
fuzzy science 模糊科学
fuzzy set 模糊集(合)
fuzzy set method 模糊集法
fuzzy set sytem 模糊集合系统
fuzzy set theory 模糊集理论;模糊集合论
fuzzy simple disjunctive decomposition 模糊简单析取分解
fuzzy space partition 模糊空间划分
fuzzy stable 模糊稳定的
fuzzy state 模糊状态
fuzzy state regulator 模糊状态整流子
fuzzy state transition matrix 模糊状态转移矩阵
fuzzy state transition tree 模糊状态转移树
fuzzy structure 模糊结构
fuzzy subgraph 模糊子图
fuzzy subset 模糊子集
fuzzy switching function 模糊开关函数
fuzzy synthetic evaluation 模糊综合评估
fuzzy system 模数体系;模糊系统
fuzzy system mapping 模糊系统映射
fuzzy terminal regulator 模糊终端调整器
fuzzy theory 模糊理论
fuzzy topological space 模糊拓扑空间
fuzzy topology 模糊拓扑;不分明拓扑
fuzzy trajectory 模糊轨道
fuzzy transition function 模糊转变函数
fuzzy transition matrix 模糊转移矩阵
fuzzy transition vector 模糊转移矢量
fuzzy truth 模糊真值
fuzzy upperbound 模糊上限;模糊上界
fuzzy utility 模糊效用
fuzzy variable 模糊变量
fuzzy vector 模糊向量
fuzzy weight 模糊权
Fülleborn's method 菲勒本法
FV 现场十字板剪切试验
Fyberstone 弗伯斯通板(一种专卖的防水墙板)
fylfot 卍字形
fynchenite 凤凰石

G

GaAs spontaneous infrared source 砷化镓自发红外光源
gab 凹节
gabarage 打包用粗布(爱尔兰)
gabardine 工作服
gabarit 净空界限;样板;外形尺寸;外廓;曲线板
gabarite 限界
gabbard 苏格兰早期的帆船
gabbart (旧时苏格兰内河用的)平底河船
gabbart scaffold 葛巴特脚手架;方木脚手架
gabble cope 山墙压顶
gabbro 辉长岩
gabbro aplite 辉长细晶岩
gabbro-copper and nickel-bearing formation 辉长岩合铜镍建造
gabbro diorite 辉长闪长岩
gabbro group 辉长岩类
gabbroic anorthosite 辉长斜长岩
gabbroic aplite 辉长细晶岩
gabbroic lens 辉长岩镜片
gabbroic texture 辉长结构
gabbroid 辉长岩状
gabbroid-basaltic rock 辉长玄武岩
gabbroitic dike 辉长岩(修成的)堤岸
gabbro nelsonite 辉长钛铁磷灰岩;辉长纳尔逊岩
gabbronorite 辉长苏长岩
gabbropegmatite 辉长伟晶岩
gabbrophyre 辉熔斑岩;辉长煌斑岩
gabbro porphyrite 辉长玢岩;辉长斑岩
gabbroporphyry 辉状斑岩
gabbro sill 辉长岩床
gabbro-syenite 辉长正长岩
gabbro texture 辉长结构
gabelle 法国盐务税
gaberdine 工作服
gabion 泥框;蔑框;填石铁丝笼;石笼;石筐
gabionade 石笼坝;堆石墙;土石筐垒成的堤;石笼组;蛇笼堤
gabion boom 石笼拦河埂;石笼挡栅
gabion cofferdam 笼网围堰
gabion dam 石笼坝
gable 山墙;三角形部分【建】
gable and hip roof 歇山
gable arch 山形(人字)拱
gable board 山墙(顶)封檐板;博风板
gable bottom car 尖底车
gable brick 山墙砖
gable column 山柱
gable coping 山墙压顶
gable cross 山墙(头)十字架
gable crown 屋脊式路拱
gabled 有山墙的
gable(d) bent 人字形排架
gable(d) dormer(window) 人字(形)老虎窗
gable(d) house 有山墙的房屋;人字屋顶房屋
gable dormer 山墙形老虎窗;山墙式构架;人字形老虎窗;花山形老虎窗;山墙头顶窗;山头顶窗
gable(d) roof dormer 有山墙的屋顶天窗;人字屋顶窗
gabled tower 山墙塔楼
gable eave board 搏风板
gable end 山墙端
gable flashing 山墙泛水
gable frame 山墙构架;山墙端框架
gable inlet ventilation 山墙进气通风
gable louver 山墙百叶窗
gable masonry wall 砖石山墙;人字山头砌筑墙
gable mo(u)lding 山墙装饰线脚;山墙线脚;山墙饰线
gable painting 山墙涂漆;山墙绘画
gable peak 山墙尖端;山尖
gable pole 山墙头桁条
gable post 封檐板顶端支撑;山墙(小)柱(山头顶点的短柱)
gable roof 双坡屋顶;山形屋顶;三角屋顶;人字(形)屋顶;双坡屋面
gable roof ridge turret 人字屋脊上小塔;人字屋顶上小塔

gable roof truss 人字(形)屋架
gable roof type 人字形屋顶(畜舍)
gable-shaped lintel 山墙形门窗楣;山墙形过梁
gable-shaped window 山墙形窗
gable shoulder 山墙压顶线脚;山墙托肩;山墙基石线脚
gable side 山墙面;侧山墙
gable slab 山墙(墙)板
gable slate 山墙石板
gable springer 山墙挑檐的托石;山墙起拱支座;山墙基石;山墙托臂;山墙拱底石;堰头【建】
gablet (壁龛或小墙洞顶上的)山墙饰;小山墙;花山头
gable tile 檐瓦;(山墙与屋面连接处的)角瓦;山形瓦;山墙顶盖瓦;半页瓦
gable top carton 山形盖顶纸盒
gable tower 人字形房;山形塔
gable tracery 山墙(窗)花格
gable type dormer(window) 山墙形老虎窗
gab-lever 凹节杆
gable wall 玻璃窑投料侧壁;人字墙;前墙;山墙(墙身);玻璃投料端墙
gable wall balcony 山墙阳台
gable wall belfry 山墙钟楼
gable window 带山头的窗;山墙窗
gable with corbel steps 阶式山墙;托阶式山墙
gab-motion 偏心轮配气
Gabon nucleus 加蓬陆核
Gabon 加蓬(红)木;加斑木
Gabon mahogany 加蓬桃花心木;加蓬红木
Gabor 伽柏
Gabor's expansion theorem 伽柏展开原理
Gabor's hologram 伽柏全息图
Gabor's in-line hologram 伽柏共轴全息图
Gabor's method 伽柏法
Gabor's zone plate 伽柏波带片
Gabriel's synthesis 加布雷尔合成
Gabriel reaction 加布雷尔反应
gabrielsonite 羟砷铁铅矿
gab rope 缩帆带
gad 小钢凿;销;錾;钢楔;钢钎;量规;键;切刀;沉胶渣;车刀;测杆
gadder 钻岩器;钻机架;钻机车;凿岩;凿孔机;流动钻眼机;穿孔器
gadding 开采块石
gadding car 钻机车
gadding machine 钻机车;开石机;钻孔机;楔劈石机
gadding plant 蔓草植物
gadget 小器具;精巧的小机械;机件
gadgeteering 小器具设计
gadgetry 小机件;小玩意
gadiometer 磁强梯度计;磁强陡度计
gadolinite 硅铍钇矿
gadolinium molybdate 钼酸钆
gadolinium ore 钆矿
gad picker 凿子;鹤嘴锄;钎子
gad pole 测杆
gadroon 圆线条装饰;刻纹圆线条装饰
gad tongs 平口钳
Gaede's high-speed metal pump 盖德高速金属扩散泵
Gaede's ionization ga(u)ge 盖德电离真空计
Gaede-type pump 盖德回转式真空泵
Gaertner spectrograph 伽特纳摄谱仪
gaff 鱼钩;钩杆;攀钩;挽钩;吊钩;带钩阀;大应力;杂抗场;斜桁
gaffer 工长;领班
gag 离合垫片;塞子;塞铁;塞盖
gagarinite 氟钙钠钇石
gagate 煤精
gagatite 煤玉
gagatization 煤玉化作用
gag bit 大口衔
gagger 校正轨距工人;型材辊式矫正机;铸模工具;造型(工具);铁骨;砂型吊钩;吊砂钩;撑子
gagger board 铁骨板
gagging 冷矫正
gaging 放射性计测

gaging thread 毛圈边线
gaglet 披水槽砌块
gag lever 止动杆
gag lever post 限位杆;挡杆
gag press 矫直压力机;压直机
gag process 矫直过程;压弯成形(法);构件成形工艺
gag rein 大勒缰
gahnite 锌尖晶石
gaidonnayite 斜方钠锆石
Gaima old land 盖马古陆
gain 腰槽;增量;增益;获得;得益;槽沟;放大
gain adjustment 增益调整;增益调节
gain admittance 增益导纳
gain advantage 获益
gain amplification 增益放大
gain amplifier 增益放大器
gain and loss 损益
gain antenna 定向天线;增益天线
gain around a feedback 反馈环路增益
gain asymptote 增益渐近线
gain balance 增益平衡
gainband merit 增益带宽指标
gain bandwidth 增益带宽
gain bandwidth factor 增益带宽乘积因数
gain bandwidth product 增益带宽积
gain calibration 增益校准;增益校正
gain calorie 获得热量
gain changer 增益变换器
gain characteristic 增益特性(曲线)
gain clipper 增益限幅器
gain coefficient 增益系数
gain coefficient of medium 媒质增益系数
gain compression 增益压缩
gain condition 增益条件
gain constant 增益常数
gain control 增益控制;增益调整;增益调节;亮度调节
gain control amplifier 增益控制放大器
gain control key 增益控制键
gain controller 增益控制器
gain control mode 增益控制方式
gain control range 增益控制范围
gain control set 增益调整器
gain crossover frequency 增益窜度频率
gain curve 增益曲线
gain curve of medium 媒质增益曲线
gain display 增益显示
gain drift 增益漂移
gaine 装饰台座;盒子
gained ability 增重能力
gained work 盈功
gain equation 增益方程
gain error 增益误差
gainesite 磷锆钠石
gain experience 获得经验
gain factor 增益因子;增益因数;增益系数;再生系数;放大效益;放大系数
gain factor of laser 激光器增益系数
gain flatness 增益平稳度
gain floor 增益下限
gain fluctuation 增益波动
gain for threshold value 阈值增益
gain frequency characteristic 增益与频率的关系特性
gain frequency relationship 增益频率关系
gain from trade 贸易收益
gain function 增益函数
gain groove 槽沟
gain inequality 增益差
gaining 开槽;单向开槽;单榫接;增益
gaining ability 增重效率;增重能力
gaining in weight 增重
gaining stream 盈水溪;盈水河;地下水补给河
gaining strength 强度增长;逐渐加强
gaining twist 变距纹
gaining valuable data 获得有价值资料
gain in mass 质量增加

gain in precision 精确度增益
gain in space 增加空间;空间增益
gain in strength 增加强度;强度增长
gain inversion 增益反转
gain in weight 重量的增加;增重量
gain in-weight batch blender 增重分批掺和机
gain joint 榫接口
gain level 增益级;增益电平
gain limited receiver 增益有限接收机
gain-loss ratio 得失比率
gain margin 增益裕度;增益余量;增益容限;增益极限;获利边际
gain matching 增益匹配
gain matrix 增益矩阵
gain measurement 增益测量
gain measuring device 增益测量装置
gain measuring set 增益测量器
gain modulation 增益调制
gain of head 水头增长;水头恢复
gain of heat 增热
gain of light 光增益
gain of night vision instrument 夜视增益
gain on conversion of investment 投资变换收益
gain on disposal 非流动资产出售收益
gain on insurance claim 保险索赔所得;保险索赔收益
gain on realization of assets 变卖资产收益;变产利益
gain on sale of investment 出售投资收益
gain or loss on sale of equipment 设备变卖损益
gain parameter 增益参数
gain phase analysis 增益相位分析
gain phase characteristic 增益相位特性
gain power 增大功率
gain-programmed amplifier 增益程序化放大器;程控增益放大器
gain ranging analog 增益变化范围模拟
gain recovery 增益恢复
gain reduction 增益衰减
gain reduction indicator 增益衰减指示器
gain returns on investment 投资收入
gains 收益
gains contribution to depreciation 应归属折价的收益
gain select 增益选择
gain selector switch 增益选择开关
gain sensitivity control 增益灵敏度控制
gain set 增音器;增音机
gain setting 增益调整
gains from proportionate stratified sampling 按比例分层抽样的得益
gain sharing plan 收益分配计划
gain slope 增益斜率
gains offset the losses 损益相抵
gain speed 渐渐增加速度
gain stability 增益稳定性
gain stability margin 增益稳定余量
gain stabilization 增益稳定
gain stage 增益级
gain step accuracy 增益台阶精度
gain step number 增益台阶数
gain switch 增益开关
gain time 增益时间
gain time control 增益时间控制
gain time controller 增益时间控制器
gain to noise temperature ratio 增益同噪声温度比
gain trace 增益曲线
gain turn-down 增益降低;增益自动下降
gain weight 体重增加
gain word of play back 回放增益字
gaitite 羟砷锌钙石
gait test 步行试验
gaize 蛋白岩;海绿云母细砂岩;脆性白土
gaize cement 蛋白土水泥;蛋白土胶凝材料;盖兹水泥
gal 伽(重力加速度单位,1 伽 = 10^{-3} 米/秒^2)
Gala beds 加拉层
galactan 半乳聚糖
galactaric acid 半乳糖二酸
galactic center 银心;银道圈;银河中心
galactic center region 银心区
galactic center source 银心区源
galactic concentration 银聚度

galactic coordinate 银道坐标
galactic coordinate system 银道坐标系
galactic corona 银冕
galactic disk 银盘
galactic equator 银道
galactic halo 银晕
galactic halo emission 银晕辐射
galactic latitude 银纬
galactic longitude 银经
galactic orbit 银心轨道
galactic pericenter 近银心点
galactic plane 银道面
galactic pole 银极
galactic system 银道坐标系
galactitol 半乳糖醇
galactogen 半乳多糖
galactoglucomannan 半乳葡甘露聚糖
galactolipid 半乳糖脂
galactomannan 半乳甘露聚糖
galactometer 乳比重计
galactonic acid 半乳糖酸
galactosaccharic acid 半乳糖二酸
galactosan 半乳聚糖
galactose 半乳糖
galactose tolerancetest 半乳糖耐量试验
galacturonic acid 半乳糖醛酸
galalith 乳石
galangal oil 高良姜油
Galanthus 雪花属
Galapagos fracture zone 加拉帕戈斯破裂带
Galapagos ridge 加拉帕戈斯海岭
Galapagos spreading center 加拉帕戈斯扩张轴
galatine glue 动物胶
galaxite 锰尖晶石
galaxy 星系;银河;天河
galbanum 白松香;阿魏脂;古蓬树脂;波斯树脂
galcio-eustatism 冰川海面升降;冰川海面进退
gale 大(暴)风;暴风;八级风
Gale's duality theorem 盖尔氏对偶定理
Gale's formula 盖尔氏公式
galea 盔瓣
galea forceps 威勒特氏钳
galea of chclicera 螯盔
gale cone 大风锥形号型;大风信号;大风号志;大风标志
gale damage 风灾
gale information 大风消息
galeite 氟钠矾
galena 硫化铅;方铅矿
galena crystal detector 方铅晶体检测器
galena detector 方铅矿石检波器
galena ore 方铅矿石
galena pulp 硫化铅矿浆;方铅矿浆
galenite 方铅矿
galenobismutite 辉铋铅矿
gale pollution 大风污染;疾风污染;风暴污染
Galerkin's finite element model 伽辽金有限元模型
Galerkin's method 伽辽金法
galerne 加勒内风
gale signal 大风信号
galet 碎石;石屑
galette silk 粗绢丝
gale urgent warning 大风紧急警报
gale warning 大风警报
Gal gravimeter 伽尔型重力仪
Galician-free fall 钢绳冲击钻进
Galician method 钢绳冲击钻进(方)法
Galilean binoculars 伽利略双筒望远镜;伽利略双目望远镜
Galilean eyepiece 伽利略目镜
Galilean glass 伽利略望远镜
Galilean number 伽利略数
Galilean satellite 伽利略卫星
Galilean telescope 伽利略望远镜
Galilean telescope system 伽利略望远系统
Galilean transformation 伽利略变换
Galilean viewfinder 伽利略取景器
galilee (哥特式教堂的西端的)门廊;(教堂的)门廊
Gal(ileo) 伽(利略)(重力加速度单位)
Galileo's law of inertia 牛顿第一定律
Galileo number 伽利略准数
Galileo slide-ga(u)ge 伽利略滑尺
galipot gum 海松树胶

Galizin seismograph 加利津地震仪
Galizin seismometer 加利津地震计
galkhaite 硫砷铊汞矿
gall 磨损处;擦伤
Gall's chain 套筒滚子链;高尔链;平环链
Gall's projection 高尔投影
galled spots 表面擦伤
galled threads 挤扁的丝扣
gallego 加耶果风
galleria forest 走廊
galleried upper stor(e)y 楼台上层
gallery 走廊;工作台;楼台;廊道;坑道地下水廊;坑道;架空过道;美术(陈列)馆;绘画展览室;横坑道;平巷;外廊;顶层楼座;灯罩架;导坑;船尾瞭望台;虫道;撑竿走道;长廊
gallery apartment building 内廊道式公寓建筑
gallery apartment house 廊式公寓住宅;连廊式公寓;外廊式公寓;通廊式公寓
gallery arcade 有拱顶的通廊;连拱廊
gallery beck 船尾下甲板
gallery break point 长廊转折点
gallery bridge 廊桥
gallery burner 喷火筒;通廊式燃烧器;通廊式喷火口
gallery cable 坑道电缆
gallery camera 反像复照仪
gallery canal 地下渠道
gallery column 廊柱
gallery corridor type 外廊式
gallery crypt 通廊式地窖
gallery documentation 坑道编录
gallery driving 坑道开凿;坑道开挖;平峒掘进;隧道掘进
gallery floor 坑道底板
gallery forest 走廊林
gallery for sediment transport 输沙廊道
gallery frame 坑道支撑;坑道支架
gallery frame timbering 框式支撑
gallery furnace 长廊炉
gallery grave 有石铺走廊的史前墓地
gallery hole 纵向润滑油孔
gallery kiln 多室窑
gallery-like cave 长洞
gallery measured 实测巷道
gallery niche 通廊式洞龛;通廊式壁龛
gallery of painting 画廊
gallery of sediment transport 输沙廊道
gallery opening 坑道口;长廊口
gallery port 通廊式喷火口;(平炉的)加料口
gallery portal 廊口
gallery practice ammunition 室内射击练习枪弹
gallery roof 通廊式屋顶
gallery system 人行廊道体系;行人道体系
gallery testing of explosive 封堵炮泥爆破试验
gallery-type 外廊式
gallery vault 通廊式拱顶
gallery ventilation 坑道通风
gallery ware 地中海各国锡釉陶器
gallery window 通廊式窗
gal(l)et 碎石块;石屑;碎石片
galleting 用碎瓦片填高屋脊瓦片的位置;嵌灰缝碎石片;填塞石缝;碎石片嵌灰缝;碎石片
galleting tile 小瓦
g-alleviation 过载减弱;加速度作用减弱
galley 中型舢板;军舰舰长用艇;地下廊道;大木船;船上厨房;长方形炉
galley dresser 厨桌
galley funnel 厨房烟筒
galley proof 长条校样
galley range 厨房炉灶
galley stroke 大幅度划桨
galley under floor 闸底廊道
gallic acid 没食子酸;五倍子酸
gallic oxide 三氧化二镓
Gallimore 镍钨锌系合金
galling 咬住;金属磨损;摩擦造成的粗糙面;塑变;大磨损
galling of rings 活塞圈磨损
gallipoli oil 劣质橄榄油
gallipot 陶罐;胆形瓶
gallite 硫镓铜矿
gallium arsenide 砷化镓
gallium-arsenide detector 砷化镓探测器
gallium-arsenide diode 砷化镓二极管
gallium-arsenide infrared emitter 砷化镓红外发

射器
gallium-arsenide-junction light source 砷化镓结光源
gallium-arsenide laser 砷化镓激光器
gallium-arsenide light source 砷化镓光源
gallium-arsenide luminescence diode 砷化镓发光二极管
gallium-arsenide optical filter 砷化镓滤光片
gallium-arsenide photocathode 砷化镓光电阴极
gallium-arsenide semiconductor 砷化镓半导体
gallium-arsenide solid-state lamp 砷化镓固态灯
gallium-arsenide surface barrier detector 砷化镓面垒探测器
gallium deposit 镓矿床
gallium halide 卤化镓
gallium nitride 氮化镓
gallium ores 镓矿
gallium oxide 氧化镓
gallon 伽仑；加仑
gallonage 加仑数
gallon jug 加仑罐
gallon mile 加仑英里
gallon-octane 加仑-辛烷值
gallon per capita per day 每人每日加仑量
gallons capita per day 加仑/人日
gallons-degree 加仑·度(美国采用的冷却单位)
gallons per day 加仑/天；加仑/日
gallons per day acre 加仑/日英亩
gallons per day per capita 加仑/日人
gallons per day per square foot 加仑/日平方英尺
gallons per hour 加仑/小时；每小时加仑数
gallons per millenary cubic feet 每千立方英尺加仑数
gallons per minute 加仑/分；每分钟加仑数；每分钟加仑量
gallons per minute per square foot 加仑/分·平方英尺
gallons per second 加仑/秒；每秒钟加仑数
galloon 缎带；花边；金银珠花边
gallop 运转不稳定；快速运送；快速运输；飞奔；发电机不正常运转
gallop bridge 立体交叉桥
galloping 弛振；运转不平稳(发动机)；跃步；点头振动
galloping form 奔马型
galloping ghost 跳动重影
galloping inflation 飞涨的通货膨胀
galloping motion 阶跃运动
galloping pattern 跳步图
gallop rhythm 奔马律
gallotannic acid 丹宁酸
gallotannin 丹宁酸；倍丹宁
Galloway boiler 加罗威锅炉
Galloway sinking and walling stage 两用吊盘
Galloway stage 多层凿井吊盘
gallow pulley 张力惰轮
gallows 构架；井架；门式卸卷机；盘条卸卷机；舱面吊架；龙门吊架
gallows arm 聚光灯吊架
gallows bit(ts) 双柱井架；船甲板中央木架
gallows bracket 木构架斜撑
gallows frame 龙门起重架；龙门吊架；井架；门形吊架；门式吊架
gallows frame derrick 门式起重架；龙门架(式)起重机
gallows plow 双轮单铧犁
gallows roll 导向滚柱
gallows stanchion 承架支柱
gallows timber 木构架；撑木
galmei 异极矿
galmins 加仑/分钟；每分钟加仑数
Galois canonical form 伽罗瓦标准形式
Galois field 伽罗瓦域
Galois group 伽罗瓦群
Galois theory 伽罗瓦理论
galosh 高筒橡胶套鞋
Galton's curve 高尔顿曲线
Galton's function 高尔顿函数
Galton's individual difference problem 高尔顿个体差异问题
Galton's law 高尔顿法则
Galton's rank order test 高尔顿等级次序检验
Galtonian curve 高尔顿曲线
Galtonian function 高尔顿函数

Galtonian individual difference problem 高尔顿个体差异问题
Galtonian law 高尔顿法则
Galtonian rank order test 高尔顿等级次序检验
Galton method 高尔顿采暖通风法
galvanic 电流的
galvanic action 电流作用；电池作用
galvanic action protector 电化腐蚀防护器
galvanic anode 流电阳极；动电阳极
galvanic bath 电镀槽
galvanic battery 蓄电池组；原电池组
galvanic cathodic protection 电阴极防腐
galvanic cell 原电池；一次电池；伽伐尼电池；化学电池
galvanic circle 导电回路
galvanic corrosion 原电池腐蚀；接触腐蚀；电蚀；电(偶)腐蚀；电流腐蚀；电解侵蚀；电化(学)腐蚀
galvanic couple 电偶
galvanic current 直流电；化电流；动电电流；电流
galvanic deposit 电(极)沉积
galvanic electricity 流电；动电
galvanic electrocoating 电沉积法
galvanic pile 电锥；电堆
galvanic protection 镀锌保护；电镀保护；电流保护
galvanic series 电化序列；电(动)势序
galvanic theodolite 电流经纬仪
galvanised iron 镀锌铁皮；白铁皮
galvanising 镀锌
galvanism 流电；电流
galvanite 加凡漆(一种钢窗锌基防锈底漆)
galvanization 镀锌工作；镀锌(作用)；电镀
galvanize 镀锌；电解沉淀
galvanized 镀锌的
galvanized and welded steel pipe 镀锌焊缝钢管
galvanized barbed roofing nail 镀锌刻花瓦楞钉
galvanized bath 镀锌池
galvanized bolt 镀锌螺栓
galvanized bunched thick pipe 镀锌束接厚管
galvanized bunched thin pipe 镀锌束接薄管
galvanized chain 镀锌链条
galvanized chainlink fence 镀锌链结栅
galvanized coating 电镀层
galvanized coil 镀锌钢板卷
galvanized common round iron wire nail 镀锌普通圆铁钉
galvanized copper wire 镀锌镜框
galvanized corrugated sheet 镀锌波纹铁皮
galvanized corrugated steel 镀锌瓦楞(薄)钢板
galvanized countersunk head machine screw 镀锌平头机器螺钉
galvanized curtain rod 镀锌窗帘杆
galvanized double-J bolt 镀锌双钩螺栓
galvanized elbow 镀锌弯头
galvanized electric(al)welded mesh 镀锌电焊网
galvanized embrittlement 镀锌脆化
galvanized fabric 镀锌(金属)网
galvanized flange 镀锌法兰盘
galvanized flexible conduit 镀锌蛇皮管
galvanized flexible iron tube 镀锌蛇皮铁管
galvanized flexible joint 镀锌活接头
galvanized flexible tube 镀锌软管
galvanized hexagonal wire mesh 六角网；镀锌六角金属丝网
galvanized hexagonal wire netting 镀锌六角(铁丝)网
galvanized insulator bolt with nut 带螺帽镀锌瓷瓶穿钉
galvanized iron 马口铁；镀锌铁皮；镀锌铁；白铁
galvanized iron barbed wire 镀锌刺铁丝
galvanized iron brace 镀锌铁撑脚
galvanized iron bucket 镀锌铁桶
galvanized iron chain 镀锌铁链
galvanized iron concealed hinge 镀锌铁暗铰
galvanized iron conduit 镀锌铁管
galvanized iron corn head screw 镀锌尖头螺钉
galvanized iron curved washer 镀锌弯铁垫圈
galvanized iron drum 镀锌铁桶
galvanized iron flat roof 镀锌铁板平屋面
galvanized iron flat wire 镀锌扁铁丝
galvanized iron hasp 镀锌搭扣
galvanized iron hinge 镀锌铁合页
galvanized iron hook 镀锌铁钩
galvanized iron hook and eye for gate and shutter 镀锌门窗钩

galvanized iron hook for lamp 镀锌灯钩
galvanized iron hoop 镀锌铁箍
galvanized iron lamp holder box 镀锌灯头箱
galvanized iron leader 白铁皮水落管
galvanized iron lined door 白铁皮防火门
galvanized iron mush-room head bolt 白铁圆头螺栓
galvanized iron oil pot 白铁油壶
galvanized iron pail 镀锌铁提桶
galvanized iron paint 镀锌铁板涂料；镀锌钢板涂料
galvanized iron pipe 镀锌(白)铁管；白铁管
galvanized iron plain sheet 镀锌铁皮
galvanized iron plate 白铁板
galvanized iron ridge capping 镀锌铁皮脊瓦
galvanized iron rivet 镀锌铁铆钉
galvanized iron rope 镀锌丝绳
galvanized iron screw 白铁螺钉
galvanized iron screw hook 镀锌螺钩
galvanized iron sheet 马口铁；镀锌铁皮；镀锌铁板
galvanized iron sheet down pipe 镀锌铁皮落水管；白铁皮水落管
galvanized iron sheet flashing 镀锌铁皮泛水；白铁泛水
galvanized iron sheeting weathering 白铁披水片
galvanized iron sheet roof 镀锌铁皮屋面；白铁皮屋面
galvanized iron sheet valley 白铁天沟
galvanized iron shingle 白铁瓦
galvanized iron slope plate 镀锌坡度铁板
galvanized iron spring 镀锌弹簧
galvanized iron square boat nail 镀锌四方船钉
galvanized iron tube 白铁管
galvanized iron wash tub 镀锌铁洗衣盆
galvanized iron water bucket 白铁水桶
galvanized iron wire 镀锌铁丝；铅丝
galvanized iron wire brush 白铁丝刷
galvanized iron wire cloth 镀锌丝布
galvanized iron wire gauze 白铁丝布
galvanized iron wire knitting ware 镀锌铁丝编织品
galvanized iron wire nail 镀锌铁钉
galvanized iron wire netting 镀锌铁丝窗纱
galvanized locknut 镀锌锁母
galvanized machine 镀锌机
galvanized malleable clip 玛钢卡头
galvanized material 镀锌材料
galvanized mesh 镀锌钢丝网
galvanized metal flexible hose 镀锌金属软管
galvanized mild steel countersunk head wood screw 镀锌沉头木螺钉
galvanized mild steel drive screw 镀锌螺钉
galvanized mild steel hexagonal bolt and nut 镀锌六角螺栓螺帽
galvanized mild steel hexagonal wire netting 镀锌六角(铁丝)网
galvanized mild steel link chain 镀锌铁链条
galvanized mild steel roofing bolt and nut 镀锌瓦楞螺栓螺帽
galvanized mild steel roofing screw 镀锌瓦楞螺钉
galvanized mild steel round head machine screw 镀锌圆头机器螺钉
galvanized mild steel round head wood screw 镀锌圆头木螺钉
galvanized mild steel square nut 镀锌四方螺帽
galvanized mushroom head bolt 镀锌圆头螺栓
galvanized nail 镀锌铁钉；镀锌钉
galvanized nipple 镀锌套管接头
galvanized pipe 镀锌(钢)管；白铁管
galvanized pipe bend 镀锌管弯头
galvanized pipe socket 镀锌管套
galvanized plug 镀锌插头
galvanized pole steps 镀锌上杆钉
galvanized pot 镀锌锅
galvanized pot sprinkler 镀锌洒水壶
galvanized primer 高锌粉底漆
galvanized reducing coupling 镀锌异径管箍
galvanized reducing elbow 镀锌异径弯头
galvanized reducing socket 镀锌异径管节
galvanized reinforcement 镀锌钢筋
galvanized rigging shackle 镀锌索具锁扣
galvanized rigging thimble 镀锌索具套环
galvanized roofing bolt 镀锌瓦楞螺栓
galvanized roofing nail 镀锌屋面钉；镀锌瓦楞钉
galvanized roofing nail twisted 镀锌螺旋瓦楞钉

galvanized screw 镀锌螺钉
galvanized screw plug 镀锌旋塞
galvanized seizing wire 绑扎用镀锌铁丝
galvanized sheet flashing 白铁泛水
galvanized sheet(iron)镀锌铁片;镀锌钢皮;镀锌(薄)铁皮;镀锌(薄)板;白铁(皮);马口铁
galvanized sheet iron pipe 镀锌铁皮管
galvanized sheet metal 镀锌薄钢板
galvanized sheet roof 马口铁屋顶
galvanized sheet steel 镀锌钢片;镀锌(钢)板
galvanized single-J bolt 镀锌单勾螺栓
galvanized square boat nail 镀锌四方船钉
galvanized square washer 镀锌方形垫圈
galvanized square(wire)镀锌方眼网
galvanized square wire netting 镀锌方眼网
galvanized stay adjustable screw 镀锌拉线调整螺钉
galvanized stay clamp 镀锌拉线夹板
galvanized stay thimble 镀锌拉线套环
galvanized steel 镀锌铁皮;镀锌钢;白铁皮
galvanized steel conduit 镀锌钢导线管
galvanized steel corrugated sheet 镀锌波纹钢板
galvanized steel decking 镀锌钢盖板
galvanized steel iron nail 镀锌钢钉
galvanized steel meeting face 镀锌钢板贴接面
galvanized steel pipe 镀锌钢管
galvanized steel plain sheet 镀锌平钢板;镀锌钢板
galvanized steel reinforcement mesh 镀锌钢筋网
galvanized steel reinforcing mesh 镀锌钢筋网
galvanized steel self tapping screw 镀锌自攻螺线
galvanized steel sheet 镀锌钢屋面板;镀锌钢墙板;镀锌钢挡水板;镀锌钢皮;镀锌(钢)板
galvanized steel strip 镀锌钢带
galvanized steel stud 轻钢龙骨;镀锌钢龙骨
galvanized steel tile 镀锌钢瓦
galvanized steel tube 镀锌钢管
galvanized steel wash boiler 镀锌钢洗衣煮锅
galvanized steel wire 镀锌铁丝;镀锌钢丝;镀锌绞线
galvanized steel wire netting 镀锌钢丝网
galvanized steel wire rope 镀锌钢丝绳
galvanized steel wire rope clip 镀锌钢丝绳卡头
galvanized steel wire strand 镀锌钢绞线
galvanized stitching flat wire 镀锌扁铁丝
galvanized straight shackle 镀锌直型卸扣
galvanized strand 铅丝绳股
galvanized stranded wire 镀锌钢绞线
galvanized strip 镀锌带材
galvanized swanneck bolt 镀锌弯钩螺栓
galvanized tee 镀锌三通管
galvanized thick pipe 镀锌厚管
galvanized transposition plate 镀锌交叉铁板
galvanized tube 镀锌管
galvanized turnbuckle 镀锌螺钉接头
galvanized U-shaped iron sheet 镀锌 U 形铁皮
galvanized vessel 镀锌容器
galvanized washer 镀锌垫圈
galvanized welded tube 镀锌焊缝管
galvanized welded wire mesh 镀锌电焊网
galvanized wire 镀锌铁丝;镀锌铅丝;镀锌电镀线材
galvanized wire netting 镀锌窗纱
galvanized wire-receiving machine 镀锌收线机
galvanized wire rope 镀锌钢丝绳
galvanized wire rope clip 镀锌钢丝绳卡头
galvanized wire rope thimble 镀锌钢索具套环
galvanized wrought-iron pipe 镀锌熟铁管
galvanizer 铅丝;电镀器;电镀工
galvanizing 镀锌
galvanizing bath 镀锌槽
galvanizing blanket 镀锌覆盖物
galvanizing brittleness 镀锌脆性
galvanizing by dipping 热浸镀锌
galvanizing embrittlement 镀锌脆性
galvanizing equipment 镀锌设备
galvanizing flux 镀锌熔剂
galvanizing preparation 电镀制剂
galvanizing primer 高锌粉底漆
galvanizing process 电镀程序
galvanizing rig 镀锌装置
galvanizing shop 镀锌车间;电镀车间
galvanizing workshop wastewater 电镀车间废水
galvannealed sheet 镀锌铁皮
galvannealing 镀锌法

galvannealing process 镀锌退火法
galvanocautery 电烙器;电烙术
galvano-chemical solution action 电化学溶解作用
galvano-chemistry 电化学
galvanograph 电镀板;电板;电流记录图
galvanography 电流记录术;电镀制版术
galvanoluminescence 电流发光
galvanolysis 电解
galvanomagnetic effect 磁场电效应;电磁效应
galvanomagnetic property 电磁性能
galvano magnetism 电磁(学)
galvanometer 检流仪表;检流计;电(流)表;电表;安培计
galvanometer amplifier 电流计放大器
galvanometer constant 检流计常数;电流计常数;电流常数
galvanometer deflection 电流计偏转
galvanometer light modulator 电流计(式)光调制器
galvanometer mirror 检流计反射镜
galvanometer oscillograph 回线示波器
galvanometer record 电流记录
galvanometer recorder 检流计记录器;电流计式记录器
galvanometer registration 电流记录
galvanometer relay 检流式继电器
galvanometer-type relay 检流计式继电器;电流计式继电器
galvanometric property 电流测定的参数
galvanometric system 电流测定方式
galvanometry 电流测定法
galvanoplastics 电铸(技)术
galvanoplasty 电镀;电铸(技)术
galvanoscope 验电(流)器;验电盘;检流仪表;检流计
galvanostat 恒流器
galvanotaxis 趋电性
galvanotropism 趋电性
galvano-voltmetter 伏安计;伏安表
Galvanum 格尔瓦纳姆铝合金
galvo 金属烟雾热
gamagarite 水钒钡石
Gamann wicket dam 加马旋转闸门活动坝;加马翻板坝
Gambia abyssal plain 冈比亚深海平原
Gambier tannin 棕儿茶丹宁;里儿茶丹宁
gambler 投机商(人)
gambling hell 赌场
gambling house 赌场
gamboge 橙黄色;藤黄
gamboge for painting 绘画用藤黄
gambrel roof 复折形屋顶;斜折线形屋顶;双折屋顶;复折(式)屋顶;复斜屋顶
gambrel vent 山形百叶窗;三角形固定(设于屋脊处的)百叶窗;复折通风;复折通气窗
game 猎物;模拟推断;博弈;比赛
game against nature 用于自然界的对策
game area 狩猎区
game court 运动场
game-cover 禁猎区
game fishes 游钓鱼类
game graph 博弈图
game land 狩猎场
game machine 博弈机
game map 禁猎区图
game of chance 机会对策
game of market competition 市场竞争对策
game pay-off matrix 博弈支付矩阵
game playing 博弈
game playing algorithm 博弈算法
game preserve 禁渔区;禁猎区;禽兽养护地;狩猎保留地
game program(me)博弈程序
game refuge 鱼猎禁区;禁猎区(策略)
game reserve 猎物禁猎区;禁猎区
game room 娱乐室;文娱室
games hall 体育馆
games hall complex 综合体育馆
game simulation 对策模拟
game theoretical model 博弈论模型
game theory 形态运筹法;对策(理)论;策略运筹法;博弈论
game tree 对策树形图;对策树;博弈树状图;比赛树
game tree of water pollution control 水污染控制

对策树
game-tree search 博弈树搜索
game-type control system 博弈型控制系统
game watch 秒表;比赛计时表
game with complete information 全信息对策
game without a value 没有值的对策
game with perfect information 全信息对策
gamfexine 更非新
gaming 对策;博弈
gaming house 赌场
gaming simulation 对策仿真;博弈模拟
Gamlan shales 加姆兰页岩
gamma 微克;反衬度;反差系数
gamma acid 伽马酸
gamma activation analysis method 伽马活化分析法
gamma amplifier 伽马放大器;灰度放大器
gamma anomaly curve area 伽马异常曲线面积
gammacerane 伽马蜡烷
gamma characteristic 灰度特性
gamma constant 伽马常数
gamma constant of isotope 同位素伽马常数
gamma control 图像灰度控制
gamma-control amplifier 非线性调节放大器
gamma correction 伽马校正;亮度等级校正;灰度校正;图像灰度校正
gamma corrector 亮度等级校正器;灰度校正器
gamma decay 伽马衰变
gammadion 万字形
gamma distribution 伽马分布
gamma documentary value 伽马编录值
gamma-emitting radioscope 伽马放射性同位表
gamma enhancement 反差增强
gamma exponent 伽马指数;传输特性等级指数
gamma field 丙种射线场
gamma field grade level 伽马场级限
gamma function 伽马函数
gamma heat releasing light method 伽马热释光法
gamma intensity measurement 伽马强度测定法
gamma log 伽马测井
gamma-log converted coefficient 伽马测井换算系数
gamma matrix 伽马矩阵
gamma-neutron log 伽马—中子测井
gamma-neutron log curve 伽马—中子测井曲线
gamma photo detector 伽马光子探测器
gamma photon 伽马光子
gamma photo source 伽马光子源
gamma polysaprobic zone 伽马强腐水性地区
gamma probability distribution 伽马概率分布
gamma radiation 伽马辐射
gamma radiation detector 伽马辐射探测器
gamma radiation level indicator 伽马射线辐射料位计
gamma radiation sampling 伽马辐射取样
gamma radioactivity 伽马放射性
gamma radiography 伽马射线照相术
gamma ray 丙种射线;丙射线;伽马射线
gamma-ray absorption fluid density meter 伽马射线吸收型流体密度计
gamma-ray counter 伽马射线计数器
gamma-ray detector 伽马射线探测器
gamma-ray emitter 伽马射线辐射器
gamma-ray energy 伽马射线能量
gamma-ray examination 伽马射线检验
gamma-ray irradiation 伽马射线照射;伽马射线辐照
gamma-ray level controller 伽马射线料位控制装置
gamma-ray log 自然伽马测井;伽马射线录井;伽马射线测井剖面;伽马射线测井
gamma ray log calibration value 自然伽马刻度值
gamma ray log calibrator 自然伽马刻度器
gamma ray log curve 自然伽马测井曲线
gamma-ray logging 伽马射线测井
gamma-ray moisture meter 伽马射线湿度计
gamma-ray observation 伽马射线观测
gamma-ray probe 伽马射线探测
gamma-ray resonance 伽马射线共振
gamma-ray scattering 康普顿散射
gamma-ray sensor 伽马射线传感器
gamma ray shielding 伽马射线屏蔽;伽马辐射屏蔽
gamma-ray source 伽马射线源
gamma-ray spectrometer 伽马射线分光计;伽马射线波谱仪

gamma-ray spectrometry 伽马射线谱法;伽马射线光谱测定法
gamma ray spectrum log 自然伽马能谱测井
gamma ray spectrum log curve 自然伽马能谱测井曲线
gamma ray survey with share 犁铧伽马测量
gamma ray survey with shield 带屏伽马测量
gamma-ray tracking 射线跟踪
gamma real data figure 伽马实际资料图
gamma sampling by uphole method 炮眼法伽马取样
gamma spectrometer 伽马谱仪
gamma spectrometry 伽马能谱法
gamma spectrum sampling 伽马能谱取样
gamma-spectrum survey in shallow sea 浅海伽马能谱测量
gamma-spectrum survey on lake bottom 湖底伽马能谱测量
gamma spectrum survey with share 犁铧能谱测量
Gamma survey method 伽马法
gamma survey on lake bottom 湖底伽马测量
gamma survey on snow with truck 雪层上汽车伽马测量
gammatia 万字形
gamma-time curve 伽马一时间曲线
gamma-trace converted coefficient 伽马一径迹换算系数
gamma transform 伽马变换
gamma transition 玻璃转化
gammeter 反差系数计
gammil 伽密
gamming chair 吊椅(大风浪中病人上下舢板用)
gamnitude 倒幅度
Gamow barrier 伽莫夫势垒
Gamow-Condon-Gurncy theory 伽莫夫一康登一哥乃理论
Gamow-Teller interaction 伽莫夫一特勒相互作用
Gamow-Teller selection rules 伽模夫一特勒选择定则
gamut 全音阶;全量程;全范围;色移
gamut of colo(u)r 色调范围
Ganadian series 加拿大统
ganat 坎儿井;(古式过滤的)集水廊道
Gandian geosynclinal region 冈底斯北槽区
Gandise-Himalaya sea 冈底斯一喜马拉雅海
Gandise low-land 冈底斯低地
Gandise old land 冈底斯古陆
gandy stick 匝道辊;捣棒
gang 一组;一队;一组;一带;作业班;组;工作班(组);工班;成套;班组
gang adjustment 组调;统调;同轴调整
gang blanking die 组合落料模
gang board 脚手架便桥;滑轮跳板;跳板;脚手架板
gang boarding 工作走道梯板;工作走道铺板;脚手架铺板;木条跳板;搁架;工作马道
gang boarding roof ladder 陡桅梯
gang-boss 组长;工长;领班;班长
gang bush breaker 分组式灌木铲除机
gang capacitor 同轴电容器
gang condenser 共轴电容器;同轴(可变)电容器;电容器组
gang control 联动控制;同轴控制;同轴调整;同轴调节
gang cultivator 多行中耕机
gang die 组合模(具);多头冲模;冲模组;顺序动作模;复式模;复合模
gang-die forming 群模成形
gang disk plough 多体圆盘犁;多行圆盘犁
gang disk plow 多组圆盘犁
gang drill 排钻;排式钻机;多头钻岩机;多头凿岩机;分组式条播机;排式钻床
gang driller 排式钻床
ganged 成组的
ganged condenser 同轴电容器
ganged control 同轴联动控制
ganged form 大型组装模板;成套预制模板
ganged form construction 组合模板施工
ganged gain control 公共增益控制
gang edger 裁边圆排锯
ganged switch 联动开关;双联开关
ganged tuning 联动调谐;同调;同轴调谐
ganger 工作队长;线路工区工长;工头;工长;领工;领班;吊锚短索
Ganges abyssal fan 恒河深海扇

gang feed 成堆料的送进
gang foreman 监工员;线路工区领工员;工长;领班;班长
gang forming 组合模板
gang form system 大模体系
gang form(work) 大模板;成组立模;组模;组合模板;成组模板;成套模板
gang grooved 组合榫
gang harrow 拖拉机耙
gang head 组合头;组合刀具;组合刀具
gang hour 工组小时;搬运计时
ganging 机械连接;同调
ganging car on cableway 索道测流缆车
ganging circuit 统调电路;同轴联动谐振电路
ganglion 残余油滴
gang maintenance 道班养路
gang mandrel 串叠心轴
gang master 工作队长;工长;领工员;领班
gang mill 直锯;立式排锯;立式框锯;框锯制材厂;框锯;排锯机;排式铣床;排锯制材厂;大排锯
gang milling 组合铣刀;铣削;排锯;排式铣削;多刀铣削
gangmillingcutter 组合铣刀
gang model 成组模板
gang mo(u)ld 多巢模板;联模;连模;成组模(板)
gang mounted 组装的
gang nail 齿板连接件;齿扎链;阻钉;群钉
gang of cavities 多槽模型;多槽模
gang of labo(u)rs 工作小组;工作(班)组;工作队;工作班
gang of wells 组合井;井组;井群
gang-operated 同轴操作;双联同轴操作
gang piece work system 班组计件制
gang pilot 领航员
gang plank 脚手架板;滑轮跳板;跳板(码头用)
gangplank principle 跳板原则
gang planning 排刨
gang plough 多铧犁;工组犁;多组犁;带座多铧犁
gang plow 联犁;多铧犁
gang potentiometer 同轴电位器
gang press 多模压机;联合压机;排式压床
gang punch 群(联动)穿孔;全套穿孔;排式冲床;多位穿孔;复穿孔
gang pusher 组长
gang reamer 组合铰刀
gangrene 坏疽
gangrenous coryza 恶性卡他热
gang rider 矿车跟车工
gang riveting 排式铆钉
gang rooter 联犁铧
gang saw 组锯;直锯;框锯;排锯(机)
gang-saw cutting 群锯切削
gang-sawing machine 排机锯;锯木架;排锯
gang-sawing technique 排锯切割技术
gang sawmill 排锯制材厂
gang-sawn 排锯的石块表面
gang selector 同轴选择器
gang shears 多圆盘剪切机;多刀剪切机
gang sheet 班组记录单
gang showers 排列成行的淋浴装置
gang slitter 多圆盘剪床
gang slitter nibbling machine 多圆盘分移剪切机
gang slitter shears 多圆盘剪床
gang slitting machine 多圆盘纵切机
gang slitting shears 多圆盘式剪切机
gang socket 连接插座
gang softening 带卷成批退火
gangspiel 锚链绞盘
gangspill 锚链绞盘
gang stirrer 成套搅拌器
gang stitcher 多头订书机
gang summary punch 总计穿孔机
gang survey 集团性调查
gang switch 转换开关;联动开关;同轴开关
gang tool 多刀机床
gang tool axle lathe 多刀切削车轴车床
gang tool operation 排刀程序
gang trapped 共用水封;共用存水弯
gang tuning 同轴调谐;同时调谐
gang tuning capacitor 同轴调谐电容器
gangue 矸石;矿渣;矿物杂质;矿石中杂质;煤矸石;脉石;尾矿
gangue brick 煤矸砖
gangue element 脉石元素

gangue froth 脉石泡沫
gangue hollow block 煤矸石空心砌块
gangue mineral 脉石矿物
gangue piles 煤矸石堆
gangue quartz 脉石类
gangue rejection 脉石剔除;矸石剔除
gangue rock 脉石岩
Ganguillet formula 甘古列特公式(计算排水管中水流速度的公式)
gang-up 机器编组;编组
gang vibrators 振器组
gangway 工作便桥;舷梯;舷门;过道;观众厅走道;工作走道;工作马道;绳梯;出闸斜坡
gangway bellows 过道折棚;通道折棚
gangway bellows cover 通道折棚盖
gangway board 搭板
gangway bridge 步桥(油船)
gangway chain 栏链
gangway connections between coaches 通过台
gangway conveyer 主运输平巷运输机
gangway diaphragm 通道折棚
gangway door 舷门
gangway falls 吊舷梯复滑车
gangway ladder 舷梯;舷门;平巷装载机
gangway man 舷门值班人员
gangway platform 舷梯平台
gangway port 舷门;梯口门;侧舷门
gangway port handling 侧舷门装卸
gangway rail 栏杆
gangway tackle 舷梯绞辘
gangway watch 舷梯值班(船靠码头时)
ganiometer 量角器
ganiometry 测向
ganister 炉衬料;硅火泥;致密(硅)岩;硅石
ganister brick 硅砖
ganister measures 致密硅岩系
ganister sand 硅粉;石英砂
ganomalite 硅钙铅矿;羟硅钙铅矿
ganophyllite 辉叶石
ganophyre 花纹岩
ganosis 欠光彩;暗淡;滞钝
gantlet 套式轨道
gantline 滑车索;桅顶吊索;定端绞辘;穿过单滑车的吊索
gantree crane 龙门吊
gantry 龙门(起重机)架;龙门吊;跨线桥;门形构架;门式起重机;门架;横动桥形台;起重机台架;起重机机架;吊机架
gantry beam 刚架横梁;龙门架梁
gantry berth 舣装码头
gantry clearance 门架净空
gantry column 构台支柱;龙门架柱;摆动柱
gantry container crane 集装箱龙门起重机
gantry conveyer 高架输送机
gantry conveying belt 高架输送带
gantry crane 门式吊机;装御机;轨道车;高架(移动式)起重机;高架吊车;龙门吊机;龙门吊(车);门(座)式起重机;门式吊车;门架吊(机);门吊;桥式起重机;桥式吊(车);岸边集装箱装卸桥
gantry crane for general use 通用门式起重机
gantry crane on board 船用集装箱起重机
gantry crane portal 龙门起重机门架
gantry crane with cantilever 单悬臂门式起重机
gantry crane with chain hoist 手动葫芦门式起重机
gantry crane with electric(al) hoist 电动葫芦门式起重机
gantry crane with hinged boom 铰接悬臂门式起重机
gantry crane with hinged shoreside boom 带铰接前伸臂的装卸桥
gantry crane with hook 吊钩门式起重机
gantry crane with jib crane 带悬臂吊车的龙门起重机
gantry crane with level luffing crane 带俯仰式吊车的龙门起重机
gantry crane with man-trolley 带驾驶员小车的龙门起重机
gantry crane with movable girder 移动主梁门式起重机
gantry crane with retractable boom 悬臂伸缩门式起重机
gantry crane with rope trolley 索控小车门式起重机
gantry crane with saddle 框架形门式起重机

gantry crane with self-propelled trolley 自行小车门式起重机
gantry crane with slewing man-trolley 带驾驶员小车的旋臂龙门起重机
gantry elevator 龙门架升降机
gantry frame 龙门架;门吊架
gantry girder 钢架横梁;龙门架大梁
gantry house 龙门起重机房
gantry lathe 两脚车床
gantry loading system 门式进料系统
gantry mounting 门架
gantry pillar 龙门架(支)柱;摆动支座
gantry platform 龙门架平台;脚手架台板
gantry post 振荡柱;龙门架柱;枢轴承
gantry running 在悬架上运转
gantry scaffold 塔架
gantry slewing crane 门框回转式起重机
gantry slinger 行车式抛砂机
gantry spiral coal unloader 门式螺旋卸煤机
gantry support for catenary 缆车龙门支柱
gantry tower 门式铁塔
gantry transfer crane 门式搬运起重机
gantry travel 龙门架行驶
gantry travel(l)er 移动桥式起重机;门式行动吊车;门架吊车;起重机移动架;移动桥式吊车;移动门(式起)机;龙门起重机上起重小车;门式移动吊车
gantry truck 运货车
gantry type jib crane 门吊
gantry unloader 装441桥式卸货机
Gantt bar chart 甘特图
Gantt chart 甘特进度表;甘特图;施工进度表;线条图(表)
Gantt chart scheduling and controlling 甘特计划进度和控制
Gantt charts for planning engineering project 甘特式计划工程项目表
Gantt layout chart 甘特布置图
Gantt load sheet 甘特负荷表
Gantt progress chart 线条进度图;甘特进行图
Gantt project planning chart 甘特工程计划表
Gantt scheduling chart 甘特日程图
Gantt shop loading chart 甘特车间负荷图
Gaoxiong Port 高雄港
gap 余隙;辊缝;龙口(打桩船);裂孔;离缝;间隙;间断;缺口;山峡;山口;山凹;地质应用计划;差距;保安隙;凹口;隘口
gap acceptance 可接受空当
gap acceptance merge control 可接受空当交汇控制
gap adjustment 间隙调整
gap admittance 间隙导纳
gap aggregate gradation 骨料间级配;间断集料级配;间断骨料级配
gap alignment 间隙对准
gap allowed for expansion 伸缩间隙;伸缩缝
gap ampere turns 气隙安匝
gap analysis 缺口分析
gap arrester 火花避雷器;气隙放电器
gap at joint 接头缝隙;缝隙
gap band 死区;不灵敏区
gap bar 接缝杆
gap bed 槽形机座
gap-bed lathe 马鞍形车床;马鞍(式)车床
gap between breakwaters 防波堤缺口
gap between imports and exports 进出口差额
gap between revenues and expenditures 收支差额
gap between town and country 城乡差别
gap between well wall and casing 井壁和套管间隙
gap breakdown 空隙击穿
gap bridge 过桥
gap butt 留间对接;开口对接;明缝
gap cavitation 缝隙空穴
gap change 空挡换相
gap character 间隙符;间隔字符
gap choke 空气隙铁芯扼流圈
gap-chord ratio 节弦比
gap clearance 接合点间隙;对接间隙
gap-clearance of piston rings 活塞环开口
gap coding 间隙编码
gap coefficient 隙缝系数;间隙系数
gap control 间隙控制

gap correction 间隙修正
gap counter 分离计数器
gap cutter 齿槽铣刀
gap diagram 轨缝图
gap digit 间隙位;间隔位;间隔数字
gape 张开的阔度;张开;给料口最大宽度;裂口;裂开;裂缝;开裂;豁口;呵欠
gape erosion 电极端腐蚀
gap eliminator 间隙消除装置
gape-to-set ratio 给料口—排料口宽度比(破碎机)
gap factor 间隙系数
gap field 气隙场
gap filled joint 填缝接头
gap filler 裂缝填充物;填缝料;填缝料;填充物;雷达辅助天线
gapfiller data 填隙数据
gap filling 隙填充
gap-filling adhesive 填隙粘合剂;空隙充填黏合剂;填缝黏结剂;填缝胶(黏剂)
gap-filling capacity 填隙嵌缝能力
gap-filling cement 填缝水泥;填缝胶泥
gap-filling glue 填缝胶
gap-filling no-heat adhesive 填缝冷黏剂;冷胶填缝
gap-filling strategy 填补差距的策略
gap financing 资金缺口
gap-frame press 马鞍(式)压床;开式压力机
gap ga(u)ge 间隙规
gap ga(u)ge for spark plug 火花塞间隙表
gap gradation 间断级配;不连续级配
gap-graded 间隙分级的;间隔分级的;间断粒级的;间级配的;粗糙分级的
gap-graded aggregate 间隙级配集料;间级配骨料;缺中径颗粒骨料
gap-graded concrete 间隙级配混凝土;不连续级配混凝土;间级配混凝土
gap-graded design 间断级配设计
gap-graded material 间断级配材料
gap grade mix(ture) 间断级配混合料
gap grading 间级配
gap grain size 不连续颗粒
gap heating 缝隙加热
gap hydraulic press 马鞍压床
gaping place 孔
gap in record 观测记录间断
gap in the frog 辙叉的有害空间;辙叉有害空间
gap in the train diagram 运行图天窗
gap joint 拼缝隙
gap junction 缝(管)隙连接
gap lane change 利用空当转换车道
gap lathe 马鞍(式)车床;凹口车床
gap length 隙宽;隙长;间隙大小;间隙长度
gapless 无间隙(的)
gapless tape 无间隔带
gapless track 无缝线路
gap loss 间隙损失;间隙损耗
gap marker 通道标志
gap measuring wedge 轨缝测量器
gap mill 凹口研磨机
gap mode 空当通行式
gap observation 裂缝观测
gap of record(ing)electric(al)pole 记录电极间隙
gap of screen strip 筛条隔隙
gap open 开路
gap packing 封气包装
gapped aggregate grading 集料间断级配;骨料间断级配
gapped bed 马鞍形床身;凹口床身
gapped grading 间断级配
gapped tape 有间隔带;间隔带
gap piece 马鞍块;床身(过桥)凹口镶块
gapping 间隙;裂缝;缩裂;不紧密接触
gapping crack 张口裂隙;张口裂缝
gap(ping)fault 张开断层
gapping fissure 张开裂隙
gapping joint-fissure 张开裂缝
gapping of the joints of a vault 拱顶接头缝隙
gapping switch 桥接开关
gap plate 补隙板
gap press 马鞍压床;马鞍式压床
gap press with fixed bed 单柱固定台压力机
gap repair 缺口修复
gap rolls 锻造辊筒
gap scanning 隙扫描

gap scatter 多路磁头缝隙离散;缝隙离散
gap shears 马鞍剪床;凹口剪床;凹口剪切机
gap site 建筑空隙地
gap-sized grading 间断级配
gap span ratio 翼隔翼展比;隔展比
gap-speed detector 车间车速检测器
gap squaring shears 空隙剪压机
gap survey 裂缝测量
gap test 爆震波传递试验;殉爆试验;空隙检验;间隙检验
gap tilt effect 隙倾效应;倾斜效应
gap time setting 空当时间配置
gap town 峡谷城市;峡口城市
gap-type filter 片式滤清器
gap weld 特殊点焊;双极单点焊;断续焊
gap welding 断续焊接
gap width 间隙值;间隙宽度;缝隙宽度
gap window 窄长窗;间隙窗
garage 汽车修配厂;汽车修理厂;汽车修理(车)间;汽车库;车库;飞机库
garage account 虚构账户
garage apartment 有停车库的公寓
garage compound 集合车库
garage construction 车库建筑
garage court 车库(小)院子
garage door 汽车库;车库门
garage door furniture 汽车间门配件;车库门配件
garage door leaf 车库门扇;车库门扉
garage door operator 车库门开启装置;车库门控制器
garage drainage 车库排水
garage drive 入车库的车道;车库入口
garage equipment 修车器具;汽车房设备
garage floor(ing)(finish) 车库地面
garage(fore)court 车库前庭
garage heating 车库采暖
garage jack 修车起重器;车库滑动门;修车千斤顶;修车起重机
garage lamp 金属护网安全灯;安全灯
garage storage 汽车储(贮)藏库
garage threshold 车库阈限;车库门口;车库门槛
garage trap 车库滤油阀
garage warming 车库采暖
garaging facility 车库设备
garancine lacquer 茜草色硝基(醇酸)清漆;茜草色清漆
garavellite 硫锑铋铁矿
garbage 垃圾;烹饪废物;无用信息;厨房垃圾内脏;废信息;废料;不用单元
garbage and waste disposal 垃圾与废物处置
garbage bag 垃圾袋
garbage bin 垃圾箱;清洁箱
garbage boat 垃圾船
garbage burner 垃圾焚烧器
garbage can 垃圾箱;垃圾筒;垃圾桶
garbage can washer 垃圾桶冲洗器
garbage chute 垃圾装卸槽;垃圾溜槽;垃圾井筒;垃圾滑槽;垃圾(管)道
garbage chute lid 垃圾滑槽盖
garbage cleaning vessel 垃圾清扫船
garbage collection 垃圾收集;无用单元的收集
garbage collection algorithm 不用单元收集算法
garbage collection routine 不用单元收集例行程序
garbage collector 无用单元收集程序;废料收集器;不用单元收集程序
garbage combustion chamber 垃圾燃烧炉
garbage compost 垃圾堆肥
garbage conservancy system 垃圾存置系统
garbage container 废料箱;垃圾箱
garbage crusher 垃圾粉碎机
garbage destructor 垃圾焚烧炉
garbage disposal 垃圾处置;垃圾处理(法)
garbage disposal chute 垃圾处理槽
garbage disposal plant 垃圾处置场;垃圾处理厂;垃圾处理场;垃圾处理厂
garbage disposal shaft 垃圾处理竖井
garbage disposal site 垃圾处理场
garbage disposal unit 垃圾处理装置
garbage disposer 垃圾处置器;废物处置器;碎污机
garbage dump 垃圾堆弃地;垃圾填土;厨房垃圾堆
garbage dumping 垃圾倾弃
garbage dumping place 垃圾倾倒区;垃圾倾倒场
garbage furnace 垃圾焚化炉
garbage grease 杂碎油脂

garbage grinder 垃圾破碎机;垃圾磨碎机
garbage grinding 垃圾磨碎
garbage grinding treatment 垃圾磨碎处理
garbage heap 垃圾堆
garbage in 无用(数据)输入
garbage incineration 垃圾焚化
garbage incineration house 垃圾焚化站
garbage incinerator plant 垃圾焚烧厂;垃圾焚化厂
garbage-in garbage-out 废进废出
garbage lighter 垃圾驳(船)
garbage odo(u)r 垃圾臭气
garbage pitch 垃圾沥青
garbage pollution 垃圾污染
garbage power 垃圾发电
garbage press 垃圾压捆机
garbage receptacle 废物箱
garbage removal 垃圾清除
garbage room 垃圾房;垃圾间
garbage shaft 垃圾筒
garbage shoot 垃圾滑槽
garbage signal 无用信号
garbage tankage 垃圾动物肥
garbage transport 垃圾运输
garbage treatment plant 垃圾处置厂
garbage truck 垃圾车;清洁车
garbage yard 垃圾倾倒区;垃圾(倾倒)场
garbin 加宾风
garble 拣选;筛拣
garbling 精选
garboard 龙骨翼板
garboard planking 逐级增厚翼板
garboard strake 龙骨翼板;龙骨邻板
garbologist 清除垃圾工人
Garbutt attachment 格氏拉杆
Garbutt rod 格氏拉杆
Garchey sink 垃圾溜槽;加奇家用垃圾坑
Garchey system 加奇垃圾管系统
garcia nutans seed oil 猩猩草子油
garcinia butter 藤黄油
garden 庭园;花园
garden aesthetics 园林美学
garden and park 园林
garden and tourism 园林绿化与旅游
garden apartment 别墅公寓;花园公寓;庭园住宅
garden architect 花园建筑家;园林建筑家
garden architecture 园林建筑学;庭园建筑学
garden area division 园林区划
garden art 园林艺术
garden balsam 凤仙花
garden bench 庭园凳子;庭园长椅;园凳
garden block planning 园林分区规划
garden bond 园墙砌法;五顺一丁或四顺一丁砌法;园墙砌合
garden bridge 园桥
garden building 庭园建筑物
garden building unit 庭园建筑单元
garden burnet 地榆
garden carpet 花园纹地毯
garden ceramics 庭园陶瓷制品
garden chair 园椅
garden city 低密度花园住宅区;花园城市;田园城市
garden city latch 花园城市住宅的门窗插销
garden city theory 花园城市理论;田园城市理论
garden clay brick wall 花园黏土墙
garden colony 花园居住区
garden construction 造园
garden cottage community 花园住宅区;花园住宅村舍
garden couplet 园林楹联
garden court 花园庭院;庭园
garden craft 造园术;园林术;庭院术
garden crop 园艺作物
garden design 园林设计;庭园设计
garden drill 园艺播种机
garden dwelling unit 花园住宅(区)
garden engine 庭园用的小型抽水机;园艺用泵
garden engineering 园标工程
gardener 园艺家;园林工人;花匠
garden facade 花园建筑物正面;花园建筑物立面
garden fence 花园栅栏;花园围篱;园篱
garden figure 花园形象;花园外貌
garden fountain 花园喷泉
garden frame 栽培植物用框架
garden front 花园正面
garden furnishing 户外用布
garden furniture 庭园小品;庭园家具
garden glass 园艺玻璃;温室玻璃
garden ground 花园场地
garden hedge 园篱
garden hose 灌溉橡胶管;花园浇水用软管;花园浇水带;花园水龙带
garden house 花园住宅;花园凉亭
garden(housing)estate 花园住宅区
garden hydrant 花园给水栓
gardening 园艺(学)
gardening and greening 园林绿化
gardening engineer 园艺技师
gardening system 园景系统
gardening up 修整铆钉梢
garden label 植物名签
garden lamp 庭园灯
garden layout 园林布局;园林施工
garden loam 园田壤土
garden machine 园林机械
garden making 造田;造园学
garden management 园林(经营)管理
garden marking 花园设施;造园
garden masonry wall 花园圬工墙
garden master planning 园林总体规划
garden mattock 园用镐
garden microclimate 园林小气候
garden mo(u)ld 园土;菜园土
garden nesting 花园住宅区
garden of acclimatization 驯化园
garden of pot plant 盆景园
garden on the Yangtze Delta 江南园林
garden operation 园圃操作
garden ornament 庭园小品点缀
garden pansy 三色堇(植物)
garden party 游园会
garden path 花园小径
garden path design 园林道路设计
garden path(paving)flag(stone)花园铺路石
garden path work 园路工程
garden perennial 多年生园艺植物
garden pink 常夏石竹
garden planning 园林规划
garden plant 园林植物
garden plot 果菜园;园地;庭园地
garden public administration 园林管理机构
garden restaurant 花园饭店
garden rocket oil 园岩油
garden roller 园艺用镇压器
garden roof 花园屋顶
gardens 园林
gardens afforestation 园林绿化
garden sculpture 庭院刻物
garden seat 园椅
garden seeder 播种器;园圃播种机
garden shed 花园棚屋;花棚;水果;蔬菜
garden-shelter forest 护园林
garden shovel 园用铲
garden site survey map 园址测量图
garden space 园林空间
garden spot 园艺场
gardens sculpture 园林雕塑
garden stair(case)花园楼梯
garden statuary 庭园雕像
garden stuff 花卉
garden-styled farming 耕作田园化
garden suburb 园林(化)郊区;成郊田园住宅区;花园市郊;田园市郊;田园郊区
garden suburban 园林化郊区
garden surround 花园周围;花园环境
garden tablet 园林匾额
garden terrace 庭园舞台
garden tile 花园瓷砖
garden tillage 园圃耕作
garden tool 园用工具
garden tool shed 花园工具棚屋;花园工具堆房
garden town 花园城
garden tractor 园艺(用)拖拉机;手扶拖拉机
garden type apartment 花园式公寓
garden unit 花园建筑构件
garden variety 常见式样;园艺品种
garden village 花园村;庭园露台
garden walks and pavement 园路工程
garden wall 园墙;花园围墙
garden wall bond 花园围墙(式)砌合;园墙砌合
garden watering demand 绿化用水
garden wicket 花园小门;庭院便门
garden work 园艺工作
garder 测油探尺
garderobe 衣柜;私室
Gardner adherometer 加德纳附着力测定仪
Gardner-Coleman method 加德纳—科尔曼吸油量测定
Gardner colo(u)r difference meter 加德纳色差计;加氏色差计
Gardner colo(u)r number 加德纳色值
Gardner colo(u)r scale 加德纳色标
Gardner colo(u)r standards 加德纳色标
Gardner crusher 加德纳型破碎机;加德纳摇锤式破碎机;锤式破碎机
Gardner drying meter 加德纳(漆膜)干燥计
Gardner drying time recorder 加德纳干燥时间记录器
Gardner gloss meter 加德纳光泽计
Gardner-Holdt bubble viscometer 加德纳—霍尔德气泡黏度计
Gardner Holdt tube 加德纳—霍尔德管
Gardner-Holdt viscosity tube 加德纳—霍尔德黏度管
Gardner mark 加德纳符号
Gardner mobilometer 加德纳淌度计
Gardner-Park's adhesion test 加德纳—帕克附着力试验
Gardner-Park's tensile meter 加德纳—帕克抗张强度试验机
Gardner straight line wash ability machine 加德纳直线式耐洗刷性试验机
Gardner viscometer 加德纳黏度计
Gardner viscosity 加德纳黏度
gare 错船系泊处
garewaite 透橄斑岩
Gargasian(stage)加尔加斯亚阶【地】
gargoyle 水柱;(怪形雕塑的)放水口;(哥特式建筑上的)滴水嘴;喷水口;排水口
gargoyle lining 出水管内衬
garigue 地中海常绿矮灌丛
garland 花彩;(矿井筒壁上的)集水圈;华饰;花环雕饰;索环;水手餐袋
garland chain 花环斜挂链
garland chain system 花环斜挂链条系统
garland circular window 饰以花彩的圆窗
garland hung chain 花环链
garland leaf 花卉彩饰
garland ornament 花环彩饰
garlic 大蒜
Garlock oil seal 加氏油封
garment 外层;涂层;覆盖物;服装
garment blank 衣坯
garment container 服装集装箱
garment for protection against fire 防火服
garment industry 制衣业;成衣业
garment of clean room 防尘服
garment piece 衣片
garment press 压力矫正机
garment tag 外表特征
garner 谷仓
garnet 石榴红色;石榴(子)石;深红色
garnet abrasive 石榴石磨料
garnet amphibole pyroxene granulite 石榴闪辉麻粒岩
garnet biotite quartz schist 石榴石黑云母石英片岩
garnet brown 石榴石棕(红棕色)
garnet compound 石榴石化合物
garnet content 石榴子石含量
garnet crystal 石榴石晶体
garnet deposit 石榴子石矿床
garnet dimicaceous schist 石榴石二云母片岩
garnet diopside amphibole plagioclase granulite 石榴透辉角闪斜长麻粒岩
garnet diopside anorthite gneiss 石榴透辉钙长片麻岩
garnet diopside sharn 石榴石透辉石矽卡岩
garnet epidote sharn 石榴石绿帘石矽卡岩
garnet felsic granulite 石榴长英麻粒岩
garnet gneiss 石榴(石)片麻岩
garnet granulite 石榴变粒岩
garnet hinge 丁字铰链;T形门铰;石榴铰链(T形门铰)

garnet hypersthene diopside plagioclase granulite 石榴紫苏透辉斜长麻粒岩
garnet hypersthene plagioclase granulite 石榴紫苏斜长麻粒岩
garnetiferous 石榴石的
garnetiferous plagioclaso orthoclase granulite 含石榴石斜长正长麻粒岩
garnetiferous skarn 石榴硅卡岩
garneting 砂浆碎石嵌缝；填塞石缝
garnetite 石榴子石岩
garnetization 石榴子石化
garnet lac 石榴色紫胶；深红色的精制紫胶
garnet laser 石榴石激光器
garnet lawsonite glaucophane schist 石榴硬柱蓝闪片岩
garnet lherzolite 石榴石二辉橄榄岩
garnet mica hypersthene light colo(u)red granulite 榴云紫苏浅色麻粒岩
garnet-mica schist 石榴石云母片岩
garnet muscovite quartz schist 石榴石白云母石英片岩
garnetoid 似石榴子石
garnet paper 墙面纸；石榴石砂纸；石榴石粉砂纸；砂纸
garnet peridotite 石榴石橄榄岩；石榴橄榄岩
garnet-phyllite 石榴石千枚岩
garnet plagioclase-gneiss 石榴斜长片麻岩
garnet purchase 装卸货绞辘；收帆绞辘
garnet pyroxene plagioclase amphibolite 榴辉斜长角闪岩
garnet pyroxenite 石榴石辉石岩
garnet-quartz schist 石榴石石英片岩
garnet rock 石榴石岩；石榴子岩
garnet sand 石榴石砂
garnet-sericite phyllite 石榴石绢云母千枚岩
garnet sharn 石榴石矽卡岩
garnet(shel)lac 石榴(石)紫胶；石榴石虫胶片
garnetting 扯松
garnett machine 打回丝机
garnett wire 钢刺条；锯齿刺丝
garnet two mica quartz schist 石榴石二云母石英片岩
garnet type ferrite 石榴石型铁氧体
garnet type pigment 石榴石型颜料
garnetyte 石榴石岩
garnierite 硅镁镍矿；暗镍蛇纹石
garnish 装饰品；扣押债务人的财产
garnish bolt 装饰螺栓；花样螺栓；雕槽螺栓
garnished mo(u)lding 装饰镶嵌
garnishee 扣押债务人的财产
garnishment 扣发；财产扣押令
garnish rail 装饰轨
garnison church 设防教堂
garniture 附属品；装饰品
garr 船底须菌
garrelsite 硅硼钠钡石
garret 阁楼；瞭望台；塞塞缝隙(用小石块)；顶楼
garret beam 屋顶层横梁
garret floor 顶楼层；阁楼地板；阁楼层；顶层楼
garretgap-graded 屋顶层
garreting 砂浆碎石嵌缝；填塞石缝；碎石片嵌灰缝
Garrett coiler 加勒特式卷取机
Garrett looping rod mill 加勒特活套式线材轧机
Garrett mill 加勒特式线材轧机
Garrett reel 加勒特式小型钢卷取机
Garret type layout 阶梯式布置
garret window 顶楼窗；老虎窗
garrison hospital 卫戍部队医院
garron 大铁钉(一端为平头，另一端呈玫瑰状)
garronite 十字沸石
gar-shift modem 变速调制解调器
Garstang sandstone 加斯唐砂岩
garter spring 夹紧盘簧；夹紧弹簧；卡紧弹簧；弹簧圈
garth 庭院；内院；场地；捕鱼用的鱼坝；鱼坝
Garth grit 加思丘陵层【地】
Garth hill beds 加思丘陵层【地】
Garton lightning arrester 加顿避雷器
garua 浓湿雾
garyansellite 水磷铁镁石
gas 煤气；燃气；气体；瓦斯
gas above oil reservoir 油层上部气
gas absorbed amount 气体吸收量
gas absorbed coefficient 气体吸收系数
gas absorbent 气体吸收剂

gas absorbent bed 气体吸收床
gas absorber 气体吸收器；气体吸附器
gas absorber-oil 气体吸收油
gas-absorbing agent 气体吸收剂
gas absorption 气体吸附；气体吸收
gas absorption cell 气体吸收元件
gas absorption equipment 气体吸收装置
gas absorption operation 气体吸收操作
gas absorption refrigerator 煤气吸收式制冷机
gas accumulation 气体聚集；天然气积聚
gas-accumulator relay 气体积聚继电器
gas-activated battery 气体激活电池
gas activation 气体活化
gas activation detector 气体激活探测器
gas active material 气体激活材料
gas activity meter 气体放射性计
gas-actuated relay 气体继电器
gas adsorbing 气体吸附
gas adsorption 气体吸着；气体吸附
gas adsorption chromatography 气相吸附色谱(法)；气相吸附层析
gas adsorption device 气体吸附装置
gas adsorption method 吸附气体法；气体吸附测定法
gas adsorption rate 气体吸收率；气体吸附率
gas adsorption trap 气体吸附阱
gas-aerosol mixture 气—气溶胶混合物
gas agent 气化剂
gas air mixer 煤气空气混合器
gas-air mixing plant 煤气空气混合站
gas alarm 气体警报器；瓦斯警报；毒气警报(器)
gas alloying apparatus 燃气合金分析仪
gas amplification 气体膨胀；气体(电离)放大
gas amplification factor 气体放大系数
gas analyzer 气体分析仪；气体分析器
gas analysis 气体分析；天然气分析
gas analysis apparatus 气体分析仪器
gas analysis instrument 烟气分析仪；烟气分析器
gas analysis meter 气体分析仪；气体分析器
gas analysis recorder 气体分析记录仪
gas analytical apparatus 气体分析装置
gas analyzed 气体指标
gas anchor 气锚
gas anchor packer 气锚封隔器
gas and coal basin 含油气煤盆地
gas and coke plant waste 煤气焦化废水
gas and gas correlation 气—气对比
gas and gasoline threeway valve 煤气与汽油三通阀
gas and its loss 气体及损失量
gas and liquid mixture 气液混合液
gas and mist sampler 瓦斯及烟雾取样仪
gas and oil pressing equipment 油气处理设备
gas and oil process and storage tanker 油气处理储存船
gas and oil separating plant 油气分离装置；油气分离站
gas and source rock correlation 气—母岩对比
gas and water expenses 煤气及水费
gas apparatus 燃气装置
gas appliance 燃(气用)具；煤气用具；燃气设备
gas appliance engineers 煤气设备工程师
gas appliance recess 煤气设备壁龛
gas appliance test 燃具试验
gas approach 烟气进口；入口烟道
gas arc lamp 煤气(喷)灯；煤气弧光灯；充气弧光灯
gas-arc welding 气体保护电弧焊；气电焊
gas-arc welding gun 气电焊焊嘴
gas area 含气面积
gas-ash concrete 加气灰渣混凝土；加气粉煤灰混凝土
gas-ash silicate concrete 加粉煤灰硅酸盐混凝土
gas associated with condensate oil 凝析油伴生气
gas associated with crude oil 石油伴生气
gasateria 自动加(气)油站；投币加(气)油站
gas atmosphere inlet 保护气体入口
gas atomization 气体喷雾
gas attack 气体腐蚀；毒气攻击
gas backstreaming 气体回流
gas baffle 烟气挡板
gas bag 气囊；气袋
gas balance 气体天平
gas ballast air 气镇空气
gas ballast control 气镇控制

gas ballast device 气镇器件
gas ballast flow 气镇流
gas ballasting 气镇
gas ballast inlet 气镇入口
gas ballast port 气镇口
gas ballast principle 气镇原理
gas ballast pump 气体压载泵(排压舱水用)；载气泵；气镇泵
gas ballast roughing-holding pump 气镇粗抽泵
gas ballast valve 气镇阀
gas balloon 气体比重瓶；称气瓶
gas barrel 煤气管
gas barrel handrail 煤气桶扶手；煤气管扶手
gas barrier 气体闭塞装置；排气装置
gas-based power generation 沼气发电
gas basin 含气盆地
gas bath 气浴
gas battery 气体电池组
gas bearing 含气；气体轴承
gas-bearing bed 气层
gas-bearing boundary 含气边界
gas-bearing depth 气层深度
gas-bearing formation 含气岩层
gas-bearing inter margin 含气内边界
gas-bearing outer margin 含气外边界
gas-bearing stratum 含气岩层
gas-bearing trap 含气圈闭
gas-bearing water bed 含气水层
gas bed 气层
gas bell 集气器钟罩；气柜钟罩；气鼓泡
gas bending stress 气流弯曲应力
gas black 烟黑；灯黑；气烟末；气黑；瓦特炭黑
gas blanket 气(体)层；气膜
gas-blast circuit breaker 气吹断路器
gas bleaching 气相漂白
gas bleed 气体冲洗
gas bleeder 放气管
gas bleed flange 气体冲洗凸缘；气体冲洗法兰
gas bleeding 气体分出；抽出气体
gas bleed valve 气体冲洗阀
gas blowby 窜气
gas blower 瓦斯泄出
gas blowing engine 煤气鼓风机
gas blow off 排气
gas blowout 气喷；气体喷出
gas blow pipe 吹气管
gas blow torch 煤气吹管
gas blue 铁蓝(旧称)
gas boat 照明灯艇；气灯灯船；灯标艇；汽艇
gas boiler 燃气锅炉；沼气锅炉；煤气锅炉
gas boiler flue 煤气锅炉烟道
gas bomb 气体钢瓶；毒气炸药；毒气弹；氧化瓶；储[贮]气瓶
gas booster 压气设备；气体输送压缩设备；气体升压机
gas-borne particles 悬浮气体中的微粒
gas bottle 洗气瓶；氧气瓶
gas bracket 煤气灯管
gas brazing 气焰硬钎焊；气焰铜焊
gas breakdown 气体击穿
gas breaker for coal combination 采煤机瓦斯断电仪
gas bubble 气泡；天然气泡
gas bubble disease 气泡病(鱼类)
gas bubble method 气体发泡法；气泡法
gas bubble protective device 布克霍尔茨保护装置
gas bubble pulsation 气泡脉动
gas bubble pump 气泡泵；气举泵
gas buffer tank 气体缓冲箱
gas-bulb sensor 气泡感受元件
gas buoy 燃气灯光浮标；气灯浮标；瓦斯浮标
gas burette 气体量管；气体滴定管
gas burner 煤气灶；煤气燃烧室；煤气燃烧器；煤气喷嘴；煤气喷燃器；煤气(喷)灯；燃气喷嘴
gas burner pliers 煤气灯头钳
gas burning appliance 煤气燃烧装置
gas burning appliance recess 煤气燃烧壁龛
gas burst 气体爆炸
gas buster 气体分离器(钻井泥浆)
gas by-pass 气体旁路
gas by-passing 天然气串流
gas by-products 煤气副产品
gas cable 充气电缆
gas cable hut 充气电缆线路储气站

gas caloricity 煤气热值
gas calorimeter 气体热量计;气体量热计
gas can 汽油桶
gas cap 气帽;气盖;气顶
gas capacity 气体容量
gas cap behavio(u)r 气顶动态
gas cap drive 气顶驱动;气驱动(油面天然气驱动使油入井)
gas-cap expansion 气顶膨胀
gas-cap reservoir 气顶油藏
gas-cap shrinkage 气顶收缩
gas carbon 气碳
gas carbon-arc welding 气保护碳弧焊
gas carbon black pigment 煤气炭黑颜料
gas carbonizing method 气体渗碳法
gas carburizing 炭化
gas carburizing furance 气体渗碳炉
gas carburization 气体渗碳
gas carburization sintering 煤气渗碳烧结
gas carburizer 气体渗碳剂
gas carburizing 煤气渗碳法;气体渗碳
gas carburizing system 气体渗碳装置
gas carrier 气体运输船
gas carrier code 气体运载法规;气体运输规则
gas catalytic oxidation equipment 气体催化氧化器
gas catalytic reduction equipment 气体催化还原器
gas cavity cell 气孔池
gas cell 气体电池;气室;气囊;充气囊;充气光电管;充气光电池
gas cell alarm 充气光电管报警器
gas cell fabric 气囊织物
gas cell frequency standard 气体池频率标准
gas cell phase shifter 充气移相器
gas cementation 气体渗碳
gas central heating 煤气集中供热
gas centrifugation 气体离心
gas centrifuge process 气体离心法
gas centrifuger 气体离心机
gas certificate 气体检验证明书;舱内油气状况检查证书(油船卸空后的)
gas chamber 煤气室;煤气柜;气室;毒气室
gas channel 气道;气槽
gas characteristic 气体特性
gas charging 充气
gas charging connection 充气接头
gas charging valve 充气阀
gas checking 晶纹;气致皱纹(饰面);气污;污气网纹;气裂;龟裂
gas check pad 紧塞垫
gas check valve 煤气止回阀;气阀;闭气阀
gas chemical industry 天然气化工
gas chlorine 气态氯
gas chromatogram 气相色谱(图);气相层析谱图
gas chromatograph 用气相色谱(法)分析;气相色谱仪;气相色谱;气相色层分离仪;色谱法分析气体
gas chromatograph coupled with Fourier transform infrared spectrometer 气相色谱/傅立叶红外光谱联机
gas chromatograph equipment with a flame ionization detector 火焰电离检测器气相色谱仪
gas chromatograph for cupola furnace 冲天炉用气相色谱仪
gas chromatograph-Fourier transform infrared spectrometer 气相色谱仪一傅立叶变换红外光谱联用
gas chromatographic(al) analysis 气体色层分析;气相色谱分析(法)
gas chromatographic (al) chemical ionization detection 气相色谱化学电离质谱仪
gas chromatographic(al) column 气相色谱柱
gas chromatographic (al) detector 气相色谱检测器;气相层析检测器;气相色谱探测器
gas chromatographic (al) flame ionization detection 气相色谱火焰电离检测法
gas chromatographic (al) head space analysis 液面上气相色谱分析
gas chromatographic(al) infrared spectrometry combination 色谱—红外光谱联机
gas chromatographic (al) mass spectrometry combination 色谱—质谱联机
gas chromatographic (al) mass spectrometry infrared spectrometry combination 色谱—质谱—红外光谱联机

gas chromatographic(al)method 气相色谱层析法
gas chromatographic(al)method with selected ion monitoring 离子选择监测气相色谱法
gas chromatographic(al)purity 气相色谱纯度
gas chromatographic(al)technique 气相色谱技术
gas chromatograph-mass spectrogram data system 气相色谱仪—质谱图数据系统
gas chromatograph-mass spectrometer 气相色谱—质谱联用仪;气相色谱仪—质谱仪联用
gas chromatograph-mass spectrometer-computer 气相色谱仪—质谱仪—计算机联用
gas chromatograph-mass spectrometer-singly ion detector 气相色谱仪—质谱仪—单离子检测器联用
gas chromatograph-mass spectrometer system 气相色谱—质谱仪装置
gas chromatograph-quadpole mass spectrometer 气相色谱四极质谱仪
gas chromatography 气相色谱分析(法);气相色谱—质谱法联用数据系统闭合环路溶出分析;气相色谱法;气相层析;气相色谱法;气体色层(分离)法
gas chromatography combined with gas spectrometry 气体色谱质谱联用
gas chromatography flame ionization detection 气相层析火焰电离检测法
gas chromatography flame photometric detection 气相层析火焰光度检测法;气相色谱火焰光度检测法
gas chromatography flame thermionic detector 气相层析火焰热离子检测器
gas chromatography-infrared technique 气相色谱—红外技术
gas chromatography-mass spectrometry 气相色谱—质谱(分析)法;气相色谱法—质谱法联用
gas chromatography-mass spectrometry-selected ion monitor detector 气相色谱法—质谱法—离子选择监测检测器联用
gas chromatography-mass spectrometry-singly ion detector 气相色谱法—质谱法—单离子检测器联用
gas chromatography-negative chemical ionization mass spectrometry 气相色谱法—负化学电离质谱法联用
gas chromatography-ultraviolet spectrometry 气相色谱法—紫外光谱联用
gas chromatography with a flame thermionic detector 气相色谱火焰热离子检测法
gas chromatography with electron capture detection 气相色谱电子俘获检测法;气相层析电子俘获检测法
gas chromatography with high efficiency packed column 高效填充塔气相层色法
gas circuit 气路
gas-circulating plant 气体循环装置
gas-circulating system 气体循环系统
gas circulation 气体循环
gas circulator 煤气供应热水器;煤气循环器;气体循环泵
gas classification 气体分级
gas cleaner 洗气机;净气器;气体净化器;气体滤清器;气体净化器
gas cleaning 煤气清洗;煤气净化
gas-cleaning cyclone 气体净化旋风器
gas cleaning device 气体净化装置;煤气净化装置
gas cleaning equipment 气体净化装置
gas cleaning pipe 用气体净化管道;气体净化管道
gas cleaning plant 气体净化装置
gas cleaning unit 气体洗涤装置
gas clean-up 气体清除;提高真空度
gas-cloth ration 气布比
gas coal 煤气用煤;气煤
gas cock 煤气旋塞;煤气龙头;煤气喷灯
gas coke 煤气焦炭;气焦
gas collecting cap 集气帽
gas collecting channel 集气罩
gas collecting dome 集气室
gas collecting jar with ground-on cover plate 集气筒
gas collecting main 煤气总管
gas collecting ring 集气环
gas collecting tube 煤气聚集管
gas collection 集气;气体收集
gas collector 集气器;集气瓶;煤气收集器;煤气聚集器

gas column 气柱
gas combustion 气体燃烧
gas company 煤气公司
gas component 气体组分;气体成分
gas composition 煤气组成;气体组成;气体成分
gas composition analysis 气体成分分析
gas composition analysis equipment 天然气组份分析设备
gas compression 气体压缩
gas compression cable 压缩气体电缆;压气电缆
gas compression cycle 气体压缩循环
gas compression pump 煤气压缩机;煤气压缩泵
gas compression tap 煤气压缩旋塞
gas compressor 燃气压缩机;压气机;空压机;空气压缩机;气体压缩机
gas compressor capacity 空压机容量
gas compressor station 天然气压气站
gas concentration 气体浓度;瓦斯浓度
gas concrete 加气混凝土;充气混凝土;产气轻质混凝土;发气混凝土
gas concrete article 加气混凝土制品
gas concrete block 加气混凝土(砌)块
gas concrete cast(ing) 加气混凝土浇制(件)
gas concrete cavity tile 加气混凝土空心砖
gas concrete coating 加气混凝土涂层
gas concrete hollow block 加气混凝土空心砖;加气混凝土空心砌块
gas concrete inside wall 加气混凝土内墙
gas concrete insulating slab 加气混凝土绝缘板;加气混凝土隔热板
gas concrete mixer 加气混凝土搅拌机
gas concrete plant 加气混凝土浇制厂
gas concrete pot 加气混凝土盆;加气混凝土槽
gas concrete product 加气混凝土制品
gas concrete purpose-made block 加气混凝土特制砖;加气混凝土特制砌块
gas concrete slab 加气混凝土板
gas concrete specimen 加气混凝土试件
gas concrete tile 加气混凝土砖;加气混凝土瓦
gas concrete wall block 加气混凝土砖墙;加气混凝土砌块
gas concrete wall slab 加气混凝土墙板
gas concrete wall tile 加气混凝土墙面砖
gas condensate field 凝析气田
gas condensate reservoir 气凝聚层
gas condensate well 凝析气井
gas condensation 气体冷凝
gas condensation purification equipment 气体冷凝净化装置
gas condenser 气体冷凝器;气冷凝器
gas conditioner 气体调节器
gas conditioning 煤气质量调整;煤气加工;煤气处理
gas conditioning tower 气体调节塔
gas conduction 气体流导
gas conductivity 气体电导率
gas conductor 气体导电体
gas conduit 煤气管道;气体导管
gas coning 气锥
gas-coning in oil reservoir 储[贮]油库中的气锥
gas constant 气体常数;气体常量
gas consumption 用气量;耗气量;燃气流量;天然气消耗量
gas consumption per second 每秒气耗量
gas consumption quota 用气定额;用气量指标
gas consumption rate 汽油消耗率
gas container 容气器;气体容器;气瓶;储气罐
gas contaminant 气体污染物
gas contamination 气体污染
gas content 含气量;气体含量
gas content barometry 气测压法
gas continuous-flow calorimeter 燃气连续流量热器
gas control instrument 气体控制仪表
gas controlled field 气驱油田
gas controller 气体调节器
gas control lever 煤气控制杆
gas control tube 充气控制管
gas control valve 气体调节阀
gas conversion process 气体转化过程
gas converter 气体转化器
gas converter in sintering 烧结气体转化器
gas cooker 燃气灶;烤箱灶;煤气(炉)灶
gas cooking stove 煤气灶

gas coolant 气体冷却剂
gas-cooled 气冷(却)的
gas-cooled electrical machine 气冷电机
gas-cooled fast breeder reactor 气冷快中子增殖堆
gas-cooled generator 气冷发电机
gas-cooled machine 气冷式电(动)机
gas-cooled monitor 气体冷却控制器
gas-cooled nuclear reactor 气冷核反应堆
gas-cooled reactor 气体冷却反应器;气体冷却反应堆;气冷反应堆
gas cooler 煤气冷却器;气体冷却器
gas cooling 煤气冷却;气冷
gas cooling coil 气冷蛇(形)管
gas core reactor 气芯堆;气体活性区反应堆
gas corrosion 气体腐蚀
gas cost 煤气费用
gas coulomb-meter 气体电量计
gas counter 煤气表;气体计数器
gas counting 气体计数
gas-coupled turbine 分轴式透平机
gas cracking 气相裂化
gas crazing 晶裂;龟裂;气裂;污气网纹
gas current 离子电流;气体电流;气流
gas curtain 气雾;气幕
gas curtain furnace 气体帘幕热处理炉
gas cushion 气体缓冲器;气垫
gas cushion cable 气垫电缆
gas cushioned hydraulic breaker 气垫式液压破碎机
gas cushioned tempering 气垫钢化;气垫淬火
gas customer penetration 用气普及率
gas cut fluid 气侵液
gas cut mud 气侵泥浆;气侵钻泥
gas cutter 氧炔切割机
gas cut(ting) 火焰切割;氧炔切割;气割
gas cutting machine 气割器;气割机
gas cutting machine with magnetic wheel 磁轮气割机
gas cutting of fluid 液体气侵
gas cutting torch 气割吹管
gas cutting torch nozzle 气割吹管嘴
gas cutting troubles 气侵事故
gas cycle 燃气循环;气体循环
gas-cycle reactor 气体循环堆
gas cycling 天然气回注
gas cylinder 压缩煤气瓶;压缩气罐;储[贮]气瓶;高压气筒;高压气瓶;钢瓶;煤气罐;气瓶;瓦斯筒;导气筒
gas cylinder cart 气体瓶车
gas cylinder room 压缩煤气瓶储藏室;气瓶室
gas damage 气体损害
gas decay tank 气态放射性废物衰变槽;气态衰变槽
gas defense 毒气防御
gas deflector cone 导风锥
gas dehydration 气体脱水
gas dehydrator 气体脱水器
gas delivery 供气量;气体输送
gas demand 燃气需用量;燃气负荷
gas densitometer 气体密度计
gas density 气体密度
gas density balance 气体密度天平;气体密度计
gas density balance detector 气体密度天平检测器
gas density ga(u)ge 气体密度计
gas density meter 气体密度计;气体密度测量器
gas density recorder 气体密度记录仪
gas depletion drive 内部气驱
gas detecting tube 气体检测管
gas detection 气体检测
gas detection hammer 气体检测锤;煤气检测锤
gas detection system 气体探测系统
gas detector 检气管;煤气检测器;气体探测器;气体检验器;气体检漏器;气体检测器;瓦斯检定器;瓦斯检测装置
gas detector method 检气管法
gas detector relay 气体检测继电器;气体继电器
gas detector system 气体探测装置
gas detector tube 气体检定管;气体检测管
gas detect reagent 气体检测试剂
gas development 气体发生;发生气泡;沼气开发
gas deviation factor 天然气偏差系数;压缩因数
gas devourer 消除油轮残余气体工具
gas dewatering 气体脱水
gas dewatering device 天然气脱水装置

gas-diesel engine 气体燃料压燃机;柴油煤气发动机
gas diffusion 气体扩散;气态扩散
gas diffusion column 气体扩散柱
gas diffusion electrode 气体扩散电极
gas diffusion process 气体扩散法
gas diffusion ring 气体扩散环
gas diode 充气二极管
gas-diode switch 充气二极管开关
gas direct combustion equipment 气体直接燃烧器
gas direct oxidation equipment 气体直接氧化器
gas-discharge 气体放电
gas-discharge analysis 气体放电分析
gas-discharge cathode 气体放电阴极
gas-discharge cell 气体放电盒;气体放电电池
gas-discharge characteristic curve 气体放电特性曲线
gas-discharge colo(u)r method 气体放电色彩检验法
gas-discharge counter 气体放电(式)计数管
gas-discharge current 气体放电电流
gas-discharge detector 气体放电检测器
gas-discharge device 气体放电设备;气体放电器(件)
gas-discharge display 气体放电显示
gas-discharge display panel 气体放电显示板
gas-discharge duplexer 气体放电天线双工器
gas-discharge etching 气体放电蚀刻
gas-discharge ga(u)ge 气体放电真空计
gas-discharge ion source 气体放电离子源
gas-discharge lamp 气体放电管;气体放电灯;放电灯
gas-discharge laser detector 气体放电激光检测器
gas-discharge noise 气体放电噪声
gas-discharge noise generator 气体放电噪声发生器
gas-discharge noise source 气体放电噪声源
gas-discharge nozzle 排气喷嘴
gas-discharge optic(al) maser 气体光学脉泽;气体放电光学微波激射
gas-discharger 充气放电器
gas-discharge relay 电离继电器
gas-discharge source 气体放电光源
gas-discharge strobe light 喷气指示灯
gas-discharge theory 气体放电理论
gas-discharge tube 气体放电管
gas-discharge tube noise generator 气体放电管噪声发生器
gas-discharge wallboard display 气体放电壁显示板
gas-discharge zone 气体放电区
gas disperser 气体弥散器;气体分散器
gas dispersion 气体扩散
gas dispersoid 气态分散体;气溶胶
gas displacement 气体置换
gas display 气体显示
gas disposer 垃圾焚化炉
gas dissolving 气体溶解
gas distillate 凝析油
gas-distributing channel 分气道
gas-distributing pipe 分气管
gas-distributing system 煤气供应方法;气体分配系统
gas distribution automatic control 气体分配自动控制
gas distribution installation 气体分配装置
gas distribution pipe line 煤气管道
gas distribution piping 煤气配气管网
gas distribution system 燃气管网系统;配气系统
gas distribution system capacity 气体分配系统流量
gas distributor 气体分配器;气体分布器
gas dome 消化池的圆顶盖;储[贮]气室;沼气室;气顶;储气筒;储气盖(用于污泥消化池)
gas-doped getter 充气消气剂
gas doping 气体掺杂
gas doping technique 气相掺杂技术;气体掺杂技术
gas dosing 气体计量
gas dosing leak 气体计量泄漏
gas downtake 下气管;下气道
gas drain 瓦斯排泄道
gas drilling 天然气钻进
gas drilling rig 气钻进机
gas drip 煤气管道凝结水
gas drippings 汽油漏油;汽油滴

gas drive 气驱
gas-driven 汽油驱动的
gas-driven generator 煤气发电机;气汽发电机;气动发电机
gas-driven hammer 煤气锤
gas-driven reservoir 气驱油藏
gas-driven roller 汽油驱动的压路机
gas-driven truck 汽油驱动载货车
gas drum 汽油桶
gas dryer 气体干燥剂
gas dry filter trap 气体干燥滤阱
gas drying 气体干燥
gas drying agent 气体干燥剂
gas drying apparatus 气体干燥器
gas drying bottle 气体干燥瓶
gas drying plant 气体干燥设备;气体干燥厂
gas drying tower 气体干燥塔
gas duct 烟道;气道;气体导管
gas dump 汽油堆放;汽油仓库
gas-dust cloud 气体尘埃云;气尘云
gas-dust complex 气尘复合体
gas-dynamic(al) 气动的
gas dynamic(al) bearing 气体动力轴承
gas dynamic(al) behavio(u)r 气体动力特性
gas dynamic(al) facility 气体动力设备
gas dynamic(al) flow 气体动态流动
gas dynamic(al) mixing laser 气动混合激光器
gas dynamic(al) mode 气动模
gas dynamics 气体动力学
gas eddy 气旋;气涡
gas efficiency 出气率
gaseity 气态
gas ejector 气体喷射器
gas electric(al) automobile 气电自动车;气电动车
gas electric(al) car 煤气机电动车
gas electric(al) drive 天然气电力驱动;煤气电力驱动;气体发动机发电机驱动
gas electric(al) rig 天然气发电驱动钻井装置
gas electric(al) submarine torch 水下气电割炬
gas electric(al) type indicator 气电式示功器
gas electric(al) welding 气电联合焊接
gas electrode 气体电极
gas electron diffraction 气体电子衍射(法)
gaselier 具有多盏煤气灯头的饰灯;煤气吊灯
gas embolism 气泡栓塞
gas emission 瓦斯散发;气体发射;瓦斯泄出;瓦斯析出
gas enclosure 夹气
gas end 气体喷出口;煤气口端墙
gas engine 燃气机;沼气发电机;内燃机;煤油机;煤气内燃机;煤气(发动)机;汽油发动机
gas engined ship 燃气发动机船
gas engineering 煤气工程
gas engineering and management 天然气工程与管理
gas engine inlet valve 煤气机进气阀
gas engine oil 气体发动机润滑油
gas engines for digester gas 消化池发动机
gas entrapment 夹气
gas envelope 气体包层
gaseous 气体的;气态的
gaseous absorbent 气态吸附
gaseous absorptivity 气体吸收性
gaseous adsorption 气态吸收;气态吸附
gaseous aggregate 气体集料(指加气混凝土中的气泡);气体骨料(指加气混凝土中的气泡)
gaseous ammonia 气态氨
gaseous anomaly 气体异常
gaseous blast 喷气流
gaseous bleaching 气态漂白
gaseous buffering techniques 气体缓冲技术
gaseous burst 气爆搅动
gaseous carbon 气态碳
gaseous cement 气体掺碳剂
gaseous chlorination 气态氯化处理
gaseous chlorine 气态氯
gaseous combustion 气体燃烧
gaseous commodities 气体矿产
gaseous conductance 气体传导
gaseous conduction 气体导电
gaseous conduction analyzer 气体导电分析器
gaseous conduction rectifier 充气管整流器;气体传导整流器

gaseous conductor 气体导体;气体导电体;导电气体
gaseous contaminant 气体性污染物;气态污染物
gaseous coolant 气体冷却剂;气态冷却剂
gaseous core 气芯
gaseous corrosion 气相腐蚀
gaseous counter 气体计数管
gaseous cushion kiln 气垫窑
gaseous detonation 气体爆炸
gaseous development 气体显影
gaseous dielectric 气体电介质
gaseous diffusion 气体扩散
gaseous diffusion cascade 气体扩散级联
gaseous diffusion cell 气体扩散槽
gaseous diffusion method 气体扩散法
gaseous diffusion plant 气体扩散设备
gaseous diffusion process 气体扩散法
gaseous diffusion separator 气体扩散分离器
gaseous diffusion source 气体扩散源
gaseous-discharge 气体放电
gaseous-discharge lamp 气体放电管;气体放电灯
gaseous-discharge tube 气体放电管
gaseous dispersion 气态分散
gaseous dispersion halo 气体色散晕
gaseous dissociation 气体离解
gaseous effluent 气体排放物;气体流出物
gaseous effluent cooler 废气冷却器
gaseous electronic detector 气态电子检测器
gaseous emissivity 气体辐射系数
gaseous envelope 气膜;气袋
gaseous exchange 气体交换
gaseous film 气态膜
gaseous film cooling 气膜冷却
gaseous fluid 气态流体
gaseous fluidization 气态流化作用
gaseous fuel 气体燃料
gaseous fuel automobile 气体燃料汽车
gaseous fuel calorific value 气体燃料的发热值
gaseous fuel commodity 气体燃料矿产
gaseous fuel dew point 气体燃料的露点
gaseous fuel specific gravity 气体燃料的比重
gaseous fuel water content 气体燃料的水分含量
gaseous guide 充气波导管
gaseous heating fuel 气体供热燃料
gaseous hydrocarbon 气态烃类
gaseous impurity 气态杂质
gaseous inclusion 夹附气体;气体夹杂物
gaseous insulant 气态绝缘料
gaseous insulation 气体绝缘
gaseous insulator 气体绝缘体
gaseous interchange 气体交换
gaseous ion 气体离子
gaseous ionization counter 气体电离计数器
gaseous jet 燃气喷流
gaseous laser 气体激光器
gaseous law of volumes 气体体积定律
gaseous losses of nitrogen 氮的气体损失
gaseous material 气态材料
gaseous medium 气体媒质;气态介质
gaseous membrane 气态膜
gaseous memory cell 气体存储元件
gaseous metal 气态金属
gaseous mixture 气体混合物;气态混合物
gaseousness 气态
gaseous oil 含溶解气石油
gaseous ore 气态矿
gaseous oxygen 气态氧
gaseous phase 气相
gaseous phase of soil 土壤气相
gaseous phase radiochromatograph 气相辐射色谱仪
gaseous plasma 气态等离子体
gaseous pollutant 气体(性)污染物;气态污染物
gaseous pollution 气体污染
gaseous polymerization 气相聚合法
gaseous product 气体产物
gaseous propellant engine 气体燃料发动机
gaseous protector tube 充气保险管
gaseous radiation 气体辐射
gaseous radiation counter 气体辐射计
gaseous radiation waste 气体辐射废物;气态辐射废物
gaseous radiation wastewater 气体辐射废水
gaseous rectifier 充气管整流器

gaseous regulator 气体调节器
gaseous scintillation counter 气体闪烁计数器
gaseous shield 瓦斯保护装置
gaseous solution 气溶体
gaseous spectrum 气体光谱
gaseous spoilage 产气腐败
gaseous state 气态
gaseous steam 气态蒸汽
gaseous sterilant 气体灭菌剂
gaseous system 气相系统
gaseous train 气体流星余迹
gaseous transfer 气体搬运
gaseous transfer differentiation 气化分异作用
gaseous tube 充气管
gaseous volcano 气体火山
gaseous waste 煤气废水;气态废物;废气
gas equation 气体方程
gas equation of state 气体状态方程;气态方程
gas eruption 气体喷发
gas escape 漏气;煤气漏失;燃气漏泄;气体逸出
gas escape tube 排气管
gas escape valve 煤气放气阀;放气阀(门)
gas etching 汽相刻蚀;气蚀
gas evolution 气体透出;气体放出
gas evolution analysis 气体放出分析
gas evolution burette 气体发生量管
gas evolution method 气体发生法
gas examination 气体查验
gas exchange 气体交换;气体电离
gas exchange action 气体交换作用
gas exchange area 气体交换面积
gas exchange loss 气体交换损失
gas exchange quotient 气体交换系数
gas exhauster 排气器;排气机
gas exhaust manifold 排气歧管
gas exhaust piping 排气管路
gas-expanded rubber 闭孔泡沫橡胶
gas expander 气体膨胀机
gas expansion 气体膨胀
gas expansion hole 气胀孔
gas expansion thermometer 气体膨胀温度计
gas expansion turbine 气体膨胀式涡轮
gas expeller 排气机
gas expelling effect 排气效应
gas exploder 气隙雷管;气枪;气爆震源;气爆引信;气爆雷管
gas explosion 煤气爆炸;气爆;瓦斯爆炸
gas explosion fire 煤气爆炸火灾
gas explosion tube 气体爆燃管
gas extraction 气体抽取;天然气的开采
gas extractor 抽气机
gas factor 油气比;气体常数
gas fading 烟熏褪色;气体褪色
gas family 石油气族
gas fat coal 气肥煤
gas-fed fuel cell 气体燃料电池
gas feeder 进气格栅
gas feed rate 气流量
gas field 油田气;气田;天然气田;天然气产区
gas field exploitation 天然气田开发
gas-filled 充气的;充有气体的
gas-filled bulb 充气灯泡
gas-filled cable 充气电缆
gas-filled capacitor 充气电容器
gas-filled counter 充气计数管
gas-filled detector 充气检测器
gas-filled diode 充气二极管
gas-filled dual coil lamp bulb 充气双绞丝灯泡
gas-filled electron tube 充气电子管
gas-filled gap 充气放电器
gas-filled hot cathode rectifier 充气热阴极式整流器
gas-filled lamp 充气灯
gas-filled lamp bulb 充气灯泡
gas-filled laser tube 充气激光管
gas filled membrane adsorption 充气膜吸附
gas-filled photocell 充气光电管
gas-filled photoemissive cell 充气光电发射元件
gas-filled porosity 充气孔隙度
gas-filled radiation counter 充气式辐射计数器
gas-filled rectifier tube 充气整流管
gas-filled relay 充气继电器
gas-filled stabilizer 充气稳压管;充气管稳定器
gas-filled thermometer 气体压力温度计;充气温度计

gas-filled thyratron 充气闸流管
gas-filled triode 闸流管;充气三极管
gas-filled tube 充气管
gas-filled tube arrester 充气避雷器
gas-filled tube rectifier 充气整流管
gas-filled tungsten filament lamp 充气钨丝灯
gas-filled type explosionproof machine 充气防爆式电机
gas-filled valve 充气管
gas-filled X-ray tube 充气X射线管
gas-filling 充气
gas-filling equipment 充气设备
gas filling maintaining optic(al) cable 充气维护电缆
gas filling of spaces 气充空间
gas filling station 汽车加油站;充气站
gas film 气膜
gas-film coefficient 气膜系数
gas-film control(ling) 气膜控制
gas film mass transfer coefficient 气膜传质系数
gas film resistance 气膜阻力
gas filter 滤(煤)气器;煤气滤器;汽油滤(清)器;气体滤清器
gas filter drain cock 煤气过滤器放水旋塞;煤气过滤器
gas filter outlet pipe 煤气过滤器输出管
gas filter shell 煤气过滤器壳
gas filter tube 气体过滤管
gas filtration 煤气过滤;气体过滤
gas fire 煤气取暖器;煤气火(焰);煤气采暖炉
gas-fired 煤气燃烧;用气体燃烧的
gas-fired air furnace 煤气反射炉
gas-fired appliance 煤气用具
gas-fired boiler 燃气锅炉;煤气锅炉;烧气锅炉
gas-fired boiler flue 燃气锅炉烟道
gas-fired burner 煤气烧嘴;煤气燃烧器
gas-fired calcining 煤气煅烧窑
gas-fired cement kiln 煤气燃烧水泥窑
gas-fired convector 燃烧煤气的对流加热器
gas-fired crucible furnace 煤气坩埚炉
gas-fired cupola 煤气化铁炉;煤气冲天炉
gas-fired equipment 燃具;煤气设备;工业煤气设备
gas-fired floor furnace 室内用煤气炉
gas-fired furnace 煤气(加热)炉;燃气窑
gas-fired heating 煤气供热;煤气采暖
gas-fired heating boiler 燃气采暖锅炉;煤气采暖锅炉
gas-fired heating installation 煤气供热装置
gas-fired heating unit 煤气加热设备
gas-fired hot water heater 煤气热水供应装置;煤气热水采暖器
gas-fired hot water system 煤气热水系统
gas-fired kiln 煤气窑;燃气窑
gas-fired plant 煤气采暖装置
gas-fired power generation 沼气发电
gas-fired range 煤气炉灶
gas-fired radiant tubes 燃气辐射管
gas-fired recess 煤气炉壁龛
gas-fired reverberatory furnace 煤气渗碳炉
gas-fired space heating 燃烧煤气的室内采暖
gas-fired stove 燃气炉;燃气的炉子;煤气炉
gas-fired thermal plant 燃气热电站;燃气热电厂
gas-fired turbofurnace unit 燃气旋风炉
gas-fired unit heater 燃气采暖设备;煤气采暖设备
gas-fired warm air heating 煤气热风供热;煤气热风采暖
gas-fired warming 煤气采暖
gas-fired washing machine 煤气洗涤机
gas-fired water heater 燃气热水器;煤气热水器;燃烧沼气的热水器
gas-fired water heating appliance 煤气热水供热装置
gas-fired water system 煤气热水系统
gas-fire extinguishing 气体灭火
gas-fire radiant 气体燃烧辐射器
gas firing 燃气供热(法)
gas fitter 煤气(装修)工;装修煤气管的工人
gas fitting 煤气装置;煤气装配;煤气装备;煤气设备;煤气配件
gas fitting work 煤气安装工程
gas fixture 煤气设备;煤气(灯)装置
gas-flame brazing 气焰钎焊
gas flame coal 易燃气煤;高级烟煤;气焰煤

gas flame singeing machine 煤气烧毛机
gas flame thermopile 气焰热电偶
gas flame welding 气烟焊接
gas flexible conduit 煤气软管
gas floor furnace 楼面煤气炉
gas flow 气体流动；气流
gas flow adjusting flange 气流调节板
gas flow calculation 气流计算
gas flow controller 气流控制器
gas flow control system 气流调节系统
gas flow counter 流动气体计数管；气体流量计
gas flow counter tube 流气型计数管
gas flow equation 气体流量方程
gas flow indicator 气体流量指示器
gas flow meter 气流计；气体流量测定仪；气体流量计
gas flow path 气体流路
gas flow radiation counter 流气型计数管
gas flow rate 气体流速
gas flow temperature 气流温度
gas flue 煤气管道
gas fluid 气态流体
gas fluidized bed 气体流化床
gas flush syringe 气加注射器
gas flux 气体焊剂；气焊焊剂
gas foamer 泡沫剂
gas-foaming admixture 泡沫剂；发泡剂
gas-foaming agent 加气剂（混凝土）；发泡剂（混凝土）
gas-focused oscillograph 气体聚焦示波器
gas focusing 气体聚集
gas force closed type 气闭式
gas forge 煤气锻炉
gas formation 沼气形成
gas former 发气剂
gas-forming admixture 加气剂；发泡剂；泡沫剂
gas-forming agent 造气剂；加气剂；发泡剂
gas-forming chemical 煤气形成的化学物质
gas-forming constituent 造气剂
gas-forming curve 发气曲线
gas-forming property 发气性
gas-forming reaction 产气作用
gas-forming styrol concrete 泡沫苯乙烯混凝土
gas for motor fuel 气态发动机燃料；动力煤气
gas for works' use 厂内自用煤气
gas fractionation unit 气体分馏装置
gas fractionator 气体分馏塔
gas-free 不含气的
gas free certificate 气体消除证书；无可燃气体证书
gas-free certificate detector 除气检验器
gas freeing 清舱内油气；气体消除
gas freezing 除气
gas from coal and coal measure 煤成气
gas from explosion 爆破气体
gas fuel 气体燃料
gas-fueled vehicle 燃气车辆；天然气汽车
gas fuel nozzle 气体燃料喷嘴
gas furnace 煤气炉；煤气发生炉
gas furnace recess 煤气炉壁龛
gas-gas interaction 气体—气体相互作用
gas gasoline 天然气液化汽油
gas gathering dome 集气室
gas gathering system 集气系统
gas ga(u)ge 煤气压力计；气体压力计；气量计
gas ga(u)ge switch 汽油表开关
gas-generating agent 加气剂
gas-generating bottle 气体发生瓶
gas-generating set 煤气发生组；煤气发生器
gas generation 气体发生
gas generation absorption method 气体发生吸收法
gas generator 煤气发生器；煤气发生炉；燃气发生器；气体发生器；气化器
gas generator ash 煤气发生炉的炉渣；煤气发生炉的煤灰
gas generator cinder 煤气发生炉的炉渣；煤气发生炉的煤灰
gas generator cycle 煤气发生炉循环
gas generator unit 煤气发生炉机组
gas geyser 气体间歇喷泉；煤气热水器；气体间歇泉
gas gouging 气隔开槽；气割槽
gas governor 煤气调压器；气体调节器
gas gradient 瓦斯梯度
gas-graphite reactor 石墨气冷反应堆

gas gravimeter 气体重力仪；气体重差计
gas gravity differentiation 气体重力分异
gas grid 气体供应阀；煤气供应阀；供气管网
gas grid line 气体供应管路
gas-guide tube 导气管
gas gun 气枪
gas guzzler 大食车（耗油量大）
gash 齿隙；齿缝；深痕
gas halo diagram 气晕图
gas halo method 气晕法
gas hand hammer(rock) drill （手动的）汽油冲击式凿岩机
gas handling system 进气系统；气体处理系统
gash angle 齿隙角；（铣刀的）齿缝角
gas-hardening treatment 气体增硬处理
gas head 气体压头
gas header 集气管；煤气取暖炉
gas headlight 煤气头灯
gas hearth system 气垫淬火系统
gas heat capacity 气体热容量
gas-heated 煤气加热的
gas-heated air 煤气加热的空气
gas-heated brooder 煤气加热式育雏器
gas-heated panel 煤气加热的板
gas-heated steamer 煤气加热式蒸煮器
gas heater 煤气（取暖）炉；煤气炉；煤气加热器；煤气供暖机（组）
gas heating 煤气加热；煤气供暖；燃气供暖（法）
gas heating boiler 煤气加热锅炉
gas heating device 煤气加热装置
gas heating installation 煤气加热装置
gas heating unit 煤气加热装置
gashed rotor 整锻转子
gas helmet 防毒面具
gash fracture 张裂缝；张口破裂
gashing cutter 齿槽铣刀
gashing tool 切槽刀
gas hoist 汽油提升机；汽油卷扬机
gas holder 储[贮]气器；煤气储[贮存]罐；煤气库；煤气柜；煤气（储）罐；储气筒；储气柜；储罐；储[贮]气柜
gas-holder bell 气柜的钟罩；储气器的钟罩
gas-holder foundation 气柜基础
gas-holder grease 气柜润滑脂
gas-holder lift 储气罐塔节
gas-holder oil 气柜油
gas-holder operation 气柜操作
gas-holder piston 储气罐活塞
gas-holder station 储配站；气体储存站；储罐站
gas-holder tank 气柜
gas hold-up 气体滞留体积；气体滞留（量）
gas hole 气眼；吹气孔
gas hood 气罩
gas horizon 含气地层
gas hose 煤气软管；气管
gas hot water system 煤气热水系统
gas house 煤气房；煤气厂；煤气站；煤气工厂
gashouse coal tar 煤气厂（煤）焦油；煤气焦油（沥青）
gashouse coal-tar pitch 煤气厂焦油沥青
gashouse waste 煤气厂废水
gash spacing error 齿隙误差
gash vein 裂缝脉；裂缝矿脉
gas hydrate 气体水合物；气水化合物；天然气水化合物
gas hydrate process 煤气水化法；气体水合过程
gas hydraulic type spring 气液压型联动弹簧；气体液压式弹簧
gas hydride atomic absorption spectrophotometry 气相氢化物原子系数分光光度法
gasifiable 可气化
gasifiable pattern 气化模
gasificating desulfurization 气化脱硫
gasification 气化（作用）
gasification ash water 气化灰水
gasification coal 气化用煤
gasification efficiency 气化效率
gasification form 气化方式
gasification gas 气化煤气
gasification latent heat 气化潜热
gasification of coal 煤的气化
gasification of refuse 城市垃圾气化
gasification strength 气化强度
gasification system 煤气化系统

gasification technical parameter 气化工艺指标
gasification technique 气化工艺
gasification wastewater 气化废水
gasification zone 气化区；气化带
gasified pattern 气化模型；气化模
gasifier 燃气发生器；气体发生器；气化器；气化炉
gasifier nozzle diaphragm 燃气发生器；透平喷射器
gasifier section 燃气发生装置
gasifier speed 燃气发生器转速
gasifier turbine 燃气发生器涡轮机
gasify 气化
gasifying agent 气化剂
gasifying medium 气化介质
gas ignition 燃气点燃；瓦斯着火
gas ignition engine 煤气（内燃）机
gas ignitor 煤气点火器
gas illumination 煤气照明
gas impermeability 不透气性
gas impermeability test 气密性试验；气密性试验；不透气性试验
gas imperviousness 不透气性
gas impregnated cable 充气浸渍电缆；气浸电缆
gas impurity 气体杂质
gas incinerater[incinerator] 气体焚烧器
gas inclusion 气体包裹物
gas in coal 煤层瓦斯
gas industry 煤气工业
gas inflation automatically operated type life-jacket 自动充气救生衣
gasing 充气
gas injection 注气；气体喷射；气体喷射；喷气
gas injection well 注气井；天然气回注井
gas inlet 气体进口；瓦斯入口
gas inlet housing 涡轮进气壳
gas inlet pipe 煤气入口管
gas in mass 气块
gas input factor 注入油气比
gas input line 进气管
gas input well 气井
gas in solution 溶解气
gas installation work 煤气管道安装工作
gas-insulated cable 气体绝缘电缆
gas-insulated switchgear 六氟化硫全封闭开关；紧凑式气体绝缘开关柜；气体绝缘开关
gas-insulated transformer 气体绝缘变压器
gas insulation 气体绝缘
gas intake 进气口
gas integral process 全气化过程
gas interceptor 截（汽）油器
gas interchange disturbance 气体交换障碍
gas interlock 气体活门
gas internal vibrator 油动插入式振捣器；内燃式振捣器
gas in water solution 天然气水溶液
gas ion 气体离子
gas ion constant 气体电离常数
gas ionization 气体离子化；气体电离
gas ionization battery 气体电离电池
gas ionization chamber 气体电离室
gas ionization counter 气体电离计数器
gas ionization detector 气体电离探测器
gas ionized display 气体电离显示
gas jet 煤气喷嘴；煤气喷头；燃气流；气体喷射；气体喷流；气体射流；气焊枪；气灯火焰
gas jet compressor 气体喷射压缩机
gas jet method 气体喷射法
gas jet propulsion 喷气推进
gas jet technique 气体喷射技术
gas jet tube 煤气喷管
gasket 热圈；捆帆绳；密封填料；密封垫圈；密封带；密封垫（衬）垫；密闭填料；填垫片；垫圈；垫片；垫衬；衬圈；衬片；衬垫
gasket cement 密封胶；衬片黏胶
gasket coating 密封涂层
gasketed holder 垫压圈
gasketed joining system 加密封垫接头法
gasketed joint 填实接缝；垫圈接头；有衬垫接合
gasketed pipe 带密封垫圈的管子
gasketed PVC pipe 加密封垫聚氯乙烯管
gasket factor 密封系数；垫片系数
gasket glazing 用密封垫镶玻璃；密封垫镶嵌玻璃；填片密封镶嵌
gasketing material 衬垫材料
gasket inserting tool 垫圈装入器

gasket iron 锤垫扁铁;填垫扁铁
gasket joint 密封垫圈接缝
gasket leakage 垫片渗漏
gasket load 垫片荷载
gasket material 密闭填料;填料;垫片材料;垫衬材料
gasket materials and contact facings 垫片材料和接触面
gasket mounted 板式连接的;底板安装
gasket mounting 填密片板式连接;板式连接
gasket packing 板式填料
gasket paper 纸垫;纸衬;垫片纸;衬纸
gasket piece-cutting machine 衬里片割机
gasket piston 带衬套活塞
gasket resiliency 垫片回弹率
gasket ring 密封片;密封圈;垫片环;垫环
gasket seal 密封垫圈;密封垫片;垫片密封
gasket sealed 衬垫密封接头
gasket seal gasket 衬垫式密封垫
gasket seat 垫圈座
gasket splicer 密封圈接头机
gasket spring 金属垫圈弹簧
gasket system 加密封垫法
gasketting tape 结构缝隙水密封条
gasket-type seal 静密封
gasket unit 垫片组
gas kick 轻微井喷
gaskin 密闭填料;气隙料;衬垫
gas kinetic collision 气体动力学碰撞
gas-kinetic diameter 气体动力学直径
gas-kinetic pressure 气体动压强
gas knock 气体爆击
gas lamp 煤气吊灯;气灯
gas laser 气体激光器
gas laser altimeter 气体激光高度计
gas laser interferometer 气体激光干涉仪
gas laser system 气体激光系统
gas law 气体定律
gas leakage 漏气;气体漏泄
gas leakage detector 漏气检查器;漏气检测仪
gas leak age indicator 漏气指示器
gas leak detector 漏风检测器
gas length 气隙长度
gas lens 气体透镜;气体浓度透镜
gasless 无气体的
gas lever 气塞杆
gas lift 气体升液器;气流提升;气力起重机;气体浮力;气举;气动(法);气升(器)
gas lift conveyer 气升运输机
gas lift flow 气升流动
gas lift flow area 气提升流区
gas lift flowing well 引气促流井;压缩机抽水井;气举出水井
gas lift gas 气举天然气
gas lift hook-up 气升挂钩;气动挂钩
gas lift intermitter 气举间隙诱导流筒;气举采油间歇调节器
gas lift pressure 气举压力
gas lift pump 气动泵;气体提升泵;气提升泵
gas lift recovery 气升回收;气举萃取;气动采取
gas lift unit 气体升液装置;气升装置
gas light 煤气灯光;煤气(吊)灯
gas lighter 煤气引燃器;煤气点火器
gas lighting 煤气照明
gas light oil 煤干馏轻油
gas light paper 感光纸
gas lime 煤气石灰
gas lime concrete 加气石灰混凝土
gas lime concrete block 加气石灰混凝土砖;加气石灰混凝土砌块
gas lime concrete tile 加气石灰混凝土砖
gas line 燃气管(道);气体管线
gas line compressor 天然气管线压气机
gas line dehydrator 气体管道脱水器
gas line material 煤气管道材料
gas line packing 管道储气
gas line pipe 煤气管
gas linking valve 燃气连接阀
gas liquefaction 气体液化
gas-liquid area ratio 气液面积比
gas-liquid chromatograph 气液色谱仪;气液层析
gas-liquid chromatograph-mass spectrometer 气液相色谱仪—质谱仪联用
gas-liquid chromatography 气液相色谱法;气液相层分析;气液相层析;气液色谱(法);气液色层分离法
gas-liquid column 气液柱
gas-liquid elastomechanics 气液弹性力学
gas-liquid equilibrium 气液平衡
gas-liquid exchange process 气液交换法
gas-liquid exchange process in wastewater treatment 废水处理的气液交换法
gas-liquid exchange treatment of wastewater 废水气液交换处理法
gas-liquid interaction 气体—液体相互作用
gas-liquid interface 气液界面
gas-liquid interphase reactor 气液相间反应器
gas-liquid mixture 气液混合液
gas-liquid partition 气液分配
gas-liquid partition chromatography 气液分配色谱(法);气液分配层析
gas-liquid partition coefficient 气液分配系数
gas-liquid phase 气体—液体相
gas-liquid reaction 气液反应
gas-liquid separation 气液分离
gas-liquid separator 气液分离器
gas-liquid-solid chromatography 气液固色谱(法);气液固层析
gas-liquid spray contactor 气液喷雾接触器
gas-liquid system 气液法
gas-liquid volume ratio 气液体积比
gas liquor 煤气液;煤气水溶液;煤气冷凝液
gas liquor tank 煤气水溶液罐;煤气水罐
gas load 燃气需用量;燃气负荷
gas loaded accumulator 充气蓄能器
gas loaded electrode 充气电极
gas loading 充气
gas locator 气体检测器;煤气探测器
gas lock 气体闸;煤气阀;气闸;气封
gas lock system 气体密封装置;锁气装置
gas log (煤气暖炉的)圆材燃烧
gas log all hydrocarbon 气测全烃测井
gas logger 气测井仪
gas logging 气体录井;气(体)测井
gas log heavy hydrocarbon 气测重烃测井
gas log interval 气测录井间距
gas log light hydrocarbon 气测轻烃测井
gas loop 气体波腹
gas-lubricated 气体润滑的
gas-lubricated bearing 气体润滑轴承
gas-lubricated rotor 气体润滑转子
gas lubrication 气体润滑
gas luminaire(fixture) 煤气照明装置;煤气灯
gas maar 喷气口
gas magnification 电离放大
gas main 燃气输送管线;输气管线煤气总管;煤气干管
gas maker 煤气工
gas making 造气;气体发生
gas making log book 煤气制造记录簿
gas making period 气化周期
gas making plant 煤气(工)厂
gas making process 煤气制造法
gas making retort 制气甑
gas manifold 气体集合管;瓦斯集合管
gas manometer 气体压力计;气体压差表
gas mantle 煤气灯(网)罩;汽灯罩
gas manufacture 煤气生产
gas manufacturing process 煤气制取过程
gas mark 气体污浊;气浊
gas maser 气态脉塞
gas mask 一氧化碳自救器;防毒面具
gas measuring apparatus 气体流量计
gas measuring flowmeter 气体流量计
gas metal-arc welding 熔化极气体弧焊;气体保护金属极电弧焊;气保护金属极电弧焊
gas metal(l)izator 金属喷涂器
gas meter 燃气表;量气计;煤气(量)计;煤气计量器;煤气(罐)表;气体流量计;气量计;气表
gas meter house 煤气表房
gas meter inclosure 煤气表罩壳
gas metering station 气体流量测定站
gas meter leather 气表革
gasmetry 气体测量
gas migration 天然气运移
gas mill wastewater 煤气厂废水
gas mixed high rate sludge digestion tank 气体混合高负荷污泥消化池
gas mixing blower 燃气混合风机
gas mixing device 气体混合设备
gas mixing tank 煤气混合箱
gas mixture 气体混合物
gas modulator 气体调节器
gas molecule 气体分子
gas-molecule-absorption spectrophotometer 气体分子吸收分光光度计
gas monitor 气体监测器
gas monitoring 气体监测;瓦斯监测
gas motor 沼气发电机;煤气机;煤气发动机
gas multiplication 气体倍增
gas multiplication factor 气体倍增系数
gas naphtha 气体石脑油
gas network 燃气管网系统
gas nipple 气体喷嘴
gas nitriding 气体渗氮
gas nozzle 焊炬喷嘴;气体喷嘴;排气孔
gas nucleus 气核
gas occlusion 气体吸住;气体吸溜;气体停止;气体隔断
gasoclastic sediment 喷气碎屑沉积物
gas odo(u)rizer 气体臭味鉴定器
gas of coal mine 矿井瓦斯
gas of combustion 废气
gas offtake 烟囱道;出气口;出气道
gas offtake borehole 煤气输出钻孔
gas offtake target 采气目的层
gas of kinds 煤气的种类
gasogene 木炭燃气;汽水制造机;小型煤气发生器
gasogenic anomaly 气成异常
gas oil 轻柴油;汽油;气体油;瓦斯油;粗柴油
gas-oil contact 油气接触(面)
gas-oil contact line 气油界面
gas-oil displacement 驱油气
gas-oil interface 气油分界面
gas-oil mixture 气态和液态石油产品混合物
gas-oil permeability 气油渗透率
gas-oil pool 气油藏
gas-oil pressure ga(u)ge 气体油料压力计
gas-oil pump 气体油料泵
gas-oil ratio 气油比;油—气比(率);生产油气比
gas-oil separator 油气分离器
gas-oil solenoid valve 气体油料螺管阀
gas-oil surface 气—油接触面;油气界面
gasol 液化石油气
gasolene 挥发油
gasolene compartment 汽油箱
gasolene tank 汽油柜
gasolene tanker 汽油运油船
gasolier 煤气吊灯
gas oline 汽油
gasoline additive 汽油添加剂
gasoline adjustment valve 汽油供给量调节阀
gasoline alarm 低油位警报器
gasoline anti-freeze mixture 汽油防冻混合物
gasoline anti-icing additive 汽油防冻添加剂
gasoline automobile 汽油(机)汽车
gasoline barge 汽油驳船
gasoline blast burner 汽油喷灯
gasoline blending stock 汽油调合料
gasoline blow pipe torch 汽油吹管
gasoline bowser 汽油加油车
gasoline buoyancy system 汽油浮力系统
gasoline burner 汽油灯
gasoline bus 汽油机公共汽车
gasoline can 汽油罐
gasoline cap 汽油箱盖
gasoline capacity 汽油容量
gasoline car 汽油车
gasoline carrier 汽油槽车
gasoline carrying 汽油输送
gasoline carrying vehicle 汽油运输车
gasoline chamber 汽油室
gasoline cistern 汽油罐
gasoline coalescer 汽油的凝结剂
gasoline cock 汽油旋塞
gasoline concrete mixer 汽油混凝土搅拌机;汽油混凝土拌和机
gasoline condenser 汽油冷凝器
gasoline consumption 汽油消耗(量)
gasoline container 汽油罐
gasoline content ga(u)ge 汽油表
gasoline content ga(u)ge with float 浮子式汽油表

gasoline corrosion cup 汽油试蚀杯
gasoline crane 汽油起重机
gasoline depth ga(u)ge 汽油油位表;汽油深度计
gasoline deterioration 汽油变质
gasoline dispenser 加油机;汽油加油车
gasoline dispensing equipment 汽油分配装置
gasoline dispensing facility 汽油分送设备
gasoline dispersant 汽油分散剂
gasoline distributing point 汽油分配站
gasoline doping 汽油掺添加剂
gasoline drippings 汽油滴
gasoline-driven 汽油驱动
gasoline-driven roller 汽油驱动压路机
gasoline-driven truck 汽油驱动载货汽车
gasoline drum 汽油桶
gasoline dump 汽油堆放;汽油堆场
gasoline economizer 节油器
gasoline-electric(al) 汽电动的
gasoline-electric(al) bus 汽油电动公共汽车
gasoline-electric(al) car 汽油电动车辆
gasoline electric(al) tractor 汽油电力牵引车
gasoline-electric(al) vehicle 汽油电力车;汽油电动车辆
gasoline engine 汽油内燃机;汽油(发动)机
gasoline-engine boat 汽油机船
gasoline engine driven 汽油机驱动
gasoline-engine driven machine 汽油动力机
gasoline engine exhaust 汽油机排气
gasoline feed pipe 汽油供给管
gasoline feed pump 汽油泵
gasoline filler neck 汽油箱加油口管
gasoline filling station 加油站;汽油加油站
gasoline filter 汽油过滤器;汽油滤(清)器
gasoline filter bowl 汽油滤杯
gasoline ga(u)ge 汽油液面指示器;汽油表
gasoline ga(u)ge dash unit 汽油表盘指示器
gasoline ga(u)ge dial bracket 汽油表盘支架
gasoline ga(u)ge line 汽油油量表传感线
gasoline ga(u)ge switch 汽油油量表开关
gasoline ga(u)ge take unit 汽油油位表传感器
gasoline ga(u)ge with float 汽油浮标表
gasoline gel 固体汽油
gasoline generator 汽油发电机
gasoline generator set 汽油发电机组
gasoline gum 汽油胶
gasoline gumming test cup 汽油胶质测定杯
gasoline heater 汽油加热器
gasoline hoist 汽油驱动提升机;汽油(驱动)卷扬机
gasoline hungry 汽油不足
gasoline injection 汽油喷射
gasoline-injection engine 汽油喷射式发动机
gasoline-injection pump 汽油喷射泵
gasoline interceptor 汽油滤(清)器
gasoline intermediate fraction 汽油的中间馏分
gasoline internal vibrator 汽油(驱动)插入式振捣器
gasoline knocking 汽油抗爆性;汽油爆震性
gasoline leak 汽油滴漏
gasoline level 汽油油位
gasoline level ga(u)ge 汽油油位表
gasoline level indicator 汽油油面指示器
gasoline(lift)pump 汽油泵
gasoline locomotive 矿用内燃机车;汽油机车
gasoline measure 汽油加注量桶
gasoline mercaptan absorber 汽油中硫醇的吸收塔
gasoline meter 汽油计量表
gasoline meter pump 汽油计量泵
gasoline mileage 汽油里程
gasoline mileage tester 汽油消耗量测定装置
gasoline motor 汽油发动机
gasoline motor car 汽油机车
gasoline naphtha 气体石脑油
gasoline octane number 汽油辛烷值
gasoline octane rating 汽油辛烷值
gasoline officer 汽油供应员
gasoline oil 汽油
gasoline operated(pile driving)hammer 汽油驱动的打桩锤
gasoline performance 汽油特性
gasoline pick-up fraction 汽油的加速馏分
gasoline pipe 汽油管
gasoline(pipe)line 汽油管路
gasoline plant 汽油压缩机
gasoline poisoning 汽油中毒
gasoline pool 汽油调合组分总和
gasoline-powered lift truck 汽油驱动的自动装卸车
gasoline powered truck 汽油机车;汽油发动机载货汽车
gasoline-powered winch 汽油驱动的卷扬机
gasoline precipitation test 汽油沉淀试验
gasoline pressure ga(u)ge 汽油压力计
gasoline pressure hand pump 汽油压力手泵
gasoline-proof 汽油振荡器;耐汽油的;防汽油
gasoline-proof grease 抗汽油润滑剂
gasoline pump 汽油泵;加油泵
gasoline pump assembly 汽油泵总成
gasoline pump case 汽油泵壳
gasoline pump diaphragm protecting washer 汽油泵膜片保护垫圈
gasoline pump glass-bowl U-clamp 汽油泵油杯U形钢丝夹
gasoline pump settling bowl 汽油泵沉淀杯
gasoline pump settling glass bowl 汽油泵沉淀玻璃杯
gasoliner 汽艇
gasoline refining 汽油精制
gasoline refining equipment 精制汽油的设备
gasoline reservoir 汽油储油器
gasoline resistance 耐汽油性
gasoline-resistant coating 耐汽油涂料
gasoline resisting test 耐汽油性试验
gasoline rock drill 带汽油凿岩机
gasoline roller 汽油压路机
gasoline scale 汽油秤
gasoline separator 汽油离析器;汽油净化器;汽油分离器;分汽油器;汽油滤(清)器
gasoline service pump 汽车加油站
gasoline sewage 汽油污水
gasoline shortage 汽油不足
gasoline shovel 内燃式挖土机;汽油铲土机
gasoline skidder 汽油集材机
gasoline splitter 汽油分馏塔
gasoline stand 加汽油站
gasoline startability 汽油起动性质
gasoline starting 汽油机起动
gasoline starting engine 汽油驱动的起动机
gasoline station 加油站;汽油站
gasoline stocks 汽油储备
gasoline storage 汽油库
gasoline storage can 汽油储[贮]存桶
gasoline storage stability 汽油储[贮]存稳定性
gasoline storage tank 汽油储存箱;地下储[贮](汽)油罐
gasoline substitutes 汽油代用品
gasoline sulfur test 汽油中硫含量的测定
gasoline supply pump 汽油泵
gasoline sweetener 汽油脱硫设备
gasoline tank 汽油箱
gasoline tank bracket 汽油箱支架
gasoline tank cap 汽油箱盖
gasoline tank capacity 汽油箱容积
gasoline tank cap chain 汽油箱盖吊链
gasoline tank car 汽油油罐车
gasoline tank dump valve 汽油箱放泄阀
gasoline tanker 汽油运输船;汽油船
gasoline tank ga(u)ge 汽油箱油位表
gasoline tank settler 汽油箱沉油器
gasoline tank truck 运汽油车;汽油油罐车
gasoline testing outfit 汽油试验装置
gasoline tetraethyl lead test 汽油中四乙铅含量测定试验
gasoline torch 汽油喷灯
gasoline torch lamp 汽油喷灯
gasoline tractor 汽油拖拉机
gasoline trap 聚汽油室;汽油凝气阱;汽油滤(清)器;汽油分离器;汽油分离阱
gasoline valve 汽油阀(门)
gasoline vehicle 汽油车辆
gasoline vibrator 汽油振荡器;汽油振捣器
gasoline yield 汽油产率
gasoloid 气溶胶;气胶溶体
gasomagnetron 充气磁控管
gasometer 蓄[气]瓶;储[贮]气柜;量气计;煤气计(量器);煤气计;煤气柜;煤气(罐)表;气体计(数器);气体定量器;气量计;气量表
gasometer(flask) 气量瓶
gasometric 气体定量的
gasometric analysis 气量分析
gasometric determination 气量测定
gasometric method 气体定量分析法
gasometric titration 气量滴定
gasometry 气体分析;气体定量法
gas on-line plant 瓦斯管线上加工厂
gas opacifier 气体乳浊剂
gas operated device 气动操纵设备
gas operated(pile driving)hammer 汽油驱动的打桩机
gas operated smoke-house 煤气加热熏房
gas orifice 引射器;喷嘴;气眼;喷气孔
gasoscope 气体检验器
gas outburst 瓦斯爆发;气喷;气体的排出;瓦斯涌出;瓦斯喷出
gas outlet 气体出口;泄气管;气体排除口;排气口;导气管;出气口
gas outlet hausing 涡轮出气壳
gas outlet hole 气体出口孔
gas outlet point 煤气出口处
gas output 煤气输出量;出气量;气量;产气量
gas oven 煤气灶;煤气烘炉
gas packer 气体封隔器
gas packing 充气包装
gas panel 气体显示板;气体放电显示屏
Gaspar 加斯帕
gas parameter 气体参数
gas partition chromatography 气相分配色谱法;气相分配层析
gas passage 煤气通道
gas path 气路
gas pedal 气动踏板
gaspeite 菱镍矿;菱镁镍矿;麦镁镍矿
gas permeability 气体渗透率;瓦斯渗透率;透气性;天然气渗透率;透气率
gas permeable glass 透气玻璃
gas-permeable membrane 透气膜;气体渗透膜
gas permeation 气体透过
gas pervious material 透气性材料
gas phase 气相;气态
gas-phase activity 气相活度
gas-phase activity coefficient 气相活度系数
gas-phase adsorption 气相吸附
gas-phase catalytic oxidation process 气相催化氧化法烟气脱硫
gas-phase chlorination 气相氯化
gas-phase chromatograph 气相色谱
gas-phase combustion 气相燃烧
gas-phase concentration 气相组分;气相浓度
gas-phase coulometry 气相电量分析
gas-phase electron adsorption 气相电子吸附
gas-phase isotherms 气相等温线
gas-phase mass concentration 气相质量浓度
gas-phase mole fraction 气相摩尔分数
gas-phase oxidation 气相氧化
gas-phase permeation 气相渗透
gas-phase pressure drop 气相压降
gas-phase radioassay 气相放射性分析
gas-phase reaction 气相反应
gas-phase region 气相区
gas-phase sedimentation 气相沉降
gas-phase separation 气相分离
gas-phase suspension process 气相悬浮过程;气相悬浮法
gas-phase thermometric titration 气相温度滴定法;气相热量滴定
gas-phase titration 气相滴定
gas phototube 气体光电管
gas pilot 煤气引燃器;瓦斯引火烧嘴
gas pin 气针孔
gas pipe 煤气管;气体管
gas pipe connector 煤气管接头
gas pipe hose 煤气软管
gas pipe line 燃气管道;煤气管道;气导管;气管线;气管线;气体管路
gas pipe line construction rig 煤气排管设备;煤气管线施工设备
gas pipeline network 气管网
gas pipeline packed cock 煤气管道压缩式旋塞
gas pipeline standards 输气管规格
gas pipeline transportation service 气管道输送
gas pipe network 煤气管网
gas pipe tap 输气管丝锥
gas pipe thread 煤气管螺纹;瓦斯螺纹;气管螺纹
gas pipe tongs 煤气管钳
gas pipet(te) 验气球管

gas piping 煤气管道;气体管路
gas piping system 煤气管道系统
gas piston ring 活塞压环
gas pitch 煤气柏油脂
gas plant 煤气发生厂;煤气厂;气体分馏装置
gas plant waste(water) 煤气生产废水;煤气厂废水
gas plasma display 等离子体气体显示器
gas platform 气井钻台
gas plating 气相扩散渗镀
gas pliers 煤气管钳;气管钳
gas pocket 气窝;气泡;气囊;气孔;气袋
gas point 煤气引入点
gas poisoning 煤气中毒;毒气中毒
gas poker 点火器
gas polarization 气体极化
gas pollutant 气体污染物
gas pollution 气(体)污染
gas pollution monitor 空气污染度监视器
gas pool 气藏
gas pore 气泡;气孔
gas porosity 疏松度
gas port 煤气(喷出)口
gas port nose 煤气烧嘴口
gas potential 气势
gas potential gradient 气势梯度
gas-powered catapult 气动弹射器
gas-powered winch 汽油驱动卷扬机
gas power engine 燃气发动机;煤气发动机
gas power locomotive 煤气机车
gas power plant 煤气发电厂;燃气发电厂
gas power station 燃气发电站;燃气发电厂
gas precharge valve 充气阀
gas precipitation flo(a)tation 气泡析出浮选
gas pressure 燃气压力;煤气压力;气体压力
gas pressure bonding 气压黏合;气体压力连接;气压胶结
gas pressure cable 气压电缆
gas pressure compacting 气压成形
gas pressure consolidation 气压固结
gas pressure drive 气压驱动
gas pressure feed system 气压输送系统;气压供给系统
gas pressure force 瓦斯压力
gas pressure ga(u)ge 气压表(测气体)
gas pressure gravimeter 气压重力仪
gas pressure head 气体压力头
gas pressure indicator 气体压力指示器
gas pressure inlet 气体压入口
gas pressure lubrication feed 气压润滑法
gas pressure lubricator 气压润滑器
gas pressure maintenance 注气保持压力
gas pressure meter 气压表(测气体);气体压力计
gas pressure protection 气体压力保护
gas pressure regulating governor 气压调节(控制)器
gas pressure-regulating station 煤气调压站
gas pressure regulation 燃气压力调节
gas pressure regulator 燃气调节器;煤气压力调节阀;气体压力调节器;燃气调节阀;燃气压力调节器;气压调节器
gas pressure regulator valve 输气管调压阀
gas pressure relay 气体继电器
gas pressure sintering 气体压力烧结;气氛加压烧结法
gas pressure supervision alarm system 气压维护通信设备;气压监警设备
gas pressure test 气压试验
gas pressure type(self) starter 气压式起动器
gas pressure weld(ing) 气压焊(接)
gas pressure welding machine 气压焊接机
gas pressurization system 气体加压系统
gas-pressurized 加压的
gas-pressurized generator 带压气体发生器
gas-pressurized system 气压输送系统
gas-press weld(ing) 气压焊(接)
gas probe 气体分析探头
gas processing 煤气处理
gas processing plant 气体加工装置
gas-produced black 瓦斯炭黑;天然气炭黑
gas producer 气化炉;煤气发生器;煤气发生器;气体发生器
gas producer coal tar 煤气发生炉焦油
gas producer man 煤气工
gas producer retort 煤气发生炉罐

gas producer vehicle 煤气发生炉汽车
gas producing bacteria 产气细菌
gas producing plant 煤气工厂
gas producing process 煤气制取过程
gas production 产气量;采气
gas production measuring 测气产量
gas production rate of sludge 污泥产气量
gas production well 煤气发生井
gas productivity 产气率
gas productivity index 采气指标
gas-proof 气密(的);防毒气(的);不透气(的);不漏气
gas-proof curtain 气幕;气帘
gas-proof machine 气密型机器;防瓦斯式电机
gas proofness 防污气性
gas-proof sealing 气密密封
gas prospection 天然气勘探
gas protection 煤气防护;气体保护;瓦斯保护;毒气防护;防化;防毒
gas protection equipment 防毒器材
gas protective clothing 防毒气服装
gas pulsation 气流脉动
gas pumice concrete 加气浮石混凝土
gas pump 汽油泵;打气筒;鼓风机;煤气吹送机
gas pump cycle 气泵循环
gas pump meter 气泵表
gas pure material 气体纯物质
gas purge 气体清洗;气体净化
gas purger 气体排放器;放气器;不凝性气体排除器
gas purge system 气体净化系统
gas purification 气体处理;煤气净化;气体净化
gas purification equipment 气体净化装置
gas purification equipment by electron beam irradiation process 气体电子照射净化装置
gas purification tower 煤气净化塔
gas purification train 气体纯化组列
gas purifier 气体净化设备;气体净化器;除气用空气过滤器
gas purifier fan 净气器风扇
gas purifying 煤气净化;气体净化
gas purifying equipment 气体净化设备
gas purifying installation 气体净化装置
gas quenching 气体冷却淬火;气冷淬火
gas quench system 气体冷却装置
gas radiant heater 煤气辐射加热器
gas radiation 气体辐射
gas radiation counter 气体辐射计
gas radiator 煤气(辐射)供暖器;煤气放热器;燃气放热器
gas radioactivity meter 气体放射性测量计
gas range 烤箱灶;煤气灶;煤气炉
gas rank of mine 矿井瓦斯等级
gas rated pressure 燃气额定压力
gas ratio 油气比;离子电流率;气体比
gas ratio control 气体比率控制
gas ration 汽油消耗定额
gas reaction 气体反应
gas reaction stage 气体反应台
gas receiver 气柜;储气桶;储气罐
gas recirculating fan 烟气再循环风机
gas recirculation 气体循环
gas recombination 气体复合
gas recombiner 气体复合器
gas recovery 气体回收;采气
gas recovery system 气体回收系统
gas recycle process 气体循环过程
gas recycle pump 气体循环泵
gas recycling 废气循环
gas-recycling plant 气体循环装置
gas reduction building 气体压缩机房
gas reduction oxidation equipment 气体氧化净化装置
gas reduction purification equipment 气体还原净化装置
gas referee's test 校表气罐
gas referees' photometer 气体检验光度计
gas reflux 气体回流;气体反流
gas reflux apparatus 气体逆流装置;气体回流装置
gas reformer 气体转化器;煤气体转化器
gas reforming 气体转化
gas refrigerating machine 气体致冷机
gas refrigeration 煤气制冷;气体冷冻
gas refrigerator 煤气冷柜;煤气冰箱
gas regenerator 煤气蓄热室

gas regenerator chamber 煤气蓄热室
gas regenerator flue 煤气蓄热室(火)烟道
gas region 含气区
gas register 气压自记器
gas regulator 煤气稳压阀;气体调节器
gas re-heater 煤气再加热器
gas re-injection 天然气回注
gas relative permeability 天然气相对渗透率
gas relay 气体继电器;瓦斯继电器
gas release 放气
gas relief valve 煤气安全阀;气体安全阀
gas removal system 除气系统
gas remover 除气用空气过滤器
gas removing capacity 除气容量
gas removing efficiency 气体过滤效率;除气效率
gas replacement testing 气体置换试验
gas repressuring 注气恢复地层压力
gas reservoir 储[贮]气罐;煤气存储器;气储[贮];气田;气体存储器;气储集层
gas reservoir model 汽油库模型
gas-reservoir truck 筒式运煤气车
gas resistant 不透气的
gas retaining property 气体保持本领
gas retention ages 气体保留年龄
gas retort 干馏甑;干馏炉
gas return 天然气返回
gas return path 气体回路
gas reversal valve 煤气交换器;煤气换向阀
gas rig 煤气驱动的钻探设备
gas ring 煤气灶
gas-ring burner 环形管进气喷燃器
gas roaster 煤气烘烤机;煤气焙烧炉
gas rock 气石
gas room heating 煤气房间采暖
gas rubbish incinerator 煤气废物焚化炉
gas rush 瓦斯冲出
gas sample 气体样品;气体取样
gas sample analysis 气体取样分析
gas sample collector 气体取样器
gas sample introduction 气体样品进样
gas sampler 气体取样器;气体采样设备;气体采样器;气体采样管
gas sample scrubber 气样洗涤器
gas sample tube 气体取样管;气体采样管
gas sampling 气体取样;气体采样
gas sampling bottle 天然气样瓶
gas sampling device 气体取样设备;气体采样设备
gas sampling equipment 气体取样设备
gas sampling system 气体取样装置
gas sampling tap 气体取样分接头
gas sampling tube 气体采样管
gas sand 气层;天然气砂(岩)
gas saturation 含气饱和度;气体饱和(率)
gas saver 节气器
gas scattering 气体散射
gas scattering chamber 气体散射室
gas scattering loss 气体散射损失
gas scavenger 煤气净化剂
gas scavenging flange 气体冲洗凸缘;气体冲洗法兰
gas schist 含天然气的片岩
gas scintillator 气体闪烁体
gas scintillator detector 气体闪烁探测器
gas scrubber 净气器;煤气洗涤器;气体洗涤塔;气体洗涤器;湿式除尘器
gas scrubbing 煤气洗涤;气体洗涤
gas scrubbing oil 气体洗涤油
gas scrubbing system 气体洗涤系统;气体洗涤设备
gas scrubbing tower 煤气洗涤塔;气体洗涤塔
gas scrubbing waste 煤气洗涤废水
gas scrubbing water 煤气洗涤水
gas seal 气密(层);气封
gas-seal bell 气封钟形盖
gas seal durability 气密件寿命
gas sealing 气封
gas sealing agent 气密封剂
gas sealing mechanism 气密封机
gas sealing system 气封系统
gas seat 煤气阀座
gas seepage 气体渗出;气苗
gas sensing electrode 气敏电极
gas sensing electrode method 气敏电极法
gas sensitive ceramics 气敏陶瓷
gas sensitive effect 气敏效应

gas sensitive leak locator 气敏查漏仪
gas sensitive metal 气敏金属
gas sensor 气敏元件;气敏传感器
gas sensory semiconductor 气敏半导体
gas separating 气体分离
gas separation 气流分离
gas separation membrane 气体分离膜
gas separation plant 气体分离装置
gas separation unit 气体分离装置
gas separator 气水分离器;油气分离器;气体分离器;分气器
gasser 喷气井;天然气井
gas service pipes 用户室内安装的煤气管;气体分配管
gas sewage 汽油污水;含汽油污水
gas shaft 气窗
gas shaft hood 气管塞;气囱罩
gas shale 含瓦斯页岩;含气油页岩;含沥青页岩
gas shield 气体保护
gas shield automatic submerged-arc welding 气体保护自动埋弧焊
gas shielded arc welding 气体保护(电弧)焊
gas shielded arc welding machine 气(体)保护弧焊机
gas shielded electrode 气包焊条
gas shielded magnetic flux arc welding 磁性焊剂气体保护(电弧)焊
gas shielded metal arc welding 气体保护金属弧焊
gas-shielded stud welding 气保护柱钉焊接
gas shield(ing) 气保护
gas shield welding 气体保护(弧)焊
gas shock 气流激波;气流冲击
gas show 天然气露头;天然气苗;气苗
gas showings 气苗
gas shut-off relief valve 煤气切断安全阀
gas shut-off valve 煤气关断阀
gassi 沙丘酸沟
gas silicate 加气硅酸盐
gas-silicate concrete 加气硅酸盐混凝土
gassiness 多气性
gassing 吹入气体;出气泡;出气;放毒气
gas singeing 煤气烧毛
gassing factor 充气系数
gassing from the wire 成串气泡
gassing machine 熏蒸机
gassing of mud cone 喷气泥锥
gassing of copper 铜气泡
gassing out 管道吹扫
gassing time 吹气时间
gassing-up 通气
gas skirt 集气罩
gas slag 煤气沉渣室
gas slag arch 煤气沉渣室拱顶
gas slag concrete 煤气炉渣混凝土
gas slippage 天然气滑升
gas slot 通气缝;气缝
gas smothering system in cargo hold 货舱惰(性)气(体)灭火系统
gas snow melter 煤气融雪器
gas soft nitriding 气体软氮化
gas soldering copper 煤气加热烙铁
gas solenoid 气管电磁阀
gas-solid adhesion 气固黏附
gas-solid chromatography 气固色谱法;气固层析
gas-solid contact 气固接触
gas-solid contact condition 气固接触条件
gas-solid contactor 气固接触设备
gas-solid equilibrium 气固平衡
gas-solid film 气固膜
gas-solid fluidization 气固流态化
gas-solid interaction 气相—固相相互作用;气体—固体相互作用
gas-solid interface 气体—固体界面;气固界面
gas-solid partition chromatography 气固分配色谱法
gas-solid reaction 气固反应
gas solubility 气体溶解性;气体溶度
gas-solubility coefficient 气体溶(解)度系数
gas-solubility factor 气体溶解度因数
gas-solution interface 气(溶)液界面
gas-solvent reaction 气体溶剂反应
gas sorption 气体吸附
gas source diffusion 气源扩散
gas-source potential 生气潜量

gas source rock 气源岩
gas space 气体空间
gas space heater 燃气供暖器
gas space heating 煤气室内采暖
gas space warming 煤气室内采暖
gas sparger 气体喷洒器;气体鼓泡搅拌器
gas specific gravity meter 天然气比重计
gas specific weight 气体比重
gas spectrometer 气体光谱分析仪
gas sphere 气界;大气
gas sphere gas loop 气圈
gas spill 气体逸出
gas spring 气压弹簧
gas spurt 气喷聚集;气丘
gas standard 气体参数
gas starter 煤气起动机
gas starting engine 燃气启动机;汽油启动机
gas station 汽车加油站;加油站;煤气站
gas-steam-water fluid 气—蒸汽—水混合流体
gas steel pipe 煤气钢管
gas stimulation 激化采气
gas stock and dies 煤气管螺丝扳牙和扳手
gas storage 燃气储存;气调储[贮]藏;充气储[贮]藏
gas storage can 汽油储[贮]存桶
gas storage heater 储气供暖器;煤气储热器
gas storage holder 气罐
gas storage reservoir 煤气罐
gas storage stability 汽油储[贮]存稳定性
gas storage tank 储气箱
gas storage unit and ballast system 储气设备和压载系统
gas stove 烤箱灶;煤气(烘)炉
gas strainer 滤气器
gas streak 夹层气;含气夹层
gas stream 气(体)流
gas stream atomizer 气流喷雾器
gas strength 气体浓度
gas stripper 气体剥离器
gas stripping 气体剥离
gas-stripping canal 气体剥离通道
gas submarine pipeline 水下气管线
gas suction plant 吸气装置;抽气装置
gas supersaturation 气体过饱和
gas supply 供气;煤气供应
gas supply agreement 供气协议
gas supply cubic meterage 供气量
gas supply line 供气管(道);煤气用户屋内设备
gas supply priorities 供气优先顺序
gas surging pressure 排气压力
gas surging pressure tide ga(u)ge 排气压力验潮仪
gas survey 气体测量
gas suspension preheater 气体悬浮预热器
gas sweetening 气体脱硫
gas sweetening unit 气体脱硫装置
gassy 气状;气体的;气态的;瓦斯矿;充满气体
gassy fermentation 气态发酵;产气发酵
gassy melt 含气金属液
gas system 供气系统
gas take 滤气阀
gas tank 煤气箱;煤气库;煤气柜;气罐;气柜;气罐;储气瓶;储气罐
gas tank bracket 汽油箱架;储气箱架
gas tank cap 油箱盖
gas tanker 汽油油罐车
gas tank filler 汽油箱加油器
gas tank float 汽油箱浮子
gas tank truck 汽油罐车
gas tank valve 汽油箱汽门
gas tap 管用丝锥;管螺纹丝锥;煤气旋塞
gas tar 煤气焦油
gas target 气体靶
gas temperature controller 燃气温度调节器
gas temperature probe 气体温度探头
gas tempering 烟气调节
gas tension 气体张力
gaster 柄后腹
gas test 空隙试验;气体分析
gas tester 气体鉴定器
gas-testing lamp 检查瓦斯灯
gas-testing safety man 煤气安全检查员
gas-tetrode relaxation oscillator 充气四极管张弛振荡器

gas thermometer 气体温度计
gas thermometry 气体测温法
gas thread 管螺纹;螺纹管
gas throttle lever 煤气节流杆
gas-tight 不透气(的);气密(的);不漏气(的)
gas-tight casing 气密外壳;气密机座
gas-tight container 气密容器
gas-tight end shield 气密端盖
gas-tight entry 气密进口
gas-tight feeding mechanism 气密加料机构
gas-tight gland 气密压盖
gas-tight housing 气密机座
gas-tight joint 不透气接合;气密接头
gas-tight lysimeter 气密液度计;气密测试计
gas-tight machine 密封式电机
gas-tight material 气密(性)材料
gas-tightness 气密性;不透气性;不透气(性)材料
gas-tight piston ring 活塞压环
gas-tight seal 气密密封
gas-tight shell 气密壳
gas-tight shielded enclosure 气密屏蔽包层
gas-tight silo 密封仓;气密青储[贮]塔
gas-tight sleeve 气闭套管
gas-tight test 气密性试验;气密(性)实验
gas-tight thread 气密螺纹
gas tip 气体喷头
gas titration 气体滴定
gas-to-dust ratio 气尘比
gas tongs 煤气管钳
gas-to-oil ratio 油气比
gas torch 气炬;气焊火焰;气焊枪;气焊焊炬
gas torch cutting 气炬切割
gas torch for mo(u)lding 造型用煤气喷灯
gas torch welding 气炬焊(接)
gas tower 洗气塔
gas tractor 内燃牵引车
gas transfer 气体转移
gas transfer model 气相转移模型
gas transfer rate 气体传输(速)率
gas transfer system 气体传输系统
gas transfer velocity 气体转移速度
gas transmission and distribution system 城市燃气输配系统
gas transmission company 煤气运输公司
gas transmission line 燃气输送管线;煤气输送管路
gas transmission pipeline 燃气输送管道;煤气输送管道
gas transmission system 输气系统
gas transport 气体输运
gas transportation machine 气体输送机
gas trap 油气分离器;浊气井;集气器;气体分离器;气体阀;气体捕集器;排气存水弯;冷凝液罐
gas trapping surface 气体捕获表面
gas travel 气体行程
gas treating process 气体净化法;气体处理法
gas treating system 气体净化装置;气体净化系统;气体处理系统;天然气处理系统
gas treatment 煤气处理;气体净化
gastriode 充气三极管
gastriode memory 充气三极管存储器
gastriode(relay) 闸流管
gastronomy 烹调法
gastropod limestone 腹足类灰岩
gas truck 汽油油罐车
gas trunking 气体围井
gas tube 煤气管;气体管(线);导气管;充气管
gas-tube boiler 废热锅炉;火管锅炉
gas-tube counter 充气电子管计数器
gas-tube handrail 煤气管扶手
gas-tube pulse generator 充气管脉冲发生器
gas-tube rectifier 充气管整流器
gas-tube switch 充气管开关
gas tungsten-arc welding 气(体)保护钨极电弧焊;钨极气体保护焊
gas tungsten-arc welding method 钨电极气体保护焊接法
gas turbine 燃气涡轮(机);燃气透平;燃气轮机;气涡轮机;气轮涡轮机
gas turbine automobile 燃气轮机汽车
gas turbine blade 燃气涡轮叶片
gas turbine booster propulsion set 燃气轮机加力推进装置
gas turbine centrifugal compressor 燃气涡轮离

心式压缩机
gas turbine compressor 燃气轮机压缩机
gas turbine control board 燃气轮机控制盘
gas turbine cycle 燃气轮循环;燃气轮机循环
gas turbine diesel locomotive 燃气柴油机车
gas turbine drier 燃气轮式干燥机
gas turbine driven generator 燃气轮发电机
gas turbine electric(al) locomotive 燃气轮机电气机车
gas turbine electric(al) power plant 燃气轮机发电厂
gas turbine engine 燃气涡轮发动机;燃气轮发动机
gas turbine engine lubricant 燃气轮机轮机润滑剂
gas turbine fuel 燃气轮机燃料
gas turbine generator 燃气轮发电机
gas turbine installation 燃气轮机装置
gas turbine jet 燃气涡轮喷气发动机;燃气轮机喷气发动机
gas turbine liquid-fuel burner 燃气轮机液体燃料燃烧器
gas turbine locomotive 燃气轮机(机)车
gas turbine locomotive with electric(al) transmission 燃气轮电力传动机车
gas turbine module 燃气轮机组装件
gas turbine nozzle 燃气轮机喷嘴
gas turbine plant 燃气涡轮机设备;燃气轮机发电厂
gas turbine powered car 燃气轮机小客车
gas turbine powered generator 燃气轮机发电机
gas turbine power plant 燃气轮机发电厂;燃气轮机动力装置
gas turbine power station 燃气涡轮发电站;燃气轮机发电厂
gas turbine power unit 燃气轮机动力装置
gas turbine principle 燃气轮机原理
gas turbine propulsion 燃气轮机牵引
gas turbine pump combination 涡轮泵组
gas turbine pump system 燃气轮机泵组
gas turbine railcar 燃气轮机(动)车;燃气轮动车组
gas turbine room 燃气轮机舱
gas turbine rotor 燃气涡轮转子
gas turbine ship 燃气轮机船;气涡轮机船
gas turbine starter 燃气轮机起动机
gas turbine supercharged boiler 燃气轮机增压锅炉
gas turbine tracked vehicle 燃气轮发动机的履带车辆
gas turbine traction unit 燃气轮牵引装置
gas turbine train 燃气轮列车
gas turbine trainset 燃气轮列车组
gas turbine train unit 燃气轮单元列车
gas turbine truck 燃气轮机载货汽车
gas turbine unit 燃气涡轮发动机
gas turbine wheel 燃气轮机叶轮
gas turbine with regenerator 回热式燃气轮机
gas turbo-alternator 燃气涡轮交流发电机
gas turbo-blower 燃气轮机鼓风机组;燃气涡轮鼓风机
gas turbo-compressor 燃气轮机压缩机组
gas turbo-electric(al) drive 燃气轮机电力驱动
gas turbo-electric(al) locomotive 燃气轮电力机车;气轮电力机车
gas turbo-generator 燃气涡轮发电机
gas type fire extinguishing system 气体型灭火系统
gas type fitting 气管接头
gas underground storage 地下储气
gas undertaking 煤气企业;煤气厂
gas unit 气举装置
gas up 充气
gas uptake 煤气上升道
gas utility 煤气供应;公用煤气事业;煤气效用;煤气公司
gas utilization 沼气利用
gas vacuole 气泡
gas valve 燃气阀;煤气阀;充气管
gas valve screw 气阀螺钉
gas valve spring 煤气阀簧
gas vane 燃气舵
gas-vapo(u)r 蒸汽燃气
gas velocity 气体速度
gas vent 气体出口;喷气口;喷气孔;排气竖管;排气口;排气孔;通气口;通气道;导气孔;放空
gas-vent 通气孔
gas vent connector 排气口接头

gas venting system 透气系统
gas vessel 汽油容器;气体容器;煤气容器
gas vibrator 燃气振动器;燃气振捣器
gas voltameter 气体伏安计;气体电压计
gas volume 气体体积
gas volume factor 气体体积修正系数
gas volumeter 气体容量计;气量表
gas-volumetric(al) analysis 气体容量分析
gas-volumetric(al) chromatography 气体体积色谱(法)
gas-volumetric(al) measuring apparatus 气体容量测定仪
gas-volumetric(al) method 气体体积分析法;气体容量分析法
gas volumetry 气体容量法
gas-warmed 煤气加热
gas-warmed air 煤气加热空气
gas warmer 煤气加热器
gas warming 煤气采暖
gas warming air heating 煤气热空气取暖
gas warming appliance 煤气采暖装置
gas warming device 煤气采暖装置
gas warming unit 煤气采暖装置
gas washer 洗气器;煤气洗涤器;气体洗涤器;湿式气体洗涤器;湿式煤气洗涤器;气体净化器
gas washing 洗气;煤气洗涤;气体洗涤
gas washing blower 气体洗涤鼓风机
gas washing bottle 洗气瓶;气体洗瓶;气泡吸收管
gas washing installation 气体洗涤装置
gas washing machine 煤气洗涤机
gas washing plant 气体洗涤设备
gas washing wastewater 洗气废水;煤气洗涤废水
gas washing water 涤气水
gas wash tower 洗气塔;煤气洗涤塔;气洗涤塔;除尘塔
gas waste incinerator 废气焚化炉
gas water 洗气用水;洗气水
gas water bed 气水同层
gas water circulation heating 煤气热水循环供热;煤气热水循环采暖
gas water contact 气水界面;气水接触(面)
gas water heater 燃气热水器;煤气热水器;煤气烧水炉
gas water interface 气水界面
gas water ratio 气水比
gas water surface 气水界面;气水结合面;气水交界面
gas water system 煤气热水系统
gas wave 气波
gas weigh(ing) balloon 称气瓶
gas weight flow 燃气重量流量
gas welded joint 气焊接头
gas welder 气焊机;气焊装置;气焊工
gas welding 乙炔焊;气焊
gas welding device 气焊设备
gas welding equipment 气焊设备
gas welding machine 气焊机
gas welding outfit 气焊机(组)
gas welding rod 气焊焊丝
gas welding rubber hose 气焊橡胶管
gas welding technique 气焊工艺
gas welding torch butt 气焊进气硬管
gas welding tube mill 瓦斯焊接机
gas welding work 气焊工作
gas well 气井;天然(气)井;(火井)
gas well gas 气井气
gas withdrawal 排气道;天然气回采
gas works 煤气(工)厂;煤气车间;煤气工程
gasworks(coal)tar 煤气焦油(沥青)
gasworks coke 煤气焦炭
gasworks liquor 煤气厂水
gasworks pitch 煤气厂沥青
gasworks tar 煤气厂焦油
gasworks tar pitch 煤气厂焦油(沥青)
gasworks waste 煤气厂废水
gas wrench 煤气管(扳)钳
gas yield 燃气产率;沼气产量;气体产率
gas-yielding polymer 释气聚合物
gas zone 含气岩层;含气带;气带
gat 海峡;港道;狭(窄)航道
gatage indicator 门开度指示器
gate 院门;整流栅;门;控制极;内浇道;门控;屏蔽闸门;片窗;跑道控制门;调节砖;时间限制电路;舌瓣;登机口;大门

gate action 闸门作用
gate-actuating rod 导叶推拉杆;导叶操纵杆
gateage 闸门开度;门叶开度;导叶开度
gateage area 导叶开口面积
gateage indicator 导叶开度指示器
gate amplifier 门放大器
gate anchorage 闸门锚具
gate-and-dam discharge 闸门排料
gate-and-dam jig 闸门排料式跳汰机
gate apparatus 大门装置;大门配件;叶片式调节阀门;导向器
gate arch 门拱
gate area 闸门面积
gate arm 支承腿架;支臂(架);栅门臂;拦(路)木;弧形闸门支承架
gate arrivals type 直达式
gate assembly 闸板组件
gate back catch 门后钩
gate bar 门闩;大门栓
gate barrier 栏木
gate bay (船闸的)闸室;闸段;船闸闸首
gate beam 闸门横梁;横肋梁;辅助横梁
gate blade 锁紧叶片
gate block 门墩;有闸门的坝段;闸首墩;闸门墩
gate brick 浇口砖
gate caisson 浮坞门
gate cell 浮坞门灌水舱室
gate chamber (船闸的)闸门室;闸门库;闸门井;传达室
gate chamber floor 闸室底板
gate chamber roof 闸室顶板
gate chamber wall 闸室墙
gate change 滑槽式换挡
gate circle 导叶分布圆
gate circuit 选通电路;闸门电路;门电路
gate clerk 门卫管理员;大门管理员
gate closer 闸门关闭器
gate closure 闸门;水闸门
gate closure type safety device 门式安全装置
gate contact 门触点
gate control 选通控制
gate control block 门控块
gate control characteristic 开关控制特性(曲线)
gate control house 闸门控制室
gate-controlled rectifier 闸控整流器
gate-controlled spillway 装有闸门的溢洪口;装有闸门的溢洪道
gate-controlled switch 键控开关;门控开关;闸控开关
gate-controlled thyristor turn-on time 闸流控制接通时间
gate-controlled turn-on time 闸流控制接通时间
gate-controlled wasteway 装有闸门的溢洪口;闸门控制的溢洪道;闸门控制的废水口;闸门控制的废水道
gate control theory 闸门控制理论
gate conveyer 平巷输送机
gate core 浇口芯
gate current 栅流;门电流
gate cutter 浇口切割机
gate cutting machine 浇口切割机
gate dam 闸门(式)堰;闸坝
gated amplifier 选通放大器;闸门放大器;门信号放大器
gated buffer 门控缓冲器
gated clock 选通时钟
gated counter 选通计数器
gated culvert 有闸门控制的涵洞
gated detector 门检波器
gated device 变速杆定位板
gate detector 选通门检测器
gated filp-flop 选通触发电路;门控触发器
gate diagram 门线图
gate diversion works 闸门分水设备
gated latch 门拉门锁
gated noise measurement 门控噪声测量
gate dog 闸门钩;闸门开启沟
gated ogee structure 装有闸门的反弧形溢流堰
gate dogging device 导叶锁定装置
gated oscillator counter 闸门式振荡计数器;控制振荡计数器
gated outlet 有闸门的泄水口;闸门泄水口
gated pattern 带浇口的模型
gated pipe 带闸门的管

gated radiometer 选通辐射计
gate driver 门驱动器
gated scaler 选通定标器
gated shipbuilding berth 有闸式船台
gated spillway 闸控溢洪道
gated squarewave generator 选通方波发生器
gated sweep 选通扫描；控制扫描
gated throttle 限动油门
gated tracking filter 选通跟踪滤波器
gated tunnel 有闸门控制的涵洞
gated weir 装有闸门的堰
gate electrode 栅电极；门电极
gate end 煤巷内端
gate end box 电缆终端套管
gate end conveyer[conveyor] 联络巷道端部转载运输机；平巷转载运输机
gate end loader 联络巷道端部装载机；平巷装载机
gate end panel 平巷配电箱
gate escape 退水闸门
gate etching 栅腐蚀
gate feed 门式进料
gate feeder 闸门式送料装置；闸门式给料器；框式给料器
gate feed hopper 闸板给料斗；框式加料器；框式加料斗
gate flap 门叶；门舌；舌瓣
gate fold 折叠式插页
gate footstep 闸门底脚
gate frame(work) 门框(架)
gate generator 选通脉冲发生器；门脉冲发生器；时钟脉冲产生器
gate groove 闸门(导向)槽；闸槽；门槽
gate ground 泄水底孔闸门
gate guard 闸栅板式安全保护装置；门式安全装置；保护栅
gate gudgeon 闸门耳轴；门轴柱(边梃)
gate guide 闸门导向槽；闸门导轨；闸槽
gate gurgle valve 泄油阀
gate handling 闸门启闭
gate hanger 闸门吊架
gate head 门道
gate hinge 门合页；门铰链；大门合页
gate hoist 闸门提升机；启门机
gate hono(u)r 正门
gate hook 门柱铰链钩片；门钩；悬挑式门枢轴；大门枢轴
gate house 机房；传达室；城楼；闸门室；闸门间；闸门操作室；门房；大门；启机室；通行检查站
gate impedance 控制极阻抗
gate interlock 运输联锁系统；门联锁(装置)
gate inverter 门脉冲放大逆变器；门反相器
gate jack 闸门千斤顶
gatekeeper 门卫；看门人；道口看守员
gatekeeper's box 道口看守房
gatekeeper's cabin 道口看守房
gatekeeper's dwelling 看门人住屋
gatekeeper's house 门房；警卫室
gatekeeper's hut 道口看守房
gatekeeper's lodge 看门人住屋；传达室
gatekeeper's office 门卫室
gate latching device 闸门锁定装置；闸门锁定设备
gate lateral position 门横向位置；波门水平位置
gate leaf 闸板；门叶；门扇
gate leakage 闸门漏水量
gate legged 折叠腿；铰链腿
gate legged table 折叠(式)桌
gate length 选通脉冲宽度；门信号宽度
gate level logic simulation 门级逻辑模拟
gate level simulation 门级模拟
gate lever 节流板杆
gate lift 闸门提升机(构)；闸门启闭装置；闸板升降机
gate lifter 闸门启闭机
gate lifting device 闸门提升设备；闸门启闭装置；闸门启闭机；门提升装置；启机；吊门设备
gate light 门灯
gate linkage 导叶传动连杆
gate lock 栏木锁闭器；门锁；导叶锁定装置
gate lodge 门房
gate logging device 闸门锁定装置；闸门锁定设备
gate lowering mechanism 下闸装置；闸门升降机；闸门降落机构；闭门(机械)装置
gateman 收票员；道口看守员
gate-man's lodge 栅栏看守房
gate mechanism 闸门机制；闸门机构

gate meeting face 门的贴接面
gate mixer 框式混合器
gate money 入场费
gate monolith 闸首墩；闸门墩
gate motor 栏木电动机
gate multivibrator 选通多谐振荡器；门信号多谐振荡器
gate niche 闸门库
gate nontrigger voltage 栅极不启动电压
gate occupancy time 门的占用时间
Gate of Herculaneum at Pompeli 蓬拜城之赫尔库兰门(古罗马)
gate of honour (庆典用的)正门
Gate of Lions 米禅勒城之狮门(古希腊)
gate on 门通
gate open(ing) 闸门开启(度)；闸门孔；闸门孔开度；导叶开度
gate-opening conveyance 门上输水
gate operating chamber 闸门操纵室；大门启闭室
gate operating control 闸门启闭操纵；大门启闭控制
gate operating deck 闸门工作桥；闸门工作平台；闸门操作平台
gate operating gear 水闸操作机械；水闸操作机构；水闸操纵机械；水闸操纵机构
gate operating machinery 闸门操作机械
gate operating mechanism 闸门操作机构；导叶操作机构；导水机构
gate operating platform 闸门操作平台
gate operating ring 控制环；调速环；导叶操作环
gate operation platform 闸门操作桥；工作桥；闸门工作桥；闸门工作平台
gate outlet 泄水闸口
gate out switch 出口开关
gate overlap 栅覆盖层
gate piece 锁块
gate pier 大门墩柱；闸门墩；门墩(柱)
gate pin 浇口棒
gate pinch-off voltage 栅夹断电压
gate pivot 闸门支枢；闸门枢轴；门枢
gate plate 闸板
gate plug 浇口塞
gate position (候机室的)站台位置；闸门布置
gate position transmitter 导叶位置传感器
gate post 门柱
gate-producing multivibrator 门信号多谐振荡器
gate pulse 选通脉冲；门脉冲
gate pulse generator 闸门脉冲发生器
gate rail 门轨
gate recess 门龛；闸门库；闸门龛；闸门槽；闸门凹座；门龛；水闸闸槽
gate resistance 栅电阻
gate restoring mechanism 导叶复位机构
gate rigging 导叶组装；导叶装置
gate ring 导环
gate road 联络巷道(采矿区内)；矿内巷道；运输平巷；煤巷；采空区内运输巷
gate road bunker 调节仓；仓式运输机
gate runner 浇道
gate saw 框锯；排锯
Gates crusher 盖茨破碎机
gate seal 闸门止水；门封；导叶密封
gate seat 门槛；闸门(门)座；闸门(门)槛；门座
gate segment 门爪
gate shaft 闸门竖井；闸门井
gate shears 铡刀剪切机；龙门剪床；双柱式剪切机
gate sheet 闸门板
gate signal 门信号
gate sill 门槛；闸门(门)槛；坞槛
gate slot 闸门门槽；闸门槽；门槽
gate spillway 设闸溢洪道
gate spring 门弹簧
gate stem 闸门杆
gate step 门墩(支承铁门门枢的铁制品)
gate stick 直浇口棒；浇口棒
gate stop 门碰；门挡；门钩
gate strainer 浇口滤渣芯片；浇口滤渣器
gate strut 闸门推拉杆；门柱
gate suppressor 控制极抑制器
gate switch 门开关
gate table 折叠式桌子
gate tensioner 栅栏式张力器；梳形张力器
gate terminal 门接线端

gate thickness 内浇道厚度
gate throttle 节流门；节流阀
gate tie 大门斜撑；大门斜拉条
gate tile 浇注管；浇注砖
gate-to-gate transportation 门对门运输
gate tower 城门；城楼；门楼；启门塔；城门楼
gate-towering mechanism 闸门升降机构
gate trap 闸门阱；闸龛；闸盒
gate travel 闸板行程
gate trigger 控制门触发器
gate trigger circuit 控制门触发电路
gate trigger current 栅极触发电流
gate trigger pulse 控制门触脉冲
gate trunnion 门锚支座
gate tube 选择管；选通管；闸门管；闸管；门电子管
gate turn off 矩形脉冲断开；门极可断开关；门电路断开；闸门电路断开
gate turnoff silicon-controlled rectifier 电子门电路断开硅可控整流器
gate turnoff thyristor 闸门关断可控硅整流器
gate type agitator 框式搅拌器
gate type gear shift 格式变速器
gate type gear shift lever 格式变速杆
gate type hydrant 阀门式消防龙头；阀门式给水栓；闸门式消火栓；闸门式消防栓；门式消防栓
gate type stall 侧出式挤奶台
gate type stirrer 框式搅拌器
gate valve 选通阀；闸(门)阀；滑阀；滑动式闸门；滑板阀；平板式阀；挡闸阀门；大阀
gate-valve chamber 闸阀室
gate valve key 闸阀钥匙
gate vault 闸门室
gate vibration 闸门振动
gate voltage 触发电压
gateway 口门；门道；采区联络煤巷；采空区平巷；采空区内运输巷；门框架建筑结构；口岸；门路连接器；人口；网间连接器；网关；途径；通路；手段；采区平巷
gateway arch 门拱
gateway bridge 门道桥
gateway of triumph 凯旋门
gateway processor 入口处理机
gateway router 网关路由器
gate wedge 斜铁；门楔
gate weir 门堰
gate well 闸门坑；闸门井；闸门槽；凹槽
gatewidth 选通脉冲宽度；内浇道宽度；门宽；门脉冲宽度
gatewidth control 选通脉冲宽度调整；门宽调整
gate winch 栏木绞车
gate winding 选通线圈；选通绕组
gather 渐增；汇集调车；采集；聚集
gather density 聚集密度
gatherer 集(茎)器；挑料工；收集器；导入装置；导入器
gatherer board 集茎器板
gathering 积聚；搜集；收皱；收集；采集业；板材粘辊；聚集
gathering after completion of dredging work 收工集合【疏】
gathering agent 搜集剂
gathering and distributing system of port 港口集疏运系统
gathering area 集水区；储集区
gathering arm 扒爪
gathering arm loader 蟹爪式装载机；耙爪式装载机；扒爪式装载机
gathering attachment 收集装置
gathering auger 收集堆运螺旋
gathering barrel 收集桶
gathering bubble 挑料(加入的)气泡
gathering cargo of port 港口集运
gathering chain 集荼夹送链
gathering comb 采集梳；分槽集束器
gathering conveyer[conveyor] 集煤输送机；集矿运输机
gathering device 集束器；集茎夹送装置
gathering drum 集草滚筒
gathering economy 采集经济
gathering end 采集端
gathering facility 集运管线；集运管道
gathering ground 集流区；流域；聚水区；集水区；集水面积；集合地域；汇流区
gathering haulage 集聚运输；汇集调车

Gaussian minimum shift key 高斯最小频移键控
Gaussian mode 高斯模
Gaussian model 高斯模型
Gaussian noise 高斯噪声
Gaussian noise generator 高斯噪声发生器
Gaussian normal distribution curve 高斯正态分布曲线
Gaussian normal epoch 高斯正向期;高斯正向磁极期
Gaussian normal polarity chron 高斯正向极性时
Gaussian normal polarity chronzone 高斯正向极性时间带
Gaussian normal polarity epoch 高斯正极性期
Gaussian normal polarity zone 高斯正向极性带
Gaussian number 高斯随机数
Gaussian optics 高斯光学(装置)
Gaussian pattern 高斯方向图
Gaussian peak 高斯峰
Gaussian plane coordinates 高斯平面坐标
Gaussian plane coordinate system 高斯平面坐标系
Gaussian plume model 高斯烟流模式
Gaussian point 高斯点
Gaussian positions 高斯位置
Gaussian principle 高斯原理
Gaussian principle of least constraint 高斯最小约束原理
Gaussian probability curve 高斯概率曲线
Gaussian probability-density function 高斯概率密度函数
Gaussian probability distribution 高斯概率分布
Gaussian probability integral 高斯概率积分
Gaussian probability process 高斯概率过程
Gaussian process 高斯过程;高斯法
Gaussian profile 高斯剖面
Gaussian projection 高斯投影
Gaussian projection plane 高斯投影面
Gaussian pulse 高斯脉冲
Gaussian quadrature 高斯求积
Gaussian quadrature formula 高斯求积公式
Gaussian random noise 高斯随机噪声
Gaussian random process 高斯随机过程
Gaussian random variable 高斯随机变量
Gaussian random vector 高斯随机向量
Gaussian rectangular plane coordinate 高斯平面直角坐标
Gaussian reduction 高斯消去法
Gaussian reduction method 高斯约化法
Gaussian reference sphere 高斯参考球
Gaussian reflectivity 高斯反射率
Gaussian region 高斯区域
Gaussian response 高斯频响特性
Gaussian roll-off curve 高斯型衰减曲线
Gaussian-shaped concentration distribution 高斯型浓度分布
Gaussian-shaped pulse 高斯型脉冲
Gaussian spread function 高斯扩展函数
Gaussian standard normal curve 高斯标准正态曲线
Gaussian stochastic process 高斯随机过程
Gaussian stress quadric 高斯应力曲面
Gaussian system of units 高斯单位制
Gaussian test 高斯判别法
Gaussian theorem 高斯定理
Gaussian transformation 高斯变换
Gaussian units 高斯制单位;高斯随机变量
Gaussian wave group 高斯波群
Gaussian wave packet 高斯波群
Gaussian wave train 高斯波列
Gaussian well 高斯位阱;高斯势阱
Gaussian white noise 高斯白噪声
Gaussian year 高斯年
Gauss integration 高斯积分
Gaussina response 高斯响应
Gauss-invariant coordinates 高斯不变坐标
gaussistor 磁阻放大器
Gauss-Jordan elimination 高斯—若尔当消去法
Gauss-Kruger coordinate system 高斯—克吕格坐标
Gauss-Kruger plane rectangular coordinate system 高斯—克吕格平面直角坐标系
Gauss-Kruger projection 高斯—克吕格投影
Gauss-Laplace curve 高斯—拉普拉斯曲线
Gauss-Legendre rule 高斯—勒让德法则

Gauss lens system 塞洛尔透镜系统
Gauss-Markov criterion 高斯—马尔可夫判别准则
Gauss-Markov corollary 高斯—马尔可夫推论
Gauss-Markov estimator 高斯—马尔可夫估计量
Gauss-Markov process 高斯—马尔可夫过程
Gauss-Markov theorem 高斯—马尔可夫定理
gaussmeter 磁感应测定计;高斯计;磁强计
Gauss-Newton method 高斯—牛顿法
Gauss objective lens 高斯物镜
Gauss-Poisson distribution 高斯—泊松分布
Gauss-Seidel iteration method 高斯—赛德尔迭代法
Gauss-Seidel matrix 高斯—赛德尔矩阵
Gauss-Seidel method 高斯—赛德尔法;赛德尔(方)法
Gauvain's fluid 戈维恩氏液
gauze 冷布;金属丝网;铜纱;纱网;纱布
gauze bag 纱布袋
gauze brush 网刷;铜丝布电刷
gauze cloth 纱罗织物
gauze cone 网状锥体
gauze effect 纱罗花纹
gauze element 滤器零件;网状滤心
gauze fabric 金属织网;纱罗织物;纱布
gauze filter 网式滤水器;网状(过)滤器
gauze filter stray 网状过滤盘
gauze filtration 网状过滤
gauze nozzle 滤网式喷管
gauze oil strainer 网式机油过滤器
gauze packing 网状填充物
gauze pad 网垫
gauze platinum electrode 铂网电极
gauze reed 纱罗筘
gauze ring 网状填充环
gauze sponge 叠片纱布
gauze strain 网状过滤
gauze strainer 滤(器)网;网状(过)滤器
gauze strainer stray 网状过滤盘
gauze top burner 网头灯
gauze wire 细目丝网;网线;细眼丝网
gauze wire cloth 金属丝网布
gauze wire cloth with cropped edges 修过边的金属丝网布
gauze wire cloth with folded edges 折边的金属丝网布
gauze wire cloth with turned up edges 边向上弯转的金属丝网布
gauze wire screen cloth 金属丝筛网布
gauzing machine 贴纱布机
gavel 双坡;山墙;人字头;(茅屋顶的)禾草束;大木槌;小槌
gavelock 斜桁;撬棍;铁杆;铁棒
Gay-Lussac's law 盖—吕萨克定律
gaylussite 针碳钠钙石
gazebo 凉亭;有玻璃窗的阳台;观景楼;瞭望亭;屋顶塔楼
gazette 公告;公报
gazetteer 地名辞典
gazetteer index 地名索引
gazogene 木炭燃气
GB factor 增益带宽因数
GB freight container 国家标准集装箱
GB-IB 通用仪器接口标准
gbyte 千兆字节
G-clamp G字形夹具
G-cramp G字形螺旋夹钳
gear 排挡数;船具,传动装置;齿轮
gear and pinion drive 大小齿轮传动;边缘传动(装置)
gear and spline rolling machines 齿轮和花键滚轧机
gear and thread cutting machine 齿轮和螺纹加工机床
gear and thread grinding machine 齿轮和螺纹磨床
gear assembly 减速器;齿轮组件;齿轮传动装置
gear backlash 齿轮啮合间隙;齿(轮)隙;齿轮(间)隙;齿(轮)侧隙
gear bank 齿轮组
gear blank 齿轮坯料;齿轮毛坯
gear box 联动盒;连接箱;传动箱;齿轮箱;变速箱;进给箱;进刀箱;传动机匣
gearbox adaptor 齿轮箱接头
gearbox body 变速箱体

gearbox breather 齿轮箱通气孔
gearbox case 齿轮箱外壳
gearbox casing 齿轮箱体;变速箱体
gearbox clutch pedal 变速箱离合器踏板
gearbox cover 减速箱盖;齿轮箱盖;变速箱盖
gearbox cover packing 变速箱盖密垫
gearbox dipstick 齿轮箱油尺
gearbox draining pump 齿轮箱泄油泵
gearbox flange 齿轮箱凸缘
gearbox head 变速箱头
gearbox housing 齿轮箱体;变速箱体
gearbox input 齿轮箱输入功率
gearbox input shaft 齿轮箱输入轴
gearbox main drive shaft 齿轮箱主动轴
gearbox oil 齿轮箱油
gearbox oil filler 齿轮箱加油口
gearbox operating mechanism 变速箱操纵机构
gearbox output 齿轮箱输出
gearbox output shaft 齿轮箱输出轴
gearbox pulley 变速箱皮带盘
gearbox rubber ring 变速箱胶圈
gearbox storage rack 减速机置台
gearbox suspension 齿轮箱悬置
gearbox torque 变速箱力矩
gearbox workshop 齿轮箱车间
gear bracket 齿轮托
gear bracket support 齿轮托支架
gear brake 齿轮闸
gear broach 齿轮拉刀
gear bronze 齿轮青铜
gear burnishing 齿面抛光法
gear burnishing machine 齿轮抛光机;齿轮滚光机
gear burnishing tool 齿轮滚光轮
gear burr machine 去齿轮毛边机
gear bush 齿轮轴套
gear case 连接箱;齿轮箱
gear case gland 齿轮箱压盖
gear case oil 齿轮箱油
gear case stuffing box 齿轮箱填密函
gear casing 齿轮箱
gear center 齿轮中心
gear chain 联动挡链系
gear chamfering 齿轮倒角
gear change 齿轮变速
gear change lever 调速杆;换挡杆;齿轮变速手柄;齿轮变速杆
gear change set 齿轮变速装置
gear change shift fork 变速杆拨叉
gear change time 换挡时间(齿轮);变速时间(齿轮)
gear changing 调挡;变速;换挡;换齿轮
gear changing by foot 脚踏换挡
gear checking device 齿轮检验器
gear checking equipment 齿轮检验设备
gear checking instruments 齿轮校正仪
gear clearance 齿轮间隙
gear cluster 齿轮组;齿轮块
gear clutch 齿(轮)式离合器
gear collar 齿轮定位环
gear compound 齿轮油;复齿轮
gear compressor 齿轮式压缩机
gear cone 齿轮锥
gear connection 齿轮连接
gear contact 齿轮齿面接触
gear contact pattern 齿轮接触斑点;齿轮工作面接触斑点
gear control 齿轮控制器
gear control lever 齿轮变速控制手柄
gear counter 齿轮计数器
gear coupling 齿式联轴器;齿轮耦合;齿轮联轴器;齿轮联轴节;齿轮联系;齿轮蹲接
gear cover 齿轮罩
gear coverseal 齿轮箱盖衬垫
gear crimper 齿轮卷曲机
gear crown 齿轮冠
gear cutter 切齿机;切齿刀;齿轮铣刀;齿轮刀具
gear cutter hob 齿轮滚刀
gear cutting 切齿加工
gear cutting consideration 切齿方法
gear cutting machine 切齿轮;切齿机;齿轮加工机床
gear cutting tool 齿轮切削刀具;齿轮加工刀具
gear cutting tool grinding machine 齿轮切削工具磨床

gear deburring and tooth pointing machine 齿轮去毛刺及齿顶成型机
gear deburring machine 齿轮去毛刺机
gear distress 齿轮损坏;齿轮故障
gear down 减速传动(装置);减速(齿轮传动);换低速挡;低速齿轮传动;齿轮减速
gear drive 齿轮传动
gear-driven 齿轮传动的
gear-driven accuracy 齿轮传动精度
gear-driven camshaft 齿轮传动式凸轮轴
gear-driven efficiency 齿轮传动效率
gear-driven fan 齿轮传动扇
gear-driven fixed mount 齿轮传动固定架
gear-driven head 齿轮驱动喷洒头
gear-driven supercharger 齿轮传动增压器
gear driving arm 齿轮传动臂
gear-driving equipment 传动装置
gear duck 齿轮帆布
gear eccentricity tester 齿轮偏心检查仪
geared bar-bender 齿轮式钢筋弯折机
geared block 齿轮滑车
geared blower 机械传动式鼓风机;齿轮传动式鼓风机
geared brake 齿轮传动手制动机
geared capstan 齿轮绞盘
geared chuck 齿轮夹头
geared container ship 带船机集装箱船
geared crane ladle 手摇吊包
geared diesel 齿轮传动柴油机
geared diesel boat 齿轮减速内燃机船
geared diesel engine 齿轮减速式柴油机
geared differential hoist 齿轮式差动卷扬机
geared-down engine 减速发动机
geared-down motor 齿轮减速发动机
geared-down speed 低挡速度
geared drill 齿轮钻床
geared-drive machine 齿轮传动机械
geared electric(al) motor 齿轮传动的电动机
geared elevator 齿轮传动升降机
geared engine 齿轮传动发动机;齿轮变速式发动机;变速发动机
geared feed 齿轮进给
geared feed pump 齿轮传动给水泵
geared foot-operated generator 齿轮传动脚踏发电机
geared gate valve 齿轮传动闸阀
geared head 齿轮变速箱;车头箱
geared-head drill 螺旋差动式钻机
geared-head engine lathe 齿轮箱车床
geared-head gap-bed lathe 马鞍车床
geared-head lathe 全齿轮车床
geared-head lathe with gap bed 马鞍式全齿轮车床
geared hoist 带减速机的提升机;齿轮传动卷扬机
geared incentive scheme 分段刺激工资制
geared inclinable crank press 齿轮传动可倾式曲柄压力机
geared ladle-hoist 浇包传动机构;齿轮传动的浇包起重机
geared limit switch 齿轮传动的限制开关
geared locomotive 齿轮传动机车
geared motor 减速电动机;带齿轮箱的电动机;带变速齿轮箱的电动机;齿轮(传)电动机
geared node turbine 齿轮节点式涡轮机
geared noncondensing turbine 齿轮传动背压式汽轮机
geared non-inclinable crank press 齿轮传动固定式曲柄压力机
geared propeller 齿轮减速螺旋桨
geared pump 齿轮传动泵
geared pumping unit 齿轮传动抽油机
geared reducing press 齿轮传动拉延压力机
geared retracter 齿轮移锭器
geared ring 齿圈
geared sleeve 齿轮套
geared system 齿轮装置
geared timing accelerator 齿轮定时加速器
geared traction machine 齿轮传动牵引机
geared turbine 齿轮降速涡轮机;齿轮传动式汽轮机
geared-up 增速的
geared valve 传动阀
geared vessel 有传动设备船
geared worm screw 传动蜗杆螺旋
gear efficiency 传动装置效率

gear end 机头;车头
gear engagement 齿轮啮合
gear engaging switch 合齿开关
gear error meter 齿轮误差测量仪
gear excessive lash 齿轮过大的齿隙
gear face 齿(轮端)面
gear feed 齿轮进刀;齿轮给进
gear feed (swivel) head 螺旋给进回转器
gear finisher 齿轮光整加工机
gear finishing machine 齿轮精加工机床
gear flow meter 齿轮式流量计
gear for feed rod 给料杠齿轮
gear form 齿形
gear form-grinding machine 成形砂轮磨齿机
gear forming 齿轮成形加工法
gear forming machine 齿轮成形加工机
gear frame 传动架
gear ga(u)ge 齿距规
gear generating 齿轮展成加工法
gear-generating machine 齿轮滚铣机床
gear generation 齿轮滚铣法
gear generator 刨齿机;刨齿床
gear graduation 齿轮分度;齿轮变速
gear grease 齿轮润滑油
gear grinder 齿轮磨床
gear grinding 磨齿;齿轮磨削
gear grinding machine 磨齿机;齿轮磨床
gear grinding machine with form wheel 成形砂轮磨齿机
gear guard 齿轮罩;齿轮传动防护罩
gearhead attendant 机工
gear head motor 齿轮减速电动机
gear head stock 齿轮床头箱
gear hob 齿轮滚刀
gear hobber 滚齿机;齿轮铣刀
gear hobbing 滚削(齿)
gear hobbing machine 滚齿机
gear hob measuring machine 齿轮滚刀检查仪
gear hoist 齿轮式起重机;齿轮传动提升机
gear hole 齿孔
gear honing machine 齿轮珩床
gear housing 齿轮箱(外壳)
gear hub 齿轮毂
gear hydraulic motor 齿轮液压马达
gear idle 空转轮
gear increaser 齿轮增速装置;齿轮增速机
gear inertia 齿轮惯量
gearing 齿轮传动
gearing attachment 联动附件
gearing bronze 齿轮青铜
gearing bronze alloy 齿轮青铜合金
gearing chain 联动链系;循环链;传动链;齿轮链系
gearing chain conveyer 循环链式输送机
gearing down 齿轮减速传动
gearing down unit 减速传动装置
gearing for mining duties 采矿工作中的齿轮传动装置
gearing(in) 齿轮啮合
gearing mesh 齿轮啮合
gearing rate 联动比率
gearing shaft 齿轮传动轴
gearing tolerance 齿轮公差
gearing toothed-wheel gearing 齿轮装置
gearing train diagram 传动系统图
gearing-up 增速传动;齿轮增速传动
gearing wheel 传动齿轮
gear input shaft 齿轮输入轴
gear integrated error tester 齿轮整体误差测量仪
gear into 啮合
gear in train 齿轮传动的轮系
gear jack 齿轮起重器
gear jammer 货运汽车驾驶员
gear knob 齿轮传动钮
gearksutite 钙铝氟石;氟钼钙矿;氟钙铝石
gear lapping 研齿
gear lapping machine 齿轮研磨机
gear lathe 齿轮车床
gear lead checker 齿轮导程校正机
gearless ball mill drive 无齿轮球磨机传动装置
gearless container ship 无装卸设备的集装箱船;无船机集装箱船
gearless differential 无行星齿轮分速器
gearless locomotive 无齿轮机车
gearless mill drive 无齿轮磨机传动

gearless mill motor 无齿轮磨机电动机
gearless reduction gyratory 无齿轮式圆锥破碎机
gearless ship 无装卸设备的船
gearless vessel 无装卸设备的船
gear level 齿轮等速
gear lever 变速杆;齿轮变速手柄
gear lever interlock 变速杆联锁装置
gear lever lock 变速杆锁
gearlex coupling 齿轮式弹性轴接
gear load (飞机的)单轮荷重
gear loading 齿轮负荷
gear lock 齿轮保险装置
gear lost 掉牙轮
gear lubrication 传动装置润滑;齿轮润滑
gearman 码头工具管理员
gear-manufacturing machine 齿轮制型机
gear mark 传动链痕迹;印刷杠子
gear mass 齿轮体
gear measuring cylinder 齿轮测量圆柱
gear measuring machine 齿轮测量机
gear measuring wires 齿轮测量线
gear mechanical static cone penetration test 齿轮机械式静力触探试验
gear mechanism 齿轮机构
gear member shaft 齿轮构件轴
gear mesh 齿轮啮合
gear meshing error meter 齿轮啮合误差测量仪
gear-meshing time limit 啮合时限
gear meter 齿轮流量计
gear milling 铣齿
gear milling cutter 齿轮铣刀
gear milling lathe 铲床
gear milling machine 齿轮铣床
gear mission 齿轮变速箱
gear motor 齿轮(降速)马达
gear motor for screw conveyer 螺旋输送器减速电动机
gear mo(u)lding machine 齿轮造型机
gear noise 齿轮噪声
gear noise testing machine 齿轮噪声检查机
gear oil 齿轮油
gear oil additive 齿轮油添加剂
gear oil filler cap 齿轮油进口盖
gear oil pump 齿轮油泵
gear oil pump with connecting bend pipe 带连接弯管的齿轮油泵
gear on worm 蜗杆齿轮
gear operating shaft 齿轮操作轴
gear output shaft 齿轮输出轴
gear pair 齿轮副
gear pair with modified centre distance 角变位圆柱齿轮
gear pair with reference center distance 标准中心距齿轮副
gear pattern 齿轮图型
gear pin 传动销轴
gear pitch measuring instrument 齿轮节距测量仪
gear planer 刨齿机
gear power transmission 齿轮传动
gear preshaving cutter 齿轮剃前刀具
gear processing 齿轮加工程序
gear profile checking machine 齿轮轮廓检验机
gear protuberance cutter 齿轮剃前刀具
gear puller 齿轮拉出器;齿轮拆卸器
gear pump 齿轮泵
gear pump for heavy oil service 重油齿轮泵
gear pump strainer assembly 齿轮泵滤网装置
gear pump with pressurized side plates 带承压侧板的齿轮泵
gear quadrant steering gear 齿扇转舵装置
gear rack 齿(轮)条
gear rail 齿轨
gear range 传动范围;传动比级;齿轮速比范围;齿耗速比范围
gear ratio 速比;传动比;齿轮速比;齿轮(传动)比;齿轮(齿数)比
gear ratio order 传速系数裁定指令
gear reduced drive 齿轮减速传动
gear reducer 减压齿轮;齿轮减速箱;齿轮减速(机)器
gear reduction 减速齿轮
gear reduction ratio 齿轮减速比
gear reduction unit 齿轮减速装置;齿轮减速设备;齿轮减速器

gear rework 齿轮返修
gear rim 大齿圈;齿轮轮缘;齿(槽)轮辋
gear ring 齿轮圈;齿轮环
gear ring chuck 齿圈夹盘
gear roller 齿轮滚柱;齿轮滚片;齿辊
gear roll-finishing machine 齿轮精滚机
gear rolling 齿轮轧制
gear rolling die 齿轮滚模
gear rolling machine 齿轮滚轧机
gear rotary pump 齿轮(回转)泵
gear run 齿轮传动
gear running test machine 齿轮跑合检查仪
gear runout 齿轮摆差
gear scoring 齿轮擦伤
gear scroll chuck 齿轮三爪卡盘
gear sector 扇形齿轮
gear segment 扇形齿轮
gear selection 齿轮选择
gear selector 齿轮选择器
gear selector lever 换挡杆;变速杆
gear set 齿轮组
gear set case 齿轮箱
gear set case suspension 齿轮箱悬架
gear shaft 齿轮轴
gear shaft bush 齿轮轴套
gear shaft roller bearing 齿轮轴滚柱轴承
gear shaft sleeve 齿轮轴套
gear shaper 插齿机;刨齿机
gear shaper cutter 刨齿刀
gear shaper cutting 刨齿
gear shaping 插齿;刨齿
gear shaping cutter 齿轮刨刀
gear shaping machine 插齿机
gear shaping recess 刨齿退刀槽
gear shaping tool 插齿刀
gear shaver 剃齿刀
gear shaving cutter 剃齿刀
gear shaving machine 剃齿机
gear shift 调挡;齿轮变速机构;变速(器);换挡(装置);变速调挡
gear shift bar selector 变速选择杆
gear shift base 变速箱座
gear shift base gasket 变速箱底垫密片
gear shift cover 变速箱盖;变速器盖
gear shift cover gasket 变速箱盖垫密片
gear shift dust cover 变速器防尘罩
gear shifter group 变速杆组
gear shifter shaft 变速拨叉轴
gear shifter shaft lock 变速拨叉轴闭锁器
gear shifter shaft lock spring 变速拨叉轴闭锁器锁簧
gear shift fork 齿轮拨叉;变速拨叉
gear shift fork shaft 变速叉轴
gear shift gate 变速滑槽
gear shift housing 变速箱
gear shifting 换挡;变速(器);变速换挡
gear shifting diagram 齿轮变速图解;变速图解
gear shifting gate 齿轮变速滑槽;变速滑槽
gear shifting lever 齿轮换挡杆
gear shifting lock ball 换挡止动球
gear shifting mechanism 变速机构
gear shifting position diagram 换挡位置图
gear shifting quadrant 变速扇形板
gear shifting time 换挡时间
gear shift lever 换挡杆;变速杆
gear shift lever ball 变速杆球端
gear shift lever ball housing 变速杆球壳
gear shift lever bracket 变速杆架
gear shift lever dust cover 变速杆防尘罩
gear shift lever housing gasket 变速杆壳垫
gear shift lever knob 变速杆捏手
gear shift lever oil seal 变速杆油封
gear shift lever pin 变速杆锁
gear shift lever rubber cap 变速杆防尘罩
gear shift lever spring 变速杆弹簧
gear shift lever spring seat 变速杆簧垫
gear shift operating rod 换挡操纵杆
gear shift pedal 变速踏板
gear shift rail interlock ball 变速杆联锁球
gear shift rail interlock pin 变速轨互锁销
gear shift rail lock ball 变速又轴锁球
gear shift reverse latch spring 变速回动闩弹簧
gear shift rod lock ball 变速杆锁球
gear shift shaft 变速杆轴

gear shift shaft bracket 变速轴托架
gear shift shaft spring 变速杆弹簧
gear shift sleeve 换挡离合器
gear shift sleeve position indicator 换挡离合套位置指示器
gear shift tower 换挡机构箱架;齿轮换挡变速机构箱架
gear side bearing 齿轮边轴承
gear side movement 齿轮端移动
gear slide sleeve 齿轮滑套
gear sliding type transmission 滑动齿轮式变速箱
gear slotter 齿轮插床
gear slotting cutter 插齿床
gear slotting machine 齿轮插床;插齿机
gear sound testing machine 齿轮噪声试验机
gear spade(r) 铲齿机
gear speeder 齿轮调速装置
gear speed reducer 齿轮减速(机)器
gear spindle 齿轴
gear spindle drive 齿轮轴传动
gear spinning pump 齿轮纺丝泵
gear-stage velocity ratio 级速比
gear stand 齿轮支架
gear stock 齿轮床头箱
gear stocking cutter 齿轮粗加工用铣刀
gear switch 变速开关
gear system 齿轮系
gear teeth 齿轮齿
gear teeth ga(u)ge 量齿微分尺;齿规
gear teeth lapping machine 轮齿研磨机
gear tester 齿轮试验机;齿轮检验器;齿轮检查仪;测齿仪
gear testing apparatus 齿轮检查仪
gear testing instrument 齿轮测试仪
gear testing machine 齿轮检查机;齿轮测试机
gear test machine 齿轮试验机
gear thicking ga(u)ge 齿厚规
gear thickness ga(u)ge 齿厚规
gear tip clearance 齿顶间隙
gear tooth 齿轮齿;轮齿
gear tooth burnishing machine 齿轮齿轧光机
gear tooth cal(l)ipers 齿轮游标卡尺;齿轮卡规
gear tooth chamfering and rounding machine 齿轮倒角和修圆机床
gear tooth chamfering machine 齿轮齿倒角机
gear tooth comparator 公法线卡尺;齿厚比较仪
gear tooth engagement 齿轮齿啮合
gear tooth forming 轮齿成形
gear tooth ga(u)ge 轮齿规;齿距规
gear tooth grinding machine 齿轮磨床
gear tooth micrometer 齿厚千分尺;测齿高千分尺
gear tooth profile 齿轮齿的外形;齿轮齿的断面
gear tooth roughing cutter 齿轮齿粗削齿刀
gear tooth rounding machine 齿轮修圆机床
gear tooth shaving machine 剃齿机
gear tooth stocking cutter 齿轮齿粗削齿刀
gear tooth tangent micrometer 齿轮切线千分尺
gear tooth vernier 齿距卡规
gear tooth vernier calipers 齿轮游标卡尺
gear tooth vernier ga(u)ge 游标齿厚仪;齿轮游标卡尺
gear train 传动轮系;齿轮系;齿轮传动链;齿轮传动机构
gear train assembly 齿轮传动
gear train diagram 齿轮系图解
gear train efficiency 齿轮系传动效率
gear transmission 齿轮传动
gear tumbler 齿轮换向器
gear type 齿轮式
gear-type balancer 齿轮式平衡器
gear-type booster pump 齿轮式增压泵
gear-type clevis end wedge-type clamp 齿形双耳楔型线夹
gear-type coupling 齿轮联轴节
gear-type hydraulic motor 齿轮式液压马达
gear-type locating ring 齿形定位环
gear-type motor 齿轮液压马达;齿轮液压电动机
gear-type pump 齿轮液压(型)泵
gear-type recycling pump 齿轮式循环泵
gear-type size pump 齿轮式输浆泵
gear-type spindle 齿型连接轴
gear unit 齿轮装置;齿轮机构
gear up 增速;高速齿轮传动;加速传动;换高速档;齿轮增速

gear wheel 大齿轮;齿(缘)轮
gear wheel drilling machine 齿轮钻机
gear wheel indexing arrangement 齿轮分度装置
gear wheel lubrication grease 齿轮润滑油脂
gear wheel metering pump 齿轮计量泵
gear wheel milling machine 齿轮铣床
gear wheel mo(u)lding machine 齿轮制型机
gear wheel pump 齿轮泵
gear wheel reversing 齿轮换向
gear wheel Rockweel tester 齿轮洛氏硬度计
gear wheel sequence 齿轮排列次序
gear wheel shaft 齿轮轴
gear wheel shaving toothed tool 齿轮剃齿刀
gear wheel with curved teeth 弧齿齿轮
gear with curved teeth 弧齿轮
gear with drawer 齿轮拆卸器
gear with equaladdendum teeth 等高齿顶齿轮
gear-within-gear motor 内啮合齿轮马达
gear-within-gear pump 内啮合齿轮泵
gear with jaw 带楔齿轮
Geber model automatic drafting system 格贝尔自动绘图系统
gebhardite 葛氯砷铅石
Gebhardt apparatus 盖勃哈特测斜仪(测定钻孔倾斜度用)
Gecalloy 盖克洛磁芯合金;铁粉磁芯用镍铁合金
Geco flo(a)tation cell 盖柯型浮选机
gecondary focus 次级焦点
gedanite 脂状琥珀;软琥珀
Gedge's alloy 格迪奇合金
Gedinian (stage) 葛丁阶【地】
gedrite 铝直闪石;铝直闪岩
gedunoha 桃花心木(红褐色硬木,产于西非和东非)
Geepound 斯勒格
geerite 吉硫铜矿
Geerz sun compass 吉尔兹太阳罗盘
Geetainer 奇泰纳(一种盘式集装箱)
geffroyite 盖硒铜矿
Geg 吉克旋风
gegenschein 对日照
gehelinite 铝方柱石
gehlenite 钙(铝)黄长石
gehlenite hydrate 水化黄长石
Gehlhoff spring 盖尔夫热点;盖尔夫泉
geic acid 赤榆酸
Geigel's reflex 盖格尔氏反射
Geiger counter 盖革氏离子计数器;盖革计数器
Geiger counter tube 盖革计数管
Geiger counting region 盖革区
Geiger-Muler counter 盖革—弥勒计数器
Geiger-Muler probe 盖革—弥勒计数器的探测设备
Geiger plateau 盖革坪
Geiger probe 盖革计数器
Geiger probing equipment 盖革计数器探测设备
Geiger steel analyzer[analyser] 盖革X射线钢分析仪
Geiger test 盖革试验
Geiger threshold 盖革阈
Geiger threshold potential 盖革阈势
Geiger-tube telescope 盖革计数望远镜
geigyne meteorology 卫生气象学
geikielite 镁钛矿
geine 土壤有机质
geison 柱顶檐悬挑;三角形檐饰悬挑;飞檐
Geissler pump 盖斯勒泵
Geissler tube 盖斯勒管
geking 电磁海流计测流
gel 胶滞体;胶质体;胶凝体;胶冻;凝胶(体);明胶;冻胶;胶化
gel addition process 凝胶加添过程
gel alumin(i)um hydroxide 氢氧化铝胶
Gelamite 吉拉买特半胶质炸药
gelasin 琼脂素
gelata 凝胶剂
gelate 胶凝
gelatification 凝胶(化)作用
gelatina 明胶剂
gelatinate 明胶合物
gelatination 胶凝作用;胶化;凝胶化
gelatin(e) 胶(质);凝胶;明胶;白明胶
gelatin(e) blasting 胶质炸药;胶体炸药
gelatin(e) board 明胶纸板
gelatin(e) capsule 胶囊
gelatin(e)-coated paper 明胶涂层纸

gelatin(e) coating 明胶涂层;明胶膜
gelatin(e)-compass (测孔斜的)凝胶罗盘
gelatin(e) culture 明胶培养
gelatin(e) culture medium 明胶培养基
gelatin(e) dry filter 明胶滤色片
gelatin(e) dry photographic(al) plate 明胶干照相底版
gelatin(e) dry plate 明胶干版
gelatin(e) duplicating 明胶版复印
gelatin(e) duplicating process paper 明胶版复印纸
gelatin(e) dynamite 黄色炸药;高强炸药;胶质(硝化甘油)炸药;胶结炸药;明胶炸药;硝化甘油炸药
gelatin(e) effect 明胶效应
gelatin(e) embedding 明胶包埋法
gelatin(e) emulsion 明胶乳剂
gelatin(e) envelope 胶质封套
gelatin(e) explosive 明胶炸药
gelatin(e) extra 特制炸胶
gelatin(e) film 明胶膜
gelatin(e) filter 明胶滤色器;明胶滤光片;胶质滤光片
gelatin(e) foam 明胶海绵
gelatin(e) foil 明胶片;明胶薄膜
gelatin(e) glaze 明胶包衣
gelatin(e) glue 明胶;动物胶
gelatin(e) image 明胶相片
gelatin(e) layer 凝胶层;明胶层;明胶膜
gelatin(e) liquefaction 明胶液化
gelatin(e) liquefaction test 明胶液化试验
gelatin(e) model 电解模型
gelatin(e) mo(u)ld 明胶石膏模
gelatin(e) mo(u)lding 明胶制模;明胶模铸线脚;胶模;胶模造型(法)
gelatin(e) network model 明胶网格模型
gelatineous 胶质的;胶状(的)
gelatin(eous) (blasting) explosive 明胶炸药;胶状的爆炸药物
gelatin(e) overcoat 明胶涂层
gelatine-pad printing 凸版印刷法
gelatin(e) paper 明胶纸
gelatin(e) plate 胶板;明胶照相干版;明胶板(照相制板)
gelatin(e) printing 明胶复印;明胶版印刷
gelatin(e) printing plate 明胶印刷版
gelatin(e) process 明胶制版法;胶版
gelatin(e) relief 明胶浮雕
gelatin(e) relief process 明胶凸版制版法
gelatin(e) silk 凝胶丝
gelatin(e) size 明胶浆
gelatin(e) solution spinning 凝胶液纺丝
gelatin(e) sponge 胶棉球;明胶海绵;海绵胶
gelatin(e) structure 凝胶构造
gelatin(e) substitute 明胶代用品
gelatin(e) supercoat 透明胶表面涂层
gelatin(e) treatment 明胶处理
gelatin(e) waste 废动物胶
gelatin(e) wastewater 明胶废水
gelating 凝胶化
gelating method (测孔斜的)凝胶法
gelating relief image 明胶相片
gelatini form 明胶样的
gelatinisation 胶化(作用)
gelatinization 胶化;胶化(作用);凝胶化;明胶化
gelatinize 胶质化;凝胶化
gelatinized cellulose 胶化纤维素
gelatinized gasoline 凝固汽油
gelatinized silk 涂胶丝
gelatinizer 胶凝剂;胶化物;稠化剂
gelatinizing agent 胶凝剂;胶化剂
gelatino-bromide paper 明胶溴化银相纸
gelatino-bromide plate 明胶溴化银干板
gelatino-chloride paper 明胶氯化银相纸
gelatinoid 明胶样的
gelatinolytic 明胶分解的
gelatinous 胶的;胶凝状的
gelatinous curd 凝胶状乳凝块
gelatinous curd deject 凝乳结块不良
gelatinous explosive 胶质炸药
gelatinous fiber[fibre] 凝胶纤维;黏质纤维;胶质纤维
gelatinous precipitate 凝胶状沉淀
gelatinous substance 胶状质

gelatinous tissue 胶性组织
gelatinum alba 白明胶
gelation 胶凝作用;胶化(作用);凝胶化;明胶化
gelation model 冻胶(电)模型
gelation of ink 油墨胶凝
gelation temperature 胶凝温度
gelation test 胶化试验;凝胶试验
gelation time 胶凝时间;胶凝期
gelatum 胶冻;凝胶
gel-cement 胶质水泥
gel chromatography 凝胶色谱(法);凝胶层析
gel coat 胶衣(层);表面涂漆;胶面;可以厚涂的耐化学的触变性聚酯涂料;凝胶漆
gel coated surface 凝胶涂刷面
gel coat(ing) 凝胶涂料
gel coat resin 胶衣树脂
gelcoat sprayer 凝胶喷射器
gel coherence 凝胶内聚力
gel column 凝胶柱
gel condition 凝凝状态;胶凝状态
gel consistency 凝胶稠度
gel deposition 凝胶沉积
gel diffusion 凝胶扩散
gel diffusion method 凝胶扩散法
gel diffusion precipitation 凝胶扩散沉淀试验
gel-dipping 胶漆浸涂法
gel disc electrophoresis 凝胶盘状电泳
gel dy(e)ing method 凝胶染色法
gel effect 凝胶效应
gelefusainization 凝胶丝炭化作用
gel electrofocusing 凝胶电聚焦
gel electrophoresis 凝胶电泳
gelemeter 凝胶时间测定计
gel exclusion chromatography 凝胶排阻色谱(法)
gel fiber 凝胶纤维
gelfilm 明胶软片
gel filtration 胶过滤;凝胶过滤
gel filtration chromatography 凝胶过滤色谱(法);凝胶过滤层析
gel formation 凝胶形成
gel forming 形成胶体;胶质形成的
gel fractionator 凝胶分段分离器
gel grained texture 胶粒结构
gel groundmass 胶体基质
gel growth method 凝胶生长法
gelid 冰冷的
gelified circleinite 凝胶化浑圆体
gelified groudmassinite 凝胶化基质体
gelified sclerotinite 凝胶化菌类体
gelifluxion 冰缘土溜;冰冻泥流
gelifraction 融冻崩解作用
gelifusinite 胶丝质类
gelifusinite-collinite 胶丝质类无结构体
gelifusinite-posttelinite 胶丝质类次结构体
gelifusinite-precollinite 胶丝质似无结构体
gelifusinite-telinite 胶丝质结构体
gelignite 硝酸甘油胶质炸药;炸胶;葛里炸药
gelinite 凝胶体;凝胶类
gelinite-collinite 胶质无结构体
gelinite group 凝胶化组
gelinite-posttelinite 胶质次结构体
gelinite-procollinite 胶质似无结构体
gelinite-sclerotinite 胶质菌核体
gelinite-telinite 胶质结构体
gel initial 初静切力(泥浆)
geliturbation 融冻泥流(作用)
gelivation 融冻泥流(作用);融冻崩解作用;冰劈(作用);冰冻作用
gel lacquer 凝胶型涂料;凝胶漆
gelled cell 胶质电池
gelled electrolyte 胶化电解液
gelled fibre 凝胶纤维
gelled gasoline 凝固汽油
gelled patterning and gilding 沥粉贴金
gel-like 胶状的
gel-like compound 凝胶状化合物;凝胶状的
gelling 胶凝;凝胶化作用;胶化
gelling agent 修补膏;胶凝剂
gelling material 凝胶物质
gelling technique 凝胶化法
gelling type rust preventive 凝胶型防锈添加剂
gel membrane 胶膜
gel mineral 无定形矿物;似矿物

gel network 凝胶网络
gelocollinite 胶质镜质体
gelodiagnosis 凝胶鉴定法
gelose 琼脂
gelosic coal 藻煤;胶质煤
gelosis 凝块
gel osmometer 凝胶渗压计
gel paint 凝胶漆
gel particle 凝胶块;胶凝粒子
gel permeation chromatography 凝胶渗透色谱(法);凝胶渗透层析
gel permeation liquid chromatography 凝胶渗透液相色谱(法);凝胶渗透液相层析
gel phase 胶相
gel point 胶凝点
gel polymer 胶凝聚合物
gel pore 凝胶(气)孔;凝胶孔隙;凝胶空隙
gel pore volume 凝胶体积
gel pore water 凝胶孔水
gel porosity 凝胶孔隙率
gel precipitation 凝胶析出
gel precipitation probability 凝胶沉淀概率
gel precipitation process 凝胶沉淀过程
gel precipitin test 凝胶扩散沉淀试验
gelquartzite 胶结石英岩
gel rubber 凝(胶)橡胶
gels 凝胶剂
gel scintillator 凝胶闪烁体
gel skeleton 凝胶骨架
gel solubilizing warmer 明胶溶化保温器
gel-space ratio 胶孔比;凝胶空间比
gel spinning 凝胶自旋;凝胶纺丝
gel sponge 凝胶海绵
gel state 凝胶状态
gel strength 凝胶强度
gel strength of drilling fluid 井液的胶凝强度
gel structure 胶状构造;凝胶结构
gel sub 凝胶底层
gel surface area 凝胶表面积
gel swelling 凝胶溶胀度
gel ten minutes 静置10分钟的泥浆静切力
gel test 胶化试验;凝胶试验
gel time 胶凝时间;凝胶时间
gel time of tung oil 桐油的胶化时间
gel treatment 凝胶处理
gel type 凝胶型
gel type binder 凝胶型基料
gel type ion exchange resin 凝胶型阳离子树脂;凝胶态离子交换树脂
gel type resin 凝胶型树脂
gel type solid 凝胶型固体
gelutong 节路顿树脂;节路顿胶
gel-void ratio 凝胶空间比
gel water 凝胶水
gem 宝石
gem crystal 宝石晶体
gem cutting 宝石琢磨
gem diamond 宝石级金刚石
gem-dimethyl 偕二甲基
gemeine hornblende 普通角闪石
gemel 铰链
gemel arch 铰链拱;对拱
gemel hinge 环钩铰链
gemel window 对窗;李窗
gem grade diamond 优质金刚石;宝石级金刚石
gem gravels 宝石砾层
geminate 加倍;成双
geminated columns 李柱;双柱;对柱
gemini 充气橡胶艇
GEM lighter 气垫驳船
gemmary 宝石学;宝石雕刻师
gemmels 铰链(旧名)
gemmho 微姆欧
gem mineral 宝石矿物
gem mounting 宝石镶嵌
gemmule 侧棘
gemnology 宝石学
Gemolite 宝石检查镜
Gemowinkel 正切规
gemstone analyzer 宝石分析仪
gemstone and jadestone commodities 宝石及玉石矿产
gemstone and jewel(l)ery industry 珍宝业
gem whisker 白宝石晶须

Genarco alloy 吉纳科合金
gencon 一般租船契约
genealogical classification 系统分类
genealogical table 系谱表
genealogical tree 系谱树;谱系树
genealogy 系统学;系谱;宗谱;谱系
genecology 物种生态学
Genelite 非润滑烧结青铜轴承合金
genemotor 电动发动机
general 总的;概述;将军俑
general, sub-contracting of construction 施工总分包
general acceptability criterion 普通验收标准;一般可接受的准则
general acceptance 一般承兑
general acceptance criterion 普通验收标准
general account 总账;一般会计;普通账户
general accountant 总会计师
general accounting 普通会计
general accounting function 总会计职能
general accounting office 总会计师室
general accounting officer 总会计师
general accounting principles 一般会计原理
general accounting system 一般会计制度;普通会计制度
general act 总协议书
general activity simulation program (me) 一般活动仿真程序
general adaption syndrome 一般适应性综合征
general addition theorem 一般加法定理
general address 综合地址;通用寄存器地址
general address reading device 通用地址阅读装置
general administration cost 一般管理费用
general administration skill 一般管理技术
general administrative and selling expenses 一般管理和销售费用;一般管理费和销售费用
general administrative expense budget 一般管理费预算
general administrative expenses 一般管理费用;普通管理费用
general adoption of the market principle 市场化
general affairs 总务
general affairs department 综合办公室
general affairs section 总务科
general afforestation 普通造林
general agency 一般代理;总经销;总代理
general agent 一般代理人;总代理行;总代理商;总代理人
general agreement on participation 参股总协定
General Agreement on Tariff and Trade 关税与贸易总协定;关贸总协定
General Agreement on Trade and Tariffs 贸易及关税总协定
general agreements to borrow 借款总安排
general air change 全面通风
General Aircraft Corporation 通用飞机公司(美国)
general air traffic 普通空中交通
general alarm switch 总警报开关
general algebraic variety 一般代数簇
general algorithm 一般算法
general ambient light 总照明
general anchorage 一般锚地
general and administrative expenses 总务及管理费用
general and conventioned tariff 普通及协定关税
general annealing glass 普通退火玻璃
general announcing system 全船广播系统
general annual reports 年度总(决算)报告
general anomaly map 异常综合图
general apparatus of logging 测井通用设备
general appearance 总外貌
general application 普通施用
general aptitude test battery 一般能力成套测验
general arrangement 总图布置;总体布置(图);总(平面)布置;总纲;总安排;通用装置;船舶总布置
general arrangement diagram 总体布置图
general arrangement drawing 总平面图;总(体)布置图
general arrangement (layout) 总图
general arrangement of feed water heater 给水加热器总图
general arrangement of port 港口总体规划
general arrangement of the apparatus 设备的总体布置

general arrangement of works 工程总体布置
general arrangement plan 总体布置图;总平面布置图
general arrangement plan of construction 施工总平面图
general arrangement to borrow 一般借款协定;借款总安排
general arrival 一般到达
general articles 总则
general aspects 一般性质;一般特性;一般情况
general assembling 总装
general assembly 总装配;总安装;全体大会
general assembly drawing 总装配图;总安装图
general assembly line 总装配线
general assignment 总任务;全部财产的移交
general astronomy 普通天文学
general atlas 普通地图集
general atmospheric circulation 大气总循环;大气总环流
general attacking 普遍发作
general audit 普通审计
general average 一般平均值;总平均值;共同海损
general average act 共同海损行为;共同海损条例
general average adjuster 共同海损理算师
general average adjustment 共同海损理算
general average and salvage 共同海损及救助
general average contribution 共同海损分摊额;共同海损分担
general average deposit 共同海损备用金
general average guarantee 共同海损担保书
general average loss 共同海损损失
general average sacrifice 共同海损牺牲
general average security 共同海损担保
general bacterial population 一般细菌数
general balance 总贸易差额
general balance equation 通用平衡方程
general balance sheet 总资产负债表
general banking 一般银行业务
general base level 总基准面
general base map 一般基本图;基本地图
general basic geologic (al) data base 通用基础地质数据库
general bathymetric (al) chart of the oceans 普通海洋等深线图;大洋水深总图;世界大洋水深图
general bed load transport 一般推移质搬运;一般底沙搬运;普通底沙输移
general benefit 一般收益;总效益
general bid 公开投标
general bill of lading 货运总单;提货单
general binding machine 通用装订机
general biofilter 普通生物滤池
general boat alarm signal 救生信号
general bonded-debt and interest group of accounts 普通债券本息账类
general bonded-debt fund 一般债券基金
general boundary condition 广义边界条件
general boundary value problem 一般边值问题
general broadcast 综合广播
general budget 总预算
general building contractor 总建筑承包者;总建筑承包商;总建筑承包人
general bulk carrier 散装杂货船
general burden 一般间接费(用)
general business environment 一般企业环境
general call 全呼
general calling 全体叫通
general call to all stations 普通呼叫
general call to two or more specific (al) stations 选台呼叫
general capital 总资本;总股本
general capital increase 普遍增加资本
general caretaking expenses 总管理费
general cargo 一般杂货;一般客货;一般货物;载重船;杂货;件杂货;普通货(物);百杂货
general cargo berth 件杂货泊位
general cargo breakbulk wharf 件杂货码头
general cargo carrier 普通货轮
general cargo conveyer 件货输送机
general cargo handling 件杂货装卸
general cargo rate 一般货物运费率;杂货运价率;杂货费率
general cargo ship 杂货船;件杂货船;普通货轮
general cargo (stack) yard 件杂货堆场

general cargo terminal 杂货码头;件杂货码头
general cargo traffic 杂货运输
general cargo vessel 件杂货船
general cargo wharf 一般客货码头;杂货码头;件杂货码头
general cargo zone 件杂货(作业)区
general cartography 普通地图制图学
general case 普通情况
general cash 一般现金;普通现金
general casher 总出纳
general catalog (ue) 总目录;普通目录
General Catalogue of Variable Stars 变星总表
general census 全面普查
general chaos in bookkeeping 财力混乱
general character 共性
general characteristic 一般特性;通性
general characteristic of reservoir 储集层一般特征
general characteristic of source rock 烃源岩一般特征
general charges 总管理费
general chart 一览图;总图;海图总图;普通海图
general chart of a coast 海岸总图
general chart of coast 沿海总图
general chart of sea 海区总图
general chemical index 综合化学指数
general chief engineer 总轮机长
general circulation 体循环;大循环
general circulation model of atmosphere 大气环流模式
general circulation of atmosphere 大气循环;大气环流
general circulation of the ocean 海洋大气环流
general claim agent 索赔总代理人
general classification test 普通入级检验
general clause 一般条款;普通条款
general clean-up 大扫除
general clerical test 一般文书工作测验
general clerk 一般事务员
general closing 普通决算
general close planting 普遍密植
general coastline 化简海岸线
general cohesive soil 一般黏性土
general colo (u) r rendering index 一般显色指数;一般显色指数
general combining ability 一般配合力(效应)
general comment 一般注释
general communication 常规通信
general commutative ring 一般交换环
general compact space 一般紧致空间
general comparator 通用比较器
general competitive analysis 一般竞争分析
general computer 普通计算机;通用计算机
general concept of mineral deposit 矿床一般概念
general concrete 一般混凝土;普通混凝土
general condition 概况;一般条件;普通保险条款;总则
general condition of contract 合同的一般条件
general conditions for the delivery of goods 一般商品交付条件
general conditions of construction 施工总则;施工概况
general conditions of contract 标准承包合同条款;通用合同条款
general conditions of delivery of goods 交货共同条件
general conditions of sale 买卖共同条件
general condition survey 一般状况检查
General Conference of Weights and Measures 国际计量大会
general confocal resonator 泛共焦共振腔
general connection diagram 总接线图;全部设备接线图
general considerations 一般问题;总体设想;总体考虑;概论
general constant 通用常数
general construction 总结构
general construction schedule 施工总进度
general contingency reserve 一般意外盈余准备;一般意外损失准备金
general contract 总包合约;总(承)包合同;综合契约
general contractor 建筑公司;总(承)包者;总承包商;总承包人
general contractor's fixed price offer 总承包商固定价格开价单

general contractor and subcontractor 总包与分包
general contractor order 总承包商订货单
general contractual liabilities 合同的一般责任
general control 一般管制;综合控制
general control area 一般管制区
general control panel 总控制盘
general coordinating budget 一般协调预算
general corrosion 一般腐蚀;全面腐蚀;普遍腐蚀
general cost 一般费用;一般成本;完全成本
general cost function 一般价值函数
general cost of administration 一般管理费;总管理费
general cost of transportation 运输总支出
General Council of Trade Unions (of Japan) [SOHYO] 工会总评议会(日本)
general counsel 总法律顾问
general credit 一般信用证
general creditor 普通债权人;一般债权人
general crossed check 一般横线支票
general crossed product 一般叉积
general crossing 普通划线;无担保项目
general custom 一般习惯
general cutting-plan 总采
general damage 普通损坏
general damage to buildings 房屋普遍破坏
general data 一般数据
general datum 基本资料
general debt 普通债务
general deduction 一般减免
general delivery 邮件的存局候领处
general density 总体谱密度
general depot 补给总库
general derivative 一般导数
general description 一般说明;总说明
general description of construction 构造说明;施工总则;施工说明(书)
general description of operation 操作方法描述
general design 一般设计;总(体)设计;综合设计;初步设计
general design of port 港口总体设计
general destiny 普遍命运
general destruction of buildings 房屋普遍破坏
general detuning 一般失调
general device 普通电极系
general device resistivity log 普通电极系电阻率测井
general device resistivity log curve 普通电极系电阻率测井曲线
general diagram 总体布置图;总图
general diffuse(d) lighting 漫射光均匀照明;一般漫射照明
general diffuse(d) ray 漫射光线
general diffuse luminaire (fixture) 普通漫射光照明装置
general diffusion equation 一般扩散方程
general digestion 普通消化
general digital computer 通用数字计算机
general dimension 总(体)尺寸;主要尺寸;概要尺寸
general direction of coast 海岸带的一般走向
general direction of traffic flow 船舶总流向
general director's room 总指挥室
general discharging instructions 卸箱指示(集装箱)
general discretized model 一般间断模型
general discussion 一般讨论
general distribution 一般分配法;一般分布
general distribution system 普通供电系统
general distributor 总批发商
general domain 一般域
general drawing 一览图;总图;总体布置图;概(要)图;全图
general dry cargo 普通干货
general dry cargo ship 普通干货船
general duty 一般用途;万能的
General Dynamics Corp. 通用动力公司(美国)
general ecology 普通生态学
general economic analysis 一般经济分析
general economic equilibrium 一般经济均衡
general economic law 一般经济规律
general economic statistics 一般经济统计
general efficiency 全体效率
General Electric Co. Limited 通用电气公司(美国)
general electric(al) image system 通用电像系统
general electric(al) synchrotron 广义电同步加速
general elliptic type 一般椭圆形
general endorsement 一般背书;不记名背书
general energy 一般能源
General Energy Research Council of Japan 日本一般能源研究委员会
general engineer 总工程师
general engineering 通用工程
general engineering contractor 总工程承包者;总工程承包商;总工程承包人
general engineering geologic(al) map 综合工程地质图
general engineering research 一般工程研究
general engineering set 总施工图;工程设计施工详图
general engineering system 通用工程系统
general environment 一般环境;总环境
general environmental condition 总体环境条件
general environmental impact 总体环境影响
general environmental message 总体环境通报
general environmental sustainability 总体环境可持续性
general equation 广义方程;一般方程;通用公式
general equation of second degree 一般二次方程
general equilibrium 一般均衡;总体均衡;全面均衡
general equilibrium analysis 一般均衡分析;全面平衡分析法
general equilibrium model 一般均衡模型
general equilibrium noise 一般均衡噪声
general equilibrium state 一般均衡状态
general equilibrium system 一般均衡系统;全面平衡体系
general equilibrium theory 一般均衡(理)论
general equipment 通用设备
general equipment package 通用设备包
general equipment package for depot 车辆段通用设备包
general equivalent in value 一般等价物
general equivalent points 一般等效点
general equivalent point system 一般等效点系
general Erlangian distribution 一般爱尔朗分布
general erosion 一般冲刷
general error 总误差
general escape 广义换码
general estimate 总概算;概算;设计概算;初步估算;初步估计
general estimate method 工程粗略总预算;概算粗估法;粗略总预算
general estimate of bridge construction 桥梁施工概算
general estimation 一般判定
general Euclidean connections 一般欧几里得联络
general evaluation 总评价
general excavation 普通开挖
general exhaust mechanical ventilation 全面抽出式机械通风
general exhaust ventilation 全面排风
general expansive cement 普通膨胀水泥
general expenses 一般(营业)费用;总管理费;综合费用;日用开支;普通费用;普通费用
general expense reappointment 一般费用再分配
general exponential smoothing 一般指数修匀
general export 总出口
general expression 一般式;普通式【数】;通式
general extension 大体的扩建;总伸长;均匀伸长;均匀拉伸;全面扩建
general fabrication method 普通装配方法;普通制造方法
general factor 一般因子;一般因素
general factory 总厂
general failure 一般性破坏
general fallow 普遍休闲
general farm 多种经营农场
general feasible solution 基本可行解
general feature 一般特性;一般特征
general features of construction 施工概要
general feedback servomechanism 一般的带有反馈的伺服机构
general field equations 一般场方程
general file 通用文件
general file translator 通用文件翻译程序
general filtering problem 一般滤波问题
general final accounts 总决算
general finishing 一般光制

general fishing tools 事故处理一般工具
general fixed assets 一般固定资产;普通固定资产
general fixed assets fund 一般固定资产基金
general fixed assets group of accounts 一般固定资产账类
general flexibility 总柔度
general flow 综合流程
general flow chart 总框图;综合流程图;总流程图;总操作程序图
General Foods Corp. 通用食品公司(美国)
general forecast 一般气象预报;综合天气预报
general forecasting 一般预报
general foreman 总领工员;总领班;总监(工);总工务员;总工头;总工长
general form 运算通式;一般形;一般格式;通式
general format 一般格式
general form of matrix 矩阵通式
general form of value 一般价值形式
general formula 一般公式;通用公式;通式
general Fourier transformation 一般傅立叶变换
general freight agent 货运总代理
general freight station 综合性货运站
general fund 普通基金
general gas law 理想气体方程;理想气体定律;普遍气体定律
general geography 普通地理学
general geology 普通地质学;地质概况
general geometry of path 一般道路几何
general geothermal gradient 总地热梯度
general goal of the macrocontrol 宏观控制的总目标
general goods 一般商品;杂货;普通商品;通用产品
general goods platform 普通站台
general goods station 综合性货运站
general government expenditures 一般行政费用
general government expenses 普通行政费用
general grant 一般性拨款;统筹财政补贴
general grouting 一般灌浆
general heat exchange 一般性热置换
general helix 一般螺旋线
general hoist rope 一般起吊绳索
general hospital 综合医院
general hyperplane 一般超平面
general idea 一般概念
general illumination 一般照明;全面照明
general illumination design 一般照明设计
general import 总进口
general improvement 一般设施;公共设施;全面好转
general impulsive force 一般冲力
general income 一般收入;总收入(量)
general income tax 一般所得税
general increase 一般涨价
general index 综合指数;总指数;总索引
general index number 总指数;普通指数
general index of retail price 一般消费物价指数
general indirect labo(u)r cost 普通间接人工费用
general industrial occupancy 常规工业建筑
general industrial statistics 一般产业统计
general industry 一般工业
general inference 一般天气预测;一般天气推断;天气形势预报
general information 一般资料;组织简介;总论
general information lossless 一般信息无损耗
general information processing system 通用信息处理系统
general inquiry 一般询问
general inspection 一般(性)检查;普遍检查
general installation drawing 总安装图
general instruction 总论
general instruction to tenders 投标须知
general insurance policy 预定保险;总保险单;长期有效保险单
general integral 一般积分;通积分
general interests 全局利益
general interference 一般干扰
general internal combustion engine plant 内燃机总厂
general interpretative program(me) 通用解释程序
general intersection theory 一般交点理论
general inventory 财产目录
general inversion 普通回返
general investigation 一般调查;普查
general isoplanatism theorem 广义等晕定理

generalist 多面手
generality 一般性;共性;概论;普遍性;通用性
generalization 一般化;归纳;广义(性);概括;全面化;普遍化;推广;通则;泛化
generalization marking 标描
generalization of binomial theorem 二项式定理的推广
generalization of pollution sources 污染源概况
generalizations on cost-volume profit analysis 成本—产量—利润分析原理
generalization transfer 标描过渡
generalize 总结;概括
generalized 广义的;通用化
generalized acceleration 广义加速度
generalized approximants 广义近似式
generalized arrangement 广义排列
generalized binomial distribution 广义二项分布
generalized binomial trials model 广义二项试验模型
generalized Bruns equation 广义布隆斯方程
generalized capillary potential 广义毛管势
generalized chart 概化曲线图
generalized circuit 广义回路
generalized compressibility factor 通用压缩因数
generalized condition 广义条件
generalized constant 通用化常数;标准化常数
generalized continuum mechanics 广义连续力学
generalized contour 综合等高线;简化等高线
generalized convexity 广义凸性
generalized convex programming 广义凸规则
generalized convolution 广义褶积;广义褶合式
generalized coordinates 广义坐标
generalized correlation 广义相关
generalized cost 广义费用
generalized Coulomb's equation 广义库仑方程
generalized Coulomb's formula 广义库仑公式
generalized country risk 一般化国别风险
generalized damping 广义阻尼
generalized data manipulation program(me) 一般化数据操作程序
generalized diffraction grating 广义衍射光栅
generalized displacement 广义位移
generalized drawing primitive 广义绘图图元
generalized effectiveness function value 广义效用函数值
generalized eigenvalue 广义特征值
generalized eigenvalue problem 广义特征值问题;广义本征值问题
generalized ensemble 广义组合;广义系统;概化系综
generalized equation 广义方程
generalized error 广义误差
generalized error-correcting tree automaton 广义的误差校正树状自动机
generalized estimate 概化估计
generalized exciting force 广义激振力
generalized experience 总结经验;综合经验
generalized expression 广义表达式
generalized fifth order aberration 广义五级像差
generalized file processing 广义文件处理;通用文件处理系统
generalized finite automaton theory 广义有限自动机理论
generalized floating 普遍浮动
generalized fluidization 广义流态化
generalized force 总合成力;广义力
generalized forcing function 广义力函数
generalized form of value general law of addition 一般价值形式
generalized Fourier analysis 广义傅立叶分析
generalized Fourier transform 广义傅立叶变换
generalized function 广义函数
generalized geoid 广义大地水准面
generalized geologic(al) section 综合地质剖面
generalized geometric(al) stiffness 广义几何刚度
generalized harmonic analysis 综合谐波分析;广义调和分析;概化调和分析
generalized heat capacity correction 通用热容校正
generalized Hooke's law 广义虎克定律
generalized hydrostatic(al) equation 广义流体静力学方程
generalized identity function 广义单位函数
generalized impedance 广义阻抗
generalized inertia force 广义惯性力

generalized inflation 全面通货膨胀
generalized information system 通用信息系统
generalized inverse matrix 广义逆矩阵
generalized job selection 广义作业选择
generalized journey fee 广义出行费用
generalized Lagrange invariant 广义拉格朗日不变量
generalized Laguerre function 广义拉盖尔函数
generalized Laguerre polynomial 广义拉盖尔多项式
generalized least square method 广义最小二乘法
generalized limit 广义极限
generalized linear system 广义线性系统
generalized list 广义(链)表
generalized load 广义荷载
generalized machine 一般化电机
generalized main sequence 广义主序
generalized map of hydrogeologic(al) condition 水文地质条件概化图
generalized marketing 广义市场
generalized mass 广义质量
generalized mathematical induction 广义数学归纳法
generalized matrix inverse 广义逆矩阵
generalized modulus 广义模数
generalized modus ponens 广义假言推理
generalized momentum 广义动量
generalized monetary 广义货币
generalized network simulator 通用网络模拟器
generalized Newtonian fluid 广义牛顿流体
generalized nodal displacement 广义节点位移
generalized nodal force 广义节点力
generalized operating procedure 一般作业程序
generalized phasor 一般化相量
generalized physical model 概化(实体)模型
generalized plane stress 概化平面应力
generalized Poisson distribution 广义泊松分布
generalized Poisson process 广义泊松过程
generalized preference 普遍优惠
generalized procedure of slices 广义条分法
generalized programming 通用程序设计
generalized programming language 通用程序设计语言
generalized projection 广义投影
generalized property 广义特性
generalized pupil function 广义光瞳函数
generalized quantity 广义量
generalized reduced gradient method 广义既约梯度法
generalized relative aperture 广义相对孔径
generalized rheological body 广义流变体
generalized routine 广义例行程序;广义例程;通用程序;标准程序
generalized scheme of preferences 一般特惠制
generalized section 一般化剖面;综合剖面
generalized series 广义级数
generalized shear failure 整体剪切破坏
generalized simplex method 广义单纯形法
generalized single-regime s-c model 广义单态速度—密度模型
generalized solid model 广义固体模型
generalized solution 广义解
generalized solution of Lagrange 广义拉格朗日解
generalized sort program(me) 通用分类程序
generalized spectral density 广义谱密度
generalized stiffness 广义刚度
generalized stock production 综合群体生产模式
generalized Stokes formula 广义司托克斯公式
generalized stress 广义应力
generalized stress condition 广义应力状态
generalized subroutine 广义子例程
generalized system of preferences 普遍优惠制
generalized trace facility 广义跟踪功能;通用跟踪功能
generalized transduction 一般换能
generalized transmission function 广义透射函数;广义传递函数
generalized two-dimensional Fourier transform 广义二维傅立叶变换
generalized variance 广义方差
generalized variational principle 广义变分原理
generalized vector grating ray-tracing equation 广义矢量光栅追迹方程
generalized velocity 广义速度

generalized vibration equation 广义振动方程
generalized wind wave spectrum 普遍风浪谱
generalize flood 普通化
generalizing 一般化;普遍化
general job 一般作业
general joiner 万能木材加工机
general journal 普通日记账
general journal entry 分录
general knowledge 一般知识;常识
general laboratory method 普通实验室方法
general labor process 一般劳动过程
general labo(u)r 壮工;普通工(人)
general law of error 一般误差定律
general law of reciprocity 一般互反律
general layout 总体布置;总平面(图);总计划;总布置(图);平面布置总图
general layout design of aids-to-navigation 航标总体配布设计
general layout of construction 施工总(平面)布置图
general layout of mine area 矿区总体设计
general layout of outdoor pipelines 室外管道综合图
general layout of outdoor water supply and drainage system 室外给排水总平面图
general layout of piping system 管道综合图
general layout of port 港口总图规划;港口总体布置
general layout of project 工程总体布置
general layout of sewage treatment plants 污水厂总体布置
general learning ability test 一般学习能力测验
general ledger 总账;清账;普通分类账;分类(总)账
general ledger accounts 总分类账
general ledger for property and commodities 财产商品总分类账
general left recurrence 一般左递归
general length of service 一般工龄
general letter of credit 一般信用证
general letter of hypothecation 一般质押权;一般押汇质权书;一般押汇担保函;押汇总质押书
general level 普通水准仪
general level of employment 一般就业水平
general level of market prices 物价总水平
general level of prices 一般价格水准
general level of retail price 零售物价总水平
general license 一般许可证
general lien 一般留置权;总留置权
general light and power distribution 一般照明及电力配电
general lighting 一般照明;全区照明
general lighting design 一般照明设计
general lighting distribution box 工作照明配电箱
general line 总路线
general linear differential equation 一般线性微分方程
general linear group 一般线性群
general linear hypothesis 一般线性假设
general linear model 一般线性模型
general linear recurrence system 一般线性递归系统
general linear regression model 一般线性回归模型
general linear transformation 一般线性变换
general list 总清单
general loan and collateral agreement 一般仿射一般放款抵押契约
general locality 总地区
general local lighting 分区一般照明;分区普通照明
general location map 平面位置图
general location plan 区域总平面图
general location sheet 总位置图;位置(总)图;地盘图
general locking 集中同步;强制同步;强迫同步系统;同步锁相;台从同步
general logarithm 一般对数
general lower derivate 一般下导数
generally 大体上
generally accepted accounting principles 一般公认会计原则;通常公认的会计原则
generally accepted auditing standards 通常公认的审计标准
generally accepted standard 通用标准
generally acknowledged 公理【数】
generally controlled 达到大致控制
generally recognized 公认的

generally recognized as safe 通常认为安全的
generally soil 一般性土
generally used dynamic(al) quantifier 通用动态计量仪
general machine tool 普通机床
general machining cell 通用加工室
general magnetic field 普通磁场;普通磁场
general maintenance 一般养护;一般维修
general maintenance organization 一般维护组织
general management 一般管理;全面管理
general management expenses 一般管理费
general management technique 一般管理技术
general management trust 一般管理信托
general manager 总经理
general manager's room 总经理室
general map 一览图;总图
general map of highway and transportation 公路图
general market 共同市场
general market risk 系统性风险
general masonry 一般砌筑工程
general mean 广义平均值
general measure of reliability 可靠性的一般测定
general measuring instrument 综合测量仪
general mechanics 一般力学
general meeting 全体会议
general meetings of the staff and workers 全体职工大会
general member 通用构件
general member system 通用构件系统
general merchandise 杂货;日用商品;百(杂)货
general merchandise wholesaler 杂货批发商
general meteorology 普通气象学
general method 一般方法
general method optimization 一般方法最优化
general migration 常规迁移
General Mills Inc. 通用面粉公司(美国)
general mine 一般项目
general model 一般模型
general model of mixed-exchanged isotopes 同位素混合—交换总模式
general monitor unit 通用监视装置
general mortgage 总抵押;一般抵押
general mortgage bond 一般抵押债券;普通抵押债券
general most favo(u)red nation clause 一般最惠国条款
general motor 通用电动机
General Motor Corporation 通用汽车公司(美国)
general motorization 全面汽车化;普通汽车化
general mo(u)lded goods 一般模制品
general navigation beacon 通用导航信标
general navigation computer 通用导航计算机
general net theory 广义网论
general noncost system 普通非成本制度
general nonlinearity 一般非线性
general normal equation 一般正规方程;通式
general objective 总目标;一般目标
general obligation bond 一般偿务债券;一般责任保证;普通责任的债务
general obligations 一般义务
general oceanography 普通海洋学
general offer 一般发盘;一般发价;一般报价;总报价
general office 总局;总管理处;总公司;总办事处;总办公室;综合室;办公厅
general office of the hospital 院部办公室
generalogical relationship 系谱关系
general operating expenses 一般业务费用
general operating provision 一般操作规定
general operating specification 一般操作规程
general operation(al) requirement 一般使用要求;一般操作条件
general operation procedure 一般操作程序
general optimal value function 一般最优值函数
general order 总订货单
general oscillating control servomechanism 广义振荡控制伺服机构
general oscillation 普遍振荡
general outline 总平面图;总纲;总布置图;概要;大纲
general overall 全面检修
general overhaul 彻底检修;一般大修理;总翻修;全面检修;全面检查;大修
general overhead 一般间接费(用)

general overhead charges 一般管理费
general overhead cost 总开支;总管理费用
general overhead expenses 一般管理费
general overhead of office on cost 一般间接费(用)
general oversupply 一般供给过剩
general packet radio service 通用分组无线电业务
general parallel language 通用并行语言
general participation clause 普遍参加条款
general partition 一般划分
general partners 全体合伙人
general partnership 一般合伙企业;普通合伙
general parts of machine 普通机件
general performance number 一般性能数据
general perturbation 普遍摄动
general petrochemical works 石油化工总厂
general picture 概貌
general placement design of aids-to-navigation 航标总体配布设计
general plan 计划概要;总图;总体布置图;总体布置;总平面;总计划;总布置图;整体规划;平面布置总图;总体规划
general plan design of port 港口总平面设计
general plane of finish construction 竣工总平面图
general plane theorem 广义平面定理
general planning 总体布局;总体方案;基本方案;总体规划
general plan of as-built works 竣工总平面图
general plan of outdoor pipelines 室外管线总平面图
general plan of port 港口总图规划
general plan of working 总经营计划
general plan restriction 总体规划限制
general points of view 一般观点
general polarity 普通极性
general policy 总政策;总方针
general policy conditions 保险单一般规定
general policy control 一般政策控制
general policy model 一般政策模型
general poll 一般轮询
general polling 通用查询
general population movement 一般性人口变动
general port 综合性港口
general port overheads 港口总管理费
general position 普通位置
general post office 邮政总局
General Post Office cable 邮政总局通信电缆
General Post Office line 邮政总局通信线路
General Post Office tower 邮政(总局)电信塔楼
general potential 一般位势
general power (委托人授予代理人的)全权;一般幂
general power of attorney 全权委托书
general precession 总岁差
general precession in longitude 黄经总岁差
general precision 一般精度
general preferential duties 一般优惠关税;普通优惠税
general preferential tariff 普通优惠关税税则
general preferential treatment 普遍优惠制待遇
general pressure 总压力
general pressure drop 总压(力)降
general price 物价
general price index 一般物价指数;总物价指数;物价总指数
general price-level adjustment 一般物价水平调整
general price level change 总物价水平的变动;一般物价水平变动
general price-level method 一般物价水平法
general principles 总则
General Principles of the Civil Law of the People's Republic of China 中华人民共和国民法通则
general probability 总概率
general problem area 一般问题的范围
general problem solver 一般问题解决者;通用问题解算机
general procedure 一般程序
General Procedures Governing the Operations of the Fund of United Nations Environment 联合国环境计划基金作业总则
General Procedures Governing the Operations of the Fund of United Nations Environmental Program(me) 联合国环境方案基金作业总则
general process 一般性处理

general processor 通用处理机
general processor program(me) 通用处理程序
general procurement 一般采购广告
general profit rate 一般利润率
general program(me) 综合程序;通用程序
general progress of construction 施工总进度
general projective geometry 一般投影几何
general property 一般属性
general property tax 一般财产税
general proposition 一般命题
general protocol 总议定书
general provisions 一般条款;总则
general prudential rule 一般谨慎条款
general public room 应接室;交际室
general public service 一般公众业务
general purpose 万向的;通用的;普通业务;多种用途的
general purpose accessory 通用附件;通用辅助设备
general purpose adhesive 通用黏结剂
general purpose aeroplane 通用飞机
general purpose amplifier 通用放大器
general purpose amplifier tube 通用放大管
general purpose analog(ue) computer 通用模拟计算机
general purpose and industrial tread tire[tyre] 通用性工业用轮胎
general purpose audit program(me) 通用审计方案
general purpose body 通用(犁)体
general purpose bonding medium 通用胶黏剂
general purpose bottom 通用犁体
general purpose box car 通用棚车
general purpose branch circuit 普通多支电路;通用多支电路
general purpose bucket 通用吊桶;普通铲斗
general purpose bulk ship 通用散(装)货船
general purpose burnt(clay) brick 通用黏土砖
general purpose cable 通用电缆
general purpose calender 通用压延机
general purpose camera 通用摄影机
general purpose cargo vessel 通用货轮
general purpose cement 普通水泥
general purpose cement(ing agent) 通用胶凝剂
general purpose computer 通用计算机
general purpose computing system 通用计算系统
general purpose container 通用集装箱;多用途集装箱;一般用途货柜
general purpose controller 通用控制器
general purpose data base 通用基本数据
general purpose data management system 通用数据管理系统
general purpose device 通用器件
general purpose digital computer 通用数字计算机
general purpose discharge standard 通用排放标准
general purpose display system 通用显示系统
general purpose entry method 通用输入方式
general purpose equipment 通用设备
general purpose fired(clay) brick 通用黏土砖
general purpose flip-flop 通用触发器
general purpose foils 通用金属箔
general purpose function 通用函数
general purpose function generator 通用函数发生器
general purpose furnace black 通用炉黑
general purpose grease 通用润滑剂
general purpose harbo(u)r 通用(性)港口;通用港(口)
general purpose hauling 通用运输
general purpose indexes 通用指数
general purpose instruction 通用指令
general purpose language 通用语言
general purpose lines of overhead 一般用途信用限额
general purpose locomotive 通用机车
general purpose machine 通用机器;通用机车;通用电动机
general purpose machine tool 普通机床
general purpose machining center 普通加工中心
general purpose macrogenerator 通用宏功能生成程序
general purpose manipulator 万用键控器;万能机械手;通用键控器;通用机械手;通用操纵器
general purpose map 普通地图

general purpose memory 通用存储[贮]器
general purpose microprogramming 通用微程序设计
general purpose middle breaker bottom 通用双壁开沟犁体
general purpose monitor 通用监察器
general purpose motor 通用电动机
general purpose mo(u)ldboard 通用型犁壁
general purpose operating program(me) 通用操作程序
general purpose operating system 通用操作系统
general purpose oscilloscope 通用示波器
general purpose overhead crane 通用桥式起重机
general purpose plane 通用飞机;全能飞机
general purpose plastics 通用塑料
general purpose pliers 通用手钳
general purpose plow 通用犁
general purpose polyester resin 通用型聚酯树脂
general purpose port 通用(性)港口;通用港(口)
general purpose Portland cement 普通硅酸盐水泥
general purpose processor 通用处理机
general purpose program(me) 一般目的程式;标准程序
general purpose pump 通用泵
general purpose radar 通用雷达
general purpose radio receiver 通用无线电接收机
general purpose reactor 通用反应堆
general purpose receiver 通用接收机
general purpose register 通用寄存器
general purpose register file 通用寄存器组;通用寄存器文件
general purpose relay 通用继电器
general purpose routine 通用例程
general purpose rubber 通用橡胶
general purpose sealant 通用嵌缝膏
general purpose ship 通用船
general purpose shop truck 通用修理工程车
general purpose simulation system 通用模拟系统;通用的模拟系统程序
general purpose software 通用软件
general purpose statement 普通决算表;通用决算表;通用结算表
general purpose storage 通用存储器
general purpose system 通用系统
general purpose system simulation 通用系统模拟(语言)
general purpose system simulator 通用系统模拟程序;一般目的系统模拟程序
general purpose terminal 通用码头
general purpose tester 通用测试器
general purpose test-signal generator 通用测试信号发生器
general purpose tipper 通用翻斗
general purpose tool 通用工具
general purpose tractor 万能拖拉机;通用拖拉机
general purpose transformer 通用变压器
general purpose transistor digital computer 晶体管通用数字计算机
general purpose trunk 通用中继线
general purpose type 通用式
general purpose van 通用篷车
general purpose varnish 通用清漆
general purpose vehicle 通用车辆;一般用途的车辆
general quadric surface 一般二次曲面
general quantity 一般量
general quarters 全船战备部署
general radiation 一般辐射
general radiocommunication 常规无线电通信
general rain 总降雨量;总降水(量);大面积降雨;大面积降水
general rainfall 一般雨量;总降雨量
general range of work speeds 工件转速范围
general rate 一般税率;普通税率;统税
general rate of interest 一般利率
general rate of profit 一般利润率
general rational fraction 一般有理分数
general reagent 类别试剂
general reciprocal 广义倒数
general reciprocity law 一般互反律
general recognized as safe 公认无害
general recommendation 一般建议
general reconstruction 大(翻)修;翻修
general record 普通记录
general recursive function 一般递归函数
general recursiveness 一般递归性
general reference to standard 一般性引用标准
general refrigerated ship 通用冷藏船
general refrigerated vessel 通用冷藏船
general register address 通用寄存器地址
general register unit 通用寄存器单元
general regulation 一般规则;通则
general regulations of port 港章
general relativistic collapse 广义相对论坍缩
general relativistic effect 广义相对论效应
general relativity theory 广义相对论
general remark 一般说明;概说;概要;概述;总论
general repair 一般修理;普通修理;大修
general repair work 大修工作
general replenishments 一般性补充资金
general requirement 一般要求;一般规格;一般规定;总的要求
general reserve 一般营业费用;一般储备
general reset 总复位
general resettlement policy 移民安置总方针
general responsibilities 一般责任;一般义务
general restoration repair 基本恢复修理
general revenue sharing 一般收入分享
general rights 一般权利
general risk analysis 一般风险分析
general route 总体路线
general routine 标准程序;通用程序
general rule 一般规则;一般规律
general rule of testing method for appropriate transport packaging 合适包装货物试验方法通则
general rules 总则;通则
general runoff zone 一般径流带
general run of things 一般趋势;一般情况
general sales agent 总经销人
general sales tax 一般销售税
general scale 主比例尺;基本比例(尺);通用比例尺
general scanning algorithm 通用扫描法
general scene lighting 舞台照明
general schedule 总表;普通等级
general schedule of construction 施工总进度
general schedule of project construction 工程施工总进度
general scour(ing) 一般冲刷;普通冲刷
general scour under bridge opening 桥下一般冲刷
general secretarial 总秘书处
general seismology 普通地震学
general selling expenses 一般销售费用
general semisimple algebraic group 一般半单代数群
general sense 一般含义
general service 一般用途(的);一般勤务;总务;公用设施;日用泵;普通业务;万用;通用;多种用途的
general service counter 总服务台
general service desk 总操作台;通用控制台
general service hose 通用胶管
general service launch 港务船;交通艇
general service oil 通用内燃机油
general service paint 普通油漆
general service pump 杂用泵;通用(排水)泵
general service-time 一般服务时间
general set 一般集合
general set theorygeneral fund 一般集论
general settlement 整体沉降
general share 通用犁铧
general shear failure 总剪切破坏;整体剪切破坏
general shear pattern 总体剪切模式
general shear slide 总体剪切滑动
general sheet of transport revenue 运输收入进款资产负债表
general ship 一般货船;零运船
general shop 百货店
general shunt 通用分流器
general signal 普通信号
general sign and symbol 通用标志和符号
general simulation test 总体模拟试验
general site consideration 一般场地考虑
general situation 概况
general sketch 总略图
general slope 总比降
general slope of basin 流域总坡降;流域总坡度
generals of surveying and mapping 测绘总类
general solution 通解
general specification 一般技术要求;一般规格;总体规定;总说明(书)
general specification of building construction 土建施工说明书
general specifications for the building of ship 通用船舶建造规范
general speed 合速度
general stability analysis 整体稳定性分析
general stability computation 整体稳定性计算
general stability criterion 一般稳定性准则
general standard 通用标准
general state of strain 一般应变状态;空间应变状态
general state of stress 一般应力状态;总的应力状态;空间应力状态;三维应力状态
general storage equation 通用存储[贮]方程
general store 杂货店;百货商店
general storm 普遍风暴;普遍暴雨;大面积暴雨
general-stratigraphic(al) column 综合地层柱状剖面图
general strike 总罢工
general strike of a deposit 矿床总走向
general stripping ratio 总剥采比
general structural concrete 普通结构混凝土
general structural steel 普通结构钢
general structural strength 总结构强度
general structure 普通结构
general subgrade 一般路基
general subscript grammar 一般下标文法
general summary of account 总决算
general superintendent 总监督员;总管(指管理人员);总负责人
general supervision 普通监督
general supervision engineer 总监理工程师
general supplies 一般性补给
general supply 总电源
general supply fund 一般物资供应基金
general support maintenance 一般保障维护
General Surface Current Circulation of the World 世界表层环流图
general surface plan 总图
general surveillance monitoring 通用监控
general survey 一般检验;群体普查;普查;总体勘测
general survey drill hole for hydrogeology 水文地质普查孔
general survey mission 全面调查团
general survey of pollution sources 污染源普查
general survey of soil and land planning 土壤普查和土壤规划
general surveyor 总测量员;海事检查人
general survey procedure 总体测量程序
general switched telephone network 通用电话(交换)网
general symbol 通用符号;常用符号
general system of preference 普通优惠制
general systems simulation 一般系统模拟
general system theory 一般系统论
general table 总表
general tailing sampler 总尾矿取样机
general tariff 一般税则;一般税率;普通运价;普通税则;通用税则
general tax reduction 一般减税
general technical office 综合技术室
general technical specifications 通用技术条件
general technology satellite 通用技术卫星
General Telephone Electronics Corporation 通用电话电子公司(美国)
general telephone network 公用电话网
general tenor 主旨
general term 一般项;普通项;通项;泛称;一般条件;统称
general terms and conditions 一般交易条件
general terns of delivery 交货共同条件
general territory 一般厂区
general test 总体调试
general testing meter 通用测试仪器
general test problem 一般试验问题
general theodolite 普通经纬仪
general theorems of dynamics 动力学普遍原理
general theory 一般理论;统一理论
general theory of perturbation 一般摄动理论
general theory of relativity 广义相对论
general theory of statistics 统计学理论
general tool 普通工具
general topological space 一般拓扑空间
general track foreman 养路总工长

general trade 一般贸易;总贸易
general training 一般培训
general transduction 一般换能
general transport administrative vehicle 一般运输后勤车辆
General Transport Co. 通用运输公司
general transport equation 一般运输方程
general transverse tracks 通用横向磁迹
general trend 一般走向;一般趋势
general trust fund 普通信托基金
general(unit)topographic(al)map 普通地形图
general upper derivate 一般上导数
general-use article 一般的制品
general-use paper 通用纸张;普通建筑用纸
general-use resin 通用树脂;普通树脂
general user privilege class 一般用户特权级
general-use snap switch 通用按钮开关
general utility 一般效用;总效用;万用;普通业务
general utility function 通用辅助操作
general utility puller 通用拉出器
general utility tool 通用工具
general utility wrench 通用扳手
general value 一般价值
general ventilation 全面通风
general vibration 全身振动
general view 总图;概要;概貌;全视图;全貌;全景(概要);外貌;外观;视察;大纲
general wagon 普通货车;通用货车
general wall 总壁
general ward 普通病室
general warranty deed 普通担保合同;全权证书;全面保证契据
general washout 一般冲刷
general waste disposal 一般废物处置
general wastes 一般废物
general water 中等水
general water quality 总体水质
general water quality vulnerability zone map 水质脆弱带总图
general weather situation 天气形势
general wind force 一般风力
general work 杂项工程;杂务;总厂
general working plan 总经营计划
general yield 全面屈服
general yielding 总变形
general yield table 总收获表
generant of the toroid 超环面(的)母圆
generated 发电的
generated address 形成地址;合成地址;生成地址
generated code 形成码;合成码;派生码
generated corrosion 均匀腐蚀
generated data 综合数据;合成数据
generated energy 发电量
generated error 生成误差
generated frequency 振荡频率
generated gas kerogen 成气干酪根
generated gear 展成法齿轮
generated hereditary class 生成的可传类
generated ideal 生成的理想
generated involute 展成渐开线
generated matrix 生成矩阵
generated monotone class 生成的单调类
generated oil kerogen 成油干酪根
generated oil zone 成油带
generated output 生成出力;发出电力
generated output power 发电机输出功率
generated pair 展成副
generated profile 展成齿廓;包络齿廓
generated ring 生成环
generated statement 生成语句
generated subgroup 生成的子群
generated symbol 生成符号
generated teeth (用滚齿法所得到的)滚铣轮齿
generated traffic 新增交通量
generated trip of traffic 发生交通量
generated voltage 电动势
generate electricity 发电
generate pressure 产生压力
generate sound project 拟定合理的项目
generating algorithm 生成算法
generating alternatives 产生备选方案
generating angle 啮合角;模孔喇叭角
generating area 起波区;风区
generating automorphism 生成自同构
generating broach 逐步成形拉刀
generating cam 展成鼓轮
generating capacity 电站容量;发电容量;发电能力
generating celestial body 生成天体
generating circle 基圆;母圆
generating code 生成代码
generating collision 振荡碰撞
generating cone 母锥;生成锥
generating contour 描绘等高线
generating cost 发电成本
generating curve 母曲线
generating cutting 滚齿切削法
generating distance 吹送距离;风区长度
generating efficiency 发电效率
generating element 生成元
generating ellipse 母椭圆
generating flank 产形齿面
generating flow 边界层增长期流动
generating function 母函数;生成函数
generating functional 生成泛函
generating function for probability distribution 概率分布生成函数
generating function of binomial coefficients 二项式系数的生成函数
generating gear 形齿轮
generating globe 地球投影图形
generating hob 齿轮滚刀
generating laser 激光发射器
generating line 母线;生成线;产生线
generating machine(ry) 发电机;发电设备
generating magnetometer 发电机式磁强计
generating mechanism 展成机构
generating method 滚切(方)法
generating milling 滚齿铣
generating motion 展成运动
generating mounting surface 加工安装基面
generating of arc 引弧
generating of steam 发生蒸汽
generating operator 生成算子
generating overhaul 大修
generating pair 生成对
generating phase 滚切阶段
generating plant 发电设备;电站;发电站;发电厂
generating power plant 发电厂
generating probability 生成概率
generating process 滚齿法
generating program(me) 生成程序
generating rack 齿条形刨齿刀
generating routine 生成例程;生成程序;编辑程序
generating run 生成运行
generating scheme 生成模式
generating set 发电装置;发电设备;发电机(组)
generating set with petrol engine 汽油机发电机组
generating standard report 编制标准报告
generating station 发电所;电厂;发电站
generating station capacity 发电站容量或额定功率
generating stroke 展成行程
generating subset 生成子集
generating system 发电系统
generating tool 展成刀具
generating transformation 生成变换
generating tube 蒸发管;拱砖管;气化管
generating unit 发电装置;发电设备;发电机组
generating voltage 发电电压
generation 一代人;世代;生成;产生;繁殖世代
generation age curve 世代年龄曲线
generational hydrocarbon product classification of kerogen 干酪根成烃产物分类
generation alternative 生成替代方案
generation area 生风区域
generation data group 相继数据组;数据组世代;世代数据组;生成数据组
generation data set 世代数据集
generation file group 世代文件组
generation gear grinding 齿轮展成法;齿轮滚磨法
generation hydrocarbon evolution stage of organic matter 有机质成烃作用演化阶段
generation index 世代指数
generation input stream 生成输入流
generation interval 世代间隔
generation length 世代长度;世代长度
generation lifetime 生成寿命
generation material for maturity 世代材料的成熟期
generation number 世代数;世代号;生成数
generation occurrence 生成出现;定值性出现
generation of alternatives 备择方案的产生
generation of a variable 变量的生成
generation of disturbances 扰动的生成
generation of electric(al)energy 发电
generation of electric(al)power by thermal power 火力发电
generation of electricity 电力产生
generation of excess heat 过热
generation of foliation 面理的世代
generation of heat 生成热
generation of lineation 线理的世代
generation of mineralization 矿代世代
generation of multiple output prime implicant 多端输出质蕴涵项的产生
generation of objective 目标的产生
generation of ocean wave 海浪生成
generation of pure oxygen 纯氧发生
generation of random numbers 随机数生成;随机数产生;乱数的产生
generation of steam 蒸汽产生
generation of storage 存储器的生成
generation outage 停止发电
generation period 世代(周)期
generation phase 生成程序段
generation printing 生成添印
generation process 滚铣过程
generation rate 产生率
generation reproduction rate 一代人再生产率
generation rule 产生规则
generation set 发电设备
generation time 世代时间
generation tree 产生树
generation type 形成方式
generative capacity 生成能力
generative center[centre] 发生中心
generative fuel 再生燃料;气体发生炉燃料
generative genesial 发生的
generative grammar 生成文法
generative graphics 生成式(计算机)制图法
generative period 世代期
generative power 生成力;发生力
generative process 生成过程
generative rule 生成规则
generative scheme 生成图式
generative software 生成软件
generative transformational grammar 生成转换文法
generative type macro 生成式宏程序
generator 振荡器;生成元;生成程序;产生者;产生器;发生器;发生炉;发电机;编制程序
generator accessories box hanger 发电机附属品箱吊
generator active power 发电机有功功率
generator adjusting arm 发电机调整臂
generator adjusting strap 发电机调整臂
generator-air cooler 发电机空气冷却器
generator and starter armature tester 发电机和起动机电枢检查器
generator and transformer room 发电机和变压器间
generator armature 发电机转子;发电机电枢
generator ash 煤气发生炉灰渣
generator body 发电机壳
generator bracket 发电机架
generator braking equipment 发电机制动设备
generator breaker 发电机断路器
generator brush 发电机电刷
generator brush arm 发电机电刷臂
generator brush spring 发电机电刷弹簧
generator building 发电机房
generator bus 发电机母线
generator busbar 发电机汇流条
generator busbar assembly 发电机组合母线
generator capability curve 发电机输出效能曲线
generator capacity 发电效率;发电机容量
generator casing 发电机外壳
generator chock coil 发电机抗流圈
generator cinder 煤气发生炉炉渣
generator circuit breaker 发电机断路器
generator clinker 煤气发生炉炉渣
generator commutation 发电机整流
generator commutator 发电机整流子

generator commutator end cap 发电机正流子端盖
generator control desk 发电机控制台
generator coupling 发电机轴节
generator cover 发电机罩
generator cover band 发电机罩带
generator cut-out 发电机断流器
generator drive 发电机驱动
generator drive axle 发电机传动轴
generator drive gear end cap 发电机传动端盖
generator driving belt 发电机拖动皮带
generator dust band 发电机防尘箍
generator effect 发电机效应
generator efficiency 发电机效率
generator end 发电机端
generator erection 发电机安装
generator excitation 发电机励磁
generator exciter 发电机励磁机
generator exciting winding 发电机励磁绕组
generator failure 发电机故障
generator field 发电机磁场
generator field coil 发电机磁场线圈
generator field control 发电机磁场控制
generator field decelerating relay 磁场调节减速继电器
generator field fuse 发电机电磁场熔断丝
generator floor 安装发电机的楼面层；发电机层
generator frame 发电机机体
generator frame through screw 发电机架长螺杆
generator gas 发生炉煤气
generator gate 产生门
generator hanger 发电机吊架
generator head band 发电机罩带
generator housing 发电机机体
generator ignition 发电机点火
generator insulation over-temperature monitor 发电机绝缘过热监控器
generator kit 发电机的整套零件
generator lead wire 发电机导线
generator locating arm 发电机固定臂
generator loss 发电机(功率)损失
generator loss of excitation protection 发电机失磁保护(装置)
generator main brush 发电机主电刷；发电机炭刷
generator main leads gallery 发电机主出线廊道
generator mark 发电机型号
generator matrix 发生器点阵
generator-motor set 发电机电动机组
generator neutral grounding equipment 发电机中性接地设备
generator of a quadric 二次曲面的母线
generator of a ruled surface 直纹形母线
generator oil thrower 发电机拨油圈
generator on charge 发电机在充电状态
generator operating conditions 发电机运行情况
generator operation 发电机运行
generator output 发电机出力
generator panel 发电机盘
generator pit 发电机坑
generator pole shoe 发电机磁极瓦
generator polynomial 生成多项式
generator potential 发生器电位
generator power 发电机功率
generator-power transformer unit differential protection 发—变组差动保护
generator protection 发电机保护
generator protective equipment 发电机保护设备
generator pulley 发电机皮带轮
generator pump 发生器泵
generator reactive power 发电机无功功率
generator reactor 发电机电抗器
generator regulator 发电机调整器；发电机调压器
generator resistance 发电机电阻
generator room 发电机房
generator rotor 发电机转子
generator set 发电机组
generator shaft 发电机轴
generator shell 发电机壳
generator shield 发电机罩
generator shielding 发电机防流罩
generator ship 供电船
generator shunt field resistor 发电机分激磁场电阻器
generator speed 发电机转速
generator stand 发电机座
generator stator 发电机定子
generator stator ground protection 发电机定子接地保护(装置)
generator step-up transformer 发电机升压变压器
generator support 发电机支座；发电机支架
generator support bracket 发电机托架
generator suspension 发电机吊挂
generator tachometer 发电机式转速计
generator terminal 发电机线接头
generator terminal stud insulator 发电机线头柱螺栓绝缘体
generator through bolt 发电机拉紧螺栓
generator-transformer bank 发电机变压器组
generator-transformer block 发电机变压器单元
generator-transformer unit connection 单元结线；发电机变压器组结线
generator-transformer unit differential protection 发电机变压器单元差动保护
generator truck 带发电装置的载货车
generator-type blasting machine 发电机式发爆器
generator-type tachometer 发电机式转速表
generator unloading point 发电机释载点
generator van 发电车
generator voltage 发电机电压
generator voltage regulator 发电机电压调整器
generator voltage relay 发电机电压继电器
generator with flywheel 飞轮(式)发电机
generatrix 母线；动线
generatrix of tank 构成储油灌的壁
generette 发电机样件
generic constant 一般常数
generic descriptor list 通称描述符表
generic efficiency 相对效率
generic element 通用部分
generic entry name 总入口名
generic family 通称族
generic function 总函数；类函数；通称函数
generic key 通称关键字
generic name 一般名称；总名称
generic phase 类分相
generic redox half-cell reaction 类氧化还原半电池反应
generic redox reaction 类氧化还原反应
generic term 通用术语；通称；地理通名；总称
generous flowing lines 大量的流线
generous profits 丰厚的利润
generous soil 沃土
genesic 发生的
genesis 起源；成因；发生
genesis classification of minerals 矿物成因分类
genesis of agricultural soil 农业土壤发生
genesis of fluoride-containing groundwater 含氟地下水成因
genesis of mineral deposit 矿床成因
genesis of minerals 矿物的成因
genesis of petroleum 石油成因
genesis of soil 土壤发生
genesis type of discontinuity 结构面成因类型
Genessee Vallley bluestone 吉恩尼斯谷青石(产于美国纽约州的一种深色灰青石)
genetical increment of strata 地层成因增量
genetically improved strain of tree 改良树种
genetical mineralogy 成因矿物学
genetical relationship 亲铁元素
genetical sequence of strata 地层成因层序
genetical type of soil 土的成因类型
genetical type of terrace 阶地成因类型
genetical types of ore deposits 矿床成因类型
genetic analysis 成因分析
genetic classification 成因分类
genetic classification of hydrocarbon gas 烃气成因类型
genetic classification of nature water 天然水成因类型
genetic classification of non-hydrocarbon gas 非烃气成因类型
genetic classification of trop 圈闭成因分类
genetic connection 演化关联
genetic engineering 遗传工程
genetic factor of natural water 天然水成因系数
genetic geologic(al) model method 成因地质模型法
genetic geomorphology 地貌发生学
genetic horizon 发生层
genetic inception 演化发生
genetic material 遗传物质
genetic method 成因法；发生法
genetic mineral 原生矿物
genetic morphology 地貌发生学
genetic polymerism 同质异重
genetic potential of kerogen 干酪根成烃潜量
genetic probability map 概率成因图
genetic relation of hydrocarbon and source 烃源成因联系
genetic relationship 亲缘关系；成因关系
genetic runoff formula 径流成因公式
genetic sequence of strata 地层的成因序列
genetic series 成因系列
genetic soil 原生土(壤)；生成土
genetic stratigraphic(al)framework 成因地层格架
genetic stratigraphic(al) unit 成因地层单位
genetic type 煤的成因类型；成因类型；成因关系
genetic type map of quaternary system 第四系成因类型图
genetic type of groundwater regime 地下水动态成因类型
genetic type of lineation 线理成因类型
genetic type of magma 岩浆成因类型
genetic type of mineral deposits 矿床成因类型
genetic type of ore 矿石成因类型
genetic type of plain 平原成因类型
genetic type of pore 孔隙成因类型
genetic type of river 河流发生型
genetic type of stream 河流发生型
genetic unit 成因单位
Genetron 基尼昂(一种致冷剂)
geneva 十字的；星形的；间歇的
Geneva 日内瓦(瑞士)
Geneva Bureau 日内瓦总秘书处(国际电信联盟)
geneva cam (十字轮机构的)；星形轮十字结构星形轮
Geneva Circular 日内瓦通函
Geneva Circular Telegram 日内瓦通电
Geneva Convention 日内瓦公约
geneva cross 十字形接头
Geneva crystal 日内瓦玻璃
geneva gear 马氏间歇机构；十字轮机构
geneva mechanism 马氏机构；十字轮机构；槽轮机构
geneva motion 间隙运动；日内瓦运动
geneva movement 间歇式送料机构；间歇工作盘装置；间隙运动；日内瓦式机心
Geneva nomenclature 日内瓦命名法
Geneva Notification 日内瓦通知书
Geneva observatory 日内瓦天文台
Geneva Round 日内瓦回合
Geneva stop 日内瓦式发条限上装置
Geneva stop work 日内瓦式发条停止上条装置
Geneva Trade Conference 日内瓦贸易会议
geneva wheel 间歇工作轮；十字形机构轮
genial weather 和暖的天气
geniculate 膝状的
geniculate body 膝状体
geniculate twin 漆状双晶【地】
genin 配质
genistein 染料木素
genius 创造能力
genius loci 地方性
genkinite 四方锑铂矿
genlock 同步耦合器
genlock equipment 集中同步设备；台从同步设备
genlocking 集中同步；同步锁相
genocide 种族灭绝
genotype 属模标本
genovertical plates 侧额区
genre 风俗画
Gen-saki 日本短期货币市场
gent's saw 绅士锯
gentes 拱门三角墙尖顶(早期英国建筑)
gentese 门尖参尖
gentex office 自通电报局
gentian violet 巴黎紫
gentle acceleration 慢加速
gentle ascent 缓坡
gentle angular unconformity 微度不整合
gentle anticline zone 平缓背斜带
gentle bank 小坡度；大转弯
gentle breeze 轻风；微风；三级风

gentle breezes and calm waves 风平浪静
gentle curve 平缓曲线
gentle dip 平缓倾斜;微倾斜
gentle fold 平缓褶皱
gentle incline 平缓的坡度
gentleman's agreement 君子协定
gentlemen's agreement on export credit 出口信贷君子协定
gentlemen's room 男厕所
gentleness 平缓(指地形)
gentle outcrop 平坦露头
gentle ramp 缓冰坡
gentle side slope 平缓边坡
gentle slope 缓倾斜;缓坡;缓和坡度;平缓坡度;平缓坡道;平缓比降
gentle slope belt 缓坡带
gentle sloping surface 缓坡面
gentle start 平稳起步;平稳开动
gentlest slope 极缓坡
gentle turn 平稳的回转
gentle wind 和风
gently 平缓地
gently dipping 平缓下降的;平缓倾斜的
gently dipping bed 缓倾岩层
gently inclined coal seam 缓倾斜煤层
gently modulated profile 逐渐转变的剖面;逐渐转变的轮廓
gently plunging fold 缓倾伏褶皱
gently rolling country 缓坡地;缓和丘陵地区;平缓丘陵地
gently rolling topography 平缓起伏地形
gently sloping bed 缓坡河底
gently sloping coast 缓斜岸
gently sloping surface 缓斜面;缓坡面
gently undulated 舒缓波状的
gently wavy fault 舒缓波状断层
gentnerite 阴铜硫铬矿;银特纳矿
gentrification 中产阶级化;社区复兴
genual 膝状的
genuine arch bridge 真正的拱桥
genuine costs of circulation 纯流通费
genuinely collective goods and services 纯公共性货物与劳务
genuine part 原配件
genuine soap 纯皂
genuine ultramarine 天然群青
genuine vitrinite 真正镜质体
genus 种类;亏格
genus of a curve 曲线的亏格
genu valgum 叉形腿
genveite 日内瓦石
genvironments 超重环境
geo 海门口;深峡海湾;陡峭海岸的狭海湾;陡壁峡口;长狭潮道
geoacoustics 地声学
geoanticline 地背斜;大背斜
geoastrophysics 地球天体物理学
geobarometer 地质压力计
geobasin 深厚水平沉积盆地;地盆
geobenthos 湖底生物
geobiology 地质生物学;地球生物学
geobiont 土壤生物;地栖生物
geobotanical anomaly 地植物异常
geobotanical chart 地植物图
geobotanical method 地植物方法
geobotanical prospecting 地植物探矿;地(面)植物勘探
geobotanist 地球植物学家
geobotany 地(球)植物学
geocarpy 地下结果性
geocartographer 地理制图工作者
geoceiver 大地接收机
geocenter 地心;地球质量中心
geocentric 地心的
geocentric altitude 地心高度
geocentric angle 地心角
geocentric apparent motion 地心视动
geocentric astronomical phenomenon 地心天象
geocentric celestial phenomenon 地心天象
geocentric colatitude 地心余纬
geocentric conjunction in right ascension 地心赤经合
geocentric coordinates 地心坐标
geocentric coordinate system 地心坐标系

geocentric datum 地心基准
geocentric diameter 地心直径
geocentric distance 地心距离
geocentric ecliptic coordinate axes 地心黄道坐标轴
geocentric ephemeris 地心历表
geocentric equatorial coordinate axes 地心赤道坐标轴
geocentric geodetic coordinates 地心大地坐标
geocentric gravitational constant 地心引力常数
geocentric horizon 地心真地平;地心地平(面)
geocentricism 地球中心说;地球中心论
geocentric latitude 地心纬度
geocentric longitude 地心经度
geocentric orbit 地心轨道
geocentric origin 地心原点
geocentric parallax 地心视差
geocentric place 地心位置
geocentric position 地心位置
geocentric radiant 地心辐射点
geocentric radius vector 地心向径
geocentric rectangular coordinate 地心直角坐标
geocentric right ascension 地心赤经
geocentric system 地心制;地心体系
geocentric unit sphere 地心单位球
geocentric velocity 地心速度
geocentric vertical 地向径向
geocentric zenith 地心天顶
geocerite 硬蜡
geochemical 地球化学的
geochemical analysis 地球化学分析
geochemical analytical method 地球化学分析方法
geochemical anomaly 地球化学异常
geochemical atlas 地球化学图册
geochemical background 地球化学背景
geochemical background value 地球化学背景值
geochemical barrier 地球化学障;地球化学垒
geochemical behavior 地球化学行为
geochemical behaviour of elements 元素地球化学行为
geochemical characteristic 地球化学特征
geochemical closed system 地球化学封闭体系
geochemical coalification 地球化学煤化作用
geochemical condition 地球化学条件
geochemical constituent 地球化学组分
geochemical controls 地球化学控制因素
geochemical correlation of oil and gas 油气地球化学对比
geochemical critical value 地球化学临界值
geochemical cycle 地球化学循环;地球化学旋回
geochemical detailed survey 地球化学详查
geochemical dispersion 地球化学扩散;地球化学分散
geochemical dispersion aureole 地球化学分散晕
geochemical distribution 地球化学分布
geochemical distribution of element 元素的地球化学分布
geochemical drainage reconnaissance 地球化学水系普查;地球化学水系勘查
geochemical effect 地球化学效应
geochemical electrovalence 极化系数
geochemical environment 地球化学环境
geochemical equilibrium 地球化学平衡
geochemical evolution 地球化学演化
geochemical exploration 化探;地球化学探矿;地球化学勘探
geochemical exploration analysis 化探分析
geochemical exploration and logging 地球化学勘探和测井
geochemical exploration for chromium and nickel 铬镍化探
geochemical exploration for copper 铜矿化探
geochemical exploration for geothermal field 地热化探
geochemical exploration for gold 金矿化探
geochemical exploration for molybdenum 钼矿化探
geochemical exploration for non-metals 非金属化探
geochemical exploration for oil and gas 油气化探
geochemical exploration for ore deposits 金属矿化探
geochemical exploration for polymetallic ore 多金属矿化探

geochemical exploration for precious metals 贵金属矿化探
geochemical exploration for tin and tungsten 钨锡矿化探
geochemical exploration for uranium 铀矿化学
geochemical exploration method 化探方法;地球化学勘探方法
geochemical exploration report 化探报告
geochemical facies 地球化学相
geochemical field operation program(me) 化探野外作业程序
geochemical flow 地球化学流
geochemical gas survey 气体地球化学测量
geochemical gradient 地球化学梯度
geochemical indicators 地球化学指标;地球化学标志
geochemical landscape 地球化学景观
geochemical logging 地球化学测井
geochemical log method 地球化学测井方法
geochemical map 地球化学图
geochemical mapping 地球化学制图
geochemical markers 地球化学标志
geochemical mechanism 地球化学机制
geochemical method 地球化学方法
geochemical mobility 地球化学活动性
geochemical model 地球化学模式
geochemical ocean section study 海洋地球化学剖面研究
geochemical open system 地球化学开放体系
geochemical operation system 地球化学工作方法;化探工作方法
geochemical partition of element 元素的地球化学分配
geochemical pattern 地球化学模式
geochemical population 地球化学总体
geochemical processing package 化探处理程序包
geochemical profile 地球化学剖面
geochemical prospecting 地球化学探矿;地球化学勘探
geochemical prospecting method 地球化学找矿法
geochemical province 地球化学省
geochemical reaction 地球化学反应
geochemical reconnaissance 地球化学普查
geochemical redox sequence 地球化学氧化还原序列
geochemical reservoir 地球化学库
geochemical rock survey 岩石地球化学测量;地球化学岩石测量
geochemical sample 化探样品;地球化学样品
geochemical section 地球化学剖面
geochemical soil survey 土壤地球化学测量;地球化学土壤测量
geochemical standard rock sample 地球化学岩石标准样
geochemical standard sample 地球化学标准样
geochemical standard soil sample 地球化学土壤标准样
geochemical standard stream sediment sample 地球化学水系沉积物标准样
geochemical stream sediment survey 水系沉积物地球化学测量
geochemical survey 地球化学调查
geochemical system 地球化学体系
geochemical water survey 水地球化学测量
geochemical zoning 地球化学分带
geochemistry 地质化学;地球化学
geochemistry and geochemical prospecting data base 地球化学及化探数据库
geochemistry data map 地球化学数据图
geochemistry derail survey 地球化学详查
geochemistry factor 地球化学因素
geochemistry of atmosphere 大气圈地球化学
geochemistry of biosphere 生物圈地球化学
geochemistry of fluoride-containing groundwater 含氟地下水地球化学
geochemistry of hydrosphere 水圈地球化学
geochemistry of hydrothermal processes 热液作用地球化学
geochemistry of landscape 景观地球化学
geochemistry of lithosphere 岩石圈地球化学
geochemistry of magmatic processes 岩浆作用地球化学
geochemistry of metamorphic processes 变质作用地球化学

geochemistry of mineral deposits 矿床地球化学
geochemistry of pegmatitic processes 伟晶作用地球化学
geochemistry of pneumato-hydrothermal processes 气化热液作用地球化学
geochemistry of rare elements 稀有元素地球化学
geochemistry of sedimentary processes 沉积作用地球化学
geochemistry of solids 固体地球化学
geochemistry of strata-bound 层控作用地球化学
geochemistry of weathering 风化作用地球化学
geochemistry primary prospecting 地球化学踏勘
geochemistry province 地球化学省
geochron 岩石地层时;地质时期间隔;地质年代;地时间过程
geochronic geology 历史地质学;地史学
geochronological unit 地质年代单位
geochronology 地质年代学
geochronometry 地质年代测定学;地质年代测定法
geoclimatic drying power 地面干燥能力
geocline 地形倾差;地斜
geocode data 地理代码数据
geo-coded topographic(al) data base 测地编码地形数据库
geocoding 地形编码;地理编码
geocoenosis 地理群落
geocole 地栖的
geocomplex 地理综合体
geocorona 地冕
geocrinite 硫砷锑铅矿
geocronite 斜方硫砷锑铅矿;砷硫锑铅矿
geocryology 多年冻土学;地球冰雪学;地冰学;冰冻地质学
geocushion 土工垫
geodata 地理数据
geode 晶腺;晶球;晶孔;晶洞
geodepression 地凹;地陷
geodesic 大地测量学的;测地线的;测大地线
geodesic altimeter 大地测量测高计
geodesic astronomy 大地天文学
geodesic azimuth 大地方位角
geodesic base line 测地基线
geodesic circle 测地圈
geodesic construction 壳罩结构;大量重复构件构造;网格状结构
geodesic control survey 大地控制测量
geodesic coordinates 大地坐标;短程线坐标;测地坐标
geodesic coordinate system 测地坐标系
geodesic cupola 网格圆顶;网格球
geodesic curvature 大地曲率;测曲率
geodesic curve 大地曲线;测地线
geodesic datum 大地基准(点)
geodesic distance 距离
geodesic distance meter 大地测距仪
geodesic dome 短程线穹顶;密网格穹隆;球形网架;网格球顶;网架穹顶;短程线坐标
geodesic formation survey 大地形变测量
geodesic head 静压力水头
geodesic isotension 短程等张力
geodesic-isotension contour 短程等张力曲面
geodesic latitude 大地纬度
geodesic lens 测地摄影镜头
geodesic level 大地基准面;大地测量水准仪
geodesic leveling 大地水准测量
geodesic line 最短线;短程线;测地线
geodesic longitude 大地经度
geodesic meridional plane 大地子午面
geodesic method 短程线法;测地线法
geodesic motion 短程线运动
geodesic network 大地控制网
geodesic normal 大地线法线
geodesic normal coordinate 测地法线坐标
geodesic ovaloid 短程卵形面
geodesic parallel 测地平行线
geodesic parameter 大地线参数;测地参数
geodesic point 大地站
geodesic position 大地位置
geodesic radius 测地半径
geodesic range finder 大地测距仪
geodesic satellite 大地测量卫星;测地卫星
geodesic spacecraft 测地宇宙飞船
geodesic structures 大量重复构架结构
geodesic theodolite 大地经纬仪

geodesic torsion 大地线挠率
geodesic triangle 测地三角形
geodesic triangulation 大地三角测量
geodesist 大地测量工作者
geodesy 大地测量(学);测地学
geodesy and cartography 测绘学
geodesy on the ellipsoid 椭球面大地测量学
geodetic 短程的;大地测量的
geodetic astronomy 大地天文学;测地天文学
geodetic astronomy survey 大地天文测量
geodetic azimuth 大地方位角
geodetic base line 大地测量基线
geodetic bearing 大地方位
geodetic benchmark 大地测量基准
geodetic book 大地测量(手)簿
geodetic boundary value problem 大地测量边值问题
geodetic calculation 大地测量计算
geodetic chain 大地测量控制锁;大地测量控制链
geodetic circle 大地圆
geodetic computation 大地测量计算
geodetic computer 大地测量计算机
geodetic connection 大地联测
geodetic connection of aiming azimuth 瞄准方位角连测
geodetic constant 大地常数
geodetic construction 最短线式结构;球模式构造;轻型受拉杆系结构;外壳承载式结构
geodetic control 大地(测量)控制
geodetic control chain 大地测量控制锁;大地测量控制链
geodetic control data 大地控制数据
geodetic control net(work) 大地控制网
geodetic control point 大地(控制)点
geodetic control survey 大地控制测量
geodetic coordinates 短程坐标;大地(测量)坐标
geodetic coordinate system 大地坐标系;大地测量坐标系统
geodetic cupola 网格圆顶;网格球
geodetic curvature 大地曲率
geodetic curve 大地线;大地测量曲线
geodetic data 大地测量资料;大地测量数据
geodetic data base 大地测量数据库
geodetic data sheet 大地(测量)成果表
geodetic datum 大地水准面;地理基准点;大地基准面;大地基准(点)
geodetic datum line 大地基准线
geodetic distance 大地距离
geodetic Doppler receiver 大地测量多普勒接收机
geodetic earth orbiting satellite 大地测量轨道卫星
geodetic engineer 大地测量工程师
geodetic engineering 大地测量(技术)
geodetic equator 大地赤道
geodetic equipment 大地测量设备
geodetic fix 大地定位
geodetic formation survey 大地形变测量
geodetic gang 大地测量队
geodetic gravimeter 大地重力测量计
geodetic gravimetry 大地重力学;大地重力测量(学)
geodetic gravity meter 大地型重力仪
geodetic head 总水头;静压水头;大地水头
geodetic height 大地(高程)
geodetic instrument 大地测量仪;大地测量器
geodetic intersection traverse 大地交会导线
geodetic laser survey system 激光大地测量系统
geodetic latitude 地理纬度;大地纬度;测地纬度
geodetic lens 大地测量透镜
geodetic level 大地(测量)水准仪;大地基准面
geodetic level(l)ing 精密水准测量;大地水准测量
geodetic line 短程线;大地线;测地线;最短线
geodetic location 大地定位
geodetic longitude 地理经度;大地经线;大地经度
geodetic measurement 大地测量
geodetic measuring technique 大地测量技术
geodetic meridian 地理纬度平行圈;大地子午线
geodetic method 大地测量法
geodetic network 大地控制网;大地测量网
geodetic normal coordinates 大地正则坐标
geodetic north pole 大地北极
geodetic origin 大地原点
geodetic origin data 大地起始数据
geodetic parallel 大地纬圈
geodetic parallel coordinates 大地平行坐标

geodetic party 大地测量队
geodetic photogrammetry 摄影大地测量学
geodetic point 大地点
geodetic point of origin 大地原点
geodetic polar coordinates 大地极坐标
geodetic position 地理位置;大地坐标;大地位置;大地定位
geodetic pressure head 自流水头
geodetic quadrangle 大地四边形
geodetic reconnaissance 大地测量踏勘
geodetic rectangular coordinates in space 大地空间直角坐标系
geodetic reference system 大地(测量)参考系(统);测地参考系统
geodetic refraction 地平大气折射
geodetic ring 大地闭合网
geodetics 大地测量学
geodetic satellite 大地测量卫星
geodetic sea level 大地测量水准面
geodetic shift 大地基准变换
geodetic station 大地点
geodetic stellar camera 大地恒星摄影机
geodetic survey board 大地测量局
geodetic survey(ing) 大地测量(学)
geodetic surveying tape 大地测量卷尺
geodetic surveyor 大地测量员
geodetic survey station 大地测量点
geodetic system 大地测量系统
geodetic team 大地测量队
geodetic theodolite 大地经纬仪
geodetic tie 大地联测
geodetic transformation 大地基准变换
geodetic transit 大地经纬仪
geodetic triangle 大地定位三角形
geodetic triangulation 大地三角测量
geodetic zenith 大地天顶
geodimeter 测距仪;光速测距仪;光电测距仪;光电测距仪
geodrain 地内排水板
geodrain method 排水板法
geodynamic(al) 地球动力学的
geodynamic(al) constant 地球动力学常数
geodynamic(al) height 动力高度
geodynamic(al) meter 动力米
geodynamic(al) method 地质动力学法;地球动力学法
geodynamic(al) phenomenon 地球动力现象
geodynamics 地质动力学;地球动力学
geodynamics satellite 地球动力测量卫星
geodyte 地上生物
geoecology 环境地质学;生态地质学;地质生态学
geoeconomy 地理经济学
geoecotype 地理生态型
geoelectric(al) anomaly 地电异常
geoelectric(al) cross section 地电断面
geoelectric(al) instrument 地电仪
geoelectric(al) measurement 地电探测;地电测量
geoelectric(al) section 地电断面图
geoelectric(al) survey 地电测量;地质电测
geoelectricity 地电(学);大地电
geoelectricity field 地电场
geofabric 土工布;土工织物
geoflex 爆炸索
geofracture 区域性大断裂;地裂隙;地裂缝;地缝合线;地断裂(带);地带断裂
geofracture observation 地裂缝观测
geogenesis 地球发生论
geogeny 地球成因学
geognosy 记录(构造)地质学;描述地质学;地知学
geogony 地球成因学
geogram 地学环境图
geographer 地理学家;地理学工作者
geographic(al) analysis 地理分析
geographic(al) area 地理区(域)
geographic(al) area call 区域性呼叫
geographic(al) area group calls 地区区域群呼
geographic(al) arrangement 地理分布;按地理排序
geographic(al) atlas 地理图集
geographic(al) attributes 地理类
geographic(al) azimuth 地理方位角
geographic(al) base 地理底图
geographic(al) base file 地理基础文件
geographic(al) botany 植物地理(学)
geographic(al) choronym 地理名称

geographic(al) circuitry 站场型网络
geographic(al) code 地名代码;地理代码
geographic(al) coding system 地理编码系统
geographic(al) concentration 生产地理集中
geographic(al) condition 地理环境
geographic(al) coordinate grid 地理坐标网格
geographic(al) coordinate net 地理坐标网格
geographic(al) coordinates 经纬度;地理坐标;地理位置
geographic(al) coordination of geologic(al) observation point 地质点地理坐标
geographic(al) correlation 地理相关
geographic(al) cycle 地貌循环;侵蚀旋回;地球旋回
geographic(al) cycle and physiographic(al) stage 地理循环及地文期
geographic(al) data 地理数据
geographic(al) data base 地理数据库
geographic(al) demarcation 分区
geographic(al) dictionary 地理词典
geographic(al) distribution 布局;地域分布;地理分布
geographic(al) distribution of disease 疾病地理分布
geographic(al) distribution of population 人口的地理分布
geographic(al) distribution of resources 资源的地理分布
geographic(al) distribution source rock 烃源岩地理分布
geographic(al) division of work 地域分工
geographic(al) element 地理要素
geographic(al) employment distribution 职业地区分布
geographic(al) environment 地理环境
geographic(al) environment determinism 地理环境决定论
geographic(al) equator 地理赤道
geographic(al) exploration 地理勘察
geographic(al) exploration traverse 地理勘察路线
geographic(al) features 地势;地理要素;地理特征
geographic(al) graticule 地理坐标网
geographic(al) graticule grid 地理坐标网
geographic(al) grid 地理网格
geographic(al) horizon 地平线
geographic(al) index 地理位置索引
geographic(al) inertia 地理惯性
geographic(al) information system 地理信息系统
geographic(al) information system digital formats 地理信息系统的数字格式
geographic(al) information system digital system 地理信息系统数字系统
geographic(al) information system digital technique 地理信息系统数字技术
geographic(al) information system laboratory 地理信息系统实验室
geographic(al) information system spatial analysis 地理信息系统空间分析
geographic(al) isolation 地理隔离
geographic(al) landscape 地理景观
geographic(al) latitude 地理纬度;大地纬度
geographic(al) layout 布局
geographic(al) limit 内图廓线
geographic(al) local time 地方时(间)
geographic(al) location 经纬度;地理坐标;地理位置;地理定位
geographic(al) longitude 地理经度;大地经度
geographic(al) longitude line 地理经度线
geographically load-shedding 分区减载
geographically referenced data storage and retrieval system 地理参考数据存储检索系统
geographically specific term 地理专名
geographic(al) map 地(理)图;地形图
geographic(al) map of volcanoes 火山分布图
geographic(al) mapping 地理图制图
geographic(al) meridian 地理子午线;地理子午圈;大地子午线
geographic(al) mesh 经纬度网格;地理网格
geographic(al) mesh numbering 经纬度编号法
geographic(al) mile 地理(英)里
geographic(al) name 地名;地理名称
geographic(al) name bank 地名库
geographic(al) name data base 地名数据库
geographic(al) names retrieval system 地名检索系统

geographic(al) name transcription 地名转写
geographic(al) nautical mile 地理海里
geographic(al) net 经纬线网
geographic(al) nomenclature 地理通名
geographic(al) north 地理北
geographic(al) number 地理编号
geographic(al) object 地理目标
geographic(al) ozone distribution 臭氧的地理分布
geographic(al) parallel 地理纬圈;大地纬圈
geographic(al) phenomenon 地理现象
geographic(al) plot 地理标绘;绝对运动图;对地运动图
geographic(al) plotting 绝对运动作图;对地运动作图
geographic(al) political elements 地理政治因素
geographic(al) position 地理位置
geographic(al) position inverse 地理位置反演
geographic(al) position locator 地理位置测定器
geographic(al) position of celestial body 天体地理位置
geographic(al) projection 地理投影
geographic(al) province 地理区(域)
geographic(al) race 地理宗
geographic(al) range 地理视距;地理射程;地理能见距离
geographic(al) range of an object 物标地理视距
geographic(al) range of light 灯光地理能见距离
geographic(al) reference 地面参照物;地标
geographic(al) region 地理区(域)
geographic(al) relationship 地形关系
geographic(al) resistant sulfate cement 普通抗硫酸盐水泥
geographic(al) scale 地理图比例尺
geographic(al) sea mile 地理海里
geographic(al) search 地理搜索
geographic(al) section signal 地名信号
geographic(al) sector search 定点展开扇形搜索;地理扇形搜索
geographic(al) sequence 评估地产位置归档法
geographic(al) sheet line 地理坐标网
geographic(al) signal 地名信号
geographic(al) silicate cement 普通硅酸盐水泥
geographic(al) speciation 地理物种形成
geographic(al) species 地理物种
geographic(al) square search 地理正方形搜索;定点展开方形搜索
geographic(al) survey 地理调查
geographic(al) symbol 地理符号
geographic(al) synecology 植物群落地理学;地理群落生态学
geographic(al) target 地理目标
geographic(al) terminology 地名学
geographic(al) tongue 地图样舌
geographic(al) unconformity 地理不整合
geographic(al) variant 地理变形
geographic(al) vertical 铅垂线;地下垂线;地球表面法线方向;地平经圈;地理垂线;大地水准面法线方向
geographic(al) vicariad 地理性代替种
geographic(al) viewing distance 地理视距
geographic(al) visibility range 地理能见距离
geographic(al) zones of climate 地理气候带
geography 地理(学)
geography of circulation 流通地理学
geography of energy 能源地理
geography of maritime transport 海洋运输地理学
geography of race 人种地理学
geography of the ocean 海洋地理学
geography paper 地图纸
geo-grating 土工格室
geogrid 网格土工布;土工格栅
geo-grille 土工格栅
geohistory 地质历史学;地史学
geohydrographic(al) unit 水文地质单元
geohydrologic(al) arkers 水文地质标志
geohydrologic(al) environment 水文地质环境
geohydrologic(al) geochemistry 水文地质化学
geohydrology 水文地质学;地质水文学;地下(水)水文学
geohygiene 地理卫生学
geoid 重力平面;地球形(体);地球体;大地体;大地水准面
geoidal distance 大地水准面距离
geoidal height 正高;大地水准面高度;大地水准面

高程;大地高(程)
geoidal height map 大地水准面高度图
geoidal horizon 海面水平面;地理水准面地平圈;大地水准面地平圈
geoidal level normal 大地水准面法线
geoidal level section 大地水准面断面
geoidal map 大地水准面图
geoidal potential 大地水准面势
geoidal rise 大地水准面隆起
geoidal slope 海平坡度;大地水准面坡度;大地水准面倾斜度
geoidal surface 大地水准面
geoidal tangent plane 大地水准面切面
geoidal undulation 大地水准面起伏
geoid contour 大地水准面等高线
geoid ellipsoid separation 大地水准面差距
geoidmeter 测地仪
geoid separation 大地水准面差距
geoid-spheroid separation 大地水准面差距
geoid surface 大地水准面
geoinformation system 地理信息系统
geoisotherm 地等温线;等地温线;等地温面
geoisothermal line 等地温线
geoisothermal map 地热等温线图
geokinetics 地球运动学
geolith 岩石地层单元;岩石地层单位
geologic(al) action 地质作用
geologic(al) aerosurveying 地质航空测量
geologic(al) aerovisual observation 空中地质观测;航空地质目测
geologic(al) age 地质时代;地质期;地质年龄;地质(年)代
geologic(al) age determination 地质年龄测定;地质年代测定
geologic(al) agent 地质作用力;地质营力;地质因素
geologic(al) age of landform 地貌地质年龄
geologic(al) airview 空中地质观测;航空地质目测
geologic(al) analogic(al) error 地质类比误差
geologic(al) and geophysical comprehensive survey 地质与地球物理综合调查
geologic(al) and mining for reasons of policy adjustment 地质矿业政策性调整
geologic(al) and ore deposition assessment 矿产地质评议
geologic(al) annotation 地质调绘
geologic(al) anomaly 地质异常;地质变态
geologic(al) aspects of soil formation 土层的地质学问题
geologic(al) atlas 地质图集
geologic(al) barometer 地质压力计
geologic(al) basemap 地质底图
geologic(al) bets 地质推断
geologic(al) block method 地质块段法
geologic(al) body 地质体
geologic(al) body cell 地质体单元
geologic(al) body stress field simulation 地质体应力场模拟
geologic(al) boundary 地质界线;地质边界
geologic(al) calamity 地质灾害
geologic(al) calendar 地质年历
geologic(al) camp 地质野营
geologic(al) casing(pipe) 地质套管
geologic(al) characteristic 地质特征
geologic(al) chronology 地质年代学
geologic(al) circle of pollutants 污染物地址大循环
geologic(al) classification 地质分类
geologic(al) classification analysis 地质分类分析
geologic(al) classification of soil 土壤的地质分类
geologic(al) classification system 地质分类系统
geologic(al) climate 古气候;地质气候
geologic(al) column 地质柱状图;地质柱状剖面;地层柱状图
geologic(al) columnar section 地质柱状(剖面)图
geologic(al) compass 矿用罗盘;地质指南针;地质罗盘
geologic(al) computing program(me) 地质计算程序
geologic(al) condition of ore deposition is bed 成矿地质条件不利
geologic(al) condition of ore deposition is well 成矿地质条件较好
geologic(al) condition 地质条件
geologic(al) conditions for prospecting 矿产普查地质条件

geologic(al) content of asbestos in the ore 石棉矿石的石棉地质含量
geologic(al) control 地质控制
geologic(al) control of mining in pit 露天开采地质指导
geologic(al) control of underground mining 地下开采地质指导
geologic(al) correlation 地质对比
geologic(al) cross-section 地质剖面图;地质横剖面图
geologic(al) cycle 地质循环;地质旋回
geologic(al) cycle of pollutants 污染物的地质大循环;污染地质循环
geologic(al) data 地质数据
geologic(al) data base 地质数据库
geologic(al) data base design 地质数据库设计
geologic(al) data base management system 地质数据库管理系统
geologic(al) dating 地质年龄测定;地质年代测定
geologic(al) deformation 地质变形
geologic(al) deposit 地质沉积
geologic(al) design 地质设计
geologic(al) development 地质展视图
geologic(al) dip of planar feature 平面地物的地质下沉
geologic(al) discontinuity 地质不连续性(面)
geologic(al) distribution 地质分布
geologic(al) documentation 地质编录
geologic(al) documentation of drill hole 钻孔地质编录
geologic(al) documentation of exploratory trenches 探槽地质编录
geologic(al) documentation of opening 坑道地质编录
geologic(al) documentation of shallow shaft 浅井地质编录
geologic(al) drawing 地质图
geologic(al) drilling 地质钻探
geologic(al) drilling rig 地质钻机
geologic(al) drill pipe 地质钻管
geologic(al) engineer 地质工程师
geologic(al) engineering 工程地质学;地质工程(学)
geologic(al) environment 地质环境
geologic(al) environment assessment 地质环境评价
geologic(al) environment monitoring 地质环境监测
geologic(al) environment protection 地质环境保护
geologic(al) environments of geothermal system 地热系统的地质循环
geologic(al) environment supervision and management 地质环境监督管理
geologic(al) epoch 地质时期
geologic(al) era 地质(时)代
geologic(al) erosion 自然侵蚀;自然冲刷;正常侵蚀;地质侵蚀
geologic(al) event 地质事件
geologic(al) examination 地质勘探;地质勘察;地质勘查;地质勘测;地质调查
geologic(al) expert system 地质专家系统
geologic(al) exploration 地质探测;地质勘探(钻进)
geologic(al) exploration at bridge site 桥位地质查勘
geologic(al) exploration for groundwater 地下水地质勘探
geologic(al) exploration instrument 地质勘探仪器
geologic(al) exploration party 地质勘探队
geologic(al) exploration team 地质勘探队
geologic(al) factor 地质因素
geologic(al) factors of ore-controlling 控矿地质因素
geologic(al) fault 地质断层
geologic(al) feature 地质特征;地形特征;地貌;地质要素;地质景观
geologic(al) field data 地质野外资料
geologic(al) field party 野外地质勘探队
geologic(al) fold 地质褶皱
geologic(al) formation 地质组成;地质建造;地质构成;地质层组
geologic(al) fracture 地质破裂
geologic(al) function 地质作用
geologic(al) genetic analysis 地质成因分析

geologic(al)-geophysical coordination 地质物探工作的配合
geologic(al) ground investigation 地质(土地)调查;地基地质调查
geologic(al) ground survey 地质地面调查
geologic(al) group 地质组
geologic(al) hammer 地质锤
geologic(al) hazard 地质灾害
geologic(al) historic(al) analysis 地质历史分析
geologic(al) history 地质历史(学);地质(发展)史
geologic(al) history of moon 月球地质历史
geologic(al) hole 地质钻孔
geologic(al) horizon 地质地层;地质层位
geologic(al) industry 地质业
geologic(al) informatics 地质信息学
geologic(al) information 地质资料
geologic(al) instrument 地质仪器
geologic(al) interpretation 地质解译
geologic(al) interpretation of aerial photograph 航片地质判读;航片地质解译
geologic(al) interpretation of photograph 像片地质判读;像片地质解译
geologic(al) interpretation of satellite photograph 卫片地质判译
geologic(al) investigation 地质勘察;地质勘测;地质调查
geologic(al) landscape 地质景观
geologic(al) legend 地质图例
geologic(al) log 地质柱状图;地质柱状剖面;地质(测井)记录;地质编录
geologic(al) logs of drill hole 钻孔地质剖面
geologic(al) longitudinal section 地质纵剖面(图)
geologic(al) map 地质图;岩石层图
geologic(al) map-making 地质图制作
geologic(al) map of bedrocks 基岩地质图
geologic(al) map of coalfield 煤田地质图
geologic(al) map of exploration area 勘探区地质图
geologic(al) map of mine district 矿区地质图
geologic(al) map of mineral deposits 矿产地质图
geologic(al) map of ore deposits 矿床地质图
geologic(al) map of ore district 矿区地质图
geologic(al) map of ore field 矿田地质图
geologic(al) map of placer deposit 砂矿地质图
geologic(al) map of quaternary system 第四系地质图
geologic(al) map of volcano 火山地质图
geologic(al) mapping 地质制图;地质填图
geologic(al) mapping area 地质测绘面积
geologic(al) mapping area of dam site 坝址区测绘面积
geologic(al) mapping area of irrigation region 灌区测绘面积
geologic(al) mapping area of reservoir region 水库区测绘面积
geologic(al) mapping area of road 线路测绘面积
geologic(al) mapping based on remote sensing 遥感地质制图
geologic(al) mapping method 地质测绘方法
geologic(al) mapping procedure based on remote sensing 遥感地质制图程序
geologic(al) market management 地质市场经营
geologic(al) market receipt charges 市场收入额
geologic(al) material 地质材料;地质资料
geologic(al) measuring instrument 地质测量仪
geologic(al) megacycle 地质大循环
geologic(al) method 地质法
geologic(al) mine radiation meter 地质矿山辐射仪
geologic(al) mineral data base 地质矿产数据库
geologic(al) minerogenetic model based on remote sensing 遥感地质成矿模式
geologic(al) model 地质模型
geologic(al) model of pool 油藏地质模型
geologic(al) museum 地质博物馆
geologic(al) noise 地质噪声
geologic(al) norm 天然地质环境;地质标准矿物分类;地质标志矿物分类
geologic(al) norm of erosion 纯地质侵蚀
geologic(al) observation point 地质(观测)点
geologic(al) observation point in bedrock 基岩地质观测点
geologic(al) observation spot 地质观察点

geologic(al) oceanography 海洋地质学;海底地质学;地质海洋学
geologic(al) orbital photography 地质卫星轨道摄影
geologic(al) oreblock method 地质块段法
geologic(al) origin 地质成因
geologic(al) period 地质时期;地质时代
geologic(al) photograph 地质照片
geologic(al) photomap 影像地质图
geologic(al) plan 水平地质剖面;地质平切面
geologic(al) plan of level of ore deposits 矿床中段地质图
geologic(al) plan of level ore body 矿体水平地质图
geologic(al) plan of mining bench 平台地质平面图
geologic(al) plan of mining level 中段地质平面图
geologic(al) plotting program(me) library 地质绘图程序库
geologic(al) point survey 地质点测量
geologic(al) prediction 地质预测
geologic(al) prediction analysis 地质预报分析
geologic(al) probing 地质探测
geologic(al) process 地质作用;地质过程
geologic(al) processing machinery 地质勘探机械
geologic(al) process of flood current 洪流地质作用
geologic(al) process of glacier 冰川地质作用
geologic(al) process of groundwater 地下水地质作用
geologic(al) process of lake 湖泊地质作用
geologic(al) process of marsh 沼泽地质作用
geologic(al) process of sheet flow 片流地质作用
geologic(al) process of surface water 地面流水地质作用
geologic(al) process of wind 风的地质作用
geologic(al) profile 地质剖面线;地质剖面(图)
geologic(al) profile along exploratory line 勘探线地质剖面图
geologic(al) profile survey 地质剖面测量
geologic(al) prospecting 地质找矿;地质勘探
geologic(al) prospecting and exploration strategy 地质勘探工作战略决策
geologic(al) prospecting engineering survey 地质勘探工程测量
geologic(al) prospecting method 地质找矿法
geologic(al) prospecting party 地质勘探队
geologic(al) prospecting team 地质勘探队
geologic(al) province 区域地质;地质省;地质区(域)
geologic(al) radar 地质雷达
geologic(al) radar method 地质雷达法
geologic(al) reasoning 地质推断
geologic(al) reconnaissance 地质踏勘;地质普查
geologic(al) record 地质记录;地质编录
geologic(al) remapping based on remote sensing 遥感修编地质图
geologic(al) remote sensing 地质遥感
geologic(al) report 地质报告
geologic(al) report for mining 采矿地质报告
geologic(al) report in exploration stage 勘探地质报告
geologic(al) report(s in prospecting and exploration) 地质报告
geologic(al) report target 地质成果指标
geologic(al) research 地质研究
geologic(al) research report 地质研究报告
geologic(al) research vessel 地质勘探船
geologic(al) reservation 地质自然保护区
geologic(al) reserves 地质储量
geologic(al) route survey 地质路线调查
geologic(al) sample 地质样品
geologic(al) sampler 地质取样器
geologic(al) sampling 地质采样
geologic(al) satellite 地质卫星
geologic(al) scale of time 地质时期
geologic(al) scheme 地质略图
geologic(al) section 地质剖面(图);地质断面
geologic(al) section map 地质剖面图
geologic(al) section survey 地质剖面测量
geologic(al) setting 地质条件;地质环境;地质背景
geologic(al) simultaneity 地质同时性
geologic(al) site condition 场地地质条件
geologic(al) sketch 地质素描

geologic(al) sketch map 地质素描图;地质草图
geologic(al) sketch map based on remote sensing 遥感概略地质图
geologic(al) sketch of exposure 露头(地质)素描图
geologic(al) stable area 地质稳定区
geologic(al) strata 地质层
geologic(al) stratum compass 地质地层罗盘
geologic(al) strike 地层走向
geologic(al) structure 地质构造
geologic(al) structure condition 地质结构条件
geologic(al) structure map 构造地质图
geologic(al) study of deep sea mineral resources 深海矿物资源地质学研究
geologic(al) succession 地质演变;地层系列;地质顺序
geologic(al) suitability 地质适宜性;地质适合性
geologic(al) suitability of site 场地地质适合性
geologic(al) supervision of mining field 采场地质管理工作
geologic(al) surprises 地质异常情况
Geologic(al) Survey 地质调查所
geologic(al) survey(ing) 地质测量;地质勘探;地质踏勘;地质勘查;地质勘测;地质调查;地质勘绘
geologic(al) surveying and mapping 地质测绘
geologic(al) survey map 地质调查图
geologic(al) survey of seismic calamity 震害地质调查
geologic(al) survey rate meter 地质勘探速率计
geologic(al) survey ship 地质勘探船;地质调查船
geologic(al) symbols 地质符号
geologic(al) technical instruction manual of drill hole 钻孔地质技术指导书
geologic(al)-tectonic map 地质构造图
geologic(al) tenor of asbestos ore 石棉矿石的地质品位
geologic(al) thematic mapping based on remote sensing 遥感地质专题编图
geologic(al) thermometer 古温标;地质温度计;地质温度表
geologic(al) thermometry 古温测量
geologic(al) thickness 地质厚度
geologic(al) time 地质时期;地质时间;地质时代;地质年代;地史时期
geologic(al) time scale 地质时间表;地质时代表;地质时标;地质年代表
geologic(al) time unit 地质年代单元
geologic(al) topographic(al) planimetric(al) map 地质地形平面图
geologic(al) transportation 地质搬运
geologic(al) treatment flow 地质加工流程
geologic(al) trend 地质走向
geologic(al) trend analysis 地质趋势分析
geologic(al) type of gravity anomaly 重力异常地质类型
geologic(al) type of strong runoff zone 强径流带地质类型
geologic(al) unit 地质单元;地质单位
geologic(al) variable 地质变量
geologic(al) vector length analysis method 地质向量长度分析法
geologic(al) working management 地质工作管理
geologic(al) work in the mine design 矿山设计中地质工作
geologic(al) work of river 河流的地质作功
geologist 地质学家;地质工作者
geologist's compass 地质罗盘
geologist's pick 地质师(手)锤
geologize 地质调查
geolograph 钻速记录仪;地质记录器
geology 地质(学)
geology mapping of canal 渠道地质测绘
geology mapping of dam site 坝址区地质测绘
geology mapping of reservoir area 水库区地质测绘
geology mapping of spillway 溢洪道地质测绘
geology mapping of tunnel 隧洞地质测绘
geology of groundwater 地下水地质学
geology of Jupiter's satellites 木星卫星地质学
geology of mars 火星地质学
geology of martian satellites 火星卫星地质学
geology of mercury 水星地质学
geology of mineral deposits 矿床地质
geology of moon 月球地质学

geology of ore body 矿体地质
geology steel tube 地质钢管
geomagnetic 地磁的
geomagnetic abnormal detection 地磁异常探测
geomagnetic activity 地磁活动
geomagnetic airborne survey system 地磁空中测量系统
geomagnetic anomaly 地磁异常
geomagnetic axis 地磁轴
geomagnetic base point 地磁基点
geomagnetic boundary 地磁边界
geomagnetic cavity 地磁穴
geomagnetic chart 地磁图
geomagnetic composition 地磁场构成
geomagnetic coordinates 地磁坐标
geomagnetic current measurement 地磁流测量
geomagnetic cutoff 地磁截止
geomagnetic cut-off latitude 地磁截止纬度
geomagnetic declination 地磁偏角
geomagnetic dipole 地磁偶极(子)
geomagnetic dipole moment 地磁偶极矩
geomagnetic disturbance 地磁扰动
geomagnetic electrokinetograph 电地磁流量计;电磁海流计;电磁测流器;地磁动电(测流)计;地磁电动测速仪
geomagnetic element 地磁要素
geomagnetic equator 地(理)磁赤道
geomagnetic field 地球(基本)磁场;地磁场
geomagnetic field reversal 地磁场反转
geomagnetic inclination 地磁倾角
geomagnetic indicator 地磁强测量仪
geomagnetic induction 地磁感应
geomagnetic inductor 地磁感应仪;地磁感应器
geomagnetic inductor compass 地磁感应罗盘
geomagnetic isoporic line 地磁图等年变线
geomagnetic latitude 地磁纬度
geomagnetic longitude 地磁经度
geomagnetic map 地磁图
geomagnetic measurement 地磁测量
geomagnetic meridian 地磁子午线;地磁子午圈
geomagnetic method 地磁勘探方法
geomagnetic micropulsation 地磁微脉动
geomagnetic moment 地球磁矩
geomagnetic nadir point 地磁底点
geomagnetic noise 地磁噪声
geomagnetic north pole position 地磁北极位置
geomagnetic observatory 地磁台
geomagnetic observatory position 地磁台位置
geomagnetic parallel 地磁纬圈
geomagnetic parameter 地磁参数
geomagnetic polarity reversal 磁极倒转反转;磁极倒转反向
geomagnetic polarity time scale 地磁极性年(代)表
geomagnetic pole 地(理)磁极
geomagnetic pole position 地磁极位置
geomagnetic reversal 地磁逆转;地磁反转;地磁反向;地磁倒转;磁极倒转反转;磁极倒转反向
geomagnetic shell 地磁壳层
geomagnetic solar daily variation 地磁太阳日变化
geomagnetic solar diurnal variation 地磁太阳日变化
geomagnetic solar lunar variation 地磁太阳月变化
geomagnetic south pole position 地磁南极位置
geomagnetic spiral field 地磁螺旋场
geomagnetic storm 地磁暴
geomagnetic strength recorder 地磁强度记录器
geomagnetic survey 地磁测量
geomagnetic tail 地磁尾
geomagnetic tide 地磁潮
geomagnetic time scale 地磁年代表
geomagnetic variation 地磁(场)变化
geomagnetism 地磁(学)
geomagnetochronology 地磁年代学
geomagnetograph 地磁记录仪
geomagnetograph survey 地磁记录测量
geomagnetometer 地磁仪
geomanetograph 地磁计
geomantic omen 风水
geomat 土工垫
geomaterial 土工(合成)材料;地质材料
geomathematical indication 地质数学标志
geomathematics 数学地质;地质数学
geomechanical model test 地质力学模型试验

geomechanics 岩土力学;地质力学;地球力学
geomechanism 地球力学
geomembrane 土工膜
geomembrane-based liner 土工膜基衬垫
geometer 几何学家;测量员
geometric(al) 几何(学)的
geometric(al) aberration 几何像差
geometric(al) accuracy 几何准确度;几何精(确)度
geometric(al) acoustics 几何声学;射线声学
geometric(al) addition 几何加法
geometric(al) albedo 几何反照率
geometric(al) analog problem 几何类比问题
geometric(al) analogy 几何图形模拟
geometric(al) analysis 几何分析
geometric(al) and physical analysis 几何物理分析
geometric(al) anisotropy 几何异向性
geometric(al) approximation 几何近似法
geometric(al) attenuation 几何衰减
geometric(al) average 几何平均(数)
geometric(al) axiom 几何公理
geometric(al) axis 几何轴(线)
geometric(al) axis diagram of member 杆件几何轴线图
geometric(al) body 几何体
geometric(al) boundary condition 几何边界条件
geometric(al) broadening 几何展宽
geometric(al) buckling 几何曲率;几何曲度常数
geometric(al) building block 几何构图块
geometric(al) canon 分度标准
geometric(al) center 几何形心;几何中心
geometric(al) center of lens 透镜的几何中心
geometric(al) characteristic 形状特性;几何特征
geometric(al) characteristic of river 河道几何特征
geometric(al) closure 几何闭合差
geometric(al) concentrating ratio 几何聚光比
geometric(al) condition 几何条件
geometric(al) condition of rectification 纠正几何条件
geometric(al) configuration 几何外形;几何构型
geometric(al) constraint 几何约束
geometric(al) construction 几何作图;几何构造
geometric(al) content 几何内容
geometric(al) continuity 几何连续性
geometric(al) convergence 几何收敛
geometric(al) correction 几何校正
geometric(al) correlation 几何相关
geometric(al) criterion 几何判据
geometric(al) cross-section 几何截面
geometric(al) crystallography 几何结晶学
geometric(al) damping 几何阻尼
geometric(al) data 几何数据
geometric(al) database 几何数据库
geometric(al) data of cascade 叶栅的几何参数
geometric(al) decorated style 几何装饰格式;几何形式样
geometric(al) decoration 几何形装饰
geometric(al) definition 几何定义
geometric(al) deflection 几何挠度
geometric(al) defocussing 几何散焦
geometric(al) deformation 几何图形变形
geometric(al) delay 几何延迟
geometric(al) depth 几何深度
geometric(al) design 线形设计;几何(形状)设计;几何纹
geometric(al) deviation 几何偏差
geometric(al) dilution of precision 几何误差放大因子(GPS定位);几何精度因子
geometric(al) dimension of axis 几何尺寸轴线
geometric(al) dimension 几何尺寸
geometric(al) dip 几何眼高差;几何倾角;几何俯角
geometric(al) dissimilarity 几何不相似性
geometric(al) distortion 几何失真;几何畸变;几何变形
geometric(al) distortion correction 几何畸变校正
geometric(al) distortion in hydraulic model 水工模型几何变态
geometric(al) distribution 几何分布
geometric(al) draft 型吃水
geometric(al) drawing 几何图
geometric(al) effect 几何效应
geometric(al) elastic stress concentration factor 几何弹性应力集中系数
geometric(al) element 几何要素
geometric(al) element of crystal 晶体的几何要素

geometric(al) ellipsoid 几何椭球
geometric(al) ergodicity 几何的遍历性
geometric(al) error 几何误差
geometric(al) factor 几何因子;几何因素;几何系数
geometric(al) factor of the adjacent bed 围岩的几何因子
geometric(al) factor of the invaded zone 浸入带的几何因子
geometric(al) factors of hole 钻孔几何因素
geometric(al) features 几何特征;几何特性
geometric(al) fidelity 几何保真度
geometric(al) figure 几何(图)形
geometric(al) focal range 几何焦深
geometric(al) focus 几何焦点
geometric(al) form 几何形状
geometric(al) freeboard 型干舷
geometric(al) free camber of spring 弹簧几何挠度
geometric(al) funicular form 几何索线性
geometric(al) gasket width 垫片几何宽度
geometric(al) geodesy 几何大地测量学
geometric(al) Gothic style 几何形哥特式格式
geometric(al) grade scale 几何粒级表;几何分级标准
geometric(al) growth 几何增大;几何增长;几何成长
geometric(al) head 几何位差;位置头
geometric(al) height 几何高(度)
geometric(al) horizon 几何(视)地平
geometric(al) horopter 几何无双像场
geometric(al) image 几何像
geometric(al) inertial navigation 几何式惯性导航
geometric(al) instability 几何形状不稳定性
geometric(al) integrity 几何完整性
geometric(al) interpolation 几何插值法
geometric(al) interpretation 几何解释
geometric(al) invariance 几何不变性
geometric(al) isomer 几何异构体
geometric(al) isomerism 几何异构
geometric(al) isotropy 几何各向同性
geometric(al) lathe 靠模车床
geometric(al) latitude 归心纬度;归化纬度;几何纬度
geometric(al) level(l)ing 几何水准测量;测高
geometric(al) libration 几何天平动
geometric(al) light-path 几何光路
geometric(al) locus 轨迹曲线;几何轨迹
geometric(al) longitude 几何经度
geometrically-distorted model 几何变态模型
geometrically normal frequency 几何正态频率
geometrically optic(al) approximation 几何光学近似
geometrically similar model 几何相似模型
geometrically similar pump 几何相似泵
geometrically unstable system 几何可变体系
geometrically true pyramid 几何真实的棱锥体
geometrically stable system 几何不可变体系
geometric(al) mean 几何平均(数);几何均数;等比中项
geometric(al) mean diameter 几何平均粒径;几何平均粒度
geometric(al) mean-diameter of silt 泥沙的几何平均粒径
geometric(al) mean error 几何平均误差
geometric(al) mean frequency 几何平均频率
geometric(al) mean grain size 几何平均粒径;几何平均粒度
geometric(al) meaning 几何意义
geometric(al) mean of U.S. weighted index 以美国为权数的几何平均指数
geometric(al) mean pitch 几何平均螺距
geometric(al) mean radius 几何平均半径
geometric(al) mean size 几何平均尺度
geometric(al) mean(value) 几何平均值
geometric(al) measurement 几何形态测量
geometric(al) metamerism 几何位变异构现象;几何条件配色
geometric(al) method 几何(方)法
geometric(al) method of satellite geodesy 卫星大地测量几何法
geometric(al) method of satellite-geodesy 几何卫星大地测量
geometric(al) method of satellite orbit 卫星轨道几何法
geometric(al) model 几何模型

geometric(al) modelling 几何模拟
geometric(al) moment of area 截面矩量;几何面积矩
geometric(al) moment of inertia 截面矩量;几何惯性矩
geometric(al) motif 几何图案花纹;几何花纹图案
geometric(al) non-linearity 几何非线性
geometric(al) number theory 几何数论
geometric(al) object 几何对象
geometric(al) optics 几何光学
geometric(al) optics aberration 几何光学像差
geometric(al) optics approach 几何光学法
geometric(al) optics approximation 几何光学近似
geometric(al) optics expansion 几何光学展开
geometric(al) optics field 几何光学场
geometric(al) optics limit 几何光学极限
geometric(al) optics ray 几何光学射线;几何光束
geometric(al) orientation 几何定向
geometric(al) ornament 几何(图)形装饰
geometric(al) parallax 几何视差
geometric(al) parameter 几何参数
geometric(al) path 几何路径
geometric(al) pattern 几何形花纹;几何图形;几何图案;几何模型;绕型(玻璃钢);几何形式
geometric(al) perturbation 几何形态干扰
geometric(al) pitch 几何螺距
geometric(al) point 几何点;图解点
geometric(al) position 几何位置
geometric(al) power diagram 功率矢量图;矢量功率图
geometric(al) probability 几何概率
geometric(al) processing 几何图形处理功能;几何处理
geometric(al) programming 几何规划(法)
geometric(al) progression 几何级数;等比级数
geometric(al) projection 几何投影(法);无穷远点射投影;点射投影
geometric(al) property 几何性质;几何特性
geometric(al) proportion 几何比(例)
geometric(al) quantity 几何量
geometric(al) reconfiguration 几何图形变形
geometric(al) reference surface 几何参考面
geometric(al) reflection 几何反射
geometric(al) relationship 几何关系
geometric(al) resolution 几何分辨率
geometric(al) satellite geodesy 卫星几何测地
geometric(al) satellite triangulation 几何法卫星三角测量
geometric(al) scanner 几何扫描仪
geometric(al) scattering factor 几何散射因数
geometric(al) scattering waves 几何散射波
geometric(al) seismology 几何地震学
geometric(al) sense 几何量
geometric(al) sequence 几何序列
geometric(al) series 几何级数;等比数列;等比级数
geometric(al) shadow 几何投影;几何阴影
geometric(al) shape 几何图形
geometric(al) signification 几何意义
geometric(al) similarity 几何相似(性)
geometric(al) similarity method 几何近似法
geometric(al) similitude 几何相似
geometric(al) solid 几何体
geometric(al) solution 几何解
geometric(al) spreading 几何散开
geometric(al) stability 几何稳定性
geometric(al) stair(case) 弯曲楼梯;圆弧形楼梯;螺旋形楼梯;盘旋楼梯;螺旋楼梯(边梁)
geometric(al) standard 几何标准
geometric(al) standard deviation 几何标准偏差
geometric(al) stiffness 几何刚度
geometric(al) stiffness coefficient 几何刚度系数
geometric(al) stiffness matrix 几何刚度矩阵
geometric(al) structure problem 几何结构问题
geometric(al) style 几何(形)式样;几何图形饰
geometric(al) symbol 几何符号
geometric(al) symmetry 几何对称
geometric(al) timing 几何定时
geometric(al) tolerance 几何公差
geometric(al) tracery 几何形花格;几何图形窗饰
geometric(al) tracery window 几何图案花格窗
geometric(al) transformation 几何变换
geometric(al) variable 几何变量
geometric(al) vertical 地向径向
geometric(al) volume of holder 储气罐几何容积

geometric(al) wave propagation 波的几何传播
geometric(al) width of gasket 垫片几何宽度
geometric(al) wireframe 几何形状线框图
geometrics 几何学图形
geometrize 用几何图形表示;作几何学图形
geometrodynamics 几何动力学
geometronics 电子制图法;地球测绘电子学
geometronics engineer 测量工程师
geometry 几何学;几何形状;几何图;几何结构
geometry coefficient 几何系数
geometry correction 几何校正
geometry correction of imagery 图像几何校正
geometry dilution of precision 精度的几何因子
geometry of basin 盆地几何形态
geometry of cable 缆索几何学
geometry of coal bed 煤层几何形态
geometry of coal seam 煤层几何形态
geometry of machinery 机械几何学
geometry of mapping 测图的几何形态
geometry of nets 网络几何(学)
geometry of parting 分歧处形态
geometry of projection 投影几何(学)
geometry of shell 薄壳几何学
geometry of sphere 球面几何(学)
geometry processor 几何信息处理机;几何处理程序
geometry test 几何畸变测试
geomonocline 地单斜(地槽边缘单斜沉积)
geomorphic(al) accident 地形突变
geomorphic(al) analysis 地貌分析
geomorphic(al) anomaly 地貌异常
geomorphic(al) architecture 地貌建筑学
geomorphic(al) area classification system 地貌区划系统
geomorphic(al) breaks 地形突变
geomorphic(al) composite profile 地貌综合剖面图
geomorphic(al) composition 地貌物质组成
geomorphic(al) condition 地貌条件
geomorphic(al) configuration 地貌外形
geomorphic(al) cycle 地貌循环;地貌旋回;冲积循环
geomorphic(al) data handling system 地貌数据处理系统
geomorphic(al) data processing system 地貌数据处理系统
geomorphic(al) data treatment system 地貌数据处理系统
geomorphic(al) description 地貌描述
geomorphic(al) disruption 地貌破坏
geomorphic(al) feature 地貌特征
geomorphic(al) geology 地貌学;地貌地质学
geomorphic(al) grade 地貌等级
geomorphic(al) legend 地貌图例
geomorphic(al) location 地貌位置
geomorphic(al) mapping 地貌填326
geomorphic(al) process 地貌过程;地貌变迁;地貌(形成)过程
geomorphic(al) profile 地貌剖面图
geomorphic(al) representation 地貌表示法
geomorphic(al) sensor 地貌传感器
geomorphic(al) shape 地貌形态
geomorphic(al) shaped line 地貌形态线
geomorphic(al) snow line 地形雪线
geomorphic(al) structure 地貌结构
geomorphic(al) unit 地貌单元
geomorphic(al) unit of mountain 山地地貌单元
geomorphic(al) unit of plain 平原地貌单元
geomorphogenic region 地貌发生区
geomorphogeny 地貌发生学;地貌成因学
geomorpho-geologic(al) map 地貌地质图
geomorphography 地貌叙述学;地貌描述学
geomorphologic(al) 地貌的
geomorphologic(al) agent 地貌力量
geomorphologic(al) condition 地貌条件
geomorphologic(al) deformation 地貌变形
geomorphologic(al) element 地貌因素;地貌要素;地貌成分
geomorphologic(al) engraving 地貌刻图
geomorphologic(al) indication 地貌标志
geomorphologic(al) landscape zoning 地貌景观分带
geomorphologic(al) map 地貌图;地形图
geomorphologic(al) mapping 地貌测绘
geomorphologic(al) marker 地貌标志
geomorphologic(al) method 地貌法

geomorphologic(al) name 地貌要素名
geomorphologic(al) remote sensing 地貌遥感
geomorphologic(al) setting 地貌特征；地貌环境；地貌背景
geomorphologic(al) survey (of river) 形态调查（河流的）
geomorphologic(al) type map 地貌类型图
geomorphologic(al) type 地貌类型
geomorphologic(al) unit of beach 海滩地貌单元
geomorphologic(al) unit of continental shelf 大陆架地貌单元
geomorphologic(al) unit of continental slope 大陆坡地貌单元
geomorphologic(al) unit of delta 三角洲地貌单元
geomorphologic(al) unit of intertidal zone 潮间带地貌单元
geomorphology 地形学；地貌学；地貌特征；地表形态学
geomorphology of abyssal fan 深海扇地貌
geomorphology of coral reef 珊瑚礁地貌
geomorphology of island 岛屿地貌
geomorphy 地形学；地貌
geomyricin 针蜡【地】
geomyricite 针蜡【地】
geon 吉纶
geonasty 感地性
geonavigation 地文航海(法)；地文导航；地标航行
geonet 土工席垫；土工网
geonomy 地(球)学
Geonor cone penetrometer apparatus 杰诺锥体贯入度仪（测定土的不排水抗剪强度等用）
Geonor settlement probe 杰诺沉陷探针（测定地基的长期沉降）
geonosy 地球构造学
Geopause 地球同步卫星
geopedology 地质土壤学
geopetal criterion 示序标志
geopetal fabric 示序组构；示顶底组构
geopetal structure 示顶底构造
geophilous plant 喜土植物
geophone 小型地震仪；听地器；地震检波器；地声探测器；地下震波检测器；地声仪；地声测听器
geophone filtering 检波器滤波
geophone model 检波器型号
geophone sensitivity 检波器灵敏度
geophone type 检波器类型
geophoto 地质照片；地质摄影
geophysical 地球物理学的
geophysical change 地球物理变化
geophysical classification 地球物理分类
geophysical condition 地球物理条件
geophysical effect 地球物理效应
geophysical electric(al) power generation 地热发电
geophysical engineering 地球物理技术
geophysical exploration 物探；地球物理探测；地球物理勘探；地球物理勘测
geophysical exploration and logging 地球物理勘探和测井
geophysical exploration company 地球物理探矿公司
geophysical exploration map 地球物理勘探图
geophysical exploration method 地球物理勘探方法
geophysical exploration report 物探报告
geophysical field 地球物理场
geophysical indication 地球物理标志
geophysical instrument 地球物理勘探仪(器)
geophysical investigation 地球物理研究；地球物理调查
geophysical jetting bit 物探(钻孔)用喷射式钻头
geophysical log 地球物理测井
geophysical magnetometer 地球物理磁强计
geophysical map 地球物理图
geophysical markers 地球物理标志
geophysical method 地球物理(方)法
geophysical method of exploration 地球物理勘探法
geophysical method of prospecting 地球物理勘探法
geophysical monitoring for climate change 气候变化地球物理监测
geophysical monitoring for climate change laboratory 气候变化地球物理监测实验室

geophysical observation satellite 地球物理观测卫星
geophysical prospecting 物探；地球物理探测；地球物理勘探；地球物理勘测
geophysical prospecting drilling vehicle 地球物理勘探车
geophysical prospecting instrument 地球物理勘探仪器
geophysical prospecting method 地球物理勘探方法；地球物理找矿法
geophysical prospecting of hydrology 水文物探
geophysical prospecting well-logging 地球物理勘探测井
geophysical refraction survey 地球物理折射波测量
geophysical satellite 地球物理卫星
geophysical search equipment 地球物理勘探设备
geophysical seismic exploration 地球物理地震探测；地球物理地震勘探；地球物理地震调查
geophysical seismic investigation 地球物理地震调查
geophysical site investigation 地球物理现场调研
geophysical subsurface investigation 地球物理地下调研
geophysical survey 地球物理勘探；地球物理勘测；地球物理调查；地球物理测量
geophysical surveying 地球物理测量
geophysical survey method 地球物理调查方法
geophysical survey ship 地球物理调查船
geophysical team 地球物理队
geophysical technique 地球物理技术
geophysical test(ing) 地球物理试验
geophysical (well-) log(ging) 地球物理测井(法)；物理测井
geophysical year 地球物理年
geophysicist 地球物理学家；地球物理工作者
geophysics 地球物理(学)
geophysiography 地理学
geophyte 地下芽植物
geophytia 地面植被群落
geopolitics 地理政治论
geopolymer 土工聚合物
geoponder 大地应答器
geoponic 耕作的
geopotential 重力(位)势；位势(的)；地重力势；地球重力势；大地位
geopotential energy 位势能
geopotential foot 位势英尺
geopotential height 位势高度；位高；地重力势高度
geopotential meter 位势米
geopotential number 位势数；位热数；地球位数
geopotential scale factor 地球重力位尺度因子
geopotential slope 位势坡度
geopotential surface 位势面；地球等位面；等重力势面；等位势面
geopotential thickness 位势厚度
geopotential unit 位势单位；地球重力位单位
geopressed geothermal electricity generation 地压地热发电
geopressured aquifer 超压水层
geopressured geothermal system 地压地热系统
geopressured reservoir 地压型热储
geopressured well 地压井
geopressure geothermal electricity generation plant 地压型地热电站
geopressure geothermal water 地压型地下热水
geopressurized geothermal system 地内密封地热系统
geoprocessing system 地理处理系统
geoproduct 土工产品
geordie 矿用安全灯
geordie turnout 方截面钢棒式道岔
georeceiver 大地接收机
geo-referenced data 与地理相关的数据
geo-referenced information 与地学有关的信息
Georgechaoite 乔治赵石
georgeite 水羟碳铜石
georgette 乔其纱
Georgia buggy 橡胶轮手推车；乔治亚斗车
georgiadesite 氯砷铅石；砷氯铅矿
georgia green 淡绿色
Georgian arch 平圆拱；乔治式拱
Georgian architecture 乔治时代(王朝)建筑
Georgian colonial architecture 乔治殖民式建筑
Georgian glass 乔治式(夹丝)玻璃

Georgian marble 乔治亚大理石
Georgian series 乔治·统【地】
Georgian style 乔治时代建筑风格
geoscience 地学；地球科学
geoscientist 地球科学家
geosere 地史学演替系列
geospacer 土工垫
geosphere 岩石圈；陆圈；陆界；壳圈；地圈；地球（岩石）圈
geostatic 土压力的；土静力的
geostatic arch 耐地压的拱；耐压拱；土压拱
geostatic curve 地压曲线
geostatic pressure 自重压力；地面压强；地(面)压力；地静压力
geostatics 刚体力学
geostatic stress 地(静)应力
geostatic stress field 地应力场
geostational satellite 同步卫星
geostationary 对地同步的；对地静止的
geostationary communication satellite 静止通信卫星
geostationary meteorologic(al) satellite 对地静止气象卫星；同步气象卫星；地球同步气象卫星；地球静止气象卫星
geostationary meteorological satellite system 同步气象卫星系统
geostationary operational environment 同步业务卫星
geostationary operational environment(al) satellite 同步环境应用卫星；地球同步环境卫星
geostationary operational meteorological satellite 地球静止业务气象卫星
geostationary orbit 静止轨道；同步轨道
geostationary satellite 赤道同步卫星；对地静止卫星；地球同步卫星；地球静止卫星
geostationary satellite launch vehicle 地球同步卫星运载火箭
geostatistical analysis 地球统计分析
geostatistic(al) computing program(me) 地质统计学计算程序
geostatistic(al) plotting program(me) 地质统计学绘图程序
geostatistic(al) program(me) library 地质统计学程序库
geostatistics 地球统计学；统计地质学；地质统计学
geostenogram 地理速测图；草测图
geostenography 地理速测
geostratigraphy 全球地层学
geostress anomaly 地应力异常
geostress survey 地应力测量
geostrophic 地转的
geostrophic acceleration 地转加速度
geostrophic amphidromic region 地转无潮区
geostrophic approximation 地转近似；地转假定
geostrophic assumption 地转近似
geostrophic contour current 地转平流
geostrophic current 地转流
geostrophic departure 地转偏差
geostrophic deviation 地转偏差
geostrophic distance 地转距离
geostrophic effect 地转影响；地转效应
geostrophic equation 地转方程
geostrophic equilibrium 地转平衡
geostrophic flow 地转流；地转(偏向)水流
geostrophic flux 地转通量
geostrophic force 地转力
geostrophic motion 地转(性)运动
geostrophic turbulence 地转湍流
geostrophic vorticity 地转涡度
geostrophic wind 地转风
geostrophic-wind level 地转风高度
geostrophic-wind scale 地转风风速标尺
geostrophy 地转
geostructure 地质构造
geosurvey 大地测量
geosuture 断裂线；地裂隙；地裂缝；地缝合线；地断裂(带)
geosynchronous operational environmental satellite 地球同步业务环境卫星
geosynchronous satellite 地球同步人造卫星
geosynclinal 地向斜的
geosynclinal anticlinorium 槽背斜
geosynclinal area 地槽区
geosynclinal axis 地槽轴(线)；大向斜轴

geosynclinal basin 地槽盆地
geosynclinal bauxite 地槽铝土矿
geosynclinal classification 地槽的类型
geosynclinal close stage 地槽封闭
geosynclinal couple 地槽偶
geosynclinal cycle 构造循环；构造旋回；构造回旋；地槽(式)旋回
geosynclinal facies 地槽相
geosynclinal folding 地槽褶皱作用
geosynclinal fold region 地槽褶皱区
Geosynclinal fold region of Island Arcs of the western pacific 西太平洋岛弧地槽褶皱区
geosynclinal inversion stage 地槽回返阶段
geosynclinal polarity 地槽极性
geosynclinal sea 地槽海
geosynclinal stage 地槽阶段
geosynclinal structural layer 地槽构造层
geosynclinal synclinorium 槽向斜
geosynclinal-type magmatic formation 地槽型岩浆建造
geosyncline 陆沉带；地向斜；地槽；大向斜；大地槽
geosyncline anticlinorium 槽背斜
geosyncline axis 地槽轴(线)
geosyncline chain 地槽(山)链；大向斜山脉
geosyncline deep fracture 地槽区的深断裂
geosyncline-diwa rebuilding type 地槽地洼再造型
geosyncline-diwa reform type 地槽地洼改造型
geosyncline evolution 地槽的演化
geosyncline-platform superimposition type 地槽地台叠加型
geosyncline-platform theory 地槽地台学说
geosyncline-platform theory for metallogenesis 槽合成矿理论
geosyncline province 地槽区
geosyncline system 地槽系
geosyncline type formation 地槽型沉积建造
geosyncline zone 地槽带
geosynthetics 土工聚合物；土工纤维织物；土工合成物；土工合成材料
geotaxis 趋地性
geotechnic(al) 土工(学)
geotechnic(al) analysis and evaluation 岩土工程分析评价
geotechnic(al) classification 岩土工程分类
geotechnic(al) engineer (岩)土工程师；土工工程师
geotechnic(al) engineering 土质工程；土力学工程(学)；土工技术工程；土工工程学；土工工程；岩土工程(学)
geotechnic(al) evaluation 土工评价
geotechnic(al) fabric 土工织物；土工布
geotechnic(al) history 土工历史
geotechnic(al) instruction manual 地质技术指导手册
geotechnic(al) investigation 岩土调研；土质调查；岩土工程勘察
geotechnic(al) investigation report 岩土工程勘察报告
geotechnic(al) map 土工图
geotechnic(al) method 土工技术方法
geotechnic(al) model 岩土力学模型
geotechnic(al) mo(u)lded bag 土工模袋
geotechnic(al) net mat 土工网垫
geotechnic(al) parameter 土工技术参数
geotechnic(al) process 土工技术(方法)；土工处理法；岩土工程方法；土基处理
geotechnic(al) property 岩土工程性质
geotechnic(al) strip 工程带(拉筋)
geotechnic(al) study 岩土工程勘察
geotechnic(al) study report 岩土工程勘察报告
geotechnic(al) survey 土质调查
geotechnic(al) test 土工试验
geotechnician 土工程师
geotechnics 岩土力学；岩土工程学；土质技术；土工学；土工技术；地质技术学
geotechnique 岩土力学；岩土技术；土质学；土质技术；土工学；土工技术；地质技术
geotechnology 岩土工程学；岩石工程学；工艺地质学；地质工艺学
geotectogene 拗陷带
geotectology 大地构造学
geotectonic 地质构造的
geotectonic cycle 造山旋回；大地构造旋回
geotectonic evolution stage 大地构造演化阶段

geotectonic geology 大地构造(地质)学
geotectonic grade 大地构造级别
geotectonic map 大地构造图
geotectonic movement 大地构造运动
geotectonic position of the ore district 矿区大地构造位置
geotectonics 构造地质学；地质构造学；大地构造(学)
geotectonic site 大地构造部位
geotectonic type 大地构造类型
geotectonic unit 大地构造单元
geotectonic valley 构造谷；大地构造谷
geotectono-geochemistry 大地构造地球化学
geotector 地音探测器
geotemperature 地下温度；地温；地热温度
geotextile 无纺布；土工(纤维)织物；土工合成物；土工合成材料；土工布；地层布
geotextile mattress 土工织物排
geotexture 地体结构；大地结构
geotherm 地热；地等温线
geothermal 地热的
geothermal and heat flow measurement 地温和热流测量
geothermal aquifer 地下热水层
geothermal-aquifer stimulation 地热含水层的激发
geothermal area 地热区；地热地带
geothermal bath 地热浴
geothermal breed 地热养殖
geothermal brine 地下热卤水
geothermal capacity 地热能容量
geothermal circulation system 地热循环系统
geothermal conversion 地热能转换
geothermal corrosion 地热腐蚀
geothermal damage predication 地热事故预测
geothermal damage threshold 地热破坏范围
geothermal data 地热资料
geothermal degree 地热增温率；地热增温级；地热增温陡度
geothermal depth 地温深度
geothermal desalination 地热脱盐
geothermal-development programme 地热开发程序
geothermal discharge water 流出地表的地下热水
geothermal distillation 地热蒸馏
geothermal disturbance 地热扰动
geothermal drilling 地热钻探
geothermal drilling rig 地热钻探设备；地热钻机
geothermal dry steam 地热干蒸汽
geothermal economics 地热经济
geothermal electric(al) power station 地热发电站
geothermal electropower station 地热发电站
geothermal energy 地热能
geothermal energy resources 地热能源
geothermal energy system 地热能利用系统
geothermal exploitation 地热开发
geothermal exploration 地温勘探；地热勘探；地热勘察
geothermal exploration stage 地热勘探阶段
geothermal field 地热田；地热场
geothermal field geologic(al) map 热田地质图
geothermal flashsteam power plant 地热骤增蒸汽发电厂
geothermal geologic(al) map 地热地质图
geothermal geologic(al) section 热田地质剖面图
geothermal grade 地温强度
geothermal gradient 地温梯度；地热增温率；地热增温级；地热增温陡度；地热梯度；地内增热率
geothermal gradient log 地温梯度测井
geothermal-gradient map 地热梯度图
geothermal greenhouse 地热温室
geothermal hazard 地下热害
geothermal heat cost 地热能成本
geothermal heat flow 地热流；大地热流量
geothermal heating 地热采暖
geothermal heating system 地热供暖系统
geothermal-hydrogeologic(al) survey 地热水文地质调查
geothermal investigation 地热调研
geothermal logging 地热测井
geothermal manifestations 地热显示
geothermal maps 地热地质图件
geothermal media 地热介质
geothermal metallogenesis east pacific 东太平洋地热金属成矿调查

geothermal metamorphism 地热变质(作用)
geothermal pollution 地热污染
geothermal power 地热能；地热动力；地热电力
geothermal power-generating capacity 地热发电装机容量
geothermal power generation 地热发电
geothermal power plant 地热(发)电站；地热发电厂
geothermal power plant condensate 地热电厂冷凝水
geothermal power station 地热电站
geothermal power system 地热发电系统
geothermal prospecting 地热普查；地热勘探；地热勘察
geothermal reserve 地热储量
geothermal reservoir 地热库；地热储量
geothermal reservoir capacity 地热储容量
geothermal resources 地热资源
geothermal resources data base 地热资源数据库
Geothermal Resources International 国际地热资源局
geothermal steam 地热蒸汽
geothermal steam plant 地热蒸汽发电厂
geothermal steam well cement 地热油井水泥；地热气井水泥
geothermal step 单位深度地温差
geothermal stream 地热流
geothermal stress 地热应力
geothermal survey 地热调查
geothermal system 地热系统
geothermal technology 地热开发利用工艺
geothermal turbodrill 地热涡轮钻具
geothermal vapo(u)r 地热蒸气
geothermal water 地(下)热水
geothermal well 地热钻孔；地热井
geothermal well cement 地热井水泥
geothermal well type 地热井类别
geothermic 地热(的)
geothermic degree 地热增温率；地热增温级；地热增温陡度；地热梯度
geothermic depth 地热深度；地热级
geothermic electropower station 地热发电厂
geothermic gradient 地温增加率；地温梯度；地热增温率；地热增温级；地热增温陡度
geothermics 地(球)热学
geothermics anomaly 地热异常
geothermic zoning 地热分带
geothermoelectric(al) plant 地热电站
geothermometer 古温标；地质温度计；地质温度表；地温计；地温表；地温标；地热计；地热测温仪；地热表
geothermometer temperature 地热温标温度
geothermometry 古温测量；地温测量
geothermy 地热(学)；地球热学
geotome 取土器
geotrichosis 地丝菌病
geotropic(al) relation 屈地关系
geotropic root 向地根
geotropism 向地性
geotumor 地瘤
geotype 地理型
geoxene 偶适地性生物；外来地栖动物
geozoologic(al) dispersion 地动物分散
geozoologic(al) method 地动物学方法
geracomium 养老院
gerade 偶态
geranial 香叶醛
geraniol 香叶醇；粗制香叶油；草叶油
geranium lake 猩红色淀
geranium oil 香叶油
geranyl butyrate 丁酸香叶酯
geranyl phenyl acetate 苯乙酸香叶酯
Gera porcelain 格腊硬瓷
Gerard's pine oil 格雷松油
Gerard reagent 吉拉试剂
gerasimovskite 钛铌锰石
Gerber's beam 葛尔培式梁；铰接悬臂梁
Gerber's beam bridge 葛尔培式梁桥
Gerber's diagram of moments 葛尔培力矩图
Gerber's girder bridge 铰接悬臂梁桥
Gerber's test 葛尔培试验
Gerber beam 悬臂连续梁
Gerber girder 铰接连梁；葛尔培大梁
Gerber hinge 葛尔培铰链

Gerber lattice(d)girder 葛尔培花格大梁
Gerber purlin(e)葛尔培檩条;葛尔培桁条
ger-bond 热塑料树脂黏合剂
ger-delay circuit 触发延迟电路
gerhardtite 铜硝石
geriatric hospital 老人医院
germ 细菌
German architecture 德国式建筑
German Baroque 德国巴洛克式建筑风格
German black(pigment)德国黑(颜料)
German brick architecture 德国式砖建筑
German cupellation 德国灰吹法
German degree 德国度
German eye splice 顺绳扭劲的眼环结
German feather board 德式鱼尾板;德式鱼鳞板
German flexible pavement design method 德国柔性路面设计法
German Gothic 德国哥特式(建筑)
germanic chloride 四氯化锗
germanic oxide 氧化锗
Germanischer Lloyd 西德船级社;德意志(联邦共和国)劳埃德船级社
Germanischer Lloyds 德国船级社
germanite 硫锗铜矿;二阶锗酸盐
germanium 锗
germanium deposit 锗矿床
germanium diode 锗二极管
germanium-doped optic(al)fiber 掺锗光学纤维
germanium-doped silica 掺锗二氧化硅
germanium-doped silica core single mode fiber 掺锗石英芯单模光纤
germanium-doped silica fiber 掺锗石英光纤
germanium halide 卤化锗
germanium monoxide 一氧化锗
germanium nitride 氮化(物)锗
germanium ores 锗矿
germanium photocell 锗光电管
germanium rectifier 锗整流器;锗二极管
germanium transistor 锗晶体管
German kieselguhr 德国硅藻土
German Late Gothic(style)德国后哥特式(建筑)
German method 核心支持法
German method of tunnel(l)ing 侧壁导坑先墙后拱法
German mounting 德国式装置
German nozzle 抛物线形喷嘴
germanous 二价锗
germanous oxide 一氧化锗
German pattern trowel 德国式纹饰镘刀;德国式花纹修平刀
German Register 德国船级社
German Renaissance 德国文艺复兴时期(的建筑风格)
German R unit 德国 R 单位
German siding 德国式墙板;德国式披叠板;德式鱼尾板;德式鱼鳞板;德国式檐板
German silver 锌白铜;钢锌镍合金;镍锌铜合金;铜镍锌合金;德国银
German silver bar 德银条;德银锭
German silver solder 铜镍合金钎料;德银钎料
German silver wire 德银线
German Society for soil 德国土壤力学协会
German soil classification 德国土壤分类法
German standard sieve 德国标准筛
German-style metal lensring 透镜垫
German tin 德国锡
German tubbing 德国式井壁丘宾筒
German turpentine 德国松节油
German type bevalled firmer chisel 德式斜边木凿
German type carpenter's pincers 德式木工钳
German type folds 日耳曼式褶皱
German type mason's hammer 德式石工锤
German type orogeny 日耳曼型造山运动
germarite 紫苏辉石
germ contamination 细菌污染
germ-free 无菌的
germ-free animal 无菌动物
germfree isolator 无菌隔离室
germicidal agent 杀菌剂
germicidal effect 杀菌效力
germicidal efficiency 杀菌效果
germicidal lamp 杀菌灯
germicidal lamp glass 杀菌灯玻璃
germicidal paint 杀菌涂料

germicidal treatment 灭菌处理;杀菌处理
germicide 杀菌剂
germifuge 抑菌剂;抗菌剂
germinal streak 初腔
germinate 萌芽体
germinate species 孪生种
germinating bed 发芽床
germinating in autumn 秋发芽
germination 颗粒生长;巨粒化;晶粒畸形长大现象;晶核化;萌发;生长;长晶粒;发芽
germination transmission of pathogen 病原种胚传播
germinative temperature 巨粒温度;晶核化温度
germinator 发芽器
germitron 紫外线灯装置
germless chamber 无菌室
germ nucleus 晶体中心;晶核中心
germ repellent paint 杀菌涂料
germ vector 病原菌传递体
gerocomium 养老院
gerontic 老年期的
gerontic stage 老衰期
gerotor 内齿轮油泵
Gerotor pump 摆线泵;盖劳特泵
gerry building 低劣房屋
gerrymander 不公正的分区规划
gersdorffite 辉砷镍矿
gerstleyite 红硫锑砷钠矿
Gerstner wave 格斯特纳波
gerwood 硬木板;墙板
Gesatamin 阿特拉通
gesellschaft 协会;公司;团体
gesometer tank 浮顶式气罐
gesso 油漆之前灰浆底层;加胶大白;石膏粉;石膏底子;雕塑用石膏粉
gestalt 格式量
gestaltism 形态论
gestalt psychology 形态心理学
gestation index 活产指数
gestation period of investment 投资酝酿阶段
get aboard 贴靠他船;登船
get accounts squared 结清账目
get a disease 染病
get ahead of 还清
get alongside 靠拢停泊
get an advance against the bill 凭汇票取得贷款
get a patent for 获得专利
get approval 得到批准
get area 占用区
get a ship afloat 把船浮起
get-away time 准备时间
get beam to wind 使船横风
getchellite 硫砷锑矿
get clear of a danger 避开险境
get close 靠拢过来
get customers 开张;开业
get damp 返潮
get flurried 发慌
get flustered 发慌
get full payment for all spending 实报实销
get good wages 拿高工资
get hot 变热
get instruction 取得指令
get in the mo(u)ld 垫平铸型
get into debt 借债
get loose 松脱
get off 驶离
get out of control 失却控制
get out of order 发生故障
get out of plumb 不垂直
get out of the way 让开航路
get quick answer 尽快答复
get raw materials from local resources 就地取材
get rid of oil 去油
getter 消气剂;吸收剂;吸气剂;吸气器;收气剂
getter action 吸气(剂)作用;除气作用
getter binder 消气剂黏合剂;消气剂的胶合剂
getter capacity 吸气总量
getter device 吸气装置
getter evapo(u)ration 吸气剂蒸发
getter flash 吸气剂快速蒸发;收气剂溅散
gettering container 吸气容器
getter-ion pump 吸气离子泵
getter mirror 金属吸气膜

getter patch 金属吸气膜
getter sputtering 吸气喷镀
getter sputtering equipment 吸气剂溅射设备
getter tab 吸气剂托
getting 开采;采矿;采掘工作
getting clay 瓷土采掘
getting-down mill 开坯轧机
getting-down roll 开坯机座轧辊
getting method 开采方法
getting of clay 黏土开采
getting-off the rails 出轨
getting rid of the source of infection 消除病源
getting work 采掘工作
get under way 准备航行;出港
Gevaet colo(u)r film 盖维特色片
geversite 锑铂矿
Geware 哥窑器
Geware glaze 哥窑釉
gewel hinge 搭扣铰链
geyser 快速热水器;快速热水炉;间歇(温)泉;间歇(式)喷泉;间歇(喷)泉;热水器
geyser action 间歇泉活动
geyser basin 间歇泉地带;间歇喷泉区
geyser column 间歇泉水柱
geyser crater 间歇泉喷出坑
geysering well 间歇泉井
geyserite 硅华;间歇泉华岩
geyser pool 间歇泉塘
geyser reservoir 间歇泉的水室
Geyser spring method 盖塞搅拌法
geyser system 间歇泉系统
geyser tube 间歇泉喉管
geyser vent 间歇泉喷口
geyser water 间歇泉水
g factor g 因数
gf-value 加权振子强度
Ghana nucleus 加纳陆核【地】
ghanat 坎儿井
Gharra 格拉风
gharry 马车(印度、埃及等)
ghatti gum 印度胶;茄替胶;达瓦树胶
gha(u)t 石阶码头;河岸台级;岸边台阶
Ghaziyah 拉格特风
ghetto 犹太人居住区;少数民族居住地区;城市少数民族区
ghizite 蓝云沸玄岩
Ghori ware 戈里瓷(中国青瓷称谓)
ghost 虚反射;影痕;幻影;幻像;双重图像;重影;重像
ghost crystal 阴影晶体
ghost echo 散乱回波;反常回波
ghosted view 幻图
ghost effect 寄生效应;幻影效应;幻像效应;幻相效应;双重图像效应;重影效应
ghost harbo(u)r 死港
ghost image 叠影;虚假脉冲;幻影;重影;重像
ghosting 吸光(无光漆表面明确不均);偷工减料(油漆作业);鬼脸纹油漆;发花;隐现花斑;无光漆发花现象;暗影
ghosting effect 怪峰效应
ghost line 幻影线
ghost marking 重影;幻影
ghost mode 重像模式
ghost peak 假峰
ghost phenomenon 重像现象
ghost pulse 虚假脉冲;寄生脉冲
ghosts 叠影;重叠影像
ghost signal 虚假信号;虚假脉冲;雷达幻影;假信号;幻影信号;幻像信号;重像信号;超幻线干扰信号
ghost stratigraphy 幻像地层学;残迹地层(学)
ghost structure 带状组织;带纹构造
ghost town 被遗弃城市的遗迹;城市遗迹
ghost wires 细波纹
Ghyben-Herzberg formula 吉本—赫兹别格公式
Ghyben-Herzberg lens 吉本—赫兹伯格扁豆状体
ghyll 峡谷
gianellaite 氮汞矾
giant 高压水枪;巨大的;水枪;大轮胎;冲矿机
giant's kettle 瓯穴;锅穴
giant's strike (公园内的)旋转秋千
giant batch winder 大卷装卷布机
giant-bedded orebody 巨厚层矿体
giant bow compass 大型测径规;大卡钳
giant brain 大型电子计算机;大型电脑

giant business 巨大企业
giant cantilever crane 巨型悬臂起重机
giant caterpillar truck 大型履带载货车
giant cave 巨洞
giant colony 巨大菌落
giant computer 巨型机
giant crane 重型起重机;巨型起重机
giant dipole 巨偶极子
giant electric(al) potential 巨大电位
giant flat truck 大型平板车
giant floe 巨型浮冰(横跨10千米以上)
giant garbage 大型垃圾
giant gas field 巨型气田
giant-grained 巨粒的
giant-grained granite 伟晶花岗岩
giant head 巨大头像
giant insulator 大尺寸棒形绝缘子
giantism 巨型
giant jet 水枪射流
giant kelp 巨藻;褐藻
giant knotweed 虎杖
giant magnetostrictive material 强磁致伸缩材料
giant micelle 巨胶束
giant mole 巨鼠型掘进机;地下旋转掘进机;大型石质建筑物(码头防波堤)
giant nozzle 大喷嘴
giant oil field 巨型油田
giant optic(al) pulsation 巨光脉动
giant optic(al) pulse 巨光脉冲
giant order 高柱式;巨柱(式);巨型柱(两层通高的柱式)
giant orebody 巨型矿体
giant panda 大熊猫
giant planet 大行星
giant plough 大型犁
giant pneumatic tire 巨型轮胎
giant pulse 窄尖大脉冲;巨脉冲
giant pulse emission 巨脉冲发射
giant pulse hologram 巨脉冲全息图
giant pulse holography 巨脉冲全息术
giant pulse laser 巨脉冲激光器
giant pulse technique 巨脉冲技术
giant rainer 大型喷灌装置
giant resonance 巨共振
giant robot brain 大型自动计算机
giant rolls 大型对辊破碎机
giant salamander 娃娃鱼;大鲵
giant-scale computer 巨型(计算)机
giant sequoia 巨杉
giant silo 巨型筒仓
giant size 特大号包装
giant straddle strafer 巨型跨立式脚手架
giant straddle truck 巨型跨坐式货车
giant stride 旋转秋千
giant tanker 巨型油船
giant telescope 巨型望远镜
giant transistor 大功率晶体管
giant tire 大轮胎;巨型轮胎
giant ultra-highspeed computer 超高速巨型计算机
giant void 巨洞
giant weight tamper 特重碓;特重夯;特重捣棒
giant white gourd knife 大冬瓜刀
gib 夹条;起重杆;扒杆;机臂;吊杆;吊车臂;扁栓;凹字楔
gib and cotter (立柱的)楔销;合楔
gib-and-cotter joint 夹条扁销连接
gib and key 合楔
gib arm 起重机臂;吊车(起重)臂
gib arm of crane 起重机悬臂;起重机斜撑;吊杆起重扒杆;吊车起重扒杆;吊车臂
Gibb's equation 吸附公式
gibber 棱石;风刻石
gibberellic acid 赤霉酸
gibberish total 控制和;校验和
gibber plain 砾漠
gibbet 吊平臂;吊机臂;撑架(托座)
gibbet(tree) 起重杆
gib block 导向条
gibbosity 凸面状;隆起
gibbous moon 凸月
gibbs 吉布斯(吸收单位,10⁻¹⁰克分子数,1吉布斯=厘米的表面浓度)
Gibbs adsorption equation 吉布斯吸附公式;吉布斯吸附方程(式)

Gibbs adsorption isotherm 吉布斯吸附等温式
Gibbs adsorption theorem 吉布斯吸附定理
Gibbs diaphragm cell 吉布斯隔膜电解池
Gibbs-Duhem equation 吉布斯—迪亨方程式
Gibbs elasticity 吉布斯弹性
Gibbs energy 吉布斯能量
Gibbs energy change 吉布斯能量变化
Gibbs energy contribution 吉布斯能量分布
Gibbs energy function 吉布斯能量函数
Gibbs energy level 吉布斯能级
Gibbs energy of electrons 吉布斯电子能量
Gibbs energy of reaction 吉布斯反应能量
Gibbs formula 吉布斯公式
Gibbs free energy 吉布斯自由能
Gibbs free energy of activation 吉布斯活化自由能
Gibbs function criterion 吉布斯函数判据
Gibbs-Helmholtz equation 吉布斯—亥姆霍兹方程(式)
gibbsite 水铝氧;水铝矿;三水氧化铝;三水铝石;三水铝矿;大水铝石
gibbsite layer 氢氧化铝八面体层
gibbsite ore 三水型铝土矿石
gibbsite refractory 三水铝耐火材料
gibbsite refractory product 水铝矿耐火材料制品
gibbsitic laterite 三水铝石红土
Gibbs-Konowalow rule 吉布斯—康诺瓦洛夫定律
Gibbs phase rule 吉布斯相律;吉布斯相法则
Gibbs phenomenon 吉布斯现象
Gibbs-Poynting equation 吉布斯—坡印亭方程式
Gibbs relation 吉布斯公式
Gibbs rule 吉布斯相律;吉布斯相法则
Gibbs system 吉布斯系统
gib clamp 扁栓制动机构
gib crane 扒杆(式)起重机;挺杆式起重机;伸臂式起重机
gib door 墙面齐平门;隐门
gib-headed bolt 钩头螺栓;扁头螺栓
gib-head(ed)key 弯头键;钩头楔键
gib head taper key 钩头斜键
gib head taper stock key 钩头键
gib hoist 悬臂(式)起重机
gibing 改变帆的方向
gib key 有头键;钩头键;凹(字)形键
giblet check 有外向槽的门框(供外开门用);门框槽头;石门框槽口
gib nut 翻边锁紧螺母
gib pole 起重扒杆
gib screw 固定螺钉;夹紧调整螺钉;调整楔用螺钉;锁紧螺旋;锁紧螺钉
gibson girl 袖珍发报机(救生艇属具)
Gibson method 惯性压力法;吉布森流量测定法
Gibson mix 吉布森混合(计算)法
giddiness 眩晕
Gidgee 相思树装饰性硬木(产于澳大利亚)
Giegy-Hardisty process 吉埃琪—哈迪斯蒂法
gieseckite 霞霞石
Gieseler plastometer test 基氏塑性测定
giessenite 针辉铋铅矿
gifblaar poison 氟乙酸
Gifford's buffer solution 吉弗氏缓冲液
Giffard's injector 吉弗德(蒸气)喷射器
gift 自动转让财产;赠与
gift and souvenir department 礼品部
gift causa mortis 死亡财产自动转让
gift coupon 赠券
gift enterprise 附送赠品的行业
gift loan 无息贷款
gift rope 牵引索;牵引绳;拖船安定绳
gift tax 赠与税;馈赠税
gig 旋转物;轻快小艇;轻便双轮马车;吊桶;单列座小舢板
giga 千周
giga billion 千兆
gigabyte 千兆字节
gigacycle 千兆周
gigacycle computer 千兆周计算机
gigacycles per second 千兆周/秒;千兆赫
giga-electron-volts 十亿电子伏特(美国)
gigaflops 吉拍
gigahertz 吉赫;千兆周/秒;千兆赫
gigahertz computer 千兆赫计算机
gigantic tanker 巨型油轮
gigantism 巨形发育;巨大症;巨大畸形
gigantomachy 以巨人与神的战斗为主题之作品

gigaohm 千兆欧(姆)
gigawatt-day/tonne 千兆瓦日/吨(燃耗单位)
giga-watte 千兆瓦
gigawatt hour 兆千瓦小时;百万千瓦小时;百万度(电)
gigging machine 刺绒起绒机
gig house 卷扬机房;绞车房
Gigli's saw 钢丝锯
Gigli's wire saw 季格利线锯
gig mill 拉绒厂;起绒厂;拉毛长
gig saw 直立往复锯;带锯床
gig stick 半径杆;绞车杆;回转棒
gilalite 水硅铜石
Gilbert 吉佰(磁通势单位)
Gilbert code 吉尔伯码
Gilbert delta 吉尔伯特三角洲
gilbertite 丝光白云母
Gilbert reversed polarity chron 吉尔伯特反向极性时
Gilbert reversed polarity chronzone 吉尔伯特反向极性时间带
Gilbert reversed polarity epoch 吉尔伯特负极性期
Gilbert reversed polarity zone 吉尔伯特反向极性带
Gilbert reverse epoch 吉佰逆磁极期
Gilbert-type delta 吉尔伯特型三角洲
Gilbert-type deltaic deposit 吉尔伯特型三角洲沉积
gild 描金;同业公会;烫金
gilded 涂金色的;贴金箔的;镀金的
gilded corbel bracket 镏金斗拱
gilded frame 镀金镜框
gilded roof 镏金顶
gilding 涂金;贴金;饰金;镀金(术);飞金
gilding alloy 装饰用铜合金
gilding machine 镀金机
gilding metal 手饰铜;仿金合金
gilding press 烫金机
gilding technique 镏金技术;贴金技术
gilflex 套装电线的非金属软管
gilgai 小起伏微地形;黏土小洼地
gilhoist 登陆艇运输车
gill 鱼鳞板;吉耳;翅片
gill cowling 发动机罩裙
gill edge 菌褶缘
gilled heater 翼片取暖器;翼片加热器
gilled heating pipe 翼片加热器
gilled pipe 肋管;带肋片的管子;翅片管
gilled radiator 腮片散热器;片式散热器
gilled ring 两肋(立窑)
gilled tube 肋形管
gilled tube heater 带有散热片的管式取暖器;带有散热片的管式加热器
gilled tube radiator 带有散热片的管式散热器;带有散热片的管式暖气片;管片式散热器
giller 马毛绳
Gillespie equilibrium still 吉莱斯皮平衡蒸馏锅
gillespite 硅铁钡矿
gill faller spreader 落叶延展机
Gilliland correlation 吉利兰关联式
gilling equipment 加水和加油设备
Gill modification 格尔修正
Gillmore apparatus 吉尔摩水泥稠度试验计
Gillmore needle 吉尔摩针(水泥稠度试验用);吉尔摩仪;吉尔摩水泥稠度试验针
Gillmore needle test 吉尔摩针凝结试验
Gillmore test 吉尔摩试验(测水泥凝结时间);吉尔摩凝结时间测定
gill net 刺网
gill netter 刺网渔船
gill preparer 预备针梳机
gills 鱼鳞板;可控(的)散热瓣
gillyflower 紫罗兰花
Gilmer Ghez Cabrera theory 体—表面联合扩散理论
gilpinite 硫酸铜铀矿
Gilsa event 吉尔萨事件
Gilsa normal polarity subchron 吉尔萨正向极性亚时
Gilsa normal polarity subchronzone 吉尔萨正向极性亚时间带
Gilsa normal polarity subzone 吉尔萨正向极性亚带
gilsonite 矿物橡胶;硬(质)沥青;黑沥青
gilt 金色涂层;涂金的;烫金;镀金材料
gilt bronze 仿金铜箔

gilt edge 烫金边
gilt roof 镏金顶
gilt stamped cover 烫金精印封面
gilt watch 镀金表
gimbal 万向支架;万向接头;常平架;平衡环
gimbal assembly 常平架组
gimbal axis 框架轴;平衡架轴;常平架轴;常平环轴
gimbaled engine 换向发动机
gimbaled hanger assembly 万向平衡式吊架组合
gimbaled motor 悬挂式电(动)机
gimbaled nozzle 万向喷嘴;万向喷头
gimbaled power plant 万向悬挂式动力装置
gimbaled rocket 万向火箭发动机
gimbal error 框架误差;常平架误差
gimbal freedom 框架自由度;万向支架自由度;常平架自由度
gimbaling 万向支架连接
gimbaling error 常平架误差
gimbaling rocket motor 万向架支座火箭发动机
gimbal joint 万向节;万向接头
gimballess gyroscope 无框架陀螺仪
gimballess inertial navigation equipment 无常平架惯性导航设备;束带式惯性导航设备
gimbal lock 框架自锁;常平架锁定
gimbal moment of inertia 框架转动惯量;常平架转动惯量
gimbal mount 万向(悬挂)架
gimbal-mounted engine 万向架支承的发动机
gimbal mounting 常平架框
gimbal pick-up 万向架位置传感器
gimbal pivot 万向节支枢
gimbal ring 常平环;平衡环
gimbals 平衡架
gimbal servo 常平架伺服机构
gimbal servo motor 万向伺服电动机
gimbal suspension 万向悬挂架;常平架
gimbal system 万向节系统
gimbal throttle system 万向节流系统
gimlet 小螺丝钻;螺丝锥;木工钻;手锥;长木钻;手钻
gimlet bit 螺旋钻(头);手钻钻头;手用螺旋尖钻
gimlet for nail 钉孔钻
gimlet point 手钻钻头
gimmer 钩扣;铰链
gimmick 绞合电容器;花样
gimp 操作噪声
gimp pin 花边钉
gin 陷阱;辘轳;绞盘;绞车;起重三角架;提升机;三脚起重机;三脚吊货架;三脚起重架;打桩机网;打桩机
gina 尖肋型止水胶垫
gina profile 尖肋型止水胶垫
gin block 单轮滑车;单饼铁滑车;差动滑车
gin compress 轧花压包机
gin fall 轧花机落棉
gingbread 华丽派建筑
Gingbread style (19世纪美国的)华丽装饰风格
gingbread work 繁俗装饰;庸俗建筑装饰;华丽俗气装修;华而不实建筑
ginger 生姜;淡赤黄色的
gingerin 姜油脂
ginger in vinegar 醋姜
gingham 方格色织布;条格平布
gingko(biloba) 银杏;白果树
gin house 轧棉厂
Gini coefficient 基尼系数
giniite 水磷铁石
gink(g)o 银杏;白果
Ginkgoales 银杏目
ginnel 狭窄通道
ginnywink 单臂人字起重机
ginorite 基性硼钙石;水硼钙石
gin pit 非机械长浅井
gin pole 中央立柱;起重桅(杆);起重机主桅;起重抱杆;起重(扒)杆;桅杆起重机;三脚起重的杆;单杆起重架;安装用起重架
gin pole derrick 把杆式起重机;桅杆(式)起重机
gin-pole truck 桅杆式起重吊车
gin pole type concrete spouting plant 混凝土浇灌塔
gin popper 三角架中间腿
gin saw 轧花机底部梳齿板
Ginsibourg's microlithotypes of coal 金兹堡显微煤岩类型
gin strap 三角起重机吊索

gin tackle 起重滑车
gin wheel 起重(单)滑轮;辘轳;滑车
gio 入门口;深峡海湾;陡峭海岸的狭海湾
giobertite 菱镁矿
giorgiosite 异水菱镁矿
Gippsland basin 吉普斯兰盆地
gipsy 锚机滚筒
gipsy sheave 链滑轮
giraffe 长颈鹿
girandole 壁上烛架;壁灯架;旋转喷水嘴;枝形烛架;多枝烛台;多枝烛架
Girard agent 吉拉得试剂
Girard turbine 吉拉德水轮机
girasol 蓝蛋白石;青蛋白石
giraudite 砷—硒黝铜矿
gird 箍梁;围梁;包带
gird bridge 路桥;梁式桥;架空桥
girded 抛前后锚时船为一链阻不能旋转;拖轮打横
girder 梁;加劲梁;桁;柁;大梁
girder action 桁梁作用;梁作用
girderage (房屋的)梁木体系
girder and beam connecting bridge 梁式桥
girder and beam connection 主次梁连接;主次梁接合;大小梁连接
girder and rail mill 轨梁轧机
girder beam 主梁;大梁
girder bearing 大梁支承
girder bearing plate 梁垫板
girder bending press 弯梁压力机
girder bent 梁式排架;带实腹横梁框架;实腹横梁框架
girder block 铁制墩木;铁枕
girder boom 大梁吊杆
girder bottom boom 桁架下吊杆;大梁下吊杆
girder bottom chord 桁架下弦杆;大梁下弦杆
girder bottom flange 桁架下翼缘;大梁下翼缘
girder bottom plate 纵桁下翼板
girder brace 桁撑
girder bridge 架空铁桥;板梁桥;梁(式)桥;桁架桥
girder bridge bearing 桁架桥支座
girder casing 梁模;梁套;大梁护面
girder chord 桁架弦杆;大梁弦杆
girder connection 梁节点;大小梁连接
girder connection at an angle 大梁在角上的连接
girder construction 大梁施工建造;梁式构;桁架结构
girder construction system 梁式结构体系
girder cross-section 大梁横截面
girder depth 大梁高度;梁高
girder design formula 大梁设计公式
girder dogs 起梁钩;吊梁钳
girder element 大梁构件
girder-erecting machine 架桥机
girder-erection survey 架梁测量
girder fabrication 梁式结构
girder face plate 桁材面板
girder fixed at both ends 两端固定的大梁
girder flange 大梁翼缘;桁材带板
girder floor 大梁楼板
girder fo one bay 单跨大梁
girder fork 梁叉
girder forms 大梁模板
girder for seat 基座纵桁
girder frame 横梁;桁架梁;梁(式)桁架
girder grid system 大梁网格体系
girder grillage 格排梁;钢梁(组成的)梁床;钢梁格床;梁式排梁;梁式承台
girder grille 大梁格栅
girder guard rail 板梁护轨
girder head 柁头
girder heating plate 大梁热板
girder hook 吊梁钩
girder interval 大梁间距
girder iron 梁铁;钢梁
girder joint 大梁接口
girderless deck 无梁面板
girderless floor 无梁楼板;无梁楼板
girderless floor construction 无梁楼盖构造;无梁楼盖结构
girderless floor system 无梁楼盖系统
girder load moment 大梁荷载力矩
girder lower boom 桁架下吊杆;大梁下吊杆
girder lower chord 桁架下弦杆;大梁下弦杆
girder lower flange 桁架下翼缘;大梁下翼缘

girder manufacture 大梁制造
girder material 大梁材料
girder moment 梁力矩
girder mo(u)ld 梁(的)覆面层
girder mo(u)ld board 大梁模板
girder pass 钢梁孔型
girder plan 纵桁布置图
girder pole 格状桅杆;桁架杆柱
girder post 大梁支柱
girder propping 安顶梁
girder radial drilling machine 梁式摇臂钻床
girder rail 宽底轨(条)
girder restraint provided by walls with openings 受有开孔墙约束的大梁
girder rolling mill 钢梁轧机
girder roof 大梁屋顶
girder section 大梁型钢;大梁断面
girder shuttering 大梁模板
girder space 横梁间隔;桁间隔;梁间距;桁间距
girder spacing 大梁间距;大梁间隔
girder span 梁跨(度)
girder stay 桁撑条
girder steel 工字钢;工字梁;钢梁
girder-stiffened 用梁加劲的
girder stiffener 大梁加劲肋
girder structural system 桁架体系;大梁结构体系
girder structure 梁式结构;大梁结构
girder subjected to bending 受弯的梁
girder support 梁支承
girder supported at both ends 两支点梁;简支梁
girder supported on two ends 简支梁
girder supported on two supports 两支点梁
girder system 桁架体系;梁系;梁(式)体系
girder test 梁试验
girder-to-beam connection 主次梁连接;主次梁接合
girder-to-cable distance 梁索间距
girder top boom 桁架上吊杆;大梁上吊杆
girder top chord 桁架上弦杆;大梁上弦杆
girder top flange 桁架上翼缘;大梁上翼缘
girder top plate 纵桁上翼板
girder truss 梁构桁架;桁架梁
girder upper boom 桁架上吊杆;大梁上吊杆
girder upper chord 桁架上弦杆;大梁上弦杆
girder upper flange 桁架上翼缘;大梁上翼缘
girder wall 桁架墙;桁构墙
girder web 大梁腹板
girder with cross bracing 交叉撑桁架梁
girder with lapped joints 搭接梁
girder without ballast 无砟梁
girder without ballast and sleeper 无砟无枕梁
girding 横拖
girdite 复碲铅石
girdle 引力带;柱带;(柱头及基座处的)环柱饰;环带;壳环;围场
girdle axis 环带轴
girdle canal 环围道
girdle sensation 紧缩感
girdle structure 环带构造
girdle texture 环带状纹
girdling 树木环腹带;环状刮皮(阻止树木生长);束腰
girekenite 天青黑云岩
girl's gymnasium 女子健身房
giro-cheque 转账性支付
Girondian(stage) 基隆德超阶【地】
giro system 直接转账系统;转账付款系统
girt(h) 木圈梁;带尺;周长;扎线;圈梁;船腹围长;鞍带;檐;加劲系梁;抛前后锚时船为一链阻不能旋转;拖轮打横;水平拉杆;方框支架横撑
girth class 干围级
girth gear 矢圈;矢轮
girth gear drive 大齿轮传动装置;边缘传动(装置)
girth joint 环形接缝;环向接头;环焊缝;环焊缝(整周焊缝)
girth limit 干围级界限
girth limit cutting 围级择伐
girth member 加固件
girth quotient 干围率
girth rail 圈栏
girth ring drive 大齿圈传动
girth rivet seam 沿圈周铆接
girth seam 周向焊缝;圈缝
girth sheets 圈板

girth strip 卧梁
girth weld 环焊缝
girth welding 环缝焊接
girtline 桅顶吊索;定端绞辘
girt strip 墙边卧材;脚手横梁
girtwise 沿走向的
Giseeler fluidity 基氏流动度
gisement 坐标(纵线)偏角
Gismo 吉斯莫万能采掘机
Gismo-jumbo 吉斯莫型钻车
gismondite 斜方钙沸石;水钙沸石
git 浇铸沟
gittinsite 硅锆钙石
give a good account of oneself 付清欠账
give a long price for 高价(购)买
give an account of 汇报
give-and-take 协商(的);折衷;折中(的);交换(的);互让(的)
give-and-take lines (计算不规形面积的)取舍线;计算不规形面积的取舍线
give-and-take trade 互让贸易;平等交换贸易
give an overall consideration 统筹安排
give a quarter turn 翻转90度
give a signal 打信号
give a wide berth to 宽让
give bank reference 提供可以征信的银行名称
give clearance to the cutting edge 铲齿
give customs clearance 海关放行
give evidence 作证
give gas 给气
give guidance to the direction of investment 引导投资方向
give in charges 托付
give more helm 加多舵角
give more rein to market forces 扩大市场作用
give more screw 松动调整螺钉
given 签订
given accuracy 给定精度
given commodity 一定量的商品
given credit 赊账
given data 已知数据
give new set 新指标
given frequency 已知频率
given grading 指定级配
given lime in soil 在土壤中施加石灰
given load 规定负荷
given mirror plane 给定镜像面
given number 已知数
given off 释放
given period 计算期
given point 已知点;给定点
given pressure 定压
given price 已知价格;计算价格;特定价
given quotation 约定汇价
given site 已知场地;特定场地
given size 规定尺寸
Giventian(stage) 吉维特阶【地】
given time 给定时间
given total cost 计划总成本
given value 已知值;给定值
given way vessel 被让路船
given year 计算年
give off vapo(u)rs 放出蒸气
give one's hand on 保证履行契约
give one the bag 解雇
give out 生产力
give overall consideration 统筹兼顾
give reasons for an award 仲裁裁决注明理由
give secret support 暗中支持
give the alarm 发出警报
give the keel 龙骨向上
give trade reference 提供可以征信的企业名称
give up 放弃
give up a business 放弃交易
give up cultivation 废耕
give-up interval 重复传输时间
give wage increases to workers and staff members 调整职工工资
give way 让路;让开航路
give-way area for ship 船舶避让区
give-way marking 让路标志
give-way sight distance 错车视距
give-way vessel 义务船;让路船;避让船
giving account 支出账目

giving great bargain 特别大减价
giving of line 定隧道准线;定隧道直线
giving quotation 应付汇价;开出价格
giving springers 设置拱脚石
giving state subsidies for price differential 返还差价款
gizmo montage amplifier 特技用混合放大器
glabrate 近无毛的;近光滑的
glabrous 无毛的
glabrous chaffed wheat 光壳小麦
glaced wall tile 瓷壁砖
glacial 极冷的;冰样的;冰期的;冰河质的;冰川越流;冰川的
glacial ablation 冰川消融
glacial ablation boundary line 冰川消融界线
glacial ablation factor 冰川消融因素
glacial ablation height 冰川消融高度
glacial ablation region 冰川消融区
glacial ablation season 冰川消融季节
glacial ablation velocity 冰川消融速度
glacial abrasion 冰蚀;冰川磨蚀
glacial accretion 冰川加积作用
glacial accumulation region 冰川积累区
glacial accumulation type 冰川推积型
glacial acetic acid 冰乙酸;冰醋酸
glacial action 冰川(作用)
glacial advance 冰川前进
glacial advance velocity 冰进速度
glacial age 冰河时代;冰河期;冰川时代
glacial age snow line 冰期雪线
glacial alluvion 冰碛层;冰川冲积层
glacial alluvium 冰川冲积层
glacial amphitheater 冰斗
glacial anti-cyclone 冰原反气旋
glacial basin 冰川盆地
glacial bed 冰川河床
glacial berg 冰山;冰川冰山
glacial boulder 冰川漂砾
glacial boundary 冰川边界
glacial breccia 冰川角砾(岩)
glacial cap 冰帽
glacial carved valley 冰蚀(河)谷
glacial cascade 冰瀑;冰川瀑布
glacial chronology 冰川年代
glacial chute 冰笕
glacial cirque 冰斗
glacial clay 冰构黏土;冰川黏土
glacial cliff 冰崖
glacial climate 冰雪气候;冰期气候
glacial coast 冰蚀海岸
glacial conglomerate 冰碛砾岩
glacial contact landform 冰川接触地貌
glacial contour 冰川等高线
glacial control theory 冰川控制说
glacial cycle 冰蚀循环;冰流过程;冰川旋回;冰川轮回
glacial debris 冰碛物;冰碛砂砾
glacial delta 冰川三角洲
glacial delta deposit 冰川三角洲沉积
glacial delta facies 冰川三角洲相
glacial denudation 冰蚀作用;冰川剥蚀作用
glacial deposit 冰碛物;冰积物;冰积土;冰积层;冰川沉积(物)
glacial deposition coast 冰积海岸
glacial derived sediment 冰川沉积(物)
glacial dirt band 冰川碎石带
glacial drift 浮冰;冰碛(石);冰河冲积层;冰川漂流;冰川冰碛
glacial drift facies 冰碛相
glacial drift map 冰碛图
glacial drift well 冰碛井
glacial environment 冰川环境
glacial epoch 冰进期;冰川世;冰川(时)期
glacial era 冰川期
glacial erosion 冰蚀;冰川侵蚀
glacial erosion cycle 冰蚀循环
glacial erosion lake 冰蚀湖
glacial-erosion landform 冰蚀地貌
glacial eustasy 冰期海面升降
glacial facies 冰川相
glacial fall 冰川瀑布
glacial feature 冰川特征;冰川地形
glacial-fed river 冰川补给河
glacial-fed stream 冰川补给河

glacial fill 冰川补给
glacial flora 冰川植物群
glacial flow 冰川流动
glacial-fluvial soil 冰河沉积土;冰水沉积土;冰川沉积土
glacial form line 冰川地形地貌形态
glacial forward movement 冰川前移
glacial geology 冰川地质(学)
glacial geomorphy 冰川地貌
glacial gravel 冰川砾石
glacial groove 冰刻沟;冰川刻槽
glacial hanging valley 冰蚀悬谷;冰成悬谷
glacial high 冰原反气旋
glacial ice 冰川冰
glacial ice drilling 冰上钻进
glacial inwash 冰川边缘沉积
glacialite 白蒙脱石
glacial lake 冰川湖;冰成湖
glacial lake deposit 冰湖沉积;冰川湖积土
glacial lake facies 冰川湖泊相
glacial landform 冰川地形
glacial layer 冰碛层
glacial length 冰川长度
glacial loam 冰川壤土
glacial lobe 冰川舌
glacial loess 冰川黄土
glacial lowering of sea level 冰期海面下降
glacial marginal lake 冰缘湖
glacial marine 冰海的
glacial maximum 最大冰川作用
glacial milk 冰川乳(浆)
glacial mill 冰川壶穴
glacial motion 冰川运动
glacial movement 冰川运动
glacial mud 冰川泥
glacial name 冰川名称
glacial number 冰川编号
glacial oscillation 冰川振荡
glacial outburst flood 冰川暴发洪水
glacial outwash 冰水沉积(物);冰川冲刷
glacial outwash gravel 冰川沉积砾石
glacial overburden anomaly 冰积物异常
glacial pavement 冰溜面
glacial period 冰(河)期;冰川(时)期;冰川纪
glacial phosphoric acid 冰磷酸
glacial plain 冰川平原;冰成平原
glacial planation 冰川削平作用;冰川均夷
glacial planning 冰川陵削
glacial plucking 冰川挖掘(作用);冰川剥蚀作用;冰川剥蚀;拔蚀作用
glacial protection theory 冰川保护理论
glacial recession 冰川退缩;冰川后退
glacial recession velocity 冰退速度
glacial record 冰川记录
glacial relict 冰河期残遗物
glacial replenishment 冰川补给
glacial reservoir 冰川补给区
glacial retreat 冰川后退
glacial river 冰源河(流);冰(川)河
glacial rock 冰成岩
glacial sand 冰川砂(层)
glacial sapping 冰川挖掘(作用)
glacial scour 冰川刨蚀;冰川冲刷
glacial sediment 冰川沉积(物)
glacial sediment sample 冰碛物样品
glacial sheet 冰盖
glacial sheet of north America 北美大陆冰盖
glacial sheet of north-mid-Europe 欧洲中北部大陆冰盖
glacial sheet of Siberia 西伯利亚大陆冰盖
glacial shelf 冰蚀陆架
glacial shrinkage 冰川收缩
glacial silt 冰川砂泥;冰川泥沙;冰川粉土
glacial sliding 冰川滑动
glacial slope 冰坡
glacial soil 冰积土;冰川土
glacial source height 冰川源头高度
glacial stage 冰(河)期;冰川阶段
glacial stagnation 冰川停顿
glacial stair(case) 冰阶
glacial stream 冰源河(流);冰(川)河
glacial striae 冰川擦痕
glacial striation 冰川擦痕
glacial table 冰川台地

glacial terrace 冰川阶地;冰成阶地
glacial theory 冰川(学)说
glacial thickness 冰川厚度
glacial threshold 冰川谷坎
glacial till 冰碛土;冰碛泥;冰砾泥;冰河冰碛物;冰川漂石黏土;冰川积层
glacial tongue 冰川舌
glacial topography 冰川地形
glacial transport 冰川搬运
glacial trough 冰蚀槽;冰川槽;冰成海槽
glacial valley 冰蚀河谷;冰(蚀)谷;冰川蚀谷;冰川(河)谷
glacial valley lake 冰谷湖
glacial varve 季候泥;冰川纹泥
glacial velocity 冰川流速
glacial width 冰川宽度
glaciated 受冰河作用的;冰川生成的;冰川覆盖的;冰冰封的
glaciated coast 冰蚀海岸
glaciated plain 冰蚀平原
glaciated rock 冰擦岩
glaciated terrain 冰蚀地面
glaciated valley 冰蚀谷
glaciation 创蚀作用;冰蚀;冰化作用;冰川化;冰川(作用)
glaciation limit 冰蚀极限
glacieluvial 冰川残积的
glacier 大冰块;冰川
glacier accumulate area ratio 冰川积累面积比率
glacier accumulate region area 冰川积累区面积
glacier advance 冰进
Glacier-antifriction alloy 格拉西减摩合金
glacier avalanche 冰崩
glacier bed 冰川河床
glacier boulder method 冰川漂砾法
glacier breeze 冰川风
glacier burst 冰川暴裂;冰川暴发
glacier cap 冰帽
glacier cover 冰川覆盖
glacier-covered 冰川覆盖的
glacier-dammed lake 冰塞湖;冰川湖
glacier delta 冰川三角洲
glacier dynamics 冰川动力学
glaciered 受冰河作用的;冰川生成的;冰川覆盖的;冰冰封的
glacieret 小冰川;二级冰川
glacier face 冰山侧面
glacier fall 冰川瀑布
glacier-fed stream 冰川补给河
glacier flow 冰川水流;冰川流动
glacier fluctuation 冰川变动
glacier glacial cap 冰冠
glacier ice 冰川流动冰
glacier(ice)berg 不规则的冰山;冰川冰山
glacier-ice-blocked lake 冰川冰塞湖
glacierized 受冰川作用的;冰川覆盖的
glacier lake 冰川湖;冰成湖
glacier mapping 冰川测图
glacier milk 冰川乳(浆);(由冰山下流出的)白浆冰水
glacier mill 破冰机;冰川瓯穴;冰川锅穴;冰川壶穴
glacier mud 冰(川)泥
glacier mud-rock flow 冰川泥石流
glacier nourishment 冰川补给
glacier oscillation 冰川的颤动
glacier outburst flood 冰融洪水;冰融泛滥;冰川暴发洪水
glacier plain 冰川平原
glacier pothole 冰川壶穴
glacier recession 冰退;冰川消退;冰川退缩;冰川后退
glacier regime 冰川情况;冰川动态
glacier region area 冰川消融区面积
glacier replenishment 冰川补给
glacier retreat 冰川消退;冰川退缩;冰川后退
glacier river 冰源河(流);冰河
glacier rock 冰擦岩
glacier silt 冰川粉
glacier snout 冰川前锋;冰川前端;冰川鼻
glacier snow 冰川雪
glacier spring 冰川泉
glacier stream 冰源河(流);冰河;冰川水流
glacier surge 冰川跃动
glacier survey 冰川测量

glacier table 冰川台地;平顶冰块;冰桌;冰(石)台;冰川基准面
glacier tongue 冰(川)舌;(伸入海中的)冰河舌
glacier total area 冰川总面积
glacier type mud-stone flow 冰川型泥石流
glacier well 冰川壶穴
glacier wind 冰川风
glacigene 冰成的
glacigenous 冰成的
glacioaqueous 冰水(作用)的
glacio-eustasy 冰川海面升降;冰川海面进退
glacio-eustation 冰川海面升降;冰川海面进退
glacio-eustatism 冰川性海面升降运动
glaciofluvial 冰水(作用)的;冰川水形成的;冰川和河流的
glaciofluvial delta 冰水三角洲
glaciofluvial deposit 冰川洪积;冰水沉积(物);冰河沉积
glaciofluvial drift 冰水碛;冰河漂积物
glaciofluvial environment 冰水环境
glaciofluvial fans 冰水扇
glaciofluvial lake 冰水湖
glaciofluvial landform 冰水地貌
glaciofluvial outwash plain deposit 冰水冲积平原沉积
glaciofluvial outwash plain facies 冰水外冲平原相
glaciofluvial terrace 冰水阶地
glaciogeology 冰川地质(学)
glaciokarst 冰川岩溶;冰川喀斯特
glaciolacustrine 冰湖的
glaciology 冰川学;冰川学;(某地区的)冰川特征
glaciology of the seas 海冰学
glaciomarine 冰海的
glaciometer 测冰仪;冰川仪;冰川计
glaciosolifluction 冰融泥流作用
glacis 斜堤;斜岸;缓斜坡
glacitex 描述布
glacon 中等冰块
glactaric acid 黏酸
glade 林中空地;林间空地;湿地;冰穴
gladite 柱硫铋铜铅矿
Gladstone Dale law 格拉斯顿—代尔定律
glairin 黏胶质;胶素
glaisher's table 湿度表格
glallzing 冲淡法
glance 硫化矿类;闪躲【船】
glance coal 光亮煤;无烟煤
glance copper 辉铜矿
glance-legibility distance 瞬间识别距离;瞬读距离
glance pitch 光泽地沥青石;纯硬化沥青;辉沥青;光泽沥青
glancing blow (船与码头)斜触
glancing collision 擦碰;擦边碰撞
glancing incidence 掠入射;水平入射
gland 压盖;塞栓;衬片
gland-bearing oak 槠栎
gland bolt 压盖螺栓;垫圈紧固螺栓
gland bonnet (轴端的)密封盖
gland box 填料箱;填料函
gland bush 压盖;密封衬套;衬片
gland casing 气封体
gland cock 封盖栓
gland cover 密封压盖;填料盖;密封套
gland follower 油封封面圈;密封压盖随件
gland housing 气封体
gland joint 密封接头;伸缩接头;伸缩管缝
gland leak-off 气封漏汽
glandless lining 压盖衬层
glandless pump 无填料泵
gland liner 压盖衬套
gland lock 压盖保险
gland neck 填料压盖颈
gland-neck bush 填料压盖轴颈衬套
gland nut 压紧螺母;压紧螺帽;压盖螺母;填料函螺母
gland oil 填料油
gland packer 压盖密填
gland packing 填料垫;压盖填料;压盖密封
gland-packing leakage 气封漏气
gland plate 填料压盖板
gland plunger 压盖柱塞
gland pocket 密封盒
gland retainer plate 轴密封盖;轴端护板
gland ring 密封环

glands 气封
gland seal 压盖密封;轴封;密封装置
gland sealing steam collecting pipe 气封蒸汽集气管
gland sealing steam control valve 气封蒸汽调节阀
gland sealing steam exhaust 气封排汽
gland sealing steam supply 气封供汽
gland sealing steam supply system 气封供汽系统
gland seal ring 压盖密封环
gland sleeve 气封套
gland spring 气封弹簧
gland steam 轴封蒸汽;气封蒸汽
gland steam condenser 气封蒸汽凝气器
gland steam connexion 气封蒸汽连接管
gland steam control regulator system 气封蒸汽调节系统
gland steam exhauster 气封蒸汽排气管
gland steam pocket 气封室
gland steam regulator 轴封供汽调器;气封蒸汽调节器
Glan-Foucault polarizing prism 格兰—傅科偏振棱镜
Glan-Foucault prism 格兰—傅科棱镜
Glan polarizer 格兰起偏镜
Glan prism 格兰棱镜
Glan-Thompson prism 格兰—汤普森棱镜
glare 眩目;眩光;晃眼;强光;水平线冰光;双占用;刺目光
glare control glass 防眩玻璃
glare effect 眩光效应;眩光效应;耀眼影响
glare-free 无眩光的;不眩目的
glare-free glass 无眩光玻璃
glare ice 光滑冰;凝水;薄冰
glare index 眩光指数;闪光指数盘;闪光度盘
glare index of window 窗眩光指数
glare of the foot lights 舞台灯光耀眼
glare-proof glass 防眩玻璃
glare-proof mirror 防眩(后视)镜;不闪光镜
glare protection 遮光保护
glare protection lens 防眩光透镜
glare-reducing 无眩光的;不眩目的
glare-reducing glass 减眩玻璃;防眩玻璃
glare screen 眩光屏;遮眩板;反眩目屏
glare shield 闪光罩;闪光屏(挡);防眩板
glare vision 眩光视力
glare zone 眩光带
glarimeter 光泽计;光泽测定器;闪光计
glaring 炫耀的;镶玻璃
glaring light 眩耀光;眩目灯光;眩光(灯)
glaserite 钾芒硝
Glasgow type generator 格拉斯哥式水煤气发生炉
glasizer's diamond 割玻璃用的金刚钻石
glasphalt 玻璃沥青
glass 镜片;玻璃(制品);玻璃无槽引上;玻璃体;玻璃品钳子
glass abrasive paper 玻璃砂纸
glass absorber 玻璃吸收器
glass agate 黑曜岩
glass aggregate 玻璃集料;玻璃骨料
glass-air interface 玻璃板—空气界面
glass ampule 玻璃安瓿
glass analysis 玻璃分析
glass and wood panel door 玻璃镶板门
glass anti-dimmer 玻璃抗雾剂
glass apparatus 玻璃仪器
glass appliance 玻璃用具;玻璃器具
glass applicator 玻璃涂药棒
glass architecture 玻璃建筑
glass area 玻璃面积
glass article 玻璃制品
glass art products 玻璃艺术制品;玻璃器器
glass-asphalt 玻璃纤维地沥青
glass-backed mica tape 玻璃布底云母带
glass bag 玻璃袋
glass bag dust collector 玻璃纤维(布)袋式集尘器
glass bag filter cloth 玻璃纤维圆筒过滤布
glass-balustrade escalator 玻璃栏杆自动滚梯
glass bar 玻璃棒
glass base laminated material 玻璃层压材料;玻璃层压制品
glass base plate 玻璃底板
glass basis 玻璃基体;玻璃底板;玻基
glass batch 玻璃配合料
glass bath 玻璃池

glass batt 玻璃丝毡
glass bead 玻璃压条；玻璃小珠；玻璃嵌条；玻璃（细）珠；玻璃微珠
glass bead cleaning 喷玻璃丸清表面
glass bead cylinder method 玻璃珠柱法
glass-beaded 玻璃熔接的
glass bead peening 玻璃丸喷丸；玻璃球喷丸处
glass bead screen 玻璃微珠屏幕
glass beads reinforced plastic 玻璃细珠增强塑料
glass beaker 玻璃烧杯
glass bed receiver 玻璃台架接收器
glass bell 玻璃钟罩
glass bell jar 玻璃钟罩
glass bending machine 玻璃弯曲机
glass bending process 玻璃弯曲法
glass binder 玻璃状黏合剂
glass black 灯黑
glass blackening 玻璃发黑
glass blank 玻璃雏形
glass blanket 玻璃纤维织物包覆层；玻璃丝毡
glass blank for astronomic(al) telescope 天文望远镜玻璃毛坯
glass block 镜片；玻璃砖；玻璃（垫）块
glass-block class-concrete block 玻璃砖类混凝土砌块
glass-block masonry 玻璃砖砌筑物
glass-block masonry work 玻璃砖砌筑工作；玻璃块砌筑工作
glass-block panel 玻璃砖镶板
glass-block partition(wall) 玻璃砖隔墙
glass-block roof-light 玻璃砖屋顶
glass-block roof-light panel 玻璃砖屋顶采光板
glass-block skylight 玻璃砖天窗
glass-block window 玻璃砖窗子
glass blower 吹玻璃机；吹玻璃工；玻璃工；玻璃吹制工
glass blower's burner 玻璃灯工用煤气喷灯
glass blower's emphyesma 吹玻璃工肺气肿
glass blowing 玻璃（的）吹制；吹玻璃
glass blowing lathe 玻璃吹制车床
glass blowing out 玻璃吹制
glass-body-demilune body 半月体
glass-bonded mica 玻璃结合云母；玻璃胶合云母；玻璃黏合云母
glass-bonded synthetic(al) mica 玻璃结合合成云母
glass bottle 玻璃瓶
glass bottom 玻璃钮扣
glass boundary 玻璃边
glass bowl 玻璃杯
glass box 玻璃盒子；玻璃盖匣
glass break detector 玻璃破碎检测器
glass break simulator 玻璃破碎模拟器
glass brick 镜片；玻璃（空心）砖；玻璃块；玻璃（地）砖
glass brush 玻璃丝刷
glass building block 建筑用玻璃砖
glass bulb 玻璃（灯）泡
glass bulb blowing machine 灯泡机
glass bulb cleaning 玻壳清洗
glass bulb compressive strength tester 玻壳耐压试验机
glass bulb forming machine 灯泡吹制机
glass bulb mercury arc rectifier 玻璃泡汞弧整流器
glass bulb mercury rectifier 玻璃泡水银整流管
glass bulb rectifier 玻璃壳整流器
glass bulb side wall type sprinkler 玻璃球边墙型喷淋头
glass bushing 玻璃套管
glass cable 玻璃丝电缆
glass cameo 玻璃浮雕
glass capacitor 玻璃电容器
glass capillary 玻璃毛细管
glass capillary column 玻璃毛细管柱
glass capillary electrode method 玻璃毛细管电极法
glass capillary gas chromatography 玻璃毛细管气相色谱法；玻璃毛细管气相层析
glass capillary tube viscometer 玻璃毛细管黏度计
glass capillary viscometer 玻璃毛细管黏度计
glass carboy 玻璃坛；玻璃酸瓶
glass carving machine 玻璃刻花机
glass case 玻璃橱（柜）
glass casement 玻璃落地窗
glass cassette 玻璃箱盒
glass catcher bars 玻璃收集推杆
glass ceiling 玻璃天花板；玻璃顶棚
glass cell 玻璃试液池；玻璃比色槽
glass cement 玻璃黏合剂；玻璃胶
glass cemented filter 玻璃夹胶滤色镜
glass-centering shim 玻璃定位填隙片
glass-ceramic 玻璃陶瓷
glass ceramic active medium 玻璃陶瓷活性媒质
glass ceramic capacitance thermometer 玻璃陶瓷电容温度计
glass ceramic capacitor 玻璃陶瓷介质电容器
glass ceramic composite 微晶玻璃复合材料
glass ceramic die 玻璃陶瓷模
glass ceramic fiber optic(al) sensor 玻璃陶瓷光纤传感器
glass ceramic fibre 玻璃陶瓷纤维
glass ceramic process 微晶化过程
glass ceramics 结晶玻璃；微晶玻璃
glass ceramics capacitor 玻璃陶瓷电容器
glass ceramic solder 微晶玻璃焊料
glass charge 玻璃配合料
glass check 天窗玻璃压边框
glass chimney 玻璃灯罩
glass church 玻璃教堂
glass-clad 玻璃封裹的；玻璃包覆的
glass clamp 玻璃夹子
glass cleaner 挡风玻璃刮水器；玻璃清洁剂；玻璃擦洗器
glass clear 透明的；明亮的；玻璃亮光的
glass cloth 玻璃织物；玻璃（纤维）布；玻璃（丝）布；玻璃（砂）布
glass-cloth for wall covering 玻璃贴墙布
glass-cloth insulation 玻璃布绝缘
glass-cloth laminate 玻璃布层压板；玻璃布层压制品
glass coating 搪玻璃；玻璃涂料；玻璃涂层；玻璃镀膜
glass cock 玻璃（旋）塞
glass coffer 玻璃天花板镶板
glass coil 玻璃蛇管
glass cold trap 玻璃冷阱
glass collection unit 玻璃收集装置
glass colo(u)r 玻璃着色用的颜料；玻璃颜料
glass component 玻璃构件；玻璃组分；玻璃成分
glass composition 玻璃成分；玻璃组成
glass concrete 玻璃纤维（增强）混凝土
glass concrete block 玻璃纤维混凝土砌块
glass concrete construction 镶空心玻璃砖建筑；玻璃纤维混凝土结构；玻璃纤维混凝土结构；带反光玻璃构造；玻璃砌块构造；嵌玻璃砖混凝土构造
glass concrete(load) bearing structure 玻璃纤维混凝土承载结构
glass concrete panel 玻璃纤维增强混凝土板
glass concrete rooflight 玻璃纤维混凝土天窗；玻璃纤维混凝土采光屋顶
glass concrete structural system 玻璃纤维混凝土结构系统
glass concrete supporting structure 玻璃纤维混凝土支承结构
glass concrete window 玻璃纤维混凝土窗子
glass condenser 玻璃冷凝器；玻璃介质电容器
glass conditioning 玻璃均化
glass constituent 玻璃组分
glass construction 玻璃结构；玻璃建筑
glass-construction vacuum pressure measuring ga(u)ge 玻璃结构真空压强测量规
glass contact area 玻璃液接触部位
glass contact block 池壁砖
glass container 玻璃容器
glass container manufacture 玻璃容器制造
glass containing clinker 含玻璃相的熟料
glass content 玻璃含量
glass cord 玻璃（纤维）绳；玻璃（帘子）线
glass cordage 玻璃纤维帘子线
glass core 玻璃棒；玻璃芯
glass core detector 玻璃条纹检验仪
glass corner 玻璃（隅角）反射器
glass corridor 玻璃走廊
glass cotton 玻璃纤维；玻璃丝；玻璃绒；玻璃棉
glass counter ceiling 玻璃吊平顶
glass cover 玻璃罩
glass covered wire 玻璃丝包线
glass cover plate 玻璃盖板
glass-crete 透明玻璃块；钢丝玻璃；玻璃混凝土
glass cross 玻璃丝网
glass culture 暖房栽培
glass culture tube 玻璃栽培试管
glass cupola 玻璃圆屋顶
glass current 玻璃液流
glass curtain 半透明织物窗帘；窗纱
glass curtain wall(ing) 玻璃幕墙
glass curved in two planes 天窗玻璃；拱形玻璃
glass cutter 玻璃切（割）工；划割玻璃刀；雕刻玻璃工；玻璃切割器；玻璃（切）刀；玻璃（割）刀
glass cutter with cutting wheels 装有切割轮的玻璃切割刀
glass cutting 玻璃刻花；玻璃切割
glass cutting fluid 玻璃切削液
glass cutting wheel 玻璃刻磨轮
glass cutting workshop 玻璃切裁车间
glass cylinder 玻璃量筒
glass-dead seal 玻璃封口
glass decoration 玻璃装饰；玻璃彩饰
glass decoration colo(u)r 玻璃色釉
glass defect 玻璃缺陷
glass delay line 玻璃延迟线
glass density 玻璃密度
glass depth 玻璃液深度
glass detector 玻璃检测器
glass dial 玻璃（刻）度盘
glass diamond 玻璃刀；划玻璃钻
glass diapositive 玻璃正片
glass dielectric(al) capacitor 玻璃介质电容器
glass diffusion pump 玻璃扩散真空泵
glass dimmer 玻璃减光器
glass diode 玻璃二极管
glass disc[disk] 玻璃度盘；玻璃皿；玻璃碟
glass-disc-laser amplifier 玻璃圆盘激光放大器
glass dish 玻璃皿
glass dish evapo(u)ration test 玻璃皿蒸发试验
glass disk laser 玻璃圆盘激光器
glass disk scale 玻璃刻度盘
glass distilled water 蒸馏水
glass dome 玻璃穹隆
glass dome light 玻璃屋顶天窗
glass dome roof-light 玻璃圆屋顶天窗
glass door 玻璃门；窗门
glass door control 玻璃门控制（机构）
glass dosimeter 玻璃剂量计
glass draining hole 放玻璃孔
glass draining tank 放玻璃液池
glass drawing machine 玻璃拉制机
glass drill 玻璃钻
glass drop 玻璃滴；玻璃珠
glass dust 玻璃屑；玻璃粉（尘）
glassed 磨光（对大理石或花岗岩表面）；打光（对大理石或花岗岩表面）
glass edge sealed glazing unit 玻璃封边的玻璃窗
glass edging machine 磨边机
glassed-in 玻璃包围着的
glassed-in area 四周镶嵌玻璃的地区；玻璃笼罩着的场地
glassed-in observation deck 装玻璃窗的瞭望台
glassed-in opening 装玻璃的开孔；装配玻璃的窗孔
glassed-in timber door 装配玻璃的木门
glassed-in veranda(h) 镶玻璃的阳台；镶玻璃的外廊
glassed-in viewing platform 装配玻璃的瞭望台
glassed-in wood(en) door 装配玻璃的木门
glassed partition 玻璃隔板；玻璃间壁
glassed steel 搪瓷钢；搪玻璃
glassed stone 磨光（的）石料
glassed surface 磨光玻璃表面；平板玻璃
glassed vessel 覆盖有玻璃的钢质容器
glass electrode 玻璃电极
glass electrode method 玻璃电极法
glass electrode pH meter 玻璃电极 pH 计
glass electrode response 玻璃电极响应
glass embossing 玻璃浮雕
glass enamel 透明釉；搪玻璃；玻璃搪瓷
glass encapsulant 玻璃封装剂
glass encapsulative 玻璃封装
glass-enclosed 周围镶有玻璃的；玻璃包盖的
glass-enclosed building 玻璃建筑物
glass-enclosed scale clinical thermometer 玻壳体温计
glass enclosure 玻璃罩

glass engraving 玻璃板刻图
glass engraving machine 玻璃雕刻机
glass envelope 玻壳
glass envelope sealing machine 屏锥封接机
glass epoxy 环氧玻璃钢板;玻璃钢板(环氧);玻璃纤维环氧树脂
glasses 双筒镜
glass etcher 玻璃蚀刻雕花机
glass etching 玻璃蚀刻法
glass-etch process 玻璃板刻图
glass extrusion process 玻璃润滑热挤压法
glass fabric 玻璃(纤维)织物;玻璃(纤维)布
glass fabric end connecting machine 玻璃布接头机
glass fabric felt 玻璃纤维布油毡
glass fabric finishing aggregate 玻璃布表面处理机组
glass fabric for copper coated laminate 涂铜板用玻璃织物
glass fabric for reinforcing plastics 增强塑料用玻璃织物
glass fabric for reinforcing rubber 增强橡胶用玻璃织物
glass fabric roofing 玻璃纤维屋面材料
glass facade 镶有玻璃的建筑正面
glass-facade block 镶饰建筑物正面的玻璃块;镶饰建筑物立面的玻璃块
glass-facade building 正面为玻璃的建筑物;玻璃立面建筑
glass face work 镶饰玻璃面工作;玻璃面工作
glass facing 镶饰玻璃面;玻璃面
glass facing tile 玻璃面砖
glass feeder 玻璃加料器
glass felt 玻璃丝毡
glass fertilizer 玻璃肥料
glass fiber 玻璃丝;玻璃(光学)纤维
glass fiber acoustic(al)ceiling 玻璃纤维吸声顶棚
glass fiber asphalt 玻璃纤维沥青
glass fiber based strip 玻璃纤维底板条
glass fiber based tape 玻璃纤维带
glass fiber battery separator 玻璃纤维蓄电池隔离片
glass fiber board 玻璃纤维板
glass fiber coating 玻璃纤维涂料;玻璃纤维涂层
glass fiber cloth 玻璃纤维布
glass fiber coating 玻璃纤维涂层
glass fiber composite 玻璃纤维复合物
glass fiber concrete 玻璃纤维混凝土
glass fiber covered wire 玻璃丝包线
glass fiber covering 玻璃丝被覆;玻璃丝包缠
glass fiber curtain 玻璃纤维帘
glass fibered plaster 加玻璃纤维的粉刷
glass fiber extrusion 玻璃纤维的喷出拉制;玻璃纤维的挤出拉制
glass fiber facing unit 玻璃纤维饰面元件
glass fiber felt 玻璃纤维毡;玻璃棉毡;玻璃丝毡
glass fiber filter 玻璃纤维(过)滤器;玻璃纤维袋集尘器
glass fiber filtering medium 玻璃纤维滤料
glass fiber filter wool 过滤用玻璃纤维
glass fiber filtration material 玻璃纤维过滤材料
glass fiber finish 玻璃纤维表面处理剂
glass fiber flooring backing 玻璃纤维地板材料背衬
glass fiber form(work) 玻璃纤维模板
glass fiber guide 玻璃光纤波导
glass fiber heat insulation rope 玻璃纤维保温绳
glass fiber insulated wire 玻璃纤维绝缘电线
glass fiber insulation 玻璃纤维绝缘
glass fiber joint runner 玻璃纤维接缝填料
glass fiber laminate 玻璃纤维层压板
glass fiber laser 玻璃纤维激光器
glass fiber lubricant 玻璃纤维润滑剂
glass fiber mat 玻璃纤维毡(毯);玻璃纤维(席)垫
glass fiber material 玻璃纤维材料
glass fiber mat wool 玻璃纤维棉垫
glass fiber mo(u)ld 玻璃纤维模(型)
glass fiber paper 玻璃纤维纸
glass fiber partition laminated 玻璃纤维层压隔板
glass fiber pipe 玻璃钢管;玻璃纤维管
glass fiber pipe section 玻璃纤维管壳
glass fiber plastic 玻璃纤维塑料
glass fiber polyester 玻璃纤维聚酯
glass fiber polyester sheet 玻璃纤维聚酯板
glass fiber pouring rope 玻璃纤维铸索

glass fiber product 玻璃纤维制品
glass fiber reinforced 玻璃纤维增强的
glass fiber reinforced cement 玻璃纤维增强水泥;玻璃纤维强化水泥
glass fiber reinforced concrete 玻璃纤维加筋混凝土;玻璃纤维增强混凝土
glass fiber reinforced concrete pipe 玻璃纤维增强混凝土管
glass fiber reinforced elastomer 玻璃纤维增强弹性体
glass fiber reinforced epoxy plastics 玻璃纤维增强环氧塑料
glass fiber reinforced epoxy resin 玻璃纤维增强环氧树脂
glass fiber reinforced gypsum 玻璃纤维增强石膏
glass fiber reinforced magnesite 玻璃纤维增强菱苦土
glass fiber reinforced nylon 玻璃纤维增强尼龙
glass fiber reinforced phenolic plastics 玻璃纤维增强酚醛塑料
glass fiber reinforced plaster 玻璃纤维增强石膏
glass fiber reinforced plastic 玻璃纤维加强塑料;玻璃纤维增强塑料;玻璃钢
glass fiber reinforced plastic bathtub 玻璃纤维增强塑料浴缸
glass fiber reinforced plastic blade 玻璃纤维增强塑料桨叶
glass fiber reinforced plastic boat 玻璃钢船
glass fiber reinforced plastic bridge 玻璃钢桥
glass fiber reinforced plastic cutter 玻璃纤维增强塑料切割机
glass fiber reinforced plastic door and window frames 玻璃纤维增强塑料门窗框
glass fiber reinforced plastic for mo(u)lding 玻璃纤维增强模压塑料
glass fiber reinforced plastic hull 玻璃钢船体
glass fiber reinforced plastic laminate 玻璃纤维增强塑料层压板
glass fiber reinforced plastic lifeboat 玻璃钢救生艇;玻璃钢救生筏
glass fiber reinforced plastic life suit 玻璃钢救生衣
glass fiber reinforced plastic pipe 玻璃纤维增强管材
glass fiber reinforced plastic pipe making process 玻璃纤维增强塑料制管工艺
glass fiber reinforced plastic pressure vessel 玻璃纤维增强塑料压力气瓶
glass fiber reinforced plastic profile 玻璃纤维增强的塑料型材;玻璃纤维增强的塑料模板
glass fiber reinforced plastic sheet 玻璃纤维增强的塑料薄板
glass fiber reinforced plastic tank 玻璃纤维增强塑料储罐
glass fiber reinforced plastic transparent sheet 玻璃纤维增强塑料透明板
glass fiber reinforced plastic valve 玻璃纤维增强阀门
glass fiber reinforced plastic vaulting pole 玻璃纤维增强塑料撑竿
glass fiber reinforced polycarbonate 玻璃纤维增强聚碳酸酯
glass fiber reinforced polyester 玻璃纤维增强聚酯
glass fiber reinforced polyester coating 玻璃纤维增强聚酯涂层
glass fiber reinforced polyester corrugated sheet 玻璃纤维增强聚酯波纹板
glass fiber reinforced polyester resin 玻璃纤维增强聚酯树脂;玻璃丝加强的聚酯树脂
glass fiber reinforced polyester sheet 玻璃纤维聚酯板
glass fiber reinforced polypropylene 玻璃纤维增强聚丙烯
glass fiber reinforced resin 玻璃纤维增强树脂
glass fiber reinforced resin panel 玻璃纤维增强的树脂墙板
glass fiber reinforced structural thermoplastic foam 玻璃纤维增强结构热塑泡沫(塑料)
glass fiber reinforced synthetic resin 玻璃纤维增强的人造树脂;玻璃纤维增强的合成树脂
glass fiber reinforced thermoplastics 玻璃纤维增强热塑性塑料
glass fiber reinforced thermoset plastics 热固性玻璃纤维增强塑料

glass fiber reinforced unsaturated polyester 玻璃纤维增强的未饱和树脂
glass fiber reinforcement 玻璃纤维增强(材料)
glass fiber resin panel 玻璃纤维树脂板
glass fiber rod 玻璃纤维杆;玻璃纤维棒
glass fiber round-straight converter 光纤维圆直变换器
glass fiber seal(ing)rope 玻璃纤维密封绳
glass fiber sheet 玻璃纤维薄板
glass fiber shrinking method 玻璃丝收缩法
glass fiber shuttering 玻璃纤维模壳;玻璃纤维模板
glass fiber sleeve-springs 玻璃纤维套筒弹簧
glass fiber steel cooling tower 玻璃钢冷却塔
glass fiber strand 玻璃纤维绳
glass fiber surfacing unit 玻璃纤维饰面元件
glass fiber tape 玻璃纤维带
glass fiber technique 玻璃纤维技术
glass fiber thread 玻璃纤维线
glass fiber tile 玻璃纤维砖
glass fiber waveguide 玻璃光纤波导
glass fiber winding machine 玻璃纤维绕丝机
glass fiber wire 玻璃纤维丝
glass fiber wound case 玻璃纤维缠绕的外壳
glass fiber wrapping material 玻璃包缠布;玻璃纤维包装材料
glass fiber yarn 玻璃纱
glass figurine 玻璃浮雕;玻璃雕塑
glass filament 长玻璃纤维;玻璃单丝;玻璃(长)丝
glass filament bushing 玻璃纤维拉丝漏板
glass filament case 玻璃纤维壳体
glass filament extrusion bushing 挤拉玻璃纤维的漏板
glass filament wind-up 玻璃丝的卷绕
glass-filled material 玻璃填料
glass film 玻璃薄膜
glass film condenser 玻璃薄膜电容器
glass film evapo(u)rator 玻璃液膜蒸发器
glass film plate 玻璃底板
glass filter 滤光镜;玻璃滤器
glass filter cloth 玻璃纤维滤布
glass filter method 玻璃滤器法
glass filtration fabric 玻璃纤维滤布
glass finish 玻璃表面处理剂
glass finishing 玻璃加工
glass fission detector 玻璃裂变探测器
glass fitter 玻璃工
glass fittings 玻璃器具;玻璃配件
glass flake 玻璃薄片;鳞片玻璃
glass flake composite 玻璃片组合件
glass flask 玻璃烧瓶
glass flexible fiber optics 玻璃纤维光学
glass floor 玻璃地板
glass floor cover(ing) 玻璃地板覆面层
glass floor finish 玻璃地板饰面层
glass flooring 玻璃地板面层
glass flooring tile 玻璃地面砖
glass floorplate 玻璃楼面板
glass flour 玻璃粉
glass flow 玻璃液流
glass flow control needle 玻璃液流量控制计
glass flume 玻璃水槽(水工模型试验用)
glass flushing tank 玻璃冲洗水箱
glass flux 玻璃状焊剂
glass foam 泡沫玻璃;玻璃泡沫
glass footed stem mo(u)ld 高脚杯桎模
glass for electrode 电极用玻璃
glass for glazing 窗用玻璃;门窗用玻璃
glass for infrared rays 红外(线)玻璃
glass for lithography 平板印刷用玻璃
glass formation 玻璃的形成
glass formation range 玻璃形成范围
glass former 成玻璃物;玻璃形成体
glass forming 玻璃成型
glass forming machine 玻璃成形机
glass forming oxide 玻璃形成氧化物;玻璃成形氧化物
glass forming press 玻璃压制成型机
glass forming system 玻璃成型系统
glass for photographic(al)plate 照相底板用玻璃
glass for sealing 封接用玻璃
glass for ultraviolet rays 透紫外线玻璃
glass for vehicle window 车窗玻璃
glass fragment 玻璃碎片
glass-frame cabinet 玻璃柜

glass-frame riser 玻璃升降器
glass framing 玻璃窗框
glass-free clinker 不含玻璃相的熟料
glass-free porcelain 无玻璃相陶瓷
glass frit 玻璃熔块
glass-fronted building 正面为玻璃的建筑物;玻璃立面建筑
glass frost 薄片碎玻璃;玻璃箔
glass funnel 玻璃漏斗
glass furnace 玻璃熔(化)炉
glass furniture 玻璃家具
glass ga(u)ge 水位玻璃管;玻璃液位表;玻璃水位指示管;玻璃管规
glass gilding 玻璃镀金
glass glaze 玻璃釉
glass-glaze condenser 玻璃釉电容器
glass-glazed 涂有玻璃釉的
glass glaze resistor 玻璃釉电阻器
glass glue 玻璃黏合剂
glass gob 玻璃料坯
glass gob delivery 玻璃料滴供给
glass gob for marble making 球坯;玻璃球坯
glass graduate 玻璃量筒
glass graduate cylinder 玻璃量筒
glass granite 玻璃花岗岩
glass grinder 玻璃研磨机
glass grinding machine 玻璃研磨机
glass grinding workshop 玻璃研磨车间
glass grit 玻璃粗粒
glass groove 安装玻璃用槽口
glass guide 导向玻璃棒;玻璃导向棒
glass gutter 玻璃檐槽;玻璃天沟
glass hard 高硬度;玻璃硬度(HRC65以上)
glass-hardened 激淬;很硬的
glass hardness 很硬淬火
glass-hard steel 特硬钢
glass-hard surface 光硬表面
glass hardware 玻璃器皿
glass head 玻璃熔接
glass head seal 玻璃熔封
glass heat exchanger 玻璃热交换器;玻璃换热器
glass heating panel (镀有导电膜的)框格玻璃电热板;玻璃电热板
glass holder 玻璃吸盘;玻璃托
glass hollow block 玻璃空心砖
glass hollow brick 玻璃空心砖
glasshouse 暖房;玻璃厂;熔制车间;温室;玻璃温室
glasshouse climate 温室气候
glasshouse culture 温室栽培;温室培养
glasshouse effect 暖房效应
glasshouse plant 温室植物
glass housing mercury rectifier 玻璃壳水银整流器
glass ice 冰壳
glass inclusion 玻璃包裹体
glass industry 玻璃装配行业;玻璃工业
glassine 耐油纸;薄半透明纸;玻璃纸
glassine bag 玻璃纸袋
glassine ink 玻璃纸用油墨
glassine paper dark-rose 黑玫瑰色半透明玻璃纸
glassine paper white 白色半透明玻璃纸
glassine paper yellow 黄色半透明玻璃纸
glassiness 玻璃质
glassing 装配玻璃;涂玻璃粉;成玻璃状
glassing burner 加工玻璃喷灯
glassing jack 磨光机;打光机
glassing machine 磨光机;打光机
glass ingredient 玻璃组分;玻璃成分
glass ink 玻璃用油墨
glass inlaying 镶嵌玻璃
glass insert 玻璃插入物
glass-insulated 玻璃绝缘的
glass insulated wire 玻璃绝缘线
glass insulation material 玻璃绝缘材料
glass insulator 玻璃绝缘子;玻璃绝缘体;玻璃绝缘板
glass ionomer cement 玻璃离子键聚合物水泥
glass iron wire 玻璃铁丝
glassivation 涂附玻璃;玻璃钝化;玻璃保护层
glass jalousie (固定式的)玻璃百叶窗;玻璃百叶窗;固定式玻璃百叶窗
glass jam 玻璃堵塞
glass jar 玻璃罐
glass knob 玻璃球形(门)把手
glass laminate 玻璃层板;安全玻璃
glass laminated sheet 玻璃层布板

glass lamp chimney 玻璃灯罩
glass lamp shade 玻璃灯具
glass laser 玻璃激光器
glass laser material 玻璃激光材料
glass laser rod 玻璃激光棒
glass laser target 玻璃激光靶
glass lathe 玻璃(加工)车床
glass lehr 玻璃退火窑
glass length scale 玻璃分划尺
glass lens polishing 玻璃透镜的抛光
glass letter 玻璃字母
glass level 玻璃液面(高度)
glass level attack 玻璃液面线侵蚀
glass level control 玻璃液面控制
glass level controller 玻璃液面控制装置;玻璃液面控制器
glass level cut 液面线侵蚀
glass lifter 玻璃升降机
glass light cupola 玻璃圆顶天窗
glass-like 玻璃样的
glass limb 玻璃度盘
glass line 玻璃液面线
glass-lined 搪瓷的;搪玻璃的;涂瓷釉;玻璃衬里的
glass-lined microcrater 搪玻璃微月坑
glass-lined pipe 内壁涂瓷釉的管道;玻璃内衬管;玻璃衬里水管
glass-lined steel 搪玻璃钢件(耐化学侵蚀的钢件);玻璃衬里钢
glass-lined steel pipe 衬玻璃钢管
glass lining 搪瓷;搪玻璃;衬玻璃;玻璃(内)衬
glass lining pipe 搪瓷管;衬玻璃管
glass liquid 玻璃液
glass liquid hydrometer 玻璃液体比重计
glass louver 玻璃百叶(窗);玻璃天窗
glass lubricant 玻璃润滑剂
glass machinery 玻璃机械
glass machining technique 玻璃加工工艺
glass magnetic material 玻璃态磁性材料
glassmaker's soap 玻璃脱色剂
glassmaker's tool 玻璃制造工具
glassmaking 玻璃制造
glassmaking equipment 玻璃制造设备
glassmaking furnace 玻璃窑
glassmaking materials 玻璃原料
glassmaking sand 玻璃生产用石英砂
glass manometer 玻璃压力计
glass manufactory wastewater 玻璃制造废水;玻璃厂废水
glass manufacture 玻璃制造
glass manufacturer 玻璃制造者
glass manufacturing waste 玻璃工业废物
glass marble 玻璃球
glass marble furnace 玻璃球窑
glass marker 玻璃刻度器
glass-marking pencil 玻璃划线笔
glass mask 玻璃掩模
glass masonry 空心玻璃砖砌体
glass mat 玻璃纤维(薄)毡;玻璃纤维板;玻璃垫
glass mat displacement 玻璃纤维毡移位
glass matrix 玻璃母体;玻璃基质
glass matrix composite 玻璃基复合材料
glass matrix porphyritic texture 玻基斑状结构
glass measure 量杯
glass measuring cylinder 玻璃量筒
glass measuring device 玻璃量具(有容量刻度的玻璃器皿)
glass melt 玻璃(熔)液;玻璃熔体
glass melted from batch only 仅用配合料熔制的玻璃;全生料熔制玻璃
glass melted from cullet 碎玻璃熔成的玻璃;碎玻璃熔制玻璃
glass melter 玻璃熔窑
glass melting 玻璃熔制;玻璃的熔融
glass melting furnace 玻璃熔窑
glass melting tank 玻璃熔窑
glass member 玻璃构件
glass membrane electrode 玻璃膜电极
glass metal 玻璃液
glass-metal seal(ing) 玻璃金属封焊
glass-mica cloth 玻璃云母布
glass-mica combination 玻璃云母复合材料
glass-mica fabric 玻璃云母布
glass-mica tape 玻璃(布底)云母带
glass microballoon 玻璃微珠

glass microelectrode 玻璃微电极
glass microsphere 玻璃微珠
glass mirror 玻璃镜
glass mirror chopper 玻璃反射镜调制盘
glass modifier 玻璃调整体;玻璃调整剂
glass monofilament 玻璃单丝
glass mortar 玻璃研钵
glass mosaic 锦玻璃;玻璃面砖;玻璃马赛克;玻璃锦砖
glass mosaic tile 玻璃马赛克;玻璃锦砖
glass mo(u)ld 玻璃(铸)型;玻璃模具
glass mo(u)ld diode 玻璃模二极管
glass needle 玻璃针
glass needle lubricator 玻璃针注油器
glass negative 玻璃底片
glass network 玻璃网络
glass network former 玻璃网络形成体
glass network intermediate 玻璃网络中间体
glass network modifier 玻璃网络形成剂;玻璃网络调整剂
glass object 玻璃品
glass observation area 玻璃瞭望区
glass observation tube 玻璃观察管
glass oil cup 玻璃油杯
glass oiler with screen 带滤网的玻璃油杯
glass opal 玻璃蛋(白)石
glass optical fiber[fibre] 玻璃光纤
glass out flow pipe 玻璃液流出管
glass oven 玻璃退火炉;玻璃熔化炉
glass ovenware 玻璃炊具
glass package 玻璃封装
glass packaging 玻璃封装
glass paint 玻璃涂料
glass painter 玻璃油漆工;玻璃上绘图者
glass painting 玻璃(涂)油漆;玻璃画;玻璃彩饰
Glass Palace 水晶宫
glass pane 窗格玻璃;玻璃板
glass panel 玻璃窗;建筑上嵌玻璃部分;玻璃镶板;玻璃嵌板
glass panel heater 玻璃嵌板加热器
glass-panel(l)ed balustrade 玻璃栏板
glass-panel(l)ed flume 玻璃水槽(水工模型试验用)
glass panel(l)ed tilting flume 活动玻璃水槽
glass paper 砂纸;玻璃(砂)纸
glass papering machine 玻璃砂纸磨光机
glass parapet slab 玻璃女儿墙板
glass parison 玻璃雏型
glass particle formation 玻璃粒的形成
glass partition 玻璃隔墙;玻璃隔断
glass partition wall(ing) 玻璃隔墙;玻璃隔板;玻璃壁
glass patio 玻璃庭院
glass pattern 玻璃图样
glass paving slab 玻璃铺地砖
glass pearl 玻璃珠
glass peep hole 玻璃检视孔
glass pellet 玻璃料块
glass pencil 特种铅笔
glass pestle 玻璃杵
glass petri disk 玻璃培养皿
glassphalt 碎玻璃沥青路面;玻璃沥青(一种以玻璃制成的铺路材料)
glassphalte 碎玻璃沥青路面
glass pH electrode 玻璃pH电极
glass pine 美国松
glass pipe 玻璃管(道)
glass pipe line 玻璃管道
glass piping 玻璃管(路)
glass plant waste 玻璃厂废水
glass plate 玻璃平板;玻璃(厚)板
glass plate capacitor 玻璃片电容器;玻璃(板)电容器
glass plate grid 玻璃网格板
glass plate level ga(u)ge 玻璃板液面计
glass plug 玻璃塞
glass pocket 无用玻璃收集器
glass polisher 玻璃抛光器
glass polygon 玻璃多面体
glass polyhedron 玻璃多面体
glass porphyry 玻(璃)斑岩
glass positioner 玻璃定位器
glass positive 玻璃正片
glass pot 玻璃坩埚

glass-pot clay 陶土；陶瓷耐火黏土
glass powder 玻璃粉
glass powder sintering 玻璃粉烧结
glass precision scale 精密玻璃刻尺
glass preparation 玻璃制备
glass press 玻璃压制（成型）机
glass pressing blank 玻璃毛坯
glass pressing machine 玻璃压制机
glass pressure tube 玻璃压力管
glass printer 玻璃印花机
glass printing 玻璃印花
glass prism 棱面玻璃砖；棱镜；玻璃棱镜
glass prism monochromator 玻璃棱镜单色仪
glass prism spectro-graph 玻璃棱镜摄谱仪
glass processing 玻璃深加工
glass products 玻璃制品
glass property 玻璃性能
glass protecting cover 玻璃护罩
glass pull 玻璃引出量
glass-pumic concrete plank 玻璃浮石混凝土板
glass pump 玻璃泵
glass putty 锡粉；嵌玻璃油灰；铅粉；窗用油灰；玻璃油灰；玻璃腻子；宝石磨粉
glass rabate 镶玻璃槽口
glass rabbet 镶玻璃槽口
glass rack 载物片架
glass rainwater gutter 玻璃天沟
glass raw material commodities 玻璃原料矿产
glass receiver 玻璃浇铸台
glass refining 玻璃澄清
glass reflector 玻璃反射器；玻璃反射镜
glass regulations 玻璃质量规定
glass-reinforced cement 玻璃纤维增强水泥
glass-reinforced concrete 玻璃纤维增强混凝土
glass-reinforced laminate 玻璃增强层压板；玻璃增强塑料板
glass-reinforced panel 玻璃增强板；增强玻璃板
glass-reinforced plastic boat 玻璃纤维增强塑料艇；玻璃钢艇
glass-reinforced plastic forms for reinforced concrete 增强混凝土用玻璃纤维增强塑料模板
glass-reinforced plastic lifeboat 玻璃钢救生艇
glass-reinforced plastic motor lifeboat 玻璃钢机动救生艇
glass-reinforced plastic oar rowing lifeboat 玻璃钢划桨救生艇
glass-reinforced plastic pipe 玻璃钢管
glass-reinforced plastics 强化玻璃钢；玻璃钢
glass-reinforced structural plastics 玻璃丝加固塑料
glass-reinforced thermoplastics 玻璃丝增强热塑塑料
glass-reinforced water-proof paper 玻璃加筋防水纸
glass remover 除玻璃碎片器
glass resistor 玻璃电阻器
glass rest （支承玻璃的）玻璃支架
glass retort 玻璃曲颈瓶
glass ribbon 玻璃条；玻璃带
glass rim 玻璃框
glass rod 玻璃杆；玻璃棒
glass-rod thermometer 玻璃棒形温度计
glass roof 玻璃屋顶
glass-roof cladding 玻璃屋顶覆盖物
glass-roof construction 玻璃屋顶建筑
glass roof gutter 玻璃天沟
glass roofing 玻璃屋面（料）
glass roof(ing) material 玻璃屋面料
glass roof(ing) tile 玻璃（屋面）瓦
glass rotameter 玻璃转子流量计
glass roundel 玻璃小圆窗
glass roving(cloth) 玻璃纤维无捻粗纱
glass run 窗玻璃槽
glass-run channel 车窗玻璃升降槽
glass runner 指示器
glass-run weatherstrip 玻璃导向器阻风雨带
glass sand 高纯度石英砂；玻璃砂
glass sand paper 玻璃砂纸
glass saucer 玻璃碟
glass saucer dome 玻璃盘状穹隆
glass saw machine 玻璃锯切机
glass scale 玻璃比例尺
glass schorl 斧石
glass scoring wheel 玻璃切割刀轮

glass screen 风挡；玻璃（荧光）屏；玻璃网目屏；玻璃滤光镜
glass screen with metal frame 带金属框玻璃屏
glass scrim for grinding wheel 砂轮用玻璃纤维网格布
glass seal 玻璃封口
glass-sealed 玻璃焊封的；玻璃密封的
glass-sealed resistor 玻璃密封电阻器
glass-sealed thermometer 玻璃密封温度计
glass-sealed transistor 玻璃密封（的）晶体管
glass sealing 玻璃密封
glass sealing alloy 玻璃密封合金
glass sealing-packing equipment 玻璃封装专用设备
glass seal type transistor 玻璃密封式晶体管
glass seam 玻璃缝
glass semi-conductor 玻璃半导体
glass semi-conductor device 玻璃半导体器件
glass semi-conductor memory 玻璃半导体存储器
glass semi-conductor read only memory 玻璃半导体只读存储器
glass severing device 玻璃掰板装置；掰板装置
glass severing machine 玻璃掰板机；掰板机
glass sewing thread 玻璃纤维缝纫线
glass shade 玻璃罩
glass shaper 玻璃成型机
glass sheet 玻璃片
glass sheet supporter 玻璃板放置架
glass sheet transfer 玻璃板传送
glass shell 玻璃（外）壳
glass-shell target 玻璃空心球靶
glass shield 玻璃护罩；玻璃护屏
glass shutter 玻璃百叶窗
glass side panel 玻璃侧板
glass sieve plate distillation column 玻璃筛板式精馏塔
glass sight ga(u)ge 玻璃水位计；玻璃观测计；玻璃观测窗
glass sign board 玻璃信号板
glass silk 玻璃丝
glass silk covered round copper wire 玻璃丝包圆铜线
glass silk wall lining 玻璃丝护墙；玻璃丝覆盖墙面
glass silk-wrapped enameled copper wire 玻璃丝漆包铜线
glass silvered 镀银玻璃
glass silvering 玻璃镀银
glass siphon 玻璃虹吸管
glass siphon trap 玻璃虹吸阱
glass size 玻璃尺寸
glass skylight 玻璃天窗
glass skyscraper 玻璃摩天（大）楼
glass skyscraper project 玻璃摩天楼设计方案
glass slab 玻璃调板
glass slab amplifier 玻璃板条条放大器
glass slate 玻璃板；玻璃（板）瓦
glass slate roof 玻璃板屋顶
glass sleeve 玻璃套管
glass sleeving 玻璃套管
glass slicker 玻璃刮刀
glass slide 载物玻片；玻璃载片
glass sliding door 玻璃滑动扯门
glass sock 玻璃套
glass solder 玻璃焊条；玻璃焊（接）料；玻璃焊剂
glass solder seal 玻璃焊接密封
glass solidification 玻璃固化
glass spacer block 玻璃模具定位块
glass spatula 玻璃勺；玻璃刮刀
glass specification 玻璃规格
glass specimen 玻璃试样
glass spectrograph 玻璃（棱镜）摄谱仪
glass sphere 玻璃球
glass spherule 玻璃球粒
glass spinning 玻璃拉丝
glass spinning machine 玻璃拉丝机
glass splicing lathe 玻璃焊接车床
glass splinter 玻璃碎屑；玻璃碎片
glass sponge 硅质海绵
glass spreading rod 玻璃涂布棒
glass stain 玻璃表面着色
glass stainer 玻璃画家
glass staining 玻璃表面着色
glass standard comparator 玻璃标准比色计
glass standoff insulator 高脚玻璃绝缘体

glass staple fibre woven fabric 定长玻璃纤维纱；定长玻璃纤维纺织布
glass staple fibre yarn 玻璃短纤维纱；不连续玻璃纤维线
glass staple yarn 定长玻璃纤维纱
glass state 玻璃（状）态
glass steel reflector 玻璃钢反射器
glass steel tile 玻璃钢瓦
glass steel tube 玻璃钢管
glass stem 玻璃杆
glass stem sealed thermometer 玻璃棒封接式温度计
glass stem thermometer 玻璃（棒式）温度计
glass stirrer 玻璃搅拌器
glass stop 玻璃挡条；玻璃压条；活动玻璃镶边；玻璃嵌条；玻璃窗挡头
glass stopcock 玻璃活塞；玻璃活门
glass stopped bottle 玻璃塞瓶
glass stopper 玻璃塞
glass stoppered bottle 带玻璃塞瓶
glass storage 玻璃品库
glass strand 玻璃（纤维）原丝
glass strand cohesion 玻璃原丝内聚力
glass strand yield 玻璃原丝支数
glass stream 玻璃液流
glass stream cutter 玻璃料股剪刀
glass strip 窗玻璃密封条；玻璃条
glass strip for outlining patterns 描绘出图形花样的玻璃条
glass structure 玻璃结构
glass substitute 玻璃代用品
glass substrate 玻璃基板；玻璃衬底
glass superconductor 玻璃超导体
glass surface 玻璃面
glass swing door 双开玻璃摆动门
glass switch 玻璃开关
glass syringe 玻璃注射器
glass tableware 玻璃餐具
glass tank 玻璃桶；玻璃熔窑；玻璃池窑；玻璃池炉
glass tank block 玻璃池窑大砖
glass tank checker 玻璃池棋盘式蓄热室；玻璃池检验器
glass tank crown 玻璃池窑顶；玻璃池冠
glass tank furnace 玻璃窑炉；玻璃熔炉；玻璃浴炉
glass tank rectifier 玻璃筒整流器；引燃管整流器
glass tape 玻璃带
glass target 玻璃靶
glass technology 玻璃工艺学
glass temperature 玻璃化温度
glass tempering 玻璃回火
glass terrace 玻璃露台
glass textile 玻璃纤维纺织品
glass textolite 层形树脂玻璃布
glass thermistor 玻璃热敏电阻器
glass thermistor thermometer 玻璃热敏电阻温度计
glass thermometer 玻璃管温度计
glass thermometer with etched stem 附有刻度的玻璃温度计
glass thickness determination 玻璃厚度的测定
glass thread 玻璃（纤维）线
glass thyratron 玻璃闸流管
glass tiff 方解石
glass tile 玻璃（面）砖；玻璃（板）瓦
glass tile for glass concrete 透光混凝土；玻璃装饰板材
glass tile roof 玻璃瓦屋顶
glass tinted by impurities 杂质着色的玻璃；非纯色玻璃
glass tissue 玻璃织物；玻璃薄纸；玻璃纤维纸
glass tissue sheet(ing) 玻璃纤维薄板
glass to ceramic seal(ing) 玻璃与陶瓷封接
glass-to-fuel ratio 玻璃燃料比
glass-to-glass seal(ing) 玻璃与玻璃封接
glass-to-metal seal(ing) 玻璃（与）金属封接；玻璃金属密封
glass tongue depressor 玻璃压舌板
glass towel 抹玻璃器皿布
glass tower 玻璃塔
glass tracing table 透写台
glass transformation temperature 玻璃转变温度；玻璃化温度
glass transition 玻璃转化；玻璃化转变
glass transition temperature 玻璃（态）转变温度；玻璃（软）化温度；玻璃化转变温度

glass triangle 玻璃三角
glass tube 氢氟酸测斜玻璃试管;玻璃(电子)管;玻管
glass tube correcting 玻壳校准
glass tube cutter 玻璃管切割机;玻璃管割刀
glass tube cutting-grinding machine 玻璃管切磨机
glass tube fenestration 玻璃管窗户布局
glass tube fuse 玻璃管熔断器;玻璃管熔断丝
glass tube joint 玻璃管接头
glass tube manometer 玻璃管(式)压力计
glass tube mercury-arc rectifier 玻璃管型汞弧整流器
glass tube protector 玻璃管保护套
glass tube separator 隔离玻璃管
glass tube welder 玻璃管的焊接器
glass tubing 玻璃管
glass tubing cutter 玻璃管切割机
glass tubing gauge 玻璃管量规
glass tumbler 大玻璃杯;玻璃滚筒
glass type 玻璃类型
glass-type ceramic coating 玻璃质陶瓷涂层
glass-type coating material 玻璃镀层材料
glass-type tube 玻壳管
glass tire cord 玻璃纤维轮胎带子线
glass uniformity 光泽均匀度
glass vacuum stopcock 玻璃真空活塞
glass vacuum system 玻璃真空系统
glass valve 玻璃阀
glass vault 玻璃拱顶
glass ventilating brick 通风玻璃砖
glass vessel 玻璃容器
glass vial 玻璃试管(测斜用)
glass wadding 玻璃填料;玻璃衬料
glass waffle 玻璃格栅
glass wall 玻璃(挡风)墙;玻璃幕墙
glass wall facing 玻璃墙面;玻璃饰面(墙)
glass wall(ing) 玻璃墙
glass wall panel 玻璃墙板
glass wall system 玻璃墙壁系统
glass wall tile 玻璃墙(面)砖;玻璃墙护板;玻璃壁瓷砖
glass ware 玻璃仪器;玻璃器皿;玻璃制品;玻璃器具
glassware 料器;玻璃制品;玻璃器皿
glassware chest 玻璃器械橱
glassware instrument 玻璃仪器
glass ware washer 玻璃制品洗涤器
glass washing machine 洗杯机
glass waste 玻璃厂废水
glass water ga(u)ge 水位计;玻璃水位计;玻璃水位管;玻璃水位表
glass water waste preventer 玻璃冲水水箱
glass weathering 玻璃发霉
glass web 玻璃毡;玻璃网布
glass wedge 玻璃楔
glass whitening 玻璃发白
glass window 玻璃窗(口)
glass window cell 玻璃窗口试验池
glass windscreen bending furnace 挡风玻璃板的弯曲炉
glass wire 玻璃纤维线;玻璃(丝)线
glass with heavy cords 有粗线道的玻璃
glass with reams 带纹的玻璃
glass with wavy cords 带波的玻璃
glass wool 玻璃纤维;玻璃丝;玻璃绒;玻璃棉
glass wool acoustic(al)board 吸声玻璃棉板
glass wool blanket 玻璃棉毡
glass wool board 玻璃棉纤维板;玻璃棉板
glass wool braided tube 玻璃丝编织管
glass wool ceiling insulation 玻璃棉顶棚保温
glass-wool felt 玻璃丝毡
glass wool filter 玻璃棉过滤器
glass wool insulation 玻璃棉隔热(材料)
glass wool lagging 玻璃棉保温外套
glass wool mat 玻璃纤维(绒)毡;玻璃纤维绒垫
glass wool pipe section 玻璃棉管段
glass wool plate 玻璃棉板
glass wool plug 玻璃纤维塞;玻璃棉塞
glass wool prefilter 玻璃棉预滤器;玻璃毛预滤器
glass wool quilt 玻璃丝毡;玻璃棉毡
glass wool roll 玻璃棉卷材
glass wool rope 玻璃棉绳
glass wool rope machine 玻璃棉绳机
glass wool slab 玻璃棉板;玻璃纤维板

glass wool strip 玻璃绒镶条
glasswork 玻璃制造业;玻璃制品;玻璃工作
glass worker 玻璃工人
glass work(ing) 玻璃加工
glass working machine 玻璃加工机械
glass works 玻璃(工)厂
glass wort 玻璃草(自其灰中可取得碱作玻璃原料)
glass woven fabric 玻璃布
glass wrapped wire 玻璃丝包线
glassy 玻璃状(的);玻璃质的
glassy aggregate 透明集料;玻璃质集料
glassy andesite 安山玻璃
glass yarn 玻璃纤维纱
glass yarn braided sleeving 玻璃纱(线)编织套管
glass yarn dying 玻璃纤维纱染色
glass yarn for wire insulation 电线绝缘用玻璃纱
glass yarn layer 玻璃纱层
glass yarn sleeving 玻璃纱线套管
glassy bond 玻璃黏结
glassy break 玻璃状断口
glassy calm(sea) 蒲福零级浪;海平如镜
glassy carbon 玻璃状炭;玻璃(态)炭;玻璃化炭黑
glassy carbon electrode 炭玻电极;玻璃炭电极
glassy electrolyte 玻璃电解质
glassy feldspar 透长石;玻长石
glassy fibrous texture 玻纤状结构
glassy fracture 玻璃状断口
glassy ice 玻璃状冰
glass yield 玻璃获得率;玻璃成品率;玻璃产量
glassy inclusion 玻璃态夹杂物
glassy lava 玻璃状熔岩
glassy layer 透明层;玻璃层
glassy lustre 玻璃光泽
glassy mass 玻璃状物质
glassy material 透明材料;玻璃状材料;玻璃材料
glassy matrix 玻璃化矩阵
glassy micro phase 玻璃态微观相
glassy millboard 光面纸板
glassy monchiquite 玻质方沸霞煌岩
glassy phase 玻璃相
glassy phosphate 玻璃状磷酸盐
glassy polymer 玻璃状聚合物
glassy porcelain 熔块瓷;玻璃态瓷
glassy rock 玻璃质岩石
glassy semi-conductor 玻璃半导体
glassy silicate host material 玻璃质硅酸盐母体材料
glassy state 玻态;玻璃态
glassy surface 光泽面;如镜面
glassy texture 玻璃状组织
glassy transition 玻璃转化;玻璃态转变
glassy tuff 玻质凝灰岩
glassy yellow 透明黄色(亮黄色)
glauberite 钙芒硝
glauberite rock 钙芒硝岩
Glauber salt 芒硝
glaucochroite 钙锰橄榄石;绿粒橄榄石
glaucodot(e) 钴硫砷铁矿;硫砷钴矿
glaucokerinite 锌铜铝矾
glauconite 海绿石
glauconite cement 海绿石胶结物
glauconite chalk 海绿白垩
glauconite deposit 海绿石矿床
glauconite mud 海绿石泥
glauconite phosphatic rock sub-formation 海绿石磷质岩亚建造
glauconite sand 海绿石砂
glauconitic anhydrock 海绿石硬石膏岩
glauconitic areritic-limestone 海绿石质砂屑灰岩
glauconitic arkose 海绿石长石砂岩
glauconitic bentonite 海绿石斑脱岩
glauconitic chalk 海绿石白垩
glauconitic limestone 海绿石灰岩;海绿灰岩
glauconitic nodular phosphorite 海绿石结核状磷块岩
glauconitic opaline 海绿石蛋白岩
glauconitic-opaline quartz sandstone 海绿石蛋白石质石英砂岩
glauconitic opaline rock 海绿石蛋白石质岩
glauconitic phosphorite 海绿石磷块岩
glauconitic quartz sandstone 海绿石石英砂岩
glauconitic rock 磷质海绿石岩;海绿石质岩
glauconitic sandstone 海绿(石)砂岩
glauconitic shale 海绿石页岩

glauconitic siliceous shale 海绿石硅质页岩
glauconization 海绿石化
glaucophane 蓝闪石
glaucophane-chlorite schist 蓝闪绿泥片岩
glaucophane-lawsonite chlorite schist 蓝闪硬柱绿泥片岩
glaucophane-lawsonite schist facies 蓝闪石硬柱石片岩相
glaucophane-mica schist 蓝闪云母片岩
glaucophane muscovite chlorite schist 蓝闪石白云母绿泥石片岩
glaucophane schist 蓝闪石片岩
glaucophane schist-green schist facies group 蓝片岩—绿片岩相组
glaucophane schist series 蓝闪石片岩系列
glaucophanite 蓝闪岩
glaucoquartzite 海绿石石英岩
glaucous 淡灰绿色的
glaucous ice 淡绿色冰(指污染过的);暗色冰
glaucous water 暗色水
glaueophane 蓝绿石
glaueophane schist 蓝绿片岩
glaueophane schist series 蓝绿片岩系
glaukosphaerite 镍孔雀石
glaur 沉底泥
Glavebel process 平板玻璃水平拉制法
glavis 格拉维斯风
glaze 雨凇(下降时呈液态,着地后就冰冻);釉料;光滑层;涂釉;上光研磨;冻雨
glaze ash 釉灰
glaze ash dealer 釉灰行
glaze-body fit 坯釉适应性
glaze-body interface 坯釉中间层
Glazebrook prism 格雷兹布鲁克棱镜
glaze bubble 釉泡
glaze capacitor 瓷釉电容器
glaze cement 火山灰(质)水泥
glaze-ceramic interface 釉坯中间层
glaze coat 罩光(半透明)涂层;上层釉;罩面清漆;透明面层;沥青光面外层;防水釉面外层
glaze coating 瓷釉涂层
glaze colo(u)r 釉色
glaze covered brick 釉面砖
glaze craze 釉裂;炸釉
glaze crazing 鳞釉;细裂纹釉
glazed aggregate 釉面骨料;光滑面集料;光滑面骨料
glazed and beveled tile 釉面斜切砖
glazed area 四周镶玻璃的区域
glazed block 玻璃砖块
glazed brick 釉(面)砖;琉璃砖;瓷砖
glazed building units 上釉建筑构件
glazed calico 摩擦轧光细布
glazed ceramic 涂釉陶瓷;施釉陶瓷
glazed ceramic mosaic tile 釉面陶瓷锦砖
glazed ceramics 上釉陶瓷
glazed ceramic tile 上釉瓷砖
glazed ceramic wall tile 釉面瓷砖
glazed chintz 摩擦轧光艳丽花布
glazed chintz finish 摩擦轧光整理
glazed clay tile 釉面陶砖;釉面陶瓦
glazed clayware 上釉陶器
glazed clayware drain pipe 釉面陶土排水管
glazed coated book paper 高光泽涂布书籍纸
glazed coated cover paper 高光泽涂布封面纸
glazed coated paper 高光泽涂布纸
glazed coat(ing) 上釉的涂层;高光泽修饰
glazed colo(u)red ceramic tile 彩釉瓷砖
glazed colo(u)red mosaic 彩釉马赛克
glazed colo(u)red pottery 彩釉陶
glazed concrete 磨光混凝土;釉面混凝土;釉面彩色饰面混凝土
glazed cornice 琉璃檐口
glazed corridor 装配玻璃的走廊
glazed cotton 上光棉线
glazed door 窗门;镶玻璃门;玻璃门
glazed earthenware 釉面陶器;上釉陶器
glazed earthenware pipe 上釉陶(土)管
glazed earthenware wall tile 釉面陶瓷墙壁砖
glazed earthware 釉面瓷器
glazed facing tile 琉璃瓦;釉瓷瓦;琉璃面砖;釉面砖
glazed finish 挂釉面;高光泽整理;上釉彩色饰面;上光涂饰剂;喷塑彩色饰面
glazed flat roof 玻璃平屋顶

glazed floor tile 上釉地砖
glazed frame 抛光框格;抛光窗框
glazed frost 雨淞冻;雨淞(下降时呈液态,着地后就冰冻);雨后凝水;凝霜;冻雨;冰雨
glazed frost indicator 雨淞指示仪;雨后凝冰指示仪
glazed greaseproof paper 高光泽防油纸
glazed hollow block 上釉空心砖
glazed ice 雨淞(下降时呈液态,着地后冻成冰)
glazed imitation greaseproof paper 高光泽仿防油纸
glazed imitation parchment 釉光仿羊皮纸
glazed interior tile 室内用釉面砖
glazed joint failure 玻璃状破坏
glazed lookout deck 四面装(配)玻璃的瞭望台
glazed material 上釉的材料
glazed mosaic 玻璃马赛克
glazed observation platform 四面装(配)玻璃的平台
glazed offset paper 高光泽胶版印刷纸
glazed opening 装玻璃的孔穴;装玻璃的洞口
glazed ornament 玻璃花饰
glazed pagoda 琉璃塔
glazed panelled door 玻璃镶板门
glazed paper 蜡光纸;釉(光)纸;高光泽纸
glazed parchment 蜡光仿羊皮纸
glazed partition 玻璃棚板墙篱;玻璃隔墙;玻璃隔断;镶玻璃隔墙
glazed partition wall 玻璃隔墙
glazed patio 玻璃庭院
glazed pig 高硅生铁;脆性生铁
glazed porcelain 施釉陶瓷;上釉瓷
glazed porcelain tile 釉面瓷砖
glazed pot 釉罐;施釉坩埚
glazed pottery 釉陶
glazed print 上光像纸
glazed printing paper 道林纸
glazed quality 玻璃(安装)质量
glazed rain 冻雨;冰雨
glazed resistance 涂釉电阻
glazed roof 嵌玻璃屋面;玻璃屋顶
glazed roof covering 玻璃屋顶覆盖层
glazed roofing tile 玻璃瓦
glazed roof sheathing 玻璃屋顶的望板
glazed roof tile 釉面瓦
glazed sash 玻璃窗框
glazed semiburnt tile 未烧透釉面砖;未烧透釉面瓦
glazed silk 上光丝
glazed slag 玻璃状炉渣
glazed sliding door 玻璃滑动门
glazed stoneware 上釉硬质陶土管;上釉陶器;釉面陶管;上釉火石器;粗釉陶
glazed stoneware pipe 上釉陶管
glazed stoneware tile 釉面陶砖;釉面陶瓦
glazed storm 雨淞暴
glazed structural unit 上釉(结构)构件
glazed terrace 平屋顶晒台;玻璃露台
glazed thread 上光线
glazed tile 釉面瓦;釉瓷瓦;琉璃瓦;瓷砖
glazed tile decorative pattern 琉璃彩画
glazed tile facing 玻璃饰面
glazed tile flooring 釉面陶楼地面;瓷砖楼地面
glazed tile kiln 琉璃窑
glazed tile pipe 釉面陶管
glazed tile roof 琉璃瓦屋顶
glazed timber door 镶玻璃的木门
glazed transparent paper 高光泽透明纸
glazed viewing platform 四面装(配)玻璃的月台【铁】
glazed wallboard 釉面墙板
glazed wall coat(ing) 墙面涂釉层
glazed wall tile 墙面(墙)砖
glazed ware 釉面(粗)陶器;上釉陶瓷器
glazed ware pipe 釉面陶瓷管
glazed window 玻璃窗
glazed wire 漆包线
glazed work 釉面砖圬工
glaze enamel 釉
glaze fault 釉面缺陷
glaze fire 釉烧
glaze firing 釉烧
glaze fit test 坯与釉适应性
glaze flow 釉的流动
glaze for electric(al) porcelain 电瓷釉
glaze formula 釉式

glaze ice 雨淞冰;透明冰
glaze kiln 釉瓷窑;搪瓷窑;上釉窑
glazeless porcelain 反瓷
glazement 平滑的表面涂料;防水釉面材料
glaze of furry appearance 丝毛釉
glaze patching 衬釉
glazer 研磨轮;光泽加工轮;抛光轮;上光轮;上光机;上边机;打光工人
glazer's pliers 装玻璃工安装用钳
glazeraise 磨光混凝土表面层的工艺
glaze removal 除釉
glaze resistor 涂釉电阻器
glaze slip 釉浆
glaze spray booth 喷釉室;喷釉柜
glaze stain 釉用着色剂;釉用色料;釉染色剂
glaze sticking 粘釉
glaze stone 釉石
glaze stress 釉应力
glaze stress tester 釉应力测定仪
glaze suspender 釉浆悬浮剂
glaze tile 釉面砖;釉面瓦;玻璃瓦
glaze viscosity 釉的黏度
glaze wheel 研磨(砂)轮;抛光轮
glaze with metallic lustre 具有金属光泽的釉
glazier 装玻璃工;轧光机;镶玻璃工人;釉工;玻璃(安装)
glazier's chisel 油泥刀;刮腻子刀;油灰刀;玻璃工用的凿子
glazier's diamond 刮玻璃用金刚钻;玻璃(割)刀;玻璃切割刀
glazier's hacking knife 玻工砍刀;玻璃割刀
glazier's hammer 玻璃安装工用锤
glazier's lead 镶嵌玻璃用铅
glazier's pick hammer 玻璃安装工用鹤嘴锤
glazier's pliers 安装玻璃用钳;玻璃钳;玻璃安装工用钳
glazier's point (镶嵌玻璃用的)三角条
glazier's putty 镶玻璃(用)油灰;镶嵌玻璃用的油灰;玻璃油灰;玻璃腻子
glazier's rack 镶嵌玻璃用支架
glazier's sprig 镶玻璃用扁头针;镶玻璃扁头钉
glazier work 装配玻璃工作
glazing 釉状光斑;釉化;装(配)玻璃;罩光;施釉;上釉;上光;玻璃装配
glazing agent 上光剂
glazing angle 镶玻璃用角状压条
glazing bar 玻璃窗棂;窗芯条;窗棂;玻璃格条
glazing bar window 有窗格条的窗子;有玻璃格条的窗子
glazing bead 活动玻璃镶边;玻璃圆嵌条;镶玻璃条;嵌玻璃条;玻璃压条
glazing bedded in washleather 衬垫鹿皮装配玻璃
glazing block 装配玻璃定位块;镶玻璃定位块
glazing booth 施釉柜
glazing brad 玻璃钉;装玻璃用钉
glazing by atomization 喷釉;喷雾法上釉
glazing by dipping 汤浸釉;沉浸法上釉
glazing by dusting 涂粉式上釉
glazing by immersion 汤蘸釉
glazing by insufflation 吹釉
glazing by pouring 浇釉
glazing by rinsing 汤釉;荡釉
glazing by salting 上盐釉
glazing by splashing 泼釉
glazing by spraying 喷釉;喷雾法上釉
glazing calender 轧光机;摩擦轧光机;擦光机
glazing clip 镶玻璃夹;嵌玻璃夹;玻璃夹;固定玻璃用的弹簧夹;玻璃卡子
glazing colo(u)r 玻璃色料;釉色;施釉色料
glazing compound 镶玻璃密封料;嵌缝材料;镶玻璃料团;玻璃油灰;玻璃腻子
glazing dimension 装配玻璃尺度
glazing felt 上毛毯
glazing fillet 玻璃嵌条;玻璃压条
glazing/floor area ratio 窗地面积比
glazing furnace 釉面炉
glazing gasket 玻璃镶嵌条;玻璃密封条;镶玻璃垫条
glazing glass 窗用玻璃
glazing gripper 上釉用的夹子
glazing groove 嵌玻璃槽;玻璃槽
glazing industry 玻璃窗装配行业
glazing kiln 釉烧窑;上釉窑
glazing knife 切玻璃刀

glazing land 牧场
glazing liquid 加色清漆;加彩清漆;釉液
glazing machine 研光机;研光机;磨光机;烘干上光机;施釉机;上光机;打光机
glazing materials 镶玻璃材料;窗玻璃;上釉材料
glazing method 上釉法;上光法
glazing mill 电子管密封玻璃管制造机
glazing mo(u)lding 镶玻璃法;镶玻璃槽口;玻璃压条;玻璃线脚
glazing of diamond 金刚石表面磨光
glazing of food 食品包冰衣
glazing paint 上光涂料;发光漆
glazing pin 固定玻璃用钉
glazing plate 上光板
glazing plate glass 厚玻璃;橱窗玻璃
glazing point 安装玻璃用钉
glazing pottery 上釉陶器
glazing product (上釉产品;上光产品
glazing profile 镶玻璃的型材;上釉的型材
glazing purlin(e) 镶玻璃的檩条;上釉的檩条
glazing putty 镶玻璃油灰;油灰;嵌玻璃油灰;玻璃油灰;玻璃腻子
glazing quality 玻璃安装质量
glazing rabbet 玻璃槽口
glazing rebate 镶玻璃裁口;镶玻璃槽口
glazing reinforced concrete 上釉加筋混凝土
glazing service map 牧业图
glazing shape 装配玻璃的型材;装配玻璃的部件
glazing sheet 上光板
glazing size 玻璃尺寸
glazing spacer block 玻璃框条
glazing spray gun 喷釉枪
glazing sprig 镶玻璃用扁头钉;玻璃钉;无头钉;嵌玻璃针
glazing stop 玻璃压条
glazing tape 玻璃密封条;封玻璃带
glazing technique 镶玻璃技术;镶玻璃方法;上釉技术;上釉方法
glazing temperature 上釉温度
glazing trade 镶玻璃行业
glazing trim 装配玻璃修饰(物)
glazing unit 镶玻璃部件;上釉部件
glazing wheel 研磨轮
glazing without frame 无窗框安装法
glazing with putty 油灰镶玻璃法
glazing work 装配玻璃工作;玻璃安装工程
gleam 微光
Gleamax 格利马克斯电解液
gleaner 割捆机;搜集者;束谷机
Gleason bevel gear cutter 格里森伞齿轮铣刀
Gleason bevel gear generator 格里森伞齿轮滚齿机
Gleason bevel gear shaper 格里森伞齿轮刨齿机
Gleason bevel gear system 格里森伞齿轮系
Gleason cutter 格里森伞齿轮刨刀
Gleason spiral bevel gear cutter 格里森螺旋伞齿轮铣刀
Gleason spiral bevel gear generator 格里森螺旋伞齿轮滚齿机
Gleason straight bevel gear cutter 格里森直伞齿轮刨刀
Gleason tooth 格氏齿
glebe 教堂土地;含矿(地)带;成矿区
glei 潜育层
glei alluvial brown soil 潜育棕色冲积土
gleisoil 潜育土
gleithretter 滑片;滑板;剪理片
gleization 潜育(化)作用
glen 峡谷;平底河谷
Glen's sedimentator 格伦沉淀器
Glen-bulk 格伦变形丝
Glengarriff grit 格伦加里夫粗砂岩
Glenkiln shales 格伦金页岩
glenmuirite 正沸绿岩
Glenn effect 格伦效应
Glenodinium sp. 薄甲藻
Glen Rose limestone 格伦罗斯灰岩
gleptoferron 庚糖酐铁
gley 潜育土;潜育层;土壤潜育层
gleyed soil 潜育化土壤
gleyey process 潜育过程
gley horizon 潜育层
gleying 潜育作用;潜育化
gley meadow soil 潜育化草甸土
gley podzol 潜育(状)灰壤

gley sod-pozolic soil 潜育生草灰化土
gley soil 潜育土
gleysolic soil 潜育土类
Gliclazide pharmaceutical wastewater 格列齐特制药废水
glidant 助流剂
glide 下滑;滑走台;滑移
glide angle 下滑角
glide band 滑移带
glide bedding 滑移层理;滑动层面;滑动层理
glide cataclastite 滑裂岩
glide chute 溜泥槽
glide direction 滑移方向;滑动方向
glide fold 剪褶皱;剪切褶曲
glide lamella 滑移层;滑动夹层
glide line 滑移线
glide mirror(plane) 滑移平面
glide off 滑动
glide path 下滑轨迹;下滑(轨)道;滑翔道
glide-path beacon 下滑道信标
glide-path indicator 降压指示器
glide-path landing system 下滑着陆系统
glide-path localizer 下滑信标
glide-path receiver 滑行着陆接收机
glide-path transmitter 下滑指向标发射机
glide plane 滑移面;滑动面
glide plane of symmetry 滑移对称面;滑动对称面
glider 滑行(快)艇;滑翔机;水上快艇
glide reflection 滑移反射
glide reflection plane 滑动反射面
glide shoe 仿形滑脚
glide slope 下滑坡度;下滑面;下滑角;下滑道
glide slope antenna 下滑道天线(机场)
glide slope beam radio transmitter 下滑道无线电波束发射机(机场)
glide slope radio beam 下滑道无线电波束(机场)
glide slope signal generator 测滑道接收机信号发生器
glide step 滑板托架脚蹬
glide tectonics 滑动构造
glide trajectory 下滑轨迹
glide-wheel 滑轮
glidewheel conveyer 滚道式输送机
gliding 下滑;镏金;滑动的
gliding angle 下滑角
gliding bacteria 滑行细菌;滑移细菌
gliding block 滑块
gliding craft 飞升航空器
gliding flow 滑移流动;滑动流(动)
gliding fracture 韧性断裂;韧性断口
gliding mark 浮游测标
gliding mark scale 滑标尺
gliding mass 滑体
gliding motility 滑行运动
gliding movement 滑走运动
gliding nappe 滑覆体
gliding plane 滑移面;滑翔面;滑动面
gliding property 平滑性
gliding slab 滑片
gliding surface 滑动面
gliding system 滑动系统
gliding tectonics 滑动构造
gliding texture 滑动构造
gliding twin 滑移双晶
gliding window 滑移式窗
Gliever bearing alloy 铅基轴承合金
Gliksten 格利克斯木纤维墙板
glim 灯笼;灯光
glime 雨雾凇
glim lamp 阴极放电管;辉光放电灯
glimmer 云母;薄光
glimmer ice 闪光冰
glimmering lustre 微闪光泽
glimmerite 云母岩
glimmerton 云母土;伊利石
glim relay tube 闪光继电管
glint 回波起伏
glint error 角闪烁误差
gliosa 胶灰质
gliosome 胶质粒
glissade 滑坡
glissette 推成曲线
glist 云母
glisten 闪烁

glister 磨砂玻璃装饰
glitch 一闪信号;自转突变;假信号
Glitsch distiilation tray 格利希蒸馏塔板
Glitsch trays 格里茨塔盘
glitter 光泽;闪耀(海面对点光源的反射);闪光剂
glitter finish 光泽罩面;闪烁罩面
g-load 过荷
global 全局的;全球的
global addition 全局加法
global address 全局地址
global address vector 全局地址向量
global air pollution 全球(性)大气污染
global analysis 整体分析;大范围分析
global approximation 整体近似
global area 公用区;全局区域
global arrangements 全球性安排
global asymptotic stability 整体渐近稳定;全局渐近稳定性
global atmosphere 全球大气
global atmosphere watch 全球大气监视网
global atmospheric(al) research program(me) 全球大气研究计划
global bandwidth 全球带宽
global beam 全球波束
global beam antenna 全球覆盖天线
global behavio(u)r 全局行为
global binding 全局结合
global bond 全部债券
global change 全球变化;全局变化
global change and biodiversity conservation 全球变化及生物多样性环境保护
global circulation 全球环流
global circulation model 全球环流模式
global climate 全球(性)气候
global climate change 全球气候变化
global climatologic change 全球气候改变;全球气候变化
global clock 全局时钟
global cloud fields 全球云区
global cluster 综合性群集
global clustered beams 综合性群梁
global code 全局代码
global coefficient of energy scattering 能量散射的球面系数
global command 全局指令
global common 全球共有物
global common subexpression 全局共用子表达式
global communication coverage 全球通信范围
global communication satellite system 全球通信卫星系统
global communication system 全球(卫星)通信系统
global concept 整体概念
global contaminant 全球性污染物
global contamination 全球性污染
global control 全局控制
global control bus 全局控制总线
global control line 全局控制线
global control section 全局控制段
global convention 全球性公约
global convergence 整体收敛;全局收敛
global convergence theorem 整体收敛性定理
global coordinate 大坐标
global coordinate system 整体坐标系
global copy operation 全局复写操作
global crisis 全球性危机
global criterion 全局性准则;球状准则
global cycle 全球循环
global data 全局数据
global data area 整体数据区
global data base 全局数据库
global data flow analysis 全局数据流分析
global data-processing system 全球资料处理系统
global deformation pattern 全球形变模型
global degree 整体次
global dictionary 全局目录
global differential geometry 整体微分几何
global dimension 整体维数
global dispersion 全球性扩散
global display address 全程区头向量地址
global distribution of elements 元素全球分布
global ecology 全球生态学
global effect 整体效应;全球性影响
global effect of pollution 污染的全球影响

global emissions 全球排放量
global energy and water cycle experiment 全球能量和水循环实验
global environment 全球环境
global environmental anomaly 全球性环境异常
global environmental monitoring system 全球环境监测系统
global environmental monitoring system for water 全球水环境监测系统
global environmental problem 全球(性)环境问题
global environmental quality assessment 全球环境质量影响评估
global equilibrium 全球均衡
global error 整体误差;全面误差;全局误差
global estimated value of recoverable reserves 可采储量的整体估计值
global estimation 整体估计
global estimation of recoverable reserves 可采储量的整体估计
global estimation value 整体估计值
global estimation variance 整体估计方差
global eustatic cycle 全球性海面升降旋回
global eustatic movement 全球性海面变化
global facility 公用部件
global fall-out 全球性散落
global floral realm 全球植物地理区系
global flow analysis 全局流分析
global fossil bulge 全球的残留膨胀
global function 全功能
global gravity anomaly 全球重力异常
global grid of fault 全球性断层网
global information system 全球情报系统
global instability 全局不稳定性
global investigation of pollution in the marine environment 全球海洋环境污染调查
globalism 全球性干涉政策
global isohyetal line 全球雨水线
globality 整体性
globalization 全球化
global knowledge 全局知识
global liquidity 全球流通手段
global location 全程单元
global lock 全局锁
global Loran navigation chart 全球罗兰导航图
global-ly-addressed header 全局编址标头
globally optimal solution 全局最优解
globally symmetric(al) Riemannian space 整体黎曼对称空间
global main process 全局主过程
global map 全局图
global mapping 整体映射
global maritime distress and safety system 全球海上遇险安全系统
global market 全球市场
global matrix 总体矩阵
global maximum 总体极大;全局极大值
global mean heat flow 全球平均热流
global mean mantle flow 全球平均地幔热流
global metallogenetic belt 全球性巨成矿带
global migration 地表迁移;表地迁移
global minimum 总体极小;全域极小;全局最小;全局极小值
global minimum point 全局极小点
global mode 全局方式
global-mode coverage 全球方式覆盖
global monitoring 全球监测
global monitoring for climate change program(me) 气候变化全球监测方案
global monitoring of air pollution 空气污染全球监测;全球性空气污染监测
global monitoring system 全球监测系统
global multiply operation 全局乘法运算
global name 全局名称
global navigation and planning chart 全球导航与计划图
global navigation chart 全球航空导航图;全球航线图
global navigation system 全球导航系统
global negotiations 全球谈判
global net(work) 全局网络;全球网
global network of research station 全球研究站网
global object 全局目标
global observation system 全球观测系统
global observing system 全球观测系统

global oil production 世界石油产量
global optimization 整体(最)优化;整体最佳化;整体优化;全局优化
global optimization theory 全局最优化理论
global optimum 总体最优值;全局最优(值)
global orbiting navigation satellite system 全球轨道导航卫星系统
global outlook of ecology 生态学世界观
global ozone distribution 全球臭氧分布
global ozone layer 全球臭氧层
global ozone observation system 全球臭氧观测系统
global ozone transport 臭氧的全球输送
global parameter 全局参数
global parameter buffer 全局参数缓冲器
global pesticide contamination 全球农药污染
global planning 全球规划
global pollution 全球性污染
global positioning satellite 全球定位卫星
global positioning system 导航星系;全球(卫星)定位系统
global positioning system receiver 全球卫星定位系统接收机
global precipitation climatology center 全球降水气候学中心
global precipitation climatology project 全球降水气候学计划
global processor 全局处理机;全局处理程序
global program(me) control 全局程序控制
global program(me) for natural reduction 减少自然灾害的全球方案
global property 整体性质
global protection 全球保护
global quality 全球质量
global quota 绝对配额;全球配额;全定额
global radiation 总辐射;环球辐射;全球辐射
global reactor 球形反应器
global real-time repeater 环球实时转发器
global receive antenna 全球波束接收天线
global recession 全球(性)经济衰退
global reference 全局引用;全程引用
global reference frame 整体参考系统
global register optimization 全局寄存器优化
global regression period 全球海退期
global regularization 全局正规化
global relative fall of sea level 全球性海面相对下降
global relative rise of sea level 全球性海面相对上升
global relative stillstand of sea level 全球性海面相对稳定
global representation 全球代理
global rescue alarm net 全球救助警报网
global reservoir system 全球热储系
global resource information database 全球资源信息数据库
global response to climate change 对气候变化的全球性回应
global restriction 全局限制;全局约束
global rifting system 全球裂谷系
global rigidity 总刚度
global routing table 全局路径选择表
global safety factor 总安全系数
global satellite system 全球卫星系统
global satellite traffic 全球卫星通信业务
global scale 全球规模;全球构造尺度
global scale of tectonic movement 构造运动全球性
global scale of transgressions and regressions 全球性海水进退
global scene 地球实况
global sea level observing system 全球海平面观测系统
global search 全局搜索;全局检索;全局查找;全程搜索
global search and replace 总体查找与替代
global sea surface temperature project 全球海面温度计划
global section 全局段;整体截影
global section data base 全局段数据基
global seismic activity 全球地震活动性;全球地表活动性
global seismicity map 全球地震活动图
global seismic monitoring 全球地震监视;全球地震监测
global selection 整体选择
global semaphore 公用信号(量)
global sensitivity 全局灵敏度
global service 全局服务
global settlement 统一结算
global shared resource 全局共享资源
global stability 全局稳定性;大范围稳定
global stack top 全局栈顶
global stack top location 全局栈顶单元
global step multiplication 全局步进乘法
global stiffness 整体刚度
global storage 全局存储(器)
global stratigraphy 全球地层学
global structure 总体结构
global subtraction 全局减法
global subtraction command 全局减法命令
global sulfur budget 全球硫收支
global sulphur budget 全球硫收支
global sum 总计
global surveillance station 全球观察站;全球跟踪站
global surveillance system 全球监视系统
global switch 整体开关
global symbol 全局符号
global symbol table 全局符号符号表
global system 全球系统;全球税制
global system for mobile communication 全球移动通信系统
global tectonic active belt 环球性构造活动带
global tectonic regime 全球构造域
global tectonics 全球构造(学);全球地质构造
global telecommunications system 全球电信系统
global temperature 全球性温度
global tendering 全球招标
global test 全局测试
global theory 整体理论;全局整体理论
global theory of curves 整体曲线论
global tide hypothesis 世界潮汐假说
global time synchronization 全球时间同步
global time synchronization system 全球时间同步系统
global tracking network 全球跟踪网
global trade 全球贸易
global traffic 全球通信业务
global transfer command 全局传送命令
global transformation 全局变换
global transgression period 全球海浸期
global transmit antenna 全球波束发射天线
global transport 全球运送
global trunk quality 全球干线通信质量
global value 全程值
global variable 总变量;全局变量;全程变量
global variable reference 全局变量引用;全程变量引用
global variable symbol 全局变量符(号);全程变量符
global vector table 全程向量表
global vibration test 全系统振动试验
global view 全局观点
global virtual protocol 全局虚拟协议
global warming potential 全球升温潜能值
global warning 全球变暖
global water budget 全球水(量)平衡
global water distribution 环球水分布
global weather 全球天气
global weather experiment 全球天气实验
global weather reconnaissance 全球天气搜索;全球天气勘察
global weather test 全球天气试验
Globar 格罗巴碳化硅电阻加热元件
Globar heating element 格罗巴硅碳棒加热元件
globate 球形的
globe 圆球灯罩;球状物;球形物;球体;地球仪
globe bearing 球面轴承
globe body 球体
globe box 密闭操作箱
globe buoy 球状浮标;球形浮标;球顶(杆)状浮标;带球杆形浮标
globe calliper 球径规
globe cam 球形凸轮
globe cartography 地球仪制图学
globe case 球形机壳
globe cased turbine 球壳式涡轮机;球壳式水轮机;灯泡式水轮机
globe chair 球形椅
globe cock 直通开关;球(形)旋塞;球形塞门;球形阀(门);球阀
globecom 全球通信系统
globefish 河豚
globefish toxin 河豚毒
globe gore 地球仪贴面条带
globe holder 球形卡座;球形灯座;球形灯罩;灯座
globe housing 球形外壳
globe insulator 拉紧绝缘子;环形绝缘子;球形绝缘子
globe joint 球窝连接;球窝接头;球窝接口;球窝接合;球节(点);球铰;球关节
globe journal 球形轴颈
globe lamp 球形灯
globe lantern 球(形)灯
globe lathe 球面车床
globe lens 球透镜
globe lighting 球罩灯照明
globe lightning 球形闪电
globe lining 球形衬
Globeloy 格洛拜洛伊耐热铸铁
globe making 地球仪制作
globe mill 球磨机
globe of diffusion 漫射球
globe photometer 球形光度计;球式光度计
globe pipe 球形管
globe pliers 球钳
globe probing method 球形探针法
globe retaining ring 球形扣环
globe rivet head 球形铆钉头
globe-roof 圆顶;球形顶
globe-roof tank 球顶储罐
globe rotary bleacher 球形旋转漂白器
globe-shaped tree 球形树
globe stop check valve 球形截止止回阀
globe stop valve 球形截流阀
globe strain insulator 球形耐拉绝缘子
globe T 球形三通
globe tap 球形水龙头;浴缸水龙头
globe tee 球形三通;球形丁字管节
globe thermometer 球形温度计;球面辐射测温计
globe tube 球形管
globe type luminescence 球形发光
globe type silent check valve 球形消声防逆阀
globe valve 球形阀(门);球(心)阀
globewide mixed organization 全球性混合组织
globigerina 球房虫;[复]globigerinae
globigerina ooze 远洋沉积物;海底软泥;抱球虫软泥
globoid 拟球体;球状体;球状的;球样的;球形体
globoidal support 弧面支座
globoidal worm 球面蜗杆
globoidal worm toothing 球面蜗杆啮合
globoid bodies 球样体
globoid cam 球形凸轮
globoid gearing 曲面轮传动
globoid worm 球形蜗杆
globoid worm gear 球形蜗轮蜗杆;球面蜗轮;球面蜗杆传动
globoid worm gearing 球面蜗轮传动装置
globose 球形的
globose nucleus 球状核
globosity 球状;球形
globular 小球状的;球状的;球形的;滴状的
globular arc 球形弧
globular bulb 球形灯泡
globular calcified lamellar bodies 球形钙化层状小体
globular carbide 球状碳化物
globular chart 球面投影地图
globular discharge 球形放电;球形放电
globular embryo 球形胚
globular flower 球状花
globular graphite 球状石墨
globular interface 球形界面
globularity 球状
globular joint(ing) 球状节理
globular light 球状灯
globular micelle 球状胶束
globular oxide inclusion 球状氧化物夹杂
globular polymer 球形聚合物
globular powder 球状粉末;球粒剂
globular projection 全球投影;球状投影;球形投影;球面投影

globular sailing 大圈航法
globular shape 球形
globular stage 球形期
globular structure 球状构造
globular texture 球状结构
globular transfer 大滴过渡；粗滴过渡
globulate 成球形
globulation 成球作用
globule 小球；熔滴；球剂；球滴；水珠
globule method of arcing 球形电弧法
globule of oil 油球
globulite 球趋晶；球雏晶
globulitic limestone 球状灰岩
globulus 球剂
globurizing 球化（退火）
glocken cell 钟式电解池
glockerite 纤水绿矾
gloia cement 高度水硬性石灰
glomerate 成球形物
glomeration 聚集成球；集块；球形物；团聚
glomeroclastic 聚合碎屑状
glomerocryst 聚晶
glomerogranulitic texture 聚合微粒结构
glomerophyric 聚合斑状
glomerular filtration 小球过滤
glomerular zone 球状带
glomerulose 团伞花序的
glomerulotubular balance 球管平衡
gloom 阴暗（天空）；干燥炉
gloomy weather 阴天；阴晦天气
GLORIA survey 大型曳航式远程旁测声呐测量
glory 彩光环
glory hole 引束孔道；落碴孔；露天矿下放溜井；露天矿坑；露天放矿漏斗；漏斗形溢洪道；看火孔；船尾中甲板的船员室计储〔贮〕藏室；杂物室；杂物柜；巨洞；炉口；大型露天矿
glory-hole method of quarrying 露天（采）矿坑采石法
glory-hole mining 露天漏斗采矿法；露天溜井采矿法
glory-hole spillway 漏斗式溢洪道；竖井式溢洪道；喇叭孔溢洪道
glory hole tunnel system 溜井平洞系统
glory scattering 辉光散射
gloss 光泽（度）；光亮度；光彩；浅饰；棒状氧化铁抛光膏
gloss agent 光泽剂；光亮剂
glossal canal 舌槽
glossary 名词汇编；汇总表；词汇（表）；词典
glossary of vessel 船舶词汇
gloss clear plastic coating 透明有光塑料涂料
gloss clear varnish 光泽清漆；光泽凡立水
gloss coat 光泽涂层；光泽膜；光泽涂层
gloss colo(u)r 有光色料；光亮色（彩）
gloss developing 显光性
gloss distinctness of image 鲜映性；投影光泽
gloss electroplating 光亮电镀
gloss emulsion 有光乳胶漆；有光乳化漆
gloss enamel 高光泽瓷漆；有光瓷漆；光亮釉
gloss finish 光泽整理；光泽罩面；光泽度；出亮
gloss galvanization 光亮镀铬
gloss haze 光泽变（混）浊；雾光
glossimeter 单向反射率计
gloss index 光泽度
glossiness 有光泽；研光度；光泽性；光泽度；光滑
glossiness of glaze 釉的光泽
glossiness test 光泽度试验
glossing 刨亮；精抛光
glossing agent 增光剂
glossing up 产生光泽
gloss ink 有光油墨
glossmeter 光泽计；单向反射率计
gloss oil 光（泽）油；光（滑）油；钙脂清漆；亮油；石灰松香清漆
glossology 命名学
glossopetra 舌形石
gloss paint 有光涂料；有光漆；光泽涂料；光泽漆
gloss paint finish 有光饰片
gloss point 釉面出现光泽的温度
glossproof 耐光泽的；不消光的
gloss rating 光泽等级
gloss reducer 消光剂
gloss reduction 光泽减少
gloss retention 光泽保持性；保光性

gloss shitting 光泽不匀；发花
gloss type of finish 光泽罩面；光泽出面
gloss varnish 有光清漆；光泽清漆；上光漆
gloss varnish for paper 纸张罩光漆
gloss ware 釉皿
gloss white 有光白；光泽白；铝钡白
glossy 研光的；整洁的；光泽的；光滑的；大光地纸
glossy black bituminous coal 黑色亮烟煤
glossy finish(ing) 最后抛光；光泽罩面
glossy glaze 光泽釉
glossy glazing 有光泽釉料；涂光泽瓷漆
glossy ink 光泽油墨；亮光漆
glossy paper 光泽纸；光面纸；蜡光纸
glossy photo 大光像纸相
glossy picture 大像纸像片
glossy print 光面照片；大光相纸相片
glossy privet 女贞【道】
glossy surface 有光泽表面；光泽面
glossy up 显现光泽；产生光泽
glossy varnish 有光泽清漆
glost 釉
glost burn 釉烧
glost fire 釉烧用火
glost firing 釉烧
glost kiln 釉（烧）窑
glost ware 上釉制品
glove anesthesia 手套式感觉缺失
glove box 干燥箱；放射物操作箱；手套箱；手套式操作箱（真空设备用）
glove-box door latch 手套箱盖锁扣
glove-box lamp 手套箱照明灯
glove box shield 手套箱屏
glove-box train 手套箱线路；手套箱串列
glove compartment 手套箱；小型工具箱；工具袋；手工具袋
glove compartment cover 手工具箱盖
glove flannel 手套棉法兰绒
glove knitting machine 手套机
glove leather 手套革
glove liner 手套衬里
glove machine 手套机
glove port 手套机
glove presser 手套压烫机
Glover tower 格洛弗塔；喷淋式气体洗涤塔
gloves 手套
gloves for protection against accident 防事故手套
gloves-leather 皮手套
glove sponge 劣质石棉
glove yarn 手套纱线
gloving 手套制作
glow 灼热；辉光
glow bulb 辉光灯炮
glow cathode rectifier 辉光阴极整流器
glow corona 辉光电晕
glow curve 辉光曲线
glow-discharge 辉光放电
glow-discharge amplifier tube 辉光放电放大管
glow-discharge anemometer 辉光放电风速计
glow-discharge cathode 辉光放电阴极
glow-discharge cleaning 辉光放电清洗
glow-discharge cold-cathode tube 辉光放电管
glow-discharge detector 辉光放电控制器；辉光放电检测器
glow-discharge display panel 辉光放电显示板
glow-discharge electron gun 辉光放电电子枪
glow-discharge lamp 辉光放电灯；辉光灯
glow-discharge lubricant 辉光放电所得润滑剂
glow-discharge manometer 辉光放电压力计
glow-discharge method 辉光放电法
glow-discharge microphone 辉光放电传声器
glow-discharge nitriding 辉光放电氮化
glow-discharge plasma 辉光放电等离子体
glow-discharge positional indicator 辉光放电位置指示器
glow-discharge potential 辉光放电电位
glow-discharge rectifier 辉光放电整流器
glow-discharge stabilizer 辉光放电稳定管
glow-discharge tube 辉光放电管
glow-discharge voltage 辉光放电电压
glow-discharge voltage regulator 辉光放电调压管
glow-discharge voltage regulator tube 辉光调压放电管
glowed-in mud 流入淤泥
glower 炽热体；炽灯丝；白炽体；白炽灯丝

glow-gap divider 放电管分压器
glow heater 辉光加热器；辉光灯丝
glowing avalanche 灼热崩落
glowing cathode 焰热阴极；辉光阴极
glowing cloud 炽热火山云
glowing colo(u)r 光亮色（彩）
glowing combustion 白热燃烧
glowing filament 灼热丝
glowing furnace 淬火炉
glowing heat 炽热
glowing red 红热
glow ion nitrogen furnace 辉光离子氮化炉
glow lamp 辉光放电管；录影灯；录像管；辉光灯
glow modulator tube 辉光调制管
glow numerating tube 辉光数码管
glow plasma 辉光等离子体
glow plug 电热塞
glow potential 辉光电位
glow priming hole 引辉孔
glow starter 荧光灯启动器；辉光启动器（日光灯）；日光灯启动器
glow switch 引燃开关；辉光（放电）开关
glow-switch starter 辉光启动器；辉光点燃器
glow tail 犁柄
glow thyratron 辉光闸流管
glow transfer indicator 辉光转移指示灯
glow tube 辉光管
glow-tube oscillator 辉光放电管振荡器
glow-tube oscilloscope 辉光放电管示波器
glow-watch 夜光表
glow wire 热灯丝
gloxinia 大岩桐属
glucic acid 丙烯醇酸
glucine 羟磷钙铍石
glucinum ethyl 二乙基铍
gluconolactone 葡萄糖酸内酯
glue 胶液；胶水；胶合剂
glue applicator 上胶（机）器
glue binding medium 胶黏剂
glue bleed through 泛胶的；渗胶的
glue block 黏结木块；胶黏合；胶合板；角木块
glue boiler 熔胶锅
glue bond 胶黏结；胶（黏）合
glue cement 胶质水泥；胶态水泥
glue coating 黏结涂层
glue colo(u)r 胶质颜料
glue connection 胶黏结
glued 胶合的
glued adhesion 胶着接合
glued and blocked joint 分段胶接
glued board 胶合板；层合板
glued boarding 胶合板
glued build-up 胶合结构
glued built-up members 胶合的组合构件
glued built-up sections 胶合的组合截面
glued circle pressure tube 胶圈压力管
glued connection 胶接（合）；胶连接
glued construction 胶合结构
glued floor system 胶合楼盖体系；胶合地板系统；胶黏地板做法
glued insulated joint 胶结绝缘接头；胶接绝缘接头
glued joint 胶合接头；胶结（接头）；胶接接头；胶接（接）合
glued-jointed two-piece panel 胶合两块的组合件；胶合两块的组合板
glued joint of steel reinforcement 钢筋黏胶连接
glued laminate 胶合层压板
glued-laminated arch 胶合叠板拱
glued-laminated beam 胶合板梁
glued-laminated board 胶合层板
glued-laminated construction 胶合叠板结构；胶合板层结构
glued-laminated girder 胶合叠梁
glued-laminated structural member 胶合层积构件
glued-laminated timber 胶合层积木；胶合板
glued-laminated timber arch 胶合叠层板拱
glued-laminated timber construction 胶合叠板结构
glued-laminated timber lattice girder 胶合板格构大梁
glued-laminated timber portal frame 胶合板龙门构架
glued-laminated timber rafter 胶合层压木椽；胶合板木椽

glued-laminated timber rigid frame 胶合木刚架
glued-laminated wood 胶合板(层积木);胶合层积材;胶合叠层木料
glued-laminated wood arched girder 胶合板拱形大梁
glued-laminated wood for structural members 结构用胶合木构件
glued laminates 胶合层压制品
glued line 胶黏接头;胶缝
glued lumber 胶合木材
glued method 黏结剂法
glued-on 胶黏住;胶上
glued plywood system 胶合板系统
glued roof truss 胶合屋顶桁架
glued sectional panel 胶合的组合板
glued slab flooring 胶合楼面;胶板楼板;胶板地面
glued structure 胶合结构
glued table 接砧台
glued timber 胶合木材
glued timber beam 胶合木梁
glued timber bridge 胶合木桥
glued timber construction 胶合木结构
glued together 胶合在一起
glued two-piece panel 胶合两块的组合件;胶合两块的组合板;胶合板组合件
glued wood 三合板;胶合(木)板
glued wood construction 胶合木建筑
glue ear 胶耳
glue-etched glass 冰花玻璃
glue-etching 胶蚀
glue fixing method 黏结固结法
glue for cold pressing 冷压黏胶
glue for tile 瓷砖胶
glue grinder 碎胶器
glue gun 黏合剂用喷枪
glue heater 胶加热器
glue in film form 胶水纸;薄膜胶
glu(e)ing material 胶合材料
glue joint 胶接(接头);胶接榫;胶接合;胶合(接头)
glue kettle 熬胶锅;脱胶罐(骨灰窑生产用)
glue-laminated 胶合(板);胶接
glue-laminated beam 胶合叠层梁;胶合层积板梁
glue-laminated construction 层胶合构造
glue-laminated structural lumber 胶合层积结构木料;胶合的薄板状结构木料
glue-laminated timber 多层胶合木(料)
glue-laminated wood 多层胶合木(料);胶合板
glue laminates 胶合(板);薄板
glue line 胶层;胶合线;胶缝
glue line heating 胶层加热
glue manufacture 胶质制造
glue medium 胶黏剂
glue mixer 黏结剂搅拌机;调胶机;拌胶器
glue nailed 钉接胶合
glue of tortoise plastron 龟板胶
glue pot 胶锅;胶桶;熬胶锅
glue powder 牛皮胶粉
glue preparation 胶结剂调制
glue press (木工的)胶夹;胶合夹
glue putty 胶质油灰
gluer 涂胶器
glue refuse 胶渣
glue resin 胶黏树脂
glue roller 胶辊
glue shear-strength 胶合剪切强度
glue-size 稀析胶(油漆用)
glue sizing 木器涂胶层
glue solution 胶液
glue spots 胶斑
glue spread 涂胶(量)
glue spreader 黏结剂摊铺机;涂胶器;涂胶机;涂胶辊
glue spreading 涂胶
glue spreading roller 涂胶滚子
glue-up 胶合
glue wastewater 炼胶废水
glue water 胶水
glue water paint 水胶涂料
Glug 格拉格
gluing 黏合;胶黏;胶合;上胶
gluing bed 胶合工作台
gluing time 胶合时间
gluino 超胶子
glulam 胶合板;胶合层积材
glulam beam 胶合板梁;胶合叠层梁
glulam column 多层胶合(木)柱
glulam timber frame 多层胶合木结构
glulam timber lattice girder 多层胶合木格构大梁
glulam timber portal frame 多层胶合木门框
glulam timber rafter 多层胶合木椽子
glulam timber raking strut 多层胶合木斜撑
glulam timber shell roof 多层胶合木薄壳屋顶
glulam timber truss(ed girder) 多层胶合木桁架梁;多层胶合板构架梁
glulam wood 多层胶合木
glulam wood arched girder 多层胶合木拱梁
glunzpech 辉沥青
gluon 胶子
glushinskite 草酸镁石
glut 楔;镶块;填缝小砖;封口砖;帆索环
glutaconaldehyde 戊烯二醛
glutamic acid 谷氨酸;氨基戊二酸
glut arch 瓶式陶瓷窑大口下面的拱
glutenous 黏的;胶质的
glutin 明胶朊;明胶蛋白
glutinousness 黏滞度
glutol 明胶甲醛
glutoscope 凝集检查镜
glut the market 市场过剩;市场存货过多;使存货过剩
glutting 镶块
glyceraldehyde 甘油醛
glyceride 甘油酯
glyceride transesterification 甘油(酯)酯交换
glycerin(e) 甘油;丙三醇
glycerin(e) attack 甘油侵蚀
glycerin(e) cement 甘油型胶泥
glycerin(e) dibromohydrin 二溴甘油
glycerin(e) dichlorohydrin 二氯甘油
glycerin(e) ester 酯胶;甘油酯
glycerin(e) kaolin slurry 甘油高岭土浆
glycerin(e) mastic 甘油油灰;甘油泥
glycerin(e) monofatty ester 脂肪酸单甘油酯;甘油单脂肪酸酯
glycerin(e) monostearate 单硬脂酸甘油酯
glycerinum zinci oxide 氧化锌甘油(剂)
glycerium phenolis 酚甘油
glycerm tri-acetate 甘油三乙酸酯
glyceroborate 硼酸甘油酯
glycerol 甘油;丙三醇
glycerol acetate 乙酸甘油酯
glycerol epoxy resin 丙三醇环氧树脂
glycerol ester 甘油酯
glycerol esterification 甘油酯化(作用)
glycerol monoacetate 单乙酸甘油酯
glycerol monobutyralte 丁酸甘油酯
glycerol monostearate 单硬脂酸甘油酯
glycerol-retention test 甘油保量试验
glycerol trinitrate 硝化甘油;三硝酸甘油
glycerolysis 甘油醇解
glyceryl alcohol 丙三醇
glyceryl diacetate 甘油双乙酸酯
glyceryl ester 甘油酯;丙三酯
glyceryl laurate ester 月桂酸甘油酯
glyceryl monooleate 单油酸甘油酯
glyceryl monophthalimide 甘油基—邻苯二甲酰亚胺
glyceryl phthalate 邻苯二甲酸甘油酯
glyceryl phthalate resin 邻苯二甲酸甘油树脂
glyceryl stearate 甘油硬脂酸酯
glyceryl triacetate 甘油三乙酸酯
glyceryl trinitrate 硝基甘油
glyceryl tristearate 三硬脂酸甘油酯
glycidamide 环氧丙酰胺
glycide 环氧丙醇;缩三甘油
glycidol 环氧丙醇;缩三甘油
glycidyl 缩水甘油基
glycidyl acrylate 丙烯酸缩水甘油酯
glycidyl allyl ether 烯丙基缩水甘油醚;缩水甘油烯丙醚
glycidyl ether bisphenoll-A epoxy 缩水甘油醚双酚A 环氧树脂
glycidyl ether novolac 缩水甘油醚树脂
glycidyl ether resin 缩水甘油醚树脂
glycidyl ethyl ether 乙基缩水甘油醚;缩水甘油乙醚
glycidyl methacrylate 甲基丙烯酸缩水甘油酯
glycidyl phenyl ether 缩水甘油苯醚;苯基缩水甘油醚
glycine-o-cresol red 氨基乙酸邻甲酚红
Glyco 哥里科合金
glycol 甘醇;二醇
glycoladehyde 羟乙醛
glycol adsorption 乙二醇吸附
glycol adsorption method 干油吸附法
glycolaldehyde 乙醇醛;羟乙醛
glycolaldehyde dimer 乙醇醛二聚物
glycolate 乙二醇盐
glycol diacetate 乙二醇二乙酸酯;乙二醇二醋酸酯
glycol dilaurate 乙二醇二月桂酸酯
glycol ester 乙二醇酯
glycol ether 乙二醇醚
glycol ether acetate 乙二醇醚乙酸酉目
glycol ether ester 乙二醇醚酯(类)
glycol ethylidene-acetal 乙二醇缩乙醛
glycoleucine 己氨酸
glycol isopropyl ether 乙二醇异丙醚
glycollate 羟乙酸酯
glycollic acid 乙醇酸
glycol maleate 乙二醇顺丁烯二酸酯
glycol monoacetate 乙二醇一乙酸酯
glycol monoethyl ether 乙二醇一乙醚
glycol nitrate 乙二醇硝酸酯
glycol phthalate 邻苯二甲酸乙二醇酯
glycophyte 甜土植物;淡土植物
glyoxal 乙二醛
glyoxalic acid 乙醛酸
glyoxylate cycle 乙醛酸循环
glyoxylic acid 水合乙醛酸
glyph 束腰竖沟;竖面浅槽饰;雕像
glyphograph 电刻版
glyphography 电刻术
glyphosate 草甘膦
glyphosate-producing wastewater 草甘膦废水
glyphosine 草甘二膦
glyptal 真空黑漆
glyptal resin 松香;甘酞树脂
glyptic 雕刻的
glyptics 雕刻术
glyptograph (宝石上雕刻的)花饰
glyptography 宝石雕刻术
glyptolith 风棱石;风刻石
gmelinite 钠菱沸石
G metal 铜锡锌合金
g-meter 加速度计
Gm-meter 跨导计;电子管互导测量仪;电子管电导测量仪
gnarl 木节;木瘤
gnarled 多疤的;多节瘤的
gnarly 多疤的;多节瘤的
gnathite 颚形附器
gnathobase 颚基
gnathodynamics 咬合力学
gnathodynamometer 咬合力计
gnathostegite 颚板
gnawed 缺刻状的
gnawing beetle 谷盗
gneiessosity 片麻理
gneiss 片麻岩
gneiss dome 片麻岩弯窿
gneiss granite 片麻花岗岩
gneissic 片麻岩的
gneissic granite 片麻(状)花岗岩
gneissic regions 片麻岩区
gneissic structure 片麻结构
gneissic texture 片麻状结构
gneissification 片麻岩化
gneiss-mica schist 片麻状云母片岩
gneissoid 片麻状的;片麻岩的;似片麻岩状
gneissose 片麻岩的
gneissose granite 片麻状花岗岩
gneissose texture 片麻状结构
gneissosity 片麻状构造;片麻(状节)理
gneiss period 片麻岩期
gneiss system 片麻岩系
gnessic 片麻状的
gnessic structure 片麻状构造
gnessose 片麻状的
gnomon 指时针(日晷);指针针;晷表;日晷指针;日晷(标竿);太阳高度指示器
gnomonic chart 球心投影海图
gnomonic map 球心投影地图
gnomonic projection 心射面切投影;心射极平投

影;极平投影;日晷投影;球心投影
gnomonic ruler 心射投影尺
gnomonics 日晷原理
gnomonic scale 心射投影尺
gnomon shadow template 晷
gnosia 认识能力;真知,直觉
gnotobasis 无菌环境
gnotobiology 限菌生物学;无菌生物学
gnotobiosis 无菌过程
gnotobiota 限菌区系
gnotobiotics 限菌生物学
go 楼梯级距;梯段长度
goa 藏原羚
go about 换帆;着手做;传开;从事
goaf 老塘;采空区;废矿;废矸堆;不含矿的岩石
goaf filling 采空区充填
goaf pack building 垒岩石带
goaf stowing 采空区充填
goaf water 老窑水
go aground 坐礁
go ahead 一直向前;前进;放行信号;向导信号
goal 目的;目标;球门
goal activity 目标行动
goal analysis 目的分析
goal attainment 目标达到情况
goal congruence 目标一致
goal congruent system 目标协调制度
goal consistency model 目标相容性模式
goal coordination 目标协调
goal-directed behavior 目标导向行为
goal-directed function invocation 目标制导功能调用
goal-post mast 双柱桅
goal object 目标事物
goal obsession 迷于目标的
goal of population control 人口控制目标
goal of water quality 水质目标
goal-orientation 以目标为中心
goal-oriented 目标指向;面向目标的
goal-oriented approach 面向目标的方法
goal-oriented inference 面向目标的推理
goal-oriented parser 面向目标的分析程序
goal-oriented recognizer 面向目标(的)识别程序
goal plan(ning) 目标规划
goal post 球门柱;龙门架;门桩;门柱;门形起重桅
goal post mast 龙门桅(杆)
goal program(me) library 目标程序库
goal programming 目的规划;目的程序设计;目标规划;目标程序设计
goal response 目标反应
goal-seeking approach 目标搜索法
goal-seeking loop 目标搜寻环
goal set 目标集(合)
goal setting 目标制定
goal setting process 目标规定过程
goals of transportation planning 交通规划目标
go-and-no-go ga(u)ge 双端规;控制质量的容许度检验器
go-and-return 两端间
go-and-return resistance 环线电阻
go-anywhere vehicle 越野车
go as course 保持航向
go ashore 上岸
go astern 倒车[船];反(方)向移动
goat 转辙机
goat's beard 假升麻属
goatstone 山羊粪石
gob 黏块;杂石;团块;料滴;塑性团块;大量;废弃物
gobbed up 炉缸冻结
gobbet 小块
gob bin 废石仓
gobbing(up) 充填
gobbinsite 戈硅钠铝石
gobbled up by flood 洪水淹没
gob bucket 吊桶;混凝土吊斗;混凝土料斗
gob cutting 料滴剪切;剪料
gob distributor 分料器
go begging 没销路
gobelin blue 暗青绿色
gobelin tapestry 壁挂
go-between 中间网络
go beyond 超出
gob feeder 滴料机;滴料供料机
gob fire 采空区着火

gob floor 木板假顶;采空区底板
gob-guide knob 导料滴旋钮
gob guiding apparatus 导料滴设备
gob heading 采空区(内)平巷
gob hopper 平衡料轮;混凝土熟料斗;混凝土出料斗;漏斗;活底料斗
gobi 戈壁;荒漠
gobi Altai 戈壁阿尔泰【地】
Gobi basin 戈壁盆地
Gobi deposit 戈壁沉积
Gobi facies 戈壁相
gob lagging 充填隔板
Goble pile-driver 戈尔布打桩机
Goble pile-driving analyzer 戈尔布打桩分析仪
goblet 饮料杯
goblet pruning 杯状修剪
goblet with Southern Song imperial kiln type glaze 仿官青釉瓯
gobo 遮光片;遮光板;亮度突然降低;镜头挡光板;透镜遮光片;传声器遮声障板
gob of glass 玻璃球坯
gobo flag 镜头遮光罩;镜头遮光器
gobpile 废石堆
gob process 滴料供料法
gob separator 分坯器
gob speed 滴料速度
gob stink 煤自燃发火臭味;采空区臭味
gob-stowing machine 充填机
gob tail 尾状条纹
go-cart 手推车
go cipher 发送密码
God's acre 墓地;公墓;坟场
God's area 墓地
Godel process 格多尔水泥砂造型法
godet 导丝辊;倒V形三角布
godet roller 导丝轮
godet wheel 导丝轮
go-devil 运名车;橡胶塞(美国俗称);手推车;清管刮刀;木材搬运橇;刮管器;防护挡板;堵塞检查器;坠撞器;管子清洁器;管道清洁器;软木塞;清管器
go-devil bullet 刮管器
go-devil launcher 清管器发送器
go-devil plane 自滑道;重力斜面
go-devil signaler 清管信号装置
go-devil station 清管器收发站
godlevskite 斜方硫镍矿
godown 栈房;港区仓库;港口仓库;货栈;货仓;后方货场;后方仓库;堆栈;仓库
godown charges 仓储费(用)
godown entry 入库单
godown rent 仓库租金;仓储费(用)
godown risk 仓库险
godown tally 仓库理货员
godown warrant 仓库保证书
godroom 串珠饰
Godschmidt's discontinuous solution 哥特施密特间断解
Godthaab nucleus 戈德霍普陆核
goe 海蚀龛;海蚀洞
goedkenite 羟磷铝锶石
Goertler parameter 戈特勒参数
goethite 针铁矿
goethite aggregate 针铁矿骨料;针铁矿集料
goethite rock 针铁矿岩
Goetz size separator 戈茨粒度分离器
gof(f)er 皱褶;起皱
gof(f)ered filter 皱纹滤纸
gof(f)ered iron 皱纹铁
gof(f)ered paper 皱纹纸
gof(f)ered plate 皱纹钢板;网纹钢板
gof(f)ered sheet iron 皱纹铁板
gof(f)ering 形成皱纹
gof(f)er machine 压纹机;皱褶机
go fishing 打捞
go forward 前进
Gogan hardness 高氏硬度
go-ga(u)ge 通过量规;通用量规
goggles 防尘目镜;护目镜;风镜;防护眼镜
go-go 最新式的;最现代化的;时髦的
go hurry ahead 急速前进
going 梯段(水平距离);踏步宽度;顺车
going aboard 上船
going aground 搁浅;船只搁浅

going alongside 靠泊
going alongside against the current 顶流靠泊
going alongside against the tide 顶流靠泊
going alongside between berthed ships 嵌档靠码头
going alongside in calm weather 无风流靠泊
going alongside removing the ice with ship's deck 船首扫冰靠码头
going alongside the steep shore 靠陡岸
going alongside with an anchor down 抛锚驶靠
going alongside with large angle 大角度驶靠
going alongside with no wind and current 无风无流靠码头
going alongside with offshore wind 吹开风靠码头
going alongside with onshore wind 吹拢风靠码头
going alongside with small angle 小角度驶靠
going alongside with the current 顺流靠泊
going alongside with the tide 顺流靠泊
going barrel 旋转鼓;发条盒
going bord 煤房运煤支巷
going concern 业务发达的商行;继续经营
going concern assumption 继续经营假设
going concern theory 继续营业企业原理
going-concern value 营业价值;经营价值
going down 下降;减少;沉没;下钻
going fire 在燃火
going free 顺风驶帆
going headway 煤房运煤支巷
going in 下钻
going off 偏斜(指钻孔钻斜)
going of the flight 梯段水平距离;一跑楼梯的长度
going of tread 踏步进深
going on without a break 继续不断
going price 现行价格;通行价格
going project 进行中的项目
going public 公共投资
going rate 现价;成交价;现行(汇)率;通行汇率
going rate pricing 现行价格法
going rod 踏步定距杆;梯段杆;定距杆
going train 运转轮系
going up 上升;建造起来;被炸毁
going value 现行价值
go into effect 施行
go into force 生效
go into liquidation 停业清理
go into operation 开工;投入生产;实施
go into service 被雇佣
go into stream 投入生产
go into the hole 下钻
go-kart 微型竞赛汽车
gola 反曲线
Golay cell 戈利辐射计;高莱探测器
Golay cell detector 高利池探测器
Golay detector 高莱探测器
Golay's column 戈雷氏柱
Golay's equation 戈雷方程式
gold 自然金;金色;黄金
goldamalgam 金汞齐
gold analyzer 黄金含量分析仪
gold and silver 货币及金银清单
gold and silver alloy 金银合金
gold and silver assay 金银鉴定
gold and silver control 金银管理
gold auctions 拍卖黄金
gold-bearing gravel 含金砾石
Goldbeck's method of pavement design 哥特培克路面设计法
Goldbeck formula 哥特培克公式(一种设计柔性路面厚度的公式)
Goldbeck(pressure)cell 哥特培克压力盒
Goldberg Mohn friction 戈德堡—莫恩摩擦力
Goldberg wedge 戈尔德贝格光楔
gold blackbody 金黑体
gold-black coating 金黑涂层
Goldblatt unit 戈德布拉特单位
gold blocking 烫金
gold bond 以黄金偿付的债券
gold-bricking 限制生产定额
gold bromide 溴化金
gold bronze 金色铜粉;金青铜
gold bronze powder 黄铜粉
gold bullion 金锭
gold bullion standard 纯金本位制
gold bushing 黄金衬套

gold carving 雕金
gold chloride 氯化金;三氯化金
goldclad wire 包金线
gold-coated 涂金的;镀金的;包金的
gold coating 镀金层
gold colloid 胶体金
gold colloidal 放射性胶体金
gold-copper deposit in volcanics 火山岩中金铜矿床
gold currency 金本位货币
gold current standard 金币本位制
gold cushion 烫金垫
gold cyanide 氰化金
gold detector 黄金探测器
gold dichloride 二氯化金
gold dioxide 二氧化金
gold-doped 掺金的
gold-doped germanium detector 掺金锗探测器
gold-doped germanium infrared detector 掺金锗红外探测器
gold-doped transistor 掺金型晶体管
gold doping 掺金
gold dredge investment 采金船投资
gold dredger 采金船
gold dust 金粉;砂金
gold electrode 金电极
gold embargo 禁止黄金出口
golden 金黄色;黄金的
golden age 黄金时代
golden colo(u)r 金色
Golden Gate Bridge 金门大桥(美国)
Golden gateway 蒂奥克莱提宫殿金门(古罗马)
golden handshake 去职补偿费
Golden House 金宫(位于意大利威尼斯)
golden larch 金钱松
golden maple 金钱槭
golden mean 黄金分割法
golden module 黄金模数
golden pheasant 金鸡
golden powder 金粉
golden rain tree 栾树
golden red 金红色
golden red glass 金红玻璃
golden rule 比例法
golden rule fallacy 谬误之推理
golden section 黄金分割
golden section method 黄金分割法
golden section search 黄金分割搜索
golden section value 黄金分割值
golde solder 金焊料
golden yellow 金黄色的
gold exchange standard 金币兑换制
goldfield 金矿区
gold filament yellow glaze 金丝黄釉
gold filigree 金丝镶嵌;金属镶嵌
gold filled 镀金;金包饰;包金
gold-film glass 涂金膜玻璃;包金膜玻璃
gold-film mercury analyzer 金膜测汞仪
gold-film resistor mercury analyzer 金膜电阻测汞仪
goldfish 金鱼
goldfish pond 金鱼池
gold fix 议定金价
gold foil 金箔
gold foil painting 贴金
gold franc 金法郎
Goldhaber triangle 哥特哈伯三角形
gold holdings 黄金储备额
gold hydroxide 氢氧化金
goldichite 柱钾铁矾
gold-imitation processing technology 仿金工艺技术
gold ingot 金锭
gold ink 金色油墨
gold lacquer 金色喷漆
gold leaf 薄金箔;金叶;金箔
goldleaf electrometer 金箔静电计
gold leafing 贴金叶装饰;贴金
gold leaves 金叶;金箔
gold luster 金色光泽
goldmanit 钙钒榴石
Goldman vapo(u)rizer 戈德曼氏气化器
goldmark 记录搜索接收机
goldmine 金矿

gold mine tunnel 金矿隧道
gold mining 金矿开采;采掘黄金
goldmining vessel 采金船
gold ocher 浅土黄(指颜料)
gold opal 金黄蛋白石;火蛋白石
gold ore 金矿石;金矿床
gold ore associated by sulfide 硫化物型金矿石
gold ore of quartz-calcite type 石英—方解石脉型金矿石
gold ore of quartz vein type 石英脉型金矿石
gold-overlaid 贴金
gold oxide 氧化金
gold paint 金漆
gold placer 砂金矿床
gold-placer sand 砂金砂
gold-plated 镀金的;包金的
gold-plated contact 镀金触点
gold-plated frame 镀金镜架
gold-plated PVC flexible hose 镀金聚氯乙烯软管
gold-plated watch bracelet 镀金表链
gold-plating 涂金;饰金;飞金;镀金面;镀金材料;镀金(术);镏金
gold plating thickness 镀金厚度
gold-platinum alloy clasp-wire 金铂合金卡环丝
gold point 金点;输金点
gold pointing 点金
gold-polymetallic vein in granite 花岗岩中金多金属矿脉
gold pool 黄金总库
gold potassium cyanide 氰化亚金钾
gold powder 黄金粉末
gold-quartz vein in gneiss-migmatite 片麻岩混合岩中含金石英脉
gold-quartz vein in greenstone belt 绿岩带中含金石英脉
gold ruby 金红
gold ruby glass 金红玻璃
gold rush 淘金热
gold sandwich contact 金夹层接触
Goldschmid's law 哥特希密定律
Gold schmidt 铝热焊(接)
Goldschmidt's alternator 哥特施密特交流发电机
Goldschmidt's law 哥特施密特定律
Goldschmidt's mineralogical phase rule 哥特施密特矿物相律
Goldschmidt's phase rule 哥特施密特相律
Goldschmidt's process 哥特施密特方法
goldschmidtine 脆银矿
goldschmidtite 针碲金银矿
gold scouring 磨光金
gold-secured loans 以黄金作抵押的贷款
gold silver 涂金胶
gold silver deposit in volcanics 火山岩中金银矿床
gold silver jewelleries 涂金油;贴金漆
gold size (高干燥剂的)嵌缝树脂清漆;描金胶;硬油清漆;金胶漆;金浆;贴金漆;短油钙脂清漆;镀金粘料
Gold slide 戈尔德计算尺
gold slide(scale)气压计误差订正滑尺
goldsmith principle 金匠原则
goldsmiths' notes 金匠收据
gold-sodium chloride 金钠氯化物
gold sodium cyanide 氰化钠金
gold sodium thiosulfate 硫代硫酸钠金
gold sol 金溶胶
gold solder 金焊
gold spring 金簧
gold stamping 烫金
gold standard 金本位(制)
gold standard act 金本位制法
Goldstone bosons 戈德斯坦玻色子
goldstone glaze 金星釉;砂金釉
Goldstone-Haystack interferometer 戈德斯通—海斯塔克干涉仪
gold stoving 金色清烘漆
gold stoving varnish 金色烤漆;金黄烤漆;描金烤清漆
gold-tin purple 金锡紫
gold toning 金调色法;调金色法
gold tooling 烫金工具
gold trioxid 三氧化二金
gold-vanadium alloy 金钒电阻合金
gold wire 金线
gold wire seal 金丝密封圈

gold-work 黄金工
gole 泄水闸;溪谷;水沟;溢流坝;闸门
golfada 戈尔法达风
golf club 高尔夫球俱乐部
golf course 高尔夫球场
golf course hazard lake 高尔夫球场障碍湖
golf course irrigation 高尔夫球场灌溉
golf links 高尔夫球场
Golgi apparatus 内网器
goliath 移动式巨型起重机;轨道起重机;(可移动的)大型起重机
goliath base 大型管座;大型管底
goliath cap 大型灯头
goliath crane 移动式巨型起重机;高架起重机;巨型(移动式)起重机;强力起重机;(可移动的)大型起重机;重型龙门起重机
go light 前进灯;绿灯
go limit 通过端极限
gombo 黏土(状)
Gombroon ware 哥布朗玲珑器皿
gome 滑润油的积炭
gomma 时序标记
Gommeson method 格姆逊法(钻孔弯曲度测量法)
gompholite 泥砾岩(杂色块状砾石)
gomphosis 钉状关节
Gomphosphaceria lacustris 湖生束球球藻
gon 哥恩(角度单位,等于直角的1%);冈;百分度(数)
gondang wax 榕树蜡;无花果蜡
gondite 锰榴石英岩;石英锰榴岩
gondola 高边敞车;料箱;气球吊篮;平底狭长小船;敞篷货车;发动机短舱
gondola car 敞篷货车【铁】;高帮敞车;无盖货车
gondola flat 铁路敞篷车辆
gondola sales 平台销售
Gondwana breakup 冈瓦纳大陆解体
Gondwana coral realm 冈瓦纳珊瑚地理区系
Gondwana crustal cupola 冈瓦纳壳块
Gondwana floral realm 冈瓦纳植物地理区系
Gondwana land 冈瓦纳大陆
Gondwana pal(a)eocontinent 冈瓦纳古陆
Gondwana-type platform 冈瓦纳型地台
gonecystolith 精囊石
gone off hole 偏斜钻孔
gone of primary deposit 原生沉积带
gone to water 钻孔涌水
gong 音簧;铃碗;皿形钟
gong buoy 装锣浮标;警钟浮标
gon grade 百分度(数)
goniasmometer 量角器;角度仪
gonidial layer 藻层
goniochromatic effect 视角闪色效应
goniochromatism 角异色性
gonioisochromatic 方向等色的
gonioisochromatism 方向等色性
goniometer 量角仪;晶体界面角测量器;角度仪;角度计;测向器;测向计;测角仪;测角器;测角计
goniometer eyepiece 测角目镜
goniometer head 测角计头
goniometer system 测向装置;测角系统
goniometer type direction finder 测角计型测向器
goniometric(al) 测角的
goniometric(al) locator 测向定位器
goniometric(al) network 测角网
goniometric(al) sight 测角瞄准镜
goniometry 量角学;测向术;测角术
goniomicroscope 测角显微镜
goniophotometer 测向光度仪;测角光度计;变角光度计
gonioradiometer 测角辐射计;变角辐射计
goniospectrophotometer 测角分光光度计
gonnardite 纤沸石
go-no-go 分检
go-no-go dosimeter 阈值剂量计
go-no-go inspection test 合格检查试验
go-no-go test 是否试验;功能试验
go-no-go test equipment 极限测试装置
go/no judg(e)ment 合格不合格判别
gonyerite 富锰绿泥石
Gooch crucible 古氏坩埚
Gooch filter 古氏滤器;古奇坩埚
Gooch funnel 古氏漏斗
Gooch valve motion 古氏阀动装置
Good's interrupted projection 古德分瓣投影

good agricultural practice 农药安全使用法
good agricultural regulation 农药安全使用规定
good angel 吉神
good apposition 对位良好
good aquifer 强含水层
good bargain 赚钱买卖
good bearing earth 坚土
good business condition 景气
good caking coal 良好黏结性煤
good circle and vicious circle 良性循环与恶性循环
good-class joinery 高级细木作
good colo(u)r 墨色均匀
good conditions for soil moisture 墒情好
good conductor 良导体
good coordination 良好的协调作业
good cycle of agroecosystem 农业生态良性循环
good delivery receipt 妥善的投送回单
good design 好设计
good design mark (工业制品的)优良设计标志
Goodeve's rotational visco(si)meter 古迪夫旋转黏计
good faith 合理;公平;诚信
good faith estimate 正确评估;实价评估
good faith money 定金;保证金
good faith purchaser 无知的买主;外行买主
good farmland 良田
good fence 良好围栏
good fish population 活鱼数目;未受污染的鱼群
good gradient 平缓坡度;平缓坡道
good ground 良好土层
good growing weather 适宜的气候
good harbo(u)r 良港口
good holding ground 好抓力锚地
good investment 有利的投资
Good King James's Gothic 英国雅可布式风格 (1603—1625年)
Goodloe packing 古德洛填料
Goodman's stress diagram 古德曼应力图
Goodman's duckbill loader 古特曼型鸭嘴铲式装载机
Goodman loader 古特门液压装载机
good merchantable 良好的商品质量
good merchantable quality 良好销售品质;适销质量;符合销售要求的质量
good mooring 良好系泊
good muscovado 上等粗糖
good neighbo(u)r policy 睦邻政策
goodness 优势
goodness of fit 拟合优度;拟合良好性;配合适度;吻合度;适(合)度;分布拟合优度
goodness of fit test for distribution 分布拟合检验
goodness of fit test(ing) 拟合优度检定;拟合优度检测;拟合优度检验;拟合良好性检验;符合优度检验;配合检验;配合度检定;适合度测定;适合度检测
goodness of geometry 结构优度
goodness of receiver 接收机品质因数
good oil 优质原油;提余液;提纯油
good one side 单面优良;一面光;单面光
good order 情况正常
good order and condition 良好状态
good paper 可靠(的)支票
good performance 良好性能
good product 合格品
good quality 质量好
good-quality undisturbed sample 质量好的原状土(样)
good quality water 优质水
good river 畅流河道
good roadbed soil 良好路基土
good rolling car 易行车【铁】
good rolling track 易行线
good running 正常运行
goods 制品;货物
good safety 安全可靠
goods afloat 未卸货物
goods and chattels 货运站;仓库
goods and depot 货运站;仓库
goods and materials 物资
goods and materials in stock 现存物资
goods and services 货物及劳务
goods area 货区
goods awaiting shipment 备运货物
goods car 运货(汽)车;载货汽车;货车

goods carried on deck 舱面货(物)
goods-carrying vehicle 运货车
goods carted into designated goods section 进货
goods category 货物品类
goods charges 货物运费
goods chassis 运货车底盘
goods clerk 收货员
goods collecting and distributing centers 货物集散地
goods collecting train 零担摘挂列车
goods consignment note 货物运单
goods consumed 被消费的货物
goods credit 货物抵押贷款
goods damaged by sea 海损货
goods damage in transport 运输途中货损
goods declaration 货物报关单;货物报告单
goods-delivered summary 发出商品汇总表
goods delivery 货物交付
goods delivery receipt 送妥回单;送货凭证
goods depot 仓库
goods dispatching[despatching] 货运调度
good seamanship 良好船艺
goods elevator 运货升降机;运货电梯
goods engine 货运机车
goods entrance 货物入口
goods exchanging trade 易货贸易
goods exported under special license 特许出口商品
goods fitting for container transport 适箱货物
goods flow 货流;财货流量
goods flow diagram 货流图
goods for everyday consumption 日用消费品
goods forwarding and receiving units 发货和收货单位
goods freight 货物运费
goods handling 物质管理
goods handling charges 货物装卸费
good sharpness 各种标志清晰度好;清晰度好
goods in bad order 货物混乱;残损货物
goods in barrels 桶装物
goods in bond 在关栈货物;保税货物;保税仓库内的货物
goods in bulk 散装货
goods-in-bulk loading track 散装货物线
goods in customs storage 海关库存货物
goods in great demand 热门货(品)
goods in hand 保税货物
goods in process 在产品
goods in short supply 短线产品
goods in stock 现货;存货
goods insurance rate 货物保险费率
goods in transit 运送中物;在途物
goods in use 使用财资
goods kept in stock 库存物资
goods label 货物标记;货签
goods lift 运货升降机;货运电梯;运货电梯;升降机;货梯;起货升降机;运货吊梯
goods lift bridge 运货吊桥
goods line 运货线;货物线
goods loaded in tank wagon 罐装货物
goods loading banks 装货箱[贮]料器
goods locomotive 货运机车
goods moved by rail 铁路货物周转量
goods movement planning 货物流通规则;货物流动规则
goods of a well-known brand 名牌货
goods of concentrated weight 集重货物
goods of first order 直接需要品
goods of particular class controller 特种货运调度员
good soil management 良好土壤管理
good solvent 强溶剂
goods on approval 看货后购买的货物
goods on hand 库存商品;盘存货(物)
goods on the way 在途货物
goods operation at destination 货物到达作业
goods operation en route 货物途中作业
goods ordering 订货公
good source rock 好烃源岩
goods package for transport 货物运输包装
good spark 强火花
goods-passenger braking system 客货车可调式制动系统
goods-passenger change-over brake 客货车

转换的制动机
goods-passenger lift 货客梯
goods platform 装卸货站台;货物站台
goods price 好价(钱)
goods rate 货物运价率
goods received note 收货票据
goods reception 接收货物
goods rejected 退货
goods rejection 废品剔除
goods returned 退货
goods section 货位
goods service 货运服务
goods service equipment 货运服务设备
goods shed 雨棚;货棚
goods shed track 货棚线
goods shipped in bulk 散装船货
goods shipped in transit 发出商品
goods sold 已销售货物
goods sold over the counter 门市商品
goods standard theory 货物本位论
goods station 货运站;货物站
goods storage 仓库
goods storage charges 货物存放费
goods straddled on two or more wagons 跨装货物
goods suited to popular tastes 产品对路
goods tariff 货物运价
goods tariff No. 货物运价号
goods temperature required at the time of acceptance of conveyance 承运温度
goods through 联运
goods through transport 货物联运
goods to arrive account 未到商品账户
goods to be transshipped 转运货物;中转货物
goods traffic 货运量;货运交通;货物运输
goods traffic plan 货运工作计划
goods traffic railway 货运铁路
goods traffic revenue 货运收入
goods train 运货列车;货运列车;货物列车;货车【铁】
goods train kilometers 货物列车(走行)公里
goods train operation equipment 货运运转设备
goods train passing track 货物会让线
goods train track 货物列车停留线
goods transfer point 货物交接所
goods transport accident 货运事故
goods transport contract 货物运输合同
goods transport facility 货运设备
goods transport implements 货运用具
goods transport office 货运室
goods transport plan 货物运输计划
goods transport record 货运记录
goods transport under escort 货物押运
goods transport within a station 站界内搬运
goods transshipment of international through transport at frontier station 国际联运货物换装
good stream shape 良好的流线型
goods trolley 货物搬运车
good structure 远景结构
goods turning rack 货物转向架
goods turnover 货运周转量
goods under customs bond 保关税货物
goods van 货车;有盖货车
goods vehicle 货运汽车;货运车辆
goods wagon 货车
goods warehouse 货仓;货栈
goods weight 货物重量
goods which sell well 热门货(品)
goods working program(me) 货运工作方案
goods yard 货物堆场;货场;堆场
goods yard road 货场道路
goods yard sheet 防湿篷布
goods yard track 货场配线
good the day 限当日
good thermal stability coal 热稳定性良好煤
good this month 限当月;本月内有效
good this week 本周内有效
good through 一直有效
good-till-cancelled 解约前有效;取消前有效;未撤消前有效
good time 正常(工作)时间;繁荣时期
good title 有效的所有权
good training condition 良好训练状态
good until cancelled 取消前有效

good value for money 物有所值
good visibility 良好能见度;好能见度
good water 好水
good weather and smooth water 风平浪静
goodwill 信誉;好信誉;商誉好;商信
goodwill entertaining 业务招待费
goodwill positive 正商誉
good working condition 完整状态
good year 丰年
Goody random model 古迪随机模式
go-off 起爆
go off the beaten track 打破常规
googol 大数【数】
Googolplex 古戈尔普勒克斯
goongarrite 纤硫铋铅矿
go on shore 上岸
go on symbol 继续符号
goop 镁尘糊块
goose barnacle 鹅颈藤壶(木船底、码头木桩上的海生物)
gooseberry 醋栗
gooseberry stone 钙铝榴石
goosecreekite 古柱沸石
goose down ware 鹅绒白
goose flesh 鹅皮
gooseneck 钢横梁;牵引拱架;弯曲管;弹簧式弯头车刀;水旋转接头;吊杆坐转轴;吊杆枢;鹅颈头;鹅颈钩;鹅颈弯
gooseneck band 轻型吊杆座
gooseneck boom 转轴式吊杆;鹅颈式吊杆;鹅颈架起重臂
gooseneck bracket 鹅颈型帘杆托架;鹅颈座
gooseneck clamp 鹅颈夹具
gooseneck claw bar 弯脖撬棍
gooseneck cleaner 桥管清扫器
gooseneck connection 鹅颈式接合;鹅颈(形)接头;鹅颈连接合;鹅颈连接管
gooseneck crane 鹅颈式起重机
gooseneck davit 转轴式吊艇柱;鹅颈式吊艇柱;鹅颈式吊艇架
gooseneck dolly 鹅颈式小车
gooseneck dumper 鹅颈式自卸车
gooseneck elbow 鹅颈弯
gooseneck faucet 鹅颈形龙头;鹅颈式龙头
gooseneck jib 鹅颈臂
gooseneck jib tower crane 折臂式塔式起重机;鹅颈式塔式起重机
gooseneck joint 鹅颈接头
gooseneck machine 鹅颈式压铸机
gooseneck pediment 鹅颈形檐饰
gooseneck pipe S 形弯管;鹅颈(形)管
gooseneck scraper 弯头刮刀;鹅颈形刮刀
gooseneck socket 鹅颈座
gooseneck tool 弹簧刀;鹅颈刀
gooseneck trailer 鹅颈挂车
gooseneck tube 鹅颈管
gooseneck tunnel (集装箱的)鹅颈槽
gooseneck-type construction 鹅颈式结构;鹅颈式建筑
gooseneck-type wagon 半挂式拖车
gooseneck ventilator 鹅颈式通风筒;弯管通风筒
goosepen 鹅圈;鹅棚;火烧洞
goose skin 鹅皮
goose type river bend 鹅颈形河弯
goosewinging 蝴蝶帆
Gooth crucible 古奇坩埚
go-out 熄灭;候潮闸
go out of business 歇业
gopher ditcher 履带式挖沟机
gopher drift 沿脉勘探巷道
gopher hole 獾洞式孔;鼠洞式孔;药室
gopher-hole type explosion 地鼠洞式爆炸法
gophering 浅井勘探
gopherman 挖土工;伐木工
gopher protected cable 防鼠咬电缆
go plug ga(u)ge 过规
go-public 挂牌
gopura(m)印度庙宇山门;山门上大塔(金字塔形);华饰庙门
goral 青羊(野羊);斑羚
gorceixite 磷钡铝石
gordian technique 关键性技术
gordian technique problem 关键性技术问题
Gordner-Holdt tube 加氏管

Gordon's equation 戈登方程
Gordon's formula 戈登公式;高氏公式(柱体破坏荷载计算公式)
gordonite 磷镁铝石
Gordon simulator 戈登模拟程序
gore 三角区
gore area 分道角区
gore lot 三角地带
gore of diverted river 转向峡谷
gore sign 分道标志
gore strake 并列板
gorge 峡(谷);河道险阻;航道中冲积的碍航物;排流峡谷;凹刻;凹弧饰;山峡
gorge area 峡区
gorge cut 小凹圆线脚
gorge district 峡区
gorgerin 柱颈
gorget 有槽引针;有槽导子
gorge type reservoir V 形水库;狭谷式水库;峡谷形水库
gorge wall 峡谷壁
gorgeyite 斜水钙钾矾
gorgoneion 妖女饰
gorgoneum 妖女饰;[复]gorgoneia 或 gorgonea
gorlic acid 环戊烯十三碳烯酸
gormanite 哥磷铁铝石
go round of form 模板周转
go-round-style garden 环游庭园
gortdrumite 硫汞铜矿
Goryaef's ruling 戈里阿耶夫氏划线
go side 通过端
Goskar dryer 哥斯卡干燥室
goslarite 硫酸锌矿;皓矾
Gospel side 教堂北端(宣读福音的地方)
gosphel ambo 福音堂的读经台
gosphel hall 福音堂
gossamer 薄雨衣;薄纱
gossan 铁帽
gossan mineral 铁冒矿物
Gossan-type iron deposit 铁冒型铁矿床
Goss process for the continuous casting of metals 高斯连续铸造
go-stop 交通指挥灯;交通信号
Gotar lens 戈塔头镜
Gothenburg reversed polarity subzone 哥特堡反向极性亚带
Gothic 哥特体;哥特式的;尖拱式的;双圆弧形
Gothic abacus 哥特式柱顶板(圆柱顶上的);哥特式冠板(圆柱顶上的)
Gothic abutment 哥特式支座;哥特式拱座
Gothic altar 哥特式祭坛
Gothic arcade 哥特式连拱廊
Gothic arch 尖券;哥特式(尖)拱;尖拱形
Gothic arch bridge 尖拱桥
Gothic architecture 哥特式建筑(公元 12~16 世纪时盛行于西欧)
Gothic arch plain tile 尖拱平砖;哥特式拱门平砖
Gothic art 哥特艺术
Gothic basilica 哥特式长方形建筑
Gothic bond 哥特式砖石砌合
Gothic brick architecture 哥特式砖建筑
Gothic(building)style 哥特式建筑风格
Gothic buttress 哥特式支墩;哥特式扶垛;哥特式扶壁
Gothic buttressing pier 哥特式扶壁墩
Gothic cathedral 哥特式教堂
Gothic cathedral St. Vitus 哥特式圣威斯斯大教堂
Gothic cathedral style 哥特式大教堂建筑风格
Gothic Charles Bridge at Prague 布拉格的哥特式理斯桥
Gothic Chippendale 哥特式家具
Gothic church 哥特式教堂
Gothic church nave 哥特式教堂的中堂;哥特式教堂的正堂
Gothic crypt 哥特式墓穴;哥特式地窖
Gothic decoration art 哥特式装饰艺术
Gothic detail 哥特式建筑元件;哥特式建筑细部;哥特式建筑零件
Gothic equilateral pointed arch 哥特式等边尖拱
Gothic groove 弧菱形轧槽
Gothicism 哥特式
Gothicist 哥特式建筑风格支持者
Gothicized 具有哥特式建筑风格的
Gothic(masonry)bond 哥特式砖石砌合

Gothic master 哥特式建筑大师
Gothic middle vessel 哥特式教堂的中堂;哥特式教堂的正堂
Gothic ornamental art 哥特式装饰艺术
Gothic palace 哥特式宫殿
Gothic pass 弧菱形孔型
Gothic pillar 哥特式独立支柱
Gothic pitch 哥特式屋顶坡度(60度);哥特式屋面坡度(60度)
Gothic pulpit 哥特式讲坛
Gothic quire choir 哥特式圣坛
Gothic raised table 哥特式柱顶板
Gothic revival 哥特式复兴时代;哥特复兴式;哥特复兴时代
Gothic roof 哥特式屋顶
Gothic section 弧边形方坯
Gothic sextant 可卸气泡六分仪
Gothic style 哥特式
Gothic tower 哥特式塔(楼)
Gothic vault 哥特式拱顶
Gothic window 哥特式窗
Gothic window tracery 哥特式石雕花格窗;哥特式窗子装饰
Gothic wing 哥特式翼
Gothite 哥赛欧脱(一种缓凝的半水化合灰浆)
Gothlandian period 戈特兰纪【地】
go through 卖完;通过
go-through machine 网眼花边机
Gotlandian 哥特兰纪【地】
got mix silo 热混合料储仓
Goto pair 高特对
go to statement 转向语句【计】
go to the sea 出海航行
got-plate bar 炉条
gotten 采完报废的巷道
gotzerite 氟硅钙钛矿
gouache 古阿颜料;水彩颜料涂白法;树胶水彩画
gouache paint 树胶水彩颜料
Goubau line 高保线
goudeyite 三水砷铝铜石
goudron 沥青;焦油;减压渣油
goudronator 沥青撒布机;沥青喷洒机;沥青喷布机
goudron highway 沥青路
gouffre 落水洞
goufing 加固墙的基础
gouge 圆凿(子);凿孔;弧口凿;断层泥;断层脉壁泥;打眼;半圆凿
gouge angle 切削角
gouge auger 匙形螺钻
gouge bit 圆弧钻头;扩孔钻头;弧口钻;勺形钻头;槽钻头
gouge carving 凿槽
gouge channel 槽状水道
gouge drill 勺形钻
gouge forceps 圆凿钳
gouge handle 弧口凿手柄
gouge hole 半圆凿穴
gouge-nippers 圆凿钳
gouge(out)凿槽;挖出;凿出
gouge slip 滑磨石;圆弧磨石;油石;弧口凿磨石
gouge spade 半圆凿
gouge work 用凿雕刻的木工
gouging 表面切割;表面吹割;刨槽;刨凹
gouging abrasion 碰撞磨损
gouging acting 刮削作用
gouging blow pipe 表面切割割矩
gouging chisel 槽凿;表面切割用凿
gouging machine 冲裁机
gouging-scrapping action 定径旋铣作用
gouging shot 掏槽眼
Gould belt 古德带【地】
Gould plotter 古德绘图仪
gourd-shaped carved lacquer vase 雕漆葫芦瓶
gour rock 菌状石
goutte d'eau 白黄玉
goux pail 带衬便桶(便桶内有吸收性衬里,当粪便倾倒后,便桶要更换衬里)
Gouy 古伊(动电学单位)
Gouy balance 古伊天平
Gouy-Chapman equation 古伊一查普曼方程
Gouy-layer 高伊层
go-valve 起动阀
govern 支配;调速
governed by law 法治

governed engine speed 发动机限速
governing 控制;操纵
governing air signal pressure 调速器空气信号压力
governing aspect 关键(问题)
governing body 管理机构
governing box 调速箱
governing characteristic 调节特性
Governing Council of the United Nations Environment Program(me) 联合国环境规划管理理事会
governing criterion 规定标准
governing device 调节装置;调节设备
governing differential equation of groundwater movement 地下水运动基本微分方程
governing equation 控制方程;基本控制方程;基本方程(式)
governing error 主要误差;支配误差;控制误差;调节误差
governing factor 支配因素;关键因素;控制因素
governing laws 依据的法律;适用的法律
governing loop 调速环节;调节回路;调节环节
governing mechanism 调节机构
governing point 控制点
governing principle 指导原则
governing response 调节响应
governing rock characteristics 岩石基本性态
governing screw 调节螺钉
governing stage 调节级
governing system 调节系统
governing time 调整时间;调速时间
governing torch 气刨枪
governing valve 调节阀
government accounting 政府会计
government(al) agency 政府机关;政府机构
government(al) building 政府大厦;政府大楼
government(al)-owned patent 政府专利
government(al) securities 国债;公债
government(al)-sponsored program(me) 政府主导计划
government(al) trade agreement 政府贸易协定
government anchor V形锚铁
government appropriation 政府公报;政府拨款
government assistance 政府援助
government band 政府通信波段
government bilateral loans 政府双边贷款
government block 政府大楼
government body 政府机关
government bond 政府债券;政府公债;公债(券)
government broker 政府经纪人
government bronze 炮铜合金
government budget 政府预算
government building 政府大楼;政府建筑
government building sign 政府建筑物标志
government bulk-buying 政府大宗采购
government call 政府电话
government capital 国有资本
government cargo rate 军公货物运价
government censor 政府检查员
government commission 政府委员会
government credit guarantee 政府信用保证
government decree order 政令
government departments 政府部门
government enterprise 国营企业
government environmental expenditures 政府环境开支
government finance 政府财政
government forces 政府因素
government form chartering 政府形式租船
government furnished equipment 国家提供的设备
government grant 政府赠款;政府拨款
government-guaranteed bond 政府担保债券
government house 官邸;政府大厦;政府大楼;政府办公楼;政府办公房屋
government intervention 政府干预
government investment 国家投资;政府投资
government land 公共用地;公地
government license 政府许可证
government loan 政府放款
government monopoly 政府专利
Government National Mortgage Association 政府抵押贷款协会(美国)
government obligations 政府债务
government offices 国家机关
government owned 国有;国营;公有;政府所有的
government-owned railroad[railway] 国营铁路
government ownership 国家所有制
government palace 政府宫殿
government paper 政府发行有价证券
government patent 由联邦或州政府赠予或转让给个人的土地(美国)
government property 国有土地
government quota 政府指标;政府限额;政府定额
government railway 国家铁路
government regulation 政府规章;政府规程
government reservation 政府保留用地
government restriction 政府管制
governments 可转让政府证券
government security 政府(有价)证券
government standards manual 政府标准手册(美国)
government stock 政府股份
government subsidy 政府补助金
government surplus 财政盈余
government survey 政府检验
government survey method 政府土地测量法(美国)
government test 政府试验
government trad emission 政府贸易代表团
government traffic 政务通报
government tug boat 国营拖船
government utility undertaker 政府公共设施承揽者
governor 燃气调压器;控制器;调整器;调速(控制)器;调速机;调节(用)变阻器
governor actuator 调速器的传动装置;调节器传动装置
governor adjusting screw 调节器调整螺钉
governor adjusting screw cap 调速器调整螺钉帽
governor arm 调速器杆
governor arm screw 调速器杆螺钉
governor assist plunger 调动器助动柱塞
governor assist spring 调速器副弹簧
governor balance gear 平衡机构调速器
governor ball 离心调速器;飞锤
governor ball arm 调速器球杆
governor body 调节器壳体
governor booster 调速助力器
governor booster piston rod 调速助力器活塞杆
governor box 调速器箱
governor bushing 调速器衬套
governor cabinet 调速器(操作)柜
governor cap 调节器盖
governor characteristic 调速器特性
governor control 调速器控制
governor control box 调速器操纵箱
governor-controlled sheave 调速器控制皮带轮
governor control safety valve 调速器控制安全阀
governor control shaft 调速器控制轴
governor deflection 调整范围;调速器偏转
governor drive 转速调节器传动;调速器驱动机构
governor droop 调速器下降特性
governor fork 调速器杠杆的叉头;调速器拨叉
governor free state 无调速状态
governor friction 调速器中的摩擦
governor gallery 调速器廊道
governor gear 调速器齿轮
governor gearing 调速器传动装置
governor gear with oil relay 带油继动器的调节装置
governor generator 调速器发电机
governor handle 调速器把手
governor high limit spring 调速器高速限制弹簧
governor house 调压站
governor housing 调速器壳体
governor hunting 调速器周期性振动
governor idle gear 调速惰轮
governor idling spring 调节器空转弹簧
governor impeller 调速叶轮泵
governor intermediate oil 调速器中间油压油
governor intermittent oil 调速器间断油压油
governor lever 调节(器)杆
governor link 调速器连杆
governor linkage 调速器连杆
governor link roller pin 调速器杆滚轮销
governor link screw 调节器连杆螺钉
governor link with roller 带滚轮的调速器联杆
governor magnet valve 调速器磁铁阀
governor manifold 调速器歧管
governor mechanism 调速器机构
governor motion 调速器传动;调节机构
governor motor 调整(机用)电动机;调速(器用)电动机;调速马达
governor movement 调速器传动装置
governor of velocity 调速器
governor oil pressure 调速器油压
governor oil system 调速器油压系统
governor overriding control 调速器超速控制
governor pipe insulating joint 调速器绝缘联接管
governor pressure oil 调速器高压油
governor pulley 调速器皮带轮
governor pump 调速器泵
governor pump gear 调节泵
governor pump gear shaft 调节泵齿轮轴
governor regulation 调速器不等率
governor response rate 调速器反应率
governor ring 调速器环
governor rod 调速器拉杆
governor roller bearing 调速器滚柱轴承
governor seal 调节器封
governor shaft 调速器轴
governor sleeve 调速器套筒
governor sleeve bushing 调节器套筒衬套
governor slide valve 调节器滑阀
governor socket 调速器套节
governor solenoid 调速器螺管
governor solenoid valve 调速器螺管阀
governor speed changer 调速器速度变换器
governor spring 调速器弹簧;调节器弹簧
governor station 调压站
governor stop arm 调速器停车手柄
governor stop block 调速器止块
governor stop solenoid 调速器停止螺管
governor sump oil 调速器油槽油(无压油)
governor switch 调速器开关
governor synchronizing system 调压器同步系统
governor test 调速器试验
governor trapped oil 调速器闭油路
governor union 调压器连接管
governor union nut 调速器连接管螺母
governor union stud 调速器连接管柱螺栓
governor valve 调速阀;调节活门;调节阀;速控压阀;速控阀
governor valve box 调节器阀盒
governor valve lock nut 调节阀防松螺母
governor valve position indicator 调节阀开度计
governor valve position recorder 调节阀行程记录仪
governor valve spring 调节器阀簧
governor weight 调速器重锤(离心式);调速器离心锤;调节锤
governor work capacity 调速器工作能力
governor work output 调速器工作能力
go wagon 用车运输
gowan 裂解了的花岗石;风化的花岗石
Gow caisson 凿井式沉箱;高氏沉箱;多级套管式沉井;波斯顿沉箱;多级套筒式沉井
Gow caisson pile 高氏沉箱桩
gowerite 戈硼钙石
gowk storm 布鸽风暴;布谷鸟风暴
gox 气态氧
goyazite 磷铝锶石
goyol 告衣醇
goz(gozes) 沙丘状积砂
gozzan 铁帽
GPS aero-triangulation GPS 空中三角测量;全球定位系统空中三角测量
GPS control network GPS 控制网;全球定位系统控制网
GPS differential correction service GPS 差分修正业务;全球定位系统差分修正业务
GPS mapping GPS 测绘;全球定位系统测绘
GPS real-time differential receiver GPS 实时差分接收机;全球定位系统实时差分接收机
GPS receiver GPS 接收机;全球定位系统接收机
grab 抓扬机;抓取;抓具;卡爪;爬杆脚扣;采泥器;采泥机;非法牟取
grab attachment 抓斗附件;抓货附具
grab bag 摸彩袋
grab bar 把手;抓条;扶手(棍);扶手(杆)
grabber 抓取器
grabbing 抓扬机装载;抓岩机抓岩
grabbing bucket 抓斗
grabbing crane 抓式起重机;抓斗(式)起重机;抓

斗吊车
grabbing crane bucket 起重机抓斗
grabbing equipment 抓斗装置;抓斗设备
grabbing excavator 抓掘机;抓斗式挖土机
grabbing floating crane 水上抓斗起重机;浮式抓斗起重机
grabbing goliath 轨道式抓斗起重机;门式抓斗起重机
grabbing load 抓起荷载;抓斗荷载
grabbing pit 装有抓斗吊车的储库
grabbing rig 抓斗起重设备
grabbing work 抓斗起重工作
grab boat 抓斗(式)挖泥船
grab bucket 抓斗;抓斗;挖土机抓斗
grab bucket conveyer 抓斗式运载机;抓斗式输送机
grab bucket crane 抓斗起重机
grab bucket dredge(r) 抓斗式挖泥船;抓斗式挖泥机
grab bucket opening and closing mechanism 抓斗启闭机构
grab capacity 抓斗
grab clamshell 抓岩机;(挖土机的)抓斗盘
grab closing circuit 抓斗闭合系统;抓斗闭合电路
grab closing drum 抓斗闭合索盘
grab closing rope 抓斗闭锁钢丝绳
grab crane 抓岩机吊车;抓斗(式)起重机;伸臂式起重机
grab crane bucket 起重机抓斗
grab discharge 抓斗卸货
grab-dredge(r) 攫斗挖泥机;抓斗(式)挖泥机
grab-dredge(r) with spud poles 钢桩抓斗挖泥船
grab-dredging 抓斗挖泥;抓斗挖掘船采矿
graben 断层槽;地堑;地沟
graben basin 地堑盆地
graben fault 地堑断层
graben faulting 断陷作用
graben-horst structure 地堑地垒式构造
graben lake 地堑湖
graben valley 地堑谷
graben zone 地堑带
Graber's organ 格氏器官
grab excavator 抓斗式挖泥机;抓斗式挖掘机
grab for excavating 挖掘用抓斗(机)
grab handle 抓柄;握柄
grab hoist 抓岩机绞车;抓斗绞车
grab hook 抓斗吊钩;起重钩;抓爪
grab hopper 装舱抓斗
grab hopper dredge(r) 装舱抓斗挖泥船;抓斗式自航挖泥船
grab hydraulic cylinder 抓斗液压缸
grab iron 抓钩;铁撬棍;铁扶手
grab jaw 抓铲
grab line 救生握索;牵索
grab loader 抓岩机
grab machine 抓斗机
grab method 攫取法
grab outreach 抓斗臂距
grab picking 粗选
grab pontoon 平底船抓斗
grab pontoon dredge(r) 平底船抓斗挖泥船
grab rail 靠墙扶手;扶手栏杆
grab rig 抓斗机具
grab rope 系索;救生握索
grab rotation 抓斗旋钩
grab rotation hydraulic motor 抓斗回转液压马达
grabs 链钩
grab sample 不定时采集的水样;定时取(集水)样;自由选取的试样;抓取样品;抓取水样;抓取试样;平均取集代表性样品;随意取样;手工取样;定时试样;定时取集的样品
grab sampler 抓斗式采样器;抓斗采样器;抓斗取样器;海底采样器;海底采样器;定时采购的样品
grab sampling 抓取法取样;拣块采样;定时取样
grab sampling machine 抓样机;简单取样机;手选取样机
grab set 急凝(混凝土);速凝
grab ship unloader 抓斗卸船机
grab slewing crane 回转式抓斗起重机
grab slewing mechanism 抓斗回转机构
grab spade 攫斗铲
grab spread 抓斗开度
grab stabilized line 抓斗稳定索
grab stabilizer 抓斗稳定器
grab strength 抓样强度

grab tensile strength 抓抗拉强度
grab tensile test 抓拉试验
grab test 抓样强力试验;布张力试验
grab tongs 起重夹钳
grab tooth 抓斗齿
grab traverse mechanism 抓斗横移机构
grab troll(e)y 抓斗小车
grab type 抓斗式
grab-type bale stock lifter 抓爪式草捆堆提升机
grab-type loader 抓斗式装载机;抓斗式装岩机
grab weight 抓斗重
grab with guide device 导板抓斗
grab with teeth 带(齿)抓斗;带牙抓斗
grace 宽限;宽惠
graceful degradation 故障弱化;机件故障降级操作;适度退化
graceful sweep 美丽的视野;精巧精整
grace payment 预定支付
grace period 优惠期;宽限期
grad(i)ometer 梯度磁强计
gradability 可分等级性;拖曳力;最大爬坡率;爬坡能力
gradable 可分类;可分级的;可分等
Gradall 挖掘平整机
gradate 顺次排列
gradated coating 过渡涂层
gradation 修坡;粒级作用;颗粒分级作用;均夷作用;渐近性;渐变;级配;级差;蔓延;多级过渡过程;等级;程度;层次;分粒作用;分类;分等(级);标度
gradational boundary 过渡边界;分级界线
gradational graduated acting 分级作用
gradational level 风化(表)面
gradation analysis test 粒度分析试验
gradation band 级配曲线范围
gradation change 级配变化;色调变化;色彩层次变化
gradation coefficient 分级系数
gradation colo(u)r 颜色的深浅程度;颜色的浓淡程度
gradation composition 粒径组成;粒度组成;级配组分;级配组成;配合成分
gradation control plant 级配控制设备
gradation curve 颗粒分析曲线;级配曲线;大小颗粒分布曲线
gradation dyeing 晕染
gradation etching 分层腐蚀
gradation factor 粒度分级系数;分级系数
gradation in size 粗细级配
gradation limit 级配限度;级配范围
gradation modulus 级配模数;级配模量
gradation of aggregate 骨料级配;骨料等级;集料级配
gradation of colo(u)r 色彩层次
gradation of gray 灰度层次
gradation of grinding media 研磨体级配
gradation of image 图像深淡程度
gradation of ore reserves 储量级别
gradation of stone 石料等级(路用)
gradation of test 级配筛分试验
gradation resistance 坡度阻力
gradation scale 颗粒分级标准;层次梯尺
gradation series 明暗层次级数
gradation sphericity factor 土的级配和球度系数
gradation test 粒度分析试验;粒度分析试验;颗粒级配筛分试验;级配(筛分)试验
gradation unit 连续投配器
gradatory 多级台阶(尤指由走廊进入教堂的台阶)
grade 整平地面;类别;径选;级别;内底高程;评级;品位;品级;牌号;室外地面;度;定坡度;电视等级;等级;程度;分级;标号
grade 4 chain 四极锚链
grade A 甲级;A级
grade ability 爬坡能力
grade adjustment 坡度调整
grade A fireprooffing door 甲级防火门
grade analysis 她别分析
grade analysis standard 粒度分析标准
grade A nickel 甲级镍
grade aqueduct 顺坡输水道
grade assistance 坡度助力
grade A wood 头等木材
grade B B级
grade bar 测坡杆;定幅范(在玻璃纤维分条整经

机上)
grade beam 斜坡梁;基础梁;合乎等级的梁;合格梁;地(基)梁
grade block 基础(墙顶层)砌块;合格混凝土块
grade braking 坡道制动
grade builder 拉坡推土机;斜板推土机;整坡机;推拉推土机
grade C C级
grade calculation 坡度计算
grade change point 变坡点
grade classification 等级分类
grade climbing 爬坡
grade climbing capacity 爬坡能力
grade collection efficiency 分级除尘效率
grade compensation 纵坡折减
grade compensation for curve 曲线折减坡度
grade computation 坡度计算
grade contour 等坡线
grade control 级配控制;坡度控制;品位控制
grade correction 倾斜(度)改正;坡度校正;坡度改正
grade course 防潮层;纵坡层;找平层;坡度层
grade crossing 铁路道口;平交(口);平面交叉;平交道(口);道口
grade-crossing-elimination structure 高架桥;道路立体交叉建筑物;立体交叉结构
grade crossing pavement 道口铺面
grade crossing predictor 道口信号预报器
grade crossing protection 平交道口防护
grade crossing signal 道口信号
grade crossing signal(l)er 道口信号机
grade crossing watchman 道口看守员
grade cross-section 平面交叉
grade D D级
graded 级配的;递级;分品;分级的;分次
graded aggregate 级配集料;级配骨料;分级粒料;分级集料
graded aggregate mixture 级配集料混合物;级配骨料混合物;级配集料混合料;级配混合集料;级配骨料混合料
graded aggregate pavement 级配路面
graded aggregate type 级配集料型;级配骨料型
graded aggregate type road mix(ture) surface 分级集料型路拌混合料路面
graded air-gap 不对称空气隙;阶梯形气隙
graded base (晶体管的)缓变基区;坡度基区
graded bed 序粒层;粒序层;粒级(递变)层;递变层
graded bedding 级配垫层;粒级层理;渐变层;级配基床;递变层理
graded bedding structure 递变层理构造
graded bench terrace 分级水平梯田
graded broken stone 级配碎石;分级碎石
graded cable 分层绝缘电缆
graded-channel fish pass 斜槽式鱼道
graded coarse sediment 级配粗的泥砂
graded coastline 均粒海岸线;均粒疾线
graded coating 颜色渐次变化的涂料;级配喷涂层;分层涂层
graded coefficient 级配系数
graded coil 分段线圈
graded colo(u)ring 分层设色法
graded column 分级桩
graded concrete aggregate 级配的混凝土骨料
graded contraction 间歇缩流
graded crossing 立体交叉
graded crushing 分级压碎;分级破碎;分段破碎
graded cyclic-sequence 递变式周期层序
graded-density filter 分级密度滤光片
graded-density particle board 密度分级刨花板;分级密度碎料板
graded-density skin 变密度表层
graded deposition 均粒沉积
graded depression 路面沉陷
graded description 质量等级说明;坡度说明;地基高度说明;等级说明
graded distribution 渐变分布;梯度分布
graded distribution method 梯形分配法
graded down 压低等级
graded dry distillation liquefaction 分段干馏液化
graded earth road 整轧土路;整形土路
graded effect 量效应
graded energy gap solar cell 分级能带宽度太阳能电池
grade design 坡度设计

graded factor 级配系数
graded filter 颗粒过滤器；级配（砂）滤池；级配（反）滤层；级配（倒）滤层；回水过滤器；滤波器；分级滤波器；分级过滤器；分级过滤池
graded filter drain 级配反滤层排水沟
graded filter of drainage 排水沟级配反滤层
graded fine sand 级配细砂
graded glass powder 筛分过的玻璃粉；分级的玻璃粉
graded glass seal 玻璃递级封接
graded gravel 砾石级配；级配砾石；分级砾石
graded grave mix(ture) 分级的砾石混合物
graded hardening 分级淬火
graded height 均衡高度
graded hourly rate 分级小时制；分级计时工资
graded hydrolysis 分段水解
grade diagram of line 线路坡度图
graded illumination 分级照明
graded-index 缓变折射率
graded-index fiber[fibre] 渐变型光纤；分级检索纤维；渐变折射率光纤
graded-index glass fiber[fibre] 渐变折射率玻璃纤维
graded-index optic(al) fiber[fibre] 集束性光导纤维；渐变折射率光学纤维
graded-index optic(al) waveguide 渐变折射率光波导
graded-index profile 渐变折射率剖面；渐变折射率分布
graded-index waveguide fibre 梯度波导纤维
graded insulated cable 分段绝缘电缆
graded insulated winding 分级绝缘绕组
graded insulating winding 分级绝缘绕组
graded insulation 分段绝缘
graded insulation transformer 分段绝缘变压器
graded insulator 分层绝缘子
grade distribution 粒径分配
grade division 分度尺
graded joint 递级接头
graded junction 缓变结
graded-junction phototransistor 缓变结光电晶体管；梯度结光电晶体管
graded-junction transistor 变速生长晶体管
graded layer 粒级层
graded lease 定期加租（按通胀率）
graded level 均衡水位；均衡高度
graded macadam 级配碎石
graded material 级配（物）料；级配泥沙
graded maturities 分类（的）满期日
graded modulus 级配模数；级配模量
graded multiple 分品复接
graded net 等级网格
graded organization 分级制度（土地、时间）
grade down 降序；按比例折减
graded particle 分级颗粒；渐变密度碎料板
graded particle board 分等刨花板
graded plain 均夷平原
graded potential 分段电势
graded potential system 坡度电位法
graded potentiometer 非线性电势计
graded-powder bonding 递级粉末封接
graded pressing 分段压制
graded product 分级产品
graded profile 均衡剖面
graded rate of products 产品等级率
graded reach （河床比降变化不大的）平缓河段；冲淤均衡段
graded response 等级应答
graded rhythmite 粒级韵律层；递变韵律层
graded riprap slope 级配抛石斜坡
graded river 均衡河流；缓达；坡度平缓的河流；冲淤均衡河流
graded river floodplain 均衡河（流的洪）泛平原
graded river section 冲淤均衡河段
graded-rubble mound with armo(u)r rock 分级堆石防波堤
graded sand 过筛砂；级配砂；分级砂
graded sand and stone 级配砂石
graded sand mix(ture) 筛过的混合砂；分级过的混合砂
graded sandstone 级配砂岩
graded sand-stone mixture 级配砂石
graded scale curve 刻度弧规；分度弧规
graded seal 过渡焊接；递级封接

graded seal glass tubing 递级封接玻璃管
graded sediment 粒度递变沉积物；均粒沉积(物)；级配泥沙；级配沉积物；级配沉积；分级沉积
graded sediments 递变沉积
graded shoreline 均夷滨线；均粒海岸线；均粒滨线
graded shoulder 整平路肩；整修路肩
graded shunt arrester 多级分路避雷器
graded sizes 骨料分级规格尺寸；集料分级规格尺寸；集料分级大小；级配大小；分级粒度；分级规格尺寸
graded slag filling 级配矿渣填料
graded slope 冲淤均衡比降；均衡坡面
graded soil 级配土
graded-soil mixture 级配(砂)土混合料
graded standard sand 渥太华砂；级配标准砂；标准级配砂
graded stream 均衡河流；缓变平衡河槽；缓变均衡河川；平缓河流；冲淤均衡河流
graded stream section 冲淤均衡河段
graded suspension 递变悬浮液
graded tasks 分等工作
graded tax 分级税；分等税
graded terrace 分级台地；分级排屋；斜坡梯田
graded texture 粒级结构；级配结构
graded thermoelectric(al) arm 分段热电臂
graded tidal-cyclic sequence 递变式潮后周期层序
graded time （冲淤的）均衡时期
graded time-lag relay 可调(整)延时继电器；分段延时继电器
graded time step 分段限时
graded time step-voltage test 按时升压试验
graded topocline 梯度地理倾差
graded trunk line 分品中断线
graded tube 刻度管
graded use of energy 能源梯级利用
graded width 修整宽度
grade easement 坡度缓和；坡度改进
grade elevation of road surface 路面标高；路面高程；坡度线高程
grade elimination 高架桥；高梁桥；立交；减缓坡度；坡度减缓
grade estimation 质量评定；等级评定
grade evaluation 质量评价；质量评定
grade factor 质量等级因素；级配因素
grade five discontinuity 五级结构面
grade five texture body 五级结构体
grade floor 有坡度的楼面
grade for butt seam inspection 对接焊缝探伤合格级别
grade for starting 起动坡度
grade four discontinuity 四级结构面
grade four texture body 四级结构体
grade frame 刻度框
grade grid 经纬线网；分度网格
grade ground level 地面高程
grade incline 倾斜度
grade indicator 坡度标
grade in longitudinal direction 纵向坡度
grade intersection 平交口
grade in tunnel 隧道倾斜度
grade labelling 商品质量的标签说明；按质分等级
grade length limitation 坡长限制
grade level 平整(后)地坪高程；地坪高程
grade-level elevator 半室外运货升降机；半地面运货升降机
grade limit(ation) 极限坡度；坡度限制
grade line 准线；定位线；巷道腰线；纵坡线；纵侧面线；坡(度)线
grade line elevation 坡度线高程
grade line level 地面线
grade line limit 坡长限制
grade line of canal 渠道纵坡线
grade location 路基设计；坡度设计；坡度测设
grade man 推土机手
grade material 级配物料
grade measurement 坡度测量
grade meridian measurement 子午线弧度测量
grade of abrasive tool 磨具硬度
grade of activity of clay mineral 黏性土的活动性等级
grade of bearing wearing 轴承磨损级别
grade of bit wearing 钻头磨损级别
grade of brick 砖标号
grade of brilliance 亮度

grade of cave stability 溶洞稳定性分级
grade of cement 水泥质量等级；水泥强度等级
grade of cement mortar 水泥砂浆强度等级
grade of cement quality 水泥质量等级
grade of city and town 城镇等级
grade of coal 煤炭(品质)分类；煤(分)级
grade of concentration 浓度等级；富集品位
grade of concrete 混凝土强度等级
grade of crude ore 出矿品位
grade of diamonds 金刚石品级；金刚石类别；金刚石级别；金刚石等级
grade of discharge 泄水坡度
grade of discontinuity 结构面等级
grade of electric(al) meter 电表等级
grade of finish 表面光洁度等级
grade of fit 配合度；配合等级；适合度；适航等级
grade of floatability 可浮性等级
grade of geosyncline-platform 地槽地台的级别
grade of goods transport accident 货运事故等级
grade of ground check on aeromagnetic anomalies 航磁异常地面查证等级
grade of hump yard tracks 驼峰调车场线路坡度
grade of hydraulic projects 水利工程等级
grade of inhomogeneity of carbide 碳化物不均匀性级别
grade of investigation district 调查地区等级
grade of levee crown 堤顶纵坡
grade of load （起重机的）负载等级
grade of lock condition 闸况等级
grade of material 材料等级
grade of metamorphism 变质程度
grade of meter 电表等级
grade of mined ore 采矿品位
grade of mortar 砂浆强度等级
grade of oil 石油品种
grade of oil and gas bed 油气层等级
grade of ore 矿石品级
grade of out-of-ga(u)ge 超限等级
grade of physical property of minerals 矿物物理性质度量
grade of pipe 管子等级
grade of prospecting of natural building material 天然建材勘察级别
grade of prospective district 远景区级别
grade of railway 铁路等级
grade of reserves 储量级别
grade of resource 资源量分级
grade of risk 危险性分级
grade of service 业务等级；工作良好度；服务范围；服务级
grade of side slope 边坡坡度
grade of slope 坡度（等级）
grade of stability of regional crust 区域地壳稳定性分级
grade of station site 站坪坡度
grade of steel 钢筋等级；钢号
grade of switch area 道岔区坡
grade of texture body 结构体等级
grade of tolerance 公差等级
grade of tooth wearing 牙齿磨损级别
grade of transmission 传输等级
grade of tubing steel 油管钢级
grade of washability 可洗选性等级
grade of water in agriculture 农业用水评价等级
grade of water temperature 水的温度分级
grade one discontinuity 一级结构面
grade one texture body 一级结构体
grade parallel measurement 平行圈弧度测量
grade parking performance 坡道驻车性能
grade peg 坡度(线)标桩；坡度桩【测】
grade planer 路基平整机；分级刨床
grade point 坡(度)点；变坡点
grade post 坡度标（桩）
grade products according to quality 按质量分等
grade profile 斜техн纵剖面（图）；坡度纵剖面；坡度剖面图（道路等）
grade protection layer 基面保护层；护坡层
grader 平土机(械)；平路机；平地(机)；分选工；分级机
grader and separator 分离分选机
grade rate 坡度率
grader blade 平地机刀片
grade record 坡度记录
grade rectification 坡度校正；坡道化直

grade reduction 坡度折减;坡度减小
grade reduction problem 降he问题
grade reference 坡度基准
grader elevator 平土升运机
grade resistance 坡度阻力;坡道附加阻力
grade ripper 耙路机
grader man 土壤分选工;平土机手
grade rod 坡度尺;水准标尺;水平尺
grader scraper 刮土平地(造坡)机
grade scale 粒级(标准);粒级标度;粒度分级标准;颗粒分级标准;土粒大小尺度;分级标准
grade scale of coal quality 煤质主要指标的分级标准
grade section 坡段
grade sensor 坡度感测器
grade-separated 立体式的;不同高度的
grade-separated bifurcation 分叉式立体交叉
grade-separated bridge 立交桥;跨线桥
grade-separated interchange 道路立体枢纽;立体交叉(道路)
grade-separated intersection 道路分层交汇处
grade-separated junction 立交枢纽;不同坡度的交汇点;立体交叉
grade-separated structure 道路立体交叉结构
grade-separating efficiency 分离级效率
grade separation 立交;立体交叉;简单立体交叉;四叶形立体交叉;等级分类;分级配
grade-separation bridge 跨线桥
grade-separation junction 立体交叉
grade-separation of pedestrian 人行立体交叉
grade-separation structure 立体交叉建筑物;立体交叉结构
grade severity rating sign 下坡速率路标
grades for building (接近住房外墙的)地面高度;住房等级
grades for penstock tunnel 水渠隧道坡度
grades for rail haulage 有轨运输坡度
grades for trucks 车行坡度
grade slab 斜坡板块;地基板
grades of airborne survey perspective area 航测远景区级别
grades of house construction 房屋建筑等级
grades of rosin 松香的颜色等级
grades of solid mineral reserves 固体矿产储量分级
grades of tax rates 税率等级
grades of water conservancy engineering 水工建筑物级别
grade specific gravity 分选比重
grade specific gravity ±0.1 rate 分选比重正负0.1率
grade stability 纵坡稳定
grade stabilization 边坡加固;纵坡稳定
grade stabilization structure 固床建筑物
grade stabilizing structure 固床设施;固床建筑物;固坡结构
grade-staff 土石方工程工长;领工员(俚语)
grade stake 填挖方高程标准桩;道路坡度桩;纵坡标桩;护桩;坡度(标)桩;平整标桩;填挖桩;水平桩;高程桩
grade standard 分级标准
grade stream 定坡河川
grade stress 等级应力
grade strip 分段条;模盒标尺板条(混凝土路坡)
grade surface 坡面
grade tariff 多级税率
grade tester 磨具硬度计
grade texture of grain 颗粒粒级结构
grade three-chain 三级锚链
grade three-discontinuity 三级结构面
grade three-texture body 三级结构体
grade trimmer 坡度整平机
gradetto 圆箍处脚;嵌条处脚;环形处脚
grade tunnel 有水力梯度的隧洞;缓坡隧道;缓坡隧道;不满流隧洞
grade two chain 二级锚链
grade two-discontinuity 二级结构面
grade two texture body 二级结构体
grade washer 倾斜垫圈
grade waste pond 磨料粉砂沉积池
grade width 路基宽度;地基宽度
grade wire 坡度控制线
Gradhov number 格拉舍夫数
gradient 斜率;增减率;阶度;倾度;坡度;陡度
gradient analysis 梯度分析
gradient and Laplacian operator 梯度和拉普拉斯算子
gradient anomaly curve 梯度异常曲线
gradient anomaly curve of mise-a-la-masse method 充电法梯度异常曲线
gradient array of single electrode 单极梯度装置
gradient board 指示坡度的路标牌;坡度牌;坡道指示牌;测斜仪;测斜板
gradient brack 坡度折点
gradient break 坡度转折(点);坡折
gradient break point 坡度转折点
gradient calorimeter 梯度形测热器
gradient centrifugation 梯度离心(法)
gradient coefficient 坡度修正系数
gradient constant 斜率常数;梯度常数
gradient corrector 梯度校正器
gradient coupling 梯度耦合
gradient covariant vector 梯度协变矢量
gradient current 坡面流;坡降流;坡度流;梯度流
gradient curve 水蚀曲线;冲刷曲线;梯度曲线
gradient descent 梯度下降
gradient determination 梯度确定
gradient device 梯度装置
gradient diagram of line 线路坡度图
gradient distribution 梯度分布
gradient drift current 梯度漂移电流
gradient eigenvector method 梯度本征向量法
gradient elevated track 坡式高架线
gradient elution 梯度洗脱;梯度洗提;梯度淋洗
gradient elution adsorbent system 梯度洗脱吸附剂系统;梯度洗提吸附系统
gradient elution analysis 梯度洗脱分析
gradient elution band press filter 梯度洗脱谱带压滤
gradient elution bandwidth 梯度洗脱谱带宽度
gradient elution chromatography 梯度淋洗色谱法
gradient elution device 梯度洗脱器
gradient elution partition chromatography 梯度洗脱分配色谱(法)
gradient elution separation 梯度洗脱分离
gradient equation 梯度方程
gradienter 斜度仪;斜度计;倾斜计;倾度测定器;水准仪;水平仪;测斜率仪;测梯度仪
gradienter screw 有刻度倾斜螺旋
gradient factor 均差化等额因数;级配系数;分选系数
gradient field of water head 水头梯度场
gradient flow 梯度(风气)流
gradient former 梯度模型
gradient freeze growth apparatus 梯度凝固生长装置
gradient freeze technique 梯度凝固技术
gradient freezing 梯度冷凝
gradient furnace 梯温炉;梯度炉
gradient grid 梯度栅
gradient hole 地热梯度井
gradient hydrophone 压差水听器
gradient improvement 坡度改善
gradient index 梯度指数
gradient index fiber 梯度折射率光纤
gradient index optic(al) fiber 梯度折射率光学纤维
gradient indicator 坡度标
gradient irrigation 沿坡沟灌
gradient layer 梯度层
gradient length 坡段长度
gradient level 梯度风高度
gradient line 梯度线
gradient mapping 梯度映射
gradient mask 梯度掩模
gradient matched filter 梯度匹配滤波器
gradient matrix 梯度矩阵
gradient measurement in self-potential method 自然电场法梯度测量
gradient meter 坡度测定仪;量坡仪;测斜器;测斜仪;高度指示器;梯度计;磁偏计;测斜仪;测坡器
gradient method 斜量法;快速下降法;梯度法
gradient microphone 压差传声器
gradient mixer 梯度混合器
gradient mixing device 梯度混合装置
gradient mixing device system 梯度混合装置系统
gradient model 梯度模型
gradient of communities and environments 群落和环境梯度
gradient of density 密度梯度
gradient of ecosystem 生态系统梯度
gradient of equal traction 等牵引力坡度
gradient of field variable 场变量的梯度
gradient of friction 摩阻梯度
gradient of gravity 重力梯度
gradient of groundwater table 地下水位坡降;地下水比降
gradient of head 水头梯度
gradient of inclined well 斜井坡度
gradient of moisture 湿气梯度;湿度梯度
gradient of piezometric head 水压比降;测压管水头比降
gradient of position function 位置函数梯度
gradient of position line 位置线梯度
gradient of potential energy 势能梯度
gradient of pressure 压力梯度
gradient of river 河流比降;河道坡降
gradient of riverbed 河床坡降;河床坡度;河床比降
gradient of side ditch 沟底坡度
gradient of slope 斜坡坡度;倾斜率;坡斜度;坡度角
gradient of station site 站坪坡度
gradient of stream 河流坡降;河流坡度;河流比降
gradient of temperature 温度梯度;温度升降
gradient of velocity 速度梯度
gradient of water head 水头梯度
gradient of water table 地下水位坡降;地下水位坡度;地下水位比降
gradient operation 梯度操作
gradient operator 梯度运算符;梯度算子
gradient overcome by momentum 动能闯坡
gradient peg 斜坡标桩;填土标桩;坡度桩【测】
gradient pitch 级配比降
gradient plate 梯度浓度培养皿
gradient plate method 倾斜平板法
gradient post 坡度标(桩)
gradient preference 梯度适应
gradient profile 坡度纵断面
gradient projection 梯度投影
gradient projection algorithm 梯度投影算法
gradient projection method 梯度投影法
gradient quenching 顶端淬火
gradient range 梯度范围
gradient ratio 比降;坡降;坡道值
gradient ratio test 梯度比试验
gradient related method 梯度相关法
gradient-related method of statistic(al) inference 统计推论的梯度相关法
gradient resistance 坡度阻力;坡道(附加)阻力
gradient run 梯度操作
gradient scale 坡度尺
gradient screen survey 梯度筛选调查
gradient screw 倾斜螺旋
gradient search 梯度搜索
gradient section 坡道段
gradient separation 梯度分离
gradient separation factor 梯度分离因素
gradient separation process 梯度分离技术
gradient separation technique 梯度分离技术
gradient series factor 等差系列换算因子
gradient shape 梯度形状
gradient sign 坡度标
gradient slope 梯度坡度
gradient solvent 梯度溶剂
gradient speed 梯度速度;定标法感光度
gradient stake 斜坡标桩;填土标桩
gradient start 梯度起点
gradient steepness 梯度倾斜度
gradient survey 梯度测量
gradient survey in mise-a-la-masse method 充电法梯度测量
gradient terrace 内倾水平梯田;坡田
gradient theory 梯度理论
gradient thin-layer chromatography 梯度薄层色谱(法)
gradient time 梯度时间
gradient tint negative 梯度色调底片
gradient tints 高程分层设色;高层分层设色;海图上表明高度或水深的着色
gradient transfer theory 梯度输送理论
gradient transfer theory for atmospheric diffusion 大气扩散梯度输送理论
gradient type 梯度形式

gradient vector 梯度向量;梯度矢量
gradient velocity 梯度速度;梯度风速
gradient volume 梯度体积
gradient waveguide fiber 梯度波导纤维
gradient wind 倾度风;梯度风
gradient wind equation 梯度风公式
gradient wind level 地转风高度
gradient within switching area 道岔区坡
gradient zone of gravity anomaly 重力异常梯度带
gradimeter 测斜仪
gradin(e) 圣台背面突出物;低踏步;阶梯的一级;梯度式座位;齿凿
grading 修坡;削坡;整坡;路基施工;路基平整;颗粒级配;级配;平整土方;平整土地;平土;配级;土方修整;土地平整;定坡度;分选;分品连接;分品法;分等(级);土方平整
grading abrasive wheels 砂轮分级
grading analysis 粒径分析;粒度分析;颗粒(级配)分析;级配分析
grading and shielding ring 均压屏蔽环
grading and washing plant 分级清洗设备
grading angle 坡角(度)
grading attachment 水准测量附件;筛选附件;分选附件
grading board 刮板
grading bracket 级配托板;分选托座
grading characteristic of sediment 泥沙级配特性
grading charts 级配图表
grading coils 分段绕制线圈
grading collapse settlement 分级湿陷量
grading control 粒度控制;级配控制
grading curve 粒径曲线;粒径(分配)曲线;粒径分布曲线;粒度(分布)曲线;颗粒级配曲线;级配曲线;土积曲线
grading curve of aggregate 集料级配曲线
grading curve representation in normal scale 正常尺度级配曲线;正常尺度分级曲线;正常尺寸筛分曲线
grading diamond 对金刚石分级
grading drain 坡度排水
grading efficiency 筛选效率;分选效率
grading elevation 路基高程
grading envelope 筛分范围;粒度范围;级配范围
grading equipment 路基整平设备;整地机械;路基施工机械;土平整设备
grading factor 粒度比;级配系数;分选因子;分选因素;分选系数
grading fraction 粒径分级
grading function 照明函数
grading group 分品群;分级群
grading instrument 斜度仪;坡度(测定)仪;测坡水准仪
grading into uniform size 按尺寸分级
grading job 路基整平工作;平土工作;整平工作;土工平整工作
grading level(l)ing 平整场地
grading limit 粒度范围;级配范围
grading limitation 粒度范围;级配范围
grading machine 土工平整机;路基整平机;整地机;整地工程;分级机;分级机
grading machinery 分级机械
grading map 土方地形整理图
grading method 分品法;分级评价法;分等级方法
grading of aggregate 集料级配;骨料级配;骨粒等级
grading of aggregation 集料级配
grading of collapse settlement 分级湿陷量
grading of core material 堤心石的分级
grading of grog 熟料过筛
grading of nodules 料球分级
grading of river 河流均夷作用
grading of(river)bank 河岸整坡;河槽整坡
grading of rock mound 抛石围坡
grading of sediment 泥沙级配
grading of soil 土壤级配;土的级配
grading of soil salinity 土壤盐渍度分级
grading of stream 河流均夷作用
grading of stream bank 河槽整坡
grading of timber 等级材;按质分类(木材);木材分级;木材等级
grading operation 筛选操作;土工平整工作;平土工作;整坡作业;平土作业;土地平整工作;土方平整
grading outfit 路基整平设备;整坡机具;平土机具;土工平整设备;土方平整设备
grading plan 坡度断面图;地形及路面高程图
grading plant 拣选装置;筛分装置;筛分机;筛分厂;分级装置
grading price 修建路基造价;路基造价
grading principle 分选原理
grading quality 筛选质量;分选质量
grading range 筛分范围;级配范围;粒度范围;分级范围
grading reel 分选圆筒筛;分级圆筛
grading requirements 路基整平标准;级配要求;级配标准
grading requirements of aggregate 骨料级配要求
grading resistance 分段电阻
grading ring 均压环;屏蔽环;分段环
grading rule 分等规则;分等标准;分级规范
grading scale 粒度范围
grading screen 分极筛;分级筛
grading shield 分段屏蔽
grading specification 级配规范;按质分等的标准
grading standard 分级标准;叙级标准
grading system 粒径分级系统
grading test 粒径分析试验;级配(筛分)试验
grading tool 筛选工具;平土工具;分选工具;手锥(测定磨具硬度工具)
grading toxicity 毒性等级
grading under pavement 路面下层整平
grading with frame offset 框架横移地进行筛分
grading work 土工整平工作;平土工作;整坡作业;平整土方;平土作业;土方平整
grading zone 级配区
gradiograph 测坡仪;测坡器;测斜仪(测量铺设排水管的坡度)
gradiomanometer 压差密度计
gradiometer 重力梯度仪;量坡仪;梯仪;梯度计;斜度仪;斜度计;倾斜(测定)器;坡度(测定)仪;测斜仪;测坡仪;测坡器
gradiometer sensitivity 梯度仪灵敏度
gradiometer survey 重力梯度计测量
gradiometry 重力梯度测量
gradocol filter 分级滤器
gradometer 坡度测定仪;磁梯度计
gradual application 阶段制动
gradual approximation 渐次近似法
gradual attenuation 平缓衰减
gradual braking 阶段制动
gradual burst 缓慢爆炸
gradual case analysis 渐次分析法
gradual change 徐变;渐变(的)
gradual construction 分阶段施工
gradual continual non-regular variation 逐渐连续不规则
gradual continual regular variation 逐渐连续有规则
gradual contraction 断面渐渐缩小;逐渐缩小;逐渐收缩;渐变收缩
gradual contraction of section 截面渐渐缩小
gradual convergence 逐渐聚焦
gradual cooling 逐渐冷却
gradual cut filter 阶梯式截止滤光片
gradual degradation 逐渐退化
gradual development 逐级展开
gradual ecotone 渐进式生态过渡带
gradual engagement 逐渐接合
gradual enlargement 逐步扩大
gradual enlargement of section 截面逐渐扩大
gradual execution 逐步实施;分阶段施工
gradual exhaustion 渐次消耗
gradual expanding 逐渐扩张
gradual failure 渐变故障;逐渐失效;逐步失效;渐变失效
gradual hydraulic jump 渐变水跃
gradual incline method 逐次倾斜法
gradualism 渐变论
gradual load 逐渐加载(法);渐加荷载;缓慢加载
gradual loss 渐次损失;渐变损失
gradually 逐渐地
gradually applied load 逐渐施加荷载;渐加荷载;缓慢加载
gradually increased the cultivation area 栽培面积逐步增加
gradually thinning 逐渐变薄
gradually varied curve widening 加宽渐变段
gradually varied flow 渐变流;缓变流
gradually varied unsteady flow 渐变不稳定流
gradual massing 逐渐集中
gradual narrowing 逐渐缩窄
gradualness 逐次
gradual pollution 渐进式污染
gradual-pressure crusher 逐渐增压式破碎机
gradual production 缓慢产生
gradual reduction of displacement 位移渐小
gradual release 逐渐释放;阶段缓解
gradual rolloff 逐渐跌落
gradual-separation 逐渐分离
gradual settlement 逐渐沉陷;逐渐沉降
gradual slope 缓坡
gradual transition 逐次跃迁;渐变过渡段
gradual variation 逐渐变化;渐变
gradual widening 逐渐扩张;逐渐放宽
graduate 刻度量器;分度;分等级
graduate braking 逐渐制动
graduated 累进的;渐进的;分级的;分度的
graduated acting 刻度控制器;刻度设备
graduated application 阶段制动
graduated arc 分度弧;渐变弧
graduated arc limb 弧形刻度板
graduated bar 分度标杆
graduated braking 阶段制动
graduated cable 测绳;分度缆
graduated calliper 刻度卡钳
graduated circle 刻度盘;度盘;分度盘;分度环
graduated coating 分层镀层
graduated collar 分度环
graduated course 渐变层;分等层;分层砖石(砌体);递减厚的石板铺设法;递减行距瓦层;迭减行距瓦层
graduated cylinder 量筒;刻度量筒
graduated dial 标度盘
graduated disk 刻度盘
graduated engineer 特许工程师
graduated eyepiece 刻度目镜
graduated filter 渐变滤光镜;分级滤光镜;分度滤光片
graduated flask 量瓶;刻度量瓶
graduated float rod 分度浮尺
graduated float tape 分度浮尺
graduated glass 量杯;刻度杯;量筒
graduated glass(ware) 玻璃量器
graduated hardening 分级淬火
graduated hopper-charging 用定量计加料;定量的斗容量
graduated horizontal circle 水平度盘
graduated income tax 分级收入税
graduated inside calliper 刻度内卡钳
graduate diploma/graduate degree's diploma 研究生毕业证/学位证
graduated lease 分级分期计租租约
graduated limb 弧形刻度板
graduated L-square 刻度角尺
graduated measure 刻度量杯
graduated measuring glass cylinder 玻璃刻度量筒
graduated measuring rod 水准测量标尺
graduated meridians 分度子午线
graduated outside calliper 刻度外卡钳
graduated parallels 分度纬线
graduated patterns 色彩渐变花纹
graduated payment adjustable rate mortgage 递增付款可调利率抵押贷款
graduated payment mortgage 累进支付抵押;分级支付抵押;递增偿还抵押贷款
graduated rate schedule 超额佣金递增办法
graduated release valve 阶段缓解的三通阀
graduated rental lease 递增租金租赁
graduated rheostat 分级变阻器
graduated ring 刻度圈;分度盘;分度圈
graduated rod 量尺;刻度尺;测深杆;分度标杆
graduated round bottom centrifuge tube 圆底有刻度离心管
graduated rule 分度尺
graduated scale 分度标;英制刻度;分度盘;分度尺;比例尺
graduated scale line 分划线
graduated sign 等级符号
graduated staff 分度标杆
graduated staff ga(u)ge 刻度水位标尺
graduated streaking chart 梯度拖影测试卡
graduated string 几段不同套管组成的套管柱;复合套管柱

graduated symbol 等级符号
graduated symbol map 分级统计图
graduated taper 刻度锥;刻度退拔
graduated tax(ation) 分级(深)税
graduated tax rates 分级税率
graduated vessel 刻度容器
graduated ware 带刻度器皿
graduate mark 分度标志
graduate payment loan 累进偿付贷款
graduate rod 分度标杆
graduate scale 分度标尺
graduate school 研究院(大学)
graduate student 研究生
graduate tax 毕业税
graduating device 刻线装置;减速装置
graduating diaphragm 分度光阑
graduating function 阶梯函数
graduating machine 分度机
graduating mark 分划线
graduating nut 递动杆螺母
graduating release position 阶段缓解位
graduating sleeve 递动弹簧套
graduating spring 节制弹簧;递动弹簧
graduating spring sleeve 递动弹簧套
graduating stem 递动杆
graduating stem nut 递动杆螺母
graduating valve 节制阀
graduating valve lever 递开阀杆
graduating valve spring 递开阀簧
graduation 修均法;加浓;分度
graduation arc 刻度弧
graduation error 刻度误差;分度误差
graduation error of micrometer 测微器分划误差
graduation in degrees 按度分刻度
graduation interval 分度间隔
graduation line 分划线;分度线;标度线
graduation mark 刻度线;分度线;分度符号;分度标记;标度线
graduation of curve 曲线修整;曲线修匀
graduation of data 数据修匀;数据(的)修整法;数据的修匀法
graduation of grinding ball 磨球分级
graduation of micrometer for vertical circle 垂直度盘测微器分划值
graduation of optic(al) micrometer 光学测微器分划值
graduation of the motor currents 电动机电流级加法
graduation scale 刻度
graduations in degrees 分度
graduation tower 梯塔
graduation value 刻度值
graduation work 校正工作
graduator 刻度员;均整线圈;分度器
Graebe-Ulmann carbazole synthesis 格雷伯—乌尔曼咔唑合成法
Graebe-Ullmann reaction 格雷伯—乌尔曼反应
Graeco-Roman Corinthian column 古希腊罗马考林辛式柱
Graeco-Roman facade 古希腊罗马建筑物正面;古希腊罗马建筑物立面
Graeffe's method 格拉夫方法
Graef rotor 平吹旋转炉;筒式旋转氧气炼钢炉
graemite 水碲铜石
Graetz number 格雷兹数
Graetz problem 格雷兹问题
graffiti 在壁画上乱涂;壁画末道浆
graffiti-marred surface 被涂鸦的表面
graffiti-out surface 被乱涂乱画的表面
graffiti removal 壁画上涂鸦清除
graffito 乱涂抹画;彩色右毛(粉刷);壁画末道浆;水泥浆露894处理;石膏图案装饰;仿雕刻装修
graffito pollution 涂写污染
graffito tile 双色涂层瓷砖;彩色毛面瓷砖
Grafmaster 格拉夫马斯特图形生成系统
Graf sea gravimeter 格拉夫海洋重力仪
graft copolymer 嫁接共聚物
grafted ternary ethylenepropylene rubber modifier 接枝三元乙丙橡胶改性剂
grafter 平锹;平铲
grafting 室内嫁接;分级复联
grafting clay 接泥
grafting knife 嫁接刀;接枝刀
grafting material 接合材料;嫁接移植物

grafting proper 枝接
grafting technique 嫁接技术
grafting tool 泥铲;平锹;平铲;土锄
grafting wax 接蜡
graft modification 接枝改性
graftonite 钙磷铁锰矿;磷锰铁矿
graft polymer 接枝聚合物
graft polymerization 接枝聚合(物)
graft rubber 接枝橡胶(高分子)
graged finish 摩擦轧光整理
gragon beam 承托脊橡梁
Graham's law 格拉海姆定律
Graham escapement 格拉海姆式擒纵机构
grahamite 葛氏脆沥青;天然脆沥青;脆沥青
Graham normal polarity hyperchron 格拉海姆正向极性巨时
Graham normal polarity hyperchronzone 格拉海姆正向极性巨时时间带
Graham normal polarity hyperzone 格拉海姆正向极性巨带
Graham salt 格拉海姆盐(可溶性偏磷酸钠)
Grahm'a law of diffusion 格拉海姆扩散定律
grail 细砾(石);盘;砂砾;鹅卵石
grain 易劈向;谷物;谷粒;谷类(货);粮食;颗粒;火药筒;纹理;地势趋向;仿木纹
grain-abrasion testing machine 磨粒磨损试验机
grain accumulation curve 颗粒累积曲线
grain adhesive 颗粒黏合剂;颗粒胶黏剂
grain aeration 谷物通风
grain aerator 谷物通风器
grainage 粒度;颗粒尺寸
grain agitator 谷物搅动器
grain alcohol 乙醇;酒精
grain and oil storage 粮油储[贮]藏
grain angle 纹理角度
grain arrangement 晶体方位;晶粒结构;晶粒方位
grainary 粮食仓
grain attack 晶粒腐蚀
grain attrition 颗粒磨损;颗粒磨耗;颗粒磨擦
grain auger 谷粒螺旋
grain base 粮食基地
grain bin 谷仓
grain bin aerator 粮箱通风器
grain bin for tipping trailer 翻斗拖车粮箱
grain blower 谷粒吹送器
grain board 谷粒滑板;木纹板
grain body 谷物运载车厢
grain bottom auger 底部谷粒螺旋推运器
grain boundary 粒径界限;颗粒界限;颗粒界;颗粒边界;晶粒间界;晶粒界限;晶界
grain boundary adhesion 颗粒边界黏附
grain-boundary area 晶界区
grain-boundary crack 晶界疏松;晶界裂纹
grain-boundary cracking 晶界间界开裂
grain-boundary energy 晶粒间界能
grain boundary eutectic 晶界共晶体
grain-boundary flow 晶界流变
grain-boundary migration 晶粒间界迁动
grain-boundary movement 晶粒间界运动
grain-boundary plane 晶粒边界面
grain-boundary precipitation 晶粒间界淀积
grain-boundary segregation 晶界偏析
grain-boundary separation 晶界分离
grain-boundary shape 晶界形状
grain-boundary sliding 晶界滑动
grain boundary slip 晶粒边界滑移
grain-boundary strength 晶界强度
grain-boundary weakness 晶界弱化
grain bounding 粒径界限
grain breakage 颗粒破碎率
grain-brush 谷物刷;清粮机刷
grain bulkhead 谷物舱壁;散装隔舱板
grain-by-grain selection 粒选
grain capacity 散装容积;散装舱容;散粮舱容
grain cargo certificate 谷物运输许可证
grain cargo hold 谷舱
grain carrier 运粮拖车;运粮船;谷物运送器;谷物运输船;粮谷专用船;散粮船
grain carrier beater 逐镐轮
grain cart 运粮拖车
grain casting 装药铸造
grain catcher 集谷器
grain chaff 谷糠
grain character 纹理特征

grain check canvas 帆布挡帘
grain checking 木纹状微裂纹
grain chute 谷粒滑槽
grain clearing fan 谷粒清选风扇
grain coarsening 晶粒变粗
grain collector 集粮器
grain collision 颗粒碰撞
grain colony 晶粒团;晶粒范围
grain column 颗粒纵列
grain complex 土粒复合体
grain composition 颗粒组成;颗粒级配(成分);晶粒组成
grain composition analysis of soil 土的颗粒成分分析
grain composition of aggregate 集料的颗粒组成
grain conduction 粒间电导;颗粒电导
grain conductor 导种管
grain connection 颗粒间连接
grain contamination 粮食污染
grain contrast 晶粒对比
grain control 颗粒控制
grain conveyer 谷物输送器;散粮输送机
grain cooling system 谷物降温系统
grain co-ordination number 晶粒配位数
grain count 颗粒计数
grain cover 谷物压送升运带
grain cracking 木纹开裂
grain crop 谷类作物
grain crusher 谷物碾碎机;碎谷机
grain cubic 散装舱容
grain-curve 粒度曲线
grain damage 穿晶损伤
grain deflector 谷物导板(用于调整割晒机的铺条宽度)
grain density 颗粒密度
grain depot 粮栈;粮库
grain deterioration 谷物变质;粮食变质
grain diameter 颗粒尺寸;粒径;颗粒直径
grain direction 木纹方向;流线方向;纹理方向
grain discharge vertical auger 立式卸粮螺旋
grain distillery 酒精厂
grain distribution 颗粒分布
grain distribution curve 颗粒分布曲线
grain distribution station 粮站
grain drag 谷物输送器
grain dryer 谷物干燥机;谷物干燥机
grain dryer burner 谷粒烘干机用加热炉
grain drying 谷物干燥设备
grain drying-and-storage equipment 谷物干燥和储[贮]存设备
grain drying machine 谷物烘干机;谷物干燥机
grain drying plant 谷物干燥装置;谷物干燥设备
grain drying system 谷物干燥系统
grain drying unit 谷物干燥设备
grain dust 谷尘
grained 粒状的;分粒的;木纹状的
grained board 木纹纸板
grained composition 粒度成分
grained half-tone plate 砂目半色调板
grained layer 颗粒层
grained marble 粒状大理石
grained metal 金属粒
grained paper 木纹纸
grained phosphoraite 粒状磷块岩
grained rock 粒状岩(石)
grained rocks 粒状岩类
grained stone facing 粒状石面;颗粒石面;米粒石面;粗粒石面
grained stone lithography 摹描石印
grained tinplate 糙面马口铁;糙面镀锡薄(钢)板
grained zinc press plate 砂目锌板
grain effect 压晶效应;穿晶效应
grain elevator 谷物输送机;谷物升运器;谷物斗式提升机;谷粒升运机;升运机;散粮提升机;斗式谷物出舱机
grain elevator terminal 散粮筒仓码头
grain embrittlement 穿晶脆化
grain end 药柱端面
grain equilibrium 颗粒平衡
grain equilibrium moisture content determination 谷物平衡的水分测定
grainer 压纹机;制纹机;刮毛刀;起纹器;漆木纹的工人;碎石机
grain failure 穿晶破坏

grain fanning-sorting machine 谷物风选机
grain feed 谷物饲料
grain feed auger 谷粒喂入螺旋
grain feeder 谷物传送器;谷舱灌补器
grain-feeding pen 精料饲喂间
grain field 农田水利工程
grain final moisture 谷物最终水分
grain fineness 颗粒细度
grain fineness number 砂(粒)细度
grain fitting 谷类防动隔舱设备;粮谷铺舱作业;粮船铺舱作业;谷物防动隔舱设备
grain flow 颗粒流;晶粒流动
grain flow line 晶粒流线
grain flow meter 谷物流量计
grain foliated structure 颗粒叶片状构造
grain formation 晶粒形成
grain form factor 颗粒形状因素
grain fraction 颗粒分级
grain fracture 穿晶断口
grain fragmentation 晶粒碎裂
grain fumigant 谷物熏蒸剂
grain gallery 粮谷输送廊道
grain glue 颗粒胶合剂
grain gradation 颗粒级配
grain grade 粒组
grain grinder 谷粒磨粉机
grain growth 颗粒生长;晶粒生长;晶粒长大
grain growth coarsening 晶粒粗化
grain growth inhibition 晶粒长大抑制
grain handling 谷物输送;谷物装卸;谷物处理
grain handling machinery 粮谷装卸机械
grain handling unit 谷物装卸运输机
grain handling wharf 粮食装卸码头;粮谷装卸码头
grain harbo(u)r 谷物港
grain hardening 晶粒硬化
grain hardness tester 谷物硬度测定仪
grain history 谷物历史
grainiform 粒状的
grain industry 谷物工业
graininess 粒性;粒度;颗粒状;颗粒度;起粒
graining 析皂;液印木纹;粒化;假木纹漆;木纹装饰;木纹涂装法;木纹饰面;起纹;起颗粒;漆绘木纹法;漆(画)木纹;漆成木纹的方法
graining board 压纹板
graining box 颗粒箱
graining brush 木纹刷
graining colo(u)r 木纹色;绘纹色料
graining comb 绘纹算
graining drawing polyester film 机械处理打毛聚脂绘图薄膜
graining lac 紫胶粒;粒状紫胶
graining liquid 木纹漆
graining machine 压纹机;胶印磨版机;磨版机
graining of crystallization 晶籽
graining oxides 滚印用陶瓷呈色氧化物
graining paint 木纹漆
graining paste 液印用颜料油浆;颜料浆
graining rag 绘纹用擦布
graining roll 印花辊
graining roller 压纹滚
graining sand 砂粒;细砂;磨版砂
graining tool 木纹描绘用具
grain interlocking 颗粒间连接
grain laden ship 谷物运输船
grain leather 粗面皮革
grainless plate 无砂目平板
grain levelling auger 谷粒分配螺旋
grain loader 装粮机
grain loading 颗粒负荷
grain loading bin 装粮箱
grain-long paper 顺纹纸
grain mass 谷堆
grain maturity 谷物成熟度
grain meter 谷物计量器
grain micrometer 谷物粒度计
grain migration 晶粒徙动
grain mill 碾房;碾坊
grain mix(ture) 颗粒混合物
grain moisture meter 谷物湿度计;粮食水分测试仪
grain noise 晶粒噪声
grain of crystallization 晶核籽;结晶中心
grain of ice 米雪;冰粒
grain of paper fibre 纸张纤维方向
grain of rice (瓷器中的)透明花纹;瓷器透明花纹

grain of sand 砂粒
grain of the country 地质结构;地势走向
grain of wood 木(的)纹理;木材纹理
grain orientation 晶粒取向
grain-oriented 晶粒定向的
grain-oriented alloy 粒子定向排列合金;晶粒取向合金
grain-oriented electrical steel 取向性硅钢片
grain-oriented silicon-iron 晶粒取向硅钢
grain-oriented steel 晶粒取向钢
grain output 粮食产量
grain-o-vator 装粮机
grain oxidation 晶粒氧化
grain packaging machine 谷物包装机
grain pan 谷粒盘
grain pan finger bar 谷粒盘指杆筛
grain pan finger rake 谷粒盘指杆筛
grain pattern 木理纹
grain per gallon 谷/加仑;格令/加仑
grain piler 装粮机
grain pipe 谷物输送管
grain plate 砂目平板
grain-polishing mill 碾米厂
grain porosity 颗粒孔隙度
grain port 谷物港埠;粮食装卸码头;粮港
grain precipitate 晶粒沉淀物
grain precleaner 谷物预净器
grain pressure 粒间压力;结晶颗粒压力
grain probe 谷粒取样器
grain processing plant 粮食加工厂
grain producing area 产粮区
grain production 谷物生产
grain production and processing 谷物生产和加工
grain propagation 晶粒扩展
grain property 颗粒特性
grain quality 谷物质量;谷粒品质
grain quay 粮谷码头
grain rain 谷雨
grain raising 木纹隆起;木纹隆起;纹理突起
grain rations 粮食定量
grain refined steel 细晶粒钢
grain refinement 细化化;细晶化;晶粒细化
grain refiner 晶粒细化剂
grain refining 细晶化;晶粒细化
grain refining inoculant 晶粒细化孕育剂
grain refining steel 细晶粒钢
grain register 收获谷物记量器
grain relaxation 晶粒弛豫
grain reserve 储备粮;粮食储备
grain returns pan 谷粒盘
grain roll (砂型的)铸铁轧辊;麻口细晶合金铸铁轧辊
grain roller 谷物压碎辊
grain roughness 砂粒粗(糙)度;砂粒糙率
grain rupture 纹理裂隙
grains 交流汇合处
grain sampler 谷物取样器;粮食取样器
grain saving pan 谷粒收集器
grain scale 谷粒秤
grain scraper 刮谷机
grain screen 粒状荧光屏
grain screw 谷粒螺旋
grain shape 粒形;颗粒形状;晶粒形状
grain shape factor 颗粒形状因数
grain shape test 颗粒形状试验
grain shear stress 泥沙颗粒剪应力
grain ship 运粮船;粮谷专用船
grain shoot 装矿石滑槽
grain shop 粮店
grain shovel 谷铲
grain shrinkage 谷物皱缩
grain side 光面
grain sieve 谷粒筛
grain silo 圆筒粮仓;谷物筒仓;谷筒;粮食筒仓;筒式粮(谷)仓
grain silo bin 谷物圆筒储[贮]仓;谷物筒仓
grain size 粒径(大小);粒度;颗粒直径;颗粒尺寸;晶粒度;晶粒大小;晶粒尺寸
grain-size 颗粒粒度;颗粒大小
grain-size accumulation 颗粒级配线
grain-size accumulation curve 粒径累积曲线
grain-size accumulation test 粒径累积试验
grain-size analysis 过筛分析;粒度分析;粒径分析;颗粒(组成)分析

grain-size analysis of sediment 泥沙粒径分析
grain-size analyzer 粒径分析仪;粒度分析仪;粒度分析器
grain-size characteristic curve 粒度特性曲线
grain-size characteristics 粒径特征;粒径特性
grain-size characteristics diagram 粒径特征图;粒径特性图
grain-size classification 粒径分级;粒径分类(法);粒度分类;粒度分级
grain-size composition 粒径组成
grain-size control 粒度控制;晶粒粒度控制
grain-size counter 谷物粒度计
grain-size curve 粒级曲线;粒径曲线
grain-size determination apparatus 粒度测定设备
grain-size distribution 粒度分布;颗粒级配;颗粒(尺寸)分布;粒度分配;粒径分配;粒径分布;粒度级配
grain-size distribution analysis 颗粒分布分析
grain-size distribution chart 粒径分布曲线
grain-size distribution curve 颗粒级配曲线;级配曲线;筛分曲线;大小颗粒分布曲线;粒径分配曲线;粒径分布曲线;粒级分布曲线;粒径级配曲线
grain-size distribution diagram 粒度分布图;颗粒细度分布图
grain-size distribution of soil 土的级配
grain-size division 粒组划分
grain-size fraction 粒径组(合);粒度级别;泥沙粒径组成
grain-size frequency curve 粒径频率曲线
grain-size frequency diagram 颗粒大小频率图
grain-size grade 粒级;颗粒粒级
grain-size grading 颗粒尺寸级配
grain-size limit 粒度极限
grain-size measurement 颗粒粒度分析;颗粒粒度测定
grain-size measuring eyepiece 粒度测量目镜
grain-size number 结晶粒度
grain-size of cave pearl 穴珠粒级
grain-size of recrystallization 重结晶颗粒大小
grain-size of sediment 沉积物粒度
grain-size of sediment discharge 输沙粒级
grain-size parameter 粒度参数
grain-size refinement 颗粒尺寸改进
grain-size scale 粒径比例;粒径分级标准;粒度仪;粒径分级标尺
grain sizing 颗粒分级
grain sizing machine 谷粒径选机;粒料分选机
grain skeleton 颗粒骨架
grain slag 粒化渣;水渣
grain slope 木纹(倾)斜度;木纹角
grain sludge 粒状淤渣
grain sluice 谷粒节流器
grains of emulsion 乳剂颗粒性
grains of equal size 细度相同的颗粒
grains of Paradise 乐园子
grains of sand 沙粒
grain sorghum 高粱
grain space 散装舱容
grain spacing 颗粒配比与分布;颗粒间距
grains per standard cubic foot 颗粒数/标准英尺
grains per stere 颗粒数/立方米
grain spout 谷粒滑槽
grain spreader 谷物匀布机
grain standard 谷物质量标准
grain-stirring device 谷物搅拌装置
grain stock 粮食储备
grainstone 粒状碳酸盐岩;粒状灰岩;颗粒石灰岩
grain storage 谷仓;粮食仓库;粮谷仓库;粮仓
grain storage silo 筒式粮仓
grain store 粮仓
grain strength 颗粒强度
grain structure 粒状组织;粒状构造;颗粒结构;晶粒构造
grain structure analysis 粒度分析;颗粒结构分析
grain-sunning ground 晒谷场
grain supply 粮食供应
grain supply centre 粮站
grain-supported 颗粒支撑
grain-supported fabric 颗粒支撑组构
grain surface 颗粒表面;粒面
grain surface area 颗粒表面
grain tank discharge lever 粮箱卸粮杆
grain terminal 谷物码头;粮食装卸码头;散粮码头
grain test 粉末法
grain testing sieve 谷粒试验筛

grain thermometer 谷物温度计
grain-tight divisional bulkhead 谷物分隔舱壁
grain tin 砂锡;粗粒锡石
grain-to-grain bearing strength 颗粒对颗粒的压力
grain to grain boundary 晶间边界
grain-to-grain contact 粒间接触压力
grain-to-grain stress 颗粒之间的应力
grain to straw ratio 谷与草之比
grain traveller 钢丝圈号数
grain trier 谷粒取样器
grain unit 谷粒容重
grain unloader 卸粮机
grain warehouse 谷物仓库;谷仓;粮食仓库;粮仓
grain washer 谷物清洗机
grain wear 颗粒磨损;颗粒磨耗;颗粒磨擦
grain wharf 粮谷码头;散粮码头
grain wood 纵材木(依木纹锯开的木材)
grainy 粒状(的)
graisenization 云英岩化
grait 卵石;砂
gramagrass 格兰马草
gram atom 克原子
gram atomic weight 克原子量
gram-calorie 克卡(热量单位)
gram-centimeter 克·厘米
gram/cm³/cm 克/立方厘米/厘米
Gram determinant 格莱姆行列式
gram element/gram rock 克元素/克岩石
gramenite 绿脱石
gram equivalent 克当量
gramer 漆绘木纹用具
gram-force gram-weight 克·力
gramineous 草绿色
gramineous 青草味的
gram-ion 克离子
gram-ion/litre 克离子浓度
grammar school 文法学校
grammar with control set 带有控制集合的文法
grammar with coordinate 带有坐标的文法
gram-mass 克质量
grammatical inference by enumeration 穷举法文法推断
grammatical inference by induction 归纳法文法推断
grammatite 透闪石
gramme equivalent 克当量
gramme moledular solution 克分子溶液
gramme moledular volume 克分子容积
gramme ring winding 环形绕组
grammes per square metre 每平方米克重
grammet 麻刀油灰
gram-meter 克·米
grammite 硅灰石
grammole 克分子
gram molecular solution 克分子溶液;摩尔溶液
gram molecular volume 克分子体积
gram molecular weight 克分子量;摩尔量
gram molecule 克分子
Gram-negative 葛兰氏阴性(水处理)
gramophone library 唱片图书馆
gram per liter 克/升
Gram positive 革兰氏阳性
Grampound grits 格兰庞德粗砂岩
grampus 大铁钳
gram-roentgen 克·伦琴(能量转换单位)
Gram-Schmidt orthogonalization 格莱姆—施密特正交化
Gram-Schmidt orthogonalization procedure 格拉姆—施密特正交化程序
Gram-Schmidt process 格莱姆—施密特正交化方法
gram-second 克·秒
grams per hour 克/小时
grams per second 克/秒
grana 质体基粒
granary 储藏室;谷仓;粮食仓库;粮仓
granary warehouse 粮谷仓库
granatohedron 菱形十二面体
granatum 石榴皮
Granby car 格兰贝式矿车
Granby type trolley 格兰贝型侧卸出碴车
grand 大钢琴
Grandagon 格朗达衮(德国的105度广角镜头)
grand arcade 大拱廊
grand average 总平均

Grand Banks 大沙洲
grand base level 主基(准)面
grand bounce 解雇;免职
grand calorie 大卡;千卡
Grand Canal 大运河
grand Canyon 大峡谷
grand canyon series 大峡谷统【地】
grand chamber 宴会厅;宴会室
grand champion 最优奖获得者
grand-circle 大剧院楼厅观众席
grand climax 最高潮;顶点
grand clock 落地大座钟
grand daughter 第三代子核
grand entrance 进口大厅;大门;正门
grand entrance court 宏大的入口庭园
grandfather 原始数据组;存档
grandfather chair 高背椅子
grandfather clause 保留条款
grandfather clock (古式)落地钟
grandfather cycle 存档期
grandfathered activities 不追溯业务
grand feu 高温釉烧
grand final 终结;结局
grand hall 豪华大厅;高级宴会厅;大厅
grand hopper 大型装料斗
grand hotel 高级旅馆
grandiderite 硅硼镁铝矿
grandifoliate 大叶的
grandite 钙铝铁榴石
grand list 税单
grandmaster key 总(万能)钥匙;万能钥匙
grandmaster-keyed lock 总钥匙锁
grandmaster-keyed series 总钥匙系列;一套总钥匙
grand master pattern 原模
grand opera house 大歌剧院
grand order 高柱式
grand order of architecture 高巨柱形建筑
grand period of growth 大生长期
grand piano 大钢琴
grand prize 特奖
grand prospects 宏伟前景
grand relief 高浮雕;隆浮雕
grand scale integration 超大规模集成电路
grand slam 优胜法
grand stair(case) 豪华楼梯;(剧场等公共场所的)正面大楼梯;大楼梯
grand stairway 豪华楼梯
grandstand 大看台;(运动场等的)正面大看台;主看台;观礼台
grandstand-type seat 大看台式座位
grand sum 总和
grand sweet 游览广场
grand swell 大增减音器
grand tide 大潮
grand tier 花坛顶层
grand total 总值;总计;共计;累计;合计
grand touring car 双座轿车
granellae 钡粒
granellarium 钡粒管系
Graney bronze 哥瑞内青铜
grange 庄园;农舍;农村住宅;谷仓;农庄;农场;田庄
grangesite 锰铁绿泥石
granide 花岗岩类(岩名)
graniphyric 花斑状的;文象斑状
granite 花岗岩;花岗石
granite aplite 花岗细晶岩
granite arch 花岗岩拱
granite ballast 花岗岩道渣
granite ballast concrete 花岗岩道渣混凝土
granite block 花岗岩块;花岗石块
granite block pavement 花岗岩块铺砌路面;花岗石块路面
granite blockwork 花岗石(块)工;花岗石块加工
granite board 花岗岩纹纸板
granite broken sand 花岗岩碎砂
granite chippings 花岗石碎片;花岗石屑;花岗石屑
granite chips 花岗石屑;花岗石碎片
granite clan 花岗岩族
granite coarse aggregate 花岗石粗骨料
granite column 花岗石柱
granite column face 花岗岩柱面
granite concrete 花岗岩碎石混凝土
granite concrete tile 花岗石混凝土砖;花岗石混凝土瓦

granite crushed sand 花岗石碎砂
granite cube 花岗石方块
granite cube floor 花岗石块地面
granite curb 花岗石缘墙;花岗石路缘;花岗石道牙
granite curtain(wall) 花岗石幕墙
granite deposit 花岗岩矿床
granite dust 花岗石粉末
granite enceinte 花岗石围墙
granite-faced 花岗石贴面的;花岗石铺面的
granite facing 花岗岩砌面;花岗石(饰)面
granite finish 花岗石修饰;花岗岩饰面
granite flour 花岗石粉末
granite for build 建筑用花岗岩
granite for ornament 狮面用花岗岩
granite-gneiss 花岗片麻岩;原生片麻岩
granite gravel 花岗石砾石
granite greisen 花岗云英岩
granite gruss 花岗岩碎石
granite-(h)aplite 花岗细晶岩
granite kerb 花岗石缘墙;花岗石路缘
granite-laterite 花岗岩红土
granite layer 花岗岩层(硅铝层)
granite lining 花岗石衬里
granite lino(leum) 花岗石油地毡
granitelle 辉石花岗石;二云花岗岩
granite masonry work 花岗石圬工工作
granite mastic 花岗石砂胶
granite-monzonitic granite group 花岗岩—二长花岗岩类
granite paper 花岗石纹纸
granite pavement 花岗石路面;花岗石铺面
granite paving sett 花岗石路面石块
granite pegmatite 花岗伟晶岩
granite plank 花岗岩板材;花岗石板材
granite plaster 石膏基快速硬化灰
granite plate 花岗石板;花岗岩石碑
granite porphyry 花岗斑岩;石英斑岩
granite powder 花岗岩粉末
granite quarry 花岗岩石矿;花岗岩采石场
granite rare earth element-bearing formation 花岗岩含稀土元素构造
granite rocks 花岗岩类
granite sand 花岗岩砂
granite screenings 花岗石筛屑
granite series 花岗岩系
granite sett 花岗石小方石;花岗石条石;花岗石小方石块
granite setter 花岗石镶嵌工
granite sett paving 花岗岩石块铺砌路面
granite slab 花岗石板
granite slab floor(ing) 花岗岩铺饰楼面;花岗石板地面
granite step 花岗石梯级;花岗石台阶
granite surface 花岗石面层
granite surface plate 花岗岩平板
granite surfacing 花岗石饰面;花岗石铺面
granite veneer facing 花岗石板饰面
granite ware 有花岗石纹的器皿;仿花岗器皿
granite wash 花岗岩冲积物;花岗砂岩
granite wash reservoir 花岗岩冲积物储集层
granitic 花岗状的
granitic amphogneiss 花岗质混合片麻岩
granitic batholith 花岗岩岩基
granitic conglomerate 花岗岩砾岩
granitic finish 花岗石状加工;花岗石面处理;细石混凝土路石饰面;假石抹面;仿花岗石面层
granitic layer 硅铝带;花岗岩层(硅铝层)
granitic magma 花岗状岩浆岩
granitic pegmatite 花岗伟晶岩
granitic pegmatite deposit 花岗伟晶岩床
granitic plaster 花岗状粉刷;洗石子粉刷;假石抹面;人造花岗石面;汰石子;水刷石
granitic-rhyolitic rock 花岗流纹岩
granitic rocks 花岗岩类
granitic sandstone 花岗质砂岩
granitic structure 花岗构造
granitic stucco coating 水刷石墙面
granitic texture 花岗结构
granitiform 花岗石状的
granite 石榴冒岩;十字石;黑云(母)花岗岩
granitization 花岗岩化(作用)
granitizational-hydrothermal solution 花岗岩化热液
granitization granite 花岗岩化花岗岩

granitization theory of metallization 花岗岩化成矿理论
granitization way 花岗岩化作用方式
granitoid 花岗岩状料;花岗(岩)状的;花岗岩类;人造花岗石(面);似花岗岩状
granitoidal texture 花岗状结构
granitoid floor 仿花岗岩地面;仿花岗岩地板
granitoid texture 似花岗岩状结构
graniton 辉长岩
granitophile element 亲花岗岩元素
granitotrachytic 花岗粗面(结构);含长(结构)
granny's knot 错误平结
granny knot 易解开的绳结;易解绳结
granny rag 涂沥青用的布;拉拉布
grano 花岗石
granoblastic 花岗变晶状
granoblastic texture 花岗变晶结构
granoclastic texture 花岗碎屑结构
grano concrete 磨石子地坪混凝土
granodiorite 花岗闪长岩
granodiorite group 花岗闪长岩类
granodiorite porphyry 花岗闪长斑岩
granodioritic amphogneiss 花岗闪长质混合片麻岩
granodolerite 花岗粒玄岩
granodraw (钢丝干式拉拔前的)磷酸锌处理
grano fibrous blastic texture 花岗纤维变晶结构
granogabbro 花岗辉长岩
grano lepidoblastic texture 花岗鳞片变晶结构
granolite 花岗状火成岩
granolith 人造铺面;人造地石;水磨石
granolithic 人造石铺面的
granolithic base 人造铺地石基底
granolithic concrete 花岗碎石混凝土;假石混凝土;磨石子地坪混凝土;人造石饰地面;水磨石;仿石混凝土
granolithic concrete aggregate 花岗碎石混凝土骨料
granolithic concrete course 人造石混凝土铺面层
granolithic concrete flooring tiling 人造石混凝土楼面砖
granolithic concrete layer 人造石混凝土铺面层
granolithic concrete paving 人造石混凝土铺路
granolithic concrete screed 人造石混凝土地面
granolithic concrete surface 人造石地面;磨石子地面
granolithic concrete tile [tiling] 人造石混凝土(铺)地砖
granolithic concrete topping 人造石混凝土铺面层
granolithic concrete tread 人造石混凝土梯级踏步
granolithic finish 花岗碎石混凝土饰面;假石抹面;人造石饰面;人造石铺面;人造石地面
granolithic floor(ing) 人造石楼面;人造石地面
granolithic layer 人造石铺面层
granolithic paving 人造石铺面
granolithic plate 浇制人造石板用的铁板
granolithic screed 人造石铺面整平板;细石混凝土面层;人造石铺面找平
granolithic sprinkle finish floor(ing) 喷撒人造石铺面楼(地)板
granolithic tread 人造石踏步
granolithic worker 花岗(岩)碎石工
granomerite 全晶粒岩
granopatic 花岗状的
granophyre 花岗斑岩;文象斑岩
granophyric 花岗斑岩的;花斑状的;文象斑状
granophyric texture 花斑岩构造;花斑结构
granoporphyritic texture 连斑结构;花岗斑状结构
granosealing 磷酸盐处理;磷化处理
granosyenite 花岗正长岩
Granox P-e-M 钼锰克混剂
grant 赠款;授与;补贴
grant a certificate 签发证书
grant-aided 受补助的
grant certificate 让与证书
grant clause 许诺条款
grant deed 转让证书;津贴证书;拨款证书
grantee 买主;产业继承人;让受人;受让者;受让人;被授予者
granter 让与人;出让人;出让方
grant for local finance balance 地方财政平衡补贴
grant-in-aid 补助金;补助拨款;拨款
granting bank credit 提供银行信贷
granting of loans 发放贷款
Grant-Manes model 格兰特—马内斯模型

grant of land 批地
Granton sandstone 格兰顿砂岩
grantor 卖主;产业转让人;让与人;授予者
grantor of credit 信用让与人
grant pension 发给补助金
grant-reeled silk 花纹绞丝
grants-in-aid fund 财政补贴基金
grants program(me) 补助计划
granula 粒剂
granular 粒状的;颗粒状的
granular activated carbon 粒状活性炭;颗粒(状)活性炭
granular activated carbon adsorption 颗粒活性炭吸附
granular activated carbon adsorption method 颗粒活性炭吸附法
granular activated carbon anaerobic fluidized bed process 颗粒活性炭厌氧流化床工艺
granular activated carbon bed 颗粒状活性炭滤床
granular activated carbon filter 颗粒状活性炭滤池
granular activated carbon filter absorber 粒状活性炭滤池
granular activated carbon filtration 颗粒活性炭过滤
granular activated carbon intensified sand filter 颗粒活性炭强化砂滤池
granular activated carbon packed bed reactor 颗粒活性炭填料床反应器
granular activated carbon process 颗粒状活性碳法
granular active anthracite 颗粒活性无烟煤
granular aggregate 粒状集料;粒状骨料
granular amphibolite 粒状角闪岩
granular area 颗粒区
granular ash 粒状碱;粒状灰
granular base 粒状基层;粒状底层;粒料基层;粒料(类)底层
granular bed filter 颗粒床除尘器;颗粒层收尘器;颗粒层集尘器;颗粒层除尘器
granular bed filter with integral cyclone 带整体式旋风筒的颗粒层集尘器
granular bed filter with separate cyclone 旋风筒分装的颗粒层集尘器
granular bed separator 颗粒床分离器;颗粒层收尘器;颗粒层集尘器;颗粒层除尘器
granular biological activated carbon 颗粒生物活性炭
granular black 粒状炭黑
granular bog iron ore 粒状沼铁矿
granular carbide 粒状碳化物
granular carbon bed 颗粒状活性炭滤床
granular carbon process 粒状活性炭法
granular cast 颗粒管型
granular cementation 粒状胶结(作用);颗粒胶结
granular cementite 粒状渗碳体
granular classification of coal 煤炭粒度分级
granular cloudiness 颗粒状混浊
granular coherer 碳粒凝合检波器
granular-cohesive soil 粒状黏性土
granular colloidal silica 粒状硅胶
granular composition 颗粒组成;颗粒结构;颗粒组分;颗粒级配
granular composition analysis 颗粒结构分析
granular cork 颗粒状软木
granular cork surfacing 颗粒状软木铺面
granular cover material 颗粒铺面材料
granular crystalline 粗晶体
granular crystalline gypsum 晶粒石膏
granular degeneration 粒状变性
granular deposit 颗粒沉积物;颗粒沉淀物
granular disintegration 粒状剥落;粒状崩解;粒化崩解;球状崩解作用
granular dust filler 粒状矿质填料
granular exfoliation 粒状剥落;粒状崩解
granular facing 颗粒铺面
granular ferric hydroxide 颗粒氢氧化铁
granular fertilizer 粒状肥料
granular fibrousblastic texture 粒状纤维变晶结构
granular filler 颗粒填充物
granular-fill insulation 颗粒填充绝缘材料;(松散颗粒填充的)隔热层;颗粒材料保温
granular filter 粒料过滤器;粒料过滤池
granular filter material 粒状过滤料;颗粒滤料
granular flux 粒状焊剂
granular form 粒形

granular formulation 颗粒制剂
granular fracture 粒状破裂;粒状破坏;粒状断口;颗粒裂面;粗粒断口
granular fuel 颗粒(状)燃料
granular gypsum 粒状(结构)石膏
granular ice 粒状冰;晶粒冰
granular insulant 颗粒绝缘材料
granular insulation material 粒状绝缘材料;粒状隔声材料;粒状隔热材料
granular insulator 粒状绝缘体
granular iron 粒铁
granularity 粒性;粒度;颗粒性;颗粒度;多粒象
granularity of emulsion 乳剂颗粒度
granularity value 粒度值
granular layer 粒层
granular lepidoblastic texture 粒状鳞片变晶结构
granular limestone 大理石;云石;粒状(石灰)岩;粒状石灰石
granular loose-fill thermal insulation 松散粒状绝热材料
granular marble 粒状大理石
granular material 颗粒(状)材料;粒状物料;粒材料;粒体;粒料
granular measurement curve 颗粒级配曲线
granular-medium filtration 粒状滤料过滤;颗粒介质过滤
granular membrane 筛网过滤膜
granularmetric analysis 粒径分析;颗粒分析
granular micrinite 粒状微粒体
granular mo(u)lding compound 粒状成形料
granular opaque matter 粒状不透明物质
granular pearlite 粒状珠光体;粒状球光体
granular pesticide 粒状农药;颗粒剂农药
granular porphyrite 粒状斑岩
granular powder 粒状粉末;粒状炸药;粒状粉料
granular product 粒状产品
granular quality 颗粒质量
granular residue 颗粒残渣
granular resin 颗粒状树脂
granular rock wool 粒状岩绵
granular rupture 粒状破坏
granular sand 粒砂
granular sensation 颗粒感
granular shoulder 粒料加固路肩
granular size 颗粒大小
granular skeleton 粒状骨架
granular sludge 颗粒(状)污泥
granular sludge membrane bioreactor 颗粒污泥膜生物反应器
granular sludge sequencing batch reactor 颗粒污泥序批间歇式反应器
granular snow 粒状雪;米雪;春天粒雪
granular soda 颗粒碱
granular sodium alginate-polyethylene oxide gel adsorbent 颗粒海藻酸钠聚氧化乙烯凝胶吸附剂
granular soil 粒状土(壤);颗粒土(壤);无黏性土壤
granular soil stabilization 碎石土面层;机械稳定土;骨粉土面层
granular solid 粒状(固)体
granular solid mixing 粒状固体拌和
granular stability 颗粒稳定
granular-stabilized 用粒料加固的
granular structure 麻面(塑料制品缺陷);粒状构造;颗粒(状)结构
granular structure analysis 颗粒结构分析
granular structure surface 麻面;粗面
granular sub-base course 粒状基层下层;粒状底基层
granular surfacing 粒状表面;粒状铺面
granular surfacing material 粒状铺面材料;粒状表面材料
granular terrace 粒料平台
granular texture 粒状(变晶)结构;团粒结构
granular texture surface 粒状结构面
granular type road 粒料路
granular water filter 粒状滤池;粒料滤池
granular wearing surface material 粒状耐磨表面材料
granular wool 粒状绵
granulate 粒化;使成粒状;成为粒状
granulated 粒状的;成粒状的
granulated ammonium nitrate 多孔粒状硝酸铵
granulated blast furnace slag 水淬(高炉)矿渣;

粒状高炉矿渣;粒化高炉矿渣;溶渣砂
granulated blast furnace slag sand 粒状高炉矿渣砂
granulated carbide 粒状电石
granulated carbon filter 粒状炭滤池
granulated cinder concrete 粒状煤渣混凝土;粒状炉渣混凝土;粒状矿渣混凝土
granulated cinder cored block 粒状煤渣空心砖;粒状炉渣空心砖;粒状矿渣空心砖
granulated cinder tile 粒状煤渣瓦;粒状炉渣瓦;粒状矿渣瓦
granulated cobalt 粒状钴
granulated cork 颗粒状软木;软木颗粒;软木粒嵌缝料;粒状软木(嵌缝材料)
granulated cork filler 软木粒嵌缝料;软木粒嵌缝
granulated cork surfacing 粒状软木面层
granulated feed 碎粒饲料
granulated fertilizer 粒化肥料;颗粒肥料
granulated finish 粒状修饰
granulated gas carburizing 固体渗碳剂渗碳
granulated gel 粒状凝胶;颗粒凝胶
granulated glass 轧碎玻璃;粒状玻璃;玻璃细粒
granulated hammer 碎石锤;麻面锤
granulated iron powder 粒化铁粉
granulated material 轧碎材料;粒状材料
granulated metal 粒状金属;粒化金属
granulated mica 粒状云母
granulated plaster 粒状粉刷
granulated rough 粗涩;粗糙不平(粘有砂粒等)
granulated screen 颗粒网目板
granulated slag 水淬渣;粒状(熔)渣;粒状(矿)渣;粒化渣;颗粒矿渣;水碎渣
granulated slag block 粒状矿渣砌块
granulated slag concrete 粒状矿渣混凝土
granulated slag cored block 粒状矿渣空心砖
granulated slag sand 粒状矿渣砂
granulated slate surfacing 粒状石板铺盖面层
granulated soil 团粒土
granulated stone facing 粒状石面;颗粒石面;米粒石面
granulated sugar 砂糖
granulated welding composition 粒状焊药
granulated wool 格状绵
granulater 制球机
granulating 颗粒化;轧碎(工作)
granulating device 粒化装置
granulating hammer 碎石锤;麻面锤
granulating of the batch 配合料粒化
granulating pan 成球盘
granulating pit 炉渣粒化池
granulating screen 水帘管;粗筛
granulation 形成颗粒;制粒法;造粒;粒化作用;颗粒;颗粒(化);金属粒化处理;成粒(作用);团粒作用
granulation column 造粒塔
granulation launder 粒化流槽;水碎流槽
granulation of pig iron 铁水粒化
granulation of slag 冲水渣
granulation of soil 土壤团粒(形成)作用
granulation size 粒度尺寸
granulator 制粒机;造粒机;粒化器;磨粉机;成球盘;成球机;成粒器;成粒机
granulators plant 造粒装置
granule 小(颗)粒;粒状斑点;粒砂;粒剂;粒斑;颗粒;微粒;团粒;砂砾
granule breccia 细砾角砾岩
granule embedment 粒状嵌入
granule hopper 颗粒料斗
granule mineral surfacing 粒状矿物撒布料
granule mo(u)lding compound 颗粒模塑料
granule of carbon 炭粒
granule ripple 细砾波痕
granule roundstone 小粒圆石
granules 生料球
granule size 粒度
granulestone 细砾岩
granule texture 细砾结构;粒状结构;麻粒结构
granulite 麻粒岩;变质麻粒岩;变粒岩;白粒岩
granulite facies 麻粒岩相
granulitic 麻粒的
granulitic texture 麻粒结构;他形花岗变晶结构;碎粒构造;等粒结构;变粒结构
granulitization 麻粒岩化(作用)
granulomere 颗粒区

granulometer 粒度计;颗粒测量仪;颗粒测定仪
granulometer analysis 粒径计法
granulometric 颗粒的
granulometric analysis 粒径分析;粒度分析;颗粒分析(法);颗粒度分析;颗粒度测定
granulometric analysis curve 颗粒分析曲线
granulometric composition 粒径组成;粒度组成;粒度成分;颗粒组成;级配组成
granulometric composition of soil 土的粒度成分;土的颗粒组成
granulometric curve 颗粒级配曲线;颗粒分析曲线
granulometric distribution 颗粒级配;颗粒分布
granulometric facies 粒度相
granulometric grading 粒径分析;粒度级配;颗粒分级;颗粒级配分析
granulometric principle 颗粒原理
granulometric property 粒度性质;颗粒性质;级配性质
granulometry 粒度分析;粒度测量(术);粒度测定(术);颗粒分析;颗粒测量;颗粒测定(方)法;测粒术
granulopexy 颗粒固定
granulophyre 微花斑岩
granuloplastic 颗粒形成的
granulose 粒状;麻粒状;淀粉粒质
granulose structure 麻粒构造
granulosis 颗粒团形成
granulosity 骨料粒质
granulous 粒状(的)
granulovacuolar degeneration 粒状空泡变性
granum 基粒
grape 深紫色
grape like clinker 葡萄状熟料
grape-myrtle 紫薇
grapery 葡萄园;葡萄温室
grape seed oil 葡萄子油
grape-shot 大钻粒
grapestone 葡萄状灰岩
grapestone lump 葡萄石团块
grape sugar 葡萄糖
grape trellis 葡萄架
grapevine 葡萄树;葡萄藤;非正式信息交流网
grapevine drainage 格状水系;蔓极排水系统;葡萄藤状水系;网式排水系统;葡萄状排水系统
grapevine stopper 高压电缆塞
graph 过程线;曲线图;图形;图像(表示);图表;标绘图
graph algorithm 图解算法
graph-analytic method 图解分析法
graph application 图应用
graph artist 图案艺术家;美术设计人
graph block 图形块
graph card 图卡
graph command 图形命令
graph data 图解数据
graph decomposition 图形分解
graph disc 图表盘
graphec(h)on 记忆管
grapher 自动记录(仪)器;图示仪;记录器
graphetic level 复制能力
graph evaluation and review technique 图解评审技术
graph follower 图形跟踪器;曲线阅读器;图形复制器;图形变示器;读图器
graph for pure dominance 纯支配的图;纯优势图
graph grammar 图文法
graphic(al) 版面;图解的
graphic(al) abstract programming language 图形抽象程序设计语言
graphic(al) access method 图形存取法
graphic(al) accuracy 图解精度
graphic(al) addition 图解加法
graphic(al) adjustment 图形平差;图解平差法
graphic(al) aggregation 图形集
graphic(al) algebra 图解代数
graphic(al) alphanumeric display 图形字母数字显示器;图示显示器
graphic(al) alphanumeric generator 图形字母数字发生器
graphic(al) ammeter 记录式安培计
graphic(al) analysis 图像分析;图解分析;图解(法)
graphic(al) analysis of three-dimensional data 三维数据图解分析

graphic(al) analytic(al) 图解分析的
graphic(al) analytic(al) method 图解解析法;图解分析法
graphic(al) appraisal method 图解评价法
graphic(al) approximation method 图解近似法
graphic(al) arch analysis 图解拱的分析
graphic(al) array 图阵;图形阵列
graphic(al) arts 图表艺术
graphic(al) arts coating 绘图标记用涂料;标牌用涂料
graphic(al) arts quality character 美术字
graphic(al) aspect 图解形式
graphic(al) atlas 图形集
graphic(al) bank 图形库
graphic(al) base 图形库
graphic(al) base design 图形库设计
graphic(al) blast 图像照相制版法;图像照相刻蚀法
graphic(al) board 图形板
graphic(al) calculate representation 图解
graphic(al) calculation 图算;图解计算(法)
graphic(al) capability 绘图能力
graphic(al) character 图形(显示)字符;美术字;图示字符;显示字符
graphic(al) character conversion 图形显示字符变换
graphic(al) character modification 图形字符修改
graphic(al) chart 曲线图;图解(表);图表
graphic(al) check 图解验证
graphic(al) clarifier design procedure 澄清池设计程序表
graphic(al) classification 图解分类
graphic(al) clipping 图形裁剪
graphic(al) code extension 图形码扩展
graphic(al) command 图形命令
graphic(al) command center for history 历史记录中心
graphic(al) command center for maps 图像命令中心
graphic(al) comparison 图形比较
graphic(al) compass traverse 罗盘仪图解导线
graphic(al) computation 图解计算
graphic(al) computation of carrying capacity at station 车站通过能力图解计算法
graphic(al) computer terminal 图解计算机终端装置
graphic(al) computing system 图解计算系统
graphic(al) console 图示控制台
graphic(al) construction 作图(法);图解作法
graphic(al) control 图解控制
graphic(al) correction 图解校正
graphic(al) correlation 图解相关
graphic(al) cursor 图形光标
graphic(al) data 绘图数据;图形数据;图像数据;图解数据
graphic(al) data analysis 图形数据分析
graphic(al) data bank 图形数据库
graphic(al) database 图像数据库
graphic(al) database management system 图形数据库管理系统
graphic(al) data input device 图形数据输入装置
graphic(al) data processing 图形数据处理;图示数据处理;图解数据处理
graphic(al) data reduction 图形数据整理
graphic(al) data structure 图形数据结构
graphic(al) data system 图形数据系统
graphic(al) dead reckoning 图解推算船只法;图解推算船位法
graphic(al) depth record 图示深度记录
graphic(al) derivation 图解推导(法)
graphic(al) design 图形设计;图示设计;图解设计;图表设计法
graphic(al) design of optic(al) system 光学系统图解设计
graphic(al) determination 绘图测定;图解测定
graphic(al) deviation of results from data 图解推求法
graphic(al) diagram 符号图;图解图表
graphic(al) diagram of hydraulic system 液压系统图解
graphic(al) differentiation 图解微分法
graphic(al) digital form 数字图解形式
graphic(al) digitiser 图形数字化器
graphic(al) display 图形显示(程序);曲线显示;图像显示器;图式显示;图表显示;图表显示;行

车图形显示器
graphic(al)display control 图形显示控制
graphic(al)display definition mode 图形显示定义方式
graphic(al)display interface 图像显示接口
graphic(al)display processor 显图处理机
graphic(al)display resolution 图形显示分辨率
graphic(al)display software 图形显示软件
graphic(al)display system 图形显示系统
graphic(al)display technics 图形显示技术
graphic(al)display terminal 图形显示终端;图像显示终端
graphic(al)display(unit) 图形显示器;图形显示装置
graphic(al)distortion 图形失真
graphic(al)documentation 图形文档资料;图形文件化
graphic(al)dot 图解点
graphic(al)drawing 图解图
graphic(al)echo 绘图回波
graphic(al)editor 图形编辑程序
graphic(al)effect 图形效果
graphic(al)element 图形元素;图素;图形要素
graphic(al)enlargement 图解放大
graphic(al)entity 图形结构;图形基本要素;图解实体
graphic(al)equaliser 多频(率)音调补偿器
graphic(al)equalizer 分段式等化器;图像均衡器
graphic(al)equation 图形方程
graphic(al)estimation 图解估计
graphic(al)estimator 图解估计量
graphic(al)evaluation 图解评价;图解计算(法)
graphic(al)evaluation network method 图解网络评核法
graphic(al)expression 图示
graphic(al)extension 图解说明;图解加密;辐射三角测量
graphic(al)extension control 图解加密控制
graphic(al)extrapolation 图解外推(法)
graphic(al)-extrapolation method 图解外推法
graphic(al)file maintenance 图形文件维护
graphic(al)follow 图形跟踪
graphic(al)form 图型;图案;图像形式;图解形式
graphic(al)formula 图(解)式
graphic(al)granite 文象花岗岩
graphic(al)graticule grid 经纬线网
graphic(al)illustration 图解说明;图解例证;图解积分法
graphic(al)illustration of the carrying capacity in various sections on a single-track district 单线区段的区间通过能力
graphic(al)image 线划图像
graphic(al)index 接图表
graphic(al)information 图形信息;图像信息
graphic(al)information processing 图像信息处理
graphic(al)input 图形输入
graphic(al)input device 图形输入装置
graphic(al)input interaction technique 图形输入交互式技术
graphic(al)input language 图形输入语言
graphic(al)input mode 图形输入模式
graphic(al)input system 图形输入系统
graphic(al)input terminal 图形输入终端
graphic(al)input unit 图形输入设备
graphic(al)instrument 自动记录仪器;自记仪器;自动记录仪表;制图仪器;图示器
graphic(al)instrument method 图示法
graphic(al)integraph 图解积分法
graphic(al)integration 图解积分
graphic(al)interface 图形接口
graphic(al)intergrowth 文象共生
graphic(al)interpolation 内插图解法;作图插值法;图解内插法
graphic(al)interpretation 图表阐释
graphic(al)interpreter 图形解释程序
graphic(al)intersection 图解交会
graphic(al)intersection method 图解交会法
graphic(al)investigation 图解研究
graphic(al)job processing 图形作业处理
graphic(al)job processor 图形作业处理程序
graphic(al)kernel system 绘图核心系统
graphic(al)language 图形语言;图像语言
graphic(al)level(l)ing 图根水准测量
graphic(al)level recorder 图示电平记录器

graphic(al)location(method) 图解定位法
graphic(al)log 柱状平面图(地);岩性录井(图);图示录井图;图解测井记录;图解测井
graphic(al)manipulation 图形处理
graphic(al)manipulation function 图形管理功能;图形处理功能;图形变换功能
graphic(al)mapping control point 图解图根点
graphic(al)mark 图示符号;图解符
graphic(al)matching 图形匹配
graphic(al)material 绘图材料
graphic(al)mean 图解平均值;图形法
graphic(al)measurement 图解量测;图解法
graphic(al)mechanics 图解力学
graphic(al)medium 制图材料
graphic(al)meter 图解记录仪;自动记录仪器
graphic(al)method 描记法;图示法;图上作业法;图解法
graphic(al)method for proportioning 图解法配料
graphic(al)method for steady flow 稳定流图解法
graphic(al)method of integration 图解积分法
graphic(al)method of land use 土地使用图
graphic(al)method of population projection 人口预测图解法
graphic(al)method of signal 信号图解法
graphic(al)method of statistic(al)inference 统计推论的图解法
graphic(al)method of water level recovery 恢复水位的图解法
graphic(al)milliammeter 毫安记录表
graphic(al)model 三维图;立体图;图解模型
graphic(al)network software 图形网络软件
graphic(al)object 图形目标;图形对象
graphic(al)optimization 图解最优化
graphic(al)output 图形输出
graphic(al)output unit 图形输出设备
graphic(al)package 图形软件包;图形程序包
graphic(al)panel (测量系统的)图示(面)板;有线路配的配电板;图形板;图示配电盘;图示控制盘;图解(式)面板
graphic(al)paper 坐标纸
graphic(al)parabola method 图解抛物线法
graphic(al)partition(ing) 图划分
graphic(al)parts 图形部分
graphic(al)pattern 图型;图案
graphic(al)performance 图解特性
graphic(al)perspective 图解透视网格
graphic(al)phototriangulation 图解像片三角测量
graphic(al)plot 制图
graphic(al)plotter 绘图仪;制图仪;制图机;绘图机
graphic(al)plotting 图解测图法
graphic(al)point 图解点
graphic(al)position finding 图解定位
graphic(al)position fixing 图解定位
graphic(al)position indicator 图示位置显示器
graphic(al)presentation 图示;图像显示;图表示法
graphic(al)presentation of frequency distribution 频率分布的图示法
graphic(al)primitive 图形原语;图形单元
graphic(al)printer 图形打印机;印图机
graphic(al)procedure 图解法(步骤)
graphic(al)processing 图表处理法
graphic(al)processing language 图形处理语言
graphic(al)processor 图形处理程序;图形处理中心;图形处理机
graphic(al)profile 图解剖面
graphic(al)programming 图形程序设计;图形程序编辑
graphic(al)progress chart 图示工程进度表
graphic(al)proportioning of aggregate 图解法级配集料
graphic(al)radical triangulation 图解辐射三角测量
graphic(al)rating scale 图表测度法
graphic(al)reactor design 图解法反应器设计
graphic(al)recognition 图形识别算法
graphic(al)record 图示记录
graphic(al)recorder 图示记录仪;图解记录器
graphic(al)recording 图示记录
graphic(al)recording instrument 图示器;自动记录仪器
graphic(al)recording unit 图式记录装置
graphic(al)recording voltmeter 自动记录伏特计
graphic(al)recording wattmeter 自动记录瓦特计
graphic(al)rectification 图形纠正

graphic(al)reduction 图解缩小转绘
graphic(al)repertoire 图形库
graphic(al)report generator 图形报告生成程序
graphic(al)representation 作图表示;图像表示;图示(法);图解(法);图解表示(法)
graphic(al)representations of structure 构造研究图件
graphic(al)reproduction 图形复制
graphic(al)reproduction technique 复照技术
graphic(al)resection 图解后方交会
graphic(al)scale 图示比例尺;图解量表;图解比例尺
graphic(al)sequence 图序列
graphic(al)sequencing 图解顺序
graphic(al)simulator 图形模拟程序
graphic(al)solution 图示解法;图解(法)
graphic(al)solution of a game 对策的图形解法
graphic(al)standards 图样规范
graphic(al)statement 图形决算表
graphic(al)statics 图解静力学
graphic(al)statistic(al)analysis 图解统计分析
graphic(al)statistics 统计图表
graphic(al)stereometer 图解立体测图仪
graphic(al)stereoplotting 图解立体测图
graphic(al)strength 图形强度;图像强度
graphic(al)structural analysis 图解结构分析
graphic(al)structure 文象构造
graphic(al)subroutine 图解子程序
graphic(al)subroutine package 图形子程序包
graphic(al)subsystem 绘图子系统
graphic(al)symbol(ization) 图解符号;图示符号;图解例图;图例(符号);图解符号
graphic(al)system 绘图系统;图形(处理)系统
graphic(al)tablet 小型绘图台;图形输入板
graphic(al)tacheometer 图解式测距仪
graphic(al)technique 图解技术
graphic(al)tellurium 针碲金银矿
graphic(al)terminal 图形(显示)终端
graphic(al)texture 文象结构;图解结构
graphic(al)time table 列车运行图
graphic(al)tolerance 图解限差
graphic(al)tracker 图形跟踪器
graphic(al)train recorder 列车运行图记录器
graphic(al)transfer 图解变换
graphic(al)transformation 图形转换;图形变换;图解转换
graphic(al)traversing 图解导线测量
graphic(al)treatment 图解法;图解处理
graphic(al)triangulation 图解三角测量
graphic(al)type machine 图解机
graphic(al)value 图解值
graphic(al)variable 图形变量
graphic(al)voltmeter 电压记录表
graphic(al)water-stage register 图示水位表;水位自记仪;水位图示仪
graphic(al)window 图形窗口
graphic-arts technique 图形法
graphic-like texture 似纹象结构
graphic-recording 自动记录式的
graphics 图形学;制图法;两维图;图(形)学;图(解)学;图解计算;图表法
graphics curve fittings program(me) 图形曲线拟合程序
graphics device interface 图形设备接口
graphics facility 制图设备
graphics field 图形字段;图形域
graphics hardware 制图硬件
graphics hold 图形同步并接
graphics input system 图形输入系统
graphics insertion 图形插入
graphics language 制图语言
graphics mode 图形方式
graphics peripheral 图形外围设备
graphics plotter 绘图仪
graphics presentation 图形表示
graphics program(me) 图形程序;绘图程序
graphics set 图形集合;图集
graphics software 制图软件;图形软件
graphics software package 制图程序组件
graphics software system 图解软件系统
graphics solution 图解
graphics standard 图形标准
graphics statement 图形语句
graphics statics 图形静力学

graphics structure input 图形结构输入
graphics subroutine system 图形子程序系统
graphics symbol 图形符号
graphics system 图形系统
graphics tablet 图形输入板
graphics terminal 图形终端
graphics text composition 图形文本组合
graphics transmission 图形传输
graphics user interface 图形用户界面;图形用户接口
graphics workstation 图形工作站
graphidox 硅钙钛铁合金
graphie file 图形文件
graph information retrieval language 图表信息检索语言
graphing 绘制图形
graphing method 作图法
graphite 黑铅;铅粉;炭精;石墨(含量)
graphite acid 石墨酸
graphite aggregate 退火石墨
graphite agitator 石墨搅拌器
graphite alloy 石墨合金
graphite annulus 石墨环(面)
graphite anode 石墨阳极
graphite anode basket 石墨阳极篮
graphite apex seal 石墨径向密封片
graphite atomizer 石墨喷雾器
graphite backing 石墨衬垫
graphite bar electric(al) furnace 石墨棒电炉
graphite-base composite material 石墨基合成材料;石墨基复合材料
graphite bearing 石墨轴承;石墨板
graphite biotite plagioclase gneiss 石墨黑云斜长片麻岩
graphite block 石墨块
graphite block air seal 石墨块密封圈;石墨块挡风圈
graphite block body 石墨块体
graphite boat 石墨舟
graphite body 石墨体
graphite bonded refractory 石墨结合耐火材料
graphite-brass composition 石墨黄铜混合料
graphite breather 石墨通气器
graphite brick 石墨(耐火)砖
graphite brick work 石墨砖结构
graphite bronze 石墨青铜
graphite bronze bearing 石墨青铜轴承
graphite brush 石墨电刷
graphite buffer 碳缓冲
graphite bush 石墨套管
graphite bushing 石墨衬套
graphite cake 石墨片;石墨块;石墨板
graphite carbon 石墨碳
graphite-ceramic fiber reinforced phenolic resin 石墨陶瓷纤维
graphite clay 石墨黏土
graphite-clay brick 石墨黏土砖
graphite coating 石墨涂层
graphite container 石墨容器
graphite-containing bearing 含石墨的轴承
graphite corrosion 石墨腐蚀
graphite crucible 石墨坩埚;石墨尘锅
graphite-cup atomizer 石墨杯雾化器
graphite cup grease 石墨杯脂
graphite cuvette 石墨槽
graphite cuvette atomizer 石墨槽原子化器
graphite cylinder 石墨圆筒
graphited 涂石墨剂的
graphited bearing metal 含石墨轴承合金
graphited braided asbestos packing 石墨处理的编织填料
graphite deflector 石墨偏转板
graphite deposit 石墨矿床
graphite-diamond-equilibrium line 石墨金刚石相平衡线
graphite-diamond-liquid triple point 石墨金刚石液相三相点
graphited oil 石墨润滑脂
graphited oilless bearing 石墨润滑的无油轴承
graphite dust 石墨粉(尘)
graphite electrode 石墨电极
graphite electrode slab 石墨电极块
graphite element 石墨化元素
graphite fabric 石墨纤维织物
graphite fiber 石墨纤维

graphite filled Teflon 聚四氟乙烯石墨;特氟隆石墨
graphite flake 片状石墨粉粒;石墨片
graphite flo(a)tation 石墨漂浮
graphite form 石墨形态
graphite furnace 石墨炉
graphite furnace atomic absorption sepctrometry 石墨炉原子吸收光谱法
graphite furnace atomizer 石墨炉雾化器
graphite gneiss 石墨片麻岩
graphite granule 粒状石墨
graphite grease 石墨(油)脂;石墨润滑脂;石墨膏;拌石墨的牛油
graphite grog crucible 石墨熟料坩埚
graphite heater furnace 石墨电阻炉
graphite heat exchanger 石墨换热器
graphite ignition anode 石墨引燃阳极
graphite in flake form 片状石墨
graphite insert 石墨衬套
graphite in sphere form 球状石墨
graphite jig 石墨模具
graphite jointing compound 石墨接合剂
graphite ladle 石墨浇勺;石墨端包
graphite ladle liner 石墨包衬
graphite layer 石墨层
graphite lead paint 石墨铅油
graphite-like material 石墨状材料
graphite-like structure 石墨状结构
graphite lime base grease 石墨钙基润滑脂
graphite liner 石墨衬套
graphite lining 石墨衬里
graphite low energy experimental pile 低功率石墨实验性反应堆
graphite lubricant 石墨润滑剂
graphite lubricating oil 石墨润滑油
graphite lubricating rod 石墨润滑棒
graphite lubrication 石墨润滑(法)
graphite lubricator 石墨润滑器
graphite marble 石墨大理岩
graphite metal 铅基轴承合金;石墨铅基合金
graphite-metal lamellar compound 石墨金属层间化合物
graphite metering pump 石墨计量泵
graphite mica schist 墨云片岩
graphite micropowder 石墨微粉
graphite-moderated lattice 石墨慢化栅格
graphite-moderated reactor 石墨(原子)减速反应堆;石墨慢化堆
graphite moderator 石墨减速剂
graphite moderator stringer 石墨慢化棒
graphite mo(u)ld 石墨模(具);石墨型
graphite mo(u)ld department 石墨模具车间
graphite nest 石墨粗大疏松组织;石墨巢孔(铸件缺陷)
graphite nodule 石墨球
graphite oil 石墨(润滑)油
graphite ore 石墨矿石
graphite oven 石墨炉
graphite oxide 石墨氧化物
graphite oxide membrane 石墨氧化膜
graphite packing 石墨密封垫;石墨衬垫;石墨填密;石墨填料;石墨盘根
graphite paint 黑铅漆;石墨(油)漆;石墨涂料;石墨防腐涂料
graphite pebble 石墨球
graphite pig iron 灰口(生)铁
graphite pipe 石墨管
graphite piston 石墨活塞
graphite piston ring 石墨活塞环
graphite plastic 石墨塑料
graphite plate 石墨板
graphite pneumoconiosis 石墨尘肺
graphite pot 石墨坩埚
graphite powder 石墨粉
graphite precipitation 石墨析出
graphite preparation 石墨剂
graphite prestressed pile 石墨增强预应力柱
graphite product 石墨制品
graphite pyrometer 石墨高温计
graphite reactor 石墨反应堆
graphite reflected 带石墨反射的
graphite reflector 石墨反射层
graphite refractory product 石墨耐火材料制品
graphite resistance 石墨电阻
graphite resistance heater 石墨片阻加热器;石墨电阻加热器

graphite resistor 石墨电阻器
graphite resistor rod 石墨电阻棒;石墨电极
graphite rod 石墨棒
graphite-rod anode 石墨棒阳极
graphite-rod cathode 石墨棒阴极
graphite-rod furnace 石墨棒炉
graphite-rod melting method 石墨棒熔融法
graphite rosette 菊花状石墨;菊花形石墨
graphite schist 石墨片岩
graphite seal 石墨密封环
graphite sealing pad 石墨填料密封环
graphite sealing products 石墨密封制品
graphite septum 石墨隔板
graphite shape 石墨形状
graphite sheath 石墨套管
graphite sheet 石墨片;石墨板材
graphite shielding 石墨屏蔽
graphite single crystal 石墨单晶
graphite size 石墨尺寸
graphite slate 石墨板岩
graphite sleeve 石墨套管
graphite slip ring 石墨滑环
graphite solid cylinder 石墨棒
graphite sovite 石墨黑云碳酸岩
graphite sparge tube 石墨喷射管
graphite spark technique 石墨火花技术
graphite spine 石墨细棒
graphite spout 石墨流槽
graphite stack 石墨堆
graphite structure 石墨结构
graphite susceptor 石墨接受器
graphite susceptor ring 石墨感应环
graphite thermal column 石墨热柱
graphite to silicon-carbide couple 石墨和碳化硅热电偶
graphite tube 石墨管
graphite-tube furnace 石墨管炉
graphite tube resistance furnace 碳管电阻炉
graphite type 石墨类型
graphite type structure fibre 石墨型结构纤维
graphite-uranium pile 石墨铀堆
graphite vessel 石墨容器
graphite washer 石墨制垫圈
graphite water 洗涤石墨;石墨淘洗水
graphite wedge 石墨楔
graphite whisker 石墨须晶
graphitic absorber 石墨吸收器
graphitic action 石墨化作用
graphitic carbon 石墨
graphitic cell 石墨电解槽
graphitic chemical equipment 石墨化工设备
graphitic clay 石墨质黏土
graphitic corrosion 留碳腐蚀;碳化腐蚀;石墨化腐蚀
graphitic mica 石墨云母
graphitic nitralloy 石墨体氮化钢
graphitic pig iron 灰口铁
graphitic pressed pipe 石墨压型管
graphitic rock 石墨岩
graphitic steel 石墨(体)钢
graphitic synthetic(al) furnace 石墨合成炉
graphitiferous 含石墨的
graphitised anode block 石墨化阳极块
graphitizable steel 石墨化钢
graphitizating carbon 石墨化碳
graphitizating furnace 石墨化炉
graphitization 石墨化(作用)
graphitization molecular sieve 石墨化分子筛
graphitization of diamond 金刚石墨化
graphitization with compression 加压石墨化
graphitized black 石墨化炭黑
graphitized bronze 石墨青铜
graphitized carbon 石墨化碳
graphitized carbon black 石墨化炭黑;导电炭黑
graphitized carbon filament 石墨化碳丝
graphitized electrode 石墨化电极
graphitized filamentary material 石墨纤维材料
graphitized packing 石墨垫料
graphitizer 墨化剂;石墨化剂
graphitizing 留碳作用;石墨化(作用);石墨化(处理)
graphitizing annealing 石墨化退火
graphitizing carbon 石墨化碳
graphitizing element 石墨化元素;促石墨化元素
graphitizing medium 墨化剂

graphitizing of diamonds 金刚石的石墨化
graph manipulation 图形处理
graph model 图解模型
graph notation 图形符号;图形表示法
graphocite 石墨质岩
graph of a correspondence 对应的图形
graph of a fuzzy switching function 模糊开关函数的图
graph of a relation 关系图
graph of equation 方程的图
graph of errors 误差曲线(图)
graph of friction(al) resistance curve 摩擦阻力头曲线
graph of function 函数图
graph of fuzzy-switching functions 模糊开关函数图
graph of molecular orbital 分子轨道图形
graph of synthetic(al) result of drill hole 钻孔综合成果图表
graph of tracer concentration change 示踪剂浓度变化过程曲线
graph of variable series 变量数列图
grapholite 石墨片岩
graphology 笔迹学
graphometer 量角仪;半圆仪
graphon 石墨化炭黑
graphoscope 电脑显示器
graphosis 石墨尘肺
graphotest 记录测微计;图示测微计
graph paper 坐标纸;计算纸;毫米纸;方格纸
graph pen 绘图笔
graph plotter 绘图仪;制图器
graph plotting 绘图
graph search control 图搜索控制
graph search strategy 图搜索策略
graph structure algorithm 图形结构算法
graph subspace 图形子空间
graph-table 图表
graph-table entry 图表项目
graph theory 图论;图解理论
graph theory model 图解理论模型(法)
graphtyper 图像电传机
grapnel 小锚;锚;探锚;四爪(小)锚;上杆脚扣;多爪(小)锚
grapnel anchor 四爪(小)锚
grapnel travel(l)ing crane 抓斗移行吊车;锚钩式行走起重机
grapnel tube 钻杆捞管;钻杆打捞爪
grappier 石灰渣
grappier cement 渣块水泥;石灰渣水泥
grapple 用锚将船固定住;抓住;抓扬机,抓泥机;夹木器;爬桩脚扣;四爪小锚;四爪(锚)
grapple dredge(r) 抓斗式挖泥机;抓斗式挖泥船;多齿抓斗挖泥船;斗式挖泥船
grapple equipped crane 抓斗起重机;锚固式起重机;多齿起重机
grapple fork 抓钩叉子
grapple hook 抓升钩
grapple iron 铁橇棍
grappler 尾孔楔钉;设置挑出式脚手架用墙上预留孔六;爬桩脚扣;抓斗;支承钩;尾孔楔形块;抓钩(器)
grappler cement 石灰水泥
grappler skidder 抓斗式集材机;抓钩式集材机
grapple travel(l)ing crane 抓斗行走吊车
grapple yarding 抓钩动力集材
grappling 小锚;锚定;捉住;锚住;拉牢
grappling fixture 抓头
grappling hook 抓升钩;四爪小锚;多爪小锚
grappling iron 小锚;抓升钩;抓机;铁橇;四爪小锚;多爪锚
grappling of arch 拱的锚定
grappling tool 带抓卡装置的打捞工具
graptolite shale 笔石页岩
graptolitic facies 笔石相【地】
Grashof's number 格拉斯霍夫数
Grashof formula 格拉肖夫公式
Grashof number 格拉肖夫数
grasp 抓住;紧握;夹紧;锚钩
grasper 抓紧器
grasping means 抓取方法
grasping reflex 抓握反射;握住反射
grasping tongs 抓取钳
grasp reflex 抓握反射;握住反射

grasp weakness 握力减弱
grass 噪声(细)条;茅草干扰;禾草;附着在船上的无柄海藻
grass airfield 草地机场
grass-arable system 草田轮作制
grass basketry 草织盛具
grass beach 草地海滩
grass belly 草腹
grass blade 草叶
grass bleaching 草地曝晒漂白法
grass building machinery 草原建设机械
grass canopy 草皮(盖)
grass carbon 草炭
grass carp 草鱼
grass carpet 草坪
grass catcher 集草器
grass cloth (壁纸或墙布的)背衬织物;夏布;(覆盖墙面用的)植物纤维疏松织物
grass-clump hybrid dwarfness 丛草型杂种矮生性
grass cock 新鲜干草垛
grass comb 轻型草地耙
grass cottage 茅舍
grass court 草地(网)球场
grass cover 草坪;草地覆盖;草场
grass-covered area 草皮覆盖地区
grass-covered ground for sunbathing or resting 供日光浴或休憩用的草坪
grass covered land 青草覆盖的土地;生草地
grass cutter 割草机;草地剪草机
grass cylinder 草籽清洗筒
grass divider 分草器
grassed area 铺草皮地区;铺草地带;铺草(皮)面积
grassed central reserve 植草的中央分隔带;(多车道的)铺草的中央分隔带
grassed central strip 中间植草带
grassed channel 植草明渠;草皮镶护明渠;草皮护坡明渠
grassed percolation area 植渗流带;草皮渗流区
grassed region 铺草皮地区
grassed slope 草皮护坡
grassed spillway 草皮溢水道
grassed surface 铺草地面
grassed swale 植草沟
grassed verge 路旁草坪
grassed water course 长草河(流)
grassed waterway 植草水道;草皮泄水道;草地泄水道
grass eliminator 除草机
grass embankment 铺草路堤
grass fabric filter 草纤维过滤器
grass farm 草地牧场
grass fen 草沼地
grass fiber 草纤维
grassfield ecology 草地生态学
grassfield farming 草地农作制
grass filtration 草地过滤
grass fire 草地火灾;气化器
grass fodder 牧草;青草饲料
grass garden 草坪庭园;草地庭园
grass glade 林中草地;林地草地
grass green 草绿色
grass harvester 牧草收割机
grass hawser 大草索
grass-heath 丛生草原
grass hopper 小吊车;蚱蜢(小型起重机的俚语);平口焊接器;转送装置;轻型单翼机
grasshopper fuse 报警熔丝;弹簧保安器
grasshopper ga(u)ge 细木工规尺;滑动接头量规
grasshopper joint 滑动接头
grasshopper pipe coupling method 导管装配夹组法
grasshopper spring 半椭圆形悬臂钢板弹簧
grassing 草上晒白法
grass killer 除草剂
grassland 草原;草地;草场;草本植被区
grassland agriculture 草田轮作法;草地畜牧业;草地轮作法
grassland body 草地犁体
grassland bottom 草地犁体
grassland climate 半湿润气候
grassland cover 草原覆盖;草地植被
grassland cultivation 草地栽培
grassland degeneration 草地退化
grassland desertification 草地沙化

grassland deterioration 草原退化
grassland ecology 草地生态学
grassland ecosystem 草地生态系统
grassland environment 草原环境
grassland establishment 建立草地
grassland farming 草地农业
grassland forest 草地森林
grassland for grazing 牧草地
grassland harrow 草地耙
grassland improvement 草原改良
grasslanding area 降落草坪;草地飞机场
grassland irrigation 草地灌溉
grassland law 草原法
grassland meteorology 草原气象学
grassland of tallgrass type 高草型地
grassland pedology 草原土壤学
grassland rejuvenator 草地松土器;草地刨土机;草地刨土机
Grassland Research Institute 草地研究所
grassland resources 草地资源
grassland science 草原学
grassland sod drill 草地播种机
grassland sod seeder 草皮地播种机
grassland soil 草地土壤
grassland type flexible harrow 草地网状耙;草地挠性耙
grassland vegetation 草原植被;草地植被;草本植被
grass like plant 禾草状植物
grass linen 夏布
Grassmann's law 格拉斯曼定律
Grassmannian 格拉斯曼流形
Grassmann manifold 格拉斯曼流形
grass margin 草地路缘;草地边缘;植草边道
grass marsh 草原沼泽
grass mat 草垫
grass matting 草席
grass meal 草粉
grass meal pellets 草粉粒
grass minimum 最低草温
grass minimum thermometer 最低草温表
grass moor 沼泽草原;禾草沼泽;草沼泽
grass mo(u)ld 腐草土
grass mower 割草机
grass mowing machine 割草机
grassot fluxmeter 动圈式磁通计;格拉索特磁通计
grass peat 草本泥炭
grass planting 植草
grass play area 草地游戏场
grass playground (供儿童游戏用的)草坪
grass plot 小草坪;小(块)草地;草地(污水处理);庭园青草地;草坪
grass plot treatment 草地处理
grass protection 植草保护;草皮护坡;草皮护岸
grass quartz 草石英
grass rejuvenator 草籽破皮机
grass-remover 锄草机
grass reseeder 草地播种机
grass resources 草场资源
grass rod 挡草杆
grass roller 滚草机;草地滚压机
grass roots 农业地带;农业区;草根
grass roots deposit 表层矿床
grass roots industry 草根工业
grass roots plant 处女地工厂
grass roots scheme 处女地的建设规划
grass roots unit 基层单位
grass rope 草(搓)绳
grass rubber 草胶;草本根胶
grass rubber plant 橡胶草
grass rug 草编小地毯
grass runway 草地跑道
grass seed 草籽
grass-seed attachment 草籽播种附加装置
grass-seed box 草籽箱
grass-seed cleaner 草籽清选机
grass-seed combine 草籽联合收获机
grass-seed drier 草籽干燥机
grass seeder 草籽播种机
grass-seed feed 草籽排种器
grass-seed harvester 草籽收获机
grass-seed hopper 草籽箱
grass-seed tube 草籽输种管
grass shoulder 草皮路肩
grass sod 草地

grass sod cutter 割草机;草地修剪机
grass spader 锄草机
grass spreader 草条撒散机
grass stalk 稻草杆
grass stick 拨草杆
grass strip 长槽地带;植草(地)带;草皮简易机场
grass surface 植草皮地面
grass-surfaced area 铺草地带
grass swamp 草原沼泽
grass swamp facies 草原沼泽相
grass table 哥特式建筑勒脚顶部;泥土台
grass temperature 草温
grass tex 沥青与植物纤维混合铺地面(运动场用)
grass thatch 茅草屋面
grass thicket 灌草丛
grass tree 禾木
grass-tree gum 禾木(树)胶
grass verge 草坪边缘;草皮路肩;路旁草皮
grass walk 铺草人行道;铺草路
grass waterway 草他泄水道;植草水道
grass wool filter 草纤维过滤器
grass work 矿坑外作业;洞外作业;坑外作业
grass woven products 草编织品
grassy chestnut 茅栗
grassy coast 多草海岸
grassy oak 茅栎
grassy plain 草原
grassy ridge 生长苔藓的(海洋)浅滩;多草的山脊;草绿色的山脉
grassy shore 草丛岸
grate 线栅;固定筛;格栅;炉栅;炉条;炉排;炉算子;铁算子;筛格;算子(板);算条筛;壁炉;壁垒;算子;栅条
grate area 炉排面积;算栅面积;火床面积
grate ball mill 格子排料式球磨机
grate bar 炉排;算条;筛条;栅条
grate bar structure 条条结构
grate bar thimble 炉条套筒
grate burning rate 炉排热强度
grate cooler 栅格冷却器
grate cooler stroke 算式冷却机冲程
grated hatch cover 格子舱(口)盖
grated inlet 算式雨水口;算式进水口;栅式进水口;格栅(式)进水口;算式雨水口
grated intake 格栅进水口
grated sandsheet 梁窝状沙地
grate feeder 算式喂料机
grate-fired furnace 层燃炉(膛);链条炉炉膛
grate firing 炉算燃烧室;层燃
grate frame 炉床架
grate frame bearer 炉床架支座
grate frame support 炉床框支架
grate funnel 炉排风箱
grate furnace 炉排炉;层燃炉
grate guard 格子安全排
grate heater 算子加热机
grate hopper 带栅格料斗
grate layer material 铺底料
grateless producer 无炉算煤气发生器
grate mill 格子型球磨机;格子式磨碎机
grate of sink 污水池格栅
grate opening 栅格孔口;栅格进口;帘格进口
grate opening of inlet 算式进水口;算式进水口
grate preheater 炉算(子)预热器;算子预热机
grater 粗齿木锉
grate rack bar screen 格筛;格栅
grate release rate 炉排热强度
grate ring 炉排座圈
grate room 火室
grate shaft 摇炉轴
grate shaker 炉排摇动器;算算摇动器
grate shoe 算板
grate side frame 炉床侧梁
grate stage cooler 阶梯形算式冷却器
grate surface 炉排面积;算床面;算算面;火床面
grate surface area 算板表面积
grate surface loading 炉算子面积负荷;算床负荷
grate-type cooler 算式冷却机;算式冷却机
grate-type gutter 算式雨水沟;算式雨水沟
grate-type inlet 帘格式进水口;算式进水口;算式进水口
grate type preheater 算式预加热机
grate type throttling 格栅式节流阀
grate water tube 炉床水管

grate with forced draught 强迫通风炉算
grate with water circulation 水冷炉算
graticulated glasses 带分划镜的望远镜
graticulation 打方格画法;方格缩放法;方格画法
graticule 量板;经纬网;经纬网格;网格;分度镜;方格图;标线片;标线板
graticule intersection 经纬网交会
graticule line 地理坐标网;方格线
graticule meridian 网格
graticule mesh 网格
graticule method for quasi two-dimensional bodies 似二度量板法
graticule method with corrected coefficient 带校正系数量板法
graticule net 经纬线网
graticule of meridians and parallels 经纬网
graticule plate 制图格网板
graticule template 制图格网展绘模片
graticule ticks 经纬仪延长短线
graticule value 坐标网格值;地理坐标值
gratin filtering 光栅滤波
grating 衍射光栅;光栅;格子底板;格子(板);格栅;滤栅;水筛;栅(板);筛条;算子(盖);算子板
grating acuity 光栅锐度
grating and dot generator 点栅信号发生器
grating anomaly equation 光栅异常方程
grating aperture 栅线隙距
grating array 栅阵
grating assembly 光栅排列
grating beam 槛木;排架座木
grating beam-divider 光栅光束分配器
grating chromatic resolving power 光栅色分辨率
grating cill 排架座木;槛木;长枕木
grating constant 光栅常数;晶格恒量;点阵常数
grating converter 光栅变换器;双线栅变频器
grating-coupled radiation 光栅耦合辐射
grating coupler 光栅耦合器
grating cover 格子盖
grating diffraction 光栅的衍射
grating digital display 光栅数字显示器
grating digital measurement milling machine 光栅数测铣床
grating digital readout 光栅数显表
grating dispersion 光栅色散
grating dividing engine 光栅刻线机
grating efficiency 光栅效率
grating element 光栅要素
grating encoding 光栅编码
grating energy measuring device 光栅能量测定仪
grating equation 光栅方程
grating filter 光栅滤光片
grating focus 光栅焦点
grating formula 光栅公式
grating frame 花格板框架
grating generator 栅形场振荡器;栅形场信号发生器;交叉线信号发生器
grating ghost 光栅鬼线
grating graduated circle 光栅度盘
grating grid 算子
grating groove 光栅(刻线)槽
grating gutter 格栅式明沟;算式雨水沟;算式雨水沟;帘格式街沟;格栅式街沟
grating hatch 格子舱口
grating hatch covering 格子舱(口)盖
grating image 光栅星像
grating incidence mount 光栅入射装置
grating infrared spectrometer 光栅红外光谱仪;光栅红外分光光度计
grating inlet 格栅(式)进水口
grating intake 格栅进水口
grating interferometer 光栅干涉仪
grating light colo(u)rimeter 光栅光比色计
grating-like hologram 类光栅全息图
grating line 光栅刻线;光栅划线
grating lobe 栅瓣;光栅波瓣
grating monochromator 光栅单色仪
grating nephoscope 栅式测云器
grating of gears 齿轮噪声
grating of sink 污水盆格算
grating of timber 木梁格栅
grating opening 栅孔
grating pair 光栅对
grating platform 格子平台
grating polychromator 光栅多色仪

grating recombiner 光栅光束重合器
grating reflector 栅状反射器;光栅反射器
grating ring 栅环
grating ruling 光栅刻线
grating-ruling engine 光栅刻线机
grating sampling method 光栅取样法
grating satellite 光栅伴线
grating scanner 光栅扫描器;光栅扫描机
grating screen 溜算子筛;除渣栅
grating selected resonator 光栅选频共振腔
grating sensor 光栅传感器
grating setting 光栅调节
grating shearing interferometer 光栅剪切干涉仪
grating sheet 栅板
grating sill 排架座木;槛木;长枕木
grating space 栅线间距;光栅间距;晶面距离
grating spacing 栅线间距;光栅间距
grating spectral order 光栅光谱级
grating spectrograph 光栅摄谱仪;光栅光谱仪
grating spectrometer 光栅光谱仪;光栅分光计
grating spectrophotometer 光栅分光度计
grating spectroscope 光栅分光镜
grating spectrum 光栅光谱
grating spectrum satellite 光栅光谱伴线
grating stereographic(al)map 光栅立体地图
grating storage target 光栅信息存储靶
grating structure 格状构造;格形构造;格栅式结构
grating texture 格(子)状结构;格形结构
grating with oblique incidence 倾斜入射光栅
gratis 免费;无偿地
gratis services 免费服务
gratonite 细硫砷铅矿
gratte-ciel 摩天楼
gratuitous 免费的;无代价的;无偿的
gratuitous contract 单方受益契约
gratuitous goods 免费商品;天然资源
gratuitous help 无偿援助
gratuitous investment 无偿投资
gratuitous service 免费服务
gratuity 退职金;赏银;小费
graunch 意外错误
graupel 雪丸;霰;软雹(霰)
graupel shower 一阵雪丸;冰暴
Gravatt level 格拉瓦特定镜水准仪
Gravatt level(l)ing rod 格拉瓦特水准尺
grave 重大的;墓(葬);清除船底污垢;低沉;坟墓
grave accident 重大事故
graved roll coating 槽辊涂装
gravel 小石;砾(石);石子;卵石;角砾;圆砾
gravel aggregate 砾石集料;砾石骨料
gravel aggregate concrete 砾石集料混凝土;砾石骨料混凝土
gravel and chip(ping)s works 砾石和碎石作业
gravel and sand filter bed 砂砾滤床
gravel and shingle foreshore 杂石滩
gravel asphalt 砾石沥青
gravel asphalt concrete 砾石沥青混凝土
gravel ballast(ing) 砾石道砟;卵石道砟
gravel band 砂砾带;砂砾层
gravel bank 砾石滩;砾石堆坝;砾石岸
gravel bar 砾石(沙)洲;砾石滩;砾石场
gravel barge 砾石驳(运)船
gravel barrow 运砾石手推车
gravel base 砾石基层
gravel base course 砾石基层
gravel basin 砾石水箱;砂砾水槽
gravel beach 砾石海滩;沙砾滩
gravel-bearing breccia 含砾角砾岩
gravel-bearing mudstone 含砾泥岩
gravel-bearing sand 含砾砂
gravel-bearing sandstone 含砾砂岩
gravel-bearing siltstone 含砾粉砂岩
gravel bed 砾石(垫)层;砾石河床
gravel bed constructed wetland 砂砾床人工湿地
gravel bedded stream 砾石床面河流
gravel bedding layer 砾石基层
gravel bed dust filter 颗粒层收尘器;颗粒层集尘器;颗粒层除尘器
gravel bed filter 颗砾层收尘器;砾石滤床过滤器;砾石滤床过滤池
gravel bed hydroponic constructed wetland 砂砾床水栽发人工湿地
gravel bed hydroponics 砾石床水栽法
gravel bed river 砾石底河床

gravel bed stream 砾石底河床
gravel blanket 砾石盖层
gravel board 围栏底板;砾石(垫)板;木栏底下的长石板;(木栏底下的)长石板
gravel bottom 砾石河底;砾石底(质)
gravel box 砾石箱;砾石笼;拦沙箱
gravel-breaking plant 轧砾石厂
gravel built-up roof cladding 砾石组成的屋顶覆盖物
gravel by nature 天然砾石
gravel cement mixture 砾石水泥混合料;水泥砾石混合料
gravel chamber filter 砾石过滤器
gravel chip(ping)s 砾石碎屑
gravel clast 砾屑
gravel coast 砾质海岸
gravel coating 砾石面层
gravel cobble 卵石
gravel coefficient 砾石系数
gravel concrete 卵石混凝土;砾石混凝土
gravel concrete slab 砾石混凝土板
gravel concrete wall panel 砾石混凝土墙板
gravel contact purification 砂砾层接触净化
gravel containing fine grains 含细粒的砾石
gravel containing lightly fine grains 微含细粒的砾石
gravel content 含砾量
gravel core 砾芯
gravel core fascine 砾石芯梢笼;砾石芯梢捆;砾石芯埽枕;砾石芯沉梢;砾石芯沉排;梢枕
gravel core roll 砾石芯梢笼;砾石芯梢捆;砾石芯埽枕;砾石芯沉梢;砾石芯沉排;梢枕
gravel course 砾石层
gravel covering 砾石盖面
gravel culture 砂培(法)
gravel cushion 砾石盖层
gravel dam 砾石坝
gravel decant line 拦沙箱
gravel deposit 砾石沉积层;砾石沉积物;砾石场;冲积矿床
gravel desert 砾质沙漠;砾漠
gravel distributor 砾石撒布器
gravel drain 砾石排水沟;盲沟
gravel dredge(r) 砾石挖泥船;砾石挖掘机;砾石疏浚船;挖石船
gravel dressing 砾石选矿
gravel dressing plant 砾石选矿厂
gravel dryer 砾石干燥器
gravel drive 砾石路
gravel dust filter 颗粒层收尘器;颗粒层集尘器;颗粒层除尘器
gravel elastic texture 砾屑结构
gravel envelope 砂砾保护井;砾石(井)壁;砾石保护层;砂砾保护层;沙砾保护层
gravel envelope well 填砂砾井;砂砾围护井
gravel extraction 砾石提炼
gravel extraction and preparation 砾石采矿与制备
gravel extraction plant 采矿厂
gravel face 砂矿工作面
gravel fill 排水暗沟;填石水沟
gravel-filled dam 砾石(填充)坝
gravel-filled drain trench 砾石(填充的)排水沟;砾石填槽排水沟
gravel filled trench 砾石填槽
gravel fillet 砾石料堆
gravel filling 砾石填充(料)
gravel filter 细卵石过滤器;砾石填料;砾石滤池;砾石滤器;砾石过滤器;护底碎石;砂砾滤池
gravel filter layer 砾石滤层
gravel filter well 砾石过滤井;砾石滤水井
gravel flag 粒状石板
gravel flood coat 砾石防洪层
gravel for garden paths 铺庭院小径的砾石
gravel for sett pavings 石块铺砌路面的砾石
gravel foundation 卵石地基;砾石基础;砾石地基
gravel fraction 砾石成分;砾石部分;砾石粒级
gravel grit 砂砾
gravel ground 砾质土;砾石地
gravel grouting 砾石灌浆
gravel hammer 碎石锤
gravel hardpan 砾石硬(土)层
gravel layer 砾石层
gravel(l)ed area 铺砾石地区
gravel(l)ed path 铺砂砾路

gravel(l)ing 铺砾石;建筑砾石路面;嵌进砾石
gravel(l)ing of road 砾石铺路(面)
gravel loader 砾石装载机
gravelly 砾质
gravelly braided-stream deposit 砾质辫状河沉积
gravelly clay 砾质黏土
gravelly ground 砾质地;砾石地
gravelly limestone 砾状灰岩
gravelly loam 砾质亚黏土;砾质壤土
gravelly phosphoraite 砾状磷块岩
gravelly sand 砾质砂土;砾砂
gravelly sand concrete aggregate 砾质砂混凝土骨料
gravelly sand loam 砾质亚砂土
gravelly sandstone 砾质砂岩
gravelly soil 砾类土;砾质土;砾石土;含砾土
gravel method 砾石找矿法
gravel mill 卵石磨;砾磨机
gravelmine 砂金矿
gravel mound 砾石堆
gravel mulch 砾质遮护料
gravelometer 抗砂(砾)试验器
gravelous sand concrete 砾石混凝土
gravel pack 砾石填料层;砾石填充层;填砾
gravel packed filter 卵石填料滤池
gravel packed water well 砂砾填充水井
gravel packed well 砾壁井
gravel packer 砾石封隔器
gravel packing 卵石填料;砾石填料;砾石回填;砾石堵塞;砾石充填(法);填砾
gravel packing completion 砂砾充填完井
gravel pack material 卵石填料
gravel pack thickness 砾石衬填厚度
gravel pass 砾石排放孔;砾石排放道;砾石排出孔
gravel path 砾石小径;砾石路
gravel pavement 砾石路面
gravel pebble 小卵石;小圆石
gravel pile 碎石桩;砾石柱
gravel pile by dry method 干法碎石桩
gravel pit 砾石(料)坑;采石场;采石坑
gravel plank 砾石板;木栏着地板;木栏底下的长石板
gravel plant 砾石筛选机;砾石筛选厂
gravel plug 砾石(堵)塞
gravel pocket 砾石填坑;蜂窝(混凝土)
gravel pontoon 砾石起重机船;砾石趸船
gravel preparation plant 砾石制备厂
gravel production 砾石生产
gravel pump 砾石泵;砂石泵;砂砾(运输)泵;沙石泵
gravel rampart 砾石埂
gravel resistance testing 抗砾石试验
gravel ridge 砾石埂
gravel river bed 砾石河床
gravel road 砾石路
gravel roll 砾石梢笼
gravel roof 砾石屋顶;组合屋面;油毡绿豆砂屋面;防水毡屋顶卵石罩面
gravel roofing 组合屋面;油毡绿豆砂屋面;砂砾面;豆石油毡屋面
gravel runway 砾石跑道
gravel sample 碎石样
gravel sand 砾石砂
gravel-sand-clay 含砂砾黏土;砾石砂土混合物
gravel-sand cushion 砂砾(石)垫层
gravel scoop 砾石铲斗
gravel screen 砾石筛;砾石滤器
gravel screening 筛石
gravel separator 卵石分离器;清除砾石装置
gravel shoulder 砾石路肩
gravel(side)walk 砾石人行道
gravel sieve 砾石筛
gravel siever 砾石筛分机
gravel silo 砾石筒仓
gravel soil 砾(质)土;砾(石类)土;砂砾土
gravel sorter 砾石筛分机;砾石分选器;砾石分选机
gravel splay 砾石冲积扇
gravel spread 卵石覆盖层
gravel spreader 砾石撒铺器;砾石撒布机
gravel sticking filter 贴砾过滤器
gravel stone 卵石;砾石;小圆石
gravel stop 挡石板;砾石挡板;挡石片
gravel stratum 砾石层;砂砾层
gravel stuffing filter 填砾过滤器
gravel substructure 砾石底层结构

gravel support 砾石承托层
gravel surface built-up roof 砾石面组合屋顶
gravel surfaced 砾石铺面的
gravel surface guilt-up roof 砾石面组合屋顶
gravel surfacing 砾石铺面
gravel terrace 砾石台地
gravel-topped macadam 砾石面路;砾石碎石路面;砾石铺面碎石路
gravel trap 砾石拦截坑
gravel trench drainage 砾石沟排水
gravel-type gully 用砾石填充的进水井;用砾石填充的集水沟;盲沟式进水井
gravel underlayer 碎石垫层
gravel walk 砾石人行道;砾石步道;砾石小路;砾石小径
gravel wall 砾石墙
gravel wall well 砾石壁井;填砾井
gravel wash 洗净砾石;冲积砾石
gravel washer 洗砾机;砾石洗涤机
gravel washing plant 洗砾石场;洗石机;砾石洗涤设备
gravel washing screen 洗砾筛;洗砾石机;洗砾筛
gravel wave altitude 砾浪高度
gravel wave space 砾浪间距
gravel well 砾壁井
gravel works 砾石作业;砾石件;砾石工
gravel yard 砾石堆场;砾石堆场
gravely sediments 砾质沉积物
grave monument 墓碑
grave-mound 坟墩
graven 雕刻的
graven image 偶像;雕像
grave of a martyr 烈士墓
graver 刻图仪;雕刻工具;雕刻刀
graver point 刻刀刃
grave sand dust 喷砂雕刻
grave slab 墓穴盖石;墓石
gravestone 墓(石)碑
grave surround 墓地
grave trap 舞台升降平台
graveyard 公墓;墓地;埋藏场;坟地;坟场
graveyard chapel 墓地礼拜堂
graveyard hours 流量稀少时间
graveyard shift 夜间60班;夜班
graveyard test 土埋耐腐试验
graveyard watch 二副夜班(指凌晨0~4时)
graviation tank 高位油罐
gravics 重力场学
gravimeter 重力计;重差计;比重计;比重测定器
gravimeter beam 重力仪杆
gravimeter drift 重力仪掉格
gravimeter measurement 容积法测流
gravimeter method 重力探矿法;重力勘探法
gravimeter with linear system 线路系统重力仪
gravimetre 液体密度测量计
gravimetric(al) 重量分析的;重量测定的;重力测量的;比重测定的
gravimetric(al) absorption method 重量吸收法
gravimetric(al) altimeter 重力高度计;重力测高计;重差高度计;重差测高计
gravimetric(al) analysis 重量分析;重量定量分析;重力分析
gravimetric(al) batching 重量配料(法);按重量配料(法)
gravimetric(al) blender 重力掺和机
gravimetric(al) concentration 重力选矿(法)
gravimetric(al) coordinates 重力坐标
gravimetric(al) correction 重力改正;重力校正
gravimetric(al) data 重力资料;重力(测量)数据
gravimetric(al) datum 重力基准
gravimetric(al) datum orientation 大地基准的重力定位
gravimetric(al) deflection(of the vertical) 重力垂线偏差
gravimetric(al) density detemination 假比重;重量密度;重量法测定
gravimetric(al) determination 重量分析测定法
gravimetric(al) dust sampler 重量分析粉尘采样器
gravimetric(al) efficiency 重量效率
gravimetric(al) factor 重量分析因数;重力测量因子
gravimetric(al) feed 重力喂料
gravimetric(al) feeder 重力喂料器;重力送料器;称量给料机
gravimetric(al) fineness 称重法细度

gravimetric(al) geodesy 重力大地测量学
gravimetric(al) geoid 重力大地水准面
gravimetric(al) map 重力场变化图
gravimetric(al) measurement 重力测量
gravimetric(al) method 重量分析法;重量法(加气混凝土配料);重力(方)法;重力测量法;称重法
gravimetric(al) microrelief 微起伏重力测量
gravimetric(al) moisture content 重量水溶度
gravimetric(al) network 重力(测)网
gravimetric(al) observation 重力观测;重力测定
gravimetric(al) point 重力点
gravimetric(al) procedure 重量分析步骤;重力分析法
gravimetric(al) reduction 重力改化
gravimetric(al) sample 重量试样
gravimetric(al) sampling 计重法采样
gravimetric(al) separation method 重力分离法
gravimetric(al) station 重力测量点
gravimetric(al) survey 重力(法)检查;重力(法)测量;重力(法)测定
gravimetric(al) system 重力系统
gravimetric(al) table 重力测量用表
gravimetric(al) traverse 重力测量导线
gravimetric(al) undulation 重力大地水准面起伏
gravimetric(al) variometer 重力变化记录计;重力变化测定仪
gravimetric(al) yield test 坍落度试验(混凝土)
gravimetrist 重力测量员
gravimetroscope 液体蒸发速度测定器
gravimetry 重量分析(法);重量测定;重力(测量)学;重力测定法
gravimetry data 重力测量资料
gravimetry instrument 密度测定仪
gravimetry point 重力点
graving 雕刻(品);船底整修;船底清扫;船底的清扫和涂油;船底除锈涂漆
graving dock 坞门干船坞;修船船坞;干船坞;(修船的)船坞
graving drydock 修船(干)船坞;(修船的)船坞
graving expenses 船底清扫费
graving flower saw blade 雕花锯条
graving piece 镶补木料
graving tool 雕刻工具;雕刻刀
gravipause 引力分界;重力分界
graviperception 重力感
graviplanation 重力夷平(作用)
gravireceptor = gravirecepter
gravirecepter 重力感受器
gravisat 重力测量卫星
gravisphere 引力层;重力圈;引力范围
gravistatic 坠积的
gravitach(e)ometer 重力速测仪;重力测速仪
gravitate 重力作用下沉;重力沉降;受重力作用
gravitate downwards 重力吸下
gravitate settling 重力地面沉降
gravitating 重力吸引;重力溜放
gravitating disk 引力盘
gravitating fluid 引力流体
gravitation 引力;重力;地心吸力
gravitational 地心引力的
gravitational acceleration 引力加速度;重力加速度
gravitational anomaly 重力异常
gravitational apparatus 重力装置
gravitational astronomy 引力波天文学;天体力学
gravitational attraction 引力吸引;重力(引力);万有引力;地球引力
gravitational attraction of moon and sun 日月引力
gravitational backwater 重力式回水
gravitational casting 重力铸造
gravitational charge 引力荷
gravitational circulation 重力循环
gravitational classifying 重力选分;重力分级
gravitational clustering 引力成团(学说)
gravitational coagulation 重力凝聚
gravitational collapse 引力坍缩
gravitational collector 重力集尘器
gravitational compaction 重力压实
gravitational condensation 重力凝聚
gravitational constant 万有引力常数;引力常数;重力常数
gravitational constant of center of the earth 地心引力常数
gravitational contraction 引力收缩;重力收缩
gravitational convection 自然对流;重力对流;热对流
gravitational cooler 重力冷却器
gravitational creep 重力蠕动;重力蠕变;重力潜移
gravitational dam 重力坝
gravitational darkening 引力昏暗
gravitational deflection 引力弯曲
gravitational determination 重力测量
gravitational differentiation 重力弥散;重力分异(作用);重力差分(作用)
gravitational displacement 引力位移;重力位移
gravitational dissipative instability 重力耗散不稳定性
gravitational disturbance 引力扰动
gravitational dust collector 重力集尘器;重力除尘器
gravitational dust precipitator 重力式集尘器;惯性除尘器
gravitational dust setting 粉尘重力沉降
gravitational dust setting chamber 粉尘重力沉降室
gravitational effect 引力效应;重力效应
gravitational encounter 引力碰撞
gravitational energy 重力能;引力能;位能
gravitational equilibrium 引力平衡;重力平衡;重力均衡
gravitational erosion 重力侵蚀;重力流失
gravitational exploration 重力勘探
gravitational exploration method 重力勘探法
gravitational feed tank 重力送料箱
gravitational field 引力场;重力场
gravitational field equations 重力场方程
gravitational field theory 引力场理论
gravitational filter 重力滤器;重力滤池
gravitational fissure 重力裂隙
gravitational flattening 引力扁率
gravitational flow 重力水流;重力流变
gravitational flow irrigation 重力流灌溉
gravitational flute instability 重力槽形不稳定性
gravitational flux 引力通量;重力通量
gravitational flux density 引力位移
gravitational force 引力;重力;地心吸力
gravitational gliding 重力滑移;重力滑动
gravitational gradient 引力梯度;重力梯度
gravitational gradient guidance 重力梯度制导
gravitational grouting 自流灌浆
gravitational hammer 重力锤
gravitational harmonic function 引力谐函数
gravitational head 重力水头
gravitational high 重力高(的)
gravitational hot water 自然流动式热水
gravitational hot water heating system 重力循环热水供暖系统
gravitational inertia classifier 重力惯性分级机
gravitational instability 引力不稳定性;重力失稳;重力不稳定性
gravitational interaction 引力作用
gravitational lens 引力透镜
gravitational-lens effect 引力透镜效应
gravitational load and geostress 重力负载和地应力
gravitational low 引力最小值;重力低(的)
gravitational magnitude 重力量值
gravitational map 重力图
gravitational mass 引力质量
gravitational measurement 重力测定
gravitational mechanism 重力机理
gravitational method 重力方法
gravitational method of exploration 重力探测法
gravitational moisture 重力水分
gravitational moment 重力矩
gravitational motion 重力运动
gravitational network 重力网
gravitational packing method 重力填充法
gravitational paradox 引力佯谬
gravitational pendulum 重力摆
gravitational perturbation 引力摄动
gravitational point 重力点
gravitational potential 引力位;引力势;重力位(势);重力势
gravitational potential energy 引力位能;重力位能;重力势能
gravitational potential function 引力位函数
gravitational potential well 引力势阱
gravitational pressure 重力压力;静水压力
gravitational processes on slope 斜坡重力作用
gravitational prospecting 重力勘探
gravitational radiation 引力辐射;重力辐射
gravitational radius of the earth 地球引力半径
gravitational receptor 重力感受器
gravitational red shift 引力红移
gravitational repulsion 万有斥力
gravitational return 重力式回水
gravitational roughing air classifier 重力式粗选风力分级机
gravitational sediment 重力沉淀
gravitational sedimentation 重力沉淀
gravitational separation 重力选分;重力分离(作用);重力分级;重力分层;密度分离法
gravitational separation processing 重力分离处理法
gravitational separator 重力(式)分离器
gravitational settling 重力沉陷;重力沉降
gravitational shift 引力偏移
gravitational similarity rule 重力相似准则
gravitational sliding 重力滑移;重力滑动
gravitational sliding basin 重力滑动盆地
gravitational sliding plate 重力滑动板(块)
gravitational slip(ping) 重力滑移;重力滑动;重力滑坡
gravitational slipping basin 重力滑动盆地
gravitational sounding 重力测探法
gravitational spreading 重力扩张;重力扩展
gravitational survey 重力调查;重力测量
gravitational synchrotron radiation 引力同步加速辐射
gravitational system 重力点网
gravitational system of units 重力单位制;工程制
gravitational tank 重力柜;重力罐;自流供料罐;自流供料储备;重力供油箱;重力供水箱
gravitational tectonics theory 重力构造学说
gravitational tensor 引力张量
gravitational tidal force of the moon 月潮引潮力
gravitational tidal force of the sun 太阳引力潮
gravitational tide 引力潮
gravitational torque 重力转矩
gravitational treatment 重选矿法
gravitational unit 重力单位
gravitational unit system 重力单位制;工程制
gravitational vector 引力向量;重力矢量
gravitational ventilation 重力通风
gravitational water 自流水;重力水
gravitational wave 引力波;重力波
gravitational wave source 引力波源
gravitational wave surface 重力波面
gravitational yard 重力编组场;驼峰调车场
gravitation-capillary pore space 重力毛管孔隙
gravitation dynamics 重力动力学
gravitation filter 重力式滤器;过滤澄清器
gravitation settler 重力沉降槽
gravitation tank 自流油罐;高架罐
gravitation transporter 重力运输机;重力输送器
gravitation type relay 重力继电器
gravitation zone of groundwater 地下水重力作用带
gravitative adjustment 重力析离;重力调节
gravitative differentiation 重力分异(作用)
gravitative faulting 重力断层作用
gravitative pressure 重力压强
gravitative water 自由地下水;重力水
gravitino 超引力子
gravitino fields 超引力子场
gravitional slide plate 重力滑动板块
gravi(to)meter 重力仪;密度(测量)计;比重计;验重器;重力计;重差计;比重测定器
graviton 引力子;重力子
gravitron 气体放电管
gravity 引力;重力
gravity abutment 实体岸墩;重力式桥台;重力式拱座;重力式墩台;重力式岸墩
gravity acceleration 重力加速度
gravity acceleration in space 空间重力加速度
gravity accumulation 重力积聚(作用)
gravity accumulation type 重力堆积型
gravity accumulator 重力储蓄器
gravity action 重力作用
gravity air heating 自动空气循环供暖
gravity allocation 重心布置

gravity ammeter 重力式安培计
gravity anchor 重力式锚碇
gravity anomaly 重力异常;重力反常
gravity anomaly configuration 重力异常形态
gravity anomaly continuation 重力异常延拓
gravity anomaly dividing 重力异常划分
gravity anomaly field 重力异常场
gravity anomaly map 重力异常图
gravity anomaly maximum 重力异常极大值
gravity anomaly minimum 重力异常极小值
gravity anomaly of continental border 陆缘重力异常
gravity anomaly of continental shelf border 陆架边缘重力异常
gravity anomaly over basin 海盆重力异常
gravity anomaly over mid-ridge 中脊重力异常
gravity anomaly over seamount 海山重力异常
gravity anomaly over trench 海沟重力异常
gravity anomaly peak 重力异常峰值
gravity anomaly profile 重力异常剖面图
gravity anomaly simulator 重力异常模拟器
gravity anomaly smoothing 重力异常圆滑
gravity anomaly strike 重力异常走向
gravity anomaly surface 重力异常面
gravity anomaly vector 重力异常向量
gravity apparatus 自溜式装置;重力仪(器);重力式装置
gravity aqueduct 重力输水管
gravity arch abutment 重力式拱座
gravity arch dam 重力(式)拱坝
gravity arc welding 重力式电弧焊
gravity-artesian well 潜水承压井
gravity aspirator bit 重力抽吸式钻头;重力吸气钻头
gravity assist 重力助推
gravity-assist pouring 重力浇注
gravity axis 重心轴(线);重力轴
gravity bag 重力储[贮]藏袋
gravity bag filter 重力袋滤器
gravity balance 重力平衡;重力秤;比重计;比重秤
gravity band 重心箍
gravity bar 重力栓;重力连接杆
gravity base(line) 重力基线
gravity base point 基本重力点
gravity base station 重力基(准)点;基本重力点
gravity base station network 重力基点网
gravity batch mixer 自流式计量给料拌和机;自动式计量给料拌和机;重力间歇式混合机;重力间歇式拌和机
gravity battery 重力电池;比重液电池
gravity bed 重力床
gravity bedding 重力层化
gravity bin 自卸仓
gravity blending 重力混合
gravity block 倒立卷筒
gravity block wire-drawing machine 倒立卷筒拔丝机
gravity bob 重力摆
gravity bottle 比重瓶
gravity bottle of mud 泥浆比重瓶
gravity breakwater 重力式防波堤
gravity breakwater with cutaway cross-section 削角式防波堤
gravity bucket conveyer[conveyor] 重力斗式输送机;斗式重力输送机
gravity bucket elevator 重力式斗式提升机
gravity bulkhead 重力式挡土墙;重力式岸壁
gravity burning appliance 重力式燃烧装置;重力式燃烧设备
gravity cable logging 重力索道集材
gravity cable(way) 重力索道;倾斜索道
gravity caging 重力装罐
gravity caisson structure 重力式沉箱结构
gravity caisson wharf 重力式沉箱码头
gravity carrier 重力投料设备;重力(式)加料装置;重力加料设备
gravity casting 重力浇铸法
gravity cell 重力电池
gravity cement texture 重力型胶结物结构
gravity center 重心
gravity center integral 重心积分
gravity central heating 重力循环采暖
gravity chance 重力作业区
gravity chute 重力溜槽
gravity chute feed 斜槽(式)重力进料

gravity circulating oil system 重力自流润滑系统;重力进给润滑系统
gravity circulation 自流循环;重力循环;重力环流
gravity circulation heater 重力循环加热器
gravity circulation heating installation 重力循环供热装置
gravity circulation heating system 重力循环供热系统
gravity circulation lubrication 重力循环润滑
gravity circulation of hot-water 热水重力循环
gravity circulation system 重力循环系统
gravity classification 重力分级;按比重分级
gravity classification yard 重力编组场
gravity classifying 自由沉降分级;重力分级
gravity clock 重力钟
gravity closed damper 重力式关闭风门;重锤式关闭风门;重锤阀
gravity closing 重力封闭
gravity collapse structure 重力塌陷构造;塌陷构造;重力崩塌构造
gravity compensation 重力补偿
gravity component 重力分量
gravity computer 重力计算机
gravity concentrate 重(力)选精矿
gravity concentration 重力浓集;重力选矿(法);重力富集法
gravity concentration apparatus 重选装置
gravity concentration method 重力选矿法;重力浓缩法
gravity concrete drum-mixer 鼓筒形自落式混凝土搅拌机
gravity conduit 自流输水管;重力管道;重力导管
gravity conjunction 重力联测
gravity constant 重力常数
gravity control 重力控制
gravity control base station 重力控制基点
gravity controlled gyro 重力控制陀螺
gravity controlled instruments 重力控制仪器
gravity control network 重力控制网
gravity control station 重力控制点
gravity conveyer 自流式输送器;重力(式)输送机;滚棒输送机(重力式);倾斜式滚道输送机;液棒运输机;倾斜辊道
gravity cooler 重力冷却器
gravity corer 重力岩芯提取器;重力垂直取样管
gravity core sampler 重力式采样管;重锤岩芯取样器
gravity correction 重力校正;重力归算
gravity coverage 有重力资料地区;重力测量(地)区
gravity crusher 重力破碎机
gravity culvert 重力排水涵洞;无压涵洞
gravity current 自流;重力流
gravity curve 抛物线
gravity dam (油水分离机的)重力屏障;重力坝
gravity dam of triangular section 三角形断面的重力坝
gravity darkening 重力昏暗
gravity data 重力资料;重力数据
gravity davit 重力式吊艇柱;重力式吊艇杆;滑轨式吊艇杆
gravity dedusting 重力除尘;惯性除尘
gravity deficiency 重力不足
gravity deposit chamber 重力沉淀池
gravity determination 重力测量;重力测定
gravity dewatering 重力疏干;重力降水
gravity dewaxing 重力脱蜡
gravity diaphragm wall 重力式地下连续墙
gravity die 硬型
gravity diecasting 硬型铸造
gravity die foundry 重力硬模铸造
gravity difference 重力差
gravity difference to base station 相对基点重力差
gravity difference to main base station 相对总基点重力差
gravity direction 重力方向;重力定向
gravity direct method 重力正演法
gravity disc 重力圆盘
gravity discharge 自流排水;重力排水;重力卸料;重力排放
gravity discharge and return of the skip 翻斗车的自动卸料和复位;翻斗车的重力卸料和复位
gravity discharge ball mill 重力排矿球磨机;重力出料球磨机
gravity-discharge chute 自流式卸料槽;重力滑料槽

gravity discharge elevator 重力式卸载升运器
gravity-displaced deposit 重力移位沉积
gravity distribution 自溜式撒布;重力分配;重力分布
gravity distributor 重力式喷洒机;自溜式喷洒机;重力喷洒车;重力喷布器
gravity disturbance 重力异常;重力扰动
gravity disturbance vector 重力扰动矢量
gravity dock wall 重力式坞墙
gravity drain 自重泄油(管路)
gravity drainage 自流排水;重力排水
gravity drainage curve 重力排油曲线
gravity drainage reservoir 重力排储层
gravity drain line 重力排水线
gravity drain plug 重力排泄堵塞
gravity dressing 重力选矿(法)
gravity drift 重力漂移
gravity drive 重力驱动;重力传动
gravity driving clock 重力钟
gravity drop 重力下降
gravity drop armature 重力式衔铁
gravity drop hammer 重力(式)落锤
gravity drydock floor 重力式干坞底板
gravity drydock wall 重力式干坞墙
gravity duct 重力管
gravity dump body 自卸车身;自重倾卸车身;重力卸料车;重力倾卸车身
gravity dumper 重力翻车机
gravity dumping 重力卸料
gravity dust collection 重力集尘
gravity dust separator 重力集尘装置;重力除尘器
gravity dynamometer 重力测力计
gravity edge effect 重力边缘效应
gravity effect 重力影响;重力效应
gravity electrode 重力焊条
gravity elevation correction factor 重力高度校正因数
gravity elevator 重力升降机
gravity energy 重力能
gravity energy of crude 原油重力能
gravity equilibrium 重力平衡
gravity erosion 重力冲蚀
gravity escapement 重力式擒纵机构
gravity exhaust ventilation 重力排气通风
gravity fault 重力断层;正(常)断层
gravity-fed 重力给料的
gravity fed carburet(t)or 重力给油气化器
gravity feed 自溜供料;重力(自动)加料;重力喂料;重力供油;重力供水;重力给料
gravity feed apparatus 重力给料装置
gravity feed circulating system 重力进给循环系统
gravity feed cup 重力进给杯
gravity feeder 重力喂料机;重力加料器;重力进料器
gravity feeding 重力进料
gravity feed line 重力进料线;重力自流进料管;重力送料管;重力给水管路
gravity-feed lubrication 重力润滑
gravity feed lubricator 重力给油润滑器
gravity feed machine 重力供料机
gravity feed oil 重力供油
gravity feed oiler 重力(式)注油器
gravity feed pipe 重力自流进料管
gravity feed stoker 重力加煤机
gravity feed system 重力式供给系统
gravity feed tank 重力送料箱;重力式供油柜;重力感;重力槽;自流式输油箱
gravity feed type boiler 自重送料锅炉
gravity feed type gun 重力送料式喷枪
gravity fender 重力式护舷;重力式防撞装置
gravity field 引力场
gravity field observation 重力场观测
gravity field of the earth 地球重力场
gravity field strength 重力(式)场强度
gravity filling 重力加料
gravity filter 自溜式过滤器;重力(式)滤池;重力(式)过滤器;重力滤器
gravity filtration 重力过滤
gravity filtration process 重力过滤法
gravity flattening 重力扁率
gravity-flo(a)tation plant 重力浮选装置
gravity-flo(a)tation technique 重选浮选技术
gravity flood 自重注水
gravity flow 重力运动;自然流动;自流;重力水流;重力流(动);无压流
gravity-flow concentrator 重力选矿机;重选机

gravity flow drier 自流式干燥机
gravity flow folding 重力流动褶皱
gravity flow gathering system 重力收集系统
gravity flow grain box 自流卸载式谷物车厢
gravity flow heating system 重力流供热系统
gravity flowing bed 重力流动床
gravity flow installation 重力流装置;重力流设备
gravity flow machine 重力流供料机
gravity flow of air 空气重力流
gravity flow screen 固定筛;自流筛
gravity flow sequence 重力流层序
gravity flowsheet 重力选矿流程
gravity flushing 自重冲洗
gravity fold 重力褶皱
gravity formula 重力公式
gravity foundation 重力基础
gravity fractionation 重力分离作用
gravity froster 重力流动式冻结机
gravity fuel feed 重力给油
gravity fuel system 重力给油系统
gravity furnace 重力暖气锅炉;重力式热风炉
gravity gasoline feed 自流加(汽)油;重力加(汽)油
gravity gasoline tank 重力汽油箱
gravity gate 自流卸料门孔;重锤阀
gravity-glide fault 重力滑动断层
gravity gliding 重力滑移(作用);重力滑动(作用)
gravity gliding tectonics 重力滑动构造
gravity grab 重力抓钩
gravity graben 重力地堑
gravity gradient 重力梯度
gravity gradient attitude control 重力梯度姿态控制仪
gravity gradient centrifugation 重力梯度离心法
gravity gradient satellite 重力梯度卫星
gravity gradient stability 重力梯度稳定
gravity gradient test 重力梯度试验
gravity gradient test satellite 重力梯度试验卫星
gravity gradiometer 重力陡度计
gravity groundwater 重力地下水;自由地下水;地下重力水
gravity grouting 自流灌浆
gravity hammer 重力(桩)锤;重力式(打桩)锤
gravity haulage 自动斜坡;重力运输;滑溜运输;滑动运输
gravity head 重心高差;重力压头;重力水头;重力高;位置头
gravity head feeder 压力送料机;重力高差给料器;重力落差进料器
gravity heater 重力循环加热器
gravity heating 重力式(循环)供暖
gravity-held check-valve 重力式单向阀
gravity high 重力高
gravity hinge 重力铰链
gravity hot water central heating 重力式热水中央供暖
gravity hot water heating system 重力热水供暖系统
gravity hump 重力驼峰
gravity hydrocyclone 重力式水力旋流器
gravity hydrocyclone settling tank 重力式水力旋流沉淀池
gravity impulse 重力驱动
gravity incline 重力斜坡道;轮子坡;驼峰溜放坡
gravity inside the earth 地球内部重力
gravity installation 重力式设备
gravity instrument 重力仪
gravity interceptor 重力截流器;重力截流管
gravity inversion method 重力反演法
gravity irrigation 自流灌溉;重力灌溉
gravity isostatic correction value 重力均衡改正值
gravity knife 重力折刀;重力弹簧刀
gravity line 自重运行路线;自流管(道);重力线;重力导管
gravity loaded accumulator 重力(锤)蓄力器
gravity load(ing) 重力荷载;自动装载;自动装料;重力装料
gravity lock 重力式锁定器
gravity lock wall 重力式(船闸)闸墙
gravity log 重力测井
gravity loss additive 降比剂
gravity low 重力低
gravity lowering 重力下降
gravity low-pressure steam heating 重力低压蒸汽供暖

gravity lubricating 重力润滑
gravity lubricating oil system 重力式滑油系统
gravity lubricating system 重力(自流)润滑系统
gravity lubrication 自流润滑法;重力润滑法
gravity main 自流管路;自流干管;重力总管;重力泄水管;重力输水管
gravity main base station 重力总基点
gravity map 重力地图;重力地图
gravity marshalling yard 重力编组场
gravity masonry wall 重力砖石墙;重砌筑墙体
gravity maximum 重力最大值
gravity measurement 重力测量;重力测定
gravity meter 重力仪;重差计
gravity method 自流法
gravity mid per cent 比重中百分曲线
gravity mill 重力式磨煤机;重力磨;捣碎机
gravity minimum 重力最小值
gravity mirror 重力镜
gravity mixer 重力拌和机;重力搅拌机;重力混合机
gravity mixing 重力拌和;重力混合
gravity mode 重力模式
gravity model 引力模式;重力模型
gravity model approach 重心法;重力模型法
gravity model method 重力模型法
gravity motion 重力流动
gravity network 重力网
gravity nut 重心铊
gravity observation 重力观测
gravity observation method 重力观测方法
gravity ocean wave 重力海波
gravity of acceleration 重力加速度
gravity of slurry 水泥浆比重
gravity of spacer 隔离液比重
gravity of the earth 地球的重力
gravity of volume 容重;体积重;重度
gravity oil feed 重力自流加油
gravity oil filling bearing 重力注油式轴承
gravity oiling 重力加油
gravity oil system 重力自流润滑系统;重力滑油系统
gravity oil tank 重力油柜
gravity oil-water separation technology 重力油水分离技术
gravity-operated haulage 自重运输
gravity-operated shutter 重力动作闸门
gravity ore pass 重力放矿溜井
gravity orogenesis 重力造山作用
gravity packing 重力填充法
gravity pawl 重力爪
gravity pendulum 重力摆
gravity percolation 重力(式)渗滤
gravity petrol feed 自流加(汽)油;重力加(汽)油
gravity petrol tank 重力汽油箱
gravity pier 重力式墩台
gravity pier and abutment 重力式墩台
gravity pipe 重力自流管
gravity(pipe)line 自流管线;自流管路
gravity piston sampler 重力活塞采样器
gravity plane 轮子坡
gravity plant 重力式拌和厂
gravity platform 重力(式)平台
gravity plenum heating 重力正压供热
gravity point of attachment effect 固定点重力效应
gravity potential 重力位;重力势
gravity pouring machine 沉积机
gravity preblending 重力预搅拌
gravity prediction 重力预测
gravity preparation 重力选
gravity prestress 在无外荷载时重力对构件产生的应力
gravity process 重力供料方法;滴料供料法
gravity profile 重力剖面
gravity prospecting 重力找矿;重力勘探
gravity pump 重力泵
gravity purity 重力纯度
gravity quay 重力式码头
gravity quay wall 重力式(驳岸)岸壁;重力式(码头)岸壁
gravity radius of the earth 地球重力半径
gravity railroad 重力缆道;重力缆(车)道(下坡重车带动上坡空车)
gravity railway 重力铁路
gravity ram 重力回程柱塞
gravity reconnaissance 重力普查
gravity reconnaissance survey 重力普查勘探

gravity reduction 重力换算;重力归算
gravity reduction table 重力归算用表
gravity reference point 重力基准点
gravity reference station 重力基准点
gravity reflux 重力回流
gravity regulator 重力调节器
gravity reinjection nozzle 重力再注入喷嘴;重力回燃喷管
gravity reiteration 重力复测
gravity relief trap 重力除水阀
gravity replenished circuit 重力式补充回路
gravity restoring moment 重力恢复力矩
gravity retaining wall 重力(式)挡土墙
gravity retaining wall with balance platform 衡重式挡土墙
gravity return 重力回水;重力回流
gravity return boiler 重力回水锅炉
gravity return feed 重力回水进给
gravity return of condenser 冷凝水重力回水
gravity return ram 重力回程柱塞
gravity return system 重力回水系统
gravity reversal 重力反向
gravity roller 重力辊道
gravity roller carrier 重力运输机
gravity roller conveyer 重力(式)滚轴输送机;重力滚道运输机;重力辊式输送机
gravity roof ventilator 重力式屋顶通风器
gravity ropeway 自溜式索道
gravity rundown lines 重力自流转油管线
gravity runway 自重运输斜坡道
gravity sack shoot 袋装货物斜槽;袋装货物滑道
gravity sampling 重力取样
gravity sand filter 重力砂滤器;重力砂滤池
gravity scale 重度标度;比重标度
gravity scavenging 重力扫选
gravity scram 重力落棒快速停堆
gravity sedimentation tank 重力沉降池
gravity seed cleaner 重力选种机
gravity segregation 依比重分层;重力偏析;重力分层;比重偏析
gravity sensing switch 重力传感开关
gravity sensor 重力传感器
gravity separation 重力选矿(法);重力分选;重力分离作用;重力分离
gravity separation method 重力选矿法
gravity separation of wastewater 污水重力分离;废水重力分离处理法
gravity separator 重力除尘器;重力(沉降)分离器;比重分离器
gravity setting chamber 重力除尘室
gravity settlement 重力沉降
gravity settler 重力沉降槽
gravity settling 重力下沉;重力沉降(澄清);密度沉降
gravity settling basin 重力沉降池;重力沉淀池
gravity settling chamber 重力稳定容器;重力沉降器;重力沉淀室;重力沉淀池
gravity settling tank 重力沉降澄清池;重力沉降槽;重力沉淀池
gravity sewer 重力下水道
gravity shaft 溜井
gravity shunting 重力调车
gravity simulation 重力模拟
gravity slide 重力滑移;重力滑坡;重力滑动
gravity sliding 重力滑移;重力滑动
gravity slipping 重力滑移;重力滑动
gravity slope 重力滑坡
gravity sludge thickener 重力污泥浓缩器;重力污泥浓缩池
gravity sluicing system 重力水力除灰系统;低压力水力除灰系统
gravity solution 重液
gravity speed fine particle board 重力铺装刨花板
gravity spillway dam 重力式溢流坝
gravity spiral conveyer[conveyor] 螺旋下落式输送机
gravity spray tower 重力喷雾洗涤器
gravity spreader 重力撒布器;重力洒布器
gravity spreading 重力扩张;重力扩展
gravity spring 下降泉;重力自流泉;重力(喷)泉
gravity stabilization system 重力稳定系统
gravity stabilized gyroscope 重力稳定陀螺仪
gravity-stabilized satellite 重力稳定卫星
gravity stabilized soil mixing plant 自落式稳定

土搅拌设备
gravity stamp 落锤捣(碎)矿(石)机;捣矿机
gravity stamp battery 重力捣碎机组;捣磨机
gravity stamp mill 捣碎机
gravity stamp shoe 捣碎机落锤
gravity station 重力点
gravity station mark 重力点标志
gravity steering control 重力操纵控制系统
gravity storage 重力位能积聚
gravity stowing 自流充填(法);重力充填
gravity stripping 重力效应逐层去除法
gravity structure 重力式结构
gravity supply 重力(式)供料;重力(式)供水
gravity supply and exhaust system 重力式进排气系统
gravity survey 重力测量
gravity suspended water 重力悬着水;重力悬水
gravity switching 重力调车
gravity system 重力供水系统;自流输送系统;自动给料系统;重力制;重力体系;重力(输送)系统;重力放矿法
gravity-system water cooling 重力水冷;热虹吸水冷却系统;环流水冷却系统
gravity table 重力测量用表;重选摇床;比重分选台
gravity tailings 重选尾矿
gravity take-up 重锤式张紧装置
gravity tank 自流储[贮]罐;自动送料箱;重力(水)柜;重力供油箱;重力供水箱;重力供料箱;高架水箱;高架水塔
gravity tank truck 重力倾尽式油槽车;重力式自动洒水车
gravity tectonics 重力构造
gravity tensioning 重力拉紧
gravity test 重力试验;比重测定
gravity thickener 重力脱水机;重力浓缩机;重力浓缩池;重力厚浆池
gravity thickening 重力增稠
gravity tide 引力潮
gravity tipping conveyer 重力倾卸式运输机
gravity tipping skip 重力倾卸小车;重力倾卸斗;翻斗车
gravity tipping valve 重力翻板阀
gravity tipple 翻斗车;重力倾翻式手推车
gravity track 重力送料道
gravity transfer 重力排注
gravity transporting cableway 重力输送索道;重力输送渠道
gravity trap 重力式凝汽阀
gravity traverse 重力测线
gravity treatment 重选
gravity tube 比重瓶
gravity tunnel 重力隧道
gravity turn 重力转向
gravity type 重型;重力刑
gravity type abutment 重力式桥台;重力式桥墩
gravity type air 重力型风力分级机
gravity type air classifier 重力空气分级机
gravity type arc welding 重力焊(接)
gravity type bank indicator 重力型倾斜指示器
gravity type central furnace 重力循环式集中炉
gravity type concrete mixing plant 自落式混凝土搅拌设备
gravity type construction 重力式构筑物
gravity type dam 重力坝
gravity type diversion weir 重力式引水堰
gravity type fender 悬挂重力式防撞装置;重力式防撞装置
gravity type fiber-bind filter 重力式纤维束滤池
gravity type filter 重力滤器;重力滤池
gravity type grab 重力式抓钩
gravity type lock 重力式船闸
gravity type lock wall 重力式(船闸)闸墙
gravity type one-pipe heating 重力式单管供热;重力单管采暖
gravity type pawl 重力式爪
gravity type pier 重力式桥墩
gravity type quay 重力式码头
gravity type quaywall 重力式岸壁
gravity type retaining wall 重力式挡土墙
gravity type retarder 重力式缓冲器;重力式减速器
gravity type separator 重力选矿机
gravity type structure 重力式构筑物;重力型结构
gravity type suspension fender 重力式悬挂碰垫
gravity type thickener 重力型增厚器;重力型增稠器;重力型增稠剂
gravity-type venting system 重力式通风系统
gravity type weir 重力型堰
gravity type welding 倚焊
gravity unit 重力单位
gravity unload with mechanical action chute 机械作用斜槽
gravity upright wall 重力式直墙防波堤
gravity value 重力值
gravity variometer 重力变化感器
gravity vector 重力向量;重力矢量
gravity ventilation 重力(自流)通风
gravity vertical 重力线
gravity voltameter 重力型电量计
gravity wall 重力(式)墙;重力式挡土墙
gravity wall of giant blocks 巨型方块防波堤
gravity wall type 重力墙式
gravity wall type of lock 重力墙式船闸
gravity warm-air heating plant 重力热风采暖装置
gravity warm water heating 重力式热风供暖(系统);重力式热风采暖(系统)
gravity water 重力(地下)水
gravity water return 重力回水
gravity water supply 重力供水系统;自流供水;重力(式)供水;重力(式)给水
gravity water system 重力排水法;重力流水系统;重力给水系统
gravity water tank 重力水箱
gravity water wheel 重力式水车;重力式水轮
gravity wave 重力波
gravity wave drag 重力波阻
gravity weight 铅锤
gravity weir 重力堰
gravity weld 倚焊
gravity welding 重力式电弧焊;重力焊(接)
gravity well 自流井
gravity wharf 重力式码头
gravity wheel 重力水轮;重力水车
gravity wheel conveyer[conveyor] 重力式滚轴输送机;重力滚轮输送机;重力辊式输送机
gravity wind 下降风;重力风;山风
gravity wing (分水闸口)重力翼墙;翼式重力桥台
gravity wing wall 重力式翼墙
gravity wirerope way 重力式缆索道;重力轻便索道;溜式索道
gravity wireway 自溜式索道
gravity yard 重力调车场;重力编组场;驼峰调车场
gravure 照相凹版(印刷术);[凹][印];凹板
gravure coating 照相凹版式涂敷
gravure cylinder polishing machine 凹印滚筒抛光机
gravure engraving 照相凹版雕刻
gravure etching process 凹版腐蚀法
gravure ink 照相凹版油墨;轮转凹印油墨;凹印油墨
gravure machine 照相凹版印刷机
gravure offset printing press 照相凹版胶印机;凹版胶印机
gravure plate-making 凹版制版
gravure press 凹版印刷机
gravure printing 照相凹版印刷;凹版印刷
gravure process 照相凹版制版法
gravure tissue 照相纸
gray 戈瑞(吸收剂量单位);灰色;灰暗;暗淡
Gray's formula 葛莱氏公式
Gray's point 格雷氏点
gray antimony 羽毛矿;辉锑矿砂
gray area 灰色面积
gray atmosphere 灰大气
gray atrophy 灰色萎缩
gray bacteria 灰色细菌
gray bed 页岩砂岩交错层
gray-black 灰黑色
gray-blue 灰蓝色
gray board 灰纸板
gray body 灰(色)体
gray brick 青砖
gray-brown desert soil 灰棕色荒漠土;灰棕漠境土
gray-brown earth 灰棕壤
gray-brown forest soil 灰棕色森林土(壤)
gray-brown podzolic soil 灰棕色灰化土;灰壤
gray-brown soil 灰褐土壤
Gray Canyon 浅蓝灰色砂岩(产于美国俄亥俄州的)
gray castings 灰口铁铸件
gray cast iron 灰口(铸)铁
gray cast-iron bollard 灰铸铁系缆柱;灰铸铁交通标柱
gray cast-iron brake shoe 灰铸铁闸瓦
gray cast-iron castings 灰口铁铸件
gray cast-iron fitting 灰铸铁配件
gray cast-iron powder 灰口铸铁粉
gray cement 灰色水泥;普通水泥
gray cinnamonic soil 灰褐土
gray clay 灰色黏土
Gray clay treating 格莱白土处理法
gray cobalt 辉砷钴矿
Gray code 葛莱(编)码
Gray column 四槽钢柱
gray contact screen 灰色接触网版
gray correction 灰度校正
gray corrosion 灰色蚀斑
gray crescent 灰质新月
Gray cyclic code 葛莱循环码
gray desert soil 灰色沙漠土;灰漠土;灰钙土
Gray desulfurization process 格雷脱硫法
gray durite 灰色微暗煤
Gray encoder 格雷编码器
gray epoxy anti-corrosive paint 灰色环氧防腐漆
gray fabric 坯布
gray face plate 灰色荧光屏
gray ferruginous soil 灰色铁质土
gray filter 中性密度滤片片;灰色滤光片;灰色滤光镜;灰色滤波器
gray forest soil 灰色森林土(壤)
gray glass 灰色玻璃
gray goods 坯布;本色布
gray-green mud 灰绿色淤泥
gray grit 灰砂岩(一种天然砂混合物);灰色砂岩
gray hair 灰发
gray induration 灰色硬结
gray infiltration 灰色浸润
graying 泛灰色;发灰色;变成灰色
Grayioc tubing joint 格雷依沃克型油管接箍
gray iron 灰铸铁;灰口铁
gray iron casting 灰口铁铸件
gray iron foundry 灰铁铸造厂
gray iron foundry wastewater 灰铸铁铸造废水
gray iron pipe 灰铸铁管
gray(ish) 灰色的
grayish brown 灰褐色
grayish-brown clay 灰棕色黏土
grayish colo(u)r 暗灰色
grayish-green 灰绿色
grayish-white 灰白色
grayish-yellow 灰黄色
grayish-yellow grit 灰黄色硬砂岩
grayite 水磷铅钍石
Gray-king assay 葛金干馏试验
Gray-king standard coke type 葛金标准焦型
gray layer 灰质层
gray leaf 灰色叶
gray level 灰度值
gray level image 灰度图像
gray level thresholding 灰度阈值化
gray manganese ore 灰锰矿;水锰矿
graymap 灰阶图
gray marble 花岗岩大理石
gray market 半黑市
gray matter 灰色物质;灰(白)质
gray micaceous iron oxide alkyd enamel 灰色云铁醇酸磁漆
gray mud 灰色软泥
grayness 灰色度
gray oil 灰色油
gray pig iron 灰生铁;灰口(生)铁
gray plate glass 灰色平板玻璃;半透明平板玻璃
gray platinum 灰色铂
gray podzolic soil 灰色灰化土
gray Portland cement 灰色硅酸盐水泥;灰色波特兰水泥
gray post 灰色砂岩
gray quick drying putty 灰色油性易干腻子
gray radiator 灰体辐射器
gray-red 灰红色
gray resistance 抗泛灰性
gray rubber 灰色橡胶
gray sandstone 灰色砂岩
gray scale 亮度色标;灰阶;灰度梯尺;灰度(色)标;灰度(级);灰标;灰比;黑白亮度级

gray scale chart 灰度级测视图
gray scale display 灰阶显示
gray-scale film 灰度胶片
gray-scale map 灰度图
gray-scale rendition 灰度再现;灰度重现
gray-scale resolution 灰度级分辨率
gray silt mud 灰色淤泥
gray softening 灰色软化
gray soil 灰(色)土
Grayson's ruling 葛莱生划线
gray speck 灰斑点
gray spot (可锻铸铁的)灰点
gray steppe soil 灰色草原土
graystone 玄武石;中粒玄武岩;灰色岩
gray substance 灰质
gray surface 灰表面
gray system 灰色系统
gray tile 青瓦
gray-tone response 灰度响应;灰度特性
gray transmission 高频混合传送
graywacke 硬砂岩;玄武土;灰瓦岩;杂(色)砂岩;灰色杂砂岩
graywacke limestone 杂砂石灰岩
graywacke sandstone 硬砂岩砂石
graywacke schist 杂砂片岩
graywacke slate 杂砂石板
gray warp soil 灰色淤积泥;灰色冲积土壤
gray waste(water) 灰色废水
gray water 灰水;无养生活污水
gray water reclamation 灰水回收
gray water septic tank 灰水化粪池污水
gray wedge knob 灰色光楔钮
gray wooded soil 灰色森林土(壤)
gray wrap 灰色包装纸
grazable woodland 可牧林地
graze 相切
grazed forest 混牧林;放牧林
grazer 放牧牲畜
grazier 畜牧业者
graziery 畜牧业
grazing 放牧
grazing acreage 牧场面积
grazing angle 掠射角;入射余角;切线角;擦地角
grazing animal 食草动物;放牧牲畜
grazing capacity 载畜量
grazing collision 擦边碰撞
grazing cycle 放牧周期
grazing density 放牧密度
grazing district 牧区;放牧区
grazing fire 低伸火力
grazing food chain 草木链;放牧食品链
grazing frequency 放牧频率
grazing ground 放牧地
grazing herbivore 食草动物
grazing incidence 掠入射;切向入射;擦地入射
grazing incidence ray 旁掠射线
grazing intensity 放牧强度
grazing land 放牧场;牧场;牧(草)地;放牧地
grazing line 擦地视线
grazing load 载畜量
grazing meadow 放牧草地
grazing method 放牧法
grazing path 掠射路径;临界视距路径
grazing preference 放牧优先权
grazing pressure 放牧强度
grazing region 放牧区
grazing season 放牧期;放牧季节
grazing sight line 擦地视线
grazing site 放牧场所
grazing succession 放牧演替
grazing system 放牧制;放牧法
grazing the cattle 放牧
grazing too closely 放牧过度
grazing trace 啮食迹
grazing unit 放牧单位
grease 油脂状物;油腻;油泥;牛油;黄油;黄矿脂;滑脂;滑油;润滑(油)脂
grease alkali 润滑脂碱
grease analysis 润滑脂分析
grease and flats 油脂
grease and oil system 干油和稀油润滑系统
grease anti-oxidant 润滑脂抗氧化剂
grease apparent viscosity 润滑脂表现黏度
grease baffle 遮油物;遮油板;挡油片

grease bag 滑脂袋
grease ball 油脂球
grease base 润滑脂基
grease basin 润滑脂盒
grease bearing waste 含油脂废水
grease-belt 油脂皮带轮
grease-belt separation 油脂带分选
grease-belt treatment 油脂带处理
grease bleeding 润滑脂分油
grease box 滑脂匣;润滑脂箱;润滑油箱
grease bucket 润滑油桶
grease burning 油脂焙烧的;焙烧脱脂的
grease can 润滑脂壶
grease cartridge lubricator 润滑脂弹筒式润滑器
grease chamber 注油器;油杯;润滑油室;油室;滑油箱
grease channeling 润滑脂的成沟现象
grease characteristic 润滑脂性质
grease chemical industry 润滑脂化工厂
grease classification 润滑脂分类
grease coating 油脂涂层;涂脂
grease cock 滑脂旋塞;润滑脂旋塞
grease collar 油脂环
grease composition 润滑脂组成
grease compounder 润滑脂调制设备
grease compounding 润滑脂组分混合
grease compounding unit 润滑脂调制装置
grease compressor 油脂枪
grease consistency 润滑脂稠度
grease consistency numbers 润滑脂稠度号
grease consistometer 润滑脂稠度计
grease contamination 润滑脂沾污
grease content 油脂含量
grease coupler 润滑脂注入嘴;润滑过的管接头
grease cover 润滑油杯盖
grease creeping 润滑脂流出
grease cup 油(脂)杯;油杯;注油器;脂油环;牛油杯;黄油杯;滑脂杯;润滑脂杯;杯状润滑器
grease cylinder feeding nozzle 润滑脂筒给油嘴
greased bar 涂脂抽芯管
grease decomposition bacteria 油脂分解菌
greased-for-life 永久性上油(润滑)
grease-dispensing system 润滑脂分配系统
grease dispensing test 润滑脂流动试验
grease distributing groove 油脂分配槽
grease dropping point 润滑脂滴点
greased ship-building berth 涂油滑道【船】
greased sleeve 涂油脂的套管
greased sluice 油选溜槽
greased-surface concentration 油脂选矿法
greased-surface concentrator 油脂选矿机
greased table 油选台
grease elasticity 润滑脂弹性
grease evapo(u)ration loss 润滑脂蒸发损失
grease extractor 润滑脂抽出器
grease feedability 润滑脂供送性能
grease filler 润滑脂填料
grease filter 油脂过滤器;润滑脂过滤器
grease(filth) 油泥
grease fitting 润滑脂油嘴
grease flo(a)tation tank 油脂浮选池
grease flow test 润滑脂流动试验
grease for belt 传动带用润滑脂
grease free 无油脂;不含油脂
grease gate 润滑脂充填头
grease gelling 润滑脂胶凝
grease grades 润滑品级;润滑油稠度等级
grease-graphite mo(u)ld 润滑脂石墨混合物
grease groove 油槽;润滑脂槽
grease gun 注油枪;牛油枪;滑脂枪;润滑脂枪
grease gun adapter 润滑脂枪接头
grease gun fitting 润滑脂枪喷嘴
grease gun lever type 横杆式润滑脂枪;拉柄式注油枪
grease gun viscometer 油枪式黏度计
grease gun with movable plunger 油塞移动增压油枪
grease hardening resistance 润滑脂抗硬化安定性
grease hole 油脂孔;润滑(油)脂孔
grease hole plug 润滑油口塞
grease homogenization 润滑脂均匀化
grease hood 排油烟罩
grease horn 牛油容器
grease ice 糊状冰;海水初冻形成的泞糊冰;油脂状冰;泥泞冰
grease insolubles 润滑脂不溶物
grease interceptor[intercepter] 油脂截留器;油脂分离器;隔油器;隔油井;截油井;除油器
grease kettle 润滑脂锅
greaseless 无润滑油的
greaseless valve 无脂密封阀
grease liquor 涂油
grease lubricant 滑脂;润滑脂
grease lubricating pump 润滑油泵
grease lubricating system 干油润滑系统
grease lubrication 滑脂润滑;润滑脂润滑
grease lubrication fitting 滑脂润滑的接头;润滑脂润滑接头
grease lubricator 油脂润滑器;油脂枪;滑脂灌注器
grease-making plant 润滑脂工厂
grease marks 油迹;油斑
grease melting point 润滑脂熔点
grease monkey 汽车修配工具;润滑工;机修工;飞机检修工
grease nipple 油脂枪喷嘴;油脂喷壶;注油嘴;牛油枪喷嘴;滑脂嘴
grease non-soap thickener 润滑脂非皂基稠化剂
grease nozzle 滑脂嘴
grease oil 润滑油脂;润滑油
grease organic filler 润滑脂有机填料
grease packing 油封;牛油填料;润滑填料
grease pad 油脂垫
grease pail 润滑脂桶
grease paint 油膏涂料;油彩
grease pan 润滑脂盘
grease passage 油脂通路
grease pencil 润滑脂铅笔
grease penetration test 润滑脂针入度测定
grease penetrometer 润滑脂针入度测定器
grease pit 油脂井;油印标记
grease plant 润滑脂装置
grease plug 润滑脂塞
grease plug lock nut 滑脂塞防松螺帽
grease pocket 润滑脂腔
grease polluted 油脂污染的
grease-pot rollers 油锅辊
grease press 压油机
grease pressure gun 注油枪;油脂压力枪;油枪
grease-proof 防油的
grease-proof anti-oxidant paper 防油抗氧纸
grease-proof ink and coating 耐油性油墨和涂料
grease-proofness 防油性;耐油
grease-proof paint 耐油性涂料
grease-proof paper 防油纸
grease-proof paper coating 耐油(食品)包装纸涂料
grease-proof wrapping 防油包装纸
grease pump 油脂泵;润滑脂(油)泵;润滑脂枪
grease pumpability 润滑脂泵送性能
greaser 加脂器;加油工;润滑脂枪;滑脂杯;注油器;润滑脂注入器;润滑器具;润滑工;润滑杯;涂油器;擦拭工人
grease rack 润滑脂注入架
grease removal 油脂去除;除脂
grease-removal tank 除油(酯)槽;除油池
grease remover 油脂去除器;除油剂;润滑脂脱除剂;除脂器
grease reservoir (轴承内的)润滑脂集储[贮]区
grease residue 油渣
grease resistance 抗油性;耐油性(能);耐润油性
grease resistant coating 耐油脂涂料
grease resistant paper coating 食品包装纸用涂料
grease retainer 护脂圈;润滑脂挡圈
grease retainer felt 护脂毡环
grease retainer remover 密封护脂圈拆卸器;护脂圈拆卸工具
grease retainer seal 储[贮]油箱密封;油脂密封圈
greaser oiler 加油工
grease seal 油膏填料;油封;润滑脂密封
grease separation 油脂分离;滑脂分离;润滑脂分油
grease separator 油脂分离器;隔油器;滑脂分离器;润滑脂分离器;除油器;除油池
grease shield 遮油物;遮油板
grease shot 油脂喷射;加油枪
grease skimmer 撇油器
grease-skimming tank 除油池;撇油池;撇沫器
grease smudge 油渍
grease-soaked lining 浸透润滑脂衬片

grease solidification	润滑脂固化
grease-spoiled oil	润滑脂污染的油料
grease spot	油渍
grease spot photometer	油斑光度计
grease spray lubrication	润滑脂喷射润滑
grease spray lubrication system	喷油润滑系统
grease squirt	注油器;润滑油枪
grease stability	润滑脂稳定性
grease stabilizer	润滑脂稳定剂
grease stable to radiation	对辐射稳定的油脂
grease stain	油渍
grease stick	棒状研磨膏
grease stripper	油脂去除器;除油器
grease structure	润滑油结构
grease substance	润滑脂
grease surface	油面
grease-surface separation	油脂选矿;油膏选矿
grease swelling	润滑脂溶胀
grease tabling	油脂摇床选
grease tank	油脂箱;润滑脂桶;油脂罐;油脂槽
grease tap	润滑嘴;润滑孔
grease tester	油脂测定器
grease testing	润滑脂试验
grease testing machine	油脂试验机
grease thickener	润滑脂稠化剂
grease thickening	润滑脂稠化
grease trap	油脂分离器;防油污管;油脂隔离器;油脂捕集器;隔油器;隔油井;截油井;润滑脂分离器;防油污存水弯
grease tube	滑脂管
grease waste(water)	油脂废水;含油脂废水
grease way	滑脂槽
grease-well lubrication	润滑脂槽润滑
grease worker	润滑脂工作器;润滑脂用搅和器
grease working machine	润滑脂搅和机
grease worms	失去润滑
greasiness	油脂性;油腻(性);渗油性;多脂
greasing	注油;脏版;润滑脂润滑;润滑(过程);污版;涂油;渗油
greasing and lubricating service truck	加油脂和润滑油的服务卡车
greasing and lubricating service vehicle	加油脂和润滑油的服务车辆
greasing apparatus	润滑器
greasing device	钢轨涂油器
greasing equipment	润滑用设备
greasing substance	润滑物质;润滑剂;润滑材料
greasing truck	润滑车
greasy	油污的;油腻;油滑的;含脂的;多油脂的
greasy dirt	油污
greasy feeling	滑感
greasy filth	油泥
greasy flame	油脂焰
greasy gold	纯金
greasy ink	多腊油墨
greasy luster[lustre]	油脂光泽;脂状光泽
greasy mor	滑腻状的森林腐殖质
greasy property	润滑性能
greasy quartz	油状石英;乳石英
greasy road	油滑道路;泥泞道路;湿滑路
greasy water	油腻水
great abundant of water quantity	水量极丰富
great altar	圣堂
great American file	半圆底三角锉
Great An Guo Temple	大安寺
great artesian basin	大自流盆地
great Australia Bay	大澳大利亚湾
great barrier reef	大堡礁
Great basin	大盆地
Great Basin high	大盆地高压
Great bell Temple	大钟寺
great bulrush	水葱
great calorie	千卡;大卡
great circle	大(圆)圈
great-circle arc	大圆弧
great-circle bearing	大圆方位;大圆弧航向;大圆方向角
great-circle belt	大圆(构造)带
great-circle chart	球心投影海图;大圆(弧)海图
great-circle correction	大圆校正
great-circle course	大圆航向
great-circle differential longitude	大圆航行分点经度;大圆航行的经度差
great-circle direction	大圆方向;大圆方位
great-circle distance	大圆(航线)距离
great-circle girdle	大圆环带
great-circle line	最大地球弧线;大圆弧线
great-circle navigation	大圆(弧)航行
great-circle path	大圆路径
great-circle sailing	大圆圈线;大圆航线;大圆航迹计算法;大圆航海术;大圆航法
great-circle sailing chart	大圆航行路线图;大圆航法图
great-circle section	大圆航行分段;大圆航线分段
great-circle track	大圆航迹
great-circle track chart	大圆航线图;大圆航路图
Great Cold	大寒
great contrast	影像反差大
great depth dive work	大深度潜水作业
great depth diving	大深度潜水
great depth diving accident	大深度潜水事故
great depth diving accident management	大深度潜水事故处理
great depth diving accident treatment	大深度潜水事故处理
great depth diving experiment	大深度潜水实验
great depth diving suit	大深度潜水服
great dipper	北斗七星
great diurnal range	大日较差;大日潮差
great diurnal tidal range	最大日潮差
great divide	主要分水岭;大分界线
great drought	大旱
great earthquake	大(地)震
great elliptic(al)arc	大椭圆弧
greater	较大
greater bunet root	地榆
greater circulation	大循环
greater coasting	近海航线
greater coasting area	沿海航线区;沿海地区;近海区(域)
greater coasting service	沿海航线
greater coasting vessel	近海航线船
greater drop in yield	显著下降
greater ebb	较大退潮流;大落潮;大落潮流
greater ebb tidal current	大落潮流
greater flood	较大涨潮流;大涨潮(流);大潮涨潮流
greater flood tidal current	大潮涨潮流
greater fool policy	大傻瓜政策;白痴政策
Greater London Council	大伦敦议会(英国)(1960年建)
greater plantain	大车前
greater-than match	大于符合
greater than or equal to	大于或等于
greater tolerance	更大耐性
greater trochanter	大转子
greatest brilliancy	最大亮度
greatest coefficient	最大系数
greatest common divisor	最大公约数;最大公因子
greatest common factor	最大公约数;最大公因子;最大公因数
greatest common measure	最大公测度
greatest common subgroup	最大公共子群
greatest difference of rates	最大位差
greatest eastern elongation	东大距
greatest elongation	大距
greatest limit	最大极限
greatest lower bound	下确界;最大下限;最大下界
greatest lower cluster set	最大聚值集
greatest mean diurnal range	最大平均日潮差
greatest north latitude	最大北黄纬
greatest south latitude	最大南黄纬
greatest variation of rates	最大日较差
great expansion	大发展
great extinction	大灭绝
Great Falls converter	竖式转炉
great glacial epoch	大冰期
great gross	十二罗(计数单位,等于1728个);大罗
great group of soils	土类
greathead shield	隧道工人安全帽
Great Heat	大暑
great hemp	博洛涅大麻
great hundred	一百二十(一种计数单位);十打
Great Ice Age	大冰期
great improvement	很大改进
great inequality	中心差
great lower boundary	最大下限;最大下界
great magnetic bight	大磁弯
greatness	巨大
great oceanic basin	大洋深海底
great oolite	大鲕状岩;大鲕石;大鲕粒;大鱼卵石
great palace of Minos at Knossos	克诺索斯的米诺斯王宫(位于克里特岛)
great picture	名画
great pluvial	大雨期
great power	巨大功率;强大功率
great pyramid of Cheops	齐奥普斯金字塔(埃及)
great pyramids	金字塔群
great radiation belt	大辐射带
great red spot	大红斑
great shock	大地震
Great slave old land	大奴古陆【地】
great soil group	土类
Great Southern Ocean	南大洋
great strikes	迅速发展
great temple of Amon at Karnak	卡纳克的阿蒙神庙(埃及)
Great Temple of Parthenon	帕提农神庙
great triangular sail	大三角帆
great trigonometrical survey station	大三角测量站(陆标)
great tropical range	回归大潮(潮)差;大潮高高潮和大潮低低潮的潮高差;大潮潮差
great variety of goods	品种齐全
Great Wall	万里长城
great wall of China	中国长城
great waterfall	大瀑布
great wheel	二轮
great wheel arbor	二轮轴
great year	大年;柏拉图年(分点绕黄道一周约为25800年)
Greaves Etchell furnace	大电极型弧阻式电炉
greben	悬岩
grece	一段楼梯(英国方言)
grecian	古希腊建筑特征
Grecian acanthus leaf	希腊式叶形装饰
Grecian architectural style	希腊建筑形式
Grecian architecture	希腊(式)建筑
Grecian column	希腊式柱
Grecian cross	希腊式十字架
Grecian Doric capital	希腊陶立克式柱头
Grecian Doric order	希腊陶立克式
Grecian Doric temple	希腊陶立克式神庙
Grecian fret	希腊回纹装饰
Grecian Ionic order	希腊爱奥尼亚柱式
Grecian key pattern	希腊回纹装饰
Grecian order	希腊柱式
Grecian ornamental art	希腊式装饰艺术
Grecian pitch	希腊式屋面坡度(15°)
Grecian-Roman(style)	希腊罗马式
Grecian temple	希腊神庙
Grecian type antenna	倒V形天线
greco	格雷科风
Greco-Latin square method	希腊拉丁方方法
Greco-Roman style	希腊罗马式
greda	漂白土
gredag	胶体油墨;石墨油膏
Greek alphabet	希腊字母
Greek and Roman tile	希腊罗马式瓦
Greek architectural style	希腊建筑形式
Greek architecture	古希腊建筑
Greek art	希腊艺术
Greek column	希腊式柱
Greek cross	希腊十字形
Greek-cross plane	正十字形平面
Greek decoration art	希腊式装饰艺术
Greek Doric capital	希腊陶立克式柱头
Greek Doric order	希腊陶立克式
Greek Doric temple	希腊陶立克式神庙
Greek fir	希腊冷杉
Greek fire	希腊燃烧剂
Greek fret	希腊回纹(装)饰
Greek Ionic order	希腊爱奥尼亚柱式
Greek juniper	希腊桧
Greek key pattern	希腊回纹装饰
Greek lace	希腊式挖花花边
Greek letter	希腊字母
Greek mo(u)lding	希腊装饰线条;希腊(装饰)线脚
Greek order	希腊柱式【建】;希腊柱型【建】
Greek ornament	希腊装饰(多在柱顶盘上用对称的树叶形和玫瑰花形装饰)
Greek ornamental art	希腊式装饰艺术
Greek ovolo mo(u)lding	希腊卵形装饰线脚
Greek Register	希腊船级社

Greek revival 希腊复兴式;希腊复古式
Greek-Roman tile 希腊—罗马式瓦
Greek roof tile 希腊式屋面瓦
Greek Senate building 希腊议院建筑
Greek stela 希腊石碑
Greek stripes 希腊彩条粗布
Greek temple 希腊神庙
Greek theatre 古希腊露天剧院
green 绿色;青色;生材的
green acid 绿酸
green agave 剑麻
green alder 绿桤木
green alga 绿藻
green algae 绿藻类
greenalite 铁蛇纹石
greenalite rock 铁蛇纹岩
green alum 青矾
green anhydrous oxide chromium pigment 无水氧化铬绿色颜料
green apex 绿基点
green area 绿化区;绿化带;绿地;草地
green area plan 绿化区总平面图
green argon illumination 绿氩光照明
green argon light 绿氩光
green ash 绿灰
Greenawalt type sintering machine 盘式烧结机
green back 美钞俗称
green background 绿色背景
green bacteria 绿色细菌
green-ball 生球
green ban 绿禁令
green band 绿波带
greenband system 绿灯带引导系统
green basic copper carbonate 石绿(颜料);孔雀石
Greenbat machine 卧式锻造机
green belt 处女地;绿色地带;绿化(地)带;绿带;无霜带;绿地
green belt around farmland 农田林网
green belt planning 城市绿地规划
green belt sprinkling 绿化用水
greenbelt town 绿地城市;绿化带的城镇;绿带城镇
green bit 烧坏的钻料
green black 墨绿;绿黑色
green black level 绿路黑电平
green block 新浇制砌块;未烧砖坯
green blue 竹绿色;翠绿色
green board 耐湿度石膏板
green body 坯体;湿坯;生坯
green bond 湿(黏结)强度
green book (英、意等国政府的)绿皮书
green bottom 污底
green break 常绿防火林
green brick 坯料;未烧砖坯
green brick size enlargement 砖坯模型放尺率
green brick work 未干的砖砌体;新砌的砖砌体;干砌的格子体
green-brown 绿褐色
green building material 绿色建筑材料
green buoy 绿色浮标
Greenburg-Smith impinger 格林伯格—史密斯撞击油尘器;格林柏格—史密斯撞击式检尘器
green casting 铸态铸件;未经(热)处理铸件;湿砂铸法
green cathode 绿枪阴极
green chain 生材输送台;生材链
green chalcedony 绿玉髓
green clause 绿色条款
green clause credit 绿色条款信用证
green clause letter of credit 绿色条款信用证;部分预支绿条款信用证
green clay brick 新制黏土砖;绿色黏土砖
Green Cloud Temple 碧云寺(北京香山)
green coating 环境适应型涂料
green compact 原坯;未烧结的压块;生坯
Green comprehensive air pollution index 格林大气污染综合指数
green compression strength 湿压强度
green concrete 新浇混凝土;新拌混凝土;绿色环保混凝土;未到龄期混凝土;生混凝土;尚未硬化的混凝土
green concrete density 新拌混凝土密度
green consumptionism 绿色消费主义
green conveyer 轮饲牧场
green copper ore 孔雀石

green core 湿砂芯
Green Country sandstone 格林镇砂岩(一种产于美国宾州的浅灰色砂岩)
green coverage ratio 绿化覆盖率
green covering 植被
green cracks 收缩开裂;生坯开裂
green crop 青饲料作物
green cut 带水斩毛
green cut clean-up 水刷石饰面法
green cutting 凿毛(混凝土浇筑层层面);冲毛
green deformation 湿变形
green density 压坯密度;生坯密度;生材密度
green diameter 原始直径
green dichroic mirror 绿二向色反射镜
green district 绿化地段
green dragon shaking tail method 苍龙摆尾法
green earth 石绿;绿泥石;绿土
green earth tere verte 海绿石
green ebony 青黑檀
green economics 绿色经济学
green efflorescence 苔藓风化;绿色风化
green electetrode 压制的炭电极
green embankment 新填路堤
green-emitting phosphor 绿色荧光粉;绿色荧光体;绿色磷光体
green enamel 绿瓷漆
green enameled iron wire netting 绿漆窗纱;绿瓷漆铁丝窗纱
green end 进料口;湿端;生材入口
greenery 草木圆
greenery conservation 树木保护
green extension period 绿灯延长时间
green facility 绿地设施
green fallow 栽种绿肥的休耕地;绿肥作物休闲地
green fambe blaze 窑变绿釉
green feed 青绿饲料
green fence 植株篱笆;绿篱;树篱笆
greenfield 绿场;暗线管道
green filter 绿色滤光器;绿色滤光片
green fingers 园艺技能
green flash 绿闪(光)
green forage 青绿饲料
green fountain 绿麻石
green fuel 新添加燃料
green gadget 无干扰雷达设备
green glass 绿(色)玻璃;瓶料玻璃
green glazing 生坯上釉
green glue stock 生胶料
green-gold 绿光金黄色
green gram 绿豆
green-grey 绿灰色
green grocery 蔬菜水果店
green hand 新手;无经验的人;生手
green hard 半干状态
green heart 新心硬木(造船用);樟属大树(产于圭亚那);绿心硬木(产于圭亚那)
green hide 生皮;粗制兽皮
green highway 草坪公路
green holder 青饲料
green hole 湿出铁口
Greenhough microscope 格林豪显微镜
greenhouse 暖房;花房;温室;玻璃温室;玻璃暖房;生坯存放室
greenhouse benches 温室木架
greenhouse cast glass 温室用玻璃
greenhouse climate 温室气候
greenhouse climate controller 温室气候控制器
greenhouse construction 温室构造
greenhouse culture 温室栽培
greenhouse culture soybean 温室栽培大豆
greenhouse culture wheat 温室栽培小麦
greenhouse effect 温室作用;温室效应
greenhouse-effect gas 温室气体
greenhouse experiment 温室试验
greenhouse facility 温室设备
greenhouse for sludge drying 污泥干化房
greenhouse gas 由温室效应气体;温室气体;二氧化碳、甲烷等导致温度效应的气体
greenhouse glass 花房玻璃;花瓶玻璃;暖房玻璃;温室玻璃
greenhouse heating 温室加温
greenhouse-induced variational of climate 温室效应引起的气候变化
greenhouse management 温室管理

greenhouse model 温室模型
greenhouse planning 温室设计
greenhouse plant 温室植物
greenhouse rot 暖房腐蚀
greenhouse sprayer 温室喷雾器
greenhouse test 温室试验
greenhouse vegetable growing 温室蔬菜栽培
greenhouse warming 温室效应引起的气温升高
greenhouse warming potential 温室升高潜能值
green indicating lamp 安全指示灯
greening 绿化
greening area 绿地面积
greening coverage 绿地率
greening cover percentage 绿色覆盖率;绿化覆盖率
greening design 绿化设计
greening lacquer 古铜色漆
greening layout 绿地布置
greening nature 绿地性质
green ink clause credit 绿色条款信用证
greenish 绿色的;浅绿色的;带绿色的
greenish black 墨绿;墨绿色的
greenish blue 绿相蓝;绿光蓝
greenish-brown 褐绿色的
greenish patina 铜绿;(金属表面的)绿苔
greenish-white marble 青白玉
greenish-white porcelain 青白瓷
greenish-yellow 黄绿色的
green jade 翡翠
green karst 绿岩溶
Green key 希腊回纹饰
green lake 绿色淀;青红色
green land 绿地
Greenland anti-cyclone 格陵兰反气旋
Greenland current 格陵兰洋流
greenlandite 铌铁矿
Greenland-north America separation 格陵兰岛与北美大陆分离
Greenland Sea 格陵兰海
Greenland shield 格陵兰地盾
Greenland spar 冰晶石
Greenland type glacier 格陵兰型冰川;格陵兰式冰川
green laver 石莼(属)
green light 绿色光;绿灯
green light relay 绿灯继电器
green line 绿线
green liquor 绿液
green liquor strainer 绿液过滤器;绿液粗滤器
green log 新伐原木
green lumber 新木料;新(伐)木材;未干燥(木)材;湿木板;湿材
green lumber storage 生材堆场
green machining 生坯加工
greenmail 绿函
green man 没有经验的工作人员
green manure 绿肥(植物)
green manure crop 绿肥作物
green manuring 压青
green masonry 新筑圬工;新筑砌体;新砌圬工;新砌砌体;未硬化砌体
green material 生材;绿色材料;青材料
green material store 生坯库
green matrix 绿化资源地
green member 新制构件
green mercury light 汞绿光
green mortar 新拌灰浆;未干灰浆;未硬化砂浆
green mosque 绿清真寺
green moss peat 绿色苔藓泥炭土
green mud 绿(污)泥;海绿泥
green network 绿地系统;绿地网
green ocher 赭绿;绿颜料
greenockite 硫镉矿
green oil 高级石油;绿(松香)油
green oil tar 绿焦油
green onyx 绿玉石
green open space 绿地空间
Green order 希腊柱型;希腊柱式
green ore 原矿石
green organic pigment 绿色有机颜料
green oxide chromium pigment 铬绿颜料
green paint 绿油漆;绿色涂料;无污染涂料
green pale 淡绿
green panel 新浇制墙板

green paper 绿色文件
green parlour 绿化客厅；陈设花卉植物的大厅
green passagaway 绿色通道
green patch distortion 绿色片痕
green pattern 绿色图案
Green Peace 绿色和平组织
Green Peace Movement 绿色和平运动
green permeability 透湿气性；湿透气性
green phosphomolybdic acid lake 磷钼酸绿色淀
green phosphomolybdic acid toner 磷钼酸绿色原
green phosphor 绿色磷光体
green phosphotungstic acid toner 磷钨酸绿色原
green pigment 绿颜料；绿色颜料
green planning 绿化规划
green plant 绿色植物
green planted city 绿化城市
green plants in environment protection 绿化在环保中的作用
green plaster 新鲜灰浆；湿稠灰浆；新鲜砂浆；未干砂浆
green plate 装甲板
green plot 块状绿地
green plot sprinkling 绿化用水
green politics 绿色政治
green power 绿色魔力
green product 新制产品
green property 湿态性能
green public space 公共绿地
green raster 绿光栅
green ratio 绿信比
green ray 绿射线；绿闪光
green reflecting dichroic 绿色反射镜
green region 绿区（可见光谱部分）
green resin 环境适应型树脂；未固化的树脂；生树脂
green revolution 绿色革命
green river ordinance 绿河条例（指环境保护）
green roasting 不完全焙烧；半焙烧
green rock 绿岩
green rod 热轧盘条
Green-Roman style 希腊罗马式
green roof 新露顶板
green room 休息室；剧院后台休息室；演员休息室；工厂内未加工产品储存室
green rot 新腐烂；绿腐朽
Green rotational viscosity 格林旋转黏度计
green run 试运转
green sailor 新水手；无（航海）经验的水手
green salt 绿盐；铬砷合剂（木材防腐剂）
green salt glaze 绿色釉
green sand 新取砂；新采砂；绿砂（钾质肥料）；海绿石砂；湿砂
greensand 绿砂；海绿砂；湿（型）砂
greensand casting 湿砂铸造
greensand core 湿（泥）芯；湿（砂型）芯
greensand facing 湿型涂料
greensand match 湿砂胎模
greensand mo(u)lding 潮砂造型；湿砂（造）型；湿（砂）型
greensand pressure filter 绿砂压滤池
greensand spinning 湿型离心铸造
green sandstone 绿砂岩；青砂岩（石）；青砂石
green saturation scale 绿色饱和度标度
green schist 绿（色）片岩
greenschist facies 绿（色）片岩相
green schools of(food) grains 青苗
green screen 绿辉荧光屏
Green's dyadic 格林重算子
green sea 连续海浪；卷到甲板上的涌浪；舱面巨浪积水
green seal 刚浇制的缝
green seaman 新水手
Green's economizer 格林式省煤器
Green's formula 格林公式
Green's formula of potential theory 位理论格林公式
Green's function 格林函数
green segment 绿闪（瞬间）
green sensitizer 绿光敏化剂
green shear strength 湿剪强度
green sheet 预算明细比较表
Greenshield's traffic flow mould 格林希尔德交通流法
Green's identity 格林恒等式
green silicon carbide 绿色碳化硅

green sized 湿加工尺寸
green sky 绿空
green slab 新浇制楼板
green slate 绿色板岩
green sludge 鲜污泥；绿污泥；生污泥；新鲜污泥
Green's law 格林定律
green snakeskin glaze 蛇皮绿釉
green snow 绿雪
green sorting deck 生材（链条）分类台
green space 绿化用地；绿化区；绿化面积；绿化（地）带；绿地；格林空间
green space conservation area 绿地保护区
green space effect 绿地效果
green space index 绿化用地指数
green space layout 绿地布局
green space of residential 居住区绿化
green spot 点状绿地；绿斑
green stability 未凝固稳定性
green staghorn sumac 绿花鹿角漆树
green stain 绿色釉点；绿色沉着斑；绿斑；绿变
green state 新制阶段；新鲜状态；绿色状态
Green's theorem 格林定理
green stock 生胶料
greenstone 绿岩；煌绿岩；软玉；深青岩；闪绿岩
greenstone belt 绿岩带
greenstone chip(ping)s 铝岩碎屑
greenstone formation of glaucophane type 蓝闪石型绿岩建造
greenstone geosyncline 绿岩地槽
greenstone-granite association 绿岩花岗岩组合
greenstone schist 绿岩片岩
greenstone tuff 辉铝凝灰岩
green straight-through arrow 绿色直行箭头
green strength （型砂的）湿态强度；湿强度；生坯强度；压坯强度
green stress （木材的）湿应力
green strip 绿（化）带
green striped heavy cees 绿线平纹袋
green structure 绿化结构
green stuff 草类；蔬菜
green sulfur bacteria 绿色硫黄细菌
green sun 绿闪（太阳）
green surface 新铺面层；未加工（表）面
Green's viscosity 格林黏度计
greensward 草坪；草皮；草地
green tack 初步黏合；成形黏着性
green tar 绿焦油；粗焦油
green tea 绿茶
green tensile strength 湿拉强度
green test 连续负荷试验；试运转；试车；格林试验；汽油中胶质测定
Green theorem 格林定理
green thumb 园艺技能
green tile 砖瓦坯；新制砖瓦；瓦坯
green timber 新木料；新（伐）木材；未干燥（木）材；湿材；生材
green tint 淡绿色
green tuff 绿色凝灰岩
green ultramarine 深蓝绿色
green unit 新制品；新制块材；新制构件
green vegetation 活植被
green verditer 绿色铜盐颜料；铜蓝
green vermilion 铁绿色
green vitriol 绿矾；七水（合）硫酸亚铁
green ware 生陶瓷；未烧器皿；生坯体；陶坯
green water 跃波；绿水
green wave 绿波
green wave band 绿波带
green wave band width 绿波（带）带宽
green wave effect 绿光波效应
green wax 绿蜡
green way 林荫（道）路；园林路；绿荫路；绿化道路
green wedge 绿楔；楔形绿地
green weight 鲜重；湿（材）重
Greenwell formula 格林韦尔公式
Greenwich 格林威治
Greenwich apparent noon 格林威治视正午
Greenwich apparent time 格林威治真恒星时；格林威治视时
Greenwich astronomical time 格林威治天文时
Greenwich civil time 格林威治平时；格林威治民用时；世界时
Greenwich hour angle 格林威治时角
Greenwich interval 格林威治时间间隔

Greenwich lunar time 格林威治太阴时
Greenwich mean noon 格林威治平正午
Greenwich mean sidereal time 格林威治平恒星时
Greenwich mean time 与格林威治的对照时间；格林威治标准时间；世界时；格林威治平时
Greenwich meridian 格林威治子午线；零度经线；基准子午线；起始子午线；本初子午线
Greenwich moon 格林威治正午
Greenwich sidereal date 格林威治恒星日
Greenwich sidereal time 格林威治恒星时
Greenwich standard time 格林威治标准时间
Greenwich time 格林威治时(间)
Greenwich zone time 格林威治区时
green willemite phosphor 绿光硅酸锌荧光体
green wood 新木料；未干燥（木）材；生材；新（伐）木材；湿材
Greenwood Airvac 格林伍德通风机（该品牌通风机的总称）
green-yellow 黄绿色
green zone 绿化区；绿化（地）带
gree of preciseness 准确度
grees 楼梯踏板
gregale 格雷大风；干冽风；（地中海中西部及欧洲沿岸的）猛烈东北风
gregarina 簇虫
gregariousness 群居本能
gregarious plant 形成群落的植物
gregaroid colony 聚生群体
gregarous 群居性
Gregorian antenna 格里天线；格里戈里天线
Gregorian calendar 格雷戈里历；公历
Gregorian mirror 格里果里反射镜
Gregorian system 格里果里折叠系统
Gregorian telescope 格里望远镜；格里果里望远镜；格雷戈里反射式望远镜
Gregorian year 格雷戈里年
Gregory's powder 格雷戈里氏散
Greibach normal form 格里巴赫范式
Greibach normal form theorem 格里巴赫范式定理
greig(e) 绸坯
greigite 硫复铁矿
greisec-type tin deposit 云英岩型锡矿床
greisen 云英岩
greisenization 云母英化作用
Greisser rolling gate 葛氏滚筒闸门
grenade 灭火弹
grenade instrumentation 发射装置
grenadine 赤色染料
grenadine twist yarn 多股加捻丝线
Grenaille 粗铝粉
grenatite 白榴石
grenet cell 杵式电池
grens 透镜棱柱
Grenville orogeny 格伦维尔造山运动
Grenville Series 格伦维尔统【地】
grenz ray 跨界辐射；境界射线
gressorial 步行适应
gretstone 土壤磨石；砂砾石；粗砂岩
Gretz rectifier 格列次全波整流器
Grevak 格雷弗克反虹吸作用的密封
Greville orogeny 格林威尔造山运动
grey = gray
Grey's beam 格雷式梁（宽缘工字钢）
Grey's girder 格雷式大梁
Grey's mill 格雷式轧机
grey alkyd primer 灰醇酸底漆
grey antimony 辉锑矿
grey atmosphere 灰色大气
grey atrophy 灰色萎缩
grey balance 灰色平衡
grey base 素图
grey-black 灰黑色
grey-black limestone powder 黑石灰石粉
grey-black limestone powder dry-ground 干磨黑石灰石粉
grey blibe 芒硝泡
grey-blue 灰蓝色
grey board 灰板纸
grey body 灰（色）体
grey-body radiation 灰体辐射
grey brick 青砖
grey-brown 灰褐色
grey-brown desert soil 灰棕色荒漠土；灰棕漠境土
grey-brown earth 灰棕壤

grey-brown forest soil 灰棕色森林土(壤)
grey-brown podzolic soil 灰棕色灰化土
grey-brown soil 灰褐土壤
grey cast iron 灰(口)铸铁
grey cast-iron brake shoe 灰铸铁闸瓦
grey cement 青水泥;灰色水泥
grey chart 灰度图
grey cinnamonic soil 灰褐土
grey clay 灰色黏土
grey cobalt ore 灰钴矿
Grey code 葛莱码;反射码
grey copper ore 黝铜矿
grey corrosion 灰色蚀斑
grey crescent 灰质新月
grey cutting 轻度刻磨
grey desert soil 灰色沙漠土
grey desert steppe soil 灰色荒原草原土
grey durite 灰色微暗煤
grey epoxy finish 灰色环氧面漆
grey experience 老练
grey face plate 灰色荧光屏
grey ferruginous soil 灰色铁质土
grey filter 灰色滤光片;灰色滤光镜;灰色滤波器
grey forest soil 灰色森林土(壤)
grey forge pig 制造锻铁用灰口铁
grey glass 灰色玻璃
grey-green mud 灰绿色淤泥
grey grit 灰色砂岩
grey hair 灰发
grey hole 灰洞
greyhound 长途公共汽车;(比钻塔短的)短立根
grey humic acid 灰色腐植酸
grey ice 灰冰
grey induration 灰色硬结
greying 变成灰色
greying resistance 抗泛灰性;抗变灰性
grey iron 灰口铁
grey iron block 铁砖
grey iron foundry wastewater 灰铸铁铸造废水
grey iron pipe 灰铸铁管
greyish 灰色的
greyish-brown clay 灰棕色黏土
greyish-green 灰绿色
greyish white 灰白的
greyish-yellow 灰黄色
greyish-yellow grit 灰黄色硬砂岩
grey Jun glaze 灰钧釉
grey layer 灰质层
grey leaf 灰色叶
grey level 灰度(等)级;灰度(电平)
grey lime 灰色石灰;青灰
grey-lime core roof 棋心盘屋顶
grey lime mortar 灰色石灰砂浆
grey lime roof deck 青灰背平屋顶
grey lime roofing 青灰背屋顶
grey line 碳线
grey Manila board 灰底白板纸
greymap 灰阶图
grey matter 灰质;灰色物质
grey micaceous iron oxide alkyd enamel 灰色云铁醇酸瓷漆
grey-mode identification model 灰色模式识别模型
grey mud 灰色软泥
greyness 灰色度;灰斑(钢板酸洗缺陷)
grey oil 灰色油
grey pale 淡灰
grey paper 灰纸版;底纸
Grey Past 太古【地】
grey pedestal 灰度台阶;灰度基线
grey photometric(al) wedge 光度学灰色光楔
grey pigiron 灰口生铁
grey plaster 青灰色灰浆
grey Portland cement 灰色硅酸盐水泥
grey post 灰色砂岩
grey quick drying putty 灰色油性易干腻子
grey receiver 灰体吸收体
grey-red 灰红色
grey room 粗布帐篷
grey rubber 灰色橡胶
grey sandstone 灰色砂岩
grey sateen 横贡坯
grey scale 灰色标志;灰阶;灰度梯尺,灰度(等)级;灰度;灰标;灰比
grey scale display 灰阶显示
grey scale display unit 灰度显示装置
grey-scale film 灰度胶片
grey scale for assessing change in colo(u)r 评定变色用灰色样卡
grey scale for assessing staining of colo(u)r 评定沾色用灰色样卡
grey scale generator 灰度信号发生器
grey scale of colo(u)r change 色泽变化灰色分级卡
grey scale value 灰度值
grey scour 初酸洗
grey shade 灰色调
grey silt mud 灰色淤泥
grey silver 银灰色
grey slate 灰色板岩
greyslick 海面油膜;平滑水面
grey soil 灰(色)土
grey speck 灰色斑
grey step wedge 灰色密度梯尺
grey stock 颜色不规则的砖;疵斑砖
greystone 灰色岩
greystone lime (白垩土烧制的)灰石灰
greystone lime mortar 灰石石灰砂浆
greystone slate 石板瓦盖;石板
grey substance 灰质
grey system 灰色系统
grey system assessment method 灰色评估法
grey tile 青瓦
grey transmission 高频混合传送
grey tropical soil 灰色热带土
grey value 灰度值;灰度等级
grey value based matching 基于灰度影像匹配
grey-violet 灰紫色
greywacke 硬砂岩;杂砂岩;灰色杂砂岩;灰玄土;灰瓦岩
greywacke reservoir 杂砂岩储集层
grey washing 坯布初洗;初酸洗
grey waste(water) 灰色废水
grey wedge 灰楔
grey wedge knob 灰色光楔钮
grey-white ice 灰白冰
greywood 银灰木(印度产的黑白两种硬木)
grey wooded soil 灰色森林土(壤)
grey zinc mush 灰色锌糊
Gr-Fe ratio 铬铁比
gribble 蛀木水蚤;吃木虫;船蛆
griblet 素图
grib stacking 三角堆垛法
grid 系统网络;(坐标网;格子(算板筛);格栅;高压输电网;滤线栅;极板网栅;网格;图网;栅条;栅格;算子
grid accumulator 栅条蓄电池
grid action 算栅作用
grid admittance 栅导纳
grid air navigation chart 坐标网航空图
grid alignment 栅极调整
grid amplitude (东西方向的)网格幅度
grid-anode ignition potential 栅阳极着火电位
grid-anode transconductance 跨导
grid area 格栅区
grid azimuth 坐标方位角;网格方位角;平面方位角;网格方位角
grid bar 格形炉栅
grid base 栅基
grid battery 栅极电池
grid beam 梁格
grid bearer 炉条格架;炉算托架
grid bearing 地岩上测量线间的方位角;坐标限角;坐标方位(角);网格象限角;网格方位(角);平面方向角;网格方位(角);方位角偏角
grid bed 制芯铁砂床
grid bell 栅铃
grid-bias battery 栅极偏压电池组
grid bias relay 栅偏压继电器
grid bias resistance 栅偏电阻
grid bias supply 栅偏压电源
grid bias voltage 栅偏压
grid blocking 栅极阻塞
grid board 坐标网板
grid boring 格点钻探
grid bypass 栅极旁路
grid cage 栅极罩
grid cantilever footing 网格状悬臂底脚
grid cap 栅极引出帽
grid-cathode cavity 栅阴空腔
grid-cathode circuit 栅阴回路
grid-cathode conductance 栅阴间电导
grid-cathode-feed amplifier 栅阴极馈电放大器
grid-cathode gap 栅阴间空隙
grid-cathode ignition potential 栅极阴极着火电位
grid-cathode resistance 栅阴电阻
grid-cathode voltage 栅阴极电压
grid-cavity tuner 栅极空腔调谐器
grid ceiling 悬挂式顶棚;井式平顶;井式顶棚;网状顶棚;网格顶棚;栅顶
grid characteristic 栅极特性
grid chart 格栅图表;网格图;网格坐标海图;网格图;方格图
grid checker packing 格子体堆砌
grid chill 栅形冷铁
grid choke 栅极抗流圈
grid circuit 栅极电路
grid-circuit clipping 栅极电路削波
grid-circuit impedance 栅极电路阻抗
grid clamping 栅极箝位
grid cleaning 电解清洗
grid clip 栅帽接线柱
grid coil 栅极线圈
grid computation 网格计算
grid conductance 栅极电导;电板电导
grid connection 电源接合;栅极引线;栅极接线
grid control 栅极控制
grid control angle 栅极控制角
grid control characteristics 栅极控制特性
grid control circuit 栅极控制电路
grid-controlled ionization ga(u)ge 栅控电离压力计
grid-controlled rectifier 栅极控制整流管;闸流管
grid-controlled rectifier tube 栅极整流管
grid-controlled relay tube 栅极继电器管
grid control method 栅控法
grid control ratio 栅控系数
grid control tube 栅控管
grid control voltage 栅极控制电压
grid convergence 坐标纵线偏角;子午线收敛角;网格偏角;平面子午线收敛角
grid coordinates 平面直角坐标;网格坐标;方格坐标
grid coordinates chart 网格坐标海图
grid coordinate system 坐标网格
grid core 栅门芯;网格芯;夹层板芯层
grid coupling 栅极耦合
grid course 网格水平航向;网格航向
grid cover 栅极屏蔽
grid current 供电网络;栅极电流
grid-current capacity 栅流容量
grid-current characteristic 栅流特性
grid-current cut-off 栅流截止
grid current cut-off voltage 栅极电流截止电压
grid-current detection 栅流检波
grid-current distortion 栅流失真
grid current impulse 栅流脉冲
grid-current modulation 栅流调制
grid cut-off separator 截止信号分离器
grid cut-off voltage 栅极截止电压
grid deck 空格桥面格式桥面
grid decking 梁格式桥面系
grid declination 格网偏角
gridded chart 有坐标网格海图
gridded data map 网格化数据图
gridded map 有坐标网格地图;方格坐标地图
gridded oblique photo 带网格的倾斜像片
gridded plane 网格平面
gridded probe 栅探针
grid detection 栅极检波
grid detection characteristic 栅极检波特性
grid detection coefficient 栅极检波系数
grid detection voltmeter 栅极检波电子管伏特计
grid detector 栅极检波器
gridding 格栅化
grid-dip meter 栅流陷落式测试表;栅陷振荡器
grid-dip oscillator 栅陷振荡器
grid-dip wavemeter 栅陷式波长计;栅流降落式波长计
grid direction 网格方向;网格方位
grid disc 栅极圆盘
grid distance 坐标网格距离;栅格间距
grid distributor 分布板
griddle 筛子;大孔筛

griddle plate 接触烤板;煎板
grid dome 网格穹隆
grid drawing 坐标网原图
grid drive 栅极控制电压
grid-drive characteristic 栅极激励特性
grid driving power 栅极调制激励功率
grid electricity 栅极电流;供电网络
grid electrode 栅极
grid electrostatic paint spray apparatus 栅极式静电喷涂装置
grid element 栅网电阻元件
grid emission 栅极发射
grid end covering 栅网端盖
grid equator 网格赤道
grider rail 槽形轨
grid excitation 栅极激励
grid-excitation winding 栅极激励绕组
grid exposure frame 网格曝光框架
grid extinguishing 栅极闭锁
grid facade 方格形房屋立面
grid failure 栅极损坏;栅极破坏;电网损坏
grid-feedback winding 栅极反馈绕组;栅极反馈绕组
grid feeding hopper 带格栅的装料斗
grid field 栅极控制场
grid floor 格栅楼面;格形地面
grid floor cover(ing) 方格形地面铺设;格栅楼面面层
grid-focused linac 栅网聚焦直线加速器
grid-focused linear accelerator 栅网聚焦直线加速器
grid-focused operation 栅网聚焦运行
grid focusing 栅网聚焦
grid focusing property 栅网聚焦性能
grid focusing system 栅网聚焦系统
grid footing 网格状底脚
grid formation 网格式结构布置;铁骨构架配置设计
grid foundation 网格基础;格栅式基础;交叉梁基础
grid frame(work) 网架(结构);(三维的)空间网架;网格框架
grid gas 管道煤气
grid gate pulse 栅极选通脉冲
grid gating 栅极选通
grid ga(u)ge 格栅量规
grid girder 梁格
grid glass plate 坐标网格玻璃板
grid-glow relay 栅极辉光放电继电器
grid glow tube 栅极辉光放电管
grid graph program(me) 栅状图程序
grid hatch cover 栅格式舱盖
grid heading 网格(坐标)航向;网格航向
grid holder 栅网架
grid hopper 栅板箱
grid hum 栅极交流声
grid impedance 栅极阻抗
grid impulse 栅极脉冲
grid impulse voltage 栅极脉冲电压
grid indication 网格坐标指示
grid indicator 可见呼叫指示器
grid induced noise 栅路感应噪声
grid inductance 栅路电感
grid induction 栅极感应
grid injection 栅极注频
grid input capacitance 栅极输入电容
grid input impedance 栅极输入阻抗
grid-interception noise 栅流分布起伏噪声
grid intersection 坐标网交点;栅格十字线;网格交点
grid interval 网格间距;栅距
grid inverse current 栅极反向电流
grid ionization chamber 网电离室
gridiron 修船架;格子船台;(格状物;栅格网;梁格结构;栅格;葡萄棚;铁框格;铁格子;筛;架;侧线
gridiron, ring and diagonal line type road network 棋盘环线加对角线形线网
gridiron and ring road network 棋盘加环形线网
gridiron city 采用棋式道路系统的城市
gridiron dock 格框船台;框格式船台
gridiron drag 铁耙路刮;铁框刮路器;方格式拖
gridiron drainage 格状水系
gridiron fashion street layout 格状道路布置
gridiron in dock 坞内梁格结构
gridiron ladder 栅形梯线
gridiron network 环状管网

gridiron pattern 方格式;棋盘格;方格形
gridiron pendulum 栅形补偿摆;杆式补偿摆
gridiron plan 方格式路网;网格规划图
gridiron road system 棋盘式道路系统;棋盘式道路网
gridiron siding 栅形调车场线路
gridiron system 格状(排水)系统;网格式体系;网格式配管系统;环状管网;网格式(给排水)系统
gridiron tile-drainage system 栅格式瓦管排水系统
gridiron town 采用棋盘式系统的城镇;棋盘式道路的城镇
grid iron type 方格式
grid iron type layout 方格式街道布置
grid(iron)type road network 棋盘形线网
grid iron type street layout 棋盘形街道布置;棋盘式街道布置;方格式街道布置
gridiron valve 格子阀
gridiron yard 栅形编组场
gridistor 隐栅管;场效应晶体管
grid junction 网格连接线
grid keying 栅极键控法
grid latitude 网格纬度
grid-lead shield 栅极引线屏蔽
grid lead(wire) 栅极引线
grid leak 栅漏
grid-leak bias 栅漏偏压
grid-leak bias separation 栅漏偏压分离
grid-leak detector 栅漏检波器;累积式栅极检波器
grid-leak modulation 栅漏调制
grid-leak rectification 栅漏整流
grid-leak resistance 栅漏电阻
grid length (应变片内的)金属栅长度;坐标网距
grid letter 坐标网注记
grid level(l)ing 面(积)水准测量;方格水准测量
grid limiter 栅极限幅器
grid limiting 栅极限幅
grid line 分格条;坐标(格)网线;栅(坐标)线;高压输电网线;网格线;方格坐标线
grid-loaded 栅网加载
grid-loaded waveguide 栅网加载波导
grid location 坐标位置
grid locking 栅极闭锁
grid longitude 网格经度
grid loss 栅极损耗
grid magnetic angle 网格磁偏角
grid magnetic azimuth adjustment 栅磁方位角改正;网格磁方位校正
grid map 网格地图
grid mark 网格标记
grid material 栅极材料
grid-mat foundation 格排基础;格筏基础
grid measuring system 网格测量系统
grid meridian 坐标网纵线;网格子午线
grid mesh 网眼;栅门;栅板网孔
grid metal 铅板合金
grid metal lath partition 网眼钢皮抹灰隔墙
grid method 网格(校正)法
grid method for strain measurement 网格法应变测量
grid model 网格模型;网格模式
grid-modulated amplifier 栅极调制放大器
grid modulation 栅极调制
grid modulation circuit 栅极调制电路
grid module 网格
grid navigation 网格航法;网格导航;网格导航
grid nephoscope 格栅测云器
grid network 电力系统;电力网
grid neutralization 栅极中和
grid neutrodyne method 栅极中和法
grid noise 栅极噪声
grid north 网格北(向)
grid note 坐标网说明
grid number 网格线号数
grid of elevation 高程网格
grid of foundation 网格基础
grid of mirrors 镜栅
grid of neighbo(u)ring zone 邻带方里网
grid of parallels and meridians 经纬网
grid of prevent sand 防沙栏栅
grid of reference 参考网格
grid of reinforcement 钢筋网格
grid of screw dislocations 螺形位错十字格
grid opening 网格孔
grid origin 坐标原点;网格原点

grid overlay 网格叠置片
grid-packed column 栅条填充柱
grid packed tower 栅条填料塔
grid packing 栅条填料;栅格填料
grid parallel 网格纬线
grid pattern 网格图形;栅格测试图;格状形式;棋盘式;网格状线
grid-pattern town 方格网城镇
grid paving 炉算板底层
grid peak-voltage 栅极尖峰电压
grid pitch (丝)距
grid plan 房屋建筑设计总体规划平面;网格平面(布置);房屋建筑设计总体规划图
grid plate 栅板;格子板;网格板;涂浆型极板;分布板;算板
grid-plate cavity 栅板空腔
grid plate characteristic 栅板特性
grid plate coupling 栅板耦合
grid plate for calibration 检校网板
grid plate oscillator 栅板振荡器
grid plate transconductance 栅板跨导;跨导
grid plenum 栅网联箱
grid plug 栅板节流塞
grid point 格子点;网(格)点;格点
grid point distribution 格子点分布
grid points 网格点
grid pool tube 栅极汞槽整流管
grid position 坐标位置;网格位置
grid potential 栅极电位
grid potentiometer 栅极分压器
grid-power supply 栅极电源
grid prime vertical 网格主垂面
grid printing 印格
grid problem 网格问题
grid proximity effect 栅极邻近效应
grid pulse 栅控脉冲
grid pulse modulation 栅极脉冲调制
grid pulsing 栅极脉冲
grid rack 算条;栅条
grid rail 铁路网
grid ratio 栅格比值
grid readings 网格读数
grid reception stations 栅式接受站
grid recognition 网格识别
grid rectification 栅极检波
grid rectification detector 栅极检波器
grid reference 网格基准;网格坐标;图网坐标;参考坐标网
grid reference box 内外图廓坐标记
grid reference point 栅网参考点
grid regenerative detection system 栅极再生检波方式
grid residual voltage 栅极侧剩余电压
grid resistance 栅极电阻
grid resistor 栅极电阻
grid resonance frequency 栅极谐振频率
grid resonance type oscillator 栅极调谐式振荡器
grid retardation 栅极减速
grid return 栅极引线
grid-return tube 反射栅电子管
grid rhumb line 网格恒向线;网格等斜线
grid road pattern 网格状道路形式
grid roller 方格状铁板压路机;网格式碾压机;网格式路碾;方格压印路滚;方格压印滚筒;格栅压路机
grid scale 坐标网尺;网格比例尺
grid scale constant 坐标网尺度常数
grid screen 栅屏
grid search technique 格点搜索技术;格点搜索法
grid section 算条
grid shadowing method 栅极屏蔽法
grid sheet 坐标网板
grid sheet system 格栅式薄板体系
grid shielding 栅极屏蔽
grid shielding can 栅极屏蔽罩
grid shortening bar 栅极短路棒
grid side-rods 栅极边条
grid signal 栅极信号
grids in building 建筑用方网格
grid slab 方格形板
grid space 坐标空间
grid-spaced contact 板形(多点)接触
grid spacer 定位格架
grid spacing 栅网间距;管网间隔;网格间距;网格间距;网格间隔;算缝间距

grid spectrometer 栅格分光计
grid square mapping 网格编图
grid stacking 格网叠加
grid steel support 格栅钢支撑
grid stiffened plate 网格加筋板
grid stopper 栅极寄生振荡抑制器
grid street layout 棋盘式街道布置(网)
grid stretcher 栅极扩मही
grid stroke 光栅线
grid structural system 网格状结构系统
grid structure 空间网架;格子式网架;网格结构
grid structure of guide rails 导轨格架结构
grid substation 电网分站
grid subtransmission 网络状二次输电
grid survey 网格法地形测量
grid suspension system 网格状悬吊顶棚
grid sweep 栅压摆幅
grid swing 栅压摆幅
grid system 直角坐标网格;格式系统;格构体系;高压电网;露龙骨吊顶系统;环(状管)网;网格式系统;输电网;方格系统
grid tank 栅极振荡回路
grid template 坐标网格模片
grid terminal 栅极出线端
grid time constant 栅极电路时间常数
grid-to-ground capacitance 栅极对地电容
grid town 方网格城镇
grid track 网格航线
grid transformer 栅极变压器
grid transmission system 高压电网
grid tray 栅条塔板;栅板塔盘
grid tray column 栅条盘分馏柱
grid tube 栅条管
grid-tuned 栅极调谐
grid-type accumulator 格栅式储蓄器
grid-type chord road 井形弦线
grid-type coil 格(栅)式盘管
grid-type concave 栅格式凹板
grid-type curtain wall 栅格式幕墙
grid-type extraction valve 栅形抽气阀
grid-type indicator 栅格形指示器
grid-type ion source 栅网形离子源
grid-type level detector 栅极式电平检波器
grid-type medium upflow filtration 格式升流过滤
grid-type radiant burner 栅格式红外线加热器
grid-type stirrer 栅式搅拌器;框式搅拌机
grid unit 栅极单元
grid value 网格值
grid value map 网格化数据图
grid valve 栅形阀
grid variation 网格偏差;网格磁差;网格变化
grid variometer 格栅式可变电感器
grid vent 栅格风口
grid voltage 栅极电压
grid-voltage cut-off 栅压截止
grid voltage lag 栅压滞后
grid voltage supply 栅偏压电源
grid wave 栅极信号波;栅决信号
grid wheel 笼式加宽轮(一种拖拉机防下陷的轮,装在轮侧)
grid winding 栅极绕组
grid window 网格窗口
grid-wire spacing 栅丝间距
grid work 网格组织
gridwork girder 格子梁
grid zone 坐标带
grid zone designation 坐标格网带编号
Griebhard's rings 格里伯哈特环
grief joint 方钻杆
grief kelly 主动钻杆;方钻杆
grief stem 钻的方套管;方钻杆
grievance 不满意见
grievance period (公众向当地政府投诉税收不当的)投诉期
griffe 爪状饰;圆柱基部的虎爪形装饰;虎爪式柱座;提刀
griffin 翼狮象
griffin fresco 鹰头怪兽壁画
Griffin mill 格里芬磨碎机
Griffin system 格里芬空气预热法
Griffin wheel 格里芬轮
Griffin wheel pressure forming process 格里芬车轮低压浇注法
Griffith's white 立德粉

Griffith casing tongs 格里菲四铰链式管钳
Griffith crack 格里菲破裂;格里菲裂纹
Griffith crack theory 格里菲微裂纹理论;格里菲破裂理论;格里菲裂隙理论
Griffith criterion of brittle failure 格里菲脆性破坏准则
Griffith equation 格里菲方程
Griffith flaw 格里菲微裂纹
Griffith fracture 格里菲破裂
Griffith-Irwin theory 格里菲—伊文理论
griffithite 水绿皂石
Griffith method 格里菲方法
Griffith microcrack model 格里菲微裂纹模型
Griffith strength criterion 格里菲强度理论
Grignard compound 格里纳德化合物
Grignard reagent 格里纳德试剂
Grignard synthesis 格里纳德合成
grike 岩(溶)沟;灰岩深沟
grillage 窑底梁;窑底格排法;格栅;格排;格架;格床
grillage analogy 格构比拟
grillage analogy method 梁格法
grillage analysis 梁格分析
grillage beam 井字梁;梁排;格子梁;格排梁;交叉梁系;网格梁;板架
grillage cantilever footing 网格状悬臂底脚
grillage column base 格排柱座;格排柱基
grillage floor 格排式楼板
grillage flooring 格栅楼面
grillage footing 格排基础;格排基础;格排底座
grillage foundation 格床基础;格形基础;格排基础
grillage girder 格排梁;格床梁;网格梁
grillage girder bridge 格床梁桥
grillage girder system 梁格系
grillage of fascine poles 柴束格床
grillage raft foundation 格排筏式基础
grillage system of bridge floor 桥面梁格系
grill(e) 格筛;格构;烤肉餐厅;栅;焙器;格栅;网屏;铁花格;铁算子;地理坐标网;风口箅子;装饰栅;格子窗
grill(e) block 多孔砌块
grill(e) burner 烤炙炉
grill(e) ceiling 格栅式天棚
grill(e) cloth 格栅布
grill(e) cover 格栅盖
grilled door 格栅门
grilled sliding door 铁栅推拉门
grilled tube radiator 加肋管散热器
grilled ventilation door 通风栅门
grill(e) fence 格式栅栏
grill(e) flocculating tank 栅网格絮凝池
grill(e) flooring 格子底板;铁格子板
grill(e) framing 格式框架
grill(e) generator 栅格信号发生器
grill-grid crusher 格栅污物破碎机
grill(e) louver 格形百叶窗
grill(e) member 花格构件
grill(e) of radiator 炉片格栅
griller 烤炙器
grill(e) room 烤炙室;烤肉餐厅
grill(e) type door 格子门
grilling 烤架;烤板;格栅
grillwork 格架
grimaldiite 三方羟铬矿
grimble ring 罗盘柜常平环
grime 烟灰;污秽物;灰尘;尘垢
Grimm discharge lamp 格里姆放电灯
Grimm discharge spectroscopic analysis 格里姆放电光谱分析
grimselite 碳钾铀矿
grimthorpe 形式重建
grin burner 烤炉
grind 磨损的绳索;磨;研碎
grindability 易磨性;可磨性;可磨度;磨削性
grindability coefficient 易磨性系数
grindability curve 可磨性曲线
grindability factor 易磨性系数
grindability index 易磨性指数;(岩石的)破碎系数
grindability of clinker 熟料(的)易磨性
grindability test 易磨性试验
grindability tester 可磨性试验机
grind away 磨样
grind-burn-leach process 磨碎燃烧浸取过程
grind crack 研磨裂纹
grind down 磨成粉

grind drier 研磨烘干机
grinder 交流干扰喀啦声;磨(碎)机;磨(矿)机;磨工;磨床;碎石机;碎木机;砂轮机
grinder's rot 磨工肺坏疽
grinder and polisher 磨平机和抛光机
grinder belt 金刚砂带
grinder bin 粉碎机装料斗
grinder buffing 粗轮打磨
grinder carrier 磨床座架
grinder feeder 研磨机的喂料器;磨床进料器
grinder jaw 磨刀夹刀板
grinder-mixer 粉碎搅拌机;粉碎混合机
grinder-polisher 磨削抛光机
grinders 磨工工人;天电干扰声
grinder shell 研磨机罩
grinder spindle 研磨机心轴;研磨机立轴
grinders pneumoconiosis 磨工尘肺
grindery 研磨车间;磨工(研磨)车间;磨坊
grind fineness 研细程度;研磨细度
grind grading 磨矿分级
grind in 磨合
grinding 研磨;磨削;磨碎;磨钢丝;磨的;摩擦的;打磨;粉磨;研磨
grinding abrasion 磨料磨损
grinding action 研磨作用;磨削作用;磨削操作
grinding agent 研磨剂
grinding aid 助磨剂;助磨料;促磨剂
grinding allowance 磨削加工余量;磨削(加工)留量
grinding amalgamator 研磨式混汞器
grinding-and-buffing attachment 研磨与抛光装置
grinding and polishing 研磨和抛光
grinding and polishing drum 研磨和抛光滚筒
grinding and polishing machine 研磨和抛光机
grinding angle 研磨角
grinding angle and colo(u)ring analysis 磨角染色分析
grinding apparatus 磨具;粉磨设备
grinding apparatus for lathe centers 车床顶尖磨具
grinding at low-temperature 低温粉碎
grinding attachment 磨削装置;磨削附件;磨钎装置;磨矿设备
grinding ball 磨球;研磨钢球
grinding ball charge 钢球装载量;钢球填充量
grinding ball loss 磨球损耗
grinding ball wear 钢球磨损
grinding base 碾盘;磨盘
grinding bed 研磨层;砂轮清理台;粉磨料层
grinding belt 研磨带;磨(削)砂带;磨带
grinding body 研磨体
grinding bowl 辊盘;磨盘
grinding building 磨矿车间
grinding burn 磨削烧伤
grinding capacity 磨削容量;粉磨能力
grinding carriage 磨床床头;砂轮架
grinding chamber 粉磨仓
grinding charge 研磨体;磨机装量;被(研)磨物
grinding chips 研磨碎片
grinding circuit 磨矿回路;粉磨流程
grinding compartment 粉磨仓
grinding composition 磨剂
grinding compound 研磨膏;金刚砂;磨料;磨剂
grinding-concentration unit 磨矿精选装置
grinding contact 磨削接触
grinding corrosion 研磨溶蚀
grinding crack 磨削裂纹;磨痕
grinding crystal 磨光(的)晶体
grinding-cum-drying plant 粉磨兼烘干设备
grinding cushion 粉磨缓冲垫层
grinding cycle 研磨循环;磨削循环
grinding cylinder 管磨机;球磨机
grinding damage 磨削损伤
grinding device 研磨工具
grinding dish 磨坑
grinding disk 砂轮;研磨盘砂轮
grinding drum 磨矿机;粉磨滚筒
grinding dust 研磨粉
grinding efficiency 研磨效率;碾磨效率;磨矿效率;粉磨效率
grinding element 研磨机组件;磨床组件
grinding energy 粉磨能
grinding equipment 研磨设备
grinding face 磨削面
grinding factor 粉磨系数
grinding fineness of cement 水泥研磨细度

grinding finish 磨光面
grinding flowsheet 磨矿流程(图)
grinding fluid 研磨液;磨削液;润磨液
grinding fluid coolant 研削润滑冷却液
grinding ga(u)ge 磨床用卡规
grinding hardness 研磨硬度
grinding head 研磨头;磨头
grinding head motor 磨头电动机
grinding hole 研磨孔
grinding housing 粉磨辊外壳
grinding in 研磨密封
grinding in closed circuit 圈流粉磨;闭路粉磨
grinding in of valve 阀门的研磨
grinding in open circuit 开路粉磨;开流粉磨
grinding installation 研磨设备;粉磨设备
grinding into fine powder 研成细末
grinding lap 磨石盘
grinding lathe 车磨两用床
grinding length 磨削长度
grinding line 研磨线;研磨伤痕;磨削(灼伤)条痕
grinding liquid 研磨液体
grinding lubricant 研磨油;研磨液;润磨液
grinding machine 研磨机;磨床;砂轮机;粉碎机
grinding machine head(stock) 磨床头
grinding machine operator 磨工
grinding machines cam numeric(al) control 数字控制凸轮磨床
grinding marks 研磨痕
grinding material 磨料
grinding material box 磨料盒
grinding mechanism 磨碎装置
grinding media 研磨体;碾磨介质;研磨介质;磨料;磨矿介质
grinding media charge 研磨体装载量;研磨介质载量
grinding media filling ratio 研磨体装填比例
grinding media grading 研磨体级配;研磨介质的级配
grinding media load 研磨体装载量
grinding media segregation 研磨体自行分级
grinding media to clinker mass ratio 研磨体与熟料重量比
grinding media wear 研磨体磨耗;研磨介质磨耗
grinding method 研磨法
grinding mill 细磨机;研磨机;辊磨;磨(碎)机;磨矿机(械);球磨机;磨坊
grinding mill ball 研磨用球
grinding mill block 研磨车间
grinding miller 研磨轮
grinding mill feed bin 磨机磨头仓
grinding mill feeder 研磨机进器;磨碎机进器
grinding milling 磨坊
grinding mill of the edge runner 砂磨机;轮碾型磨机
grinding of base (玻璃、陶瓷器皿的)磨底;底座磨平
grinding off 研磨掉;磨耗掉
grinding of rail 钢轨的研磨
grinding of terrazzo 水磨石磨光
grinding oil 金属研磨用冷却油乳液;研磨油;润磨液
grinding operation 研削操作
grinding ore 磨矿
grinding out 磨内圆
grinding pan 研磨盘;磨盘;盘转磨机
grinding paper 磨光纸
grinding paste 研磨膏;磨削(用)冷却剂
grinding pattern 齿面磨纹
grinding pebble 磨球;磨砾(砾磨机用)
grinding performance 研磨性能
grinding plant 研磨设备;磨碎设备;粉磨设备
grinding plate 研磨平台;研磨平板;磨碎砥石
grinding point 磨平头的钉
grinding position 研削磨位置
grinding powder 气门砂;磨光粉
grinding power 研磨功率
grinding process 水磨工序;粉磨过程
grinding product 粉磨产品
grinding property 磨碎能力
grinding pump 研磨泵
grinding rate 研磨速度;磨细度;磨碎速度;磨矿速度;粉磨速度
grinding ratio 研磨比;磨削比
grinding relief 研磨用凹槽
grinding rest 磨床刀架

grinding ring 环形磨槽;研磨环
grinding rod 研磨(用钢)棒;钢磨棒
grinding roll 研磨辊;细碎对辊机
grinding roller 磨辊
grinding runner 研磨头的转动部分;转动磨石
grindings 磨屑
grinding sand 研磨用砂
grinding scheme 粉磨流程
grinding screen 粉碎筛;分碎筛
grinding seal 研合密封
grinding section 磨矿段;粉磨车间
grinding segment 研磨瓦
grinding sensitivity 研磨敏感性;磨削裂纹敏感性
grindings hardness determination 抗磨硬度测定
grinding size 研磨尺寸
grinding skin 磨削变质层
grinding slab 研磨(基)板
grinding sleeve 研磨套
grinding slip 研磨石;油石;磨刀凿用的油石
grinding spindle 研磨主轴;研磨轴梗
grinding stage 研磨台;研磨架;研磨程度
grinding stock 磨削留量
grinding stone 砂轮;细砂质磨石;磨石(砂轮)
grinding stress 研磨应力;磨削应力
grinding surface 磨(削)面
grinding system 粉碎系统;粉磨系统
grinding table 磨盘;研磨台
grinding technique 磨削技术
grinding teeth 磨齿
grinding temper 磨削时高温回火
grinding test 磨损试验
grinding to ga(u)ge 磨削到标准尺寸
grinding tolerance 磨削裕度;磨削留量
grinding tool 磨削工具;研磨工具
grinding tools electrolytically 电解磨削刀具
grinding tooth 臼齿
grinding track 粉磨轨道
grinding treatment 磨碎处理
grinding type resin 研磨型树脂
grinding undercut 磨削空刀槽;磨削越程槽
grinding unit 研磨装置;研磨头;磨头
grinding valve 磨阀
grinding wheel 砂轮;磨损;磨盘;磨轮
grinding wheel bearing 砂轮轴承
grinding wheel bearing protecting cover 砂轮轴承护盖
grinding wheel bonding material 砂轮黏结材料
grinding wheel cloth 增强砂轮布
grinding wheel conditioning unit 砂轮修整机组
grinding wheel contact surface 磨轮接触面
grinding wheel dresser 砂轮整形器;砂轮研磨机;砂轮磨钎机
grinding wheel dressing 砂轮整形;砂轮修整
grinding wheel drive motor 砂轮电机
grinding wheel driving box 砂轮传动箱
grinding wheel flange 砂轮法兰盘;砂轮垫盘
grinding wheel for pneumatic bit grinder 风动磨钎机用砂轮
grinding wheel for steel balls 磨钢球砂轮
grinding wheel grade 砂轮等级
grinding wheel head 磨头;砂轮座
grinding wheel shape 砂轮形状
grinding wheel slide 砂轮滑座
grinding wheel spectacles 磨轮眼镜;砂轮护目镜
grinding wheel speed 砂轮转速
grinding wheel spindle 砂轮主轴
grinding wheel spindle extension 砂轮主轴伸出
grinding wheel spindle sleeve 砂轮轴套
grinding wheel stand 砂轮座
grinding wheel structure 砂轮结构
grinding wheel testing machine 磨轮试验机;砂轮试验机
grinding wheel truing 砂轮整形
grinding wheel unit 砂轮装置
grinding work 磨工工作;粉磨功;粉磨功
grind into fine powder 研成粉末状
grindle 小沟;沟渠
grind-leach process 磨碎浸取过程
grind master 磨削靠模
grind off 磨掉
grindometer 细度计
grind on 研磨
grind per pass 一次通过的研磨物料量
grind size 研磨粒度

grindstone 研磨石料;磨(刀)石;砂轮
grindstone dresser 磨石刻槽器
grindstone dust 磨石屑
grindstone pit 磨石井
grindstone swarf 磨石屑
grindstone-trough 磨刀石水盘
grindstone with chest 带罩壳砂轮
grinning 旧墙纸表面涂料显示失调的缺点;底色外透(底层油漆色彩透现在外);露底
grinning effect 露底现象
grinning through 露出底层;底色外透(底层油漆色彩透现在外);露底;透出;露油漆层;露板条
griotte 棱角大理岩;红纹石灰岩;红纹大理岩;大理岩;大理石
grip (机械手的)抓手;抓牢;执手;辊拉边夹出的波纹;螺栓有效长度;紧握;夹子;钳取机构;握把;手柄;扼;把手;把数;夹(紧)
grip alignment 夹具对准
grip anchorage 握裹锚固
grip anchorage system 夹锚体系
grip between concrete and steel 钢筋与混凝土间握裹力;钢筋与混凝土间握固力;钢筋混凝土握裹力;混凝土钢筋间的握力
grip block 夹持块;夹持板;断绳防坠器
grip bolt 夹持螺栓
grip brake 手制动;夹紧制动
grip cheek 夹紧颊板
grip chuck 套爪卡盘;套爪夹头
grip clip 弹簧线夹
grip coat 搪瓷底层;底层(搪瓷)
grip concrete 混凝土握裹力;混凝土握固力
grip design 夹具设计
grip die 夹紧模
grip disc 凸缘链轮
grip dynamometer 普通握力计
gripe 抓住;固定索
gripe in 缚稳舢板
gripe lashing 缚带系绳
grip end 抓柄;机械手握物端
grip-face belt 皮704运输机中带肋皮带
gripfase 墙面板的固定板条
grip finger 夹钳爪;夹钳钳口
grip gear 升降机保险装置
grip groove 夹持槽
griphand 置景工
grip handle 抓柄;执手;门执手;把手;拉手架
griphite 疑难榴石;暧昧石
grip hold arc 夹弧;包围弧
grip holder 固定架;夹头;夹圈固定器;夹持器
grip hole 倾斜炮眼
grip hook 轮爪
griphtie 暧昧石
grip iron 铁夹
grip jaw 夹紧卡瓦;夹紧颚爪;颚形夹爪
grip length 锚固长度;钢筋握裹;锚着长度;握裹长度;握固长度
grip length of reinforcement 钢筋握裹长度;钢筋握固长度
grip mark 卡爪印;夹爪印
grip module 夹钳组件;手爪模件
grip nut 夹紧螺母;防松螺母
grip of bolt 螺栓茎长;螺栓柄;夹紧螺栓
grip of concrete 混凝土握裹力;混凝土握固力
grip of hole 炮眼倾斜
grip of rivet 铆钉头距离;铆钉深入度;铆钉茎长
grip of testing machine 试验机夹具
gripped end 夹持端
gripper 牙板夹头;抓爪(器);抓器;夹坯机;夹具;夹持器;叼纸牙
gripper Axminster carpet 夹片式阿克明斯特地毯;机织阿克明斯特地毯
gripper brake 手制动器;夹紧制动器
gripper die 夹紧模
gripper edge 夹紧边;夹持边;叼口
gripper feed 夹提给料;夹钳送料;夹持进给;夹持给料
gripper-feed mechanism 抓爪进给机构
gripper fork 叉形抓爪
gripper jaw 夹持爪
gripper mark 叼爪规
gripper mechanism 抓爪机构;夹爪机构
gripper-spool Axminster carpet 机织阿克明斯特地毯
gripper system 抓爪系统

gripping 过于逼风驶帆；夹紧
gripping action 矢锥扭录；矢锥吃口
gripping apparatus 抓扎；扣紧工具
gripping appliance 捕集设备(夹具)
gripping belt 夹送皮带
gripping clutch 握式离合器
gripping conveyer 夹持输送器
gripping device 抓手凹槽；抓取装置；抓取器；固定器；夹具；夹紧装置
gripping dies 卡瓦
gripping force 握裹力
gripping fork 垫叉；叉形件
gripping head 夹头；夹持帽；握固头
gripping jack 夹紧千斤顶
gripping jaw 夹爪；卡爪；卡瓦
gripping mechanism 夹钳装置；夹持机制
gripping pattern 轮胎防滑花纹；防滑花纹
gripping pliers 抓爪钳；夹(管)钳
gripping plough 开沟犁
gripping range 卡紧范围
gripping roll 拉料辊；擒料辊
gripping surface 抓取面；攫取面
gripping tongs 鸭嘴钳；两用钳；平口钳；扁钳
gripping tool 抓取工具
grip plate 折边板
grip resistance 抗滑阻力
grip ring 夹圈；夹环；夹持环
grip roll 夹持转子
grips 抓爪
grip safety 握柄保险机
grip seal joint 夹持密封接头
grip shot 斜炮眼；斜炮孔；斜炮洞
grip slide 夹紧滑块
grip slipper 夹持座
grip socket 压环
grip spread 栅形排列
grip spreader 夹挂吊架；夹挂吊车
grip strength 握力
grip-tie fitting 夹紧式配件
grip tire 防滑轮胎
grip-tite push-in mechanical joint 紧密推入式机械接头
grip type tilting manipulator 夹辊式翻钢推床
grip vise 夹紧虎钳；压夹虎钳
grip wedge 止动楔；保险楔；安全楔
grip wrench 管子钳
griquaite 透辉石榴岩
griqualandite 青石棉
grisaille 不透明玻璃；灰色装饰画法；浮雕式彩绘(法国)
grisaille painting 墨彩(陶瓷装饰法)；灰彩
grischunit 砷钙锰石
griseous 浅灰色的；白色带棕斑的
grisounite 硝酸甘油
grisoutite 硝酸钾炸药
grissaille 浮雕装饰画法
gristmill 磨坊
grit 硬渣；杂粒；棱角粒料；金属锯屑；磨光粉；砂(岩)；带孔纸；粗砂
grit arrestor 收尘器
grit basin 沉砂池
grit bin 石屑料仓；砂石料仓
grit blast 喷粒处理；喷丸清理；喷丸器；喷射清理；喷砂
grit blasted 用硬渣喷射的；用砂砾喷射的；喷砂(处理)
grit blasting 喷丸；喷砂碎面(法)；喷砂打磨(法)；喷砂(处理)；喷丸处理；喷砂消除法；喷砂清除法；喷砂除锈；喷软粒；喷软(砂)；喷粒处理；吹砂(处理)
grit blasting method 喷砂清理法
grit-blinded 铺石屑的
grit blinding 喷石屑
grit blow-off 排砂管；吹砂
grit box 集污砂箱
grit bucket 集砂桶
grit carborundum 金刚砂砾
grit catcher 捕砂器；除砂池；沉砂池；集砂器
grit chamber 渣室；杂粒池；砂砾(生物)滤池；除渣室；沉渣池；沉渣池；沉渣槽
grit chamber agitator 沉砂池搅拌器
grit chamber channel 沉砂池渠道；沉砂池槽道
grit chamber design 沉砂池设计
grit channel 沉砂池；沉砂渠；沉砂槽

grit collector 集渣池；集沙室；集砂器
grit compartment 沉沙室；沉渣池
gritcrete 砂石混凝土；砂砾屑混凝土
grit discharge by gravity 重力排沙
grit discharge pocket 杂粒排放坑
grit discharge with hydraulic elevator 水力提升器排沙
grit discharge with spiral pump 螺旋泵排沙
grit disposal 沉沙处置
grit dredge(r) 挖沙机；挖沙船；除沙机
grit duct 沉沙管
grit emission 磨料飞散
grit erosion 风沙侵蚀
grit finish 磨砂处理
grit-flume 粗粉空气输送料槽
grit-free 没有砂砾的
grit grard 抛粒护板
grit gravel 细粒砾石；细砾；砂砾；粗砂岩
grit gulley 沉沙沟
grit hopper 杂粒斗；沉沙斗
grit incineration 砂泥焚化；沙泥焚化
grit number 磨料筛目数；磨料粒度数；筛网号
grit operation 除砂操作
grit paper 粗砂纸
grit pile-up site 堆砂场
grit-proof 隔岩屑的
grit removal 除砂
grit-removal basin 除砂池
grit-removal efficiency 除砂效率
grit-removal eliminator 除砂设备
grit-removal facility 除砂设备
grit-removal operation 除砂操作
grit-removal tank 除砂池
grit reservoir 沉渣池；沉沙水库；沉沙(砾)池
gritrock 粗砂岩
gritrock reservoir 粗砂岩储集层
grit sandstone 粗砂岩
grit separator 煤灰分离器；灰砂分离器；粗粉分离器
grit size 磨料粒度
grit spreading 铺撒石屑(路面)；铺撒砂砾(路面)
grit spreading vehicle 石屑铺撒车；砂砾铺撒车
grits storage hangar 石屑堆放棚；磨料堆放棚
gritstone 天然磨石；砂石；细砂岩；天然砥石；砂岩(石)；粗砂岩
grit stratum 砂砾层
grit tank 沉沙池；沉沙槽
gritted 铺石屑的
gritter 铺砂机
grittiness 含砂量；砂性
gritting 铺砂工程；铺砂工作
gritting blasting 喷砂处理
gritting lorry 砂砾铺撒车；石屑铺撒车
gritting machine 铺砂机
gritting material 研磨料；砂砾材料
gritting material bin 石屑储[贮]仓；磨料储[贮]仓
gritting material hangar 石屑堆放棚；磨料堆放棚
gritting material store 石屑堆栈；磨料堆栈
gritting roll attachment 石屑碾辊附件
gritting sand 铺盖砂；铺砂
gritting service 砂石铺撒作业
gritting surface dressing 石屑铺面
gritting truck 砂石铺撒车
gritty 含砂砾的；粗砂质的
gritty consistence 含砂(砾)度
gritty coverstone 面层石屑；道路面层石屑
gritty dust 石粉；砂砾屑
gritty finish 石屑铺面
gritty scale 带砂表皮
gritty soil 含细砾的土；砂砾质土；粗砂(砾质)土
gritty(surface)finish 石屑铺面
grit washer 洗砂装；洗砂机
grit washing 杂粒沉沙池洗涤(污水处理厂操作)；洗砂冲洗
grit washing mechanisms 杂粒冲洗机械
grivelle 斑点花纹
grizzle 高硫劣煤；硫劣煤；劣质煤；欠火建筑砖
grizzle brick 次等砖；欠火砖
grizzled 有灰斑的；灰色的；灰白的；斑白的
grizzly 栅筛；铁格筛
grizzly bar 格筛条；铁栅(筛栅)条；栅条
grizzly chamber 格筛室；棒条筛室
grizzly chute 带格筛的溜道
grizzly crusher 格筛破碎机；带栅破碎机
grizzly feeder 污水厂入流格筛；滚筒筛喂器；格筛

grizzly grate 箅条筛
grizzly level 格栅水平
grizzlyman 格筛工
grizzly screen 格筛；格栅；铁栅筛
grizzly with depressed crossbar 凹下横杆式格栅
Grobal environmental assessment program(me) 格洛保环境评价方案
grocer 食品商
grocer's store 副食店
grocery 杂货店；食品杂货店；副食店
grocery store 食品杂货店；副食店
groceteria 自选副食品；食品杂货店
grog 黏土熟料；耐火熟料；陶渣；熟料；烧粉
grog brick 耐火(熟料)砖
grog-containing whiteware 含熟料的白色陶瓷器
grog fireclay mortar 耐火黏土砂浆；熟料耐火黏土
groggery 低级酒吧
grog mill 熟料磨碎机；熟料磨料机
grog refractory 耐火砖；耐火熟料；熟料(耐火材料)
grogshop 低级酒吧
groin 折流坝；交叉拱；交叉坝；海堤；排流坝；丁坝；腹股沟；防沙堤
groin basin 丁坝间水区
groin body 丁坝坝体
groin ceiling 穹棱天花板
groin centering 肋拱撑架；交叉拱架
groind arch 穹棱屋顶；两筒相贯穹顶；交叉拱肋式
groind point (无肋交叉拱顶的)交接线
groin(e) 穹棱
groined arch 交叉拱；交叉拱肋；对角线拱；穹棱屋顶；穹棱拱；两筒相套穹顶
groined ceiling 交叉拱顶天棚；交叉穹顶棚；穹隆天花板；穹棱顶棚
groined point 穹棱交点
groined ribs 交叉肋
groined roof 穹棱屋顶；交叉拱棱屋顶；交叉拱屋顶；穹隆屋顶
groined slab 带拱肋的板；井字梁楼板
groined vault 交叉筒拱；穹棱拱顶；两筒正交相贯穹顶；两筒正交相套穹顶
groined vaulting 交叉拱顶(筑法)
groin end slope 丁坝端坡
groin engineering 丁坝工程
groin field 丁坝场；丁坝坝田
groin head 防波堤头；突堤堤头；丁坝坝头
groining 筑防波堤；做交叉拱；做成穹棱；筑丁坝；造防波堤；穹棱工作
groin of training works 导流丁坝
groin pointing downstream 倾向下游的防波堤；倾向下游的丁坝
groin pointing slightly upstream 略斜向上游的丁坝
groin pointing upstream 倾向上游的防波堤；倾向上游的丁坝
groin rib 拱肋；穹棱肋
groin root 防波堤脚；丁坝脚；岸边接头(船舶用水、电、油等)；丁坝根部
groin slab 井字形梁板
groin spacing 丁坝间距
groin system 丁坝系统；丁坝群
groin training 丁坝整治
groin vault 正交穹顶；棱拱
groin works 丁坝设施；丁坝工程
grommet 橡胶密封圈；螺栓垫圈；金属孔眼；金属封油环；麻丝垫环；环管；护孔环；索眼；索环；垫环
grommet fender 碰垫球；绳圈碰垫
grommet punch 帆布打洞机
grommet ring 油灰麻绳填料环；索环；帆布金属眼环
groom 擦洗和加油
grooming behavio(u)r 清整动作
groove 细槽沟；轧槽；沟；环槽；切口；企口；剖口；坡口；膛线；槽；擦痕沟；凹线；凹槽
grooveability 沟纹耐久性
groove and tongue 企口；槽舌(榫)；凸凹榫
groove angle 轧槽边斜角；焊缝坡口角度；(焊接的)坡口角(度)；齿槽角；槽角(度)；凹槽角
groove bit 槽形钻
groove bit nose 在钻头上开槽
groove-bottom diameter 内槽径
groove cast 沟铸型；沟纹模型；沟(脊)模
groove connection 榫式接合；企口连接；榫槽接合
groove contour 槽形
groove control valve 槽式节流阀
groove corrosion 焊接部位腐蚀；槽腐蚀

groove cutter	刻纹机;切槽口刀
groove-cutting chisel	槽刨
groove-cutting machine	开槽机
grooved	开槽的
grooved and hyperboloidal composite liners	槽沟双面衬板
grooved and tongued	企口接合的
grooved and tongued block	企口砌块
grooved and tongued floor(ing)	槽舌地板;企口地板
grooved and tongue(d) joint	企口接合;槽舌接合
grooved and tongued panel	企口墙板
grooved armature	有槽电枢
grooved-back brick	沟纹砖
grooved ball bearing	凹槽式滚珠轴承
grooved ball seat	有槽球座
grooved barrel	有槽卷筒;螺旋槽面卷筒
grooved bit	带槽钻头;打孔钻头
grooved brick	沟波砖;带槽(耐火)砖
grooved butt ca(u)lking tool	刮平刀
grooved cam	凸轮槽盘;槽凸轮
grooved casting	有槽(止动)铁柱;开沟浇铸;开沟浇注
grooved connection	槽式接合
grooved construction joint	开槽伸缩缝
grooved contraction joint	(混凝土路面的)槽式缩缝;开槽伸缩缝;槽形伸缩缝
grooved core swirlspray nozzle	槽芯漩涡式喷嘴
grooved couplings	槽纹套管;槽管套
grooved cylinder	有槽滚筒
grooved cylindric(al) pin	带槽圆柱形销
grooved dead centre	带槽死顶尖
grooved director	有槽导子
grooved disk	有槽圆盘
grooved disk cam	有槽盘形凸轮
grooved drive stud	锯齿形传动键
grooved drum	缠索筒;沟面盘车;螺旋槽卷筒;绳沟滚筒;缠索筒;槽筒
grooved end-grain wood block	横切面企口木板块
groove depth	坡口深度;坡口高度;槽深
grooved expansion joint	沟槽胀接
grooved facing tile	背面有凹槽的面砖
grooved fiber-alignment connector	槽形光纤定位连接器;槽形对接光纤连接器
grooved friction wheel	槽形摩擦轮
grooved gearing	三角皮带轮传动装置
grooved glass	槽形玻璃;槽纹玻璃;槽舌玻璃;凹槽玻璃
grooved heart cam	有槽心形凸轮
grooved hexagonal nut	带槽内六角螺母
grooved hexagonal screw	带槽六角螺钉
grooved hoist drum	绳槽卷筒
groove digitizing method	槽纹跟踪数字化法
grooved iron roll	有槽铸铁轧辊
grooved joint	槽式接合;槽缝;凹缝;槽舌接(缝)
grooved liner plate	有槽形衬板
grooved match ceiling boarding	企口顶棚板条
grooved metal gasket	齿形垫片
grooved metallic gasket	透镜垫
grooved milling cutter	槽形铣刀
grooved mount	槽形底托
grooved needle	槽针
grooved needle holder	钳口带槽持针钳
grooved nut	槽顶螺母;槽顶螺帽
grooved panel	槽纹镶板;槽舌镶板
grooved pile	企口板桩;槽口板桩
grooved pin	带槽的销子
grooved pipe	丝牙口管子;槽口管子
grooved piston	带槽活塞
grooved piston ring	有槽活塞环
grooved plunger	带槽的柱塞
grooved plywood	开槽胶合板;企口胶合板
grooved probe	有沟探子;有沟探针;有槽引针
grooved pulley	三角皮带轮(滑轮);槽轮
grooved rail	有槽钢轨;有导轨,槽(形)轨
grooved ring	槽环
grooved roll	有槽轧辊;带槽轧辊;槽形辊压机;纹滚筒;凹槽滚筒
grooved roller	有槽滚轴;压路机;纹路碾;槽滚压路机;槽辊
grooved roller inking unit	槽辊上墨装置
grooved roll washing machine	槽纹辊洗涤机
grooved rotor	有槽转子
grooved round nut	带槽圆螺母
groove drum	槽筒
grooved runway	槽纹道面跑道
grooved seam	(金属板的)折缝接合
grooved setting-up table	槽面装置台
grooved shaft	槽轴
grooved shank drill	槽柄钻头
grooved shed insulator	茶托绝缘子
grooved sheet pile	企口板桩
grooved shell	有沟壳
grooved slab	企口板;槽形板;槽纹板
grooved slab insulation	绝热槽形板
grooved soil sampler	开槽黏土取样器
grooved specimen	开槽试样
grooved spring bar	有槽弹簧杆
grooved spring base	带槽的弹簧底板
grooved spring steel	带槽弹簧钢
grooved steel plate pile	槽形钢板桩
grooved stud	有槽双端螺栓;定位销钉
grooved table roller	辊道的带槽辊
grooved taper pin	有槽锥形销;带槽锥形销
grooved tile	有凹槽面砖;有槽瓦;槽形瓦
grooved tire	槽纹式轮胎
grooved track	槽形轨线路
grooved trolley wire	沟纹滑接线;槽触轮架空线
grooved-type contraction joint	槽形伸缩缝
grooved upland	沟切高原;镂空高地;槽蚀高地
grooved valve stem	有槽阀杆
grooved water piston	带槽水泵活塞
grooved waveguide	槽纹波导
grooved wheel	槽轮
grooved wire	槽线
groove end	坡口端
groove face	坡面;坡口(表)面;槽面
groove finisher	表面刻槽机(自动筑路机)
groove flange	槽面法兰
groove for coupling	联结用槽
groove form diffraction grating	槽形衍射光栅
groove for sash	吊窗槽
groove grinder	轴承沟道磨床;轴承槽床;槽磨床
groove insert	活塞槽嵌入物(顶活塞环用)
groove joint	假缝;槽线;槽缝;半缝;凹(圆)缝;凹(槽)缝;槽式结合
grooveless flange gasket	无槽凸缘垫圈
groove made by back chipping	铲根并坡口
groove mark	沟痕
groove markings	槽形纹
groove mask	影格板
groove muller	槽形碾砂机
groove nut	有槽螺母
groove of insulator	绝缘子槽
groove of pulley	滑轮槽
groove of sheave	滑轮槽
groove of the notch	切口槽
groove of thread	螺纹谷;螺(纹)槽
groove of trolley wheel	接地轮槽
groove out of square	沟侧摆
groove part	型槽
groove pin	槽针
groove pipe	带槽形接口的管
groove plane	槽刨
groove profile	槽剖面
groover	开槽机;(混凝土路面的)切缝机;切槽装置;挖槽机;切缝机
groove radius	槽半径
groove-root diameter	槽底内径
groove sash	吊窗槽
groove seam	起槽缝;企口缝
groove shape	槽形
groove shed insulator	复檐绝缘器
groove steel	槽钢
groove-type contraction joint	凹槽形收缩缝;凹槽形伸缩缝
groove type optic(al) cable	骨架型光缆
groove wear	沟槽磨损
groove weld(ing)	坡口焊(接);有坡口焊缝;开坡口焊接;开槽焊;槽焊(接);凹焊(接)
groove welding seam	坡口焊缝
groove-weld joint	有坡口焊接接头
groove width	槽宽
groove with variable pitch	变距槽
grooving	沟蚀;刻槽;开槽;切槽;温差膨胀裂纹;套槽;槽蚀;凹凸榫接
grooving and tonguing	槽舌接合;企口榫接;企口接合
grooving chisel	槽凿
grooving cutter	铣槽刀;切槽刀;槽铣刀
grooving for sleeve bearing	套筒轴承槽
grooving machine	刻槽机;开槽机;(混凝土路面的)切缝机
grooving method	刻槽法
grooving of roll	轧辊孔型设计
grooving of valve seat	阀座磨损
grooving plane	开沟刨;开槽刨;槽刨
grooving rotor	有槽转子
grooving runway	槽纹道面跑道
grooving saw	铣槽锯;开槽锯;锯槽锯
grooving tool	铣槽刀具;刻沟器;开槽工具;切槽工具;切槽刀
groovy	槽的
groovy fit seal face	榫槽密封面
groped ends	索绪
groping brush	索绪帚
groping end basin	索绪锅
groping end efficiency	索绪效率
groping end machine	索绪机
groping end though	索绪槽
groping-reflex	摸索反射
grorudite	霓细花岗岩
gros forts	厚实家具用布
grosgrain	茜丽绸;罗缎
grospydite	辉榴蓝晶岩
gross	总量;罗(计量单位,等于12打);基本质量;粗的
gross ablation	总消融量(冰川)
gross absorption	总吸收(量);毛吸收量
gross activity	总活度;总放射性
gross activity measurement	总活度测量;总放射性测量
gross amount	总量;总计;总额;毛额
gross amount of structural steel	钢材数量表
gross analysis	全量分析
gross and composite method	分组总计法
gross annual income	年总收入;全年总收入
gross annual output	年总产量
gross annual production	年总产量
gross apartment unit area	公寓单元总面积
gross appreciation	涨价总额
gross area	总面积;建筑占地面积;毛面积;全面积
gross area of section	毛截(面)面积
gross assessed value of property	财产估价总值
gross assets	资产总额;投资总额
gross assimilation	总同化
gross-austausch	大型交换
gross available head	总有效水头;总水位差
gross available income	现有总收入
gross average	共同海损;救助及特别费用;平均毛额
gross average deposit	共同海损备用金
gross benefit	毛效益;总收益
Grossberg unit	格罗斯伯格单位
gross blockage factor	堵塞系数
gross blow hole	大气孔
gross bonded debt	债券负债总额
gross book value	账面总价值;账面原值
gross brake horsepower	总制动马力
gross brake power test	总制动功率试验
gross budget	总预算
gross building area	总建筑占地面积;建筑(占地)总面积
gross building cost	建筑毛造价
gross building density	总建筑面积;总建筑密度
gross buoyancy	总浮力
gross business expenses	总费用
gross business product	商业总产值
gross calorie value	总热值
gross calorific power	总发热量
gross calorific value	总热值;高位发热量
gross calorific value in moist ash free basis	恒湿无灰基高位发热量
gross calorific value of geothermal energy	地热能总热值
gross capacity	总容量;总能力
gross capacity of reservoir	总库容;毛库容
gross capital formation	资本形成总值
gross cash flow	现金总流量;总现金流量
gross cave	毛洞
gross change method	变动毛数法
gross charges	总支出
gross charter	总租船契约;总承付租赁;总包租

gross circulation 总发行额
gross collector area 集热器总面积
gross column area 总柱面积
gross combination weight 车辆总重(量)
gross commercial area 总商业面积
gross composition 基本成分
gross constant 总常数
gross cost of manufacture 总造价
gross cost of merchandise sold 销货总成本
gross count 总计数
gross-count rate 总计数率
gross coverage 总占地面积
gross credit 总信贷;总收入(量)
gross cross-sectional area 总截面面积;总横截面积;毛截(面)面积;断面总面积
gross culture 总体培养
gross cut 总挖方量
gross damaged value 残损总价
gross deadweight 总载重量
gross declination 跌价总额
gross decontamination factor 总去污系数
gross demonstrated capacity 总容量
gross density 总密度;毛密度
gross discharge 总流量;毛流量
gross dividend 国民总所得
gross domestic investment 国内投资总额
gross domestic product 国内生产总值;国内产品总值;境内生产总值
gross domestic product per year 国内年生产总值
gross draft rate 毛取水率
gross drilling time 总钻孔时间;总钻进时间
gross dry weight 总净重
gross duty(of water) 毛用水率;毛灌水定额;渠首用水率;总用水率;总灌溉率
gross dynamics 普通动力学
gross earning depreciation method 总收入提成折旧法
gross earnings 毛收入;总收入(量)
gross earth pressure 总土压力
gross effect 总效应
gross efficiency 总效率
gross empty weight 总皮重
gross energy 总能量
gross energy intake 采食总能
gross erosion 总侵蚀量;总冲刷(量)
gross error 粗(误)差;显著误差【测】;严重错误;总误差;过失误差
gross error detection 粗差检验
gross estimate 毛估;毛估
gross evapo(u)ration 总蒸发量;毛蒸发(量)
gross examination 肉眼检查
gross exhaust gas 总排气量
gross expenditures 总支出;支出总额
gross export value 出口总额
gross fall 总水头;总的降落(指水头、水压)
gross features 外貌特征;宏观特征
gross filter area 收尘总面积
gross fire 毛火
gross fission product 总裂变产物
gross fixed assets formation 固定资产投资总额
gross fixed capital formation 固定资本形成总值
gross floor area 建筑总面积;建筑毛面积
gross floor space index 总占地指标
gross focusing 总聚焦
gross fog 浓雾
gross forecasting time 总预见期;总预报时间
gross for net 以毛作净;以毛重作净重
gross freeboard 总净空;总干舷高;总出水高度;总超高(坝顶)
gross freight 总运费
gross fuel capacity 总容量
gross generation 总发电量
gross hauled tonnage 总的牵引吨数
gross head 总水头;总落差;毛水头
gross head of dredge pump 泥泵总扬程
gross heat budget 总热量
gross-heat-conductivity 总热传导
gross heat consumption 总热耗
gross heating value 总热值;总发热量
gross heat of combustion 总热值;总燃烧热
gross heat rate 总热耗率;毛热耗率
gross hole sampling 全孔法
gross horsepower 总马力;总功率
gross hypothesis 粗略假设

gross import 进口总值;进口总额
gross import value 进口总值
gross income 总收益;总收入(量);毛收入;收入总额
gross income for distribution 可分配总收入
gross income from sales 销售的毛收入
gross income multiplier 总收入系数
gross income tax 总所得税;毛收入税
gross index 粗索引
gross industrial and agricultural value of output 工农业总产值
gross industrial output value 工业总产值
gross information content 总信息量
gross inspection 肉眼检查
gross interception loss (植物对降水的)总截留损失(量)
gross interest 总利息;纳税前的投资利息收入;毛利息
gross investment 总投资(额);建设资金;(未扣除设备折旧的)毛投资;投资总(金)额
gross investment income 投资毛收入
gross investment in constructional period 建设期投资总额
gross investment of construction project 建设项目总投资
gross irrigation requirement 毛灌溉需水量
gross laden weight 总载货车重
gross leak test 总漏泄试验
gross leasable area 可出租毛面积
gross lease 毛额租金
gross liability 总负债
gross life 总寿命
gross lift (水泵的)总扬程
gross line 总保额
gross liquid mixed cargo samples 总装载液体混合样品
gross lithology estimates 总体岩性估计
gross load 总热量;总装载量;总荷重(量);总荷载;总负载;总负荷;全负载;全负荷
gross load hauled 牵引总重
gross loading intensity 总荷载强度;总负载强度;基底压力
gross loading strength 总荷载强度
gross loss 毛损
gross main 总的
gross margin 销货毛利;毛利(润)
gross margin percentage 毛损率;毛利率
gross margin pricing 毛利定价法
gross margin ratio 毛利比率
gross margin recognition on installment sales 分期付款销售毛利的列计
gross mass 总质量
gross mass of freight container 集装箱额定质量
gross merchandise margin 商品销售总差距
gross misconduct 严重渎职
gross model 有效模型
gross moment of inertia 总惯性模量
gross morphology 一般形态;总体形态学
gross national consumption 国民消费总值
gross national farm product 国民农业生产总值
gross national income 国民收入
gross national product 国民(生产)总值
gross national product per capita 人均国民生产总值
gross national supply 国民总供给
gross negligence 显著的疏忽;重大过失
gross net premium 净保险费总额
gross oil fuel capacity 燃油舱总容量
gross operating head 总运用水头;总运行水头;总使用水头;毛工作水头
gross operating income 总营业收入
gross outlay 总支出
gross output 总产量;总产值;毛出力;毛产量
gross output of cattle breeding 畜牧业总产量
gross output of railway transport industry 铁路运输业总产值
gross output of trade 商业总产出
gross output(value) 总产值
gross parameter 粗参数
gross pay 薪工总额
gross payload 总有效载重量;总有效荷载
gross payroll 工资总额
gross performance characteristic 全节流发动机特性

gross photosynthesis 总光合作用
gross pit sampling 全坑法
gross population density 总人口密度;人口毛密度
gross possible rent 租金总收入
gross power 总功率
gross precipitation 总雨量;总降水量;毛雨量
gross prediction 粗略(的)预计
gross premium 毛保(险)费
gross pressure 总压力
gross price 总价(格)
gross price method 总价(入账)法
gross price method of discount recognition 确定折扣的总价法
gross primary production 总第一级生产量;总初级生产量
gross primary productivity 总初级生产力
gross proceeds 销售总额;总货价收入;售价总数
gross product 总产物;总产品;总产量;矢量积
gross production 总产量;毛产量
gross production rate 总产率
gross profit 总利润;利润总额;毛利(润)
gross profit analysis 销货毛利分析;毛利分析
gross profit margin 毛利率;毛利额
gross profit method 销货毛利法;毛利率法;总利润法
gross profit on sales 销售毛利;销货毛利
gross profit percentage 毛利百分率
gross profit rate 毛利率
gross profit ratio 毛利率
gross profit ratio test 毛利率测试
gross profit test 毛利测算法
gross profit variation 毛利差异
gross project cost 建设项目总成本
gross project income 建设项目的总收入
gross property 总财产
gross pump head 水泵总扬程
gross pumping head (水泵的)总扬程
gross purchases 进货总额
gross rack cross section 拦污栅毛断面
gross radioactivity 总放射性;全放射性
gross radioactivity measurement 总放射测量
gross rail load on axle 轴重
gross rainfall 总雨量;毛雨量
gross rate 总收益率;毛沸率
gross rated capacity 总额定出力
gross rate of increase of domestic animals 牲畜总增率
gross rating 总定额;毛定额
gross raw material 毛量原料
gross receipt 总收入(量);总经收益;收入总额
gross recoverable value of ore 总回收价值
gross reduction ratio 总破碎比
gross register(ed) ton(nage) 总登记吨(位);总注册吨(位);登记总吨位
gross rent multiplier 租金总收入系数;按租金估算地价的乘数
gross reserves 总储(存)量;总储藏量
gross reservoir capacity 总库容;总蓄水量;毛库容
gross residential area 总居住面积
gross residential density 居住(区)毛密度
gross return 毛利
gross revenue 总收益;总收入(量);总收益额;收入总额
gross revenue earning 年总收入;年毛收入
gross room number 总箱位数(集装箱等)
gross rubber 充油充炭黑胶料
gross salaries 薪资总额
gross sales 销售总额;销货总额
gross sample 总样(品);总试样
gross section 总截面;总断面;毛截面;毛断面;全部截面
gross sectional area 断面总面积;总断面面积;毛截(面)面积
gross section yielding 总截面变形
gross segregation 宏观偏析
gross selling price 总售价
gross shipping weight 总载货量;装运总量;毛装运重量
gross shrinkage 总收缩(量);集中性缩孔
gross site area 总(建筑)基地面积
gross soil loss 土壤流失总量
gross solid separator 大块固体分离器
gross spread 毛差额
gross stage efficiency 总级效率

gross start-stop distortion 总起止畸变
gross storage 总库容;总储(存)量;总储藏量;毛库容
gross stress 毛应力
gross stress concentration factor 总应力集中系数
gross stress reaction 总体应力反应
gross structure 总体结构;粗视构造
gross terms 总承付租货;船方不承担装卸费
gross thermal efficiency 总热效率
gross thermal value 总热值
gross thickness 毛厚度
gross thrust 总推力
gross tolerance 总公差
gross ton 总吨;长吨
gross ton-kilometer 总吨公理;总重吨公里
gross ton-kilometer hauled 牵引的总重吨公里
gross tonnage 总吨位;总吨数;毛吨数;登记总吨位;船的总吨数
gross tonnage measurement 总吨位
gross tonnage of freight locomotive hauled 货运机车牵引总重
gross tonnage of ship 船舶总吨数
gross tonne-kilometre worked 工作的总重吨公里
gross torque 总转矩;总力矩
gross total 总计
gross tractive effort 总牵引(能)力
gross tractive force 总牵引(能)力
gross tractive power 总牵引(能)力
gross trade turnover 商业周转总额
gross trading profit 销售毛利;销货毛利;贸易毛利
gross traffic hauled 总拖运量;总牵引运输量
gross trailing load 牵引总(载)重
gross train weight 车辆总重(量)
gross tunnel 毛洞
grossular 钙铝榴石
grossularite 钙铝榴石
gross unit value of ore 矿石单位毛值
gross up 返计还原;补偿费
gross value 毛值;总值
gross value of industrial output 工业总产值
gross vehicle load 车辆总负载(自重在内);车辆总装载量;车辆总重(量);车辆总载重量;总载重;车辆全部载重
gross vehicle weight 总车重;汽车总重;车辆毛重
gross volume 搅拌机总容量;搅拌槽容量;总容量;毛体积;粗材积
gross volume of concrete mixers 混凝土搅拌机总容积
gross wages 工资总额
gross water requirement 总需水量
gross water use 总用水量
gross weight 总重(量);毛重;全重(量)
gross weight distribution 总重分布
gross weight hauled 牵引总重
gross weight of car 车辆总重(量)
gross weight of equipment 设备总重
gross weight of vehicle 车辆总重(量)
gross weight range 总重范围
gross weight terms 毛重条件
gross working capital 总周转资金;总营运资金;流动资本总额;毛流动资本
gross world product 世界生产总值
gross yards 总码数
gross yield 毛收益率
grotesque 怪诞的装饰图形;奇形装饰
grotesque ornament 奇形装饰(品)
Grotthus-Draper law 哥罗特苏斯—德рейpe伯定律
grotto 穴洞;岩穴;岩洞;人工洞室;洞室;洞室(用于避暑或娱乐的人造洞穴);峒室
grotto column 洞穴支柱
grotto of Pythagoras in Cartone 卡尔东的毕达哥拉斯墓穴
grotto plan 溶洞平面图
grotto work 山洞工程
grouan 粗砂
ground 陆地;木针;磨碎的;坯料;土(壤)底(材);场地
ground 210Po survey method 地表钋210法
ground absorption 地面吸收(率);大地吸收
ground acquisition 地面探测;地面检测
ground adhesion 研磨黏附(作用)
ground adhesion capacity 地面附着能力
groundage (船泊的)进港费;停泊费;泊船费
ground air 地面气体;陆空的;地下空气;地空的

ground-air radio frequency 地空通信频率
ground alumin(i)um sulphate 粉状硫酸铝
ground amplification 土(壤)放大作用
ground amplitude 地面振幅;地层幅度
ground anchor 系机地锚;锚杆;土锚;地下锚定装置;地锚;地层锚杆
ground anchorage 地牛
ground-anchorage tendon 地锚预应力钢丝束
ground-anchored 锚着于地面的
ground anchor of bridge cable 桥缆地锚
ground anchors 活动住房泊驻
ground and polished piston 研磨活塞
ground-and-washed chalk 重质碳酸钙
ground apples 石块(俚语)
ground application 地面喷药
ground arc 接地电弧
ground arch 平衡拱;地拱
ground area 房屋及外围占地面积;底层面积;接地面积;建筑工地;地面面积
ground artery 地下水脉
ground asbestos 石棉屑;石棉粉末
ground asphalt storage 地上沥青储仓
ground auger 土钻;地钻
ground avalanche 全层雪崩;大坍方
ground aviation radio exchange 地面航空无线电交换机
ground bar 接地棒;隧道支撑的横档木;大横梁
ground base 土基;地基
ground based 地面的
ground-based astronomy 地面天文学
ground-based beacon 地面雷达信标
ground-based computer 地面计算机
ground-based duct 地沟管道
ground-based electronic omnidirectional satellite communications antenna 地面式电子全向卫星通信天线;地基式电子全向卫星通信天线
ground-based navigation(al) aids 地面导航设备
ground-based observation 地面观测
ground-based optic(al) receiver 地面光学接收器;地面光学接收机
ground-based ozone station 陆基臭氧站;地面臭氧站
ground-based radar 地面雷达
ground-based repeater 地面转发站
ground-based scanning 地面扫描
ground-based scatterometer 地面散射计
ground-based spraying 地面喷药
ground-based station type 地面基站类型
ground-based system 陆基系统;地面系统
ground-based telescope 地面望远镜
ground-based terminal 地面终端(设备)
ground basic slag 碱性炉渣粉;磨碎碱性炉渣
ground batch 粉碎配合料
ground batching plant 泥浆拌和机
ground beam 卧木;基础梁;地(基)梁;枕木;槛木;地(脚)梁
ground beam sleeper 地梁枕木
ground bearing pressure 地基承载力;地基承压力
ground bearing test 地基承载试验
ground bed cable 地床电缆
ground-bedded 放在地上的
ground beetle 爬行甲虫;土鳖虫;地面甲虫
ground blast furnace slag 磨细高炉矿渣
ground bolt 锚栓;地脚螺栓;板座栓
ground-borne vibration 地面产生的振动
ground brace 卧木;槛木;地木;枕木;支撑
ground bracing 杆底加固
ground breaking 开工(尤指土木工程方面);破土(动工)
ground-breaking ceremony 破土典礼;开工典礼;动工仪式
ground break-point 地面倾斜变换点
ground brick 砖粉;碎砖
ground bridge 圆木铺成的路(沼泽地上);(沼泽地上的)圆木路
ground brush 涂底刷;接地电刷;圆头(油漆)刷;底涂刷;粗涂刷
ground bunker 地下储(贮)料槽
ground bus 接地母线;接地总线
ground cable 接地线;锚定链;卧底链;地线;地下电缆
ground cable buoy 水底锚链浮筒;水底锚链浮标
ground calcium carbonate 重质碳酸钙;磨细碳酸钙
ground camera 地面摄影机

ground capacitance 接地电容;对地电容
ground car 高通过性汽车
ground cascade 地下水瀑布
ground casing 百页窗窗框
ground caustic 碱粉;粉状腐蚀剂
ground cementation 地面稳定化;地面固化;岩层灌浆;土壤灌浆
ground cementing 地面浇水泥;地面固化
ground chain 制动链;固定在水底的锚链(供系船用);接地线;接地锚链;锚定链;海底锚链;地链
ground chain buoy 水底锚链浮筒;水底锚链浮标
ground chalk 地白垩;细磨白垩;重质碳酸钙;纯白垩
ground characteristic 地面特性
ground check 地面校正;地面检查
ground check chamber 地面校正室
ground checkout system 地面检查系统
ground check result map of survey area 测区地检结果图
ground circle 基圆
ground circuit 接地回路
ground clamp 地线夹子(电焊时用);接地夹(子);接地端子;地线接线柱;地线夹(头)
ground class 地面类别;地面级别
ground clay brick 砖屑;砖粉
ground clearance 林地清除;离地距离;离地净空;离地(净)高;离地间隙;离地高(度);车架净空;车底净空
ground clearance measurement 离地间隙测定
ground clearance under chassis 底盘净空
ground clearing 田地清除
ground clutter 地物回波;地物反射波;地面杂乱回波;地面反射波
ground clutter area 地面杂乱回波区;地面反射区
ground coat 底涂料;底涂(层);底漆;底层涂料
ground coat(enamel) 底釉
ground coat paint 底涂油漆
ground coefficient 地基系数
ground collapse 地表塌陷
ground colo(u)r 背景颜色;土色;底(涂)色
ground colo(u)r light signal 矮型色灯信号机
ground communication 地面通信;地面交通
ground communication equipment 地面通信设备
ground communication facility 地面通信设施
ground communication system 地面通信系统
ground communication tracking system 地面通信跟踪系统
ground compaction 地基压实
ground compliance 基础柔量;地层适从效应
ground concentration 地面浓度
ground concentration of pollutant 污染物质的地面浓度
ground concentration of pollution 地面污染浓度
ground concentration of pollution material 污染物质的地面浓度
ground condition 地面条件;地基条件
ground condition factor 场地条件系数
ground conductivity 地面电导率;大地电导率;大地导电率
ground conductor 地线
ground conduit 地线管道
ground configuration 地形
ground connection 磨口接头;接地(连接)
ground connection traverse 地面联测导线
ground connector 接地器;地面电源插头
ground consolidation 地面加固;地基固结
ground contact 地线接触点;车轮与地面的接触面
ground contact area 接地面积
ground contacting area of track 履带支面
ground-contour-following device 地面仿形装置
ground control 外业控制;地面制导设备;地面指挥;地面控制
ground control approach 地面控制进场(飞机着陆)
ground control approach system 地面控制渐近系统
ground control center 地面控制中心
ground control equipment 地面控制设备
ground-controlled approach 地面控制进场
ground-controlled approach minimums 地控进场最低数值
ground-controlled space system 地面控制空间系统
ground control point 地面控制点;大地控制点
ground control point survey 地面控制点测量
ground control post 地面控制站

ground control station 地面控制站
ground control survey 地面控制测量
ground control system 地面控制系统
ground control work 地面控制作业
ground coordinates 地面坐标
ground coordinate system 地面坐标系统
ground cork 软木屑
ground correction 地面校正
ground count 路上行车计数
ground course 地面(圬工)层
ground course converter 航向变换器
ground cover 路堑植物护坡;植被(护坡);土地植物;地面覆盖物;地表盖层;地被含量;地被(植)物;地被(地面覆盖);地面覆盖(层)
ground coverage 制图区(域);萌蔽地;航摄范围;地面覆盖(范围);地面覆盖(层);成图区域;成图情况;成图面积
ground cover areas 种植区
ground cover plant 地被植物
ground crack 地面裂缝
ground crew 地勤人员
ground curvature 地面曲率
ground cushion vehicle 气垫车
ground cylinder effect 土坏作用
ground dampness 地基含水量
ground data 围岩的物理力学性质;地面数据
ground data equipment 地面数据设备
ground data handling system 地面数据处理系统
ground data processing system 地面数据处理系统
ground data system 地面数据系统
ground deformation 地面形变,地壳形变
ground deformation anomaly 地形变异常
ground deformation measurement 地形变测量
ground department/national remote sensing center[centre] 国家遥感中心地面部
ground deposit(ion) 地面沉积
ground depression 洼地;地面凹陷;地表下沉
ground-derived navigation(al) data 地面提供的导航数据
ground detector 接地指示器;接地探测器;通地指示器
ground device 接地装置
ground dimension 实地尺寸
ground directional relay 接地定向继电器
ground discharge 云地间放电
ground disinfection 土壤消毒;地面消毒
ground dispensing point 地面配料点
ground displacement 地面位移
ground displacement measure 地中变位测量
ground display system 地面显示系统
ground disposal 埋地处理;地下处置
ground distance 地面距离
ground drain 地下排水;地面排水沟
ground drainage 地面排水
ground drift indicator 地速偏航指示器
ground drilling 地面钻探;地基钻探
ground drive mechanism 行走轮驱动机构
ground-driven 地轮传动的
ground-driven mower 地轮驱动式割草机
ground-driven pickup 地轮驱动的捡拾器
ground-driven rotary harrow 地轮传动旋转耙
ground dry 干打基础
ground dust 土尘
ground echo 地物回波;地面回波;地面反射波
ground echo pattern 地面回波图形
grounded 接地线;接地的;搁浅的
grounded-anode amplifier 阴极输出器;阳极接地放大器
grounded antenna 接地天线
grounded base 共基极;基极接地
grounded-base circuit 基极接地电路
grounded-base connection 基极接地连接
grounded-base equivalent circuit 基极接地等效电路
grounded bridge 接地电桥
grounded cable 地线
grounded circuit 接地电路
grounded collector (接地的)共集(电)极;集电极接地
grounded-collector connection 集电极接地连接
grounded concentric wiring system 接地同轴电缆制
grounded conductor 接地导线
grounded counterpoise 接地平衡网络;接地地网;地网

grounded dock 坞组
grounded emitter 接地发射极;发射极接地
grounded emitter amplifier 共射放大器;发射极接地放大器
grounded emitter connection 发射极接地连接
grounded grid 接地栅极
grounded grid amplifier 接地栅极放大器
grounded hummock 固冰丘
grounded neutral 接地中(性)点;接地中线
grounded neutral conductor 接地中线导体
grounded neutral system 中线接地制
grounded parts 接地部件
grounded plate amplifier 阳极接地放大器;阴极输出器
grounded resistance 接地电阻
grounded rubble 浆砌块石
grounded screen 接地电网;接地屏栅
grounded shield 接地护罩
grounded ship 搁浅船(舶)
grounded system 接地制;接地系统
grounded work 与木砖装连的装修木工;固定在地面上的细木工作;与预埋件连接的细木工作
ground effect 地面效应;大地效应
ground effect boat 冲翼艇
ground effect machine 气垫车,地效机;地面效应器;地面效应机
ground effect vehicle 地面效应气垫运载工具;地面效应车
ground electric(al) locomotive 地面电力机车
ground electrode 地电极
ground elevation 地面高程;室外地坪标高;地面标高
ground elevation point 高程点
ground emitter transistor amplifier 发射极接地晶体管放大器
ground-engaging wheel 驱动行走轮
ground engineer 地质工程师;地勤工程师(飞机)
ground engineering 地面工程;地基工程
ground environment 地面环境;地电设施
ground equalizer 接地均衡器
ground equalizer inductor 接地均衡器线圈;接地均衡电感器
ground equipment 地上设备
ground equipment system 地面设备装置
ground equipment test shop 地下设备测试站
ground experiment 地面试验
ground exploration 地基勘探
ground exploration of high construction 高层建筑物地基勘察
ground fabric 地布
ground facility 地面设施;地面设备
ground failure 地面破坏
ground failure opportunity map 地面破坏机率图
ground fall 地表崩坍
ground fallout plot 地面沉降图
ground fan 落地风扇
ground fault 接地故障
ground fault circuit interrupter 接地故障电路中断
ground fault circuit protection 接地故障线路保护
ground fault interrupter 接地故障断路器
ground fault neutralizer 接地事故消除器
ground fault personnel protection 接地故障人身保护
ground feature 地面特征
ground fertilizer 基肥
ground field 基本域
ground fill 填土
ground filling point 地面装料点
ground fine pitch 地面小距
ground finish 精整加工;磨削(精整)加工;磨光;抛光
ground fire 地下火;地表火
ground fish 底栖鱼
ground fissuration 地裂
ground fissure 地裂除;地裂缝
ground flat 研磨平面
ground floor 一楼;埋地板;地面层;底层(英国楼层叫法);首层(英国)
ground floor area 地面层面积;底层面积
ground-floor building 平房
ground-floor construction 底层构造;底层(地面)构造;低层建筑(物)

ground-floor entrance 底层入口
ground-floor floor 底层地板
ground-floor(ground) plan 底层平面(图)(英国);首层平面(英国)
ground-floor height 底层高度
ground-floor masonry wall 底层砖石墙
ground-floor slab 底层楼板
ground flora 植物群;地表植物区系;地被区系
ground flutter 地面杂乱回波
ground flux 熔剂粉
ground fog 地面雾;低雾
ground form 地形
ground fracturing 地裂
ground frame 地面控框架
ground freezing 地面冻结;地层冰结
ground freezing method 冻结法【岩】地基冻结法
ground frequency 基本频率
ground friction 地面摩擦
ground frost 地面霜
ground frost indicator 地面霜冻指示仪;地面霜冻指示器
ground gained side ways 旁向航摄范围
ground gamma-ray spectrometry 地面伽马能谱测量
ground garbage 磨碎垃圾
ground gate 底孔闸门
ground gear 地面设备(机场);降落设备
ground geophysics prospecting 地面调查
ground glass 磨砂玻璃;磨口玻璃;毛玻璃(屏);玻璃粉
ground glass apparatus 磨口玻璃仪器
ground glass appearance 毛玻璃样
ground glass diffuser 毛玻璃漫射器
ground glass finder 毛玻璃寻像器;方框式取景器
ground glass focusing 毛玻璃片对光
ground glass joint 磨口玻璃接头
ground glass-like 毛玻璃样
ground glass prism 毛玻璃棱镜
ground glass rangefinder 毛玻璃测距仪
ground glass screen 毛玻璃网目片;毛玻璃投影屏
ground glass screen image 毛玻璃屏像
ground glass sounding tube 毛玻璃测管
ground glass stopper 磨砂玻璃塞;磨口玻璃塞
ground glass stoppered bottle 磨口玻璃瓶
ground glass stoppered flask 磨口玻璃瓶
ground glass viewing 毛玻璃检影
ground glaze 底釉;底涂层
ground grain hulls 磨碎的谷壳
ground granulated furnace slag 高炉矿渣粉
ground graphite 粉状石墨;石墨粉
ground-grip tire 高通过性轮胎;带有防滑链的轮胎;钩土轮胎
ground grouting 岩层灌浆;土壤灌浆
ground guard 接地保护
ground guidance system 地面制导系统
ground gypsum 石膏粉
ground handling crane 地面使用起重机
ground handling equipment 地面保障设备
ground handling staff 地勤职员
ground hardening 地基硬化
ground hay 碎干草
ground heat 地(面的)热
ground heaving 地面上升
ground height 地面高度;地面高程;地基高度
ground hob 磨齿滚刀
ground-hog 火车上的制动手;牵引车;上坡牵引车
ground-hog kiln 斜坡地洞窑;艺术瓷窑
ground hold 锚泊属具;锚泊索具
ground hopper 进料沟
ground hose 灌浆软管
ground humidity 土的湿度
ground hydrant 地面消火栓;地面消防栓
ground ice 锚冰【水文】;河底冰;土内冰;地下冰;地面冰;底冰
ground ice mound 土底冰丘;冰丘
ground identification of satellite 卫星地面识别
ground impact 着陆冲击
ground improvement 地基加固
ground inclination 地基坡度;地(面)倾斜
ground indicator 地面指示器
ground information 地面信息
ground information processing system 地面信息处理系统
grounding 底子;搁浅;接地;使船搁浅;船只搁浅

grounding accident 搁浅事故
grounding apparatus 接地设备
grounding bus 接零母线
grounding cable 接地电缆
grounding conductor 接地导体
grounding contact 接地触点
grounding current 接地电流
grounding device 接地装置;接地设施
grounding electrode 接地电极
grounding elevation 地面高程
grounding facility of broadcast continuity 播控中心的地线系统
grounding for lightning 防雷接地
grounding height 地面高程
grounding insulation 主绝缘
grounding jumper 接地跳线
grounding keel 坞座龙骨
ground in glass stopper 磨砂玻璃塞
grounding lead 接地引线
grounding level 地面高程
grounding material 接地材料
grounding metal sheet 接地金属片
grounding network 接地线路;接地系统
grounding of network 电力网接地装置
grounding of pole 磁极接地
grounding on ground floor 一层接地平面图
grounding outlet 接地式电源插座
grounding pad 接地板;接地垫
grounding path 接地通道
grounding percentage 接地百分率
grounding plate 接地板
grounding plug 接地(式)插头;接地插座
grounding reaction 搁座力
grounding receptacle 接地插座;接地插孔
grounding relay 接地继电器
grounding resistance 接地电阻
grounding rod 接地棒
grounding steady tube for car body 车体接地预埋固定管
grounding steady tube for trunk body 车体接地预埋固定管
grounding surge 接地过电压
grounding switch 接地开关
grounding system 接地系统;地线系统;地面系统
grounding terminal 接地线端
grounding transformer 接地变压器
grounding type receptacle 接地型插座
grounding washer 接地垫圈
ground injection test 地下压注试验
ground in joint 磨口接头
ground in oil 油中研细(涂料)
ground in picking up 地面回升值
ground installation 地面装置;地面设备
ground instance 基本例子
ground inversion 地面逆温
ground investigation 地面勘查;地质勘探;地面调查
ground irrigation 地面灌溉
ground isocenter 地面等角点
ground jack 起重机
ground jammer 地面人为干扰发射机
ground joint 金属紧密配合连接;接地连接;磨口接头;磨光接合;无浆连接;砖石砌体干砌;管道阀门接头
ground joist 落地格栅;底层地表格栅;地板栅
ground key valve 光键阀;栓塞阀
ground layer 近地层;土被;贴地层;地表边界层;地被层;底土层
ground laying 上底色
ground laying machine 上色打底机(陶瓷制品);地面铺设机
ground lead 工件引线;接地引线;地面导索滑车
ground lead cable logging 全拖式钢索集材
ground leaf cover 枯枝落叶层
ground leak 地层漏水
ground leakage resistance 漏地电阻
ground lease 土地租约
groundless 无基础的
ground level 接地电平;基态;基(能)级;地平线;地平高度;地面找平;地面高(度);地面测平;地面标高;地电平;地水准面;地平(面)
ground-level car park 地面停车场
ground-level concentration 基级浓度;落地浓度
ground-level concentration distribution 基级浓度分布
ground-level concentration of contaminant 污染物的近地面浓度
ground-level concentration of smoke and soot 煤烟地面浓度
ground-level floor(ing) 地面高程铺设地板;底层地板面层
ground-levele freeway 地平式高速公路
ground-level heliport 地面直升机场
ground-level in mining 矿山地平面
ground-level(l)ing 平整场地;地面水准测量
ground-level loader 落地装载车
ground-level ozone 地面臭氧
ground-level parking lot 地面停车场
ground-level pollution 地面污染
ground-level railroad 地面铁道
ground-level reservoir 地面水库
ground-level source 地面源
ground-level station 地面站
ground-level storage 地面蓄水
ground-level wall 底层墙壁
ground limestone 重质碳酸钙;石灰石粉
ground line 地面线;原地面线;轴线;干线;基线;基础;天然地平线;地平线;地面轨迹
ground-line area 地际表面积
ground-line gradient 自然坡度;地面(自然)坡度;地面(天然)坡度
ground-line treatment 地际处理
ground litharge 黄铅丹
ground litter 枯枝落叶层;地被物
ground load 地面荷载
ground loading point 地面装料点
ground location 地面位置;地面探测;地面测位
ground locomotive 地面用电力机车
ground log 测速锤
ground loop 接地回路
ground lug 抓地爪;抓地板
ground maintenance 地面维修
ground making 整地
groundman 挖土工;铺轨工;接地线电工;地面工作人员;装岩工
ground map 地形图
ground mapping 地图测绘;地图标记
ground mapping radar 测描地面雷达
ground mass 金属基体;基质;基体
ground mast 地面信号机柱
ground mat 接地网;地网;地垫
ground material 磨细材料;粉屑状材料;原始资料;磨细的物料;粉状物料
ground material types studied 研究的地物类型
ground means of transportation 地面运输设备
ground-measured value 地面测量值
ground measurement 地面测量
groundmeter 接地电阻测量仪
ground method 地面方法
ground mica 粉状云母
ground mine 海底水雷;潜埋水雷;沉底雷
ground minute trace gamma survey 地表微迹伽马测量
ground mist 地面薄雾
ground moistening 土壤湿润;土地湿润;地下湿润;地下浸润
ground moisture 土壤湿度(等)级
ground moraine 地堆石;地(冰)碛;底碛(层)
ground motion 地面运动;地(震)动
ground motion array 强震观测台阵
ground motion characteristic 地面运动特征
ground motion parameter 地面运动参数
ground motion strength 地面运动强度
ground mo(u)ld 排水沟模型;地模;倒置隧道模型;土堤堆土模;堆土木模;地面铸型;地坑造型
ground movement 岩层移动;围岩移动;地面形变;地面塌陷;地动;地层运动
ground nadir 地面(天)底点
ground nadir point 地底点
ground navigation 地面导航
ground navigation(al) aids 地面导航系统
ground navigation(al) system 地面导航系统
ground net 固定鱼网;地线网;曳网;拖网
ground net system 地线网络系统
ground noise 背景噪声;原底噪声;地噪声;大地噪声;本底噪声
ground nutshell 碎果壳(堵漏用)
ground object 地物;地面目标;方位物
ground object detection 地面目标探测
ground-object identification 地面(物体)识别
ground oblique-angle 地面倾斜角
ground observation post 地球观测站
ground observation station 地面观测站
ground off saw 磨边锯
ground orientation 地面定向
ground outlet 接地引出线;接地出线座;穿墙出线
ground over 土被
ground oyster shell 牡蛎壳粉;天然碳酸钙;贝壳粉
ground paper 厚纸;纸坯
ground parallel 地面纬圈;地横线
ground-particle velocity 地面质点速度
ground paste 研磨过的色浆;色漆料浆
ground patch area 地面小块面积
ground pattern 底样
ground peg 小桩子【测】;量地木栓
ground penetrating blade 切土部件
ground penetrating radar technique 穿土雷达技术
ground perlite 珍珠岩粉;膨胀珍珠岩粉
ground phenomenon 地面现象
ground phosphorite 磷矿粉
ground photogrammetric survey 地面摄影测量
ground photogrammetry 地面摄影测量(学)
ground photograph 陆摄相片;地面照片
ground photography 地面摄影
ground pigment 粉末填充剂
ground pipe 接地导管;地下管道
ground plan 闸底平面图;零米标高图;平面布置图;水平投影图;地面图;地层平面图;底面图;底层平面(图)(英国);大体方案;初步计划;草案;下层平面图;底层图样
ground plan dimension 地层平面尺寸;底层平面尺寸
ground plane 接地(平)面;接地层;屏蔽面;透视地平面;地线层;地平面
ground-plane antenna 地面天线;接地平面天线;水平极化天线
ground plane plot 水平距离图;地平面图;水平距离显示
ground plan of mine workings 矿区巷道底面图
ground plastering 粉刷基底;粉刷底层
ground plate 卧木;木地板;槛木;地梁;接地导板;埋地板;底板;柱脚垫木;接地铜板
ground plate glass 研磨的平板玻璃
ground plot 平面(地形)图;航测标绘图;地形平面图;地面图
ground plumb point 地面铅锤点
ground point 地面点;地爬式转辙器
ground pole 接地柱;接地极
ground polyonometry 地面导线测量
ground position 地面位置
ground position indicator 对地位置指示器;地面位置指示器
ground position light signal 矮型位灯信号;矮型幻灯信号机
ground potential 大地电位;地电势
ground powder 研磨粉末
ground power house 地面厂房
ground power unit 地面动力装置
ground preference relay 接地保护继电器
ground preparation 整地
ground pressure 岩层压力;接地压力;接地比压;土压或地压;土壤压力;对地压力;地压(力);地面压强;地面(接触)压力;地层压力
ground pressure per unit area 单位面积地基压力
ground price 土地价格
ground probe 地面探测
ground processing equipment 地面处理设备
ground product 研磨产品
ground product gas 地下气化煤气
ground prop 地下(开挖)支撑;地层支撑
ground propagation 地面传播
ground protection 接地保护
ground protection installation 接地保护装置
ground pumice 浮石粉
ground pyramid method 地面锥形法
ground quartz 石英粉
ground quick lime 地面生石灰
ground radiation 地面辐射
ground radiometer 地面辐射仪
ground rainfall 地面雨量;地表降雨量
ground rain(fall) ga(u)ge data 地面雨量资料
ground range 地面距离
ground range image 平距图像

ground range scan 地面距离扫描
ground range sweep 地面距离扫描
ground ray 直接辐射波；地面射线
ground readout station 地面读出站
ground reamer 磨齿铰刀
ground receiver 地面接收装置；地面接收站；地面接收器；地面接收机
ground receiving equipment 地面接收设备
ground receiving station 地面接收站
ground receiving telemetering station 地面接收遥测站
ground reception bunker 地下煤斗
ground recharge 大地再充电
ground reconnaissance 初步踏勘；初步勘察；地面选点
ground recording telemetry station 地面记录遥测站
ground rectangular coordinates 地面直角坐标
ground-referenced navigation(al)data 地面基准导航数据
ground reference navigation 地文导航
ground reflected wave 地面反射波
ground reflecting 地面反射
ground reflection 地面反射
ground refuse 地面垃圾
ground relaying 继电接地保护
ground relief 地貌；地面起伏
ground remote sensing 地面遥感
ground rent 地租；地面租金
ground requirement list 总需求清单
ground research 地面勘查；地基研究
ground resistance 大地电阻
ground resistivity gradient log curve 接地电阻梯度法测井曲线
ground resistivity log 接地电阻法测井曲线；接地电阻测井
ground resolution 地面分辨率；地面分辨力
ground resolution area 地面可分辨面积
ground resonance 地面共振
ground response 地面运动反应
ground return 接地回路；地面(杂乱)回波；地面返回；地面反射；地回路
ground return circuit 大地回路；地回路电路
ground-return system 大地回路制
ground ring 接地环
ground river pollution 地下河流污染
ground rock 磨细岩石；石屑
ground rock powder 石粉
ground rod 接地(金属)棒
ground roll 滑行；地滚波
ground roller (水流的)底部旋滚
ground rule 程序；基础性准则；基本准则；基本规则；基本规定
ground rupture 地裂
grounds 木砖
ground safety lock 防收锁
ground sample 地基试样；研磨样品
ground sample process 磨样法
ground sand 磨细砂
ground satellite station 地面卫星站
ground satellite terminal 地面卫星终端装置
ground scan system 地面扫描系统
ground scatter propagation 地面散射传播
ground scene 地面景物
ground screen 地网；地面屏蔽
ground screw 地脚螺钉
ground sea 长涌浪
ground seat union 密封油任
ground seismic motions 地层动力地震运动
groundsel 岩床；作基础的木材；急湍；急滩；木结构最下部分；潜堰；石滩；地槛；底槛
ground sensor 地温探测器；地貌传感器
ground servicing equipment 地勤设备
ground servicing of aircraft 飞机地勤
ground servitude 地面工务
ground set 地面台
ground set flagpole 竖放旗杆
ground setting 地面沉降
ground settlement 地面沉降；基土沉降；地面下沉；地面陷落；地层陷落；地表塌陷
ground settlement method of artificial control 人工控制地面沉降方法
ground settlement observation 地表沉降观测
ground shade 地色

ground shaker 落地式振动运输机；底板簸动运输机
ground shaking 地振动
ground shaping 地面整形；地面整平
ground sheet 防潮布
ground ship way 固定滑道【船】
ground shoe 仿形滑板
ground sign 地面符号
ground signal 地面信号；矮型信号
ground signaler 矮型信号机
ground signal projector 信号枪；地面信号发射器
ground sill 卧木；槛木；基础板；地梁；地槛；底板；岩床；急湍；急滩；木结构最下部分；潜堰；石滩；地槛；底槛；作基础的木材
ground sill rocked in the bank 潜丁坝
ground sill rooted in the bank 伸入岸中的土坝
ground site of automatic telemetry 地面自动遥测站
ground site of remote sensing 地面遥感试验站
ground sketch 地形略图；地志资料；地面草图
ground skidder 绞盘机
ground skidding 全拖式集材
ground slab 地板
ground slag 磨碎炉渣；渣粉；磨细矿渣
ground slide 载玻片
ground slope 地面坡角；地面坡度
ground slope factor 地面倾斜因素；地面倾斜系数
ground sluice 地沟；地层砂金洗选槽；底部泄水闸
ground-sluice gate 底部泄水闸门
ground sluicing 水力运土(方)
ground slurry 研磨浆料
grounds of railway economic efficiency audit 铁路经济效益审计依据
ground soil 地面土壤
ground solid foundation 地下实地
ground solidification 土壤固化
ground sonic meter 地音仪
ground sources 地面水源
ground-space coordinate system 地面点位坐标系统
ground specific gravity of crude oil 地下原油比重
ground spectral data value 地物波谱数值
ground spectral measurement data 地物坡谱测量数据
ground spectral measurement techniques 地物波谱测量方法
ground spectrum studies 地物波谱研究
ground speed 对地速度；对岸速度；地速；地面上空速度
ground speed control 前进速度调速；前进速度调节
ground speed meter 地速计
ground speed of plane 飞机速度
ground speed regulator 前进速度调节器
ground sprinkling 地面洒水
ground squirrel 地松鼠
ground stability 地面稳定(性)
ground stabilization 围岩稳定处理；地基加固
ground stabilization by cement grouting 用水泥灌浆法稳定土壤
ground starter 地上起动机
ground state 基态
ground state level 基态能级
ground state relaxation 基态弛豫
ground station 无线电岸台；地球站；地面探测站；地面电台
Ground Station Committee 地面站委员会
ground station equipment 地面站设备
ground station identification code 地面站识别码
ground-station interrogator 地面站问答机
ground steel wire 磨光钢丝
ground stereophotogrammetric survey 地面立体摄影测量；地面立体摄影测绘
ground stereophotogrammetry 地面立体摄影测量；地面立体摄影测绘
ground stereoplotter 地面摄影立体测图仪
ground stiffness 地基劲度
ground stone 基(础)石；磨光石料；磨细石料
ground storage 露天堆放场；地下埋存；地下储量；地下储藏；地上仓库
ground storage of water 水的地下储存；地下水储存量；地下水储藏量；地下储水量；地下储水
ground storage pool 地下储放池
ground storage tank 地面储[贮]水池
ground stor(e)y 一层(楼)；地被层；底层(楼)
ground straight-shank twist drill 磨制直柄麻花钻

ground strap 接地母线；接地汇流排
ground streamer 地面流光
ground strength 地层强度；地基强度
ground strengthening 地基加固
ground stress 地应力；地基中应力
ground stress field 地应力场
ground stress in measurement indicator 地应力测量仪器
ground stress measuring hole 地应力测量孔
ground stress measuring method 地应力测量方法
ground strip 狭条工地；接地片
ground subsidence 地面下陷；地面陷落；地面塌陷；地面沉陷；地面沉降因素；地基下沉；地基沉降；地层塌陷；地层塌落；地表沉陷
ground subsidence measuring 地面坍陷量
ground substance of bone 骨基质
ground suitable for mechanical excavation 适用于机械挖掘的土壤
ground supply socket 地面电源插座
ground support 地层支护
ground-support device 地面维护设备
ground-supported insulator 地上绝缘子
ground support equipment 地面保障设备；地面辅助设备
ground surface 地面；研磨面；路面；零位面；毛砂面；地表(面)
ground surface acceleration 地面加速度
ground surface distress 地面开裂
ground surface elevation 地面高程
ground surface line 地面线
ground surface measurement 地面度量
ground surface pulse 地面脉动
ground surface subsidence over tunnel 隧道地表沉陷
ground surface symbolization of salinization 盐渍化地表标志
ground surveillance radar 地面监视雷达
ground survey control 地面测量控制
ground survey(ing) 地形测量；地面测量
ground swell 暴涌；激涌；海涌；浅水长涌；地面隆起；地隆；底涌；底隆；长涌(浪)；岸涛；岸浪
ground swell sea 底涌浪
ground swing 地面反射变化
ground swing error 地面反射误差
ground switch 接地线开关
ground table (墙基、柱基、台基等的)贴地层；地坪高程；基床；地面高程；接地石基层；地基面
ground tackle 锚具；锚定区；锚泊装置；锚泊属具；锚泊索具；海底固定具
ground tackling 锚具
ground tap 磨牙丝锥
ground tap for radiator 磨制暖气片丝锥
ground target 地面目标
ground target detection 地面目标探测
ground target identification 地面目标识别
ground telemetering equipment 地面遥测装置；地面遥测设备
ground telemetering station 地面遥测站
ground temperature 地下温度；地温；地面温度
ground term 基项
ground terminal 照明接地线；接地(终)端；接地螺钉；接地端子；地线接线柱；地面终端(设备)
ground terminal system 地面终端系统
ground test 静态试验；接地试验；土壤试验；地面试验
ground test equipment 地面测试设备
ground-testing plant 地面试验设备
ground test pieces 土样(试块)
ground thermometer 接地温度计；地温计；地温表
ground the tanks 油罐接地
ground throw lever 地面扳道握柄
ground throw stand 地面板道座
groundtight 不漏土的
ground tile 地砖
ground tilling mill 旋耕碎土机
ground tilt 地倾斜
ground timing system 地面定时系统
ground tint 底色调；地纹
ground to air 地对空
ground to air radio station 地空通信无线电台
ground to be filled 还填土
ground-to-cloud discharge 地云间放电
ground to ground 地对地

ground-to-plane radio station 地空通信无线电台
ground torpedo 海底鱼雷
ground-to-sea missile 岸对舰导弹
ground-to-ship missile 岸对舰导弹
ground track 地面航迹;地面轨迹;地面跟踪
ground track camera 对地摄影机
ground tracking equipment 地面跟踪设备
ground track point lock 轨道电路控制的道岔闭锁器
ground track scan 地面航迹扫描
ground track sweep 地面航迹扫描
ground traffic (机场的)地面交通
ground transmitter 地面发射机;地面传感器
ground transportation 地面运输
ground treatment 围岩处理;地基处理;地层处理
ground triangle 地面三角形
ground triangulation 地面三角测量
ground Trinidad épuré with rock flour 含石粉的特立尼达沥青粉
ground truth 地面景物;地面真相;地面验证资料;地面实况(资料);地面实测资料;地面实测数据;地面调查资料
ground truth investigation 地面实况调查
ground truth system 地面实况系统
ground-tunnel structure interaction 地层—隧道结构相互作用
ground-type drinking fountain 立式饮水泉
ground-type Koepe winding 地面式戈培提升
ground-type plug 接地式插头
ground-type twin-rope friction winder 地面式双绳摩擦提升机
ground vegetation 植被;地面绿化;土地植物;地面植被
ground vegetative cover 地面植被
ground vehicle 陆上车辆
ground velocity 对地速度;地速
ground vibration 地面振动;地层振动
ground vibrational level 基础振动水平;地面振动级
ground vibration load 地面振动荷载
ground-viewing satellite 地面观察卫星
ground viscosity of crude oil 地下原油黏度
ground visibility 地面视域;地面能见度
ground wall 基础墙;地墙
ground wall paper 背景墙纸
ground washer 接地垫圈
groundwater 潜水(指地下水);地下水;地层水
groundwater accretion 地下水增量;地下水增加;地下水增长;地下水增补
groundwater accumulation 地下水积储[贮]
groundwater age 地下水年龄
groundwater age measurement 地下水年龄测定
groundwater aquifer 地下水含水层
groundwater artery 地下水(干)脉;地下水干道
groundwater attack 地下水侵蚀
groundwater balance 地下水(量)平衡;地下水均衡
groundwater balance calculation 地下水均衡计算
groundwater barrier 地下水隔水层;地下水阻挡层
groundwater basin 地下蓄水库;地下水蓄水池;地下水盆地;地下水流域
groundwater behavio(u)r observation 地下水动态观测
groundwater blow 地下涌水
groundwater body 地下水体
groundwater brine 地下水盐水
groundwater budget 地下水资源总表;地下水量平衡计算(表);地下水(量)平衡;地下水均衡;地下水出入计算
groundwater capacity 潜水储量;地下水蕴藏量;地下水储量
groundwater capture 地下水袭夺
groundwater cascade 地下水梯级;地下水瀑布;地下水落差;地下水跌落;地下水跌水;地下水跌差
groundwater catchment 地下水流域
groundwater chart 地下水位图
groundwater chemical index 地下水化学指数
groundwater chemical quality 地下水化学质量
groundwater chemistry 地下水化学
groundwater chemistry map 地下水化学图
groundwater circulation metallogenic model 地下水循环成矿模式
groundwater circulation system 地下水循环系统
groundwater collecting construction 地下集水建筑物
groundwater compartment 潜水含水层

groundwater component 地下水成分
groundwater condition 地下水状况
groundwater connectivity test 地下水连通试验
groundwater conservation 地下水保持
groundwater consumption rate 地下水消耗量
groundwater contaminant 地下水污染物(质)
groundwater contaminant sources 地下水污染源
groundwater contamination 地下水沾污;地下水污染
groundwater contour 地下水位等值线;地下水(位)等高线;地下水等水位线
groundwater control 地下水控制;地层水控制
groundwater corrosivity 地下水侵蚀性;地下水腐蚀性
groundwater dam 地下水堰;地下水坝
groundwater dating 地下水年龄鉴定
groundwater decrement 地下水减退
groundwater deflector 地下水变流装置
groundwater depletion 地下水消落;地下水消耗;地下水量枯竭;地下水亏耗;地下水耗损;地下水耗竭;地下径流衰竭
groundwater-depletion curve 地下水消退曲线;地下水耗损曲线;地下水亏耗曲线
groundwater deposit 地下水沉积;地下水沉淀
groundwater depression 地下水位下降
groundwater depth 地下水埋深;地下水深度
groundwater development 地下水开发
groundwater direction 地下水流方向
groundwater disaster in mine 矿井地下水灾害
groundwater discharge 涌水量;地下水涌水量;地下水溢出量;地下水排泄;地下水排出;地下水出流(量);地下水喷分
groundwater discharge area 地下水溢出区;地下水排泄区;地下水流出面积;地下水出流区;地下水过水面积
groundwater disinfection rule 地下水消毒规则
groundwater divide 潜水分水岭;潜水分界线;潜水分界面;地下水分线;地下水分岭;地下水分界;地下分水线;地下分水岭
groundwater drain 地下水排水管;地下水排水沟
groundwater drainage 地基排水;地下水排除;地下水排出
groundwater drainage pump 地下水抽水泵
groundwater drain pipe 地下水排水管
groundwater drawdown 地下水水位泄降;地下水位下降;地下水水位降深;地下水水位降落;地下水面降落
groundwater drawdown continuously 地下水位持续下降
groundwater drawdown pump 地下排水泵
groundwater dynamics 地下水动力学
groundwater dynamics formula 地下水动力学公式
groundwater dynamics method 地下水动力学法
groundwater elevation 潜水位;潜水标高;地下水(水)位;地下水高程
groundwater environment 地下水环境
groundwater environmental background value 地下水环境背景值
groundwater environmental quality assessment 地下水环境质量评价
groundwater equation 地下水(平衡)方程
groundwater evapo(u)ration 地下水蒸发
groundwater ever development 地下水超采
groundwater exploration 地下水勘测
groundwater extraction 地下水提取
groundwater fall line 地下水瀑布线;地下水坡降线
groundwater-fed lake 地下水补给湖
groundwater feed 地下(水)补给
groundwater floor 地下水层
groundwater flow (河流流量中的)地下水流量;潜流;地下水流(动);地下(水)径流
groundwater flow-rate 地下水流量
groundwater fluctuation 地下水位升降变化;地下水面波动
groundwater forecast 地下水情预报
groundwater for use in building 建筑用地下水
groundwater geology 地下水地质学
groundwater gradient 地下水面坡降;地下水(面)坡度
groundwater hardness 地下水硬度
groundwater head in curtain grouting heavy 帷幕内地下水头
groundwater head outside the grouting heavy curtain 帷幕外地下水头

groundwater hill 潜水丘;地下水丘;地下水脊(线)
groundwater hole 地下水坑
groundwater horizontal zoning 地下水水平分带
groundwater hydraulics 地下水水力学
groundwater hydrograph 地下水位图;地下水基流过程线;地下径流过程线
groundwater hydrologist 地下水水文学家
groundwater hydrology 地下水水文学
groundwater hydrothermal solution 地下水热液
groundwater hygiene 地下水卫生学
groundwater hypersalinity 高盐度地下水
groundwater in alluviall-pluvial piedmont plain 山前冲洪积平原地下水
groundwater in alluvial plain 冲积平原地下水
groundwater in apron plain 冰积平原地下水
groundwater in bedrock 基岩地下水
groundwater increment 灌注量;地下水增量;地下水增加;地下水增长;地下水增补;地下水补给量
groundwater in desert 沙漠地下水
groundwater index 地下水指数
groundwater indicator 地下水指示器
groundwater in eluvial-cliff debris 残坡积地下水
groundwater infiltration 地下水渗流
groundwater inflow 地下水入流;地下水流入量
groundwater in intermontane basin 山间盆地地下水
groundwater in lacustrine plain 湖积平原地下水
groundwater in littoral plain 滨海平原地下水
groundwater in loess platform 黄土台塬地下水
groundwater in Mesozoic-Cenozoic basin 中新生界盆地地下水
groundwater in old course 古河道地下水
groundwater in red bed 红层地下水
groundwater intake 地下水进水口
groundwater inventory 地下水储量;地下水资源总表;地下水(水量)平衡计算表;地下水库存清单;地下水地量
groundwater investigation 地下水调查;地下水勘测
groundwater irrigation 井水灌溉;地下水灌溉
groundwater isobath 地下水位等深线
groundwater isopiestic line 地下水等位线
groundwater isotopic composition 地下水同位素成分
groundwater law 地下水法
groundwater level 潜水面(指地下水);潜水位;地下水(水)位;地下水(水)面;地下水高程
groundwater level fluctuation 地下水位波动;地下水位变动
groundwater level line 地下水位线
groundwater line 潜水位线;地下水位线;地下水流线
groundwater lowering 降低水位(指地下水);降低地下水法;降低地下水位;地下水位下降;地下水位降低;地下水面降落;地下水降低
groundwater lowering installation 降低地下水位的装置
groundwater lowering works 地下水水位降低工程
groundwater management 地下水管理
groundwater management system 地下水管理系统
groundwater map 地下水分布图
groundwater mapping 地下水分布图
groundwater mineralization 地下水矿化
groundwater mining 地下水抽取;地下水超采
groundwater model 地下水模型
groundwater modelling 地下水模拟
groundwater monitoring 地下水控制;地下水监测
groundwater monitoring network 地下水监测网
groundwater monitoring system 地下水监测系统
groundwater mound 潜水丘;地下水壅高;地下水丘;地下水面穹起;地下水岗
groundwater movement 地下水运动
groundwater movement determination 地下水运动测定
groundwater net 地下水网
groundwater observation station 地下水观测站
groundwater observation well 地下水观测井
groundwater of the upper zone 上层地下水
groundwater organic contamination 地下水有机污染
groundwater organic pollution 地下水有机污染
groundwater origin 地下水成因
groundwater origin type 地下水成因类型

groundwater outcrop 地下水露头
groundwater outcrop type on image 影像地下水露头类型
groundwater outflow 地下水溢出;地下水流(出)量;地下水出流(量);地下出流
groundwater overdevelopment 地下水过量开发
groundwater overdraft 过度抽取地下水;地下水过度抽取;地下水超抽;地下水超采
groundwater permeation fluid mechanics 地下水渗流力学
groundwater physics 地下水物理(学)
groundwater piracy 地下水截夺;地下水夺流
groundwater plane 地下水面
groundwater plume 地下水污染带
groundwater podzol 潜水灰壤
groundwater podzolization 潜水灰壤作用
groundwater pollutant 地下水污染物(质)
groundwater pollution 地下水污染
groundwater pollution assessment 地下水污染评价
groundwater pollution cause 地下水污染原因
groundwater pollution degree 地下水污染程度
groundwater pollution fashion 地下水污染方式
groundwater pollution forecasting 地下水污染预测
groundwater pollution map 地下水污染图
groundwater pollution model 地下水污染模型
groundwater pollution monitoring 地下水污染监测
groundwater pollution pathway 地下水污染途径
groundwater pollution potential 地下水污染势
groundwater pollution source identification 地下水污染源识别
groundwater pollution sources 地下水污染源
groundwater pollution transport 地下水污染迁移
groundwater pollution type 地下水污染类型
groundwater pollution zone 地下水污染带
groundwater power plant 地下水(力)发电厂
groundwater pressure 地下水压力
groundwater pressure head 地下水压力水头;地下水压头
groundwater profile 地下水位剖面图
groundwater protection 地下水保护
groundwater protection district 地下水保护区
groundwater protection law 地下水保护法
groundwater protection policy 地下水保护政策
groundwater protection standard 地下水保护标准
groundwater province 水文地质分区;地下水区(域);地下水(分)区
groundwater quality 地下水水质;地下水水质
groundwater quality assessment 地下水质量评价
groundwater quality deterioration 地下水质恶化
groundwater quality management 地下水质管理
groundwater quality map 地下水质图
groundwater quality model 地下水(水)质模型
groundwater quality modeling 地下水质模拟
groundwater quality monitoring 地下水质监测;地下水水质监测
groundwater quality monitoring network 地下水质监测网
groundwater quality prediction 地下水质预测
groundwater quality protection 地下水水质保护
groundwater quality standard 地下水质(量)标准
groundwater quality suitable 适用地下水水质
groundwater quality testing method 地下水质检验法
groundwater quality variable 地下水质变化
groundwater quantity 地下水量
groundwater recession 地下水消退曲线;地下水消落;地下水消耗;地下水下降;地下水退水;地下水亏损
groundwater recession curve 地下水消耗曲线;地下水位下降曲线;地下水亏损曲线;地下水退水曲线
groundwater recession hydrograph 地下水消落过程线;地下水亏耗过程线
groundwater recharge 灌注量;地下水重蓄;地下水(再)补充;地下水回灌(量);地下水灌注;地下水补给(量)
groundwater regime 地下水系统;地下水情(况);地下水机制;地下水动态
groundwater regime elements 地下水动态要素
groundwater regime in mountain area 山区地下水动态

groundwater regime in permafrost region 多年冻土区的地下水动态
groundwater regime in plain area 平原区地下水动态
groundwater regression 地下水消落
groundwater-related hazard 地下水公害
groundwater replenishing 地下水补充
groundwater replenishment 灌注量;地下水重蓄;地下水(再)补给
groundwater reserve 地下水储量
groundwater reserve condition 地下水赋存条件
groundwater reservoir 地下水蓄水库;地下水蓄水池;地下水(水)库;地下水储水层;地下含水层
groundwater resource development 地下水资源的开发
groundwater resource management 地下水资源管理
groundwater resource management model 地下水资源管理模型
groundwater resource model 地下水源模型
groundwater resources 地下水资源;地下水蕴藏量;地下水储藏量
groundwater resources evaluation 地下水资源评价
groundwater restoration 地下水恢复
groundwater rice soil 潜育水稻土
groundwater ridge 脊形地下水面;地下水穹丘;地下水隆起;地下水脊(线);地下水分水线
groundwater run 潜水源水河
groundwater runoff 基流;地下水流(量);地下水径流;地下水出流
groundwater sample 地下水样本
groundwater sampling 地下水采样
groundwater sampling point 地下水采样点
groundwater saturation zone 地下水饱和带
groundwater seep 地下水出口
groundwater seepage 地下水渗漏;地下水渗透
groundwater shed 地面水流域
groundwater soil 潜育土;潜水土
groundwater solution 地下水溶解
groundwater source amount 地下水资源量
groundwater sources 地下水(水)源
groundwater stage 地下水水位
groundwater station 地下水站
groundwater steady flow 地下水稳定流
groundwater storage 潜水储量;地下蓄水量;地下水储[贮]存量;地下水蕴藏量;地下水蓄水量;地下水储(水)量;地下水储存量;地下水储藏量;地下储水量
groundwater storage curve 地下水蓄水曲线;地下水储量曲线
groundwater storage element 地下水储[贮]存要素
groundwater stream 阴河;地下水流(河)
groundwater supply 地下水供应;地下水(水)供;地下水供给;地下给水
groundwater surface 潜水面(指地下水);地下水位(表面);地下(水)水面
groundwater surface lift 地下水位抬高
groundwater survey 地下水调查
groundwater system 地下水系统
groundwater table 潜水面(指地下水);地下水位;地下水面
groundwater table(contour) plan 地下水位等高线图
groundwater table fluctuation 地下水位升降;地下水波动
groundwater table rise 地下水上升
groundwater tank 地下水箱
groundwater tapping 地下水流出口
groundwater temperature 地下水温
groundwater tracer 地下水示踪剂;地下水流动示踪物
groundwater tracer test 地下水示踪试验;地下水跟踪试验
groundwater transmissibility 地下水导水系数
groundwater trapping 地下水截流
groundwater treatment 地下水处理
groundwater treatment plant 地下水处理厂
groundwater treatment system 地下水处理系统
groundwater trench 地下水位槽陷;地下水沟;地下水槽
groundwater trenching 地下水面凹陷
groundwater turbidity 地下水浑浊度

groundwater turbulent flow 地下水紊流
groundwater type after intrusion 入侵后的地下水型
groundwater type before intrusion 入侵前的地下水型
groundwater unsteady flow 地下水非稳定流;地下水不稳定流
groundwater vein 地下水脉
groundwater velocity measurement 地下水流速测定
groundwater vertical zoning 地下水垂直分带
groundwater wave 地下水波
groundwater works 地下水工程
groundwater yield 地下水出水量;地下水产(水)量
groundwater zone 地下水区;地下水层
groundwater zoning in alluvial-pluvial fan 冲洪积扇潜水分带
ground wave 直接辐射波;地震波;地面电声波;地面电光波;地面(电)波;地(层)波
ground wave operating distance 地波作用距离
ground wave propagation 地波传播
ground wave radio signal 地波无线电信号
ground wave to sky wave correction 地波配合天波改正量
ground ways 下水滑道;造船滑道;底滑道
ground weather radar station 地面天气雷达站
ground weight sinker 沉块
ground wheel driven hay loader 地轮驱动的干草装载机
ground-wheel driven machine 地轮驱动的机具
ground white lead(in oil) 铅白粉(调油漆用)
ground whiting 重质碳酸钙
ground wind 地面风
ground wind indicator 地面方向仪
ground wire (喷射混凝土施工时的)基准钢丝;定位线;照准地线;接地(电)线;地线;避雷(地)线
ground wire carrier communication 地线载波通信
ground-wire gradient 架空地线电位梯度
ground-with-shirt mounting 带裙板地面固定件
ground-with-supports mounting 基座嵌固
groundwood decker 磨木浆浓缩机
groundwood fiber 磨木浆纤维
groundwood filled board 磨木浆芯层纸板
groundwood paper 木浆(制)纸
groundwood printing paper 磨木浆印刷纸
groundwood pulp 磨木浆
ground wood waste-water 磨木纸浆废水
ground work 铺设挂瓦条;垫层铺设;基础(工程);基本工作;挂瓦条;基本成分;土方工程
ground-working disk blade 切土圆盘刀
ground-working equipment 土壤耕作机具
groundwork management 土方调配
ground yarding 地曳式集材机
ground zero 地面零点
group 组;族;集团;群;团体;班组
group 3 demountable cargo container 第三组拆卸式货物集装箱
group accounts 集团账户;分类账户
group action 桩作用;组合作用
group action of piles 桩群作用;群桩作用
group active power regulating device 有功功率成组调节装置
group activity 分组活动
group additive utility function 群可加效用函数
group address 组地址
group addresses 群地址
group addressing 成组访问;成组寻址
group address message 群地址信息
group advance send 成组超前发送
group after group at different time 分期分批
groupage 零星货混装运输
groupage bill of lading 拼箱提单;成组提单
groupage shipment 集装运货
groupage system 拼箱制(航运);编组制【铁】
groupage traffic 拼箱运输
groupage wagon 直达整装零担货车;分组装车
group agglutination 群凝集
group algebra 群代数
group allowance 群体限额;分组限额
group alternating light 联互光
group ambient temperature 群环境温度
group amplifier 组合放大器;群放大器
group analysis of asphalt 沥青组分分析
group-and-cluster housing 成组成团住宅

group annuity 团体年金;分组年金
group antenna 群天线;分组天线
group arrival time 群到时
group assembly parts 组装零件
group atmosphere 团体气氛
group automatic callback 组群回叫
group automatic operation 组合式自动操纵
group average 组平均值
group averaging 群上平均
groupband modem 宽带调制解调器
group bandpass filter 群路带通滤波器
group banking 集团银行业
group battery 电池组
group behavio(u)r 群体行为;团体行为
group bidding 团体投标
group block 复式滑车
group bonus 团体奖金
group-bonus wage plan 小组奖金工资制
group booking 团体预订
group bus 组合母线
group busying facility 群忙设备
group busying hour call 群忙时呼叫
group button 成组按钮
group calculation 分群计算
group call 群呼(叫)
group calling 组呼;群呼(叫)
group carrier frequency 群载频
group carrier supply bay 群载频供给架
group carry 小组进位;成组进位;分群进位
group casting 底铸(法)
group center 中心局;长途电话局
group character 组分隔符;群特征;分组标识
group charger 电池组充电器
group check 成组检查
group circle graticule 接目象限标线
group classification code 归组分类码
group code 组符号;群码;分组码
group coded record(ing) 成组编码记录
group code record 群码记录
group cohesiveness 团体内聚力
group comparison 分组比较;分类比较
group compensation 组补充
group component 组元
group connection 成组连接;群终接
group connector 群终接器;群连接器
group constant 群常数
group control 群控(制);分组控制
group control change 群控变换
group control panel 群控制屏
group converter 合群转换器
group costing 分类成本计算
group cost system 分组成本制度
group cross-section 群截面
group cutting 群状母树作业法
group data 分组数据;分析资料
group decision-making 群决策;集体决策
group degenerate modes 群简并波
group delay 群延迟;群时延
group delay characteristic measuring equipment 群延迟特性测量装置
group delay correction 群时延校正
group delay data 群时延数据
group delay distortion 群延迟失真;群时延失真
group delay distortion correction 群时延失真校正
group delay equalization 群时延均衡
group delay equalizer 群延迟均衡器;群时延均衡器
group delay frequency 群延迟频率;群时延频率
group delay-frequency characteristic 群时延频率特性
group delay measuring equipment 群延迟测量设备
group delay measuring set 群时延测试仪
group delay response 群时延响应
group delay spread 群时延展宽
group delay time 群延迟时间;群时延(时间)
group delay variation 群时延变化
group demodulation 群解调
group demodulator 群解调器
group demodulator filter 群解调滤波器
group demultiplexer 群多路分解器
group depreciation 分组折旧;分类折旧
group detector 群检测器
group determinant 群行列式
group dialing 群拨号

group diffusion method 群扩散法;分群扩散法
group digital check 成组数字检查;分群数字检查
group-directional characteristic 群方向特性
group disability insurance 团体残废(保)险
group discount 团体折扣
group discussion 分组讨论
group distress alerting 群遇险报警
group divisible design 分组设计
group divisible incomplete block design 分类不完全区组设计
group divisible rotable design 分组旋转设计
group drilling 多孔底钻进;丛式钻进
group drive 组合传动;分组运转方式
group driving 屏风式打桩法;成组传动
group dwelling 住宅群
group dynamics 集团动态;集体动力学;团体动态学;分组动态特性
group dynamics theory 团体动力学理论
group dynamics training 团体动力训练
group echo 群回波
grouped 化合的
grouped bar chart 分组条状图
grouped bar graph 分组条状图
grouped columns 集束柱;群柱
grouped commercial district 商业集中区;商业密集区
grouped comparison design 组间比较设计
grouped controls 组合控制装置
grouped data 分组资料;分组数据;分类资料;分类数据
grouped distribution 分组分布
grouped docks 码头群;坞组;成组港池
grouped financial statement 财务报表的分组合并
grouped fire zone 分组防火区
grouped frequency 分组频率
grouped-frequency operation 组合频率制
grouped frequency series 分组频率数列
grouped joint 会接
grouped meters 集中设置用户水、电表
grouped pass 分组穿综法
grouped pilasters 组壁柱
grouped pulse generator 脉冲群振荡器
grouped record 成组记录;分组记录
grouped regression model 分组回归模型
grouped sequential inspection 分组序列检查
grouped sequential sampling inspection 分组逐次抽样检查
grouped site 住宅组群用地;连片用地【建】
group effect 群体效应
group effect of anchors 群锚效应
group efficient 群桩效率
group energy 群能
group energy interval 群能间隔
group enterprise 集体事业(单位);集体企业
group equation 群方程
group extension 群扩张
group factor 群体因子
group fan layout 扇形集束布置
group filter 群滤波器
group financial statement 分组合并财务报表
group-flashing light 组闪灯;连续闪光灯;群合闪光灯;闪光组;联闪光;短联闪光
group forming criterion 分组原则
group foster home 儿童领养所
group frequency 基团频率;群频(率)
group garage 集合停车库
group goal 团体目标
group grilling pumping 群孔抽水
group heating 联片供暖;分区供暖
group house 联立住宅群;集体宿舍;合租住宅;住宅组群;成组住宅
group housing development 群体住房建设
group housing project 合租住房建设项目
group hunting 自动寻找
group I demountable cargo container 第一组拆卸式货物集装箱
group incentive plan 小组奖励计划;团体奖工计划
group in charge of organizing the sources of goods traffic at station 车站货源组织小组
group index 组指数(路基土壤分类);组指标(路基土壤分类);组合指数;群折射率;分组指数;分类指数;分类指标;分层指数
group index method 分组指数法;分组指标法
group indication 组号;分类标志

groupiness factor 群性因子
grouping 集团;车群;组群方式;归组;集聚;集合法;使成组;成组;分组;分群;分类
grouping and internationalization of the economy 经济集团化
grouping by destination 按目的地编组【铁】
grouping error 分组误差;分群误差
grouping for homogeneity 同质性分类
grouping information 分组信息
grouping interval 组距;分组区间
grouping key 并席键
grouping lattice 分组格
grouping method 分组法
grouping of commodities 商品组合
grouping of discriminant 判别归组
grouping of equations 方程分组;方程分群
grouping of hump yard 驼峰调车场头部;驼峰编组场头部
grouping of population 人口分组
grouping of radio beacon 无线电航标台组
grouping of records 记录编组
grouping of township enterprises 乡镇企业集团化
grouping of tracks 线路分组
grouping of variables 变量的分组
grouping of wells 井群布置;丛式钻井
grouping piles 群桩
grouping probability 归组概率
grouping procedure 分组法
grouping sample 分类样本
grouping selection 选组
grouping sheet 分组认表;分类单
grouping system 分组体系
grouping tile drainage system 集组瓦管排水系统
group in periodic(al) table 周期表属
group instruction 分组教学
group insurance 团体保险
group insurance premiums 团体保险费
group integration 团体积分
group interaction 群体交叉
group interrupted quick flashing light 联断急闪光
group interval 道间距;分组段
group interview 小组访谈
group item 组合项;成组项
group iteration 群迭代法
group iterative method 群迭代法;成组迭代法
group knife 组合闸刀
group lead track 溜放线
group life insurance 团体人寿险
group light 组灯(航);集体照明;分组照明
group line 成组生产线
group line switch 区组预选器;分组预选器
group mark 信息组标记;组符号;群标志;组标记
group marking relay 群信号继电器
group-matrix 群矩阵
group mean 组平均值
group mean latitude 组平均纬度
group method 组丛法(分析沥青用);组合加工法;群状采伐
group method of data handling 成组数据处理法
group method of depreciation 分组折旧法;分类折旧法
group milling cutter 组合铣刀
group mind 集团心理
group model 多家厂商共同设计的类型;群体模型
group modem equipment 群调制反调制设备
group modification 群体变异
group modulation 群调制
group modulator 群调制器
group moral 集团道德
group motion 群动
group multilinear utility function 群体多线性效用函数
group multiplexer 群多路复用器
group name 组名
group nozzle governing 分组喷嘴调节
group number 组号;潮群编号;分类指标
group number of analysis and experiment samples 分析试验样品组数
group number of test 试验组数
group number of testing soil sample 试验土样组数
group number of test well 试验孔组号
group occulting light 群体隐蔽照明;遮蔽组灯;联明暗光;明暗灯(光组);联顿光

group of bars 钢筋束
group of basin 盆地群
group of beds 岩层群
group of bins 筒仓组;筒仓群;储[贮]仓组
group of buildings 建筑组群
group of bulk plants 罐区
group of conductors 导体组
group of contaminants 污染物类
group of depot 罐区
group of drawing 冲压级别
group of economic experts on air pollution 空气污染问题经济专家组
group of experts on the scientific aspects of marine pollution 海洋污染科学专家组
group of faults 断层组
group of heads 磁头组
group of hump yard tracks 驼峰调车场线群
group of invariants 常数组;不变量组
group of islands 群岛
group of large-diameter vertical piles 大直径直柱群
group of layers 层组(路面结构)
group of lines 线束【铁】;线群
group of loads 一组荷载
group of locks 船闸组
group of phenotype 表现型群
group of piles 桩组;桩群;群桩
group of pulleys 滑轮组
group of ranking piles 斜桩群
group of silos 储[贮]仓组
group of spheric(al)tanks 球形储[贮]罐组群
group of springs 泉群
group of stations 罐区
group of tank farms 罐区
group of tanks 槽组
group of tracks 线束【铁】;线群
group of transformation 变换群
group of turnouts 道岔组
group of two observations 双观测组
group of waves 波群
group of wells 井群;群井
groupoid 广群
group order 分类定货
group out 成组输出
group passenger 团体旅客
group payment system 集体付款制
group phenomenon 群现象
group piece work system 班组计件制
group pile effect 群桩效应
group pilot frequency 群导频
group planting 成群丛植;组植;群植
group precipitant 组沉淀剂
group precipitation 组沉淀
group preference assumption 集团偏好假设
group printing 成组印刷;成组打印
group processes 团体相互影响过程
group propagate 组传送
group propagation time 群传播时间
group property 组特征;群体特性;团体特征
group purchase 组合购买
group quarantine 集体检疫
group quarters 集体住宅;集体住所
group quick light 联快闪光
group rapid transit 集体快速交通
group rate 分区运费率
group rate system 分组运率制度
group rationality 群体合理性
group reaction 组反应
group reactive power regulating device 无功功率成组调节装置
group reagent 类别试剂;分组试剂
group record 组记录;组合记录
group redundant technique 分组冗余技术
group refractive index 群折射率
group regulation 成组调节
group relationship 集体关系
group relaxation 群松弛;成组松弛
group relay 群继电器;群继电器
group relief 团体弥补免税法
group repetition frequency 组重复频率
group repetition interval 组重复周期;组重复间隔;群重复周期
group repetition period 组重复周期
group replacement 组换

group representation 群表示
group reservation 团体预约
group response 群体反应
group retarder 线束减速器;线束缓行器
group ride 集合乘坐
group sales company 集团销售公司
group sampling 分层取样;分组抽样;分群抽样;分层抽样;分层采样
group selection 组选择;群选(择);群体选择
group-selection cutting 群状择伐
group-selection felling 群状择伐
group-selection system 群状择法作业法
group selectivity 类选择性
group selector 选组器;组选择器;群选择器
group separation 组分离
group separator 组分隔符;成组分离符;分组分隔符
group separator character 群分离符
group sequential sampling 分组序贯抽样
group sequential sampling plan 分组连续抽样法
group settlement 群体沉陷
group settling 群体沉降
group shelterwood cutting 群状择伐
group shelterwood felling 群状择伐
group short flashing light 群短闪光
group shot 群摄;全摄;全景
group signal(l)ing equipment 组信号设备
group spring 弹簧组
group staggered parking 成组交错停车
group stand 展示板组合
group standard 团体标准
group standing mo(u)ld 成组立模
group starter 分组起动器
group starting singal(l)er 线群出站信号机
group study 群组调查研究
group suppression 群抑制
group survival 集群生存
group switch 组开关
group symbiosis 群体共生
group synchronization 群同步
group system 组合制;组合系统;群状采伐方式作业
group technology 组合工艺学;群组技术;成组技术;成组工艺学
group terminal equipment 群终端设备
group term life insurance 团体定期人寿保险
group theory 群论
group time 群时
group tourism 团体旅游
group transfer polymerization 基团转移聚合反应
group translating equipment 群变换设备
group translation 群频转译
group travel 团体旅游
group type venting system 分组泄气系统
group utility function 集团效用函数;集体效用;群体效用函数
group value function 集团值函数;群体值函数
group valve 组合阀
group variation 集群差异
group velocity 群速(度);波群(传播)速度
group velocity dispersion value 群速度频散值
group velocity method 群速度法
group velocity of waves 波群速度
group vent 组合排气管;组合透气管
group very quick light 联甚快闪光
group vibration 基团振动
group visa 团体签证
group ware 群件
group wave velocity 群波速度
group work 小组集体工作
grouser 轮爪;履刺;履齿;临时(定位)桩;锚定桩;定位桩;防滑凸纹(轮胎);防滑齿片(履带)
grouser bar 链条抓地板;履带销
grouser boxes 履带式盖板
grouser plate 履带轮爪;履带齿片;履板
grouser shoe 履带链板;带履刺履带板
grouser track 履带
grout 修饰涂料;注浆材料;浆液;灰浆;砂浆;薄胶泥;薄(灰)浆
groutability 可灌性
groutability ratio 可灌比
grout acceptance 吸浆量;吃浆量
grout-acceptance rate 水泥浆吸入速度
grout agent 注浆材料
grout agitator 泥浆搅拌器;泥浆搅拌机
grout anchored deformed bar 灌浆锚定的螺纹钢杆

grout blanket 灌浆层
grout box 灌浆盒;压力灌浆盒
grout cap 灌浆帽
grout car(t) 运灰浆车;灰浆(小)车
grout casing 水泥浆封固套管
grout cell 灌浆盒
grout compartment 灌浆(区)段
grout concrete 灌浆混凝土
grout consistency 压浆稠度;注浆稠度;水泥浆稠度;砂浆稠度
grout consistency meter 灌浆用砂浆稠度仪;砂浆稠度仪
grout consumption 耗浆量;吃浆量
grout core 水泥黏结岩芯
grout curtain 灌浆帷幕;灌浆屏幕;(在一排孔中灌水泥形成的)水泥屏障
grout duct 灌浆孔道
grouted aggregate 集料灌浆
grouted aggregate concrete 预填集料混凝土;灌浆混凝土;预填骨料混凝土;骨料灌浆混凝土
grouted alluvium 用灌浆方法固结的冲积层;灌浆(固结的)冲积层
grouted alluvium cut-off wall 冲积层灌浆截水墙;灌浆冲积土式截水墙
grouted anchor 灌浆锚杆
grouted annulus 灌浆的环形筒体;灌浆的环形空间
grouted area 灌浆区
grouted asphalt macadam 灌沥青碎石;灌地沥青碎石(路面);沥青灌浆碎石路(面)
grouted asphalt revetment 沥青灌浆护岸
grouted block pavement 灌浆式块料路面;灌缝式块料路面
grouted bolt 灌浆锚杆;灌浆螺栓
grouted brick 浆砌砖;砌砖灌浆
grouted brick masonry 灌浆砖砌体
grouted butt-type 灌浆对接式
grouted column pile wall 柱列灌注桩
grouted concrete 压浆混凝土;预填集料混凝土;灌浆混凝土
grouted cut-off wall 灌浆帷幕;灌浆截水墙;灌浆防渗墙
grouted foundation 灌浆基础
grouted frame 砂浆填充门框
grouted-in 已灌浆(的)
grouted in their conduits 管道压浆
grouted joint 灌浆接头;灌浆缝;浆缝
grouted macadam 灌沥青碎石路;灌浆碎石路
grouted masonry 灌浆圬工;灌浆砌体
grouted pile 灌浆桩
grouted pitching 灌浆块石护岸;浆砌块石护岸
grouted prepacked aggregate 预埋集料
grouted procedure 灌浆工序;灌浆(方)法
grouted riprap 灌浆抛石;浆砌乱石(块);抛石灌浆
grouted riprap arch bridge 浆砌乱石拱桥
grouted rock bolt 注浆型锚杆;灌浆岩石锚杆;灌浆(的)岩石锚栓
grouted rockfill dam 浆砌堆石坝
grouted roof-bolt 洞顶灌浆锚杆
grouted rubble 灌浆砌块;浆砌块石
grouted scarf joint 砂浆嵌缝接头
grouted stone 浆砌块石
grouted tendon 灌浆固结的钢丝束;灌浆固结的钢筋束
grout ejector 灌浆注射器
grouter 灌浆机;灌浆工;水泥喷补枪
grout fabrication 浆液配制
grout feed line 灌浆输送线
grout-filled fabric mat 灌浆土工布护排
grout filler 压浆填料;灌浆填料缝;灌缝用(薄)砂浆;灌缝料;灰浆填料
grout flow cone 水泥浆渗散试验锥
grout fluidifier 灰或水泥浆流化剂;灌浆用增塑剂;灰浆流化剂
grout groove 灌浆槽
grout head 灌浆头
grout header 灌浆集管
grout hole 灌浆孔;压浆孔;注浆孔
grout-hole drilling 注水泥浆钻孔钻进;灌浆孔钻进
grouting 压浆;注浆;灌浆(方)法;灌浆;浆液;浆砌;涂薄胶泥;生浆制备
grouting admixture 灌浆外加剂
grouting agent 黏结剂;灌浆剂;灌浆材料;胶结剂;浆料;浆液

grouting aid 成浆剂
grouting air vent 灌浆排气口
grouting all the depth once 全深一次注浆
grouting and sealing technique 灌浆止水技术;灌浆封孔技术;灌浆封堵技术;灌浆防渗技术;灌浆堵水技术;灌浆闭气技术
grouting apparatus 灌浆设备
grouting at back of shaft wall 井壁后注浆
grouting(bore)hole 灌浆孔
grouting cement 灌浆水泥
grouting chamber 灌浆箱
grouting chemicals 注入化学药剂;灌注用的化学药剂
grouting cofferdam 灌浆围堰
grouting compound 填灌混合料;灌浆混合料
grouting cup 灌浆漏斗
grouting curtain 灌浆帷幕;灌(形成的)拦(浆)墙
grouting cut-off wall 灌浆(形成的)拦(浆)墙
grouting depth 灌浆深度
grouting device 灌浆设备
grouting dolly 灌浆筒;灌浆棒
grouting drill hole space 灌浆孔距离
grouting effect 灌浆效果
grouting equipment 灌浆设备
grouting filling 抹平料
grouting fluid 薄浆
grouting gallery 灌浆廊道
grouting gantry 灌浆工作台
grouting gun 灌浆枪;灌浆器;灌水泥浆枪
grouting height ratio 灌入高度比
grouting hole 灌浆孔;灌浆漏斗;衬背注浆孔
grouting hole depth 灌浆孔深度
grouting hose 灌浆软管
grouting injection apparatus 灌浆设备
grouting injector 灌浆机
grouting inlet 灌浆口
grouting installation 灌浆装置
grouting job 灌浆工作
grouting joint 灰浆灌缝
grouting lance 灌浆喷枪
grouting laying 灌浆垒砌
grouting level height 灌浆段高程
grouting level length 灌浆段长度
grouting machine 洋灰搅拌机;灌浆机(械);水泥搅拌机
grouting material 灌浆材料
grouting material in urgent need 应急注浆材料
grouting method 注浆方法;灌浆(方)法
grouting mortar 灌填砂浆;灌(浆)灰浆;水泥砂浆
grouting nipple 灌浆管接头
grouting of annular space 周边空隙内灌浆
grouting of ducts 管道灌浆
grouting operation 灌浆操作
grouting pan 灌浆机;灌浆泵
grouting pile hole construction 灌注桩施工法;灌注桩施工
grouting pipe 灌浆管
grouting plant 灌浆装置
grouting pressure 注入压力;注浆压力;灌浆压力
grouting procedure 灌浆工序;灌浆法
grouting pump 灌浆泵;压浆泵
grouting quality 灌浆质量
grouting radius 灌浆半径
grouting rate 衬背注浆率
grout ingredient 灌浆成分;浆料成分
grouting resin 灌注树脂
grouting rod 灌浆枪;灌浆杆
grouting sand 灌浆用砂
grouting screen 灌浆帷幕
grouting socket 灌浆承窝;灌浆承口
grouting structure 灌浆结构
grouting suspension 灌注悬浮液;灌浆悬凝
grouting system 注浆方法
grouting technique 灌浆技术
grouting technology 灌浆工艺
grouting test 灌浆试验
grouting time 灌浆时间
grouting tool 灌浆工具
grouting type 灌浆类型
grouting up 灌浆
grouting with asphalt 灌沥青
grouting with cement slurry 灌薄水泥浆
grouting with plastics 塑料堵漏
grouting work 灌浆工作

grout injection 喷浆;注水泥(浆)
grout injection apparatus 压浆设备;注浆设备
grout injection method 灌浆法(加固土壤)
grout injection pipe 灌浆管
grout injector 水泥浆注入器
grout inlet tube 进浆管
grout-intruded concrete 挤浆混凝土(向集料挤入水泥浆而成的混凝土)
grout intrusion 强制灌浆
grout in with cement 用水泥填充
groutite 斜方水锰矿
grout joint 灌浆缝
grout laying 灌浆铺砌;灌浆砌筑
grout life 灌浆高度;灌浆量;表面高度
grout line 灌浆管道
grout-lock brick 灌浆槽孔砖
grout loss 表面起砂
grout machine 灌浆机
grout mix computation chart 水泥浆计算诺模图
grout mixer 灰浆拌和机;灌浆混合机;灰浆拌和器;水泥浆搅拌机;水泥浆拌和机;砂浆拌和机;薄浆搅拌机;拌浆器;拌浆机
grout mixer and placer 砂浆拌灌机;砂浆搅拌喷射器;砂浆搅拌灌注器
grout mixing plant 灌浆拌和设备
grout mix proportion 浆液配合比
grout mix(ture) 薄浆混合料;灌浆混合料;浆液混合料
grout nick 灌浆凹槽(圬工接缝处)
grout nipple 灌浆喷嘴
grout nozzle 灌浆喷嘴
grout off 注水泥止水
grout outlet 出浆口
grout outlet hose 水泥排出管
grout packer 灌浆封隔器
grout pile 灌注桩
grout pipe 灌浆管;压力灌浆管;注水泥管
grout pipe system 灌浆管灌系;灌浆管系统
grout pocket 灌浆孔隙
grout pressure 压浆压力
grout proportion 浆液配比
grout pump 注浆泵;水泥灌浆泵;砂浆泵
grout pumping rig 灌浆泵送设备
grout recess 灌浆槽
grout remixing 砂浆重新拌和
grout repairing 灌浆修补
grout retaining cap 挡浆帽
grout return pipe 回浆管
grout return tube 回浆管
grout sealing 水泥封闭
grout slope 斜坡法灌浆;灌浆坡度;灰浆坡度;薄浆流动坡度
grout socket 压浆套管
grout spreader 泥浆铺撒器;灰浆铺撒器
grout stop 止浆片
grout take 耗浆量
grout to seal water 注浆堵水
grout trough 泥浆槽
grout valve 灌浆阀
grout vent 压浆孔;灌浆(排气)孔;灌浆排气管
grout volume 注浆量
grove 园林;小丛林;树林;树丛
Grove's synthesis 格罗夫合成
groveling of road 砾石路面
grove shape meter 槽形仪表
grow 增长;栽种;栽培;生长
grow-back 带材的厚度差
grower 栽培者
grower washer 弹簧垫圈
grow fat 变肥
growing 长大
growing area 种植面积
growing automaton 增长性自动机
growing building 建造中的建筑物
growing conditions in North China 华北地区生长条件
growing directions 生长方法
growing district 产区
growing equity mortgage 定息递增偿付抵押贷款
growing house 温室
growing in competition 竞争中增长
growing needs 不断增长的需要
growing of rope 接长钢丝绳
growing on hill-side 山地栽培

growing pains 前进中的困难;发展过程中的问题
growing pattern of fault 断层生长型式
growing period 生长期
growing period of fault 生长断层活动期
growing plant without soil 无土栽培
growing point 生长点
growing rate of fault 断层生长速度
growing season 生长期;生长季节
growing stock 立木蓄积
growing stock of forest 森林存量增长
growing wave 增幅波;生长波
growing wave tube 生长波管
grow in pace with 同步增长
grow in phase with 同步增长
grow in step with 同步增长
growl 轰鸣
growler 小冰山;葫芦形小冰山(露出水面不超过一米);短路线圈检查仪;短路线圈测试仪;残碎冰山
growler board 千斤顶木垫块
growlery 私室
grown 天然helm材
grown form 基本形态
grown in situ 就地生长
grown junction 生长结
grown-junction photocell 生长结光电池
grown land 王室领地
grown sea 暴风涛
grown spar 现成杆柱
grow out of nothing 从无到有
grow perceptibly 见长
growth 增长;生长;长大
growth and decline 消长
growth anticline 生长背斜【地】
growth area 增长区
growth by Robertson's law 按罗伯逊定律的增长
growth center 增长中心;成长中心
growth coefficient 增殖系数
growth coil 生长管盘
growth condition of glacier 冰川发育条件
growth constant 增殖常数;增长常数;生长常数
growth-controlled fracture 控增长断裂
growth curve 增长曲线
growth curve of activated sludge 活性污泥增长曲线
growth curve of bacteria 细菌生长曲线
growth cycle 生活周期
growth defect 生长缺陷
growth dynamics 生长动力学
growth effect 生长效应
growth equation 生长方程
growther 经济大发展地区
growth fabric 生长组构
growth factor 增长因子;增长系数
growth factor method 增长指数法;增长系数法;增长因素法
growth fault 生长缺陷;同生断层;生长断层
growth fault-roll-over anticlinal trap 生长断层-滚动背斜圈闭
growth-framework pore 生物骨架孔隙
growth fund 迅速发展投资基金;发展基金
growth habit 生长习性
growth hillock 生长小锥;生长丘
growth in demand 需求增长
growth index 生长指数
growth index of fault 断层生长指数
growth industry 新兴行业;新兴工业;发展特快的新行业
growth inhibitor 生长抑制剂
growth interface 生长分界面
growth in thickness 加厚生长
growth in volume 材积生长
growth junction transistor 生长结型晶体管
growth lamellae 生长层
growth lattice 生长管盘;生长格子
growth layer 生长层
growth limiting factor 生长限制因素
growth line 生长线
growth management 社区增长管理
growthmanship 增大增长速度
growth media 生长基
growth of algae 海藻的生长
growth of cast-iron 铸铁生长
growth of concrete 混凝土膨胀

growth of fungi 霉菌的生长
growth of load 负载增长;负荷增长
growth of mechanical power 机械动力的发展过程
growth of population 人口增长
growth of quartz 石英结晶;石英的生长
growth of the crystal 晶体生长
growth of the market 市场发育
growth operator 增量算子
growth order 增长序
growth-oriented environment 面向扩展的环境
growth parameter 生长参数
growth peak 生长峰
growth period 生长期
growth phase 生长相
growth pit 生长凹坑
growth plan(ning) 发展计划
growth pole theory 增长极论
growth potential 增长潜力
growth production 发展生产
growth rate 增长率;生长(速)率;生长速度
growth rate fluctuating 生长速率起伏
growth rate method 增长率法
growth rate of earmarked fund 专用基金增长率
growth rate of population 人口增长率
growth rate of port traffic 港口吞吐量增长率
growth rate of real assets 固定资产增长率
growth ratio 生长比率
growth regulation 生产调节剂
growth response 增长感应;生长反应
growth retardation 增长停滞;生长迟缓
growth-retarding effect 阻碍生长作用
growth rhythm 生长韵律;生长节律;生长规律
growth ring 树木年轮;生长轮
growth sector boundary 生长扇面边界
growth semiconductor junction 生长结
growth shake 木心环裂;木材心裂
growth site 生长位置
growth stage 增长时期
growth step 生长阶(梯)
growth striation 生长条纹;生长辉纹;生长层
growth structure 生长构造
growth surface 生长面
growth theory of the firm 厂商的增长理论
growth times 发展倍数
growth tube 生长管
growth twin 生长双晶
growth variable 生长变量
growth vector 生长向量
growth zone of reef body 礁体生长带
growth zoning 生长环带
grow up 长大
grow very heavy 茂盛
groyne 折流坝;交叉拱;排流坝;丁坝;防沙堤;防波堤
groyne basin 丁坝间水区
groyne(d) vault 正交穹隆
groyne engineering 丁坝工程
groyne head 防波堤头;丁坝坝头
groyne pointing downstream 倾向河流下游的防波堤;倾向河流下游的丁坝;指向河流下游的丁坝
groyne pointing upstream 倾向河流上游的防波堤;倾向河流上游的丁坝;朝向河流上游的丁坝
groyne root 防波堤脚;防波堤根部;丁坝脚;丁坝根部
groyne spacing 丁坝间距
groyne with short training wall 有防潮矮墙的防波堤;有防潮矮墙的丁坝
groyne work 折流坝工程;丁坝工程
groyning 筑防波堤;筑丁坝
grozing iron 烙铁;铅管接口抹刀
grub 清理场地;挖根
grub axe 掘根斧;锄头
grubber 掘土机;掘根机;挖树根工具;挖根犁;碎土机;树根清理工;深耕中耕机;除根机;除草机
grubber plow 掘根犁
grubber rope 除根索;拔除树桩的绳索
grubbing 清除树根;除根
grubbing axe 斩根斧;斩除树桩的斧头;除根斧
grubbing harrow 除草耙;圆盘犁
grubbing hoe 斩除树桩的斧头;挖掘铲;挖根锄
grubbing implement 拔除树根的工具;挖除树桩的工具
grubbing machine 拔除树根的机器;拔除木桩的机器
grubbing mattock (除根用的)鹤嘴锄
grubbing tool 除根工具
grubbing winch 除根机
grub-breaker 掘根开荒犁
grub-breaker plow 挖掘开荒犁
grub felling 连根伐
grub hoe 鹤嘴锄
grub hook 挖根钩
grub out 连根挖
grub saw 锯石用手锯;大理石(手)锯;石锯
grub screw 固定螺钉;木螺丝;木螺钉;平头螺钉;无头螺丝;无头螺钉
grudging post 立根【岩】
gruenlingite 碲硫铋矿
grume 小堆
grumeaux 灰泥凝块
grummet 堵头;垫环;金属孔眼;麻丝垫环;垫圈
Grum recording spectroradiometer 格卢姆自动记录光谱辐射计
grunauite 辉铋镍矿
grunching 无自由面爆破
grundite 伊利黏土
grundsill 木框架底部横档;槛木;地槛
grunerite 铁闪石
grunerite schist 铁闪石片岩
grunnching 无自由面爆破(除工作面外)
grunt 架线学徒工;工人
grunter 电工助手
Grupo Matarazzo 马塔拉佐财团(巴西)
grus 碎砂砾堆积
grush 花岗岩碎砾
gruss 粒状岩(石);花岗岩碎砾;碎砂砾堆积
Grusz process 格鲁兹双金属铸造法
gruzdevite 顾硫锑汞铜矿
gryke 岩沟
Gryphite grits 贝壳粗砂岩
gryposis 异常弯曲
gryptcrete 纤维质灰浆混凝土
G-system 英国双曲线无线电近程导航系统
g-tolerance 加速度耐力;G容限
GTO thyristor 闸门关断可控硅整流器
guag 采空矿穴
guaiacel 邻甲氧基苯酚
guaiaci lignum 愈创木
guaiacol 愈创木酚
Guaiacum officinale 愈创木
guaiac-wood 愈创木
guaiac wood oil 愈创木油
gualac 愈创木树脂
gualacum 愈创木树脂
guambo 深黄褐色硬木(产于中美洲)
guanajuatite 硒铋矿
guanamine 鸟粪胺
guanamine resin 鸟粪胺树脂
guanazole 胍唑
Guang colo(u)rs 广彩
Guang Jun glaze 广钧釉
Guang kingfisher blue 广翠
guanidine aldehyde resin 胍醛树脂
guanine 鸟嘌呤石
guanite 鸟粪石
guano 鸟粪石;鸟粪(层);海鸟粪
guano phosphorite 鸟粪磷钙土
guano-type deposit 鸟粪石型矿床
Guan ware 官窑(器)
guarantee 作保证(人);货物保证书;担保品;保证书;保证(人)
guarantee acceptance 承兑保证
guarantee against defects 质量保证书
guarantee agreement 担保协定;担保协议
guarantee bond 担保书
guarantee both quality and quantity 保质保量
guarantee by manufacturer 制造商担保
guarantee circuit 保证电路
guarantee clause 保证条款
guarantee contract 担保合同
guarantee cost 担保费用
guaranteed accuracy 保证精度
guaranteed annual wage 年保证工资;保证年度工资
guaranteed annual wage plan 最低年工资保证制
guaranteed annuity interest 保证年(金)利息
guaranteed annuity option 保证年金选择权
guaranteed bill 保证票据
guaranteed bond 保证债券;信用保证债券;附保证的公司债券
guaranteed by indorsement 背书担保
guaranteed calculation for regulation 调节保证计算
guaranteed capacity 保证容量
guaranteed cobalt-blue kiln 包青窑
guaranteed cost 担保费用
guaranteed credit 有担保的信贷
guaranteed deadweight, bale-space 保证载重量和包装容积
guaranteed debt 保证债务
guaranteed deposit 保证存款
guaranteed depth 保证深度
guaranteed dividend 保证红利;保证股息;保证股利
guaranteed efficiency 保证效率
guarantee deposit 押金;保证金
guaranteed freight 保付运费
guaranteed fund 保证基金
guaranteed horsepower 保证马力
guaranteed issue 保证发行
guaranteed letter of credit 凭保证开发的信用证;保兑信用卡
guaranteed liability 保证负债
guaranteed loan 有担保的贷款;保证借款
guaranteed market investment 有保证市场投资
guaranteed maximum continuous output 保证最大连续出产
guaranteed maximum contract 保证最高价合同
guaranteed maximum cost 合约规定工程总费用最高限额;工程总费用最高限额
guaranteed maximum cost contract 最高限价担保合同
guaranteed mortgage system 保证抵押制度
guaranteed output 保证功率;保证出力
guaranteed payment 保证支付
guaranteed performance 保证性能;保证合同履行
guaranteed period 保证期;保用期
guaranteed precision 保证精度
guaranteed price 保证价格
guaranteed purchase 保证价格购买
guaranteed quality 保证质量;保证品质
guaranteed rate 保证(工资)率
guaranteed rate of channel depth 航道水深保证率
guaranteed rate of depth datum 深度基准面保证率
guaranteed rate of lock's navigable time 船闸通航时间保证率
guaranteed rate of navigable water level 通航水位保证率
guaranteed rate of navigable water stage 通航水位保证率
guaranteed rate of the highest navigable stage 最高通航水位保证率
guaranteed rate of the highest navigable water level 最高通航水位保证率
guaranteed rate of the lowest navigable stage 最低通航水位保证率
guaranteed rate of the lowest navigable water level 最低通航水位保证率
guaranteed reserve 保证准备(金)
guaranteed share 保证股票;保证股份
guaranteed speed 保证航速
guaranteed steam consumption 保证汽耗
guaranteed stock 保证股票
guaranteed suitable 包用
guaranteed system 保证制度
guaranteed turbine efficiency 水轮机保证效率
guaranteed vacuum 保证真空
guaranteed value 保证值
guaranteed wage plans 保证工资计划
guaranteed wage system 保证工资制
guaranteed warranty 保单
guaranteed water depth 保证水深
guaranteed weight 保证重量
guaranteed working week 保证工作周
guaranteed yield strength 保证屈服强度
guaranteed yield stress 保证屈服应力
guarantee for loans 信贷担保
guarantee for repair 包修
guarantee fuel 备用燃料
guarantee fund 保证金

guaranteeing delivery 保证交货
guaranteeing of promissory notes 本票的保证
guarantee insurance 保证保险
guarantee letter 保函
guarantee letter of credit 担保信用证
guarantee load 额定安全重量;额定安全荷载;额定安全负载
guarantee of a bill 票据保证
guarantee of bills 票据保证
guarantee of delivery 交货保证
guarantee of export credits 出口信用担保;出口信用保证
guarantee of good working order 保证良好操作
guarantee of insurance 保险担保书
guarantee of payment 付款保证
guarantee of performance 履约担保;履约保证书;履行义务保证
guarantee of quality 质量保证(书)
guarantee of repayment 借款偿还保证
guarantee of shipper 货主保证
guarantee of title 买方对所有权ების证
guarantee of weight 保证重量
guarantee on first demand 保证债权人提出立刻履行
guarantee on the first demand 首次要求即付保函
guarantee output 保证出力
guarantee pay 保证付款
guarantee performance 保证履行合同
guarantee period 保证期;担保期限;保用期;保险期
guarantee point 保证点
guarantee rate of designed discharge 设计流量保证率
guarantee rate of designed flow 设计流量保证率
guarantee rate of designed lowest stage 设计最低水位保障率
guarantee rate of designed minimum discharge 设计最小流量保障率
guarantee rate of designed water-level 设计水位保证率
guarantee reagent 优级纯;保证试剂
guarantee slip 保单
guarantee stock 保息股票
guarantee sum 保证金额
guarantee test 保证试验
guarantee the payment of the debts 保证偿还债务
guarantee time 保修期限
guarantee to keep something in good repair 保修
guarantee to pay compensations 包赔
guarantee upper bounds 确保的上界
guarantee work 保修工作
guarantor 票据担保人;担保人;保(证)人
guarantor company 担保公司
guaranty bond 押标金;投标保证金;履约保证(书);工料支付保证(书);标价契约;担保契约;标价契约;使用契约;劳动力和材料支付契约;保证契约
guaranty money of tender 投标保证金
guaranty principal 被保证本人
guaranty travellers draft 保证旅行汇票
guard 限程器;列车长;警戒;守卫;防物罩;防护
guard's valve 紧急制动阀;停开关;车长阀
guard's van 守车
guard's van valve 守车制动阀
guard against corrosive influence 反腐蚀
guard against damp 谨防潮湿;防潮
guard against error 防止错误
guard amplifier 保护放大器
guard and alarm system 安全防范报警系统
guard aperture 护孔
guard area 警戒区
guard arm 护臂;保护横杆
guard balanced by a counter weight 重锤平衡卫板
guard band 防护频带;保护间距;保护带
guard band filter 保护频带滤波器
guard bar 扶手;栏杆;护墙栏(杆);护栏
guard bead 木档条;滑窗挡轨(杆);护栏;护轨;滑窗内轨;护条(吊窗);护木条
guard bee 监视蜂
guard belt 护轮带
guard bit 保护位
guard block 防护块体;保护部件
guard board 指路牌;(起重机的)构台;导板;栏板;护板;挡(边)板
guard boat 巡逻艇;巡逻船;警戒船
guard boom 防护屏;护栏木;防护栅
guard byte 保护字节
guard cable 钢索护栏;防护索;安全索;安全防护用钢丝绳
guard camber 禁闭室
guard car 缓冲车
guard catalyst 保护催化剂
guard cell 保卫细胞
guard chamber 警卫室
guard circle 保护圆;保护圈
guard circuit 保护电路
guard column 保护柱
guard cradle 保护网
guard curb 防护路缘;防护侧石
guard dam 防护坝
guard digit 保护位;保护数(位)
guard duty 警卫岗
guarded command 保护命令
guarded crossing 有人看守道口
guarded hot box method 保护热箱法
guarded plate method 保护平板法
guarded switch 保险开关
guard electrode 保护电极;屏蔽电极
guard electrode log 聚流电极测井记录
guard enable 允许保护
guarder 保护装置
guard facility 警卫设备
guard fence 围墙;栏栅;护篱;护栅(栏);护栏;防护栅
guard for belt pulley 皮带轮防护罩
guard front lock 防护面锁
guard gate 应急闸门;防护闸(门);备用闸门;安全闸(门)
guard gating 保护栅
guard grating 安全栅
guard heater 外加热器
guard house 警卫宿舍;警卫室;门卫室
guardian angel 吉神
guardian block 防护块体
guardian by statute 法定监护人
guardianship process 守护进程
guarding 消除表面漏泄法;渔网加强边缘;警卫岗;帆布滚边
guarding a target 守卫一个目标
guarding counter 屏蔽计数器
guarding equipment 警卫用具
guarding figure 保险数位
guarding outfit 警卫用具
guard kerb 防护路缘;防护侧石
guard lamp 告警灯;保护灯
guard latch 保护锁键
guardless type axle box 无导框式轴箱
guard line 保护线
guard lock 入口船闸;挡潮闸(门);潮汐船闸;防护闸(门);防洪闸(门);保安联动
guard log 屏蔽测井
guard magnet 防护磁铁;保险磁铁
guard mask 防护面罩
guard method 保护方法
guard mode 保护状态
guard net 安全网;防护网;保护网
guard pennant 值勤旗
guard period 防护期间;保护时段
guard pile 护桩;防护桩;防冲桩
guard pin 叉头钉;保护销
guard planting 门卫种植
guard plate 防护板;保护板;安全(挡)板;紧固托板
guard plot 防护区
guard position 保护位
guard post 护柱;围护桩;防撞柱;防护桩;标柱
guard rail 舷栏杆;钢轨护栏;护墙栏(杆);护轮机;护栏;护轨;防舷木;防护栏(杆);护轨;扶杆
guard rail bolt 护轨螺栓
guard rail brace 护轨撑
guard rail check ga(u)ge 查照间隔
guard rail clamp 护轨夹
guard rail crossing 设栏平交道;有栅栏的平交道(铁路上)
guard rail face ga(u)ge 护轨间隔
guard rail groove 护轨轮缘槽
guard rail of bridge 桥梁护轨
guard rail tie plate 护轨垫板
guard rail washing machine 护轨洗涤机
guard relay 防护继电器;保持继电器;保安继电器
guard ring 止推护环;止环;警戒圈;护圈;色保护环;保险环;保护环;安全环
guard ring detector 保护环探测器
guard ring type capacitor 保护环式电容器
guard room 警卫室
guard row 保护带;保护行
guards 粘插图纸条
guard screen 屏蔽;护网;防护网
guard sealed method 保护屏蔽法
guard shield 防护屏蔽;保护屏蔽
guard ship 巡逻艇;巡逻船;警戒舰
guard signal 保险信号;保护信号;防护信号;安全信号
guard sill 护坎
guard sleeper 护轨枕木
guard space 保护距离;保护间隔
guard stake 护桩;保护桩【测】
guard stanchion 栏杆柱
guard stand 岗亭
guard staple 卡套
guard stone 护石;排衔石;侧石
guard timber 护木
guard timber bolt 护木螺栓
guard timber of bridge 桥梁护木
guard time 防护时间
guard track 保险间距
guard tree 保护林
guard unit 防火责任区
guard vacuum 防护真空;保护真空装置
guard valve 速动闸门;速动事故阀;速动安全阀;事故(保险)阀;事故(安全)阀;防护阀
guard wagon 缓冲车
guard wall 护墙;墙式护栏
guard window 护窗;防范窗
guard wire 安全线;保护导线
guard wood 护木
guard zone 保护区
guar gum 瓜耳(树)胶
guarinite 片榍石
Guastavino 古斯塔维诺法(穹隆顶内壳砌筑用)
Guatacre 淡褐色硬木(产于巴西)
Guatemala 危地马拉
guatta-percha 马来乳胶
guba 居巴雨飑(新几内亚海上)
gubernaculum 引神;副刺
Gucker and Rose aerosol counter 古克—罗斯悬浮粒计数器
Gudden-Pohl effect 古登—波尔效应
Gudermannian 古德曼函数
gudgeon 轴头(销);增强衬圈;悬挂式合页;铁销;门轴;轴柱;螺栓;托架;枢轴;枢轴;耳轴
gudgeon bearing 端轴承
gudgeon block 轴承
gudgeon hole hone 活塞销孔搪磨头
gudgeon journal 轴颈
gudgeon of universal joint 万向节销
gudgeon pin 轴头(销);杆头销;活塞销;十字头销;舵针;舵栓;耳轴销
gudgeon pin bearing 活塞销轴承
gudgeon pin boss 耳轴销壳
gudgeon pin bush 活塞销衬套
gudgeon pin cover 活塞销盖
gudgeon pin hole cap 活塞销孔盖
gudgeon pin plug 活塞销塞
gudmundite 硫锑铁矿
Gudumlund faience 古都隆锡釉陶
guerinite 格水砷钙石
Guerin press 格林式橡胶模
Guerin process 格林橡胶模冲压法;金属薄板成形法
guerite 古碉堡塔楼;卫兵哨亭
guess 推测
guessed average 假定平均数
guessed mean 假定平均数
guess rope 攀索;辅助缆索;系艇缆;拖船安定绳
guess stick 计算尺
guesstimate 概算;粗略估计;猜测
guess value 推测值
guess warp 系艇缆
guesswork 推测
guest bedroom 客人卧室;客房
guest car ratio 宾客拥车比率

guest chamber 客房
guest cottage 别墅
guest-crystal 客晶
guest hall 大客厅
guest-host display 宾主效应彩色液晶显示
guest-host effect 宾主效应
guest-host interaction 宾主作用
guest house 招待所;宾馆
guest lounge 宾客休息室;宾馆休息室
guest machine 备用机
guest mineral 客矿物
guest room 客房;会客厅
guest rope 系艇缆;牵引索;牵引绳;辅助缆索;攀索;拖缆辅助绳
guests 客串用户【计】
Guest unit 格斯特单位
guest warp 系艇缆;牵引索;牵引绳;拖船安定绳;攀索;辅助缆索
guest warp boom 舷侧系艇杆;系艇杆
guest worker 客籍工人
guettardite 格硫锑铅矿
gug 下坡道(采矿、隧道、坑道的)
gugiaite 顾家石
guglia 方尖碑;方尖柱
gugout 地穴
guhr 硅藻土
guhr dynamite 硅藻土炸药
guichet 办事处询问台;办事处询问窗口;办公室询问台;办公室询问窗口;售票窗
guidance 引导;制导;导星;导航;导承
guidance accuracy 引导精度
guidance and navigation computer 制导与导航计算机
guidance and navigation control system 制导与导航控制系统
guidance apparatus 制导装置
guidance axis 瞄准轴;导引轴;导向支承
guidance code 引导代码
guidance equipment 制导设备
guidance error 制导误差
guidance formula 波导公式
guidance information 制导信息
guidance key 导向键
guidance misdistance 制导误差
guidance of planning 计划指导
guidance of table 平台导槽
guidance panel 操作面板
guidance plan 指导性计划
guidance quality 导向性
guidance radar 制导雷达;导航雷达
guidance section 仪表舱
guidance sensor 制导传感器
guidance sign 导航标志;导航标志
guidance site 制导站;制导场
guidance station 制导站;导航站;导航台
guidance system 引导系统;制导系统;导向装置;导航系统
guidance tape 导向带;导板带
guidance tracking receiver 导引跟踪(装置)接收器
guidance unit 制导设备;瞄准器
guidance value 准值
guidance network 制导网
guide 销套;引导;指引;指导;滑槽;入门(指说明书等的入门知识);导子;引导器;导向;导承;波导
guide adjusting wedge 导轨调整楔
guide angle 导角;导向角(钢)
guide apparatus 制导装置;导向装置
guide arm 导向臂;导杆
guide arrangement 引导装置
guide baffle 导向挡板
guide bank 导流护岸;导堤
guide bar 支承梁;滑杆;导(向)杆
guide bar bolt 导杆螺栓
guide bar oil cup 导杆油杯
guide bar receiver 导杆支座
guide barrel 导风管
guide bar set bolt 导杆防松螺栓
guide bar support 导杆支架
guide bar unit 导杆组
guide bar yoke 导杆轭
guide base 井口盘;导向座;导向架
guide bead 窗滑轨;内导条(垂拉窗);活页导条;提拉窗导轨;导条

guide bearing 定位轴承;导轴轴承;导轴承;导引轴承;导向轴承;导向支承
guide bearing bracket 导轴承托架;导向轴承支架
guide bend test 控制弯曲试验
guide blade 异流片;隔声板;导叶(片);导向叶片;导向器叶片;挡板
guide blade carrier 导叶持环;导片持环
guide blade loss 导叶损失
guide blade rim 导向叶片出口边
guide blade row 导叶列
guide blade segment 导叶组
guide blade support ring 导叶持环
guide block 导块;滑块;导向块;导向滑车;导瓦;导丝器;导丝板;导流砖;八字砖
guide block bearing 导向块轴承
guide block brass 导向块铜衬
guide board 路(名)牌;标板;指路牌;路标
guide body 导向架
guide bolt 导向螺栓
guidebook 指南;指导书;入门(指说明书等的入门知识);参考手册
guide borehole 导向钻孔
guide box 导箱
guide bracket 固定罐道钢夹子;导座;导卡(人字闸门);导架
guide building 导航建筑物
guide bund 导水墙;导流墙
guide bush 导(轴)衬
guide bushing 导向轴衬;(模具的)导套
guide cage 导板盒
guide cam 导向凸轮
guide canal 导汽槽
guide cap bolt 导向盖螺栓
guide card 引导卡片;指引卡片;导卡片
guide carriage 导轮
guide casing 导箱;导架
guide channel 导流河渠;导沟;导槽
guide characteristic wave impedance 波导管特性波阻抗
guide chase 导框
guide clamp 导向线夹
guide clearance 导向部间隙;导承间隙
guide coat 刮粗层;打底层;标记涂层
guide column of gasholder frame 储气罐导柱
guide comb 导梳;箕形板
guide consumption 引导消费
guide copy 底图图形
guide core 岩芯导向胶塞
guide coupling 扩孔用导向杆;导管连接器
guide crest 导流堰顶
guide cross registration 规格线套合
guide curve 导向曲线
guide cylinder 导向筒
guide dam 导水坝
guided automatic welding 导向自动焊接
guided bend test 型板弯曲试验;靠模弯曲试验;定形弯曲试验
guided bend test for butt welded joint 对接接头靠模弯曲试验
guided discharge 导致放电
guided earth satellite 可控地球卫星
guide derrick 导向钻塔
guided ground transport 引导式地面运输
guide disc 导盘
guided light tramway 轻型导向电车线路
guided light transit 轻型导向交通系统
guided missile armed destroyer 导弹驱逐舰
guided missile base 导弹基地
guided missile counter-measure 防导弹措施
guided missile lauching site 导弹发射场
guided propagation 导行传播;导向传播;波导传播
guided radar 导向雷达
guide drum 导向轮;导向滚筒;导向鼓
guided transit 导向交通
guided transmission 波导传输
guided transport system 导向运输系统
guide duct 导风管
guided wave 循轨波;遵导波;被导波;导波
guided-wave radio 波导无线电
guided way 导向轨
guide edge 引导边缘;横向导向;导向边;导向板
guide elbow 定向弯头
guide error sensor 导柱误差传感器
guide extension 导向延伸(部分)

guide face 导面
guide field 引导场
guide finger 指针;指示箭头;导向销
guide flow cylinder 导流筒
guide flume 导流槽
guide force 导向力
guide for container equipment inspection 集装箱设备检查指南
guide fork 导叉
guide for navigation 导航
guide for the valve rod 阀门杆导路
guide fossil 主导化石;指示(性)化石;指导性化石;标准化石;标志化石
guide frame 丝架;导承框;导(承)架
guide-framed gas holder 直升式储气罐
guide-framed holder 直升式储气罐
guide frame system 导架减摇方式
guide funnel 导向漏斗
guide gear 导向齿轮
guide gib 导向扁销;导向夹
guide gip 导承扁栓
guide girder 导梁
guide grid 定向栅(格)
guide groove 导(向)槽
guide holder 导向座;导向架
guide hole 导向(钻)孔;导孔;超前炮眼;超前孔
guide horizon 测量标志水平面
guide housing 导向罩;导向框
guide image 底图图形
guide inside micrometer 支承式内径千分尺
guide jetty 导航排架;导堤
guide key 导(向)键
guide lamp 导向灯
guide ledge 导板
guideless die 无导向模
guide levee 导堤
guide lever 转向横拉杆臂;基准杠杆;导向杆
guide lift 向左(手势)
guide lifter 带导向槽提升器
guideline 须知;引导线;准线;控制线(路);导线【测】;导引线;导(向)绳;导向标线;标(志)线;导向图表;准则;指南;方针
guideline for design 设计的指导方针
guideline for firm's training plan 公司培训计划指南
Guideline for Procurement of I-BRD 世界银行采购指南
Guideline for Procurement under International Bank for Reconstruction and Development loans and International Development Association 国际复兴开发银行贷款国际开发协会信贷采购指南
guideline for training in the packing of cargo in freight container 集装箱货物装载培训指南
Guideline for Withdrawal of Precedes of International Bank for Reconstruction and Development loans and International Development Association 国际复兴开发银行贷款国际开发协会信贷提款指南
guideline level 指导性残留量
guideline life 标准使用年限
guideline of largest of largest value 最大最大准则
guideline of largest of smallest value 最大最小准则
guide liner 键槽划线盘;导轴衬;导槽划线盘;导板衬
guidelines for design 设计准则
guidelines of management of smokedust regions of cities 城市烟尘控制区管理办法
guidelines of pollutant control for industries 工业企业污染物排放控制指标
guidelines of supervision and management of sewage treatment facility on environment (al) protection 污水处理设备环境保护监督管理办法
guide line specification 规格准则
guide line tensioner 导向绳张紧器;导绳张力器
guide link 平动连杆;导向滑块
guide lip 导向挡边
guide lug 履带板齿;导缘;导耳
guide margin 引导余量;引导边缘;导向边宽度;导边(数据孔)间距
guide mark 导板划痕;标记
guide marker 导向标(志)

guide mast 导桩;导柱;导桅
guide mast type 导向门架式
guide mechanism 导向机构
guide member 导引引件;导向构件;导轨;导板
guide meridian 导子午线
guide microscope 测量显微镜
guide mill 围盘轧机
guide mineral 指示矿物;标型矿物
guide nose 导尖
guide number 闪光指数
guide nut 导向螺母;导向螺帽
guide of rod 杆导承
guide pad 导向块
guide pass 导轨面
guide passage 导向通道(离心泵);引道;导路
guide path 导(向)槽
guide peg 导桩;导向杆
guide pile 定位桩;导桩
guide pillars 导柱
guide pin 定位销;导钉;定位钉;滑槽销;导柱;导针;导(向)销
guide pin bushing 导套
guide pin bushing boss 导套的凸台
guide pipe 导(向)管
guide pipe with outer insulation 外隔热导流管
guide piston 导向活塞
guide plane 导面
guide plank 导向木桩
guide plate 导流片;支承板;导(向隔)板;导(向)板;导(流)板
guide plate die 导板模
guide plate of gear rack 齿条导向板
guide plug 导向插头
guide point 导向点;导杆节点
guide pole 导杆
guide position 行导位置
guide positively 使具有正位移
guide post 方向标;指向柱;指路标(志);路标;导柱;导向路标;导标;标柱;控制指标
guide-post bushing 导套
guide post die 导柱式冲模
guide post typo die 导柱模
guide price 指导价格
guide principle 指导原则
guide pulley 压带轮;导向皮带轮;导向轮;导向滑轮;导向滑车;导轮;导辊
guider 自准直悄镜;自调位托辊;导星望远镜;导向器
guide rail 护栏;护轨;竖井导轨;导轨
guide rail of gasholder 储气罐导轨
guide reamer 导向铰刀
guide rib 导肋;导向棱;导向挡边
guide right 向右(手势)
guide ring 导圈;准环;控制环;调速环;导向环;导(料)环
guide ring coupling 带导向环的长接头
guide rod 导辊;滑杆;投影导杆;导向钻杆;导杆
guide rod type diesel pile hammer 导杆式柴油打桩机
guide roll 转向轮(三轮滚轧机);驾驶三轮辊;导辊;转向滚轮;主动辊;拉引辊;导向碾轮(压路机);导向辊;导滚
guide roller 导向滚轴;托辊;导柱;导向滚轮;导辊;导筒;导轮;挡轮
guide roller bolt 导辊螺栓
guide roller of ejector 卸土板导向滚轮
guide roller of gasholder 储气罐导轮
guide rolling 导卫轧制
guide rope 导(向)绳;导索
guide rope hoist 钢丝绳式升降机
guide rope winch 导索绞车
guider servo 导杆伺服
guide rule 准则;导引
guide runner 导向滑车;导(向)板;导滑槽
guide sample 标样
guides and guards 导卫装置
guide score 导卫板划痕
guide scratch 导板痕
guide screw 丝杆;导螺杆
guide seam 主导层;标准(矿)层;标志层(位);标型矿层
guide section 导向部分
guide shaft 导轴
guide shearing 导卫板划伤
guide sheave 导向滑轮;导向滑车

guide shell (凿岩机的)托盘;(凿岩机的)托架;导流筒
guide shoe 引鞋;导瓦;导块;转盘托板
guide sign 指向标志;指路标(志);路标;导向标志
guide size 导槽尺寸
guide sleeve 导向套筒;导(向)套
guide slot 定向槽;导(向)槽;导缝
guide socket 定向插座
guide specification 指导性规范
guide spindle 导轴
guide spindle bearing 导轴轴承
guide stem 导杆
guide step 滑板托架脚蹬
guide strap 抱钩
guide strap of balance weight 坠砣抱箍
guide structure 导流建筑物;导航建筑物
guide sub 导向接头
guide suction vane 导流叶(片)
guide surface 导轨面
guide system 导向系统;导向装置
guide tap 导向丝锥
guide-tape rod 导带柱
guide thimble 导向套管
guide to international ocean graphic(al) data exchange 国际海洋数据交换简介
guide to sailing 航行指南
guide track 轴承环;导轨
guide trench 导水沟;导(流)沟
guide tube 套管;导向套管;导管
guide twist drill 有导径的深孔麻花钻
guide type 导轨型
guide type jack 导向式千斤顶
guide value 参考值;标准值
guide valve 导向阀
guide valve boot 导阀罩
guide vane 导向叶片;导流叶片;向翼;导翼;导叶;导向叶轮;导向叶片;导叶板;导片;导流器;导流片;导流板;导板
guide vane apparatus 叶片式调节阀门;导向器
guide vane arm 导叶(拐)臂
guide vane bearing 导叶轴承;导柄轴承
guide vane control valve 导叶控制阀
guide vane cover 导叶上环
guide vane lever 导叶臂杆
guide vane of turbine 水轮机导叶
guide vane opening 导叶开度
guide vane packing 导叶(止水)填料;导叶密封条
guide vane ring 座环;导叶(固定)环;导(向)叶环;导水机构
guide vane servomotor 导向叶片式伺服发动机
guide vane stem 导叶叶柄
guide vane wheel 导向叶轮
guide waling for pile driving 导桩架
guide wall 导流岸壁;翼墙;顺坝;导(水)墙;导水堤;导(流)墙;导流堤;导(航)墙;导航墙
guide wave 导波
guide wave demultiplexer 导波分路器
guide wave impedance 导波管(波)阻抗
guide wave length 波导(管)波长
guide way 导向体;导向槽;导流;导沟;导路;导轨;定向线路
guideway grinder 导轨磨床
guide wheel 导叶组;导(向)轮;导流叶轮
guide wheel blade 导轮叶
guide wheel bucket 导向叶片
guide wheel core ring 导轮芯环
guide wheel disk 导向轮盘
guide wheel impeller 导轮叶片
guide wheel rim 导向轮轮缘
guide wheel shaft 导轮轴
guide wheel shaft bush 导轮轴套
guide wing 滑板翼片
guide wire 导向钢丝绳;导索;准线
guide work 导航设施;导航建筑(物);导堤
guide yoke 滑板托架
guiding 制导;指导性;导星
guiding accuracy 制导精确度
guiding aircraft 导航机
guiding air in spirality 螺旋形导气
guiding arrangement 引导装置;导向装置
guiding attachment 导向附件;导向附加装置
guiding axle 导向轴
guiding beater 导向击送轮
guiding bed 指示薄煤带;标志层(位)

guiding board 导向牌
guiding bush 导轴衬
guiding center 导向中心
guiding clearance 导向间隙;导隙
guiding collar 导向轴环;导向垫圈
guiding device 导星装置;导向装置;导出装置
guiding dike 导航堤
guiding discharge 导致放电
guiding-diving dike 导航隔流堤
guiding dolphin 导向系船墩;导向浮标
guiding edge 导杆边
guiding edge strip 导线条(路面上)
guiding effect 导向效用
guiding error 导星误差
guiding eyepiece 导星目镜
guiding field 导向场
guiding fin 导向滑板
guiding force 导向力
guiding groove 导槽
guiding guard 导向护板
guiding gutter 导(水)沟;导水槽
guiding handle 导向柄
guiding hole 中导孔;导孔
guiding jetty 导航排架;导(航)堤
guiding lamp 导向灯
guiding landmark 方向标
guiding light 灯光导标;导灯;指示灯
guiding line of safety in production 安全生产方针
guiding mechanism 导向装置
guiding microscope 引导显微镜;导星显微镜
guiding nut 导螺母
guiding ocular 引导目镜
guiding pile 定位桩;导桩
guiding pipe 导管
guiding plan 指导性计划
guiding price 指导价格
guiding principle 方针
guiding principles for environmental protection 环境保护工作指导方针;环境保护工作方式
guiding prism 引导棱镜;导引棱镜
guiding rafter 量规;导橡;标准计;标准橡
guiding rod 导辊
guiding rule 样板;规准;规板;准则;靠尺;木准尺
guiding runner 导向滑道
guiding shears 导剪
guiding shoe 导瓦
guiding sign 导向牌
guiding slot 导向槽
guiding star 引导星
guiding system 制导系统;导向系统
guiding telescope 导远镜;导星望远镜;导星镜
guiding tongue 导向尖轨
guiding unit 导向装置
guiding value 参考值;标准值
guiding valve 滑阀
guiding vane 导向
guiding wall 导流墙;导堤
guiding wheels device 导向轮装置
Guier coordinate plane 吉尔坐标平面
Guignet's green 基尼绿;酸性绿;碱式氧化铬绿;翠铬绿
guijiao 圭角
Guijo 褐红色硬木(产于菲律宾)
guild 行业协会;同业公会;同业工会
Guild colo(u)rimeter 吉尔德比色计
guilder rose 绿球花
guildhall 会馆;同业会所
guildite 四水铜铁矾
guild-ship 同业公会
Guillaume's metal 铜铋合金
guillaume alloy 铁镍低膨胀系数合金
Guillaume balance 积分摆轮
Guillaume metal 铜铋合金
guilleminite 硒钡铀矿
Guillemin line 基勒明仿真线
Guillery cupping test 吉莱利拉延成型性能试验;吉莱利杯突试验
Guillery ruler 吉莱利玻璃尺;吉耳列利玻璃尺
guilloche 扭索饰【建】
Guilloche work 花纹皮辊
guilloching machine 雕彩纹机
guillotine 铡除刀;轧刀;截流器;截锯机;截断机;剪断机;剪床;横切机;切纸机;切断器;裁切机
guillotine attenuator 刀形衰减器

guillotine bar shear 铡刀式钢材剪切机
guillotine blade-type damper 闸刀式叶片型风机
guillotine cutter 闸刀式切纸机
guillotine cutting machine 闸刀式切纸机
guillotine factor 截断因子
guillotine machine 截断机
guillotine operation 挤切法
guillotine plate shears 闸刀式剪切机；剪板机
guillotine shears 剪切机；剪板机；闸刀式剪切机
guillotine shutter 闸门式快门
guillotine type gate 起落式闸门；垂直吊升闸门
guillotine window 吊升窗
Guimet's blue 合成群青
guinea 觇标【测】；整坡标桩
Guinea architecture 几内亚建筑
Guinea basin 几内亚海盆
guinea corn 高粱
Guinea current 几内亚海流
Guinea grains 乐园子
Guinea green 几内亚绿；古纳特绿
guinea-pig 实验材料
Guinier focusing camera 纪尼埃聚焦照相机
guitermanite 块硫砷铅矿
Gujarat architecture 古吉拉特建筑(15~17 世纪印度的伊斯兰建筑)
Gukhman number 库克曼数
gula 反曲线；S 形线；空心铸模；葱形饰
gulch 峡谷；沟壑；沟谷；干谷；深沟；冲沟
gulch gold 砂金
Guldberg Waage group 古尔脱伯格—瓦格数
gulf 湾；大矿体；海湾
gulf area 海湾地区
gulf-Atlantic coast bivalve subprovince 海湾—大西洋沿岸双壳类地理亚区
gulf coast geology 港湾海岸地质学
gulf coast geomorphy 港湾海岸地貌学
Gulf coasts 海湾沿海地区
gulf current 湾流
Gulf-European Freight Association 海湾欧洲货运协会
gulf flow 湾流
gulfinning 海湾精制法
Gulf intracoastal canal 墨西哥沿岸运河
Gulf intracoastal waterway 墨西哥沿岸航路
Gulf mining coastal oil 海湾开采的石油
gulf ocotea 湾奥寇梯木
Gulf of Aden 亚丁湾
Gulf of Alaska 阿拉斯加湾
Gulf of California 加利福尼亚湾
Gulf of Guinea 几内亚湾
Gulf of Mexico 墨西哥湾
Gulf of Thailand 泰国湾
gulf port 海湾内港埠
Gulf ports 墨西哥湾各港
gulf red 铁红
gulf science 海湾科学
Gulf states 海湾国家
Gulf stream 北大西洋暖流；墨西哥湾流；墨西哥暖流；墨西哥海流
gulf stream 湾流
gulf stream eddy 湾流环
gulf stream front 湾流锋
gulf stream meander 湾流蛇曲
gulf stream system 湾流系统
gull 狭谷；(地滑造成的)张裂缝；扩大节理；海鸥；充满碎石的裂隙
gullet 锯齿高度；狭窄的航道入口；锯齿间空隙；水落管；齿室；齿槽
gullet depth 齿槽深度
gulleting 修整锯齿；狭路出土挖方法；切割锯齿；(结合狭路出土的)挖土力；舵柱承槽；舵板栓槽
gulleting of rudder 镶舵入座
gulleting saw file 圆锯锉
gullet of saw 锯齿间凹槽
gullet saw 钩齿锯；齿槽锯
gullet tooth 偏锋齿
gull(e)y 排水沟；冲刷沟
gull(e)y 排水沟；冲刷沟；沟渠；沟壑；街沟；海底狭沟(伸入陆地)；雏谷；冲沟
gull(e)y bed neighbouring to culvert 涵洞沟床
gull(e)y control 冲沟控制；沟蚀控制；沟蚀防治；沟壑控制
gull(e)y control dam 谷坊；水土保持坝
gull(e)y control planting 沟蚀防治种植；防治沟蚀植树造林法
gull(e)y correction 沟蚀校正
gull(e)y cover 沟盖(板)；窨井盖
gull(e)y cutting 沟蚀冲切
gull(e)y drain 下水道；雨水口连接管；排水渠(道)
gull(e)y drainage 排水沟
gull(e)y emptier 雨水口吸泥车；雨水口除泥车；吸去沟渠中泥沙的卡车；排空式油槽车
gull(e)y emptying 雨水口除泥
gull(e)y erosion 冲沟侵蚀；沟状侵蚀；沟状冲刷；沟形侵蚀；沟蚀；沟壑冲蚀
gull(e)y erosion form 沟蚀形态
gull(e)y grate 雨水口栅盖
gull(e)y grating 进水井盖；箅栅；雨水口；雨水栅盖
gull(e)y gravure 沟坡冲蚀后退；山坡沟蚀后退
gull(e)y grid 雨水口栅盖
gull(e)y head erosion 沟头侵蚀；沟头冲蚀
gull(e)y head tile 沟头瓦
gull(e)y hole 集水孔；沟渠集水孔；集水口
gull(e)ying 沟状侵蚀；沟状冲刷；沟蚀(作用)；开沟；冲沟作用
gull(e)y of loessland 黄土区冲沟
gull(e)y opening 雨水口孔隙
gull(e)y plugging 闸沟
gull(e)y pot 排水井；雨水井
gull(e)y reclamation 填谷造地；填沟造地
gull(e)y squall 峡谷飑
gull(e)y stabilization structure 沟壑防护工程；沟壑稳定建筑物
gull(e)y stabilization work 沟壑稳定工程
gull(e)y structure 沟状构造
gull(e)y sucker 路沟窨井吸淤车
gull(e)y surround 雨水口四周铺砌；进水口四周铺砌
gull(e)y trap 阴沟存水弯；进水口防臭设备；进水口防臭阱；集水井(水隔间)；下水道入口截污井；雨水口截流井；雨水口水封；雨水管道沉砂井；阴沟头；水沟排污阱；水沟防臭水封；沉泥井
gull(e)y washer 大暴雨
gullied area 沟蚀地区
gullied surface 沟壑地表；地面冲沟发育
gulls 气球假目标雷达反射器
gulp 抑制；字群；字节组
gulp valve 补气阀
gum 橡胶；胶；木焦油；煤气胶；树胶；刺槐树胶；桉树属
gum acacia 金合欢胶；阿拉伯胶
gum accroides 禾木树胶
gumacum 愈创木脂
gum arabic 阿拉伯(树)胶
gum asafetida 阿魏胶
gum band 橡胶带
Gumbel distribution 甘倍尔分布
gumbelite 含镁伊利石
gum belt 树胶带
gum benzoin 安息香(树)胶；安息香
gum bichromate 重铬酸盐光敏树胶
gum bloom 漆斑；树胶致雾
gum blushing 树脂致白
gumbo 坚硬黏土；黏土(状)的；强黏土；肥黏土
gumbo bank 淤泥河岸；泥滩
gumbo bit 钻进黏土层用钻头
gumbo(clay) 碱性黏土；重黏土；肥黏土
gum-boots 长统胶靴
gumboti 硬紧黏土
gum brush 胶水刷
gum canal 树胶管
gum cement 橡胶黏结剂；橡胶黏合剂；橡胶结合剂；树胶结合剂
gum compound 纯胶料
gum content 胶质含量；树胶含量
gum deposit 胶质沉淀物；积胶
gum duct 树胶管
gum dynamite 胶质火药；黄炸药
gum elastic 橡胶；弹性树胶
gum emulsion paint 合成橡胶(系)乳化漆
gum filler 橡胶填料
gum formation 树脂形成
gum forming 形成胶态；胶质形成；胶化
gum hose 树胶软管
gum inhibitor 阻胶剂；胶质抑制剂
gum kino 吉纳树胶；奇诺胶；桉树胶
gum lac 橡胶树脂；紫胶的别名
gumlactree 树脂油桐
gum level 胶质含量
gum like material 类胶物质
gum loader 除粉器
gummase 漆酵素
gum mastic 玛树胶；乳香树脂；乳香胶
gummed 涂胶的
gummed label 涂胶标签
gummed paper 胶纸
gummed sealing tape 树胶密封层
gummed tape 胶纸带
gummer 修锯机；煤粉清除工；涂胶机；除粉器
gummer machine 锯齿板磨锐机
gummet 麻刀油灰
gummic acid 阿拉伯酸
gummiferous 有树胶的；产树胶的
gumminess 含有树胶；树胶状；树胶质
gumming 浸油；浸胶；结胶；胶接；加树脂含黄；涂胶；树胶分泌；树胶的采集；除粉
gumming dirt 胶泥
gumming of piston rings 活塞圈烧焦
gumming test 黏合试验
gummi pertscha 马来乳胶
gummi plasticum 马来乳胶
gummosis 流胶现象
gummosity 黏性
gummous reaction 流胶反应
gummy 含树胶；拖漆刷的；树胶制
gummy appearance 树胶状的；黏稠的
gummy formation 黏性岩石；树胶状结构
gummy oil 胶质油；含大量胶质的油
gummy precipitation 胶质沉淀
gummy residue 胶质状残渣；胶质残余；树胶状沉淀
gummy sand 黏砂；含黏土多的砂
gum naval stores 树胶树脂产品；树胶产品
gum of pine 松树脂
gum passage 树胶道
gum plug 树胶塞
gumpnon 调和颜料的溶剂
gum pocket 树胶囊
gum pole 圆木
gum process 树胶印画法
gumption 颜料调和法
gum residue 树脂残渣
gum resin 橡胶树脂；脂松香；胶胶(树)脂；树胶
gum rubber 纯胶胶料
gum running 树胶熔炼
gum Sandarac 山达脂；香松树胶
gum set 黏凝
gumshoe 橡胶套鞋
gum-solution 树胶熔炼
gum spirit 松节油
gum-spirit cement 松节油灰泥；防水灰泥
gum spirits of turpentine 树胶精油
gum stock 纯胶料
gum stock solutions 纯胶料溶液
gumstower 除粉器
gum streak 树胶条纹
gum strip calender 条胶压延机
gum tape 树胶黏条
gum thus 结晶松香
gum tolerance 最高容许的胶质含量；胶质容许量
gum tree 桉树(木材)；桉木
gum turpentine 脂松节油；松节油；树胶精油
gum turpentine oil 橡胶用溶汽油；汽蒸松节油
gum vein 树胶纹线；树胶斑纹
gum waste 制胶废水；废树胶
gum water 阿拉伯胶溶液；胶水
gum window 橡胶透声窗
gumwood 橡胶树；枫木；含树脂木材
gun 注射器；火炮；喷射机；喷镀枪；喷补(机)
gun accessory 火炮附件
gun adapter 油枪嘴
gun application 喷涂；喷射
gun-applied 喷涂作业
gun-applied concrete 喷射混凝土
gun-applied furnace insulation 喷涂的加热炉绝热材料
gun-applied mortar 喷铺砂浆
gunar 舰用电子射击系统
gun asphalt 喷涂沥青
gun barrel 旋流分离器；枪管；枪筒；炮筒；沉淀罐；气体分离器；炮管

gunboat 自动卸载船;自动卸载车;炮艇;炮舰
gun brass 炮铜
gun bridge 枪桥
gun burner 枪式喷燃器
gun camera 照相枪;手提式摄影机
gun carriage 炮架
gun carrier 炮架
gun cement 喷涂水泥
gun consistence 喷浆用水泥;可喷射稠度
guncotton 硝化纤维素;硝化棉;火棉
guncotton magazine 火药库
gun cradle 炮架
guncreting 压力灌浆混凝土;压灌混凝土;喷(射灌)浆混凝土
gundeck 炮台甲板
gun deflection board 方向修正板
gun drill 枪孔钻;单槽钻床
gun drilling 深钻孔
gun drilling machine 枪孔钻床;深孔钻床
gun-driven rivet 枪铆铆钉
gun-driven riveting 枪射铆接
gun elevation 炮的仰角
gun-feed reverberator 料枪给料反射炉
gun finish 喷射混凝土涂层;喷射混凝土饰面;喷枪修整;喷枪罩面;未经手工抹光的喷浆面层;表面喷枪修整
gun fore-end 枪前托
gunge 油污
gun grade 宜喷射稠度
gun-grade asphalt 喷洒沥青;喷涂沥青
gun-grade cork 喷涂软木(屑)
gun-grade form 喷涂泡沫材料
gun-grade rendering 喷枪粉刷
gun-grade sealant 枪用密封剂
gun-grade vermiculite 喷涂蛭石(屑)
gun hose 喷射器管;喷枪软管
guningite 水锌钒
gun-injection 枪注
gun iron 炮铁;半钢;钢性铸铁
gunis 采空矿穴
Gunite 钢性铸铁;冈奈特可锻铸铁;压力浆;喷涂水泥砂浆;喷浆;喷射水泥(砂)浆;喷浆法;喷枪;喷浆;素喷法;水泥(喷)枪
gunite coat(ing) 喷浆涂层;喷浆保护层;喷浆涂层
gunite covering 喷浆面层;压力喷浆面层;喷浆(罩)面;喷浆涂层;喷浆(保)护层
gunited concrete 喷浆混凝土;喷射(浇灌)混凝土
gunited concreting 喷射浇灌
gunited material 喷涂材料
gunite gun 喷水泥枪;喷浆器
gunite jacket 喷枪套
gunite layer 喷浆(保护)层
gunite lining 压力喷浆法衬砌;喷浆支护;喷浆衬砌;水泥喷射灌浆;水泥喷浆衬砌
gunite lining of tank 喷浆罐池
gunite machine 压力喷浆机
gunite material 喷浆材料
gunite method 喷浆法
gunite mortar 喷浆砂浆
gunite plant 喷浆装置;喷浆车间
gunite process 喷浆法;涂水泥法
guniter 灌浆机操作工
gunite slab 喷浆混凝土板
gunite work 混凝土喷灌;喷浆工作
guniting 压力喷浆;喷射水泥;喷射灌浆(法);喷射法;喷浆(法)
guniting aid 喷浆辅助器
guniting machine 喷浆器;喷浆机
guniting thickness 喷浆厚度
gunjet 喷水器;喷枪
gunjet nozzle 喷枪式喷嘴
Gunk 冈克(商品名)
gunk 泥状料;预混料;油腻料
gunlock 集中同步系统;闭锁机
gunman 喷浆工;喷射混凝土工
gun metal 锡锌青铜;锡铜合金;青铜;炮筒;炮铜
gunmetal 锡铜合金;炮铜
gunmetal bearing 炮铜轴承
gunmetal bronze 炮铜
gunmetal bush 炮铜轴衬
gunmetal cage 青铜保持架
gunmetal gray 铁灰色
gunmetals motor carriage 机械化炮车
gunmetals mount 炮架

gun mix bond 喷浆黏合料;喷浆黏合层
gun mount 炮架
gunnable 可喷涂的;可喷浇的
Gunn altimeter 电容式测高计
Gunn diode 耿氏振荡器
gunned asphalt 喷洒的沥青
gunned castable 喷射水泥;喷浆
gunned coat 喷涂层
gunned coat of paint 喷漆层
gunned film 喷涂薄膜
gunned insulation 喷涂隔热隔声材料
gunned ladle 喷补盛钢桶
gunned lining 喷补的内衬
gunned mastic 喷浆玛瑞脂;喷涂沥青
gunned plastics 喷涂塑料
gunned putty 喷灌油灰
gunned refractory 喷补用耐火材料
gunned repair 喷浆修理
gunned varnish 喷清漆
Gunn effect oscillator 耿氏效应振荡器
Gunn effect power amplifier 体效应功率放大器
Gunn effect solid state frequency source 耿氏效应固态频率发生器
gunnel 舷边;护舷材;船舷上缘
gunnel tank 舷侧水柜
gunner's quadrant 火炮象限仪
gunnery base 射击基地
gunnery range 射击场;靶场
gunnery ship 炮舰
gunning 压力喷浆;喷浆;喷补(炉衬)
gunning and slinger process 喷投成型法
gunning concrete 喷涂混凝土
gunning crew 喷射混凝土施工队
gunning material 喷射材料;喷浆材料
gunning mix 喷射(混合)料;喷补料
gunning pattern (喷射混凝土形成的);圆锥外形喷射浇模;喷浆直径
gunning process 喷射成型
gunning refractory 耐火喷射料;喷补料
gunning rig 喷浆设备;喷浆枪
gunning technique 喷涂技术;喷浇技术
gunnite 压力喷浆
Gunn mode 耿氏模式
Gunn oscillation element 耿氏振荡元件
Gunn oscillator 耿氏振荡器
gunny 麻布;粗(黄)麻布
gunny-and-paste pretreatment for painting 披麻捉灰(中国在木结构上刷油漆的传统做法)
gunny bag 麻(布)袋;粗麻袋
gunny cloth 麻袋片;麻布
gunny fiber 粗黄麻纤维
gunny needle 麻包针
gunny sack 粗麻(布)袋;麻袋
gunny sackbag 粗麻布袋
gun parallax 方位差
gun park 停炮场
gun pendulum 弹道摆;冲击摆
gun-perforate 用打孔枪穿孔机;放炮穿孔
gun perforator 射孔器(深钻钻工具);穿孔器(油井)
gun-perforator loader 射孔枪装药器
gun performation 射孔
gun platform 炮座
gunpowder 有烟火药;火药;黑色炸药
gunpowder residue 火药残留物
gun power 火药
gun pressure 腔压
gun putty 喷灌油灰
gun rubber 天然橡胶
gun sampler 冲击式(侧壁)取样器
gun sealant 喷射嵌缝膏
gunshot 射程
gunsight 瞄准具
gunsight aiming point camera 瞄准照相枪
gunsight computer 瞄准具计算机
gunsight device 瞄准装置
gun sprayer 喷枪式喷雾机
gun spraying 喷射;喷枪喷雾;喷枪喷涂;喷枪喷洒;喷浆
gun spring 扁方弹簧
gun stock 枪托状门窗边梃;倒向器
gunstocking 船首尼木甲板边梃
gunstock stile 枪托状门窗边梃;不等宽门窗边梃
gun tackle 起重滑车;神仙葫芦

gun tap 枪式丝锥
gun template (由板条组成的)三角形样板
Gunter's chain 英尺测链;甘特测链;66尺测链;测链(相当于66英尺长)
gunter iron 挂帆钩
gunter yard (小艇的)三角帆滑动上桅
gun tube 炮管
gun turret 炮塔
gun-type burner 喷射式燃烧器
gun-type coring machine 枪式取芯器
gun-type oil burner 枪式油炉枪;枪式油燃器
gun type soot blower 枪式吹灰机
gunwale 舷缘;舷边(缘);甲板边缘;护舷材;船舷上缘
gunwale(angle)bar 舷边角钢
gunwale down 舷缘着水;舷边和水面相平
gunwale plate 舷缘边板
gunwale rail 舷边栏杆
gunwale strake 舷边列板
gunwale tank 舷侧水柜
gunwale to 舷缘着水;舷边和水面相平
gunwale under 舷边没入水面以下
gun welder 半自动电弧焊接机
gun welding machine 手提式点焊机;点焊枪
Gunz glacial stage 贡兹冰期【地】
Gunz glaciation 贡兹冰期【地】
Gunz-Mindel interglacial stage 贡兹—民德间冰期
Guo Biao[GB] 国标
gupaiite 古北矿
Gupta's long arm centrifugal pipette 哥普塔长臂离心管
gurdy(gurgite) 卷绳车
Gurevich effect 古雷维治效应
Gurgan 红褐色硬木(产于印度、缅甸、安达曼群岛等处)
gurgling well 脉动喷井
gurgolye (哥特式建筑上的)滴水嘴
gurhofite 雪白云石
Gurjun 红褐色硬木(印度、缅甸、安达曼群岛等处)
gurjun 古芸香胶
gurjun balsam 古芸香脂
gurlet 丁字斧
Gurney-Lurie chart 葛尼—鲁利传热图
gush 泉涌;喷出
gusher 自喷井;喷油井;喷出物;间歇喷泉
gusher hole 喷泉口
gusher type well 自流钻井
gusher(well) 自喷(油)井
gushing spring 涌泉;火山泉;喷泉
gushing water 涌水;突水
gushing water hole 涌流孔
guss asphalt 流态地沥青;摊铺地沥青
guss-asphalt surfacing 摊铺地沥青(混凝土)面层
guss concrete 流态混凝土;摊铺混凝土
gusset 连接板;角撑板;加力片;加力板;结点板;加固板
gusset angle 扣角钢;节点角钢
gusset box 接点板
gusset connection 结点板连接;节点板连接
gusset connector 节点板接头
gusseted base 角板底座
gusseted connection 角撑连接
gusseted joint 节点板接头
gusset felt 箱衬纸板
gusset piece 金属屋面连接片;屋顶坡槽间隙盖板;连接板
gusset plate 加固板;缀板;连接板;扣板;结点(连接)板;角撑板
gusset plate connection 节点板联结
gusset plate splice 接点板拼接
gusset stay 节点拉条;结节撑;扣牵条;结点牵条;结点(角)撑;角板撑条
gussetted multiwall paper bag 装有角撑板的多层纸袋
gust 骤风;突发
gustavite 辉铋银铅矿
gust duration 阵风时间
gust factor 阵风系数
gust frequency 阵风频数
gust-gradient distance 突风梯度距离
gustiness 阵(风)性;阵风风速;阵(发)性;大气湍流度
gustiness components 阵风分量
gustiness factor 骤风因素;阵风系数

gust load 骤风荷载;阵风荷载
gust load factor 骤风荷载系数
gust of rain 骤雨;阵雨
gust of wind 阵风;疾风;风流;风暴
gust peak speed 阵风最大风速
gust room 客房
gustsonde 阵风探空仪
gust spectrum 阵风谱
gust speed 阵风速度;聚风风道
gust tunnel 风洞;突风风洞
gust velocity 阵风速度
gusty wind 阵风
gut 狭水道;狭海峡;窄峡
gut burglar 工地炊事员
Gutenberg discontinuity 古登堡不连续面
Gutenberg magnitude relation 古登堡能量—震级关系式
Gutenberg-Richter's law of magnitude 古登堡—里克特震级定律
Gutenberg-Richter magnitude scale 古登堡—里克特分级表
Gutenberg-Richter relation 古登堡—里克特关系式
Gutenberg-Richter table 古登堡—里克特走时表
gut hammer 铁磬
Gutherie cutch extract 格思里树皮胶
guthion 谷硫磷
Guti weather 古蒂天气
gut line 缆车用粗钢缆
gutsevichite 水磷钒铝矿
gutta 钉头饰;滴状装饰;圆锥饰;固塔坡胶;[复] guttae
gutta-balata 巴拉塔树胶
gutta-jelutong 节路顿胶
gutta percha 马来树胶;杜仲(橡)胶;马来亚胶;树胶汁;古塔胶
gutta percha cable 杜仲树胶绝缘电缆;杜仲胶(海底)电缆
gutta-percha ring 树胶圈
guttation 吐水作用【给】
gutted 火毁的
gutted structure 遭火灾的房屋;火毁的房屋
gutte pipe filter 水落管滤池
gutter 中沟;流槽;空页【计】;街沟;角形火焰稳定器;明沟;排水(边)沟;排除雨水的檐沟;通风口;水沟;水槽;出料槽
gutter angle 边沟转角
gutter application 雨水管工程
gutter apron 平自【道】
gutter area 飞边槽面积
gutter beam of U section 槽形檐沟梁
gutter bearer 檐槽托(架);天沟托板;天沟托木
gutter bed 铅制檐沟床;披水板;天沟挡水板
gutter block 边沟砌块
gutter board 屋顶天沟挡雪板;天沟托板;檐槽挑口板;挑檐板;挑口板(檐槽);槽檐板;挡雪板
gutter bolt 天沟螺栓;檐槽螺栓
gutter bracket 檐槽托(架);天沟托架
gutter cast 小沟铸型
gutter cleaner 清沟机
gutter corner tile 天沟边角瓦
gutter drainage 明沟排水;渠道排水;边沟排水;明渠排水
gutter end stop 排水沟终端
gutter end tile 排水沟终端瓦
gutter expansion joint 沟渠的伸缩节
gutter fitting 排水沟配件
gutter grade 边沟底坡
gutter grade line 边沟纵坡线
gutter gradient 沟槽坡度;排水沟坡降
gutter grate 街沟水箅;街沟进水格栅;边沟进水格栅
gutter grating inlet 平箅式进水口;格栅式进水器
gutter hanger 檐沟托;檐沟挂钩;檐槽吊钩;天沟支脚
gutter hook 檐沟挂钩
guttering 开沟(法)
gutter inlet 雨水口;路沟进水口
gutter man 摊贩;清阴沟的工人
gutter member 水落饰件
gutter of filter 滤池洗水槽
gutter offtake 路沟泄水口;马路边沟;马路边排水口
gutter of roadside 路旁阴沟;路面阴沟
gutter outlet 天沟排水口
gutter paving 沟槽铺面;沟槽面
gutter plank 槽盖;檐槽(木)板

gutter plate 檐槽(竖)板;天沟侧板;天沟下托梁板
gutter-plough 犁式挖沟机
gutter plow 开沟犁
gutter-receiving stone 雨水管下承石
gutter section 路沟断面;街沟断面
gutter sett 砌沟石
gutter sheet 檐沟薄板
gutter slope 排水沟边坡
gutter sole 沟底
gutter spout funnel 落水斗;水落斗
gutter stabilizer 窄槽稳定器
gutter steel bearer 天沟钢托
gutter steel hanger 天沟钢吊
gutter stone 街沟石;沟(底)石
gutter support 天沟支承
gutter tile 天沟瓦;沟瓦
gutter tool 开沟工具;排水沟整修器;排水沟建造工具
gutter way 排水沟;雨水沟
gutter work 沟槽工程
guttings 路面用细石屑
Guttman process 谷特曼化学加固法
guttra 格特拉飚
Gutzeit's test 古特泽特砷检测法
guy 牵索;天线杆拉线
Guyana basin 圭亚那海盆
guyanaite 圭羟铬矿
Guyana old land 圭亚那古陆
Guyana shield 圭亚那地盾
guy anchor 拉索桩;系索桩;拉线桩;拉线地锚;拉索锚(固);拉索锚(碇);锚碇;地锚
guy cable 缆风;缆绳;拉索
guy chain 拉链;牵链
guy clamp 拉线夹(板);锚接绳固定线夹;绑绳卡子
guy clevis eye-bolt 双耳连接头
guy clip 线卡子
guyed antenna mast 拉线式天线杆
guyed derrick 牵索起重机;拉索式起重机;索桅式动臂起重机;摆转起重架
guyed derrick crane 动臂式人字形起重机;牵索(式)桅杆起重机;牵索转臂起重机;牵索人字起重机;索桅式动臂起重机;缆绳式起重机;动臂式桅杆起重机;桅缆式动臂起重机
guyed iron chimney 牵索铁烟囱
guyed mast 桅杆;拉线式塔架;拉线式井架;拉线式电杆;绑绳稳定的轻便桅杆
guyed pole 拉线杆
guyed steel stack 用风缆拉紧的钢烟囱;牵索钢烟囱;牵索钢烟囱
guyed structure 牵索结构;拉索结构
guyed tower 曳索塔;牵索塔;带缆塔架;带拉线塔
guy hook 拉线钩
guying 牵索的;牵索调位
guy insulator 拉线绝缘子
guy line 缆风;缆绳;拉线;牵绳;堵漏毡吊索;绷索;绑绳
guyot 桌状海丘;海底平顶山;平顶海山
guy ring 集索圈;绑绳环
guy rod 支杆;拉线桩;拉杆
guy rope 系缆;支索;张索;钢缆;绳索;拉索;绑绳;牵索;稳索
guy span 吊长件货物索具
guy stake 系索桩;索系桩
guy strain insulator 拉线耐拉绝缘子
guy stub 拉线桩
guy table 钢丝绳转台(桅杆起重机);钢丝绳转盘(桅杆起重机)
guy tackle 稳索绞辘
guy thimble 牵索套管
guy tightener 紧索轮
guy winch 张索绞车
guy wire 牵索;拉线;钢缆;支索;张索;绑绳
guy-wire insulator 拉线绝缘子
guy-wire slide (起火时用)紧急下塔滑绳
guzebo 阳台
gybe 改变航道
gybing 改变帆的方向
Gyffin shale 吉芬页岩
Gylden method 吉尔当方法
gylot 陀螺罗经操舵仪
gym 健身房;体育馆
Gymbals 杰姆巴万向支架
gym finish 健身房地板面漆
gym-horse 木马

gym mat 运动场用垫
gymmer 一对
gynaeceum 妇女专用房间(古希腊、古罗马建筑中)
gymnasium 体育馆;练身房;健身房;[复]gymnasia
gymnasium equipment 体育馆设备
gymnastic hall 健身房
gymnastic room 健身房;体操房
gymnite 水蛇纹石
Gymnophiona 两栖类
gymnosperm 裸子植物
Gymnospermae 裸子植物门
gym room 练身房
gynaeceum 闺房(回教的);阁闱(希腊人住宅中的)
gynecological examining table 妇科诊察床
gynecological surgery 妇外科
gynocardia oil 大风子油
gynospore 大孢子
gyotaku 鱼拓
gyp 绝缘检漏仪
gyphosate solution 草甘膦水剂
gypklith 刨花(丝)建筑板材
gyppo (实力不足的)小公司
Gyproc 建筑产品的商业总名称
gyprock 石膏岩;石膏为主要成分的岩石
gyps(e)ous 石膏状的;含石膏的
gypseous alabaster 雪花石膏
gypseous horizon 石膏聚积层
gypseous karst 石膏岩溶
gypseous marl 土状石膏;石膏状灰泥
gypseous sand 石膏屑
gypseous scale 石膏冰垢
gypseous solid rock 石膏状坚石
gypseous spar 石膏晶石
gypsey 间歇河(流)
gypsiferous 含石膏的
gypsification 石膏化
gypsite 土(状)石膏
gypsography 石膏雕刻术
gypsoide 石膏状矿物
gypsophila patrini 巴氏丝石竹
Gypstele system 吉普斯塔拉体系(一种合金边框石膏板构造体系)
gypsum 石膏(块);二水石膏
gypsum acoustic(al)board 石膏吸声板
gypsum activation 石膏激发
gypsum adhesive 石膏胶黏剂
gypsum aluminate expansive cement 石膏矾土膨胀水泥
gypsum and hydrite deposit 石膏和硬石膏矿床
gypsum anhydrite 无水石膏;硬石膏
gypsum-anhydrite ore 石膏硬石膏矿石
gypsum article 石膏制品
gypsum backerboard 石膏衬板
gypsum backing board 石膏衬板
gypsum backing(mixed)plaster 石膏打底抹灰
gypsum base board 石膏灰泥纤维板;石膏底板
gypsum basecoat(mixed)plaster 石膏打底抹灰
gypsum based 以石膏为基料的;石膏打底的
gypsum based mortar 石膏基料砂浆
gypsum bin 石膏仓
gypsum block 石膏材料;石膏墙板;石膏砌块;石膏块材
gypsum block cement 石膏块材
gypsum block medium 石膏块培养基
gypsum block partition 石膏砌块隔断
gypsum board 石膏板顶棚;石膏(条)板
gypsum board enclosure system 石膏板围护系统;石膏板隔墙系统
gypsum board lath 石膏板条
gypsum board sheathing 石膏板衬板
gypsum board wall 石膏板墙
gypsum bond(ing)plaster 黏结用石膏灰;石膏打底灰;黏贴石膏
gypsum breaker 石膏粉碎器
gypsum brick 石膏砖
gypsum building material 石膏砌块;石膏建材;石膏板
gypsum calcinating kettle 石膏炒锅
gypsum calcination 石膏煅烧
gypsum cassette slab 石膏夹芯板
gypsum cast building unit 预浇石膏建筑单元
gypsum ceiling 石膏吊顶
gypsum ceiling board 石膏天花板;石膏吊顶板
gypsum cement 石膏水泥;石膏胶结物;石膏灰

泥;石膏粘结剂;金氏水泥(干固水泥)
gypsum cement plaster 净石膏灰;石膏胶凝灰
gypsum-coated wall fabric 石膏浸渗墙布
gypsum composite panel 石膏复合板
gypsum concrete 石膏混凝土
gypsum concrete mill mixture 石膏混凝土拌和料
gypsum content 石膏含量;含石膏成分
gypsum continuous kettle 石膏连续炒锅
gypsum core 石膏芯
gypsum core board 石膏芯板
gypsum crusher 石膏粉碎器;石膏破碎机
gypsum crushing plant 石膏破碎机;石膏破碎车间
gypsum crystal 石膏晶体
gypsum cutting 石膏侵
Gypsum deep 吉普瑟姆海渊
gypsum deposit 石膏矿床
gypsum-dolomite formation 石膏白云岩建造
gypsum-dolomite sub-formation 石膏白云岩亚建造
gypsum drywall 石膏板墙
gypsum earth 石膏土;土(状)石膏
gypsum expansion on setting 凝结时石膏膨胀
gypsum fiber concrete 石膏纤维混凝土
gypsum filler(block) 石膏填块;石膏填孔料
gypsum filling 石膏填孔
gypsum finish 石膏抹灰饰面
gypsum fireproofing 石膏防火盖面;石膏防火盖板
gypsum floor 无缝石膏地面
gypsum floor pot 空心石膏地砖
gypsum form board 石膏模板
gypsum-free 不含石膏
gypsum free cement 无石膏水泥
gypsum ga(u)ging plaster 石膏掺和浆;调凝石膏浆;石膏罩面灰泥
gypsum gauze 石膏纱布
gypsum hand plaster 手工粉刷石膏
gypsum heated in kettle 锅制石膏
gypsum heated in pan 盘制石膏
gypsum hemi-hydrate 半水石膏
gypsum hollow wall panel 石膏空芯墙板
gypsum hopper 石膏仓
gypsum horizon 石膏层
gypsum insulation 石膏保温
gypsum iron slag cement 石膏化铁炉渣水泥
gypsum job-mix(ed) basecoat plaster 现拌石膏底灰灰浆
gypsum joint filler 石膏嵌缝料;石膏嵌缝膏
gypsum kettle 石膏烧锅
gypsum lath board base 穿孔石膏板底层
gypsum lath(ing) 石膏抹灰底板;纸面石膏板;石膏(条)板
gypsum lath nail 石膏条板钉;石膏板(条)钉
gypsum lath partition 石膏板隔断
gypsum layer 石膏层
gypsum-less 不含石膏
gypsum-lime mortar 石膏(石)灰浆;建筑用石膏;雪花石膏
gypsum-lime plaster 石膏石灰灰浆
gypsum-lime stuff 石膏石灰拌料
gypsum machine plaster 机械粉刷石膏
gypsum marble 石膏大理石
gypsum marl 石膏泥灰
gypsum masonry 石膏圬工;石膏砌体
gypsum meal 石膏粉
gypsum model 石膏模(型)
gypsum mortar 石膏灰浆
gypsum mo(u)ld 石膏型;石膏模(型);石膏外模
gypsum mo(u)lding plaster 石膏线条粉刷;模型石膏粉刷;造型石膏;铸模石膏;石膏线脚灰
gypsum mud 石膏泥浆
gypsum neat plaster 净石膏灰;纯石膏灰;净纯石膏灰泥;石膏净粉饰
gypsum ore 石膏矿石
gypsum pan 石膏(烧)盘
gypsum panel 石膏板;石膏(芯)墙板;石膏隔墙板
gypsum partition block 石膏隔墙砌块
gypsum partition tile 石膏隔墙砖
gypsum paste 石膏浆
gypsum pat 石膏试块
gypsum pattern 石膏模
gypsum perlite plaster 珍珠岩石膏灰泥;石膏珍珠岩灰浆
gypsum perlite wall board 石膏珍珠岩墙板
gypsum plank 石膏抹灰底板;石膏(条)板
gypsum plank with reed 芦苇石膏板
gypsum plant 石膏工厂;石膏植物
gypsum plaster 石膏抹料;石膏胶凝材料;石膏;石膏灰(泥);粉饰用石膏
gypsum plasterboard 石膏壁板;石膏装饰板;石膏灰胶纸柏板;纸面石膏板;石膏(墙)板
gypsum plasterboard panel 石膏抹灰板
gypsum plaster ceiling 石膏抹灰顶棚;石膏天花板
gypsum plaster for building purpose 建筑石膏
gypsum plaster(ing) 石膏粉刷;石膏粉饰
gypsum plaster mixture 石膏抹灰混合料
gypsum plate 石膏片;石膏板
gypsum pottery plaster 陶瓷用熟石膏
gypsum powder 石膏粉
gypsum precast building unit 预制石膏构件
gypsum product 石膏制品
gypsum putty 石膏浆
gypsum quarry 石膏开采
gypsum ready-mixed plaster 预拌石膏灰;预拌石膏砂浆
gypsum ready-mixed stuff 现成石膏抹灰拌料
gypsum ready-sanded plaster 预掺砂石膏灰
gypsum requirement 石膏需用量
gypsum-retarded cement 石膏缓凝水泥;石膏缓冷水泥
gypsum retarder 石膏缓凝剂
gypsum retarding agent 石膏缓凝剂
gypsum ribwall wallboard partition 带肋石膏墙板隔墙
gypsum road 石膏(三合土)路
gypsum rock 天然石膏;石膏岩
gypsum roof(ing) board 石膏屋面板
gypsum roof(ing) slab 石膏屋面板
gypsum roof(ing) tile 石膏瓦
gypsum roof(ing) plank 石膏屋面板
gypsum-salt red beds association of continental rift type 大陆裂谷型膏盐红层联合
gypsum sand 碎屑石膏;石膏粒
gypsum-sand mortar 石膏砂浆
gypsum-sand plaster 石膏砂抹灰
gypsum sheathing 石膏盖板;石膏衬板;石膏望板
gypsum sheathing board 石膏墙面板;防水石膏衬板;石膏望板
gypsum sheathing plasterboard 石膏墙面板
gypsum sheet 石膏薄板
gypsum sheet ceiling 石膏薄板顶棚
gypsum sheet partition 石膏薄板隔墙
gypsum slab partition 石膏板隔断
gypsum slag board 石膏矿渣板
gypsum slag cement 石膏矿渣水泥;石膏块材
gypsum slurry 石膏泥浆;石膏稠浆
gypsum spar 石膏晶石;透石膏
gypsum sphere 石膏球
gypsum stains 石膏斑
gypsum stor(ag)e 石膏堆栈;石膏仓库
gypsum store 生石膏存放
gypsum stud 石膏墙筋;石膏龙骨
gypsum stuff 石膏拌料
gypsum-sulfuric acid process 石膏制造硫酸(和水泥)的方法
gypsum tile 石膏板;石膏瓦;石膏(面)砖
gypsum transformation 石膏变形
gypsum trowel finish 石膏抹面装饰;石膏粉刷饰面
gypsum undercoat(mixed) plaster 石膏底抹灰
gypsum unit 石膏构件;石膏单元;石膏墙板;石膏砌块
gypsum veneer plaster 石膏镶饰灰
gypsum-vermiculite plaster 石膏蛭石灰浆
gypsum waffle slab 石膏双向密肋板
gypsum wall baseboard 石膏墙板
gypsum wallboard 纸面石膏板;石膏墙板;石膏壁板
gypsum wall plaster 石膏墙面抹灰
gypsum wood board 石膏刨花板
gypsum wood-fibered plaster 木纤维石膏板;夹木纤维石膏抹灰;石膏木丝灰泥;石膏木丝灰浆
gypsum wood slab 石膏刨花板
gypsy (绞车上的)铰绳筒;绞缆筒;绞车筒;锚机滚筒;卧式绞盘
gypsy capstan 铰绳起锚机
gypsy head 绞绳筒;绞缆筒
gypsy spool 绞车辘轳
gypsy wheel 锚链轮;锚机持链轮;副卷筒
gypsy winch 小型手绞车
gyracone crusher 旋回圆锥破碎机
gyractor 旋流池
gyradisc crusher 转盘式细轧机;转盘式破碎机;转盘式粉碎机;转盘式轧机;转盘式碎石机
gyrasphere crusher 旋回球面破碎机
gyrate 旋转;回转
gyrating breaker 回转(式)破碎机
gyrating crusher 回转(式)粉碎机
gyrating current 旋转流;螺旋流;涡流;八卦流
gyrating mass 旋转质量
gyrating screen 平面旋回筛
gyration 旋转(运动);回转;环动
gyration radii 回转半径
gyration release 回转释放;回转解除
gyrator 旋转子;旋转器
gyrator filter 回转器滤波器
gyratory 回旋破碎机
gyratory ball-mill 回旋球磨机
gyratory breaker 旋回(式)碎石机;回转(式)轧碎机;回转(式)碎石机;回转(式)破碎机;环动(式)碎石机;环动(式)轧碎机;环动(式)破碎机
gyratory compactor 旋转式压实机;回转式压实机;回转式夯实机;回转式夯具
gyratory cone breaker 回转式圆锥破碎机
gyratory cone crusher 锥形破碎机
gyratory crusher 旋转式碎石机;旋转式破碎机;旋回(式)破碎机;圆锥碎石机;回转压碎机;回转式轧碎机;回转式碎石机;回转(式)破碎机;环动碎石机;环动破碎机
gyratory fixed-spindle crusher 定轴式回旋碎矿机
gyratory intersection 环形交叉;转盘式交叉(道路)
gyratory junction 转盘式交叉(口)
gyratory lift bridge 旋转升降桥;回旋升降桥
gyratory mill 锥形磨机
gyratory motion 旋回运动;回转运动
gyratory movement 旋回运动;回转运动
gyratory pillar shaft crusher 定轴式回旋碎矿机
gyratory press 回转式压力机
gyratory primary crusher 旋回式一次破碎机
gyratory riddle 偏心振动筛;偏心吊筛
gyratory screen 旋转筛;回转(振动)筛;平面旋回筛;偏心振筛
gyratory shaker 旋转搅拌机
gyratory shaker table 偏心振动台
gyratory shear method 旋转剪力法
gyratory sifter 回转式筛
gyratory stabilizing system 陀螺稳定系统
gyratory testing machine 回转式试验机
gyratory traffic (交叉口的)环行交通
gyre 环行洋流;环流圈;涡旋;大漩涡
gyro 陀螺仪
gyroaccelerometer 陀螺加速计
gyro adjuster 罗盘校正器
gyroantihunt 陀螺仪减摇振装置
gyroattachment 陀螺仪附件
gyroautomatic navigation system 陀螺自动导航系统
gyroaxis 陀螺轴;陀螺仪(主)轴;回转轴
gyroazimuth 陀螺方位角
gyroazimuth theodolite 陀螺经纬仪
gyroazimuth transit 陀螺经纬仪
gyroball 陀螺球
gyrobearing 陀螺仪方位;陀螺方位(角)
gyrobus 飞轮车
gyrocamera 陀螺仪照相机
gyrocar 单轨车
gyrocase 陀螺组合件的外壳
gyroclinometer 回转式倾斜仪;回转式倾斜计;陀螺测斜仪
gyrocompass 回转罗盘;陀螺仪;陀螺罗盘;陀螺罗经;电罗经
gyrocompass alignment 陀螺罗经校准指北;陀螺罗经对准
gyrocompass bearing 陀螺罗方位
gyrocompass course 陀螺罗航向
gyrocompass error 陀螺罗经(误)差
gyrocompassing 陀螺平台指北;陀螺罗经状态
gyrocompass log 陀螺罗经日志
gyrocompass north 陀螺罗北
gyrocompass repeater 回转罗盘转发器;陀螺罗经复示器;电罗经复示器;辅助罗盘;辅助罗经
gyrocompass room 陀螺罗经室
gyrocon 高功率微波放大器
gyrocontrol 陀螺控制;陀螺操纵
gyrocontrol installation 陀螺仪控制装置

gyrocontrol unit 陀螺罗经控制箱
gyroconverter 陀螺转换器
gyrocourse recorder 陀螺罗经航向记录器
gyrodozer 铲斗自由倾斜式推土机
gyro-driving 回转传动
gyrodynamics 陀螺动力学
gyrodyne aircraft 直升(飞)机
gyroelectric 回转电路
gyroelectric(al) medium 旋电介质
gyro emission 回转发射
gyroerectional navigation system 陀螺光学导航系统
gyroerror 陀螺仪误差;陀螺罗经差
gyro flux-gate compass 磁通闸门陀螺罗经;陀螺感应同步罗盘;陀螺感应罗经
gyroflywheel 陀螺转子
gyro frequency 旋转频率
gyrogain 陀螺增益
gyro-granulator 细碎圆锥破碎机;陀螺成球机
gyrograph 旋转测度器;转数指示器;转数图示器;转数记录器;陀螺仪曲线图;陀螺船首偏荡指示图
gyro gunsight 回转式瞄准器;陀螺仪瞄准器
gyrohorizon 陀螺水平仪;陀螺地平
gyrohorizon indicator 回转地平仪;陀螺地平仪
gyrohousing 陀螺仪外壳
gyrohydraulic steering control 陀螺液压操舵系统
gyrohydraulic steering system 陀螺液压操舵装置
gyroid 五角三八面体
gyroindication 陀螺仪指示
gyroinstrument 陀螺仪
gyrointegrating accelerometer 陀螺积分加速度计
gyrointegrator 陀螺积分器
gyrolaw of precession 陀螺运动规律
gyrolevel 陀螺水准仪;陀螺水平(仪);陀螺测斜仪
gyrolite 吉水硅钙石;白钙沸石
gyrolog 陀螺仪记录簿;陀螺船首偏荡记录器
gyromachine 陀螺电机
gyromag 陀螺磁罗盘;陀螺磁罗经
gyromagnetic 回转磁
gyromagnetic compass 陀螺磁罗盘;陀螺磁罗经
gyromagnetic constant 回转磁常数
gyromagnetic coupler 回转磁耦合器
gyromagnetic effect 旋磁效应;回转磁效应
gyromagnetic ferrite 旋磁铁氧体
gyromagnetic frequency 旋磁频率;回旋频率
gyromagnetic material 旋磁材料
gyromagnetic medium 旋磁介质
gyromagnetic radius 拉莫尔半径
gyromagnetic ratio 旋磁比
gyromagnetic resonance parameter 回转磁谐振参量
gyromagnetics 旋磁学
Gyromax balance 基隆麦克斯摆轮
gyromechanism 陀螺机构
gyromeridian 陀螺仪子午线
gyrometer 陀螺测速仪;陀螺测试仪
gyromixer 回转(式)拌和机;回转(式)搅拌机;环动(式)拌和机
gyro-moment 回转力矩;陀螺力矩
gyromotor 陀螺马达;陀螺电动机
gyroon-off switch 陀螺开关
gyroorientation error 陀螺定向误差
gyrooutput-axis pick-up 陀螺输出轴信号传感器
gyropackage 陀螺仪组
gyropanel 陀螺平台
gyro-pen 回转笔
gyropendulum 陀螺摆
gyropilot 自动驾驶仪;陀螺自动操舵(装置);陀螺罗经自动舵;陀螺驾驶仪
gyropilot steering 陀螺驾驶仪操舵
gyro pitch and roll recorder 纵横摇舵自动记录仪
gyroplane 旋翼机;自转旋翼机
gyroplatform 陀螺平台
gyroprecession 陀螺进动性
gyroprobe 陀螺探管
gyropter 旋转式飞机;旋翼飞机
gyrorake 立轴回转式搂草机
gyro-reciprocating engine 旋转往复式发动机
gyroreference package 陀螺基准组件
gyroreference system 陀螺基准系统
gyrorelaxation effect 回旋弛豫效应
gyrorelaxation heating 碰撞加热
gyrorepeater 电罗经复示器;陀螺复示器
gyroresonance 旋磁共振
gyroroom 陀螺主罗经室
gyro-rotor 回转体;陀螺转子
gyrorotor end play micrometer 陀螺转子端隙测量计
gyrorudder 陀螺自动驾驶仪
gyrorudder control 陀螺自动驾驶仪控制
gyroscope 旋转仪;回转仪;回转器;环动仪;陀螺仪
gyroscope autopilot 陀螺自动驾驶仪
gyroscope equipment 陀螺仪装备
gyroscope integrator 陀螺积分器
gyroscope meridian line 陀螺仪子午线
gyroscope position pick-up 陀螺仪位置传感器
gyroscope reference system 陀螺基准系统
gyroscope rotor 陀螺转子
gyroscope wheel 回转轮
gyroscopic accelerator 陀螺加速度表
gyroscopic action 回转作用;陀螺作用;陀螺动作
gyroscopically balanced flowmeter 陀螺作用的流量计
gyroscopic apparatus 陀螺仪
gyroscopic bearing 陀螺方位
gyroscopic camera 陀螺仪照相机
gyroscopic-clinograph method 陀螺测斜仪法
gyroscopic clinometer 陀螺倾斜仪
gyroscopic compass 回转(式)罗盘;陀螺罗盘;陀螺罗经;电罗经
gyroscopic compass repeater 电罗经复示器
gyroscopic compass room 陀螺罗经室
gyroscopic couple 回转偶力;陀螺力偶;陀螺力矩
gyroscopic deflection 陀螺罗盘偏转
gyroscopic drift 陀螺(仪)漂移
gyroscopic drift indicator 陀螺航向指示器
gyroscopic dynamics 回转仪动力学
gyroscopic effect 回转效应;陀螺效应
gyroscopic equilibrium 陀螺平衡
gyroscopic equipment 陀螺仪
gyroscopic erected navigation 陀螺光学导航
gyroscopic force 回转力;环动力;陀螺力
gyroscopic horizon 陀螺水平仪;陀螺地平
gyroscopic inclinometer 回转倾斜仪;回转倾斜计;陀螺式倾斜仪;陀螺测斜仪;陀螺倾斜计
gyroscopic inertia 回转仪惯性;陀螺惯性;定轴性
gyroscopic instrument 陀螺仪表
gyroscopic integrator 回转积分仪;回转积分器;陀螺积分仪
gyroscopic intertia 陀螺定轴性
gyroscopic level 陀螺水准仪
gyroscopic moment 回转力矩;陀螺力矩
gyroscopic motion 回转运动
gyroscopic orientation 陀螺仪定向
gyroscopic orientation survey 陀螺定向测量
gyroscopic pendulum 回转摆;陀螺摆
gyroscopic pitching couple 陀螺俯仰力偶;陀螺俯仰力矩
gyroscopic precession 进动性;陀螺(仪)进动
gyroscopics 陀螺仪学;陀螺力学
gyroscopic seismometer 陀螺地震计
gyroscopic sextant 陀螺六分仪
gyroscopic sight 陀螺瞄准具
gyroscopic stability 陀螺稳定性
gyroscopic stabilization 回转器稳定法
gyroscopic stabilizer 回转稳定器;陀螺稳定器;陀螺防摇装置
gyroscopic theodolite 陀螺经纬仪
gyroscopic torque 回转转矩;陀螺转矩
gyroscopic turn meter 陀螺自转指示器;陀螺回转指示计
gyroscopic well surveying device 回转油井测量装置
gyro-sextant 回转式六分仪;陀螺六分仪
gyroship stabilizer 陀螺船首防摇器
gyrosight 陀螺瞄准器;陀螺瞄准具
gyrosphere 陀螺球
gyrosphere crusher 回转破碎机;回转轧碎机
gyrospin axis 陀螺自转轴
gyrostabilization 陀螺仪的稳定
gyrostabilization unit 陀螺稳定装置
gyrostabilized antenna 陀螺稳定天线
gyrostabilized compass 陀螺稳定罗盘
gyrostabilized gravimeter 陀螺稳定重力仪
gyrostabilized instrument 陀螺稳定仪表
gyrostabilized magnetic compass 陀螺稳定磁罗经;陀螺磁罗盘
gyrostabilized mount 陀螺稳定装置
gyrostabilized platform 陀螺稳定(平)台;陀螺平台
gyrostabilized solar satellite 陀螺稳定太阳卫星
gyrostabilized system 陀螺稳定系统
gyrostabilized unit 陀螺稳定装置
gyrostabilizer 陀螺稳定器;陀螺防摇装置
gyrostat 回转仪;回转轮;回转仪;陀螺稳定器
gyrostatic compass 回转式罗盘;陀螺罗盘
gyrostatic effect 陀螺稳定效应
gyrostatics 陀螺学
gyrosteering gear 陀螺操舵装置
gyrosyn 陀螺同步罗盘;陀螺同步罗经;陀螺感应罗盘
gyrosyn compass 陀螺同步罗盘;陀螺同步罗经
gyrosystem 陀螺系统
gyrotedder hay conditioner 旋转式干草摊晒机
gyrotheodolite 回转经纬仪;陀螺经纬仪
gyrotheodolite method 陀螺经纬仪法
gyrotiller 旋耕机
gyrotor 回转器
gyrotor classifier 旋转分级机;离心分级机
gyrotory crusher 环动式粉碎机
gyrotransit 陀螺经纬仪
gyrotransit method 陀螺经纬仪法
gyrotraverse mechanism 转塔回转机构
gyrotraverse moment 回转力矩
gyrotron 振动陀螺仪;回旋振荡管;陀螺振子
gyrotron maser 陀螺振子微波激射器
gyrounit 陀螺组件;陀螺部件
gyrovector axis 陀螺自转矢量轴
gyrovibration absorber 陀螺减振器
gyrowheel 陀螺(仪)转子
gyrus dentatus 齿状回
Gysi's articulator 盖西氏咬合架
gyttja 软泥;湖底沉积物;湖底软泥;腐(殖黑)泥;[复]gyttjor
gyttja soil 湖底腐殖质土;湖成腐殖质

H

H₃ content in water 水中氚含量
Haanel depth rule 哈内尔深度法则
haapalaite 叠镁硫镍矿
haar 哈雾(苏格兰东部一种海雾)
haarlem oil 亚麻仁油
Haase system 哈泽管桩凿井法
Habann magnetron 分辨阳极磁控管
Habbord type booster 哈邦尔德式调压机
habeas corpus 人身保护权
habendum clause 地契的持有条款
haberdasher's shop 针线铺
haberdashery 针线铺
habilitation for mines 矿山投资
habit 习性;习惯
habitability 可居住性;居住适应性;居住适宜性;适于居住;生境习性
habitable 可(居)住的
habitable area 居住面积;居住空间
habitable attic 可居住的屋顶阁楼
habitable house 住房;居室;住宅;住所;可居住房屋
habitable land area 可居住地面积
habitable room 住所;住房;居室;可居住用房
habitable space 适宜居住的场地
habitable stor(e)y 可住人的楼层
habitacle 宿舍;雕像壁龛
habitancy 居住地
habitant 居民;居住者
habitat 海底探测船;习生地;(动植物生长的)自然环境;住地;聚集处;栖息地;栖息场所;生境;生长环境
habitat analysis 生境分析
habitat complex 生境总体;生境混合作用
habitat condition 生境条件
habitat condition index 生境条件指数
habitat conservation 生境保护
habitat degradation index 生境恶化指数
habitat destruction 生境破坏
habitat deterioration 生境变化
habitat dislocation 扰乱生境;生境破坏
habitat diversification 生境多样化
habitat diversity 生境多样性
habitat evaluation procedure 生境评价法
habitat factor 生境因子;生境因素
habitat form 生境形式;生境形态
habitat gradient 生境梯度
habitat group 生境(类)群
habitat grouping 生境集合
habitat indicator 栖所指示
habitation 住宅;住所;居住(所);居民地;栖息(习性);生活环境
habitation representation 居民点表示法
habitation-shortage household 居住困难户
habitation-uniment system 房屋构造体系(加拿大、美国的一种房屋构造体系,采用混凝土薄壳、内部轻质隔墙)
habitat isolation 生境隔离
habitat maintenance of fish 鱼类生境维持
habitat management 生境管理
habitat mapping 生境图
habitat modification 生境改善
habitat of deposition 沉积场所
habitat preservation 生境保护
habitat protection 生境保护
habitat quality 生境质量
habitat quality index 生境质量指数
habitat range 生境范围
habitat segregation 生境分离
habitat selection 生境选择
habitat stability 生境稳定性
habitat status 生境状况
habitat suitability index 生境适宜指数
habitat type 生境型
habitat variability 生境变化性
habitat volume 生境容量
habit face 习性面
habit formation principle 习惯形成原理
habit modification 习性变化
habit modifier 改晶剂;调晶剂

habit of growth 生长习性
habit persistence hypothesis 习惯持续性假设
habit persistence model 习惯持续性模型
habit plane 习性面;习惯面;惯习面;惯态平面
habits and customs 习惯与习俗
habitual loudness level 习惯响度级
habitual residence 习惯居所
habituation 习惯作用;习惯性;习惯形成;习惯化
habitude 习俗
haboob 哈布沙暴
habotai 电力纺
habutae 电力纺
hachure 晕翳线;用影线表示;痕迹影线
hachured map 晕滃图
hachure(line) 影线;晕滃线
hachure map 影线图
hachure method 晕渲法
hachuring 晕翳表示法;晕滃法;刻线;地貌晕线法;地貌晕滃法
hacienda (美国南部的)庄园住宅;种植园
hack 斫开;鹤嘴锄;丁字斧;碎土
hack barrow 运坯手推车;运砖手推车;运砖架;砖坯车
hackberry tree seed oil 朴树子油
hack cap 砍工具的棚盖;存放壁;晒砖场木栅
hack chronometer 停读时表
hacked bolt 棘螺栓;锚栓;底脚螺栓;粗尾脚螺栓;凹痕螺栓
hacked soffit 括有糙痕的拱底面;用拱顶石连住的拱腹
hacked-type hammer 斧式锤片;(粉碎机的)楔形锤片
hacker 硬件通;砍伐工
hacket 木工用斧
hack file 刀锉;手锯锉;菱形手锉
hack hammer 劈石斧;琢石斧;斧形锤
hackia 砍斧木
hackie 出租汽车驾驶员
hacking 凿毛;流沙槽;拉毛【建】;刻痕;砍;开槽;花砌石层;砌体断线;劈;堆码坯;打毛(混凝土);碎土
hacking knife 油灰刮刀;砍刀;刮灰刀;刮腻刀(安装玻璃用)
hacking machine 码垛机
hacking off 打毛;墙面打毛
hacking(-out)knife 油灰刮刀
hack iron 十字镐
hackle 砍光;手工梳麻法;齿状突起
hackle bench 梳麻台
hackle drum 梳齿滚筒
hackle guide 梳麻导板
hackle mark 锯齿纹
hack lever 架杆
hackling machine 梳理机
hackly 锯齿状;粗糙不平的
hackly fissures 锯齿状裂纹
hackly fracture 锯齿(状)断口
hackly surface 粗糙不平表面
hackmanite 紫方钠石
hackmatack 西方落叶松;北美落叶松;美洲落叶松
hack saw 弓形(钢)锯;弓锯;钢锯
hacksaw blade 弓锯条;弓锯片;钢锯条
hacksaw cutter 钢锯铣刀
hacksaw frame 固定式钢锯架;弓锯架;钢锯架
hacksaw frame bow saw frame 弓锯框
hacksaw frame with plastic handle solid type 固定式塑柄钢锯架
hacksaw frame with wood(en)handle 木柄钢锯架
hacksaw(ing)锯床
hacksaw(ing)machine 弓锯机;弓锯床
hacksaw structure 锯齿状构造
hack stand 出租汽车停车处
hack the sea 破浪前进
hack watch 船表;航行表
hack work 苦工
Hackworth valve gear 哈克沃斯阀动装置
Hacoba bulking system 哈科巴连续膨化变形法
haco oil 鲸鱼油

Hadacidin 氢乙钠
hadal 深渊底;超深渊(的)
hadal depth 超深渊
hadal sedimentation 超深渊沉积作用
hadal zone 超深渊带
Hadamard's evaluation 哈马德评价式
Hadamard's inequality 哈马德不等式
Hadamard transform 哈马德变换
Hadamard transformation 哈达马德变换
Hadamard transformed image 哈达马德变换图像
Hadamard transform spectroscopy 哈达马德变换光谱
Hadang 橄榄色硬木(产于印度)
hade 倾斜余角;偏垂角;伸角;断层余角;断层伸向;垂直倾斜
hade of fault 断层倾角
hadeslip fault 倾向滑断层
Hadfield manganese steel 哈氏高锰钢;奥氏体锰钢
Hadfield steel 哈德菲钢;哈德菲尔德高锰钢;奥氏体高锰钢
hading 倾斜的;偏垂的
Hadley cell 哈得来环流;哈得来环流
Hadrian's villa at Tivoil 蒂奥里的哈德良离宫
Hadsel mill 海德赛尔磨矿机
Haefely process 卷烘处理
haematite 赤铁矿
haematite rock 赤铁矿岩
h(a)emorrhagic fever 登革热
haff 淡泻湖
Haffield breccia 哈菲尔德角砾岩
haffit 教堂长板凳的一端;盖子的固定和开启部分有铰链
Haffotty Shales 哈弗特页岩
Hafner ware 哈夫内陶器(16世纪中期德国的陶器,常制成炉瓦和厚壁容器)
hafnium boride 硼化铪
hafnium deposit 铪矿床
hafnium ion 铪离子
hafnium metal powder 金属铪粉
hafnium-neodymium isotope correlation 铪钕同位素相关性
hafnium nitride 氮化铪
hafnium ores 铪矿
hafnium oxide 氧化铪;二氧化铪
hafnium-strontium isotope correlation 铪锶同位素相关性
hafnium titanate 钛酸铪
hafnon 铪石
haft 旋钮;装柄;柄;把手
haft duplex operation 半双工运用
haftplatte 粘板
hag 沼地;砍伐树林
Hag's tooth 不好看的成品
hagatalite 波方石
Hagedorn equation of state 哈格多恩状态方程
Hagen's hypothesis 哈根假说
hagendorfite 黑磷铁钠石
Hagen-Poiseuille's equation 哈根—伯肃叶方程
Hagen-Poiseuille's law 哈根—伯肃叶定律
Hager disc 海格离心盘
Hagg's principle 海格准则
Haggen-Macher equation 哈根—马歇方程式
haggite 三羟钒石
haggle 讨价还价
haggle over a bargain 讲价
haggle over prices 讲价
Haggloader 立爪式装载机;赫格装载机
haggy 潮湿和崎岖的
Hagia Sophia at Constantinople 君士坦丁堡的圣莎菲亚教堂
hagiasterium 洗礼池;圣地
hagioscope (中世纪教堂的)壁上窄窗;(十字形教堂的)斜孔小窗;窥视窗;内墙小窗
Hagley ashes 哈格雷火山灰
Hague Arbitration Convention 海牙仲裁公约
Hague Conference(1907) 海牙国际法会议(1907)
Hague Convention 海牙公约
Hague Rules 海牙规则

Hague-Visby Rules 海牙—威斯比规则
ha-ha 围护界沟；拦家畜用壕沟；暗墙；矮篱墙；隐蔽围墙；沉篱
hah-hah fence 隐篱；沟中边篱
Haidinger fringe 海丁格尔干涉条纹
haidingerite 砷钙石
Haidinger rings 海丁格尔干涉条纹
Haigh fatiguetesting machine 黑格疲劳试验机
Haigh kiln 黑格窑
Haigh machine 黑格疲劳试验机
haikal （哥普特教堂三个圣所的）中心附属礼拜堂
Haik tectonic segment 哈尔克构造段
hail 船籍港；冰雹；雹（子）
Hailar block 海拉尔地块
hail damage 雹灾
hail day 雹日
hail embryo 冰雹坯胎
hailer 高声信号器；汽笛；手提扩音器
hailfall 降雹
hail forecaster 冰雹预报器
hail imprints 冰雹印痕
hailing station 船闸指挥站
hail insurance 雹害保险
hail lobe 雹瓣
hail mark 雹痕
hail mitigation 消雹
hail proof 防冰雹的
hail shooting 降雹
hailsquall 雹飑
hail stage 成雹阶段；雹期
hailstone 冰雹；雹子；雹块
hail storm 下雹；夹雹暴风雨；大冰暴；雹暴
hail suppression 防雹
haily 冰雹的
Hainan craze 海南热
hair 金属丝；麻丝；麻刀
hairball 毛粪石
hair beater 麻刀打散器
hair-bonded rubber 毛毡橡胶
hair bracket 船首旋花装饰
hair breadth 发差；微差；发隙距离
hair-breadth tuning 高精度调谐
hair brush 发刷；毛刷
hair bulb 毛球
hair canal 毛管
hair catcher 集毛器；除毛器
hair check 发裂纹；细裂纹
hair checking 细裂缝；发丝裂缝；发状辐射；发状辐裂；发裂
haircloth 毛布
hair colo(u)r 毛色
hair compass 微调两脚规；微调圆规
hair copper 毛赤铜矿
haircord carpet 发毛起圈地毯
hair crack 发状裂缝；细裂（纹）；毛细状裂纹；微裂纹；发丝裂纹
hair cracking 发裂
hair crepe 发辫
hair cross 十字丝；叉线
hair cruces 发交叉
hair cut finance 修剪融资；削发融资（美国术语）
hair cuticle print 毛小皮印痕
haircuts 黏丝
hair cystolith 毛内钟乳体
hair disc 毛盘
hair divider 微调分规
hairdresser's parlo(u)r 女子理发店
hairdressing saloon 美发厅；理发店
haired cement mortar 加纤维的水泥砂浆；麻丝水泥灰浆
haired fairing 毛辫整流
haired gypsum 掺麻丝石膏
haired lime mortar 麻丝石灰砂浆
haired mortar 麻丝砂浆；麻刀灰；含毛发砂浆
hairen 毛制的
hair felt 油毛毡；毛毡；发毡
hair felted fabric 毛毡
hair-fibered cement mortar 麻丝水泥砂浆；麻刀水泥泥浆
hair-fibered gypsum 麻丝石膏
hair-fibered lime mortar 麻刀石灰泥浆
hair-fibered mortar 麻丝砂浆
hair-fibered plaster 麻刀灰泥
hair-fields 毛区

hair grease 毛填料润滑脂
hair gypsum 麻刀石膏
hair gypsum-lime mortar 麻刀石膏灰泥
hair hook 灰泥中拌麻丝器具；麻刀灰拌和钉耙；抹灰工具
hair hydrograph 毛发湿度仪（表）；毛发湿度计
hair hygrometer 毛发湿度仪（表）；毛发湿度计
hairiness reducer 毛丝消除器
hair interceptor 滤毛发网（下水道中）
hair interlining 黑炭衬
hair kiln 纤维灶窑
hair lacquer 发（喷）漆
hair-like 像毛的
hair-like crack 发裂
hair line 细长条纹；照准（十字）丝；极微细的线；瞄准线；气泡线；十字线；十字丝；发丝（机）；发线；发纹裂；发际；细线条；游丝；毛筋；黑线纹
hairline crack 发丝(状)裂缝；毛细裂缝；细裂缝；塑性收缩裂缝；毛细裂缝；发样裂纹；细纹；毛发丝状裂隙；发状裂纹；发裂
hairline cracking 细微裂纹
hairline finishing 精细直线打磨法
hairline joint 对接细缝；发丝接缝
hairline pointer 瞄准器
hairline register 套合规矩线
hairline seams 发纹
hair-lock 毛发塞垫
hair manganese 发锰
hair matrix 毛母质；毛基质
hair mercury 发汞
hair mortar 麻刀灰泥；麻刀灰（浆）
hair orchid 毛兰
hair ornament 发饰品
hair papilla 毛乳头
hair pencil 画笔；毛鬃
hairpin 发针形楔入物；细销
hairpin bed 发卡形路基；发针形（曲线）路基
hairpin bend 马蹄形弯道；发针形弯段；U形转弯；回头弯道；回头曲线；U形存水弯
hairpin bend dune 马蹄形沙丘；抛物线形沙丘；发针形沙丘
hairpin conductor 发夹型导体
hairpin cooler U形冷却器
hairpin coupling loop 发夹式耦合环
hairpin curve 发针形曲线；回填曲线；U形曲线；回头曲线（道路）
hairpin curve location 回填曲线测设
hairpin furnace 马蹄窑；U形窑
hairpin lace 毛工饰边花边
hairpin reinforcement 胡子筋（大板接缝用）
hairpin structure 发夹结构
hairpin tube bundle U形管束
hairpin turn 回头曲线；发针形弯曲
hairpin winding 发针形绕组；发卡式绕组
hair plate 毛板
hair pounder 马车夫(俚语)；驾驶工
hair powder 发粉
hair-processing machine 毛绒处理机
hair product wastewater 发制品废水
hairs 十字叉线
hair salon 美发廊
hair salt 发盐
hair-scales 毛区
hair scratch 毛状痕迹；毛状刮纹
hair seam 发（丝）裂缝；细裂缝；毛发缝；发缝
hair shaft 毛干
hair-shaped 毛细的
hair side of belt 皮带的毛面
hair sieve 细网筛；细孔筛；密眼筛；粗毛滤网；马尾筛
hair slat 毛矾石
hair space 最小的空铅
hair spring 细弹簧；游丝
hairspring cleaner 游丝除垢剂
hairspring collet removers 游丝内柱拆除器
hairspring collet tool 游丝内柱工具
hairspring dividers 细弹簧分规；细弹簧两脚规
hairspring holder 游丝夹
hairspring inside stud 游丝内柱
hairspring lever 游丝校平器
hairspring pin 游丝销
hairspring screw 游丝螺钉
hairspring set lever 游丝调整杆
hairspring setting 游丝定长
hairspring stud 游丝外柱
hairspring stud screw 游丝外柱螺钉

hairspring support 游丝架
hairspring tweezer 游丝镊子
hair stone 毛发水晶
hair-streams 毛流；毛顺向
hairtail 带鱼
hair-thin 细如毛发的
hair trap 集毛器
hair trigger 微力扳机；微力触发器
hair-trigger method 发状触发方法
hair tube U形管
hair wastes 发渣
hair wire 游丝
hairwork 发制品
hairy 毛（状）的；毛发状
hairy caterpillar 晕嗡
hairy roving 毛纤维
hairy sheath 毛鞘
hairy surface 毛表面；粗表面
haitcloth 马尼衬
Haitienne 埃蒂安绸
haiweeite 多硅钙铀矿
Haizman plug 海茨曼炮塞
hake 干燥瓦坯的格栅；砖坯干燥槽架；牵引调节板；瓦坯干燥器
hakite 硒汞铜矿
hakuchi 港外锚地
halarch succession 盐生演替系列
halation 晕影；晕光作用；光晕；充晕；成晕现象
halation by diffusion 扩散晕影
halation by reflection 反射晕影
halation effect 晕电荷
Halco 哈尔科铬钢
Halcomb 哈尔库姆合金钢
Halcut 哈尔卡特铬钨钢
halcyon days 平安时期；平安时间
halcyon seas 风平浪静
Haldane's apparatus 哈尔丹空气分析器；霍尔丹氏呼吸气体分析器；耳丹分析仪
Haldane scale 霍尔丹氏标度
haldenhang 崖脚缓坡；重力坡（脚）；剥蚀基坡
Haldi 哈尔迪铬钢
Haldu 黄褐色硬木（产于印度）
hale-hale water 大雨
haleite 二硝基乙胺
Hale telescope 海尔望远镜
half 一半；模瓣；半个
half-absorption layer 半吸收层
half-absorption thickness 半吸收厚度
half-add 半加
half adder 半求和器；半加（法）器
half-adder binary 二进制半加器
half-adder-subtractor circuit 半求和求差电路；半加减电路
half adjust 四舍五入；舍入
half a gale 半飓风
half ahead 前进二【航海】
half alloy dead center 半缺合金顶尖
half amide of malonic acid 丙二酸半酰胺
half-amplitude duration 半幅值持续时间；半幅宽度
half-amplitude point 半幅度点
half amplitude value of log 测井曲线半幅值
half-anchor ear 半桩环；半锚杆环
half anchor pole 半锚杆
half-anchor tower 半锚塔；半拉杆塔
half-and-half 一半一半；等量；各半；一比一
half-and-half bearing 半轴承
half-and-half joint 对拼接头
half-and-half lap scarf 木材两端各切一半
half-and-half solder 锡铅（各半的）焊料；铅锡各半软钎料
half angle 半角
half-angle cone 半角锥
half-angle formula 半角公式
half-angle of projection 圆锥半角
half-angle of spread 半扩散角
half-angular breadth 半角宽度
half a point 半成点
half-arc 半弧光灯
half-arc angle 半弧角
half-arch 拱背
half arid area underground water area 半干旱地区潜水区
half astern 后退二【航海】；半速后退

half-automatic ignition 半自动点火
half-auto residual static correction 半自动剩余静校正
half-axle 半轴
half axle gear 半轴齿轮
half-backed-bench saw 半压台锯（锯条的一半被压住）
half-bag mixer （每次拌和半袋的）小型混凝土拌和机；小型混凝土搅拌机
half-baked 半焙烧
half-baked brick 半（焙）烧砖
half balance 半平衡；不完全平衡
half balk 顶板支梁
half-ball value 半球形活门；半球形阀
half-baluster 半伸出栏杆柱
half bandwidth 谱带半宽度；半谱带宽度；半带宽
half bank 单坡防洪堤
half basin 半水槽
half bat 半（块）砖
half-bath toilet room 无浴缸的盥洗室
half batten 半板条
half-bat thick 半砖厚
half-bat wall 半砖墙
half-bay 半跨
half-beam 半（横）梁；从舱口通过船半边的梁
half-bearing 半轴承
half bearing thickness 轴瓦厚度
half-bed 半接头；半底层；砌半石基底
half black 半处理的；半加工的；未完成的
half blank 半制品
half-blind dovetail 半盲燕尾榫；半盲鸠尾榫；盖头燕尾榫；抽头燕尾榫
half block 半字组；半（数据）块；半瓷砖
half block model 半边船模
half-body of revolution 半旋转体
half bog area 半酸沼区
half bog soil 半沼泽土
half boiling 半脱胶
half bond 半键
half bordered pit pair 半具缘纹孔对
half-bound 半精装的
half bow-knot 半缩帆结
half box 无盖轴箱
half braking 局部制动
half breadth 半宽（度）；（船的）中轴距离；半值宽度
half-breadth plan 半宽（水线）图
half breast water wheel 胸射式（中射式）水轮（机）
half brick 半砖
half-brick partition 半砖隔断
half-brick wall(ing) 半砖（厚）墙
half-bridge 半电桥
half bridge circuit 半桥式电路
half-bright 半光制的
half-bright bolt 半光制螺栓
half-bright nut 半精致螺母；半精制螺母
half-bright screw 半光制螺钉
half-bright washer 半光制垫圈
half-broken 半（碾）碎的；半破坏的；半驯服的
half-bubble 半磁泡
half buckler 锚孔盖
half built-up crank shaft 半组合式曲轴
half-buried tank 半埋式（箱）罐；半埋藏式水箱
half-burnt brick 未烧透砖；半烧透砖
half-burnt clay brick 未烧透黏土砖；半烧透黏土砖
half-byte 半字节；半位组
half-cap crown 露面冠；开面冠
half car 半履带式机车（美国）
half-card 大光钡地纸
half cardinals 象限点；隅点
half cargo condition 半载状态
half carry 半进位
half-carry flag 半进位标志
half-cell 单电极系统；半电池；半单元
half-cell potential 半电池电势
half-cell reaction 半（电）池反应
half-cell reaction energy change 半电池反应能量变化
half-center 半缺顶尖
half-centimeter division 五毫米分划
half-chair 半椅式
half cheese antenna 半盒形天线；半饼形天线
half chimney block 半烟囱砌块；对分烟囱砖
half-chord 半翼弦

half chronometer 半精密记时表
half circle protractor 半圆分度器
half-circular drain 半圆形排水沟
half circular system of locomotive running 半循环运转制【铁】
half clock period 半时钟脉冲周期
half closed and half open 半睁半闭
half-closed layer 半封层；单向排水土层
half cloverleaf intersection 半苜蓿叶形交叉
half coating 半覆盖膜；半镀膜
half cock 半击发状态
half-cocked 机头半张开；处于半击发状态的
half cofferdam 半围堰
half coil 半（圈）线圈；半（节）线圈
half-coil spacing 半圈间距；半节线圈间距；半加感节距
half column 半（露）柱；半（身）柱
half-commission man 半经纪人
half-completely differentiated intrusive body 不完全分异岩体
half compression cam 减压凸轮；半压缩凸轮
half-concentric resonator 半共心谐振腔
half convergency 半收敛差；半聚合差；转换角
half cooked 半热的；准备不够的；半熟的
half corrosion water 半腐蚀性水
half-countersunk 半埋头
half-countersunk head 半埋头
half-countersunk rivet 半埋头式铆钉
half-countersunk rivet head 半埋头式铆钉头
half-coupling 半联轴节；半靠背
half-crawler tractor 半履带式拖拉机
half-cream 半膏
half-cross(ed) belt 直角挂铰皮带；半交皮带
half cross-section 半剖面
half crown seizing 半冠形合扎
half-crushed 半（碾）碎的
half crystal 半晶玻璃；半成晶
half-crystal can relay 半晶体密封继电器
half-crystal glass 半晶体玻璃；半结晶玻璃
half-crystallization time 半结晶期
half cupola 半球形顶棚大楼；半圆顶
half cupped fracture 半杯形断裂；半杯形断口
half current 半选电流
half-current pulse 半激励电流脉冲
half-curved needle 半弯形缝合针
half cut 半路堑
half cutoff 半截止；半切断
half cutwork 半雕绣
half cycle 半周期；半循环
half-cycle displacement input 半周期位移输入
half-cycle displacement output 半周期位移输出
half-cycle transmission 半周传输
half-cylinder 半圆柱体
half davit 收锚杆；半吊杆
half-day system 半日制
half dead centre 半缺顶尖
half dead escapement 半反冲擒纵机构
half deck 后部甲板；平台甲板；半甲板
half decked boat 靴式舢板；半甲板艇
half decker 靴式舢板；半甲板艇
half decode 半译码
half-deflection method 半偏转（测）法
half deformation 半变形；半畸形
half degummed silk 半脱胶丝
half desert 半沙漠【地】；半荒漠
half-diamond indention 倒三角形
half diesel engine 半柴油发动机
half dislocation 半位错
half-dock type shipbuilding berth 半坞式船台
half dollar 美元的半元银币
half-dome action 半穹顶作用
half done 半成的；半熟的
half door 半（截）门
half double 半重瓣的
half-dressed quarry stone 粗琢块石
half-dressed stone 粗琢石
half-drill strip method 半斜带法
half-dry spinning frame 半干精纺机；半干纺细纱机
half-duplex 半双用；半双向；半双工
half-duplex basis 半双工制
half-duplex channel 单向通道；半双向通道；半双工信道
half-duplex circuit 半双工线路；半双工电路

half-duplex communication 半双工通信
half-duplex communication line 半双工通信线路
half-duplex mode 半双工方式
half-duplex operation 半双向操作；半双工通信；半双工操作
half-duplex repeater 半双工中继器
half-duplex service 半双工服务
half-duplex transmission 半双工传输
half-duty 半税
half-ebb 半落潮
half echo suppressor 半回波抑制器
half elbow 双锚单交叉
half element 单电极系统
half elevation 半立面
half-elliptic 半椭圆形的
half elliptic spring 弓形弹簧；半椭圆形弹簧
half embankment 半路堤
half emulation mode 半仿真方式
Halfen anchor channel 哈尔芬锚固槽钢（混凝土预埋件）
Halfen concrete insert 哈尔芬预埋件
half-engaged column 半（露）柱
halfer 单脚
half-ester 半酯
half-ever green 半常绿树
half-excited core 半激励磁芯
half excretion time 半排出期
half exterior and half interior syndrome 半表半里
half face brick 半块面砖
half-faced 显出一个侧面的
half fading time 半褪色时间
half fare 半票；半价运费
half-fat 半黑体字；半粗体字
half-field angle 半视场角
half field-of-view 半视场
half-fife of activation product 活化产物半衰期
half-filled end-slots 半填充端部槽
half-finished 半成品；半加工的
half-finished goods 半成品；半制品
half-finished material 半制品；半成品
half-finished part 零配件半成品
half-finished product 半成品
half-finished washer 半光制垫圈
half-finishing nut 半光制螺母
half-fired brick 未烧透砖；半烧透砖
half-fired clay brick 未烧透黏土砖；半烧透黏土砖
half-flat 半平的；半套房间；半层
half flight 半跑（梯）；半跑段
half flood 半潮汛；半涨潮
half floor 楼梯（休息）平台；半层
half-fork strut 半叉形支柱
half frame 半框架；一梁一柱式支架
half-frame camera 半幅照相机
half frequency signal 半频信号
half-frequency spin waves 半频旋转波
half-frog brake-shoe escapement 半辙叉脱轨器
half-front view 半前视图
half frosted lamp bulb 半磨砂灯泡
half-full 半满
half gable 半山墙
half gantry crane 单支腿龙门起重机；单脚高架起重机；半龙门起重机；半龙门吊
half-gas fired furnace 半煤气燃烧炉
half gas firing 半煤气烧成
half gate 半门；补偿栅
half-gateway 半信关
half-glass door （上部安装玻璃的）半（截）玻璃门；上半截装配玻璃的门；上部全玻璃的门
half-glazed door 半玻璃门
half-glazed partition 半截玻璃隔断
half-groove 半陇；半槽；半企口
half halved joint 半缺接合
half-hard 中硬的
half-hard metal 中硬金属
half-hard rubber 半硬质胶；半硬橡胶
half-hard steel 中硬钢；半硬钢
half-hardy annual 半耐汉一年生（植物）
half-hardy plant 半耐寒植物
half hatchet 可起钉的木工斧；羊角锤；短柄（半）斧；半边小斧头
half header 镶墙边的半砖砌；收尾的半砖；收头半砖；半顶砖；半圆木大顶梁
half height block 半高砌块；半厚砌块；4英寸厚混凝土块体

half-height container 半高集装箱
half-Hertz transmission 半赫兹传输
half-high container 半截货柜半高集装箱
half high width of peak 峰的半高宽
half hip 部分斜脊;半斜脊
half-hipped roof 半斜脊屋顶
half hitch 简单结子;半结
half-hour rating 半小时额定值
half-hour synthetic(al) finishes 半小时合成漆
half-hunter 半双盖表
half image 半影
half-in-and-half-out of the water 半浮半沉的
half-inch scale 半英寸比例尺
half-inch survey 半英寸比例尺测图
half-integer 半整数
half-integer angular momentum 半整数动量矩
half-integer synchronization 半整数时差
half intensity 半强度
half-intensity width 半强度宽度
half-interval contour 中插等高线;半距等高线
half-interval search 区间分半检索
half-intracavity laser 半内腔激光器
half island 半岛
half-joint 半接头;半底层
half-joist T-截面格栅;T形(截面)格栅
half keystone ring 单面梯形活塞气环
half-kilometer sign 半公里标
half landing 中间休息板;楼梯平台;全宽梯台;半层休息板
half-landing 楼梯平台;半空间平台
half-lap 半折叠;半叠盖
half-lap coupling 半搭接联轴节
half-lap joint 半叠接(合);对搭缝;对搭接(接头)
half-lapped joint 对搭缝;对搭接;半叠接
half-lap scarf 半叠接(合)
half-lap scarf joint 半叠斜接;半叠楔接(合);半叠拼接;半搭接
half-lap winding 半叠包绕组
half-lattice(d) girder 华伦式大梁;格构(大)梁;半格组(大)梁;半格构架
half-lattice(d) frame 半(柜)格式构架;半格组构架;半格(构式)构架;半框格式构架
half-length 半身梁
half lethal dose 半致死(剂)量;50%致死量
half life 半衰期;半减期;半寿期;半寿命;半反应期
half life of recurrent selection process 轮回选择过程的半衰期
half life period 半衰期
half life-size 实物一半大小的
half-light 暗淡光(线)
half line 半直线
half-linear width 半线宽
half-line frequency 半行频率
half-line pulse 半直线脉冲
halfliner 半轴瓦
half load 半荷载;半负载
half-loaded 半装填
half-loading section 半节线圈区段;半负载段
half log 对开(圆)木;对开材;半原木
half main rafter 半主椽;(伸不到屋脊的)半椽条;半人字木
half-mark 实测标
half martensite 半马氏体
half-mask respirator 半面罩式呼吸防护器
half mast 半旗
half-matte 半光泽的
half-matt gloss 半光(泽)
half-maximum line breadth 半峰线宽
half-Maxwell lens 半麦克斯威尔透镜
half mean spring range 大潮坡半潮幅;大潮均半潮差
half measure 折中办法;妥协方案;折衷办法;姑息手段;权宜办法;双栏式排版
half-mechanized shield 半机械化盾构
half-metal 半金属
half method 分半(方)法
half mile rule 半海里法则
half-mirror 半透明膜;半(透明)反射镜
half-miter joint 半斜接
half model 半边船模;半模型
half-modular cable 半模块式电缆
half module 半模(数)
half monitor 锯齿形屋顶

half moon 半月(形);上下弦;(城堡的)半月形外堡
half-moon break (钢轨的)半裂
half-moon crosstie 半月形轨枕
half-moon dike 半月形堤
half-moon hoe 半月形锄
half-moon shape 半月形
half moon-shaped 新月形的
half mortise hinge 半嵌铰链
half mo(u)ld 半型
half-normal bend 弯曲管(弯角135°);半标准弯头
half nut 开合螺母;开缝螺母;对开螺母
half-nut cam 开合螺母凸轮
half-nut lever 对开螺母手柄
half-octagon ring 半八角形密封环
half of mo(u)ld 半模
half-open 半开
half-open corner joint 半开口角接头
half-open cube 半开正方形;半开立方体
half-opened corner joint 半开角焊接
half open interval 半开区间
half-open-jointed 半开式接缝(的)
half-open position 半开状态
half-open switch 不密贴道岔
half ordinary rate 普通价目的一半
Halford window 哥特式金属窗
half-outdoor power plant 半露天式电厂
half-outdoor power station 半露天式发电站
half-oval controller 半椭圆控制器
half-oval edge 半椭圆立轧边
half-oval pass 半椭圆孔形
half-oval profiled wax 半卵圆铸造蜡条
half-oval wire 半卵形(截面)金属丝
half overall model 半整体模型
half-oxidizing roasting 半氧化焙烧
half pace 踏步平台;凸窗高台;梯口;半梯台
halfpace landing 半中平台;楼梯平台
halfpace stair(case) 转身楼梯
half parameter 半参数
half-part mo(u)lding box 半分砂箱
half-path 半通路
half-pay 半薪
half-peak breadth 半峰宽(度)
half-peak width 半峰宽(度)
half period 半衰期;半寿期;半周期
half-period average current 半周平均电流
half-period average value 半周(期)平均值;半周期均价值
half-period fluctuation 半周期波动
half-period zones 半周期带
half peripheral length 轴瓦半周周长
half-picture storage tube 半像存储管
half-pier 半窗间壁
half-pig 半清管器
half pike 短柄锄
half pile length 半桩长
half pilotage 半引航费
half-pipe coil jacket 半管螺旋式夹套
half-pipe section 半管套
half-pitch roof 半斜屋顶
half-plain work 石面半砍平工作;石面半砍平细工;半琢平工作(石面)
half plan 半平面图
half plane 半(平)面
half plane in the pattern space 模式空间中的半平面
half-plate 半波片
half point 半罗经点
half polished plate glass 半磨(平板)玻璃
half polished spring 半抛光弹簧
half poop 短船尾楼
half port 小舷门;舷孔半盖;半舵
half-portal 半门架
half portal crane 半门式起重机(双门柱不等长)
half-portal scraper 半门式刮板取料机
half power 半幂;半功率
half-power angle 半功率角
half-power bandwidth 半功率带宽
half-power beamwidth 半功率束宽
half-power frequency 半功率频率
half-power method 半功率法
half-power point 半功率点
half-power width 半功率宽度
half-prestressed beam 半预应力梁
half price ticket 半(价)票

half principal 半主椽;(伸不到屋脊的)半椽(条)
half prism 半棱镜
half profile 半剖面
half-profit system 利润平分制
half pulse recurrence rate delay 半脉冲重复频率延时
half pulse repetition-rate delay 脉冲重复频率半延迟
half quantum number 半量子数
half quarter cut 径锯材
half quartered wood 径切对开木材
half rabbeted lock 半槽口门锁;半榫插销
half-race 半宗
half-range of tide 半潮差
half-range series 半幅级数
half-reaction 半反应
half-reaction potential 半反应电势
half reading pulse 半读出脉冲
half rear axle 半轴
half rear axle gear gasket 半轴承轮
half-rear view 半后视图
half-recessed element 半下沉的部件;半缩进的部件;半凹部件
half-reflecting mirror 半反射镜
half reflection 半反射
half relief 半(凸)浮雕
half residence time 半留时间;半存期
half restrained 半约束的
half reverse 半反向
half-rib 半肋
half ripper 细木锯
half rip-saw 细(齿)木锯
half rod 半圆木杆(放在炮眼内,保证按平巷轮廓炸掉岩石)
half roofing tile 半屋面瓦
half roof tile 半屋面瓦
half-rotary cut 半圆旋切的
half-rough 半粗糙
half-rough nut 半粗制螺母
half-round 半圆材(的);半圆形(的)
half-round apse 半圆室
half-round apsis 半圆室;半圆形前廊
half-round arch 半圆拱
half-round bar 半圆钢(筋);半圆杆;半圆形钢条;半圆条
half-round barrel vault 半圆(筒形)拱顶
half-round bar steel 半圆棒钢
half-round bastard file 粗齿半圆锯
half-round bastion 半圆形堡垒
half-round batten 半圆狭板
half-round bent scraper 半圆弯头刮刀
half-round ceiling 半圆形顶棚;半圆形天花板
half-round channel pipe 半圆沟管
half-round chisel 半圆凿;半圆錾子
half-round clamping ring 半圆卡环
half-round coarse file 半圆粗锉
half-round concave milling cutter 凹半圆铣刀
half-round countersunk head bolt 半圆沉头螺栓
half-round cross-vault 半圆形交叉拱顶;半圆形交叉筒拱
half-round cut-water 半圆分水尖
half-round cylindric(al) roof 半圆柱形屋顶
half-round drill 勺钻;半圆钻
half-round eaves-tile ornament 半瓦当
half-rounded edge 半圆边
half-round edge 圆弧棱边;半圆边缘
half-round exedra 半圆形(露天)室
half-round file 弯面锉;半圆(平)锉
half-round fillet 半圆线饰
half-round flatter 半圆踩锤
half-round floor channel 半圆形明沟
half-round gutter 半圆(天)沟;半圆(形)檐沟;半圆(形)檐槽
half-round hammer 半圆头锤
half-round head file for carpentry 半圆头木工锉
half-round head rivet 半圆头铆钉
half-round head screw 半圆头螺钉
half-round head square neck bolt 半圆头方颈螺栓
half-round iron 半圆铁(杆);半圆(形截面)型钢;半圆钢
half-round iron-bar 半圆铁条
half-round iron wire 半圆铁线
half-round middle file 半圆中锉
half-round mo(u)lding 半圆线条;半圆线脚
half-round niche 半圆形壁龛

half-round nip(pers) 半圆钳
half-round nose bit 半圆唇金刚石钻头
half-round nosing 半圆(形)突边
half-round oil stone 半圆油石
half-round pliers 半圆钳
half-round pointing 半圆(勾)缝
half-round profile 半圆形截面;半圆形轮廓
half-round rasp 半圆锉
half-round reamer 半圆铰刀
half-round rib 半圆形肋
half-round ridge covering 半圆脊帽
half-round ridge tile 半圆形脊瓦
half-round rigid receiving tray 半圆形刚性接收盘
half-round rivet 半圆(形)铆钉
half-round(roof)gutter 半圆形天沟
half rounds 各种直径的半圆钢;半圆钢
half-round saw 方形锯;半圆锯
half-round saw file 半圆锯锉
half-round scraper 半圆(形)刮刀
half-round screw 半圆头螺栓;半圆头螺钉
half-round section 半圆形型材;半圆形截面;半圆剖面
half-round section duct 半圆形管道
half-round smooth file 半圆细锉
half-round spade 半圆铲
half-round steel 半圆(形截面)钢材
half-round steeper 半圆枕木
half-round stretcher 半圆形露侧砖
half-round table 半圆形桌
half-round termination 半圆形端头;半圆形界限;半圆形地界
half-round timber 对开圆木
half-round tower 半圆形塔(架)
half-round transverse arch 横向半圆拱
half-round trip fare 海上部分来回票价
half-round tunnel vault 筒形拱顶;半圆形隧洞拱顶
half-round veneer 半圆单板;(胶合板的)半圆旋切单板
half-round veneer slicing 半圆单板刨切
half-round wagon vault 筒形拱顶
half-round window 半圆形窗
half-round wire 半圆金属丝
half-round wood rasp 半圆木锉
half-round wood screw 半圆头木螺旋;半圆头木螺钉
half-sample 半样本
half-saturated 半饱和的
half-saturation 半饱和
half-saturation time 半饱和时间
half-saturation tissue 半饱和时间组织
half-saturation unit 半饱和时间单位
half sawing 一开(木料锯成两半)
half-sawn 半锯产品
half-sawn stone 半锯石;锯开石
half scale 二分之一比例;半足尺;半标度
half-scale model 半尺寸模型
half scarf 半搭接
halfscoop 抓斗
half-scraper 半清管器
half seas over 在航海途中
half-second pendulum 半秒摆
half section 320 英亩土地(根据美国政府测量法);半剖视;半剖面;半截面;半节;半断面
half sectional dock 半分段式浮船坞
half sectional side elevation 半剖视侧图
half sectional view 半剖视图;半剖面图
half section duct 半截管道
half section excavation 半断面开挖
half section excavation method 半断面隧道开挖法
half section gutter 半截水沟
half section network 半节网络
half section pipe 半截管
half-selected core 半选磁芯
half-selected pulse 半选脉冲
half-selected voltage 半选电压
half-serve brake 半制动闸;半继动闸
half-shade analyzer 半影分析器
half-shade angle 半遮角
half-shade compensator 半影补偿器
half-shade device 半影装置;半影器件
half-shade eyepiece 半影目镜
half-shade plate 半影板
half-shade polariscope 半阴旋光计
half-shade polarizer 半影起偏振镜

half-shadow 半影
half-shadow analyzer 半影分析器
half-shadow angle 半影角
half-shadow device 半影装置
half-shadow polarimeter 半阴旋光计
half-shadow prism 半影棱镜
half shaft 带法兰的半轴;半轴;半井状通道
half shape rubber bobbin 半圆形橡胶滚轮
half sheet 半页图
half shell 半壳
half shell pressure vessel 瓦片式压力容器
half-shift register 半移位寄存器
half shovel 单翼平铲
half shroud 齿轮的半罩壳;半齿高加强板
half-shrub 半灌木
half side formula 半边公式
half side milling cutter 半边靠模铣刀
half side-view 半侧面
half-silvered mirror 半面涂银镜;半镀银镜
half-silvered surface 半涂银面
half sine pulse 半正弦脉冲
half sine shock pulse 正弦半波冲击脉冲;半正弦冲击脉冲
half sine wave 半正弦波
half-sinusoid 正弦半波
half-sip saw 细齿锯
half-size 半粒径(的);尺寸缩小一半的;原尺寸之半;缩小一半的
half-sized detail 半足尺详图
half-size factor (筛分的)半数;(筛分的)半大颗粒系数
half-size ratio 小于筛孔半径的粒度比
half-size scale 半尺寸
half slating 疏铺石板
half slot 半圆槽
half small column 壁柱;半柱
half-socket pipe 半节接管;半插套管;下层土排水沟
half sole 半圆金属焊垫
half solid floor 空心肋板
half soling 半圆形衬垫
half-space 楼梯休息平台;休息平台;梯台;半叠接;半空间;半(空)格;转弯休息平台;楼梯转身平台;半层休息板
half space analysis 半空间分析
half-space axial symmetry 半空间轴对称
half-space key 半(空)格键
half-space landing 楼梯平台;转弯休息平台;中间休息板;半层休息板
half-space model 梯台模型;半空间模型【数】
half-space stair(case) 转弯平台式楼梯;转身楼梯
half-span house 半屋面温室
half-span loading 半跨荷载
half-span roof 一面坡屋顶;单坡屋顶;半坡屋顶
half speed 半速前进;半速(度)
half-speed gear 降速齿轮传动;半速齿轮(传动)
half-speed prism 半速棱镜
half-speed shaft 半速轴
halfsphere 半球
half-spheric(al)steel bobbin 半圆形钢质滚轮
half-spheric(al)washer 半球形垫板
half-split flow 半削裂流;半截流;半分流
half-split pipe 半截管;半圆管道;半圆(形)檐槽
half-split pipe duct 半圆槽
half staff 半旗
half-standard pitch auger 半标准螺距螺旋
half steel 钢性生铁
half step 半音
half stock 半股;半额股票
half stopped 半阻塞的
half stor(e)y 屋顶层;半层;走廊层;夹楼层;回廊层;阁楼;顶楼
half structurization decision 半结构化决策
half-stuff 半料浆
half-subtracter 半减(法)器
half sunk 半下沉;半埋沉;半埋入的
half sunk rivet 半埋头铆钉
half sunk roadway 半穿式公路桥
half-supply voltage principle 半供电压原则;半电源电压原理
half supporting ring 半支持环
half-surface hinge 半露铰链
half sweep 单翼平铲
half-symmetric(al)resonator 半对称谐振腔
half-symmetric(al)unstable resonator 半对称非稳定谐振腔

half-tap 不割断主导线的中心抽头
half tenon 粗短榫
half the number 半数
half-thickness 半(值)厚度;半吸收厚度
half thread 半螺纹
half-through arch 半穿式拱
half-through arch bridge 中承式拱桥
half-through bridge 下承矮墙桁架;半下承桥;中承式桥;中承矮梁桥;半下承桥;半穿式(梁);矮桁梁下承桥
half-through girder 下承板梁
half-through span 半穿式桥跨
half-through truss 中承式桁架;半穿式桁架
half tidal basin 半感潮港池;半潮港(池)
half tidal harbo(u)r 半潮港(池)
half tidal level 半潮面
half tidal lock 半潮闸
half tide 半潮
half tide basin 半感潮港池;半潮港(池);半潮船坞;半潮船渠;半潮池;半潮差港池
half tide canal 半潮船渠
half tide cofferdam 部分挡潮围堰;半潮围堰(较低的挡水并能溢流的建筑物)
half tide landing 半潮起岸
half tide level 平均潮位;平均潮面;半潮(水)位;中潮面;半潮面;半潮高程
half tide rock 半潮干出礁石
half tide wall 半潮高护墙;半潮挡浪墙
half-tile 半瓷砖
half tile(d)wall 瓷砖台度;半砖墙;半截贴面(砖)墙
half-timber 半木结构;砖木混合结构;外角用的短木材;半方材
half-timber beam 露明木架墙;半扁平木梁
half-timber building 半露木架房屋;半露木(构架房)屋
half-timber construction 砖木(混合)结构;半(露)木结构
half-timbered 砖木结构的;露明木架;半木结构的
half-timbered building 露明木骨架建筑;半露(明)木架建筑
half-timbered member 半木构件
half-timbered wall 露明木(骨)架墙
half-timbered work 半木结构建筑
half-timber house 半露木(构架房)屋;半露木架房屋
half-timber wall 半露木墙
half time 半排出期(同位素);半工半薪;半衰期;半寿期
half-time emitter 中间时间发送器;中间发送器;半时发送器;半脉冲发送器
half-time shrinkage 半期收缩
half-time survey 中间检验;半期检验;半期检测
half-tine grab 半爪抓斗;半齿抓斗;短齿抓斗
half tint 中间色调
half-title 副标题
half-tone 网目铜版;中间色调;照相铜版;半音(度);半色调
half-tone black ink 照相凸版黑墨
half-tone block 网目凸版;网线凸版
half-tone characteristic 半色调特性;灰度特性
half-tone charge pattern 半色调电荷图
half-tone distortion 半色调失真;半音失真
half-tone dot 半色调点;网点
half-tone dot reproduction 半色调网点再现
half-tone engraving 半色调版
half-tone etching 网纹版
half-tone image 半色调(图)像
half-tone information 半色调信息;半浓度信息
half-tone ink 照相铜版印刷油墨
half-tone method 浓淡色调法;半色调法
half-tone modulating screen pseudocolo(u)r encoding 半色调屏假彩色编码
half-tone negative 半色调底片
half-tone news 铜纸
half-tone original 半色调原版文件
half-tone original drawing 半色调原图
half-tone paper 铜纸;网目印相纸
half-tone photograph 挂网片
half-tone photography 半色调摄影
half-tone picture 半色调(图)像
half-tone plate 半色调版
half-tone positive 半色调正片;网目阳图铜版
half-tone processing 半色调处理

half-tone reading 半色调显示
half-tone reproduction 灰度再现;网目复制
half-tone response 半色调特性
half-tone-rich original 纯中间色调原稿
half-tone screen 半色调网目片;照相制版网屏;网版
half-tone signal 半色调信号
half-tone storage tube 半色调存储管
half top 半桅合板
half top view 半剖俯视图
half track 半履带式(车);半轨;重型卡车(前后轮推动的载货车);半履带(驱动);半磁迹
half-track car 半履带式汽车;半履带车
half-track carrier 半履带式战车
half-track combine 半履带式联合收获机
half-tracked 半履带式的
half-tracked vehicle 半履带式货车
half-track motor lorry 半履带式货车
half-track recorder 半迹记录器
half-track tractor 半履带式牵引车;半履带式拖拉机
half-track truck 半履带运货车;半履带式货车
half-track unit 半履带式行走部分;半履带式行走装置
half-track vehicle 半履带式汽车;半履带式车辆
half-train 半拖车【铁】
half-transmitting mirror 半透射镜
half trap 半圆形存水弯
half trip 单项工作(提升或下放钻具)
half truss 半屋架;半桁架
half tuck 奇形怪状的凸嵌(一种嵌砖缝的方法)
half tunnel 半隧洞;半隧道;半山洞
half tunnel vault 半隧洞拱块
half turn 双跑楼梯
half-turn coil 半匝线圈;半圈线圈
half-turn coil winding 半匝绕组
half-turn plow 翻转铧式犁
half-turn socket 半圆形打捞器
half-turn stair(case) 两次直角转弯楼梯;两次直角拐弯楼梯
half-turn stair(case) with landings 带平台直角式楼梯
half-turn stair(case) with winders 带扇步直角式楼梯
half-twisted bit 半螺旋钻;半麻花钻头
half-twisted auger 半螺旋钻
half-type grab 短齿抓斗
half value 半值
half-value angle 半值角;半亮度角
half-value layer 半值层;半吸收层;半衰减层;半价层;半层值;半值厚度
half-value period 半值期;半衰期;半寿命
half-value thickness 强度减半厚度;半值层;半吸收厚度;半值厚度
half-value width 半值宽度
half-volume 半容积
half vortex blading 半涡流形叶片
half waistband 侧腰带
half warmed turbine 半热体汽轮机
half water gas 半水煤气
half-wave 半波
half-wave amplifier 半波放大器
half-wave antenna 半波天线
half-wave character 半波特性
half-wave circuit 半波电路
half-wave conductor 半波导线
half-wave detector 半波检波器
half-wave dielectric(al) layer 半波介质层
half-wave diode rectifier 半波二极管整流器
half-wave dipole 半波振子;半波偶极子
half-wave doubler 半波倍压器
half-wave doublet antenna 半波对称天线
half-wave element 半波振子;半波辐射器;半波单元
half-wave layer 半波层
half-wave-length 半波长(度)
half-wave-length dipole 半波偶长极子
half-wave line 半波线
half-wave linear radiator 半波线状辐射器
half-wave mercury vapo(u)r rectifier 半波贡气整流器
half-wave mercury vapo(u)r rectifier tube 半波汞气整流管
half-wave plate 半波偏振片;半波(晶)片
half-wave polarization plate 半波长偏振片
half-wave potential 半波电势
half-wave radiation 半波辐射
half-wave radiator 半波辐射器

half-wave rectification 半波整流
half-wave rectifier 半波整流器
half-wave rectifier tube 半波整流管
half-wave rectifying circuit 半波整流电路
half-wave retarder 半波延迟器
half-wave suppressor coil 半波抑制(感应)线圈
half-wave thyristor rectifier 半波可控硅整流器
half-wave transformer 半波变压器
half-wave transmission line 半波输电线;半波传输线
half-wave vibrator 半波振子;半波振动器
half-wave voltage 半波电压
half-wave voltage doubler 半波倍压整流电路
half-wave zone 半波区;半波带
halfway 中途的
halfway around 半转
halfway house 中途旅馆;中途康复站
halfway housing 半途停工的住房建设
halfway rectification 半波整流器
halfway unit 半工业装置
halfway up the mountain 山腰
half weathered rock 半风化岩石
half weathering feldspar 半风化岩石
half-white glass 半白玻璃
half-white oil 半亮油
half width 半宽(度)
half width at half-maximum 半峰半宽度
half width construction 半宽施工
half width of spectrum line 谱线半高宽
half width work 半边宽度筑法
half-window 夹层楼面砖;半窗
half winterness 半冬性
half-wool 棉毛交织物
half-word 半字
half-word block 半字组
half-word boundary 半字界
half-word input 半字输入
half-worn wheel 半磨耗车轮
half wrap drive 包围轮一半的皮带轮传动;绕半个轮轴的绳索传动
half wraps 半圈
half-write pulse 半写脉冲
half-wrought timber 半成品木材
half-yearly account 半年账单
half-yearly audit 半年查账
half-yearly closing 半年决算;半年结账
half-yearly economic report 半年经济报告
half-yearly premium 半年保险(费)
half-yearly settlement 半年决算
halibut 星鲽;大比目鱼
Halicarnassos Mausoleum 哈里卡纳苏斯陵墓
halicole 盐土植物
halide 卤素;卤化物
halide acid 卤化物酸
halide crystal 卤化物晶体
halide glass 卤化物玻璃
halide ion 卤根离子
halide lamp 卤素灯;卤化物灯;检卤灯
halide torch 卤素管;卤化物检漏灯;检卤漏灯
Halifax Port 哈利法克斯港(加拿大)
haline water 高咸水
haliplankton 咸水浮游生物
halite 星云母;岩盐;卤酸酯;天然氯化钠;石盐;大青盐
halite rock 石盐岩
halite water 高咸水
hall 礼堂;会堂;厅;大厅
hall access type 厅式平面(各室入口开向大厅的)布置方式
Hallade recorder 哈拉德记录仪
hall-and-corridor type apartment 复合式公寓
Hall angle 霍尔角
Hall angular displacement transducer 霍尔角位移传感器
hall bedroom 隔厅卧室;门厅卧室
hall building 厅堂建筑
hall carpet 大厅地毯
hall choir 排练厅;唱诗班坐席
hall-church 哥特式教堂
hallcist 古代作坟墓用的以石板筑成的地;坟墓地下巨廊
hall clock 大厅钟
hall closet 门厅壁橱;大ँ内的套间
Hall coefficient 霍尔系数

Hall constant 霍尔常数
hall construction 大楼式建筑
hall crane 行车
Hall current 霍尔电流
Hall deep cell 霍尔深型浮选机
hall dome 半球形顶棚大楼
hall-donjon 城堡主楼
hall door 礼堂(大厅)门
hall-dungeon 城堡下层厅堂;城堡的大厅
Halle's point 阿累氏点
Hall effect 霍尔效应
Hall effect amplifier 霍尔效应放大器
Hall effect compass 霍尔效应式罗盘
Hall effect device 霍尔效应器件
Hall effect linear detector 霍尔效应线性检波器
Hall effect magnetometer 霍尔效应磁强计
Hall effect mixer 霍尔效应混频器
Hall effect multiplier 霍尔效应乘法器
Hall effect sensor 霍尔效应传感器
Hall effect transducer 霍尔(效应)传感器
halleflinta 长英角岩
Hallen derrick system 哈轮吊杆方式【机】
Haller's law 哈勒定律
hallerite 锂钠云母
Hallett table 哈列梯床
Hall-flowmeter 霍尔流动性测量仪
hall form 大楼模壳;大楼类型
hall furniture 大厅家具
Hallian stage 哈利阶
halliard 扬索;旗绳;(旗帆等的)升降索;升降绳
Halliburton test 哈利伯顿试验
Halliburton thickening time tester 哈利伯顿(水泥浆)增稠时间测试仪;哈利伯顿(水泥浆)变硬时间测试仪
Halligan tool 哈利根铁铤
Hallikainen capillary viscometer 哈利凯南毛细管黏度计
Hallikainen rotating-disk viscometer 哈利凯南转盘式黏度计
Hallikainen sliding-plate viscometer 哈利凯南滑板式黏度计
hallimomdite 三斜砷铅铀矿
Hallinger shield 哈林格盾构(法)
hallite 星蛭石
hall-keep 城堡内居室;城堡的大厅
hall lantern 站厅信号灯
hall mark 品质证明;检验刻记纯度;纯度(检验印记)
hall mirror 门厅镜
Hall mobility 霍尔迁移率
hall-nave 教堂坐席;(车站的)中央广场
hall noise 厅堂噪声
Hall of Abencerrages 狮子院南厅(西班牙阿尔罕伯拉宫)
Hall of Ambassadors 使节厅(西班牙阿尔罕伯拉宫)
hall of arts and crafts 工艺品馆
hall of columns 柱厅;圆柱大厅
Hall of Fame 贤良名人祠
hall of heavenly phenomenon 天象厅;天象馆
Hall of Judg(e)ment 评判厅(西班牙阿尔罕伯拉宫)
Hall of Mysteries at Eleusis 埃莱夫西斯的神秘厅
hall of pottery and porcelain 陶瓷管
Hall of Prayer for Good Harvests 天坛祈年殿
Hall of the Double Axes 双轴厅(克里特的克诺索斯宫中的)
Hall of the Hundred Columns (波斯波利斯的)百柱厅
Hall of Two Sisters 姐妹厅(西班牙阿尔罕伯拉宫)
hallostachys belangeriana 盐穗草
Hallowell granite 哈洛威尔石(产于美国缅因州的一种浅灰色花岗岩)
halloysite 叙永石;准埃洛石;多水高岭土;多水高岭石;埃洛石
halloysite clay 埃洛石黏土
Halloysitum Rubrum 赤石脂
Hall process 霍尔方法;电解还原制铝法
hall quire 有唱诗班席位的大厅(教堂);唱诗班坐席;排练厅
hall roof 厅堂屋顶
hall roof truss 厅堂屋(顶桁)架
halls kiln 环形多室窑;长廊窑
halls of ivy 名牌大学的古老建筑(美国)
hall stand 大厅衣帽架;衣帽架
hall steel door 厅堂钢门

hall system apartment 厅式公寓
Hall tablemount 霍尔平顶海山
hall transept (教堂的)交叉通道;厅堂翼部;(教堂的)交叉甬道
hall tree 落地衣帽架;大厅衣帽间;衣帽架
Hall-type automobile electronic igniter 霍尔型汽车电子点火器
hall-type block 厅堂式房屋
hall-type building 厅堂式房屋;厅式建筑;厅式大楼
Hall unit 霍尔元件
hall-ventilation 大厅通风
Hall voltage 霍尔电压
Hallwach's effect 光电效应
hallway 回廊;过道;门厅;穿堂
Halmahera Sea 哈马黑拉海
Halman 铜锰铝合金电阻丝
halmeic deposit 海成沉积物
halmeic mineral 海生矿物
halmyrolysis 海解作用;海底风化(作用)
halmyrolytic deposit 海解沉积
halmyrolytic Mn nodule 海解锰结核
halo 晕圈;晕轮;光晕
haloacetic acid 卤代乙酸
haloacetonitrile 卤代乙腈
haloalkane 卤代烷
halo-anhydrite 盐硬石膏
halobiont 咸水生物
halobios 咸水生物;盐生生物;海洋(水)生物;海产生物
halobolite 海底锰块
halocarbon 卤烃;卤化碳
halocarbon plastic 碳卤塑料
halocarbon resin 碳卤树脂
halochromism 卤色化(作用);加酸显色
halochromy 加酸显色现象
halocline 盐跃层;盐度跃(变)层
halococci 嗜盐球菌
halocyanophytes 盐生蓝藻类
haloduric 耐盐的
halo effect 光晕效应;光圈效应;光环效应
haloeremion 盐生漠境;盐生荒漠
halo fiber 光晕纤维
haloform formation potential 卤仿生成势
halogen 卤素
halogen analyzer 卤素分析仪
halogenate 卤化
halogenated alkane 卤烷
halogenated anthanthrone 卤化蒽烯蒽酮
halogenated aralkyl-aryl ether 卤化芳烷基芳基醚
halogenated compound 卤代化合物
halogenated derivative 卤化衍生物
halogenated (fire) extinguishing agent 卤化物灭火剂
halogenated hydrocarbon 卤代烃
halogenated hydrocarbon degrading 卤代烃分解
halogenated hydrocarbon pollution 卤代烃污染
halogenated isoviolanthrone 卤化异紫蒽酮
halogenated methane 卤甲烷
halogenated paraffin 卤蜡
halogenated polymer 卤代聚合物
halogenated solvent 卤化溶剂
halogenating agent 卤化剂
halogenation 卤化(作用)
halogenation process 卤化法
halogen cell 卤素电池
halogen compound 卤素化合物
halogen-containing 含卤素的
halogen-containing organsim 含卤素有机物
halogen-containing substance 含卤物质
halogen counter 卤素计数管
halogen disinfectant 卤素消毒剂
halogenerated organic solvent 卤代有机溶剂
halogen ether 卤代醚
halogen family 卤族
halogen family element analysis 卤族元素分析
halogen family elements 卤族元素
halogen free 无卤
halogenic deposit 盐(类)沉积
halogenic soil 盐成土
halogenide 卤化物;含卤物
halogen incandescent lamp 卤素白炽灯
halogen lamp 卤素灯
halogen leak detector 卤素检漏器
halogen leak locator 卤素查漏仪

halogen organics 卤代有机物
halogenosilane 卤代硅烷
halogenosilicic acid ester 卤代硅酸酯
halogenous 含卤素的
halogenous organism 含卤素有机物
halogenous reaction 卤化反应
halogenous substance 含卤物质
halogen poisoning 卤族元素中毒
halogen process 卤化法
halogen quenched counter 卤素猝灭计数器
halogen rock 盐岩
halogen-specific detector 卤素专用检测器
halohydrin 卤代醇
halohydrocarbon 卤烃类
halo-hydrocarbon degrading 卤代烃分解
halo-hydrocarbon pollution 卤代烃污染
haloid fire-extinguishing system 卤化物灭火系统
haloidite 卤化物岩
haloid oxyacid 卤素含氧酸
halokainite 钾盐钾镁矾
halo-ketone 卤代酮
halolite 盐屑岩;盐类岩屑;盐鲕
halomask 晕光蒙片
halometer 盐量计
halomethylation 卤甲基化(作用)
halomorphic soil 盐(碱)土;盐渍土(壤);盐生土壤;盐成土
halomorphism 盐生形态
halon 碳、溴或卤的气体化合物
halonereid 海洋浮游生物
halon fire extinguisher 卤盐灭火器
Halon Management Center 哈龙管理中心
Halon management plan 哈龙管理计划
halon smothering system 卤化物灭火系统
halo of radio-burst region 射电暴晕
halopelite 盐质泥岩
halophenol 卤代酚
halophile 喜盐植物;喜盐微生物;亲卤素的;嗜盐微生物;适盐微生物
halophilic 喜盐的
halophilic microorganism 嗜盐微生物
halophilism 好盐性
halophilous bacteria 嗜氧细菌;嗜盐细菌
halophilous plant 好盐植物
halophily 盐质适应
halophitic vegetation 盐生植被
halophobe 嫌盐植物;避盐植物
halophobe plant 避盐植物
halophobic 避盐的
halophobous 避盐的
halophyte 盐土植物;盐生植物
halophytic vegetation 盐土植被;盐生植物
haloplankton 咸水浮游生物
halo ring circular features 晕圈环形体
halosere 盐生演替系列;海洋演替系列
halo-sylvite 钾石盐
halothane 三氟溴乙烷;氟烷
halotrichite 铁明矾;铁铝矾
halo-type anomaly 晕环异常
halo water 晕水
halpace 室内升高的地面;楼梯转弯处的平台
Halphen-Hicks test 哈尔芬—希克斯松香定性检验法
halt 暂停;停机;停顿
halt circuit 停止电路
halt command 停机命令
halted state 暂停状态;停止状态
halter 马笼头;平衡棒
halt indicator 停机指示器
halting place 休息场地;野餐场所
halting problem 停机问题
halting problem of flowchart schema 框图模式停机问题
halt instruction 停止指令;停机指令
halt mode 停机状态;停机方式
halt number 停止号码;停机号
halt problem 停机问题
halt sign 停车标志
halt signal baton 停车信号杆
halt switch 停止开关;停机开关
halurgite 哈硼镁石
halvans 杂质多的矿石;贫矿(石)
Halvan tool steel 铬钒系工具钢
halve 平分
halved belt 交叉带

halved door 半截门
halved joint 半搭接合;相嵌接合;互嵌接合;对(开)搭接;单裁口接接;重接
halved splice 半叠接;半搭接
halved tie 对束;两分枕木
halver 二等分
halver operator 等分算子;二等分算子
halves pass 减半传递
halve the production cost 生产成本降低一半
halving 开平接合;锁接;对分;等分;二等分;半开叠接;半迭接
halving adjustment 二等分校正
halving circuit 平分电路
halving joint 嵌接;边叠接;半搭接;对开搭接;单裁口搭接
halving line 平分线;半像线
halving method 二等分法
halving register 平分寄存器
halvings 贫矿
halving saw 切片锯
halyard 扬帆缩;升降索;扬索;旗绳;升降绳
halyard block 篷索滑车;升降绳滑车
ham 小村庄
hamada 石漠
hamartomatous 错构的
hamate 钩头状的
hamber 细绳
hambergite 硼铍石
hambrolin(e) 细绳
hambrough 细绳
Hamburg blue 汉堡蓝
Hamburg Port 汉堡港(德国)
Hamersley-type iron deposit 哈默斯利型铁矿床
hames 马颈轭
Hametag impact mill 漩涡冲击式磨机;哈梅塔格漩涡冲击式磨机
Hametag mill 漩涡磨机
Ham Hill stone 哈姆丘陵岩
Hami block 哈密地块
Hamilton-Cayley theorem 凯莱—哈密顿定理
Hamilton group 汉密尔顿群【地】
Hamiltonian characteristic function 哈密顿特征函数
Hamiltonian coordinate 哈密顿坐标
Hamiltonian cycle 哈密顿循环
Hamiltonian function 哈密顿函数
Hamiltonian graph 哈密顿图
Hamiltonian method 哈密顿法
Hamiltonian operator 能量算符
Hamiltonian principle 哈密顿原理
Hamiltonina equations of motion 哈密顿运动方程
Hamilton metal 哈密顿合金
hamisyncope 半晕厥
Hami-Tarim block-zone 哈密—塔里木地块带
hamlet 小村(庄);居民点;村落;部落
hammada 岩漠;石漠;石质荒漠
hammam 澡堂;土耳其式浴室
hammarite 哈硫铋铜铅矿
Hammatoidea sinensis 中华双尖藻
hammer 榔头;手持式凿岩机;锻锤;锤(子);冲击式凿岩机;凹印
hammer and sickle 带有锤子和镰刀圆形的旗帜
hammer anvil 铁砧;锤砧;锤垫
hammer apparatus 机械打桩机;机械锤;锤击设备;机动桩锤;打桩设备;打桩机
hammer automatic operating device 自动司锤装置
hammer axe 锤斧
hammer bank 击打部件;锤头组
hammer beam 悬臂托梁;木拱脚悬臂托梁;悬臂梁;锤柄;橡尼梁
hammer-beam roof 托臂梁屋顶;椽尼梁屋顶;悬臂梁屋顶
hammer-beam roof truss 托梁木屋架(晚期哥特式的一种)
hammer-beam truss 悬臂(托)梁桁架
hammer bit 冲钻
hammer blade mill 锤式粉碎机
hammer block 锤体
hammer blocked 用锤分块的
hammer blow 水击;水锤;液压撞击;锤击状花纹;锤击(噪声);锤打
hammer blow tamper 冲击夯;锤击夯
hammer-blow type pulse-velocity measuring device

锤击式脉冲进度检测仪
hammer body 锤机柱架
hammer brace 椽尾梁斜撑
hammer break 锤形衔铁断路器
hammer breaker 锤式破碎机;锤式轧碎机;锤式碎石机;锤(击破)碎机
hammer burr 击锤辊花
hammer butt 总木
hammer cable (击碎机上的)落锤缆
hammer carrier 锤架
hammer circle 锤头循环打击圈
hammer circle diameter 锤圆直径;锤头打击圈直径
hammer-claw 爪心锤
hammer coating 锤纹涂层
hammer cog 粗锻
hammer cogging 锻造开坯
hammer crane 锤头式起重机
hammer crane head 锤式起重机头
hammer crusher 锤碎机;锤式轧碎机;锤头螺栓;锤式碎石机;锤式破碎机
hammer crusher with feed rolls 带进料滚筒的锤式破碎机
hammer cushion 桩锤垫
hammer cylinder 气锤汽缸
hammer die 锻模
hammer disk 锤片转盘
hammer down 用锤钉上
hammer-down-the-hole machine 冲孔机
hammer dressed 锤琢的
hammer dressed ashlar 锤琢石;斩面条石;斩面方石;锤斩方石
hammer dressed ashlar masonry 锤琢石砌体;锤琢石圬工
hammer dressed method 锤凿法
hammer dressed quarry stone 锤琢毛石;锤琢块石
hammer dressed slab 锤击石板材
hammer dressed stone 锤琢(毛)石;锤琢石面
hammer dressing 用锤整修;锤琢;锤整;锤修(整)
hammer drifter 架柱式风钻
hammer drill 冲击(式)钻机;凿岩机;取芯凿岩机;锤钻;冲钻机;冲击凿岩机;风钻
hammer-drill hole 凿岩机钻孔
hammer drilling 冲击式钻进;冲击式凿岩
hammer-drill stop(p)er 伸缩式凿岩机
hammer driver 锤驱动器
hammer-drive screw 钉状螺钉;锤击螺钉;锤入式螺钉
hammer driving 锤击驱动;锤击打桩
hammer drop 桩锤落高;落锤高度
hammer drum 锤片滚筒
hammered bottle surface 凹凸不平的瓶面
hammered cathedral 有突起圆边的玻璃
hammered finish 锤琢面层;锤琢石面
hammered finish of stone 锤琢石面
hammered finish tile 锤琢瓦面
hammered glass 锻玻璃;有锤痕的玻璃;压花玻璃;锤痕玻璃;凹印玻璃
hammered granolithic finish 斩假石面;剁斧石面
hammered granolithic flooring 剁斧石楼地面
hammered iron 锻铁
hammered piston ring 锤打活塞环
hammered point 铆钉锤端
hammered resistance welding 电阻加热锻焊;锤锻电阻焊
hammered rivet 锤铆钉
hammered sheet iron 被锻铁板
hammered steel 锻钢
hammered work 锻造工作
hammer eye 钻孔凿;冲击钻(头);锤柄眼
hammer face 锤面
hammer faced 用锤凿面的
hammer-faced stone 锤琢石
hammer finger 锤状指
hammer finish 锤琢石珐琅饰面;锤纹漆涂法装;锤琢修整;锤琢饰面;锤纹漆饰面
hammer finish enamel 锤纹瓷漆
hammer flattener 平锤
hammer forging 自由锻造;锤锻(件)
hammer foundation grill 锤机基础格层
hammer foundation structure 汽锤基础结构
hammer frame 锤机柱架
hammer gate 落锤式闸门
hammer grab 桩锤垫;冲抓斗;冲击式钻孔抓具;冲击式抓斗
hammer-grab pile boring 冲抓式钻孔桩法
hammer-grab pile boring machine 冲抓式钻机
hammer-grab pile drilling 冲抓式钻孔桩法
hammer-grab work 冲抓钻孔作业(灌注桩)
hammer grinder 锤式研磨机;锤碎机;锤式粉碎机;锤击式碎石机
hammer grinding mill 锤式研磨工场;锤击式研磨机
hammer guide 锤导座;锤导承
hammer hack 劈石斧;斧形锤
hammer hand crane 塔式悬臂吊车
hammer hand drill 手提风钻
hammer hand-held drill 手持式风钻
hammer handle 锤柄
hammer harden 冷作硬化;锤硬
hammer-hardened 锤锻硬化的
hammer hardening 锤硬法
hammerhead 锤头;反尖削机翼
hammerhead bolt 大头螺栓;锤头螺栓
hammerhead-boom 塔式起重机(美国)
hammerhead chisel 槽榫錾;锤头凿
hammerhead crane 塔式(悬臂)起重机;塔式悬臂吊车;锤(头)式起重机
hammerhead(ed) key 锤头形键
hammerhead groin 锤(头)形防波堤;T形丁坝
hammerhead groyne 锤形防波堤
hammerhead jib 锤头支架;塔式起重机摇臂;塔式起重机旋臂;塔式起重机悬臂;锤臂
hammerhead key 锤头形键;锤头式防波堤;(硬木制的)银锭榫
hammerhead oil pier T形石油码头
hammerhead pier T形码头;塔式起重机支柱
hammerhead slewing crane 锤头式旋转起重机
hammer hook 击锤锁扣;击锤卡钩
hammer in 打入
hammering 噪声;敲击;锻造;锤击;锤锻
hammering action 冲击作用
hammering block 砧
hammering cut of dents 凹部校直
hammering device 锤打装置
hammering energy analyzer 锤击能量分析仪
hammering machine 锤锻机;锤锻
hammering press 锻造压力机;锻压机
hammering spanner 单头开口爪扳手
hammering test 锤击试验
hammer interrupter 锤形衔铁断路器
hammer latch 击锤卡榫
hammerless 无冲击的
hammer lever 锤击杆
hammer line 桩锤吊索
hammerlock moor 止荡锚泊法
hammer machine 锤击试验机
hammer magnet 锤击磁铁
hammer man 铁匠;铁工;锻工;打桩工
hammer mark 锤击伤痕;锤击钢印标记
hammer mill 舂碎机;离心破碎机;锻工场;锤机;锤式研磨机;锤式破碎机;锤式磨;锤式粉碎机;锤磨机
hammer-mill cage 锤碎机筛笼;锤碎机筛格
hammer-milled fiber 锤磨纤维
hammer-mill grate 锤碎机固定筛;锤碎机算条
hammer milling 锤磨作用;锤碎
hammer-mill sieve 锤式粉碎机筛片
hammer-mill type stalk shredder 锤式碎茎机
hammer-mill with air-classifier 风选锤式粉碎机
hammer-mill with predryer 带预烘干装置的锤式磨
hammer nylon 尼龙锤
hammer off 锤截断
hammer out 锤平
hammer paint 锤纹漆
hammer path 撞击轨迹;撞击轨迹;锤头轨迹;锤程
hammer peening 锤头打铃
hammer pick 鹤嘴镐;尖锤;锤尖
hammer piling 锤击沉桩
hammer pin 锤头销;锤片轴销
hammer pin body 锤击销体
hammer pinching 锤琢石面
hammer piston 凿岩机活塞;气锤活塞
hammer plate 夹锤板
hammer plugger 取芯凿岩机
hammer post 椽尾支柱;椽梁支柱;椽横梁支墩;悬臂托柱
hammer pulse 锤击脉动
hammer ram 夯锤;锤头;冲击活塞;风动铆(钉)锤

hammer rating 锻锤的吨位
hammer refining 晶粒锻压纯化;锻压纯化
hammer retaining pin 锤片保持轴销
hammer reversing drill 锤式可逆钻
hammer riveting 锤击铆接;锤打铆接
hammer riveting machine 锤击铆接机;锤铆机
hammer rock drill 凿岩机;风钻;冲击式钻机;冲击石钻机
hammer rolls 锤碎机
hammer rope 桩锤吊索
hammer rotor 锤头转子
hammer scale 锻铁屑;锻鳞;锤鳞
hammer shaft 锤把;锤柄
hammer sharpener 锤磨器
hammer-shears 手工剪
hammer shock 水锤冲击
hammer sinker 向下冲击式凿岩机;冲击式向下凿岩机
hammer-size formula 桩锤尺寸公式
hammer slag 锤渣
hammer sledge 大锤
hammer smith 锻工;铁匠
hammer spring 锤簧
hammer stand 锤架;桩锤台;桩锤架;锤机(栓)架
hammer standard 锤机柱架
hammer stem 锤杆
hammer stock machine 槌心缩绒机
hammer striking mechanism 打锤机构
hammer stroke 桩锤冲程;锤程
hammer-stroke slag crusher 锤击碎渣机
hammer stud 锤杆桩
hammer support 锤杆座
hammer swing mill 锤磨机;摆锤粉碎机
hammer swing sledge mill 锤击磨
hammer tail 锤杆
hammer tamper 桩锤夯;打夯机
hammer tamping 锤夯
hammer tamping finisher 打夯修整工
hammer test 锤击试验
hammer test machine 锤击试验机
hammer toe 锤状趾;锤形趾
hammer tone 锤琢饰面;锤面状涂层;锤纹(饰面)
hammertone emulsion paint 锤纹乳化漆
hammertone enamel 锤纹瓷漆;锤纹瓷漆
hammertone finish 锤纹饰面
hammertone hardboard 锤纹硬纸板
hammertone silver paint 锤纹银粉漆;锤纹铝粉漆
hammer tongs 锻锤夹钳
hammer tool 锤锻工具;锤打工具
hammer track 锤形径迹
hammer trigger 锤机扳柄
hammer tup 打桩锤(的)撞头;撞锤
hammer-type holder-on 锤式气顶
hammer-type pneumatic hand-held drill 锤式手持风钻
hammer-type roof 露明木构架屋顶
hammer-type sprinkler 锤式喷灌器
hammer-type vibrator 锤式振动器
hammer valve 锤形阀
hammer valve gear 锤机阀动装置
hammer weight 冲击锤重量
hammer welding 锻接;锻焊(接);锤焊
hammer weld pipe 锻(焊)接管
hammer with tubular shaft 管柄锤
hamme ton silver 龟裂花纹银色涂料;龟裂花样银花涂料
Hammett equation 哈梅特方程式
Hammett indicator 哈梅特酸指标指示剂
Hammett scale 哈梅特酸标度
hamming 加重平衡
Hamming check 汉明校验
Hamming code 汉明码;误差检测及校正码
Hamming distance 汉明距离;代码间距
Hamming weight 码重
Hamming window 哈明窗
hammock 圆丘;硬木产区(美国南部);冰丘;吊铺;吊床
hammock chair 帆布椅
hammocked ice 堆积冰
hammock forest 内陆常绿阔叶林
hammock hitch 吊铺捆结
hammock shaped distortion 吊床形畸变
hammock shroud 海葬用的帆布床
Hamm tip 橄榄形喷嘴(喷混凝土用);哈姆喷嘴

hampdenite 硬蛇纹石
hamper 有盖大篮;障碍船具;(平时需用但在紧急时会妨碍操作的)甲板上船具
hampered vessel 受阻船
hamper navigation 妨碍航行
hamrock 暗灰色砂石(产于欧洲爱尔兰克来尔)
Hamstead beds 哈姆斯特德层
hamster 仓鼠
hamula 小钩
hamulus 翅钩
hamulus laminae spiralis 螺旋板钩
hamulus of spiral lamina 螺旋板钩
hanaways 杂质多的矿石
hance 托架小拱;承托小拱;矮拱;舷墙等急转折部分;拱腋;拱腰;梁腋
hance arch 三心拱;平圆拱;矮矢高椭圆拱【建】
H-anchorage 非张拉端锚具;H形锚具
hancockite 铅黝帘石
Hancock jig 联杆凸轮传动跳汰机
hand-actuated 手动的
hand adjustable reamer 手调铰刀
hand adjustment 手(工)调整;手调(节)
hand adjustment device 手调装置
hand advance 手动提前点火
hand air pump 手动气泵
hand air valve 手控气阀
hand anemometer 轻便风速仪;轻便风速计;轻便风速表;手提风速表;手持风速表
hand anvil 手砧
hand aperture card mounter 手动窗孔卡片安装机
hand appliance 手动装置
hand applicator 手工涂布机
hand applied(mixed)plaster 手工抹灰;人力拌制的灰浆
hand a project to client 工程移交给甲方
hand-arbo(u)r press 手扳压床
hand-ashed producer 手工除尘气体发生炉
hand ashing 手工除尘
hand aspirated type 手捏吸气式
hand assembly 手工装配
hand atlas 便携式地图集
hand auger 手钻;手提螺钻
hand auger work 人力打炮眼
hand axe 手斧
hand back saw 手锯
hand bag 手提包
hand baggage 手提包
hand bailing 手动捞水
handball court 手球场
handball playground 手球场
hand banisters 扶手栏杆
hand bar cutter 手动钢筋切断器;人工钢筋切断机
hand barrow 提架;担架;两轮手推车;塌车;双轮手推车;小车
hand basin 洗手盆
hand bearing compass 手提罗经
hand beetle 手夯(锤)
handbell 手摇铃
hand bellowed duster 手摇风箱式喷粉器
hand bellows 手风箱;皮老虎;手用吹风器;手工吹风箱
hand belt sander 手操纵带式打磨器
hand belt sanding machine 手操纵砂带磨机
hand bench 手工台
hand bench grinder 手动台式磨床
hand bender 手操纵弯折机;手动弯曲机
hand bending 手弯
hand bending machine 手工弯筋机;手工弯折机;手工弯曲机
hand bilge pump 手摇污水泵;手摇舱底水泵;手摇舱水泵
hand-bill 剪枝器
handbills and waterspouts 喷砂冒水
hand bit 手钻子;手凿子;手钎子
hand block(and tackle)手动滑轮吊车;手动起重机滑轮组
hand blocked 手工压模的(壁纸);手工印制的
hand-blocked fabric 手工模版印花织物
hand-blocked(wall)paper 手工印花墙纸
hand blocking 手工模版印制;人工印花;木刻版印刷(术)
hand block printing 手工模版印花
hand block sander 人工进料砂光机
hand blower 皮老虎(锻工用);手摇(鼓)风机;手风箱

hand blowing 人工吹制
hand blown glass 人工吹制玻璃
hand blown process 人工吹制法
handbook 指南;说明书;手册
handbook of inspection requirement 检验要求手册
handbook of overhaul instruction 大修须知
hand boom 手动喷水器
hand booster 手摇升压器
hand bore 手摇钻
hand borer 便携钻孔器;便携打眼器;手动钻头;手动钻孔
hand boring 手钻孔;手工钻探;手工凿孔
hand boring bit 手钻钻头
hand boring machine 手(摇)钻机
handbow 手弓
hand brace 人力钻;手摇(曲柄)钻
hand brace drill 手摇曲柄钻
hand brake 手动制动装置;手制动(器);手刹车;手(动)闸
hand brake adjusting bolt 手闸调准螺栓
hand brake adjusting bolt bracket 手闸调整螺栓托架
hand brake adjusting bolt spring 手闸调准螺栓簧
hand brake band 手闸带
hand brake band anchor bar adjusting screw 手闸带锚定杆调整螺钉
hand brake band anchor bracket 手闸箍端托架
hand brake band anchor clip spring 手闸带碇夹弹簧
hand brake band bracket 手闸箍端托架
hand brake bracket 手制动器支架
hand brake cam 手闸凸轮
hand brake cam lever 手闸凸轮杆
hand brake chain guide 手闸动链导板
hand brake chain wheel 手闸动链轮
hand brake compression spring 手闸压缩弹簧
hand brake control 手闸控制
hand brake cross-shaft 手闸横轴
hand brake disc[disk] 手制动盘
hand brake drive rod adjusting fork 手制动器驱动杆调整叉
hand brake drum 手力制动鼓
hand brake guide plate 手制动轴卡板
hand brake latch rod 手制动爪杆
hand brake lever 手闸杆;手刹车操纵杆
hand brake lever adapter 手闸杆接头
hand brake lever arm 手闸杆
hand brake lever bracket 手闸杆托架
hand brake lever grip spring 手闸杆握柄弹簧
hand brake lever latch 手闸杆闩
hand brake lever latch rod knob 手闸杆弹键杆钮
hand brake lever operating spring 手闸杆操作弹簧
hand brake lever operating stop 手闸杆操作止点
hand brake lever pawl 手闸杆爪
hand brake lever pawl pin 手闸杆爪销
hand brake lever pawl rod 手闸棘爪杆
hand brake lever pawl spring cup 手闸杆爪弹簧盖
hand brake lever pull rod yoke 手闸杆拉杆轭
hand brake lever sector 手闸杆扇形齿轮
hand brake lever yoke 手闸杆轭
hand brake lining rivet 手闸衬片铆钉
hand brake locating bracket 手闸定位托架
hand brake master step 手制动轴托
hand brake mast support 手制动轴导架
hand brake on 手制动器已接合
hand brake operating lever 手闸操作杆
hand brake operating lever link 手闸操作杆联杆
hand brake operating lever link pin 手闸操作杆联节销
hand brake operating lever shoe 手闸操作杆瓦
hand brake pawl rod 手动棘爪拉杆
hand brake pull rod chain 手制动拉杆链
hand brake pull rod guide 手制动拉杆导架
hand brake pull rod guide anchor 手制动拉杆吊
hand brake release spring 手闸动放松弹簧
hand brake shaft rigging 手制动轴装置
hand brake shoe pin 手制动瓦销
hand brake support 手闸支架
hand brake support tie rod 手闸棘爪杆
hand brake telltale 手制动接合指示灯

hand brake transfer lever 手闸传力杆
hand brake valve 手制动阀
hand braking 手制动
hand breaking 人工破碎
hand brick 手工砖
hand Brinell's hardness tester 手提布氏硬度计
hand-broken metal 人工敲碎的石块;小碎石料
hand-broken stone 人工碎石;小碎石子
hand-broom 扫帚
hand brush 洗手刷;手刷
hand brushing 手工刷涂;手工刷漆
hand bucket 手提桶;手动戽斗;吊桶
hand buddle 手工淘汰盘
hand buggy 手(推)车;手(拉)车
hand-built product 手工品
hand bumping 手工锤击
hand bumping tool set 成套手锤
hand by-pass pump (内燃机的)手动增压泵
hand by-pass valve 手动旁通阀
hand cable 辅索
hand cable winch (绞盘上的)手摇缆索;手摇缆索绞车;手动绞车
hand caging 人工装罐
hand calciner 手工锻炉
hand calculation 手算
hand calculator 手摇计算装置;手摇计算器;手提计算器;手控计算器
hand camera 手提(式)摄影机
hand capacity 接触电容
hand capstan 人力绞盘;人力绞磨;手摇(式)绞盘;手推绞盘
hand car 手拉车;挂车;手摇(轨道)车;手动车
hand carding 手工梳理
hand-carry centrifugal pump 手提离心式水泵
hand carrying type rail tongs 手抬式钢轨钳
hand cart 手拉小车;两轮手推车;手推车;单轨小车;小车
handcarving 手雕
hand case high pressure acetylene generator 手提箱式高压乙炔发生器
hand ca(u)lking 手工凿密
hand centrifuge 手摇离心机
hand chain 手(拉)链
hand chain block 手动链滑车;手动神仙葫芦
hand chain block hoist 手动链滑车卷扬机
hand chain drive 手拉链驱动;人力链传动(装置)
hand chain lever hoist 环链手扳葫芦
hand chain sounding machine 手动链式触探机
hand channel 人工渠道
hand charging 人工装料
hand chaser 手动螺纹梳刀
hand chipping 手工凿平;手工修整;人工铺撒石屑;人工铲削清理;人工凿平;人工修整;手工铲削清理
hand chipping and scraping 手工敲铲除锈
hand chippings spreader 石屑铺撒机;石屑铺撒工
hand choke 手动阻风阀
hand chuck 手动卡盘
hand chucking hole 卡盘扳手孔
hand churn drill 手动钢绳冲击钻机;手动冲钻
hand churn drilling 手动钢绳冲击钻进
hand clamp 螺丝夹钳
hand-cleaned rack 人工清除格栅
hand-cleaned screen 手工清除筛;人工清洗滤网
hand cleaning 人工清除;人工清理;手选;手拣;手工除锈;重涂时表面处理等级
hand cleft wood shingle 手劈木瓦片;手工劈制的木质屋顶板
hand cleft(wood)siding shingle 手劈外墙面板
hand clipping 手工剪修
hand clutch lever 手动离合器杆
hand coal cutting machine 手扶采煤机
hand coat 手工刷浆
hand coating job 手工装饰工作
hand coded analyser 手编分析程序
hand-coded compiler 手编编译程序
hand cold chisel 手冷錾
hand cold header 手动冷镦机
hand-colo(u)red 手工着色的
hand-colo(u)red map 手工着色地图
hand-colo(u)ring 手工着色
hand-colo(u)r separation 手工分色
hand combination set 手持送受话器
hand combing 手工梳理

hand-compacted 人工捣实的
hand-compacted concrete 人工捣实混凝土
hand compacting beam 手工捣棒
hand compaction 人工夯实；人工捣实
hand compass 手提罗经
hand completion 手工修改
hand computation 手提计算；手算
hand computer 手摇计算机
hand concrete-cart 混凝土手推车
hand concrete-cart mixing 混凝土手推车拌制；混凝土手推车拌和
hand concrete mixing 人工混凝土拌和
hand contamination monitor 手部污染监测器
hand control 人力控制；人工控制；人工操纵；手工控制；手工操纵；手动控制
hand-controlled 用手控制的
hand-controlled branch 手控式水枪
hand-controlled grass cutter 手动剪草机
hand-controlled manipulation 手控操纵
hand-controlled nozzle 手控式水枪
hand-controlled regulator 手动调整器
hand-controlled sprinkler 手操纵喷灌器
hand-controlled target 手控靶
hand controller 手动控制器
hand controls 手控制系统
hand control signal 手控信号
hand control signal box 手控信号台
hand control system 人工控制系统；手动控制系统
hand control unit 人工控制机构
hand control valve 手动控制阀；手动阀（门）；手操纵阀
hand control valve body 手控制阀体
hand control valve cover 手控阀盖
hand control valve piston 手控制阀活塞
hand counter 手控计数器
hand crab 人工起重小车
hand-craft 手工业；石棉—水泥制品；手艺
handcraft roofing 大块石棉水泥瓦楞屋面板
hand crane 手操纵的吊车；手摇起重机；手动起重机；手动吊车
hand crank 起动摇把；起动曲柄；手摇曲柄；手动曲柄
hand crank grinding tool 手磨气门摇臂
hand cranking 手摇起重；手摇起动
hand cranking device 摇手装置
hand cranking leg 手摇腿
hand crew 作业组
hand crew boss 手工灭火队领班
hand crusher 手摇轧碎机；手摇破碎机；手动破碎机；手摇碎矿机
hand-cut overlay 手工垫版
hand-cut peat 人工粉碎泥炭；手工采掘泥炭
hand-cut random rectangular ashlar 手工加工的矩形石
hand cutter 用手切割者；手工切削刀具；(坯面上的)手工冲印
hand cutting 手工切割
hand cutting tool 手工刀具
hand cylinder method 人工吹筒摊平法（平板玻璃）；手工吹筒法
hand cylinder process 人工吹筒摊平法（平板玻璃）
hand deck-pump 手摇舱面泵
hand decoration 手工彩饰
hand diamond drill 手动金刚石钻机
hand die 手锻模
hand digging 人工挖掘；手工挖掘
hand digitized 手工数值化
hand direction indicator 手动示向器
hand director 手指规
hand distress signal 手操纵式危急信号
hand distributor 人工撒机；手压撒布机；手喷枪；手喷器
hand dog 钻杆（螺钉）扳手；变形法兰式扳手
hand door 手动闸门
hand doping 人工绝缘和包缠
hand drawing 手工描绘
hand-drawing method 手扯长度法
hand-drawn coarse fiber 手拉粗丝
hand-drawn original 清绘原图；手工绘原图
hand-drawn roller 人工喂料辊；手拉压路机；手拉滚筒
hand dredge(r) 手动挖泥机；人力挖泥机；人工挖泥耙；(人工的)挖泥耙
hand dressed 手工琢石；手动修整
hand dressing 手工梳麻

hand drier 干手器；烘手箱；烘手器
hand drill 钻孔器；手持式凿岩机；手持式风钻；手持式凿岩钻
hand driller 手力钻机
hand drilling 手工钻孔；土钻孔；手动钻进
hand drill(ing machine) 手摇钻；手钻机
hand drill steel 手持式钢钎
hand drive 人工开动；手传动
hand-driven 手动的
hand-driven batten 手动吊杆
hand-driven crab 手摇起重绞车
hand-driven gear transmission bit 手摇研磨器
hand-driven gear transmission drill 手摇研磨器
hand-driven generator 手摇发电机
hand-driven grinder 手摇砂轮机
hand-driven pottery's wheel 手动陶车
hand-driven rivet 手铆铆钉
hand-driven screw press 手动平板机
hand-driven starter 手摇起动机
hand driver 手工打桩机；手工打入工具
hand drum winch 手工筒式卷扬机；手工筒式绞车
hand dryer 干手器；烘手箱；烘手器；手吹风机
hand drying apparatus 人工干燥装置
hand-dug caisson 人工挖孔桩
hand-dug pit 手掘井筒
hand-dug shaft 手掘(竖)井
hand-dug well 人力挖井；手工挖掘井
hand duster 手摇喷粉器；手摇喷粉机
hand-dynamo 手摇发电机
hand-dynamometer 握力计
hand earth auger 手摇钻土器；人力土钻
handedness 旋向性
hand effect 手效应
handel 镜臂
hand electric(al) drill 手提式电钻
hand electric(al) lamp 矿用手提灯
hand electric(al) tool 手持电动工具
hand elevator 手动升降机
hand emergency signal(l)ing apparatus 手摇报警器
hand-engaged twist-lock 手动扭锁
hand engagement 手工接合
hand engraving 手工刻图；手工雕刻
hander 座；支持器；架；夹头
hand erected scaffold 人工架设的脚手架
hand excavation 手工开挖；人工土石方；人工开挖
hand-excavation shaft 人工开挖井孔
hand expansion valve 手动膨胀阀
hand extinguisher 手提式灭火器；手式灭火机
hand extinguishing appliance 手提式灭火装置
hand-fabricated 人工制造的
hand face shield 手提面罩(焊工用)
hand facility 搬运设施
hand-fed producer 手工加料气体发生炉
hand-fed transplanter 手置式移栽机
hand feed 人工给料；人工加料；手摇推进；手力进料；手动给进钻机；手操纵进刀；人工馈送
hand feed band-saw 手推带锯
hand feed core drill 手把给进岩芯钻机
hand feeder 手动续纸器
hand feed grinder 手送碎木机
hand feeding 用手进料；用手供料；人工加料
hand feeding mechanism 手动进给机构
hand feed punch 人工馈送穿孔机；手进给穿孔机；手工穿孔机
hand feed punch 手工输入穿孔机
hand feed surfacer 手进刀平面刨床
hand feed wheel 手进给轮
hand feel and drape 手感舒适
hand feeling 手感
hand felling 手工缝边
hand fettle 人工修坯；人工补炉
hand fid 插接木笔
hand file 锉刀；平锉；手夹
hand-fill 人工装载
hand-filled face 人工装载工作面；手工装载工作面
hand filling 用手填实；人工填充；人工回填；人工包装；人工灌装
hand finish 手工整修；手工最后加工；手工修整；手工精整加工；手工精修；手工成型
hand-finish concrete 人工抹面的混凝土
hand-finished 手工整修的
hand finisher 混凝土手工修整器；手工修整器；手工整修机

hand finishing 手工修整；人工修整；手工整修；手工精削；手工加工；手工成型
hand finishing noise 手工作业噪声
hand fire alarm(device) 手动火警报警器
hand-fired 人工加煤的
hand-fired boiler 手烧锅炉
hand-fired furnace 人工加煤炉；手烧炉膛
hand fire extinguisher 手携灭火器；手动式灭火器；手提式灭火器；手提(式)灭火机
hand fire pump 手摇消防泵
hand-firing 手工烧火；人工燃烧；手工燃烧
hand fit 压入配合；人工配合
hand fitting 手工组装；手工装配；手工安装；人工修配
hand fixed 人工加煤的
hand flag 手旗
hand flame cutter 手提火焰切割器；灭火机
hand flare 手提火号
hand float 木抹子；抹子；镘刀；单纹锉刀；抹灰工具；手镘(板)
hand float finish 手镘抹面；手镘抹面
hand floating dry concrete 手镘干硬性混凝土
hand float light 手提泛光照明灯
hand float trowel 抹子；镘刀；手镘刀
hand follow-up steering 手控随动操舵
hand force pump 手摇泵；手压泵
hand-forged ironwork 手工锻制铁工
hand forging 手工锻制的；手制锻件；手工锻(造)；手锻
hand fork 手叉
hand-fork truck 手动叉式装卸车
hand form block 手工模型
hand-formed brick 手制砖
hand-formed joint 手制节点
hand forming 手工捏塑；手工成型
hand frame 抬斗；担架
hand fuel pump 手摇燃料泵
hand fuller 套柄铁锤
hand furnace 人工炉
hand gear 手拨轮
hand gear control 手捏变速操纵装置
hand gear shift lever 手动换挡杆
hand generator 手摇发电机
hand-ginned cotton 手轧棉
hand glass 沙漏
hand glassing 手工推光
hand gobbing 人工充填
hand goniometer 手提式测向器；手提式测角器；手动测角计；接触测角仪；手持测角仪
hand-go shower set with soap holder 带皂托手动淋浴器
hand-got 手工采掘
hand gouge 手圆凿
hand governing 人工调节
hand grab 扶手(棍)
hand grader 人工平路机；手扶平路机
hand graining 手工搓纹
hand gravure 手工凹印
hand grease gun 人工注油枪；手持润滑油枪；手控黄油枪
hand grinder 手工(研)磨机；手工磨床；手摇砂轮；手工磨光器；手持式砂轮机
hand grinding 手工磨削
hand grinding machine 手摇砂轮
hand grip 手柄；有力的握手；紧握
hand grip control 手柄控制
hand gritter 人工石屑撒布机；手持式岩粉撒布器
hand gritting 手工铺撒石屑
hand grooving and tonguing plane 手用槽刨和舌刨
hand guard 护手；扶手
hand guard bar 扶手栏杆
hand-guided 人工控制的；用手掌握的；用手引导的
hand-guided power-propelled vibrating roller 手控电动振动压路机
hand-guided roller 人力压路机；人力路碾；手拉滚筒
hand-guided snow remover 手扶式除雪机
hand-guided two-wheel dozer 手控双轮推土机
hand-guided vibrating concrete finisher 手控振动混凝土抹面机
hand gun 手(持)喷枪；手用喷枪(喷雾时用)；手压注油枪；手动油枪
hand gun duster 手动喷粉机
hand gun loader 手压油枪注油器

hand hackle 手工梳麻
hand hack saw blade 手弓锯片
hand hack saw frame 手弓锯框
hand hammer 手锤;手持式凿岩机
hand hammer drill 手冲钻机;手冲击钻;手持式风钻
hand hammer drilling 手工凿岩;手锤钻眼
hand hammer rock drill 钻凿手锤
hand handle 手感
hand-handling 手工生产;手工操作
hand haulage 人力运输;人工运输;人工搬运
hand hay cutter 铡刀
hand-HB tester 手锤布氏硬度计
hand-headset assembly 手持头戴送受话器
hand-held 手持式(的);手提式报话器
hand-held aerial camera 手提航拍摄影机
hand-held anemometer 手持式风速仪
hand-held binocular 手持式双目镜
hand-held calculator 手摇计算器;手持计算器
hand-held camera 手提摄影机;手持摄影机
hand-held colo(u)r TV camera system 手持式彩色电视摄像系统
hand-held computer 手提式计算机;手持(式)计算机
hand-held console 手动控制台;手提操作台
hand-held(design) 手提式
hand-held drifter 重型手持式凿岩机
hand-held drill 手提式风钻;手持风钻;手持式凿岩钻;手持式凿岩机
hand-held drilling machine 手持式凿岩机
hand-held electric(al) drill 手提电钻;手持式电钻
hand-held exposure 手持摄影
hand-held extinguishing appliance 手提式灭火设备
hand-held flood light 手持探照灯;手持泛光灯
hand-held hammer drill 手持式冲击凿岩机
hand-held knotting device 手持式打结器
hand-held meter 手持式曝光表
hand-held microphone 手持式传声器
hand-held moisture meter 手持湿度测量仪
hand-held photography 手持摄影
hand-held pointer 手感应笔
hand-held portable electric(al) tool 手提式轻便电力工具
hand-held power tool 手扶动力工具
hand-held(rock) drill 手持式凿岩机
hand-held self-rotating air-hammer drill 手持式回转风动凿岩机
hand-held sinker 手凿
hand-held thermal imager 手持热像仪
hand-held thermocouple probe 手持数字热电偶高温计;手持热电偶探测器
hand-held TV camera 手提电视摄像机
hand-held welder 手动焊接机
hand hemming 手工缝边
hand high 手高(堆货高度)
hand hoe 手锄
hand hoist 手持提升器;手持卷扬机;人力卷杨机;手摇卷扬机;手摇绞车
hand hoisting 人工提升;手动提升
hand hold 手把;手柄;握把;拉手【建】;紧握;手持
handhold at driver's cab doors 驾驶室门旁扶手
handhold at the pocket sliding door car end 2 二号车端头滑动处扶手
handhold drill 手提凿岩机
handhold infrared alarm 手持红外报警器
handhole 手工掏槽;检查孔;探孔;手孔
handhole and manhole openings 手孔和人孔开口
handhole cover 检查孔盖;手孔盖;观察孔盖
handhole door 探孔门;检查用手孔门;储[贮]水槽门;注油管;手洞门
handhole door for inspection 检查用手孔门
handhole flange nipple 手孔凸缘螺纹接套
handhole nipple cap 手孔接头盖
handhole plate 手孔板;手孔盖
handhole plate bolt 手孔板螺栓
handhole plate gasket 手孔板垫密片
handhole with butt welded flange 对焊法兰手孔
handhole with flat cover 平盖手孔
handhole yoke 手孔轭
hand holing 手工掏槽;人工打眼
hand hook 手钩
hand hose line 小口径水带线
hand-hydraulic steering gear 手动液压操舵装置
handicap 障碍;残疾人

handicap door opening system 便开门(残疾人用)
handicaped elevator 残疾人电梯
handicapped people 残废者
handicapped person 残疾人
handicapped persons use of public buildings 残疾人用的公房
handicap person's center 残疾人中心
handicap persons use of public buildings 残疾人用的公房
handicap water cooler 杠杆冷水器(残疾人用)
handicraft 工艺品;手工艺品;手工艺;手工(业);特种工艺
handicraft art 手工艺
handicraft article 工艺品
handicraft forging 手工锻制的
handicraft industry 手工业
handicraftsman 手艺(工)人;手工艺工人
handicraft-type metal grille 人工制作的金属格栅;手工艺型金属栅;手工艺型花格窗
handie-lookie 手提照相机
handie-talkie 携带用小型收发报机;舰对岸手提无线电;手提式无线电话机;手提(式)步话机;手持式步话机;步谈机;步话机;报话机;微型双工电台
hand impact screwdriver 击松螺丝刀
hand-in 交发
hand inertia starter 手摇惯性起动机
handiness 易操纵性;轻便;操纵方便
hand-in-glove 密切合作
handing nipple in tubing 油管工作筒
handing object 悬吊物
handing of door 门扇开关方向
handing of window 窗扇开关方向
handing over 移交;交付
handing-over certificate 移交证书
handing-over point 搬移点
handing-over post 交接所
handing profits over-to the state 财政上交
handing sampling 掌子面采样
handing/taking-over 交接
handing time 辅助工作时间
handing tool 提放工具
handing traverse 支导线【测】
hand-in-hand chain 环形交叉链
hand injector 手提灌注器
hand insertion 人工装入;手工下线;手工嵌线
hand inspection 手检查
hand interruptor 手动断续器
hand in the business license for cancellation 缴销营业执照
hand introduction 人工装入
hand iron 铁镫;攀登铁爬梯;烙铁
handiwork 工艺品;手工(制品)
hand jack 手摇千斤顶;手压千斤顶;手力起重器;手动千斤顶;手动起重器
hand jack(screw) 手力千斤顶
hand jack type bar shears 手动液压式钢筋切断器
hand jig 手摇跳汰机;手动淘汰机
hand jigger 手工磨光器
hand jigging 人力淘汰;人工跳汰(选)
hand-kept 手工记录
hand kneading 手工捏练
hand knitred hosiery 手工针织品
hand knitting 手工针织
hand-knitting yarn 手编绒线
hand knob 球形捏手;手柄
hand-knotted pile carpet 手织簇绒地毯
hand-knotted rug 手织地毯;手结地毯
hand labo(u)r 人工劳动;手工制作;手工劳动
hand labo(u)r method 人工方法;手工方法
hand ladle 手勺;手浇包;端包;长柄手勺
hand-laid 手砌(的)
hand-laid foundation 手砌基础
hand-laid stone-filled asphalt 人工铺筑填石沥青
hand lamp 手(提)灯
hand lamp transformer 手提灯变压器
hand lance 手提式吹灰器;手喷枪;喷水器;喷枪;手压泵
hand lantern 提灯
hand laser night vision device 手持激光夜视仪
hand lashing 手工出渣;手工出砟
hand lathe 手摇车床
hand lawnmover 手提式剪草机
hand laying 手撒的;手铺的;人工敷设

hand laying gang 人工安装队;手工铺路组;手工铺路队
hand laying up 人工绝缘和包缠
hand lay-up 手糊成形;手工制造增强塑料膜
hand lay-up method 手工积层法
hand lay-up process 手铺法
handle 执手;拉手【建】;句柄;经营;手感;处理;铲柄;操作;办理;把手
handleability 可运用性;可操作性
hand lead 木砣;手锤(球);手测深锤;测深手锤;测深(铅)锤
hand lead bend 测锤结
hand lead line 水砣绳
handlead sounding 水砣测深;手锤测深
handlead survey 水砣测深;手锤测深
handle air blower 皮老虎;手用吹风器;手风箱
handle bar 车把;操纵杆;把手;手柄杆
handle bar bolt 手杆螺栓
handle bar bushing 手杆衬套
handle bar clamping ring 把手杆夹圈
handle bar end nut 把手端螺母
handle bar grip 手柄杆;车把手握柄
handle bar guide tube 手柄杆导管
handle bar plug 把手塞
handle bar plunger 把手柱塞
handle bar spiral 把手螺旋
handle bar stem 车把手杆
handle blank 手柄木;宜于做把手或手柄的一段木材
handle change 远距离控制;远距离操纵;转向盘柱换挡;转向盘柱变速
handle contractual disputes 处理合同争议
handle control 遥控
handle credit 经营信贷
handle extension 手柄伸长部
handle eye 把手孔眼
handle file 针座
handle for cover close 机盖压紧手轮
handle for die block 模具把
handle for pan position 旋转锅体用手轮
handle for regulating vibration speed 振动速率调节手柄
handle for valve stem 阀杆手柄
handle gouge 带柄圆凿;带柄半圆锉
handle grip 手柄
handle guide plate 手柄导板
handle hay rake 小型草耙
handle-insulated 柄绝缘自动剥线器
handle knob 把手;捏手
handle lever 手柄杆
handle lever fastening bolt 手柄杆螺栓
handle lock 把手锁件
handle locking 手柄锁紧
handle lock nut 手柄锁紧螺母;手柄防松螺母
handle magneto 手摇磁电机
hand length 手长
hand lens 简单显微镜;手持放大镜;放大镜
handle of axe 斧柄
handle operation 手柄操纵
handle pin 手柄销
handle-plate 门执手护板
handle position 手柄位置
handler 装卸装置;管理者;机械人;堆垛机;处理程序;处理器;处理机;操纵者
handler routine 处理程序的例行程序
handle set screw 车辆制动螺钉
handle setting knob 手柄设定旋钮
handle setting plate 手柄固定板
handle setting shaft 手柄固定轴
handle setting spring 手柄固定弹簧
handle shears 圈把剪刀
handle socket 手柄窝
handle stickling machine 贴柄机
handle strength 手扯强力
handle-talker 手提无线电步话机
handle tornob 握手
hand lettering 手写注记
handle-type fuse 手柄型熔断丝;把手式熔断丝
hand level 手提(式)水准仪;手用水准仪;手用水平仪;手持水准仪;手持水平仪;手水准(仪)
hand level with vertical angle 带角手水准
hand lever 手(控)杆;手动拉杆;手柄
hand lever adjustment 手柄调整
hand lever brake 制动手柄;手力杠杆闸
hand lever latch 手杆闩

hand lever pivot shaft 手杆枢轴
hand lever shearing machine 手柄剪机
hand lever shears 手动切板机
hand lever shifter 手杆式开关
hand lever type grease pump 手摇滑脂增压器
hand lever tyre expander 手动杠杆扩胎器
handle with care 小心轻放;小心搬运
hand lift 手摇起重机;手动启闭机
hand-lift device 手摇起重装置
hand lift for-mounted mower 中间悬挂割草机的手动起落机构
hand-lifting 手拔
hand lifting lever 手起落杆
hand lifting winch 手摇式起重机;手摇卷扬机
hand-lift plow 手起落式犁
hand-lift scraper 手提升式刮土铲运机
hand-lift truck 手动装卸车;手摇起重机车
hand-lift unit 手摇起重装置
hand light 手提灯
hand light static cone penetration test 手摇式轻型静力触探
hand line 操纵索;操作索;手执的钩丝;手钓丝
handline boat 手钓艇
hand-line ga(u)ge 手工线测仪
hand-liner 手钓船
handling 装载;装卸;加工;货物搬运;吊运;处理;搬运
handling ability 处理能力
handling and loading 装运;搬装
handling and shipping equipment 操作及装运设备
handling and storage equipment 装卸运输与储藏设备
handling and transportation 装卸与运输
handling a very large tanker alongside 巨型油船靠泊作业
handling a well 钻孔钻进规程的调整
handling bay 装卸港池
handling breakage 装运中的破裂;操作时破碎
handling bridge 桥式装卸机;桥型装卸机;桥形装卸机
handling capacity 处理数量;处理能力;处理容量;装卸能力;吞吐能力;吞吐量;调车场作业能力
handling capacity of berth 泊位吞吐量
handling capacity of radar transponder 雷达应答器容量
handling cask 装卸容器
handling characteristic 操纵特性;加工性能
handling charges 装卸费(用);输水费用;搬运费(用);搬货费用
handling command 作业指令;操作指令
handling condition 装卸条件;搬运条件
handling container 集装箱
handling cost 手续费
handling crew 管理人员
handling device 运转设施;装卸装置;操作装置
handling ease 加工容易;便于加工
handling efficiency 装卸效率
handling equipment 加工设备;吊运设备;运输设备;装卸设备;起重(运输)设备;搬运设备
handling essentials 操纵要领
handling expenses 处理费(用)
handling facility 起重(运输)设备;操纵灵便;装卸设备;搬运设施
handling failure 处理失效;处理故障
handling fee 手续费
handling frame 操作架;操纵架
handling from storage 货物出入仓库;从仓库装卸
handling grab 装卸抓钩
handling guy 搬运索
handling hole 吊环螺钉孔
handling hook 搬钩
handling in machinery 机械操作工序吨
handling installation 输水设施;搬运设施
handling instruction 操作说明
handling labo(u)r 保养工
handling length 起吊长度
handling length of piles 桩的起吊长度
handling life 储[贮]运寿命
handling line 牵引索;牵引绳;搬运线
handling loaded 人力加载的;手工上料的
handling loss 装卸损失
handling machinery 装卸设备
handling mechanization 装卸作业机械化

handling misfiring 瞎炮处理
handling of blasting caps 爆破雷管处理
handling of bulk cargo 散货装卸
handling of bulk freight 散货装卸
handling of bulk goods 散货装卸
handling of bulk materials 散装材料装卸;散装材料处理;大量材料装卸;大量材料处理
handling of cartographic(al) data in digit 地图数字信息处理
handling of goods and materials 物资管理;物料管理
handling of labo(u)r 劳工管理;劳动力管理;劳动力调配
handling of land 土地开发
handling of load(s) 装载处理;重物(的)装卸;加荷处理
handling of map data in digit 地图数字信息处理
handling of materials 材料控制;散货装卸;搬运材料
handling of plane surface 平面交会的处理
handling of plants 田间管理
handling of sample 样品处理;样本处理
handling of shield 盾构操作
handling of special cargo 特殊货物搬运
handling of the camera 摄影机操纵
handling operated pump 手摇泵
handling operation 运输作业;作业;装卸作业;管理工作;连续操作;手动控制;服务工作;搬运作业
handling plant 装卸设备;搬运设施
handling problem 处理问题
handling pump 手摇泵
handling radius 悬臂伸距;工作半径;活动半径;起重机伸距;起重半径;吊车伸出臂半径
handling radius of crane 起重机臂工作半径
handling rail 地面操作扶手
handling rate 处理费;装卸费(率)
handling reinforcement (混凝土构件的)预埋件
handling safety 操作安全
handling scratch 装运擦伤;操作时擦伤
handling shock 装卸振动
handling slot 搬物扣手
handling speed 周转速度
handling stone 卵石
handling strain 装运变形;搬运变形
handling strength 运输强度;起操强度
handling stress 装卸应力;起吊应力;搬运应力
handling system 装卸;装卸(运输)系统;处理系统;搬运系统
handling technique 装运方法;装卸技术
handling technology 装卸工艺
handling term 装卸期限
handling tight 轧紧;旋紧;箍紧;拧紧管接头;丝扣管接头
handling time 装卸时间;处理时间;搬运时间
handling tonnage of handling machinery 装卸机械起运量
handling tons 操作量(指装卸)
handling tool 装卸工具
handling volume 装卸量
handling work 手工
handling yard 装卸场
hand lining 手工画边;手钓
hand-loaded 人力发料的;人力加载的;手工加载的
hand loading job 手工装载工作
hand log 手控测程仪;手持计程仪;扇板测速仪;扇板测程仪
hand logger 手工采运机
hand log line 测速绳
hand looping 人工造成活套
hand lotion 擦手洗剂
hand lubrication 人工润滑法
hand lubricator 手油壶;手提润滑器;手动润滑器
hand luggage 手提行李
hand machine 手摇机器;手动机;手力机
hand-made 手制;手工(制)的
hand-made article 手工品
hand-made bentonite 人造膨润土
hand-made bentonite ball 人造膨润土粒
hand-made brick 手制砖;手工砖
hand-made glass 人工制造的玻璃
hand-made nail 手制钉
hand-made paper 手工纸;手抄纸
hand-made paper mill 手工造纸作坊
hand-made shingle 手工木制瓦

hand magnet 手磁铁
hand magnifier 手提放大镜;手持放大镜
hand-mallet 手木槌
hand-manipulated vibrator 手动振动器;手动振捣器;人工操作振捣器
hand mapping camera 手提(式)测绘摄影机;手提航拍摄影机;手提测图摄影机
hand mark 侧面标志
hand mast 短桅
hand-measured staple length 手扯纤维长度测定法
hand measurement 手测量
hand-me-down 现成的;廉价劣质的;旧事物
hand method 手工方法
hand micrometer 手工测微器
hand microphone 手持式传声器
hand microtelephone 手提(式)微型电话机;手持送受话器
hand microtome 哈氏切片机;手工切片机
hand mill 手磨机
hand miller 手动铣床
hand milling 手铣
hand milling machine 手动铣床
hand-mined face 手工采矿面;人工采掘面
hand miner 手工采掘工
hand mix 手拌(料)
hand-mixed cement grout 人工搅拌水泥浆
hand-mixed concrete 手拌混凝土;人工拌制的混凝土
hand mixing 人工搅拌;人工混合;人工拌和(法);手拌;手工拌法
hand mix procedure 人工拌和(法);手拌法
hand mixture 手拌和料
hand money 预付款;定金
hand monitor 手监测器
hand mortising machine 手工制榫眼机
hand motion 手(开)动;手带动
hand motor 手电动机
hand mo(u)ld 手工模制;手工压模;手工铸模
hand mo(u)lded 手工压模的;手工成型的
hand mo(u)lded brick 手工制砖;手工砖
hand-mo(u)lded grey iron casting 手工造型灰铁铸件
hand mo(u)lding 手工线脚;手工线条;手工造形(法);手工印坯;手工成型
hand-mo(u)lding press 手工模压机;手工成型法
hand-mo(u)ld tile 手工(制)型瓦
hand movement 手推
hand mower 手推割草机;手力割草机;手推剪草机
hand mucker 人工土方工
hand mucking 人工挖掘;人工清理;人工掘土;人工出渣;人工挖运软土;人工挖掘;手工装岩
hand nut 手螺母
hand of door 门的左右手向
hand off 手动切断
hand of rotation 旋转方向
hand of spiral 螺线方向
hand oiler 手注油器;手加油器
hand oiler body 手注油器体
hand oiler cap 手加油(器)盖
hand oiler spring 手注油器簧
hand oiler valve 手注油器阀
hand oiling 手工加油
hand oil lamp glasses 手提油灯玻璃罩
hand oil pump 手摇油泵;手动油泵
hand-on background 工作经验
hand operate 手动运算
hand-operated 人工驱动的;手动控制(的);手动的;人工操作的;手摇的;手工操作的;手操纵的
hand-operated air pump 手动打气泵
hand-operated auger 手摇螺旋钻;手动螺旋钻
hand-operated bar bender 手工操作的弯钢筋器
hand-operated bar cutter 钢筋手工切割器
hand-operated barrow mounted sprayer 推车式手压喷雾器
hand-operated bending roll 手动弯曲辊
hand-operated block (and tackle) 手动神仙葫芦;手动滑轮组
hand-operated boring machine 手摇钻孔机
hand-operated brake valve 手动操纵制动阀
hand-operated camera 手控摄影机
hand-operated ca(u)lking gun 手动嵌缝枪
hand-operated chain drive 手工操作传动链
hand-operated chippings spreader 手工石屑铺撒器

hand-operated chopper 手摇切碎机
hand-operated compacting beam 手持捣实棒
hand-operated crane 手动吊车
hand-operated cutting machine 手动切割机
hand-operated direction valve 手动换向阀
hand-operated distribution panel 手动操作配电盘
hand-operated door 手动启闭门
hand-operated drill 手摇钻
hand-operated driver 手动驱动器；手动打桩机
hand-operated dual horn 手动双雾角
hand-operated duster 手动喷粉器；手摇喷粉器
hand-operated electric(al) drill 手动电钻
hand-operated elevator 手工操纵的货运升降机；手工操纵的货运电梯
hand-operated feed 手动给进
hand-operated fire alarm installation 手提式火灾报警装置
hand-operated fruit press 手摇水果榨汁机
hand-operated ga(u)ge 手动液面计
hand-operated gear 手动驱动；手(传)动装置
hand-operated grinder 手摇研磨机
hand-operated gritter 手动石屑铺砂机；手动石屑铺撒器
hand-operated gun 手动水泥喷枪
hand-operated hoist block 手工操作的提升机滑轮
hand-operated hydraulic press 手动液压机
hand-operated jack 手操作千斤顶
hand-operated jaw crusher 手动颚式破碎机
hand-operated jig 手动跳汰机
hand-operated knapsack sprayer 背负式手动喷射器
hand-operated level ga(u)ge 手动液面计
hand-operated lifting machine 手动提升机
hand-operated machinery 人工操作机器
hand-operated mortar gun 手动灰浆喷枪
hand-operated overhang crane 手动单梁桥吊
hand-operated piston pump 手动活塞泵
hand-operated press 手压机
hand-operated propeller 手操推进器
hand-operated pulley block 手动滑轮组；手动滑车组
hand-operated pump 手摇泵；手压泵；手动泵；手抽泵
hand-operated push feed 手动推斯送进
hand-operated reel 手摇卷线车
hand-operated regulator 手动调节器
hand-operated rocker arm 手动摇臂
hand-operated rolling shutter 手动卷百叶
hand-operated rotary cutting tools 手动旋转式切削工具
hand-operated screed 手工整平板
hand-operated screen 焊工面罩
hand-operated screw press 手动螺旋压力机
hand-operated service lift 手动杂务梯
hand-operated slide feed 手动滑板送料
hand-operated soil disinfector 手动土壤消毒器
hand-operated spray(ing) pump 手动喷雾泵；手动喷射泵
hand-operated spreader 手动撒布器；手动喷射器
hand-operated sprinkler 手压喷洒器
hand-operated starter 手摇起动机；手动起动器
hand-operated stirring gear 手动搅拌齿轮(装置)
hand-operated suspended monorail system 手动悬挂式单轨系统
hand-operated switch 手动开关；手扳道岔
hand-operated tamping beam 手持捣实棒
hand-operated tipping drive 手动倾翻传动装置
hand-operated travel(l)ing bridge crane 手动桥式行车；手动桥式吊车
hand-operated traverser 手工操纵转车台
hand-operated troll(e)y hoist 手动小车升降机
hand-operated valve 手控阀；手动阀(门)
hand-operated vertical hydraulic press 手动立式液力压榨机
hand-operated vertical screw press 手摇立式螺旋压榨机
hand-operated vibrating finisher 手动振动抹面机；手动振动磨光机
hand-operated vibrating screed 手工振动式整平板
hand-operated vibration screed 手工振动式整平板
hand-operated winch 手动绞车；手摇绞车
hand-operating 人工操作；徒手操纵
hand-operating chuck 手动夹盘
hand-operating crank 手摇起动柄

hand-operating device 手动操作装置
hand-operating electrostatic sprayer 手动式静电喷漆机
hand-operating lever 手柄杆；手操作杆
hand-operating mechanism 手动机构
hand operation 手(工)操作；手动操作
hand operation equipment 手工操作设备
hand operator 手动开关器
hand out (免费奉送的)广告品；分发资料
handover 交接；图像拖尾；移交；交出
hand-over against payment of the price 凭付款交付
hand-over discharge 交卸
hand-over foreign exchange 支付外汇
hand-over hand 两手交替拉绳法
hand-over item by item 点交
hand-over of the works 工程接管
hand-over to the next shift 交班
hand-over word 交换代码
hand-packed 人工包扎的；人工夯实的；手工夯实的；手工包装的
hand-packed bottoming 人工夯实的基础；人工夯实的基层；人工毛石铺底
hand-packed hardcore 人工装填石块；人工夯实的石填料
hand-packed rock facing 人工压紧的岩石面层；人工砌筑的岩石面层
hand-packed rockfill 人工砌石；人工堆石；人工抛石；手工砌石
hand-packed rubble 人工砌石；人工铺砌块石；人工抛石
hand-packed stone 人工堆石；人工装填石块
hand-packed Telford type subbase 人工夯实的泰福式基层
hand packing 人工夯实；人工充填；人工包扎
hand-packing strip 人工(充)填带
hand-painted carpet 手工描花地毯
hand-painted picture tile 手工绘制瓷砖
hand painting 手工彩绘
hand pallet truck 手摇平板车
hand paper cutter 手摇切纸机
hand peening 手锤敲击硬化
hand penetrometer 手动贯入仪
hand percussion boring 人力冲击钻进
hand-photogrammetric camera 手控航空摄影机
hand pick 精选；手工挑选
hand-picked 精选的
hand-picked reject 手拣矸石
hand picker 手选工
hand picking 手捡；手工凿掘；人工凿掘；拣选；人工挑选；手选
handpicking stone 手选矸石
handpick miner 手工采掘工
handpiece 机头；手持件
hand pile driver 人工打桩机
hand pinching 手捏
hand pipe cutter 人工管子工具
hand pipe tool 手工管子切割器；人工修管工具
hand piston pump 人力操纵活塞泵
hand pit 人工掘井
hand-pitched 人工敲碎的
hand-pitched base 人工碎石块铺底层；手铺小块石底层
hand-pitched broken stone 手(铺)碎石
hand-pitched foundation 人工铺砌的基础
hand-pitched stone 人工琢石
hand-pitched stone subbase 人工铺砌碎石基层
hand-placed 手放的；人工铺设的；人工砌的
hand-placed concrete 人工浇筑混凝土
hand-placed riprap 人工堆筑的抛石护坡；人工抛石；手铺片石
hand-placed rock 手砌块石；干砌块石；人工砌石
hand-placed rubble 人工铺置乱石
hand-placed stone 人工铺置石块
hand-placed stone riprap 人工铺设的防冲乱石
hand placement 手工置放；手工铺设；手工摊铺
hand placing 手工置放；手工铺设
hand-plaited straw articles 草编织品
hand plane 手刨
hand planer 手刨床
hand planing machine 手刨床；手动送料刨板机
hand-plant 手栽幼苗
hand plastering 手工抹灰
hand plate 门推板；门上推手板；推板

hand plate saw 板锯
hand plate shears 手动板材切割机；手动板材剪切机；手工平板切割机
hand plate working 手工板金加工
hand plotting 人工标图
hand plow 手推犁
hand plug ga(u)ge 手塞规
hand plug tap 手丝锥
hand plunger grease pump 手动杠杆式上油泵
hand pointer 手动压尖机
hand-poked producer 手工出灰气体发生器
hand poking 人工通炉；手工出灰
hand-polished 手工抛光的
hand polisher 人工抛光盘
hand polishing 手磨；手工磨光
hand portable extinguisher 轻便灭火器
hand portable riveter 人力轻便铆钉器；轻便手力铆接机；手力轻便铆钉机
hand pot 手提(喷油)壶
hand potato cutter 手摇马铃薯种切块机
hand pouring pot 手提灌油壶
hand power 手力
hand-powered 手摇的
hand-powered capstan 手动起锚绞盘；手动起锚机
hand-powered cutter 手动切断机；手动切刀
hand-powered emergency transmitter 手摇供电式紧急发信机
hand-powered knapsack sprayer 手动背负式喷雾器
hand power elevator 手动升降机；人力升降机
hand power press 手压机
hand power screw press 手动螺旋压力机
hand power screw punching press 手动螺旋压孔机
hand power service lift 手动杂务梯
hand power steering gear 手动操舵装置
hand power tool 手提动力工具
hand power track crane 手力移动式起重机
hand power travelling crane 手移起重机
hand power trolley 手摇车
hand power truck crane 手拉行车
hand precision reamer 精密手铰刀
hand preparation 手选
hand press 手(工)压机；手压床；手动操作印刷机
hand-pressing 押手
hand press-packed bale 人力压装的包；人力紧压捆包
hand press-packing 人工打包
hand primer 手动注油装置；手动起动泵；手打油杆
hand priming device 手动装置
hand priming lever spring 手摇杆弹簧
hand priming of pump 手起动泵
hand print 手纹
hand-printed(wall)paper 手工印花墙纸；手工印刷的(糊)墙纸
hand printing 手工印制；手工印刷
hand proof 手工打样样张
hand-propelled 手推的
hand-propelled compacting machine 手推压实机；手推夯实机
hand-propelled generator 手推式(火焰)发生器(路面加热用)
hand-propelled travel(l)ing crane 手控桥式吊车
hand-propelled tripper 手摇推进式倾料器
hand-propelled vibrating roller 手推振动压路机；手推式振动碾(压机)
hand-propelled vibro-finisher 手推振动抹面机；手推振动磨光机
hand propelling boat 手摇螺旋桨舢板
hand property 手感
hand puddling 人工捣férence
hand pulled 人工提升的
hand pulled template 手拉样板
hand pulley block 手拉滑车组
hand pump 手摇泵；手压泵；手力唧筒；手动抽水机；手动泵；手(揿)泵
hand pump for tyre inflation 车胎充气手泵
hand punch 手动穿孔(机)；手动冲压机；手动冲孔(机)；手冲床；手扳冲刀
hand punched card 人工穿孔卡
hand punching and shearing machine 手力压剪机
hand punching machine 手动冲孔机
hand punning 人工捣实
hand push cart 手推车

hand-pushed indicator diagram 手拉示功图
hand push grease gun 手推润滑油枪
hand putter 手摇绞车
hand putting 人力推车
hand pyrometer 手提式高温计
hand-rabbled furnace 人工(搅拌)炉
hand rabbling 人工搅拌
hand radar 便携式雷达
hand rail 栏杆(扶手);扶手带
handrail baluster 栏杆柱;扶手栏杆
handrail banister 栏杆柱
handrail bolt 双头螺栓;扶手螺栓
handrail bracket 扶手托架
handrail conveyer 扶手橡胶带
handrail for fan deck 排风机平台栏杆
handrail height 扶手高度
handrail(ing) 栏杆;扶手
handrail iron 铁栏杆;铁扶手
handrail of stair(case) 楼梯扶手
handrail plane 扶手顶面
handrail profile 扶手外形;扶手断面
handrail punch 扶手螺栓打眼钻;扶手螺栓板子
handrail screw 扶手螺栓;插销螺钉;楼梯扶手螺栓
handrail scroll 栏杆末端的蜗形;栏杆卷饰;扶手涡饰;扶手卷形饰;扶手涡卷端;楼梯涡旋端
handrail section 栏杆型材;扶手断面
handrail shape 栏杆式样;扶手形状
handrail stanchion 栏杆柱
handrail standard 栏杆支架;栏杆柱
handrail support 栏杆托座;扶手托座
handrail trim 扶手形状
handrail wreath 螺旋扶手端头;扶手涡旋冠端;转弯扶手
hand raising 人工天井掘进
hand rake 手耙(清理拦污栅用)
hand raked 手耙的
hand-raked furnace 人工炉
hand raking 人工扒动
hand ram 人工夯;手夯实;手夯锤;手夯槌
hand rammer 人力夯;人工打夯机;手夯;手工捣锤
hand ramming 手工捣实
hand rapids-warping 人力绞滩
hand ratchet 手动制轮机;手动制杆机;手动棘轮
hand reach 手操作空间范围
hand reamer 手(动)绞刀;手动扩孔钻
hand reamer with pilot 带导向柱的手用铰刀
hand reamer with spiral flutes 带螺线槽的手用铰刀
hand reaming 手铰孔
hand reaper 人力收割机
hand receiver 听筒;手持听筒;手持式受话器
hand reeling machine 手摇纺车
hand refinishing kit 手用修理工具袋
hand refractometer 手提析光计
hand-regulated arc lamp 手调弧光灯
hand-regulated warp let-off device 手调式送经装置
hand-regulating expansion valve 手调节膨胀阀
hand regulation 子动调节;人工调节;手动调整;手调节
hand regulator 手动调节器
hand release 手动释放装置
hand remover 分针拆卸工具
hand rendering 手工涂刷;手工抹灰;手工粉刷
hand reset 人工重调;手工重调;手动复原;手动调整;手动复位
hand reset relay 手动还原继电器;手动复位继电器
hand reset system 人工复原式
hand resetting 手动复归
hand rest 手摇刀架;手工工具架
hand-restoring 手力复原;手动复原
hand-restoring indicator 手动复归信号器
hand reverberatory furnace 人工操作反射炉
hand reversing gear 手动反向机构
hand revolution counter 手持转数计
hand riveted hammer 手铆(钉)锤
hand riveter 拉铆枪;手动铆(钉)机
hand rivet(ing) 手铆;手工铆接
hand rivet(ing) hammer 手铆(钉)锤;手动铆钉锤
hand rivet(ing) machine 手铆钉枪;手铆(接)机
hand rock facing 人工砌石面层
hand rodding 人工(用棒)捣实;手工捣实;手工捣固
hand roller 手压碾;手推墨辊;手拉镇压器
hand rope 手缆;扶手绳

hand rope operation 手绳操作
hand rope winch 手缆绞车
hand rotary drilling 手转钻井(法)
hand rotary drilling machine 手转钻井机
hand rotary electric(al) planer 手扶旋转电刨
hand rotation stoper 手转伸缩式凿岩机;手摇上向凿岩机
hand round pass 手工喂料轧制的圆孔型
hand rounds 手轧圆钢
hand rubble 手工砌筑毛石;手工砌筑块石
hand rubble fill 碎石填充
hand rule 手定则(右手或左手定则)
hand run 机械制成而用手工修整的
hand-run tucks 手打褶裥
hand rust removing 手工除锈
hands 水手
hand sample 手工样品;小样品;小样本;手工取样
hand sample cutter 手工取样器
hand sampler 人工取样器
hand sampling 人工取样;人工抽样;人工采试样;手工取样;手工采样
hand sander reach rod 手拉撒砂器传动杆
hand saver 手护;保护手套
hand saw 手锯
hand saw blade 手锯条
hand sawing plywood 手锯胶合板
hand saw with wood(en) handle 木柄手锯
hand scalloping 手工分离牙边法
hand scarfing 手动火焰清理
hand scoop 手勺
hand scrape 手动刮板
hand scraper 手控刮铲小车;人力拖铲;手刮刀;手扶机械铲;手动刮板
hand scraper for unloading railway cars 火车卸货用的手控刮铲小车
hand scraper unloader 手控刮铲卸货车
hand scraper winch 手控刮铲车的绞车;手控铲运车的绞车
hand scraper with belt 有皮带的手控铲运车
hand screen 手动筛;焊工面罩;手摇筛
hand screening 人工清筛;丝网遮蔽法;手工丝网印刷法
hand screen-printed fabric 手工筛网印花织物
hand screw 手旋螺钉;手力千斤顶;手动千斤顶;手动螺杆;手旋螺丝夹(木工夹具);木框螺栓夹板;手动起重器
hand screw clamp 手螺旋夹
hand screwdown gear 手动压下装置
hand screw holdfast 手螺旋夹
hand screwing tool 方板牙扳手
hand screw press 手转压力机;手动螺旋压力机
hand screw tap 手用丝锥
hand-screw vulcanizing press 手旋硫化机
hand scribing 手工刻划
handscroll of painting 画卷
hand scrubber 手刷
hand scudding 手工推挤
hand scutching 手工打麻
hand seamer 手摇封罐机
hand sediment sampler 手提式泥沙采样器
hand seed dresser 手提式种子拌药机
hand seeder 手摇播种机
handsel 定金;初试;保证金
hand selection 手选
hand semi-rotatory pump 手摇半转泵
handset 手工排字;手工排版;手工排;手(提步话)机;手持收发话筒;手持的小型装置;手柄;手提电钻
hand set bit 手镶(金刚石)钻头;手工镶嵌金刚石钻头
hand setter 金刚石镶嵌工
hand set(ting) 用手调整;人工铺筑;人工安装;手工镶嵌;手工推平;手调
hand set transmitter 手持送话器
hand-sewn seam 手缝接缝
handshake control 信号交换输入输出控制
handshake cycle 交换过程处理周期
handshake protocol 握手协议
hand shaker 手摇抖动器
handshaking 信号交换;符号交换;联络;交接过程;交接处理;符号变换
handshaking protocol 交接规约程序
hand shank 抬包架;手铸铁水包;手柄;端包
hand shaving 手工刨皮

hand shear cutting machine 手扳铡断机
hand shearer 手工剪板机
hand shears 手(工)剪;手动剪切机
hand shield 手持式焊工面罩;(焊工的)手持护目罩;手持焊接护目罩;手持电焊面罩
hand shifting 手换挡
hand shovel 手铲
hand shovel(l)ing 手工铲装
hand shovel(l)ing of coal 人工投煤
hand shower 手持式(淋浴)莲蓬头;人工喷雾器;手持式淋浴器
hand shutter 手工控快门
hand sieve 手(工)筛
hand sieving 手工过筛;手工筛分
hand signal 手(势)信号;手(旗)信号;手(动)信号
hand signal(l)ing device 手动信号设备;手动信号设施
hand sinker 手冲钻;手沉锤
hand sinking 人工凿井;人工挖深;人工掘进
hand skidding 人力集材
hand sleeves 手套
hand-slewing crane 手旋起重机;手推旋转起重机
hand slicking 手工推平
hand slide rest 手动刀架
hand slitting 手工剖幅
hand smoothing 陡手平滑
hand snips 手力剪机;铁丝钳;白铁剪子
hands-off 手动断路;手工闸
hands-off control 放手操纵
hands-off economics 不干预经济学
hands-off operation 不开放式操作
hands-off policy 不干涉政策
hands-off speed 离手速度
hand soil auger 手动土壤螺旋(麻花)钻
hand soldering bit 手焊烙铁
handsome price 好价(钱)
hands-on 交互性工作;实践经验;实际训练;动手上机
hand sonar 便携式声呐
hands-on background 实际工作能力
hands-on operation 内行操作
hand-sort 手选;手检;手拣
hand-sort card 手动分类卡片
hand sorting 手选;手拣;手工分选(的)
hand sorting method 手选法
hand sorting separation 人工分选;手选(分离)
hand-sorting table 手选台
hand sower 手摇撒播机
hand spading 手工铲土;手工铲掘
hand specimen 人工试样;手工样品;手大小的矿石标本;手采岩样;手标本
hand spectrophotometer 手提式分光光度计
hand spectroscope 手提(式)分光镜
hand speedometer 手持测速器
hand spike 撬棒;绞盘杠杆;人工打桩;撬棍;推杠;铁笔;杠;铁柄;手杆
hand spindle brake 手动主轴制动器
hand split 手工劈裂;手工劈开
hand split and resawn 手劈再锯木瓦板
hand split wood shingle 手劈木瓦片;手工劈制的屋顶板;手工劈制的木片瓦
hand split wood siding shingle 手劈外墙面板;手工劈制的木质板壁
hand spoke wheel 手辐轮;操纵辐轮
hand spray 手持式喷枪;喷笔
hand sprayer 手喷器;人力喷雾器;人控撒布机;手摇喷雾器;手提喷洒器;手动喷雾器;手动喷水器;手动喷洒器;手喷枪
hand sprayer with platform mounted pump 泵装在平板上的手动喷雾器
hand spray gun 手(提式)喷枪
hand spraying 手工喷淋;手工喷雾
hand spreader 手工铺撒器;手工铺撒机;手工铺撒工
hand spreading 手工铺砂;手工铺撒;手工摊铺
hand spring balance 手提式弹簧吊秤
hand spring scale 手提式弹簧秤;手拉弹簧秤
hand sprinkler 手洒水器;手喷雾器
hand square passes 手工喂料轧制的方孔型系统
hand squeezer 人力压榨机
hand stacker 手推堆垛机;手工堆垛工
hand-stand tool 倒立器
hand stapling 手扯纤维长度测定法
hand starter 手工启动器;手摇起动机
hand starting 手工启动;人力起动;手摇起动

hand starting arrangement 手起动装置
hand starting button 手控起动开关
hand starting magneto 手摇起动磁电机
hand steel cable winch 手动钢索卷扬机；手动钢索绞车
hand steel rope winch 手动钢索卷扬机；手动钢索绞车
hand steel shears 手动钢筋剪断机
hand steel snap 手用钢铆钉模
hand steered 用手驾驶的；手工操纵的
hand steer(ing) 用手操纵的；用手驾驶的
hand steering gear 手动转向装置；手动操舵装置
hand steering room 手动操舵室
hand stirrer 手工搅拌器
hand stirring 人工搅拌
hand stitch 手缝线迹
hand stoked 手工加煤
hand stoker 手工加煤机
handstone 手用油石；小石子；用手操作的研磨石；带柄手用油石；鹅卵石
handstone subbase 小石子底座
handstop 手动挡水板
hand stop valve 手动截止阀
hand stowage 人工码垛
hand stowing 手工充填
hand straw cutter 人力切草机；人力锄草器
hand-stuff 人工填料
hand stuffing 手工加脂法
hand suspended monorail system 人工高架单轨系统
hand swabbing 手工刷色；手工刷浆
hands wanted 招工
hand sweeping 人工清扫
hand switch box 手动搬道机
hand syringe 手动喷射器
hand tacheometer 手提转速器；手提转速计
hand tachometer 手携式转数计；手持转速表
hand tally （手摇的）计数器；手摇计算器；手揿点货机；手摇计数器；手揿点货机
hand tamp 人力夯；手(工)夯；手锤
hand-tamped 手筑；手捣
hand-tamped pipe 手捣管
hand tamper 手扶打夯机；手(力)夯
hand tamping 人工捣实；人工夯实；手工夯实；手工夯击
hand-tank grease gun 手压罐式润滑油枪
hand tap 手用丝锥；手用螺纹攻；手扳丝锥；手扳螺纹攻
hand taper pin reamer 手用锥(形)销孔铰刀
hand taper reamer 锥形手铰刀
hand taper tap 手用锥形丝锥
hand tap for pipe thread 手用管牙丝锥
hand tapping 手工套扣；手动攻丝
hand tar spraying machine 手动焦油喷洒机
hand taut 用手拉紧
hand tear test 手撕试验
hand teem(ing) 人工浇铸；手工浇铸
hand telephone 手持受话器
hand telephone set 手持电话机
hand template 手拉样板
hand-templet radiation plot 手置模片辐射三角测量
hand-templet radiation triangulation 手置模片辐射三角测量
hand test 手上试验；手感试验
hand tester 手感测试仪
hand testing sieve 人工试验筛
hand threshed 人力脱粒
hand threshing machine 手摇脱粒机；打谷机
hand throttle 手动节气阀；人工节流阀；手拉加速杆；手风门；手(动)油门
hand throttle button 手动油门按钮
hand throttle lever 手动油门杆；手动风门杆；手节流杆
hand throttle link cover bracket 手油门拉杆盖支架
hand throttle wire 手油门拉线
hand throw crossover 人口操纵渡线
hand throwing 手工拉坯
hand tight 手揿加紧；手拉紧；手工管接头
hand tiller 手动舵柄；手操舵柄
hand tilting device 手动倾动设备；手动翻转装置；人力倾覆装置
hand tip-cart 手卸车；手推倾卸车；手倾车
hand tipping-barrow 手卸车；手倾车
hand-tipping type ladle 人工倾斜式金属桶

hand-to-mouth buying 现买现卖；零星采购；即用即购方式
hand-to-mouth operation 小本经营
hand tongs 手钳
hand tool 手工(操作)工具；手(动)工具；手持工具
hand tooled finish 手工整修；手工凿面；手工錾面
hand tool lathe 手摇工具车床
hand tool rest 丁字形刀架
hand torch 手动焊割炬；手持焊枪
hand touch 手感
hand towel 手巾
hand tracing 手工跟踪
hand tracking 人工跟踪；手动跟踪
hand tractor 手扶拖拉机
hand traffic count 人工车辆统计
hand traffic counter 人工运量计数器
hand trammel 木工规
hand trammer 推车工
hand tramming 人力运输；人力推车
hand transferring 手工转绘
hand transmitter receiver 便携式收发两用机
hand transposition 手工换位
handtrap 手掷靶装置
hand travel(l)ing crane 手移起重机；手推起重机
hand travel(l)ing gear 手动齿轮(装置)
hand traverse 手摇转
hand traverse gear 手动横向进给机构
hand traverse pinion 手摇方向机小齿轮
hand traverse shaft 手摇方向机轴
hand traversing brake linkage 手摇方向机制动联动装置
hand traversing mechanism 手摇方向机
hand-trimming method 手修法
hand trim wheel 手控配平轮
hand trip control 手动跳闸把手
hand trip gear 手动脱扣器
hand trip(ping) 手动跳闸；手动脱扣
hand trolley 手推运货车
hand trolley hand winch 手摇车
hand trowel 手(工)镘刀
hand truck 平板车；手推(运货)车；手推(货)车；搬运小车
hand tub 老式唧筒
hand tube cutter 手工管子切割器
hand tuning 手调谐
hand turf cutter 手动泥炭采掘机；手动泥煤采掘机；手动割草机
hand turning gear 手动盘车装置
hand turning tool 手工车刀
hand twiner 手工拈线机；手动拈线机
hand tying 手工打结
hand ultrasonic detecting cart for rails 手推式钢轨超声波探伤机
hand valve 手动控制阀；手动阀(门)
hand valve grinder 手动气门研磨器
hand vibrating finisher 手动振动抛光机；手动振动抹面机
hand vibrating screed 手控振动刮板
hand vibrator 手持振动器
hand vice 老虎钳；手(虎)钳
hand viewer 袖珍放大镜；便携式缩微阅读器；便携式观察器
hand vise 手虎钳；老虎钳
hand wagon 手推车
hand washbasin 洗手盆；洗手器
hand water pump 手动水泵
hand water sprayer 手动喷水器；手动喷洒器
hand wedging 用楔块塞瓷；手工揉泥
hand welder 手电焊器；手持电焊机；人工焊机
hand welding 手焊
hand welding installation 手焊设备
hand wharf crane 手摇码头起重机
hand-wheel 转向轮；转向盘；驾驶盘；手动(转)轮；操作盘；操纵(手)轮
hand-wheel-dressing tool 手动砂轮整形工具
hand-wheel handle 手轮把手
hand-wheel nut 手轮螺母
handwheel propelled vibrating finisher 手轮推进的振动抹面器；手轮推进的振动抹面机
hand wheel screw down 手轮压下装置
hand wheel shaft 手轮轴
hand wheel shaft cap 手轮轴盖
hand wheel shaft cover 手轮轴盖
handwheel switch 手轮开关

handwheel valve 手轮操纵阀；手轮操作阀
hand winch 手摇绞盘；手(摇)绞车
hand-winch take-up 手摇绞车拉紧装置
hand winding 手绕法
hand winding device 手动车装置
hand winding watch 手动上条表
hand windlass 杠杆起锚机；人力起锚机；手摇卷扬机
hand window(control)gearing 手动窗开关器
hand wiper 手动刮水器
hand wire cable winch 手动钢丝索绞车
hand won face 手工开采工作面
hand work 精细工艺；手工作业；手工制品；手工(加工)
hand-worked 手工加工的
hand-worked lift 手动升降机
hand-worked roaster 人工操作焙烧炉
hand-work hand labo(u)r 手工操作
handwork hood 手工操作通风柜
hand-wound 手绕的
handwriting 手迹；笔迹
hand-written compiler 手编编译程序
hand-written form 手写体
hand-written signature 亲笔签字
hand-wrought 手工制成的
handy 轻便的；手边；便于使用的；便携的；便当的
handy billy 小绞辘；轻便(抽水)泵；手摇舱面泵；轻便消防泵
handy concrete mixer 手动混凝土搅拌机
handy connection 手动易拧紧接头
handy lift hoist block 手动葫芦
handy lift unit 轻便升降装置
handy man 杂工；巧手工；多面手
handyman's special 需要大翻修的房产
handy microtome 轻便切片机
handy mixer 手动式拌和机
handy-size bulk carrier 集装箱散货船；轻型散装货轮
handy-size bulker 灵便型散货船
handy-size clean carrier 轻型轻油船
handy-size dirty carrier 轻型重油船
handy tackle 轻便绞辘
handy-talkie 步话机；手持式步话机；步谈机
handy-type computer 便携式台式计算机；便携式台式机
hang 悬挂；下垂状态；卡住；上弯线；垂
hangar 跨线雨棚；机库；棚厂；飞机(修理)库；飞机棚
hangar block 机库区
hangar deck 机库平台；机库甲板；棚厂甲板
hangarette 小型机库；小飞机库
hangar floor 飞机库地坪
hang-arm electronic belt scale 悬臂式电子皮带秤
hangar tunnel 棚洞
hang bolt 吊螺栓
hang dry 悬干
hanged extinguishing equipment 悬挂式灭火装置
hanged pipe 悬吊管
hanged powder extinguishing equipment 悬挂式干粉灭火器
hanged tube 悬吊管
hangelite 弹性藻沥青
hanger 悬架；悬挂器；衣架；梁托；起旋器；起吊扁担；上盘；陡坡林地；吊弦；吊具；吊挂件；吊钩；大库棚；垂饰；飞机库
hanger abrator 悬挂式抛丸清理机
hanger attachment 悬吊装置；吊车
hanger beam 悬吊梁
hanger bearing 悬挂支承；吊轴承
hanger bearing assembly 悬管轴承组合
hanger bolt 吊(架)螺栓；吊(挂)螺栓
hanger bolt head 吊架螺栓头
hanger box 悬挂轴承盒
hanger bracket 悬架
hanger brick 挂耳砖；吊顶砖
hanger cable 电缆吊索
hanger clamp 吊弦夹
hanger container 挂衣集装箱
hanger coupling 悬挂接头
hanger crack 悬挂裂纹
hanger for ceiling 风渠隔板吊杆
hanger frame 吊架
hanger girder 吊梁
hanger insert 吊环预埋件；吊钩预埋件
hanger iron 挂铁
hanger lever 吊架杆

hanger method 悬钩法;悬杆法
hanger of air duct 风管吊架
hanger-on 缠附物
hanger-pack clothing container 挂衣集装箱
hanger pin 挂钩销
hanger plank 吊架板
hanger plate 吊板
hanger pocket 挂钩套;挂钩槽
hanger rail 吊轨
hanger ring 吊环
hanger rod 悬(物)杆;悬挂棒;吊杆
hanger rope 吊索;吊缆(索)
hangers (电解板上的)挂耳
hanger shafting 吊轴系
hanger spring 吊挂弹簧
hanger supported shaking screen 悬挂式振动筛
hanger winch 吊杆顶牵索定位绞车
hanger wire 吊绳;悬挂线;吊缆
hanger wood 垂枝
hangfire 滞火;迟发火
hang-fire behavio(u)r 迟缓点火状态
hang-fire tire 防击轮胎
hang five 五趾吊
hang gliding 悬挂式滑翅运动
hanging 悬空;悬吊;下垂的;挂料;固定不动;工作吊架;倾斜;帷幔;顶盖;吊挂;壁挂
hanging adit 悬帮平峒;上盘平峒
hanging alluvial fan 悬冲积扇
hanging and power-driven crane with single-beam 电动单梁悬挂起重机
hanging arch 悬拱;悬顶;炉上的吊顶;吊窟;吊拱
hanging armo(u)r jacket 吊挂式炉喉保护板
hanging baffle 悬吊隔声板
hanging bar 吊杠
hanging basket 悬篮;吊篮
hanging batten 吊灯排
hanging beam 悬臂梁;吊梁
hanging bearing 悬吊轴承
hanging block 悬挂滑车
hanging bolt 悬挂螺栓;悬杆;挂钩;吊架
hanging bolt method 悬吊螺栓方法
hanging brace 悬式撑臂;吊撑
hanging bracket 悬托架;悬臂支架;吊架
hanging bridge 悬桥;吊桥
hanging-bucket conveyer 吊斗式输送机
hanging buttress 悬(式)扶垛;悬扶壁;悬臂扶垛;悬挑扶垛;悬扶支墩
hanging cabinet 吊柜
hanging cable 吊索;吊缆
hanging case 吊柜
hanging ceiling 吊顶
hanging chain 吊链;悬链
hanging-chain excavator 挂链式多斗挖掘机
hanging chart 挂图
hanging cirque 悬冰斗
hanging clamp 船用固定夹
hanging clinometer 悬式测斜仪;悬式测斜器;悬挂式倾斜仪
hanging clock 挂钟
hanging coal 悬空煤
hanging colter 立柄犁刀;吊柄犁刀
hanging compass 悬式罗盘;悬挂罗经;倒置罗经
hanging container 衣架货柜
hanging cupboard 立柜
hanging curtain 悬挂式帷幕;帘
hanging cutter 直犁刀
hanging dam 浮冰坝
hanging drier 悬挂式干燥器
hanging drop 悬滴
hanging drop atomizer 悬滴雾化器
hanging drop cell 悬滴槽
hanging drop preparation 悬滴标本;悬滴装置
hanging drop technique 悬滴术
hanging drop test 悬滴试验
hanging dryer 悬挂式烘燥机
hanging effect 悬吊作用
hanging elastic member 悬挂弹性件
hanging explosion-proof and power-driven crane with single beam 防爆电动单梁悬挂起重机
hanging fan 吊扇
hanging fascia 挂落
hanging feeder 吊挂式饲槽
hanging fender 悬挂式护舷
hanging fiber 悬吊纤维板

hanging filter 吊式过滤器
hanging flashing 吊挂防雨板;吊挂防漏板
hanging floor 悬吊楼板;吊式楼面;吊式楼板
hanging garden 悬园;(古巴比伦建的)空中花园;架空庭园
hanging garment container 衣架货柜
hanging glacier 悬冰川
hanging guard 上卫板
hanging guide 悬吊护板(上轧辊出口侧的);悬挂导架;悬挂导杆;吊挂护板
hanging gutter 悬吊槽;吊挂檐沟;吊挂檐槽
hanging harness 悬吊工具
hanging head 铰链横档
hanging ice 悬浮冰
hanging indention 首行满排余均缩排
hanging it out 怠工
hanging junction 悬垂汇流
hanging keel 倒挂龙骨
hanging knee 倒挂龙骨
hanging ladder 悬梯;吊梯
hanging lamp 挂灯;吊灯
hanging latticed architrave 吊挂眉子;倒挂眉子
hanging layer 悬着水层;上盘岩层
hanging leader 起重机转轴钢架;悬式导架;悬立式打桩导架
hanging-leave ornament 垂叶饰
hanging leaves 垂叶饰
hanging level 悬挂水准管;悬式水准仪;悬式水平仪;挂水准器
hanging load 悬挂荷载(钻塔);悬锤
hanging measuring instrument 悬式测量器
hanging mercury drop electrode 悬(滴)汞电极
hanging method 间断卸料法
hanging nozzle drier 悬挂式喷气烘燥机
hanging of a charge 挂料
hanging of lining 砌井壁
hanging ornament 吊饰;垂饰
hanging over 叠涩;拨礁
hanging paste 贴墙纸糨糊;悬胶
hanging pendulum 悬摆仪(测定坝、桥梁和基础等的水平位移用)
hanging pillar 悬柱;吊柱
hanging pipe 悬吊管
hanging pipe clamp anchor 吊管卡座
hanging plate 挂盘
hanging post 吊柱;旋门柱;铰接门柱;挂门柱
hanging post truss 吊柱桁架
hanging power hay mower 悬挂式牵引割草机
hanging rainwater gutter 悬式雨水的檐槽
hanging ring 提手
hanging river 悬河
hanging rock 危岩;危石
hanging rod 悬式标尺
hanging rod for pipe 管子拉杆
hanging roof 悬索屋盖;悬挂(式)屋盖
hanging roof gutter 悬挂式天沟
hanging roof truss 悬挂式屋架;吊桁架
hanging rooms 悬化室
hanging rudder 吊舵
hangings 吊帘;窗帘
hanging sash 垂直拉窗;吊窗(格框)
hanging scaffold(ing) 悬空脚手架;悬挂(式)脚手架;悬臂脚手架;悬吊脚手架;挂脚手架;吊架
hanging scoop scale 吊盘秤
hanging screen 帘;吊墙
hanging scroll 吊挂卷轴
hanging set 悬挂垛盘的木框;基础垛盘
hanging sheave 悬吊滑车;悬吊滑轮
hanging shelf 吊架;悬式格架
hanging shingling 悬挂式木瓦;悬吊墙面板;吊屋面板
hanging side 悬崖;上盘;顶板(矿体)
hanging socket 悬垂插座;悬式灯座
hanging sprayer 悬挂式喷雾机
hanging stage 悬臂脚手架;工作吊板
hanging stair(case) 悬挑(式)楼梯;悬空楼梯
hanging steps 悬挑楼梯;悬空式梯台;悬挑踏步;悬踏步;半悬梯级
hanging stick 悬杆
hanging stile 铰链窗边框;(门窗的)铰链挺;铰链栅门;提拉窗滑车槽
hanging stor(e)y 悬式楼面;悬式楼板
hanging strip 装铰链条
hanging support 悬式支承;悬挂支承;吊杆
hanging telephone set 挂式电话机

hanging test 挂重法断裂试验
hanging theodolite 悬(挂)式经纬仪
hanging tie 悬吊系杆
hanging tile 挂瓦
hanging top roller 悬吊式拉边机
hanging transit 悬式经纬仪
hanging truss 悬式桁架;悬挂桁架;悬吊(式)桁架;吊(柱)桁架;吊挂桁架
hanging truss bridge 悬挂式桁架桥;吊挂式桁架桥
hanging truss roof 悬挂屋架
hanging type of mooring ring 垂悬式系船环
hanging-up 悬挂;挂起;工作期;挂料
hanging-up of round 低效爆破
hanging valley 悬谷;悬沟
hanging valve 悬阀;吊阀;翻板阀
hanging wall 悬垂墙;悬帮;绝壁;上盘;顶壁;顶板
hanging wall active upthrow 上盘主动上投
hanging wall case 吊柜
hanging wall downthrow 上盘下降
hanging wall drift 上盘平巷
hanging wall leg 上盘立柱
hanging wall panel 挂墙板
hanging wall paper 糊墙纸
hanging wall upthrow 上盘上升
hanging wall waste 上盘废石
hanging water 悬着水;悬挂水
hanging window 吊窗
hanging zone 悬着水带
hang inside 下套管
Hangman grits 杭曼粗砂岩
hang off 拉开
hang over 释放延迟;拖尾
hang over delay 释放延迟时间
hangover fire 潜燃火
hangover time 闭锁时间
hang paper 假造支票(美俚)
hang plate 倾斜板
hang retention water 悬着水
hangrod 水平挂衣杆
hang roll(er) 上辊筒
hang scoop scale 盘秤
hang spring 吊丝
hangtag 使用保养说明标签
hang ten 十趾吊
hang tiling 挂瓦
hang tongs 吊钳
hang-type drinking fountain 挂式饮水泉
hang-type urinal 挂式小便器
hang-up 悬起;意外停机;搁置;拖延;挂起;起;错乱;放矿溜道堵塞;中止(操作);悬料
hang-up prevention 死机预防;防错;意外停机预防
hang-up signal 挂机信号
hang-up telephone 挂断电话
hang wall 挡烟垂壁
hang wall paper 糊墙纸
hang wind 坡风
hank 线束;一绞;纵帆前缘上的环;归龙;工字形框子;卷线轴
Hankel function 亨克尔函数;汉克尔函数
Hankel integral 亨克尔积分
Hankel inversion theorem 亨克尔反演定理
Hankel transform 亨克尔变换;汉克尔变换
hank for hank 两船并列逆风驶帆
Hankow willow 旱柳
hank reel 摇绞机
hank reeling machine 成绞机;复摇机
hank scouring machine 散纱洗涤机
hank shellac 绞状漂白紫胶
hanksite 碳酸芒硝;碳钾钠矾
hank spreading device 绷绞机
hannayite 水磷铵镁石
Hanning window 哈宁窗
hansa orange 耐晒橙
hansa scarlet 耐晒猩红
Hansa yellow 汉撒黄;耐晒黄
hanse 梁腋
hanse arch 加腋拱;多心拱
Hansen's parameter 汉森参数
Hansen's problem 汉森问题
Hansen's theory 汉森理论
Hansgirg process 高温碳素还原制镁法
Hanusiodine number 哈纽斯碘值

Hanus method 哈纽斯碘值测定法
haphazard building 无规划建造的房屋;不规则建筑
haphazard ignition 偶然发火
haphazard parking 不规则停车
haphazard parking arrangement 不规则停车排列
haphazard plan 盲目的计划
haphazard selection 任意抽选
haphazard walking 无秩序行走
haphazard weight 任意除数
haplite 细晶岩;简单花岗岩
haplobasalt 人造玄武岩
haploid 单倍的
haplont 单元体
haplophase 单元期
haplophyll 初始叶
haplopore 单孔
haploscope 视轴测定立体镜
haplosis 减半作用
haplostele 单中柱
happenchance 偶然事件
happened place of debris flow 泥石流发生地点
happened time of debris flow 泥石流发生时间
happening shanty town 自发建设的贫民窟
happenstance 偶然事件
hapten 半抗原
hapteron 附着器
haptobenthos 接触水底植物
haptogen 凝膜
haptophore 结合簇
haptophore group 结合簇
haptophyte 黏着植物
haptor 吸附盘
hap-type rudder 襟翼舵
Hapug amplifier 浮动载频放大器
haradaite 硅钒锶石
haram 闺房(回教的);后宫
harbinger 前兆
harbolite 硬辉沥青
harbortite 钠磷铝石
harbo(u)r 港口吞吐量;港口;港;海港
harbo(u)r accommodation 港湾设备;港内泊船能力;港口设施;港口设备
harbo(u)r administrative office 港湾行政办公室
harbo(u)r advisory radar 港口咨询雷达
harbo(u)r advisory system 港口咨询系统
harbo(u)rage 港湾;港内锚地;停泊处;遮蔽
harbo(u)r agency 港务机构
harbo(u)r ago 停泊处
harbo(u)r anchorage 港湾锚地;港内锚地
harbo(u)r anchorage atlas 港湾锚地图集
harbo(u)r and dock appliances 港口码头设施
harbo(u)r approach 进港处;港口航道;进港航道
harbo(u)r approach chart 进港图
harbo(u)r area 港区
harbo(u)r assets 港口资产
harbo(u)r authority 港务机关;港务(管理)局;港务管理机构;港湾管理局;港内当局;港口管理局;港口当局;海港当局
harbo(u)r bar 港口沙洲;港口沙坝;港口拦(门)沙(坝)
harbo(u)r barge 港口驳船;码头驳船
harbo(u)r basin 港坞;港塘;港池
harbo(u)r block 港区;港口作业区
harbo(u)r board 港务局;港务机关;海港当局
harbo(u)r board president 港务局长
harbo(u)r boat(launch) 港务船;港口作业船
harbo(u)r boat service 港务船服务处
harbo(u)r bottom 海港底部
harbo(u)r bound 港区范围
harbo(u)r boundary 港界(线)
harbo(u)r breakwater 港湾防波堤
harbo(u)r building 海港建筑;港口建筑物
harbo(u)r bureau 港务局
harbo(u)r camber 港内盆地
harbo(u)r capacity 港口容量;港口能力
harbo(u)r cargo handling appliance 港口货物装卸机械
harbo(u)r chart 港域海图;港湾(海)图;港区图;港内水深图;港口图;港泊图;海港(航)图
harbo(u)r charter 港内租船(口头)协议
harbo(u)r city 海港城市
harbo(u)r commerce statistics 港口业务统计资料
harbo(u)r commission 港务(管理)委员会

harbo(u)r congestion 港口拥塞;港口拥挤
harbo(u)r construction 海港建设;港湾建筑;港口建筑;港口建设;港口工程施工
harbo(u)r construction engineer 海港建设工程师
harbo(u)r construction plant 筑港设备;筑港机具
harbo(u)r construction site 海港施工现场;海港建筑现场;港湾建筑工地
harbo(u)r craft 港作拖轮;港作船;港口作业船;港口运输船;港口工作船
harbo(u)r crane 港口起重设备;港口吊车;港湾起重机;岸上起重机
harbo(u)r culture 港养
harbo(u)r dam 港堤
harbo(u)r deepening 海港挖深;海港加深;港口浚深;港口水域浚深
harbo(u)r deposit 湖泊沉积(物)
harbo(u)r depot 港站
harbo(u)r depth 港域水深;港口水深
harbo(u)r design 港口设计
harbo(u)r desilting 港口清淤
harbo(u)r development 海港发展;港湾开发;港口改建
harbo(u)r disbursement 港务杂费
harbo(u)r district 港湾地区;港(内)区
harbo(u)r diving 港口潜水
harbo(u)r dock (有闸的)港池
harbo(u)r dues 港务费;港(口)税;入港税;入港费;停泊费;船舶港务费
harbo(u)r dues on cargo 货物港务费
harbo(u)r duty 港口关务
harbo(u)r echo-ranging and listening device 海港超声波测听器
harbo(u)r efficiency 港口效率
harbo(u)r engineer 港口工程师
harbo(u)r engineering 海港工程学;港湾工程;港口工程(学);港工学;海港工程(学)
harbo(u)r engineering model 港工模型
harbo(u)r engineering model test 港工模型试验
harbo(u)r engineering service 港口工程部
harbo(u)r engineering survey 港口工程测量
harbo(u)r entrance 港湾入口;港口入口(处);港(口)门;港的口门;进港口门;港口入口;防波堤口门
harbo(u)r entrance control post 海港进口检查站
harbo(u)r entrance control vessel 进港口检查船
harbo(u)r entrance jetty 港湾进口导堤;港口入口导堤;港口进口导堤;港口(门)导堤;海港入口导堤
harbo(u)r entrance light 港口入口灯塔;港口门灯标;进港口门灯标;进港口(门)灯塔;进港口灯船
harbo(u)r entry obstacles 进港口障碍物
harbo(u)r equipment 港口设备;码头装卸设备
harbo(u)r extension 港口扩建(工程)
harbo(u)r facility 港口设施;港口设施;海港设施
harbo(u)r fairway 港区
harbo(u)r fees 港务费
harbo(u)r ferry 港内交通船;港口交通船;港口渡轮;港口渡船
harbo(u)r fireboat 港口消防船
harbo(u)r fleet 港口船队
harbo(u)r floating crane 港湾浮式起重机
harbo(u)r for lagging up river craft 小船停泊处;小船过冬停泊水域
harbo(u)r for laying-up craft 避风停泊水域
harbo(u)r frontage 港区临水面;港口临时面;港岸线;港岸
harbo(u)r generating set 停泊发动机组
harbo(u)r generator 港口发电机组
harbo(u)r greve 港口市市长(英国)
harbo(u)r guide 港务手册;港口指南
harbo(u)r handling appliance 港口装卸机械
harbo(u)r handling efficiency 港口装卸效率
harbo(u)r handling equipment 码头装卸设备
harbo(u)r handling machinery 港口装卸机械
harbo(u)r handling technology of port 港口装卸工艺
harbo(u)r health officer 港口检疫(人)员
harbo(u)r hydraulic reclamation 港区吹填
harbo(u)r illumination 港口照明
harbo(u)r improvement 海港改进;港口整治;港口改建
harbo(u)r installation 港口设施;港口设备;港口设施;海港设备
harbo(u)r investment 港口投资额

harbo(u)r land area 港口陆域
harbo(u)r launch 港务艇;港作(业)船;港用汽艇;港湾工作交通艇;港口作业船
harbo(u)r layout 港口布置
harbo(u)rless 无港的
harbo(u)r light 港口灯塔;港口灯(标);港灯;入港灯
harbo(u)r lighter 港口驳船
harbo(u)r lighting 港口照明
harbo(u)r light tower 港口灯塔
harbo(u)r limit 港区边界;港区界线;港界(线)
harbo(u)r line 港区(界)线;港口(专用)线;港口水域建筑界线;港界(线);海港线;码头线
harbo(u)r lock 港口水闸
harbo(u)r log 驻港日记
harbo(u)r manager 港务主任;港务局长;港务经理;港口经理
harbo(u)r map 港湾图
harbo(u)r marshalling park 港区调车场
harbo(u)r marshalling station 港湾编组站
harbo(u)r master 驻港船长;港务(监督)长
harbo(u)r masterplan 港湾布置图
harbo(u)r mechanical handling 港口机械装卸
harbo(u)r mechanical technique 港口机械装卸技术
harbo(u)r model 港湾模型;港口模型
harbo(u)r model test 港湾模型试验
harbo(u)r mole 港口防波堤;海港防坡堤
harbo(u)r mooring 港湾碇泊;港内系泊(设备);港湾系泊缆;港湾系船缆
harbo(u)r mouth 港口口门
harbo(u)r of export 运出港
harbo(u)r office 港务办公室
harbo(u)r officer 港务人员;港口官员
harbo(u)r of refuge 避风港;安全港(口)
harbo(u)r oil pollution problem 港区石油污染问题
harbo(u)r oil spill problem 港区石油污染问题
harbo(u)r operating organization 港口营运机构
harbo(u)r oscillation 港域水面振动;港内水面波动
harbo(u)r painter 轻便(的)艇首缆
harbo(u)r passage time 船舶卸货所需时间
harbo(u)r pilot 领港员;港口引航员
harbo(u)r plan 港湾(平面)图;港区图;港口(平面)图;港泊图
harbo(u)r planner 港口规划人员
harbo(u)r planning 港口规划
harbo(u)r police 港务警察;水上公安机关
harbo(u)r pollution 港口污染
harbo(u)r portal crane 港口门式吊车;港口龙门吊车;港口门式起重机
harbo(u)r practices 港口业务手续
harbo(u)r project 港湾工程
harbo(u)r promotion group 港口业务发展组
harbo(u)r promotion section 港口业务发展组
harbo(u)r quarter 港务局及附属建筑物所在区域
harbo(u)r radar 港口雷达;海港雷达
harbo(u)r radar system 港口雷达系统
harbo(u)r radio communication 港口无线电通信
harbo(u)r radio telephone facility 港口无线电话设施
harbo(u)r railing road 港区铁路;港区铁道
harbo(u)r railroad 港内铁路;港口铁路(美国)
harbo(u)r railway 港内铁路;港口铁路(美国)
harbo(u)r railway junction terminal 港湾铁路枢纽
harbo(u)r rates 港税
harbo(u)r reach 港区(河段);港内河段;港口河段;通到港口的河道
harbo(u)r regulation 港务规章;港口规则;港口管理条例;港章
harbo(u)r representative 港方代表
harbo(u)r resonance 港湾谐振
harbo(u)r response model experiment 港口响应模型实验
harbo(u)r risks 港口险
harbo(u)r rule 港口规则;港口管理条例
harbo(u)r service 港务;港口业务;港口通信业务;海岸电台业务
harbo(u)r shelter 港口避风泊地
harbo(u)r side 港边;港旁;临港面
harbo(u)r side walking 港侧人行道
harbo(u)r sign 港口号亭
harbo(u)r signals 港口信号
harbo(u)r signal station 港口信号台

harbo(u)r silo 港口筒仓
harbo(u)r silting 港内淤积
harbo(u)r site 港址
harbo(u)r site selection 港址选择
harbo(u)r siting 港址选择;港口选址
harbo(u)r sounding map 港内水深图
harbo(u)r speed 港内速度;港内航速;港口装卸效率
harbo(u)r station 港湾站;港区车站;港口(车)站
harbo(u)r statistics 港口统计资料
harbo(u)r structure 海港结构;港湾建筑物;港口建筑物
harbo(u)r superintendency administration 港务监督
harbo(u)r superintendent 港务监督长
harbo(u)r superintending officer 港务监督员
harbo(u)r supervision office 港务监督(办公)室
harbo(u)r surging 港域水面振动;港内水面波动
harbo(u)r surveillance control radar 港口监视指挥雷达
harbo(u)r survey(ing) 港湾测量;港口测量;港湾调查
harbo(u)r tanker 港口油轮
harbo(u)r tender 港口供应船
harbo(u)r terminal 港湾枢纽;港口码头
harbo(u)r tonnage dues 吨税【港】
harbo(u)r towage 港内牵引
harbo(u)r town 港埠
harbo(u)r track 码头线
harbo(u)r traffic 海港运输;海港交通;港内交通;港口交通
harbo(u)r traffic capacity 港口通过能力
harbo(u)r traffic facility 港内运输工具
harbo(u)r traffic tools 港口运输工具
harbo(u)r transload 港口吞吐量
harbo(u)r transport 港内运输;港口运输
harbo(u)r tug 海港拖船;海港拖驳;港作拖船;港口拖船
harbo(u)r user 港口使用者
harbo(u)r utility craft 港作船;港湾工作交通艇
harbo(u)r volume 港域容积
harbo(u)r wall 海港围墙;港口岸壁
harbo(u)r warehouse 港口堆栈;港口货栈;港口仓库
harbo(u)r watch 驻港值班;港泊值班【船】
harbo(u)r water depth 港口水深
harbo(u)r water fluctuation 港内水面波动
harbo(u)r waters 港区水域;港口水域
harbo(u)r with parallel jetties 有平行码头的海港;有平行防堤的海港
harbo(u)r workboat 港口作业船;港作船
harbo(u)r workboat wharf 港作船码头
harbo(u)r worker 码头工人
harbo(u)r working 海港作业;海港工作
harbo(u)r works 筑港工程;港湾工程;港口设施;港口工程;海港工程;港口建筑物;港工
Harcus 哈卡斯铬锰钢
hard 埋硬;坚硬;反复强的
hard acid 硬酸
hard adder 硬加法器
hard adhesive 硬质黏结剂
hard aggregate 硬质骨料;硬质集料
hard alee 下风满舵
hard alloy 硬(质)合金;高强度合金
hard alloy bit 硬质合金钻头
hard alloy cutter 硬质合金刀具
hard alloy grinding block 硬质合金磨块
hard alloy steel 硬(质)合金钢
hard alumin(i)um 硬铝
hard and fast ashore 牢固搁浅
hard-and-fast rule 硬性规则;硬性规定;固定规则;严格的规则;严格规定
hard anodic oxidation coating 硬质阳极氧化膜
hard arc 强电弧
hard area 硬区
hard ash coal 硬灰煤
hard asphalt 硬(质)沥青;硬地沥青(针入度10以下)
hard asphalt determination 硬沥青检定
hard asphalt mix 硬地沥青混合料(采用硬地沥青粉作黏结料的混合料)
hardas process 硬质氧化铝膜处理法
hard aweather 上风满舵
hard axis 难轴;难轴

hardback 硬皮书;粗糙背面木板
hard baked slab 素烧瓷板
hard bark 硬皮
hard base 硬碱
hard-base rock 硬基岩
hard bath 硬性显影液
hard beach 硬底海滩;海滨拉船硬斜道
hard bed 硬质河床;有岩石河床
hard-bed material 硬底质
hard-bed river 石底河流
hard-bed stream 岩床河流;石底河流;石底河川
hard bilge 尖舭
hard black 硬烟末;补强炭黑
hard blister 玻璃内部气泡
hard blue limestone 硬质青石灰岩
hard board 硬(质)板;高压板;刚纸;人造板;硬质纤维板
hardboard door 硬质板门
hardboard-faced plywood 硬质纤维板贴面(胶合)板;贴面胶合板
hardboard finish 硬质板铺面
hardboard floor(ing) 硬质板地面
hardboard machine 硬纸板机;高压板机
hardboard mill waste 木材厂废料
hardboard mill wastewater 硬纸板厂废水;木材厂废水
hardboard panel 硬质纤维板
hardboard siding 硬质纤维板板壁;硬质纤维板护墙板
hardboard underlayer 硬质纤维板垫层
hardboard underlayment 硬质纤维板垫层;硬质纤维板垫板
hard boiled 强硬的
hard-bonded furnace deposit 炉内硬块沉积
hard borosilicate glass 耐火玻璃;硬质硼硅玻璃
hard bottle 硬质玻璃瓶
hard bottom 硬底;坚固底板
hard boundary 硬质边界
hard braking 紧急制动
hard brass 硬黄铜
hard brazing 硬焊
hard bread 压缩饼干(救生艇用)
hard breakdown 硬击穿;刚性击空;刚性击穿
hard brick 硬砖
hard brittle material 脆性材料
hard brittle rock 硬脆岩石
hard broken ground 硬脆地层
hard bronze 硬青铜
hard bubble 硬泡
hard bubble suppression 硬泡抑制
hard-burned 硬烧的;过烧的;死烧的;过火的;烧透的
hard-burned brick 硬砖;炼砖;过火砖
hard-burned clinker 过烧熟料
hard-burned free lime 生石灰
hard-burned gypsum 高温焙烧(的)石膏;炼石膏
hard-burned magnesia 死烧氧化镁;硬烧氧化镁
hard-burned plaster 高温焙制石膏;煅烧过度的石膏
hard-burned refractory ware 硬烧耐火制品;硬烧耐火器材
hard burning 硬烧;急烧
hard burning driven kiln 强化煅烧窑
hard burning mix 难烧生料
hard burnt 硬烧;无水石膏;烧石膏;高温焙烧的;过火(砖)
hard-burnt clayware 素烧陶瓷
hard-burnt facade brick 过烧面砖
hard-burnt plaster 无水石膏;烧石膏
hard-burnt stoneware 素烧陶器
hard calcareous slate 硬质石灰岩石板;硬石灰板
hard carbide industry 硬质合金工业
hard carbon 硬炭
hard carbon black 硬质炭黑
hard carbon film 硬炭薄膜
hard cargo berth 活动板泊位
hard case meter 铸壳煤气表;硬壳煤气表
hard cash 现款;现金;硬币
hard casting 硬铸铁;硬铸件
hard cast iron 硬铸铁
hard cast steel 硬铸钢
hard cemented soil 硬结土(壤)
hard center nozzle 硬芯喷嘴;耐磨插芯喷嘴
hard center 硬的夹心
hard chalk 硬白垩

hard china clay 硬质陶土
hard chine boat 直框架结构船
hard chrome 硬铬
hard-chromed 镀硬铬的
hard chrome facing 表面镀硬铬
hard chrome plating 镀硬铬
hard chromium 硬铬(镀层);厚层镀铬
hard chromium plated tool 硬质镀铬工具
hard chromium plating 硬铬电镀;硬铬镀层;镀硬铬
hard clay 硬质(黏)土;硬黏土
hard clay ore 硬质黏土矿石
hard clod 硬土块
hard coal 硬煤;无烟煤
hard coal international classification 硬煤国际分类
hard coal plough 硬煤层刨煤机
hard coat 硬膜;硬罩;硬盖
hard coated 涂有硬膜的
hard coat system 硬涂层系统
hard coke 焦炭
hard colo(u)r 硬彩
hard-combustible component 难燃烧体
hard compact clod 坚硬土块
hard compact soil 稳定土;压实土;坚实土;密实土(壤);夯实土(壤)
hard component 硬成分
hard concrete 干(硬)性混凝土
hard condition 刚性条件
hard contact 硬触点;金属触点
hard contrast 高反差
hard conversion 计量单位转换
hard copal 硬库巴树脂
hard copper 硬铜;冷加工铜
hard copy 硬副本;硬拷贝;印刷记录;可读复制资料;复制件;复印文本;裱粘底图;裱糊底涂
hard-copy enlarger 放大印机
hard-copy log 硬拷贝记录;复印记录
hard-copy material 硬拷贝材料
hard-copy output 硬拷贝输出;复制件输出
hard-copy peripherals 硬拷贝外围设备
hard-copy plotting board 硬拷贝图板
hard-copy terminal 硬拷贝终端
hard core 硬芯;硬路基;硬核;路基碎料;硬垫层;硬底层;骨干;坚实基础;核心硬件;(用作路面或基础下的底基层的)天然岩石碎块;碎石垫层;碎石路基;碎块混凝土块;石填料
hard core cusp 硬芯会切
hard core drain 填石排水沟
hard core pinch 硬芯箍缩
hard core potential 硬芯势
hard cosmic ray 硬宇宙射线
hard cost 硬费用
hard court 硬地网球场
hardcover 精装的
hard crown 硬冕玻璃
hard cure 硬焙烘
hard currency 硬通货;硬币
hard cutting 硬岩掏槽
hard data 硬数据
hard-degradable organic compound 难降解有机化合物
hard-degradable organic wastewater 难降解有机废水
hard detergent 硬(性)洗涤剂
hard diamond like material 硬的金刚石类材料
hard direction 难磁化方向
hard disk 硬盘【计】
hard disk cartridge 硬盘盒
hard disk cylinder 硬盘柱(面)
hard disk drive 硬盘驱动器
hard disk pack 硬盘组
hard disk system 硬盘系统
hard dollars 硬通货;现金
hard dot 实点
hard dough 硬蜡熟
hard-dough stage 硬黄熟期
hard down 下风满舵;贴近下风向
hard draw 硬冷披;硬拉
hard-drawing 冷拔
hard drawing brass 冷拉黄铜
hard drawn 冷拉;冷拔
hard-drawn alumin(i)um wire 硬铝线
hard-drawn copper contact wire clip 硬拉铜接触线线夹
hard-drawn copper stranded conductor 硬铜绞

合线
hard-drawn copper strand wire 硬铜绞线
hard-drawn copper wire 硬铜线;冷拉铜丝;冷拉铜线
hard-drawn mild steel wire 冷拔低炭钢丝
hard-drawn steel wire 硬拉钢丝
hard-drawn wire 冷拉丝;硬拉线;硬拔线;冷拔丝
hard-drawn wire reinforcement 冷拉钢筋
hard drill 硬钻
hard drilling 硬层钻进
hard driven kiln 强化煅烧窑;强化操作窑
hard driving 硬岩打进;猛力驾驶;猛力传动;艰难打桩
hard dry 硬干;干透的;实干(涂层)
hard-drying 硬结;干硬;干结
hard drying time 硬干时间;实干时间
hard-earned capital 血本
hard-edge 硬边
hard edged hammer 硬缘锤
hard edges 铸件毛刺
hard elastic fibre 硬弹性纤维
hard elastic region 硬弹性区域
hard electron 高能电子
harden 变硬
hardenability 硬化性能;硬化特性;硬化(程)度;可硬化性;可硬化度;可淬硬性;淬硬性;淬透性(值);淬火性
hardenability band 硬化带;硬锻带;可淬硬性带;淬透性带
hardenability characteristic 淬透性
hardenability chart 淬透性图表
hardenability limits 淬透性极限
hardenability line 淬透性直线
hardenability test 淬透性试验
hardenability value 硬化指数
hardenable 可硬化的
harden and grind 硬化与研磨
harden depth 淬硬深度
hardened 硬化的
hardened and ground worm 淬硬磨光蜗轮
hardened and tempered 淬火与回火(的)
hardened and tempered steel 调质钢
hardened case 表面渗碳硬化
hardened cement 硬化水泥;水泥石
hardened cement paste 硬化水泥浆(体);水泥石
hardened circuit 硬化电路
hardened concrete 硬结混凝土;硬化混凝土;已硬化(的)混凝土
hardened cose 表面渗碳硬化
hardened coupling 淬硬接头
hardened distribution 硬化谱分布
hardened dust 硬化填料
hardened emulsion 硬化乳化层
hardened explosive 硬化炸药
hardened face 硬化面;淬硬表面;淬火表面
hardened fat 硬化脂
hardened glass 钢化玻璃;强化玻璃
hardened in air 在空气中硬化的;气硬的
hardened layer wood 硬化层积(木)材
hardened links 防辐射无线电通信线路
hardened material 硬化材料
hardened mortar 硬化砂浆;硬化泥浆;硬化灰浆
hardened oil 硬化油
hardened outside verge 已硬化的外缘
hardened plate 淬硬钢板
hardened PVC 硬化聚氯乙烯
hardened resin 硬树脂;凝固树脂
hardened rosin 硬化松香;钙脂(松香);石灰松香
hardened rubber 硬化橡胶
hardened soil 板结土
hardened spectrum 硬化谱
hardened steel 硬化钢;淬硬钢;淬火钢
hardened steel bushing 淬火钢衬套
hardened steel guideway 淬火钢导轨
hardened steel inserts (在卡盘上、钳上的)淬火镶嵌钢
hardened steel nail 淬火钢钉
hardened steel roller 淬火钢滚轮
hardened steel tire 硬化钢环箍
hardened surface 淬硬表面
hardened verge 硬路肩;加固的路边
hardened way 淬硬导轨;淬火导轨
hardener 硬化剂;硬化机;硬化工;硬化成分;增硬剂;固化剂;坚膜剂

hardener dust 硬化粉剂
hardener filler 硬化填充料
hardener of surface skin of concrete 混凝土表面硬度增强剂
harden extent 板结程度
Harden fast 哈登法斯特混凝土块速凝结剂
harden grout film 水泥结石
hardening 硬结;硬化(混凝土);淬硬;淬火;板结
hardening accelerating admixture 固化促进剂;促硬剂;硬化加速剂
hardening accelerating agent 催硬剂
hardening accelerator 早强剂;固化促进剂;硬化加速剂;速凝剂;促硬剂
hardening action 硬化作用
hardening admix(ture) 助凝剂;硬化剂
hardening age 硬化期;硬化时效
hardening ag(e)ing 硬化时效
hardening agent 固化剂;硬化剂;坚膜剂;强化剂;淬火介质;淬火剂
hardening alumin(i)um alloy 硬化铝合金
hardening and tempering 调质
hardening asphalt 硬化沥青
hardening at subcritical temperature 低温硬化
hardening babbit 硬质巴比合金;低锡巴氏合金
hardening bath 硬化浴;硬化液;坚膜液;淬火浴;淬火池
hardening behavio(u)r 硬化过程
hardening bilinear system 硬化型双(曲)线性系统
hardening bitumen 硬化沥青
hardening by cooling 冷却硬化
hardening by hammer 锤击硬化
hardening by hammering 锤击硬化;锤锻硬化
hardening by high frequency current 高频淬火
hardening by isothermal heat treatment 等温处理加硬化
hardening by itself 自身硬化
hardening cabinet 硬化箱
hardening capacity 硬化性能;硬化能力;硬化量;硬化本领;可淬性
hardening carbon 硬化炭
hardening catalytic agent 硬化催化剂
hardening chamber 硬化室;淬火室
hardening characteristic 硬化特征;硬化特性
hardening coat 硬化面层
hardening compound 淬火剂
hardening constituent 强化
hardening crack 骤冷裂纹;淬致裂纹;淬致裂痕;淬火裂纹
hardening crack(ing) 淬裂
hardening curve 硬化曲线;淬火曲线
hardening degree 硬化(程)度
hardening effect 硬化效应
hardening energy 硬化能量;硬化能力
hardening filler 硬结填料
hardening film preventives 硬膜防锈剂
hardening flaw 淬火裂纹
hardening floor paint 地板漆
hardening furnace 硬化炉;淬火炉
hardening gloss paint 硬光漆
hardening heat 硬化热;淬火温度
hardening in boiling water 沸腾水淬硬化
hardening in ice 冰淬硬化
hardening in lead bath 铅淬硬化
hardening in running water 水的硬化
hardening kiln 硬化窑
hardening kinetics of cement 水泥硬化动力学
hardening lens 安全镜片
hardening liquid 硬化液;淬火液
hardening machine 淬火机
hardening mechanism 硬化机理
hardening medium 淬火剂
hardening metal 硬化金属
hardening modulus 硬化模量
hardening of cement 水泥硬化
hardening of concrete 混凝土硬化
hardening of gelatin(e) 明胶硬化
hardening of loan terms 货款条件紧缩
hardening of oils 油类硬化
hardening of soil 土壤板结
hardening of steel 钢的硬化
hardening oil 淬火油;硬化油
hardening-on kiln 烤烧釉下颜料窑
hardening period 硬化期
hardening plant 淬火车间

hardening point 硬化点
hardening point of asphalt 沥青硬化点
hardening process 硬凝过程;硬化过程;硬化法
hardening rate 硬化速度;硬化速率
hardening ratio 硬化率
hardening rule 硬化规则;硬化保护层;强化规律
hardening shop 淬火车间
hardening strain 硬化应变;淬火应变
hardening stress 硬化应力;淬火应力
hardening tank 淬火槽
hardening temperature 硬化温度;淬火温度
hardening test 硬化试验;凝固试验
hardening test of concrete 混凝土硬化试验;混凝土脱模强度试验
hardening time 硬化时间
hardening tongs 硬化钳
hardening treatment 硬化处理;淬火处理
hardening yield value 硬化屈服值
hardening zone 硬化区
hardenit 细马氏体
hardenite 硬(化)体
harden polymerization 聚合硬化
harden quenching 淬硬
harden right out 使完全硬化
harder 压帽坯机
hard eraser 硬橡皮擦
hard error 硬错误
hard error rate 硬错误率
hard error status 硬错状态
hardest formation 坚硬地层
hardest water 极硬水
hard facade 坚硬表面
hard face 硬(表)面
hard-faced 堆焊硬质合金的;有硬层的;硬表面的;正面硬化
hard-faced bit 镶有硬质合金(的)钻头;堆焊硬质合金钻头
hard-faced share 硬面犁铧
hard facility 硬设施
hard-facing 覆硬层;硬质焊敷层;表面耐磨堆焊;表面硬化;表面渗碳(处理);表面淬火(处理);加焊硬面(法);硬面处理;堆焊硬质合金;耐磨堆焊
hard-facing alloy 硬面合金;表面加硬用合金
hard-facing electrode 耐磨堆焊焊条
hard-facing material 表面补强材料
hard-facing metal 镀焊用硬质合金;表面加硬用合金
hard-facing of tool joint 钻头接头表面硬化
hard-facing rod 镀硬面焊条
hard-facing welding 耐用堆焊
hard failure 硬失效
Hardfast 哈登法斯特混凝土硬化剂
hard fat 硬脂
hard fat lubrication 滑脂润滑
hard feed water 硬给水
hard ferrite 硬磁铁氧体
hard fiber 钢纸;硬(质)纤维
hard fiber bearing 硬刚纸轴承
hard fiber board 硬质纤维板
hard fiber core 硬纤维芯
hard fiber plug 硬纤维塞
hard fiber washer 硬纤维垫圈
hard filler 硬填缝料
hard film 硬性底片;硬膜
hard filth water 硬垢水
hard fine pottery 硬质精陶
hard finish 硬饰面;石膏抹面;(用细砂灰浆抹成的)光滑硬面层;粉刷的面层;硬质罩面;干细抹面;精细磨光;精磨;精加工;抹灰压光面;多砂粉刷
hard finished 压光的
hard-finished plaster 需加催硬剂的无水煅烧石膏;硬面灰浆;无水煅烧石膏
hard-finished plastering 压光饰面;压光磨面
hard finish floor 硬装修地面
hard finishing 硬挺整理
hard finish plaster 缓硬石膏;硬饰面石膏;马丁水泥
hard fire 高温烧成
hard-fire colo(u)r 高温颜料
hard-fired 烧透的
hard-fired brick 烧透的砖;焙烧适度的砖
hard-fired ware 硬质烧成陶瓷;高温烧成的窑具;高温烧成的瓷器;高温烧成陶瓷
hard flame 高能火焰
hard floor cover(ing) 地面硬质面层

hard floor finishes 硬质地面层
hard floor paint 硬地板漆
hard flour 硬质面粉
hard flow 低流动性
hard-flowing material 流散性差的材料
hard flux 硬掺和料(提高沥青稠度)
hard focus lens 硬调透镜
hard fold 硬折痕
hard format 硬格式
hard formation 硬(岩)层;硬(地)层;坚硬岩层
hard formation bit 钻硬岩钻头
hard formation cutting head 坚硬岩石钻凿钻头;坚硬岩石钻凿刀盘;硬岩层钻头
hard freeze 黑冻
hard frost 严重冰况;酷寒;黑霜;黑冻
hard-frozen soil 坚硬冻土;硬冻土
hard fusite 硬微丝煤
hard gamma 硬量子
hard genuine babbitt alloy 巴比硬合金
hard glass 硬(质)玻璃;耐高温玻璃;高强度玻璃;硬(化)玻璃
hard glass bulb 硬质玻璃灯壳
hard glaze 高温釉
hard glazed coat(ing) 盐釉质面层;硬珐琅(质)涂层
hard gloss 硬珐琅(质)
hard gloss paint 硬光泽;高光泽涂料;硬光漆
hard glossy film 光泽硬质薄膜层
hard goods 耐用(消费)品
hard-grade bar 硬钢筋
hard-grade billet 硬钢坯段(料)
hard grades of bitumen 硬质沥青
hard grain 硬粒
hard grained 粗粒状的
hard-grained roll 硬铸铁滚筒
Hardgrave grindability index 哈氏可磨性指数
hard grazing 强度放牧
hard grease 硬(润滑)脂
hard grid facing 表面喷镀
hard grinding stone 硬磨石
hard ground 坚固地基;硬地(基);硬防蚀剂;硬底;岩底;坚实地基;不整合面
hard-ground man 硬岩开采工人;硬土操作工人
hard ground structure 硬底构造
Hardgrove grindability 哈德格罗夫易磨性
Hardgrove grindability index 哈德格罗夫可磨度指数
Hard grove number 哈德格罗夫可磨度指数
Hardgrove sandstone 哈德格拉夫砂岩
hard gum 硬树胶;化石树胶
hard gummed skein 硬胶丝
hardhat 硬壳帽;防护帽;矿工、建筑工人、潜水帽;安全帽
hard hat diver 带潜水盔的潜水员
hard hat diving 重装潜水
hard hat dress 重潜水服
hard head 硬质巴比合金;硬头
hard-headed hammer 硬头榔头
hard heading 岩石平巷;岩石工作面;石巷
hard-heading man 硬岩掘进工(人)
hard heading work 硬岩掘进工作
hardhead sponge 硬海棉
hard helm 扳满舵
hard ice 坚固冰
hard ice condition 严重冰况
hardie 尖锐底模
hardie-hole 底模砧孔
hard image 硬(调)图像;高反差图像;黑白鲜明图像;黑白分明图像;强对比度图像
hard inclusion 硬质点夹杂
hardiness 抵抗性
hard inflammable material 难燃材料
Hardinge ball mill 锥形球磨机;哈丁奇型圆锥球磨机
Hardinge cascade mill 哈丁奇型泻落式磨机
Hardinge conic(al) mill 哈丁奇型圆锥球磨机
Hardinge countercurrent classifier 圆筒形交流分级机;哈丁奇型逆流分析
Hardinge disk feeder 哈丁奇型水平转盘给料机
Hardinge feeder-weigher 哈丁奇型称量给料器
Hardinge hydroclassifier 哈丁奇型水力分级机
Hardinge mill 哈定格磨机;哈丁奇型球磨机;哈丁磨
Hardinge superfine classifier 哈丁奇型超分级机
Hardinge thickener 哈丁奇型控制
Hardinge thickner 哈丁奇型沉降槽
Hardinge type loop classifier 哈丁奇型回路分级机

hard insulating material 硬质保温材料
hard intensity 大强度
hard iron 硬铁;冷硬铸铁
hardish 较硬的
Hardite 哈代特耐热镍铬硅合金
hard kaolin deposit 硬质高岭土矿床
hard knot 死结
hard lac 硬虫胶
hard lacquer 硬喷漆
hard lac resin 硬紫胶树脂;硬虫胶
hard laid 硬搓(绳)
hard laid wire rope 硬绳
hard landing 硬着陆
hard landscape 硬质景观;硬风景
hard lay 硬捻;硬搓绳
hard layer 硬岩层;硬地层
hardle 下型锤
hard lead 铅锌合金;硬铅;锑铅合金
hard lead alloy 硬铅合金
hard lead pipe 硬铅管(道)
hard lead pressure pipe 锑铅压力管
hard lead sheet 锑铅薄板;硬铅皮
hard leather 硬革
hard light 强光
hard lime 硬质石灰层
hard limestone 硬(质)石灰岩
hard limited integrator 硬性受限积分器;硬限量积分器
hard limiter 硬(性)限幅器
hard limiting 硬限制;硬极限;硬限幅
hard limiting transponder 硬限幅转发器
hard line fibre 硬质长纤维
hard lining 硬质衬砌
hard loan 需用硬通货偿还的贷款;硬贷款;严厉条件贷款;条件苛刻的条款
hard lumber 硬质木材
hard lump 硬块;硬土堆
hard lump of cement 整袋水泥僵块
hard machine check 硬设备效验
hard magnesia 硬烧镁氧
hard magnetic material 硬磁(性)材料
hard magnetization 强磁化
hard maple 硬槭木;糖槭;硬枫木
hard mass 人造宝石
hard mastic asphalt 硬沥青玛琋脂;硬质砂胶沥青
hard material 硬质材料;坚硬物料
hard matrix 硬金属胎体
hard metal 硬(性)金属
hard metal alloy 硬质合金钻头;高强度合金
hard metal article 硬质合金制品
hard metal bit 硬合金钻头
hard metal compound 硬质合金化合物
hard metal disease 硬金属病
hard metal drilling 硬质合金钻进
hard metal facing 表面堆焊硬合金
hard metal insert 硬质合金镶嵌物
hard metal sheathed cable 硬金属铠装电缆;硬金属套电缆
hard metal tip 硬质合金片
hard metal(tipped) bit 硬质合金钻头
hard metal tool 硬质合金刀具
hard mica 弯曲时不分层主云母
hard money 现金;可兑换货币
hard money mortgage 现金抵押
hard mortar 硬性灰浆;干硬(性)砂浆;干硬性灰浆
hard mosaic floor 硬木块镶嵌地板
hard mo(u)lding plaster for ceramic industry 硬质模具石膏
hard mo(u)lding plaster for foundry industry 硬质翻砂石膏
hard mo(u)lding plaster for medical use 硬质医用模具石膏
hardness 硬性;硬度;刚度;难解
hardness ag(e)ing 加工时效;硬度老化
hardness analysis 硬度分析
hardness analyzer 硬度分析仪
hardness Barcol 巴氏硬度
hardness breakthrough 硬度泄漏
hardness change 硬度变化量
hardness class 硬度(等)级
hardness control 硬度控制
hardness conversion 硬度换算
hardness crack 硬度裂缝

hardness creep 硬度蠕变
hardness curve 硬度曲线
hardness degree 硬度(标度)
hardness degree of rock 岩石坚硬程度
hardness-depth curve 硬度深度曲线
hardness determination 硬度判定;硬度测定(法)
hardness drop tester 肖氏硬度计;回跳硬度计
hardness factor 硬度值
hardness ga(u)ge 硬度计;硬度标
hardness increase 加强硬度
hardness index 硬度指数
hardness ions 硬性离子
hardness limit 硬度极限
hardness loss 硬度消失;硬度衰减;硬度损失
hardness measurement 硬度测量
hardness method 硬度法(非破损性混凝土强度测定法)
hardness number 硬度(指)数;硬度指标;硬度值
hardness number conversion 硬度值转换
hardness of aggregate 集料硬度
hardness of concrete 混凝土干硬性;混凝土硬度
hardness of grain 磨粒的硬度
hardness of grease 润滑脂的硬度
hardness of stone 石材硬度
hardness of water 水(的)硬度
hardness of wood 木材硬度
hardness penetration (钢材的)硬化深度;硬化度;淬硬深度;淬火深度;淬化深度
hardness point 硬度标
hardness process(ing) of wood 木材硬化法
hardness range 硬度极限;硬度范围
hardness ratio 硬度比
hardness reducer 硬度软化剂
hardness removal 水的软化;除硬度
hardness resin 固体树脂
hardness rubber 硬橡胶
hardness scale 硬度(计);标度;硬率;硬度计表盘刻度;硬度尺寸;硬度表;硬度标
hardness scale(degree) 硬度标度
hardness self-excitation 硬自励
hardness solidness 硬度
hardness test 硬度实验;硬度测试
hardness tester 硬度(试验)计;硬度试验机;硬度测试仪
hardness test(ing) 硬度试验
hardness testing automation 硬度测试自动化
hardness testing device 硬度计;硬度测试仪
hardness testing machine 硬度测试机;硬度试验机;硬度实验机;硬度计;硬度测试仪
hardness value 硬度值
hardness vector 硬度矢量
hardnester 锉式硬度试验器
hard of hearing 听觉不良
hard oil 慢干油;硬油
hard oil finish 干性油涂饰;干性油饰面
hard oil putty 硬质油性腻子;干硬性油腻子;干硬性油灰
hardometer 硬度计
hard ore 硬矿石
hard oscillation 强振荡
hard over 满舵;扳满舵
hard over position 最大转舵位置
hard-packed snow 压实的雪
hard-packed snow remover 坚雪铲除工
hard packing 硬垫
hardpan 硬质地层;硬岩盘;坚固基础;不透水层;硬盘层;硬土(层);硬磐;坚固基础;灰质壳;铁磐;底价
hardpan spring 硬岩盘泉
hard paper 硬性感光纸
hard paraffin 硬石蜡;固体石蜡
hard parking-place 硬地停车场
hard particle 硬颗粒
hard parton 硬部分子
hard paste 硬坯;硬质瓷(旧用名词)
hardpaste 硬瓷
hard paste porcelain 瓷器
hard patch 铆接补板
hard pavement 硬(质)路面
hard pedal 笨踏板
hard phase 硬质相
hard phenolic resin 硬酚醛树脂
hard physical labo(u)r 重体力劳动
hard picture 硬图像;强对比度图像

hard pine 硬松
hard pitch 硬沥青;硬焦油脂
hard plant 耐寒植物
hard plaster 无水石膏;烧石膏;硬质灰膏;墙的水泥抹面
hard plastic tube 硬塑料管
hard plating 低廉的镀铬;镀硬铬
hard-point 硬(化)点;结构加固点
hard polyvinyl choride pipe 硬聚氯乙烯管
hard polyethylene 硬性聚乙烯
hard polymer 硬聚合物
hard porcelain 硬质瓷器;硬瓷;刚性瓷器
hard potential 硬势
hard pressed fiberboard 高压纤维板
hard-pressed particle board 高压刨花板
hard-pumped 难抽;打足气的
hard putty 硬质油性腻子;硬(质)油灰
hard radiation 硬(性)辐射
hard rain 暴雨
hard ray 硬性射线
hard regular card financial statement system 常备客票结账管理系统
hard remanent magnetization 硬剩磁磁化
hard residue 硬性残留物
hard resin 硬(性)树脂;固体树脂
hard rime 霜凇
hard road 硬道路面
hard rock 硬质岩石;硬岩石;硬岩;硬石岩;坚岩;坚石
hard rock boring 硬岩钻进
hard rock boring program(me) 硬岩钻探程序
hard rock chip(ping)s 硬岩石屑;硬碎石
hard rock concrete 硬石岩骨料混凝土;硬石混凝土;硬石岩集料混凝土
hard rock driller 硬岩钻工
hard rock drilling 硬岩钻进
hard rock flour 硬石粉末;硬岩粉末
hard rock jaw crusher 硬岩颚式碎石机;颚式硬石碎石机
hard rock mine 硬岩矿
hard rock miner 硬岩(采)矿工;硬岩开采机
hard rock mining 硬岩开采
hard rock phosphate 硬岩磷酸盐
hard rock product 硬石产品
hard rock quarry 硬石开采(场);硬岩采石场
hard rock raise boring machine 硬岩天井钻进机
hard rock reaction crusher 硬岩反击式破碎机
hard rock slab 硬石石板;硬岩石板
hard rock tunnel boring 硬岩平巷掘进机钻进法
hard rock tunneling 硬岩隧道施工
hard rock tunneling machine 硬岩石掘进机联合机
hard-rolled 冷轧的
hard rolling (轴颈圆角等的)硬化滚压;滚压硬化
hard rolling car 难行车【铁】
hard rolling track 难行线
hard roof 坚硬顶板;稳固顶板
hard rot 干腐;烂疤
hard roving 硬粗纱
hard rubber 硬质(橡)胶;硬橡胶;胶木
hard rubber article 硬质胶制品
hard rubber base 硬橡胶垫
hard rubber battery box 硬橡胶蓄电池壳
hard rubber bead 硬橡胶撑轮圈
hard rubber board 硬橡胶板
hard rubber box 硬橡胶壳
hard rubber circular rod 硬橡胶圆辊
hard rubber meeting face 硬质橡胶贴接面
hard-rubber pipe 硬橡胶管
hard rubber roller 硬橡胶滚轮
hard rubber round rod 硬橡胶圆辊
hard rubber scraps 硬质胶屑
hard rubber sheet 硬橡胶片;硬橡胶薄板;硬胶板
hard-rubber steering wheel 硬橡胶转向盘
hard rubber tire 硬橡胶轮胎;硬橡胶轮胎
hard rubber tube 硬橡胶管;硬橡胶管
hard rule 硬性规定
hard running track 难行线
hards 麻屑
hard sandstone 硬砂岩
hard saturated fat 硬化饱和脂
hard schist 硬片岩
hard science 硬科学
hard scientist 硬科学家
hard Scope 袖珍式(肖氏)硬度计

hard seat 基岩;底岩
hard sectoring 硬分段
hard seeds 硬粒
hard segment 硬链段
hard self excitation 硬自励;硬自激
hard self-excited oscillation 硬自激振荡
hard sell 强行推销
hard service 重负荷工作状态;超负荷工作(状态);不良使用
hard servo 硬伺服
hard set 牢牢固定的;装牢的;凝固的
hard-setting cement 铁矿水泥
hard-settling 硬结;沉底结块
hard shadow 硬性阴影;清晰阴影
hard shell clam 硬壳蛤;硬壳蚌;蛤蜊
hard-shelled 硬壳的
hard shell sphere model 硬壳球体模型
hardship 艰难困苦
hard shore 基岩岸
hard shoulder 硬(质)路肩
hard shower 硬射流;硬簇射;穿透射流
hard shut-down 突然停机
hard site 加固基地防御;地下场;地下设施;坚固发射场
hard sludge 硬(淤)渣;硬污泥
hard snow 压实的雪;压坚硬的雪
hard snow remover 坚雪铲除机;坚雪铲除工;除雪机
hard soap 硬皂
hard-soft acid-base principle 硬软酸碱原理
hard solder 含铜多的焊锡;硬钎料;硬焊药;硬焊料;(含铜较多的)焊锡;钎焊料;铜焊
hard soldered 硬焊接的
hard soldered joint 硬焊接缝;硬焊接合
hard soldering 硬钎焊;硬焊剂;硬焊(的)
hard solder joint 硬焊料焊接
hard solid rock 坚硬岩层
hard solid rubber 硬质橡胶
hard sphere 硬球
hard-sphere collision 刚性环碰撞
hard-sphere fluid 硬球形流体
hard-sphere model 硬球模型
hard-sphere scattering 刚球散射
hard spot 麻点;硬(麻)点;部分过硬
hard stand 停机场;硬地面;停机坪;可停放车辆、飞机等的硬质地面
hard standing 硬面停车场;硬面层;路路面;坚实停车场;停机坪
hard starting load 重起动负荷
hard start of oscillation 振荡突起
hard steel 硬钢;高碳钢
hardstem 速凝石膏炮泥
hard still enable 硬静止起动
hard stock 过烧变形硬;硬浆
hard stock brick 耐火砖;普通硬砖
hard stone 坚硬岩石;坚石
hard stone coast 基岩岸
hard stop 硬停;立即停机
hard stopper 硬质填塞料
hard stopping 硬质填(塞)料;硬质充料;快干水性石膏腻子
hard-stoved enamel finish 烘漆饰面;烘干瓷漆饰面
hard strip 硬带材
hard structure 硬式结构
hard subsoil 硬下层土;硬基土;硬地层;硬底土
hard suction 硬质吸水管
hard sundry particle 硬质杂粒
hard super-conducting alloy 超导电性硬合金
hard surface 淬火表面;硬质路面;硬表面;壳体;硬面
hard surface in-contact mask 表面坚固内接触掩膜
hard surface mask 表面坚固的掩膜
hard surface plate 硬表面板
hard surface road 硬质路面的道路;硬面(道)路
hard surface runway 高级道面跑道
hard surface share 表面硬化犁铧
hard surfacing 硬质面层;硬质路面;硬质合金覆面;硬质焊敷层;表面硬化
hard surfacing material 表面补强材料
hard surrounding rock 坚硬围岩
hard swell 硬膨罐
hard swell can 硬胀罐(用手指压不回去)
hard swell spoilage 硬膨罐腐败
hard-switch modulator 刚性(开关)调制器

hard system 硬系统
hard tap 出渣口凝结
hard target 硬目标
hard technique of decision 决策硬技术
hard technology 工程技术
hard technology of decision 决策硬技术
hard-temperature 硬化温度
hard temper sheet 硬回火的薄钢板
hard terminal end 夹持端
hard texture 硬件刻画膜
hard-textured chipped wood concrete 硬质木屑混凝土
hard-textured wood 坚实木材
hard-textured wood fibre concrete 坚木纤维混凝土
hard through-dry 实干
hard timber 硬质木材
hard tin alloy 硬锡合金
hard-to-break 难切断
hard-to-burn-refuse 难燃垃圾
hard-to-cut material 难加工材料
hard-to-cut rock 难凿岩石
hard to dry aggregate 难干集料
hard to enumerate group 难于查点的人群
hard-to-get data 难以取得的数据;难得数据
hard to manage the small peasant economy 不好管小农经济
hard tooth combination saw blade 硬齿多用锯片
hard top 硬质路面;硬质地面的空地或道路;硬顶
hard top container 硬顶式集装箱(一种顶开式集装箱)
hard-top minibus 硬顶小客车
hard-topped highway 硬路面的公路
hard-to-reach 难以达到的
hard-to-screen material 难以过筛的材料
hard-to-scribble symbol 难刻符号
hard-to-sinter 难烧结
hard-to-start 难起动
hard-troweled surface 硬镘光表面
hard tube 高真空(电子)管
hard-tube modulator 硬管调制器;高真空管调制器
hard-tube pulser 电子管脉冲发生器;硬管脉冲发生器;高真空电子管脉冲发生器
hard turn bilge 尖舭
hard twist 硬捻;硬拈
hard-type modulator 真空管调制器
hard-type stay cable 竖琴形拉索
hard type surfactant 硬性表面活性剂;硬型界面活性剂
hard up 贴近上风向
hard up of rudder 上风满舵
hard usage 强烈使用;粗略利用
hard vacuum 高度真空
hard valve 高真空电子管
hard varnish 硬漆
hard vector surface 硬矢面
hard wait 硬等待
hard wall 石膏抹底墙;石膏抹灰基层;石膏打底抹灰
hardwall board 硬墙纸板
hardwall hose 水带
hardwall of mill 磨机盖板
hardwall plaster 墙壁水泥抹面;水泥粉饰;水泥粉;灰浆层罩面
hardwall plaster wall 水泥粉饰墙
hardware 小五金;硬设备;硬件;重兵器;金属元件;金属器具;金属构件;金属附件;五金
hardware address control 机器地址控制
hardware and electric(al) appliance 五金电料
hardware arbiter 硬件判优器
hardware architecture 硬件体系结构
hardware assembler 硬件汇编器;硬件汇编程序
hardware base address 硬件基地址
hardware check 硬件检验
hardware cloth 钢丝网
hardware compatibility 硬件兼容性;设备兼容(性)
hardware component 硬件成分;硬件部件
hardware configuration 硬件配置;设备配置
hardware construction 硬件构造
hardware context 硬件关联
hardware control 硬件控制
hardware deficiency 硬件缺陷
hardware description language 硬件描述语言

hardware design 硬件设计
hardware development 硬件开发
hardware digital processing 硬件数字处理
hardware emulator 硬件仿真程序
hardware error recovery management system 硬件错误恢复管理系统
hardware factory 五金零件工厂；五金厂
hardware fault 硬件故障
hardware finish 特光洁铸造表面
hardware firmware software trade-off 硬件固件软件权衡
hardware fitting electric(al) porcelain insulator 电器绝缘瓷座
hardware fittings 小五金配件
hardware industry 五金制品工业
hardware instruction 硬件指令
hardware interface 硬件接口
hardware interlock register 硬件互锁寄存器
hardware interrupt facility 硬件中断设备
hardware interrupt level 硬件中断级
hardware item 硬件项目；小五金项目
hardware layer 硬件层
hardware lockout 硬件封锁
hardware logic 硬件逻辑
hardware malfunction 硬件故障
hardware man 五金商人
hardware management 硬件管理
hardware manufactory 五金工厂
hardware mesh 金层丝网
hardware modularity 硬件模块化
hardware module 硬构件；硬件模块
hardware monitor 硬件监视器；硬件监督程序；设备监督器
hardware monitoring 硬件监视
hardware mounting machine 五金零件装配机
hardware net 硬件网络
hardware of automated cartography 自动制图硬件
hardware padding sequence 硬件填充序列
hardware pattern generator 硬件图形发生器
hardware priority interrupt 硬件优先(权)中断
hardware process control block 硬件进程控制块
hardware product factory 小五金制品厂
hardware redundancy 硬件冗余法
hardware representation 硬件表示
hardware reset 硬件复位
hardware resource 硬件资源
hardware resource control 硬件资源控制
hardware selection criteria 硬件选择标准
hardware selector 硬件选择器
hardware simulator 硬件模拟程序
hardware-software harmony 硬软件协调
hardware-software support 硬件软件支援；硬件软件支持
hardware stack 硬件堆栈
hardware store 五金店
hardware subassembly 硬件部件
hardware support 硬件支援；硬件支持
hardware synchronization mechanism 硬件同步机制
hardware system 硬件系统
hardware system availability 硬件系统可用性
hardware system reliability 硬件系统可靠性
hardware technology 硬件技术
hardware telemetry 硬件遥测
hardware timer 硬件计时器
hardware work 小五金安装工程
hard washing 难选
hardwaste breaker 开回丝机
hardwaste dump 废料堆放场
hardwaste recovery 废料回收
hard water 硬水
hard water inlet 硬水入口
hard water soap 硬水肥皂
hard water well 硬水井；苦水井
hard wax 硬蜡
hardway 难劈向
hardwearing 耐磨的
hardwearing coat 耐磨面层
hardwearing floor cover(ing) 耐磨地面面层
hard weather 恶劣天气
hard-welding 硬质合金熔焊
hard white coat 硬质白灰罩面
Hardwick conveyor loader head 哈德威克型运输机集尘装置；经济餐馆
hard-wire 高碳钢丝；硬(拔)线；实线
hard-wired 硬连接；电路的；线路固定的；硬连线；硬接线
hard-wired circuit 硬连线电路
hard-wired controller 硬连线控制器
hard-wired index 变址电路
hard-wired instruction 硬连线指令
hard-wired interconnection 硬连线互连
hard-wired logic 布线逻辑；硬连线逻辑；硬连线逻辑
hard wired logic(al) system 硬线逻辑系统
hard-wired numerical control 硬连线数控器；硬连接数控装置
hard-wired system 硬连线系统
hard-wire logic 硬(布)线逻辑
hard-wire-oriented engineer 硬连线系统工程师
hard-wire telemetry 有线遥测术
hard-won achievements 来之不易
hard wood 硬(质)木材；硬木；硬材；阔叶树林
hardwood basket weave parquet flooring 硬木席纹地板
hardwood bearing 硬木轴承
hardwood block flooring 硬木块楼地面
hardwood board 硬材板
hardwood charcoal 硬材木炭
hardwood creosote 硬木杂酚油
hardwood cutting 硬木插；硬材扦插；硬插条
hardwood cutting method 硬木插条法
hardwood decorative carveddoor 雕饰硬木木门
hardwood distillation 硬材干馏
hardwood dowel 硬木键
hardwood fender 硬木护(舷)木
hardwood finish 硬木装修；硬木表饰
hardwood floor block 阔叶材木地板块
hardwood floor(ing) 硬木地板
hardwood flour 纤维质木粉
hardwood forest 阔叶树林
hardwood furniture 硬木家具
hardwood guide 硬木导块
hardwood mosaic floor 硬木镶块地板；阔叶树林镶嵌地板
hardwood parquet flooring 硬木拼花地板
hardwood plywood 阔叶材胶合板；硬木胶合板
hardwood rail 硬材扶杆
hardwood rough board flooring 硬木毛地板
hardwood shear key 硬木剪力键
hardwood sleeper 硬木轨枕
hardwood straight edge 硬木直尺
hardwood strip 硬木板条
hardwood strip floor 硬木条形地板；硬木板条地板
hardwood strip flooring 硬木条形(企口)地板
hardwood tar 硬材焦油；硬木焦油沥青
hardwood tar pitch 硬材沥青；阔叶材焦油沥青；硬木焦油脂
hardwood tie 硬木轨枕
hardwood timber block 硬木垛墩
hardwood tongue-groove parquet floor strips 硬木拼花企口板
hardwood wedge 硬木楔
hard-wrought 冷加工；冷锻
hard X-ray 硬性(X)射线；高透力X射线
Hardy's thin cross-section(al) device 哈氏切片机
hardy annual plant 耐寒一年生植物
hard yarn package 硬卷装
Hardy Cross 哈地·克罗斯弯矩分配法
Hardy Cross method 哈地·克罗斯法【给】
hardy hole 葫芦孔；铁砧插模孔；锤柄孔
Hardy joint 哈代式弹性万向联轴器
Hardypick drifting machine 哈迪皮克型平巷电钻机组
Hardy plankton indicator 哈迪浮游生物指示器
hardy plant 耐寒植物
hardy stock 耐寒砧木
hardystonite 锌黄长石
hard zinc 硬锌
hard zone 硬化区
Hare's apparatus 黑尔仪器
hare's fur 兔毫釉
Hare's hydrometer 黑尔比重计
harebell 钩钟柳；经济餐馆
harem (伊斯兰教的)闺房；后宫
hargus 哈古斯锰钢
harie 毛墙面
harkerite 硼硅钙镁石；碳硼硅钙镁石
Hark formula 哈克公式
Harkins and Jura method 哈—焦气体吸附法(测粉末比表面)
Harkondale foot 澳洲木材体积标准(1HDF=1.27立方米)
Harkort test 哈可试验法
harl 冷雾；粗灰泥墙
Harlech dome 哈莱奇穹丘
Harlechian 哈莱奇阶【地】
Harlech series 哈莱奇统
harlequin opal 猫眼石
harling 干粘石或喷黏石的施工操作
harm 损害；伤害
H-armature 梭形电枢
harm component of ore 矿石有害组分
harmel 骆驼蓬
Harmet process 钢锭浇注法
harmful admixture 有害杂质
harmful agent 有害溶剂
harmful agents control 控制有害因素
harmful alagal bloom 有害藻类水华
harmful bacteria 有害细菌
harmful carbon deposition 有害的炭沉积
harmful chemic(al) monitoring 化学有害物监测
harmful component 有害组分
harmful constituent 有害组分
harmful district 有害坡段
harmful dust 有害粉尘
harmful effect 有害作用；有害影响；有害效应；不良影响
harmful element 有害元素
harmful gas 有害气体
harmful gas poisoning 有害气体中毒
harmful geologic(al) phenomenon displayed on image 影像显示不良地质现象
harmful habit 不良习惯
harmful impurity 有害杂质；有害混杂物
harmful ingredient 有害配料；有害成分
harmful insect 有害昆虫
harmful interference 有害干扰；危害性干扰
harmful marginal water 有害边缘水
harmful material 有害物(质)
harmful microelement 有害微粒元素
harmfulness 有害(性)；危害性
harmful pitch 有害树脂
harmful pollutant 有害污染物质
harmful radiation 有害辐射
harmful refuse 有害垃圾
harmful substance 有害物(质)
harmful trend 不良倾向
harmful variation 有害变异
harmful waste 有害物(质)；有害废物
harmful water 有害(的)水；有毒水
harmless 无害的
harmless carbon deposition 无害的炭沉积
harmless-depth theory 无害深度假说
harmless district 无害地段
harmless flaw detector 无损探伤术
harmless grade 无害坡度
harmless test 无损检验
harmodotron 电子束管
harm of immersion 浸没危害
Harmon hinge 旋转铰链
harmonic 谐音；谐波的
harmonic absorber 谐波吸收器
harmonic acceleration 谐波加速
harmonic aerial 谐波天线
harmonical 和谐的
harmonically balanced crankshaft 谐波平衡曲轴
harmonically conjugate points 调和共轭点
harmonically conjugate system 谐波共轭体系
harmonic vase 谐音容器；共鸣箱；共鸣罐；调和(花)瓶饰
harmonic amplifier 谐频放大器；谐波放大器
harmonic amplitude 谐(波)振幅
harmonic amplitude function 调和振幅函数
harmonic analysis 谐波分析；调和分析
harmonic analysis of tide 潮汐谐和分析；潮汐调和分析
harmonic analyzer 调和分析仪；调和分析机；谐波分析器
harmonic antenna 谐波天线
harmonic approximation 谐波近似
harmonic asynchronous torque 谐波异步转矩

harmonic attenuation 谐波衰减
harmonic average 调和平均(值);调和(平)均数
harmonic average index 调和平均指数
harmonic balance 谐波平衡
harmonic balancer 谐波平衡仪
harmonic bias 谐波偏置;谐波偏流;偏置谐波电流
harmonic blocking relay 谐波闭锁继电器
harmonic boundary 调和边界
harmonic bunching 谐波聚束
harmonic cam 谐和凸轮
harmonic cancellation 谐波消除
harmonic center 调和中心
harmonic circulating current 谐波环流
harmonic coal mining method 谐和地压采煤法
harmonic coefficient 谐波系数;调和系数
harmonic compensation 谐波补偿
harmonic complex 调和线丛
harmonic component 谐频分量;谐波分量;谐和(部分);调和分量
harmonic component generator 谐波分量发生器
harmonic compressor 谐波压缩器
harmonic configuration 调和构形
harmonic conjugate 调和共轭
harmonic conjugate line 调和共轭直线
harmonic conjugate rays 调和共轭射线
harmonic constant 谐和常数;谐波系数;谐波常数;调和常数
harmonic constituent 谐波分量;调和分潮
harmonic content 谐波含量
harmonic continuation 调和开拓
harmonic conversion 谐波变频
harmonic conversion transducer 谐波变换换能器
harmonic converter 谐波变频器;谐波变换器
harmonic coordinates 调和坐标
harmonic correction 谐波校正
harmonic correlation 调和相关
harmonic crystal 谐波型晶体
harmonic current 谐波电流;正弦电流
harmonic current constant 谐调潮流常数;周期性潮流常数
harmonic curve 谐波曲线;调和曲线
harmonic cycling 谐波振荡
harmonic decline 谐和衰减
harmonic detector 谐波指示器;谐波检波器
harmonic development 谐波展开
harmonic diagram 调和曲线图
harmonic dial 谐波标度盘
harmonic differential 调和微分
harmonic differential form 调和微分形式
harmonic dimension 调和维数
harmonic displacement 谐波位移
harmonic distortion 谐波失真;谐波畸变;振幅失真;非线性失真;非曲线畸变
harmonic distortion attenuation 谐波失真衰减
harmonic distortion factor 谐波失真因数
harmonic distortion level 谐波失真电平
harmonic distortion measuring set 谐波失真测量设备
harmonic distortion method 谐波畸变法
harmonic distribution 调和分布
harmonic disturbance 谐波扰动
harmonic disturbing force 谐波干扰力
harmonic division 调和分割
harmonic drive 谐和传动;谐波激励
harmonic echo 谐和回声;谐波回声;和谐回声
harmonic effect 谐波效应
harmonic element 谐和元素;调和元素
harmonic eliminator 谐波消除器
harmonic emission 谐波发射
harmonic equation 调和方程
harmonic excitation 谐波励磁;简谐激振
harmonic excited generator 谐波励磁发电机
harmonic exciting torque 谐波扰动力矩
harmonic excluder 谐波消除器
harmonic expansion 谐波(级数)展开;级数谐波展开;调和展开;傅立叶级数展开
harmonic expression 调和式
harmonic factor of inverter output voltage 变压器的输出谐波系数
harmonic fields 谐波场
harmonic filter 谐波滤波器;去谐滤波器
harmonic filtering circuit 谐波滤波电路
harmonic filtration 谐波过滤
harmonic flow 调和流

harmonic fold 谐和褶皱;协调褶皱
harmonic folding 谐和褶皱作用
harmonic force 谐和力;简谐力
harmonic forms 谐波形
harmonic formula 调谐公式
harmonic frequency 谐振频率;谐频;谐和频率;谐波频率
harmonic frequency characteristic 谐波频率特性
harmonic frequency converter 谐波频率转换器
harmonic frequency response 谐波频率响应
harmonic function 谐和函数;谐波函数;调和函数
harmonic gear 谐波齿轮
harmonic generation 谐波产生
harmonic generator 谐波发生器
harmonic group 谐波群
harmonic habitat 谐调生境;协调生境;和睦生境;融洽生境
harmonic index 谐波指数
harmonic index number 调和指数
harmonic integral 调和积分
harmonic interference 谐波干涉;谐波干扰
harmonic ion 谐振离子
harmonicity 调和性
harmonic lag 调和滞后
harmonic leakage reactance 谐波漏电抗
harmonic line (三次曲线的)调和线
harmonic linearization 谐波线性化
harmonic load 谐(和)荷载
harmonic loss 谐波损耗
harmonic majorant 调和强函数
harmonic mapping 调和映射
harmonic mean 调和中项;调和平均值;调和平均(数);调和均数
harmonic mean diameter 调和平均直径
harmonic measure 调和测度
harmonic membrane 弹簧膜压力计
harmonic method 调谐推潮法;调和法
harmonic mixer 谐波混频器
harmonic mode crystal 谐波型晶体
harmonic motion 谐(和)运动;谐波运动;正弦运动;简谐振动;简谐运动;和谐运动
harmonic motion cam 谐波运动凸轮
harmonic motion drive 谐波传动
harmonic motor 谐波电动机
harmonic mountain range 谐和山脉;协和山脉
harmonic number 谐波级次
harmonic of field errors 场不均匀性谐波
harmonic operation 谐波运行
harmonic order 谐波级次
harmonic order number 谐波阶数
harmonic oscillation 谐震;谐(和)振荡;谐和摆动;谐(波)振荡
harmonic oscillator 谐振子;谐(波)振荡器;正弦波发生器;简谐振子
harmonic output power 谐波输出功率
harmonic peak 谐振峰(值)
harmonic pencil 调和束
harmonic pencil of lines 调和线束
harmonic pencil of planes 调和面束
harmonic period 谐振周期
harmonic phase 谐波相位
harmonic point 谐和点;调和点
harmonic power 谐波功率
harmonic prediction 调和预报
harmonic prediction (潮汐的)调和推算
harmonic producer 谐波发生器
harmonic progression 谐波级数;调和数列;调和级数
harmonic proportion 调和比(例)
harmonic pulsation 谐波脉动
harmonic quantity 谐量;周期量
harmonic radar 谐波雷达
harmonic radiation 谐振辐射;谐波辐射
harmonic range of points 调和点列
harmonic ratio 调和比(例)
harmonic receiver 谐波接收器
harmonic reduction 谐波分析;调和分析
harmonic relaying 谐波继电保护
harmonic resonance 谐波共振
harmonic response 谐波响应
harmonic response characteristic 谐波响应特性(曲线)
harmonic response diagram 谐波响应图
harmonic response method 谐波响应法

harmonic ringing 选频振铃;调谐信号
harmonics 谐频;和声学
harmonic scale 调和音阶
harmonic section 调和分割
harmonic separation 调和分隔;调波间隔
harmonic sequence 调和序列
harmonic series 谐(波)级数;谐波系(列);调和数列;调和级数
harmonic series of sound 谐音系列
harmonic shutter 谐波抑制器
harmonic single-force point source 谐单力点源
harmonic source 谐波源;谐波发生器
harmonic speed changer 谐波减速装置
harmonic stop 谐音音栓
harmonic subflow 调和子流
harmonic suppression 谐波抑制
harmonic suppression network 谐波抑制网络
harmonic suppressor 谐波抑制器;谐波滤除电路
harmonic synthesis 谐波综合;谐波合成
harmonic synthesizer 谐波综合器;调和退算机
harmonic tensor field 调和张量场
harmonic test 谐波试验
harmonic tidal constant 潮汐调和常数
harmonic tide function 潮汐调和函数
harmonic tideplane 平均大潮低低潮(面)
harmonic tide prediction 调和潮候推算;潮汐调和预报
harmonic tolerance 谐波容限;非线性失真容限
harmonic tone 谐音
harmonic torque 谐波转矩
harmonic trap 谐波抑制器
harmonic tremor 谐震;谐和颤动(火山喷发时发生的地壳轻微持续的振动)
harmonic trend analysis 谐和趋势分析
harmonic tuning 谐波调谐
harmonic variation 调和变化
harmonic vector field 调和向量场
harmonic vibration 谐振;谐和振动;谐波振动;简谐振动;和谐振动
harmonic vibration-rotation band 简谐振转动谱带
harmonic vibrator 谐振子
harmonic voltage 谐波电压
harmonic wave 谐波;调和波
harmonic wave analyzer 调和波分析器
harmonic wave factor 谐波因数;谐波系数
harmonic waveguide 谐波波导管
harmonic winding 谐波绕组
harmonic wire projector 谐波定向天线
harmonious body lines 车身平滑线
harmonious equapotentiality 调和等能性
harmonious term 调和项
harmonization 谐调
harmonization of standard 标准的协调
harmonize 调准;调停
harmonized commodity description and coding system 商品统一分类和编码办法
harmonized standard 协调标准
harmonized system 调和关税制度
harmonizing colo(u)r 调和色
harmonograph 谐振记录器
harmony 谐和;谐调;协调性;和声(学);调和
harmony of alignment 线形协调;调和线向
harmony of economic returns and contribution to society and environmental protection 经济效益、社会效益和环境效益的统一
harmony of interest 利益协调
harmony of lines 线条协调;线条一致
harmony of population and land 人一地相称
harmony of scale 比例尺上的协调
harmony planting 调和绿化
harmophane 刚玉
harmotome 重十字沸石;交沸石
harmus 盖缝瓦片;瓦管接头盖
Harnack's first convergence theorem 哈拿克第一收敛定理
Harnage Shales 哈尔纳格页岩
harness 装具;捆束;吊带;导线系统;导线系列
harness an animal to a cart 套车
harness assembly 捆束组件
harness a well 压住井喷
harness building machine 编综机
harness hitch 攀踏结
harness horse 挽车马
harnessing 整治

harness mountain 山区治理
harness room 马具室
harness run 捆束主干
harnmodotron 毫米波振荡管
harntail 角尾
Haro-Herbig object 阿罗—赫比格天体
Haroltonian operator 哈密顿算子
harp 形似竖琴的物件；集电器滑轴夹；切泥弓；竖琴式管子结构加热炉
harp antenna 扇形射束天线；扇形(波束)天线
harp arrangement 平行弦式布置；竖琴式(布置)
harp configuration 竖琴形
harped tendons (偏离构件重心轴配置的)弯起钢丝束；偏斜钢丝束
harpes 抱握器
harpin(g)s 临时牵条；木桁船首向外张的最顶部的厚外板；船首部外侧腰板
harp longitudinal cable arrangement of cable-stayed bridge 斜拉桥竖琴形纵向布置
harp mesh 单向纱网；平行啮合
harp mesh cloth 单向纱筛布
harpolith 岩镰
harp-shaped shackle 竖琴形卸扣
harp-shaped stay cables bridge 竖琴形斜拉桥；竖琴式斜拉桥
harp-shaped shackle 联锚卸扣；联锚卸机
harp-type cable-stayed bridge 竖琴形斜拉桥；竖琴式斜拉桥；竖琴式斜缆桥；平行弦式斜拉桥；平行弦式斜缆桥
harp type screen 经丝筛；单向丝弦筛；竖琴式筛
harr 冷雾；门柱铰链；车铰链门柱
harranca 冰川或大陆架的大裂缝
harrier 推车工
Harrington's bronze 哈林顿耐摩轴承青铜
Harrington's bronze alloy 哈林顿青铜合金
Harrington's water-jet pump 哈林顿喷水抽气泵
"Harris, Ullman's multi-nuclei model" 哈里斯—乌尔曼复核心模式
harrisite 钙长橄榄岩；方辉铜矿
harrisitic texture 网纹斑杂状结构
Harrison cone 哈里森圆锥体
harrow 旋转式碎土机；路耙；料浆池耙子；耙(路机)；草耙
harrower 耙土机
harrowing 耙掘；耙地
harrow mark 耙痕
harrow plough 垂直小圆盘犁
harrow plow 垂直圆盘犁
harrow sole 耙底层
harrua 浓湿雾
harsh 粗糙的；不光滑的
harsh aggregate 粗集料；粗骨料
harsh arc 刚性弧
harsh clutch 刚性离合器；急剧接合器
harsh concrete 干硬性混凝土；粗(颗)粒混凝土；粗糙混凝土
harsh contract 苛刻的合同
harsh desert 荒漠
harsh environment 苛刻环境；恶劣环境
harsh grain 粗粒革
harsh image 强对比度图像；色调鲜明图像；粗糙影像
harsh mix(ture) (无和易性，不均匀的)粗糙搅拌混凝土；不良级配；干(硬)性混合料；干硬性拌和料；干性混合物；粗糙搅拌(料)；粗拌料；粗颗粒混凝土；粗糙混合料
harshness 轮胎耐(路面)粗糙性；粗糙性
harshness 粗糙度(混凝土)；干硬度(混凝土)
harsh running 不稳定运转
harsh sand 棱角砂；粗砂
harsh terms 苛刻(的)条件
harsh water environment 苛性水环境
harstigite 硅铍锰钙石；铍柱石
hart 毛墙面
Hart condenser 哈特冷凝器
Hartfell shales 哈特费尔页岩
Hartford No.28 press-blow machine 哈特福特28型成型机
Hartham Park (英国威尔特郡出产的)深奶色石
Hart impact tester 哈特式冲击强度试验机
hartite 晶蜡石；晶粒石
Hartley formula 哈特莱公式
Hartley oscillator 哈脱莱振荡器
Hartley principle 哈特利原理

Hartmann criterion 哈特曼判据
Hartmann diaphragm 哈特曼光阑
Hartmann dispersion formula 哈特曼色散公式
Hartmann flow 哈特曼流动
Hartmann formula 哈特曼公式
Hartmann lines 哈特曼线
Hartmann number 哈特曼数
Hartmann oscillator 哈特曼哨
Hartmann solution 哈特曼溶液
Hartmann test 哈特曼试验；哈特曼检验
Hartmann unit 哈特曼单位
Hartree equation 哈特里方程
Hartree method 哈特里方法
hartsalz 钾石盐；硬盐
Hartshill quartzite 哈特谢尔石英岩
Hartshone bridge 互感测量电桥
hartung 哈通钨钼钢
Hartwell Clay 哈特韦尔黏土
Harvard classification 哈佛分类法
Harvard miniature compaction test 哈佛小型压实试验
Harvard Standard Region 哈佛标准天区
Harvard twill 二上二下斜纹
harvest 收获；收成
harvestable fish 可捕捞鱼
harvest aid 落叶剂
harvest cooler 收获用冷藏装置
harvest cutting 主伐；成熟伐
harvest duck 收获机用输送器帆布带
harvester 收获机；采集装置
harvester ants 农田蚁类
harvester conveyer 收获机的输送器
harvester oil 农业机械用润滑油
harvesters 采集联合机
harvester stacker 收获堆垛机
harvester thresher 联合收获机；割脱两用机
harvester-tracker 堆垛机
harvesting 林产采收；收割；采运
harvesting apparatus 收获装置
harvesting dryer 割晒机
harvesting equipment 收获机具
harvesting mechanism 收获装置
harvesting strategy 捕捞策略
harvesting the ocean 海洋食物资源开发
harvest ladder 收获用附加栏板
harvest method 收获(测定)法
harvest of forest trees 采伐林木
harvest strategy 丰收策略
harvest table 长折面桌(供丰收时聚餐用)
harvest yield 收获量
harveyizing 甲板渗碳硬化
Harvey process 哈维法
Harvey steel 固体渗碳硬化钢
Harwich cement 哈威奇水泥(一种罗马水泥)
Harwit's imaging spectrometer 哈维特成像分光计
harzburgite 斜方辉石橄榄岩；斜方辉橄岩；方辉橄榄石
Harz jig 哈兹型活塞跳汰机
Hascrome 铬钼钢
hash 杂乱脉冲干扰；茅草干扰；无用数据；散列；电噪声
hashab 优质浅色阿拉伯树胶
hash class 散列表类
hash code 散列码
hash coding 无规则编码；随机编码；散列编码
hashed value 散列值
hasheesh 印度大麻
hashemite 铬重晶石
hash function 散列函数
hash house 经济餐馆；经济小吃店
hash index 散列索引
hashing 散列法
hashing addressing 相近寻址；相近选址
hashing method 散列法
hashing total 散列总和
hash mark 条痕
hash noise 干扰噪声；杂乱噪声
hash table 散列表
hash table bucket 散列表元
hash table entry 散列表项
hash table method 散列数据表法
hash total 混列总量；数位总和
hash type 散列型
Haskinizing 用防腐剂处理木材(一种专利方法)

has lamp 煤气灯
hasp 钩；铁扣；搭扣
hasp and staple 搭扣(和锁环)；铁钩和钩环
hasp iron 铁钩
hasp lock 狱门锁；搭扣锁；搭扣销
Hass 弯管内曲度；哈斯铝合金
Hassall's joint 哈索尔接头(一种排水管上用的双重沥青密封接头)
hassock 蒲团；拜垫；软钙质砂岩；草垫；草丛
hastarium 拍卖行
Hastelloy 耐盐酸镍基合金；哈司特莱镍(基)合金
Hastelloy electorde 哈司特合金焊条
hasting of maturity 加速成熟
hastingsite 绿钠闪石；绿钙闪石
Hastings sand 铁质砂岩；哈斯丁砂；哈斯丁沙滩(英国南部一个城市)
hastite 白硒钴矿
hasty map 速成地图
hasty mine field 快速布雷区
hasty profile 速测断面图
hasty road 简易公路
hasty soil survey 快速土质调查
hasty survey 快速测量；简易测量；速成测量；速测
hasty trench 简易堑壕
haswelite 毫威炸药
hat 随机编码
hat-and-coat hook 衣帽(挂)钩
hat band 帽圈
hat-block 帽模
hat box 污水阱；污水井
hat brush 帽刷
hatch 窗口；舱口；阴影线；升降口；下舱孔
hatch area 舱口面积
hatch awning 舱口雨篷；舱口天幕；舱口防雨罩
hatchback 有尾窗的；后活动顶盖汽车
hatch bar 舱盖压条；封舱闩条
hatch batten 舱围压条；舱口压条【船】
hatch batten clip 舱口楔耳
hatch battening arrangement 封舱设备
hatch battens 舱位压条
hatch beam 舱口(活动)梁
hatch beam carrier 舱口活动梁座
hatch beam shoe 舱口梁座
hatch beam sling 舱口梁吊索
hatch bearer 舱口盖承
hatch board 舱盖(板)
hatch boom 小关(双杆装卸舱口吊杆)；固定在舱口上空的吊杆；舱口吊杆
hatch canvas 盖舱帆布
hatch capacity 货舱容积；舱容(量)
hatch cargo list 分舱单
hatch carling 舱口纵梁；舱口纵桁
hatch carrier 舱口梁座
hatch checker 舱口理货员
hatch clamping batten 舱口压条【船】
hatch clamping beam 舱盖压条
hatch cleat 锁条；舱口楔耳
hatch coaming 舱口拦板
hatch companion 舱口天窗
hatch control position 舱口操纵部位
hatch conveyer 船用装卸输送机
hatch cover 舱口盖
hatchcover buffer 舱盖板缓冲器
hatchcover for refrigerated hold 冷藏舱盖
hatchcover gasket 舱盖板水密装置
hatchcover handling winch 舱盖开关绞车
hatchcover jacking device 舱盖板起升装置
hatchcoverless container ship 无舱盖集装箱船
hatchcover weather-tight sealing arrangement 舱盖板防浪装置
hatch crane 舱口起重机
hatch cross beam 舱口横围板；舱口横梁
hatch deck 舱口
hatch dog 舱口盖夹扣
hatch door 仓门；舱门
hatched area 打斜线部分；加影线部分；书有阴影线的面积
hatched drawing 细线条图
hatched fry 初孵鱼苗
hatched grafting 镶接
hatched knot 系紧的绳
hatched ornamental 绞纹装饰
hatch end beam 舱口端梁
hatch end coaming 舱口端围板

hatch end protection 木舱盖端护铁
hatcher 船舷门
hatchery 幼猪养殖场;孵化场
hatchery pond 孵化池
hatchery residue 孵化厂废品
hatchet 小斧(救生艇配置品);短柄(小)斧;带把斧;斧头
hatchet iron 斧形烙铁
hatchet planimeter 斧状面积仪
hatchet stake 曲铁桩砧
hatchet stone 手斧石
hatchettite 伟晶蜡石
hatch feeder 舱口装货器
hatch foreman 货舱装卸组长;货舱装卸领班
hatch for entry and exit 出入舱口
hatch grating 升降口花格;格栅式舱盖;格子舱口;人孔格栅
hatch hook 吊货钩
hatch hour 舱时;各舱装卸作业总时间
hatch-hour rate 舱时量
hatching 晕滃;影线(图);制图衬影;画影线;剖面线;打斜线;阴(影)线
hatching and grinning 应用于不规则的砖砌墙面
hatching apparatus 晕线描绘器
hatching effect 阴影效果
hatching house 孵化室
hatching pond 孵化池
hatching power 牵引力
hatching winch 绞车
hatchite 细硫砷铅矿;硫砷铊银铅矿
hatch ladder 舱口梯
hatch ledge bar 舱口盖承;舱口盖Z型承铁
hatch list 船舱装货明细表;舱(口)单;分舱货物明细表;分舱单
hatch list telegram 分舱货物电报
hatch locking bar 带销封舱门
hatchman 看关手;舱口装卸指挥
hatch mast 舱口起重杆
hatchment 盾形纹章(建筑饰面)
hatchminder 看关手;舱口装卸指挥
hatch money 船长资金
hatch mo(u)lding 舱口护缘
hatch number 舱口号码
hatch opening 舱口
hatch rest bar 舱口盖承
hatch rest section 舱口盖承
hatch ring 舱口环(系绳用);舱口盖板握环
hatch roller 舱口滚轮
hatch securing bar 压条
hatch side coaming 舱口围板【船】
hatch side girder 舱口边桁
hatch signal man 舱口装卸指挥;看关手
hatch socket 舱口梁座
hatch speed 舱口装卸速度
hatch stanchion 舱口支柱
hatch stiffener 舱口围板扶强材
hatch stopper 舱口边桁
hatch survey 货舱检验;货舱检查;舱口检验;舱口检视
hatch survey report 货舱检查报告;舱口检验报告;舱口检查报告
hatch tackle 舱口吊货绞辘
hatch tarpaulin 盖舱油布;盖舱帆布;舱盖布
hatch tender 看关手;舱口装卸指挥
hatch tent 舱口雨篷;舱口天幕;舱口防雨罩
hatchty 斧状的
hatchures 阴影线;囊状线
hatchway 升降口;下舱孔;闸门;平面出入口;天花板出入口;地窖口;舱口;舱井
hatchway beam 舱口活动梁
hatchway coaming 舱口围板【船】
hatchway cover 舱盖(板)
hatchway door 升降口门
hatchway door interlock 升降口门连锁装置
hatchway enclosure 升降口围栏
hatchway gate 升降口闸门
hatchway tonnage 舱口吨位(即舱口超额吨位)
hatch way trunk 舱口井
hatch web 舱口活动梁
hatch wedge 舱口楔;封舱(木)楔
hatch whip 舱口吊货绞辘
hatch window 舱口窗
hat felt 帽坯
hatfield time yield 短期蠕变试验准则;短期徐变试验准则
hat flange 帽形法兰;带帽法兰
hat-forming machine 帽坯机
hat front 帽盾
Hathaway oscillograph 十二回线示波器
hathet face 斧头面
hat hook 挂帽钩
Hathor capital 爱神柱头
Hathor head 哈托尔头像(古埃及女神)
Hathor-headed 爱神刻像的
Hathor headed capital 哈托尔头像柱头(古埃及女神)
Hathoric 有关哈托尔女神及其头像柱的
Hathoric column 爱神柱
Hathoric type of capital 爱神柱
Hathor mask 哈托尔面具(饰)
Hathor temple 哈托尔神庙(古埃及女神)
hat money 额外酬金
hat orifice 圆柱形锐孔;帽形孔板
hat packing 帽形轴封
hat pass 立轧梯形孔型;帽形孔型
hat peg 挂帽钩;帽钉
hat pin 挂帽钩
hat rack 帽架
hatrurite 哈硅钙石
Hatscek process 哈谢可法
hat seal 帽形密封
hat section 帽形截面
hat shape 初制帽坯
hat shield 帽盾
hat stand 帽架
hatted code 随机码
hatten 条板
hatter's plush 帽绒
Hatteras abyssal plain 哈特勒斯深海平原
Hattersley dobby 哈特斯莱复动式多臂机
hat tree 衣帽架
hat type 帽式
hat type foundation 帽式金属底盘(安装小型机器用)
hat type seal 帽形止水
hat washer 帽形圈
hauchecornite 硫锑铋镍矿
hauckite 羟碳铁镁锌矾
hauerite 褐硫锰矿;方硫锰矿
haugh 河边的冲积低地(苏格兰)
haughland 泛滥平原
haughtonite 富铁黑云母
haul 改变路线;改变航向;改变航线;改变方向;此处吊起;风向逆时针方向改变;运输;牵引;拖(运)
haulabout 煤(驳)船;供煤船;运煤驳船;(装有输送机的)加煤驳船
haulage 运输(量);运费;牵引量;拖曳费;输运;搬运费(用);搬运
haulage and ventilation drift 运输及通风平巷
haulage angle 拖运角;牵引角(皮带运输机上);运输角度
haulage appliance 拖运工具
haulage berm 运输平盘
haulage box 运输巷道装车溜井口;装车漏口
haulage business 搬运(事)业;运输(事)业
haulage by rail-mounted vehicles 铁道车辆运输
haulage by truck 卡车拖运
haulage cable 拖运缆索;牵引索
haulage capacity 牵引能力;拖运力
haulage capacity of locomotive 机车牵引能力
haulage car 运输车
haulage chain 牵引链
haulage clip 牵引夹
haulage company 搬运公司
haulage conveyer 采区中间运输机
haulage cross-cut 运输石门
haulage curve 运输平巷弯道
haulage distance 拖运距离
haulage drift 运输巷道;运输平巷
haulage drum 提升滚筒;牵引滚筒
haulage efficiency 牵引效率
haulage engine 卷扬机;绞车
haulage entry 运输平巷
haulage equipment 搬运设备
haulage facility 船台滑道拉曳设备
haulage fleet 运输车队;车队
haulage gantry 运输栈桥
haulage gear 牵引机构;牵引传动(设备)
haulage grade 运输道坡度
haulage heading 运输巷道;运输平巷;运输导洞
haulage horizon 运输水平
haulage level 运输水平
haulage-man 运输工
haulage means 运输工具
haulage motor 机车【铁】;牵引车;电机车
haulage of muck 岩渣托运
haulage plant 运输设备;牵引装置
haulage road 拖运道路
haulage roadway 运输巷道
haulage rope 运输钢丝绳;牵引索;牵引(钢丝)绳;拖缆
haulage speed 牵引速度
haulage stage 运输阶段
haulage system with endless pulling device 有循环拖拉装置的牵引体系
haulage tariff 公路运价表
haulage time 运输时间
haulage tractor 液引拖拉机
haulage truck 载货车
haulage tunnel 运输巷道;运输隧洞;运输隧道;运输平巷;运输平硐
haulage vehicle 运输容器;运输车辆
haulage way 运输巷道;运输(路)线;运输道路;运输坑道(连接地下开挖的出入坑道);煤矿中的运输巷道;主隧道
haulage winch 运输绞车;上坡机;绞盘;拖运绞车
haul away 启航;运输挖方法;出渣;出土
haul-away equipment 拖离设施;拖开设施
haul-away vehicle 拖去车辆
haul-away wagon 运送货车
haul back 拉线;拉回
haul-back block 回索滑轮;回空索滑车
haul-backline 回空索
haul block 导向滑车
haul cycle 行程;运输周期;运输循环
haul-cycle time 运输周期
haul distance 行程长度;run距;运程;动距
haul down 卸下;拉下
hauled element 牵引构件
hauled load 牵引质量;牵引荷载
hauled vehicle 被牵引车辆
hauled weight 运输重量
hauler 运输行;运输机;运输工;拉线;牵索;起动机;拖运机;拖曳者;承运人
haul forward 向前斜拉;风转向正横前
haul her wind 船首向上风靠近
haulier 运输行;运输工
haul in 引船进坞;拉进
haul in equipment 引船进坞设备
haul 收紧;拖运
hauling away 货车运货;运出
hauling cable 牵引索;牵引绳
hauling capacity 牵引力;牵引量;运输能力;牵引(能)力;拖运能力
hauling capstan 绞车
hauling chain 牵引链;拖链
hauling charges 运费
hauling container 运输容器
hauling container of concrete 混凝土运输车;混凝土运料斗
hauling contractor 运输承包者;运输承包商;运输承包人
hauling crushed rock 出渣
hauling cycle 运输周期
hauling distance 运距
hauling drum 牵引卷筒;牵引滚筒
hauling engine 卷扬机;牵运机;牵引机(车)
hauling entry 运输平巷
hauling equipment 运输设备;运输工具;牵引设备;拖运设备
hauling force 牵引力
hauling gallery 运输平巷
hauling line 引缆索;引缆绳;引火绳;撇缆;吊工具上桅的小绳
hauling machine 牵引机;运输机;拖运机
hauling-off anchor 退滩锚
hauling-off pull 脱浅拉力
hauling operation 运输作业;运输工作
hauling plant 运输设备
hauling power 牵引力;牵引力;牵引功率;拖力

hauling road 运料路
hauling rod universal joint bushing 拖杆万向节衬套
hauling rod wheel bushing 拖杆轮套
hauling rope 拖拉绳索；牵引索；牵引(钢丝)绳
hauling scraper 运土刮土机；牵引式刮运装置；牵引式铲运机；牵引刮土机；拖曳刮土机
hauling speed 牵引速度；拖运速度
hauling stress 搬运应力
hauling truck 载货车；牵引车；拖车
hauling unit 运土车；运输设备；运输工具；牵引车
hauling-up device 起重装置；起重设备；提升装置
hauling-up slip 修船滑道
hauling-up the slip(way) 拉上滑道；拉上船台
hauling vehicle 拖曳车
hauling velocity 牵引速度
hauling winch 绞车；牵引绞车；升降绞车；电引绞车
hauling wire rope 运输钢丝绳
haul in with 改向驶近
haul length 运距
haulm 盖顶稻草
haul manual 大修手册
haulm cutter 茎叶切碎器
haulm divider 分茎器
haulm hook 分茎刀
haulm pulling machine 拔茎机
haulm pulverizer 茎叶切碎机
haulm separator 除茎叶器
haul off 改向驶离；(螺杆的)引出装置
haul of fresh concrete 新拌混凝土的运送
haul road 运输道(路)；运料道路；运材道路；工地运料路；进出道
haul round 风向逐渐改变
hauls 捕获数
Haultain infrasizer 豪尔顿细粒淘析器
haul to wind 掉转向风；船首向风
haul up 使船头更近风向(帆船)
haul vehicle 运载工具
haul wagon 小型运输货车
haulway 运输路线
haulyard (旗帆等的)升降索
haul yardage 运渣方数；运土量；运土方数
haulyard shackle 升降索卡环
haunch 拱腋；梁腋；(道路的)厚边；榫腋脚
haunch arch 多心拱
haunch boards 加腋模板；梁腋板；混凝土梁侧模板
haunched 加腋的；起拱的
haunched arch 加腋拱；护拱
haunched beam 加腋梁；起拱梁；托臂梁；变截面梁
haunched board 加腋模板
haunched concrete 基脚上突起的混凝土
haunched floor 加腋楼板
haunched member 加腋构件；突起构件
haunched mortise-and-tenon joint 有榫腋脚的榫接头；加腋榫接
haunched slab 加腋板
haunched soffit 加腋拱腹
haunched tenon 加腋榫；腋脚凸榫；变截面榫
hauncheon 榫腋脚
haunching 加腋；加托臂；护拱；(道路上的)水泥混凝土镶边
haunching coefficient 加腋系数
haunching concrete 增厚混凝土
haunch of arch 拱的托臂；拱腋(托)；拱背圈；梁腋
haunch up 拱起
Hauner ratio 豪纳比值
Hausdorff maximal principle 豪斯多夫极大原理
Hauser's alloy 赫氏易熔合金
Hausler type roof cladding 豪斯勒屋面材料
hausmannite 黑锰矿
hausmannite ore 黑锰矿矿石
Hausner process 高频镀铬法
hautefeuillite 水钙磷镁石
Hauterivian 欧特里阶(早白垩世)【地】
haut-pass 部分抬高的地板(大厅内)；半梯台
hauyne 蓝方石
Havana dock 哈瓦那式浮船坞
Havana wood 哈瓦那杉木；西印度杉木
have a good command of 精通
have an accident 出事故
have and to hold (地契中的)持有条款的开头语
have a spell 接班
have holidays by turns 轮休
have jurisdiction over 管辖

haven 港；泊船处；避难所；安全地方
haven master 港务长
haven of refuge 避风港
haversack 帆布背包
Haversian canal 哈佛(氏)管
haversine 半正矢
haversine formula 半正矢公式
have several times the life 超保险期使用
have the same educational level 同等学历
have tiny funds and very small earnings 本小利微
have to 必须
have trade relations 通商
havgul 豪古尔风
having a moderate climate 气候温和
having dentils 具齿状的
Havoc 哈佛克硅钼钒钢
havoc 严重灾害；严重破坏；大破坏
Havre trough 哈佛海槽
Hawaiian activity 夏威夷式火山活动
Hawaiian Islands 夏威夷群岛
Hawaiian ridge 夏威夷海岭
Hawaiian type eruption 夏威夷式喷发
hawbill snips 鹰嘴钳；弯口钳
haw-haw 沉篱；隐蔽围墙；拦家畜用壕沟；暗墙；矮篱墙
hawi 河谷泛滥低地
hawiite 中长玄武岩
hawk 镘灰板；灰浆板；托灰板
hawk's eye 鹰眼石
hawkbell 鹰铃
hawkbill 焊钳；铁钳
hawkbill pliers 鹰嘴钳；弯头钳
hawkbill snips 曲刃剪；鹰嘴剪；弯口剪
hawk-boy 泥水匠助手
hawk eagle 鹰雕
Hawkers Act 小贩法例
Hawkins cell 铁镍电池
Hawk pug 豪克真空练泥机
hawksbeak 鸟啄装饰线脚；鸟嘴装饰线脚
Hawksley's formula 霍克斯累公式(用于计算水库浪高)
hawleyite 方硫镉矿
hawse 小峡谷；小埠；锚链孔所在的船首部分；锚孔；山口；船与锚之间水面部分
hawse bag 锚链筒下端塞(防浪软包)；锚链孔下端塞(防浪软包)；锚链孔塞包
hawse block 锚链筒上端塞
hawse bolster 锚链枕垫；锚链孔唇口
hawse boy 锚链筒下端塞(防浪软包)
hawse buckler 锚链孔盖
hawse flap 锚链孔盖
hawse full 在颠簸中的锚泊
hawse-hole 锚链孔
hawse hook 链孔肘板；尖蹼板；锚链孔肋板
hawse jackass 锚链筒下端塞(防浪软包)；锚链孔下端塞(防浪软包)
hawse piece 锚链孔肋材
hawse pipe 锚链筒；锚链孔
hawse pipe flange 锚链筒凸缘；锚链筒滚口
hawse plug 锚链筒上端塞
hawser 系船索；系泊缆(索)；钢丝索；缆索；缆(绳)；巨缆(周径在10英寸以上的)；大索(一般周长在6英寸以上)；大缆；粗绳；安全索
hawser apparatus 抛缆器
hawser bend 缆索捆绑结头；互穿两个半结
hawser board 拖缆托板
hawser clamp 缆索制止器
hawser cutter 钢索切割器
hawser fastening 粗索拴牢
hawser hole 带缆孔
hawser laid 正搓
hawser laid rope 正搓绳；普通大缆
hawse rope 解链引绳
hawser port 导缆孔；导缆板
hawser reel 卷索车；绳车
hawser rudder 拖绳应急舵
hawser thimble 大号钢索端嵌环
hawser winch 卷索绞车
hawsing beetle 扩缝锤
hawsing iron 捻缝凿
hawsing mallet 捻缝(木)槌
hawthorn 山楂
haxonite 陨碳铁矿；碳铁矿

hay 小额款项；干草
hay-and-straw elevator 干草与茎秆升运器
hay baling 干草打捆
hay band 草绳
hay band braider 干草带编织机
hay band spinning machine 打草绳机
hay belly 草腹
hay blower 干草吹送器
haybox (保暖用的)干草箱
haycap 干草堆顶覆盖物
hay carrier 干草运输拖车
hay chopper 干草切碎机
haycock 孤立锥形冰堆；浮冰块
hay cocker 干草堆垛机
hay cocking machine 干草堆垛机
haycockite 斜方硫铁铜矿
hay condenser 干草压捆机压实器
haycook 侧立冰
hay crimper 干草压扁碾折机
hay crop 牧草
haycruiser 自走式干草捡拾压捆机
hay crushing roll 干草压扁辊
hay cube 干草垫块
hay curing 晒制干草
hay cutter 铡草机
haydite 页岩陶粒；陶粒
haydite aggregated concrete 陶粒集料混凝土；陶粒骨料混凝土
haydite concrete 陶粒混凝土
haydite concrete wall panel 陶粒混凝土墙板
haydite type aggregate 陶粒集料；陶粒骨料
hay drying 人工制干草
hay dust 干草尘
hay elevator 干草升运机
Hayem's solution 阿杨氏溶液
hay equivalent 干草等价；干草当量
hayesite 水硼钙石
Hayes pegboard test 海斯栓钉入板测验
hayfield 干草地
hay filter 干草过滤器
Hayford-Bowie method of isostatic reduction 海福特—鲍伊均衡归算法
Hayford-Bullard method of isostatic reduction 海福特—布拉德均衡归算法
Hayford effect 海福特效应
Hayford ellipsoid 海福特椭球
Hayford gravity anomaly 海福特重力异常
Hayford spheroid 海福特椭球
hay fork 干草叉；草叉
hay grab 干草抓斗
hay grab with monorail trol(l)ey 单轨吊车式牧草抓爪
hay grinder 干草粉碎机
hay harvester 干草收获机
hay harvest(ing) 干草收割
hay holder 草捆止退板
hay knife 割(干)草刀
hayland 牧草种植地
haylaye 干草料
hay liner 自走式干草捡拾压捆机
hay loader 干草装车机；装载机；干草装载机
hayloft 干草架；秣仓
hay maker 晒草机；干草摊晒机
hay-making chute 干草铺条导向槽
hay-making equipment 干草收获制备机具
hay-making machinery 干草制备机械
hay-making range 制备干草成套机具
hay-making system 干草制备分系统
hay manger 干草槽
hay measuring 干草测量
haymow 干草堆；干草棚
hay mower 割草机
Hayne's alloys 哈氏钴铬钨合金
Haynes alloy 海钠合金
Haynes stellite 哈氏钨铬钴合金
hay packer 干草装填器
hay pelleter 干草制颗粒机；干草压饼机
hay pick-up 干草捡拾
hay pick-up-press 干草拣拾压捆机
hay pole 干草堆垛柱
hay press 压干草机；干草压缩机；干草压捆机；干草捆压机
hayrack 装载耙架；干草槽；雷达信标；导向式雷达指标台；带有传动装置的雷达指示器；草料槽；有

传动装置的雷达信标
hayrack boom loader 梯臂式装车机
hay rake 靠叉(塔上钻杆)
hay retainer 干草止退板
hayrick 干草垛
hay ricking machine 干草堆垛机
hay ridger 侧向搂草铺条机
hay road 乡村道路;农村道路
hay rope 草绳
hay scatterer 干草铺条撒布器
hay shed 干草棚
hay spreader 草条撒散机;翻草机
haystack boiler 草堆形锅炉
hay stacking cable 堆草用钢缆
haystack temperature probe 干草垛温度传感器
hay sweep 干草堆集器
hay tank 过滤罐
hay tedder 干草翻晒机;摊草机;翻草机
hay thermometer 测草温度计
Hayton floor 哈顿地板(一种玻璃弹簧跳舞地板)
hay value 干草等价
hay wafer 干草压块机
hay wafering machine 干草压饼机
Hayward grab bucket 海沃德抓斗
Hayward orange peel 海沃德四颗板抓斗
haywire 临时电线
haywire wiring 临时布线
hayyesenite 水硼钙石
hay yield 干草产量
hazard 冒险(性);危险;危害;风险
hazard analysis 危害(性)分析
hazard area 危险区(域)
hazard assessment 危害性估诂;危害评价
hazard associated with fire 火灾隐患
hazard beacon 障碍物信标;危险(警告)信标;濒危标志
hazard bonus 事故补偿金;公害津贴
hazard buoy 危险浮标
hazard classification 易燃等级;公害等级;火灾等级;危险分类
hazard classification system 危险性分类制度
hazard clean-up 消除隐患
hazard code 危险等级标号
hazard concentration limit 危险浓度极限
hazard content 建筑内危险物品
hazard control 危险控制
hazard control zone 危险控制区
hazard diamond mark 危险品金钢标志
hazard evaluation 危害性估计;危害评价
hazard forecast 灾害预报
hazard-forming stage 成险水位
hazard function 危害函数;风险函数
hazard geology 灾害地学
hazard identification 危害性识别;危害性鉴定
hazard index 危险符号;危害指数
hazard label 危险标记
hazard level 危险程度;危害度
hazard light beacon 危险灯标
hazard limit 危险限度
hazard mapping 绘制危险地区示意图
hazard marker 危险标志(牌)
hazard marking light 危险标灯
hazard materials 危险物品
hazard mitigation measure 减灾措施
hazard of aquatic organism 水生生物受害
hazard of content 建筑物内物品的危险性
hazard of crops 农作物受害
hazard of debris flow 泥石流的灾害
hazard of forest 森林受害
hazard of organism 生物受害
hazard of potential pollution 潜在污染危害
hazard of vibration 振动危害
hazard of well engineering 水井工程病害
hazardous 危险的
hazardous act 危险行为
hazardous air pollutant 有害(的)空气污染物;有害的大气污染物
hazardous and noxious substance 危险与有毒物质
hazardous area 危险品区;危险场所;危险区间;危险面积
hazardous atmosphere 有危险的空气
hazardous behavio(u)r 危险状态
hazardous bend 危险弯段

hazardous cargo 危险货物
hazardous cargo clause 危险品条款
hazardous cargo container yard 危险货物箱堆场
hazardous cargo house 危险品(仓)库
hazardous characteristic 危险特性
hazardous chemical data manual 危险化学品数据手册
hazardous chemical reaction 危险性化学反应
hazardous chemicals 有害化学物;化学有害物;危险性化学药品;危险的化学品
hazardous classification 危险性分类
hazardous commodity 危险商品
hazardous compound 有害化合物
hazardous concentration of gas 气体危险浓度
hazardous condition 危险状态;危险情况
hazardous earthquake 危险地震
hazardous element 有害元素
hazardous failure 危险性损坏
hazardous goods 危险品;危险(的)货物
hazardous goods terminal 危险品码头
hazardous health effects of water pollution 水污染对健康的影响
hazardous identification number 危险识别号
hazardous index 危险指数
hazardous industrial waste 工业有害废渣
hazardous industrial wastewater 工业有害废水
hazardous industry 危险工业
hazardous insurance 高率保险
hazardous location 危险路线;危险地段;危险场所;危险部位
hazardous marker 危险指示标
hazardous material 有害物(质);危险物料;危险材料;危害物
hazardous material disposal 有害三废处理
hazardous material incident 危险品事故
hazardous material management 危险品管理
hazardous material release 危险物品泄漏
hazardous material rescue 危险品的救援
hazardous material response team 危险品事故响应队
Hazardous Materials Advisory Council 危险品咨询理事会
hazardous materials team communication 危险品处理工作队通信
hazardous material storage 危险品(仓)库
hazardous material transportation 危险货物运输
Hazardous Material Transportation Act 危险品运输条例
hazardous matter 危险物质;危险物品
hazardous medical waste 危险医疗废物
hazardous noise 危害性噪声
hazardous occupation 具有危险性的职业;危险性职业
hazardous operation 危险(的)作业
hazardous organic chemicals 有害有机化合物
hazardous particle 有害微粒;有毒微粒
hazardous passage 危险水道
hazardous passage of boil-eddy type 泡漩险滩
hazardous passage of over-ledge-flow type 滑梁水险滩
hazardous passage of reef type 礁石险滩
hazardous passage of sharp bend pattern 急弯型险滩
hazardous passage of sharp bend type 急弯险滩
hazardous production material 危险性生产材料
hazardous rapids 险滩
hazardous rating 危害等级;防爆等级
hazardous reaction 危险反应
hazardous rock 危石
hazardous roof 危险顶板
hazardous roofing materials 危害性屋面材料
hazardous shoulder 危险路肩
hazardous site 危险场所
hazardous site and spot 危险场所和地点
hazardous solid waste 危险固体废物
hazardous spill 危险溢出物
hazardous storage 危险品储藏室
hazardous substance 有害性物质;危险物质;危险物品
hazardous substance mark 危险物品包装标贴
hazardous top 危险顶板
hazardous waste 有害垃圾;有害废液;危险废物;危险废(弃)物
hazardous waste disposal 有害废物处理

hazardous waste management 有害废物管理;危险废物管理
hazardous waste manifest system 危险废料表示系统
hazard pay 危险津贴
hazard point 危险点
hazard potential 潜在危害能力;危害潜力;危害可能性
hazard prevention 防灾
hazard rate 冒险率;危害率;风险率
hazard rating 危害分级
hazard reduction 减少火险
hazard report 危害报告书
hazard-resistant building 防灾建筑;耐震建筑物;防险建筑物
hazards 危险灾害
hazard segregation 危害隔绝法
hazards of electromagnetic radiation 电磁辐射危害
hazards of new construction 新建工程的危险
hazards of pollutant 污染物的危害
hazard threshold 危险阈值
hazard to aquatic organism 水生生物受害
hazard to crops 农作物受害
hazard to forest 森林受害
hazard to organism 生物受害
hazard variation 危险性变化
hazard warning light 危险警告灯
hazard way light 危险信号闪光
hazchem 危险化学品标签系统
hazchem code 化学危险品标志符号
hazchem protective suit 危险化学品防护服
haze 光雾;模糊度;雾斑;薄雾
haze and fog 雾霾
haze dome 气室
haze factor 霾因子
haze filter 消雾滤光器;去雾滤光片
haze finish 蒙雾修饰;表面朦雾处理
haze horizon 霾层顶
hazel 页岩质砂岩;浅棕色
haze layer 霾层
hazel earth 砂黏土
haze line 霾线
hazelnut oil 榛子油
hazel pine 苏合香木材
hazel sandstone 淡褐色砂岩
hazel wattle hurdle 榛树枝条篱栏
hazemeter 烟柱测距仪;能见度测量仪;混浊度测量仪;大气透射计;薄膜混浊度测量仪
Hazen's equation 黑曾方程
Hazen's formula 黑曾公式(经验频率绘点位置公式)
Hazen's law 黑曾定律(估算土壤渗透系数用)
Hazen's method 黑曾法(水文频率计算)
Hazen's number 黑曾数;黑曾色值
Hazen's uniformity coefficient 黑曾均匀度系数
Hazen-colo(u)rimeter 黑曾比色计
Hazen probability graph paper 黑曾几率格纸
Hazen-Williams diagram 黑曾—威廉斯曲线图
Hazen-Williams equation 黑曾—威廉斯方程
Hazen-Williams formula 黑曾—威廉斯公式
Hazen-Williams rough coefficient 黑曾—威廉斯粗糙系数
Hazen-Williams roughness coefficient 黑曾—威廉斯粗糙度系数
hazeometer 漆膜混浊度测量仪
haze penetration 雾穿透
haze surface finish 蒙雾表面修饰
haze tallow 日本蜡;漆脂;皮油
haze test 蒙雾试验;表面雾化试验
haze value 漆膜光雾值
haze wax 皮油
haziness 模糊度;混浊性
hazing 混浊;清漆轻度发浑;起雾
hazle 砂页岩
hazmat management 危险品管理
haz-mat unit 处理危险物品的车辆
hazy 有薄雾的;烟雾弥漫的;多烟雾的
hazy atmosphere 被烟雾污染的大气
hazy picture 模糊图像
hazy weather 烟雾天气
H-bar 宽缘工字钢;宽翼工字钢
H-bar control 水平条控制
H-beam 宽翼(缘)工字梁;H 字梁;工字梁
H-beam bunton 工字梁(横)撑;工字钢罐道梁

H-beam cap 工字钢顶梁
H-beam grillage 工字梁格排
H-beam pile H 形桩;工字桩;工字(形)桩;工字钢桩;宽缘工字钢桩;宽翼(缘)工字桩
H-beam section 工字钢梁
H-beam steel 工字梁钢
H-beam web 工字梁腹板
H-block 工字砌块;调整间隙垫板
H-broad beam 宽缘工字梁
H-clay 氢黏土
H-column 工形柱;工字(形)柱;宽缘工字柱;宽翼缘工字杆;宽翼工字柱
H-column with covers 宽翼缘加板工字形钢柱
H-display 分叉点显示
He abundance 氦丰度
head 固有头;高岬;刻纹头;前部;顶端;船首;标题
head access 磁头存取
head access window 磁头存取窗
headache 头痛;安全帽
headache ball 破碎球;撞球(拆除建筑物用);拆房重锤
headache post 摇臂下保险立架(钢绳冲击钻机)
headache rack 安全保险架(俚语)
head amplifier 前置放大器;前级视频放大器
head anchor winch 艏锚绞车
head and chest set 戴挂式电话机
head and front 紧要部分
head and shell opening 封头和壳体开孔
head and stern leading marks 首尾导标
head and stern mooring 船首尾系泊
head arm 磁头臂
head assembly 磁头组件
head attachment 柱头构造;头部连接(法);头部附件
head azimuth 磁头方位角
head ball 重力冲击球
head band 拱ібіл门;拱石圈饰;装饰顶带;头饰带;头环
head banding 磁头条带效应
head basin 上游池
head bay 渠首;前池;上闸首;上游河段;上游池;船闸上游河段
head bay floor 上游池底板
head bay of lock 船闸前首;船闸前港
head bay wall 上游墙
head beach 岬角滩
head beam 顶梁;露梁
head bearing 止推轴承;上轴承
head block 辙尖枕木;挡推力木块;天车;上滑车;吊货滑车;垫块;导滑车;制动轮;前副垫块
head block tie 辙尖枕木;尖轨长枕
head board 系兽板;床头板;护顶木板;端面板;顶板横梁副卡楔;横架;推出板;立式配电盘
head bond 顶层接合
head book 主要账簿
head box 压力盒;流料箱;绝热帽;压头箱
head breadth 头宽
head bridge 高架桥
head bulk depot 首站
head bulk plant 首站
head cap 头帽;(精装书的)书脊护舌
head-capacity curve 压头排量曲线;水头—容量曲线
head casing 门窗头线条板;门额;门窗头框;门头线(条板)
head-center 破碎机头中轴(圆锥破碎机)
head chainman 前点测工
head checks (in rail) 钢轨头部龟裂
head chip 磁头工作间隙
head clamping lever 进刀夹紧固手柄
head cleaner 磁头清洁剂
head clearance 头顶净空;头顶空隙;顶部净空
head clogging 磁头堵塞
head coefficient 压头系数
head conduit 压力水管;压力(输送)
head cone 龙头内锥
head construction 柱头构造;头部连接(法)
head contour 封头外形
head-control gate 标头控制闸门
head controlling gate 水头控制闸门
head core 磁头芯
head corner of rail 钢轨踏面角
head count 人口普查
head counterbore 顶头扩孔(钻);顶头锥口孔
head cover 端盖;顶盖

head crash 磁头碰撞;磁头划碰
head crescent 带隔板端盖(内啮合齿轮泵)
head cup 铆钉机顶具
head current 急流头部
head cut 沟头切割
head cut in gully 沟头冲蚀
head cylinder 磁头柱体
head demagnetizer 磁头消磁器
head depot 首站
head design 封头设计
head difference 压头差;水头差
head difference between neighbouring drawdown 相邻落程的水头差
head differential 压头差;水头差
head-discharge relation 水头流量关系
head-discharge relation curve 水头流量关系曲线
head distribution 水头分布
head ditch 进水渠;斗渠
head-drag coefficient 头部阻力系数
head drain 地头沟
head driller 顶部钻孔机
head drive 机头传动
head driving 巷道掘进(工程);开凿导洞;导坑掘进(工程);导洞掘进(工程)
head drum 磁头鼓
head duration curve 水头历时曲线
headed bolt 撑帽式锚杆;露头螺栓;撑帽式杆柱
headed drip 铁皮泛水圆形卷边
headed test specimen 突头试件
head effect 对地电容效应
head end 装料端;机头;机车;头端;头部;上游端;端载端;端头;收发中心
headend equipment 前端设备
headend ignition 前端发火;前端点火
headend of hump yard 驼峰编组场头部
headend operation 掉头(车辆)
headend pressure 活塞头压力
headend revenue 邮政行旅车收益
headend system 机车端供电系统
headend type tractor-loader 前端型拖拉装载车
headend unit 头端单元
head equation 水头方程(式)
header 管箱;露头;露头石;联箱;掘进护板;集管;帽木;活挡头;横楞木;平巷掘进机;头沟;头部;首部(结构);端板(柱);顶盖;丁砖;丁砌;标题;报头;半端梁格栅
header and stretcher 横直砌合
header area 标题区
head eraser 磁头消磁器
header assembly 管座装置
header band 端横带
header bar 顶端钢筋
header bland 镦锻坯料
header block 露头砌块;丁头接缝砌块;标题字组;顶头接缝砌块
header bolt 冷锻螺栓
header bond 满丁砌合;全丁砌合(法);顶砖砌法;丁砖砌体;丁砌砌合法;丁砌合(法);丁砖层
header box 卸粮拖车
header brick 丁(头)砖;露头砖;顶砖
header buffer 首部缓冲器
header card 首(号)卡;首标卡;标题卡(片)
header cell 标题单元
header channel 进水槽
header clay brick 全丁砖
header course 全丁砖皮(层);丁头行;露头层;满丁砌行;丁砌层;顶砌层;丁砖层;丁头层;丁砌砌层
headere 露头砖;露头石;掘进护板;集管;顶梁;丁砖
header entry 标题项目
header error control 信头差错控制
header face 端面
header field 标题字段
header for cleaned air 净化空气的集管
header fork 叉式摘穗器
header format 标题格式
header-high 到第一全丁砖层的高度
header-index 标题索引
header information 标头资料
header joint 丁砖缝;砖砌合;满丁砌合;丁砖砌合
header joist 过梁格栅(支承短格栅的梁);格栅端头大梁;活挡头;横楞木;丁头格栅
header label 首标;首部标签
header line 主要管线(指水管);表头线;标题行
header list 标题表

header maker 镦锻机制造厂
header message 首部信息;标题消息
header of the gin pole 钻塔上支架水平横梁
header of thermal medium inlet 热载体入口总管
head erosion 沟头侵蚀;头侵蚀
header pipe 总管(气、水等);集(气)管;送风总管
header record 标题记录
header segment 标题段
header sheet 标题表
header statement 标题语句
header stone 丁(头)石;露头石
header-subcode 索引副码
header support 总管管架;集箱支座
header table 标题表
header tank 高位槽
header tank of pumping house 泵房出水池
header-terminal capacitance 顶端电容
header tile 露头瓷砖;丁砖;带槽面砖
header type boiler 直管锅炉
header type feed heater 联箱式给水加热器
header word 标题字
head face 端面
head facility 首站设施
head fall 压力降落;纵向倾斜;纵向坡度;水头落差;水头降落;端面;跌水
head fast 头缆;艏(系)缆;船首(系)缆;船首系索
head field 磁头场
head file 剑形锉
head fire 顺风火
head flag 起始标记;磁头标记
head flashing 屋顶凸起物周边的泛水;顶边;突出部分防雨板;屋顶边泛水;头顶防水板
head flow 脉动式喷出
head flow chamber 闸道流量计
head flowmeter 压头式流量计
head flume 头部水槽;导流槽
head for 船向某地航行
head for boring bar 镗孔头
head fork weldment 顶叉焊接件
headform 头模
head formed by spinning 旋压封头
head-form test 人头模型试验
head fraction 头馏分;初馏物
headframe 支架架;井架(浅井、竖井);隧洞进洞支撑构架;顶架
headframe hoist 塔式提升机
headframe of shaft 竖井(的)顶架;竖井井架
headframe pulley 井架天轮
head friction 水头摩擦
head gap 磁头空隙;磁头间隙;磁头缝隙
head gasket 盖垫密片
head gate 引水闸门;总水(闸)门;进水闸(门);渠首闸门;首部闸门;上闸闸门
head-gate duty of water 毛灌水率;首部闸门灌水率;引水闸灌溉率;毛灌水定额;渠首闸门灌水率
head gear 井塔;井架;船首帆缆装置;安全帽;帽子;头饰;头盔;钻塔
headgear bin 井口矿仓
head grain 垂直于纹理方向
head groove 头部槽
head guard 墙挡板;(墙洞上边防止木材腐蚀的)铅制排水沟;门窗顶泛水;门窗头泛水
head-hardened rail 头部热处理轨
head harness 装带
head heave 导滑车
head height 头高
head helmet 头兜
head-house 工作楼;闸门控制室(渠首等);候车室;顶架室;井口建筑(物);隧道洞口棚架
head housing 磁头罩
head in bend 弯头水头
head increaser 水头增流器
heading 预锻;引水点;露天砖石;坑道;进向;浇口布置法;镦头;航向;前进方向;平巷;桶盖桶底材;题目;艏向;首标;上端沉淀精矿;镦头;导坑;导洞;船首(方)向;表头(标题);标题;标目;报头首部;报表表首
heading advance 巷道掘进;平巷掘进;导坑掘进;导洞掘进
heading and bench blasting 梯段(超前)掘进爆破(法);台阶工作面爆破(法)
heading and benching 上部半断面超前台阶施工法
heading and bench method 巷道梯段开挖法;巷道梯段掘进法;短台阶法;导坑巷道梯段开挖法

导坑巷道梯段掘进法;导洞梯段开挖法;导洞梯段掘进法;正台阶采矿法
heading and bench tunnel(l)ing method 导洞与层阶法
heading and bending blasting 梯段掘进爆破
heading and cut 正台段掘进
heading and footing 加标题和页码;标题和页码
heading and stope mining 正台阶回采
heading and stope system 正台阶回采法
heading angle 航向改变角
heading area 标题区
heading-area coating 标题区涂层
heading-back 截短枝条
heading beam 顶梁
heading blast(ing) 坑道爆破;工作面爆破;掘进爆破;峒室爆破;梯段爆破
heading blasting charge 峒室装药;工作面爆破装药
heading bond 全丁砖砌合;一砖(厚的)墙;满丁砌合;丁砖砌合
heading card 方位分度盘
heading character 标题字符
heading chisel 樺凿
heading code 识别码
heading collar 巷道口;横盖木;平巷口;顶木;导坑口;导洞口
heading control loop 方向控制系统
heading course 一皮丁砖;露头砖层;满丁砖行;丁砖层;丁砌砌层
heading crew 为首班组;领班船员
heading device 航向仪
heading die 粗镦模
heading driver 掘进工;导洞掘进机;导洞掘进工
heading driving 导洞掘进
heading face 掌子面;工作面前壁;开挖面;掘进工作面
heading flash 船首线标志;船首闪灯;船首标志线
heading for advancing 掘进巷道;掘进平巷;掘进导坑;掘进导洞
heading geologic(al) control 采准地质指导
heading gyroscope 首向陀螺仪
heading index 首向标线
heading indication 首向指示
heading indicator 航向指示器
heading joint 直角(接)合;两块板的对接线;横缝;端缝;端(接)合;顶头接
heading light 前灯;头灯
heading line 艏向线;首向指示;船首线标志;方位线;标题(行)
heading line off 船首标志线消失
heading machine 压头机;导坑掘进机;封头机
heading marker 航向指示器;航向标志器;船首(线)标志;船首标志(线)
heading marker alignment 船首标志调准
heading marker contact 船首标志接触器
heading marker off 船首标志线消失
heading message 标题信息
heading method of tunneling 导洞法掘进隧道;导洞法隧道掘进
heading monitor 艏向监测器
heading of moving vehicle 运动体航向
heading overhand bench 下梯段超前掘进
heading overhand bench method 超前台阶式掘进法
heading per standard compass 按标准罗航向
heading per steering compass 按操舵罗航向
heading printing 标题印刷
heading prop 导坑支柱
heading record 标题记录
heading selector 航向选择器
heading side 下盘;底盘
heading slope 仰坡
heading-stage 抽穗阶段
heading statement 标题语句
heading stone 丁头石
heading stope 采矿工作面
heading the sea 迎浪行驶
heading tool 头部镦锻工具;钉头型模
heading-to-steer mark 操舵首向标志
heading up 壅水;抬高水位
heading upward plan position indicator 相对方位平面位置指示器;航向平面位置显示器;船首线向上的平面位置指示器
heading upward presentation 相对方位显示;船首线向上的显示

heading ventilation 巷道面通风;掌子面通风;工作面通风;掘进通风;掘进工作面的通风
heading volume 标题卷宗
heading wall 矿脉下盘
heading work 拱顶石饰(工);掘进工作
heading working 拱顶石饰细工
head in parking 垂直停车
head installation 首站设施
head into shell 封头接入壳体
head intrusion 磁头插入度
head jamb 框架顶木;门帽;门顶框;门(樘)上槛;门上框
head joint 砌体竖缝;横缝;端缝;垂直灰缝;竖(砖)缝;端部接缝;直角接合;端接
head lagging 顶板背板
head lamp 前照灯;头灯;车头灯;额灯;照明灯
head lamp adjusting screw 前灯调节螺钉;前照灯调节螺钉
head lamp body 前灯壳;前照灯壳
head lamp bracket 前灯支架;前照灯支架
head lamp bright viewing distance 前照灯明视距
head lamp brush guard 前灯栅罩;前照灯栅罩
head lamp bulb 头灯灯泡;前照灯泡
head lamp case 头灯罩;前照灯罩
head lamp casing 头灯壳;前照灯壳
head lamp connector 前灯接头;前照灯接头
head lamp deflector 头灯回光罩;前照灯回光罩
head lamp dimmer 前灯减光灯;前照灯减光灯
head lamp door 头灯框;前照灯框
head lamp door latch 头灯框闩;前照灯框闩
head lamp door rocket 头灯框座;前照灯框座
head lamp door spring 头灯框闭锁弹簧;前照灯框闭锁弹簧
head lamp driving bulb 头灯近光灯泡;前照灯近光灯泡
head lamp holder 头灯架;前照灯框闭锁弹簧
head lamp housing 头灯壳;前照灯壳
head lamp lens 头灯玻璃;前照灯玻璃
head lamp mask 前灯遮光;前照灯遮光
head lamp of locomotive 机车头灯
head lamp parking bulb 近光灯泡(汽车)
head lamp reflector 前灯反射罩;前照灯反射罩
head lamp support 前照灯头灯架
head lamp swivel 头灯活节;前照灯活节
head lamp tester 头灯试验器;前照灯试验器
head lamp upper beam 前灯上光线;前照灯上光线
head lamp with anti-vibration device 带减振器头灯
headland 河源;陆岬;岬;海角;前沿地
headland head 陡岬
headland light 陡岬灯塔
head lap 端接接;木瓦接接;(屋面板之间的)搭接长度;搭接板上下边沿的最短距离;搭接(重叠)部分
head lead 水砣
head lease 开头租赁
headledge 舱口端围板
head length 头长
headless 无头的
headless bolt for welding 焊接单头螺栓
headless nail 无头钉
headless package 无边盘卷装
headless plug 无头塞
headless rivet 无头铆钉
headless screw 无头螺钉
headless set screw 无头定位螺栓
headless shoulder screw 无头轴肩螺钉
headless slotted plug 无头槽塞
headless system 直送方式【机】
head letter 标题字
headlight 樘;前樘灯;车头灯;照明灯;前照灯;头灯;大灯;前照灯(汽车)
headlight adapter 前照灯接头;前照灯适配器
headlight beam indicator 头灯光度指示器
headlight body 头灯壳;前照灯壳
headlight bracket 头灯架;前照灯架
headlight bulb 头灯灯泡;前照灯泡
headlight bulb mounting seat 头灯灯泡座;前照灯泡座
headlight bulb retainer ring 头灯灯泡护圈;前照灯泡护圈
headlight cable 头灯电线;前照灯电线
headlight cable terminal block 头灯接线板;前照灯接线板

headlight casing 头灯壳;前照灯壳
headlight casing stand 头灯壳座;前照灯壳座
headlight clip 前照灯卡
headlight control 头灯控制;前照灯控制
headlight dazzle 前照灯眩光(汽车)
headlight dimmer 前照灯调光器;前照灯变光器;大灯调光器;大灯变光器;前灯减光器
headlight door 前灯盖;头灯框
headlight glare 前照灯眩光;前照灯闪光
headlight guard 头灯护架;前照灯护架
headlight high beam filament 远光灯丝
headlight hood 前照灯护罩
headlighting 前照灯照明
headlight lens 前照灯玻璃
headlight lens locator 前照灯玻璃固定器
headlight mo(u)lding 头灯框;前照灯框
headlight oil 头光油
headlight reflector 头灯反射镜;前照灯反射镜
headlight reflector seal 头灯反射镜封垫;前照灯反射镜封垫
headlight shell 头灯壳;前照灯壳
headlight sight distance 夜间视距;头灯视距;前照灯视距
headlight socket 头灯座;前照灯座
headlight step 头灯级;前照灯级
headlight switch 头灯开关;前照灯开关
headlight switch knob 头灯开关按钮;前照灯开关按钮
headlight turbine 前照灯汽轮机
headlight visor 头灯防眩罩;前照灯防眩罩
headlight wiring conduit 前灯导线管;前照灯导线管
headline 压缩头线;头缆;艏缆;首缆;首页标题;船首缆
headline main winch 船首主锚绞车
headline of bucket loader 链斗挖泥船艏锚缆
headliner 磨头衬板;池炉前墙内衬;池炉端墙内衬;标题制作者
head line scow 艏缆方驳
head load 磁头装入;磁头加载
head loading mechanism 磁头加载机构
head load pad 磁头加载垫
head load solenoid 磁头加载螺线管
head load system 磁头加载系统
head-lock tile 头部交锁砖
head log 前缘木;设置在滑道前端底部的圆木
head loss 扬程损失;压头损失;落差损失;水头损失
head loss due to enlargement 扩大段水头损失
head loss in bend 弯头水头损失;弯管水头损失;弯段水头损失
head loss indicator 水头损失指示器
head loss of discharge pipeline 排泥管线水头损失
head loss of pipeline 管线水头损失;管道水头损失
head loss of suction pipe 吸泥管水头损失
head loss recorder 水头损失记录仪
headman 监工;工头;工长;斜井口把钩工;运煤工
head marker 磁头标志
head mast 前柱;柱顶;缆索塔顶;前桅;桅顶;吊塔
head metal 冒口
head meter 水柱压测量仪;水头计;压差流量计;落差流量计
headmeter 压力计;落差流量计
head misalignment loss 压头偏差损耗
head money 人头税
Headmore 黑德莫尔铬钒钢
head mortgage 开头抵押
headmost 队头的
head motion 摇动机构;机头运动;传动机头
head motor 头部发动机
head mo(u)ld(ing) 门窗额饰;门窗线条;门窗上面的线脚;墙孔上部的线脚;拱圈线脚;敝口线脚;过梁线条;拱石圈线条;(孔口的)顶部装饰线脚
head nail 头排钉;钉头
head nailing 挂瓦钉;端头钉;前部钉法;瓦头钉法;上部钉法;钉端部
head nurse's room 护士长室
head of a family 当家人
head of a groin 防波堤端头;丁坝端头
head of anchor 锚冠
head of approach 行进水头;行近水头
head of a river 源流;河源;水源
head of a well 井顶
head of axe 斧背
head of bay 湾顶

head of canal 渠首
head of casing 窗框上槛
head of column 柱头
head of culvert 涵洞端墙;涵洞头
head of cutter 刀头
head of cylinder 汽缸头
head of delivery 扬程;压头
head of diving team 潜水长
head of drain 排水系统中的最高点;排水渠水头;排水渠首;排水渠落差;排水孔口;排水管端;排水沟头
head of dredge pump 泥泵扬程
head of expenditures 支出项目
head off 开头;(船头先离泊位)
head office 总局,总机构;总管理处;总公司;总店;总部;总办事处;指挥部;厂部办公室;本部
head office cost 总部费
head office of railway material 铁路物资总公司
head office organization 承包者的公司组织;承包商的公司组织;承包人的公司组织
head office overheads 总公司经费
head office staff 总公司工作人员
head-of-fleet 驳船队前端
head of frame 上槛
head of glass 玻璃液压头
head of groyne 防波堤端头;丁坝端头;丁坝坝头
head of hump 驼峰编组场头部
head of ingot 锭头
head of ladder 梯端
head of liquid 液位差;液柱压力;水柱高度
head of mast 杆塔顶部
head of mill 磨头
head of navigation 通航终点
head of nozzle 喷嘴头
head of oil fuel 燃油压头
head of pile 桩头
head of pressure 水头;压头
head of pump 泵压头
head of reservoir 库尾
head of river 源流;河源
head of spur-dike 丁坝坝头
head of the clay room 坯房头
head of the department of medical administration 医务部主任
head of the mast 电杆顶部
head of the queue 队首
head of the reactor vessel 反应堆外壳的顶盖
head of tide water 感潮区上游段
head of tripod 三角架头
head of water 水柱高度;水位差;水头
head of water over weir 堰顶水头
head of water to be retained 挡水高度
head on 迎头对遇;对遇
head on approach 顶头靠近【船】;顶头靠泊【船】
Headon beds 赫登层【地】
head-on collision 正面冲突;迎头撞击;迎面碰撞;直接碰撞;正碰;两船对撞;对撞
head-on course 迎头航线
head-on cross-section 正截面
head-on impact 迎头碰撞;正冲
head-on pump 水泵压头
head-on radiation 迎面辐射;直接定向辐射;正面辐射
head-on the spillway 溢流堰顶水头
head-on view 观测落差;前视图;正视图
head-on wind 逆风;迎面风
head opening 封头开孔
head oriented equation 按水头列的方程
head oscillation curve 水头波动曲线;水头摆动曲线
head over weir 堰顶水头
head packing 柱帽垫;头部填密
head phone 头戴式耳机
head phone 记录电话;头戴受话机;收发话器
headphone unit 头戴式收音器
head piece 横梁;顶梁;流口;木隔断的压顶木;顶部物;井口油水分离器
head pin 烧针
headpin 头柱
head pivot point 头的转动中心
head plate 木骨架顶板;顶板;封头板;木骨架顶头木条
head plate of stanchion 支柱顶板
head pond 前池;上游池
head pool 上游池;河源地
head position 领头位置

head positioner 磁头定位器
head positioner logic 磁头定位器逻辑
head positioning mechanism 磁头定位机构
head positioning time 磁头定位时间
head post 畜舍中紧靠饲料槽的柱
head post office 邮政总局
head pressing 封头冲压
head pressure 压位差;压(力)头;机头压力;入口压强;排出压力;水头压差;水头(压力);输送压力
head product 初馏物
head protection 头部防护
head pulley (皮带运输机的)头上滑轮;主滑轮;主动皮带轮;前鼓轮;头轮;天轮;首轮;上鼓轮;端部滚筒
head pulley drive 头上滑轮驱动;天轮驱动
head-pulley-drive conveyer 端部滚筒驱动的传送带
head pump 甲板冲洗泵;水头泵;船首泵
headquarters 总局;总管理处;总店;设总部;总部;指挥部;司令部;本部
headrace 引水渠(槽);前渠;引水渠道;进水渠道;上游水道;上游渠道;上游;渠首
headrace bay 进口河湾;前池
headrace channel 引水渠
headrace conduit 进水渠;上游渠道;上游进水渠(电站)
headrace gallery 引水渠坑道
headrace surge tank 上游调压井;首部调压井;上游调压水槽
headrace system of hydropower station 水电站引水系统
headrace tunnel 引水隧洞;引水隧道
head rail 上框;上槛;机头导轨;上冒头;船首斜帆;船首栏杆装置;门框(上)冒头
head range 扬程范围;压头范围;水头范围
head ratio 不同水位比
head reach 向前冲程;直线冲程;闸首区;河源区;渠首部分;上游段;冲程;正车惯性滑距
head recovery 水头恢复
head regulator 渠首调节建筑物;上游节制闸
head requirements 水头要求
head reservoir 压力水池;上水库
head resistance 迎面阻力;正面阻力;头部阻力
head rest 顶座;额头发垫;靠头之物;头靠
head-rest for vehicle seat 车辆座椅头垫
head restraint 头部保护装置
head rig 主备备;主锯机
head-rigging 前桅索具
head river 河源段;上游河流
head rod 头杆;第一连接杆
head roll 驱动滚筒
head roller 首轮
headroom 净高;无障碍空间;巷道高度;开采高度;净空(空间);净空高(度);上游闸室;顶部空间;高度净空;楼梯净空;甲板下净空;头上空间
headroom of flight 梯段净高
headroom of landing 平台净高
headroom under a bridge 桥下净空;桥梁出水净高
head rope 头缆;首绳;重载绳
head rotation 磁头走向
head running effect 磁头工作效应
heads 头留分;富集矿物
head sampling 选矿取样
head saw 板材锯;圆材锯;原木锯;主锯;剖料锯;头锯
head scarp 滑坡后壁
head schedule 标题字字体表
head screen 焊工罩面;焊工面罩
head sea 迎浪;逆浪;顶风浪;船头浪
head section 机头部分
head select signal 磁头选择信号
headset 头戴受话器;头戴式耳机和送话器;停筒
head shaft 顶轴;驱动轴;首端轴;主动轴
head sheave 卷扬机滑轮;天轮;天车;首轮;端滑轮;主滑轮
headsheets 艇首座板
head shield 护目罩
head shifter 磁头移行器
head shot 拍摄头部
head shot number 首炮号
head side 缸盖端
head sill 门窗框顶梁;上槛(门或窗的框顶上的横梁)
head sink 压头沉落
head slab 柱头板;盖顶板

head slotted plug 顶端带槽螺塞
headsman 推车工
head sod revetment 沟头
head space 液上气相分离;空容积;膨胀缝;上部;顶空(法);预留空间;顶端空间;顶部空间
headspace analysis 液面上气体分析;顶空分析
headspace gas chromatograph 顶空气相层析
headspace gas chromatography 顶空气相色谱法
headspace response 液面上响应
headspace sampling 顶空进样
headspace tester 顶隙测定计
head span 弹性软横跨;软横跨
headspan bracket 软横跨固定底座
headspan cross messenger wire 软横跨承力索
headspan mast 横跨支柱
headspan suspension 软横跨
headspan wire 横撑力索
head spar 主集材架杆
head spar(tree) 首柱木
head spring 源泉;前倒缆;起源;水源
head sprocket 主(动)链轮;首部链轮
head sprouting season 抽穗季节
head stack 磁头组(件)
head staffman 前司尺员【测】
headstall 马笼头
headstamp 标印
head stay 前支(架)索;桅顶前方支索
head steel 附加螺旋钢筋;附加构造钢筋
head step 起步踏步
headstep settling time 磁头步进稳定时间
head stevedore 装卸组长
headsticks 三角帆顶角撑木;井架
head still 门楣
headstock 悬挂架;主轴箱;(旋转部件的)支架;矿井架;机头架;机头;头座;头架;床头箱;车床箱;车头;车床头(座);测长机机头;刨床头座;刨床刀架;轮轴头架;底架端梁
headstock center 头座顶尖
headstock center collet 床头顶心套筒
headstock cone 床头锥轮
headstock cone pulley 床头塔轮
headstock gear 启闭机
headstock housing 床头箱盖
headstock rack pinion 床头箱齿条小齿轮
headstock spindle 机头座轴
head stone 拱心石;奠基石;墓石;础石;墙角(基)石
head stream 源流;河源(段);上游河流;源头;发源地
head support arm 磁头(支持)臂
head surge basin 上游调压井;上游调压池
head surge chamber 上游调压室;上游调压井
head swell 迎面涌浪;逆涌
head switch 读头开关
head-tail instability 首尾不稳定性
head-tail structure 头尾结构
head tank 压头箱;原料罐;压力水箱;压力柜;压力槽;高位水箱;高位水池;落差储[贮]水池;进料桶;进料罐;水塔
head tank farm 首站
head tapeman 前司尺员【测】
head tax 人头税
head tenon 木榫顶方榫
head tide 逆潮(流)
head tin 上端沉淀精矿
head to head 头接头;对遇
head to head structure 头对头结构
head to tail 头尾相接
head to tail addition 头尾加成
head to tail connection 头尾(系统)连接
head to tail linkage 头对尾键合
head to tail structure 头(对)尾结构
head to the wind 正顶风
head touch 磁头接触
head tower (架空索道或缆索起重机的)塔架;首塔(缆式起重用);端塔;顶塔
head track (推拉隔断上的)顶轨
head traverse of radial drilling machine 摇臂钻进给箱横切
head tree 柱顶木托块;分叉柱;支柱横木;顶木托块;柱头横木
head trim 前倾度;艏倾;船首倾度
head-turned angle 头的转动角度
head unload 磁头卸下
head up 艏向向上(雷达显示)

head-up display 平视显示器;车头显示
head-up display windscreen 屏显挡风玻璃
head upper drum 磁头上鼓
head value 首次分析价值
head valve 头阀;顶(置)阀
head valve of CO_2 cylinder 二氧化碳瓶头阀
head variation 压头变化
head variation in glass capillary viscometer 玻璃毛细管黏度计压头变化
head vat 头池
head wall 拱面墙(桥拱平面内的墙);胸墙;山墙;端墙;端面墙;陡壁;面墙(洞道进出口);斗壁
head wall abutment 一字形桥台
headward 朝向头部的
headward end of dock 坞首
headward erosion 向源侵蚀;沟头冲刷;溯源侵蚀
headwater 涌水区;河源头;河源(段);河流源头;水源;上水流;上层水;上游水域;上游水面;上游河段
headwater area 上游地区
headwater area of protection 上游保护区
headwater basin 水源盆地
headwater channel 源头水渠;引水渠;前渠;上游进水渠(电站);上游河槽
headwater control 源头控制;水源控制;水头控制;水头调节;上游水位控制;上游防洪措施
headwater elevation 水头高程;水源高程;上游水位(高程)
headwater erosion 向源侵蚀;向源冲刷;溯源侵蚀;溯源冲刷
headwater forest 水源(树)林
headwater-level 上游水位
headwater-level ga(u)ge 上游水位计
headwater pond 蓄水前池
headwater pressure 上游水压力
headwater prospect map 水源勘察图
headwater protection area 源头河保护区
headwater region 源头区;河源区;上游区
headwater retention 源头保水性
headwaters area 河源地区
headwater section 上游段
headwaters of river 上游
headwater storage reservoir 上游水库
headwater survey 河源调查;水源调查
headwater tunnel 进水隧洞;进水隧道
head wave 前波;首波;顶头波
headway (楼梯踏步的)垂直净空;车头距;行车间隔;钻进;主到理(方向);连发列车间隔时分/间隔;净空高(了);进展;进航速度;进尺;横巷;前进运动;通航高度;车间时距;(前后两车船之间的)安全距离
headway calculation 列车时间间隔计算
headway control system 车间距离控制系统
headway density controller 车头距密度控制机
head weather mo(u)lding 窗顶披水条,(门窗的)顶部泻水线脚;顶部披水
head-web separation 轨头轨腰分裂
head well 高产(油)井
head winch 首端绞车
head wind 逆风;顶(头)风;超前卷绕
head winding 磁头绕组
head wind velocity 迎面风速
head wire 头缆;船首缆
head wire pontoon 头缆抬缆方艇
head wire scow 头缆抬缆方艇
headworks 准备工程;渠头控制水量建筑物;准备工作;头形构饰;渠首构筑物;拱(顶)石饰;装有绞关索具的平台;掘进工作;掘进工程;井架;脑力劳动;进水口工程;渠首工程
headworks structure 渠首工程结构
head yard 前桅上的横桁
heal 用土覆盖
healant 修补剂
heal bite 弹簧车刀
heald-braiding machine 编综丝机
heald frame hook 综钩
healed joint 愈合节理
healing 铺屋面;裂缝愈合;裂缝合拢
healing agent 修补剂;修补膏
healing effect 弥复效应
healing of concrete 混凝土裂缝愈合;混凝土裂缝合拢
healing power 遮盖能力
healing property 复合性

healing rate constant 恢复速率常数
healing spring 医疗(矿)泉
healing stone 合拢铺石料;屋顶岩板;石(板)瓦
healing time 恢复时间
HE alloy 硅镁铝青铜
health 卫生
health administrative science 健康管理科学
health agency 卫生机构
health allowance 保健津贴
Health and Sanitary Regulation 卫生检疫规定
health and welfare fund 保健和福利基金
health and well-being 健康和福利
health authorities 卫生主管部门;卫生当局
health building 医院;疗养所
health care 卫生保健;保健
health-care facility 保健设施;卫生保健设施;疗养所
health-care measures 卫生保健设施;保健措施
health-care system 保健制度
health center[centre] 保健中心;医疗站;卫生所;保健站;保健所
health center for women and children 妇幼保健站
health certificate 健康证书;健康证明(书)
health chackup 体格检查
health check 正常检查
health check program(me) 正常检查程序;健康检查程序
health city 正常的城市
health club 健身俱乐部
health consultation 健康诊断
health control 健康管理
health department 卫生处
health development industry 健康开发产业
health education telecommunication 卫生教育电信
health effect 健康影响
health effect appraisal 健康效应评价
health effects of atmospheric pollution 大气污染对健康的影响
health effects of polychlorinated biphenyl pollution 多氯联苯污染对健康的影响
health examination 健康检查
health examination survey 健康调查
health facility 卫生设施
health filling method by leaching cycle 滤沥循环卫生填埋法
health fund 保健基金
health guard 检疫(人)员;保健
health guidance 健康指导;保健指导
health hazard 有害于健康;健康危害;人身事故;对健康有害的事物
health hazards in paint making 制漆过程中的健康危害
healthier trend 更有利的趋势
health index 健康指数;保健指标
health indicator 卫生指标;保健指标
health inspection 卫生监督
health insurance 健康(保)险
health insurance expenses 健康保险费(用)
health insurance plan 健康保险计划
health insurance premium 健康保险费(用)
health legislation 健康立法;卫生法
health level 健康状况;健康水平
health management 保健管理
health monitor 环境保护监视器;保健检测仪
health-monitoring installation 剂量控制装置;剂量测量装置
health officer 检疫人员;卫生人员
health physicist 保健物理学家
health physics 有害辐射防护学;放射卫生(学);保健物理学
health planning 卫生事业规划
health problem 健康问题
health protection 卫生防护
health protection forest 卫生防护林
health quarantine 卫生检疫
health quarantine service 卫生检疫中心;卫生检疫站;卫生检疫所
health regulation 卫生法规
health-related monitoring 与健康有关的监测
health requirements and standards for surface water 地面水卫生要求和标准
health resort 休养station;疗养地
health resort hotel 休养区旅馆
health resort park 休养区公园

health risk 有害于健康
health risk assessment 健康风险评价;健康风险评估
health satellite 完好卫星
health screening 健康普查
health service 医务处;卫生所;卫生(保健)事业;保健(事业)
health sociology 社会卫生学
health spa 休养所;减肥中心
health state 完好状态
health station 保健站
health statistics 卫生统计学
health status 健康状态
health stream 卫生河流
health supervision 健康监护;卫生监督;保健管理
health surveillance 健康监护
health survey 健康调查
health system 医疗卫生系统
health work 保健工作
healthy building 健康建筑物
healthy carrier 健康带菌者
healthy city 健全的城市;清洁城市
healthy material 健康材料
healthy residence 健康住宅
healthy ship 健康船舶
healthy stream 无污染河流;清洁河流
healthy worker effect 健康工人效应
heam pass 钢梁孔型
heap 料堆,堆(阵);堆积;堆叠;大量
heap capacity 装载容量;加载重
heap cloud 直展云
heap coking 土法炼焦
heaped capacity 最大容量;装载容量;加载量;堆积容量;堆装斗容量;堆载量;堆尖斗容量;铲装容量
heaped concrete 未摊铺混凝土;成堆混凝土
heaped load 堆积荷载;堆集荷载;堆货荷载
heaped measure 堆量
heaped-up island 堆积岛
heaper dump leaching 堆浸
heap fermentation 堆积发酵
heaping 堆(量);堆阵操作
heaping up of water by wind 风增水
heaping weight 堆积重
heap leaching 堆摊浸滤;堆摊浸出;堆滤法
heap load 堆积荷载
heap of boulders 巨砾堆
heap of clay 黏土堆场
heap of debris 碎片堆,垃圾堆;残渣堆
heap of granular material 颗粒材料堆
heap of rubble 毛石堆;乱石堆;块石堆;石堆
heap of tripod 三脚架头
heap roasting 堆摊焙烧
heaps 炼焦堆
heap sampling 堆锥缩样法
heap sand 铸造用砂;填(充)砂;堆砂
heap sort 群分类;堆叠分类
heap stead 井口房;井口建筑(物);井架
heap symbol 堆叠符号;大堆阵符号
heap together 堆积
heap type 堆叠式
hearing aid 助听器
hearing aid amplifier 助听放大器
Hearing cell 希尔令型浮选槽
hearing clerk 接待员
hearing conservation 听力保护;听觉保护
hearing defect 难听
hearing loss 听力损失
hearing protection 听力保护
hearing protector 听力保护器
hearing threshold 听力域
Hearne current 赫恩角海流
hearse (重要人物墓上的)铁框架;木制或金属烛台架;墓架
heart 索芯;髓心部
heart and dart 叶与箭线脚装饰
heart and square 平勺;心形镘刀
heart block 心形单眼木拼
heart board 髓心板;带心板材
heart bond 心形砌合法;中心结合;墙心顶块砌接
heart cam 心形凸轮;桃轮
heart carrier 桃子夹头
heart center[centre] 髓心;中心
heart check 髓心干裂;内部裂缝;心裂;中心裂纹;

木心环裂;木心辐裂
heart cut 中心切割;中心馏分
heart face 去掉边材的木料正面
heart-face board 心材面板
heart gear 中心齿轮
hearth 炉膛;炉床;炉边;锻造炉;船上厨房;敞炉;壁炉地面
hearth accretion 炉瘤;炉结块;炉缸冷结
hearth area 炉底面积
hearth block 炉底砖;炉床砖;座板;熔矿炉底石;炉板
hearth bottom 炉底
hearth brush 炉刷
hearth casing 炉缸外壳
hearth chill 炉缸冻结
hearth cinder 炉渣;熟铁渣
hearth construction 炉膛构造
hearth electrode 炉底电极
hearth extension 炉床前的砖石地面
hearth fettling 炉底衬
hearth furnace 高壁炉;墙壁高炉;平炉;膛式炉;床炉
hearth jacket 炉套;炉缸外壳
hearth layer 底料层;底层炉料
hearth level 炉底标高;炉缸高程
hearth line 炉缸水平线
hearth lining 炉膛衬层;炉床内衬
hearth melting furnance 床式熔炉
hearth number 熔炼炉号
hearth of pot furnace 坩埚窑作业室
hearth plate 炉底座板
hearth refining 炉床精炼
hearth ring machine 炉圈翻砂机
hearth roaster 床式焙烧炉
hearth-roaster gas 多膛焙烧炉炉气
hearth roasting 炉膛焙烧;多层焙烧法
hearth rug 火炉口毡
hearth soaking zone 炉内均热带
hearth sow 炉缸结块
hearth stone 火炉石;炉(膛)石;壁炉块石
hearth trimmer 炉前格栅;壁炉前的托梁
hearth wall 炉床内壁
hearting 心墙;心部;填筑心墙
hearting concrete 填心混凝土
hearting wall 不透水心墙
heart land 心脏地带;中心地区
heart of the town 城市中心;城市商务区
heart piece 桃轮
heart plank 心材板;厚心板(厚5~15厘米,宽大于23厘米);髓心厚板
heart rot 木心腐烂;心材腐朽
heart shackle 心形卸扣
heart shake 心裂;中心裂纹;木心环裂;木心辐裂
heart-shaped shackle 心形卡环
heart shape reception 心形方向图接收
heart shape thimble 牛眼环;绳端心环
heart stain 心材变色
heart substance 心材含有物
heart tie 心材枕木
heart wall 不透水心墙;心墙
heart wall dam 心墙坝
heartwood 心木;心材
heartwood rot 心材腐朽;心材腐蚀
heartwood tree 心材树(木)
heart yarn 麻芯索条
heat 炉次;加热;每炉熔炼量;熔炼序次;熔次;热量;取暖照明及动力成本
heatable 可加热的
heatable stage microscope 热显微镜
heat absorbed in evapo(u)ration of water 蒸发水分吸收的热量
heat absorbent 吸热剂;热吸收剂
heat-absorbent surface 吸热面;冷却面
heat absorber 吸热器;热吸收器
heat-absorbing 吸热
heat-absorbing action 吸热作用
heat-absorbing filter 吸热滤光片
heat-absorbing glass 吸收红外线玻璃;吸热玻璃;保温吸热玻璃;隔热玻璃
heat-absorbing reaction 吸热反应
heat-absorbing shield 吸热罩
heat-absorbing surface 受热面
heat-absorbing windshield 吸热挡风玻璃
heat absorptance 热吸收率
heat absorption 热吸收;吸热量

heat absorption capacity 吸热能力;吸热量;热容量
heat absorption efficiency 吸热效率
heat absorption rate 吸热速率
heat absorptive glass 吸热玻璃
heat absorptivity 吸热性
heat abstraction 热的散失
heat abstractor 散热器;散热装置
heat accepting medium 受热介质
heat accepting surface 受热表面
heat acclimatization 热适应
heat account 热平衡(计算)
heat accumulating of revolution type 旋转型蓄热式
heat accumulating type 蓄热式
heat accumulating type of shift change 阀门切换型蓄热式
heat accumulation 热储积;累积吸热量;热积聚
heat accumulation capacity 热储积能力
heat accumulation power 热储积能力
heat accumulation property 热储积性能
heat accumulation quality 聚热质量
heat accumulator 蓄热体;蓄热器
heat action 热作用;热影响
heat action zone crack 热影响区裂纹
heat-activable tanning agent 热激活性鞣剂
heat-activated adhesive 热熔胶合剂;热活化黏结剂
heat-activated battery 热激活电池
heat activation 加热激活;热激活
heat-actuated 热动的
heat-actuated device 热温控装置;热(启)动装置
heat-actuated fire door 自动防火门;热(驱)动防火门
heat-actuated means 热启动装置
heat adaptation 热适应(作用)
heat addition 加热;供热
heat-adsorbent 热吸附剂
heat adsorption 吸热
heat adsorption capacity 吸热(容)量;吸热能力
heat adsorption efficiency 吸热效率
heat adsorption surface 吸热面
heat-affected zone 热影响区;温度影响区域;受热区
heat-affected zone crack(ing) 热影响区裂缝
heat affecting 热影响
heat-aged discolo(u)ration 热老化变色
heat ag(e)ing 热养护;热老化;热时效
heat ag(e)ing inhibitor 防热老化剂
heat ag(e)ing test 加热老化试验;热老化试验
heat-agglomerating 加热烧结
heat air ag(e)ing 热空气老化
heat air delivery 输送热风
heat alarm 过热信号;高温警报器;热警报(信号);温升报警信号
heat-altered coal 热力变质煤
heat-altered low volatile steam coal 热变质低挥发分蒸汽煤
heat-altered medium-volatile coal 热变质中挥发分煤
heat analysis 熔炼分析
heat and corrosion resistant steel 耐热防锈钢
heat and humidity ratio 热湿比
heat and humidity testing equipment 潮热试验设备
heat and material balance 热量与物料平衡
heat and moisture transfer 热湿交换
heat and power cogeneration 热电联产
heat and power plant 热电厂
heat and power station 热电站
heat and sweat damage 潮热蚀
heat and vent system 加热与通风系统
heat apparatus 加热装置
heat application 供热;热灌筑;热供应
heat-applied coating 热敷层
heat availability 热效性;热利用率;热的有效性
heat baffle plate 挡热板
heat balance 热(量)平衡;热量对照表
heat balance data 热量平衡数据
heat balance diagram 热平衡图
heat balance equation of aquifer 含水层热量均衡方程
heat balance flowmeter 热平衡流量计
heat balance in sludge digestion 污泥消化热平衡
heat balance of dryer 干燥器热平衡
heat balance of groundwater 地下水热量均衡
heat balance of kiln 窑炉热平衡

heat balance of soil 土壤热量平衡
heat balance of the ocean 海洋热平衡
heat balancer 热平衡器
heat balance table 热平衡表
heat balance test 热平衡试验
heat balancing 热平衡
heat band 热波段
heat-banking area 封热辐射区域;热辐射区域
heat barrier 热障;热屏(蔽)
heat barrier coating 隔热涂层;隔热保护层;热绝缘涂层
heat barrier material 隔热材料
heat batch number 炉号
heat bath 热浴(室)
heat bath furnace 热浴炉
heat bitumen grouting method 热沥青灌浆法
heat blanket 隔热毡
heat blast 热鼓风
heat-blocking 热黏着性
heat-blocking action 隔热作用;绝热作用;热封闭作用
heat blower 热风发生器;热风吹送机
heat board 保温板
heat-bodied oil 热稠化油;厚油;热聚合油
heat-bodied tung oil 熟桐油
heat body 热稠化
heat bodying 热炼
heat bodying of oil 油的热聚合
heat bondable fibre 热黏合纤维
heat bonded fabric 热黏合织物
heat bonding 热黏结;热黏合;热胶合
heat booster 助热器;增热器;加热器;热力增强器;升热器
heat box 加热室
heat bridge 热桥;热量外流通道
heat budget 热总量;热预算;热量收支;热量平衡;热量计算;热量概算
heat budget method 热量平衡法;热预算方法
heat budget of the earth 地球热量平衡
heat-bulb electric(al) velometer 热球式电风速计
heat cable 热缆
heat calculation 热(量)计算
heat calorie 热质
heat capacitivity 热容率
heat capacity 热容(量);供热能力
heat capacity at constant pressure 恒压热容
heat capacity at constant volume 恒容热容
heat capacity lag 热容量滞后
heat capacity mapping 热容成像
heat capacity mapping mission 热容量制图卫星;热容量成像卫星;热能勘测任务
heat capacity mapping radiometer 热容量成像辐射计
heat capacity of water 水的热容量
heat capacity per unit volume 单位体积热容
heat capacity per unit weight 单位重量热容(量)
heat capacity ratio 热容比
heat carrier 吸热体;载热体;载热介质;热载体;导热体;传热器
heat carrying agent 载热体;热媒;载热剂;热载体
heat carrying capacity 吸热容量;吸热能力;载热能力;热容量;导热能力
heat carrying fluid 载热(流)体
heat catalysis 热催化作用;传热媒介
heat cell 热辐射指示器
heat change 热交换
heat channel 热桥;热沟;热槽
heat character 热学性质
heat characteristic 加热特性
heat check(ing) 热裂
heat chemical treatment 化学热处理;热化学处理
heat circulation 热循环
heat cleaned 热清除
heat cleaning 热清洗;热清洁法
heat cloud 热成云
heat coagulant 热凝固剂
heat coil 热熔线圈;加热线圈
heat collapse 热破坏;热变形
heat-collecting zone 热量积聚带
heat colo(u)r 火色
heat compensating jacket 热补偿夹套
heat compensation 热补偿
heat compensator 热平衡器

heat condition 热状态;热条件;热量状况
heat conditioning 热调节
heat conductance 热(传)导;导热性
heat-conducting 热传导的;导热(的)
heat-conducting fin 导热片
heat-conducting flow 热传导流
heat-conducting fluid 载热流体;流体载热剂;导热的流体
heat-conducting material 导热材料
heat-conducting medium 导热介质
heat-conducting property 导热性能
heat-conducting strip 导热片
heat conduction 热传导现象;热传导;导热(作用);导热(现象)
heat conduction problem 热传导问题
heat conductivity 热导率;热传导性;导热性;导热率;导热度
heat conductivity coefficient 导热系数
heat conductivity equation 热传导方程
heat conductivity factor 热导系数
heat conductivity of rock 岩石的热导性;岩石的导热性
heat conductivity pressure ga(u)ge 热传导压力表
heat conductivity vacuum ga(u)ge 热导式真空规
heat conductor 热(传)导体;导热体;导热材料;传热导体
heat conservation 热量守恒;热量保存
heat conservation shutter 保热罩;热量保存启闭器
heat consumer 用热户;热用户
heat consumption 耗热量;含热量;热消(耗)量;热气消耗;热量消耗;热耗量
heat consumption of clinker 熟料热耗
heat consumption quota 耗热定额
heat consumption rate 热耗率
heat consumption test 热耗试验
heat content 焓;含热量;热容量(系数);热函;热含量
heat content of solid waste 固体废物含热量
heat control 温度控制;热(量)控制;加热调节
heat controller 热控制器
heat control multilayer filter 热控多层滤光器
heat control switch 热控(制)开关
heat control valve 热控制阀
heat control valve shaft 暖气阀轴
heat convection 热对流;热传送;对流换热
heat convection coefficient 热对流系数
heat conversion factor 热换算因数
heat-convertibility 热转化性;热固性
heat-convertible resin 热固(性)树脂
heat counter 热度计
heat crack 热裂纹;高温裂纹;热致裂缝;温度裂缝
heat cracking 热裂
heat craze 加热裂纹
heat cumulus 热成积云
heat cure 热硫化
heat cured 热固性的;加热养护的
heat cured insulation 热固性绝缘
heat curing 热硫化
heat curing accelerator 热硫化促进剂
heat curing cycle 加热养护周期
heat curing instrument 热养护设备
heat curing methyl methacrylate 加热固化型塑料
heat curing plastics 加热固化型塑料
heat curing system 热硫化法;热固化系统
heat current 热流
heat cycle 加热周期;机热模塑周期;热周期;热(力学)循环
heat cycle effect 热循环效应
heat cycle of a welding operation 焊接的热循环
heat cycling 循环加热
heat cylinder 热汽缸
heat dam 热障
heat damage 热破坏;受热变质
heat deairing 加热脱气
heat death 热寂
heat debt 热亏损
heat decomposition 热分解
heat-decrepitation extraction method 热爆浸取法
heat deflection temperature 热挠曲温度
heat deflector 挡热板
heat deformation 热变形
heat degree 热度
heat delivery surface 供热面;散热面;传热面;放热(表)面

heat demand 用热要求
heat denaturation 热变性
heat denaturation test 热变性试验
heat density 热能密度;热量密度;表面放热率
heat desizing 热脱浆
heat destortion point 热扭变点
heat detecting equipment 热探测设备
heat-detection system 感温探测系统
heat detector 热(源)探测器;热量探测器;热检测器;热检波器;热辐射自动导引头灵敏元件;温感(探测)器
heat deterioration 加热劣化变质;热老化
heat development 热显影
heat-die pressing process 热压法
heat diffusion 热散逸;热扩散(作用);导热
heat diffusivity 热扩散率;热导率
heat discharge 热排放(物);热量排放
heat discolo(u)red 受热变色
heat dispersion 热分散
heat dissipating ability 热消散能力
heat dissipating capacity 散热量
heat dissipating capacity of equipment 设备散热量
heat dissipating tile 散热砖
heat dissipation 耗热量;热消散;热损失;热散逸;热扩散(作用);热散;散热
heat dissipation factor 热消散系数
heat dissipation rate 热量耗散率
heat distance 热辐射距离
heat distortion 热致扭曲;热扭变;热畸变;热变形
heat distortion point 热变形点
heat distortion temperature 加热变形温度;热扭变温度;热畸变温度
heat distributing network 热网
heat distribution 热(量)分布;热分配
heat dried 热干的
heat drift 热漂移
heat drilling 热钻孔;热钻(法)
heat-driven oscillation 热致振动;热振荡
heat drop 焓降;热(落)差;热量变化率;热降(低);温降;温度梯度;温度落差
heat dry 热力干燥的
heat drying (污泥的)加热干燥;烘干;热干化
heat drying of sludge 污泥加热干燥;污泥加热干化
heat dump 吸热器
heat durability 耐热度
heat duty 热负荷
heat duty of checker 格子体热负荷值
heat dynamometer 热测力计
heat economising 热量节约;热量节省
heat economizer 节热器;废气预热器
heat economizing 热量节约;热量节省
heat economy 热经济
heated absorption tube 热吸收管
heated air 热风
heated air circulation 热风循环
heated air drier 热风式干燥机
heated air drying 热风干燥
heated air intake system 加热空气调温进气系统
heated air output 热风输出;暖气输出量
heated air psychrometer 热空气干湿表
heated air screen 热风幕
heated area 受热面积
heated area of checkers 格子砖受热面积
heated asphalt 热沥青
heated bar 测热杆;测热棒
heated bit 加热的钻头;加热的烙铁;加热的刀头
heated board 加热保温板
heated body 加热的物体
heated cathode 间热阴极;热阴极
heated ceiling 热顶棚;采暖顶棚
heated cleaning screen 加温的清选筛
heated concrete 加热混凝土;热拌混凝土
heated container 加热集装箱
heated core box 热芯盒
heated crusher 烘干破碎机
heated-die pressing process 热模压法
heated digester 加热消化池
heated digestion chamber 加热消化室
heated doorway 保温门道
heated dry air 干热空气
heated effluent 热废液;热废水
heated electron evapo(u)ration rate sensing device 热电子蒸发速率测量器件

heated energy 热能
heated equivalent work 热功当量
heated funnel 加热漏斗
heated garage 采暖车库
heated glass 加热玻璃
heated graphite atomizer 加热石墨原子化器
heated hung ceiling 加热吊顶棚;采暖吊顶棚
heated iron 加热的钻头;加热的烙铁;加热的刀头;焊铁
heated launder 热流槽
heated metal suspended ceiling 采暖金属吊顶棚
heated mortar 热砂浆;加热的砂浆
heated mo(u)ld 热模
heated-orifice storage ga(u)ge 加热式累加降水量
heated plate 加热板
heated roller 加热滚筒;火滚筒
heated skin detection 蒙皮热辐射探测
heated soil 变热的土壤
heated spot 热迹
heated stream 渗入热水的溪流
heated succession 热顺序
heated suspended ceiling 加热吊顶棚;采暖吊顶棚
heated tank 加热油罐
heated thermometer anemometer 热风速仪
heated-tool welding 热烙铁焊接
heated treatment rate 热处理率
heated value 热值
heated waste 热废水
heated water discharge 热水排放
heated wire 加热丝
heat effect 热效应
heat effect of solid laser 固体激光器热效应
heat efficiency 热效率;热吸率;热效应
heat efficiency of boiler operation 锅炉运行热效率
heat efficiency of cooler 冷却机热效率
heat efficiency of furnace 熔窑热效率
heat efficiency of kiln 窑炉热效率
heat efficiency of the dryer system 干燥系统热效率
heat egress 热输出
heat electric(al) couple 热电偶
heat-eliminating medium 冷却介质
heat elimination 热气排除;热量消除(法);排热(法);散热(法);放热(法)
heat embrittlement 热硬化;热脆化;受热变脆
heat emission 热(气)放射;热(量)发射;热辐射;热发散;散热;放热
heat emission coefficient 散热系数
heat emission equipment 散热设备
heat emission factor 放热系数;放热因数
heat emission rate 热量排放率;散热率
heat emissivity 散热;热辐射;放热;热发射率
heat emissivity coefficient 热辐射系数;热放射系数;热发射系数
heat emitting 放热
heat emitting apparatus 放热器
heat emitting surface 放热表面
heat endurance 耐热性;耐热度;热稳定性
heat energy 热能
heat engine 热气机;热力(发动)机;热机
heat engine ceramics 热机陶瓷
heat engineering 热力工程;热工(学)
heat engineering calculation 热工计算
heat-engine plant 火电站;火电厂
heat-engine traction unit 热力牵引发电机组
heat entering mill 进磨热量
heat enthalpy 焓;热容量;热函;热含量
heat equation 热流方程;热力方程
heat equator 热赤道;温度赤道
heat equilibrium 热平衡
heat equivalent 热当量
heat equivalent of work 热功当量
heater 炉;硫化机;硫化罐;加热装置;加热元件;加热器;加热炉;加热工;加热导体;暖器;暖气设备;热源;热丝;燃烧室;取暖器;采暖机;放热器;发热器
heater air flow pressure switch 加温气流压力开关
heater and blower unit 加热和吹风装置;加热和吹风器件
heater assembly 加热设备
heater band 加热带
heater battery 放热器组
heater bias 灯丝偏压;灯丝偏压
heater block 加热部件

heater box 加热箱
heater brick 加热器格子砖
heater button 加热器开关
heater car 加热车箱
heater carrying fluid 载热流体
heater case 电热箱
heater casing 加热器箱盒；暖气设备罩
heater cathode 直热式阴极
heater-cathode connection 直热阴极接线
heater-cathode leakage 灯丝阴极间漏泄
heater-cathode leakage current 灯丝阴极间漏电电流
heater chain 热丝电路；灯丝电路
heater choke 加热器扼流圈；灯丝扼流圈
heater circuit 加热电路；热丝电路；电热电路；灯丝电路
heater coil 加热盘管；盘管加热装置
heater cooler 火焰管冷却器
heater cord 电炉花线
heater current 热丝电流；灯丝电流
heater discharge pipe 热水器放水管
heater drain valve hole 加热器放水阀孔
heater-drip pump 加热器疏水泵
heater element 加温器；加热元件
heater emission 热丝极发射
heater exchange 热交换
heater for vehicle 车辆暖气装置
heater friction 加热器阻力
heater hose 加热器软管
heater housing 加热器(壳)罩；加热器罩；加热器外壳；加热炉罩
heater hum 热极交流声；丝极交流声
heater intermittent test 灯丝断续试验
heater lead 灯丝引线
heater lower cover 加热器下盖
heater mica plate 电热器用云母板
heater mixer 加热拌和机；加热搅拌机
heater oil 加热用油
heat erosion 高温削蚀；高温腐蚀；热侵蚀；热腐蚀
heater outlet couple 炉出口处热电偶
heater output 加热器出量
heater planer 加热整平机；烫平机
heater platen 加热压模板
heater plug 加热器塞；热源消防栓
heater power 加热器功率；热丝功率；灯丝功率
heater power circuit 热丝电源电路
heater reflector 加热反射器
heater resistance 加热器电阻；灯丝电阻；发热器电阻
heater screen 热源屏障
heater selector 加热选择器
heater shield 挡热板
heater shut-off valve 加热器断流阀
heater storage space 加热器水空间
heater supply 加热电源；灯丝电源
heater surface 加热器换热面
heater switch 加热器开关；电热开关
heater system 加热器系统
heater temperature 加热器温度；加热炉温度
heater thermometer 加热器温度计
heater transformer 热丝变压器
heater transformer windings 热丝变压器绕组
heater type 旁热式；旁热型
heater-type thermistor 旁热式热敏电阻器
heater unit 加热装置；暖风机
heater upper cover 加热器上盖
heater valve balance 加热阀平衡器
heater valve cover 加热阀盖
heater valve shaft bushing 加热器阀轴套管
heater valve thermostat 加热阀恒温器
heater volt 热丝电压
heater voltage 热丝电压
heater washing hole 加热器洗孔
heater well 加热器套管
heater winding 加热线圈；灯丝线圈
heater wire 加热线
heat etching 热侵蚀
heat evolution 热析出；放热；发热
heat exchange 换热；热(量)交换
heat-exchange area of checker 格子砖换热面积
heat-exchange by convection 对流热交换
heat-exchange by radiation 辐射热交换
heat-exchange chamber 换热室；传热室
heat-exchange coefficient 热交换系数
heat-exchange coefficient of interior surface 内表面换热系数
heat-exchange coil 热交换蛇形管
heat-exchange cycle 回(流)热循环；热交换循环
heat-exchange dynamics 热交换动力学
heat-exchange efficiency 换热效率
heat-exchange equipment 换热设备；热交换器
heat-exchange facility 换热器
heat-exchange fluid 热交换流体；载热流体；流体载热剂
heat-exchange in digestion tank 消化池热交换器
heat-exchange instability 热交换不稳定性
heat-exchange method 热交换法
heat-exchange of heat pipe 热管换热器
heat-exchange of the ocean 海洋热交换
heat-exchange plate 换热板片
heat-exchanger 冷却器；换热器；热交换器；散热器
heat-exchanger circuit 换热器回路
heat-exchanger coil 热交换器线圈；热交换器蛇形管
heat-exchanger effectiveness 热交换器有效性
heat-exchanger grate 热交换箅子板
heat-exchanger lag 热交换器延迟(反应堆)
heat-exchanger leak detection probe 热交换器探漏器
heat-exchanger pipe assembly 热交换器管组装
heat-exchanger principle 热交换器原理
heat-exchanger pump 热水泵
heat-exchanger rig 换热装置
heat exchangers in parallel 换热器的并联
heat exchangers in series 换热器的串联
heat exchangers of the plate-type 板式热交换器
heat-exchanger thermal ratio 热交换器回热度
heat-exchanger tower 热交换塔
heat-exchanger tube 换热器管
heat-exchanger type regenerator 换热式回热器
heat-exchanger wall 换热器壁
heat-exchanger with internal heat source 有内部热源的热交换器
heat-exchange station 热交换站
heat-exchange surface 热交换面
heat-exchange surface area 换热面积
heat-exchange system 热交换系数
heat-exchange tank 热交换池
heat-exchange tube 换热管；热交换管
heat-exchange tube bundle 热交换管束
heat-exchange type 换热型
heat-exchanging apparatus 热交换器
heat-exchanging process 热交换工作程序
heat-exchanging system 热交换装置
heat-exchanging unit 换热器
heat excluding glass 隔热玻璃
heat exhaustion 热衰竭；热量消耗
heat expansion 热(膨)胀
heat expansion coefficient 热膨胀系数
heat expenses 热支出；热消耗
heat exposure threshold limit value 热接触阈限值
heat extraction 热抽提；排热；除热；抽出热量
heat-extraction system 排热系统
heat facility 热装置
heat fade 热衰减
heat fading 热衰退
heat fastness 热稳定性；耐热性
heat fatigue 热疲劳
heat filter 滤热片
heat-filter glass 滤热玻璃
heat-filtering glass 热防护玻璃
heat fire detector 感温火灾探测器
heat flash 强热
heat flexibility test 热柔韧性试验；热挠曲试验
heat flow 热流(动)
heat flow anomaly 热流异常
heat flow anomaly zone 热流异常带
heat flow balance 热流平衡
heat flow diagram 热能流程图
heat flow equation 热流方程
heat flow measurement 热流测量
heat flow meter 热流(量)计
heat flow problem 热流问题
heat flow property 热流特性
heat flow rate 热流量；热流(速)率
heat flow rate per unit area 热流强度
heat flow remote measurement 热流远距离测量
heat flow sensor 热流传感器
heat flow survey 热流测量
heat flow transducer 热流传感器
heat flow unit 热流(量)单位
heat flow value 热流值
heat flow variation 热流量变化
heat flow vector 热流向量
heat flush 热冲洗
heat flux 热流量；热(流)通量；热流(动)；热焊剂；热负荷
heat flux density 热流密度
heat flux differential scanning calorimetry 热流式差示扫描量热法
heat flux distribution 热流分布
heat flux equation 热通量方程
heat flux transducer 热流传感器
heat flux vector 热通量向量
heat for disinfection 加热消毒
heat forging drawing 热锻件图
heat forming process 热成型过程；热成型工艺；热成型(方)法
heat from arcing 电弧放电热
heat function 焓；热焓；热函(数)
heat-fusion-bonded epoxy coating 热熔黏合环氧涂层
heat fusion joint 加热熔压接头
heat gain 增热量；热增量；热增加
heat gain and loss 热量获得和损失
heat gain from lighting 照明散热量
heat gain from occupant 人体散热量
heat ga(u)ge 温度计；热压力计；热量计；热感应塞
heat generated by equipment 设备产热
heat generated by lighting 照明产热
heat generated by people 人身产热
heat generating agent 发热剂
heat generating appliance 热发生器
heat generating capacity 热值；发热量
heat generating device 热发生设备
heat generating reaction 放热反应
heat generation 产热；热量生产；热发生；释热；生热(性)；发热
heat generator 回热器；热风炉；热发生器
heat gradient 热(量)梯度；热量变化率
heat grain 谷物加热
heat guard 绝热体
heat gun 煤气喷枪；热风器
heath 石南
heat hammer drill 热锤钻孔
heat-hardenable resin 热固性树脂
heat hardening 加热硬化；淬火
heat hardiness 抗热力；耐热力
heat hazard of mining area 矿区地下热害
heat haze 热霾
heath bell 石南花
heath forest 石南林
heath fruit 荒地果树
heath gravel 荒地砾石
heat history 热历史
heath land 荒地；荒野；石南荒原
heat-homer 热自动瞄准头
heat homing guidance 热辐射寻的制导
heath peat 灌木泥炭；石南泥炭
heath podzol 石南灰壤
heat imaging 热成像
heat imaging device 热成装置
heat impact resistance 耐热冲击性
heat impulsive method 热脉冲测定法
heat increment 热增耗
heat increment rate 热增耗率
heat index 热(量)指数
heat indicating paint 示温涂料；示温漆；变色漆
heat-indicating pigment 示温颜料
heat indicator 热量指示器；热力指示器；温度(指示)计
heat-induced collapse 热融滑塌
heat influence 热影响
heat influx 注入热量；热补给(量)
heating 供热；供暖；加温；取暖；受热；采暖；发热
"heating, ventilation and air-conditioning" 暖通调控
heating accessories 加热器具
heating adhesion 加热黏合
heating ag(e)ing 热老化
heating agency 载热体
heating agent 加热剂
heating air 热风；暖气
heating allowance 煤火补贴

heating alloy 合金电热丝
heating alternator 加温用交流发电机
heating and cooking equipment 供暖与烹调设备
heating and cooling cost 热处理和冷却费
heating and lighting wagon 暖气照明组合车
heating and plumbing 暖卫
heating and power center 热电站
heating and power plant 热电厂
heating and power producing centre 热电联合中心
heating and ventilating 供暖与通风;暖通;取暖与通风;采暖通风
heating and ventilating discipline 暖通专业
heating and ventilating engineer 暖通工程师
heating and ventilation 采暖通风;暖通设计
heating and ventilation system 供暖与通风系统
heating apparatus 加热装置;供热装置;加温装置;加热器;暖气装置;取暖器
heating apparatus for defrosting window of vehicle 车窗除雾用加热器
heating appliance 供热设备;加热设备;取暖用具;采暖设备
heating area 供热面积;供暖面积;加热面积;受热面积;传热面积;采暖面积
heating area of boiler 锅炉的受热面积
heating arrangement 加热装置;供暖装置
heating attachment 加热装置
heating basement 供热地下室
heating bath 加热槽
heating blowpipe 加热炬
heating body 受热体
heating boiler 供热锅炉;供暖锅炉;采暖锅炉
heating by circulating air 循环空气加热
heating by condensation 凝结增温
heating by convection 对流采暖
heating by exhaust gases 排气加热
heating by proximity effect 邻近效应加热
heating by radiation 辐射增温;辐射采暖
heating by town gas 管道煤气取暖;城市煤气供热
heating by waste heat 废气供热
heating cabinet 加热箱
heating cable 加热电线;加热电缆
heating cable unit 加热电缆单元
heating calculation 热力计算;供热计算
heating capacity 供热能力;热容量;供暖能力;给热能力;热含量;发热效能
heating chamber 加热室;暖气室;燃烧室
heating circuit 供热电路;加热电路;电热电路
heating circulating pump 供热循环泵
heating circulation 供热系统;供热范围;供暖系统
heating coefficient 加热系数
heating coil 供暖盘管;加热旋管;加热蛇管;加热盘管;暖(气盘)管;热盘管;采暖盘管;加热线圈
heating coil drain valve 加热盘管排泄阀
heating coil inlet valve 加热盘管入口阀
heating coil outlet system 加热盘管排出系统
heating coil set 加热盘管组
heating coils in digester tank 消化池加热盘管
heating compensating jacket 热补偿夹套
heating computer 加热计算机
heating concrete 热拌混凝土
heating conductor 加热导体
heating connect order 电热接电单
heating consumption 热耗
heating contactor 采暖接触器
heating control 加热控制
heating controller 加热调节器
heating convector 对流加热器
heating-cooling draw die 加热—冷却拉延模
heating-cooling mixer 加热—冷却混合机
heating cord 电热丝
heating cure 加温固化
heating curing 加温养护
heating current 加热电流
heating curve 加热曲线;热加曲线
heating-curve determination method 加热曲线测定法
heating cycle 加热循环;加热工况
heating cylinder 加热汽缸
heating degree-day 增温度/日
heating depth 加热深度
heating design 供暖设计;暖气设计;采暖设计
heating device 加热装置;加热器;发热器;供暖装置;供暖设备

heating drum 加热滚筒
heating duct 暖气管道;供热导管;加热导管
heating duration 供暖时间;加温时间
heating effect 热效应;供暖效应
heating effect indicator 热值测定器
heating efficiency 供暖效率;加热能力
heating efficiency of the dryer 烘干机热效率
heating electrode 加热电极
heating element 暖气片;加热部件;供暖部件;加热元件;加热器;热体;热媒;电热元件;电热丝;发热元件
heating emission 传热
heating energy 热能
heating engineer 暖气工程师;供热工程师
heating engineering 供暖工程
heating environment 加热环境
heating equipment 供暖设备;加热设备;采暖设备;暖气设备
heating facility 采暖设备
heating field 加热场
heating firm 供热公司;供暖公司
heating floor 保温地板
heating flow pipe 采暖供水管
heating flue 供暖烟道;热烟道
heating for building operations 房屋建造所需的加热
heating for construction operations 施工操作中的加热
heating form maitenance 热模养护
heating-freezing stage 冷热两用台
heating fuel 供暖燃料;加热用油
heating furnace 加热炉
heating furnace for blooms 大铜坯加热高炉
heating fuse 热用熔断丝
heating gas 燃气;加热燃气;加热煤气;热煤气
heating gate 加热(孔)口
heating generator 高频加热器;热发生器
heating glass pane 加热玻璃片
heating grid 加热栅栏;热格
heating hardness 加热硬性;加热硬度
heating heat 暖气热量
heating heat pump 供暖热泵
heating heat requirement 暖气要求
heating hood 加热罩;暖气装置外壳;暖气装置防护套;暖气装置挡板
heating hose coupling 暖气软管接头
heating index 热指标
heating inductor 加热电感器
heating industry 供热企业;供热工业
heating in salt bath furnace 浴炉加热
heating inset 电炉加热元件
heating installation 加热装置;加热器;供热装置;暖装置;暖气设备;取暖设备
heating-insulating material 隔热材料
heating insulation 热绝缘
heating isolation 热隔绝
heating jacket 加热(夹)套;保温套;暖气装置外套
heating jumper 采暖连接线
heating kettle 加热壶;加热锅;加热釜;暖水壶
heating lamp 焊接灯;喷灯
heating layout 采暖平面图
heating lead 加热引线
heating limit 发热限制
heating line 采暖管线
heating line of train 列车采暖线路
heating load 供暖负荷;热负载;热负荷;供热量
heating loa diagram 热负荷图
heating load duration graph 热负荷延续时间图
heating loss 加热损失
heating machine 加热机;供热机
heating main 供暖总管;供热干管;暖气总管
heating mantle 加热(外)罩;保温套;暖气罩
heating means 加热方法
heating medium 供热介质;载热体;加热介质;热媒
heating medium for high temperature 高温载热体
heating medium inlet 热载体入口
heating medium outlet 热载体出口
heating medium parameter 热媒参数
heating microscope 高温显微镜
heating mixer 加热拌和机
heating muff 加热套筒
heating network with condensating pipe 有凝结水管热网
heating network without condensating pipe 无凝结水管热网

heating of building and home 房屋取暖
heating of contact 触点发热
heating of turbine 暖机
heating oil 加热用油;燃料油
heating oil barrier 油加热栅栏;油加热隔阻器
heating oil-fired central heating 燃油集中供热
heating oil interceptor 油加热截流器
heating oil pumping station 燃料油泵送站
heating oil-resistant 加热油阻
heating oil storage 燃料油库
heating oil storage tank 燃料油罐;燃油储[贮]罐;燃油储[贮]存柜
heating oil store 燃油库;燃料油商店;燃料油房;燃料油堆栈;燃料油仓库
heating oil stripper 燃料油卸料器;燃料油洗提器
heating oil supply system 燃料油供应系统
heating oil tank 燃料油罐
heating oil-tight 燃油密封;加热燃油密封
heating operation 加热操作
heating or cooling medium 加热或冷却介质
heating oscillator 高频加热器;加热振荡器
heating output 热输出力;热输出功率;热输出(量)
heating panel (埋入墙内的)盘管取暖板;加热板;加热板;散热板;玻璃电热板;板式取暖器
heating passage 传热;暖气通道
heating peak 放热峰
heating period 加热期;加热阶段;烘窑期;升温期;采暖期
heating period of the engine 发动机加热延续时间
heating pipe 供暖管;加热管;加气管;暖气管;热(力)管
heating pipeline 采暖管道
heating piping 供热管
heating piping system drawing 采暖管道系统图
heating plan 暖气布置图
heating plant 集中供暖系统;供热装置;供热设备;供暖设备;加热设备;加热机房;暖气设备;暖气厂;采暖装置
heating plant room 暖气设备室;加热设备室;锅炉房
heating plant stack 加热设备烟囱;锅炉房烟囱
heating plate 加热板
heating platen 热板
heating plug 火花塞
heating pot 加热坩埚;加热壶;加热锅
heating power 供暖能力;热力;发热能力
heating process 加热过程
heating pump 供暖输送泵
heating radiator 热散射器;暖气(炉)片;供暖散热片;热辐射器;采暖器
heating radius 供热半径
heating rate 耗热率;加热速率;加热速度;升温速率;升温速度;单位热流;比热率;热耗
heating-rate curve 升温速率曲线
heating reflow coating 热再流平涂层
heating region 采暖地区
heating residue 加热涂料的残余
heating resistance 耐热值
heating resistant enamel 耐热瓷漆
heating resisting metal 耐热金属
heating resistor 加热电阻(器)
heating return pipe 采暖回水管
heating riser 采暖系统立管;采暖的下供立管
heating rod 加热棒
heating roller 加热辊
heating schedule 加热制度;加热规范
heating season 供暖季节;采暖期;采暖季节;供暖期
heating sink 热耗
heating sludge-digestion tank 加热污泥消化池
heating source 加热源
heating spiral 加热盘管;暖气盘旋管
heating spraying 热喷涂
heating stage 加热台
heating steam 加热(用)蒸汽;热蒸汽
heating stove 火炉
heating strip 取暖带
heating stylus 加热针
heating substation 热力点
heating supply area 受热面面积
heating supply pipe 供热管
heating supply tube 供热管
heating supply tubing 供热管

heating surface 蓄热面积;加热(表)面;受热面; 放热面
heating surface area 加热面积;受热面面积;受热面积;散热面积
heating surface area rating 加热表面积额定能力
heating surface of boiler 锅炉受热面
heating surface of the firebox 火箱受热面
heating surface of tubes 管受热面
heating system 加热装置;供暖系统;加热系统;暖气系统;采暖(供热)系统
heating tape 绝热胶布;加热带
heating tape unit 加热带单元
heating technician 采暖技术员;采暖工程师;供暖技术员;供暖技师;供暖技工
heating temperature of phosphor 磷光体加热温度
heating test 加热试验;加热检查
heating time 加热时间;增温时间;供暖时间;受热时间
heating time of phosphor 磷光体加热时间
heating tongs 加热钳;火钳
heating torch 加热炬;气焊枪
heating training 热锻
heating transfer 传热;热交换
heating transfer medium 传热介质
heating transfer property 传热性能
heating transfer rate 传热率
heating transfer resistance 传热总热阻
heating transfer velocity 传热速度
heating transformer 灯丝变压器
heating transmission 传热;热交换
heating transmission coefficient 传热系数
heating treating film 热处理膜
heating trench 暖气管沟
heating trough 供暖管道
heating tube 热管;加热管
heating tube in section of convection(al) chamber 对流段炉管
heating tube in section of radiation chamber 辐射段炉管
heating tube support 炉管支承架
heating type 供暖形式
heating unit 加热装置;热单位;供热装置;采暖单元;供暖装置;供暖机组;电热元件;发热体
heating-up 加热;烤窑;升温
heating-up crack 升温裂纹
heating-up curve 升温曲线;烘窑曲线
heating-up flue 加热烟道
heating-up from cold 从冷态升温
heating-up loss 升温热损失
heating-up period 加热期;点火期
heating-up temperature 加热温度
heating-up time 加热时间;烤窑时间;升温时间
heating-up zone 加热区;加热段
heating value 发热量;热值;供热值
heating value determination 热值测定
heating value of digester gas 沼气热值
heating value of gas 气体热值
heating-ventilating and cooling 加热通风和冷却
heating-ventilating assembly 供暖通风(两用)机组;采暖通风两用机组
heating-ventilating unit 供暖通风机组
heating-ventilation 供暖通风
heating voltage 加热电压
heating wall 火墙;燃烧室墙
heating warpage test 热畸变试验;热移定性试验
heating water 取暖用热水;加温用热水
heating water pipe 取暖热水管
heating water return 采暖回水管
heating water supply 采暖供水管
heating wire 高电阻丝;电热丝;电炉丝
heating wire method 电热丝法
heating with humidifying 增湿加热
heating with recirculated air 再循环暖气供暖
heating with waste heat 余热采暖;废热采暖
heating works 采暖工程
heating zone 加热区;加热段;加热带
heat injection noise 热喷射噪声
heat injury 灼伤;热损伤
heat input 热负荷;供热;热输入(量);热量输入;热量补给
heat instability 热不稳定性
heat insulant 隔热材料;保温材料
heat-insulated 隔热的;绝热的
heat-insulated container 隔热容器

heat-insulating 绝热的;热绝缘的
heat insulating belt 带状保温材料
heat insulating blanket 隔热毡
heat insulating block 隔热砌体;隔热块;绝热块体
heat insulating board 隔热(纸)板
heat insulating brick 隔热砖
heat insulating bucket 保温水桶
heat insulating capacity 隔热性能;保暖性能
heat insulating coat(ing) 热绝缘涂层;隔热涂层;隔热保护层
heat insulating concrete 隔热混凝土;绝热胶泥;绝热混凝土
heat insulating course 绝缘层;隔热层
heat insulating cover 保温管套
heat insulating efficiency 隔热效率
heat insulating glass 隔热玻璃;绝热玻璃
heat insulating glazing unit 隔热玻璃构件
heat insulating hanging 隔热帷幔
heat insulating hatchcover 绝热舱盖
heat insulating layer 绝缘层;隔热层;绝热层
heat insulating mass 隔热材料
heat insulating material 隔热材料;绝热材料;热绝缘材料;保温材料
heat insulating mat(t) 隔热毡
heat insulating mortar 隔热砂浆;保温砂浆
heat insulating paper 隔热墙纸
heat insulating plaster 隔热抹灰
heat insulating power 隔热能力
heat insulating property 隔热性能;绝热性能;保温性能
heat insulating quality 隔热质量
heat insulating quilt 隔热毡
heat insulating refractory 隔热耐火材料
heat insulating rope 保温绳
heat insulating sheet 隔热薄片
heat insulating shield 绝热板
heat insulating slab 隔热板;绝热板
heat insulating structural panel 隔热结构墙板
heat insulating tile 隔热瓦;隔热面砖
heat insulating washer 绝热垫圈
heat insulating work 保温工程
heat insulating zone 隔热区;保温区
heat insulation 隔热(保温层);绝热(保温层);热的绝缘;保温
heat insulation and heat control 隔热降温
heat insulation and refractory material 绝热及耐火材料
heat insulation board 保温板
heat insulation brick 隔热砖
heat insulation coefficient 隔热系数
heat insulation course 隔热层
heat insulation efficiency 隔热效率
heat insulation glass 绝热玻璃;隔热玻璃
heat insulation-grade material 隔热材料
heat insulation-grade plaster 隔热抹灰
heat insulation material 隔热材料
heat insulation material of asbestos fibre 石棉纤维绝热材料
heat insulation measures 隔热措施
heat insulation tape 绝热带
heat insulator 隔热物;绝热体;热绝缘子;热绝缘体;热绝缘材料;保温材料
heat intensity 热强度
heat intensity of hearth 炉底热强度
heat-intercepting glass 遮绝热玻璃;隔热玻璃
heat interchange 热交换;互换
heat interchanger 换热器;回热器;热交换器
heat in the exterior 表热
heat-intolerant 不耐热的
heat intra vane pump 内叶片热泵
heat inverter unit 热转换器
heat island (城市的)热岛现象;热岛
heat island effect 热岛效应
heat joining 热接合
heat kill 热杀
heat killed 热杀的
heat label 热标记;测温纸
heat-labile 不耐热的;热不稳定的
heat-labile factor 不耐热因素
heat-labile test 热不稳定试验
heat lag 热延迟
heat lamp 电热灯
heat-latent 潜热
heat law 热学定律

heat leak(age) 热漏泄;热漏失;热渗透
heat leakage test 隔热试验
heat leaving mill 离磨热量
heatless lehr 无光玻璃退火炉
heat liberation per unit furnace volume 单位炉容放热量
heat liberation per unit heating surface 单位加热面放热量
heat liberation rate 放热(速)率
heat lightning 热闪;无雷声闪电
heat load 热负荷;载热量
heat loading 热负荷
load of furnace 熔窑热耗;熔窑热负荷
heat localization 热积聚
heat loss 热消耗;热损耗;热(量)损失
heat loss by infiltration 冷风渗透耗热量
heat loss calculation 热损失计算
heat loss computation 热损耗计算
heat loss estimate 热损失估计
heat loss function 热损失函数
heat loss in exit gas 废气热损失
heat loss of pipeline 管道热损失
heat loss rate 散热率
heat loss through walls 墙壁热损失
heat low 热低压【气】
heat luminescence parameter 热放光参数
heat luminescent method 裂变径迹方法
heat machine 热机
heat map 热红外分布图
heat mark 加热痕
heat mat 隔热垫
heat medium 热介质
heat medium oil 热载体油
heat melting-down 热熔化
heatmeter 量热计
heat meter with micro-processor 带有微处理器的热量表
heat mixed paper 隔热纸
heat mixed value 热对流值
heat moisture ratio 热湿比
heat motion 热运动
heat movement 热量传动;热量移动
heat nuisance 热干扰
heat number 热指数
heat of ablation 消融热;烧蚀热
heat of absorption 吸(收)热;吸(附)热
heat of absorption detector 吸附热检测器
heat of activation 活化热
heat of adhesion 附着热
heat of aggregation 聚集热
heat of association 缔合热
heat of carbonization 碳化热
heat of clinker formation 熟料生成热
heat of combination 化合热
heat of combustion 热值;热当量;燃烧热;发热量
heat of compression 压缩热
heat of condensation 冷凝热;凝结热
heat of content 燃烧热
heat of conversion 转化热
heat of cooling 冷却热
heat of crystallization 析晶热;结晶热
heat of decomposition 分解热
heat of dehydration 脱水热
heat of denaturation 变性热
heat of desorption 退吸热
heat of dilution 稀释热;冲淡热
heat of dissociation 离解热;分解热
heat of dissolution 分解热
heat of emission 发射热
heat of evapo(u)ration 汽化热;蒸发热
heat of fermentation 发酵热
heat off hinge 抽心铰链;抽心合页
heat of formation 形成热;生成热
heat of freezing 凝固热
heat of fusion 聚变热;熔融热;熔解热;熔化热;溶化热
heat of fusion of snow 融雪热
heat of gasification 气化热(液)
heat of hardening 硬化热;凝结放热量;凝固热
heat of heat 熔热
heat of hydration 水化热;水合热
heat of hydrogenation 氢化热
heat of immersion 浸渍热
heat of initial set(ting) (混凝土的)初凝热

heat of ionization 电离热
heat of irreversible reaction 不可逆反应热
heat of mixing 混合热
heat of molecular orientation 分子定向热
heat of oxidation 氧化热
heat of polymerization 聚合热
heat of radiation 辐射热
heat of reaction 反应热
heat of setting 硬结热;凝结热
heat of solidification 固化热;凝固热
heat of solution 溶解热
heat of solution method 溶解热法
heat of sorption detector 吸着热检测器
heat of stirring 搅拌热
heat of sublimation 升华热
heat of superheat 过热热量
heat of superheating 过热热量
heat of the earth's interior 地热
heat of transformation 相变热;转变热
heat of vapo(u)rization 汽化热;气化热(液);蒸发热
heat of vapo(u)rization for free water 自由水面汽化热
heat of vapo(u)rization for grain 谷物汽化热
heat of wetting 润湿热;湿润热;湿化热
heat only boilers 纯供热锅炉
heat-operated refrigerating system 热风制冷系统
heat-operated refrigeration 热力制冷
heat outlet 热气出口
heat outlet washer 导热垫圈
heat output 供热量;热值;热(量)输出;热功率;输出热量;传热量;发热量
heat output of a reactor 反应堆的热功率
heat pack 热装罐
heat passage 热道;热传递;传热的通路
heat pattern 加热曲线
heat penetration 加热深度;热透(入)
heat perception coefficient 热感系数
heat permeability 透热性;热渗透(性)
heat pick-up 热敏传感器
heat picture 热图像(红外线暗视器摄取的图像)
heat pipe 热(导)管;导热管
heat-pipe motor 热управление冷电动机
heat pipe of capillarity 毛细热管
heat pipe preheater 热管式空气预热器
heat plant 供热设备;供热车间
heat plasticization 热塑炼
heat plug 热插头(马达上的)
heat pollution 热污染
heat pollution effect 热污染效应
heat pollution monitoring 热污染监测
heat pollution source 热污染源
heat polymer 热致聚合物;热聚物
heat polymerization 热聚合(作用)
heat polymerization rubber 热聚橡胶
heat polymerized oil 热聚合油
heat pour method 热法法
heat power 热功率;热动力
heat power engineering 热工学
heat power plant 火力发电站;热电站
heat power station 火力发电站;热电站
heat power supply 热力供应
heat precontrol 预控热
heat pressure 热压
heat pressure method 热压法
heat preventing coating 防热涂层
heat price 热能价格;热价
heat probe 探热器
heat process 热加工
heat producer 热源
heat producing capability 发热能力
heat-producing chemical reaction 放热化学反应
heat production 热(能)生产;热能产量;热的产生;生热(性);产发热(量);放热;发热(量)
heat production rate 热能生产率
heat productivity 热产率;发热量
heat productivity of K 钾的热产率
heat productivity of K-40 钾40的热产率
heat productivity of Rb 铷的热产率
heat productivity of Rb-87 铷-87的热产率
heat productivity of Th-232 钍232的热产率
heat productivity of U-238 铀238的热产率
heat productivity of uranium 铀的热产率
heat-proof 不透热的;不传热的;隔热的;保温的;隔声;抗热;耐热(的);热稳定的;防热(的)
heat-proof glass 隔热玻璃;耐热玻璃
heat-proof laminated rubber hose 耐热夹布胶管
heat-proof material 耐热材料
heat-proof porcelain 耐热陶瓷;耐热瓷器
heat-proof quality 隔热性能
heat prostration 热衰竭
heat protecting filter 隔热滤光镜
heat protection 抗热;防热
heat-protection layer 热保护层;防热层
heat-protection shield 隔热罩
heat-protection system 防热系统
heat-protective clothing 热防护服
heat-protective coating 隔热保护层
heat-protective surface 热防护面;保温面
heat prover 废气和排出气体分析器;废气和排出气体分析仪
heat pugmill(mixer) 热捏和机;热搅拌机
heat pulse 热(能)脉冲
heat pump 供热泵;热(力)泵;气泵
heat pump air conditioner 热泵式空(气)调(节)器
heat pump drier 热泵式干燥机
heat pump driven by combustion engine 内燃机驱动的热泵
heat pump driven by diesel engine 柴油机驱动的热泵
heat pump driven by gas engine 煤气内燃机驱动的热泵
heat pump using exhaust air 利用废气的热泵
heat pump using outside air 利用外部空气的热泵
heat quantity 热量
heat quilt 隔热毡
heat radiated from the flame 火焰热辐射
heat radiating by conduction 传导散热
heat radiating by evapo(u)ration 蒸发散热
heat radiating equipment 热辐射设备
heat-radiating imaging 热辐射成像
heat-radiating wave 热辐射波
heat radiation 热辐射;散热;辐射传热
heat radiation device 热回收装置
heat radiation intensity 热辐射强度
heat radiation pyrometer 光学测温计;光测高温器;光测高温计;热辐射高温计
heat radiation sensing device 热辐射传感设备;热辐射敏感元件
heat radiator 热辐射体;热辐射器
heat radiometer 热辐射计
heat range of heat-supply service 供热半径
heat rate 加热率;热消费率;耗热率;热耗
heat rating 热功率
heat ray 热线;红外光;红外(线);热射线;放射线
heat reactive phenolic resin 热反应性酚醛树脂
heat reactive resin 热反应性树脂
heat reactivity 热反应性
heat reactor 供热反应堆;核供热堆;热化反应堆
heat receiver 吸热体;受热器
heat reclaim 热(量)回收
heat reclamation system 热量回收系统
heat recovering tower 热量回收塔
heat recovery 余热回收
heat recovery boiler 余热锅炉;废热回收锅炉;发热锅炉
heat recovery box 热回收箱
heat recovery system 热(量)回收系统
heat-recovery unit 热回收装置
heat-recovery wheel (空调装置中的)空气冷热调节轮;热回收轮
heat recuperating 余热利用;蓄热;热回收
heat recuperator 换热器;热回收装置
heat-reducing filter 滤热片;滤热玻璃
heat refining 加热细化
heat reflectance 热反射率
heat-reflecting 热反射的
heat-reflecting film 光热分离膜
heat-reflecting glass 热反射玻璃
heat-reflecting material 热反射材料
heat reflecting paint 热反射漆
heat reflection 热消耗;热反射(能力)
heat reflection filter 热反滤光器
heat-reflective coating 热反射涂层;热发射层
heat-reflective glass 热反射玻璃
heat reflectivity 热反射性;热反射率
heat refused joint 热熔接缝
heat regeneration 回热;热量再生
heat regenerator 蓄热室;热量回收
heat regime 热量状况;热动态
heat regulating lever 调热杆
heat regulation 热(量)调节
heat regulator 热(量)调节器;调器
heat rejection 热损失;排热;废热排放
heat-rejection circuit 排热回路
heat-rejection equipment 废热排放设备
heat-rejection load 暖热量
heat-rejection temperature 废热温度
heat release 热释放;热放出;释放热(量);散热量;放热
heat release from flame 火焰放热(量)
heat release kinetics 热释放动力学
heat release luminescent method 热释发光法
heat release rate 炉膛容积热强度;放热率
heat release rate calorimeter test 释热容量热计试验
heat release rate of grate area 炉排面积放热率
heat releasing fluid 流体载热剂
heat releasing medium 放热介质
heat removal 热量排除;排热
heat-removal equipment 排热设备
heat-removal mechanism 排热机理
heat-removal system 排热系统;除热系统
heat-removing agent 排热剂
heat-repellent 抗热;防热
heat-repelling 抗热的;防热的
heat requirement 热需要量;热能需求量
heat reserve 热储备
heat reserving curing 保温养护
heat reserving work 保温工程
heat reservoir 蓄热器;储[贮]热器;热源;热库;储热器
heat resistance 耐热(性);抗热性;热阻;热强性;耐热(度)
heat-resistance cement plate 耐热水泥板
heat-resistance concrete 抗热型混凝土
heat-resistance current meter 耐热流速仪;耐热流速计
heat-resistance index (沥青混凝土的)抗热指数
heat-resistance of ca(u)lk 油膏耐热度
heat-resistance of malthoid 油毡耐热度
heat-resistance paint 耐高温漆
heat-resistance test 热稳定性试验;耐热试验
heat-resistant 抗热(的);耐温;耐热的;防热的
heat-resistant alloy 耐热合金
heat-resistant bulb 耐热玻璃瓶
heat-resistant cast iron mo(u)ld 耐热铸铁模
heat-resistant concrete 抗热性混凝土;耐热混凝土
heat-resistant core 耐热芯子
heat-resistant enamel 耐热瓷釉;耐热搪瓷
heat-resistant enamel paint 耐热瓷漆
heat-resistant explosive 耐热炸药
heat-resistant fabric 耐热织物
heat-resistant finish 耐热面漆
heat-resistant furnace 耐热炉
heat-resistant glass 耐热玻璃
heat-resistant lubricant 耐热润滑剂
heat-resistant material 耐热材料
heat-resistant motor 高温电动机;耐热电动机
heat-resistant oil 耐热油品
heat-resistant paint 耐热(油)漆;耐热涂料;耐高温漆
heat-resistant pigment 耐热颜料
heat-resistant plant 耐热植物
heat-resistant plastic nozzle 耐热塑料喷管
heat-resistant plastics 耐热塑料
heat-resistant polymer 耐热聚合物;耐高温聚合物
heat-resistant quality 耐热特性
heat-resistant resin matrix 耐热树脂基体
heat-resistant rubber 耐热橡胶
heat-resistant steel 耐热钢
heat-resistant steel chain 耐热钢链条
heat-resistant steel electrode 耐热钢焊条
heat-resistant unit 耐热玻璃设备
heat-resisting 耐热(的);耐高温的;防热的
heat-resisting aggregate 耐热集料
heat-resisting alloy 耐热合金
heat-resisting alloy steel 耐热合金钢
heat-resisting alumin(i)um alloy 耐热铝合金
heat-resisting alumin(i)um paint 耐热铝粉漆
heat-resisting cable 绝缘耐热电缆
heat-resisting casting 耐热铸件

heat-resisting cast iron 耐热铸铁
heat-resisting cement plate 耐热混凝土板
heat-resisting coating 耐热涂层
heat-resisting coefficient 热阻(力)系数
heat-resisting concrete 耐热混凝土
heat-resisting durability 耐热持久性
heat-resisting enamel 耐热搪瓷
heat-resisting exhauster 耐热风机
heat-resisting fabric 耐热布
heat-resisting gasket 耐热衬垫
heat-resisting glass 抗热玻璃;耐热玻璃
heat-resisting glass-ceramic coating 耐热微晶玻璃涂层
heat-resisting grease 耐热滑脂
heat-resisting iron 耐热铸铁
heat-resisting material 耐热材料
heat-resisting paint 耐热油漆;耐热涂料;耐热漆
heat-resisting plant 耐热植物
heat-resisting porcelain 耐热瓷
heat-resisting property 耐热性能
heat-resisting quality 耐热特性
heat-resisting rubber 耐热橡胶
heat-resisting steel 耐热钢;热强钢
heat-resisting test 耐热试验
heat-resisting tile 耐热贴面
heat-resisting tire 耐热轮胎
heat-resistivity 耐热性;热阻率
heat-resist metal 耐热金属
heat-resistor 耐热器
heat responsive detector 热敏探测器
heat responsive element 热响应元件;热敏元件
heat responsive time 热响应时间
heat retainer 蓄热体;保温器;保体;保热器
heat retaining 储[贮]热能力;保温的;热保持
heat retaining capacity 保温能力;保热容量;保温能力
heat retaining mass 储热物质;蓄热体
heat retaining property 保温性
heat retention 蓄热
heat retention capacity 蓄热能力
heat return 回热
heat rise in mass concrete 大体积混凝土水化热;大体积混凝土内热量升高;大块混凝土水化热
heat riveting 热铆
heatronic 高频电热的
heatronic mo(u)lding 高频电热模塑法
heatronic welding 高频电热焊接
heat run 老炼试验;加热试验;热运转;热试转;热试车;发热试验
heat run test 长期运行发热试验
heat-rupture test 高温强度试验;热破坏性试验
heat-saving 节热(的)
heat saving apparatus 省热器
heat scatter(ing) 热散射
heat screen 隔热屏(障);绝热屏
heat-seal 熔焊(接);热封
heat sealability 热封性能(塑料薄膜袋)
heat-sealer 热封缄器
heat-seal film over-wrapping machine 包装薄膜热封机
heat seal(ing) 热灌(缝);热封(合);热封接;热焊接
heat sea(ling) adhesive 热熔性胶黏剂
heat seal(ing) coating 热封涂层
heat seal(ing) machine 熔接机;热封机
heat seal method 热合法
heat seeker 热导引头;热跟踪头
heat-seeking 热辐射导引
heat-seeking guidance system 热辐射导引系统
heat seeling coating 热瓷涂布
heat select 加热选择
heat sensimeter 热感测定仪
heat-sensing 热灵敏
heat sensing device 感温器;热(敏)传感器;热灵敏装置;热灵敏器件
heat-sensitive 热敏的
heat-sensitive adhesive agent 热敏黏合剂
heat-sensitive cable 热敏电缆
heat-sensitive detector 感温探测器
heat-sensitive device 热敏器件
heat-sensitive element 热敏元件
heat-sensitive eye 热敏元件;热辐射探测器
heat-sensitive ferrite 热敏铁氧体
heat-sensitive latex 热敏性胶乳
heat-sensitive material 热敏性材料;示温物质

heat-sensitive paint 热变色(油)漆;热敏油漆;热变颜料;示温漆
heat-sensitive paper 热敏纸
heat-sensitive sensor 热敏传感器
heat-sensitive transducer 热敏传感器
heat-sensitive transfer sheet 热敏转印纸
heat-sensitive varnish 热敏油漆
heat sensitivity 热敏性;热灵敏度
heat sensitivity of explosive 炸药热感度
heat sensitization 热敏化(作用)
heat sensitized mixing 热敏混炼
heat sensitizing agent 热敏(化)剂
heat sensor 热敏元件;热传感器
heat separation 热选法
heat service 高温下使用
heat set 热固着
heat set ink 热固着油墨;热干油墨;速干印油;热干漆
heat settability 热定形性
heat setter 热定形机
heat set(ting) 热定形;快干;热固性的;热凝结;热凝(固);热定形;加热凝结
heat-setting bonding 热定形黏合
heat-setting jointing material 热硬性黏接材料
heat-setting machine 热定形机
heat-setting mortar 热硬性水泥砂浆;热固性火泥
heat-setting refractory 热固性耐火材料
heat-setting refractory mortar 高温硬耐火砂浆
heat shield 隔热层;隔热罩;隔热屏(障);隔热板;热防护屏;热防护层;挡热罩;防热罩;防热层;避热罩
heat-shielded cathode 热屏蔽阴极;保温阴极
heat shielder 热屏(蔽)
heat shield(ing) 热屏(蔽);防热
heat-shielding performance 隔热性能
heat shield panel 防热板
heat shimmer 热流闪烁
heat shock 热振(荡);冲击;热骤变
heat-shock resistance 抗热振性;耐热冲击
heat shortness 热脆性
heat shrinkable 热缩的
heat shrinkable fiber 热收缩纤维
heat-shrinkable tubing 热缩性管材
heat shrinkage material 热收缩材料
heat-shrink plastic shield 热缩塑性防护层
heat-shrink plastic tubing joint 热缩塑性套管接头
heat shroud 热遮板;防热套
heat shunt 热分流器
heat sidewise 侧向热运动
heat sink 吸热装置;吸热设备;吸热器;吸热部件;整流器散热片;冷源;热沉;受热器;散热装置;散热器;散热片吸热;散热片
heat-sink compound 散热器填料
heat-sink cooling 散热冷却
heat-sink copper 吸热铜;铜吸热剂
heat-sink design 热沉系统设计
heat-sinking capability 散热能力
heat-sinking capacity 吸热容量
heat-sink shell 吸热层;热沉层
heat-sink shield 消热罩;吸热罩;热沉罩
heat-sink strip 散热条
heat-sink system 热沉系统
heat size 炉容量
heat skin detection 热表皮探测
heat soak-back 热回放
heat soak-back temperature 热回放温度
heat-soaked turbine 热态汽轮机
heat softened resin 热软化树脂
heat softening 热塑炼
heat sonic screening machine 热声筛选机
heat-sound automatic recorder 热声自动记录仪
heat-sound method 热声法
heat source 供热源;热源
heat-source drone target 热源靶机
heat source movement 热源移动
heat source rock 热源岩体
heat spall 热力剥落
heat spot 过热点
heat stability 热稳定性
heat stability of malthoid 油毡热稳定性
heat stability test 热稳定性试验
heat stabilization test 热稳定(性)试验
heat-stabilized bearing 热稳定轴承
heat-stabilized compressed 加热压缩材

heat-stabilized compressed wood 加热稳定压缩木材
heat-stabilized wood 热塑(木)材
heat stabilizer 热稳定剂
heat stable 热稳定的
heat stable anti-oxidant 热稳定防老剂;防热老化剂
heat stable inhibitor 热稳定剂
heat stable material 耐热材料
heat stable plant 耐热植物
heat stagnation 热滞
heat standing wave ratio 热驻波比
heat steadiness 热稳定性
heat sterilization 热消毒;热灭菌;热力灭菌法
heat storage(capacity) 热储量;热容量;蓄能(能力);储热能力;储[贮]热能力;供热能力;热储[贮]藏;热仓库;(材料的)热的储[贮]存;热蓄积;储热
heat storage characteristic 蓄热特性
heat storage index 蓄热指数
heat storage maintenace 蓄热养护
heat storage material 蓄热材料
heat storage power 蓄热能力
heat storage property 蓄热性能
heat storage quality 蓄热性能
heat storage quilt 保暖垫
heat storage unit 储热器
heat storage well 储热井
heat-stored curing 蓄热养生;蓄热养护(混凝土)
heat stored in the country rocks 围岩储[贮]存的热量
heat storing capacity 蓄热能力
heat straightening 热矫正
heat stream 热流
heat strength 钢化
heat-strengthened 钢化的;热处理的;淬火的
heat-strengthened glass 热强化玻璃;钢化玻璃
heat stress 热应力;热应激
heat stress index 热应力指数;热应力指标;热强度指数
heat stress resistance factor 热应力抵抗因子
heat stretch 热拉
heat stretched fibre 热拉伸纤维
heat-stretch factor 热膨胀系数
heat stretching machine 热拉伸机
heat stretch zone 热延伸区(直接成条机的)
heat stroke 热射病;冒暑
heatstroke prevention 防暑(降温)
heatstroke preventive 防暑药
heat summation 热量总和
heat supply 供热;热源;热能供应;热补给
heat supply distribution and control station 热网站
heat supply installation 供热装置
heat supply network 热能供应网;热(力)网
heat supply pipe(line) 供热管道;供热管线;供热导管
heat supply program(me) 供热方案
heat supply set 供热机组
heat-supply system 供热系统
heat-supply system based upon geothermal energy 地热供热系统
heat-supply system based upon heating plant 区域锅炉房供热系统
heat-supply system based upon heating power cogeneration plant 热电厂供热系统
heat-supply system based upon industrial waste heat 工业余热供热系统
heat-supply system based upon low temperature nuclear reactor 低温核供热系统
heat supply unit 供热机组
heat surge 温度突然上升
heat switch 供热开关;热开关
heat system 热量系统;热力系统
heat temperature 热量温度
heat test(ing) 耐热试验;加热试验;热试验
heat test method 热试验法
heat the slide over an alcohol hurner 在酒精灯上将载玻片加热
heat throughput 发热量
heat thunderstorm 热雷暴
heat-tight machine 耐热电机
heat tint 氧化膜色;回火色
heat tinting 热染色;氧化着色;加热(氧化)着色;烘染;升温着色

heat tolerance threshold 耐热阈
heat tone 加热时的颜色
heat-to-power efficiency 热能变成电能过程的有效作用系数
heat-toughened 热韧化(处理)
heat transfer 换热;热量输送;热交换器;热传输;热传递;热传导;导热
heat-transfer agent 载热剂
heat-transferal 热传导
heat-transfer area 传热面积
heat-transfer arrangement 热传递装置
heat-transfer by conduction 导热;传导(传)热
heat-transfer by convection 对流传热
heat-transfer by radiation 辐射散热;辐射传热;散热
heat-transfer capacity 传热能力
heat-transfer character 传热特性
heat-transfer coefficient 热转移系数;水面综合散热系数;输热系数;传热系数
heat-transfer coil 传热蛇管;传热盘管
heat-transfer computer 热传递计算机
heat-transfer conductance 热导
heat-transfer control 传热控制
heat-transfer cycle 传热循环
heat-transfer element 热传元件
heat-transference 热传导
heat-transfer equation 热传导方程;传热方程
heat-transfer equipment 传热设备
heat-transfer factor 传热因数;热传输系数
heat-transfer fluid 载热流体;流体载热剂;传热媒质;传热流体
heat-transfer head 传热(测量)头
heat-transfer index 传热指数
heat-transfer in fluidized bed 流动床传热
heat-transfer intensification 传热的强化
heat-transfer intensification of heat exchanger 换热器传热的强化
heat-transfer intensity 传热强度
heat-transfer labeling equipment 印制标签的传热装置
heat-transfer level 传热水平
heat-transfer loop 传热回路
heat-transfer machine 传热机
heat-transfer material 载热体;热转移介质
heat-transfer mechanism 热量传递机制;传热机理
heat-transfer medium 传热介质;载热剂;传热媒质
heat-transfer medium inlet 传热介质入口
heat-transfer medium inlet and outlet 热载体出入口
heat-transfer medium outlet 传热介质出口
heat-transfer metal 热传导金属
heat-transfer of geothermal energy 地热能的热量传递
heat-transfer oil 载热油;传热油
heat-transfer passage 导热孔道
heat transfer printing 热转移印花
heat-transfer process 传热过程
heat-transfer property 传热性
heat-transfer rate 热流量;热负荷;比热流;导热速度;传热速率
heat-transfer reactor 换热反应器
heat-transfer salt 热传导盐;传热盐
heat-transfer surface 传热面
heat-transfer system 热转换系统
heat-transfer theory 换热理论;热交换理论;热传导理论;传热理论
heat-transfer through curved wall 非平壁导热
heat-transfer through panel wall 平壁导热
heat-transfer through wall 墙壁传热
heat-transfer unit 传热装置
heat transform 热的交换
heat transmissibility 传热性
heat transmission 热流;热交换;热传递;热传导
heat transmission coefficient 传热系数
heat transmission cost 热能输送成本
heat transmission efficiency 传热效率
heat transmission factor 传热因子
heat transmission in soil 土中热传导
heat transmittance 热透射率;传热系数;传热性
heat transmitter 热传递器;传热器
heat transmitting 热透射
heat transmitting glass 透红外线玻璃;透热玻璃;导热玻璃
heat transport 热输运;热量输送
heat transport(ing)fluid 载热流体;流体载热剂

heat transport system 热传输系统
heat trap 吸热器;热收集器
heat-trapping ability 致温室效应气体的吸热能力;截热能力
heat-treatability 可热处理性;热处理性
heat-treatable 可热处理的
heat-treatable alumin(i)um alloy 热处理铝合金
heat-treatable steel 热处理钢
heat-treated 热处理的;经过热处理的
heat-treated alloy steel 热处理的合金钢
heat-treated condition 热处理状态
heat-treated flat glass 热处理平板玻璃
heat-treated frog 热处理辙叉
heat-treated glass 热处理玻璃;钢化玻璃
heat-treated metal 热处理金属
heat-treated metal foil 热处理金属箔
heat-treated nail 热处理钉
heat-treated plate glass 钢化平板玻璃
heat-treated polished plate glass 钢化镜面平板玻璃
heat-treated rail 热处理钢轨
heat-treated safety glass 钢化安全玻璃
heat-treated steel 淬火钢;热处理钢
heat-treated steel bar 热处理钢筋
heat-treated steel track 热处理钢履带
heat-treated structure 热处理组织
heat-treated surface 热处理过的表面
heat-treated track pin 热处理履带销
heat-treated track shoe 热处理履带瓦
heat-treating 热处理
heat-treating condition 热处理条件
heat-treating department 热处理车间
heat-treating facility 热处理设备
heat-treating film 热处理(氧化)膜
heat-treating furnace 热处理炉
heat-treating machine 热处理机
heat-treating oil 热处理(用)油
heat-treating regime 热处理制度
heat-treating waste 热处理废物
heat-treatment 加热处理;热处理
heat-treatment after forming 成型后的热处理
heat-treatment between working procedures 工序间热处理
heat-treatment condition 热处理状态;热处理条件
heat-treatment crack 热处理裂纹
heat-treatment cycle curve 热处理工艺曲线
heat-treatment department 热处理车间
heat-treatment fixture 热处理设备
heat-treatment for shrinkage relief 消除收缩的热处理
heat-treatment furnace 热处理炉
heat-treatment furnace for stress relieving 消除应力热处理炉
heat-treatment line 热处理作业线
heat-treatment liquor 热处理液
heat-treatment method 热处理法
heat-treatment of basalt paving setts 玄武岩铺路小方石的热处理
heat-treatment of glass 玻璃热处理
heat-treatment of optic(al)crystal 光学晶体热处理
heat-treatment of sludge 污泥(加)热处理
heat-treatment of steel 钢的热处理
heat-treatment oil 热处理用油
heat-treatment practice 热处理施行
heat-treatment process 热处理过程
heat-treatment protective coating 热处理保护涂料;热处理保护涂层
heat-treatment regime(n) 热处理规范
heat-treatment shop 热处理车间
heat-treatment technics 热处理技术
heat-treatment temperature 热处理温度
heat-treatment verification test 热处理验证试验
heat-treatment with laser beam 激光束热处理
heat-treatment zone 热处理区
heat-trickling filter 热滴滤器
heat under reflux 回流加热
heat unit 热(量)单位;热辐射体
heat unstability test 热不稳定试验
heat-up burner 烤窑用燃烧器
heat up rate 升温速率;升温速度
heat up time 增温时间
heat utilization(efficiency) 热利用(率);热利用
heat-vacuum method 热真空法
heat value 卡值;热值;热当量;发热量

heat value formula 热值公式
heat value meter 热值仪
heat value of sludge gas 污泥气热值
heat variable 热敏的
heat variable resistor 热敏电阻器;热变电阻器
heat venting facility 排热设施
heat vibration 热量变化
heat vibration method 热捣法
heat vulcanize 热硫化胶
heat vulcanization 热硫化
heat wall 火墙
heat waste 热损(失);热损(耗);热量浪费
heat water pollution 热水污染
heat wave 红外(线)辐射波;热浪;热波
heat-welding adhesive 热熔性胶黏剂
heat well 热井
heat wheel 转轮式换热器
heat withdrawal 排热
heat work 热修
heat-wrapping 热封塑膜包装
heat writing oscillograph 热电式示波器
heat writing recorder 热写记录仪
heat yield 产热量
heat zone 高温层
heave 隆胀;隆起;横断距;平错;水平移动;水平距;垂荡
heave aback 顶流停船;顶着风浪停船
heave a cable short 缩短锚链
heave ahead 绞进(吊车口令)
heave a line taut 绞紧
heave and cast 举起抛出
heave and set 起伏
heave a ship about ahead 使船急转前进
heave a ship ahead 收链使船前进
heave away the rope 绞缆
heave away the wire 绞缆
heave compensation 起伏补偿
heave compensation equipment 升沉补偿设备
heave compensator 起伏校正仪;升沉补偿器(海上钻探时,钻探船随波浪升沉而钻压不变);浮沉补偿器
heaved block 脊形断块
heave down 使船倾斜
heaved pile 回升桩
heave fault 挏断层;横推断层;平移断层
heave fishing 滥捕
heave force 隆胀力
heave force of foundation 基础隆胀力
heave ga(u)ge 隆起监测计(沟底土层);土地隆起监测器
heave here 从此吊起;此处提起
heave-ho 起锚
heave hook 转环提引钩;提引水龙头
heave in 绞进
heave in lively 迅速绞缆
heave lift 重型起重机
heave meter 升沉仪
heave motion 法向移动
heave motivator 法向操纵装置
heavenly bamboo 天竹子;南天竹
heavenly body 天体
heavenly sphere 天穹
heavenly stem 天干
heave of base 基底隆起
heave-off hinge 抽芯铰链;抽芯合页
heave of the sea 波浪起伏
heave out 扯起
heave out and lash up 捆起吊桶
heaver 重量;小铁梃;钩键;杠杆;大秤;杠杆;临时搬运工;举物人;举起物;绞棒;叉簧;移重工具
heave ratio 冻胀比
heave root 密封
heave stake 隆起标桩
heave taut 绞紧
heave the lead 打水砣(用手砣测深)
heave tight 绞紧
heave to 顶风漂泊;顶风(浪)停船
heave up 起锚
heave up anchor 起锚
heavier component 较重组分
heavier direction traffic 交通重负荷方向
heavier-duty 重型;加强式
heavier-duty contact 重负载接点
heavier polluted area 较重污染区

heavier textured soil 黏重土壤
heaviest piece weight 最重件重量
heaviest seas 最大波浪
heaviest wheel load 最重车轮荷载
heavily burdened river 多沙河流
heavily burdened stream 多沙河流
heavily coated electrode 厚涂层焊条
heavily cracked 深度裂化的
heavily damped circuit 强阻尼电路
heavily developed populated area 人口众多发达地区；高密度地区
heavily dope 重掺杂
heavily doped crystal 高掺杂晶体
heavily doped layer 重掺杂层；高掺杂层
heavily doped seed 重掺杂籽晶
heavily etched surface 重腐蚀面
heavily faulted crystal 高层错晶体
heavily feathered cable 高强度羽状电缆
heavily filtered X-ray beam 重度过滤 X 射线束
heavily involved in debts 债台高筑
heavily loaded line 重载铁路
heavily loaded track 重载轨道
heavily milled iron powder 过磨铁粉
heavily moistureladen steam 高饱蒸汽
heavily polluted area 严重污染区（域）
heavily polluted river 重污染河流
heavily polluted source 严重污染水源
heavily polluted tidal river 重污染感潮河道
heavily polluted water body 重污染水域
heavily polluted waters 重污染水域
heavily reinforced 超倍加强；大量配筋的；大大加强的；超配钢筋的；重重加强的；超倍增强的
heavily reinforced concrete 超配筋（的）混凝土
heavily reinforcement 大量配筋
heavily salted 重盐渍的
heavily sediment-laden river 多沙河流
heavily sediment-laden stream 多沙河流
heavily shaken area 严重受震区
heavily silt-carrying river 多沙河流
heavily silt-carrying stream 多沙河流
heavily stressed 高应力的
heavily supported 坚牢固定的
heavily traffic 交通频繁
heavily trafficked highway 交通繁忙的公路
heavily trafficked road 繁重交通道路
heavily trafficked track 运量繁忙的线路
heavily traffic reach 交通频繁河段
heavily travel(l)ed 交通频繁的
heavily travel(l)ed river reach 交通频繁河段
heavily travel(l)ed road 交通繁重的道路；交通繁忙路段；交通繁忙的道路；繁密交通道路；繁忙交通道路
heavily water area 灌溉区域
heavily watered ground 大量浸水地层
heavily weathered rock layer 风化严重岩层
heavily wooded area 树木很密的地
Heavimet 重合金的商名
heaviness of felling 采伐强度
heaving 隆胀；隆起；拉起；上下起伏
heaving a beached wreck to higher position 搁滩沉船绞至较高位置
heaving action 抛射作用；爆破作用
heaving and hauling line 撇缆
heaving bottom 冻胀土基；冻胀地基
heaving exploaive 低速炸药
heaving foundation 冻胀地基
heaving line 引（缆）索；引缆绳；引火绳；撇缆；抛缆；吊工具上桅的小绳
heaving line bend 撇缆接结
heaving line knot 撇缆接结
heaving line slip knot 撇缆活结
heaving mallet 绞绳木槌
heaving motion 船舶摇力；拥升动作（船舶）
heaving of bottom 坑底隆起
heaving of floor 底板鼓起
heaving of road 道路的冻胀
heaving of tunnel 隧道的底臌
heaving oscillation 起伏振荡
heaving property 冻胀性质
heaving sand 涌沙；胀沙；流沙（也称流沙）
heaving shale 易塌页岩；黏土页岩的冻胀；膨胀页岩；崩塌页岩
heaving soil 冻胀土
heaving value of frozen soil 冻土的冻胀量

Heaviside's calculus 海维西特运算微积
Heaviside's expansion theorem 海维西特展开定理
Heaviside's function 海维西特函数；阶跃函数
Heaviside layer 海氏层
heavy 重物；厚大的；笨重
heavy(1) dynamic(al) sounding 重一型动力触探
heavy(2) dynamic(al) sounding 重二型动力触探
heavy abutment 重型桥台
heavy accident 重大事故
heavy accumulation of snow 大量积雪
heavy advance 大涨价
heavy aggregate 重骨料；重集料
heavy-aggregate concrete 重（骨料）混凝土
heavy-aggregate concrete shield 重混凝土防护层
heavy air 压缩空气；重空气
heavy alloy 高密度合金；重合金
heavy anchor arm 重型锚臂
heavy and crude oil engine 重油原油发动机
heavy and lengthy rates 特重特长运费率
heavy and medium plate mill 中厚板轧机
heavy anode 不分瓣阳极
heavy application 加厚涂层
heavy ash 重苏打灰
heavy asphalt 重石油沥青；重（质）沥青；稠（地）沥青
heavy asphaltic base oil 重沥青基油
heavy asphaltic crude 重质沥青基原油
heavy asphaltic oil 重质沥青油
heavy asphaltic residues 重质沥青残渣油
heavy-atom method 重原子法
heavy-at-one-end desk 一头沉桌
heavy axle load 重轴荷载
heavy bacteria 密集菌
heavy ballast 沉重的压载物
heavy base layer 重掺杂基区层
heavy beam seas 大横浪
heavy-bedded 厚铺的
heavy bending 重型钢筋弯曲
heavy benzene 重质苯
heavy benzol 重苯
heavy bitumen 重（质）沥青；稠沥青
heavy blasting 强力爆破
heavy block 重型滑车（有三、四个滑轮的）
heavy bob 重锤
heavy-bodied 浓稠度；黏稠（质）的；浓稠的；高黏度的；稠厚的
heavy-bodied oil 重体油
heavy-bodied paint 高稠度油漆；稠涂料
heavy body 高黏度；高稠度
heavy body oil 重体油；高黏性油
heavy-bonded paint 高黏性涂料
heavy boring 粗镗
heavy break 密集损伤
heavy bridge 重型桥
heavy bridge deck panel 重型桥梁桥面板
heavy building 构造坚固
heavy bulk(y) cargo 重散（装）货；比重大的重散装；长大重件
heavy burden 重负荷料
heavy burdened river 挟带大量泥沙的河流
heavy burdened stream 挟带大量泥沙的河流
heavy burder 重料
heavy-burned magnesia 重质氧化镁；低活性氧化镁
heavy buy(ing) 大量购买；大量采购
heavy calcium carbonate 重质碳酸钙
heavy caliber 大口径；大孔径
heavy camber 大弯度
heavy cap 重压盖
heavy capital goods 大型资本财产
heavy car 重载货车
heavy cargo 重（件）货；重货物；特重货（物）；特重货；笨重货物
heavy cargo carrier 重货船；重件运输船
heavy cargo ship 重货船
heavy carrier gas 重载气
heavy cast red brass 重铸红铜
heavy-centered pattern 厚心喷射图形
heavy ceramics 建筑陶瓷；重质陶瓷
heavy chain 重载链；重型链
heavy charge 重质原料
heavy charged particle 重带电粒子
heavy chemical 重化学品；粗制化学药品；化工原料
heavy chemical industry 重化学工业
heavy clamshell 重型抓斗

heavy clay 重（质）黏土；致密黏土
heavy clay article 重黏土制品
heavy clayey loam 重亚黏土
heavy clay flooring tile 重黏土地砖；重质黏土铺地瓷砖
heavy clay industry 砖瓦工业；重黏土工业；粗陶工业
heavy clay kiln 砖瓦窑；重黏土（砖瓦）窑
heavy clay loam 重黏壤土
heavy clay product 重黏土产品；砖瓦（制品）；粗陶制品
heavy clay tile 重黏土砖；重质黏土瓦
heavy clay ware 黏土质建材制品
heavy coat 厚涂料
heavy coated electrode 厚（药）皮焊条；厚涂层焊条
heavy coating aggregate for glass fabrics 玻璃布重涂层处理机组
heavy coil-loading 重加感
heavy-colo(u)r 深色
heavy commodity 大宗物资
heavy communication satellite 重型通信卫星
heavy compaction 强夯
heavy compactor 重型夯具
heavy concrete 重质材料制成的混凝土（防辐射用）；重（骨料）混凝土；高密度混凝土（通常用于防辐射）；稠混凝土
heavy concrete block 重混凝土砌块
heavy condensation products 重质冷凝产物
heavy constituent 重组分
heavy construction 重型结构；重型建筑；高强度施工；大型工程
heavy consumption 大量消费
heavy contact 重负载接点；重负载触头
heavy cord 粗绳索；粗条纹
heavy corner 过厚壁角
heavy counter 大圆形船尾
heavy cover 厚（层）覆盖岩层
heavy crawler dozer 重型履带（式）推土机
heavy cross section gasket 大尺寸截面垫片
heavy crude 重（质）原油；重质石油
heavy crude asphalt petroleum 重沥青原油
heavy crude oil 重质原油
heavy cruiser 重型巡洋舰
heavy current 强（电）流；大电流
heavy current busbar 大电流母线
heavy current cable 强电流电缆
heavy current connector 强流接头
heavy current engineering 强电工程
heavy current feedthrough 强电流馈送
heavy current insulator 强电流绝缘子
heavy current lead-in 强电流引入
heavy current line 强电流线路
heavy current slow-speed generator 大电流低转速发电机
heavy curtain 帷幕
heavy curtain type 帷幕式
heavy cut 重切削；强力切削；深挖方；深挖掘；深挖沟；深挖槽；深开挖；重留分
heavy cutting 深挖工作；重研磨；厚件切割；深挖作业；深切；大开挖
heavy cycle oil 重循环油
heavy damage 严重破坏
heavy damping 强阻尼
heavy deck 重型甲板；厚面板
heavy decker 重甲板船
heavy deck vessel 重甲板船
heavy-density medium 重密度介质
heavy-density recovery 重悬浮液再生
heavy deposition 严重淤积；严重淀积；严重电积；严重沉积
heavy derrick 重（型）吊杆
heavy derrick step 重吊杆基座
heavy derrick stool 重型吊杆座
heavy diesel oil 重柴油
heavy digester 重载消化池
heavy discolo(u)ration 严重变色
heavy discount 大折扣
heavy distillate 重馏分
heavy disturbed 严重扰动
heavy dividing top strip 表面厚分隔条
heavy doubling frame 重型捻线机
heavy down gradient 陡下坡；长大下坡道
heavy downpour 倾盆大雨
heavy draft 深吃水

heavy draught 大压下
heavy draught vessel 深吃水船(舶)
heavy drawn binder 重型割捆机
heavy drawn lead 重拉制铅
heavy drifter 重型架式凿岩机；重型风钻
heavy drill 重型凿岩机
heavy drip 大量滴下
heavy drop 重型空投
heavy drop-hammer compaction 强夯(法)
heavy dumper 重型自卸载货汽车
heavy dunnage 厚货垫
heavy-duty 有特殊抗力的；重质(的)；重载；重(的)；重级工作制；经得起损耗；强有力，重负荷的
heavy-duty acetylene generator 大型乙炔发生器
heavy-duty air vibrator 重型气动振动器
heavy-duty anti-corrosive coating 重防腐蚀涂料
heavy-duty anti-foam bearing 重型防泡沫轴承
heavy-duty apparatus 重型消防炮车
heavy-duty apron feeder 重型链板喂料机
heavy-duty armored conveyer 重型铠装运输机
heavy-duty balance 大型天平
heavy-duty ball bearing 重型球轴承
heavy-duty bearing 重载轴承
heavy-duty bench 重型作业工作台
heavy-duty bogie 重载平板车
heavy-duty bogie wagon 载重平板车
heavy-duty boiler 大容量锅炉
heavy-duty brake 重型制动器
heavy-duty bridge 承担繁重交通的桥梁
heavy-duty car 重型汽车；重级轨道车
heavy-duty cast-iron well cover 重型铸铁井盖
heavy-duty cast-iron well cover seating 重型铸铁井盖座
heavy-duty chain 重型链
heavy-duty chassis 重型车辆底盘
heavy-duty chuck 重载夹盘
heavy-duty compressor 重型压缩机
heavy-duty concrete pipe 混凝土重型管
heavy-duty cone crusher 重载圆锥式破碎机
heavy-duty contacts 重负载电接触器材
heavy-duty converter 重型变矩器
heavy-duty crane 重型吊车；重型吊车；重级工作制(起重)吊车；巨型起重机
heavy-duty crane rail 重型起重机轨道
heavy-duty crate 搬运箱
heavy-duty cross-section device 重型切片器
heavy-duty crusher 重型破碎机
heavy-duty cutaway disc harrow 大型缺口圆盘耙；重型缺口圆盘耙
heavy-duty cutter 重型切割器；粗加工铣刀
heavy-duty cutting blade 重型(平地机)刮刀
heavy-duty design 重型设计(方案)
heavy-duty diagonal cutting nippers 大型斜嘴偏口钳
heavy-duty diesel oil 重型柴油
heavy-duty diesel vehicle 重型柴油车辆
heavy-duty digester 重载消化池
heavy-duty disc harrow 大型圆盘耙；重型圆盘耙
heavy-duty disk plow 重型圆盘犁
heavy-duty drill 重型钻头
heavy-duty drilling machine 重型钻床
heavy-duty drive 重载传动
heavy-duty engine 重型发动机；重负荷工作的发动机；负荷工作的发动机
heavy-duty engine lathe 重型车床
heavy-duty engine oil 重型发动机油
heavy-duty equipment 重型设备
heavy-duty final drive 重负荷侧传动
heavy-duty flexible(swing)door 重型平开门
heavy-duty floor(ing)受重负的楼板；使用频繁楼板；重载(型)楼板地面
heavy-duty frame 重型构架
heavy-duty freight car 重型载货车
heavy-duty gas turbine 大功率燃气轮机；重型燃气轮机
heavy-duty gear 重载齿轮
heavy-duty geared head lathe 重型普通车床
heavy-duty generator 大型发电机；大容量发电机
heavy-duty glass rope 玻璃纤维承力索
heavy-duty guard 重型护刃器；加强护刃器
heavy-duty harrow 重(型)耙
heavy-duty highway 交通繁忙的公路；重型公路；承担繁重交通的公路
heavy-duty hoisting equipment 重型起重设备

heavy-duty hook 重型钩
heavy-duty hose 高压软管
heavy-duty industrial floor cover(ing)厚实工业地面
heavy-duty lathe 重型车床
heavy-duty lifting machine 重型起重机械
heavy-duty(loa)重负载
heavy-duty loader 重型装载机
heavy-duty lock 强力锁
heavy-duty lubricating grease 重型润滑脂；重负荷润滑脂
heavy-duty machine 重型机器
heavy-duty machinery 重型机械；重载机械；重型机床
heavy-duty machinery plant 重型机械厂
heavy-duty machinery shop 重型机械厂
heavy-duty machine tool 重型机床
heavy-duty magneto 强力磁电机
heavy-duty mainline 繁忙干线
heavy-duty mainline railroad 繁忙干线铁路
heavy-duty manipulator 重型机械操纵器
heavy-duty mechanic 重型车辆技工
heavy-duty millinng cutter 粗齿铣刀
heavy-duty mirror bracket 大型镜架
heavy-duty multiple disk steering clutch 重负荷多片式转向离合器
heavy-duty oil 重型(发动机)油；重负荷机油
heavy-duty oil bath air cleaner 重油浴式空气滤清器
heavy-duty operating condition 重负荷的使用条件
heavy-duty paste mixer 厚浆混合机
heavy-duty pavement 重级路面(用于繁重交通要道)；(承担繁重运输的)重负路面；高级铺面；高级路面
heavy-duty plain cutter 重型普通铣刀
heavy-duty plastic bag 大型塑料袋
heavy-duty plastic sack machine 大型塑料袋机
heavy-duty preselect radial drilling machine 重型预选摇臂钻床
heavy-duty press 重型压力机
heavy-duty pressure tunnel 高压隧洞
heavy-duty radial drilling machine 重型摇臂钻床
heavy-duty rectifier 大功率整流器；强力整流管；强功率整流器
heavy-duty regulation work 重型整治(工程)
heavy-duty revetment 厚护坡；厚护岸
heavy-duty road 重载公路；重型公路
heavy-duty roller bearing 重型圆柱
heavy-duty rope 承力索
heavy-duty roughing tool 重型粗加工刀具
heavy-duty rubber spring 重负荷橡胶弹簧
heavy-duty runway 重型飞机跑道；重级飞机跑道
heavy-duty samming and setting machine 重型平展匀湿两用机
heavy-duty scaffold 频繁使用的脚手架；重型脚手架
heavy-duty seabed vehicle 重型海底作业潜水器
heavy-duty semi-mounted disk harrow 半悬挂重型圆盘耙
heavy-duty service 重荷条件下使用
heavy-duty shears 强力修枝剪；重型剪切机
heavy-duty side-dumping rock loader 大型侧卸装岩机
heavy-duty slab milling cutter 重型平面铣刀
heavy-duty slurry pump 重型泥泵
heavy-duty snow plough 重型雪犁
heavy-duty spiketooth harrow 大型钉齿耙；重型钉齿耙
heavy-duty spring 重型弹簧
heavy-duty stretcher 大型绳索伸长器
heavy-duty supporting screen pipe 重型支承筛管
heavy duty surface treatment 加厚表面处理工
heavy-duty switch machine 大功率转辙机
heavy-duty tail pipe supported packer 支承式封隔器
heavy-duty test 重载试验
heavy-duty thread 重型螺纹
heavy-duty tipper 重型倾卸车
heavy-duty tire 重载轮胎；重型轮胎
heavy-duty tool 重型机床；粗切刀
heavy-duty tool block 单臂刀架
heavy-duty track 重型轨道
heavy-duty tractor 重型拖拉机
heavy-duty traffic 重型货车交通
heavy-duty traffic railroad 重载铁路

heavy-duty trailer 重型挂车
heavy-duty transfer case 重负荷分动箱
heavy-duty transport machine 重载运输机；重型运输机
heavy-duty travel(l)ing type radial drilling machine 可移式重型摇臂钻床
heavy-duty trawler 重型拖网渔船
heavy-duty tripod 重型三角架
heavy-duty truck 重型自卸汽车；重型载货汽车；重型运货车；重型载货车
heavy-duty turning lathe 重型车床
heavy-duty type 重型
heavy-duty type brake fluid 重负荷型制动液
heavy-duty type vernier caliper with fine adjustment 带微调重型卡尺
heavy-duty vehicle 重型车辆
heavy-duty vibrating grizzly 重型振动格筛
heavy-duty vibrating screen 大号振动筛
heavy-duty wagon 重载平板车
heavy-duty water suction and discharge hose 大号排吸水胶管
heavy-duty winch 重型绞车
heavy earth 重晶石(钡白)；重土
heavy earthwork 大量退鹅工程；大量(的)土方工程
heavy edge 过缘加厚的；边缘加厚的
heavy edge reinforcement 路缘加强钢筋；边缘加密钢筋；边缘加强钢筋
heavy-edge work 边缘加厚工作
heavy electrical plant 重型电气设备
heavy electron 重电子(即介子)
heavy element 沉重元件；沉重构件；重元素
heavy enamel 厚瓷漆
heavy-enamel single cotton 厚漆单层纱包的
heavy-enamel single glass 厚漆单层玻璃
heavy-enamel single ink 厚漆单层玻璃纸包的
heavy-end power distillate 重质拖拉饥煤油
heavy ends 重质馏分；重组分；残液
heavy equipment manufacturer 重型设备制造厂家
heavy equipment transport 重装备运输工具
heavy excavation 大规模挖方工程；大(规模)开挖
heavy explosive charge 大量装药
heavy fall 大跌价
heavy fall of rain 雹暴
heavy feed 重进刀
heavy feeder 优质饲料；高负载馈电线；高负荷馈电线
heavy feed stock 重质原料
heavy filing 重锉法
heavy film 厚膜；厚层
heavy filtering medium 重滤料
heavy filter material 重滤料
heavy filtration 重过滤
heavy finishing 重浆整理
heavy fission fragments 重裂变碎片
heavy fixed hand hack saw frame 重型固定手弓锯框
heavy flannelette checked 厚格绒
heavy flint glass 重燧石玻璃；重火石玻璃
heavy floe 厚浮冰(超过2英尺厚)
heavy flow 强烈油流
heavy flow of water 强水流
heavy fluid 有重量流体；重液；重(量)流体
heavy fluid separation 重液分离
heavy fluid washer 重液洗选机
heavy fog 重雾；浓雾；大雾
heavy foliage 密叶
heavy force fit 重压紧配合
heavy forging 大型锻件
heavy foundation 重型工业厂房基础；重型基础
heavy fraction of oil 原油的重馏分
heavy frame 承力架；承力隔框
heavy freight 重质货物
heavy freight train 重货物列车
heavy frost 严寒；浓霜
heavy fuel 重质燃料；重质料油；慢燃物
heavy fuel booster pump 重油升压泵
heavy fuel burning engine 重油发动机
heavy fuel engine 重油发动机
heavy fuel equipment 重油设备
heavy fuel feeding pump 重油供油泵
heavy fuel oil 重柴油；重质燃烧油
heavy fuel oil clarifier 重油分杂机
heavy fuel oil purifier 重油分水机
heavy fuel powerplant 重油发动机

heavy fuel rotating combustion engine 重油转子发动机
heavy fuel separator 重油分离器
heavy gas model 重气体模型
heavy gas oil 重质瓦斯油
heavy gasoline 重质汽油
heavy ga(u)ge 大型量规;大剖面;大口径的;大尺寸
heavy ga(u)ge copper tube for general purposes 通用大口径铜管
heavy ga(u)ged 大断面的
heavy ga(u)ge metal (按英国标准线规规定的)厚级板材
heavy ga(u)ge nail 大号钉
heavy-ga(u)ge roofing sheet 重型屋面板;加厚屋面板
heavy-ga(u)ge rubber sheet 厚胶片
heavy ga(u)ge steel plate 厚钢板
heavy ga(u)ge wire 粗导线
heavy-going 难于通行的
heavy-going road 难以通行的道路
heavy goods 件杂货;重货物
heavy goods vehicle 重型货车
heavy grade 大坡度;陡坡
heavy grade stone protection 重型块石护坡
heavy gradient 陡坡(度);大坡度
heavy gradient section 陡坡段
heavy grading 大量平整土地
heavy grazing 过牧;过度放牧
heavy grease 滑脂
heavy ground 难于通行的地带;危险悬帮
heavy hail 大雹
heavy hammer 重锤
heavy hammer fall distance 重锤落距
heavy hand-held rock drill 重型手持式凿岩机;重型手持式风钻
heavy handy dead weight 装卸最重限额
heavy haul 重载运输
heavy hauler 重型牵引车
heavy haul freight line 重载货物线路
heavy haul railway 重载铁路
heavy haul railway transport 铁路重载运输
heavy haul trailer 重型平板挂车
heavy haul train 重载列车
heavy hilly area 重丘区
heavy hitter 有实力的房地产投资者或开发商;房地产大款
heavy hole 重空子;重空穴
heavy hole band 重空穴带
heavy hole state 重空穴态
heavy horizontal pendulum 重水平摆
heavy horsedrawn steerage hoe 操向式重型马拉中耕机
heavy hull repair ship 重型船体修理舰
heavy hydraulic press 重型水压机
heavy hydrocarbon gas 重烃类气体
heavy hydrocarbon oil 重质烃油
heavy hydrocarbons 重烃
heavy hydrogen 重氢;超重氢
heavy ice 重水冰;重冰(重水);固态重水;厚堆积冰(厚度10英尺以上的海水);厚冰
heavy ice target 重冰靶
heavy idler 加重的传送带的托轮
heavy impulse 高冲量
heavy industrial district 重工业区
heavy industrial landscape 重工业景观
heavy industry 重工业
heavy initiation 强力起爆
heavy injury 密集损伤
heavy in section 大截面;大断面的
heavy insulated tape 高等级绝缘加热带
heavy insulation 强化绝缘
heavy intensity 大强度
heavy intermediate submarine cable 重型中继海底电缆
heavy intermittent duty 重间歇负载
heavy intermittent test 重荷间歇试验;断续重负载试验;重负载断续试验;大负载断续试验
heavy internal 加重的内部零件
heavy ion 重离子;大离子
heavy iron 厚锌层(镀锌薄)钢板
heavy irrigation 足量灌溉;过量罐溉;过冷灌溉;过度灌溉;大量灌溉
heavy isotope 重同位素
heavy isotope build-up 重同位素积累

heavy jackstay (两船间传送用的)重型牵索
heavy joist 大尺寸木格栅;重型托梁;重型格栅
heavy jute mats 厚麻布
heavy keying fit 固定配合
heavy label 重标记
heavy-laden 重载的
heavy layer 厚层
heavy layer of fluid mud 厚浮泥层
heavy leather 厚革
heavy lesion 密集损伤
heavy lift (可移动的)大型起重机;超重货(物);特重货物;重件
heavy lift additional 超重附加费
heavy lift boom 重吊杆
heavy lift carrier 重件专用船
heavy lift charges 超重货物装卸费;超重(附加)费
heavy lift cum container berth 重货及集装箱泊位
heavy lift equipment 强力起重设备;重型装卸设备(起重能力大于50吨);重型起重设备;重件装卸设备
heavy lift ship 重型起重货船
heavy lift vessel 重吊船;重件专用船;重货船
heavy line 黑线;粗(实)线
heavy liquid 重质液体;重液
heavy liquid bubble chamber 重液体气泡室
heavy liquid concentration 重液分选
heavy liquid inlet 重液入口;重相入口
heavy liquid method 重液法
heavy liquid petrolatum 重质液体蜡膏
heavy liquid phase 重液相
heavy liquid process 重液选矿法
heavy liquid residuum 重质液体残油
heavy liquid separation 重质液体分离(法);重液选矿
heavy liquor 重液
heavy list 大横倾
heavy load adjustment 重负载调整
heavy-load-carrying column 重载柱;重型柱
heavy-load compensating device 重负载补偿装置;重负荷补偿装置
heavy-loaded plain bearing 大负荷普通轴承
heavy loader 重型装车机;重型装车机(泥浆)
heavy load(ing) 重负载;重(加)载;重加感;重负荷;重件;重负(载)
heavy loading district 重加感区;重负荷区
heavy load nominal rating 重负荷标准定额
heavy load relay 重负载继电器
heavy load ship 重件专用船
heavy load truck 重型卡车
heavy loam 重亚黏土;重砂质黏土;重壤土;重垆姆
heavy lock 加强的闭锁装置
heavy lodging 极度倒伏
heavy long swell 八级涌浪
heavy lorry 重型自动车;重型载货汽车
heavy losses 严重亏损
heavy lube stock 重质润滑油原料
heavy lubricating oil 重润滑油;厚黏度润滑油
heavy machine building industry 重型机器制造工业
heavy machinery shop 重型机械厂
heavy machinery works 重型机器厂
heavy machine shop 重工程车间
heavy machine tool 重型机床
heavy machine tool plant 重型机床厂
heavy magnesia 重镁氧
heavy magnesium carbonate 重质碳酸镁
heavy magnesium subcarbonate 重质碱式碳酸镁
heavy maintenance 大修(工程)
heavy mallet 木夯;大锤;槌
heavy mark 粗纬
heavy market 市场呆滞
heavy masonry 重质圬工
heavy medium 重悬浮液;重介质
heavy medium circuit 重介质回路
heavy medium cyclone 重介质旋流器
heavy-medium cyclone separator 热介质旋流分离器
heavy medium jigging 重介质跳汰
heavy medium method 重液悬浮法
heavy-medium operation 重液浮选法
heavy-medium process 重介质选矿(法)
heavy-medium separation 重介选矿法;重液选洗法
heavy medium separation method 重液分离法
heavy-medium separator 重介质分选机

heavy-medium suspension 重介质悬浮液
heavy melting scrap 重废钢
heavy merchant mill 大型轧钢机;大型轧(钢)机
heavy meson 重介子
heavy metal 重金属
heavy metal adsorption 重金属吸附
heavy metal bioavailability 重金属生物可利用性
heavy metal biotoxicity 重金属生物毒性
heavy metal chelating agent 重金属捕集剂
heavy metal chromium 重金属铬
heavy metal commodities 重金属矿产
heavy metal concentration 重金属浓度
heavy metal-containing wastewater 重金属废水
heavy metal contamination 重金属(元素)污染
heavy metal content 重金属含量
heavy metal index 重金属指标
heavy metal index method 重金属指标法
heavy metal input 重金属进入
heavy metal ion 重金属离子
heavy metal ion adsorbent 重金属离子吸附剂
heavy metal ion-containg wastewater 含重金属离子废水
heavy metal ion pollution of water body 水体重金属离子污染
heavy metal ion removal 除重金属离子
heavy metal ion source of water pollution 水污染重金属离子源;水(的)重金属离子污染源
heavy metallic form 重金属形态
heavy metallic salt 重金属盐
heavy metal model of river 河流重金属模型
heavy metal plate 厚重金属板
heavy metal plate lining 厚钢板内衬
heavy metal plate rigidity 厚钢板刚度
heavy metal plate stiffness 厚钢板刚度
heavy metal poisoning 重金属中毒
heavy metal pollutant 重金属污染物
heavy metal pollution 重金属污染
heavy metal pollution of water body 水体重金属污染
heavy metal precipitation 重金属沉淀
heavy metal removal 除重金属
heavy metals-containing wastewater 含重金属废水
heavy metals entering sewage treatment works 重金属进入污水处理厂
heavy metal soap 重金属皂
heavy metal sorption 重金属吸附
heavy metal specification 重金属表征
heavy metal wastewater purifying agent 重金属废水净化剂
heavy mild sand 重亚砂土
heavy mineral 重矿物
heavy mineral anomaly 重矿物质异常
heavy mineral concentrate 重砂样品
heavy mineral oil 重溶剂油
heavy mineral prospecting 跟踪找矿
heavy mineral prospecting method 重砂找矿法
heavy mineral province 重矿物区
heavy mineral sampling 重砂采样
heavy mineral soil 重矿物质土壤
heavy mineral spirit 重质溶剂油;重石油醚
heavy mobile machine shop 重型活动修理车间
heavy money 大笔款
heavy mortar 稠砂浆;重砂浆;浓灰浆;稠水泥砂浆;稠灰浆
heavy motorcycle 重型摩托车
heavy motor truck 重型载货汽车
heavy mud 重钻泥;重泥浆
heavy naphtha 重石脑油
heavy natural gasoline 重质天然汽油
heavy needle holder 粗针持针钳
heavy neutral oil 重质中性油
heavy neutral stock 重质中性油料
heavy non-ferrous metal 重有色金属
heavy nuclear reaction 重核反应
heavy nucleus 重核
heavy oil 原油;重质石油;重油;杂酚油;高粘度油
heavy oil burner 重油燃烧器
heavy oil carburet(t)or 重油化油器
heavy oil cracking 重油裂化
heavy oil desulfurization 重油脱硫
heavy oil engine 重油发动机
heavy oil fraction 重油馏分
heavy oil gasification 重油气化

heavy oil heater 重油加热炉
heavy oil pump 重油泵
heavy oil sand 重油砂
heavy oil service tank 重油储[贮]罐
heavy oil wastewater 稠油废水
heavy order 大量订货
heavy ore 重矿石
heavy organic muck 重质有机垃圾
heavy overcast 阴沉
heavy oxygen 重氧
heavy package 超重货(物)
heavy pack ice 重浮冰
heavy padder 重型浸轧机
heavy panel 高强(度)墙板
heavy paper 纸板
heavy paraffinic hydrocarbon 重质烷烃
heavy parking 大量停车
heavy particle 重质子;重颗粒
heavy parts 重型零件
heavy passing shower 大阵雨
heavy paste 稠浆
heavy payment 巨额支出
heavy pelting rain 连绵大雨
heavy penalty tax 重罚税款
heavy petrol 重汽油;低级汽油
heavy petroleum oil 重石油
heavy petroleum spirit 重石油醇;重质溶剂油;重石油醚
heavy petroleum wax 重石油蜡
heavy phase 重相
heavy phase weir 重相堰
heavy physical labo(u)r 重体力劳动
heavy piece 重件
heavy pinching 重摘心
heavy pine 美国西部黄松
heavy pinus pondarosa 美国西部黄松
heavy placer anomaly 重砂异常
heavy placer mineral 重砂矿物
heavy placer mineral analysis 重砂矿物分析
heavy placer mineral association 重砂矿物组合
heavy placer survey 重砂测量
heavy plate 厚板
heavy plate mill 厚板轧机
heavy plate shears 厚板剪切机
heavy plate straightener 厚板矫正机
heavy platinum metal 铂族重金属;铂类重金属
heavy plough 重型犁
heavy plow 重型犁
heavy pneumatic drifter 重型风动式钻机
heavy point 粗黑点
heavy pollutant 重质污染物
heavy pollution 严重污染;重(度)污染
heavy polymer 高分子量聚合物;重聚合物
heavy pot 高锅
heavy power 大功率
heavy precasting 重型构件预制
heavy pressure 高压(力)
heavy price 高价;昂价
heavy primary ray 强初级射线
heavy-producing 高产的
heavy profile 重型构件断面;大号型材
heavy pruning 重剪
heavy pumping 强抽运;大量抽水
heavy punch 重锤
heavy purse 富裕;富有
heavy rail 重型(轨)轨;重轨
heavy rail rapid transit 重轨快速交通
heavy rail track bolting wrench 重轨螺栓扳手
heavy rail transit 城市重交通
heavy rain 大雨;暴雨
heavy rainfall 大降雨量
heavy ram 重锤夯
heavy rare earth 重稀土
heavy rare earth deficiency 重稀土亏损
heavy rare earth ores 重稀土矿
heavy rare earths 重稀土类
heavy recoil oil 重反冲油
heavy recovery section 重型装备抢救排
heavy recycle stock 重质循环油
heavy reinforcement 大量配(钢)筋
heavy relay 大电流切换继电器
heavy repair 大修
heavy repairing funds 大修基金
heavy repair manual 大修手册

heavy rescue company 有重型救援设备的消防队
heavy residual fuel oil 重质残渣燃料油
heavy residual stocks 重质残油
heavy residue 残液;重(质)残渣
heavy resin oil 重质树脂油
heavy resurfacing 路面大修
heavy ring 承力环
heavy ring segments 重型环
heavy ripper 重型松土机
heavy road oil 重质铺路油
heavy rod 粗盘条
heavy roller check 深碾裂缝
heavy rolling 重型碾压;重吨碾压
heavy rolls 重型辊碎机
heavy Romanesque(style) 盛饰罗马风
heavy roof 重屋顶
heavy root system 庞大根系
heavy roughing cut 强力粗切削
heavy route 重路由
heavy route circuit 重路由电路
heavy route earth station 重路由地面站
heavy route station 重路由站
heavy route telephone message 重路由电话通信
heavy route toll telephone 重路由长途电话
heavy rubber cot 厚胶指套
heavy sail 下桅帆
heavy saline-alkaline water 重盐碱水
heavy salinization 严重盐碱化
heavy sand 重砂
heavy sand mineral identification 重砂矿物鉴定
heavy sand mineral quantitative determination method 重砂矿物定量测定法
heavy sand mineral separation and identification instrument 重砂矿物分离及鉴定仪器
heavy sandy loam 重砂壤土
heavy satellite 重型卫星
heavy satellite platform 重型卫星平台
heavy scale 厚氧化皮
heavy sea 汹涌海面;汹涛;狂风大浪;破浪;大浪;风涛海面;波涛汹涌的海面
heavy seal coat 路面加厚封层;加强保护层;(路面的)加厚封层
heavy section 大型型钢;重型型钢;重型材;厚断面;厚壁;大型剖面;大型材;大厚度
heavy section car 重型轨道车
heavy-section casting 大型铸件;厚壁铸件
heavy section mill 大型钢轧机
heavy section prop 重型支柱
heavy section steel 大型钢
heavy section stringer 大截面桁条;粗桁条
heavy sediment-laden flow 走沙水流;高含沙水流
heavy sediment laden river 多沙河流;大量挟沙河流
heavy sediment laden stream 多沙河流
heavy seed 密集小气泡;密布灰泡
heavy seeding 密播
heavy segmented flat roller 重型分组光面镇压器
heavy selling 抛售
heavy service 重负载运行;重负荷运行
heavy service car 大型服务车
heavy set 密镶的(金刚石)
heavy-set bit 高布齿密度钻头
heavy sewage purifier 高浓度污水净水器
heavy shade 饱和色
heavy shape 重型构件断面
heavy shear line 重型剪切机作业线
heavy sheet 加厚窗玻璃;厚窗玻璃(厚度为5~6毫米);橱窗玻璃
heavy sheet glass 厚玻璃
heavy shirting 重磅平布
heavy shock load 强烈冲击荷载
heavy shot 猛烈射孔;猛烈爆炸
heavy shoulder 厚(轮)胎缘
heavy shower 大阵雨;大暴雨;暴雨
heavy side (壁厚不均匀管子的)重侧
heavy silt 重质粉土
heavy silty loam 重粉壤土
heavy sizing 重浆
heavy slag 重炉渣
heavy slash 重废材;粗采伐剩余物
heavy sludge 稠污泥
heavy smooth wheel roller 重型平碾
heavy snow 大雪
heavy soda 重苏打

heavy soda ash 重苏打灰
heavy soil 重质土;重土壤;黏重黏土(壤);黏质土;黏土质
heavy solid 机械杂质粗粒
heavy solution 重液
heavy solvent naphtha 重质溶剂油
heavy spar 重晶石
heavy-spring standard 重型弹性铲柄;高强度弹性铲柄
heavy squall 烈飑;暴风
heavy stain (试验沥青时的)浓油斑;深色斑
heavy-stained sapwood 深色斑边材
heavy starting duty 重载启动负载;重型起动工况
heavy steel plate 厚钢板;厚钢板
heavy steel rolling mill 大型轧钢厂
heavy steel rolling plant 大型轧钢厂
heavy steel slab 厚钢板
heavy still bottoms 重质残渣
heavy storm 强风暴;大风暴;大暴雨;暴风雨
heavy straw board 厚草纸板
heavy stream 集水射流
heavy structural system 重型结构系统
heavy structure 重型结构
heavy sugar 繁忙运输线
heavy suite 重矿物组分;重矿物颗粒
heavy surf 拍岸大浪
heavy swell 汹涌的长浪;六级涌;狂浪;七级涌;大涌
heavy swell long 八级涌
heavy tails 重质尾部馏分;重尾馏分
heavy tamping 重夯;重锤夯实
heavy tamping method 重锤夯实法
heavy tap 斜底
heavy tar 重质柏油;厚柏油
heavy tar distillate 重质焦油馏出物
heavy tar oil 重质焦油
heavy taxation 课征重税
heavy test 满载试验
heavy-textured soil 重黏质土;重黏结土;质地的严重土壤
heavy timber 重木;大木;耐火木结构
heavy timber construction 大(型)木结构;重(型)木结构;重型木建筑;重型木建筑
heavy-tined harrow 重型钉齿耙
heavy tipper 重型自(动倾)卸车
heavy-tonnage lines of communications 重吨位车辆的交通线
heavy tower bolt 粗芯插销
heavy tower crane 重型塔式起重机
heavy track 重轨道
heavy traction vehicle 重型牵引车
heavy tractor 重型拖拉机
heavy traffic 大流量交通;繁重交通;繁忙运输;繁忙交通;服务繁忙
heavy traffic artery 繁忙交通干道
heavy traffic bridge 承担繁重交通的桥梁
heavy traffic line 繁忙运输线
heavy traffic period 运输繁忙期间;业务繁忙时间
heavy traffic reach 交通频繁河段
heavy traffic road 繁密交通道路;繁忙交通道路
heavy traffic route 繁忙运输线
heavy traffic stream 拥挤车流
heavy traffic volume 繁密交通量
heavy traffic zone 交通繁忙区
heavy trailer 重型拖车
heavy triangular blunt band saw file 粗重直三角带锯锉
heavy triangular blunt saw file 粗重直三角锯锉
heavy triangular taper saw file 粗重斜三角锯锉
heavy truck 重型载货汽车
heavy trucking 重型货车货运
heavy truss 重型桁架
heavy turbulence 严重紊流
heavy turning movements 强烈的转向运动;繁密的转弯交通
heavy twist 强捻
heavy type 重型
heavy-type cab-tire cable 重型橡套电缆
heavy type grab 重型抓斗
heavy type sectionalized shield 重型分节式掩护支架
heavy-type truck 重型自卸汽车
heavy undercarriage 重型起落架
heavy unemployment period 严重失业时期
heavy unit 重型构件

heavy unit shop 重型器材修理厂
heavy vector meson 重矢量介子
heavy vehicle 重型车(辆)
heavy version 重型设计(方案)
heavy vessel 重型容器
heavy viscous mud 高黏泥浆
heavy viscous oil 高黏度油
heavy wall 厚(壁)墙
heavy-walled 厚墙的;厚壁的
heavy wall pipe 厚壁管
heavy ware 重型车辆;重型物件
heavy water 重水
heavy water boiling reactor 重水沸腾堆
heavy-water cooled reactor 重水冷却反应堆
heavy water moderated reactor 重水减速反应堆
heavy water moderator 重水慢化剂;重水减速剂
heavy water plant 重水工厂
heavy water probe 重水靶探头
heavy water reactor 重水反应堆
heavy water reflector 重水反射层
heavy water system 重水系统
heavy wax 重质蜡
heavy wear 严重磨损
heavy weather 阴天;坏天气;大风浪天气;恶劣天气
heavy weather condition 恶劣气候条件
heavy weathering 严重风化
heavy weather vessel 适航性良好的船
heavy-web straight shank drill 厚钻心直柄钻头
heavy weed growth 杂草丛生
heavy weight 重件
heavy weight aggregate 重骨料;重集料
heavy weight concrete 高密度混凝土;重(质)混凝土;重骨料混凝土;特重混凝土
heavy weight concrete block 重混凝土砌块
heavy weight diving 重潜水(服)
heavy weight drilling 加重泥浆钻进
heavy weight drilling fluid 重质钻液;重质钻孔泥浆;重质钻井泥浆
heavy weight helmet 重潜水盔
heavy weight hide 重磅皮
heavy weight load 重荷载
heavy weight rubber product 重型制品
heavy weight volumetric(al) component 特重物件
heavy welding 大断面焊接
heavy welding electrode 粗电焊条
heavy wheel-load traffic 重载汽车运输;重轮载(汽车)交通
heavy wire 粗钢丝
heavy-wire screen 重型金属丝筛网(用于筒筛)
heavy wood 重(木)材
heavy wool grease 粗毛润滑脂
heavy work 重作业;繁重的工作
heavy yielder 高产作物
heazlewoodite 六方硫镍矿;硫镍矿;黄镍铁矿
Heberden's node 希柏登结
Heberden-Rosenbach node 希柏登结
Hebraic granite 文象花岗岩
Hebridean shield 赫布里底地盾
hecatolite 月长石
hecatompedon 百尺庙(古希腊雅典城)
hecatonstylon 有百柱建筑
heck 拦鱼格栅;衬锭;水下独立开合双扇门;格构大门
Heckenham metal 黑肯哈姆铜镍合金
heckling hards 梳亚麻粗纤维
heckling tow 梳亚麻短纤维
heck sample 校核试样;检查用试样
hectarage 公亩数
hectare 公顷(一公顷=100公亩,等于10^4平方米)
hectar service 高负荷运行
hectastyle 用六柱的;有六柱的
hectic 潮热的
hectic fever 消耗热;潮热
hectobar 百巴(气压单位)
hectogamma 百微克
hectogram(me) 百克;公两
hectograph 胶版(印)
hectoliter 公石;百升
hectolitre 公升;百升
hectometer 百米;百公尺
hectometer 100米
hectometer post 百米标
hectometer stake 百米桩【测】
hectometer wave 百米波
hectometric wave 百米波

hectonewton 百牛顿
hectorite 锂皂石;锂蒙脱石
hectostere 100立方米;百立方米(法国通用)
hectowatt 百瓦
hectowatt hour 百瓦时
heddur 黑杜尔铝金合金
hedenbergite 钙铁辉石
hedenbergite sharn 钙铁辉石矽卡岩
hedeoma oil 海地油
Hedera 常春藤属
hedge 篱笆(栅栏);树篱;绿篱
hedge against inflation 避免通货膨胀损失的套期保值
hedge bill 长柄镰
hedge buying 套买保值
hedge clause 免责条款;套头交易条款;套期保值条款;避险条款;避免责任条款
hedgecutter 绿篱修剪器
hedge fund 避险基金
hedgehog 棱形拒马;(钢纤维混凝土搅拌不当时产生的)刺猬团;键钮
hedgehog spin anchorage (压力钢管的)刺猬式锚固
hedge plant 绿篱植物
hedge planting 篱笆式栽植
hedge planting system 篱笆式栽植法
hedge purchase 套买
hedger 筑篱工人;避险者(外汇市场)
hedgerow 绿篱;栅篱;灌木树篱;树篱(笆);矮树篱
hedgerow system 篱形整枝法
hedge sale 套卖
hedge selling 卖期保值;套买保值
hedge trimmer 树篱修剪器;修剪机;修边机
hedge-type assets 保障资产
hedging 套头交易;套期保值;避险(外汇市场);包围
hedleyite 碲铋矿
hedonal 氨基甲酸-2-戊酯
hedonic price 内涵价格
hedreocraton 大陆克拉通
hedrites 多角晶
Hedvall effect 海德华效应
hedyphane 钙砷铅矿;铅砷磷灰石
heel 出水面坡脚;柱脚;跟面;跟部;结趾;焊跟;倾侧,桅的下部或下桅甲板下部分;垫块;底结;刀跟;挡土墙(土侧)底部;背棱;横倾【船】
heel-and-toe watches 轮流值班
heel-and-toe wear 胎面边缘磨损
heel angle of switch rail 尖轨跟角
heel bar 对接贴板
heel bead 脚条;打底密封剂;装玻璃垫条
heel block 垫板;辙叉跟间铁块;间隔铁(护轮机);垫块;导向滑车
heel block of switch 尖轨跟间隔铁
heel-board 踵板
heel boom 底脚吊杆
heel brace 下舵枢;舵踵钮;舵托
heel bracket 尾柱肘板
heel burner 烧边器
heel chair 轨跟枢
heel contact 踵形接触;动接点;大端接触
heel cutting 锤状插条
heel disc 舵踵盘座
heel distance 辙叉跟距;跟距
heel dolly bar 铆钉托
heeled water plane 倾侧水线面【船】
heel end slug 跟端缓动铜环
heeler 快速帆船
heel extension 辙叉跟延长轨
heel fishplate 辙叉跟端鱼尾板
heel gudgeon 下舵枢;舵踵钮;舵托
heeling 倾斜飞行;倾斜【船】;侧倾;横倾【船】
heeling adjuster 倾斜自差器;倾差校正仪
heeling angle 横倾角;倾角;侧倾角
heeling coefficient 倾差系数
heeling compensation 倾斜调整
heeling condition 倾斜状态;倾侧状态
heeling corrector 倾差校正磁铁
heeling deviation 倾斜自差;倾差
heeling error 倾斜自差;倾斜(误)差;磁罗经倾斜自差;侧倾误差
heeling error instrument 倾斜自差调整器;倾差仪(磁罗经)
heeling experiment 倾斜试验
heeling in 浅沟斜栽苗木;假植;埋植

heeling lever 倾侧力矩的力臂
heeling magnet 校正倾斜磁铁;倾差校正磁铁
heeling moment 横倾力矩;倾斜力矩;倾侧力矩;侧倾力矩
heeling resistance 横倾阻力
heeling tank 倾侧柜;摆动水柜(破冰船上产生强制摆动用)
heel-jar 跟震
heel joint bar for switch point 尖轨跟尖板
heel joint bolt 辙叉趾端连接螺栓
heel-knee test 跟膝试验
heell-and-toe 顺序接班
heel length 辙叉跟长度
heel length of frog 辙叉跟长度
heel line 坝踵线;横倾线;倾侧线
heel man 握梯手
heel mark 锥印;履痕
heel nailing machine 钉跟机
heel of angle bar 角铁的角部
heel of a rafter 椽脚;椽端
heel of a ship 横倾角;船桅与地面垂线的夹角
heel of a shot 炮眼离装药最远部分;炮眼口
heel of blade 辙叉跟
heel of dam 坝踵
heel of frog 辙叉跟
heel of metal 熔金属面
heel of pillar 支柱踵
heel of switch rail 尖轨跟(端)
heel of tool 刀头跟面
heel of tooth 齿根面
heel pintle 舵踵栓
heel pivot 轨跟枢
heel plane 圆刨(床)
heel plate 消防梯脚板
heel plate seat 侧支承块底座;背靠块座
heel post 柱脚;间壁门柱;闸门侧立柱;螺旋桨柱;门轴柱(边框);门侧铰链柱;推进器柱;承重柱;侧立柱
heelpost 船尾柱
heel pressure 跟压力
heel quoin 门桄石
heel radius 肘形弯管的外半径;弯头的外半径
heel ring 加固圈
heel riser(block) 辙叉跟高间铁块
heel rope 杆底绳
heels 残余料
heel sheave 张紧滑轮;(冲击钻的)后滑轮
heel side 背缘
heel slab 后部底板
heel socket 吊艇栓承座
heel spread 尖轨跟距
heel spread clip 跟距扣板
heel spread of frog 辙叉跟宽;辙叉跟开口距
heels seat 跟座
heels slab 柱脚板
heels stay 防滑垫
heel stone 门桄臼石;门桄石座;门桄插石
heel strap 扒头铁板;椽梁系板
heel tackle 张紧滑车组
heel tap 斜底
heel teeth 边缘齿(牙轮)
heel tenon 底梳下端接榫
heel tooth notch 边缘齿折断(牙轮)
Heenan dynamic dynamometer 希南动态测功计
Heenan hydraulic torque meter 希南液力力矩计
heermatical seal 真空密封接头;气密封接头
Hefner candle 赫夫纳烛光
Hefnerkerze 赫夫纳烛光
Hefner lamp 赫夫纳灯
heft 拉手;把手;手感重量
Hegeler furnace 赫格勒炉
Hegman fineness of grind ga(u)ge 赫格曼细度计
Hegman ga(u)ge 汉克门规;赫格曼细度计
heideite 硫钛铁矿
heidornite 氯硫硼钠钙石
Heigedaling orogeny 黑疙瘩岭运动
heighbo(u)ring track 相邻轨道
height 高度;高程;海拔;长宽高
height above airport 机场上高度
height above datum 基准面以上高度;基(准)面以上高程;海图零点以上高程;水尺零点以上高度;水尺零点以上高程;标高
height above ground level 地面以上高度
height above ground surface 地面以上高度

height above lowest foundation of dam 坝基最底部以上高度
height above sea level 海拔(高度);海拔(高程);海拔高度
height above the hull 船体以上高度
height above touch-down 着陆点上的高度
height accuracy 高程精度
height adjustment 高度调节
height altitude 地平纬度
height analyser 波高振幅分析器
height anomaly 高程异常
height-area curve 高度面积曲线
height arm 高程臂
height at withers 体高
height-azimuth-range position indicator 高度方位距离位置显示器
height balanced tree 高度平衡树;深度平衡树
height below datum 海图零点以下高程
height between stories 层高(指楼层);楼层高(度)
height board 高度板;量层杆;皮数杆【测】;梯级样板;楼梯踏步高度样板
height bridge 高程托桥
height calipers 高度卡尺
height card 高度绘图;高度测绘板
height-change chart 高度变化图
height check micrometer 测高千分尺
height class 树高级木
height clearance 净高
height coefficient of resource abundance 资源丰度系数高
height compensator 爬车器
height component 垂直方向性
height computation 高程计算
height computer 高度计算机;高程计算机
height control 高程控制;图像高度调整;高度限制
height control cylinder 高度调节油缸
height controller 高度控制器
height control lever 高度控制杆
height correction 高度校正;高程校正;高程改正;航高校正
height counter 高度计数器;高程计数器
height cut-out 开孔高度
height data 高程资料;高程数据
height datum 高程基准(面);高;水准零点
height density 限制房屋高度的区划规定
height determination 高程测定
height-diameter ratio 高径比;高度直径比
height difference 高度差;高(程)差
height difference interpretoscope 高差判读仪
height difference of benchmarks and height list 水准点高差及高程表
height difference of two air stations 两摄影站高差
height dimension 高度尺寸
height displacement 高程位移
height distortion 高程变形
height district 建筑高度限制区
height driver 高度传动器
heightened block 加高的房屋
heighten(ing) 增级;升高;增加;增大;加强;加高;提高;增高
heightening of crest 堤的加高
heightening of dam 坝顶加高;坝的加高
heightening of dike 坝的加高
heightening of fill 堤的加高
heightening of submerged bar 潜滩加高
heighten one's vocational level 提高业务水平
height equation 高度方程;高程方程
height equivalent 等效高度
height equivalent to an effective theoretical plate 有效理论塔板等效高度
height equivalent to a theoretic(al) plate 理论塔板等效高度
height equivalent to theoretic(al) plate 等板等效高度
height equivalent to transfer unit 当量传质单元高度
height error 高度误差
height factor 高度系数
height finder 高度测定器;测高仪;测高器;测高计
height-finder radar 测高雷达
height finding 测高
height-finding instrument 高度测定器;测高仪
height-finding microscope 测高显微镜
height-finding radar 测高雷达

height flip-flop 高度双稳(态)多谐振荡器;高度双稳态触发器
height from observation point to surface 测点地面高度
height from observation point to water table 测点至水面高度
heightgage 测高器
height gain 高度增益;高度增量
height-gain curve 高度增益曲线
height-gain factor 高度增益系数
height ga(u)ge 高度游标卡尺;高度计;高度规;测高仪
height ga(u)ge with magnifier 带放大镜高度尺
height grid 高程网格
height index circuit 标高电路;高程电路
height indicator 高度表;高度(指示)器;高度计;测高仪
height indicator of gyrosphere 陀螺球高度位置指示器
height information 高程资料
heighting 测定高程
height input 高度数据输入
height limitator 高度限位器
height limit sign 高度限制标志
height mark 高度标记;高程注记;标高
height marker 高度标识器
height-marker-intensity compensation 测高标记亮度补偿
height master 高度游标卡尺;高度规
height measurement 高度测量;高程测量;测高
height measurer 高程计
height measuring device 测高装置
height measuring error 测高误差
height measuring ga(u)ge 高度测量器
height measuring sextant 测高六分仪
height micrometer 高度测微仪
height money 高空作业费;高空作业津贴
height of adsorption zone 吸附区高度
height of a fuzzy set 模糊集的高
height of a hump 峰高【铁】
height of arc 弧高;矢高
height of arch 拱矢
height of a transfer unit 传质单元高度
height of backwater 壅水高(度)
height of baffle plate 挡板高度
height of beam 梁的高度
height of bench 阶梯高度
height of bridge superstructure 桥梁建筑高度
height of bubble cap 泡罩高度
height of building 建筑(物)高度;房屋高度
height of burst 炸高
height of camber 拱矢高;路拱高度
height of capillary rise 毛细提升高度;毛细(水)上升高度
height of capillary water 毛管水高度
height of center of gravity 重心高(度)
height of center of buoyancy 浮心高度
height of chassis above ground 车架高度(指底座离地高度)
height of chimney 烟囱高度
height of cloud base 云底高(度)
height of collapse arch 坍落拱高度
height of collapse zone 冒落带高度
height of collimation 视准高度
height of component tide 分潮高度
height of confined water 承压水(水)头
height of contour 外形凸点
height of course 层高【建】
height of court 天井的高度
height of cover 覆土厚度;管道埋深;顶部覆盖高度
height of curve 曲线高度
height of cut 泥层厚度;挖泥厚度
height of cut fill at centre stake 中桩填挖高度
height of dam 坝高
height of dammed water due to spur dike 丁坝壅水高度
height of damming 回水高度
height of delivery 水泵扬程;输送高度
height of diapir core 底辟核高度
height of downcomer 下液管高度
height of drop 落距;落锤高度;落差
height of dry dock cope 干坞坞墙顶高
height of elevation 拔起高度【铁】
height of embankment 路堤高度;堤岸高度

height of extended part of shuttle 货叉伸出高度
height of eye 眼高;观测高度
height of eye correction 眼高差;伏角差;视高校正;视高改正
height of fall 坠落高度;落锤高度;落程;锤落高度
height of fault scarp 断层崖高度
height of fence 栅栏高度;篱笆高度;围墙高度
height of fill 路堤高度
height of floor 楼面高度
height of floor above rail level 地板致轨面距离
height of fractured zone 破裂带高度
height of free fall 自由落程
height of freshwater table above sea level 淡水面在海平面以上高度
height of gas cap 气顶高度
height of gas column 气柱高度
height of geologic(al) observation point 地质点高程
height of gravel stuffing 填砾高度
height of gravitational centre 重心高(度)
height of groundwater parting 地下分水岭高程
height of high tide 高潮高水位;高潮潮高;满潮高
height of high water 高潮面高;高潮高水位;高潮高
height of hitting wave 击浪高度
height of hump 驼峰高度
height of hydraulic jump 水跃高度
height of hydrocarbon column 烃柱高度
height-of-image adjustment 像高修正
height of incidence 入射高(度)
height of instrument 水平仪高;仪器高(度)
height of jet 射流高度
height of jig bed 跳汰机床层厚度
height of jump 水跃高度
height of jump-movement 跳移高度
height of landslide cliff 滑坡臂高度
height of layer 料层厚度
height of leading marks 导标高度
height of level 水平面标高;水平面高程
height of lift 一次铺成的厚度;混凝土浇筑的高度;扬程;浇筑层厚度;起吊高度;铺筑路面厚度;提升高度;分段高度;泵的吸升高度
height of lift(ing) 克服高度;提升高度;拔起高度【铁】
height of light 灯光高度;灯高
height of lighthouse 灯塔高度
height of load removal 卸载高度
height of low-water 枯水位高程;低潮线;低潮高;低潮潮高
height of mast 桅顶高度
height of mean high water of neap tide 小潮平均高潮位;小潮平均高潮面
height of mean high water of spring tide 大潮平均高潮位;大潮平均高潮面
height of mean low-water of spring tide 大潮平均低潮位;大潮平均低潮面
height of metacentric above keel 稳心在龙骨上的高度
height of nozzle 接管高度
height of nozzles and manhole from outside vessel 接管与人孔离容器外侧高度
height of oil column 油柱高度
height of oil ring 油环高度
height of orbit centre above still stage 波浪中心高度
height of overflow 溢流高度
height of pencil control 铅笔升降控制器
height of portal clearance 门架净高度
height of pour 浇注高度;浇筑高度
height of projection 凸起高度
height of proscenium 台口高度
height of protrusion in flange joint 法兰连接凸出部分高度
height of rail 钢轨高度
height of reaction unit 反应装置高度
height of reinforcement of weld 加强焊缝高度
height of release 排放高度
height of reservoir water table when earthquake is induced 地震时库水位标高;地震时水位高程
height of rise of water-level 水位上涨高度
height of rising flood 涨水高度
height of roof above rail level 车顶至轨面距离
height of room 居室高度

height of runoff 径流深度
height of sample 试件高度
height of sand flush from well 孔内涌砂高度
height of seat of fifth wheel 第五轮高度(汽车)
height of segment 管片高度
height of shelter belt 防护林带高度
height of sighting line 视线高度
height of sinter terrace 泉华阶地高度
height of site 测站高
height-of-site correction 测站高程校正
height of slope 边坡高
height of smoke outlet 排烟口高度
height of spring gushed above the surface 泉水涌出地面高度
height of stalagmite 石笋高度
height of station 测站高
height of steady head 稳定水头高度
height of step 踏步高度
height of stone fall 石瀑高度
height of stor(e)y 层高【建】
height of structural closure 构造闭合高度
height of subgrade 路基高度
height of summit 山峰高程
height of summite 山顶高度
height of support 支座高度
height of surveying sextant 六分仪高度
height of swell 涌浪高度;涌高;壅水高(度);涨水高度
height of the maximum storage 最大蓄量水位
height of thread 螺纹齿高
height of tide 高潮位;潮(位)高
height of tooth 齿高
height of top and bottom nozzles flanges from tangential line 顶部与底部接管法兰至切线的高度
height of transfer unit 传递单元高度
height of transverse section 横剖面高度
height of tree 树高
height of vadose overflow 渗流溢山点高程
height of vision 视线高
height of wall 墙高
height of water 水柱高度;水位落差
height of water flush at well mouth 孔口涌水高度
height of water injection 注水高度
height of water-level 水位高程
height of wave 浪高;波浪高
height of wave climbing 浪爬高度
height of wave erosion 波浪侵蚀高度
height of weir 堰高
height of weir plate 溢流堰高度
height overall 全高
height overlap coverage 高度重叠覆盖范围
height pattern 高度形式;垂直方向性;垂直方向图
height peak 高峰
height point 高程点
height pole 皮数杆【测】
height-position indicator 高度位置显示器
height potentiometer 高度分压器;高度电位器
height power factor 高度功率因数
height profile of ozone 臭氧铅直分布图
height range finder 测高测距仪
height range indicator 高度距离显示器
height ratio 高比
height ratio of tide 潮高比
height reduction 高程归算
height regulation 高度规定(房屋)
height restriction 建筑物高度限制
height scale 高度标尺;高度比例尺;垂直比(例)尺
height sensor 高度传感器
height servo 高度伺服机构
height setting 高度调节
height shoot 顶枝
height slide 高程滑尺
height survey 高程测量
height system 高程系统
height table 高程表
height to a theoretic(al) plate 理论塔板高度
height-to-distance ratio 感光特性曲线
height tolerance 高度公差;垂直公差
height-to-width ratio 高宽比
height traverse 高程导线
height traverse survey 高程导线测量
height under hook 钩下高度;吊钩高度
height variation coefficient of wind pressure 风压高度变化系数
height vernier calliper 高度游标尺
height vernier ga(u)ge 高度游标尺
height-volume curve 高度体积曲线
height warning device 高度警告装置
height zone 建筑容积分区
height zoning (of building) 建筑高度分区(规划);建筑高度区划;建筑容积分区
heikolite 青钛闪石;平康石
heiligenschein 露面宝光
heiligtag effect 干扰波引起的误差;波干扰误差
heilosphere 氦层
Heim's hypothesis 海姆静水压假说
Heimitian form 埃尔米特型
Heimlich maneuver 西姆林清操纵法
heinrichite 钡砷铀云母
heir 继承人
heir at law 法定继承人
heir by operation of law 法定继承人
heir in tail 限定继承人
heirloom 相传动产
heirs and assigns 继承人和受让人
Heisenberg's matrix mechanics 海森伯矩阵力学
Heisenberg's uncertainty principle 海森伯测不准原理
Heisenberg's uncertainty relation 海森伯测不准关系
Heising modulation 恒流调制;板极定流调制
Heister's valves 螺旋瓣
hekistotherm 适寒植物
helarch succession 海洋演替系列
held covered 已按每月保比例保险
held covered at the discretion of the underwriter 根据保险公司意愿继续承保
heldiver 联合掘进机(斜井用)
held over 延搁
held retention water 支持水
held terminal 挂起终端;截听终端
held water 吸着水;支持水;黏滞水;结合水;束缚水;薄膜水
helenite 弹性地蜡
heleoplankton 池沼浮游生物
heleox saturation diving 氦氧饱和潜水
helerobar 异量元素
Hele-Shaw motor 径向活塞式液压电动机
heliacal cycle 太阳周
heliambulance 救护直升机
Helianthus annuus 向日葵
heliarc cutting 氦弧切割(法)
heliarc unit 氦弧焊具
heliarc welder 氦弧焊机
heliarc welding 氦弧焊(法);氦电弧焊
heliare 氦弧
helibase 直升(飞)机基地
helibase manager 直升(飞)机基地总管
helicab 出租直升机
helical 螺旋形的;螺线
helical accelerator 螺旋加速器
helical aerial 螺旋形天线
helical angle 螺旋角
helical antenna 螺旋(形)天线;螺旋(线)天线
helical architecture 螺旋形建筑
helical auger 螺旋钻;螺纹钻
helical band 螺旋带
helical band friction clutch 螺旋带摩擦离合器
helical barrel vault 螺旋形筒拱
helical beam antenna 螺旋束天线
helical bevel gear 螺旋伞齿轮
helical binding 螺旋箍筋
helical blade 螺旋叶片
helical blade stirrer 螺旋桨式搅拌机
helical blower 叶轮式鼓风机;轴向鼓风机;螺旋鼓风机
helical Bourdon tube 螺旋弹簧管
helical burr 螺纹
helical car unloader 螺旋式卸车机
helical casing 螺旋壳
helical cast 螺旋排绕;反盘
helical chain 螺旋斜挂链
helical coil 螺旋状电圈;螺旋形线圈;蛇管
helical compression spring 压力盘簧;螺旋压簧
helical configuration 螺旋构型
helical contact 螺旋弹簧接触体
helical conveyer 螺旋(式)输送机;螺旋输送器
helical conveyer centrifugal 螺旋卸料离心机
helical conveyer feeder 螺旋运输给料机
helical conveying feeder 螺旋输送给料机
helical conveyor feeder 螺旋运输给料机;螺旋喂料机
helical convolute 螺旋面
helical current meter 旋桨式流速仪
helical curve 螺旋(曲)线
helical cutting 螺旋切槽
helical delay cable 螺旋延迟电缆
helical diagram 螺旋取向照像
helical digger 螺旋挖掘机
helical dislocation 螺旋位错;螺线位错
helical drill 螺旋钻
helical extruded shape 螺旋形挤压型材
helical eyepiece 螺旋(形)目镜
helical fan 螺旋式通风机;螺旋风扇
helical fibre 螺旋形光纤
helical field 螺旋场
helical fin 螺旋状肋片;螺旋翅片
helical fin section 螺翅
helical flash lamp 螺旋形闪光泡;螺旋形闪光灯;螺旋闪灯
helical flash tube 螺旋形闪光管
helical flow 漩流;螺旋(水)流,涡流
helical-flow turbine 回流式汽轮机
helical forging 螺旋轧制
helical form container 螺旋形盛料器
helical funnel bit 漏斗状螺旋钻
helical gas lens 螺旋气体透镜
helical gear 斜齿轮;螺旋齿轮;螺线轮
helical gear cutting 螺旋齿轮切齿法
helical gear hub 螺旋齿轮毂
helical gearing 螺旋(齿轮)传动
helical gear pump 斜齿轮(式)泵
helical gear reducer 螺旋齿轮减速器
helical gear shaft 螺旋齿轮轴
helical gear with helix angle 带螺旋角的螺旋齿轮
helical gear with jaw 带楔斜齿轮
helical gravity chute 螺旋重力溜子;螺旋重力溜槽
helical groove 螺旋槽
helical heater 螺旋状灯丝
helical hinge 双向弹簧铰链;双开式旋转铰链
helical instability 螺旋不稳定性
helical knife cutter head 螺旋滚刀式切碎器
helical lamp 螺旋形灯管
helical lens 螺旋透镜
helical linac 螺旋线慢波结构直线加速器
helical line 螺旋线
helical line type filter 螺旋线形滤波器
helical lobe compressor 螺杆压缩机
helically 成螺旋形
helically bladed screw conveyor 螺旋叶片式输送器
helically grilled tube 连接弯管;伸缩管
helically reinforced column 螺旋箍筋柱
helically threaded nail 螺丝钉
helically welded case 螺旋焊接壳体
helically welded tube 螺旋(缝)焊接管
helically wound 螺旋绕组;螺旋绕法
helical manometer 螺旋式压力计
helical master 螺旋主盘
helical mill cutter 螺齿铣刀
helical milling 螺旋铣(削);螺旋铣(法)
helical mixer 螺旋式混合器
helical mixer type aerator 螺旋混合型曝气器
helical motion 螺旋运动
helical mount 螺旋支架
helical oil groove 螺旋油槽
helical out lobe 螺旋外叶
helical pinion 螺旋小齿轮
helical pinion cutter 螺旋插齿刀
helical plating stress tester 螺旋式镀层应力测试仪
helical plow 螺旋式犁
helical polarization 螺旋偏振
helical potentiometer 螺旋线电位器
helical pressure tube 螺旋形压力弹簧管
helical propelling sleeve 螺旋推进套
helical rack 螺旋齿条
helical rake angle (铣刀片的)轴向刀面角;螺旋前角;螺旋截尖角
helical ramp 螺旋(形)坡道;螺旋坡线
helical recording 螺旋扫描记录法;螺线扫描记录法
helical reinforcement 螺旋(形)钢筋;螺纹钢筋;螺旋箍筋

helical resonator 螺旋形谐振器;螺旋共振器
helical ribbon mixer 螺旋带式搅拌机
helical runner 螺旋叶轮
helical scan 螺线扫描
helical screw pulp conveyer 螺旋纸浆传送器
helical screw spreader 螺旋形(粒料)撒布机
helical shaped bit 螺旋形钻
helical shell 螺形薄壳
helical-shell conic(al)shaped bowl 锥形实壳体转筒
helical slit drum 螺旋槽形磁鼓
helical sounding borer 螺旋形触探钻
helical spline broach 螺旋花键拉刀
helical spool 螺旋槽阀芯
helical spring 螺旋(形)弹簧;盘簧
helical stair(case)旋梯;螺旋(式)楼梯;盘梯
helical steel support 螺旋形井筒支架
helical strake 螺旋形箍
helical strand 螺旋钢丝索
helical structure 螺旋结构
helical stump cutter 螺旋式树桩挖掘机
helical surface 螺旋面
helical symmetry 螺旋状对称;螺旋对称性
helical tank 螺旋形油箱
helical tension spring 张力盘簧
helical tip speed 螺线桨尖速度
helical tooth 斜齿;螺旋齿
helical tooth(milling)cutter 螺旋齿铣刀
helical tube conveyer 螺管输送机
helical type pump 螺旋式泵
helical vault 螺旋拱顶
helical vibration feeder 螺旋振动上料器
helical-welded tube 螺旋焊接管
helical welding 螺旋焊接
helical weld pipe 螺旋焊管
helical wheel 螺旋轮
helical winding machine 螺旋绕丝机
helical worm gear 螺旋(形)蜗轮
helical wound spring 螺旋形盘簧
helical xenon-arc lamp 螺旋形氙(电)弧灯
heliciric texture 残缕结构
helicitic phenocryst 旋转变斑晶
helicline 螺旋形匝道;渐升坡;螺旋坡线;螺旋坡道;盘旋(式)斜坡道
helicograph 螺旋规
helicoid 漩涡形;螺旋体;螺旋面;螺旋纹;螺圈
helicoidal 螺旋形的
helicoidal anemometer 螺旋桨式风速仪;螺旋桨式风速计;螺旋桨式风速表
helicoidal bevel final drive 螺旋伞齿轮终端传动
helicoidal conveyer 螺旋输送机
helicoidal flow 螺旋状水流;螺旋(水)流
helicoidal pump 螺旋泵
helicoidal ramp 螺旋桨式坡道
helicoidal saw 石工锯
helicoidal shell 螺状薄壳
helicoidal step configuration 螺旋阶梯状
helicoidal surface 螺旋面
helicoidal wheel shaft 螺状轮轴
helicoid conveyer 螺旋(式)输送机;螺旋输送器
helicoid cyme 螺状聚伞花序
helicoid dichotomy 螺状二岐分枝
helicoid inflorescence 螺纹花序
helicoids drill 螺杆钻具
helicone 螺旋形极化天线
Helicon gear 交叉齿轮
helicon wave 螺旋波
helicopod gait 螺旋形步态
helicopodia 螺旋形步态
helicopter 直升(飞)机
helicopter autopilot 直升(飞)机自动驾驶仪
helicopter-borne survey 直升(飞)机普查
helicopter deck 直升(飞)机升降甲板
helicopter-dunked sonar 直升(飞)机吊放式声呐
helicopter for crane work 供起重用的直升飞机
helicopter ground 直升飞机场
helicopter landing-pad 直升(飞)机降落垫板
helicopter muffler 直升(飞)机消声器
helicopter pad 直升(飞)机起落坪
helicopter platform 直升(飞)机起落平台
helicopter rescue 用直升机救援;直升(飞)机援救
helicopter skid 直升(飞)机滑道
helicopter sonar 直升(飞)机声呐
helicopter sower 直升(飞)机撒播(飞)机
helicopter sprayer 直升(飞)机喷雾(飞)机
helicopter traffic 直升(飞)机运输
helicotrema 蜗孔
helicotron 螺旋质谱仪;螺旋质谱计;螺线质谱仪
helictic structure 残缕构造
helictite 卷曲石
helidrome 直升(飞)机(飞)机场
helier 潮洞
helihome 直升汽车(房)屋
heliocentric 螺旋心;日心的
heliocentric angle 日心角
heliocentric coordinates 日心坐标
heliocentric distance 日心距
heliocentric ephemeris 日心历表
heliocentric gravitational constant 日心引力常数
heliocentricism 太阳中心说
heliocentric latitude 日心纬度
heliocentric longitude 日心经度
heliocentric orbit 日心轨道
heliocentric parallax 周年视差;恒星视差;日心视差
heliocentric phenomenon 日心天象
heliocentric place 日心天体位置
heliocentric position 日心位置
heliocentric radial velocity 日心视向速度
heliocentric system 日心系统;日心体系
heliocentric theory 日心说
heliocentric velocity 日心速度
heliochrome 天然(彩)色照片;彩色照片
heliochromy 天然色照相术
heliodon 日影仪
heliodor 金绿柱石
helioelectric power plant 日光发电厂
heliofilter 滤镜
heliogram 回光信号;日光反射信号;太阳照相图
heliogramma 日照纸
heliograph 回光通信;日照计;日光摄影机;日光胶版;日计;日光反射(信号)仪;日光反射信号器;日光反射信号机;太阳照相仪;太阳照相机;太阳摄影机;太阳光度计;反光通信
heliographic(al)chart 日面图
heliographic(al)coordinates 日面坐标
heliographic(al)distribution 日面分布
heliographic(al)latitude 日面纬度
heliographic(al)longitude 日面经度
heliographic(al)paper 晒图纸
heliographic(al)pole 日面极
heliographic(al)print 日光晒印
heliography 照相制版法;日光摄影术;日光摄影法;日光反射信号法
heliogravure 凹版照相(术);凹版摄影
heliogreenhouse 日光温室
heliogyro 直升(飞)机
Helio-Klischograph 凹印电子雕刻机
heliolamp 日光灯
heliolatitude 日面纬度
heliolite 日长石
heliology 太阳学
heliolongitude 日面经度
helio magnetosphere 日磁层
heliometer 量日仪;测日仪
heliometry 量日仪测量术
heliomicrometer 太阳测微计
helion 氦核
helion filament 太阳灯丝
heliopathia 日光病
heliopause 太阳风层顶
heliophile 喜阳植物;喜光植物;适阳植物
heliophilous plant 喜阳植物;适日植物
heliophobe 嫌阳植物;避阳植物;避光植物
heliophobous plant 阴性植物
heliophyllite 斜方氯砷铅矿;日叶石
heliophysics 太阳物理学
heliophytes 阳生植物
heliophytia 阳生植物群落
helioplant 太阳能利用装置
helioscene 可折遮阴棚架
helioscope 观日望远镜;量日镜;回照器;太阳望远镜;太阳(目视)观测镜;太阳观察镜
helioscope eyepiece 目镜
helioscopic ocular 天文回照目镜
helioseismology 日震学
heliosensitivity 日光敏感性
heliosphere 日光层;太阳层
heliostat 追日镜;定日镜
heliotaxis 趋阳性;趋日性
heliotechnics 日光能技术;太阳能技术
heliothermometer 日温(测量)计;日光温度表
heliotrope 血滴石;淡紫色;紫红色;鸡血石;回照器;回光仪;日光回照仪;日光反射信号器;青莲色;太阳反射仪
heliotrope-carnation 丁香紫兰
heliotropic(al)wind 日转风;日成风
heliotropism 向日性
heliotype 照相胶版;摄影制版
heliozincograph 摄影制锌版
heliozincography 摄影制锌版术
helipad 直升(飞)机升降场
heliport 直升(飞)机站;直升(飞)机场
heliport deck 直升(飞)机停机坪
helipot 螺旋线圈电位计
helipot bridge circuit 交流电阻电桥电路
helipumper 靠直升飞机搬运的轻型手台泵
helispot 直升(飞)机临时降落场;直升(飞)机降落场;直升(飞)机(临时)降落场;直升(飞)机停机坪
helitack 直升(飞)机和空降灭火
helitack foreman 直升(飞)机灭火领班
helitank 直升(飞)机吊箱
helitanker 直升灭火飞机;灭火直升(飞)机
helitron 电子螺旋管;电子螺丝管
helitron oscillator 螺旋管振荡器;螺线管振荡器
helium 沼泽群落
helium age method 氦年代测定法【地】
helium ages 氦法年龄
helium arc welding 氦弧焊(法)
helium-atmosphere 氦保护
helium-atmosphere box 充氦盒
helium backscattering analyzer 氦反向散射分析仪
helium cadmium laser 氦镉激光器
helium charging unit 充氦装置
helium compressor 氦(气)压缩机
helium content 氦含量
helium coolant 氦冷却剂
helium-cooled reactor 氦冷反应堆
helium dating 氦年代测定法【地】
helium detector 氦检测器
helium diving bell 充氦潜水钟
helium family element 氦族元素
helium gas 氦气
helium ionization detector 氦电离检测器
helium lamp 氦灯
helium layer 氦气层
helium leak detector 氦探漏仪;氦检漏器
helium leaking rate 氦漏失率
helium luminous tube light 氦气管灯
helium magnetometer 氦磁力仪;氦磁力计
helium mass spectrometer leak detector 氦质谱检漏仪
helium method 氦(测)法
helium neon laser 氦氖激光器
helium nitrogen oxygen mixture 氦—氮—氧混合气
helium oxygen decompression 氦氧减压
helium oxygen decompression table 氦氧减压表
helium oxygen dive(diving)氦氧潜水
helium oxygen diving equipment 氦氧潜水装具
helium oxygen diving inter-communicator 氦氧潜水对讲机(器)
helium oxygen diving telephone 氦氧潜水电话
helium oxygen heavy weight diving apparatus 氦氧重潜水装具
helium oxygen mixture 氦氧混合气
helium oxygen subsaturation diving 氦氧亚饱和潜水
helium reclamation unit 氦气回收装置
helium refrigerator 氦制冷机
helium-rich core 富氦核
heliumshield 氦保护
helium ship 氦气艇
helium spectrometer 氦分光计
helium spectrum lamp 氦谱灯
helium survey 氦气法;氦气测量
helium survey apparatus type 氦测量仪类型
helium syncope 氦昏厥
helium tank 氦气瓶
helium tube 氦管;充氦管
heliweld 赫利焊接
helix 螺旋形天线;螺旋线饰【建】;螺旋(线);螺旋

弹簧
helix accelerator 螺旋波导直线加速器
helixal inlet 蜗卷式(螺旋线)入口
helix angle 螺旋(升)角;节面角
helix angle of tooth 轮齿倾斜角
helix angle tester 螺旋角检查仪
helix array 螺旋天线阵
helix Bourdon tube 螺旋弹簧管
helix-coupled-vane circuit 螺旋线耦合叶片慢波线
helix element 螺旋辐射元
helix flash tube 螺旋形闪光管
helix heater 螺旋状阴极;螺旋形灯丝;螺旋丝电热器
helix impedance 螺旋线阻抗
helix linac 螺旋波导直线加速器
helix-milling 螺旋线铣削
helix mixer 螺旋混合器;螺旋混合机
helix pair 螺旋副
helix ribbon type agitator 螺带式搅拌器
helixseal 螺旋密封
helix-to-coaxial-line transducer 螺旋同轴线匹配交换器
helix travel(l)ing wave tube 螺线形行波管
helix tube 螺旋管
helix voltage 螺旋线电压
helix waveguide 螺旋状导管;螺旋形导管
Helizarin pigment colo(u)rs 海立柴林涂料
hell 废物焚化炉
hellandite 硼硅钇钙石
Hellas 希腊盆地
helldriver 联合掘进机
Hellenic architecture 希腊建筑
Hellenic Register 希腊船级社
Hellenic Register of Shipping 希腊船级社
Hellenic Shipyards Company 海伦尼克造船公司
Hellenistic 古希腊建筑式的
Hellenistic Baroque 希腊化巴洛克
Hellenistic Basilica 希腊化长方形会堂;希腊化巴西利卡
Hellenization 希腊化
Heller's test 赫勒试验
Heller system 赫勒系统
hell fence 木桩围栏
Hellige colo(u)rimeter 海氏比色计
Hellige meter 黑利格比色计
Hellinger-Reissner principle 赫格林—赖斯纳原理
hello program(me) 试用程序
hell raiser 磁铁打捞器
Helly's metric(al) system 海利米制
Helly's theorem 海利定理
hellyerite 水碳镍矿
helm 掌舵;艄;舵(轮);舵(机)
helm alee 下风舵
helm amidship 正舵
helm angle 舵角;操舵角度
helm bar 舵柄
helmcloud 山头云
Helmert's blocking technique 赫伯特分区平差法
Helmert's condensation reduction 赫伯特压缩改正
Helmert's formula 赫伯特公式
Helmert's formula of gravity 赫伯特重力公式
Helmert's height 赫伯特高程
Helmert's transformation 赫伯特变换
helmet 钢盔;盔;机罩;焊工帽;头盔;防护帽
helmet air supply 头盔供气
helmet and breast plate 头盔与领盘
helmet attachment 头盔附属物
helmet crane 帽形起重机
helmet cushion 头盔垫子
helmet diver 潜水员
helmet diving 头盔潜水
helmet diving apparatus 头盔式潜水器
helmeted 戴头盔
helmet-hose diving 头盔—软管式潜水
helmet liner 钢盔内帽
helmetmounted binoculars 头盔式双目望远镜
helmet radio 钢盔式无线电设备
helmet radiophone 头盔无线电话机
helmet roof 盔顶
helmet safety lock 头盔安全锁
helmet-shaped 盔形的
helmet-shaped roof 盔形屋顶
helmet shield 工作帽;盔式护罩;焊工罩;安全帽
helmet transceiver 钢盔式无线电收发机
helmet-type diving apparatus 头盔式潜水装具

helm gear 转舵装置;舵机装置;操舵装置
Helmholtz coil 亥姆霍兹线圈;探向线圈
Helmholtz contraction time 亥姆霍兹收缩时间
Helmholtz equation 亥姆霍兹方程
Helmholtz free energy 亥姆霍兹自由能
Helmhatz resonator 亥姆霍兹共振器
helm indicator 驾驶指示器;舵角指示器
helminth 寄生虫;蠕虫;肠虫
helminthes 蠕绿泥石
helminthic disease 蠕虫病
helm kick 舵跳动(海水冲击所致)
helm order 舵令
helm pintle 舵销;舵栓
helm port 舵轴筒;舵杆孔;舵轴孔
helm roof 四坡攒角屋顶;顶层;尖(盔)屋顶
helm rudder gear 转舵装置
helm signal 舵角信号
helmsman 操纵机构;摄像车操纵员;舵手;舵工
helmsman's room 舵工室
helm starboard 右舵
helmstock 舵柄
helmutwinklerite 水砷锌铅石
helm wind 舵轮风(英国北部河谷寒冷东北强风)
helobios 沼泽生物;沼生生物;池沼生物
helobious 沼泽地生的
helodium 沼泽疏林区
helodric 沼泽植丛群落
helohylium 沼泽森林群落
helolochmium 草甸植丛群落
helophyte 沼生植物
helophytia 涨落沼泽生物控制
heloplankton 沼泽浮游生物
helotism 菌藻共生;奴役现象
help 协助;支援;扶助;帮助
help command 帮助指令
helper 助手;辅助机车;辅助工;帮工
helper engine 辅助机车
helper grade 推挽坡度;双机坡度【铁】;多机坡度;补机坡度
helper hold track 补机停留线
helper hole 辅助炮眼
helper locomotive 助推机车
helper post 辅助支柱
helper roll 辅助辊
helper spring 上层钢板弹簧(汽车);附加弹簧;副钢板;辅助弹簧
helper station 补机站;补机牵引始终点站
helper stringer 肋纵梁;辅助纵梁
helper tug 助航拖轮
helper-up 推车工助手
helping earth-scraping process 助铲法
helping professions 辅助性专业
helping wind 顺风
helpmate 助手;助理(人)员
help program(me) 求助程序
help slow down the pace of global warning 帮助减缓全球变暖速度
help with farm work 帮工
help-yourself-system 自助制度
helraser 磁铁打捞器
helsinkite 绿帘钠长岩
helve 斧柄;柄
helve hammer 摇锤;杠杆锤
helve ring 柄圈
Helvetian orogeny 海尔欧(造山)运动
Helvetian stage 海尔微阶
helvine 日光榴石
helvolus 赤黄色
helying 用瓦盖屋面;用石板盖屋面
hem 折边;卷边;蜗缘饰
hemafibrite 红纤维石;水羟砷锰石
hemartine extract 苏木浸膏
hemastrontium 苏木精锶染剂
hematine 苏木精
hematin extract 苏木浸膏
hematite 红铁矿;赤铁体;赤铁矿
hematite alvikite 赤明矿方解石碳酸岩
hematite beforsite 赤铁矿镁云碳酸岩
hematite cement 赤铁矿胶结物
hematite-chlorite oolite 赤铁矿-绿泥石鲕状岩
hematite iron 低磷生铁;赤铁矿生铁
hematite ironstone 赤铁矿铁岩
hematite mud 赤铁矿粉泥浆
hematite oolite 赤铁矿鲕状岩

hematite ore 赤铁矿矿石
hematitepig iron 低磷生铁
hematite rauhaugite 赤铁矿白云(石)碳酸岩
hematite rock 赤铁岩
hematite slag 赤铁渣
hematite sovite 赤铁矿黑云碳酸岩
hematite type macrocrystalline glaze 赤铁矿型巨晶釉
hematitization 赤铁矿化
hematolite 红砷锰矿;红砷铝锰石
hematophanite 红铁铅矿
hematoxylon 苏木
hemeralopia 昼盲
hemeraphotometer 昼光光度计
hemerocology 人为地面形态;栽培生态学;人工环境生态学
hemerophyte 栽培植物;人工引种植物
hemi-acetal 半缩醛
hemi-acetal group 半缩基酸
hemi-arid fan 半旱地扇(半干旱地区冲积扇)
hemi-ascomycetes 半子囊菌类
hemi-azygous vein 半奇静脉
hemi-base 半底面
hemi-bel 半贝(尔)
hemi-biotrophs 半活养寄生物
hemi-block 半支阻滞
hemi-cellulose 半纤维素
hemi-c material 半分解纤维质
hemi-colloid 半胶质;半胶体
hemi-colloid structure 半胶体结构
hemi-concentric resonator 半共心谐振腔
hemi-cone 半锥体
hemi-cone wash 冲积锥
hemi-continuous 半连续的;强弱连续
hemi-continuous rolling mill 半连续式轧机
hemi-cryptophyte 半隐芽植物;地面芽植物
hemi-crystalline 半晶状(的);半晶质(的);半结晶的
hemi-crystalline-porphyritic 半晶斑状
hemi-crystalline rock 半晶质岩
hemi-crystalline rocks 半晶质岩类
hemi-crystalline texture 半晶质结构
hemi-cube algorithm 半立方体算法
hemi-cyanine 半菁
hemi-cycle 半圆形(结构);半圆形(建筑物);半圆室;半循环
hemi-cycle arena 半圆竟技场
hemi-cyclic double bond 半环状双键
hemi-dome 半圆(屋)顶;半(圆)穹隆;半坡面
hemi-epiphyte 半附生植物
hemi-formal 半缩甲醛
hemi-glyph 半竖陇;半三槽板的边槽;半竖槽
hemi-hedral crystal 半面型晶体;半晶
hemi-hedral field 半面像场
hemi-hedral form 半面型形
hemi-hedral symmetry 半面型对称
hemi-hedrism 半对称性
hemi-hedron 半面体
hemihedry 半对称(多指晶体)
hemi-hihedrite 硅铬锌铅石
hemi-holohedral 半全面型
hemi-hyaline texture 半玻璃质结构
hemi-hydrate 半水化(合)物;熟石膏
hemi-hydrate gypsum 半水石膏
hemi-hydrate gypsum plaster 半水化合石膏抹灰
hemi-hydrate of calcium sulfate 半水石膏
hemi-hydrate plaster 半水(合)石膏;熟石膏灰泥
hemi-hydraulic lime 半水硬(性)石灰
hemi-hydraulic lime mortar 半水硬石灰砂浆
hemi-ketal 半缩酮
hemi-mercaptol 半硫代半缩醛
hemi-metabola 半变态类
hemi-morph (晶体的)异极像
hemi-morphic crystal 半形晶体
hemi-morphism 半形性;半对称形;异极性(晶体);异极性;异极像(晶体)
hemi-morphite 异极矿
Hemi-ngfordian 亥明佛德阶【地】
hemi-parasite 半寄生物
hemi-pelagic abyssal sediment 近海—深海沉积物
hemi-pelagic deposit 半远洋沉积;近海沉积(物)
hemi-pelagic environment 半远洋环境
hemi-pelagic region 半远洋区
hemi-pelagic sediment 半远洋沉积
hemi-pinacoid 半轴面

hemi-plankton 半浮游生物
hemi-polymer 易溶聚合物
hemi-prism 半棱柱
hemi-prismatic 半棱晶
hemi-pyramid 半锥;半(棱)锥体
hemi-receiver 半球波束接收机
hemi-reflector 半球波束反射器
hemi-saprophyte 兼性腐生植物
hemi-sphere 半天球;半球形头部汽缸;半球形;半球体;半球(面);半地球
hemi-sphere map 半球(地)图
hemi-sphere of sky of uniform brightness 均匀亮度的天空半球
hemi-sphere temperature of ash T_2 灰半球(软化)温度
hemi-spheric(al) 半球形(的);半球的
hemi-spheric(al) absorptance 半球状吸收比
hemi-spheric(al) bowl 半球形钵体
hemi-spheric(al) candle power 半球(形)烛光
hemi-spheric(al) circulation 半球环流
hemi-spheric(al) combustion chamber 半球形燃烧室
hemi-spheric(al) coverage 半球形视野
hemi-spheric(al) distribution 半球分布
hemi-spheric(al) dome 半圆体屋顶;半球形圆顶
hemi-spheric(al) dome structure 半球形圆顶结构
hemi-spheric(al) emittance 半球形辐射;半球发射率
hemi-spheric(al) fusion temperature 半球(熔融)温度
hemi-spheric(al) head 半球形封头;球形封头
hemi-spheric(al) immersion lens 半球状油浸透镜
hemi-spheric(al) integration 半球面积分
hemi-spheric(al) lamp 半球形灯
hemi-spheric(al) lens 半球形透镜
hemi-spheric(al) mirror 半球形反射镜
hemi-spheric(al) (moud) 半球形窣堵坡
hemi-spheric(al) nose 半球形头部
hemi-spheric(al) projection 半球形投影
hemi-spheric(al) pyrheliometer 半球式日射强度表;半球式日射强度计
hemi-spheric(al) reflection 半球状反射
hemi-spheric(al) resonator 半球面谐振腔
hemi-spheric(al) rotor pole 半圆形转子磁极
hemi-spheric(al) shell 半球形壳
hemi-spheric(al) stupa(moud) 半球形浮图
hemi-spheric(al) substrate 半球形基底
hemi-spheric(al) thermodynamic(al) model of climate 气候的半球热力模式
hemi-spheric(al) transmittance 半球状透射比
hemi-spheroid 半球体;滴形油罐;半球形储罐
hemi-symmetrical 半对称的
hemi-symmetry 半对称
hemi-systole 半(减)收缩
hemi-teratic 半畸形的
hemi-triglyph 半三槽板间距(陶立克建筑)
hemi-trope 半体双晶
hemi-tropic(al) winding 半线圈绕组
hemi-variate 半变量
hemi-vertebra 半脊椎畸形【地】
hemi-zygosity 半合子状态
hem joint 钩接
hemline 贴边线
hemlock 铁杉
hemlock bark extract 北美铁杉树皮提取物(稀释泥浆用)
hemlock fir 铁杉属;铁杉木材
hemlock needle oil 铁杉油
hemlock spruce 铁杉属
hemlock tannin 铁杉丹宁
hemming machine 折缘机;缝口机;缝边机
hemming press 折边压力机
Hemmingway 沉积碳酸钙的石材防腐法
hemo-core wire rope 麻芯钢丝绳
hemoglobinuric fever 黑水热
hemo packing 麻线填料
hemo tallow packing 麻脂填料
hemo yarn 麻线
hemp 苎麻;麻丝;麻类;大麻
hempa 六甲磷
hemp-bag 麻袋
hemp-breaked and scutcher 碎茎打麻机
hemp burlap mat 苎麻布垫
hemp cable 大麻缆;麻绳;大麻绳
hemp canvas 麻帆布
hemp ca(u)lking 麻丝嵌填
hemp center 钢丝绳麻芯;麻芯
hemp-centre cable 黄麻芯钢索
hemp cloth 大麻织物
hemp combings 大麻短纤维
hemp cord 麻绳;索芯
hemp core (钢缆的)麻芯;索芯
hempcore cable 标准钢丝绳
hemp cored wire rope 麻心钢索;麻心钢丝绳
hemp cut 麻丝;麻筋;麻刀
hemp-cut plaster finishing coat 麻刀灰罩面
hemp dust 大麻尘
hempen core 麻芯线
hemp fiber 苎麻纤维;麻筋;麻刀(纤维);大麻纤维;麻絮
hemp-fibered plaster 麻刀灰
hemp-fibered plaster base 麻刀灰打底
hemp-fibered plaster finishing coat 麻刀灰罩面
hemp floor mat 麻地垫
hemp gasket 麻垫片
hemp gasket joints for pipes 管道麻绳垫圈接缝
hemp hards 麻屑;短麻
hemp hawser 麻绳缆
hemp Hessian mat 大麻布垫
Hemphillian 亥姆菲尔阶【地】
hemp hose 麻织水龙带
hemp jointing 麻填料;麻丝嵌填;麻垫料
hemp lashing 麻绳捆扎
hemp line 大麻长纤维
hemp packed piston 麻填密活塞
hemp packing 麻填密;麻丝填料;麻丝封填;麻绳填料;麻垫(料)
hemp palm 棕榈
hemp products 大麻制品
hemp rope 麻绳
hemps 麻屑
hemp sack 麻袋
hemp sacking 制麻袋
hemp scutcher 扩大麻机
hempseed 大麻籽
hemp-seed oil 大麻(籽)油;小麻籽油;线麻籽油
hemp shives 麻布头尖
hemp strap 麻织系带
hemp tallowed packing 涂油麻填料
hemp thread 麻丝
hemp tow 麻屑;短麻;大麻短纤维
hemp-twist 麻绳
hemp wall covering 大麻纤维墙布
hemp wrapped 大麻布包裹的
hemp yarn 大麻纱
hem shoe 闸瓦
hemusite 硫钼锡铜矿
hench 拱高;烟囱筒腰
hen-coop 鸡笼
hen-cooping 鸡笼式填石木笼;三角形填石木笼护岸
hendecagon 十一角形;十一边形
hendecahedron 十一面体
hendecanal 十一醛
hendecane 十一烷
hendecanoic acid 十一烷酸
hendecaploidy 十一倍性
hendecene 十一烯
hendecyl 十一基
hendecyne 十一碳炔
hendersonite 复钒钙石
Hendre Shales 亨德里页岩
hendricksite 锌云母
heneicosane 二十一烷
heneicosane dicarboxylic acid 二十一烷二羧酸
heneicosanedioic acid 二十一烷二酸
heneicosanic acid 二十一酸
heneicosanoic acid 二十一烷酸
heneicosene 二十一碳烯
heneicosoic acid 二十一酸
henequen 龙舌兰纤维
Hengstebeck approximation 亨斯得贝克近似计算
henhouse 鸡舍
henkelplush 毛圈长毛绒织物
Henkel process 亨克尔法
henna 棕红色;红褐色
Hennebique pile 韩涅布克柱
hennery 养鸡场;家禽饲养场
hennibique 钢筋混凝土中由扁钢箍定位的圆钢筋
henostyle 独柱式
henry 亨利(电感单位)
Henry's constant 亨利常数
Henry's equation 亨利方程
Henry's law 亨利定律
Henry's law constant 亨利定律常数
Henry's unit 亨利单位
Henry emmenopterys 香果树
henryite 碲银铜矿
Henrymeter 电感表;亨利计;电感计
henryptite 水钙锰榴石
Henry wilsontree 山白树
Henschel mixer 亨舍尔混合机
Hensen's line 亨森线
hens house 鸡舍
heofucoxanthin 新叶黄素
heory duty face lathe series 落地车床
heotastylos 有7根柱子的建筑物(如古典式寺庙)
hepar sulfuris 硫化钾
hepaticae 苔类(苔藓植物);苔纲
hepatic cinnabar 朱砂一硫化汞
hepatic iron ore 肝铁矿
hepatic pyrite 肝铁矿
hepatic water 含硫矿水
hepatitis 肝炎
hepatitis A 甲型肝炎
hepatitis B 乙型肝炎
Hepex process 海派克斯过程
Hephaisteion 赫费斯提翁神庙(也称提修斯神庙,古希腊)
Hepple Whilte-Gray lamp 海柏尔怀特一格雷火焰安全灯
hepronicate 癸烟酯;灭酯灵
heptabarbital 庚巴比妥
heptacene 庚省;并七苯
heptachlor 七氯
heptachlor epoxide 七氯环氧化物
heptachlorinated dibenzofuran 七氯化二苯并呋喃
heptachlorobiphenyl 七氯联苯
heptachlorofuran 七氯呋喃
heptacodium 七子花属
heptacosane 二十七烷
heptacosanoic acid 二十七酸
heptadecane 十七烷
heptadecanoic acid 十七酸
heptadecanol 十七醇
heptadecanoyl 十七酰
heptadecene 十七碳烯
heptadecyl 十七基
heptadecylamine 十七烷胺
heptadecyl phosphoric acid 十七烷基磷酸
heptadiene 庚二烯
heptadiyne 庚二炔
heptaepoxide 庚环氧化物
heptagon 七角形;七边形
heptagonal prism 七角棱镜
heptahedron 七面体
heptaldehyde 庚醛;庚醛
heptalin acetate 乙酸甲基环己酯
heptaline 甲基环己醇
heptamethylene glycol 庚二醇
heptanal 庚醛
heptandiol 庚二醇
heptane 庚烷
heptane diacid 庚二酸
heptanedicic acid 庚二酸
heptane number 庚烷值
heptanoic anhydride 庚酸酐
heptanol 庚醇
heptanone 庚酮
heptanthiol 庚硫醇
heptantriene 庚三烯
heptastyle 七柱式
heptastyle building 七柱式建筑
heptastylos 七柱式建筑
heptatriacontane 七十烷
heptene 庚烯
heptenoic acid 庚烯酸
heptenone 庚烯酮
heptenyl 庚烯基
heptine 庚炔
heptobarbital 苯甲基比妥
heptorite 兰方煌沸岩
heptose 庚糖

heptulose 庚酮糖
heptyl 庚基
heptyl acetate 乙酸庚酯
heptyl aldehyde 庚醛
heptyl amice 庚胺
heptyl amine 庚胺
heptyl benzene 庚基苯
heptylene 庚烯
heptylene diacid 庚烯二酸
heptyl ether 庚醚
heptyl heptylate 庚酸庚酯
heptylic acid 庚酸
heptylic anhydride 庚酸酐
heptyne 庚炔
heptynoic acid 庚炔酸
H-equation 水头方程(式)
Heraion at Olympia 奥林匹亚的赫拉神殿
Heraklith 希拉克利绝缘材料(一种厚度不大于3英寸的检验材料)
Heraklith magnesite-bound excelsior building slab 希拉克利菱苦土木屑建筑板
HERALD equipment 海港防御用声呐和测听装置
heraldic china 纹章装饰瓷器
herb 草本植物;草本的
herbaceous 草本的
herbaceous aquatic vegetation 草本水生植物
herbaceous border 草本植物的花坛;草皮护面花坛
herbaceous cover 草皮护面;草皮覆盖;草皮覆被
herbaceous cutting 草本的杆插
herbaceous flower bed 草药花坛
herbaceous fruit 草本果
herbaceous fuel 草本可可燃物
herbaceous layer 草本层
herbaceous perennid 多年生草本植物
herbaceous plant 草本植物
herbaceous soil covering 草皮覆盖层;草本地被(物)
herbaceous stage 草本群落阶段
herbaceous stem 草质
herbaceous steppe 草本干草原
herbaceous swamp 草本沼泽
herbaceous type 草本型
herbaceous weed 草本杂草
herbage 草本植物(群);放牧权
herbage cover 草皮覆被
herbage intake 采草量
herbage with dwarf grass 矮收草放牧
herbal medicine shop 药铺
herban 草完隆
herbarium 植物标本室
herbarium material 标本室植物
herbary 菜园
herberge 小旅店;小客栈
Herbert's cloud burst test 赫伯特喷丸试验
Herbert's hardness 赫伯特硬度
Herbert's pendulum hardness 赫伯特摆式硬度
Herbert's pendulum hardness test 赫伯特摆锤硬度试验
Herbert's pendulum hardness tester 赫伯特摆式硬度计;赫伯特摆锤硬度试验器;赫伯特摆锤硬度试验机
Herbert's pendulum method 赫伯特摆法
Herbert's test 赫伯特腐蚀试验
herb garden 草药园;药用植物园
herbicidal chemcial 除莠化学药品;除莠剂;除草剂
herbicide 灭草剂;杀草剂;除草剂
herbicide action 除草效力
herbicide damage 除草剂药害
herbicide placement 定点施药
herbicide poisoning 除莠剂中毒
herbicide-processed film sheet 避光灭草膜
herbicide sprayer 除莠剂喷雾机
herbicide test 除莠剂试验
herbivorous animal 食草动物
herb-lane 路侧车道
herb layer 草层
herbosa 草木植被;草甸植被;草丛;草本群落
herb rubber 草胶
herbshop 草药店
herbstore 草药店
hercoal F 赫尔科尔 F 炸药
Hercogel 赫尔科爆胶
Hercolt aluminum process 赫科特制铝法
Hercomite 赫尔科炸药
Hercules 重型机器;大力神
Hercules crane 重型塔式起重机
Hercules duster 赫尔科利斯除尘机
Hercules foundation at Augsburg 奥格斯堡赫尔科利斯泉
Hercules graphic adaptor 彩色图形接口板
Hercules metal 赫尔科利斯合金
Hercules powder 赫尔科利斯炸药;矿山用炸药
Hercules Powder process 赫尔科利斯火药公司法
Hercules press 大力神压机
Hercules stone 极磁铁矿
Hercules trap 赫尔科利斯阱
Hercules wire rope 赫尔科利斯多股钢丝绳;大力神(多股)钢丝绳
herculite 钢化玻璃;赫尔克莱特厚平板玻璃
Herculoy 锻造铜硅合金
Hercynian basement 海西期基底【地】
Hercynian cycle 海西旋回【地】
Hercynian epoch geosyncline 海西期地槽
Hercynian geosyncline 海西地槽【地】
Hercynian marine trough 海西海槽
Hercynian mountain 海西山地
Hercynian movement 海西运动【地】
Hercynian orogency 海西(造山)运动【地】
Hercynian period 海西期【地】
hercynide 海西褶皱带;海西构造带【地】
hercynite 铁(铝)尖晶石
herd 放牧
herder 牧者
herderite 磷铍钙石
herd infection 集体传染
herding 集中畜群
herding the cattle 放牧
herd instinct 群集本能
herdman's hut 牧人小屋
herd number 注册号码
herd to dwarf herbage 矮草放牧
hereafter 下文;此后
here-by 特此
hereditament 继承;可继承的财产
hereditary characteristic (弹塑性的)后效特性
hereditary engineering 遗传工程
hereditary estate 世袭财产
hereditary property 世袭财产
here-in 此中
heretical imputation 异端分配
heretical set 异端集
Hering furnace 赫林电炉
heritable 可转让
heritable obligation 可继承债务
heritable rights 可继承权利
heritable rond 附以继承权的契约
heritage 遗产;世袭财产
heritage coast 自然遗产海岸
heritage preservation 自然遗产维护
herkinghead 交叉支撑
herm 古希腊刻有汉密士神象的石碑
hermannite 蔷薇辉石
hermaphrodite 两性体
hermaphrodite cal(l)ipers 单边卡钳;复式卡钳;内外卡钳
hermaphroditical contact 中性接触体
hermaphroditic connector 鸳鸯插头;阴阳插头;第一型插座;单一型插头;等同接合连接器
hermaphroditic coupling 内扣式接口
hermaphroditic polyacrylamide 两性聚丙烯酰胺
hermatopelago 沉溺海脊
hermatype 造礁珊瑚
hermetex 修理屋面用防水塑性石棉合成物
hermetical 密封的
hermetically sealed 密封的;全封闭的;气密封口;气密的
hermetically sealed cable 密封电缆
hermetically sealed casing 密封机罩
hermetically sealed chamber 全密封室
hermetically sealed condensing unit 密封式冷凝机
hermetically sealed construction 密封结构
hermetically sealed container 密闭容器
hermetically sealed dehydrated captive air space 气密干燥空腔
hermetically sealed door 密封门
hermetically sealed double glazing 密封双层玻璃
hermetically sealed edge 密封边
hermetically sealed electromagnetic relay 密封电磁继电器
hermetically sealed instrument 密封仪器;密封式测量仪器
hermetically sealed magnetic drive pump 封闭式电磁泵
hermetically sealed motor 密封(式)电动机;密封型电动机
hermetically sealed pulse transformer 密封脉冲变压器
hermetically sealed relay 密封式继电器
hermetically sealed switchbox 密封式(继)开关箱
hermetically sealed type transformer 密封(式)变压器
hermetic cabin 气密室
hermetic centrifuge 密封式离心机
hermetic closure 密封;密闭
hermetic compressor 密封压缩机;封闭式压缩机
hermetic cover 密闭盖
hermetic door 密封门
hermetic enclosure header 密封体引线头
hermetic hood 密闭罩
hermetic integrating gyroscope 密封式积分陀螺仪
hermeticity analysis 密封性分析
hermetic machine 密封式电机
hermetic material 气密性材料;不透气材料
hermetic microwave digestion 密封微波溶出
hermetic motor 密封型电动机;密封式电动机
hermetic package 气密封装
hermetic pressure-feed type centrifuge 密闭加压进料式离心机
hermetic pscakge 密封包装
hermetic reciprocating compressor 封闭往复式压缩机
hermetic refrigerant motor-compressor 封闭式压缩机
hermetic screw pump 密封式螺杆泵
hermetic seal 真空密封;密封止水;密封(接头);气密封接;气密封(闭)
hermetic seal furnace 气密封炉
hermetic separator 密封分离器
hermetic turbocompressor 密封式透平压缩机
hermetic type bowl 封闭型转筒
hermetic unit 密封装置;密封部分;密闭部件
hermetization 密封
hermitage 修道院;隐庐;僻静的住处
Hermite 艾米插值
Hermite's polynomials 厄密多项式
Hermite-Gauss beam 厄密—高密光束
Hermite-Gauss modes 厄密—高斯波形
Hermitian form 厄密型
Hermitian kernel 厄密核
Hermitian matrix 厄密矩阵
hermitian operator 厄密算符
hermosa pink 浅粉色
hernial protrusion of membrane 膜膨出
heroin(e) 海洛因
herol 赫罗尔胶
heron 苍鹭
Heroult furnace 赫劳特电弧炉
Heroult process 赫劳尔特电解炼铝法
heroum 英雄墓舍
herpolhode cone 空间锥面
Herreshoff furnace 窄轴式多腔焙烧炉;赫雷肖夫炉
herring 青鱼;鲱
herring boat 鲱鱼捕劳队
herringbone 鱼脊形;交叉缝式;十字缝;鲱骨式的;人字形
herringbone arrangement 鱼脊形排列(集装箱)
herringbone ashlar 人字形琢石墙面;人字形琢石面墙
herringbone bearing 人字形曲槽轴承
herringbone bond 斜置丁砖皮;人字形面砖排列;对角接合;人字(形)砌合;人字(式)砌合
herringbone bracing 斜十字撑;人字撑
herringbone brickwork 交叉缝砌砖;人字形砖砌体;人字形砌砖;错缝砌体
herringbone bridging 人字撑
herringbone convex ring 人字凸环
herringbone crack 人字形裂缝
herringbone cross bedding 人字形交错层理
herringbone cross bedding structure 鱼骨状交错层理构造
herringbone cross lamination 鲱骨式交错层理
herringbone distortion 人字形扭曲变形

herringbone drain 鱼脊式排水管道;人字形排水沟;人字布置排水管
herringbone drainage 鲱骨式排水;人字形排水
herringbone drain(age) system V 形排水系统;人字形排水系统;人字形排水法;梳式排水系统
herringbone dressing 人字纹修琢(石面)
herringbone earth 鱼骨形ेत地
herringbone effect 人字花形
herringbone face 鲱骨式割面
herringbone fashion 人字式
herringbone gear 人字(形)齿轮;人字(齿)轮;双螺旋齿轮
herringbone gear cutting machine 人字齿轮切齿机
herringbone gear milling machine 人字齿轮铣床
herringbone gear planer 人字齿轮刨床
herringbone gear pump 人字齿轮泵
herringbone gear tooth 人字齿
herringbone joint 人字榫;人字缝
herringbone marking 鱼骨状流痕
herringbone masonry 人字铺砌砌体;人字式铺地;人字砌圬工
herringbone masonry bond 人字形砌合
herringbone masonry course 人字形砖层
herringbone matching 木纹拼花;人字形企口;人字拼接;人字拼花
herringbone mesh 人字形网眼
herringbone mesh opening 人字形网孔
herringbone method 人字形矿房采矿法
herringbone milking bail 鱼刺形挤奶装置
herringbone parquet floor 人字拼花地板
herringbone parquetry 席纹(木)地板;人字拼木地板;人字拼花地板
herringbone pattern 人字形(图案);人字形(布置);人字焊纹;梳式;席纹式
herringbone pavement 人字式铺地;人字形铺(砌)路)面;人字路面;人字式(铺砌)路面
herringbone paving 人字形(铺砌)路面;人字式铺砌法
herringbone planking 人字形排列的板
herringbone reducer 人字齿轮减速器
herringbone reflector 鱼骨形反射器
herringbone roller conveyer 人字滚轴运输机
herringbone room arrangement 人字式矿房布置
herringbone seal ring 人字密封环
herringbone stitch 人字形缝
herringbone stoping 人字形采矿法;鲱骨式采矿法
herringbone structure 鲱骨式构造;人字形结构
herringbone strut(ting) 格栅撑;剪刀撑;人字撑;人字形支撑;人字形斜撑
herringbone system 鱼骨形排水系统;人字形(排水)系统;人字形排水管系;人字形灌溉系统;鲱骨式排水系统;鲱骨式(灌溉)系统
herringbone system of drains 人字形排水系统
herringbone texture 人字形结构
herringbone tile drainage system 人字形瓦管排水系统
herringbone timbering 人字支架(隧道);人字形支架;人字形支护法
herringbone track 箭翎线【铁】
herringbone track layout 箭翎形线路布置
herringbone twill 人字形斜纹;山形斜纹
herringbone weave 人字形组织
herringbone wharf layout 鱼骨式码头布置;人字式码头布置;梳式码头布置
herringbone wheel 人字齿轮
herringbone winding 人字形绕组
herringbone work 之字形砌合工程;鱼骨形砌合工程;人字铺面工程;人字铺砌工;人字形砌工;人字形砌法
herringboning 紧缝;十字缝法;人字形采矿法
herring-gear 人字形齿轮
herring knot 拖网结
herring oil 青鱼油;鲱鱼油
Herschel' telescope 赫歇尔望远镜
Herschel effect 赫歇尔效应
Herschel formula 赫歇尔公式
herschelite 碱菱沸石
Herschel-type Venturi tube 赫歇尔型文丘里管
herse 吊闸;(城门的)吊门
Hersey detector meter 赫西检测器水表
Hertz 赫(兹);每秒周数
Hertz(ian) antenna 赫兹天线
hertz(ian) contact strain formula 赫兹接触应变公式
Hertz(ian) dipole 赫兹偶极子
Hertz(ian) doublet 赫兹偶极子
Hertz(ian) doublet antenna 赫兹偶极天线
Hertz(ian) effect 赫兹效应
Hertz(ian) fracture 皱形裂纹
Hertz(ian) law 赫兹定律
Hertz(ian) oscillator 赫兹振荡器
Hertz(ian) radiator 赫兹辐射器
hertz(ian) rays 赫兹氏射线
Hertz(ian) resonator 赫兹振荡器;赫兹共振器
Hertz(ian) stress 赫氏应力
Hertz(ian) theory 赫兹理论
Hertz(ian) vector 赫兹矢量
Hertz(ian) wave 赫兹波;电磁波
Hertzsprung gap 赫兹伯伦空隙;赫氏空隙
herzenbergite 硫锡矿
hesitating relay 缓动继电器
hesitation 暂停;暂时停机
hesitation pumping 间歇泵送
hesitation set(ting) 假凝(现象);虚凝
hesitation squeeze 间歇挤水泥
hesperetol 橙皮酚
hesperidene 橙皮烯
hesperidine 橙皮碱
hesperitinic acid 橙皮酸
Hess's diagram 赫斯图
Hess's law 赫斯定律
Hessbil 赫斯比衬垫(一种沥青衬垫)
Hesselman engine 赫塞尔曼发动机
hessian 麻布;打包麻布;粗麻布罩(养护混凝土用);粗麻布
hessian-based bituminous felt 麻布油毡
hessian-based bituminous sheeting 麻布油毡
Hessian canvas 大麻帆布
hessian cloth 粗麻布;浸沥青的麻绳;麻袋片
hessian fire hose 麻布水龙带
hessian rope 麻绳
hessian sacking 粗麻布袋装;粗麻袋(养护用)
hessian sheeting 粗麻布沥青油毡
hessite 天然碲化银;碲银矿
Hess-Ives colo(u)r meter 赫斯—艾夫斯比色计
Hessle boulder clay 赫塞尔泥砾
hessonite 桂榴石;钙铝榴石
hest face 最优材面
hetaerolite 锌锰矿
hetegony 伴星起源说
heteracumulate texture 异补堆积结构
heter-aerobic pond 稳定塘—兼性厌氧塘
heteresthesia 差异感觉
heteroacid 杂酸
heteroagglutination 异种凝集
heteroagglutinin 异种凝集素
heteroaromatic compound 芳香杂环化合物
heteroaromatics 杂芳族化合物
heteroatomic bond 杂原子链
heteroatomic ring 杂环
heteroazeotrope 多相非共沸混合物
heterobaric heterotope 异量异序元素
heterobasidium 不定形担子
heteroblastic 异源发生
heteroblastic texture 不均匀变晶结构
heterobrochantite 异水胆矾
heterobrochate 异形网状
heterocatenary polymer 杂链聚合物
heterocercal 歪形尾
heterochain 杂链
heterochain polymer 杂链聚合物
heterocharge 混杂电荷
heterochiral 左右异向的
heterochromatic 异色的;多色的;非单色的
heterochromatic magnitude 混色星等
heterochromatic paint 多彩涂料
heterochromatic photometer 多色光度计
heterochromatic photometry 异色光度学;多色光(度)学
heterochromatic stimuli 异色刺激
heterochromous 异色的;不同色的
heterochromous digital signal 不均匀数字信号
heterochronia 异时性
heterochronic restitution 不同时恢复
heterochronic rhythm 异时节律
heterochronism 异时(生成)性
heterochronogenous soil 次生土(壤)
heterochronous 异时生成的;差时步
heterochronous convergence 异时趋同
heterochrony 异时性;异时发生
heterocoagulation 异质凝结;杂凝聚
heterocollinite 不均匀镜质体
heterocomplex 杂络物
heterocompound 杂化合物
heterocondensation 杂缩合
hetero conjugation 杂共轭
heterocrystal 异种晶体;异质晶体
heterocycle 杂环
heterocyclic 杂环的
heterocyclic(al) compound 杂环化合物
heterocyclic(al) dye 杂环(结构)染料
heterocyclic(al) nitrogen compounds 杂环氮化物
heterocyclic(al) oxygen compound 氧杂环化合物
heterocyclic(al) ring 杂环
heterocyclic(al) system 杂环的系统
heterodesmic structure 多型键构造
heterodisperse 杂分散;非均相分散
heterodispersity 杂分散性;非均相分散度
heterodromy 异向旋转式
heterodynamics 异态动力学
heterodyne 外差;成拍
heterodyne action 外差作用
heterodyne analyzer 外差式分析仪;外差分析器
heterodyne buzzer oscillator 外差蜂鸣振荡器
heterodyne converter 外差变频器;多差式变频器
heterodyne detection 外差检波
heterodyne detection method 外差检波法
heterodyne detector 外差检波器
heterodyne filter 外差滤波器
heterodyne frequency 外差频率
heterodyne frequency meter 外差式频率计
heterodyne harmonic analyzer 外差谐波分析仪;外差式谐波分析器
heterodyne interference 外差干扰
heterodyne marker adder 外差标志叠加器
heterodyne measurement 外差法测量
heterodyne method 外差法
heterodyne modulation 外差调制
heterodyne modulator 混频器
heterodyne oscillator 外差振荡器
heterodyne receiver 外差(式)接收机
heterodyne reception 外差接收法
heterodyne relay system 外差中继方式
heterodyne repeater 外差中继站
heterodyne repeating 外差转发;外差中继
heterodyne slave filter 外差伺服滤波器
heterodyne sound analyzer 外差式声分器
heterodyne stage 外差级
heterodyne system 外差式
heterodyne-type frequency meter 外差频率计
heterodyne voltmeter 外差式电压计
heterodyne wave analysis 外差式谐波分析
heterodyne wave analyzer 外差式波形分析仪
heterodyne wavemeter 外差式波长计
heterodyne whistle 外差啸声
heterodyne whistling 外差啸鸣
heterodyning 外差作用;差拍变频
heteroepitaxial film 异质外延膜
heteroepitaxial optic(al) waveguide 异质外延光波导
heteroepitaxy 异质外延
hetero fiber 异质纤维
heterofilm 异相薄膜
heterogel 杂凝胶
heterogeneity 异质(性);异性质;异类;多相性;非均质;非均匀性;不统一性;不均匀性
heterogeneity error 不均匀误差
heterogeneity of rock 岩石的非均质性
heterogeneity of variance 方差不齐
heterogeneity test 间杂性检验;多相性测定(沥青物质);非均匀性试验
heterogeneous 异质的;多相的;多相催化湿空气氧化工艺;参差;非均质的;非均匀的
heterogeneous alloy 多相合金;不匀合金
heterogeneous anisotropic aquifer 非均质各向异性含水层
heterogeneous anisotropic medium 非均质各向异性介质;非均质各向异性介质
heterogeneous aqueous system 不均匀含水体系

heterogeneous aquifer 非均质含水层
heterogeneous atmosphere reaction 多相大气反应
heterogeneous azeotrope 非均匀共沸混合物
heterogeneous blend 不均匀混和
heterogeneous body 非均质体;非均匀体;不均匀体
heterogeneous broadening 非均匀加宽
heterogeneous cargo 异质货物
heterogeneous catalysis 多相催化;非均相催化
heterogeneous catalyst 多相催化剂
heterogeneous catalytic oxidation 多相催化氧化
heterogeneous catalytic oxidation method 多相催化氧化法
heterogeneous catalytic ozonation 多相催化臭氧氧化
heterogeneous catalytic ozonation process 多相催化臭氧氧化工艺
heterogeneous catalytic ozone decomposition 多相催化臭氧分解
heterogeneous chain compound 杂链化合物
heterogeneous chemical reaction 多相化学反应
heterogeneous circuit 不均匀电路
heterogeneous classification 不同质分类
heterogeneous colo(u)rs 多色性染料
heterogeneous combustion 异相燃烧;非均质燃烧
heterogeneous decomposition 多相分解
heterogeneous deposit 非均匀沉积层
heterogeneous die 异形冷拔模
heterogeneous differentiation 不均匀分异作用
heterogeneous distortion 不均匀失真
heterogeneous distribution 非均匀分布;非纯一分布;不均匀分布
heterogeneous effect 非均匀效应
heterogeneous elastic medium 非均匀弹性介质
heterogeneous equilibrium 多相平衡;复杂平衡;复相平衡;非均相平衡;不均匀平衡
heterogeneous equilibrium state 非均匀平衡状态
heterogeneous error 不均匀误差
heterogeneous fabric 不均匀组构
heterogeneous flow 非均匀流;非均相流动
heterogeneous fluctuation 非均匀涨落
heterogeneous fluid 非均匀流体;非均值流体;非均匀流体
heterogeneous foundation 非均匀地基;不均匀地基
heterogeneous ground 非均质基础
heterogeneous group 异类组;非同质组
heterogeneous isotropic aquifer 非均质各向同性含水层
heterogeneous layer(stratum) 非均质层
heterogeneous light 杂色光
heterogeneously fluidized bed 非均一流化床
heterogeneous map coverage 有各种地图资料地区
heterogeneous material 各向异性材料;多相材料;非均质材料;不均匀材料
heterogeneous medium 非均匀介质
heterogeneous membrane electrode 异相膜电极
heterogeneous mixture 非匀质混合物;非均匀混合物;不均匀混合物;不均匀混合料
heterogeneous multiplexing 非均匀复用
heterogeneous multiprocessor 多机功能型多处理机
heterogeneous network 多机种网络【计】;多机型网
heterogeneous nucleation 异相成核;多相核化;非均匀成核;不均匀生核
heterogeneous output 非同质产出
heterogeneous oxide 异质氧化物
heterogeneous plastic body 非均质塑性体
heterogeneous polymerization 非均相聚合
heterogeneous population 异质性群体;非同质总体;非纯一总体
heterogeneous radiation 杂辐射;多频(率)辐射;非均匀辐射;不均匀辐射;不均匀放射
heterogeneous radiator 多频(率)辐射器;非均匀辐射器
heterogeneous ray 异形射线
heterogeneous reaction 多相反应;复相反应;非均质反应;非均相反应;非均向反应
heterogeneous reactor 非均匀反应堆;非均相反应堆
heterogeneous reactor type 不均匀堆型
heterogeneous reservoir 非均质储集层
heterogeneous restriction 非齐性限制
heterogeneous ring compound 杂环化合物
heterogeneous rock mass 非均质岩体
heterogeneous sample 不均匀试样
heterogeneous sand 非均质砂

heterogeneous settling 不均匀沉落
heterogeneous shear layer 非均匀剪力层
heterogeneous sintering 多元系烧结;多相烧结
heterogeneous soil 非均匀土;不均质土
heterogeneous spheric(al)earth 非均匀球状地球
heterogeneous state 非均匀态
heterogeneous strain 非均匀应变;非均匀应变
heterogeneous stratum 非均质地层
heterogeneous stress 非均质应力;不均匀应力
heterogeneous structure 异类结构;多相组织;多相结构;不均匀组织
heterogeneous summation 异质相加
heterogeneous surface 多相表面
heterogeneous system 多相(体)系;复相系;非均质系(统);非均匀体系;非均相体系;非纯一系统;不均匀(体)系;不均(相)系
heterogeneous thermoelectric(al) effect 非均匀温度差电效应
heterogeneous turbulence 非均相湍流
heterogeneous ultraviolet-Fenton reaction 非均相紫外线—芬顿反应
heterogeneous variance 异质性的方差
heterogeneous waste 混杂废物
heterogeneous zeotrope 多相非共沸混合物
heterogenesis 偶然发生
heterogenetic 多相的
heterogenite 羟氧钴矿;水钴矿
heterogenous body 多相体
heterogenous composition 多相组成
heterogenous diffusion model 多相扩散模型
heterogenous dispersion 多相分散
heterogenous oxidation 多相氧化作用
heterogenous photocatalysis 多相光催化
heterogenous photocatalyst 多相光催化剂
heterogenous photocatalytic decomposition 多相光催化分解
heterogenous photocatalytic degradation 多相光催化降解
heterogenous photocatalytic oxidation 多相光催化氧化
heterogenous photocatalytic ozone oxidation 多相光催化臭氧氧化
heterogenous photocatalytic treatment 多相光催化处理
heterogenous photochemical electron transfer 多相光化学电子传输
heterogenous photochemical reaction 多相光化学反应
heterogenous soil mass 非均质土体
heterogeny 异质
heterojunction 异端连接
heterojunction infrared detector 异质结红外探测器
heterojunction material 异质结材料
heterojunction photocathode 异质结光电阴极
heterojunction photovoltaic cell 异质结光电池
heterojunction solar battery 异质结太阳电池
heterojunction structure 异质结构
heterokinesis 异化分裂
heteroleaf false panax 异叶果玉茶
heterolipid 杂脂
heterolysis 异种溶解;异溶化作用
heteromerism 异数
heteromerite 符山石
heteromery 异数
heterometric titration 比浊滴定
heterometry 比浊滴定(法)
heteromixis 异源融合
heteromorphic alternation 异性交替
heteromorphism 异形(性);异形现象;异像;异态性;异态现象;结晶同质异像;同质异象;多晶(性)现象
heteromorphite 异硫锑铅矿
heteromorphosis 异形;异形再生;异形形成
heteromorphous rock 异形岩
heteromorphy 异态性
heteropathy 反应性异常
heterophagy 异溶化作用
heterophase material 多相材料
heterophase polymerization 多相聚合
heterophyte 异养植物;异形植物
heteropic 无均性的
heteropic(al) 非均匀性的
heteropic(al) deposit 异相沉积;同时异相沉积
heteroplanobios 河流漂浮物

heteropolar alternator 异极交流发电机
heteropolar bond 有极键;极性键;电价键
heteropolar class 异极对称型
heteropolar compound 异极化合物
heteropolar direct current linear motor 多极直流直线电动机
heteropolar dynamo 异极电机
heteropolar field magnet 异极场磁铁
heteropolar generator 异极发电机
heteropolar inductor generator 异极感应子发电机
heteropolarity 异极性(晶体)
heteropolar lattice 异极点阵
heteropolar liquid 异极性液体
heteropolar machine 异极电机
heteropolar type of axial symmetry 异极型轴对称
heteropole 异极
heteropoly acid 杂多酸
heteropoly blue 杂多蓝
heteropoly compound 杂多化合物
heteropolymer 杂聚(合)物
heteropolymerization 异相聚合;杂聚合(作用)
heteropoly molybdic acid 杂多钼酸
heteroscedasticity 方差不齐
heteroscedastic variances 非齐性方差
heteroscian region 日影异向区
Heterosigma akashiwo 赤潮异弯藻
heterosiloxane 杂硅氧烷
heterosite 异磷铁锰矿;磷铁石
heterosphere 非均质层;非均匀(大)气层
heterostatic(al) 异位差的
heterostatic(al) method 异位差连接法
heterostrobe 零拍闸门
heterostrophy 以与常规方向相反的方向卷绕所得的品质或状态
heterostructure 异质结构;异晶结构;非结构性
heterosuggestion 外源暗示
heterosynapsis 异形联会
heterotactic 不均匀有规立构
heterotactic fabric 异变组构
heterotactic polymer 杂同(立构)聚合物;杂规聚合物
heterotaxy 地层变位
heterothermic animal 异温动物
heterotope 异序元素;同量异序元素
heterotroph 异养生物
heterotrophic aerobic biofilm 异养好氧生物膜
heterotrophic bacteria 异养(细)菌
heterotrophic bacterial plate count 异养菌平板计数
heterotrophic cycle 异养循环
heterotrophic microbe 异养微生物
heterotrophic microorganism 异养微生物
heterotrophic nanoplankton 异养微浮游生物
heterotrophic nitrification 异养硝化
heterotrophic organism 异养生物
heterotrophic plant 异养植物
heterotrophism 异养性
heterotropic 斜交的
heterotype 异形;同类异性物
heterotypic(a)effect 异形作用
heterozone organism 多生境生物
heterozygosis 杂合
heterozygous 杂合的
heterpolar attachment energy 异极附着能
het flux 热通量
Hettangian 赫特唐阶【地】
hettocyrtosis 轻度弯曲
heulandite 片沸石
Heulinger equation 休林格方程
heuristic 启发式(的);试探
heuristic algorithm 探试算法
heuristic approach 试探步骤
heuristic function 启发函数
heuristic method 直接推断法;启发性方法;探索法;发展法
heuristic model 启发模式
heuristic non-preemptive algorithm 探试非抢先算法
heuristic power 启发能力
heuristic procedure 启发式程序;发展程序
heuristic program(me) 探索程序;助解程序;启发式规划;启发程序;试探程序
heuristic programming 启发式程序设计;试探规划法

heuristic pruning of game tree 博弈树(的)启发式修剪
heuristic routine 探索程序;试探程序
heuristic rule 启发式规则;试探规则
heuristics 直观推断;启发式;试探法
heuristic search 启发式(的)搜索;探试搜索;试探搜索
heuristic search technique 启发式查找技术;试探查找技术
heuristics robot 职能机器人
Heuslars alloy 惠斯勒磁性合金
Heusler's magnetic alloy 铜基锰铝磁性合金
Heusler alloy 锰铝铜强磁性合金
hevelian halo 淡晕
hew 劈
hew down 砍倒
hewed tie 斧砍枕木;斧砍沉排
hewer 采煤工人;采伐者
hewettite 针钒钙石
hewn 斧砍的
hewn foliage 经修整的树形
hewn natural stone 经斩琢的石块;经斩琢的方石
hewn square 劈枋
hewn stone 方石;毛石;粗凿石;粗屑石
hewn stone bonder 方石横向砌块
hewn stone bonding header 方石横向砌块
hewn stone masonry(work) 方石(墙)工
hewn stone slab 方石板
hewn stone vault 方石拱顶
hewn stone window 方石窗
hewn stonework 方石(墙)工
hewn timber 砍平原木;粗制木材;被砍劈的木材
hexaboron silicide 硅化六硼
hexabromide value 六溴化物值
hexacalcium aluminate 六钙铝酸盐
hexacalcium dialuminoferrite 铁二铝酸六钙
hexacene 并六苯
hexachalcocite 六方辉铜矿
hexachlorobenzene 六氯(代)苯;六氯丙烯
hexachlorobutadiene 六氯丁二烯
hexachloro-cndomethylene tetrahydrophthalk anhydride 氯桥酸酐
hexachlorocyclohexane 六氯环己烷;六氯化苯;六六六
hexachlorocyclohexane poisoning 六氯化苯中毒
hexachloro endomethylene tetrahydrophthalic acid 氯桥酸;氯茵酸
hexachloroendomethylene tetrahydrophthalic anhydride acid 六氯代内亚甲基四氢邻苯二甲酸酐
hexachloroethane 六氯乙烷
hexachlorophene 六氯酚
hexachlorophenol 六氯(苯)酚
hexacode 六位代码
hexacontane 六十(碳)烷
hexacosandiacid 二十六烷二酸
hexacosandienoic acid 二十六碳二烯酸
hexacosane 二十六烷
hexacosanol 二十六醇
hexacosenoic acid 二十六碳烯酸
hexacosoic acid 二十六烷酸
hexacosyl alcohol 二十六醇
hexactin 六放体
hexacyclic compound 六环化合物
hexad 六价元素
hexad axis 六次对称轴
hexadecadienedioic acid 十六碳二烯二酸
hexadecadienoic acid 十六碳二烯酸
hexadecafluorooxy-cyclooctane 十六氟氧基环辛烷
hexadecandioic acid (hexadecane diacid) 十六烷二酸
hexadecane 十六烷
hexadecane dicarboxylic acid 十六烷二甲酸
hexadecanoic acid 十六酸
hexadecanol 十六醇
hexadecanolide 十六内酯
hexadecapole moment 十六极矩
hexadecatrienoic acid 十六碳三烯酸
hexadecendioic acid 十六碳烯二酸
hexadecimal 十六进制的
hexadecimal byte 十六进制字节
hexadecimal code 十六进制代码
hexadecimal constant 十六进制常数
hexadecimal counter 十六进位计数器
hexadecimal digit 十六进制数字;十六进制数位

hexadecimal display format 十六进制显示格式
hexadecimal format 十六进制格式
hexadecimal loader 十六进制装入程序
hexadecimal multiplication 十六进制乘法
hexadecimal notation 十六进制记数法;十六进制表示法;十六进位法
hexadecimal number 十六进制数位
hexadecimal numbering system 十六进制记数系统
hexadecimal number system 十六进制数系统
hexadecimal numeral 十六进制数字
hexadecimal pad 十六进制输入键盘
hexadecimal point 十六进制小数点
hexadecimal program(me) 十六进制程序
hexadecimal system 十六进制
hexadecimal value 十六进制值
hexadecine 十六碳炔
hexadecoic acid 十六酸
hexadecyl 十六基
hexadecyl alcohol 十六醇
hexadecylamine 十六烷基胺
hexadecyl amine-hydrochloride 十六胺盐酸盐
hexadecylene diacid 十六碳炔二酸
hexadecylene dicarboxylic acid 十六碳烯二甲酸;十六碳炔二羧酸
hexadecylenic acid 十六碳烯酸
hexadecyl hydrosulfate 十六烷基硫酸
hexadecylic acid 十六酸
hexadecyl mercaptan 十六基硫醇
hexadecylol amine 十六醇胺
hexadecyl palmitate 十六酸十六酯
hexadecyl phosphate 十六基磷酸盐
hexadecyl sulfenamide 十六基亚磺酰胺
hexadecyne 十六碳炔
hexadecynoic acid 十六碳炔酸
hexadeeene 十六碳烯
hexadeeene diacid 十六碳烯二酸
hexadeeene dicarboxylic acid 十六碳烯二羧酸
hexadiene 己二烯
hexadienoic acid 己二烯酸
hexadienol 己二烯醇
hexadiine 己二炔
Hexadiline 海沙地林
hexadiyne 己二炔
hexaethyl disiloxane 六乙基二(甲)硅氮烷
hexaethylditin 三乙基锡
hexafluoride 六氟化物
hexafluoroethane 全氟乙烷
hexafluoropropylene 六氟丙烯
hexagon 六角形;六角体;六方形;六边形
hexagonal 六角(的);六方(晶体);六边形的
hexagonal aperture 六角形孔径
hexagonal axis 六重轴
hexagonal bar(iron) 六角钢(条);六角形铁杆
hexagonal bastion 六角形堡垒
hexagonal bipyramid 六方双锥
hexagonal bit 六角钻头
hexagonal bolt 六角形螺栓
hexagonal bolts and nuts 六角闩
hexagonal close-packed structure 六角密集(堆积)结构;六方最紧密堆积结构
hexagonal closest packing 六方紧密堆积;六方密集堆
hexagonal column 六方柱状冰晶
hexagonal crystal 六角晶体;六边形晶体
hexagonal crystal system 六角晶系
hexagonal diamond 六方金刚石
hexagonal dipyramid 六方复双锥
hexagonal dipyramidal class 六方复双锥类
hexagonal dome 六角(形)穹顶
hexagonal double end tubular wrench 六角双头套筒扳手
hexagonal element 六角形部件
hexagonal ferrite 六角形铁氧体
hexagonal ground plan 六角形平面
hexagonal head 六角(形)头
hexagonal head bolt 六角头螺栓
hexagonal head screw 六角头螺钉;六角头螺栓
hexagonal head tap bolt 六角头螺栓
hexagonal head wrench 六角头扳手
hexagonal interior joint 六角内接头
hexagonal iron bar 六角钢条
hexagonal kelly 六方钻杆
hexagonal key 六角形扳手

hexagonal lattice 六角点阵;六方晶格
hexagonal lenticulation 六角(形)透镜光栅
hexagonal liner 六角衬板
hexagonal mesh 六角形网格;六边形网格
hexagonal mesh wire net 六角形网格
hexagonal mesoporous silica 六方间隙多孔硅石
hexagonal mirror 六角反射镜
hexagonal mosaic 六角形马赛克
hexagonal mosaic tile 六角形马赛克砖
hexagonal net 六角形网络
hexagonal nomogram 六角线图
hexagonal nut 六角(形)螺母;六角(螺)帽
hexagonal packing 六方堆积
hexagonal pattern 六角形纹样
hexagonal pavilion 六角亭
hexagonal pile 六角桩
hexagonal platelet 六方形片晶
hexagonal pressing die 六方压接模
hexagonal prism 六角棱镜;六角导轮;六方柱
hexagonal profile 六角形断面
hexagonal pyramid 六角锥体;六角棱锥;六角单锥
hexagonal rod 六角钢条
hexagonal scalenohedron 六方偏三角面体
hexagonal screw 六角螺钉
hexagonal shaft 六角钻杆
hexagonal shape 六角形
hexagonal socket 六角套筒
hexagonal socket screw 六角凹头螺钉
hexagonal socket screw key 六角凹头螺钉键
hexagonal socket wrench 六角套筒扳手
hexagonal space grid 六角空间网架
hexagonal spanner 六方扳手
hexagon(al) steel 六角钻杆;六角钢
hexagonal steel bar 六角形钢筋
hexagonal structure 六角形结构;六角结构
hexagonal system 六角系;六角晶系;六方晶系
hexagonal threading nut 六角形扣纹螺母
hexagonal tile 六角形面砖
hexagonal trapezohedral class 六方偏方面体类
hexagonal trapezohedron 六方偏方面体
hexagonal trommel 六面滚筒筛
hexagonal turret 六角形塔楼
hexagonal washer 六角垫圈
hexagonal weight 六方坠陀
hexagonal wire(d)glass 六角形铁丝网玻璃
hexagonal wire netting 六角形铁丝网
hexagonal wood screw 六角形头木螺钉
hexagon bar 六角钢(材);六角形钢筋
hexagon bar iron 六角钢
hexagon bar key 六角杆键
hexagon bar steel 六角条钢
hexagon bed turret 六角转塔刀架
hexagon bolt 六角头螺栓
hexagon cap bolt 六角头螺栓
hexagon cap screw 六角帽螺钉
hexagon check nut 六角防松螺母
hexagon die nut 六角扳牙
hexagon drive socket set 成套六角螺钉套头
hexagon ga(u)ge block 六角轨距块
hexagon-headed bolt 六角头螺栓
hexagon iron 六角铁
hexagon nipples 六角外螺钉;六角螺纹管子扳头
hexagon nut 六角螺帽
hexagon pin punch 六角形弹簧冲头
hexagons 六角钢
hexagon shank 六角钎尾
hexagon socket spinner wrench 有柄插入式六角套筒扳手
hexagon spanner 六角扳手
hexagon steel 六角钢
hexagon thin nut 六角扁螺母
hexagon voltage 六角形电压
hexagon wrench key 六角扳手键
hexagram 等(边)六角形;六线形;六芒星形;等六边形
hexahedral bar 六角钢
hexahedrite 六面体陨铁;方陨铁
hexahedron 正六面体;六面体;六方体;立方体
hexahydrite 六水泻盐
hexahydrobenzene 六氢化苯;环己烷
hexahydroborite 六水硼钙石
hexahydrophenyl-naphthalene 六氢化苯基萘
hexahydrophthalate acid 六氢化邻苯甲酸

hexahydrophthalic acid 环己二酸
hexahydrophthalic anhydride 六氢化邻苯二甲酸酐
hexakismethoxymethyl melamine 六甲氧基三聚氰胺
hexakisoocta-hedron 六八面体
hexalaminar membrane 六板层膜
hexaldehyde 己醛
hexaleg block 六角块(体)
hexamer 六聚物
hexamethylene 己烯
hexamethylene diamine 六亚甲基二胺;己二胺
hexamethylene diisocyanate 六亚甲基二异氰酸酯
hexamethylene tetramine 六亚甲基四胺,乌洛托品
hexamethylol melamine 六羟甲基三聚氰胺
hexamine 乌洛托品
hexanal 己醛
hexandial 己二醛
hexane 己烷
hexane-dinitrile 己二腈
hexanedioic acid 己二酸
hexane-extractable material 可萃取的乙烷物质
hexangle 六角形
hexangular 六角(的)
hexanoic acid 己酸
hexanol 己醇
hexaoctahedral class 六面体类
hexaoctahedron 六八面体
hexapartite vault 六肋穹顶
hexapetalous flower 六瓣花
hexaphase 六相
hexaphenylethane 六苯乙烷
hexaplanar 六角平面
hexaplanar system 六角(晶)系
hexaploid 六瓣体
hexapod 六角锥体;六角块(体)
hexapole 六极
hexastar 六放星
hexastyle 六柱式(柱廊);六柱式(门廊)
hexastyle building 六柱式建筑
hexastyle portico 六柱式门廊
hexastyle temple 六柱式神庙
hexastylos 有六根柱子的建筑物(古代寺庙);六柱式建筑
hexasymmetric(al) faces 六对称面
hexatetracontanoic acid 己四十烷酸
hexatetrahedral class 六四面体类
hex belt 六角形皮带
hex bolt 六角(头)螺栓
hex crystal 六角晶体
hexene 己烯
hexenoic acid 己烯酸
hexenoic aldehyde 己烯醛
hexenone 己烯酮
hexenyl 己烯
hexetetrahedron 六四面体
hex head screw 六角头螺钉
Hexi-Corridor depression in front mountains 河西走廊山前拗陷
Hexi-Corridor seismotectonic zone 河西走廊地震构造带
hexin 己炔
hex inverter 六位反演器
Hexi tectonics system 河西构造体系
hexitol 己六醇
hex key 六角键
hex nut 六角螺母;六角螺帽
hexoctahedral class 六八面体类
hexocyclium 己环铵
hexode 六极管
hexogen 黑索金炸药
hexone 甲基异丁基(甲)酮
hexpsan 半纤维素
hex roofing 六角屋顶
hex socket 六角套筒
hex stop nut 六角止动螺母
hex wrench 六角扳手
hexyl 己(烷)基
hexyl acetate 乙酸己酯
hexylamine 己胺
hexylene 己烯
hexylene glycol 己二醇
hexyl mercaptan 己硫醇
hexynic acid 己炔酸
hexynoic acid 己炔酸

Heyd separator 海德选粉机
heyite 钒铁铅矿
Heyn stress 晶粒间应力;海恩应力
heyrovskyite 富硫铋铅矿
Heywood 海意伍德法(一种玻璃装配法)
H flume H形(测流)槽
HF process 氟氢酸法
H frame H形框架;H形电杆;H形车架
Hg conversion 水银法转换
Hg delay line 汞延迟线
H-girder 宽翼缘工字梁;H形梁;工字大梁
Hg placer ore 砂汞矿石
H-hinge 工字铰链;工字合页;H形铰链
H-horizon H层;腐殖质层
hial shower 雹阵
hiatal 非等粒的
hiatus 裂孔;缺失;沉积间断
hiatus break 间断
hiba oil 桧叶油;日本罗汉柏叶油
hibar prepacked column 高压预填柱
Hibbert cell 希伯特电池
Hibbert standard 希伯特标准
hibernaculum 越冬场所;冬芽;冬眠场所(越冬巢)
hibernal annual 冬性一年生植物
hibernal annual plant 一年生越冬植物
hibernant amphibious animal 冬眠两栖动物
hibernating state 越冬状态
hibernation 越冬;深度冬眠;冬眠
Hibernian orogeny 伊里亚造山运动
Hiberno-Romanesque style 爱尔兰—罗马风格
Hiberno-Salairian Articulate brachiopod region 希伯尼—沙莱里有铰腕足动物地理大区
hibiscus root powder 木槿根粉
hibonite 黑铝钙石
hi-bottom 高速犁体
hibschite 水榴石;八面硅钙铝石
Hi-cap fuse 大容量熔断丝;大切断功率熔断丝
hiccough 电子放大镜
hickey 弯管器;螺纹接合器
hickey bar 弯钢筋;板筋成形工具
hick joint 平缝砌;平(灰)缝
hick joint pointing 平勾缝
hickory 山核桃木
hickory bow 胡桃木弓
Hickson visco(si)meter 希克森黏度计
hicktown 远离大都市的乡镇;小镇
hicky 弯管器;螺纹接合器
hicore 不锈钼铬钢;不锈铬钼钢
hicrogranitic texture 微花岗结构
hidage 每一百英亩的土地税
Hidaka orogeny 日高造山运动
Hidaka tectonic zone 日高构造带
hidalgoite 砷铝钒
Hidamet 高阻尼合金
hidden abutment 埋式桥台
hidden anomaly 隐伏异常
hidden arc furnace 埋弧电炉
hidden arc welding 埋弧焊;潜弧焊
hidden assets 账外资产;账外物资
hidden axis 隐轴;暗轴
hidden bit 隐匿位
hidden camera 隐蔽照相机
hidden conductor 隐蔽导线;暗线
hidden corner 隐蔽角
hidden cost 隐藏成本;隐蔽费用;隐蔽成本;不可预见费(用)
hidden damage 暗伤
hidden danger 隐患
hidden defect 隐患(提防);内在瑕疵;潜在缺陷
hidden determinant 隐藏决定簇
hidden discount 暗扣
hidden ditch 暗渠
hidden domain 隐式域
hidden dumping 隐蔽倾销
hidden economy 隐性经济
hidden fault 隐伏断层
hidden fire 隐火
hidden flaws 暗伤
hidden framing glass curtain wall 隐框玻璃幕墙
hidden gutter 暗沟
hidden harbo(u)r 陆封港
hidden heat 潜热
hidden installation 隐蔽装置
hiddenite 翠绿锂辉石

hidden layer 隐蔽层;掩盖层;缺失地层
hidden line 虚线;隐(藏)线;隐蔽线
hidden line algorithm 隐藏线算法
hidden line elimination 隐(藏)线消除
hidden line plot 隐线绘图
hidden line removal 隐(藏)线消除
hidden line rendering 隐线浓淡处理
hidden ore body anomaly 隐伏矿体异常
hidden outcrop 隐蔽露头;掩蔽露头
hidden peril 隐患
hidden periodicity 潜周期性;暗隐周期性
hidden-periodicity model 暗隐周期性模型
hidden pipe 暗管
hidden plastic 潜塑
hidden price increase 变相涨价
hidden reefs and dangerous shallows 暗礁与险滩
hidden reserve 隐蔽准备;秘密盈余
hidden resources 地下资源
hidden rock 水下岩石;暗礁
hidden shoal 暗滩
hidden stair(case) 隐蔽楼梯
hidden surface 隐(藏)面;非显见面
hidden surface algorithm 隐藏面算法
hidden tack 暗泡灯
hidden tax 间接税
hidden terrace 暗阶地
hidden trouble 隐患
hidden tube 暗管
hidden valley 暗天沟;埋藏谷
hidden valve 隐蔽阀
hidden variable 隐变数;隐变量
hidden wealth 隐蔽财产
hidden wiring 暗线
hide 隐棚;遮盖;生皮
hide and seek 隐藏与寻找
hide-away 隐蔽处;偏僻小吃店;僻静娱乐处
hide container 厚皮集装箱;通风集装箱;兽皮集装箱;生皮专用集装箱
hide finish 皮革涂饰剂;皮革面漆
hide glue 皮胶
hide house 生皮仓库
hide inspection bureau 生皮检验局
hideland 土地分配
hide-marking hammer 生皮打号锤
hide option 隐藏选择操作
hide-out 掩蔽所
hide pulling machine 剥皮机
hide rope 皮带;皮绳
hide skinning knife 剥皮刀
hide stamping machine 生皮打印机
hiding economy 遮盖经济性
hiding effect 隐蔽作用
hiding-in earthquake 潜伏地震
hiding pigment 盖底颜料
hiding place 躲藏处
hiding power (油漆的)不透明性;掩盖力;(油漆等的)遮盖力;遮盖本领;盖底力;被覆力
hiding power black and white checker board 遮盖力黑白格板
hiding power chart 遮盖力试验纸
hiding power value 遮盖力值
hiding strategy 隐蔽策略
hidumimium 铝铜镍合金
hielmite 钙铌钽矿
hiemal aspect 冬季相
hiemal plant 雨绿植物
hiemeliogosa 雨绿木本群落
hiemifruticeta 雨绿灌木群落
hiemiherbosa 雨绿草本群落
hiemisilvae 雨绿乔木群落
hieracosphinx 鹰狮头雕像
hierarchic(al) 层次的;分级的;分层的
hierarchic(al) approach 等级剖析法;等级逼近法
hierarchic(al) classification 分层;系统分组
hierarchic(al) common bus 分层公共母线
hierarchic(al) computer network 分级计算机网络
hierarchic(al) control 多级控制;递阶控制;层次控制;分级控制;分层控制
hierarchic(al) control system 分级控制系统;分层控制系统
hierarchic(al) cosmology 等级式宇宙论
hierarchic(al) data base 层次数据库;分级数据库;分层数据库
hierarchic(al) database 层次型数据库

hierarchic(al) data model 分级数据模型
hierarchic(al) data structure 分级数据结构
hierarchic(al) decision making 分层决策
hierarchic(al) decomposition 层次分解
hierarchic(al) dendrogram 谱系树枝图
hierarchic(al) design 分级设计
hierarchic(al) design method 层次设计法
hierarchic(al) diagram 谱系图
hierarchic(al) diffusion of information 信息按等级层次扩散
hierarchic(al) direct 层次直接;分级直接存取法
hierarchic(al) direct access method 层次直接存取方法;分层直接存取法;分级直接存取法
hierarchic(al) direct organization 分级直接存取组织结构;分层直接结构
hierarchic(al) distributed subloop control 分层分布回路控制
hierarchic(al) distributed system 分层分布系统
hierarchic(al) division 层次划分
hierarchic(al) file 分级文件
hierarchic(al) file structure 分级资料结构;分级文件结构
hierarchic(al) file system 分级文件系统
hierarchic(al) geometry design 分级几何设计
hierarchic(al) group divisible design 分层分组设计
hierarchic(al) indexed direct access method 层次索引直接存取方法;分级索引直接存取法;分层索引直接存取法
hierarchic(al) indexed sequential access method 分层索引顺序存取法
hierarchic(al) interactive graphics system 层次型交互图形系统
hierarchic(al) Kalman filter 多级卡尔曼滤波器
hierarchic(al) level 分级级数
hierarchic(al) level approach 分级结构法
hierarchically sequential 分极顺序的
hierarchically structure distributed 分级结构分散控制系统
hierarchic(al) memory storage 按级分类项存储[贮]器
hierarchic(al) method 分段法
hierarchic(al) model 等级式模型;递阶模型;层次模型;分层模型
hierarchic(al) model following controller 多级模型跟踪控制器
hierarchic(al) modeling of water quality control 分层水质控制模拟
hierarchic(al) monitor 分级监督程序
hierarchic(al) multiprogramming 分级多道程序设计
hierarchic(al) network 分级网络;分层网络
hierarchic(al) number 层次数
hierarchic(al) operating system 分级(结构)操作系统
hierarchic(al) organization 等级结构
hierarchic(al) path 层次路径;分层路径
hierarchic(al) pointer 分层指示字
hierarchic(al) rank 等级
hierarchic(al) responsibility 分层负责
hierarchic(al) revenue-sharing methods 分类分成(办)法
hierarchic(al) routing 分级路由选择
hierarchic(al) sampling 谱系采样
hierarchic(al) segment 层次片段
hierarchic(al) segment theorem 层次片段定理
hierarchic(al) sequence 层次顺序;分级顺序
hierarchic(al) sequence key 分层序列关键码
hierarchic(al) sequential 层次顺序;分级顺序的
hierarchic(al) sequential access method 层次顺序存取方法;分级顺序存取法
hierarchic(al) sequential organization 分级顺序组织;分层顺序结构
hierarchic(al) set 层次集合;分级集合
hierarchic(al) structure 等级式结构;递阶结构;层次结构;分级结构;分层构造
hierarchic(al) structure of directory 目录的层次结构
hierarchic(al) switching 分级交换
hierarchic(al) system 分层结构系统
hierarch of function 机能层次
hierarchy 级别;次序;层级;层次;分级制度(土地、时间);分级结构;分层结构;分层接合;谱系
hierarchy access time 分级存取时间

hierarchy configuration 分级配置;分级结构
hierarchy data structure 分级数据结构
hierarchy-distributed system 分级式分布系统
hierarchy manager 分级管理程序
hierarchy model 分级结构模型
hierarchy of authority 权责层次
hierarchy of block 分程序层次(结构)
hierarchy of consumption 消费层次
hierarchy of control 控制体系;控制层次
hierarchy of human need 人类需要层次
hierarchy of layer 分层层次
hierarchy of memory 分级存储器系统;分层存储器系统
hierarchy of needs theory 需求层次理论
hierarchy of operation 操作层次
hierarchy of risk 技术层次
hierarchy of roads 道路网主干线;道路干线系统
hierarchy submodel 次级模型
hierarchy system 分级系统;分级多机控制系统;分级的多计算机控制系统;分层控制系统;谱系
hieratite 方氟硅钾石
hierogloph 像形印痕
hierograph 象形文字
hieron 神位圣舍;庙舍
Hi-Fix 高精度定位系统
HiFix navigation 哈菲克斯导航
Higbee cut 斜削接口螺纹
Higbee indicator 接口螺纹开始标记
Higbie model 希格比模型
higganite-(Yb) 镱兴安石
Higgins furnace 希金斯炉(一种熔制耐火材料的电炉)
higgle 讨价还价
Higgs bosons 希格斯玻色子
Higgs mechanism 希格斯机制
high abnormal pressure 高异常压力
high abrasion furnace black 高耐磨炉黑;高耐磨炉法炭黑
high abrasion-resistance polyurethane varnish 高耐磨聚氨酯清漆
high-abrasive material 耐磨材料
high absolute photocathode 高纯度光电阴极
high absorber 高吸收体
high absorption spectrophotometry 高吸光度法
high-accident location 交通事故多发地段;事故多发地段(交通)
high accuracy 高准确度;高精(确)度
high accuracy air flow measuring system 高精度气流测试系统
high accuracy and fast edit plotter 高精度快速绘图机
high accuracy arithmetic 高精度运算
high accuracy centerless grinding machine 高精度无心磨床
high accuracy control point 高精度控制点
high accuracy formula 高精(确)度公式
high accuracy gravimeter 高精度重力仪
high accuracy hob sharpening machine 高精度滚刀刃磨床
high accuracy position system 高精度定位系统
high accuracy radar data transmission 高精度雷达数据传输
high accuracy radio-fix 高精度无线电定位
high accuracy survey 高精度测量
high accuracy tuning 高精度调谐
high acid leaching 浓酸浸出
high activity area 高活性区
high activity data processing 高效数据处理
high activity liquor 强放射性废液
high activity processing 高效处理
high activity waste container 高放射性废物容器
high acuity 高镜度
high acutance developer 高锐度显影液
high address 高位地址
high aggregate/cement ratio 高集料水泥比;高集灰比
high air 高层空气;高压空气
high air entry piezometer tip 高进气孔隙压力测头;高进气测压管测头
high air pollution potential 高层空气污染潜力
high alarm 高位警报;上限信号器
high albedo 高反照率
high albite 高钠长石
high alcohol 高碳醇;高级醇

high algae-laden 高藻
high algae-laden raw water 高藻原水
high alicyclic solvent 高脂环溶剂
high alkali cement 高碱(度)水泥
high alkali cement clinker 高碱水泥熟料
high alkali clinker 高碱熟料
high alkali glass 高碱玻璃
high alkali glass fiber 高碱玻璃纤维
high alkalinity 高碱性;高碱度
high alkalinity mud 高碱泥浆
high alloy 高合金
high alloy martensite 高合金马氏体
high alloy steel 高(质)合金钢
high alloy steel vessel 高合金钢容器
high aloft 高空反气旋
high altar 主坛
high altitude 高空;大高度
high altitude accident 高空失事
high altitude aerodrome 高原机场
high altitude air-conditioning equipment 高空空气调节设备
high altitude aircraft photograph 高空航空相片
high altitude airport 高原机场
high altitude anoxia 高原缺氧
high altitude balloon 高空气球
high altitude colo(u)r photograph 高空彩色像片
high altitude diving 高海拔潜水
high altitude environment 高空环境
high altitude infrared sensor system 高空红外传感器系统
high altitude laboratory 高空试验室;高空实验室
high altitude localities 高海拔地区
high altitude machine 高原电机
high altitude magnetic field 高空磁场
high altitude method 高空高度法
high altitude nuclear burst 高空核爆炸
high altitude observatory 高山观测台;高空观测站
High Altitude Observatory 高山天文台
high altitude orbit 高轨道
high altitude passage 高仰角通过
high altitude photograph 高空像片
high altitude photography 高空摄影(术)
high altitude reconnaisance 高空搜摄
high altitude reconnaissance 高空侦察
high altitude rocket (气球带到高空发射的)高空探测火箭
high altitude sampling program(me) 高空取样计划
high altitude satellite 高空卫星
high altitude sickness 高原病
high altitude station 高山
high alicutude strip chart 高空航线图
high altitude suspension 高空悬浮物
high altitude syncope 高原晕厥
high altitude test 高空试验
high altitude test vehicle 高空试验飞行器
high altitude transformer 高原变压器
high altitude very high frequency omniranger 高空甚高频全向信标
高铝氧的
high alumina basalt 高铝(氧)玄武岩
high alumina brick 高铝(氧)砖
high alumina castable 高铝耐火浇注料
high alumina cement 高铝水泥;高矾水泥;矾土水泥;矾土泥浆
high alumina cement clinker 高铝水泥熟料
high alumina cement concrete 高铝水泥混凝土
high alumina ceramic 高氧化铝陶瓷
high alumina clay 高铝黏土
high alumina clay chamot(te) 高铝矾土熟料
high alumina electric(al) porcelain 高铝电瓷
high alumina fiber 高铝质耐火纤维
high alumina fireclay brick 高铝耐火砖
high alumina insulation firebrick 高铝绝热砖
high alumina products 高铝制品
high alumina refractory 高铝(质)耐火材料
high alumina refractory fiber and its products 高铝耐火纤维及其制品
high alumina refractory product 高铝耐火材料制品
high alumina sand 高铝砂
high alumina slag 高铝渣
high alumin(i)um brick 高铝砖
high alumin(i)um grinding ball 高铝球磨球

high ammonia nitrogen chemical industry wastewater 高氨氮化工废水
high ammonia nitrogen wastewater 高氨氮废水
high amperage 高安培数;大电流量
high amplitude 高振幅
high amplitude continuous reflection mode 强振幅连续反射波模式
high amplitude socket 高振幅锚座
high-amplitude wave 强波
high analysis fertilizer 浓缩肥料
high and dry 露底搁浅
high and low amplitude reflection mode 强弱振幅反射模式
high and low energy wet scrubber 高能及低能湿式洗涤器
high and low pass filter 高低通滤波器
high and low points method 高低方法;高低点法
high and low pressure area 高压区及低压区
high and low pressure turbine 高低压汽轮机
high and low railway drawbridge 高低道式吊桥
high and low temperature Brinell hardness tester 高低温布氏硬度计
high and low temperature shift 高低温变换
high and low tension porcelain insulator 高低压瓷绝缘子
high and low voltage power capacitor 高低压电容器
high and low water alarm 水位报警(器)
high and new technology 高新技术
high and stable yield 高产稳产
high angle 高角(度)
high angle boundary 高角度边界
high angle fault 高角断层;陡角断层;陡倾断层
high angle fracture 陡角裂隙;陡角断裂
high angle grain boundary 大角度晶粒间界
high angle reverse fault 陡角逆断层
high angle shot 高角度拍摄
high angle smooth bottom steel bench plane 高角平底钢刨
high angle strafing 大角度扫射
high angle thrust 高角度冲断层;陡冲断层
high angular momentum state 高角动量态
high annealing 高温退火
high anti-knock fuel 高抗爆燃料
high antiquity 远古
high apartment 高层公寓
high aperture beam 大孔径光束
high aperture lens 强光透镜
high arch tractor 高地隙拖拉机
high area 反气旋
high area limit of gamma 伽马高区限
high area storm 高压区风暴
high aromatic solvent 高芳烃溶剂
high aromatic white spirit 高芳烃矿石油溶剂
high art 纯艺术
high ash 灰分高的;高灰分的
high ash coal 高灰分煤
high ash content 高含灰量
high ash fuel 高灰分燃料
high assay 高指标样品
high atmospheric pressure 高气压
high attack angle 大冲角
high attenuation 高衰减量
high attenuation cable 大衰减电缆
high attenuation zone 高衰减带
high audit risk 高审计风险
high azimuthal observation 高方位观测
high back 高靠背
high background contamination 高本底污染
high background level 高本底水平
high background noise 高本底噪声
high background pollution 高本底污染
high back jumbo 高工作面钻车
high back pressure 高回压
high back pressure foam maker 高背压泡沫发生器
high back pressure turbine 高背压涡轮机
high bake 高温烘烤
highball 高速火车;(火车的)全速前进信号
high band 高频段;高(频)带
high band standard 高带标准
high band system 高频带方式
high bandwidth 高带宽
high bank 高边坡;高堤;高岸
high bank flow 高岸流量

high bar 高单杠
high barometric(al) maximum 高气压(的)最高值
high barometric(al) pressure 高气压
high Baroque effect 高巴洛克式影响
high Baroque Rome 罗马高巴洛克式
high basic clinker 高碱熟料
high basin 高库
high bay lighting 高架灯具
high bay warehouse 高货架仓库
high beam 远光灯;高光束;车前灯(的)远距离光束
high beam of head lamp 前灯远光
high beam plow 高架犁
high bed 浅滩;沙洲
high benefit field 高效益田
high beta stellarator 高比压仿星器
high bidders 标价高的投标人
high biomass concentration 高生物质密度
high birefringence optic(al) fiber 高双折射光纤
high bit rate 高比特率
high bit rate digital subscriber line 高比特率数字用户线
high blasting explosive 高爆炸药
high block 高层大楼
high block of flats 高层公寓
high blower 高增压器
high bog 高地沼泽
high boiler 高沸溶剂;高沸点(化合)物
high boiling 高温沸腾;高沸点的
high boiling component 高沸点组分
high boiling compound 高沸化合物
high boiling hydrocarbons 高沸点烃类
high boiling petroleum products 高沸点石油产品
high boiling phenol 高沸点苯酚
high boiling point 高沸点
high boiling point impurity 高沸点杂质
high boiling point naphtha 高沸点石脑油
high boiling point solvent 高沸点溶剂
high-bond bar 高握裹力钢筋;竹节钢筋;变形钢筋
high-bond concrete reinforcing bar 高粘着力钢筋
high-bond reinforcing bar 高握裹力钢筋
high boost 高频提升;高频补偿
high boron steel 高硼钢
high bottom phase 高滩地
high-bow rescue boat 绞接船头救援艇
high boy 高五斗橱;高脚七斗橱
high brass 优质黄铜;高锌黄铜;高级黄铜
high breaking capacity fuse 高断力熔断器
high bright contrast 高反差
high brightness 高明度;高亮度
high brightness beam 高亮度光束
high brightness coating 高光泽涂层
high brightness concentration factor 高亮度聚光因子
high brightness indicator 高亮度显示器
high brightness slide projector 高亮度幻灯机
high brightness stationary cine projector 高亮度固定式放映机
high-brown 鲜明棕色的
highbrow publication 高学术水平刊物
high brush 高灌木丛
high Btu gasification process 高发热量气化过程
high Btu oil-gas process 高发热值油气法
high Btu pipeline gas 高热值管道气
high build 高固体性;高成膜性;成厚涂层性
high build asbestos bituminous anti-corrosive paste 厚浆型沥青石棉防腐浆
high build coating 厚涂层涂料;厚膜涂料;厚膜涂层
high build coating 多余涂层
high build epoxy bituminous anti-corrosive paint 环氧沥青厚浆型防腐漆
high building 高层大楼
high-build paint 厚(涂)层涂料;厚涂层漆;稠涂料
high build system 厚涂层系统
high-bulk book paper 高松厚度书籍纸
high bulk fiber 高膨体纤维
high bulking 高膨化
high burned 高温焙烧过的
high burnup fuel 高燃耗燃料
high by-pass ratio 高涵道流量比
high-calcium concentration 高钙浓度
high-calcium-containing wastewater 含高钙废水
high-calcium flyash 高钙粉煤灰
high-calcium flyash cement 高钙粉煤灰水泥
high-calcium lime 高钙石灰;富石灰

high-calcium lime paste 高钙石灰浆
high-calcium lime putty 高钙石灰浆
high-calcium limestone 富钙灰岩;高钙石灰岩
high-calcium lump lime 高钙石灰块
high-calcium magnesite clinker 高钙镁砂
high-calcium quicklime 高钙生石灰
high calorific power 高热值
high calorific value 高热值;高发热量
high calorific value coal 高发热量煤
high cam 高镶条
high camber 高路拱
high capacitance cable 高容量电缆
high capacitance transmitter 高容量发射机
high capacity 高功率(的);高负载;大容量
high-capacity and high-speed electronic computer 大容量高速电子计算机
high-capacity blower 大能力鼓风机
high-capacity bucket elevator 大功率斗式提升机
high-capacity coach 大容量客车
high-capacity communication system 大容量通信系统;大容量交通系统
high-capacity condenser 大容量电容器
high-capacity container 大容量容器
high-capacity deep foundation 承载力大的深基础
high-capacity filter 高容量滤池
high-capacity fuse 大容量熔断丝;大切断功率熔断丝
high-capacity heating equipment 高容量采暖设备
high-capacity hydroelectric(al) plant 大容量水电站
high-capacity machine units 大容量机组
high-capacity marshalling yard 大能力编组场【铁】
high-capacity meter 大容量煤气表;大流量煤气表
high-capacity motor 高功率发动机;高功率电动机;大型电动机
high-capacity power plant 大容量电厂
high-capacity power station 大容量电站
high-capacity reservoir 高产油层
high-capacity silo 大容量储仓
high-capacity squeezer 高效轧车
high-capacity turbine 大容量水轮机
high-capacity tire[tyre] 载重轮胎
high-capacity unit 高容量送风系统;高功率装置;高功力装置;大容量送风系统
high-capacity wagon 长大货物车;大型货车
high-capacity water power station 大容量水电站
high capped pile foundation 高承台桩基;高桩承台
high-carbon acid 高碳酸
high-carbon alkyne 高碳炔
high-carbon alloy steel 高碳合金钢
high-carbon chromium 高碳铬
high-carbon chromium steel 高碳铬钢
high-carbon coke 高碳焦炭;高炭焦
high-carbon content 高含碳量
high-carbon ferro-chrome 高碳铬铁
high-carbon high-tensile steel 高碳高张力钢
high-carbon high-tensile steel wire 高碳高张力钢线
high-carbon iron 高碳生铁
high-carbon iron alloy 高碳铁合金
high-carbon iron powder 高碳铁粉
high-carbon martensite 高碳马氏体
high-carbon steel 高碳钢(板)
high-carbon steel heat treated 热处理过的高碳钢
high-carbon tar 高碳焦油
high-carbon wire 高碳钢丝
high ceiling 高顶棚
high-ceilinged 高顶棚的
high celerity wash of rapid filter 快滤池高速冲洗
high-cement concrete 富灰混凝土
high-center bit 中部凸出的钻头;中部凸出的钎头
high center of gravity 高重心
high centralized managing pattern 高度集团管理模式
high chair 钢筋支座;钢筋支架;高脚椅
high channel 高通道
high channel antenna 高频信道天线
high chaotic situation 高混沌态
high chloride level 高氯化物水平
high chlorine coal 高氯煤
high chopper 高削器
high chrome 高铬钢
high chrome alloy steel 高铬合金钢

high chrome ball 高铬球
high chrome steel 高铬钢
high chromium alloy 高铬合金
high chromium cast-iron ball 高铬铸铁球
high chromium content cermet 高铬金属陶瓷
high chromium iron 高铬生铁
high chromium stainless steel 高铬不锈钢
high chromium steel 高铬钢
high class 高质量的;高精度级;高精度等级的;优质的
high class cargo 高级货物
high class carpenters' work 高级大木作
high class fit 高级精度配合
high class furniture brushing lacquer 高级家具手刷漆
high class gear 高级齿轮
high class joiners' work 高级细木作
high class product 优等品
high class residential zone 高级住宅区;高尚住宅区;高等阶层住宅区
high clearance duster 高地隙喷粉机
high clearance sprayer 高地隙喷雾机
high clearance tractor 高地隙拖拉机
high cliff coast 高崖海岸
high climber 高空作业工
high closure reservoir 高闭合度储集层
high coast 悬崖陡岸;陡峭的海岸;陡(崖)岸
high coastline 陡岸线
high coercive 高矫顽磁性的
high coercivity 高矫顽磁性;高矫顽磁力
high coincident circuit 高符合电路
high cold desert 高寒荒漠
high cold meadow 高寒草甸
high cold scrub 高寒灌丛
high cold steppe 高寒草原
high colo(u)r 深色
high colo(u)r rendering fluorescent lamp 高显色荧光灯
high combustion rate burner 高速喷嘴
high compressibility 高压缩性
high compressible soft soil 高压缩性软土
high compressible soft soil foundation 高压缩性软土地基
high compressible soil 高压缩性土
high compression combustion chamber 高压燃烧室
high compression cylinder head 高压缩比汽缸盖
high compression engine 高压缩比发动机
high compression lean burn engine 高压缩比稀混合气体发动机
high compression motor 高压缩发动机
high compression ratio 高压缩比
high compression ring 高压环
high concentrated organic cyanide wastewater 高浓度有机氰化物废水
high concentrated organic pharmaceutical wastewater 高浓度有机制药废水
high concentrated organic wastewater 高浓度有机废水
high concentrated organic wastewater pesticide wastewater 高浓度有机农药废水
high concentrated organic wastewater pollutant 高浓度有机污染物
high concentrated organic wastewater treatment 高浓度有机废水处理
high concentrate dust 高浓度粉尘
high concentrate dust agent 高浓度粉剂
high concentrate formulation 高浓度制剂
high concentration coal oil mixture 高浓度煤油混合燃料
high concentration oxygen 高浓度氧
high concentration sewage 高浓度制污水
high conductivity 高电导性;高电导率;高导电性;高导电率
high conductivity compound 高导电化合物
high conductivity copper 高导电性铜
high conductivity for electricity 高导电率
high confining pressure test 高侧限压力试验
high connection supervision 高阶连接监控
high consistency pulp 高浓纸浆
high consistency rotational visco(si)meter 高黏度用(转子)黏度计
high consistency visco(si)meter 高稠度黏度计
high constructive delta 高建设性三角洲

high contact use 高度接触类用途
high content alloy 高合金
high content of corundum product 高刚玉制品
high content of phosphorus 高磷含量
high contrast 高对比度;高调;高对比;硬调
high contrast anomalies 高对比度异常
high contrast developer 高反差显影液
high contrast edge 高反差边缘
high contrast film 高反差胶片
high contrast image 高对比度图像;对比度强的图像;硬图像
high contrast imagery 高对比成像
high contrast pan film 高反差全色软片
high contrast plate 高衬度底片
high contrast target 高对比目标;高衬度靶
high control survey 高程测量
high copper 高铜
high corona 日冕高层
high correlation 高度相关
high cost 费用高
high cost energy resource 高成本能源
high coupon 高收益投资
high coverage 大有效区;大视野
high crack edge 高绽边
high credit 信贷限额
high-creep strength steel 高蠕变强度钢
high cristobalite 高温型方石
high critical pressure 高临界压强
high crown 高路拱;横坡大的路拱
high-crowned 高路拱的
high-crowned arch 高顶拱
high cube box car 大容量棚车
high cube cargo 轻(泡)货;松泡货
high cube container 高集装箱
high curie point alloy 高居里点合金
high current 高速流;强流;大电流
high current arc welding 大电流(电)弧焊
high current injection test set 大电流注入测试仪
high-current microtron 强流微波加速器
high current plasma arc welding 大电流等离子体弧焊
high current rectifier 大电流整流器
high current switch 大电流开关
high cut 向上切割;深挖;上挖
high cut filter 高阻滤波器
high cut frequency 高截频率
high cut frequency slope 高截频徒度
high cutting bar 高割型切割器
high cycle 高频的
high cycle fatigue 高周次疲劳;高频疲劳;高循环疲劳
high cylindrical valve 高圆筒阀
high dam 高坝
high damage factor 高损伤因素
high damping metal 哑金属;高阻尼合金
high dam with large reservoir 高坝大库
high data rate laser system 高数据速率激光系统
high data rate storage system 高速资料存储系统;高数据率存储系统
high day 节日;假日
high debt-service obligation 高额偿付债务款项
high deck of bridge 上层桥面
high definition 高清晰度
high definition detector 高分辨率探测器
high definition developer 高分辨力显影剂
high definition image(ry) 高清晰度图像
high definition laser radar 高分辨率激光雷达
high definition pick-up 高分辨力摄像
high definition picture 高分辨率图像
high definition range finder 高分辨率测距仪
high definition scanning system 高分辨力扫描系统;高分辨率扫描系统
high definition surveillance radar 高分频监视雷达;高分辨率监视雷达
high definition television 高清晰度电视
high deflection coupling 高偏转接头
high deformation 高度形变
high degree cable shielding 高级电缆屏蔽
high degree of accuracy 高精度
high degree of autonomy 高度自治权
high degree of scarcity 稀缺程度高
high demand for investment 投资过热
high denier-fiber 高支度纤维;粗纤度纤维

high-dense concrete 重骨料混凝土
high density 高密度
high-density aggregate 重集料;重骨料;高容重集料;高容重骨料
high-density alloy 高比合金
high-density ammonia dynamite 高密度铵爆炸药
high-density assembly 高密度组装;高密度装配
high-density bag blaster 高密度吹袋机
high-density bale 高密度紧包
high-density baling 高密度压捆
high-density bipolar code 高密度双极性码
high-density bleacher 高浓度漂白池;高密度漂白器
high-density cargo 比重大的货物;重量货(物);重货
high-density carrier 高密度载波机
high-density cement 高比重水泥
high-density charge 高密度炉料
high-density concrete 重骨料混凝土;高容重混凝土;高密度混凝土(通常用于防辐射)
high-density current 高密度值
high-density data system 高密度数据系统
high-density development 高密度发展;稠密建筑区
high-density digital recording 高密度数字记录
high-density digital storage 高密度数字存储器;高集成度数字存储器
high-density economy and society 高密度经济
high-density electron beam optics 高密度电子束光学;强流电子光学
high-density explosive 高密度炸药
high-density foam 高密度泡沫
high-density funnel 高压风洞
high-density grading 高密度实度级配;高密度级配
high-density graphite 高密度石墨
high-density hard board 高密度硬质纤维板
high-density high-conductivity charge 高密度高导热性炉料
high-density line 高运输密度线路
high-density method 高密度法;高电流密度锌电解法
high-density overlay 高密度贴面胶合板
high-density packaging technique 高密度组装技术
high-density particle board 高密度刨花板
high-density pick-up baler 高密度捡拾压捆机
high-density plywood 高压胶合板;高密度(贴面)胶合板
high-density point of interest 高密度特征点
high-density polyethylene 高密度聚乙烯
high-density polyethylene glass paper 高密度聚乙烯玻璃纸
high-density polyethylene pipe 高密度聚乙烯管
high-density polythene 高密度聚乙烯
high-density powder 高密度粉末
high-density pressing 高密度坯块
high-density propellant 高密度燃料
high-density PVC tangle film 高密度聚氯乙烯扭结膜
high-density PVC tangle film glassine 高密度聚氯乙烯扭结膜玻璃纸
high-density residential district 高密度住宅区
high-density storage 高密度存储器
high-density traffic 高密度通信业务;高密度交通
high-density tungsten alloy 高密度钨合金
high-density weighing materials 高密度加重剂
high-density weighting material 高密度加重材料
high-density wood 压实木材;高密度木材
high-density wood chipboard 高密度碎木板
high-density wood door 压实板木门
high-density zone 高密度区
high depression angle 高俯角
high destructive delta 高破坏性三角洲
high detectivity detector 高效探测器;高探测率探测器
high diamond 优质金刚石
high diffusivity path 高效扩散路径
high dilution 高度稀释
high dimensional pattern 高维模式
high dip 陡坡
high dip angle 高倾角;陡倾角
high dipping 倾斜度大的;急倾斜(大于15度)
high directional antenna 锐方向性天线
high discharge mixer 高效能混凝土搅拌机;搅拌机车
high discharge rate accumulator 高放电率蓄电池
high discharge skip 高卸料(小)车

high disinfection byproduct formation potential 强消毒产物生成势
high dispersing rosin emulsion 高分散松香胶乳
high dispersion 大色散度
high dispersion crown glass 高色散冕玻璃
high dispersion glass 高色散玻璃
high dispersion nozzle 高雾化细度喷嘴；细雾化喷嘴
high dispersion spectrum 高色散光谱
high-dissolved oxygen 高溶解氧
high distillation thermometer 高温蒸馏温度计
high distortion 高度失真
high dither 高频抖动
high-domed 高拱的
high dose tolerance 高剂量耐受性
high drag 高阻力
high drift angle 高偏角
high drop 高水头；高空投放
high drop dam 高水头坝
high drop lock 高水头船闸
high dropping point grease 高滴点润滑脂
high drop ship lift 高水头升船机
high drop ship passing structure 高水头过船建筑物
high drop ship passing through facility 高水头过船设施
high dry(ing)shrinkage concrete 高干缩率混凝土
high dry objective 高倍干燥接物镜
high dry strength 高干强度；干强度高
high ductile alloy 高塑性合金
high ductility alloy 高韧性合金
high ductility steel 高塑性钢；高柔性钢；高韧性钢
high dudy fireclay 高重负耐火砖
high dump box 高倾卸式车厢
high dump wagon 高位倾卸(式)拖车
high duty 有特殊抗力的；重质的；高生产率(的)；高能率(的)；高负荷的；重型的；重型；税率高的
high-duty boiler 高压锅炉；高效率锅炉
high-duty brick 高强度砖
high-duty cast iron 高级(优质)铸铁
high-duty detergent 高效洗涤剂
high-duty die 高负荷模
high-duty fireclay 高耐火度黏土
high-duty(fire)clay brick 高级黏土砖；高效耐火砖；重级耐火砖
high-duty iron 高质量铸铁
high-duty lubricating 高温高压用润滑油
high-duty malleable cast iron 高强度可锻铸铁
high-duty of water 高效用水率；高灌溉率
high-duty pigiron 优质生铁
high-duty pump 大排量泵；高效能泵；高能率泵
high-duty refractory 高级耐火材料
high-duty running 高负荷运行
high-duty service 高负荷运行
high-duty shaft kiln 大型立窑
high-duty silica 特级硅氧材料
high-duty steel 高强钢
high dynamic(al)response 高动态响应
high dynamic(al)strain indicator 超动态应变仪
high-early cement 早强水泥；快硬水泥
high-early concrete 早强混凝土；快硬混凝土
high-early Portland cement 早强波特兰水泥
high-early strength 早高强
high-early strength agent 早强剂
high-early strength cement 快硬水泥；早强(型)水泥；高早强水泥
high-early strength concrete 早强型混凝土；早高强混凝土
high-early strength Portland cement 早强硅酸盐水泥；早强波特兰水泥
high echo 高波
high efficacy heat conduction grease 高效导热脂
high efficiency 高效率
high efficiency air filter 高效空气滤清器；高效空气过滤器
high efficiency air filter unit 高效空气过滤器设备
high efficiency blower 高效鼓风机
high efficiency clarifier 高效澄清池
high efficiency classifier 高效选粉机
high efficiency coagulation sedimentation filtration 高效混凝过滤
high efficiency computer number control turning machine 高效能计算机数控车床
high efficiency cyclone 高效旋风筒

high efficiency deoxidiser 高效脱氧剂
high efficiency driling system for small and deep hole 高效小直径深孔钻削系统
high efficiency dust collection equipment 高效吸尘设备
high efficiency dust collector 高效吸尘器
high efficiency electrode 高效率焊条
high efficiency electrostatic precipitator 高效电集尘器
high efficiency energy conversion technique 高效能量转换技术
high efficiency filter 高效过滤器
high efficiency gear hobbing machine 高效滚齿机
high efficiency hydrogen purifier 高效氢纯化器
high efficiency liquid chromatography 高效液相色谱(法)
high efficiency modulation 高效率调制
high efficiency packing tower 高效填料塔
high efficiency particle filter 高效粒子过滤器
high efficiency particular air filter 高效微粒空气过滤器
high efficiency particulate absolute air filter 超高效空气过滤器
high efficiency particulated air filter 高效过滤材料
high efficiency submerged automatic arc welding process 高效埋弧自动焊接法
high efficiency water reducing agent 高效减水剂
high efficient air separation 高效空气选粉
high efficient air separator 高效空气选粉机
high efficient band width 高效带宽
high efficient bioaugmentation technique 高效生物强化技术
high elastic deformation 高弹形变；高弹变形
high elasticity 高弹性
high elastic limit steel 高弹性限度钢；高弹性限；高弹性极限钢
high elastic rubber 高弹性橡胶
high electric(al)field effect 强场效应
high elevation 高地势
highe-level radioactive liquid waste 强放射性废液
high elongation acetate yarn 高伸长醋酯纤维
high elongation strain ga(u)ge 大应变应变计
high emission photocathode 高发射光电阴极
high emissivity 高发射率
high end 高档
high end microprocessor 高档微处理机
high end point 高终沸点
high energy 高能率；高能(量)
high energy accelerator 高能加速器
high energy accelerator physics 高能加速器物理学
high energy activation 高能活化
high energy astronomy 高能天文学
high energy astronomy observatory 高能体温观测站
high energy astrophysics 高能天体物理(学)
high energy aviation fuel 高能航空燃料
high energy balance 高能量平衡
high energy baryon 高能重子
high energy battery 高能电池
high energy beam 高能束
high energy betatron 高能感应加速器
high energy bond 高能键
high energy bremsstrahlung 高能轫致辐射
high energy chemistry 高能化学
high energy coast 高能海岸
high energy combination 高能结合
high energy component 高能成分
high energy-consuming enterprises 高耗能企业
high energy-density 高比能量
high energy detector 高能探测器
high energy disintegration 高能蜕变
high energy electron 高能电子
high-energy electron beam irradiation 高能电子束辐照
high energy electron diffraction 高能电子衍射
high energy electronics 高能电子学
high energy environment 高能环境
high energy event 高能核转变
high energy flashlamp 高能闪光灯
high energy foam generation system 高能泡沫发生系统
high energy foam generator 高能泡沫发生器
high energy fuel 高能燃料
high energy gamma photon generator 高能伽马光子发生器

high energy gamma-ray 高能伽马射线
high energy geophysics 高能地球物理学
high energy geothermal field 高能位地热田
high energy harmonic oscillator 高能谐振子
high energy heavy ion nuclear reactor 高能重离子反应堆
high energy igniter 高能点火器
high energy ignition 高能点火
high energy ignition system 高能点火系统
high energy impact pressing 高能(束)冲压
high energy instrument 高能仪表
high energy ion analyzer 高能离子分析仪
high energy irradiation 高能照射
high-energy irradiation treatment 高能辐照处理
high energy laser 高能激光器
high energy laser pump 高能激光泵
high energy laser system 高能激光系统
high energy light(weight)propellant 轻重量高能推进剂
high energy linear collider 高能直线碰撞机
high energy liquid laser 高能液体激光器
high energy liquid oxidizer 高能液体氧化剂
high energy momentum 高能动量
high energy neutron 高能中子
high energy neutron detector 高能中子探测器
high energy nuclear magnetic resonance detector 高能核磁共振仪
high energy nuclear physics 高能核物理学
high energy particle 高能粒
high energy particle radiography 高能粒子摄影
high energy phosphate 高能磷酸盐
high energy phosphate bond energy-rich phosphate bond 高能磷酸键
high energy physics 高能物理(学)；高能粒子物理学；粒子物理学
high energy positron 高能正电子
high energy propellant 高能燃料
high energy proton 高能质子
high energy proton detector 高能质子探测器
high energy pulse laser 高能脉冲激光器
high energy radiation 高能射线；高能辐射
high energy rate 高能速
high energy rate deformation 高能速变形
high energy rate forging 高能高速(锤)锻造；高能快速锻造
high energy rate forging hammer 高速锤
high energy rate forging machine 高能高速锻(造)机
high energy rate forming 高能形成；高能率成形；高能高速成形；高能快速成形
high energy rate forming machine 高速成型机；高速成形机
high energy resonance 高能共振
high energy rig 高能装置
high energy scattering 高能散射
high energy scrubber 高效洗涤器；高能洗涤器
high energy short-pulse flashlamp 高能短脉冲闪光灯
high energy silage 高能青储[贮]料
high energy silverzinc battery 高能银锌电池
high energy spark forming 高能率电火花成形
high energy state 高能态
high energy tail 高能尾；高能端
high energy throughput echelle spectrometer 高能红外光栅光谱仪
high energy total absorption detector 高能全吸收探测器
high energy underwater power source 高能水下动力源
high energy vibration 高能振动
high enriched fuel 高浓缩燃料
high enriched uranium 高度浓缩铀
high enrichment 高度浓缩
high enrichment fuel 高浓缩燃料
high enrichment leacher 高加浓燃料浸取器
high enrichment reactor 高浓缩性反应堆；高浓缩燃料反应堆；高加浓反应堆
high enthalpy 高热焓
high enthalpy fluid 高焓流体
high enthalpy well 高焓井
high entry porous stone 高进气透水石
high environmental risk area 高环境风险区
high environment test system 高空环境试验系统

high equilibrium concentration 高平衡浓度
higher 较高
higher algebra 高等代数(学)
higher aliphatic alcohol 高脂族醇
higher ambient transistor 高温度稳定晶体管
higher area limit of gamma 伽马偏海限
higher aromatic hydrocarbon 高级芳烃
higher aromatics 高级芳烃
higher array 高维阵列
higher authorities 上级(机关)
higher availability 利用率高
higher bacteria 高等细菌
higher bleaching powder 高效漂白粉
higher body 上级机关
higher bronze 高级青铜
higher bulk 高膨松体
higher calorific value 高卡值;高位热值;高发热值
higher categories 高级类别;高级储量
higher category 高等项目;较高的分类级别
higher center cradle 高顶托架
higher composition of capital 高位资本构成
higher-cost region 成本较高的区域
higher court 高级法院
higher critical velocity 高临界流速;超临界速度;超临界流速
higher cut-off frequency 上限截止频率
higher decimal 小数的整数部分
higher degree curve 高次曲线
higher derivative 高阶微商;高阶导数;高级衍生物
higher derivative method 高阶导数法
higher difference 高阶差分
higher dimentional space 高维空间
higher education 高等教育
higher energy orbital 高能轨道
higher energy phosphate 高能磷酸化物
higher energy state 高能量状态
higher energy thioester bond 高能硫酯键
higher fatty acid 高级脂肪酸
higher fungi 高级真菌;高等真菌
higher geodesy 高等大地测量(学)
higher geometry 高等几何学
higher goods platform 高站台
higher grade deposit 高品位矿床
higher grade industry 高层
higher-grade pay 高工资
higher harmonic mode 高次模
higher harmonic resonance 高谐波共振
higher harmonics 高次谐波
higher harmonic voltage 高次谐波电压
higher harmonic wave 高次谐波;较高次谐波
higher hazard area 高危险区
higher heating value 高位发热量;高发热值
higher high tide 高高潮(位)
higher high water 较高高水位;较高高潮;高高潮(位);高潮高水位
higher high water interval 高高潮间隙
higher high water level 历史较高水位
Higher Himalayan geodome system 高喜马拉雅地穹系
higher homologue 高级同系物
higher hydrocarbon 重烃;高级烃
higher layer line synthesis 高层线综合法
higher layers of the atmosphere 比平流层更高的大气层
higher level hierarehies 更高级的层次结构
higher level of cultural practices 高水平的栽培措施
higher levels 较高层次
higher level skill 高技术技能
higher low tide 高低潮
higher low water 较高低水位;较高低潮;高低潮位;低潮高水位
higher low water interval 高低潮间隙(月亮过中天和下一高高潮的时间间隔)
higher mathematics 高等数学
higher mature source rock 高成熟源岩
higher mature stage 高成熟阶段
higher-method of mining 矿业开发技术高
higher mode 高振型
higher-mode coupling 高次波形耦合
higher-mode effect 高振型影响
higher-mode excitation 高次谐波励磁
higher-mode interference 高阶干扰
higher-mode of vibration 高振型振动;高振型

higher-mode Rayleih wave 高振型瑞利波
higher modes 高阶模;高次模式
higher-mode surface wave 高振型表面波
higher order 高阶;高次
higher-order algebraic equation solver 高次代数方程式解算器
higher-order approximation 高阶逼近
higher-order attenuation 高次波的衰减
higher-order autoregressive scheme 高阶自回归形式
higher-order coherence effect 高阶相干效应
higher-order coherence function 高次相干函数
higher-order compound 高级化合物
higher-order derivative of gravity potential 重力位高阶导数
higher-order diffracted neutrons 高次衍射中子
higher-order geodesy 高等大地测量(学)
higher-order harmonics 高次谐波
higher-order lag structure 高阶滞后结构
higher-order language 高级语言
higher-order mode 高次模;高阶振型;高阶模
higher-order of minuteness 高阶微量
higher-order of oscillation 高频振荡
higher-order phase transition 高阶相变;高级相变
higher-order precedence grammar 高阶优先文法
higher-order residual aberration 高次残余像差
higher-order software 高级软件
higher-order software technology 高级软件技术
higher-order structure 高级构造
higher-order surface wave 高阶表面波
higher-order survey 高等测量
higher-order transformation 高次变换
higher-order transitions 高级跃迁
higher-order triangulation 高等(级)三角测量
higher-order vibration 较高次振动
higher-order wave 高次波
higher oxygen consumption water 耗氧量稍高的水
higher pair 线点对偶;高副
higher partial derivative 高阶偏微商
higher phenols 高级酚
higher plane curve 高次平面曲线
higher plant 高等植物
higher polyolefins 高级聚烯烃
higher potential barrier 高遮挡
higher pressure deaerator 高压除气器
higher priority 较重要优先项目
higher priority goal 较高优先级目标
higher priority item 比较优先考虑的项目
higher residual pesticide 高残留农药
higher slice 上部分层
higher source luminance 高亮度光源
higher space 高维空间
higher specific speed Francis turbine 高比速混流式水轮机;高比速辐向轴流式水轮机
higher state 高能态
higher surveying 高等测量学
higher tar acid 高级焦油酸
higher taxon 高等单元
higher techniques of mineral commodity processing 矿产加工技术高
higher than average crustal heat flow 高于地壳平均值的热流
higher than the average cost 高于平均数额费用
higher transition probability 高阶转移概率
higher-up 上级
higher vibration 高次振动
higher water 高潮;高水位;洪水
highest 最高的
highest adsorption density 最大吸附密度
highest and best use 最佳用途
highest and best use study 地产的最佳使用研究
highest annual hourly traffic volume 年最高小时交通量
highest annual hourly volume 年度最高小时(交通)量
highest astronomical tide 最高天文潮
highest attained vacuum 极限真空;极度真空
highest bid 最高出价;最高标价;最高递价;最高报价;竞买的最高价(指投标的)
highest bidder 最高价投标人;最高价竞买人;买方出价最高者;递价最高的投标人;出价最高的投标人;出高价的报标人
highest bit 竞卖最高价

highest common divisor 最高公约式;最大公约数;最大公因子
highest discharge 最大流量
highest elevation 最高点;最大高程
highest energy 最高能量
highest ever-known discharge 已知最大流量;历史最大流量
highest ever-known tide level 历史最高潮位
highest ever-known water level 历史最高水位;历史最高潮位
highest flood discharge 最大洪水流量
highest flood level 最高洪水位
highest flood peak 最高洪峰
highest flood period 最高洪水期
highest frequency 最高频率
highest high discharge 历史最大流量
highest high tide level 最高高潮位
highest high water(level) 最高洪水位;最高水位;最高高潮;历史最高(高)潮位;历史最高水位;最高(高)潮位
highest high water level on record 历史最高水位;历史最高潮位
highest high water spring tides 春潮的最高高水位
highest ice floe water level 最高流冰水位
highest ineffective concentration tested 测试的最大无影响浓度
highest-in first-out 高入先出法
highest limit 最大限度
highest low tide level 最高低潮位
highest low water 最高低水位;最高低潮位
highest low water of neap tides 小潮的最高低潮位
highest mode 最高振型
highest navigable flood level 最高通航水位
highest navigable stage 最高通航水位
highest nuchal line 最上顶线
highest observed frequency 最高观测频率
highest order 最高位;最高序列
highest organic soil 高有机土
highest oxygen consumption water 耗氧量很高的水
highest possible frequency 最高可能的频率
highest possible high water 可能最高水位;可能最高高潮位
highest possible price 可能得到的最高价;尽可能的最高价
highest power factor 最高功率因数
highest price method 最高价格法
highest priority 最重要优先项目
highest-probable frequency 最高可及频率
highest quality 超级;超等质量
highest quotation 最高报价
highest recorded stage 最高记录水位;实测最高水位
highest recorded(water) level 最高记录水位;历史最高水位;历史最高潮位;记录最高水位;实测最高水位
highest relief 极高突起
highest reservoir level 最高水库水位
highest resonance of energy 最大能量的共振
highest runoff 最大径流(量)
highest significant position 最高有效位
highest speed 最高速度;最高航速;最高车速
highest stage 最高水位
highest stage with drift ice 流冰最高水位
highest temperature 最高温度
highest tender 最高报价
highest tide level 最高潮位
highest upper pool elevation 上游最高库水位
highest upsurge 最高(上)涌浪
highest useful compression ratio 最高有效压缩比
highest valence 最高价【化】
highest water level 最高水位
highest water mark 最高水位线
highest water table 最高水位
highest wave 最大波浪
highest wave crest 最高波峰
highest world standard 世界先进水平
highest yield 最高收益
high-expansion alloy 高膨胀合金
high-expansion cement 高膨胀水泥;快膨胀(性)水泥;强膨胀水泥
high-expansion cement concrete 高膨胀水泥混凝土

high-expansion cement mortar 高膨胀水泥砂浆
high-expansion concrete 高膨胀混凝土
high-expansion foam 高级膨胀泡沫体
high-expansion foam agent 高倍数泡沫液
high-expansion foam extinguisher 高爆塑性装药弹
high-expansion foam extinguishing agent 高倍数泡沫灭火剂
high-expansion foam extinguishing system 高倍数泡沫灭火系统
high-expansion foam fire-extinguishment system 高倍数泡沫灭火系统
high-expansion foam-generating equipment 高倍数泡沫发生装置
high-expansion foam generator 高膨胀泡沫发生器灭火装置;高倍数泡沫发生器
high-expansion foam system 高倍数泡沫系统
high-expansion foam truck 高倍数泡沫车
high-expansion material 高膨胀(系数)材料
high expansive cement 高膨胀水泥
high exploration area 勘探程度高区
high explosion 强爆炸
high explosive 高效炸药;高能炸药;高级炸药;高爆炸药;烈性炸药;猛炸药;高威力炸药
high explosive compacting 高爆炸力成形
high-explosive injury 爆炸伤
high-explosive plastic 塑性猛烈炸药
high-explosive plastic projectile 高爆塑性装药弹
high expression roll 高效压榨辊筒
high extensibility fiber 高延伸度纤维
high face 高割面
high face jumbo 高掌子面钻车;用于高工作面的钻车
high face room-and-pillar mining 高工作面房柱式开采
high feed 快速给进
high feed gear 快速给进齿轮
high ferric cement 高铁水泥
high ferrite cement 高铁水泥
high fidelity 高逼真度;高保真(度);高度保真
high fidelity amplifier 高保真度放大器
high fidelity deconvolution 高保真度反褶积
high fidelity dynamic(al)correction 高保真度动校正
high fidelity receiver 高保真度接收机
high fidelity stack 高保真叠加
high fidelity system 高保真度系统
high fidelity treatment 高保真度处理
high field 高磁场;强电场
high field booster 高磁场蓄电池升压器
high field distribution map of survey area 测区高场分布图
high field domain luminescence 强场畴发光
high field effect 高场效应
high field electroluminescence 强场电致发光
high field emission 场致发射
high field intensity 高场强
high field mobility 高场迁移率
high field superconductor 第二类超导体
high field type booster 强场式升压机
high fill 高填土;高路堤
high filter 高频噪声滤波器
high finance 高明筹资;复杂的财务
high fine semi-auto universal cylindrical grinder 高精度半自动万能外圆磨床
high finish 光制;高级整理;高度光泽;粘磨;研磨
high finished 精制的
high fired 高温焙烧过的;强煅烧的
high fired glaze 高温釉
high fired pitch fibre 高温焙烧沥青纤维
high fired porcelain 高温瓷器
high fire hazard 高度火灾危险
high-fire-risk 高度火灾危险
high fix 高精度定位系统
high flash 高闪点
high flash fuel 高闪点燃料
high flash naphtha 高闪点石脑油
high flash oil 高闪点油
high flash point 高闪点
high flats 高层公寓
high flood 高水位
high flood level 高洪水位
high flood plain 高河漫滩
high flow 高流动性

high flow capacity 高流量
high flow gas laser 高速流气体激光器
high flow rate 高流率
high flow season 汛期;洪水期;洪水季(节)
high flow speed 高流速
high flow year 多水年;丰水年
high fluidity 高流动性
high fluid potential belt 高流体热能带
high fluid pressure 高液压
high-fluoride drinking water 高氟饮用水
high flush cistern 高冲洗水池
high flush tank 高水箱;高架水塔;高架(冲洗)水箱
high flutter magnet 高颤磁铁;大调变度磁铁
high flux 最大密度流;强力流
high flux beam reactor 高通量(中子)束堆
high flux core 高磁密铁芯
high flux hollow fiber 高通量中空纤维
high flux isotope reactor 高通量同位素反应堆
high flux reactor 高通量反应堆
high flux research reactor 高通量研究反应堆
high-flying glamor stocks 高价位魅力股
high-flying highway 上跨道路;高架公路
high focal plane buoy 高焦面浮标;高灯芯浮标(一般在10米高以上)
high fog 高雾
high forest 高山林;乔木林
high format 高规格;大凯本;大尺寸
high format(clay)brick 大型砖;大尺寸砖
high fouling potential 高污着势
high framed portal crane 高架门(式)起重机
high freeboarded 干舷高的
high freeboard ship 高干弦船
high free rosin size 高游离松香胶
high-freggquency saturable reactor 高频饱和扼流圈
high-frequency 高周率;高周波;高频(率)
high-frequency absorber 高频吸收装置;高频吸收器
high-frequency absorption section 高频吸收剖面
high-frequency air-leg rock drill 高频气腿式凿岩机
high-frequency alternate current generator 高频交流发电机
high-frequency alternate current welding machine 高频交流电焊机
high-frequency alternator 高频交流(发)电机
high-frequency ammeter 高频电流表
high-frequency amperemeter 高频电流表
high-frequency amplification 高频放大
high-frequency amplification receiver 高放式接收机
high-frequency amplifier 高频放大器
high-frequency amplifier tube 高频放大管
high-frequency amplitude 高频振幅
high-frequency and low noise transistor 高频低噪声晶体管
high-frequency antenna 高频天线
high-frequency arc stabilizer 高频稳弧器
high-frequency auto centrifugal casting furnace 高频自动离心铸造炉
high-frequency backscatter 高频返回散射
high-frequency band 高频带
high-frequency barrier 高频位垒
high-frequency bias 高频偏磁
high-frequency bipolar transistor 高频双极晶体管
high-frequency brazing 高频钎焊
high-frequency brazing machine 高频钎焊机
high-frequency bridge 高频(测量)电桥
high-frequency bus 高频总线
high-frequency buzzer 高频蜂鸣器
high-frequency cable 高频电缆
high-frequency capacitor 高频电容器
high-frequency carrier 高频载波
high-frequency carrier channel 高频载波通道
high-frequency carrier telegraphy 高频载波电报
high-frequency cauterization 高频电灼
high-frequency centrifugal casting machine 高频(电流)离心铸造机
high-frequency ceramic element 高频陶瓷元件
high-frequency ceramic filter 高频陶瓷滤波器
high-frequency ceramics 高频陶瓷
high-frequency ceramic soldering machine 高频陶瓷焊接机
high-frequency change over switch 高频转换开关
high-frequency channel 高频信道
high-frequency choke 高频抗流圈;高频扼流圈

high-frequency circuit 高频电路
high-frequency coaxial cable 高频同轴电缆
high-frequency coil 高频线圈
high-frequency coil with core 高频磁芯线圈
high-frequency communication cable 高频通信电缆
high-frequency communication engineering 高频通信工程
high-frequency compensation 高频补偿(电路)
high-frequency component 高频分量
high-frequency concentration meter 高频浓度计
high-frequency confinement 高频约束
high-frequency connector 高频接插件
high-frequency control 高频操纵
high-frequency converter 高频变换器
high-frequency conveyer 高频输送机
high-frequency core 高频铁芯
high-frequency cracks test 高频噪声试验
high-frequency crystal filter 高频晶体滤波器
high-frequency crystal oscillator 高频晶体振荡器
high-frequency curing 高频硬化
high-frequency current 高频电流
high-frequency current transformer 高频变流器
high-frequency deformation test 高频率变形试验
high-frequency demodulator 高频解调器
high-frequency densimeter 高频密度计
high-frequency diathermy electrocoagulation device 高频透热电凝器
high-frequency diathermy heating 高频电热
high-frequency dielectric 高频电解质
high-frequency dielectric(al)separation method 高频介电分离法
high-frequency dielectric(al)separator 高频介电分离仪
high-frequency dielectric(al)welding 高频介电焊
high-frequency direction finder 高频方向测定仪;高频测向仪;高频测向器
high-frequency direction finding 高频测向
high-frequency direction finding station 高频测向台
high-frequency discharge 高频放电
high-frequency discharge lamp 高频放电管
high-frequency distortion finder 高频失真仪
high-frequency doublet antenna 高频偶极天线;高频对称天线
high-frequency drier 高频加热干燥机
high-frequency drill 高频凿岩机
high-frequency drilling 高频冲击钻进
high-frequency dryer 高频烘燥机
high-frequency drying 高频干燥(法)
high-frequency drying stove 高频干燥炉;高介质烘干炉
high-frequency effect 高频影响;高频效应
high-frequency electric(al)current sterilization 高频电流灭菌法
high-frequency electric(al)drilling 高频电热钻进
high-frequency electric(al)drying 高频电干燥
high-frequency electric(al)field 高频电场
high-frequency electric(al)furnace 高频电炉
high-frequency electrically heated mo(u)ld 高频加热模
high-frequency electric(al)machines 高频电机
high-frequency electric(al)welded steel pipe 高频电焊钢管
high-frequency electrolysis 高频电解
high-frequency electromagnetic field 高频电磁场
high-frequency engineering 高频工程学
high-frequency erasure 高频消磁
high-frequency error 高频误差
high-frequency error function 高频误差函数
high-frequency excitation 高频激发
high-frequency fatigue testing machine 高频疲劳试验机
high-frequency feeder 高频馈路
high-frequency field strength measuring meter 高频场强测量仪
high-frequency figure of merit 高频品质因数
high-frequency filtering image with partial coherent light 部分相干光高通滤波图像
high-frequency fine 高频微调
high-frequency finisher 高频整修器;高频平整器
high-frequency flaw detection method 高频探伤法
high-frequency fluctuation 高频波动

high-frequency frequency-spectrum analyser 高频频谱分析仪
high-frequency furnace 高频(电)炉
high-frequency ga(u)ge 高频仪
high-frequency generator 高频发生器;高频(发)电机
high-frequency geophone 高频检波器
high-frequency gluing 高频黏结
high-frequency hacksaw blade painted 高频淬火涂漆钢锯条
high-frequency hardening 高频淬火
high-frequency hardening machine 高频淬火机
high-frequency harmonic oscillator 高频谐振子
high-frequency head tester for TV 电视机高频头测试仪
high-frequency head video recorder 高频头录像机
high-frequency heat 高频电炉
high-frequency heating 高频(电感)加热
high-frequency impact loads 高频冲击力
high-frequency impedance 高频阻抗
high-frequency impulse 高频脉冲
high-frequency impulse electro-erosion machine 高频脉冲电蚀机床
high-frequency induction coil 高频感应(线)圈
high-frequency induction furnace 高频感应(电)炉;高频电感(加热)炉
high-frequency induction generator 高频感应发生器
high-frequency induction hardening 高频淬火
high-frequency induction-heated cell 高频感应加热电解槽
high-frequency induction heating apparatus 高频感应加热设备
high-frequency induction heating coating equipment 高频感应加热蒸镀设备
high-frequency induction heating equipment 高频感应加热设备;高频感应加热淬火式炉
high-frequency induction melting furnace 高频感应熔化炉
high-frequency induction quenching 高频感应加热淬火
high-frequency induction soldering 高频感应焊接
high-frequency induction welder 高频焊机;高频感应(式)焊机
high-frequency induction welding 高频焊接;高频感应焊(接)
high-frequency instantaneous variation 高频瞬变
high-frequency instrument 高频仪表
high-frequency insulator 高频绝缘子
high-frequency interference 高频干扰
high-frequency internal vibration 高频内振动
high-frequency internal vibrator 高频内振动器;高频内振捣器
high-frequency ionization chromatographic(al) detector 高频电离色谱检测器
high-frequency iron 高频铁
high-frequency iron core 高频铁芯
high-frequency iron dust core 高频铁粉芯
high-frequency jig 高频跳汰机
high-frequency knotter 高频打捆机
high-frequency line and patching bay 高频线路及调度架
high-frequency loss 高频损耗
high-frequency magnetic anomaly 高频磁异常
high-frequency magnetic core 高频磁芯
high-frequency magnetic mirror 高频磁镜
high-frequency magneto 高频率永磁发电机
high-frequency measuring instrument 高频测量仪器
high-frequency medium electric(al) separator 高频介电分离仪
high-frequency medium hot combination equipment 高频介质热合设备
high-frequency melting process 高频熔炼法
high-frequency microvoltmeter 高频微伏表
high-frequency mid-voltage standard apparatus 高频中电压标准仪
high-frequency millivolt meter 高频毫伏表
high-frequency millivoltmeter division calibrator 高频毫伏表定度仪
high-frequency modulation 高频调制
high-frequency modulator 高频调制器
high-frequency motor 高频马达;高频电动机
high-frequency motor generator 高频电动发电机

high-frequency mo(u)lding technology and equipment 高频注模技术及设备
high-frequency multiplex communication 高频多路通信
high-frequency noise 高频噪声
high-frequency office cable 高频局用电缆
high-frequency orientation protector 高频方向保护装置
high-frequency oscillation 高频摆动
high-frequency oscillator 高频振荡器
high-frequency oscillograph 高频示波器
high-frequency percussion 高频冲击
high-frequency pipe bending machine 高频弯管机
high-frequency plasma 高频等离子体
high-frequency plasma generator 高频等离子发生器
high-frequency plasma melting method 高频等离子熔融体法
high-frequency plasmatron 高频等离子管
high-frequency plastic hot-bonding machine 高频塑料热合机
high-frequency plastic welder 高频塑料热合机
high-frequency plug 高频塞
high-frequency poker vibrator 高频插入式振捣器
high-frequency polarograph 高频极谱仪
high-frequency polarography 高频极谱
high-frequency porcelain 高频用瓷料
high-frequency potentiometer 高频电位器
high-frequency power 高频功率
high-frequency power generator 高频功率发生器
high-frequency prediction 短波频率预测
high-frequency preheater 高频预
high-frequency preheating 高频预热
high-frequency pressure indicator 高频压力指示器
high-frequency probe 高频探针;高频探头
high-frequency process 高频法
high-frequency propagation 高频传布
high-frequency protection 高频保护(装置)
high-frequency protector 高频保安器
high-frequency protector panel 高频保安器盘
high-frequency pulp screen 高频筛浆机
high-frequency pulse 高频脉冲
high-frequency pulse generator 高频脉冲发生器
high-frequency push-pull fatigue tester 高频拉压疲劳试验机
high-frequency pyrolysis 高频热解
high-frequency Q meter 高频 Q 表
high-frequency quenching 高频加热淬火(法);高频淬火
high-frequency quenching machine 高频淬火机(床)
high-frequency radiation 高频辐射
high-frequency radio physics 高频无线电物理学
high-frequency radio transmitter 高频无线电发射机
high-frequency radio wave 高频电波
high-frequency receiver 高频接收机
high-frequency recombination 高频重组
high-frequency recombination strain 高频次重组菌株
high-frequency relay 高频继电器
high-frequency repeater distribution frame 高频中继配线架
high-frequency resistance 高频电阻
high-frequency resistance hardening 高频电阻淬火
high-frequency resistance welding 高频接触焊;高频电阻焊
high-frequency resistor 高频电阻器
high-frequency response function 高频响应函数
high-frequency response pick-up 高频响应传感器
high-frequency rotating switch 高频旋转开关
high-frequency rotational transition 高频转动跃迁
high-frequency sandwich composite 高频夹心式复合材料
high-frequency screen 高频筛
high-frequency sealing 高频封接
high-frequency seal machine 高频熔接机
high-frequency seasoning 高频干燥
high-frequency section 高频段
high-frequency seism 高频地震
high-frequency seismic survey 高频地震勘探
high-frequency seismograph 高频地震仪

high-frequency selectivity 高频选择法
high-frequency separation of diamond 金刚石高频选矿
high-frequency shale shaker 高频振动筛
high-frequency signal generator 高频信号发生器
high-frequency single sideband 高频单边带
high-frequency single sideband radio telephone 高频单边带无线电话
high-frequency sky wave field strength 高频天波场强
high-frequency soldering 高频钎焊;高频焊接
high-frequency sounder 高频测深仪
high-frequency sound insulation 高频声隔
high-frequency spark chamber 高频电火花室
high-frequency spark gap oscillator 高频电火花隙发生器
high-frequency speaker 高频扬声器
high-frequency spectrum 高频光谱
high-frequency spraying 高频感应喷涂
high-frequency sputtering equipment 高频溅射设备
high-frequency stage 高频级
high-frequency stamper 高频捣实机
high-frequency steel furnace 高频炼钢炉
high-frequency straight seam welded pipe 高频直缝焊接钢管
high-frequency substitution method 高频替代法
high-frequency surface finishing machine 高频表面平整机
high-frequency surface hardening 高频表面淬火
high-frequency susceptibility 高频磁化率
high-frequency symmetrical cable 高频对称电缆
high-frequency telephone 高频电话
high-frequency telex radio 高频电传无线电(设备)
high-frequency terminal bay 高频终端架
high-frequency test bay 高频测试架
high-frequency testing apparatus 高频测试装置
high-frequency tetrode 高频四极管
high-frequency thin film thickness meter 高频薄膜厚度计
high-frequency thyristor 高频可控硅
high-frequency titration 高频滴定(法)
high-frequency titrimeter 高频滴定计
high-frequency tool 高频工具
high-frequency torque fatigue tester 高频扭转疲劳试验机
high-frequency track circuit 高频轨道电路
high-frequency transceiver 高频收发两用机
high-frequency transduction 高频传导
high-frequency transfer 高频转移
high-frequency transformer 高频变压器
high-frequency transformer silicon steel sheet 高频变压器硅钢片
high-frequency transistor 高频晶体管
high-frequency transmitter 高频发射机
high-frequency treatment 高频处理
high-frequency trimmer 高频微调电容(器)
high-frequency triode oscillator 高频三极管振荡器
high-frequency tube 高频管
high-frequency tuner 高频调谐器
high-frequency ultrasound 高频超声
high-frequency ultraviolet lamp 高频紫外线灯
high-frequency unit 高频部件
high-frequency vacuum-meter 高频率真空测定计
high-frequency valve oscillator 高频电子管振荡器
high-frequency vibrated concrete 高频振动混凝土
high-frequency vibrating screen 高频振动筛
high-frequency vibration 高频振动
high-frequency vibration technic 高频振捣技术
high-frequency vibration transducer 高频振动传感器
high-frequency vibrator 高频振动器
high-frequency vibratory compactor 高频振动捣实器
high-frequency voltmeter 高频伏特计;高频电压表
high-frequency watch 高频手表
high-frequency wattmeter 高频瓦特计
high-frequency wave 短波;高频波
high-frequency wave analyzer 高频波分析仪
high-frequency welded spiral fin tube 高频焊接螺旋翅片管
high-frequency welding 高频电焊;高频焊接

high-frequency welding magnetic bar 高频焊接磁棒
high-frequency wire communication 高频有线通信
high-frequency wired distribution system 高频有线分配系统
high-frequency wire-wound resistor 高频线绕电阻器
high-frequency work sampling 高频率工作取样
high frigid cement 高寒水泥
high front manger 高前沿饲槽
high front shovel 长臂挖掘机
high fundamental frequency blade 高基频叶片
high furnace 高炉
high fusibility ash 高熔灰分
high fusible ash 高熔灰分
high fusion-ash coal 高熔灰煤
high gain 高增益
high gain amplifier 高增益放大器
high gain and low-drift operational amplifier 高增益低漂移运算放大器
high gain antenna 高增益天线
high gainband-width differential amplifier 高增益宽带差动放大器
high gain control loop 高增益控制回路
high gain gas laser 高增益气体激光器
high gain image intensifier 高增益像增强器
high gain imaging array 高增益成像阵列
high gain laser 高增益激光器
high gain loop 高增益环路
high gain maser 高增益微波激射器
high gain operational amplifier 高增益运算放大器
high gain oscillation 高增益振荡
high gain oscilloscope 高增益示波器
high gain repeater 高增益转发器;高增益增音机
high gain seismograph 高增益地震仪
high gain transition 高增益跃迁
high gain tube 高增益管
high-gantry cable bulldozer 高台架缆索推土机;高台架缆索粗碎机
high-gap compound 宽禁带化合物
high gas-oil ratio 高气油比
high gas pressure switch 高气压开关
high gas pressure test 高气压试验
high gate drive 控制极强驱动
high gear 高速挡;高速传动
high-geared capital 高速资本
high-geared operation 高速操纵;快速施工
high gel rubbers 高冻橡胶
high glass content mo(u)lding compound 高强模塑料
high glaze 高光泽
high gloss 高光洁度
high gloss enamel 高光瓷漆
high gloss enamel tile 高光泽釉瓷砖
high gloss finish 高光泽表面处理;高光泽面漆
high gloss ink 高光泽墨
high gloss lacquer 高光泽清漆
high gloss offset printing ink 高光泽胶印油墨
high gloss paint 高光泽涂料;全光漆;强光泽涂料
high gloss paper 高光泽纸
high gloss varnish 高光泽清漆
high gloss wall slab 高度光滑表面墙板
high Gothic 高哥特式
high Gothic tracery 高哥特式装饰花纹
high grade 优质;优等(的);高质量(的);高坡;高品位(的);高精尖;高级的
high grade alloy 高级合金
high grade alloy steel 高级合金钢
high grade alloy structure steel 高级合金结构(钢)
high grade carbon steel 高级碳素钢
high grade cargo 高级货物
high grade cast iron 高级铸铁
high grade cataclasis way 高级碎裂作用方式
high grade cement 优质水泥;高级水泥;高标号水泥
high grade cement mortar 高强度等级水泥砂浆
high grade clay 高级黏土
high grade coating 超级镀膜
high grade concrete 高强度等级混凝土;高强度混凝土
high grade consumer goods 高级消费品
high grade corporate bonds 高级公司债券
high grade dyestuff 高级染料

high grade dynamite 高级炸药
high grade electrode 高级焊条
high grade energy 高(级)能
high grade energy welding 高能焊
high grade finish 高级整理
high grade fuel 高级燃料
high grade hotel 高级旅馆
high grade insulating materials 高级绝缘材料
high grade melting scrap 高品位废金属
high grade metamorphic bituminous 高变质烟煤
high grade metamorphism 高级变质作用
high grade metamorphosed 高变质
high grade mill 高品位矿石选矿厂
high grade mineral 高品位矿物
high grade of transparency 高透明度
high grade ore 高品位矿石;富矿(石)
high grade paper 高级纸(张)
high grade pressure ga(u)ge 高精度压力计
high grade processed goods 高级加工产品
high grade product 优质产品;高品位产量
high grade reinforcement steel 高强钢筋
high grade residence 高级住宅
high grade steel 优质钢;高级钢
high grade structural steel 高强结构钢
high grade timber 优质木材
high grade timber board 优质木板
high grade time relay 高级时间继电器
high grade white interference colo(u)r 高级白干涉色
high gradient 高梯度;陡比降;大坡度;大比降
high gradient magnetic separation 高梯度磁选;高梯度磁分离
high gradient magnetic separation process in wastewater treatment 废水处理高梯度磁分离法
high gradient magnetic separation technology 高梯度磁分离技术
high gradient magnetic separator 高梯度磁选机;高梯度磁分离器
high gradient separator 锥度大的选粉机
high grading 选择性开采
high grass savanna 高草稀树草原;热带高草草原
high gravity environment 高重力加速环境
high gravity filler 重质填料
high gravity fuel 高比重燃料
high ground 高地
high grounding 高阻接地方式
high handle chain saw 高把链锯
high hardness 高硬度
high hardness and low-alkalinity wastewater 高硬度低碱性废水
high hat 遮光罩;灯伞
high hat kiln 高帽窑
high hazard building 高度危险房屋
high hazard content 危险品;高度危险储[贮]物;高火灾隐患
high hazard group 高危险品房屋群;高危害性组群
high hazard industrial buildings 含高危险品的工业房屋;含高危险品的工业建筑物
high hazard industry occupancy 高火险工业用房
high hazard occupancy 高火险用房
high head 高压头;高水头
high head dam 高水头(水)坝
high head development 高水头开发项目;高水头电站
high head gravity flow system 高位自流系统
high head(hydraulic) turbine 高水头水轮机
high head hydropower plant 高水头水力发电厂
high head hydropower station 高水头水力发电站
high head plant 高水头电站
high head pump 高扬程(水)泵
high headroom grinding plant 高净空粉磨车间
high head scheme 高水头开发项目;高水头方案
high head ship passing structure 高水头过船建筑物
high head ship passing through facility 高水头过船设施
high head water power plant 高水头(水)电站
high head water power station 高水头水力发电站
high heat 高温的;高热量
high heat appliance 高温设备
high heat calorie gas 高热值煤气
high heat capacity 高热容量
high heat cement 高热水泥
high heat coat(ing) 耐热面层

high heat conductivity 高热导率
high heat duty firebrick 高耐火黏土砖;高温耐火砖
high heat exchange rate 高热交换率
high heat flow zone 高热流带
high heating radiation 高热辐射
high heating value gas 高热量煤气
high heat interchange rate 高热互换率
high heat level evapo(u)rator 高热面蒸发器
high heat method 高热法
high heat oil well cement 高热油井水泥
high heat paint 高温漆
high heat release burner 高热负荷燃烧器
high heat solder 高热焊条
high heat treatment 高温热处理
high heat value 高热值
high hedge 高篱
high helix drill 高速螺旋钻;高螺旋钻头
high-helix straight shank drill 大螺旋角直柄钻头
high hiding colo(u)r 高遮盖力着色料
high holding power anchor 大抓力锚
high holding power drag embedment anchor 大抓力埋入式海锚
high hot mineral spring 高热矿泉
high humic acids coal 高腐殖酸煤
high humidity 湿度大
high humidity and condensation test 高湿度结露试验
high humidity test chamber 高湿度测试箱
high humidity treatment 高湿度养护;高湿度处理
high hump 高驼峰【铁】
high hydraulic valve 高压液压阀
high-hysteresis type 高滞变型;高磁滞型
high idle 无功效的;高空载的
high impact 高冲力;高冲击(强度)
high impact acrylic mo(u)lding powder 耐冲击丙烯酸模塑料
high impact grade 高抗冲级
high impact interlayer 高耐冲击夹层;高抗穿透夹层
high impact plastics 耐冲击塑料
high impact polystyrene 高抗冲聚苯乙烯;耐冲性聚苯乙烯
high impact rayon 高(冲击)强度人造丝
high impact-velocity forging 高速锻造(法)
high impedance 高阻抗
high impedance cable 高阻抗电缆
high impedance coil 高阻抗线圈
high impedance electrode 高阻抗电极
high impedance nuclear magnetic resonance spectrometer 高阻抗核磁共振波谱仪
high impedance photoresistor 高阻抗光敏电阻
high impedance preamplifier 高阻抗前置放大器
high impedance relay 高阻抗继电器
high impedance rotor 高阻抗转子
high impedance sampling oscilloscope 高阻取样示波器
high impedance shunt 高阻抗分流器
high impedance state 高阻(状)态
high impedance voltmeter 高阻伏特计
high incendiary explosive 高燃炸药
high inclination satellite 高倾角卫星
high index 高指数
high index coupling medium 高折射率耦合媒质
high index glass rod 高指数玻璃棒
high index material 高折射率材料
high index of refraction 高折射率
high indicator 高值指示符
high induction furnace 高感应电炉
high industrial block(of flats) 预制高层公寓;工业化高层公寓
high inerts coal 高灰分煤;地劣质煤
high information delta modulation 高信息增量调制
high initial Portland cement 早强波特兰水泥
high initial strength Portland cement 早强(硅酸盐)水泥;早强波特兰水泥
high input impedance 高输入阻抗
high input-impedance stage 高阻抗输入级
high insulating property 高绝缘性能
high integrity casting 高致密铸件
high intensity 高声强;高烈度;强烈
high-intensity approach lights 高亮进场照明
high-intensity arc 高强电弧;强光电弧
high-intensity arc lamp 高强度弧光灯
high-intensity atomizer 高强度雾化器

high-intensity carbon arc 高强度炭弧灯
high-intensity carbon arc lamp 高强度炭弧灯;高光强炭弧灯
high-intensity cathode-ray tube 高亮度阴极射线管
high-intensity cold light supply 高强度冷光源
high-intensity combustion 强烈燃烧
high-intensity cone aerator 高强度锥形曝气器
high-intensity current 高强度电流
high-intensity discharge lamp 高强度放电电源;高强度放电灯;钠光灯;弧光灯;高压金属卤化物灯;高压汞灯
high-intensity emission 高强度发射
high-intensity lamp 高强度灯
high-intensity light 高强光
high-intensity light source 高强度光源
high-intensity magnetic field 高强度磁场
high-intensity microphone 高声强传声器
high-intensity mixer 高强度搅拌机
high-intensity noise 高强度噪声
high-intensity projector lamp 强光投射灯
high-intensity pulsed laser illumination 高强度脉冲激光照明
high-intensity rainfall 高强度降雨
high-intensity retro-reflect sheeting 高亮度后向反射板栅
high-intensity separation 高强度分选
high-intensity sound room 高声强实验室
high-intensity turbulent flame 强紊流火焰
high-intensity wastewater treatment 高强度废水处理
high interest 高息;高利
high interest rate 高利率
high interfacial tension 高界面张力
high interpretability 可判程度高
high inversion fog 高逆温雾
high investment 高额投资
high iron cement 高铁水泥
high iron metal heat 大量生铁熔铁
high iron Portland cement 高铁硅酸盐水泥;高铁波特兰水泥
high irradiance 高辐照度
high irrigated emulsion 高含水量乳化液
high island 高岛
high joint (混凝土路面因冻胀而形成的)梯级状抬高;凸缝
high-joint pointing 原浆勾缝;凸勾缝
high key 高调;亮色调;亮色调图象调节键
high key image 高调图像
high key light(ing) 强光照明;高调照明
high killing rate 高度杀伤率
high kilovoltage radiography 高千伏 X 射线照相术
high knock rating gasoline 高辛烷值汽油
high lag thermometer 高滞后温度表
high land 高原;高地(的);山地
highland basalt 高地玄武岩
highland boundary fault 高地边界断层
highland climate 山岳气候
highland ecosystems 高原生态系统
highland glacier 高地冰川
highland ice 高地冰川;山地冰
highland lake 高原湖沼;高地湖泊
highland moor 高原湖沼
highland range 高地山脉
highland rock 高地岩石
high lanthanum rare earth chloride 高富镧氧化稀土
high lateral order 高侧序
high lateral wall 高侧墙
high latitude 高纬度
high latitude climate 高纬度气候
high latitude region 高纬度地区
high layer protocol interworking 高层协议互通
high lead-antimony 高铅锑
high lead bearing alloy 高铅轴承合金
high lead bronze 高铅青铜
high lead cable logging 半拖式钢索集材
high lead crystal glass 高铅晶质玻璃;全高铝质玻璃
high leaded brass 高铅黄铜
high leaded brass alloy 高铅黄铜合金
high lead skidding 半拖式集材
high leakage transformer 高泄漏变压器;高漏泄变压器
high least 高级的

high level 高平地;高(能)级;高级(的);高电平
high-level anti-cyclone 高空反气旋
high-level bridge 高标准的桥;高架桥
high-level cave 高水平热室;高放热室
high-level cell 高放射性物质工作室;高放热室;强放射性热室
high-level cistern 高(位)水箱;高位水槽
high-level cleaner-loader 高置式清理装载机
high-level cloud 高云
high-level command 高级命令
high-level compiler 高级编译程序
high-level computer 高级计算机
high-level condenser 高位冷凝器
high-level counting 高强度计数
high-level crossings 立体交叉;高架跨越
high-level cyclone 高空气旋
high-level data link 高级数据链路;高级数据链接
high-level data link control 高阶数据链路控制;高级数据链路控制规程;高级数据链接控制程序;高级数据链路控制程序
high-level data link control procedure 高级数据通信线路控制过程;高级数据链路控制规程;高级数据链接控制程序
high-level data link control protocol 高级数据链路控制规程
high-level deck 高桩平台;高架台面
high-level delivery tipping trailer 高位翻卸挂车
high-level flush toilet 高水箱抽水马桶
high-level forecasting 高空预报
high-level graphic language 高级图形语言
high-level graphics programming instruction 高级图形编程指令
high-level graphics software 高级绘图软件
high-level heat recovery 高位热回收
high-level housing 高层住房
high-level index 高级指数
high-level indicator 高级指示器
high-level instrumentation 高功率检测仪表;高功率测量仪器;强放射性测量仪器
high-level intake 浅孔式进水口
high-level interface 高级接口
high-level irradiator 高强度辐照器;强放射性辐照器
high-level job control language 高级作业控制语言
high-level language 高级语言
high-level limit indicator 高粒位极限指示器
high-level mechanical extract system 高位机械排烟系统
high-level micro-programming language 高级微程序设计语言
high-level mill 高液面排矿磨碎机
high-level modulation 高电平调制
high-level motorway 高级机动车道
high-level network 高级网络
high-level network service 高级网络服务
high-level nuclear waste 强放射性废料
high-level of available aluminum 高含量有效铅
high-level of cave water 洞穴高水位
high-level of financing 筹集高额资金
high-level of liquidity 清偿能力高
high-level of phosphorus 高含磷水平
high-level of solar radiation 高能级太阳辐射
high-level opening light 高级顶灯
high-level overflow gate 高位溢流门
high-level ozone 高空臭氧
high-level platform 高承台
high-level platform supported bearing piles 高桩码头
high-level platform supported on pile 高桩承台
high-level process control language 高级过程控制语言
high-level profile 高标高断面
high-level programming interface 高级编程接口
high-level programming language 高级程序设计语言
high-level protocol 高级协议
high-level radiation 高能级辐射
high-level radio 高强度放射
high-level radioactive liquid waste 高品位放射性废液
high-level radioactive waste 强密度放射性废水;高含量放射性废水
high-level radioactive waste 强放射性废物
high-level radioactive wastewater 高放射性废水
high-level railway bridge 高位铁路桥梁

high-level relief 高能位释放
high-level reservoir 高位水库
high-level ridge 高空脊
high-level ring interchange 高标准环形交叉路
high-level round-about 高位迂回路线;高位环形交叉;高标准环形交叉路
high-level scheduling 高级调度
high-level slewing tower crane 上回转塔式起重机
high levels of phosphorus 高磷水平
high-level software 高级软件
high-level storage bin 高位储[贮]仓;高标准储[贮]料仓
high-level switch 高液位开关
high-level system 高程系统
high-level talks 高级谈判
high-level tank 高位罐;高位(水)箱;高水槽;高架水塔;高架水塔;高架水柜
high-level testing 高级静水压力试验(英国);屈服试验
high-level track 高线路
high-level trough 高空槽
high-level waste 高放射性强度废物;高放废物;强放射性废物
high-level waste container 高放射性废物容器
high lever granite 高位花岗岩
high lift 高扬程;高块(混凝土浇筑层);叉式装卸机;叉车;高举;大升程;铲运机
high lift blooming mill 大升程初轧机
high lift boom 接长伸臂
high lift concrete construction method 混凝土高块浇筑法
high lift concreting 高块浇筑
high lift construction for mass concrete 高块施工
high lift dam 高水头坝
high lift fork stacking truck 叉式升降堆积车
high lift grouting 高层段灌浆法;高升程灌浆
high lift lateral and front stacking truck 高货位三向叉车
high lift lock 高水头船闸;高水级船闸
high lift orderpicker 高货位拣选车
high lift placing 高块浇筑
high lift pump 高扬程(水)泵;高压泵;二级泵
high lift pumping station 高扬程抽水站;二级泵站
high lift shiplift 高水头升船机
high lift slabbing mill 大升程板坯初轧机
high-lift track 高举式起重机
high lift truck 高举升车
high lift valve 高升程阀
highlight 高亮度;高度光泽;著部分;要点;重点;高光(绘画或摄影中物体的最亮部分);辉亮部分;强光;强调;图像最亮部分;突出
highlight area 高光区域;明亮部分
highlight brightness 强光(激)亮度;最大亮度
highlight contrast 高光反差
highlight control 反差检查
highlight details 高光层次
highlight dot 高光点子
highlight equalizer 高亮度均衡器
highlight exposure 高光曝光;亮部曝光;强光曝光
highlight flux 强光通量
highlight gradation 高调层次
highlighting 加光;光线最强的;浅色处理面;精彩纹理面
highlight intensity 强光强度
highlight laser 强光激光器
highlight mask 高光蒙片
highlight retouching 高光修版
highlight signal 最亮信号
highlight test 强光试验
highlight-to-lowlight ratio 反差系数
highlight window 高窗
high lime 高石灰含量
high lime ash 高钙粉煤灰
high lime clinker 高钙熟料
high limed slag 高钙矿渣
high lime treatment 高氧化钙处理
high limit 上限
high limit control 高限控制
high limit of size 上限尺寸
high limit of tolerance 容许上限;上限公差;上差
high limit of water stable area 水稳定场上限
high limit register 上限寄存器
high limit relay 高限压力中断器;高精度减压阀
high line 高压(电)线;高压电缆;公用动力线;架

空索道;架空线;索道高压线;天线
high line conduit 高线引水道;高线引水渠(道);高线管渠;无压管道
high line eliminator 高压线干扰消除器
high line logging 高缆索运木法
high link 高链节
high liquid-gas ratio 高液气比
high liquid level 高液面
high liquid limit clay 高液限黏土
high liquid limit soil 高液限土
high living unit 高层公寓单元
high load 高荷载
high loaded process 高负荷运转
high load factor 高负荷因素
high load-factor operation 电站在高负载系数下工作
high loading 高填充(剂用)量
high-loading aeration method 高负荷曝气法
high-loading algae pond 高负荷藻塘
high-loading biological filter 高负荷生物滤池
high load-lifting speed 起升高速度【机】
high load lowering speed 下降高速度【机】
high load pot bearings 大荷载盆式支座
high localized stress 局部高应力
high located peat 高位泥炭
high loft fabric 高蓬松度织物
high loss fibre 高损耗光纤
high loss polyiron material 高(损)耗铁粉材料
high low 高低桩;多层桩
high low bars 高低杠
high low bay roof system 高低跨屋顶体系
high low bias check 高低偏压校验
high low bias test 高低压偏值测试;高低限测定;高低偏压检测;边缘校验;边缘检查
high low circuit 高低压回路
high lower water 高低潮位
high low extending/retracting speed of shuttle 货叉伸缩高(低)速度
high low firing appliance 高低两速点火装置
high low flow alarm device 高低流量警报装置
high low frequency connector with mixed contact 高低频混装接插件
high low generator 高低频发电机;变速定压发电机
high low graph 高低图;顶端—深底图形
high low horizontal travel(l)ing speed 水平运行高低速度【机】
high low ice level control 高低冰位控制器
high low index 高低指数
high low inversion 高低温型转化
high low lamp 变光灯泡;变烛电灯
high low level bin signal 高低仓位指示信号器
high low level control 高低水位调节器;双位电平调节器
high low lever 高低液位面(指控制器上);高低挡手柄
high low limit 高低限值;上下极限
high low method 高低点法
high low method of cost estimation 成本估算的高低法
high low pad 高低垫
high low points method 高低点法
high low pulper 高低速双转盘水力碎浆机
high low pump 高低压泵
high low rail 高低轨
high low-range switch 高低量程转换开关
high low safety system 高低压安全切断系统
high low signal alarm 高低位报警
high low signal selector 高低信号选择器
high low technique 高低法
high low temperature spectrum 高低温光谱
high low temperature tester 高低温试验机
high low tide 高低潮位
high low transformation 高低温型转变
high low voltage relay 高低压继电器
high low voltmeter 高低压(警报)电压表
high low walled batch drier 高低壁式分批
high low water alarm 高低水位报警器
high luminance colo(u)r 高亮度色;亮色
high luminosity spectrometer 高亮度分光计
high luminosity star 高光度星
high luster 强光泽
high luster polishing 高光泽抛光
high lustre coating 高光泽涂层
highly aberrated laser beam 高像差激光束

highly abrasion-proof 高度耐磨的
highly acid 高酸的
highly acidic 高度酸性
highly acid slag 高酸渣
highly active 高效
highly amorphous cellulose structure 高度非晶态纤维素结构
highly basic 高碱的
highly basis slag 高碱矿渣
highly capitalized 大量投入资本的
highly charged particle 多电荷粒子
highly coherent beam
highly collapsible 强烈湿陷
highly collimated beam 高准直光束
highly colo(u)red river water 浓色河水
highly combustible 高度可燃的
highly combustible construction 高度易燃性构造
highly complicated 高度复杂的
highly compressed 高度受压的;高压的;高度压缩的
highly compressed steam 高压蒸汽
highly concentrated activated sludge process 高浓度活性污泥法
highly concentrated ammoniacal wastewater 高浓度白色废水
highly concentrated ammonial sewage 高浓度氨污水
highly concentrated ammonia-nitrogen 高浓度氨氮
highly concentrated ammonia-nitrogen industrial wastewater 高浓度氨氮工业废水
highly concentrated dye intermediate wastewater 高浓度染料中间体废水
highly concentrated dye wastewater 高浓度染料废水
highly concentrated lipidic wastewater 高浓度油脂废水
highly concentrated nitrogen-containing starch wastewater 含高浓度氮淀粉废水
highly concentrated oil-containing wastewater 含高浓度油废水
highly concentrated phenol wastewater 含高浓度酚废水
highly concentrated printing and dyeing wastewater 高浓度印染废水
highly concentrated refractory organic wastewater 高浓度难降解有机废水
highly concentrated saline water 高浓度咸水
highly concentrated salt-containing dyestuff wastewater 含高浓度盐染料废水
highly concentrated silt discharge 高含沙洪流
highly concentrated solution 高浓度溶液
highly concentrated tetrazoline production wastewater 高浓度四氮唑生产废水
highly concentrated wastewater 高浓度(化工)废水
highly concentrated wool scouring wastewater 高浓度洗毛废水
highly conductive fluid 高导电流体
highly conductive synthetic (al) fiber 高导电性合成纤维
highly contaminated wastewater 高污染废水
highly convergent series 高度收敛级数
highly coupled 高度耦合的
highly damped instrument 高阻尼仪表
highly dangerous cargo 烈性危险货
highly decorated 装饰华丽的
highly detergent oil 高级去垢油
highly deviated 高度偏离
highly diluted 高稀释
highly diluted principle 高度稀释原理
highly directional antenna 强方向性天线
highly directional device 高精度定向装置
highly directional system 强方向性系统
highly dispersed 高度色散(的);高度弥散的
highly dispersed system 高级分散体系
highly dispersivity 高分散性
highly doped 高度掺杂
highly ductility 韧性高
highly eccentric orbit satellite 高偏心轨道卫星
highly effective 高效的
highly effective anti-foaming agent 高效消泡剂
highly effective anti-mo(u)ld 高效防霉剂
highly effective complex microorganism 高效复

合微生物菌群
highly effective compound fluidized bed bio-reactor 高效复合流化床生物反应器
highly effective degrading bacteria 高效降解菌
highly effective degrading bacteria flora 高效降解细菌菌群
highly effective distribution 高效分布
highly effective flame retardant synergist 高效阻燃增效剂
highly effective levelling agent 高效匀染剂
highly effective scouring agent 高效炼助剂
highly effective scouring penetrant 高效精炼渗透剂
highly effective strain 高效菌株
highly efficient 高效率的
highly efficient 4-stroke low-speed diesel engine 高效四冲程低速柴油机
highly efficient biological aerated filter 高效曝气生物滤池
highly efficient diffuser 高效率扩压器
highly efficient equipment 高效率设备
highly efficient regeneration 高效再生
highly elastic 高弹性的;富有弹性的
highly elastic alloy wire 高弹性合金线
highly electronegative element 高电负元素
highly enriched uranium 高浓缩铀
highly faired hull 流线型船身
highly finished 满膛
highly fire-resisting 高度耐火的
highly fire-resisting component 高度耐火构件
highly fissured 高度裂隙化的;裂隙(非常)发育的
highly flammable material 高度易燃性材料
highly flexible 高度挠性的
highly fluorescing waters 强荧光水域
highly folded structure 褶皱强裂的构造;强褶皱构造
highly fractured region 强破裂区
highly gassy mine 高瓦斯矿井
highly hydrated lime 高度水化石灰
high lying resonance 高位共振(能级)
highly intensified corrugated paper 高强度瓦楞纸
highly ionized noble gas 高电离惰性气体
highly jointed 裂隙(非常)发育的;节理发育的
highly liquid asset 高度流动性资产
highly loaded diffuser 高负荷扩散器
highly matched 高度匹配
highly mineralized geothermal brine 高矿化地热卤水
highly mineralized water treatment 高矿化度处理
highly olefinic gasoline 富烯烃汽油
highly organic soil 有机质含量高的土;富有机质土
highly oriented fiber 高(度)定向纤维
highly oriented pyrolytic graphite 高定向热解石墨
highly oscillatory mode 强振荡形态
highly permeable ground 高渗透性土层
highly pigmented 色彩浓艳的
highly plastic 高塑性的
highly plastic clay 高塑(性)黏土
highly plastic soil 高塑性土
highly polarized light 高度偏振光
highly polished 高度抛光的
highly polluted 高度污染
highly porous 高度疏松的;高孔隙度的
highly pressed 受高压的;高度压缩的
highly prospective area 富有远景的地区
highly pure indium antimonide single crystal 高纯锑化铟单晶
highly purified gas purifying equipment 高纯气体净化设备
highly qualified specialist 高水平专家
highly radiating gas 强辐射气体
highly random distribution 高度无规分布
highly reactive hydrocarbons 高度活性碳氢化合物
highly refined 高度精制的
highly refined oil 高度精制油
highly reflecting film 高反射膜
highly reflective foil 高反射金属薄片
highly refracted ray 高折射光线
highly refractory 高度耐火的
highly reliable and intelligent ship 高可靠性智能化船舶
highly relief 强烈起伏
highly resistant calcium silicate brick 高抗性硅酸盐砖

highly resistant compass engineering brick 高抗性弧形(工程)砖
highly resistant compass sand-lime 高抗性弧形灰砂砖
highly resistant engineering brick 高抗性工程砖
highly resistant material 高阻材料
highly resistant radial engineering brick 高抗性扇形砖;高抗性径向(工程)砖
highly resistant radial lime sand brick 高抗性扇形灰砂砖;高抗性径向灰砂砖
highly resonating 高共振的
highly seismic zone 高烈度区
highly selective resin 高选择性树脂
highly sensitive 高度灵敏的
highly shock resistant 高度抗震的
highly significant 高度显著
highly skewed distribution 高度偏斜分布
highly skilled worker 高度熟料的技工
highly specialized 高度专业化的
highly specialized personnel 高级专业人才
highly statistical significance 高度统计显著性
highly stressed 高度受力的
highly stressed zone 高应力区
highly swing 高度摆动
highly symmetric(al)graph 高度对称图
highly toxic 高毒性的
highly toxic compound 高毒性物
highly toxic material 高毒性物质;剧毒材料
highly toxic substance 高毒性物质
highly turbid 重浊
highly variable chimeric 高度易变的嵌合体
highly viscous oil 高黏石油
highly viscous sand-laden fluid 高黏携砂液
highly volatile 易挥发的
highly volatile liquid 高度挥发性液体
highly volatile product 高挥发性油品
highly volumetric loading 高容积负荷
highly wear-resistant 高度耐磨的
highly weathered 强风化的
highly weathered rock 强风化岩石
highly weathered soils 高度风化土壤
highly weathered stratum 强风化层
highly weather resistant 高度耐老化的;高度防风雨的
highly wrought goods 精加工制成品
highly yielding 高产性的
highly yielding ability 高产性能
high Mach number engine 高马赫数发动机
high magnesia cement 高镁水泥
high magnesian aragonite 高镁文石
high magnesian calcite 高镁方解石
high magnesia Portland cement 高镁硅酸盐水泥;高镁波特兰水泥
high magnesium lime 镁质石灰;高镁石灰
high magnetic arcade 高磁拱
high magnetic field intensity 高磁场强度
high magnetic metal rod 强磁金属棒
high magnification 高倍放大
high magnification seismograph 高倍率地震计
high magnified glass 高倍放大镜
high magnified image 高倍放大影像
high maintainability 高维修性
high manganese carbon steel 高锰碳钢
high manganese cast steel 高锰铸钢
high manganese pig iron 高锰生铁
high manganese stainless steel 高锰不锈钢
high manganese steel 高锰钢
high manganese steel casting 高锰钢铸件
high manganese steel frog 高锰钢辙叉
high manganese steel rail and guard rail 高锰钢辙叉及护轨
high manganese zinc alloy 高锰锌合金
high margined item 高利产品;高贡献边际产品
high margin product 高利产品
high mark cement 高强度等级水泥
high mast lighting 高杆(式)照明
high mast lighting system 高杆灯
high meander belt deposit 高蛇曲带沉积
high mechanical 高度机械化的
high mechanical construction 高度机械化施工
high mechanization 高度机械化
high medium horizontal travel(l)ing speed 水平运行高中速度【机】
high medium mountain 高中山(绝对高度3500~1000米)
high megohmmeter 超绝缘测试仪;超高阻表
high megohm resistance comparator 高阻比较器;超高阻比较器
high melt(ing) 难熔的;高温熔炼
high melting casting metal investment 高熔铸金包埋料
high melting fiber 高熔点纤维
high melting glass 难熔玻璃
high melting glaze 高温釉
high melting metal 高熔金属
high melting plastics 高熔点塑料
high melting point 高熔点
high melting point alloy 高熔点合金
high melting point asphalt 高熔点沥青
high melting point metal 高熔点金属
high melting point refractory 高熔点耐火材料
high melting wax 高熔点蜡
high memory area 高端内存区
high methane gas 富甲烷气
high methane mine 高沼气矿井
high mica 高级云母
high milling 精磨
high mineralization 高矿化(作用);矿化度大
high mirror finishing sheet 镜面薄板
high mixture ratio pneumatic conveyer 高浓度气力输送机
high mobility 高迁移率
high mode coupling 高次波型耦合
high mode vibration 高频振动
high modification-ratio 高变性比
high modulus 高模量
high modulus carbon fiber 高模量碳纤维
high modulus composite sheet pile 高模量组合板桩
high modulus fabric 高模量织物
high modulus fiber 高模数纤维;高模量纤维
high modulus furnace black 高模数炉黑
high modulus glass 高模量玻璃;高弹玻璃
high modulus glass fiber 高模量玻璃纤维
high modulus glass fiber 高模数玻璃纤维
high modulus inclusion 高模量插入体;高模量包体
high modulus polyester yarns 高模数聚酯纱
high modulus polymer fiber 高模量聚合物纤维
high modulus polyvinyl alcohol fiber 高模量聚乙烯醇纤维
high moisture and condensation test 高湿度结露试验
high moisture content 高含水量
high moisture grain 高湿度谷粒
high molecular 高分子的
high-molecular coagulant 高分子凝结剂
high molecular compound 高分子化合物
high molecular flocculant 高分子吸附剂
high molecular oxiacid 高分子含氧酸
high molecular polymer 高分子聚合物;高分子聚合体
high molecular surfactant 高分子表面活性剂
high molecular synthetic(al)material 高分子合成材料
high molecular waste 高分子废物
highmolecular weight 高分子量
high molecular weight copolymer 高分子量共聚物
high molecular weight demulsifier 高分子破乳剂
high molecular weight high density polyethylene 高分子量高密度聚乙烯
high molecular weight hydrocarbon 高分子量烃
high molecular weight material 高分子量物质
high money 高空作业费(英国)
high moor(bag) 藓类沼泽;高地沼泽;高位沼泽;高位湿原;高地酸泽;贫养沼源;贫氧酸沼地;高沼地
highmoor peat 高沼泥炭
high mortality 高死亡率
high moss 高地沼泽
high mountain 高山(绝对高度3500~5000米)
high mountain high plateau desert soil zone 高山高原荒漠土带
high mountain meadow soil and timber soil area 高山草甸土及草原土区
high mountainous karst 高山岩溶
high mountainous terrain 高山地
high mountain station 高山测站
high mountain timber soil illimerized soil zone 高山草原土灰化土带
high mountain underground water area 高山潜水区
high multifilament yarn 多纤维复丝
high myopia 高度近视
high natural frequency accelerograph 高固有频率自记加速计
high necked 高领的
high negative correlation 高度负相关
high nickel 高镍
high nickel alloy 高镍合金
high nickel-chrome steel 高镍铬钢
high nickel-copper-tin alloys 高镍铜锡合金
high nickel iron-base alloy 高镍铁基合金
high nickel steel 高镍(合金)钢
high nitrate high-salinity wastewater 高硝酸盐高盐度废水
high nitrate wastewater 高硝酸盐废水
high nitrogen 高氮量
high nitrogen oil 高氮石油
high nitrogen pressure sintering 高氮压烧结
high noise immunity logic 高抗扰度逻辑
high noise immunity logic interface device 高抗扰度接口器件
high noon 正午
high northern latitude 北半球高纬度
high note buzzer 高音(调)蜂鸣器
high nutrient 高营养
high nutrient sludge and compost 高营养污泥与堆肥
high oblique 高倾斜的;深倾
high oblique air survey 深倾航空摄影测量
high oblique convergent photograph 深倾交向摄影像片
high oblique photograph 大角度倾斜的航空照片
high oblique photography 深倾摄影
high observatory 高觇标
high-occupancy vehicle 高载客量车辆;多乘客车辆
high-occupancy vehicle lane 高容量车辆车道;合用车专用车道;多乘客车辆专用车道
high octane 高辛烷值(的)
high octane gasoline 高辛烷值汽油
high octane number component 高辛烷值组分
high octane rating 高辛烷值
high of reservoir level 热储的水位高度
high ohmic resistance 高欧姆电阻
high ohmic resistance alloy 高欧姆电阻合金
high oil 高油量
high-oil-content circuit breaker 多油式断路器
high oil-level tank 高位油罐
high oil pressure cable 高油压电缆
high oil recovery 高出油率
high oil temperature switch 高油温开关
high-opacity foils 不透明金属箔
high open circuit voltage 高开路电压
high opening trawl 高网口拖网
high operand 上界操作数
high order 高位;高序;高次
high order aberration 高次像差
high order add circuit 高阶加法电路
high order approximation 高阶近似
high order assembler 高阶组装器
high order bit 高部位
high order byte 高位二进制数组
high order column 高位列
high order component 高次分量
hIgh order derivative 高次导数
high order detonation 高级爆轰
high order digit 最高位数位;高位数;高数位;高阶数字;高部位
high ordered equation 高次方程
high order effect 高级效应
high order elasticity 高阶弹性力学
high order elastic model 高阶弹性模型
high order elastic modulus 高阶弹性模量
high order end 最左端;高位端
high order fission product 高次裂变产物
high order focusing 高阶聚焦;高次聚焦
high order focusing spectrometer 强聚焦分光仪
high order function 高阶函数;高级功能
high order goal 高阶目的
high order grating 高次光栅
high order interface 高阶接口
high order interpolation 高次差内插;非线性内插

high order logic 高阶逻辑
high order loop 高阶环路
high order manufacturing 高级工业
high order merge 高阶合并
high order mode 高阶模
high order mode filter 高次模滤除器
high order moment 高次多项式
high order multiples 高阶多次反射波
high order multipole radiation 高次多极辐射
high order oscillation 高次振荡模
high order parabola 高阶抛物线
high order path adaptor 高阶通道适配器
high order path connection 高阶通道连接
high order path overhead monitoring 高阶通道开销监视
high order path termination 高阶通道终端
high order path unequipped generator 高阶通道未装载发生器
high order(position) 高(数)位；高阶位
high order reservoir 高位水库
high order term 高次项
high order transverse mode 高阶横波型；高次横向波型；高次横模
high order transverse wave 高阶横波
high order virtual container 高阶虚容器
high order white 高级白
high organic loading rate 高有机负荷率
high output 高出力；高产量；丰产
high output fluorescent lamp 大功率荧光灯；高输出日光灯
high output gas generator 高输出燃气发生器
high output laser system 高输
high output machine 高产量机械；高产机
high output three-phase induction motor 高功率三相感应电动机
high overcast 高云密布；高密度云天空
high overvoltage anode 高超压阳极
high oxide 高价氧化物
high oxygen consumption water 耗氧量高的水
high ozone-depletion potential 高臭氧消耗潜能值
high parallel light rays 高度平行光
high parapet 高女儿墙
high partition 高隔墙
high pass 高通
high pass and low-pass filter 高低通滤波器
high pass filter 高通滤光片；高通滤波器
high pass filtering 高通滤波
high pass main frequency 高通主频
high pass network 高通网络
high pay 高空作业津贴；高空作业费(美国)
high peak current 高峰(值)电流
high-peaked roof 陡坡屋顶
high peaker 高频增益提高电路；高频峰化器；高频补偿电路
high peaker circuit 高频峰化电路
high peaker test 高频峰化器测试
high peaking network 高频峰化网络
high peak power 高峰值功率
high peak power laser 高峰值功率激光器
high pedestal jib crane 门座式悬臂起重机
high penetration bitumen 高渗透性沥青
high penetration power 高穿透本领
high penetration resistance 高抗穿透性
high penetration resistance glass 高抗贯穿性玻璃
high percentage 百分率很高
high performance 性能优良；性能良好；优越性能；高准确度；高指标；高性能(的)；高效能；高效率
high performance airscrew 高效螺旋桨
high performance air separator 高效空气选粉机
high performance alloy 高导磁率合金
high performance anti-static fiber 高性能抗静电纤维
high performance ceramics 高性能陶瓷；高技术陶瓷
high performance characteristic 性能良好
high performance chromatograph 高效色谱仪
high performance chromatography 高效离子层析
high performance column 高性能柱
high performance computer system 超高性能计算机系统
high performance concrete 高性能混凝土
high performance engine 高性能发动机
high performance equipment 高性能设备；高效能设备

high performance horizontal machining centre 高性能卧式加工中心
high performance hydraulic pump 高性能液压泵
high performance ion chromatography 高效离子色谱法
high performance ion exclusion chromatography 高效离子排斥色谱法
high performance laser 高性能激光器
high performance lens 高性能透镜
high performance liquid chromatograph 高效液相色谱仪
high performance liquid chromatography 高效液相色谱(法)
high performance material 高性能材料
high performance memory 高性能存储器
high performance microwave amplifier 高效微波放大器
high performance navigation system 高性能导航系统
high performance objective 高性能物镜
high performance open back power press 后开式高性能压力机
high performance optic(al) cable connector 高性能光缆连接器
high performance piping 高技术铺管法
high performance propellant 高热值燃料
high performance protective coating 高性能防护涂层
high performance resin 高性能树脂
high performance resin matrix 高性能树脂基体
high performance reverse phase liquid chromatography 高效倒相液相色谱法；高效倒相液相层析
high performance sealant 高功能密封膏
high performance size exclusion chromatography 高效体积排阻色谱法；高效体积排阻层析
high performance superheteodyne receiver 高性能超外差式接收机
high performance tape 高性能纸带
high performance thin-layer chromatography 高效薄层色谱法；高效薄层层析法
high performance unit 高性能部件
high performance vane pump 高性能叶片泵
high performance vehicle 高性能漆料
high performance weld flux 高性能焊剂
high performing car 高性能汽车
high permeability alloy 高导磁率合金；高磁导合金
high permeability Co-Fe alloy 高导磁率钴铁合金
high permeability material 高磁导率材料
high permeability nickel-iron alloy 高导磁率镍铁合金
high permeability silicon alloy 高导磁率硅合金
high permeability zone 高渗透层
high persistent pesticide 高残留农药
high pH mud 高碱度泥浆
high phosphor brake shoe 高磷闸瓦
high phosphorous(cast)iron 高磷铸铁
high phosphorous(pig)iron 高磷生铁
high phosphorus coal 高磷煤
high phosphorus concentration 高磷浓度
high pH value cement 高碱度水泥
high pigment level 高颜料浓度；高颜料含量
high piled combustible material 高堆可燃性材料
high piled jetty 高桩码头
high piled pier 高桩码头
high piled storage 高堆储[贮]藏物
high pile(d) wharf 高桩码头
high pile feeder 高纸堆自动给纸机
high pitch 高音调；陡坡
high-pitched 坡度陡的；高坡的(屋顶等)；急坡的；急倾斜的；屋顶坡度陡的
high-pitched alternating current 高声频交流电
high-pitched buzzer 尖声蜂鸣器；高频蜂鸣器；高音气笛
high-pitched roof 陡屋面；陡坡屋顶；高坡屋顶
high-pitched tone 高音调
high pitch regulator 高音调节器
high place 高殿
high-placed window 高窗
high plain 高平原
high planting 堆植法
high plant population 密植
high plastic clay soil 高塑性黏质土
high plasticity 高塑性

high plasticity soil 高塑性土
high plastic limit 高塑限
high plastic silt 高塑性粉土
high plateau desert soil area 高原荒漠土区
high plateau zonation 高原地带性
high platform 高(站)台
high platform building 高台建筑
high point of gutter 天沟分水线
high point of structure 构造高点
high point of the slope 坡的高端
high polar glacier 高极地冰川
high-pole 大圆材
high polish 高度抛光；高度磨光
high polymer 高聚物；高(分子)聚合物；高分子；大分子聚合物
high polymer anti-wear strip 高分子抗磨导轨软带
high polymer chemistry 高分子(聚合)化学
high polymer coagulant 高分子絮凝剂
high polymer electrolyte 高聚物电解质
high polymeric compound 高聚化合物
high polymerised pulp 高聚合度浆粕
high polymer microphone 高聚物扬声器；高聚(合)物传声器
high polymer piezoelectric(al) material 高分子压电材料
high polymer synthesis industry 高分子合成工业
high polymer waste 高聚物废物
high pool level 高库水面
high population capacity 高人口容量
high population density 高人口密度
high porosity gel-state fiber 高孔隙率凝胶态纤维
high porosity grinding wheel 大气孔砂轮
high porosity media 高孔隙率填料
high portal-framed crane 高架门(式起重)机
high Portland 多水泥的
high Portland cement concrete 富混凝土；硅酸盐水泥用量的混凝土
high positioned defect 高位缺损
high-positive correlation 高度正相关
high-positive indicator 高正指示器；正值指示字符；大于零指示位
high potency 高效能
high potential 高势能；高电位
high potential test 高压试验【电】
high potential test for electric(al) equipment 电气设备高压试验
high potting 高压试验
high pour-point crude 高倾点原油
high power 高功率(的)；高倍；强功率；大功率
high-power acquisition radar 大功率探测雷达
high-power amplifier 高功率放大器；大功率放大器
high-power binoculars 高倍双筒望远镜
high-power biological microscope 高倍生物显微镜
high-power bolometer 大功率测辐射热计
high-power cable 大功率电缆
high-power capacity 大功率电容；高容量
high-power circulator 大功率环行器
high-power connector 大功率接插件
high-power density 大功率密度
high-power drilling machine 高功率钻床
high-powered drill 大功率电力驱动
high-powered engine 大功率发动机
high-powered equipment 大功率设备
high-powered low-speed marine diesel engine 大功率低速船用柴油机
high-powered money 高效货币；强力货币
high-powered vessel 大马力船舶
high-power electric(al) capstan 大功率电动绞盘
high-power engineering 大功率工程
high-power explosive 高强度炸药
high-power factor 高功率因素
high-power field 高倍视野
high-power forward-wave tube transmitter 大功率前向波管发射机
high-power gas laser 高功率气体激光器
high-power gear reducer 大功率减速机
high-power generator 大功率发电机
high-power graphite electrode 高功率石墨电极
high-power homing beacon 高功率归航信标台
high-power laser 高功率激光器；大功率激光器
high-power laser bonding 高功率激光焊接
high-power lens 高功率透镜

high-power lithium battery 高能锂电池
high-power magnetic relay 高功率电磁继电器
high-power magnetron 大功率磁控管
high-power microscope 高倍显微镜
high-power microwave assembly 高功率微波装置
high-power modulation 高功率调制
high-power objective 高倍率镜头;高倍(接)物镜
high-power objective 强光物镜
high-power object lens 强光力物镜
high-power ocular 高倍目镜
high-power operation 高功率操作;全负荷运行
high-power optic(al)fiber 大功率光纤
high-power parabola 高次抛物线
high-power periscope 高倍潜望镜
high-power phenomenon 高功率现象
high-power photoobjective 高倍照相物镜
high-power projector lamp 强光投射灯
high-power pulse 强脉冲
high-power pulsed klystron 大功率脉冲速调管
high-power pulser 大功率脉冲发生器
high-power radio station 强功率无线电台
high-power relay 大功率继电器
high-power screw press 高能螺旋压力机
high-power semi-conductor element 大功率半导体元件
high-power sensitivity 高功率灵敏度
high-power station 大电厂
high-power stereoviewer 高倍立体观测镜
high-power sweep generator 大功率扫频信号发生器
high-power switching device 大功率转换设备
high-power telescope 高倍率望远镜
high-power term 高幂项
high-power transistor 大功率晶体管
high-power transmitter 大功率发射机
high-power transmitter energy 大功率发射能
high-power travel(l)ing wave tube 大功率行波管
high-power valve 大功率管
high precedence 高优等级
high precise current source 高精度恒流源
high precision 高精(密)度;高度精密
high precision AC servo number-control mechanical slide unit 高精度交流伺服数控机械滑台
high precision aeromagnetic survey 高精度航空磁测
high precision and multifunction speed meter 高精度多功能转速计
high precision arithmetic 高精度运算
high precision centerless grinder 高精度无心磨床
high precision collet chuck 高精度弹簧夹头
high precision components 高精密度部件
high precision computation 高精度计算
high precision coordinate measuring instrument 高精度坐标测量机
high precision cutting tool 高精密度切削工具
high precision cylindrical grinder 高精度外圆磨床
high precision depth recorder 高精度深度记录仪
high precision digital multimeter 高精度数字式万用表
high precision face grinder 高精度平面磨床
high precision gear 高精度齿轮
high precision gear hobber 高精度滚齿机
high precision geodesy 高精度大地测量学
high precision granite surface plate 高精度花岗岩平台
high precision grinding machine 高精度磨床
high precision hard alloy cutter grinder 高精度硬质合金工具磨床
high precision harmonic micrometer 高精度谐波测微仪
high precision hob sharpening grinding machine 高精度滚刀刃磨床
high precision kelmet set 高精度油膜轴承组
high precision laser 高精度激光器
high precision laser installation 高精度激光装置
high precision lead-screw cutting lathe 高精度丝杠车床
high precision level(instrument)高精度水准仪
high precision level(l)ing 高精度水准测量;精密水准测量
high precision low frequency signal generator 高精度低频信号发生器
high precision master gear 高精度标准齿轮
high precision micro computer-controlled jig borer 高精度微机控制坐标镗床
high precision number-control boring head 高精度数控镗头
high precision oil filter 精密油滤
high precision optic(al)plate 高精度光学平板
high precision photogrammetry 高精度摄影测量学
high precision plastic connector 高精度塑料连接器
high precision probe and its system 高精度测头系统
high precision shoran 高精度肖兰(导航系统)
high precision short range equipment in electronic navigation 近程高精度无线电导航设备
high precision small-module horizontal gear hobber 高精度小模数卧式滚齿机
high precision stereometer 高精度立体量测仪
high precision sun sensor 高精度太阳传感器
high precision surface grinding machine 高精度平面磨床
high precision survey 高精度测量
high precision track scale 高精度轨道衡
high precision travel-control straightening press 高精密行程控制校直压力机
high precision tube 高精管
high precision universal lathe 高精度万能车床
high precision vertical circular sawing machine 高精度立式圆盘锯床
high premium 高度保险费
high pressed particle board 高压碎料板
high-pressrue connecting rod 高压连杆
high-pressure 高压力;高压(的)
high-pressure absorber 高压吸收器
high-pressure accumulator 高压蓄力器
high-pressure acetylene generator 高压乙炔发生器
high-pressure admission 高压进汽;高压进气
high-pressure air 高压空气;压缩空气
high-pressure air atomization 高压空气喷雾法
high-pressure air atomizer 高压空气喷雾器
high-pressure air compressor 高压空气压缩机
high-pressure air-conditioning system 高压空(气)调(节)系统
high-pressure air floating fender with pressure relief valve 附有或不带减压阀的高气压浮动式护舷
high-pressure airgun 高压空气枪
high-pressure airless spray 高压无(空)气喷涂
high-pressure airless spraying machine 高压无(空)气喷涂机
high-pressure air-outlet pipe 高压空气管
high-pressure air pipe 高压空气导管
high-pressure air rifle 高压气枪
high-pressure air system 高压(空)气系统
high-pressure ammonia receiver 高压储[贮]氨器
high-pressure and low pressure turbine 高低压汽轮机
high-pressure apparatus 高压设备
high-pressure aqueous environment 高压水环境
high-pressure arc technique 高压电弧技术
high-pressure area 反气旋;高压区域;高(气)压区
high-pressure asbestos rubber board 高压石棉橡胶板
high-pressure atomizer 高压雾化器
high-pressure autoclave 高压釜
high-pressure axial piston pump 高压轴向活塞泵
high-pressure baler 高压打捆机;高压打包机
high-pressure beck 高压绳状染槽
high-pressure bellows manometer 高压波纹管式压力计
high-pressure belt 高压带
high-pressure bicycle pump 高压气筒
high-pressure blast 高压风
high-pressure blower 高压鼓风机
high-pressure blowpipe 高压焊炬
high-pressure boiler 高压锅炉
high-pressure boiler-feed pump 高压锅炉给水泵
high-pressure boiler tube 高压锅炉管
high-pressure bottle 高压容器
high-pressure burner 高压烧嘴;高压燃烧器
high-pressure cage 高压罩
high-pressure caisson operation 高压沉箱作业
high-pressure capillary rheometer 高压毛细管流变仪
high-pressure casting 高压铸造
high-pressure catalytic unit 高压催化设备
high-pressure cell 高压室;高压腔;高压传感器
high-pressure cementation 高压黏结;高压胶(黏)结
high-pressure center 高压中心
high-pressure centrifugal blower 高压离心鼓风机
high-pressure centrifugal ventilator 高压离心通风机
high-pressure chamber 高压室
high-pressure charge 孔内爆破药包
high-pressure charging 高压装料;高压进料
high-pressure chemistry 高压化学
high-pressure cloud chamber 高压云室
high-pressure coal hydrogenation 煤高压加氢
high-pressure cock 高压水龙头;高压开关
high-pressure cold water 高压冷水
high-pressure combustion 高压燃烧
high-pressure combustion chamber 高压燃烧室
high-pressure combustion process 高压燃烧过程
high-pressure comminution 高压粉磨
high-pressure complex master and auxiliary vane pump 高压子母叶片泵
high-pressure compression 高压压实
high-pressure compressor 高压压缩机;高压压气机;高压压风机
high-pressure condenser 高压冷凝器
high-pressure connection 高压接头
high-pressure consolidometer 高压固结仪
high-pressure controller 高压控制器
high-pressure cooker 高压锅
high-pressure counter 高压计数管
high-pressure cross head 高压十字头
high-pressure crystal growth furnace 高压晶体生长炉
high-pressure cut-out 高压断路装置;高压保险装置;高压断路器
high-pressure cylinder 高压汽缸;高压气瓶
high-pressure desander 高压除砂器
high-pressure diamond 高压金刚石
high-pressure diaphragm pump 高压隔膜泵
high-pressure differential thermal analysis apparatus 高压差示热分析仪
high-pressure differential thermal analysis unit 高压差示热分析仪
high-pressure direct spinning technology 高压直接纺丝技术
high-pressure discharge 高压气体放电
high-pressure discharge hose 高压送水管
high-pressure discharge lamp 高压强放电灯
high-pressure distributor 高压分配器
high-pressure district 高压区
high-pressure double-cylinder oil pump 高压双缸油泵
high-pressure draft 高压气流
high-pressure draught 高压气流
high-pressure dredge pump 高压泥浆泵
high-pressure drilling 高压钻进
high-pressure drilling hose 高压钻探胶管
high-pressure dry gas flooding 高压干气驱
high-pressure duct 高压管路
high-pressure duct system 高压管路系统
high-pressure dy(e)ing 高压染色法
high-pressure eccentric 高压偏心轮
high-pressure economizer 高压节煤器
high-pressure electrode 高压电极
high-pressure electrolysis 高压电解
high-pressure engine 高压发动机
high-pressure erosion bit 高压冲蚀钻头
high-pressure evapo(u)ration 加压蒸发
high-pressure evapo(u)ration tower 高压蒸发塔
high-pressure extinction 高压灭火
high-pressure extraction cycle 高压回热循环
high-pressure extruding machine 高压涂药机
high-pressure fan 高速排气扇;高速风扇;高压风扇
high-pressure feed-water heater 高压给水加热器
high-pressure field 高压油田
high-pressure filter 高压失水仪;高压滤油器;高压过滤器
high-pressure fire service main 高压(水)消防总管
high-pressure fire system 高压消防系统;高压水消防系统
high-pressure fixed displacement vane pump 高压定量叶片泵

high-pressure flange 高压法兰
high-pressure float valve 高压浮球阀
high-pressure fluorescent mercury lamp 高压水银灯
high-pressure fog 高压喷雾
high-pressure fog extinction 高压喷雾灭火
high-pressure fog stream 高压喷雾射流
high-pressure forming 高压成形
high-pressure fuel injection pump assembly 高压油泵总成
high-pressure fuel pump 高压燃料泵
high-pressure fumarole 高压喷汽孔
high-pressure furnace 高压炉
high-pressure gas 压缩煤气;高压(气体)
high-pressure gas chromatography 高压气相色谱法
high-pressure gas container 蓄压器
high-pressure gas control law 高压气体管制法
high-pressure gas drive 高压气驱
high-pressure gas dynamic(al) laser 高压气动激光器
high-pressure gaseous source 高压气态源
high-pressure gasholder 高压储气罐
high-pressure gas injection 高压气体注射法;高压注气;高压进汽
high-pressure gas pipe 高压煤气管
high-pressure gas pipeline 高压煤气管线
high-pressure gas producer 高压煤气发生器;高压煤气发生炉
high-pressure gas system 高压煤气系统
high-pressure gate 高压闸门;高水头闸门
high-pressure ga(u)ge 高压(压力)表;高压计;排出压力表
high-pressure gear type pump 高压齿轮型泵
high-pressure generator 高压发生器
high-pressure grease gun 高压润滑油枪
high-pressure grease lubrication 高压油脂润滑
high-pressure grinding roll 辊压机;高压磨辊
high-pressure grouting 高压注浆;高压灌浆
high-pressure head 高压位差;高压头
high-pressure heater 高压加热器
high-pressure heating 高压供暖
high-pressure heating system 高压供暖系统
high-pressure helium gas scintillation counter 高压氦气闪烁计数器
high-pressure holder 高压堵气罐
high-pressure hole 高压钻孔
high-pressure horizontal autoclave 高压卧式消毒器
high-pressure hose 高压龙头;高压软管;高压胶管
high-pressure hose line 高压软管线(路);高压蛇形管线
high-pressure hose reel 高压暖管卷盘
high-pressure hot water heating 高压热水供暖
high-pressure hot water heating system 高压热水供暖系统
high-pressure housing 高压传送器
high-pressure hydraulic 高压液压系统
high-pressure hydraulic fire system 高压水力灭火系统
high-pressure hydraulic gear pump 高压液压齿轮泵
high-pressure hydraulic hand pump 高压液压手泵
high-pressure hydraulic jet machine 高压水射流器
high-pressure hydraulic pump 高压液压泵;高压水泵
high-pressure hydraulic system 高液压系统
high-pressure hydrogenation 高压氢化;高压加氢
high-pressure incandescent lamp 高压白炽灯
high-pressure inflater 高压气筒
high-pressure injection 高压喷水
high-pressure injection method 高压注浆法
high-pressure injection pile 高压旋喷桩
high-pressure injection pump 喷射(油)泵;高压注射泵;高压射流泵
high-pressure injector 高压注射器
high-pressure inlet 高压进水口;高压进气口
high-pressure installation 高压装置
high-pressure intake 高压入口;高压进水设备;高压进水口;高压进气口
high-pressure ion exchange 高压离子交换
high-pressure ionization chamber 高压电离室
high-pressure ionization ga(u)ge 高压强真空计;高压力电离真空计;高压电离规

high-pressure ionization vacuum ga(u)ge 高压电离真空规
high-pressure ionizer-sustainer laser 高压电离维持激光器
high-pressure jet grouting 高压喷射注浆;高压喷射灌浆
high-pressure jet nozzle 高压喷嘴
high-pressure jig 高压卷染机
high-pressure jointing 高压接合
high-pressure kier 高压精炼锅
high-pressure laminate 高压层(压)板;高压胶合板;高压层积材
high-pressure laminate door 层压塑料门
high-pressure laminating 高压叠层
high-pressure lamp 高压灯
high-pressure leak detector 高压系统测漏器;高压系统测漏气装置
high-pressure level controller 高压液面调节器
high-pressure line 高压线;高压管路
high-pressure liquid chromatograph 高压液相色谱仪
high-pressure liquid chromatography 高压液相色谱分析;高压液相色谱(法);高压液相层析;高压液体色谱法
high-pressure liquid encapsulation pulling method 高压液封直拉法
high-pressure liquid gas volume 高压液化气容器
high-pressure liquid level ga(u)ge 高压液面计
high-pressure liquid system 高压流体系统
high-pressure live steam 高压直接蒸汽;高压新蒸汽
high-pressure load reducing spring 高压减荷弹簧
high-pressure/low temperature metamorphic belt 高压/低温变质带
high-pressure lubricating system 高压润滑系统
high-pressure lubrication grease 高压润滑油脂
high-pressure lubricator 高压润滑器;高压润滑泵
high-pressure magneto-ignition 高压磁点火
high-pressure main (水、电、煤气、下水道等的)高压总管线;高压总管
high-pressure mangle 高压轧液机;重型轧车
high-pressure manometer 高压力表;高压力计
high-pressure medium 高压介质
high-pressure melting method 高压熔融法
high-pressure mercury discharge lamp 高压汞放电灯
high-pressure mercury lamp 高压水银灯
high-pressure mercury vapo(u)r lamp 高压汞(气)灯;高压水银灯
high-pressure meter 高压煤气表;高压计
high-pressure mica-paper capacitor 高压云母纸介电容器
high-pressure mo(u)lding 高压造形;高压模塑
high-pressure mo(u)lding machine 高压造形机
high-pressure natural gas-water leakage 高压天然气—水渗出
high-pressure nitrogen liquid-gas truck 高压液氮气化车
high-pressure nucleus 高压核心
high-pressure oil burner 高压油(喷)灯;高压燃油炉
high-pressure oil-gas well cementing 高压油气井注水泥
high-pressure oiling pump 高压注油泵
high-pressure oil jack 高压油千斤顶
high-pressure oil pump 高压油泵
high-pressure oil pump tester 高压油泵试验机
high-pressure oil pump testing platform 高压油泵试验台
high-pressure oil saver 高压防油器
high-pressure oil tube 高压油管
high-pressure orchard sprayer 高压果园喷雾机
high-pressure overlay 高压面板
high-pressure oxygen 高压氧
high-pressure pan 高压罐
high-pressure penstock 高压输水管道
high-pressure/performance liquid chromatography 高压高效液相色谱法
high-pressure phenomenon 高压现象
high-pressure physics 高压物理学
high-pressure piezoelectric(al) ceramic hydrophone 高压电陶瓷水听器
high-pressure pipe 高压管
high-pressure pipe flange 高压管法兰

high-pressure pipeline 高压管线;高压管道
high-pressure pipe pump 高压管道泵
high-pressure piping 高压管道
high-pressure piston 高压活塞
high-pressure piston rod 高压活塞杆
high-pressure plastic overlay 高压塑料面板
high-pressure pneumatic conveyer 高压气流式输送器
high-pressure pneumatics 高压气动技术
high-pressure polarimetric sensor 高压偏振型传感器
high-pressure polyethylene 高压聚乙烯
high-pressure polyethylene film 高压聚乙烯薄膜
high-pressure polyethylene packing 高压聚乙烯盘根;聚乙烯高压盘根
high-pressure polymerization 高压聚合
high-pressure positive blower 高压(正鼓)风机
high-pressure preheater 高压预热器
high-pressure process 高压注气工艺;高压法
high-pressure producer 高压发生器
high-pressure propellant tank 高压燃料箱
high-pressure pulse pump 高压脉动泵
high-pressure pump 高压抽水机;高压泵
high-pressure pumper 高压泵车
high-pressure purge 高压气体吹洗
high-pressure radiation-chemical laser 高压辐射化学激光器
high-pressure reactor 高压反应器
high-pressure receiver 高压气体储[贮]罐
high-pressure regional metamorphic facies series 高压区域变质相系
high-pressure regional metamorphism 高压区域变质作用
high-pressure regulator 高压调节器
high-pressure relief valve 高压释放阀
high-pressure ridge 高压脊
high-pressure rock triaxial testing machine 高压岩石三轴试验仪
high-pressure rotating pile 高压旋喷水泥土桩
high-pressure rubber hose 高压橡胶软管
high-pressure safety cut-out 高压安全切断器
high-pressure safety valve 高压安全阀
high-pressure sampler 高压取样器
high-pressure science 高压科学
high-pressure scouring 高压冲洗
high-pressure screw pump 高压螺旋泵
high-pressure scrubber 高压洗涤器塔
high-pressure section 高压部分
high-pressure selling 高压推销;倾力推销
high-pressure separator 高压分离器
high-pressure service 高压运转
high-pressure service fire pump 高压消防泵
high-pressure side 管道弯头外侧;高压侧;制冷系统高压边
high-pressure slide valve 高压滑阀
high-pressure sluice valve 高压闸阀
high-pressure sodium lamp 高压钠灯
high-pressure sodium vapo(u)r lamp 高压钠汽灯
high-pressure solid jet 高压密集射流
high-pressure solution growth 高压溶液生长
high-pressure spectroscopy 高压光谱学
high-pressure spectrum 高压光谱
high-pressure spot 高压点
high-pressure spray 高压射流;高压喷雾
high-pressure spraying set-up 高压喷雾装置;高压喷射装置;高压喷洒装置
high-pressure spray machine 高压喷雾机
high-pressure sprinkler 高压喷淋器;高压(头)喷灌机;高压洒水器;高压洒水车;高压人工降雨装置
high-pressure squeeze method 高压挤水泥法
high-pressure stage 高压极;压风机的高压级
high-pressure stainless steel hose 高压不锈钢软管
high-pressure steam 高压蒸汽
high-pressure steam boiler 高压蒸汽锅炉
high-pressure steam cured 高压蒸汽养护的
high-pressure steam curing 高压蒸汽养护(法);高压蒸汽消毒法;蒸汽养护;蒸釜养护
high-pressure steam curing process 高压蒸汽养护法
high-pressure steam digestion 高压蒸汽溶出
high-pressure steam engine 高压蒸汽机
high-pressure steam etching 高压蒸汽腐蚀
high-pressure steam heating 高压蒸汽加热;高压蒸汽供暖;高压蒸汽采暖

high-pressure steam heating system 高压蒸汽供暖系统
high-pressure steam installation 高压蒸汽装置
high-pressure steam jet 高压蒸汽喷嘴
high-pressure steam line 高压蒸汽管线
high-pressure steam oxidation 高气压水汽氧化
high-pressure steam plant 高压蒸汽设备
high-pressure steam power plant 高压火力发电厂
high-pressure steam process 高压蒸汽法
high-pressure steam sterilizer 高压蒸汽消毒器;高压蒸汽消毒柜;高压蒸汽灭菌器
high-pressure steam sterilizer of portable type 手提式高压蒸汽消毒器
high-pressure steam system 高压蒸汽系统
high-pressure steam treatment 高压蒸汽处理;高压蒸汽养护
high-pressure steam turbine 高压蒸汽涡轮
high-pressure sterilization 高压蒸汽消毒灭菌法
high-pressure sterilizer 高压(蒸汽)消毒器
high-pressure superheated steam 高压过热蒸汽
high-pressure system 高压系统
high-pressure tail cutter shower 高压切边水针
high-pressure tank 高压气瓶;高压气柜;承压箱
high-pressure tap 高压龙头
high-pressure technology 高压技术
high-pressure test 高压试验
high-pressure testing pump 高压试压泵
high-pressure three-throw pump 高压三冲程泵;高压三向射流泵
high-pressure thrust bearing 高压推力轴承
high-pressure tire 高压轮胎
high-pressure tower 高压塔
high-pressure triaxial apparatus 高压三轴仪
high-pressure tube 高压管
high-pressure tube calibration 高压管(法)校准
high-pressure tunnel 高压隧洞
high-pressure turbine 高压涡轮(机);高压透平;高压汽轮机
high-pressure turbine stage 涡轮高压级
high-pressure turbocharger 高压涡轮增压器
high-pressure turn 高压转换管
high-pressure twin hydraulic pump 高压双缸液压泵;高压双(向)水泵
high-pressure twin-roll mill 辊压机;高压双辊磨
high-pressure tire 高压轮胎
high-pressure unit 高压装置;高压送风器
high-pressure valve 高压活门;高压阀(门)
high-pressure vane pump 高压叶片泵
high-pressure vapo(u)r oxidation method 高压水汽氧化法
high-pressure variable vane pump 高压变量叶片泵
high-pressure varistor 高压压敏电阻器
high-pressure vessel 高压容器
high-pressure washing 高压冲洗
high-pressure water 高压水
high-pressure water blast 高压喷水除锈法
high-pressure water blasting 高压喷水清理
high-pressure water cleaning 高压水清洗
high-pressure water derusting 高压水除锈
high-pressure water feeding pump for boiler 高压锅炉给水泵
high-pressure water hose 高压水(软)管;高压输水胶管
high-pressure water-injection pump 高压注水泵
high-pressure water jet 高压喷射嘴;高压水射流;高压水枪;高压水射器;高压水舱;高压射水;高压喷水器
high-pressure water main 高压总水管
high-pressure water mist system 高压水水雾系统
high-pressure water or air 高压水或空气
high-pressure water pump 高压水泵
high-pressure well 高压井
high-pressure wind tunnel 高压风洞
high-pressure wire braided hose 高压钢丝编织胶管
high-pressure xenon arc lamp 高压氙弧光灯
high-pressure xenon lamp 高压氙灯
high-pressure zone 高压区;高压层
high price 高价
high priced durable consumer goods 高价耐用消费品
high priced service 高价劳务
high priority communication 最优先等级的通信
high priority investment project 最优先投资项目
high priority project 主要优先项目;重要优先项目
high priority record queue 高优先级记录队列
high prob sizer 高效率概率筛
high producing 高产的
high product 优质产品
high production 高生产率
high production area 高产区
high production central-mixing plant 高产中央搅拌厂(混凝土)
high production kiln 高产(砖)窑;高产瓦窑
high production long-set-up machines 高效率调整时间长的机器
high production oil-gas bed 高产油气层
high production press 大生产率压力机
high production rate 高生产率
high productive 高生产率的
high productive capacity 高生产能力
high productive soil 高产土壤
high professional qualification 高技能
high profile 高型的
high profit item 高利项目
high profit product 高利产品
high protrusion barb 高突刺
high pruning 修荫枝
high pulling torque 大力矩
high pulse repetition laser 高脉冲重复(率)激光器
high pumping rate 高抽速
high punch 高位穿孔
high puritied magnesia brick 高纯镁砖
high purity 高纯度
high purity alumin(i)um 高纯铝
high purity alumin(i)um foil 高纯铝箔
high purity antimony 高纯锑
high purity arsenic 高纯砷
high purity beryllium oxide 高纯氧化铍
high purity bismuth 高纯铋
high purity boron 高纯硼
high purity cadmium 高纯镉
high purity cellulose 高纯度纤维素
high purity cerium dioxide 高纯氧化铈
high purity chlorine dioxide 高纯二氧化氯
high purity copper 高纯铜
high purity detergent 高纯洗净剂
high purity ferrosilicon 高纯硅铁
high purity gallium 高纯镓
high purity gas 超纯气体
high purity germanium detector 高纯锗探测器
high purity gold 高成色黄金
high purity graphite products 高纯石墨制品
high purity hydrogen generator 高纯氢气发生器
high purity indium 高纯铟
high purity indium antimonide single crystal 高纯锑化铟单晶
high purity lead 高纯铅
high purity lithium 高纯锂
high purity material 高纯度材料
high purity metal 高纯金属
high purity mullite ceramics 高纯莫来石陶瓷
high purity nickel 高纯镍
high purity niobium oxide 高纯氧化铌
high purity oxygen 高纯氧
high purity oxygen aeration activated sludge process 高纯氧曝气活性污泥法
high purity oxygen aeration system 高纯氧曝气系统
high purity phosphorus 高纯磷
high purity powder 高纯粉末
high purity reagent 高纯试剂
high purity silicon tetrachloride 高纯四氯化硅
high purity sulphur 高纯硫
high purity tantalum oxide 高纯氧化钽
high purity tellurium 高纯碲
high purity tin 高纯锡
high purity titanium 高纯钛
high purity vanadium 高纯钒
high purity water 高纯水
high purity zinc 高纯锌
high purity zirconium 高纯锆
high purity zirconium hydride powder 高纯度氢化锆粉
high quality 质地优良;高质量;高品质;高级的;高标号的;优质(的)
high-quality cast iron 高级铸铁;优质铸件
high-quality cement 优质水泥;高级水泥;高强度等级水泥
high-quality concrete 高质量混凝土;高级混凝土
high-quality drinking water 优质饮用水
high-quality effluent 优质出水
high-quality factor 高品质因素;高品质因数
high-quality finishing 高级整理
high-quality grade 优质(等)级
high-quality grey iron 高级灰口铁
high-quality inductor 高品质电感
high-quality laser rod 高质量激光棒
high-quality measuring 高品位测量
high-quality nickel steel tube 高级镍钢管
high-quality primary mirror 优质主镜
high-quality product 优质产品
high-quality project 优质工程
high-quality rectified image 高质量校正图像
high-quality renovated water 优质再生水
high-quality reuse water 优质再用水
high-quality shale oil 高质页岩油
high-quality steel 优质钢;高质量钢;高级钢
high-quality thermometer 高质量温度表
high-quality vane pump 高性能叶片泵
high-quality water 优质水
high-quality well-known goods 名优产品
high quantity item 高定额项目
high quantum number 高量子数
high quartz 高温型石英
high quartz gravel 高石英砾石
high quartz solid solution 高温石英固溶体
high rack storage 高架仓库
high rack storage system 高架储存系统
high radiating gas 高辐射气体
high radiation area 高辐射区
high radiation field 强度射场
high radiation flux 高辐射通量
high radioactive area 高度放射性区域
high radioactivity 高放射性
high rail 曲线外轨【铁】
high rail side of curve 曲线外轨一侧【铁】
high rainfall area 多雨区;雨量大的地面
high range 大射程;大量程;大刻度;大范围;高山脉;高量程;大比例
high range water reducer 高含量减水剂;高效减水剂
high range water-reducing admixture 高效减水剂
high rank coal 高级煤
high rank fuel 高级燃料
high rank graywacke 高级杂砂岩;长石杂砂岩
high rank hydrothermal alteration 高级水热蚀变
high rank metamorphism 高级变质作用
high rate 高效率;高速
high rate activated sludge process 高负荷活性污泥(处理)法;高效活性污泥法;高速活性污泥处理法
high rate activated sludge waste(water) treatment 高效活性污泥污水处理
high rate adsorption bio-oxidation treatment 高效吸附生物氧化处理
high rate adsorption bioxidation 高负荷吸附生物氧化处理
high rate aeration 高效曝气;高速通气法;高速曝气;高负荷曝气
high rate aeration basin 高效曝气池;高负荷曝气池
high rate aeration process 高速通气法;高负荷活性污泥(处理)法
high rate aeration settling process 高负荷曝气沉降法;高效曝气沉淀工艺;高速曝气沉淀法
high rate aeration settling tank 高速曝气沉淀池
high rate aeration tank 高负荷曝气池
high rate aerobic algal ponds 高负荷好氧藻类塘
high rate aerobic treatment 高效好氧处理;高负荷需氧处理;高负荷曝气处理;高负荷好氧处理
high rate anaerobic digestion 高effi厌氧消化
high rate anaerobic treatment 高效厌氧处理
high rate biofilter 高效生物滤池;高负荷生物滤池
high rate biofiltration system 高曝气过滤系统;高负荷生物过滤系统
high rate biological filter 高效生物滤池;高负荷生物滤池
high rate cathode 高倍率阴极
high rate degrading bacteria 高效降解菌
high rate digester 高速消化池;高负荷消化池
high rate digestion 高效消化;高速率消化;高负荷

消化(法)
high rate digestor 高效消化池
high rate discharge tester 负荷检查器
high rate filter 高效率滤器;高效(率)滤池;高速(生物)滤池;高速过滤器;高速(度)滤池;高负荷(生物)滤池;滤池
high rate filtration 高效过滤(结构);高速过滤;高负荷过滤
high rate fixed film 高效固着膜
high rate forming 高速成形
high rate freezing 急速冷却(法)
high rate of advance 高推进速度
high rate of rainfall 高强降雨量
high rate of reusability 资源再生利用率高
high rate of success in competition 竞争效应高
high rate oxidation pond 高效氧化塘
high rate pond 高效塘
high rate pressing 高能速压制
high rate rapid filter 高速快滤池
high rate settling tank 高效沉淀池;高负荷沉淀池
high rate silver zinc battery 高效银锌蓄电池
high rate single-stage digestion tank 高负荷一级消化池
high rate sprinkling filter 高效喷滤池;高负荷喷滤池
high rate telemetry 高速遥测
high rate treatment 高负荷处理
high rate trickling filter 高效滴滤池;高速滴滤器;高负荷滴滤池
high rate trickling filter plant 高效滴滤池装置
high rate trickling filtration 高效滴滤;快速滴滤法
high rate water level 高水位
high rate water wash 高效水冲洗;高强度冲洗
high rate water washing 高强度冲洗
high-rating generator 大型发电机
high ratio gear 高速比传动装置
high ratio of modulus 高模量比岩石
high ratio of rainfall 高强度降雨
high ratio transformer 大变比变压器
high ratio zoom lens 高倍变焦距镜头
high reaction force 高反力层;高反力(护舷)
high reactivity resin 高反应性树脂
high recombination rate contact 高复合率接触
high recycled content 高再循环材料含量
high reduction 高缩微率
high reference point 高基准点
high reflectance zone 高反射带
high reflecting dielectric(al)film 高反射电介质膜
high reflecting glass 高反射玻璃
high reflection 高反射
high reflection coating 高反射膜
high refractive glass beads 高折射玻璃微珠
high refractive index and low dispersion optic(al)glass 高折射率低色散光学玻璃
high refractive index optic(al)glass 高折射率光学玻璃
high refractive low dispersive glass 高折射低色散玻璃
high refractive low dispersive optic(al)glass 高折射率低色散光学玻璃
high refractory 高耐火(的)
high refractory brick 高级耐火砖
high refractory oxide 高耐火性氧化物
high regger 高索架设者
high relaxation 高弛豫
high reliability 高可靠性
high reliability item 高可靠性项目
high relief 高(凸)浮雕;高峻地形;隆浮雕;深浮雕
high-relief area 起伏大的地区
high remanence 高顽磁性
high Renaissance 高文艺复兴(式)
high rented 高租金的
high residual-phosphorus copper 高残磷铜
high residue level 高残留度
high resistance 高电阻
high resistance alloy 高阻合金
high resistance barrier 高电阻阻挡层
high resistance electrothermic alloy 高电阻电热合金
high resistance ground 高电阻土壤
high resistance joint 高强度连接;高电阻接头
high resistance measurement 高阻测量
high resistance metal 高阻金属
high resistance meter 高阻计

high resistance potentiometer 高阻电位计
high resistance pyrometer 高阻高温计
high resistance refractory rod 高电阻耐火杆
high resistance resistor 高阻电阻器
high resistance soil 高电阻土壤
high resistance transmitter 高阻式送话器
high resistance voltmeter 高(电)阻伏特计
high resistance wind 高电阻绕组
high resistant sulfate cement 高抗硫酸盐水泥
high resisting material 高电阻材料
high resistive interesting point 高阻交点
high resistive shielding disturbance 高阻屏蔽干扰
high resistive shielding layer 高阻屏蔽层
high resistivity alloy 高电阻率合金
high resistivity dust 高比电阻粉尘
high resistivity layer epitaxy 高阻层外延
high resistivity shielding 高阻屏蔽
high resolution 高清晰度;高分辨率;高分辨力;高分辨(能力)
high resolution aerial camera 高分辨率航空照相机
high resolution aerial photograph 高分辨率航摄照片
high resolution aeromagnetic survey 高分辨率航空磁测
high resolution atomic absorption spectrometry 高分辨原子吸收光谱测量术
high resolution Bragg spectrometer 高分辨率布雷格光谱计
high resolution camera 高分辨率摄影机
high resolution capillary column gas chromatography 高分辨率毛细管柱状气相色谱法;高分辨率毛细管柱状气相层析
high resolution cathode ray tube 高分辨(能力)显示管;高分辨率喷管
high resolution chromatographic(al)separation 高分辨色谱分离
high resolution colo(u)r graph control board 高分辨率彩色图形控制板
high resolution colo(u)r graphic(al)controller board 高分辨率彩色图形控制板
high resolution colo(u)r picture 高分辨率彩色图像
high resolution column 高分辨柱
high resolution column gas chromatography 高分辨率气相色谱法;高分辨率气相层析
high resolution column gas chromatography/high resolution mass spectrometry 高分辨率气相色谱法—高分辨率质谱法联用
high resolution column gas chromatography/high resolution mass spectrometry/computer 高分辨率气相色谱法—高分辨率质谱法—计算机联用
high resolution column gas chromatography/mass spectrometry 高分辨率气相色谱法—质谱法联用
high resolution computer graphic system 高分辨率计算机图形系统;高分辨度计算机图形系统
high resolution deep-level transient spectrometer 高分辨率深能级瞬态谱仪
high resolution densilog 高分辨率密度测井
high resolution detector 高分辨(率)探测器;高分辨率检波器
high resolution diffraction 高分辨率衍射
high resolution dip log 高分辨率地层倾角测井
high resolution dipmeter 高分辨倾角仪;高分辨率地层倾角仪
high resolution dipmeter tool 高分辨地层倾角测井仪
high resolution electron microscope 高分辨(率)电子显微镜
high resolution fused fiber optics component 高分辨熔融光纤元件
high resolution gamma spectrometry 高分辨伽马射线测谱学
high-resolution gas chromatography with electron capture detection 电子俘获检测高分辨气相层析
high-resolution gas chromatography with flame photometric(al)detection 火焰光度检测高分辨率气相色谱法
high resolution geophone streamer 高分辨率检波器拖缆
high resolution graph 高分辨度图形
high resolution graphics 高分辨图形

high resolution graphic terminal 高分辨率图形终端;高分辨度图形终端
high resolution grating 高分辨率光栅
high resolution image 高分辨图像
high resolution image sensor 高分辨图像传感器
high resolution imaging radar 高分辨率成像雷达
high resolution infrared detector 高分辨率红外探测器
high resolution infrared radiation sounder 高分辨率红外辐射探测器;高分辨力红外辐射探测器
high resolution infrared radiometer 高分辨(率)红外辐射仪;高分辨率红外辐射计
high resolution infrared sounder 高分辨率红外探测仪;高分辨率红外探测器
high resolution infrared spectroscopy 高分辨率红外光谱学
high resolution instrument 高分辨率仪器
high resolution interferometer 高分辨率干涉仪
high resolution interferometry 高分辨率干涉量度学
high resolution laser radar 高分辨率激光雷达
high resolution laser spectroscopy 高分辨率激光光谱学
high resolution lens 高分辨率透镜
high resolution liquid chromatography 高离析液相色谱法;高分辨液相色谱(法)
high resolution liquid mass spectrometry 高分辨液相质谱法
high resolution log curve 高分辨率密度测井曲线
high resolution low frequency spectrum analyzer 高分辨率低频谱分析仪
high resolution mass spectrometer 高分辨率质谱仪;高分辨质谱计
high resolution mass spectrometry 高分辨质谱(测定)法
high resolution Michelson interferometer 高分辨率迈克尔逊干涉仪
high resolution multispectral scanner 高分辨率多光谱扫描器
high resolution nuclear-magnetic-resonance spectrometer 高分辨率核磁共振波谱仪
high resolution optic(al)mask 高分辨率光学掩膜
high resolution optic(al)memory 高分辨率光存储器
high resolution optics 高分辨光学
high resolution photographic(al)film 高分辨率胶片
high resolution photographic(al)particles 高分辨率成像微粒
high resolution photoresist 高分辨率光致抗蚀剂
high resolution picture 高分辨率图像
high resolution picture transmission 高分辨率图像传输
high resolution plate 高清晰度底片;高分解力底片;高分辨板;超微粒干版
high resolution pulse Doppler radar 高分辨率脉冲多普勒雷达
high resolution pulser 高分辨率脉冲发生器
high resolution radar 高分辨率雷达
high resolution radar image 高分辨率雷达影像
high resolution range finder 高分辨率测距仪
high resolution seismic 高分辨率地震
high resolution seismic data processing 高分辨率地震资料处理
high resolution seismic profile 高分辨率地震剖面仪;高分辨地震剖面
high resolution sensing survey 高分辨率地震测量
high resolution spectrometer 高分辨率分光计
high resolution spectroscopy 高分辨(率)光谱学
high resolution surveillance radar 高分辨率监视雷达
high resolution surveillance radar station 高分辨率监视雷达站
high resolution telescope 高分辨率望远镜
high resolution television camera 高分辨率电视摄像机
high resolution thermometer 高分辨率温度计
high resolution visible sensor 高分辨率可见光传感器
high respond 快速响应
high return 高反射
high return high risk 高效益高风险
high Reynola number flow 高雷诺数水流
high Reynola number wind tunnel 高雷诺数风洞

high rib lath 高肋金属网；瓦棱网眼钢皮
high riffle table 高格条摇床
high rigidity plastics 高刚度塑料
high rise 高耸的；高层大楼
high-rise apartment 高层公寓
high-rise apartment building 高层公寓建筑
high-rise block 多层高楼街区；高层建筑（街坊）
high-rise(building) 高层建筑（物）；高层房屋
high-rise building construction 高层建筑施工
high-rise capacity 高生产能力
high-rise city 空间发展城市
high-rise dwelling 高层住宅
high-rise erection 高层建筑安装
high-rise flats 高层公寓
high-rise garage 多层停车库
high rise/high density 高层高密度
high-rise hotel 高层旅馆；高层饭店
high-rise housing 高层住宅
high-rise low density 高层低密度
high-rise manifold 超高度歧管
high-rise operation 高空作业
high-rise parking building 高层（停车）车库
high-rise parking garage 高层（停车）车库
high-rise pile cap 高桩承台
high-rise platform pile foundation 高承台桩基础
high rise(r) 高层住宅或办公楼
high-rise rescue 高层建筑救援
high-rise residence sketon plan 高层住宅方案
high-rise residential area 高层住宅区
high-rise roof 高屋顶
high-rise slums 高层贫民窟
high-rise structure 高耸结构（物）；高层结构；高层建筑（物）
high risk area 高风险区
high risk decision 高风险决策
high risk investment 高风险投资
high river flow 高流速河道水流
high road 公路；干路；大路
high rocky ridge 高多岩山脉；高多岩山岗；高岩质山脉；高岩质山岭；高岩质山脊
high rod 高水准尺
high roof mining system 高矿房回采
high-run production 大量成批生产
high runway 高原跑道
high rupture strength 高断裂强度
high rupturing capacity fuse 高遮断容量熔丝
high safety glazing 高安全度窗玻璃
high saline wastewater 高盐度废水
high salinity 高矿化度；高含盐量
high salinity aniline-cotaining wastewater 高盐度苯胺废水
high salinity bump 高盐跃值
high salinity mud 高矿化度泥浆
high salinity reservoir 高盐度热储
high salinity waste 高盐量废水
high salinity water 高盐（度）水
high salt content 高含盐量
high salt waste 高盐量废水
high sand content production 高含沙采油
high sanidine 高温型透长石
high saturation 饱和度大
high scale 刻度上段；上部刻度；辅助刻度
high school 中学
high sea 远洋；高浪；狂浪；开敞海面；猛浪；外海；道格拉斯六级浪；大浪；公海
high seam 厚煤层；厚矿层
high sediment discharge 高输沙量
high seismic risk 高地震危险性
high seismic-wave velocity zone 高地震波速度带
high selectivity 高选择性
high selector relay 高压选择阀
high sensibility tester 高灵敏度试验器
high sensitive chromogenic 高灵敏钍显色剂
high sensitive chromogenie reagent for thorium 高灵敏度钍显色剂
high sensitive relay 高灵敏（度）继电器
high sensitive smoke detector 高灵敏度感烟探测器
high sensitive vacuum tube voltmeter 高灵敏度真空管电压表
high sensitivity 高灵敏度
high sensitivity demodulator 高灵敏度解调器
high sensitivity detector 高灵敏度检测器
high sensitivity flowmeter 高灵敏度流量计

high sensitivity galvanometer 高灵敏度电流计
high sensitivity laser detection system 高灵敏度激光探测系统
high sensitivity oscilloscope 高灵敏度示波器
high sensitivity seismograph 高灵敏度地震仪
high sequioxide soil 高倍半氧化物土
high service 高压供水
high service system 高压供水系统
high service works 高区供水厂
highset detector 高值指示装置；高值检测器
high set setting 高值整定；高定值
high severity cracking 高苛刻度裂化
high shaft furnace 高竖炉
high shear 高剪切
high shear continuous mixer 高剪切连续搅拌机
high shear grinding 高切力研磨；高剪切研磨
high shearing stress 高切应变
high shearing-type mixer 高剪切型混合器
high shear mixer 高剪切混料机
high shear modulus 高切变模量
high shear viscosity 高剪切黏度
high sheet cloud 高层云
high shielded 高度屏蔽
high shoulder 超高路肩
high shrinkage fiber 高（收）缩纤维；高缩率纤维
high shrinkage oil 高收缩率原油
high shrinking aggregate 高收缩率集料；高收缩率骨料
high shrink staple 高缩率短纤维
high side 制冷系统高压边；管道弯头外侧；傍山的一边（道路）
high-sided gondola car 高边敞车
high-sided open goods wagon 高边敞车
high-sided pack 上部充填带
high sided plate conveyer 高侧板式输送机
high-sided truck 高栏板卡车
high sided wagon 高边车
high side float valve 高压浮球阀
high side light 高侧窗
high side(of road) 傍山的一边（道路）
high side roller mill 高侧轮碾机
high side window 高侧窗
high signal 高柱信号；强信号
high silica brick 高硅（质）砖
high silica cement 高硅水泥
high silica-content cement 高硅水泥
high silica content fiber 高硅石量纤维；高硅含量纤维
high silica content glass 高硅含量玻璃
high silica electric(al) porcelain 高硅电瓷
high silica flux 高硅焊剂
high silica glass 高硅玻璃
high silica glass fabric 高硅氧玻璃布
high silica glass fiber 高硅氧玻璃纤维
high silica glass wool 高硅氧玻璃棉
high silica iron 高硅铁
high silica limestone 高硅石灰岩
high silica magnesite brick 高硅镁砖
high silicate glass 高硅酸盐玻璃
high siliceous cement 高硅水泥
high silicon bronze 高硅青铜
high silicon bronze alloy 高硅青铜合金
high silicon cast iron 高硅铸铁
high silicon cast-iron anode 高硅铸铁阳极
high silicon chromium iron 高硅铬铁
high silicon iron 高矽铁；高硅铸铁；高硅生铁
high silicon Monel 高硅蒙奈尔合金
high silicon pig iron 高硅生铁
high silicon sheet iron 高硅钢片
high sill spillway 高槛溢洪道
high silver crust 高银壳
high sintered parts 高温烧结零件
high sintering 高温烧结
high-sintering dolomite 难溶白云石
high sinuosity channel 高弯度河道
high slip induction motor 高滑差感应电动机
high slip motor 高转差率电机；高滑率电机
high slope 陡坡
high slovency naphtha 高溶解度石脑油
high slump concrete 高坍落度混凝土；高塑性混凝土；大坍落度混凝土
high sodium high sulfur fuel oil 高钠高硫燃料油
high softening point resin 高软化点树脂
high solid 高固体分

high solid coating technology 高固体粒子涂料技术
high solid content sludge 坚硬污泥
high-solidity cascade 大调修叶栅；大稠度叶栅；稠叶栅
high solid lacquer 高硬性喷漆；高固体漆
high solid mud 高固相泥浆
high solubility binder glass mat 高溶解度黏合剂玻璃纤维毡
high solution bacteria 高分解细菌
high solvating plasticizer 高溶合性增塑剂
high solvency naphtha 高熔解度石脑油；高溶解力石脑油
high sounding horn 高音喇叭
high sound insulation 高度隔声
high source temperature 高热源温度
high southern latitudes 南半球高纬度地区
high spaces 高隔空铅
high specific activity 高比放射性
high specific speed pump 高比（转）速泵
high specific speed turbine 高比速透平
high-specific thrust 最大比推力
high-speed 高速度；高速（的）
high-speed(accounting)computer 高速计算机
high-speed accounting machine 高速计账机；高速会计机
high-speed adder 高速加法器
high-speed adjustment 高速调整
high-speed aerodynamics 高速气体动力学；高速空气动力学
high-speed aeronautics 高速航空学
high-speed agitator 高速搅拌器
high-speed aircraft 高速飞机
high-speed air switch 高速空气开关
high-speed alphanumeric printer 高速字符印字机
high-speed amplifier 高速放大器
high-speed analog(ue)computer 高速模拟计算机
high-speed analog(ue)digital converter 高速模数转换器
high-speed and high-tension filter 高压高速滤波器
high-speed and multitool cutting method 高速多刀复刃切削法
high-speed auger 高转速螺旋
high-speed automated highway 自动化高速公路
high-speed automatic fine puncher 高速自动精密冲床
high-speed automatic press 高速自动压床
high-speed automatic program(me)unit 高速自动程序装置
high-speed automatic transmitter 高速自动发射机
high-speed balancing 高速动平衡
high-speed ball mill 高速球磨机
high-speed bar code printer 高速条码打印机
high-speed barrier curb 高速车道的栏式路缘石
high-speed bar tacking machine 高速套结机
high-speed bed 高速带
high-speed belt conveyer 高速皮带输送机
high-speed bench drilling machine 高速台钻
high-speed bending moment 高速弯曲力矩
high-speed bit 高速钻头
high-speed blanking press 高速压力机
high-speed boat 高速艇
high-speed boring head 高速镗头
high-speed boring machine 高速镗床
high-speed bottom 高速犁体
high-speed brake 高速制动器；快速制动器
high-speed brass 高速切削黄铜
high-speed brittleness 高速脆性
high-speed buff 高速抛光轮
high-speed buffer 高速缓冲器
high-speed buffer register 高速缓冲寄存器
high-speed building hoist 高速施工吊车；高速建筑吊车
high-speed bus 高速总线；高速母线
high-speed cage 高速尘笼
high-speed calculator 高速计算器
high-speed camera 高速照相机；高速摄影机
high-speed car 高速汽车
high-speed card punch 高速卡片凿孔机；高速卡片电传打字机穿孔机
high-speed card reader 高速读卡机
high-speed card-teletype 高速卡片电传打字机
high-speed card-teletype terminal 高速卡片电传

打字机终端
high-speed cargo liner 快速货轮
high-speed carry 高速进位(方式)
high-speed cement 早强水泥;快硬水泥;快凝水泥
high-speed centerless grinding machine 高速无心磨床
high-speed centrifuge 高速离心机
high-speed channel 高速信道;高速通道
high-speed check-row valve 高速方形穴播阀
high-speed cinecamera 高速电影摄影机
high-speed cine theodolite 高速电影经纬仪
high-speed circuit 高速油路;主油系
high-speed circuit breaker 高速熔断器;高速断路器;高速断路开关
high-speed circuit breaker arrangement 高速断路器配置
high-speed circuit runway 高速环形跑道
high-speed clutch 高速离合器
high-speed coagulative precipitation unit 快速混凝土沉淀装置;速效凝结沉淀装置;速效混凝土沉淀装置
high-speed coefficient 高速系数
high-speed cold Pilger mill 高速冷轧管机
high-speed compaction 高速压实(用于粉末冶金);高速成形
high-speed comparator 高速比较器
high-speed compound terminal 高速混合终端;高速复式终端机
high-speed computational capability 高速计算能力
high-speed condenser 高速冷凝器
high-speed conditioner 高速搅拌筒
high-speed cone winder 高速络筒机;络筒机
high-speed conical refiner 高速圆形精磨机
high-speed controller 高速控制器
high-speed controller and lever safety valve 高速调节器及柄式安全阀
high-speed conveyer 高速输送机
high-speed copier 高速复印机
high-speed copying machine 高速复印机
high-speed correlator 快速相关器
high-speed counter 高速计数器
high-speed counting 高速计数
high-speed counting circuit 高速计数电路
high-speed craft 高速艇
high-speed cropping 高速剪断
high-speed crushing mill 高速粉磨机
high-speed cutterbar 高速切割器
high-speed cutting 高速切削;快速切削
high-speed cutting machine 高速砂轮切断机
high-speed cutting nozzle 高速割嘴
high-speed cutting-off wheel 高速切割砂轮
high-speed data 高速数据
high-speed data acquisition 高速数据收集;高速数据集合
high-speed data acquisition system 高速数据采集系统
high-speed data collection system 高速数据采集系统
high-speed data rate 快速数据率
high-speed data transfer 高速数据传送
high-speed data word 高速数字
high-speed decision 快速决策
high-speed deformable mirror 高速可变形反射镜
high-speed developer 高速显影液
high-speed development 高速显影
high-speed diesel engine 高速柴油机
high-speed diesel fuel 高速柴油
high-speed die winder 高速染色络筒机
high-speed differential protection 高速差动保护
high-speed digital computer 高速数字计算机;快速数字计算机
high-speed digital counter 高速数字计算机
high-speed digital recorder 高速数字记录器
high-speed directional overcurrent relay 高速定向过流继电器;高速定量过载继电器
high-speed disc atomizer 高速碟式雾化器
high-speed disintegrator 高速破碎机;高速粉碎机
high-speed dispersion 高速分散
high-speed dissolver 高速溶解器;高速分散机
high-speed distance relay 高速(远)距高继电器
high-speed diverter switch 高速分流器开关
high-speed document communication terminal 高速文件通信终端

high-speed drill 高速钻机
high-speed drilling attachment 高速钻孔附件
high-speed drilling machine 高速钻床
high-speed drivage 快速掘进
high-speed drive 高速(度)传动
high-speed drum 高速磁鼓
high-speed drying press 高速压干机
high-speed duct 高速管道
high-speed dump 快速转储
high-speed dy(e)ing machine 高速染色机
high-speed dynamic(al) pressure bearing 高速动压轴承
high-speed earth compactor 高速土壤压实机
high-speed electrified line 高速电气化铁路
high-speed electronic bandknife splitting machine 高速电子带刀削薄机
high-speed electronic computer 高速电子计算机
high-speed electronic printer 高速电子打印机
high-speed electrospark perforating electrode tube 高速电火花打孔电极管
high-speed electrostatic printer 高速静电打印机
high-speed elevator 高速电梯
high-speed emulation memory 高速仿真存储器
high-speed emulsion 高速感光乳剂
high-speed encryptordecryptor 高速加脱密机
high-speed-energy forging machine 高速锤
high-speed engine 高速发动机;高速发电机
high-speed excitation 高速励磁;快速励磁;强行励磁
high-speed excitation system 高速励磁系统
high-speed exciter 高速励磁机
high-speed exit taxiway 高速脱离滑行道;高速滑出道;高速出口滑行道
high-speed external cylindrical grinder 高速外圆磨床
high-speed extruder 高速挤压机
high-speed extrusion 高速挤压
high-speed facsimile apparatus 高速传真机
high-speed fan blender 高速风扇式混和器
high-speed fascimile 高速传真
high-speed fibre optics strander 光纤高速绞合机
high-speed figure 高速数字
high-speed film 高速感光胶片;高感光度胶片
high-speed filter 高速滤池
high-speed flat bed sewing machine with blade 高速带刀平缝机
high-speed flat bed sewing machine with puller 高速滚筒平缝机
high-speed flat-screen washer 高速平网洗涤机
high-speed flip-flop 高速双稳态触发器
high-speed floating aerator 高速浮动式曝气器;高速浮动式曝气机
high-speed flow 高速水流
high-speed fluid filler 高速灌装机
high-speed forging 高速锻造(法)
high-speed forming 高速成型
high-speed forming machine 高速成形型
high-speed Francis turbine 高转速混流式水轮机;高转速辐向轴流式水轮机;高转速法兰西斯式水轮机
high-speed freezing centrifuge 高速冷冻离心机
high-speed freight vehicle 高速货车
high-speed gas chromatography 高速气相色谱(法)
high speed gear 高速齿轮
high-speed gear coupling 高速齿轮联轴节
high-speed gear hobber 高速滚齿机
high-speed gear shaper 高速插齿机
high-speed generator 高速发电机
high-speed grease 高速润滑脂;高速牛滑脂
high-speed grinding wheel 高速砂轮
high-speed guideway 高速导轨
high-speed hacksaw blade 高速钢锯条
high-speed hammer mill 高速锤磨机
high-speed hand grinder 高速手摇砂轮
high-speed handling 高速操纵
high-speed handpiece 高速机头
high-speed hardening furnace 高速淬火炉
high-speed head 高速头
high-speed heat 高速加热
high-speed heating furnace 高温快速加热炉
high-speed high resistance counter 高速高电阻计数器
high-speed highway 高速公路
high-speed hobbing 高速滚削

high-speed holographic(al) camera 高速全息摄影机
high-speed hot press 高速热压机;高速垫压机
high-speed hydraulic cylinder 高速液压缸
high-speed hydraulic hammer 高速液压锤
high-speed hydraulic hose coupling machine 高速液压软管接管机
high-speed imaging device 高速成像装置
high-speed increment plotter 高速增量绘图机
high-speed indicator 高速指示器
high-speed industrial flat bed sewing machine 高速工业平缝机
high-speed infrared detector 高速红外探测器
high-speed infrared film 高速红外胶片
high-speed infrared radiometer 高速红外辐射计
high-speed interactive retrieval system technique 快速交互检索系统技术
high-speed intercity rail 市际高速铁路
high-speed intercity traffic 高速城市间运输
high-speed internal grinder 高速内磨床
high-speed isostatic pressing 高速等静压制
high-speed jet 高速射流;主量孔
high-speed job selection 高速作业选择
high-speed joint 高速旋转接头
high-speed knife 高速切割器
high-speed label loom 高速标签织机
high-speed lane 高速车道
high-speed laser print 高速激光印刷
high-speed laser printer 高速激光打印机
high-speed laser recording equipment 高速激光记录设备
high-speed lathe 高速车床
high-speed lens 高速镜头
high-speed level recorder 高速水平记录器
high-speed lifeboat 救生快艇
high-speed lift 高速电梯
high-speed line 高速线路
high-speed line electric(al) machine 高速直线电机
high-speed line mill 高速线磨
high-speed line printer 高速宽行打印机
high-speed liquid chromatograph 高速液相色谱仪;高速液体色层分析仪
high-speed liquid chromatography 高速液相色谱(法)
high-speed liquid-liquid chromatography 高速液—液色谱法
high-speed locomotive 高速机车
high-speed look ahead carry generator 高速先行进位发生器
high-speed loop 高速环
high-speed low-noise synchronous motor 高速低噪声异步电动机
high-speed machine 高速钻机
high-speed machine tool 高速机车
high-speed maglev system 高速磁悬浮系统
high-speed magnetic amplifier 高速磁放大器
high-speed magnetic levitation 高速磁浮
high-speed mechanism 高速机构
high-speed membrane osmometer 高速膜渗透计
high-speed memory 高速存储器
high-speed metal cutting 高速切削法
high-speed microhole electron discharge machining tool 高速电火花小孔加工机床
high-speed microwave switch 高速微波转换
high-speed mill 高速轧机;高速磨机
high-speed milling attachment 高速铣工附件
high-speed milling machine 高速铣床
high-speed mixer 高速搅拌机;高速混合机;高速拌和机
high-speed mixer grinder 高速混合研磨机
high-speed mixing 高速混合;高速拌和
high-speed motion-picture camera 高速电影摄影机
high-speed motor 高速马达;高速电动机
high-speed multiplexer 高速多路复用器
high-speed multiplication 高速乘法
high-speed multiplier 高速乘法器
high-speed multiply 高速乘
high-speed needle valve 高速针阀
high-speed network 高速网络
high-speed numerical counter 高速数字计数器
high-speed operational amplifier 高速型运算放大器
high-speed optic(al) switch 高速(光学)开关

high-speed oscillograph 高速示波器
high-speed oscilloscope 高速(记录)示波器;快速扫描示波器;快速记录示波仪
high-speed paper cutter 高速切纸机
high-speed paper cutting machine 高速切纸机
high-speed paper machine 高速造纸机
high-speed paper sack bottomer 高速纸袋封底机
high-speed paper tape punch 高速纸带穿孔机
high-speed paper tape reader 高速纸带输入机
high-speed passenger elevator 高速载人电梯;高速载客电梯
high-speed passenger lift 高速载客电梯;高速客梯
high-speed passenger line 高速客运线路
high-speed passenger train 高速旅客列车
high-speed patching method 路坑快速修补法;快速修补法
high-speed perforator 高速凿孔机;快速凿孔机
high-speed peripheral 高速外围设备
high-speed photodetector 高速光电探测器
high-speed photograph 高速照相
high-speed photographic(al) technique 高速照相技术
high-speed photography 高速照相术;高速摄影;快速摄影
high-speed photometer 高速光度计
high-speed photometry 高速测光
high-speed planer 高速刨床
high-speed planing machine 高速刨床
high-speed plankton sampler 高速浮游生物取样器
high-speed plasma chromatography 高速等离子色谱
high-speed plate drier 高速制版机
high-speed plotter 高速绘图机;高速描绘器;高速绘图仪
high-speed plough 高速刨煤机;高速犁
high-speed pneumatic tool 高速气动工具
high-speed polishing 高速抛光
high-speed powderless etching machine 高速无粉腐蚀机
high-speed precision lathe 高速精
high-speed press 高速压机
high-speed pressing machine 高速压合机
high-speed printer 高速印刷装置;高速印刷机;高速打印机
high-speed printing press 高速印刷机
high-speed projector 高速放映机
high-speed pulverizer 高速粉磨机
high-speed pump 高速泵
high-speed pumping installation 高速抽水装置
high-speed pumping unit 高速抽气装置
high-speed radial drilling machine 高速径向钻孔机
high-speed railway 高速铁路
high-speed railway network 高速铁路网
high-speed raster photography 高速网络摄影术
high-speed reader 高速读出器;高速读出机
high-speed reading system 高速读数方式
high-speed read/punch diagnostic 高速读卡/穿孔诊断
high-speed reamer 高速钢铰刀
high-speed reception 高速收报;高速接收
high-speed reciprocating friction testing machine 高速往复式摩擦试验机
high-speed reclosing 高速重合
high-speed record 高速记录
high-speed recorder 高速记录器
high-speed reducing valve 高速减压阀
high-speed register 高速寄存器
high-speed regulator 高速调节器;快动调节器
high-speed relay 高速继电器;快速继电器
high-speed repair 快速检修
high-speed repair method 快速检修法
high-speed repetitive operation 高速重复操作
high-speed road towing 高速道路拖曳;高速道路牵引
high-speed roller mill 高速辊磨
high-speed rolls 高速辊式破碎机
high-speed rotary head milling machine 高速回转头铣床
high-speed rotary press 高速轮转印刷机
high-speed rotary printing ink 高速轮转油墨
high-speed rotary table 高速(旋)转台
high-speed rotating dish spray drier 高速转盘喷雾干燥机

high-speed runner 高速转轮
high-speed sand filtration 高速砂滤法;快速砂(过)滤
high-speed saw blade 高速钢锯条
high-speed schlieren photograph 高速纹影摄影
high-speed scintillation autoradiography 高速闪烁放射自显影术
high-speed scrubber 高速洗涤器;高速涤气器
high-speed section 高速区段
high-speed selector 高速选择器
high-speed selector channel 高速选择通道
high-speed separating spindle 高速分离锭子
high-speed serial printer 高速串行打印机
high-speed series tripping 高速串联断路法
high-speed servo motor 高速伺服电动机
high-speed shaft 高速轴
high-speed shaping machine 高速牛头刨床
high-speed ship 高速船舶
high-speed shutter 高速快门
high-speed simulation 高速仿真;快速仿真
high-speed slot-puncher 高速冲槽机
high-speed slow-scanning oscillograph 高速慢扫描示波器
high-speed small copy milling machine 高速小仿形铣床
high-speed solar wind 高速太阳风
high-speed spectrograph 高速摄谱仪
high-speed spectroscopy 高速分光学;高速波谱学
high-speed spindle oil 高速锭子油
high-speed spinner 高速分配轮
high-speed spray 高速喷雾
high-speed sprayer 高速喷雾器
high-speed starting 高速起动
high-speed static pressure bearing 高速静压轴承
high-speed steam curing 快速蒸汽养
high-speed steam curing process 快速蒸汽养护法
high-speed steel 硬钢;风钢;锋钢;高速钢
high-speed steel bit 高速钢钻头;高速钢车刀
high-speed steel cutter 高速钢刀具
high-speed steel die milling cutter 高速钢模具铣刀
high-speed steel drill 高速钢钻头
high-speed steel end mill 高速钢端铣刀
high-speed steel geared tap 高速钢磨牙丝锥
high-speed steel ground groove drill 高速钢槽钻
high-speed steel ground tap 高速钢磨削丝锥
high-speed steel ingot 高速钢锭
high-speed steel lathe tool 高速钢车刀;白钢车刀
high-speed steel machine saw blade 高速钢机用锯条
high-speed steel power hacksaw blade 高速钢机用锯条
high-speed steel sawblade 高速钢锯片
high-speed steel tool 高速钢刀具
high-speed steel tool bit 高速钢刀具
high-speed steel turning bit 高速钢车刀条
high-speed steel turning tool 高速钢车刀
high-speed steel twist drill 高速钢麻花钻
high-speed steel wire 高速钢丝
high-speed step-out relay 高速失步继电器
high-speed sterilizer 高速消毒器
high-speed storage 高速存储器
high-speed streak camera 高速条纹摄影机
high-speed stretch 高速拉伸
high-speed survey 高速巡检
high-speed sweep 高速扫描
high-speed switch 高速寻线机;高速开关;快速开关
high-speed synchroniser 高速同步器
high-speed tabletop centrifuge 高速台式离心机
high-speed tablet press 高速压片机
high-speed tape perforator 高速纸带穿孔机
high-speed tape punch 高速纸带穿孔机
high-speed tape reader 高速纸带读出器
high-speed target 高速目标
high-speed telegraph 快速电报
high-speed telemetering device 快速遥测设备
high-speed telemetry 高速遥测
high-speed telescope 高速望远镜;强光力望远镜
high-speed test 高速;快速试验
high-speed test controller 高速测试控制器
high-speed tester 高速试验机
high-speed testing technique 高速试验技术;高速试验方法
high-speed test track 高速试验跑道;高速试验轨道

high-speed texture meter 高速构造深度仪
high-speed thread book sewing machine 高速穿线订书机
high-speed thread cutting attachment 高速螺纹切削附件
high-speed three roller mill 高速三辊磨
high-speed tool 高速工具
high-speed tool bit 高速刀具
high-speed tool steel 高速工具钢;高工钢
high-speed track 高速线路;高速示踪;高速轨道
high-speed tracking laser 高速跟踪激光器
high-speed track inspection car 高速轨道检查车
high-speed trade-mark printing machine 高速商标印刷机
high-speed traffic 高速交通
high-speed train 高速列车
high-speed train prior than low speed train 先客后货
high-speed transit 高速交通
high-speed trenching machine 高速挖沟机
high-speed trunk line 高速干线
high-speed tunnel(l)ing 隧道快速掘进(法);快速隧道施工;高速隧道施工;隧洞快速掘进(法)
high-speed turbine 高速涡轮机
high-speed turbine aerator 高速涡轮机曝气器;高速透平曝气器
high-speed turbine stirrer 高速涡轮式搅拌器
high-speed turning tool 高速刀具;高速车刀
high-speed turnouts 高速道岔
high-speed type 高速型
high-speed type end mill 高速端铣刀
high-speed underground railroad 高速地下铁道
high-speed underground railway 高速地下铁道
high-speed universal lathe 高速万能车床
high-speed vacuum seamer 高速真空封罐机
high-speed valve 高速阀
high-speed vehicle lane 高速车道;快车道
high-speed vertical external broacher 高速立式外拉床
high-speed vertical miller 高速立式铣床
high-speed vertical universal milling machine 高速立式万能铣床
high-speed vessel 高速船舶
high-speed vibrator 高速振动器;高频振动器
high-speed vibratory screen 高速振动筛
high-speed voltage comparator 高速电压比较器
high-speed wash 高速冲洗
highspeed welding 快速焊接
high-speed whirly crane 高速回转式起重机
high-speed winder 高速缠绕机
high-speed winding machine 高速缠绕机
high-speed wind tunnel 高速风洞
high-speed wind-up 高速卷取
high-speed wire rod finishing block 高速线材精轧机
high-speed xerographic(al) printer 高速电摄影打印机
high-sphericity 高球度
high spiker 补漏钉工人(俚语)
high-spin 高自旋
high-spin complex 高自旋络合物
high-spin state 高自旋态
high-spot 高点;强放射性点;突出部分;凸起处(路面);高光强聚光灯;路隆
high-spot reform 重点改造
high-spot review 重点审查;重点考察;重点考查
high stability 高稳定性;高稳定度
high stability crystal oscillator 高稳定度晶体振荡器
high stability quartz crystal oscillator 高稳定度石英晶体振荡器
high stability temperature variation meter 高稳定度变温仪
high stabilized photomultiplier 高稳定光电倍增管
high stable laser 高稳定激光器
high stable oscillator 高稳定度振荡器
high stacked storage 高架储[贮]藏
high stack effect 烟囱效应
high stage compressor 高压压缩机;第二级压风机;二段式压风机
high stage microscope 高镜台显微镜
high stake rack truck 高侧板车厢
highstand 高水位期
high stand hose reel 高位水带卷

high steam 高压蒸汽
high steam pressure switch 高蒸汽压开关
high steel 硬钢;高碳钢
high stop filter 高截止滤波器
high strain-rate forming 高应变率成形;高压变率成形
high strain-rate phenomenon 高速应变现象
high strain-rate testing 高应变率试验
high strand integrity fibre 高度集束的原纱纤维
high stratosphere 平流层上部
high street 大街;正街
high strength 高强度
high-strength alloy 高强合金
high-strength alloyed steel 高强合金钢
high-strength alumina-chrome brick 高强铝铬砖
high-strength alumin(i)um alloy 高强度铝合金
high-strength and high-modulus glass 高强高弹玻璃
high-strength bar 高强钢筋;高强度粗钢筋
high-strength bevel steel washer 高强度(钢)斜垫圈
high-strength bolt 高强螺栓;高强度螺栓
high-strength bolt-welded 栓焊
high-strength brass 高强度黄铜
high-strength brick 高强砖
high-strength bronze 高强度青铜
high-strength canvas for conveyor belts 高强传送带帆布
high-strength casting pipe 高强度铸铁管
high-strength cast iron 高铁力铸铁;高速铸铁;高强度铸铁
high-strength cement 高强(度)水泥;高标号水泥
high-strength china 高强度瓷
high-strength cinder sand block 高强(度)煤渣砂砌块
high-strength coal 高强度煤
high-strength composite 高强度复合材料
high-strength concrete 高强(度)混凝土
high-strength construction steel 高强结构钢
high-strength deformed bar 高强异形钢筋
high-strength diamond 高强度金刚石
high-strength diatomaceous brick 高强硅藻土砖
high-strength drill pipe 高强度钻探管
high-strength driving shaft 高强度传动轴
high-strength electrode 高强度焊条
high-strength enamelled wire 高强度漆包线
high-strength expanding grout 高强膨胀性浇灌;高强度膨胀(性)水(泥)浆;高强膨胀性灌浆
high-strength explosive 高威力炸药
high-strength fiber 高强度纤维
high-strength field 高场
high-strength flame resistant fiber 高强(度)耐火纤维
high-strength friction grip bolt 高强摩擦夹紧螺栓;高强度摩阻螺栓;高强度摩擦紧固螺栓
high-strength glass disc 高强度玻璃盘
high-strength glass fiber 高强度玻璃纤维
high-strength granulated cinder tile 高强(度)颗粒煤渣地砖
high-strength granulated slag tile 高强(度)颗粒炉渣地砖
high-strength gypsum 高强(度)石膏
high-strength gypsum ga(u)ged plaster 高强石膏调麻灰浆;高强石膏罩面灰泥
high-strength hexamine-containing wastewater 含高浓度六胺废水
high-strength high-modulus polymer 高强度高模量聚合物
high-strength hot rolled reinforced bar 高强度热轧钢筋
high-strength hull steel 高强度船体结构钢
high-strength hydraulic cylinder 高强度液压缸
high-strength insulator 高强度绝缘子
high-strength integrated wastewater 高浓度综合废水
high-strength lightweight concrete 高强轻质混凝土
high-strength low-alloy steel 高强度耐低温钢;高强(度)低合金钢;低合金高强度钢
high-strength material 高强度材料
high-strength metallic alloy 高强度金属合金
high-strength micro-crystalline glass 高强度微晶玻璃
high-strength mortar 高强(度)砂浆

high-strength nitrogeneous wastewater 含高浓度氮废水
high-strength nut 高强度螺帽
high-strength oil well cement 高强油井水泥
high-strength optic(al)fiber 高强度光纤
high-strength organic waste 高强度有机废水;高含量有机废物
high-strength parts 高强度部件
high-strength pipe 高强度管
high-strength plain steel washer 高强度平钢垫圈
high-strength plain washer 高强度平垫圈
high-strength plaster 高强度石膏
high-strength porcelain bushing 高强瓷套管
high-strength porcelain bushing shell 高强度瓷套壳
high-strength Portland cement 高强硅酸盐水泥;高强波特兰水泥
high-strength quality steel wire 高强优质钢丝
high-strength radio 高强度放射
high-strength rail 高强度钢轨
high-strength reinforcement 高强钢筋
high-strength reinforcing 高强度钢筋
high-strength reinforcing steel 高强钢
high-strength rivet 高强铆钉
high-strength rock 强度高的岩石
high-strength screw 高强螺钉
high-strength semi-killed steel 高强度半镇静钢
high-strength semi-lightweight concrete 高强半轻质混凝土
high-strength shotcrete 高强喷散混凝土
high-strength silicone rubber 高强度硅橡胶
high-strength small concrete block 高强度混凝土小块
high-strength steel 高强钢;高强度钢(材)
high-strength steel electrode 高强度钢焊条
high-strength steel plate 高强度钢板
high-strength steel washer 高强度(钢)垫圈
high-strength(steel)wire 高强钢丝
high-strength structural adhesive 高强度结构黏合剂
high-strength structural steel 高强度结构钢
high-strength superabrasive wheel 高强度超硬砂轮
high-strength suspension insulator 高强度悬式绝缘子
high-strength synthetic(al)fiber 高强度合成纤维
high-strength thermal resistant alloy 高强度耐热合金
high-strength tube 高强度管
high-strength wastewater containing high-level of sulfate 含高浓度硫酸盐的高强度废水
high-strength weldable alumin(i)um alloy 高强度可焊铝合金
high-strength welded pipe 高强度焊接管
high-strength welded steelpipe 高强度焊接钢管
high-strength yellow brass 高强度黄铜
high-stress 高应力
high stress fiber 高应力纤维
high stroke 高冲程
high stroke friction press 高冲程摩擦压机
high styrene resin 高苯乙烯树脂
high styrene rubber 高苯乙烯橡胶
high sudsing detergent 高泡沫洗涤剂
high sulfate-resistant cement 高抗硫水泥
high sulfate-resistant oil well cement 高抗硫油井水泥
high sulfur coal 高硫煤
high sulfur combustor 高硫燃料
high sulfur content fuel 高含硫燃料
high sulfur crude 高硫原油
high sulfur crude oil 高硫原油
high sulfur fossil fuel 高硫矿物燃料
high sulfur fuel 高硫燃料
high sulfur gas oil 高硫瓦斯油
high sulfur oil 高硫(石)油
high sulfur residues 高硫残液
high sulfur steel 高硫钢
high sulphur coal 高硫煤
high sulphur coke 高硫焦碳
high sulphur content 高硫的
high sulphur fuel 高硫燃料
high sulphur fuel oil 高硫燃料油
high sulphur heavy crude 高含硫重质原油
high sulphur light crude 高含硫轻质原油

high sulphur residues 高硫残渣
high summer 盛夏
high supercharged engine of low compression ratio 低压缩比高增压发动机
high swell 狂涌;强涌;高膨胀
high symmetry 高(度)对称
high table 大宴会中较其他桌子高的餐桌
high-talc body 高滑石含量瓷坯
high tar coal 高油煤
High Tech 高技派(在建筑造形风格上着重表现高度生产技术);高科技;高技术
high tech cement based composite 高技术水泥复合材料
high tech circle 高技术圈
high-technical industrial development area 高新技术产业区
high technology 高(等)技术
high technology ceramics 高技术(陶)瓷
high technology deterrence 高技术威慑力
high technology for civilian use 民用高技术
high technology industry 高科技产业;高技术产业
high technology robot 高级技术机器人
high technology sector 高技术部门
high tech products 高技术产品
high temperate controlling 高温调查
high-temperature 高温
high-temperature absorption heat pump 高温型吸收式热泵
high-temperature adhesive 高温黏合剂;高温胶黏剂
high-temperature aerodynamics 高温气体动力学
high-temperature ag(e)ing 高温时效
high-temperature age 高温年龄
high-temperature alloy 高温合金;耐热合金
high-temperature and high-pressure cake 高温高压泥饼
high-temperature and high-pressure consistometer 高温高压稠度仪
high-temperature and high-pressure cut-off valve 高温高压截止阀
high-temperature and high-pressure dy(e)ing 高温高压染色
high-temperature and high-pressure dy(e)ing machine 高温高压染色机
high-temperature and high-pressure filter loss 高温高压失水
high-temperature and high-pressure filter press 高温高压失水仪
high-temperature and pressure electrochemistry 高温高压电化学
high-temperature and special inorganic coating 高温及特种无机涂层
high-temperature annealing 高温退火
high-temperature anti-sticking coating 高温防黏涂层
high-temperature baked enamel finish 高温烤漆饰面
high-temperature baking 高温烘烤
high-temperature behavio(u)r 高温性能
high-temperature belt 高温带
high-temperature blower 高温鼓风机
high-temperature box-type furnace 高温箱形电炉
high-temperature brazed joint 高温铜焊接头;高温硬焊接合
high-temperature brazing gas cooling furnace 高温钎焊气冷炉
high-temperature brine 高温热卤水
high-temperature brittleness 高温脆性
high-temperature burned anhydrite 高温煅烧石膏
high-temperature cabinet 高温箱
high-temperature calcinations of(raw)gypsum 高温煅烧石膏
high-temperature camera 高温照相机
high-temperature carbonization 高温碳化;高温干馏
high-temperature carburizing 高温渗碳
high-temperature cellulose-decomposting bacteria 高温纤维素分解菌
high-temperature cement 高温水泥;耐高温水泥
high-temperature cemented carbide 耐高温硬质合金
high-temperature ceramics 高温陶瓷
high-temperature chemical reaction engineering 高温化学反应工程

high-temperature chemistry 高温化学
high-temperature chimney 高温烟囱
high-temperature circuit 高温循环回路;高温循回路线
high-temperature circulation path 高温循环路线
high-temperature coal tar 高温焦油沥青;高温煤焦油沥青(或煤沥青)
high-temperature coating 耐高温涂料;耐高温涂层
high-temperature coaxial cable 高温同轴电缆
high-temperature coke 高温焦炭
high-temperature colo(u)r-changing porcelain 高温变色釉瓷
high-temperature colo(u)red glaze 高温颜色釉
high-temperature components 高温元件
high-temperature concrete 耐火混凝土;耐高温混凝土
high-temperature conductive paint 高导热性涂料
high-temperature conversion catalyst 高温变换催化剂
high-temperature corrosion 高温腐蚀
high-temperature cracking 高温裂化
high-temperature creep 高温蠕变
high-temperature creep and rupture testing machine 高温蠕变及持久强度试验机
high-temperature crucible 耐高温坩埚
high-temperature crystal growth 高温晶体生长
high-temperature curing 高温养护法;高温固化
high-temperature curing under pressure 高温加压养护
high-temperature cut-off valve 高温闸阀
high-temperature decorating colo(u)r 高温色料
high-temperature deflection test bed 高温曲绕试验台
high-temperature deformation 高温变形
high-temperature dehydration 高温去水
high-temperature demo(u)lding agent 高温脱膜剂
high-temperature detector 高温探测器
high-temperature dielectric(al) coating 高温介电涂层
high-temperature diffusion 高温扩散
high-temperature digestion 高温煮解;高温溶解
high-temperature digestive treatment 高温消化处理
high-temperature dilatometer 高温膨胀仪
high-temperature disc feeder 高温圆盘给料机
high-temperature discolo(u)ration 高温变色
high-temperature dispersant 高温分散剂
high-temperature dispersing furnace 高温扩散炉
high-temperature distillation 高温蒸馏
high-temperatured photomultiplier 高温光电倍增管
high-temperature drier 高温干燥机
high-temperature drying 高温干燥
high-temperature dy(e)ing machine 高温染色机
high-temperature effect 高温效应
high-temperature electric(al) heating panel 高温电热板
high-temperature electric(al) insulating coating 高温电绝缘涂层
high-temperature electric(al) resistance furnace 高温电炉
high-temperature electric(al) wire 高温电线
high-temperature electrochemistry 高温电化学
high-temperature electrostatic precipitator 高温静电沉淀器
high-temperature electrothermal alloy wire 高温电热合金丝
high-temperature embrittlement 高温致脆
high-temperature enamel 高温搪瓷
high-temperature engineering material 高温工程材料
high-temperature environment 高温环境
high-temperature equilibrium method 高温平衡法
high-temperature exhaust 高温排放
high-temperature exposure test 高温曝晒试验
high-temperature extensometer 高温引伸计
high-temperature fatigue testing machine 高温疲劳试验机
high-temperature field 高温场
high-temperature fired magnesite chrome brick 高温烧成镁铬砖
high-temperature flare 高温耀斑
high-temperature flexural strength tester (high-temperature MOR tester) 高温抗弯强度测试仪
high-temperature flue gas 高温烟气
high-temperature fuel cell 高温燃料电池
high-temperature furnace 高温炉
high-temperature gain fall 高温增益下降
high-temperature gas chromatography 高温气相色谱(法)
high-temperature gas-cooled reactor 高温气体冷却式反应器;高温气冷反应堆
high-temperature gas filter 高温燃气过滤器
high-temperature gas physical mechanics 高温气体物理力学
high-temperature gel 高温凝胶
high-temperature glaze 高温釉
high-temperature grease 高温润滑油
high-temperature heat 高温熔炼
high-temperature heater 高温加热器
high-temperature heat pump 高温热泵
high-temperature heat-resisting aluminum paint 耐高温银灰漆
high-temperature heat source 高温热源
high-temperature-high-pressure apparatus 高温高压设备
high-temperature hook-up wire 高温安装线
high-temperature-hot-water 高温(热)水
high-temperature hot water pump 高温热水泵
high-temperature hydrofluorination 高温氢氟化作用
high-temperature hydrogen furnace 高温氢气炉
high-temperature indicator 高温指示器
high-temperature induction heating 高温感应加热
high-temperature infrared transmitting glass 高温红外透射玻璃
high-temperature infrared window 高温红外窗口
high-temperature inorganic coating 高温无机涂层
high-temperature insostatic pressing 热等静压
high-temperature insulant 隔高温材料
high-temperature insulating coating 高温隔热涂层
high-temperature insulating material 高温隔热材料
high-temperature insulation 高温(度)绝缘;耐热绝缘
high-temperature insulation coating 高温隔热涂层;耐高温绝缘涂层
high-temperature investment 高温包埋料
high-temperature isostatic pressing 高温等静压
high-temperature kiln drying 高温窑干
high-temperature lacquer 高温漆
high-temperature lasting tester 高温持久强度试验机
high-temperature levelling agent 高温匀染剂
high-temperature life 高温下使用寿命
high-temperature limit switch 高温限制开关
high-temperature long loop ager 长环高温蒸化器
high-temperature long loop steamer 长环高温汽蒸机
high-temperature/low-pressure metamorphic belt 高温低压变质带
high-temperature lubricant 高温润滑剂
high-temperature lubricating coating 高温润滑涂层
high-temperature machining 高温切削;高温加工
high-temperature manganese steel 高温度锰钢
high-temperature martensite 高温马氏体
high-temperature material 高温材料
high-temperature material technology 高温材料工艺学
high-temperature metallographic(al) microscope 高温金相显微镜
high-temperature metallography 高温金相学
high-temperature metering pump 高温计量泵
high-temperature microscope 高温显微镜
high-temperature millivoltmeter 高温毫伏计
high-temperature mill scale 高温轧制铁鳞
high-temperature modification 高温型
high-temperature MOR tester 高温抗弯强度测试仪
high-temperature nuclear particle detector 高温核粒子探测器
high-temperature oil furnace 高温油炉
high-temperature oil well cement 高温油井水泥
high-temperature operation 高温作业;高温操作
high-temperature operation life test 高温工作寿命试验
high-temperature optic(al) fiber sensor 高温光纤传感器
high-temperature orogenic zone 高温造山带
high-temperature oscillating crucible visco(si)meter 高温振动坩埚黏度计
high-temperature oxidation 高温氧化
high-temperature oxidation method 高温氧化法
high-temperature oxidation-resistant coating 高温抗氧化涂层
high-temperature oxide 高温氧化物
high-temperature package dy(e)ing machine 高温筒子染色机
high-temperature pasteurization 高温巴氏灭菌法
high-temperature permeability 高温透气性
high-temperature phenomenon 高温现象
high-temperature phosphate 高温磷酸盐
high-temperature physics 高温物理学
high-temperature pitch 高温硬沥青
high-temperature polymerization 高温聚合
high-temperature preservation 高温保藏法
high-temperature pressing 高温压制
high-temperature-pressure beck 高温高压绳状染色机
high-temperature-pressure dy(e)ing machine 高温高压染
high-temperature-pressure jigger 高温高压卷染机
high-temperature-pressure sample dy(e)ing machine 高温高压样品染色机
high-temperature process equipment 高温处理设备
high-temperature processing 高温处理
high-temperature pyrolytic cracking 高温热解法
high temperature quartz 高温石英
high-temperature radioactive substance 高温放射性物质
high-temperature radio-frequency cable 高温射频电缆
high-temperature rapid firing 高温快烧
high-temperature rapid firing stains 高温快烧斑点
high-temperature reaction 高温反应
high-temperature reactor 高温反应堆
high-temperature refractory material 高温耐火材料
high-temperature region 高温区
high-temperature reservoir 高温储热器
high-temperature resistance 耐高温性
high-temperature resistant fiber 耐高温纤维
high-temperature resistant material 耐高温材料
high-temperature resistant polymer 耐高温聚合物
high-temperature resisting refractory concrete 耐高温耐火混凝土
high-temperature roasting 高温焙烧
high-temperature salt bath electrode furnace 高温电极盐浴电阻炉
high-temperature sample dy(e)ing machine 高温样品染色机
high-temperature scaling 高温剥落
high-temperature science 高温科学
high-temperature sensor 高温传感器
high-temperature series expansion 高温级数展开
high-temperature service 高温作业;高温设备(装置)
high-temperature setting 高温定形
high-temperature setting glue 高温固化胶
high-temperature shift 高温变换
high-temperature short-time 高温短时灭菌法
high-temperature short-time pasteurization 高温快速灭菌法;高温短时(巴氏)灭菌法
high-temperature side 高温测
high-temperature sintering 高温烧结
high-temperature sludge 高温氧化淤渣
high-temperature solar energy collector 高温太阳能收集器
high-temperature solution 高温溶液
high-temperature solution growth 高温溶液生长
high-temperature solution method 高温溶液法
high-temperature spectrum 高温谱
high-temperature sprinkler 高温洒水喷头;高温喷水器
high-temperature stability 高温稳定性
high-temperature stability test 高温稳定性试验
high-temperature state equation 高温状态方程
high-temperature steam curing 高温蒸汽养护;蒸釜养护
high-temperature steaming 高温汽蒸法

high-temperature steel 耐热钢;耐高温钢
high-temperature sterilization 高温灭菌法
high-temperature sterilizing tunnel oven 高温灭菌隧道烘箱
high-temperature storage life test 高温储[贮]存寿命试验
high-temperature strain ga(u)ge 高热应变仪;张线式高温传感器
high-temperature strength 高温强度;热强度
high-temperature strength stabilizer 高温强度稳定剂
high-temperature strength test 高温强度试验
high-temperature structural material 高温结构材料
high-temperature sulfur-containing solution thermal water 高温含硫热水
high-temperature superconducting ceramics 高温超导体陶瓷
high-temperature superconducting material 高温超导体材料
high-temperature superconductor 高温超导体
high-temperature surface 高温受热面
high-temperature synthesis under atmosphere 常压高温合成法
high-temperature tar 高温(煤)焦油
high-temperature technology 高温技术;高温工艺(学)
high-temperature test 高温试验;耐高温试验
high-temperature test chamber 高温试验箱;高温试验室;高温测试箱
high-temperature textile 高温度纺织品
high-temperature textile printing wastewater 高温印染废水
high-temperature thermal resistor ceramics 高温热敏陶瓷
high-temperature thermal system 高温地热系统
high-temperature thermistor 高温热敏电阻器
high-temperature thermocouple 高温热电偶
high-temperature thermodynamics 高温热力学
high-temperature thermoelectric(al) furnace 高温热电炉
high-temperature thermogravimetric(al) apparatus 高温热重力测定仪
high-temperature thermomechanical treatment 高温形变热处理
high-temperature thermometer 高温温度计
high-temperature thermosetting 高温热定形
high-temperature top burning hot blast stove 高温顶燃式热风炉
high-temperature torsional strength 高温扭曲强度
high-temperature total hydrocarbon analyzer 高温总烃量分析仪
high-temperature treatment 高温处理
high-temperature tunnel kiln 高温隧道窑
high-temperature ultraviolet microscope 高温紫外显微镜
high-temperature vacuum furnace 高温真空炉
high-temperature vacuum furnace microscopy 高温真空炉显微术
high-temperature varnish 高温清漆
high-temperature visco(si)meter 高温黏度计
high-temperature visible column thermometer 视柱式高温计
high-temperature waste(water) 高温废水
high-temperature water 高温水
high-temperature water boiler 高温热水锅炉
high-temperature water heating system 高温水采暖系统
high-temperature water pump assembly 高温水泵总成;高温水泵结合组
high-temperature water reservoir 高温热水储
high-temperature wax 高温蜡
high-temperature well 高温井
high-temperature winch-beck 高温高压绳状染色机
high-temperature workshop 高温车间
high-temperature X-ray diffraction analysis 高温 X 射线衍射分析
high-temperature zone 高温区
high tempering 高温回火
high temper steel 高温回火钢
high tenacity 高韧性;高韧度
high tenacity cord 高强力帘线
high tenacity fiber 高强力纤维
high tenacity nylon filament 高强力耐纶丝

high tenacity rayon 高强度人造丝
high tenacity staple fiber 高强力黏胶纤维
high tenacity viscose fiber 高强力黏胶纤维
high-tensile 高张拉的
high-tensile alloy 高张拉合金;高强合金;高抗拉合金;耐拉合金
high-tensile alloy steel bar 高强合金钢钢筋
high-tensile bolt 高强(度)螺栓
high-tensile brass 高强度黄铜
high-tensile bronze 高强度青铜
high-tensile cast iron 高强度铸铁
high-tensile deformation bar 高张力异形钢筋
high-tensile deformed bar 高强异形钢筋
high-tensile deformed steel bar 高拉力异形钢筋
high-tensile electric(al) welding rod 高强焊条;高强度电焊条
high-tensile quality 高抗拉性能
high-tensile reinforcement 高强(受拉)钢筋
high-tensile reinforcing steel 高强力钢筋;高强钢筋
high-tensile round-link chain 高强度圆环链
high-tensile steel 高张拉钢;高强(度)钢;高拉力钢筋;高抗拉钢;高级钢;耐拉钢
high-tensile steel bar 高强钢;高强(度)钢筋;高强度粗钢筋
high-tensile steel derrick 重型钻塔
high-tensile steel plate 高强钢板
high-tensile steel reinforcement bar fixture 高强钢筋夹具
high-tensile steel tendon 高强(预应力)钢丝束
high-tensile steel tube 高强度钢管
high-tensile steel wire 高强力钢丝
high-tensile strength 高抗拉强度;高拉力;高抗张强度
high-tensile strength wire 高强力钢丝
high-tensile structural steel 高抗拉结构钢;高强度结构钢
high-tensile thin steel plate 高强薄钢板
high-tensile wire 高强(度)钢丝
high tension 高压(电);大拉力;强大拉力;高张力;高张拉;高强度;高拉力;高电压
high-tension ammeter 高压电流计;高压安培计
high-tension apparatus 高压设备
high-tension arc 高压电弧
high-tension band 拉力钢皮带
high-tension battery 高压电池组
high-tension bolt 高强度螺栓;高强抗拉螺栓
high-tension boost 高压提升
high-tension bridge 高压电桥
high-tension busbar 高压母线;高压汇流排;高压电线
high-tension bushing 高压套管
high-tension cab(in) 高压室;高压柜
high-tension cable 高压电缆
high-tension cell 高压间隔
high-tension ceramics 高压陶瓷
high-tension circuit 高压线圈;高压电路
high-tension coil 高压线圈
high-tension compartment 高压间
high-tension coupling condenser 高压耦合电容器
high-tension current 高压电流
high-tension current inductor 高压电流互感器
high-tension current transformer 高压电流互感器
high-tension damper 高压阻尼器
high-tension delay time 高压迟加时间
high-tension detonator 高压引燃剂;高压(引爆)雷管;高压引爆雷管;高压起爆雷管;高压起爆剂
high-tension direct current generator 高压直流发电机
high-tension direct current plant 高压直流装置
high-tension direct current potentiometer 高压直流电位差计
high-tension distribution box 高压配电箱
high-tension distribution center 高压配电中心
high-tension distribution panel 高压配电盘
high-tension distributor 高压分电器
high-tension drop type fuse 高压跌落式熔断器
high-tension egg insulator 高压蛋形绝缘子
high-tension electric(al) porcelain insulator 高压电瓷绝缘子
high-tension electricity 高压电
high-tension fuse 高压熔断器
high-tension ignition 高压点火;火花点火
high-tension ignition coil 高压发火线圈
high-tension indoor wall entrance bushing sleeve 高压户内穿墙套管

high-tension insulating gloves 高压绝缘手套
high-tension insulator 高压电瓷;高强绝缘子
high-tension kilowatt-hour meter 高压电度表
high-tension laboratory 高压试验室
high-tension lightning arrester 高压避雷装置
high-tension line 高压线路
high-tension loop 高压回路
high-tension magneto 高压永磁(发)电机;高压磁电机
high-tension main 高压馈电线
high-tension measuring instrument 高压测量仪器
high-tension megger 高压兆欧表
high-tension mercuric bulb 高压水银灯泡
high-tension mercury lamp ballast 高压水银灯镇流器
high-tension mercury search light 高压水银投光灯
high-tension mercury street light 高压水银路灯
high-tension mercury-vapo(u)r spotlight 高压水银投光灯
high-tension motor 高压电动机
high-tension network 高压电力网
high-tension oil filled cable 高压充油电缆
high-tension oil switch 高压油开关
high-tension overhead line 高压架空线路
high-tension pin-type insulator 高压针式绝缘子
high-tension porcelain insulator 高压电瓷绝缘子;高压瓷瓶
high-tension porosity tester 高张拉孔隙度试验机
high-tension power line 高压输电线
high-tension power plant 高压发电厂
high-tension power supply 高压电源
high-tension pulse 高压脉冲
high-tension radio-graphic(al) apparatus 高压 X 线摄影机;高压摄影 X 线机
high-tension regulating transformer 高压调节变压器
high-tension relay 高压继电器
high-tension rod 高强度粗钢筋
high-tension room 高压室;高压间
high-tension rubber gloves 高压橡皮手套
high-tension selenium rectifier 高压硒整流器
high-tension shackle insulator 高压绝缘瓷瓶
high-tension shielding 高压屏蔽
high-tension side 高压侧
high-tension spark 高压火花
high-tension spark method 高压火花法
high-tension spark plug 高压火花塞
high-tension stator coil 高压定子线圈
high-tension steel 高强度钢
high-tension steel wire 高强钢丝
high-tension supply main 高压馈电干线
high-tension switch 高压开关
high-tension switchboard 高压开关板
high-tension switch cabinet 高压开关柜
high-tension switch gear 高压开关装置;高压开关设备;高压开关柜
high-tension terminal 高压接柱
high-tension test pencil 高压测电笔
high-tension thread rolling machine 高压螺纹滚轧机
high-tension transformer 高压变压器;高压变换器
high-tension transmission line 高压输电线(路)
high-tension valve 高压阀(门)
high-tension valve type lightning arrester 高压阀式避雷器
high-tension voltmeter 高压伏特计
high-tension warm-up time 高压预热时间
high-tension winding 高压绕组;高压绕线
high-tension wire 高压线
high-tension wire to 1st cylinder 第一汽缸高压
high terrace building 高台建筑
high terraced 台榭
high test 优质的;经严格试验的;严格考验(的)
high test bleaching powder 高级漂白粉
high test cast iron 高级铸铁
high test cast steel 高级铸钢
high test cement 优质水泥;高级水泥
high test chain 高强度链条
high test hypochloride 优质次氯酸盐
high test hypochlorite 高级次氯酸盐
high test metal 高质量金属
high thermal conductivity 高导热性
high thermal water 高温热水

high thickness emulsion 高稠度乳胶
high threshold 高阈值
high threshold logic circuit 高抗电路
high thrust drill 高钻压钻机;高推力钻机
high thrust rig 高钻压钻机;高推力钻机
high tide 高潮;满潮
high tide alarm 高潮报警器
high tide forecasting 高潮预报
high tide high water 最高高潮;最高高水位
high tide level 高潮(水)位;高潮面
high tide line 高水位;高潮线
high tide mark 高潮线
high tide of water 水的高潮
high tide shoreline 高潮滨线;高潮岸线
high tide slack(water) 高潮憩流
high tide stage 高潮水位
high tide surface 高潮面
high tied static crane 高杆固定起重机;高杆固定吊车
high time 高空作业津贴;正合时机;时机成熟
high time allowance 高空作业津贴;高空作业费
high tin babbit 高锡巴氏合金
high tin bronze 高锡青铜
high titanium oxide type electrode 高(氧)化钛型焊条
high titanium slag 高钛渣
high tomb 塔形墓
high tone loudspeaker 高音喇叭
high top heat 高温炉顶
high top knurled nut 滚花高螺母
high-top-pressure capability 炉顶高压能力
high torque 高转矩
high torque alternate current motor 大转矩交流电动机
high torque and low-speed motor 大转矩低速电动机
high torque at low-speed 低速大转矩
high torque capacity 高转矩
high-torqued bolt 高抗矩螺栓
high torque motor 高起动转矩电机
high torque on the drill string 钻杆柱受高扭矩负荷
high torque rotary drill 高转矩回转钻机;大转矩回转钻机
high torque test 高扭矩试验
high-to-width ratio 高阔比;高宽比
high traffic area 大业务量地区(通信)
high traffic ship 高通信业务灵船舶
high transconductance tube 高跨导管;高互导管
high transmission glass 高透光玻璃;高透射玻璃;高透明玻璃
high transmittance optic(al)glass 高透射光学玻璃
high transportation volume 高运量
high troposphere 对流层上部
high-trust sprinkler 高穿透洒水喷头
high tulization limit refractory 高使用极限值耐火材料
high turbidity water 高浊度水
high turbidity water sedimentation 高浊度水沉淀
high turbidity zone 高浊带
high turbine 大功率透平
high-turbulence combustion 高紊流燃烧
high turbulence combustion chamber 高紊流燃烧室
high-type 高级的
high-type highway 高级公路
high-type layout 高式布置(机电)
high-type pavement 高级铺面;高级路面
high-type road 高级道路
high-type surface 高级路面
high usage 利用率高
high usage group 高利用率干线组
high usage line 利用率高的线路
high usage route 高效路由
high usage trunk 高利用率传输线;高度使用的传输线
high vacuum 高(度)真空
high-vacuum apparatus 高真空设备
high-vacuum asphalt 高真空沥青
high-vacuum brazing material 高真空钎料;高真空焊料
high-vacuum breakdown 高真空击穿
high-vacuum coating machine 高真空镀膜机

high-vacuum coating unit 高真空镀膜机
high-vacuum connector 高真空接插件
high-vacuum distillation 高真空蒸馏
high-vacuum electron-beam fusion welding 高真空电子束熔焊
high-vacuum electron tube 高真空电子管
high-vacuum evapo(u)ration technique 高真空蒸镀技术
high-vacuum evapo(u)rator 高真空蒸发器
high-vacuum film plating table 高真空镀膜台
high-vacuum fitting 高真空组件
high-vacuum furnace 高真空炉
high-vacuum ga(u)ge 高真空计;高真空测量压力表
high-vacuum grease 高度真空油脂
high-vacuum induction heater furnace 高真空感应加热炉
high-vacuum installation 高真空装置
high-vacuum insulation 高真空绝热
high-vacuum jar 高真空瓶
high-vacuum melting furnace 高真空冶炼炉
high-vacuum metallizing 高真空敷镀金属
high-vacuum oil diffuse pump 高真空油扩散泵
high-vacuum orbital simulator 高真空轨道运行模拟器
high-vacuum phototube 高真空光电管
high-vacuum pump 高(度)真空泵
high-vacuum rectifier 高真空整流器
high-vacuum rectifier tube 高真空整流管
high-vacuum seal 高真空密封
high-vacuum superinsulation cryogenic container 高真空超绝热低温容器
high-vacuum switching tube 高真空开关管
high-vacuum system 高真空系统
high-vacuum technology 高真空技术
high-vacuum television tube 高真空显像管
high-vacuum tube 高真空管
high-vacuum valve 高真空整流管;高真空阀(门)
high valence 高价
high valence ion 高价离子
high valley 高天沟;高山谷
high value 高位值
high-value company 商业中心区消防队
high value scrap 高值废料
high vanadium steel 高矾钢
high-vaulted 成高拱顶的
high veld 高韦尔得草原
high-velocity 高速(度)
high-velocity air conditioning system 高速空调系统
high-velocity aircraft rocket 高速航空火箭
high-velocity beam scanning image pick-up tube 高速射束扫描摄像管
high-velocity blasting agent 高速炸药;烈性炸药
high-velocity boat 高速艇
high-velocity booster 高速引爆剂
high-velocity brittleness 高速脆性
high-velocity burner 高速燃烧器;高速烧嘴
high-velocity camera tube 高速摄像管
high-velocity channel 高速渠道
high-velocity combustion 高速燃烧
high-velocity computer 高速计算机
high-velocity conduit 高速水道
high-velocity convection burner 高速对流燃烧器
high-velocity craft 高速艇
high-velocity double-chamber bed 高流速双室床
high-velocity duct 高速管道
high-velocity duct system 高速风道系统
high-velocity electron beam 高速电子束;快速电子束
high-velocity electron camera tube 高速电子摄像管
high-velocity explosive 高速炸药
high-velocity flow 高速水流;高速水道
high-velocity fluidized bed 高速流化床
high-velocity fog 高速喷雾
high-velocity forging 高速锻造(法)
high-velocity forging machine 高速锻造机
high-velocity forming 高速成形
high-velocity forming die 高速成形模具
high-velocity ground roll 高速地滚(波)
high-velocity gunite 高速喷浆
high-velocity head 高速头
high-velocity impact 高速碰撞

high-velocity jet 高速射流;高速喷射
high-velocity jet flow 高速射流
high-velocity layer 高速带
high-velocity liquid jet machining 高速射流机加工
high-velocity loop 高速度回路
high-velocity ram machine 高速冲压机;高速冲床
high-velocity scanning 高速(电子)扫描
high-velocity shooting flow 高速射流
high-velocity shooting jet 高速射流
high-velocity star 高速星
high-velocity supply air system 高速送风系统
high-velocity thermo couple 快速反应热电偶
high-velocity tool 高速工具
high-velocity wash 高速冲洗
high-velocity water 高速流水
high vertical development cloud 直展云
high-very-high frequency 上限甚高频
high viscosity 高黏性;高黏度
high-viscosity asphalt(ic)cutback 高黏(滞)度液体沥青
high-viscosity crude oil field 高黏度油气田
high-viscosity fluid 高黏性流体;高粒性流性
high-viscosity fracturing fluid 高黏度压裂液
high-viscosity glue spraying machine 高黏度涂胶机
high-viscosity index 高黏性指数;高黏度指数
high-viscosity index lubricating oil 高黏度指数润滑油
high-viscosity index oil 高黏度指数油液
high-viscosity nitrocellulose 高黏度硝化纤维素
high-viscosity oil 高黏性油;高黏石油
high-viscosity pad 高黏前置垫
high-viscosity paint 高黏性涂料
high-viscosity solvent refined oil 高黏度溶剂精制油
high-viscosity tar 高黏性焦油
high-viscosity triple screw pump 高黏度三螺杆泵
high visibility 高能见度
high visibility paint 高可见度漆
high visibility pigment 高可见度颜料
high visibility reflective fabrics 高可见反射织物
high volatile 高挥发性的
high volatile bituminous coal 高挥发烟煤;高挥发性沥青煤
high volatile coal 高挥发煤
high volatile cutbacks 高挥发性稀释剂
high-voltage 高压【电】;高电压
high-voltage accelerator 高压加速器
high-voltage annunciator 过电压报警器
high-voltage anode 高压阳极
high-voltage arc 高压电弧
high-voltage arc technique 高压电弧技术
high-voltage bleeder 高压分压器
high-voltage block 高压堆
high-voltage board 高压柜
high-voltage bridge 高压电桥
high-voltage bushing 高压套管
high-voltage cab(in) 高压室;高压柜
high-voltage cable 高压线;高压电缆
high-voltage cable end box 高压电缆终端盒
high-voltage capacitor 高压电容器
high-voltage ceramic capacitor 高压陶瓷电容器
high-voltage ceramic vacuum switch tube 高压陶瓷真空开关管
high-voltage circuit 高压电路
high-voltage circuit breaker 高压断路器;高压开关
high-voltage circuit tester 高压电路测试器
high-voltage coil 高压线圈
high-voltage combination starter 高压综合启动器
high-voltage compartment 高压箱;高压部分
high-voltage condenser 高压电容器
high-voltage condenser paper 高压电容纸
high-voltage conductivity 高压导电性;高压导电率
high-voltage conductor 高压导线
high-voltage connection 高压引线;高压接线
high-voltage controlling transformer 高压控制变压器
high-voltage converter 高压变流机
high-voltage corona 高压电晕
high-voltage corona device 高压电晕装置
high-voltage cubicle 高压配电盘;高压开关柜
high-voltage current 高压电流
high-voltage damper 高压阻尼线圈
high-voltage direct current 高压直流线;高压直

流电
high-voltage direct current cable 高压直流电缆
high-voltage direct current substation 高压直流变电站
high-voltage direct current transmission 高压直流输电
high-voltage direct current transmission line 高压直流输电线
high-voltage discharge 高压放电
high-voltage discharge electrode 高压放电极
high-voltage distribution equipment 高压配电装置
high-voltage drive tube 高压激励管
high-voltage dropout explosion fuse 高压脱落式熔断器
high-voltage drop-out type fuse cutout 高压跌落式熔断器
high-voltage earth discharge 高压接地放电
high-voltage electric(al) apparatus 高压电器
high-voltage electric(al) porcelain 高压电瓷
high-voltage electrode 高压电极
high-voltage electron microscope 高压电子显微镜
high-voltage electron microscopy 高压电子显微术;高压电子显微法
high-voltage electrophoresis 高压电泳;高电压电泳(法)
high-voltage electrostatic precipitating system 高压静电除尘系统
high-voltage electrostatic separator 高压静电分离仪
high-voltage electrostatic vacuum cleaner 高压静电吸尘器
high-voltage electrostatic voltmeter 高压静电电压表
high-voltage fence 高压电网(围篱)
high-voltage field 高压(电)场
high-voltage flash-over tester 高压闪络测试仪
high-voltage fuse 高压熔断器
high-voltage generator 高压发生器;高压发电机
high-voltage glow discharge lamp 高压辉光放电灯
high-voltage glow tube 高压辉光管
high-voltage ground protection 高压地保护
high-voltage/high frequency 高压/高频
high-voltage high-speed oscilloscope 高压高速示波器
high-voltage holding test 耐压试验【电】
high-voltage indicator 高压指示器
high-voltage insulating boots 高压绝缘靴
high-voltage insulating gloves 高压绝缘手套
high-voltage insulating oil 高压绝缘油
high-voltage insulation 高压绝缘
high-voltage insulator 高压绝缘子
high-voltage insulator compartment 高压绝缘子室
high-voltage interference 强电干扰
high-voltage isolator 高压隔离开关
high-voltage lamp 高压电路试验灯;高压灯
high-voltage line 高压导线;高压线路;高压电线
high-voltage line suspension insulator 高压线悬式绝缘子
high-voltage load disconnecting switch 高压负荷切断开关
high-voltage load switch 高压负荷(切)开关
high-voltage magnet dynamo 高压磁石发电机
high-voltage meter 高压计
high-voltage micaceous condenser 高压云母电容器
high-voltage motor 高压(交流)电动机
high-voltage negative-ion generator 高压负离子发生器
high-voltage neutral 高压中性点
high-voltage nuclear battery 高压核电池
high-voltage oil circuit breaker 高压油断路器
high-voltage oscilloscope 高压示波器
high-voltage out-door disconnector 高压户外隔离开关
high-voltage overhead transmission line 高架高压输电线
high-voltage paper electrophoresis 高压纸电泳;高电压纸电泳(法)
high-voltage photovoltaic converter 高伏特光电转换器
high-voltage pin insulator 高压针式绝缘子
high-voltage porcelain insulator 高压电瓷绝缘子
high-voltage potential transformer 高压电压互感器
high-voltage pothead 高压电缆头
high-voltage power line 高压输电线
high-voltage power supply 高压电源
high-voltage power unit 高压电源装置
high-voltage probe 高压探针
high-voltage protection 高压防护;高压保护
high-voltage pulse 高压脉冲
high-voltage pulse corona 高压脉冲电晕
high-voltage pulse discharge-ozonation process 高压脉冲放电-臭氧氧化法
high-voltage pulse generator 高压脉冲发生器
high-voltage radiography 高压射线摄影术
high-voltage rectifier 高压整流器
high-voltage regulation factor 高压调整系数
high-voltage relay 高压继电器
high-voltage resistor 高压电阻器
high-voltage side terminal 高压侧端子
high-voltage silicon rectifier stack 高压硅堆
high-voltage slow wave 高波幅慢波
high-voltage source box 高压电源箱
high-voltage spark 高压火花
high-voltage stillwater separation method 高压静水分离法
high-voltage substation 高压变电所(站)
high-voltage supply system 高压供电系统
high-voltage support frame 高压支架
high-voltage switch 高压开关
high-voltage switch and protection element 高压开关和保护元件
high-voltage switchboard 高压(开关)板
high-voltage switch cabinet 高压(开关)柜
high-voltage switchgear 高压开关设备;高压配电装置
high-voltage synchronous motor 高压同步电动机
high-voltage system 高压系统
high-voltage terminal 高压电极;高压端子
high-voltage test 高压试验【电】
high-voltage test for insulation 高压绝缘试验
high-voltage testing device 高压测试器;高电压试验器
high-voltage testing equipment 高压试验设备
high-voltage testing transformer 高压试验用变压器
high-voltage time test 高压时间试验
high-voltage tongs 高压钳
high-voltage transformer 高压变压器;主变压器
high-voltage transmission line 高压输电线(路);高电压线
high-voltage(transmission) mast 高压输电电线塔;高压输电(电)线干
high-voltage tube 高压管
high-voltage tube fuse 高压管式熔断器
high-voltage unit 高压设备;高压发生器
high-voltage video coupler 高压视频耦合器
high-voltage winding 高压绕组
high-voltage wire 高压线(路)
high-voltage wound asynchronous motor 高压卷线异步电动机
high volume 高运量;高容量
high volume air-borne dust sampler 大流量飘尘采样器
high volume air sampler 高容量空气采样器
high volume air sampling system 高容量空气采样装置
high volume cascade impactor 高强度串联冲击器
high volume electric(al) speaker 高音电动扬声器
high volume filter 高容量过滤器
high volume method 高容量法
high volume product 大量生产制品
high volume pumping 大排量泵送
high volume road 高交通量道路
high volume run 大容量运行;大量运用
high volume sampler 高容量取样器;高容量采样器;大容量采样器
high volume spray(ing) 高容量喷雾;高容量喷布
high volume stack sampler 高容量烟囱采样器
high volume through-tubing flowmeter 大容量流量计
high votalge converter 高压变流器
high wage 高工资
high wall (新开挖土石方的)土石方露明面;高边坡(基)坑;露天矿未开采工作面;采矿工作面;(开挖的)深堑侧壁
highwall-drilling machine 立式钻机;边坡钻机
highwall drilling rig 边坡钻机
highwall slope 高边坡;边坡坡度
high water 高水位;满潮;大水;高潮
high water alarm 最高水位指示器;高水位警报器;高水位报警器(器);高潮警报器;洪水位警报器
high water base vane pump 高水基叶片泵
high water bed 洪水河床
high water berth 高水位泊位
high water channel 高水槽
high water consumption 高耗水量
high water-content crude oil 高含水原油
high water control 高水位控制
high water-cut crude oil 高含水原油
high water-cut well 高含水井
high water dam 高水坝
high water difference 高水位差
high water dock 高水位码头
high water embankment 防洪堤
high water equinoctial spring tide 分点大潮高潮位
high water flow 洪水流量
high water full and change 平均朔望潮高潮间隙;朔望高潮间隙
high waterhead hydraulic system 高水头水力枢纽
high water inequality 高潮不等
high water interval 高水位间隔;高潮间隙
high water(level) 高潮位;高水位;高潮面;洪水位
high water level bridge 高水位桥
high water level of ordinary spring tide 大汛平均高水位
high water level of spring tide 大潮高潮位;大潮高潮面
high water limit 高潮线;洪水水位线
high water line 高水位线;高潮(水位)线;最高潮位线
high water lunitidal interval 月潮高潮间隙;月潮间隙;高潮月潮间隙;高潮间隙;潮候时
high water mark 高水位注记;高水位(线);高水位标志;高潮线;高潮痕;高潮标记;洪水(水)位线;洪水痕迹;洪水标记
high water neaps 小潮高潮(位);小潮高潮面
high water neap-tide 小潮高潮(位);小潮高潮面
high water of ordinary neap tide 小潮平均高位;小潮平均高潮面;一般小潮高潮
high water of ordinary spring tide 一般大潮高水位;一般大潮高潮(位)
high water period 高水期;高潮期;洪水期;汛期;丰水期
high water plane 高水面
high water platform 高潮浪蚀台;浪蚀台地;浪蚀平台
high waterproof plywood 高耐水性胶合板
high water quay 高水位码头
high water revetment 高水位护岸
high water(river) bed 洪水(期)河床;河漫滩;高水位河床
high water rock 高潮石;高潮礁
high water season 洪水(汛)期;洪水季(节);汛期
high water slack 高水平潮;高潮憩流;高潮平流
high water span 高水位跨度
high water spring 子午高潮;朔望高潮(位);大潮高潮(面)
high water spring tide 高水位春潮;大潮高水位
high water stage 高潮水位
high water stand 高潮停潮;高潮憩流;高潮平潮
high water table 高潜水位;高地下水位
high water temperature switch 高水温开关
high water time 高潮潮时
high water training 高水导治;洪水治导
high water wharf 高水位码头
high wattage lamp 大功率电灯
high watt-density-swaged cartridge heater 高瓦特密度型锻圆筒加热器
high wave 大波
high wave condition 大浪海况
high wave echo 高波
high wave energy 高波能
high wax containing asphalt 高蜡沥青
high wax content oil 高含蜡原油
high wax oil 高蜡石油
highway 公路;汇流通道
Highway, Bridges and Engineering Works 公路与桥梁和工程建筑物(英国期刊名)
highway accident 公路事故

highway act 公路条例;公路法
highway administration 公路管理(局)
highway administrative organization 公路管理机构
highway aerial photogrammetry 公路航空摄影测量
highway aesthetics 公路美学
highway alignment 公路线形;公路线路
highway and rail transit bridge 公铁两用桥
highway and railway double use bridge 公路铁路两用桥
highway appearance 路容
highway approach 公路引道
highway appurtenance 公路附属设施
highway authorities 公路管理局;公路管理机构
highway beam bridge 公路梁式桥
highway beautification 公路美化
highway binder 路用结合料;公路铺路沥青(黏结料)
highway boundary line 公路界线
highway bridge 道路桥;公路桥(梁)
highway bridge approach 公路桥引道
highway broom 公路路帚;公路路刷
highway building technique 公路施工技术
highway bus station 公路客运站
highway capability 公路通过能力
highway capacity 公路通行能力;公路容量;交通通过能力
highway capacity manual 公路通行能力手册
highway chip(ping)s 筑路石屑
highway circuit 信息主通路
highway classification 公路等级;公路分类
highway code 公路条例;公路规程;公路法令;公路法规
highway communication 公路交通
highway compaction machine 公路压缩机
highway concrete 道路混凝土
highway concrete sand 道路混凝土用砂
highway construction 公路施工;公路建筑;公路建设;公路工程
highway construction and maintenance 公路建设和维护
highway construction budget 公路施工预算;公路建筑预算
highway construction cost 公路造价
highway construction financing 公路修建投资
highway construction project 公路修建计划
highway construction site 公路施工现场
highway construction soil 公路修建用土壤
highway construction technique 公路施工技术
highway construction tester 公路施工试验者;公路施工试验机
highway construction work 公路施工工作
highway contractor 公路承建者;公路承建商;公路承包人;公路包工
highway crane 公路起重机
highway crossing 公路交叉(口);公路道口;公路穿越
highway crossing annunciating device 道口通知设备
highway crossing flashing signal 道口闪光信号
highway crossing mono-indication obstruction signal 道口遮断信号
highway crossing outdoor audible device 道口室外音响器
highway crossing protection 公路交叉口防护
highway crossing sign 道口标志
highway crossing signal 公路交叉口标志;道口信号
highway crossing signal control panel 道口信号控制盘
highway crossing signal indicator 道口信号指示器
highway crossing signal(l)er 道口信号机
highway curve 公路曲线;公路弯尺
highway curve table 公路曲线测设用表
highway cut 公路开挖
highway data bank 道路数据库
highway deficiency rating 公路缺陷分级
highway deicing salt 公路除冰盐
highway delivery 公路输送(量)
highway department 公路(管理)处
highway depot 公路仓库;公路补给站;公路段车站储[贮]藏所;公路段车站仓库
highway design 公路设计
highway design engineer 公路设计工程师

highway design standards 公路设计规范;公路设计标准
highway discharge 公路排水
highway dispatch 公路交通调度
highway ditch 公路排水沟
highway division 公路分段
highway drainage 公路排水(系统)
highway drainage pipe 公路排水管
highway economics 公路经济
highway embankment 路堤;公路路堤
highway embankment built in loess plateau 黄土路堤;土堤
highway emulsion 筑路乳液
highway engineer 公路工程师;道路工程师
highway engineering 公路工程(学);道路工程(学)
highway engineering progress 公路工程进度;公路工程发展
highway equipment 公路设备
highway erosion control 公路防冲措施
highway facility 公路设施
highway fill 公路填土;公路填方;公路路堤;公路回填
highway finance bond 公路投资公债
highway fork 公路分叉
highway freight transportation 公路货运
highway frontage 朝向公路的土地
highway fuel economy test 公路燃油经济性试验
highway furniture 公路设施
highway geometric(al) design 公路几何设计
highway geometry 公路线形
highway glare 公路眩光;公路光亮表面的眩光
highway grade crossing 公铁平交道口;公路平面交叉;平交道
highway grade crossing surface 平交道口路面
highway grade intersection 公路斜坡断面
highway grade separation 公路立体交叉;公路立交道口
highway grading 公路(土方)整平;公路平整;公路路基整型
highway gravel 筑路用砾石
highway greening 公路绿化
highway grinder 公路碎石机
highway grooving machine 公路开槽机
highway heating 公路加热
highway heating installation 公路加热装置
highway hopper 筑路斗式运料车;筑路用底卸式自卸汽车
highway ice-control 公路冰冻控制;公路防冰冻措施;防止路面结冰措施
highway illumination 公路照明
highway improvement 公路改善;公路改良
highway industry 公路工业
highway interchange 公路立体交叉
highway intersection 公路交叉(口);道路交叉
highway investment and expenditures 公路投资及支出
highway investment criteria 公路投资标准
highway joint 公路接合处;公路接缝
highway junction 公路连接点;公路交叉口
highway landscape 公路风景;公路景色
highway landscape design 公路景观设计
highway landscaping 公路景色;公路风景;公路景观
highway lane 公路行车道
highway lantern 公路路灯
highway layout 公路设计;公路布置
highway life 公路使用年限
highway lighting 公路照明;道路照明
highway lighting mast 公路路灯杆
highway live load 公路活载
highway load 公路负荷
highway location 公路定线;公路布置
highway machinery 筑路机械
highway maintenance 公路养护
highway maintenance cost 养路费
highway maintenance fee 公路养路费
highway maintenance tax 养路税
highway map 公路(路线)图;公路交通图
highway mark(er) 公路标志;公路指示器;公路路标;公路里程碑
highway marking 公路路标;公路标志线
highway marking machine 公路标志机
highway material 筑路材料;公路材料

highway materials survey 公路材料检测;公路材料调查
highway mover 公路割草机
highway mower 公路除草机
highway needs 公路用品
highway net(work) 公路网;高速公路网
highway noise 公路噪声
highway on stilts 高架公路
highway over crossing 公路上跨(立体)交叉;公路立体交叉;公路旱桥立体交叉
highway over crossing bridge 公路立交桥
highway-overpass 上跨公路桥
highway pavement 公路路面
highway pavement coring machine 公路路面钻样机
highway pavement foundation 公路路面基础
highway pavement tester 公路路面试验机
highway planing machine 公路平整机
highway planning 公路计划;公路规划
highway planning section 公路设计组;公路设计科;公路设计处;公路规划组;公路规划科;公路规划处
highway planning survey 公路规划调查
highway planting 公路绿化
highway pollution sources 公路污染源
highway programming 公路计划;公路规划
highway project 公路设计;公路规划;公路方案
highway pump 公路用泵
highway rail bridge 公路铁路桥
highway railroad crossing 公铁平交道口;公路铁路(平)交叉(道口)
highway-railway bridge 公路—铁路桥梁
highway-railway crossing 公铁平交道口
highway-railway grade crossing 公铁平交道口
highway-railway level crossing 公路—铁路平面交叉
highway relocation 公路改线
Highway Research Abstracts 公路研究文摘(美国期刊名)
Highway Research Board 公路研究委员会(美国)
highway research information service 道路研究资料服务中心
highway road marker 公路路牌
highway route 公路路线
highway safety 公路安全
highway safety research 公路安全研究
highway salt 公路防冻盐;公路用防冻剂
highway service 公路服务(设施)
highway setting-out 公路定线
highway show 公路展览;公路外观
highway sign 公路标志
highway skid resistance 公路防滑性
highway source 源于公路的污染
highway speed 公路车速(限制)
highway spiral 公路螺旋曲线;公路缓和曲线
highway sprayer 路用材料洒布机
highway stilts 高架公路
highway stone chip(ping)s 筑路石屑
highway structure 公路构筑物;公路建筑物
highway subgrade 公路路基
highway subgrade soil 公路路基土
highway surfacing 公路铺面;公路路面
highway surfacing foundation 公路路面基础
highway surfacing tester 公路路面试验机
highway survey(ing) 公路测量;公路勘测
highway switching 多路交换
highway system 公路系统;公路网
highway tax 公路使用税
highway test 公路试验
highway tonnage capability 公路载重能力
highway tractor 筑路用拖拉机;运输车;公路牵引索
highway traffic 公路交通
highway traffic control 公路交通控制;公路交通管制;公路交通管理
highway traffic engineering 公路交通工程
highway traffic fuel 公路交通用燃料
highway traffic law 公路交通法(规)
highway traffic noise 公路交通噪声
highway traffic safety 公路交通安全
highway traffic sign 公路交通标志
highway traffic signal 公路交通信号
highway traffic survey 公路交通观测
highway trailer 公路拖车;公路牵引车
highway transition curve 公路缓和曲线;公路渐

变曲线;公路过渡曲线
highway transition spiral 公路缓和螺旋线;公路缓和过渡线
highway transportation 公路运输;公路客运量
highway transportation means 公路运输手段;公路运输方式
highway trust fund 公路信用基金;公路信托基金
highway tube 公路隧道;公路管道
highway tunnel 公路隧洞;公路隧道
highway type 公路型式;公路类型
highway under crossing 公路下穿交叉
highway underdrainage facility 公路排水设施
highway user 公路使用者
highway vehicle 公路车辆
highway vehicle noise 公路车辆噪声
highway widening 公路拓宽
highway width 总线宽
highway with rising gradient 上坡公路
highway work 公路工作
high wearing feature 高耐磨性;高抗磨性
high-webbed tee iron 宽腰T形钢
high weight mud 重的泥浆;超重泥土浆
high weight solids content 高比重固相含量
high weir 高堰
high weir type classifier 高溢流堰式分级机
high weir type screw classifier 高堰式螺旋分级机
high wet modulus 高湿模量
high wet modulus fiber 高湿模量纤维
high wet modulus rayon 高湿模量人造丝
high wet modulus viscose fiber 高湿模量黏胶纤维
high wet strength paper 高湿强纸
high wheeled car 高轮车
high-wide load clearance detector 货车装载高宽限界检测器
high-wide load detector 货车装载高宽检测器
high-wide load indicator 货车装载高宽指示器
high wind 劲风;狂风;疾风;七级风;大风
high wind diagram 强风图
high window 高窗
high window workshop 高窗车间;高窗厂房
high wing furrower 高翼开沟铲
high wire 高钢索
high-wire chair 高凳式钢筋座
high workability 高和易性;优良和易性;高流动性
high workability concrete 高工作度混凝土;高和易性混凝土
high writingspeed socilloscope tube 高记录速度示波管
high yield 更大收获;高产率;高产额;丰产
high yield clay 高膨胀黏土
high yield cold-rolled 高屈服强度冷扎的
high yield field 高产田
high yield generator 大功率振荡器
high yielding 高产的
high yielding strain 高产品系
high yielding well 高产(油)井
high yield low-alloy steel 高强度低合金钢
high yield plot 丰产田
high yield potential 高产潜力
high yield strength 高屈服强度;高屈服点
high yield strength reinforcement 高屈服强度钢筋
high yield strength steel 高屈服强度钢;高屈服点钢
high yield stress steel 高屈服强度钢
highy potent herbicide 高效除草剂
high zirconium content 高锆含量
high zonal recirculation 强西风环流;强纬向环流
high μ tube 高放大系数管
hike 加价;突然移动
hike in prices 涨价
hiking map 游览图
Hilaire method 希莱尔定位法;高度差法
hilairite 三水钠锆石
Hilbert's basis theorem 希尔伯特基定理
Hilbert's characteristic function 希尔伯特特征函数
Hilbert's cube 希尔伯特立体
Hilbert's invariant integral 希尔伯特不变积分
Hilbert's matrix 希尔伯特矩阵
Hilbert's modular form 希尔伯特模形式
Hilbert's modular function 希尔伯特模函数
Hilbert's modular group 希尔伯特模群
Hilbert's nullstellensatz 希尔伯特零点定理
Hilbert's parallelotope 希尔伯特超平行体
Hilbert's polynomial 希尔伯特多项式

Hilbert's scheme 希尔伯特概型
Hilbert's transform 希尔伯特变换
Hilbert's transformer 希尔伯特变压器
Hilbertian basis 希尔伯特基
Hilbert matrix 希尔伯特矩阵
Hilbert space 希尔伯特空间
Hilbert transformation 希尔伯特变换
Hilbert-type problem 希尔伯特型问题
Hilborn detector 希尔伯恩探测器
Hild differential drive 赫尔德钻头给进差动机构
Hildebrand cell 希尔德布兰德电解池
Hildebrand function 希耳德布兰特函数;希尔德布兰德函数
Hildebrandt extractor 希得波兰特萃取机
Hiley's formula 希列(打桩)公式
hilgardite 氯羟硼钙石;水氯硼钙石
hiliocentricism 日心说
hi-lite 图像最亮部分
hill 小山(丘);陵
Hill's equation 希尔方程式
Hill adams 希尔阿丹斯止门滑轮
hill-and-dale 深度
hill-and-dale recording 垂直录声
hill-and-dalero 横越分水线的道路
hill-and-dale route 横越分水线的道路
hill-city 山城
hill-climbing 登山
hill climbing capacity 上坡能力
hill climbing method 爬山法;登山法
hill-creep 山坡土爬;山(坡)蠕动;滑坡;山坡物质缓慢下滑
hill crest 山顶
hill culture 垅作;坡地栽培
hill depression 溶丘洼地
Hill determinant 希尔行列
hill diarrhoea 高山泄肖
hill drainage plough 坡地排水沟犁
hill-drop plate 多粒槽口式排种盘
hillebrandite 针硅钙石
hiller 培土器
hiller hoe 培土锄铲
hill features 山地地貌
hill garden 筑山庭园
hill hachures 山丘的影线
hill holder 停车防滑器(坡道上汽车);(汽车的)下坡防滑机构;上坡防退器
hilliness coefficient 起伏系数
hilling bottom 培土犁体
hill making 掇山
hillock 山岗;小丘;土坡;土墩;废石堆;异常析出
hillock plain 低丘平原
hill peat 高地泥炭;丘陵泥炭
hill plane 物面
hill planting 穴植
hill point 山顶
hill pump station 山坡泵站
hill reaction 堆积反应
hill river 丘陵区河流
hills 丘陵
hill shading 晕渲(画木线的阴影);晕渝;直照晕渲【测】;地貌晕线法;地貌晕渝法;阴影法
hill-shading(method) 晕渲法
hill shape 山形
hillside 山腰;小山坡;丘陵坡;山坡;山脚
hillside architecture 山坡建筑
hillside bin 山腰料仓;山坡斜坡道
hillside borrow 山坡采场;山坡取土坑
hillside brick 铺坡砖
hillside bridge 傍山桥
hillside combine 山地联合收获机
hillside contact spring 山麓接触泉
hillside covering works 山坡保护工程;山坡覆盖层
hillside creep 山坡坍方;山坡滑坡;山坡物质缓慢下滑;山坡蠕滑;山坡潜移;滑坡
hillside cultivated 坡耕地
hillside cut and fill 山腰上半挖半填;山坡半挖半填
hillside cut-off trench 山坡截水沟
hillside dam 山坡挡水坝;山麓小坝
hillside diggings 山坡砂矿
hillside ditch 山坡排水沟;山脚沟;单侧沟;傍山沟
hillside field 坡地
hillside fill 山坡填土
hillside flanking 护坡;加固山坡
hillside gravel 山腰砂砾

hillside line 山坡(腰)线
hillside location 山坡现场部位;山坡勘测;山坡定位;山坡场所
hillside orchard 山地果园
hillside placer 山腰砂矿
hillside plot 山田
hillside plough 双向犁
hillside quarry 山坡采石场
hillside ramp 山坡斜坡;山坡坡道;山坡台阶
hillside road 山坡路;傍山路
hillside section 山坡区段
hillside seepage 山坡渗流
hillside silo 山坡筒仓
hillside slope 山坡斜度
hillside surface water 山坡地表水
hillside town 山坡城镇
hillside tractor 山(坡)地拖拉机
hillside waste 山麓碎石
hillside work 山坡土方
hillside workings 山坡开采区
hill sign 山坡标志
hillslope 坡面;山坡(面)
hillslope erosion 山坡冲刷
hillslope evolution 坡画演化
hillslope hydrology 山麓水文学
hillslope landform 坡地地貌
hillslope process 坡面过程
hillslope runoff 坡面径流;山坡径流
hillslope shape 坡地形态
hill spacing 穴潲
hill station 山中避暑地(印度等地)
hill stream 丘陵(地)区河流;山溪;山区河流;山涧
hill to full of bumpy and holes 凹凸不平的山地
hill toning 缔约程序;地貌晕线法;地貌晕渝法
hill top 山峰;小山;小山山顶;丘顶
hilltop reservoir 高山水库;山顶水库;山顶水池
hill torrent 山溪;山区河流;山涧
hill up 培土
hillwash 坡地坍滑;坡地冲刷(作用);坡底冲刷
hillwork 地貌晕渣
hilly 多(小)山的;多丘陵的
hilly and plain karst 丘陵平原岩溶
hilly area 丘陵区
hilly city 山区城市
hilly coast 丘陵海岸
hilly country 丘陵地区;丘陵地(带)
hilly county with low mountains 低山丘陵区
hilly grassland 丘陵草地
hilly ground 丘陵地(区)
hilly land 丘陵地
hilly land location 丘陵地区选线
hilly landscape 丘陵景观
hilly road 丘陵地区道路
hilly terrain 丘陵地带
hi-load 高荷载
hi-load nail 高荷载钉
hi-lo-check 高低端检查;计算结果检查
hi-lo-circuit 高低压回路
hi-lo signal alarm 高低位报警
hilt 刀柄
Hilt's law 希尔特定律
Himalaya ammonite region 喜马拉雅菊石地理大区
Himalaya cedar 喜马拉雅山杉木
Himalaya fault-fold swell 喜马拉雅继褶隆起
Himalaya inheriting fault-fold zone 喜马拉雅继承性断褶带
Himalaya marine faunal province 喜马拉雅海相动物地理区
Himalaya mountains 喜马拉雅山
Himalaya-Nanling tectonic zone 喜马拉雅—南岭构造带
Himalayan birch 糙皮桦
Himalayan brachiopod region 喜马拉雅腕足动物地理大区
Himalayan cycle 喜马拉雅旋回
Himalayan fir 喜马拉雅冷杉
Himalayan fold system 喜马拉雅褶皱系
Himalayan geosynclinal fold region 喜马拉雅地槽褶皱区
Himalayan geosyncline 喜马拉雅期地槽
Himalayan mountains 喜马拉雅山山脉
Himalayan movement 喜马拉雅山运动【地】
Himalayan Nappe fracture zone 喜马拉雅辗掩断

裂系
Himalayan orogeny 喜马拉雅造山运动
Himalayan period 喜山期
Himalayan platform folded belt 喜马拉雅台褶带
Himalayan southern foot fault belt 喜马拉雅山南麓断裂构造带
Himalayan spruce 喜马拉雅云杉
Himalayan subcycle 喜马拉雅亚旋回
Himalaya region 喜马拉雅大区
Himalaya regression 喜马拉雅海退
Himalayas 喜马拉雅山区;喜拉玛雅山山脉
Himalaya sea 喜马拉雅海
Himalaya tectonic zone 喜马拉雅构造带
Himet 碳化钛硬质合金
hind axle 后(轮)轴
hind body 后体
hind carriage 拖车
hindcasted wave 推算波浪
hindcasting 后报
hindcasting technique 追算技术
hindered amine 位阻胺;受阻胺
hindered contraction 受阻收缩
hindered phenol 位阻酚;受阻酚
hindered phenol anti-oxidant 受阻酚抗氧剂
hindered rotation 受阻旋转;受碍转动
hindered sedimentation 拥挤沉淀
hindered-setting tank classifier 槽式干扰下沉分级机
hindered settling 阻滞沉落;制约沉降;干扰沉降;受阻沉降;干涉沉降
hindered-settling column 干扰下沉柱;干扰下沉室
hindered-settling hydraulic classifier 干扰下沉式水力分级机
hindered settling ratio 受阻沉降比例;阻滞沉降率
hinder land 海岸后面陆地;海岸后面地区
hinder navigation 妨碍航行
hind head of lock 船闸尾端
Hindley hob 亨德莱滚刀
Hindley worm 欣德利球面蜗杆;亨德莱蜗杆;球面蜗杆
Hindley worm gear 欣德利蜗轮;球面蜗轮蜗杆;球面齿轮
Hindley worm toothing 球状蜗杆啮合;球面蜗杆啮合
Hindolet worm 球面蜗杆
Hindoo architecture 印度建筑
Hindoo temple 印度庙宇
hindrance 开关阻抗;障碍物
hindrance factor 曳力系数
hindrance to navigation 碍航
hindrance to traffic 交通障碍物
Hindu temple 印度庙宇
hind wheel 后轮
hinelight 高强度荧光灯
hinge 折叶;关键;铰片;铰链;胶水纸;活节;合叶;枢纽
hinge area 铰链区
hinge armature 枢轴衔铁
hinge backset 门铰链收进
hinge bar 铰接杆
hinge bolt 铰链栓
hinge bound 铰链装得不好门面被卡住
hinge-bound door 安装铰链不当的门
hinge bracket 铰链托架
hinge bracket bolt 铰链托架螺栓
hinge carrying mullion 铰链式承重竖框
hinge column 铰链柱
hinge connected beam method 铰接梁法
hinge crack 冰隙
hinged 铰接的
hinged action 铰接作用
hinged and folding hatchcover 铰接折叠式舱盖
hinged apron 平板运输机
hinged arch 铰接拱
hinged arch bridge 铰接拱桥
hinged arch(ed girder) 铰接拱形大梁
hinged arm 枢杆
hinged armature 枢轴衔铁
hinged bar 枢杆;铰接型钢;铰接杆件
hinged barrier 悬栅门;铰式栅门
hinged beam 铰接梁
hinged bearer 铰支座;铰支承
hinged bearing 铰支座;铰接支座;铰承座
hinged bolster 铰接支撑;铰接承梁;铰接板;铰垫板
hinged bolt 可拆卸螺栓;铰式螺栓
hinged boom 折叶式悬臂;铰接式悬臂;绞链吊臂
hinged bottom hopper 底开式料斗;带铰接底的排种筒
hinged bridge 活动引桥
hinged bulkhead door 绞链舱壁门
hinged cantilever 旋转腕臂【电】;铰接式腕臂
hinged cantilever beam 铰接悬臂梁
hinged cantilever beam bridge 铰接悬臂梁桥
hinged cantilever bracket 旋转腕臂底座
hinged casement 铰链平开窗
hinged casing 铰接框(架)
hinged cockpit hood 铰接座舱盖
hinged column 铰接柱
hinged connection 铰接
hinged coupling 铰接
hinged cover 连盖容器;铰链盖
hinged crankcase guard 铰接曲柄箱保护器
hinged door 折门;合页门;铰链门;铰装门
hinged end 铰链端;铰接端
hinged firedoor 铰接的防火门
hinged flap 铰链拍落式闸板
hinged flash-board 铰固闸板
hinged flash gate 舌瓣
hinged flask 可拆式砂箱
hinged floating gate 侧开式浮箱门
hinged folding hatch cover 折叠舱(口)盖
hinged foot valve 铰接式底阀
hinged fork 铰链式货叉
hinged frame 铰接框(架)
hinged funnel 可放倒烟囱
hinged gantry 绞接台架
hinged garage shutter door 铰接的车库门
hinged gate 铰链门;旋开式闸门;绞链闸门;回转式闸门;合页门
hinged girder 铰接式桁架(梁);铰接梁
hinged grid 折叠式炉栅
hinged guide bit 链接式导向钻头
hinged guide fitting 绞链导块
hinged handle 铰接摇柄;铰接曲柄
hinged hatch cover 折页式舱盖;铰式盖舱板;绞链式舱盖
hinged-hinged bar 两端铰接杆
hinged-hinged end condition 两端铰接
hinged hopper 铰式料斗;铰式斗车;翻转式装料斗
hinged immovable support 铰接固定支座
hinged impost 铰支拱基
hinged jaw tongs 铰接钳
hinged joint 铰接节点;企口缝;销接头;铰接头;铰接合;铰(链)接合;绞接口
hinged latch bolt 悬舌式弹簧销
hinged lattice(d) girder 铰链格状梁
hinged leaf gate 平开门
hinged leg 铰接支柱
hinged length 绞接长度
hinged lid 铰链盖
hinged lift bridge 铰接吊桥
hinged loading ramp 铰接的装载斜道;铰接的装料滑台
hinged mast 铰式桅杆
hinged mechanism 铰接机制
hinged movable support 可动铰接支座;铰接活动支座
hinged paddle 铰接轮叶
hinged patch 活页堵漏板
hinged pedestal 铰接柱
hinged pier 铰接支墩;摆座
hinged pin 铰接销
hinged pipe 旋转接头管
hinged pipe vise 管子台虎钳
hinged pivot 铰支轴
hinged plate 铰折板;铰链(装合)板
hinged point 铰接点
hinged pool 串联水库
hinged post 铰接墩柱
hinged radiator guard 铰式散热器防护罩;铰式暖气片防护罩;铰式散热器罩
hinged ramp 铰接坡道
hinged rudder 悬式舵;吊舵
hinged seat 铰接座
hinged shoe 铰支座;铰支承块
hinged shutter 铰接遮光门窗
hinged side 铰接侧面
hinged skylight 旋转天窗;铰接天窗
hinged steel plate 铰接钢板
hinged structure 铰接构件;铰接结构
hinged support 铰支承;铰(接)支座
hinged support of sheet piling 板桩铰接支承
hinged system 铰接系统
hinged temporary deadlight 绞链舷窗盖
hinged tie bar 铰接系杆
hinged tool holder 活节刀柄
hinged truss 铰接桁架
hinged tube 旋转接头管
hinged type contraction joint 铰式收缩缝
hinged type entry guide 铰链式导口;绞链式导箱装置
hinged unloading trap 铰接卸车地板门
hinged watertight door 绞链水密门
hinged window 铰链窗
hinged wood(en) slat 铰接的木板条
hinge end 转折端
hinge fault 挟转断层;枢纽断层
hinge forming die 铰链成形模
hinge impost 铰支拱基
hinge in the key 拱冠铰;拱顶铰
hinge jack 铰接夹头
hinge jamb 门框梃;安铰链的门边框
hinge joint 铰接(节点);铰链;铰式接合;铰式接缝;铰链(接合);活节连接
hinge-jointed column 铰接柱
hinge joint section 铰接部件
hinge knuckle 铰接转向节;铰关节
hingeless 无铰链的
hingeless arch 无铰拱
hingeless arch bridge 无铰拱桥
hingeless frame 无铰刚构架
hingeless support 无铰支点
hinge ligament 蝶铰韧带
hinge line 铰合线;枢纽线;蝶铰线
hinge mechanism 铰机制;塑性铰机构
hinge moment 铰接力矩
hinge mount 铰接座;铰接架
hinge of anticline 背斜枢(纽)
hinge of spring 簧节套
hinge of syncline 向斜枢(纽)
hinge pedestal 铰柱脚;铰接支柱;摇杆;摆柱
hinge pillar 铰链柱
hinge pin 铰接销;合叶心
hinge pivot 铰支轴
hinge plate 铰链页;铰链门框;铰合板
hinge point 铰(链)点
hinge post 铰接桥墩;铰链柱
hinge pressure 铰链压力
hinge reinforcement 铰链板
hinge-rod 铰链杆
hinge rotation 铰接转动;塑性铰转动
hinges 合叶铰链
hinge stile 铰链(门窗)梃
hinge stone 门臼石
hinge strap 装铰链板条;铰链柱;合叶
hinge strap flap 铰链板
hinge template 铰链定位板
hinge tube 管子芯骨;铰链管
hinge type contraction joint 铰式缩缝
hinge type faulting 挟转断层作用
hinge type ripper 链式松土机;链式耙路机
hinging length 铰接长度
hinging post 带铰门柱;旋门柱;铰装门柱
hingle 门铰链
Hino 日野
hinokitiol 扁柏油酚
Hinsberg reaction 兴斯堡反应
hinsdalite 磷铅锶矾;磷铅铝矾
hinterland 内地;后置地;后陆;(海岸或河岸的)后方地带;海岸或河岸的后方地区;穷乡僻壤;腹地;背地
hinterland basin 后陆盆地
hinterland connections 港口集疏运系统
hinterland-dipping duplex 后倾双冲构造
hinter surf bed 陆架沉积层;近岸沉积层
hiortdahlite 希硅锆钠钙石
hip 斜脊;角梁;戗脊
hip and flat roof 四坡平屋顶
hip and gable roof 歇山式屋顶;人字坡墙
hip and hip roof 四坡屋顶
hip and ridge finishing pieces 屋脊木瓦
hip and ridge shingle 脊(木)瓦
hip and staple 搭扣

hip-and-valley roof 四坡排水管屋顶;有交叉的四坡屋顶;四坡一阴敛屋顶
hip bath 浅浴盆;坐浴(浴)盆;坐浴
hip bevel 戗脊(椽)角
hip bevel stone 戗脊上斜石板
hip capping 屋脊盖瓦;戗脊盖瓦;戗脊盖条;交叉铺叠屋脊瓦;屋脊压盖;屋脊盖条
hip cross 十字部
hip dormer 有戗脊的老虎窗;屋脊窗
hip end 端坡
Hiperco 海皮考合金
Hipercon alloy 海波可合金
Hiperloy 高导磁率合金
hipernik 高导磁率镍铁合金;铁镍磁性合金;高导磁铁镍合金;高导磁率镍钢
Hipersil 高导磁率硅钢
hip gable 双坡顶山墙
hip hook 屋脊挂瓦钢条;脊瓦挂钩;戗脊挂瓦钩
hip iron 屋脊挂瓦铁;戗脊挂瓦钩
hip jack 端坡椽;(屋顶的)支顶木
hip jack rafter 脊椽;端坡椽
hip joint 上弦与斜端的结点(桁架)
hip knob 脊端饰;戗脊端饰
hip mansard roof 四坡孟莎屋顶;复折四坡屋顶
hip mo(u)ld(ing) 屋脊线脚;脊线脚;脊饰;戗脊线脚
hip-mounted animal ornament 戗兽【建】
hip nail 鹅头接骨钉
Hipolan 海玻琅高湿模量黏胶纤维
hipot 高压绝缘试验;高压通电试验
hipotype 亚型
Hipparion 三趾马属
hipped 三坡老虎窗
hipped dormer(window) 四面落水的天窗;有斜脊的天窗
hipped end (四坡屋顶的)端坡;四坡顶山墙端;(屋顶的)斜坡端
hipped gable 歇山屋顶山墙;半山头【建】
hipped gable roof 小戗角屋顶;半山头屋顶
hipped mansard roof 折线形斜坡屋顶(英国)
hipped-plate construction 折板结构
hipped-plate structure 折边无梁板结构;折板结构
hipped roof 四面落水屋顶;斜脊屋顶;庑殿式屋顶;四坡屋顶
hipped skylight 无框天窗
hipped slab roof 四坡面板屋顶
hipping 模式化
hippodrome 赛马的椭圆形运动场(古希腊);马技表演场;杂技场;竞技场;马戏场
hip point 上弦与斜端的结点(桁架)
hippopotamus 河马
hippuric acid 苯甲酰氨基酯酸
Hippurite limestone 马尼蛤灰岩
hip rafter 斜面梁;角梁;斜屋脊椽;脊;脊小椽;戗脊椽木;四坡屋面面坡椽
hip roll 筒形戗脊;四坡屋顶中脊上的瓦或盖条;半圆脊帽;屋脊铅皮圆筒;屋脊卷筒形装饰;筒形脊包层
hip roof 四坡顶
hip sheet 屋脊盖板
hip starting tile 脊瓦
hip stone 戗脊石
hip stress 戗脊应力
hip tank 凹面双层底舱
hip tile 斜脊瓦;肩瓦;戗脊盖瓦;屋脊盖瓦
hip token 模式化记号
hip vertical 竖吊杆;后部直拉杆件;肩竖杆;戗架拉杆
hi-rail(car) 公铁两用车
hiran 高精度肖兰(导航系统)
hire 租用;租赁
hire and payment clause 运费支付条款;租船支付条款
hire base 租船基价
hire charges 租费
hire charges for wagon 货车租用费
hired carriage 租车
hired labo(u)r 计日工作;雇工;临时工
hired labo(u)r rate 计日工资;工资率
hired plant 租用机械;租用(的)设备
hired wagon 租用货车
hire fleet 租借的船队;租借的车队
hireling 租用的;被雇佣的;被雇佣者;被雇者
hire purchase 赊购;租赁;分期付款购买;分期付款购货

hire purchase agreement 分期付款购买合同
hire purchase company 租购信贷公司;分期付款购买公司
hire purchase contract 分期付款合同
hire purchase credit 分期付款购货信贷
hire purchase credit insurance 租购信用保险
hire purchase price 分期付款价格
hire purchase sale 分期付款出售
hire purchase system 租购法
Hire Purchase Trade Association 分期付款贸易协会(英国)
hirer 租赁人
hire system 分期付款购买法;分期付款(法)
hiring 租金
hiring and layoff cost 聘用及解雇成本
hiring cost 租赁成本;租赁费用
hiring hall 职业介绍所
hiring of labo(u)r 招工;雇佣工人
hiring out depot 出租仓库
hiring rate 就业增长率
hiring system 分期付款购买法
hi-rise 高层大楼
H-iron 工字铁;氢还原铁粉
Hiroshima and Nagasaki nuclear explosion 广岛和长崎核爆炸
Hiroshima orogeny 广岛变动
hirst 河道中沙滩;砂堆
hirth coupling 直线齿形鼠牙盘离合器
Hirth minimeter 海司单杠杆比较仪
hirudin 水蛭素
hirudiniasis 水蛭病
hirudo 水蛭
Hirudo aegyptiaca 埃及水蛭
hisingerite 硅铁石;水硅铁石
hislopite 海绿方解石
Hispano-Morresque architecture 西班牙摩尔建筑
hi-spot 高光强聚光灯
hiss 电子管噪声
Hissar reversed polarity chron 希萨尔反向极性时
Hissar reversed polarity chronzone 希萨尔反向极性时间带
Hissar reversed polarity zone 希萨尔反向极性带
hissing arc 啸声电弧
histidine 组氨酸
histoautoradiography 组织放射自显影术
histochemistry 组织化学
histogram 组织图;柱状(频率)图;柱形图;柱式图解;直方图;矩形(统计)图;频率分布图;梯级频率图;长方形图
histogram computer 直方图计算机
histogram correction 直方图修正;直方图修改
histogram current meter 直方图海流计
histogram equalization 直方图平直化;直方图均衡(化)
histogram flattening 直方图平坦化;直方图平直化
histogram modification 直方图修正;直方图修改
histogram normalization 直方图正态化
histogram of age frequency distribution 年龄频率分布图
histogram of beneficial mark of uranium 铀有利性指标直方图
histogram of binomial distribution 二项分布直方图
histogram of disjunctive Kriging estimator value 析取克立格估值直方图
histogram of groundwater development 地下水开采直方图
histogram of meteorologic(al) element 气象要素直方图
histogram of size distribution 粒度分布直方图
histogram program(me) 直方图程序
histogram record 高速记录;摄影记录
histogram recorder 高速记录器;遥测记录器;无线电遥测摄影机
histogram specification 直方图规格
histogram thresholding 直方图阈值化
histograph 排水历时图
histograph of drainage 排水历时图
histography 组织图谱
histolytic 溶组织的
histophotometry 组织光度测定法
historiated 有图案的;用人物装饰的;人兽图像装饰的
historiated capital 带历史雕像的柱头

historic(al) age 建筑物的实际年龄
historic(al) American building survey 美国历史性建筑调查
historic(al) analogy method 历史比拟法
historic(al) archaeologic(al) method 历史考古法
historic(al) area 历史性建筑区
historic(al) atlas 历史地图集
historic(al) background 历史背景
historic(al) best cost of the entity 本单位历史先进水平
historic(al) break rate 历史(的)破损率
historic(al) building 古建筑(物);历史(性)建筑
historic(al) building preservation 古建筑保护
historic(al) building styles 历史性建筑类型
historic(al) chart 古海图
historic(al) city 古城市;历史名城
historic(al) climate 历史(上的)气候
historic(al) consistency 逐期一致性
historic(al) context 历史条件;历史背景
historic(al) cost 原始成本;历史成本;上期成本
historic(al) cost accounting 历史成本会计
historic(al) cost assumption 历史成本假设
historic(al) cost basis 历史成本基础
historic(al) cost convention 按历史成本计算的惯例
historic(al) costing 历史成本法
historic(al) cost method 历史成本法
historic(al) cost rules (for revenue recognition) 实际成本惯例
historic(al) cost statement 按原始成本编制的报表
historic(al) cultural city 历史文化名城
historic(al) data 履历数据;原始数据;历史资料;历史数据
historic(al) data approach 历史资料法
historic(al) data extrapolation 历年资料外推法
historic(al) demography 历史人口学
historic(al) district 历史保护区;历史性市区
historic(al) earthquake 历史地震
historic(al) earthquake survey 历史地震调查
historic(al) environment 历史环境
historic(al) event 历史事件
historic(al) factor 历史因素
historic(al) flood 历史(记载的)洪水
historic(al) flood damage 历史洪水灾害;历史洪水损失
historic(al) flood investigation 历史洪水调查
historic(al) flood level 历史洪水位
historic(al) flood survey 历史洪水调查
historic(al) flow 历史(最大)流量
historic(al) garden 古园林
historic(al) geology 地史学;历史地质学
historic(al) geomorphology 历史地貌学
historic(al) heritage 历史遗产;历史传统
historic(al) highest tidal level 历史最高潮位
historic(al) highest water level 历史最高水位;历史最高潮位
historic(al) inertia 历史惯性
historic(al) landmark 历史的(土地)界标;历史性标志;历史建筑
historic(al) least depth 历史最小水深
historic(al) lowest water level 历史最低水位;历史最低潮位
historically active fault zone 历史上断层活动区
historic(al) map 历史地图
historic(al) maximum flood 历史最大洪水
historic(al) method 历史法
historic(al) models 历史模型
historic(al) monument 古址;古迹;历史遗迹;历史古迹
historic(al) museum 历史博物馆
historic(al) park 历史区
historic(al) peak flood 历史最大洪峰
historic(al) pedology 历史土壤学
historic(al) preservation 古迹保存;历史性建筑保护
historic(al) preservation program(me) 历史性建筑保护计划
historic(al) process of nature 自然历史过程
historic(al) product-grade method 历史产品品位法
historic(al) prospective study 历史前瞻性调查
historic(al) records 历史资料;历史记载;历史记录
historic(al) relic preservation 历史文物保护
historic(al) relic preservation area 历史文物(遗址)保护区

historic(al) relics 文物;历史文物
historic(al) remains 遗迹
historic(al) remark about randomization 关于随机化的历史见解
historic(al) ruin 历史遗迹
historic(al) seismologic(al) activities 历史地震活动
historic(al) simulation 历史模拟
historic(al) site 古址;古迹;历史遗址
historic(al) site park 古迹公园
historic(al) spot 古迹
historic(al) structure 历史性建筑
historic(al) survey 历史调查
historic(al) time 有史时期
historic(al) town 古镇
historic(al) track 历史航迹
historic(al) value 历史价值
historic(al) viewpoint 历史观点
historic(al) village 古村落
historic(al) wall 古城墙
historic(al) weather map 历史天气图
historicism 历史主义
historiography 编史工作
history card 沿革记录片
history flood 历史记录洪水
history museum 历史博物馆
history of aesthetics 美学史
history of architecture 建筑(历)史
history of art 美术史
history of astronomy 天文学史
history of available space list 可利用的空间表的历史
history of binomial coefficients 二项式系数的历史
history of cartographic(al) work 图历表
history of cartography 地图制图学史
history of Chinese Architecture 中国建筑史
history of coroutine 共行程序的历史
history of environment(al) science development 环境科学发展史
history of folk custom 民俗史
history of nationality 民族史
history of project 项目历史
history of study 研究历史
history of systematics 分类学史
history of tunnel(l)ing 隧道施工历史
history of Western architecture 西洋建筑史
history register 经历寄存器
histosol 有机土;沼泽土(壤);泥炭土
hi-strength 高强度
histrionics 戏剧表演;舞台艺术
hit 碰撞;瞬断;点击;残缺性线干扰
hit a bridge 钻具中途遇阻
Hitachi 日立公司(日本)
Hitachi Lease Corporation 日立租赁公司(日本)
Hitachi Limited 日立制作所(日本)
Hitachi network architecture 日立网络体系结构
Hitachi Shipbuilding and Engineering Co. Ltd. 日立造船工程公司
hit-and-miss 湿刨
hit-and-miss effect 时现时隐效应
hit-and-miss governor 断续调节器
hit-and-miss method 尝试法;断续法
hit-and-miss ventilator 可调风量通风器;不定向通风;钻孔通风板
hit-and-miss ventilator 活动通风板
hit-and-run 开车肇事后逃跑的
hit an obstruction 钻具遇阻
hit a target 达到定额
hit bottom 触底
hitch 系结;故障;联结装置;急推;猛拉;套挂断错;索结
hitch and step 梯段式长壁工作面;阶梯式长壁采矿法
hitch angle 套钩角铁
hitch arm 活力臂;动力臂
hitch ball (悬挂装置拉杆的)连接球端;球形接头
hitch ball socket 挂挂球窝
hitch bar 连接杆;拖挂装置
hitch clevis 连接卡;连接环;挂钩
hitch crossbar 牵引横板
hitch cutter 刨电工;挖沟工;连接装置;连接器
hitched tandem paver 拖带式串联摊铺机;拖带式串联铺路机
hitcher 钓杆;撑篙

hitch feed 夹持送料
hitch frame 连接装置主架
hitch head 牵引杆接头
hitchhiker satellite 母子卫星
hitching 系留连接;结绳;打系结
hitching of rope 结绳法
hitching on 改用平衡梁钻进
hitching post 系留柱
hitch lifting jack 联结装置起落千斤顶
hitch on 改变钻法
hitch over 重新卡紧
hitch pin 连接销;牵引杆联结销
hitch pin hole 牵引杆销孔
hitch plate 索结板
hitch pole 牵引杆;辕杆
hitch timbering 打梁窝安装顶梁法
hitch to the beam 套上游梁(冲击钻动作)
hitch up 准备钻进
hitch wheel 牵引卷筒
hitch yoke 连接叉
hi-temp hi-effective thinner 高温高效稀释剂
hit file 命中文件
hither plane 近平面
hit indicator 碰击指示器
hit noise 打噪声
hit-on-the-line 短暂开路
hit-or-miss control 断续供油控制法
hit probability 击中概率;命中概率
hit probability payoff 击中概率支付
hit rate 命中率
hitter 铆(钉)枪
hit the pay 钻机进入矿层
hit the target 完成指标
hitting method 撞击法
hitting power 打击力
Hittite architecture 赫梯建筑(公元前14~公元前13世纪)
hive 巢箱;蜂巢状物;闹市
hive-off 分散经营
hi-visibility vest 透明背心
hi-vi vest 反光背心
hi-volume sampler 悬浮颗粒量测器;悬浮污泥质取样器;大流量取样器
H-layer 腐殖质层
H line 电离钙H线
hoar 地面积霜的;斑白的;灰白的
hoar crystal 霜晶
hoard 囤积
hoarder 囤积居奇者;囤户
hoarding (建房时的)临时围篱;临时板围;临时板篱;囤积;栅墙;板围
hoarding and speculation 囤积居奇
hoarding stone (建房时的)临时围篱;界石
Hoar Edge Grits 胡尔埃杰粗砂岩
hoar-frost 白霜
hoarse 噪声
hoarstone (自太古存在的作界标用的)弧石柱;界标石
hoary 久远的;灰白的;古老的;陈旧的;被灰白毛的
hoary antiquity 远古
hoary hair 灰发
hob 螺杆;截齿头;挤压母模;齿轮滚刀;滚铣
Hobann magnetron 分瓣阳极磁控管
hobbed gear 滚铣齿轮
hobbing 铣刀;滚切;挤压制模法;齿轮滚绞法;滚铣
hobbing bit 滚刀盘钻头
hobbing cutter 滚(齿)刀
hobbing machine 滚齿机
hobbing method 滚切(方)法
hobbing press 沉模压力机
hobby 铆顶;铆钉抵棒
hobby farmer 业余农民
hobby room 业余爱好工作室
hobby shop 玩物商店
hob checking equipment 滚刀检查仪
Hobda tide ga(u)ge 本多式潮位计
hob grinding machine 滚齿刀磨床
hob head 滚刀架
hob master 挤压制模原模
hob milling cutter 齿轮滚刀
Hob mouthed oven 陶瓷瓶式窑
hobnail 平头钉;大头钉
hob relieving grinding machine 滚刀铲磨床
hob resharpening machine 滚齿刀磨机

hob sharpener 滚刀磨床
hob sharpening grinding machine 滚刀刃磨机床
hob sharpening machine 滚刀磨床
hob tap 标准丝锥;标准螺纹攻;标准螺丝攻;板牙丝锥
hob tester 滚刀检查仪
hob testing machine 滚刀检查仪
hocaritite 银黄锡矿
Hochstadter's cable 霍赫斯德特高压多芯电缆
hock 齿爪
hockey area 曲棍球场;冰球场
hockey ground 曲棍球场
hockey puck 冰球
hockey stick 冰球杆
hockey stick incision 弯形切口
hock glass 大酒杯
hock joint 飞节
hockle 绳索的纽结
hock shop 当铺
Ho-coefficient 何氏系数
hoctonspheres 球形罐
hocton spheroid 球形罐
hod 砖斗;砂浆桶;煤斗;灰(砂)斗;灰(浆)桶
hod carrier 壮工;小工;搬运灰泥砖瓦工;建筑小工;灰泥砖瓦搬运工;辅助工
hoddy-doddy 灯塔的旋转灯
Hodectron 磁脉冲汞气放电管
hodgkinsonite 硅锰锌矿;褐锌锰矿
hodman 小工;砌砖工人的助手;搬运灰泥、砖瓦的工人;壮工;建筑小工;灰泥砖瓦搬运工;辅助工
hodograph 高空风(速)分析图;高空分析图;速矢端线;速矢端迹;速端图;速端曲线;速度(矢量)图;时距曲线;潮流图
hodograph equation 速矢端线方程
hodograph method 速矢端线法;速度图法
hodograph plane 速端平面;速度图平面
hodograph transformation 速矢(端线)变换;速度图变换;速度矢端曲线变形
hodometer 轮转计;路程计;里程计;里程表;计距器;车距计;车程计
hodoscope 锥光偏振仪;描迹仪
hodoscope system 速矢描迹系统
hodrushite 贺硫铋铜矿
hoe 摊平锹;锄
hoe boom 耕耘机伸臂;耕耘机构架;锄杆
hoe bucket 耕耘机铲斗;锄式铲斗;铲斗
hoe coulter 锄式开沟器
hoe dipper 耕耘机铲斗;挖土机铲斗
hoe dipper for drainage work 排水作业用长柄铲;排水作业用长柄杓
hoed mo(u)ld 披水饰
hoe excavator 反铲
hoe-fork 叉锄
hoe furrow opener 锄式开沟器
hoe gang 锄铲单组
hoe handle 锄头柄
hoe head 锄头
hoeing 用耙子装载;掘松;挖土
hoeing and ridging tool 锄地和培垄农具
hoeing machine 中耕机
Hoekestra balance plastometer 何氏可塑计
hoel of piston crown (制动主缸内的)活塞出油孔
hoe point 锄式开沟器尖
hoernesite 砷镁石
hoe scaler 剥离刀;剥离铲
Hoesch 郝氏建筑体系
Hoesch section 郝氏剖面;郝氏轮廓
hoe-trac disc harrow 锄车圆盘耙
hoe-trac plow 锄车犁
hoe-trac sickle bar 锄车镰刀排
hoe-type trenching machine 反铲挖沟机
hoe wheel 锄铲圆盘
Hoffmann's continuous kiln 霍夫曼连续窑
Hoffmann's electrometer 霍夫曼静电计
Hoffmann's kiln 霍夫曼窑;轮窑
Hofmann's reaction 霍夫曼反应
Hoffmann's rule 霍夫曼规则
Hofmann-Neill rule 霍夫曼-奈尔法则
hog 拱起;粒化机;船底长扫帚;玻璃挠曲;拱曲
hog's back ridge 半椭圆截面屋脊;链式脊
hog's back(tile) 元宝瓦;半圆形脊瓦
hog back 拱起物;陡峻的拱地;底板突出部分;等倾线岭;山脊;拱背;猪背岭
hog-backed 向上拱曲的

hog-backed beam 弓形梁;上拱梁
hog-backed floor 波浪形底板
hog-backed girder 上缘弯曲的梁;拱曲大梁;弓形大梁
hogback tile 拱背(脊)瓦
hogback truss 弓形桁架;拱背桁架
hog barge 流动挖泥船(俚语);挖泥船(俚语)
hogbomite 黑铝镁铁矿
hog box (水力冲填的)泥水混和箱;水填坝水池;沉淀箱
hog chain 倒拱式悬杆;倒拱式悬链
hog chain truss 链式桁架;双柱下撑式桁架
hog deformation 挠曲变形
hog depilating device 刮毛机
hog feeding 猪饲料
hog frame 加强架;防梁架;弓形构架;凸形构架
hogged fuel 细碎燃料
hogger 磨;碎木机
hoggery 养猪场
hoggin 夹砂历史;级配碎石混合料;级配砾石混合料;过筛碎石;粗砂;细石渣;夹砂砾石;泥拌混合料【道】;含砂砾石
hogging 中拱【船】;中段拱起;级配碎石混合料;级配砾石混合料;挠度;翘曲;扫船底;道砟;船首尾下垂;船舶中拱;拱曲
hogging bending 向上弯曲;负弯
hogging condition 中拱状态
hogging deformation 挠曲变形
hogging line 过底绳
hogging moment 中拱弯矩;中拱力矩;挠曲力矩;起拱力距;凸曲力矩;负弯矩
hogging of beam 梁的拱度
hogging stress 中拱应力
hoghead 水泵缸套
hoghorn 寻形喇叭辐射器;平滑匹配装置
hoghorn antenna 喇叭天线
hog house 猪舍
hog manure 猪厩肥
hog moment 挠曲力矩
hog nose 单刃钻
hogpen 猪圈
hog-pit 网部浆坑
hog ranch 猪饲养场
hog ring 扭曲环;拱形卡环
hogsback(tile) 元宝瓦
hogshead 液量单位;大(木)桶
hogshead hook 吊桶钩
hog shred 松散机
hog still 蒸馏塔
hogwallows 暴雨造成的坑洼
hogwash 泔水
hog weed 杂草;豚草
hog wire 倒钩铁丝
hohmannite 褐铁矾;羟水铁矾
hoick 急升
Hoimester 泥条软度计
hoist 一吊货;起动机;吊升机械;吊车;电葫芦;扯起;升起;吊装
hoist and bucket conveying 混凝土料斗提升机
hoist and drill head clutch lever 升降机及回转器分动手把
hoist and hold wires 抓认升降启闭钢缆
hoist a signal 挂信号旗
hoist-away 起重机
hoist back-out switch 吊车退回开关
hoist beam 升降机大梁
hoist bell crank 起重机直角杠杆
hoist belt 斜井皮带提升机
hoist block 起重葫芦;起重机滑轮;起重车
hoist brake 提升机制动器;卷扬机制动柄;卷扬机制动器;起重机闸
hoist bridge 卷扬斜桥;绞车桥;升降桥
hoist bucket 提升吊桶;吊罐;吊斗
hoist cable 提升机缆索;卷扬(机缆)索;起重索;启门索;升降机电缆;吊索
hoist cable scoket 卷扬机缆索卷筒;提升机缆索卷筒
hoist carriage 绞升吊车
hoist chain 提升机链条;卷扬机链条;启门链
hoist chamber 卷扬机室
hoist clutch 卷扬机离合器;启门机离合器
hoist control valve adaptor 升降机控制阀接头
hoist control valve bracket 起重机控制托架
hoist crane 绞绳葫芦
hoist cylinder 举升液压缸;起重机油缸;升降机油缸

hoist cylinder filler plug 起重机油缸注入塞
hoist cylinder gasket 起重机油缸垫密片
hoist cylinder head 起重机油缸盖
hoist cylinder oil relief valve 起重机油缸油压安全阀
hoist cylinder packing 起重机油缸填线圈
hoist cylinder piston spring 起重机油缸活塞簧
hoist down a cargo 卸下船货
hoist drum 起重鼓轮(机器卷索用)
hoist elevator 家用电梯
hoist engine 绞车发动机;卷扬机
hoist-engine house 提升机房
hoist equipment 国产设备
hoister 卷扬机;绞车驾驶员;起重机驾驶员;吊车驾驶员
hoist for mo(u)ld board carrier 模板车升降机
hoist frame 起重机架
hoist gate lifting device 卷扬式闸门启闭机
hoist girder 天车横梁
hoist guy (rope) 卷扬机拉索
hoist hole 卸料口(货物);提升间
hoist horsepower 绞车功率
hoist house 卷扬机罩
hoist incline 斜桥
hoisting 起重;吊起;提升;吊货;升起;吊装
hoisting and drilling load cycles 起下钻和钻进时间负荷比
hoisting angle 提升牵引角
hoisting apparatus 起重设备
hoisting appliance 提升设备;提升机构
hoisting barrel 绞车滚筒;吊车滚筒
hoisting beam 起重梁
hoisting beam for repairs 检修起重梁
hoisting block 绞辘下滑车;起重葫芦;起重动滑轮;起重机滑轮;起重滑车
hoisting box 起重机驾驶室;起重机操纵室
hoisting bucket 高炉吊罐;翻斗车;提升斗
hoisting by hand-operated gear 手摇绞车提升
hoisting by shallow shaft elevator 浅井提升机提升
hoisting cable 吊索;钢丝绳;起重索;起重机绳;起升索;提升钢丝绳
hoisting cage 提升(起重)笼;提升罐笼
hoisting capacity 绞车起重量;起重能力;起重量
hoisting chain 起重链
hoisting chamber 卷扬机室;启门机室
hoisting compartment 提升格间
hoisting controller 起重控制器
hoisting crane 起重机;升降起重机
hoisting cycle 提升循环时间
hoisting depth 提升深度;提升高度
hoisting device 升降装置;起重设备;起重装置;起卸方法;启闭设备
hoisting draw works drum 牵引工作圆筒
hoisting drum 卷扬机滚筒;卷扬机鼓轮;绞车(缆)筒;起重机卷索鼓轮;起重鼓轮;起重筒;提升滚筒
hoisting drum brake 提升卷筒制动器
hoisting duty 起重量;提升能力
hoisting engine 绞车机;起重机;提升机;绞车(发动机);吊重机器(起重机、绞车等)
hoisting equipment 起重设备;启闭设备;起重机备;吊装设备
hoisting equipment in pumping house 泵房起重设备
hoisting eye 吊环
hoisting fall 吊绳
hoisting frame 卷扬机架;提升机机架;启门机架;吊车架;提升架
hoisting gear 提升设备;绞升车;闸门启闭机构;绞车齿轮箱;起重装置;提升机构;吊重索具
hoisting gear brake 提升机制动杆;提升齿轮制动器
hoisting gear for cutter 铰刀起落装置
hoisting gear for draghead ladder 耙头架起落装置
hoisting gear for glide chute 溜泥槽起落装置
hoisting gin 三脚吊货架
hoisting height 起吊高度【机】;提升高度
hoisting hole 提升孔
hoisting hook 起重钩;提升钩;吊钩;提引钩
hoisting installation 起重设备
hoisting jack 千斤顶;起重千斤顶;起重器
hoisting lever 升降机把手
hoisting limiter 提升高度限位器
hoisting line 起重索;起重绳;吊索

hoisting load 提升荷载
hoisting load when jib in the highest position 吊臂在最高位置时吊起的重量
hoisting load when jib in the lowest position 吊臂在最低位置时吊起的重量
hoisting lowering 升降钻具
hoisting-lowering speed 升降速度
hoisting machine 提升机械;提升绞车;吊车
hoisting machinery 起重机械
hoisting mast 提升杆;提升柱;起重杆
hoisting mechanism 起重机构;起升机构
hoisting motion 提升动作
hoisting motor 升降电动机
hoisting muck 起吊污物;起吊垃圾
hoisting muck car 石渣升降机
hoisting operation 提钻作业
hoisting pad 有钩吊板;吊艇眼板
hoisting period 提升周期;提升循环时间
hoisting pipe 套管
hoisting plant 起重装置;提升设备
hoisting plant depreciation apportion and overhaul charges 提升设备折旧摊销及大修费
hoisting plug 起重机吊钩;提引栓;提升栓
hoisting power 起重能力;起重量;吊车能力
hoisting power of crane 起重机起重能力
hoisting production quota 提升生产定额
hoisting productivity 提升生产率
hoisting pulley 起重滑轮;提升天轮
hoisting pulley-block 提升车组;提升复式滑轮
hoisting ram 起重推杆;起重顶杆
hoisting reel 绞车滚筒
hoisting ring 吊环;起吊环
hoisting rod 吊艇铁杆
hoisting room 吊车间
hoisting rope 钢丝绳;起重索;起重钢绳;提升索;吊索;吊货钢索
hoisting rope equalizer 吊车钢索平衡器
hoisting rope safety factor 提升钢丝绳安全系数
hoisting shackle 捞船圆眼
hoisting shaft 起重竖井;提升井
hoisting sheave 天轮
hoist(ing) sheave jenny wheel 起重滑轮
hoisting sling 起重系索
hoisting speed 起重速度;起升速度【机】;提升速度
hoisting speed and loads 提升速度与起重量
hoisting spoil 起吊垃圾;起吊废料
hoisting steam engine 起重蒸汽机
hoisting steel bar 吊装钢筋
hoisting system 卷扬(机)系统;提升系统;升降系统
hoisting tackle 辘护;起重滑车(组);提升滑轮;提升滑车
hoisting tongs 夹钳;起重钳
hoisting tower 吊机(起重)塔;起重塔(架);提升塔(架);吊装塔架
hoisting type water swivel 提引水龙头
hoisting unit 绞盘;起重设备;提升设备;提升机
hoisting velocity 提升速度
hoisting velocity of winch 绞车提升速度
hoisting weight 提升重量
hoisting well 提升井道
hoisting winch 卷扬机;绞盘;绞车;提升设备;提升绞车
hoisting works 卷扬机
hoist lever 提升机杠杆
hoist lifting capacity 绞车起重能力
hoist limit gear 卷扬机限程齿轮;绞车限制齿轮
hoist line 提升缆索;起重缆索
hoist line boom point sheave 提升绳索臂端滑轮
hoist line socket 吊索管套;吊索承窝
hoist load 提升荷载
hoist machinery 起重机械;起重机装置;起重机工具
hoistman 起重机驾驶员;提升机驾驶员
hoist mechanism 起重机理;起吊机构;起重机械
hoist moment 起重力矩
hoist motor 卷扬机马达;卷扬机发动机;吊车电动机
hoist of flag 挂信号旗
hoist overspeed device 吊车超速防止装置;提升超速保护装置
hoist overwind device 吊车过头防止装置;钢丝绳绕过极限保安装置;钢绳过卷保护装置
hoist pennants 挂三角旗
hoist piston 起重机活塞

hoist piston ring 起重机活塞环
hoist piston rod 起重机活塞杆
hoist piston rod head block 起重机活塞杆块
hoist piston rod tie bar 起重机活塞柱夹杆
hoist platform 启闭台
hoist portal frame 龙门扒杆
hoist pump 起重机油泵
hoist pump body 起重机油泵体
hoist pump body bushing 起重机油泵体衬套
hoist pump bracket 起重机油泵托架
hoist pump check valve 起重机油泵止回阀
hoist pump check valve body 起重机油泵止回阀体
hoist pump control rod 起重机油泵控制杆
hoist pump control valve 起重机油泵操纵阀
hoist pump cover 起重机油泵盖
hoist pump cover gasket 起重机油泵盖垫密片
hoist pump drive gear 起重机油泵主动齿轮
hoist pump drive shaft 起重机油泵主动轴
hoist pump drive shaft bushing 起重机油泵传动轴衬套
hoist pump intermediate gear 起重机油泵中间轮
hoist pump intermediate shaft 起重机油泵中间轴
hoist pump manifold 起重机油泵歧管
hoist pump operating lever 起重机油泵操作杆
hoist reduction gear 绞车提升机减速器;绞车起重机减速器;起重(机)减速器
hoist rope 卷扬绳索;起重机钢丝索;起重机钢丝绳;起升钢丝绳;升降绳索
hoist slack-brake switch 吊车制动器过松开关
hoist station 卷扬机站
hoist tackle 起重辘轳;起重葫芦;起重滑车
hoist tower 吊机塔;起重塔(架)
hoist trolley 吊重行车
hoist-type induction coil 升降式感应圈
hoist types and capacities 升降类型及其载重量
hoist unit 提升机
hoist universal joint 起重机万向节
hoist up 挂起;绞起
hoistway 起卸口(货物);提升间;升降机井;升降机间;吊运井;电梯井;提升井
hoistway access switch 提升井检修通道入口开关;起卸口开关
hoistway biparting door 提升井门对开拉门
hoistway door 升降机井道的门;电梯厅门
hoistway door interlock 提升井门联锁装置
hoistway door locking device 提升井门锁定装置
hoistway enclosure 提升井围栏;竖井间壁;竖井墙
hoistway entrance 提升井入口;电梯通道进口
hoistway gate 提升井闸门
hoistway telescoping gate 提升井伸缩式闸门;提升井搭叠式闸门
hoistway unit system 提升井门联锁装置
hoist winch 卷扬机绞盘
hoist winch engine 卷扬机绞盘发动机
hoist wire 吊斗缆
hoist with lifting tongs 带提升钳夹的起重机
hoist works for rotary 转动的起重工作
Hoke ga(u)ge 福克块规(量块中间有孔,组合时用连接杆穿行)
Holarctic region 泛北植物区
holard 土壤水;土壤总含水量
Holbric 外墙空芯砖
holcodont 沟齿系
hold 止住;夹持;货舱;容纳;认为;认定;同期;船舱;柄榫头;保存;持有;保持
hold account procedure 封存程序
hold acknowledge 保持应答
hold acknowledge signal 保持确认信号
hold-all 工具箱;工具袋;手提箱
hold amplifier 保持放大器
hold-back 遮掩;退缩;车轮牵制器;压住;制动装置;逆止器;停止器
hold-back carrier 反载体
hold-back device 逆止器;防逆行装置
hold-back hook 速脱钩
hold-back ratchet 止回棘轮机构
hold-back winch 锁定绞车;夹紧绞车
holdbat 管箍
hold batten 货舱壁护条
hold batten cleat 舱壁护条夹
hold beam 舱(内)梁
hold bolt 固定螺栓
hold bunker 底煤舱
hold button 保持按钮

hold cable 握持缆索
hold capacitor 保持电容器
hold capacity 货舱舱容;容纳能力;舱容(量);保存能力
hold cargo 装船货物;装舱货物;船舱货物;舱内货(物);舱货
hold ceiling 舱内衬板
hold circuit 吸持电路;保持电路
hold cleaning 清舱(作业)
hold cleaning machine 清舱机
hold-closed mechanism 保持合闸机构
hold condition 持恒状态
hold control 同步调整
Holdcraft bar 霍氏测温杆
hold crew 舱内装卸工(人)
hold depth 舱深
hold-down 压(紧)板;压型;夹板;压具;夹爪
hold-down bars 固定棒
hold-down bolt 地脚螺栓;压缩(螺)栓;固定栓
hold-down clip 压紧夹板;固定夹
hold-down fixture 压牢装置
hold-down gag (剪切机的)压紧装置
hold-down gear 压紧装置
hold-down grid 压填料栅板;填料压板
hold-down groove 固定槽
hold-down mechanism 压紧机构
hold-down nut 固脚螺帽;固定螺母;压紧螺母;固定螺母
hold-down-or-up 摆动锚定装置
hold-down packer 悬挂式封隔器
hold-down plate 挡块;压紧盖板
hold-down ring 压紧环
hold-down roller 压紧辊;张紧轮
hold-down screw 压紧螺旋
hold-down slips (夹持器的)牙板
hold-down spring 压紧弹簧
hold-down strip 压条
hold-down support 固定支座
hold-down tabulator key 保持按下制表键
holdenite 红砷锌锰矿
Holden permeability bridge 导磁率测量电桥
holder 储[贮]罐;支架;支持物;支持器;刻刀;焊条钳;焊把;容器;票据持有人;持有者;保持架
holderbat 管子箍;管卡;裂环管夹;卡环
holder-cup 铆钉托
holder for die 板牙架
holder for tube 定位管卡子
holder for value 已付价持票人;价值持票人
holder in bad faith 恶意占有人
holder in due course 正当持票人;法定持有人
holder of contract 合同持有人
holder of license 牌照持有人
holder of the bill of lading 提单持有人
holder of the patent right 专利(权)持有者;专利(权)持有人
holder-on 铆工;铆钉顶棒;(压气铆钉的)气顶
holder plane 框面
holder spring 夹持器弹簧
holder steel 夹具钢
holder up 铆工;铆钉托;夹铆钉杆
hold facility 设备占用
hold facility 保持能力;保持(现场)能力
holdfast 固定底板(墙上);系锭座;吸盘;支架;固着器;固定器;钩子;紧握物;夹(钳);锚碇;平头大铁钉;地锚;保持
hold-fast coupling 固接联轴节
hold file 保存文件
hold for delivery on request 保存到请求传输
hold frame 舱内肋骨
hold fumigation 熏舱
hold gang 舱内装卸班
hold ground 锚泊地
hold harmless 赔偿责任转移;转移责任;使另一方不受损害
hold harmless agreement 一方承担义务的协议;免责约定;转移责任之约定;不受危害协议书
hold harmless amount 往年平均基金总额
hold harmless grant 往年平均基金总额超额拨款
hold-in 保持同步
hold in check 阻止
hold in custody 扣押
holding 吸持;储[贮]藏;支承;土地;所有物;存储;保温的;持(的);持有
holding ability 抓底能力

holding action 牵制作用;保持作用
holding aircraft 等待空域的飞机
holding anode 保持阳极
holding apparatus 压紧装置
holding area 等候区
holding armature 吸持衔铁;保持衔铁
holding bar 吸持棒
holding basin 储[贮]存池
holding beam 固定射束;保持电子束
holding boom 拦污硬
holding bottle 存储瓶
holding brake 减速制动器;恒速制动;防松闸;制动闸;止动闸;保压制动
holding braking 恒速制动
holding bulkhead 结构舱壁;耐压舱壁
holding capacitor 存储电容器
holding capacity 容量;容积;抓力;可蓄(容)量;锚抓力;保持能力
holding catch 止挡
holding chamber 气压保持器
holding charge 保持装填量
holding chuck 夹盘;卡盘
holding circuit 保护电路
holding coil 吸持线圈;保持线圈
holding company 股权公司;控制公司;控股公司;母公司
holding control 同步调整;同步控制
holding corporation 控股公司
holding cost 储存保管费用;保持费用;保藏费
holding current 保持电流
holding detent 制动爪
holding device 夹紧装置;吸持装置;夹具;夹紧器;夹持装置;定位装置
holding dog 楔形夹持器
holding dog with foot release pedal 脚踏式夹持器
holding-down bolt 压栓;锚固螺栓;定位螺栓;地脚螺栓;系紧螺栓;锚栓;压紧螺栓;固定螺栓;地脚螺钉;底脚螺栓
holding-down clip 锚固压板;盖条固定夹(用于金属薄板螺栓);压力板;卡子
holding-down latch 挡器;锁键
holding-down position 放平位置
holding-down rod 压紧杆;腿圈杆
holding-down screw 压紧螺旋
holding escrow 由中介持有地契的时期
holding factor 保持因数
holding flange 固定凸缘
holding force 握持力;吸持力
holding forceps 把持钳
holding frequency 调频运行方式;保持频率
holding function 动作功能
holding furnace 混合炉;保温炉
holding girder 支托大梁
holding ground 承载土;可锚泊地;锚地底质;水上储[贮]木场
holding gun 保存电子枪;保持电子枪
holding heat 空运转热消耗;维持热
holding jaw 夹爪
holding jib 支承悬臂
holding key 保持电键
holding latch 擒纵器;卡子;挡器;掣子
holding layer for filtration 过滤承托层
holding line 起重机;吊索;升降绞索;缆索;载重钢丝绳;起重索(抓斗用);停机线;提升钢丝绳;定位线
holding magnet 吸持磁铁;保持磁铁
holding maker 度假者
holding moment 约束转动力矩
holding nut 支承螺母
holding-off 瞄准点修正
holding of the court 法院裁决
holding-on 咬合
holding-on bar 顶铆杠
holding-operating company 控股兼营业公司
holding(out)coil 锁定线圈
holding pawl 固定爪
holding period 混凝土升温持续期;吸持时间;静置期;财产持有时间;保存时间
holding pin 紧固销
holding plate 垫板
holding point 暂留点;等待点
holding pond 废水滞流池;废水蓄水池;蓄养池;储[贮]池;泉塘;水塘;水池;盛液池;存储[贮]池
holding position 保持位置

holding position accuracy 定位精度
holding power 吸持功率;阻力;装载力;支持力;支撑力;固着力;静抓力;握(裹)力;维持功率;负荷力
holding power of anchor 锚抓力;锚泊驻留力
holding power of anchoring 锚泊系留力
holding power of ground 锚设备抓力
holding power to weight ratio 抓重比
holding primer 临时性底漆
holding pump 前支架
holding range 同步范围
holding register 保持寄存器
holding relay 吸持继电器
holding ring 夹环;调整环;定位环;绑环
holding rod 握杆
holding room 产品储存室
holding rope 吊索
holding screw 紧固螺钉;紧定螺钉
holding stock 保持库存
holding strength 保持强度
holding strip 夹条
holding stud 支架柱螺栓
holding tank 接受器;收集槽;存储槽;储料囤;储存池
holding tank of water 储水罐
holding temperature 保温温度
holding the line 控制物价
holding thread 紧固螺纹
holding-through 贯通的;联络小巷
holding-through survey 贯通测量
holding time 占用时间;占线时间;钎焊时间;通报时间;保温时间;保留时间;吸着时间
holding time at test pressure 试压保压时间
holding time of an international circuit 国际电路占用时间
holding time setter 保持时间给定装置
holding torque 保持转矩;保持力矩
holding track 车辆停留线;停车线
holding-up hammer 双动蒸汽锤;挡锤;圆边击平锤;铆钉顶锤;铆钉抵锤;铆钉撑锤
holding-up tool 铆顶工具
holding valve 平衡阀;闭锁阀;保持阀
holding voltage 吸起电压;吸持电压
holding yard 停车站线;停车场【铁】
holding zone 保温段;保留区
hold in hand 控制住
hold in range 保持范围;同步保持范围
hold instruction 保存指令;保持指令
hold-in winding 保持线圈
hold keelson 舱内内龙骨
hold key 保持键
hold ladder 舱口梯
hold lamp 占线指示灯;占线信号灯
hold lantern 货舱手提灯;舱内灯
hold line 制火线
hold list 保持信息表
hold machinery 舱内机械
hold man 货舱装卸工;舱内卸工(人)
hold mark 占有标记;保持符号
hold mode 保留式;保存式;保持状态;保持(方)式
hold off 脱出同步;隔开;释抑
hold-off arm 定位器
hold-off buoy 外拉浮筒
hold-off circuit 释抑电路
hold-off diode 闭锁二极管
hold-off interval 失步间隔
hold of pile 桩的打入深度
hold on 抓住
hold on coil 吸持线圈
hold on the chain 带住锚链
hold on the land 保持与陆地接触
hold on to our favorable position in 保持优势
hold-open catch 固定卡销
hold-open door 常开门
hold-open operating lock 舱盖打开时的锁挡
holdor 全息数据存储器
holdout 不下渗性;滞洪流量(水库);保持性
hold-out area 为等候进入医院而暂时停留的面积
holdout device 制止装置;维持装置
hold-over 延期;保温材料;保残木;绝缘材料;蓄冷;残留木;保持故障(法)
hold-over command 保存命令
hold-over fire 长时期隐续的火
hold over mode 保持模式

hold-over storage 逾年蓄水量;多年调蓄(量)
hold-over system 遗留体制;蓄冷系统;保持故障系统
hold over tennant 租约到期后仍不归还房产的承租户
hold page queue 保持页面队列
hold paint 耐蚀漆;货舱漆;货仓面漆;船舱漆
hold pattern 保持模式
hold pillar 舱内支柱
hold point 停止点
hold positively 联锁
hold queue 保持排队;保持队列
hold range 陷落范围;牵引范围
hold register 保持记忆装置
hold relay 保持继电器
hold request 保持要求;保持请求
hold reset 保持复位
hold reset switch 复位保持开关
hold route for shunting 非进路调车
hold shift 换舱
hold signal 持续信号
holdsman 舱内装卸工(人)
holds maximum length 船舱最大长度
hold space 货物容积;货舱容积;货舱舱容
hold space efficiency 舱位利用率
hold stanchion 舱内支柱
hold stringer 舱内侧桁
hold stripping 清舱
hold sweep(ning) 扫舱
hold switch transistor 保持开关晶体管
hold the line 等着不挂断
hold the particles apart 疏松土粒
hold the trump-card in the export of 拥有出口优势
hold time 保持时间;维持时间;吸着时间
hold together 结合在一起
hold ton 船舱(容积)吨
hold track 停车轨道
hold track for breakdown train 救援列车停留线
hold track for reserved locomotive 储[贮]备机车停留线【铁】
hold-up 扬料装置;停滞;支持;拦截;停顿;停车;滞留量
hold-up hook 运输钩
hold-up rate 扬料率
hold-up tank 储[贮]存罐
hold-up time 滞留时间;停留时间
hold-up vessel 储[贮]存容器;储料罐
hold-up volume 滞留体积
hold-up weight 滞留量
hold ventilator 舱内通风筒
hold washing 洗舱
hold water 浆挡水;挡水
hold water ballast 下水柜或下舱装水的压载
hold wire 闭斗缆【疏】
hold yard 车辆停留场;车辆停车场
hole 线道;漏洞;孔(洞);空穴;坑;海底学;洞
hole-and-slot resonator 孔隙型谐振器;槽孔型谐振器
hole angle 炮眼角度
hole annulus 环状空间
hole aperture 孔径;窗口宽度
hole application 穴施
hole armature 带通风槽电枢
hole array 钻孔排列
hole axis 钻孔轴线
hole azimuth angle 钻孔方位角
hole back 钻孔底;钻孔底
hole bailer 捞砂筒;抽砂筒(清孔用)
hole base assembly system 基孔制装配系统
hole base system 基孔制
hole basis limit system 基孔极限制
hole block 有孔的砖块;有孔的砌块;空心砌砖
hole blow 井喷
hole blowing process 通孔喷吹法
hole-bored axle from end to end 空心车轴
hole borer 钻洞机
hole boring 镗孔
hole boring cutter 镗孔刀
hole bottom 钻孔底;炮眼底
hole bottom cleaning 孔底清洗
hole bottom pressure ga(u)ge 孔底压力计
hole bubble 孔穴状气泡
hole burning 烧孔

hole-burning effect 空穴燃烧效应;烧孔效应
hole-burning model 烧孔模型
hole carrier 空穴载流子
hole caving 孔坍
hole cementation 钻孔灌浆
hole circle 孔圆;孔圈
hole cleaning 清孔
hole clearance 套管柱与井壁间的间隙
hole collar 钻孔口;炮眼口
hole concentration 空穴浓度
hole-conductor current access device 穿孔导体电流存取器
hole coordinates 钻孔坐标
hole cost 钻孔成本
hole count check 孔数检验;计孔检验
hole counter 孔计数器
hole count error 孔数错
hole-coupling 小孔耦合
hole cover 钻孔盖(板)
hole coverage 井眼环形空间
hole current 空穴电流
hole curvature 钻孔偏斜度
hole cutter 孔锯;打眼凿
holed 穿孔的;多孔的
holed by forklift hands 被铲车碰成洞
hole deflection 钻孔偏斜
hole deflexion 钻孔偏斜
hole density 空穴密度
hole depth 孔深[岩];炮眼深度
hole depth (by percussive drilling) 凿孔深度
hole depth error 孔深误差
hole depth ga(u)ge 孔深计
hole deviation 钻孔弯曲;钻孔偏斜;钻孔偏差;孔斜;井斜
hole deviation and surveying 钻孔偏斜与测量
hole deviation angle 钻孔偏斜角;井斜角
hole deviation distance 钻孔偏距
hole deviation intensity 钻孔偏斜强度
hole deviation plane 钻孔弯曲平面
hole deviation survey 钻孔偏斜测量
holed fibreboard 穿孔纤维板
hole diameter 孔径【岩】;井径;炮眼直径
hole diameter log curve 井径曲线
hole diameter measurement 井径测量
hole diameter of sieve (perforated) plate 筛板孔径
hole-diffusion length 空穴扩散长度
hole digger 钻孔器;打孔器
hole dilating drill 扩孔钻头
hole director 钻孔定向器;钻孔定位支架;炮眼导向架
hole-down packer 悬吊式堵塞器
hole drift current 电穴漂移电流
hole drill 螺纹底孔钻;螺孔钻
hole drilling 钻孔工作
hole drilling and blasting 钻眼爆破;凿眼爆破
hole eaten by insect (木材的)虫眼
hole effect 空穴效应
hole effect model 空穴效应模型
hole expanding boring machine 钻扩机
hole extruding 孔变薄翻边;变薄翻孔
hole face 孔壁;井壁
hole flanging 孔翻边
hole flushing 洗孔;洗井
hole for alignment 瞄准条校孔(测斜仪)
hole for hoist 吊装孔
hole-forming core 形成孔洞的型芯(混凝土构件内)
hole forming pin 穿孔针;穿空针
hole ga(u)ge 孔规;井径仪;井径规;内孔规;内测微计;内径规
hole grout pile 钻孔灌注桩
hole horizontal departure 钻孔水平位移
hole-in 开孔钻进
hole in aircraft floor 飞机底部摄影舱孔
hole inflow 孔内涌水
hole injection 空穴注入
hole injection rate 空穴注入速率
hole in soaking pit 均热炉坑
hole-in-the-center effect 中空效应
hole irrigating 穴灌
hole irrigation 穴灌;点浇
hole lapping 内擦准法
hole layout 炮眼布置
hole length 炮眼长度
hole loading 炮眼装药;炮孔装药

hole locator 钻孔定位器;孔定位器
hole man 爆破工;放炮工;爆炸手
hole mark 洞孔测标;峒孔测标
hole-method 穴植法
hole migration 空穴徙动;空穴迁移
hole mobility 空穴迁移率
hole noise 井口干扰
hole number 孔号
hole numbers at every circuit 每周孔数
hole opener 扩孔器
hole overall deviation distance 钻孔累积偏距
hole pattern 钻孔排列;孔模式;穿孔模式
hole-per bit 一个钻头进尺数
hole phenomenon 空穴现象
hole pincers 打孔钳
hole pitch 钻孔间距
hole pitch(distance) in sieve tray 筛板孔中心距
hole pitch of perforated tray 筛板孔距
hole placement 炮眼布置
hole placing 钻孔布置;布置炮眼
hole-planted method 穴植法
hole planting 穴状栽植;穴植
hole position 钻孔定位
hole pressure 孔压力
hole problems 钻孔复杂情况
hole production method 基准孔装配法
hole protection method with mud 泥浆护壁法
hole punched device 冲孔装置
hole punching 凿通孔
hole-punching machine 打孔机
holer 凿岩工;打眼工
hole reaming 扩孔;扩大钻孔
hole resistance 钻孔阻力
hole saw 管钻;孔锯;筒形锯
holes conduction band 空穴密集带
hole screen 网孔状屏;多孔筛
hole sealing 封孔
hole seeding 穴播
hole set 一组炮眼
hole setting 炮眼布置
holes for screw 螺栓孔
hole shooting 井中爆炸
hole shrinking sticking 缩径卡钻
holes in joints of welded vessels 焊接容器接头处的孔
holes in pattern 排孔
hole site 钻孔位置;孔位
hole size 炮眼直径;孔的尺寸;钻孔直径;孔径;孔大小
hole sizer 扩孔器;孔筛
hole slash 扩大钻孔
hole sort 孔分类;穿孔分类(法)
hole sowing 穴播;点播
hole spacing 钻孔间距;孔距;洞距
holes per face 掌子面上的炮眼数
holes per kilometer 每公里孔数【岩】
holes per square kilometer 每平方公里孔数【岩】
holes processing in tubesheet 管孔加工
hole state 空穴态
hole steel propeller 钢制空心螺旋桨
hole storage 储孔
hole structure 钻孔结构
hole survey 钻孔测量
hole survey by centerline method 中线法洞内测量
hole table 空白表
hole televiewer 钻视探头
hole-tester 钻孔试验器
hole through 贯通;穿孔
hole tile 空心砖
hole to be provided 预留洞
hole to core barrel annylus 孔壁与岩芯管环状间隙;井壁与岩芯管环状间隙
hole toe 孔底;井底
hole tolerance 孔径公差
hole top 孔口;井口
hole transplanting 穴植
hole trap 空穴陷阱
hole trouble 孔内事故【岩】
hole type 钻孔类型;空隙类型
hole volume 孔体积
hole wall 孔壁;井壁
hole-wall-protecting mud 护孔壁泥浆
hole wall protection and loss shut-off 护壁堵墙
hole wall sampling 孔壁取样;井壁取样

hole wall stability 孔壁稳定性【岩】;井壁稳定性【岩】
holey 有孔的
Holfos bronze 霍尔福斯高强度青铜
holiday 工休日;节(假)日;假期;无胶区;漏涂(点);绝缘层漏孔
holiday bungalow 度假住宅;度假别墅
holiday camp 度假营地;假日野营
holiday chalet 度假住宅;度假别墅
holiday cottage 度假住宅;度假别墅
holiday detection 空胶检查
holiday detector 针孔检测仪;漏涂点检测仪;绝缘检漏仪
holiday dwelling 假日住房
holiday flat 假日公寓
holiday for working an extra shift 调休
holiday hotel 假日旅馆
holiday house 假日住宅
holiday inn 假日旅馆
holiday locator 绝缘检漏孔
holiday pay 休假工资;假日工资
holiday premium pay 假日奖金
holiday shift 假日调班
holiday ship 度假船
holiday village 度假村
holiday with pay 带薪假日;带工资的假日
holing 钻孔;石板瓦钻孔;打孔;掏槽落煤法;打眼;穿孔
holing machine 板岩钻孔机
holing(of slab tile) 石板瓦钻孔
holing of slates 石板瓦钻孔
holings loader 除粉器
holing through 穿孔;贯通;掘通(巷道)
holism 整体主义
holistic approach 机能整体方法
holistic plan(ning) 全面论规划
holland 洁白亚麻布;荷兰布;窗帘棉布
hollander 漂打机;打浆机
Hollander beater 荷兰式打浆机
Holland formula 荷兰公式
hollandite 锰钡矿
Hollerith code 霍尔瑞斯代码
Hollerith number 何勒里斯数
hollingworthite 硫砷铑矿
Hollocast floor 霍洛卡斯特楼板(一种箱形孔钢筋混凝土楼板)
Hollock 灰褐色硬木(印度)
Hollong 红褐色硬木(印度)
hollow 中空的;开槽刨;挖空;凹陷的;凹地;凹部
hollow abutment 空心支座;空心桥台
hollow anode 空心阳极
holloware 底柱设备
hollow area 空心区域
hollow area coefficient 开口面积系数
hollow article 中空制品;空心制品
hollow axle 空心(车)轴
hollow back 凹背
hollow backed 凹背板;凹背的
hollow-backed flooring 架空地板(法);底空楼板
hollow-backed saw 凹背锯
hollow ball 空心球
hollow ball bearing 空心球轴承
hollow bead 空心颗粒
hollow beam 预制钢筋混凝土空心梁;管梁;空心射束;空心梁;空心电子注
hollow bed 接缝空穴底层;空心平缝层;底层有空穴的接缝
hollow billet 空心(钢)坯
hollow bit 岩芯钻头;空心钻头;空心钎头
hollow bit tongs 空心钳
hollow blade 空心叶片;空心桨叶
hollow block 空心砖;空心石块件;空心砌块;空心块(体);空心混凝土砌体;空心混凝土块件;空心方块(一种防波堤块体)
hollow block density 空心砌块密度
hollow block floor 空心砌块楼板
hollow block machine 空心砌块成形机
hollow block making 空心砌块制造
hollow block masonry wall 空心砌块墙
hollow block mo(u)ld 空心砌块模型;空心砌块模板
hollow-block tester 空心砌块试验机
hollow block wall 空心(砌)块墙
hollow bloom 空心管坯

hollow body 空心块(体)
hollow body wall 空心块墙
hollow bonded wall 空心砌合墙;空斗墙
hollow-bored 钻成空心的
hollow boring bit 空心钻头;空心镗刀油
hollow boring rod 空心钻杆
hollow bowl 中空内管
hollow bowl centrifuge 空心转筒离心机;空杯离心机
hollow bowl clarifier 空心转筒离心机;空心转筒澄清机
hollow box 空心石块件;空心混凝土块件
hollow box beam 箱形梁
hollow box foundation 井筒基础;沉井基础;空箱基础
hollow box girder 箱形大梁
hollow brick 空心砖
hollow brick for arches or shells 拱壳砖
hollow brick masonry 空心砖墙
hollow brick partition 空心砖板墙
hollow brick wall 空心砖墙
hollow brick with interrupted joint 间断接头的空心砖
hollow brickwork wall 空心砖墙
hollow building block 空心建筑砖;空心大砖;空心石块件;空心砌块;空心混凝土块件
hollow building brick 空心建筑砖;空心大砖
hollow buttress 空心支墩;空腹支墩
hollow buttress dam 空腹支墩坝
hollow cable 空心电缆
hollow calcium silicate brick 空心硅酸钙砖
hollow carbon microphere 中空碳微珠
hollow casting 中空铸型法;空心铸塑;空心铸件;空心注浆;糊膏中空浇铸法
hollow casting method 空心浇铸法
hollow cast-iron column 空心铸铁柱
hollow cast piston 空心铸造的活塞
hollow cast prestressed concrete floor 预应力混凝土空心楼板
hollow cast wheel 空心铸轮
hollow cathode 空心阴极
hollow cathode discharge 空心阴极放电
hollow cathode discharge spectroscopic analysis 空心阴极放电光谱分析
hollow cathode discharge tube 空心阴极放电管
hollow cathode lamp 空心阴极灯
hollow cathode tube 空心阴极管
hollow cement flag 空心水泥板
hollow center 经纬仪竖直角
hollow chamfer 斜削角;凹圆削角;凹圆线;凹(圆)槽;圆凹线脚;凹斜面
hollow charge 锥形装药;空底装药;聚能削装药;管状装药
hollow chisel 圆鼻錾;凹形凿
hollow cinder block 空心煤渣砖
hollow circular cylinder 中空圆柱
hollow cladding tile 空心镶面砖
hollow clay block 空心黏土砌块;空心陶土块;黏土空心砌块
hollow clay brick 空心黏土砖;黏土空心砖
hollow clay building block 空心黏土砌块
hollow clay building block with keyed underside 地面带扣的空心黏土砌块
hollow clay tile 空心黏土砖;空心陶土砖;黏土空心砖
hollow clay tile lintel 黏土空心砖过梁
hollow clinker block 空心炉渣砖;空心烧结砖;空心溶渣砖
hollow closing plug 空心关闭塞
hollow coil 空心线圈
hollow column 空心柱
hollow component 空心构件
hollow composite slab 空心组合板
hollow concrete 空心混凝土;大孔混凝土;蜂窝(状)混凝土
hollow concrete beam 空心混凝土梁
hollow concrete block 空心混凝土砌块;空心土块(体);混凝土空心(方)块
hollow concrete brick 空心混凝土砖
hollow concrete floor 空心混凝土楼板
hollow concrete method 空心混凝土方法
hollow concrete wall 空心混凝土墙
hollow concrete with less sand 少砂大孔混凝土
hollow conductor 空心导线

hollow-conductor-cooled rotor 内冷发电机转子
hollow cone 空心圆锥体;空心光锥
hollow cone type nozzle 空心雾锥喷嘴
hollow conical beam of light 空心光锥
hollow copper wire 空心铜线
hollow core 中空形芯
hollow core cable 空心电缆
hollow core construction 空心构造;空心结构
hollow cored building block 空心建筑砌块
hollow core door 空心门
hollow core floor slab 空心楼板
hollow core flush door 空心平面门;空心夹板门
hollow core member 空心构件
hollow core method 空心纤维法
hollow core plank 空心板
hollow core rock bolt 空心岩栓
hollow core sandwich panel 空心夹层墙板
hollow core slab 空心(楼)板
hollow core steel door 空心钢门
hollow core valve 空心阀
hollow crank pin 空心曲柄销
hollow crank shaft 空心曲柄轴
hollow cross section 空心断面
hollow cutter shaft 空心铰刀轴
hollow cutting 中空冲裁
hollow cylinder 圆筒体;空心圆柱体
hollow cylinder test 空心圆柱体试验
hollow cylindric(al) cast-in-place pile 空心圆筒就地浇注混凝土桩
hollow cylindric(al) (diametrically loaded) fender 空心圆筒(径向受力)护舷
hollow cylindric(al) fender 中空圆柱形护舷
hollow dam 空心坝
hollow derrick 管状吊杆;空心吊杆
hollow dielectric(al) waveguide 空心介质波导
hollow dome 空腔圆顶
hollow draw-in spindle 空心内柱轴
hollow drawn shape 空心冷拔型材
hollow drill 空心钻
hollow drill bit 岩芯钻头
hollow drilled stone 空心石块
hollow drilling steel 空心钻探钢材
hollow drill rod 空心(钻井)钻杆
hollow drill shank steel 钎子钢;空心钻钢;空心钻杆
hollow drill steel 中空钻探钢,六角钢;空心钻杆;炮钎钢
hollow dry process rotary kiln 中空干法回转窑
hollowed fiber membrane 中空纤维膜
hollowed fiber membrane bioreactor 中空纤维膜生物反应器
hollowed fiber membrane contactor 中空纤维膜接触器
hollowed fiber microfiltration 中空纤维微滤
hollowed fiber ultra-filtration membrane 中空纤维超滤膜
hollow-edged equalling file 凹边齐头平锉
hollow-edged file 凹边锉
hollowed microfiltration fiber membrane 中空纤维微滤膜
hollowed-out enamel 镂空陶瓷;镂花珐琅
hollowed-out work 镂空雕刻
hollow ended 中陷船首尾;水线端点凹陷
hollow end roller 端穴滚子
hollow expanded cinder concrete block 泡沫炉渣混凝土砌块
hollow extruded shape 空心挤压型材
hollow extrusion gasket 空心垫缝带
hollow fabric 管状织物
hollow face 凹脸
hollow faced wire cup brush 凹面杯形钢丝刷
hollow fiber 空心纤维
hollow fiber membrane 空心纤维膜
hollow fiber permeator 空心纤维渗透器
hollow fiber ultra-filtration 空心纤维超滤
hollow filler 空心砖;空心填块
hollow filler block floor 空心肋楼板(混凝土密肋)楼板
hollow filler brick (密肋楼板用的)空心砖
hollow filler floor 空心砖(密肋)楼板
hollow filler system 空心砖密肋(楼板)系统
hollow fillet 空心线脚;空心嵌条
hollow first-stage plug 空心第一级塞
hollow flight conveyer 带有中空叶片可通水冷却的螺旋输送机
hollow float 空心浮标;空心浮子

hollow floor 空心轻楼板;空心地板
hollow floor beam 空心楼板梁
hollow floor filler 空心地板填充料
hollow floor slab 空心(楼)板
hollow forging 中空锻造;空锻;荒管
hollow foundation pier 空心基础
hollow fraise 筒形外圆铣刀;套筒形铣刀;套料铣刀
hollow frame section 空心构件断面
hollow furred ceiling 空透性吊顶
hollow fusee 空心均力圆锥轮
hollow gasket 空心密封垫;空心垫缝带
hollow ga(u)ged brick 陶瓷空心砖
hollow ga(u)ged brick with right angle cut 直角切割的陶瓷空心砖
hollow girder 空心(大)梁
hollow glass 瓶罐玻璃;凹窝载玻片
hollow glass beads 空心玻璃细珠
hollow glass block 空心玻璃砖;空心玻璃砌块
hollow glass block partition(wall) 空心玻璃砖隔墙
hollow glass block sealing machine 空心玻璃砖封接机
hollow glass brick 空心玻璃砖
hollow glass fiber 空心玻璃纤维;中空玻璃纤维
hollow glass microspheres 空心玻璃微珠
hollow glass sphere 中空玻璃微珠
hollow glass tile 空心玻璃砖
hollow glass ware 空心玻璃制品
hollow granule 空心粒子
hollow gravity arch dam 空腹重力拱坝
hollow gravity dam 空心重力坝
hollow grip 抓手凹槽
hollow ground slide 凹载玻片
hollow ground tool 空心凿
hollow gypsum(building) block 空心石膏砌块
hollow gypsum plank 空心石膏板
hollow head set-screw 空心头固定螺钉;内六角头固定螺钉
hollow hinge pin 空心合页心
hollow inclusion 空心包体
hollowing 拉坯成型;空鼓;淘空
hollowing file 圆锉;槽形锉
hollowing knife 弧刮刨
hollow ingot 空心坯
hollowing out 挖空
hollow jet 空心喷嘴
hollow jet needle valve 空心喷嘴针形阀;空注针形阀
hollow jet valve 空注阀
hollow joint 留空缝;空心结合;空缝;凹缝
hollow keel 中陷龙骨;带槽龙骨
hollow key 空心键;空底键
hollow knot 空心节;空孔节;脱落节;树节洞;空洞
hollow lead 滚刀的沟槽导程
hollow lens 空心棱镜
hollow light(weight) slag concrete 空心轻质炉渣混凝土砌块
hollow lime-sand brick 空心灰砂砖
hollow line 波谷线
hollow lintel 空心过梁
hollow load-bearing concrete masonry unit 空心承重混凝土砌块
hollow load cell 空心压力盒
hollow mandrel 空心心轴
hollow mandrel lathes 空心心轴车床
hollow mason arch bridge 空腹石拱桥
hollow masonry unit 空心砌块;空心砌体
hollow masonry wall 虚圬墙;空芯墙;空斗墙
hollow masonry(work) 空心圬工
hollow mast 空心桅杆
hollow member 空心构件
hollow metal 空心金属(件)
hollow metal door 空腹金属门;包铁门;空心金属门
hollow metal fire door 空腹金属防火门;空心金属防火门(中垫绝缘层的)
hollow metal frame 空心金属框架
hollow metal window 空心钢窗;空腹金属窗
hollow microsphere target 空心玻璃壳微球靶
hollow mill 空心(外圆)铣刀;筒形外圆铣刀
hollow milling cutter 空心铣刀
hollow milling tool 空心铣刀
hollow mo(u)ld 空心模具
hollow mo(u)ld floor 空心模制楼板
hollow mo(u)lding 空心模;凹圆形线脚;凹饰
hollow neck 瓶颈过薄

hollowness 空心(度)
hollow newel 旋梯的筒状中空柱;无栏杆旋梯的井孔;旋梯柱井孔;装旋梯柱的井孔;螺旋梯梯井;空心楼梯栏杆柱
hollow newel stair(case) 无中心柱螺旋形楼梯;露明中柱旋梯;露明井梯;无中柱螺旋梯
hollow nose tongs 空端夹子
hollow nozzle 空壁喷管
hollow of resistance 阻力变弱
hollow out 挖空;掏空
hollow packing 空心填料
hollow panel 空芯墙板
hollow partition 空心隔断;空心隔墙;空心块间壁墙;空心夹墙
hollow partition block 空心隔墙砌块
hollow partition slab 空心隔墙板
hollow partition tile 空心隔墙砖
hollow partition wall(ing) 空心隔墙
hollow party(manory) wall 空心分户(共用)墙
hollow party wall 空心共用隔墙
hollow pier 空心桥座;空心桥台;空心(桥)墩
hollow pile 空心桩
hollow pillar 空心壁柱;空心支柱
hollow pinion 空心齿轴
hollow pin knocked out reversing valve 断销式反循环短节
hollow pipe 空心管子
hollow pipe-type breakwater 管式透空防波堤
hollow pipe waveguide 空腔波导管;空管波导
hollow piston 空心活塞
hollow piston jack 空心活塞千斤顶;空心活塞起重器
hollow piston pump 空心活塞泵
hollow piston rod 空心活塞连杆
hollow pivot 空心枢
hollow plane 槽刨;凹陷面
hollow plank 空心楼板;预应力混凝土预制空心楼板
hollow plaster panel 空心抹灰墙板;空心灰泥板
hollow plate 空心板
hollow plug 空心胶塞
hollow plunger 空心柱塞;空心滑阀;空心冲杆
hollow plunger pump 空心柱塞泵
hollow plywood 空心胶合板
hollow pot 空心砌块
hollow precast prestressed floor slab 空心预制预应力楼板
hollow preformed pile 预制空心桩
hollow prism 空心棱镜
hollow profile 空心型材;空心断面
hollow prop 空心金属支柱
hollow punch 空心冲头;冲孔器;冲孔机
hollow pyramid 中空方柱体
hollow quoin 枢轴凹座(船闸闸门);闸隅槽;闸门墩;闸臼(坞闸转动的座);空芯墙角基石
hollow ram pump 空心柱塞泵;空心水锤泵;空心活塞泵
hollow reamer 扩孔钻头;空心铰刀
hollow relief 凹雕
hollow rib 空心肋
hollow-ribbed 有空心肋的
hollow-ribbed bridge 空心肋拱桥
hollow-ribbed floor 肋形楼板;空心砖楼板
hollow rim 空心轮辋
hollow ring 空心环
hollow rivet 空心铆钉
hollow rod 空心棒;空心钻杆
hollow rod churn drill 空心杆冲钻
hollow rodding 钻杆柱
hollow rod drilling 空心钻杆钻进
hollow roll 咬口(接缝);空心辊;卷筒接头;卷边(接缝)
hollow rope 空心钢索
hollow rounded skirting 凹圆踢脚线;凹圆踢脚板
hollow round pier 空心圆形桥墩
hollows 圆角面
hollow sand-lime brick 空心石灰砂砖
hollow screw 空心螺钉
hollow screw pile 空心螺旋桩;钢管螺旋桩
hollow sea 卷浪;非常陡深的海
hollow section 空心型材;空心截面
hollow section strand 空心股线
hollow shaft(ing) 空心轴;中空轴;管轴

hollow shape 空心型材
hollow shell 空心管坯
hollow shell pile 空心(带)壳桩
hollow slab 空心板
hollow slab bridge 空心板桥
hollow slab extruder 空心楼板挤出机
hollow slab floor 空心板楼板
hollow slab frame 空心板框
hollow sliding block 联轴节型滑块
hollow slug 空心块(体)
hollow space 空穴区;空洞
hollow space in cupola 圆顶中的空间
hollow space oscillator 空腔振荡器
hollow spar 红柱石;组合杆子;空心帆桁
hollow sphere 空心球
hollow spinning block 空心旋压模
hollow spinning chuck 空心旋压模
hollow spoke 空心轮辐
hollow spool 空心阀芯
hollow spot 洼坑;凹陷地点
hollow square 空心四方体块;空心方块(一种防波堤块体);消波混凝土块体
hollow square beam 空心方钢犁梁
hollow square block 四脚空心方块;四角空心方块
hollow square mo(u)lding 空心方线脚;凹入的金字塔方阵线脚
hollow square quay wall 空心方块码头
hollow square steel 空心方钢
hollow stalk 空心茎
hollow stay 空心撑杆
hollow stay bolt 空心螺撑
hollow steel 中空钢;空心钢(材)
hollow steel door 空心钢门
hollow steel frame 空腹钢构架
hollow steel mast 钢管制桅杆;钢管轻便钻塔
hollow steel pile 空心钢管桩;空心钢桩
hollow steel propeller 中空钢制螺旋桨
hollow steel rim 空心轮辋
hollow steel roller 空心钢辊
hollow steel section 薄壁型钢
hollow steel spoke wheel 空心钢辐轮
hollow steel window 空腹钢窗
hollow stem auger 空心杆螺旋钻
hollow step 空心踏步
hollow sting 空心支架
hollow stirrer 空心搅拌器
hollow stone 空心砌块;空心石
hollow structural section 空间结构型材
hollow swage 陷型模;甩子
hollow swell 陡涌
hollow tetrahedron 空心四面体
hollow tile 空心砖;空心瓦;空心(楼)板
hollow tile density 空心砖密度;空心砖孔隙度
hollow tile floor 空心砖楼板;空心楼盖;空心板楼面;空心板楼板
hollow tile floor construction 空心砖楼板建筑
hollow tile floor slab 空心砖钢筋混凝土楼板;空心砖楼板
hollow tile masonry wall 空心砖墙
hollow tile mo(u)ld 空心砖模
hollow tile partition 空心砖隔墙;空心板砖墙
hollow tile roofing 空心砖屋顶
hollow tile slab 空心砖楼板
hollow tile wall 空心砖墙
hollow transmission line mast 空心输电线杆
hollow trip plug 空心打开塞
hollow trunnion 中空轴颈;空心轴颈
hollow trunnion bearing 中空轴轴承;中空耳轴
hollow truss 空腹桁架
hollow-trussed beam 虚构梁;花隔梁;花格梁
hollow tube fiber 空管纤维
hollow tube mode 空心管模
hollow tube pile 管柱桩
hollow tube pile foundation 管柱桩基础
hollow turbine drill rod 空心涡轮钻杆
hollow type neoprene(compressible) seal 多孔橡胶嵌缝条
hollow unit 空心石块件;空心砌块;空心混凝土块件;空心构件
hollow unit masonry 空心块砌体;空心砌块加工
hollow unit masonry construction 空心块砌体结构
hollow vacuous 空的
hollow vortex 空心涡
hollow wall 空心(夹层)墙;空斗墙;夹墙

hollow wall block 空芯墙块
hollow wall construction 空芯墙构造
hollow walling 空芯墙体
hollow wall insulation 空芯墙隔绝(性能)
hollow wall of masonry 空芯墙砌体
hollow ware 中空器皿;空心器皿;凹形器皿
hollow ware glass 空心器皿玻璃
hollow water line 波形水线
hollow waveguide 中空波导管;空心波导;空腔波导管
hollow way 沿谷道路
hollow web girder 空心腹板梁;箱形大梁
hollow weir 空心堰
hollow wire 管状线
hollow wire rope 空心钢丝绳
hollow wood construction 空心木(板)结构;空心胶合板结构;空心夹板构造;空心木结构;空腹木结构
hollow zone 凹地带
holly 冬青属植物
holm(e) 河旁低地;小岛(河或湖中)
Holme's light 救生火号
Holme's signal 荷尔米信号
Holme double scoop 荷尔米双斗挖泥器
Holme mud sampler 荷尔米泥浆取样器
holm(e) oak 圣栎
Holmes-Stewart phenomenon 回缩现象
holmite 云辉黄煌岩;富辉黄斑岩
holmium 钬
holmium ores 钬矿
Holmpress pile 霍尔姆普莱斯桩(适用于粉质土和非黏性土的现浇钢筋混凝土桩)
holmquistite 锂(蓝)闪石
holoaxial 全轴的
holoblastic cleavage 全裂
holocamera 全息照相机;全息摄影机
holocaust 大毁灭
holocellulose 综合纤维素;全纤维素
Holocene 全新世【地】
Holocene epoch 全新世【地】
Holocene series 全新统【地】
Holocenic crust movement 全新世地壳运动
holocentric 单心
holochrome 全色素
holoclastic rock 全碎屑岩
hololococcolith 全颗石
holocoenosis 综合生态影响
holocrystalline 全晶质(的);全(结)晶
holocrystalline-porphyritic 全晶斑状
holocrystalline rock 全晶质岩
holocrystalline texture 全晶质结构
holodiastolic 全舒张期的
holo film 全息胶片;全息底片
holoframe 全息帧
hologamy 整体配子;成体配合
hologenesis 完全发生
hologony 全发生
hologram 综合衍射图;全息照片;全息相片;全息图
hologram aberration 全息像差
hologram emulsion shrinkage 全息照片乳胶收缩
hologram grating 全息摄影栅
hologram illuminating wave 全息照明波
hologram image 全息(照相)图像;全息照相图像
hologram imaging 全息成像
hologram interferometer 全息图;全息干涉仪
hologram master 全息原版片
hologram material 全息照相材料
hologram matrix radar 全息照片矩阵雷达
hologram memory system 全息存储系统
hologrammetry 全息摄影测量;全对称
hologram optics 全息摄影光学
hologram page 全息图面
hologram photography 全息摄影(术)
hologram photoplate 全息照相底片
hologram picture 全息图像
hologram radar 全息雷达
hologram reconstruction 全息图再现
hologram recording material 全息(照相)记录材料
hologram technique 全息摄影技术
hologram television 全息照像电视;全息(照相)电视
hologram transmission 全息传输
hologram transmittance 全息片透射比
hologram view 全息图

holograne grating 全息光栅
holograph 全息照像;全息摄影;亲笔证书
holographic(al) 全息的
holographic(al) aberration 全息像差
holographic(al) anaglyph 全息立体互补图
holographic(al) antenna 全息天线
holographic(al)-base system 全息系统
holographic(al) cinematography 全息电影摄像术
holographic(al) coding plate 全息编码板
holographic(al) colo(u)r storage 全息彩色储存
holographic(al) concave reflection-type grating 全息凹面反射光栅
holographic(al) conjugate relation 全息共轭关系
holographic(al) contouring 全息轮廓法;全息等高线
holographic(al) coordinate system 全息摄影坐标系统
holographic(al) correlation filtering 全息相关滤波
holographic(al) coupler 全息耦合器
holographic(al) data storage 全息数据存储器
holographic(al) deconvolution 全息解卷积
holographic(al) deformation 全息畸变;全息变形
holographic(al) device 全息照相装置
holographic(al) diffraction grating 全息衍射光栅
holographic(al) display 全息显示
holographic(al) display system 全息显示系统
holographic(al) doublet 全息胶合件
holographic(al) element 全息元件
holographic(al) encoding 全息编码
holographic(al) fibre 全息纤维
holographic(al) filter 全息滤光片;全息滤波器
holographic(al) filtering 全息滤波
holographic(al) grating 全息照相栅;全息光栅
holographic(al) grating scanner 全息光栅扫描器
holographic(al) headup display 全息平视显示器
holographic(al) image 全息图像
holographic(al) image restoration 全息图像再现
holographic(al) imaging 全息成像
holographic(al) information accumulation 全息信息储存
holographic(al) information storage 全息信息存储;全息信息储存
holographic(al) interference measuring 全息干扰计量
holographic(al) interference pattern 全息干涉图
holographic(al) interferometer 全息干扰仪
holographic(al) interferometric evaluation method 全息干涉估价法
holographic(al) interferometry 全息干涉度量学;全息干涉测量(法);全息干扰测量法
holographic(al) laser beam scanner 全息激光束扫描器
holographic(al) laser visor 全息激光护目镜
holographic(al) lateral shear interferometer 全息照相横向剪切干涉仪
holographic(al) lens 全息透镜
holographic(al) lens system 全息透镜系统
holographic(al) map display 全息地图显示
holographic(al) mapping system 全息测图系统
holographic(al) mask technology 全息掩模技术
holographic(al) matched filter 全息匹配滤波器
holographic(al) memory 全息照相存储器;全息(照相)存储器
holographic(al) memory device 全息存储装置
holographic(al) method 全息摄影法
holographic(al) microscope 全身显微镜;全息显微镜
holographic(al) microscopy 全息(照相)显微术
holographic(al) mirror 全息反射镜
holographic(al) Moire technique 全息莫尔法;全息波纹法
holographic(al) Moire technology 全息莫尔法
holographic(al) movies 全息电影
holographic(al) multiplication 全息多重像
holographic(al) nondestructive testing 全息术无损检验
holographic(al) optic(al) elements 全息光学元件
holographic(al) optics 全息光学
holographic(al) panoramic stereogram 全息全景立体照片
holographic(al) parameter 全息参数
holographic(al) photoelasticity 全息光弹(方)法
holographic(al) photoelastic test 全息光弹试验
holographic(al) ray-tracing equation 全息光线

追迹方程式
holographic(al) recording device 全息摄影记录装置
holographic(al) recording parameter 全息记录参数
holographic(al) scan 全息扫描
holographic(al) scan aberration correction 全息扫描像差校正
holographic(al) scanning method 全息扫描法
holographic(al) seismograph 全息地震仪
holographic(al) simulation 全息模拟
holographic(al) spectroscopy 全息光谱学
holographic(al) stereo display 全息立体显示
holographic(al) stereomodel 全息立体模型
holographic(al) storage 全息照相存储器;全息摄影存储器;全息存储
holographic(al) storage and access system 全息存储与存取系统
holographic(al) subtraction 全息减去法
holographic(al) system 全息照相系统;全息摄影系统
holographic(al) technique 全息摄影技术
holographic(al) television 全息电视
holographic(al) three dimensional display 全息立体显示
holographic(al) tomography 全息分层照相术
holographic(al) tranducer 全息照相换能器
holographic(al) transducer analyzer 全息照相换能器分析仪
holographic(al) video disk 全息录像盘
holographic(al) viwer 全息观察仪
holograph processing 全息处理
holograph will 无见证人的手书遗嘱
holography 全息照相术;全息(照相)术;全息学;全息摄影(术)
holography camera 全息照相机
holohedral 全面型的;全对称晶形的
holohedral form 全面晶形
holohedral hemimorphism 全面异极性
holohedral symmetry 全(面)对称
holohedrism 全对称性
holohedron 全面体
holohedry 全面像;全(面)对称;全晶形
holohyaline 全玻质的
holohyaline texture 玻璃质结构
holokarst 全岩溶;完全喀斯特
hololaser 全息激光器
hololens 全息透镜
hololeucocrate 全白岩
holomagnetization 全磁化
holomelanocrate 全黑岩
holometabola 完全变态类
holometabolic 完全变态的
holometabolous 完全变态的
holometer 测高计
holometry 全息照相干扰测量术;全息(光)照相干涉测量术
holomicrography 全息显微照相术
holomictic condition 完全性条件
holomictic lake 完全混合湖;完全对流湖
holomixis 全混合(作用);完全混合作用
holomorph 全型
holomorphic 正则的
holomorphic folds 全形褶饰
holomorphic function 正则函数;全线函数;全纯函数
holomorphism 全面型;全对称形态
holonomic constraints 完整约束
holonomic system 完整系统
holontozone 全生物带
holoparasite 全寄生植物
holophane lumeter 全辉照度计
holophankton 全浮游生物
holophone 全息录音机
holophotal 全光反射的
holophotal system 全反射聚光体系
holophote 全射镜;全光反射装置;全反射系统;全反射聚光镜
holophytic nutrition 全植物式营养
holophytic protozoa 植物型原生动物
holoplankton 终生浮游生物
holoscan 全息扫描
holoscope 全息照相机
holoscopic 近复消色差

holoseismic 全息地震
holoseismic method 全息地震法
holosteric 全部固体的
holosteric barometer 固体气压表;空盒气压表
holostratotype 正层型
holostrome 全层
holostylic 全接型
holostyly 全接型
holosymmetric 全面型的
holosymmetry 全对称
holosystemic 全面型的
holosystolic 全收缩期的
holosystolic high plateau type 全收缩高平顶型
holotactic 全规整
holotactic polymer 全规速聚合物
holotape 全息录像带
holotape frame 全息带帧
holotopy 全局关系
holotype 正模标本;全型;完模标本
holoviewer 全息观察仪
holozoic nutrition 全动物型营养
holsom 耐巨浪的船体
Holsteinian interglacial epoch 霍尔斯坦间冰期
holster 轧辊堆放架
holt 林地
holtedahlite 三方羟磷镁石
holtite 锑线石
Holtz's electrical machine 霍尔兹静电起电机
Holtz's machine 霍尔兹感应电机
holy city 圣城
Holy Cross church 神圣十字架教堂
holy door 教堂圣门
Holy Grail 圣盘;圣杯
Holy of the Holies 万圣之圣
Holy Sepulchre at Jerusalem 耶路撒冷的圣墓
holystone 磨石
holy-water basin 圣水盆
holy-water stone (教堂门口的)圣水石盆
holy-water tank 循环冷却水柜
homalographic(al) 等面积的
homalographic(al) projection 等积投影
homaloidal curves 统一曲线系
homalopisthocranius 扁平后头
Homan and Rodger's floor T形钢加强的混凝土楼板
homansote 用板组成的预制木屋
Homburg's alloy 洪堡合金
hom dyne reception 零差接收
home 回原;起始位置;停打阻力;出发点;返回始位;返回
home accessories 房间内装饰品
home address 住所地址;内部地址;标识地址
home affairs 民政
home airport 终点航(空)站
home-area toll 国内长途电话
home automation 家庭自动化
home bank 家乡堤岸;国内银行
home-based 基于家庭的
home-based trip 家庭出行
home-based work trip 家庭工作出行
home block 标识块
home booking 卸货港订运
homebound 返程货物
homebound voyage 回程航次;返航航次
home bread-baking oven 家用面包炉
home brew 家庭制造
home broadcasting 国内广播
homebuilder 住宅建筑商;住宅建造商
home building 住宅建设;住宅建造
home car 本路车【铁】
home construction activity 住宅建设
home consumption 厂用消耗
home depot 机务本段
home-driven pile 打到底的桩
home drive system call 驱动器磁头回零系统调用
home earth dam 家乡土坝
home elevator 家用电梯
home environment 本地环境
home equipment 自制设备;国产设备
home equity loan 住宅资产抵押贷款
home exchange 国内汇兑
home extension intercom 内部分机通话
home fire extinguisher 家用灭火器
home fire safety 家庭防火安全

home for the aged 养老院;老人之家
home for the homeless 收容所
home freezer 家用冰箱
home freight 国内运费;回程运费
home furnishing 家用设备
home garden 庭院花园;家庭花园;宅园
home-garden type 手扶式农具
home grinder 家用研磨机
home ground garden 住宅花园
home-grown 土生的
home-grown timber 国产木材;本国木材
home handyman 家庭零杂工;受雇做(家庭)杂事的人
home heating oil 家庭取暖用油
home improvement 住房改善
home improvement loan 住房改善贷款
home industry 家庭工业
home inspection service 住宅检测
home inspector 住宅检测员
home insurance building 家庭保险公司大楼
home interview method 登门调查法
home interview survey 家访调查
home in the holder 插入管座
home key 原位键
homeland scape 家庭园景
home laundry 家用洗衣房
homelessness 无家可归者
home lift 家用电梯
home loading and unloading charges 路内装卸费
home loan 住房抵押贷款
home loan banks 住宅贷款银行
home loop 内部回路;本地回路
home loop operation 本地回路操作
home-made 手工制的;自制的;本地制的;自装仪器;国产的;当地造的
home-made rig 国产钻具;国产设备
home manufacturer 工厂生产构件的房屋制造商;预制房屋商
home mortgage loan 房屋抵押贷款
home motor 家用电动机
homenergic 等能量的
home-nursing-convalescent 康复疗养院
homeoblastic 等粒变晶状
home occupation signs 房屋占用标志;房屋使用标志
homeochromatic 颜色相似的
homeocrystalline 等粒的
home office (银行的)总行;总公司;国内(总)公司
home office switch 防电击开关
homeohormous 恒温动物的
homeohydric plant 恒水植物
homeomorph 异物同形
homeomorphism 异物同形(现象)
homeomorphy 异质同晶(现象)
home on 瞄准
home-on-jam 干扰自动跟踪
home ore 国产矿石
homeosis 异形再生
homeostasis 自稳机制;自稳机理;自动动态平衡;自动调节;内稳定状态;内环境稳定;内环境平衡;动态静止
homeostat 同态调节器
homeostatic control 稳态控制
homeostatic fabric 均变组构
homeostatic mechanism 自稳机理;自动稳态机制;稳态机制;适应性机能
homeosynapsis 同型联会
homeotype 同模(式)标本
Home Owner's Association 业主协会
homeowner's insurance 房主保险
Home Owner's Loan Corporation 房主贷款公司(美国)
homeowner's once-in-a-lifetime tax exclusion 房主一次性免税
homeowner's policy 房主保险单
homeowner's tax exemption 住宅免税
homeowner's warranty program(me) 房主保修计划
home ownership 房产所有权;住宅所有权
home ownership scheme 居者有其屋计划
home page 主页
home paint 家用油漆
homepitaxy 同外延;等外延
home port 国内港口;船籍港

home position 原始位置;原来位置;静止位置;标识位置
home-position contact 起始触点
home-produced 国内生产的
home products 本国产品
home projector receiver 家用投影电视接收机
homepycnal flow deposit 等密度流沉积
homer 自动导航设备;归航指示标;归航台(机场)
home radar chain 地面雷达网
home range 牧场;栖息区;巢区
home record 引导记录;住所记录;内档;起始记录
home registration certificate 国内注册证书
Homerian(stage) 霍梅尔阶(阿拉斯加的植物阶)
home roll 主辊
home scrap 生活垃圾;本厂垃圾;家庭垃圾
home screwed 极度拧紧的
home signal 进站信号(机);场内信号灯
home site 住宅基地
home smoke detector 家用感烟探测器
homes outs 国内外
homespun 手工纺织呢;钢花呢
homestall 一家的房地产
homestat 同态调节器
home station 本地台
homestead 住宅;自耕农场;宅基;家园
Homestead Act 移居法(美国);份地法
homestead area 居住区
homesteader 农场所有权人
homestead exemption 破产家园豁免
homestead exemption law 宅基豁免法
homesteading 定居耕种;公有房地产转让
homestead tax exemption 宅基免税
hometaxial output 轴向均匀输出
home television 民用电视
home terminal 家庭用终端设备
home town 家乡;故乡
home-to-work peak hour 上班高峰时间
home trade 国内贸易
home trader 国内运输船
home treatment 家庭处理
home treatment device 家用水处理装置
home treatment equipment 家用净水器
home unit 出售公寓(澳大利亚);家庭单元
home visit 家庭访视
home ward bound 驶回本国;回国的船
homeward cargo 返航货物
homeward rate 回程货运价
homeward return voyage 归航
homeward sailing 返航
homeward stitches 不整的针脚
homeward voyage 归国航次;回程航次;返航(航次)
home warranty 住宅担保
home warranty insurance 房产质量合格保险
home warranty program(me) 住宅担保计划
home waters 宅边水域;国内水域;领海
home water softener 家用软水剂;家用水质软化器
home water treatment 家庭水处理
home water treatment device 家用水处理装置
home water treatment system 家用水处理系统;家庭水处理系统
home wear 便服
home wind 回港风
home workshop 业余爱好劳作室
home-yard orchard 宅旁果园;房周果园
homilite 硅硼钙铁矿
homing 引导;自动引导;归位;归航;回复原位
homing action 还原动作;导引作用
homing adapter 归航仪(飞机);归航附加器
homing aid 归航设备;辅助导航设备;归航台(机场)
homing antenna 自引导天线;航向(接收)天线;方位天线;导引天线
homing beacon 归航信标;进场指标;机场导航台;导航着陆无线电信标
homing control 自动瞄准
homing device 自动跟踪装置;归航设备
homing finder 归位式选择器
homing fishes 返回鱼类
homing guidance 自动导引
homing intelligence 自动导引信息
homing mechanism 自动导引装置
homing movement 还原动作
homing on 瞄准
homing range 归航有效距离
homing relay 复位式继电器

homing route of fish 鱼类洄游路线
homing sensor 自动寻的传感器
homing sequence 引导序列;起始序例
homing station 归航台(机场)
homing system of guidance 自动跟踪装置
homing transponder 归航应答器
homing tree 导引树
homing type line switch 归位式导线机
hommock 小丘
homoaxial folds 同轴褶皱
homobium 共生体
homoborneol 高冰片
homobrochate 同形网状
homobront 雷暴等时线【气】
homocamphor 高樟脑
homocentric 共心的;同心的
homocentric axial rays 共轴复心射线
homocentric beam 共心光束
homocentric illumination 共心照明
homocentricity 共心性
homocentric lens 共心透镜
homocentric objective 共心物镜
homocentric pencil 共心光锥
homocentric pencil concentric(al)beam 同心光束
homocentric pencil of rays 共心光束;同心光束
homocentric rays 共心光线
homocentric sphere 同心球
homocentrisity 共心性
homocharge 同号电荷;纯号电荷
homochromatic 均色;等色的
homochromatic photometry 同色光度学;同色光度测量;同色测光
homochromatism 同色性
homochromatography 同系色谱法
homochromic 同色异构体
homochromo-isomer 同色异构体
homochromo-isomerism 同色异构
homochromy 同色
homochronous 类同步
homoclimate 相同气候
homoclime 相同气候区
homoclinal 均斜的
homoclinal river 单斜河
homoclinal valley 单斜谷
homocline 均斜;同斜层;单斜褶皱;单斜(层)
homocollinite 均质镜体;无结构腐殖体
homoconjugation effect 同共轭效应
homocycle 碳环
homocyclic compound 同素还化合物
homodesmic 纯键的
homodesmic structure 单型键构造;纯键结构
homodimer 同型二聚体
homodisperse 均匀分散;均相分散
homodont 同型齿
homodromy 同向旋转
homoduplex 同质双链
homodyne 零拍;零差
homodyne detection 零差探测;零差检波
homodyne detector 零拍检波器;零差检波器
homodyned holography 零差全息术
homodyne mixer 零拍混频器
homodyne receiver 零拍接收机
homodyne receiving 零拍接收法
homodyne reception 零拍接收
homodyning 零差探测
homoemorphy 同形
homoenergetic 均能的
homoeosis 同源异形
homoeothermic animal 同温动物
homoeothermic spring 等体温热泉
homoeothermic water 等体温水
homoepitaxial diamond 同质外延金刚石
homoepitaxy 同质外延
homoetric(al)pair 同效对
homofil 单组分纤维
homofocal hyperbola 共焦双曲线(网)
homofocal hyperbolic grid 共焦双曲线网格
homogamy 同形花现象
homogen 均质
homogenate 匀浆
homogeneity 匀质性;均质(性);均匀(性);均一性;均相性;绝对均匀性;齐(同)性;同质性;均次性;等质性

homogeneity beam 均匀射束

homogeneity broadening 均匀展宽
homogeneity in glass marble 玻璃球均匀性
homogeneity of concentrate 匀相乳油;均相乳油
homogeneity of drafting 制图均匀性
homogeneity of reservoir 储集层均一性
homogeneity of sample 样品的均一性;样本一致性
homogeneity of variance 方差齐性
homogeneity test 均质性测定;均匀性试验;均匀性测定;均一性测定;结构均一性测定;同质性检验
homogeneity test of variance 方差齐性检验
homogeneity to the eye 目测均匀性
homogeneous 均质的;均匀(相)的;均一的;齐次的;同名像点;同次
homogeneous accretion 齐次增大
homogeneous alignment 均匀排列;平行排列
homogeneous alloy 均匀合金
homogeneous anisotropic aquifer 均质各向异性含水层
homogeneous anisotropy 均匀各向异性
homogeneous atmosphere 均质大气(层);均匀大气(层)
homogeneous band 均匀带
homogeneous batholith 均质岩基;匀质岩基;均一岩基
homogeneous beam 均匀射束;均匀电子束
homogeneous bedding 均匀层理
homogeneous body 均质体
homogeneous boundary condition 齐次边界条件
homogeneous broadening 均匀展宽
homogeneous broadening effect 均匀增宽效应
homogeneous carburizing 均匀渗碳
homogeneous cargo 统一种类的货物;同种类货;同一规格的大宗物品;单货种货物;同质货物
homogeneous cargo bagged for stowage purpose 同类散货压舱包装
homogeneous casting process 均质铸锭法
homogeneous catalysis 均相催化
homogeneous catalyst 均相催化剂
homogeneous catalytic reaction 均相催化反应
homogeneous catalytic wet-air oxidation 均相催化湿空气氧化
homogeneous chemical equilibrium 均相化学平衡;单相化学平衡
homogeneous classification 同质分类
homogeneous cloud 均云
homogeneous coating 均匀涂层;均匀镀层
homogeneous coefficient 齐次系数
homogeneous combustion 均质燃烧
homogeneous compression 各向均匀压缩
homogeneous concrete 匀质混凝土;均质混凝土
homogeneous conductivity 均匀电导率
homogeneous coordinates 齐次坐标
homogeneous coordinates system 齐次坐标系统;均匀坐标系统
homogeneous cover(ing) 均匀涂层;均质护壁板
homogeneous crust 均匀地壳
homogeneous crystallization 单相结晶
homogeneous dam 均质(土)坝
homogeneous deformation 均质变形;均匀形变;均匀变形
homogeneous degree 均匀程度
homogeneous degree of mineralization 矿化均匀程度
homogeneous dielectric(al)film 均质介质膜
homogeneous differential equation 齐次微分方程
homogeneous dike 均质岩墙
homogeneous distribution 均匀分配;单一分布
homogeneous disturbance 均匀扰动
homogeneous earth dam 匀质土坝;均质土坝
homogeneous earth(fill)dam 均质填筑坝
homogeneous elastic solid 均质弹性体
homogeneous ellipsoid 均匀椭球
homogeneous embankment dam 均质填筑坝
homogeneous equation 齐次方程;同次方程式
homogeneous equilibrium 均相平衡;单相平衡;单纯平衡
homogeneous equilibrium state 均匀平衡状态
homogeneous excitation 均匀激发
homogeneous fabric 均匀组构
homogeneous fiber wallboard 均质纤维墙板
homogeneous field 均匀场
homogeneous flame 均相火焰
homogeneous fleece 同质套毛
homogeneous floor 同质楼板

homogeneous flow 均匀流动
homogeneous fluid 均质流体;均匀液体;均匀流体
homogeneous fluidization 均相流化
homogeneous foundation 均质地基
homogeneous function 齐次函数
homogeneous goods 单货种货物
homogeneous group 同类组
homogeneous half-space 均匀半空间
homogeneous hard rock 均质硬岩
homogeneous immersion 均匀浸润;均匀浸没
homogeneous immersion fluid 均匀浸液
homogeneous integral equation 齐次积分方程
homogeneous invariant 齐次不等式
homogeneous ionization chamber 均质电离室
homogeneous isotropic aquifer 均质各向同性含水层
homogeneous isotropic elastic half-space 均质各向同性弹性半空间
homogeneous isotropic elastic mass 均质各向同性弹性体
homogeneous isotropic soil 各向同性均质土
homogeneous isotropic turbulence 均匀各向同性紊流
homogeneous landslide 无层滑坡
homogeneous layer 均质层
homogeneous light 均匀光(线);单色光
homogeneous line 均匀谱线
homogeneous linear equation 齐次线性方程
homogeneous linear equations 齐次线性方程组
homogeneous linear restriction 齐次线性约束
homogeneous line broadening 均匀谱线增宽
homogeneous liquid membrane 均质液体膜
homogeneously fluidized bed 均匀流化床
homogeneous magnetic field 均匀磁场
homogeneous Markov chain 均匀马尔可夫链
homogeneous Markov process 均匀马尔可夫过程
homogeneous mass 均质体
homogeneous material 均质材料;均匀物料;均匀材料
homogeneous medium 均质介质;均匀媒质;均匀介质
homogeneous medium filter 均质介质滤池
homogeneous member 同类构件;均质构件
homogeneous membrane 均质膜;均匀膜;均相膜
homogeneous mixture 匀质混合物;均质混合物;均匀混合物
homogeneous multiplexed circuit 均匀复用线路
homogeneousness 均质性
homogeneous network 齐次型网络;同机种网络;同等级网;同等大地网
homogeneous nonstationary process 齐次非稳定过程
homogeneous nonstationary time series 齐次非平稳时间序列
homogeneous nucleation 均质成核;同质成核
homogeneous ocean 均匀海洋
homogeneous optic(al)waveguide 均匀光学波导
homogeneous period 同周期
homogeneous perturbation 均匀扰动
homogeneous phase 均(匀)相;同相
homogeneous phase Fenton oxidation-coagulation process 均相芬顿氧化-混凝工艺
homogeneous phase measurement 均相测量
homogeneous photocatalysis 均相光催化
homogeneous point 同名点
homogeneous Poisson process 均匀泊松过程
homogeneous polymerization 均相聚合(作用)
homogeneous polynomial 齐次多项式
homogeneous polyvinyl chloride cover(ing) 均质聚氯乙烯涂层
homogeneous porous medium 均匀多孔介质
homogeneous precipitation 均匀沉淀;均相沉淀
homogeneous precipitation method 均相沉淀法
homogeneous process 齐次过程;平稳过程
homogeneous product 同质产品
homogeneous production functions 齐次的生产函数
homogeneous program(ming) 齐次规划
homogeneous propellant 均质推进剂
homogeneous radiant energy 均匀辐射能;单色辐射能
homogeneous radiation 均匀辐射;均匀放射
homogeneous radioactivity layer 均匀放射层
homogeneous ray 均匀射线;同形射线

homogeneous ray tissue 同形射线组织
homogeneous reaction 均匀反应;均相反应;单相反应
homogeneous reactor 均质反应堆
homogeneous river 均质河
homogeneous rock 均质岩石
homogeneous roof 相似屋顶;匀质屋顶;匀称屋顶;同质屋顶;同质房顶;对称屋顶
homogeneous seal 均质密封
homogeneous section 均匀线段
homogeneous series 同系列;同类型地图
homogeneous shallow water 均匀浅海
homogeneous shear 均匀切变
homogeneous sintering 单相烧结
homogeneous social group 同质社会集团
homogeneous soil 匀质土;均质土(壤)
homogeneous solid 均匀固体
homogeneous solid solution 均匀固熔体
homogeneous solution 齐次解
homogeneous solution-type reactor 均匀溶液反应堆;水锅炉
homogeneous space 均质空间;齐性空间
homogeneous state 均态
homogeneous state of stress 均匀应力状态
homogeneous steel 均质钢
homogeneous steel plate 均质钢板
homogeneous strain 均匀应变
homogeneous stratum 均质层
homogeneous stream 均质河
homogeneous stress 均匀应力
homogeneous stress field 均匀应力场
homogeneous structure 均一状结构
homogeneous system 匀质系统;均质系统;均匀系;均相(体)系;齐次组;齐次系;同机种系统;单相系
homogeneous target 均匀目标
homogeneous temperature 均化温度
homogeneous temperature field 均一温度场
homogeneous texture 均质结构;均匀(排列)结构;均一结构;平行排列结构
homogeneous tile 均质面砖
homogeneous tissue 同形组织
homogeneous tone 色调均匀
homogeneous traffic 单一运输
homogeneous transformation 齐次变换
homogeneous tube (非焊接的)整管
homogeneous turbulence 均匀紊流;均匀湍流
homogeneous waste 同质垃圾
homogeneous wave equation 齐次波动方程
homogeneous X-ray 单色 X 射线
homogenesis 同质
homogenetic association 同源配对
homogenic 均质的
homogenising of sludge 污泥均化
homogenization 匀质性;匀化作用;均匀作用;均(匀)化;均化(作用);搅匀;同质化;等质化
homogenization of glass 玻璃的匀化
homogenization of raw material 原料均化
homogenization of raw meal 生料(粉)均化
homogenization pressure 均一压力
homogenization temperature 均一温度;均化温度
homogenize 匀化;均化;搅匀;使均匀
homogenized alloy 均质合金
homogenized grease 均化润滑脂
homogenizer 匀浆器;匀化器;高速搅拌器;高速搅拌机;匀质器;均化器;乳化机
homogenizing basin 搅拌池
homogenizing coefficient 均化系数(道路车辆与人通量)
homogenizing furnace 均热炉组;均化炉
homogenizing store 均化储库
homogenizing treatment 匀化处理;均匀热处理
homogenizing trough 均化槽
homogenous distribution 均匀分布
homograde distribution 同度分布
homographic(al) solution 对应解
homography 单应性
homohalin layer 均匀含盐层
homoharringtonine 高粗榧碱
homohopane 升藿烷
homo-ion 同离子
homoionic soil 同离子土
homo-ionic solution 同离子溶液
homoiosmotic 恒渗压的

homoiothermic animal 定温动物
homoiothermism 保持恒温
homoiothermy 恒温状态;恒温动物;恒温层;等温层;保持恒温
homojunction 同质结;单质结
homojunction laser 同质结激光器
homojunction structure 单质结结构
homologate 同意
homological algebra 同调代数
homological dimension 同调维数
homological figures 同源图形
homological functor 同调函子
homologically connected space 同调连通空间
homologically finite filtration 同调有限滤子
homological mapping 同调映射
homological property 同调性质
homologisation = homologization
homologization 同源;均裂作用;对应线对
homologize 同系
homologous 同源的;同系的;对应的
homologous angles 同调角
homologous compound 同系(化合)物
homologous dimension 对应尺寸
homologous edges 同调棱
homologous faces 同调面
homologous field 同系场
homologous gaseous sphere 同模气体区
homologous impeller 对立叶轮
homologous line pair 均称线对
homologous lines 均称谱线;同调线
homologous machines 同系机械
homologous pair 同系线对;对应线对
homologous pairing 同源配对
homologous photograph 相应像片
homologous point 同调点
homologous radio burst 同系射电暴
homologous series 同源系列群落;同系列
homologous sides 同位边;对应边
homologous star 同系星
homologous terms 同调项
homologous theory 同源学说;同源论
homologous to zero 同调于零
homologous transformation 同调变换
homologous turbine 同型水轮机;同系列水轮机
homologous variation 同源变异
homologous waves 同系波
homolographic(al) map 等面积投影地图
homolographic(al) projection 相应投影
homologue 相应物;同源包体;同系;相应;对应物
homologues compound 同系物
homologus 相应的
homology 相互影射;同源(现象);同系(现象);同调;导体同形
homology base 同调基
homology boundary 同调边缘
homology boundary homomorphism 同调边缘同态
homology class 同调类
homology covering surface 同调的覆盖面
homology dimension 同调维
homology exact sequence 同调正合序列
homology functor 同调函子
homology group 同调群
homology manifold 同调流形
homology module 同调模
homology ring 同调环
homology spectral sequence 同调谱序列
homology sphere 同调球
homology structure 同调结构
homology system 同调系
homology theory 同调论
homology transformation 同调变换
homology type 同调型
homolosine projection 等面积投影
homolysis 均裂
homolytic reaction 均裂反应
homomartonite 高马炸药
homomergic flow 均能流
homometrical 同度量
homometric(al) pair 同效对
homo-mixer 高速搅拌机
homomixis 同源融合
homomorph 同态象
homomorpha 同形变态

homomorphically irreducible graph 同胚不可约图形
homomorphic deconvolution 同态反褶积
homomorphic filter 同态滤波器
homomorphic filtering 同态滤波
homomorphic function 同态函数
homomorphic image 同态象
homomorphic inverse filtering 同态反滤波
homomorphic mapping 同形映射;同态映射
homomorphism 异质同晶(现象);同态;同晶形现象
homomorphism group 同态群
homomorphism interpolation theorem 同态内插定理
homomorphism module 同态模
homomorphism ring 同态环
homomorphs 同形体
homomorphy 同形性
homonicotinic acid 高烟酸
homonomous segmentation 同律分节
homonomy 同律性
homonuclear molecule 共核分子
homonym 异物同名;同名物
homopause 均质层顶;湍流层顶
homoperiodic 齐周期的
homophase 同相
homophoric acid 高樟脑酸
homopicnal inflow 等密度入流
homoplasmon 同质体
homoplasy 趋同性;同型
homopolar 同极(器)
homopolar alternator 单级同步发电机
homopolar bond 均极键;无极(性)键
homopolar colloid 无极性胶体;同极性胶体
homopolar component 零序分量;单极性分量
homopolar compound 无极化合物
homopolar crystal 同极晶体
homopolar dynamo 同极发电机
homopolar field magnet 同极场磁铁
homopolar generator 同极发电机;单极发电机
homopolar generator electromagnetic pump 同极发电机式电磁泵
homopolar high-voltage direct current system 同极高压直流输电系统
homopolar link 无极性链
homopolar machine 单极电机
homopolar magnet 同极磁铁;单极磁铁
homopolar molecule 单极分子
homopolar motor 单极电动机
homopolar polymer 同极聚合物
homopolar power 零序功率
homopolar tachogenerator 单极测速发电机
homopolar type of axial symmetry 同极型轴对称
homopolar valency 无极价
homopolymer 均聚(合)物;同聚物
homopolymerization 均聚(作)
homopolymer resin 均聚(合)物树脂
homopore membrane 均孔膜
homopycal inflow 等密度入流
homopycnal flow 等密度流
homoretene 高惹烯
homoscedasticity 同方差性;方差齐
homoscedastic table 同方差表
homoseis 同震线
homoseismal line 同震线
homospecificity 同种特异性
homosphere 均质层;均匀气层;混成层
homostachydrine 高水苏碱
homostasis 同态
homostrobe 零拍闸门
homosulfanilamide mafenide 高磺胺
homosyndesis 同型联合
homotactic bed 等效的路基;等效的地层
homotaxial 等列
homotaxis 排列类似【地】
homotelinite 结构腐殖体
homoterpenylic acid 高萜酸
homotherm 恒温动物
homothermal 恒温的
homothermal condition 恒温条件
homothermism 稳定调节;保持恒温
homothermous 恒温的
homothermy 恒温状态;恒温层;同温层;等温层
homothetic axis 同位相似轴
homothetic centre 相似中心;同位相似中心

homothetic conics 同位相似二次曲线
homothetic correspondence 同位相似对应
homothetic curves 同位相似曲线
homothetic figures 同位相似图形
homothetic paths 同位相似道路
homothetic ratio 同位相似比
homothetic solutions 同位相似解
homothetic triangle 同位相似三角形
homothetic work-hardening 同位相似加工硬化
homotopically 同伦地
homotopic chain-mapping 同伦链映射
homotopic critical point 同伦临界点
homotopic geodesic line 同伦测地线
homotopic mapping 同伦映射
homotopic paths 同伦道路
homotopic to zero morphism 同伦于零射
homotopy 伦移;同伦
homotopy addition 同伦加法
homotopy addition theorem 同伦加法定理
homotopy axiom 同伦公理
homotopy boundary homomorphism 同伦边缘同态
homotopy boundary operation 同伦边缘运算
homotopy class 同伦类
homotopy classification 同伦分类
homotopy classification problem 同伦分类问题
homotopy coboundary operation 同伦上边缘运算
homotopy cochain 同伦上链
homotopy exact sequence 同伦正合序列
homotopy extension property 同伦扩张性
homotopy extension theorem 同伦扩张定理
homotopy groups 同伦群
homotopy invariance theorem 同伦不变性定理
homotopy invariant 同伦不变量
homotopy obstruction 同伦障碍
homotopy operations 同伦运算
homotopy sequence 同化序列
homotopy set 同伦集
homotopy sphere 同伦球面
homotopy system 同伦系
homotopy theorem 同伦定理
homotopy theory 同伦(理)论
homotopy type invariant 同伦型不变量
homo-treatment 均匀热处理
homotropy 同向弯曲式
homotype 同型;似型;等模标本
homotypic(al) effect 同型作用
homotyposis 同型原理
homotypy 同型性
homozygosis 纯质性
homozygote 同质接合子;同型合子;纯合体
Hompress pile 霍尔姆普斯桩
honcho 头子;老板;工头(俚语)
Honda alloy 本多镍钴钛系磁钢;本多磁钢
Honda steel 本多磁钢
hondrometer 粒度计;微粒特性测定计
Honduras cedar 洪都拉斯雪松属
Honduras mahogany 洪都拉斯桃心木
hone 细磨石;油石;刮器;极细砂岩;磨(刀)石
honecomb duct 导流管(道)
honed finish 磨制表面;珩磨面;细磨加工面;搪磨后的表面(光洁度)
honed finishing 搪磨
honed slipper-holder 角形滑接片保持器
hone out 去毛刺;抛光;磨光;去毛刺
honer 搪磨机
hone spacing block 磨刀石隔块
honessite 铁镍矾
honest material 好料
honestone 磨刀石;均密砂岩;极细砂岩磨
honey 蜂蜜
honey butter 加蜜黄油
honey centrifuge 甩蜜机
honeycomb 整流栅;内裂缝;梳状;蜂窝状;蜂形;蜂窝(器);蜂房(式);蜂巢
honeycomb aggregate 蜂窝结构的集料;蜂窝结构的骨料
honeycomb and scale 蜂窝麻面(指混凝土)
honeycomb and surface voids 蜂窝麻面(指混凝土)
honeycomb base 蜂窝底托
honeycomb blister 蜂窝气泡
honeycomb block 蜂窝状砖块;蜂窝状砌块
honeycomb board 蜂窝夹芯胶合板

honeycomb bonded structure 蜂窝胶合结构
honeycomb brick 蜂窝砖
honeycomb briquet 蜂窝煤
honeycomb cellular structrue 蜂窝结构
honeycomb ceramics 蜂窝陶瓷;多孔陶瓷
honeycomb check 蜂窝裂缝
honeycomb clinker 蜂巢状熟料;蜂巢状烧结块
honeycomb coil 网格线圈;蜂房(式)线圈
honeycomb collimator 蜂房型准直器
honeycomb concrete 蜂窝(状)混凝土
honeycomb coral 蜂窝珊瑚
honeycomb core 蜂窝芯(材);夹芯纸板;夹芯镶板;蜂窝式夹芯墙板
honeycomb core material 蜂窝夹层材料
honeycomb core panel 蜂窝填芯板
honeycomb core plywood 蜂窝夹芯胶合板
honeycomb core sandwich wall board 蜂窝夹芯墙板
honeycomb crack 龟裂;网状裂缝;蜂窝状裂纹
honeycomb duct 导流管道;多孔管道
honeycombed 蜂窝状的;蜂窝形;蜂窝结构的
honeycombed block 密孔砌块
honeycombed casting 多气泡铸件
honeycombed check 蜂窝状裂纹;蜂窝状黏合
honeycombed(clay) brick 多孔砖
honeycombed coffering 六角形平条;蜂窝形格顶棚
honeycombed coke 蜂窝状炉焦炭
honeycombed coke process 蜂窝式炼焦法
honeycombed concrete 蜂窝形混凝土
honeycombed core 蜂窝状芯;蜂窝状夹层
honeycombed coring 蜂窝芯;成蜂窝状芯
honeycombed fascine raft 蜂窝状柴(排)筏
honeycombed fissure 蜂窝状裂缝
honeycombed ice 蜂窝状冰
honeycombed masonry wall 蜂窝形砖石墙
honeycombed paper 蜂窝状纸板
honeycombed paper core 蜂窝纸芯
honeycombed sandwich radome 蜂窝形夹层圆顶
honeycombed slating 蜂窝形铺石板
honeycombed solution surface 蜂窝状溶蚀表面
honeycombed structure 蜂窝状结构
honeycombed texture 蜂窝状结构
honeycombed tile 六角面砖;蜂窝形面砖
honeycombed wall 蜂窝状地垄墙;蜂窝式墙(壁);花墙(水工模型试验用);多孔墙
honeycombed window 蜂窝形窗
honeycomb electrostatic storage 蜂房式静电存储器
honeycomb fashion 蜂窝式
honeycomb filter 蜂窝式过滤器
honeycomb fire damper 蜂窝式防火门
honeycomb furnace crown linings 蜂窝式大窟砖
honeycomb ground 六角锦砖地面;蜂窝形底
honeycomb in concrete 混凝土中蜂窝;混凝土蜂窝现象
honeycombing 形成蜂窝(表面);内裂;木材龟裂;蜂窝状孔洞;蜂窝的;蜂窝现象(指混凝土);蜂窝裂
honeycombing of brush 电刷中形成细孔
honeycombing of wood 木材龟裂
honeycomb labyrinth seal 蜂巢状迷宫式密封
honeycomb like 蜂巢状的
honeycomb(line) plate 蜂窝板
honeycomb material 蜂窝材料
honeycomb memory 蜂房形存储器;蜂窝式存储器
honeycomb metal 蜂巢状金属
honeycomb model 蜂巢模型
honeycomb module 蜂巢微型组件
honeycomb pattern 蜂窝形图形;蜂窝形图案
honeycomb pipe 窝管
honeycomb pitting 蜂窝蚀孔
honeycomb plastics 蜂窝塑料
honeycomb plate 配水孔板
honeycomb plywood 蜂窝夹芯胶合板
honeycomb process 蜂窝成形
honeycomb radiator 蜂窝式散热器;蜂窝式散热片
honeycomb rot 蜂窝腐朽
honeycomb sandwich 多孔夹
honeycomb sandwich construction 蜂窝夹层结构
honeycomb sandwich panel 蜂窝夹层板
honeycomb sandwich structure 蜂窝夹层结构
honeycomb screen 整流栅;整流格
honeycomb seal 多孔密封
honeycomb-shaped dune 蜂窝状沙丘

honeycomb-shaped plastic medium 蜂窝状塑料滤料
honeycomb-shaped rust stain 蜂窝状锈斑
honeycomb slating 蜂窝状铺砌；蜂窝状铺石板
honeycomb sleeper wall 蜂窝式地龙墙
honeycomb storage 蜂窝式存储器
honeycomb structure 蜂窝（状）构造；蜂窝（式）构造；蜂房构造
honeycomb structure laminate 蜂窝状结构层压板；蜂窝状层压板
honeycomb structure transducer 蜂窝结构换能器
honeycomb super silica brick 蜂窝状高级硅砖
honeycomb texture 蜂窝结构
honeycomb tile 蜂窝状砖块
honeycomb tube 窝管
honeycomb-type geo-composite 蜂窝形土工复合物
honeycomb-type house 蜂窝式房屋
honeycomb vault 一种立体悬桃的蜂窝状装饰
honeycomb wall 蜂窝墙；漏ত（地垄）墙；蜂窝状（地垄）墙；地垄墙；地龙墙；蜂窝式地龙墙
honeycomb weathering 蜂窝状风化；蜂窝式风化
honeycomb weave 蜂巢组织
honeycomb winding 蜂房式绕组
honeycomb window 蜂窝窗
honeydew 蜜露
honey extractor 甩蜜机
honey flow 蜜流季节
honey glaze 蜂蜜色釉
honey-gold 蜜黄色
honey poisoning 蜂蜜中毒
honeystone 蜜蜡石
honeysuckle 金银花草纹；忍冬（金银花）
honeysuckle ornament 刃冬饰；忍冬饰；希腊建筑花饰（以棕叶饰为基本装饰）
honey-wag(g)on 垃圾车
honey wall 蜂窝墙
Hong Kong currency 港币
Hong Kong Dollar 港元
Hong Kong Seamen's Union 香港海员工会
Hong Kong stone 香港石色油漆
hongquiite 红旗矿
hongshiite 红石矿
Honigmann method of shaft lining 赫涅曼竖井衬砌法
Honigmann shaft-boring process 赫涅曼矿井钻进法
honing 金属表面磨损；磨石磨孔法；珩磨；搪磨；镗磨
honing clearance grooves 磨头退出槽
honing finish 研磨加工
honing ga(u)ge 石料凿磨量规；刃口样板；磨刀夹具
honing head 搪磨头
honing machine 油石珩磨机；条石磨孔机；搪磨机
honing oil 磨刀油
honing stroke 镗磨行程
honing tool 镗磨头
honing wheel 珩磨轮
honitest 计尘试验
Honolulu/Japan route 火奴鲁鲁（檀香山）—日本航线
hono(u)r a bill 支付期票
hono(u)r a claim 承付索赔；承兑赔偿金额
hono(u)r a contract 履行合同
hono(u)r a debt in advance 提前还债
hono(u)r agreement 履行协议；名誉契约
hono(u)r an agreement 履约
hono(u)r an application 批准申请
hono(u)rarium 酬金
hono(u)rary 名誉团体；名誉的
hono(u)rary advisor 名誉顾问
hono(u)rary director 名义董事
hono(u)rary doctor 名誉博士
hono(u)rary member 名誉会员
hono(u)rary office 名誉职务
hono(u)rary post 名誉职位
hono(u)r cheque 承兑支票
hono(u)r claim 承付赔偿要求
hono(u)r contract 守合同；名誉契约
hono(u)red 已承兑
hono(u)ring contract and acting in good faith 守约重义
hono(u)r one's liability 承担赔偿责任
hono(u)r policy 信用保（险）单；信誉保险单；名誉保险单

hood 发动机罩；遮（光）板；罩布；局部排风罩；机罩；帽盖；排烟罩；排风罩；无底坩埚；通风柜；兜帽；厨房拔气罩；车篷；舱口罩棚；拔气罩
Hood's rule 虎德公式（计算房间采暖所需管长的公式）
hood anti-squeak 发动机罩减声片
hood bump 发动机罩挡
hood catch 机罩卡钩
hood connecting rod 车篷连杆
hood cover 机罩套
hooded ball casters 家用脚轮
hooded crane 羽冠鹤；白头鹤
hooded dry drilling 加吸尘罩的干式钻进；封闭式干钻凿岩
hooded impeller 闭式叶轮
hooded jacket 风雪衣
hooded roof ventilator 有罩的屋顶通风管
hooded shield 加罩盾构
hooded ventilator 有罩通风机；有罩通风管
hood end 木板镶槽端
hood face 曲面
hood fastener 盖锁扣；罩子挂钩；机罩卡钩；外罩钩扣
hood for fume 烟罩；集气罩
hood for vehicle 车篷
hood for vehicle engine 车辆引擎罩
hood gate 漏斗闸门；料仓闸门
hood hinge 机罩铰链
hood hinge rod bracket 机罩的铰链拉杆支架
hood inlet 虹吸进水口
hood jettison 座舱盖抛放
hood lacing 机罩衬带
hood latch 机罩（锁）闩；机罩锁扣
hood ledge lacing 机罩减振垫片
hood lock 机罩锁
hood lock handle 机罩关锁手柄
hood lock handle spring 机罩锁把弹簧
hood lock hook 罩锁钩
hood lock latch 机罩锁闩
hood lock plate 机罩锁板
hood loss 排气管损失；排汽缸损失
hood mo(u)lding 拱檐线脚；滴水罩饰；披水线脚；发动机罩嵌线饰；门窗头的出檐线脚；出檐垂饰
hood mo(u)lding fastener 罩嵌条钩扣
hoodoos 土柱；石柱；石林；峰林
hood operating cylinder 带有罩的操作汽缸
hood pad 罩垫
hood pile cap 套式桩帽
hood plate 船尾材外板
hood poop 鲸背形船尾楼
hood side 机罩侧面
hood side panel 车头罩侧板
hood side under panel 机罩下侧板
hood side upper panel 机罩上侧板
hood support 机罩支架
hood support bumper 发动机罩托架防撞器
hood test 罩盖法试验
hood top panel 车头罩顶板
hood-type annealing furnace 罩式退火炉
hood webbing 机罩折边
hood wrap 罩盖程度
hoof foot 蹄形家具腿
hoof print 蹄印
hoof shovel 蹄形松土铲
Hoogoven cement 矿渣硅酸盐水泥
hook 钩（子）；钩（状刃口）；钩（形物）；扣钩；河湾钩形物；弯曲沙嘴；弯钩；弯成钩形；沙钩；吊钩；挂钩
Hook's generalized stress-strain law 广义虎克应力—应变定律
Hook's spring 虎克弹簧
hook accessories 钩的附件；吊钩附件
Hookah type diving apparatus 水面供气式潜水装具
Hookah type diving device 水面供气式潜水装具
Hookah type diving equipment 水面供气式潜水装具
Hookah type diving unit 水面供气式潜水装具
hook allowance 吊钩余量
hook anchor 钩状锚
hook anchorage 带钩锚具；带钩锚固；钩形锚具
hook-and-band hinge 大门铰链；门窗头的出檐线脚；束带式铰链；T形铰链

hook and butt eye 钩与扣眼
hook and butt joint 钩扣连接
hook and dee ring 金属钩子和D形环
hook and eye 钩与螺旋扣眼；钩扣铰接；风钩和羊眼圈
hook and eye hinge 钩扣铰接；钩和铰链
hook and hook turnbuckle 双钩伸缩螺扣；双钩法兰螺扣
hook and ladder truck 钩梯形载货车
hook and loop fastener 钩环扣
hook and torque cap insulator 钩子和单耳连接绝缘子
hook and triangular plate 吊钩和三角形板
hook angle 前角
hook assembly 吊钩滑轮组
hook bar 曲杆；弯钩钢筋
hook belt 带安全钩的安全带
hook bending 弯成
hook block 挂钩滑轮；钩吊；吊杆滑轮组；带钩滑车
hook bolt 钩头螺栓；丁字头螺栓；地脚螺栓；带钩螺栓
hook bolt block 钩头插销锁
hook bottom block 有钩的下滑轮；钩的底座
hook cable 钩索
hook capacity 钩的负载量
hook cargo 钩吊货物
hook chain 钩链
hook chisel 钩形凿
hook clip 吊线钩
hook collector 钩状集电极
hook conveyer[conveyor] 钩式运输机；钩式输送机
hook crane 带钩吊车
hook cycle 吊钩周转率
hook damage 钩伤；货物造成的货损
hook damage and contamination 钩损及沾污险
hook down 弯钩朝下
Hooke's body 虎克体
Hooke's coupling 虎克接头；万向联轴节
Hooke's law 虎克定律
Hooke's model 虎克模型
Hooke's solid 虎克固体
Hooke's solid model 虎克固体模型
Hooke's universal joint 虎克万向联轴器；万向联轴节
Hookean body 虎克体
hookean coupling 虎克接头
Hookean deformation 虎克变形
Hookean effect 虎克效应
Hookean elastic body 虎克弹性体
Hookean law 虎克定律
Hookean model 虎克模型
Hookean solid 理想弹性（固）体；虎克固体
Hookean solid model 虎克固体模型
Hookean spring 线性弹簧；虎克弹簧
Hookean substance 虎克物质
Hookean universal joint 虎克万向联轴器
hooked 钩状的
hooked bar 弯筋；曲杆；曲柄；带弯钩钢筋；弯钩钢筋
hooked bolt 钩（头）螺栓；弯钩螺栓
hooked end of bar 钢筋弯（钩）端
hooked ends 尾端弯钩
hooked end steel fiber 端部带弯的钢纤维
hooked foundation bolt 弯钩地脚螺栓
hooked ga(u)ge 钩规
hooked groin 钩形（丁）坝；弯丁坝；勾头丁坝；钩连丁坝
hooked groin contraction work 钩形丁坝束水工程
hooked groyne 钩形（丁）坝；弯丁坝；钩连丁坝
hooked joint 密缝接头；密缝接合
hooked key 钩形扳手
hooked line 钩状线道
hooked nail 钩（头）钉；钩钉
hooked plates 单页插图
hooked rod 弯头钢筋
hooked rug 钩针编结地毯
hooked rule 钩规
hooked scraper 钩形刮刀
hooked sole plate 钩形底板
hook(ed) spanner 弯钩的开脚扳手；有钩扳手；钩形扳手
hooked spike 钩头道钉
hooked spit 钩状岬；钩形沙嘴；钩形沙咀；弯（曲）沙嘴
hooked tommy 圆头帽钩头扳手；圆螺帽钩头扳手

hooked wrench 弯钩的开脚扳手;钩形扳手
hook end clamp 钩头鞍子;垂直定位钩
hook end fitting 定位钩
Hooke number 柯西数
hooker 挂钩
Hooker's diaphragm cell 胡克隔膜电解池
Hooker's extrusion method 胡克挤压法
Hooker's law 胡克定律
Hooker's process 胡克薄壁管挤压法
hooker-on 挂钩工人
Hooker process 正向冲击挤压法
hooket 小钩
hook evapo(u)ration ga(u)ge 钩形测针蒸发器
hook eye 钩眼
hook face 曲面
hook feed 钩式送料
hook feeder 钩式送料装置
hook file 钩形信件整理器
hook for cleaning chips 切屑钩;刨花钩
hook for pipe fixing 装管用钩
hook for starting up 上炉钩子
hook foundry nail 型箱用钩头钉
hook gantry crane 吊钩龙门吊
hook ga(u)ge 管压力表;钩形水表;钩形表规;钩(规)尺;钩形(水位)测针;钩形水位计;钩(形水)尺;钩形测针;钩(形标)尺
hook-ga(u)ged method 管压力计法(测定新拌混凝土中含气量);钩形表法(加气混凝土配料法的一种)
hook goods 钩吊货物
hook grab magnet gantry crane 三用门式起重机(吊钩—抓斗—磁铁)
hook groin 钩状丁坝
hook groyne 钩状丁坝
hook guard 熔丝架;保护器架
hook hasket 钩筐
hook-headed spike 钩头道钉
hook height 车前端挂钩高度;吊钩(提升)高度;吊钩
hook hitch 挂钩
hook hoist speed 吊钩起升速度
hook hole damage 钩伤
hook housing 钩架
hooking 屏幕变形
hooking iron 清缝凿
hooking up 连结;用成组吊钩吊起
hook iron 清缝凿;钩状铁
hook joint 钩式接头;钩扣连接;钩接(合);钩环接头;钩形接缝;S形接缝
hook joint chain 链接链
hook joint strength 勾接强度
hook junction 钩结
hook knife 弯刀
hook ladder 挂梯;带钩梯
hook ladder belt 带钩安全带
hooklet 小钩(子)
hook lever 翻木机
hook lifting rate 吊钩提升速率;吊钩吊升速度
hook-like halving 钩状对搭接头
hook-like structure 似钩状构造
hook link chain 钩链
hook load capacity 大钩负荷能力
hook load rating 大钩负荷定额
hook lowering speed 吊钩下降速度
hook mark 吊钩痕
hook nail 集索环
hook off 摘开
hook-off joint 卸开接头
Hook of Holland 荷兰角
hook of rod 钢筋弯头;钢筋钩
hook on 钩住;钩车;钩上;钳形的
hook-on bucket 钩接式料斗;钩接式吊斗;钩接式铲斗;钩形挑流鼻坎;钩形弧段
hook-on grab 钩接式抓斗;钩接式夹具
hook-on instrument 悬挂式仪表;钩接式仪表
hook-on type meter 钩接式仪表
hook opening width 吊钩张开宽度;钩口宽度
hook operation 线路中继操作
hook-out blind 离窗卷帘;离窗遮帘
hook-out carved lacquer article 彩绘勾刀漆器
hook phototransistor 钩形光电晶体管
hook pin 钩头栓;钩销;绝缘子弯脚钉
hook plate 钩头垫板;钩板
hook pole 舢板钩杆
hook position 吊钩位置

hook pulley 吊钩滑轮
hook reach 吊钩工作半径
hook rebate 门窗梃上闭合槽;钩头企口槽;钩形凹凸榫;S形凹凸榫
hook release circuit 钩释放电路
hook ring 箱柜钩眼;门窗钩眼;吊环
hook roller 旋转式挖土机;吊钩滚轴;吊钩滚轮
hook rope 带钩绳(索)
hook rope knife 切断井下钢丝绳的割刀
hook rule 钩尺
hook scale 带钩尺
hook scarf 钩形嵌接;钩式嵌接
hook screw 钩头螺钉;带钩螺钉
hook seam 咬口接缝
hook separation point 脱钩点
hooks for piling of anchor's chain 排锚链的钩
hook-shaped double leaf gate 钩形双扇闸门
hook-shaped rod 钩形杆
hook-shaped structure 钩形曲构造
hook spanner 钩(形)扳手
hook speed 提引钩提升速度
hook strip 挂钩板条(衣服)
hook switch 挂钩开关;叉簧
hook tackle 带钩绞辘
hook tender 杂工工头;斜井上下把钩(采矿);工头助手(俚语)
hook tool 钩形刀具
hook tooth 棘轮齿
hook-tooth file 平行齿半圆锉
hook-tooth sprocket gear 不可逆链轮
hook travel 吊钩行程
hook turn-over device 钩式翻钢机
hook type airless shot blasting machine 单钩真空抛丸清理机
hook type bottom block 钩式下滑轮;钩式底座
hook type crane 钩式起重机
hook type current transformer 钩形变流器
hook type double leaf gate 钩形双扇闸门
hook type outside locking device 钩式外锁闭
hook type stripper 钩式卸料板
hookup 弯钩朝上;钩住;专用布置;联结装置;接续图;试验线路;试验电路;挂钩
hookup mechanism 连结装置;连接装置
hookup wire 架空电缆;架空电线;单连线;布线用电线;安装线
hook valley 逆向支流
hook wall flooding packer 悬挂式可拆卸封隔器;注入孔用悬挂式孔壁封隔器;孔壁封隔器;井壁封隔器
hook wall packer 悬吊式堵塞器;壁钩型封隔器
hook water ga(u)ge 带钩形水测针
hook weight 吊钩重量
hook well floating packer 注水井用悬挂式堵塞器
hook winch 吊钩绞车
hookworm 钩虫病
hook wrench 钩形把手;带钩扳子(螺钉钻子);带钩螺钉钻子
hoop 箍(圈);卡箍;环(箍);水泥的横向张骨
hoop antenna 圆柱形天线
hoop back 围圈靠背
hoop bamboo 箍篾
hoop chain 环形链
hoop cone system 锚环张拉法
hoop cutter 环状割刀
hoop deflection 环向变形
hoop deflexion 环向变形
hoop driver 紧箍器
hoop-driving machine 打箍机
hoop-drop relay 落弓式继电器
hooped bar 环筋
hooped column 箍筋(筋)柱;环柱
hooped concrete 箍筋混凝土;环筋混凝土;配环筋混凝土;配箍筋混凝土
hooped penstock 加箍压力水管
hooped pile 箍筋桩;配箍(筋)桩
hooped reinforcement 箍筋;环形钢筋;环状钢筋;环状箍筋;环筋
hooped steel 环形钢筋
Hooper jib 胡波型风力跳汰机
hooper shape 漏斗形状
hooper water closet 斗式马桶
hoop fastener 紧箍钉
hoop force 环向力
hoop guard 防夹子

hooping 箍筋;箍环;加箍筋;加箍筋;环筋;打箍
hooping dike 月堤;环形围堤
hoop interlock 环向锁口
hoop iron 箍铁(条);箍钢;铁箍
hoop-iron band 箍铁皮;箍铁结合;卡箍连接;砌体铁拉件
hoop-linked chain 连环链
hoop load 环向荷载
hoop mill 打包用的钢带轧机;箍铁轧机;箍钢轧机;带扎机;带钢压延机
hoop missing 货筒箍失落
hoop-net 圈网
hoop pin anchorage 环销锚
Hoop pine 澳大利亚软木
hoop reinforcement 环状箍筋;环筋;箍筋;环箍钢筋
hoop rolling mill 带钢轧机
hoop soil method 箍土法
hoop steel 箍钢带;箍带钢
hoop stick 车棚轻型构件
hoop strain 环向应变
hoop stress 圆周应力;周向应力;周线应力;箍应力;链应力;环形电压;环向应力;围压
hoop tension 周张力;箍张力;箍拉力;环向拉力;环筋张力;环箍张力;环箍拉力
hoop tension strength 环向抗拉强度
hoop type heater 环形管式加热器
hoop winding 周向缠绕
hoop wood 光滑冬青;桶材
hoos(e)gow 警卫室;监狱;厕所
Hoosier pole 平衡杆
hoot 汽笛响声
hootch 茅屋
hoot-collector 开关式集电极
hootenanny 机器配件
hooter 喇叭;警报器;汽笛;气笛
hoot owl 夜班
Hoover 真空吸尘器
Hoover Dam 胡佛坝(美国)
Hoover muller 平磨机
hoow key 鞍形键
hop 跳跃;传输部分;免费搭乘;免费搭车;船身上弯
hopcalite 霍布卡特(一氧化碳消灭剂);浩networking炸药
hop clover 蛇麻草
hope 小湾;小盲谷【地】;小河口;沼泽中干地;河口附近的锚地;期望
Hope's destroyer 虎佩木材防腐剂
Hope's patent glazing 虎佩窗(几种形式特殊的窗和天窗)
hopeite 磷锌矿
hope line 希望路线
hope line draft 希望路线图
Hope sapphire 人工尖晶石
hop-fibre cloth 蛇麻草纤维织物
hop-field 蛇麻草田
hopfnerite 透闪石
hop-garden 蛇麻(草)园
hophornbeam 铁木
Hopkin's bioclimate law 霍普金生物气候法则
Hopkinson test 背靠背试验
hopped-up engine 加力式发电机;强力发动机
hopper 装料斗;料[贮]液槽;储[贮]水槽;储[贮]斗;供料器;给料器;漏斗;料斗;开ధ式泥斗;接受器;接收机;集尘箱;泥舱;席斗;定量给料器;储片盒;储斗箱;储卡箱;仓斗
hopper ash level indicator 灰斗料位指示器
hopper ash pan 斗形灰盘
hopper barge 底卸式泥驳;运泥船;自卸砂驳;自卸驳(船);开底泥驳(船);泥舱船;泥驳
hopper barge winding gear 底卸式泥驳卷绕装置
hopper barge with bottom door 开底泥驳
hopper barge without bottom door 满底泥驳
hopper base 箱底;料斗支架
hopper bin 储[贮]料漏斗
hopper blower 带储[贮]料斗的抛送器;带储[贮]料斗的吹送器
hopper body 活底车
hopper bottom 种子箱底;种子筒底;料斗底;粮箱底;储存设备底
hopper bottom car 漏底车;底卸(式)手推车;漏斗式底卸车;漏卸车;底卸式车辆
hopper bottom door 泥门
hopper-bottomed bin 锥底库;漏斗式料斗;底卸式料斗;底卸式料仓
hopper bottom freight car 底卸式货车

hopper bottom rail car 漏斗形铁路卸货车；底卸式铁路货运车
hopper bottom tank 斗底沉淀池
hopper bucket dredge(r) 装舱链斗挖泥船
hopper capacity 漏斗容量；料斗容量；泥舱容量；泥驳容量
hopper capacity curve 泥舱容量曲线
hopper car 戽斗车；漏斗式底卸车；漏斗车；料斗车；斗推车；底卸式车(辆)；底卸车
hopper car body 自动卸料车体；活底车身
hopper car unloader 链板卸车机
hopper changing delay factor 换舱延迟系数
hopper charging bucket 活底料罐；活底斗
hopper chute 溜槽；漏斗形溜槽；漏斗式斜槽；漏斗式滑槽；漏斗车；滑槽
hopper clearance 漏斗离地高度
hopper closet 带储[贮]水槽的抽水马桶
hopper compartment 漏斗分隔间；漏斗分隔舱；储[贮]料仓
hopper container 漏斗集装箱；漏斗货柜；底卸式集装箱
hopper conveyer 漏斗式运输机；戽斗(式)输送机
hopper-cooled 连续水冷(却)的
hopper discharge 漏斗卸料
hopper door 进料斗盖；泥(舱)门；(开底泥驳的)底开门
hopper door control device 泥门启闭装置
hopper door fill valve 料斗盖装料阀；料斗底装料阀
hopper door hydraulic ram 泥门液压柱塞
hopper drawing channel 料斗卸料槽
hopper drawing device 料斗卸料装置
hopper dredge(r) 吸扬(式)挖泥船；有泥舱的挖泥船；自航式挖泥船；自航耙吸式挖泥船；开底挖泥船；泥舱式挖泥船；吸扬式挖泥船；仓式挖泥船
hopper drier[dryer] 箱式干燥器；料斗干燥器；加料斗干燥器
hoppered bottom 锥形底；斗底
hoppered bottom basin 斗底水池
hoppered bottom tank 斗底水池
hoppered floor 装有卸料漏斗的楼板
hoppered outlet 漏斗形出口
hopper feed 漏斗式上料；料斗供料机
hopper feedback 选料斗
hopper feeder 给料仓；料斗送料器；料斗供料机；斗式加料器
hopper feeding height 料斗喂料高度；喂料高度(料斗)
hopper filling 漏斗装料
hopper filter 锥形过滤器；钟式滤水器；斗式过滤器
hopper flange 料斗法兰
hopper for the Unionmelt 埋弧焊漏斗；合熔焊漏斗
hopper frame 内开下旋窗扇
hopper gate 漏斗闸门
hopper grinder 漏斗式磨床；斗式研磨机；斗式碾磨机
hopper gritter 漏斗式铺砂机；斗式铺砂机
hopper head 雨水斗；水落斗
hopper installation 漏斗装置；漏斗设备
hopper lead 单管接合管
hopper light 内开下旋气窗；侧铰中悬窗；内推下旋式窗扇；外推式窗扇；内推上悬气窗；倒开气窗
hopper lining 装料斗衬里
hopper lite 侧铰中旋窗；内开下旋气窗
hopper loader 料斗式装载机；料斗加料器；斗式装载机；斗式装运机；斗式装料机
hopper-loading efficiency 装舱效率
hopper-loading operation 装舱作业
hopper-loading piping 装舱管系
hopper-loading rate 装舱速度；装舱率
hopper-loading valve 装舱阀
hopper magnet 料斗磁铁
hopper measurement 泥舱计量
hopper mill 斗式粉碎机；斗式碾磨机
hopper mixer 漏斗状混合器
hopper-on rails 行车式漏斗；行车式料斗
hopper-on-rails type 轨承行斗式
hopper opening 漏斗口；料斗卸料口
hopper ore car 漏斗矿石车
hopper outlet 漏斗出口；料仓出口
hopper overflow device 泥舱溢流装置
hopper overflow gate 泥舱溢流门
hopper overflow weir 泥舱溢流堰；泥舱溢流门
hopper-pressing and vibrocasting process 挂斗振浆法

hopper sash 倾斜式窗框；斗窗
hopper scale 自动戽斗定量秤；料斗式秤；料斗电子秤；库秤
hopper scow 底卸式平底船；底卸式方驳
hopper slide 漏斗滑板
hopper slope 斗壁倾角
hopper spreader 漏斗式粒料撒布机；漏槽撒布机；漏斗式撒布机；斗式(粒料)撒布机
hopper storage 漏斗储[贮]藏舱
hopper structure 漏斗构造
hopper suction dredge(r) 吸扬装舱挖泥船；装舱吸扬挖泥船；开底吸扬式挖泥船
hopper suction pipe 泥舱吸泥管；抽舱管
hopper suction valve 抽舱阀
hopper throat 漏斗卸料口；斗式卸料孔
hopper trailer 斗式牵引车(铺黄砂、石子用)；料斗(式)拖车
hopper truck 带式载货车；底卸式载货车；漏斗形底卸车；斗式载货车；斗式载货车
hopper type 漏斗式；煤仓式
hopper type clarifier 倒锥形澄清池
hopper type concrete spreader 斗式混凝土布料机
hopper type feeder 箱斗式饲槽
hopper type gritter with feed roll 有喂料滚筒的斗式铺砂机
hopper type portal crane 带斗门式起重机
hopper type single opening shaft kiln discharge 漏斗型单闸门卸料装置
hopper vent 下旋内开气窗；漏斗式通气孔；漏斗式通风口
hopper ventilator 漏斗式通风口；漏斗式空气调节器
hopper ventilator window 外推上旋窗；通风翻窗
hopper vibration method 挂斗振浆法
hopper vibrator 斗式振动筛
hopper wagon 自动倾卸车；漏斗式料车；漏斗式底卸车；漏斗(底)车；带头推车
hopper water closet 斗式马桶
hopper W.C. 斗式马桶
hopper weigher 料斗秤
hopper weigh(ing) batcher scale 漏斗自动秤量计；料仓自动秤量计
hopper well densimeter 泥舱密度计
hopper window 上悬外推窗；底铰内开窗；下悬窗；外推上旋窗
hop pillow 蛇麻草枕头
hopping 免费搭乘；免费搭车；船身上弯
hopping conduction 漂移电导
hopping model 跳跃模型
hopping photoconduction 跳动光电导
hopping process 跳动过程
hoppit 提升大吊桶；凿井用吊桶
Hoppler(falling ball) viscometer 郝勒(落球)黏度计
Hoppler rolling ball viscosimeter 郝普勒落球黏度计
Hoppler visco(si)meter 郝普勒落球黏度计
hop pocket 蛇麻草袋
hoppus ton 哈柏士吨
hops 琵琶桶
hopsack 方平织物
hopsacking 粗织席纹呢
hop skywave 反射天波
Hopton wood (英国德比郡产奶油色、灰色或褐色的)郝铺顿石灰石
horadiam drilling 水平辐射状钻孔
horary 每小时的；时间的
horepipe 蛇皮管
horeshead 超前纵梁
Horinger tectonic segment 和林格尔构造段
Horiz 卧式
horizocardia 横位心
horizon 土层；水平；地(平)线；地层；层位
horizon autonavigator 水平自动导航仪
horizon bar 水平棒
horizon camera 水平(式)摄影机；地平(线)摄影机
horizon circle 地平经仪
horizon closure 全圆闭合差；水平闭合(差)
horizon cloth 粗帆布舞台天幕
horizon communication 视距通信
horizon dial 水平标度盘
horizon dip 地平俯角
horizon distance 测者视地平距离
horizon glass 地平镜；水平镜；水平玻璃观察孔
horizon light 天幕灯槽；水平光；地平线标志灯

horizon line 水平线；地平线；合线
horizon marker 标准层位
horizon mining 水平开采；多水平开采(法)
horizon mirror 水平镜
horizon of roof rock 盖层层位
horizon of soil 土壤层次；土层
horizon parallax 地平视差
horizon peak 层位嶂
horizon photograph 地平线像片
horizon picture 地平线像片
horizon plain 地平面
horizon plane 水平面；地平面
horizon range 水平距离；直视距离；视线距离；测者视地平距离
horizon range from an object 物标视地平距离
horizon ring 地平刻度环
horizon sample 水平试样；地层试样
horizon scan 水平搜索
horizon scanner 水平线跟踪系统
horizon-seeking stabilization 水平瞄准稳定
horizon sensor 水平传感器；地平仪；地平传感器
horizon shade (定镜前的)地平遮光镜
horizontal 横(平)的；水平线；水平的；卧式
horizontal acceleration 横向加速度；水平加速度
horizontal adit 横洞；水平通道；水平洞
horizontal adjustment 水平校正；水平调正；水平调整
horizontal aerotriangulation 平面空中三角测量
horizontal air collector 卧式集气罐
horizontal air elevator 卧式气力提升机
horizontal air pump 卧式气泵
horizontal alignment 平面线型；平面线路；平面定线；水平校准；水平线向；水平线位；水平线路；水平线定向；水平定线；水平线形
horizontal alignment coil 水平校线线圈
horizontal aligument 水平对准
horizontal alinement 水平定线
horizontal amalgamation 横向合并
horizontal amplifier 水平放大器
horizontal amplitude 行幅度；水平幅度
horizontal amplitude adjustment 水平幅度调整
horizontal amplitude control 水平幅度控制
horizontal amplitude section 振幅水平剖面
horizontal ampoule forming machine 转筒式安瓿机
horizontal anaerobic filter 平流式厌氧滤池
horizontal analysis 横向分析；水平分析
horizontal anchorage 水平锚具；水平锚固；水平锚碇
horizontal and elevation control point 平高点
horizontal and elevation picture control point 像片平高控制点
horizontal and vertical clearance above bridge deck 桥面净空；桥面限界
horizontal and vertical control point 平高控制点
horizontal and vertical control point of photograph 像片平高控制点
horizontal and vertical dividers 横竖分隔条
horizontal and vertical extension 平面控制高程加密
horizontal and vertical parity check code 阵码；水平和垂直奇偶检验码
horizontal and vertical photo control point 像片平高控制点
horizontal angle 水平角；方位角
horizontal angle brace 水平隅撑
horizontal angle observation 水平角观测
horizontal angle of book 水平方向观测手薄
horizontal angle of deviation 水平偏向角
horizontal angles of first break 初至波的水平角
horizontal antenna 水平天线
horizontal aperture correction 水平孔径校正
horizontal application 水平安装
horizontal apron 横挡板；水平护垣；水平海墁
horizontal arbor 水平柄轴
horizontal arch element 拱环；(拱坝的)水平拱元件
horizontal array 水平阵
horizontal assembly 地上装配；水平组装；水平装配
horizontal association 横向联合
horizontal astigmatism compensator 水平像散补偿器
horizontal astigmatism corrector 水平像散校正器
horizontal asymptote 水平渐近线
horizontal auger 卧钻；卧式土钻机
horizontal axis 水平轴(线)

horizontal axis concrete mixer 水平轴混凝土拌和机
horizontal axis current meter 水平轴流速仪；横轴流速仪
horizontal axis mixer 卧式(混凝土)搅拌机；非倾斜鼓筒式搅拌机；水平轴式搅拌机；横轴式拌和机
horizontal axis of graph 运行图横轴
horizontal axle concrete mixer (即非倾倒式或筒式的)水平轴混凝土拌和机；横轴混凝土搅拌机
horizontal axle current meter 横轴流速仪；水平轴流速仪
horizontal backfill 水平回填
horizontal baffle 水平隔板
horizontal balanced reciprocating compressor 对称平衡往复式压缩机
horizontal bands (浮标的)横条纹；横带
horizontal band saw 水平带锯
horizontal band sawing machine 卧式带锯机
horizontal bandwidth 水平带宽
horizontal bar 横杆；横棒；水平条；水平杆；单杠；防擦龙骨；水平钢筋；横向钢筋
horizontal bar, parallel bars 单双杠
horizontal bar-chart 横条图
horizontal bar control 水平条信号控制
horizontal bar generator 横条信号发生器；水平条信号发生器
horizontal bar induction 水平软铁磁感应
horizontal bar keel 背板龙骨；保护龙骨
horizontal barrel distortion 水平桶形失真
horizontal base-line method 水平基线法
horizontal beam 横梁；水平束；水平梁
horizontal beam deflection 电子注的平面偏移
horizontal beam hole 水平束孔
horizontal beam size 水平束的截面大小
horizontal beam width 水平约束宽度；水平波束宽度；射束水平宽度
horizontal bed 水平岩层；水平基床
horizontal bedding 水平层理
horizontal bedding structure 水平层理构造
horizontal-bed glazing jack 水平式打光机
horizontal bed type milling machine 水平床身式铣床
horizontal belt conveyer 卧式胶带运输机；平带运输机；卧式运输带；水平胶带输送机
horizontal belt filter 水平带式滤机
horizontal bench drill 卧式台钻
horizontal bending vibration 水平挠曲振动
horizontal bicycle stand 水平自行车架
horizontal blackout period 水平消隐脉冲周期
horizontal blanking 水平消隐；水平熄灭；水平回描消隐；水平电子束熄灭
horizontal blanking impulse 水平消隐脉冲
horizontal blanking interval 水平消隐间隔；水平扫描消隐时间
horizontal blanking level 水平消隐电平
horizontal blanking multivibrator 水平消隐多谐振荡器
horizontal blanking period 水平回扫消隐时间；水平回描消隐时期
horizontal blanking pulse 水平消隐脉冲
horizontal blanking signal 行扫描消隐信号；水平消隐信号
horizontal blanking time 水平消隐时间
horizontal blasting 水平爆破
horizontal blind 水平软百叶；水平排水暗管
horizontal block 卧式卷线筒
horizontal blowing process 水平喷吹法
horizontal boarding wall 水平木板壁
horizontal boat 水平舟
horizontal boat-shaped crucible 水平舟形坩埚
horizontal boat zone refining 水平舟区熔提纯
horizontal boiler 卧式锅炉
horizontal boom 水平吊杆(美语)
horizontal boom tower crane 水平杆塔式吊车
horizontal borehole 水平钻孔
horizontal borehole drain 水平钻孔排水
horizontal borehole pressure recorder 水平钻孔压力记录器
horizontal borehole profile 水平钻孔断面
horizontal borer 卧式钻孔机；水平钻孔机
horizontal boring 水平钻孔
horizontal boring and milling machine 龙门铣床；卧式铣床；卧式镗(铣)床
horizontal boring machine 卧式钻孔机；卧式镗床

horizontal boring rig 水平钻机
horizontal boring unit 卧式镗床；水平钻孔机
horizontal bottom furnace 平炉底炉膛
horizontal bow 横弓
horizontal box furnace 卧式箱形炉
horizontal brace 横拉条；水平(横)撑；水平拉梁；脚手横撑；横板水平撑
horizontal bracing 水平(支)撑；横拉条；水平联结式
horizontal branch 水平排水支管；排水支管；水平枝；水平支(管)；水平管(道)；水平根；排水支渠；排水支沟
horizontal branch pipe 水平支管
horizontal branch star 水平支恒星
horizontal breadth 水平厚度
horizontal break switch 水平断路开关
horizontal brick kiln 水平砖窑
horizontal bridging 平面加密；水平桥接；水平格栅撑；水平支杆
horizontal bridging operation 平面控制加密；水平桥接作业；水平桥接操作
horizontal Bridgman method 水平布里季曼法
horizontal broaching machine 卧式拉床
horizontal buffer 横向缓冲器
horizontal building berth 水平船台
horizontal bulkhead 水平隔墙
horizontal bus 水平母线
horizontal calciner 卧式煅烧炉
horizontal call loan 横向拆借资金
horizontal camera 水准复照仪；水平照相机
horizontal cavity 水平空腔
horizontal cell 水平空腔
horizontal cell tile 平砌多孔砖
horizontal centering 水平中心调整；水平合轴
horizontal centering alignment 水平中心调整
horizontal centering amplifier 水平定心放大器
horizontal centering control 水平中心控制；水平居中调整；水平对中控制
horizontal centering potentiometer 水平中心位置电位器；水平定心电位计
horizontal center line 水平中心线；水平(的)中线
horizontal center-pivot-hung sash window 水平中轴旋窗
horizontal centrifugal casting machine 卧式离心浇铸机
horizontal centrifugal grinder 卧式离心碾磨机
horizontal centrifugal oil pump 卧式离心油泵
horizontal centrifugal pump 卧式离心泵
horizontal centrifugal screen 卧式离心筛
horizontal centrifuge 卧式离心机
horizontal chaining 水平丈量
horizontal chamber furnace 卧式炉
horizontal chamber oven 卧式炉
horizontal channel 水平通道
horizontal charge 平放锭料
horizontal chassis 水平底盘
horizontal check 横向校验；横向检验；水平检验
horizontal checkout 水平测试
horizontal check valve 水平逆止阀；水平递止阀
horizontal chromatography 水平色谱法
horizontal chucking multi-spindle 卧式多轴夹盘车床
horizontal circle 水平刻盘；水平度盘；地平圈
horizontal circle level 水平度盘水准仪
horizontal circle setting screw 水平度盘定位螺丝
horizontal circular chromatography 水平圆环色谱
horizontal city 水平式城市
horizontal cladding 水平外挂墙板
horizontal clamping 水平钳位；水平脉冲箝位
horizontal clamping screw 水平制动螺钉
horizontal cleaner 卧式滑净装置
horizontal clearance 横向净宽；横向净空；水平净空；水平间隙；净宽
horizontal clearance plane 平劈面；横劈面
horizontal cleavage 水平节理
horizontal closed head bead mill 卧式密闭砂磨机
horizontal closure(method) 平堵法截流(法)
horizontal coal seam 水平煤层
horizontal coherent enhancement 水平相干加强
horizontal coil 水平线圈
horizontal collecting water plant 水平集水布置
horizontal collimation 水平照准线
horizontal collimation error 水平视准差
horizontal column 按图幅编号排列存放的地图
horizontal combination 横向企业合并；横向联合；同业联合；同业合并

horizontal combustion 水平燃烧
horizontal common-plane array 水平共面装置
horizontal communication 平行沟通
horizontal comparator 水平比较仪；水平比测器；水平比测计
horizontal compartment 由外墙及防火楼层包围部分房屋
horizontal competition 同业竞争
horizontal component 水平分量；水平分力；水平部分
horizontal component of electric(al) field 水平电场
horizontal component of magnetic field 磁场水平分量
horizontal component of secondary field 二次场水平分量
horizontal component of the cable tension 缆索拉力的水平分力
horizontal component of the earth's magnetic field 地磁水平分力
horizontal component seismogram 水平分量记录
horizontal component seismograph 水平向地震仪；水平分量地震仪；水平地震仪
horizontal compressional effect 水平压缩效应
horizontal compressional stress 水平压缩应力
horizontal compressor 卧式压缩机
horizontal condenser 卧式冷气器；卧式冷凝器
horizontal cones 卧式宝塔筒子
horizontal consolidation pressure 水平固结压力
horizontal constitution 水平垒结
horizontal continuous-acting auger press 卧式连续动作螺旋压榨机
horizontal contour corrector 水平轮廓校正器
horizontal contour signal 水平轮廓信号
horizontal contraction 水平收缩
horizontal control 平面控制；水平控制
horizontal control accuracy 平面控制精度
horizontal control base 平面控制基线
horizontal control data 平面控制数据
horizontal control monument 水平面控制界牌；水平面控制标石；平面控制标石
horizontal control network 平面控制网；水平面控制(测量)网；水平控制网
horizontal control operator 平面控制点坐标计算员
horizontal control photography 平面控制地形摄影
horizontal control point 平面控制点
horizontal control section 水平面控制面
horizontal control station 平面控制点
horizontal control survey 平面控制测量
horizontal control survey net(work) 平面控制测量网
horizontal convergence 水平会聚
horizontal convergence circuit 水平会聚电路
horizontal convergence coil 水平会聚线圈
horizontal convergence control 水平收敛调整
horizontal convergence current 水平会聚电流
horizontal convergence measurement 水平收敛位移量测
horizontal conveyer screen 卧式运输筛
horizontal conveyer 水平传送机
horizontal coordinates 横轴坐标；地平坐标；平面坐标；水平坐标
horizontal coordinate system 平面坐标(统)；地平坐标系；水平坐标系统
horizontal coordination 横向协调
horizontal coplane 像对平共面
horizontal cordon 水平单干形
horizontal core 水平内芯
horizontal coring 水平内芯
horizontal coring block 横向空心砌块
horizontal coring brick 长孔空心砖；水平孔空心砖
horizontal cornice 水平挑檐
horizontal corrosion zone 水平溶蚀带
horizontal counter-balanced action shaking screen 具有水平平衡作用的摇筛机
horizontal counterblow hammer 卧式无钻座锤
horizontal coupling 水平联轴节
horizontal course 平砌层(方块)
horizontal covergence shape control 水平会聚形状控制
horizontal creep 水平向徐变
horizontal crest 平顶波峰
horizontal crossbar 水平横梁；水平横杆

horizontal cross brace 水平剪力撑
horizontal crosscut 水平运输平巷掘进
horizontal crossing 水平交叉口
horizontal cross-section 水平截面
horizontal crusher 卧式破碎机
horizontal crystal growth 横拉法
horizontal current(air) classifier 水平(气)流分级机
horizontal current density component 水平电流密度分量
horizontal curvature 水平曲率
horizontal curve 平(面)曲线;水平(断面)曲线
horizontal curve method 水平曲线法
horizontal cut 横向缺口;水平掏槽;水平开挖
horizontal cut and fill stoping 水平分层充填采矿法
horizontal cutter 横切机
horizontal cyclone water-film scrubber 旋筒式水膜除尘器
horizontal cylinder 卧式汽缸;水平圆柱
horizontal cylinder dryer 卧式圆筒烘燥机
horizontal cylindric(al) tank 卧式圆形油罐;卧式圆筒形储罐
horizontal damping 水平阻尼
horizontal damp-proof course 水平防潮层
horizontal danger angle 水平危险角;水平方向危险角
horizontal decision matrix 水平决策矩阵
horizontal deephole drilling machine 卧式深孔钻床
horizontal definition 水平清晰度;水平分辨(力)
horizontal deflecting circuit 水平偏转电路
horizontal deflection 横向挠曲(度);行偏转;水平偏转;水平挠度
horizontal deflection amplifier 水平致偏放大器;水平偏转放大管
horizontal deflection amplifier tube 水平偏转放大管
horizontal deflection coil 水平偏转线圈
horizontal deflection control 水平偏转调整
horizontal deflection electrode 水平偏转电极
horizontal deflection oscillator 水平偏转振荡器
horizontal deflection output circuit 水平偏转输出电路
horizontal deflection plate 水平偏转板
horizontal deflection stage 水平偏转级
horizontal deflection system 水平偏转系统
horizontal deflection tube 水平偏转输出管
horizontal deflection waveform 水平偏转波型
horizontal deflection yoke 水平偏转线圈
horizontal deformation 水平形变;水平扭曲
horizontal delivery 水平流动
horizontal dentoalveolarfiber 水平位牙槽纤维
horizontal departure 钻孔水平间距;水平偏差
horizontal detail 行清晰度;水平清晰度
horizontal development 水平展开
horizontal deviation 水平偏差
horizontal deviation accuracy 水平偏准确度
horizontal dial feed 水平转盘送料
horizontal diamond boring machine 卧式钻石镗床
horizontal diaphragm 水平隔板
horizontal die slotting machine 卧式冲键插床
horizontal difference table 水平式差分表
horizontal diffuser 卧式扩散器;水平扩散器
horizontal diffusion 水平扩散
horizontal diffusion coefficient 横向扩散系数;水平扩散系数
horizontal diffusion furnace 卧式扩散炉
horizontal digester 卧式蒸煮器
horizontal dilution of precision 水平误差放大因子[GPS定位];水平精度几何因子
horizontal dimensioning 水平标注
horizontal dipole 水平偶极子
horizontal dipole curtain antenna 水平偶极天线阵
horizontal dipole sounding 水平偶极测深
horizontal dip slip 水平倾滑
horizontal direction 行偏转;水平方向
horizontal direction formation 水平方向成层法
horizontal direction or coordinates 水平方向或座标
horizontal directive tendency 水平方向性
horizontal discharge tube 水平放电管
horizontal dish crusher 卧式圆盘破碎机
horizontal disk separator 卧式圆盘种子清选机
horizontal dislocation 水平位错;水平错动;水平断层;水平错动
horizontal dispersion 水平色散
horizontal dispersion of interference colo(u)rs 干涉色(的)水平色散
horizontal displacement 走向滑距;平面位移;水平移位;水平位移;水平变位
horizontal displacement measurement 水平位移测量
horizontal displacement observation 水平位移观测
horizontal displacement observation for bridge substructures 墩台水平位移观测
horizontal displacement of bridge substructures 墩台水平位移
horizontal displacement of crown 冠顶的水平位移;拱顶的水平位移
horizontal displacement of the bottom of the hole 井底水平移位
horizontal displacement stage 水平位移阶段
horizontal disposition 横排列;横列排
horizontal distance 平距;水平距离;水平间距
horizontal distance or coordinates 水平距离或座标
horizontal distribution 横向分布;水平配光曲线;水平分布
horizontal distribution pipe 水平配管
horizontal distributor 水平配管
horizontal disturbing acceleration 水平干扰加速度
horizontal ditch 水平沟
horizontal divergence 水平散度;水平发散(度)
horizontal divergent-nozzle pump 水平发散喷嘴扩散泵
horizontal division 水平划分;建筑立面横线条处理;横向分割
horizontal double action oill hydraulic press 卧式双动油压机
horizontal double antenna 水平对称天线
horizontal-doublet antenna 水平偶极子天线
horizontal draft 横吸式
horizontal draft carburetor 横吸式气化器
horizontal drain 水平排水沟
horizontal drainage blanket 横向排水垫层;水平排水铺盖
horizontal drainage branch 水平排水支管
horizontal drainage gallery 水平排水垫层
horizontal drainage pipe 水平排水管
horizontal drawing machine 水平拉制机
horizontal draw-out metal-clad switchgear 水平抽出的铠装开关装置
horizontal drift 水平漂移
horizontal drift of a(bore)hole 钻孔的水平位移;孔的水平位移
horizontal drill(er) 水平孔钻机;水平钻孔钻机
horizontal(drill)hole 水平钻孔
horizontal drilling 水平钻探;水平钻孔;水平(孔)钻进
horizontal drilling machine 水平钻孔机
horizontal drilling pit 水平钻坑;水平探坑
horizontal drilling rig 卧式钻机;水平钻机
horizontal drilling system 水平钻探系统
horizontal drill press 卧式钻床
horizontal drive 水平运输平巷掘进;水平推动;水平激励;水平传动
horizontal drive control 水平推动控制
horizontal driver 水平偏转激励器
horizontal drive screw 水平微动螺旋
horizontal drive signal 水平同步信号
horizontal drive stage 水平推动级
horizontal driving impulse 水平推动脉冲
horizontal driving signal 水平推动信号
horizontal drum mixer 卧式混凝土搅拌机;卧筒式搅拌机;卧(筒)式拌和机
horizontal drying machine 卧式烘干机
horizontal duct 水平导管
horizontal ductile shear zone 水平韧性剪切带
horizontal duplex drill 卧式双轴钻床
horizontal dynamic amplitude 水平动态幅度
horizontal dynamic amplitude control 水平动态振幅控制
horizontal dynamic convergence 水平动态会聚校正;水平动态会聚
horizontal dynamic focusing 水平动态聚焦
horizontal earthboring machine 横向钻孔机
horizontal earth friction force 水平土摩擦力
horizontal earth pressure 水平土压力
horizontal earth rate 水平地球速率
horizontal echograph 横向回声测距计
horizontal edger 卧式轧边机
horizontal efficiency 水平扫描效率
horizontal electrode 水平电极
horizontal electrofilter 卧式电集尘器
horizontal electrostatic precipitator 卧式电集尘器
horizontal elutriator 卧式淘析器
horizontal emittance 水平发射度
horizontal engine 卧式发动机
horizontal epitaxial furnace 卧式外延炉
horizontal epitaxial reactor 水平外延反应管
horizontal equilibrium 水平平衡
horizontal equity 纳税横向均等
horizontal equivalent 斜坡的换算水平距离;换算水平距离;水平等效(值)
horizontal erection 水平建造;地上装配;水平组装;水平装配
horizontal error 水平(测量)误差
horizontal escapement 工字轮式擒纵机构
horizontal evapo(u)rator 卧式蒸发器
horizontal exaggeration 水平放大
horizontal excavation 水平开挖
horizontal exit 横向疏散口;水平安全出口
horizontal exploration 水平勘探
horizontal exploration method 水平勘探法
horizontal exploratory tunnel 水平勘探巷道支护
horizontal extension 平面控制加密
horizontal extent 水平引伸;水平范围
horizontal extruder 卧式挤压机;水平挤出机
horizontal extrusion presses 卧式挤压机
horizontal face plate 水平旋台
horizontal fault 平移断层;水平断层
horizontal faunal area 水平动物区系范围
horizontal feed 水平馈送;水平进给
horizontal feed chipper 水平进料削片机
horizontal fendering 横护(舷)木
horizontal fibre 水平纤维
horizontal field balance 水平磁力仪
horizontal field magnetometer 水平场强磁力仪
horizontal field of fire 水平射击区
horizontal field of view 水平视角
horizontal fieldstrength diagrams 水平场强图
horizontal fillet weld(ing) 横向角缝焊接;横角焊;水平角焊
horizontal film evapo(u)rator 卧式膜蒸发器
horizontal filter 卧盘式真空过滤器;水平式滤池;水平滤池
horizontal filter paper chromatography 水平滤纸色谱
horizontal filter-well 水平排水井;水平滤井;导流孔;水平渗(流)井
horizontal fin 横翅;水平散热筋;水平安定面
horizontal fineboring machine 卧式精镗床
horizontal fine motion screw 水平微动螺旋
horizontal fire grate 横放炉条;水平炉箅
horizontal firewall 水平防火壁
horizontal firing 水平燃烧
horizontal fish finder 水平鱼探仪;水平探鱼仪
horizontal flail slasher 水平甩链式茎杆切碎机
horizontal flame pot furnace 平焰式坩埚窑
horizontal flange 水平中分面;水平法兰面
horizontal flange angle iron 水平翼缘角钢
horizontal flat screen 卧式平板筛;水平式平板筛
horizontal flight 水平飞行
horizontal flight indicator 水平飞行指示器
horizontal floor 水平楼板
horizontal flow 水平(水)流;水平径流
horizontal flow chart 水平流程图
horizontal flow compensation 水平整流设备
horizontal flow electrostatic precipitator 卧式电集尘器
horizontal flow grit chamber 平流式沉砂池
horizontal flow nozzle 水平流动型喷管
horizontal flow roughing filter 平流式粗滤池
horizontal flow roughing filtration 平流粗滤
horizontal flow sand filter 平流砂滤池
horizontal flow sedimentation basin 平流沉淀池
horizontal flow sedimentation tank 平流(式)沉淀池
horizontal flow settling basin 水平流沉淀池
horizontal flow tank 水平沉淀池;平流池;平流(沉降)槽

horizontal flue 水平烟道
horizontal-flue coke oven 水平烟道炼焦炉
horizontal-flued oven 卧囪炉;水平烟道炉
horizontal fluid layer 水平流层
horizontal flyback 水平回扫
horizontal flyback period 水平回扫周期
horizontal flyback pulse 水平回扫脉冲
horizontal flyback sweep 水平回扫
horizontal flyback time 水平回扫时间
horizontal flyback transformer 水平回扫变压器
horizontal focus 水平聚焦
horizontal focusing 水平聚焦
horizontal fold 水平褶皱;非倾伏褶曲
horizontal force 横向力;水平力
horizontal force instrument 水平磁强仪;摆动磁针
horizontal force magnetometer 水平场强磁力仪
horizontal forging machine 卧式锻造机
horizontal format 横向格式
horizontal forming shoe 水平摆锻锤头
horizontal form(work)support 水平模板支撑
horizontal fracture 水平裂缝
horizontal frame 横向构架;水平框架
horizontal frame gate 横架式闸门
horizontal frame saw 横框锯;卧式框锯
horizontal frame saw mill 卧式框锯制材厂
horizontal free face 水平临空面
horizontal frequency 行频;水平扫描频率
horizontal frontal line 正平线
horizontal frontal plane 正平面
horizontal furnace 卧式炉
horizontal gallery 水平通道;水平集水廊道
horizontal gallery camera 卧式制版照相机
horizontal gas coke 卧炉焦
horizontal gas-water interface 水平的气水界面
horizontal gear 水平齿轮;平齿轮
horizontal generator 卧式静电加速器
horizontal geodetic base 平面大地探测基础
horizontal geodetic datum 大地测量平面基准
horizontal geothermal gradient 水平地温梯度
horizontal girder 横梁;水平桁
horizontal girder gate 横梁式闸门
horizontal glass 地平镜;定镜(六分仪)
horizontal glazing 水平玻璃(安装)
horizontal glazing bar 玻璃横格条
horizontal glued-laminated timber beam 水平胶结叠层木梁;水平胶合木梁
horizontal gradient 水平梯度
horizontal gradient of gravity 重力水平梯度;重力水平坡度;水平重力梯度
horizontal gradient of gravity anomaly 重力异常水平梯度
horizontal gradient of gravity profile 重力水平梯度剖面图
horizontal gradient of magnetic anomaly 磁异常水平梯度
horizontal gradient of ZZ 的水平梯度值
horizontal gradient technique 水平温差法
horizontal grate cooler 水平算式冷却器
horizontal gravity take up 水平重锤拉紧装置
horizontal grid 水平方格
horizontal grill(e) 横芯
horizontal grinding dispersion machine 卧式研磨分散机
horizontal grit settling tank 平流式沉砂池
horizontal grizzly 水平格筛
horizontal ground motion 水平向地面运动
horizontal grout joint 水平浆缝
horizontal guiding wheels 水平轮组【机】
horizontal guy 水平拉线
horizontal hair 横丝
horizontal hammer mill 卧式锤碎机
horizontal handle 横执手
horizontal handling type 水平装卸方式
horizontal heart 横位心
horizontal heave force 水平冻胀力
horizontal hobbing machine 卧式滚齿机
horizontal hold 行同步;水平同步
horizontal hole 平孔;水平孔
horizontal hone 卧式磨床
horizontal hum bars 水平哼声干扰条纹
horizontal hunting 图像水平摆动;水平摆动
horizontal hydraulic jack 卧式油压千斤顶
horizontal hydraulic press 卧式液力压榨机
horizontal hydrogen resistance furnace 卧式氢气电阻炉
horizontal illumination 水平照明;水平照度
horizontal image 水平图像
horizontal impact crusher 卧式冲击破碎机;水平冲击破碎机
horizontal impervious bottom bed 水平隔水底板
horizontal impervious layer 水平不透水层
horizontal indexing 水平指引
horizontal induction 水平磁感应
horizontal induction furnace 卧式感应加热炉
horizontal induction motor 卧式感应电动机
horizontal ingot 水平锭料
horizontal inscribed board 匾;额匾
horizontal integration 横向联营;同业联合;同业合并
horizontal intensity 水平强度
horizontal intensity chart 水平强度分布图
horizontal intensity isoporic charts 水平强度等磁变线图
horizontal intensity variometer 水平强度磁变仪;水平磁强变感器;水平磁强计
horizontal-interlace technique 隔行扫描技术;水平扫描隔行扫描技术;扫描技术
horizontal intersection 水平交会
horizontal interval 水平间隔
horizontal inverter 水平偏转换向器;水平反向器
horizontal isomagnetic chart 水平强度等磁力线图
horizontal isotherm 水平等温线
horizontality 水平状态;水平性质;水平度
horizontality of support ring 支承圈水平度
horizontality of the rim 轮缘的水平状态
horizontalization 置平【测】
horizontalizing model 模型置平
horizontal jet 水平射流
horizontal jib 水平吊杆;水平臂;水平起重杆
horizontal jib tower crane 水平吊杆塔吊
horizontal jitter 水平晃动
horizontal joint 横缝;水平中分面;水平联轴节;水平节理;水平接头;水平(接)缝;层节理
horizontal joint flange 水平中分面法兰
horizontal karst 水平层岩溶
horizontal keying pulse 水平键控脉冲
horizontal keystone correction 水平梯形畸变校正
horizontal kier 卧式煮炼锅
horizontal kiln 卧式窑;水平窑
horizontal kneeand-column type milling machine 卧式升降台铣床
horizontal labor union 同业工会
horizontal ladder 卧式梯;水平梯
horizontal ladder gasket 水平梯形密封垫
horizontal laminar flow 水平层流
horizontal lamination 水平纹层
horizontal landing 水平着陆
horizontal lap 水平搭接(钢筋)
horizontal lateral 水平侧枝
horizontal lateral bracing 水平横向支撑系
horizontal lathe 卧式车床
horizontal lattice 水平帘子
horizontal launching 水平发射
horizontal layer 水平层
horizontal layer stacker 水平层堆料机
horizontal layer stacking 水平层堆料
horizontal laying 水平敷设
horizontal length of photo-baseline 摄影基线水平长度
horizontal length of tee 三通横长
horizontal lens angle 透镜水平角
horizontal level(l)ing 水平测量
horizontal level luffing crane 水平俯仰起重机
horizontal lever 水平杠杆
horizontal lifting beam 水平提升杆
horizontal light beam adjusting screw 水平光调节螺钉
horizontal light component 光波水平分量;水平光分量
horizontal lighting 水平照明
horizontal limb 水平度盘
horizon(tal)line 水平线;地平线;环纹;横线;横条;平线
horizontal linear element 水平线状元素
horizontal linearity 水平线性
horizontal linearity coil 水平线性调整线圈
horizontal linearity control 水平线性调整
horizontal line frequency 水平扫描频率
horizontal line of sight 水平视线
horizontal line-transfer 水平线传导
horizontal linkage 横向联系;水平式联系
horizontal live load 水平活荷载
horizontal load 横向负载;横向负荷
horizontal load carrying member 水平受力构件
horizontal load(ing)水平荷载;水平加荷;横向荷载
horizontal loading test for pavement 路面水平荷载试验
horizontal load reaction 水平荷载反作用;水平荷载反应;水平荷载反力
horizontal lock 平锁;横向锁;水平锁相
horizontal lode 水平矿脉
horizontal long-tube evapo(u)rator 卧式长管蒸发器
horizontal loop method 水平线圈法
horizontal louvers 水平百叶窗
horizontal low-pass filtering action 水平低通滤波作用
horizontally driven grate cooler 水平推动算式冷却器
horizontally dump truck 横式倾卸车;侧面式倾卸车
horizontally focusing quadrupole 水平聚焦四极透镜
horizontally focusing sector 水平聚焦扇块
horizontally framed mitre gate 横架式人字闸门
horizontally interlaced dot 水平交错点
horizontally interlaced pattern 水平隔行扫描图形
horizontally laminated wood 水平层积木
horizontally layered medium 水平层状介质
horizontally loaded pile 承受水平荷载桩
horizontally opposed flat-four engine 水平对置式四缸发动机
horizontally pivoted hung window 中悬式窗
horizontally pivoted window 水平铰接窗;水平(方向)旋窗;横向铰接窗;翻窗
horizontally pivot hung 中悬
horizontally pivoting window 翻窗;中悬窗
horizontally polarized antenna 水平偏振天线;水平极化天线
horizontally polarized array 水平极化天线阵
horizontally polarized electromagnetic wave 水平偏振电磁波
horizontally polarized non-directional antenna 水平极化无方向性天线
horizontally polarized receiving antenna 水平极化接收天线
horizontally polarized wave 水平偏振波
horizontally projected jet 水平射流
horizontally sliding sash 平拉窗;横向滑动窗
horizontally sliding window 平拉窗;横向滑动窗
horizontally sliding(window)sash 横向窗扇
horizontally split 水平剖分
horizontally split casing 水平剖分式机壳
horizontally split cross culvert 水平剖分式横向(分流)廊道(船闸)
horizontally stratified sandstone 水平成层砂岩
horizontal machine 卧式电机
horizontal machining center with pallet 带工作台的卧式加工中心
horizontal magnetic force 水平磁力
horizontal magnetic intensity 水平磁力强度
horizontal magnetic needle 水平磁针
horizontal magnetization 水平磁化
horizontal magnetometer 水平磁强仪;水平磁力仪
horizontal magnification factor 水平放大率
horizontal manhole 水平人孔
horizontal map of heights 等高线平面图
horizontal map with heights or with contour 等高线平面图
horizontal mark computing system 横向标记计数方式
horizontal material seal 水平料封
horizontal measurement 平面位置测量
horizontal member 横构件;水平构件;水平杆件
horizontal merger 同业合并
horizontal microinstruction 横向微指令
horizontal micrometer 卧式测微仪
horizontal microprogram(me)水平微程序;直接控制微程序
horizontal microprogramming 水平微程序设计
horizontal miller 卧式铣床
horizontal milling 水平铣
horizontal mirror 地平镜

horizontal mixer 卧式搅拌机
horizontal model 水平模型
horizontal moment 水平力矩
horizontal momentum 水平分动量
horizontal monitoring control network 平面监测网
horizontal motion 水平运动
horizontal motion seismograph 水平运动地震仪;水平向地震仪
horizontal motor 卧式电动机
horizontal mo(u)lding 横线条
horizontal movement 横向移动;水平运动;水平移动;水平位移
horizontal movement associated with vertical displacement 伴有垂直位移的水平运动
horizontal movement ga(u)ge 水平位移计
horizontal multicompartment fluidized bed dryer 卧式多室流化床干燥器
horizontal multiple layering 水平复式压条;水平多式压条
horizontal multivibrator 水平扫描多谐振荡器
horizontal mutin 横棂条
horizontal navigation 平面导航;水平面航行
horizontal nogging piece 木骨架平顶镶板
horizontal nozzle 水平接管
horizontal objective 水平物
horizontal obstacle sonar 水平障碍声呐
horizontal offset 水平错开
horizontal of technology 横向技术
horizontal oil-firing boiler 卧式燃油锅炉
horizontal oil groove 水平油槽
horizontal oil hydraulic 卧式油压机
horizontal oill-water interface 水平的油水界面
horizontal one-pipe system 水平单管系统
horizontal one-stage pump 卧式单级泵
horizontal opposed engine 卧式对置发动机
horizontal opposed reciprocating compressor 卧式往复压缩机;对称平衡往复式压缩机
horizontal ordinate 横坐标
horizontal orifice 水平口;水平孔
horizontal or inclined rollstack 水平或倾斜的滚辊叠片
horizontal oscillating transformer 水平扫描振荡器用变压器
horizontal oscillator 水平扫描振荡器
horizontal oscillator circuit 水平扫描振荡电路
horizontal oscillator tube 水平振荡器
horizontal output 水平输出
horizontal output circuit 水平输出电路
horizontal output deflection circuit 水平输出偏转电路
horizontal output stage 水平输出级
horizontal output transformer 水平输出变压器
horizontal output tube 水平输出管
horizontal overhung 水平遮阳板;水平出挑
horizontal overlap 水平掩覆
horizontal overscan 水平过扫描
horizontal ozone transport 臭氧的水平输送
horizontal palmette 水平棕叶
horizontal panel 横向嵌板;墙上横板;水平镶板;水平区格板
horizontal pan mixer 盘式混合机
horizontal parabola 水平抛物波
horizontal parabola control 水平抛物控制
horizontal parabola dc control voltage 水平抛物波直流控制电压
horizontal parallax 左右视差【测】;水平视差;地平视差;X—视差
horizontal parallax correction 水平视差校正
horizontal parallax difference 左右视差较
horizontal parallax screw 左右视差螺旋;水平视差螺旋
horizontal parallax slide 左右视差滑尺;水平视差滑尺
horizontal parallel strap 水平并联板
horizontal parameter 水平参数
horizontal parity check 纵向奇偶性检验;水平检验
horizontal particle 水平运动粒子
horizontal partition wall 扇面板
horizontal passage 水平通道
horizontal pass point 平面过渡点
horizontal path 水平轧线
horizontal path transmittance 水平光程透射率
horizontal peak acceleration 水平加速度峰值
horizontal pendulum 水平摆

horizontal pendulum factor 水平振摆因子
horizontal pendulum footing 水平向摆式基础
horizontal pendulum starter 水平摆式起动器
horizontal permeability 水平渗透率
horizontal permeability coefficient 水平渗透系数
horizontal phase 水平相位
horizontal phase section 相ం水平剖面
horizontal phase shifter 水平移相器
horizontal phasing control 水平相位控制
horizontal photo-control 相片平面控制
horizontal photo-control point 相片平面控制点
horizontal photograph 水平摄影;水平摄像片;地平线像片
horizontal pick-up 水平检波器
horizontal picture 水平像片;水平图像
horizontal piling reaction 桩的水平反力;水平打桩反力
horizontal pincushion correction 水平枕形畸变校正
horizontal pipe 横管;水平管(道)
horizontal pipe jacking 水平顶管
horizontal pipe well 水平管井
horizontal piping 水平管桶
horizontal pitman crusher 水平摇杆型破碎机;水平连杆型破碎机
horizontal pitman jaw crusher 水平连杆型颚式破碎机
horizontal plan 平面图;水平投影
horizontal plane 地平面;横剖面;水平板;水平面
horizontal plane machine 卧式刨床
horizontal planer 卧式刨床
horizontal plank free standing silo 自立横板式密仓
horizontal plank silo 横板式密仓
horizontal planning 水平面布置图
horizontal planning machine 横刨机
horizontal plastic injection mo(u)lding machine 卧式塑料注射成形机
horizontal plate 水平盘
horizontal plate anchorage 水平锚碇板
horizontal platebending machine 卧式弯板机
horizontal plate ga(u)ge 板式水平位移计;板式水平测头
horizontal plough mixer 水平犁头混合机
horizontal pneumatic press 卧式气力压榨机
horizontal polarization 水平偏振;水平极化
horizontal polarized shear wave 水平偏振横波
horizontal porcupine opener 卧式豪猪开棉机
horizontal porosity 水平孔隙度
horizontal position 横焊位置;平焊位;水平位置
horizontal position accuracy 平面位置精度
horizontal position dial down 水平位置面下
horizontal position dial up 水平位置面上
horizontal positioning 水平位置调整
horizontal position welding 横焊;平焊;水平(位置)焊接
horizontal precipitation 水平析出;水平沉淀
horizontal press 卧式压床
horizontal pressure 横压力;水平压力
horizontal pressure foot 水平压板
horizontal pressure generator 卧式充压型静电加速器
horizontal pressure gradient 水平压强梯度
horizontal pressure leaf filter 水平加压叶滤机
horizontal prevention leakage 水平防渗
horizontal principal plane 水平主平面
horizontal prism 水平棱柱;水平棱镜
horizontal profile 水平断面;横剖面(图);水平剖面(图);水平截面
horizontal profiling 水平剖面测量
horizontal projected area 水平投影面积
horizontal projected window 上悬窗;上旋窗
horizontal projection 水平投影
horizontal projection area 水平投影面积
horizontal projection calculating reserve 储量计算水平投影图
horizontal projection echo sounder 水平回声探测仪
horizontal projection of orebody 矿体水平投影图
horizontal projection of orebody for reserve calculation 矿体储量计算水平投影图
horizontal projection plane 水平投影面
horizontal property 房产
Horizontal Property Act 共有公寓住宅法案

horizontal prop leg 卧式支脚
horizontal pull 水平拉力
horizontal pulling on machine 卧式套皮壳机
horizontal pulling technique 水平拉晶技术
horizontal pulp-current classifier 水平矿浆流分级机
horizontal pulse 水平脉冲
horizontal pulse extractor 卧式脉冲萃取器
horizontal pulse timing 水平脉冲同步;水平脉冲定时
horizontal pulse width set 水平脉冲宽度调整
horizontal pump 卧式(水)泵
horizontal punching machine 卧式冲床
horizontal pushing jack 横推千斤顶
horizontal-push shear test 水平推剪试验
horizontal radial diamond drilling 水平辐射状钻孔金刚石钻进
horizontal radiation pattern 水平面内的辐射图
horizontal range 广度;水平射程;水平距离
horizontal reciprocating-table surface grinder 卧式往返工作台平面磨床
horizontal redundancy check 横向冗余校验
horizontal redundancy check character 横向冗余校验字符
horizontal reel 卧式卷取机
horizontal reference 水平基准(面)
horizontal reference plane 基准水平面
horizontal refraction 地平(大气)折射;水平折射;地平折光差;地平蒙气差
horizontal refraction error 水平折光差
horizontal regenerator 卧式蓄热室
horizontal registration control 水平重合控制
horizontal reinforcement 水平钢筋
horizontal reinforcing 水平加强;水平加筋
horizontal relevance tree 水平关联树
horizontal repetition rate 行扫描频率;水平扫描频率
horizontal resection 平面后方交会(法);水平后方交会(法)
horizontal reset 水平复位
horizontal reset clock 水平复位时钟
horizontal resolution 行分辨能力;行分辨力;水平析像能力;水平清晰度;水平分解力;水平分辨率;水平分辨(力)
horizontal resolution bars 水平分解力测试条
horizontal resolution limit 水平分辨率极限
horizontal resolution wedges 水平分辩力检验楔
horizontal resultant 水平合力
horizontal retort 横式罐;平罐;卧式釜;水平炭化炉;水平罐
horizontal retort tar 卧式蒸罐制焦油
horizontal retrace 水平回描
horizontal retrace period 水平回归周期
horizontal retrace ratio 水平回扫时间比
horizontal return tubular boiler 卧式回管锅炉
horizontal reversing apparatus 水平轴转换装置
horizontal rhombic aerials 水平菱形天线
horizontal rib 水平肋板
horizontal ribbon blender 水平螺条掺和机
horizontal ring 水平辐射炮眼组圈
horizontal ring drilling 水平辐射环形钻进;水平辐射状钻孔
horizontal ripple 横波纹织物
horizontal roll 水平轧辊
horizontal-rolled position welding 滚动水平焊;水平转动焊接
horizontal roller(grinding)mill 水平滚轴碎石机;水平滚轴碾磨机
horizontal roller guiding open washing machine 横导辊平洗机
horizontal roller mill 水平辊磨;水平环辊式磨机
horizontal roll forging mechanical hand 水平辊锻机械手
horizontal roll stand 水平(轧)辊机座
horizontal root 水平根
horizontal rotary disk type vacuum filter 水平回转圆盘式真空过滤机
horizontal rotary dryer 卧式回转干燥器;卧式烘干滚筒
horizontal rotary engine 卧式星形发动机
horizontal rotary furnace 卧式转炉
horizontal rotary-knife shredder-mower 水平转刀式割草切碎机
horizontal rotary pump 卧式回转泵

horizontal rotary pyrites burner 卧式硫铁矿回转炉
horizontal rotating autoclave 卧式回转高压釜
horizontal rubber sawing machine 卧式锯胶机
horizontal rudder 潜水舵；水平舵
horizontal saddle 水平鞍座
horizontal sampler 水平式采样器
horizontal sandalwood plane 红木一字刨
horizontal sand grinder 卧式砂磨机
horizontal saw mill 卧式锯床
horizontal sawtooth 水平锯齿波
horizontal sawtooth buff 水平锯齿波缓冲
horizontal sawtooth current 水平锯齿波电流
horizontal sawtooth former 水平锯齿波发生器
horizontal scale 水平标尺；横坐标；横向比例尺；平面尺度；平面比例(尺)；水平(线)比例尺；水平度盘比(例)尺
horizontal scale of apparent resistivity 视电阻率曲线的横向比例尺
horizontal scanner 水平扫描器
horizontal scan(ning) 水平扫描；行扫描
horizontal-scan(ning) circuit 水平扫描电路
horizontal-scan(ning) device 水平扫描装置
horizontal-scan(ning) frequency 水平扫描频率；行扫描频率
horizontal-scan(ning) generator 水平扫描发生器
horizontal-scan(ning) interval 水平扫描时间；水平扫描间隔
horizontal-scan(ning) line 水平扫描线；平面同步扫描线
horizontal scan nonlinearity 水平扫描非线性
horizontal screen 平面筛；躺筛；水平筛
horizontal screen chassis 水平筛架
horizontal screen well 水平筛井
horizontal screw pump 卧式螺旋泵
horizontal seam 水平岩层；水平煤层；水平矿层；水平接头；水平接缝；水平(焊)缝
horizontal searching mode 水平搜索方式
horizontal seat(ing) 水平支座
horizontal section 横剖面(图)；平截面；水平剖面(图)；水平断面；水平段；水平截面
horizontal section method 水平断面法
horizontal section of coal bed 煤层水平断面图
horizontal section of coal seam 煤层水平断面图
horizontal section of orebody 矿体水平断面图
horizontal section of ore deposit 矿床水平断面图
horizontal section warper 卧式分条整经机
horizontal sector 水平光弧
horizontal sedimentation 水平沉淀
horizontal sedimentation basin 水平沉淀池
horizontal sedimentation tank 平流沉降槽；水平沉淀池
horizontal seepage test method 水平渗透试验法
horizontal segment 水平段
horizontal seismic coefficient 横向震动系数；水平地震系数
horizontal seismograph 横向震动仪；水平地震仪
horizontal seismometer 横向震动仪
horizontal sensor 水平传感器
horizontal separation 走向滑距；水平间距；水平分离；水平错开；水平错距
horizontal separation fault 水平错位断层
horizontal separation indicator 水平间隔指示器
horizontal separator 水平同步脉冲分离器；水平分离器
horizontal-series lattice spacing 水平序列晶格空间
horizontal-setting 水平砌筑法；水平安装
horizontal settling tank 平流式沉淀池；水平沉淀池
horizontal sextant angles method 水平夹角法
horizontal shading 水平阴影
horizontal shading signal 水平阴影信号
horizontal shaft 水平杆；横八轴；水平轴
horizontal shaft arrangement 卧轴式布置(水轮机)；横轴式布置(水轮机)
horizontal shaft centrifugal fire pump 水平轴离心消防泵
horizontal shaft centrifugal pump 水平轴离心泵；卧轴式离心泵
horizontal shaft complete 横轴总成
horizontal shaft current meter 水平轴流速仪；横轴流速仪
horizontal shaft disc crusher 平轴盘式碎矿机；水平盘式破碎机
horizontal shaft double turbine 卧式双水轮机；横轴式双水轮机

horizontal shaft generator 卧式发电机
horizontal shaft mixer 横轴搅拌机；横轴拌和机；卧式拌和机；水平转轴搅拌机
horizontal shaft pump 卧式水泵
horizontal shaft type power plant 水平轴型发电厂；水平轴型动力装置；水平轴型动力厂
horizontal shaft water turbine 水平轴式水轮机
horizontal shaper 卧式牛头刨床
horizontal shear 水平剪切；水平剪力；卧式剪床
horizontal shearing 水平剪切作用
horizontal shearing stress 水平剪应力
horizontal shear machine 水平剪切机；水平剪力仪；水平剪力机；卧式剪床
horizontal shear method 水平剪切法；水平剪力法
horizontal shear strength 水平抗剪强度
horizontal shear stress 水平剪(切)应力
horizontal sheeting 横木排板护壁；长横木撑板(横向排列的木板)；水平挖方支撑；水平护板
horizontal sheeting for excavation 水平挖方支撑
horizontal Shemens diffractometer 卧式西门子衍射仪
horizontal shift 行偏移；水平位移
horizontal shift control 水平偏移校正
horizontal shift(ing) 水平移动
horizontal shore 水平支梁；水平支承架；水平支承桁架；水平撑(混凝土梁模板的)跨间支撑；可调跨度的梁或桁架式构件；集合水平支撑；横撑；悬空顶撑；水平顶撑
horizontal shoring 水平(顶)撑；横撑；大跨水平撑
horizontal shunting yard 水平调车场
horizontal shuttering support 水平支撑
horizontal siding 水平外墙面板；水平披叠板
horizontal sight 水平瞄准
horizontal sight distance 水平视距
horizontal sight distance inside of curve 弯道视距
horizontal signal 水平信号
horizontal silo 筒式筒仓；青储[贮]窖
horizontal simultaneous 水平同步的
horizontal single action presses 卧式单动冲床
horizontal sintering of strip 粉带平放烧结
horizontal situation indicator 水平位置指示器
horizontal size 水平尺寸
horizontal size control 行幅度调整；水平幅度调整
horizontal size press 水平式施胶压榨
horizontal skew 水平斜撑；水平失真；水平偏斜；水平畸变
horizontal slice 水平分层
horizontal slice of layer amplitude 层位振幅水平切片
horizontal slice of layer velocity 层位速度水平切片
horizontal slice of migration amplitude 偏移振幅水平切片
horizontal slice of migration phase 偏移相位水平切片
horizontal slicing 水平分层开采法
horizontal slide door 横拉门
horizontal slider 平推拉窗
horizontal slide type entrance 水平滑动式入口
horizontal sliding 水平滑动
horizontal sliding at base 基底水平滑动
horizontal sliding bar 水平滑杆
horizontal sliding door 横向滑动门；平推(拉)门；推拉门；水平推拉门
horizontal sliding fire door 平推防火门
horizontal sliding-frarne hot saw 水平滑座式热锯
horizontal sliding of slab 路面板的水平滑移
horizontal sliding wheel 平滑轮
horizontal sliding window 平推拉窗；水平推拉窗；水平滑动窗
horizontal slip 水平滑距；水平滑动
horizontal slipforming 水平向滑模
horizontal slope 平坡
horizontal slotting machine 卧式插床
horizontal slow motion device 水平微动装置
horizontal slow motion knob 水平微动钮
horizontal soft iron 横向软铁
horizontal soil friction force 土的水平摩擦力；水平土摩擦力
horizontal solar telescope 水平太阳望远镜；地平式太阳望远镜
horizontal sole leather roller 卧式底革滚压机
horizontal solid-bowl centrifuge 沉降式离心脱水机
horizontal spacer 水平定距器；水平定距块；隔片；垫块；水平定位片

horizontal spacing 水平间距；水平间隔
horizontal spectrograph 地平摄谱仪
horizontal spindle 水平(主)轴
horizontal spindle press 水平卧式压榨机
horizontal spindle pump 卧式水泵；水平轴泵
horizontal spindle rotary table grinding machine 卧式回转工作台磨床
horizontal spindle surface grinding machine 水平轴平面磨床
horizontal spinning extruder 卧式挤压纺丝机
horizontal spiral type dry water meter 水平螺翼式干式水表
horizontal split 水平中分面；水平分裂
horizontal-split multi-stage pump 水平剖分式多级泵
horizontal spout chipper 水平进料削片机
horizontal spray chamber 横式喷雾室
horizontal spray chamber type cooling tower 卧式喷雾凉水塔
horizontal spring 水平簧
horizontal spring hinge 水平弹簧铰链
horizontal sprinkler 水平型洒水喷头
horizontal square-set method 水平方框支架采矿法
horizontal square-set system 水平分层方框支架采矿法
horizontal square wave 水平正方波
horizontal stability 水平稳定(性)
horizontal stabilizer 水平稳定器；水平安定面
horizontal stacking 水平叠加
horizontal stack section 水平叠加剖面
horizontal stadia 横基尺视距法
horizontal stand 水平台
horizontal starter 水平起动器
horizontal static convergence 水平静态会聚
horizontal static convergence magnet 水平静会聚磁铁
horizontal static load displacement 水平向静态位移量
horizontal stay 水平拉线
horizontal steam engine 卧式蒸汽机
horizontal steering magnet 水平导向磁铁
horizontal sterilizer 卧式杀菌锅
horizontal stiffener 水平支肋；水平加强筋；水平加劲肋；水平加劲杆；水平防挠材
horizontal stiffening 水平加劲
horizontal stiffness 水平刚度
horizontal stock chest 卧式储[贮]浆池
horizontal storage 平放储[贮]存
horizontal storage tank 卧式储[贮]罐
horizontal straight edge 水平直规
horizontal straight line 水平直线
horizontal strain 水平应变
horizontal strain modulus 水平应变模量
horizontal stratum 水平岩层
horizontal stratification 水平层理；水平分层
horizontal stratigraphic(al) separation 水平地层离距
horizontal stress 水平应力
horizontal strip borer 水平钻孔钻机(露天采矿用)；水平钻孔穿孔机
horizontal stripe (浮标的)横条(纹)
horizontal structure of community 群落的水平结构
horizontal strut 水平支杆；水平撑
horizontal strutting 水平柱；水平支撑
horizontal subheading 横标目
horizontal subsurface flow constructed wetland 水平潜流人工湿地
horizontal suction 犁体水平间隙
horizontal sundial 地平(式)日规
horizontal sunshade 水平遮阳
horizontal sunshading board 水平遮阳板
horizontal supplying branch 水平供水支管
horizontal supply pipe 水平供水支管
horizontal support 卧式支座
horizontal surface 水平面
horizontal surface grinder 卧式平面磨床
horizontal surface grinding machine 卧式平面磨床
horizontal survey 平行测量；水平测量
horizontal suspension 水平悬挂；水平摆动悬架
horizontal sway 水平侧倾
horizontal sweep 水平扫描
horizontal sweep amplifier 水平扫描放大器
horizontal sweep driver 水平扫描激励器

horizontal sweep generator 水平扫描振荡器
horizontal sweep selection 水平扫描选择
horizontal sweep system 水平扫描系统
horizontal sweep voltage 水平扫描电压
horizontal symmetry correction 水平对称校正
horizontal sync bit 水平同步比特
horizontal sync circuit 水平同步电路
horizontal sync control 水平同步控制
horizontal sync discriminator 水平同步鉴别器
horizontal sync generator 水平同步信号发生器
horizontal synchronization 水平同步
horizontal synchronization pulse separator 水平同步脉冲分离器
horizontal synchronizing circuit 水平同步电路
horizontal synchronizing device 水平同步装置
horizontal synchronizing generator 水平同步信号发生器
horizontal synchronizing impulse 水平同步脉冲
horizontal synchronizing pulse 水平同步脉冲
horizontal synchronizing signal 水平同步信号
horizontal sync marker 水平同步标志
horizontal sync pedestal 水平同步脉冲基底电平
horizontal sync potentiometer 水平同步调节电位器
horizontal sync pulse 水平同步脉冲
horizontal sync regulator 水平同步调节器
horizontal sync time 水平同步时间
horizontal system 横向系统
horizontal system of combined exploratory opening drilling 水平坑钻结合系统
horizontal system of coordinates 地平坐标系
horizontal system of exploring opening 水平坑道系统
horizontal table filter 平面过滤机
horizontal tabulation 横向制表；横向表列；水平制表
horizontal tabulation character 横向制表字符；横向列表字符
horizontal take-off 水平起飞
horizontal tandem generator 卧式串列静电加速器
horizontal tangent 水平切线
horizontal tangent screw 水平微动螺旋
horizontal tank 水平箱柜；水平池槽；卧式(油)罐
horizontal taping and plumbing 卷尺持平垂准置距法
horizontal telescope 地平望远镜
horizontal tempering 水平钢化
horizontal thickness 水平厚度
horizontal thread 横丝
horizontal through-draught kiln 横焰窑
horizontal throw 水平行程
horizontal thrust 横向推力；平冲式冲断层；水平推力
horizontal thrust borer 水平掘孔机；水平冲击式凿眼机；水平冲击钎子
horizontal thrust fault 平冲断层；水平冲断层
horizontal tie 水平拉杆；横向联系
horizontal tieback 水平拉索；水平拉杆
horizontal tie beam 水平拉杆
horizontal tilt 行倾斜；水平倾斜
horizontal timber 横木
horizontal timbering 水平支撑(法)
horizontal timber plank 水平木板
horizontal timber sheeting 水平木板栅；水平木板挡板
horizontal timebase 水平扫描信号；水平时基
horizontal timebase generator 水平扫描信号发生器
horizontal tolerance 容许超宽；水平允许偏差；超宽
horizontal tool head 水平刀架
horizontal total stress 水平总应力
horizontal trace 水平迹线
horizontal traction shaft 行走横轴
horizontal trade competition 同业竞争
horizontal traffic 水平装卸法
horizontal traffic segregation 平面交通分隔
horizontal training 水平整枝
horizontal transfer 横向转移；水平转移；水平传送
horizontal transit circle 水平子午环；地平子午环
horizontal transit instrument 地平中星仪
horizontal translation 横向摆动
horizontal transmission 横向传播；水平传送；水平传播
horizontal transportation 水平运输
horizontal trapezoid generator 水平梯形波发生器
horizontal travel(l)ing acceleration 水平运行加速度
horizontal travel(l)ing formwork 水平滑移模板
horizontal travel(l)ing mechanism 水平运行机构
horizontal travel(l)ing shaft 行走横轴
horizontal travel(l)ing speed 水平运行速度【机】
horizontal travel(l)ing time 水平运行时间
horizontal trench 水平沟
horizontal trench sheeting and bracing 水平沟槽的挡板和支撑；水平沟槽的挡板和拉条
horizontal trigger 水平触发器
horizontal trim(ming) 水平调整；水平修边
horizontal trunnion mount 水平耳轴安装架
horizontal tube 水平管(道)
horizontal tube bundle units 水平管束单元
horizontal tube condenser 横管冷凝器
horizontal tube evaporator 平管蒸发器
horizontal tube evapo(u)rator 横管蒸发器；卧式管形蒸发器；卧式管式蒸发器；卧管蒸发器；水平管式蒸发器
horizontal-tube multiple-effect 水平管多效
horizontal tube multistage evapo(u)ration 水平多级蒸发
horizontal tube type vibrator 水平管式振动器
horizontal tube zonation generator 卧管式臭氧发生器
horizontal tubular absorber 水平管吸收器
horizontal tubular boiler 卧式火管锅炉
horizontal tunnel 水平坑道
horizontal turbine 横轴式涡轮机；卧式涡轮机；卧式水轮机
horizontal turbulent diffusion 水平紊动扩散；水平湍流扩散
horizontal twin-roller machine 卧式双滚延压机
horizontal-type 横置型；横式；卧式(的)；水平式
horizontal-type accelerator 卧式加速器
horizontal type alkaline cleaning 卧式碱洗
horizontal type central furnace 卧式集中(采暖)炉
horizontal type compressor 水平式压缩机
horizontal type evapo(u)rator 卧管蒸发器；横管蒸发器
horizontal type feed water heater 卧式给水加热器
horizontal type generator 卧式发电机
horizontal type high pressure steak sterilizer 卧式高压消毒器
horizontal type high pressure sterilizer 卧式高压蒸气消毒柜
horizontal type motor 卧式发电机
horizontal type pump 卧式泵
horizontal type reciprocating compressor 卧式往复活塞压气机
horizontal type relay 水平(式)继电器
horizontal type sampling mill 卧式试样磨机
horizontal type shading device 水平式遮阳装置
horizontal under-scan 水平扫描不够
horizontal union 同业工会
horizontal untwining of station bottleneck 车站咽喉的平面疏解
horizontal vaccum filter 卧式真空过滤机
horizontal vacuum resistance furnace 卧式真空电阻炉
horizontal valve 水平阀
horizontal variation 水平变化
horizontal variometer 水平磁强变感器；水平磁力仪；水平磁秤
horizontal vector loop 横截面量环
horizontal vee antenna 水平 V 形天线
horizontal velocity 水平速度；水平流速
horizontal velocity distribution curve 水平速度分布曲线
horizontal velocity gradient 水平速度梯度；水平流速梯度
horizontal velocity of the scan(ning) spot 水平扫描点的水平速度
horizontal veneer slicer 平板刨切机
horizontal-vertical bridging 平面高程加密
horizontal-vertical deflection coil 水平竖直偏转线圈
horizontal-vertical drilling system 水平和垂直钻探系统
horizontal-vertical fillet weld 横向垂直角焊缝
horizontal-vertical system exploring opening-drilling 水平和垂直坑钻结合系统
horizontal-vertical system of exploring opening 水平和垂直坑道系统
horizontal vessel 水平储[贮]罐；水平容器；卧式容器
horizontal vibrating centrifuge 卧式振动离心机
horizontal vibrating needle 水平磁强仪；摆动磁针
horizontal vibrating screen 水平振动筛
horizontal vibrating screening centrifuge 卧式振动离心脱水机
horizontal vibration 水平振动
horizontal view(ing) angle 水平视角
horizontal visibility 水平能见度
horizontal volute propeller pump 卧式蜗壳旋桨泵
horizontal wall anchorage 水平锚碇墙
horizontal wall slab 横向墙板
horizontal warper 卧式整经机
horizontal washing machine 卧式清洗机
horizontal water-collecting layout 平面集水布置
horizontal water-film cyclone dust cleaning 卧式旋风水膜除尘
horizontal water-tube cooler 横管冷却器
horizontal wave 水平波
horizontal weather-boarding 横向鱼鳞板
horizontal wedge-cut 水平式楔形陶槽
horizontal wedge wipe 水平楔形消除
horizontal weld 横向焊缝；水平焊
horizontal welding 横焊；平焊(接)；水平焊接
horizontal well 平井；水平井
horizontal wheel 导线平轮
horizontal wheel assembly 导线平轮组
horizontal whirler 卧式烘干机
horizontal width of outcrop 露头水平宽度
horizontal wind bracing 水平抗风撑架
horizontal wind direction 水平风向
horizontal winding machine 卧式缠绕机
horizontal wind system 水平风系
horizontal wipe 水平消除
horizontal wire 横丝；水平照准线
horizontal wire rod reel 卧式线材卷取机
horizontal yoke current 水平偏转电流
horizontal zenith telescope 水平天顶仪；地平(式)天顶仪
horizontal zero 水平零位
horizontal zone 水平地带
horizontal zone melting 水平区域熔炼
horizontal zoning 水平分带(性)
horizon trace 像片地平线；水平迹线；地平迹线
horizon tracker 水平线跟踪系统
horizon transmission 直接视距传输
hormic theory 策动理论
hormites 海泡石组
hormogonium 连锁体
horn 弓角；杆端耳圈；杆端叉口；喇叭(天线)；角状物；角状容器；角峰；角(柄)；机臂；号角；舵托；电极臂
horn aerial 喇叭形天线
horn angle 号形角
horn antenna 喇叭天线；号角天线
horn anvil 角砧
horn arrester 角形避雷器；带有喇叭裂口的避雷器
horn balance 空气动力平衡；突角补偿
horn base 喇叭座
hornbeam 鹅耳枥
hornblende 角闪石；普通角闪石
hornblende andesite 角闪安山岩；普通闪石安山岩
hornblende anorthosite rock 角闪斜长岩
hornblende basalt 角闪玄武岩
hornblende diopside anorthite gneiss 角闪透辉钙长片麻岩
hornblende diopside leptynite 角闪透辉变粒岩
hornblende diorite 角闪闪长岩；普通角闪石闪长岩
hornblende dolerite 普通角闪石粒玄岩
hornblende gabbro 角闪石辉长岩
hornblende gneiss 角闪片麻岩
hornblende-gneiss-granite 角闪片麻花岗岩
hornblende-gneiss-granite-schist 角闪片麻花岗片岩
hornblende granite 角闪花岗岩
hornblende hornfels facies 角闪角页岩相
hornblende latite 角闪安粗岩
hornblende miascite 普通角闪石云霞正长岩
hornblende mica peridotite 普通角闪石云母橄榄岩
hornblende mica picrite 普通角闪石云母苦橄岩
hornblende mica pyroxenite 普通角闪石云母辉石岩
hornblende olivine gabbro 普通角闪石橄榄辉长岩
hornblende orthorhombic and monoclinic pyrox-

ene plagioclase gneiss 角闪二辉斜长片麻岩
hornblende picrite 角闪苦橄岩;普通角闪苦橄岩
hornblende plagioclase eptynite 角闪斜长变粒岩
hornblende plagioclase gneiss 角闪斜长片麻岩
hornblende porphyrite 角闪玢岩
hornblende porphyry 角闪斑岩
hornblende propylite 角闪青盘岩
hornblende pyroxene plagioclase gneiss 闪辉斜长片麻岩
hornblende pyroxenite 普通角闪辉石岩
hornblende schist 角闪片岩
hornblendes granoblastite 角闪变粒岩
hornblendes orthorhombic and monoclinic pyroxene granulite 角闪二辉麻粒岩
hornblendes potash-feldspar and plagiocolase gneiss 角闪二长片麻岩
hornblende syenite 角闪正长岩
hornblende trachyte 角闪石粗面岩
hornblendite 角闪石岩;南闪石岩;普通角闪石岩
hornblendite group 角闪石岩类
horn block 角块
horn brand(ing) 角上烙印
horn-break fuse 号角形开关
horn-break switch 锥形开关;角形开关;灭弧角开关
horn buoy 雾角浮标;雾号浮标
horn button 喇叭按钮
horn button base plate 喇叭按钮座板
horn cable contact cup 喇叭线接触杯
horn cupping 角法
horn-cyclide 角形圆纹曲面
horn diaphragm 喇叭膜
horn diaphragm washer 喇叭膜护圈
horn disk 喇叭盘
horned(screw)nut 有羊角螺母;有角螺母;花螺母;冠形螺母;齿形螺母
horn equation 喇叭方程
Horner's method 何纳计算方法
horn-fed paraboloid 喇叭馈电抛物面天线
horn feed 喇叭(形)天线馈电;喇叭(形)馈源
horn feed antenna 喇叭形天线
hornfels 角页岩
hornfelsicated mudstone 角岩化泥岩
hornfelsicated siltstone 角岩化粉砂岩
hornfelsicated texture 角岩结构
hornfelsicated tuff 角岩化凝灰岩
horn fire alarm 火灾报警(器)
horn for vehicle 车辆喇叭
horn fuse 角形熔断丝;角式保险器;角式熔断丝
horn gap 角(火花)隙
horn-gap arrester 角隙避雷器;号角式避雷器
horn-gap switch 角隙开关
horn gate 角形闸门;角形水门;角形水口;角状浇口;牛角浇口
horn hood 喇叭盖
horn housing 喇叭壳
horn-hunter 侦声器
horning 竖立
horning hammer 开槽锤
horning tool 槽模
horn knob 喇叭钮
horn-lens 喇叭透镜
horn-lens antenna 喇叭天线
horn lightening arrester 角隙避雷器
horn membrane 喇叭膜;风笛膜片
horn of bank 沙洲突端
horn of the anvil beak 铁砧角
horn of variation of flare angle 变张角喇叭
horn parabolic(al) antenna 喇叭抛物线天线
horn press 柱状工作台压力机;悬臂式压力机
horn protractor 量角器;角规;分度器
horn push 喇叭按钮(车辆)
horn quicksilver 角银矿;汞齐
horn radiator 喇叭(形)天线;喇叭形辐射器
horn reflector antenna 喇叭形反射器天线;喇叭抛物线天线;喇叭反射面天线
horn ring 角轮
horns 门窗框突角
horn-shape 喇叭形的
horn-shaped pole 兽角形杆
horn shavings 角屑
horn shell 喇叭壳
horn silver 角银矿
hornskin 漆皮
horn socket 锥形管套;喇叭打捞器;角锥形打捞器

horn socket with bowl 导向漏斗打捞器
horn spacing 悬臂距离
horn spheric(al)back cover 喇叭球背罩
horn spoon 牛角匙
hornstone 硅质页岩;角岩;角石;黑硅石;燧石
horn structure 喇叭结构
horn support plate 喇叭支板
horn switch 喇叭开关
horn throat 喇叭喉
horn throat and mouth 喇叭喉部和口
horn toothed wheel 喇叭齿轮
horn tube 喇叭管
horn-type antenna 喇叭形天线
horn-type transducer 喇叭换能器
horn valve 角形阀
horn wire 喇叭线
horn work 角制品;角堡(防城用)
horny button 角状按钮
horn zone 霍恩区
horobetsuite 辉锑铋矿
horocycle 极限圆
horologe 钟表
horological 定时花的
horological gear 时计齿轮
horologium 钟表
Horologium of Andronikos Cyrrhestes 安德隆尼柯·塞拉斯特钟塔(古瑞典)
horology 测时术
horometer 测时器
horopter 无双像场
horopter curve 无双像场曲线
horoscope 昼夜长短表;天宫图
horosphere 极限球面
horotelic evolution 常速进化
hor propagation 跳跃传播
Horrebow level(attachment) 赫瑞鲍水准器
hors concours 不合格的
horse 高凳;干燥瓦片用托架;夹石;滑绳;桅座挂滑车铁杆;断夹块;船架模型
horse's head 挂轮架
horseback 马脊岭
horseback mountain 马背岭
horse block 下马石;手砣站台
horse boat 牲口船
horse box 运马车棚
horse brush 马刷(子)
horse chestnut 七叶树;朱红色
horse chestnut oil 马栗油
horse dam 临时坝
horsed joint 圬工的踏步形接合;鞍形接头;咬口接头
horsedrawn 马拉的
horsedrawn binder 马拉(式)割捆机
horsedrawn cart 马拖车;马拉车
horsedrawn expanding harrow 马拉展式钉齿耙
horsedrawn finger-wheel rake 马拉指轮式搂草机
horsedrawn motorized duster 马拉(式)机动喷粉机
horsedrawn rake 马拉搂草机
horsedrawn ridging plow 马拉(式)起垄犁
horsedrawn roller 马拉路碾;马拉滚筒
horsedrawn side delivery rake 马拉侧向搂草机
horsedrawn traffic 马车交通;马车运输
horsedrawn van 运货马车
horsedrawn vehicle 马拉消防车
horseflesh 马肉色木材
horseflesh ore 马肉矿
horse gang plow 马拉式多铧犁
horse gear 马拉传动装置
horse gin 马拉提升滚筒
horse hack 马拉(式)碎土锄
horsehair 马毛织的家具罩;马毛
horse hair broom 马鬃扫帚
horse hair brush 马鬃刷
horse hair fibre 马毛纤维
horse hair hand brush 马鬃手刷
horse hair lamp brush 马鬃灯刷
horse harrow 马拉耙
horsehead 马头门;吊装框架;隧道临时支撑;隧道超前纵梁;轻型框架;滑轮支架;(升降机的)罐笼
horse head beam 马头梁
horse head girder 马头门横梁
horseheading 前部支撑(采矿隧道木支撑)
horsehide gloves 马皮手套
horse iron 长柄捻缝凿

horse latitude high 副热带高压
horse latitude 马纬度;回归线无风带(南北纬30～35度之间);副热带无风带(纬度)
horsemand frieze 骑像饰檐壁
horse market 马市
horse motor 马拉传动装置
horse mo(u)ld 砖坯模;挑檐抹灰样板;灰线模具
horse mouth rasp 马口铁皮粗锉
horse of barren rock 脉石夹层
horse parlor 马厅
horse path 马道
horse pot 马厩的排水坑(在污水进入排水沟管前)
horse power 畜力驱动;功率;马力;马拉传动
horsepower curve 马力曲线
horsepower formula 功率公式
horsepower hour 马力·小时
horsepower input 马力输入
horsepower loading 马力载荷;马力负荷
horsepower nominal 标准马力;登记马力;公称马力
horsepower of equipment 设备功率
horsepower of transmission 传动功率
horsepower output 马力输出
horsepower per liter 公升功率
horsepower petrol engine 马力汽油机
horsepower plow 马拉犁
horsepower rating 功率;马力标定
horsepower requirement 功率要求;需用马力
horse power spectrum 马力谱
horsepower to weight ratio 马力重量比
horse race 赛马
horse rail 驶帆架
horserake 马拉集草机;马拉(搂)耙
horse ranch 养马牧场
horse rasp 大号粗锉
horse roller 马拉(路)碾
horse saddle 马鞍石
horse scaffold 木马架;轻便脚手架
horse scraper 马拉刮路机
horseshaped section 马蹄形断面
horseshoe 楼厅;马蹄形物;马蹄形;蹄铁;二楼楼厅
horseshoe apsis 马蹄形圆室
horseshoe arch 马蹄形拱(圈);马蹄拱
horseshoe base 马蹄形底座
horseshoe bend 马蹄形弯道;U形存水弯
horseshoe channel 弯形航道;蹄形航道
horseshoe collar 马蹄环
horseshoe conduit 马蹄形(排)水管;马蹄形管渠;马蹄形管道
horseshoe curve 马蹄形曲线
horseshoe curve development 马蹄形展线
horseshoe drain 排水用U形管
horseshoe electromagnet 马蹄形电磁铁
horseshoe filament 马蹄形灯丝
horseshoe flame 马蹄(形)焰
horseshoe furnace 马蹄形窑炉
horseshoe ga(u)ge 马蹄规
horseshoe holder 马蹄形夹具
horse-shoeing 装马蹄铁
horseshoe iron 马蹄铁
horseshoe lake 马蹄形湖
horseshoe magnet 马蹄形磁铁;马蹄铁;马蹄形磁体
horseshoe magnet(iron) 马蹄形磁铁
horseshoe main 马蹄形总管;马蹄铁管
horseshoe manifold 马蹄形管道
horseshoe meander 马蹄形河曲
horseshoe mining method 马蹄形开采法
horseshoe mixer 马蹄式搅拌机
horseshoe nail 马掌钉
horseshoe quality 马蹄铁质量
horseshoe reef 马蹄形旋礁;马蹄礁
horseshoe reef pool trend 马蹄礁油气藏趋向带
horseshoe riveter 马蹄形铆钉机
horseshoe section 马蹄形截面;马蹄形断面
horseshoe sectional tunnel 马蹄形隧洞;马蹄形隧道
horseshoe section sewer 马蹄形断面阴沟
horseshoe set 马蹄形支架
horseshoe sewer 马蹄形排水管;马蹄形下水道
horseshoe-shaped 马蹄形的
horseshoe-shaped betwixtoland 马蹄形盾地
horseshoe-shaped curves of zero velocity 马蹄形零速度线
horseshoe-shaped glacier 马蹄形冰川
horseshoe-shaped lining 马蹄形衬砌

horseshoe shaped section 马蹄形截面
horseshoe-shaped sewer 马蹄形阴沟;马蹄形排水管
horseshoe-shaped shield 马蹄形盾构
horseshoe shaped tunnel 马蹄形隧洞;马蹄形隧道
horseshoe stair(case) 马蹄式楼梯
horseshoe stirrer 马蹄式搅拌机
horseshoe trip 马蹄形切刀
horseshoe trip knife 马蹄形切刀的棕绳切割器
horseshoe tunnel 马蹄形断面隧道;马蹄形隧道
horseshoe type 马蹄式
horseshoe type brake 蹄式制动器
horseshoe type furnace 马蹄形炉膛
horseshoe type mixer 马蹄形混合器;马蹄形混合机
horseshoe type thrust bearing 马蹄形推力轴承
horseshoe vortex 马蹄形漩涡
horseshoe vortex system 马蹄形涡流系
horseshoe washer 开口垫圈;蹄形垫圈
horse shore reef 马蹄礁
horse stall container 马厩集装箱
horses teeth 长石
horse stone 马石
horse sweeper 马拉扫路机
horsetail 马尾丝线;马尾式
horsetail fault 马尾状断层
horsetail fissure 马尾状裂隙
horsetail ore 马尾状矿脉;马尾矿
horsetail shape 马尾状
horsetail splitting 马尾状分岔
horsetail structure 马尾丝状构造;马尾构造
horsetail vein 马尾丝状矿脉
horse tender 放马员
horse track 马车道
horse traction 马力牵引;马拉
horse traffic 马车运输
horse tram-car 有轨马车
horse vehicle 马车
horse wagon 运货马车;马车
horse way 马行道;马车道
horse with ommel 鞍马
horsfordite 锑铜矿
horsing 槌击捻缝;临时挑出支撑
horsing beetle 扩缝锤
horsing iron 捻缝凿;长柄捻缝凿
horsing-up 房屋抹灰线脚;安装挑梁檐饰线
horsing-up mo(u)ld 组装线脚木模
horst 悬垂体(洞顶);地垒
horst fault 垒断层;地垒(状)断层
horst mountain 垒断山;地垒山
horst uplifting 地垒隆起
horsy survey beacon 马架标
horticultural building 园艺房;温室
horticultural cast glass 温室玻璃
horticultural cathedral glass 温室用的镶嵌玻璃
horticultural crop 园艺作物
horticultural glass 温室玻璃;园艺用玻璃
horticultural machine 园艺机器
horticultural machinery 园艺机械
horticultural plant 园艺作物
horticultural scheme 园艺规划
horticultural sheet glass 园艺用平板玻璃
horticultural spray 园艺喷雾用油
horticultural structure 园艺建筑物
horticultural tractor 园艺拖拉机
horticulture 园艺学;园艺
horticulturist 园艺家;花卉学家
Horton number 霍顿数;赫顿数
hortonolite 镁铁橄榄石
Horton sphere 霍顿型球形压力储罐;球状气体储[贮]罐;可加压球状气罐
Horton spheroid 水滴形油罐
hose 胶(皮)管;软管;柔性管;皮带管;水龙管;水龙带
hose accessory 水带附件
hose adapter 水带异径接口
hose and belts duck 软管和传送带用帆布
hose angle valve 软管的折角阀
hose-apparatus 软管呼吸器
hose armouring wire 铠装胶管用钢丝
hose assembly 软管接头
hose balancer 软管平衡器
hose bandage 水带包布
hose becket 水带吊带
hose bed 水带箱

hose belt 水带提带
hose bib 软管接嘴;软管(小)龙头;水龙带旋塞;水龙带龙头
hose box 水龙带箱
hose box with spray-jet fire nozzle 带喷雾直流水枪的水带箱
hose braider 水龙带编结机
hose branch holder 带架水枪支架
hose bridge 水带护桥
hose bundle 软管束
hose cabinet 水龙带箱;消火栓箱
hose cap 管帽;水带堵头
hose car 水管车(消防用)
hose carcass 水带编织层
hose carrier 软管运输小车
hose cart 消防队水管车;软管车
hose casing 水带覆盖层
hose clamp 软管支架;软管夹(子)
hose clip 软管夹(子)
hose cock 洒水车;软管旋塞;软管龙头;软管钩;皮(式)龙头;水带旋塞;水龙带龙头
hose company 消防队
hose compartment 水带箱
hose connection 软管连接;软管接头;皮带(式)龙头
hose connection valve 软管连接阀
hose connector 软连续管;软管连接器;软管接头
hose control 水带三脚架
hose coupler 软管接头
hose coupling 软管联轴节;软管连接(器);软管接头
hose coupling nipple 软管(用)接头
hose coupling spanner 水带接口扳手
hose coupling wrench 水带接口扳手
hose davit 装油软管吊柱;软管吊柱
hose director 软管喷嘴
hose down 用软管洗涤;用水龙冲洗;软管放油
hose dryer 水带烘房
hose drum 软管绞盘;水龙带卷筒
hose-drying tower 消防水龙带晾干塔
hose duck 软管帆布
hose elbow 软(管)弯头
hose-end fitting 水带接头
hose-end pressure 软管末端压力
hose end sprinkler 管端喷灌器
hose expansion 管体膨胀
hose extension 软管加长节
hose-fed diving 管供式潜水;软管供气潜水
hose fitting 软管配件;软管附件
hose flushing 软管冲洗
hose gaiter 水带包布
hose globe valve 软管球形阀
hose guard 软管防护罩
hose handling frame 水龙管架;水带举升架;软管操纵架
hose handling gantry 软管操纵架
hose hanger 软管吊板
hose hawser 软管吊绳
hose hoist 水带吊钩
hose house 消防水龙带屋;水带房
hose-in-hose 重叠花
hose jacket 水带外层
hose joint 软管接头
hose joint fastener 胶管紧固卡
hose jumper 水带护桥
hose-lay 铺放水龙管
hose laying 水带(的)铺设
hose laying lorry 水带泵车
hose layout 铺设水带线
hose levelling instrument 软管水准器
hose line 软管管路;软管管道;水龙管;软管线
hose line foam nozzle 空气泡沫枪
hose liner 软管套头
hose load 水带负载
hose machine 软管机
hoseman 软管喷水工;消防人员
hose manifold 水带分水器
hose mash 胶管呼吸器
hose mask 软管面罩;防毒呼吸器
hose mender 修水龙带用套夹;软管修理工
hose nipple 软管嘴;软管螺纹接头;软管螺纹接套
hose nozzle 软管喷嘴;水枪喷嘴;水龙带射口;水龙带喷嘴
hose oiler 软管加油器

hose operator 水带操作员
hose outlet 消火栓出水口
hose packing 软管包装
hose pipe 软管;挠性管;水龙(软)管;蛇形管;水龙带
hose pipeline 软管管路;软管管道
hose pipe tire 软管轮胎
hose plug 水带螺塞
hose-proof generator 防水发电机
hose-proof machine 防水式电机
hose protector 软管防护罩;软管防护装置;软管防护套
hose pulsation 水带跳动
hose rack 软管架子;软管夹(子)
hose ramp 水带护桥
hose reel 胶管卷筒;软管卷绕轮;软管卷盘;软管绞车;水龙带卷筒
hose reel branch 胶管卷水枪
hose reel nozzle 胶管卷水枪
hose reel pack 软管卷组
hose reel system 软管卷盘系统
hose reel tubing 卷盘胶管
hose reel unit for subsea lubrication valve 水下钢丝防喷阀管缆绞车
hose reel unit for subsea test tree 海底测试树管缆绞车
hose roll 水带卷盘
hose roller 水带滚筒
hose shaft 软管轴
hoseshoe hose load 水带的马蹄形叠装
hose siphon 软管虹吸;虹吸软管
hose sling 水带吊索
hose spanner 软管扳手;皮龙扳手
hose station 墙式消火栓
hose storage rack 水带储藏架
hose storage room 水带储藏室
hose strangler 水带断流器
hose strap 水带挂钩
hose-stream test 水流冲击试验;水龙射水试验(对热的隔墙或门作耐力试验)
hose support 软管支架
hose supporting clip 软管挂环
hose tank 水带箱
hose tap 消防水龙头
hose tender 水带泵车
hose test 软管试验
hose testing 冲水试验
hose thread 水龙带接头螺纹;软管接头螺纹;水带螺纹;水带接口螺纹
hose tools 水带附件
hose tower 水带塔
hose tube 软管
hose type pump 软管式泵
hose union 软管联管节;软管接合器;水龙带接头;水龙带管接(头)
hose valve 水龙带阀门
hose with coated plastics 包塑软管
hose wrench 软管扳手
hosiery 针织厂;针织品
hosiery yarn 毛纱
Hoskold factor 霍斯科尔德因素(用于估价年薪的因数)
hospice 招待所;(教会办的)旅客招待所;救济院
hospitable grant 交际费
hospital 医院;卫生院
hospital architecture 医院建筑
hospital arm pull 医院的带臂门拉手;医院肘臂执手
hospital bank 医院银行
hospital barn 病畜舍
hospital bed 医院用床;医院病床;病床
hospital bed lift 医院病床提升器
hospital busbar 旁路母线;故障备用汇流条;备用汇流条
hospital car 医务车
hospital door 平门;光平门;医院门
hospital for gynecology and obstetrics 妇产医院
hospital furniture 医院家具
hospitality expenses 招待费
hospitalization cost 医疗费
hospitalization insurance 医疗保险
hospitalization premiums 医疗保险费
hospital partition 病床围帘
hospital relay group 故障线收容继电器群
hospital service contract 医疗合同
hospital service system 医疗设施系统

hospital sewage treatment 医院污水处理
hospital ship 医院船;医务船;医疗船
hospital solid waste 医院固体废物
hospital stop 止门器(医院)
hospital switch 自动转换开关;故障转换开关
hospital tip 安全铰端
hospital transport 医院输送舰
hospital ward 医院病房
hospital wastewater 医院污水;医院废水
hospital wastewater disposal 医院污水处理
hospital window 上悬外推窗;外推上悬窗(医院);下悬内开窗(医院)
hospitium 旅店
host 寄生;基质;承办单位
hostage 抵押品
host area 移民安置区
host attachment 主系统连接装置
host bakery 圣餐面包房;圣饼烘房;主面包房
host based support program(me) 供主机(用的)支援程序
host choice 宿主选择
host community 原有公众;老居民
host compound 主体化合物
host computer 虚拟计算机;主计算机;主机
host country 侨居国;所有国;东道国
host crystal 主晶;主剂;结晶核;基质晶体
host digital termination 局用数字终端
hostel 招待所;旅舍;小旅馆;寄宿舍;宿舍
host element 主元素
hostel for seamen 海员招待所
hostel for the elderly 老年人住房
hostel-like process 类似站进程
hostel occupant 招待所住客
hostel residence 寄宿舍
hostelry 旅馆;客栈
host exchanger 主交换机
host farming community 原有村社
host flora 寄生植物群
host formation 宿主地层
host glass 晶核玻璃;基质玻璃
host grain 宿主颗粒
host-host protocol 传输协议
hostile 坏天气
hostile attack 恶意攻击
hostile environment 有害环境
hostile identification 敌意认同
hostile witness 恶意证人
hostility 敌对行动
hostility between capital and labo(u)r 劳资冲突
host insect 寄主昆虫
host-ion interaction 基质离子互作用
host language data base 主语言数据库
host language system 主语言系统【计】
host lattice 主晶格;主点阵
hostler 机车维修人
hostless system 分布系统
host machine 主机
host material 晶核材料;基质材料
host media 基底介质
host medium 基质媒质
host mineral 主矿物;载体矿物
host name 主机名称
host node 主节点
host number 主机号
host-parasite interaction 宿主寄生物间相互作用
host-parasite relationship 虫主关系
host party 东道缔约方
host plant 寄主植物;寄生植物
host population 原有人口;老居民
host preparation facility 主系统准备功能
host process 主进程
host processing system 主处理系统
host processor 主处理机
host-pyrolite 主火成岩
host race 寄主族
host rock 主岩;主体岩石;容矿岩;围岩
hostry 旅店
host snail 宿主钉螺
host specificity 寄生专一性
host-stone 成矿母岩
host subscriber 主机用户
host system 主系统
host tap 主接头
host-to-host protocol 主机与主机协议

host unit 主办单位
host user 主机用户
hot 炎热;强放射性
hot acetic acid treatment 加热醋酸处理
hot acid 热酸
hot acid polymerization process 热酸聚合过程
hot acid system 热酸回收系统
hot acid treatment 热酸处理
hot adhesive 热熔性胶黏剂
hot adhesive composition 热黏结材料
hot aggregate 热粒料;热骨料;热集料
hot aggregate bin 热料仓
hot aggregate conveyer 热集料输送机
hot aggregate storage bin 热集料(储)仓;热骨料(储)仓
hot aggregate vibration screen 热料振动筛
hot-air 热空气;热风
hot-air apparatus 热风装置
hot-air bath 热汽浴
hot air blast 热风
hot-air blast hole 热风鼓风口
hot-air blower 热风鼓风机
hot-air box 热空气箱
hot-air-blown oven 热风炉
hot-air cabinet 热风烘箱
hot-air chamber 热风干燥室
hot-air circular drying machine 热风循环干燥机
hot-air circulating oven 干热循环箱
hot-air circulation 热风循环
hot-air circulation system 热风循环系统
hot-air cocoon drying machine 热空气干燥烘茧机
hot-air collector 热气集合器
hot-air control valve 热空气调节阀
hot-air conveying type dryer 热风输送式干燥机
hot-air coupleciton 热气集合体
hot-air cure 热空气熟化;热空气硫化
hot-air curing 热风养护
hot-air deicer 热气防冰设备
hot-air distributor 热风分布器
hot-air drier 热风干燥器
hot-air drying 热空气干燥;热气烘干;热风干燥
hot-air drying kiln 热气干燥室
hot-air duct 暖气管道;热气导管;热风管(道)
hot-air engine 热(空)气发动机
hot-air fan 热风机
hot-air funnel 热气漏斗
hot-air furnace 热空气炉;热风炉
hot-air generator 热气炉;热风机
hot-air gun 热气喷枪
hot-air heater 热气加热器;热气供暖器;空气供暖器;热风供暖器;热空气取暖设备
hot-air heater-defroster 暖气除霜器
hot-air heating 热空气(式)取暖;热空气(式)加热;热空气(式)供暖;热风供暖;热风采暖
hot-air heating flue 热风供暖烟道
hot-air heating standpipe method 立管热风供暖法
hot-air heating system 热空气供暖系统;热气供暖装置;热风供暖系统;热风采暖系统
hot-air horn 热空气喇叭
hot-air intake 热气进口
hot-air jacket 热风加热套
hot-air jet dryer 热风喷射烘燥机
hot-air jig tenter 热风摆动布铗拉幅机
hot-air oven 热风炉;热风烘箱
hot-air plant 热风装置
hot-air retraction 热空气收缩
hot-air roll 热风辊
hot-air seasoning 窑干;热气烘干;热气干燥(法);热空气烘干;热风干燥(法)
hot-air setting 热空气定形
hot-air shrinkage 热风收缩
hot-air sizing machine 热风式浆纱机
hot-air slasher 热风式浆纱机
hot-air soldering 热空气钎焊
hot-air staking machine 热空气拉软机
hot-air standard cycle 热气标准循环
hot-air steam cure 热气蒸汽硫化
hot-air stenter 热风烘燥拉幅机
hot-air sterilization 干热(空气)灭菌法;热汽灭菌法;热空气消毒;热空气灭菌法
hot-air sterilizer 干热消毒器;干热灭菌器;热气灭菌器
hot-air stripper 热空气脱漆器;热气剥漆器
hot-air stuffing mill 热风加脂转鼓

hot-air system 热气供暖系统;热风系统
hot-air test 热气试验
hot-air treatment 热气养护;热(空)气处理
hot-air-turbine 高温空气透平
hot-air type heater 热空气加热设备;热空气取暖设备
hot-air unit 热风炉
hot-air vulcanization 热空气硫化
hot-air welding machine 热空气焊接机
hot-air welding method 热空气焊接法
hotal dining room 旅馆餐厅
hotal laundry 旅馆洗衣房
hot-alumin(i)um mill 铝材热轧机
hot and cold bath (e)热冷槽(浸注)法(木材防腐处理)
hot-and-cold load technique 热冷负载法
hot-and-cold open tank 热冷开口槽
hot-and-cold open-tank treatment 热冷槽处理
hot-and-cold soak process 热冷浸泡法(木材防腐处理);热冷槽浸渍(处理)
hot and cold steeping 热冷槽浸渍(处理)(木材防腐处理)
hot and cold water immersion test 冷热水浸渍试验
hot and cold water mixer 冷热水管混合器
hot and cold water system 冷热水系统;冷热水管系
hot and dry rolling 热干碾压法
hot application 热灌筑;热敷
hot application of concrete 混凝土热天浇灌;热天浇筑混凝土;热天浇注混凝土;热天浇灌混凝土;暑天浇灌混凝土
hot application tar 热用焦油(沥青)
hot application type rust preventive 热施型防锈剂
hot-applied 热用的;热涂
hot-applied bitumen 热铺沥青;热施工沥青
hot-applied coal tar coating 热煤焦油覆盖
hot-applied coating 加热敷层;热施工涂料;热敷层
hot-applied(joint)pouring compound 热注封缝材料
hot-applied sealant 热施工密封膏
hot arc 热弧
hot area 加热区;加热段;受热面
hot arid zone 炎热干旱区
hot ascending flow 上行热流量
hot asphalt 热沥青
hot asphalt adhesive composition 沥青热黏合料
hot asphalt adhesive compound 沥青热黏合料
hot asphalt bitumen impregation 热地沥青灌注
hot asphalt bonding composition 沥青热黏合料
hot asphalt cement 热沥青胶泥;沥青热黏合料
hot asphalt cementing compound 沥青热黏合料
hot asphalt-coated gravel 热拌沥青砾石
hot asphalt-coated sand 热拌沥青砂
hot asphalt concrete 热铺沥青混凝土;热拌沥青混凝土
hot asphaltic bitumen 热沥青
hot asphaltic concrete 热沥青混凝土
hot asphaltic concrete pavement 热铺(地)沥青混凝土路面
hot asphaltic mortar 热沥青砂浆
hot asphalt impregation 热地沥青灌注
hot asphalt plant 热拌沥青集料厂
hot atom annealing 热原子退火
hot atom reaction 热原子反应
hot baffle 挡热板
hot ball type anemometer 热球式风速表
hot-banded steel pipe 热箍式钢管
hot band meter 热条式电表
hot bank 热备用
hot bar method 热棒测定法
hot bar shears 条钢热剪机
hot bath 热水浴
hot-bath quench aging 热浴淬火时效
hot-bath quenching 热浴淬火
hot bay 高温段
hot bearing grease 热的轴承油脂
hot bed 热床;轧钢机台架;暖床;温床
hotbed combine 联合温床
hotbed of carbon 热炭层
hot belt 热的地带
hot bending 热弯(曲)
hot bend(ing)test 热弯(曲)试验;高温弯曲试验
hot bent bend 热弯的弯头

hot bin 高温箱
hot binder 热黏合料;热黏合剂
hot-bin starvation (沥青拌和中的)空仓
hot bitumen 热沥青
hot bitumen cement 热沥青胶泥;热膏体沥青
hot bitumen-coated gravel 热拌沥青砾石
hot bitumen-coated sand 热拌沥青砂
hot bitumen concrete 热沥青混凝土
hot bitumen grout 热浇灌沥青;热用沥青
hot bitumen impregnation 热沥青灌注
hot bituminous mortar 热铺沥青砂浆
hot blast 热鼓风;热空气喷流;气动力加热;炽热空气流
hot-blast air 热风
hot-blast cupola 热风冲天炉;热风化铁炉
hot blast fan high pressure system 热风高压(暖气)系统
hot-blast furnace 热风炉
hot blast heater 热风炉
hot-blast heating 鼓风加热供暖;热风暖气;热风供暖;热风采暖
hot-blast main 热风总管
hot-blast outlet 热风出口
hot-blast stove 高炉热风器;热风炉
hot-blast stove gas burner 热风炉燃烧器
hot-blast system 热风系统;热风暖气系统
hot blast technique 热风技术
hot-blast valve 热风阀
hot bleaching 加热漂白;热法漂白
hot blow 热吹
hot blown iron 热风炼铁
hot body 热体
hot bonding composition 热黏合料
hot bonding compound 热黏合料
hot bonding roofing cement 屋面热黏合料
hot box 加热箱;热轴;热芯盒;过热的轴颈箱
hot-box detector 热轴检测器
hot-box process 热芯盒造型法;热盒制芯法
hot breakdown 热灯浆
hot brick 保温帽砖
hot brine 热盐水;热卤水
hot brine deposit 热卤水沉积
hot brine deposition 热卤水沉积作用
hot brine field 热卤水田
hot brine mineral survey 热卤水矿调查
hot brine ore-forming theory 热卤水成矿说
hot briquetting 加热成形;热压;热团矿
hot brittle 红脆的;热脆的
hot brittle iron 热脆铁
hot brittle material 热脆材料
hot brittlement 热脆
hot-bulb 热球
hot-bulb anemometer 热球式电风速计
hot-bulb arrangement 热球装置
hot-bulb engine 热球(式)柴油机;烧球式发动机
hot-bulb ignition 热球点火
hot calendering 热轧光
hot carbonate process 热碳酸盐过程
hot-cast 热铸
hot cathode 热阴极钨丝荧光灯;强光荧光灯;热(离子)阴极
hot cathode discharge 热阴极放电
hot cathode discharge tube 热阴极放电管
hot cathode gas filed rectifier tube 热阴极充气整流管
hot cathode gas-filled rectifier 热阴极充气管整流器
hot cathode gas filled valve 闸流管
hot cathode gas rectifier tube 热阴极充气管整流管
hot cathode glow tube 热阴极辉光放电管
hot cathode ionization ga(u)ge 热阴极电离计;热阴极电离规
hot cathode ion source 热阴极离子源
hot cathode lamp 热阴极灯;热阴极(电子)管
hot cathode magnetron ga(u)ge 热阴极磁控管真空计
hot cathode magnetron ionization ga(u)ge 热阴极磁控管电离规
hot cathode mercury-arc rectifier 热阴极汞弧整流器
hot cathode mercury discharge lamp 热阴极水银放电灯
hot cathode mercury vapo(u)r rectifier 热阴极汞汽整流器
hot cathode preheat lamp 热阴极预热灯
hot cathode rapid-start lamp 热阴极快速启动灯
hot cathode rectifier 热阴极整流器
hot cathode type magnetron ga(u)ge 热磁控超高真空计
hot cathode X-ray tube 热阴极 X 射线管
hot cave 热室;热洞
hot cell 热室;热单位
hot cement 未适当冷却的水泥;热水泥
hot cement elevator 热混凝土提升机
hot cement(ing) composition 热黏合料
hot chamber 烘房;室室;热试验容器
hot-chamber die casting 热室压铸法;热箱压铸
hot chamber die casting machine 热室压铸机
hot chamber machine 热室压铸机
hot channel factor 热管系数
hot charge 热原料
hot charging 热装(料)
hot chassis 接地底盘
hot check 热裂纹
hot chemical gas flame 热化学气体火焰
hotching 跳汰机选矿;跳汰机产物
hot chip(ping)s-bonding roofing cement 屋面热拌石屑黏合料
hot chisel 热凿;热錾;切热金属凿刀
hotchpot(ch) 掺杂物;杂捡;混合岩层;混合物;财产混杂
hot chromatography 热色谱(法)
hot clean criticality 热净临界
hot cleaner 热水清洗装置
hot clean loading 热净装载
hot clean reactor 热净反应堆
hot clear end-point 热透明点
hot climate cell 高温试验室
hot closed-die forging 热模锻
hotcloset 保温箱(保存菜肴用)
hot cloud 热云
hot coal gas 热煤气
hot coal-tar coating 热煤焦油覆盖
hot-coated-stone plant 石子热涂(沥青)车间;石子热涂(沥青)厂
hot coating 热涂(装);热涂层
hot coil 热线圈;热盘管
hot coil conveyer 热带卷运输机
hot coiler 热卷取机
hot-coining 热精压
hot-cold lysis 热冷溶解
hot-cold work(ing) 中温加工;热冷加工
hot colo(u)r 热色
hot compress 热敷
hot compression 热压
hot concentrated alkali 热浓碱
hot concrete 热混凝土
hot condition 热态
hot consolidation 热致密化;热固结
hot constant 热态常数
hot corebox 热芯盒
hot crackability 热裂性
hot-crack(ing) test 热裂(纹)试验
hot creek 热水沟
hot creep 热致蠕动
hot crimping 热卷边
hot critical reactor 热临界堆
hot cropping 热剪下料
hot cross-section 热截面
hot cupboard 保温菜橱;食物保温箱
hot cure 热养护;热固化;热处治
hot curing 热养护;热硫化;热固化
hot curve 热态有效性曲线
hot cut flare machine 热切扩口机
hot cut method 热切法
hot cutter 热作用具
hot cutting 热切削;热切割
hot cycle 热循环
hot cyclone 高温旋风筒
hot cylinder 热汽缸
hot cylinder setting 热滚筒定形
hot deformation 高温变形;热(态)变形
hot densification 热致密化
hot desert 热沙漠;热漠
hot desert test 热砂试验
hot die 热(压)模
hot die-face cutting 热模横切割
hot die forging 热模锻
hot die steel 热作模具钢;热锻模钢;高温模具钢;热变形模具钢
hot diffusion pump 热扩散泵
hot dimpling machine 热压曲面机;热压波纹机
hot dip 热液浸渍;热镀;浸渍;热浸(镀)
hot-dip alloying 热浸合金过程
hot-dip alumin(i)um coating 热浸镀铝层
hot-dip aluminizing 热(浸)镀铝
hot-dip coating 热浴镀层;热浸涂;热镀
hot-dip coating process 热浸镀法;热浸镀层法
hot-dip compound 热浸渍型防锈剂
hot-dip galvanization steel 热浸镀锌钢
hot-dip galvanized 热浸渍镀锌的
hot-dip galvanized steel 热浸镀锌电镀钢
hot-dip galvanized steel wire 热镀锌钢丝
hot-dip galvanizing 热浸渍镀锌;热浸镀锌
hot-dip metallic coated product 热浸金属涂层制品
hot-dip metallic coated sheet steel 热浸金属涂层钢板
hot-dipped galvanized sheet steel 热浴镀锌薄板钢材
hot-dipped terne 热浸镀铅锡合金层
hot-dipped tinplate 热浸镀锡薄板
hot dipping 浸热度;热镀(锡)
hot dipping method 热浸法
hot-dip plating 热渍电镀;热浸涂镀
hot-dip process 热浸镀锌法
hot-dip sheet galvanizing 薄钢板的热镀锌
hot-dip striping compound 热浸渍脱模剂
hot-dip tinning 热浸镀锡
hot-dip tinning plant 热镀锡车间
hot-dip tinning stack 热镀锡装置
hot-dip type rust preventive 热浸渍型防锈剂
hot-dip zincing 热浸镀锌
hot dozzle 保温帽
hot drawing 热拉;热拔(的)
hot drawing wire 热拉钢丝
hot-drawn 热拉;热拔(的)
hot-drawn tube 热拉管
hot drilling water 预热的钻进用水
hot drip gum test 热滴胶质试验
hot driven rivet 预热铆钉
hot driving 热排放;热铆
hot drop saw 杠杆式热锯;热锯
hot dry climate 干热气候
hot dry environment 干热环境
hot dry rock 热干岩
hot dry rock reservoir 干热岩储
hot dry rocks stimulation 干热岩体激发
hot dry rock system 干热岩系统
hot ductility 热塑性(混凝土);热延性
hot-ductility test 高温延性试验;热塑性试验
hot dy(e)ing exhaust process 高热竭染法
hot dyer 高温染色工
hotel 大型旅馆;旅馆;酒店
hotel bedroom 旅馆卧室;旅馆客房
hot elbowing by inductive heating 中频加热弯管
hot elbowing with wrinkle 折皱加热弯管
hotel building 旅馆建筑
hotel car 带餐室卧车;带餐室的轿车;餐卧车厢
hotel construction 旅馆建设;旅馆建筑(物)
hotel de ville 巴黎市政厅
hot electric(al) pressing 电热压
hot electrode 热电极
hot-electron emission 热电发射
hot electrostatic precipitator 高温电除尘器
hotel entrance 旅馆进口
hotel entrance lounge 饭店门厅;宾馆休息大厅
hot elevator 热混凝土升降机;热混凝土提升机;热料提升机
hotel foyer 旅馆门厅
hotel glass ware 玻璃餐具
hotel lobby 饭店门厅;宾馆休息大厅
hotel lock 组锁;旅馆锁
hotel-motel-motor hotel 汽车旅馆
hotel restaurant 旅馆餐厅
hotel towel holder 旅馆毛巾架
hotel use porcelain 旅馆用瓷
hotel vestibule 宾馆接待厅
hotel ware 旅馆用品
hot embrittlement 加热使脆;热脆(性)
hot enamel coating 热搪瓷涂层
hot end 热焊点;热端
hot end coating 热端镀膜

hot end dust 热端灰泡;热端灰尘
hot end operation 热端操作
hot engine 热发动机
hot environment 高温环境
hot epitaxy 热壁外延
hot etching test 热浸蚀试验
hot extrude 热挤
hot extrusion 热(挤)压;热挤加工;热挤
hot extrusion forming 热挤出成形
hot extrusion mo(u)lding 热挤模塑
hot face 耐熔面;热面
hot face cutting system 耐火切割系统;热面切割系统
hot face pelletizing system 热面造粒系统
hot face zone 热面层
hot facility 强放射性设备
hot feed 热加料;热进料;发热饲料
hot-feed remotecontrolled pump 热馈遥控泵
hot filament 热灯丝
hot filament chemical vapo(u)r deposition 热丝化学气相沉积法
hot filament ionization ga(u)ge 热阴极电离真空计;热灯丝电离真空计;灯丝电离真空计
hot filling 热灌装
hot film anemometer 热膜风速计;热膜风速表;暖膜风速表
hot film current meter 热膜流速仪
hot filter 加热油滤器
hot finished 热光制
hot-finished material 热轧材料
hot-finished rod 热轧盘条
hot-finished steel pipe 热轧钢管
hot-finished steel tube 热轧钢管
hot-finished tubing 热轧钢管;热加工管
hot finisher 热轧精轧机
hot finishing mill 热轧精轧机座
hot firing 热试车
hot fixtre soldering technique 热夹具焊接技术
hot flame 高温火焰
hot flo(a)tation 热浮选
hot flo(a)tation circuit 热浮选回路
hot floor 干燥坑;平底干燥器
hot floor dryer 热地板干燥器;地坑(式)干燥器
hot flue 热烟道
hot flue dryer 热风烘燥机
hot flue gas 热烟道气体
hot fluid 热流体
hot fluid injection 注入热液
hot fluid jet texturing process 热气流喷射变形工艺
hot flush 热潮红
hot foiling 热包工艺
hot food server 食品保温器
hot food table 电热饭桌
hot forge rolling 热辊锻;锻工热滚轧法
hot forging 热锻
hot former 热成型机
hot forming 高温成形;热压成形;热加工;热成形
hot forming process 热成形过程;热成形工艺;热成(方)形
hot-forming property 热成形性
hot-galvanize 热镀锌
hot galvanizing 高温镀锌;热浸镀锌;热镀锌(法);热电镀
hot gas 高温废气;热气
hot-gas bearing 热气轴承
hot-gas by-pass 热气旁通管
hot-gas defrosting 热气体除霜;热气融霜
hot-gas driven generator 热气驱动发电机
hot gas electrostatic precipitator 高温气体电吸尘器
hot-gas flame 热气体火焰
hot-gas line 热气管(线)
hot-gas line for defrosting 热气融霜管路
hot gas main 热气总管
hot-gas multi-stage booster 热气体多级增压机
hot-gas outlet duct 热空气出口管
hot-gas plant 热气发生器
hot-gas pressure compaction 热气压成形
hot-gas purification 热气净化
hot-gas recycle process 热气循环过程;热气循环法
hot-gas reflow soldering 热气流焊接
hot-gas servo 热气伺服机构
hot-gas servosystem 热气伺服系统

hot-gas soldering 热气流焊接
hot-gas squeeze 热气压结
hot-gas stabilizing system 热气稳定系统
hot-gas system 热气装置
hot-gas welding 气焊;热气焊接;热空气焊接;热风焊接
hot glass wire 火线夹子
hot glass wire cutting 火线夹子切割(法)
hot glue 预热黏合剂
hot gluing 热胶合;热压黏结
hot grinding 热(研)磨
hot-groove rolling 热辊轧制
hot ground cement 热磨水泥
hot ground pulp 热法磨木浆
hot groundwater 地下热水
hot gunning 热喷涂
hot hardness 热硬性;热硬度
hot hardness property 热硬性质
hot-head engine 热头发动机;热头引擎
hot-head ignition 热头点火
hot-head press 热压头压烫机
hot heat 热炉次
hot heating 热水供暖
hot hobbing 热挤压制模法
hot hole 热孔;热空穴
hot-hole blasting 热鼓风炮眼爆破
hot house 温室;烘干室;浴室
hot-house effect 温室作用;温室效应
hot-house plant 温室植物
hot humid climate 湿热气候
hot hydraulic system 热液压系统
hot ice 高融冰点
hot idle 热空转
hot impact 热冲击
hot ingot peeling machine 热钢锭剥皮机床
hot injection mo(u)lding 热压铸成形
hot injection mo(u)lding machine 热压铸机
hot inspection 热检验
hot insulation mastic 绝热玛琋脂
hot intrusion 热侵入体
hot intrusive 热侵入
hot investment casting 精密铸造
hot iron 高温熨烫;铁水
hot ironing 热熨平;热压薄
hot iron mo(u)ld blowing 热模吹制
hot iron mo(u)ld ware 热模制品
hot isopressing 热等静压
hot isostatic apparatus 热等静压设备
hot isostatic bonding 热等静压(黏结)
hot-isostatic-bonding and pressing process 热等静压制法;等静压制法
hot isostatic compaction 热等压成形;热等静压
hot isostatic press(ing) 热等静压(机);高温等静压;等静压机
hot isostatic pressing treatment 热等静压处理
hot issue 热门股票
hot items 热门货(品)
hot jacket 热套
hot jet welding 热风焊接
hot job 紧急任务
hot joining 热接
hot joint 热接缝
hot junction 热结;热接点;热端
hot junction temperature 热接点温度
hot kerosine drying 热煤油干燥
hot key 热键
hot knife 热切刀
hot lab(oratory) 放射性实验室;原子核实验所;热实验室;强放射性物质实验室;强放射性物质(研究)实验室
hot laboratory cave 热室
hot lacquering 热喷漆
hot lahar 热泥流
hot-laid asphaltic concrete 热铺沥青混凝土
hot-laid bituminous concrete 热铺沥青混凝土
hot-laid bituminous macadam 热铺(沥青)碎石路
hot-laid coarse tar concrete 热铺粗粒焦油混凝土
hot-laid fine tar concrete 热铺细粒焦油混凝土
hot-laid method 热铺法
hot-laid mixture 热铺混合料;热拌混合料;热铺沥青
hot-laid pavement 热铺路面
hot-laid process of construction 热铺法施工
hot-laid rolled asphalt 热铺碾压沥青

hot-laid surfacing 热铺面层
hot landing 高速降落(飞机)
hot laying 热铺
hot level(l)ing machine 热平整机
hot light 主光;热光
hot lighting 主光照明
hotline 直通电话线路;热线
hot-line job 热线作业;带电作业;带电工作;带电操作
hot-line service 热线通信业务;热线服务
hot-line support 热线支持
hot-line tool 带电操作工具
hot-line work 带电操作
hot link 热链接
hot lip K-cell 热唇分子束炉
hot list 热表
hot listing 热销房产
hot loading test 高温强度试验
hot load test (耐火材料的)热压试验;加热剪切变形试验;热荷载试验;热负荷试验
hot load testing furnace 热负荷测试炉
hot logging 流水集材作业;流水式采运
hot loop 热回路
hot-maceration method 热浸法
hot machine 高温机器
hot machining 高温切削;热加工
hot mangle 板材热矫直机
hot mastication 热塑炼
hot material 热材料;强放射性材料
hot melt 易熔色料;热软化剂;热熔性;热熔体;热熔融;热熔料;热熔(化);热熔黏接剂
hot-melt adhesive 热熔(性)胶黏剂;热熔胶;热熔黏接剂;热熔胶合剂
hot-melt application 热熔施工
hot-melt applicator 热熔体涂装器
hot-melt coater 热熔体涂布机
hot-melt coating 热熔性涂布;热熔融涂装法;热浸涂可剥涂层
hot-melt extruder 热软化挤出装置;热熔挤出机
hot-melt gun 热熔性喷枪
hot-melt method 热熔黏结法
hot-melt painting 热熔涂装
hot-melt plastic stripe 热熔塑料条
hot-melt road marking paint 热熔型路标漆
hot-melt sealant 热熔密封膏
hot-melt strength 热熔强度
hot metal 液态金属;高熔金属;熔融金属
hot-metal bottle 铁水罐
hot-metal break-out 铁水穿漏
hot-metal car 铁水车
hot-metal charge 熔融金属加料
hot-metal composing machine 铸排机
hot-metal ladle 液态金属盛桶
hot-metal ladle and carriage 铁水罐车
hot metalliferous brine 含金属的热卤水
hot metallurgy laboratory 热冶金实验室
hot-metal pouring stank 兑铁水处
hot-metal receiver 熔(融)金属接受器
hot-metal sawing machine 热锯机;热金属锯机
hot mill 热轧厂;热轧成形;热轧设备
hot mill motor 热轧电动机
hot mineral spring 热矿泉
hot mirror 热反射镜
hot mix 热拌和面材料;热拌(和);热拌混和料
hot-mix asphalt 热拌沥青混合料;热沥青混合物
hot-mix asphalt base course 热拌沥青基层;热拌沥青底层
hot-mix asphalt lining 热拌沥青衬砌
hot-mix asphalt plant 沥青热拌厂
hot-mix asphalt surface course 热拌沥青面层
hot-mix bituminous pavement 热拌沥青铺面
hot-mix-cold-laid 热拌冷铺
hot-mix construction of roads 热拌混合料铺路;路面的热拌铺筑法
hot-mixed asphalt 热拌沥青混合料
hot-mixed asphalt(ic) concrete 热拌沥青混凝土
hot-mixed asphalt plant 热拌沥青骨料厂
hot-mixed bituminous concrete 热拌沥青混凝土
hot-mixed bituminous macadam 热拌沥青碎石
hot-mixed cold-lay type 热拌冷铺式(沥青混合料)
hot-mixed concrete 热拌混凝土
hot-mixed construction 热拌铺
hot-mixed construction of roads 路面的热拌铺筑法;热拌混合料铺路

hot-mixed macadam 热拌碎石
hot-mixed plant 热拌设备
hot-mixed recap method 热拌修复路面法
hot-mixed recycling 热拌再生利用
hot-mixed surface 热拌路面
hot-mixed topping 热拌面层
hot mixing 热拌
hot-mix(ing) method 热拌法;加热搅拌法
hot-mix plant 热拌设备;热拌厂
hot-mix plant mixture 工厂热拌材料;工厂热拌混合料
hot-mix recap method 热拌(混合料)修复路面法
hot-mix surface 热拌路面
hot mixture 热拌混合料;热拌和料
hot model test 热态模型试验
hot-moist environment 湿热环境
hot money 国际游资;游资
hot money volatile money 游资移动
hot motoring method 热机马达法
hot motor part detection 发动机发热部分探测
hot mo(u)ld 热模型;热模塑
hot mo(u)ld blowing 热模吹制
hot mo(u)lded blank 热压型毛坯
hot mo(u)lding 热榨;热塑;热模制;热模压
hot mo(u)lding mica 热塑云母
hot mud 热泥
hot mud flow 热泥流
hot-neck grease 热轴颈用润滑油;滚棒轴头润滑脂(热轧机)
hot needle-leaf forest 热性针叶林
hot nodule 热结节
hot number 热门货(品)
hot oil centrifugal pump 热油离心泵
hot oil distillation 热油蒸馏
hot oil duct 热油导管
hot oil(dy(e)ing) process 热油轧染
hot oil expression 热压油法
hot oiling 热油洗井
hot oil machine 热油机
hot oil pump 热油泵
hot oil quenching 热油淬火
hot operation 高温运行
hot oven slag 鼓风炉渣;高炉渣
hot override switch 热断开关
hot pack(ing) 热囊法;热包裹法
hot-pack method 热装罐法
hot-pack mill 叠板热轧机
hot padding 热压印
hot particle 强放射性粒子
hot pass 焊珠
hot patch 热补;热拌冷补的
hot patching 热法补坑;热补
hot patch maintenance 热补养护
hot patch outfit 热补设备
hot peening 高温喷砂处理;热喷丸;热打铁;热锤击
hot peening marquenching 热锻分级淬火
hot penetration 热贯;热灌
hot penetration bitumen 热灌沥青
hot penetration bituminous macadam 热灌沥青碎石路;热贯沥青碎石
hot penetration construction 热浸沥青施工;热灌沥青路面施工
hot penetration construction method 热浸施工法
hot penetration method 热浸渗法;热灌法;热贯法
hot permeability 高温透气性;热透气性
hot piercing 热穿孔
hot pin 热棒
hot piping 热管道输送
hotpit 热窖
hot plant 热拌沥青;热拌厂
hot-plant mixing 热的厂拌(沥青材料)
hot-plant mixing method 热的厂拌法
hot plastic anti-fouling paint 热塑性船底防污漆
hot plasticity 热塑性(塑性)
hot plastic paint 热涂塑性涂料
hot plate 燃气板;灶面板;炉台;厨房炉灶;菜肴保温板;加热板;热板;电热板;保温盘;铁板灶
hot plate apparatus 热板设备
hot plate bar 炉台条
hot plate drier 热板式干燥机
hot plate magnetic stirrer 热极磁扰动器
hot platen press 热压机;层积热压机
hot plate press 热板压呢机
hot plate spalling test 热板式崩裂性试验

hot plate straightening machine 钢板热矫正机
hot plate tempering 烘干回火;热钢板回火
hot plate veneer redrier 热单板重干机
hot plug 热型火花塞;热点火塞
hot plug engine 热塞柴油机
hot plugging 热插入
hot plume 热烟云
hot pneumatic type portable sprayer 热气式轻便弥雾机
hot point 摩擦发热点
hot polymer 热聚物
hot polymerization 热聚合(作用)
hot polymerization drying 热聚合干燥
hot pond 蒸木料槽;蒸木池
hot pool 热水塘;温泉池
hot potassium carbonate process 热碳酸钾过程
hot pourable 热态时可浇的
hot-poured 热灌的;热浇(的)
hot-poured compound 热浇混合料
hot-poured crack filler 热灌填(缝)料
hot-poured joint 热灌接口
hot-poured sealant 热浇注密封膏
hot-poured type 热灌式
hot pouring compound 热浇混合料
hot precipitator 高温除尘器
hot precoated sand 热覆膜砂
hot preparation 热制备
hot press 热压机;高温压块
hot-press approach 热压法
hot-press arrangement 热压装置
hot-press bed 热压机座
hot pressboard 热压板
hot-press cermet 热压金属陶瓷
hot-pressed alloy 热压合金
hot-pressed alloy powders 热压合金粉坯
hot-pressed aluminium nitride 热压氮化铝
hot-pressed and sintered product 热压烧结制品
hot-pressed brass 热压黄铜
hot-pressed bronze 热压青铜
hot-pressed bronze compact 热压青铜块坯
hot-pressed cadmium telluride ceramics 热压碲化镉陶瓷
hot-pressed ceramics 热压(陶)瓷
hot-pressed chippings 热嵌石屑法
hot-pressed compact 热压块坯
hot-pressed gold 热压金
hot-pressed iron compact 热压铁坯块
hot-pressed lanthanum fluoride ceramics 热压氟化镧陶瓷
hot-pressed mica 热压云母
hot-pressed oil 热榨油
hot-pressed parts 热压零件
hot-pressed process 热压过程;热压法
hot-pressed product 热压制品
hot-pressed silicon nitride 热压氮化硅
hot-pressed sintering 热压烧结
hot-pressed strontium fluoride ceramics 热压氟化锶陶瓷
hot-presser 热压机操作工
hot-press forge 热模锻;热锻压机
hot-press forging 热压锻造
hot-press gluing 热压胶合
hot-press(ing) 热压;热压成形
hot-pressing infrared element 热压红外元件
hot-pressing mo(u)ld 热压模
hot-pressing pressure 热压压力
hot-pressing temperature 热压温度
hot-pressing tool 热压工具
hot-press loading time 热压机装板时间
hot-press method 热压法
hot-press mo(u)lding 热压模制;热压模塑
hot-press printing 热压箔
hot-press ram 热压冲杆
hot-press resin 热压树脂
hot-pressure 热压
hot-pressure casting 热压铸成形
hot-pressure resin 热压树脂
hot-pressure welding 预热压力焊;热压(接)
hot-press welding 热压焊(接)
hot primary air 一次热空气
hot probe method 热探针法
hot procedure 热加工过程;热加工工序
hot process 加热过程;热铺法
hot process plywood press 热压胶板机

hot process purifier 加热净化器;热法净化器
hot process softening 热法软化
hot property 热性能
hot pulp flo(a)tation 热浆浮选
hot quenching 高温淬火;热淬(火);热脆
hot rack 轧制机台;热钢材冷却台架
hot rail 禁火标志
hot ray 热线
hot reaction 热反应
hot reactor 热态反应堆
hot recycle pump 热循环泵
hot reducing 热压缩
hot reduction 热轧
hot reeling machine 旋进式热轧机;热卷取机;热管材整径机
hot-refined pulp 热精制浆粕
hot refining 热法精制
hot reflux 热回流
hot reflux condenser 热回流管
hot reheat 高温级再热
hot repair 热修(补);热态检修;热补;热拌冷补的
hot repress(ing) 再热压
hot reserve 暖机预备
hot residue 热残渣
hot resistance 热电阻
hot ring 热环
hot riser 热冒口
hot-riveted 烧红铆钉铆固;热铆固
hot riveting 热铆
hot roll 热轧(辊)
hot-roll annealing 热轧(后)退火
hot-rolled 热辊压榨的;热轧的
hot-rolled asphalt 热铺碾压沥青
hot-rolled asphalt base layer 热碾压沥青基层
hot-rolled band 热轧带材;热轧带带
hot-rolled bar 热轧钢筋;热轧条材;热轧圆钢
hot-rolled bronze door 热轧铜门
hot-rolled channel steel 热轧槽钢
hot-rolled coil 热轧带卷
hot-rolled copper 热轧铜
hot-rolled deformed(steel)bar 热轧异形钢筋;热轧变形圆钢;热轧变形钢筋
hot-rolled drill 热滚麻花钻
hot-rolled finish 热轧表面加工;热轧光洁度
hot-rolled finished sheet 热轧成品薄板
hot-rolled flat product 热轧扁钢
hot-rolled joist 热轧钢龙骨
hot-rolled long product 热轧线材
hot-rolled mild steel 热轧软钢
hot-rolled narrow strip 热轧窄钢条
hot-rolled pipe 热轧管
hot-rolled plate 热轧板
hot-rolled reinforced bar 热轧钢筋
hot-rolled round steel 热轧圆钢
hot-rolled sand carpet 热压沥青砂毡层
hot-rolled section 热轧型钢
hot-rolled sheet metal 热轧金属板
hot-rolled sheet(steel) 热轧钢板
hot-rolled sillcon steel sheet 热轧硅钢板
hot-rolled steel 热轧钢
hot-rolled steel bar 热轧钢筋
hot-rolled steel joist 热轧钢龙骨
hot-rolled steel section 热轧型钢
hot-rolled strip 热轧钢条
hot-rolled structural section 热轧结构型钢
hot-rolled tube 热轧管
hot-rolled uncoated flat product 热轧无镀扁钢
hot-rolled window sash 热轧钢窗框
hot-rolled wire 热轧钢丝;热轧线材
hot rolling 加热发光;热辗压;热卷;热轧
hot-rolling arrangement 热轧装置
hot rolling mill 热轧机;热轧厂
hot-rolling roll 热轧用轧辊
hot-rolling wastewater 热轧钢废水;热轧(厂)废水
hot roll neck grease 热轧辊颈润滑脂
hot room 烘房;热室
hot rubber 热聚合橡胶
hot runner 热流道
hot-runner mo(u)ld 热流道模;热浇道模
hot-running 热滑
hot run table 热金属辊道
hots 热门货(品)
hot saline solubility test 热盐水溶解试验
hot salt quenching 盐浴淬火

hot sate 热固化;热定形
hot saw(ing) 热锯;热轧钢切削;热态锯切
hot scarfer 热轧件火焰清理机
hot scarfing 高温修切边缘;热轧件的火焰清理;烧剥
hot screed 热的抹平;热的刮平
hot screen 热料筛
hot screen separation 热料筛析
hot sealing compound 热封缝材料
hot sealing wax 热封蜡
hot-seal packing machine 热封包装机
hot-seal pressure vessel 热封式高压釜
hot season 热季
hot set 热固(化);热定形;热变定;热硬化
hot set hammer 热作锤
hot-setting adhesive 热凝黏合剂;热固(化)胶黏剂;热固性黏结剂;热固性胶黏剂
hot-setting glue 热凝结胶;热补用胶水;热固树脂
hot-setting phenol resin adhesive 热固性酚醛树脂黏结剂
hot-setting resin 热固性树脂
hot shake-out 高温打箱(落砂用);热打箱
hot shaped 热轧的;热成形的
hot shaping 加热成形;热轧;热态变形;热锻压
hot shearing 热剪(切)
hot shears 热剪(切)
hot shoe 闪光灯插座;触点式插座
hot shop 热(工)车间
hot short 快速货运;红脆的;热脆(性)的;熟练工(人);不耐热的
hot-short iron 热脆铁
hot shortness 红脆性;热脆性
hot shot 淬沥过度;(钢铁的)脆性状态;过热;快机;快船;快车;向热气流发射试验;高速车
hotshot wind tunnel 热射式风洞
hot shower 热水淋浴
hot shrinkage 热缩
hot shrinkage-tension 热缩拉力
hot side 高压边;高放射性侧;热面;热端;玻璃带热面
hot size 热模精压;热定径
hot-sizing press 热校正压力机
hot skidway 流水垫木楞场
hot smoking 热法熏制
hot soarfing 热剥
hot soil 热土
hot soluble extract 热水溶解的浸膏
hot solvent bath 热溶剂浴
hot solvent tank cleaning 热溶剂箱清洗
hotsonite 水羟磷铝矾
hot spa 温泉
hot space 高温区;热区
hot spark 强火花
hot spark plug 热型火花塞
hot spell 热浪;长热期
hot spinning 热旋压
hot spinning temperature chart 热旋压温度图
hot splicer 热接片机
hot spot 局部过热;过热点;过热部位;烈火点;亮斑;局部加热;加热点;摩擦发热点;红窑点;热节;热斑;热点;多灾区
hot spot application 热面预热
hot spot indicator 灼热点温度指定器
hot spot intake manifold 借热进气管
hot spot manifold 热点式进气歧管
hot spot model 热斑模型
hot spot of kiln shell 红窑
hot spot of plaster 粉刷面上起斑泡;粉刷面上起泡斑点
hot spot temperature 热点温度
hot spot temperature of transformer 变压器热点温度
hot spotting 热点;烈火点控制
hot spray 热喷涂
hot spray apparatus 热喷涂装置
hot spray coating 热喷涂涂料
hot spraying machine 热喷雾机;热喷淋机
hot spraying pistol 热喷雾枪
hot spray paint 热喷漆
hot spray process 热喷涂法
hot spray traffic paint 热喷型路标漆;热喷涂道路划线漆
hot spring 热泉;温室;温泉
hot spring anomaly of contact gone 接触带温泉异常

hot spring gas 热泉气
hot spring marsh 热泉沼泽
hot spring sanitarium 温泉疗养院
hot spring scenic spot 温泉风景区
hot spring sinter 热泉泉华
hot spring soil residual 温泉余土
hot spring vapo(u)r 热泉汽雾
hot spring water 温泉水
Hotspur 郝司波铬镍钢
hot stability of bitumen 沥青热稳定性
hot stack 热堆积
hot stage 发热期
hot stage electron microscope 高温电子显微镜
hot stage microscope 热台显微镜;高温载物台显微镜;温台显微镜
hot stage photomicrograph 高温显微照相
hot stall 热车熄火
hot stamping 热压箔;热冲压;烫印
hot stamping brass 热压黄铜合金
hot stamping foils 热冲压箔
hot stamping shop 热压车间
hot standby 双机热备;热备用
hot standing wave ratio 热驻波比
hot star 热星
hot start lamp 热启动灯;热启动灯
hot-start-up time 热态起动时间
hot stenter setting 热拉幅定形
hot stick 带电操作杆
hot stock treatment 热浆处理
hot storage (材料的)热储[贮]存;热藏;蓄热
hot storage bin 热储[贮]料斗
hot stove burner 热风炉烧嘴
hot straightening 热矫直
hot straining 热应变
hot straining embrittlement 热应变脆性
hot strength 高温强度;拉伸强度;热(态)强度
hot strength index 热强度指数
hot stretch 热拉伸
hot-stretched cord 热拉伸合成帘子线
hot stretching 热拉直法
hot-strip 热轧钢材;热轧带钢;热轧带材
hot-strip mill 带钢热轧机;扁钢热扎机;热轧带钢机;带材热轧机
hot-strip reels 热轧带钢卷取机
hot stuff 煤焦油晶碱;热沥青
hot stuffing 热加脂
hot stuff man 热料工
hot stylus 热记录笔
hot surface 高吸收表面;高碱性表面;碱性表面;热表面
hot surface ignition engine 热球(式)柴油机
hot surface treatment 热敷(路)面处理;热法表面处理;表面热处理
hot-swage 热旋锻
hot swaging 热型锻
hot swelling 热胀罐
hot tack 热黏性
hot tap 热水龙头
hot tapping 带气连接;铁板热焊
hot tar 热焦油
hot tar concrete 热拌焦油沥青混凝土
hot target 热目标
hot tarring 热沥青铺路;热浇柏油;热灌柏油
hot tear(ing) 形成热裂缝;热(撕)裂
hot tear crack 拉裂
hot tensile strength 热张力;热拉伸强度
hot tension test 高温拉力试验
hot test for soundness 沸水安装性试验
hot test(ing) 热(态)试验;高温试验
hottest spot 热斑
hottest-spot temperature allowance 最热点温度容差
hottest water 吸热的水
hot-tinned strip 热镀锡薄钢板
hot tinning 热镀锡
hot-tinting 氧化膜色;火色;回火色
hot top 保温台;热炉顶;热顶;低温帽;保温帽
hot topped ingot 帽钢锭
hot-topped mo(u)ld 带保温帽钢锭模
hot-transfer operation 热转移操作
hot trepanning 中空凸模冲孔
hot trephining 中孔凸模冲孔
hot trichloroacetic acid 热三氯乙酸
hot trimming 高温清整(铸件用);热切毛边;热精整

hot trimming die 热修边模
hot tube expanding machine 热扩管机
hot tunnel climatic test 热风洞试验
hot-type plug 热型火花塞
hot vacuum 热真空
hot vacuum treatment 高温真空处理
hot vapo(u)rizer 热蒸发器
hot-vapo(u)r line 热蒸汽线
hot varnish 热漆
hot venetian type ship bottom paint 热喷涂溶解型船底漆
hot vinyl 热塑性乙烯树脂
hot vulcanization 热硫化
hot vulcanizing 热补
hot wall 热爆墙;火墙;暖墙(设采暖烟道的墙壁);热损墙面
hot wall reactor 热壁反应器
hot-wall tube furnace 热壁管式炉
hot wash 热水洗涤
hot waste 热废物;强放射性废物
hot wastewater 高温废水
hot-water 热水
hot-water accumulator 热水聚集器
hot-water apparatus 热水设备;热水器;家用热水器;热水装置
hot-water bag 热水袋
hot-water basin 热水池
hot-water boiler 热水锅炉
hot-water bottle 热水袋;热水瓶
hot-water branch 热水支管
hot-water calorifier 热水供暖器
hot-water central heating 集中热水供暖
hot-water channel 热水通道
hot-water check valve 热水止回阀
hot-water circulating 热水循环
hot-water circulating flow 热水循环流
hot-water circulating pump 热水循环泵
hot-water circulating rate 热水循环流量
hot-water circulation system 热水循环系统
hot-water circulator 热水循环器
hot-water consumption 耗热水量
hot-water convector 热水对流器
hot-water corrosion 热水溶蚀
hot-water cost 热水成本
hot-water cups 热水注
hot-water cure 热水养护(法)
hot-water curing 热水养护
hot-water curtain 热水幕
hot-water cylinder 热水筒;热水罐
hot-water deposit 热水沉积
hot-water distribution basin 池式配水系统
hot-water drying tank 热水干燥槽
hot-water extraction pump 热水抽吸泵
hot-waterfall 热水瀑布
hot-water filter 保温滤器
hot-water flashing system 热水扩容系统
hot-water flo(a)tation 热水浮选
hot-water funnel 热水漏斗
hot-water generator 热水器;热水炉;温水发生器
hot-water gravity system 热水重力循环系统
hot-water heat 热水热量
hot-water heater 热水锅炉;热水加热器;热水加热炉;热水采暖装置
hot-water heating 热水(式)供暖;热水取暖;热水加热;热水采暖
hot-water heating design 热水供暖设计
hot-water heating equipment 热水供暖设备
hot-water heating installation 热水供暖装置
hot-water heating load 热水供暖负荷
hot-water heating network 热水供热网
hot-water heating pipe 热水供暖管
hot-water heat-supply network 热水热网
hot-water heating system 热水加热装置;热水供暖系统;热水采暖系统
hot-water heating with forced circulation 压力循环式热水供暖
hot-water injector 热水喷射器
hot-water inlet 热水口
hot-water inlet valve 温水入口阀
hot-water line 热水管线
hot-water main 热水总管
hot-water network 热水系统;热水(管)网
hot-water open circuit diving suit 开式回路热水潜水服

hot-water permeation method 热汤渗透法
hot-water pipe 热水管
hot-water(pipe)line 热水管线
hot-water preparer 热水器;热水炉
hot-water pressure heating 高压热水供暖;压力热水供暖
hot-water pump cylinder cover 热水泵缸盖
hot-water pump cylinder liner 热水泵缸衬
hot-water pump piston 热水泵活塞
hot-water pump piston follower 热水泵活塞填密盖
hot-water radiation 水暖散热器
hot-water radiator 热水散热器
hot-water resistance test 耐热水测定
hot-water return 热水回水;热水回路
hot-water return pipe 热水回水管
hot-water rinse 热水清洗
hot-water rinsing bath 热水淋浴;热水淋洗
hot-water service 热水供应;热水供给
hot-water service room 开水间;热水供应室
hot-water steam radiator 热水蒸汽散热片;暖气片
hot-water storage plant 热水储备设备
hot-water storage system 热水储存系统
hot-water storage tank 热水(储水)箱;热水储[贮]槽
hot-water storgae 热水槽
hot-water stove 热水(加热)炉;热水加热器;水暖器
hot-water suction pipe 热水吸管
hot-water supply 热水供应
hot-water supply line 热水供给管线
hot-water supply load 热水供应负荷
hot-water supply main 热水供给干线;热水干管
hot-water supply room 开水间;热水供应室
hot-water supply station 热水供应站
hot-water supply system 热水供应系统
hot-water surface 热水表面
hot-water system 热水系统
hot-water tank 热水箱;热水柜;热水池;热水槽
hot-water test 热水试验
hot-water treatment 热水处理(法)
hot-water tube 热水管
hot-water type 热水式
hot-water upper cover 热水筒上盖
hot-water vase 温壶
hot-water vessel 热水箱;热水容器
hot-water washing 热水洗
hot-water wash scheme 热水冲洗方案
hot-water well 热水井
hot-water wet suit 湿式水加热潜水服
hot wave 热浪
hot wax ink 耐热蜡油墨
hot-weather concreting 酷热气候混凝土施工;夏季浇注混凝土;炎热气候下混凝土浇注;热天浇筑混凝土;热天浇注混凝土;热天浇灌混凝土;暑季混凝土施工
hot-weather construction 热天施工;夏季施工
hot-weather trial 热带气候试验
hot weld encapsulation 热焊封袋
hot welding 热焊(接)
hot well 高温井;凝结水井;热(水)井;温井;天然温井
hot well cement 高温油井水泥
hot-well depression 凝结水过冷(却)
hot well pump 凝结水泵;热水池泵;热井泵
hot well(tank) 热水箱
hot wetting 热湿润
hot wind 热风
hot wire 热丝;热(电阻)线;热电阻丝;带电电线;波拉尼真空计
hot-wire ammeter 热线式电流计;热线式电流表;热线安培计
hot-wire analyzer 热电阻丝型分析仪
hot-wire anemometer 热线风速计;热敏电阻风速计;热线风速表;热丝风速计
hot-wire bridge 热线式电桥
hot-wire cell 热丝池
hot-wire coil 热线圈
hot-wire current meter 热线流速仪
hot-wire cutting 热金属丝切割法
hot-wire detector 热丝检测器;热线式检波器
hot-wire flowmeter 热线流量计
hot-wire fuse lighter 热线点火器
hot-wire galvanometer 热线式电流计
hot-wire ignition 热线点火;电阻丝点火
hot-wire instrument 热线仪表;热线式仪表;膨胀式仪表
hot-wire ionization ga(u)ge 热丝电离规
hot-wire lighter 热线点火棒
hot-wire manometer 热线压力表;热线压力计
hot-wire measurement 热线测量
hot-wire meter 热线式仪表
hot-wire method 热线法
hot-wire microphone 热线式传声;热线传声器
hot-wire noise generator 热线噪声发生器
hot-wire oscillograph 热线(式)示波器
hot-wire pressure ga(u)ge 热丝压强计;热线压力计
hot-wire pressure meter 热线压力计
hot-wire probe 热线测针
hot-wire relay 热线式继电器
hot-wire resistance seismometer 热线电阻地震计
hot-wire saw 电热线锯
hot-wire scalloping 热熔分离制牙边法
hot-wire technique 热线测试技术
hot-wire telephone 热线电话
hot-wire type anemometer 热线式风速仪;热线式风速表
hot-wire vacuum switch 热线式真空开关
hot-wire wattmeter 热线式瓦特计
hot-wire welding 热丝焊
hot work 热作(业);玻璃制品的热成形;热加工
hot work ability 热加工性
hot worked 热作的;热加工的
hot working 热加工;热成形;热作
hot-working character 热加工性能
hot-working die steel 热锻模钢
hot-working face 高温工作面
hot-working treatment 热加工处理
hot work mo(u)ld steel 热作业用模具钢
hot workshop 热工车间
hot work tool steel 热锻模具钢;热镦模具钢
hot zirconia 热氧化锆
hot zone 热区;热带
hot zone of shaft kiln 立窑高温带
hot zone ring 熟料圈
hound 桅肩
hound band 桅肩箍
hounding 桅肩下
hound of a mast 桅肩
hound rod 驱动杆
hourage 计时数
hour angle 相位角;子午圈角;时角
hour angle-declination axis mount 时角赤纬轴座架
hour angle difference 子午圈角差;子午角差;时角差
hour angle hyperbola 时角方位图
hour angle of apparent sun 视太阳时角
hour angle of heavenly body 天体时角
hour angle system 时角系统
hour angle system of coordinates 第一赤道坐标系;时角坐标系
hour circle 子午线;时圈;时刻度环;时标盘;赤经度盘
hour cost 小时成本
hour counter additional intermediate wheel 时计数附加中间轮
hour counter cam 时计数凸轮
hour counter column wheel 时计数柱轮
hour counter column wheel jumper 时计数柱轮跳杆
hour counter coupling stud 时计数器连接桩
hour counter driving pinion 时计数器传动齿轴
hour counter driving wheel 时计数器传动轮
hour counter hand 时针
hour counter indicator 时计数指示器
hour counter intermediate wheel 时计数器中间轮
hour counter jumper 时计数器跳杆
hour counter jumper eccentric 时计数器跳杆偏心轴
hour counter jumper seating 时计数器跳杆垫
hour counter lock 时计数器锁杆
hour counter lock spring 时计数器锁杆簧
hour counter operating lever 时计数器活动杆
hour counter operating lever hook 时计数器活动杆构压片
hour counter operating lever spring 时计数器活动杆簧
hour counter setting wheel 时计数器过轮
hour counter spindle 时计数器轴
hour-counting wheel 时计数轮
hour-counting wheel bridge 时计数轮夹板
hour-counting wheel friction spring 时计数轮摩擦簧
hour-counting wheel stud 时计数轮桩
hour-counting wheel tube 时计数轮套
hour curve 子午线
hourdis 大型建筑砖
hourdis stone 肋间砌块
hour-glass 沙漏
hour-glass contraction 葫芦状收缩
hour-glass fracture 沙漏状断口
hour-glass piston 沙漏形活塞
hour-glass reflector 沙漏型反射器
hour-glass texture(or structure) 沙钟结构(或构造)
hour-glass worm 欣德利球状蜗杆;球面蜗杆
hour-glass worm wheel 球面蜗轮
hour hammer 时锤
hour hammer cover 时锤杆压片
hour hammer intermediate lever 时锤中心杆
hour hammer operating lever eccentric 时锤操纵杆偏心销
hour hammer operating lever spring 时锤操纵杆簧
hour hammer spring 时锤杆簧
hour hammer stud 时锤杆桩
hour hand 时针
hour heart 时桃轮
hour heart spring 时桃轮簧
hour-index 时标
hour indicator 时盘
hour indicator driving wheel 时指示器传动轮
hour indicator driving wheel cover 时指示器传动轮盖
hour indicator lock 时指示器锁杆
hour indicator maintaining plate 时指示器托板
hour indicator spring driver 时指示器传动簧
hour indicator support 时指示器座
hour indicator support cover 时指示器座压片
hour indicator support seating 时指示器座垫
hourless 无时间限制的
hour locking lever 时锁杆
hourly 每小时的
hourly amount 逐时量
hourly average 小时平均;小时均值
hourly capacity 小时市场能力;小时(生产)能力;每小时生产率;时用水量;时交通量
hourly consumption 小时用水量;每小时用水量;每小时消耗量;每小时耗用量
hourly cooling load 逐时冷负荷
hourly earnings 小时工资收入;每小时工资
hourly earnings average 平均小时收益
hourly efficiency 小时生产率
hourly fluctuation 每小时变化;时变化
hourly growth rate 小时增值率;时增值率
hourly heights 逐时潮高;每小时潮高
hourly load 逐时负荷;时负荷
hourly loss rate 小时损失率;每小时损失率;时损率
hourly machine rate 机器每小时生产率
hourly maximum water consumption 小时最大用水量;每小时最大用水量
hourly mean value 时均值
hourly motion 每时运动
hourly observation 逐时观测;记录观测
hourly output 小时制功率;每小时产量;时出力
hourly pay 小时工资
hourly period worker 靠工资维生的工人;计时工人
hourly precipitation 逐时降水量;每小时降水量
hourly premium system 小时奖金制
hourly production 小时生产率
hourly production capacity 小时市场能力
hourly rainfall(depth) 小时降雨量;每小时降雨量;每小时降水量
hourly rate 小时费率
hourly rate of discharge 时流量
hourly rate wage plan 小时固定工资率计划
hourly sol-air temperature 逐时综合温度
hourly tonnage 小时产量
hourly tractive effort 小时制牵引力
hourly traffic pattern 交通量时变(化)图
hourly traffic volume 小时交通量;每小时交通量

hourly value 小时值;一小时值
hourly variation 小时变化(量);每小时变化
hourly variation coefficient 逐时变化系数;每小时变化系数;时变化系数
hourly variation factor 时变化因数
hourly volume variation 交通量时变(化)图
hourly wages 小时工资;计日工资
hourly wages rate 小时工资率;计时工资率
hourly water consumption 每小时用水量;时用水量
hourly yield 逐时量
hour mark 时号
hour mechanism support banking ring 时机心托限制环
hour meter 小时表;计时仪;计时表(指仪表);小时计
hour minute 时分
hour norm 工时定额
hour of labo(u)r 工时
hour of mean solar time 平太阳时角
hour of work and output 工作时数与产出
hour-out line 时间曲线
hour plate 时钟字盘
hour rack 时齿条
hour rate 时差
hour rate wage 计时工资
hour rating 小时功率
hour recording 时记录
hours fished 捕鱼作业时数
hour snail 时蜗形凸轮
hours of daylight 日照小时;日照时数
hours of digging 控坑工时
hours of labo(u)r 劳动时间
hours of service 业务工作时间
hours of sunshine 日照小时;日照时数
hours of use 使用时间
hours on stream 连续工作时数
hours per hour 按工作时数付给计时的工资
hours run (延续)工作时间;操作小时
hour star 时星形轮
hour star jumper 时星形轮跳杆
hour star jumper eccentric 时星形轮跳杆偏心销
hour star ring 时星轮衬圈
hour star support 时星轮座
hours to stream 操作时间
hours underway 航行时间
hour target 小时定额
hour totalizer 累计计时器
hour variation coefficient 小时变化系数
hour wages system 计时工资制
hour wheel 时(针)轮
hour wheel cutter 时轮刀具
hour wheel driving wheel 时轮传动轮
hour wheel friction spring 时轮摩擦簧
hour wheel module 时轮组件
hour wheel seating 时轮垫
hour wheel spring-clip 时轮簧夹
hour wheel stud 时轮桩
hour wheel washer 时轮垫圈
hour zone 时区
house 房子;住宅;宫;屋;收藏;舍;房屋
house account 房屋账户
house agent 房屋经理者;房地产经纪人;房屋经纪人
house air bill 空运货物单据;空运分提单
house air waybill 空运货物单据;空运分提单;分提单
house allowance 住房津贴
house alteration 住房改建;房屋改建
house alternator 厂用发电机
house and building depreciation expenses 房屋建筑物折旧费
house automatic refuse compactor 家用自动垃圾压实机
house automation 住宅自动化
house bill 公司内部账单;内部汇票
house bill of laden 运输代理人提单
house bill of lading 分提单
house board 一般照明配电盘(剧院)
house boat 营房船;住宿船;水上住家;水上住房;可供住家的船;有铺位的游艇
house bote 修屋下料;修屋木料
house branch 户线管;入户管线
house-brand gasoline 普通品和汽油;汽油正规品种
housebreaker 拆屋专门设备;承包拆屋者
housebreaking 拆房;房屋拆除

house builder 施工人员;房建造者;住房营造业者;房屋承建商;住宅建筑商;住宅建造商
house building 住宅建筑;住宅建设
housebuilding industry 住房建筑工业
housebuilding program(me) 住宅建设计划;住宅建设规划
housebuilding sector expenses 房建部门支出
house cable 用户电缆;室内电缆
house-car 箱式车;箱车;冷藏车
house cistern 家用储[贮]水器;家用蓄水池
house clock 房屋上的时钟
house coaming 甲板舱室围板
house complex 住宅群
house connecting box 用户分线箱
house connection 用户接管;住宅污水管;住户连接线;户内连接管(水、气、污水管等);室内管系;房屋(连接)管;家庭污水管道
house connection passage 房屋连接走廊
house connection pipe 住宅连接管
house construction 房屋建造;房屋建筑
house correction 厂内校对
house curtain 剧院大幕;大幕
housed 封装的
house dead end line 用户管
house decorator 室内装潢设计者
house designer 室内装潢设计者
housed-in lifeboat 遮蔽式救生艇
housed joint 嵌槽接头;藏式接头;封装接头;镶嵌接头;嵌入连接
housed lot 畜棚
house document 内部文件
house door 宅门
house drain 住宅排水管;住宅排水沟;排水管户线;生活污水排水管;生活污水管(道);房屋排水管;房屋排水沟;房屋排水道
house drainage 房屋下排水;室内排水;房屋排水装置
house drainage pump 泵房排水
house drainage system 房屋排水系统
house drain pipe 住宅排水管
house(d) stair(case) 封闭式楼梯
housed string 封闭式楼梯斜梁;嵌入式楼梯斜梁;暗楼梯基
housed stringer 嵌入楼梯斜梁
house duty 房捐;房产税
house dweller 房屋居民
house exchange system 室内(电话)交换系统
housefall method 单船吊杆传送法
house fall system 联动装卸方式;联动装卸法
house famine 房荒
house filter 家用滤器;家用滤池
house fire 房屋火灾
house fixture 家庭卫生器具
house flag 公司旗;船首旗
house for rent 出租房屋;房屋出租
house for sale 出售房屋
house for the aged 老人住宅
house foundation 房屋基础
housefront 房屋(的)正面
house furnishings 家具
house garden 庭园
house gas piping 住宅煤气管道
house generating set 厂用发电机组
house generator 自备发电机;厂用发电机
house heating 住户供热;住房供暖;房屋供暖
household 住户;家用;户
household appliance 家用电器
household check 户口查对
household chemical 家用化学品
household china 家用陶瓷;日用瓷
household cinder 生活炉灰
household cleaner 家庭清洁剂
household cleaning equipment room 家庭清洁工具室
household cleanser 去污粉
household clinker 生活炉灰
household consumption 家庭消费;住户消费
household demand 生活用电需量
household detergent 家用洗涤剂;家庭洗涤剂
household disposal system 民用垃圾处理系统;生活垃圾处置系统
household double-tube solar energy water heater 家用双筒式太阳能热水器
household effects 家用物品

household electric(al) appliance 家用电器
household electric(al) appliance pollution 家用电器污染
household equipment 家庭设备
householder 住户;户主
householder's comprehensive policy 房主综合保险单
household filter 家用(过)滤器;家用滤池
household finance corporation 家庭金融公司
household fuel 家用燃料;民用燃料
household fuel gas 民用煤气;民用燃料气
household garbage 生活垃圾;家庭垃圾
household garbage grinder 民用垃圾粉碎机
household geyser 家用热水器
household god 灶神;家神
household habitation-shortage 居住困难户
household hatchet 家用斧头
household hot water 家用热水
household industry 家庭工业
household mortgages 家庭抵押债务
household necessities 日用必需品
household oil 家庭用润滑油
household paint 家用涂料;家用漆
household pesticide 家庭用农药;卫生用农药
household porcelain 日用瓷
household refrigerator 家庭冰箱
household refuse 家庭垃圾;生活垃圾
household refuse disposal 生活垃圾处理
household refuse processing plant 家庭垃圾处理厂
household scissors 家用剪刀
household scourer 家庭用的洗擦物
household separation 分户
household sewage 生活污水;家庭污水
household sideline production 家庭副业
household spray 家用杀虫液剂;卫生用杀虫液剂
household store room 家用储[贮]藏室
household stuff 家庭家具陈设
household synthetic detergent 民用合成洗涤剂
household to be relocated 拆迁户
household trip 家庭出行
household unit 家庭单元
household vaccum packer 家用真空包装机
household vice 家用台钳
household wares 家用器具;家用品
household waste 家庭垃圾;家庭洗涤废水
household wastewater 家庭废水;生活污水
household wastewater irrigation 生活污水灌溉
household water 家庭用水
household water filter 家用用水过滤器;家用热水;民用滤水器
household water meter 家用水表
household water pipe 家用水管
household water piping system 家用输水系统
household with inconvenient space distribution 居住不方便户
house hunter 寻者;寻户
house improvement 住房改善
house inlet 进水管
house insurance 房地产保险
house-interview survey 登门访问调查
house joint 套接;承插连接
house journal 室内装潢杂志;室内纸张工具箱;家庭杂志
housekeeper 房屋管理人
housekeeping 整理工作(房内);内务工作;内务处理;内务操作
housekeeping depot 房产建筑段
housekeeping information 内务辅助信息
housekeeping instruction 辅助指令;管理指令;内务指令
housekeeping operation 整理操作;内务操作
housekeeping package 家庭整洁(工作);整理组装
housekeeping program(me) 内务程序;辅助程序
housekeeping routine 内务例行程序;内务处理程序;内务操作程序;辅助程序
housekeeping software 整理软件;内务软件
house lateral 用户管
house lead-in 进户线
house lease 房契
house let 小屋;小房子
houselight 观众席灯光(剧场);观众席照明;观众大厅照明
house line 室内线;三股油麻绳

house machine 厂用机组
house magazine 室内装潢杂志;室内纸张工具箱;家庭杂志
housemaid's sink 民用洗涤槽;女佣用的水槽;落地污水盆;洗拖把池
house main 干线(配电)
houseman 控制室操作人员
house microclimate 居室微小气候;室内小气候
house moss 室内角落尘埃;室内尘埃
house mouse 家鼠
house number 门牌号
house of cards 不牢靠的计划;不切合实际无法实现的计划;筹划不周难以实现的计划
house of correction 教养院;教养所
house of culture 文化宫
house of detention 拘留所
house of God 教堂;寺庙
house of mercy 慈善堂
house of Pansa at Pompeil 庞贝的潘萨住宅
house of prayer 教堂
house of refugee 养育院;难民收容所
House of the Future 未来展览馆(1929年雅各布森与拉尔森所设计的展览馆,呈圆形,屋顶可升降直升机)
house of worship 街区内做礼拜的地方
house organ 公司内部刊物;内部刊物
house orientation 房屋方位;房屋朝向
houseowner's comprehensive policy 房主综合保险单
house paint 建筑用漆;民用漆;房屋漆;民用房屋漆
house painter 房屋油漆工
house paper 内部汇票
house phone 内线电话
house pipe system 户内管网
house place 农场(宿舍)起居室
house plant 室内植物;室内装饰植物
house plumbing 住宅卫生设备;房屋卫生设备安装;房屋卫生设备安装;房屋管道安装
house plunder 家庭用品
house-poor homeowner 因购房而致贫的业主
house preservative 房屋防腐材料;房屋防腐剂
house property 房产
house purchase price 住房零售价格
house raising 邻里助建房
house reference 商业备咨
house refuse 生活垃圾
house refuse disposal 生活垃圾处理(法)
house remodel(l)ing 住房改建
house rent 房租
house rent insurance 房租保险
house riser 室内立管;主管(指供气供水管)
house room 卧室
house scow 具有舱面室的平底驳
house service 室内配线
house-service circuit 厂用电路
house-service consumption 厂用动力损耗;厂用电损耗
house service cutout 家用断流器
house-service equipment 厂用电设备
house-service gas pipe 厂区管道
house-service meter 家庭用仪表;家用水表;家用电表;普通用户电度表
house-service network 厂用电力网
house service pump 厂用水泵
house-service turbo-generator 厂用汽轮发电机组
house-service wire 进户线
house sewage 生活污水;用户污水;家庭污水
house sewage system 户内排水系统
house sewer 用户污水管;住宅污水管;家庭污水管;家庭污水(管)系统;污水管进户线
house sewerage 房屋的排水系统
house sewer system 房屋下水系统
house silhouette 房屋轮廓
house sitter 看房人
house slant 室外下水三通接头
housesmith 建房铁工
house stuff 家具和陈设品
house subdrain 住宅地下排水管;房屋地下排水管
house substation 专用变电所;厂用变电所
house substructure 房屋基础
house supply 厂用动力;厂用电源;厂用电(力)
house tank 房屋水箱
house tax 房(屋)税;房捐;房产税
house telephone 内部电话

house telephone system 室内电话系统;内部电话交换系统
house the anchor 收锚进孔
house to be let 房屋招租
house-to-curb piping 户外管道安装
house-to-house 用户到用户;门至门;门到门
house-to-house service 门对门运输
house-to-house transportation 门对门运输
house-to-let 出租房屋;招租
house top 房顶;屋顶
house to pier 门至港;门到码头
house track 站线【铁】;仓库轨道
house trailer 供居住的拖(挂)车
house trailer for construction sites 施工现场用的居住拖(挂)车
house transformer 所用变压器;厂用变压器
house trap 住宅下水出口存水弯;住宅存水弯;室内排水管存水弯
house turbine 自备涡轮机;厂用涡轮机
house urban sweeping 城市住宅的垃圾
house wall 房屋墙壁
house ware 家庭用品
house wastewater 生活废水;住宅废水
house waybill 空运货物单据;空运分提单
housewife 水手袋
house-wife program(me) 内务程序
house wiring 房屋(电)线路;室内配电线;室内布线
house with court(yard) 庭院住宅
house wrecker 拆屋工人;拆屋承包人
house wright 建房工人
housing 住宅建设;遮蔽物;遮蔽物;壳(体);架;机器外壳;锚进入锚床的状态;屏蔽套;桅杆的甲板下部分;外套;外壳;套罩;房屋;房地产;凹部
housing accommodation benefit 房屋福利
housing act 住房法规
housing adjusting screw 压下调整螺钉
housing administration 住房管理
housing affordability 购房能力指数
housing allocation 住房分配
housing allowance 住房津贴;房屋租津贴
housing anchor 无杆锚
housing and environment(al) design 住宅和环境设计
housing and home finance agency 房地产、住房贷款机构
housing and land 房地产
housing and land tax 房地产税
housing area 小区;住宅区;住宅区域;居住区
housing assistance payment 住房补助付款
housing assistance plan 住房援助计划;住房补助计划;住房资助计划
housing association 住房协会
housing authority 住房主管部门;住房管理局;住房管理部门
housing bearing seat 箱体轴承座
housing betterment 住房改善;房屋修缮与扩建;房屋修缮与改良;房屋修缮与改进
housing bill 住房议案
housing bonds 住房债券
housing box 轴箱;壳箱
housing bushing 箱体轴套
housing cap 机架盖
housing case 机壳
housing census 住房普查
housing center 居住中心
housing chain stopper 航行时固定锚的制动链
housing cluster 住宅组团
housing code 住房建筑规范;住房法规;建房法规
housing colony 居住区;居民新村;住宅小区;聚居地
housing commercialization 住房商品化
housing completions 住房竣工量
housing complex 住宅群;住宅团地
housing condition 住房状况;居住条件
housing construction 宿舍结构;住宅建设
housing consultant 住房顾问
housing cooperative 住房合作公司;住房合作社
housing corporation bond 住宅建设公司债券
housing cost 保管费用
housing demand 住房需求
housing density 住宅密度;住房(稠)密度;房屋密度
housing depreciation 住房折旧
housing development 住宅建设;开辟住宅区;住房开发

housing development corporation 住房开发公司
housing development plan 住宅建设计划
housing dispute 房屋纠纷
housing district 居民点
housing economics 住宅经济学
housing estate 住宅用地;住宅群;住宅区;居住小区;居住点;居民点
housing estate development 新村开发;新村建设;居住区建设
housing estate planning 小区规划
housing estate road 居民点道路;住宅区道路;居住区道路
housing estate unit 居住区单元住房
housing expenses 住房开支;住房费用
housing feet 机架的水平支脚
housing filler elbow 箱体加油口弯管
housing finance 住房金融;住房集资;住房费用
housing finance agency 住房筹资机构
housing financing 住房财务
housing financing supply 住宅资金供应
housing for blind and deaf people 聋哑人住房
housing funds 住房资金;住宅建设基金;住宅公债
housing grinder 柱式磨床;有外壳的研磨机;有外壳的磨床
housing group 住宅组团;住宅群
housing improvement 住房改善;住宅改善
housing improvement and betterment 房屋修缮与扩建;房屋修缮与改良;房屋修缮与改进
housing improvement program(me) 住房改善计划
housing industry 房屋建筑业
housing inventory 现有住房量
housing investment 住宅投资
housing investment guarantees 房屋投资保证
housing joint 隐蔽接头;嵌槽接头;藏式接头
housing land 住宅建设用地
housing law 住房法规;居住法(律)
housing layout 住宅布局
housing legislation 住房立法
housing line 天幕下斜的压绳
housing loan 建房贷款
housing management 住房管理
housing management bureau 房管局
housing management office 住房管理所
housing market 住宅市场;住房市场;房屋买卖市场
housing market analysis 住房市场分析
housing mast 伸缩桅
housing material 建筑住房用材料
housing mobility 住址变动
housing mortgage 住房抵押
housing need 住房需要;住房需求
housing note 入仓通知
housing obsolescence 住宅老化
housing of a grouped site 住宅组团
housing of a mast 甲板下的桅杆
housing of eyepiece 目镜箱
housing of family 家属宿舍
housing of pump 泵体
housing of the closed top 闭口式机架
housing of the open top 开口式机架
housing ordinance 住房法规
housing output 住房建设(指建设过程);住房产出量;住房建造量
housing permit 建房批准书
housing pin 压紧螺旋;压紧螺钉
housing pipe 套管
housing plan 住房(建造)计划
housing planning 住房规划
housing plate 箱体端盖
housing plug 箱体塞子
housing policy 住房政策
housing preference 住房选择
housing problem 住宅问题;住房问题;房屋问题
housing production 住房建造量
housing program(me) 住房计划;住房规划;计划住房;住房建设规划
housing project 住宅规划;住宅工程;住房(建筑)计划;住房(建设)工程
housing property taxes 房产税
housing quality 住房质量
housing quality standard 住房质量标准
housing questionnaire 住房调查表
housing rehabilitation grant 住房复兴拨款
housing rehabilitation loan 住房复兴贷款

housing renewal 住宅更新
housing requirement 住房要求
housing scheme 住房方案;住宅兴建方案;住宅建筑方案
housing screw 压下螺钉;压紧螺旋;压紧螺丝;压紧螺钉
housing seat 箱体座
housing shortage 住宿短缺;住房短缺;房荒
housing shortage ratio 房荒率
housing site 住房基地
housing situation 住房状况;居住状况
housing sleeve 箱体套筒
housing society 房屋建筑协会;住房建筑协会
housing space 住房空间
housing standard 住宅建筑标准;住房(建筑)标准;居住标准;房屋建筑标准
housing standardization 住房标准化
housing starts 住房开工量
housing statistics 住房统计
housing stock 现有住房量;住房总量
housing stopper 航行时固定锚的制动链
housing subsidy 住房补贴
housing supply 住房供应量
housing survey 住房调查
housing system 住房体系;房舍设备
housing tenure 住房产权;住房占用权
housing topmast 桩栓上桅
housing tower 住宅塔楼;塔式住房
housing tradition 住房传统
housing transition 周转房
housing type 住房类型
housing undertaking 住房建设实施
housing unit 住房单元;住宅单元;住房单位;房屋构件
housing voucher 住房补贴凭证
housing washer 座圈;外圈
Housner's spectral intensity 豪斯纳谱烈度
houttuynium syntheticum 癸酰乙醛亚硫酸氢钠
hoved 陡岬(丹麦)
hove down 使船向一侧倾斜(小船修理时)
hove in sight 逐渐靠拢到看得见
hovel 壁龛;小屋;窑的圆锥形外壳;遮蔽物;杂物间;陋宅;陋屋;茅舍;棚子
hovel kiln 锥罩式窑;瓶子窑
hoveller 无执照海员
hovelling 烟囱砌筑成茅舍状(四周开窗);茅舍状烟囱
hovel shelter 临时暖房
hover 悬浮;停悬;覆盖
hoverbarge 气垫驳船
hovercar 气垫车
hovercraft 气垫式运输机;悬浮运载工具;气垫艇;气垫飞行器;气垫船;气垫车;腾空艇
hovercraft seakeeping capability 气垫船海上行驶能力
hover dredge(r)气垫式挖泥船
hoverferry 气垫轮渡;气垫渡船
hovergem 民用气垫船
hover ground 松散地面;松散地基(基);松软地基;不坚硬土壤
hover height 气垫举升高度
hovering 悬停
hovering act 禁止船只在领海内逗留法;海岸法规
hovering endurance 悬停持续时间
hovering phase 悬停阶段
hovering pressure 滑动压力
hovering turn 悬停转弯
hovering vessel 偷入领海的走私船只
hoverliner 巨型核动力气垫船
hovermarine 气垫船
hoverpad 气垫底板
hover pallet 气垫气行器;气垫飞行器;气垫起重移位器
hoverplane 直升(飞)机
hover platform 气垫平台
hoverport 气垫船码头;气垫船港口
hover pulley 气垫滑轮
hovership 气垫船
hovershow 气垫飞行器展览
hover skirts 气垫围裙;气垫围板
hovertrain 悬浮式列车;气垫列车;气垫火车
hovertruck 气垫载重车
hove taut 拉紧
hove to 顶风漂泊

hovite 铝钙石
Howard diagram 霍华德图(求解有侧向荷载压杆的极坐标图解法)
howardite 古铜钙长无球粒陨石;钙长紫苏辉石无球粒陨石
Howe's factor 豪威系数
Howe's formula 豪威公式
Howe's string truss 豪威式龙骨桁架
Howe's truss 豪氏桁架;豪威(式)桁架
Howell's unit 豪厄尔氏单位
Howell-Bunger valve 锥形阀
Howe roof truss 豪威式屋架
Howe truss 豪威桁架;平弦再分析架;威式桁架
howieite 硅铁锰钠石
howl 啸声
howler 啸鸣器;高声信号器
Howley park 约克郡砂岩(英国)
howling 振鸣
howlite 羟硅硼钙石
hoy 平底船;大货驳;沿海岸单桅小船
hoya (崎岖山区中的)高河谷;河床;盆地
Hoyer beam 霍耶梁
Hoyer effect 霍耶效应(预应力筋摩阻力效应)
Hoyer girder 霍耶大梁
Hoyer hollow(floor)filler 先张预应力混凝土空心楼板的填料
Hoyer method 霍耶法
Hoyer method of prestressing 预拉伸
Hoyer process 霍耶法
Hoyer slab 先张预应力混凝土板
Hoyle's alloy 侯氏合金
Hoyt's metal 锡基轴承合金;候特合金
Hoyt's sector theory 霍伊特扇形理论
hp and lp mixed economizer 高低压联合式省煤器
H-pile 工字(形)桩;宽缘工字钢桩;H 形钢桩
H-post 工字杆;宽翼缘工字杆;H 形柱
H-ridge roof 工字脊屋顶
H-runner 吊用宽翼缘工字梁
H-scope 分叉点显示器
H-section 工字形剖面;工字形断面;工字截面;宽翼缘工字截面
H-section anchor pile H 形锚桩
H-section column 宽缘工字柱;宽翼工字柱
H-section iron 宽翼缘工字钢
H-section Peine girder 宽翼缘工字钢梁
H-section pile 宽翼缘工字桩;H 形桩
H-section sash 工字钢窗扇
H-section steel 宽翼缘工字钢;H 形钢
H-section steel pile 宽翼缘工字桩;H 形钢桩
H-shape 工字形
H-shaped column 工字形柱
H-shaped forehearth 前炉前床;H 形通路
H-shaped iron 宽翼缘工字钢
H-shaped isotherm 工字形等温线
H-shaped pile 工字形桩
H-shaped plan 工字形平面
H-shaped steel pile H 形钢桩
hsianghualite 香花石
H-steel 宽翼缘工字钢
H-stretcher H 形撑杆
H-theorem of Boltzmann 波尔兹曼 H 定理
H-type balanced valve 工字形无压阀门
H-type gasket H 形密封条;H 形垫条;H 形密封垫
hualage allowance 可捕量
hualage standard 可捕标准
Huanghai mean sea level 黄海平均海平面
Huanghai Vertical Datum 1956 1956 年黄海高程系统
huanghoite 黄河矿
Huangpu River 黄浦江
Huayi map 华夷图
hub 中枢;毂盘;高颈;轮毂;集线器;套节;枢(纽);承口;插孔
hub and spigot 套节及套管
hub and spigot joint 中心向联轴节;中心插口接合;突口啮接;套筒接头;套塞接头
Hubard-Field mix-design method 哈巴—费尔(沥青)混合料配合设计法
Hubard-Field stability 哈巴—费尔稳定度
Hubard-Field stability test 哈巴—费尔稳定性试验;哈巴—费尔稳定度试验
Hubbard glacier 哈巴德冰川
Hubbard tank 练力浴池
hub bearing 轴套轴承;轴柄轴承

hubbeck pale boiled linseed oil 哈明漆油
hubbed flange 高颈法兰;带颈法兰
hubber 冲压工
hubbing 压制阴模(法);拉拔法
hubbing steel 冲模钢
Hubble radius 哈勃半径
hub bolt 轮毂螺栓
hub bore 毂孔
hub borer 毂孔镗床
hub brake 轴制动器;轮毂制动器;轮闸
hub cap 毂盖
hub city 中心城市
hub cone 泄水锥
hub cover 毂盖
hub-deep 木桩深度;车辙深度;标桩深度
hub docking 毂套对接
hub dynamometer 套筒式测力计
hub eddy 浆毂涡流
hub fillet 毂肩
hub flange 毂缘
hub flange fitting 承口法兰配件
hub for vehicle wheel 车轮轮毂
hub insulation 套筒绝缘
hub key 毂键
hub lateral 承口分叉管
hubless cast-iron pipe 无套节铸铁管
hub liner 毂衬(垫)
Hubl solution 胡伯溶液
hub micrometer 中心千分尺
hub of commerce 商业中心
hub of communication 交通枢扭
hub of gear 齿轮毂
hub of industry 工业中心
hub of pipe 管子承口;管子突口
hub of piping 管子突口
hub plate 毂衬
hub plug 端塞;堵头
hub polling 中枢查询;传递探询
hub ratio 轮毂比
hub reduction 轮毂减速
hub sleeve 毂套;承口套管
hub spider 轮毂
hub splines 花键轴套
hub stake 中心(标)桩
hub station 枢纽站
hub wheel 毂轮
hub wheel motor 轮毂电机
huckaback 浮松布
huckle 背斜顶尖
huckleberry 越桔(北美灌木);黑果木
huddling chamber 混合室
hudrogen cyanide 氰化氢
Hudsen formula 赫德森公式
Hudsonian Canyon 哈德逊峡谷
Hudsonian orogeny 哈德逊运动【地】
Hudsonian Strait 哈德逊海峡
hudsonite 黑铁辉石
Hudson River Bluestone 赫德森河蓝灰砂岩
hue 主色;色相;色彩
hue adjust circuit 色调调整电路
hue angle 色调角;色彩角
huebnerite 钨锰矿
hue circle 色相环;色调环
hue circuit 色相环
hue content 色相鲜明度;色调饱和度
hue control 色调控制;色调调整;色调调节
hue controller 色调控制器
hue difference 色调差
hue-difference angle 色差角
hue error 色调误差
huegelite 砷铅铀矿
hue information 色彩信息
hue level 色调电平
huemulite 水钒镁钠石
hue of colo(u)r 色相
hue range 色调范围
hue range control 色调范围调整;色彩变化范围调整
hue removal rate 色度去除率
hue-saturation and intensity 色度章度和透明度
hue-saturation and intensity system 色彩饱和度和强度系统
hue selection 色调选择
hue sensibility 辨色敏感度

hue shift 色调偏移
Huey test 晶间腐蚀试验；不锈钢耐蚀试验
huff-buff 无线电高频测向仪
huffler 小河港引航员
Huff separator 赫夫分选机
huge 巨大的
huge 550-millimetre high l-beams 高度 550 毫米的大型工字钢
huge blast 大爆破
huge block 大型砌块
huge building 大型建筑(物)
huge chert nodule 巨型燧石结核
huge coal 特大块煤
huge concrete block 大型混凝土砌块；大型混凝土块(体)；简体结构
huge project 大型工程
huge ship 巨轮
huge size 特大型
huge-size ore deposit 特大型矿床
huge speculative profits 巨额投机利润
huge stock 大量储备
huge vessel 巨轮
Huggenberger tensometer 惠更伯格伸长计
Huggins' viscosity equation 惠金黏度方程式
Hughes' metal 休斯铅基轴承合金
hug the coast 靠岸航行；紧靠海岸航行
hug the land 靠岸航行
hug the left bank 紧抱左岸航行
hug the right bank 紧抱右岸航行
hug the shore 靠岸航行；紧靠海岸航行
hug the wind 靠风航行
hula shirt 防擦网
Hulett electric(al) unloader 休利特电动卸料机
Hulett ore unloader 胡列特式矿石卸船机
Hulett unloader 胡列特式重型抓斗卸船机
hulk 仓库船；废船(体)；笨重船
hull 去皮；船体；船身；船壳；车盘
hull and machinery 船体与机械设备
hull and materials 船壳和设备
hull appendage 船体附属物；船体附件
hull appendage ratio 船体附属物比率；船舶附属物比率
hull appurtenances 船体属具
hull armo(u)r plate 船身甲板；车座甲板
hull assembly 船体装配
hull assembly shed 船体装配车间
hull auxiliaries 船用辅具
hull batten plate 船身镶板
hull bending moment 船体弯矩
hull block division plan 分段划分图
hull-borne (气垫船等的)排水状
hull bottom 船底
hull cell 薄膜电池
hull certificate of class 船体入级证书
hull classification and inspection record 船体入级记录和检验报告
hull construction 船体结构；船体构造
hull construction technology 船体建造工艺
hull conveyer[conveyor] 壳荚输送器
hull core structure 简体结构
hull cross member 车座横梁
hull damage 船体破损；船壳损伤
hull deflection 船体挠度
hull deformation 船体变形
hull-deformed vessel 船体变形船
hull design 船体设计
hull dimensional ratio 船体尺度比
hull displacement 船壳排水量
hull down 船体下隐(在水平面下看不见)
hulled grain 脱壳谷物
hull efficiency 船体效率；船身效率；船壳效率
huller 去皮机；脱壳机
huller gin 带铃壳籽棉轧花机
huller rib 脱壳肋条
hull fabrication 船体加工
hull factor 船体附属物比率；船舶附属物比率
hull fitting 船体舾装；甲板舾装；舱面属具
hull-fitting fender 与船体线型相适应的碰垫
hull form 船体型线；船体形状；船体线型
hull foundation 壳体基础
hull frame 船体肋骨；船身构架
hull girder 相当桁；船桁体
hull girder vibration 船体梁振动
hull grillage strength 船体板架强度

hull horizontal bending strength 船体水平弯曲强度
hulling cylinder 脱壳滚筒
hulling disk 脱壳盘
hulling drum 脱壳滚筒
hulling machine 脱壳机；脱谷机
hulling mill 脱壳机
hulling separator 脱壳机
hull insurance 机身保险；船体保险；船身保险；船壳保险；车身保险
hull intermediate plate 船身中间板
hull lift hook 车盘吊钩
hull line 船体线；船体外形；船体轮廓
hull lofting 船体放样
hull longitudinal 船体纵骨
hull machinery 船体机械
hull maintenance 船体保养
hull-mounted hydrophone 船壳水听器
hull-mounted sonar 船壳声呐
hull natural frequency 船体固有振动频率
hull number 船(壳)号；船号；船舶(建造)序号
hull out 船身露出海平线
hull outfitting 船装
hull paint 船壳漆
hull plate 船身板
hull plating 船壳板
hull policy 船体保险单；船壳保险单；船舶保险单
hull pressure 船体压力
hull principal dimension 船体主尺度
hull process(ing) equipment 船体加工设备
hull projections 船体突出部
hull reinforcement plate 船身加强板
hull repair 船壳修理
hull repair ship 船体修理船
hull resistance 船壳阻力
hull resistance characteristic 船舶阻力特性
hull-return system 船体回路系统
hull risk 船舶险
hull rubber 薄膜橡胶
hull scantling 船体构件尺寸
hull shop 船体车间
hull side area 船体侧面投影面积
hull sloping plate 船身斜板
hull specification 船体说明书
hull steel 造船钢；船用钢
hull stiffness 船体刚度
hull strained 船体变形
hull-strained ship 船体变形船
hull strength 船体强度
hull stress measurement 船体应力测量
hull structural stability 船体结构稳定性
hull structure 船体结构；船体构造
hull structure similar model 船体结构相似模拟
hull sunk 全船下沉；船身下沉
hull torsional vibration 船体扭转振动
hull trawler 双拖网渔轮；双拖网渔船
hull turret plate 船身转塔板
hull underwriter 船舶保险人
hull up 船身露出海平线
hull valve 舷侧阀；通海阀
hull ventilation 全船通风
hull ventilator 舷侧阀；通海阀；车盘通风器
hull vibration 船体振动
hull vibration behavio(u)r 船体振动性态
hull vibration damping 船体振动阻尼
hull vibration logarithmic decrement 船体振动对数衰减率
hull weight 船体重量
hulsite 黑硼锡铁矿
hum 灰岩残丘；蜂音
hum adjustment 交流声调整
human acoustic measurement 人类声学测量
human acoustics 人类声学
human activity 人为活动；人类活动
human adjustment 人为调节
human aging 人的老化
human and material resources 人力物力资源
human asset accounter 人力资产会计师
human asset accounting 人力资产会计
human behavio(u)r 人类行为
human being 人类
human bias 人为偏倚
human bioclimatology 人体生物气候学
human biogeography 人类生物地理学

human biology 人类生物学；人口生物学
human capital 人力资源；人力资本
human capital approach 人力资本法
human capital flow 人力资本流动
human capital formation 人力资本成形
human capital(investment) 人力投资
human care 人口的管理
human-caused error 人为误差
human check 人工检查
human comfort 人的舒适(性)
human community 人类群落
human component characteristic 人环节特性
human contamination 人为污染
human-control characteristics 人操纵特性
human counter 人体计数器
human decision-making 人的决策
human development 人才的成长
human dimensions 人体尺度
human disease 人类疾病
human dummy impact test 人体模型冲击试验
human ecology 人类生态学；人口生态学
human efficiency 人的效率
human element 人为因素；人工地物
human element accident 责任事故；人为因素事故；人为事故
human engineering 运行工程学；人事管理；环境工程学；工效学；工程心理学；机械设备利用学；人因工程(学)；人体工程(学)；人力工程(学)；人类工程(学)；人机工程(学)；人的因素工程学
human environment 人生环境；人类环境
human equation 人为因素
human erosion 人为侵蚀；人为冲刷；人为冲蚀
human error 人为误差；人差【测】
human excrement 人排泄物；人类排泄物(如汗、粪便、尿等)；人粪尿
human excrement use 人类粪便的利用
human excreta 人类排泄物(如汗、粪便、尿等)
human exposure to pollutants 人体接触污染物
human factor 人为因素；人的因素；人类因素
human-factor engineering 人类工程学；人因工程(学)
human factors handbook 人类工程手册
human factors society 人的因素学会
human failure 人为事故
human force 人力
human geography 人文地理学；人生地理
human habitant 人类生活环境
human habitat 人类生境；人类生活环境
human habitat model 人类生境模型；人类生活环境模式
human health 人体健康；人类健康
human health risk assessment 人体健康风险评价
human history 人类历史
human impact 人类活动影响
human-impacted resources 受人类冲击的资源
human incurring error 人为误差
human induced degradation 人类引起的环境退化
human influence 人类活动影响
human information input channel 人的信息输入通道
human information output channel 人的信息输出通道
human intervention 人类干预
human investment 人力投资
humanisation 人格化
humanism 人文主义；人本主义
humanitarian management 人道主义管理
humanity 人类
humanity geography 社会地理学
humanization of nature 自然的人化
humanized 人化的
human laboratory 工程心理实验室
human labor in general 一般人类劳动
human labo(u)r power 人的劳动力
human landscape 人文景观
human living space 人类生活空间；人类生存空间
human longevity 人类寿命
human machine adaption 人机配合
human-made block dike 人造方块堤
human-made compacted embankment 人工压实土
human-made disaster 人为灾难
human-management 人道主义管理
human-memory 人的记忆

human-motivation 人力激发
human obsolescence 人力的过时
humanoid 具有人类特点的;人形机
humanoid robot 人型机器人
human operation 人工操纵
human parameters 人体参数
human period 人类时期
human physiological reaction 人体生理反应
human pollution 人体污染
human pollution burden 人体污染负荷
human potential 人力资源的潜力;人的潜力
human power 人力
human quality 资质
human reaction of earthquake 人对地震反应
human relations 人事关系
human relations management 人事关系管理
human relay working 人工转接
human resources 劳动力资源;人文资源;人力资源
human resources accounting 人力资源会计
human resources capital plan 人力资源资本计划
human resources cost 人力资源成本
human resources department 人力资源部
human resources development 人力资源开发
human resources development and management 人力资源开发管理
human resources information 人力资源情报
human resources management 人力资源管理;人力管理;人才管理
human resources model 人力资源模式
human resources office 人事室
human resources value 人力资源价值
human risks 人为风险
human safety engineering 人类安全工程学
human scale 人类尺度
human sensation level 人的感觉级
human sensitivity of vibration 人的振动敏感性
human sensitivity to vibration 人对振动的敏感性
human settlement 居民区;人类(居)住区
human settlement pattern 人类居住模式;人类居住类型
human settlements management 人类住区管理
human settlement system 居住点组群体系
human simulation 人工智能模拟;人的模拟
human socioeconomic system 人类社会经济系统
human survival limit 人的减速度死亡极限
human testimony 人证
humanthracite 无烟煤级腐植煤
humanthracon 烟煤级腐植煤
human tolerance (指对放射性等的)人体耐受量
human tolerance to noise 人对噪声的容忍度
human value 人的价值
human vision 人的视觉
human waste 人类废物;人粪尿
human wealth 人力财富
humate 腐殖酸酯;腐殖酸盐
hum balancer 交流声平衡器
hum bar 交流带;图像波纹横条
humberstonite 水硝碱镁矾
Humble relation 亨勃尔关系
Humboldt ball mill 洪堡球磨机
Humboldt Glacier 亨博尔特冰川
humboldtilite 硅黄长石
humboldtine 草酸铁矿
humboldtite 硅硼钙石;草酸铁矿
hum bucking 抵消交流声
hum distortion 交流声失真
humectant 含水加入剂;湿润剂;持湿剂;保湿剂
humectation 润湿;湿润
Hume duct 休谟管道;混凝土冷却管道;钢筋混凝土管
Hume gold flow equilibrating mechanism 休谟黄金流动平衡作用
hum eliminator 交流声抑制器
Hume pipe 钢筋混凝土管;休谟管(离心混凝土管)
humeral angle 肩角
Hume-Rothery compound 电子化合物
hum field 交流声场
hum filter 交流声滤除器;平滑滤波器
hum free 无交流声
humic acid 腐殖酸;腐蚀酸
humic acid acetamide 腐殖酸酰胺
humic acid attack 腐殖酸侵蚀
humic acid chlorination 腐植酸氯化
humic acid combined fertilizer 腐殖酸复合肥

humic acid decomposing bacteria 腐殖酸分解细菌
humic acid in brown coal 褐煤腐植酸
humic acid in combustible base 可燃基腐植酸
humic acid in peat 泥炭腐植酸
humic acid in weathered coal 风化煤腐殖酸
humic acid waters 腐殖酸水体
humicap 湿敏电容
humic-cannel coal 腐殖烛煤
humic carbonated soil 腐殖质碳酸岩土
humic coal 泥煤;腐殖煤
humic colloid 腐蚀胶体
humic complex 腐蚀络合物
humic compound 腐殖质;腐蚀质化合物
humiceram 湿敏陶瓷
humic fertilizer 腐殖肥料
humic gel 腐蚀凝胶
humic matter 腐殖(质)物质;腐殖质
humic mud 腐殖泥
humic mud lake 腐殖泥湖
humic mulch 腐殖质覆盖层;腐殖覆盖物
humic organic matter 腐蚀有机质
humics 腐殖质
humic-sapropelic type 腐殖腐泥型
humic sewage lake 污水湖
humic soil 腐殖土;腐殖泥
humic substance 腐殖(物)质
humic sulfur 腐殖硫
humic type 腐殖型
humic type source rock 腐殖型源岩
humic water lake 腐殖质水湖
humid 湿气重的;湿的;潮湿(的)
humid ag(e)ing resistance 耐湿气老化性
humid air 湿空气;潮(湿)空气
humid air plasma 湿式等离子体
humid analysis 湿法分析
humid area 湿润(地)区;潮湿地区
humid climate 湿润气候;潮湿气候
humid-dry cycling 湿干循环
humid-dry cycling resistance 耐干湿交替性
humid euphorbia 地锦【植】
humidex 湿润指数
humid fan 湿地冲积扇
humid farming 灌溉农业
humid gas 湿气
humid heat 湿比热
humid heat capacity 湿热容
humid heat capacity of air 湿空气热容
humid heat-resistance 耐湿热性
humidification 增湿(作用);加湿(作用);湿润(作用)
humidification agent 润湿剂
humidification by mixing 混合增湿
humidification-dehumidification 增湿减湿法
humidification-dehumidification technique 增湿减湿技术
humidification spray 喷湿剂;喷湿器
humidification system 增湿系统
humidified 腐殖的
humidified blast 加湿鼓风
humidified combustion 增湿燃烧
humidified greenhouse 湿润温室
humidifier 增湿器;增湿剂;加湿器;调湿装置;调湿器;输水装置
humidifier installation 纸型加湿机
humidifier section 加热段
humidifier tower 增湿塔
humidifying 增湿;给湿;加湿
humidifying and air conditioning equipment 空调湿调设备;水淋式空调设备
humidifying chamber 加湿室;加湿器
humidifying effect 湿(润)效应;加湿作用
humidifying equipment 增湿设备;加湿设备;调湿设备
humidifying installation 湿润装置
humidifying machine 增湿机
humidifying plant 增湿装置;增湿车间
humidifying ventilation 加湿通风
humidiherbosa 湿地草本群落
humidin 潮菌素
humidiometer 湿度计
humidistat 恒湿箱;恒湿器;恒湿计;湿度调节仪;湿度调节器
humidity 水气含量;水分含量;湿(气);潮气
humidity absorption 吸潮;潮湿

humidity ag(e)ing 潮湿老化
humidity analyzer 湿度分析器
humidity barrier 防潮层
humidity blushing 蒙湿气;湿气致白(漆病);湿气模糊
humidity box 湿度盒
humidity cabinet 加湿室;湿度箱;湿度室;潮湿箱
humidity cabinet test 调温调湿箱试验;潮湿箱试验
humidity cell 湿度试验室;潮湿箱
humidity chamber 湿度试验室;可调湿度室;潮湿箱;雾室;湿度试验容器
humidity chart 湿度图
humidity coefficient 湿润系数;湿度系数
humidity content 含水量
humidity control 水分控制;水分调节;湿度控制;湿度调节
humidity controller 湿度控制器;湿度调节器
humidity controlling 湿度控制;湿度调节
humidity correction factor 湿度校正系数
humidity corrosion 湿气腐蚀
humidity cover 保温覆盖(物)
humidity damage 潮湿损坏
humidity densimeter 湿度密度计
humidity density 湿度密度
humidity density meter 湿度密度计
humidity detector 湿度仪;湿度检测器
humidity determination 湿度测定
humidity drier 湿式干燥器
humidity entry 潮气侵入(处)
humidity equilibrium 湿度平衡
humidity factor 湿度效应;湿润系数;湿度因子;湿度因素
humidity flowmeter 潮湿气体流量计
humidity free material 防潮材料
humidity gradient 湿度梯度;湿度比降
humidity increase 湿度增加
humidity index 湿润指数;湿度指数
humidity-indicating paint 示湿涂料
humidity indicator 湿度指示器
humidity limit 湿度界限
humidity measurement 湿度检测
humidity meter 湿度计
humidity migration 湿气移动;湿气迁移
humidity mixing rate 湿度混合率
humidity mixing ratio 含湿比例;湿度混合比
humidity movement 湿气移动
humidity of air 空气湿度
humidity passage 湿气移动途径
humidity penetration 潮湿侵入
humidity permeability 渗湿性
humidity potential 湿度势
humidity proof roof(ing)sheet(ing) 防潮屋面板
humidity protection 湿度防护
humidity province 湿度区
humidity quantity 湿度量
humidity ratio 含湿量;湿含量;湿度比;比湿(度)
humidity ratio of moist air 湿空气的湿含量
humidity ratio of moist air in the free stream 自由气流湿空气的湿含量
humidity recorder 湿度记录器
humidity regulation 湿度调节
humidity regulator 湿度调节器
humidity removal from roofs 屋顶除潮
humidity resistance 抗潮;防潮;耐湿性;防潮性(能)
humidity seal 防潮缝
humidity sensitive ceramics 湿敏陶瓷
humidity sensitive effect 湿敏效应
humidity sensitive element 湿敏元件;湿度受感元件
humidity sensor 湿度传感器
humidity slide-rule 湿度计算尺
humidity stop 止潮层;防潮物
humidity strip 测湿片;测湿带
humidity test 耐(潮)湿试验;湿度(腐蚀)试验
humidity-testing cabinet 湿度试验柜
humidity transducer 湿度传感器
humidity ventilation 润湿通风
humidizer 增湿器;增湿剂;加湿剂
humid lithogenesis 湿成岩石成因论
humidly cured sealant 湿气固化密封膏
humid mesothermal climate 湿温气候;湿润气候;湿暖气候
humid microenvironment 潮湿微环境

humidness 湿度
humid operative temperature 湿作用温度
humidor 蒸汽饱和室;保湿装置
humidostat 自动空气调湿装置;恒湿仪;恒湿度调节仪;湿度调节仪;湿度调节器;湿度调节板
humid period 湿润期
humid process 加湿法
humid refuge 潮湿垃圾
humid region 湿润区;潮湿区
humid rock 湿成岩
humid room 潮湿房间;雾室;湿室;保湿室
humid room dampproofing 潮湿房间防潮
humid room luminaire(fixture) 潮湿房间照明灯
humid room services 潮湿房间设施
humid rubbish 潮湿垃圾
humid soil 湿润土(壤)
humid subarctic climate 亚寒带温湿气候
humid subtropical to tropical subzone 潮湿亚热热带亚带
humid subtropical to tropical zone 潮湿亚热带热带
humid temperate climate 湿润温和气候
humid temperate zone 潮湿温带
humid temperature climate 湿润温和气候;湿暖气候
humid test 湿度试验
humid transition life zone 湿润过渡生物带
humid tropical belt 湿热带
humid tropical condition 湿热状态
humid tropical lowland 潮湿热带低地
humid tropical region 湿热地区;湿热带
humid tropical to subtropical zone 潮湿热带亚热带
humid tropics 湿热带
humid tropics climate 湿热气候
humid tropics climate zone 湿热气候带
humid volume 湿体积;湿容积;湿(空气)比容
humid warm temperate zone of northern China type 北方型潮湿暖温带
humid waste 潮湿垃圾
humid weight 湿重
humid zone 湿润区;湿润带;潮湿带
humification 腐殖(质)化;腐殖(化)作用
humification modulus 腐殖化模量
humified 腐殖化的
humified organic matter 腐殖化有机质
humify 腐殖化;变成沃土
humilis 淡云
humin 胡敏素;腐黑物;腐黑酸
huminite 腐殖组
humistor 湿敏电阻
humit 温湿
humite 矽镁石;硅镁石
humite phlogopite sharn 硅镁石金云母矽卡岩
humiture 温湿度
Humm's concrete probe 赫姆探针
hum measurement 背景噪声测量
hummeler 除芒器
hummer 蜂音器;蜂鸣器
hummerite 水钒镁矿
hummer screen 电动簸动;电磁振动筛;电磁筛;冲击式电磁振动筛;哈马型冲击式电磁振动筛
hummer tone 蜂音
humming 蜂鸣音
humming sound 蜂鸣器
hummock 沼泽中的高地;小丘;圆(形)丘;堆积冰群;波状地
hummock-and-hollow topography 鼓盆地形;丘状盆地;起伏地形
hummocked ice 积冰;丘状浮冰;冰丘冰
hummocking 冰丘的形成
hummocky 浮冰拥塞;波丘地
hummocky and quiet chaotic mode 波状和平稳混乱模式
hummocky clinoform reflection configuration 波状斜坡反射结构
hummocky cross bedding structure 丘状交错层理构造
hummocky cross stratification 波状交错层理
hummocky floe 堆积浮冰块;浮冰群
hummocky surface 丘形地面
hummocky surface topography 波状地形
hum modulation 交流调制;哼声调制
hum noise 交流声

humocoll 泥炭级腐殖煤
humodetrinite 碎屑腐殖体
humodil 褐煤级腐殖煤
humodite 亚烟煤级腐殖煤
humodurite 腐殖微暗煤
humolite 腐殖煤
humosapropelic coal 腐殖腐泥煤
humovitrinite 腐殖镜质体
hump 小丘;圆丘;撞轨器;浪峰;拉坯用泥段;曲线顶点;驼峰【铁】;凸起;铁路驼峰;山岗;峰(值)
hump and kickback 驼峰回头线
hump and marshalling facility 驼峰及调车设备
hump area 驼峰区
hump avoiding line 驼峰迂回线
hump-back bridge 弓形桥;罗锅桥;驼峰式桥;驼背形桥
hump-back conveyor furnace 驼峰式退火炉
hump cabin 驼峰控制楼
hump classification yard 驼峰调车场
hump conductor 驼峰作业员
hump control tower 驼峰控制楼
hump crest 驼峰峰顶;峰顶【铁】
hump crest coupler's cabin 驼峰连接员室
hump distribution zone 驼峰溜放部分
hump drag 最大阻力;峰值阻力
humped yard 驼峰调车场
hump effect 驼峰效应
hump end of yard 驼峰头部
hump-end retarder 驼峰头部缓行器
humper 变形盘子;土方工人(俚语)
humper year 丰年
hump gradient 驼峰坡度
hump height 驼峰高度;峰高【铁】
Humphrey's spiral concentrator 汉弗莱型螺旋选煤机;汉弗莱型螺旋选矿机
Humphrey gas pump 内燃水泵
Humphries equation 汉弗里斯方程
humping automation 驼峰自动化
humping back signal 后退信号【铁】;后退溜放信号
humping bead 焊接凸起;焊接隆起
humping capacity 驼峰作业能力;驼峰解体能力
humping fast signal 加速推送信号;加速留放信号
humping gradient 驼峰推送坡;驼峰溜放坡
humping gradient section 驼峰推送坡段
humping operation 推峰作业
humping part 驼峰推峰部分
hump in gravity yard 重力驼峰调车场驼峰
humping section 推送部分【铁】
humping slow signal 减速推送信号
humping speed 溜放速度;推峰速度
humping track 驼峰线路;驼峰调车线
humping-type stilling basin 驼峰式消能池
humping velocity head 驼峰初速能高
humping zone 驼峰作业区
hump keel 中凸龙骨
hump lead 驼峰溜放线;驼峰溜放坡;推送线
hump locomotive 驼峰机车;驼峰调车机车
hump master 驼峰值班员
hump mo(u)ld 压凸用阳模;压坯用阳模;上凸模具
hump operator 驼峰作业员
hump profile 驼峰纵断面
hump rail brake 驼峰制动器;驼峰缓行器
hump resonance 共振峰
hump retarder classification yard 驼峰缓行器编组场
hump scale 驼峰轨道衡
hump-shaped pad block 驼峰【铁】
hump shunting 驼峰调车
hump signal 驼峰信号
hump signal(l)er 驼峰信号机
hump signal repeater 驼峰复示信号机
hump speed 过渡航速(气垫船);界限速度;极限速率;断阶滑水速度
hump summit 驼峰峰顶
hump switching 驼峰调车法
hump track 驼峰轨道;驼峰推送线;驼峰溜放线;驼峰调车轨道
hump trimming signal 下峰信号【铁】
hump-type stilling basin 弓背消能池;驼峰式消能池;驼峰式消力池
hump voltage 驼峰电压
humpy 小棚屋;凹凸冰磺丘
hump yard 驼峰调车场;驼峰编组站;驼峰编组场
hump yard classification throat 驼峰调车场头部

hump yard neck 驼峰调车场咽喉
hum reduction factor 滤波因数;交流声降低因数
humulite 腐殖岩
humulith 腐殖岩
humus 腐殖质;腐蚀货
humus acid 腐殖酸
humus bucket 腐殖土挖掘铲斗
humus carbonate soil 腐殖质碳酸岩土;腐殖质土
humus-clay complex 腐殖质黏土复合体
humus coal 腐殖质煤;腐殖煤
humus content 腐殖质含量
humus content of topsoils 表土的腐殖质含量
humus-formation 腐殖质的形成
humus horizon 腐殖质层
humus layer 腐殖质层;腐蚀层
humus plant 腐殖质植物
humus-rich layer 腐殖质层
humus sewage 腐污水
humus sludge 腐殖质污泥;腐蚀污泥;腐木污泥
humus soil 腐殖质土;腐殖土
humus stratum 腐殖层
humus tank 二次沉淀池;化粪池;腐殖质沉淀池;腐殖(沉淀)池
humus tank effluent 腐殖质沉淀池出水
hunch 厚片;大片;大块;圆形建筑物
hunch pit 渣坑
hundred call seconds 百秒呼叫
hundred flowers 百花
hundredfold 百倍的
hundred-fold crackles 百级碎
hundred-meter post 百米标
hundred million stere 亿立方米
hundred million tonne 亿吨
hundred-percent 百分之百
hundred-percent barchart 百分比条图
hundred-percent location 最佳地区
hundred-percent rectangle 全数必中界【数】
hundred-point emulation 小指标竞赛
hundreds dial 百分度盘
hundred weight 1/20 吨;英担
hundred-year frequency flood 百年一遇频率洪水
hundred-year once flood 百年一遇洪水
hundred-year storm condition 百年不遇的风暴
Hung's modified filtration counting method 洪氏浮集法
hung acoustic(al) ceiling 吸声吊顶棚;悬式吸声天花(板)
Hungarian architecture 匈牙利(式)建筑
Hungary architecture 匈牙利(式)建筑
Hungary flexible pavement design method 匈牙利柔性路面设计法
hung ceiling 吊顶
hung ceiling incorporating services 吊顶整套设备
hungchaoite 章氏硼镁石
hung fender 悬式护舷;悬式护(舷)木;悬挂式碰垫
hung floor 悬挂楼板;悬式楼面;悬式楼板
hung floor construction 悬挂式楼板构造
hung glazing 悬式玻璃窗
hung gutter 悬挂落水槽
hung partition(wall) 吊隔墙;悬式隔墙
hungry 强吸面;贫瘠;低塑性的;无价值的
hungry joint 凹缝;未胶牢接头;凹式接缝
hungry joint pointing 深凹勾缝
hungry spots (新浇混凝土表面的)小孔;贫瘠地区;麻面(混凝土)
hungry surface 不丰满的表面
hung sash 悬吊窗扇(用于双悬窗);垂直推拉窗扇
hung scaffold(ing) 悬挂式脚手架;吊脚手架;架空脚手架
hung shell 悬吊壳体;悬式骨架
hung-sifter swing 吊筛式摆动
hung slating 墙面上吊挂的石板;钢丝夹固定的石板瓦;石板瓦贴面
hung-span beam 悬跨梁
hung story 悬式楼面;悬式楼板
hung tile 挂瓦
hung tilework 贴墙面砖工作;悬式排水瓦管
hung tiling 墙面上吊挂的瓦片;挂瓦
hung window 吊窗
hung wood fender 悬木缓冲器;悬木防护器;悬挂式护舷木;防撞悬木
hunk 岩块;厚片;大片;大块
hunk of cable 电缆盘
hunt 摆振

Hunt continuous filter 亨脱连续过滤机
Hunt conveyer 亨氏输送机
hunt effect 摆动效应
hunter 寻觅器；搜寻器；搜索器；双盖表
Hunter's screw 差动可逆式螺旋桨
Hunter colo(u)r and gloss meter 亨特比色光泽计
Hunter difference equation 亨特色差方程
Hunter gloss meter 亨特色泽计
hunter green cloth 墨绿色布
hunter kill 射杀量
hunter-killer 猎潜艇
hunter-killer aircraft carrier 反潜航空母舰
Hunter multipurpose reflectometer 亨特万能反射(白度)计
Hunter whiteness 亨特白度
hunt group 寻线群；寻热阻
huntilite 杂砷银矿
hunting 寻线；摇头；绕一定点左右摆动；搜索；查寻；不规则(的)振荡
hunting action 搜索动作
hunting angle 振荡角
hunting arrow 打猎箭
hunting bow 打猎弓
hunting box (打猎时住的)窝棚；猎舍
hunting case watch 密面表
hunting circuit 闭锁电路
hunting effect 猎振现象
hunting field 猎场；狩猎区；猎区
hunting frequency 振荡频率；摆动频率
hunting gear 跟踪装置；从动装置
hunting ground 猎场
hunting leather 粗绒面革
hunting lodge 猎舍；打猎住处
hunting loss 摆动损失
hunting motor 从动电动机
hunting movement 蛇形运动
hunting of governor 调节器的摆动；摆节器的摆动
hunting of load 负载振荡
hunting period 寻线期间；振荡周期；摆动周期
hunting power 振荡功率
hunting probe 振动探示器；摆动探头
hunting quota 猎捕量限额
hunting range 振荡范围；摆动范围
hunting saddle 猎鞍
hunting speed 寻线速度；寻觅速度；振荡速度
hunting switch 寻线开关；寻线机
hunting system 自摆系统；猎захt系统
hunting time 寻线时间；振荡持续时间
Huntington mill 亨廷顿型磨碎机；摆式破碎机
hunting tooth 追逐齿
hunting vibration 不规则振动；不规则摆动
hunting zone 猎振带；猎区；搜索范围
huntite 高镁白云石；碳酸钙镁石；碳钙镁石
Huntland rail 幕轨
Huntonit 木纤维板
Huntsman process 亨茨曼坩埚炼钢法
hunt table 半圆形矮桌
Huon pine 塔斯马尼高大乔木(材色淡黄，有波浪纹，用于雕刻和造船)
Huowaan cloth 火浣布
hurdle dike 栅栏透水堤
hurdled ore 粗筛过的矿石
hurdle fence 篱笆；矮围墙
hurdle groin 栅栏透水丁坝
hurdle groyne 栅栏透水丁坝
hurdle scrubber 栅格式气体洗涤器
hurdle-type washer 栅格式气体洗涤器
hurdle work 河堤网格工；编条制品；编柳条制品；编篱工
hurds 粗亚麻
hurdy-gurdy 卷筒车
Hurdy-Gurdy wheel 带辐射轮叶的水轮机；赫尔第一格尔第式水轮机(水流直接冲击辐射式桨叶)
hureaulite 红磷锰矿
hurl 用力投(掷)
hurl barrow 双轮手推车
hurlbutite 磷钙铍石
hurley 跳汰机
hurley house 坍坏房屋
Hurlhinge 自动对正平接铰链
hurlinge 平铰链；平铰接
hurling pump 旋转泵
Huronian 休仑统【地】
Huronian System 休仑系

Hurrell colloid mill 赫里耳胶体磨
hurricane 龙卷风；十二级风；大旋风
hurricane alley 飓风路径
hurricane analogue 飓风模拟
hurricane area installation 抗风面的安装
hurricane barrier 飓风阻挡物；飓风屏蔽结构；飓风屏障
hurricane center 飓风中心
hurricane damage 飓风造成的损失
hurricane deck 小艇甲板；最上层甲板；天幕甲板；上层轻甲板
hurricane-deck vessel 轻甲板船
hurricane delta 飓风三角洲
hurricane dryer 风干室
hurricane force 飓风力
hurricane-force wind 强飓风
hurricane globe 防风罩
hurricane-induced surge 飓风潮
hurricane-induced tide 飓风潮
hurricane lamp 气灯；风(暴)灯；防风灯
hurricane lantern 抗风灯；风灯
hurricane oil lamp 风灯
hurricane parameter 飓风参数
hurricane path 飓风路径
hurricane radar band 螺旋带
hurricane rain 飓风雨
hurricane roof 小艇甲板；天幕甲板；上层轻甲板
hurricane seas 风暴潮海浪
hurricane storm surge 飓风潮
hurricane storm wave 飓风浪；飓风潮
hurricane surge 飓风涌浪；飓风浪波；台风狂涌；风暴潮
hurricane test 风动试验
hurricane tide 飓风狂浪；飓风涌浪；飓风涌波；台风狂涌；风暴潮；暴风潮
hurricane track 飓风路径
hurricane tracking 飓风跟踪
hurricane warning 飓风警报
hurricane warning system 飓风警报系统
hurricane wave 飓风(涌)浪；飓风涌波；台风狂涌
hurricane wind 飓风；台风
hurried provisional edition 临时急用版
hurrock 一堆石子
hurry-durry 多风多雨的
hurry gum 筛过的细料
hurry-up 快递
hurry-up fall 快递吊货索
hurry up hole 方钻杆鼠洞
hurry-up wagon 应急修理车；工程救险车；抢修车；抢险车
hurst 有树林的高地；沙岸；残丘
Hurst formula 赫斯特公式
Hurst method 赫斯特法
Hurst phenomenon 赫斯特现象
hurter 支撑杆；墙角防护桩石；墙角防护短石柱；(设置在入口处的)防护短柱；支持杆；加强物；护角桩石
Hurter and Driffield curve 曝光特性曲线
hurtle 冲撞
Hurwitz's generalization of binomial theorem 二项式定理的赫尔茨推广
Hurwitz polynomial 赫维茨多项式
Hurwitz stability criterion 霍尔维茨稳定性判据
husbandager 船舶代理费
husbandry 农业；饲养
husband-wife shop 夫妻店
huse 以油漆房屋为职业的人
hush 冲刷找矿法
hush-hush 本底噪声降低
hushing 冲刷找矿法
hush money 贿赂费；封嘴钱；封口钱
hush pipe 消声管
Husid plot 胡西德图
husk 外皮；脱壳
husked rice 加工米；极粗糙米
husker 带铃壳籽棉轧花机
husk garland 果壳形花饰
husking 剥壳
hustiement 家庭用具；小屋；棚屋；茅舍
hut(ch) 小屋；棚屋；跳汰机筛下室；分流板控制跳汰机水流；茅舍；棚屋；舍
hutching 细筛料
Hutchinson's tar tester 赫氏焦油黏度计
Hutchinson's teeth 锯齿形牙

hutchinsonite 硫砷铊铅矿；红铊铅矿
Hutchinson viscosity 赫金生黏滞度
hutch table 柜桌
hutch water 洗矿水
hutchwork 跳汰机筛下产品
hutment 临时办公室；临时营房；棚屋群
hutted 由棚屋建成的；具有棚屋的
Hutting equation 赫蒂格方程
Hutton's formula 休特顿公式(计算斜面上风压的公式)
huttonite 斜钍石；硅钍石
HU-Washizu principle 胡海昌一鹫津原理
Huwood loader 休乌特型装载机
Huwood slicer 休乌特振荡剥煤机
huydrocone crusher 液压式锥形破碎机
Huygens approximation 惠更斯近似
Huygens eyepiece 惠更斯目镜
Huygens ocular 惠更斯目镜
Huygens principle 惠更斯原理
Huygens source 惠更斯源
Huygens wavelet 惠更斯子波
Hveem cohesiometer 维姆黏聚力仪(测定沥青混合料及稳定土等黏聚力用)
Hveem cohesion test 维姆黏聚力试验
Hveem design method 维姆设计法
Hveem design method of flexible pavement 维姆柔性路面设计法
Hveem method 维姆稳定度仪法(一种设计柔性卤莽厚度的方法)
Hveem stability 维姆稳定度(一种测定沥青混凝土等稳定性的指标)
Hveem stability test 维姆稳定度试验
Hveem stabilometer 维姆稳定度仪
Hveem stabilometer method 维姆稳定度仪法(一种设计柔性卤莽厚度的方法)
Hvorslev parameter 沃斯列夫参数
Hvorslev soil model 沃斯列夫土模型
Hvorslev surface 沃斯列夫面
hvpermetamorphism 超变质作用
H-wave 流体动力波；水力动力波；H波
hyacinth 紫蓝色；锆石；红锆石；风信子
hyacinthin 苯乙醛
hyacinth oil 风信子油
hyacinth orinntal 白兰花
hyacinth pond 风信塘
hyaline 透明素；透明的；透明蛋白；玻璃状的；玻璃质；玻璃样的；玻璃(般)的；晶状结构
hyaline cap 透明冠
hyaline cast 透明管型
hyaline degeneration 透明变性；玻璃样变性
hyaline droplet 玻璃样小滴
hyaline hyalin 透明素
hyaline layer 透明层
hyaline membrane 透明膜
hyaline-quartz 玻璃石英
hyaline smooth 透明光滑的
hyaline substance 玻璃样物质
hyaline tube 透明管
hyalinization 透明化作用；玻璃样变；玻璃纤维样化
hyalinize 透明化
hyalinosis 透明变性；玻璃样变性
hyalite 玻璃蛋白石；玉滴石
hyaloallophane 透铝英石
hyalobasalt 玄武玻璃；玻质玄武岩
hyaloclastic agglomerate 淬碎集块岩
hyaloclastic breccia 淬碎角砾岩
hyaloclastic fragmental facies 玻质碎屑岩相
hyaloclastic lava 淬碎碎屑熔岩
hyaloclastic tuff 淬碎凝灰岩
hyaloclastic vitreous fragment 淬碎玻璃碎屑
hyaloclastite 玻质碎屑岩
hyalocrystalline 透明结晶质；玻晶质
hyalocrystalline calcium silicate hydrate 半晶质水化硅酸钙；半晶态水化硅酸钙
hyalocrystalline texture 玻晶质结构
hyalodacite 玻质英安岩
hyalograph 玻璃雕刻仪器
hyalography 玻璃蚀刻术
hyaloid 玻璃状；玻璃样的
hyaloid canal 透明管；玻璃体管
hyaloid membrane 玻璃体膜
hyalomitome 透明质
hyaloophitic 玻璃质辉绿岩的
hyaloophitic texture 玻基辉绿结构；玻质辉绿岩结构

hyalophane 钡冰晶;钡冰长石
hyalophyr 玻基斑岩
hyalopilitic texture 玻晶交织结构
hyaloplasm 透明质(浆)
hyalopolitic texture 玻璃交织结构
Hyalospongia 玻璃海绵纲
hyalotekite 硼硅钡铅矿
hyalotome 透明质
hyalurate 透明质酸盐
hyaluronate 透明质酸盐
hyaluronidase 透明质酸
Hyatt roller bearing 挠性滚柱轴承;弹簧圆柱滚子轴承
Hyblum alloy 海布拉姆铝镁硅合金
hybnickel 改良不锈钢
hybrid 杂化;混杂的;混频环;混合物;混合的;桥接岔路;等差作用
hybrid aircushion vehicle 混合式气垫船
hybrid anaerobic baffled reactor 复合式厌氧折流板反应器
hybrid anaerobic reactor 复合式厌氧反应器
hybrid analog(ue)computer 混合模拟计算机
hybrid analog(ue)-digital simulation 模数混合模拟;模拟数字混合模拟;模拟数字混合仿真
hybrid beam 组合梁
hybrid biofilm reactor 复合式生物膜反应器
hybrid biological reactor 复合式生物反应器
hybrid biological treatment system 复合式生物处理系统
hybrid bond 杂化键
hybrid bridge 组合桥(梁);混合结构桥
hybrid bridge circuit 差动式桥式电路
hybrid cable-supported bridge system 斜拉—悬索组合体系桥
hybrid channel assignment 混合信道指配
hybrid circuit 混合电路;波导管 T 形接头
hybrid coating 混合涂料
hybrid coil 混合线圈;混合变压器;桥接岔路线圈;等差作用线圈;等差作用变量器;差动线圈
hybrid complex 杂化配合物
hybrid composite 混杂复合材料
hybrid computation 混合(物)计算
hybrid computer 混合型计算机;混合(式)计算机(模拟—数字)
hybrid construction 混合结构;混合建筑
hybrid container ship 两用集装箱船;滚吊船
hybrid cooler 组合式冷却器
hybrid cooling system 混合式冷却器
hybrid decoding 混合译码;混合解码
hybrid diagnostic program(me)混合诊断程序
hybrid digital analogy computer 混合数字模拟计算机
hybrid drive 混合动力驱动
hybrid effect 混杂效应
hybrid element 杂交元;混合有限元
hybrid-element method 混合单元法
hybrid engine 复合发动机
hybrid error control 综合差错控制;混合差错控制
hybrid fastenings 半分开式扣件
hybrid fiber 混杂纤维
hybrid fibre/coax 混合光纤/同轴网
hybrid fibre reinforced composite 混杂纤维增强复合材料
hybrid flowsheet 混杂流程
hybrid function generator 混合式函数产生器
hybrid fusion-fission reactor 聚变—裂变混合反应堆
hybrid gas insulated substation 混合式气体绝缘开关站;混合式气体绝缘变电站
hybrid gas insulated switchgear 混合式气体绝缘开关
hybrid gas-liquid electric(al) discharge 混合气—液放电
hybrid girder 不同强度钢材组成钢板梁
hybrid hinge nickel-plated 镀镍复合式铰链
hybrid-hollow-fiber membrane bioreactor 复合式中空式纤维膜生物反应器
hybridigation 杂化
hybrid integrated circuit 混合(式)集成电路
hybrid interface 混合接口
hybrid ion 兼性离子
hybridism 杂化作用;混杂作用;混合性
hybridization 杂化
hybridization sedimentation 混杂沉积

hybridized orbital 杂化轨道
hybrid library 混合程序库
hybrid main program(me)混合主程序
hybrid mapping 混合成像法
hybrid mapping system 混合式测图系统
hybrid material 混杂材料
hybrid matrix 混杂基体
hybrid membrane aerated biofilm reactor 复合式膜曝气生物膜反应器
hybrid membrane bioreactor 复合式膜生物反应器
hybrid metal 石墨化钢
hybrid method 混合法
hybrid model 杂交模型;杂化模型;混合模型
hybrid modulation 混合调制
hybrid mortgage 混合抵押贷款
hybrid multiplex modulation 复合调制
hybrid multiplex modulation system 混合多重调制系统;复合调制方式
hybrid multiplier 混合倍增器
hybrid multivibrator 复合式多谐振荡器
hybrid operating system 混合操作系统
hybrid operation 混运
hybrid optic(al)-digital processing system 光学数字混合处理系统
hybrid orbital 杂化轨道
hybrid parameter 杂系参数;混合参量
hybrid place name 混杂语地名
hybrid platform 混合式平台
hybrid plotter 混合绘图仪
hybrid process 杂化流程
hybrid processing 混合处理(法)
hybrid propellant 固体液体燃料
hybrid radar-infrared system 雷达红外混合系统
hybrid reaction pond 混合反应池
hybrid receiver 混合式接收机
hybrid reinforcement 混杂增强材料
hybrid repeater 混合变压器
hybrid repeating coil 差动转电线圈
hybrid reserve 混合储备金
hybrid rock 混合岩;混染岩
hybrid rock filter 滤池
hybrid ro-ro container ship 滚装吊装两用集装箱船
hybrid ro-ro/lo-lo vessel 滚装/吊装兼用船
hybrid scanning radar system 混合扫描雷达系统
hybrid sedimentation 混合沉积
hybrid set 二线四线变换装置
hybrid software 混合软件
hybrid solar house 混合式太阳房
hybrid stereoplotter 混合立体测图仪
hybrid structure 杂交结构
hybrid system 混杂系统
hybrid telescope 混合式望远镜
hybrid transformer 混合变压器
hybrid treatment system anaerobic pond and trickling filter 复合式处理系统—厌氧塘和滴滤池
hybrid-type 混合型的;桥接岔路型
hybrid wave 混合波
hybrid wave function 混合波函数
hybrid waveguide junction 混合波导管连接
hycar 合成橡胶
Hycomax 铝镍钴系永久磁铁
hydantoin propionate 丙酸乙内酰脲
hydathode 排水器
hydatogenesis 液成作用;热液成矿作用;水成作用
hydatogenic 水成的
hydatogenous 液成的;热液成矿的;水成的
hydatogenous mineral 水成矿物
hydatogenous rock 水成岩;液成岩
hydatogen sediment 液成沉积物
hydatomorphic 水成的
hydatophytia 淹没植物群落
hydatopyrogenic 火水成的
Hyde Park 海德公园(伦敦)
hydoxyl proline 羟基脯氨酸
Hydra 水螅(属)
hydrabarker 水力剥皮机
hydrabeating 水力打浆
hydrability 水化性;水合性
hydrabrusher 水力碎浆机
hydracyclone 水力旋风器
hydradephaga 水生食肉类
hydra-electric(al)system 液压电气系统
hydraeroperitoneum 水气腹

hydrafiner 水化精磨机
hydraguide 油压转向装置
hydral 水合氯醛
hydra-leg 液压凿岩机腿
hydralime 熟石灰
hydralsite 水硅铝石;高铝叶蜡石
hydramatic 油压自动式;液压自动式
hydramatic transmission 液力传动;液压自动控制传动装置
Hydra metal 海德拉合金钢
hydramine 醇胺
Hydramoeba 水螅变形虫
hydrangea 八仙花
hydrant 消火栓;消防柱;消防栓;给水栓;救火龙头;灭火龙头
hydrant barrel 消火栓筒;消防筒;消防龙头套筒
hydrant bend 消防龙头;给水弯头
hydrant bonnet 消火栓帽;消防栓帽;消防龙头护罩
hydrant cabinet 消防箱
hydrant cap 消火栓管嘴帽;消防栓管嘴帽
hydrant connector 消防栓连接器
hydrant drain 消火栓泄水阀
hydrant drain valve 消防栓泄水阀
hydrant-flow calculation 消防流量计算
hydrant-flow test 消防流量测定
hydrant for passenger train 客车给水栓
hydrant outside a building 屋外消防栓;屋外给水龙头
hydrant pipe 消防水管
hydrant pit 水龙头坑
hydrant pitometer 消防栓流速计;消火栓毕托计;消防栓毕托计
hydrant size 消火栓尺寸;消防栓阀座口径;消防栓尺寸
hydrant standpipe 消防栓竖管
hydrant stem 消防栓;水龙头管颈
hydrant street 街道消防水井
hydrant tee 消防水栓三通
hydrant valve 消防龙头阀
hydrant well 消防井
hydrarch succession 水生演替;水生演变
hydrargillite 水铝氧;三水铝石;三水铝矿
hydrargyria 汞中毒;水银中毒
hydrargyrism 汞中毒;水银中毒
hydra-rib bearing 流体预过盈轴承
hydrashear cutter 液压切断器
Hydra steel 海德拉热模具钢
hydratability 水合能力
hydratable 能水合的
hydrate 水化(物);水合物
hydrate alkalinity 氢氧化物碱度;水合物碱度;水合物碱量
hydrate conversion 水化物转化
hydrated 水化的
hydrated alumina 氧化铝水合物;水合氧化铝;三水合氧化铝
hydrated alumin(i)um oxide 三水合氧化铝
hydrated alumin(i)um silicate 水合硅酸铝
hydrated aluminium sulfate 水合硫酸铝
hydrated basic 水合碱
hydrated basic carbonate of copper 水化碱性铜碳酸盐
hydrated calcium uranyl phosphate 水合磷酸铀铣钙
hydrated cation 水合阳离子
hydrated cellulose 水解纤维素;水合纤维素
hydrated cement 水化水泥
hydrated cement paste 已水化水泥
hydrated chromium green 水合铬绿
hydrated chromium oxide 水合氧化铬
hydrated chromium oxide green 水合氧化铬;铬绿
hydrated clay 水合黏土
hydrated concrete 水化混凝土
hydrated dolomitic lime 水化碳酸钙镁;水化白云石灰
hydrated electron 水合电子
hydrated electron dosimeter 水合电子剂量计
hydrated electron dosimetry 水合电子剂量测定法
hydrated ferric oxide 水化氧化铁
hydrated form 水合式
hydrated grease 水合润滑脂
hydrated halloysite 水埃洛石
hydrated hydraulic lime 水化水硬石灰
hydrated hydrazine 水合肼

hydrated ion 氢氧离子;水合离子
hydrated ion radius 水合离子半径
hydrated iron oxide 水化氧化铁
hydrated layer 水化层
hydrated lime 消石灰;氢氧化钙;水化石灰;熟石灰
hydrated lime powder 水化粉石灰
hydrated lime process 水化石灰法
hydrated lime putty 水化石灰膏泥
hydrated lime silicate 水化石灰硅酸盐
hydrated magnesium alumin(i)um silicate 水合硅酸铝镁
hydrated magnesium lime 水化碳酸钙镁;水化白云石灰
hydrated matter 水合物
hydrated mercurous nitrate 水合硝酸汞
hydrated metallic ion 水合金属离子
hydrated micelle 水化胶束
hydrated mineral 含水矿物;水化矿物
hydrated natural aggregate 水化的天然集料;水化的天然骨料
hydrated nitrocellulose 水化硝酸纤维素
hydrated oxide 氢氧化物
hydrated peroxide 水合过氧化物
hydrated potash 水合碳酸钾
hydrated proton 水化质子
hydrated radius 水化半径
hydrated salt 水合盐
hydrated sheath 水化层
hydrated silica 硅酸;水化硅氧
hydrated soap 水合皂
hydrated specifically adsorbed species 水合特性吸附物种
hydrated stock 水化浆料
hydrated sulphate of lime 石膏
hydrated titanium dioxide 水合二氧化钛
hydrated white lime 水合白石灰
hydrate heat 水合热
hydrate inhibitor 水合抑制剂;水合物抑制剂
hydrate of alumin(i)um 氢氧化铝
hydrate of lime 消石灰;石灰的水化物;熟石灰
hydrate of sodium 苛性钠;烧碱
hydrate process 水合过程
hydrate putty 水化石灰膏
hydrate storage 水化法储[贮]存
hydrate water 结合水;水合水
hydrating 水化的;水合的
hydrating capacity 水化性能
hydration 水化(作用);水合(作用)
hydration activity 水化活性
hydration at negative temperature 负温度水化
hydration catalyst 水合催化剂
hydration chemistry 水化化学
hydration classifier 水力分级机
hydration dating 黑曜岩水化年代测定法;水化层法年代测定
hydration degree 水化程度
hydration depth 水化深度
hydration disclo(u)ration 水化变色
hydration energy 水化能
hydration heat 水化热(气);水合热
hydration heat of lime 石灰水化热
hydration induced lump 水硬性大块
hydration isomerism 水合同分异构
hydration mantle 水合衣
hydration number 水合数
hydration of cement 水泥水化(作用);水泥水合作用
hydration of olefines 烯烃的水化作用
hydration period 水化期
hydration plant 水合装置
hydration polishing 水合抛光;水合精磨法;水合光泽处理
hydration process 水化过程
hydration product 水化产物
hydration rate 水化率;水合率
hydration reaction 水化反应
hydration reactor 水合反应器
hydration retarder 水化阻滞剂
hydration sheath 水化膜
hydration sphere 水合层
hydration structure 水化构造
hydration temperature 水化温度
hydration tendency 抗水化性
hydration value 水合值

hydration water 水化水;吸附水;结合水;水合水
hydration weakening 水弱化作用
hydratisomery 水合同分异构
hydratogensis 液成
hydratopneumatic 气水生成的
hydrator 水化器;水合器
hydratruck 液压传动起重车
hydraucone 锥形尾水管;喇叭口;水力圆锥体
hydraucone draft pipe 喇叭形尾水管
hydraucone draft tube 喇叭形尾水管
hydraucone tube 水力锥式尾水管
hydraucone type draft tube 水力锥式尾水管;锥形尾水管
hydrauger 水力钻具;水力螺旋钻
hydrauger hole 水冲钻孔;水力钻具钻进的钻孔
hydrauger(hole)method 水冲钻孔法;水力钻探法
hydraugillite 水铝矿
hydraulic 液压的;水压的;水力(学)的
hydraulic accumulator 蓄液器;蓄水池;水压蓄能器;液力蓄能器;水力储蓄器
hydraulic action 液压传动,水压作用;水力作用
hydraulic activity 水硬性(水泥);水凝性
hydraulic actuated excavator 液压挖土机;液压挖掘机
hydraulic actuating cylinder 液压缸;液压动作筒
hydraulic actuation 液压作动;水力开动
hydraulic actuation system 液压作动系统
hydraulic actuator 液压作动器;液压执行元件;液压执行机构;液压驱动器;液压动作筒;液压促动器
hydraulic additive 水硬性外加剂
hydraulic admixture 水硬性混合料;水硬性添加剂;水硬性掺和料
hydraulic agent 水硬剂
hydraulic agitation 液力搅拌
hydraulic agitator 液力搅拌器
hydraulic air compressor 水力压气机
hydraulic air pump 射水空气泵
hydraulic alarm 水力警铃
hydraulically actuated brake 液压制动器
hydraulically actuated disk type brake 液压驱动盘式制动器
hydraulically actuated wet disk type brake 液压驱动湿式盘式制动器
hydraulically actuated wet multiplate clutch 液压作动湿式多片离合器
hydraulically assisted steering 液压控制;液压操纵
hydraulically automatic 液压自动式
hydraulically available energy 有效水能
hydraulically balanced spindle 液压平衡式接轴
hydraulically balanced valve 液压平衡阀
hydraulically circulated clarifier 水力循环澄清池
hydraulically controlled brake 液压控制式制动器
hydraulically controlled bulldozer 液压操纵推土机
hydraulically controlled clutch 液力操纵离合器
hydraulically controlled plow 液压控制犁
hydraulically controlled scraper 液压控制式铲运机
hydraulically cushioned protection 液压减振装置;液压减振设施;液压减振防护
hydraulically damped seat 液压减振座椅
hydraulically designed sprinkler system 液压(结构的)喷水灭火系统
hydraulically driven baler 液压传动的捡拾打捆机
hydraulically driven mid-mounted mower 液力传动中间悬挂式割草机
hydraulically driven mower 液力传动式割草机
hydraulically driven pump 液压传动泵
hydraulically driven transverse propeller 液压驱动横向推进器
hydraulically equivalent particles 水力等效颗粒
hydraulically equivalent size 水力等效粒径;水力等价粒径;等容粒径;等沉速粒径
hydraulically expanded rubber die 液压成形用橡胶模
hydraulically filled 水力填筑的;吹填(的)
hydraulically operated 液压操作的;液压驱动的;液压操纵的;液力操纵的
hydraulically operated clutch 液力驱动的机械式离合器
hydraulically operated connector 液压连接器
hydraulically operated control 液压控制
hydraulically operated controller 液压控制器
hydraulically operated direction control valve 液压操纵的方向控制阀
hydraulically operated equipment 液压(操作)设备;液压驱动装置

hydraulically operated fixture 液压夹具
hydraulically operated inter-locking device 液压联锁装置
hydraulically operated lift 液压起落机构
hydraulically operated ramps 液压跳板
hydraulically operated rear-dump truck 液压操纵后卸货车
hydraulically operated side-tipping wagon 液压侧倾卸车
hydraulically operated valve 液动阀
hydraulically powered pump 液压驱动泵
hydraulically powered side thrust 液力侧推力器
hydraulically powered side thruster 液力转向装置
hydraulically powered silage grab 液压传动青储[贮]料抓斗
hydraulically powered suspension leveling system 液力悬挂自动调平系
hydraulically powered winch 液力传绞车;液力传动绞车
hydraulically pressed 液压的
hydraulically profitable section 水力上经济的断面;水力经济断面
hydraulically raised plow 液压起落犁
hydraulically rough 水力粗糙的
hydraulically rough zone 水力粗糙区
hydraulically smooth 水力光滑的
hydraulically smooth pipe 水力光滑管
hydraulically smooth zone 水力光滑区
hydraulic amplifier 液压放大器
hydraulic analog(ue) 水力学模拟;水力相似;水力比拟;液压模拟
hydraulic analog(ue)model 水力(学)模拟模型
hydraulic analog(ue)table 水力模拟盘
hydraulic analog(ue)water channel 水力模拟水槽
hydraulic analogy 流体动力模拟;水(力)模拟
hydraulic analogy water channel 水力模拟水槽
hydraulic analysis 水力分析
hydraulic and circular clarifier 水力旋式澄清池
hydraulic and mechanical drive 液力机械传动
hydraulic architecture 水工建筑技术;水工建筑(学)
hydraulic arm 液压臂;液压机构工作臂
hydraulic ash conveyer 水力输灰装置
hydraulic ash ejector 水力冲灰沟器
hydraulic ash removal 水力除灰
hydraulic-ash sluicing 水力除灰
hydraulic assist(ance) 液压辅助(装置);水力辅助(装置);水力助推
hydraulic atomization 液力喷雾
hydraulic attachment 液压附加装置
hydraulic auger 液力传动螺旋
hydraulic autocrane 液压汽车起重机
hydraulic auto-lift 液压汽车升降机
hydraulic automatic 液压自动(装置)
hydraulic automatic leveling device 液压自动找平装置
hydraulic automatic mortising machine 立式榫槽机
hydraulic automatic weld 液压自动焊接
hydraulic average depth 平均水力深度
hydraulic average radius 平均水力半径
hydraulic axis 坡降线;水力轴线
hydraulic backacter 液压反铲挖土机
hydraulic backfilling 水砂充填;水力回填
hydraulic(back)hoe 液压反(向)铲
hydraulic backhoe excavator 反铲液压挖掘机;液压式反铲挖掘机
hydraulic back-pressure valve 液压止回阀;水压逆止阀
hydraulic baker 液压剥树皮机
hydraulic balance 液压式平衡;水力平衡;测功水力秤
hydraulic balance system 液压平衡系统
hydraulic balancing disc 液压平衡盘
hydraulic bale tension control 草捆密度液压调节
hydraulic baling press 水力打大包机
hydraulic bar bender 液压钢筋弯曲机
hydraulic bar cold header 液压钢筋冷镦机
hydraulic barker 水力剥皮机
hydraulic barking 液压剥皮法;水力剥皮
hydraulic barrel lifter 水力桶状升降机
hydraulic barrier 折流挡板
hydraulic beamer 水力卷轴机

hydraulic bedplate 液压操纵的底板
hydraulic behavior 水力性能
hydraulic behavior of flow 水流水力特性
hydraulic bellows 鼓液皮老虎
hydraulic belt 扬水带;水力传送带
hydraulic bender 液压弯折机;水力弯管机;液压弯曲机;液力压弯机;水力折弯机
hydraulic bending machine 液力弯曲机;水力弯机
hydraulic bent sub 液力斜接头
hydraulic bias 液压锁紧器
hydraulic binder 水硬性胶结剂;水硬性黏结料;水硬性结合料;水硬性水凝材料;水硬性胶合剂;水力结合器
hydraulic binding agent 水硬性胶凝材料;水硬性胶结剂;水硬性黏结料;水硬性胶结料;水硬性胶合剂
hydraulic binding material 水硬性胶凝材料
hydraulic binding medium 水力黏合剂
hydraulic biofilm load 水力生物膜负荷
hydraulic bit horsepower 液压钻头功率
hydraulic blast 水曝清砂;水力清砂
hydraulic blasting 水力爆破
hydraulic block 水压台;水压块
hydraulic block-making machine 液压制砌块机
hydraulic blow 水力冲击
hydraulic blow out preventer 液压防喷器
hydraulic boiler test 锅炉水压试验
hydraulic bolt forcer 液力螺栓机
hydraulic bolting device 液压上紧(螺钉)装置
hydraulic bolting machine 液压锚杆安装机
hydraulic bond 水硬性结合剂
hydraulic bonding agent 水硬性胶结料
hydraulic boom 液压吊臂
hydraulic boom machine 液压支臂切削掘进机
hydraulic booster 液压助力器;液压增力器
hydraulic booster steering 液压助力转向
hydraulic booster system 液压辅助系统
hydraulic bore 明渠涌浪;河口涌潮;水力(激)波
hydraulic borehole ga(u)ge 液压钻孔应变计
hydraulic borehole stressmeter 液压钻孔应力计
hydraulic boring lathe 液压镗床
hydraulic boring machine 液压镗床;水力镗床
hydraulic bottomhole pump 水力井下泵
hydraulic boundary condition 水力边界条件;边界水力条件
hydraulic brake 液压制动(器);液压式制动装置;液力制动(器);液动闸;闸式水力测功器;水压闸;水力制动器;水力制动闸
hydraulic brake booster 液压制动加力器
hydraulic brake controls 液压制动操纵机构
hydraulic brake drain tube 液压制动系放气软管
hydraulic braked vehicle 液压制动车辆
hydraulic-brake filler 液压制动系统加油器
hydraulic brake fluid 液压制动液
hydraulic-brake hose 液压制动软管
hydraulic brake oil 制动液
hydraulic brake operating valve 液力制动操纵阀
hydraulic brake shaft 液力制动轴
hydraulic brake system 液力制动系统
hydraulic braking 液压制动
hydraulic braking cylinder 液压制动缸
hydraulic breaker 液压破碎机;水压破碎机
hydraulic breaking 水力破碎
hydraulic breakwater 喷水式防波堤;水力防波堤
hydraulic bronze 耐蚀(铅)锡黄铜
hydraulic bucket loader 液压斗式装载机
hydraulic buffer 液压减振器;液压缓冲器;液力缓冲器;水力消能器;水力缓冲器
hydraulic bulge forming 液压膨胀成形
hydraulic bulge test 液压胀形试验
hydraulic bulldozer 水力开土机
hydraulic bullgrader 水力平路机
hydraulic bumper 液压减振器;液压缓冲器;液力保险杠
hydraulic bumper cylinder 液压保险杠液力缸
hydraulic burster 液压破碎机
hydraulic bursting test 水压爆破试验
hydraulic cabinet 液压操纵箱
hydraulic calcium silicate 水硬性硅酸盐
hydraulic calculating 水力计算
hydraulic calculation 水力(学)计算;水工计算
hydraulic calculation for channel regulation 航道整治的水力计算

hydraulic calculation for dredge-cut 挖槽水力计算
hydraulic calender 水压轧光机
hydraulic calibrating press 水力定型机
hydraulic capliper 液压孔眼测径器
hydraulic capstan 液压绞盘;液力绞盘;水力起锚机;水力绞盘
hydraulic capsule 液压传感器;液压测力仪;液压测力计;液压薄膜测力计;膜式水力测压器;水力囊
hydraulic cargo winch 液压起货机
hydraulic car retarder 液压式缓行系统
hydraulic cartridge 液体炸药筒;水力爆破筒
hydraulic casing 液压下套管(法)
hydraulic casing extractor 液压压拔套管机
hydraulic casing jacking unit 液压套管千斤顶
hydraulic casing machine 液压装箱机
hydraulic cavitation 水压空化
hydraulic cell 液压压力盒;液压传感器;液力测压计;液力测力计
hydraulic cellar 液压系统地下室
hydraulic cement 水硬性胶凝材料;水硬(性)水泥;水凝水泥
hydraulic cement and lime composition 水化水泥和石灰成分
hydraulic cement concrete 水硬水泥混凝土
hydraulic cementing 液压灌浆
hydraulic cementing agent 水凝黏合剂;水力黏合料
hydraulic cementing material 水硬性胶凝材料;水力黏合料;水硬性黏结料
hydraulic cementitious material 水硬性胶凝材料
hydraulic cement mortar 水硬水泥砂浆
hydraulic centrifugal cleaner 水力离心除渣器
hydraulic chamber 水压室
hydraulic change-over 液力换向机构
hydraulic character for rock 岩土的水理性质
hydraulic characteristic 水力(学)特性;水力特征
hydraulic characteristic curve 水力特征曲线;水力特性曲线
hydraulic chart 水力传动图
hydraulic checking cylinder 液压制动缸
hydraulic chock 液压支垛
hydraulic chuck 液压卡盘;自动钻头夹盘;水力夹头
hydraulic circuit 液压(循环)管路;液压回路;液压传动系统图;液动循环管路
hydraulic circuitry 液压阀
hydraulic circular sawing machine 液压圆锯床
hydraulic circulating clarifier 水力循环澄清池
hydraulic circulating head 循环头(冲击钻);水力循环给水器
hydraulic circulating system 液压循环系统
hydraulic clamp 液压(式)夹紧装置;液动压板;液动夹板
hydraulic clamping device 液压夹具;液力夹具
hydraulic clamshell trenching machine 液压抓斗挖槽机
hydraulic classification 液压分类;水力分选;水力分级
hydraulic classifier 水筛机;水力分选器;水力分粒机;水力分级机
hydraulic cleaning 水力清洗;水力清除法
hydraulic clutch 液压离合器;水力联轴器
hydraulic clutch assy 液力离合器总成
hydraulic coal mining 煤矿水采;水力采煤
hydraulic coefficient 水力系数
hydraulic cog 液压嵌齿;液压轮牙
hydraulic cold-drawing machine 液压冷拉机
hydraulic cold header 液压冷镦机
hydraulic compacting press for magnetic material 磁性材料制品液压机
hydraulic compartment 液压舱
hydraulic component 液压机件;液压元件
hydraulic compressor 液压压气机;水力压缩机
hydraulic computation 水力学计算;水力计算
hydraulic computation for channel regulation 航道整治的水力计算
hydraulic computer 液压计算机
hydraulic concrete 水工混凝土;水凝混凝土
hydraulic concrete pump 液压混凝土泵;水工混凝土泵
hydraulic condition 水力条件;水力工况
hydraulic conductivity 水力传输系数;透水性;水渗导性;水力传导性;水力传导系数;水力传导率;水力传导度;渗透系数;导水性;导水率
hydraulic conductivity of soil 土壤水分导电性

hydraulic conductivity unsaturated soils 不饱和土壤水分导电性
hydraulic conduit sections 过水断面;液压管道部件;水工管道部件
hydraulic conduit transport 管道水力运输
hydraulic connection 水力连系
hydraulic connector 液压连接器
hydraulic consolidation cell 液压固结盒
hydraulic constituent 水力要素
hydraulic control 液压控制;液压空气液压控制;液压操纵;液力控制;水力(传动)控制;水力操纵
hydraulic control apparatus 液压操纵装置
hydraulic control box 液压控制箱
hydraulic control circuit 液压控制管路
hydraulic control component 液压控制元件
hydraulic control cylinder 液压操纵油缸
hydraulic control device 液压控制装置;液压操纵装置
hydraulic control lathe 液压控制车床
hydraulic-controlled bulldozer 液压操纵推土机
hydraulic-controlled motor 液压控制马达
hydraulic-controlled scraper 液压式铲运机
hydraulic controller 液压控制器;液压控制装置;水力控制器
hydraulic control lever 液压控制手柄
hydraulic control of disk clutch 液压操纵盘状离合器
hydraulic control pump 液压控制泵
hydraulic control system 液压控制系统;液压操纵系统
hydraulic control valve 液压控制阀
hydraulic conversion coefficient 水功率转换系数
hydraulic converter 液力变矩器
hydraulic conveyer 液力输送系统;液力输送机;水力输送机
hydraulic conveying 液压输送;水力输送
hydraulic conveying device 水力输送装置
hydraulic copying attachment 液压仿形刀架
hydraulic copying miller 液压仿形铣床
hydraulic copy shaper 液压仿形牛头刨床
hydraulic core 水填心墙
hydraulic core extractor 液压岩芯退取器;水力岩芯提取器;水力岩芯提取器
hydraulic core pump 液压退芯泵
hydraulic coupler 液压系统管接头;液力耦合器;液力联轴器
hydraulic coupling 液压联轴器;液压联轴节;液力联轴节;液力联接器;液力管接头;水压联轴器;水力联轴器
hydraulic crane 液压起重机;液力起重机;水压起重机;水力起重机
hydraulic crawler drill 液压履带式钻机
hydraulic crawler excavator 液压履带式挖土机;液压挖掘机;液压履带式挖掘机
hydraulic crawler loader 液压履带式装载机
hydraulic crowd(ing) 液压皱纹形成
hydraulic crusher 液压破碎机
hydraulic cupola hoist 化铁炉用液压起重机
hydraulic current 落差流
hydraulic cushion 液压缓冲
hydraulic cushion guardrail 液压缓冲护栏
hydraulic cutter 水力切割器
hydraulic cyclinder 液压圆筒式启闭机
hydraulic cyclinder feed 液压缸给进
hydraulic cyclone 旋液分离器;水力旋转分粒机;水力旋流(分离)器
hydraulic cyclone elutriator 液力回旋淘析器;水力旋流淘析器
hydraulic cylinder 液压缸;水压缸
hydraulic cylinder bleeder screw 液压缸放出管螺钉
hydraulic cylinder gasket 液压缸垫密片
hydraulic cylinder hoist 液压圆筒式启闭机
hydraulic cylinder piston 液压缸活塞
hydraulic cylinder piston stop rod 液压缸活塞止动杆
hydraulic cylinder plug 液压杆塞
hydraulic cylinder return spring 液压缸回动弹簧
hydraulic cylinder seal 液压缸油封
hydraulic cylindric(al) grinding machine 液压外圆磨床
hydraulic dam 水坝;水力冲积堤
hydraulic damage 水力破坏
hydraulic damper 油压减振器;液压减振器

hydraulic damper test stand 液压减振器试验台
hydraulic damping 液压阻尼
hydraulic damping device 液压缓冲装置
hydraulic decoking 水力清焦
hydraulic defibering machine 水力解棉机
hydraulic depth 水力深度
hydraulic derrick 液压吊货杆
hydraulic derricking ram 液压改变起重臂倾角的推杆;液压改变起重臂倾角的顶杆
hydraulic descaling system 水力除鳞系统
hydraulic design 液压计算;水利工程设计;水力设计
hydraulic design criteria 水力设计准则
hydraulic detention time 水力停留时间
hydraulic development 水利枢纽;水利开发;水力枢纽;水力开发
hydraulic development project 水电开发工程
hydraulic device 液压装置;液压设备
hydraulic diagram for heat supply 热网水压图
hydraulic diameter 液力直径;水力直径
hydraulic-diaphragm micropipetter 液压隔膜微量移液管
hydraulic die cushion 液压垫
hydraulic die press 模锻水压机
hydraulic diesel locomotive 液力传动内燃机车
hydraulic differential 液压差速器
hydraulic differential analyzer 液压微分分析器
hydraulic differential drive 液压差速传动
hydraulic diffusibility 液压渗透率
hydraulic digging attachment 液压挖掘附件
hydraulic dilatation 液压扩张术
hydraulic discharge 流量;水力排料;地下水出流量;产水率
hydraulic disc-type brake 液压盘形制动器
hydraulic disk brake 液压盘式制动器
hydraulic disorder 水力失调
hydraulic display panel 液压操纵显示板
hydraulic ditcher 液压挖沟机
hydraulic ditching shovel 液压挖沟铲
hydraulic dock (旧式的)液压升降船坞;水压升降船坞;水力(起重)船坞
hydraulic dolenoid valve 液压螺旋管阀
hydraulic double action jack 液压双动千斤顶
hydraulic down-stroke press 水力下行压机
hydraulic drag 水阻力;水流力;水力挖泥机;水力牵引
hydraulic drag bucket 液压污泥斗
hydraulic drag-shovel 液压拖铲挖泥机
hydraulic drawbar control 牵引装置液压操纵
hydraulic dredge(r) 水力冲泥机;吸扬式挖泥船;吸扬式疏浚机;挖泥船;水力吸式挖泥机;水力挖泥船;水力疏浚机
hydraulic dredging 用吸扬式挖泥船挖泥
hydraulic dressing machine 液压锻钎机
hydraulic drift net capstan 液压流网绞盘
hydraulic drift net capstan fluid connector 液压连液压流网绞盘
hydraulic drill 液压钻(眼)机;液压凿岩机;水力钻(机)
hydraulic drill collar 液压钻机套圈;液压钻机套环;液压钻机套管
hydraulic drilling 液压钻孔;液压凿岩;泥浆洗孔钻进;泥浆冲洗钻进;水力钻探;湿法凿岩;湿法打眼
hydraulic drilling control 液压进给调节器
hydraulic drill jumbo 液压钻臂台车
hydraulic drill rig 液压钻探设备;液压钻车
hydraulic drill ring 液压凿岩台车
hydraulic drive 液力传动;水力传动
hydraulic drive circuit 液压传动回路
hydraulic drive for cutter 铰刀液压传动
hydraulic driven driller 全液压钻机
hydraulic driver for cutter 铰刀液压传动
hydraulic driving machine 液压传动机械
hydraulic drop 跌水;水力落差
hydraulic drop hammer 液压落锤
hydraulic dual brake 液力双制动器
hydraulic duckham 液压吊货称
hydraulic dump 水力排土场
hydraulic dump rake 液压横向搂草机
hydraulic dump truck 水力自动倾卸车
hydraulic dynamometer 液压测力仪;液压测力计;液压测功器;液力测力计;水力功率计;水力测力计;水力测功器;水力测功率;水力测力计;水力测功机

hydraulic earth-fill cofferdam 水力填土围堰
hydraulic efficiency 液压效率;流体效率;水力效率
hydraulic ejector 排砂泵;液压脱泥机;液压排泥管;喷射式排泥管;排沙泵;水力喷射器;水力排泥泵;射流式喷射装置
hydraulic-electric(al) lift 液压电力起落机构
hydraulic element 液压元件;水力元件;水力因素;水力要素
hydraulic element test 液压元件试验
hydraulic elevating table 液压升降台
hydraulic elevator 液压提升机;液压(驱动)电梯;液压升降机;液压顶升机;水压升降机;水力提升机
hydraulic elongator 液压拉伸器
hydraulic elongator car 液压拉伸器小车
hydraulic embankment 水工堤坝
hydraulic emptying system 泄水系统;排水系统
hydraulic end loader 液压端部装载机
hydraulic energy 水能;水力能源
hydraulic engine 液压发动机;水压机;水力(发动)机
hydraulic engineer 水利工程师
hydraulic engineering 水利工程(学);水力工程(学);水工学
hydraulic engineering complex 综合利用水利工程;水利枢纽
hydraulic engineering concrete 水工混凝土
hydraulic engineering laboratory 水工(工程)试验室
hydraulic engineering research 水工研究
hydraulic engineering structure 水工建筑物;水工结构
hydraulic engineering study 水工研究
hydraulic engineering system 水利系统
hydraulic engineering works 水工建筑物
hydraulic equilibrium 水力平衡
hydraulic equipment 液压设备;液压装置;水力设备
hydraulic equipment for tipping 倾斜用液压设备
hydraulic equivalence 水力等效
hydraulic erosion 水力磨蚀
hydraulic erosion dredge(r) 水力冲刷挖泥船;水力冲刷挖泥船
hydraulic evapotranspirometer 水力式土壤蒸腾计
hydraulic excavation 水力挖掘;水力开挖;水力开采;水力冲挖;疏浚
hydraulic excavator 液压单斗挖土机;吸泥泵;液压冲采机;水力挖凿机;水力挖土机;水力挖掘机;疏浚机
hydraulic excavator digger 水力挖掘机
hydraulic excavator-loader 液压挖土一装载机
hydraulic expansion wall scraper 液压伸张式井壁刮刀
hydraulic exploration rig 液压钻探机
hydraulic exponent 水力指数
hydraulic extension 液压扩建
hydraulic extension pipe 液压系统加长软管
hydraulic extraction 脱水作用;水力提取;水力开采;水力冲挖
hydraulic extruder 油压挤塑机;液压式压出机;液压式挤出机;水压挤出机
hydraulic extrusion 液压挤压
hydraulic extrusion press 液压挤压机
hydraulic factor 水力因素;水力因数
hydraulic fan 液力风扇
hydraulic fast-travel excavator 液压快速移动挖泥机
hydraulic feed 液压推进;液压进料;液压进刀;液压给进;水力进刀;水力传送
hydraulic feedback 液压反馈
hydraulic feed core drill 液压给进式钻机
hydraulic feed cylinder 液压给进缸
hydraulic feed drill 油压钻机;液压钻机
hydraulic feed equipment 液压给进装置;液压给进式钻机
hydraulic feed mechanism 液压推进机构
hydraulic feed pulpstone doctor 液压式磨石刮刀
hydraulic feed rig 液压给进式钻机
hydraulic feed swivel head 液压给进回转器
hydraulic feed system 液压给进系统
hydraulic feed unit 液压进给机构
hydraulic felling wedge 液压伐木楔
hydraulic fender 液压式防护板;液压护舷;液压防撞装置
hydraulic fill 水力充填(物);淤填;水沙冲填料;水力填筑;水力填方;水力吹填;水力冲填;吹填;冲填土;充填材料

hydraulic fill curve 水力填筑曲线
hydraulic fill dam 淤填坝;水力填筑坝;水力冲填坝;水力充填坝
hydraulic fill embankment 水力填筑路堤
hydraulic filling 水沙充填;水力填土法
hydraulic filling system 灌水系统
hydraulic fill material 水力回填材料
hydraulic fill method 水力冲填法;水力充填法
hydraulic fill pipe 排泥管
hydraulic fill pipeline 淤填管线;排泥管线;水力土管道;输泥管;吹泥管线
hydraulic fill process 水力填筑法;水力吹填法;水力淤填法;吹填法
hydraulic fill(soil) 水力(冲)填土
hydraulic filter 液压滤油器;水力滤器
hydraulic finish 水刷表面处理
hydraulic fitting 液压装置;液压零配件
hydraulic fixed riveter 固定式水力铆机
hydraulic flange press 液压折边机
hydraulic flanging machine 液压折边机;液压弯边机;液压翻边机
hydraulic flanging press 液压翻边压力机
hydraulic flo(a)tation 水力浮选
hydraulic flocculation 水力絮凝
hydraulic floor jack 液压落地千斤顶
hydraulic flow 液压油流量;液压水流量;湍流;水(压)流;水力流(动)
hydraulic flow net 流网;水力流网
hydraulic fluid 液压油;液压用液体;液压液;液压流体;液压机液体;水力刹车油;工作液体
hydraulic fluid discharge plug 放油塞
hydraulic fluid distributor 液压油分配器
hydraulic fluid-lift platform 液动升降台
hydraulic fluid side 液体侧
hydraulic fluid tank 液压箱
hydraulic flume 水力试验槽
hydraulic flume transport 水槽运输
hydraulic flushing 水力冲洗;水力冲沙;水沙充填
hydraulic flushing of sediment 水力冲泥
hydraulic flushing of sewers 水力通沟
hydraulic flute transport 溜槽水力输送
hydraulic foam tower 水力泡沫塔
hydraulic folding hatchcover 液压折叠式舱盖
hydraulic folding marker 折叠式液压划行器
hydraulic foot brake 液压脚制动器
hydraulic force feedback 水力反馈
hydraulic forging 水压锻造
hydraulic forging press 液压锻(压)机;水压锻机;锻造水压机
hydraulic form 水力形状
hydraulic former 水力圆网成形器
hydraulic forming 水压成型
hydraulic forming press 液压成形压力机
hydraulic formula 水力学公式
hydraulic fracture 水力破坏;水力断裂
hydraulic fracture test 水压致裂试验;水力压裂试验;水力断裂试验
hydraulic fracturing 水力劈裂;水力破裂;水力断裂;液压致裂(法);高压水砂破裂法;水压致裂(法);水压胀裂(法);水压致裂;水力断裂作用;水力冲裂法
hydraulic freight elevator 液压货梯
hydraulic friction 水头摩阻;水力摩阻;水力摩擦
hydraulic friction coefficient 水力摩阻系数
hydraulic front-end loader 液压前端装载机
hydraulic front excavator 正铲液压挖掘机
hydraulic fuse 液压保险器;膜片式液压保险
hydraulic gantry crane 液压高架起重机;水力桥式起重机;水力龙门起重机
hydraulic gate 液压闸门;液压封闭器
hydraulic ga(u)ge 液压计;水压计;水力表
hydraulic gear 液压机构;液压传动装置;液压传送装置;液压传动装置;液压传动(机构);液压传递装置
hydraulic gear pump 液压齿轮泵
hydraulic generator 水轮发电机
hydraulic geometry 河相;水力几何学;水力几何形态
hydraulic giant 水力冲射器;高压水枪;水枪;水力冲洗机;水力冲矿机
hydraulic giant nozzle 水枪喷嘴
hydraulic gland 水封套
hydraulic glue 水硬(性)胶;防水胶;防湿胶;水硬

黏结剂
hydraulic gobbing 水力充填
hydraulic governing 液压调节;水力调节
hydraulic governor 液压调速器;液压调节器;水力调节器
hydraulic grab 挖掘机;液压抓斗;液压操纵抓爪
hydraulic grade 水压线;水力坡降;水力坡度;水的坡度
hydraulic grade line 水压线;水面线;水力坡降线;水力坡度线;测压管水头线
hydraulic gradient 液压梯度;水压坡度;水位差;落差梯度;水压坡降线;水压陡度;水力梯度;水力坡降;水力比降
hydraulic gradient analysis 水力梯度分析
hydraulic gradient line 水力梯度线;水力坡降线;水压线
hydraulic grading outfit 水力筛分设备
hydraulic granulation 水淬成粒(法)
hydraulic gravel 水力冲刷砾石
hydraulic gravel fill 流沙充填;水砂充填
hydraulic gravity gradient 水力重力梯度
hydraulic grease gun 牛油枪;黄油枪;润滑脂枪
hydraulic grid 水力网
hydraulic grinder 液压式磨木机
hydraulic gripper feed 液压爪式送料
hydraulic ground-testing machine 水力地面试验机;水工地面试验机
hydraulic guide 液压导向器
hydraulic guillotine cutter 液压截切机
hydraulic gun 水枪;水力冲射器
hydraulic gypsum 水硬石膏
hydraulic hammer 液压锤;水压锤;水(力)锤;水力冲击
hydraulic hand tipping gear 液压手动倾卸装置
hydraulic hanger 液压式悬挂器
hydraulic hardening 水硬硬的;(水泥)的水硬作用
hydraulic haulage 水力运送;水力运输
hydraulic head 液压头;水头;水(力)压头;水力水头
hydraulic head of flow velocity 流速水头
hydraulic height control 高度液压调节机构
hydraulic helicoids drill 液动螺杆钻具
hydraulic high-pressure bit 高压水头钻头
hydraulic hitch 水力式联结器
hydraulic hoe 液压耕耘机
hydraulic hoist 液压启闭机;液压提升机;液压起重机;水力提升机;水力起重机
hydraulic hoisting 水力提升
hydraulic holddown 水力压具
hydraulic hole calipers 液压井径仪
hydraulic hook 液压挂钩;液力钩
hydraulic horsepower 液压马力
hydraulic hose 液压软管;水力软管
hydraulic hose connection 液压软管连接套
hydraulic HP bit 喷射式钻头
hydraulic hydrant 消防水栓
hydraulic hydrated lime 水硬性化石灰;水硬性熟石灰
hydraulician 水利工程师;水力学家
hydraulic impact 水力冲击;水力冲击;水力冲击力
hydraulic inclination 水面坡降;水力坡度
hydraulic index 水硬性指数;水硬率
hydraulic index method 水力指数法
hydraulic indicator 水压计
hydraulic injection mo(u)lding machine 液压注模机
hydraulic installation 液压设备
hydraulic integral controller 液压积分控制器
hydraulic integraph 水力积分仪
hydraulic intensifier 液压增强器;液压放大器;水力增压器
hydraulic interlocking 液压联锁
hydraulic(inter)relation 水力联系
hydraulic inverted press 上压式液压机
hydraulicity 水硬性;水凝性
hydraulicity Vicat index 水凝性指数;水泥凝结指数
hydraulic jack 液压油缸;液压起重臂;油压千斤顶;液压升降车;液压千斤顶;液压起重器;水力千斤顶;水力起重器;立式千斤顶
hydraulic jack separating method 液压千斤顶分离法
hydraulic jack with separate pump 分离式油压千斤顶;分离泵式液压千斤顶
hydraulic jar 液动震击器【岩】
hydraulic jaw crusher 液压颚式破碎机

hydraulic jet 水力射流
hydraulic jet barker 水力冲击剥皮机
hydraulic jet descaler 水力除鳞设备
hydraulic jet mining 水力喷射开采
hydraulic jet mixer 喷射式搅拌器;水力射流搅拌机
hydraulic jet piling 水冲沉桩
hydraulic jet propulsion 喷水推进
hydraulic jetting 水力喷洗
hydraulic jib-type loader 转臂式液压装载机
hydraulic jig 水力跳汰机;水簸机
hydraulic jigger 水力绞车
hydraulic jigging 水力淘选
hydraulic joint 液压接合
hydraulic jumbo 液压台车
hydraulic jump 水跃
hydraulic jump mixing device 水跃混合设备
hydraulic jump property 水跃特性
hydraulic jump wave 水跃波
hydraulic jump wave height 水跃波高
hydraulicker 水力机操作者
hydraulicking 水力挖土;水力开采;水力掘进挖土;水力掘进开采;水力(掘进)冲挖;水掘法;水采法
hydraulic knock-out mechanism 水力出坯机构
hydraulic laboratory 水力试验室;水力实验室;水工试验室
hydraulic lathe 液压车床
hydraulic leak 液压系统的漏油
hydraulic leather 液压装置皮碗
hydraulic leg 液压柱;液压(支)腿
hydraulic levee 水流冲(积)堤;水力淤填堤;水力冲填堤;水力冲积堤
hydraulic leveler 液压平衡器
hydraulic leveling jack 液压调平千斤顶
hydraulic level(l)ing device 流体测沉计
hydraulic lift 液压升降装置;液压升降机;液压起重机;液压顶升机;液压电梯;水压提升机;水压升降机;水力提升机;水力提升;水力升降机;水力起重机
hydraulic lift cover 液压升降机构盖
hydraulic lift dock 液压提升船坞;水力升降船坞;(旧式的)液压升降船坞
hydraulic lift hitch 液压悬挂装置
hydraulic lifting capacity 液压提升能力;水力提升能力
hydraulic lifting cylinder 液压提升油缸
hydraulic lifting jack 液压起重器
hydraulic lifting platform 液压提升平台;水力提升平台
hydraulic lifting system 液压提升系统
hydraulic lifting truck 液压升降车;水力升降车
hydraulic lifting work platform 液压升降平台;水力提升工作台
hydraulic lift linkage 液压联动装置
hydraulic lift linkage drawbar 液压悬挂系统联结装置
hydraulic lift mechanism 液压悬挂机构;液压起落机构
hydraulic lift piler 液压升降堆垛台
hydraulic lift scraper 液压提升式铲运机
hydraulic lift system 液压悬挂系统
hydraulic lift tailgate 液压提升的船闸下游闸门
hydraulic lift truck 液压升降车
hydraulic lime 水硬性石灰;水硬石灰
hydraulic lime mortar 水硬石灰砂浆
hydraulic limestone 水硬性石灰岩;水硬石灰石
hydraulic link 液力测力环
hydraulic lip packing 液压带唇环形密封件
hydraulic liquid 压力液;水力液
hydraulic liquid tank 水力液体箱;水力液体罐
hydraulic load capsule set 液压荷载膜盒式装置
hydraulic load cell 液压传感器;液压测力仪;液压测力计;液压测力盒;水力岩压测量传感器;水力岩石压力测量传感器;水力岩压力盒
hydraulic loader 液压装载机;液压装车机;水力装载机
hydraulic load(ing) 水力荷载;水力负荷;水力装载
hydraulic loading of filter 过滤器(的)水力负荷;过滤池的水力负荷
hydraulic loading rate 水力负荷率
hydraulic loading shovel 液压装载铲
hydraulic loading system 水力荷载系统
hydraulic lock 液压锁;液压卡紧;液压封闭;液压闭锁装置;液压阻塞
hydraulic locomotive 液压机车

hydraulic log splitter 液压劈木机
hydraulic longline gurdy 液压延绳钓机
hydraulic loosening 水力崩落
hydraulic lorry-mounted excavator 装在车上的液压挖土机;装在车上的液压挖泥机
hydraulic loss 液压损失;水力损失
hydraulic luffinghydraulic luffing 液压变幅;水力变幅
hydraulic machine 液压机;液力机
hydraulic machinery 水压机械;液压机械;水力机械
hydraulic machinery of filling and emptying valves 灌水泄水阀液压启闭机械
hydraulic machinery of miter gate 人字门液压启闭机
hydraulic machinery of wedge roller gate 三角闸门液压启闭机械
hydraulic main 液压总管;总水管;水压主管;水管干线
hydraulic maintenance panel 液压维护板
hydraulic mangle 水压轧水机;水压轧机
hydraulic manifold 液压系统分路阀箱
hydraulic manifold block 液压油路板
hydraulic manipulator 液压机械手;水力操纵器
hydraulic mass 流体动力质量
hydraulic material 水硬性建筑材料;水硬性材料
hydraulic matrix 水力黏合料
hydraulic mean depth 平均水深;平均水力深度;水力平均深度;水力半径
hydraulic mean path 水力半径
hydraulic mean radius 平均水力半径;水力平均半径
hydraulic mean ratio 水力平均比例
hydraulic measurement 水力学测量;水力测量学
hydraulic mechanical efficiency 液力机械传动效率
hydraulic mechanism 液压机构
hydraulic medium 液压油;液压介质;工作液体
hydraulic method 液压法
hydraulic method of coke removal 水力清焦法
hydraulic mine 水力采矿
hydraulic mine filling 水力充填
hydraulic miner 水采工人
hydraulic mining 水力开采;水力掘进挖土;水力掘进开采;水力掘进挖沟;水力采矿
hydraulic mining giant 水力开采水枪
hydraulic mining operation 水采作业
hydraulic mining system 水力开采(法)
hydraulic misadjustment 水力失调
hydraulic mixer 水力混合器
hydraulic mix(ture) 水力混合料
hydraulic mobile crane 液压移动起重机;液压汽车起重机
hydraulic mobile excavator 液压移动挖掘机
hydraulic mockup 水工大模型
hydraulic model 水力(学)模型;水力学试验模型;水工模型
hydraulic model of port 港口水工模型
hydraulic model research 水工模型研究
hydraulic model study 水力学模型研究;水工模型研究
hydraulic model test(ing) 水力(学)模型试验;水力型试验;水工模型试验
hydraulic model test of port 港口水工模型试验
hydraulic module 水硬系数
hydraulic modulus 水硬(性)模数;水硬模量
hydraulic moldboard side shift 液压模板尖端;液压模板侧移器;液压模板侧移机
hydraulic moment variator 液压变矩器
hydraulic monitor 水流控制设施;高压水流控制设施;高压水枪;水枪
hydraulic mortar 水硬(性)砂浆;水硬性灰浆;水下速凝砂浆;水凝砂浆;水凝灰浆
hydraulic motor 油压马达;液压马达;液压发动机;液压电(动)机
hydraulic motor drive 液力马达传动
hydraulic motor SAE stall pressure 液压马达的SAE制动压力
hydraulic motor winch 液力马达驱动的绞车
hydraulic mo(u)lding machine 液压造型机;液力造型机
hydraulic mo(u)lding press 水力压型机
hydraulic mucking 水力出碴
hydraulic mule 液力供桩机
hydraulic navvy 水力挖凿机;水力挖土机;水力挖泥机

hydraulic net 水力网
hydraulic nozzle 液压喷嘴
hydraulic nut 液压螺母
hydraulic of sewers 管道水力学
hydraulic of soils 土壤的水压力
hydraulic oil 液压油;高压油
hydraulic oil accoumulator 水压蓄力器
hydraulic oil engine 液压油引擎;液压油发动机
hydraulic oil jack 液压油千斤顶
hydraulic oil press 液压榨油机
hydraulic oil pump 液压油泵
hydraulic oil pump discharge line 液压油泵泄放管路
hydraulic oil pump driving device housing 液压油泵传动装置壳体
hydraulic oil ram 液压撞锤;液压活塞
hydraulic oil-recharge pipeline 液压进油管路
hydraulic oil system 液压油路
hydraulic oil tank 液压油箱;液压油柜
hydraulic oil temperature ga(u)ge 液压油温度表
hydraulic operated 液压控制的;液压操作的;液压操纵的;液力驱动的;水力操作的;水力操纵的
hydraulic operated check valve 液控单向阀
hydraulic operated gate lifting device 液压式闸门启闭机;液压启闭机;水力启闭机
hydraulic operating condition 水力操作条件
hydraulic operating gear 液压千斤顶;液压传送装置;液压传动装置;液压传递装置
hydraulic operation 液压动作;液压操作;水力操作
hydraulic oscillating cylinder 液压摇摆油缸
hydraulic oscillating doctor 液压往复式刮刀
hydraulic oscillator 水力振荡器
hydraulic overflow settlement cell 溢水式沉降仪;溢水式沉降计
hydraulic overflow settlement ga(u)ge 溢水式沉降仪;溢水式沉降计
hydraulic overhead crane 液压桥式起重机
hydraulic overload 水力超载
hydraulic overload device 液压超载保护装置
hydraulic overload door gear 门上液压关门器
hydraulic overloading 水力过负(荷);水力超载
hydraulic package 液压组件
hydraulic packer 水力式封隔器
hydraulic packer holddown 水力封隔器定位器
hydraulic packing 液压装填;液压密封
hydraulic panel 液压系统控制盘
hydraulic parameter 水力参数
hydraulic parameter of rapids abating 消滩水力指标
hydraulic parameters of flow 水流水力参数
hydraulic passenger elevator 液压乘人升降机;液压乘人电梯;液压客梯
hydraulic peeler 水压剥皮器
hydraulic percussion(drilling)method 水力冲击(钻探)法;水力冲击(钻进)法;水力冲填法
hydraulic perforating 液压穿孔
hydraulic performance 水力(学)特性;水力性能
hydraulic permeability 透水性;水压渗透性;水力透水性;水力渗透性;渗透性能
hydraulic permeability tester 液压式渗透性试验仪
hydraulic permeation rate 水力渗透率
hydraulic piercing 液压穿孔
hydraulic piercing press 液压冲孔机
hydraulic piezometer 液压式测力计;孔隙水压力计;水压计;水力(式)测压计
hydraulic pile chuck 液压桩夹
hydraulic pile driver 水力打桩机
hydraulic pile driving 液压打桩;水力打桩
hydraulic pile hammer 液压桩锤
hydraulic pile head splitter 液压桩头破碎器
hydraulic pilot control 辅助液力控制
hydraulic pin puller 液压去栓器;液压去钉器
hydraulic pipe 液压管子截割机;液压管道;水力管(道)
hydraulic(pipe)bender 液压弯管机
hydraulic pipe bending machine 液压弯管机
hydraulic pipe catcher 液动捞管器【岩】
hydraulic pipe cleaner 水力洗管器
hydraulic pipe cutter 液压切管器;水力刮管器
hydraulic pipe line 液压管路;水输管道;水力管线;水力管路;水力管道;水管系;输水管路
hydraulic pipe line dredge(r) 水力管线开挖机;吸扬(式)挖泥船;吸泥式挖泥船;水力排管式挖泥船;水力管线冲泥;水力管路冲泥;水力管道冲泥

hydraulic pipe testing machine 管材水压试验机
hydraulic pipe transport 管道水力运输
hydraulic piping 水管系
hydraulic piston 液压活塞;水力活塞
hydraulic piston actuator 液压活塞式驱动器
hydraulic piston sampler 水压活塞取土器
hydraulic pitch control system 液压螺距操纵系统
hydraulic placement of sand 水力填砂;水力灌砂
hydraulic planer 液压刨床
hydraulic plant 液压设备
hydraulic plaster 装饰石膏;水硬性石灰水泥;水硬性抹灰
hydraulic plate bender 液压弯板机;液力弯板机
hydraulic platform 液压升降塔架
hydraulic plunger 液压柱塞
hydraulic plunger elevator 液压柱塞直顶式升降机;液压柱塞升降机
hydraulic plunger shaft 液压活塞轴;唧筒活塞轴
hydraulic plywood press 胶合液压机
hydraulic pneumatic panel 液压气动控制台
hydraulic pneumatic power hammer 液压空气锤
hydraulic posistioner 液压定位调整装置
hydraulic positional servomechanism 液压位置伺服机构
hydraulic positioner 液压定位器
hydraulic post 液压柱;水力098柱
hydraulic potential 水力位势;水力位能;水力势能
hydraulic potential gradient 水势梯度
hydraulic power 液压功率;液压动力;水能;水力(功率)
hydraulic power assistance 水力辅助设备
hydraulic power control 水力发电控制
hydraulic power development 水电开发
hydraulic power drive 液压传动
hydraulic-powered wiper motor 液压式风窗刮水器的液力驱动器
hydraulic power lift 液压起落机构
hydraulic power line 液压系统
hydraulic power oil 液压油
hydraulic power pack 液压动组件;液压动力机组
hydraulic power plant 水力发电厂;水电站;水电厂
hydraulic power pool 水力联合系统;水力发电储[贮]水池
hydraulic power project 水力发电工程方案;水力发电工程;水电工程项目
hydraulic power pump 液压泵
hydraulic power scheme 水力发电枢纽;水力发电开发方案
hydraulic power site 水力发电厂址
hydraulic power supply 水力的动力供应
hydraulic power system 水力动力系统
hydraulic power take-off 液压驱动功率输出轴
hydraulic power transmission 液压传动
hydraulic power transmitter 水力发送器
hydraulic power unit 液压动力装置;液压动力设备;液压动力机构;液压泵站;液动压力机构
hydraulic power unit for steering 液压转向装置
hydraulic preservation oil 液压防护油
hydraulic press 油压机;液压式压机;液压器;液压冲床;水压机
hydraulic press accumulator 水压机积储[贮]器
hydraulic press brake 水压压弯机
hydraulic press duck 水压机过滤帆布
hydraulic press for free forging 自由锻造水压机
hydraulic pressing machine 液压找包机
hydraulic pression extruder 液压柱塞式压出机
hydraulic press oil 水压机用油
hydraulic press packing 水压机打包;打包机
hydraulic pressure 油压(力);液压(力);水压(力)
hydraulic pressure cell 油压室;液压元件;液压室;液压盒;水压室
hydraulic pressure compensating governor 液压补偿调节器
hydraulic pressure elevator 液压提升装置
hydraulic pressure engine room 油压动力室
hydraulic pressure ga(u)ge 液压计;液压表;液体压力计
hydraulic pressure gradient 液压梯度
hydraulic pressure head 液压压头;水压(压)头;水头
hydraulic pressure indicator 液压表
hydraulic pressure manifold 液压歧管
hydraulic pressure nozzle 液压喷嘴

hydraulic pressure piston coring 液压活塞取样器取样
hydraulic pressure regulator 液压调节器;水力调节器
hydraulic pressure snubber 液压减震器;液压减压器;水压力缓冲器
hydraulic pressure stabilizer 液压稳定器
hydraulic pressure switch 水压开关
hydraulic pressure system 液压系统
hydraulic pressure test 液压试验;水压试验;油压试验
hydraulic pressure tester 液压式压力试验机
hydraulic pressure testing 水压试验
hydraulic pressure treatment 液压处理
hydraulic pressurizing 液压密合
hydraulic press with manometer 带压力计的水压机
hydraulic prime motor 水力原动机
hydraulic prime mover 水力原动力
hydraulic profile 水力(纵)剖面;水力断面
hydraulic project 水利项目;水利工程
hydraulic project management 水利工程管理
hydraulic project operation 水利工程运行
hydraulic prop 液压撑脚;顶杠;液压支柱
hydraulic prop compensating gear 液压支柱补偿装置
hydraulic propeller 喷水推进器;水力推进器
hydraulic property 水力性质;水力特征;水硬性;水力(学)特性
hydraulic property of aquifer 含水层水力性质
hydraulic property of current 水流的水力学特性
hydraulic propulsion 液力驱动;喷射驱动
hydraulic propulsion system 液压推进系统
hydraulic propulsion unit 喷水推进装置
hydraulic prospecting 水力勘探
hydraulic protector 液压保护装置
hydraulic pulldown 液压强制给进机构
hydraulic puller 液压张拉器;液压式拔卸器;液压拉拔器;液压拔杆器
hydraulic pull-shovel 液压索铲;液压拉铲
hydraulic pull tester 液压拉力检定器
hydraulic pump 液压泵;水压泵;水(力)泵
hydraulic pump aggregate 液压泵组
hydraulic pump discharge 液压泵泄放
hydraulic pumping unit 液压泵装置;液压泵汲设备
hydraulic pump lever 液压泵接合杆
hydraulic pump placer 液力泵喷浆机
hydraulic pump sprayer 水力泵喷雾器
hydraulic pump(unit) 液压泵
hydraulic punching machine 液压冲床;水力冲压机;水力冲孔机
hydraulic punching press 冲压水压机
hydraulic purse seine winch 液压围网起网机
hydraulic pusher 液压推车机;液压顶车机
hydraulic pusher for transfer car 液压传送顶车机
hydraulic quicklime 水化生石灰
hydraulic race track device 水力跑道装置
hydraulic rack limiter 液压支架限程器
hydraulic radius 水力半径
hydraulic radius scale 水力半径比例
hydraulic rail bender 液压弯轨机;水力弯轨机
hydraulic rail gap adjuster 液压轨缝调整器
hydraulic rake 水力拖耙
hydraulic rake classifier 水力耙式分级机
hydraulic ram 液压油缸;液压作动筒;液压拉杆;液压千斤顶;液压缸(柱塞);液压撞头;压力扬程机;水压扬汲机;水压头;水力夯锤;水力冲压机;水力冲锤;水锤扬水机;水锤泵;冲击起水机
hydraulic ram actuator 液压绞链开闭装置;液压顶重器开闭装置
hydraulic ram cylinder 液压提升油缸
hydraulic ram for spud 定位桩液压柱塞
hydraulic ram lift 液压悬挂装置;带液压油缸的悬挂装置
hydraulic rammer 液压压头
hydraulic ram press 液压平板机
hydraulic ram pump 水锤泵
hydraulic rapids-heaving 水力绞滩
hydraulic rapids-warping 水力绞滩
hydraulic rate 水硬率
hydraulic ratio 水力比率
hydraulic ratio changer 液压比率变换器;液压比变换装置

hydraulic rear end loader 后置式液力装载机
hydraulic recirculation clarifier 液压循环澄清池
hydraulic reclamation 吹填
hydraulic recoil mount 液压制退的炮架
hydraulic reduction gear 液压减速装置;液压减速装置
hydraulic reference tool 静液压参照工具
hydraulic refractory 水硬性耐火材料
hydraulic refractory cement 水硬性耐火水泥
hydraulic regime(n) 水力状况
hydraulic regionalization 水力区域化
hydraulic regulator 液压调节器;水力调节器
hydraulic reinforcement bar cutter 液压钢筋切断机
hydraulic relay 液压替续器
hydraulic relay valve 液压继动阀
hydraulic relay valve body 水力继动阀体
hydraulic relay valve diaphragm 水力继动阀膜
hydraulic relay valve piston 液压继动阀活塞
hydraulic relay valve piston cup 液压继动阀活塞皮碗
hydraulic relay valve return spring 液压式继动阀回动弹簧
hydraulic relief valve 液压安全阀
hydraulic remote control of cargo valve 货油阀液压遥控装置
hydraulic remote transmission 液压远距离传输
hydraulic research station 水力研究所
hydraulic reservoir 液压(系统)油箱;液压蓄力器;液压储油器
hydraulic residence time 水力停留时间
hydraulic resistance 水力损失;流体阻力;水(力)阻力;水力摩阻
hydraulic resistance balance 水阻力平衡
hydraulic resistance coefficient 水力阻力系数
hydraulic resistance of flow 水流水力阻抗
hydraulic resistivity 水力阻抗性
hydraulic retaining structure 挡水水工结构;挡水(水工)建筑物;挡水坝
hydraulic retarder 水力滞后器;水力迟缓器;液力减速器
hydraulic retention time 水力停留时间
hydraulic retractable slide base 液压进退式滑撬机架
hydraulic retraction cylinder 液压后退缸
hydraulic retraction system 液压进退系统;液压后退方法
hydraulic reversing gear 液压回动装置
hydraulic revolving-pack press 液力转塔式压榨机
hydraulic ripper 液压松土机;水力松土机
hydraulic riveter 液压铆接机;液压铆钉枪;液压铆钉机;水压铆机;水力铆接机
hydraulic riveting 液压铆接;水压铆接
hydraulic riveting machine 水力铆钉机;液压铆接机;液力铆接机
hydraulic road simulator 液压道路模拟
hydraulic robot 液压机器人
hydraulic rock breaker 液压岩石破碎机;液压碎石机;水力碎石机
hydraulic rock crusher 液压碎石机
hydraulic rock cutter 液压凿岩机
hydraulic rock cutting 液压凿岩;水力采石工作;水力凿岩
hydraulic rock drill 水力凿岩机
hydraulic rock fill 水力岩石回填
hydraulic rock grab 液压抓岩机
hydraulic rocking 水力摇床洗选
hydraulic rod break-out kit 液压卸扣装置
hydraulic rod extractor 液压钻杆提升器
hydraulic roll-balance system 轧辊液压平衡系统
hydraulic roll balancing 油压式滚筒定位
hydraulic roller 滚滚流;横卷流;水旋窝
hydraulic roof bolting equipment 液压式顶板锚栓安装设备
hydraulic roof bolting machine 液压式顶板锚栓安装机
hydraulic roof support 液压屋顶支座
hydraulic rooter 液压松土机;液压拔根机
hydraulic ropegeared elevator 液压绳轮升降机
hydraulic rotary drill 液压回转式钻机
hydraulic rotary drilling 水力驱动旋转钻进;液压旋转式钻进;液压回旋钻进;水力旋转钻探;水力回旋钻探
hydraulic rotary drilling machine 液压旋转钻井机;水力旋转钻井机
hydraulic rotary drilling rig 水力旋转凿井机组;水力旋转钻探机组
hydraulic rotary method 液压旋转方法;液压回转方法;水力旋转方法;水力回转方法
hydraulic roughness 水力糙率;水力糙度
hydraulic routing 用水力学方法演算洪水;水力学法流量演算;水力学(方法)洪水演算
hydraulics 液压系统;水硬性物质;水力学
hydraulic safety valve 液压安全阀
hydraulic sand filling 水力填砂;吹填砂
Hydraulics and Pneumatics 水力学与气体力学(期刊)
hydraulic sand sizer 水力分砂器
hydraulic sanitation 水利卫生
hydraulic scaffold(ing) 液压脚手架;液压搭脚手架
hydraulic scale 水压表;水力表;液压秤
hydraulic scale model 水工缩尺模型
hydraulic scale model research 水力缩尺模型研究;水力缩尺模型试验;水工缩尺模型研究;水工缩尺模型试验
hydraulic scaling rig 液压找顶机;液力挑顶刷帮机
hydraulic scarifier 液压松土机;液压翻地机
hydraulic scheme 液压系统图
hydraulic scour 水力冲刷
hydraulic scraper 液压刮削器;液压刮土机
hydraulic screen 液压油滤网;压力油滤网
hydraulic screw press 液压螺旋压力机
hydraulic seal 液体密封;水封;防水层
hydraulic section 水力断面
hydraulic selector 液压选择阀;液压开关
hydraulic selector valve 液压选择阀
hydraulic semi-dump trailer 液压半倾倒式拖车
hydraulics engineer 水利工程师
hydraulics engineering practice 水利工程实践
hydraulics engineering project 水利工程计划
hydraulic sensing pad 液压传力垫块
hydraulic separation 水力离析;水力分离;水力分级
hydraulic separator 液压分离器;水力离析器;水力分离器
hydraulic sequence control 水工程序控制
hydraulic sequence valve 液压顺序阀
hydraulic servo 液压助力装置;液压伺服机构
hydraulic servo-actuator 液压伺服执行器;液压伺服执行机构
hydraulic servomechanism 液压伺服机构
hydraulic servomotor 液压伺服马达;液压伺服电动机;液压接力器
hydraulic servo steering 液压伺服转向
hydraulic servo system 液压伺服系统;液压伺服机构
hydraulic set(ting) 水硬;水凝(固);水中硬化;水力装置;水力设备
hydraulic setting device 液压支架装置
hydraulic setting mortar 水硬性砂浆;水硬性耐火泥浆
hydraulic settlement ga(u)ge 水力沉降计
hydraulic shaking table 液压振动台
hydraulic shape 水力形状
hydraulic shaper 液压牛头刨;液动牛头刨
hydraulic shattering 水力破裂法
hydraulic shear 液压剪床;水力剪切力
hydraulic shear aerator 水力切变充气机
hydraulic shear diffuser 水力剪切扩散设备
hydraulic shearing machine 液力剪机
hydraulic sheet metal drawing press 薄板拉伸液压机
hydraulic sheet metal forming press 板材成型液压机
hydraulic shifter unit 液压移动装置
hydraulic shiplift 液压升船机;水力升船机
hydraulic shock 液压冲击;液体冲击波;水力冲击;水力冲击
hydraulic shock absorber 液压减振器;液压减振器;水力减振器
hydraulic shock eliminator 液压减振器;液压除振器
hydraulic shock strut 液压减振支柱
hydraulic shock vibration 液压冲击振动
hydraulic shovel 液压单斗挖土机;液压铲;液压挖掘机
hydraulic shovel loader 液压铲运装载机
hydraulic shutoff valve 液压截流阀;水力断流阀
hydraulic shutter 液压闸门
hydraulic shutter door operator 百叶门液压开关器
hydraulic side bearing 液压旁承
hydraulic sideboom 液压侧臂
hydraulic side-wall coring tool 液压侧壁取芯工具
hydraulic silting 水力泥砂充填;水沙充填
hydraulic similarity 流体动力相似;水力相似(性);水力相似律
hydraulic similitude 水力模拟;水力相似
hydraulic single column bench press 单柱台式液压机
hydraulic single column press 单柱液压机
hydraulic sizer 水力分粒机
hydraulic slab debarker 水力剥皮机
hydraulics laboratory 水利实验室
hydraulic slave cylinder 液压随动油缸
hydraulic slave motor 液压马达
hydraulic slewing crane 液压旋臂起重机;水力旋臂起重机;水力悬臂起重机
hydraulic slide rule 水力学计算尺
hydraulic sliding form 液压滑动模板
hydraulic slip 液压损失;液压传动输入输出转速差
hydraulic slip coupling 液压滑动联轴节;液压补偿联轴节
hydraulic slip loss 液力打滑损失
hydraulic slope 坡降;水力坡降;水力坡度;水力比降
hydraulic sloughing 水力脱流
hydraulic sludge withdrawal pipe 水力污泥提升管;水力污泥排放管
hydraulic sluicing 冲挖;水力开采(法);水力法排土;水力法掘进;水力冲淤(法);水力冲沙(法)
hydraulic slushing 水力灌浆;水力冲ม
hydraulics of alluvial streams 冲积平原河流水力学
hydraulics of groundwater 地下水水力学
hydraulics of open channel 明渠水力学
hydraulics of pipeline 管线水力学
hydraulics of seawater purging 海水净化水力学
hydraulics of well 井的水力学;管井水力学
hydraulic soil transportation 水力运土
hydraulic sorting 水力分选
hydraulic speed controller 液压速度控制器
hydraulic speed governor 液压式调速器
hydraulic spindle core drill 液压立轴钻机
hydraulic spinner 传动撒布机
hydraulic splitter 液压劈石机;水力劈木机
hydraulic splitting 水力劈裂
hydraulic spoil bank 水力冲毁堤岸
hydraulic spoiler 液压式扰流片
hydraulic sprayer 液压喷雾器;水力喷雾器
hydraulic spraying 液压喷漆;液压喷涂
hydraulic spray nozzle 水雾喷嘴
hydraulic spring grease cup 液压弹簧
hydraulic spud lifting gear 液压定位起吊设备
hydraulics research 水力学研究
hydraulics research station 水力学研究站
hydraulic stability 液压稳定性;水力稳定性
hydraulic stabilizer 水力稳定器
hydraulic stabilizing cylinder 液压稳定缸
hydraulic stacker 液压堆垛机
hydraulic starter 液压启动器;液压起动机
hydraulic starting 液压起动;液压启动
hydraulic starting system 水力启动系统
hydraulic static cone penetrometer 液压静力触探仪
hydraulic steel construction 水工钢结构
hydraulic steel holder 液压持纤器
hydraulic steel structure 水工钢结构
hydraulic steel support 液压钢支架
hydraulic steering 水力操纵;液压转向装置;液压传动控制;水力传动控制
hydraulic steering booster 液压转向加力器
hydraulic steering brake 液力转向加力器
hydraulic steering control system 液压式转向操纵系统
hydraulic steering gear 液压操舵装置
hydraulic steering linkage 液压转向联动装置
hydraulic steering mechanism 液压转向机构
hydraulic steering pump 液压转向助力泵
hydraulic steering unit 液压转向装置
hydraulic stepping motor 液压步进马达
hydraulics test 水力学试验
hydraulics theory 水力学理论
hydraulic stirring device 水力搅拌设备
hydraulic stop 液压止动器
hydraulic storage 液力存储器
hydraulic stowing 水沙充填;水力充填

hydraulic stowing partition 沙门子
hydraulic stowing pipe 水沙充填管道
hydraulic strength 水硬强度;水力强度
hydraulic stretcher 液压拉伸矫直机
hydraulic stripper 液压卸除机构
hydraulic stripping 水力清基;水力清除;水力开采;水力剥离
hydraulic stroker 液压加载油缸
hydraulic structure 水工构筑物;水工构造物;水工结构(物);水工建筑物
hydraulic structure for shipyard 修造船水工建筑物
hydraulic structure for water borne traffic 航运水工工程
hydraulic structure load 水工建筑物荷载
hydraulic structure model 水工结构模型;水工建筑物模型
hydraulic structure model test 水工结构模型试验
hydraulic study 水力研究
hydraulic study of bridge site 桥址的水力研究
hydraulic subsiding value 水力下沉值;水力沉降值
hydraulic suction dredge(r) 吸扬(式)挖泥船;水力挖泥船;水力吸泵挖矿机;吸泥器;吸泥机;吸泥船
hydraulic supply 液压供应
hydraulic surface grinding machine 液压平面磨床
hydraulic surface loading 表面水力负荷
hydraulic surface loading of biologic(al) filter 生物滤池表面水力负荷
hydraulic suspension 水中悬移质;悬移质;悬沙;水力悬浮
hydraulic suspension seat 液压悬架座椅
hydraulic suspension system oil cylinder 液压悬挂系统油缸
hydraulics weight 液压系统重量
hydraulic switch machine 液压转辙机
hydraulic swivel head 液压转盘
hydraulic system 水压系统;液压系统;水力系统
hydraulic system control handle 液压系统操纵手柄
hydraulic system supply and vent tank 液压系统日用油柜
hydraulic system tank 液压系统柜
hydraulic system working pressure 液压系统工作压力
hydraulic tachometer 液压转速计;液压测速仪;速头转速计
hydraulic tailing fill 水力尾沙充填
hydraulic tamping machine 液压捣固机
hydraulic tank 液压箱;水压柜
hydraulic tappet 液压挺杆
hydraulic tapping 液压系统导出孔
hydraulic technology 液压技术
hydraulic telemetry 水力遥测术
hydraulic telemotor 液压遥控马达;液压操舵器;液压操舵机
hydraulic telescopic spreader 液压伸缩吊具
hydraulic telescopic tower 液压式套管塔
hydraulic temperature control 液压温度控制器
hydraulic tensioning 液压拉伸;水力张拉
hydraulic tension regulator 液力张紧调整器
hydraulic tensor 液压张力器
hydraulic test 水力试验;水工试验
hydraulic test bench 液压试验台
hydraulic test box 水力试验箱
hydraulic test cart 液压测试车
hydraulic test chamber 液压试验室
hydraulic tester 液压试验机
hydraulic testing 水力试验;静水试验
hydraulic testing model 水力学试验模型
hydraulic test machine 液压试验机
hydraulic test method 耐水压力试验方法
hydraulic test pressure 液压试验压力
hydraulic test pump 试压泵;水压试验泵
hydraulic test stand 液压试验台
hydraulic theory 水压假说
hydraulic thrust 水推力
hydraulic thrust boring machine 液压冲击式钻机;水压钻探机;水力钻具
hydraulic thrust loss 水推力损耗
hydraulic tidal stream 水力潮汐河
hydraulic tide ga(u)ge 水压式自记验潮仪
hydraulic tightening 液压夹紧
hydraulic tilting system 液压倾卸系统;液压翻车系统

hydraulic tipper 液压翻斗车;液压自卸车
hydraulic tipping gear 液压倾卸机构
hydraulic tipping ram 液压切断(液压倾翻)机构油缸;倾卸液压油缸
hydraulic tipping trailer 液力倾卸挂车
hydraulic tire expander 液压张胎器
hydraulic toggle press 液压杠杆压砖机
hydraulic tongs 液压钳
hydraulic tool 液压工具
hydraulic top cover 液压机构顶盖
hydraulic torque 液力扭矩;水力转矩
hydraulic torque converter 液压扭矩变矩器;液变矩器
hydraulic torque converter driving wheel 液力变矩器主动轮
hydraulic torque converter with axial-flow runner 轴流式涡轮液力变矩器
hydraulic torque convertor 液压变矩器
hydraulic torquemeter 液力扭矩计
hydraulic tracing control 液压仿形控制
hydraulic track jack 液压起道机
hydraulic track lining device 液压拨道器
hydraulic track tensioning 液压履带张紧
hydraulic traction 水中推移质;水力曳引;水力拖曳;水力推移;水力牵引
hydraulic tractor 液压传动式拖拉机
hydraulic tractor-shovel 液压拖铲铲土机
hydraulic transformer 液压变速器
hydraulic transient 液流不稳定状态;水力瞬变过程;水力过渡过程
hydraulic transmission 液压变速;油压传动;液压传递;液力传动;水压传动;水力传动
hydraulic transmission box 液力变速箱
hydraulic transmission control 液压传动控制
hydraulic transmission fluid 液压传动液
hydraulic transmission gear 液压操舵装置;液力传动装置;水力传动装置
hydraulic transmission oil 液力传动油
hydraulic transmission oil heatexchanger 液力工作油热交换器
hydraulic transmitter 液压联轴节
hydraulic transport(ation) 水力输送(法)
hydraulic-transporting device 水力输送设备;水力运输设备
hydraulic transport system 水力运输设施
hydraulic trap 水力混汞捕汞器;水力捕金器
hydraulic traverse cylindric(al) grinder 液压外圆磨床
hydraulic traversing control handle 液压方向机操纵手把
hydraulic traversing electric(al) motor 液压方向机电动机
hydraulic traversing guide control 液压式往复导丝控制
hydraulic traversing mechanism 液压方向机
hydraulic traversing mechanism pump 液压方向机油泵
hydraulic traversing oil reservoir 液压方向机油箱
hydraulic trawl winch 液压拖网起网机
hydraulic trencher 液压开沟机;水力开沟机
hydraulic trench-forming shovel 液压开沟铲;水力开沟铲
hydraulic trenching 液压开沟耙;液压开沟锄;水力开沟耙;水力开沟锄
hydraulic trigger 液压触发器
hydraulic trip assembly 液压脱扣装置
hydraulic tripping device 液压脱钩装置;液压跳闸装置
hydraulic truck-mounted excavator 液压卡车挖掘机
hydraulic tube 水力管;液压管
hydraulic tube transport 管道水力运输
hydraulic tubing 液压管
hydraulic tunnel 液压隧道;水工隧洞;水工隧道
hydraulic tunnel shield jack 液压隧道掘进千斤顶;液压隧道盾构千斤顶
hydraulic turbine 液力涡轮;水轮机;水力涡轮机;水力透平
hydraulic turbine generator 水力涡轮发电机
hydraulic twin tube piezometer 水力双管压力计
hydraulic type bulldozer 液压操纵推土机
hydraulic type dust removal 水力除尘
hydraulic type vibration actuator 液压式振动驱动器

hydraulic underreamer 液压扩孔器
hydraulic unit 液压装置;液压机构
hydraulic universal crawler type excavator 液压万能履带式挖土机
hydraulic universal machine 液压万能试验机
hydraulic universal mobile excavator 液压万能汽车挖土机
hydraulic universal staking machine 液压立式刮软机
hydraulic unlock 液压松开
hydraulic uplift 静水浮托力
hydraulic uplift pressure 扬压力;静止上托力;静水扬压力;水扬压力
hydraulic up-stroke press 水力上行压机
hydraulic vacuum brake 液力真空联合制动器
hydraulic vacuum valve 液压真空阀
hydraulic value 水力值
hydraulic valve 液力阀;液压阀;调水活门;水压阀;水力阀
hydraulic valve control system 液压阀控制系统
hydraulic valve lifter 液压式气门挺杆;液压瓣升降杆
hydraulic valve lifter oil passage 液压起阀器的油孔
hydraulic vane pump 水力叶轮泵
hydraulic variable-speed coupling 液压变速联轴器
hydraulic variable speed drive 液压变速传动;液力变速传动
hydraulic variable speed gear 液压变速箱;液压变速传动装置
hydraulic vent filter 储油器通气管过滤器
hydraulic vertical shiplift 水压式垂直升船机
hydraulic vibration 水力振动
hydraulic vibration generator 液压式振动发生器
hydraulic vibrator 液压振动器;液力振动器
hydraulic vulcanizing press 水压硫化机
hydraulic wake 水力伴流
hydraulic walking chock 液压活动制动器
hydraulic wall sampler 液压井壁取样器
hydraulic wall scraper 液压井壁泥饼清除器;液压井壁刮刀
hydraulic water 水压用水
hydraulic wear 水力磨损
hydraulic wedge 液压楔
hydraulic weld process 液压焊接法
hydraulic wheel 水轮
hydraulic wheel brake 水轮制动器
hydraulic wheel press 水力压轮机
hydraulic wheel scraper 水轮刮削机
hydraulic wheel truck 液压轮卡车
hydraulic winch 液压绞车;水力绞车
hydraulic windlass 液压起锚机
hydraulic window(control)gearing 液压窗开关器
hydraulic window opening device 液压开窗装置
hydraulic wire cold header 液压钢丝冷镦机
hydraulic working platform 液压升降平台
hydraulic works 水利(工程)设施;水利工程
hydraulic wrench 液压扭力扳手
hydraulic yield 水源产量
hydraulic yielding prop 液压让压支柱
hydraw 静水压挤拔
hydrazine hydrate 水合肼
hydrazine hydrate wastewater 水合肼类废水
hydrazine perchlorate 高氯酸肼
hydrazine sulfate spectrophotometry 硫酸肼分光光度法
hydrazinobenzene 苯肼
hydrazoates 叠氮化物
hydrazobenzene 氢化偶氮苯
hydrazoic acid 叠氮酸
hydria 水罐
hydric 氢的
hydride 氢化物
hydride generation 氢化物发生
hydrideion-transfer 氢负离子转移
hydride process 氢化处理法
hydride separation-atomic absorption method 氢化物分离-原子吸收法
hydriding 氢化
hydriding failure 氢化破损
hydrigraphic(al) datum 水深基(准)面
hydriodate 氢碘酸盐
hydriodic acid 氢碘酸
hydrionic 氢离子的

Hydrmel 可洗含油色料
hydroabietyl alcohol 氢化松脂醇
hydroabrasion 液体研磨
hydroabrasive wear 流体磨料磨损
hydroacoustic(al) amplitude analyser 水声振辐分析器
hydroacoustic(al) bearing indicator 水声测向仪
hydroacoustic(al) contact 水声接触点
hydroacoustic(al) range finder 水声测距仪
hydroacoustic(al) rate 水声率
hydroacoustic(al) receiving base 水声接收基阵
hydroacoustic(al) speed 水声速度
hydroacoustic(al) station 水声站
hydroacoustic(al) velocity 水声速度
hydroacoustic(al) vibrator 水声振动器
hydroacoustics 水声学;水声
hydroagriculture 水体农业
hydro(air)plane 高速汽船;水上飞机;水上滑行艇
hydroair unit 液压气压联动装置
hydroalternator 水力发电机
hydroamphibole 水闪石
hydroapatite 磷钙土;水磷灰石
hydroarch 水生演替的
hydroarch sere 水生演替系列
hydroaromatic compound 氢化芳族化合物
hydroaromatic series 氢化芳香系
hydroauger 水力螺旋钻
hydroauger hole 水冲钻钻孔;水冲钻孔
hydroauger hole method 水冲钻孔法
hydrobarker 水力剥皮机
hydrobarometer 水深尺;测水深计;测深器;测深管
hydrobasaluminite 水羟铝矾石;水基性矾
Hydrobel 海特罗伯尔炸药
hydrobiolite 生物堆积岩
hydrobiologic(al) boundary condition 水生生物边界条件
hydrobiologic(al) characteristic 水生生物特性
hydrobiologic(al) monitoring 水生生物监测
hydrobiologic(al) phenomenon 水生态现象
hydrobiologic(al) regime 水生生物状况
hydrobiologic(al) research 水生生物研究
hydrobiologic(al) response 水生生物反应
hydrobiologic(al) sample 水生生物样本
hydrobiologic(al) simulation 水生生物模拟
hydrobiologic(al) simulation model 水生生物模拟模型
hydrobiology 水生生物学
hydrobiology of lake 湖泊水生生物学
hydrobiont 水生生物
hydrobiontic alage 水生藻类
hydrobios 水生生物
hydrobiotic constituent 水生生物组分
hydrobiotite 水黑云母
hydrobiplane 双翼水上飞机
hydrobismutite 水泡铋矿
hydroblast barrel 液压清理回转筒
hydroblast(ing) 水力清理;高压水喷射清洗;加水爆破法;水力清砂;水爆法
Hydrobon 催化加氢精制
hydroboom jumbo 液压钻架钻车;带液压钻架的钻车
hydroboracite 水硼钙石;水方硼石
hydroboring pump 水力冲洗泵(钻探用)
hydroborocalcite 水硼钙石
hydroboron 硼氢化物
hydrobowl 水力浮槽(分级机)
hydrobowl classifier 水力浮槽分级机;浮槽式水力分级机
hydrobromate 氢溴酸盐
hydrobromic 氢溴的
hydrobromic acid 氢溴酸
hydrobromic acid immersion test 氢溴酸浸渍试验
hydrobromic acid neutralization test 氢溴酸中和试验
hydrobromide 氢溴化物
hydrobrusher 水刷洗机
hydrocal 流体动力模拟计算器
hydrocalcite 水方解石
Hydrocal plaster 高强度石膏水泥
hydrocalumite 水铝钙石
hydrocancrinite 水钙霞石
hydrocarbon 烃
hydrocarbon analog(ue) 碳氢化合物类似物
hydrocarbon analysis 烃类分析

hydrocarbon analyzer 碳氢化合物分析仪
hydrocarbonate 碳酸氢盐;酸性碳酸盐
hydrocarbonate water zone 重碳酸水带
hydrocarbon automatic recorder 烃类分析自动记录仪
hydrocarbon base course 碳氢基层
hydrocarbon binder 碳氢结合料
hydrocarbon binder cooker 碳氢黏结剂蒸煮器
hydrocarbon binder film 碳氢黏结层
hydrocarbon binder melting kettle 碳氢黏结剂融化壶
hydrocarbon binder pump 碳氢黏结剂泵
hydrocarbon binder spraying machine 碳氢黏结剂喷雾机
hydrocarbon black 碳氢黑;烃黑;炭黑
hydrocarbon compound 碳氢化合物
hydrocarbon contamination 烃污染
hydrocarbon content in sediment 沉积物中碳氢化合物含量
hydrocarbon conversion 烃转化
hydrocarbon correction 油气校正
hydrocarbon cracking 烃裂解
hydrocarbon degradation bacteria 烃降解菌
hydrocarbon drying oil 烃合成干性油;石油系合成干性油
hydrocarbon fuel cell 碳氢化合物燃料电池(烃燃料电池)
hydrocarbon gas 烃气;碳氢气体
hydrocarbon group analysis 碳氢基团分析
hydrocarbon indication technique 烃类检测技术
hydrocarbon indicator section 烃类检测剖面
hydrocarbonism 碳氢化合物中毒
hydrocarbonize 碳氢化
hydrocarbon miscible flooding 烃混溶驱动
hydrocarbon mixture 混合烃
hydrocarbon oligomer 烃齐聚物
hydrocarbon-oxidizing microorganism 烃氧化微生物
hydrocarbon plastics 烃类塑料
hydrocarbon pollution 碳氢化合物污染
hydrocarbon polymer 碳氢聚合物
hydrocarbon production 碳氢化合物产品
hydrocarbon productivity curve of artificial simulate 人工模拟产率曲线
hydrocarbon productivity curve on natural section 自然剖面烃产率曲线
hydrocarbon productivity of genetic oil stage 成油阶段产烃率
hydrocarbon productivity of overmature stage 过成熟阶段产烃率
hydrocarbon productivity of wet-gas stage 湿气阶段产烃率
hydrocarbon reforming cell 碳氢化合物改质处理燃料电池
hydrocarbon resin 烃(类)树脂;石油树脂
hydrocarbon saturation 烃饱和度
hydrocarbon series 烃系
hydrocarbons monitor 碳氢有害排出物监测器
hydrocarbon soluble dyestuff 碳氢可溶性染料
hydrocarbon solvent 烃类溶剂
hydrocarbon source rock and its evolution 烃源岩及其评价
hydrocarbon spectrum 烃质谱
hydrocarbon surfactant type foam 碳氢活化剂泡沫
hydrocarbon survey 碳氢化合物测量方法
hydrocarbon synthesis 烃类合成
hydrocarbon type analysis 烃族组分分析
hydrocarbon utilizing bacteria 烃类分解细菌
hydrocarbon-utilizing microorganism 烃利用微生物
hydrocarbon with condensed rings 稠环烃
hydrocarbonylation 氢化法
hydrocarbyl cyanide 氰化烃
hydrocassiterite 水锡石
hydrocellulose 水解纤维素
hydrocellulose nitrate 水解纤维素硝酸酯
hydrocenosis 毛细管引流法;导液法
hydrocerussite 水白铅矿
hydrochart 水文图(表)
hydrocheck 制动缸
hydrochemical 水化学的
hydrochemical analysis 水化学分析
hydrochemical anomaly 水化学异常

hydrochemical basis 水化学基
hydrochemical boundary condition 水化学边界条件
hydrochemical characteristic 水化学特征
hydrochemical chart 水化学图
hydrochemical columnar section 水化学柱状图
hydrochemical component 水化学组分
hydrochemical constituent 水化学组分
hydrochemical data 水化学资料
hydrochemical dynamics 水化学动力学
hydrochemical environment 水化学环境
hydrochemical evolution 水化学演化
hydrochemical facies 水化学相
hydrochemical formation in mine area 矿区水化学形成作用
hydrochemical horizontal zoning 水化学水平分带
hydrochemical investigation 水化学调查
hydrochemical logging 水化学测井
hydrochemical map 水化学图
hydrochemical map for oil field 油田水化学图
hydrochemical map of groundwater 地下水化学图
hydrochemical mark of mineral deposits prospecting 水化学找矿标志
hydrochemical model 水化学模型
hydrochemical network 水化学监测网
hydrochemical profile 水化学剖面图
hydrochemical property 水化学特征
hydrochemical prospecting 水化学勘探
hydrochemical regime 水化学状况;水化学指标
hydrochemical research 水化学研究
hydrochemical resultant plot 水化学成果图
hydrochemical runoff modulus method 水化学径流模数法
hydrochemical sampling 水化学采样
hydrochemical stratification 水化学分层
hydrochemical stratification sampling 水化学分层采样
hydrochemical survey 水(文)化学调查;水化学测量
hydrochemical type 水化学类型
hydrochemical variation 水化学变化
hydrochemical vertical zoning 水化学垂直分带
hydrochemical zonality 水化学分带性
hydrochemical zoning 水化学分带
hydrochemistry 水质化学;水(文)化学
hydrochemistry profile for oil field 油田水化学剖面图
hydrochimous 适雨天植物
hydrochinone 对苯二酚
hydrochloric acid 氢氯酸;盐酸
hydrochloric acid cleaning 盐酸清洗
hydrochloric acid manufacture 盐酸生产
hydrochloric acid pickling wastewater 盐酸酸浸废水
hydrochloric solution of cupreous chloride 氯化亚铜的氢氧化物溶液
hydrochloride 盐酸化物
hydrochloroborite 多水氯硼钙石
hydrochlorofluorocarbon 氟氯烃化合物
hydrochore 水布植物
hydrochory 水媒传布
hydrocinnamic alcohols 苯基丙醇
hydrocinnamic aldehyde 苯基丙醛
hydroclamp 液压抓斗;液压夹钳;液压夹具
hydroclassification 水力分级
hydroclassifying 水力分类法
hydroclassifying of particle 颗粒水力分级
hydroclast 水成碎屑
hydroclastic 水成碎屑的
hydroclastic rock 水成碎屑岩
hydroclastics 水成碎屑岩
hydrocleaning 水力清洗
hydroclimate 水中生物的物理及化学环境;水文气候;水生气候;水面气候
hydroclimatic factor 水面气候因素;水文气候因素
hydroclimatology 水文气候学
hydroclinohumite 水斜硅镁石
hydroclipper mower 液力剪切式割草机
hydroclone separator 水力旋流分离器
hydrocoal mining 水力采煤
hydrocodimer 氢化共叠化合物
hydrocoele 水系腔
hydrocole 水生动物

hydrocolloid 水(解)胶体
hydrocompaction 湿陷;渗水压实作用
hydrocomplex 水利系统
hydrocomputation simulation program(me) 水文计算模拟程序
hydrocondensation 加氢缩合反应
hydrocone breaker 液压锥形破碎机
hydrocone crusher 液压圆锥破碎机;液压锥形破碎机
hydrocones gyratory crusher 液压圆锥旋回破碎机
hydrocone type 吸管式;虹吸式(水工建筑物)
hydroconion 喷雾器;喷洒器
hydroconite 水方解石
hydroconsolidation 水固结(作用);湿陷
hydroconversion 加氢变换
hydrocooler 水冷却器
hydrocooling 用水冷却;水冷(处理)
hydrocope 液压平衡
Hydrocorallina 水螅珊瑚目
hydrocore-knock-out machine 水力型芯打出机
hydrocortisone butyrate propionate 丁酸丙酸氢可的松
hydrocoupling 水压联轴器
hydrocracker 加氢裂化器
hydrocracking 加氢裂解;加氢裂化;氢压下裂化;氢化裂解
hydrocrane 液压起重机;水力起重机
hydrocratic eustacy 水动型海面升降
hydrocultivation 水栽培
hydrocuprite 水赤铜矿
hydrocushion 液压减振装置;液压缓冲(器);液压衬垫;液压平衡
hydrocushion type relief valve 液压平衡式安全阀
hydrocyanic acid 氰化氢;氢氰酸
hydrocyanic ester 氰酯
hydrocyanite 铜靛石;水蓝晶石
hydrocyclone 水力旋流器;旋液分离器;锥形除渣器;离心除尘机;水力旋流分离器;水力离心分粒器;水力离心分离器;水力离心分级器
Hydrocyclone process 水力旋流分离过程
hydrocylinder 液压(油)缸
hydro dam 水坝
hydro dam body 水坝体
hydro dam crest 水坝顶
hydrodealkylation 加氢脱烃
hydrodehazer 脱水器
hydrodensimeter (土的)含水密实度测定仪
hydrodesulfurization 加氢脱硫
hydrodesulphurisation 氢化脱硫作用
hydrodesulphurization 加氢脱硫;氢化脱硫作用
hydrodevelopment 水力开发
hydrodiffusion 水中扩散
hydrodis 船底附生物
hydrodist 海上电子测距仪
hydrodiuresis 水性多尿
hydrodolomite 水白云石
hydrodresserite 多水碳铝钡石
hydrodrill 水力钻具
hydrodrilling 液压钻进
hydrodrill rig 液压钻车
hydroduct 水下冲压式喷射发动机;水汽波导
hydrodynamic(al) 流体动力(学)的;水动力的;水动水力的;水压的
hydrodynamic(al) analogy 液动模拟;流体动力(学)模拟;水动力模拟;水动比拟
hydrodynamic(al) axis 水流动力轴线
hydrodynamic(al) barrier 水动力遮挡
hydrodynamic(al) behavior 流体动力特性
hydrodynamic(al) brake 液力制动(器)
hydrodynamic(al) cavitation 水力空化
hydrodynamic(al) characterisitic 流体动力学特性
hydrodynamic(al) chromatography 液相流动色谱(法)
hydrodynamic(al) coefficient 水动力系数
hydrodynamic(al) compressive forming 动水压成形
hydrodynamic(al) computation 流体动力计算;水动力计算
hydrodynamic(al) condition 水动力条件
hydrodynamic(al) conductivity 水动力传导性
hydrodynamic(al) coupling 液力耦合器;液力联轴节
hydrodynamic(al) damping 流体动力阻尼
hydrodynamic(al) differential equation 流体动力(学)微分方程
hydrodynamic(al) diffusion 流型扩散;流体动力学扩散;水动力扩散
hydrodynamic(al) digital model 流体动力数字模型
hydrodynamic(al) dispersion 水动力弥散
hydrodynamic(al) dispersion coefficient 水动力扩散系数
hydrodynamic(al) dispersion tensor 水动力扩散张量
hydrodynamic(al) drag 流体动力(力)阻力;水动力阻力;动水阻力;动水拖曳力
hydrodynamic(al) drive 液压传动装置;流体动力驱动;液力传动;水力传动
hydrodynamic(al) effect 流体效应;流体动力效应;水动力影响
hydrodynamic(al) equation 液体动力学方程;流体动力学方程;水动力(学)方程
hydrodynamic(al) flow field 水动力流场
hydrodynamic(al) fluid drive 水动力传动
hydrodynamic(al) force 水动力;动水(作用)力
hydrodynamic(al) form 流线型;良流线体
hydrodynamic(al) ga(u)ge 动水压力计
hydrodynamic(al) governor 液压调速器
hydrodynamic(al) head 流体动压头;水动力压头;动水压头;动(力)水头
hydrodynamic(al) impact 水动力冲击
hydrodynamic(al) instability 流体力学不稳定性;流体动力不稳定(性)
hydrodynamic(al) intensity of pressure 动水压强
hydrodynamic(al) investigation 水动力调查
hydrodynamic(al) lag 水力延缓;水动力阻滞;动水滞后
hydrodynamic(al) lift 水动力升力;水动力升举力;动升举力
hydrodynamic(al) load 流体动力荷载;动水荷载
hydrodynamic(al) lubrication 液体润滑;液动体润滑;流体动力润滑(作用)
hydrodynamic(al) lubrication model 液体箔膜模型
hydrodynamically 水动力地
hydrodynamically rough surface 水动力粗糙面
hydrodynamically smooth surface 水动力光滑面
hydrodynamic(al) machine 液压机械
hydrodynamic(al) mass 流体动力质量;水动力质量
hydrodynamic(al) mass coefficient 水动力质量系数;动水质量系数
hydrodynamic(al) means 水动力法
hydrodynamic(al) mixing characteristic 水动力混合特性
hydrodynamic(al) model 水动力学模型
hydrodynamic(al) moment 水动力力矩
hydrodynamic(al) moment transformer 水力变矩器
hydrodynamic(al) mo(u)lding press 液压模压机
hydrodynamic(al) noise 流体动力噪声;流动噪声;水动力噪声
hydrodynamic(al) operation 液力机械操作;液力操作
hydrodynamic(al) oscillator 流体动力振子
hydrodynamic(al) parameter 水动力参数
hydrodynamic(al) performance 流体动力学性能;水动力性能
hydrodynamic(al) phenomenon 流体动力现象;水动力现象
hydrodynamic(al) piezometer 水力测压计;动水式测力计
hydrodynamic(al) pool 水动力油藏
hydrodynamic(al) power transmission 液动式动力传输;动力式液压传动
hydrodynamic(al) pressure 液体动压力;流体动(力)压力;水动(力)压力;水动压(力)
hydrodynamic(al) process 液动体冲压法
hydrodynamic(al) profile 水力翼型
hydrodynamic(al) research 液体动力(学)研究;流体动力学研究;水动力(学)研究
hydrodynamic(al) resistance 流体动阻力;水动(力)阻力
hydrodynamic(al) shear force 流体动力剪切力
hydrodynamic(al) shock 液压冲击;水动力冲击
hydrodynamic(al) similitude 水动力相似
hydrodynamic(al) smooth surface 水动力平滑面
hydrodynamic(al) spectrum 流体动力谱
hydrodynamic(al) stability 流体力学稳定性;流体动力稳定性;水动力稳定性;动水稳定性
hydrodynamic(al) stress 水压应力;流体动力的应力
hydrodynamic(al) theory 水动力理论
hydrodynamic(al) time scale 流体动力学时标
hydrodynamic(al) torque-converter 液力变矩器
hydrodynamic(al) torque transformer 液动力矩变换器
hydrodynamic(al) transducer 流体动力型换能器
hydrodynamic(al) transmission 液力传动
hydrodynamic(al) trap 水动力圈闭
hydrodynamic(al) turbulence 流体力学湍动
hydrodynamic(al) water-resistance test 水力式防水试验
hydrodynamic(al) wave 流体动力波;水力动力波;水力波;H波
hydrodynamic(al) zonality of karst water 岩溶水动力分带
hydrodynamics 液体动力学;水能动力学;水动力学
hydrodynamics of equatorial ocean 赤道海洋流体动力学
hydrodynamist 流体动力学家
hydrodynamometer 流速计;水速计;水力测力计;水力测功器;水力测功率;水力测功计;水力测功机;水动力计
hydroecologic(al) factor 水生生态因素
hydroecology 水生生态学
hydroeconomic analysis 水利经济分析
hydroejector 水力喷射器;水抽子
hydroelasticity 液体弹性;液动弹性力学;流体弹性;流动弹性
hydroelasticity theory 流体弹性理论
hydroelastic model 水力弹性模型
hydroelastic stability 水力弹性稳定性
hydroelastic vibration 流体弹性振动
hydroelasto-plastic body 流体弹塑性体
hydroelectric(al) 水电的;水力发电的
hydroelectric(al) bath 电解槽
hydroelectric(al) board 水电局
hydroelectric(al) complex 水力发电联合企业
hydroelectric(al) construction 水力发电施工;水发电建筑;水电施工;水电建筑
hydroelectric(al) crushing 液电破碎
hydroelectric(al) dam 水力发电坝;水电坝
hydroelectric(al) development 水力发电开发;水力发电(建设);水电开发;水电(站)建设
hydroelectric(al) development project 水电开发工程
hydroelectric(al) elevator 电动液压升降机
hydroelectric(al) energy 水力发电能;水电能
hydroelectric(al) engineer 水力发电工程师;水电工程师
hydroelectric(al) engineering 水力发电工程
hydroelectric(al) exploitation 水力发电开发(利用);水电开发(利用)
hydroelectric(al) generating set 水轮发电机组;水力发电机组
hydroelectric(al) generating station 水力发电厂
hydroelectric(al) generator 水力发电机
hydroelectric(al) generator plant 水力发电厂
hydroelectric(al) generator set 水力发电机组
hydroelectric(al) installation 水力发电设备;水力发电装置
hydroelectric(al) mine-car conveyer 液电式推车机
hydroelectric(al) plant 水电站;水电厂;水力发电站
hydroelectric(al) plant of moderate head 中水头水电站
hydroelectric(al) potentiality 水电蕴藏量;水电潜力
hydroelectric(al) power 水电;水电蕴藏量;水力发电
hydroelectric(al) power development 水力发电工程;水力发电开发
hydroelectric(al) power facilities and related works 水电站和有关工程
hydroelectric(al) power generation 水力发电
hydroelectric(al) power houses 水电站厂房
hydroelectric(al) power planning 水力发电规划
hydroelectric(al) power plant 水电站;水力发电厂
hydroelectric(al) power plant in river channel 河床式水电站
hydroelectric(al) power project 水电工程;水力发电工程方案
hydroelectric(al) power station 水电站;水力发电站;水力发电厂
hydroelectric(al) power station operation 水电

站运行
hydroelectric(al) practice 水力发电实践
hydroelectric(al) project 水力发电项目；水力发电计划；水力发电工程；水电项目；水电计划；水电工程
hydroelectric(al) protocol 水电项目议定书(条约草案、会谈记录)
hydroelectric(al) resources 水力资源；水电资源；水电来源；水能资源
hydroelectric(al) scheme 水电开发计划；水电规划枢纽；水电规划方案
hydroelectric(al) servocontrol system 液电伺服系统
hydroelectric(al) steering engine 电动液压操舵装置
hydroelectric(al) steering gear 电液式操舵装置；电动液压操舵装置
hydroelectric(al) synchronous machine 水轮同步发电机
hydroelectric(al) use of water 水力发电用水
hydroelectric(al) wave generator 波力发电机；波浪水力发电机
hydroelectricity 水电
hydroelectricity generation 水力发电
hydroelectrometer 水静电计
hydroenergy 水能
hydroenergy utilization 水能利用
hydroengineering 水力工程(学)；水电工程；水利工程
hydroengineering complex 综合利用水利工程；水利枢纽；水利设施综合体
hydroengineering design 水利工程设计
hydroengineering geologic(al) sampler 水下工程地质取样管
hydroenvironmental simulation 水环境模拟
hydroeuxenite 水黑稀金矿
hydroexhauster 抽水机
hydroextracting cage 甩水机
hydroextraction 滴干；水力提取；水力冲挖；脱水作用；水力开采
hydroextractor 离心机；挤压机；脱水器
hydroextractor model 离心脱水机
hydrofacies 水相
hydrofall mill 湿法无介质磨
hydrofeed 液压进给装置
hydrofeeder 液压进给装置；液压给进装置
hydroferricyanate 氰高铁酸盐；氢高铁氰酸盐
hydrofiller 含沥青填充物
hydrofilter 水滤器；湿法除尘器
hydrofin 稳定板
hydrofine 氢化提纯
hydrofining 加氢精制
hydrofinish 流体抛光法
hydroflaker 水力纤维离解机
hydroflap 潜水舵；水(下)翼；水下舵；水襟翼
hydroflated tire 充水轮胎
hydrofluidic 液体射流
hydrofluoric acid 氢氟酸
hydrofluoric acid bottle inclinometer 氢氟酸测斜仪
hydrofluoric acid burn 氢氟酸灼伤
hydrofluoric acid etch 氢氟酸蚀刻
hydrofluoric acid wash 氢氟酸洗涤
hydrofluoric ether 氟代烃
hydrofluoride 氢氟酸盐；氢氟化物
hydrofluorinator 氢氟化器
hydrofluorocarbon 氢氟碳化合物
hydrofluosilicate 氢氟硅酸盐
hydrofluosilicic acid 氟硅酸
hydrofoil 水翼艇的水翼；水翼(船)；水叶；浮筒
hydrofoil amphibian 水翼式水陆两用车
hydrofoil boat 水翼艇；水翼船
hydrofoil cascade 水力翼栅
hydrofoil control system 水翼操纵系统
hydrofoil craft 水翼艇；水翼船
hydrofoil ferry 水翼渡船
hydrofoil rudder 流线型舵
hydrofoil ship 水翼船
hydrofoil stabilization device 水翼船稳定装置
hydrofoil subduer 水翼消波器
hydrofoil unit 水翼装置
hydrofoil vessel 水翼艇
hydrofoil weir 水翼式堰
hydrofoil wing 水翼

hydroforging 水力应变法
hydroform 液压成形
hydroformer 水压式成形器
hydroforming 液压成型；液压成形(法)；加氢重整
hydroforming machine 液压成形机
hydroform method 液压成形法
hydrofracture 水力压裂；水力破碎；水力断裂
hydrofracturing 水压致裂(法)；水压胀裂(法)；水力破碎法；水破碎
hydrofracturing method 水压致裂法；水力破裂法
hydrofraise 液压双轮掘削机；双轮反循环钻机
hydrofraise method 液压双轮掘削造孔法；双轮反循环钻机造孔法
hydrofreat 水力处理器
hydrofuge 拒水的；防水的；防湿的；不透水的
hydroful 液满
hydrogap rudder 水隙舵
hydrogarnet 水石榴石；水榴石
hydrogasification 加氢气化
hydrogasification of coal 煤加氢气化
hydrogasification of pretreated coal 煤加氢气化预处理
hydrogasifier 加氢化炉
hydrogasoline 加氢汽油
hydrogel 水性凝胶；水凝胶
hydrogen 氢气
hydrogen acceptor 氢受体
hydrogenant agent 氢化剂
hydrogen arc lamp 氢弧灯
hydrogenate 氢化
hydrogenated amorphous carbon 氢化非晶碳
hydrogenated castor oil 加氢蓖麻油；氢化蓖麻油
hydrogenated fuel 氢化燃料
hydrogenated naphtha 氢化石脑油
hydrogenated naphthalene 氢化萘
hydrogenated polybutene 氢化聚丁烯
hydrogenated propylene tetramer 氢化丙烯四聚物
hydrogenated rosin 氢化松香
hydrogenated rubber 氢化橡胶
hydrogenated terphenyl 氢化三联苯
hydrogenated toluene dlsocyanate 六氢化甲苯二异氰酸酯
hydrogenated vegetable oil 氢化植物油
hydrogenating 氢煤法
hydrogenating agent 氢化剂
hydrogenation 加氢作用；加氢过程；氢化(作用)；液相氢化
hydrogenation apparatus 氢化装置
hydrogenation catalyst 氢化催化剂
hydrogenation of coal 煤的氢化
hydrogenation of oil 油加氢；油的氢化
hydrogenation plant 加氢装置
hydrogenation reaction chamber 加氢反应器
hydrogenation solvent 氢化溶剂
hydrogenation unit 加氢装置
hydrogen atomic clock 氢原子钟
hydrogen atomic oscillator 氢原子振荡器
hydrogenator 氢化器
hydrogen attack 氢侵蚀；氢腐蚀；氢脆
hydrogen autunite 氢钙铀云母
hydrogen bacteria 氢细菌；产氢菌
hydrogen balance 氢平衡
hydrogen balloon 氢气球
hydrogen-based membrane biofilm reactor 基于氢膜生物膜反应器
hydrogen blistering 氢爆皮
hydrogen bond 氢键
hydrogen bonding 氢键键合
hydrogen bonding force 氢键力
hydrogen brazing 氢(钎)焊
hydrogen brazing furnace 氢气钎焊炉
hydrogen bridge 氢桥
hydrogen brittleness 氢脆(性)
hydrogen brmide 溴化氢
hydrogen bubble chamber 氢气泡室
hydrogen bubble technique 氢气泡技术；氢气泡法
hydrogen burner 氢燃烧器
hydrogen burning 氢燃烧
hydrogen-carbon atomic ratio 氢碳原子比
hydrogen carrier 氢载体
hydrogen cation exchange capacity 氢离子交换容量
hydrogen cation exchanger 氢阳离子交换剂

hydrogen cell 氢燃料电池
hydrogen chloride 氯化氢
hydrogen chrome 铬化氢
hydrogen clock 氢(原子)钟
hydrogen cloud 氢云
hydrogen-colled generator 氢气冷却发电机
hydrogen-containing 含氢的
hydrogen-containing gas 含氢气体
hydrogen content 氢含量
hydrogen content of hydrocarbon 烃类中的氢含量
hydrogen controlled covering 低氢型药皮
hydrogen convection layer 氢对流层
hydrogen coolant 氢冷却剂
hydrogen-cooled machine 氢冷式电机
hydrogen-cooled rotary converter 氢冷式旋转变流机
hydrogen-cooled rotor 氢冷转子
hydrogen-cooled synchronous condenser 氢冷式同步调相机
hydrogen-cooled turbine-generator 氢冷式涡轮发电机
hydrogen-cooled turbo-alternator 氢冷式汽轮发电机
hydrogen cooler 氢冷却器
hydrogen cooling 氢冷却
hydrogen cooling system 氢冷却装置
hydrogen corrosion 氢腐蚀
hydrogen cracking liquefaction 加氢裂解液化
hydrogen cryogenic upgrading process 氢气深冷提纯过程
hydrogen cyanide poisoning 氢氰酸中毒
hydrogen cycle 氢循环
hydrogen cylinder 氢气瓶；氢气筒
hydrogen damage 氢损伤
hydrogen demercuration 加氢去汞
hydrogen desulfurizaion 加氢脱硫
hydrogen dioxide 过氧化氢
hydrogen discharge lamp 氢灯；氢放电灯；氢气放电管
hydrogen distribution 氢分布
hydrogen efficiency 氢效率
hydrogen electrode 氢焊条；氢电极；氢标准电极
hydrogen embrittlement 氢蚀致脆；氢脆性；(钢的)氢脆性；酸脆性
hydrogen embrittlement failure 氢脆破坏
hydrogen embrittlement of metals 金属受氢蚀致脆
hydrogen emission region 氢发射区
hydrogen energy 氢能
hydrogen engine 液氢发动机；氢气发动机
hydrogen equivalent 氢当量
hydrogenerating unit 水轮发电机组
hydrogeneration 加氢作用
hydrogenerator 水力发电机；水轮发电机
hydrogenesis 氢解(作用)；水生成；水分凝结作用；水成作用；成水作用
hydrogenetic 水成的
hydrogenetic sediment 水成沉积(物)
hydrogenetic soil 水成土
hydrogen-evolution type of corrosion 放出氢气型的腐蚀
hydrogen excess pressure 氢气过压
hydrogen exponent 氢指数
hydrogen-filled rectifier 充氢整流管
hydrogen firing 氢气焙烧
hydrogen flame detector 氢火焰检测器
hydrogen flame ionization 氢火焰离子化
hydrogen flame ionization detector 氢焰离子化检测器；氢(火)焰电离检测器；氢(火)焰离子检测器
hydrogen flame photometer 氢焰曝光表；氢焰光度计
hydrogen flame temperature detector 氢火焰温度检测器
hydrogen fluoride 氟化氢
hydrogen fluoride chemical laser 氟化氢化学激光器
hydrogen fluoride poisoning 氟化氢中毒
hydrogen fluoride recovery plant 氟化氢回收设备
hydrogen free electrode 低氢焊条
hydrogen fuel 氢燃料
hydrogen-fueled system 氢燃料系统
hydrogen furnace 氢气炉
hydrogen gas 氢气

hydrogen gas automatic blow down system 氢气放空系统
hydrogen gas buffer 氢气缓冲罐
hydrogen gas compressor 氢气压缩机
hydrogen gas cooling system 氢冷系统
hydrogen gas cooling tower 氢气冷却塔
hydrogen gas cylinder 氢气钢瓶
hydrogen gas-filled thyratron 充氢闸流管
hydrogen gas fire arrester 氢气阻火器
hydrogen gas generator 氢气发生器
hydrogen gas processing 氢气处理
hydrogen gas system 氢气系统
hydrogen generation station 制氢站
hydrogen generator 氢气发生器
hydrogen geocorona 氢地冕
hydrogen halide 卤化氢
hydrogen halo 氢晕
hydrogen header 氢集管
hydrogenic 水成的
hydrogenic model 氢模型
hydrogenic rock 水生岩;水成岩
hydrogen index 含氢指数;氢指数
hydrogen index of fluid 液体的含氢指数
hydrogen index of gas 天然气的含氢指数
hydrogen index of hydrocarbon 烃类的含氢指数
hydrogen index of oil 原油的含氢指数
hydrogen index of shale 泥质的含氢指数
hydrogen index of water solution 水溶液的含氢指数
hydrogen induced cracking 氢致裂纹
hydrogen inner cooling 直接氢冷;氢内冷
hydrogen inner-cooling generator 氢内冷发电机
hydrogen inner-cooling system 氢内冷系统
hydrogen iodide 碘化氢
hydrogen ion 氢离子;水合氢离子
hydrogen ion activity 氢离子活度
hydrogen ion back diffusion 氢离子回渗
hydrogen ion concentration 氢离子浓度;氢碱度
hydrogen ion concentration meter 氢离子浓度计
hydrogen ion concentration recorder 氢离子浓度自记仪
hydrogen ion concentration value 氢离子浓度值
hydrogen ion determination apparatus 氢离子测定器
hydrogen ion exchanger 氢离子交换器
hydrogen ion exponent 氢离子指数
hydrogen ion index 氢离子指数
hydrogen ion indicator 氢离子指示剂;酸碱指示剂
hydrogen ionization condensation region 电离氢凝聚区
hydrogen-ionized 含有氢离子的
hydrogen-ion potentiometer 氢离子电位计
hydrogen-ion transfer 氢离子移转
hydrogenising 氢化
hydrogen isotope geothermometer 氢同位素地温计
hydrogenization 加氢作用;氢化
hydrogenize 氢化
hydrogen lamp 氢灯
hydrogen laser 氢激光器
hydrogen loss 氢损失
hydrogen maser 氢微波激射器
hydrogen microflare detector 氢焰温度检测器;氢微闪检测器
hydrogen migration 氢转移
hydrogen nitrate 硝酸
hydrogenolysis 加氢分解
hydrogenotrophic denitrification 氢自养反硝化
hydrogenous 含氢的;富氢的
hydrogenous coal 高水分煤;氢化煤;褐煤
hydrogenous material 水成物质
hydrogenous Mn nodule 水成锰结核
hydrogenous rock 水成岩
hydrogen outer cooling 间接氢冷;氢外冷
hydrogen outer cooling turbogenerator 氢外冷汽轮发电机
hydrogen overvoltage 氢超电压
hydrogen oxides 氢的氧化物
hydrogen oxygen cell 氢氧电池
hydrogen oxygen isotope correlation 氢氧同位素相关性
hydrogen oxygen saturation diving 氢氧饱和潜水
hydrogen peroxide 过氧化物;过氧化氢;双氧水

hydrogen peroxide catalytic oxidation 过氧化氢催化氧化
hydrogen peroxide decomposition 过氧化氢分解
hydrogen peroxide enhanced photocatalytic oxidation 过氧化氢强化光催化氧化
hydrogen peroxide isopropylbenzene 过氧化氢异丙苯
hydrogen peroxide oxidation 过氧化氢氧化
hydrogen peroxide oxygen 过氧化氢氧
hydrogen peroxide solution 过氧化氢溶液
hydrogen peroxide wastewater 过氧化氢废水
hydrogen phosphide 磷化氢
hydrogen plasma 氢等离子体
hydrogen plasma arc process 氢等离子弧过程
hydrogen plasma jet 氢等离子流
hydrogen platinum electrode system 氢铂电极系统
hydrogen polarization 氢极化
hydrogen-powered bus 氢气动力大客车
hydrogen-powered vehicle 氢燃料汽车
hydrogen pressure 氢压(力)
hydrogen pressure regulating curve 氢压力调整曲线
hydrogen processing 氢加工
hydrogen production 氢的制造
hydrogen production and distribution system 氢气发生和储配系统
hydrogen pump 氢泵
hydrogen purity 氢气纯度
hydrogen recombination line 氢复合线
hydrogen recombiner 氢气复合器
hydrogen recovery process 氢回收过程
hydrogen-reduced iron 氢还原铁粉
hydrogen-reduced powder 氢还原粉末
hydrogen reduction 氢还原
hydrogen reduction autoclave 氢还原高压釜
hydrogen relief annealing 预防白点退火
hydrogen-rich gas 富氢气体
hydrogen-rich recycle 富氢的循环气
hydrogen scale 氢温标;氢标度
hydrogen seal 氢气密封
hydrogen selenide 硒化氢
hydrogen shift polymerization 氢转移聚合(作用)
hydrogen sintering 氢气烧结
hydrogen soil 氢质土
hydrogen soldering 氢焊
hydrogen spectrum 氢光谱
hydrogen storage beam tube 氢存储束管
hydrogen storage material 储[贮]氢材料
hydrogen stream 氢气流
hydrogen sulfate 硫酸氢(盐)
hydrogen sulfide 硫化氢
hydrogen sulfide automatic analyzer 硫化氢自动分析仪;硫化氢自动分析计
hydrogen sulfide barrier 硫化氢障
hydrogen sulfide-forming bacteria degradation 产硫化氢菌降解
hydrogen sulfide mud 硫化氢淤泥
hydrogen sulfide poisoning 硫化氢中毒
hydrogen sulfide precipitation method 硫化氢沉淀法
hydrogen sulfide water 硫化氢水
hydrogen sulphide 硫化氢
hydrogen sulphide poisoning 硫化氢中毒
hydrogen test 测氢试验
hydrogen thermometer 氢(气)温度计
hydrogen thyratron 氢闸流管
hydrogen-tight 氢密封;不漏氢
hydrogen-tight house 氢密封机座
hydrogen transfer 氢转移;氢传递
hydrogen transference 氢转移
hydrogen transfer reaction 氢转移反应
hydrogen tritide 氚化氢
hydrogen tube 充氢管
hydrogen value 氢值
hydrogen water chemistry 氢水化学
hydrogen wave 氢波
hydrogen-welding 氢焊
hydrogeochemical 水文地球化学的
hydrogeochemical abnormality types 水文地球化学异常种类
hydrogeochemical anomaly 水圈地球化学异常
hydrogeochemical cycle 水文地球化学循环
hydrogeochemical district 水文地球化学区

hydrogeochemical division 水文地球化学分区
hydrogeochemical environment 水文地球化学环境
hydrogeochemical index 水文地球化学环境指标
hydrogeochemical map 水文地球化学图
hydrogeochemical prospecting 水文地球化学找矿;水文地球化学勘探
hydrogeochemical province 水文地球化学省
hydrogeochemical sub-district 水文地球化学亚区
hydrogeochemical zone 水文地球化学带
hydrogeochemical zoning 水文地球化学分带
hydrogeochemical zoning and division 水文地球化学分带和分区
hydrogeochemistry 区域水文地球化学;水文地球化学
hydrogeography 水文地理学
hydrogeologic(al) and water well drilling rig 水井钻机
hydrogeologic(al) appraisal 水文地质评议
hydrogeologic(al) (bore) hole 水文地质钻孔
hydrogeologic(al) boundary of mineral deposit 矿床水文地质边界
hydrogeologic(al) characteristic 水文地质特征;水文地质特性
hydrogeologic(al) chart 水文地质图
hydrogeologic(al) condition 水文地质条件
hydrogeologic(al) control 水文地质控制
hydrogeologic(al) drill-hole 水文地质钻孔
hydrogeologic(al) drilling 水文地质钻探;水文钻探
hydrogeologic(al) exploration 水文地质勘探
hydrogeologic(al) exploration type of open to-air mine 露天矿床水文地质勘探类型
hydrogeologic(al) information 水文地质资料
hydrogeologic(al) investigation 水文地质勘察
hydrogeologic(al) map 水文地质图
hydrogeologic(al) map for bedrock 基岩水文地质图
hydrogeologic(al) map for mineral spring 矿泉水文地质图
hydrogeologic(al) map for oil field 油田水文地质图
hydrogeologic(al) map in tunnel 坑道水文地质图
hydrogeologic(al) map of coalfield 煤田水文地质图
hydrogeologic(al) map of exploration area 勘探区水文地质图
hydrogeologic(al) map of mine district 矿区水文地质图
hydrogeologic(al) map of ore deposit 矿床水文地质图
hydrogeologic(al) map of ore district 矿区水文地质图
hydrogeologic(al) map of reclamation 土壤改良水文地质图
hydrogeologic(al) map of tunnel spreading 坑道水文地质展示图
hydrogeologic(al) map on agricultural water supply 农田供水水文地质图
hydrogeologic(al) map on irrigation 灌区水文地质图
hydrogeologic(al) map on water supply 供水水文地质图
hydrogeologic(al) mapping 水文地质测绘
hydrogeologic(al) materials 水文地质资料
hydrogeologic(al) nature 水文地质特性
hydrogeologic(al) observation data in drilling 钻探水文地质观测数据
hydrogeologic(al) observation hole 水文地质观测孔
hydrogeologic(al) observation items in drilling 钻探水文地质观测项目
hydrogeologic(al) parameter 水文地质参数
hydrogeologic(al) parameter of aquifer 含水层水文地质参数
hydrogeologic(al) phenomenon in mines 矿坑水文地质现象
hydrogeologic(al) problem of artificial environment 人为环境水文地质问题
hydrogeologic(al) problem of primary environment 原生环境水文地质问题
hydrogeologic(al) profile 水文地质剖面图
hydrogeologic(al) prospect hole 水文地质勘察孔
hydrogeologic(al) survey 简易水文地质观测;水文地质调查;水文地质测绘
hydrogeologic(al) survey of agriculture water

supply 农田供水水文地质调查
hydrogeologic(al) survey of concentrative water supply 集中供水水文地质调查
hydrogeologic(al) survey of disease caused by soil and water 水土病因水文地质调查
hydrogeologic(al) survey of mineral deposit 矿床水文地质调查
hydrogeologic(al) survey of mineral spring 矿泉水文地质调查
hydrogeologic(al) survey of oil field 油田水文地质调查
hydrogeologic(al) survey of soil improvement 土壤改良水文地质调查
hydrogeologic(al) survey of water supply 供水水文地质调查
hydrogeologic(al) test 水文地质试验
hydrogeologic(al) test hole 水文地质试验孔
hydrogeologic(al) test in field 野外水文地质试验
hydrogeologic(al) well drilling machine 水文地质水井钻机
hydrogeologist 水文地质学家；水文地质工作者
hydrogeology 水文地质(学)
hydrogeology and engineering seismic survey 水文工程地震勘探
hydrogeology data base 水文地质数据库
hydrogeology survey 水文地质调查
hydroglauberite 水钙芒硝
hydroglider 水上滑翔机
hydroglockerite 水基铁矾
hydrognosy 水体史；水体历史描述
hydrogoethite 水纤铁矿
hydrogoethite rock 水针铁矿岩
hydrogovernor 水轮机调速器
hydrogracture 水压裂缝
hydrograph 自记水位仪；自记水位计；过程线；流量图；流量速度计算仪；流量流速计算仪；水文图；水文曲线；水文过程线；水位图；水流测量图；水的过程线；水道(测量)图；湿度记录器
hydrograph analysis 过程线分析；水文过程线分析
hydrograph ascending limb 过程线上升段
hydrograph curve of discharge 流量过程线
hydrographer 流量测量员；海道测量员；水文(地理)学家；水文测量员；水文测量者；水道测量员；水道测量学家
hydrographic(al) 水道测量的；水路的
hydrographic(al) airborne laser sounder 航空激光测深仪
hydrographic(al) annotation 水文调绘
hydrographic(al) automated system 水道自动测量系统
hydrographic(al) basin 流域
hydrographic(al) bulletin 水道测量公报
hydrographic(al) cartography 水文制图学
hydrographic(al) cast 重锤测深；水文施测；断面测量
hydrographic(al) chart 海道图；海道测量局刊印的海图；水系图；水文(地理)图；水路图【测】；水道图
hydrographic(al) charting 海图制图
hydrographic(al) cruise 海测航次
hydrographic(al) curve 水文曲线
hydrographic(al) data 水文(地理)资料；水道测量资料
hydrographic(al) data acquisition 水文资料采集
hydrographic(al) data acquisition system 水道测量资料收集系统
hydrographic(al) data analysis 水文资料分析
hydrographic(al) database 水文资料库
hydrographic(al) data collection 水文资料采集
hydrographic(al) data compilation 水文资料整理；水文资料整编
hydrographic(al) data gathering 水文资料采集
hydrographic(al) data interpolation 水文资料插补
hydrographic(al) data processing 水文资料处理
hydrographic(al) data processing system 水道测量数据处理系统
hydrographic(al) data station 水文资料站
hydrographic(al) datum 海图基准(面)；水文基准(面)；水文测量基准；水位零点；水位过程线零点；潮高计算基准
hydrographic(al) department 海道测量局；水文局
hydrographic(al) digital positioning and depth recording system 水道测量数字定位与深度自记系统

hydrographic(al) engineering survey 水利工程测量
hydrographic(al) factor 水文因素
hydrographic(al) feature 水文要素；水文性质；水文(地理)特征；水文特性；水文特点
hydrographic(al) forecast(ing) 水文预报
hydrographic(al) map 海(道)图；水系图；水文地理(分区)图；水文(测验)图；水道图
hydrographic(al) markers 水文标志
hydrographic(al) measurement 水文测验；水文测量
hydrographic(al) net(work) 水文网(站)；水(道)网；河系；河网；水系
hydrographic(al) observation 水文观测
hydrographic(al) office 海道测量局；水文(测量)局；水道测量局；水道测量部
hydrographic(al) parameter 水文参数
hydrographic(al) practice and symbols 海道测量作业与符号
hydrographic(al) precision scan(ning) echo sounder 水文精密扫描回声测深仪；水道测量精密扫描回声测深仪
hydrographic(al) process 水文变化过程
hydrographic(al) prognosis 水文预象
hydrographic(al) publications 水路出版物
hydrographic(al) reconnaissance 水文勘测
hydrographic(al) record 水文测量记录
hydrographic(al) regime 水文情势；水文情况
hydrographic(al) relief globe 水域浮雕地球仪
hydrographic(al) satellite tracking and recording 水文卫星跟踪与记录
hydrographic(al) section 水文断面
hydrographic(al) service 水文局；水道测量局
hydrographic(al) sextant 水文地理测量六分仪；测深六分仪
hydrographic(al) ship 海洋考察船；海道测量船；水文测验船
hydrographic(al) sonar 水文地理声呐
hydrographic(al) sounding 水道测深
hydrographic(al) station 海洋观测站；水文(观测)站；定点观测站
hydrographic(al) survey 海道测量
hydrographic(al) survey and charting system 水道测量与成图系统
hydrographic(al) survey baseline 海道测量基线
hydrographic(al) survey(ing) 河海测量；水文调查；水文测量；水道测量
hydrographic(al) surveying sextant 水道测量六分仪
hydrographic(al) survey of channel 航道测量
hydrographic(al) surveyor 海道测量员
hydrographic(al) (survey) ship 水道测量船
hydrographic(al) (survey) vessel 水道测量船
hydrographic(al) table 水文表
hydrographic(al) trawl 扫海具
hydrographic(al) vessel 海道测量船；水文测验船；水文测量船
hydrographic(al) wire 水文测验缆索
hydrographic(al) wire slope and azimuth indicator 水文测绳倾斜角和方位角指示器
hydrographic(al) year 水文年(度)
hydrograph of demand 需用的水文图
hydrograph of flow 流水的水文图
hydrograph of inflow 流入水文图
hydrograph of river stages 河流水位过程线
hydrograph of surface runoff 地面径流过程(曲)线
hydrograph of water(tide) level 水位过程线
hydrograph recession limb 过程线下降段
hydrograph separation 过程线分析；过程线分割；水文(过程线)分析；水文过程线分割；按补给分割过程线
hydrograph shape 水文过程线形状
hydrograph synthesis 过程线综合；过程线合成；单位过程线综合
hydrography 海道测量学；水象学；水文(地理)学；水文测量学；水路测量术；水道(测量)学；水道测量术
hydrography feature 水文要素；水文特征
hydrogrossular 水绿榴石；水钙铝榴石
hydrogrossularite 水绿榴石
hydrogymnasium 水中运动场
hydrogyro 流体浮悬陀螺仪
hydrohaematite 水赤铁矿

hydrohalite 冰盐；冰石盐
hydrohalloysite 水埃洛石；水合多水高岭土
hydrohammer 液动冲击器
hydrohausmannite 水黑锰矿
hydrohauyne 水蓝方石
hydroherderite 水磷铍钙石
hydrohetaerolite 水锌锰矿
hydrohonessite 水铁镍矾
Hydroidea 水螅虫类
Hydroides 盘管虫属
hydroindication 水指示现象
hydrointegrator 水力求积仪
hydroiodic ether 乙基碘
hydroisobaric line 等水压线
hydroisobath 等水深线；等埋深线；地下水等水深线；地下水位等值线；地下水位等高线；等水位线
hydroisohypse 等(水)深线；等埋深线；地下水等高线
hydroisopiestic line 等水压线；等承压水头线
hydroisopleth chart 水文等值线图
hydroisopleth map 水文等值线图；地下水位变化图；等水文等值线图
hydroites 含水矿物
hydrojack control 液力举升器操纵
hydrojet 液力喷射；喷液；水力喷射
hydrojet cleaner 液压清洁器
hydrojet craft 喷水船
hydrojet dredge 水力(喷射式)挖泥船；水力(喷射式)挖掘机；水力采砂船
hydrojet nozzle 水力喷嘴
hydrojet propelled ship 喷水推进艇；喷射推进船
hydrojet propulsion 射水推进
hydrojunction 水利枢纽
hydrojunction layout 水利枢纽布置
hydrojunction of navigation canal 运河枢纽
hydrokaolin 水埃洛石
hydrokaolinite 水高岭土
hydrokeel sidewall craft 侧壁式气垫艇
hydrokinematics 流体运动学；水运动学
hydrokineter 炉水循环加速器
hydrokinetic 流体动力的
hydrokinetic crushing 液压动力破碎；流体动力破碎(法)
hydrokinetic fluid drive 水动力传动
hydrokinetic machine 液压动力机械；液力机械
hydrokinetic pressure 流体动力压力
hydrokinetics 液体动力学；流体动力学；水动力学
hydrokinetic transmission 动液传动
hydrokinetic-type hydraulic transmission 流动动力式液压传动
hydrol 防水剂；二聚水分子
hydrolability 水分不稳定性
hydrolaccolith 水岩盖
hydrolapse 水分直减率
hydrolastic 液力补偿悬挂；液力平衡悬架
hydroleg 液压柱；液压腿
hydrolevel(l)ing device 液压平衡装置；液压调平装置
hydrolift system 水力提升系统
hydroline 吹制油
hydrolith 氢化钙；水生岩；水成碳酸盐碎屑岩；水成沉淀岩
Hydrolithic 海多罗里提克防水材料
hydroliths 水生岩类
hydrolocation 水中定位；水声定位
hydrolocation chronoscope 水中定位测时计
hydrolocator 水中定位仪；水声定位仪
hydrolock(retort) 水封式杀菌机
hydrologic(al) 水文的
hydrologic(al) abstraction 水文损失量
hydrologic(al) accounting 水文计算；水量平衡
hydrologic(al) almanac 水文年鉴
hydrologic(al) analog 水文模拟
hydrologic(al) analogy 水文相似；水文情态模拟；水文比拟
hydrologic(al) analysis 水文分析
hydrologic(al) and engineering geology 水文工程地质学
hydrologic(al) and geologic(al) condition 水文地质条件
hydrologic(al) and meteorologic(al) condition 水文气象条件
hydrologic(al) and meteorological fixed station 水文和气象固定电台

hydrologic(al) and meteorological mobile station 水文和气象移动电台
hydrologic(al) and meteorologic(al) survey 水文与气象调查；水文与气象测验
hydrologic(al) annotation 水文调绘
hydrologic(al) anomaly 水文异常
hydrologic(al) apparatus 水文仪器
hydrologic(al) atlas 水文图集
hydrologic(al) balance 水文过程量平衡；水量平衡；水分平衡
hydrologic(al) basin 水文盆地；水文流域
hydrologic(al) benchmark 水文用水准点；水文参证
hydrologic(al) budget 水均匀；水文预算；水量平衡(表)；水均衡(预测)
hydrologic(al) calculation 水文计算
hydrologic(al) characteristic of lake 湖泊水文特征
hydrologic(al) characteristics 水文特征值；水文特性
hydrologic(al) characteristic survey 水文特征调查
hydrologic(al) circle 水文周期；水文循环
hydrologic(al) civilization 水文文明
hydrologic(al) computation 水文计算
hydrologic(al) condition 水文状况；水文条件
hydrologic(al) cycle 水(文)循环；水文周期；水分循环
hydrologic(al) data 水文资料；水文数据；水分资料
hydrologic(al) data acquisition 水文资料采集
hydrologic(al) data analysis 水文资料分析
hydrologic(al) data bank 水文资料库；水文数据库
hydrologic(al) database 水文数据库
hydrologic(al) data collection 水文资料采集
hydrologic(al) data compilation 水文资料整理；水文资料整编
hydrologic(al) data gathering 水文资料采集
hydrologic(al) data interpolation 水文资料插补
hydrologic(al) data processing 水文资料处理
hydrologic(al) data restoration 水文资料修复；水文资料还原
hydrologic(al) data sampling 水文资料选样；水文资料抽样
hydrologic(al) data station 水文资料站
hydrologic(al) decade 水文十年
hydrologic(al) design 水文设计；水文计算
hydrologic(al) divide 水文分水岭；水文分水界；水文分界线；地下水分水界
hydrologic(al) effect 水分效应
hydrologic(al) element 水文要素
hydrologic(al) engineering method 水文工程方法
hydrologic(al) equation 水文方程；水量平衡方程
hydrologic(al) equilibrium 水量平衡
hydrologic(al) event 水文事件
hydrologic(al) experiment 水文实验
hydrologic(al) experiment(al) station 径流实验站；水文实验站
hydrologic(al) exploration 水文勘查
hydrologic(al) extreme 水文极值
hydrologic(al) extreme events 水文极端事件(洪涝旱碱等)
hydrologic(al) extremes 极端水文现象
hydrologic(al) factor 水文因素；水文要素
hydrologic(al) forecast(ing) 水文预报
hydrologic(al) forecasting center[centre] 水文预报中心
hydrologic(al) forecast of lake 湖泊水文预报
hydrologic(al) forecast of spring flood 春汛预报
hydrologic(al) frequency analysis 水文频率分析
hydrologic(al) frequency calculation 水文频率计算
hydrologic(al) frequency computation 水文频率计算
hydrologic(al) frequency curve 水文频率曲线
hydrologic(al) front 水文锋面
hydrologic(al) ga(u)ge 水位计；验潮仪
hydrologic(al) geology 水文地质(学)
hydrologic(al) homogeneity 水文一致性
hydrologic(al) impact of urbanization 城市化对水文的影响
hydrologic(al) information 水文情报；水情；报汛
hydrologic(al) inventory 水文清点
hydrologic(al) investigation 水文研究；水文调查
hydrologic(al) laboratory 水文实验室
hydrologic(al) line 水涯线
hydrologic(al) manual 水文手册
hydrologic(al) map 水文图；水文等值线图

hydrologic(al) measurement 水文检测
hydrologic(al) method 水文学方法；水文地质方法
hydrologic(al) model 水文模型
hydrologic(al) modification 水文要素变化；水文情态变化
hydrologic(al) network 水文站网
hydrologic(al) network planning and design 水文站网规划设计
hydrologic(al) observation 水文观测；水文测验
hydrologic(al) period 水文周期
hydrologic(al) post 水文观测点
hydrologic(al) process 水文(变化)过程
hydrologic(al) process modelling 水文过程模拟
hydrologic(al) process modelling test 水文过程模型试验
hydrologic(al) profile 水文地质剖面图
hydrologic(al) prognosis 水文预测；水文预报
hydrologic(al) project 水利工程
hydrologic(al) property 水文性质；水文特性
hydrologic(al) record 水文记录；水文实测资料
hydrologic(al) regime 水文状况；水文情势；水文情况；水情
hydrologic(al) regime automatic monitoring and forecasting 水情自动测报
hydrologic(al) region 水文区域；水文地区
hydrologic(al) regionalization 水文区划；水文分区
hydrologic(al) remote sensing technique 水文遥测技术
hydrologic(al) research 水文研究
hydrologic(al) response 水文响应(曲线)；水文特性(曲线)
hydrologic(al) response unit 水文响应单元
hydrologic(al) routing 用水文学方法演算洪水；洪水演算；水文演算
hydrologic(al) section 水文剖面
hydrologic(al) sensor 水文探测设备；水文传感器
hydrologic(al) separation 水文过程线划分
hydrologic(al) service 水文所
hydrologic(al) services 水文服务事业
hydrologic(al) simulation 水文模拟
hydrologic(al) station 水文(观测)站
hydrologic(al) station of river 河流水文站
hydrologic(al) statistical characteristics 水文统计特征(值)
hydrologic(al) statistical parameter 水文统计参数
hydrologic(al) statistics 水文统计(学)
hydrologic(al) storm 水文风暴
hydrologic(al) survey 水文勘测；水文调查水文查勘；水文查勘
hydrologic(al) survey of glaciers 冰川水文调查
hydrologic(al) survey of lake 湖泊水文调查
hydrologic(al) survey of river basin 流域水文调查
hydrologic(al) survey of swamp 沼泽水文调查
hydrologic(al) system 水文系统；液压逻辑系统
hydrologic(al) type 水文型
hydrologic(al) warning 水文警报
hydrologic(al) well logging 水文测井
hydrologic(al) winch 水文绞车
hydrologic(al) year 水文年(度)
hydrologic(al) yearbook 水文年鉴
hydrologic(al) zonality 水文区划；水文分区
hydrologist 水文学家；水文工作者
hydrology 水文学；水文物理现象；水理学
hydrology and water resources calculation 水文水资源计算
hydrology of groundwater 地下水水文学
hydrology of lake 湖泊水文学
hydrology of land 陆地水文学
hydrolube 氢化润滑油
hydrolysate 水解液
hydrolysis 水解(作用)
hydrolysis acidification 水解酸化
hydrolysis acidification pool 水解酸化池
hydrolysis-aerobic loop sequencing batch reactor process 水解—好氧循环序批间歇式反应器工艺
hydrolysis-aerobic process 水解—好氧工艺
hydrolysis basin effluent 水解池出水
hydrolysis-biological aerated filter nitrification process 水解—曝气生物滤池硝化工艺
hydrolysis butyl flo(a)tation aids 水解丁基浮选剂
hydrolysis-contact oxidation-air flo(a)tation 水解—接触氧化—气浮—混凝—生物炭工艺
hydrolysis current 水解电流

hydrolysis equilibrium 水解平衡
hydrolysis nuclei 水解晶种
hydrolysis of ethyl acetate 乙酸乙酯水解
hydrolysis polyacrylonitrile 水解聚丙烯腈
hydrolysis polymerization 水解聚合反应
hydrolysis product 水解产物
hydrolysis rate 水解速率
hydrolysis reactor 水解反应器
hydrolysis resistance 抗水解性
hydrolysis yield 水解率
hydrolyst 水解催化剂
hydrolyte 水解质
hydrolytic 水解的
hydrolytic acidification 水解酸化
hydrolytic acidification-anaerobic-aerobic biochemical process 水解酸化—厌氧—好氧生化工艺
hydrolytic acidification-anoxic biological process 水解酸化—缺氧生物工艺
hydrolytic acidification basin 水解酸化池
hydrolytic acidification-biological contact oxidation process 水解酸化—生物接触氧化工艺
hydrolytic acidification-contact oxidation-coagulation sedimentation process 水解酸化—接触氧化—混凝沉淀工艺
hydrolytic acidification-oxic oxidation process 水解酸化—好氧氧化工艺
hydrolytic acidification-oxic process 水解酸化—好氧工艺
hydrolytic acidification-sequencing activated sludge process 水解酸化—序批式活性污泥法
hydrolytic acidification-upflow biological aerated filter process 水解酸化—升流式曝气生物滤池工艺
hydrolytic acidity 水解(性)酸度
hydrolytic action 水解作用
hydrolytic adsorption 水解吸附
hydrolytic bacteria 水解细菌
hydrolytic constant 水解常数
hydrolytic current 水解电流
hydrolytic decomposition 水解
hydrolytic degradation 水解降解
hydrolytic dissociation 水解作用；水解离解；水解电离
hydrolytic effect 水解效应
hydrolytic enzyme 水解酶
hydrolytic oxidation 水解氧化
hydrolytic-polymeric reaction 水解聚合反应
hydrolytic process 水解过程
hydrolytic reaction 水解反应
hydrolytic reactor 水解反应器
hydrolytic reagent 水解剂
hydrolytic resistance 抗水解能力
hydrolytic scission 水解分裂
hydrolytic slaked lime 水解石灰
hydrolytic spoilage 水解变质
hydrolytic stability 抗水解稳定性；水解稳定性
hydrolytic tank 水解池；水解槽
hydrolytic treatment 水解处理
hydrolyzable 可水解的
hydrolyzable nitrogen 水解性氮
hydrolyzate 水解液；水解沉积物；水解产物
hydrolyzate effluent 水解产物污水
hydrolyzation 水解
hydrolyzation and biological contact oxidation process 水解—生物接触氧化工艺
hydrolyze 水解
hydrolyze acidification 水解酸化
hydrolyzed fat 水解脂肪
hydrolyzed polyacrylamide 水解聚丙烯酰胺
hydrolyzed polymaleic anhydride 水解马来酸酐
hydrolyzed shellac 水解紫胶
hydrolyzed starch 水解淀粉
hydrolyzed straw 水解秸秆
hydrolyzer 水解器
hydromagnesite 水碳镁石；水菱镁矿
hydromagnetic 水磁
hydromagnetic brake 液压电磁制动器
hydromagnetic DC converter 磁流体直流变流器
hydromagnetic engine 磁流体(动力)发动机
hydromagnetic oscillation 磁流体振荡
hydromagnetic oscillations of the plasma 等离子体的磁流体振荡
hydromagnetics 磁流体动力学

hydromagnetic wave	磁流体动力(学)波;磁流波
hydromagnetism	水磁学;磁流体力学
hydromagnetite	水磁铁矿
hydroman	液压操作器;水力控制器
hydromanometer	液压计;液压表;流体压强计;流体压力计;测压计
hydromaster tester	液压助力器试验器
hydromatic	液压自动传动
hydromatic brake	液压制动器
hydromatic drive	液压传动
hydromatic forming	直接液压式板料成形法
hydromatic jaw clutch	液动式牙嵌离合器
hydromatic process	液压自动工作(方)法;双作用液压成形法
hydromatic propeller	液压自动变距螺(旋)桨
hydromatic retort	水静压式杀菌机
hydromatic system	液压自动传动系统
hydrombobomkite	水硫硝镍铝石
hydromechanical cell	水力机械间
hydromechanical deep drawing	机械液压深拉延
hydromechanical drawing	充液拉深
hydromechanical drive	液压机械传动
hydromechanical drive system	液压机械传动系统
hydromechanical injection mo(u)lding machine	液压机械注模机
hydromechanical leg	液压机械支柱
hydromechanical mining method	水力机械采矿法
hydromechanical model	流体力学模型
hydromechanical press	液压机械式压机
hydromechanical process	流体力学过程
hydromechanical tipper	液压翻车机
hydromechanics	液体力学;流体力学;水力学
hydromechanics theory	流体力学理论
hydromechanisation	水利机械化;水力机械化
hydromedusa	水螅水母
hydromelanothallite	水黑氯铜矿
hydromelioration	水利土壤改良
Hydroment	海若蒙黏合剂(商品名)
hydrometallurgical extraction	水冶提取
hydrometallurgical plant	水冶设备
hydrometallurgy	水力冶金;水冶;湿法冶金
hydrometamorphism	水热变质;水变质作用
hydrometeor	降水;水文气象;水气现象;水气凝结物;水气凝结体
hydrometeorologic(al) forecasting centre	水文气象预报所;水文气象预报中心
hydrometeorologic(al) instrument	海洋气象仪;水文气象仪
hydrometeorologic(al) network	水文气象(观测)站网
hydrometeorologic(al) relationship	水文气象关系
hydrometeorologic(al) service	水文气象站
hydrometeorologic(al) ship	水文气象调查船
hydrometeorologic(al) station	水文气象站
hydrometeorologic(al) survey	水文气象调查
hydrometeorologist	水文气象学家;水文气象工作者
hydrometeorology	海洋气象学;水文气象学;水文气象
hydrometer	原油比重计;液体比重计;流速计;海水比重计;土壤比重计;石油密度计;湿度比重计;比重表(指仪器名称)
hydrometer analysis	液体比重分析;密度计分析;比重计分析;比重计法
hydrometer analysis method	比重计分析法
hydrometer calibration	比重计校准
hydrometer degree	比重计标度
hydrometer jar	液体比重计瓶;液体比重测量缸
hydrometer method	比重计法;比重法
hydrometer method of analysis	比重计分析法
hydrometer of constant immersion	定浮比重计
hydrometer of variable immersion	变浮比重计
hydrometer set	液体比重计组
hydrometer syringe	比重计吸液器
hydrometer system	水文气象预报系统
hydrometric(al)	测定比重的
hydrometric(al) aerial cableway	测流架空缆道
hydrometric(al) aerial ferry	水文测验架空缆道;测流架空缆道
hydrometric(al) basic station	水文基本测站
hydrometric(al) boat	水文测验船;测船
hydrometric(al) cable car	水文缆车
hydrometric(al) cableway	水文缆道
hydrometric(al) cable-way with cable suspension	悬索式水文缆道
hydrometric(al)(cross-)section	水力测量截面
hydrometric(al) current meter	流速仪;水文流速仪
hydrometric(al) dingey	水文测艇
hydrometric(al) float	水文测验浮子;水文测验浮标;测流浮子;测流浮标
hydrometric(al) flowmeter	流速仪;水文流速仪;水文测速仪
hydrometric(al) ga(u)ge	水位(测)站;水尺
hydrometric(al) measurement	水文测验;水文测量
hydrometric(al) method	比重计测定法
hydrometric(al) network	水文测量网络;流体比重测定网络;水文测验站网;测流站网
hydrometric(al) pendulum	流速摆(测量河道流速用);摆式流速仪
hydrometric(al) profile	水文测验断面
hydrometric(al) propeller	流速仪;水文流速仪;水文测速仪
hydrometric(al) records	水文测验实测资料;水文测验记录
hydrometric(al) scheme	水文图表
hydrometric(al) section	水文观测断面;水文断面图;水文测验断面;测流断面
hydrometric(al) staff float	测流浮杆
hydrometric(al) station	水文站;水文(观)测站;测流站
hydrometric(al) station control	水文测站控制
hydrometric(al) station datum(plane)	水文站基面
hydrometric(al) station for special purposes	专用水文站;水文专用测站
hydrometric(al) surveying	水文测量
hydrometric(al) table	液体比重表
hydrometric(al) tube	测流管(测定水速的器械)
hydrometry	液体比重测定法;流体测定;水文测验学;水文测验;水文测量学;水流测定;水力测量学;测流法;测定液体方法;测比重法;比重测定法
hydrometry for reservoir	水库水文测验
hydromica	水云母
hydromica clay	水云母黏土
hydromica mudstone	水云母泥岩
hydromill	液压双轮掘削机;双轮反循环钻机
hydromilling trencher	液压碾磨造孔机
hydromill machine	液压碾磨机
hydromine	水力开采矿井;水力机械化矿井
hydromining	水力开采(法)
hydromodulus	流量模数;流量模量
hydromolysite	水铁盐
hydromonitor	水力测功器;水力检查计
hydromonoplane	水上单翼机;单翼水上飞机
hydromorphic anomaly	水成异常
hydromorphic dispersion	水成分散
hydromorphic horizon	水成层
hydromorphic soil	水成土
hydromorphism	水生形态
hydromorphous process	水成过程
hydromorphy	水成形态
hydromotive equipment	水上运输设备
hydromotor	水压发动机;射水发动机;液压马达;液压发动机;液压电(动)机;水力发动机
hydromotor jig	流体传动跳汰机
hydromotor propeller	水力推进器
hydromucker	水力装岩机;液压装岩机
hydromuscovite	水白云母
Hydronalium	海德罗纳留姆耐蚀铝镁合金;海庄铝合金
Hydronalium alloy	海德罗镁铝合金
hydronapper	液压传动起绒机
hydronasturan	水沥青铀矿
hydronaut	深水潜航器驾驶员
hydronautics	水航学
Hydron blue	海昌蓝
Hydrone	海德隆铅钠合金
hydronephelite	水霞石
hydrone theory	缔合水理论
hydroniccite	羟镍矿
hydronic heating	热水供暖
hydronic heating system	循环加热系统;循环采暖系统;热水供暖系统
hydronics	循环加热冷却系统
hydronitchblende	水沥青铀矿
hydronitric acid	叠氮酸
hydronium ion	水合氢离子;水合氢
hydronium jarosite	水合氢离子铁矾
hydronymy	水系名称;水体名词学
hydroodelhayelite	水片硅碱钙石
hydrooptics	水域光学
hydropaste	含铝涂模浆
hydropeening	喷洗;喷水清洗;喷水清理
hydropenic	缺水的
hydropercussive drilling tool	液动冲击钻具
hydropercussive tool	液动冲击器
hydroperiod	淹水期;水文周期
hydroperoxide	过氧化氢物;氢过氧化物
hydroperoxyl radical	氢过氧基游离基;过氧氢自由基
hydrophane	水蛋白石
hydrophase diagram	水文相变图
hydrophil	亲水的
hydrophilae	水媒植物
hydrophile	喜水物;亲水物;亲水胶体
hydrophile-lipophile balance	亲水物—亲脂物平衡;亲水亲油平衡
hydrophile-lipophile balance value	亲水亲油平衡值
hydrophilia	亲水(性)
hydrophilic	亲水(性)的
hydrophilic adsorbate	亲水吸附质
hydrophilic adsorbent	亲水吸附剂
hydrophilic aggregate	吸水骨料;亲水集料;亲水骨料;吸水集料
hydrophilic bacteria	喜湿细菌
hydrophilic chain	亲水链
hydrophilic channel	亲水管
hydrophilic coefficient	亲水系数
hydrophilic colloid	亲水胶体
hydrophilic contact lens	亲水性接触镜
hydrophilic dust	亲水性粉尘
hydrophilic filler	亲水填料
hydrophilic fraction	亲水组分;亲水分数
hydrophilic gas electrode	亲水气体电极
hydrophilic gel	亲水胶
hydrophilic group(ing)	亲水基(团)
hydrophilicity	亲水性;亲合性
hydrophilic-lyophobic balance	亲水疏水平衡值
hydrophilic material	亲水材料
hydrophilic method	亲水法
hydrophilic micelle	亲水微胶粒
hydrophilic modification	亲水改性
hydrophilic nature	亲水性
hydrophilic ointment	亲水性软膏
hydrophilic oleomicelle	亲水非极性芯胶束
hydrophilic organic compound	亲水性有机化合物
hydrophilic organogel	亲水有机胶体
hydrophilic particle	亲水性颗粒
hydrophilic plant	亲水植物
hydrophilic pole	亲水极
hydrophilic porous carrier	亲水性多孔载体
hydrophilic powder	亲水(性)粉末
hydrophilic sol	亲水溶胶
hydrophilic solute	亲水胶质
hydrophilic species	亲水物种
hydrophilic substance	亲水物质
hydrophilic system	亲水系统
hydrophiling	亲水化
hydrophilite	氯钙石
hydrophilization process	亲水化法
hydrophilous	亲水的
hydrophilous nature	亲水性质
hydrophily	亲水状态
hydrophite	含水石;水铁蛇纹石
hydrophlogopite	水金云母
hydrophobe	疏水物;疏水胶体
hydrophobic	憎水的;疏水的;输水的
hydrophobic activation	疏水活化
hydrophobic admixture	憎水剂;憎水性外加剂;疏水剂
hydrophobic adsorbate	疏水吸附质
hydrophobic adsorbent	疏水吸附剂
hydrophobic aggregate	厌水骨料;憎水性集料;憎水骨料;不吸水集料;不吸水骨料
hydrophobic bond	疏水键
hydrophobic bonding	疏水键合
hydrophobic cement	憎水水泥;防潮水泥
hydrophobic chain	疏水链
hydrophobic chemical pollutant	疏水化学污染物
hydrophobic colloid	憎水胶体;疏水胶体
hydrophobic compound	疏水化合物
hydrophobic core	疏水内芯

hydrophobic dust 疏水性粉尘
hydrophobic expanded pearlite 憎水膨胀珍珠岩
hydrophobic expansive cement 憎水膨胀水泥
hydrophobic fraction 疏水分数
hydrophobic gas electrode 疏水气体电极
hydrophobic gel 疏水胶体
hydrophobic glass beads 憎水玻璃细珠
hydrophobic grid membrane filter 疏水性网格膜滤池
hydrophobic group 憎水基;疏水基
hydrophobic hollow fiber membrane 疏水中空纤维膜
hydrophobic interaction 疏水性相互作用;疏水基相互作用
hydrophobic ionizable organic compound 可电离疏水有机化合物
hydrophobicity 憎水性;疏水性
hydrophobicity index 疏水性指数
hydrophobic material 憎水材料;疏水物
hydrophobic membrane 疏水膜
hydrophobic microporous membrane 疏水微孔膜
hydrophobic nature 疏水性
hydrophobic oil 疏水油
hydrophobic organic chemicals 疏水有机化学物
hydrophobic organic compound 疏水有机化合物
hydrophobic organic contaminant 疏水有机污染物
hydrophobic organic pollutant 疏水有机污染物
hydrophobic particle 疏水性颗粒
hydrophobic Portland cement 防潮波特兰水泥;憎水水泥;憎水硅酸盐水泥
hydrophobic property 憎水性能
hydrophobic pumice 疏水浮石
hydrophobic rubber 疏水橡胶
hydrophobic sol 憎水溶胶;疏水溶胶
hydrophobic solute 疏水溶质
hydrophobic species 疏水物种
hydrophobic substance 憎水物质;疏水物质
hydrophobic surface 疏水表面
hydrophobic treatment of masonry(work) 砖石抗水处理
hydrophobing agent 疏水剂
hydrophobization 憎水化
hydrophone 水中噪声测向仪;水中听声器;水中感声;水中地震检波器;水下听声器;水下拾音器;水下扩音器;水下测声器;水下测听器;水听器
hydrophone array 水听器列;水听器基阵
hydrophone bearing 水声方位
hydrophone calibrator 水下听声器校准器
hydrophone noise 水下听声器噪声
hydrophone response 水下听声器响应
hydrophone system 水采样系统
hydrophonic detector 水下探测器;水听器;水声探测器
hydrophonic plant 水栽植物
hydrophonic plant growth 水栽植物培养
hydrophonics 水栽法
hydrophore 水样取样器;采水样器
hydrophorograph 液流描记器
hydrophotometer 水下光度计
hydrophysical 水文物理的
hydrophysical law 水文物理定律
hydrophysical process 水文物理过程
hydrophysics 水文物理学
hydrophysics property of rocks 岩石水理性质
hydrophyte 水生植物
hydrophyte adnata 固着水生植物
hydrophyte natantia 漂浮水生植物
hydrophyte pond 水生植物塘
hydrophyte radicanta 有根水生植物
hydrophytic vegetation 水生植被
hydropite 蔷薇辉石
hydroplane 滑行艇;潜水舵;水翼船;水翼;水上滑行艇;水上滑行;水面滑走快艇;水上飞艇;水上飞机
hydroplane speedboat 水上飞行快艇
hydroplaning 液面滑行;水滑现象;落地前平飘(飞机);水滑;漂浮现象
hydroplaning phenomenon 水滑现象
hydroplaning speed 液面滑行速度;水滑速度
hydroplankton 水生浮游生物
hydroplant 水电站;水力发电厂;水电厂
hydroplanula 水浮浪幼体
hydroplastic 水塑的
hydroplastic corer 氢化塑料取芯器
hydroplasticity 含水塑性;水塑性

hydroplastic sediment 含水塑性沉积物
hydroplat 液压测力仪
hydroplutonic 液水深成的;水火成的
hydroplutonic rock 水火成岩
hydropneumatic 液压气动的;水气并用;水气并动
hydropneumatic accumulator 液压气式蓄能器;水气储压器
hydropneumatic brake 液压空气制动机
hydropneumatic device 液气装置
hydropneumatic die cushion 液压气动式模具缓冲器
hydropneumatic leveler 液压气动调平器
hydropneumatic recoil system 液气后座系统
hydropneumatic riveter 液压气动式铆钉机;液压气动式铆钉锤;液压气动式铆钉枪
hydropneumatics 液压气动学;液压气动技术
hydropneumatic side bearing 汽液旁承
hydropneumatic spring 液压空气弹簧;带液压枕垫的板簧
hydropneumatic storage tank 气压水箱
hydropneumatic surge arrester 液压气动涌浪制动器
hydropneumatic suspension 液压气动式空气悬架
hydropneumatic system 液压气动系统
hydropneumatic tank 液压气动水箱
hydropneumopericardium 水气心包
hydro policy 水力发电方针;水力发电策略
hydropolis 水上城市
hydropolylitionite 水多硅鳞云母
hydropolymer 氢化聚合物
hydropolymerization 氢化聚合作用
hydroponic culture 溶液培植
hydroponics 溶液培植;溶液栽培;溶液培养;水栽培;水栽法
hydropost 液压支柱;液压支架
Hydropotes inermis 河麂
hydropower 水电;液压机具;液力(源);水力
hydropower development 水能利用;水电开发
hydropower development project 水电开发工程
hydropower house 水电站
hydropower plant(with reservoir) 水电站;水力发电厂
hydropower potential 水电蕴藏量
hydropower resources 水力(发电)资源;水利资源;水电资源
hydropower site 水力发电厂址
hydropower station 水力发电站
hydropower tunnel 水电站隧洞;水力发电隧洞
hydropress 液压机;水压机
hydropressure water gun 液压水枪
hydroprocessing 氢加工
hydroprocessing technology 氢加工工艺
hydropropyl cellulose 羟丙基纤维素
hydrops 积液;积水
hydropsis 海洋预报学
hydropulse 水下脉动式喷射发动机
hydropyrochlore 水烧绿石
hydropyrometer 水测高温计
hydroquinol 对苯二酚
hydroquinone 氢醌;对苯二酚
hydroquinone monobenzyl ether 对苄氧苯酚;对甲氧酚
hydroregime 水情;水文情势;水文情况
hydroresorcinol 氢化间苯二酚
hydrorheostat 水变阻器
hydroromarchite 羟锡矿
hydrorrhea 液溢
hydrosamarskite 水铌钇矿
hydrosaturnism 水铅中毒
hydroscience 水科学;水文科学
hydroscillator 水力振荡器
hydroscillator classifier 水力式分级机
hydroscope 液压测试器;潜望镜;水中望远镜;水气计;水力测试器;湿度计;深水探视仪;深水探测仪
hydroscopic 吸水的;吸湿的;潜望镜的;深水望远镜的
hydroscopic capacity 吸水量;吸湿容量
hydroscopic coefficient 吸湿系数
hydroscopic coefficient of soil 土的吸水系数
hydroscopicity 水湿性
hydroscopic material 吸湿材料
hydroscopic moisture 吸湿水;吸湿含水率
hydroscopic moisture content 吸湿含水量

hydroscopic property 吸湿性
hydroscopic salt method 吸湿盐法
hydroscopic water 吸(收)水;吸湿水;吸附水;结合水;薄膜水
hydroscopic water content 吸湿含水量
hydroseal 水封
hydroseal(sand)pump 水封(式)泵
hydroseeder 水力播种机
hydroseim 板垂震源
hydroseparation 水力分粒作用;水力分粒法;水力分离作用;水力分离法;水力分级
hydroseparator 水力分离器;水力分选机;水力分离器;水力分离机;水力分级器;沉降槽;分水机
hydrosequence 水文系列;水文序列
hydrosere 水生演替系列;水生序列
hydroserpentine 水蛇纹石
hydroshield 泥水盾构
hydrosiderite 褐铁矿
hydrosilation 硅氢化作用
hydrosilicarenyte 水成纯砂岩
hydrosilicate 含水硅酸盐;氢化硅酸盐;水化硅酸盐;水合硅酸盐
hydrosilicofluoric acid 氟硅酸;氢化硅氟酸
hydrosilicon 硅氢化合物
hydrosilylation 氢化硅烷化
hydrosizer 水力分级机;上升水流分级机;浮选机
hydroski 水橇
hydroski craft 滑行艇
hydroskimmer 气垫船;水面掠行艇;水面滑行艇
hydroski vehicle 水面滑行艇
hydrosodalite 水方钠石
Hydrosoft 海特罗佐夫特柔软剂
hydrosol 液悬体;脱水溶胶;水溶体;水溶胶
hydrosolvent 水溶剂
hydrosonde 海底地貌测定仪;水下地震勘探仪
hydrosound sub-bottom profiler 震波水下回声地质剖面探测仪
hydrosound underwater seismic profiler 震波水下回声地质剖面探测仪;水下回声地震剖面仪
hydrospace 海洋水界
hydrospace detection 水中探测
hydrospace vehicle 深潜器
hydrospark forming process 水中放电成形法
hydrosphere 海洋水界;水圈;水气;水界;水层;地水层;地球水圈;地球水面
hydrospin(ning) 液力旋压
hydrospire 水旋板
hydrosporae 水布植物
hydrospring 水上安定器
hydrostabil 对水不稳定
hydrostability 抗水稳定性
hydrostabilizer 水上安定器;水面稳定设备
hydrostable 抗水的;对水稳定的
hydrostat 恒湿仪;恒湿箱;恒湿器;恒湿计;水压调节器
hydrostatic 静水(力学)的;水压式的;流体静力学的;静流体
hydrostatic accelerometer 流体静力加速计;静水加速仪
hydrostatic analysis 液体静压分析
hydrostatic approximation 流体静力学近似
hydrostatic arch 静水力拱(曲线);静水压力拱
hydrostatic arch curve 静水压(力)拱形曲线
hydrostatic assumption 流体静力学假设
hydrostatic balance 液压比重计;液体比重计;流体静力学平衡;比重计
hydrostatic bearing 液体静力轴承;浮动轴承;液体静压轴承;静压轴承
hydrostatic bed 静水压力基床;水褥
hydrostatic bell 水压钟
hydrostatic burst testing 水压爆破试验
hydrostatic calculation 静水力计算
hydrostatic catenary 静水压悬链线;静水压力悬链(曲)线
hydrostatic column 静水柱
hydrostatic compression 静水压缩;围压压缩
hydrostatic confining pressure 流体静力围压
hydrostatic cooker 水静压式杀菌机
hydrostatic curve 流体静力曲线;静水压(力)曲线;静水力曲线
hydrostatic cyclic fatigue test 水压循环疲劳试验
hydrostatic depth ga(u)ge 静水深度计;水静压测深棒
hydrostatic design stress 静水压力设计应力

hydrostatic dilatation 水压扩张术
hydrostatic distribution of the earth pressure 土压力的静水压分布
hydrostatic drive 液压静力传动;液压传动;静液压传动;液压驱动
hydrostatic drive bulldozer 液压传动推土机
hydrostatic earth 流体静力学地球
hydrostatic equation 液体静力(学)方程;流体静力方程;静水(压)力方程
hydrostatic equilibrium 流体静力平衡;静水压力平衡;静水(力)平衡
hydrostatic excess pressure 过剩静水压力;超流体静压力;超静水压力
hydrostatic extrusion 液力静挤压;静液压挤压;静(水)压挤压
hydrostatic force 静水压力
hydrostatic forging 液压锻造
hydrostatic fuse 水压式引信
hydrostatic ga(u)ge 液度计;液体压力计;液体静压(压力)计;流体静压(力)计
hydrostatic head 静扬程;静压头;静压高差;静水压力水头;静水头
hydrostatic head ga(u)ge 流体静压头计
hydrostatic inspection certificate 水压试验合格证
hydrostatic instability 静水不稳定(性)
hydrostatic intensity of pressure 静水压强
hydrostatic isostatic forming 等静压成形
hydrostatic joint 承水压接口;承压接口;水压接头;水压承接头
hydrostatic level 流体静压水准(测量)仪;静水压力水位;静水平面
hydrostatic level(l)ing 流体静力水准测量;液体静力水准测量
hydrostatic level(l)ing apparatus 流体静力水准(测量)仪;水管式水准仪
hydrostatic level(l)ing instrument 液体静力水准仪
hydrostatic lift 静水扬力
hydrostatic lifting 静压提升
hydrostatic line 静水压力线
hydrostatic line of pressure 静水压力线
hydrostatic load 静水(压力)荷载
hydrostatic load distribution 静水荷载分布
hydrostatic lock 静液锁闭;水力冲击
hydrostatic lubrication 液体静压润滑;流体静力润滑
hydrostatic lubricator 液压润滑器
hydrostatic motor 静发动机
hydrostatic organ 水中平衡器
hydrostatic overpressure test 静水超压试验
hydrostatic paradox 静水疑题;静水奇像
hydrostatic piezoelectric strain constant 等静压压电应变常数
hydrostatic pile press-extract machine 静力压拔桩机
hydrostatic press 液压机;水压机;等静压机
hydrostatic pressing 等静压成型
hydrostatic pressure 液压;液体静压力;静流体静压(力)强压强;流体静压(力)强压强;水压
hydrostatic pressure apparatus 静态高压设备
hydrostatic pressure cell 静水压力盒
hydrostatic pressure coefficient 静水压力系数
hydrostatic pressure distribution 静水压力分布;流体静压分布
hydrostatic pressure head 流体静压头;静水压头
hydrostatic pressure of reservoir 库水静压力
hydrostatic pressure on dock floor 坞底静水压力
hydrostatic pressure ratio 静水压力系数;静水压(力)比
hydrostatic pressure test(ing) 静水压(力)试验;流体静压试验
hydrostatic process 水压法
hydrostatic profile ga(u)ge 静水位移监测计
hydrostatic proof test 水压验证试验
hydrostatic pull 静水拉力
hydrostatic sludge removal 流体静压污染排除(法);静水(压)排泥;清除水污泥
hydrostatic stability 静水稳定性
hydrostatic state of stress 流体静力应力状态;静水应力状态
hydrostatic static of stress 静水压力状态
hydrostatic sterilization 水静压式杀菌
hydrostatic sterilizer 水静压式杀菌机
hydrostatic strength 流体静力强度;静水压(力)强度

hydrostatic stress 流体静应力;静水应力
hydrostatic stress condition 静水应力状态
hydrostatic structure 水工结构
hydrostatic swing bridge 水力驱动的平旋桥;水力传动的平旋桥
hydrostatic tension 静水张力;静水拉力;三轴向张力
hydrostatic test(ing) 静水压试验(常用于检查排水系统中有无渗漏);液压试验;流体静力(学)试验;静水(压)试验;水压试验;水力试验
hydrostatic test(ing) machine 水压试验机;液体静力试验机
hydrostatic transmission 静液传动;液压静力传动;液压传送装置;液压传动装置;液压传递装置
hydrostatic trigger 液压起重机
hydrostatic under-pressure 坞底静水压力
hydrostatic uplift 静水止托力;静水上托力;静水浮(托)力
hydrostatic valve 静水压力阀;排泥阀
hydrostatic vehicle 水静力运动器
hydrostatic water level 静水水位
hydrostatic weighing 流体静力称重
hydrostatic weighing method 静水力学称量法
hydrostatic weighting 液体比重天平测定比重
hydrostatics 液体静力学;流体静力学;水力学;湿度比学
hydrosteam ejector pump 水蒸气喷射泵
hydrosteam system 水火电(力)系统
hydrostone 多孔石膏制品
hydrostop cylinder 液压闭锁式油缸
hydrostratigraphic(al) unit 水文地层单元;水层单位
hydrostructure 水工结构;水工建筑物
hydrostructure model 水工建筑物模型;水工结构模型
hydrostructure model test 水工结构模型试验
hydrosubstitution 水置换法
hydrosulfite 亚硫酸氢盐;连二硫酸盐
hydrosulfuric acid 氢硫酸
hydrosulphuric thermal water 含硫化氢热水
hydrosynthesis 水合成
hydrosystem 水文系统;水利系统;水力系统;水电系统
hydrotalcite[hydrotalkite] 菱水碳铝镁石;水滑石
hydrotasimeter 电测水位指示器;电测水位计
hydrotator 水转分选机;水升水流分级机
hydrotator classifier 水转式分选机;水转式分级机
hydrotator-thickener 水力浓缩槽
hydrotaxis 趋水性
hydrotechnical 水工的
hydrotechnic(al) research 水工研究
hydrotechnical study 水工研究
hydrotechnic bureau 水利局
hydrotechnic research 水利科学研究
hydrotechnics 水学;水利技术;水利工程学;水力工程(学);水工(学);水工技术
hydrotechny 供水技术
hydrotenorite 水黑铜矿
hydrotephroite 水锰橄榄石
hydroterpine 氢化萜品;氢化萜二醇
hydroterpins 氢化松节油
hydrotest 水压试验
hydrotherm 热液
hydrothermal 热水的
hydrothermal action 热液作用
hydrothermal activation 水热活化
hydrothermal activity 热液活性
hydrothermal alteration 热液蚀变;水热蚀变
hydrothermal alteration halos 水热蚀变晕
hydrothermal alteration minerals 水热蚀变矿物
hydrothermal alteration zone 水热蚀变带
hydrothermal altered ground 水热蚀变地面
hydrothermal altered reservoir 水热蚀变热储
hydrothermal area 水热区
hydrothermal circulation 热液循环
hydrothermal coefficient 水热系数;水热反应
hydrothermal condition 水热状况;水热条件
hydrothermal convection(al) systems 地热对流系统
hydrothermal crater bottom 爆炸坑底部
hydrothermal crater group 爆炸坑群
hydrothermal crater lake 水热爆炸口湖
hydrothermal crater let 小火口

hydrothermal crater-pit 爆炸穴
hydrothermal crater system 水热爆炸口系统
hydrothermal crater terrace 爆炸口阶地
hydrothermal crater vent 爆炸口孔道
hydrothermal crater wall 爆炸口壁
hydrothermal crystal growth 热液晶体生长
hydrothermal curing 水热养护
hydrothermal delustring 热液消光
hydrothermal deposit 热液矿床;热水沉积
hydrothermal emissions 水热活动喷气
hydrothermal explosion 水热爆炸
hydrothermal explosion crater 水热爆炸口
hydrothermal factor 水热因素
hydrothermal fault 导水导热断裂
hydrothermal features 水热显示点
hydrothermal filling 热液充填
hydrothermal fumarole 水热喷汽孔
hydrothermal gas 水热伴生气体
hydrothermal geothermal field 水热型地热田
hydrothermal growth 水热生长
hydrothermalism 热液作用
hydrothermal isotope composition 热液同位素成分
hydrothermal metamorphism 热液变质(作用);水热变质
hydrothermal metasomatism 热液交代作用
hydrothermal method 水热法
hydrothermal method of curing 湿热养护(法)
hydrothermal mineral 热液矿物
hydrothermal mineral survey 热水矿床调查
hydrothermal Mn nodule 热液锰结核
hydrothermal model 冷热水异重流模型
hydrothermal ore deposit 热液矿床
hydrothermal orefield structure 热液矿田构造
hydrothermal plumes 热水柱
hydrothermal process 热液作用;水热工艺;鞭液作用
hydrothermal rocks 水热蚀变岩石
hydrothermal solution 热液溶解作用;热水溶液;水热溶液
hydrothermal stage 热液期
hydrothermal steam 水热蒸汽
hydrothermal sulfide system 水热硫化物系统
hydrothermal summit crater 顶峰喷口
hydrothermal synthesis 热液合成(法);水热合成
hydrothermal system 水热系统
hydrothermal treatment 水热处理;湿热处理
hydrothermal veined antimony deposit 热液脉状锑矿床
hydrotherm figure 温度雨量曲线
hydrothermic 热水的
hydrothermograph 水热仪
hydrothermomagnetic 热磁流体
hydrothermomagnetic wave 热磁流体波
hydrothermostat 水柱式恒温器
hydrotherophyte 水生一年生植物
hydrothorite 水钍石
hydrotimeter 水硬度计
hydrotimetric burette 水硬度滴定管
hydrotimetric flask 水硬度量瓶
hydrotrack drill 水力导轨钻机
hydrotrain 火车轮渡;火车渡轮;火车渡口
hydrotrain terminal 火车轮渡码头
hydrotransport 水力运送;水力运输;水力输送
hydrotransporting device 水力运输设备
hydrotreater 加氢器
hydrotreating 加氢处理;氢化处理
hydrotrencher 液压挖壕机;液压挖沟机;液力挖沟机
hydrotroilite 水陨硫铁;水单硫铁矿
hydrotrope 助水溶物;水溶助剂;水溶物
hydrotropic action 促水化作用
hydrotropic agent 水溶助剂
hydrotropic extraction 水溶抽提
hydrotropic lignin 水溶性木素
hydrotropic solubilizer 水溶增溶剂
hydrotropic solution 水溶溶液
hydrotropism 向水性
hydro tube 压力输水隧洞
hydrotungstite 水钨华
hydrotunnel 压力输水隧洞;压力隧洞;压力隧道
hydroturbine 水涡轮;水轮机
hydroturbine generator unit 水轮发电机组
hydroturbine lubrication 水轮机润滑
hydroturbojet 水下涡轮喷射发动机;水涡喷射发

动机
hydrougrandite 水钙榴石
hydrourushiol 氢化漆酚
hydrous 含水；水化的；水合的
hydrous altered mineral filling 含水蚀变矿物充填
hydrous antimony pentoxide 水合五氧化二锑
hydrous borax 含水硼砂
hydrous calcium silicate 水合硅酸钙
hydrous calcium sulfate 水合硫酸钙；含水硫酸钙
hydrous calcium sulphate 含水硫酸钙；水合硫酸钙
hydrous iron oxide 水合氧化铁
hydrous kaolin 水合高岭土
hydrous lime 熟石灰
hydrous magnesium silicate 含水硅酸镁；水合硅酸镁
hydrous manganese oxide 水合氧化锰
hydrous material 含水材料；含水物
hydrous mica 水(化)云母；含水云母
hydrous ointment 含水软膏
hydrous oxide 水合氧化物
hydrous oxide membrane 水合氧化膜
hydrous oxide precipitate 水合氧化沉淀物
hydrous salt 含水盐；水合盐
hydrous silicate 水合硅酸盐
hydrous titanium oxide 水合氧化钛
hydrovac power brake 液压真空制动器
hydrovacuum 油压真空
hydrovacuum brake 油压真空制动器；液压真空制动器
hydrovacuum cylinder 液压真空缸
hydrovalve 液压开关；液压阀；水龙头；水阀
hydrovane 水翼；水平舵
hydrowollastonite 雪硅钙石；纤硅钙石
Hydrox 粉状水泥防水剂
hydrox blaster 水蒸气爆破筒
hydrox cartridge 水蒸气爆破筒
hydrox cylinder 水蒸气爆破筒
hydroxide 氢氧化物
hydroxide alkalinity 氢氧化物碱度
hydroxide ion 氢氧离子；羟离子
hydroxide ion activity 氢氧离子活度
hydroxides and hydrousoxides 氢氧化物含水氢氧化物
hydroxide sedimentation method 氢氧化物沉淀法
hydroxide solid 氢氧化物固体
hydroxonium 水合氢
hydroxonotlite 水硬硅钙石
hydroxy 羟基
hydroxyacid 醇酸
hydroxy-aldehydic acid 醇醛酸
hydroxyapatite 羟磷灰石
hydroxyapophllite 羟鱼眼石
hydroxybenzene finder phenol detector 酚检测器
hydroxybenzylsutfonic acid formaldehyde resin 羟基苄磺酸甲醛树脂
hydroxy-ethyl acrylate 丙烯酸羟乙酯
hydroxy ethyl cellulose 羟乙基纤维素
hydroxyhydroquinone 苯三酚
hydroxyimino 异亚硝基
hydroxylamblygonite 羟磷铝锂石
hydroxylamine perchlorate 高氯酸羟胺
hydroxylapatite 羟基磷灰石
hydroxylaptite 羟磷灰石
hydroxylated contaminant 羟基污染物
hydroxylated fatty acid 氢氧化脂肪酸；羟基脂肪酸
hydroxylated oil 羟基化油
hydroxylated wood resin 羟基木质树脂
hydroxylation theory 羟化理论
hydroxylbastnaesite 羟碳铈矿
hydroxylellestadite 羟硅磷灰石
hydroxyl equivalent 羟基当量
hydroxyl group 氢氧化物；氢氧基
hydroxyl group vibration 羟基振动
hydroxyl ion absorption 氢氧根离子吸收
hydroxyl organic acid 羟基有机酸
hydroxyl radical 羟基游离基
hydroxyl solvent 羟基溶剂
hydroxyl value 羟基值；羟基数
hydroxymalonic acid 丙醇二酸
hydroxymethyl 羟甲基
hydroxymethylation 羟甲基化
hydroxymethyl cellulose 羟甲基纤维素
hydroxymethyl hydroperoxide 羟甲基化过氧氢
hydroxyphenyl-acetic acid 苯乙醇酸

hydroxy propyl acrylate 丙烯酸羟丙酯
hydroxypropyl methyl cellulose 羟丙基甲基纤维素
hydroxyquinol 苯三酚
hydroxy unsaturated acid 不饱羟基酸
hydrozincite 羟碳锌石；水锌矿
hydrozing 氢气热处理；氢气保护热处理
hydrozinkite 水锌矿
hydrozinzite 羟碳锌石
hydrozircon 水锆石
Hydrozoa 水螅(虫)纲
hyelology 降水学
hyenblenditewebsterite 角闪二辉岩
hyetal 雨量的；降雨
hyetal coefficient 雨量系数；降雨系数
hyetal equator 雨量赤道；雨赤道
hyetal interval 雨量差
hyetal province 降雨区域
hyetal region 雨水区域；雨量区域；降雨区域
hyetogram 雨量自记曲线
hyetograph 雨量图；雨量记录表；雨量计；雨量过程线；雨量分布图；自记雨量计；降雨历时线；降雨历时计
hyetographic(al) curve 雨量过程线；雨量计曲线
hyetography 雨量图法；雨量学；雨量分布学；降雨分布学；降雨地理分布学
hyetology 雨学；降雨(量)学；降水量学
hyetometer 雨量计；雨量仪；雨量表(测量用)
hyetometry 雨量测定法
hygeian rock (用作防潮层的)防水组合物
hygiene 卫生(学)；保健学；保健法
hygiene monitoring 卫生监测
hygiene of military construction 军事施工卫生
hygiene of regimen 作息制度卫生
hygiene research center 卫生研究中心
hygiene standard 卫生标准
hygienic(al) pavement 无尘耐磨路面；无尘耐磨路面
hygienic bacteriology 卫生细菌学
hygienic basis 卫生基础
hygienic characteristic 卫生特征
hygienic chemistry 卫生化学
hygienic evaluation 卫生(学)评价
hygienic feature 卫生特性
hygienic finishing 卫生整理
hygienic measure 卫生措施
hygienic requirements of sewage irrigation 污水农灌卫生要求
hygienic standard 卫生标准
hygienic standard for design of industrial enterprise 工业企业设计卫生标准
hygienic standardization 卫生标准化
hygienization 卫生化
hygieno-meteorology 卫生气象学
hygienic standard for toxic substances 毒物卫生标准
hygral change 湿度变化
hygral equilibrium 湿度平衡
hygral expansion 湿膨胀
hygral expansion index 湿膨胀指数
hygral shock 湿冲击
hygral shrinkage 润湿收缩
hygristor 湿敏电阻；湿变电阻器
hygro 湿
hygroautometer 自记湿度计；自记湿度表
hygrochase 湿裂
hygrocole 湿生动物
hygrodeik 图示湿度计
hygroexpansivity 湿膨胀率；吸湿膨胀性
hygrogram 湿度自记曲线；湿度图
hygrograph 自记湿度计；自记湿度表；湿度记录仪；湿度计
hygro-instability 吸湿不稳定性
hygrokinematics 水物质运动学
hygrokinesis 湿动态
hygrology 湿度学
hygrometabolism 湿代谢作用
hygrometer 湿度计；湿度表(指仪表)；测湿度计
hygrometer recorder 自记湿度仪
hygrometric 吸湿性的；测湿的
hygrometric continentality 降水大陆度
hygrometric deficit 湿度差
hygrometric formula 湿度公式
hygrometric instrument 测湿仪器
hygrometric moisture meter 吸湿计

hygrometric movement 吸湿运动
hygrometric sensitive ceramics 湿敏陶瓷
hygrometric state 湿态
hygrometric table 湿度表
hygrometrograph 温度描记器
hygrometry 温度测定学；湿度测定法；测湿度学；测湿(度)法
hygronom 空气湿度(参数)测定仪；温度仪
hygropetrical fauna 湿岩生动物区系
hygropetrobios 湿岩生物
hygrophile 喜水(生)的
hygrophilite 湿块云母
hygrophilous 喜水(生)的
hygrophilous plant 喜湿植物；好湿植物
hygrophobe 嫌水植物
hygrophyte 湿生植物
hygroscope 验湿仪；验湿器；温度计；湿度器；测湿计
hygroscopic 吸水的；收湿的
hygroscopic absorption 吸水(过程)；吸湿(作用)
hygroscopic agent 吸湿剂
hygroscopic capacity 吸湿(容)量；吸湿能力
hygroscopic cargo 吸湿性货物
hygroscopic degree 吸湿度
hygroscopic depression 收湿性降低
hygroscopic desiccation 吸湿干燥
hygroscopic effect 吸湿性；吸湿效应；吸湿现象；吸湿(作用)
hygroscopic equilibrium 吸湿平衡
hygroscopic expansion 吸湿膨胀
hygroscopic humidity 吸湿度
hygroscopicity 吸水性；吸湿性(能)；吸湿含水量；吸湿性；吸湿性；收湿性
hygroscopic material 吸湿(性)材料；收湿性材料
hygroscopic matter 吸湿性物质
hygroscopic mechanism 吸湿机制
hygroscopic moisture 吸湿水(分)；吸湿含水量；吸着水；湿存水
hygroscopic moisture content 湿吸量
hygroscopic moisture correction factor 吸着水校正系数
hygroscopic moisture meter 吸湿型水分测定计
hygroscopic nuclei 吸湿性核素
hygroscopic number 吸湿数
hygroscopic particle 吸湿性粒子
hygroscopic phenomenon 吸湿现象
hygroscopic potential 吸湿势；吸湿潜力
hygroscopic pressure 吸湿压力
hygroscopic property 吸湿性能
hygroscopic salt 潮解盐
hygroscopic softener 吸湿柔软剂
hygroscopic soil water 吸湿土壤水
hygroscopic substance 吸湿物质
hygroscopic test 吸湿试验
hygroscopic water 吸着水；吸湿水；吸附水；结合水；束缚水；湿存水；薄膜水
hygroscopic(water) coefficient 吸水系数
hygroscopic water content 吸着水含量；吸湿量
hygroscopy 潮解性；吸水性；湿度测定法
hygrosensor 湿度探测器
hygro-stability 湿稳性
hygrostat 恒湿器；湿度检定箱；湿度恒定器；湿度测量控制仪；测湿计
hygrostatics 湿度比学
hygrotaxis 趋湿性
hygrothemograph 温湿度计；温湿自记仪
hygrothermostat 恒湿恒温器
hygrothermal effect 湿热效应
hygrothermograph 温度湿度曲线图；湿温计；湿度温度计
hygrothermoscope 温湿仪
hygrotropism 向湿性
hylaeion 雨林
hylayion hypotropicum 副热带雨林
hylea 热带雨林
hylean 被森林覆盖
hylid 雨蛙
Hylleraas coordinates 海勒喇斯坐标
hylotropy 恒质变形
Hymag 海马格短纤维状石棉
Hyman 海曼铝合金
hymatomelenic acid 棕腐植酸
hymeniderm 膜皮
hymenology 膜学
hymenotome 膜刀

Hymm 海姆合金
hymn board 教堂唱诗班布告板
hymonette 毛衬布
hympanum 门楣中心
Hynack steel plate 海纳克护膜耐蚀钢板
Hynico 海镍钴磁性合金
hyolithid extinction 软舌螺绝灭
hyp 亥普(衰减单位)
hypabyssal 浅成的;半深成的
hypabyssal intrusive body 浅成侵入体
hypabyssal rock 浅成岩;半深成岩
hypaethral (古典无屋顶的)寺庙建筑;屋顶开孔房屋;四合庭院;露天房屋;院子;天井的
hypaethral temple 露天神殿
hypaethron 天井;天窗;院子
hypaethrus 庙宇院子;露天房屋
Hypalon ca(u)lking 海帕伦填缝材料;氯磺化聚乙烯填缝材料
Hypalon cement 氯磺化聚乙烯胶泥;海帕伦胶泥
Hypalon coating 氯磺化聚乙烯涂料;海帕伦涂料
Hypalon roofing 海帕伦屋面材料;氯磺化聚乙烯屋面材料
Hypalon rubber 海帕伦橡胶
Hypalon sealant 氯磺化聚乙烯密封膏;海帕伦密封膏
hypaque sodium 泛影钠
hypar shell 双曲抛物面薄壳;扭壳
hypautochthonous granite 半原地花岗岩
hypautochthony 微异地生成煤;微异地堆积
hypautomorphic 半自形的【地】
hypautomorphic granular 半自形粒状(的)
hypautomorphic granular texture 半自形(晶)粒状结构
hypautomorphic texture 半自形结构;半自形变晶结构
hypenpycnal flow deposit 高密度流沉积
hyperabduction 外展过度
hyperabelian group 超阿贝耳群【地】
hyperacidite 超酸性岩(石)
hyperacidity 酸过多
hyperacoustic 超声
hyperacoustic(al) 超声波的
hyperacoustic quenching 超音(频)猝熄;超声波(淬火)
hyperacoustic zone 超声波区
hyperactive fault 超工作故障
hyperacute rejection 超急性排斥反应;超急排斥
hyperalgebra 超代数
hyperalkaline 超碱性的
hyperalkaline rock 超碱性岩
hyperalkalinity 超碱性
hyperalternative 超候选者
hyperaltitude photography 超高摄影术
hyperarc 超弧
hyperbar 下降气流区;高气压
hyperbaric 高压的
hyperbaric cabinet 高压室
hyperbaric chamber 超压箱;高压舱
hyperbaric diver rescue lift boat 高压潜水员救生船
hyperbaric diving 大深度潜水
hyperbaric environment 高压环境
hyperbaric lifeboat 密封增压救生圈
hyperbaric oxygen 高压氧
hyperbaric oxygenation 高压氧法
hyperbaric oxygen chamber 高压氧舱
hyper baric oxygen drenching 高压纯氧浸透法
hyperbaric vehicle 超气压深潜器;深潜器
hyperbaric welding 水下超压焊接
hyperbarism 高压症;高气压病
hyperbaropathy 高气压病
hyperbase 超基
hyperbola 双曲线
hyperbola function 双曲线函数
hyperbola of fit 拟合双曲线
hyperbola of higher order 高阶双曲线
hyperbolic 双曲
hyperbolical grill 双曲线栅格
hyperbolical zone 双曲线区
hyperbolic amplitude 双曲线幅度
hyperbolic antenna 双曲线天线
hyperbolic arch 双曲线拱;双曲拱桥
hyperbolic arch dam 双曲拱坝
hyperbolic area coverage 双曲线覆盖范围
hyperbolic Bessel functions 双曲线贝塞耳函数
hyperbolic bundle 双曲丛
hyperbolic catenary 双曲悬链线
hyperbolic chamber 双曲线(形)室
hyperbolic characteristic 双曲线特性
hyperbolic chart 双曲线图;双曲面图
hyperbolic collineation 双曲性直射
hyperbolic conchoid 双曲螺线
hyperbolic cone 双曲线铁炮
hyperbolic congruence 双曲性线汇
hyperbolic contact 双曲线弹性插孔
hyperbolic cooling tower 双曲线冷却塔;双(面)冷却塔
hyperbolic coordinates 双曲坐标
hyperbolic coordinate system 双曲坐标系
hyperbolic cosecant 双曲余割
hyperbolic cosine 双曲余弦
hyperbolic cosine function 双曲余弦函数
hyperbolic cotangent 双曲余切
hyperbolic cotangent function 双曲余切函数
hyperbolic curve 双曲线
hyperbolic cylinder 双曲(线)柱面
hyperbolic cylindrical coordinates 双曲柱面坐标
hyperbolic cylindrical surface 双曲柱面
hyperbolic decay law 双曲线衰减定律
hyperbolic decline 双曲型衰减
hyperbolic deflector 双曲偏转板
hyperbolic differential equation 双曲型微分方程;双曲线微分方程
hyperbolic distance 双曲线距离
hyperbolic domain 双曲型域
hyperbolic electronic navigation system 双曲线电子导航系统
hyperbolic element 双曲元
hyperbolic equation 双曲线方程
hyperbolic fix 双曲线无线电船位;双曲线定位
hyperbolic flareout 双曲线拉平
hyperbolic flow equation 双曲线流动方程;双曲流(运动)方程
hyperbolic function 双曲(线)函数
hyperbolic gear 双曲线齿轮
hyperbolic geometry 双曲几何
hyperbolic grading 双曲线形梯度
hyperbolic homology 双曲透射
hyperbolic horn 双曲线形喇叭筒;双曲线式喇叭;双曲线喇叭天线;双曲线号筒
hyperbolic horn antenna 双曲线喇叭天线
hyperbolic inverse 双曲线倒数
hyperbolic inversor 双曲线变换器
hyperbolic involution 双曲对合
hyperbolicity 双曲率
hyperbolic lattice chart 双曲线网格海图
hyperbolic law 双曲线定律
hyperbolic lens 双曲透镜
hyperbolic line of position 双曲(线)位置线
hyperbolic lines 双曲性直线
hyperbolic locus 双曲线轨迹
hyperbolic logarithm 自然对数;双曲(线)对数
hyperbolic magnetic axis 双曲线形磁轴
hyperbolic magnetic lens 双曲磁镜透
hyperbolic manifold 双曲流形
hyperbolic measure 双曲测度
hyperbolic method 双曲线法
hyperbolic metric 双曲度量
hyperbolic metric geometry in a plane 双曲平面度量几何
hyperbolic mirror 双曲面镜
hyperbolic model 双曲线模型
hyperbolic motion 双曲(线)运动
hyperbolic navigation 双曲线导航
hyperbolic navigation aids 双曲线无线电导航设备;双曲线导航(法)
hyperbolic navigational aids 双曲线导航
hyperbolic navigation chart 双曲线导航图
hyperbolic navigation system 双曲线(无线电)导航系统;双曲线导航制
hyperbolic non-Euclidean geometry 双曲型非欧几里得几何学
hyperbolic operator 双曲型算子
hyperbolic orbit 双曲线轨道
hyperbolic outlet 双曲线卸料口
hyperbolic parabola 双曲抛物线
hyperbolic paraboloid 双曲(线)抛物面
hyperbolic paraboloidal roof 双曲抛物面壳顶
hyperbolic paraboloidal shell 双曲抛物面壳体;双曲抛物面(薄)壳;扭壳
hyperbolic paraboloid conoid 双曲线抛物面圆锥体
hyperbolic paraboloid roof 双曲抛物面屋顶
hyperbolic paraboloid umbrella shell 双曲抛物面伞形薄壳屋顶
hyperbolic partial differential equation 双曲线型偏微分方程(式)
hyperbolic pattern 双曲线族
hyperbolic pattern lattice 双曲线网格
hyperbolic pen 双曲线笔
hyperbolic pencil of circles 双曲性的圆束
hyperbolic plane 双曲平面
hyperbolic plano-convex lenses 双曲平凸透镜
hyperbolic point 双曲点
hyperbolic point on a surface 曲面上的双曲点
hyperbolic position 双曲线定位
hyperbolic position finder 双曲线定位仪
hyperbolic position finding 双曲线定位
hyperbolic position fixing system 双曲线定位系统
hyperbolic positioning 测距差定位
hyperbolic positioning system 双曲线定位系统
hyperbolic position line 双曲线定位线;双曲船位线
hyperbolic projectivity 双曲性直射
hyperbolic pulsed radar system 双曲线脉冲雷达系统
hyperbolic quadratic surface 双曲型二次曲面
hyperbolic radar system 双曲线雷达系统
hyperbolic radio navigation 双曲线无线电导航
hyperbolic radio positioning 无线电双曲线定位
hyperbolic region 双曲区域
hyperbolic Riemann surface 双曲型黎曼曲面
hyperbolic screw pump 双曲线式螺杆泵
hyperbolic secant 双曲正割
hyperbolic sector 双曲扇形
hyperbolic shell 平移双曲线壳;双曲线薄壳
hyperbolic shell roof 双曲线薄壳屋顶
hyperbolic sine 双曲正弦
hyperbolic sine function 双曲正弦函数
hyperbolic singular point 双曲奇点
hyperbolic space 双曲(线)空间
hyperbolic spiral 双曲螺线
hyperbolic stress distribution 双曲线应力分布
hyperbolic substitution 双曲代换
hyperbolic surface of position 双曲位置面
hyperbolic sweep generator 双曲线扫描发生器
hyperbolic system 双曲线系统;双曲线网格
hyperbolic tangent 双曲线切线
hyperbolic trace 双曲线轨迹
hyperbolic trajectory 双曲线轨道
hyperbolic transformation 双曲变换
hyperbolic trigonometry 双曲三角学
hyperbolic-type 双曲(线)型
hyperbolic velocity 双曲线速度
hyperbolic wave 双曲线波
hyperbolic wheel dirve 双曲线齿轮传动
hyperbolograph 双曲线规
hyperboloid 双曲线体;双曲面
hyperboloidal gear 双曲线齿轮
hyperboloidal mirror 双曲面反射镜
hyperboloidal secondary mirror 双曲面次镜
hyperboloidal shell 双曲面薄壳;扭壳体;双曲面壳体
hyperboloid arched dam 双曲拱坝
hyperboloid aspheric(al) surface 双曲非球面
hyperboloidic position 双曲面位置
hyperboloid of one sheet 单叶双曲面
hyperboloid of revolution 旋转双曲面
hyperboloid of rotational symmetry 旋转对称双曲面
hyperboloid of two sheets 双叶双曲面
hyperboloid of two sheets of revolution 双叶旋转双曲面
hyperboloid straightening roll 双曲线矫直辊
hyperbromic acid 高溴酸
hypercardioid microphone 超心形传声器
hypercenter 超中心
hypercentric perspective 超远心透视
hypercharacteristic 超特性
hypercharge 超(电)荷力;超荷
hyperchimaera 超嵌合体
hyperchloridation 供盐过多
hyperchromatic 着色特深的;过染色的;染深色的;多色差的

hyperchromatic lens 多色差透镜
hyperchromatism 着色过度
hyperchromatosis 着色过深
hyperchrome 增色团
hyperchromic 超色性的;增色的
hyperchromic effect 增色效应;浓色效应
hyperchromicity 增色性;增色效应;减色效应
hypercinnabar 六方辰砂
hypercircle method 超圆法
hypercoagulability 高凝结状态;凝固性过高
hypercoagulable 凝固性过高的
hypercoagulative state 高凝状态
hypercohomology 超上同调
hypercolo(u)r 超色
hypercommutator 超换位子
hypercompact 超紧
hypercomplex 超复数
hypercomplex number 结合代数;四元数
hypercomplex system 超复数系统
hypercompressor 超压压力机
hyperconcentrated flow 高含沙水流
hyperconcentration 超浓缩
hyperconcentration flow 高含沙水流
hypercone 超圆锥
hyperconical 超锥
hyperconic(al) function 超锥函数
hyperconic(al) horn 超锥喇叭
hyperconjugation 似共轭效应;超联结;超结合;超共轭效应;超共轭(作用)
hyperconnected 超连通的
hypercritical 超临界(的)
hypercritical flow 超临界(水)流
hypercrosslinked polymeric adsorbent 超交联高分子吸附剂
hypercrosslinked resin 超交联树脂
hypercube 超正方体;超立方体
hypercubes method 超方体法
hypercurve 超曲线
hypercyanotic 高度青紫的
hypercycle 超循环
hypercycle theory 超循环理论
hypercyclothem 超旋回层
hypercylinder 超柱体
hyperdiploid 超二倍体
hyperdiploidy 超二倍性
hyperdispersion 过渡散布
hyperdistention 膨胀过度
hyperdop 双曲线多瓦浦测位测速器
hyperdynamic 高动力性的
hyperdynamic(al) circulation 高动力性循环
hyperdynamic(al) circulatory state 高动力循环状态
hyperedge 超边
hyperelastic 超弹性的
hyperelastic body 超弹性物体
hyperelastic deformation 超弹性变形
hyperelasticity 超弹性
hyperelastic law 超弹性规律
hyperellipsoid 超椭圆体
hyperelliptic 超椭圆的(曲线)
hyperelliptic function 超椭圆函数
hyperemesis gravidarum 恶阻
hyperendemic 高度地方性的
hyperenvironment 超高级环境
hyperenvironment test system 超环境试验系统
hyperenzootic 超地方性兽病的
hyper Erlang distribution 超埃尔兰分布
hyperesin 贯叶连翘树脂
hypereutectic alloy 过共晶合金;高低成分的低共熔合金
hypereutectic cast iron 过共晶铸铁
hypereutectic overeutectic 过共晶
hypereutectoid 超低共析体;过共析(体)
hypereutectoid alloy 过共析合金
hypereutectoid steel 超共析钢;过共析钢
hypereutrophication 高富营养
hypereutrophic lake 高富营养湖泊
hypereutstoid 超易熔体
hyperexponential distribution 超指数分布
hyperexponential service model 超指数服务模型
hyperextension 过度伸展;过伸
hyperfiltrate 超滤(液)
hyperfiltration 超滤(法);超过滤;反渗透(法)
hyperfiltration application 超滤应用

hyperfiltration desalination 超过滤脱盐
hyperfiltration economy 超滤经济学
hyperfiltration inorganic membrane 超滤无机膜
hyperfiltration membrane 超滤膜
hyperfiltration method 超滤法
hyperfiltration process 超过滤法
hyperfiltration system 超滤系统
hyperfiltration theoretic(al) consideration 超滤理论设计
hyperfiltration treatment 超过滤处理
hyperfine 超细的;超精细;极微小的
hyperfine coherence 超精相干性
hyperfine coupling 超精细结构耦合
hyperfine field 超精细场
hyperfine interaction 超精细相互作用
hyperfine magnetic field 超精细磁场
hyperfine magnetic line 超精细磁线
hyperfine quantum number 超精细量子数
hyperfine relaxation 超精细弛豫
hyperfine resonance 超精细共振
hyperfine spectrum 超精细光谱
hyperfine structure 超(精)细结构;超微细结构;高度精细结构
hyperfine structure multiplet 超精细结构多重线
hyperfine structure parameter 超精细结构参数
hyperfine structure spectrum 超精细结构谱
hyperfine transition 超精细跃迁
hyperfinite element 超有限元
hyperflexion 过度屈曲;屈曲过度
hyperflow conveying method 密相气升输送法
hyperfluid 超流体;超流动的
hyperfluidity 超流体性;超流动性
hyperfluidized bed 超流化床
hyperfocal 超焦的
hyperfocal chart 超焦距测定表;双焦距测定表
hyperfocal distance 超像距;超焦距;超焦距(离);超大景深
hyperfocal table 焦深表
hyperfocus 超焦距
hyperformer reactor 超重整反应器
hyperforming 超重整
hyperforming process 超重整法
hyperfragment 超子裂片;超碎片;超裂片
hyperfrequency 超高频
hyperfrequency waveguide 超高频波导管
hyperfrequency waves 微波;超高频波
hyperfuchsian group 超辐赫斯群
hyperfunction 超函数
hypergame 超对策
hypergence 浅成的
hypergenesis 表生蚀变;表面蚀变作用
hypergene structure 表生构造
hypergeometric(al) 超比;超几何
hypergeometric(al) curve 超几何曲线
hypergeometric(al) differential equation 超几何微分方程
hypergeometric(al) distribution 超几何分布;超比分配
hypergeometric(al) equation 超几何方程;超比方程;高斯微分方程
hypergeometric(al) function 超几何函数;超比函数
hypergeometric(al) polynomial 超比多项式
hypergeometric(al) progression 超几何级数
hypergeometric(al) series 超(越)几何级数;超比级数
hypergeometric(al) tail-area probability 超几何尾区概率
hypergeometric(al) waiting time distribution 超几何等待时间分布
hypergeometry 超几何学;多维几何学
hypergiant 超超巨星
hypergol 自燃燃料发动机;自燃火箭燃料
hypergolic fuel 自燃火箭燃料
hypergolic igniter 自燃点火器
hypergolic mixture 自燃混合气
hypergolic propellant 自燃推进剂
hypergon 拟球心阑透镜组
hypergonar lens 柱面压缩透镜
hypergranulation 超米粒组织
hypergranule super-granule 超米粒
hypergraph 超图
hypergravity 超重(力)
hypergroup 超群
hypergroupoid 超广群
hyperhaline 超盐水

hyperhemispheric(al) lens 超半球透镜
hyperhigh-frequency 超高频
hyperhomology 超下同调
hyperhydric content 超度含水量
hyperimmersed bolometer 超浸辐射热测定器
hyperinflation 超级膨胀;恶性通货膨胀
hyperinvariant subspace 超不变子空间
hyperjump 超跃度
hyperkeratosis 过度角化
hyperkinematic granite 深成动力花岗岩
hyperkinetic 高动力性的
hyperleptene 特狭上面型
hyperleptoprosopic 特狭面型
hyperlethal 超致死量的
hyperline 超线
hyper link 超级链
hyperloop 超回路
Hypermal 海坡马尔高导磁率铁铝合金
Hypermalloy 海坡马洛伊高导磁率铁镍合金;高导磁率镍铁合金
hypermanganate 高锰酸盐
hypermarket 超级市场;超大型商场;巨型超级市场;特级市场
hypermatic 过黏的
hypermature 成熟过度的
hypermedia 超媒体
hypermelanic 深暗的
hypermetamorhism 复变态
hypermetamorphosis 复变态(期)
hypermetria 辨距过大
hypermetric 超度量
hypermicroscope 超级显微镜
hypermineralization 矿质过多
hypermodel 超模型
hypermorph 超(效)位基因
hypermultiplet 超多重(谱)线;超多重态
hypermutability 超突变性
hypernetted-chain 超网链
hypernetted-chain approximation 超网链近似
hyperneutral symmetry 超中性对称
hypernic 红色天然染料
hypernilpotent 超幂零
hypernorm 超范数
hypernormal 超常态;超常规;超常的
hypernormal dispersion 超正态离差;超常态离差
hypernormality 超正态性
hypernotion 超概念
hypernucleus 超核
hypernutrition 营养过度
hyperon 超子
hyperon atom 超子原子
hyperon gas 超子气
hyperonic atom 超原子
hyperon liquid 超子液体
hyperon pair production 超子对产生
hyperon production 超子产生
hyperoon 古希腊住宅的楼层
hyperosculation 超密切(性)
hyperosmotic 高渗的
hyperosmotic anhydration 高渗脱水
hyperosmotic solution 超渗溶液;高渗溶液
hyperosmotic state 高渗状态
hyperoxide 超氧化物;过氧化物
hyperpanchromatic 超全色(胶片);高泛色
hyperparaboloids 超抛物体
hyperparallelepiped 超平行六面体
hyperparallels 超平行
hyperparasite 过寄生物;重寄生物;重(复)寄生
hyperparasitic 重寄生的
hyperparasitism 重寄生(现象)
hyperparasitization 重复寄生现象
hyperphysical 超自然;超物质
hyperpiestic water 超承压水
hyperpigmentation 着色过度
hyperplane 超平面
hyperplane decision boundary 超平面判定边界
hyperplane method 超平面方法
hyperplane theorem 超平面定理
hyperploid 超倍体;超倍的
hyperploid state 超倍体状态
hyperploidy 超倍性
hyper-Poisson distribution 超泊松分布
hyperpolarizability 超极化率
hyperpolarizability tensor 超极化张量

hyperpolarization 超极化;过极化
hyperpolyhedron 超多面体
hyperpresbyopia 高度远视
hyperpressure 超压(力);超高压
hyperpressure gas chromatography 超后气相色谱法
hyperpseudoscopy 超幻视术
hyperpure 超纯的
hyperpure gas 超纯气体
hyperpycnal flow 高密度流
hyperpycnal inflow 超重(入)流;高密度入流;重入流
hyperpyrexia 温度过高
hyperpyrexial temperature limit 过热温度界限
hyperquadric 超二次曲面的
hyperquantivalent idea 超价值观念
hyperquantization 超量子化;二次量子化
hyper Raman gain 超拉曼增益
hyper Raman polarizability 超拉曼极化率
hyperreactive 反应过度的
hyperrectangle 超矩形
hyperreducible triangle algebra 超可约三角代数
hyperreflexia 反射过强
hyperresolution 超消解法;超分解
hyperresonance 反响过强
hyperrule 超规则
hypersaline brine 超浓卤水
hypersaline fluid 超盐度流体
hypersaline reservoir 超高盐度热储
hypersaline sink 高浓度盐池
hypersaline water 超咸水
hypersalinity 高矿化度
hypersaturated state 过饱和状态
hyperscaling 超标度
hyperscope 壕沟用潜望镜;潜望镜
hyperseal 超密封型
hypersemotactics 超义素结构学
hypersensibility 高敏性
hypersensible 高敏(感)性的
hypersensitive 超高灵敏度的;过敏的;高敏(感)性的
hypersensitivity 超敏性;超敏反应;超(高)灵敏度;超感光度;高敏性;变态反应
hypersensitization 超增感作用;超增感;超敏感(作用);超灵敏化
hypersensitized 超高灵敏度
hypersensitized film 超高感光胶片
hypersensitizing 超增感
hypersensitizing bath 超敏化浴
hypersensization 超敏化(作用)
hypersensor 超敏感元件;超敏断路器
Hypersil 海泼斯尔合金
hypersimple 超单的
hypersolvus 超熔线;全熔岩浆
hypersonic 超音速的;超声;高超音速;特超音速的;特超声速
hypersonic aerodynamics 超声波空气动力学;高超音速空气动力学;高超声速空气动力学;特超声速空气动力学
hypersonic aerospace aircraft 超音速航天飞机
hypersonic analyzer 超声波探伤器
hypersonic flight 高超音速飞行
hypersonic flow 超声波(速)流;高超音速气流;高超音速流动
hypersonic flow condition 高超音速绕流条件;特超声速流状态
hypersonic frequency 高超声频(率);特超声频(率)
hypersonic heating 高超音速加热
hypersonic inlet 高超音速进气口
hypersonic model 特超声速模型
hypersonic nozzle 高超音速喷管
hypersonic research device 高超音速研究装置;特超声速研究装置
hypersonics 超声波能学;高超音速空气动力学;特超声学
hypersonic similarity law 特超声速相似定律
hypersonic similarity parameter 特超声速相似参数
hypersonic similitude 特超声速相似律
hypersonic sound 特超声
hypersonic speed 超声波速;高超声速;特超声速
hypersonic speed flow 高超声速流
hypersonic stability 特超声速稳定性
hypersonic study 特超声速研究

hypersonic wake 特超声尾流
hypersonic wind tunnel 超声波风洞;高超声速风洞
hypersorber 超吸器;超吹器;活性吸附剂
hypersorption 超吸附
hypersorption process 超吸附过程;超吸法
hypersound 特超声
hyperspace 超(越)空间;多维空间
hyperspecialization 高度专门化
hyperspeed 超高速
hypersphere 超球(面);多维球
hypersphere geometry 超球面几何学
hyperspheric(al) coordinates 超球面坐标
hyperspheric(al) function 超球面函数
hyperspheroidal function 超球体函数
hyperstability 超稳定性
hyperstable 超稳定(的)
hyperstatic 超静定的
hyperstatic bridge 超静定桥梁
hyperstatic calculation (超静定的)静力学计算
hyperstatic frame 超静定构架
hyperstaticity 静不定性;超静定性
hyperstatic pressure 孔隙水压(力)
hyperstatic structure 超静定结构
hyperstatic system 超静定系统;超静定(体)系
hyperstereography 超远距立体摄影
hyperstereoscopic distortion 超立体畸变
hyperstereoscopic holographic(al) image 超体视全息像
hyperstereoscopic image 超体视像
hyperstereoscopic viewing 超立体观察
hyperstereoscopy 超体视;超立体观察法
hypersthene 紫苏辉石
hypersthene andesite 紫苏辉石安山岩
hypersthene chondrite 紫苏辉石球粒陨石
hypersthene diopside amphibole plagioclase granulite 紫苏透辉角闪斜长麻粒岩
hypersthene diopside plagioclase granulite 紫苏透辉斜长麻粒岩
hypersthene diorite 紫苏辉石闪长岩
hypersthene dolerite 紫苏耀石粒玄岩
hypersthene gabbro 紫苏辉长岩
hypersthene garnet biotite plagioclase gneiss 紫苏石榴黑云斜长片麻岩
hypersthene granite 紫苏辉石花岗岩
hypersthene granular porphyrite 紫苏辉石安山玢岩
hypersthene granulite 紫苏辉石麻粒岩
hypersthene light-colo(u)red granulite 紫苏辉石浅色麻粒岩
hypersthene olivine gabbro 紫苏辉石橄榄辉长岩
hypersthene peridotite 紫苏辉石橄榄岩
hypersthene picrite 紫苏辉石苦橄岩
hypersthene porphyry 紫苏辉石二长岩
hypersthenes basalt 紫苏辉石玄武岩
hypersthenfels 苏长岩
hypersthenite 紫苏石;苏长岩;紫苏岩;紫苏辉石岩
hyperstrange particle 超奇异粒子
hyperstress 超应力
hyperstructure 超级结构
hypersurface 超曲面
hypersurface contour map program(me) 超曲面等直线图程序
hypersurface trend analysis 超曲面趋势分析
hypersynchronous 超同步的
hypersynchronous braking 超同步制动
hypertape 高速带
hypertape control unit 带控制器
hypertape knit 自动上带装置
hypertelorism 距离过远
hypertensin 增压素
hypertension 压力过高;曲池
hypertensor 加压剂
hypertext markup language 超文本标记语言
hypertext transfer protocol 超文本传输协议
hyperthemamophilic microorganism 极端嗜热微生物
hypertherium 门头线上突出的檐口
hypertherm 人工发热器
hyperthermal 超热(的);超高温
hyperthermal bath 高温浴
hyperthermal material 超热材料;超高温材料
hyperthermal spring 过热泉
hyperthermic (土温的)超热状况
hyperthermic treatment 升温处理

hyperthermocouple 超温差电偶;超热热电偶
hyperthermometer 超高温温度计;超高温温度表;超高温(度)计
hyperthermy 超常温
hyperthyrum 门框线条板
hypertonia 压力过高;高张力;高渗压
hypertonic 高张的
hypertonic dehydration 缺水性脱水
hypertorus 超锚环;超环面
hypertoxicity 剧毒性
hypertraceable graph 超可描画图
hypertraction 过度牵引
hypertriangular noise 超三角形噪声
hypertron 超小型电子射线加速器
hypertrophy 营养过度
hypertropic environment 高营养环境
hyperutility function 超效用函数
hypervapotron 特超蒸发器
hypervariable 超变量
hypervariable codon 超可变密码子
hypervariable region 高变区
hypervelocity 超(高)速;特超声速
hypervelocity aerodynamics 超高速空气动力学
hypervelocity free flow 超高速自由流
hypervelocity free stream 超高速自由流
hypervelocity impact 超高速碰撞
hypervelocity particle 超速粒子
hypervelocity wind tunnel 特超声速风洞
hyperventilation 过度换气;换气过度
hyperviable region 超变位
hyperviscosity 黏滞性过高
hypervisor 管理程序
hypervolume 超体积
hypethral 无屋顶的;露天的
hypha(e) 菌丝
hyphen 联廊;带廊;连字符
hyphenated sequences 不完善顺序
hyphopodium 附着枝
hypidiomorphic 半自形的【地】
hypidiomorphic crystal 半自形晶
hypidiomorphic texture 半自形结构
hypin 缅甸本色粗布
hypo 硫代硫酸钠;海波
hypo acid 次酸
hypoalimentation 营养不足
hypo-alum toning 海波明矾调色法
hypobaric 低气压的
hypobaropathy 高空病;低气压病
hypobasic chamber 低压舱
hypobasis 次要的基层基础;柱基线脚的最下座
hypo bath 海波浴槽;海波定影液;大苏打浴槽
hypobatholitic zone 岩基深部带;深成岩基带
hypobromination 次溴酸化
hypobromite 次溴酸盐
hypobromous acid 次溴酸
hypocalcic cement 低钙水泥
hypocalcification 钙化不全;不全钙化
hypocapnia 低碳酸血
hypocardia 低位心
hypocaust 火坑式采暖装置(古罗马);地板下蓄热式采暖;火坑供爱
hypocenesthesia 存在感觉减退
hypocenter 震源
hypocenter computing process 震源计算方法
hypocenter data file 震源资料档案
hypocenter parameter 震源参数
hypocentral data file 震源资料档案
hypocentral distance 震源距(离)
hypocentral location 震源位置;震源定位;震源测定
hypocentral parameter 震源参数
hypocentral plot 震源图
hypocentral shift 震源迁移
hypocentre of explosion 爆炸源
hypocentrum 震源
hypochlorate-orcinol test 次氯酸盐苔黑酚试验
hypochloric acid 次氯酸
hypochlorite 次氯酸盐;次氯酸根
hypochlorite method 次氯酸盐法
hypochlorite of lime 次氯酸石灰
hypochlorite of soda 次氯酸苏打
hypochlorite oxidation method 次氯酸盐氧化法
hypochlorite sweetening 次氯酸盐(法)脱硫
hypochlorous anhydride 次氯酸酐
hypochromatic 染浅色的

hypochromatism 着色不足
hypochrome 减色团;淡色团
hypochromic effect 减色效应
hypochromicity 减色性;减色现象
hypochromic shift 浅色移动
hypochromism 缺色性;少色性
hypo clearing acid 海波消除剂
hypocoagulability 凝固性过低
hypocrateriform 高脚碟形的
hypocritical legal act 虚伪的法律行为
hypocrystalline 次晶质;半晶质(的);半结晶(的)
hypocrystalline porphyritic 半晶斑状
hypocycloid 圆内旋轮线;内摆线;次摆圆
hypocycloidal 内圆滚线
hypodermic syringe 注射器
hypodigm concept 种型群概念
hypodign 种型群
hypodispersion 平均分布
hypodynamia 力不足;乏力
hypoelastic 亚弹性的;准弹性的;次弹性(的)
hypoelasticity 亚弹性;次弹性
hypoelastic law 准弹性定律
hypo elimination 海冰消除
hypo eliminator 海波消除液;海波消除剂
hypoenuston 下漂浮生物
hypoepistephanine 次表千金藤碱
hypoergia 低反应性
hypoeutectic 低共低熔体
hypoeutectic alloy 亚共晶金;低级低共熔合金
hypoeutectoid 亚共析;亚晶融质;低低共熔体
hypoeutectoid steel 亚共析钢
hypofiltration 深成渗透作用
hypofocus 震源
hypofunction 功能减退;功能低下;机能衰退
hypog(a)eum 地下墓室;[复]hypog(a)ea
hypogeal 岩洞建筑;地下建筑;山洞建筑;上升生成的;地下生(长)的;地下的
hypogee 岩洞建筑;山边建筑;地下室;地下建筑
hypogene 深成的;上升生成的;上升的;地下生成
hypogene action 内生作用;内力作用;深成作用
hypogene deposit 内生矿床;深成矿床
hypogene magma 深成岩浆
hypogene mobility 深成活动性
hypogene rock 深成岩
hypogene spring 上升泉
hypogene vein deposit 深成脉状矿床
hypogene water 上升水
hypogenous cotyledon 地下子叶
hypogeum 岩洞建筑;古代地下墓室;地下室;地下建筑;地窖
hypogeum tomb 地下墓室
hypohalous 一氧化卤素;次卤酸
hypo-hurricane 假设飓风
hypoid 准双曲面的
hypoid axle 双曲面齿轮传动轴
hypoid bevel 准双曲面伞齿轮传动;偏轴伞齿轮传动
hypoid bevel gear 准双曲线伞齿轮;准双曲面锥齿轮;偏轴伞齿轮
hypoid bevel wheel 准双曲面伞齿轮
hypoid drive pinion 主动准双曲线伞齿轮
hypoid gear 准双曲面齿轮;准双曲齿轮;直角交错轴双曲齿轮;海波齿轮;偏轴伞齿轮
hypoid gear and pinion 准双曲面伞齿轮副
hypoid gear oil 双曲线齿轮油
hypoid generator 准双曲面齿轮加工机床
hypoid lubricant 耐特高压润滑剂;准双曲面齿轮润滑剂
hypoid pinion 准双曲线小锥齿轮
hypoid pinion offset 螺旋传动的偏心距
hypo indicator 定影液指示剂
hypoiodite 次碘酸盐
hypokinematic metamorphism 深成动力变质作用
hypolemmal compact layer 膜下致密层
hypolimnetic 湖下层的
hypolimnetic aeration 深冷水等温层曝气;深层曝气
hypolimnetic anoxia 深冷水等温层缺氧
hypolimnetic current 深冷水等温层水流
hypolimnetic dissolved oxygen consumption rate 深冷水等温层溶解氧消耗率
hypolimnetic layer 深冷水等温层
hypolimnetic oxygen deficit 深冷水等温层缺氧量
hypolimnetic oxygen demand 深冷水等温层需氧量
hypolimnetic oxygen depletion 深冷水等温层缺氧量
hypolimnetic region 深冷水等温层区
hypolimnetic waters 深冷水等温层水体
hypolimnion 下层滞水带;下部滞水层;均温层;湖底静水层;深冷水等温层
hypolimnion aeration 深冷水等温层曝气
hypolimnion aeration system 滞水带曝气系统
hypolimnion of lake 湖泊深冷水等温层
hypolimnion region 深冷水等温层区
hypolimnions(of lake) 湖下层水
hypolodous acid 次碘酸
hypomagma 深岩浆
hypomanganate 次锰酸盐
hypometamorphic 深变质的
hypomicron 亚微粒;亚微子
hypomigmatization way 深成混合岩化方式
hypomineralization 矿质过少
hypomonotectic 亚偏晶
hypomonotectic alloy 亚偏晶合金
hyponeuston 水面下漂浮生物;次漂浮生物
hyponome 水漏斗
hypo-osmoticity 低渗性
hypopeltate 下面盾状
hypoperfusion 灌注不足
hypoperistalsis 蠕动迟缓
hypophosphate 连二磷酸盐
hypophosphoric acid 连二磷酸
hypophosphorous acid 次磷酸
hypophyge 凹曲线脚(陶立克柱头)
hypopiestic water 次承压水;半自流水
hypopigmentation 着色不足
hypopistephanine 海冰表千金藤碱
hypoplankton 下层浮游生物;低生浮游生物
hypopnea 呼吸不足;呼吸表浅
hypopodium 次要的基层基础
hypoprosexia 注意不足
hypopsia 视力减退
hypopycnal flow deposit 低密度流沉积
hypopycnal inflow 轻入流
hyporeactive 反应不足的
hyporeflexia 反射减退
hyposaline 低盐度的
hyposcenium (古希腊剧场舞台前部下面的)矮墙
hyposcope 蟹眼式望远镜;军用潜望镜
hyposmosis 低渗透
hypo soda 大苏打
hyposphene 下楔
hypophosphate 次磷酸盐
hypo stain 定影斑点
hypostasis 下沉物
hyposteel 亚共析钢
hypostereography 超近距立体摄影
hypostoichiometry 次化学计量
hypostracum 内壳层
hypostratotype 次层型
hypostroma 下子座
hypostyle 多柱式建筑;多柱式(的)
hypostyle column 内柱;内楹柱;金柱
hypostyle edifice 多柱式建筑
hypostyle hall 连柱厅;多柱厅;多柱式大厅
hypostypsis 轻度收敛
hypostyptic 轻度收敛的
hyposulfite 连二亚硫酸盐;次亚硫酸盐
hyposulfite of soda 大苏打
hyposulphite 次亚硫酸盐
hyposulphuric acid 连二硫酸
hyposulphurous acid 连二硫酸
hyposynchronous 低于同步的;次同步的
hyposynergia 协同不足
hypo tank 定影灌
hypotaxia 控制力减弱
hypotelorism 距离过近;间距缩短
hypotension 压力减低;压力过低;张力减低
hypotenuse 弦;斜边
hypotenuse face 弦面
hypo test 海波检验;残余海波测定
hypothec 抵押权;不转移占有权的抵押
hypothecate 不转移占有权的抵押
hypothecated account 抵押账户
hypothecated assets 抵押资产
hypothecated goods 抵押品;担保品
hypothecation 抵押;担保契约
hypothecium 下盘层
hypothermal 高温热液;深成高温热液的;低温的
hypothermal deposit 高温热液矿床;深成高温热液矿床;低温热液矿床
hypothermal process 架成高温热液作用
hypo thermal regional metamorphism 高温区域变质作用
hypothermal spring 低温温泉
hypothermy 低温
hypothesis 假说;假设;假定
hypothesis for concrete fracture 混凝土断裂假说
hypothesis of bar deposition 沙洲沉积学说
hypothesis of isostasy 地壳均衡说
hypothesis of land-bridge 陆桥说
hypothesis of molecular motion 分子运动假说
hypothesis of quasi-stationarity of order 2 准二阶平稳假设
hypothesis of stationarity of order 2 二阶平稳假设
hypothesis of strict stationarity 严格平稳假设
hypothesis of the crustal-wave mosaic structure 波浪状镶嵌构造说
hypothesis test approach 假设测算法
hypothesis test(ing) 假设验证;假设检验;假设测验
hypothesize-and-test 假设和测试
hypothetical 假想的;假设的
hypothetical axis 假想轴
hypothetical combination 假定组合
hypothetical condition 假设条件
hypothetical cost 利息;假定成本
hypothetical diagram 理想线图
hypothetical discharge 假设流量;假定流量
hypothetical earthquake 假想地震
hypothetical electric(al) potential 假设电势
hypothetical equilibrium tide 假设平衡潮
hypothetical exchange 过渡交换机
hypothetical flood 臆拟洪水;假设洪水
hypothetical flood hydrograph 假想洪水过程线;设计洪水过程线
hypothetical frequency 假设频数
hypothetical gas 假设气体
hypothetical global climate 设想全球气候
hypothetical ground plane 假设地平面;假定高程起始面
hypothetical hinge (建筑或桥面上的)假铰
hypothetical hurricane 假设飓风
hypothetical hydrograph 假想水文过程线;设计水文过程线
hypothetical ideal jet 假设理想射流
hypothetical interest 假定利息
hypothetical interfacial structure 假设界面结构
hypothetical machine 理想机器
hypothetical memory 虚拟存储器
hypothetical model 假想模型
hypothetical operating system 理想操作系统;假想操作系统
hypothetical parallax 力学视差;理想视差
hypothetical path 假定途径
hypothetical pendulum 虚摆
hypothetical plane method 假定平面法
hypothetical price 假定价格
hypothetical reaction 假设反应
hypothetical reactivity 假设反应性
hypothetical reference circuit 假设参考电路
hypothetical reference connection 假设参考连接
hypothetical reference digital link 假设参考数字链路
hypothetical reference digital section 假设参考数字段
hypothetical reserve 假定储量;推测储量
hypothetical resources 假定资源
hypothetical river 假设河流
hypothetical route 假定途径
hypothetical scenario 设想方案
hypothetical single slope 臆拟单坡
hypothetical slurry 假设的泥浆
hypothetical solution 假设溶液
hypothetical solution concentration 假设溶液浓度
hypothetical standard state solution 假设标准溶液
hypothetical storm 臆拟暴雨;假设暴雨
hypothetical stream 假设河流
hypothetical structure 假设结构;假想结构
hypothetical syllogism 假设三段论
hypothetical system 假设系统
hypothetical testing 假设检验
hypothetical theory 假说理论
hypotonic 低渗(的)

hypotonic solution 低渗溶液
hypotrachelium （柱端与拇指圆饰环纹间的）柱颈；（陶立克式柱身与柱颈之间的）环槽；古希腊柱颈
hypotraction 牵引不足
hypotrematic 下侧条
hypotrochoid 次内摆线；长短辐圆内旋轮线
hypotrophy 发育障碍；不足生长
hypotype 亚型；补模标本
hypovanadate 次钒酸盐
hypovanadic oxide 二氧化钒
hypoventilation 换气不足
hypoxemia 低氧
hypoxia 氧过少；低氧
hypoxyloid 次木质的
hyprelastic deformation 超弹性变形
hypsicephalic 高头型
hypsithermal interval 气候最适宜期（9000～2500年前）；变热期；冰后温暖期；高温期
hypsobathymetric(al) map 水深地形图；水底地形图
hypsochrome 向紫增色基；浅色团
hypsochromic 向蓝移（的）
hypsochromic effect 紫移；浅色效应
hypsochromic shift 浅色位移
hypsochromy 浅色团作用
hypsocline 特种高程曲线
hypsodont 高冠齿
hypsograph 测高仪
hypsographic(al) chart 地形图；地势图；分层设色（地）图
hypsographic(al) curve 等高线；高程线；高深曲线；等深线；陆高海深曲线
hypsographic(al) feature 高度地势；高程注记；高程要素
hypsographic(al) map 等高线地形图；地势图；高程地图
hypsography 地形起伏；地形测绘学；地形测绘（法）；地貌表示法；测高学；分层设色法；比较地势学
hypsol 环氧树脂类黏合剂
hypsometer 树高测定器；三角测高仪；沸点气压计；沸点气压表；沸点测高器；沸点测高计；沸点测高表；沸点测定器；沸点测定计
hypsometric(al) 分层设色的
hypsometric(al) chart 分层设色图
hypsometric(al) contour 分层设色等高线
hypsometric(al) curve 分层设色等高线；高程面积（分布）曲线；高程线；陆高深曲线；湖海等深线；等高线
hypsometric(al) curve method 等高线法
hypsometric(al) data 高程资料
hypsometric(al) feature 地貌
hypsometric(al) formula 测高公式
hypsometric(al) integral 标高整体容积
hypsometric(al) layer 色层
hypsometric(al) level(l)ing 沸点气压高程测量
hypsometric(al) map 分层设色（地）图；立体地形图；地势图；等高线地图
hypsometric(al) measurement 树高测定
hypsometric(al) method 分层设色法
hypsometric(al) tinting 分层设色法
hypsometric(al) tinting chart 分层设色图

hypsometric(al) tints 高程分层设色；分层设色表；海图上表明高度或水深的着色
hypsometric(al) unit 分层设色层
hypsometry 高程测量学；高程测量术；高程测量（法）；测高学；测高术；测高法；沸点测定法
hypsophobia 高处恐怖
hypsothermometer 沸点测高器；沸点测高计；沸点测高表
hypsothermometry 沸点测高法
hyrated shell 水化层
Hy-rib 折板式金属网（商品名）
hy-rib reinforcement floor 折边金属网混凝土楼面；折板式钢筋混凝土楼面
hy-rib steel sheet 折边钢板网；折板式钢片
hyrogeologic(al) condition of ore district 矿区水文地质条件
hyro-vac 油压真空制动器
Hysol 海索尔（环氧树脂类黏合剂）
hyssop 圣水；海索草
hyster arch 履带式拱钩
hysteresigraph 磁滞曲线记录仪；磁滞回线记录仪
hysteresimeter 滞后计；磁滞测定器
hysteresis 滞后（作用）；滞变；平衡阻碍；磁滞（现象）；迟滞性
hysteresis active current 磁滞有功电流
hysteresis advance 磁滞提前；磁滞超前
hysteresis amplifier 磁滞放大器
hysteresis characteristic 滞后特性；滞变特性；磁滞特性
hysteresis clutch 磁滞离合器
hysteresis coefficient 滞后系数；磁滞系数
hysteresis compaction 滞回压实
hysteresis comparator 磁滞比较器
hysteresis constant 滞后常数；磁滞系数
hysteresis constant of magnetic material 磁性材料磁滞常数
hysteresiscope 磁滞回线仪
hysteresis curve 滞回曲线；滞后曲线；滞变曲线；磁滞曲线
hysteresis curve recorder 磁滞曲线记录器；迟滞曲线记录器
hysteresis cycle 磁滞周期；磁滞循环；磁滞回线
hysteresis damper 滞后阻尼器；滞变阻尼器；磁滞阻尼器
hysteresis damping 滞后阻尼；滞后缓冲作用；滞变阻尼；磁滞阻尼
hysteresis damping coefficient 滞变阻尼系数
hysteresis damping factor 滞变阻尼因数
hysteresis diagram 磁变（曲线）图
hysteresis distortion 磁滞失真
hysteresis effect 滞后作用；滞后效应；滞变效应；磁滞效应
hysteresis elasticity 滞后弹性
hysteresis element 磁滞成分
hysteresis energy dissipation curve 滞变能量消散曲线
hysteresis envelope 滞变包线
hysteresis error 滞环误差；磁滞误差
hysteresis form factor 磁滞形状系数
hysteresis free 灭磁滞
hysteresis graph 磁滞曲线图
hysteresis heat 磁滞热
hysteresis heat build-up 滞后生热性

hysteresis lag 磁滞
hysteresis loop 滞后周线；滞后回线；滞后环；滞变回线；磁滞回线；磁滞曲线；滞回环
hysteresis-loop displaying equipment 磁滞回线显示装置
hysteresis loss 滞后损失；滞后损耗；滞变损失；磁滞损失
hysteresis loss coefficient 磁滞损耗系数
hysteresis loss resistance 磁滞（损耗）等效电阻
hysteresis meter 磁滞计；磁滞测定计
hysteresis module 滞后模量
hysteresis modulus 滞回模量；磁滞模数
hysteresis motor 磁滞（式）电动机
hysteresis of the bubble 气泡滞后
hysteresis of transformation 相变滞后
hysteresisograph 磁滞回线记录仪
hysteresis phenomenon 滞后现象
hysteresis property 滞后性能；磁滞性质
hysteresis quality 滞后性
hysteresis response zone 滞后响应区
hysteresis set 滞后定形；滞后变形；磁后变形
hysteresis steel 磁滞钢
hysteresis strain energy 滞变应变能
hysteresis suppression 灭磁滞
hysteresis synchronous motor 磁滞（式）同步电动机
hysteresis tester 滞后试验机
hysteresis torque 磁滞转矩
hysteresis-type resonance curve 滞变型共振曲线
hysteresis voltage 滞后电压
hysteretic 滞后的
hysteretic angle 滞后相（移）角；磁滞（后）角
hysteretic behavio(u)r 恢复力特性；滞变性能
hysteretic characteristic 滞回特性
hysteretic curve 滞回曲线
hysteretic damping 滞后阻尼
hysteretic effect 滞后效应
hysteretic lag 磁滞
hysteretic memory cell 滞后存储元件
hysteretic nature 滞后性
hysteretic property 滞后性能
hysteretic state 滞后状态
hysterisis error 回程误差
hysterocrystalline 次生结晶
hysterocrystallization 次生结晶作用
hystero crystallization way 次生结晶作用方式
hysterogenetic 岩浆末期的
hysterosystole 过晚收缩；期后收缩
hysterothecium 船形囊壳
hysterset 功率电感调整
hystersis 滞变
hystersis error 滞后误差
hystersis loop 滞回环
hystool 高凳式钢筋座
Hyswing ball mill 海斯温球磨机；四滚筒球磨机
Hytensyl bronze 海坦西尔黄铜
Hythe beds 海茨层
hythergraph 温湿（曲线）图；温度与湿度关系图
hytherograph 温雨图；温度雨量图
Hytor 抽压机；海托尔抽压机
Hy-Tuf steel 海图夫低合金高强度钢
hyvac oil pump 高真空油泵
hyvac pump 真空泵

I

ianthinite 水(斑)铀矿
iarderellite 硼铵石
Iarimer column 十字形工字钢组合柱
iatron 投影电位示波器
I-bar 工字铁;工字钢;工字杆;工形材
I-beam 工字(形)梁
I-beam bridge 工字梁桥
I-beam steel 工字钢
Ibeco 预制牛皮纸
Ibo 预制木屋
Ibus 墙板
icand register 被乘数寄存器
ice abrasion 冰蚀;冰磨蚀
ice accretion 结冰;积冰
ice-accretion indicator 积冰指示器
ice accumulation 积冰
ice accumulation on ship 船舶积冰
ice action 冰作用
ice admixture 冰掺料(用于减少混凝土放热)
ice age 更新世冰期;冰期;冰河时代
ice airfield 冰上机场
ice amount 冰量
ice anchor 冰锚
ice and frost detector 冰霜探测仪
ice annulus 冰环
ice anvil 冰砧
ice apparatus 测冰器
ice apron 破冰体;挡冰板;防冰装置;冰裙;冰栏;冰覆面;冰挡;冰川舌
ice apron of pier 桥墩破冰体
ice arena 冰上运动场
ice atlas 冰图(集);冰情图册;冰情图(例)
ice auger 凿冰器;冰钻
ice avalanche 冰坠;冰崩
ice axe 破冰斧;冰镐
ice bag 冰袋
ice ball method 冻土移植法
ice band 冰(夹)层;冰带
ice bank 冰原;冰场;冰岸
ice bank evapo(u)rator 结冰式蒸发器;冰箱式蒸发器
ice bar 冰塞;冰堤;冰坝
ice barchan 新月形冰丘
ice barge 冷藏驳船
ice barrage 冰障;冰坝
ice bar(rier)冰阻;冰障(碍);冰堰;冰缘;冰堤;冰坝
ice barrier flood 冰阻造成的洪水
ice bath 冰浴器
ice beam 抗冰梁
ice-bearing current 挟冰水流
ice belt 浮冰带
iceberg 流冰;冰山;大冰块
iceberg alarm 冰山警报
Iceberg City 冰山城(美国安克雷奇的昵称)
iceberg furrow mark 漂冰沟痕
ice bight 冰凹湾
ice bin cooling unit 冰仓冷却装置;冰仓冷却设备;冰仓冷却器;冰仓冷却单元
ice blaster 喷水冰层
ice blasting 冰爆破
ice blink 冰源反光;海岸冰崖;冰映射;冰(映)光
ice blister 喷水冰层;冰泉;冰堆
ice blower 风力送冰机
ice blue 冰染蓝
iceboat 破冰设备;破冰船;冰上滑艇
ice body 装运冰块用的车箱;冰体;冰块
ice bond 冰联结
ice boom 河流解冻;冰轰声;挡冰栅;防冰栅
ice bore 冰钻孔
ice-borne sediment 冰夹带的沉淀物;冰积物
ice boulder 冰川巨砾
icebound 冰冻的;被冰封阻的;封冻(的);冰封区;冰封的
icebound harbo(u)r 冰封港;冻港
icebound season 冰封期;封冻季节;冻封期
ice box 电冰箱;冰箱;厨用冰箱
icebox effect 冰箱效应;冰室效应
ice box type air-conditioner 冰箱式空调器

ice break 开冻
ice breaker 破冰拦;破冰设备;破冰器;破冰机;破冰船;冰挡
icebreaker cell 拦冰墩
ice breaker of bridge pier nose 桥墩破冰结构
icebreaker ship 破冰船
ice breaking 破冰
icebreaking blasting 破冰爆破
icebreaking cable repair ship 破冰电缆修理船
icebreaking fleet 破冰船队
icebreaking machine 碎冰机
icebreaking pier 破冰墩
icebreaking ram of vessel 船用破冰锤
icebreaking tanker 破冰油船
icebreaking tool 破冰工具
icebreaking tow 破冰船型船头
ice breaking tug 破冰拖轮
ice break-up 开冻;破冰(开冻);冰裂;解冻
ice breccia 角砾冰;冰角砾
ice bridge 河川坚冰;冰桥;冰川坚冰;冰坝;碍航的河面冰
ice bulletin 冰情广播
ice bunker 冰仓
ice-bunker refrigerated wagon 冰冷藏车
ice buoy 抗冰浮标;冰(区)浮标
ice cake 浮冰块;冰块;板冰
ice calving 裂冰
ice calorimeter 冰卡表;本生冰量热器
ice can 冰桶
ice can crane 吊冰行车
ice-canoe 滑冰艇
ice canopy 流冰群
ice can truck 吊冰桶车
ice cap 陆冰;冰帽;冰冠;冰盖
ice-cap climate 冰冠气候;冰盖气候;永冻气候
ice carapace 封冻冰层;冰帽;冰冠;冰被
ice cascade 冰瀑
ice cave 冰穴;冰窖;冰洞
ice cavern 冰洞
ice cellar 冰窖
ice chain 冰链
ice chamber 冷藏库;冰箱
ice channel 冰区水道;冰间航道
ice chart 海冰图;冰情图(例);冰情分布图;冰况(分布)图;冰(海)图;冰分布图
ice chest 冷箱;冰箱;冰库;冰柜
ice chisel 冰凿;冰錾
ice chute 泄冰道;泄冰槽;运冰滑道
ice class 冰级
ice classification 冰级
ice classification of ships 冰区航行船级
ice clause 冰封条款;冰冻条款
ice clearing 浮冰穴;冰裂;冰间穴;冰间水面;冰间湖;冰湖的
ice cliff 冰崖
ice climate 冰雪气候
ice clogging 流冰壅塞;冰塞;冰块拥塞
ice cloud crystal 冰晶云
ice cluster 大冰堆
ice coating 结冰(层);积冰;冰膜衣
ice code 冰情电码
ice cofferdam 冰围堰
ice cold 冰冷(的)
ice cold sense 冰冷感
ice colo(u)r 冰染料
ice colo(u)r process 冰染法
ice compress 冰敷
ice concentration 海冰密集度;覆冰量;冰的疏密度
ice concrete 冰混凝土;冰混凝体
ice condition 冰情;冰况
ice condition forecast 冰情预报
ice condition observation 冰情观测
ice cone 冰锥
ice cone number 冰锥数
ice conglomerate 冰砾岩
ice contact delta 冰界三角洲;冰接三角洲
ice contact deposit 冰界沉积
ice contact slope 冰接坡

ice contact stratified drift 冰界层状冰碛
ice content 含冰率;含冰量
ice content of frozen soil 冻土的含水量
ice control 冰流控制;冰冻防护措施
ice control chemicals 化学抗冰剂
ice control gate 冰流控制闸门
ice controlling material mixing plant 防冻剂拌和机
ice controlling material spreader 防冻剂洒布机;防冻剂喷洒机
ice control measures 防冰措施
ice control salt 防冻盐
ice control structure 防冰构筑物
ice cover 海冰密集度;覆冰(量);冰壳;冰盖;冰层;表面结冰
ice covered 冰封的;冰(川)覆盖的
ice-covered channel 冰覆盖渠道;冰封河道;冰封的渠道;封冻水道
ice-covered lake 封冻湖
ice-covered sonar 冰下声呐
ice-covered stream 封冻河流
ice covering 表面结冰;冰盖(层)
ice cover rate 冰层覆盖率
ice cover water-rating curve 冰盖水位—流量关系曲线;结冰影响的水位流量(关系)曲线
ice crack 冰间水路;冰裂(纹)
ice cream 冰淇淋
ice cream cabinet 冷饮柜;冰淇淋柜
ice cream freezer 冰淇淋冷冻机
ice cream mix 冰淇淋粉
ice cream parlo(u)r 冷饮室
ice cream shell 冰淇淋碟
ice creep 冰蠕动;冰滑动
icecrete 冰混凝体
ice crevasse 冰隙
ice criteriaon 冰况标准
ice crusher 碎冰机
ice crust 冰壳
ice crystal 结晶冰屑;冰晶(体)
ice crystal chemistry 冰晶化学
ice crystal cloud 冰晶云
ice crystal effect 冰晶效应
ice crystal fog 冰晶雾
ice crystal haze 冰晶霾
ice crystal imprint 冰晶印痕
ice crystal mark 冰晶痕
ice crystal theory 冰晶学说;冰晶理论
ice cube 小冰块
ice cuber box 冰柜
ice culmination zone 分冰岭;分冰界
ice cutter 切冰机;冰切削器
iced 用冰冷却的;冰冷的;冰冰封的
ice dam 冰坝;冰壅塞
ice damage 冰灾;冰害
ice-dammed lake 冰塞湖;冰川湖;冰堰湖
ice day 结冰日;冰日
iced chamber 冷冻房
ice decorating 冰花装饰
ice-deflecting boom 拦冰栅
ice-deposited material 冰积物
ice-deposited moraine 冰碛物
ice desert 冰漠
ice designations 冰情图(例)
ice-detecting equipment 测冰仪
ice-detecting set 测冰仪
ice-detector 测冰仪
iced firn 冻结粒雪
iced glass 冰纹玻璃
ice dike 冰墙;冰脉
ice discharge 泄冰;冰流量
ice disintegration 冰崩
ice dissolving tank 化冰柜
ice divide 海冰分界线;分冰岭
ice dock 冰坞(在冰面开一船池保护过冬船不受压伤)
ice dome 冰穹
ice doubling 抗冰衬板;劈开的冰山;船首防冰加强材料;防冰船首加强材

ice drag 冰锚
ice-drift(ing) 冰流;浮冰;漂冰;流冰
ice-drill 冰钻
iced tea 冰茶
ice dye 冰染料
ice dy(e)ing 冰染
ice edge 冰缘;冰刃;冰区边界
ice engine 制冰机
ice engineering 冰工学
ice erosion 冰蚀;冰川侵蚀;冰川剥蚀作用
ice escape channel 泄冰渠
ice evacuation 除冰
ice evapo(u)ration 冰蒸发量
ice evapo(u)ration level 高空冰汽转相高度
ice fall 悬冰;垂冰;冰瀑(布);冰崩
ice fat 泥泞冰;冰脂
ice fathometer 回声测冰仪
ice feather 冰羽
ice fender 抗冰护舷;冰栏;冰挡
ice field 巨大的冰川体;一片巨大的冰川体;冰原;冰场
ice field belt 冰原带
ice field climate 冰原气候
ice filling point 加冰所
ice film 冰膜
ice fin 破冰鳍
ice firn 粒雪冰;万年雪冰;冻粒雪
ice flaker 刨冰片机
ice flat 油脂状冰
ice float 浮冰;冰盘
ice floe 巨大的冰川;流冰;淌凌;大浮冰体;浮冰(块);冰盘;冰排;冰源
ice floe velocity 浮冰流速;冰块流速
ice flood 凌汛
ice flow 冰水径流;冰流
ice flow control 冰流控制
ice flower glass 冰纹玻璃;冰花玻璃
ice flower-like structure 冰花状构造
ice flowers 霜花;冰花
ice flushing 冲冰(船闸冲水操作)
ice fog 冻雾;冰雾
ice foot 冰脚;冰壁
ice force 冰力
ice forecast 冰情预报
ice forecasting service 冰情预报工作
ice formation 结冰(过程);成冰(过程);冰凝
ice-formation condition 结冰条件
ice-formed rock 冰岩;岩状冰
ice forming 结冰
ice framing 抗冰骨架
ice-free 不冻的;无冰的
ice-free area 无冰区域
ice-free harbo(u)r 不冻港
ice-free period 不冻期;无冰期
ice-free port 不冻港
ice-free season 不冻季节;无冰期;无冰季节
ice-free waters 不冻水域
ice-free waterway 不冻水道;无冰水道
ice fringe 沿岸冰带;冰条线;冰条带
ice front 冰崖;冰锋;冰壁
ice gang 流冰;融冰流;淌凌
ice gate 泄冰闸门;排冰闸(门);排冰门
ice ga(u)ge 量冰尺;测冰仪;测冰尺
ice gland 冰腺
ice glass 冰状玻璃;冰花(状)玻璃
ice-glazed 涂冰
ice glow 冰耀光
ice gorge 流冰壅塞;流冰堆积;凌壅;冰阻;冰峡;冰塞;冰谷;冰坝
ice grouser 防滑链条;冰锚桩
ice gruel 粥状冰
ice guard 桥墩的破冰构造;拦冰栅;破冰设备;挡冰栅;防冰装置;防冰屏;冰栏;冰挡
ice gush 冰涌
ice hail 小雹;冰雹
ice hammer 冰锤
ice harbo(u)r 冰港;封冻港
ice hockey bat 冰球拍
ice hockey rink 冰球场
ice hockey timer 冰球计时器
ice hold 冰舱
ice hook 冰锚
ice house 冷藏库;冷藏室;制冰机房;制冰厂;冰库;冰房;冰窖

ice hummock 冰丘;冰堆积;冰堆
ice impact pressure 流冰冲击压力
ice in sheet 片状冰;条冰
ice island 浮冰岛;(浮冰的)冰岛
ice island iceberg 岛状冰山
ice jam 冰障;流冰壅塞;流冰堆积;冰阻;冰塞;冰凌堆集;冰块拥塞;冰坝
ice jam flood 冰壅凌汛;凌汛
ice jamming 冰障(碍)
ice jam prevention 防凌
ice jam stage 冰壅水位;冰塞水位;冰坝水位
ice knee 抗冰材
ice laid deposit 冰川沉积
Iceland 冰岛
Iceland agate 黑曜岩
iceland crystal 冰洲石
Icelandic 冰岛式
icelandic low 冰岛低压
icelandic low zone 冰岛低压区
icelandite 冰岛岩
Iceland moss 冰岛苔藓(含胶质)
Iceland spar 透明方解石;冰洲(晶)石;冰岛晶石矿
iceland spar deposit 冰洲石矿床
ice lane 冰隙;冰间水道;冰间航道
ice layer 冰层
ice laying 封冻期;冰封期
ice lead 冰间水道
ice ledge 贴岸冰;残冰堆积;冰栅;冰瀑;冰壁
ice lens (河流或浅海中的)底冰;冰透镜体;冰晶体;冰扁豆体
ice limit 结冰区界;冰限;冰区(边)界
ice load(ing) 冰压力;冰荷载;冰负荷
ice lobe 冰川舌
ice lolly freezer 棒冰机
ice machine 制冰机
ice machine oil 冷冻机润滑油
ice make capacity 制冰能力
ice maker 制冰机
ice making machinery 制冰机
ice mantle 冰雨;冰帽;冰冠;冰盖;冰盾
ice margin 冰缘
ice marginal lake 冰边湖
ice mass 致冷物体;制冷物质;冰块
ice massif 冰山丛
ice master 冰区引航员
ice mechanics 冰力学
ice melting agent 融冰剂;化冰剂
ice melting and filling system 冰的融解热量
ice melting current 熔冰电流
ice melting equivalent 冰的融解热当量
ice melting point 冰的融点
ice melting salt 融雪岩
ice mill 磨冰机;冰川砾磨蚀地
ice mine 破冰layer
ice mosaic 混冻冰
ice motion 冰的流动;冰流
ice mo(u)ld 冰铸
ice mound 积冰丘;土底冰丘;冰丘;冰堆
ice mountain 冰山
ice movement 冰移动;冰川运动
ice navigation 冰区航行
ice needle 冰针
IC engine spark arrester 内燃机火星捕捉器
ice nip 两块大浮冰相遇;密集冰;冰挤压;冰的挤压力
ice nipped 被冰夹住
ice nuclei 冰核
ice nuclei in atmosphere 大气冰核
ice nucleus 冰核
ice observation service 冰情观察组织
ice ocean power plant 冰洋发电站
ice of land origin 陆源外
ice pack 流冰群;积冰;大块浮冰(群);冰群;浮冰量;冰排;冰裹法;冰袋
ice palace 溜冰馆;室内溜冰场
ice pan 荷叶冰;饼状冰;冰盘;冰饼
ice paper 透明纸;冰纹纸
ice particle 冰粒
ice pass 冰水泄放道
ice pass structure 泄冰建筑物
ice patch 小面积流冰群
ice patrol 冰情巡逻
ice patrol vessel 冰情巡逻船
ice patterned glass 冰花玻璃
ice pellets 小冰球;小冰雹;冰丸

ice period 结冰(时)期;结冰季节;冻冰季节;冰冻(周)期;冰期
ice-phobic coating 防冰涂层
ice pick 冰凿;冰鏨
ice pieces 冰屑
ice piedmont 山麓冰坡;冰麓
ice pillar 霜柱;冰蘑菇
ice pilot 冰区引航员
ice pit storage 冰窖储[贮]藏
ice plank 冰区航行瞭望台
ice plant 制冰设备;制冰厂;冰厂
ice plate 冰盘
ice-ploughing device 除冰装置
ice-plowing device 除冰装置
ice pneumatic conveying system 气力输冰系统
ice point 冰点
ice pole 冰极
ice port 封冻港;冰港(由冰崖所围成的临时自然港)
ice prediction 冰情预报
ice pressure 冰压力
ice prevention 防凌
ice prevention measure 防凌措施;防冰措施
ice prisms 冰针
ice profile sonar 冰轮廓探测声呐;冰断面探测声呐;测冰声呐
ice protection 流冰保护;防凌
ice push 流冰壅塞;流冰推拥;流冰推力;冰胀;冰壅;冰推(力)
ice pushed ridge 冰脊
ice pushed terrace 冰壅阶地;冰(堆)阶地
ice quake 冰震;冰崩
ice radar 测冰雷达
ice raft 浮冰排;筏冰;冰排;冰筏
ice rafted debris 冰携碎屑
ice rain 霰;雨夹雪;冻雨;冰雨
ice rake 耙冰机
ice ram 撞冰船头
ice rampart 岸边堆积的冰脊;湖冰脊;冰河堤
ice reconnaissance 冰情侦察;冰情勘查;冰况侦察
ice recovery screw 冰回取螺旋输送机
ice reef 浮冰群;冰礁
ice reefing 冰群堆积
ice refrigerator 冰箱
ice regime 冰况;冰情
ice regime forecast 冰情预报
ice regime observation 冰情观测
ice region 冰区
ice removal 除冻器;除冰(器)
ice removal agent 除冰剂
ice removal apparatus 除冰装置
ice removal salt 防冻盐
ice removal salts 防冰盐类
ice removal system 除冰系统
ice removing wheel 除冰轮
ice report 冰情报告;冰况报告
ice resistant bow 抗冰船首
ice resistant offshore structure 抗冰近海结构物
ice ribbon 冰条纹;冰带
ice ridge 冰脊
ice rind 冰皮;冰壳
ice ring 致冷环;冰环
ice rink 溜冰场;滑冰场;冰上运动场
ice rise 冰隆
ice river 冰河
ice road 冰道;滑行道
ice run 流冰;融冰流;漂冰;淌凌;浮冰(流);冰流
ice runoff 融冰径流;冰水径流
ice sample 冰样
ice saw 冰锯
ice scoured plain 冰蚀平原;冰掘平原
ice scouring 冰挖作用;冰蚀作用
ice scouring machine 锯冰机
ice scrape 冰蚀
ice scraper 刮冰器;刮冰机
ice screwing 浮冰漩涡(形成冰丘)
ice scum 初冰;冰碴
ice sea deposit 冰海沉积
ice seamanship 冰区航行船艺
ice seamanship manual 冰区船艺手册
ice season 结冰时期;结冰季节
ice segregation 冰析;冰隔作用;冰的分凝作用
ice shed 分冰界
ice sheet 陆上的冰;陆冰;大陆冰川;薄冰层;冰原;冰盖;冰盾;冰层

ice sheet disposal 冰层处置
ice shelf 陆架冰;冰架
ice shell plating 冰带外板
ice shield 挡冰板
ice ship 冰区航行船
ice shore 向岸冰涌
ice shove 冰壅
ice shutter 挡冰板;薄冰片
ice signal station 冰况信号台
ice situation 冰情
ice-skating rink 溜冰场
ice sky 冰映射;冰映光
ice-slide 滑冰道
ice-sliding conveyer 滑冰槽
ice slip 冰滑动
ice sludge 海绵冰;碎冰
ice sluice 泄凌道;泄冰闸
ice slush 夹冰水;冰针团;冰泥;冰凌
ice slush concretion 冰泥钙结核
ice snow pressure 冰雪压力
ice spar 透长石;玻璃长石
ice spicule 冰针
ice stadium 冰上运动场
ice stone 冰晶石
ice storage(bin)冰库;冰藏库
ice storage bin rating 储[贮]冰仓容量
ice store 冰库
ice-stored refrigeration 冰蓄冷
ice store-house 冰藏库
ice storm 冰暴
ice stream 冰流;冰河;冰川
ice-strengthened ship 冰区加强船
ice-strengthened tanker 冰区加强油船
ice-strengthened tug 破冰拖轮
ice strengthening 抗冰加强(材)
ice stringer 冰带舷侧纵桁
ice strip 浮冰带
ice structure electroplating technique 电镀冰花工艺
ice supply quay 供冰码头
ice swimming 冰泳
ice system 冰川系
ice table 平面大冰块;平顶冰块
ice thickness 冰厚
ice thickness measurement 冰厚测量;冰厚测定
ice thrust 冰壅;冰压力;冰推力
ice ton 冰吨
ice tongs 冰钳;冰块夹;冰夹具
ice tongs traction bow 冰钳式牵引弓
ice tongue 伸入海中的冰山舌;冰(川)舌
ice tongue afloat 浮冰舌(延伸到海中的狭窄半岛形浮冰)
ice topography 冰貌
ice trap 冰栏;冰挡
ice trouble 冰害
ice tub 冰淇淋桶
ice undercurrent 冰川潜流;冰川底流
ice up 被冰覆盖;结冰;全面冻结
ice uplift 冰浮托
ice vein 冰脉
ice velocity 冰移动速度
ice voered island 冰覆岛
ice wall 冰墙;冰壁
ice warning 冰情警报
ice-warning indicator 结冰警告器
ice water 冰水
ice water installation 制冰水设备
ice water interface 冰水界面
ice water pipe 冰水管
ice water test 冰水试验
ice web frame 冰带加强肋骨
ice wedge 片状冰;冰楔;冰脉
ice worn 冰蚀的;被冰擦伤
ice yacht 带帆冰橇;冰上滑行游艇
ice zone 冰区
ichnite 印痕;遗迹;化石足迹
ichnofacies 遗迹(化石)相;化石(岩)相
ichnofossil 遗迹化石;足迹化石;踪迹化石;痕迹化石
ichnograph 地图;平面图;平面布置
ichnography 平面图法
ichnolite 印痕;遗迹;化石足迹;含化石足迹的岩石
ichnology 遗迹化石学;足迹化石学;痕迹学
ichography 径迹图
ichor 岩精

ichthyiod 鱼形体
ichthyoacanthotoxin 鱼刺毒
ichthyocolla 鱼胶
ichthyodont 鱼牙化石
ichthyofauna 鱼类区系
ichthyoglyptus 鱼形石英
ichthyoid 流线形的
ichthyol 鱼石脂
ichthyolite 鱼化石
ichthyologist 鱼类学者
ichthyology 鱼类学
ichthyol oil 鱼石油
ichthyophthalmite 鱼眼石
ichthyophthirius 白点病
ichthyoplankton 鱼食浮游生物
ichthyotoxic fishes 有毒鱼
ichthyotoxin 鱼毒
ichuite 动物足迹
icicle 垂冰;冰柱
icicle prevention device 防冻设备
icicle shaped 冰柱形的
icing 结冰;积冰;覆冰;冰泉
icing beater 冰糖搅拌机
icing facility 冷冻设备
icing index 结冰标志
icing indicator 结冰指示器;结冰信号器
icing level 积冰高度
icing of screen 筛网结冰
icing protection 防止结冰
icing-rate meter 积冰速率表
icing tower 碎冰楼
icing track 加冰线
I-column 工字(形)柱;前炉床
icon 图标
iconic(al) model 形象模型;图像模型
icon menu 图标菜单
iconcenter 像中心
iconocope 光电摄像管
iconography 影像学;平面图绘法;图像学;图像表示法;插图;插画
iconolog 光电读像仪
iconology 图像学
iconometer 光像测定器;量影仪;测影仪;测距镜;返光镜;反光镜
iconometry 量影学
iconoscope 光电显像管;光电析像管
iconoscope camera 光电摄像管摄像机
iconoscope film camera 光电摄像管式电视片摄像机
iconoscope tube 光电摄像管
iconostasis 圣像屏;圣障;圣壁
iconotron 移像光电摄像管
icos 地下隔墙法
icosagon 二十角形;二十边形
icosahedral 二十面体的
icosahedron 二十面体
icosane 二十烷
Icos-Flex system 艾科斯一弗莱克斯体系
icositetrahedron 二十四面体
Icotype 象模标本
ICSPRO 海洋科学计划秘书处间委员会
icy 盖着冰的;结冰的;覆盖着冰的;冰的
icy pavement 结冰的路面
icy shower 冰(阵)雨
icy snow 冻结雪
icy soil 冻土
icy wind 寒风;冰冷风
Idaglass 艾达玻璃纤维
Idaho white pine 美国山地白松;爱达荷白松
idaite 铁铜蓝
iddingsite 褐绿泥石;伊丁石
idea 主意;构思(建筑师或工程师的初步设想);概念
idea bank 意见库
idea design 概念设计
idea-finding 研讨方案
ideal 标准的
ideal aberration-free image point 理想的无像差像点
ideal absorbed solution 理想吸收溶液;理想附溶液
ideal absorbed solution theory 理想吸附溶液理论
ideal articulation 理想传声清晰度
ideal behavio(u)r 理想变化过程
ideal biocide 理想杀生剂

ideal bioclimate 理想生物气候
ideal blackbody 理想黑体
ideal boundary 理想边界
ideal buckling load 理想压曲荷载
ideal building 理想房屋;标准建筑物
ideal burning 理想燃烧
ideal circulation 理想环流
ideal city 理想城市
ideal clay 理想黏土
ideal climate 理想气候
ideal coarse-grained soil 理想粗粒土
ideal coarse-grained soil of no plasticity 无塑性理想粗粒土
ideal column 理想柱;理想塔
ideal combustion 理想燃烧
ideal communication channel 理想信道
ideal condition 理想条件
ideal constraint 理想约束
ideal coordinates 理想坐标
ideal crystal 理想晶体
ideal cupola 理想圆屋顶;理想圆穹顶
ideal cycle 理想循环
ideal deformation 理想变形
ideal detector 理想探测器
ideal diameter 理想直径
ideal dielectric 理想电介质
ideal direct current machine 理想直流电机
ideal dispersion 理想分散
ideal dome 典型圆顶
ideal drag 理想流体中的阻力
ideal efficiency 理想冷冻循环;理想效率
ideal elastic body 理想弹性体
ideal elastic-plastic 理想弹塑性的
ideal elastic solid 完全弹性固体
ideal energy gradient 理想能线
ideal engine 理想发动机
ideal engine cycle 理想发动机循环
ideal exhaust velocity 理想排气速度
ideal filter 理想滤波器
ideal flow 理想流动;无黏流动
ideal fluid 理想流体;理想流动
ideal focusing field 理想聚焦场
ideal focusing magnetic field boundary 理想聚焦磁场边界
ideal frame 理想构架
ideal framework 理想构架
ideal frequency distribution 理想的次数分配
ideal frequency domain filter 理想频域滤波器
ideal fuel 理想燃料;标准燃料
ideal function 理想函数
ideal gas 理想气体;理想流体
ideal gas constant 理想气体常数;理想气体常量
ideal gas equation 理想气体方程
ideal gas law 理想气体定律
ideal gas law equation 理想气体定律方程式
ideal grading 理想粒度;理想级配
ideal grading curve 理想土积曲线;理想(颗粒)级配曲线
ideal grain size 理想颗粒尺寸;理想粒径
ideal graphite 理想石墨
ideal horizon 理想地平
ideal image point 理想像点
ideal imperfect crystal 理想非完美晶体
ideal indicator 理想指示剂
idealine 糊状黏土
idealisation 理想化
idealism 理想主义
idealizal search service 理想的检索服务
idealization 理想化
idealize 理想化
idealized 理想化的
idealized characteristic 理想化特性
idealized earthquake ground movement 理想化地震动
idealized elastic continual mass 理想弹性连续体
idealized elastic member 理想弹性构件
idealized elastic solid 理想弹性(固)体
idealized form 理想化形式
idealized fragmental mass 理想碎屑体;理想粒状体
idealized model 理想模式;理想化模型
idealized pattern 理想化模式
idealized screen 理想筛网;理想屏幕
idealized seismologic(al) model 理想震源模型
idealized source model 理想化震源模型

idealized structure 理想化结构
idealized system 理想(化)系统
idealized value 理想(化)值
ideal jet velocity 理想射流速度
ideal line 理想直线;理想线路
ideal liquid 理想液体;理想流体
ideal load heat consumption 理想运转热消耗
ideally perfect crystal 理想完美晶体
ideally plastic soil 理想塑性土
ideally reflecting 理想反射
ideal main stress 理想主应力
ideal middle-size(d) grain 理想中等大小粒子;理想中值粒径;理想中等粒径
ideal model 理想模型;理论模型
ideal network 理想网络;无损耗网络
ideal nozzle 理想喷嘴
ideal of perfection 鉴定标准
ideal orbit 理想轨道
ideal orientation 理想取向
idealoy 理想坡莫合金
ideal package of sensors 传感器的理想组件
ideal parallel 理想并联
ideal paralleling 理想并车
ideal particle 理想粒子
ideal particle size 理想颗粒尺寸
ideal particle size distribution 理想颗粒大小分布
ideal pendulum method 理想摆法
ideal photon receiver 理想光子接收器
ideal photopair 理想像对;标准像对
ideal plan 理想方案
ideal plane 理想平面
ideal plasma 理想等离子体
ideal plasticity 理想塑性
ideal plastic mechanism 理想塑性机制
ideal plastic theory 理想塑性理论
ideal plate 理想塔板
ideal point 理想点
ideal principal stress 理想主应力
ideal productivity index 理想采油指数
ideal profile 理想剖面(图);理想横剖面
ideal propellant consumption 理想燃料消耗量
ideal quench(ing) 理想淬火
ideal radiator 理想辐射体
ideal refrigeration cycle 理想冷冰循环
ideal regenerative cycle 理想回热循环
ideal resonance 理想共振
ideal sand 理想砂(土)
ideal scan(ning) 理想扫描
ideal schematic diagram 理想的作业顺序示意图
ideal sea level 理想海平面
ideal section 理想剖面(图);理想断面
ideal sensitivity 理想灵敏度
ideal setting basin 理想沉淀池
ideal setting basin efficiency 理想沉淀池效率
ideal setting tank 理想沉淀池
ideal shear strength 理想剪切强度
ideal soil 理想土壤
ideal solid solution 理想固溶体
ideal solution 理想溶液
ideal specific impulse 理想比冲量
ideal standard cost 理想标准成本
ideal state 理想状态
ideal strain 理想应变
ideal strength 理想强度
ideal stress 理想应力
ideal structure 理想结构
ideal system 理想(的)系统
ideal theory 理想理论
ideal time domain filter 理想时域滤波器
ideal town 理想城市
ideal tracer 理想示踪物
ideal trajectory 理想轨道
ideal transducer 理想换能器
ideal transformer 理想变量器
ideal truss 理想桁架
ideal type 理想类型
ideal type elevator 筒状铰链式提引器
ideal velocity 理想速度
ideal Venturi(tube) 理想文丘里管
ideal wave 理想波
ideal weeder 标准型中耕除草器
ideal working substance 理想工质
idea man 谋士
Idea of Pacific Economic Sphere 太平洋经济圈设想
idea of style 风格思想;样式概念
ideas about business operation 经营思想
ideas competition 设计立意赛;设计思想竞赛
ideas of influence 受控观念
Idel Sonde 艾德测头
idem factor 幺并矢
idemfactor 幂等因子;等幂因子;等幂矩阵
Idemitsu Kosan Co. Ltd 出光矿山公司(日本)
idempotence 幂等性;等幂性
idempotent 幂等;幂等
idempotent element 幂等元
idempotent law 幂等(定)律
idempotent matrix 幂等矩阵
idempotent property 幂等性质
idempotent transformation 幂等变换
identical argument list 相同变元表
identical component 相同成分
identical computation 相同计算
identical congruent 相同
identical element 恒等元素;单位元素
identical entry 相同项
identical equation 恒等式
identical figure 恒等形
identical function reference 相同函数引用
identical goods 同样货物
identical graduation 同一刻度
identical ligand 同一配位体
identically equal 恒等的;全等的
identically true formula 永真式
identically varnishing 恒等于零
identical map 等角投影(地)图
identical matrix 恒等矩阵
identical network 相同网络;恒等网络
identical object 同名地物
identical operation 相同运算
identical parts 可互换零件
identical period 等同周期
identical permutation 元排列;恒等置换;恒等排列;同等排列
identical point 同名点
identical points 对合点
identical price 价格相同
identical processor 相同处理机
identical product 同一产品
identical quantity 恒等量
identical relation 恒等式;恒等关系;全等式
identical sand 统一砂
identical sections 相同片段
identical standard 相同标准
identical sublist 相同子表
identical substitution 恒等代换
identical subtree 相同子子树
identical table 相同表
identical text 同文
identical transformation 恒等变换
identifiability 可识别性;可辨识性
identifiability criterion 能识性原则
identifiable cost 可辨认成本
identifiable economic factor 可识别的经济因素
identifiable point 易辨认点;可辨认点
identifiable signal 识别信号
identifiable target 可识别恒等目标
identification 工程试验鉴定;鉴定(法);鉴别(反应);认同作用;认定;识别;证认;等同;标识;辨识;辨认
identification address 识别地址
identification algorithm 识别算法
identification and classification of subsidability of loess 黄土湿陷性的鉴定和分类
identification beacon 识别信标;识别标志
identification blink 识别闪光
identification burst 记录密度标记;判断标识信号;识别段
identification call letter 识别呼号
identification card 鉴定卡;说明卡;识别卡(片);身份证;标签
identification card reader 识别卡阅读器;标识卡片阅读器
identification character 标识(字)符
identification chart 证认图
identification code 识别码;标识码;标识符
identification code number 数码代号
identification colo(u)r 标志颜色
identification condition 识别条件;标识条件
identification data 识别数据
identification device 识别装置
identification division 识别部分;标识部分
identification equipment 识别装置
identification error 识别误差
identification field 标识字段
identification form 识别单
identification generator 识别信号发生器
identification group of card 卡片标识组
identification impulse 识别脉冲;辨认脉冲
identification index 鉴定指标;鉴别指数;识别标志
identification instrument 识别仪器
identification instrument of minerals and rocks 岩矿鉴定仪器
identification item 识别项;标识项
identification key 识别关键字;辨认标志
identification lamp 识别灯
identification leader 识别带头
identification letter 识别字母
identification light 信号灯光;指示光;识别灯;标志灯
identification mark (检验机构认可的)确认标志;识别符号;识别标志;辨认标志;鉴别标志
identification marker 鉴定标志;识别标记;鉴别标志
identification marking 鉴别标志;认识标记
identification method 识别方法;辨认法(指各项存货计价)
identification method of sample 样品鉴定方法
identification name 识别名;标识名
identification number 机器编号;识别数(码);成套设备编号;标识号
identification of anomaly boundary 异常界线划定
identification of bacteria 细菌鉴定
identification of brick 砖的鉴定
identification of castings 铸件的鉴定
identification of colo(u)r test 颜色识别试验
identification of construction contract 施工合同的签证
identification of contaminant source locations 污染源位置识别
identification of control points 控制点辨认
identification of curve type 曲线类型的鉴别
identification of drawing 图形识别
identification of environmental problem 环境问题辨识
identification of forgings 锻件鉴定
identification of ground and sky-waves 天地波识别
identification of heavy mineral sand 重砂鉴定
identification of law 法律的确认
identification of measured values 观测值鉴定
identification of mineral inclusion 矿物包裹体鉴定
identification of noise source 噪声源鉴别
identification of plates 平板鉴定
identification of podzols and podzolised soils 鉴定灰壤土和灰化土
identification of pollutant 污染物鉴定;污染物鉴别
identification of polluted ecosystem 污染生态系统识别
identification of pollution source locations 污染源位置识别
identification of pollution sources 污染源识别
identification of position 位置识别;位置的确定
identification of project 立项
identification of seams 岩层鉴别;岩层对比
identification of site intensity 场地烈度鉴定
identification of skill 技术等级
identification of soils 土壤识别;土的识别
identification of structural equation 结构方程的识别
identification of swelling soil 膨胀土的鉴别
identification of the parties 当事人各方
identification of tool 工具编号
identification of user 用户标识
identification of variety 品种鉴定
identification of welds 焊接鉴定
identification paint 标志用涂料
identification parameter 鉴定参数
identification phase 识别期
identification plate 识别板;标志板
identification point 识别点;标识点

identification post 指示标桩
identification probability 识别概率
identification problem 识别问题
identification pulse 识别脉冲
identification receiver 识别接收机;辨识接收器
identification record 识别记录
identification report of fossils 化石鉴定报告
identification report of heavy placer 重砂鉴定报告
identification report of rock and mineral 岩矿鉴定报告
identification resolution chart 清晰度测试卡;分解力测试卡
identification satellite 识别卫星
identification sequence number 识别序列号
identification set 识别无线电发射机
identification sign 识别记号;识别符号
identification signal 识别信号
identification site 识别部位
identification sonar 识别声呐
identification source 识别信号源
identification string 标识串
identification symbol 识别符号
identification system 识别系统
identification tag 识别标签
identification tape 识别带
identification technique 识别技术
identification term of sample 样品鉴定项目
identification test 鉴定试验;鉴别试验
identification thread 识别绳条
identification trailer 识别带尾
identification type of sample 样品鉴定类型
identification yarn 标志纱
identified amount of resources 查明资源量
identified cost method 识别成本法
identified distribution 特定分布
identified photograph 调绘相片
identified reserve 查明储量
identified resources 查明资源
identifier 鉴定员;鉴定(用试)剂;鉴定人;鉴别器;名称;名标;识别符;标志符;标识符(号);辨识器
identifier attribute 识别属性;标识符属性
identifier circuit 识别电路
identifier count 识别符计数;标识符计数
identifier declaration 识别符说明;标识符说明
identifier length 识别符长度;标识符长度
identifier list 识别符表;标识符表
identifier name 标识符名
identifier pointer 识别符指示字;标识符指针
identifier table 标识符表
identifier word 标识(符)字
identify 等同;标识;辨认;识别
identify alternatives 鉴定备选方案;辨识备选方案
identify anomalous reflection coefficients 鉴别异常反射系数
identify circuit 识别电路
identify code 识别码
identify control section 识别控制段;标识控制段
identify-disc 证明牌
identify dummy section 识别空段;标识哑段;标识空段
identifying 标识;识别
identifying call 识别调用
identifying code 验证码;参考代码;标识码;标定码
identifying cost of alternatives 鉴别备选方案成本
identifying date of sample 样品鉴定日期
identifying dye 识别染料
identifying feature 识别标志
identifying mark 识别标记
identifying number 识别数码
identifying of human resources 人力资源的确定
identifying on the ground 外业判读
identifying operator 鉴定员;鉴定人
identifying plate 识别牌;标号牌
identifying signal 识别信号
identifying the payee of a check 验明支票收款人
identify processes number 判识地质过程数
identify unit 标识单元
identity 验明;恒等(式);同一性;识别性;身份;等同
identity by the map 根据地图判读
identity card 工作证
identity certificate 身份证明
identity crisis 认同的转折点
identity declaration 等同说明
identity-definition 等同定义

identity distance 面间距
identity element 幺元;鉴别元件;恒等元素;全同元件;同一元素;单位元素
identity equation 恒等方程
identity gate 恒等门;全同门;同门;符合门
identity graph 幺图
identity law 同一律
identity mapping 恒等映射
identity matrix 幺阵;单位矩阵;纯量矩阵
identity of a group 单位元素
identity of operation 等同运算
identity operation 恒等运算;全同运算;全同操作
identity operator 恒等算子
identity property 等同性
identity relation 同等关系;等同关系
identity relator 等同关系符
identity symbol 恒等号
identity theorem 同一性定理
identity transformation 恒变换
identity unit 全同单元;全同部件;同门
identometer 快速校验钢条成分仪;材料鉴别仪
ideogram 意表符号;表意文字
ideograph 模式图;商标;意表符号;表意文字
ideograph chart 象形图
ideomotor action 动念动作
ideotype 非典型标本;表意标本
idexing slot 标引槽
idigbo 非洲伊地泡木(黄褐色条纹,用于高级细木工)
idioadaptation 特殊适应
idioadaptive evolution 特殊适应性演化
idioagglutinin 自发凝集素
idiobiology 个人生物学
idioblast 自形变晶
idioblastic 自形变晶的
idioblastic series 自形变晶系列
idioblastic texture 自形变晶结构
idiochromatic 自色的
idiochromatic crystal 本质色晶体;本质光电晶体
idiochromatic photoconductor 本质光电导体
idiochromatism 本质色性
idiocrasy 特异品质;特异反应性
idioelectric 非导体的
idiogenite 同成矿床
idiogenous 同成(因)的
idiogeosyncline 山间地槽;独地槽
idiographic 特殊规律研究法
idiometer 人差计
idiomorph 自形晶
idiomorphic 自形的
idiomorphic crystal 自形晶体;整形晶体
idiomorphic granular 自形晶粒状的
idiomorphic granular texture 自形晶粒状结构
idiomorphic pore 自发孔
idiomorphism 自形作用
idio-morphosis 特殊变形
idiophanism 自现干涉圈
idiophase 分化期
idioreflex 自发反射
idiosome 胶粒
idiostatic 同势差;等位差(的)
idiostatic circuit 等位差电路
idiostatic method 同位差连接法;同势差连接法;等电位法
idiosthenia 自生力
idiosyncrasy 特质;特有的风格;特异品质;特异反应性
idiosyncratic exchange 特质交易
idiosyneratic reaction 特发性反应
idiotopic 自形组构
idiot stick 挖土器;铲;计算尺;计数尺(俚语)
idiotype 个体型
idle 空运转;空(热)耗;无用的;窝工;不工作的
idle adjuster 怠速调整器
idle assets 闲置资产
idle balance 闲置余额
idle bar 闲导条;无载铜条;死导条
idle battery 闲置电池;无载电池;无负荷电池
idle block 空载程序块
idle boiler 压火锅炉
idle call 空调用
idle capacity 闲置(生产)能力;空转功率;空闲生产能力;剩余能(量);储存容量;备用容量;备用功率;备用电容
idle capacity cost 闲置能量成本;闲置能力成本

idle capacity loss 闲置能量损失;闲置能力损失
idle capacity variance 闲置生产能力差异;闲置能量差异
idle capital 闲置资金;闲置资本;闲散资金;呆滞资本
idle car 游车
idle carrier 托滚座
idle cash 闲置现金
idle channel 闲线路;空闲信道
idle channel detection 空闲信道检测
idle channel loading 空信道加荷
idle channel state 空闲通道状态
idle character 空闲字符
idle circuit 空闲电路;无效电路
idle coil 闲圈;空置线圈
idle communication mode 空闲通信方式
idle component 虚部;无功分量;无功部分;电抗分量;分量元件
idle condition 空载条件;空载工况;空闲状态
idle contact 空接点;闲接点;闲触点;空触点;间隔触点
idle cost 无效成本;停工损失
idle current 无功电流
idle-current wattmeter 乏表;无功电流瓦特计
idle cut-off 慢关闭油路
idle cycle 空转循环
idle equipment 闲置设备
idle equipment report(list) 闲置设备清单
idle facility 闲置设施
idle frequency 闲频;中心频率;未调制频率
idle fund 闲置资金;闲散资金
idle gear 换向齿轮;空转轮;惰齿轮
idle hand 闲置人员
idle hole 空转孔
idle hours 空闲时间;窝工时间;停机时间
idle indicating signal 空闲指示信号
idle interval 闭锁期间
idle jack 空插孔
idle labo(u)r 闲散劳动力;失业工人
idle labo(u)r cost 闲置人工成本
idle land 闲置地;休闲地;休耕地;空闲地;熟荒地
idle line 闲线;空线;空闲线
idle line termination 空闲线路终接负载
idle link 空闲链路
idle load heat consumption 池窑空运转热消耗
idle locomotive track 配机线
idle loop 空循环;空环
idle machine 闲置机器;闲置的机器
idle machine time 停歇机器时间
idle machine time report 闲置机器时间报告
idle man-hours 闲置人时数
idle man-year 闲置人年
idle member 零杆;无效构件;无负荷构件
idle microphone 空闲传声器
idle mode 空闲模式
idle money 闲置资金;闲置货币;闲散资金;闲散现金;游资
idle money-capital 闲置的货币资本
idle motion 无荷载下工作;空行程;空转;空载运转;空动
idle needle valve 怠速针阀
idleness 空闲(率)
idleness expenses 停机费用
idle off-stream unit 停运装置
idle operator lamp 空位表示灯
idle pass 空轧孔型
idle pattern 空组合
idle pendulum 闲摆
idle period 空转时间;无功周期;窝工时间;停运时间;停车时间
idle plant 休业工厂
idle position 空转位置;空载位置;无负位;停止位置
idle process 空闲过程
idle producer 低产井
idle pulley 空转(滑)轮;惰轮
idler 空车【铁】;空转(齿)轮;空载;引导轮;过桥轮;闲频信号;闲轮;支承轴承;张紧皮带轮;托辊
idler arm 空转轮臂
idler axle 中间心轴
idler center flange (履带的)导轮中心凸缘
idler circuit 空载电路
idle register 空闲记录器
idle resources 未利用资源
idle resources effect 闲置资源效果

idler flange 惰轮凸缘
idler gear 中间齿轮;空转(齿)轮;换向齿轮
idler hub (履带的)导带毂
idler jet 导轮连接管
idle roll(er) 空转(轧)辊;空转辊(柱);从动(轧)辊;传动轧辊
idle roller arm 从动辊支架
idler oscillation (履带的)导轮振荡
idle route indicator 路由示闲器
idle routine 空闲例行程序;空闲程序
idler pulley 空转(滑)轮;惰轮;张紧轮;支承滚筒;滚轮;皮带张紧轮;托轮;导向轮
idler pulse 无效脉冲
idler reverse gear 倒转惰齿轮
idler revolution 空转轮转速
idler roll 无载托辊
idler roller 张紧轮;惰辊;导辊;托辊
idler set (输送带的)转辊组
idler shaft (输送机的)空转辊轴;空转轮轴;导轮组
idler shaft bearing 空转轴轴承
idler shaft bearing cap 空转轴轴承盖
idler shaft bearing cone 空转轴轴承锥
idler shaft bearing cup 空转轴轴承杯
idler shaft driven gear 空转轴从动齿轮
idler shaft gear 空转轴齿轮
idler shaft low speed gear 空转轴低速齿轮
idler sheave 惰轮
idler side plate (履带的)导轮侧板
idler solenoid 空转轮螺旋线圈
idler spring 空转轮弹簧
idler sprocket(wheel) 中间齿轮;履带张紧轮
idler stand 托辊支承
idler support beam 惰轮支承梁;(履带的)导轮支承梁
idler tilt 惰轮倾斜(度);(履带的)导轮倾斜(度)
idler toe-in 惰轮前束
idler toe-out 导轮外倾;惰轮外倾
idler track 导轮轨
idler travel 惰轮移动距离;(履带的)导轮移动距离
idler tread (履带的)导轮链板
idler tumbler 张紧轮
idle run 无载起动;空转;空回次;空程
idle running 空转;空载运行;无负载运转
idle running idle stroke 空行程
idle runway 闲置跑道
idle rwheel 空转轮;惰轮
idle search 备用搜索
idle shift 停班
idle ship 停航闲置船;停泊船
idle shipping 闲置船(舶)
idle signal 空闲信号
idle space 不工作位置
idle speed 空转速度;空载转速
idle speed adjustment 空转速度调节;空转调整
idle stand 空转机座
idle state 空闲状态
idle status of resources 资源闲置情况
idle stock 呆料
idle stroke 空(冲)程;慢行程
idle time 闲置时间;中断运转时间;空走时间;空载时间;空闲时间;窝工时间;停机时间;停工时间;不生产时间
idle time cost 闲置费(用)(指设备);无效工时成本;停机费用;停工(时间)成本;停工费用
idle-time expenses 闲置时间开支
idle time pay 停工时期工资;停工期工资
idle time supplementary rate 闲置时间补充率;空闲时间补充(工资)率
idle time variance 闲置时间差异
idle track 非驱动履带板
idle trunk lamp 空中继线指示灯
idle two-way selector stage indicator 双向选组级示闲器
idle unit 闲置设备;闲置机组;空转装置;空转机组;停用设备;备用机组
idle valve 怠速阀
idle vehicle 闲置的车辆
idle wheel 惰轮;摩擦传动轮
idle wire 空线
idle work 无用功;无功
idling 闲置;空载;慢速的;无负载运转;低速轧制;怠速
idling adjusting needle 空转调整计
idling adjustment 空转调整;空转调节装置;慢速喷嘴调节装置;慢速调节;无效调整
idling beam 无用束
idling carburet(t)or 空转气化器
idling charges 闲置费(用)(指设备)
idling condition 空转状态;空载状态
idling consumption 空转消耗
idling current 无效电流
idling cut-off 空转切断;空载切断
idling device 空转装置
idling distance 空走距离
idling frequency 无效频率
idling gear 空转轮;空转齿轮
idling grade 惰行坡度
idling jet 慢速喷嘴;怠速喷嘴
idling loss 闲置损失;闲置损耗;空转损失;空转损耗;空载损失;无效损耗;低速损失
idling period 停运时间;停产时间;停产期
idling run 空转
idling signal 空闲信号
idling speed 空转转速;空转速度;空行速度(机车);怠速转速
idling system 慢车系统;怠速系统
idling time 空载时间
idocrase 符山石
idol 偶像
idonic acid 艾杜糖酸
idosaccharic acid 艾杜糖二酸
idreaction 自身反应
idrialine 绿地蜡
idrialite 绿地蜡;晨砂地蜡;辰砂地蜡
idrizite 镁铝铁矾
iduronic acid 艾杜糖醛酸
ldwall stone 爱德华尔石
lecnum 铜镍合金
ieonite 钾镁矾
lewis's base 路易斯碱
if and only if 当且仅当
ifenprodil 艾芬地尔
IF preamp 中频前置放大器
if statement 条件语句
lgatalloy 伊格塔洛伊钨钴硬质合金
l-ga(u)ge 工字形极限卡规
igdanit 硝酸铵与柴油燃料混合物
igdantin (光弹模型材料)依格达胶
lgedur 伊盖杜尔铝合金
igelite 聚氯乙烯塑料
lgel Monument near Treves 特雷沃附近的伊格尔塔形墓碑
igepons 依格泡(表面活性剂)
l-girder 工字梁;单腹板梁;工字(形)大梁
iglesiasite 锌白铅矿
igloo 爱斯基摩人的冰屋;圆顶建筑;圆顶集装箱;冰屋型货柜
igneous 火成的
igneous accumulate 火成堆积
igneous activity 火成活动(与火成岩侵入和形成有关的所有作用)
igneous barrier 岩浆岩体遮挡
igneous complex 火成杂岩
igneous cycle 火成旋回
igneous discontinuity 火成结构面
igneous drilling 火力钻进
igneous intrusion 火成侵入(体)
igneous magma 岩浆
igneous markers 岩浆标志
igneous metallurgy 高温冶金学
igneous metamorphism 火成变质(作用)
igneous mineral 火成矿物
igneous petrology 火成(岩)岩石学
igneous plutonic 熔融的;火成的
igneous plutonic rock 火成岩
igneous province 火成岩区
igneous reservoir 侵入岩热储
igneous rock 岩浆岩;火成岩
igneous rock diapir trap 岩浆岩刺穿圈闭
igneous rock reservoir 火成岩储集层
igneous rock series 火成岩系
igniextirpation 烙除法
ignifluid boiler 半沸腾锅炉
ignimbrite 熔结凝灰岩
ignimbrite plateau 熔灰岩台地
ignisation 人工热源照射法
ignitability 易燃性;着火性
ignitable 可点燃
ignitable waste 可燃废物
ignited flame length 火焰长度
ignited sample 灼烧样品
igniter 引燃器;引燃极;引爆装置;引爆剂;点火器;点火剂;打火装置;触发极;发火器(装置)
igniter body 点火装置
igniter chamber 引燃点火室
igniter circuit tester 点火器电路测试器
igniter cord 点燃导火线;点火线
igniter fuse 引信;点火引信;导火线;导爆线
igniter gas 点火气
igniter material 点火材料
igniter motor 点火发动机
igniter nozzle 点火器喷嘴
igniter pad 引爆药包;点火药包
igniter pellet 点火雷管
igniter pressure 点火压力
igniter stick 点火棒
igniter tube 点火筒;点火管
igniting 点火
igniting burner 点火用燃烧器
igniting circuit 引爆电路;点火电路
igniting composition 起火剂
igniting flame 点火火焰
igniting fuse 传爆信管
igniting magneto 点火式磁电机
igniting powder 点火药
igniting primer 引燃药;引爆器;雷管;起爆药包;传火管
ignition 起爆;着火;点燃;点火
ignitionability 可点燃性;可触发性
ignition accelerator (柴油的)着火加速剂;点燃加速器
ignition accumulator 点火蓄电池
ignition advance 点火提前
ignition advance mechanism 提前点火机构
ignition and lighting magneto 点火及照明永磁发电机
ignition anode 点火阳极;触发阳极
ignition arch 点火拱
ignition battery 点火电池
ignition booster 点火升压器
ignition burner 点火燃烧器;点火燃烧室;点火喷嘴
ignition button 点火按钮
ignition by compression 压缩发火
ignition by friction 摩擦发火
ignition by magneto 磁电机点火
ignition cable 引爆电线
ignition cam 点火凸轮
ignition cap 雷管
ignition chamber 点火室;燃烧室
ignition charge 引火药;引爆炸药;点火药;导火炸药;导火头
ignition circuit 引燃电路;点火线路
ignition circuit tester 点火电路测试仪
ignition coil 点火线圈;发火线圈
ignition coil primary cable 点火线圈低压线
ignition coil resistor 点火线圈电阻器
ignition coil secondary cable 点火线圈高压线
ignition coil tester 点火线圈试验器
ignition control 发火控制
ignition controller 提前引燃防止剂
ignition control relay 点火控制继电器
ignition cut-out 点火断电器
ignition delay 点火延迟;迟发点火
ignition delay tester 点火滞后测试仪;点火迟延时间测试器
ignition detector 点火检测器
ignition device 点火装置;点火设备
ignition distributor 点火分电器;分电器
ignition electrode 点火极;点火电极
ignition exciter unit 点火激励器
ignition explosive 导火炸药
ignition failure 点火系故障
ignition free 不计烧失量
ignition fuse 点火引线
ignition governor 点火调节器
ignition harness 引火线;点火线
ignition hazard 火灾危险;着火危险;发火危险
ignition head 感应配电帽
ignition heat 着火热;燃烧热
ignition injection 引燃点火喷射
ignition interference 点火干扰
ignition interrupter 点火断续器

ignition key 点火开关
ignition lag 延迟着火;延迟发火(装置);延迟点火;着火延迟;迟发点火
ignition limits 燃烧极度
ignition lock 点火闩;点火开关
ignition loss 灼失量;灼烧损失;灼烧减量;烧失量
ignition loss curve 烧失量曲线
ignition loss free 无灼烧损失
ignition loss of soil 土的烧失量
ignition magneto 点火永磁发电机;发火永磁发电机
ignition mixture 燃烧混合物
ignition muffle 高温灼烧炉
ignition noise 点火系噪声
ignition of gas 沼气点火;气体点火
ignition of gas generator 燃气发生器点火
ignition of precipitate 沉淀灼烧;沉淀物灼烧
ignition pad 点火药包
ignition paper 点火纸
ignition pattern 点火次序;放炮顺序
ignition performance 点火性能;点火操作
ignition pilot 点燃小火(燃气热水器)
ignition pin 发火针
ignition plug 火花塞;电嘴
ignition point 着火点;燃烧点;燃点;发火点
ignition point tester 燃点试验器;着火点试验器
ignition powder 起爆药;引燃剂;起爆剂;点火药
ignition precipitate 灼烧沉淀
ignition primer 引发剂
ignition quality 燃烧性质;发火性能
ignition quality improver 发火性增进剂
ignition rate 引爆速率;着火速率(炸药);导爆速度
ignition rating 点火定额
ignition rectifier 引燃管;均匀点火器;点火整流器
ignition residue 灼烧残渣;燃烧残渣;烧(余残)渣
ignition resistance test 耐燃试验
ignition retard 迟发点火
ignition scope 点火检查示波器
ignition sequence 点火次序
ignition spanner 点火扳手
ignition spark 点火火花;点火电火花
ignition switch 点火开关;发火开关
ignition system 点火装置;点火系统;发火系统
ignition temperature 着火温度;着火点;燃烧温度;燃烧湿度;燃点温度;点火温度
ignition test 灼烧试验;灼燃试验
ignition time 着火时间;点火时间
ignition timer 点火定时器
ignition time scattering 起爆时间离散
ignition timing 定时点火;点火定时;发火定时
ignition tube 点火管
ignition unit 点火装置;点火设备
ignition voltage 引燃电压;点火电压
ignition wire 引燃线;点火线
ignitor 引燃电极;引爆装置;点火器;点火剂;点火极;发火器装置
ignitor cord 引燃线;点火线
ignitor current 点火极电流
ignitor discharge 点火极放电
ignitor drop 起弧极电压降
ignitor fuse detonator firing 火雷管起爆法
ignitor oscillation 引燃极振荡
ignitor pad 点火药包
ignitor train 导火素
ignitron 引燃管;水银半波整流器;点火管;放电管;发火器装置
ignitron contactor 引燃管接触器
ignitron control 引燃管控制
ignitron leads 引燃管引线
ignitus 火红色(的)
ignorable coordinates 可忽视坐标;可遗坐标
ignorance factor 不可知因素
ignorant end 工具或设备最重的一头
ignorant end of tape 钢卷尺活动端
ignore 空点
ignore character 作废符;取消符;无(作)用(字)符
ignore character block 抹除符号组
ignore code 无作用(密)码;无作用符号;无作用代号
ignore data 从略数据
ignore instruction 无效指令;无动作指令
ignore specification 舍弃说明
ignore warming 无视警告
ignoring residual value 不计残值
I-groove 工形槽
IHC-Holland type draghead 荷兰 IHC 耙头

I-head cylinder 工形头汽缸;工形汽缸头;顶置气门汽缸
I-head engine 工形头发动机
IHI system 石川岛播磨装卸方式
ihleite 黄铁矾
ihrigized iron 高耐酸性硅铁
ihrigizing 固体渗硅;硅化法;渗硅
iimoriite 羟硅钇石
I-iron 工字铁;工字钢
I-iron shearing machine 工字铁剪切机
ijolite 霓霞石
ijolite group 霓霞岩类
ikaite 六水碳钙石
IK process 固体渗铬法
ikunolite 脆硫铋矿
ilesite 集晶锰矾;四水锰矾
ileus duplex 重复肠梗阻
Ilfracombe beds 伊尔弗拉库姆层
Ilgner flywheel 伊尔格纳型飞轮
ilimaussite 硅铈铌钡矿
Ilkovič equation 伊尔科维奇方程
ill-advised 失策的
ill afford 难以提供;难以负担
illam 含宝石砾层
ill-balanced soil 不平衡土壤
ill condition 不良情况;病态
ill conditioned 情况恶劣的
ill-conditioned equation 病态方程
ill-conditioned matrix 病态矩阵
ill-crystallized 结晶不良的
ill defined 粗略的;不清晰的;不明确的;不精确(的)
ill developed 不发育的
ill-effect 坏作用;不良作用
illegal act 违法活动;非法行为;不法行为
illegal action 非法行为
illegal activity 非法活动
illegal acts by audited entity 被审单位的违法行为
illegal address 非法地址
illegal address check 非法地址检查
illegal addressing mode 非法寻地址方式
illegal building 违章建筑;违法建筑
illegal character 禁用字符;不合法字符
illegal client acts 客户的不法行为
illegal code 非法代码;不合法代码
illegal combination 非法组合;非法联合;非法合并
illegal command 无效指令;非法指令【计】;非法命令
illegal-command check 非法命令检查;不合法指令检验
illegal consideration 非法约因;非法报酬
illegal contract 违法合同;非法合同
illegal control-message error 非法控制消息的错误
illegal element 非法元素
illegal element name 非法元素名
illegal encroachment 非法侵占;非法侵入
illegal entry count 非法入口计数
illegal function name 非法函数名
illegal guard mode 非法防护方式
illegal housing 非法占建
illegal immigrant 非法移民
illegal infiltration 非法渗透
illegal instruction 禁用指令;非法指令【计】
illegal interest 非法高利
illegal interrogation 非法讯问
illegal interrupt 非法中断
illegal intervention 非法干涉
illegality 非法(行为)
illegally obtained evidence 非法获得的证据
illegal means 非法手段
illegal nature 非法性质
illegal omission 非法遗漏
illegal operation 错误运算;非法操作
illegal organization 非法组织
illegal packet 非法(信息)包
illegal partnership 非法合伙
illegal payment 非法支付;非法付款
illegal possession 非法占有
illegal profit 非法所得;非法利润;不当利润;不当得利
illegal report 非法报表
illegal request 非法要求;非法请求
illegal restraint 非法限制
illegal search and seizure 非法搜查与扣押

illegal state 非法状态
illegal strike 非法罢工
illegal structure 非法建筑物
illegal tipping 非法轻弃
illegal trade 非法贸易
illegal traffic 非法运输
illegal transaction 非法交易
illegal trust 非法信托
illegal vote 非法表决
illegible mark 模糊印记
illegitimate 不正常重组
illegitimate copulation 不正常接合
illegitimate crossing over 不正常交换
illegitimate error 不合理误差
illegitimate income 非法收入;不正当收入
illegitimate instrumental variable 不合理的工具变量
illegitimate pollination 不正常授粉
illegitimate voyage 违例的航次
ill-health 不健康;健康不良;健康不佳
illicit 禁止的
illicit income 非法收入
illicit market 黑市
illicit pact 非法协定
illicit trade 违禁贸易;非法贸易
illicium oil 莽草油
Illinoian 伊里诺冰期
Illinois Central Gulf Railroad Company 伊利诺斯中央海湾铁路公司(美国)
illiquid assets 非现金的资产;非流动资产;不能立即变现的资产
illiquid fund 非流动资金
illiquid holding 不动产
illiquidity 无流动资金;非现金;非流动性;非兑现性
illite 伊利石
illite clay 伊利石黏土
illite hydromica 伊利(石)水云母
illite structure 伊利石结构
Illium 伊利姆合金;镍铬合金
ill-judged 判断失当的
ill-lit 光照暗淡
ill-management 管理不善
ill-nourished 营养不良
illogic(al) 不合逻辑的
illogic(al) thinking 非逻辑性思维
ill-shaped casting 畸形铸铁
ill sorted 不配对的
ill-structured problem 结构不完善的问题
ill-thriven 不健康的
ill-timed 不适时的
illuderite 黝帘石
illuminance 照(明)度;光照度;亮度;施照度
illuminance of ground 地面照度
illuminance of skylight 天窗照度
illuminance uniformity 照度均匀度
illuminant 照明体;照明剂;照明的;施照体;施照器;发光物
illuminants for colo(u)rimetry 比色表照明器
illuminate 照亮
illuminated area 照明区(域);照亮面积
illuminated barrier 照明式围栏
illuminated batten 照明吊棒
illuminated body 受照体
illuminated bollard 光亮的标志柱;发光的标志桩;照明系缆柱
illuminated ceiling 照明天平;照明式天花板;照明顶棚
illuminated ceiling system 顶棚照明系统
illuminated circuit diagram 照明电路图
illuminated contour 明暗等高线
illuminated contour method 明暗等高线
illuminated court 灯光球场
illuminated diagram 亮灯显示图;发光图
illuminated diagram switch board 照明配电盘
illuminated dial 照明表盘
illuminated dial ammeter 照明度盘式安培计
illuminated dial instrument 刻度盘照明仪表;发光标度盘仪器
illuminated disk 照亮圆面
illuminated duplex air ga(u)ge 照明双针气压计
illuminated globe 内照明地球仪
illuminated graphic(al) panel 照明图示板;照明控制板
illuminated guard-post 照明护柱

illuminated hemisphere 照亮半球
illuminated indicator 照明式示向标志
illuminated indoor fountain 室内灯光喷泉;照明式室内喷泉
illuminated inspection machine 照明验布机
illuminated mimic diagram 照明模拟图
illuminated mirror 受光镜;照明镜
illuminated mirror slide 照明式镜面推拉窗
illuminated part of a surface 阳面
illuminated quard-post 照明护柱
illuminated rectangular reader 照明长方形阅读放大镜
illuminated relief 地貌光影立体表示法
illuminated rocket 照明弹
illuminated scale 照明度
illuminated sectional directory 照明式区域位置示意图
illuminated sign 灯光照明标志;照明(式)标志
illuminated switch 发光开关
illuminated system 照明系统
illuminated target 照亮的目标
illuminated track diagram 照明式轨道显示图
illuminated track model 照明轨道模型
illuminated wall 受光墙;照明墙
illuminate second flashing 照秒发亮
illuminate time display 照时显示
illuminating 照明;照亮;彩饰
illuminating apparatus 照明器
illuminating beam 照射光束
illuminating carbon 照明炭棒
illuminating circuit 照明线路
illuminating cone 光锥(体)
illuminating device 照明装置
illuminating effect 照明效应;照明效果
illuminating efficiency 照明效率
illuminating engineering 照明工程(学)
Illuminating Engineering Society 照明工程学会(美国)
illuminating equipment 照明设备
illuminating flare 照明弹
illuminating gas 照明(煤)气
illuminating glass 照明玻璃;照光玻璃
illuminating glassware 玻璃灯具;照明仪器;照明器皿;照明玻璃制品
illuminating kerosene 照明灯油
illuminating lamp 照明灯
illuminating lens 照明透镜
illuminating line 照明线
illuminating mark 光点测标;光标
illuminating mirror (经纬仪上的)(反光镜)
illuminating mouth mirror 照明口镜
illuminating oil 照明用油
illuminating power 照明功率;照明本领;亮度;照度;发光力
illuminating projectile 照明弹
illuminating quality 照明质量
illuminating ray 照明光线
illuminating rocket 照明火箭
illuminating ship 灯彩船
illuminating source 光源
illuminating system 照明系统
illuminating value 照(明)度
illuminating wall 照明墙;反光墙
illuminating window 照明窗
illumination 映光;照明(学);照明法;照度;采光;灯饰;彩灯
illumination angle 照射角
illumination at low-powers 低倍照明
illumination beam 照射光束
illumination bundle 照明光束
illumination cable 照明(用)电缆
illumination calculation 照明计算;照度计算
illumination circuit 照明线路
illumination climate 光照气候
illumination component 照明组件;照明分量;照明元件
illumination condition 照明条件
illumination control 照明控制
illumination control room 照明控制室
illumination current 照明电流
illumination curve 照明曲线;照度曲线
illumination design 照明设计
illumination desk 调光台
illumination device 照明设备

illumination distribution 照明分配;光照分布
illumination effect 照射效应;光线照明效果
illumination efficiency 照明效率
illumination engineer 照明工程师
illumination engineering 照明工程(学)
illumination equipment 照明设备
illumination factor 照明因数;照明系数;照度系数
illumination fatigue effect 光照疲劳效应
illumination from skylight 天窗采光;天窗照明
illumination glass 照明玻璃
illumination height 照明高度
illumination installation 照明装置;照明设备
illumination intensity 照明强度
illumination level 照明高度;照度(级)
illumination load 照明负荷
illumination measurement 照明测量
illumination meter 照度计
illumination mirror 照明镜
illumination of deck 甲板照明
illumination of exploratory tunnel 勘探坑道照明
illumination of highway tunnel 公路隧道照明
illumination photometer 勒克司(照度)计;测光表
illumination photometry 照明光度学;照度测定
illumination power 照明功率
illumination sensitivity 照射灵敏度
illumination sign 照明信号
illumination source 照明光源
illumination standard 照明标准;照度标准
illumination surface 照射面
illumination switch 照明开关
illumination system 照明系统
illumination taper efficiency 照射递减效率
illumination tariff 照明收费表
illumination tower 照明塔
illumination value 照明值
illumination zone 照明范围
illuminative level 照度级
illuminator 反光镜;舷窗;映光器;照明装置;照明体;照明器;照明灯;施照体;照射器;反光板;发光器
illuminator lamp 照明用灯
illuminator level 照明高度
illuminator of reflected light 反射光照明器
illumine 照亮
illuminometer 流明计;照度计;亮度仪表;亮度计
illuminophore 发光基团
illumi yarn 光亮丝
illusion 幻影;幻景;错觉
illusional 错觉的
illusory offer 虚报价
illustrated book 有插图的书
illustrated catalogue 图示型录;图示目录;插图目录表
illustrated edition 插图本
illustrated parts catalog 带图解的零件的目录表;插图部分目录
illustrated parts catalogue 零件图解目录
illustration 用图表说明;注解;例证;例示;举例说明;说明;实例;插图
illustration board 插图纸板
illustration map 写景地图
illustration printing paper 插图印刷用纸
illustration with figures 用图表说明
illustrative 直观;说明性
illustrative diagram 直观图;示意图
illustraton of model 模型图
illuvial 沉积的
illuvial clay 沉积黏粒
illuvial halo 淋滤晕
illuvial horizon 土壤淀积层;淀积层;沉积层
illuvial humus 淀积腐殖质
illuvial layer 淀积层
illuvial soil 淀积土
illuviated anomaly 淋积异常
illuviation 淀积作用
illuvium 淀积物;淀积层;淋积物;[复] illuvia
ill-ventilated 通风不良的
ill weed 为害大的杂草
ilmajokite 伊硅钠钛石
Ilmenau porcelain 依曼劳硬瓷(德国)
ilmenite 钛铁矿含量;钛铁矿
ilmenite aggregate 钛铁矿集料;钛铁矿骨料
ilmenite black 钛酸亚铁;钛铁黑
ilmenite loaded concrete 钛铁混凝土
ilmenite ore 钛铁矿矿石

ilmenite-rich rock 富钛铁矿岩
ilmenite type electrode 钛铁矿型焊条
ilmenitite 钛铁岩
ilmenomagnetite 钛铁磁铁矿
ilmenorutile 黑金红石;钛铁金红石
ilmentite type granite 钛铁矿型花岗岩
ilsemannite 蓝钼矿
iluminite 铝电解研磨法;电解抛光氧化铝制品
ilvaite 黑柱石
image 映像;影像;景像;图像;表象
image acceleration 图像转移加速器
image accelerator 图像加速电极
image accelerator voltage 图像加速极电压
image activator 映像激活程序
image acutance 影像锐度
image admittance 对等导纳
image aerial 镜像天线
image alteration pattern map 影像蚀变类型图
image amplifier 图像放大器
image amplifier iconoscope 图像放大光电摄像管
image-amplifying device 图像增强器
image analysing computer 影像分析
image analysis 映像分析;影像分析;图像分析
image analysis method 影像分析法;图像分析法
image analyzer 图像扫描器
image and waveform monitor 图像和波形监视器
image angle 像角;图像角
image antenna 镜像天线
image area 图像面积;成像面积
image area coverage 图像覆盖面积
image area test 图像面积测量
image array 图像阵列
image aspect 像方位
image aspect ratio 图像宽高比
image attenuation 影像衰减;镜像衰减;图像衰减
image attenuation coefficient 镜频衰减系数;图像衰减系数
image attenuation constant 影像衰减常数;图像衰减常数;对等衰减常数
image averaging 图像平均
image background 图像背景
image band 图像波段
image bandwidth compression 图像带宽压缩
image basis 图像基础
image beam 像束
image-bearing 载像
image blurring 图像模糊
image border 图像边缘
image boundary 图像边界
image brightness 图像亮度
image brightness control 影像亮度控制
image burn 图像损缺;图像烧伤
image by inversion 反演象
image card 图像卡片
image carrier 图像载波
image carrier suppression 图像载波抑制
image-carrying fiber 传像纤维
image-cartography 影像地图制图学
image center[centre] 影像中心
image channel 图像通道
image channel gain limited sensitivity 图像通道最大灵敏度
image channel noise limited sensitivity 图像通道的有限杂波灵敏度
image channel synchronizing sensitivity 图像通道的同步灵敏度
image characteristics extraction 图像特征提取
image charge 像电荷;镜像电荷;图像电荷
image chart 象形图
image circle 像圈;图像圈
image circular feature map 影像环形体图
image circular features 影像环形体
image classification 图像分类
image coding 影像编码;图像编码
image coding devices 图像编码器件
image coding technique 图像编码技术
image-combining mirror 成像交联反射镜
image composition 图像合成
image compression 图像压缩
image conduit 传像管
image confusion 像弥散
image construction 影像作图法;图像构成
image-contained data 像容数据;像含数据
image contrast 像衬比;影像反差;图像反差;图像

对比度;摄像反差
image contrast technique 像衬比技术
image control 图像调整
image control coil 图像调整线圈
image conversion 图像转换;图像变换
image converter 像转换器;图像光电变换管;变象器
image converter camera 图像变换摄像机;变象相机;变象管摄像机;反象复照仪
image converter high-speed camera 变象管高速摄影机
image converter streak camera 像转换器扫描照相机;变象管高速扫描照相机
image converter tube 像转换管;图像光电变换管;变象管
image coordinates 像面坐标;像点坐标
image copy 影像副本
image corrector 图像改正器
image correlation 影像相关
image correlator 图像相关器
image counter 形象专柜
image criteria of geothermal feature 影像地热显示标志
image criteria of oil-gas indication 影像油气显示标志
image current 像电流;影像电流
image current density 像电流密度
image curvature 像弯曲
image curve 像曲线
image data 图像数据
image data acquisition 影像数据采集
image database 图像数据库
image database system 图像数据库系统
image data compression 图像数据压缩
image data network 图像数据网
image data page 图像数据页面
image data processing 图像数据处理
image data structure 图像数据结构
image deblurring 图像模糊消除
imaged edge 成像界限
image defect 图像缺陷;成像缺陷
image definition 图像清晰度;反差度
image degradation 图像位移;图像递降
image degradation model 图像退化模型;图像劣化模型
image density 影像深浅度
image description 图像描述
image detail 影像细部;图像细节
image detail factor 图像细节因数
image detail region 图像细节区
image detection 图像检测;图像检波
image deterioration 图像退化;图像变坏
image device 成像器件
image diagonal 图像对角线
image difference 图像差
image diffusion 图像模糊
image digital converter 图像数字转换器
image digitization 图像数字化;图像数值化
image digitizer 图像数字转换器
image digitizing 图像数字化
image dimensions 图像尺寸
image direction 像面朝向
image displacement 像(点)位移
image display 影像显示;图像显示
image display definition mode 图形显示定义方式;图像显示定义方式
image dissection 析像;图像分解
image dissection camera 析像摄像机
image dissection photography 图像分解摄影
image dissector 析像器;图像分解器
image dissector camera 析像管摄像机
image dissector camera system 图像分析照相机系统
image dissector multiplier 析像倍增管
image dissector tube 析像管;光电析像管
image dissimilarity 图像不同性
image distance 像距
image distortion 像畸变;影像畸变;图像失真;图像畸变
image disturbance ratio 图像干扰比
image divider 分像器
image dividing optic(al) system 光学分像系统
image dividing relay optic(al) system 分像转向光学系统
image drainage network map 影像水系图

image drop-out 图像信号失落
image duration 帧周期
image element 像素
image encoding 图像编码
image energy 图像能
image enhancement 影像增强;图像增强
image-enhancing equipment 图像增强设备
image enlargement 影像放大
image enlarger lamp 图像放大灯
image erection 影像扶正
image error 影像误差;成像误差;像差
image evaluation 图像估价
image factor 映射要素
image fault 图像失真
image fidelity 图像逼真度
image field 像场;镜像场
image file 影像文件;图像文件;图像档案
image filtering 滤像
image flattening lens 像场修正透镜;图像场修正信号
image flicker 图像闪烁
image flyback 帧回描
image focal point 像方焦点
image focus 图像聚焦
image focusing 图像聚焦
image focusing electrode 影像聚焦电极
image fog 影像模糊
image force 像力;镜像力
image force model 像力模型
image form 图像幅面
image formation 构像;图像形成;成像
image formation by rays 光线成像
image-forming 成像
image-forming block 成像部件
image-forming component 成像元件
image-forming condition 成像条件
image-forming electron 成像电子
image-forming infrared receiver 成像红外线接收机
image-forming system 成像系统
image-forming ultrasonic microscope 成像超声显微镜
image frequency 像频;帧频;镜像频率;镜频;图像频率
image frequency interference 像频干扰;镜频干扰
image frequency rejection ratio 像频抑制比
image frequency selectivity 像频率选择性
image frequency signal 镜频信号
image frequency spectral analysis 图像频谱分析
image function 像函数
image fundamental 图像基础
image furnace 光聚焦加热炉;聚焦炉;弧像炉
image generation 图像生成
image geometric(al) error 图像几何误差
image geometry 影像几何(性质);图像几何(性质)
image geomorphologic(al) map 影像地貌图
image graphics 非编码图形;映象图形学
image guide 传像束
image guide tube 导像管
image height 祯面高;图像高度
image histogram 图像直方图
image hologram 像全息图;影像全息图
image horizon 像片地平线
image iconoscope 移能式光电摄像管;图像光电摄像管
image identification correlation method 图形识别对比法
image illumination 图像照度
image illumination uniformity 像面照度的均匀性
image impedance 影像阻抗;镜像阻抗;对等阻抗
image index 图像索引
image information 图像信息
image information preprocessing 影像信息预处理
image information processing 图像信息处理
image information processing system 图像信息处理系统
image input 图像输入
image integrating tracker 图像累积跟踪器
image intensification 图挚增强;图像强化
image intensifier 像亮化器;光放大器;图像增强器;图像亮化器;变象管
image intensifier isocon 像增强式分流直像管
image intensifier-microchannel plate 像增强器微通道板
image intensifier orthicon tube 像增强析像管

image intensifier tube 像增强管;图像亮度放大管
image intensifying photosensitive film 像增强光敏薄膜
image intensifying system 像增强系统
image intensity 图形强度;图像强度;图像亮度
image interference 虚源干涉;影频干扰;镜像干扰;图像干扰
image interference ratio 镜频抑制比;镜频抗拒比
image intermediate frequency transformer 像信号中频变压器
image interpolation 图像内插;图像插行
image interpretability 影像判读性
image interpretation 影像判译;图像判读
image interpretation criterion 图像解译标志
image interpretation equipments 图像解译设备
image interpretation in detail 图像详细解译
image interpretation method 图像解译方法
image inverse process 影像反转过程
image inversion 像倒置;图像转换;倒像
image inverter 倒像器;倒像板
image-inverter prism 倒像棱镜
image isocon 分流直像管;分流正析像管
image jitter 图像跳动
image lag 影像延迟;成像滞后
image lens 图像透镜
image library 图像库;形象图库
image line 镜像线路;图像扫描线
image lineament map 影像线性体图
image lineaments 影像线性体
image line-amplifier output 图像线路放大器输出
image load 镜像负载
image lock 图像同步
image log 拟测井记录
image-loss 影像损失
image luminance 图像亮度
image magnification 影像放大(率)
image manipulation 图形处理
image manipulation language 图像处理语言
image map 影像地图;印象图;图像映射
image mapping apparatus 像片测图仪
image match 影像匹配;镜像匹配
image matching system 图像匹配系统
image matching technique 图像匹配法
image measurability 图像可测性
image measurement 像片量测
image measuring apparatus 像片量测仪;图像测量装置
image method 映射法;图像法
image mode 像方式
image model 图像模型
image modulation 像频调制
image monitor 图像监视器;图像监控器
image mosaic 影像镶嵌图
image motion 像移;图像移动
image motion compensation 像动稳定补偿;像移补偿;图像移动补偿
image motion compensator 像移补偿器
image motion factor 像移因素
image motion function 像移动函数
image movement 像移
image-movement compensation 像移补偿
image multiplication 图像放大
image multiplier 像倍增器;图像放大器
image name 映像名
image of a saint 圣者形象
image of a variate 变量的像
image of spheric(al) center 球心像
image of tangential circle 切圆图形
image operation 图像运算
image optic(al) reconstruction 图像光学恢复
image optics 成像光学
image orbiting facility 图像旋转设备
image orientation 像片定向;图像取向
image-oriented 面向图像的
image orthicon 超正析像管
image-orthicon assembly 图像正析像管装置;超正析像管摄像机;超正析像管装置
image-orthicon camera 图像正析像管摄像机;超正析像管摄像机
image orthiconoscope 图像正析像管;超正析像管
image output 图像输出
image overlaying 图像复合
image pairs 像片对
image parallax 影像视差

image parameter 镜像参数;图像参数
image parameter design 镜像参数设计
image patch 像斑
image pattern 像图
image pattern recognition 图像模式识别
image persistance 影像暂留;图像持续时间;图像持久性
image phase-change coefficient 图像相位变化系统
image phase constant 像相常数;影像相位常数;对等相位常数
image photocell 光电摄像管;图像光电管
image photomap 影像地图
image pick-up 摄像
image pick-up device 摄像器
image picture 相片;图像
image plane 成像面;像平面;图像平面
image plane coordinate system 像平面坐标系
image plane holography 像平面全息摄影
image plane irradiance 像面辐照度
image plane scanning 像面扫描
image plate 图像板;像信号板
image plumbicon 超氧化铅视像管
image point 像点;同名像点
image point setting 像点定位
image position 像位置
image position specification 像位规定
image power amplifier 图功率放大器
image preprocessing 影像预处理
image press 图像压缩
image printing 图像打印
image printout 图像输出
image process(ing) 图像处理;映像处理;图像加工
image processing digital system 图像处理数字系统
image processing equipment 图像处理设备
image processing in real-time 实时图像处理
image processing interpretation method 图像处理解译方法
image processing package 图像处理程序包
image processing products 图像处理产品
image processing system 图像形成装置;图像处理系统
image processing technique 图像处理技术
image processing technology 图像处理技术
image processor 图像处理机;映像处理程序
image-projected method (超声波检测的)图像投影法
image projection form 图像投影方式
image quality 像片质量;影像质量;图像质量;图像品质
image quality criterion 图像质量标准
image quality evaluation 图像质量评价
image quality indicator 像质计
imager 成像器
image raster 图像光栅
image ratio 像频抗拒比;镜像比;镜频相对增益
image ray 像点投影线
image recognition 图像识别
image reconstruction 像重建;影像重建;图像重显;图像重建
image reconstructor 图像重显装置
image record(ing) 图像记录;影像录制;图像录制;录像
image recording medium 图像记录媒体
image recording model 全图像记录方式
image recovery technique 图像恢复技术
image redundancy 图像多余度
image refresh database 影像更新数据库
image registration 影像重合;影像配准;图像重合;图像配准
image rejection 镜像干扰抑制;镜像抑制;图像载频抑制;图像载波抑制
image rejection ratio 镜频抗拒比
image reject mixer 镜像抑制混频器
image rejector 镜频抑制器
image repetition 图像重复
image representation 图像表决
image reproducer 像重现装置;图像重显装置
image reproduction 图像重显;图像再现
image resolution 像片分辨率;影像分辨力;图像清晰度
image resolution ratio 图像分辨率
image response 像频响应;镜频响应;镜道响应
image restoration 图像恢复;图像复原

image retaining panel 影像保留幕板
image retention 图像保留;残像;残留影像
image retention time 图像保留时间
image reversal 镜像
image roll 图像滚动
image rotation 影像旋转
image rotation picture 旋像
image rotation prism 转像棱镜;成像旋转棱镜
imagery 雕刻;刻像;群像;摄像
imagery radar 成像雷达
imagery recognition 图像识别
image sampling 图像取样
image scale 像片比例尺;像标;镜像比例尺;图像比例尺;成像比尺
image scale figure 像片比例尺数字
image scanner 图像扫描器;图像扫描输出仪
image-scanning device 图像扫描装置
image screw 映像旋量
image scrolling 图像滚动;图像卷起
image section 映像段;图像部分
image section descriptor 映像段描述符
image segmentation 图像分段;局部图像分析法;图像分割(法)
image sensing 图像传感
image sensing array 成像遥感器阵列
image-sensing panel 图像检测板
image sensor 影像传感器;摄像传感器
image-separating prism 分像棱镜
image separation 影像分离
image set 像集
image shape 镜像形状
image shape converter 像形状变换器
image shape picture 像幅
image sharpening 图像清晰化
image sharpness 像清晰度;影像清晰度;图像清晰度
image shield 镜像屏蔽
image shift 图像位移;图像偏移
image shift iconoscope 移像光电摄像管
image shotpoint 虚炮点
image side 像面
image signal 映像信号;镜像信号;图像信号
image signal amplifier 图像信号放大器
image signal amplitude 图像信号幅度
image signal distribution amplifier 图像信号分配放大器
image signal generator 图像信号发生器
image signal polarity 图像信号极性
image simulation 图像模拟
image size 像幅;图像尺寸
image size change 像场变化
image smear 影像污点
image smoothing 图像平滑
images of identical parallel 同纬圈图形
image source 像源;图像源;图像出处
image space 图像空间;像空间;像方
image space coordinate system 像空间坐标系(统);图像空间坐标系(统)
image spacing 画面空间
image spliting optic(al) system 影像分光系统
image splitting eyepiece 分像目镜
image spot 像点
image spot size 像点尺寸
image spread 像扩散
image stabilization 图像稳定
image stage 图像传输部分
image-stone 像石
image storage 图像存储
image storage array 图像存储阵列
image storage device 录像设备;图像存储器件
image storage space 图像存取空间
image storage translation and reproduction 图像存储变换和再生
image storage tube 储像管
image store 映像存储区
image storing tube 图像存储管;储像管
image strip 图像条(带);图带;条带
image subsection 小图像
image subtraction 影像相减
image suppression 镜频抑制
image surface 像面;焦面
image surface curvature 像面曲率
image sweep frequency 图像扫描频率
image switching 图像切换

image symbolization 图像符号化
image symmetry 影像对称
image synchronization 影像同步
image synthesis 像合成;图像合成
image synthesis array 图像综合天线阵
image system 映像系统
image termination 图像终端设备
image texture 影像结构
image texture analysis technique 图像纹理分析处理
image theory 镜像原理
image throughput 图像通过量
image tone 影像色调
image transcription 像录制;影像录制;录像;图像录制
image transfer 影像转制;影像转绘;图像传输
image transfer characteristics 图像转移特性
image transfer constant 影像转移常数;镜像转移常数;镜像传输常数
image transfer converter 图像转换设备
image transfer exponent 图像转换指数
image transfer ratio 图像转换比
image transfer system 图像传输系统
image transfer technique 图像转换技术
image transformation 图像变换
image transformer 图像变换器
image translator 图像变换器
image transmission 图像传真;图像传输
image transmission and processing 图像传输与处理
image transmission scanning 图像传输扫描
image transmission system 图像传输系统
image transmitter 图像发射机
image transmitter power 图像发射机功率
image-transmitting bundle 传像束
image treatment 图像处理
image tube 显影管;显像管
image type 图像类型
image value 图像值
image velocity sensor 图像速度传感器
image vericon 超正析像管
image-viewing tube 图像光电变换管
image visibility 图像可见度
image volume 图像容积形状变换器
image warping 图像卷曲
image wave function 图像波函数
image well 虚井;推测井;反射井
image white 图像白色
image width 像宽;图像宽度
imaginal type 表象类型
imaginary 假想的
imaginary accumulator 虚数累加器
imaginary argument 虚自变数
imaginary axis 虚轴;假想轴
imaginary beam 虚梁
imaginary circle 虚圆
imaginary circle at infinity 虚球圆
imaginary circular point 虚圆点
imaginary combustion 假想燃烧
imaginary component 虚(数)部分;虚分量
imaginary component of complex function 复(变)函数的虚分量
imaginary component of secondary field 二次场虚分量
imaginary cone 虚锥
imaginary constant 虚常数
imaginary cross-section 假设断面
imaginary cube root 虚立方根
imaginary curve 虚曲线
imaginary cyclic(al) field 虚循环域
imaginary demand 虚假需求
imaginary eccentricity 假想偏心度
imaginary elastic body 假想弹性体
imaginary element 虚元素
imaginary ellipse 虚椭圆
imaginary ellipsoid 虚椭圆面
imaginary elliptic cylinder 虚椭圆柱面
imaginary exponent 虚指数
imaginary factor 虚因子
imaginary field 虚域
imaginary focus 虚焦点
imaginary frequency 虚频率
imaginary fund 假想基金
imaginary hinge 虚铰;假铰

imaginary hypersphere 虚超球面
imaginary infinite prime divisors 虚无限素因子
imaginary intersection 虚相交
imaginary intersection point 虚交点
imaginary line 虚线;假想线
imaginary load 虚载(荷);假想荷载;假想负荷
imaginary load factor 假想荷载因数;虚荷载因素
imaginary loading 虚荷载;虚负载
imaginary mass 假想质量
imaginary number 虚数
imaginary part 虚数部分;虚部(复数)
imaginary part operation 虚部运算
imaginary part operator 虚部算子
imaginary plane 虚平面
imaginary point 虚点
imaginary potential 虚势
imaginary prestress 虚预加应力
imaginary prestressing 假想预应力
imaginary prime divisor 虚素因子
imaginary profit 预期利益
imaginary quadratic field 虚二次域
imaginary quadric hypersurface 虚二次超曲面
imaginary quantity 虚数;虚量
imaginary quaternion 虚四元数
imaginary-real component method 虚实分量法
imaginary refractive index 虚数折射率
imaginary representation 假想图示
imaginary root 虚根
imaginary sample points 虚样本点
imaginary semi-axis 虚半轴
imaginary source 假胀源
imaginary space 虚空间
imaginary sphere 虚球
imaginary stereoscope 虚拟立体模型
imaginary stressing 虚加应力
imaginary stretching 虚拉伸
imaginary surface 虚曲面
imaginary tensioning 虚张拉
imaginary terms 虚数项
imaginary time coordinate 虚时间坐标
imaginary transformation 虚数变换
imaginary unit 虚数单位;假想单位
imaginary variable 虚变量
imaginary wave front 虚波阵面
imagination 假想;创造力
imaginative geomorphologic figuration 造型地貌
imagine 设想
imaging 成像
imaging beam 成像光束
imaging component instrument without referring wire 无参考线虚分量仪
imaging deformer 成像变形器
imaging detector 成像探测器
imaging device 成像装置;成像设备
imaging equation 构像方程;成像方程
imaging fiber 成像光纤
imaging focal length 成像焦距
imaging infrared 成像红外
imaging infrared guidance 成像红外制导
imaging infrared receiver 成像红外接收机
imaging infrared system 成像红外系统
imaging intersection point 虚交点
imaging lens 成像透镜
imaging mirror 成像反射镜
imaging mosaic 成像感光镶嵌幕
imaging process 成像过程
imaging quality 成像质量
imaging radar 影像雷达;摄像雷达;成像雷达
imaging sea 迎浪
imaging sensor 图像传感器;成像传感器
imaging sonar 影像声呐
imaging system 成像系统
imaging technique 成像技术
imaging theory 成像理论
imaging time 成像时间
imaging tube 显像管
imaging with coherent light 相干光成象
imaging with incoherent light 非相干光成像
imaging with partial coherent light 部分相干光成象
Imagon lens 伊梅冈镜头
imandrite 铁钙钠石
IMA process 结合移动平均数过程
imaret (土耳其招待朝圣者的)旅店

imbalance 偏重;失(去)平衡;不稳定性
imbalance factor of locked traffic 过闸运量不均衡常数
imbalance factor of traffic 运输不均衡系数
imbalance of consumption 消费失调
imbalance of developments 发展不平衡性
imbat (地中海东部的)冷季节风
imbed 灌封;埋置;包埋
imbedded chips 嵌入(的)石屑
imbedded code 嵌入码
imbedded fin tube 镶片式翅片管
imbedded prime ideal 嵌入素理想
imbedded steel 埋置钢筋
imbedded temperature detector 埋入测温器
imbedded temperature-detector insulation 埋入测温器绝缘
imbedded(wave)guide 埋封波导
imbedded winding 埋入式绕组;嵌入绕组
imbedding 埋入
imbedding method 嵌入法
imbedding tool 埋线器
imbedment 埋置;嵌入
imbex 盖瓦
imbibant 吸胀体;吸涨体
imbibe 吸液膨胀;吸液膨润;浸透;透入
imbibing 吸收作用;吸液
imbibition 吸液;吸水膨胀;吸收;吸入作用;吸取(液体);吸滤作用;浸渗;染色传递法;润润作用;膨化;透入;渗化
imbibitional pressure 吸涨压
imbibitional water 吸胀水
imbibition moisture 吸入水
imbibition pressure 吸胀压;浸渗压
imbibition process 吸液印相法
imbibition water 饱和水;渗吸水
imbow 作拱;用拱跨越
imbowment 穹隆;拱
imbrex 鳞状饰片;筒(子)瓦;槽瓦;波形瓦;半圆形盖瓦
imbricate 叠瓦状;搭盖;覆瓦状;复瓦状的
imbricate aestivation 复瓦状花被卷叠式
imbricated dune 叠瓦沙丘;叠置沙垄
imbricated fault 叠瓦断层
imbricated fault zone 叠瓦断层带
imbricated leaf 覆丘状叶
imbricated ornament 叠覆饰;鳞状饰物;鳞状装饰
imbricated plate 鳞状板;搭盖板
imbricated roof 覆瓦屋顶;叠瓦屋顶
imbricated structure 叠瓦构造;叠覆构造
imbricated texture 鳞状结构;叠瓦状结构
imbricated tilt against current 逆流倾斜呈叠瓦状
imbricated tracery 鳞甲状窗(花)格;鳞状窗花格;鳞甲窗格
imbricated winding 链形绕组
imbricate fault 叠瓦状断层
imbricate structure 鳞片状结构;叠瓦结构
imbricate thrusts 叠瓦式冲断层
imbricate veins 迭瓦状矿脉
imbricating delta 叠瓦三角洲
imbrication 鳞甲饰;瓦状叠覆;叠瓦作用
imbrice 半圆形脊瓦
imbue 浸染;染透
Imbuia 巴西黄褐色硬木
imcomplete cleavage 部分分裂
imequentvalley 斜向谷;非顺向谷
imerina stone 散光闪石
imernite 钠透闪石;散光闪石
imgreite 伊碲镍矿
Imhoff cone 殷霍夫(式)锥形管
Imhoff cone test 殷霍夫圆锥管试验(污水沉淀);殷霍夫锥形测定
Imhoff pit 殷霍夫井
Imhoff sedimentation funnel 殷霍夫沉淀漏斗
Imhoff solid 殷霍夫固体
Imhoff tank 隐化池;殷霍夫式化粪池;殷霍夫(式)沉淀池;双层沉淀池;沉淀隐化池
imhofite 硫砷铜铊矿
imidazole[iminazole]间二氮茂;咪唑
imidazoline 咪唑啉
imidazolinone 咪唑啉酮
imide modified alkyd 亚胺改性醇酸树脂
imide resin 酰亚胺树脂
imine 亚胺
iminocarbonic acid 亚氨基碳酸

iminocarbonic ester 亚氨基碳酸酯
iminodicarboxylic acid 亚氨基二羧酸
iminophenimide 苯乙哌酮
imipramine hydrochloride 托法尼
imitant 赝品;代用品
imitate 摹写;仿效
imitate a trademark 冒牌
imitating kang colo(u)r vase with floral design 仿康彩花卉瓶
imitation 临摹;模拟;模仿;仿制(品);仿造(品)
imitational Song typeface 仿宋体
imitation antique 仿古瓷器
imitation antique carpet 仿古地毯
imitation antique porcelain 仿古瓷器
imitation art paper 高灰分印刷纸;仿铜版纸;仿美术纸
imitation art printing paper 充粉纸
imitation brand 影ескому商标;冒牌
imitation brick 仿砖;仿砖
imitation copy 摹写本
imitation curios 仿古玩
imitation effect 模仿效应
imitation glassine paper 仿玻璃纸
imitation gold 装饰用铜合金;装饰用的铜铝合金(含铝3%~5%)
imitation gold ink 仿金油墨
imitation gold yarn 充金线
imitation haircloth 充马尼衬
imitation handmade printing paper 仿手工印刷纸
imitation horse hair lining 充马尼衬
imitation Japanese paper 仿日本和纸
imitation jewel 人造宝石
imitation kraft 仿牛皮纸
imitation lag 仿制时滞
imitation leather 假皮;假革;人造革
imitation leather paper 仿革纸
imitation manganese bronze 仿造锰青铜
imitation marble 假大理石;人造大理石
imitation natural wool 充原色羊毛
imitation of style 形式抄袭;形式模仿
imitation parchment paper 仿羊皮纸
imitation parts 仿造品
imitation pearl paint 仿珠光漆
imitation platinum alloy 仿造铂合金
imitation porcelain 仿制瓷;仿古瓷
imitation red lead 假红丹
imitation silk 充丝
imitation silver 假银;仿银
imitation silver yarn 充银线
imitation stone 假(宝)石;人造石;仿造石
imitation stone block 假石砌块
imitation stone finish 假石饰面
imitation travertine 仿石灰华;人造石灰华
imitation wood 人造(木)材
imitation woolen paper 仿羊皮纸
imitative colo(u)r 模仿色
imitative growth 仿效性增长
imitative pricing 模仿价格
imitator 模拟设备;模拟器;模拟程序;模拟器;仿真器;仿真品
immaculate 无斑点的
immadium 高强度黄铜
immalleable 无韧性的;不可锻性;不可锻的
immanent cause 内因
immaterial 非物质的
immaterial assets 名义资产;无形资产
immaterial capital 无形资本
immateriality wasting 无形损耗
immaterial labo(u)r 无形人工;非物质劳动力
immaterial safety circuit 非本质安全电路
immaterial value 非物质价值
immature 未成熟的;不成熟的
immature concrete 新鲜混凝土;新拌混凝土;未终凝的混凝土;未凝结混凝土
immature cotton waste 废元棉
immature fiber 未成熟纤维
immature form 亚成熟形态
immature region 幼年地区
immature residual soil 新残积土
immature sediments 未成熟沉积物
immature segment 未成熟节片
immature soil 未熟土;未成熟(的)土;生土
immature source rock 未成熟源岩
immature stage 未成熟阶段

immature technology 不成熟技术
immature wood 未成熟木材
immaturity 幼年期;未熟阶段;未成熟性
immdediate maintenance 紧急性维修
immeasurability 不能测量;不可计量性;不可测量性
immeasurable 不能量测的;不可测量的
immedial green 直达绿
immedial pure blue 直达纯蓝
immedial sky blue 直达青
immediate 直接的
immediate acceptance 即期承兑
immediate access 立即存取;快速访问;快速存取;即时存取
immediate-access memory 立即访问存储器;即时存取存储器
immediate-access storage 直接存储器存取;立即访问存储器;快速存储器;即时存取存储器;瞬时存储器
immediate-access store 立即存取存储
immediate acidity 直接酸度
immediate acknowledgement 即时应答
immediate action 立即行动;快速;速动
immediate address 直接地址;零级地址;立即地址
immediate addressing 零级定址;立即寻址;快速寻址
immediate addressing mode 立即寻址方式
immediate address instruction 直接地址指令;立即地址指令
immediate appreciation percentage 即时可懂百分比
immediate assets 流动资产
immediate background 最近背景
immediate bank 紧靠岸边;紧靠水边的堤岸
immediate-bearing 直接受荷轴承;直接支承
immediate biochemical oxygen demand 直接生化需氧量;瞬时生化需氧量;表观生化需氧量;标准生化需氧量
immediate bottom 直接底板
immediate call 即时呼叫
immediate cancel 立即作废;立即取消
immediate capture 瞬时俘获
immediate-capture efficiency 瞬时俘获效率
immediate cash payment 立即付现;即时付现;即期付现
immediate cause 直接原因;近因
immediate checkpoint 立即检验点
immediate command 立即命令
immediate compensation 立即赔偿
immediate compression 瞬时压缩
immediate consequence 直接后果
immediate contact 直接接触
immediate crazing 早期开裂
immediate danger 紧迫危险
immediate danger area 直接危险地区
immediate data 直接数据;立即数据
immediate data dependency 直接数据相关性
immediate death 立即死亡
immediate deformation 瞬时变形
immediate delivery 直接交付;立即交货;立即发货;即时交货;即期交货
immediate development 近期发展
immediate device control block 立即设备控制块;快速设备控制块
immediate diagnosis 即刻诊断
immediate economic benefit 近期经济效益
immediate effect 直接作用;瞬时效应
immediate elastic deformation 瞬时弹性变形;瞬间弹性变形
immediate environment 直接环境
immediate environment termination 立即环境终止
immediate execution mode 立即执行方式
immediate experience 即时体验
immediate following telegram 紧急后续电报
immediate fund 即期资金
immediate hanging wall 直接上盘
immediate heave 瞬时隆起
immediate inference 直接推理
immediate installation of tunnel support 随挖随支护
immediate instruction 立即指令
immediate labo(u)r 直接劳工
immediate liability 流动负债
immediate load instruction 立即送数指令
immediate losses of prestress 瞬时预应力损失

immediately addressing 直接寻址;直接选址;直接定址;直接编址
immediate manoeuvre 随时操纵
immediate means 直接方法
immediate mission 直接任务
immediate mode 快速方式;即时方式
immediate movement 瞬时位移
immediate offshore area 近岸海域
immediate operand 立即操作数
immediate operation 立即操作
immediate operation instruction 立即操作指令
immediate oxygen demand 直接需氧量
immediate packing 直接包装
immediate passivation 瞬时钝化
immediate payment 即时付款
immediate predecessor 直接先趋块;紧接前项活动
immediate prediction 临震预报
immediate processing 立即处理;快速处理;即时加工;即时处理;按需处理
immediate production 直接生产
immediate profit 目前利润
immediate reaction 即现反应;即时反应
immediate reading 瞬间读数
immediate record 立即记录
immediate recourse 立即追索
immediate release signal 立即释放信号
immediate-request mode 立即请求方式
immediate response 即时应答
immediate-response mode 立即响应方式
immediate roof 直接顶板
immediate runoff 直接径流;地表径流(量)
immediate set 瞬间形变;瞬间变定
immediate settlement 瞬时沉降;初始沉降
immediate shipment 立即装运;立即装船
immediate significance 现实意义
immediate skip 立即跳动;立即跳令
immediate source 直接污染源;直接来源
immediate strain 瞬时应变
immediate subcommand 立即型子命令
immediate subordinate 直接从属
immediate successor 直接后继块;紧接后元
immediate superior 直接上级
immediate support 直接支援
immediate surface 直接表面;接触面
immediate task 立即任务
immediate term approach for earthquake prediction 临震预报法
immediate train distance point without siding 无配线分界点
immediate transportation 即运
immediate transportation entry 中途报关手续;途中报关进口
immediate vicinity 相邻地区
Immedium sand filter 伊曼迪姆砂滤池
immense growth 大发展
immensurability 无法计量的性质或状态
immerged coil heat exchanger 沉浸式蛇管换热器;沉浸式盘管式换热器
immerged crack 深埋裂纹
immergence 埋入;沉浸
immerse 浸入;浸泡;浸没;沉浸
immerseable finger 浸入式深度规
immerse body 潜没体
immersed 浸没式
immersed body 潜体;水下物体
immersed bolometer 浸没式热辐射计
immersed cross-section(al) area 浸水断面面积
immersed density 浸没容重;潜容重;水中密度
immersed detctor 浸没式探测器
immersed detector element 浸没探测元件
immersed electrode furnace 浸极式炉
immersed focal-plane lens 浸没焦平面透镜
immersed guide roll for coating 涂装浸没导辊
immersed impeller aerator 淹没式叶轮曝气装置;浸没式叶轮曝气器
immersed membrane 浸没式膜
immersed membrane activated sludge process 浸没式膜活性污泥法
immersed membrane bioreactor 淹没式膜生物反应器
immersed method 水浸法(一种探伤方法)
immersed nozzle 潜入式喷嘴
immersed oil pump 浸没式油泵
immersed pump 沉没泵

immersed rheostat 液体变阻器
immersed section 浸水剖面;浸水密度;浸水断面;水下断面
immersed sidewall air cushion vehicle 侧壁式气垫船
immersed sidewall hovercraft 侧壁浸水式气垫船
immersed system 浸水系统
immersed torque motor 湿式力矩马达;湿式力矩电动机
immersed transducer 浸没式换能器
immersed tube 沉管
immersed tube method 沉埋法
immersed tube tunnel 水底管形隧道
immersed tunnel 水底隧道下;水底隧道;沉埋式隧道;沉管隧道
immersed tunnelling method 沉埋法
immersed tunnel trench method 沉埋法
immersed volume 水下体积
immerse precipitation 浸没沉淀
immersible concrete vibrator 插入式混凝土振捣器
immersible electric(al) pump 电动潜水泵
immersible motor 浸入型电动机;潜水电动机
immersible pump 潜水泵
immersible sewage pump 潜水排污泵
immersible switchgear 潜水开关装置
immersing cooling 沉浸冷却
immersing heater 沉浸式加热器
immersing medium 浸没介质
immersion 浸水;浸入;浸泡;浸没;埋入;沉入;沉浸;沉放
immersion(al) wetting 浸入润湿;浸润作用;浸入湿润
immersion angle 入水角
immersion apochromat 浸没复消色差透镜
immersion bell 钟罩
immersion burner 浸没式燃烧器
immersion casting 浸没铸造
immersion cleaning 浸洗
immersion coating 浸没镀层;浸渍涂布;化学涂层
immersion colo(u)rimeter 浸没式比色计
immersion combustion 浸没燃烧
immersion-compression test 浸水抗压试验;浸水压缩试验(测定沥青混凝土水稳定性使用)
immersion condenser 浸没式聚光器
immersion cooler 浸没式冷却器;沉浸式冷凝器
immersion cooling 浸没冷却;浸冷
immersion counter 浸没式计数管
immersion couple 浸入式热电偶
immersion-cured 浸水养护的;水中养护的
immersion-curing 浸水养护
immersion depth 浸水深度;浸没深度;潜没深度
immersion electric(al) heater 浸没式电热器
immersion electrode 浸液电极
immersion electron lens 浸没式电子透镜
immersion filter tube 浸液过滤管
immersion fluid 浸液
immersion foot 泡浸足
immersion freezer 浸液制冷器;浸入式冷冻器;沉浸式制冷器
immersion furnace 沉渍式保温锅
immersion gain 光学浸没增益
immersion gas burner 浸没式煤气喷灯
immersion gilding 浸入镀金法
immersion grating 浸没光栅
immersion heater 浸液加热器;浸入式热水器;浸入式加热器;浸没(式)加热器
immersion heating 潜没加热(法)
immersion heating appliance 浸没式加热器
immersion heating element 浸没式加热元件
immersion hot water heating 浸没式热水加热
immersion hot water heating appliance 浸没式热水加热器
immersion in water 浸入水中
immersion jig 全浸式卷染机
immersion length 浸没长度
immersion lens 浸没透镜
immersion liquid quenched fuse 液体灭弧熔断器
immersion liquiol 浸液
immersion magnifier 浸没放大镜
immersion method 液浸法;浸渍法
immersion needle 埋入式振动棒
immersion objective 油镜;浸渍物镜;浸液物镜
immersion objective lens 浸没物镜
immersion of the draw bar 引砖深度

immersion of thermometer 温度计浸入深度
immersion oil 浸(渍)油;浸没油;浸镜油
immersion oil method 油浸法
immersion operation 沉没作业
immersion optics 浸没光学
immersion pH 工业用浸入式酸碱计
immersion pipe 淹没管;浸入管;浸没管(子)
immersion plate 浸镀层
immersion plating 浸镀
immersion poker 埋入式振捣棒;埋入式振动棒
immersion proof 抗浸水性;防浸器
immersion pump 沉没式泵
immersion pyrometer 浸入式高温计;埋入式高温计
immersion reflectometer 浸没反射计
immersion reflow soldering 浸流焊接
immersion refractometer 油浸折射计;浸式折光仪;浸没折射计
immersion refractometry 浸没式折射分析法
immersion resistance 耐浸渍性
immersion scale 载重标尺;水标尺;吃水标(尺)
immersion scan(ning) method 水中扫描法
immersion scan(ning) technique 水中扫描术
immersion series 油浸镜头组
immersion service 浸没条件
immersion sintering 浸没烧结
immersion stability test 水浸安定度试验(评价沥青混合物耐水性的试验方法)
immersion suit 潜水服
immersion surface 浸水面
immersion system 浸没装置;浸没系统
immersion tank 浸柜
immersion technique 油浸法
immersion test 浸渍(法)试验;浸水试验;浸没式试验;探漏试验
immersion thermocouple 埋浸式热电偶
immersion thermometer 浸入式温度计;埋入式温度计
immersion-tray test 浸盘试验
immersion treatment 浸渍处理
immersion trench 沉管基槽
immersion type 埋入式
immersion-type rotary water meter 浸没式旋转水表
immersion-type thermocouple 埋入式热电偶
immersion-type vibration method 插振法
immersion-type vibrator 插入式振动器;棒式内部振捣器;插入式振捣器;振动钎杆;振捣杆;振捣棒;浸入式振捣器
immersion-type water heater 浸没式热水器
immersion unconfined compression test 浸水无侧限压缩试验
immersion vessel 浸渍罐;浸渍容器
immersion vibration 浸没式振动;浸没式振动
immersion vibrator with handle 有柄浸入式振动器;有柄插入式振动器;有柄浸入式振捣器;有柄插入式振捣器
immersion water heating 浸入式加热
immersion weighing 浸入称重;水中称重
immigrant 移入者;移民;移来的
immigrantin 移居者
immigrant remittance 侨汇;移民汇款
immigrants admitted 被批准的移民数
immigration 移民;移居(入境);外来(的)移民
immigration and naturalization office 移民局
immigration law 移民法
immigration office 移民局;入境签证处
immigration officer 移民局官员
immigration procedure 入境手续
imminence 濒危
imminent danger 急迫危险;迫近的危险
immiscibility 不溶混性;不可混合性;不混溶性
immiscibility gap 不相混间隙;不混间隙
immiscible 不溶混的;不可混的;不混溶的;不互溶的
immiscible butane 不溶混丁烷
immiscible displacement 不混溶驱替
immiscible droplets 非混和的液滴
immiscible gas injection 注不混溶气
immiscible liquid 不溶混液体;不混溶液体;不互溶液体
immiscible liquid-liquid pair 不溶混液—液对
immiscible matter bearing inclusion 不混溶包裹体
immiscible metal 难溶熔金属
immiscible phase 不溶混相;不混溶相

immiscible range 不混溶区;不混合区
immiscible region 不混溶区
immiscible solution 不溶混溶液;不混溶溶液
immiscible solvent 不溶混溶剂;不混溶溶剂;不混合溶媒剂
immiscible with water 和水不溶混的
immiserizing growth 贫困化增长
immission 排入
immittance 导抗
immixable 不能混合的
immmutability 不易性
immobile 固定的;不动的
immobile block 定滑车
immobile dislocation 不可移动的位错
immobile element 不活动元素
immobile estate 不动产
immobile ice 固定冰
immobile liquid 固定液
immobile phosphorus 固定性磷
immobile property 不动产
immobile temperature 稳定温度
immobilisation factor 固定因子
immobility 不流动性;不动性
immobilization 制动术;固定术;固定化;降低流动性;缩小迁移率;变流动资本为固定资本
immobilization of wastes 废物固定
immobilization test 制动试验
immobilization with adhesive tape 胶布固定
immobilize 停航
immobilized activated sludge cell 固定化活性污泥槽
immobilized ammonia-oxidizing bacteria 固定化氨氧化菌
immobilized bentonite 固定化膨润土
immobilized biological activated carbon 固定化生物活性炭
immobilized biomass 固定化生物体
immobilized enzyme 固定化酶
immobilized flocculant 固定化絮凝剂
immobilized microbe pellets 固定化微生物小球
immobilized microboe 固定化微生物
immobilized microorganism technology 固定化微生物技术
immobilized nitrobacteria 固定化硝化菌
immobilized photocatalyst 固定化光催化剂
immobilized powdered biological activated carbon 固定化粉末生物活性炭
immobilized pure culture microorganism 固定化纯培养微生物
immobilized spindle 备用轴
immobilized tannin 固定化丹宁
immobilized technology 固定化技术
immobilized titanium dioxide catalyst 固定化二氧化钛催化剂
immobilize waste 固定废物
immovability 固定性;不动性;不可移动性
immovable 固定的;不可移动的;不动的;不动产
immovable bed 定床
immovable bed flow model testing 定床水流模型试验
immovable bed model 定床模型
immovable bed model testing 定床模型试验
immovable bed river engineering model 定床河工模型
immovable bed river model 定床河道模型
immovable bed sedimentation model 定床淤积模型
immovable estate 不动(地)产
immovable fitting 紧密配合
immovable fixture 固定装备
immovable grate 固定格栅
immovable landed property 不动的地产
immovable pleasure boat 不击舟
immovable property 不动(资)产
immovable restraint 固定约束
immovable roll 固定滚筒;不可移动的滚筒
immovable support 固定刀架
immoveable end fixity 端部固定性
immoveable restraint 不变的约束;不变的限制
immune from taxation 免除捐税
immune to all pleas 不容申辩
immunity 高耐蚀状态;抗扰性;抗扰度;免疫;豁免(权);不敏感性
immunity from execution of judgement 执行判决的豁免
immunity from judgement 判决的豁免
immunity from prosecution 免于公诉
immunity from taxation 免税
immunity from trial 不受审判
immunization plate 防虫处理板
immunizing 耐腐蚀处理;使钝化;高耐蚀处理;免疲热处理;不敏感处理
immunizing dose 50 半数免疫剂量
immunoprecipitation 免测沉淀法
immunosuppression 免投抑制
immuration 墙里工事;砌墙墙洞
immure 围在墙内;镶在墙上
immutable 不变的
impact 效果;撞击;碰撞;突加;输入;打击;冲突;冲力;冲击
impact abrasion 撞擦伤;冲击磨损
impact acceleration 撞击加速度;碰撞加速;冲击加速度
impact acceleration process 碰撞加速过程
impact accelerometer 碰撞加速计;冲击加速计;冲击加速度表
impact action 冲击作用
impact action bit 冲击钻头
impact adhesive 接触型黏结剂;接触型胶结剂;触压胶黏剂;冲击黏合剂
impact agitation 碰撞激励
impact air pressure 气流冲击压力
impact allowance 容许冲击量;容许冲击负荷值;冲击容许量;冲击留量
impact allowance load 允许冲击荷载;容许冲击荷载
impact analysis 影响分析
impact angle 冲击角
impact anvil 冲砧
impact area 命中面;碰撞面积
impact assessment 效果鉴定
impact assessment of groundwater environment 地下水环境影响评价
impact assessment of soil environment 土壤环境影响评价
impact-atomizing 冲碎;冲击粉碎
impact-atomizing pile driver 冲击—雾化打桩机
impact attenuation device 减冲设施
impact bar 冲击试杆
impact beam 防撞梁;防冲梁;反击板
impact bending 冲击弯曲
impact bending strength 冲击弯曲强度;冲击挠曲强度
impact bend(ing) test 冲弯试验;冲击弯曲试验
impact blaster 冲击起爆器
impact block 耐磨板;冲击块
impact blow 动载冲击;冲击作用
impact break 冲击断裂
impact breaker 轧碎机;碰撞式破碎机;双转子锤式破碎机;锤式破碎机;冲击式碎石机;冲击式破碎机;冲击式粉碎机;反击式破碎机
impact breaking 反击破碎
impact breccia 撞击角砾岩
impact brittleness 撞击脆性;碰撞脆性;冲击脆性
impact broadening 碰撞致宽
impact buddle 冲击式淘汰盘
impact burner 冲击燃烧器;射焰燃烧器
impact cast 撞击铸型;撞击模;冲击模
impact cast-in-situ pile 现场打桩
impact center 冲击中心
impact clam 冲击夹钳;冲击式抓斗
impact cleaning 喷丸清理;抛丸(喷射)清理;抛丸冲击清理
impact coal mill 冲击型碎煤机;冲击式碎煤机
impact coefficient 冲量系数;冲击系数
impact comminution 冲击粉碎
impact compacter 冲击式夯具;冲击式压实工具
impact compaction 夯击压实;冲击压密
impact compactor 冲击式压实工具;冲击式压实器;冲击(式)夯(具)
impact compression test 冲击压缩试验
impact compressive failure 冲击挤压法破碎
impact crash switch 撞击应急开关
impact crater 撞击坑
impact crusher 冲击式碎石机;冲击式破碎机;冲击式粉碎机;反击式破碎机
impact crusher dryer 反击式破碎烘干机
impact crushing 碰撞破碎;冲击破碎;冲击粉碎;

反击破碎
impact crushing drying 冲击破碎兼烘干
impact current 冲击电流
impact cutter 冲击式切割器
impact cutting 冲击式切割
impact damage 冲击破坏
impact damper 减振器；缓冲器
impact deceleration 冲击减速
impact detector 碰撞检测器
impact detonation 冲击爆破
impact device 冲击装置；击打装置
impact disintegration 碰撞蜕变；碰裂反应
impact disintegration reaction 碰撞蜕变反应
impact drill 冲(击)钻；冲击式钻机
impact-driven cased pile 冲击夯打的套管桩
impact-driven cast in-situ pile 冲击夯打就地浇注桩；打入现浇桩
impact-driven pile 打入桩
impact-driven prestressed concrete pile 冲击夯打预应力混凝土桩
impact-driven reinforced-concrete pile 冲击夯打钢筋混凝土桩
impact drive rotary head 撞击式驱动旋转喷洒头
impact-driven shell pile 冲击打入的壳套桩
impact-driven steel pile 冲击打入的钢桩
impact-driven wood(en) pile 冲击打入的木桩
impact driver 敲击式螺丝刀；撞击式螺丝刀
impact ductility 撞击韧性；冲击延性；冲击韧性
impact duration 击打持续时间
impact dust collector 冲击式除尘器
impact dynamics 冲击动力学
impact earthquake 冲击地震
impacted 嵌塞的
impacted medium 击打介质
impact effect 碰撞作用；冲击作用；冲击效应；冲击效果
impact elasticity 冲击韧；冲击弹性
impact elasticity tester 冲击弹性试验机
impact electric(al) drill 冲击电钻
impact electric(al) wrench 冲击电动扳手；电动套筒扳手
impact endurance test 抗冲击试验；耐冲(击)试验；冲击疲劳试验
impact energy 撞击能；碰撞能；冲击能(量)；冲击动能
impact energy transition curve 冲击能转变曲线
impact environment 碰撞环境
impact epoch 冲击期
impacter 冲击器；冲击机；无砧座锻锤
impacter forging hammer 无砧座锻锤
impacter for retractor 牵开器压紧器
impact erosion 冲击碎屑；冲击侵蚀；冲击破裂；冲击腐蚀
impact evaluation 影响评价
impact excavation 冲击开挖
impact excitation 碰撞激励；碰撞激发；冲击励磁；冲击激励
impact excited transmitter 脉冲激励发射机
impact exciter 强行励磁机；冲击励磁机
impact extruding press 冲挤压力机；冲挤压(压力)机
impact extrusion 冲挤；冲击挤压
impact extrusion press 冲击挤压机
impact face 撞击面
impact factor 动静载弯拉应力比；冲击因素；冲击因数；冲击系数；冲击荷载系数；动力系数
impact failure 碰撞破损；冲击损坏
impact fatigue 冲击疲劳
impact fatigue test 冲击疲劳试验
impact fatigue testing machine 冲击疲劳试验机
impact fender 撞击性护舷
impact figure 撞击图像
impact finisher 锤碾式去麸机
impact flexibility 冲击柔性
impact flexural strength 冲击弯曲强度
impact fluorescence 碰撞荧光
impact force 撞击力；碰撞力；冲击力
impact force of train 列车冲击力
impact forging 模锻；冲击成型；冲锻
impact formula 冲击公式
impact fracture 冲击断裂；冲击断口
impact fuse 着发引信；碰炸导火索
impact fuze 碰炸引信
impact generator 冲击发电机；冲击电压发生器

impact glass 撞击玻璃
impact-grab boring machine 冲抓式钻孔机；冲击式钻孔机；冲抓(锥)成孔机
impact grinder 冲击研磨机；冲击式研磨机
impact grinding 冲击研磨(法)；冲击式粉磨；冲击破碎；冲击磨
impact grinding mill 舂粉器；冲击器；冲击研磨机；冲击式磨机
impact hammer 回弹仪
impact hammer crusher 反击锤式破碎机
impact hardness 冲击硬度
impact hardness test 冲击硬度试验
impact hardness tester 冲击硬度试验机
impact hardness testing machine 冲击硬度试验机
impact hypothesis 撞击假说
impact idler 防冲托辊
impact indentation 冲击刻痕
impacting block oil nozzle 冲击块油嘴
impacting gear 冲齿齿轮
impacting shaft 冲击轴
impact injury 冲击伤
impact insulation 冲击噪声隔离层；撞击隔声
impact insulation class 冲击声级
impaction 撞击；嵌塞；碰冲
impact ionization detector 碰撞电离探测器
impact ionization switch 碰撞电离开关
impaction loss 碰撞损失
impaction of vitreous 玻璃体嵌塞
impaction separator 惯性分离器
impact isolation 撞击隔声
impactite 碰撞岩
impact jaw crusher 颚式冲击破碎机
impact jet flow 冲击射流
impact lava 撞击熔岩
impact law 碰撞定律；冲击定理
impact-lightflash sensor 碰撞闪光传感器
impact line flowmeter 冲量式流量计
impact load(ing) 碰撞荷载；碰撞负载；突加负载；冲击荷载；冲击负载；冲击负荷；冲淡负荷
impact loading condition 冲击加荷状态
impact load of drift 漂流物撞击力
impact load of ship 船只撞击能力
impact load of ship or drift 船只或漂流物的撞击力
impact load stress 冲击荷载应力
impact loan 冲击性贷款
impact loss 碰撞损失；冲击(水头)损失
impact machine 锤击机；冲击(试验)机
impact matrix 碰撞矩阵
impact measurement 冲击测量
impact mechanism 触发机构
impact melt 撞击熔融物
impactment 碰撞
impact metamorphic rock 撞击变质岩
impact metamorphism 冲击变质
impact meter 冲击计
impact method 冲击法
impact microphone 撞击传声器
impact mill 竖井式磨粉机；锤击式磨煤机；冲击式研磨机；冲击式轧机；冲击式铣刀；冲击式碾压机；冲击锤磨机
impact mixer 冲击式混合机；冲击混合机
impact modified grade 冲击性改进级
impact modulator 对冲型元件；冲流调制器；冲击调制器
impact modulator amplifier 冲击调制放大器
impact momentum 碰撞量
impact multiplier 冲击乘数
impact noise 撞击噪声；撞击声；碰撞噪声；冲击噪声
impact noise rating 撞击隔声级；撞击噪声级；冲击噪声率
impact number 冲击值读数
impact of collision 纵向冲击
impact of freeway noise 高速公路的噪声影响
impact of information flow 信息流的影响
impact of overflows 溢流影响
impact of recoil 反冲
impactometer 撞击取样计；撞击计
impact on the quality of the water and the air 对水质和空气质量的影响
impactor 撞击取样器；撞击计；碰撞器；打桩机；冲击器；冲击机；冲锤；反击式破碎机
impactor crusher 冲击式破碎机
impactor extractor for trifin nails 三翼钉打拔器

impact oscillation graph 冲击振荡图
impact paper 压敏纸
impact parameter 碰撞参数；冲击系数；冲击参数
impact penetration test 冲击穿透试验
impact penetration tester 砂型皮下强度测定器；冲击刺入度测试仪
impact penetrometer 锤击贯入仪
impact piece 冲击部件
impact pile 冲击桩
impact pile driver 冲击式打桩机
impact pile driving 锤击打桩
impact piston 冲击活塞
impact pits 碰撞坑
impact plate 击碎凹板
impact plate engraving system 冲压制版系统
impact plate flowmeter 冲板式流量计
impact point 撞击点；课征点
impact polarization 碰撞极化
impact potential 冲击势位；冲击电压；冲击电位；冲击电势
impact press 冲击式压力机
impact pressing 冲击式加压
impact size pressure 滞点压力；碰撞压强；碰撞压力；速度头；动压强；动压力；冲击压重；冲击压力
impact pressure probe 全压测针；冲压测量管
impact pressure tube 皮托管；冲压力管
impact pressure velocity meter 冲压式流速仪；冲击式流速仪
impact probability 碰撞概率
impact probe 冲击探头
impact proof board (护舷的)防冲板
impact property 抗冲击性能；冲击性能；冲击特性
impact pulverizer 锤击式磨煤机；冲击式磨(碎)机；冲击式粉碎机
impact pulverizing 冲击磨细；冲击粉碎
impact radiation 碰撞辐射
impact radius 碰撞半径
impact rammer 冲击锤
impact rate 冲击速率
impact reactance 冲击电抗
impact recombination 碰撞复合
impact recombination coefficient 冲击复合系数
impact recorder 冲击记录器
impact reduction 冲击粉碎
impact register 冲击(自动)记录器
impact resilience 抗冲冲性；冲击回弹性
impact resilience tester 抗冲(击)试验机
impact resiliometer 抗冲击强度试验机
impact resistance 抗冲撞能力；冲击抗性；抗冲强度；抗冲击力；耐冲击性；冲击阻力；抗冲力
impact resistance test 抗冲击试验
impact resistance value 冲击值
impact resistant 抗冲击的
impact resistant plastics 耐冲击塑料
impact-resisting 抗冲击
impact retention 冲击保持率
impact roll 冲击滚轮
impact roller 振动式碾压机；振动式碾碾；冲击式压路机；冲击式碾压机
impact rotor crusher 反击式转子破碎机
impact rupture 冲击断口
impact sampler 冲击式采样器
impact screen 撞击筛；振动筛；冲击式振动筛；冲击筛
impact screwdriver 敲击式螺丝刀；撞击式螺丝刀
impact sensitivity 撞击敏感度
impact separator 固定电极静电选矿机
impact shear strength 冲击剪切强度
impact shell pile 冲击式陶管桩
impact shock 冲击震动
impact simulator 冲击模拟器
impact size reduction 冲击粉磨；反击破碎
impact slag 撞击熔渣
impact soil-compaction device 夯土设备；冲击式土壤捣实设备
impact sound 撞击声；碰撞声；冲击声
impact sound index 撞击声指数
impact sound insulation 撞击声隔绝；撞击声隔声；冲击声隔绝；冲击声隔声
impact sound insulation material 撞击声隔声材料
impact sound insulation reference contour 撞击声隔声参考曲线
impact sound intensity 撞击声强度
impact sound level 撞击声级

impact sound transmission 冲击声传播;撞击传声;冲击声传输
impact sound transmission level 撞击声透射级
impact spanner 冲击(式)扳手;气压扳手;冲击扳钳
impact specimen retest 冲击试样复试
impact speed 碰撞速率;冲击速度
impact spot welding 冲击点焊(法)
impact spray test 冲击喷射式渗透性试验
impact sprinkle 冲击式喷洒器
impact statement 影响报告
impact strength 抗冲击强度;耐撞性;碰撞强度;冲击强度
impact strength index 冲击强度指数;冲击强力指数
impact strength modifier 冲击强度改进剂
impact strength test 冲击强度试验
impact stress 冲击应力
impact striae 撞击痕
impact stroke 冲击行程;冲击冲程
impact structure 撞击构造
impact study (工程项目的)影响及后果研究;影响的研究
impact surface 撞击面
impact swaging 旋转模锻
impact switch 碰撞式开关
impact tamper 冲击式捣土机
impact temperature 滞止温度;碰撞温度;冲击温度
impact tensile test 冲击张力试验
impact tension 冲击拉伸
impact tension machine 冲击拉伸试验机
impact tension test 冲击拉伸试验
impact test certification 冲击试验证明
impact tester 冲击试验仪;冲击试验机
impact test(ing) 动载试验;冲击试验;碰撞试验
impact testing apparatus 冲击试验装置;冲击试验设备;冲击试验机
impact testing machine 冲击试验机
impact test piece 冲击试片
impact test property 冲击试验性能
impact test specimen 冲击试件
impact test temperature 冲击试验温度
impact theory 撞击理论
impact tool 冲击工具;夯
impact toughness 冲击韧性
impact toughness tester 冲击韧度试验仪;冲击韧性试验机
impact transmission 冲击噪声传播
impact transmitter 脉冲发射机;冲击发射机
impact tube 迎水面承压管;全压管;皮托管;冲压管;冲击(压力)管
impact type 冲击型
impact type energy dissipator 冲击式消能工
impact type flowmeter 冲量式流量计
impact type hammer crusher 冲击锤式破碎机
impact type insole stapling machine 冲击式钉内底机
impact type pneumatic tool 冲击类风动工具
impact type rotary cutter 冲击式回转切割器
impact type screen 拍振筛;振动筛;冲击式筛
impact value 冲击值
impact vane anemometer 冲击叶式风速计
impact velocity 撞击速度;碰撞速度;冲击速度
impact viscometer 冲击黏度计
impact viscosity 冲击黏性
impact wave 冲击波
impact wear 击打磨损
impact wear test 冲击磨损试验
impact weight 冲击荷重
impact wheel 冲击式水轮(机);冲击式涡轮
impact work meter 冲击测功仪
impact wreckage 碰撞破坏
impact wrench 冲头;冲击扳钳;机动扳手;套筒扳手;冲击(式套筒)扳手
impact zone 碰撞区;碰撞带;冲击区(域)
impages 门栏杆
impair 不成对的;奇数
impaired 被损坏的
impaired area 损坏面积
impaired development 损害的发育
impaired loci 损伤点
impairment 减损;缺损
impairment grade 劣化度
impairment of capital account 亏损账户
impairment of vision 视力损害
impalement 刺穿

impaler 插入物
impalpability 细微
impalpable flour 细磨材料
impalpable structure 极细粒构造;无定形结构
impalpable texture 极细粒结构;微粒状结构
impar 无对的
imparipinnate 奇数羽状的
imparipinnate leaf 奇数羽状复叶
impartial 公正的
impartiality 公正性
impartial taxation 公平课税
impassable 不能通行的;不可通行的
impassable barrier 不可逾越的障碍
impassable gulf 不可逾越的鸿沟
impassable road 不(能)通行(的)道路
impassable trench 不通行地沟
impasse (路的)尽头;独头巷道;死路;死胡同
impasse blind alley pass 尽头路
impaste 用浆糊刷
impasto 厚涂颜料
impatiency 闭阻
impedance 阻抗;全电阻;电(阻)抗
impedance adapter 阻抗适配器
impedance amplifier 阻抗放大器
impedance balance 阻抗平衡
impedance bond 阻抗联结器;阻抗结合;扼流变压器
impedance bridge 阻抗电桥;抗阻电桥
impedance change frequency test 电阻抗变换频率试验
impedance characteristic 阻抗特性
impedance chart 阻抗圆图
impedance checker 阻抗测试器
impedance circle 阻抗圆
impedance circle diagram 阻抗圆图
impedance coil 电抗线圈;电抗线圈;扼流圈
impedance compensator 阻抗补偿器
impedance component 阻抗分量;阻抗部件
impedance conversion 阻抗变换
impedance converter 阻抗匹配变换器
impedance correction network 阻抗校正网络
impedance corrector 阻抗校正器
impedance coupled amplifier 阻抗耦合放大器
impedance coupling 阻抗耦合
impedance curve 阻抗曲线
impedance diagram 阻抗图
impedance dial(l)ing 阻抗拨号
impedance drop 阻抗(电压)降
impedance-drop test 阻抗压降试验
impedance electrometer 阻抗静电计
impedance exploration 阻抗探索;阻抗法勘探
impedance factor 阻抗因数;阻抗系数
impedance frequency 阻抗频率
impedance function 阻抗函数
impedance inverter circuit 阻抗变换电路
impedance inverting network 阻抗倒置网络
impedance irregularities 阻抗不规则性
impedance kilovolt-amperes 阻抗千伏安培
impedance loss 阻抗损耗
impedance match(ing) 阻抗匹配;感应淬火
impedance matching coil 阻抗匹配线圈
impedance matching load box 阻抗匹配负载箱
impedance-matching transformer 阻抗匹配变压器
impedance matrix 阻抗矩阵
impedance measurement 阻抗测量
impedance measuring equipment 阻抗测量设备
impedance meter 阻抗仪;阻抗计
impedance method 阻抗法
impedance method calibration 阻抗法校准
impedance mismatch 阻抗失配
impedance mismatching 阻抗失配
impedance muffler 阻抗消声器
impedance normalization 阻抗归一化;阻抗标准化
impedance of listening 收听干扰
impedance ohm 阻抗欧姆
impedance parameter 阻抗参数
impedance part 阻抗部件
impedance plethysmogram 阻抗体积描记图
impedance plethysmograph 阻抗脉波计
impedance plethysmography 阻力体积描记法
impedance plotter 阻抗图示仪
impedance protection 阻抗保护装置
impedance protective system 阻抗保护系统
impedance ratio 阻抗比

impedance relay 阻抗继电器
impedance scale 阻抗标度
impedance screw 节流螺钉
impedance silencer 阻抗消声器
impedance sound absorber 阻挠式消声器;阻抗式消声器
impedance stability 阻抗稳定性
impedance stabilization method 阻抗稳定法
impedance standard 阻抗标准
impedance starter 阻抗起动器
impedance tapered line 阻抗渐变线
impedance tester 阻抗试验器
impedance time relay 阻抗延时阻抗继电器
impedance titration 阻抗滴定
impedance transducer 阻抗转换器
impedance transformer 阻抗变压器;阻抗变换器
impedance transformer network 阻抗变换网络
impedance transforming filter 阻抗变换滤波器
impedance transposition method 阻抗换置法
impedance triangle 阻抗三角形
impedance tuner 阻抗调谐器
impedance type analogy 阻抗类型比
impedance unbalance 阻抗失衡
impedance unbalance measuring set 阻抗失衡测定器;阻抗失衡测定装置
impedance variation method 阻抗变化法
impedance voltage 阻抗电压
impedance voltage drop 阻抗压降
impeded diffusion 弥散障碍
impeded discharge of cuttings 阻碍岩粉排出
impeded drainage 不良排水;排水不良
impede navigation 妨碍航行
impede the passage 阻碍航道
impediment 障碍物
impedimenta 辎重
impedimeter 阻抗计
impedimetric cell 阻抗滴定池
impedimetric titration 阻抗滴定
impedimetry 阻抗滴定法
impedimetry cell 阻抗滴定池
impedin 阻抑素
impedometer 阻抗计;波导阻抗测量仪
impedor 阻抗器;二端阻抗元件
impel 推进
impellent 推进器;推动力
impeller 旋转混合器;压缩器;(水泵的)转子;转轮;末级叶轮;推进器;风机叶轮
impeller agitator 叶轮搅拌器
impeller approach 叶轮入口
impeller assembly 叶轮组
impeller assisted thermosiphon 叶轮推动的热虹吸管
impeller bar (破碎机转子上的)冲击杆
impeller bearing 叶轮轴承
impeller bearing house 叶轮轴承壳
impeller blade 叶轮(叶)片;叶泵;桨叶;涡轮叶片
impeller-blower 叶轮式鼓风机
impeller breaker 叶轮(式)轧碎机;叶轮(式)破碎机;叶轮(式)粉碎机;反击式破碎机
impeller casing 叶轮箱;叶轮壳;抛砂机壳体
impeller centrifugal fan 叶轮离心式风扇
impeller channel 叶轮(流)道
impeller channel velocity 叶轮通道流速
impeller chimney 叶轮罩
impeller clearance 叶片间隙;叶轮间隙;导向叶片与工作叶片的间隙
impeller collector 风扇除尘器
impeller crusher 叶轮破碎机
impeller cup 叶轮润滑油杯
impeller cut 叶轮切割
impeller diameter 叶轮直径
impeller diameter reduction 车削叶轮直径
impeller-diffuser clearance (浮选机的)叶轮扩散器间隙
impeller disk 叶轮盘
impeller drive 叶轮驱动机构
impeller exciter 叶轮激动器
impeller eye 叶轮入口
impeller eye area 叶轮中孔面积
impeller finger 叶轮爪
impeller head 抛丸头;抛丸器;抛(砂)头
impeller hub 叶轮柄;叶轮轴;叶轮轮毂
impeller impact breaker 叶轮冲击式破碎机
impeller impact crusher 叶轮冲击式破碎机

impeller impact mill 叶轮冲击磨机;叶轮反击式粉磨机
impeller inducer 导流叶轮;导风叶轮
impeller inlet 叶轮进口
impeller inlet area 叶轮进口面积
impeller inlet guide vane 叶轮进口导向叶片
impeller labyrinth 叶轮迷宫密封
impeller locking screw 叶轮端螺钉
impeller mixer 叶轮混合机
impeller mounting ring 叶轮固定环
impeller nominal diameter 叶轮名义直径
impeller nut 叶轮螺母
impeller of dredge pump 泥泵叶轮
impeller of pump 泵叶轮
impeller outer diameter 叶轮外径
impeller output 叶轮排出量
impeller packer 叶轮包装机
impeller passage 叶轮流道
impeller plate 叶轮板
impeller pump 叶片泵;叶轮泵;转子泵;离心泵;泵轮
impeller ramming 抛实;抛砂造型
impeller screw 挤压螺旋输送机
impeller selection 叶轮选择
impeller sensor 叶轮传感器
impeller shaft 叶轮轴
impeller shaping 修锉叶轮
impeller side plate 转子侧板
impellers of pumps 水泵叶轮
impeller-sparger system 叶轮曝气器系统
impeller suction 叶轮吸水口
impeller tip speed 叶轮外缘速度
impeller turbine 叶轮式水轮机
impeller type chopper 叶轮式切碎机
impeller type crusher 叶轮式破碎机
impeller type flow transmitter 叶轮式流量传感器
impeller type log 叶轮式计程仪
impeller type pulverizer 叶轮式粉磨机
impeller type root chopper 叶轮式块根切碎机
impeller type turbine 叶轮式水轮机;桨(叶)式水轮机
impeller vane 叶轮叶片;转轮叶片;桨叶;水斗
impeller wheel 叶轮;涡轮
impeller with two side discs 闭式叶轮
impelling 推进
impelling force 推进力;推动力
impelling force on particles 颗粒上的推动力
impelling ratio 推进比
impellor wheel 推轮
impending cliff 悬崖
impending earthquake 即将来临的地震
impending earthquake prediction 临震预报
impending motion 临界传动
impending plastic flow 急端塑性流动
impending skidding 紧急制动滑行
impending slough 临界稠度;稠度;(喷射混凝土的)临界流淌稠度
impenetrability 不(可)渗透性;不可(贯)入性;不可穿入性;碍性
impenetrable 不能穿过的;不可贯入的
impenetrable by water 不透水的
impenetrable layer 不渗透层
imperative 绝对必要的;紧急;命令式
imperative duty 紧急任务
imperative instruction 执行指令
imperative macro instruction 执行型宏指令
imperative operation 命令性操作;强制性操作
imperative statement 无条件指令;动作语句
imperfect 不完善的;不完全的
imperfect adhesion 黏结不良
imperfect amber 不纯琥珀
imperfect arch 斜圆拱;平圆拱
imperfect bonding 黏结不良
imperfect cleavage 不完整解理;不完全解理【地】
imperfect combustion 不完全(的)燃烧
imperfect contact 不良接触;不良接点;不良触点
imperfect contact detector 不良接触探测器
imperfect covariance 不完全协方差
imperfect crystal 真实晶体;不完整晶体
imperfect crystal formation 不(完)全晶型
imperfect dielectric 非理想(电)介质;不完善电介体
imperfect dislocation 不(完)全位错
imperfect earth 接地不良;不良接地
imperfect elasticity 不完全弹性

imperfect equilibrium 不完全平衡
imperfect evidence 不完全证据
imperfect field 不完全域
imperfect flower 不完全花;不具备花
imperfect flower-form 不具备花形的
imperfect fluid 不完全流体
imperfect frame 静不定结构;不稳定框架;不完整构架;不完全框架
imperfect framing 不稳定框架
imperfect gas 实际气体;非理想气体
imperfect gift 不完善赠与
imperfect grain 不完整粒;不饱满的籽粒
imperfect hedge 不完全对冲;不完全避险
imperfect impression 不符合质量印页
imperfect information 不完全信息
imperfect inversion 不完全反转
imperfection 不完整性;不完善度;不完全(性);不完美性;残次品;不合格折贴
imperfection in workmanship 操作缺陷
imperfection of elasticity 不完全弹性
imperfections in markets 市场不完全性
imperfect isomorphism 不完全类质同像
imperfectly drained 排水不良的
imperfectly elastic 非完全弹性的
imperfectly elastic analysis 非完全弹性分析
imperfectly elastic medium 非完全弹性介质
imperfect magnetic circuit 不全磁路
imperfect maintenance 维修不良
imperfect manufacture 加工缺点;不良加工(木材)
imperfect market 不完全竞争的市场
imperfect mixing 不完全混合
imperfect mixing flow 非理想混合流动
imperfect neutrality 不完全中立
imperfect obligation 不完全的义务;不受法律约束的责任
imperfect ownership 不完全(的)所有权
imperfect polarization 不完全极化
imperfect real right 不完全的物权
imperfect ream 不足令纸
imperfect scattering 不完全散射
imperfect sparking 不完全点火
imperfect specification of model 模型设定不完善
imperfect stage 不完全期;不完全阶段
imperfect tape 缺陷带
imperfect thread 不完全螺纹
imperfect trust 不完全信托
imperfect tuning 不全调谐
imperfect vacuum 部分真空
imperfect well 不完善井
imperforate 无气孔的
imperforation 无孔的状态;闭锁畸形
imperial 特号斜纹布
Imperial Ancestral Temple 太庙
Imperial Apartments (在斯匹拉托的)戴克里先宫;帝国公寓
imperial basilica 皇宫教堂
imperial cathedral 皇家大教堂;帝国大教堂
Imperial Chemical Industries Ltd. 帝国化学工业公司(英国)
Imperial College of Science and Technology 皇家科技学院(英国)
imperial cupola 摩尔式圆顶;洋葱式圆顶;大圆顶
imperial dome 葱头形圆顶
imperial gallon 英制加仑;英国标准加仑(1加仑=4.546升);伽仑
imperial green 翡翠绿;巴黎绿
imperial kiln 关窑(器)
Imperial Kiln's sagger 官窑匣体
Imperial Mahogany 浅红色花岗岩(产于美国明尼苏达州)
Imperial Mansion 帝国大厦(美国)
imperial ottoman 粗横棱柔软织物
imperial palace 宫城;内务府;皇家花园;皇宫;王宫
imperial paper 特大绘图纸
imperial purple 深紫红色
imperial quart 英夸(脱)
imperial red 皇家红;铁红;朱红色
Imperial Rome style 罗马王朝式;罗马帝国(建筑)风格
imperial roof 葱头式屋顶;大屋顶
imperial scale 特大规模;英制
imperial sizing 英制尺寸
imperial slate 板岩瓦片
imperial tape 厚棉带

imperial tomb 陵
imperial wire 英制标准电线规格
imperial yellow 鹅黄;杏黄
Imperical bushel 英国法定蒲式耳
imperil 危及
imperious layer 不透层
impermanence[impermanency]暂时性
impermeability 抗渗(透)性;气密性;非渗透性;防水性;不透(水)性;不透过性;不渗透性;不侵透性
impermeability coefficient 抗渗系数
impermeability factor 抗渗系数;径流系数;不透水系数;不渗透系数
impermeability of concrete 混凝土的密实性
impermeability of joints 密缝性
impermeability pressure 抗渗压力
impermeability test 抗渗(性)试验;抗渗透性试验;不渗透试验
impermeability test for concrete 混凝土的抗渗试验
impermeability test(ing) 不透水性试验;抗渗(性)试验;抗渗透性试验;不渗透试验
impermeability to gas 不透气性
impermeability to rain 防雨性
impermeability to sound 隔声性
impermeability to water 防水性
impermeability to wear 抗渗水性
impermeable 非渗透性的;不透水(带)的;不(可)渗透的;不被渗透的
impermeable barrier 隔水岩层;隔挡层;非渗透性遮挡;防渗层;不透水(隔)层;不透(水)层;不渗透挡层;不渗透屏障;不可渗层
impermeable base 不透水基底;不渗地基
impermeable bed 隔水层;不透水(岩)层;不渗透层
impermeable bed enclosing 非渗透性层包围
impermeable boundary 不渗透边界
impermeable break 不渗透夹层
impermeable breakwater 不透水防波堤;不透浪防波堤
impermeable clay 防渗黏土;不渗透黏土
impermeable coating 不透水涂层
impermeable concrete 防渗混凝土
impermeable confining bed 不透水隔层;不渗封闭(地)层
impermeable course 防透层;不透水层
impermeable diaphragm 不渗透挡板;不渗透的膜层;阻水层
impermeable diaphragm of the continuous wall type 连续墙型不渗透(水)层
impermeable dike[dyke] 不透水堤
impermeable film 不渗透薄膜
impermeable formation 不透水岩层
impermeable foundation 不渗水基础;防渗地基
impermeable graphite 不透性石墨;不渗石墨
impermeable groin 不透水排水坝;不透水丁坝;不透水防波堤;不透水折流坝;不透水挑水坝
impermeable ground 不透水围岩;不透水土;不透水地面
impermeable ground layer 不透水土层
impermeable groyne 不透水排水坝;不透水丁坝;不透水折流坝;不透水挑水坝;不透水防波堤
impermeable layer 防渗层;不透水膜;不透水膜
impermeable liner 不透水涂层;不透水衬里
impermeable material 隔绝材料;防渗材料;不透水材料
impermeable pore 不渗透孔隙
impermeable reservoir 非渗透性热储
impermeable rock 隔水岩体;不透水岩;不渗透岩(石)
impermeable rock bed 不透水岩层
impermeable rock stratum 不透水岩层
impermeable sand paper 不透水砂纸
impermeable seam 不透水接缝;不透水层
impermeable soil 不透水土(壤);不渗透土
impermeable stratum 隔水层;不透水地层
impermeable substance 不透水物质
impermeable to air 气密的;不透气
impermeable to chemicals 抗化学品的
impermeable to gas 气密的;不透气
impermeable to gasolene 不渗汽油
impermeable to oil 不透油的
impermeable to petrol 不渗汽油
impermeable to sound 隔声性
impermeable to surface water 地面水不渗透的
impermeable to vapo(u)r 不透蒸汽的
impermeable to water 不透水的;不透水带

impermeable wall 隔水墙；防渗墙；不透水墙
impermeable zone 不透水区；不透水带
impermephane 透明保护敷料；透明保护涂料
impersonal 非个人的
impersonal account 无人名账号；非人名账户
impersonal assets 不记名资产；非个人资产；非对人资产
impersonal astrolabe 超人差等高仪
impersonal corporation 非个人的公司
impersonal entity 非个人实体；法人单位
impersonal forces 非人力
impersonality 非人格性
impersonalized market mechanism 非人格化市场机制
impersonal ledger 非人名总账
impersonal micrometer 无人差测微器；超人差测微器；超人差测微计
impersonal organizational structure 非人际的组织机构
impersonal security 非个人的担保
impersonal tax 间接税；非人身税
Impervion 液态和膏状材料
impervious 抗渗的；水透不过去的；不透水(带)的；不渗水性；不能透过的；不能渗透的；不可渗透的
impervious area 防渗面积
impervious area runoff 防渗水区径流
impervious asphalt(ic) concrete layer 不渗透的沥青混凝土层；防水沥青混凝土层
impervious barrier 防渗层
impervious base 不渗水基层；防渗水基础
impervious bed 不透水河床；隔水层；防渗水河床；不透水(岩)层
impervious belt 不透水带
impervious blanket 不透水(面)层；防渗铺盖；防渗盖层；不透水铺盖；不透水覆盖层
impervious bottom 不透水基层；不透水底板；防渗水基底
impervious boundary 防渗水边界；不透水边界
impervious boundary in three-sides of an aquifer 三面隔水边界
impervious boundary of apical layer 顶板隔水边界
impervious boundary of bottom layer 底板隔水边界
impervious boundary of fan-shaped intersection 扇形相交的隔水边界
impervious break 不透水夹层；不渗透夹层
impervious carbon 不透水炭；不渗透性炭
impervious clay 不透水黏土
impervious clay blanket 防渗水黏土覆盖层；不透水黏土铺盖；不透水黏土层；防渗水黏土铺盖
impervious clay layer 不透水黏土层
impervious clay stratum 不透水黏土层
impervious compound 不透水化合物
impervious concrete diaphragm 不透水混凝土面层；混凝土防渗墙
impervious condition 不透水条件
impervious confining bed 不透水隔层
impervious construction 不透水结构
impervious core 不透水芯墙；防渗芯墙
impervious core wall 不透水芯墙
impervious course 防渗层；防渗水层
impervious curtain 防渗帷幕；不透水帷幕
impervious cut-off 不透水齿墙；防渗齿墙；不透水截水墙
impervious dam 防渗水坝；不透水坝
impervious deposit 不透水岩层
impervious diaphragm 不渗透水(面)层；不渗透挡板；不透透隔膜；防渗水芯墙
impervious diaphragm wall 不渗透的阻挡墙；防渗墙
impervious dike 不透水堤；不透堤
impervious element 不渗透构件；防渗设施
impervious facing 防渗面层；不透水盖面
impervious factor 不透水系数
impervious fault 不透水断层
impervious foundation 防渗水地基；不透水地基
impervious ground 不透水围岩
impervious layer 防渗(水)层；不透水层
impervious lens 不透水透镜体
impervious liner 不透水衬体；不透水衬层
impervious lining 不透水性衬里；不透水衬砌；不透水衬里
impervious machine 密封式电机；密封式电动机
impervious material 防渗(水)材料；不透水物质；不透水材料
impervious membrane 防渗帷幕；防渗层；防渗薄膜；不透水膜
imperviousness 密封性；不透水(性)；不透过性；不通透性
imperviousness coefficient 防渗水系数；不透水系数
imperviousness factor 防渗水因素
imperviousness of concrete 混凝土密实性
imperviousness of joints 密缝性
imperviousness sound 隔声性
imperviousness to rain 防雨性
imperviousness to water 防水性
impervious paint 防渗油漆
impervious pavement 不透水铺面
impervious river bed 不透水河床；不渗水河床
impervious rock 防渗水岩石；不透水岩(石)；不渗透岩(石)
impervious rolled fill 防渗水的碾压填土；不透水的碾压填土
impervious sandy clay core 砂质黏土防渗芯墙
impervious section 防渗截面
impervious sheathed wire 不渗透铠装线
impervious skin 不透水膜；不透水层
impervious soil 黏性土(壤)；渗透性不良土壤；不透性土；不透水土(壤)；不可渗透的土壤
impervious stratum 不透水地层；不透水地层；不透气层；不渗透地层
impervious stream bed 不渗透河床
impervious structure 不透水结构
impervious subsoil 不渗透底土
impervious surface 抗渗面；非渗透性地表；防渗水地面；不透水面(层)
impervious to air 不透空气的
impervious to chemicals 抗化学品的
impervious to gasolene 不渗汽油的
impervious to heat 不透热的
impervious to moisture 防潮的
impervious to oil 不透油的
impervious to petrol 不渗汽油的
impervious to sound 隔声的
impervious to surface water 地面水不渗透的
impervious to vapo(u)r 不透蒸汽的
impervious to vermin 防虫的
impervious wall 防渗墙；不透水墙
impervious wall of dam 不透水的坝芯墙
impervious zone 防渗水区；不透水区；不透水带
impervious-zone runoff 不透水区径流
impet dam (小型的)开口沉箱
impetuous torrent 湍流水流；湍急水流
impetus 动力；动冲量
impinge 冲击
impingement 撞击；空气采样法；紧密接触；气雾捕集器；碰撞；冲撞；冲击；冲打
impingement aerator 撞击式曝气器；冲击式曝气器；冲击曝气器
impingement angle 碰撞角；冲击角
impingement area 撞击范围；冲击面积
impingement attack 浸蚀；滴蚀；冲击侵蚀；冲击腐蚀
impingement baffle 缓冲挡板
impingement black 烟道炭黑；接触法炭黑；燃气炭黑；槽法炭黑
impingement collector 撞击式除尘器
impingement corrosion 冲击腐蚀
impingement dust collector 碰撞除尘器；冲击式除尘器
impingement particle 碰撞粒子
impingement plate 冲击板
impingement plate scrubber 板式冲击除尘器
impingement point 碰撞点
impingement rate 碰撞率；冲击率
impingement scrubber 冲击涤气器
impingement separator 撞击式气液分离器；撞击式分类器；冲击(式)分离器
impingement-type unit 冲击式装置
impingement washing scrubber 冲击水浴除尘器
impinge plate process 冲击板法
impinger 撞击式检尘器；撞击(滤尘)器；撞击集尘器；空气采样器；接触器；搅拌汽车；碰撞(取样)器；冲击式采样吸收管；冲击(滤尘)器；尘埃测定器；采尘器
impinging corrosion 水锤腐蚀；冲击腐蚀
impinging entraining scrubber 自激喷雾洗涤器
impinging hole injector 冲击射流式喷头
impinging injector 流束互击式喷射器
impinging jet 冲击性喷枪；冲采射流
impinging radiation 冲击辐射
impinging stream reactor 撞击流反应器
implacement 指定位置
implant 注入物；插入
implantation 移植法；注入；植入；埋入法
implantation of silicone 硅胶植入术
implantation period 着床期
implantation site 置入位置
implanted core 空心门芯
implanted electrode 埋藏电极
implant test 试验注入；插销试验
implastic 不可塑的
implement 用具；执行；工具；机具；航海用具；器械；器具；实施；施行
implementation 执行(过程)；供给器具；履行；实现
implementation and supervision 执行和监督
implementation classes 实际应用课程
implementation courses 实际应用课程
implementation details of worktrain running and operation 工程车运输实施细则
implementation language 工具语言
implementation management 建设管理
implementation module 实现性模块
implementation of contract 履行合同；合同的履行
implementation of normative document 标准文件的贯彻
implementation of strategies 策略的实施
implementation of the reference language 参考语言的实现
implementation period of new technique 新技术采用期
implementation period of new technology 新技术采用期
implementation phase 实施阶段
implementation plan 实施计划
implementation procedure 实现过程；实施程序
implementation project 治理工程
implementation regulations 实施细则
Implementation Rules for Law of the People's Republic of China on Prevention and Control of Water Pollution 中华人民共和国水污染控制实施细则；中华人民共和国水污染防治实施细则
implementation schedule 进度表
implementation scheduling 建设进度
implementation steps 实施步骤
implementation technique 实现技术
implement by stages 分期实施
implement carrier 通用机架
implementer 制订人
implementer name 设备名(称)
implementing a contract 履行合同
implement lowering control 农具下降速度调节
implementor 实现者
implementor name 实现者名；设备名(称)
implement porter 自动底盘；通用机架
implement shed 农具棚；工具棚；工具仓库
implements of labo(u)r 劳动工具
implements of production 生产工具
implexed 内皱
implicant 隐含数
implication 显微共生；蕴含；隐含(式)；牵连
implication function 隐含操作
implication graph 隐含图
implication imply 隐含
implication relation 蕴含关系；隐含关系
implication routine 隐含程序
implication table 隐含表
implication texture 同期显微共生结构
implicit 蕴含
implicit address 隐式地址
implicit attribute 隐式属性
implicit backward finite difference 隐函数反向差分法
implicit cash cost 内在付现成本
implicit computation 隐式计算；隐函数法计算
implicit contract 默认契约
implicit cost 隐含成本；内在成本；内含成本；潜在成本
implicit decision 隐判定
implicit declaration 隐式说明
implicit definability 隐含可定义性
implicit definition 隐定义

implicit difference equation 隐式差分方程
implicit difference method 隐式差分方法
implicit difference scheme 隐式差分格式
implicit differentiation 隐微分法;隐式微分(法)
implicit enumeration 隐数法;隐含枚举法;间接列举法
implicit estimation technique 隐式估计法
implicit expression 隐式
implicit finite difference 隐式有限差分
implicit formula 隐式公式
implicit function 隐(含)函数
implicit function equation 隐函数方程
implicit function generation 隐函数变换
implicit function theorem 隐函数定理
implicit identification 隐式标识;隐式辨识
implicit integration 隐式积分
implicit integration formula 隐式积分公式
implicit interest 内在利息;内含利息
implicit interest charges 内含利息的开支
implicit interest revenue 内含利息收入
implicit iterative method 隐式迭代法;隐含迭代法
implicitly defined array 隐定义阵列
implicit mathematical model 隐式数学模型
implicit method 隐式方法;隐含法
implicit mode 隐方式
implicit opening 隐(式)打开
implicit parameter 隐参数
implicit policy 内含政策
implicit precedence 固有优先性
implicit price deflator 综合价格换算系数
implicit procedure 隐过程
implicit procedure parameter 隐过程参数
implicit quadratic equation 隐二次方程
implicit qualification 隐式限定
implicit reference 隐式引用
implicit rental value 隐含租金价值
implicit scheme 隐(式)格式
implicit smoothing 隐式平滑
implicit solution 隐函数解;隐含解
implicit storage management 隐式存储管理
implicit structure 隐结构
implicit synchronizing signal 隐同步信号
implicit system 隐式系统
implicit tax 默示税收
implicit time-difference method 隐函时间差分法
implicit type 隐式类型
implicit type association 隐式类型结合
implicit type conversion 隐式类型转换
implicit variable 隐变量
implicity 隐含性
implied acceptance of risk 默示承担风险
implied addressing 蕴含选址;隐(式)访问;隐式编址;隐含寻址
implied agreement 默示协定;默认协议
implied and induced output 隐输出和诱导输出
implied association 隐式结合
implied authority 默许代理权;默示的授权
implied cash cost 内在付现成本
implied circulation 隐循环
implied condition 隐含的条件;默许条件;默认条件
implied consent 隐含同意;默许;默示同意
implied contract 隐含的契约;默示合同;默认契约
implied decimal point 隐含小数点
implied deed 默示契约;默示合同
implied goodwill 隐藏商誉
implied growth rate equation 内含增长率公式
implied indemnification 法律所默认的赔偿
implied linear constraint 隐式线性约束
implied listing 默许上市委托
implied objection 默示异议
implied partnership 默认合伙
implied power 默示权力
implied promise 默示承诺
implied rate of interest 内含利率
implied ratification 默示批准
implied recognition 默示承认;默示承诺
implied representation 默示代理
implied reservation 默示权益保留
implied term 默示条件;默示条款;默认条件
implied trust 内在信托;默许信托;默示信托;法定信托
implied undertaking 默示义务
implied waiver 默示放弃
implied warrant 默示担保

implied warranties in sale by sample 样品买卖的默认条件
implied warranty 内在保证;默认保证
implode 内向爆炸;内爆炸
implosion 内向爆炸;内裂;内爆(波);从外向内的压力作用;爆聚;暴缩
implosion drilling 内爆凿岩;内爆压碎钻进
implosion test 刚性试验
implosion wave 内爆波
impluvium 院内蓄水池(古罗马);房(顶)采光井;住宅内室存积雨水的方形储[贮]水池(古罗马);天井
impolarizable electrode 去极化电极
impolder 海边围垦;海滨围垦地(英国)
imponderability 无重量
imponderable 不能称的
imponderable fluid 无重量流体
imponderableness 无重量
imporosity 无孔性;无孔隙;非多孔性;不透气性
imporous 无孔的;无孔隙的
import 进口【贸】;输入品,输入(指进口物品);导入
importable merchandise 可进口的商品
import advanced technology 吸取先进技术
import agent 进口代理商;进口代理
import airway bill of lading 空运进口货物提单
importance 重要地位
importance function 重要性函数
importance of a flare 耀斑级别
importance sampling 重要性抽样;重点抽样
importance value 重要值
importance weighting 非等量加权;按重要性大小加权
import and export corporation 进出口公司
import and export duties 进出口税
import and export licence 进出口许可证
import and export license system 进出口许可制(度)
important 重要的
important agricultural base 重要农业基地
important factor 重要因素;重要性系数
important key 重要关键
important meteorite 重要陨石
important resonance 重要共振
important spare-parts program(me) 重要备件计划
important structure 重要结构
import ban 进口限额
import bill 进口汇票
import bounty 输入奖励金
import cargo 进口货(物)
import charge 进口费用
import collateral 进口担保品
import collection 进口托收
import commission 进口代理佣金;进口代办行
import commission agent 进口代理商
import commission house 进口代理商
import commitment 进口承诺
import-competing goods 进口竞争货物
import content 进口商品内容
import control 进口管制
import credit 进口信用证
import credit account 进口收入账户
import credit insurance 输入信用保险
import curtailment 削减进口;缩减进口
import customs clearance 进口通关
import debit account 进口支出账户
import declaration 进口申报单;进口报关单
import department 进口部
import deposit 进口押金;进口存款;进口保证金
import deposit scheme 进口押金制;进口存款制
import document 进口单证
import document arrival notice 进口单据到达通知书
import dumping 输入倾销
import duty 进口(关)税
import duty clause 进口关税条款
import duty memo 进口税缴纳证;海关进口税缴纳证
import duty memo of customs 海关进口税缴税单
imported and exported products 进出口货物
imported commercial goods 进口商品
imported commodity 进口商品
imported empty 空箱进口
imported equipment 引进设备;进口设备
imported labo(u)r 输入劳动力

imported materials 进口材料
imported materials and parts 进口料件
imported price 进口价格
imported product 进口产品
imported raw material 进口原料
imported style 外来形式
imported transshipment cargo 进口转口货
imported transshipment goods 进口转口货
imported water 输入水
imported weed 带进来的杂草;非乡土性杂草
import employment 输入就业
import entitlement account 进口头寸账户
import entry 进口报关手续;进口报关
importer 进口人;进口(贸易)商
importer quota 进口商配额
importer's currency 进口国货币
importer's entry of goods 进口商报关单
importers statement and guarantee bond 进口商声明及保证书
import exchange 进口外汇;进口汇兑
import-export commodity inspection 进出口商品检验
import-export industry 进出口业
import-export license system 进出口许可证制(度)
import-export link system 输出入连锁制;进出口连锁制
import-export offset agreement 输出入相抵协定
import-export operation 进出口作业
import-export operations right 进出口经营权
import-export ports 进出口口岸
import-export quota system 进出口配额制
import financing 进口资金融通;进口融资
import for expanding export 以进养出
import freight and insurance 进口运费及保险费
import freight and insurance account 进口运费用保险费账户
import gold point 进口黄金点
import goods 进口货(物)
import house 进行行;进口商行
import import link system 输入入连锁制
import import offset agreement 输入入相抵协定
importing market 进口市场
importing unemployment 输入失业
importing wholesaler 进口批发商
import labour power 输入劳动力
import letter of credit 进口信用证;进口货信用证
import levy 进口捐
import licence system 进口许可证制(度)
Import Licencing Board 进口许可委员会
import license 进口许可证
import license system 进口许可证制(度)
import licensing bank 进口签证银行
Import Licensing Code 进口签证规约
import licensing system 进口许可证制(度);进口签证许可制
import list 海关进口货物分类表
import loan repayment 偿还进口借款
import manifest 进口载货清单;进口舱单
import mark-up 进口加价
import merchant 进口商;进口贸易商
import of advanced technology 引进先进技术
import of complete machines 整机进口
import of technique 技术进口
import of technology 技术输入
import of volume index 进口物量指数
import order 进口订货单;进口订单
import permit 准进口证;进口许可证;进口(批)准单;进口批件
import price 进口价格
import procedure 进口程序
import profit indices 进口盈利指数
import quantum 进口量
import quarantine 进口检疫
import quota 限额输入;进口限额;输入限额;进口配额
import quotas system 进口配额制
import rate 进口货运费率
import relief 进口救济
import requirement 进口需求量
import restriction 进口限制;进口管制;输入限制
imports 进口商品;进口货
import shares 进口份额
imports of good and material services 货物和物

质性服务的进口
import state 进口国
import structure 输入结构
import subsidy 进口补贴
import substitute and export substitute 进口替代与出口替代
import substituting economy 进口替代型经济
import substitution 进口替代
import-substitution strategy 进口替代战略
import sundry charges 进口杂费
import sundry charge account 进口杂费账户
import surcharge 进口附加税
import surplus 逆差;入超
import surtaxes 进口附加税
import target 输入目标制
import tariff 进口税则;进口税(率)
import tax 进口税
import tender 进口招标
import trade 进口贸易
import traffic 进口运输
import value 进口值;输入价值
import volume 进口量
import water 引进水源
impose 征收
impose(a duty) on 征收
imposed budget 自上而下强制预算
imposed deformation 强加的变形
imposed fabric 强置性组构
imposed fill 加填土
imposed floor load 楼面活荷载
imposed load 施加荷载;作用荷载;强加荷载;外加荷载
impose fines on someone 罚款罚金
impose punishment in accordance with the law 依法惩处
impose sanction 进行制裁
impose tax 征税
imposing discharge fee 征收排污税
imposing stone 装版台
imposite (一种不溶于松节油的)天然沥青
imposition 课税
imposition of surcharge 收取附加费
impositor 幻灯放映机
impossibility theorem 不可能性定理
impossible 不可能的
impossible condition 不能履行的条件
impossible event 不可能事件
impossible stereoscopic effect 零立体法
impost 拱脚;拱墩;拱端托;拱端石;进口税;起拱点;窗框的竖框
impost block 拱端托块;柱顶托块;柱头拱墩;拱墩块体
impost capital 拱墩柱头;拱墩帽
imposter 冒名顶替者
impost hinge 拱趾铰;拱墩柱头铰(链);拱端铰(链)
impost joint 起拱接点;起拱接缝;拱墩柱头接缝
impost level 起拱标高
impost mo(u)lding 起拱处线脚;拱墩柱头造型
impost pressure 拱墩压力
impost section 拱趾断面;拱墩断面
impost springer 拱脚石;起拱石
impotable water 非饮用水;不宜饮用水
impotence 无力
impound 蓄水;扣押
impoundage 蓄水量;蓄水
impoundage of surface water 地表水聚集
impound body 静止水体
impounded aggregate 集水面积
impounded area 集水面积;蓄水面积;围堤区域;围堤面积
impounded basin 蓄水池;有闸港池;水库
impounded body 蓄水池;储[贮]水池;静止水体
impounded dam 蓄水坝
impounded dock 通闸港坞;恒定高水位的港坞;有闸港池;灌水式船坞;港内码头;港内船坞;封闭式港池;闭合式港口
impounded harbo(u)r 封闭式港坞
impounded harbo(u)r basin 封闭式蓄水港池
impounded level 蓄水位
impounded port 封闭式港口
impounded reservoir 蓄水库;蓄水池
impounded surface water 聚积的地表水;地面积(滞)水

impounded water 储[贮]水;蓄水;储[贮]积水;水库
impounded water level 蓄水面标高
impounding 蓄水池
impounding basin 溢液池
impounding clear water and discharging sediment-laden water 蓄清排浑
impounding culvert 灌水涵管;灌水涵洞
impounding dam 蓄水坝;拦水坝;尾矿池
impounding dock 潮水船坞;封闭式港池
impounding elevation 蓄水高程
impounding pond 蓄水池;水库
impounding reservoir 蓄水库;蓄水池;水库
impounding scheme 蓄水方案;筑坝方案
impounding water level 蓄水高程
impoundment 拦蓄;蓄水(量);集水
impoundment stage 蓄水期
impoundment water 蓄水
impound water 静止水体;水库
impoverished area 贫穷地区
impoverished fuel 贫化燃料
impoverished rubber 失去弹性橡胶
impoverished soil 贫脊土;贫脊地;贫瘠的土壤
impoverishment 贫脊化;贫化
impoverishment of anomaly 异常贫化
impracticability 难驾驭;不实用性;不切合实际性
impracticable 不切合实际的;不能实行的
imprecise information risk 不精确信息风险
imprecise interruption 不精确中断
imprecise terms 不明确的条文
imprecision 精密度不够;不精密度
impreg 浸渍木;浸渍材;树脂浸渍木材;树脂浸注木材
impregnability 浸透性能;浸渍能力
impregnant 浸渍液;浸渍剂
impregnate 浸渍;浸透;浸润;浸染
impregnated 浸渍的
impregnated and compressed wood 浸渗树脂压缩木
impregnated anodizing 浸渍阳极处理
impregnated apparatus 浸渍设备
impregnated asbestors web 浸石棉布
impregnated bit 孕镶钻头
impregnated cable 浸渍电缆
impregnated carbon 浸渍炭棒
impregnated casing bit 孕镶套管钻头
impregnated casing shoe 孕镶套管靴
impregnated cathode 浸渍式阴极
impregnated cement 浸渍水泥;饱和水泥
impregnated cloth 浸透树胶的布;浸渍布
impregnated coil 浸渍线圈
impregnated coke 镁焦
impregnated concrete 浸渍混凝土
impregnated core bit 孕镶岩芯钻头
impregnated diamond bit 孕镶金钢石钻头
impregnated diamond reaming shell 孕镶金钢石扩孔器
impregnated dust 浸渍粉剂
impregnated expanded cork brick 经浸渍的膨胀软木砖
impregnated fabric 浸渍织物
impregnated felt 浸渍(毛)毡
impregnated flax felt 沥青浸渍亚麻油毡;沥青油毡
impregnated foam 浸渍泡沫
impregnated glass colth 浸渍玻璃布
impregnated graphite materials 浸渍石墨材料
impregnated kraft paper 浸渍牛皮纸
impregnated material 浸渍材料
impregnated matrixes 孕镶胎体
impregnated on activated carbon 在活性炭上的浸渍;活性炭浸渍剂
impregnated paper 绝缘浸渍纸;浸渍(树脂)纸;浸染纸;合成纸
impregnated polyurethane 浸渍聚氨酯
impregnated polyurethane foam 浸渍聚氨酯泡沫
impregnated resin 浸渍树脂
impregnated rock 浸渍岩石;浸染岩;饱和岩(石)
impregnated seal 浸渍封接
impregnated sleeper 油浸枕木
impregnated structure 浸染状构造
impregnated system 浸渍系统
impregnated-tape metal-arc welding 浸焊条金属弧焊
impregnated timber 浸渍木

impregnated varnish 浸渍清漆
impregnated with oil 浸油
impregnated wood 防腐剂浸渍木材;浸渍防腐木材;浸灌防腐(剂的)木材
impregnate tie 浸油轨枕
impregnate with creosote 用杂酚油灌注;杂酚油浸渍
impregnating 浸染
impregnating agent 浸渍剂
impregnating apparatus 浸渍设备
impregnating asphalt 浸渍沥青
impregnating bath 浸渍浴;浸渍池;浸渍槽
impregnating composition 浸渍液成分
impregnating compound 防腐剂;浸渍剂;浸渍化合物;浸注剂;渗入剂
impregnating depth 浸渍深度
impregnating fluid 浸注液;防腐液
impregnating installation 浸渍设施
impregnating insulating varnish 浸渍绝缘清漆
impregnating liquid 浸注液
impregnating mangle 浸轧机;浸染机
impregnating mix(ture) 浸渍混合料
impregnating monomer 浸渍单体
impregnating oil 浸渍防腐油;浸透油
impregnating plant 浸渍设施;浸渍厂
impregnating resin 浸渍树脂
impregnating scumble 浸染暗彩;防腐涂料
impregnating solution 浸渍溶液
impregnating temperature 浸渍温度
impregnating test 浸渍试验
impregnating varnish 浸渍清漆
impregnating vessel 浸渍罐;浸漆罐;浸胶罐
impregnating wood 浸灌防腐木材
impregnation 注入;浸渍;浸透;浸渗;化学防护剂;防腐剂浸渍
impregnation bath 浸渍槽;浸渍池;浸渍浴;浸泡池;浸泡槽
impregnation catalyst 浸渍催化剂;浸染催化剂
impregnation compound 浸染化合物
impregnation degree 浸渍程度;浸渍度
impregnation deposit 浸染矿床
impregnation installation 浸渍装置;浸渍设施
impregnation method 浸渍法
impregnation of timber 木材浸渍(处理)
impregnation of zone 区段灌注
impregnation on dried basis 全干浸渍
impregnation on partially dried basis 含水浸渍
impregnation period 浸渍期;浸渍时期
impregnation plant 浸渍设施;浸渍厂
impregnation speed 浸渍速度
impregnation structure 浸染状构造
impregnation tank 浸渍槽
impregnation technology 浸渍工艺
impregnation test 浸渍试验
impregnation time 浸渍时间
impregnation varnish 浸渍清漆
impregnation with creosote 木材防腐油浸制
impregnator 浸渍器;浸渍机;浸胶体
imprescriptible right 不受时效限制的权利
impress 盖章;盖印;刻记号;记号;强行征用;压痕;压印
impressafiner 单动盘磨机
impressed-charge density 外加电荷密度
impressed current 强制电流;外加电流
impressed current anode 外加电流阳极
impressed current cathodic protection 外加电流阴极保护
impressed-current protection system 外加电流保护系统
impressed decoration 滚压花纹
impressed distortion 强迫歪曲;强迫扭曲
impressed electric(al) current 外加电流
impressed electromotive force 外加电动势
impressed frequency 外加频率
impressed pressure 外加压力
impressed stoneware 印纹炻器
impressed stress 外加应力
impressed vapo(u)r pressure 外加蒸气压力
impressed variation 强制变异
impressed voltage 外加电压
impressing 压印;刻痕;盖印
impression 型腔;印图;印刷(版);印数;印模;印记;印痕;压印;压迹;压花;压痕;感应;模腔;痕迹;凹陷

impression block 印痕测定器;压印模板
impression box 打印器
impression control 压印密度控制
impression-cup 印模盘
impression cylinder 压印滚筒;压花(滚)筒
impression diameter 压痕直径
impression die 模锻模
impression eccentric 压印偏心调节法
impressionism 印象派
impression material 模槽材料
impression method 印模法
impression of depth 景深效果
impression of space 立体效应;立体效果
impression packer 印痕收集器
impression per hour 每小时印数
impression plaster 印牙齿用烧石膏
impression roller 加压辊
impression system 压印系统
impression taking 取模
impression-tray 印模盘
imprest 预借;预付款;垫款
imprest account 定额储备金账户;预付款项
imprest bank account 定额备用金银行账户
imprest cash 定额备用现金
imprest cash account 定额备用金账户
imprest fund 定额备用金
imprest fund system 定额备用金制度
imprest level 定额备用金数量
imprest method 定额备用法
imprest replenishment 定额备用金的补充
imprimitive matrix 非本原矩阵
imprint 印记;印痕;盖章;痕迹
imprinter 印刷器;压印器;刻印机;戳印机;标记盖印机
imprinter fraction 可约分数;假分数
imprint impression 出版说明
imprinting 印迹作用
imprint method 印痕法
imprisoned lake 堰塞湖
imprisoned radiation 束缚辐射
improbability 非概率性
improductive forest 不成材林
improper 非正常;不正当的;不适当的;不合适的;不合格的
improper accumulation 非法收益留存
improper character 禁用字符;非正式符号;非正常特征
improper code 禁码
improper command 非法命令
improper command check 非法命令检查;不适当指令检查
improper conic 退化二次曲线
improper convex function 非正常凸函数
improper eigenfunction 异常本征函数;非正常特征函数
improper fraction 可约分数;假分数
improper Gaussian process 反常高斯过程
improper integral 异常积分;广义积分;奇异积分;非正常积分;反常积分
improper maximum 非正常最大
improper minimum 非正常最小
improper operation 误操作
improper orthogonal matrix 非正常正交矩阵
improper packing 包装不适当;包装不良
improper random variable 非正常随机变量;非合适随机变量
improper rotation 异常旋转;非正常旋转;反射旋转
improper stowage 装载不良;码垛不良
improper symbol 非正常符号
improper ventilation 通风不良
improvability 可改良性
improve 改良;改进
improved 改进的
improved channel 整治的水道;改良航道;经整治航道;经整治的水道;经整治河槽;经整治航道;疏浚航道;疏浚航槽
improved cost efficiency 改善成本—效益
improved damping least square method 改进的阻尼最小二乘法
improved design 改进设计
improved discharge 已改良的排放
improved drying oil 改良干性油
improve defense works 加固工事
improve desertic soils 改良荒漠土

improved estimate 矫正估计
improved estimation on variance 矫正方差估计
improved extraction procedure 改良提取程序
improved Fischer-Tropsch method 改良费-托法
improved grading stabilization 改善级配加固法
improved land 基本设施已建成地段;改良地段;有设施用地
improved light-weight aggregate 改良轻集料;改良轻骨料
improved lime process 改良石灰法
improved mass selection 改良混合选择
improved method of satellite orbit 卫星轨道改进法
improved nail 改良钉
improved oxidation ditch 改进型氧化沟
improved paper 改良纸
improved pasture 改良牧地
improved recovery 改进可采储量
improved road 改善道路;铺面路
improved rubber 改性橡胶;改良橡胶
improved runway surface 改进跑道道面
improved soil texture 改良土壤结构
improved soil tilth 改良土壤耕性
improved solid wood 经处理的实木;改良实心木
improved subgrade 改善路基
improved tannin 改性丹宁
improved technology 改进过的技术;改变技术
improved type 改进的型号
improved Venturi flume 改良的文丘里槽;改进的文丘里槽
improved wood 改性木材;胶合板;积层木板;改良木材
improve economic performance 提高效益
improve markedly 明显好转
improvement 整治;改善;改良(投资);改进;好转
improvement and extension 改进及扩充
improvement and extension program(me) 改良与扩充计划
improvement area 改建区;改建地区
improvement authorized 核准修缮改进
improvement by soil removing 排土改良
improvement cutting 改进伐
improvement district 改善地区;改建区(域)
improvement district bonds 改建区域债券
improvement dredging 整治性疏浚
improvement expenses 改建费;固定资产改进费用
improvement factor 噪声改善系数;提高因素
improvement felling 改进伐;抚育伐
improvement in process 在建改良工程
improvement line 道路扩建用地线;道路改建路界线;房屋改建线;房屋建筑线
improvement method of soil 土壤改良方法
improvement mortgage bond 建设抵押债券
improvement of agrarian structure 土壤改良
improvement of business operation 经营的改善
improvement of commodity quality 提高商品质量
improvement of contaminated soil 污染土壤的改良
improvement of existing lines 既有线改建与改造;现有线路改善
improvement of impact sound 撞击声改善量
improvement of orbit 轨道改进
improvement of photograph 照片质量改善
improvement of plaster bond 抹灰黏结力的提高
improvement of polder 洼地改造
improvement of reclaimed soil 填筑土的改良
improvement of river 河流治理
improvement of riverbed 河道整治;河床治理;河床改善
improvement of riverbeds in mountain areas 山区河床改造
improvement of river channel 河道整治
improvement of saline-alkali soil 盐碱土改良
improvement of soft ground 软基加固
improvement of soil 土壤改良
improvement of stream 河流治导;河流整治
improvement of stream bed 河床治理
improvement of terms 条件改革
improvement of tools 工具改革
improvement of water quality 水质改善
improvement planting 改进栽植
improvement programs 改良方案
improvement ratio 建造设施比率
improvement research 加工技术研究

improvement tax 都市设施改良税
improvement thinning 抚育性疏伐
improvement threshold 信噪比改善阈值;改良限度;改进限度;门限改善
improvement time 矫正时间
improvement to plant and equipment 厂房和设备改良
improvement trade 加工贸易
improvement trade for import 加工进口贸易
improvement value 改良价值
improvement works 整治设施;整治工程;改善工程;改良工程;技术改造工程
improve quality 改善质量
improver 增进剂;改正剂;改性剂;改进者;改良剂;促进剂
improve soil structure 增进土壤作用
improve the operational efficiency through modern technology and training 通过新技术和培训来改善管理效能
improve the quality of products 提高产品质量
improve the quality of varieties 提高品种质量
improve the utility of equipment 提高设备的效能
improving agent 改正剂
improving dirt road 泥路强固
improving environmental sanitation 改善环境卫生
improving food conversion ratio 提高饲料转化率
improving furnace 精炼炉
improving investment environment 改善投资环境
improvised darkroom 临时暗室
improvised facility 临时设施
improvised housing 临时住房
improvised installation 临时设施
improvised makeshift 临时凑合的办法
improvised plant 临时设施
impsonite 英普逊焦沥青;焦油沥青;脆沥青岩
impson stone 焦油沥青
impules regenerator 脉冲再生器
impulsator 脉冲发生器;脉冲调制器
impulse 脉冲;冲量;冲激;冲击;冲动
impulse accepting relay 脉冲接收继电器
impulse accumulator 脉冲累加装置
impulse action 脉冲作用;冲动作用
impulse ammunition 抛缆筒火药
impulse amplitude 脉冲振幅;脉冲幅度
impulse analyzer[analyser] 脉冲分析仪
impulse approximation 脉冲近似值;冲量近似
impulse arc welding 脉冲电弧焊(接)
impulse at kiln head 回风
impulse attenuation 脉冲衰减
impulse blade 冲击式叶片;冲动式叶片
impulse blaster 脉冲引爆器
impulse blocking tube 脉冲阻塞管
impulse blow 动载冲击
impulse breaker 脉冲开关
impulse broadening 脉冲展宽
impulse buoy 脉冲浮标
impulse burner 脉冲烧嘴
impulse buy 不在事先计划内的购置
impulse cam 脉冲凸轮
impulse cascade 冲动式叶栅
impulse change 脉动变化
impulse charge 推力炸药
impulse circuit 脉冲线路;脉冲电路
impulse circuit breaker 脉冲断路器
impulse clock 脉冲(时)钟;冲击钟
impulse code 脉冲码;脉冲电码
impulse code modulation 脉(冲编)码调制
impulse coding 脉冲编码
impulse compaction 脉冲压实;冲击成型
impulse condensing turbine 冲击式凝气透平
impulse condition test 脉冲状态试验
impulse conduction 冲动传导
impulse contact 脉冲接点;短时闭合触点
impulse control 脉冲控制
impulse controller 脉冲调节器;脉冲控制器
impulse correction 脉冲校正
impulse corrector 脉冲校正器
impulse counter 脉冲计数器
impulse crowding 脉冲拥挤
impulse current 脉冲电流;冲击电流
impulse current automatic butt welding machine 脉冲电流自动对焊机
impulse current generator 脉冲电流发生器;冲击电流发生器

impulse current meter 脉冲电流测量仪
impulse current relay 脉冲电流继电器
impulse current seam welding machine 脉冲电流缝焊机
impulse current semiautomatic butt welding machine 脉冲电流半自动对焊机
impulse current shunt 冲击分流器
impulse current system 脉冲电流系统
impulse cutting machine 脉冲剪切机
impulse cylinder 脉冲缸
impulse delay 脉冲延迟
impulse delay time 脉冲延迟时间
impulse density 脉冲密度
impulse discharging voltage 冲击放电电压
impulse distance measuring instrument 脉冲测距仪;脉冲测距器
impulse driven clock 脉冲驱动钟
impulse drying 脉冲干燥
impulsed spot welding 脉冲点焊
impulse duct engine 脉动式(空气)喷气发动机
impulse duration 脉冲宽度;脉冲持续时间
impulse duration system 脉冲时间系统
impulse electromagnetic automatic welding machine 电磁脉冲自动焊机
impulse element 脉冲元件
impulse equation 脉冲方程式
impulse excitation 脉动激励;脉冲激励;脉冲激发;冲击激励
impulse excited circuit 冲击激励电路
impulse excited oscillator 脉冲激励振荡器;冲击激励振荡器
impulse exciter 脉冲激励器;冲击激励器
impulse face 冲面;冲击面
impulse fan 推力风机;冲力式通风机
impulse firing 脉冲烧成;脉冲煅烧
impulse flashover voltage 脉冲闪络电压
impulse focused time-of-flight mass spectrometer 脉冲聚焦飞行时间质谱仪
impulse force 冲力;冲击力
impulse forced response 脉冲响应
impulse form 脉冲形状;脉冲形式
impulse forming tube 脉冲形成管
impulse frequency 脉冲频率
impulse frequency reponse 脉冲频率特性
impulse frequency system 脉冲频率制
impulse frequency telemetering 脉冲频率遥测
impulse front 脉冲面
impulse function 脉冲函数;冲击函数
impulse function generator 脉冲函数发生器
impulse generator 脉冲振荡器;脉冲发生器;脉冲发电机;射击电动协调装置;冲击电压发生器
impulse governed oscillator 脉冲控制振荡器
impulse hammer 冲击锤
impulse heat sealer 脉冲瞬间热封机
impulse homonuclear decoupling method 脉冲同核去耦法
impulse hunting 脉冲振荡;脉冲摆动
impulse hydrolysis 脉冲水解
impulse impeller 推进叶轮;冲力叶轮
impulse inertia 脉冲惯性;冲击惯性
impulse input 脉冲输入
impulse interference 脉冲干扰
impulse jammer 脉冲干扰机
impulse laser amplifier 脉冲激光放大器
impulse-length ratio 冲击长度比
impulse level 脉冲电平
impulse load(ing) 脉冲加载;冲击加载;冲击荷载
impulse machine 脉冲发送器;脉冲发送机
impulse magnet 脉冲电磁铁;脉冲磁铁
impulse maneuver 冲击式操纵
impulse measurement 脉冲测量
impulse measurement facility 脉冲测量设备
impulse measuring device 脉冲测量仪
impulse mechanical strength 冲击机械强度
impulse meter 脉冲计(数器);脉冲积算器
impulse method 脉冲法
impulse method calibration 冲量法校准
impulse method electric(al) finding 脉冲法电波探测
impulse microphone 脉冲传声器
impulse modulated radar 脉冲调制雷达
impulse modulated telemetering 脉冲调制遥测
impulse modulation 脉冲调制

impulse modulator tube 脉冲调制管
impulse momentum equation 冲量动量方程
impulse momentum method 冲量法
impulse momentum relationship 脉冲动量关系
impulse momentum theory 冲量理论
impulse motor 脉冲马达;脉冲电动机
impulse movement core 脉冲驱动的心
impulse noise 脉冲噪声;冲激噪声
impulse noise analysis 脉冲噪声分析
impulse noise blanking 脉冲噪声消除
impulse noise frequency 脉冲噪声频率
impulse noise generator 脉冲噪声发生器
impulse noise inverter 脉冲噪声倒置器
impulse noise level 脉冲噪声电平
impulse noise limiter 脉冲噪声限制器
impulse noise measurement 脉冲噪声测量
impulse noise processing 脉冲噪声处理
impulse noise signal 脉冲噪声信号
impulse noise susceptibility 脉冲噪声敏感性
impulse nozzle 冲式喷嘴
impulse of jet 射流冲击
impulse of motion 冲动
impulse onset 脉冲初动
impulse oscillator 脉冲振荡器
impulse oscilloscope 脉冲示波器
impulse oscilograph 脉冲示波器
impulse over-voltage 冲击过电压
impulse pallet 圆盘钉;推力钻
impulse peak 脉冲峰值
impulse peak power 脉冲峰值功率
impulse phase-locked loop 脉冲锁相环
impulse pin 冲击销
impulse plasma welding machine 脉冲等离子焊机
impulse polarography 脉冲极谱
impulse potential gate 脉冲电位门
impulse power 脉冲功率
impulse precision sound level meter 脉冲精密声级计
impulse preselection 脉冲预选
impulse pressure 冲击压力
impulse protection level 冲击保护水平
impulse pump 脉冲泵;冲击(式)泵
impulser 脉冲发送器;脉冲传感器
impulse radar transmitter 脉冲雷达发射机
impulse radiation 脉冲辐射
impulse ratio 脉冲系数;脉冲比;冲击系数
impulse-reactance 冲击电抗
impulse-reaction turbine 冲击反击式水轮机;冲动—反动式汽轮机
impulse recorder 脉冲(自动)记录器;脉冲(分频)计数器;脉冲定标器
impulse recording paper 脉冲记录纸
impulse rectifier 脉冲整流器
impulse recurrence frequency 脉冲重复频率
impulse recurrence rate 脉冲重复率
impulse recurrent timing control 脉冲再现计时控制
impulse reflection process 脉冲反射法
impulse reflection 脉冲反射
impulse regeneration 脉冲再生
impulse register 脉冲寄存器;脉冲计数器
impulse regulator 脉冲调节器
impulse relay 脉冲继电器
impulse repeater 脉冲转发器;脉冲重发器;脉冲中继器
impulse repeating 脉冲重发
impulse repetition 脉冲重复
impulse repetition rate 脉冲重复率
impulse reporter 脉冲记录器
impulse response 脉动反应;脉冲应答;脉冲响应;脉冲(频率)特性;冲击过渡函数;冲击响应
impulse response analyzer 脉冲特性分析仪
impulse response characteristic 脉冲响应特性曲线
impulse response function 脉冲响应函数;脉冲反应函数
impulse response matrix 脉冲响应矩阵
impulse response model 脉冲响应模
impulse response technique 脉冲响应技术
impulse scaler 脉冲(分频)计数器;脉冲定标器
impulse sealer 高频熔接机;脉冲热封机
impulse sealing 脉冲热合;脉冲焊接;脉冲封口
impulse section blade 冲动式叶型叶片
impulse seismic device 震源装置;激震装置
impulse seismogram 脉冲地震记录

impulse sender 脉冲发送器
impulse sending machine 脉冲发送机
impulse separator 脉冲分离器
impulse sequence 脉冲序列
impulse set 冲击式机组
impulse shaper 脉冲形成器
impulse signal 脉冲信号;脉冲系统
impulse signal generator 脉冲信号发生器
impulse signaling 脉冲通信
impulse solenoid 脉冲螺线管;脉冲电磁线圈;脉冲电磁铁
impulse sound level meter 脉冲声级计
impulse source 脉冲源;冲击源
impulse sparkover 脉冲闪络
impulse sparkovervoltage 冲击闪络电压;冲击放电电压
impulse speed 脉冲速度
impulses per minute 每分钟脉冲数
impulse spot welder 脉冲点焊机
impulse spreading circuit 脉冲展宽电路
impulse spring 脉动簧;脉动弹簧
impulse spring break 脉动簧断开
impulse springmake 脉冲簧片闭合
impulse starter 脉动起动机;脉冲起动器;冲击起动机
impulse starting 脉冲起动
impulse steam trap 脉冲式蒸汽疏水器
impulse steam turbine 冲力式蒸汽轮机;冲击式汽轮机
impulse stepless gear box 脉冲无级变速器
impulse stepping motor 脉冲步进电动机
impulse strength 冲击强度
impulse stretcher 脉冲伸张器
impulse stroke 工作行程;脉冲行程
impulse summation 脉冲相加
impulse switch 脉冲开关;冲击式断路器
impulse system 脉冲系统
impulse tachometer 脉动转速计;脉冲转数计;脉冲式转速计;冲击式转速计
impulse target diagram 脉冲速度界限曲线图;脉冲目标图
impulse technique 脉冲技术
impulse telemeter 脉冲遥测计
impulse test(ing) 脉冲(电压)试验;冲击试验
impulse testing apparatus 冲击试验设备
impulse testing machine 脉冲试验机;脉冲测试机
impulse time delay 脉冲时延
impulse time division system 脉冲多路通信制;时间分隔脉冲多路通信制
impulse time margin 脉冲时间裕度
impulse timer 脉冲时间继电器;脉冲计时器;脉冲定时器
impulse track circuit 脉冲轨道电路
impulse train 脉冲系列;脉冲群;脉冲链
impulse train pause 脉冲序列间歇
impulse transfer 脉冲变换
impulse transformer 脉冲变压器
impulse transmission 脉冲发射;脉冲传输
impulse transmitter 脉冲发送器;脉冲发射机
impulse transmitter tube 脉冲发射管
impulse transmitting relay 脉冲发送继电器
impulse transmitting tube 脉冲发射管
impulse trap 脉冲式疏水器;脉动式排水阻气阀;冲力气水阀
impulse turbine 脉冲透平;冲力式轮机;冲击式涡轮机;冲击式透平;冲击式汽轮机;冲动式涡轮机;冲动式透平;冲动式水轮机
impulse turbine with velocity stages 速度级冲动式汽轮机
impulse type 脉冲型;脉冲式;冲击式
impulse type bag filter 脉冲袋式除尘器
impulse type load(ing) 脉冲型荷载
impulse type multiplier 脉冲乘法器
impulse type output amplifier 脉冲输出放大器
impulse type relay 脉冲式继电器
impulse type stage 冲动级
impulse type steam turbine 冲动式汽轮机
impulse type telemeter 脉冲式遥测装置;脉冲式遥测仪;脉冲式遥测计
impulse type telemetering system 脉冲式遥测系统
impulse type voltage regulator 脉冲式调压器;脉冲式电压调整器
impulse unit 冲击式机组
impulse valve 脉冲阀

impulse ventilator 冲力式通风机
impulse voltage 脉冲电压;冲击电压
impulse voltage characteristics 脉冲电压特性
impulse voltage divider 冲击分压器
impulse voltage generator 脉冲电压发生器;冲击电压发生器
impulse voltage oscilloscope 脉冲电压示波器
impulse voltage test 脉冲电压试验;冲击电压试验
impulse volume ratio 冲量容积比
impulse water turbine 水戽式水轮机;冲击式水轮机
impulse water wheel 冲击式水轮;冲击式水车
impulse wave 脉冲波;冲击波
impulse wave form 脉冲波形
impulse wave generator 脉冲波振荡器
impulse weight ratio 冲量重量比
impulse welding 脉冲焊接;脉冲焊
impulse wheel 冲击(水)轮;冲击式涡轮机;冲击式水轮机;冲斗式水轮;冲动式叶轮;冲动式水轮
impulse wheel meter 脉冲轮计
impulse wheel turbine 冲击式水轮机
impulse width 脉宽;脉冲宽度
impulse width modulation 脉宽调制
impulse withstand voltage 冲击耐压
impulsing 振荡冲击;脉冲激励;发送脉冲;发生脉冲
impulsing oscilloscope 脉冲示波器
impulsing power source 脉冲电源
impulsing relay 脉冲继电器
impulsion 冲力;冲挤;冲动
impulsion fan 冲击式通风机;脉冲式通风机;冲击式通风机
impulsive 冲击的;冲动的
impulsive acceleration 脉冲加速
impulsive action 冲动作为
impulsive bandwidth 脉冲带宽
impulsive burst 脉冲暴
impulsive concretescope 脉冲式混凝土检测示波仪
impulsive control problem 冲击控制问题
impulsive current 脉冲电流
impulsive discharge 脉冲放电;冲击放电
impulsive force 脉冲力;冲力;冲击(力)
impulsive force of mudflow 泥石流冲击力
impulsive function 脉冲函数
impulsive hard phase 脉冲急速相
impulsive height 冲起高度
impulsive impedance 脉冲阻抗
impulsive load(ing) 脉冲荷载;动荷载;突加荷载;冲击荷载
impulsive measurement 脉冲测量
impulsive moment 冲量矩
impulsive noise 脉冲(噪)声;脉冲杂波;脉冲干扰;碰撞噪声;冲击噪声
impulsive noise signal 脉冲杂波信号
impulsive onset 脉冲初动
impulsive response 脉冲响应
impulsive rotatory motion 脉冲旋转运动
impulsive sound 脉冲声;冲击声
impulsive sound equation 脉冲声方程
impulsive source 脉冲源
impulsive test 脉冲试验
impulsive testing of concrete beams 混凝土梁的冲击试验
impulsive torque 冲击扭矩
impulsive variation 脉动变化
impulsive warming 爆发性增温;爆发性加温
impulsiving 脉冲激振
impunctate 无细孔
impunity 免责;免罚
impure 含杂质的
impure amber 不纯琥珀
impure clay shale 不纯黏土页岩
impure code 非纯代码
impure public goods 非纯粹的公共货物
impure sandstone 不纯砂岩
impure water 不洁水;不纯水
impurities in water 水中杂质
impurity 沾染;杂质;夹杂物;混杂物;掺杂质;不纯
impurity absorption 杂质吸收
impurity activation 杂质激活
impurity activation energy 杂质激活能
impurity activity 杂质活性
impurity and greasy dirt 杂物及油污
impurity band 杂质能带
impurity-band conduction 杂质带传导
impurity-band conduction theory 杂质带传导理论

impurity-band mobility 杂质带迁移率
impurity center 杂质中心
impurity compensation 杂质补偿
impurity concentration 杂质浓度
impurity conduction 杂质导电
impurity contamination 杂质沾污
impurity content 杂质含量
impurity control 杂质控制
impurity damage 杂质损伤
impurity defect 杂质缺陷
impurity diffusion 杂质扩散
impurity distribution 杂质分布
impurity effect 掺杂效应
impurity element 杂质元素
impurity energy level 杂质能级
impurity gas 掺杂气体
impurity gradient 杂质梯度
impurity ion instability 杂质离子不稳定性
impurity ionization energy 杂质电离能
impurity level 杂质能级;杂质量;不纯度
impurity limitation 杂质限度
impurity mineral 杂质矿物
impurity mobility 杂质迁移率
impurity penetration 杂质渗入
impurity photoconductor 杂质光电导体
impurity profile 杂质分布图
impurity profile measurement 杂质分布测量
impurity scattering 杂质散射
impurity segregation 杂质分凝
impurity selection 杂质选择
impurity source 杂质源
impurity state 杂质态
impurity striation 杂质条纹
impurity water 含杂质的水
imputation 推算的价值量;超ình向量
imputation system 扣除利润分配中重复课税的制度
imputed capital flow 被转嫁的资本流量
imputed capital value 被转嫁的资本价值
imputed cost 应归成本;推算成本;设算成本;附加成本
imputed cost vector 归属价值向量
imputed income 推算收入
imputed interest 应归利息;应付利息;估算利息;推算利息;设算利息;被转嫁的利息
imputed price 估计价格
imputed rent 估算房租;应计租金;推算租金
imputed service charges for life insurance 人寿保险服务费用估计数
imputed value 推算价值
imputed value of fuel 燃料的估算价值
iputrescibility 不腐性
imputrescible 能防腐的;不易腐烂的
imputrescible material 不易腐烂的材料
imputresciblity 不易腐烂性
imref 倒费密
imtemperate wind 烈风(九级风)
imtermediate train spacing point 分界点【铁】
imtermittent 断续的
imvite 蒙脱土
inaccessibility 难接近;不易接受性;不可达性;不可接近性
inaccessible 不通达的
inaccessible area 偏僻之处;荒野;偏僻地区
inaccessible point 不可到达点
inaccessible pore 不可入孔隙
inaccessible region 难接近地区;不可达区
inaccessible roof 高顶板
inaccessible site 不可达到的地段
inaccessible state 不可达状态
inaccessible stationary point 不可达稳定点
inaccessible value 不可达值
in accordance with 根据;按照
in accordance with agreement 按照协议
inaccuracies of fabrication 装配上的误差
inaccuracy 偏差;误差;不准确性;不准确(度);不正确;不精确;不精密(性)
inaccuracy erection 装配误差
inaccuracy in dimension 尺寸不准
inaccuracy of erection 施工误差
inaccurate 不精确(的)
inaccurate contour 草绘曲线;草绘等高线
inaccurate date 不准确数据
inaccurate drop hammer 不精确落锤
in a cool climate 在凉爽气候条件下

inact 非活性的
inactinic glass 非光化玻璃
inaction 故障;静止;钝化;不活跃;不活泼;不活动
inaction period 非作用期;无作用期间
inactivaor 灭能剂
inactivated efficiency 灭活效率
inactivated ratio 灭活率
inactivated sludge 灭活污泥
inactivating 钝化
inactivation 灭活(作用);失效;钝化(作用)
inactive 不活动的;不活泼的;静态的;停止的;惰性的;怠惰的;待用的;非现用的;非活跃的;非活化的;非活(动)性的
inactive account 不活动账户;静止账户
inactive addition 非活性混合料
inactive air 非搅拌空气
inactive basin 稳定海盆
inactive black 非活性炭黑
inactive block 静态分程序;无效程序块;待用程序块
inactive coil 端簧圈
inactive component 无源设备
inactive constraint 无效约束;不起作用的约束
inactive door 被动门;不动门
inactive echo 迟钝波
inactive element 不活动元素
inactive entry 待用入口
inactive event variable 静态事件变量;待用事件变量;非活动事件变量
inactive face 停用的工作面
inactive fault 非活动断层;不活动断层
inactive field 非搅拌区
inactive file 无效文件;待用文件;非现用文件;非现用档案;非现行文件
inactive filled line 停运油管线
inactive filler 钝性填料;非活性填充剂
inactive front 不活跃锋
inactive gas 惰性气体;惰性气氛;不活泼气体
inactive glacier 固定冰川;不活动冰川
inactive leaf 固定门扉
inactive level 非活动程序
inactive line 虚描线;无效行;停留线路;待用线(路)
inactive link 待用链路;非活动链
inactive marginal basin 稳定边缘海盆;不活动边缘盆地
inactive mode 被动式
inactive molecule 非活性分子
inactive money 闲散资金;呆滞资金
inactive needle 不工作织针
inactive node 待用节点;不活动节点
inactive optically 非旋光的
inactive optics 非能动力学
inactive page 不活动页
inactive plate 静止板块;不活动板块
inactive population 非职业人口
inactive porosity 无效孔隙度
inactive portion 不活动部分
inactive program(me) 非现用程序
inactive queue 待用队列
inactive queue of task 待处理任务队列
inactive receptor 非活性型受体
inactive region 不活动区
inactive runway 停用的跑道
inactive segment 非活跃段
inactive solvent 惰性溶剂
inactive spring 死泉
inactive state 待用状态;非活性状态;非活动状态;不活泼状态;不活动状态
inactive station 待用站;不活动站
inactive stock report 呆滞材料报告单
inactive storage 最低库容;无效库容;死库容;垫底库容;备用库容
inactive storage capacity 静存储容量
inactive substance 非活性物质
inactive test 非放射性试验
inactive thermal seepage 无热活动渗眼
inactive time 无效时间
inactive unemployed 没有收入来源的失业者
inactive volcano 死火山
inactive volume 待用卷
inactive water drive 不活跃水驱
inactive well 不出油井
inactive workings 废巷道
inactivity 休止状态;无功率;不旋光性;不活动性;

不放射性
inactivity cost 停工成本
inactual fact 事实上
in actual fire conditions 在实际火情中
inadaptability 无适应性;不适用性;不适应性;不适配性
inadaptation 不适应
inadequacy 机能不全;不适用;不适宜(性);不适当
inadequate 不足够的;不适宜;不适合的;不充分的
inadequate capacity 不敷用生产能力
inadequate consumption 消费不足
inadequate degree of geologic(al) investigation 地质研究程度不够
inadequately mapped areas 测图资料不全的地区
inadequate of data 数据不足
inadequate packing 包装不适当
inadequate penetration 未焊透
inadequate perfusion of microcirculation 微循环灌流不良
inadequate soil tillage 不合理的整地
inadequate stimulus 不适宜刺激物
inadherent 不黏结的
inadhering 不附着的
inadhesion 不黏着;不黏性
inadhesive 不黏附的;不能黏结的
inadmissibility 不允许
inadmissible 不能允许的
inadmissible assets 不可税资产
inadmissible harmonics 禁用调和函数
inadmissible strategy 不宜取策略
inadmissible structure 非批准结构
in-a-door bed 门内折叠床
in advance 提前
in advance of a signal 信号机前方
inadvertence[inadvertency]不留心的
inadvertent reuse 非计划再用
inagglutinable 不凝集的
in a humid environment 在潮湿的环境中
inalienable 不可分割(的)
inalienable right 不可让与的权利
in alignment 成一直线
Inalium 因阿铝合金
in alkaline soils 在碱性土壤中
in all 总共
in all weather 不论任何天气
in a mess 零乱;凌乱
in ancient times 在古代
in-and-in 折插包;互锁
in-and-out 自由出入;即刻买卖;进入和驶出口;内外重叠;短期结算
in-and-out bolt 贯通螺栓;贯穿螺栓
in-and-out bond 丁砖层与顺砖层竖向交错砌合;丁顺砖逐层交替的砌合法
in-and-out channel 嵌入河道
in-and-out movement 抓片机构;往复运动
in-and-out plating 内外搭板;内外叠板法
in-and-out rates 进出费率;内外差异费率
in-and-out reheating furnace 分批加热炉
in-and-out stepwise method 分批进入和退出的逐步回归法
in-and-out strakes 内外叠板(法)
in-and-out system 内外叠板法
in-and-out time card 进出时间卡
in-and-out traffic schematic drawing 出入交通量示意图
in-and-out type furnace 分批装出料的室式加热炉
in and over-clause 舱内及甲板上条款
inanimation 无生命
in antis 正面双柱式;双柱门廊
in antis temple 壁角柱之间的神庙
in a number of ways 在许多方面
inapparent 不显性的
in apparent good order and condition 外表良好
inapplicability 不能适用
inapplicable 不适用的
inappreciable 微不足道的;不足道的
inappreciable error 不显著误差
inapproachable 难接近;不易接近的
inappropriate 不恰当的;不合适的
in aqueous 在水中
inarching 靠接
inarching by cleaving 割裂靠接
inarching with a branch 单枝靠接
inarmo(u)red cable 未铠装电缆;非铠装电缆;铅皮电缆

inarray 内部数组
in arrears 拖欠
inarticulate 无关节的
inarticulation 声言模糊
inartistic 非艺术的
in assessing terms 评定(贷款)条件
inassimilable 不能同化的
in association with specific conditions 结合具体情况
inaugural exhibition 开幕展览会
inauguration 落成典礼;开幕式;开工典礼;典礼
inaugurator 主持落成仪式者;主持开幕仪式者
inaurate 黄金光泽
in bad repair 失修(的)
in ballast 压载状态
inband 同带信号传输
inband distortion 带内失真
inband diversity 带内分集
inband frequency assignment 带内频率分配
inband noise 带内噪声
inband rybat 丁石边框
inband signal(l)ing 带内信令;同带信号传输
inband signal(l)ing pattern 带内信号方式
inband system 带内制
inbank capacity 平槽泄水能力
inbank flow 平槽水流
inbark 树穴;(木材的)夹皮
in-base illuminator 显微镜座内装置式显微镜灯
in batches 分批地;分批的
in batch mode 以成批方式
in-bath measurements 熔池测试
in-batter pile driving 俯打
in bay dump 海湾抛泥
inbed 压实(回填料);夯实
in behalf of 代表……
in-between position 中间位置
inbex 盖瓦
in-bin drying 仓内干燥
in black and white 白纸黑字
in block 块状;单块;整块;滑轮机构;成组的
inblock cast 整体铸造的;整体铸造
in-block-cast member 整体铸造构件
inblock cylinder 汽缸排
inboard 内侧的;机上的;在内舱;在船内;机内;内纵的;船内(的)
inboard boom 舱口吊杆
inboard cabin 内侧舱事
inboard cargo 舱内货(物)
inboard channel 内侧信道
inboard discharge pipeline 船上排泥管
inboard engine 内侧发动机
inboard extension 内伸臂
inboard guy 里舷稳索
inboard profile 纵剖面图【船】;船内纵剖面图
inboard screw 内侧螺旋桨
inboard shaft 内侧轴
inboard shot 在船上的锚链节
inboard turning 内旋式;内向旋转;向内转动
in bond 货物在关栈中;丁头砌合
inbond brick 内砌石;丁砖
inbond rybat 丁石边框
inborn reflex 固有反射
in bottles 瓶装的
inbound 进入的;回程;返程
inbound cargo 来港货物;进口货(物)
inbound engine lead track 机车入库线
inbound freight 回程运费;到达货物
inbound lane 入境车道
inbound line 入站线路
inbound pacing 接收速度
inbound platform 到达站台
inbound ship 归航船只
inbound track 进站线(路);到达线
inbound traffic 入境交通
inbound train 到达列车
inbow 筑穹隆或拱;弧形的
in boxes 箱装
inbreak 侵入;崩落
inbred line 纯系
inbred-variety cross 顶交
in bridge 跨接;加分路;桥接;旁路;并联
in-built 内装;埋设的
in-built anti-splash 固定的防溅装置;固定挡水板;

固定挡泥板
in-built ashtray 嵌入式烟灰缸;嵌入式灰缸
in-built(bath)tub 固定装置的浴缸;埋设式浴盆
in-built cabinet 嵌入式柜橱
in-built cupboard 嵌入式碗橱
in-built furniture 固定式家具
in-built garage 内装的车库;内部车库
in-built jack 内装式起重器
in-built kitchen 内设的厨房;内部厨房
in-built kitchen cabinet 固定的厨房碗柜
in-built shower stall 嵌入式淋浴间;内部淋浴间
in-built tub 嵌入式浴缸
in-built unit 嵌入式固定单元
in-built wardrobe 固定衣柜;埋入式衣橱
in bulk 整体的;整批;桶装;散装(的);大批;大量;成块;成堆
inbye compressor 坑道用空压机;井下压风机
Inca architecture 印加建筑;秘鲁印第安式建筑
incaite 硫锑锡铅矿
in cake 成块
incalculability 无数;无量
incalculable 难以数清的;无数的
incandescence 热光源;白(炽)热;白炽(光)
incandescent 极亮的;白炽的
incandescent arc lamp 白炽弧光灯
incandescent bulb 白炽灯泡
incandescent cathode 热阴极;炽热阴极;白炽阴极
incandescent daylight lamp 白炽日光灯
incandescent electric(al) lamp 白炽电灯
incandescent filament lamp 白炽灯
incandescent flame 白热焰
incandescent globe 球形白炽灯
incandescent lamp 光热灯泡;磨砂白炽灯
incandescent lamp base 白炽灯灯头
incandescent light 白炽光;白炽(光)
incandescent lighting 白炽(灯)照明
incandescent lighting fixture 白炽(灯)照明设备
incandescent readout 白炽灯数字显示装置
incandescent special-service lamp 专用白炽灯
incandescent spot 炽点;白炽点
incandescent state 白热状态;白炽状态
incandescent time 白炽时间
in-can preservative 罐内防腐剂
incapability of meeting obligation 无法履行义务的情况
incapacitated shareholder 丧失能力的股东
incapacitation 失效;失去能力
incapacity 无能力
incarbonization 成碳作用
incarceration 箝闭
incarnadine 肉红色
incase 镶在框子内;装在箱内;箱装
in case of emergency 在紧急的情况下
incavation 洼地;空心的东西
in cavetto 反浮雕(凹雕饰)
in-cavity melt pressure 腔中熔化压力
incavo 反浮雕挖成凹形部分
incease of power 提高功率
incendiarism 纵火
incendiary 纵火者;燃烧作用
incendiary agent 燃烧剂
incendiary effect 燃烧作用
incendiary file destroyer 纵火器
incendiary fire 纵火
incendiary material 纵火材料;燃烧材料
incendiary oil 纵火油料
incendiary pencil 纵火管
incendiary rocket 燃烧火箭弹
incendiary spark 引燃火花
incense cedar 北美翠柏;香杉;香肖楠
incense juniper 香刺柏
incense wood 香木
incenter 内心;内切圆圆心;(四面体的)内切球心;内接圆心
incentive 诱因;物质刺激
incentive bonus 激励性奖金
incentive export system 奖励出口制;出口奖励制度
incentive force 鼓动力
incentive industry 奖励工业
incentive pay 奖金津贴
incentive pay plan 奖励工资计划
incentive plan 奖励计划
incentive price 鼓励性价格
incentive reduction 鼓励性减免;激励性退税

incentive scheme 奖励计划;物质刺激制度
incentive standard 奖励标准
incentive-suggestion system 奖励建议制度
incentive system 奖励制度
incentive system for rational energy utilization 合理使用能源奖励制度
incentive system of wages 奖励工资制度
incentive wage 奖励工资
incentive wage payment plan 奖励工资计划
incentive zoning 鼓励性区划
incentre 内心;内切圆圆心
inception 开始;起始;初始位置
inception of insurance cover 保险开始期
inception report 初始报告
inceptisol 始成土
inceptive 初始
inceptor 拦截设施(污水处理用);秽污分离器
incertae sedis 未定地位;分类位置未定
incertitude zone 未定带
incertum 虎皮墙;毛石坞工墙
incertum opus 砂浆小毛石墙
inch 英寸(1 英寸 = 2.54 厘米)
in-channel signal(l)ing 随路信令
in charge 主管(的)
in-charge accountant 主任会计师
in-charge of diving 潜水主管
inch board 一英寸板
incher 小管;口径以英寸计的东西
inches of head 水头英寸数
inches of water 英寸水柱
inches penetration per year 每年腐蚀深度
inches per second 每秒英寸数
inch-foot charge for water service 英寸英尺消防水费
inch for nominal size 公称直径以英寸表示
inching 小量位移;精密送料;渐进;渐动;极低速运行;缓动;瞬时断续接电;点动
inching arm 微动臂
inching button 微动按钮
inching control 渐动控制;微动控制;微调
inching drive 慢速传动
inching engine 低速转动发动机
inching motion 微动
inching motor 慢速电动机;辅助传动电动机
inching operation 微动操作
inching power 低速航行功率
inching starter 渐动起动器;低速起动器
inching switch 微动开关
inching valve 微动阀
inchmaree clause 疏忽条款
inch module 英制模数
inchoate bill 缺项汇票
inchoate instrument 不完整的票据;未登记的契约
inch of mercury 英寸汞柱
inch of water 英寸水(水头压力单位,水深单位)
inch plank 一寸尺;英寸板
inch-pound 英寸·磅
in chronological sequence 按时间顺序
inch rule 英制尺
inch scale 英制比例
inch screw die 英制螺纹板牙
inch screw tap 英制螺纹丝锥
inch screw thread 英制螺纹
inch size 英制尺寸
Inch-standard 英制标准
inch strength 英寸强度
inch stuff 英寸材
inch system screw thread 英制螺纹
inch thread 英制螺纹
inch trim moment 每英寸倾斜力矩
Inchun Port 仁川港(朝鲜)
incidence 落下;入射;冲角
incidence angle 入射角;冲角
incidence(angle) yaw 迎角
incidence axiom 关联公理
incidence beam 入射光束
incidence direction 入射方向
incidence electron-beam current 入射电子束电流
incidence formula 关联公式
incidence function 关联含数
incidence indicator 倾斜指示器;倾角指示器
incidence loss coefficient 入射损失系数
incidence matrix 关联矩阵
incidence measuring gear 测攻角机构

incidence meter 倾角计;倾角表
incidence number 关联数
incidence of hot spots 产生热点
incidence of light 光线入射角
incidence of poverty 贫困率
incidence of stabilizer 稳定器的倾斜角
incidence of taxation 纳税负担
incidence of unemployment 失业范围
incidence plane 入射面
incidence point 入射点
incidence rate 入射率;发生率;发射率;发病率
incidence relation 关联关系
incidence wire 倾角线
incident 关联的;偶发事件;入射的;事件;事故
incidental 偶然的
incidental appeal 附带上诉
incidental charges 杂费;从属费用
incidental civil action 附带民事诉讼
incidental claim 附带请求
incidental consequence 附带的结果
incidental cost 杂费;附加费用;附带成本
incidental cutting 临时采伐
incidental cutting and drilling 附带的切削和钻孔(因吊装需要)
incidental defect 偶然缺陷;偶然缺点;非主要的缺陷
incidental discharge 意外排放
incidental effect 偶发作用
incidental emission 意外倾卸;意外排放
incidental expenses 杂费;临时费用;偶然费用
incidental fee 杂费
incidental fees tariff 追加运费费率
incidental felling 临时采伐
incidental frequency modulation 临时频率调制
incidental information 偶然信息
incidental liabilities, expenses 附带责任和费用
incidental loss 附带损失;非经常性损失
incidental observation station 临时性观测站
incidental parasite 偶然寄生物
incidental phase modulation 寄生调相
incidental release 意外释放;意外排放
incidental reserve 事故备用
incidental revenue 临时(性)收入;偶然收益
incidental 临时费用
incidental service 伴随服务
incidental time 非主要工作时间
incidental transaction 附带交易;非经常性交易;额外交易
incident angle 入射角
incident angle of the entering ray 入射光线的入射角
incident beam 入射线;入射(光)束
incident blast wave 入射冲击波
incident current 入射电流
incident current velocity 入射水流速度;入射潮流速度
incident defect 轻微缺点
incident detection 事故检查;事故检测
incident direction 入射方向
incident dose 入射剂量
incident earthquake 入射地震
incident electron beam 入射电子束
incident energy 入射能(量)
incident face 入射面
incident field intensity 入射场强
incident field strength 入射场强
incident flow 迎面流;迎流;入射流
incident flux 入射流;入射(光)通量
incident gamma photon energy 入射伽马光子能量
incident gravity wave 入射重力波
incident height 入射高(度)
incident illumination 入射照明
incident image 入射影;入射(图)像
incident intensity 入射强度
incident light 入射光
incident light beam 入射光束
incident light illuminator 入射照明器
incident light intensity 入射光强度
incident light meter 入射光式曝光表
incident light microscope 反光显微镜
incident light reading 入射光测定读数
incident line 入射线
incident medium 入射介质
incident mode 入射模式
incident motion 入射运动

incident nodal point 入射节点
incident noise 入射噪声
incident normal 入射法线
incident nucleon 入射核子
incident path 入射路径
incident photon 入射光子
incident point 入射点
incident power 入射功率;入射波功率
incident pressure 入射压力
incident pulse 入射脉冲
incident radiation 阳光辐射能;入射辐射
incident rate 事件率
incident ray 入射(射)线;入射光线;表面入射束的中心线
incident record 事件记录;伴随记录
incident report 事件报告
incident signal 入射信号
incident sound 入射声
incident summary 事故摘要
incident wave 浸入波;入射波
incident wavefront 入射波前
incident wavelength 入射波长
incinderjell 凝固燃烧剂
incinerarium 火化场
incinerate 烧成灰;焚烧;灰化
incineration 火葬;化煅烧;烧成灰;焚烧(法);焚烧(法)除溶剂;焚化
incineration areas 焚化区
incineration areas at sea 海上焚化区
incineration ash 焚化灰
incineration by pyrolysis 高温分解焚烧
incineration corrosion 焚化炉腐蚀
incineration deodo(u)rization 焚烧脱臭;焚化脱臭
incineration house 垃圾焚化站
incineration method 焚化法
incineration of garbage 垃圾焚化
incineration of municipal refuse 城市垃圾焚化
incineration of refuse 垃圾焚烧
incineration of solid wastes 固体废物焚烧;固体废物焚化
incineration of wastes 废物焚烧
incineration plant 垃圾焚化厂;焚化装置
incineration scrubber water 焚化洗涤水
incineration system 焚化系统
incineration unit of garbage 垃圾焚化装置
incinerator 垃圾焚化炉;垃圾焚化炉;锻烧炉;煅烧装置;焚烧炉;焚化(垃圾)炉
incinerator collector 焚化炉集尘器
incinerator for plastics 塑料焚烧炉
incinerator jet 焚化炉喷嘴
incinerator of batch operation 间歇式垃圾焚化炉
incinerator of half-mechanized operation 半机械化垃圾焚化炉
incinerator of semi-mechanized operation 半机械化垃圾焚化炉
incinerator residue 焚化炉残渣
incinerator room 垃圾焚化间;焚化间
incinerator scrubber water 焚化炉洗涤水
incinerator stoker 焚化炉加料机;垃圾焚烧炉
incipience 起始
incipient 初发的
incipient bar 初期沙洲
incipient calcination 初期煅烧
incipient cataract 初期内障
incipient cause 远因;始因
incipient cavitation 空化初生;起始气蚀;初始气蚀;初始空蚀
incipient cavitation number 初生空穴数
incipient combustion 起始燃烧
incipient continental rift 大陆初始裂谷
incipient contraction crack 初期收缩裂缝
incipient crack 起始裂纹;初始裂纹;初始裂缝;初期裂纹;初期裂缝;初期发裂;初裂(纹)
incipient curing 初期养生;初期养护(混凝土)
incipient decay 初期腐朽;初期腐烂;初期腐败
incipient deformation point 变形软化温度;变形点
incipient detonation 初爆
incipient erosion 早期冲蚀
incipient expenditures 初期支出
incipient failure 开始破坏;起始故障;初始破坏;初期破坏;初期故障;初发故障
incipient fault 萌芽断层;潜在故障
incipient flood plain 雏形河漫滩
incipient fluidized bed 初始流化床

incipient fluidizing velocity 初始流化速度
incipient fold 萌芽褶皱;初期褶皱
incipient fusion 垂熔
incipient gullying 初期沟蚀
incipient impulse 初始脉冲
incipient inhibition 初期抑制
incipient joint 萌芽节理;初始节理
incipient LC30 起始半数致死浓度
incipient lethal concentration 起始致死浓度
incipient lethal level 立即致死量;起始致死水平;半数致死量
incipient lethal temperature 起始致死温度
incipient low 初生低压
incipient melting 初熔
incipient metamorphism 初级变质作用
incipient motion 起始运动;初始运动;初期运动
incipient movement 初期运动
incipient peneplain 初期准平原;初期侵蚀平原
incipient period 初期
incipient piping 初始管涌;初期管涌
incipient plastic flow 早期塑性流动;初塑性流动
incipient podzolization 初期灰壤化;初期灰化作用
incipient point 始动点
incipient rupture 初期破坏
incipient scour 初始冲刷;初期冲刷
incipient sediment motion 初始泥沙运动;初始沉淀运动
incipient sintering 起始烧结
incipient skidding 开始滑动
incipient spoilage 初期腐败
incipient stability 初始稳定
incipient stage 开始阶段;初(生)始阶段;初期
incipient stage of decay 初期腐朽阶段
incipient tendency 苗头
incipient velocity 起动流速
incipient wilting 初萎
incircle 内切圆;内接圆
in-circuit emulator 电路内仿真器;内部电路仿真器
in-circuit test 电路内测试
in circuit testing 内部电路测试
incisal guidance 切导
incisal guide 切导
incise 流切;切开;切割
incised 雕刻的
incised decoration 刻花
incised decoration in bronze 青铜蚀刻装饰;铜雕饰
incised inscription 阳刻
incised meander 切入河曲;刷深河弯;深切曲流;深切河湾
incised ornament 雕刻装饰;雕刻饰物
incised river 高滩曲折河道;陵削河;深切河流;深切河(谷);高槽河道;地下河(流);地割流
incised slab 雕成石板;雕刻(石)板;雕成木板
incised stream 深切河流
incised tile 雕刻瓦;雕刻砖
incised work 雕刻工作
incised wound 割伤;割切伤
incising 刻缝;雕刻
incising and filling wit colo(u)red lacquer 雕填
incising and filling with gold dust 戗金
incising of wooden tie 木枕刻痕
incision 下切作用;切口;切开;切割;深切
incision and threaddrawing therapy 切开挂线法
incision of river 河流刷深
incision of stream 河流下切;河流刷深
incision scissors 切开剪
incisive tooth 切齿
incisura angularis 角切迹
incitogram 冲动发放
in cladding and applied linings 在覆层及所用衬里
in-class training 课堂培训
in clear 净额(离)
in clearing 交换中的本行票据;经过票据所的票据交换总额
inclemency 狂风暴雨
inclement weather 严酷天气;狂风暴雨天气;寒冷天气;恶劣天气
inclinable 可倾斜的
inclinable press 可斜式压力机
inclination 斜度;眼高差;倾斜;倾度;倾侧度;坡度;偏转;偏斜;水平差
inclination and direction of boreholes 钻孔顶角与方位角
inclination angle 斜角

inclination angle of a windscreen 前风挡玻璃的安装角
inclination angle of flame jet 火焰喷射角
inclination borehole 斜孔
inclination change line 倾斜变换线
inclination compass 倾角罗盘
inclination compensator 倾斜补偿器
inclination correction 倾斜校正;倾斜改正
inclination correction factor 荷载倾斜因数
inclination drilling 偏斜(孔)钻进
inclination error 倾角误差
inclination factor 倾斜因子;倾斜因素;倾斜因数;倾斜系数
inclination ga(u)ge 倾斜计
inclination indicator 偏斜指示器
inclination joint 倾斜关节
inclination limit 极限倾角
inclination line 比降线
inclination maximum 最大倾角
inclination measurement method 钻孔测斜方法
inclination measurement with drilling 随钻测量
inclination method 倾角计算法;倾度计算法(坝工)
inclination needle 倾角指针
inclination observation 倾斜观测
inclination of aquifer 含水层产出状态
inclination of bigger rat hole 大鼠洞斜
inclination of bucket ladder 斗架倾角
inclination of caldron 炒锅倾角
inclination of deck structure 面板结构的坡度
inclination of ecliptic 黄赤交角
inclination of load 荷载倾(斜)角
inclination of mouse hole 小鼠洞斜
inclination of needle 磁(针)倾角
inclination of nozzles flange 接管法兰倾斜
inclination of orbit 轨道倾角;轨道交角
inclination of planetary orbits 行星轨道倾角
inclination of radial pipe 辐射管倾斜度
inclination of ramp 进出车道坡度角(高速公路)
inclination of roof 屋顶倾度;屋顶坡度
inclination of satellite orbit 卫星轨道倾角
inclination of stratum 地层倾斜度;地层倾斜
inclination of the borehole 钻孔倾斜
inclination of the earth's axis 地轴倾斜度
inclination of the magnetic body 磁性体倾向
inclination of the wind 风的偏差角
inclination of trunnion grid 横轴倾斜
inclination of wavefront 波前倾斜
inclination of wind 风向与等压线的交角
inclination pitch 倾斜度
inclination resistance 坡道阻力;坡道附加阻力
inclinator 电梯座椅;斜垫块;倾斜器;倾角仪;倾倒器
inclinatorium 测斜器;倾斜仪
inclind staple shaft hoisting 暗斜井提升
incline 斜坡;斜面;偏向
incline batter 内倾
incline bedding 倾斜层理
incline bogie 斜井罐笼
incline conveyer 倾斜式输送机
incline cut and fill stoping 倾斜分层
inclined adit 斜洞
inclined allowance 倾斜裕量
inclined anchor 斜锚
inclined angle 倾角
inclined antenna 倾斜天线
inclined anticline 倾斜背斜
inclined approach 斜坡道
inclined apron 斜护坦
inclined arch 斜拱;倾斜拱
inclined area 斜面积
inclined armor 斜置装甲板
inclined axis mixer 混凝土斜轴搅拌运输车;斜轴搅拌机;斜轴拌和机;高速卸料搅拌机
inclined axis truck mixer 斜轴搅拌机;斜轴拌和机
inclined ball mill 倾斜式球磨机;倾斜球磨机
inclined bar 斜钢筋
inclined barrel arch (连拱坝的)斜筒拱;斜拱筒
inclined barrel vault 斜筒形穹顶;斜筒穹顶
inclined beam 斜梁
inclined bearing 斜支承
inclined bed 倾斜岩层
inclined bedding 斜层理;倾斜层理
inclined belt conveyer 倾斜运输带;倾斜皮带输送机

inclined body wave 斜体波
inclined boom (起重机的)倾斜吊杆
inclined boom type trench excavator 斜臂式挖沟机
inclined(bore)hole 斜钻孔
inclined boring 斜向钻孔
inclined bottom car 漏底车;底卸式矿车;斜底车
inclined breakwater 斜坡式防波堤
inclined bridge 坡桥;有坡度的桥
inclined bucket elevator 倾斜斗式提升机
inclined building berth 倾斜船台
inclined buoyant jet 斜浮射流
inclined cable 斜缆(索)
inclined cable plane 斜索面
inclined cableway 斜坡道;倾斜索道;单索缆道
inclined casting 倾斜浇(注)法;倾斜浇注
inclined catenary 斜链形悬挂
inclined-catenary construction 斜链线结构
inclined catenary structure 斜链线结构
inclined chain-and-bucket trench digger 倾斜式链斗挖沟机
inclined chord 斜弦杆
inclined chord of truss 桁架斜弦杆;斜弦杆桁架
inclined chute 斜溜槽
inclined clarifier 斜式沉降器
inclined cleaner 梯形开棉机
inclined clerestory 倾斜天窗
inclined coal seam 倾斜煤层
inclined coil type 倾斜线圈型
inclined column grain drier 倾斜筒式谷物干燥机
inclined contact 倾斜接触
inclined continuous bucket elevator 倾斜连接斗式提升机
inclined conveyer 倾斜运输机;斜坡式输送机
inclined conveyer separator 倾斜输送带式清选机;倾斜输送带式分离机
inclined course 斜砌层
inclined crack 斜向裂缝;斜裂缝
inclined cut 倾斜工作面
inclined cut-and-fill 倾斜分层充填(法)
inclined cut-and-fill stope 倾斜分层充填采场
inclined cutting 倾斜切削
inclined cylinder block 斜轴式柱塞泵缸体;倾轴式柱塞泵缸体
inclined cylinder screen cleaner 倾斜式圆筒筛网清花机
inclined damper 斜插板阀
inclined derrick 斜塔(指钻塔)
inclined disk centrifuge 倾斜盘式离心机
inclined dispersion 倾斜色散
inclined dispersion of interference colo(u)rs 干涉色倾斜色散
inclined distance 斜距
inclined dock 斜坡式码头
inclined drain 倾斜排水层
inclined drift 倾斜巷道;倾斜导坑;倾斜坑道
inclined drilling 倾斜钻进
inclined drop 斜面跌水;斜槽跌水;陡槽迭水
inclined drum leacher 倾斜的鼓式浸出器
inclined earthquake zone 倾斜地震带
inclined elevator 倾斜提升机;倾斜升料机
inclined end post 端斜压杆;端斜撑;桁架端受压斜构件
inclined engine 斜置发动机
inclined escalator shaft 自动楼梯斜道;斜自动楼梯通道
inclined extinction 斜消光
inclined fault 倾斜断层
inclined fibrous texture 倾斜片状结构
inclined field 倾斜场
inclined field tube 斜场管
inclined fill 倾斜分层充填
inclined floor 斜坡式护坦;倾斜地面;倾斜底层;倾斜底板
inclined flow 偏斜水流
inclined flow precipitation tank 斜流式沉淀池
inclined flow turbine 斜流式水轮机
inclined fold 斜歪褶皱;斜歪倾伏褶皱;倾斜褶皱
inclined force 斜内作用力;斜力;倾斜力
inclined gallery 斜楼座;斜廊
inclined ga(u)ge 斜(坡)水尺;斜坡水标尺;倾斜水尺;倾斜水标尺;倾斜水尺;倾斜高计;倾斜标尺
inclined gib 斜起重臂
inclined glass curtain wall 斜玻璃幕墙

inclined glazing 装斜玻璃;斜玻璃窗
inclined grate 斜炉箅;倾斜炉栅;倾斜炉箅;倾斜的壁炉铁栏
inclined grate cooler 倾斜炉栅冷却器;倾斜式炉箅子冷却器;倾斜式炉箅子冷却机
inclined grate incinerator 斜炉箅垃圾焚化炉
inclined grizzly 斜置栅筛;倾斜格筛
inclined guide 斜置导轨;斜导杆
inclined guide plate 斜面导板
inclined hanger cable 斜吊索
inclined haulage or drift 缓倾斜巷道
inclined haulage road 缓倾斜巷道
inclined haulageway 倾斜运输道
inclined haunch (梁的)放大端部
inclined haunched beam 端部放大的梁;斜变截面梁
inclined height 斜高
inclined hoist 倾斜起重机
inclined hole 倾斜钻孔
inclined hole cut 斜孔掏槽;倾斜眼掏槽
inclined horizontal fold 斜歪水平褶皱
inclined impact test 倾斜冲击试验
inclined impervious bottom bed 倾斜隔水底板
inclined instrument 测斜仪
inclined intake 倾斜入口;倾斜进水口
inclined jib 倾斜伸臂;倾斜吊杆;斜起重臂
inclined joint 斜接口;斜接合;斜接(缝);斜缝
inclined ladder 倾斜梯
inclined lamellar thickener 倾斜浓缩箱
inclined lattice 倾斜帘子
inclined ledged frame bridge 斜腿刚架桥
inclined leg frame 斜腿刚架
inclined legged frame structure 斜腿刚构
inclined length of orebody 矿体斜长
inclined letter 斜体字
inclined lift 倾斜升降机
inclined load 斜向荷载
inclined magnetization 倾斜磁化
inclined manometer 倾斜压力计
inclined meander 倾倒曲流
inclined mirror 斜交镜;倾斜反射镜
inclined motor stair(case) shaft 自动楼梯通道;倾斜汽车螺旋形楼梯通道
inclined moving stair(case) 倾斜的正面大楼通道
inclined moving stair(case) shaft 自动楼梯通道
inclined nappe 溢流水舌
inclined opening 斜井;倾斜坑道
inclined orbit 倾斜轨道
inclined overhead contact line 斜链形悬挂的架空接触线
inclined paddle type agitator 斜桨式搅拌器
inclined parallelopiped 斜角平行六面体
inclined parking 斜列停车
inclined parking arrangement 斜向停车布置(停车场)
inclined parking lane 斜列停车道
inclined penstock intake 斜卧管式进水口
inclined piece 斜撑
inclined pile 斜桩
inclined piston pump 倾斜柱塞泵
inclined plane 缆车道;斜坡道;升船斜面;斜面;倾斜面;升船斜坡
inclined plane aerator 斜板曝气器
inclined plane clock 斜面钟
inclined plane fish way 斜面式鱼道
inclined plane method 斜面法(测试用)
inclined plane visco(si)meter 斜面黏度计
inclined plane with traversally travel(l)ing caissons 横向斜坡式升船机
inclined plank isolating-oil pool 斜板式隔油池
inclined plank settler 斜板式沉淀池
inclined plank settling tank 斜板式沉淀池
inclined plank settling tank for horizontal-direction flow 平向流斜板式沉淀池
inclined plank settling tank of downflow 下流斜板式沉淀池
inclined plank settling tank of same-direction flow 同向流斜板式沉淀池
inclined polarization 斜极化
inclined position 倾斜位置
inclined position welding 倾斜焊
inclined principal axes 倾斜主应力轴
inclined projection 斜投影
inclined pusher bar elevator 斜推杆式升料机

inclined rack 斜齿轨
inclined ramp 斜坡
inclined range 斜距(离)
inclined ray 倾斜射线
inclined ray paths 倾斜射线途径
inclined reaction 斜向反力
inclined reel 斜丝框
inclined reflectivity 倾斜反射率
inclined reinforcement 斜钢筋
inclined retort 斜式甑;倾斜式碳化炉
inclined rill system 斜工作面上向采矿法
incline driving 斜井掘进
inclined roadway 斜坡道
inclined rock face 倾斜岩面
inclined rock slope 岩石斜坡
inclined ropeway 斜缆道;斜索道;(运重物的)斜坡的空中索道
inclined screen 斜格栅;斜网;斜筛;倾斜栅网;倾斜式滤网;倾斜筛
inclined screw mixer 倾斜螺杆混合机
inclined seam 倾斜矿层
inclined segment 斜面弧形板
inclined seismic zone 倾斜地震带
inclined sewage screen 斜污水网筛
inclined sewer screen 倾斜的排水(拦污)栅
inclined shaft 斜筒;斜井;斜式导井;斜坑;倾斜轴
inclined shaft fish lock 斜井式鱼闸
inclined shaft mucking apparatus 斜井装岩机
inclined shaft pump 斜式水泵
inclined shaft skip 斜井箕斗
inclined shaft timbering 斜井木支护
inclined shaft tubular turbine 斜轴贯流式水轮机;斜轴管式水轮机
inclined shear bar 弯起钢筋
inclined ship lift 斜面升船机
inclined shoot 斜槽
inclined shore 斜支撑木;斜撑
inclined side slab 倾斜支承板(采矿巷道的钢筋混凝土)
inclined skip charger 爬式加料机
inclined skip hoist 斜井箕斗提升机;爬式加料机
inclined slicing 倾斜分层崩落回采法
inclined slide 倾斜装车台
inclined slip 斜坡主缆车道;斜坡滑道
inclined stacking 斜堆法
inclined staff ga(u)ge 斜坡水尺;斜标尺;倾斜水尺
inclined steel 斜钢筋
inclined stem globe valve 斜杆球阀
inclined step 倾斜梯级
inclined stirrup 斜向箍筋;斜(面)箍筋;斜钢箍
inclined stone 斜石
inclined stone drift 倾斜岩石巷道
inclined stratum 倾斜地层;倾斜层
inclined stress 倾斜应力
inclined subway 斜地地道
inclined suspension 倾斜悬吊
inclined tee joint 斜角丁字接头
inclined terminal 斜坡式码头
inclined test bed 倾斜试验台
inclined throat shears 斜刃剪切机
inclined timber shore 斜木撑
inclined T joint 斜接丁字接头
inclined to puddle 容易胶黏
inclined track 有坡线路;倾斜轨道
inclined trajectory 倾斜轨迹
inclined traverser 斜面转车台
inclined tray drier 倾斜盘式干燥机
inclined trough 斜槽
inclined trough chute 斜槽
inclined tube bundle unit 斜顶式管束单元
inclined tube evapo(u)rator 斜管蒸发器
inclined tube ga(u)ge 斜管气压
inclined tube manometer 斜管(式)压力计;斜管式测压计;倾斜管式压力计
inclined tube micromanometer 斜管微压计
inclined tube precipitation 斜管沉淀
inclined tube precipitation tank 斜管式沉淀池
inclined tube settler 斜管沉降器
inclined tube settling tank 斜管沉淀池
inclined tube visco(si)meter 斜管黏度计
inclined twin 斜轴双晶
inclined undercut gate 倾斜下开式闸门
inclined upstream face 倾斜上游面
inclined valve 斜置气门

inclined way for boat 船用登陆板
inclined web 斜腹板
inclined wedge 倾斜楔体
inclined wedge method 斜楔体法
inclined weir 斜背堰
inclined weld 斜焊
inclined wharf 斜坡(式)码头;斜码头
incline engine 斜井卷扬机
incline escarpment 陡坡
incline grab 斜井抓岩机
incline grade 斜坡;倾度
incline gradient 斜坡
incline hoist 斜井提升机
incline hole 斜眼;斜孔
incline inset 斜井井底车场
incline landing 斜井装卸台;下山车场
incline level 倾斜计;倾计
incline overcome by forward impetus 动能闯坡
incline-plank settling tank of lateral flow 侧向流斜板式沉淀池
incline rope 斜井提升钢丝绳
incline ship lift 斜面升船机
incline sight 倾斜观测
incline skip hoist 斜井箕斗提升机
incline stop 斜井挡车器
incline trackman 斜井轨道维修工
incline variometer 倾角可变电感器
incline weld 斜焊接(缝)
inclining angle of buoy 浮标总倾斜角
inclining experiment 倾斜试验
inclining moment 倾斜力矩;倾侧力矩
inclining test 倾斜试验
inclinometer 井斜仪;倾斜仪;倾斜计;倾角仪;倾角计;磁倾仪;磁倾计;测斜机;测斜仪;测斜仪;测斜器
inclinometer box 测斜仪盒
inclinometer case 测斜仪盒
inclinometer readout 测斜仪读数装置
inclosed 密闭的
inclosed meander 环形河湾
inclosing crystal 包晶
inclosing route 圈定路线
inclosure 围场;罩;壳;包裹体【地】
inclosure act 圈地法
inclosure of space 围封空间
inclosure wall 围墙;大围墙
include 包括;包含(物)
included 包括在内
included angle 夹角;(焊接的)坡口角(度)
included angle of arch 拱的夹角;竁拱中心角;拱圈中心角;拱的包角
included angle of crest 堰顶(圆弧)包角;拱圈中心角
included angle of thread 螺纹夹角
included aperture 内涵纹孔口
included gas 内含气体
included geographic(al) region type 所属地理区类型
included pitapeture 内含纹孔口
included poikiloblastic texture 包含嵌晶变晶结构
included sapwood 内涵边材;内含边材
included tectonic cycle 所属构造旋回
included variable 列入变量
include file 内含文件
include member 内含成员
including export packing 包含出口包装费
including particular average 单独海损在内;包括单独海损
inclusion 蕴含;夹杂(物);内含物;充填物;掺杂(物);包体【地】;包含(物)
inclusion body 内含体;包含体;包含体
inclusion character 包体性质
inclusion characteristic 包体特征
inclusion complex 包络物;包合配合物
inclusion composition determination 包体成分测定
inclusion compound 包络物
inclusion count 夹杂计数
inclusion determination 包体测定
inclusion dike 包裹岩脉
inclusion fee 土地增值税
inclusion filling 包裹体【地】
inclusion-forming substance 包合形成物
inclusion gate 蕴含门
inclusion ga(u)ge 包体仪

inclusion genesis of earthquake 包体成因
inclusion geobarometry 包裹体测压法
inclusion granule 内含颗粒
inclusion in steel 钢内夹杂
inclusion isotope composition 包裹体同位素成分
inclusion migration 包合物的迁移
inclusion of combustible material 加入可燃物法
inclusion of igneous rocks 火成岩包裹体
inclusion of metamorphic rocks 变质岩包裹体
inclusion of multicrystal face growth 多晶面生成包裹体
inclusion of parameters 包裹体参数
inclusion of peat 泥炭夹层；泥炭包体
inclusion of sand 砂的夹进；砂的混入
inclusion of skeletal growth 骸晶生成包裹体
inclusion principle 包容原理；包含原理
inclusion reaction 包合反应
inclusion relation 包含关系
inclusion relation for sets 集的包含关系
inclusion shape 包体形状
inclusion temperature 包体温度
inclusion texture 包(裹)体结构
inclusion thermometry 包裹体测温法
inclusion thermometry method 包体测温法
inclusion trail 包裹体痕迹
inclusion type 包体类型
inclusion water 包裹体水
inclusive 计算在内的；包括在内；包括的
inclusive-or-gate 或门
inclusive process 内含过程
inclusive reaction 内含反应
inclusive routine 相容程序；内含程序
inclusive segment 相容段；可兼段；集容程序段；内存(程序)段
inclusive spectrum 内含谱
incoagulability 不凝结性
incoagulable 不(可)凝(结)的；不能凝固的
incoagulable gas 不凝(性)气体
incoalation 煤化作用；成煤作用
incobation 潜育
Incochrome nickel 因科镍合金；镍铬耐热合金
incoercibility 不可压缩性
incoercible 不能压缩的
incoherence 松散；支离破碎；无黏性；不凝聚性；松散性；不相干性；不连贯(性)；不共格
incoherent 非相干的，不相关的，不相参的，不连贯的，不共格的
incoherent alluvium 无黏性冲积层；未胶结冲积层；松散沉积层
incoherent analog(ue) modulation 非相干模拟调制
incoherent averaging 非相干平均
incoherent bundle 非相关束；非相干光纤束
incoherent circular source 非相干环形辐射源
incoherent disturbance 非相干扰动；不相干扰动
incoherent emission 非相干发射
incoherent fiber bundle 非相关光纤束
incoherent fiber optics 不相干纤维光学
incoherent generation 非相干振荡
incoherent grain boundary 非共格晶粒间界
incoherent hologram 非相干全息图
incoherent holography 非相干全息术
incoherent illumination 非相干照明
incoherent imaging 非相干成像
incoherent imaging system 非相干成像系统
incoherent infrared arrangement 非相干红外装置
incoherent inter-phase boundary 不相干相间边界；非相干相间边界
incoherent light 非相干光；不相干光
incoherent light diode 非相干光二极管
incoherent light holography 非相干光全息术；非相干光全息摄像；不相干光全息术
incoherent light source 非相干光源
incoherent light system 非相干光学系统
incoherent material 未胶结物质；无黏性材料；松散材料；非固结材料
incoherentness 无黏结性；不相干性
incoherent noise 不相干噪声
incoherent optic(al) arrangement 非相干光学装置
incoherent optic(al) converter 非相干光转换器
incoherent optic(al) information processing 非相干光信息处理
incoherent optic(al) radar 非相干光学雷达
incoherent phase 非相干相位

incoherent point source 非相干点光源
incoherent quasimonochromatic source 非相干准单色光源
incoherent radiation 非相干辐射
incoherent receiver 非相干接收器
incoherent reception 非相干接收
incoherent reception system 非相干接收系统
incoherent reflecting 不相干反射
incoherent rock 不黏结岩石；不胶结岩石
incoherent scatter(ing) 非相干散射；不相干散射
incoherent scattering cross section 非相干散射截面
incoherent scattering function 非相干散射函数；不相干散射函数
incoherent signal 非相干信号
incoherent slump 松散滑塌
incoherent source 非相干光源
incoherent superposition 非相干重合
incoherent system 非相干系统
incoherent transfer function 非相干传递函数
incoherent wave 非相干波
incohesion 不黏结性；无内聚性
incohesive 无黏聚力的；不黏结的
incohesive property 不黏结性
Incoloy 因科洛伊(一种耐高温的镍铬铁合金)；耐热镍铬铁合金
incombustibility 非燃性；不燃性；不可燃(烧)性
incombustible 不燃物；不燃烧的；不能燃烧的
incombustible building material 不可燃建筑材料；阻燃建筑材料
incombustible component 不燃体
incombustible construction 耐火结构；耐火构造；非燃性构造；不燃(烧)结构
incombustible construction type 防火的结构型
incombustible dust 不可燃尘末
incombustible fabric 不燃性织物
incombustible lining 耐火支架；不燃支架
incombustible material 非燃性材料；不燃性材料；不燃物质；不燃(烧)材料；不燃物质；耐火材料
incombustible mixture 不能燃烧的混合物；不(可)燃混合物
incombustible paint 不燃性涂料
incombustible paper 耐火纸；不燃纸
incombustible refuse 不可燃废物
incombustible reguse 不可燃垃圾
incombustible sign material 不可燃标志材料；不易燃标志材料
incombustible substance 不燃烧物
incombustible transaction 耐火处理；防火处理
income 进项；进款；入款；所得；收益；收入
income account 损益计算书；收益账(户)
income account audit 损益计算书审计
income account principle 损益计算书原则
income admission limits 申请住房收入限额
income after taxes 税后所得
income analysis 所得分析
income and cost audit of railway transportation 铁路运输收支审计
income and expenditure statement 收支清单
income and expenses 收支
income and surplus accounts 收入和盈余账户
income apportionment 净利的分配；收益分配
income at liquidation 清算所得
income audit 收益预算；收益审计
income audit of railway transportation 铁路运输收入审计
income averaging 收入平均数
income before extraordinary item 未付特殊项目前的收益
income before tax 税前所得
income beneficiary 收益受益人
income benefit insurance policy 遗益享受保险单
income bond 营业收入债券；收入债券
income bone 收益债券
income bracket 收入分类
income capital conditions sheet 进款资金状况表
income ceiling 收入额高限
income charges 收益支出
income consumption curve 收入消费曲线
income cycle 收入周期
income deduction 所得扣除项目；收益减除数
income demand elasticity 需求收入的弹性
income density of demand 需求的吸入弹性
income determination 收益决定

income differential 收入差距
income distribution 利润分配；收益分配；收入分配
income duty 所得税
income effect 所得效果；收入效应
income elasticity 收入弹性
income elasticity of demand for import 进口需求的收入弹性
income exempted 免税所得
income exempted from tax 免税的所得
income expenditures 收益支出费用
income-expense ratio 收支比
income for consumption 消费收入
income form investment 投资收益
income from completed contract 已完工合约的收益
income from forfeited deposits 没收定金收益
income from immovable property 不动产收入
income from investment 投资收入
income from leasing property 财产租赁所得
income from penalties 罚金收入
income from recoveries of bad debts 收回坏账收益
income from royalties 版税收益
income from sales of assets 出售资产收入
income from sales of products 销货收入
income from transferred property 财产转移收入
income from undertakings 事业收入
income from variable yield investment 可变收益投资收入
income from work 劳动收入
income gradient 收入梯度
income group 收入分类
income in advance 预收收益
income in kind 实物收入
income level 收入水平
income management information system 收入管理信息系统
income multiplier 收入因子
income offset by depletion 冲销折耗的收益
income of transportation 运输收入
income of unincorporated enterprises 非公司企业收入
income on construction job 工程收益
income on investment 投资收益；投资收入
income option policy 可选择保险额的保险
income originating from agricultural industry 来源于农业的收入
income participation 参与收入分成的权利
income participation loan 分享收益的贷款
income per capita 人均收入
income per head 人均收入
income policy 所得政策；收入政策
income-producing equipment 生产设备
income-producing expenditures 生产所得支出
income projection 收入预测
income property loan 以房地产收入作抵押的贷款
incomer 来港船只
income ratio 收益比率
income received method 所得收入法
income redistribution 收益重分配
income sheet 收益表
income sheet ratio 损益率；收益表比率
income statement 损益计算书；损益表；收益表；百分率收益表
income statement accounts 收益表账户
income statement analysis 损益表分析
income statement and profit appropriation statement 损益表及利润分配表
income stream 收入流量
income subject to withholding 应扣缴预提税收入
income summary 收入汇总
income summary account 损益汇总账户
income tax 所得税
income tax allocation 所得税分配；分摊所得税
income taxation 征收所得税
income tax bracket 所得税等级
income tax credit 所得税减免
income tax cushion 所得税缓冲器
income tax deduction 课税所得减除额；所得税扣款
income tax expenses 所得税费用
income tax for individual 个人所得税
income tax law 所得税法
income tax leverage 所得税杠杆作用
income tax liability 所得税额；所得税负债

income tax on intercompany profit 内部利润所得税
income tax on salaries and remunerations 薪给报酬所得税
income tax payable 应付所得税
income tax rate 所得税率
income tax rates table 所得税率表
income tax reserve 所得税准备
income tax return 所得税申报表
income tax returnblank 空白所得税申报表
income tax self-assessed 自报所得税额
income tax surcharge 所得税附加额
income tax withheld at source 从源扣缴的所得税
income tax withholding 扣缴所得税
income tax withholding table 所得税预扣表
income terms of trade 贸易收入条件
income to net worth 纯益与资本净值比
income-to-total assets 收益与资产总价之比
income value 所得值
income velocity 收入速度
incoming 进料；进来的；进入的
incoming air 进入的空气；进气
incoming and outgoing feeder 进出线
incoming and outgoing line bay 进线和出线间隔
incoming barrel 收集阱
incoming beam 入射束
incoming bit stream 输入位流
incoming cabinet 进线柜
incoming call barring 禁止呼入
incoming call identification 输入呼叫识别
incoming carrier 进入载波
incoming charges 进料（费用）
incoming circuit 输入电路
incoming compartment trap 收集阱
incoming current 输入电流
incoming data 输入数据
incoming divisor 收图室
incoming end 入口端
incoming feeder 进入馈线；输入馈线
incoming feeder cubicle 进线开关柜
incoming first selector 输入第一选择器
incoming flow 进水量；来水；入流
incoming gas 进气
incoming generator 准备并车的发电机
incoming group 引入组
incoming information 输入信息
incoming inspection 进厂检验
incoming junction 输入连接点
incoming laser beam 入射激光束
incoming level 接收电平；输入电平；受电电平
incoming line 引入线；进线；进入管线；输入线（路）
incoming line reperforator 入线复凿岩机；入线复凿孔机
incoming loading platform 地图图库装卸车台
incoming manifold 进站汇管
incoming map detail 地图内容要素
incoming material control 购进材料控制
incoming multipoint unit 入局多点设备
incoming of foreign capital 输入外资
incoming panel 进线配电盘
incoming product 到达产品
incoming pulse 输入脉冲
incoming radiation 日照；入射辐射；射入辐射
incoming ray 入射光
incoming register 输入寄存器
incoming routine test 输入例行测试
incomings and outgoings 收支
incoming sediment 来沙量；来沙
incoming signal 入射信号；输入信号
incoming slab 轧前板坯；来料板坯
incoming solar irradiation 日光入射
incoming solar radiation 日照
incoming station 到达站
incoming stock 进入原料；进料
incoming stone （轧石机的）进给石料
incoming symbol 进入符号；输入符号
incoming sync 输入同步脉冲
incoming sync pulse 输入同步脉冲
incoming task 输入任务
incoming test 入库检验
incoming tide 涨潮；进潮
incoming traffic 进区交通；输入通信量
incoming train 进站列车

incoming trajectory 进入轨道
incoming transmission 输入传输
incoming trunk 入口干线；入局中继线；输入中继线
incoming vessel 进港船
incoming waste 流来的废水；进入的废水
incoming waste water 入流废水
incoming water 来水
incoming water and sediment 来水来沙
incoming wave 来（袭）波；入射波；输入（电）波
incoming well 收集井
incommensurability 无公度；不相称；不可通约性；不可比性
incommensurable 不能比较的；不可通约的
incommensurable number 不可通约（的）数
incommensurable quantity 不可通约量
incommensurate phase 不通约相
in commercial industry 工业规模（的）
in commercial quantity 工业规模地
incompact 松散的，不紧密的；不结实的
incompactible 难压实的
incomparability 非可比性，不可比性
incomparability of consequence 结果的不可比性
incomparable 不能比较的，不可比（较）的
incomparable element 不可比元素
incomparable inconsistent 非一致的
incomparable product 不可比产品
incompatibility 配伍禁忌；非兼容性；不协调(性)；不相容性；不能配合；不可混用性；不兼容性；不共存性
incompatibility problem 不兼容问题
incompatible 不能共存的；不兼容
incompatible action 不相容动作
incompatible computer 不兼容计算机
incompatible crosspoints 不兼容交叉点
incompatible data 不兼容数据
incompatible displacement model 不协调的位移模型
incompatible elements 不相容元素
incompatible equations 不协调方程；不相关方程；不能通约；不相容方程组
incompatible events 互斥事件；不相容条件；不相容事件
incompatible parameter 不兼容参数
incompatible pollutant 禁忌污染物
incompatible salts 不相容盐类
incompatible waste 不相容废物
incompetence[incompetency]关闭不全；功能不全；机能不全；不称职；无能力
incompetent attendance 不合格的维护
incompetent bed 弱（岩）层；软（弱）岩层；软弱夹层；松软岩层；不稳固岩层
incompetent fold 弱褶皱
incompetently intercalated bed 软弱夹层
incompetent plastic 软塑（的）
incompetent rock 塑性岩；软岩；弱岩石；软质岩（石）；不稳固岩层；不坚实岩层
incompetent rock stratum 弱胶结岩层
incompetent witness 无资格的证人
incomplete 残缺；不完善的；不完全的
incomplete absorption 不完全吸收
incomplete aquifuge 隔水层不完整
incomplete beta function 不完全贝塔函数
incomplete block 不（完）全阻滞；不完全区组
incomplete block design 不完全（小）区组设计；不完全配伍组设计；不完全块体设计
incomplete Bouger reduction 不完全布格改正
incomplete boundary condition 不完全边界条件
incomplete branch 不完全分支
incomplete breakdown 不完全击穿
incomplete call 未完成呼叫
incomplete census 不完全普查
incomplete chemical combination 不完全化合
incomplete closing of tongue rail 尖轨不密贴
incomplete combustion 不完全燃烧
incomplete combustion loss 不完全燃烧损失
incomplete compaction 未完全捣实；捣实不足；不完全捣实
incomplete compaction concrete 未压实混凝土；未捣实混凝土
incomplete compensatory pause 不完全代偿间歇
incomplete consolidation 不完全固结
incomplete contraction 不完全约束；不完全收缩
incomplete contraction orifice 不完全收缩口；不完全收缩孔

incomplete data 不完全资料；不完全数据
incomplete detonation 不完全起爆；不完全爆炸
incomplete dialing 未完拨号
incomplete diffusion 不完全漫射
incomplete dominance 不完全显性
incomplete double circulation 不完全双循环
incomplete ecosystem 不完全生态系统
incomplete elliptic integral 不完全椭圆积分
incomplete equation 不完全方程
incomplete equilibrium 不完全平衡
incomplete evapo(u)ration 不完全蒸发
incomplete expansion 不完全膨胀
incomplete expansion cycle 不完全膨胀循环
incomplete fertilizer 不完全肥料
incomplete figures 不完全统计资料；不完全统计数字
incomplete filling 不完全充填
incomplete flower 不完全花
incomplete fold 不完全褶皱
incomplete function 不完全函数
incomplete fusion 拟裂变；未熔合；未焊透；不完全熔融；不完全熔接
incomplete fusion zone 不完全熔化区；半熔化区
incomplete gamma function 不完全伽马函数
incomplete hiding 不完全遮盖
incomplete inter-run fusion 焊道间未熔合
incomplete inter-run penetration 层间未焊透
incomplete Latin-square 不完全拉丁方
incomplete latin square experiment 不完全拉丁方试验
incomplete leaf 不（完）全叶
incomplete linear system 不完全线性系统
incomplete linkage 不完全连锁
incomplete list 不完全目录
incomplete lubrication 不完全润滑
incompletely contracted orifice 不完全收缩口；不完全收缩孔
incompletely filling 未焊满
incompletely mixed aerated lagoon 非完全混合曝气塘
incompletely saturated 不完全饱和的
incompletely specified Boolean function 非完全指定的布尔函数；不完全说明布尔函数
incompletely specified function 不完全确定函数
incompletely specified unicode assignment 非完全指定的单码分配
incomplete mixing 搅拌不足；搅拌不匀；混合不匀；不均匀拌和；拌和不匀
incomplete mixing system 不完全混合系统
incomplete moment 不完全矩
incomplete observation 不完全观测值
incomplete overflow 不完全溢流
incomplete oxidation 不完全氧化
incomplete oxidation system 不完全氧化系统
incomplete packing 不完全充填
incomplete penetrance 不完全外显率
incomplete penetration 未钎透；未焊透
incomplete performance 不完全履行
incomplete phase operation 非全相运行（状态）
incomplete polynomial 不完全多项式
incomplete prior information 不完全先验信息
incomplete program(me) 不完全程序
incomplete quench 部分淬火
incomplete quenching 不完全淬火
incomplete randomization 不完全随机化
incomplete ranking 不完全秩评定
incomplete rarefaction wave 不完全稀疏波
incomplete reaction 不完全反应
incomplete recrystallization section 不完全的再结晶区
incomplete reduction 不完全还原
incomplete reflection 不完全反射
incomplete regeneration 不完全再生
incomplete root penetration 根部未焊透
incomplete round of direction observation 不完全方向观测
incomplete routine 不完全例行程序
incomplete secondary sewage treatment 不完全二级污水处理
incomplete segmentation 不全分裂
incomplete separation 不完全分离
incomplete set of directions 不完全方向组
incomplete sex-linkage 不完全（性）连锁
incomplete shear crack 不完全剪切裂隙

incomplete shell structure 不完全壳结构
incomplete shrinkage 不完全收缩
incomplete specification 不完全说明
incomplete statistical observation 不完全统计观测值
incomplete survey 非全面调查
incomplete system of equation 不完全方程组
incomplete task log 未完成任务记录
incomplete test 未完成的试验
incomplete tetanus 不全强直
incomplete thread 不完整螺纹;不完全螺纹
incomplete tillage 非全套耕作
incomplete treatment 不完全处理
incomplete well 浅井;未完成的井;不完善井
incomplete β function 不完全贝塔函数
in compliance with 遵照
incompressibility 不可压缩性;不可压缩条件
incompressibility condition 不可压缩性条件
incompressibility factor 不可压缩性系数
incompressibility modulus 不可压缩(性)模量
incompressible 不易压缩的;不能压缩的;不可压缩的
incompressible body 不可压缩物体
incompressible boundary layer 不可压缩的附面层;不可压缩的边界层
incompressible condition 不可压缩条件
incompressible deformation 不可压缩变形
incompressible energy equation 不可压缩能量方程
incompressible field 不可压缩场
incompressible flow 不可压缩水(流);不可压缩流动
incompressible fluid 非压缩性流体;不可压缩性流体;不可压(缩的)流体
incompressible fluid boundary layer 不可压缩流体附面层
incompressible hydrokinematics 不可压缩流体动力学
incompressible ideal fluid mechanics 不可压缩理想流体力学
incompressible jet 不可压射流
incompressible liquid 不可压缩液体
incompressible material 不可压缩(性)材料
incompressible medium 不可压缩介质
incompressible stratum 不可压缩层
incompressible viscous fluid 不可压缩黏性流体
incompressible visco(us)-fluid mechanics 不可压缩黏性流体力学
incompressible volume 不可压缩体积
incompression 压缩状态;受压
incomputable 不能计算
incondensable 不能凝缩
incondensable gas 不(冷)凝气体
incondensible 不能冷凝的
inconductivity 无传导性;不电导性
inconel 铬镍铁合金;因康镍合金
inconel weld 单面焊缝
inconformity 不一致
in conformity with design specifications 符合设计要求
in conformity with the law of nature 符合自然规律
in conformity with the specifications 符合规范规定
incongeable 不凝结的
incongruence 不交合
incongruent 不符合的
incongruent articulation 不等面关节
incongruent compound 非一致熔融化合物
incongruent drag fold 异向拖褶皱
incongruent melting 异无熔化;异成熔化;异成分熔融;非同分熔融;不一致熔融
incongruent melting compound 不一致熔融化合物
incongruent melting point 异元熔点;非一致熔融点;不一致熔点
incongruity 不一致;不协调(性);不相称;不交合;不调和
incongruous 不适合的
Inco nickel 因科镍;可锻镍
inconnector 内接符;流线内接符
inconsecutive 不连续的
inconsequence 前后不符
inconsequent 不连贯的
inconsequent drainage(system) 不顺向水系

inconsequential 无关紧要的事物
inconsequent river 非顺向河
inconsequent stream 非顺向河
inconsistency 不一致性;不一致;不相容性
inconsistent 不一致的;不相容的;不调和的
inconsistent bias 非一致性偏倚
inconsistent data 非一致数据
inconsistent equation 矛盾方程;不相关方程;不相容方程
inconsistent estimator 不一致估计量;不相容估计量
inconsistent fiducial distribution 不一致置信分布
inconsistent merchandise assortment 不配套品种
inconsistent shared data 不相容的共享数据
inconsistent statements 不相容语句
inconsistent statistic 不一致统计量
inconsistent statistor 非一致统计量
inconstancy of volume 体积膨胀
in constant dollar value 按不变价值美元计算
inconstant element 不稳定元素
inconstant fold 不稳定褶皱
inconstant value 变数
inconsumable 非消费性的
Inconterms 国际贸易条件解释通则
incontestable clause 不可抗辩条款
in control 在控制内的
incontrollable 失控;不能控制的
inconvenience 麻烦;不方便;不适合
inconvenient 不方便的
inconvertibility 不能兑换;不可转换性;不可逆性;不具兑换性
inconvertible 不能交换的;不能变换(的)
inconvertible currency 不能自由兑换的货币;不兑换通货
inconvertible paper 不能转换契约;不兑现契约
inconvertible paper currency 不可兑换纸币
incoordinate movement 不协调动作
incoordination 协调不能;不协调(性);不配合;不等同
incoordination form 不协调式
incoordination from of prospecting by mining 探采工作失调
incoordination load 不匹配负载
in-core compiler 磁芯内的编译程序;常驻(内存)编译程序
in-core detector 堆芯内探测器
in-core loop 堆芯回路
incorporate 结合;混用;存入;包括
incorporated 合并的;订入;编入
incorporated appliance 组合灶具
incorporated company 股份(有限)公司
incorporated liability 有限责任
incorporated pocket-books 皮包公司
incorporated radionuclide 结合态放射性核素
incorporated scanning 插入扫描
incorporate in 并入
incorporating foreign direct investment 吸收外商直接投资的方式
incorporation 有限公司;注册公司;公司;混入;合并;成立公司;掺入;搀入
incorporation for trailing of faults 追踪断层的归并
incorporation device 混合装置
incorporation of piles into a concrete footing 桩基与混凝土基脚的锚固
incorporeal assets 无形资产
incorporeal capital 无形资本
incorporeal chattel 无形动产
incorporeal property 无形财产
incoporeity 无形体
incorrect 错误的;错误;不正确的
incorrect data 不正确数据
incorrect grinding of tool 刀具不正确刃磨
incorrect in size 尺寸不准
incorrect length 错误长度;不正确长度
incorrect marking 标志不清
incorrect match 不合色样
incorrectness 不正确
incorrect operation 不正确动作
incorrect switching 误操作(开关)
incorrect trip 误跳闸
incorrect weld profile 焊缝形状不对
incorrect weld size 焊缝尺寸不对
incorrelate 不相关的;不相关
incorrigible 难以纠正的
incorrodible 不(受)腐蚀的

incorrosive 不腐蚀的
incorruptibility 坚固性;耐用度;不腐败性
in couples 成对的
in course of construction 在施工中
increase accumulation 增加积累
increase amount 增加金额
increase by degrees 递增
increase by wide margin 大幅度增长
increase core recovery 提高岩芯采取率
increased competition 竞争扩大
increased contrast 增强反差
increased cost 费用增加
increased counts 计数增量
increased cruising range 增大的巡航航程
increased current metering 增流计算
increased/decreased reserve by exploration 勘探增减储量
increased/decreased reserve by recalculation 重算增减储量
increased demand 需求增加
increased density 增加密度
increased factors of safety 增大的安全系数
increased incoming longwave radiation reflection 增加射入长波辐射反射
increased logarithmic phase 对数增长期
increased outgoing longwave radiation 增加射出长波辐射
increased pulse 频脉
increased rate of fire 提高的发射速度
increased resistance invasion 高侵
increased resonance 叩响增强
increased-safety motor 增安型电动机
increased unit 增产量
increase during the year 全年增长
increased value 增加的价值
increased value insurance 增值保险
increased value policy 增值保单
increased water use efficiency 增加水分利用率
increased yield value 增加的屈服值
increase economic of efficiency 提高经济效应
increase gear 增速(传动)装置
increase global radiation reflection 增加环球辐射反射
increase in absolute figures 增长的绝对数
increase in amounts 增加额
increase in capital 资本增值
increase in cost 费用的增加
increase in demand supply 需求供给量的增加
increase in density 密度增高
increase in dip 倾斜角增大
increase in exposure 曝光延长
increase in groundwater level 地下水水位增高
increase in inventories 库存增加
increase in rates 费率升高
increase in speed 增速
increase in stock of inventories 库存存货的增加
increase in the water-holding capacity of soil 增加土壤保水力
increase in value 价值的增长
increase in yield 增产
increase of accuracy 精度提高
increase of budget 增加预算
increase of contrast 反差增大
increase of cross-section 断面增大
increase of demand 增加需求
increase of efficiency 效率提高;增加效率;提高效率
increase of frequency 提高频率
increase of loading 荷载增加
increase of number of runs 增加班次或线路
increase of parallel interval with latitude 渐长纬度
increase of power 功率增加
increase of premium 增加保费
increase of pressure 气压上升
increase of production 增加生产;产量增长
increase of risk 增加危险;增加风险
increase of service life 延长使用寿命
increase of speed 增速;速率增长
increase of stocks 增加库存
increase of supply 增加供给
increase of threshold 阈域增长
increase of value 增值
increase operation rate 提高利用率
increase or decrease clause 增减条款
increase plot 繁殖区

increase position 加码
increase production and practice economy 增产节约
increase production without increasing the work force 增产不增人力
increase productivity 提高生产率
increase progressively 递增
increaser 锥形管;异径接头;联轴节缩小套节;连轴齿套;扩径(水)管;渐扩管
increase rate of radon 氡增长率
increase resistant system 增加抵抗力系统
increase the acid content of soil 提高土壤酸含量
increase the par value 提高平价;提高票面价值
increase wage 增加工资
increasing absorbing power 提高土壤吸收性能
increasing amount 增长量
increasing annuity 定期增加的年金;递增年金
increasing coefficient of allowable stress 容许应力提高系数
increasing cost 生产成本增加;递增成本;成本递增
increasing coupling 扩径管;渐扩管
increasing dosage effect 提高剂量效应
increasing exchange power 提高代替能力
increasing expenses 递增费用
increasing failure rate 单调增加故障率
increasing forward wave 增大前进波
increasing function 增(量)函数;上升函数;递增函数;单调非减函数
increasing gear 增速装置
increasing in yield 增产
increasing irrigation efficiency 提高灌溉效率
increasing of seismic activity 地震频度增大
increasing of soil fertility 提高土壤肥力
increasing oscillation 增幅振荡;升幅振荡
increasing pitch 递增螺距
increasing power 升幂
increasing process 增过程
increasing rate of economic effect 经济效益提高率
increasing rate of geologic(al) market receipt 市场收入增长率
increasing rate of management profit 经营盈利增长率
increasing rate of result value 成果价值增长率
increasing rate of salinity 矿化度的增长率
increasing return 递增收益;报酬渐增;增加利润;收益递增;递增报酬
increasing risk aversion 递增厌恶风险
increasing runoff 增强流量
increasing sequence 增序列;递增序列
increasing series 递增级数
increasing soil acidity 增加土壤酸性
increasing soil moisture content 提高土壤水分含量
increasing soil permeability 提高土壤渗透性能
increasing specific surface area 提高比表面积
increasing strength 增强能力
increasing technique 增量法
increasing the total yield 提高总产量
increasing titer gradient 效价递增
increasing traffic volume 增长交通量
increasing value 增值
increasing velocity 递增速度
increasing water quality 提高水质
increasing wave 增幅波;增长波
incredible 不可信的
increment 增值;增益;增加(值);增加物;增大;增(长)量;增长;灌注量;递增;补给量
increment adding 增量加载
increment addressing 增量寻址
incremental address 增量地址
incremental analysis 增量分析;阶段增量分析
incremental application of load 荷重增载
incremental assignment 增量分配
incremental attenuation 增量衰减
incremental backup 部分备份
incremental benefit 增量利润;边际利益
incremental benefit-cost ratio 增量效益费用比;边缘受益—费用比;边际利益费用比
incremental calculation 增量计算
incremental capacitor 精确校正电容器;精确调整电容器
incremental capital output ratio 边际资本产出率
incremental cash flow 资金流量净增额;增加的现金流动

incremental-charge polarography 增量电荷极谱法
incremental coder 增量编码器
incremental collapse 增量失稳;发展性破坏
incremental compaction 增量压缩;增量精简数据法
incremental compiler 增量编译程序
incremental computation 增量计算
incremental computer 增量计算机
incremental connector 加长连接件
incremental construction 建筑增量
incremental control 增量控制
incremental controller 增量控制器
incremental coordinates 相对坐标;增量坐标
incremental cost 增值(成本);增量成本;增量成本
incremental cost analysis 增量成本分析
incremental cost saving 增支成本节约额
incremental creep 增加的蠕变
incremental data 增量数据
incremental delivered power 输出功率增量
incremental depth 增量深度
incremental development 渐增性开发
incremental digital computer 增量数字计算机
incremental digital recorder 增量(式)数字记录器
incremental digitizer 增量数字化器
incremental discrepancy correction 增量闭合差改正数
incremental displacement 增量位移;位移增量
incremental displacement sensitivity 增量位移灵敏度
incremental display 增量显示
incremental dump 增量转储
incremental duplex 增流双工
incremental encoder 增量编码器
incremental execution 部分执行
incremental expenses 增支费用
incremental extrusion 渐近挤压
incremental fixed cost 增支固定成本
incremental flow 增支流
incremental frequence shift 增量频移
incremental frequency control 增量频率控制
incremental implementation 渐增实现
incremental inductance 增量电感
incremental inductance tuner 增量电感调谐器
incremental induction 增量感应;增量电感;增量磁感应强度
incremental installment allocation method 分期递增分配方式
incremental integrator 增量型积分器;增量积分器
incremental investment 补充投资
incremental iron loss 增量铁损
incremental launching method 顶推法(施工)
incremental line 增生线
incremental loading 渐增负荷;分级加荷
incremental loading test 逐级加荷试验
incremental load method 渐增荷载法
incremental load technique 增量荷重技术
incremental losses 增量损耗
incremental magnetic susceptibility 增量磁化率
incremental manner 递增方式
incremental method 增量法;阶段增量法;渐增法;递增法
incremental mode 增量方式
incremental model 增量模型
incremental motion 增量运动
incremental movement 增量运动
incremental net benefit 增得的净效益
incremental noise 增量噪声
incremental numerical solution 增量数值解
incremental permeability 增量噪声;增量导磁率;增量磁导率;磁导率增量
incremental plastic theory 增量塑性理论
incremental plotter 增量绘图仪;增量绘图器;绘图机;不连续曲线描绘器
incremental plotter control 增量绘图仪控制
incremental plotting rate 增量绘图速率
incremental portion 增长部分
incremental position 增量位置
incremental power 递增的电力
incremental pressure 增量压力
incremental profit 增量利润
incremental pulse 增量脉冲
incremental quantity 增量
incremental rate 增量率
incremental rate of ion concentration 离子浓度的增长率

incremental ratio 增量比
incremental recorder 增量记录器;连续记录器;级进记录器;步进记录器
incremental recording 增量记录
incremental regulation 增量调节
incremental-repetitional loading 反复递增荷载
incremental representation 增量表示法
incremental resistance 增量电阻;交流电阻与直流电阻之差;动态阳极电阻
incremental resistivity measurement 增量电阻率测量
incremental rigidity 增量刚度
incremental sampling 递增取样
incremental servo-drive 增量式伺服传动
incremental speed regulation 增量速度调整
incremental starting 分级增压起动
incremental stiffness 增量劲度;增量刚度
incremental strain 增量应变
incremental strain theory (锻压的)(增量理论)
incremental stress 增量应力
incremental system 增量系统
incremental technique 增量法
incremental ternary representation 增量三元表示法
incremental theory 增量理论
incremental theory of plasticity 塑性增量理论
incremental transducer 增量式传感器
incremental transmission loss 输电损失增量
incremental tuner 增量式调谐器;步进式调谐器
incremental value of throw with depth 落差随深度增大值
incremental vector 增量向量
increment and correction 增量与改正量
increment and correction table 增量与改正量表
increment angle 齿端角
incrementary ratio 增量比
incrementation 增量
incrementation by two 按二增量
incrementation parameter 增量参数
incrementation processing 增量处理
increment borer 取木钻;探木钻;生长钻
increment budgeting 增量预算法
increment by two 按二增量
increment distance of outer track build-up 外轨抬高递增距离
increment distance of widening on curve 弯道加宽递增距离
increment-drop test 落锤增量(冲击弯曲)试验
increment dump 增量转储
increment equation of motion 增量运动方程
increment equilibrium equation 增量平衡方程
increment felling 生长伐
increment groundwater 地下水增量;地下水增加;地下水增长;地下水增补
increment in land 土地增值税
increment internal pressure tester 耐内压试验机
increment interrupt 增量中断
increment investment 投资增加额
increment list 增量表
increment load 增量装入;递增荷载
increment load method 荷载增量法;递增荷载法
increment load procedure 增量荷载方法
increment mode display 增量式显示
increment motion 附加运动
increment of altitude 高度增量
increment of aquifer 含水层补给量
increment of azimuth 方位角增量
increment of coordinate 坐标增量
increment of decrease 递降率
increment of effective stress 有效应力增值
increment of face advance 工作面推进量;扭曲量增量;螺旋量增量(斜齿轮)
increment of foundation strength 地基强度增长
increment of function 函数增量
increment of ground strength 地基强度增长
increment of groundwater 地下水补给量
increment of height 高程增量
increment of hour angle 时角增量
increment of load(ing) 荷载增量;增载
increment of net recharge for back-pumping 回扬增加的净增量
increment of plastic strain 塑性应变增量
increment of pumping cost 抽水费用增加
increment of roll 摇摆增量

increment of speed 速度增量
increment of sun's right ascension 太阳赤经的增量
increment operation 增量运算
increment pack up 增量备份
increment percent 增量百分率
increment pulse 增量脉冲
increment rate 增加率
increment reaction 渐增反应
increment register 增量寄存器
increment routine 增量程序
increment sampling 小样取样;增量取样(法)
increments of settlement 沉降增量;沉降增量
increment speed 增长速度
increment tax 增值税
increment tax on land value 土地增值税
increment to the stock 资源增长量
increment value 增值
increment value duty 增值税
increment value tax 增值税
increment velocity 增量速度
increscent 渐增的
incriptions 记名证券
incrop 遮蔽露头
incrust 结壳;覆以硬壳;不易剥落
incrustant 建筑物表面装饰的;建筑面饰;长污垢的;垢壳;水垢
incrustate 结壳;表面硬化
incrustated well 结垢的井孔
incrustation 锈疽;锈疤;镶贴;硬渣;渣皮;结皮;结壳;锈垢;水锈;水碱;水垢;饰面细工;沉积;表面结皮
incrustation composition of casing well 管井结垢物成分
incrustation in pipes 管内积垢
incrustation sheath 外皮
incrustation texture 被壳结构
incrusted ballast 道砟板结
incrusted glass 镶嵌的玻璃;镶嵌玻璃
incrusted soil 结壳土壤
incrusting 起壳;积垢
incrusting substance 结壳物质
incrustion 黏结窑皮;表面装饰
incrystallizable 不能结晶的
incrystate 硬皮
incubater[incubator]培育箱;培养箱;恒温箱;恒温器;孵育箱;孵卵器;孵化器;定温箱;保温箱
incubation 保温培养
incubation medium 培养基
incubation period 诱导期;培养期;潜育期;潜伏期
incubative temperature 定温箱温度
incubator for eggs 孵卵器
incubator oil 孵化箱燃料油
incubator thermometer 恒温箱温度计
incubator tray 蛋盘
incumbator test 稳定度试验
incumbent 弯垂下来的;覆盖在上面的;压在上面的;负有义务的
incumbent cotyledons 背倚子叶
incumbrance (不动产的)财产抵押权;不动产置留权;保留数
incuneation 嵌入;楔入
incurable depreciation 不可弥补的贬值
incur debt 借债;负债
incur loss 蒙受损失
incur obligation 负有义务
incurred losses 已发生损失
in current price 按现价计算
incursion 侵入
incurvate 凹入的
incurvation 内弯;内曲
incurvature 内曲率
incurve 内曲球;内曲
incus 砧状云
incycle 进入循环
indalloy 英达洛依焊料;铟银焊料(含铟90%,银10%的合金)
in damaged condition 已受损
indane 茚满
indanthrene 阴丹士林;标准还原蓝
indanthrene blue 还原蓝;阴丹士林蓝
indanthrene blue-green 阴丹士林蓝绿
indanthrene bordeaux 阴丹士林枣红
indanthrene brilliant blue 阴丹士林亮蓝

indanthrene brilliant green 阴丹士林亮绿
indanthrene brilliant orange 阴丹士林亮橙
indanthrene brilliant rose 阴丹士林亮玫瑰红
indanthrene dark blue 阴丹士林暗蓝
indanthrene direct black 阴丹士林直接黑
indanthrene dyes 阴丹士林染料
indanthrene golden orange 阴丹士林金橙
indanthrene olive 阴丹士林橄榄绿
indanthrene printing blue 阴丹士林印染蓝
indanthrene red violet 阴丹士林红紫
indanthrene steamer 阴丹士林蒸箱
indanthrene vat colo(u)r 阴丹士林瓮染色
indanthrone 阴丹酮
indanthrone blue 靛蓝
in dash installation 内部仪表装配
indazole 吲唑
inddor design temperature 室内设计温度
indebted 负债的
indebtedness 负债
in decades to come in 在未来数10年内
indeciduous 常绿的;不落叶的
indecision 决定不能
indecomposability 不可分性
indecomposable code 不可分码;不可分割解码
indecomposable matrix 不能分解矩阵;不可分解矩阵
in decuplicate 一式10份
in-deep filtration 深层过滤
in-deep sampling 深层采样
indefatigability 不易疲劳
in default 违约
in default of payment 不付款
in-defined boundary 不清晰边界
indefinite 不明确的
indefinite ceiling 不定云幂高度
indefinite decision-making 非确定型决策
indefinite equation 不定方程(式)
indefinite form 不定型;不定(形)式
indefinite integral 不定积分
indefinite integration 不定积分法
indefinite limitation of time 不确定期限
indefinitely variable transmission 无级变速传动(装置)
indefinite matrix 不定矩阵
indefinite metric 不定度规
indefinite point 不定穴
indefinite projects 未定工程项目
indefinite right 没有限定的权利
indefinite scale 自由比例尺;任意比例尺
indefinite standard 不定标准
indefinite storage 无定期储存
indefinite summation 不定总和
indefinite suspension 无限中止
indefinite term 不确定期限
indefinite thing 非特定物
indefinite variability 不定变异性
indeformable 不易变形的;不变形的
in-degree of vertex 顶点的入度
indehiscent 不开裂的
indehiscent air 闭气
indelible copy 开印印样
indelible ink 擦洗不掉的墨水;防弊油墨;不灭墨水;不变墨水
indelible marking ink 不灭墨印
indelible pencil 保迹铅笔
indelta 河流分流内陆区;内陆三角洲;内河三角洲
in demand 顾客需要
indemnification 赔偿费;免遭损失;赔偿物;赔偿金
indemnify entertained 准予赔偿
indemnify for the loss incurred 赔偿所受损失
indemnifying measure 补偿措施
indemnify refused 拒绝赔偿
indemnitor 赔偿人
indemnity 免罚;赔款;赔偿(金);损害赔偿;补偿物
indemnity agreement 赔偿协议;赔偿协定;补偿协议
indemnity bond 赔偿契约;补偿保证
indemnity by employer 雇主提供的保障
indemnity clause 赔偿条款
indemnity contract 赔偿契约;赔偿合同
indemnity for area loss 用地补偿
indemnity for damage or loss 赔偿毁损
indemnity for damages 赔偿损失
indemnity for defamation 名誉损害赔偿;赔偿名誉

indemnity for injury 工伤赔偿费
indemnity for risks 风险赔偿
indemnity in kind 赔偿实物
indemnity insurance 损失补偿保险
indemnity of accident 事故赔偿
indemnity period 索赔时效
indemnity refused 拒绝赔偿
indemonstrability 无法表明
indene resin 聚茚树脂
indenometer 井身直度自动记录仪
indent 压印;压牌;给文件加边;刻痕;锯齿形;进口订货单;缺划;委托购买书(国外订货单);双联订单;使凹进;订(货)单;凹砖牙
indent account 代购账户
indentation 印压;压坑;压痕;压刻;刻痕;礧磙;海岸线弯入处;缺刻;曲折形;曲折岸;低凹;齿状饰;成穴作用;凹穴;凹陷;凹入;凹痕;凹槽
indentation anisotropy 压痕各向异性
indentation creep 压痕蠕变
indentation cup 半球形压痕
indentation depth ga(u)ge 压痕深度计
indentation hardness 印痕硬度;压痕硬度;胀压硬度;刻痕硬度;成穴硬度
indentation hardness number 压痕硬度指数;布氏硬度指数
indentation hardness of rock 岩石压入硬度
indentation hardness test 压痕硬度试验
indentation hardness tester 压痕硬度试验机;压痕硬度计
indentation index 压痕指数
indentation machine 凹凸轧印机;压痕硬度计;硬度机;压痕硬度机
indentation method 压入法;刻痕法;凹痕法
indentation of durometer 硬度计针入度
indentation ratio 压深率
indentation test 刻槽压痕试验;压入试验;压痕(硬度)试验;压凹硬度试验;球印硬度试验
indented 齿状的
indented bar 竹节钢;螺纹钢筋;齿印钢筋;刻痕钢筋
indented beam 锯齿式组合梁;错口式组合梁
indented bolt 锯齿式锚固螺栓;螺纹锚杆;刻痕螺栓;带纹螺栓;齿纹螺栓
indented chisel 牙轮钻头;锯齿錾;齿状钻头
indented coast 曲折海岸
indented coast line 海湾海岸线;曲折海岸线
indented coast of erosion 海蚀锯齿形海岸
indented fiber 刻痕纤维
indented girder 锯齿式组合梁;错口式组合大梁
indented hammer 锤头
indented joint 齿槽连接;齿(状)接合;齿合接缝;犬牙交错缝;茬口接缝;错口接合
indented mo(u)lding 齿纹线脚;有齿形装饰的线脚
indented ramp 礧磙坡道
indented ribbed bar 竹节钢;锯齿形竹节钢(筋)
indented roller 凹纹压路机;凹纹路碾;凹纹滚子
indented sill 齿形底槛;齿槛
indented steel wire 刻痕钢丝;齿纹钢丝
indented wire 轧痕钢丝;齿纹钢丝;变形钢丝;刻痕钢丝;麻面钢丝;齿痕钢丝
indentee 受托代办人
indenter 硬度计压头;压入头;压痕硬度计的压头;刻压机;刻痕压路机
indent for space 舱位订单
indent house 进口订货行;进口代办行
indenting 压窝;直荏;留荏;马牙荏【建】;齿纹线脚;成穴的;订单
indenting apparatus 凹进仪
indenting ball 硬度计球;钝钻头
indenting course 凹凸砖层;锯齿形水道
indenting hammer 锤头
indenting roller 凹纹压路机;齿状滚子;刻痕滚子
indent invoice 托购清单
indention 缺刻;成穴
indent number 订货单号
indentometer 直读压痕硬度计
indentor 压痕硬度计的压头;委托商
indent tab character 空格制表符;空格标记字符
indenture 凭单;缺划;契约;双联合同;成穴;在船实习契约
indenture deed 规定双方权利义务的契约
indenture defaulting 违约
indenture labo(u)r 合同工;契约劳工
indenture trustee 契约受托人
indenture worker 合同工

independence 自主性;无关性;独立性
independence accountant 独立会计师
independence contractor 独立承包商;独立承包人
independence in analysis of variance 方差分析中的无关性
independence of fuzzy event 模糊事件独立性;模糊事件的独立性
independence of random variables 随机变量的独立性
independence test 独立性检验;独立性测验
independency 无关
independent 独立的;单独的
independent accounting 独立核算
independent accounting system 独立核算制
independent accounting unit 独立核算单位
independent adjustment 单独调整;单独调节
independent analysis 独立分析
independent appraisal 独立估价
independent architect 独立开业建筑师
independent arch method 纯拱法
independent assortment 独立分类
independent astronomical data 独立天文测量数据
independent audit 独立审计
independent axle 半轴;独立轴;自动轮
independent axledrive 单独轴驱动
independent beam plow 组式犁
independent block pavement 独立块铺面
independent boiler 独立锅炉
independent brake 独立制动;单独制动
independent brake valve 独立制动阀;单独制动阀
independent catalogue 独立星表
independent chuck 四爪卡盘;单独移爪卡盘;分动夹盘
independent clutch controlled drive 离合器控制的传动装置
independent compilation 独立编译
independent component 独立组分;独立成分
independent component release 独立元件释放;独立部件释放;独立部件分离
independent condition 独立条件
independent construction unit 独立建筑单位
independent contact 独立触点
independent contractor 独立经营的承包商;独立合同人;独立承包商;独立承包人
independent control 独立控制;单独控制
independent controller 独立控制器
independent coordinate system 独立坐标系
independent criterion 独立准则
independent crowd 独立式推压
independent crystallization 独立结晶
independent cutting method 分别剔除法
independent cylinder 分置式油缸
independent data communication 独立数据通信
independent day number 独立日数
independent differential operators 无关微分算子
independent digit 独立数字
independent double bonds 孤双链
independent double linkages 孤双链
independent drive 独立驱动;单独驱动;单独激励
independent driven exciter system 独立励磁制
independent drive oscillator 主振式振荡器;他激振荡器
independent economic growth 经济自行增长
independent effect 独立作用效应
independent elements 无关元
independent enterprise 独立自主的企业
independent entry 独立元;独立项
independent equation 独立方程
independent equatorial coordinate system 恒星赤道坐标系
independent event 独立事件
independent excitation 单独励磁
independent failure 独立失效;独立故障
independent fastening 分开式扣件【铁】
independent fee appraiser 独立收费地产评估员
independent folds 独立褶皱
independent footing 独立基脚;独立基础;独立底脚
independent foundation 独立基础;独立地基
independent function 独立函数
independent girder 悬吊大梁;独立大梁;独立主梁
independent hearth fire 不带después板的壁炉
independent heating 单室采暖
independent heat supply 单独供热
independent height system 独立高程系

independent hotplate 板式灶
independent image-pairs 独立像对
independent increment 独立增量
independent indexing mechanism 独立变址机构
independent input/output channel 独立输入输出通道
independent inspection 独立检查
independent inspector 独立检查员
independent integral 无关积分
independent investment 独立投资;单项投资
independent joint action 独立的联合作用
independent landscape 自成景观
independent leg jackup 独立腿自升平台
independent linear equation 无关线性方程
independently mounted 独立悬挂式的
independent measurement 单独测量
independent median 独立式中央分隔带
independent mesh 独立回路
independent mineral 独立矿物
independent mobile home 独立活动式家庭挂车;独立旅游居住车
independent model 独立模型
independent model aerial triangulation 独立模型法空中三角测量
independent model aerotriangulation 独立模型法空中三角测量
independent model block triangulation 独立模型法区域空中三角测量
independent model triangulation 独立模型三角测量
independent molecule 独立分子
independent motor drive 单独电动机驱动;单独电动机传动
independent navigation 自主导航
independent network 独立网【测】
independent observation 独立观测
independent oiler 单独加油器
independent operation 独立操作;分列运行
independent overflow area 独立溢出区
independent parameter 独立参数
independent partial tide 独立分潮
independent partition scheduling 独立区分调度
independent periods 独立周期
independent physical-chemical treatment 单独物理—化学处理
independent pole 独立式脚手架(双排支柱的脚手架)
independent-pole scaffold(ing) 独立双排立柱脚手架;双排脚手架
independent probability 独立概率
independent processor 独立处理机
independent program(me) loader 独立程序的装入程序
independent pump 单动泵
independent random variable 独立随机变量
independent realm 独立王国
independent reference 独立引用
independent release 单独缓解
independent resource 独立资源
independent resurvey 独立重测
independent rotation mechanism 单独回转机构
independent sample 独立样本
independent scaffold 独立(式)脚手架(双排支柱的脚手架)
independent screw chuck 分动螺旋卡盘
independent seconds watch 每秒一跳秒表
independent sector 独立扇形装置;独立扇面
independent segment 独立段;独立程序段
independent segregation 独立分离
independent sequence of partitions 无关的分割序列
independent set 无关集
independent shipowner 独立船东
independent sideband 独立边带
independent side-band receiver 独立边带接收机
independent side band transmitter 独立边带发射机
independent solution 无关解
independent space 无关空间
independent spirit level 独立水准器
independent stair(case) 独立式楼梯;不靠墙楼梯
independent suspension 独立悬挂
independent switch 独立开关
independent synchronizing system 独立同步制

independent tank 舷侧独立舱;独立舱
independent tank center 中间独立舱
independent test 独立性检验
independent tide 独立潮
independent time lag protection 定时限保护
independent time lag relay 定时限继电器
independent trailer 全挂车
independent trailer coach 独立拖挂式客车
independent triangulation 独立三角测量
independent triangulation network 独立三角网
independent type boiler 独立式锅炉
independent unit 独立单元
independent unknown 独立未知数
independent utility 独立实用程序
independent variable 自发量;自变数;自变量;独立变数;独立变量
independent variable depth sonar 独立可变深度声呐
independent ventilation 独立通风
independent wave form 单independent波形
independent wire rope core 钢丝绳内芯
independent wire rope core wire rope 绳式股芯钢丝绳
independent wire strand core wire rope 钢芯钢丝绳
independing mode 独立方式
indepenent time-lag 定时限
in depth 深入的;彻底的
inderborite 多水硼镁钙石;变水方硼石
inderite 多水硼镁石
indescribable mathematics 模糊数学
indestructibility 不损坏性;不灭性;不可毁性
indestructible 不能破坏的;耐久的;不灭的
indeterminable 超静定的
indeterminacy 模糊;测不准;不确定(性)
indeterminacy phenomenon 不确定现象
indeterminate 未知数;待定的;不定式;不定的
indeterminate analysis 不定解析;不定(解)分析
indeterminate appropriation 无定额经费分配;不定拨款
indeterminate beam 超静定梁
indeterminate boundary 不定边界
indeterminate case 不定情况
indeterminate circle 不定圆
indeterminate cleavage 不定裂
indeterminate coefficient 未定系数;待定系数;不定系数
indeterminate constant 不定常数
indeterminate coupling constant 不定耦合常数
indeterminate equation 不定方程(式)
indeterminate equation model 非确定性系统模型
indeterminate error 未定误差;不(可)定误差;不可测误差
indeterminate fault 不确定性故障
indeterminate form 未定(形)式;不定形式
indeterminate frame 超静定构架;超静定刚架
indeterminate framing 超静定构架
indeterminate function 不定函数
indeterminate lay days 未确定装卸货日期
indeterminate line of curvature 不定的曲率线
indeterminateness 超静定性;不(确)定性;不确定度
indeterminate operation 不定操作
indeterminate plant 中小植物
indeterminate principle 测不准原理
indeterminate range 不定范围
indeterminate solution 不定解
indeterminate stem 无限型茎
indeterminate stress 不定应力
indeterminate structure 超静定结构;不定结构
indeterminate-term liability 不定期负债
indeterminate truss 超静定桁架
indeterminate zone 不确定带
indetermination 待定;不确定
indeterminism 测不准论
index 索引;指引;针盘指针;归位;幂;附标;标引;变址;编索引;指数;指标;[复] indices
indexable 可加索引
index access method 变址存取法
index adder 变址加法器
index area 索引区
index-area method 指示面积法;(径流预报的)指标—面积法
index arm 指针;指示(杆);指臂;六分仪游标;六分仪标志杆

index arm head 指臂头
index array 变址(数)寄存器组
indexation 指数化;指数法
indexation of price 物价指数化
indexation of tax allowances and taxation 课税减免指数化与税收
index bar 指标杆;指臂
index barricade register 变址隔离寄存器
index basin 指示流域;参证流域
index beam 引示射束;指引射束
index bed 指示层;标准层;标志层(位)
index bit 索比比特;变址位
index board 指
index book 目录书
index bristol 索引光泽纸板
index buffer 引导缓冲器;变址缓冲器
index card 索引卡(片)
index center[centre] 分度虎钳;分度顶尖
index change gears 分度交换齿轮架
index chart 指示图;指标图;接图表;海图索引图;索引图
index chuck 分度卡盘
index compound 母体化合物;索引化合物
index-consistency test 指数稠度试验
index contour 注有数字等高线;指示等高线;指标等高线;计曲线;基本等高线;首曲线;示量等高线;标注等高线
index correction 仪表指针校正;仪表刻度校正;仪表读数校正;指标改正(量);指标订正(量);刻度校正
index counter 指数计数器;变址计数器
index crank 分度摇柄;分度头曲柄
index curve 指数曲线
index cycle 指数循环;变址周期
index data item 标引数据项;变址数据项
index definition 指数定义;变址定义
index device 调车定位装置;分度装置
index diagram 接图表
index dial 刻度盘;表盘
index disc 指示盘
index domain 索引区
index drawer 索引抽屉
indexed access 变址访问;变址存取
indexed address 已变址地址;结果地址;变址地址;被变址地址
indexed addressing 索引寻址;变址寻址;变址选址;变址访问;变址地址;变址编址
indexed addressing mode 变址寻址方式
indexed bond 指数化债券;物价指数债券
indexed data name 加标数据名;索引型数据名;变址数据名
indexed document 索引文献
indexed family 加标族
indexed file 索引文件;加索引的文件
indexed file structure 定位文件结构
index edge 索引栏
indexed list 加下标表;索引表
indexed loan 指数化贷款
indexed mode 变址方式
indexed plan 标高平面
indexed plane 标高平面;高程平面
indexed rate 与指数挂钩的兑换率
indexed recovery rate 加标回收率
indexed sequential 变址序列;变址顺序
indexed sequential access method 索引顺序存取法
indexed sequential access mode 按索引顺序访问方式
indexed sequential data set 加下标顺序数据集;索引顺序数据集
indexed sequential file 加下标顺序文件;索引顺序(编制)文件;顺序检索文件
indexed sequential organization 加下标顺序结构;索引顺序结构;顺序编制
indexed set 加下标组;索引集
indexed system 加标系
indexed zero page addressing 变址零页寻址
indexer 索引器;分度器
index error 指数误差;指示误差;指标(误)差;分度误差
index error of sextant 六分仪指示差
index error of staff 标尺零误差
index error of vertical circle 竖盘指示差
index factor 指标因子

index feed 分度进给
index field value 变址域值
index finger 指针;食指
index flood 指示洪水
index forest 指标林
index formation 标准建造
index formula 指数公式
index fossil 标准化石
index ga(u)ge 指示计;指示表
index gear 分度齿轮
index glass (测角器的)指示镜;六分仪动镜;动镜;分度镜;标镜
index gradient 指数梯度
index hand 指针;指示针
index handle 分度柄
index head slide 分度头溜板
index hole 定位孔;分度孔
index horizon 标准层位
index indicative 指数
index indirect addressing 变址间接寻址
indexing 转位;转换角度;加下标;换档;索引;分度;标引;标行;标刻度;标定指数;变址;编制指数;指数;指标
indexing addressing 变址型寻址
indexing applications unit 变址器;变址单元
indexing attachment 分度附件
indexing center 分度顶尖
indexing circuit 指引电路
indexing feed 间歇进给;分级进给
indexing fixture 分度夹具
indexing gear constant 分度机构常数
indexing head 分度头
indexing hole 分度孔
indexing jig 分度钻模
indexing language 标引语言
indexing machine 转位机
indexing mechanism 转位机构;间歇机构;分度机构
indexing method 索引法;变址法
indexing of document 文件编索引
indexing operation 分度操作;变址操作
indexing plate 刻度盘;分度盘;标牌;标度盘;说明牌
indexing plunger 分度销
indexing register 指数寄存器
indexing set 指标集
indexing shaft 分度轴
indexing system 指数调整制度;标引系统;编目系统
indexing table 转位工作台
indexing transfer 间歇传送
indexing type milling fixture 分度式铣夹具
index instruction 变址指令
index intensity 标记强度;标记亮度
index in terms of material products 实物指标
index jumbo 带钻机定向的钻车
index law 指数律
index level 指示水准器;变址级
index lever 分度杆
index limestone 标准石灰岩
index limnograph 水位标记仪
index line 等值线;指示线;指标线;分度线
index-linked loan 保值借贷
index linking 同物价指数相联系的
index liquid 折射率液;浸液
index machine 索引机
index major contour 加粗等高线
index manual 牵引手册;索引手册
index map 接合图;航线标示图;位置图(建筑物);索引图
index mark 指标;指数;指示器;参见号;标线
index marker 索引标记
index-mark for zeros 零位标线
index matched crystal 折射率匹配晶体
index matching 折射率匹配
index mechanism 索引机构
index memory 变址存储器
index method 指数法
index microscope 指标显微镜;读图显微镜
index mineral 指示矿物;标准矿物;标志矿物
index mirror 指标镜;六分仪动镜;目标定位镜;动镜;标(度)镜
index model 指数模型
index modification 索引修改;变址修改
index modification addressing 变址修改型寻址
index modify 变址修改
index mosaic 镶嵌索引图

index movement 指数变化
index name 附标名;下标名;索引名
index number 指数;索引号(码);索引编号;分度数
index number of all commodities 总物价指数
index number of arithmetic(al) average 算术平均指数
index number of average indicatrix 平均指标指数
index number of building industry production 建筑业生产指数
index number of capacity utilization 开工率指数
index number of construction activity 建筑活动指数
index number of construction cost 建筑价格指数
index number of construction output 建筑生产指数
index number of cost of living of staff and workers 职工生活费指数
index number of export prices 出口物价指数
index number of food production 食品生产指数
index number of import prices 进口物价指数
index number of industrial production 工业生产指数
index number of merit 优质指数
index number of nominal wages 名义工资指数
index(number) of price 物价指数
index number of real wages 实际工资指数
index number of sensitive prices 敏感(性)物价指数
index number of wage 工资指数
index of abbreviation 缩写词典
index of abnormality 非正态指数
index of absorption 吸收指数
index of abundance 丰度指数
index of acidity 酸度指数
index of acousto-optic(al) property 声光性能指数
index of activity 活性指数;活度指数
index of adjoining chart 邻图索引
index of adjoining sheet 接图表
index of adsorption 吸收指数
index of affinity 亲和指数
index of alteration 蚀变系数
index of a radical 根指数
index of aridity 干燥指数;干燥率;干旱指数
index of atmospheric purity 大气纯度指数
index of average unit price paid 平均单位支付价格指数
index of basicity 碱性指数
index of bioactivity 生物活性指标
index of biologic(al) activity 生物活性指标
index of biologic(al) integrity 生物完整性指数
index of blasting action 爆破作用指数
index of business 商业指数
index of business condition 商情指数
index of coincidence 符合指数
index of complexity 复合指数
index of conformity 一致性指数
index of consumption 消费指数
index of cooperation 合作指数;合作系数
index of correlation 相关指数;相关指标
index of critical point 临界点指数
index of damage 破损指数
index of determination 可决指数
index of difference 差量指数
index of diffraction 衍射指数
index of discharge 流(出)量指数
index of dispersion 离中趋势指数;离差指数
index of dynamic(al) energy 动能量指数
index of employment cost 就业费用指数
index of eutrophication 富营养化指数
index of export price 出口价格指数
index of flow state 流态指数
index of fold 褶曲指数
index of friction 摩擦指数
index of fuzziness 模糊性指数
index of heading indicator 主要指标的指数
index of heating effect 加热指数
index of hole quality 钻孔质量指标
index of igneous and metamorphic petrology 火成岩石学和变质岩石学的指数
index of import and export 进出口指数
index of import volume 进口物量指数
index of imprimitivity 非本原指数
index of inconsistency 不一致性指数
index of inertia 惯性指数;惰性指数

index of inventors and patentees 发明人与专利人索引
index of irrigation need 灌溉需水指数
index of liquidity 液性指数;流性指数;流动性指标
index of living 生活指数
index of living environment 生物环境指数
index of living standard 生物水平指标;生活水平指数;生活水平指标
index of locomotive operation 机车运用指南
index of maximum variation of ore body 矿体最大变化标志
index of meandering 曲流指数;弯曲指数
index of mechanical property 力学性质指标
index of metallurgic(al) process 冶金工业指标
index of metamerism 条件配色指数
index of mining technique economics 采矿技术经济指标
index of modulation 调制系数
index of moisture 湿润指数
index of number employed 就业人数指标
index of open pit mining technology 露天开采技术指标
index of operating rates 开工率指数
index of partial production 局部开工生产指数
index of performance 性能指数;性能指标;效能指数;效率指数
index of places 地名索引
index of plasticity 塑性指数
index of pollution 污染指数;污染指标
index of pore in rock 孔隙指标
index of porosity 孔隙指数
index of potential inhalation toxicity 吸入中毒几率系数
index of precision 精确(度)指数;精(密)度指数
index of production capacity 生产能力指数
index of production cost 生产成本指数
index of purchasing power of exports 出口购买力指数
index of quality 质量指标;品质指标
index of radicals 根式的指数
index of reference 引用索引;访问的索引
index of reflection 反射率
index of refraction 折射系数;折射率;折光指数;折光率
index of regional difference of consumer prices 消费者地区差价指数
index of regional remotely sensed images and geologic(al) results 区域遥感图像及遥感地质资料索引
index of reliability 可靠性指数
index of ripple 波痕指数
index of routine analysis for atmospheric pollution 大气污染常规分析指标
index of routine analysis for water pollution 水污染常规分析指标
index of seasonal variation 季节变动指数
index of sedimentary petrology 沉积岩石学的指数
index of seismic damage 震害指数
index of selection 选取指数
index of sensitivity 灵敏度指数;灵敏度指标
index of sewage strength 污水强度指数
index of social purchasing power 社会购买力指数
index of specialty 特性指数
index of speculation 投机指数
index of stability 稳定(性)指数;稳定率
index of stabilization 稳定指数
index of streets 街道索引
index of suboptimality 次优性指数
index of symbols 符号表
index of terms of trade 进出口商品比价指数;贸易条件指数
index of terms of trade for exports 出口贸易条件指数
index of terms of trade for imports 进口贸易比价指数
index of thermal interia 热惰性指标
index of the scan(ning) helix 扫描螺线指针
index of tilth 耕性指数
index of transportation cost 运输费用指数
index of tumbler test for crucible non-volatile residue 焦渣转鼓指数
index of turbulence 紊流指数
index of variation characters 变化性质指数
index of wage 工资指数

index of water environment quality 地下水环境质量指数
index of water pollution 水污染指数;水污染指标
index of water quality 水质指数
index of wetness 年雨量指数;年水量指数;湿润指数
index of wholesale prices 批发物价指数
index of wholesales 批发销售额指数
index of workability 可加工性指数
index of working cost 成本指标
index organism 有机物指数;指示生物;指标生物
index organism for pollution 污染的指标生物
index part 变址部分
index path 光路图
index percent 指数率
index phrase 标引短语
index pin 指针;指度针;定位销;分度销
index pin knob 分度销钮
index plan 索引图;索引平面;标准层面【地】
index plane 标准面;标准层位
index plate 分度板
index plate housing 分度盘体
index plunger 分度销
index point 标(志)点;标定点
index prism 指示棱镜;动棱镜
index property 指示特征;指标性能;指标特性;特征指标;特征性质;特性指标
index property test 指标特性试验
index pulse 指引脉冲;标志脉冲
index qualifier 变址修改
index qualifier 变址限定符
index ratio 趋肤指数比
index realm 变址区域
index record 变址记录
index reference line 指数参考线
index register 指数寄存器;变址寄存器
index register pointer 变址寄存器指示器
index return character 变址返回符
index ring 刻度环;分度圈
index rod 指标杆
index sample 标准试样;标准试块;标准标本
index-sector separation circuitry 标志区段分离电路
index sediment concentration 含沙量指标;标准含沙量
index sequential file 索引顺序文件
index series 指标序列;指标体系
index series of environmental protection 环境保护指标体系
index set 指标集;变址组
index setting 指数调整
index shade 指标镜色片;动镜色片;动镜滤光片(六分仪部件)
index signal 指引信号;变址信号
index signal amplifier 指示射线位置
index sleeve 分度筒
index slip 标志条
index slit 变址缝口
index-smoothing forecasting 指数平滑预测法
index space 指标空间
index species 指标生物
index standard cost 指数标准成本
index state 标志状态
index station 指标站;参证站;参考站
index stereomark 立体测标
index stock 索引卡片纸
index stress 指示应力
index strip 指示带
index substance 指示物质
index system 指标系统;指标体系
index system of railway statistics 铁路统计指标体系
index table 水平分度头;分度工作台
index table machine 转位工作台机床
index tag 分度标志;标(志)牌;标签;标记
index term 索引术语;标引词
index test 折光率;分度试验;指标特性试验
index theorem 指数定理
index thermometer 指标温度计
index time 转位时间
index to boundaries 行政区划示意图
index to compilation map 编图资料示意图
index to names 地名索引
index to notation 记号索引

index tool 转位刀具
index to photograph 照片索引图
index to sheet 接图表
index to subjects of invention 发明主题索引
index track 变址道;变址磁道
index transducer 标志传感器
index tube 指引罐
index type 变址类型
index-type test 指标特性试验
index unit 指示装置;提升装置;分度装置
index value 指数;指示值;指标值;规定值;变址值
index variable 下标变量
index weight 指标全重;标志权
index window 指标窗
index word 修改量;指标字;索引字;索引词汇;变址字
index word register 变址字寄存器
index yarn 标志纱线
index zero 指标零
index zone 标准带
India-Eurasia collision 印度大陆与欧亚大陆碰撞
India-Gondwana separation 印度大陆与冈瓦纳大陆分离
India hemp 印度大麻
indialite 印度石;六方堇青石
Indiana 印第安纳高级细布(经防水处理可用作橡胶布)
Indiana cloth 印第安纳高级细布(经防水处理可用作橡胶布)
Indiana isopentane process 印第安纳异戊烷法
Indiana lamp 印第安纳灯(测定烟点用)
Indian alfalfa 印度苜蓿
Indiana limestone 贝德壳灰岩;微壳灰岩;印第安纳鲕状灰岩
Indian almond 橄仁树
Indian Antarctic basin 南极印度洋海盆
Indiana oxidation test 印第安纳氧化法
Indian architecture 印度(式)建筑;印度原始建筑
Indian blanket 印安花纹手织毛毡
Indian Buddhist architecture 印度佛教建筑
Indian cedar 印度松木
Indian Concrete Journal 印度混凝土杂志(期刊)
Indian corn 印度玉米
Indian delta 印度三角洲
Indian ebony 印度乌木
Indian era 印度纪元
Indian-fig 仙人掌
Indian gum 印度树胶;茄替胶
Indian ink 黑墨(汁)
Indian Journal of Power and River Valley Development 印度动力和流域开发杂志(期刊)
Indian kino 印度吉纳树胶
Indian laurel oil 印度月桂油
Indian linen 印度亚麻平布
Indian mahogany 印度红木
Indian monsoon 印度季风
Indian monsoon current 印度洋季风海流
Indian Ocean 印度洋
Indian Ocean area 印度洋地区
Indian Ocean coverages 印度洋区域的覆盖范围
Indian Ocean equatorial current 印度洋赤道洋流
Indian Oceanic-Canada crustal-wave system 印度洋—北美洲波系
Indian Ocean plate 印度洋板块
Indian Ocean Region 印度洋地区
Indian Ocean region 印度洋区域
Indian Ocean Satellite 印度洋卫星
Indian ochre 印度(赭)红;印度赭石;天然氧化铁
Indian okra 印度洋麻
Indian paper 印度纸;竹浆纸;凸版纸
Indian peninsula nucleus 印度半岛陆核
Indian pipe 印第安烟管
Indian pipestone 烟斗泥
Indian platform 印度地台
Indian purple 印度紫红色
Indian red 三氧化铁;印度红
Indian red ochre 印度红赭石;天然氧化铁
Indian red pigment 印度红颜料;黄红色铁矿矿砂颜料
Indian redwood 印度红木
Indian Register 印度船级社
Indian remote sensing satellite 印度遥感卫星
Indian ridge seismotectonic zone 印度洋海岭地震构造带

Indian rock-fill dam 印度式堆石坝
Indian rose chestnut 铁力木
Indian rubber 印度橡胶
Indian rubber cable 橡胶绝缘电缆
Indian rubber hose 橡胶水龙带;涂胶软管
Indian rubber tree 印度橡皮树
Indian shield 印度地盾
Indian shovel 印度铲
Indian silver greywood 印度银灰木
Indian spring low water 印度洋大潮低潮面;平均大潮低潮(面)
Indian spring water 印度大潮低潮面
Indian stone 油石;印度石
Indian style 印度式建筑;印度式
Indian subregion 印度亚区
Indian summer 印第安夏;秋老虎
Indian summer loadline 印度洋夏季吃水线
Indian tide plane 印度大潮低潮面;平均大潮低潮(面)
Indian tragacanth 卡拉雅胶
Indian transfer paper 版画印样用薄纸
Indian yellow 印度黄;酸性黄
India oil stone 印度油石(修表工磨雕刻刀用)
India-Pakistan/Japan route 印度、巴基斯坦—日本航线
India plate 印度板块
India proof paper 版画印样用薄纸
India rubber 橡皮;天然橡胶;弹性橡胶
India-rubber cable 橡胶绝缘电缆;印度橡胶电缆
India-rubber cushion 橡胶缓冲器
India-rubber hose 橡胶软管
India-rubber packing 橡胶填料
India-rubber ring 橡胶圈
India-rubber spring 橡胶簧
India-rubber wire 橡胶绝缘线;橡胶电线;印度橡胶电缆
India spring water 印度大潮低潮面
India tide plan 印度大潮低潮面
indicant 指示符
indicate 表明;标明
indicated 指示的;公称的;合计的;标明的
indicated air speed 表速;指示空速
indicated air temperature 指示气温
indicated altitude 仪表高度;计示高度;表高
indicated board 指示板
indicated course bend 指示航线弯曲
indicated course curvature 指示航线曲率
indicated course directional error 指示航线方向误差
indicated course error 指示航向误差
indicated course line 指示航线
indicated course sector 指示航向象限
indicated deposit 暂定按金
indicated device 指示器
indicated diagram 指示图
indicated direction 指示的方向
indicated displacement error 指示位移误差
indicated efficiency 指示效率
indicated ga(u)ge 指示计
indicated glass column 指示玻璃柱
indicated glide path 指示滑翔道
indicated glide path bend 指示滑翔道弯曲
indicated glide path curvature 指示滑翔道曲率
indicated glide path sector 指示滑翔道象限
indicated heat specific consumption 指示热消耗率
indicated horse-hour 指示马力小时
indicated horse power 指示功率;指示马力;理论马力;视在拉马力
indicated horsepower hour 指示马力小时
indicated instrument 指示仪表
indicated lamp 指示灯
indicated linearity sector 指示线性象限
indicated mean effective pressure 指示平均有效压力;平均指示有效压力
indicated mean effective stress 指示平均有效应力
indicated meter 指示器;指示流量计
indicated micrometer 指示测微计
indicated noise meter 噪声指示计
indicated number 指示数
indicated object 表示对象
indicated ore 指示矿量
indicated output 指示马力;指示产量;额定产量
indicated power 指示功率
indicated pressure 指示压力

indicated pressure ga(u)ge 指示压力表
indicated range 指示范围
indicated recorder 指示记录器
indicated relay 指示继电器
indicated reserve 指示储量;指明储量;推论储量;标示储量
indicated resources 推定资源
indicated slant course directional error 指示倾斜航线(方向)误差
indicated slant course line 指示倾斜航线
indicated speed 指示速率;指示车速
indicated strain 指示应变
indicated thermal efficiency 指示热效率
indicated thrust 指示推力
indicated tractive effort 指示牵引力
indicated tractive power 指示牵引力
indicated value 指示值
indicated water level 指示水位
indicated wattmeter 指示瓦特计
indicated weight 标重;标明重量
indicated weight meter 指重表
indicated work 指示功
indicate equation 指数方程
indicating accuracy 指示正确性;指示精度;示值精度
indicating arm 指示杆
indicating bell 指示铃
indicating bit 指示位
indicating bolt 门栓指示器(室内有无占用);显示门栓
indicating calipers 指示卡规
indicating circuit 指示电路
indicating controller 指示控制器;标尺式控制器
indicating desiccant 指示干燥剂
indicating device 指示装置;指示器(件)
indicating device bridge 指示装置夹板
indicating diagram 指示图
indicating dial 指示盘
indicating end 指示针
indicating equipment 指示设备
indicating error 指示误差;示值误差
indicating finger 指示针
indicating floor stand 指示地轴架
indicating flowmeter 指示流量计;带刻度流量计
indicating fuse 指示熔断器
indicating ga(u)ge 指示器;指示计;千分表;百分表
indicating head 仪表刻度盘
indicating instrument 显示仪表;指示仪器;指示仪表
indicating lamp 讯号灯;信号灯;指示灯
indicating length 指示长度
indicating light 指示灯
indicating liquid level ga(u)ge 液面指示计;指示液面计
indicating lock 指示锁
indicating measuring instrument 指示式量测仪器
indicating mechanism 指示装置
indicating meter 指示计
indicating micrometer 指示测微计
indicating needle 指针
indicating organ for the surveying values 测量数值指示机构
indicating organism 指示生物
indicating organism for water pollution 水污染指示生物
indicating oscillograph 显示式示波器;指示式示波器
indicating panel 指示仪表板;表示盘
indicating paper 指示剂;试纸
indicating plant 指示植物
indicating plant of groundwater 地下水指示植物
indicating plug ga(u)ge 内径精测仪
indicating potentiometer 指示电位计
indicating pressure ga(u)ge 指示压力计;指示压力表
indicating range 指示范围
indicating recorder 指示记录器
indicating relay 信号继电器;指示继电器
indicating selsyn 指示式自整角机;指示式自同步机
indicating snap ga(u)ge 内径精测仪
indicating switch 指示式开关
indicating system 指示系统
indicating test method 指示试验法
indicating thermocouple 指示热电偶
indicating thermometer 指示温度计;温度指示计

indicating type 指示式
indicating unit 指示装置;指示器
indicating value 示值
indicating wattmeter 瓦特指示表
indicating wattmeter method 指示瓦特计法
indication 指示
indication applied occurrence 指示应用性出现
indication board of cylinder storage 气瓶储存显示板
indication circuit 指示电路;表示电路
indication code 表示电码
indication cycle 表示周期
indication-defining occurrence 指示定义性出现
indication error 显示误差;指数误差;指示误差;示值误差;读数误差
indication information 表示信息
indication lag 指示延迟
indication lamp 指示灯;表示灯
indication magnet 指示磁铁
indication marks on goods package for transport 货物运输包装标记
indication mechanism 显示机构;指示机构
indication of analysis results 分析结果的表示法
indication of bioquality 生物质量指标
indication of depressed button 按钮指示;按钮表示
indication of fulfillment of plan 计划完成指标
indication of measuring instrument 测量器示值
indication of position of container stowage 集装箱装载位置的表示
indication panel 表示盘
indication period 表示周期
indication plug unit 表示插件
indication price 指示价格
indication range 显示范围;指示范围;示值范围
indication relay 指示继电器
indication requisition 表示条件
indication rod 表示杆
indication sign 指示标志;导向牌
indication symbol 指示符
indication transformer 表示变压器
indication unit 指示仪器
indicative abstract 要点抄录;指示摘要;指示性简述;指示简介;内容提要
indicative character 指示特征
indicative curve 指示曲线
indicative diagram of standard plan 标准计划指示图表
indicative dump 简易转储
indicative figure 指示性数字
indicative financial plan 指示性财务计划
indicative function 示性函数
indicative geobotany 指示地植物学
indicative growth 指示性增长率
indicative mark 指示性标志;指示标志
indicative phytogeography 指示植物地理学
indicative planning 指示性计划;指示式计划
indicative price 指示性价格;指示价
indicative reference 指示性引用标准
indicative sign 指示标志
indicative species 指示物种
indicative structural plane 标引性结构面
Indicative World Plan for Agricultural Development 世界农业发展指示计划
indicator 显示器;指示物;指示器;指示牌;指示剂;指示计;指示(符);找矿标志;计量仪器;示功器;表示器;标志器
indicator animals for atmospheric pollution 大气污染指示动物
indicator assemblage 指示植物群丛
indicator bacteria 指示菌
indicator bacteria of water fecal pollution 水体粪(便)污染指示菌
indicator balance 指秤天平
indicator basic colo(u)r 指示剂碱性色
indicator benthos 底栖指示生物
indicator board 指示器盘;指示(器)板;指示屏
indicator bolt 显示门闩(显示厕所有没有人);有人无人锁;厕所锁
indicator buoy 指示浮标
indicator button 门钮指示器;有人无人钮;指示器按钮
indicator card 指示图;指示卡;示功图
indicator case 指示板;指示(仪表刻度)盘
indicator-channel 示功器通道

indicator chart 指示(字)图;指示符图;示功图
indicator circuit 指示电路
indicator clip 指示器固定簧片;夹盘弹簧片
indicator clutch 指示器离合杆
indicator clutch operating lever 指示器离合操作杆
indicator clutch spring 指示器离合簧
indicator cock 指示阀;示功器旋塞
indicator colo(u)ring 指示性着色
indicator community 指示(植物)群落
indicator connection plug 指示器接头塞
indicator constant 指示剂常数
indicator cord 指示表拉绳;指示标拉绳
indicator culture 指示培养
indicator current 指示电流
indicator curve 指数曲线
indicator device clutch wheel 指示装置离合轮
indicator device driver 指示装置驱动轮
indicator device driver maintaining washer 指示装置传动轮保护隔片
indicator diagram 指示(线)图;指示器图表;示功图
indicator diagram of work 示功图
indicator dial 指示表刻度盘;指示(仪表刻度)盘
indicator dial cover 指示器标度盘壳
indicator disc 指示盘
indicator display 指示器显示
indicator distance piece 指示器隔片
indicator drive 示功器传动机构
indicator driving wheel 指示器传动轮
indicator driving wheel core 指示器传动轮座
indicator driving wheel friction spring 指示传动轮摩擦簧
indicator driving wheel ring 指示器传动轮衬套
indicator drum 指示器卷筒;记录器卷筒
indicator electrode 指示电极
indicator element 显示元件;指示元素;示踪元素
indicator equation 指示器方程
indicator for total credit 信贷总额指标
indicator function 指示函数
indicator gate 显示选通门;指示器闸门;指示器门电路
indicator ga(u)ge 指示(压力)计;指示表
indicator gear 指示器传动装置;指示器传动机构;指示表传动机构
indicator hole 指示(钻)孔
indicator horizon 指层
indicator lamp 指示灯;监视灯;度盘灯
indicator lamp cover 指示灯罩
indicator lamp terminal 指示灯接线柱
indicator length 指示长度
indicator light 指示器信号灯;指示灯;监视灯;度盘灯
indicator lock 指示锁
indicator maintaining plate distance piece 指示器定位板隔片
indicator maintaining small plate 指示器压片
indicator medium 指示性培养基;鉴别培养基
indicator microorganism 指示微生物
indicator mineral 指示矿物
indicator module 指示器组件
indicator motion 指示器动作
indicator needle 指针
indicator needle shaft 指针轴
indicator of distribution 分配指标
indicator of divergence 离散性指标;发散指标
indicator of economic development 经济发展指标
indicator of environment(al) statistics 环境统计指标
indicator-off 指示指令断开;指示器关闭
indicator of inclination 倾斜指示器
indicator of physical output 实物量指标
indicator of pollution 污染指示物;污染指示器;污染指示剂;污染指标物
indicator of potential contamination of water 潜在水污染指示物
indicator of product variety 品种指标
indicator of sewage pollution 污水污染指示物
indicator of standard physical output 标准实物产量指示
indicator of subdividing organic matter type 有机质类型指标
indicator of swing 旋角指示器
indicator of utilization of machine-hour 台时利用(程度)指标
indicator-on 指示指令接通;指示器开启

indicator organism 指示生物;指标生物
indicator organism method 指示生物法
indicator organisms for atmospheric pollution 大气污染指示物
indicator organisms for water pollution 水污染指示生物
indicator organisms of polluted water 污泥水指示有机体
indicator panel (警报器的)显示板;信号箱;信号显示板;指示屏
indicator paper 试纸;示功器用纸
indicator pencil 指示器记录头;示功器笔
indicator pinion 指示器齿轴
indicator pipe 指示器管
indicator piston 指示器活塞
indicator plankton 指示浮游生物
indicator plant 指示植物;指示设备
indicator plant of bitumen 沥青指示植物
indicator plant of boron 硼矿指示植物
indicator plant of copper 铜矿指示植物
indicator plant of gypsum 石膏指示植物
indicator plant of iron 铁矿指示植物
indicator plant of lead 铅矿指示植物
indicator plant of phosphate 磷的指示植物
indicator plant of selenium 硒矿指示植物
indicator plant of selenium and uranium 硒与铀矿指示植物
indicator plant of silver 银矿指示植物
indicator plant of soil 土壤指示植物
indicator plant of zinc 锌矿指示植物
indicator plants for atmospheric pollution 大气污染指示植物
indicator plate 指示板
indicator pointer 指示针;指示器指针
indicator post 指示器支杆(闸门);指示(标)柱;开关指示柱
indicator range 指示幅度
indicator register 指示寄存器
indicator-rod 漏箭
indicator safety 指示器保险装置
indicator scale 指示器;定尺指示器
indicator screen 指示器荧光屏
indicator seating 指示器垫
indicator setting wheel driving wheel 指示器过轮传动轮
indicator signal 指示信号
indicator sign post 指示牌
indicators of demographic(al) trend 人口趋势指标
indicators of landslide 滑坡识别标志;滑坡(鉴定)标志
indicator species 指示物
indicator spring 示功器弹簧
indicator strain 指示菌
indicator switch 指示器开关
indicator system 指示系统;指标体系
indicator system for economic results 经济效果指标体系
indicator system for moderately well-off society 小康社会指标体系
indicator test 指示剂试验;绘示功图
indicator tube 显示管;指示管
indicator valve 阀口指示器;指示阀;带(开度)指示器的阀
indicator vein 指示脉
indicator wheel 指示轮
indicator wheel bridge 指示轮夹板
indicator wheel jumper 指示轮定位杆
indicator work 指示功
indicatrices indicatrix 的复数形式
indicatrix 指示线;指示面;指示量;指标线;特征曲线;指标图;折射率椭圆体;[复] indicatrices
indicatrix ellipse 变形椭圆
indicatrix of cartography 地图投影变形椭圆
indicatrix of diffusion 漫射指示量
indicatrix of optic(al) biaxial crystal 二轴晶光率体
indicatrix of optic(al) mono-axial crystal 单轴晶光率体;一轴晶光率体
indicatrix of scattering 散射指示量
indices index 的复数形式
indices of atmospheric purity 空气清洁指数
indices of corrosion evaluation 溶蚀评价指标
indices of crystal face 晶面指数
indices of crystallographic plane 晶面指数

indices of general property 一般性指标
indices of goods traffic 货物运输指标
indices of locomotive operation 机车运用指标
indices of mineral dressing 选矿技术经济指标
indices of self-purification of water body 水体自净指数
indices of wagon utilization 货车运用指标
indices system of economic efficiency audit of engineering system 工务系统经济效益审计指标体系
indices system of economic efficiency audit of maintenance system 机务系统经济效益审计指标体系
indices system of economic efficiency audit of railway bureau 铁路局经济效益审计指标体系
indices system of economic efficiency audit of railway subbureau 铁路分局经济效益审计指标体系
indices system of economic efficiency audit of rolling stock system 车辆系统经济效益审计指标体系
indices system of economic efficiency audit of station and division 基层站段经济效益审计指标体系
indices system of economic efficiency audit of station system 车站系统经济效益审计指标体系
indicia 邮戳;标记
indicial admittance 过渡导纳;单位阶跃导纳
indicial equation 指数方程
indicial motion 示性运动
indicial response 指数特性;过渡响应
indicial transfer function 指数传递函数
indicolite 蓝电气石;蓝碧硒
indictable offence 可检控书
indiction 小纪(15年)
indictment 检控
indifference 无差别
indifference analysis 无差异分析
indifference curve 中性曲线;无差异曲线;无差别曲线;同好曲线
indifference interval 无差别区间
indifference level 无差异水平
indifference region 无差别区域
indifference surface 无差别曲面
indifferent 不偏的
indifferent air-mass 变性气团;中性气团
indifferent business 附带业务
indifferent electrode 惰性电极
indifferent electrolyte 协助电解物
indifferent equilibrium 中性稳定(度);中性平衡;随遇平衡
in different forms 不拘形式
indifferent gas 中性气体;惰性气氛
indifferentism 无差别
indifferent stability 中性稳定(度)
indifferent tension 无差异拉力
indiffused crystal waveguide 非漫射晶体波导
indiffusible 不扩散的
indiffusible ion 非扩散离子;不扩散离子
indiffusion 向内扩散;不扩散
indigenization 国产化
indigenous 乡土的;固有的;土产的;本地的
indigenous availability 当地具备
indigenous blast furnaces 土高炉
indigenous breed 本地品种
indigenous cover 乡土覆盖植物
indigenous equipment 土设备
indigenous expert 土专家
indigenous flora 乡土植物群;自然菌群
indigenous forest 原始森林
indigenous fuel 当地燃料
indigenous graphite 析出石墨
indigenous ground stress 固有地应力
indigenous groundwater 当地地下水
indigenous inhabitant 本地居民
indigenous knowledge 本地知识
indigenous materials 乡土材料
indigenous method 土办法
indigenous plant 乡土植物
indigenous raw material 当地原材料
indigenous river 本地河
indigenous soil 原生土(壤);原地土(壤);定积土
indigenous species 乡土(品)种;当地种;本地(品)种

indigenous stream 本地河
indigenous technology 土技术
indigenous value 固有价值
indigenous variety 乡土品种;本地品种
indigenous village 原居村落
indigenous wood 本地木材
indigent capital 贫困的资本
indigestibility 难消化;无法消化
indigirite 水碳镁铝石
indigo 蓝草;靛蓝(色);靛蓝类染料
indigo auxiliary 靛蓝还原助剂
indigo blue 靛蓝色
indigo carmine 靛蓝胭脂红
indigo copper 铜蓝
indigo disulphonate 二磺酸靛蓝
indigo disylvine disulfonate 靛蓝二磺酸二钾盐
indigo dyeing 靛蓝染色
indigo dyeing wastewater treatment 靛蓝染色废水处理
indigo-heliographic paper 靛蓝晒图纸
indigoid dye 靛属染料;靛类染料
indigo lake 靛蓝色淀
indigolite 蓝电气石
indigo process 靛蓝法
indigo pure powder 靛蓝粉
indigo red 靛玉红
indigosol golden yellow 溶靛素金黄
indigosol printing black 溶靛素印染黑
indigosol printing blue 溶靛素印染蓝
indigosol red-violet 溶靛素红紫
indigo vat 靛瓮
indigo vatting 靛蓝的瓮化
indigo white 靛白
indirect 间接(的)
indirect a. c. convertor 间接交流变流器
indirect acting carcinogen 间接致癌物
indirect acting factor 间接影响因素
indirect acting mutagen 间接致突变物
indirect action 间接作用
indirect action of noxious substance 有害物质间接作用
indirect activation 间接生效;间接活化
indirect activity 辅助业务;间接业务
indirect actuating relay 间接作用式继电器
indirect address 间接地址;间接选址
indirect address code 间接地址码
indirect addressing 间接寻址;间接定址;间接编址;间接选址
indirect address mode 间接地址方式
indirect adjustment 间接平差
indirect aerological analysis 间接高空(气象)分析
indirect agglutination 间接凝集反应
indirect agglutination test 间接凝集试验
indirect air-conditioning system 间接空调系统
indirect analog 间接模拟
indirect analog(ue)/digital converter 间接模/数转换器
indirect analysis 间接分析
indirect application of international standard 国际标准的间接应用
indirect arc rocking furnace 摇动式间接电弧炉
indirect assignment 间接赋值;间接地址分配
indirect atomic absorption method 间接原子吸收法
indirect attractive range 间接吸范围
indirect bank protection 间接护岸工程
indirect benefit 间接效益;间接利益
indirect benefit of project 项目间接效益
indirect binary n-cube network 间接二进制 N 方体网络
indirect bio-catalysis 间接生物催化剂
indirect branch 间接转移
indirect business 间接交易;间接买卖
indirect call 间接呼叫;间接调用;间接调入
indirect calorimetry 间接量热法;间接测热
indirect carcinogen 间接致癌物
indirect casting 下铸;间接浇铸
indirect catchment 间接集水(面积)
indirect causation 间接因果关系
indirect cell 间接环流
indirect chaining method 间接链接法
indirect charges 间接费(用)
indirect circulation 逆环流
indirect clutch 间接离合器

indirect colo(u)rimetric method 间接比色法
indirect command file 间接命令文件
indirect componential movement 间接组分运动
indirect condenser 间接冷凝器
indirect connected steering gear 间接联动舵机
indirect connection 间接连接;与房屋排污系统不直接连接
indirect construction cost 间接建设成本;工程间接费;间接工程费(用)
indirect consumption 间接消耗
indirect contact 间接接触
indirect contract 间接合同
indirect contribution 间接税
indirect control 间接控制
indirect cooking process 间接蒸煮法
indirect cooled desuperheater 中间载热体降温器;间接冷却减温器
indirect cooler 间接冷却器;间接冷却机
indirect cooling 间接冷却
indirect cooling evapo(u)rator 间接冷却式蒸发器
indirect cooling generator 间接冷却发电机
indirect cooling water 间接冷却水
indirect correlation 间接相关
indirect cost 间接费(用);间接成本
indirect cost distribution 间接成本分配
indirect cost of footage of tunneling 单位进尺间接费
indirect coupling 间接耦合
indirect cycle 间接周期;间接循环
indirect cylinder 间接热水箱
indirect damage 间接损失;间接损害
indirect data address 间接数据地址
indirect daylight factor 间接采光系数
indirect daylighting 间接采光
indirect deactivation 间接释放
indirect demand 间接需求
indirect department 辅助部门
indirect departmental cost 间接不满费用
indirect descent 旁系
indirect desulfurization 间接脱硫
indirect determination 间接测定
indirect digital frequency synthesizer 间接式数字频率合成器
indirect direct current convertor 间接直流变流器
indirect discharge measurement 间接流量测量;间接测流法
indirect diving 间接潜水
indirect division 间接分区;间接分裂
indirect drainage 间接排水;内排水
indirect drainage system 内排水系统
indirect drain pipe 间接下水管;间接排水管
indirect draught 侧压下
indirect dryer 间接烘干机
indirect-drive machine 间接驱动机械
indirect dye 间接染料
indirect echo 间接回波
indirectly heated rotary calciner 间接加热回转煅烧窑
indirect effect 间接作用;间接影响;间接效应;非直接效应
indirect effect on gravity 重力间接效应
indirect effect on height anomaly 高程异常间接效应
indirect effect on the deflection 垂线偏差间接效应
indirect electrochemical oxidation 间接电化学氧化
indirect electrostatic process 间接静电处理
indirect emission 间接发射
indirect energy-saving 间接节能
indirect eruption 间接喷发
indirect exchange 间接汇兑
indirect exchange quotation 外汇间接报价
indirect excitation 间接磁给;间接激励
indirect execution 间接执行
indirect expenditures 分配列入支出;间接支出
indirect expenses 间接工程费(用);间接费(用)
indirect expense audit 间接费审计
indirect expense of project 项目间接费用
indirect expense sheet 间接费用表
indirect experience 间接经验
indirect extrusion 间接挤压
indirect factor 间接因素
indirect factory cost 间接工厂成本
indirect fertilizer 间接废料
indirect file 间接文件

indirect finance 间接融通
indirect financing 间接供应资金
indirect fire appliance 间接燃烧用具
indirect-fired 间接加热
indirect-fired dryer 间接燃烧烘干机;间接加热烘干机
indirect-fired furnace 间接加热炉;单独燃烧室炉
indirect fire heater 水套加热炉
indirect fire suppression 间接扑火;间接灭火
indirect firing 间接烧成
indirect-firing system 间接燃烧系统
indirect flood damage 间接洪水损害;洪水(造成的)间接损失
indirect floodlight 间接散光灯
indirect flow 间接流通
indirect flushing 反向冲洗;反循环冲洗
indirect footlight 反射型脚灯;间接脚光
indirect freezing process 间接冷冻法
indirect frequency modulation 间接调频
indirect glare 间接眩光;反射眩光
indirect harvester 分段式收获机
indirect heat 旁热
indirect heater 间接加热器
indirect heat exchanger 间接式热交换器
indirect heating 集中供暖;间接加热;间接供暖;外部加热
indirect heating surface 间接受热面
indirect heating system 间接供暖系统
indirect high measurement 间接高程测量
indirect holding fastening 分开式扣件【铁】
indirect illumination 泛光照明;间接照明;无影照明;反射照明
indirect importance 间接重要性
indirect improvement of soil 土壤的间接改良
indirect incidental expenses 间接附带费用
indirect indexing 间接分度法
indirect indicator 间接指标
indirect industrial discharge 间接工业排放者
indirect infection 间接传染
indirect influence 间接影响
indirect inhibitor 间接抑制剂
indirect input 间接输入
indirect instruction 间接(地址)指令
indirect interpretation 间接解译
indirect interpretation key 间接解译标志
indirect investigation 侧面调查
indirect investment 间接投资
indirection 间接
indirectional approach 不定向逼近法
indirect isotope effect 间接同位素效应
indirect judgement 间接判定
indirect labo(u)r 间接劳动;辅助生产人员;辅助人工;辅助工人的劳动
indirect labo(u)r cost 间接人工成本
indirect labo(u)r efficiency variance 间接劳动效率差异
indirect laying 间接瞄准
indirect level(l)ing 间接水准测量;间接高程测量
indirect liability 间接负债;间接责任
indirect light 间接光
indirect lighting 泛光照明;间接照明;间接采光;反射(光)照明;反光照明
indirect lighting component 间接照明器具
indirect light reflex 间接光反射
indirect liquefaction of coal 煤(的)间接液化
indirect load 间接荷重;间接荷载
indirect loss 间接损失
indirect luminaire 间接光源
indirectly coupled system 间接耦合系统
indirectly fed antenna 无源天线
indirectly heated oven 间接式烤箱;外部式烤箱
indirectly heated thermistor 间接加热热敏电阻;旁热式热敏电阻
indirectly illuminated sign 间接照明标志
indirectly lighted sign 间接光照标志
indirectly seating 间接承垫
indirectly visible area 间接能见区
indirect maintenance expenses 间接维修费用
indirect manufacturing cost 间接制造成本
indirect mark 间接标志
indirect material 辅助材料;间接材料
indirect material cost 间接材料成本
indirect materials or supplies 间接材料或办公用品
indirect material use variance 间接材料用量差异

indirect measurement 间接量度;间接量测;间接度量;间接测量;间接测定
indirect method 间接法
indirect mode 间接状态
indirect mortgage loan 间接抵押放款
indirect mouth 河口(河口与海湾相接之处)
indirect mutagen 间接致突变物
indirect national tax 间接国家税
indirect observation 间接观测;间接测量
indirect observation with conditional equation 条件间接观测方程
indirect open-spray system 间接敞开式喷淋系统
indirect operating expenses 间接营业费用
indirect operation 间接传动;间接操作
indirect optic(al) holography 间接光学全息摄影
indirect orientation 间接定向法
indirect output 间接输出;脱机输出
indirect oven 间接式烤箱;外热式烤箱
indirect overhead 间接管理费
indirect oxidation 间接氧化
indirect oxidation process 间接氧化工艺
indirect oxidation reaction 间接氧化反应
indirect photo chemical oxidation 间接光化学氧化
indirect photography 间接摄影
indirect photometric chromatography 间接光度色谱法;间接光度层析
indirect pointer 间接指针;间接指示字;间接指示器
indirect pollution of groundwater 间接地下水污染
indirect pollution source 间接污染源
indirect possession 间接占有
indirect pouring 间接浇铸
indirect private investment 私人间接投资
indirect probability 间接概率
indirect problem 反问题;反命题
indirect process 间接法
indirect proof 间接证明
indirect prospecting indication 间接找矿标志
indirect quadruple 间接四元组
indirect quotation 间接行市;间接汇价;间接标价;间接报价
indirect quotation of exchange rate 间接报价法
indirect radiation 间接辐射
indirect radiator 间接散热器;间接放热器
indirect rail fastening 分开式扣件【铁】
indirect rate 间接汇兑率
indirect ray 间接射线;无方向射线;电离层反射波;反射射线
indirect reaction 间接反应
indirect reading 间接读数
indirect recharge 间接补给
indirect recombination 间接复合
indirect rectification 间接法纠正
indirect recursion 间接递归
indirect reference address 间接参考地址
indirect reflection 间接反射
indirect refrigerating system 间接冷冻系统
indirect refrigeration 间接制冷;间接冷却
indirect respiration 间接呼吸
indirect retainer 间接固位体
indirect reuse of wastewater 废水的间接回用
indirect river mouth 间接河口
indirect rolling action 轧制侧向压下;侧压下
indirect route 迂回道路;间接路由;间接进路;非直达线
indirect scanning 间接扫描
indirect scheme of digit rectification 间接法数字纠正
indirect screening 间接加网
indirect search 间接搜索片
indirect selection 间接选择
indirect service water heater 给水管间接加热器
indirect sharing 间接共享
indirect shop labor expenses 辅助工人工资支出
indirect sign 间接信号
indirect signal 间接信号
indirect solution 间接解法
indirect sound 间接声
indirect sounding 间接声测法
indirect spectrophotofluorometry 间接荧光分光光度法
indirect spectrophotometric method 间接分光光度法
indirect spot welding 单面点焊
indirect staining 间接染色法

indirect standardization 间接法标准化
indirect stratification 间接层理;次生层理;次级层理
indirect stress 合成应力;间接应力
indirect stroke 间接雷(击)
indirect survey 间接测量
indirect system 间接系统
indirect system of refrigeration 间接制冷系统
indirect tax 间接税
indirect tax credit 税收间接抵免
indirect temperature lamination 间接温度层压(法)
indirect template theory 间接模板学说
indirect tensile test 间接拉力试验;间接拉伸试验
indirect titration 间接滴定
indirect toxicity 间接毒性
indirect trade 间接贸易
indirect transformation 间接转变
indirect transition 间接跃迁
indirect transmission 间接传动
indirect transshipment 间接换装
indirect triple 间接三元组;三重间接
indirect-use of water 水的间接再用
indirect utility 间接效用;非直接效用
indirect vapo(u)r cycle 间接蒸气循环;间接水汽循环
indirect-view thermal imager 间视热像仪
indirect vision 周围视觉;间接视
indirect wage 间接工资
indirect waste 间接污水
indirect waste pipe 迂回排泄管;间接排水管;间接废水管;废水间接排水管
indirect water reuse 间接水再用
indirect wave 间接波;电离层反射波
indirect welding 单面点焊
indirect working men 间接生产人员
indiscrete value 连续值
indiscriminate 无差别的
indiscriminate dumping 野蛮卸料;违章卸料
indiscriminate transfer of resources 平调
indiscriminate use of chemicals 滥用化学品
indiscriminate use of pesticide 农药的滥用
in disorder 零乱;凌乱
indispendable condition 必备条件
indispensable obligation 不可推卸的责任
indisposed 不舒服的
indisposition 不适
indisputable 无可争辩的
indissolubility 不溶解性
indissoluble 难溶的;不溶的
indissoluble-resin content 不溶性树脂含量
indissolvable 难溶的;不溶解
indistinct boundary 不清晰边界
indistinct image 模糊影像
indistinctness 模糊;不清晰度
indistinguishability 不可区分性;不可分辨性
indistinguishable 难区别的;不可区别的
indite 硫铁铟矿
inditron 字码管;指示管;氖灯
indium 自然铟
indium antimonide 锑化铟
indium arsenide detector 砷化铟探测器
indium bromide 三溴化铟
indium monobromide 一溴化铟
indium monochloride 一氯化铟
indium monoiodide 一碘化铟
indium monosulfide 一硫化二铟
indium ores 铟矿
indium oxide 氧化铟
indium tribromide 三溴化铟
indium trifluoride 三氟化铟
indium triiodide 三碘化铟
indium trioxide 三氧化二铟
indiun deposit 铟矿床
individed right 不可分割的权利
individual 专用的;个体;特殊的;独立;单独的
individual account 个体计算
individual action 单独作用
individual addressable module 控制模块
individual adjustment 单独调整;分别调整
individual adsorbate molecule 单个吸附质分子
individual analysis 分别分析
individual-assignment 分片包队法
individual axle drive 单轴驱动;单轴轴驱动
individual bargaining 个别谈判

individual base 独立基础
individual baseline 独立基线
individual batch bucket 单装料罐
individual batcher 单个混凝土拌和机;分批计量器
individual blast 单孔通风
individual blasting 单孔爆破
individual block 单个大型组件;单独的块体
individual branch circuit 独立分支电路
individual building 个体建筑
individual business 个体户
individual button 单独操纵按钮
individual calling 个别呼叫;单呼
individual camera 单景航空摄影机
individual capital 个人资本
individual cast 分块铸造;分开铸造
individual casting 分割铸造(法)
individual cellar 独立地下室
individual character 个性品质
individual check 个人支票;私人支票
individual chemicals 单个化学品
individual chemical species 单个化学物种
individual chip package 单片组装
individual clutch 单独离合器
individual coach lighting 各车厢分离式照明
individual coat 一度(油漆、抹灰等);特殊涂层
individual coefficient of heat 个别热系数
individual colo(u)r 单色的
individual community density 个体群密度
individual comparisons 个体比较
individual compass error 罗盘固有误差
individual constant 个体常量
individual construction 个体建筑;单件结构
individual consumer 个体用户;特殊用户;单独用户
individual contactor equipment 单个式接触器控制牵引装置
individual control 分级控制
individual control system 分级控制系统
individual conveying duct 单独输送管道
individual correction 个别改正
individual cylinder 分块铸造汽缸;分割铸造汽缸;分段汽缸
individual death 个体死亡
individual decision making 个体决策
individual depreciation 个别折旧(法)
individual depreciation rate of fixed assets 固定资产个别折旧率
individual derivative 物质导数
individual design of rapids/shoal regulation 分滩设计
individual development 单独开发;单独发展
individual distortion 单独畸变
individual dosimeter 个人照射量计;个人剂量计
individual drafting 一版清绘
individual drawing system 单独制图制
individual drinking water supply 分设的饮水供应(系统)
individual drive 独立传动装置;单独驱动(装置);单独传动(装置)
individual-drive motor 单独传动电动机
individual earnings 个人收入
individual economy 个体经济
individual element 单个构件
individual elevation 高程注记点;高程点
individual enterprise 个体企业;私人企业
individual ergodic duration 单个各态历程
individual error 个体误差;个别误差;人差【测】;单项误差
individual escape 单人脱险
individual escape apparatus 单人脱险呼吸器
individual escape equipment 单人脱险设备
individual eyepiece focus 单目镜聚焦
individual fabrication 特殊装潢
individual farm 个人经营农场
individual feeding bail 个体饲喂栏
individual field 个别地块;私人田地
individual filament 单纤维;单丝
individual footing 独立基础;单个基础;单独基础
individual footing foundation 单独底脚基础
individual form of art 艺术的个人形式;艺术的独特风格
individual frame 单张像片
individual girder part 单独的大梁部分
individual habitat 个体生境;个体栖息地
individual heating 独立供暖;单独采暖

individual hole 单独孔
individual house 独立房屋
individual household operation 分户经营
individual hygiene 个人卫生
individual image 分色原图
individual import licence 个别进口许可证
individual income 个人所得
individual integer 单个整数
individualistic 单个的
individual item 个别项目;单项工程
individual labo(u)r 个体劳动者
individual labo(u)r movement 个别劳动力流动
individual lead 单独引入线
individual leaflet 单个小叶
individual letter sign 个人信函符号
individual lever type all-relay interlocking 单独操纵继电式电气集中联锁
individual lever type all-relay interlocking for a hump yard 单独操纵式驼峰电气集中
individual lighting 独立供电照明
individual line 专用线路;专用路线
individual line subscriber 专线用户
individual liquid sample 个别液体样(品)
individual load 单独荷载
individual lots 个人营业用地
individually driven stand 单独传动机座
individually screened cable 分析屏蔽电缆
individually tailored 单独配制的
individual management 个别管理
individual manufacture 单独制造
individual mark 分类标牌(加工后钢筋)
individual maximum wave height 单一最大波高
individual merit testing 个体生产性能测定
individual microscopic grain 单体微粒
individual micro-submersible 单人微型潜水器
individual model 单模型
individual modulation 个别调制
individual monitoring 个人监测
individual morphology 单体形态
individual mortgage 个人抵押
individual mounting 单独安装;特殊装置;特殊装配
individual negotiation 个人转让
individual office 独用办公室;个人办公室
individual one-site treatment system 单独就地处理系统
individual on-site system 就地单独系统
individual ownership 个体所有制
individual package delivery 分别包装交货
individual packing 单件包装
individual part 个别零件
individual particle 单个颗粒
individual-particle model 单粒子模型
individual pen 单圈
individual perception threshold 个人感觉阈
individual personality 个人人格
individual piece production 单件生产
individual pile 单独桩;单柱
individual plant 单株植物;单株
individual plasma 孤立等离子体;单个等离子体
individual point 独立点
individual point source 单个点污染源
individual pollutant 单个污染物
individual pollution source 单个污染源
individual population succession 单一种群演替
individual portal frame 单独门框
individual preparation 单独加工;单独制备
individual prepared-roofing shingle 个别预制的屋顶盖板
individual pressure chamber 单人压力舱
individual priority 个别优先
individual processing 单独加工
individual processor system 独立处理系统
individual producer 个体生产者
individual project 单项工程;单位工程;不可分项目
individual prophylaxis 个人预防
individual proprietorship 个人所有权;独资(经营)
individual protection 个体防护;个别防毒
individual pump-injection system 分缸供油泵式发动机
individual rain 一场雨;单场雨
individual rationality 个体合理性
individual record 个体记录
individual register 分级记发器
individual register system 分级记发制

individual relief letter 浮雕式独立字母
individual request for inventor 发明人单项服务
individual rhythm 个体节律
individual room 独用房间;单独房间
individual's account 分户账目
individual sample 个别试样
individual sculpture 独立雕塑
individual section machine 分列式制瓶机;分部式制瓶机
individual section of pipe 一段管子
individual selection 个人选择
individual sensitivity 个体敏感性
individual service 个体服务行业;个人劳务
individual service pipe 单户用户管;单用户管
individual sewage disposal system 专用污水处理系统;单个污水处理系统
individual sewage disposal system act 单个污水处理系统法令
individual shingle 整张整张屋面板
individual shot 单孔爆破
individual sleeper 单根轨枕
individual source 个别来源
individual source control measure 单个污染源控制措施
individual sovereignty 独立主权
individual species 个人种;单个物种
individual specificity 个体特异性
individual speed 单车车速
individual stall barn 单圈牛舍
individual stall housing 单栏牛舍
individual strip 单航线
individual strip triangulation 单航带三角测量;单航带法
individual structure 单件结构
individual stud clip 专用立筋卡
individual supply 单独供应
individual support points 单独点支承
individual system costing 单系统费用估计
individual task 单个任务
individual test 个体测验
individual thickness 分层厚度
individual tie 单根轨枕
individual tile 专用砖;专用瓦
individual timing device 个别定时设备
individual tolerance 个体耐受性
individual traffic 个体交通
individual transmission 个别传动
individual trip 个人出行
individual trunk 专用中继线
individual variable 个体变项;个体变量
individual variation 个别变化
individual vehicle delay 每车延误
individual vent 专用通风管;单独透气管;单独通气管;单独通风管;单独排气管
individual warm water preparation 单独烧热水
individual water distribution 单独配水
individual water quality parameter group map 单个水质参数组图
individual water quality vulnerability zone map 单个水质脆弱带图
individual water supply 单独给水
individual water supply installation 单独供水设备
individual wave 单个波
individual well 单口井;单独孔
individual well capacity 单井出水能力;单井出水量
individual well yield 单井开采量
individual working people 个体劳动者
individual workshop 小作坊
individual work submersible 单人作业潜水器
individuation 个体化
indivisibility 不可分性
indivisible 除不尽的;不可分的;不可除尽的
indivisible credit 不可分割债权
indivisible letter of credit 不可分割信用证
indivisible obligation 不可分债务
indivisible thing 不可分物
Indo-Australian realm 印澳界【地】
Indo-Chian movement 印支运动【地】
Indochina geosyncline 印支(期)地槽
Indochina massif 印支地块
Indochina orogeny 印支运动【地】
Indochina subcycle 印支亚旋回【地】
Indo-Chinese epoch 印支期【地】
indocyanine green 绿靛花青

indoleacetic acid 吲哚乙酸;吲哚基醋酸
indole test 吲哚试验
in dollar 按美元
Indo-Mediterranean bivalve region 印度一地中海双壳类地理大区
Indonesia geosyncline 印度尼西亚地槽
Indonesia/Japan route 印度尼西亚一日本航线
Indonesian Register 印尼船级社
indoor 户内;室内(的)
indoor aerial 室内天线
indoor air cooler 室内冷气机
indoor air design condition 室内空气计算参数
indoor air flow 室内气流
indoor air pollutant 室内空气污染物
indoor air pollution 室内空气污染
indoor air quality 室内空气质量
indoor air quality standard 室内空气质量标准
indoor air velocity 室内空气流速
indoor air ventilation 室内通风
indoor Alpha card survey 室内阿尔发卡法
indoor and outdoor bus bar bushing 户内外母线式穿墙套管
indoor antenna 室内天线
indoor arena 体育馆;室内斗技场
indoor bicycle parking area 自行车库
indoor building board 室内建筑板
indoor building sheet 室内建筑钢板
indoor cable 户内电缆
indoor chassis rolls dynamometer 室内滚筒式底盘测功器
indoor chlorinated rubber paint 室内氯化橡胶涂料
indoor clear varnish 室内透明清漆;室内抛光油漆
indoor climate 室内气候
indoor coating 室内涂层;室内涂料
indoor combined load switch 户内式组合负荷开关
indoor condition 室内条件
indoor court 室内庭院
indoor court-yard 内天井
indoor decorating 室内装饰
indoor decoration 室内装饰
indoor decorative plant 室内装饰植物
indoor design 室内设计
indoor design temperature 室内设计温度
indoor décor 室内装饰
indoor distribution equipment 户内配电装置
indoor emulsion paint 室内乳剂油漆;室内乳化油漆
indoor environment 室内环境
indoor equipment 户内设备;室内设备
indoor facility 室内设备
indoor film 内业胶片
indoor finish(ing) paint 室内饰面油漆
indoor fire extinguishing system 室内消防系统
indoor fire hydrant 室内消火栓;室内消防栓
indoor fittings 室内照明器附件
indoor fixture 室内装置
indoor fountain 室内喷泉
indoor garden 室内庭园;室内花园
indoor glass door 室内玻璃门
indoor gloss varnish 室内抛光油漆
indoor golf range shelter 室内高尔夫球场棚
indoor growing 室内栽培
indoor heat 室内热量
indoor heat gain 室内热增量
indoor high-voltage isolating switch 户内高压隔离开关
indoor humidity 室内湿度
indoor hydro electric(al) station 室内式水电站
indoor ice rink 溜冰馆;室内溜冰场
indoor illumination 室内照明
indoor installation 室内设备;室内装置
indoor insulator 户内绝缘子
indoor lacquer 室内漆器
indoor lamp 室内灯
indoor layout of plumbing system 室内给排水平面图
indoor learner's pool 室内初学者泳池
indoor light fitting 室内照明设备
indoor lighting 室内照明
indoor line 室内线
indoor location 室内装置;室内安装
indoor luminaire 室内照明装置
indoor marble 室内大理石
indoor market 室内市场
indoor masonry wall 室内圬工墙

indoor measuring technique 室内测量法
indoor microclimate 适内微小气候
indoor moisture 室内湿度
indoor noise 室内噪声
indoor noise criterion curve 室内噪声标准曲线
indoor oil circuit breaker 户内油断路器
indoor-outdoor carpet(ing) 室内外两用地毯
indoor-outdoor thermometer 户内外寒暑表
indoor paint 内用漆;室内油漆
indoor pane 室内侧玻璃
indoor parking area 室内停车场
indoor parking space 室内存车场
indoor photography 室内摄影
indoor pigmented varnish 室内有色清漆
inddor pipe 室内燃气管道
indoor pipe system 室内管系;室内管道系统
indoor pipe trench 室内管沟
indoor piping 室内敷管
indoor plant 室内植物;温室植物
indoor planting 室内绿化
indoor plumbing system drawing 室内给排水系统图
indoor pollution level 室内污染级
indoor pool 室内游泳池
indoor post insulator 户内柱形绝缘子
indoor processing shop 室内加工车间
indoor proving ground 室内试验场
indoor recording 室内录音
indoor recreation center 室内康乐中心
indoor reference for air temperature and relative humidity 室内温湿度基数
indoor regulator 户内式电压调整器
indoor relative humidity 室内相对湿度
indoor reveal 室内外抱
indoor road tester 室内转鼓试验台;室内道路模拟试验装置
indoors 在室内
indoor sample 室内样品
indoor scene 室内景物
indoor set 室内装置
indoor shot 室内摄影;室内拍摄
indoor single-pole isolating switch 户内单极隔离开关
indoor skating rink 室内溜冰场
indoor slatted blind 室内木条百叶窗
indoor space 室内空间
indoor stadium 体育馆;室内体育馆
indoor stair(case) 室内楼梯
indoor stock keeping 家畜舍饲
indoor storage elevator 仓内升运器
indoor stroke 吸入冲程
indoor substation 室内变电站;室内变电所;室内变压器
indoor swimming pool 室内游泳池
indoor teaching pool 室内教练池
indoor temperature 室内温度
indoor temperature control 室内温度控制;室内温度调节
indoor temperature sensor 室内温度探测器
indoor tennis court 室内网球场
indoor tennis facility 室内网球设施
indoor tennis stadium 网球馆
indoor termination end 户内终端头
indoor test 台架试验;室内试验
indoor theater 室内戏院;室内电影院
indoor thermal environment 室内热环境
indoor thermal power plant 室内火电厂
indoor thermostat 室内恒温器
indoor tile 室内铺地砖
indoor track 室内跑道
indoor transformer 室内型变压器
indoor transportation system 室内运输系统
indoor trench 室内管沟
indoor turbine 室内汽轮机
indoor type 户内型
indoor type circuit breaker 户内式断路器
indoor type transformer 户内式变压器
indoor varnish 室内清漆
indoor ventilation system 室内通风系统
indoor water supply system 室内供水系统
indoor weather 室内气候
indoor window cill[sill] 室内窗台
indoor wintering 室内越冬
indoor wiring 内线;室内布线

indoor wiring optic(al) cable 户内电缆
indoor work 内业;室内工作
Indo-Pacific belemnite region 印度太平洋箭石地理大区
Indo-Pacific bivalve region 印度太平洋双壳类地理大区
Indo-Pacific realm 印度太平洋界
Indo-Pacific trilobite realm 印度太平洋三叶虫地理区系
indophenol blue 靛酚蓝
indophenol method 靛酚法
indophenol reaction 靛酚反应
indorse 支票背书
indorsed bond 认可债券
indorsee 受让者;受让人
indorsement = endorsement
indorsement of claim 索偿的背书
indorsement of service 送达的背书
indorser[indorsor] 背书人
Indo-sinian cycle 印支旋回
Indosinian movement 印支运动【地】
Indosinian orogeny 印支造山运动
Indo-sinian period 印支期
Indo-West Pacific foraminigeral realm 印度—西太平洋有孔虫地理区系
indraft 向岸水流运动;内向流;向内流;向岸流;吸入;引入
indraught 吸入;引入
indrawing 凹入
induce 诱发;导出
induced 诱导的
induced absorption 诱导吸收
induced absorption band 感应吸收带
induced action 感应作用;感应辐射
induced activity 诱导活动性
induced adaptation mechanism 诱导适应机制
induced air 诱导空气;诱导风;引风
induced air flo(a)tation 感应气浮
induced air oxidation 诱导空气氧化
induced angle of attack 诱导迎角;次生仰角
induced break point 诱导折点法
induced catalysis 诱导催化(作用)
induced charge 感生电荷;感应电荷
induced chemical reaction 感应化学反应
induced circuit 被感应电路
induced circulation 感应环流;诱导(式)循环;强制循环
induced cleavage 诱生劈裂;诱生劈理;压力裂隙;压力裂缝;压力节理;开采裂隙
induced codeposition 诱导共沉积
induced coefficient 感应系数
induced colo(u)r 诱导色;衍生色;感应色
induced combination scattering 受迫组合散射
induced Compton scattering 受迫康普顿散射
induced consumption 由投资引起得消费增加
induced cracking 感应裂缝
induced cross-linking 诱导交联
induced crystallization 诱导结晶
induced current 诱导电流;感应电流;感生电流
induced curvature 诱导曲率
induced decomposition 诱导分解
induced detonation 远距离起爆;感应起爆
induced dipole 诱导偶极;感应偶极子
induced dipole radiation 受迫偶极辐射
induced disease 诱发疾病
induced disintegration 诱发衰变
induced draft 吸风(风扇);诱导(式)通风;诱导风扇;引风;喷气抽水;负压送风
induced-draft boiler 引风锅炉;诱导通风锅炉
induced-draft circular sinter cooler 诱导通风式烧结矿回转冷却机
induced-draft cooling tower 诱导通风式冷却塔;诱导通风式凉水塔;风式凉水塔;引导通风式凉水塔;吸风冷却塔
induced-draft crossflow cooling tower 抽风横流冷却塔
induced-draft engine 抽气通风机;抽风通风机
induced-draft fan 人工通风扇;强制通风风扇;诱导通风风扇;引风机;抽(气通)风机;排烟机
induced-draft mechanical cooling tower 抽风式机械通风冷却塔
induced-draft plant 诱导通风装置
induced-draft water-cooling tower 排风式冷却塔;引风水冷却塔

induced drag 诱导阻力
induced drag coefficient 诱导阻力系数
induced draught 诱导通风;强制通风
induced draught burner 强制通风燃烧器
induced draught fan 诱导风扇;鼓风机;引风机
induced draught filter 抽风集尘器
induced draught system 强制通风系统;抽风系统
induced draught water cooler 排风式凉水器;吸风水冷却器
induced earthquake 诱发地震
induced effect 诱发效应
induced electrical polarity 诱导极性
induced electromotive force 感应电动势;感生电动势
induced electrooptic(al) axis 感应电光轴
induced emission 感应发射
induced emission of light 感应光发射
induced environment 外界感应环境
induced equation 诱导方程
induced failure 诱发故障;诱导故障
induced fan 诱导风扇;引风机
induced fission 诱生裂变;诱发裂变
induced fission track density 人工诱发裂变径迹密度
induced-fit 诱导契合
induced flow 强迫水流;诱导水流
induced-flow heater 强制流动加热器
induced flux 感应通量;感应磁通量
induced grid noise 栅流感生起伏噪声
induced hardness 感应硬度
induced highwater 感应高潮位
induced infiltration 诱导入渗
induced inner content 诱导内容度
induced investment 诱致投资;劝诱投资;被诱导的投资
induced item 诱发项目
induced iterative method 诱导的迭代法
induced lift 诱导升力
induced lightning over voltage 感应雷击过电压
induced lysis 人工溶解
induced magnetic variation 诱发磁性变化
induced magnetism 感应磁性;感生磁
induced magnetization 感应磁化
induced magnetization strength 感应磁化强度
induced matrix norm 导出的矩阵范数
induced measure 诱导测度
induced metric space 诱导度量空间
induced mission 受迫发射
induced module 诱导模
induced moment 感应矩
induced motion 诱导运动
induced motor 感应电动机
induced movement 诱发运动
induced moving magnet instrument 感应(移)动磁(铁)式仪表
induced mutation 诱发突变
induced noise 感应噪声;感生噪声
induced norm 导出范数
induced nuclear disintegration 诱发核衰变
induced nuclear reaction 诱发核反应
induced nucleation method 诱导成核法;外加晶种法
induced operator 诱导算子
induced outer measure 诱导外测度
induced overvoltage 感应过电压
induced overvoltage withstand test 感应耐压试验
induced oxidation 诱导氧化(作用)
induced permeability 诱发渗透率
induced photon 感应光子
induced polarity 诱导极性
induced polarization 感应极化;激发极化(法);诱导极化
induced polarization charging array 激发极化充电装置
induced polarization combined profiling array 激电联剖
induced polarization dipole-dipole sounding array 激电偶极测深
induced polarization instrument controlled by microcomputer 微机激电仪
induced polarization instrument in frequency domain 频率域激电仪
induced polarization instrument in time domain 时间域激电仪

induced polarization log curve 激发极化测井曲线
induced polarization longitudinal mid-gradient array 纵向激电
induced polarization method 激发极化法
induced polarization method in bore hole 井中激发极化法
induced polarization orthogonal sounding array 激发极化正交测深
induced polarization potential log 激发电位测井
induced polarization quadripole sounding array 激电四极测深
induced polarization sounding method 激发极化测深法
induced polarization survey 感应偏振测量
induced polarization test methods 石油学会试验法(英国)
induced polarization transverse mid-gradient array 横向激电中梯阵列
induced porosity 次生孔隙
induced power supply 感应电源
induced precession 实际进动
induced precipitation 诱导沉淀(作用)
induced probability measures 诱导概率测度
induced radiation 感应辐射
induced radioactivity 诱发放射性;感应放射性;感生放射性;感光放射性
induced radioisotope 感生放射性同位素;人造放射性同位素
induced reaction 诱导反应;感应
induced recharge 诱导回灌;诱导补给量
induced recharge of surface water 地表水诱导补给
induced recombination 诱导复合;受迫复合
induced remission stage 诱导缓解期
induced resistance 诱导阻力;感应阻力
induced resources 诱导资源
induced reverse osmosis 诱导反渗透
induced rolling moment 诱导滚动力矩
induced room air intake 诱导风入口
induced seismicity 诱导震动;感应地震;诱发地震活动
induced sequence 诱导序列
induced shot firing 松发爆炸
induced sinusoid 感应正弦电压
induced spark 感应火花
induced stress 诱发应力;诱导应力;感生应力;次生应力
induced subgraph 诱导子图;导出子图
induced substance 诱导物质
induced surface hardening 感应加热表面淬火
induced surface runoff 诱导地面径流
induced test 感应试验
induced topology 诱导拓扑
induced traffic 诱增交通量(因道路交通设施改进所增加的交通量,包括新增交通量、导增和变增交通量)
induced traffic volume 诱增交通量;诱增运量;诱发交通量
induced transition 感应跃迁;受激跃迁
induced transition cross section 感应跃迁截面
induced transmission filter 感应透射滤光片
induced uniaxial anisotropy 诱导单轴各向异性
induced unitary operator 诱导酉算子
induced unitary representation 诱导酉表示
induced variation 诱导变异
induced velocity 诱导速度;诱导流速
induced velocity field 诱导速度场
induced voltage 感应电压
induced wave 诱发波
inducement 诱导
inducement of investment 投资诱因
inducement to invest 投资诱因
inducer 诱导者;诱导物;诱导轮;诱导剂;感应器;进口段;电感器;导流轮;导风轮
inducing charge 加感电荷;施感电荷
inducing colo(u)r 诱导色
inducing combustion 诱导燃烧
inducing current 施感电流
inducing factor 诱发因素
inducing field 加感场
inducing magnet 施感磁化
inducing winding 加感绕组
inductance 感应性;感应现象;感应系数;电感应;电感(器);发动机进气;线圈

inductance amplifier 电感耦合放大器
inductance approach switch 电感式接近开关
inductance attenuator 电感型衰减器
inductance balance 感应平衡;电感平衡
inductance box 电感箱
inductance box with plugs 插塞式电感箱
inductance bridge 电感电桥
inductance-capacitance filter 电感电容滤波器
inductance circuit 施感电路
inductance coefficient 电感系数
inductance coil 感应线圈;电感线圈;线圈
inductance comparator and measuring probe 电感比较器及测量头
inductance connecting three point type oscillator 电感三点式振荡器
inductance constant of crystal unit 晶体振子电感常数
inductance coupled amplifier 电感耦合放大器
inductance coupling 电感耦合
inductance effect 电感效应
inductance factor 电感因数
inductance filter 电感滤波器
inductance ga(u)ge 感应仪
inductance measurement 电感测量
inductance meter 亨利计;电感(测量)计;电感测定器
inductance mutual 互感
inductance settlement probe 电感沉降探头
inductance standards 电感标准
inductance strain ga(u)ge 电感应变计
inductance type liquidometer 电感式液位计
inductance type transducer 电感式传感器
inductance unbalance 电感性失衡
inducted draft 人工通风;抽(气通)风;负压通风
inductile 低塑性的
in-duct installation 管沟敷设
induction 诱导(作用);归纳;感应(现象)
induction accelerator 感应加速器;电子回旋加速器
induction air-conditioning system 诱导式空(气)调(节)系统
induction alternator 感应(交流)发电机
induction amplifier 感应放大器
induction-arc furnace 感应电弧炉
induction balance 感应平衡;电感平衡
induction balancer 电感平衡器
induction block 感应器
induction brake 感应制动器
induction brazing 高频钎焊;感应(硬)钎焊;感应(加热)钎焊
induction bridge 感应电桥
induction burner 感应燃烧器
induction chamber 吸气室;诱导式进气室
induction charging 感应充电
induction coil 感应线圈
induction compass 感应罗盘;感应罗经;地磁感应罗盘
induction cooker 感应炉;电磁灶
induction coupling 感应耦合器
induction crucible furnace 感应坩埚(熔)炉
induction current 诱导电流
induction degassing 感应去气法
induction density 感应密度
induction development 感应显影
induction disc type directional power relay 感应圆盘式定向功率继电器
induction disc type directional relay 感应圆盘式定向继电器
induction disc type inverse-time overcurrent relay 感应圆盘式反比时限过载流继电器
induction disc type ratio differential relay 感应圆盘式比率差动继电器
induction disc type relay 感应(圆)盘(式)继电器
induction disc wattmeter 感应圆盘瓦特计
induction door 进气门
induction drag 感应阻滞
induction drill 电动感应钻机
induction drilling 感应法钻进
induction-driven wind tunnel 诱导风洞
induction dynamometer 异步测功机;感应式功率计;感应式测力计
induction effect 诱导效应;感应作用;感应效应
induction electric(al) furnace 感应电炉
induction electrolog 感应电测井
induction electron accelerator 感应式电子加速器

induction energy 感应能
induction equation 诱导方程
induction experiment 诱导试验
induction exploration 感生探索(方法);感应探索(方法)
induction factor 诱发因素;诱导因素
induction fan 感应扇
induction field 感应场
induction field locator 感应场定位器
induction flame damper 吸气管阻焰器
induction flowmeter 感应流量计
induction fluid amplifier 引流型放大元件
induction flux 感应通量
induction force 诱导力;感应力
induction frequency conventer 感应式变频机
induction frequency meter 感应频率计
induction furnace 感应(电)炉
induction generator 异步发电机;感应(式)发电机;感应电动机
induction gum test 诱导胶质试验
induction hardening 高频淬火;感应硬化;感应淬火
induction hardening equipment 感应淬火装置;感应淬火设备
induction hardening machine 感应淬火机
induction heat 感应热
induction-heated screen 感应加热筛分机
induction heater 感应加热器;电感加热器
induction heater and control panel for rolling bearing 滚动轴承感应加
induction heater of electromagnetic cooker 电磁灶感应加热器
induction heating 感应加热;电感加热
induction heating apparatus 感应加热装置;感应加热器
induction heating coil 感应加热线圈
induction heat sterilization 感应热灭菌法
induction heat-treatment 感应热处理
induction height 导入高程
induction height survey 导入高程测量
induction hum 感应交流声;感应哼声
induction influence 感应干扰
induction instrument 感应系仪表;感应式仪表;感应式测试仪器
induction interrupter 感应断续器
inductionless 无感应的
induction lightning stroke 感应雷击
induction log 感应法测井图;感应测井记录
induction log calibration value 感应测井刻度值
induction log curve 感应电测井曲线
induction logger 感应测井仪
induction logging 感应测井
induction log resistivity 感应测井电阻率
induction log tool 感应测井仪
induction loudspeaker 感应(式)扬声器
induction machine 感应(式)电机;感应(式)加速器
induction magnetometer 感应(式)磁强计
induction manifold 吸水管;吸气歧管
induction melting 感应熔化
induction meter 感应式电度表;感应计
induction method 感应法;归纳法
induction motor 异步电动机;感应电(动)机
induction motor controller 感应电动机控制器
induction motor meter 感应式电度表
induction motor type synchronous motor 感应电动机式同步电动机
induction noise 进气噪声
induction period 诱发期;诱导期;感应(周)期
induction period of drying 干燥诱导期
induction period test 诱导期试验
induction permeability characteristic 感应导磁率特性曲线
induction permeability curve 磁通导磁率曲线
induction phase 诱导期
induction phase shifter 感应移相器
induction phenomenon 感应现象
induction pipe 吸入管;吸气管;进水管;进气管;进口管;导入管
induction port 进气门
induction potentiometer 感应式电位计
induction power supply 感应电源
induction pressure ga(u)ge 吸入压力计
induction principle 归纳原理
induction problem 感应问题
induction process 感应过程

induction pump 交流电磁泵;电磁感应泵
induction quenching 感应淬火
induction ratio 诱导比
induction receiver 感应接收机
induction regulator 感应式转数调节器;感应调压器;电感调节器
induction relay 感应式继电器
induction-ring type high-speed relay 感应环式高速继电路
induction rotor 感应转子
induction salinometer 感应盐分计
induction sensor 感应传感器
induction sheath 感应屏蔽层;磁感应屏蔽层
induction shield 感应屏蔽
induction silencer 感应消声器
induction sintering 感应烧结
induction soldering 感应(软)钎焊;感应焊接
induction spark coil 电火花感应线圈
induction spot welder 感应点焊机
induction starter 感应式起动器;感应起动器
induction stirred furnace 感应搅拌电炉
induction stirrer 感应搅拌器
induction stroke 吸入冲程;进气冲程;进气行程
induction system 诱导系统;感应装置
induction tachogenerator 感应测速发电机
induction telephone 感应式电话
induction tempering 感应回火
induction tensor 感应张量
induction time 诱导期;感应时间
induction torque 异步转矩;感应转矩
induction torquemeter 感应式转矩计
induction training 就业培训
induction turbine 中间充气汽轮机
induction type acceleration mechanism 感应加速机制
induction type alternator 感应式交流发电机
induction type ammeter 感应式安培计
induction type distance relay 感应式远距离继电器
induction type frequency converter 感应式换频机
induction type frequency meter 感应式频率计
induction type furnace 感应式电炉
induction type instrument 感应式仪表
induction type integrating watt-meter 感应式累计功率表;感应式累计瓦特计
induction type magneto 感应式磁电机
induction type overcurrent relay 感应式过流继电器
induction type overvoltage relay 感应式过电压继电路
induction type power directional relay 感应式功率定向继电器
induction type reactive power relay 感应式无功功率继电路
induction type strain ga(u)ge 电感式应变仪;电感式应变片
induction type synchronous alternator 感应式同步交流发电机
induction type undervoltage relay 感应式(过)低电压继电器
induction type voltage regulator 电感式电压调节器
induction type voltmeter 感应式伏特计
induction type watt-hour meter 感应式瓦特小时计
induction type wattmeter 感应式瓦特计
induction unit 诱导器
induction valve 进气阀;吸入阀
induction variable 归纳变量
induction variometer 变感仪
induction vector 感应矢量
induction vehicle detector 感应式车辆检测器
induction velocity 诱导速度
induction ventilation 导入式通风;诱导(式)通风
induction voltage 感应电压
induction voltage regulator 感应(式)调压器
induction watt-hour meter 感应式瓦时计
induction wattmeter 感应式瓦特计
induction welding 感应熔焊;感应焊接
induction weld mill 感应焊管机
induction wind tunnel 诱导风洞
induction zone 感应区
inductive 归纳的;电感(性)的
inductive action 感应作用
inductive aero-electromagnetic system 感应式航空电磁系统

inductive approach 归纳法
inductive assumption 归纳假设
inductive balance 电感平衡
inductive bond 阻抗接合
inductive branch 电感支路
inductive cab signal(l)ing 感应式机车信号
inductive character 电感性
inductive choke 电感线圈
inductive choke coil 电感扼流圈
inductive circuit 电感电路
inductive coil 有感线圈;感应线圈
inductive compensator 感应补偿器
inductive control 感应控制
inductive coordination 感应协调
inductive coupler 电感耦合器
inductive coupling 感应作用;电感耦合
inductive definition of sets 集合的归纳定义
inductive demonstration of argumentation 归纳论证
inductive detector 电感式车辆检测器;电感式检测器
inductive displacement transducer 感应式位移传感器
inductive divider 电感分配器
inductive drag 感应阻力
inductive drop 感应降落;电感电压降
inductive edge 归纳边
inductive effect index 诱导效应指数
inductive feedback 电感反馈
inductive filter 感应滤波器
inductive flowmeter 感应式流量计
inductive formula 归纳公式
inductive generator 感应电动机
inductive grounding 电感性接地
inductive hypothesis 归纳假设
inductive impedance 感抗;电感性阻抗
inductive inference 归纳推理
inductive interference 感应干扰
inductive internal surface analyzer[analyser] 感应式内表面分析仪
inductive iris 电感性膜片
inductive line pair 感应线对
inductive load 有感负载;感性负载;电感性负荷
inductive loading 加感
inductive logic 归纳逻辑
inductive loop 感应涡流
inductive loop detector 感应式框形检测器;电感回路检测器
inductive loop traffic detector 感应线圈交通探测器
inductively coupled plasma 电感耦合等离子体
inductively coupled plasma atomic emission spectrometry 电感耦合等离子体原子发射光谱法
inductively coupled plasma emission spectrometry 电感耦合等离子体发射光谱法
inductively coupled plasma mass spectrometry 电感耦合等离子体质谱法
inductively coupled plasma optical emission spectrometry 电感耦合等离子体光发射光谱法
inductively coupled plasma spectrometer 电感耦合等离子体光谱仪
inductively coupled plasma spectroscopy 电感耦合等离子体光谱法
inductively loaded antenna 加感天线
inductively loading 加感
inductive meter 感应式仪表
inductive method 归纳法;感应电磁勘探法
inductive micrometer 感应测微计
inductive microphone 感应式话筒
inductive neutralization 感应中和
inductive phase 诱导相
inductive pick-off 感应发送器
inductive pick-up 感应传感器
inductive post 电感柱
inductive pressure transducer 电感压力传感器
inductive probe 感应探头;感应探测器
inductive proof method 归纳证明法
inductive reactance 感抗
inductive reaction 电感反应
inductive reasoning 归纳推理
inductive relay 电感继电器
inductive resistor 有感电阻器;感应电阻器
inductive salinometer 感应盐液密度计;感应式盐度仪;感应式盐度计
inductive sawtooth generator 感应锯齿波发生器

inductive selection 感应选择
inductive sender 电感发送器
inductive shunt 感应分流器
inductive spacing 电感性间隔
inductive statistics 归纳统计学
inductive strain ga(u)ge 感应式应变仪
inductive surge 电感性电涌
inductive susceptance 电感性电纳
inductive switch 感应开关
inductive transducer 感应(式)传感器;电感传感器
inductive tuning 电感调谐
inductive ventilation 诱导通风;诱导式通风
inductive ventilation system 诱导通风系统
inductive voltage transformer 电压互感器
inductive waveform 电感性波形
inductive wave probe 电感式浪高仪
inductive winding 电感线圈
inductive window 电感性窗口
inductivity 诱导性;诱导率;感应性;感应率;介电常数
inductometer 电感计;电感表
inductor 线圈;诱导者;诱导物;感应体;感应器;手摇发电机;电感线圈;电感器
inductor blocks 感应器
inductor filter 电感滤波器
inductor form 线圈骨管;线圈架
inductor generator 感应子电机;感应体发电机
inductorium 火花感应线圈
inductor machine 感应子电机
inductor microphone 感应子传声器;感应式传声器
inductor motor 作为发电机的感应电动机;感应子电动机
inductor tachometer 感应式转速计
inductor type alternator 感应式交流发电机
inductor type generator 感应式发电机
inductor type synchronous alternator 感应式同步交流发电机
inductor type synchronous motor 感应子同步电动机
inductor type tachometer 电感式转速计
inductor voltage regulator 感应式电压调整器
inductosyn 感应同步器;感应式传感器
inductosyn droop meter 感应式传感器倾斜计
inductosyn scale 感应同步尺
inductuner 电感调谐设备;电感调谐器
induflux method 疲劳电磁测定法
induit 代购契约
indulge 容许延期付款
indulgence 付款延期
indulin 纯松木素
induline 引杜林(染料);对氮蒽蓝
induline bases 引杜林色基
in duplicate 一式两份;一式二份
indurated 固结的
indurated clay 固结黏土;硬化黏土
indurated fiber 硬纤维
indurated rock 固结岩石;硬化的岩石
indurated soil 硬磐土;固结土(壤);固结土(壤)
indurated talc 滑石板岩;硬滑石
induration 硬结(作用);硬化作用;固结(体)
Induroleum 无接缝铺地板法
Indus river basin 印度河盆地
industrial absorption tower 工业吸收塔
industrial abstraction 工业抽水
industrial accident 工业事故;工伤事故
industrial accident fund 工伤抚养基金
industrial accident fund reserve 工伤抚养准备基金
industrial accounting 工业会计
industrial acidity meter 工业酸度计
industrial activited carbon 工业活性炭
industrial activity 工业活动
industrial added value 工业增加值
industrial adhesive 工业黏合剂
industrial administration 工业经营学;工业部门
industrial-agricultural enterprise 工农联合企业
industrial air conditionine 工厂空气调节
industrial air conditioniner 工艺性空调
industrial air conditioning 工艺性空气调节
industrial air pollution 工业大气污染;工业空气污染
industrial alcohol 工业乙醇;工业酒精
industrial aluminum sulphate 工业硫酸铝
industrial analysis 工业分析
industrial and agricultural products 工农业产品
industrial and commercial administration 工商行政管理

industrial and commercial consolidated tax 工商业统一税
industrial and commercial income tax 工商所得税
industrial and commercial responsibility insurance 工商业责任保险
industrial and commercial tax 工商业税
industrial and mining area 工矿区
industrial and mining establishment 工矿企业
industrial anthropometry 工业人体测量
industrial anti-pollution technique 工业抗污染技术
industrial appliance 工业用具
industrial applicability of an invention 发明的工业实用性
industrial arbitration 劳动仲裁
industrial architect 工业建筑师
industrial architecture 工业建筑(学)
industrial area 工业区
industrial aromatic hydrocarbons 工业芳香烃
industrial art 工艺美术;工业美术
industrial asphalt(ic) tile 工业沥青瓦
industrial atmosphere 工业气氛;工业区环境;工业区空气
industrial autoclave 工业用蒸压器
industrial automation 工业自动化
industrial automation instrument 工业自动化仪表
industrial bacteriology 工业细菌学
industrial balance 工业天平
industrial ball 工业钢球
industrial bank 实业银行
Industrial Bank of Japan 日本工业银行
industrial base 工业基地
industrial battery 工业电池
industrial bearing 工业轴承
industrial belt 工业地带
industrial boiler 工业锅炉
industrial boiler make-up water 工业锅炉补给水
industrial bond 工业债券
industrial boom 工业繁荣
industrial borescope 工业光学孔径计
industrial broker 工业房地产经纪人
industrial brokerage 工厂用地交易的经纪业务
industrial brush 工业用刷
industrial building 工业馆;工业用房;工业建筑(物);工业大厦;厂房
industrial buildings allowances 工业建筑物免税额
industrial burning appliance 工业燃烧器
industrial capital 工业资本
industrial car 车间窄轨运输车;车间机动运输车;工业机动车
industrial carrier 厂矿专用运输(工具)
industrial caster 工业用脚轮
industrial caterpillar tractor 工业履带式牵引车
industrial cathode ray tube 工业用阴极射线管
industrial cell 工业电解槽
industrial center 工业中心
industrial centralization 工业集中化
industrial ceramics 工业陶瓷
industrial chain 工业链条
industrial chemical muriatic acid 工业化学盐酸
industrial chemical plant 化工厂
industrial chemicals 工业化学品;化工原料
industrial chemistry 工业化学
industrial chimney 工业烟囱;工厂烟囱
industrial chromatograph 流程色谱仪
industrial chromatography 工业色谱
industrial circulating fund 工业流动资金
industrial circulating water system 工业循环水系统
industrial city 工业都市;工业城市
industrial-civil foundation ga(u)ge method 工民建地基规范法
industrial classification 工业分类
industrial cleaning 工业清洗
industrial climatology 工业气候学
industrial clinker 工业熟料
industrial coal briquette 工业型煤
industrial coating 工业涂装
industrial code 工业规范;工业法规
Industrial Communication Association 工业通信协会
industrial company 工业公司
industrial complex 工业综合体;工业联合企业;大工业中心
industrial computation 工业计算

industrial concern 工业部门
industrial condensing unit 工业用压缩冷凝机组
industrial construction 工业建设;工业建筑(物)
industrial construction project 工厂建设计划;工厂建筑方案;工业建筑工程
industrial construction site 工厂施工现场;工业施工现场;工业施工工地
industrial consumer 工业用户
industrial consumer goods 日用工业品
industrial consumption 工业生产消费
industrial contaminant 工业污染物
industrial contamination 工业污染
industrial control 工业控制;工业管理;工业调整
industrial control analog(ue) module 工业控制模拟设备
industrial control communication 工业控制通信
industrial control component 工业控制部件
industrial control computer 工业控制(用)计算机
industrial control computer language 工业控制(用)计算机语言
industrial controlling 工业控制用
industrial control module 工业控制组件
industrial control push button 工业控制按钮
industrial control unit 工业控制(用)计算机;工业控制器
industrial cooling 工业冷却
industrial cooperation 工业合作
industrial corrosion 工业腐蚀
industrial cost control 工业成本控制
industrial cost recovery 工业成本回收
industrial crawler tractor 工业履带式拖拉机
industrial crop 工业(原料)作物;经济作物
industrial crystallization 工业结晶;大量结晶
industrial curtain wall 工业幕墙;工业用帷幕墙
industrial cut stock 工业用材
industrial data processing 工业数据处理
industrial daylighting 工业采光
industrial decentralization 工业分散(化);工业疏散
industrial density 工业密度
industrial department 工业部门
industrial dermatitis 工业性皮炎
industrial dermatosis 职业性皮肤病;工业(性)皮肤病
industrial design 工业(制品)设计
industrial detergent 工业洗涤剂
industrial detonator 工业雷管
industrial development 工业发展
Industrial Development Center for Arab States 阿拉伯国家工业发展中心
industrial development park 工业发展区
industrial development zone 工业发展区
industrial diamond 工业钻石;工业金刚石;碎粒金刚石;黑金刚石
industrial diamond stone 工业用金刚石
industrial diaphragm pressure ga(u)ge 工业膜片式压力表
industrial diesel engine 工程机械用柴油机
industrial disaster 工业灾害
industrial discharge 工业排放(物);工业排出物;工业废物
industrial disease 职业病;工业(疾)病
industrial dispersion 工业离散
industrial distribution 工业布局
industrial district 工业区;商业区
industrial district road 工业区道路
industrial dust 工业粉尘;工业尘
industrial dust discharge 工业粉尘排放
industrial dust pollution 工业粉尘污染
industrial dynamics 工业动态;工业动力学
industrial economics 工业经济学;经济性工学
industrial economy 工业经济
industrial effluent 工业污水;工业排水;工业排放物;工业排出物;工业流出物;工业废液;工业废水;产业废水
industrial effluent transport 工业污水输送
industrial electric(al) conduction meter 工业电导仪
industrial electric(al) drying 工频电干燥
industrial electric(al) equipment 工业电器
industrial electric(al) furnace 工业(用)电炉
industrial electric(al) heating 工业电加热;工频电加热
industrial electricity 工业用电
industrial electrolyzer 工业电解槽

industrial electronics 工业电子学
industrial electron tube 工业用电子管
industrial embroidery machine 工业绣花机
industrial emission 工业污水排放;工业排放(物)
industrial emission source 工业排污源;工业排放源
industrial emmition source 工业排出源
industrial engineer 工业(管理)工程师
industrial engineering 工业工程(学);经营管理(工程)学;经营工程(学);企业管理学;生产组织技术
industrial engineering method in cost estimation 成本估计的工程法
industrial enterprise 工业企业
industrial environment 工业环境
industrial environment(al) health 工业环境卫生
industrial establishment 工业企业
industrial estate 工业区;工业用地;工业地产
industrial estate development plan 工业区发展计划
industrial estate for small and medium size industry 中小工业用地
industrial ethyl alcohol 工业用乙醇
industrial exhaust heat 工业余热
industrial exhaust system 工业排风系统;工业废气排放系统
industrial exhibition 工业展览会
industrial exposure 工业接触;工业辐照
industrial extraction unit 工业通风机
industrial extractor 工业通风机
industrial extract ventilation unit;industrial fan 工业通风机;工业吸风装置
industrial fabric 工业织物
industrial fabrics in wet filtration 湿法过滤中的工业纤维
industrial facility 工业设施;工业设备
industrial facility discharge 工厂污水排放
industrial factors for ores 矿石工业指标
industrial factory building 工业厂房
industrial fair 工业展览
industrial fait 工业博览会
industrial fan 工业鼓风机
industrial fiber 工业用纤维
industrial fiber scope 工业纤维镜
industrial finish 工业建筑面层处理;工业加工精洁面;工业用面漆
industrial finishes or coatings 工业用涂料
industrial fire 工业加热炉
industrial fixture 工厂(照明)设备
industrial flat bed sewing machine 工业平缝机
industrial floor 工厂地面;工业楼面;工业楼板
industrial floor brick 工业用铺地砖;工厂铺地砖
industrial floor finish 工业楼面修饰;工业楼面整修
industrial floor(ing) tile 工厂地面铺块;工业用铺楼面砖
industrial foils 工业用金属箔
industrial forestry 工业性林业
industrial frequency 工业频率
industrial frequency alternate current electric (al) traction system 工频交流电力牵引制
industrial frequency current 工频电流
industrial frequency traction 工频牵引
industrial fuel consumption 工业燃料消耗
industrial fume 工业烟尘;工业性烟气
industrial function 工业功能
industrial fund 工业资金;工业基金
industrial furnace 工业加热炉;工业(窑)炉
industrial furnace centrifugal blower 工业炉离心鼓风机
industrial garage 工厂车库;工业停车房
industrial gas 工业气体;工业煤气
industrial gas appliance 工业燃(气用)具
industrial gas chromatograph 工业气相色谱仪
industrial gas chromatographic method 工业气相色谱
industrial gaseous waste 工业废气
industrial gas furnace 燃气工业炉
industrial gasoline 工业汽油
industrial gas turbine 工业用燃气轮机;工业燃气轮机
industrial gelatine 工业明胶
industrial geography 工业地理学
industrial glass 工业玻璃
industrial glass fabric 工业用玻璃织物
industrial glazing 工厂玻璃窗(安装);工业磨光

（机）
industrial glue 工业黏合剂；工业黏胶
industrial goods 工业品；生产资料
industrial grade 工业用；工业品位；技术等级
industrial grade benzene 工业用苯
industrial grade diamond 工业品位金刚石
industrial grade of ore 矿石工业品位
industrial grade toluene 工业级甲苯
industrial grease 工业润滑油脂
industrial ground 工业场地
industrial grouping 工业组合
industrial growth 工业增长
industrial growth potential 工业增长潜力
industrial harbo(u)r 工业港(口)
industrial hazard 工业灾害；工业危害性；工业事故；工业公害
industrial health 工业保健；工业卫生
industrial health statistics 工业卫生统计
industrial heating 工厂取热；工业加热
industrial heating equipment 工业供热设备
industrial heating facility 工业加热设施；工业供暖装置
industrial heat pump 工业热泵
industrial housing 工业化住房(建设)；工业用房；工业房屋
industrial hygiene 工业卫生
industrial hygiene regulation 工业卫生条例
industrial hygiene standard 工业卫生标准
industrial hygienics 工业卫生学
industrial hygienist 工业卫生学家
industrial hygienology 工业卫生学
industrial illness 职业病；工业病
industrial illumination 工业照明
industrial incinerator 工业灰化炉；工业焚化炉
industrial indices of mineral resources 矿产工业指标
industrial inertia 工业惯性
industrial information center [centre] 工业情报中心
industrial infrastructure 工业基础结构
industrial inhalant 工业性吸入物
industrial injury 工业外伤；工业损伤；工业(生产)性伤害；生产性外伤
industrial installation 工业设备安装；工业生产装置；工业生产设备
industrial instrument 工业测量仪器；工业测量仪表；技术仪表
industrial instrumentation 工业测量仪表
industrial interference 工业干扰
industrial investment 工业投资
industrialist and businessman 职能资本家
industrialization 工业化；产业
industrialization emigration 因工业化人口外迁
industrialization of building 建筑工业化
industrialization of construction industry 建筑工业的工业化
industrialization of rural 郊区工业化
industrialization of rural areas 农村工业化
industrialization of the country 国家工业化
industrialized apartment building 工业化公寓建筑
industrialized building 工业化房屋；预制部件装配式建筑；工业化建筑(物)；大型预制件装配式建筑法
industrialized building construction 工业化施工
industrialized building method 工业化建造法；工业法建造法
industrialized building system 工业化(房屋)建筑体系；工业化建筑系统
industrialized building technique 工业化建筑技术
industrialized concrete construction 工业化混凝土工程
industrialized concrete house building factory 混凝土房屋预制厂
industrialized construction 工业化施工
industrialized construction industry 工业化的建筑工业
industrialized construction method 工业化建造方法
industrialized country 工业化国家
industrialized facade 预制构件立面
industrialized housing 工业化住房
industrialized housing construction 工业化住房建设
industrialized production of houses 工业化房屋建造

industrialized reinforced concrete (system) construction 工业化钢筋混凝土体系施工
industrialized structure 工业化结构
industrialized system building 工业化系统建筑法；工业化体系建筑法
industrialized unit 工业化单元
industrialized wall 预制墙壁；工业化墙壁
industrial jewel 工业宝石
industrial kiln 工业窑
industrial kilns and furnaces 工业炉窑
industrial kitchen 工厂厨房
industrial laminate 工业用层压板
industrial laminate(d) board 层压板
industrial land 工业用地
industrial landscape 工业景观
industrial land use 工业土地利用
industrial layout 工业配置；工业布局
industrial legislation 工业立法
industrial lift 工业电梯
industrial lift truck 工业升降式装卸车；工业汽车吊
industrial light 工厂照明器
industrial light fitting 工业照明装置
industrial lighting 工业照明；工厂照明
industrial light(ing) fixture 工厂照明设施；工业照明设备
industrial lime 工业用石灰
industrial line 专用线；专用铁路线；工业支线
industrial linkage 工业联系
industrial liquid waste 工业液体废物
industrial load 工业用电量；工业(用电)负荷
industrial loader 工业装载机
industrial loan 工业贷款
industrial localization factor 工业择地因素
industrial location 工业选址；工业(所在)地点；工业区位
industrial location factor 工业布点因素
industrial location theory 工业布局理论
industrial logic-sequence controller 工业用逻辑顺序控制器
industrial lubricant 工业润滑油
industrial luminaire(fixture) 工厂照明设施；工业照明设备
industrially built block(of flats) 工业化公寓
industrially manufactured house 工业生产住房
industrially pure titanium 工业纯钛
industrial maintenance paint 工业维护用涂料
industrial management 工业管理学
industrial management program(me) 工业管理程序
industrial marshalling station 工业编组站
industrial mastic 工业胶脂地板
industrial material 工业材料
industrial materials from solid wastes 固体废物废物工业材料
industrial material talc 工业原料滑石
industrial measurement instrument 工业测量仪表
industrial medicine 工业医学
industrial melanism 工业黑化现象
industrial meteorology 工业气象学
industrial meter 工业仪表
industrial methylated spirit 工业用甲基化酒精；工业用甲醇变性酒精
industrial microbiology 工业微生物学
industrial microcomputer 工业(用)微计算机
industrial microscope 工业用显微镜
industrial migration 工业迁徙
industrial mineral 工业矿物
industrial mobile crane 工业用自走式起重机
industrial modernization investment 更新改造措施投资
industrial modernization project 更新改造措施项目
industrial muriatic acid 工业用盐酸
industrial naphtha 工业石脑油
industrial naphthalene 工业萘
industrial noise 工业噪声
industrial noise criterion 工业噪声标准
industrial noise management 工业噪声管理
industrial nuclear power 工业用核动力
industrial nuisance 工业公害
industrial occupancy 工业区
industrial oil 工业用油
industrial oil-gas bed 工业油气层
industrial or commercial 工业的或商业的
industrial ore ratio 工业含矿率

industrial organic wastewater 工业有机废水
industrial organization 工业体制
industrial oscillograph 工业用示波器
industrial outlet water 工业排水
industrial output value 工业产值
industrial oxygen blow-pipe 工业用吹氧管
industrial packaging 工业包装
industrial paint 工业用漆；工业涂料；工业漆
industrial park 工业园；工业区停车场；工业(区)公园；工业区；花园工厂
industrial parquetry block 工业镶木地板块料
industrial partition(wall) 工厂隔墙；工业隔墙
industrial pattern 工业形式
industrial performance standard 工业作业标准
industrial personal computer 工业控制微机
industrial photogrammetry 工业摄影测量
industrial physician 工业卫生医师
industrial pipe 工业用管
industrial plant 工厂
industrial plant equipment 工业设备
industrial plaster for fillers and decoration 填料和装潢用工业石膏
industrial plaster for mo(u)lding 工业模具石膏
industrial pneumoconiosis 职业性矽肺病；工业肺尘埃尘着病
industrial poison 工业毒物
industrial poisoning 工业中毒
industrial policy 工业政策
industrial pollutant 工业污染物
industrial polluter 工业污染肇事者
industrial pollution 工业污染
industrial pollution control 工业污染控制
industrial pollution control planning 工业污染控制计划；工业污染控制规划
industrial pollution of water body 水体工业污染
industrial pollution source 工业污染源
industrial pollution source management 工业污染源管理
industrial pollution source rating 工业污染源评价
industrial pollution threshold limit 工业污染阈限值
industrial population 工业人口
industrial porcelain 工业瓷
industrial port (工厂企业所属的)专用港；工业港(口)
industrial portable track 工业轻便轨道
industrial potential 工业潜力
industrial potentiometer 工业用电位计
industrial power 工业动力
industrial power plant 企业自备电厂
industrial power supply 工业供电
industrial premises 工厂不动产；工业房产连地基；工业厂房
industrial preservative 工业防腐剂
industrial pretreatment standard 工业预处理标准
industrial process 工业生产过程；工业生产方法
industrial process control 工业过程控制；工业工程控制；生产过程控制
industrial processing system 工业加工系统
industrial process water 工业生产用水
industrial produce 工业生产
industrial producing waste 工业废物
industrial production 工业生产
industrial production index 工业生产指数
industrial production statistics 工业生产统计
industrial productivity 工业生产能力
industrial products 工业制品；工业产品
industrial project 工业项目
industrial property 工厂资产；工业特性；工业产权；产业所有权
industrial property rights 工业产权
industrial prototype cell 工业原型电解槽
industrial provision 工业设施
industrial psychiatry 工业精神病学
industrial psychology 工业心理学
industrial public nuisance 工业公害
industrial pump 工业泵
industrial pump factory 工业泵厂
industrial pumping 工业抽水；工业泵送
industrial pure iron 工业纯铁
industrial quality 产业素质
industrial quality control 工业质量管理
industrial quarter 工业区
industrial radioactive waste 工业放射性废物

industrial radiology 工业放射学
industrial railcar 轻轨料车
industrial railroad 工矿企业专用铁路线
industrial railway 专用铁路线；工业(专用)铁路；工矿企业专用铁路线
industrial railway junction terminal 工业铁路枢纽
industrial raw ground water 工业原料用地下水
industrial raw medical 工业原料
industrial raw mica 工业原料云母
industrial reactor 工业用原子反应堆
industrial recycling and reuse 工业用水的回用和再用
industrial reduced model 工业缩小模型
industrial reflector 工厂照明反射器
industrial refrigerating plant 工业制冷装置
industrial refrigerating system 工业制冷系统
industrial refuse 工业垃圾；工业固体废物
industrial relations research association 工业相互关系研究协会
industrial relocation 工业区位转移
industrial renovation 工业革新
industrial requirements for minerals 矿物工业要求
industrial requirements for ores 矿产工业要求
industrial reserve army 工业后备军
industrial reserves 工业储量
industrial reservoir 工业储[贮]池；工业储[贮]罐；工业蓄水池；工业水箱
industrial residue 工业残渣
industrial residue pollution 工业废渣污染
industrial resistance furnace 工业用电阻炉
industrial restructuring 工业结构改革
industrial reuse 工业再用
industrial revenue bond 工业债券
industrial revolution 工业革命；产业革命
industrial rheology 工业流变学
industrial rights of man 人类产业权利
industrial river 工业河流
industrial road 工业道路
industrial robot 工业(用)机器人
industrial rolling needle 工业滚针
industrial roof 工业屋顶
industrial roofing 工业屋面铺盖
industrial royalties 工业特许使用费
industrial safety colo(u)r 工业安全颜色；工业安全保护色
industrial salt 工业用盐
industrial sanitary chemistry 工业卫生化学
industrial sanitary facility 工业卫生设施
industrial scale 工业磅秤；工业规模
industrial school 工艺学校；工业学校
industrial screen 工业用筛
industrial sealants bonds 工业密封结合体
industrial sector 工业部门
industrial securities 工业证券
industrial security 工业安全
industrial security manual 工业安全手册
industrial separation 工业分离
industrial service 工业供气
industrial sewage 工业污水；工业废水
industrial sewage effluent 工业污水处理后出水
industrial sewage treatment works 工业污水处理厂
industrial sewer 工业污水管
industrial sewerage 工业污水工程；工业排水工程
industrial sewerage system 工业污物排放系统；工业排污管网
industrial sewing machine 工业缝纫机
industrial sewing machine and spare parts 工业缝纫机及零件
industrial sewing machine stand 工业缝纫机机架
industrial shift 工业转移
industrial shutter door 工业百叶门
industrial siding 专用铁路线；工业专(用)线；厂矿专用铁路线
industrial silt 工业淤泥；工业粉砂
industrial site 工厂所在地
industrial skip 轻轨料车
industrial slag 工业渣；工业矿渣
industrial sludge 工业污泥
industrial soap 工业皂
industrial society 工业社会
industrial sociology 工业社会学
industrial sodium densimeter 工业钠度计
industrial solid waste 工业废渣(路用)；工业固体废物
industrial solvent 工业溶剂
industrial sorting of ore 矿石工业品级
industrial sources of water pollution 水污染工业源；水工业污染源
industrial space-heating 工厂空间采暖；工业环流供暖
industrial square 工业广场
industrial stack 工业烟囱(群)
industrial stair(case) 工业楼梯(间)
industrial standard 工业标准
industrial station 工业站
industrial steam turbine 工业汽轮机
industrial steel window 工业厂房钢窗
industrial stethoscope 工业探伤仪
industrial stoichiometry 工业化学计算
industrial storied building 工厂楼房；工业高层建筑
industrial stream 工业河流
industrial structure 工业厂房；工业(部门)结构
industrial suburb 工业市郊；市郊工业区
industrial sulphuric acid 工业硫酸
industrial sundry expenses 生产用杂费
industrial system 工业系统；工业体系；实业制度
industrial talc 工业用滑石粉
industrial tale 工业滑石
industrial tank 工业油箱；工业水箱
industrial tax exemption 工业税豁免
industrial technology 工业技术
industrial telemetering 工业遥测术
industrial telemetry 工业遥测
industrial television 工业电视
industrial television camera 工业电视摄像头；工业电视摄像机
industrial television monitoring 工业电视监视
industrial television monitoring system 工业电视监控系统
industrial television set 工业(用)电视机
industrial terminal 工业枢纽
industrial test 工业试验
industrial threshold limit 工业(排污)阈限值
industrial tile 工业砖(瓦)
industrial tissue 碳素相纸
industrial town 工业市镇；工业都市；工业城镇；工业城市
industrial toxicant 工业毒物；工业毒剂
industrial toxicology 工业毒物学；工业毒理学
industrial track 工业专(用)线；工业专用铁路
industrial tractor 工业(用)拖拉机；工矿牵引车；牵引车
industrial training center 工业训练中心
industrial trial 工业性实验
industrial truck 工业用车(辆)
industrial trunk route 工业干线
industrial turbine set 工业汽轮机组
industrial TV camera 工业电视摄像机
industrial type employment distribution 按产业类型的职业分布
industrial type fluid 工业用液体
industrial type of ore 矿石工业类型
industrial types of (mineral) deposits 矿床工业类型
industrial tyre 工业车辆轮胎
industrial undertakings 工业企业
industrial union 产业工会
industrial unit 工业装置；工业设备
industrial urbanization 工业城市化
industrial use of water 工业用水
industrial user 工业用户
industrial utilization of flyash 粉煤灰工业利用
industrial utilization of gas 工业用气
industrial vapo(u)r 工业蒸气
industrial varnish 工业用清漆；工业油漆
industrial vegetable oil 工业用植物油
industrial ventilation 工业通风
industrial visco(si)meter 工业黏度计
industrial wall tile 工业墙砖
industrial waste 工业废渣；工业废物；工业废水；工业废弃物；工业废料；工业废液
industrial waste base(course) 工业废渣基层
industrial waste disposal 工业废物处置；工业废物处理；工业废水处理
industrial waste fluid 工业废液
industrial waste gas 工业废气；工业废料管理
industrial waste heat 工业余热；工业废热
industrial waste law 工业用水法规
industrial waste mixture base course 工业废渣混合料类基层
industrial waste residue 工业废渣
industrial waste residue treatment method 工业废渣处理方法
Industrial Wastes Committee 工业污水委员会
industrial waste steam 工业废汽
industrial waste stream 工业污水流
industrial waste surcharge 工业废液附件罚款
industrial waste survey 工业废水调查
industrial wastewater 工业污水；工业废水；产业废水
industrial wastewater discharge 工业污水排放
industrial wastewater disposal 工业废水处理
industrial wastewater flow 工业污水流
industrial wastewater line 工业废水管线
industrial wastewater treatment 工业污水处理；工业废水处理
industrial wastewater treatment plant 工业污水处理厂
industrial wastewater variation 工业废水变化
industries wastewater works 工业污水处理厂
industrial water conditioning 工业水处理
industrial water consumption 工业通水量；工业耗水量；工业用水量
industrial water drainage 工业排水
industrial water law 工业用水法规
industrial water law and regulation 工业用水法规
industrial water pollution 工业(用)水污染
industrial water quality 工业用水水质
industrial water quality assessment 工业用水质量评价；工业用水水质评价
industrial water quality requirement 工业用水水质要求
industrial water quality standard 工业用水水质标准
industrial water receive tank 工业水储[贮]存箱
industrial water requirement 工业水需求量
industrial water reservoirs 工业水场
industrial water service 工业给水；工业用水设施；工业供水
industrial water supply 工业(用)供水；工业给水；生产供水
industrial water supply standard 工业供水标准
industrial water system 工业用水系统
industrial water tower 工业水塔
industrial water treatment 工业水处理
industrial water use 工业用水
industrial water works 工业用水工程；工业水厂；工业供水工程
industrial weighing machinery 工业衡器
industrial weights 工业砝码
industrial wheeled equipment 轮式工程机械(设备)
industrial wheel tractor 工业用轮式牵引车
industrial window 工业型窗
industrial withdrawal 工业排放
industrial worker 产业工人
industrial X-ray apparatus 工业用X射线机
industrial X-ray film 工业X光胶片
industrial X-ray radiographic equipment 工业用X射线照相设备
industrial yarn 工业纱
industrial yeast 工业酵母
industrial zone 工业区
industries close to consumption areas 消费趋向性工业
industries fair 工业展销会
industries in the non-material sphere 非物质领域中的产业部门
industries that are monopolistic in nature 天然垄断性的行业
industrist 实业家
industry 工业；行业；产业
industry accounting 工业会计
industry airfield 工厂机场
industry and environment office 工业和环境办事处
industry and factory railway 工业企业铁路
industry and side-line business 工附业
industry and subordinary industry expenses 工附业支出
industry application program(me) 工业应用程序

industry-based student 以工业为基地的学员
industry brand 煤的工业牌号
industry code 工业代码
industry construction 工业建设
industry crossing 工业道口
industry-education marriage 工教结合
industry effect 产业效果
industry gas turbine 工业燃气轮机
industry hygiene standard 工业企业卫生标准
industry in the coastal regions 沿海工业
industry in the interior 内地工业
industry law 工业法
industry molybdenum trioxide 工业级三氧化钼
industry of explosive material 爆炸物工业
industry-owned 工矿企业所属的
industry plotting subroutine 工业用设计子程序；工业绘图子程序
industry refuse 工业垃圾；工业固体废物
industry response time 工业反应时间
industry standard architecture 工业标准体系结构
industry standard architecture bus 工业标准结构总线
industry standard item 工业标准项目
industry standard number 工业标准数
industry standard plotting package 工业标准绘图软件包
industry standard specifications 工业标准规格；工业标准规范
industry track 专用线
industry type 煤的工业类型
industry waste 工业废物
industry wastewater 工业废水
industry-wide econometric model 全工业经济计量模型
induvia 集中层体
indxeable tool holder for turning tools 机械夹固式车刀杆
in echelon 成梯形
in economy 按经济原则
inedible oil 非食用油
ineeterminacy 不定性
in effect 实际上
ineffective 无效的
ineffective anomaly 无效异常
ineffective call 无效呼叫
ineffective call attempt 无效试呼
ineffective dose 无影响剂量；无效量；无效剂量
ineffective drawdown 无效降深
ineffective length of anchorage 锚定的无效长度；无效锚固长度
ineffectiveness 无效
ineffective point 无效点
ineffective precipitation 非有效降水量
ineffective rise 无效上升
ineffective strain 无效菌株
ineffective temperature 无效温度
ineffective tillering 无效分蘖
ineffective time 无效时间
ineffective traffic 无效运输
inefficiency 效率低；无效率；无效(力)；低效(率)；不称职
inefficient 效率很低的；无能的；低效的
inefficient combustion 不良燃烧
inefficient insecticide 无效杀虫剂
inefficient pump 低能泵
inefficient type 无效类型
inelastic 非弹性的
inelastic activity 非弹性活动
inelastic analysis 非弹性分析
inelastic behavio(u)r 塑性状态；非弹性特性；非弹性变形；不适应性的；非弹性性状；非弹性性能；不能变通的非弹性性状
inelastic bending 非弹性弯曲
inelastic buckling 非弹性压曲；非弹性失稳；非弹性屈曲；非弹性翘曲
inelastic closure 非弹性闭合
inelastic collision 非弹性碰撞
inelastic collision of ions 离子非弹性碰撞
inelastic consolidation 非弹性固结
inelastic construction 非弹性结构
inelastic deflection 非弹性变形；非弹性挠曲
inelastic deflexion 非弹性变形；非弹性挠曲
inelastic deformation 非弹性变形
inelastic demand 无弹性需求

inelastic densification 非弹性致密
inelastic design response spectrum 非弹性设计反应谱
inelastic drift 非弹性侧移
inelastic dynamic(al) analysis 非弹性动力分析
inelastic earthquake response 非弹性地震反应
inelastic fluid 非弹性流体
inelastic hinge 非弹性铰
inelasticity 缺乏弹性；无适应性；无弹性
inelastic lateral buckling 非弹性侧向压曲
inelastic loading 非弹性荷载
inelastic problem 不能通融的问题
inelastic property 非弹性性质
inelastic range 非弹性范围；非弹性阶段
inelastic region 非线性区；非弹性区(域)
inelastic reinforced concrete frame 非弹性钢筋混凝土框架
inelastic resistance 非弹性阻力；非弹性抗力
inelastic scattering 非弹性散射
inelastic stage 非弹性阶段
inelastic strain 非弹性应变
inelastic supply 无弹性供给
inelastic system 僵硬的系统；不灵活系统；非弹性体系；非弹性系统
inelastic torsion 塑性扭曲；非弹性扭曲
inelastic vibration 非弹性振动
inelastic volumetric(al) strain 非弹性体积应变
inelastic wave 非弹性波
inelegance 粗糙
ineligible acceptance 不合格的银行承兑汇票
ineligible bills 不合格票据
ineligible commercial paper 不合格商业票证
in empty 空箱(指集装箱)
inenucleable 不能剔出的
inequal heating 不均匀加热
inequalities between means 平均值间之不等式
inequality 不平坦；差别；不相同；不相等；不平等；不等性；不等式；不等量
inequality coefficient 不等式系数
inequality constraint 不等性约束；不等式约束
inequality of variance 方差不等式
inequality phenomenon of daily tide 潮汐日不等现象
inequality proportion 不等比例
inequality-restricted estimator 不等式约束估计式
inequality restriction 不等式限制
inequant fold 不等翼褶皱
inequation 不等式方程；不等式
in-equidistant grid 不等距网格
inequigranular 不等粒状
inequigranular fabric 不等粒组构
inequigranular psammitic texture 不等粒砂状结构
inequigranular texture 不等粒结构
inequilateral 不等边(的)
in equilibrium 处于平衡
inequilibrium system economics 非平衡系统经济学
inequitable exchange 不公平交换
inequities 不公平待遇
inequivalent site 不等效晶位
inequivalved shell 不等瓣壳
ineradicable 不能根绝的
inerconversion 互换
inerlocking quality 联锁性
inerrancy 无错误
inert 无自动力的；惰性(的)；不活泼的
inert addition 惰性混合料；惰性添加料
inert additive 惰性添加剂
inert aggregate 矿物集料；矿物骨料；无效拌和料；惰性集料；惰性骨料；非活性集料；非活性骨料
inert alumina silicate bead 惰性矾土硅酸盐珠；惰性矾土硅酸盐球
inert amorphous mineral silicate 惰性非晶形矿物硅酸盐
inertance 惰性
inert anode 惰性阳极
inert atmosphere 惰性气氛；惰性大气层
inert atmosphere processing cell 惰性气氛处理热室
inert base 惰性填(充)料；惰性基质；惰性基料
inert carbon arc welding 惰性气体保护碳弧焊
inert cell 注水电池
inert chromosome 惰性染色体
inert coating 惰性涂层

inert complex 惰性络合物
inert component 惰性组分
inert diluent 惰性稀释剂
inert diluent effect 惰性稀释剂效应
inert dust 岩粉；惰性尘末
inert electrode 惰性电极
inert element 惰性元素
inert explosive 惰性炸药
inert filler 中性填料；惰性炸药；惰性填(充)料；惰性掺和料
inert filling material 惰性填料
inert food 无效食物
inert free gas 无惰性气体
inert gas 惰性气体；惰性气氛；不凝性气体
inert gas arc spot welding 惰性气体电弧点焊
inert gas arc welding 惰性气体电弧焊；气弧保护焊；惰性气体保护焊
inert gas backing 惰性气体衬垫
inert gas blanket 惰性气体覆盖层
inert gas carbon arc welding 惰性气体保护碳极电弧焊
inert gas distribution system 惰性气体分配系统
inert gas element analysis 惰性气体元素分析
inert gas extinguishing system 惰性气体灭火系统
inert gas-filled 充有惰性气体的
inert gas generator 惰性气体发生器
inert gas hand welding installation 惰性气体手焊设备
inert gas lubrication 惰性气体润滑
inert gas main 惰性气体总管
inert gas-metal arc welding 惰性气体保护金属极电弧焊
inert gas root pass welding 惰性气体根部焊缝焊接；惰性气体第一层焊接
inert gas seal 惰性气体密封
inert gas-shielded arc cutting 惰性气体保护电弧(切割)
inert gas-shielded arc welding 惰性气体保护电弧焊
inert gas shielding 惰性气体保护
inert gas spot welding 惰性气体保护电弧点焊
inert gas system 惰性气体装置；惰性气体系统
inert gas welding 惰性气体焊接法；惰性气体保护焊
inert humus 惰性腐殖质
inertia 惯性；惯量；无力；惰性；不活动
inertia area 惯性面积
inertia balance 惯性平衡
inertia bar 惯量棒
inertia block 惯性减振块；惯性砌块
inertia brake 小制动器
inertia break-off 惯性折断
inertia characteristic 惯性特征；惯性特性
inertia coefficient 惯性系数
inertia comparator 摆轮转动惯量比较仪
inertia compensation 惯性补偿
inertia constant 惯性常数
inertia couple 惯性力偶；惯性矩
inertia coupling 惯性耦联；惯性
inertia current 惯性流
inertia damper 惯性阻尼器
inertia effect 惯性效应
inertia element 惯性元件
inertia equalization 惯性平衡
inertia factor 惯性因数
inertia flight 惯性飞行
inertia flow 惯性流
inertia force 惯性力；惰力
inertia force classifier 惯性(力)分级机
inertia force coefficient 惯性力系数
inertia force separator 惯性力分选器；惯性力除尘器
inertia governor 惯性调速器；惯性调节器
inertia grade 惯性坡度；动力坡度
inertia grate stoker (垃圾的)惯性输送机
inertia-gravitational wave 惯性重力波
inertia-gravity wave 惯性重力波
inertia head 惯性水头
inertia immunity sync circuit 惯性同步电路
inertia interaction analysis 惯性相互作用分析
inertia knock-out grid 惯性落砂机
inertial angular acceleration sensor 惯性角
inertial autonavigator 自动惯性导航仪；惯性自动导航仪
inertial axis 惯性轴

inertial block 惯性块
inertial boundary layer 惯性边界层
inertial centrifugal separator 惯性离心式分离器
inertial circle 惯性圆
inertial component 惯性分量
inertial confinement 惯性约束
inertial confinement fusion 惯性约束聚变
inertial control 惯性控制
inertial control system 惯性控制系数
inertial coordinates 惯性坐标
inertial coordinate system 惯性坐标系
inertial damping 惯性阻尼
inertial delay 惯性延迟
inertial directional survey 惯性定向测量法
inertial Doppler navigation equipment 惯性多普勒导航设备
inertial dust collection 惯性除尘
inertial dust collection equipment 惯性除尘装置
inertial dust collector 惯性吸尘装置
inertial dust separator 惯性除尘器；惰性触尘器
inertial effect 惯性作用
inertial ellipse 惯性椭圆
inertial ellipsoid 惯性椭球
inertial error 惯性误差
inertial flight 惯性飞行
inertial flow 惯性流动
inertial force 惯性力
inertial frame 惯性架
inertial grade 惯性坡度
inertial guidance 惯性制导；惯性导航
inertial guidance computer 惯性制导计算机
inertial guidance information 惯性制导信息
inertial guidance integrating gyro 惯性制导积分陀螺
inertial guidance system 惯性制导系统
inertial gyrocompass 惯性陀螺罗盘
inertial impact 惯性冲击力
inertial impaction 惯性撞击；惯性碰撞(法)
inertial instability 惯性不稳定性
inertial instrumen system 惯性仪表系统
inertial laser sensor 惯性激光传感器；激光惯性传感器
inertial load 惯性荷载
inertial loading component 惯性荷载分量
inertial lock 惯性自锁
inertially balanced stabilized platform 惯性平衡稳定平台
inertially stabilized platform 惯性稳定平台
inertial mass 惯性质量
inertial matrix 惯性矩阵
inertial measurement 惯性基准法
inertial mist separator 惯性雾滴分离器
inertial moment 惯性矩
inertial navigation 惯性导航
inertial navigation and guidance 惯性导航与制导
inertial navigation computer 惯性导航计算机
inertial navigation equipment 惯(性)导(航)设备
inertial navigation set 惯性导航设备
inertial navigation system 惯性导航系统
inertial navigator 惯性导航仪
inertia load 惯性荷载
inertial-omega integrated navigation system 惯性欧米伽组合导航系统
inertial oscillation 惯性振荡
inertial period 惯性周期
inertial platform 惯性平台
inertial positioning system 惯性定位系统
inertial pressure 惯性压力
inertial pressure head 惯性压头
inertial pressure method 惯性压力法
inertial rate integrating gyroscope 惯性速率积分陀螺仪
inertial reactance 惯性抗力
inertial reference frame 惯性参考坐标系
inertial reference integrating gyroscope 惯性基准分陀螺仪
inertial reference system 惯性参照系；惯性参考系
inertial resistance 惯性阻力
inertial sensor 惯性传感器
inertial sensor system 惯性传感系统
inertial separator 惯性集尘器；惯性分离器
inertial setting 离心沉淀
inertial settling 惯性沉降
inertial shakeout machine 惯性落砂机

inertial size 空气动力学粒径
inertial sizing device 惯性粒度分级装置
inertial space 惯性空间
inertial space reference system 惯性空间基准系统
inertial starter 惯性起动器
inertial surveying device 惯性测量装置
inertial surveying system 惯性测量系统
inertial surveying unit 惯性测量装置
inertial switch 惯性开关
inertial system 惯性装置；惯性系(统)
inertial tensor 惯性张量
inertial theory 惯性理论
inertial time 惯性时
inertial type brake tester 惯性式制动试验台
inertial velocity meter 惯性速度表
inertial vibrating screen 惯性振动筛
inertial vibration 惯性振动
inertial wave 惯性波
inertial welding 惯性焊接
inertia mass 惯性质量
inertia mass bearing 惯性体轴承
inertia matrix 惯性力矩阵
inertia measurement 惯性测量
inertia modulus 惯性模量
inertia motion 惯性运动
inertia of energy 能量的惯性
inertia of flowing water 流水的惯性
inertia of photo 感光惰性
inertia of population 人口惯性
inertia oscillation 惯性振荡
inertia pole 惯性极
inertia pressure 惯性压力
inertia radius 惯性半径
inertia ratio 惯性矩比；惯性比
inertia reaction 惯性反作用力
inertia reactionary torque testing instrument 惯性反作用力矩测试仪
inertia reaction moment tester 惯性反作用力矩测试仪
inertia resistance 惯性阻力；惰性阻力
inertia revolution counter 惯性式转速表
inertia seismograph 惯性地震仪
inertia seismometer 惯性地震计
inertia separator 惯性分离器
inertia shaking force 摇动惯性力
inertia shear 惯性剪力
inertia stabilizer 惯性稳定器
inertia starter 惯性启动器
inertia starting switch 惯性起动开关
inertia strain instrument 惯性应变仪
inertia switch 延时开关
inertia tensor 惯性张量
inertia test 惯性试验
inertia time constant 惯性时间常数
inertia torque 惯性力矩
inertia trial 冲程试验
inertia type air classifier 惯性风力分级机；惯性空气分级机
inertia type shaker 惯性抖动器
inertia type starter 惯性起动器
inertia wave 惯性波
inertia welding 储[贮]能摩擦焊；惯性焊接；储能摩擦焊
inertia wheel 惯性轮
inerting concentration 惰化浓度
inert ingredient 农药助剂
inertinite 惰性组
inertinite kerogen 惰质组干酪根
inertinization 丝炭化作用
inertite 微惰性煤
inert material 惰性物料；惰性材料；非活性材料
inert matter 惰性物质
inert medium 惰性介质
inert mineral 惰性矿物
inertoderinite 惰屑体
inertodetrimite 碎屑惰性体
inertodetrite 微惰屑煤
inert particle agglutination test 无关颗粒凝集试验
inert pigment 惰性颜料；无活性颜料；惰性颜料（油漆中不起化学反应的颜料）
inert primer 惰性火帽
inert region 惰性部分
inert salt 惰性盐(安全炸药用)
inert salt anion 惰性盐阴离子

inert solid 惰性固体；非活性固体
inert solid pollution 惰性固体污染
inert solvent 惰性溶剂
inert tracer 惰性示踪剂
inert type 安定型
inert vibrating shakeout 惯性振动落砂机
inert water 没有腐蚀性的水；惰性水
inert zone 死区；非灵敏区
inerversion 倒置
inescapable cost 不可避免的成本；必然发生的成本
inesite 红硅钙锰矿
inessential cooperative game 非本质合作对策
inessential general game 非本质一般对策
inessentials 非必需品
inestimable 难估量
inevitable 不可避免的；必然的
inevitable accident 不可避免的(意外)事故
inevitable outcome 必然结果
inevitable result 必然结果
inevitable succession 必然演替
inexact 不精确(的)
inexact concept 不确切概念
inexact data 不准确数据；不精确数据
inexact environment 模糊环境
inexact function 不确切函数
inexactitude 不精确
inexactness 不确切性；不精确(性)
inexact regression analysis 不确切回归分析
inexact statement 不确切命题
in excess of production capacity 生产能力过剩
in excess of the quota 超过限额
inexhaustible resource 无限资源
inexorable 不可抗矩的
inexorable law 必然规律
inexorable trend 必然趋势
inexpansibility 不可膨胀性
inexpensive 廉价的
inexperience 缺乏经验
inexperienced 没有经验的
inexperienced operator 生手
inextensibility 非延伸性；不可延展性
inextensible 不能延伸的
inextensible cable 不伸长索
inextensile 不能延伸的
inextensional buckling 非延伸压屈
inextensional deformation 非延伸性变形；非伸缩变形；不可开拓性变形
inextractable 不可萃取
in extremis 极端危险状态
inextricable mixture 不可混合物
in-face 重点养护；悬崖坡
infall process 沉降过程
infancy 幼年
infancy period of erosion(al) cycle 侵蚀旋回幼年期
infant 法定未成年人
infant(bath)tub 儿童浴缸
infant farming 原始农业；原始耕作
infantile landforms 幼年地形
infant industry 新兴工业；新建工业；新办工业
infant marketing 外销初始期
infant mortality 幼儿死亡率；早期损坏率
infant's bath 儿童浴缸
infant school 幼儿园
infared image tube 红外(线)图像转换器
infauna 内栖动物；海底动物；潜底性动物
infaunal organism 内生动物群生物
infeasibility 不可行性
infeasible 不可行的
infeasible linear program(me) 不可行线性规划
infeasible method 不可行法
infeasible program(me) 不可行规划
infeasible region 不可行区域
infeasible solution 不可行解
infected ship 有疫情的船
infected water 污染水；含菌水
infection 侵染现象；侵染
infection aerialis 空气污染
infection by air or water 通过空气或水的传染
infection center 发病中心
infection disease 传染病
infections focus 疫源地
infectious diarrhea 传染性腹泻
infectious hepatitis 传染性肝炎

infectious hospital 传染病医院
infectious room 隔离病室
infective articles 感染性物品
infeed 横进给
infeed accumulator 进料活套塔
infeed cam 横进给凸轮;横给进凸轮
infeed conveyer 给料运输机
infeed grinding 横向进给磨削;横磨法
infeed method grinding 横磨法
infeed rate 横进给率
infeed set-up 横切装置
infeed stroke 横向行程
inference engine 推理机
inference method 推理法
inference network method 推理网络法
inference of causation 因果关系推定
inference rule 推理规则
inference technique 推论技术
inferencial statistics 推断统计学
inferential flow meter 间接流量计
inferential liquidlevel meter 间接液面计
inferential measurement 推理估测;独立测定
inferential meter 涡轮式水表
inferential statistics 推论统计学
inferior 下位;劣质;低劣
inferior advantage 较小利益
inferior angle 下线角
inferior arc 劣弧
inferior cloud 低云
inferior coal 劣质煤;低质煤
inferior conjunction 下合
inferior field 无穷域
inferior figures 下附数字
inferior fuel 低质燃料
inferior geodesy 初等大地测量学
inferior goods 低档货;次等商品
inferior horn 下角
inferiority 下级;劣等(感);次等
inferiority gradient 劣势差
inferiority of quality 质量低劣
inferiority rate 次比
inferiority rate control 次比控制
inferior limit 下(极)限;最小限度;最低限度;最小尺寸
inferior limit event 下极限事件
inferior meridian 下子午线
inferior mirage 下现蜃景;下蜃景
inferior nuchal line 下项线
inferior perforated spot 下筛斑
inferior planet 内行星
inferior purlin(e) 劣质桁条;下桁条
inferior quality 质量低劣;质量不高;劣品;劣等;品质较差;品质低劣;低质量
inferior quality-high price 质次价高
inferior solution 劣解
inferior tide 下位潮;下合潮;朔潮
inferior to none 最优;第一
inferior transit 下中天
inferior trunk 下干
inferolateral surface 下外侧面
inferred contact 推断接触带
inferred fault 推断断层
inferred ore 推断的矿石;推定储量;推测矿量
inferred reserves 推算储(备)量;推定储量;推测储量
inferred resources 推测资源
inferred temperature 推算温度
inferred variable 推测变量
inferred zero 刻度外的零点
inferred-zero instrument 刻度非零开始仪器;无零点仪器
infertile soil 贫瘠的土地;瘦土
infestation 侵染;侵害现象
infestation by termites 白蚁侵扰
infidelity 无保真性;不精确(性);不保真
infield 可耕地;可耕土;安装地点
in file for reference 存档备查
infill block 填充块(体)
infill(bore)hole 加密孔
infill brick 填充砖(块)
infill development 填空性建设;填空性建筑
infilled wall 填充(的)墙
infiller block 填空砌块
infiller brick 填空砖

infiller concrete panel 填空混凝土(墙)板
infiller material 填充建筑材料;填空材料
infiller wall 填空墙
infill factor 回淤率
infill frame 填实框架
infill housing 填空性住房
infilling 空隙填料;填隙作用;填实空隙
infilling block 填空砌块
infilling brick 填空砖
infilling concrete 填充混凝土
infilling concrete panel 填空混凝土(墙)板
infilling masonry 填充圬工
infilling material 填充(建筑材)料;填空材料
infilling wall 填空墙
infilling work 填充工程
infill masonry work 填充的圬工工作
infill method 加密法
infill panel 内嵌板;内镶(嵌)板;填充板
infill slab 填充板
infill system 填充体系
infill tile 填充砖
infill walling 填充(的)墙
infill well 加密井;插补井
infiltrant 浸渍剂
infiltrate 共渗;浸润;渗入物
infiltrated air 漏入空气
infiltrated dampproof course 渗排水防潮层
infiltrated flow 渗流
infiltrated water 下渗水
infiltrated wetland-biology ecology system 渗滤式湿地生物生态系统
infiltrating area 渗透面积
infiltrating effect 渗滤效应
infiltrating rainfall 渗入雨水;入渗雨量
infiltration 下渗;浸透;浸渍;浸润;入渗(作用);渗透(作用);渗入(物);渗滤(作用);地表渗入
infiltration air 渗透空气
infiltration and redistribution 渗透作用和再分配
infiltration area 下渗区;渗透区;渗透面积;渗水面积;渗入面积;地下水补给区
infiltration basin 渗水池;渗入盆地;渗滤池
infiltration bed 渗滤床
infiltration canal 渗水渠;渗渠
infiltration capacity 下渗容量;下渗能力;入渗能力;入渗量;透过能力;渗透能力;渗透量;渗入能力;入渗量;渗漏能力
infiltration capacity curve 下渗容量曲线;入渗能力曲线;渗透量曲线;入渗量曲线
infiltration capacity equation 下渗容量方程;入渗能力方程
infiltration channel 渗(流集)水渠
infiltration coefficient 下渗系数;浸渍系数;渗透系数;渗入系数
infiltration coefficient of annual precipitation 年降水入渗系数
infiltration coefficient of precipitation each time 每次降水入渗系数
infiltration control 渗透控制
infiltration curve 渗透曲线
infiltration deposit 淋积矿床
infiltration depth of annual precipitation 年降水入渗深度
infiltration design allowance 设计允许渗入量
infiltration dilution 渗滤稀释
infiltration dilution activity coefficient 渗滤稀释活度系数
infiltration ditch 盲沟;入渗渠;排水暗沟;渗(水)渠;渗(水)沟;渗入渠道
infiltration ditch of sewage 污水渗渠
infiltration diversion 渗透取水;渗水管引水(河底)
infiltration electric(al) field 过滤电场;渗透电场
infiltration experiment 渗透试验
infiltration field 渗流场
infiltration fissure 渗水缝
infiltration flow 入渗流量;渗入流量;渗流
infiltration from spreading surface land 地面入渗法
infiltration from water basin 水盆地渗入补给
infiltration gallery 集水管道(在地下埋设的有孔渗流管道);渗水通道;渗水廊道;渗(入通)渠;廊道;渗流集水廊道
infiltration ground water 渗滤地下水
infiltration gutter 渗水沟

infiltration halo 渗滤晕
infiltration head 渗透压头;渗水水头
infiltration heat loss 冷风渗透耗热量
infiltration index 下渗指数;入渗指数;入渗率;渗透指数;渗入指数;渗滤指数;渗流指数
infiltration intensity 下渗强度;入渗强度;渗透强度;渗入强度
infiltration land treatment system 渗滤法土地处理系统
infiltration line 浸润线
infiltration loss 下渗损失;入渗损失;渗失量;渗入损失;渗漏损失;渗流损失
infiltration metasomatism 渗滤交代作用
infiltration metasomatism way 渗滤交代作用方式
infiltration method 浸渍法
infiltration observation 降水渗入量观测
infiltration of precipitation 大气降水渗入
infiltration parameter 渗透参数
infiltration path 入渗路径
infiltration pipe 渗水管;渗滤管
infiltration pit of sewage 污水渗坑
infiltration point 入渗点
infiltration pond 渗水池;渗入池
infiltration process 下渗过程;浸渍法;入渗过程
infiltration process with cold pressing 冷压浸渍法
infiltration process with vibration 无压浸渍法
infiltration rate 吸收速度;入渗(速)率;渗透率;渗入(速)率;渗入速度;渗滤速率;地表渗入率
infiltration rate curve 渗透率曲线
infiltration rate of precipitation 降水入渗速率
infiltration rate to sewer 污水渗滤速率
infiltration recharge 渗滤补给
infiltration recharge on unit area 单位面积入渗补给量
infiltration reservoir of sewage 污水渗库
infiltration routing 下渗演算;入渗推算;渗滤路线
infiltration sintering 浸渗烧结
infiltration slit 透水缝;渗水缝
infiltration stage 浸润期
infiltration strength 渗入强度
infiltration stress 渗透应力
infiltration tank 入渗池
infiltration test 渗水试验
infiltration test apparatus 渗水试验仪
infiltration theory 下渗理论;入渗理论
infiltration tunnel 渗水隧道;地下渗水道
infiltration type 渗入型
infiltration value 渗滤值;下渗值
infiltration vein 渗成脉
infiltration velocity 下渗率;入渗速度;渗入速率;渗入速度;渗滤速度
infiltration volume 渗透水量;渗入容量
infiltration water 下渗水;过滤水;入渗水;渗透水;渗入水;渗滤水
infiltration water source 渗入型水源地
infiltration well 入渗井;渗滤井
infiltration well of sewage 污水渗井
infiltration zone 下渗区
infiltrative type 浸润型
infiltrator 渗入者
infiltrometer 透水性测定仪;渗透仪;渗透(测定)计;测渗仪
infimum 下确界;最大下界
infinite 无限的;无穷大
infinite aggregate 无穷尽集;无穷集
infinite amount of vaporization 无限蒸发
infinite anisotropic source 无限各向异性源
infinite aquifer 无限含水层
infinite aquifer zone 无限含水带
infinite attenuation 无限衰减
infinite attenuation network 无限大衰减器
infinite automata 无限自动机
infinite automaton 无限自动机
infinite baffle 无限障板
infinite-baffle loudspeaker 无限障板式扬声器
infinite bath 无限浴比
infinite beam 无限(长)梁
infinite body 无限(物)体
infinite boundary 无限边界
infinite branch 无穷分枝
infinite cardinal number 无限基数
infinite chain 无限链
infinite classical group 无限典型群
infinite complex 无穷复合形

infinite conical horn 无限长锥形号筒
infinite continued fraction 无限连分数
infinite convergence 无限收敛
infinite cyclic group 无限循环群
infinite dam 巨坝
infinite decimal 无穷小数;无尽小数
infinite degree of stability 无限稳定度
infinite degree of freedom 无限自由度
infinite dilution 无限稀释;无限稀度;无限冲淡
infinite dilution activity coefficient 无限稀释活度系数
infinite dilution reference state means 无限稀释参比状态法
infinite dilution resonance integral 无限稀释共振积分
infinite dilution worth 无限稀释价值(材料)
infinite dimensional control 无限维控制
infinite dimensional linear systems theory 无限维线性系统理论
infinite dimensional probability distribution 无限维概率分布
infinite direct product 无限直积
infinite discontinuity 无穷不连续点
infinite discrete group 无穷离散群
infinite displacement 无限小位移
infinite distance 无穷远
infinite domain 无限域
infinite duration 无限持续时间
infinite elastic body 无限弹性体
infinite elasticity 无限弹性
infinite elastic solid 无限弹性体
infinite elastic strip 无限弹性带
infinite elastic wedge 无限弹性楔子;无限弹性楔型
infinite emanation medium 无限射气介质
infinite expansion 无限扩展
infinite extension 无限扩张
infinite fluid viscometer 无限流体黏度计
infinite game 无限对策
infinite gap 无限间断
infinite graph 无限图
infinite Grassmann manifold 无限格拉斯曼流形
infinite group 无穷群
infinite height 无限高
infinite hierarchy model 无限层次模型
infinite impedance detection 无限阻抗检波
infinite impedance detector 无限阻抗检波器
infinite increment 无穷增量
infinite induction 无限归纳
infinite integral 无穷积分
infinite interval 无穷区间
infinite isotropic source 无限各向同性源
infinite journal bearing 无限滑动轴承
infinite lattice 无限栅格
infinite length 无限长
infinite life 无限寿命
infinite lifetime 无限寿期
infinite limit 无穷限
infinite line 无限长线
infinite loop 无限循环;无穷(循)环;死循环
infinitely differentiable function 无限次可微函数
infinitely dilute solution 无限稀释溶液
infinitely divisible distribution 无穷可分分布
infinitely divisible law 无穷可分律
infinitely fast control system 无惯性控制系统
infinitely great 无限大;无穷大
infinitely large quantity 无限大值
infinitely long line 无穷长线
infinitely narrow gap 无限窄间隙
infinitely near 无穷近
infinitely near points 无限接近点
infinitely reflected reactor 无限反射层堆
infinitely regulation 无级调节
infinitely rigid 无限刚性的
infinitely rigid pile 无限刚性桩
infinitely safe geometry 无限安全几何条件
infinitely sharp leading edge 无限陡前沿
infinitely slow process 无限慢过程
infinitely small 无限小;无穷小
infinitely small quantity 无穷小值
infinitely small resultant 无限小合力;无穷小结果
infinitely thick sample 无限厚样品
infinitely thick target 无限厚靶
infinitely variable 平滑调整(的);无穷变量
infinitely variable adjustment 无级调整

infinitely variable control 无级调节
infinitely variable hydrodynamic(al) transmission 无级液体静压传动器
infinitely variable speed 无级变速
infinitely variable speed adjustment 无限变速调整
infinitely variable speed drive 无级变速传动(装置)
infinitely variable speed gearing 无级变速传动机构
infinitely variable speed transmission 无级变速传动装置
infinite magnetic field 无限磁场
infinite mass 无限质量
infinite matrix 无穷矩阵
infinite medium 无限媒质;无限介质
infinite medium spectrum 无限介质谱
infinite membrane 无限大膜
infinite memory filter 无限存储过滤器
infinite multiplets 无限多重谱线
infinite multiplication constant 无穷介质倍增常数
infinite multiplication factor 无限介质倍增因子;无限倍增系数
infinite networks 无限网状结构
infinite number 无穷大数
infinite open interval 无穷开区间
infinite order 无限阶
infinite orthogonal group 无限正交群
infinite-pad method 无穷反衬法;无限插入法
infinite pile 无限桩
infinite plane source 无限平面源
infinite plate 无限平面;无限大板
infinite point 无穷远点
infinite population 无限总体
infinite power network 无限大电网
infinite prime divisor 无限素因子
infinite product 无穷(乘)积
infinite product representation 无穷乘积表示
infinite programming 无限规划
infinite quantity 无穷量
infinite range approximation 无穷范围逼近
infinite ratio 无限大比
infinite rays 平行射线
infinite reactor 无限堆
infinite reflux 全回流;无限回流
infinite reflux operation 无限回流操作
infinite reflux rate 无限回流率
infinite region 无穷区域
infinite regress 无穷回归
infinite regular continued fraction 无限正则连分数
infinite rejection filter 无穷衰减滤波器
infinite reservoir 无限大容器
infinite rest mass 无限静止质量
infinite return 无限回流
infinite rigidity 无限刚性
infinite rod bundle 无限棒束
infinite semicomputable set 无穷半可计算集
infinite sequence 无限序列;无穷序列
infinite series 无限级数;无穷数列;无穷级数
infinite set 无限集;无穷集
infinitesimal 极小量;无限小的;无穷小
infinitesimal analysis 微元分析;微积分
infinitesimal area 无穷小邻域
infinitesimal balance equation 微元平衡方程
infinitesimal calculus 无穷小计算;微分运算;微积分学
infinitesimal canonical transformations 无穷小典型变换
infinitesimal change 无限小变化
infinitesimal character 无穷小特征标
infinitesimal circle 无穷小圆
infinitesimal constant 无穷小常数
infinitesimal deformation 无限小形变;无穷小形变;微小形变
infinitesimal dipole 无穷小偶极子
infinitesimal dislocation 无限小位错
infinitesimal displacement 无限小位移;无穷小位移;微小位移
infinitesimal disturbance 微扰
infinitesimal element 无穷小元素;无穷小单元;微元
infinitesimal error probability 无限小误差概率
infinitesimal force 无限小力
infinitesimal generator 无穷小生成元
infinitesimal geometry 微分几何(学)
infinitesimal group 无穷小群
infinitesimal increment 无穷小增量
infinitesimality 无穷小

infinitesimal linear transformation 无穷小线性变换
infinitesimally small upper limit 无穷小上限
infinitesimal mapping 无限小映射
infinitesimal method 无穷小法
infinitesimal motion 无穷小运动
infinitesimal neighbo(u)rhood 无穷小邻域
infinitesimal nucleus 无穷小核
infinitesimal operators 无穷小算子
infinitesimal radius 极小半径
infinitesimal real number 无穷小实数
infinitesimal relative displacement 无穷小相对位移
infinitesimal rotation 无穷小旋转
infinitesimal strain 无限小应变;无穷小应变
infinitesimal time 无限短时间
infinitesimal transformation 无穷小变换
infinitesimal unitary transformation 无穷小幺正变换
infinitesimal value 无穷小数
infinite slab 无限(大)平板
infinite slab reactor 无限平板堆
infinite slab source 无限平面源
infinite slope 无限边坡
infinite solution 无限解
infinite source 无限源
infinite span 无限跨度
infinite speed variation 无级变速;无级调速
infinite state automata 无限状态自动机
infinite static stability 无限静稳定性
infinite stiffness 无限刚性
infinite strip 无限长板条;无限(地)带
infinite sum 无穷和
infinite symplectic group 无限辛群
infinite term 无限词项
infinite tree 无穷树
infinite trigonometrical series 无穷三角级数
infinite tube bundle 无限管束
infinite type 无限型
infinite unitary group 无限酉群
infinite universe 无限全域
infinite value logic 无限多值逻辑
infinite variable speed mechanism 无级变速装置
infinite wave train 无限波列
infinite weight 无限大权
infinitive progression 无穷级数
infinitive stage transmission 无级变速
infinitive variable gear box 无级变速齿轮箱
infinitude 无限量;无穷
infinity 无限;无穷(大);超位数
infinity bar (测距仪的)校正杆
infinity focus 无限远焦点
infinity plug 无穷大插塞
infinity point 无限大点
infirmary 医务所;附属医院
infirmary for lay brethren 修士医院
infitrometer plot 下渗实验区
infix 镶边;中缀;中加成分;穿入;插入
infix expression 中缀表达式;插入表达式
infix extract 中缀摘录
infix form 插入形式
infix notation 中缀表示法;插入记号;插入表示法
infix operation 插入操作
infix operator 中缀运算符;插入算符
inflame 引燃;着火(的)
inflame furnace 火焰直接加热炉
inflamer 燃烧物;燃烧器
inflammability 易燃性;可燃性
inflammability limit 着火极限;可燃极限
inflammability limiting concentration 可燃极限浓度
inflammability test 易燃性试验;可燃性试验
inflammable 易燃的;可燃的
inflammable cargo 易燃货物;易燃货品
inflammable compound 易燃剂
inflammable constituent 易燃成分;可燃组分
inflammable constituent of petroleum 石油可燃性组分
inflammable freight 易燃货物;易燃货品
inflammable gas 易燃气体;可燃气体
inflammable gas detector 可燃气体探测器
inflammable goods 易燃货物;易燃货品
inflammable limit 易燃限度
inflammable liquid 易燃液体
inflammable material 易燃(性)材料;易燃品
inflammable matter 可燃物(质)

inflammable mixture 可燃混合物
inflammables 易燃物
inflammable solid 易燃固体
inflammable solid waste 易燃性固体废物
inflammable store 易燃品储[贮]藏库
inflammable substance 易燃物;易燃品;可燃物质
inflammable vapo(u)r 易燃蒸气
inflammation 着火;燃烧;发炎
inflammation point 燃点
inflammation temperature 燃点
inflatable 可膨胀的;可吹胀的;可充气的
inflatable aircraft 充气式飞机
inflatable boat 橡皮艇;充水式小舟;充气橡皮艇
inflatable building 充气房屋;充气建筑
inflatable closure 充气型密封
inflatable cushion 气垫
inflatable dam 充气坝
inflatable diaphragm 气力缓冲器
inflatable dry suit 充气式干式潜水服
inflatable fabric 充气织物
inflatable fender 可膨胀橡胶防冲材;膨胀式碰垫
inflatable gasket 充气垫圈
inflatable gripper 充气夹具
inflatable habitat 充气式居住舱
inflatable life belt 气胀式救生带
inflatable lifeboat 充气式救生筏;橡胶救生艇;气胀式救生艇;充气救生艇
inflatable life jacket 充气救生衣
inflatable life-raft 充气式救生筏
inflatable mattress 充气垫
inflatable packer 膨胀型封隔器;膨胀式封隔器;充气止浆塞;充气封隔器;充气垫
inflatable pneumatic platen cushion 压片充气锤
inflatable raft 橡皮筏;充水式救生筏
inflatable rubber boat 气胀式橡胶艇
inflatable rubber craft 充气橡胶艇
inflatable rubber dinghy 气胀式橡胶工作艇
inflatable rubber seal 充气橡胶密封
inflatable rubble tube 充气橡胶管
inflatable seal 膨胀式密封;充气式密封
inflatable seat 充气座椅
inflatable shed 充气式仓库
inflatable skeleton 充气骨架;充气房屋
inflatable structure 充气(式)结构
inflatable tire 充气轮胎
inflatable tube 充气芯管
inflatable void former (混凝土的)气孔形成剂
inflatable warehouse 充气式仓库
inflate 使通货膨胀
inflated 膨大的
inflated appraisal 过高估价
inflated bridge 充气桥
inflated budget 膨胀的预算
inflated drag cone 充气制动锥
inflated-hose 充气软管
inflated inner form 充气内模
inflated lava 膨胀熔岩
inflated price 腾贵的物价
inflated profits 虚抬利润
inflated rubber bag 充气胶囊
inflated slag 有孔渣;多孔(熔)渣
inflated stock 过多库存
inflated store 过多库存
inflated structure 充气结构
inflatee 通货膨胀受害人
inflater 充气机
inflating medium 生气剂
inflating pump 打气筒;打气泵
inflation 膨胀;通货膨胀;吹胀;吹张法;充气(膨胀)
inflation accounting 通货膨胀会计
inflationary cost 通货膨胀费用
inflationary gap 膨胀缺口;通货膨胀缺口
inflationary high prices 膨胀性高价
inflationary pressure 通货膨胀压力
inflationary spiral 膨胀螺旋;通货膨胀的恶性循环
inflationary trends 通货膨胀趋势
inflationary universe 暴涨宇宙
inflation control 控制通货膨胀;通货膨胀控制
inflation film manufacturing machine 吹膜机
inflation inlet 充气(进)口
inflation of consumption 消费膨胀
inflation of consumption demand 消费需求膨胀
inflation of cost 成本膨胀
inflation of prices 涨价;物价上涨

inflation pressure 充气压力
inflation pressure thermometer 充气压力温度计
inflation process 充气法
inflation pump 打气筒;打气泵
inflation rate 通货膨胀(年)率
inflation receptor 充气感受器
inflation sleeve 充气套筒
inflation tax 通货膨胀税
inflation tube 充气管
inflation valve 充气阀
inflatoplane 充气橡胶飞机
inflator 增压泵;气筒;送压泵;打气筒;吹胀器;充气器;充气机;充气泵
inflator pump 充气筒;充气泵
inflect 向内弯曲;使弯曲
inflected arch 仰拱;弯成拱;反弯拱;反曲拱
inflecting point 拐点
inflection 衍射;偏转;反弯(曲);反曲;反挠
inflectional asymptote 拐渐近线
inflectional normal 拐法线
inflectional tangent 拐切线
inflection angle 拐角;偏转角
inflection point 转折点;转向点;拐点;偏转位置;反弯点;反曲点
inflection point emission current 拐点发射电流
inflection point method 拐点法;反弯点法
inflection stretch 反弯曲段
inflection-tangent-intersection method 拐点正交切线法
inflection temperature 拐点温度
inflector assembly 偏转装置
inflector plate 偏转板
inflecture point 偏转位置;拐点;转化点;反弯点
inflexibility 硬性;刚性;无适应性;非柔性;非挠性;不挠性;不(可)弯曲性
inflexible 刚性的;不能弯曲的;不灵活的;不可弯曲
inflexible burden 固定间接费
inflexible steel rope 硬钢丝绳
inflexible steel wire 硬钢丝绳
inflexible tile 刚性砖
inflexible wheel base 固定轴距
inflexion 偏转;反弯曲;反变
inflexion point 拐点;转折点;反曲点
inflexion point of seismic fault 地层断裂拐点
inflexion strath 反曲大河谷
inflexion stretch 拐折伸展
in-flight 飞行中的
inflight processing 飞行中处理
infloat switch 带浮子开关
inflow 流入(量);进水(量);进气;入流;垂直补给
inflow angle 流入角;入流角
inflow channel 进水渠(道);进水槽
inflow current 进流;入流
inflow device 流入设备
inflow discharge 进水流量
inflow discharge curve 来水流量曲线;入流曲线;进水流量曲线
inflow face 入口断面(岩芯筒)
inflow forecast 来水预测
inflow forecasting criterion 入流预报准则
inflow from the cross-section(al) of aquifer 含水层断面流入
inflow hydrograph 来水过程线;进水过程线;入流过程线
inflowing reach 入流段
inflowing river 入流河
inflowing sediment 入库泥沙;入湖泥沙;入海泥沙
inflowing stream 入流河
inflow into reach 河段入流
inflow level 流入标高
inflow Mach number 进口马赫数
inflow of air 空气对流
inflow of capital 资本流入
inflow of fund 资金的流入
inflow of ground water 地下水流量;地下水流入;地下水来水量
inflow of phreatic water 地下水流量;地下水流入;潜水来水量
inflow of surface water 地表水的流入量
inflow of water 涌水量;进水量
inflow of water vapo(u)r 水蒸气的入流
inflow pattern 到达车流图式
inflow pipe 流入管
inflow rate 涌水率;涌水量;流入速率;进水率;入流量

inflow ratio 入流比
inflow splitting 进水分流
inflow-storage-discharge curve 进水蓄水泄水曲线;进水蓄水排水曲线;入流储量出流曲线
inflow-storage-discharge method 入流蓄水出流法(洪水追踪)
inflow-storage-outflow curve 入流蓄水出流曲线
inflow-storage-outflow method 入流蓄水出流法(洪水追踪)
inflow tee-piece 入流三通
inflow to reach 河段来水
inflow T-piece 入流三通
inflow tube 流入管
inflow variability 进流可变性
inflow velocity 进水流速
inflow wastewater quality 入流污水水质
inflow wheel (液力传动的)向心式叶轮
influence 感应;势力
influence aeration of the soil 影响土壤的通气
influence area 影响面积
influence basin 浸水面积(洪水);影响区;影响盆地;影响范围
influence characteristic 影响特性
influence chart 影响图;感应图
influence circle 影响圆;影响圈
influence coefficient 影响系数;干扰系数
influence coefficient method 影响系数法
influence coefficient of load 荷载影响系数
influence coefficient of weakly pervious layer 弱透水层影响系数
influence coefficient value 影响系数值
influence diagram 影响图;感应图
influence diagram of stress 应力感应图
influence electricity 感应电
influence factor 感应系数
influence factor of groundwater regime 地下水动态影响因素
influence factor of strain 应变感应系数
influence field 影响场
influence function 影响函数
influence fuse 感应信管
influence fuze 感应引信
influence line 影响线
influence line analysis 影响线分析
influence line diagram 影响线图
influence line for an arch 拱的影响线
influence line for reactions 反力影响线
influence line of moments 弯矩影响线
influence line of reactions 反力影响线
influence machine 感应起电机
influence mine 感应水雷
influence number 影响数
influence of aero-elevation 航高影响
influence of aero-velocity 航速影响
influence of air pollution 空气污染影响
influence of colo(u)r 色彩影响
influence of forest on soil 森林对土壤的影响
influence of gravity 重力影响
influence of heat 热影响;热效应
influence of inert time 惰性时间影响
influence of local geology 局部地质影响
influence of the noise and vibration to the neighbo(u)rhood in pile driving 打桩对邻居的噪声和振动影响
influence of topography 地形影响
influence of ultraviolet ray 紫外线影响
influence of wall thickness 壁厚的影响
influence of water-level 水位影响
influence of water table 水位影响;地下水位影响
influence ordinate 影响纵坐标
influence quantity 影响量
influence radius formula Vinikin 维里金影响半径公式
influence range 影响范围
influence scope of settlement 沉降影响范围
influence surface 影响面
influence value 影响值;感应值
influence vector 影响向量
influence zone 影响区;影响带
influencing factor 影响因素;作用因素
influent 影响者;流入液
influent action 渗水作用
influent analysis 入流液分析

influent channel 进风道;进水渠
influent flow 入流量
influent gradient 流入物梯度
influent header 入流联管箱;入流集(水)管
influential action 感应作用
influential force of concrete shrinkage 混凝土收缩的影响力
influential force of temperature change 稳定变化的影响力;温度变化的影响力
influential zone of pathogen 病原感应圈
influent impounded body 拦蓄水量;补给地下水的水体
influent lake 与地下水有联系的湖;渗流湖
influent line 进水管线
influent mixing and particle stabilization stage 进水混合与微粒稳定阶段
influent pipe 进水管;渗流管
influent river 支汊;(补给地下水的)亏水河;入渗河流;渗水河;渗漏河;出渗河(流)
influent seepage 向内渗漏;入渗;渗透;渗入补给;渗入;渗漏
influent stream 补给河流;(补给地下水的)亏水河;入渗河流;入流河川;潜水注入河;渗水河;渗漏河;出渗河(流);补给地下水的河流
influent substance 流入底物
influent substrate 渗入物
influent water 流入(的)水;渗流水;渗漏水
influent water quality 进水水质
influent water quality index 进水水质指数
influent weir 进水堰
influenza (马、猪等的)流行性热病;流行性感冒
influenza virus 流感病毒
influx 注入;流注;内流;汇集;河流汇合处;输入通量
influx of cold air 冷空气流入(量);漏入冷风
influx of order 订货单涌至
influx of traffic 交通汇流;交通汇合处
influx of visitors 参观(访问)者入流(量)
influx population 流入人口
influx rate 涌入速度
infolding 内折
info quote 参考报价
inform 通知;报告
informal 非正式的;非形式的
informal agreement 非正式协商;非正式协定;非正式契约
informal application 非正式申请
informal audit report 非正式审计报告
informal axiomatics 非形式公理学
informal coalition 非正式联盟
informal contract 非正式合同
informal design 不规则式设计
informal documentation 非正式文件
informal environmental education 非正式环境教育
informal garden 不规则庭园;非规则式庭园;不规则园林
informal garden style 非规整园林
informal group 非正式组织
informal liaison roles 非正式联络职位
informal meeting 非正式会议
informal method 非形式法
informal organization 非正式组织
informal organization norm 非正式组织的规范
informal record 非正式记录
informal report 非正式报告
informal restriction 非正式限制
informal road system 自然道路系统
informal sector 非正规行业;非正式部门
informal state diagram 非形式状态图
informal test 非正式试验
informal training cost 非正式训练费用
informal unit 非正式地层单位
informant 信息提供者
informant representation 信息来源表示法
informatics 信息控制论;信息(科)学;计量情报学;情报学
information 信息;资料;情报;通知;辅助信息
information abstract 信息摘要
information access 信息访问
information accumulation 信息积累;信息储存
information acquisition 信息获取;信息采集;情报收集
information acquisition of surveying and mapping 测绘信息获取
information additivity 信息可加性

information age 信息化时代
informational activity 信息获取活动
informational capacity 信息总量;信息容量
informational committee 情报委员会
information algebra 信息代数
informational linkage 信息联系
informationally efficient 信息功效
informational rate 信息率
informational role 信息方面的角色
informational sign 导向标志
informational strategy 信息策略
information and drawing 图纸资料
information and reception desk 服务台
information annex 资料性附录
information architecture 信息体系结构
information at source 所得额来源资料
information audit 情报审计
information bandwidth 信息带宽
information bank 信息库;资料库;情报所
information bank system 信息库系统;情报库系统
information bit 信息位
information block 信息块
information block length 信息块长度
information board 情报板;问询台;查询台
information bus 信息通道
information capacity 信息容量
information carrier 信息载体
information center 情报中心;信息中心
information channel 信息信道;信息通路;信息通道;信息波道
information checklist for design 设计数据校核表
information circuit 信息电路;信息传送线路
information circular 资料通报;情报交流
information city 信息城市
information classification 信息分类
information clearing house 信息交流中心
information code 信息码
information collection 信息收集;情报收集
information collector 信息收集器
information communication system 信息通信系统
information community 信息共同体
information compression 信息压缩
information computing center 情报计算中心
information concerning submarine 有关水下资料
information content 信息量;信据特征;负平均信息量
information control 信息控制
information control program(me) 信息控制程序
information control system 信息控制系统
information conversion 信息转换;信息变换
information conversion function 信息转换功能
information cost 信息成本
information cost function 信息价格函数
information counter 问询台;问讯台;服务台
information cycle 信息周期;情报周期
information data 信息数据
information data set 信息数据集
information decoding 信息译码
information density 信息密度
information department 情报部(门)
information department/national remote sensing center 国家遥感中心资料部
information desk 问讯台;问询台;查询台;查号台
information display 信息再现;信息显示(器)
information display rate 信息再现速率;信息显示速度
information display system(s) 信息显示系统
information dissemination 资料分发
information distributor 信息分配器
information document 情报资料
information drawing 条件图
information drop-in 信息混入
information dropout 信息丢失
information economics 信息经济学
information encoding 信息编码
information engineer 信息工程师;情报工程师
information engineering 信息工程(学)
information environmentalist 信息环境保护者
information environment science 信息环境学
information error 信息误差
information excess 信息冗余
information exchange 情报交流;信息交换
information explosion 信息爆炸
information extraction 信息提取

information extraction ability 信息提取能力
information feedback 信息反馈
information feedback center 信息反馈中心
information feedback control loop 信息反馈回路
information feedback system 信息反馈系统
information field 信息域;信息段;信息场
information firms 信息企业
information flag 信息标记
information flow 信息流;情报交流
information flow analysis 信息流分析
information flow chart 信息流程图
information flow-rate 信息流速
information form 信息形式
information format 信息格式
information functional unit 信息功能单元
information function of a partition 划分的信息函数
information fusion 信息融合
information gain 信息增益;情报增益
information gate 信息闸门
information gathering 信息收集
information generator 信息源;信息发生器
information handling 信息处理;情报处理
information-handling apparatus 信息处理装置
information-handling capacity 信息容量
information-handling center 情报处理中心
information-handling machine 信息处理机
information-handling program(me) 信息处理程序
information-handling rate 信息处理速率
information-handling system 信息处理系统
information-handling technique 情报处理技术
information heading 信息首部;信息标头
information hiding 信息隐蔽
information image 信息图像
information industry 信息产业
information infrastructure 信息基础
information input 信息输入
information inquiry facilitiy 信息查询设施
information-intensivetypes of industries 信息密集工业
information interchange 信息交换
information kit 信息工具包
information language 信息语言
information lattice 信息点阵结构
information limitation 信息极限
information link 信息链路;数据自动传输装置
information logical machine 信息逻辑机
information loss 信息损失
information lossless machine 信息无损计算机
information machine 信息机
information management 信息管理检索和传播系统;情报管理
information management system 信息管理系统;情报管理系统
information mark 信息标志
information market 信息市场
information material 信息材料
information matrix 信息矩阵
information measure 信息测度
information measurement 信息测量
information medium 信息媒介;信息媒体
information model 信息模型
information network 信息网络;情报网(络)
information network system 信息网络系统
information number 信息数
information of a set system 集系信息
information offering 信息提供商
information office 情报室;情报处;情报部(门);问讯处
information of heat-treatment 热处理资料
information of the chart 海图说明栏
information operator 查询台
information-oriented language 信息语言
information output 信息输出
information package 资料程序包
information packet 信息包
information path 信息通道
information pattern 信息模式;信息方式
information pollution 信息污染
information process analysis 信息处理分析
information processing 信息加工;信息处理;情报整理;情报处理
information processing apparatus 信息处理装置
information processing capability 信息处理能力

information processing center 信息处理中心
information processing code 信息处理编码
information processing element 信息加工元件
information processing for satellite remote sensing 卫星遥感信息处理
information processing language 信息处理语言
information processing machine 信息处理机
information processing of surveying and mapping 测绘信息处理
information processing program(me) 信息处理程序
information processing standards for computers 用于计算机的信息处理标准
information processing system 信息处理装置;信息处理系统;人对信息的处理系统
information processor 信息处理机
information programme 信息程序
information pulse 信息脉冲;光学信息脉冲
information quantity 信息量
information queue 信息排队
information rate 信息速率;信息接收速度;数据接收速度
information readout time 信息读出时间
information-read-wire 信息读出线
information region 信息区
information register 信息寄存器
information report 情况报告
information representation 信息表示
information requirement 信息要求
information resources 信息资源
information resources sharing 信息资源共享
information retrieval 信息检索;信息恢复;资料检索;情报检索
information retrieval and documentation of experiments 试验质料检存
information retrieval display 情报检索显示器;情报检索显示
information retrieval language 信息检索语言;情报检索语言
information retrieval system 信息检索系统;情报检索系统
information retrieval technique 信息检索技术;情报检索技术
information revolution 信息革命
information room 问讯处
information science 信息(科)学;资料学;情报(科)学
information science institute 信息科学讨论会
information science technology 信息科学技术
information search 信息检索
information security 信息保密;信息安全;情报安全性
information selection system 信息选择系统
information sensing 情报读出
information sensing capability 信息感受能力
information separator 信息分离器;信息分离符;信息分隔符
information separator character 信息分隔符
information sequence 信息序列
information service 信息服务;情报业务;情报工作;情报服务;查询业务;查询台
information service center 资料服务中心
information set 信息集
information sheet 信息表;资料表;单页资料
information show 样品展示
information sign 导向标志;路标;指路标(志)
information signal 信息信号;强风警报旗
information sink 信息接收器
information society 信息社会;信息化社会
information source 信息源;信号源;资料来源;情报(来)源
information source entropy 信息源的平均信息量
information space 信息空间
information stability 信息稳定性
information storage 信息存储;信息储存;情报储[贮]存;数据存储(器)
information storage and retrieval 信息存储和检索;情报存储和检索
information storage and retrieval system 信息储存和检索系统
information storage means 信息存储方式
information storage system 信息存储系统
information storage tube 信息存储管
information storage unit 信息存储器

information storing device 信息存储器;信息存储装置
information stream 信息流
information structure 信息结构
information structure design 信息结构设计
information structure model 信息结构模型
information subsystem 信息子系统
information super highway 信息高速公路
information symbol 信息符号
information system 信息系统;指示系统;情报系统
information system of inventory control 库存管理信息系统
information system of personnel and payroll 人事工资管理信息系统
information system of remote sensing and environmental analysis 遥感与环境分析信息系统
information table 问讯台
information tailoring 信息加工
information technique 信息技术
information technology 信息技术;信息工程;情报科学
information theory 信息论;信息理论;情报理论
information to aid in decision 决策辅助信息
information track 信息道
information transfer 信息传递
information transfer efficiency 信息传输效率
information transmission 信息传输
information transmission equipment 信息传输设备
information transmission speed 信息传输速度
information transmission system 信息传输系统
information treatment 信息处理
information treatment machine 信息处理机
information treatment program(me) 信息处理程序
information trunk 查询线
information unit 信息量单位;信息单元
information user 信息用户
information vector 信息向量;信息矢量
information weight of index 标志信息权
information wire 信息线
informationwise 信息方式
information word 信息字;信息元
information work of standard 标准情报工作
information-write-wire 信息写入线
informative abstract 信息(性)摘要;内容提要;情报文献
informative drawing 参考图纸
informative sign 指示标志
informis 无定状云
informofer 信息子;信息转换体
informosome 信息体
infra-acoustic(al) 声下的;亚声;次声频的
infra-acoustic(al) frequency 亚声频率;次声频
infra-acoustic(al) telegraphy 次声频电报
infra-acoustic(al) vibration 次声振动
infra-audible 亚声
infra-audible sound 次声;超低声
infrabar 低气压
infrablack 黑外
infrablack synchronizing signal 黑外同步信号
Infracambrian 先寒武系的;始寒武统【地】
infracapillary space 毛管间孔隙
Infraclass 附纲
infracord spectrophotometer 红外(线)记录分光光度计
infracrust 内地壳层
infracuticular region 小皮板下区
infradeposit 远岸沉积
infradyne 低外差机;低外差法;低层差法
infradyne receiver 低外差接收机
infradyne reception 低外差接收法
infraglacial 冰底的
infraglacial deposit 底碛层;底碛(冰);底积层;冰底碛
infragranitic 花岗岩层下的
infra-gravity wave 亚重力波;长周期重力波
infra-littoral 远离岸的
infra-littoral deposit 远岸沉积
inframarginal externlities 边际内的外部效应
inframedian zone 水深在300~600英尺间的海底
inframundance 地表下
infraneritic 浅海的
infraneritic environment 浅海外环境;浅海环境

infraneuston 表膜下漂浮生物
infrapermafrost water 冻结层下水;冰冻层下水
infraplacement 向下移位
infra-ray diameter ga(u)ge 红外(线)测径仪
infra-ray width ga(u)ge 红外(线)测宽仪
infrared 红外(线)的
infrared absorber 红外(线)吸收体;红外(线)吸收器
infrared-absorbing gas 吸收红外线气体
infrared-absorbing glass 红外(线)吸收玻璃;吸收红外线玻璃
infrared absorption 红外(线)吸收
infrared absorption coefficient 红外(线)吸收系数
infrared absorption hydrogen chloride gas monitor 红外(线)法氯化氢监测仪
infrared absorption hygrometer 红外(线)吸收湿度测定仪
infrared absorption method 红外(线)吸收分析法
infrared absorption spectra hygrometer 红外(线)吸收光谱湿度计
infrared absorption spectrometry 红外(线)吸收光谱法
infrared absorption spectroscopy 红外(线)吸收光谱分析
infrared absorption spectrum 红外(线)吸收光谱
infrared acquisition 红外(线)探测
infrared acquisition system 红外(线)搜索系统
infrared activity 红外(线)活性
infrared aerial camera 红外(线)航空照相机
infrared aids 红外(线)装置
infrared aimed laser 红外(线)瞄准激光器
infrared airborne radar 红外(线)机载雷达
infrared albedo 红外(线)反照率
infrared analysis 红外(线)分析
infrared analyzer 红外(线)检偏镜;红外(线)分析仪
infrared and thermographic(al) system 红外(线)和热红外成像系统
infrared and ultraviolet adsorbing Vycor glass 吸红外和紫外高硅氧玻璃
infrared angle tracker 红外(线)角跟踪
infrared arrangement 红外(线)装置
infrared array detector 红外(线)阵列检测器
infrared astronomical satellite 红外(线)天文卫星
infrared astronomy 红外(线)天文学
infrared astronomy telescope 红外(线)天文望远镜
infrared autonavigator 红外(线)自动导航仪
infrared axial temperature detecting system 红外(线)轴温探测系统【铁】
infrared baking 红外(线)烤干
infrared band 红外(线)谱带;红外(线)频带;红外(线)波段
infrared beacon 红外(线)信标
infrared beam 红外(线)射束;红外(线)光束
infrared beam control 红外(线)射线控制
infrared beam splitter 红外(线)分光镜
infrared beam spread 红外(线)光束发散
infrared binoculars 红外(线)望远镜;红外(线)双筒(望远)镜
infrared blasting 红外(线)引爆
infrared bolometer 红外(线)辐射热测量计;红外(线)测辐射热计
infrared brazing 红外(线)钎焊
infrared burner 红外线燃烧器
infrared calibration device 红外(线)校准装置
infrared camera 红外(线)照相机;红外(线)摄影机
infrared camera tube 红外(线)摄像管
infrared camouflage-detection camera 假目标红外探测机
infrared carbon dioxide analyser 红外(线)二氧化碳分析器
infrared carbon monoxide monitor 红外(线)法一氧化碳监测仪
infrared catastrophe 红外(线)灾难
infrared cell 红外(线)试池
infrared chromatography 红外(线)色谱法
infrared cloud mapper 红外(线)云层成像仪
infrared cloud picture 红外(线)云层图
infrared colo(u)r film 红外(线)彩色胶片
infrared colo(u)r index 红外(线)色指数
infrared communication 红外(线)通信
infrared communication link 红外(线)通信线路;红外(线)通信链路
infrared communications set 红外(线)通信装置
infrared components 红外(线)元件

infrared contrast 红外(线)对比度
infrared converter tube 红外(线)变像管
infrared correlation chart 红外(线)相关图
infrared counter 红外(线)计数器
infrared curing 红外(线)养护
infrared detecting satellite system 红外(线)卫星探测系统
infrared detection 红外(线)探伤;红外(线)探测;红外(线)检测
infrared detection system 红外(线)探测系统
infrared detection technique 红外(线)探测技术
infrared detection unit 热探头
infrared detector 红外(线)探测器;红外(线)检测器
infrared detector cell 红外(线)探测元件
infrared diagnosis 红外(线)诊断
infrared dichroism 红外(线)二向色性
infrared digital tachometer 红外(线)数字转速表
infrared diode laser 红外(线)二极管激光器
infrared display 红外(线)显示器
infrared distance measuring instrument 红外(线)测距仪
infrared distancer 红外(线)测距仪;红外(线)测距计
infrared dome 红外(线)整流罩
infrared dry curing ink 红外(线)干燥油墨
infrared dryer 红外(线)干燥器;红外(线)干燥机
infrared drying 红外(线)烘干;红外(线)干燥
infrared drying aids process 红外(线)炉法
infrared drying by electric lamps 电红外线管干燥
infrared drying by gas heated panels 煤气加热板红外线烘干
infrared drying equipment 红外(线)烘干设备
infrared drying stove 红外(线)烘干炉;红外(线)干燥炉
infrared early-warning 红外(线)预警
infrared electronic distance measuring instrument 红外(线)测距仪
infrared electronic range-measurement 红外(线)测距仪
infrared emanator 红外(线)辐射源
infrared emission 红外(线)发射
infrared emitter 红外(线)发射体
infrared emitting ability 红外(线)发射率
infrared emitting diode 红外(线)发射二极管
infrared engineering 红外(线)工程
infrared equipment 红外(线)设备
infrared excess 红外(线)超
infrared excess object 红外(线)超天体
infrared excitation 红外(线)激励
infrared eye 红外(线)摄像装置
infrared facility 红外(线)装置
infrared fast dryer 红外(线)快速干燥器
infrared film 红外(线)摄影胶片;红外(线)片;红外(线)胶片
infrared filter 红外(线)滤片片;红外(线)滤波器
infrared fire 红外(线)火;红外(线)炉
infrared food warmer 红外(线)食品加热器
infrared Fourier spectrometer 红外(线)傅立叶光谱仪
infrared frequency 红外(线)频率
infrared gas analyzer 红外(线)气体分析仪;红外(线)气体分析器
infrared gaseous pollution meter 红外(线)气体污染分析仪
infrared gas-oven with electron lighter 电子打火红外燃气灶
infrared ga(u)ge 红外(线)测厚仪
infrared gear 红外(线)仪器
infrared generation 红外(线)振荡
infrared generator 红外(线)发生器
infrared glass 红外(线)玻璃
infrared grating 红外(线)光栅
infrared head-phone 红外(线)接收耳机
infrared heat alarm 红外(线)热报警
infrared heater 红外(线)取暖器;红外(线)加热器;红外(线)供暖器
infrared heating 红外(线)加热;红外(线)供暖
infrared heating appliance 红外线加热设备
infrared heating furnace 红外(线)加热炉
infrared heating treatment 红外(线)热处理
infrared heat lamp 红外(线)加热灯
infrared helioseismology 红外(线)日震学
infrared heterodyne detector 红外(线)外差探测器
infrared heterodyne radiometer 红外(线)外差式辐射仪
infrared high precision thermometer 红外(线)高精度测温仪
infrared homing 红外(线)自动寻的;红外(线)自动跟踪
infrared horizon scanner 红外(线)水平扫描器
infrared horizon sensor 红外(线)地平仪
infrared humidifier 红外(线)湿度计
infrared illumination 红外(线)光照射;红外(线)光照明
infrared illuminator 红外(线)发光器
infrared image 红外(线)影像;红外(线)图像
infrared image converter 红外(线)像变换器;红外(线)图像转换器
infrared image processor 红外(线)成像处理器
infrared imagery 红外(线)成像
infrared imaging 红外(线)成像
infrared imaging spectroscopy 红外(线)成像光谱学
infrared imaging system 红外(线)热像仪;红外(线)成像系统
infrared index 红外(线)指数
infrared indicator 红外(线)指示器
infrared inspection 红外(线)探伤;红外(线)检验;红外(线)检查;红外(线)检测
infrared instrument 红外(线)仪器
infrared intensity 红外(线)辐射强度
infrared interference filter 红外(线)干涉滤光片
infrared interference method 红外(线)干涉法
infrared interferometer 红外(线)干涉仪
infrared interferometer spectrometer 红外(线)干涉分光计
infrared interferometric(al) spectrophotometer 红外(线)干涉分光光度计
infrared interferometry 红外(线)干涉测量
infrared introscope 红外(线)内壁检验仪
infrared jammer 红外(线)干扰器
infrared jamming 红外(线)干扰
infrared joint heater 红外线缝道加热器
infrared lamp 红外(线)电灯;红外(线)灯
infrared laser 红外(线)激光器;红外(线)激光
infrared laser methane alarm 红外(线)激光甲烷报警器
infrared lead 红外(线)导引头
infrared leak detector 红外(线)检漏器
infrared light 红外(线)光照射;红外(线)光
infrared light diagnosis instrument 红外(线)光诊断仪
infrared light supply 红外(线)光源
infrared limit 红外(线)极限
infrared line 红外(线)谱线
infrared line-scanner 红外(线)行扫描仪
infrared line-scanning system 红外(线)行扫描系统
infrared locator 红外(线)探测器;红外(线)定位器
infrared lock-on 红外(线)锁定;红外(线)跟踪
infrared mapping 红外(线)测绘
infrared mapping system 红外(线)测绘系统
infrared maser 红外(线)脉塞;红外(线)激射器
infrared material 红外(线)感光材料
infrared measurement 红外(线)测量
infrared measuring radar 红外(线)光测雷达
infrared microgas analyzer 红外(线)微量气体分析仪
infrared microscanner 红外(线)显微扫描器
infrared microscope 红外(线)显微镜
infrared microscopic thermographer 红外(线)显微热像仪
infrared microspectrometer 红外(线)显微分光仪
infrared modulation technology 红外(线)调制技术
infrared modulator 红外(线)调制器
infrared moisture meter 红外(线)水分测定仪
infrared multifunctional massager 红外(线)多功能按摩器
infrared navigation 红外(线)导航
infrared non-activity 红外(线)非活性
infrared object 红外(线)天体
infrared objective 红外(线)物镜
infrared oil analyzer 红外(线)法油份分析仪
infrared optic(al) fiber 红外(线)光纤
infrared optic(al) fused silica 红外(线)光学石英玻璃
infrared optic(al) material 红外(线)光学材料
infrared optic(al) measuring system 红外(线)光学测量系统
infrared origin 红外(线)辐射源
infrared oven 红外(线)炉;红外(线)烤箱;红外(线)烘箱;红外(线)干燥炉
infrared oven stand 红外线烘干装置
infrared over-the-horizon communication 超视距红外通信
infrared pavement heater 红外辐射路面加热器;红外(线)铺路(用)加热器
infrared permeability 红外(线)穿透性
infrared phosphor 红外(线)磷光体
infrared photocell 红外(线)光电管
infrared photoconductivity detector 红外(线)光电导探测器
infrared photoconductor 红外(线)光电导体
infrared photodetector 红外(线)光电探测器
infrared photodiode 红外(线)光电二极管
infrared photo frequency 红外(线)光频率
infrared photographic(al) material 红外(线)片
infrared photographic(al) film 红外(线)照相胶片
infrared photographic(al) plate 红外(线)照相底片
infrared photography 红外(线)照相;红外(线)摄影(技术);红外(线)摄影(术)
infrared photomacrography 红外(线)低倍放大摄影
infrared photometer 红外(线)光度计
infrared photometry 红外(线)光度测量
infrared photomicrograph 红外(线)显微照片
infrared photo multiplier 红外(线)光电倍增管
infrared pick-off 红外(线)传感器敏感元件
infrared picture 红外(线)图像
infrared plate 红外(线)光谱感光板
infrared powder 对红外线敏感的粉末
infrared prism interferometer-spectrometer 红外(线)棱镜干涉光谱仪
infrared problem 红外(线)灾难
infrared projector 红外(线)信号发射器
infrared proximity action 红外(线)近距引爆
infrared pyrometer 红外(线)高温计
infrared radar 红外(线)雷达
infrared radiant heater 红外线辐射器;辐射采暖炉;辐射板;红外(线)辐射加热器
infrared radiation 红外(线)照射;红外(线)辐射
infrared radiation ceramics 红外(线)辐射陶瓷
infrared radiation curve 红外(线)辐射曲线
infrared radiation damage 红外(线)辐射损伤
infrared radiation detection system 红外(线)辐射探测系统
infrared radiation detector 红外(线)辐射探测器
infrared radiation heater 红外(线)辐射加热器
infrared radiation heating 红外(线)辐射加热
infrared radiation pattern 红外(线)辐射图形
infrared radiation pyrometer 红外(线)光电高温计;红外(线)辐射高温计
infrared radiation sounder 红外(线)辐射探测器
infrared radiation thermometer 红外(线)辐射温度计;红外(线)辐射式测温仪
infrared radiation wave 红外(线)辐射波
infrared radiator 红外(线)辐射器
infrared radiometer 红外(线)辐射仪;红外(线)辐射计
infrared radiothermometer 红外(线)辐射温度计
infrared radistion 红外(线)无线电导航系统
infrared radome 红外(线)整流罩
infrared range 红外(线)范围
infrared range and detection equipment 红外(线)探测设备
infrared range-measurement system instrument 红外(线)测距仪
infrared ranging and directing 红外(线)测距(与)测向(仪)
infrared ranging and directing system 红外(线)测距与测向系统
infrared ranging system 红外(线)测距系统
infrared ratio image 红外(线)比影像
infrared ray 红外(线)
infrared ray absorbent glass 吸收红外线玻璃;吸红外线玻璃
infrared ray bulb 红外(线)灯泡
infrared ray dryer 红外(线)烘干装置;红外(线)烘干机

infrared ray drying oven 红外(线)干燥炉
infrared ray insulating glass 隔绝红外线玻璃;隔红外线玻璃
infrared ray lamp 红外(线)灯
infrared ray reflecting coating 红外(线)反射涂层
infrared ray transmitting glass 透红外线玻璃
infrared receiver 红外(线)探测器;红外(线)接收机
infrared receiving set 红外(线)接收机;红外(线)辐射接收器
infrared reconnaissance system 红外(线)侦绘系统
infrared reflectance analyzer 红外(线)反射率分析仪
infrared reflecting camouflage paint 红外(线)反射伪装涂料
infrared reflection method 红外(线)反射法
infrared reflow soldering 红外(线)回流焊接
infrared refractometer 红外(线)折射仪
infrared region 红外(线)区;红外(线)范围
infrared remote controller 红外(线)遥控器
infrared remote receiver 红外(线)遥控接收机
infrared remote sensing 红外(线)遥感
infrared remote sensing of atmosphere 红外(线)大气遥感
infrared remote sensing technique 红外(线)遥感术
infrared remote sensing technology 红外(线)遥感术
infrared remote transmitter 红外(线)遥控发射机
infrared scanner 红外(线)扫描装置;红外(线)扫描仪;红外(线)扫描器
infrared scanning device 红外(线)扫描装置
infrared scanning imager 红外(线)扫描成像仪
infrared scanning imagery 红外(线)扫描图像
infrared scanning imaging 红外(线)扫描成像
infrared scanning radiometer 红外(线)扫描辐射计
infrared search 红外(线)搜索
infrared search apparatus 红外(线)搜索装置
infrared searchlight 红外(线)探照灯
infrared search set 红外(线)搜索装置;红外(线)定向仪
infrared search system 红外(线)搜索系统
infrared seeker 红外(线)探索器
infrared sensing 红外(线)传感
infrared sensing element 红外(线)灵敏元件
infrared sensing head 红外(线)灵敏头
infrared sensitive emulsion 红外(线)感光乳剂
infrared sensitive film 红外(线)感光胶片
infrared sensitive material 红外(线)光敏材料
infrared sensitive plate 红外(线)底片
infrared sensor 红外(线)传感仪;红外(线)传感器
infrared shell scanner 红外(线)筒体扫描仪
infrared signature identification 红外(线)目标识别
infrared slot reader 红外(线)槽式阅读器
infrared slow-scan TV camera 红外(线)慢扫描电视摄像机
infrared sniperscope 红外(线)瞄准镜
infrared solar radiation 太阳红外辐射
infrared soldering 红外(线)软钎焊
infrared sound transmitter 红外(线)声发射器
infrared source 红外(线)源;热辐射源
infrared space astronomy 红外(线)空间天文学
infrared spectral energy 红外(线)谱能
infrared spectral qualitative analysis 红外光谱定性分析
infrared spectral sensitivity 红外(线)光谱灵敏度
infrared spectrogram 红外(线)光谱图
infrared spectrograph 红外(线)摄谱仪
infrared spectrographic technique 红外(线)光谱摄像技术
infrared spectrography 红外(线)分光摄谱学
infrared spectrometer 红外(线)光谱仪;红外(线)分光计;红外(线)分光(光度)计
infrared spectrometer paper 红外(线)光谱分析纸
infrared spectrometric(al) analyzer 红外(线)光谱分析仪
infrared spectrometry 红外(线)光谱法
infrared spectrophotometer 红外(线)干涉光度计;红外(线)分光(光度)计
infrared spectrophotometry 红外(线)干涉光度学;红外(线)分光光度测量;红外(线)分光光度(测定)法
infrared spectroscope 红外(线)分光镜
infrared spectroscopy 红外(线)光谱学

infrared spectrum 红外(线)谱;红外(线)光谱
infrared spectrum analysis 红外(线)光谱分析
infrared spin-scan radiometer 红外(线)旋转扫描辐射计
infrared spotlight 红外(线)点光源
infrared static sensor 红外(线)静态传感器
infrared stealth 红外(线)隐身
infrared stealth material 红外(线)隐身材料
infrared stove 红外(线)炉
infrared stoving 红外线干燥
infrared stoving by electric lamps 电红外管干燥
infrared stoving finish 红外(线)烘干涂料
infrared surveillance equipment 红外(线)监控器
infrared surveillance system 红外(线)监视系统
infrared survey 红外(线)测量
infrared system 红外(线)系统
infrared tableware sterilizing equipment 红外(线)餐具消毒设备
infrared target capture 红外(线)目标俘获
infrared target contrast 红外(线)目标衬比
infrared target seeking device 红外(线)目标自动探寻器
infrared technique 红外(线)技术
infrared technique for remote sensing 红外(线)遥感术
infrared technology 红外(线)技术
infrared telescope 红外(线)望远镜
infrared television 红外(线)电视
infrared television camera tube 红外(线)电视摄像管
infrared temperature 红外(线)温度
infrared temperature profile radiometer 红外(线)温度剖面辐射计;红外(线)温度轮廓辐射计
infrared temperature sounder 红外(线)温度探测仪
infrared thermogram 红外(线)热谱图
infrared thermometer 红外(线)温度计;红外(线)测温仪
infrared trace 红外(线)跟踪
infrared tracker 红外(线)跟踪器
infrared tracking equipment 红外(线)跟踪设备
infrared tracking radar 红外(线)跟踪雷达
infrared transmission 红外(线)透射
infrared transmission characteristic 红外(线)透射特性
infrared transmitter 红外(线)发射机
infrared transmitting ceramics 透红外陶瓷
infrared transmitting fiber 红外(线)透射光纤;透红外纤维
infrared transmitting filter 红外(线)透射滤器
infrared transmitting glass 红外(线)光透玻璃;红外(线)玻璃;透红外线玻璃
infrared transmitting material 透红外线材料;透红外材料
infrared transmitting set 红外(线)发射机
infrared-transparent glass 透红外线玻璃
infrared transparent material 红外(线)透明材料
infrared transparent nose 红外(线)头整流罩
infrared transparent substrate 红外(线)透明衬底
infrared video camera 红外(线)摄像仪
infrared vidicon 红外(线)视像管;红外(线)摄像管;红外(线)光导摄像管
infrared viewer 红外(线)观察镜;红外(线)观测器
infrared vulcanization 红外(线)硫化(作用)
infrared warmer 红外(线)取暖器
infrared warming appliance 红外取暖器
infrared warning device 红外(线)报警装置
infrared warning receiver 红外(线)警戒接收机
infrared wave 红外(线)波
infrared wave spectrum 红外(线)波谱
infrared welding machine 红外(线)焊接机
infrared window 红外(线)窗口
infrarefraction 红外(线)折射
infrasil 红外硅材料(其中的一种)
infrasizer 空气粒粒器;微粒空气分级器
infrasizing 超微粒空气分级
infrasonic 亚音速的;亚声速的;次声波的
infrasonic frequency 亚声频(率);次声频(率)
infrasonics 次声学;次声
infrasonic sound 次声
infrasound 亚声;次声
infrasound generator 次声发生器
infrasound receiver 超声接收器
infrasound source 次声源
infra-specific diversity 种下多样性

infrastructural upwelling 下构造层物质上涌
infrastructure 下面结构;下层结构;下部结构;下部建筑;永久性防御设施;构造基底;军事建筑工程;基础设施;基础结构;基本装卸设施;基本设施;基本经济建筑;内壳构造;深层构造;地下建筑物;地下建设;底层结构;底层构造
infrastructure charges 下部建筑费用
infrastructure construction 基础设施建设
infrastructure contract 基础设施承包合同
infrastructure cost 下部建筑成本
infrastructure facility 基础设施
infrastructure for daily needs 日常需求的基础结构
infrastructure investment 基础设施投资
infrastructure map 深层构造图
infrastructure project 基础设施项目
infrastructure rehabilitation 基础设施更新
infrastructure support 基础结构支助项目
infrequent comprehensive inquiry 非经常性综合调查
infrequent flooding 非常洪水
infrequent types 稀有类型
infrernce network 推理网络
infringe contract 违反契约;违反合约;违反合同
infringement 冒用商标;侵权行为
infringement of contract 违反契约;违反合约;违反合同
infringement of map copyright 侵犯地图版权
infringement of patent 侵权
infringement of patent rights 专利权的侵犯;侵犯专利权
infringement of title 侵犯所有权
infringement of trademark 侵犯商标(专用)权
infringe upon patent right 侵犯专利权
in full 全数;全费
in full charge 负全部责任
In full premium 全部保险费
in full sail 满帆
in full swing 全面开展
infundibular 漏斗状的
in funds 有资本
infuscate 烟褐色
infuse 灌注;浸渍
infused oil 油浸剂
infusibility 难熔性;不熔性
infusible 难熔的;不熔(化)的
infusible compound 难熔化合物
infusible material 难溶建筑材料;不溶建筑材料
infusible precipitate 不熔性沉淀
infusion 注入物;灌输;灌液;浸入;浸剂
infusion in seam 注入到油层;煤层注水
infusion jar 浸剂罐
infusion jet method 水压松散冲米法
infusion of desiralbe character 导入理想性状
infusion pump 灌入泵;灌流泵
infusion time 注入延续时间
infusion tube 高压注水管
infusodecoction 浸煎剂
infusorial earth 硅藻土
infusorians concrete 硅藻土混凝土
infusum 浸剂
ingate 浇注入口孔;内浇口;内浇道;码头门;入口孔;输入门
in-gate core 内浇口心
in gay colo(u)rs 色彩鲜艳
in gear 开动的;啮合
in general 大体上
in general use 普遍采用
Ingenhausz's experiment 印尖霍丝试验;金属棒相对导热率试验
ingenite 内成岩
ingenue 淡黄绿色
ingenuity 创造性;创造才能
in geographic(al) rotation 按地理次序
Ingersoll glarimeter 英格索尔光泽测定仪;英氏光泽计
Ingersoll gloss meter 英格索尔(偏光)光泽计
ingigo blue 靛蓝
in-glaze colo(u)r 釉中彩
in-glaze decoration 釉中彩装饰;釉中彩饰
ingle 炉膛;密抹;壁炉;壁垒
ingle channel data 单道数据
inglenook 炉边安乐椅;壁炉墙角;炉边(凹处)
ingle recess 壁炉旁凹座;炉边凹处
ingle rotor impactor 单转子反击式破碎机
ingleside 炉边

ingo 门窗侧墙；门窗侧边；门窗侧壁
ingodite 因硫碲铋矿
ingoing 入渗；墙上凹面；洞口侧面；门窗橙子
ingoing control post 轧机进口侧操纵台
ingoing flood 进潮流
ingoing ga(u)ge 轧前厚度
ingoing power 进入功率
ingoing side 进料侧
ingoing stream 涨潮流；进入水流；进潮流
ingoing tenant 新租户
Ingold cutter 英戈尔德刀具
ingold fraise 整形铣刀
ingoldsby car 重型卸料车
in good condition 完好的
in good condition rate of equipment 设备完好率
in good order 情况良好
in good preparations 在良好的制片中
in good repair 维修良好
in good working order 良好的工作状态
ingo plate 洞口侧；进面板；洞口侧面板
ingot 铸铁；结晶块；坯料；锭(坯)；初轧坯
ingot adapter 移锭装置连接器
ingot bar 铸块
ingot bay 钢锭仓库；钢锭跨
ingot blank 锭坯
ingot bogie 送锭车
ingot breaker 锭块轧断机；钢锭折断机；碎锭机
ingot-breaker press 钢锭压断机
ingot buggy 送锭车；运金属的车辆；翻锭车
ingot butt 钢(锭)头
ingot cap 钢锭压盖
ingot car 锭车
ingot case 钢锭型；锭型
ingot casting 铸锭；钢锭浇铸
ingot casting bay 铸锭跨
ingot chair 送锭车的锭座
ingot charging crane 装锭吊车；铸锭起重机；钢锭装料吊车
ingot charging gear 装锭设备
ingot clamp 锭料夹头
ingot cooling chamber 锭料冷却室
ingot copper 铜锭；锭铜
ingot crack 钢锭裂纹
ingot crane 钢锭起重机；钢锭吊车；锭块起重机
ingot crop end 锭截头
ingot cropping 晶锭修整
ingot crusher 锭块压碎机
ingot cylinder 结晶圆筒
ingot dogs 锭钳
ingot drawing crane 脱锭吊车
ingot drawing machine 脱锭机；锭块牵引机
ingot ejector 锭块顶推机；排锭器
ingot furnace 铸锭炉
ingot gold 金
ingot gripper 钢锭夹钳
ingot handling equipment 钢锭处理设备
ingot heat furnace 热锭炉
ingot ingot 铁锭
ingot iron 工业纯铁；铁锭；低碳钢
ingot iron cover 铸铁盖
ingot iron pipe 铸铁管
ingotism 树枝状巨晶；树枝状结晶(钢锭结构缺陷)
ingotism macrocrystalline 巨晶
ingot lathe 铸锭车床
ingot manipulator 铸造用器具
ingot metal 金属铸块；金属锭
ingot metallurgy 铸锭冶金法
ingot mo(u)ld 钢锭型；铸锭模；钢锭模；铁模；锭模
ingot mo(u)ld milling machine 钢锭模铣床
ingot pattern 钢锭试样
ingot piler 锭块堆垛机；垛锭机
ingot pit 铸锭均热坑
ingot pit crane 均热炉钳子吊车
ingot process 熔块法
ingot production process 金属锭生产法
ingot puller 拉锭器
ingot pusher 推锭机；钢锭推出器；钢锭推出机
ingot retracter 取锭器
ingot run 铸锭车过道
ingot scales 钢锭秤
ingot scalping machine 钢锭剥皮机床；钢锭表面修整机
ingot segregation 浇锭分凝
ingot shape 锭形
ingot skin 锭皮
ingot slab 扁钢锭
ingot slab mo(u)ld 扁钢锭模
ingot steel 铸钢；钢锭；锭钢
ingot stirrup 锭钳
ingot stripper 脱模机
ingot stripping crane 钢锭脱模吊车；脱模吊车
ingot structure 铸锭组织
ingot surface 锭表面
ingot tilter 垫块倾卸车
ingot tin 锡锭
ingotting 铸锭
ingot tipper 翻锭机
ingot transfer car 运锭块的车；锭块转运车
ingot tumbler 翻锭机
ingot weigher 锭秤
ingot yard 储锭场
ingrafted river 接流河
ingrafted stream 接流河
ingrain 原纱染色的
ingrain art paper 花纹铜版纸
ingrain carpet 双面提花地毯
ingrain dye 显色染料
ingrained 纱染的
ingrain jute carpet 双面提花黄麻地毯
ingrain lining paper 染色衬纸
ingrain wallpaper 染色墙纸；双面提花墙纸
ingrain wallpaper cover 双面提花墙纸覆盖层
in grant 凭证件才能让渡的财产
ingredient 组分；组成部分；拌料(成分)；拼分；成分
ingredient of concrete 混凝土成分
ingredient ore 配矿
ingredient ore of blasting 爆破配矿
ingredient ore of incoming deposit 入仓配矿
ingredient ore of load car 装车配矿
ingredients of concrete 混凝土组分
ingredient supply 配料供应品
ingress 进入；进口处；初亏
ingress and egress 进出权
ingressed lightning wave overvoltage 侵入雷电波过电压
ingression 内移；海侵；海进；海泛；侵入
ingression of groundwater 地下水侵入
ingression sea 进侵海
ingression speed 海侵速率
ingress of air 进风；接触空气
ingress of dirt 尘埃侵入
ingress of groundwater 地下水侵入
ingress of liquids 液体侵入
ingress of moisture 潮气侵入
ingress of water 水(分)渗入；水的进入
ingress pipe 导入管
ingress reflector 入射反光镜
ingress transition 侵入过渡层
in-ground storage for liquefied gas 液化气地下储存
in-group rating 类集指数；分组指数；分组指标
ingrown bark 夹皮；树木夹皮
ingrown meander 内生曲流；内生河湾；内切河湾
ingrown valley meander 深切河谷曲流
ingrowth 向内生长
Ingson diagram 因格松图解法；因格松图解
inguinal reflex 盖格尔氏反射
ingyin 银金木(印度和东印度产的坚强的深褐色硬木)
inhabitable 适于居住的
inhabitant 住户；栖居的动物；常住居民
inhabitation 住处；居住
inhabited area 居民区
inhabited environment 居住环境；栖居环境
inhabited locality 居民点
inhabited oasis 有人烟绿洲
inhabited place 居民点
inhabited satellite 载人卫星
inhalable particles 可吸入微粒
inhalant 吸入剂
inhalant siphon 进水管；入水管
inhalant syphon 入水管；进水管
inhalation 吸入物；吸入法
inhalation apparatus 吸入器械
inhalation dose 吸入剂量
inhalation hazard 吸入危害(性)
inhalation hazard test 吸入危害试验
inhalation of radioactive material 放射性物质的吸入
inhalation of silica particles 矽尘吸入
inhalation of siliceous dust 硅尘吸入
inhalation route 吸入途径
inhalation test 吸入试验
inhalation toxicity 吸入中毒度
inhalation tube 吸气管
inhalation valve 吸气阀
inhalator 吸入器；人工呼吸器
inhaler 吸入器；吸气器；滤气器；呼吸面具
inhaling hose 吸气软管
inhaling method 吸入法
in hand 正在进行
in-hand collections 交入款项
inharmonic 不协调的
inharmonic frequency 不调和频率
inharmonious fold 不调和褶皱
inharmony 不协调(性)
inhaul 引索；卷帆索；牵收索；拖铲索
inhaul cable 拉索；(缆索开挖机的)牵收缆索；卸载拉绳
in hauling 在搬运中
inhaul line (缆索开挖机的)牵收缆索
in-hearth out-hearth furnace 旧砂再生炉
inherence [inherency] 固有形状；内在(性)
inherent 固有的；本来的
inherent addressing 固有寻址
inherent ambiguity 固有歧义性；固有二义性
inherent ash 原生灰分；固有灰分；内在灰分
inherent Austenitic grain size 奥氏体本质晶粒度
inherent balance (发动机等本身结构决定的)固有平衡
inherent burst 固有岩爆
inherent capability 固有能力
inherent cause 附带原因
inherent character 本性
inherent characteristic 固有特性
inherent colo(u)r variation 固有色差
inherent command 固有命令
inherent concrete heat 内在混凝土热量；混凝土发热
inherent contradictions 内在矛盾
inherent damping 固有阻尼
inherent defect 固有缺陷；内在缺陷
inherent erodibility 天然易蚀性
inherent error 固有误差；固有错误；内在误差；内含误差
inherent feedback 内反馈
inherent filtration 自过滤
inherent fine grain 本质细晶粒
inherent flaw 固有裂纹
inherent floatability 固有可浮性；天然可浮性
inherent grain size 本质晶粒度
inherent heat 固有热量
inherent impurity 固有杂质；内有杂质
inherent instability 固有的不稳定性；内在不稳定性
inherent interference 内部电磁干扰
inherent law 内在规律
inherent loss 固有损失
inherently ambiguous language 固有多义语言
inherently safe 自我调控的；自身安全的
inherent memory 固定存储器
inherent mineral mater 内在矿物质
inherent moisture 内在湿度；原有湿气；原有湿度；固有水分；固有湿度；内在水分；内在灰分
inherent moisture of aggregates 集料原有湿度；骨料原有湿度
inherent nature 固有性质；固有特性；本性
inherent noise 固有噪声
inherent noise level 固有噪声级
inherent oscillation 固有振荡；内部振荡
inherent porosity 固有孔隙；内在孔隙
inherent property 固有性能
inherent protection 固有保护
inherent pulsed source 内生脉冲源
inherent quality 内在质量
inherent regulation 固有调节；固有变动率；内部自动调节
inherent regulation rate 内调节速率
inherent reliability 固有可靠性；内在可靠性
inherent resistance 内阻(力)
inherent restriking voltage 固有再击穿电压
inherent rigidity 固有刚度
inherent road right 固有路权

inherent sensitivity 固有灵敏度
inherent settlement 固有沉陷
inherent shear strength 固有抗剪强度
inherent shortcircuit protection 固有的短路保护
inherent stability 原有稳定性;自稳定性(能);自身稳定(性);固有稳定性;固有稳定度
inherent stiffness 固有刚度
inherent storage 固有存储器
inherent strain 内在应力
inherent strength 固有强度
inherent stress 无外荷时的应力;自重应力;预应力;固有应力;内主应力;内在应力;初应力;初始应力
inherent transparency 固有(的)透明性
inherent value 原本价值
inherent vibration 自振;固有振动
inherent vice 自然变质;固有瑕疵;固有缺陷;内在缺陷;内在缺点;(铸件橡胶制品等的)内部缺陷
inherent viscosity 固有黏度;特性黏度
inherent viscosity number 特性黏(度)数
inherent voltage regulation 内部自动电压调节
inherent weakness failure 固有弱点带来的损坏;内部缺陷引起的损坏
inherit 继承
inheritable basin 继承盆地
inheritable movement 继承性活动
inheritance 法定遗产;遗产;继承物
inheritance attribute 继承属性
inheritance by operation of law 法定继承
inheritance code 继承代码
inheritance hierarchies 继承分层
inheritance hierarchy 继承等级
inheritance of protective reaction 保护反应遗传
inheritance tax 继承税
inheritance tax law 遗产税法
inherited argon 继承氩
inherited attribute 继承属性
inherited drainage 遗留水系
inherited error 原有误差;遗留误差;固有误差;固有错误;继承误差;承袭误差
inherited fabric 继承性构造
inherited geosyncline 继承地槽
inherited meander 嵌入曲流
inherited river 遗留河
inherited stream 遗留河
inherited valley 遗留谷
inheritor 继承者;继承人
inhibitant concentration 50 半抑制浓度
inhibit circuit 抑制电路;禁阻电路;禁止电路
inhibit command 禁止指令
inhibit current pulse 阻流脉冲
inhibit driver 禁止驱动器
inhibited 禁止的
inhibited activity 抑制活动
inhibited brass 耐蚀黄铜
inhibited effect 抑制作用
inhibited oil 抑制油;阻化油;抗氧化油
inhibited oxidation 抑制氧化;受抑制氧化
inhibited pulse 被禁止脉冲
inhibited reaction 受抑制反应
inhibit element 禁止元件
inhibit enable input 禁止启动输入
inhibiter 禁止器
inhibit function 抑制作用
inhibit gate 禁止门
inhibiting 加阻化剂
inhibiting action retarding effect 抑制效应
inhibiting agent 抑制剂;阻聚剂;缓蚀剂;缓凝剂;防锈剂
inhibiting alkali-aggregate expansion 抑制碱集料膨胀
inhibiting degradation 抑制降解
inhibiting factor 抑制因子
inhibiting film 抑制膜
inhibiting input 禁止输入
inhibiting oil 防锈油
inhibiting pigment 缓蚀颜料(底漆用);防锈颜料;防锈漆(金属涂用);缓凝颜料
inhibiting signal 禁止信号
inhibiting value 抑制值;安定性增高值
inhibition 抑制(作用);阻滞;阻止;制止(令);禁止
inhibition-action balance 抑制行动平衡
inhibition choke 扼流
inhibition constant 阻抑常数

inhibition mechanism 阻抑机理
inhibition of complement 补体抑制
inhibition of corrosion 抑制腐蚀;腐蚀的阻力;防腐蚀
inhibition of nitrification 硝化作用的抑制
inhibition of oxidation 氧化的抑制
inhibition period 抑制期
inhibition phenomenon 阻抑现象
inhibition ratio 抑制比率
inhibition test 抑制试验
inhibition time 抑制时间
inhibition zone 抑制带
inhibitive tissue 防氧化薄页纸
inhibit line 禁止线
inhibitor 抑制因子;抑制物;抑制器;抑制剂;阻抑剂;阻聚剂;阻化剂;抗聚剂;禁止器;防锈蚀剂;禁制因素
inhibitor adsorption 抗吸附剂
inhibitor desulfurization 抑制剂脱硫
inhibitor dye 抑制剂染料
inhibitor effectiveness 抑制效能;抑制剂效力
inhibitor for nitrification 硝化抑制剂
inhibitor of ignition 延爆剂;阻爆剂
inhibitor response 抑制剂感应
inhibitor susceptibility 抑制剂感应
inhibitor sweetening 抑制剂脱臭法
inhibitor valve 限制阀
inhibitory 禁止的
inhibitory action 抑制作用
inhibitory analogues 抑制类似物
inhibitory area 抑制区
inhibitory coating 防护涂料;防护(涂)层(油漆);保护层
inhibitory connection 抑制联系
inhibitory effect 抑制效应
inhibitory input 抑制输入
inhibitory kinetic spectrophotometry 阻抑动力学分光光度法
inhibitory phase 抑制相
inhibitory reflex 抑制反射
inhibitory stage 抑制期
inhibitory state 抑制状态
inhibitory toxicity 抑制性毒性
inhibitory type 抑制型
inhibit pulse 禁止脉冲
inhibit set 禁置位
inhibit signal 抑制信号;延迟信号;保持信号
inhibit stage 抑制级
inhibit winding 禁止绕阻
inhibit wire 禁止线
in hold 在舱内
in hole perforator 潜孔钻机
inhommogeneous soil 非均匀土
inhomogeneity 多相性;非均质性;不匀(一)性;不统一
inhomogeneity scale 不均匀性尺度
inhomogeneos semi-conductor 不均匀半导体
inhomogeneous 非均质的
inhomogeneous anisotropic(al) medium 不均匀各向异性介质
inhomogeneous anisotropy 非均匀各向异性
inhomogeneous bank material 不均匀河岸物质
inhomogeneous boundary condition 非齐次边界条件
inhomogeneous boundary value problem 非齐次边值问题
inhomogeneous broadening 非均匀展宽;不均匀增宽
inhomogeneous broadening effect 不均匀(增宽)效应
inhomogeneous coating 非均匀镀层
inhomogeneous coordinates 非齐次坐标
inhomogeneous deformation 不均匀变形
inhomogeneous difference equation 非齐次差分方程
inhomogeneous dilatancy 非均匀膨胀
inhomogeneous dispersion 非均匀色散
inhomogeneous displacement 不均匀位移
inhomogeneous excitation 不均匀激励
inhomogeneous fiber 非均匀光纤
inhomogeneous field 非均匀场
inhomogeneous flow 非均相流动;不均匀流动
inhomogeneous force system 不均匀力系
inhomogeneous glide 不均匀滑移

inhomogeneous kiln feed 成分不均齐的入窑生料
inhomogeneous lattice 非均匀空格
inhomogeneous linear difference equation 非齐次线性差分方程
inhomogeneous line broadening 不均匀线展宽
inhomogeneous Lorentz group 不均匀洛伦兹群
inhomogeneous medium 非均匀介质;不均匀媒质;不均匀介质
inhomogeneous mesh calculation 非均匀网络计算
inhomogeneous plane wave 不均匀平面波
inhomogeneous plastic body 非均匀塑性体
inhomogeneous plastic flow 不均匀塑性流变;不均匀范性流变
inhomogeneous pumping 非均匀抽运;不均匀抽运
inhomogeneous settlement 不均匀沉降
inhomogeneous sizing 浸润不良;吃油不均
inhomogeneous soil 非均质土
inhomogeneous strain 不均匀应变
inhomogeneous system of equations 非齐次方程组
inhomogeneous tide 不均匀潮汐
inhomogeneous turbulence 非均匀(性)紊流
inhomogeneous universe 非均匀宇宙
inhomogeneous wave 不均匀波
inhomogeneous wave equation 非齐次波动方程
inhomogenous 不匀的
inhomogenous accretion 非齐次增大
inhour 反时针
inhour equation 倒时数公式
in-house 本身的
in-house colo(u)r 初色
in-house design 自办设计
in-house facility (企业的)内部设备
in-house line 内部线路
in-house literature 内部刊物
in-house monetary accounting 厂内货币核算
in-house network 内部网络
in-house planned price 厂内计划价格
in-house price 厂内价格
in-house recycling 厂内回用
in-house sale 单一代理人代理房地产的上市和出售
in-house skill training program(me) 厂内技术训练计划
in-house system 近距离系统;近距系统;厂内系统
in-house test 实验室试验
in-house transaction 安全交易
in-house treatment 厂内处理
in-house utilization coefficient 厂内利用系数
in-hull suction pipeline 船内吸泥管
in-hull suction valve 船内吸泥阀
inhumation 埋葬;土葬
in humid climates 在气候潮湿的地区
inicarone 吡香豆酮
inimical impurity 有害杂质
in installments 分期地
in issue 有争议的;争论中的
initial 最初的;首字母;初始(的);初期的;草签
initial a agreement 草签协议;草签协定
initial absolute pressure 绝对初压力
initial absorption 原吸收量;初级吸收;初步吸收
initial abstraction 初始吸收(渗失)量;初始损失;初始排除;初(期)损(失)
initial abstraction retention 初始滞留雨量;(降水的)初始截留
initial acceleration 初始加速度;初步加速度
initial acceptance 初步验收
initial access time 初始访问时间;初始存取时间
initial acidity 起始酸度
initial a contract 草签合同
initial acquisition 初始捕获
initial action 初始作用
initial activity 初活性
initial actuation time 起始激励时间
initial address 起始地址
initial address message 起始地址信息
initial adhesion 初期黏附力
initial adhesive strength 初始黏结强度;初期黏合(强)度
initial adjustment 起始调整
initial a document 草签文件
initial age 初期年龄
initial age distribution 初始年龄分布
initial aiming point 最初瞄准点
initial air content 初始空气含量

initial alignment 起点定向;起点调整;初始对准
initial allowance 机械加工裕度;机械加工余量;机械加工留量;机械加工公差;初步折扣
initial amount of pollutant 初始污染物量
initial amplitude 初振幅
initial angle 初始角
initial appearance 初始状态
initial application 最初申请
initial appropriation 首次拨款;初期拨款;初步拨款
initial approximate value 初始近似值
initial approximation 初始近似
initial argon 初始氩
initial assumption 初始假设
initial astronomical point 起始天文点
initial attribute 初始属性
initial audit 初期审计;初级审计;初次审计;初步审计
initial axial ratio of markers 标志体原始轴率
initial azimuth 起始方位角
initial back bevel 初后斜面
initial backfill 前回填;起始回填料
initial balance 期初余额;期初差额;初始余额
initial ball charge 磨球初装量;初始装球量(球磨机的)
initial basic feasible solution 初始基本可行解
initial basic solution 初始基本解
initial beating-up period 打浆初期
initial bed 底料层
initial behavio(u)r regime 初始特性状态
initial bending 初始弯曲;初始弯沉
initial bias 初始偏置
initial biochemical oxygen demand 初始生化需氧量
initial blast 初爆破
initial blasting 早期爆破;初次爆破;初步爆破
initial boiling point 始沸点;初馏点;初沸点
initial boiling point of petroleum 石油的初沸点
initial bond 押标金
initial bonding strength 初期黏合(强)度
initial bond strength 初期结合强度
initial bottomhole flowing pressure 初始井底流压
initial boundary value problem 初始边值问题
initial breakdown 初次压轧;初步击穿
initial buckling 初始压曲
initial budget 最初预算
initial budget estimate 初步概算;初步框算
initial burning 开始着火
initial byte synchronization 初始字节同步
initial calibration 初始校准
initial camber 起始拱度;起始弯度
initial capacity 初始容量
initial capital 创办资本;创业资本;初始资本
initial capital cost 创办资本
initial capital investment 创业资本投资;创办投资;初始投资;初期投资
initial capitalization 最初投资
initial carriage return 起始回车
initial carrier 最初承运人;出发道路
initial cathode emission 阴极起始发射
initial chaining value 初始链值
initial charge 启动注水;期初费用;初始进料;初电荷;初充电
initial closing 草签结算
initial closure loop 初期截流围堰
initial closure works 初期截流工程
initial coastal profile 原始海岸剖面
initial collapse pressure 湿陷初始压力
initial collision 首次碰撞
initial colo(u)r 最初颜色
initial commitment 草签约定
initial community 先锋群落
initial completion 初次完成
initial component 最初成分
initial compression 初始压缩
initial compression curve 初始压缩曲线
initial compressive strength 初始压力
initial concentration 最初浓度;起始浓度;初始浓度;初(期)浓度
initial concentration distribution 初始浓度分布
initial concrete 初始混凝土;初凝混凝土
initial condensation 初始凝结;初凝
initial condition 原始条件;起始状态;起始条件;初值条件;初始条件;初期条件
initial condition adjustment 起始条件调整;初始条件调整

initial conditional code 初始条件码
initial condition circuit 起始条件电路
initial condition code 起始条件码
initial condition hierarchy 初始条件层次
initial condition mode 初始状态方式;初始条件状态;初始条件方式;复位方式
initial condition setter 起始条件调节器
initial conductor tension 线路起始张力
initial configuration 初始配置;初始结构;初始构形;初始格局
initial confining pressure 初始围压;初始侧限压力
initial consolidation 初始固结
initial consolidation pressure 初始固结压力
initial construction cost 初期建设投资
initial containment 初始容量
initial content 初含量
initial contraction crack 原始收缩裂缝;初期收缩裂缝
initial control 起始控制
initial convergence 初始闭合;初始下沉;初始收敛
initial cooling 预冷却(法);初始冷却;初冷却
initial cost 原始成本;原价;最初费用;最初成本;开办费;开办成本;基(本)建(设)费(用);生产成本;创业成本;创办费;初始费用;初次费用
initial course 起始航向;始航向
initial crac 初期裂纹;初期裂缝
initial crack 早期裂缝;初(始)裂缝;初裂纹
initial cracking 初裂
initial cracking stress 初裂应力
initial creep 初始徐变;初(始)蠕变
initial critical flux 初始临界通量
initial criticality test 初次临界试验
initial cross-section 初始截面
initial crushing 初压碎
initial crushing strength 初始压碎力;初始抗碎强度
initial curing 初期养生(混凝土);初期养护(混凝土);初期固化
initial current 起始电流;初电流
initial curve 初始曲线
initial cut 切槽
initial daily production 最初日产量
initial daily traffic 初期日交通量
initial data 原始资料;原始数据;起始数据
initial data base 初始数据库
initial data base description 初始数据库描述
initial data error 起始数据误差
initial day 起息日;起算日
initial decay 初始衰减
initial definition of rights 权利初始界定
initial deflect error 初偏差
initial deflection 初始弯沉;原有挠度;(初)始偏差;初始挠度;初期弯沉
initial deflection of derrick 吊杆初挠度
initial deformation 初始形变;初始变形
initial deformation modulus 初步变形模量
initial dehydration 初始脱水
initial delay 起始延迟;初始延迟
initial demand 初始需(要)量
initial density 起始密度;初始密度
initial deposit 初始沉积
initial depth 原有水深;初始水深;初始深度
initial depth of drilling coal 见煤深度
initial design 原设计;初步设计
initial design parameter 初步设计参数
initial design stage 初始设计阶段
initial detention 地面滞留(水量);地面蓄水;地面截流水量;初始滞水量;初始滞留;初始持水量
initial detonating agent 起爆剂;初始起爆剂
initial development 初步发育
initial deviation 初始偏差
initial dewatering 初期排水
initial diagenesis 初始成岩作用
initial diameter of drilling bit 开孔钻头直径
initial diameter of injection hole 注浆孔直径
initial differential capacitance 初始微分电容
initial digitalization 初期洋地黄化
initial digitizing 第一次数字化
initial dilution 初始稀释;初始稀释度
initial dip 原始倾斜
initial directed line 起始方向线
initial direction 起始方向;初始方向
initial discharge 初始放电
initial displacement 初始位移;初位移
initial disposal 初步处置

initial dissolved oxygen demand 初始溶解需求量
initial distortion 初始畸变
initial distribution 初始分布
initial disturbance 初始扰动
initial dose 首次(剂)量;初剂量;初级剂量
initial dredging 基建(性)挖泥;基建性疏浚
initial dry(ing) shrinkage 初始干缩度(新浇混凝土试块与干燥后长度之差);原始干缩(率);最初干缩;初始干缩(率)
initial earthquake 初始地震
initial earth stress 初(始)地应力
initial edge 射出边
initial effective stress 初有效应力
initial effluent degradation 初始出水坡降
initial elasticity 初始弹性
initial elevater 第一升运器
initial endorsement 草签保证;草签背书
initial energy 初能量
initial enrichment factor 初始增浓因子
initial environmental assessment 初始环境评估
initial environmental examination 初始环境分析
initial ephemeris data 初始历书数据
initial equation 初始方程
initial erection 初始竖立;初始安装
initial error 原始误差;起始误差;初始误差
initial eruption 初始喷发
initial estimate 初始估计(值);初步测算
initial examination 初次检查
initial excitation response 起始励磁响应
initial expenditures 开办费;创业费;创办费
initial expenses 开办费;创办费;初次费用
initial explosive 起爆药;起爆剂
initial explosive ratio 一次爆破单位炸药消耗量;初次爆破耗药比
initial external procedure 初始外过程
initial face 爆破后的边坡面
initial factor of safety 初期安全系数(提升钢丝绳)
initial failure 早期失效;初期故障
initial fiber structure 初始纤维结构
initial filling 早期注水;第一次充料;初期蓄水;初次充水
initial fill material 电泳原漆;电泳初投涂料
initial filtrated water 初滤过的水
initial financial difficulties 初期财务困难
initial financing 创业资本筹借
initial firing 初烧
initial firing current 起动电流;点火电流
initial fissure 初裂缝
initial flaw 原始裂纹(混凝土骨料周围的)
initial flow 起涨量;初始流量
initial flowing pressure 初始流压
initial flow value 初始流量值
initial foam height 起始泡沫高度
initial form 原有形状;原(始)形;原始类型;原来形状
initial formation pressure 原始地层压力
initial fracturing 初步破碎
initial fragmentation 一次爆破;初次爆破
initial free space 初始自由空间
initial frequency 起始频率
initial fuel-conversion ratio 初始燃料转换比
initial function 初始函数
initial fund 开办资金;开办基金
initial gain 起始增益
initial gap 初始间隔
initial gas production 初产气量
initial gas saturation 原始含气饱和度
initial gel 初切(应)力(指泥浆)
initial geodetic data 起始大地数据
initial geodetic point 起始大地点
initial geologic(al) logging 原始地质编录
initial geologic(al) logging in exploratory well 探井原始地质编录
initial geologic(al) logging in galley 坑道原始地质编录
initial geologic(al) logging in geologic(al) mapping 地质填图原始地质编录
initial geologic(al) logging of geologic(al) section 地质剖面原始地质编录
initial geologic(al) logging of sampling 采样原始地质记录
initial gloss 初始光泽
initial gradient 原坡高;初始梯度;初始坡降
initial great-circle course 初始大圆航路
initial guarantee 投标保证金

initial handling 预处理;初加工
initial hardening 初步硬化
initial hardness 起始硬度;初始硬度
initial heading 初始航向
initial heat of hydration 初始水化热
initial height of water 开始水头高度;初始水头高度
initial hog(ging) 初始拱曲(变形)
initial hole diameter 开钻直径;开孔直径
initial hydraulic condition 初始水力条件
initial hydraulic gradient 起始水力梯度
initial ignition voltage 初始点燃电压
initial ignorance 起始未知
initial imperfection 初始缺陷;初始不完善性
initial impoundment 初次蓄水
initial impulse 起始脉冲;起始冲量;初始脉冲;初期脉冲
initial incidence angle 初始入射角
initial inclined bed 原始倾斜岩层
initial index 初始变址
initial index of refraction 初始折射率
initial infiltration capacity 初始下渗容量;初容量
initial infiltration rate 初始下渗量;初始入渗量;初渗率;初渗量;初期浸透强度
initial information 初始信息
initial input 原始进料量;最初投入
initial input program(me) 初始输入程序;起始输入程序
initial input routine 起始输入程序;初始输入例行程序
initial inspection 初验
initial installation 初期装机
initial installation expenses 最初装备费
initial instruction 起始指令
initial interface velocity 初始界面沉速;初始交接速度
initial interval 初始间隔
initial intrusion 开始入侵
initial inventory 初期存货
initial inverse voltage 起始反向电压;初始反向电压
initial inversion 初始反转
initial investment 最初投资;开始投资;开办费;建设投资;起始投资;期初投资;创办投资;创办费;初始投资;初期投资;初次投资
initial ionization 起始电离;初步电离
initial ion pair 初始离子对
initialism 缩写词;首字母缩略词
initial isobaric period 初始等压时间
initial isochrone 初始瞬时曲线;初始等时线
initial isotope composition 初始同位素成分
initial iteration 初始迭代
initialization 预置;初置;初值发送;初始准备;初始状态;初始(化);初始工作;初始发送
initialization command 起始命令
initialization directive 初始化指令
initialization mode 初始化方式
initialization of construction 开始施工
initialization of program(me) 开始执行程序
initialization statement 预置语句;初始化语句
initialize 起始;初始化
initialize a variable 给变量赋初值
initialized 起始步骤
initialize data 预置数据
initialize format 预置格式
initialize operation 预置操作
initialize program(me) 预置程序
initializer 初始化程序;初始程序
initialize routine 预置程序;起始程序;初始化程序
initializer routine 起始程序
initialize signal 初始信号
initialize utility 预置实用程序
initialize verification 预置校验
initializing data 起始数据
initial lag 起始滞后;初始滞后
initial landform 原始地形;原地形
initial landmass 原始陆体;原始陆块
initial lane insert 初始巷置入
initial leaching rate 初(始)释毒速度;初(始)渗毒速度
initial lead 提前量;初始铅
initial length 初长度
initial length of markers 标志体原始长度
initial lesion 初期损害
initial level 零米高程;基准面高程;起始水准面;初始水平;初始能级

initial life-stage 原始生物期
initial line 开始行;极轴;起线;起始线;始线;初始行;初始线
initialling 临时签署
initial liquefaction 初始液化
initial load 预加荷载;始点荷载;初始装入;初始荷载;初始负载;初始调入;初负荷
initial load-bearing capacity 初始承载力;荷载承受能力
initial loader 初始装入程序
initial loading 初始加荷;初始装入
initial lock mechanism 初始闭锁机构
initial loop 初始环路
initial loss 开办损失;初始损失;初期损失
initial loss estimated 预计损失费
initial low point 初始低点
initial lubrication 起动润滑
initial luminosity function 初始光度函数
initial lump sum 最初付费;逐渐提成
initially 起初
initially complex model 最初复杂模型
initially simple model 最初简单模型
initially slow-setting 初始慢凝
initially twisted beam 扭曲线型梁
initial machine code 初始机器码
initial machine load 初始机器加载
initial magnetic field 初始磁场
initial magnetic permeability 起始磁导率
initial magnetization 初磁化
initial magnetization characteristic 起始磁化特性
initial magnetization curve 起始磁化曲线;初步磁化曲线
initial magnification 初始放大率
initial main sequence 初始主序
initial main sequence star 初始主序星
initial maintenance 初期养生(混凝土);初期养护(混凝土)
initial management 初期管理
initial margin 暂定按金
initial market share 初期市场占有率
initial mass concentration 初始质量浓度
initial mass function 初始质量函数
initial materials 原始物料
initial measurement 初始量测
initial medium 初始介质
initial melting temperature 初熔温度
initial memory address 初始存储器地址
initial meridian 起始子午线;本初子午线
initial metacentre 初稳心
initial microprogram(me) loading 初始微程序装入
initial migration 初始运移
initial mining district 首采区
initial mixing 初期混合
initial mobility period 初动期
initial mode 起始方法;初始方式
initial model 初始模型
initial modulus 起始模量;初始模量
initial moist curing 初期湿养护
initial moisture(content) 初期水分;初始水分;原始含水量;初含水量
initial moisture deficiency 初始水分不足量;初期水分缺失;初期水分缺缺量;初期水分缺差;初期湿度差
initial moment 原力矩;初力矩
initial momentum 初始动量
initial motion 初始运动;初始运动;初动
initial movement 初期运动
initial Newtonian viscosity 起始牛顿黏度
initial node 初始结点
initial nonzero stress state 初始非零应力状态
initial nuclear radiation 早期核辐射;初始核辐射
initial observation date 初期观测日期
initial-of-message signal 开始通信信号;启动信号
initial oil production 初产油量
initial oil saturation 原始含油饱和度
initial operating period 初始建期(指公共住宅区中的1/4可达到95%有住户)
initial operation 初始运转;初始运行;初始操作
initial orbit 初轨
initial order 起始指令;起始程序;期初订货;首批订货;初始顺序
initial organic concentration 初始有机物浓度
initial ornamental 饰花字头
initial oscillation 原有波动

initial outlay 初期投资;工程施工投资;创办费;初始投资
initial output 初期产量
initial output test 起始出力试验
initial overburden pressure 起始超载压力
initial parameter 最初参数;初值参数;初始参数
initial parameters method 初参数法
initial payment 最初付款;入门费;首(期)付款;定金;定价金;初付费;分期付款中的首期付款
initial penetration rate 初始钻进;初始钻速
initial performance 新品性能;初始性能
initial period 起始阶段;初始周期;初始时间;初期
initial period of construction 施工初期阶段
initial permeability 初磁导率
initial perturbation 初始微扰
initial pH 初始 pH
initial phase 初相(位);初位相;初始相;初始阶段;初起阶段
initial phosphorus precipitate 初始磷沉淀物
initial photon density 初始光子密度
initial pile 原堆
initial pitch 初始斜度
initial pitting 初期点蚀
initial placement 初始填筑
initial placement condition 初始填筑条件
initial plan(ning) 原规划;初步规划
initial plastic state 初始潜塑状态
initial play 初始空隙
initial point 原点;起始点;始点;初始位置;初始点
initial point of pipeline 管线始点
initial pollutant 初始污染物
initial pollutant mass 初始污染物质量
initial polluted zone settling velocity 初始污染带沉降速度
initial population 原始群体
initial population inversion 初始粒子数反转
initial pore pressure 初始孔隙压力
initial pore water 初始孔隙水
initial pore water pressure 初始孔隙水压力
initial porosity 起始孔隙率;初始孔隙度
initial portion 初始部分;初(期)绿(灯)时间
initial position 起始位置;起点(位置);初位;初始位置
initial potential flowing 初始自喷产量
initial power 起动功率;启动功率;初始功率
initial power generation 初期发电
initial precipitation 初始沉淀;径流出现以前的降水;初期降水
initial premium 初期保险费
initial pressure 起始压力;初压(力);初始压(力)
initial pressure distribution 初始压力分布
initial pressure of subsidability of loess 黄土的湿陷起始压力
initial prestress 初预应力;初始预应力;初施预应力
initial price 牌价;初步价格
initial problem 初始问题
initial procedure 初始过程
initial processing 初步加工;初步处理
initial process name 初始进程名;初始过程名
initial production 原始产品;初期产量
initial production check 开始生产检验
initial program(me) 初始程序
initial program(me) budget proposal 初步方案概算
initial program(me) loader 初始程序装入程序
initial program(me) loading 初始程序装入【计】;初始程序调入【计】
initial program(me) load mode switch 初始程序装入方式开关
initial proposal 初步概算
initial prosperity 初步繁荣
initial protection 初期支护
initial protion 原始部分
initial pulse 起始脉冲
initial quality 原有的质量
initial quantity 起算量
initial rain(fall) 初期降雨;径流出现以前的降雨;初期雨量
initial range 原始距离
initial ranking 初步排列顺序
initial rate 初速率
initial rate of absorption 吸收初速;初始吸水率
initial rate of dyeing 初染速率
initial rate of penetration 初始钻速

initial rating of well 井的开始产量
initial ray 原始射线
initial reading 起始读数;初(始)读数;初次读数
initial record 起始记录
initial relative permeability 初始相对渗透性
initial relief model 原始地形模型
initial removal 起始去除;初始去除
initial request 初始请求
initial reserve 初期准备金;初期储备
initial reservoir pressure 原始油层压力
initial reset 初始复位
initial residual stress 初始残余应力
initial resistance 初阻力
initial resistance of filter 过滤器初阻力
initial retention 初期滞流;初保持量
initial reverberation time 初始混响时间
initial rise 初升
initial rolling 初(次)碾压;初步碾压
initial row 初始行
initial rubber 原料胶
initial rudder adjustment 初始舵角调整
initial run 开始运转
initial run-in 初次磨合运转
initial running-in 初次试转
initial run welding 根焊缝焊接
initial sample 原始样品;原始样本;原始抽样
initial saturation 原始饱和度
initial scour 初冲刷
initial section 起始段;初始截面;初始断面
initial segment 初始线段;初始段
initial selection of line (公路、铁路的)路线出选
initial sensitivity 初始灵敏度
initial sentential form 初始句型
initial separation 初选
initial separatory cell 粗选槽
initial series of production 新产品开发
initial set 起始沉陷;初凝;初变定
initial set of concrete 混凝土初凝
initial set of petroleum 石油的初凝点
initial setter 起始置位器
initial setting 起始调整;初调;初变形(弹簧的);起始置定;初整定;初始位置
initial setting capacity 初凝能力
initial setting curve 初凝曲线
initial setting energy 初凝能量
initial setting heat 初凝热(量)
initial setting period 初凝阶段;初凝期
initial setting power 初凝能力
initial setting process 初凝过程
initial setting property 初凝性质
initial setting quality 初凝质量
initial setting rate 初凝速度
initial setting reaction 初凝反应
initial setting shrinkage 初凝收缩
initial setting time 初凝时间
initial settlement 原有沉降;瞬时沉降;初始沉陷;初始沉降;初(期)沉(降)
initial settling rate 初始沉淀速度
initial shear resistance 初始剪阻力
initial shear stiffness 初始剪切劲度
initial shear strength 初始抗剪强度
initial shear stress 初切应力
initial shear stress ratio 初始剪应力比
initial shear thrust 原始剪冲断层
initial shock spectrum 初始冲击谱
initial-short circuit current 起始短路电流
initial shot start pressure 初始发射压强
initial shrinkage 早期收缩;初始收缩(量);初期收缩
initial shut-in 初关井
initial shut-in pressure 初始关闭压力
initial side 初算边;起始边
initial side of triangulation 三角锁网起始边
initial signal 起始信号
initial signal unit 起始信号单元;初始信号装置
initial sinking stage 初始沉降阶段
initial skid resistance 原始抗滑力
initial slope 初始斜率
initial slow decay 初始慢衰减
initial softening 初始软化
initial soil 原始土壤
initial soil moisture 初始土壤水分;初期土壤水分
initial soil storage 初期土壤储水量
initial solution 初始溶液;初始解
initial special depreciation 初期特别折旧

initial species 原始种
initial speed 起始速率;起始速度;初速率;初速(度);初始速度
initial spin 初始自旋
initial spring tension 弹簧起始张力
initial stability 初稳(定)性;初始稳定性
initial stability period 初定期
initial stage 初态;开始阶段;起始水位;创办时期;初始阶段;初期;初创阶地;草创时期
initial stage of production 开采初期
initial stages of selection 选择的初期阶段
initial star 初始恒星
initial start 初始启动
initial start dam 初期坝
initial starting 起步
initial start-up 初次起动
initial start-up period 最初起动期
initial state 原始状态;起始状态;初(始)状态
initial state function 初态函数
initial state manifold 初态流形
initial statement 开始语句
initial state of stress 应力初始状态
initial static pressure 起始静压力
initial station 起始点;始发站
initial stationary phase 最初稳定期;最初静止期
initial steady discharge 原稳定流量;初始稳定流量
initial steady level 原稳定水位;初始稳定水位
initial steam pressure 新汽压力;进汽压力
initial steam pressure regulator 新汽压力调节器
initial steam temperature 新汽温度
initial step 起步
initial stiffness 初始刚度
initial stock 初期存货
initial storage 初始库容
initial storage gas weight 压缩气体最初重量
initial strain 初应变;初始应变
initial strain stage 初始蠕变阶段
initial strength 早期强度;初始强度;初始浓度;初期强度
initial stress 原应力;预(加)应力;初(始)应力
initial stress condition 初始应力状态
initial stress method 初(始)应力法
initial stress state 初始应力状态
initial stress tensioning 预应力张拉
initial strong component 初始强分量
initial structure 先期结构;初始构造
initial subscription 初期认缴金
initial substance 初始物质
initial sum 起始和;初始和
initial sun 初始太阳
initial support 隧道初期支护;初期支护
initial surface 起始面
initial surface absorption 表面初始吸水率
initial surge 初生涌浪
initial surge voltage 初始浪涌电压
initial surplus 期初盈余;初期盈余
initial survey 初(始)检(验);初测;初步检验
initial survey and test of cargo gear 起货设备初次检验和试验
initial susceptibility 初步磁化率;起始磁化率;初(始)磁化率
initial symbol 初始符号
initial synchronization 初始同步;初并网
initial system 初始系统
initial table 初始表
initial tableau 初始表
initial tangential modulus 初始切线模量
initial tangent modulus 原始切线模量;原切模量;最初切线模量;初始切线模量
initial tangent stiffness 初始切线刚度
initial tape 起始带
initial task index 起始任务指标
initial tax on land per-mu 初税亩
initial temperature 最初温度;起始温度;初(始)温(度)
initial temperature difference 初温差
initial tension 原张力;初(始)张力;初拉力
initial term 初项;初始项
initial terminal 始发枢纽
initial thermal strain 初始热应变
initial thickness of soil stratum 原始土层原度
initial throughput 初始输运能力
initial thrust 初始推力;初始推力
initial time 初始时间;出发时间

initial time of test 试验开始时间
initial torque 初始转矩;初始力矩
initial toxicity 初期毒性
initial traffic number 初期交通量
initial transient 初始瞬值
initial transmission 初始透过率
initial traverse 初始横断
initial treatment 初步处理
initial treaty 初期协议;初期契约
initial tremor 初期脉动
initial trial 初步实验
initial trim 龙骨倾斜;空船首尼吃水差;设计吃水差;初纵倾(度)【船】
initial turbidity breakthrough 初始浑浊度穿透
initial turbulence 初始湍流
initial unloaded sag 起始无载垂度
initial update 首次更新
initial upper bound solution 初始上界解
initial utility 最初效用
initial vacuum 初期减压
initial value 原始(价)值;起始值;输入值;始值;初(始)值;初期值;基础资料
initial value declaration 初值说明
initial value estimator 初值估计器
initial value method 初值方法
initial value of groundwater pollution 地下水污染起始值
initial value of parameter 参数初值
initial value of slope angle 边坡角起始值
initial value parameter 初值参数
initial value problem 初值问题
initial value theorem 初值定理
initial vehicle interval 初始车辆
initial velocity 起始速度;起始流速;初速(度)
initial velocity model 初始速度模型
initial velocity of reaction 反应初速度
initial vertex 起始顶点;始点;初始顶点
initial viscosity 原黏度;起始黏度;初始黏度
initial void ratio 原始孔隙比;初始孔隙比
initial volatile rate 初始挥发率
initial voltage response 起始电压响应
initial volume 原始体积;初始体积
initial wastewater concentration 初始废水浓度
initial water 初生水
initial water breakthrough 开始水串
initial water content of frost heaving 起始冻胀含水量
initial water deficiency 初始水分不足量;初始亏水量
initial water level 初刻水位
initial water level in aquifer 含水层初见水位
initial water-loss shrinkage 初始脱水收缩
initial water production 初产水量
initial water ratio 原始水灰比
initial water-resistance 初期耐水性
initial water saturation 原始含水饱和度
initial water table 初始水位
initial wave 初生波
initial wear 起始磨损
initial wearing rate 初始磨速
initial weight 初重;初始重量
initial weight of sample 样品原始重量
initial wet curing 初期湿养护
initial width of bank ruin of reservoir 初期坍岸带宽度
initial workpiece 首件
initial yaw 初始摇摆
initial yield load 初始屈服荷载
initial yield value 初始屈服值
initial zone 零时区
initiate 首创
initiate key 启(动)键
initiate log-on request 启动请求·
initiate mode 启用状态
initiate potential 启动电位
initiate stage of the engineering system 工程系统的开发阶段
initiate statement 初值化语句
initiate trigger 启动触发器
initiating 起动触发;启动
initiating ability 引发作用;起爆能力
initiating additive 引发添加剂
initiating agent 起爆药
initiating charge 起爆药(包)

initiating composition 起爆药
initiating element 起动元件
initiating equipment 起爆器材
initiating explosive 引炸药;起爆药
initiating laser 引爆激光器
initiating mechanism 引发机构
initiating particle 原始粒子
initiating process 起始过程
initiating pulse 起动脉冲;启动脉冲;触发脉冲
initiating relay 始动继电器
initiating signal 起始信号;启动信号
initiating task 启动任务;初始任务
initiating trigger 起始触发器
initiating unit 起动元件
initiation 引发(作用);起始;起燃;起爆;产生
initiation area discriminator 起始区判别器;初始区域鉴别器
initiation complex 起始复合物
initiation control device 起始控制装置
initiation device 起爆装置
initiation fee 入会费
initiation material 起爆材料
initiation of construction 开始施工
initiation of flame 火焰的发生
initiation of freezing 开始冰冻
initiation of grain motion 颗粒起动
initiation of motion 开动;起动
initiation pattern 起爆方式
initiation point 起爆点
initiation signal 起始信号
initiation site 起始位点
initiation system 起爆系统
initiative 积极性;能动性;首创精神;初步
initiative reaction 起步反应
initiative wastewater treatment plant 初级废水处理厂
initiator 引发剂;励磁机;起动物;起始器;起爆器;起爆剂;始发量;发起人;爆炸雷管
initiator codon 起始密码
initiator program(me) 初始程序;起始程序
initiator-terminator 启动终止程序
initil output 初始产量
inject 注射;注满;喷射
injectability 注射能力
injectable pressure 灌浆压力
injectable soil 地基土的可灌性
injectant 喷入物
injected annulus 注入环
injected body 贯入体
injected dike 侵入岩脉
injected foundation 注射基础;注液加固基础;灌浆基础
injected frequency 注入频率
injected fuel spray 注射燃料雾体;燃料喷雾
injected gas 注入气体
injected hole 注浆孔;灌浆凝固钻孔;水泥封孔
injected mass 贯入体
injected orebody 贯入矿体
injected pulse 注入脉冲
injected rock 贯入岩(层)
injected stream 注入流
injected tendon anchor 灌浆锚杆
injected value of current 注入电流值
injected water 射出水;注入水
injecting electrode structure 注入电极结构
injecting grout 灰泥注浆
injecting hole 注入孔
injecting nozzle 喷嘴
injecting particle 注入粒子
injecting paste material 浆灌材料
injecting ridge 注水脊
injecting system 注入系统
injecting voltage 注入电压
injecting water sectionally 分段压水
injectio 注射剂
injection 引射;铸入;注射剂;注入(法);注入;注频;注浆(施工);灌浆;进入轨道;入射;喷射;射入轨道;单射
injection adjusting apparatus 注射液调节器
injection advance 喷射提前
injection advance device 喷油提前装置
injection aerator 喷射曝气装置;喷射曝气器
injection aerator design 喷射曝气器设计
injection agent 喷灌器;喷灌剂;注入剂;灌浆剂

injection aid 喷灌剂
injection air 喷射空气
injection air vent 注射式通风道
injection amplifier 注入式放大器
injection angle 喷射角(度)
injection apparatus 注入器械;灌浆设备;喷射装置
injection band spreading 注射谱带扩展
injection block 注射块
injection blow mo(u)lding 注射吹塑(成型);注坯吹塑
injection burner 引射式燃烧器;喷射(式)燃烧器
injection cam 喷油凸轮
injection camshaft 喷油凸轮轴
injection capacity 注射能力;注入能力
injection carburet(t)or 喷射式汽化器;喷射(式)化油器
injection catheter 注射导管
injection cement 喷浆水泥
injection chamber 压射室
injection channel 注入通道
injection cock 喷嘴;注射旋塞;喷管
injection column 灌浆管柱
injection complex 贯入杂岩
injection condenser 喷水凝气器;喷射冷凝器
injection condition 注入条件;喷射条件
injection cooking 注液蒸煮法
injection cooler 喷射式冷凝器
injection cooling 注入冷却
injection current 注入电流
injection cylinder 压射(汽)缸;注塑缸;喷射汽缸
injection delay 喷射滞后
injection density 喷射密度
injection depth 注射深度
injection device 注入装置
injection differential pressure 喷嘴的压降
injection domain 注入区
injection drain pipe 喷射器放油管
injection drawing process 注拉法
injection drill 注水孔钻机;注水孔凿岩机
injection drill-bit 湿式凿岩钻头
injection drop 喷嘴压降
injection duration 喷射持续时间
injection earthquake relationship 注水地震关系
injection efficiency 注入效率
injection electroluminescence 注入式电致发光
injection equipment 喷射机构
injection error 注射误差
injection feeder 喷射进料器
injection fiber 注入光纤
injection field 注入场
injection flow 注入流
injection flow method 注射流动方法
injection fluid 注入液;注入流体
injection-fluid front 注入剂前缘
injection follow-up pressure 注塑后续压力
injection force 注射力
injection gallery 灌浆平巷;灌浆坑道;灌浆廊道
injection gas-fluid ratio 注入气油比
injection gear 喷射装置
injection gneiss 贯入片麻岩;注入片麻岩
injection grease gun 高压润滑油枪
injection grid 注入式栅极
injection grout 压力灌浆;注入(浆)液;灌浆
injection grouting 注射灌浆;压力灌浆
injection grout pile 压力式灌注桩
injection gun 注入器;注射;喷枪
injection head 注入头(使钻液注入钻杆的水龙头接箍);喷头
injection heater 注射加热器
injection heating 注入加热
injection hole 喷灌孔;压浆孔;注浆孔;灌浆孔
injection hose 喷射软管
injection hotspot 注射加热部位
injection installation 注灌设备;喷射装置
injection instant 注入瞬间
injection interval 喷油间隔
injection lag 喷油滞后;喷射迟后
injection lane 喷射枪;喷枪
injection laser 注入型激光器
injection length 注入时间
injection life 注射寿命
injection line 注入管线;高压燃油管;喷油管路
injection locking 注入锁相;注入锁定
injection luminescence 注入发光

injection machine 喷射机
injection magnet 注入磁铁
injection material 喷注材料;注浆材料;灌浆材料
injection mechanism 喷射机构
injection metamorphism 贯入变质(作用)
injection metasomatism 贯入交代作用
injection method 灌注法;注入法;注气法;灌入法;灌浆(方)法;喷射法
injection method in tunnel construction 注射法隧道施工
injection migmatite 注入混合岩
injection mix 压入料
injection mode 注入方式
injection modulator 注频调制器
injection mortar 喷注砂浆;注入灰浆;喷射灰浆
injection mo(u)ld 注塑模具;注射模;注模
injection mo(u)lded part 喷射成型零件
injection mo(u)lded plastics 喷射成型塑料
injection-mo(u)lded item 注压制品
injection mo(u)lder 注射模型成型机
injection mo(u)lding 压射成型;压力铸造法;注塑铸造法;注射成型;注射塑制;注射模塑;挤压出线脚;热压铸成型;喷射铸造(模型)法;喷射成型
injection mo(u)lding machine 注射成型机;注塑成型机;喷射模铸机
injection mo(u)lding pressure 注射模塑压力
injection neutral beam 注入中性束
injection noise 喷注噪声;喷射噪声
injection nozzle 注射喷嘴;注射管嘴;喷(射管)嘴
injection nozzle carrier 喷嘴架
injection nozzle tester 喷油嘴试验器
injection of ammonia 喷氨处理
injection of water 水喷射
injection oil engine 喷油发动机
injection optics 注入光学
injection optics diagram 注入光学系统图
injection orbit 注入轨道
injection orbit radius 注入轨道半径
injection orifice 注射孔;喷嘴孔;射孔
injection parameter 注入参数
injection parameters for grouting 注浆参数
injection period 喷射期
injection phase 注入位差
injection phase angle 注入相位角
injection phase locking 注入锁相
injection pipe 注射管;灌浆管;喷射管
injection pipe jack 注浆管千斤顶
injection piston 压塞
injection plant 喷射装置;喷射设备
injection point 注入位置
injection press 注塑机;注射成形机
injection pressure 压射压力;注射水压力;注入压力;注浆压力;灌注压力;喷注压力;喷射压力
injection procedure 灌入法
injection process 注液法;注射法;灌浆(方)法;喷射法
injection process of clay 黏土灌浆法
injection profile 注水断面
injection project 注水计划
injection pump 注射泵;注入泵;注塑泵;灌入泵;喷油泵;喷油(油)泵;射流泵
injection pump assembly 喷油泵总成
injection pump barrel 喷油泵套筒
injection pump housing 喷油泵体
injection pump lever 喷射泵操纵杆
injection pump plunger 喷油泵柱塞;喷射泵柱塞
injection pump stroke 喷油泵柱塞行程
injection quadrant 注入象限
injection ram 注射活塞
injection rate 注入速度;注入率;喷吹率
injection ratio 注入系数
injection regime 注入方式
injection resin 压注树脂
injection revolution frequency 注入绕转频率
injection rinsing machine 喷洗机
injection rod 喷射枪;喷射棒;喷杆
injection sandstone 贯入砂岩
injection scheme 注入图
injection schlieren 贯入异离体
injection semiconductor laser 注入形半导体激光器
injection slush grouting 灌泥浆
injection socket 喷注插口;喷射插口
injection source 喷射源
injection speed 注射速度

injection splicer 注压接头机
injection spray pipe 喷雾管
injection spread 喷射分散
injection starter 注射起动机
injection station (色液测流法的)注入站
injection straight section 注入直线段
injection stream 喷(嘴)射流
injection syringe 注射管
injection system 喷油系统;喷射系统;喷浆系统
injection tank 灌注槽
injection technique 注射技术;注入技术;喷浆技术
injection technology 注入工艺
injection temperature 注射温度;进口温度
injection test 压水试验;注水试验
injection time 注入时间;喷射时间
injection timing 喷油时限;喷油定时;喷射定时
injection tower 注射塔
injection trajectory 注入轨道
injection tube 注射管;注入管
injection type engine 喷射式发动机
injection type steam heating system 蒸汽喷射热水系统
injection valve 注水阀;喷射阀
injection-valve closing pressure 喷油阀关闭压力
injection-valve opening pressure 喷油阀开启压力
injection water 注射水;注入水;灌注水;喷射水
injection water cooling 喷水冷却
injection wear 喷射磨损
injection well 注水井;注入井;灌注井;回灌井;补给井
injection well plugging 注入井堵塞
injection with asphalt 沥青注射;沥青喷射
injection work 喷射功
injective cochain complex 内射上链复形
injective dimension 内射维数
injective endomorphism 内射自同态
injective envelope 内射包络
injective function 内射的函数
injective homomorphism 内射同态
injective module 内射模
injective resolution 内射分解
injectivity 注入性;内射性
injectivity index 吸水指数
injectivity index test 注水系数试验;注水系数测试;注水系数测定
injectivity survey 注入量测量
injectivity test 注入井试井
injector 引射器;注水器;注射器;注入器;注入机;注模机;灌浆机;进样器;喷嘴;喷注器;喷油器;喷射器;喷射泵
injector accelerator 注入器加速器
injector anode 注入器阳极
injector assembly 喷油嘴总成
injector blow pipe 喷射式焊炬;喷射式吹管;低压喷焊器
injector body 注吸器体;喷射器体
injector burner 引射式燃烧器
injector cathode 注入器阴极
injector circulation 喷射循环
injector column 注入器柱
injector combining nozzle 喷射器联合嘴
injector condenser 喷射冷凝器;注射冷凝器
injector configuration 喷嘴排布
injector cooling pump 喷油器冷却泵
injector current 注入器束流
injector delivery 喷嘴器输出管;喷射器输出管
injector dribble gallery pipe 喷射器滴油总管
injector efficiency 注入器效率
injecto-regenerative diving outfit 喷射再生式潜水装置
injector emittance 注入器发射
injector face 喷嘴头部面积;喷射头部表面
injector fuel pipe 注射燃料管
injector gas burner 高压煤气喷烧器
injector grid 注入栅极;喷嘴网
injector gun 注入器枪
injector head 喷头;喷射头
injector hole area 喷嘴孔的通过截面面积
injector hot-water lifter 引射式热水提升器;热水器(能自动送水至高处)
injector kier 注射式煮布锅
injector manifold 喷雾器集流腔;喷雾器管道
injector method 注浆法
injector mixer 喷射混凝土搅拌机;注射混合器;喷射搅拌机;喷射混合器
injector modulator 注入器调制器
injector nozzle 喷射器嘴
injector nozzle holder 喷射器嘴座
injector orifice 喷油嘴孔
injector output 注入器输出
injector overflow 喷油器溢流孔
injector overflow pipe 喷射器溢油管
injector port 注射口
injector pump 注射泵;高压油泵
injector section 注入器部分
injector set 喷嘴组
injector setting (喷射器的)喷油嘴安装
injector slot 喷射器缝口
injector spacing 喷嘴间距
injector spray tip 喷头;喷射头;喷射器喷头
injector stage 注入器级
injector staging 喷嘴配置
injector stop valve 喷射器断流阀
injector stream 射流
injector stud 喷嘴固定螺柱
injector system 喷射系统
injector terminal 注入器高压电极
injector testing 喷射器试验
injector torch 喷射式焊炬;低压吹管
injector tube 注入器管道
injector tunnel 注入孔道
injector-type mill 喷射式磨机
injector valve 喷油嘴针阀
injector ventilation 注射通风
injectron 高压转换管;高压开关管
injunction 禁制令;禁(止)令
injurant 伤害剂
injure-accident rate 伤害事故率(交通);负伤事故率
injured party 受害方
injured powder 变质炸药
injured state 受害国
injured well 损坏的井
injurious amount 有害含量
injurious animal 害兽
injurious defect 有害缺陷
injurious deformation 有害变形
injurious ingredient 有害成分
injurious material 有害物(质)
injurious weed 有害杂草
injury 工伤;损伤;伤痕
injury accident from collision 撞伤
injury accident from falling 砸伤
injury accident from hitting each other 殴斗
injury accident of passenger falling from train 旅客坠车致伤
injury accident of passenger jumping from train 跳车
injury by falling 坠落伤
injury by frost 霜害
injury by hail 雹害
injury by nuclear weapon 核武器损伤
injury done to the fields 对用地的损害
injury-free 无工伤的
injury(suffered) on the job 工伤
injustice 不公平
Inka aeration 因卡曝气(法)【给】;浅层曝气
Inka aeration system 因卡式曝气系统
ink absorptivity of paper 纸张吸墨性
ink adjustment key 墨斗键
ink and wash 水墨画
Inka process 因卡法
ink-ball 擦墨球垫
ink base 油墨色浆
ink black 墨黑色
ink bleed 墨水渗迹;墨水渗出;渗墨
inkblot 墨迹
inkblot test 墨迹测验
ink blue 油墨蓝;墨水蓝
ink bottle 墨水瓶;黑色厚玻璃瓶
ink-bottle pore 墨水瓶状孔
ink box 墨盒
ink cartridge 墨水盒
ink case and paper-weight 墨盒和镇尺
ink compounds 调墨油
ink crocking 油墨擦掉或弄污
ink density 墨水浓度
ink diffusivity of paper 纸张施胶度
ink disease 黑水病
ink distribution 打墨;传墨
ink dot 墨点
ink drafting 清绘
ink drawing 墨线图;墨水画;清绘
ink droplet 墨滴
ink dry back 油墨干后变暗
ink duct roller 墨斗辊
ink dyestuff 墨水色料
inked 修补着色;墨水涂染
inked drawing 上墨图
ink-embossed paper 压凸印刷墙纸
ink emulsification 油墨乳化
inker 油墨辊;油黑印码器;印字机;墨滚
ink eraser 擦墨药
inker wash-up machine 墨辊冲洗机
ink essence 墨水精
ink film 染色胶片;彩色胶片;彩色胶卷
ink-film breakdown 油墨膜断裂
ink-film thickness 油墨层厚度
ink flo(a)tation test 墨水飘浮试验法
ink fog printer 墨水雾式打印机
ink fountain 墨斗
ink fountain agitator 墨斗搅拌器
ink gloss 油墨光泽;印刷面光泽
ink grade 油墨浓度
ink grinding mill 轧墨机
in kind 以货代款
in-kind payment 实物支付
inkiness 墨黑;漆黑
inking 墨迹画图;涂抹墨;上墨水线
inking ball 擦墨球垫
inking chart 记录图表
inking mechanism 上墨装置
inking print 油墨印件
inking roller 墨辊
inking roller casting machine 墨辊浇铸机
inking system 输墨系统;上墨系统
inking up 上墨;加油墨
inking wheel 油墨轮
ink intermediate 油墨半成品
ink jet console 喷墨式控制台打字机
ink-jet image 喷墨图像
ink-jet method 喷墨法
ink-jet plotter 喷墨绘图机
ink-jet printer 墨水喷墨打印机;喷墨印刷机;油墨喷射打印机
ink knife 油墨刮刀;油墨刮板
ink lay 油墨层
ink lever 上墨手柄
ink line 墨线
ink manuscript 清绘原图
ink mileage 油墨展开面积
ink mist printer 墨水雾化打印机
ink mist recording 印迹记录;喷墨记录
inkometer 油墨黏性计;油墨拉力计;墨水黏度计
ink pad 打印台;印泥
ink particle 油墨颗粒
ink pen 墨水笔
ink penetration 油墨渗透
ink plotter 墨水描迹器;墨水绘图机
inkpot 墨水瓶
ink powder 墨粉
ink pump 墨泵
ink rail 墨斗轨
ink-receptive 亲油墨的
ink receptivity 吸墨性;油墨吸收能力
ink recorder 油墨记录器;墨水记录仪;带涂料记录仪;带墨水记录仪
ink recording 油墨记录
ink reflectance 墨水反射率;墨色反射特性;墨迹反射
ink resistance 抗墨性
ink rest 墨床
ink ribbon 墨带;色带;墨水带
ink roller cleaning device 墨辊清洗装置
ink roll mill 油墨用三辊磨
Inkrom process 固体渗铬法
ink slab 调墨台
ink slice 油墨刮板
ink smudge 墨水污迹
ink speck 墨斑
inkspot 墨水点
ink squeezeout 墨水挤出
inkstand 墨水台

inkstandish 墨水瓶架
ink stick 墨锭
ink stone 砚台
ink-stone screen 砚屏
inkstone with water annulus 辟雍砚
ink system 墨斗装置
ink tank 墨水容器
ink taste 墨水味
ink test 品红试验(陶瓷吸水性试验);吸墨水试验;吸红试验
ink transfer 油墨转移
ink uniformity 墨水均匀性;墨迹均匀性
ink-up table 油墨台
ink vapo(u)r recorder 喷墨记录器
ink vapo(u)r recording 喷墨记录法
ink vibrator interrupter 墨辊摆动中断器
ink well 墨水水池;油墨缸
ink writer 印字机;油墨印码器
ink writing recorder 墨水笔绘记录器;笔录器
inky 有墨迹的
inlaid 镶嵌的
inlaid brick 路面标示砖;平砌砖;平埋砖
inlaid decoration 镶嵌装饰
inlaid flood plain 内叠河漫滩
inlaid flooring 镶嵌地面;镶嵌地板
inlaid flooring brick 镶嵌铺面砖
inlaid lacquer ware 镶嵌漆器
inlaid linoleum 镶嵌油毡;花纹油地毯;嵌花油地毯
inlaid material 镶嵌物
inlaid model method 镶嵌模型法
inlaid mo(u)lding 镶嵌线条
inlaid panel 嵌花镶板
inlaid panel mo(u)lding 嵌花镶板线脚
inlaid parquet 席纹地面;镶木地板;镶块拼花地板;镶嵌地板;席纹地板
inlaid relationship 内叠关系
inlaid strip 嵌花板条
inlaid-strip floor cover(ing) 拼条地面铺饰
inlaid strip flooring 镶条地面;镶嵌地板
inlaid structure 镶嵌构造
in-laid terrace 内叠阶地
inlaid unconformity 嵌入不整合
inlaid vinyl goods 镶嵌乙烯制品;镶嵌乙烯材料;乙烯嵌花(装饰)品
inlaid wood 镶嵌木
inlaid work 镶嵌细工;镶嵌加工;镶嵌工程;镶嵌作业;镶嵌制品
in-lake channel 湖泊航道
inland 内陆;内地
inland arc 内陆弧
inland barge 内河驳船
inland basin 内陆盆地;内流盆地
inland bill 境内票据
inland bill of lading 陆运提单;内陆提单
inland bill of lading clause 内陆提单条款
inland BL 内陆提单
inland boat 内河船(舶)
inland canal 内陆运河
inland channel 内河航道
inland city 内陆城市;内地城市
inland clearance depot 内陆通关基地;内地验关站;内地货物集散地;内地保税货站
inland consolidation depot 内陆拼装站
inland container 内陆集装箱
inland container depot 内地集装箱站(栈);内地集装箱货场
inland country 内陆地区
inland craft 内河船(舶)
inland customs dues 国内关税
inland customs house 内地海关
inland dam 内堤;大陆沙坝
inland delivery price 内陆交货价格
inland delta 内陆三角洲
inland depot 内陆装卸站;内地货运站
inland depression plant 内陆洼地电站
inland desert 内陆沙漠
inland dike 内堤
inland distribution depot 内地调运站
inland drainage 内陆(排)水系
inland drainage basin 内陆流域盆地
inland(dry)dock 内河船坞
inland dune 内陆沙丘;大陆沙丘;风积沙丘
inland duty 国内税;内地税
inland earthquake 内陆地震

inland fishery 内陆渔业
inland fleet 内河船队;内河船(舶)
inland fog 内陆雾
inland forwarding expenses 内陆运费用;内地运费
inland freight 国内运费;内陆运输;内陆运费;内地运输
inland freight depot 内地货运站
inland freight haulage 内地货运
inland harbo(u)r 内陆港(口);内河港口;内地港(口);河港
inland haulage 内陆运费用
inland ice 内陆冰(川);大陆冰盖;冰原;冰盾
inland inundation 内涝
inland lake 内陆湖;内流湖
inland lock 内河水闸
inland marine insurance 陆运险;内陆水运保险;内陆(水上)运输保险
inland-moving dune 内陆移动沙丘
inland native country 内地
inland navigation 内河航运;内河航行;河运
inland navigation canal 内陆运河
inland navigation channel 内河航道
inland navigation projects 内河航运工程
inland navigation regulation 内河航行规章
inland navigation rights 内河航行权
inland navigation system analysis 内河航行系统分析
inland orientated industry 内地工业
inland place of discharge 内陆卸货地点
inland plain 内陆平原
inland port 内陆港(口);内河港口;内河港(埠);内地港(口);河港
inland province 内地省份
inland radiotelegraph 国内无线电报
inland region 内地地区
inland revenue 国内税收;内陆税收
inland river 内陆河(流);内河
inland rule 内河航行规章
inland Sabkha deposit 内陆萨巴哈沉积
inland Sabkha facies 内陆萨巴哈相
inland Sabkha sedimentation model 内陆萨巴哈沉积模式
inland sand 内陆砂
inland sea 陆缘海沟;内(陆)海;内河
inland sea dike 内陆海堤
inland sea pilot 内海引航员
inland ship 内河船(舶)
inland shipment 内陆运输
Inland Steel 内陆钢铁公司(美国)
inland stream 内河河(流);内河
inland telegraph 国内电报
inland telephone network 国内电话网
inland terminal 内河码头
inland terminal depot 内陆枢纽站;内河码头
inland town 内地城市
inland trade 国内贸易
inland transport(ation) 内陆运输;内地运输
inland transportation charges 内陆运费用
inland trunk call 国内长途电话
inland vessel 内河船(舶)
inland water 内河水
inland water-borne transport 内地水运
inland water-borne transportation 内陆水运
inland watercourse 内陆水道
inland water gripper 内河运煤船
inland waters 领水;内陆水域;内陆水系;内陆水体;内河水域;内河;内海;内地水域
inland waters fishery 内陆水域渔业
inland waters pollution 内陆水域污染
inland water survey 内陆水文测量
inland water transportation 内河水运
inland waterway 内河水路系统;国内水道;内河水道;内河水道;内河航道
inland waterway bill of lading 内水水运提单;内河航运提单
inland waterway bulletin 内河航道公报
inland waterway chart 内河航行图
inland waterway classification standard 内河航道分级标准
inland waterway fleet 内河船队
inland waterway for sea-going vessel 海船进江航道
inland waterway gripper 内河运煤船
inland waterway message 航道通信

inland waterway network 内河航道系统;内河水道网;内河航道网
inland waterway port 内河港口;内河港(埠);内陆水道港口
inland water(way) transport(ation) 内河运输
inland waterway vessel 内河船(舶)
in-law suite 所增建的小套间
inlay 镶入;镶嵌(物);镶嵌饰;镶嵌工艺;工艺镶嵌;里层;嵌体;嵌入图案;嵌入法;嵌衬;插入
inlay camera 插入摄像机
inlay carver 嵌体雕刻刀
inlay carving 嵌雕
inlay cladding 镶嵌金属包覆
inlay clad plate 双金属层板
inlay contact 插入接点
inlayer 镶嵌工;镶嵌者
in layers 分层地
inlaying 镶嵌
inlaying grafting 插接
inlaying saw 镶嵌锯
inlay mask 插入罩片
inlay pattern 镶嵌花纹
inlay thread 衬垫纱线
inlay wax 嵌体蜡
inlay window 窗嵌体
inlay work of colo(u)red marble 彩色大理石镶嵌
inleakage 吸入;漏泄;漏入;漏电;入渗;不密封
inleakage of air 空气渗入;空气漏入
inlet 小湾;镶嵌物;雨水口;供料口;进线口;进水(道);入入(量);进汽道;进口;入孔;投料口;通入;通海小河;插入物;分水口
inlet absolute pressure 进口绝对压力
inlet air 进入空气;入风
inlet air block 进气口块体
inlet air temperature 入口气温
inlet air tile 进气口面砖
inlet amount 进料量
inlet and outlet losses 进出口损耗
inlet and outlet pipe of gasholder 储气罐进气兼出气管井
inlet angle 进口角(度);入射角;入(口)角
inlet annulus 进气环室
inlet aperture 入口
inlet attack 进口侵蚀
inlet baffle plate 进口挡板
inlet bell 入口铃形绝缘子
inlet bend 进口弯头
inlet blade angle 叶片进口角
inlet box 吸入箱;滤油箱;进口箱
inlet bush 引入套筒;引入衬管
inlet cam 进气凸轮
inlet cam inlet-valvecam 吸进凸轮
inlet cam roller lever 进气阀凸轮滚轴杠杆
inlet camshaft 进气阀凸轮轴
inlet capacity 吸入容量;吸入能力
inlet casing 进汽缸
inlet cave 地下河消蚀作用
inlet chamber 吸入室;进水室;进水井(倒虹吸);进气室;输入腔
inlet chamber of inverted siphon 倒虹管进水井
inlet channel 进水渠;进水沟;进水槽;进气道
inlet characteristic 进气道性能
inlet chest 吸入箱
inlet chute 进料溜子;进料溜槽
inlet circuit 输入端电路
inlet clack 吸入瓣
inlet close 进气停止
inlet coefficient 入口系数
inlet compartment 进气室
inlet component 进气道组件
inlet compressor stage 压缩器的第一级
inlet cone 入口锥(体)
inlet connecting pipe 雨水口连接管;街道进水连管
inlet connection 进口连接(件);入口连接;入口接管;输入连接;进气管接头
inlet control 进气道调节;进口调节;进口(水位流量)控制;入口控制
inlet-control water heater 进口控制式热水器;敞口式热水器
inlet culvert 进水涵管;进水涵洞;进水暗渠
inlet device 入口装置
inlet diameter (入口的)进口直径;入口直径
inlet diffuser 进口扩压器
inlet distribution channel 进水配水渠

inlet duct 进水道;进气装置;进气道;进风道;送风道
inlet-duct area 进口(截面)面积
inlet duct ventilation 入口管道通风
inlet elbow 进口弯头
inlet end 进料端;入口端;喂料端
inlet end ring 料浆圈;挡砖圈;挡料圈
inlet energy dissipation 进口能量消耗
inlet fan 进气风扇
inlet filter 进水滤网;进水滤器;进口过滤器;入口过滤层
inlet flange 入口法兰
inlet flange diameter 进口法兰直径
inlet flow 吸入流
inlet flow rate 进口流量
inlet flushing system for pipe and cone 管道和导风锥的冲洗装置
inlet for storm water 雨水(排水)口
inlet gas 进气;入口气体
inlet gate 进水闸(门);进气闸门
inlet grate 进水口格栅;进气口格栅
inlet grating 雨水口算;雨水进水算;进水口帘格;进气炉算;进气算子
inlet grid 进水口格栅;进气口格栅
inlet grille 进水口格栅;进气口格栅
inlet head 入口压头;吸入压头;进料端
inlet header 多路进浆管
inlet hole 吸入孔;入口
inlet hood 进料端
inlet horn 入口喇叭口
inlet-in screen 进气滤网
inlet line 吸油管路;吸入管线;输入线(路);输入管线
inlet line filter 输入线路滤波器
inlet loading 进气负荷
inlet loss 进气损失;进口损失;入口损失;入口损耗
inlet louver 入风网
inlet Mach number 进口马赫数
inlet manhole 进入孔;检修孔;人孔入口
inlet manifold 进气歧管;进气多头管;进口汇管;入口集合管
inlet method 进水口法
inlet nipple 进口螺纹接套
inlet nominal size 进口公称尺寸
inlet non-return valve 入口止回阀
inlet nozzle 吸入喷嘴
inlet of branch channel 汊道进口
inlet of culvert 涵洞进水口
inlet of dust contained gas 含尘空气入口
inlet of grate-in-gutter type 帘格式进水口
inlet of pass 孔型入口侧
inlet of sewer 下水道入口;下水道口
inlet of surface water 地面水入口;地面入水口
inlet of ventilating system 通风系统进口
inlet of weep hole 流水槽入口
inlet open 开放吸气;进气阀开
inlet opening 雨水口进水孔;进水孔;进水井;入口
inlet operating temperature 入口操作温度
inlet orifice 入口
inlet outlet 进出口
inlet part 埋入件;埋入部分
inlet passage 进气通道
inlet pipe 引入管;进油管;进水管;入管;进气管;进油管廊道;进气管廊道;进水管廊道
inlet pipe gallery 进水管坑道;进油管坑道;进水管坑道;进油管廊道;进气管廊道;进水管廊道
inlet plate 入口盖
inlet port 入口;进油口;进气口;吸油口;进水口
inlet pressure 进口压力;入口压力
inlet pressure control 入口压力控制
inlet pressure of column 柱入口压力【化】
inlet ram 进气挺杆
inlet region 入口区
inlet regulator 进油调节器;进气调节器;进水调节器
inlet relief valve 进口减压阀;进口安全释放阀
inlet reservoir 入口蓄水池
inlet rocker 进气摇杆
inlet scoop 吸气口;吸风口;进水斗
inlet scroll 进口涡管
inlet section 进口段
inlet shaft 进水井;进水口竖井
inlet side 吸入端;真空侧;进口侧;入口侧
inlet sill 进水潜坝;进水口底槛;进口底槛
inlet sleeve 吸气套筒
inlet sluice 吸入闸门;进水闸(门);进气水闸;渠首闸

inlets of grit chamber 沉沙池进口
inlet steam pressure 进汽压力
inlet steam temperature 进汽温度
inlet strainer 进水滤器;进水连蓬头;进水(口)滤网;入口滤器
inlet stroke 吸入冲程;进气冲程
inlet structure 进水建筑物;进口建筑物;入口构造
inlet submerged culvert 半压力式涵洞
inlet surface 进水面
inlet surround 进水口周围
inlet swirl 进口旋流
inlets with horizontal gratings 平箅式雨水口
inlets with vertical gratings 立箅式雨水口
inlet temperature 吸入温度;入口温度
inlet temperature of air 空气入口温度;进气温度
inlet time 进口时间;进口集流时间;集水时间;集流时间;地面集水时间
inlet time to storm drains 雨水管进水时间
inlet training wall 进口导墙;进口导堤
inlet transition 进口渐变段
inlet trunnion 进料端空心轴
inlet tube 进水管;入口管
inlet union nut 进口接管螺母
inlet unsubmerged culvert 无压力式涵洞
inlet valve 吸入阀;压力水进水阀;供给阀;进油阀;进水阀;进气阀门;进料阀(门);进(给)气阀;入口阀
inlet valve bank 进气活门栅
inlet valve cap 进给阀盖
inlet valve chest 进气阀柜
inlet valve deck 进给阀盖
inlet valve spring 进给阀簧
inlet vane 进口叶片;进口导叶片
inlet velocity 进水速度;进入速度;进气速度;进口速度;进口流速;入口速度
inlet velocity head 入口速度压头
inlet volute 进口蜗壳
inlet water 进水;入口水柱
inlet well 吸水井;雨水井;进水井;集水井;入口井;导水井
inlet wing wall 进水口翼墙
inlet wire 进线
inlet wire diameter 进线直径
inlet works 进水泵站
inlet zone 进水区
inlier 构造窗;内围层;内露层;内窗层【地】
in lieu of 代替
in lighter 装驳船
in light of instruction 按照指示
in light weight 重量轻
in-line 上线;成一直线;液压进油管路;轴向(式)的;直列的;在线;在管线内;联机;进油(管)路;并行
in-line antenna 八木天线
in-line arrangement 直列式;顺排;顺列布置
in-line assembly 顺序组装;成行装配;插入组装
in-line assembly machine 一列式组装机;流水装配机
in-line bank 顺列布置管束
in-line barge transportation 一列式拖带法
in-line blender 管线混浆器;管道混合器
in-line blending 在管道中进行掺和;管道混合
in-line booster 序列式增压器;轴向加速器;串联助推器;串列(式)助推器
in-line check-valve 直通单向阀
in-line coagulation 在线混凝
in-line coagulation-ultra-filtration process 在线混凝超滤工艺
in-line coding 直接插入编码;联机编码;内部编码;成簇(数据)编码
in-line combination 直列组合
in-line combine 直列型联动收割机
in-line computer 在线计算机
in-line continuous production 流水线连续生产
in-line control 直接控制;在线控制
in-line coupling 排式接头
in-line crown block 单轴天车
in-line cylinder arrangement 直列汽缸排列
in-line data processing 成簇数据处理
in-line diagnosis 串行诊断;成簇诊断
in-line drive 共轴传动;联机传动;串联传动;直列式传动(装置)
in-line engine 直排发动机;直列式发动机;单列汽缸发动机
in-line engine crankshaft 直列式发动机曲柄

in-line equalization 在流程线上调节
in-line equipment 位于一线上的设备
in-line filter 管路过滤器;串联过滤器
in-line filtration 直流过滤
in-line five-cyl engine 直排五缸发动机
in-line flocculation 管道絮凝
in-line four-point probe 四点一列探针
in-line gears 同轴齿轮
in-line hologram 一列式全息图;同轴全息图
in-line holography 轴向全息摄影;同轴全息摄影
in-line hydraulic measurement 连线水力测量
in-line installation 直连安装
in-line linkage 直连式动力转向装置
in-line macro 在线宏指令
in-line meter 在线仪表;在股道中的仪器;在管道中的仪器
in-line mixer 液体连续搅拌机
in-line mixing 管线中混合
in-line motor 直列式(柱塞)马达
in-line needle valve 管道针状阀
in-line oscillation 顺流振荡
in-line package 直插式封装
in-line packing machine 嘴子排成直列的固定式装包机
in-line pallet transfer machine 板架轴向式连续自动工作机床
in-line panel grinding and polishing machine 管屏连续细磨抛光机
in-line plunger motor 直列式柱塞电动机;直列式柱塞马达
in-line plunger pump 直列式柱塞泵
in-line position 顺列布置
in-line price 与市场一致的价格
in-line procedure 直线程序;直接插入子程序;插入过程;联机程序
in-line processing 直接处理;在线处理;内部处理;直接插入处理;联机处理;成簇处理
in-line production 流水作业
in-line pump 直列式柱塞泵;管线泵;连续泵;管道泵
in-line pumps 顺列布置泵组
in-line quick coupling 准直快速接头
in-line readout 在线读出;即时读出
in-line square pitch 正方形直列;正方形排列
in-line stationary mixer 管线静态混合器
in-line stockpile 直线堆场;顺列堆料
in-line subroutine 直接插入子程序;在子程序;联机子程序;内子程序
in-line system 成簇数据处理系统
in-line transfer 直进式传送
in-line triangular pitch 正三角形直立;三角形排列
in-line type thermoforming machine 生产线中热成形机
in-line valve 管路上活门
in-line vibration 纵向振动;轴向振动
inliquidity 难以转为现金的资产
in-list 内目录
in lump 块状
Inman's Nautical Tables 应氏航海表
inmarsat 国际海事(通信)卫星
INMARSAT back-up distress altering network 国际海事卫星备用遇险报警系统
INMARSAT ship earth station 国际海事卫星船舶地面站
in mass 大量的;成批的
in merchantable condition 适合商销
in metric 用公制
in-migration 移来;迁入
in-milling 横向铣削
in-mine seismic prospecting 矿井地震勘探
in money terms 按货币计算
in most cases 大多数情况下
in most of varieties 在多数品种中
inmost shadow 最深阴影
in-motion viewing 动观
in-mo(u)ld coating 模内涂覆
in-mo(u)ld coating of plastics 塑料模塑涂装法
in-mo(u)ld coating system 模中涂布系统
in-mo(u)ld decorating system 模中装饰系统
in-mo(u)ld decorating with coatings 模内喷涂装饰
in-mo(u)ld decorating with foils 模内箔装饰
in-movement 横向进磨运动;横向进给运动
inn 小旅馆;小酒店;旅馆;旅店;客栈;酒馆
innage 油水柜内实际油水量的测量;装料量;剩油量

in name 名义上
innards 内部结构
innate 固有的;底着的
innate character 本质
innate-releasing mechanism 本能释放机制
in native state 天然状态
innavigable 不通航的;不能通航的
in need of repair 需要修理的
innelite 硅钛钠钡石
inner 内部的;内部
inner absorption 内吸收
inner-adjustable focus 内调焦
inner-adjustable focus collimator 内调(焦)平行光管
inner air cooler 内部空气冷却器
inner and outer contact of associated small intrusion 派生小侵入体内外接触
inner and outer contact of large intrusion dome 大型侵入体穹隆内外接触带
inner and outer strake 瓦叠式列板
inner and outer tube 内外管(岩芯管)
inner angle 内角
inner angle number of closed figure 闭合图形内角个数
inner anterior quadrant 内前四分体
inner arc 内弧
inner-arc basin 内弧盆地
inner arch 内部发券;内拱
inner auroral zone 内级光带
inner automorphism 内自同构
inner automorphism group 内自同构群
inner band brake 内带外张式制动器
inner bank 内岸;凸岸
inner bar 内沙洲;内沟坝;隔板;挡板;内滩
inner bark 内皮
inner barrel 内岩芯筒,内(岩芯)管;内取心管
inner basin 内港池
inner basin of levee 堤内集水区
inner battery housing 内电池盒
inner bead 内线条;灰饼(标志抹灰厚度);滑窗内轨;扯窗滚珠;抹灰弯点;提拉窗导轨
inner beam 内梁
inner bearing 内轴承
inner belt 内环路;内环
inner bending moment 内弯矩
inner berm 内护道;内戗道
inner blanket 内部再生区
inner block 内型
inner body 内部零件
inner border 内图廓【测】
inner bottom 内底;双层底顶部
inner bottom plating 内底板
inner boundary line of ore body 矿体内边界线
inner bracket 内肘板
inner brake 内制动器;内闸
inner brake hub 内制动器毂
inner brake shoe 内制动瓦
inner breakwater 内(部)防波堤;内港小型防波堤
inner building board 内部建筑板
inner bulwark 内舷墙
inner cage (汽轮发电机的)内机座
inner cal(l)ipers 内卡钳
inner cap 内盖
inner capacities 内容量
inner capillary water 内毛细水
inner capsid 内衣壳
inner capsule 内膜
inner carrier 内活塞
inner casing 内罩;内(层)缸;内套(管);内汽缸;内门窗框;内壳;内隔板
inner casing of joint (万向接头的)内活动接头
inner cavity wall 内部空芯墙
inner center 内心
inner chamber 内室
inner check valve guide 内止回阀导座
inner-chill 内冷铁
inner circle 内圈;内环路
inner circular layer 内环层
inner circumference highway 内环路
inner city 旧城;城内衰落区;内城(区)
inner city decay 内城衰落
inner clearance ratio 内净空比;内净径比;内间隙比;内间距比
inner clinch 内旋结

inner closure dike 背水侧截梳戗堤(围堰)
inner cluster sets 内部聚值集
inner clutch plate 内鼓摩擦片
inner coat 内层
inner coil 内蛇管;内层线圈
inner collar 凹辊环
inner column 内柱
inner combustion chamber 内燃烧室
inner common tangent 内公切线
inner complex 内络合物
inner complex salt 内络盐
inner computer 内部计算机
inner condenser 内冷凝器
inner condensing tube 冷凝器内管
inner conductor 中心线;零线;内(部)导体
inner cone 焰芯
inner cone of flame 内焰
inner consistency 内部符合
inner container 内容器
inner content 内容度
inner continental basin red beds association 内陆盆地红层组合
inner continental shelf 内陆架;大陆棚内半部
inner contour of tunnel lining 隧道衬砌内轮廓
inner contradictions 内在矛盾
inner control limits 内(在)管理界限
inner conversion 内在转换
inner-cooled machine 内冷电机
inner-cooled stator coil 内冷定子线圈
inner-cooling 内冷
inner coordination sphere 内配位层
inner core 内筒;内核
inner core barrel length 内岩芯筒长度
inner core barrel size 内岩芯筒尺寸
inner core tube 内管(指岩芯管)
inner corona 内冕
inner corridor 内走廊
inner couette 内筒
inner court 室内球场;内院
inner court width 内院宽度
inner cover 内罩
inner cross-section 净断面
inner curtain (wall) 内层幕墙;内隔板
inner curve 内曲线
inner cutting angle 内刃角
inner cutting shoe (双层岩心钻的)内靴
inner cycloid 内摆线
inner cylinder 心管;内圆筒
inner dead center 内止点
inner dead point 内止点
inner decorating 内装修
inner deformation 内在形变
inner deforming loss 内变形损耗
inner derivation 内部求导
inner diameter 内径
inner diameter ga(u)ge 内径标准尺
inner diameter ga(u)ge stone 内径边刃金刚石
inner diameter saw 内径锯
inner diameter tolerance 内径公差
inner die surface 模内面
inner diffusion 内扩散
inner dike 内堤
inner dimensions 内(部)尺寸
inner door 内针门;内门
inner drive joint 内传动接合
inner dump 内部排土场
inner ear 内耳
inner-eaves corbel bracket 内檐斗拱
inner edge 内缘;内图廓【测】;内边
inner edge of frame 肋骨内缘
inner electrolysis 内电解法
inner element 内部元素
inner emulsion paint 内乳胶漆
inner enceinte(wall) 内层幕墙;内围墙
inner end 内端
inner end of tooth 齿的内端
inner energy 内能
inner envelope 内包叶
inner envelope rotor flank (转子发动机的)内包络线转子工作面
inner equilibrium 内平衡
inner ester 内酯
inner ether 内醚
inner evapo(u)ration of soil 土壤内蒸发

inner expanding brake 内胀式制动器
inner extension tube 内管短节
inner face 正面;里面;内面;内表面
inner face tubbing 内面带法兰盘的丘宾筒
inner facing 内衬
inner fan 内扇
inner fan deposit 内扇沉积
inner fender 内挡泥板
inner fiber board finish 内部纤维板装修;内部纤维板装饰
inner film coefficient 内膜系数
inner finish 内装饰
inner finish(ing) paint 内装饰涂料
inner fire box 内火箱
inner fish-plate 内(侧)鱼尼板
inner fissure 内壳缝
inner fixtures 室内装置
inner flag register 内标记寄存器
inner flame 内焰
inner flange 内法兰
inner flank 内侧
inner flat keel 内平板龙骨
inner flue 内烟道
inner focusing telescope 内对光望远镜
inner force 内力
inner form 反面版
inner formwork 内模板
inner frame 内框架;内骨架
inner friction 内摩擦;内耗
inner function 内函数
inner gallery 内走廊
inner gallery apartment building 内廊道式公寓建筑
inner garden 内园;内花园
inner gate 内(闸)门;船闸上游闸门;内人门
inner(-ga(u)ge) stone (金刚石岩心钻头的)内缘钻石;(钻头的)内侧刃保径金刚石
inner gear 内齿轮
inner gearing 内啮合;内齿轮
inner gimbal 内万向悬挂支架
inner gimbal axis of gyro 陀螺仪内框架轴
inner gimbal ring 内平衡环
inner gimbal suspension of gyro 陀螺仪内框架
inner girder 屋内大梁
inner glass door 内部玻璃门
inner glazing 内部门窗镶嵌玻璃
inner gloss(clear) varnish 内部上光清漆
inner glume 内颖
inner grease retainer 内油挡
inner grid 内栅板;调制栅
inner guard rail 内侧护轨
inner guide 内导杆
inner hall 内殿
inner handrail (栏杆的)内扶手
inner harbo(u)r 港池;内港
inner harbo(u)r basin 内港池;内港
inner harbo(u)r line 内海港线;内港界线
inner hardboard finish 内侧硬质板饰面
inner harmonic measure 内调和测度
inner hearth 壁炉床;建筑内部得热;壁炉内地面
inner heat 内部热量;内部供暖;内部发热
inner heating 内加热
inner heating surface 内加热面
inner hexagon spanner 内六角扳手
inner hinge plate 内铰合板
inner hip 内肩节(平旋桥)
inner hull 内壳;耐压壳(体)
inner humidity 内部湿度
inner hydration sphere 内水合层
inner hyperboloid 内双曲面
inner illumination 内部照明(度)
inner insulation 内绝缘;内部绝热;内部隔声
inner iodine value 内碘值
inner isomorphism 内同构
inner iteration 内迭代
inner jar 内容器
inner jetty 内突堤;内防波堤
inner jib 内侧船首三角帆
inner joinery 室内细木工
inner keel 内平板龙骨;内龙骨
inner Lagrangian point 内拉格朗日点
inner lane 内侧车道
inner-lapping pluvial fan 内叠式洪积扇
inner layer 内层

inner layout 内部布置;室内规划;室内布置
inner lead 岩石间航道
inner lead bonder 内部引线接合器
inner lead bonding 内部引线接合法
inner leak 内部泄漏
inner leakage 内部泄漏量
inner lens cone 内镜筒;航摄仪内镜筒
inner level memory 内级存储器
inner lift arm and tappet assembly 内提升臂及挺杆总成
inner lighting 室内采光
inner linearization 内线性化
inner liner 内部衬垫
inner lining 内壳(垂直扯窗窗框内重锤箱);内衬;内部衬砌
inner lining board 内衬板
inner lining ring puller of motor bearings 电机轴承内套拔出装置
inner link 内在联系
inner lip 内唇
inner-lock train 直达列车
inner loop 内(循)环;内环路;内侧道;副环;辅助环;(立体交叉中的)内转匝道
inner loop fluidized bed 内循环流化床
inner loop sequencing batch reactor 序批式内循环反应器
inner lot line court (一边有土地界线的)内庭院
inner macroinstruction 内部宏指令
inner magnetosphere 内磁层
inner management 内部管理
inner mantle 下地幔,内地幔【地】;地幔深部
inner margin 内缘
inner marker 内指点标;内指标台(机场)
inner masonry dividing wall 内围砌分隔墙
inner mast 内门架【港】
inner measure 内测度
inner memory 内存储器
inner-mesh 内网
inner mixing air type burner 内混合式烧嘴
inner model 内模型
inner moisture 内部湿度;内部潮气
innermost 最内部的;最里面部分
innermost isoseismal line 最内地震线
innermost layer 最内层
innermost loop 最内层循环
innermost part 圣所;深闺;(寺庙、教堂的)内殿;最深部分
innermost suburbs 近郊
inner mo(u)ld 内模
inner multiplication 内乘法
inner nave aisle 教堂座位间走道
inner normal 内法线
inner nozzle 内水口
inner nuclear layer 内核层
inner oil seal 内油封
inner oil seal ring 内油封环
inner oil sump 内油槽
inner opposite angle 内对角
inner orbit 内层轨道
inner-orbital configuration 内轨型
inner orientation 内方位;内定向
inner-outer core boundary 内外核界面
inner packaging 内包装
inner partitioning 内部分隔
inner passage 内部通道
inner perforated bowl 带孔内转鼓
inner pericycle 内生中柱鞘
inner periphery of lining for vertical wall 直墙衬砌内轮廓线
inner photoeffect 内光电效应
inner pipe nozzle 管程接管口
inner piping(system) 内部管道系统
inner piston 内活塞
inner pit aperture 纹孔内口
inner pivot ring 内导环
inner planet 内行星;带内行星
inner planet(ary) gear 内行星齿轮
inner planking 内层(木壳)板
inner plate 内墙板
inner plexiform layer 内网织层
inner plug 内塞盖
inner point 内点
inner-pole type alternator 内极式交流发电机
inner port line 内港界线

inner posterior quadrant 内后四分体
inner post type container 内柱式集装箱
inner potential 内电位
inner pressure relief 内部减压
inner pressure test 内压试验
inner prestress 内预应力
inner primer 涂料底度;内部湿度;内底漆
inner product 内积
inner product control 内积控制
inner product feedback control 内积反馈控制
inner product of tensors 张量的内积
inner product of vector 矢量内积
inner product space 内积空间
inner proscenium 内部舞台;舞台前部
inner pump housing 内泵壳
inner punch 内冲头
inner quality 内在质量
inner quality standard 内控标准
inner quantum number 内量子数
inner race 内座圈;内圈
inner radiation belt 内辐射带
inner radiation zone 内辐射带
inner rail 里轨;内围栏;内轨
inner redecoration 内部重新装修;内部重新装饰
inner reflected reactor 内反射反应堆
inner reflector 内部反射层
inner reflux 内回流
inner region 内区域
inner region of boundary layer 边界层内区
inner regularity 内正则性
inner regular set 内正则集
inner relief road 市内疏散通路
inner residential area 市内住宅区
inner reveal (门窗框的)内侧壁;内部外抱
inner ring 内环
inner ringless roller bearing 无内圈滚子轴承
inner ring road 内环路
inner ring spacer 内隔圈
inner roadstead 内锚泊地
inner room 里屋;套间
inner rope 内绳
inner rotor 内转子
inner row teeth (牙轮的)里面的一排齿;内排齿
inner saddle 内鞍
inner sanctum 私室;书斋;圣所;至圣处
inner scaffold 室内脚手架
inner sea 内海
inner sealed box 内部密封箱
inner seal(ing) 内部填缝;内部封闭;内部封接
inner set 内集
inner shafts 内侧轴
inner shell 内壳(层);耐压壳(体)
inner-shell ionization 内壳电离
inner shield 内屏蔽
inner shoe 内支承块;内刃脚
inner side 内侧;内壁
inner(side) aisle (座位间的)走道;内部侧廊;教堂座位与列柱间的走道
inner side of a vertical surfaces 内旋
inner sill panel 内门坎板(车身)
inner skin 内表层;内层壳板;内部罩面层
inner skirt 内撑裙
inner slatted blind 窗内百叶遮阳;内层百叶遮阳;内部木条百叶窗
inner sleeve 内套筒
inner slide 内滑道
inner slope 内坡
inner solar system 内太阳系
innersole 内底
inner space 内层空间;内部空间
inner span 内跨
inner sphere complex 内层络合物
inner sphere surface complex 内层表面络合物
inner spindle 内轴
inner spiral bundle 内螺旋束
inner splice bar 内鱼尾棒
inner split ring 内隔环
inner spring 内弹簧
innerspring cotton-filled mattress 棉塞弹簧垫
inner stair(case) rail 内楼梯拦杆
inner stator 内定子
inner stay cone 座环内圈(贯流式水轮机)
inner steered angle 内侧车轮的回转角
inner stone (钻头的)内侧刃金刚石

inner stope 内回采工作面
inner storage 内存储器
inner strain 内应变
inner strake 内列板
inner stress 内应力
inner string 屋内楼梯斜梁;内楼梯斜梁
inner string cementing 内管注水泥
inner stringer 内纵梁
inner stripping material 内剥离物
inner structure 内部结构
inner structure of nodule 结核内部构造
inner stud 隔墙竖根;内门窗框
inner suburban district 近郊区
inner suction pipe machine 内真空制管机
inner suction pipe process 内真空制管法
inner surface 拱券内表面;心材面;内表面
inner surface area 内表面积
inner surface inwall 内壁
inner surface of vault(ing) 拱顶内表面
inner surface of wall 内墙面
inner surfacing 室内墙面处理;内衬
inner tangent common 内公切线
inner temperature control 内部温(度)控(制)
inner term 内项
inner terminal 内端子
inner terminal curve 内末端曲线
inner tidal delta 内潮汐三角洲
inner tile 内瓦
inner tire 内胎
inner topology 内部拓扑
inner total reflection 内全反射
inner tower 内架
inner tracery 内窗(花)格
inner track 内磁道;内部路径
inner track roller frame bearing 内滚珠框架轴承
inner transformation 内部变换
inner tube 内岩芯管;内取芯管
inner-tube adapter 内管导向套(岩芯管);内岩芯管连接管
inner-tube assembly 内管组件
inner-tube core lifter 内岩芯管岩芯提断器;内管岩芯卡取器
inner-tube deflation 内胎放气
inner-tube extension 内岩芯管接长管;伸长内管
inner-tube head for wire line 绳索取芯内管接头
inner-tube of pneumatic tyre 充气轮胎内胎
inner-tube of type 内胎
inner-tube projects 内管凸出体
inner-tube shoe 内岩芯管靴;内管管鞋
inner-tube stabilizer 内管稳定器
inner-tube valve 内胎气门
inner-tube valve cap 内胎气门嘴帽
inner-tube wrench 内管扳手
inner tyre 内胎
inner vacuum vessel 内部真空室
inner vale 单面山前低地;封闭盆地
inner valve 内瓣
inner veil 内菌幕
inner vertical shaft 内垂直轴
inner viscosity 结构黏度;内在黏度
inner volume 内体积
inner vortex 内螺旋线(旋流集尘器)
inner wall(ing) 内(衬)墙;内壁
inner wall spacing 内壁间距
inner warp 内缠绕层
inner warping 内缠绕层
inner waterbreak 内(部)防波堤;内港小型防波堤
inner waterproofing 内防水
inner waters 内陆水域
inner wedge 内楔形炮孔(掏槽用)
inner wheelpath 内轮迹带;内轮迹线;内侧车道
inner width 净宽
inner window 内窗
inner window cill 室内窗槛;内窗槛
inner window frame 内层窗樘;内窗框(架)
inner window sill 室内窗槛;内窗槛
inner work 室内装修与设施;内部工作;内部工程
inner work function 内功函数
inner yoke 内磁轭
inner zone 内带
inner zone of southwest Japan 西南日本内带
inning 涨出地;排干的沼泽地;围垦土地;围垦(地)
inninmorite 斑玄武岩
innocent mispresentation 非蓄意的虚报

innocent party 无优先权的一方
innocent purchase for value 无知的高价购买
innocent third party 善意第三方
innocent treatment 无害化处理
innocuity 无害
innocuity test 安全试验
innocuous 无害的
innocuous effluent 无害出流水;无害排出物;无害废水;无毒流出物
innocuousness 无害性
innocuous substance 无害物质
innocuous zone 无害区
innominatal 无名的
innominate 无名的
in nonuplicate 一式九份
innovation 革新;改革;合理化建议;发明
innovation and creation 发明创造
innovation cost 革新成本;技术革新费用
innovation life cycle 技术创新的寿命周期
innovation shoot 更新枝
innovation system 创新体系
innovation theory 革新理论
innovative and alternative technique 革新代用技术
innovative approach 新型方法
innovative business 技术革新业务
innovative delivery method 革新的交货方法
innovative financing technique 提供贷款的新方法
innovative method 新型方法
innovative technology 新工艺;创新的技术;新技术
innovator 革新者
innoxious 无害的
innoxious substance 无害物质
innumerability 无数
innyard 客栈的庭院
incohcerent rotation 不相干转动
in octuplicate 一式八份
inoculate 培植接种
inoculated cast iron 变性铸铁
inoculating crystal 晶种
inoculating river 合并河
inoculating seed 晶种
inoculating stream 合并河
inoculation 变质处理;预防注射
inoculation of the soil 土壤接种(对土壤注射造氮菌)
inoculation sludge 接种污泥
inoculation sludge basin 接种污泥池
inodorous felt 作衬垫的石棉毡;无气味油毡
inodo(u)rous 无气味的
in off position 在关闭位置
inoganic nitrogenous fertilizer 无机氮肥
inoglia 纤维胶质
inolith 纤维石
in one piece nozzle 单喷嘴
in-opening door 内开门
in-opening window 内开(式)窗
inoperable 不能实行的;不能工作的;不能操作
inoperable time 不能操作时间;不可操作时间
inoperating contact 停用触点
in operation 正在操作;在操作中
inoperative period 停机期;非运行期
in order 顺次(的);按顺序
inorder for binary tree 二叉树的中根次序
in order to minimize overdredging 为使超挖最小
in order to promote patronage or sales 以广招徕
inordinate wear 异常磨损
inorfil 无机纤维
inorganic 无机的
inorganic accelerator 无机促进剂
inorganic acid 无机酸
inorganic acid ester 无机酸酯
inorganic acidity 无机酸酸度
inorganic acidotrophic lake 无机酸营养湖
inorganic active filler 活性无机填料
inorganic adhesive 无机胶黏剂
inorganic adsorbent 无机吸附剂
inorganic agent 处理剂
inorganic aggregate 无机集料;无机骨料
inorganic alumin(i)um salt waterproofing agent 无机铝盐防水剂
inorganic ammonia nitrogen wastewater 无机氨氮废水
inorganic analysis 无机分析

inorganic architecture coating 无机建筑涂料
inorganic arsenic chemicals 无机砷化合物
inorganic base 无机碱
inorganic base grease 无机物稠化的润滑脂
inorganic bentonite 无机膨润土
inorganic binder 无机黏结剂;无机黏合剂;无机漆基;无机结合料
inorganic bond 无机结合剂
inorganic-bonded composite 无机胶凝复合材料
inorganic-bonded heat resistant fluorophlogopite plate 无机黏合耐热氟金云母纸层压板
inorganic building material 无机建筑材料
inorganic byproduct 无机副产物
inorganic carbon 无机碳
inorganic carbonic acid 无机碳酸
inorganic carbon system 无机碳体系
inorganic carcinogen 无机致癌物
inorganic cement 无机胶结剂
inorganic cementing material 无机胶凝材料
inorganic ceramic membrane 无机陶瓷膜
inorganic chemical dosimeter 无机化学剂量计
inorganic chemical industry 无机化工
inorganic chemical pollutant 无机化学污染物
inorganic chemical pollution 无机化学物污染;无机化学品污染
inorganic chemical reagent 无机化学试剂
inorganic chemicals 无机化学物
inorganic chemistry 无机化学
inorganic chert 无机燧石
inorganic clay 无机黏土
inorganic coagulant 无机絮凝剂;无机凝结剂;无机混凝剂
inorganic coating 无机涂料;无机涂层;无机敷层
inorganic colloid 无机胶体
inorganic colloid-gelled grease 无机胶体凝胶润滑脂
inorganic colloid in soil 土壤无机胶体
inorganic complex 无机综合物;无机络合物;无机复合物
inorganic complex type coagulant 无机物复合型混凝剂
inorganic component 无机物组分
inorganic compound 无机化合物
inorganic concentration 无机物浓度
inorganic concrete aggregate 无机混凝土骨料
inorganic constituent 无机物组分;无机成分;非有机物成分
inorganic constructional material 无机建筑材料
inorganic crystal scintillator 无机晶体闪烁器
inorganic deposit 无机沉积(物);无机沉淀(物)
inorganic dissolved component 溶解的无机组分;无机溶解组分
inorganic dissolved substance 溶解的无机物;无机溶解物
inorganic dust 无机(性)粉尘
inorganic environment 无机环境
inorganic fertilizer 无机肥料
inorganic fiberboard 无机纤维板
inorganic fiber 无机纤维
inorganic fiber board 无机纤维板
inorganic filler 矿质填(充)料;无机(物)填料
inorganic flocculant 无机絮凝剂
inorganic fluorine compound 无机氟化合物
inorganic forms of nitrogen 氮的无机形态
inorganic formulation 无机制剂
inorganic foulant 无机污垢
inorganic fungicide 无机杀菌剂
inorganic gas 无机气体
inorganic gel-thickened grease 无机稠化润滑脂
inorganic heat insulating material 无机隔热材料
inorganic hydrochemistry 无机水化学
inorganic index 无机物指数
inorganic insecticide 无机杀虫剂
inorganic insoluble substance 无机不溶物
inorganic insoluble test 无机不溶试验
inorganic insulation 无机绝缘
inorganic insulation material 无机绝缘材料
inorganic ion content 无机离子含量
inorganic ion exchange 无机离子交换
inorganic ion exchange membrane 无机离子交换膜
inorganic ion exchange paper 无机离子交换纸
inorganic ion exchanger 无机离子交换剂
inorganic ions 无机离子

inorganic limestone 无机碳酸钙
inorganic mass spectrometry 无机质谱法
inorganic material 无机物质;无机材料
inorganic materials which exist in the earth 地球中存在的无机质
inorganic matter 无机物质;无机物
inorganic membrane 无机膜
inorganic membrane-bioreactor 无机膜—生物反应器
inorganic mercury 无机汞
inorganic mercury chemicals 无机汞化学物;无机汞化学品
inorganic mercury compound 无机汞化合物
inorganic mercury poisoning 无机汞中毒
inorganic microconstituent 无机微组分
inorganic microgel grease 无机微胶粒润滑脂
inorganic microparticle 无机微颗粒
inorganic monophosphate 无机单磷酸盐
inorganic nitrogen 无机氮
inorganic nitrogenous wastewater 含无机氮废水
inorganic non-metallic material 无机非金属材料
inorganic nutrient 无机营养物
inorganic nutrition 矿质营养;无机营养
inorganic-organic polymer composite flocculant 无机有机高分子复合絮凝剂
inorganic origin theory 无机成因学说
inorganic paint 无机漆
inorganic particle 无机颗粒
inorganic particulate matter 无机颗粒物
inorganic peroxide 无机过氧化物
inorganic pesticide 无机农药
inorganic petrochemicals 无机石油化学品
inorganic phosphate 无机磷酸盐
inorganic phosphorescent paint 无机夜光漆;无机磷光漆
inorganic phosphorus 无机磷
inorganic pigment 无机颜料
inorganic pollutant 无机污染物
inorganic pollution 无机物污染
inorganic pollution of water body 水体无机物污染
inorganic polymer 无机聚合物;无机高分子聚合物
inorganic polymer coagulant 无机高分子混凝剂
inorganic polymer composite flocculant 无机高分子复合絮凝剂
inorganic polymer flocculant 无机高分子絮凝剂
inorganic radio iodine 无机放射性碘
inorganic refuse 无机垃圾
inorganic resin 无机树脂
inorganic rubber 无机橡胶
inorganic salt 无机盐
inorganic salt pollution 无机盐污染
inorganic salt production 无机盐生产
inorganic scintillation crystal 无机闪烁晶体
inorganic sediment 无机沉积(物);无机沉淀(物)
inorganic separation membrane 无机分离膜
inorganic silt 无机淤泥;粉砂;矿质泥沙
inorganic sludge 无机污泥
inorganic soil 无机土(壤)
inorganic solid 无机固体
inorganic solute 无机溶质
inorganic solvent 无机溶剂
inorganic sorbent 无机吸着剂
inorganic sphere 无生界;无机界
inorganic structural material 无机建筑材料
inorganic substance 无机质;无机物
inorganic sulfur 无机硫
inorganic suspended sediment 无机悬沙
inorganic synthetic(al) dye 无机合成颜料;无机合成染料
inorganic thin film 无机薄膜
inorganic tin 无机锡
inorganic toxic material 无机有毒物(质);无机毒物
inorganic waste 无机废弃物
inorganic wastewater 无机污水;无机废水
inorganic weak acid 无机弱酸
inorganic world 无机界
inorganic zinc rich coating 无机富锌涂料
inorganic zinc rich paint 无机富锌漆
inorganic zinc rich primer 无机富锌涂料;无机富锌打底料
inorganization 无组织
inornate 无华丽装饰的
inosilicate 链硅酸盐
inosine wastewater 含肌苷废水

inositol 环己六醇
inotropic(al) 变力性的
inotropic(al) action 变力作用
in-out register 出入寄存器
inoxidability 不可氧化性
inoxidable 不可氧化
inoxidizability 抗氧化性;不可被氧化性
inoxidizable 抗氧化的;不可被氧化的
inoxidizing coating 防蚀面层;(油漆的)防护层
inoxidzable coating 防氧化涂层
inpaint 修复漆面;补涂
in parallel 相平行;并联
in particular cases 特殊情况下
in passing 顺便;附带(提及)
in passive position 处于被动
in-patient department 住院部
in payment for 付价款
in payment of 付款额
in per capita terms 按人口平均计算
in perpetuity 永远;无限期
in phase 同相;同时协调的
in-phase angle 同相角
in-phase beam 同相束
in-phase chrominance signal 同相彩色信号
in-phase component 同相分量;同相部分
in-phase compression ratio 同相压缩比
in-phase cross correlation 同相交叉相关
in-phase current 同相电流
in-phased horizontal antenna array 同相水平天线阵
in-phase energy spectrum 同相能量谱
in-phase filter 同相滤波器
in-phase loss 欧姆损耗;同相损失
in-phase mixer 同相混频器
in-phase operation 同相运行;同相运用
in-phase opposition 反相
in-phase overlap 同相垫纱
in-phase particles 同相粒子
in-phase potentiometer 同相电位计
in-phase rejection 共相抑制;同相抑制
in-phase signal 同相信号
in-phase state 同相
in-phase terminal 同相终端;同相端子
in-phase vibration 同相振动
in-phase voltage 同相电压
in-phase yarn 同相纱
in-pile 堆内
in-pile loop 堆内回路
in-pile test 堆内试验
in place 现场的;在原位;就位;当地;原状
in-place bedrock 原状基岩
in-place cleaning 现场清洗
in-place concreting 混凝土现场浇注
in-place construction 现场建造
in-place density 原状密(实)度
in-place drowning 就地沉没
in-place gravel 就地的砾石
in-place leaching 原地浸出
in-place measurement 现场量测;现场测定
in-place mix combine 路拌联合装置
in-place mixer 路拌机
in-place mixing 现场拌和
in-place mo(u)lding 原位模塑
in-place nuclear density 用核子密度计在现场测得的重度;核子测定的原状密实度
in-place pipe cleaning 就地清洗水管;管线原地清洗
in-place push test 原位推压试验
in-place regeneration 原地再生;就地再生
in-place slump test 现场坍落试验
in-place soil 就地的土壤
in-place soil-shearing test 原状土剪切试验
in-place stress field 环境应力场
in-place test 现场试验;实地试验
in-place testing 原地试验
in plane 平面内
in-plane force 在平面力;平面内力
in-plane headshape 平面缠绕封头曲面
in-plane loading 平面内荷载;在平面内荷载
in-plane rigidity 平面内刚度;在平面刚度
in-plane shear 水平剪切;水平剪力
in-plane shear crack 平面内剪力裂缝;在平面剪裂缝
in-plane shear test 平面抗剪试验
in-plane vibration 面内振动

in-plant assay 厂内测定;厂内装配
in-plant atmosphere 厂内空气
in-plant control 近距控制;厂内控制
in-plant control system 近距离控制系统
in-plant handling 厂内输送;厂内搬运
in-plant mixer 厂内搅拌机
in-plant noise 厂内噪声
in-plant over-all economic calculation system 厂内全面经济核算制
in-plant recycling 厂内循环;厂内回用
in-plant recycling of waste streams 厂内重复使用
in-plant recycling of wastewater stream 废水的厂内循环使用
in-plant reuse 厂内再用
in-plant road 厂内道路
in-plant survey 厂内调查
in-plant system 近距离系统;厂内系统
in-plant track 厂内轨道
in-plant training program(me) 工厂培训计划;厂内培训计划
in-plant transport 厂内输送
in-plant treatment 厂内处理
in-plant water recycling 厂内水循环
in-plate permeability 顺面渗透性
inpolar 内极点
inpolar conic 内极二次曲线
inpolygon 内接多边形
inpolyhedron 内接多面体
in portable form 便携式的
in position 已在仓;正确布置;正常位置;进入位置;到达位置
in-position viewing 静观
in powder form 移动式的;粉状(的)
in preference to 先于……;优于……;比……好
in preparation 筹备中
in-process defect 生产过程产生的缺陷
in-process inspection 工序间检验;生产过程中的检验
in-process inventory 过程中投入量;加工清单;加工过程中的产品清单
in-process material 在制品
in-process order 在制订货
in-process product 中间产物;中间产品
in-process quality control 过程内质量控制;工序内质量管理
in-process raw material 加工用原材料
in-process stock 加工中的材料
in-process storage 半成品保管
inproper alignment 不合适的对准
in production 在生产中
in proportion to the expenditures of living labo-(u)r 比例于活劳动的支出
in prospect 预料;在望;即将到来;期望
in-pulp electrolysis 矿浆电解
in pursuance of 依照……;执行;在执行……时;按照……
input 进料;进量;输给;投入;输入(量);输入额
input absolute noise level 输入端绝对噪声电平;输入端等效噪声电平
input acoustic(al) power 输入声功率
input action 输入作用
input adapter 输入衔接器
input adjustment 投入调整
input admittance 输入导纳
input air 风量
input amount 投入量
input amplifier 输入放大器;输入端放大器
input amplitude 输入信号振幅
input analysis 投入分析法;输入分析
input and output analysis 投入产出分析
input and output operation 输入/输出操作
input angle 输入角
input aperture 入射孔径
input architecture 输入设备结构
input area 输入区
input argument 输入变元
input attenuation 输入衰减
input attribute 输入属性
input axis 输入轴
input back-off 输入补偿
input bars 输入条
input beam 输入束;输入光束
input block 输入信息组;输入信息块;输入缓冲区;输入分程序;输入(存储)区;输入部件

input borrow 输入借位
input buffer 输入缓冲器
input buffer register 输入缓冲寄存器
input bundle 输入束
input bus 输入总线
input button 输入按钮
input capability 输入能力
input capacitorless 无电容输入
input capacity 输入能力
input card 输入卡片
input carry 输入进位
input category 输入分类
input cavity 输入空腔谐振器;输入空腔
input channel 输入电路;输入信道;输入通道;输入流道;输入端
input character 输入字符
input characteristic 输入特性
input checking equipment 输入校验设备
input chip 输入片
input choke 输入扼流圈
input circuit 输入电路
input clock 输入时钟
input code 输入代码
input coefficient 投入系数
input combination 成本投入组合
input common-mode range 输入共模范围
input common-mode rejection ratio 输入共模抑制比
input computer 输入计算机
input concentration 进料浓度;输入浓度
input conductance 输入电导
input connection 吸入口接头;输入连接;输入接线
input control 输入控制
input controllability 输入可控性
input controller 输入控制器
input control shift register 输入控制移位寄存器
input control unit 输入控制器
input control valve 输入调节阀
input converter 输入转换器
input correlation 输入相关
input cost 投入成本
input coupler 输入耦合器
input coupling factor 输入耦合因数
input crystal 输入晶体
input current 输入电流
input curve 输入量曲线
input data 输入数据
input data bus 输入数据总线
input data instruction 输入数据指令
input data processor 输入数据处理机
input data proof 输入数据的检验
input data set 输入数据集
input data strobe 输入数据选通(脉冲)
input data structure 输入数据结构
input data system 输入数据系统
input data translator 输入数据转换器
input data validation 输入数据合法性确认
input decoupling zero 输入解耦零点
input demodulator 输入解调器
input derivative feedback 输入微分反馈
input detector 输入探测器
input device 输入装置;输入设备
input digit 输入数(字);输入数位
input digital data 输入数字数据
input display unit 输入显示设备
input distributed processing 输入分布处理
input distribution 投入分布
input drift 输入信号变化
input duct 输入流道
inputed income 估算收入
input editing 输入编辑
input efficiency 输入效率
input electrode 输入电极
input element 输入环节
input end 输入端
input energy 输入能量
input energy stability 输入能量稳定度
input ensemble 输入集
input equation 输入方程
input equipment 输入装置;输入设备
input error 输入误差;输入错误
input expander 输入扩展器;输入扩展电路
input face 输入面
input factor 输入系数

input file 输入文件
input file code 输入文件代码
input file label handling 输入文件标号处理
input filter 输入滤波器
input flow 吸入流;输入流量
input flow range 输入流量范围
input flyback 输入回描
input format 输入格式
input format control 输入格式控制
input gap 输入隙
input gap voltage 输入隙电压
input gas 注入气体
input gate 输入门
input generator 输入(信号)发生器
input head 输入头
input header 输入标题
input hopper 输入漏斗
input horsepower 输入马力;输入功率
input horsepower of mud pump 泥浆泵输入功率
input hum level 输入交流声电平
input identification 输入识别
input image 输入图像
input impedance 输入阻抗
input impedance of detector 检波器输入阻抗
input-impulse 输入脉冲
input in cash 现金投入
input information 输入信息;输入数据
input inhibitor 输入禁止器
input in kind 实物投入
input instruction code 输入指令码
input intensity 输入强度
input interface 输入接口
input interface adapter 输入(端)接口器
input interface in a compiler 编译程序中的输入接口
input interleaving control 输入交叉控制
input jack 输入塞孔
input job flow 输入作业流
input job queue 输入作业队列
input job stream 输入作业流
input keyboard 输入键盘
input leakage 输入漏泄
input level 输入电平
input light beam 输入光束
input limit 输入极限
input limiter 输入限制器
input line 输入线(路)
input linear group 输入线性部分
input link 输入网络节;输入连线
input list 输入表
input listing format 输入列表格式
input loading 输入加载
input loading factor 输入加载因数;输入负载率
input loading type preamplifier 输入加载式前置放大器
input logic 输入逻辑
input logic variable 输入逻辑变量
input loop 输入环
input machine variable 输入计算机变量
input magazine 卡片输入箱;输入库;输入储卡箱
input mask 输入框
input material 原料
input matrix 输入矩阵
input measure 输入测度
input medium 输入媒体
input mirror 输入镜
input mirror transmission 输入镜透射率
input modulation 输入调制
input module 输入组件;输入模块
input module valve 输入模块阀
input motion 输入运动
input network 输入网络
input node 输入节点
input nozzle 输入喷嘴
input number 输入数
input objective 输入物镜
input offset current 输入失调电流;输入补偿电流
input offset current drift 输入补偿电流漂移
input offset voltage 输入补偿电压
input offset voltage drift 输入补偿电压漂移
input operation 输入操作
input option 输入任选
input orientation 投入导向
input-oriented budgeting 侧重投入的预算法;编

制侧重投入的预算
input-output 投入产出;输入输出
input-output accounting 投入产出会计;投入产出核算
input-output analysis 投入产出分析;投产出产分析;输入输出分析
input-output analysis method 投入产出分析法
input-output approach 投入产出法
input-output channel 输入输出通道
input-output characteristic 输入输出特征;输入量与输出量的关系特性
input-output circuit board 输入输出电路板
input-output coefficient 投入产出系数
input-output command 输入输出指令
input-output control program(me) 输入输出控制程序
input-output control system 输入输出控制系统
input-output data channel 输入输出数据通道
input-output device 输入输出装置;输入输出设备
input-output device controller 输入输出设备控制器
input-output equipment 输入输出装置;输入输出设备
input-output executive 输入输出执行程序
input output forecast model 投入产出预测模型
input-output framework 投入产出结构
input-output interface system 输入输出接口系统
input-output interrupt 输入输出中断
input-output library 输入输出程序库
input-output limited 受输入输出限制
input-output mathematical model 投入产出数学模型
input-output model 投入产出模型;输入输出模型;输入产出模型
input-output operation 输入输出操作
input-output optimization model 投入产出优化模型
input output price 投入输出成本
input-output process 投入产出过程;输入输出过程
input-output processor 输入输出处理器
input-output programming system 输入输出程序系统
input-output ratio 投入产出比率;投入产出比
input-output recording medium 输入输出记录介质
input-output register 输入输出寄存器
input-output relation 输入输出关系
input-output relationship 投入产出关系
input-output request word 输入输出请求字
input-output standard interface 输入输出标准接口
input-output storage 输入输出存储区
input-output study 路段投入驶出调查
input-output table 进料出料表;投入产出表
input-output table of environment(al) resources 环境资源投入产出表
input-output technique 投入产出技术
input-output technology 投入产出技术
input-output teletype unit 输入输出电传机
input-output termination 输入输出终端
input-output unit 输入输出(控制)装置;输入输出控制器;输入输出部件
input-output variable 投入产出变量
input-output vector 投入产出矢量
input-out table of environmental protection 环境保护投入产出表
input-out unit 输入输出装置
input pad 输入衰减器
input parameter 输入参数
input pass-band 输入通带
input passenger flow 输入客流
input path 入射光路
input peripheral control 输入外围控制
input pick-up (发动机示波器等的)输入传感器
input picture 输入图像
input pinion 输入小齿轮;传动小齿轮
input point 输入点
input polarization 输入偏振
input port 输入口;输入端
input potentiometer 输入电位计
input power 轴功率;输入功率
input power winding 输入电源绕组
input pressure 输入压力;进口压力
input primitive 输入图元;输入基元
input process(ing) 输入处理;输入加工;输入进程;输入过程
input processor 输入处理机

input profile of a well 井的注入剖面
input program(me) 输入程序
input program(me) tape 程序输入带
input pulse 输入脉冲
input pulse amplitude 输入脉冲幅度
input pulse frequency 输入脉冲频率
input pulse width 输入脉冲宽度
input quantity 进料量;输入量
input queue 输入排队;输入队列
input queue overflow 输入队列溢出
input rank 输入队列
input rate 注入速度;输入速度
input rating 额度输入量;额度热负荷;输入率;额定热负荷
input ratio 输入率
input ray pencil 入射光锥
input reactance 输入电抗
input reactivity spectrum 输入反应性谱
input rearrangement 输入重新排列
input reference 输入基准值;输入参考值
input reference axis 输入基准轴
input reflection coefficient 输入反射系数
input register 输入寄存器
input relay 输入继电器
input repeat time 投入重复期
input requirement 投入需求量
input resistance 输入电阻
input resistor 输入电阻
input resolution 输入分辨能力
input resonant circuit 输入谐振电路
input resonator 输入共振器
input ripple 输入脉动
input routine 输入例行程序;输入例程;输入程序
input scan 输入扫描
input screen 输入屏
input sensing 输入感测
input sequence number 输入顺序号
input service 输入工作
input shaft 输入轴
input shaft forward reverse 前进后退传动轴
input side 输入侧
input signal 输入信号
input signal voltage 输入信号电压
input slit 输入狭缝
inputs of services 服务投入
input space 输入空间
input species 输入物种
input specification 输入规格
input speed 注入速度;输入速率;输入速度
input stability 输入稳定度
input stacker 输入接卡箱
input stage 输入级
input-step 输入阶跃
input storage 输入存储器
input store 输入储存
input string 输入串
input stroke 进气冲程;输入冲击脉冲
input structure 投入结构
input subsystem 输入子系统
input symbiont 输入共存程序
input symbol 输入符号
input system 进风系统;输入数据系统;吸气通风系统;输入(管道)系统;鼓风通气系统
input table 投入表;输入台
input tape 输入带
input tape density 输入带密度
input tape format 输入带型号
input tape sorting 输入带排序
input time 投入期;输入时间
input time constant 输入时间常数
input time series 输入时间系列
input torque 输入转矩;输入力矩
input total sediment inflow 输入泥沙总量
input traffic control 输入信息量控制;输入流量控制
input transaction processing 输入事务处理
input transfer 输入传送
input transformer 输入变压器
input transformer type 输入变压器式
input translator 输入翻译程序;输入变换器
input-truncated condensation 模式被输入截入缩合
input tube 输入管
input unit 投入单位;输入装置;输入设备;输入机
input value 输入值
input value of system 系统输入值

input variable 输入变数;输入(变)量
input variable name 输入变量名
input vector 输入向量
input velocity 输入速度
input velocity gradient 速度梯度输入
input voltage 输入电压
input voltage filtration 输入电压滤波
input voltage range 输入电压范围
input wave 输入波
input well 输入源(信息);注水井;输入井
input white noise 输入白噪声
input winding 输入线圈;输入绕组
input window 输入窗口
input wiring diagram 输入端接线图
input with open collector 集电极开路的输入
input work 指示功(发动机);机器的总功;输入功;输入工作
input work queue 输入作业队列;输入排队
in quadruplicate 一式四份
inquiline 寄居动物
in quintuplicate 一式五份
inquire by file 按文件询问
inquire by unit 按部件询问
inquire response 询问应答
inquiry 询问;询价;门诊;问价;探价;调查;查询
inquiry agency 征信机构
inquiry and communication system 询问与通信系统
inquiry and transaction processing 查询和事务处理
inquiry application 咨询应用
inquiry character 询问符;查询符
inquiry circuit 查询电路
inquiry desk 询问台
inquiry display terminal 询问显示终端
inquiry form 查询单
inquiry language 查询语言
inquiry list 询价单
inquiry logical terminal 查询逻辑终端
inquiry office 问讯处
inquiry of luggage and parcel accident 行包事故查询
inquiry processing 询问处理;查询处理
inquiry-response communication 查询响应通信
inquiry-response operation 查询响应操作
inquiry session 查询会话;查询对话
inquiry sheet 询价单;查询表格
inquiry station 询问站;询问栈;询问用终端设备;询问台;查询站
inquiry system 询问系统
inquiry terminal display 询问终端显示器
inquiry test 探询反应测验
inquiry transaction 查询事项;查询交往
inquiry unit 询问设备
inquiry with competitiveness 竞争性询价
inquiry-with-update 查询更新处理;查询并更新
inquisition 探询
inquisitor 询问雷达信标辅助装置
in ramp 入口坡道
in range 串视
in ratio 成比例
in readiness 准备就绪
in rear of a signal 信号机后方
inreducible graph 不可约图
in regime 在规定的制度下;(河床的)冲淤均衡;在制定规定下;(河道的)可以控制情况
in regular turn 依常例顺序
in regular turn clause 停泊条款
in relays 轮班
in response to 根据
irreversible 不可逆
irreversible colloid 不可逆胶(质)体
inroad 侵入;侵害;侵犯
in roll damping 滞滚作用
in round number 以整数计
inrudder 内侧舵
inrush 溃堤;(堤坝的)决口;起动功率;起动冲量
inrush current (通电瞬间的)冲击电流;合闸电流;涌流
inrush of water 水流涌入;涌水;来水
inrush water 涌流的水;涌入水
in-sack dryer 袋装农产品干燥机
in-sack drier platform 成袋粮食干燥台
in-sack drying 袋装农产品干燥

in sand cooling 砂内冷却
ins and outs 详细情节;迂回曲折;来龙去脉;底细
insane asylum 精神病院;疯人院
inscattering 内散射
inscattering correction 内散射校正
inscribable 可雕刻的
inscribe 写上;注册;内接
inscribed 内接的
inscribed angle 内接角
inscribed circle 内切圆;内接圆
inscribed circle of a triangle 三角形的内切圆
inscribed cone 内接圆锥
inscribed control region 内接控制区
inscribed cylinder 内切圆柱
inscribed figure 内接(图)形
inscribed in a circle 圆内接
inscribed polygon 内接多边形
inscribed polyhedron 内接多面体
inscribed prism 内接三棱形;内接棱柱(体)
inscribed pyramid 内接棱锥
inscribed quadrilateral 内接四边形
inscribed securities 记名证券
inscribed sphere 内切球
inscribed square 内接四边形
inscribed stock 记名股票
inscribed triangle 内接三角形
inscriber 记录器
inscribe to curves 曲线内接
inscription 记入;铭文;铭刻;碑文
inscription of a tablet 碑文
inscription on a tablet 碑文
insculptate 雕空的
inseam 内缝
in-seam driving 脉内成巷
in-seam sealant 缝隙密封剂
in season 应时的;合时宜的;恰当时刻;时令的
in seat starting 就地启动
insect 害虫
Insecta 昆虫纲
insectarium 养虫室
insectary 养虫室
insect attack 虫害;虫蛀
insect attractant 诱虫剂;引虫剂
insect bite 虫咬
insect cloth 蚊帐纱
insect damage 虫害
insect disease prevention 防治病虫害
insect farm 杀虫剂车间(石油炼厂内)
insecticidal lacquer 杀虫漆
insecticidal oil 杀虫油
insecticidal paint 杀虫涂料;杀虫漆
insecticide 杀虫剂
insecticide for timber 木材用杀虫剂
insecticide kerosene 杀虫剂用煤油
insecticide net 防虫网
insecticide paint 杀虫涂料;杀虫漆
insectifuge 驱虫剂;杀虫剂
insect infestation 遭受虫害
insect injury 虫害
insection 齿纹
insectivorous plant 食虫植物
insect net 防虫网
insectofungicide 杀虫灭菌剂
insect parasite 寄生天敌
insect-paste 杀虫胶
insect pest 虫害
insect pests and plant disease 植物病害
insect population 虫口
insect powder 杀虫粉
insect prevention 防虫
insect proof 驱虫的;防虫的
insect repellant 驱虫剂;杀虫剂
insect-repellent finishes 防蛀整理(如地毯、织物)
insect screen 纱窗门;窗纱;纱窗
insect-trap light 灯光捕虫器
insect wax 虫蜡;虫白蜡;白蜡
insect wire screening 防虫网;纱窗
insect with sucking mouth part 刺吸式口器害虫
insecure 易坍的;不牢靠的;不安全的;不可靠的
insecure investment 无保障的投资;不保险的投资
insecure work 危险工作
insecurity 不安全(性);不安全感
inselberg 岛状山;岛山;残山
insensibility 人事不省;不灵敏性

insensible perspiration 不显汗
insensitive 无感觉的;不敏感的;不灵敏的
insensitive clay 非灵敏黏土
insensitive interval 不灵敏时间
insensitiveness 不灵敏(性);不灵敏度
insensitivity 不敏感(性);不敏感;不灵敏(度)
inseparability 不可分离性
inseparable 不可分离的;不可分割的;不可分(的)
inseparable action 不可分动作
inseparable bearing 不可分离型轴承
inseparable element 不可分元;不可分割元
inseparable field 不可分域
inseparable graph 不可分图
inseparable polynomial 不可分多项式
in septuplicate 一式七份
in sequence 相继;按序;按顺序
insequent 斜向的
insequent drainage(system) 斜向水系;无定向水系;非顺向水系
insequent landslide 斜向滑坡;切层滑坡
insequent river 斜向河(流);偶向河
insequent slide 切层滑动
insequent stream 斜向河(流)
insequent valley 斜向谷
in series 串联(地)
insert 镶入;介入;接入;内埋件;嵌入式标志;嵌片;刀片;衬垫;插页;插入(物)
insertable impact bar 插入式反击锤
insert bearing 镶嵌式轴瓦;精密轴承
insert bit 镶刃刀片;镶嵌钻头;镶刃钻头;嵌镶钻头;嵌入式钻头
insert blank 镶刃刀胚
insert block 预埋砌块
insert boom 插入臂
insert camera 插入摄像机
insert chip 镶装刀片
insert core 插入型芯;插入泥芯
insert cover 嵌入盖
insert cursor 插入光标
insert depth 嵌入深度
insert die 镶入式模具;镶块模
insert earphone 小型耳机;插入式耳机
inserted arch 嵌入窟;反窟
inserted arch brick 反窟砖
inserted blade 镶入齿;镶齿刀头;插入式叶片
inserted blade core drill 镶齿扩孔钻
inserted blade end mill 镶齿端铣刀
inserted blade reamer 镶齿铰刀
inserted bolt 预埋螺栓;预埋螺杆
inserted broach 镶齿拉刀
inserted carbide 镶齿硬质合金刀片
inserted casing 插入套管
inserted ceiling 嵌入平顶
inserted chaser tap 镶齿螺纹梳刀丝锥
inserted column 嵌入墙内的柱;嵌墙柱;附墙柱
inserted crystal 镶嵌晶体
inserted die 镶入板牙
inserted drill 镶刃钻头;硬质合金钻头
inserted forging die 镶块锻模
inserted fraise 镶齿铣刀
inserted grill(e) 嵌入式格子窗;嵌入格栅
inserted implement 镶嵌机具;(拖拉机上的)悬挂式机具
inserted into concrete 插入混凝土
inserted joint casing 嵌接套管
inserted line 嵌线
inserted liner 镶入式缸套
inserted link 插杆
inserted magnet 嵌入的磁铁
inserted mode 插入方式
inserted num 嵌入螺母
inserted piece 预埋件(管子或电线等);嵌(入加强)块;砂型骨
inserted pin 插销
inserted reamer 镶齿铰刀
inserted shim 嵌入垫片
inserted side milling cutter 镶齿三面刃铣刀
inserted sleeve 镶入的衬套
inserted spandrel 嵌入式路肩
inserted strips 镶嵌条带
inserted stuffing box 衬垫压盖
inserted subroutine 插入子程序
inserted tap 镶齿丝锥
inserted tenon 假榫;嵌入榫头

inserted tongue 嵌入舌片
inserted tool 嵌入式工具;硬质合金刀具
inserted tooth 镶嵌齿;镶齿
inserted tooth broach 镶齿拉刀
inserted tooth cutter 镶齿铣刀
inserted tooth facing milling cutter 镶齿平面铣刀
inserted tooth hob 镶齿滚刀
inserted tooth saw 嵌齿锯;镶齿锯
inserted trick (镶嵌针槽的)镶钢片
inserted type of screw casing 嵌接式套管;插入式套管
inserted type thermometer 插入式温度计
inserted valve seat 嵌入阀座
inserted wedge 嵌入楔块
inserted wooden plate method 插板掘进法
insert electrode tip 电极头
insert equipment 插入设备
inserter 插入器;隔板;插件
insert film 插入胶片
insert fitting joint 插入连接接头
insert glove 镶边手套
insert hinge 嵌入式铰链
insert hole 嵌入孔
inserting 插补
inserting bottom rail 下侧梁嵌补(集装箱)
inserting contour 插绘等高线
inserting control 插入控制
inserting film 装片
inserting gooseneck tunnel longitudinal rails 鹅颈槽纵梁嵌补(集装箱)
inserting individual point 插点
inserting network 插网【测】
inserting optic(al) cable connector 插入光缆连接器
inserting tool 下线工具;嵌线工具
inserting top rail 上侧梁嵌补(集装箱)
inserting winding 嵌线
insertion 夹入;嵌装物;嵌装件;嵌填物;衬垫;插页
insertional growth 插入生长
insertional translocation 插入易位
insertion breakage 插入断裂
insertion character 插入字符
insertion class 插入类型
insertion coefficient 插入系数
insertion compound 插入化合物
insertion connection 插入连接
insertion criterion 插入规则
insertion editing 插入编辑
insertion funnel 插入式漏斗
insertion gain 介入增益;接入增益;插入增益
insertion group 嵌入集团
insertion head 插入装置
insertion into deque 插入双排队;插入到双向队列中
insertion into thread tree 插入到线索树;插入到穿线树
insertion into tree 插入到树
insertion lace 镶边花边
insertion length 插入长度
insertion liner 镶入式缸套
insertion loss 介入损耗;介入衰减;接入损耗;插入损失;插入损耗
insertion machine 插入机
insertion mutation 插入突变
insertion of node 节点的插入
insertion of plug 装入堵头
insertion of stack 堆栈的插入
insertion operation 插入操作
insertion picture character 插入图形字符
insertion piece 垫块;嵌块;插入件
insertion point 插入点
insertion potential 插入电位
insertion power function 介入功率函数
insertion ratio 介入比;插入系数
insertion reaction 插入反应
insertion sequence 插入顺序;插入序列
insertion sort 插入分类;插入排序
insertion spanner 镶接扳手;接头扳手
insertion switch 引入开关;插入开关
insertion test signal 插入测试信号
insertion thermostat 浸入式恒温计;插入式恒温计
insertion turbine flowmeter 插入式透平流量计
insertion-type thermocouple 插入式热电偶
insertion-type vibrator 插入式振捣器

insertion-type vortex flow transmitter 插入式涡流变送器
insertion waves 细波纹;网波纹
insert jib section 中间臂节
insert joint casing 嵌接套管;插接套管
insert length 嵌入长度
insert liner 镶嵌式轴衬
insert map 插图
insert mo(u)lding 镶嵌造形
insert network method 插网法
insert nozzle 镶装喷管
insert pattern 组合模板
insert pin 插销
insert plate valve 插板阀
insert point method 插点法
insert pump 杆式泵
insert punch 嵌入冲头
insert reaming shell 镶有金刚石条的扩孔器
insert ring 可溶镶块;嵌条
insert rock bit 镶嵌钎头
inserts 镶嵌件
insert set bit 镶切刃岩芯钻头
insert stressmeter 插入式应力计
insert sub 插入接头
insert subcommand 插入子命令
insert subroutine 插入子程序
insert the casing 下套管
insert tube 插入管
insert-type reaming shell 金刚石条带嵌入型扩孔器
insert-type turbine flowmeter 插入式涡轮流量计
insert valve 嵌入式阀;插入阀
insert vane pump 插入式叶片泵;插入式叶轮泵
insert width 嵌入宽度
insert with gate valve 带阀的管接头
in-serve behavior 使用品质
in service 运行中;在使用中;运转中
inserviceable 不合用的
in-service behavior 工作状态;工作性能;使用情况
in-service carbonation 进行中的炭化作用
in-service date 开始使用日期;投产日期
in-service freight car maintenance depot 装卸检修所【铁】
in-service installation 在运行中安装
in-service life 使用寿命
in-service personnel 在编人员
in-service test(ing) 实用试验
in-service trainine 在职训练;在职培训
in-service treatment 原地处理
in-service use 实际使用
inset 插页;插入晶;斑晶
inset balcony 内阳台;嵌入式阳台
inset fire 嵌入壁炉采暖炉
inset grate 嵌入式格栅;嵌入式帘格
inset hinge compensator 内移铰链轴补偿器
inset inserting machine 插页机
inset light 嵌入灯
inset sink 嵌入式洗涤盆
insetting machine 插页机
in setting safety valves 在安装安全阀时
inset transmitter 插入式话筒
inset type 镶入式;嵌入式
in sextuplicate 一式六份
inshave 弧刮刨
in shear 受剪
in shifts 轮班
inshore 沿海的;近岸;水下岸坡
inshore bottom contour 沿海地形;近海海底地形
inshore breeze 向岸风
inshore current 向岸水流运动;向岸流;近海海流;近滨流;近岸流
inshore environment 沿海环境
inshore fishery 近海渔业;沿海渔业
inshore fishes 沿岸渔业
inshore fishing 沿海渔业;沿海渔业
inshore fishing ground 沿海渔场
inshore flow 近岸流
inshore hydrographic(al) survey 沿海水文测量;近岸水文测量
inshore hydrography 近岸水文测量学
inshore ice 近岸冰
inshore longitudinal girder (减海台式码头的)内岸纵主梁
inshore marine environment 沿海海洋环境;近海海洋环境

inshore navigable zone 沿岸通航区;沿岸通航带
inshore navigation 沿海航行
inshore oceanographic(al) survey 沿岸海洋调查
inshore sample 近岸试样
inshore shoal 近岸沙洲
inshore spawning ground 沿海产卵场
inshore station 沿海台站
inshore survey 近岸测量
inshore traffic zone 沿岸通航区;沿岸通航带
inshore trawler 近海拖网渔船
inshore water 近岸水
inshore waters 沿岸水域;近岸水域
inshore wind 向岸风
inshore zone 近海带;近岸区;外滩地段
in short supply 供不应求
inshot 跟进装置
inshot burner 水平火焰燃烧器
inside 内舷;内侧;内部的;内部
inside admission 内进汽(蒸汽机)
inside air cooler 室内冷气机
inside air temperature 内部空气温度
inside and outside grinding 内外磨削
inside-angle tool 修饰阴角的镘板;阴角抹子;内角抹子
inside antenna 内部天线
inside arch 内拱
inside arm 固定在舱口上空的吊杆
inside axle box 内轴箱
inside back 里封底
inside bag 单人呼吸器的呼吸囊
inside band 活动内托条;砂型加固圈
inside bank 内坡
inside bank slope 河岸内坡
inside bath(room) 内浴室
inside bevel(led) chromium piston ring 内倒角镀铬活塞环
inside bevel(led) piston ring 内斜面活塞环
inside bevel(led) taper-face piston ring 内倒角锥形活塞环
inside blade 内侧刀齿
inside brake 内制动器;内制动
inside bridge wall 内桥砖
inside broach 内拉刀
inside broker 正式经纪人
inside building board 内部建筑板
inside butt strap 内对搭接板;平接内贴板
inside cabin 内侧舱事;不靠船舷的舱位
inside cable 内部电缆
inside cal(l)ipers 内径卡尺;测孔器;内卡钳;内径卡规;内(侧)径规
inside cal(l)ipers ga(u)ge 内径千分(卡)尺
inside cal(l)ipers micrometer 内卡钳千分表
inside cap 内轴承盖
inside casing (门窗的)内贴脸;内压条;内框
inside ceiling panel 内顶板
inside cellar wall 内部隔墙
inside centering ring 内定心环
inside chaser 内螺纹梳刀
inside circumference 内圆周
inside clearance 内余隙;内径隙规;内侧尺寸
inside clearance ratio 内净径比;内间隙比;内间距比;内(部)净空比
inside clinch 内旋结;内绳扣
inside coat 室内墙顶涂层;内层;里层
inside coating 内壁涂层
inside column 内柱
inside conductor 内导线
inside connecting rod 内连杆
inside-controlled system of railway enterprise 铁路企业内部控制制度
inside cord 内绳
inside core 内芯
inside corner 室内墙隅;阴角;内圆角子(修型工具)
inside corner mo(u)lding 阴角压条;内角镶条;内角线条
inside corner trowel 阴角抹子
inside corner weld(ing) 内角焊(接)
inside counting 内计数
inside-coupled casing 接头连接套管
inside court 内院
inside courtyard pipe network 院内管网
inside cover 里封面
inside covering 内套(管)

inside crank 内曲柄
inside crankshaft 内曲轴
inside cross-section 衬砌断面
inside cross-section of tunnel 隧道净断面
inside curve 内曲线
inside curved lead rail 内曲合拢轨
inside cutter 切管器;切管机
inside cutting blade 内切刀齿
inside damper 内部阻尼器
inside dam power station 坝内式水电站
inside decorating 内部装修
inside dial indicator 内径千分表;内径千分(卡)尺
inside diameter 内直径;内径;去皮直径
inside diameter character 内径代号
inside diameter kicker 内径掏槽钻
inside diameter of collars 钻铤内径
inside diameter of drill pipe 钻杆内经
inside diameter of kelly 方钻杆内径
inside diameter of vessel 容器内径
inside diameter saw 内圆锯
inside diameter slicer 内圆切片机
inside dimension 内尺寸;内部尺寸
inside director 内部董事
inside ditch 内侧沟
inside door 内门
inside-door lock 房门锁
inside dozer 内侧推土机;内侧刮铲
inside dozer blade 内侧推土机刮铲
inside drainage 内排水
inside drainage system 内排水系统
inside drill 侧孔电钻
inside-drum filter 内滤式过滤器;内鼓式过滤器
inside embankment toe 内侧堤坝坡脚;内侧坝趾
inside-emptied brick 中空砖
inside engaged gear 内啮合齿轮
inside exhaust pipe 内排气管
inside expanding brake 内胀式制动器
inside face (钻头的)底唇面
inside facing 内衬;室内饰面
inside fastening 内钉法
inside fiber board finish 内部纤维板装修;内部纤维板装饰
inside fillet 内角焊缝
inside fillet weld 内角焊
inside fillet welding 内角焊接
inside film coefficient 内部传热膜层系数;内表面散热系数
inside finish 内(部)装修
inside fixtures 室内装置
inside flange 内法兰
inside force 内侧力
inside-frosted bulb 内磨砂灯泡
inside-frosted lamp 内磨砂玻璃灯;乳白(色)灯泡;内阁光(毛玻璃)灯泡;内表面阿光灯泡
inside gallery 内走廊
inside gallery apartment building 内走廊式公寓建筑
inside ga(u)ge 内径量规;内径规
inside-ga(u)ge stone 内侧保径金刚石
inside gear 内啮合
inside gear asphalt pump 内啮合齿轮沥青泵
inside gearing 内啮合
inside glass door 内玻璃门;室内玻璃门
inside glazing 内部安装玻璃法;内镶玻璃
inside gloss(clear) varnish 内部上光清漆
inside groove 内槽
inside guard rail 内护轨
inside handle 室内门执手;内把手
inside handrail 内扶手(栏杆)
inside hardboard finish 内侧硬质板饰面
inside heat 内部供暖
inside height 内高
inside humidity 室内湿度
inside illumination 内部照明;室内照明
inside indicator 内径测微指示计
inside info 内部情报
inside information 内部资料
inside insulating building 内绝热建筑
inside insulation 内绝缘;内部绝热;内部隔声
inside jamb block 内侧墙基砖
inside jaw chuck 内爪卡盘
inside joinery 室内细木工
inside kicker (金刚石岩心钻头的)内缘钻石;(钻头的)内侧刃金刚石

inside kitchen 屋内厨房;内厨房
inside lag 决策时间滞差;内在时间滞差;内部滞后
inside lap 内余面;内搭接;排气余面;乏气余面
inside layer 内层
inside layout 室内规划;室内布置
inside lead 内螺纹导程
inside lead circle of pile 板桩内导向圈
inside leader 内水落管
inside lead ga(u)ge 内螺纹导程(螺距)仪
inside lighting 室内采光
inside lining 内框;内(部)衬砌;内衬;衬里
inside lining board 内衬板
inside link 内拉杆
inside lip surface 内唇面
inside lot 内侧地块;不沿城市道路的地块
inside market 内部市场
inside masonry dividing 内部围砌分隔墙
inside measurement 内尺寸
inside micrometer 内径千分(卡)尺;内径测微器;内径测微计
inside micrometer cal(l)ipers 内测千分卡尺;内径千分(卡)尺
inside micrometer ga(u)ge 千分内径规
inside mix nozzle 内部混合喷嘴
inside money 内部货币
inside noise 室内噪声
inside noise criterion curve 室内噪声标准曲线
inside nominal diameter 公称通径
inside of a fouling point 警冲标内方【铁】
inside of a fouling post 警冲标内方【铁】
inside of blades 叶片嵌装
inside of curve 曲线内侧
inside of the levee 堤内边
inside or outside water channels 内外水槽
inside or outside water ways 内外水槽
inside-out 里面向外翻
inside-out filter 外流式过滤器
inside-out motor 旋转电枢式同步电动机;反结构同步电动机
inside-out redrawing 反向再拉延;反拉延
inside-outside effect 内外效应
inside-packed type piston pump 内部填塞式活塞泵
inside packing 内包装
inside paint 底漆
inside paint coat 底漆层
inside pane 双层玻璃窗的室内侧玻璃;室内侧玻璃
inside partitioning 内部分隔
inside passage 内线航道
inside pitch line length 齿根高度
inside plant 同一建筑物内的;内站;内部通信装置;室内线缆;室内设施
inside pressure ga(u)ge 舱内压力表
inside pressure pneumatic system reinforced by cable 钢索加固的内气压系统(充气建筑)
inside pressure pneumatic system reinforced by membrane ribs 膜肋加固的内气压系统(充气建筑)
inside pressure pneumatic system with interior anchor points 带内部锚固点加固的内气压系统(充气建筑)
inside price 内部价格;批发价(格)
inside race 内套圈
inside radius 内半径
insider dealing 内幕交易;幕后交易;秘密交易
inside reamer (钻头的)内侧刃金刚石
inside rear view mirror (车身的)内部后视镜
inside recess 内凹座
inside recessing tool 内凹槽车刀
inside redecoration 内部重新装修;内部重新装饰
inside reveal (门窗的)内侧壁;内部外抱;内部窗侧
inside rolling hitch 内转结
insider trading 内线交易;内幕交易
inside scaffold 室内脚手架
inside screw 螺母
inside screw and non-rising stem 暗杆内螺纹
inside screw and rising stem 明杆内螺纹
inside screw rising stem 内螺纹明杆
inside seal(ing) 内部填塞;内部封闭
inside shape 内部净空断面形状
inside shedding 内侧开口运动
inside shoe 内履
inside shoe brake 内蹄式制动器
inside skin 内皮;内部罩面层

inside slatted blind 内层百叶遮阳;窗内百叶遮阳;内部木条百页窗
inside slip 内滑
inside slope (水库、水池的)内侧斜坡;暗斜井;井下斜井;内坡
inside spider 内支撑片
inside spin 内螺旋
inside spline ga(u)ge 内花键量规
inside square-corner slick 内直角光子
inside stair(case) rail 内楼梯拦杆
inside step design 内台阶设计
inside stone (钻头的)内侧刃金刚石
inside stop 内压条;墙内木桩;墙内木砖;门窗上动条;窗框贴脸板
inside strake 内列板;内刮板
inside stud 隔墙竖筋;内门窗梃
inside surface 内表面;木材向心面
inside surface area 内表面积
inside surface temperature 内表面温度
inside system layout 室内系统布置
inside tap 打捞公锥
inside taper ga(u)ge 内锥度量规
inside tappet 内侧踏盘
inside thread 内螺纹
inside tile 内砖
inside tire 内轮胎
inside tracery 内窗(花)格
inside track 有利地位;里圈;内轨
inside treadle motion 内踏盘运动(指开口运动)
inside trim (门窗的)额楣;内部整修;门窗内镶边;内框;内部镶边
inside tube 内管
inside tube diameter 内管直径
inside tube wall 内管壁
inside tunnel controlling survey 峒内控制测量
inside turn 内转弯;内盘旋
inside turning radius 内部回转半径;内转弯半径
inside union 内部工会
inside upset 内加厚
inside vapo(u)r phase oxidation process 管内气相氧化法
inside view 内视图
inside wall 内墙
inside wall frame 内墙框架
inside wall of cylinder 缸内壁
inside wall sill 内墙基石
inside wall surface 内墙表面
inside waterproofing 内防水
inside waters 内陆水域
inside weld 内焊缝
inside wheel turning angle 内转向轮转向角
inside width 内宽
inside window 内窗
inside window cill[sill] 内窗槛
inside wiring 室内线路
inside work 坑道内作业;塔内作业;室内工作
insight 理解;见识;洞察力
in sight of one another 互见中
insignia 证章;徽章;标志
insignificant 无意义的
insignificant error 不显著误差
insignificant variable 不显著变量
insinkability 不沉性
insinkable test 不沉性试验
insipid 无味的
insistent 固立的
in-site 在施工现场
in-site analysis 现场分析
in-site data 现场数据
in-site oxidation 就地氧化
in-situ 原位置;在原位;在原地;当地;原位;原地;在(施工)现场
in-situ accessibility 就地修理
in-situ activation analysis 现场活化分析
in-situ aerated concrete 现场浇注加气混凝土
in-situ aeration concrete 现浇加气混凝土
in-situ aeration well 就地曝气井
in-situ analysis 现场分析
in-situ artistic terrazzo 现制美术水磨石
in-situ assimilation and contamination 就地同化混染作用
in-situ bio-denitrification system 就地生物滤池脱氮系统;就地生物滤池反硝化系统
in-situ biofilm 原位生物膜

in-situ biological groundwater denitrification 就地生物地下水脱氮;就地生物地下水反硝化
in-situ biological treatment 原位生物处理
in-situ bioremediation 原位生物修复
in-situ brickwork 现场砌筑砖工;就地砌砖
in-situ cable duct 工地电缆槽;现浇电缆沟
in-situ California bearing ratio test 原位加州承载比试验
in-situ cap 现浇桩帽
in-situ-cast 现浇;现场浇筑
in-situ-cast aerated light concrete 现浇加气轻质混凝土
in-situ-cast aerated mortar 现浇加气砂浆
in-situ-cast aerated reinforced concrete 现浇加气钢筋混凝土
in-situ-cast concrete 现浇混凝土
in-situ-cast concrete floor 就地浇注混凝土楼板
in-situ-cast construction method 现浇施工方法
in-situ-cast foaming 现场模制泡沫
in-situ casting work 现场浇筑工程;现场浇注工程;现场浇灌工程
in-situ-cast mortar 现灌砂浆
in-situ-cast pile 就地浇注桩
in-situ-cast prestressed concrete pile 现浇预应力混凝土桩
in-situ-cast stair(case) 现浇浇注楼梯间;现浇混凝土楼梯
in-situ-cast string 现浇混凝土楼梯斜梁
in-situ-cast topping 现浇面层
in-situ CBR test 原位加州承载比试验
in-situ check and test method 现场检测方法
in-situ chemical treatment 原位化学处理
in-situ coal gasification 煤地下气化;煤的地下气化
in-situ combustion 地下燃烧法;(采掘石油时的地下)现场燃烧
in-situ composite 定向凝固的共晶体复合材料;原位复合材料
in-situ concrete 原地混凝土;就地混凝土;现浇混凝土;现场(浇注)混凝土
in-situ concrete filling 混凝土灌缝
in-situ concrete pile 现浇混凝土桩;就地灌注桩;现场钻孔灌注桩
in-situ concrete rib 现浇混凝土肋
in-situ concrete shell 现浇混凝土薄壳
in-situ concrete slab 现浇混凝土板
in-situ concrete structure 现浇混凝土结构
in-situ concreting 混凝土现场浇注
in-situ condition 天然条件
in-situ conservation 就地保护
in-situ construction 现浇建筑(工程)
in-situ conversion 就地改建
in-situ cube strength 钻芯折算混凝土立方体强度
in-situ curb beam 现浇(混凝土)侧石梁;就地(浇筑)路缘梁
in-situ denitrification activity 就地反硝化活力;就地生物脱氮活力
in-situ density 现场密(实)度;原state密(实)度;原位密度;原地密度(不采样测定的密度);实体密度
in-situ determination 就地确定
in-situ developed dyes 原地显色染料
in-situ differentiation 就地分异作用
in-situ direct shear test 原位直剪试验
in-situ dynamic(al) compaction procedure 原位动力压实法
in-situ effective overburden pressure 原位有效上覆压力
in-situ effective pressure 原位有效压力
in-situ field measurement 原地野外测量
in-situ filling 现浇(混凝土)填料;现场填充;就地装填
in-situ floor 现浇地面
in-situ floor finishes 现浇筑地板
in-situ flooring 现浇地板;现制楼地面
in-situ foaming 就地发泡;现场发泡
insituform 薄管套衬里
in-situ formation 原地形成
in-situ gravel 原地砾石
in-situ impulse test 原位激振试验
in-situ inspection 现场检验
in-situ landing 现浇平台;就地着陆
in-situ leaching 原地浸出
in-situ lightweight-aggregate concrete 现浇轻骨料混凝土
in-situ lining 现场衬砌;现场衬里

in-situ loading test 现场荷载试验
in-situ measurement 原位测量;现场量测;现场测量;下方(的)测定;原位测试
in-situ mixer 就地拌和机
in-situ modulus 现场测量模量
in-situ moisture meter 现场用湿度计
in-situ monitor(ing) 现场监测;原位监测
in-situ mortar 现浇灰浆;就地(制)灰浆
in-situ overburden pressure 原位上覆压力
in-situ oxidation 就地氧化
in-situ permeability test 原位渗透(性)试验;渗水试验
in-situ pier 现浇桥台;现浇墩子;现浇防波堤
in-situ pile 现浇混凝土桩;现浇桩;就地灌注桩
in-situ pile cap 现浇桩帽
in-situ piling machine 钻孔灌注桩机;钻孔灌注桩打桩机
in-situ pipeline 现浇管道
in-situ pipe lining 就地敷设管道
in-situ placed wall 地下连续墙
in-situ plastics flooring 现制塑料楼地板
in-situ polymerization 原地聚合;现场聚合
in-situ porosity 原始孔隙率;原始孔隙度
in-situ pouring 混凝土现场浇注
in-situ precipitation 原地沉淀
in-situ precipitation method 原位沉淀法
in-situ preparation 就地制造
in-situ pressiometer 原位压力计;现场压力计
in-situ pressure meter 现场压力计
in-situ production 下方产率
in-situ quantification 现场定量
in-situ quantitation 原位定量
in-situ recovery 原地回收
in-situ reinforced concrete 现场浇混凝土;现浇钢筋混凝土
in-situ reinforced concrete bridge 现浇钢筋混凝土桥梁
in-situ reinforced concrete floor 现浇钢筋混凝土楼板
in-situ reinforced concrete slab 现场浇混凝土板
in-situ resource 实地资源
in-situ rib 就地浇注肋
in-situ rock mass 原岩体
in-situ rock mechanics 岩石力学现场研究
in-situ rock stress 原地岩石应力
in-situ sampling 就地采样
in-situ sand 现场制沙
in-situ shear sounding apparatus 现场剪切触探仪
in-situ shear test 现场剪切试验
in-situ shell pile 现浇包壳(混凝土)桩
in-situ soil 现场土;原位土(壤);原地土(壤)
in-situ soil mass test 土体原位测试
in-situ soil-shearing test 原状土剪切试验
in-situ soil strength 现场土强度
in-situ soil test 原地土壤试验;现场土壤试验;原位土工试验
in-situ solid concrete floor 现浇实心混凝土楼板
in-situ stabilized column 原位加固土桩
in-situ strain 现场测定应变
in-situ strength 现场强度;原状强度;原位强度;自然条件下的强度
in-situ stress 原岩应力;现场(测定)应力;原位应力
in-situ stress measurement 现场应力测试;现场应力测量
in-situ terrazzo 现制水磨石;现浇水磨石
in-situ terrazzo border 现制水磨石镶边
in-situ terrazzo facing 现制水磨石面
in-situ terrazzo flooring 现制水磨石楼地面
in-situ terrazzo skirting 现制水磨石踢脚
in-situ test(ing) 原位试验;工地试验;现场试验;现场测试;原位测试;就地试验
in-situ test of rock mass 岩体原位试验
in-situ theory 原地生长说
in-situ thrust test 原位推裂试验
in-situ topping 就地封顶
in-situ tracer 原位示踪剂
in-situ treatment 原位处理
in-situ tunnel 现浇隧道
in-situ uniaxial compressive test 原位单轴抗压试验
in-situ unit weight 原位容重
in-situ vitrification 就地玻璃化法
in-situ void ratio 原位孔隙比
in-situ volume 下方【疏】量

in-situ water content 原位含水量
in situ weathered material 原地风化物质
insizwaite 等轴铋铂矿
insley plant 现场混凝土构件厂;(提升和浇灌混凝土的)机械装置
inslope 内坡
in-slot signal(l)ing 时隙内信号方式
in snatches 断断续续地
insoak 未饱和土壤对地表自由水的吸收
insolation 日照;日射能量;日射率;日射;曝晒;晒干
insolation area 日照面积
insolation duration 日照时间;日照持续时间
insolation weathering 日晒风化
insolilith 日晒风化砾石
insolubility 非熔性;不溶(解)性;不可溶(解)性
insolubilize 不溶解
insolubilizer 不溶黏料
insolubilizing process 不溶性处理
insoluble 难溶的;不溶解(的);不可溶的
insoluble anode 不溶性阳极
insoluble azo colo(u)r 不溶性偶氮颜料;不溶偶氮染料
insoluble azo dye 不溶性偶氮染料
insoluble barite 不溶性重晶石
insoluble colo(u)rant 不溶性色料
insoluble compound 不溶性化合物;不溶化的物质
insoluble film 不溶性膜
insoluble granule 不溶性颗粒
insoluble humic acid 不可溶腐殖酸
insoluble hydroxide 不可溶性氢氧化物
insoluble impurities 不溶性杂质
insoluble in water 不溶于水的
insoluble material 非溶解质;不可溶性物质;不可溶物质
insoluble matrix type anti-fouling paint 基料不溶型防污漆
insoluble matter 不溶物(质);不溶解物(质);不可溶性物质
insoluble mineral 不可溶性无机物
insoluble monolayer 不溶性单层
insoluble organic dyestuff 不溶性有机染料
insoluble pigment 不溶性颜料
insoluble polymeric contact disinfection 不可溶性高分子接触消毒剂
insoluble quaternary ammonium salt cation ion resin bactericide 不可溶性季铵盐阳离子树脂杀菌剂
insoluble residue 不溶性残渣;不溶(性)残余物;不溶(解)残渣;不溶残积;不可溶残渣;不可溶残余物
insoluble residue content 水不溶物含量
insoluble residue in acid 酸不溶物
insoluble residue in water 水不溶物
insoluble residue line 不溶积线
insoluble resins 不溶树脂
insolubles 不溶物(质)
insoluble salt 不可溶性盐
insoluble salts 不溶盐类
insoluble scale 不溶性屑垢
insoluble sludge 不溶残渣;不可溶性污泥
insoluble soil 不溶性沾污物
insoluble starch 不溶性淀粉
insoluble substance 不溶性物质
insoluble substance content 酸不溶物含量
insoluble suspension 不可溶性悬浮物体
insoluble tar 沉淀焦油
insoluble toxin 不溶性毒素
in solution environment 在溶解环境中
insolvable 不能偿还
insolvency 破产;无力偿还;无力偿付;无偿付能力
insolvency clause 破产条款
insolvency guaranteed fund 破产担保基金
insolvent 破产者;无力偿还者;无偿还能力的
insolvent debtor 无偿债能力的债务人
insolvent laws 破产法
in some detail 稍为详细的
insonification 声照射;声透射;声穿透
insonified zone 有声区;水下声音传播区;声音传播区
insonify 声穿透
in sound condition 情况正常
in soundings area 在有测量水深的地区
inspect and repair only as needed 仅在必需时进行检修

inspected 已检查
inspected approval 业经验货同意
inspected area 探伤范围
inspected point 检查点
inspecting chart 检查表
inspecting commission 检查费
inspecting engineer 验收工程师
inspecting eye 检查孔
inspecting hole 检查孔;探孔
inspecting item 检验项目
inspecting of construction 施工检查
inspecting period for monitoring control network 检测周期
inspecting pit 探坑;检查坑
inspecting port 检视口;检查孔
inspecting shaft 检查孔
inspecting stair(case) 检查梯
inspecting stand 检查台
inspecting tag 检查标签
inspection 考察;校对;检验;检查;调查
inspection acceptance of finished work 竣工验收
inspection after construction 施工后检查;成品检验
inspection agency 检验代理商
inspection and acceptance 验收
inspection and analysis of the plan 计划的检查分析
inspection and certificate fee 检验证明费用
inspection and checkout 检查与测试
inspection and claim 检查和索赔
inspection and claim clause 检查和索赔条款
inspection and maintenance 检查和维修
inspection and maintenance equipment 检修设备
inspection and quality control 检查与质量控制
inspection and repair as necessary 按需进行检修
inspection and repair of buildings and structures 房建检修
inspection and repair shed for bogie 转向架检修间
inspection and repair trouble 检修不良
inspection and testing 检查和检验
inspection and testing company 检验公司
inspection and testing of imports and exports 进出口商品检验
inspection and test of cargo gear 起货设备检验和试验
inspection assorting process 检查分选法
inspection before assembling 组装之前检查
inspection before delivery 交货前检验
inspection bench 检验台;检查台;产品检验台
inspection bench for bogie 转向架检修台
inspection between process 工序间检验
inspection body 检验机构
inspection(bore)hole 检查钻孔
inspection box 观察箱
inspection bureau 检查所;检查局
inspection by attributes 品质检验;按固定指标检验
inspection by certificate 合格证检查
inspection by sampling 取样检验
inspection by variable 变量检验
inspection card frame 检查卡插座
inspection car of railway track 铁路轨道检查车
inspection certificate 检验合格证书;检验证(明)书;检查证(明)书
inspection certificate of origin 产地检验书
inspection certificate of quality 质量检验证书;品质检验(证)书
inspection certificate of quantity 数量检验书
inspection certificate of weight 重量检验书
inspection certificate on damaged cargo 验残检验书;残损货物检验证书
inspection chamber 检验室;检查室;检查井;人孔
inspection charges 检验费
inspection chart 检验表
inspection clause 检查条款
inspection code 检修指示书;检修规范
inspection code for special equipment 特殊设备检验规范
inspection cover 检查孔盖;检查孔板
inspection cover for nozzle 喷嘴检查盖
inspection culvert 检查廊道;检查涵洞
inspection delegation 考察团
inspection department 检验部门
inspection device 检验装置
inspection directorate 稽查人(职务)

inspection door 检查孔;观察孔;观察门;窥视口;检查门;检查口
inspection during construction 在建工程检验
inspection during fabrication 在制造期间检查
inspection enlarger 检查放大机
inspection equipment 检验设备
inspection examination 查验
inspection expenses 检查费
inspection eye 检查装置;管道清扫孔;检查口;检查孔
inspection fitting 检查口;检查接头;活接头;检验装置;检修用的接头;检查孔(板)
inspection for dredge pump wearing 泥泵磨损检查
inspection for leakage 渗漏检查
inspection free 免费检验
inspection gallery 检验廊;检查廊道;树查廊道
inspection ga(u)ge 验收规;检验量规;检查规
inspection glass 看火玻璃
inspection glasses 看火目镜
inspection hatch 检查口盖;检查窗
inspection hole 观室孔;检查口;检查孔;测视孔;检修孔
inspection hole cap 窥视孔盖
inspection hole flag 检查孔盖板
inspection hole for exhaust gas pressure 排气压力检查孔
inspection hole of ceiling 顶棚检查孔
inspection hole on roof 屋顶检查孔
inspection instrument 检验仪器
inspection item 检验项目
inspection jig 检验样板;检验模具;检验夹具
inspection junction 检查井;检查口;检查接管件(下水道支管汇接井)
inspection ladder 检查梯
inspection ladder of abutment 桥头检修梯
inspection lamp 修车灯;检验灯;检查灯
inspection launch 检查艇
inspection level 检验等级;抽检百分比
inspection light 检查灯
inspection list 承包工程项目清单
inspection machine 校正机
inspection manhole 进入孔;检查孔;检修人孔
inspection manual 检查手册
inspection map of anomalies 异常检查平面图
inspection map of mineralization point 矿点检查图
inspection memorandum 检查记录
inspection of anomaly 异常检查
inspection of blank 毛坯检验
inspection of chemical equipment 化工设备检验
inspection of construction 施工检查
inspection of contracted-out parts 协作件检验
inspection of edge preparation 坡口检查
inspection office at the port 口岸检查机关
inspection of fixed assets 固定资产检查
inspection of foundation subsoil 验槽
inspection of goods 买方检查货物
inspection of hull lofting 船体放样检验
inspection of line 线路检查
inspection of lining 衬砌检查
inspection of materials 材料检验
inspection of mineralization point 矿点检查
inspection of optic(al) crystal 光学晶体检验
inspection of plate edges 板边检查
inspection of quality conformity 质量一致性检验
inspection of quantity 数量检验
inspection of rail 钢轨检验
inspection of risks 被保险物件检查
inspection of site 视察现场;现场勘察
inspection of the boiler 锅炉检验
inspection of track 线路检查
inspection on damaged cargo 货物残损检验
inspection on spot 现场检验
inspection opening 检查孔
inspection outline 检验大纲
inspection pane 窥视孔;窥视玻璃孔
inspection peep hole 检查视孔
inspection percentage 抽检百分比
inspection photoelectric(al) quality 产品质量的光电检查
inspection pig 管道内检测器
inspection pipe 检查管
inspection pit 检验井;检修坑;检查坑;检查井;探坑;探井

inspection pit track 机车检查坑线
inspection plan 检验大纲
inspection plate 人孔盖;检查孔盖(板);监视孔板
inspection platform 检验平台;检查用走廊
inspection plug 检查孔塞;观察孔塞
inspection pocket 检查袋(铸铁管上可取下的盖板便于疏导)
inspection point 检查点
inspection port 观察口;检查孔
inspection pressure 检查压力
inspection procedure specification 检查规程
inspection provision 检验准备;检验规定
inspection rack 汽车检修台
inspection report 巡检报告;鉴定报告;检验报告
inspection requirements 检查要求
inspection requirements manual 检查要求手册
inspection result 检查结果
inspection result of main dimension 主要尺寸检查结果
inspection result of main item 主要项目检查结果
inspection routine 检查程序
inspection schedule 检查图表
inspection screw 检查螺钉;检查孔螺旋盖
inspection shaft 检查(竖)井;检查孔;检查坑
inspection sheet 检验样;检验单
inspection sieve 检验筛
inspection skid 检查台
inspection staff 检验人员
inspection stair(case) 检查梯
inspection stamp 检验印记
inspection standard 验收标准;检验标准;检修标准
inspection table 检验台;检查表
inspection tap 检查室
inspection technique 检查技术
inspection test 复核试验;收货检验
inspection ticket 检验单
inspection trolley (桥梁、线路的)查道车;检查用吊车;检查手摇车
inspection truck for construction ga(u)ge 建筑限界检查车
inspection tube 检查管(道)
inspection tunnel 检查隧道
inspection well 观测井;检查井;探井
inspection window 观察窗;检验窗;检察窗;检查窗
inspector 验收员;检验员;检查员;监工;施工检查员;产品检查员
inspectorate 调查处;检查员职责
inspector-general 总检查长
inspector-in chief 总检查长
inspector of material 材料检验员
inspectors trunk 监测线
inspectoscope 金属裂缝探伤器;检验镜;检查境;探伤镜;(射线透视的)违禁品检查仪
inspersed 渗入的
inspiration 吸气;汲入
inspirator 吸入器;吸气器;注射器;注入器;呼吸器;喷注器;喷射器;喷气注水器
inspirator burner 注射燃烧器
inspiratory 吸入的;吸气的
inspiratory capacity 吸气量
inspiratory flow rate 吸气流速
inspiratory force 吸气力
inspiratory phase 吸气相
inspiratory reserve volume 吸气储(备)量
inspiratory standstill 吸气停顿
inspiratory valve 吸气阀
inspire 吸入
inspired air 吸入气
inspirited burner 自动雾化喷嘴
inspirometer 吸气计;吸气测量计
inspissate 浓厚的
inspissated deposit 浓缩油藏
inspissated oil 浓缩石油
inspissation 蒸浓(法);浓缩作用
inspissator 蒸浓器
in spite of 不管
instability 不稳固性;不稳定性;不稳定(度);不安定性
instability at resonance 共振不稳定性
instability coefficient 不稳定性系数;不稳定度系数
instability condition 不稳定条件
instability constant 不稳定常数;不安定常数

instability due to sliding 滑移导致不稳定性
instability due to uneven settlement 不均匀沉降导致不稳定性
instability effect 不稳定性效应；非稳定效应
instability factor 不稳定因子
instability index 不稳定(性)指数
instability in pitch 纵向不稳定性
instability line 不稳定线
instability mode 不稳定型
instability neutrons 不稳定性中子
instability of gold market prices 市场金价的不稳定
instability rain 不稳定雨
instability shower 对流性阵雨；不稳定降雨
instability stress 不稳定应力
instability strip 不稳定带
instability theory 不稳定理论
instable 不稳固的；不稳定的
instable buoyancy equilibrium 不稳定的浮力平衡；浮选游不定平衡
instable frame 不稳定框架
instable governor 不稳定调节器
instable grout 不稳定浆液
instable layer 不稳定层
instable shower 不稳定性阵雨
instable slope 不稳定边坡
in stages 分阶段的
in staggered pattern 左右交错
install 设置；安设
installaion recess 安装孔洞
installation 装置(设备)；装备；计算站；机械设备；埋设；设置；设施；设备；安装
installation accuracy 安装精度
installation allowance 安置津贴
installation and checkout 安装与测试；安置测试
installation and debugging 安装与调试；安装调试
installation and equipment 设施和设备
installation and maintenance 安装和维修；安装和维护
installation and maintenance condition at field 现场安装和维修条件
installation and operation 安装和运转
installation and overhaul specification 安装及检修规范
installation and plumbing 安装和装管
installation and trial run 安装调试
installation capacity 装置容量；设备容量；安装能力
installation charges 装机费(用)；安装费(用)
installation cost 装置费；设备投资；设备费(用)；安装费(用)；安装成本
installation cost as per machine capacity 单位容量安装成本
installation damages 安装损伤
installation data 安装数据
installation data of air compressor 空压机装置数据
installation date 装竣日期
installation design standard 设备设计标准
installation deviation 安装偏差
installation diagram 装备图；安装图
installation dimension 设备尺寸；安装尺寸
installation drawing 装配图；装置图(样)；装备图；安装图
installation engineer 安装工程师
installation equipment 安装设备
installation error 安装误差；安置误差
installation exercise 安装作业
installation exit routine 安装出口程序
installation facility 安装设施
installation failure 安装故障
installation fee 开办费；安装费(用)
installation floater 安装临时工(人)
installation for centrifugally cast concrete 混凝土旋转浇灌设施
installation for manufacturing prepared roofing 组合屋面材料制造设备
installation for spun concrete 混凝土旋转浇灌设施
installation grant 安置津贴
installaion insurance 安装保险
installation kit 装置工具；装配工具；零配件
installation material 安装材料
installation method 装置法
installation microscope 安装显微镜
installation of bus line 母线安装
installation of cable 电缆敷设；电缆安装

installation of christmas tree 采油树安装
installation of cost accounting system 成本会计制度的设置
installation of equipment 设备安装
installation of ga(u)ging station 测站装置
installation of heating 暖气装置
installation of machine and equipment 机器和设备的安装
installation of pile 桩的设置
installation of track 轨道敷设
installation of well-points system 井点系统安装
installation on vessels in service 容器在役时的安装
installation parts list 安装零件清单
installation pipe 用户管道
installation pipework 用户屋内设备
installation power 安装功率
installation process drawing 安装过程图
installation processing control 计算站处理控制；设备处理控制
installation production order 设备生产订货
installation productivity option 计算中心生产率选件
installation project 安装项目
installation rate 安装速度
installation riser 户内主管(道)
installation rule 安装规则
installation site 安装位置；安装地点；安装场所
installation size 安装尺寸
installations of pump 水泵安装
installation specification 安装说明书；安装检修规程；安装规范；装配指导书
installation squadron 安装中队
installation stress 安装应力
installation supplying 设施供应
installation survey 安装测量
installation system 铺装方式
installation tape number 装配带号；装带号
installation temperature 安装温度
installation time 装配时间；安装时间
installation tools 安装工具
installation to pantograph 受电弓安装
installation unit weight 设备单位重量
installation utilization rate 设备利用率
installation verification procedure 安装验证过程
installation weight 装置重量；安装重量
installation work 安装工作；安装工程
install drawing 安装图
installed accounting 分期付款会计
installed account payable 应付分期账款
installed and interest charges 分期还本付息
installed bond 分期偿还债券
installed book 分期付款簿
installed business 分期付款业务
installed capacity 安装容量；安装能力；已安装能力；装置容量；装料容量；装机容量；设备容量；发电厂设备容量；额定功率；安装的设备能力
installed capacity of power station 电站装机容量
installed cash credit 分期付款的现款信贷
installed cost of project 新投资添置成本
installed distribution liquidation 分期分配结算
installed engine 装机发动机
installed engine power 装机容量
installed fittings 已安装配件；安装配件
installed generating capacity 装机容量
installed gross capacity 总装机容量
installed height of lifting bridge 吊桥安装高度
installed horse power 装置功率；装机功率力数
installed load 接入荷载；安装荷重；安装荷载
installed net horsepower 装机净功率
installed net torque 装机净力矩
installed plant capacity 已安装的工厂生产能力；电站装机容量
installed position and shapes 安装位置与形式
installed power 装机功率；安装功率
installed power per employee 全员动力准备水平；全员动力准备程度
installed user program(me) 安装的用户程序；用户安装程序
installed vacuum cleaner 固定式真空吸尘器
installed weight 安装后的重量
installed wheel capacity 涡轮机设备容量；装机轮容量
installer 安装者；安装工

install for centrifugally cast concrete 离心力浇筑混凝土的装置
installing 装置；阵列
installing brick 砌砖
installing column of sand 铺填砂井
installing pile shaft by excavation 挖孔灌桩法
installing remedial measure 安装补救措施
installing replacement pipe 安装更换管道
install machinery 安装机器
install machinery and equipment 安装机器和设备
install winnower 安装风选机
instalment 分期；分期信贷；分期付款
instalment account receivable 应收分期账款
instalment and interest charges 分期还本付息
instalment and margin purchase 分期付款和保证金购买
instalment basis 分期付款标准；分期付款法；分期付款偿还法
instalment bond 分期付款债券；分期偿还公司债
instalment business 分期支付营业
instalment buying 分期付款购买；分期付款购货
instalment certificates 分期付款收据；分期付款股票；分期付款单据
instalment contract 分期履行的合同；分期付款(售货)合同；分期分批履行的合同；定期弃行合同；分期付款契约
instalment credit 分期付款信贷；分期付款信贷；分期偿还贷款
instalment credit control 分期付款信贷控制
instalment credit selling 分期付款赊销
instalment delivery 分期交货
instalment distribution liquidation 分期派款清算
instalment financing 分期还款信贷；分期付款资金融通
instalment guarantee 分期付款保证函
instalment house 分期付款商店
instalment in full 全额分期付款
instalment insurance 分期缴费的保险
instalment loan 分期偿还的借款
instalment method 分期付款(偿还)法
instalment of arrears 分期偿还欠款
instalment paid 每期应付款项付讫
instalment payment 分期付款的支付；分期付款
instalment payment account 分期付款账户
instalment payment clause 分期支付条款；分期付款条款
instalment payment credit 分期付款信用证
instalment payment of tax 分期缴税
instalment plan 分期付款(购货)法
instalment plan system 分期付款销货
instalment premium 分期支付的保险费；分期缴付的保险费
instalment purchase 分期付款进货；分期付款购货
instalment receipt 分期付款收据
instalment receivable account 分期应收账款
instalment reducing depreciation method 分期递减折旧法
instalment reporting 分期付款报告
instalment sale 分期付款销售
instalment sales basis 分期付款销货法
instalment sales contract 分期付款销售合同；分期付款销货合同；分期分批的销售合同
instalment scrip 分期付款收据
instalment scrip book 分期付款收据簿
instalment selling 分期付款销售法；分期付款销货；分期付款出售
instalment shipment 分期装运；分批装运；分批装船
instalment store 分期付款商店
instalment system 分期缴付制度；分期付款销货
instalment terms 分期付款条款；分期付款条件
instalment trade 分期付款交易
instalmenttrading 分期付款交易
instal paid 付讫的分期支付款项
instanataneous speed regulation 瞬时速度调节
instance 事例；实例；场合
instant 速溶的；瞬间
instant acting 瞬时动作(的)；有效瞬间
instant adjustment 炉前调整
instantaneity 即时性；瞬时性
instantaneous 瞬时的
instantaneous acceleration 瞬时加速度
instantaneous access 立即访问；立即存取
instantaneous acoustic(al) pressure 瞬时声压
instantaneous acting relay 瞬时继电器；瞬动继电器

instantaneous action 瞬时动作
instantaneous adhesive 瞬间黏结剂
instantaneous aeration 瞬时曝气
instantaneous aerator 瞬时曝气器
instantaneous air speed 瞬时空气速度
instantaneous amplifier 瞬时放大器
instantaneous amplitude 瞬态振幅;瞬时振幅;瞬时幅度
instantaneous amplitude section 瞬时振幅剖面
instantaneous analyzer 瞬态分析器
instantaneous angular frequency 瞬时角频率
instantaneous angular speed 瞬时角速度
instantaneous angular velocity 瞬时角速度
instantaneous annealing point 瞬时退火点;瞬时熟炼点
instantaneous applied load 突然施荷载
instantaneous assembly 瞬时临界系统
instantaneous astronomical north pole 瞬时天文北极
instantaneous attracted armature relay 瞬动吸铁继电器
instantaneous automatic frequency control 瞬时自动频率控制
instantaneous automatic gain control 瞬时自动增益控制
instantaneous availability 瞬时可用性
instantaneous axis 瞬时轴
instantaneous axis of rotation 瞬时转动轴
instantaneous band 瞬时频带;瞬时带宽
instantaneous behavio(u)r 瞬时特性
instantaneous bending strength 瞬时抗弯强力
instantaneous billet energy 瞬时毛坯能量
instantaneous blast 同时放炮;瞬时爆破;瞬发爆破
instantaneous braking power 瞬时制动力
instantaneous braking ratio 瞬时制动率
instantaneous break 瞬时切断;瞬时断流
instantaneous breaker 速断开关
instantaneous calendar 瞬跳日历机构
instantaneous cap 即发雷管;瞬时或同步起爆雷管;瞬发雷管;顺发雷管
instantaneous capacity 瞬时功率
instantaneous carrying current 瞬时载流;瞬时极限电流
instantaneous center 瞬时中心
instantaneous center of acceleration 瞬时加速度中心
instantaneous center of rotation 瞬时旋转中心
instantaneous center of velocity 瞬时速度中心;速度瞬心
instantaneous circle 瞬时轨道圆
instantaneous closure of valve 阀的瞬时关闭
instantaneous code 立即码
instantaneous cohesion of frozen soil 冻土的瞬时内聚力
instantaneous combustion 瞬时燃烧
instantaneous communication 瞬时通信
instantaneous companding 瞬时压扩
instantaneous compandor 瞬时压缩扩展器;瞬时压扩器
instantaneous complete rejection of load 瞬时全部弃荷
instantaneous compliance 瞬时柔量
instantaneous compressor 瞬时压缩机
instantaneous concentration 瞬时浓度
instantaneous condition 瞬时条件
instantaneous contact 瞬时接触
instantaneous control 瞬间控制
instantaneous conversion 瞬时转化
instantaneous correlation matrix 瞬时相关矩阵
instantaneous corrosion rate 瞬时腐蚀(速)率
instantaneous couple 瞬时力偶
instantaneous coupling modulus 瞬时偶合模量
instantaneous current 瞬时电流
instantaneous current efficiency 瞬时电流效率
instantaneous cut 瞬时掏槽
instantaneous daily rate 瞬时日差
instantaneous damage 瞬时破坏
instantaneous damper current 瞬时阻尼电流
instantaneous data-transfer rate 瞬时数据传输率
instantaneous date and day mechanism 瞬跳双历机构
instantaneous date mechanism 瞬跳日历机构
instantaneous deflection 瞬时变位
instantaneous deformation 瞬时形变;瞬时变形

instantaneous delivery 瞬时流量
instantaneous depth of cut 瞬时切入深度
instantaneous description 瞬时描述
instantaneous detector 瞬时探测器
instantaneous detonator 激光雷管;瞬发雷管;顺发起爆管
instantaneous deviation control 瞬时偏移控制;瞬间偏移控制
instantaneous deviation indicator 瞬时频偏指示器;瞬时偏移指示器
instantaneous deviator stress 瞬时偏应力
instantaneous dipole moment 瞬时电极矩
instantaneous discharge 瞬时排放;瞬时流量;瞬时放电
instantaneous distance 瞬时距离
instantaneous distribution 瞬时分布
instantaneous drawdown 瞬时水位降低;瞬时水位下降
instantaneous effect 瞬时效应
instantaneous elastic deflection 瞬时弹性挠度
instantaneous elasticity 瞬时弹性
instantaneous elastic strain 瞬时弹性应变
instantaneous electric(al) detonator 瞬时电起爆器;瞬时(发)电雷管;瞬时电子引信;瞬时电子引爆剂;瞬发电雷管
instantaneous electric(al) water heater 快速热水器;瞬时电热水器
instantaneous electromotive force 瞬时电动势
instantaneous electronic scan(ning) 瞬时电子扫描
instantaneous elements 瞬时根数
instantaneous emf 瞬时电动势
instantaneous erection 瞬时安装(法)
instantaneous error 瞬时误差
instantaneous error measurement 瞬时误差测定
instantaneous exciter 瞬时激励器
instantaneous expander 瞬时扩展器
instantaneous exposure 自动快速曝光;瞬时曝光
instantaneous extension 瞬时延伸
instantaneous fading 瞬时衰落
instantaneous failure 瞬时破坏
instantaneous failure rate 瞬时失效率;瞬时故障率
instantaneous field 瞬时场
instantaneous field of view 瞬时视场
instantaneous firing 瞬时引爆;瞬时起爆;瞬发;瞬发爆发
instantaneous flash 瞬时闪火
instantaneous flow rate 瞬时流速
instantaneous(flow) velocity 瞬时流速
instantaneous Fourier spectrum 瞬时傅立叶谱
instantaneous freezing 瞬时冻结
instantaneous frequency 瞬时频率
instantaneous frequency-indicating receiver 瞬时频率指示接收机
instantaneous frequency section 瞬时频率剖面
instantaneous frequency stability 瞬时频率稳定度;瞬时频率度
instantaneous fulcrum 瞬时支点
instantaneous fuse 瞬发引信
instantaneous fusing 瞬熔;瞬发引线
instantaneous gas-oil ratio 瞬时油气比
instantaneous gas water heater 水流式煤气热水器;瞬时煤气热水器;快速煤气热水器
instantaneous geyser 直通式洗澡用锅炉;间歇泉;瞬时加热器;瞬时热水器
instantaneous ground resolution 瞬时地面分辨率
instantaneous head 瞬时水头
instantaneous heater 瞬时加热器
instantaneous heating 瞬时加热
instantaneous ignition 瞬时起爆;瞬发点火
instantaneous image 瞬时像
instantaneous impact fuze 瞬发撞击引信
instantaneous(im)pulse 瞬时脉冲;瞬时传冲
instantaneous intensity 瞬时强度
instantaneous killing factor 瞬时杀伤因素
instantaneous latitude 瞬时纬度
instantaneous linear velocity 瞬时线速度
instantaneous load(ing) 瞬时载重;瞬时荷载;瞬时负荷;瞬间载荷;瞬变荷载
instantaneous loading capacity 瞬时装载能力
instantaneous longitude 瞬时经度
instantaneous loss of shunting 瞬时分路不良
instantaneously changing 瞬跳
instantaneously changing simple calender 瞬变单日历

instantaneously operating apparatus 瞬时运行电器
instantaneously repeating communication satellite 即时转发型通信卫星
instantaneous magnitude 瞬时值
instantaneous mass 瞬时质量
instantaneous maximum current of feeding section 供电臂瞬时最大电流
instantaneous maximum wind 瞬时最大风
instantaneous maximum wind velocity 瞬时最大风速
instantaneous measurement 瞬间测量;瞬变测量
instantaneous measuring apparatus 瞬态测试仪器
instantaneous model 瞬时模型
instantaneous modulus 瞬时模量
instantaneous modulus of elasticity 瞬时弹性模量
instantaneous molal concentration 瞬时摩尔浓度
instantaneous movement 瞬跳机构
instantaneousness 瞬间
instantaneous neutron 瞬发中子
instantaneous noise 瞬时噪声
instantaneous number average degree of polymerization 瞬时数均聚合度
instantaneous operation 瞬时动作
instantaneous orbit 瞬时轨道
instantaneous outburst 瞬时喷出;瞬时爆发
instantaneous output 瞬时输出
instantaneous overcurrent relay 瞬动过电流继电器
instantaneous overload 瞬时过载
instantaneous overvoltage 瞬时过电压
instantaneous overvoltage capacity 瞬时耐压容量
instantaneous oxygen demand 瞬时摄影片
instantaneous peak current 瞬时峰值电流
instantaneous peak discharge 瞬时洪峰流量
instantaneous peak value 瞬时峰值
instantaneous penetration rate 瞬时钻速
instantaneous phase 瞬时相位
instantaneous phase current 瞬时相位电流
instantaneous phase indicator 瞬时相位指示器
instantaneous phase section 瞬时相位剖面
instantaneous photograph 瞬时需氧量
instantaneous point-source 瞬时点源
instantaneous pole 瞬时极
instantaneous polymerization 瞬时聚合
instantaneous pore pressure 瞬时孔隙压力
instantaneous position 瞬时位置
instantaneous power 瞬时功率
instantaneous power output 瞬时功率输出
instantaneous power spectrum 瞬时功率谱
instantaneous pressure 瞬时压力
instantaneous probability of detection 瞬时检测概率
instantaneous profile (潮波引起的)瞬时纵断面
instantaneous pulsation 瞬时脉动
instantaneous punch energy 瞬时冲头能量
instantaneous radiation 瞬时辐射
instantaneous radius 瞬时半径
instantaneous rate 瞬时速率;瞬时日差
instantaneous rate of deformation 瞬时形变速率
instantaneous rate of flow 瞬时流量
instantaneous rate of increase 瞬时人口净长率
instantaneous reactance 瞬时电抗
instantaneous reaction yield 瞬时反应收率
instantaneous reactivity 瞬时反应性
instantaneous reading 瞬时读数
instantaneous-reading tape 直读卷尺
instantaneous readout 瞬时读出
instantaneous reclosing 瞬时重合闸
instantaneous record 瞬时记录
instantaneous recorder 瞬态记录器
instantaneous recovery 瞬时回变;瞬时恢复
instantaneous recovery time 瞬时恢复时间
instantaneous relay 速动继电器;瞬动继电器
instantaneous relay blast 瞬迟发爆破
instantaneous release 瞬时释放
instantaneous removal of forms 立即脱模
instantaneous reoperate time 瞬时再动作时间
instantaneous resonance 瞬态共振
instantaneous response 瞬时响应;瞬时反应
instantaneous response analysis 瞬态响应分析
instantaneous response function 瞬时响应函数
instantaneous restoring force 瞬时恢复力
instantaneous return time 瞬时复原时间
instantaneous reverse 瞬时换向

instantaneous rotation 瞬时旋转
instantaneous rupture 瞬时破坏
instantaneous safety gear 瞬时式安全钳
instantaneous safety gear with buffered effect 具有缓冲作用的瞬时式安全钳
instantaneous sample 瞬时取样
instantaneous sampler 快速采样器;瞬时采样器
instantaneous sampling 瞬时取样;瞬时采样
instantaneous set 瞬时变定
instantaneous short circuit 瞬时短路
instantaneous short-circuit test 瞬时短路试验
instantaneous shot 瞬时摄影
instantaneous shunt 瞬时分路
instantaneous shutter 瞬时快门;瞬间快门
instantaneous(sink) water heater 快速热水器;瞬时热水器
instantaneous slope 瞬时斜率
instantaneous sound energy density 瞬时声能密度
instantaneous sound pressure 瞬时声压
instantaneous source 瞬时工作源
instantaneous specific heat 瞬时比热;瞬间比热容
instantaneous spectrum 瞬时光谱
instantaneous speed 瞬时车速
instantaneous speed change 瞬时速度变动率
instantaneous squib 瞬时电点火管
instantaneous stability 瞬时稳定性
instantaneous stand-by 瞬时备用;短时备用
instantaneous state 瞬时状态
instantaneous stiffness 瞬时劲度
instantaneous stopping 瞬时停顿
instantaneous storage 立即存取存储;瞬时存储器
instantaneous strain 瞬时应变;瞬时间应变
instantaneous strain stage 瞬时变形阶段
instantaneous stream density 瞬时流密度
instantaneous streamline 瞬时流线
instantaneous strength 瞬时强度
instantaneous stress 瞬时应力
instantaneous stress-strain curve 瞬时应力应变曲线
instantaneous surface profile observation 瞬时水面线观测
instantaneous surface tension 瞬时表面张力
instantaneous surge 瞬时涌浪
instantaneous suspension sampler 瞬时悬移质取样器
instantaneous switch 瞬时开关
instantaneous switching 瞬时转换
instantaneous system 瞬时系统
instantaneous tangent modulus 瞬时正切模量
instantaneous terrestrial system 地球瞬时坐标系
instantaneous theoretic(al) output 瞬时理论产量
instantaneous time 瞬时时间
instantaneous torque 瞬时转矩
instantaneous total mortality coefficient 瞬时总死亡系数
instantaneous traffic load 瞬时交通荷载
instantaneous transmission rate 瞬时传输率;瞬变传输率
instantaneous trip 瞬间自动跳闸;瞬间切断;瞬间断路
instantaneous trip circuit breaker 瞬时跳闸断路器
instantaneous type water heater 快速式热交换器
instantaneous unit hydrograph 瞬时单位水过程线
instantaneous unlocking mechanism 瞬跳机构
instantaneous value 瞬时值
instantaneous velocity 瞬时速度
instantaneous velocity of reaction 瞬时反应速度
instantaneous velocity vector 瞬时速度向量
instantaneous video check 瞬时视频检验
instantaneous voltage 瞬时电压
instantaneous warm water 瞬时热水;快速暖水
instantaneous water heater 流水加热器;快速热水器;快速热水炉;瞬时热水器
instantaneous water-level 瞬时水位
instantaneous water stage 瞬时水位
instantaneous wave elevation 瞬时波面;波面瞬时值
instantaneous weight 瞬时重量
instantaneous wind speed 瞬时风速
instantanous center[centre] 瞬心
instant bandwidth 瞬时带宽
instant bathroom(unit) 预装配整套浴室;快速浴室
instant blasting 瞬时爆破

instant bridge 瞬时建成桥(一种快速建成的新型预应桥)
instant camera 一步成像照相机
instant center[centre] 瞬心;瞬时中心
instant-change calendar 瞬时变换日历
instant chromatography 瞬息色谱法
instant city 当代城市
instant data setting 瞬时换日
instant data setting calendar 瞬跳换日
instant dispatching 迅速发运
instantiation 示例
instant insanity 顿时错乱
instantizing 速溶化
instant-jump calendar 瞬跳日历
instant lock 碰锁;暗锁
instantly acting time-limit resetting system 瞬动时限复归制
instant map 瞬变地图
instant noodle 方便面
instant of exposure 曝光瞬间
instant of failure 断裂瞬间;破裂瞬间;故障瞬间
instantograph 快照
instant-on switch 瞬动开关
instant on system 即时接通制;瞬间接收制
instant payment 立即付款
instant photography 一步成像摄影(术);瞬时摄影
instant program(me) access 瞬时程序存取
instant reply 可即时放送的录像
instant reset 瞬时复位
instant return mirror mechanism 瞬时回镜机构
instant reverse 瞬时换向
instant-set polymer 速凝聚合物
instant-start 瞬时起燃;瞬时接入
instant-start fluorescent lamp 瞬时起动荧光灯
instant starting boiler 快速启动锅炉
instant stop type 立即停止方式
instant talk-in system 即时对话系统
instant tender 现场投标
instant thin-layer chromatography 瞬时薄层色谱法
instant vapo(u)rization 瞬间气化
instant vision 瞬间可视图像;瞬间接收图像
in star connection 按星形接法
instaseal 防漏粗黏粉
instauration 重建;修复
instead of 代替
in-step 同相;同步(的);同级
in-step condition 同相条件;同步条件
instillation 滴注法
instillator 滴注器
instinct 本能
instinctive 本能的
instinctive activity 本能活动
instinctive behavio(u)r 本能行为
instinctive behaviour disturbance 本能行为障碍
instinctive centre 本能中枢
instinctive migratory journey 本能的长途洄游
instinct of workmanship 提高艺术的本能
institute 学院;学术会议;学会;协会;研究院;研究所;专科学校;公会;机关;设计院
institute a claim 提出索赔
Institute Air Cargo Clause 协会航空货物保险条款
institute an inquiry 开始调查
Institute Cargo Clauses 协会货物平安险条款;协会全险条款;协会货运保险条款
institute clause 学会条款;协会保险条款;公会条款;附带条款
institute cost reduction program(me) 制订成本降低计划
Institute Dangerous Drugs Clause 协会危险药品条款
institute for nationalities 民族学院
Institute for Telecommunication Science 电信科学协会
institute of architectural design 建筑设计院
institute of biologic(al) products 生物制品研究所
institute of chartered ship brokers 租船经纪人公会
institute of civil engineering 建筑工程学院
institute of design 设计院
institute of economics 经济研究所
Institute of Electric(al) and Electronic Engineers 电机及电子工程师学会(美国)
Institute of Electric(al) electronic Engineers 电气与电子工程师协会
Institute of Electric(al) electronic Engineer standards 电气与电子工程师协会标准
Institute of Experimental Biology 实验生物研究所
Institute of Export 出口学会
Institute of Glaciology 中国科学院兰州冰川冻土沙漠研究所
institute of human relations 人群关系研究所
Institute of Information Scientists 信息科学家协会
Institute of International Container Lessors 国际集装箱出租者协会;国际集装箱出租商协会
institute of labo(u)r health 劳动卫生研究所
institute of medical instruments and apparatus 医疗仪器及器械研究所
Institute of Mining and Metallurgy sieve 矿业与冶金协会规定用筛(美国)
Institute of Mining Engineers 采矿工程师学会
institute of navigation 航海学会
institute of oceanography 海洋研究所
institute of physiology 生理研究所
Institute of Real Estate Management 房地产管理协会
institute of science and technology 理工学院
Institute of Traffic Administration 交通管理学会
Institute of Traffic Engineer 交通工程师学会(美国)
Institute of Transport 运输学会
Institute of Water Engineers 水工程师协会
Institute of Water Pollution Control 水污染控制学会;水污染控制协会
institute of water research 水研究所
institute of zootechnics and veterinary science 畜牧兽医科学研究所
Institute policy 协会保险单
institute precautions 制订防范措施
Institute Replacement Clauses 协会重置条款
Institute Strikes Clause 协会罢工险条款(货物)
institute ten percent disbursement clause 公会百分制之十费用条款
institute warranties 公会保证条款
institution 学会;研究所;机构;事业单位;设置;设立
institutional accounting 机关会计;事业机构会计
institutional advertising 厂商广告
institutional adviser 法律顾问
institutional approach 惯例方法
institutional arbitration 机构仲裁
institutional building 学院建筑;学会建筑;社团建筑
institutional consideration 法律研讨
institutional credit 机构贷款
institutional economics 制度经济学
institutional factor 体制因素;人文因素
institutional food 规格食品
institutional framework 相关机构网;组织机构;体制
institutional function 体制函数
institutional influence 制度影响
institutional investment 机构投资
institutional investors 有组织的集体投资者;团体投资人
institutional issues 法律问题
institutional land use 社团土地利用
institutional lender 金融机构
institutional loan 金融机构贷款
institutional network 机构交易
institutional occupancy 公共机关用房
institutional organization 事业机构
institutional property 社团地产
institutional purchase 厂商间购买
institutional sector of origin 来源机构部门
institutional society 社会慈善事业协会
institutional structure 体制结构
institutional uses 非营利机构用房
institutional waste 公共场所垃圾
institutional wastewater 公共废水;机关废水;社会团体废水
institution of action 民事诉讼的起诉
Institution of Civil Engineers 土木工程师学会(英国);土木工程师协会
Institution of Electrical Engineers 电气工程师学会;电气工程师协会
Institution of Environmental Engineers 环境工程师协会
Institution of Naval Architects 造船工程师协会
Institution of Railway Signal Engineers 铁路信号工程师学会(英国)
Institution of Water and Power Engineers 水电工程师协会

Institution of Water Engineers 自来水工程师学会;给排水工程师协会;水利工程师协会
institutions of higher learning 高等院校;高等学校
in stock 现成的;有库存的;定型的
in-storage drying 储藏干燥
instoscope 目视曝光计
in-stream aeration 河水复氧
in-stream dissolved oxygen concentration 河水溶解氧浓度
in-stream flow consideration 河道水流条件
in-stream flow requirement 河道水流要求
in-stream habitat 河道生境
in-streaming 内流的
in-stream non-point source nutrient model 河道非点污染源营养物模型
in-stream phosphorus concentration 河水含磷浓度
in-stream waste concentration 河道污水浓度
in-stream water quality measurement 河道水质测定
in-stream water quality monitoring station 河道水质监测站
instroke 排气行程
instron 拉伸强度试验机
Instron apparatus 英斯特朗电子强力机
Instron capillary rheometer 英斯特朗毛细管流变仪
instron tensile(strength) tester 拉力测定仪;英斯特朗电子强力测试仪
instruct a bank 授权银行
instructed grading 指定级配
instructed mix 指定配合比
instruction 指示(书);指令
instruction address 指令地址
instruction address register 指令地址寄存器
instruction address source 指令地址源
instructional charts, diagrams 教学海图及图表
instructional design 教学设计
instructional film 指导片
instructional gaming program(me) 教学游戏程序
instructional investment 公共团体投资
instructional management system 教育管理系统
instructional map 教学地图
instructional mode 教学形式
instructional objectives 教学目标
instructional software 教学软件
instructional strategy 教学策略
instructional system 教学系统
instructional television 教育电视;教学电视
instructional terminal 教学终端
instructional testing 教学测试
instructional(work)shop 培训车间
instruction architecture 指令体系结构
instruction area 指令存储区
instruction array 指令阵列
instruction bank 指令存储体
instruction-based architecture 以指令为基础的结构
instruction book 指南;养路工作手册;说明书
instruction book of spike tooth harrow 钉齿耙使用说明书
instruction buffer 指令缓冲器
instruction buffer register 指令缓冲寄存器
instruction bus 指令总线
instruction cache 指令缓存
instruction card 指导卡片;说明卡
instruction carriage 试验车
instruction catalogue 指令表
instruction character 指令字符;指令符号;控制符(号);操作符
instruction check indicator 指令检验指示器
instruction classification 指令分类
instruction clerk 工作指导
instruction code 指令码
instruction command 指令形式的命令
instruction compatibility 指令兼容性
instruction constant 指令常数
instruction control 指令控制
instruction control unit 指令控制器
instruction counter 指令计数器
instruction counter register 指令记数寄存器
instruction deck 指令卡片组
instruction decode network 指令译码网络
instruction dependency 指令相关性
instruction display 示教显示器
instruction distribution channel 指令分配通道

instruction element 指令元素;指令元件
instruction execute 指令执行
instruction execution 指令执行
instruction execution control 指令执行控制
instruction execution logic 指令执行逻辑
instruction execution time 指令执行时间
instruction fetch 指令取出;取指令
instruction fetch microoperation 取指令微操作
instruction fetch operation 取指令操作
instruction fetch phase 取指令周期;取指令阶段
instruction fetch routine 取指令程序
instruction fetch time 取指令时间
instruction field 指令字段
instruction flow 指令流
instruction for assembly 装配技术规程
instruction for bidding 招标须知;投标须知
instruction for despatch of goods 空运货物委托书
instruction for erection work 装配指导;建造指导;装配技术规程
instruction for installation 安装说明书
instruction for maintenance 维修说明书
instruction format 指令形式;指令格式
instruction forming 指令形成
instruction for mounting 安装说明书
instruction for planning construction project 建设项目计划任务书
instruction for safe use 安全使用说明书
instruction generation 指令生成程序
instruction generator 指令发生器
instruction ignore 指令无效;指令无动作
instruction in draughtsmanship 制图质量技术规程
instruction information 指令信息
instruction interval 指令作用距离
instruction length 指令长度
instruction length code 指令长度码
instruction level multiprogram(me) 指令级多道程序
instruction level parallel processing 指令级并行处理
instruction list 指令表
instruction location 指令地址
instruction location register 指令地址寄存器
instruction machine code 指令机器码
instruction main line 指令主线
instruction manual 用户说明;作业细则;指令手册;说明手册;使用说明(书);使用手册;操作说明书;安装手册
instruction map 指令变换
instruction mark for dangerous cargo 危险货物标志
instruction memory 指令存储器
instruction mix 指令混合比例
instruction modification 指令(字)修改;指令改变
instruction modifier 指令修改量
instruction number 指令数号;指令数
instruction of design 设计说明书
instruction of technical operation 技术操作规程
instruction operation code 指令操作码
instruction packet 指令包
instruction path 指令通路
instruction pipeline 指令流水线
instruction processing unit 指令处理器
instruction processor 指令处理机
instruction program(me) 指令程序
instruction queue 指令队列
instruction receiver 指令接收机
instruction regarding explosive picked up at sea 有海面上打捞爆炸物的说明
instruction register 指令寄存器
instruction repertoire 指令系统;指令代码;指令程序
instruction repertory 指令系统
instruction replica 教学模型
instruction retry 指令重复执行;指令复执
instruction routine 指令常规
instruction segment 指令段
instruction sequence 指令序列
instruction set 指令系统
instruction set component 指令系统成分
instruction set expandability 指令系统扩展性
instructions given in one's presence 当面指示
instruction sheet 工序卡;说明图表;说明书
instruction signal 指令信号
instruction space 指令空间
instruction step 指令步

instructions to bidder 投标指南;投标人须知
instructions to contractor 向承包商下达指令
instruction storage 指令存储[贮]器
instructions to tenderer 投标者须知
instructions to tendering 投标指南;投标说明
instructions to the user 用户使用说明
instruction stream 指令流
instruction subsystem 指令子系统
instruction system 指令系统
instruction table 指令表
instruction tape 指令带
instruction terminal 指令终端
instruction time 指示时间;指令执行时间;指令(取出)时间
instructions to bidders 投标指南;投标人须知
instructions to contractors 向承包商下达指令
instructions to tenderers 投标者须知
instruction to the structure 结构说明
instruction tracing 指令追踪
instruction type 指令形式;指令类型
instruction unit 指令部件
instruction word 指令字
instruction words of location 位置指令字
instructive card 工作指导卡
instructive theory 指导学说
instructor 指导者
in-structure ventilation 建筑内通风
instrument 仪表;仪器;证书;工具;器械;票据;放射性同位素监测仪
instrument adjustment 仪器调整
instrument air 仪表用风;仪表气源
instrument air system 仪表用空气系统
instrumental activation analysis 仪器活化分析
instrumental aerial triangulation 仪器空中三角测量
instrumental analysis 仪器分析;物化分析
instrumental angle 仪器角
instrumental azimuth 仪器方位
instrumental bearing 仪器方位
instrumental bias 仪器偏差
instrumental broadening 仪器致宽
instrumental capital 生财资本
instrumental charged-particle activation analysis 仪器带电粒子活化分析
instrumental conditioning 操作性条件反射
instrumental constant 仪器常数
instrumental contour 实测等高线;仪器实测等高线;仪器轮廓
instrumental correction 仪器校正;仪器改正
instrumental defect 仪器缺陷
instrumental degradation 仪器递降
instrumental drawing 仪器制图;仪器绘图
instrumental drift 仪器零点漂移
instrumental epicenter[epicenter] 仪器(测定的)震中;微观震中;实测震中
instrumental error 仪器误差;仪表误差;仪器视差
instrumental function 仪器功能
instrumental gallery 观测廊道
instrumental height 仪器高
instrumental instability 仪器不稳定(性)
instrumental interpretation method 仪器解译方法
instrumentalist 仪表专业人员
instrumentality 债券;媒介
instrumental landing system 降落仪表系统(飞机)
instrumental latitude 仪器纬度
instrumental limits 仪器使用范围
instrumental line profile 仪器谱线轮廓
instrumental longitude 仪器经度
instrumental mannequin 仪器式人体模型(耐燃试验用)
instrumental match prediction 仪器配色测示法
instrumental measurement 仪器测定(法);仪器量测
instrumental method 仪器分析法;仪器法
instrumental music 器乐
instrumental neutron activation analysis 仪器中子活化分析
instrumental observation 用仪器观测
instrumental operation 自动操作
instrumental optics 仪器光学
instrumental parallax 仪器视差
instrumental parameter 仪器参数
instrumental photon activation analysis 仪器光子活化分析
instrumental range 仪器使用范围

instrumental record 用仪器记录
instrumental refraction 仪器折射;仪器内大气折射
instrumental resolution 仪器分辨率;仪器的鉴别力
instrumental response 仪器响应
instrumental segment 仪器测定的音节
instrumental seismology 仪器地震学
instrumental sensitivity 仪器灵敏度
instrumental shaft 仪器轴
instrumental shaft plumbing 用仪器进行竖井铅锤测量
instrumental straggling 仪器误差
instrumental switch 仪器开关
instrumental technique 仪表技术
instrumental thermal neutron activation analysis 仪器热中子活化分析
instrumental tracking and telemetry ship 跟踪遥测船
instrumental variable method 仪器变量法
instrumental variables 仪器变量
instrumental variable technique 辅助变量法
instrument and meter plant 仪表厂
instrument and meter workshop 仪表间
instrument approach 仪表进场着陆
instrument approach chart 盲目降落临场导航图
instrument approach procedure chart 仪表进场导航图
instrumentarium 全套器械
instrument assembly 仪器装配
instrumentation 仪器(装备);仪器配备;仪表学;仪表设备;仪表化;仪表工业;装设仪器;测试设备;测量仪器;测量仪表;测量设备
instrumentation amplifier 仪器放大器;测量放大器
instrumentation analysis 仪器分析
instrumentation buoy 带仪器浮标
instrumentation coefficient 测量系数
instrumentation complex 全套测量设备
instrumentation console 仪表台;仪表板;操纵台
instrumentation control rack 操纵台
instrumentation diagram 仪表系统图
instrumentation engineering 仪表工程学
instrumentation for control 控制仪表
instrumentation grounding 仪表接地
instrumentation operation station 仪表操作台
instrumentation panel 控制台
instrumentation payload 仪器有效荷载
instrumentation plan 测量示意图
instrumentation sensor 仪表装置传感器;测试设备传感器
instrumentation symbol 测量仪表符号
instrumentation system 仪表测量系统;测试设备系统;测量系统
instrument autotransformer 仪表用自耦变压器
instrument axis 仪器轴
instrument background 仪器本底
instrument base 仪器基座
instrument bay 仪器舱;仪表舱
instrument bezel (仪器的)玻璃框
instrument block valve 仪表截止阀
instrument board 仪器板;仪表盘;仪表板
instrument-board bracket clamp 仪表板托架夹
instrument-board clock 仪表板时钟
instrument-board dial glass 仪表板玻璃字盘
instrument box 仪器箱;仪表箱
instrument branch 仪表支线
instrument cabinet 仪器舱
instrument calibration 仪表校正
instrument capsule 仪表容器
instrument carriage 测试仪表托架
instrument case 仪器箱;仪器柜;仪表盒;工具角色
instrument chart 仪表飞行图
instrument circuit 仪表电路
instrument cluster 仪表组;仪表板
instrument cluster lamp 仪表板灯
instrument collection 票据托收
instrument combination 仪器组合
instrument compartment 仪器舱
instrument computer 仪表用计算机
instrument connection 测试孔;测试接管
instrument constant 仪表常数
instrument container 仪器箱
instrument contour 实测等高线
instrument control 仪器控制;仪表控制
instrument controller 仪表控制器
instrument converter 仪表用变流器

instrument coordinates 仪器坐标
instrument cord 仪表电绳
instrument correction 仪器修正;仪器误差改正(量);仪器订正
instrument cubicle 仪表室;仪表间
instrument data 仪器资料
instrument deconvolution 仪器反褶积
instrument defect 仪器缺陷
instrument desk 仪表台
instrument detection limit 仪器检出限
instrument device 仪器装置
instrument dial 仪器刻度盘;仪器标尺;仪表标度盘
instrument digital technique 仪表数字技术
instrument direction 仪器导向
instrument discriminator 仪表鉴别器
instrument drawing 工具图;仪器图
instrument drift effect 仪器漂移效应
instrumented 装有仪器的
instrumented pile 装有量测元件的桩;装有量测器的桩
instrumented pile driver 装有仪器的试桩机
instrumented satellite 测量卫星
instrument elevation (测量时的)仪器高程
instrument error 量具误差
instrument face panel 仪表面板
instrument fascia bezel 仪表饰带卡环
instrument feeding 仪器供电;仪器电源
instrument flight 仪器操纵飞行
instrument flying chart 仪表飞行图
instrument for automatic sequence control 自动顺序控制仪
instrument for cryotherapy 深低温治疗器
instrument for determining surface finish 表面光洁度测定仪
instrument for induced polarization 激发极化仪
instrument for magnetic property measurement 磁性测量仪
instrument for measuring oxygen in molten steel 钢液定氧仪
instrument for measuring water temperature 水温测量仪器
instrument for quality analysis of water 水质分析仪(器)
instrument for wax pattern preparation 制作蜡型器械
instrument frequency response 仪器频率响应
instrument gallery 仪表廊;观测廊道
instrument gears 仪表齿轮
instrument glass 仪表玻璃
instrument glass dial 仪器玻璃标盘;仪表玻璃标度盘
instrument grease 仪表脂
instrument head 测量头
instrument headed screw 半埋头螺钉
instrument identification 仪器标记
instrument indication 仪器显示
instrument industry 仪表工业;仪器仪表制造业;仪表制造业
instrument intensity scale 仪器测定烈度表
instrument lag 仪表惰性;仪表的滞后
instrument lamp 仪表指示灯;仪表(板)灯
instrument landing 仪表着陆;盲目着陆;盲目降落
instrument landing approach 盲目着陆方式
instrument landing station 仪表着陆无线电台
instrument landing system 仪表着陆系统;盲目着陆系统;盲目降落临场设备
instrument landing system glide path 仪表导降系统滑道
instrument landing system localizer 盲目着陆系统信标
instrument lathe 仪表车床
instrument leads 仪表导线
instrument level (测量时的)仪器高程
instrument light 仪表信号灯;仪表盘灯;仪表灯
instrument limits 仪器使用范围
instrument line 仪器线
instrument lubricant 仪器用润滑剂
instrument luster 仪表饰框
instrument magnitude 仪器测定震级
instrument maintenance code 仪器维护编号;仪表维修规程
instrument-making industry 仪器制造工业
instrument man 测量员;仪表操作人员
instrument manual 仪器手册

instrument mask 仪表罩
instrument-matching 仪器配色
instrument measurement 用仪器测量
instrument meteorological condition 仪器气象状况;仪表飞行气象条件
instrument microcomputer 仪表用微计算机
instrument mode 仪表工作状态
instrument model 仪器型号
instrument monitoring 仪器监测
instrument motor 仪表电动机
instrument mounting plate 仪表座板
instrument multiplier 仪表扩程器;仪表附加电阻
instrument name 仪器名(称)
instrument network 仪表网络
instrument observation 仪表观测
instrument of certification 证明文件;批准书
instrument of credit 信用证书;商业证券;商业信用凭证
instrument of credit control 信用管制手段
instrument of electric(al) prospecting 电法勘探仪器
instrument of expression 表达工具
instrument of flow measurement 流量测仪器
instrument of ga(u)ge observation 水位观测仪器
instrument of gravity measurement 重力测量仪器
instrument of labo(u)r 劳动工具
instrument of magnetic survey 磁测仪器
instrument of modeling test 模拟试验仪器
instrument of payment 支付手段
instrument of precipitation observation 降水观测仪器
instrument of production 生产工具
instrument of ratification 批准证书;批准书
instrument of sediment measurement 泥沙测验仪器
instrument of surveying and mapping 测绘仪;测绘器
instrument of use 使用仪器
instrument oil 仪器用油;仪表油
instrument operation 仪器工作
instrument optics 仪器光学
instrument panel 操纵板;操纵盘;仪器表面板;仪表屏;仪表盘;仪表(控制)板;仪表操纵板
instrument-panel bracket 仪表板支架
instrument-panel gasket 仪表板衬
instrument-panel lamp 仪表板照明灯
instrument-panel light 仪表板灯
instrument-panel support 仪表板支架
instrument-panel warning light 仪表板信号灯
instrument parameter 仪器参数
instrument payload 仪表有效荷载
instrument pen 仪表笔;记录器笔
instrument photo triangulation 平面照片三角测量;仪器照片三角测量
instrument piping 仪表管路
instrument plant 仪器厂
instrument platform 仪器平台
instrument plug 仪表插头
instrument plumbing 仪器检测垂直度
instrument power supply 仪器电源
instrument precision 仪表精确度
instrument programmability 仪表程序控制性
instrument program(me) 仪表程序
instrument range 仪器量程;仪器测量范围;仪表区段;仪表量程
instrument reading 仪器读数;仪表读数
instrument reading time 仪器读出时间
instrument record 仪器记录
instrument recording camera 仪表记录摄影机
instrument recording photography 仪器记录照相
instrument relay 仪表继电器;计数继电器
instrument resistor 仪表电阻器
instrument room 仪器室;仪器舱;仪表室;仪表间
instrument route survey 仪表线路测量
instrument runway 仪表跑道;盲目起降跑道;仪器飞行起落跑道
instruments and manpower requirements 仪器和人力需求
instruments and meters 仪器仪表
instrument screen 仪器百叶箱
instrument service division 仪表服务部
instrument servo 仪表伺服系统
instrument servo mechanism 仪表伺服机构
instrument setting 仪器上输入数据

instrument shed 工具棚
instrument shelter 仪器罩;仪器(百叶)箱
instrument shunt 仪表分流器
Instrument Society of American 美国仪器协会
instruments of crystal goniometry 晶体测角仪器
instruments of pollution prevention of vessel 船舶防污设施
instrument stand 仪器座;仪器架;仪表架
instrument station 仪器站;量测站;测站
instrument straggling 仪器歧离
instrument survey 仪表测量
instrument-survey method 仪器测量法
instruments weight 仪器重量
instrument switch 仪表开关
instrument symbols 仪器符号
instrument system 仪器系统
instrument table 仪器桌
instrument technician 仪表技术员
instrument technology 检验仪表工艺学
instrument terminal 仪表接线端子
instrument test repair laboratory 仪器检测修理实验室
instrument transformer 仪表变压器;仪表(用)互感器
instrument transformer for all purpose 仪表用综合互感器
instrument truck 仪器车;仪表车
instrument type 仪器类型
instrument-type of cementing 固井仪表分类
instrument-type relay 仪表式继电器
instrument unit 仪器组
instrument use 仪器使用
instrument weather 仪表天气
instrument with electrostatic screening 静电屏蔽仪表
instrument workshop 仪表修理间
instrument zero 仪器零位置
instrument zero correction 仪器零点校正
instrumentation 仪器仪表
insubmersibility 抗沉性;不沉性
insubmersibility test 不沉性试验
in substance 实质上
in substantial 无实体的
in succession 顺次
insufficiency 关闭不全;功能不全;机能不全;闭锁不全
insufficient 不足的;不够的
insufficient compression 压缩不足
insufficient funds 资金不足
insufficient gloss 光泽不足
insufficient impregnation 浸渍不足
insufficient leg 焊脚厚度不足
insufficiently burnt 未烧透的;生烧的
insufficiently compacted concrete 非完全密实混凝土
insufficiently prepaid 预付款不足
insufficient mineralization 不足的矿化作用
insufficient or excess reinforcement of weld 加强焊缝不足或过量
insufficient orientation 概略定向
insufficient packing 包装不足;包装不良;包装不牢固
insufficient penetration 熔深不足
insufficient reason rule 同等几率原则
insufficient section 不足部分
insufficient span of bridge 桥梁孔径不足
insufficient V-belt tension 三角皮带张力不足
insufflation 注气法;吹送粉料;吹色法;吹入剂;吹入法
insufflation dust 喷窑灰
insufflator 吹药器;吹入器
insula 多层式公共住宅(罗马建筑);建筑群;岛
insulae 群屋;(古罗马的)建筑群
insulance 绝缘电阻;介质电阻
insulant 绝缘物质;绝缘材料
insular 独立地段(周围为道路)
insular building 独立式房屋;独立建筑物
insular climate 岛屿性气候
insular motif(pattern)岛状基型
insular shelf 岛架;岛基台
insular shoulder 岛坡
insular slope 岛坡
insular state 岛国
insular structure 岛状结构

insular talus 岛坡
insular water bottle 绝缘采水瓶
insulate 绝缘
insulated 绝缘的;保温的
insulated annealed copper 绝缘韧铜线
insulated body 被绝缘体
insulated bolt 绝缘螺栓
insulated boxcar 保温车箱
insulated brush 绝缘电刷
insulated cable 绝缘电缆
insulated cambric pipe 麻布绝缘管
insulated capacity 保温能力
insulated catenary wire support clamp 绝缘子组合承力索座
insulated ceiling 隔热平顶
insulated clip 绝缘夹线板
insulated column 隔立柱;隔离柱;独立柱;保温蒸馏塔
insulated conductor 绝缘导体;绝缘导线
insulated conduit 绝缘导管
insulated container 隔热集装箱;绝热集装箱;保温集装箱;保温货柜
insulated cooling 隔热缓冷
insulated cover 绝热覆盖棚
insulated curing 保温养护
insulated dowel 绝缘定位销
insulated electric(al) conductor 绝缘电线;绝缘电导体
insulated evapo(u)ration pan 隔热蒸发器;隔热蒸发皿;挡雨式蒸发皿
insulated eye 绝缘孔眼
insulated feeder 保温冒口
insulated flange 绝缘法兰盘
insulated flexible pipe 绝缘软管
insulated floor 保温地板
insulated flue pipe 隔热烟道管
insulated form 保温模板
insulated furnace 热处理绝热炉
insulated ga(u)ge bar 绝缘轨距杆
insulated ga(u)ge rod 绝缘轨距杆
insulated handle 绝缘手柄
insulated hanger 绝缘吊线
insulated hatch cover 隔热舱盖;密封舱盖;密封舱盖
insulated heater wire 加热丝绝缘
insulated hold 隔热舱;保温舱
insulated hook 绝缘钩
insulated hook end clamp 垂直绝缘定位钩
insulated hot water heater 隔热热水加热器
insulated impound body 隔离水体;隔离水库
insulated impounded body 隔离水体
insulated joint (钢轨的)绝缘节;隔热接头;绝缘接头,绝缘接缝
insulated joints located within the clearance 侵入限界绝缘
insulated joints within a turnout 岔中绝缘
insulated lake 隔水湖;隔离湖
insulated large container 大尺寸绝缘集装箱
insulated line hanger 绝缘吊弦
insulated lorry 保温汽车
insulated material 隔热材料;绝缘材料;保温材料
insulated metal roofing 加热金属屋面;保温金属屋面
insulated middle wire 不接地(的)中线
insulated neutral (对地的)绝缘中线;不接地中性线;不接地中点
insulated overlap 绝缘锚段关节
insulated overlapped section 绝缘关节
insulated paint 绝缘漆
insulated paper 绝缘纸
insulated pipe 绝缘管;保温管
insulated pipe line 绝缘管线
insulated plate for protection 绝缘防护板
insulated pliers 绝缘钳
insulated produce container 隔热集装箱;保温集装箱;保冷集装箱
insulated rail 绝缘钢轨
insulated rail joint 绝缘钢轨接头;绝缘导轨接头
insulated-return power system 绝缘回路供电制
insulated-return system 绝缘回流制;绝缘回流系统
insulated riser 保温冒口
insulated river 隔离河流
insulated roof 保温顶盖
insulated single point water heater 有栓塞活门的热水加热器
insulated sleeve 绝缘护套
insulated spring 绝缘弹簧
insulated steam curing cloche 蒸汽养护保温罩
insulated steel sheathed conduit 绝缘钢套导线管
insulated storage container 隔热储[贮]箱
insulated stream 隔水河;隔离河流;隔底水流;(和地下水饱和区隔开的)隔底河流;地表水流
insulated supply system 不接地电源
insulated switch section 道岔绝缘段
insulated swivel clip holder 绝缘支持器;分体绝缘支持器
insulated system 不接地制
insulated tank 绝热罐
insulated tape 绝缘胶带
insulated tarpaulin 绝热盖布
insulated tent 绝热帐篷
insulated thermistor 绝缘型热敏电阻器
insulated thimble 绝缘护环
insulated track 绝缘轨道
insulated track section 绝缘轨道区段
insulated transition mast 绝缘转换柱
insulated transport 保温运输
insulated truck 隔热车
insulated tube 绝缘管
insulated wagon 铁路保温车
insulated wall 保温墙
insulated wire 绝缘(导)线
insulater 绝缘物质
insulate the organizational management system 封闭化
insulating 绝缘的
insulating ability 绝缘能力;保暖性能
insulating article 绝缘制品
insulating asbestos board 绝缘石棉板
insulating asbestos board sheet 隔热石棉板
insulating asbestos paper 石棉绝缘纸
insulating back-up material 外部隔热材料
insulating bar 绝缘杆;绝缘棒
insulating barrier 绝缘隔板
insulating base 绝缘基座;绝缘地基;绝缘底座;绝缘(底)板
insulating beads 绝缘小珠
insulating blanket 绝缘毡;绝缘垫层;绝热毯
insulating blanket of digester 消化池保温层
insulating block 绝缘砌块;绝热砖
insulating board 隔热板;隔声板;绝缘板;保温板
insulating boot 绝缘罩
insulating brick 隔热砖;绝缘砖;绝热砖;保温砖
insulating buffer layer 绝缘缓冲垫层
insulating buffer plate 绝缘缓冲垫板
insulating building material 隔声绝缘材料;隔热(隔声)绝缘材料;绝热建筑材料
insulating bulk-head 隔热间壁
insulating bushing 绝缘套筒;绝缘套管
insulating capacity 隔热隔声能力
insulating cardboard 绝缘纸板;绝缘厚纸
insulating castable 隔热铸块;保温铸块;绝热浇注件
insulating ceiling 隔热顶棚;隔热顶棚;隔热天花板
insulating cell 绝缘隔间;绝缘隔板;绝缘衬垫
insulating cement 隔热水泥;绝缘水泥;绝缘胶(结材料);绝热胶(泥);耐温水泥
insulating ceramics 绝缘陶瓷
insulating clamp 绝缘线夹
insulating cloth 绝缘布
insulating coating 隔声面层;隔热面层;绝缘外套;绝缘涂层;绝缘漆;绝热涂层;保温涂层
insulating coefficient 绝热系数
insulating compound 绝缘物质;绝缘剂;绝缘膏
insulating concrete 隔热混凝土;绝缘混凝土;绝热混凝土;保温混凝土
insulating coolant 绝缘冷却剂
insulating core 绝缘芯
insulating cork 隔热隔声软木
insulating corkboard 隔热隔声软木板
insulating cork sheet(ing) 隔热隔声软木(薄)片;隔热隔声软木(薄)板
insulating cork slab 隔热软木(厚)板;绝缘软木(厚)板
insulating corrugated cardboard 隔热隔声瓦楞纸板;绝缘波纹卡(片)纸板
insulating coupling 绝缘联轴节
insulating course 隔声层;隔离层;隔断层;绝缘层

insulating covering 绝缘套
insulating detector 绝缘检漏仪
insulating disk 绝缘盘
insulating door 隔声门;保温门
insulating efficiency 保温效率;绝缘性能
insulating facing 绝缘涂料
insulating felt 隔热隔声毡;绝缘油毛毡
insulating fiber 绝热纤维
insulating fiber board 隔热隔声纤维板;隔热防寒纤维板;绝缘纤维板;绝热纤维板
insulating fiber board lath 抹灰用隔热纤维板条;绝热纤维板条(墙)
insulating fiber material 隔热纤维材料
insulating firebrick 隔热(耐)火砖;绝热耐火砖;耐火隔热砖
insulating fire-retarding glass 隔热防火玻璃
insulating fishplate 绝缘鱼尾板
insulating flange 绝缘法兰
insulating floor 绝缘楼板
insulating flue brick 隔热烟囱砖
insulating foam 隔声泡沫;隔热泡沫;绝缘泡沫
insulating foil 隔热箔;绝缘金属薄片
insulating formboard 绝热模板
insulating glass 隔热隔声玻璃;隔层玻璃;绝热玻璃
insulating glaze 绝缘釉;绝缘瓷釉
insulating glazing 上绝缘釉;隔热隔声玻璃窗(安装);窗用玻璃
insulating glazing unit 绝缘釉料元件
insulating gloves 绝缘手套
insulating gum 隔电胶
insulating gypsum 隔声石膏;隔热石膏
insulating gypsum board 隔热石膏板
insulating gypsum lath 绝热石膏条板
insulating gypsum wallboard 绝热石膏墙板
insulating insert(ion) 隔热隔声填充(物);绝缘垫片
insulating jacket 隔热夹套;隔热挡板;绝热夹套;保温套
insulating joint 绝缘接头
insulating joint bar 绝缘夹板
insulating laminated product 绝缘层压制品
insulating layer 绝缘层;隔温层;隔离层;绝热层;防寒层;保温层
insulating lead felt 含铅保护毡
insulating lining 隔热衬料;隔热衬里;绝热里衬;绝缘覆面;绝缘衬里;绝缘衬垫;绝热衬里;绝热衬里;保温内衬;保温里衬;保温衬里
insulating liquid 绝缘液体
insulating locator 绝缘检漏孔
insulating machine 绝缘机
insulating masonry(work) 绝缘圬工工程
insulating mat 绝缘垫
insulating material 隔热材料;绝缘材料;绝热材料;保温材料
insulating matter 绝热材料
insulating medium 绝缘介质
insulating metal roofing 隔热金属屋面;隔热金属屋顶;绝热金属屋面
insulating mixed plaster 绝缘混合灰泥
insulating molded product 绝缘模压制品
insulating mortar 保温砂浆
insulating mo(u)ld coating 隔热涂料
insulating oil 绝缘油;变压器油
insulating oil flash point 绝缘油闪点
insulating oil saponification number 绝缘油皂化值
insulating packing 隔热填料;绝缘填料
insulating pack-up block 绝热垫砖
insulating pad 绝缘垫片
insulating panel 隔热隔声墙板;保温板
insulating paper 绝缘纸;保温纸
insulating partition(wall) 绝热隔板
insulating paste 绝缘胶;绝缘膏
insulating peat 隔热泥炭
insulating pipe 绝缘管
insulating plank 绝热厚板
insulating plaster 保温抹灰;绝缘灰泥
insulating plasterboard 隔热石膏板;隔热灰泥板;铝箔塑料绝缘板
insulating plasterboard(foil back) (铅箔衬里的)隔热抹灰板
insulating plastic foam board 泡沫塑料保温板;绝缘泡沫塑料板;保温泡沫塑料板
insulating plastics 绝缘塑料
insulating polymer 绝缘聚合物
insulating powder 粉状保温材料;泡沫粉末;保温粉

insulating power 隔热能力;绝缘能力
insulating property 绝缘性质;绝缘性能
insulating pumice 保温浮石
insulating pumice gravel 保温浮石粒;保温浮砾石
insulating quality 隔热隔声质量
insulating refractory 隔热耐火材料;绝热耐火材料
insulating refractory brick 隔热耐火砖;绝热耐火砖
insulating refractory concrete 保温耐火混凝土;绝热耐火混凝土
insulating refractory furnace lining 隔热耐火炉衬
insulating resistance 绝缘电阻
insulating resistance test 绝缘电阻试验
insulating riser sleeve 保温冒口套
insulating roof deck 隔热屋面板;隔热防寒屋面板
insulating roof deck plate 绝缘屋面板
insulating roof fill 屋顶保温填充料;隔热保温屋面充填(材料);绝湿屋面填充(材料)
insulating roofing 隔热屋顶
insulating roof(ing) material 隔热屋面材料
insulating roof(ing) tile 隔热瓦;绝缘屋瓦
insulating rubberized fabrics 绝缘胶布
insulating screed 隔热找平层
insulating screed material 隔热找平材料
insulating section 隔热隔声型材料
insulating sheath 绝缘层;隔离层;保温层;绝缘护套
insulating sheathing muff 保温套
insulating sheet 绝缘薄板
insulating shoes 绝缘鞋
insulating skin 保温面层;绝缘外壳
insulating slab 绝缘石板;绝缘厚板
insulating sleeve 导线护套;绝缘套管;保温套;绝缘套筒;隔离套筒
insulating stand 绝缘台;绝缘架
insulating storage surface 绝缘靶
insulating straw board 隔声纤维板;隔热纤维板;隔热麦杆板;隔热稻草板;绝缘麦杆板;绝缘稻草板
insulating strength 绝缘强度
insulating strip 绝缘条;隔离条;伸缩条
insulating structural material 隔热隔声结构材料
insulating structural panel 隔热保温建筑板
insulating substance 绝缘物质
insulating substrate 绝缘衬底
insulating system 绝缘系统;隔热隔声系统
insulating tape 电线胶布;电绝缘胶带;绝缘(用)胶带;绝缘胶布;绝缘带
insulating test 绝热试验
insulating tile 保温瓦;保温地砖;隔热瓦
insulating transformer 隔离变压器
insulating trim 绝缘镶边;绝缘贴面
insulating trip 绝缘片
insulating tube 绝缘管;保温管
insulating-type particle board 软质刨花板
insulating unit 隔热隔声单元
insulating value 绝缘值;保温值
insulating varnish 绝缘清漆
insulating varnish and enamel 绝缘涂料
insulating varnished silk 绝缘漆绸;黄蜡绸
insulating varnished sleeving 绝缘漆管
insulating wall 隔热隔声墙;保温墙;绝缘墙
insulating wallboard 绝缘墙板
insulating wallpaper 保温墙纸;绝缘墙板
insulating washer 绝缘垫圈
insulating wax 绝缘蜡
insulating window 隔热隔声窗;隔热保温窗
insulating wire 绝缘线
insulating wool 保温棉
insulating work 隔热保温工程
insulation 保温;绝缘
insulation against heat 热的绝缘
insulation against oscillation 隔震
insulation against vibration 隔振
insulation ag(e)ing 绝缘老化
insulation ag(e)ing rate 绝缘老化速度
insulation allowance 绝缘容差;绝缘公差
insulation batt 绝缘合成板
insulation between plates 片间绝缘
insulation blanket 隔热衬垫;绝缘毯;绝缘合成板;保温毯
insulation blocking 隔离保护
insulation board 绝热板;隔声板;绝缘板;保温板
insulation breakdown 绝缘破损
insulation brick 隔热隔声砖;隔热砖;绝热砖;保温砖
insulation building material 隔声建筑材料;隔热(隔声)建筑材料

insulation by aerated concrete 加气混凝土保温
insulation can 隔热容器;保温箱
insulation class 绝缘等级
insulation coating 绝缘涂层
insulation control 绝缘控制;绝缘监督
insulation coordination 绝缘协调;绝缘配合
insulation core 绝缘芯子
insulation cork board 隔热软木板
insulation course 隔离层;保温层
insulation distance 绝缘距离
insulation facing 隔热隔声面层
insulation failure 绝缘失效;绝缘破坏;绝缘击穿;绝缘故障;保温失效
insulation fault 绝缘破损;绝缘不良
insulation fiber 绝缘纤维;绝热纤维;保温纤维
insulation figure 保温值
insulation film 绝缘薄膜
insulation for wall 墙身保温
insulation gap 绝缘间隙
insulation glass 隔热隔声玻璃
insulation glass unit 中空玻璃
insulation-grade block 绝缘等级块体
insulation hatch 防热舱盖
insulation heat fast attain method 绝热快速到达法
insulation heat fast pass method 绝热快速通过法
insulation in bags 袋装保温隔声材料
insulation in guy 拉线绝缘
insulation inspecting device 绝缘监视装置
insulation installation 隔热隔声装置
insulation integrity 全绝缘
insulation jacket 保温隔声罩
insulation joint 绝缘承口;绝缘节;绝缘接头;绝缘接缝
insulation lamina 绝热带
insulation lath 绝缘灰板条
insulation layer 隔热隔声层;保温层;隔离层;绝缘层
insulation level 绝缘水平;绝缘等级
insulation lining 隔热内衬;绝缘内衬;隔热隔声内衬
insulation material 隔热隔声材料;隔热隔声材料;绝缘材料;保温材料;绝热材料
insulation material against structure-borne sounds 隔热结构传声材料
insulation material of vibration 防振材料;防震材料
insulation measurement 绝缘电阻测量
insulation medium 绝缘介质
insulation of digester 消化池保温
insulation of furnace wall 窑体保温
insulation of impact sounds 撞击声选择
insulation of pipes in heating network 管道保温
insulation of reinforcement 钢筋保护层
insulation paint 绝缘漆
insulation paper 隔热纸
insulation paste 绝缘胶;绝热胶
insulation plate 绝缘板
insulation porcelain 绝缘瓷(瓶)
insulation powder 绝缘粉;保温粉
insulation product 绝缘产品;隔热隔声产品
insulation profile 隔热隔声型材
insulation protection 绝缘保护
insulation pumice 保温浮石
insulation puncture 绝缘击穿
insulation quilt 保温垫子;保温毡
insulation resistance 绝缘电阻
insulation sampler 绝缘采样器;保温取样器
insulation section 隔热隔声型材
insulation sheath 绝缘外皮
insulation shielding 绝缘屏蔽
insulation sleeve 绝热套;保温套
insulation slit 绝缘缝
insulation space 隔热处所
insulation space factor 绝缘占空系数
insulation spacing 绝缘间距
insulation straw board 隔热隔声纤维板
insulation strip (把变速车道和公路本身分隔开的)分隔带
insulation structural material 隔热隔声结构材料
insulation support ring for vertical vessels 立式容器的保温支承圈
insulation tape 绝缘包布
insulation test 绝缘(层电阻)试验;绝缘测试
insulation thickness 绝缘厚度;保温层厚度
insulation to falls 保温层找坡
insulation value 绝缘值;保温值
insulation wall 绝缘壁;绝热壁

insulation wallboard 隔热隔声墙板
insulation window 保温窗
insulation wool 保温棉
insulation work 保温
insulation workshop 保温车间
insulative curtain 绝热幕帘
insulative water paint 水性绝缘漆
insulativity 绝缘性;绝缘度;比绝缘电阻
insulator 隔离物;绝缘子;绝缘物质;绝缘体;绝缘器;绝缘块
insulator bolt 瓷瓶穿钉
insulator bracket 绝缘子托架
insulator chain 绝缘子串
insulator failure 绝缘体失效
insulator flashover 绝缘子闪络
insulator glass 绝缘子玻璃
insulator leakage distance 绝缘子泄漏距离
insulator pin 绝缘体销
insulator series 绝缘子串
insulator shed 绝缘子裙部
insulator string 保温线;保温绳
insulband 绝缘绑带
insulcrete 绝缘混凝土
insulcrete concrete floor 绝缘混凝土地板
insulcrete slab 绝缘混凝土板
insulex (隔墙等用的)石膏绝热材料;平顶
insulite 木纤维墙板
insulite floor 胶料地板
insullac 绝缘漆;电绝缘漆
insulosity 岛屿度;岛(屿)率(即岛屿面积与总水面之比)
insulwood 木纤维墙板
insuperable barrier 不可逾越的鸿沟
insurable 应保险;可以保险的
insurable industry 保险业
insurable interest 可保利益
insurable liability 保险责任
insurable portfolio 保险业务总量
insurable property 可接受保险的财产;可保财产
insurable risk 可受保的风险
insurable title 可以投保的所有权
insurable value 可保价值
insurance 卸货费在内的到岸价格条款;船舱底交货的到岸价格;舱底交货的到岸价格;保险加空运费价格;保险;安全保障
Insurance Accounting and Statistic(al) Association 保险会计和统计协会
insurance against accident 事故保险
insurance against accident to workmen 人员事故保险;工人的事故保险;人工伤害保险
insurance against all-risks 保综合险;保全险;全能保险
insurance against breakage 保破碎险
insurance against litigation 诉讼保险
insurance against loss or damage 遗失或损坏保险
insurance against theft 防失窃保险
insurance agent 保险代理人
insurance amount 投保金额;保险总额
insurance and freight 保险费和运费
insurance and freight net 净到岸价格
insurance application 保险申请
insurance appraiser 保险损害评价人
insurance broker 保险代理人
insurance business 保险业业务;保险事业
insurance by way of indemnity 补偿保险
insurance cable 保险缆索
insurance canvasser 保险推广员
insurance certificate 保险凭证;分保单;保险单
insurance charges 保险费
insurance cheater 安全带(井架工用)
insurance claim 保险索赔
insurance clause 承保条款;保险条款
insurance company 保险公司
insurance compensation 保险赔偿金;保险赔偿费
insurance condition 保险条件
insurance conglomerate 保险承保集团
insurance contract 保险契约;保险合同
insurance cost 保险费
insurance cover 保险总额;保险责任;保险范围;保险保障
insurance coverage 保险金额
insurance documents 保险单据
insurance estimating 保险费预算
insurance expenses 保险费(用)

insurance fee 保险费
insurance for impaired lives 伤残保险
insurance for land transportation 陆运保险
insurance for liability 对责任的保险
insurance for life 终身保险;人寿(保)险
insurance for medical care 医疗保险
insurance for old age 老年保险
insurance for the transportation of imported and exported goods 进出口物资运输保险
insurance for unemployment 失业保险
insurance for war risks 兵险;战争(风)险
insurance freight liner terms 班轮到岸条款
insurance fund 保险基金
insurance fund reserve 保险基金准备
insurance hawser 应急拖缆
insurance indemnity 保险赔偿金;保险赔偿费
insurance instruction 投保通知
insurance item 保险项目
insurance law 保险法
insurance law of labo(u)r accident 职工意外事故保险法
insurance liability 保险责任
insurance lock 保险门锁
insurance mechanism 保险机制
insurance money 保险赔偿金;保险金
insurance notification 保险通知
insurance of client's property 业主财产保险
insurance of credit guarantee 信用保险
insurance of hull and appurtenances 船壳保险
insurance of property 财产保险
insurance of workmen 劳工险
insurance of works 工程保险
insurance on goods 货物保险
insurance period 保险期限
insurance policies in force 生效保险单
insurance policy 保险凭证;保险单
insurance policy holder 投保方
insurance portfolio 保险业投资组合
insurance premium 保险奖金;保险费
insurance premium deduction 保险费扣除
insurance premium of commercial enterprises 商业企业保险费用
insurance premium unexpired 未过期保险费
insurance prepaid 预付保险费
insurance proceeds 保险赔偿费
insurance rate 保险费率
insurance ratio 保费率
insurance rebate 保险回扣
insurance reserve 保险准备
insurance risk 保险危险
insurance slip 投保单;保险申请
insurance subject 保险标
insurance surveyor 保险损害鉴定人
insurance till death 人寿(保)险
insurance title 保险项目名称
insurance trust 保险信托
insurance under a life policy 人寿保险单下的保险
insurance value 保险值;保险价值;保险额
insurant 受保人;投保人;投保方;被保险者;被保险人
insure 投保
insure against death 保寿险
insure against fire 保火险
insured 被保险人;保价邮件
insured advances 保险预付款
insured amount 投保金额;保险金额
insured bank deposits 已保险银行存款
insured closing letter 保险成交文件
insured goods 被保险货物
insured interests 被保险物
insured letter 保价信
insured loan 有保险的贷款
insured mortgage 保险抵押
insured parcel 保价包裹
insured peril 专项保险
insured savings 保险储蓄
insured unemployment rate 已保险失业率
insured value 投保价值;保险价值
insured value clause 投保价值条款
insurer 承保人;保险商;保险人;保险公司;保险方
insurer security 保险人保证书
insure works 投保工程险
insuring to value 按实价保险
insurmountable barrier 不可逾越的障碍;不可逾越的鸿沟

insurrection 暴动
insusceptibility 不易感受性;不敏感性;不灵敏性
insusceptible 不灵敏的
insusceptible to moisture 耐潮的
in suspension 悬浮着的;悬浮状态的
insweep 下半部瓶身缩小
inswept 前端窄
inswept frame 窄式车架;流线型机架;前端窄缩式车架
in-swinging casement 内开扇;向内开的窗扇
in-swinging casement window 内开(式)窗
in-swinging window 活动窗;旋转窗
in-swing side hinged 内开边梃铰接的
in switch 输入开关
insymbol 内符号;内部符号
in synchronism 协调;同步
intack 无损
intact 未扰动的;完整无损的;完整的
intact bulkhead 整体舱壁
intact buoyancy 无损浮力;完整浮力
intact clay 无裂隙的黏土;无损伤的黏土;原状黏土;原地黏土;未扰动黏土;完整黏土
intact condition of karst 洞穴顶板岩体完整情况
intact ecosystem 完整的生态系统
intact gravel framework 完整砾石格架
intact ground 未扰动地基
intact leaves 完整叶
intactness coefficient of rock mass 岩体完整性系数
intactness index of rock mass 岩体完整性指数
intact overconsolidated clay 未扰动超固结黏土
intact overconsolidated soil 未扰动超固结土
intact rock 完整岩石
intact rock bridge 完整岩桥
intact sample 原始样品;原始样本;完整样品
intact shear strength 原地剪切强度;未扰动剪切强度;完整剪切强度
intact ship 未受损船
intact specimen 原状试件
intact stability 无损稳定性;完整稳性
intact wall 整体墙
intagliated 压纹的;凹雕的
intagliated ornament 凹刻装饰
intaglio 阴雕;玻璃刻花;凹刻;凹雕;凹板
intaglio engraved process 凹刻制版术
intaglio ink 凹版油墨;凹板墨
intaglio plate 凹版印版
intaglio printing 铜板移印法;凹版印刷
intaglio relievo 平面浮雕
intaglio rilevato 凹浮雕
intaglio tile 压纹瓦;凹雕瓦
intakage of air 空气渗入;空气入渗
intake 进入口;吸入(量);引入量;装船的货物数量;灌注量;管道入口;井口;进气巷道;进口;进给;进风量;取水口(取水头部);输入能量;摄取;垂直补给
intake adit 进风巷道
intake advance 进气提前
intake air 进入空气
intake area 回灌区;进水面积;进口(截面)面积;受水区;垂直补给区
intake area of aquifer 含水层受水区;含水层来水区;含水层补给区
intake auxiliary air 辅助进气口
intake basin 压力前池;进水池;进入池;取水池
intake belt course 入口线脚砌筑层;缩窄带
intake brine 压入盐水
intake bulk station 进料站
intake canal 进水渠
intake capacity 吸收能力(钻孔);容纳量
intake casing 进气口外室
intake chamber 引入线室;进水(闸)室;进水室;进水间
intake channel 进水渠;进水槽
intake check ball 进口止球阀;进口止回球
intake check valve 进水止回阀
intake condition 进口参数
intake conduit 进水渠;进水管(道);进(口)水道
intake conduit pipeline 进水管线
intake cone liner [upper, lower] 料斗衬板(上、下)
intake dam 引水坝;进水口挡水墙;分水坝
intake deck 进水口平台
intake depot 进料站
intake duct 进水渠;进入道;进气装置;进气管

intake end 进口端
intake energy 进口能
intake facility 进料设施
intake fish baffle 进水口挡鱼结构;进水口挡鱼隔板
intake flange 吸入口法兰
intake flow 吸入流动
intake gallery 引水隧洞;进水隧洞;进水廊道
intake gate 进水闸(门);进料闸(门)
intake gate road 进水闸门巷道
intake grate 进水(口护)栅
intake guidance 吸气导管
intake guide 吸气导管
intake guide unit 导流装置;吸气导管装置
intake head 吸入高度;进风压头;引水渠首;渠首工程
intake header 进水头部;进水干管;取水口(取水头部)
intake heading 引水渠首;拱顶石饰;进水渠首;取水首部
intake hopper 进料漏斗;进料仓
intake installation 快速接头;进料设施
intake line 吸油管路;进入管道;进口管线
intake loss 进水损失;进水损耗;进口损失
intake loss coefficient 吸入损失系数
intake main 进水总管;进水干管;取水干管
intake main of lock chamber 闸室进水总管;船闸闸室进水总管
intake manifold 吸入歧管;进水汇管;进气歧管
intake manifold stud 进气歧管柱螺栓
intake measure 装船尺寸
intake nozzle 吸气管嘴
intake of groundwater 地下水补给
intake of hydropower station 水电站进水口
intake of mill 磨机进料口
intake of the hole 钻孔吸收(泥浆)能力
intake of water works 水厂取水口;自来水厂进水口;水厂进水口
intake of well 取水戽斗
intake opening 吸入口;进风口
intake period 吸气期
intake pipe 吸油管;进水管;进气管;取水管
intake pipe gallery 进水管线廊道
intake pipeline 进口管线
intake plant 进料站
intake port 进水口
intake pressure 进气压力;进口压力
intake project 进料工程
intake rate 下渗率;进水速率;入流率;取水率;土壤吸水速率
intake recharge (含水层的)引入回灌量;含水层补给量;垂直补给
intake resistance 进气阻力
intake screen 吸入滤网;莲蓬头;进水筛网;进料滤网;进口滤网;进口拦污栅;进口拦栅;取水口格栅
intake shaft 输入轴;进风井;进水竖井
intake shutter 进气活门片
intake silencer 进口消声器;入口消声器
intake sluice 进水闸(门)
intake strainer 吸入滤网;莲蓬头
intake stroke 吸入冲程;进气冲程;进给冲程;进气行程
intake structure 引水建筑物;引水构筑物;进水(口)建筑物;进水建筑物;取水建筑物;取水构筑物;通风建筑物
intake system 吸入系统
intake tank farm 进料罐区
intake terminal 入口端
intake tower 进水塔;排水塔
intake tunnel 进水隧道;进水隧洞;引水隧洞;引水隧道;取水隧洞
intake valve 进水阀;进汽阀;进气阀(门);进口阀(门)
intake valve cage 进气阀罩
intake velocity 进水流速;进气速度;进口速度
intake weight 运入重量
intake well 取水井;进水井;注入井
intake well head pressure 注入井井口压力
intake with automatic flushing 自动冲沙进水口
intake works 进水建筑物;引水建筑物;引水构筑物;进水工程;进料工程;取水建筑物;取水构筑物;取水工程
intaking without dam 无坝取水
intalio press 凹版印刷机
intalox saddle 矩鞍形填料;槽鞍形填料

in tandem 相互合作地;一前一后的;纵列(的);串列(的);串联的;成纵列的;串联(轧钢机)
in-tandem model 串联模型;串联模式
intangible 不能触摸的
intangible analysis 无形因素评价
intangible assets 固定无形资产;无形资产
intangible assets acquisition audit 无形资产取得审计
intangible assets amortization audit 无形资产摊销审计
intangible assets audit 无形资产审计
intangible assets investment audit 无形资产投资审计
intangible assets transfer audit 无形资产转让审计
intangible benefit 无形效益;不可计算的利益
intangible cost 无形费用;无形成本
intangible depreciable property 无形应折旧财产
intangible development cost 无形开发成本
intangible drilling cost 无形钻探费
intangible factor 难以确定的因素
intangible fixed assets 无形固定资产
intangible flood damage 难确定的洪水灾害
intangible goods 无形商品
intangible hand 无形手
intangible investment 无形投资
intangible item 无形资产项目
intangible merchandise 无形商品
intangible property 无形财产
intangible property rights 无形财产权
intangible resources 无形资源
intangible standard 无形的标准
intangible tax 对某些无形资产的捐税
intangible value 无形价值
in-tank 罐内
in-tank mixing 罐内混合
intarometer 盲孔千分尺
intarsia 镶嵌饰;细木镶嵌装饰;木镶嵌饰;嵌木制品;嵌木细工;嵌花
intarsia flat knitting machine 嵌花横机
intarsia pattern 嵌花花纹
intarsist 细木镶嵌工
intecell 星形仓
intectate-reticulate 无覆盖层
integer 整体;整数【数】
integer angular momentum 整数角动量
integer attribute 整数属性
integer constant 整(数)常数
integer conversion 整数转换;定点数据转换
integer data 整型数据
integer expression 整型表达式
integer form 整型形式
integer harmonics 整数倍谐波
integer item 整数项
integer linear programming 整体线性规划;整数线性规划
integer multiple 整数倍(数)
integer overflow trap 整数溢出自陷
integer part 整数部分
integer programming 整体程序设计;整数规划
integer programming problem 整数规划问题
integer quantity 整型量
integer representation 整数表示
integer-slot winding 整槽绕组
integer solution 整数解
integer spin 整数自旋
integer type 整型
integer value 整数值
integer variable 整型变量
integer variable dimension 整型变量维数
integer variable name 整型变量名
integer variable reference 整型变量引用
integer vector 整型向量
integrability 可积性;可积分性
integrability condition 可积(分)条件
integrable 可积分的
integrable function 可积函数;可积分函数
integral 总体;整体积分;整体的;全悬挂
integral abrasive edging 整体式防滑条(浇筑前嵌入楼梯踏板的防滑条);整体式耐磨镶边
integral absorbed dose 积分剂量
integral accelerator 积分加速器
integral action 整体作用;积分作用;积分演算;积分动作
integral action coefficient 积分作用系数;积分演

算系数
integral action limiter 积分作用限幅器;积分动作限制器;积分动作限制程序
integral air pump 整装抽气泵
integral air-scour 整体气洗
integral algebraic(al) number 代数整数
integral algebraic(al) function 整体代数函数
integral approach 积分近似
integral approach to the solution 积分求解法
integral armband 整体式表带
integral atomic time 累积原子时
integral basis 整体基础;整基
integral bearing 浇铸轴承
integral bearing housing 整体轴承箱体
integral bell joint 整体承口接头
integral bilinear form 整双线性型
integral binary cubic forms 整二元三次型
integral bit 整体式钎头
integral blade 整体叶片;带叶冠叶片
integral body 整体车身
integral boiling reactor 整体沸腾式反应堆
integral boundary 整数界限
integral bracelet 整体式表带
integral bridge abutment 整体式桥台
integral bumper 整体组合式减振器;整体挡板
integral cage 整体保持架
integral calculus 积分学
integral cam 整体凸轮轴
integral cam shaft 整体(铸造的)凸轮轴
integral capacitance 总电容
integral cap nut 整体帽螺母
integral case 整体外壳
integral casing 无接箍套管
integral cast 整体浇灌
integral cast handle 固定把手
integral cast(ing) 整体(浇)铸;整体浇注;整体浇筑;整体浇灌
integral cast-iron dished head 整体铸铁碟形封头
integral chart recorder 积分图记录器
integral check valve 内装式单向阀;内装单向阀
integral choice pattern 整选择模式
integral circuit 积分电路
integral circulating pump 整体式的循环泵
integral closure 整闭包
integral coefficients 整系数
integral colo(u)r 总的色泽;整体颜色;内色
integral colo(u)ring 内部着色;全部着色;整体着色
integral colo(u)ring anodizing[anodising] 整体阳极色彩处理
integral colo(u)ring anodizing process 整体阳极上色程序
integral compensation 积分补偿
integral composite structure 整体复合材料结构
integral compound 灰浆防水掺和料;混凝土防水掺和料
integral computer 整数计算机
integral condenser 整体式凝气器
integral constant 整常数;积分常数
integral control 总体控制;整体控制;积分控制;积分调节
integral control action 整体控制作用
integral controlled action 整体调整作用
integral controller 积分控制器;积分调节器
integral convergence 整数收敛
integral convergence test 积分收敛判别法
integral convolution 积分褶合式
integral cooler 装在回转窑上冷却器;多筒冷却器;多筒冷却机
integral cooling system 整体化冷却系统
integral coordination 一条龙协作
integral cosine 余弦积分
integral counterweight of crankshaft 整体式曲轴平衡重
integral counterweight 整体式平衡重
integral coupling 整体联轴器
integral cover 整体围带
integral curb (与路面整体连接的)整体路缘;整体道牙
integral current 全流
integral curve 累积曲线;积分曲线
integral denotation 整数标志;整标志
integral density 总电谱密度;累积密度;积分密度
integral design 总体设计;综合设计
integral detector 积分检测器

integral-differential equation 积分微分方程
integral differential operation 积分微分运算
integral digital map 积分式数字地图
integral discriminator 积分鉴别器
integral-disk rotor 整锻转子
integral distribution curve 累积分布曲线;积分分布曲线
integral divisor 整因子
integral domain 整域;整环;积分区域
integral dose 积分剂量
integral dot 积分点
integral drill bit 整体式钻头
integral drill steel 整体钻钎
integral drive 整体式驱动装置;整体式传动装置
integral electrical heating 整体电加热
integral enclosure 整体密封罩
integral equalizer 积分稳定环节
integral equation 积分方程
integral equivalent 等价整数
integral erection 整体安装
integral experiment 整体实验
integral exponent 整指数
integral exposure meter 积分曝光表
integral expression 整式;积分表达式
integral factor 积分因数
integral feasibility solution 整数可行解
integral fin 整体散热筋;集成散热片
integral fixedbed reactor 固定床积分反应器
integral flow curve 径流累积曲线;径流积分曲线
integral form 整型
integral formula 积分公式
integral formwork 整体模板
integral frame 整体门框;集装门框
integral fuel cell 整体燃料箱
integral-function 整函数
integral-furnace boiler 水冷炉膛锅炉
integral gear rotor 整体带齿旋转活塞
integral glow curve 积分辉光曲线
integral governing 积分调整
integral group algebras 整群代数
integral gyro 积分陀螺仪
integral hardener 增硬剂;整体母合金
integral hardening agent 整体硬化剂
integral heat 变浓热
integral heat-exchange unit 内装热交换装置
integral heat of solution 溶解热
integral height indicator 积分测高仪
integral hoisting 整体吊装
integral hoisting process 整体吊装法
integral horsepower motor 整数马力电动机;大于1马力的电动机
integral housing 整体式壳体;整体桥壳
integral hub bearing 整体式轮毂轴承
integral ideal 整数理想子环
integral inner ring 整体内圈
integral insulation 整体绝缘
integral intake manifold 整体进气歧管
integral intensifier system 总体式增压器
integral invariant 积分不变式
integrality 整体性;整体
integralization 整化
integral joint 整铸接头
integral-joint casing 整体连接套管
integral joint tubing 整体连接水管
integral key 花键
integral key shaft 花键轴
integral kiln 整体窑
integral lath(ing) 抹灰钢丝网
integral lighting switch 整体照明开关
integral linear mapping 整线性映射
integral linear polarization 累积线偏振
integral linear programming 整数线性规划
integral line-breadth 积分谱线宽度
integral lining 整体式衬砌
integral linkage power steering system 整体杆系式动力转向系统
integral lock 整体锁;固定锁
integrally annealed 整体退火
integrally cast 整体浇注
integrally closed ring 整闭环
integrally colo(u)red 整体上色的
integrally finned tubes 整体翅片管
integrally forged impulse wheel 整锻冲动式转轮
integrally forged monolayered cylinder 整体锻造式单层圆筒
integrally heated 整体加热的
integrally raising of fractured wreck 断船整捞
integrally stiffened plating 倒L形铝板加劲板
integral mesh 结合网
integral mesh tube 结合网管
integral metal 整体轴承;浇铸轴承合金
integral method 积分法
integral method of waterproofing 整体防水法;混凝土内加入特种外加剂防水方法
integral-mode controller 积分型控制器
integral motor 积分电动机;内装式电动机
integral mo(u)ld 整模型
integral mullion 窗间立柱
integral multiple 整数倍(数)
integral norm 整范数
integral number 整数【数】
integral number of times 整倍数
integral of angular momentum 角动量积分
integral of energy 能量积分
integral of force vive 活力积分
integral of living force 活力积分
integral one-piece mainshaft (多缸转子发动机用的)整体式偏心轴
integral operation 积分运算
integral operator 积分算子;积分算符
integral outer ring 整体外圈
integral packaging 集成包装;完整包装
integral part 组成部分;主要部分;整体(组成)部分;积分部分;不可分割部分
integral pattern 整模式
integral photometer 积分光度计
integral pilot 整体超前型钻头
integral pitch winding 整距绕组
integral plate 整体垫板
integral plotting surface 整体测图面
integral point 整体式探头;整数点
integral procedure decomposition temperature 积分过程分解温度
integral process 整体方法(混凝土中加入粉、液或浆以填充孔隙提高防水性能);整数法
integral programming 整体规划
integral proofer 整体防护料
integral quadratic forms 整二次型
integral quantity 整数值;积分值
integral rational function 整有理函数
integral rational invariant 整有理不变式
integral rational operator 整有理算子
integral reactor 一体化反应堆
integral refractometer 积分折射率仪
integral reinforcement 整体补强
integral relation 积分关系式
integral relief valve 整体式减压阀
integral representation 整数表示;积分表示
integral resonance 整数共振
integral restrictor 积分限制器;积分开关
integral ring 整体托圈;整数环
integral rotor 整体转子
integral sampling 累积采样
integral scattering 积分散射
integral schematic description 总概要说明
integral sensitivity 积分灵敏度
integral shaft 整体轴
integral slot winding 整数槽绕组
integral solution 整数解
integral sound suppressor 内装式消声器
integral spectrum 积分谱
integral spin 整数自旋
integral square error 平方误差积分
integral square-error approximation 积分平方误差近似
integral square-error method 误差平方积分法
integral stability of boom 吊杆总体稳定性
integral steel 整体钎子;整体钢钎
integral steel casting 整体铸钢件;整体钢铸件
integral structure 整体结构
integral surface 积分曲面
integral system 混凝土掺和防水材料工艺;灰浆掺和防水材料工艺
integral tank 整体式水箱;必备水箱
integral temperature 积分温度
integral test 综合测试;积分判别法;积分检验法
integral testing 整体试验
integral test system 综合测试系统;综合测试设备
integral time 积分时间
integral time scale 时间积分比尺
integral(tip) shroud 整体围带
integral tone 积分音调
integral tool joint 整体钻杆接头
integral transcendental function 整超越函数
integral transform 积分变换
integral transformation 积分变换
integral transform method 积分变换法
integral transmission package 整体式传动装置
integral transmittance 积分透过率
integral treatment 多合处理
integral tripack 累积彩色底版
integral type 整体式
integral type flange 整体法兰;整体凸缘
integral type powersteering 整体型转向加力装置
integral unit 联合机组;全套部件;成套机组
integral unit drive 直接传动(的)
integral value 整数值;积分值
integral valued polynomial 整值多项式
integral water jacket 整体式水套
integral water-proofer 整体防水剂;总体防水剂
integral waterproof(ing) 刚性防水;整体防水;总体防水
integral waterproofing of concrete 混凝土整体防水
integral water repelling agent 整体防水剂
integral weight distribution curve 积分重量分布曲线;积分加权分布曲线
integral wreck-raising 沉船整体打捞
integrand 被积函数
integrant 要素;成分
integraph 积分仪;积分曲线仪;积分描图仪
integrate 结合;集成
integrate control 计划防治
integrated absorption 积分吸收
integrated absorption coefficient 累积吸收系数
integrated absorption cross-section 积分吸收截面
integrated absorption method 积分吸收法
integrated adapter 集成转接器;集成适配器
integrated aeration-filtration reactor 一体式曝气过滤反应器
integrated agriculture 综合农业
integrated analysis 综合分析
integrated appraisal 综合评价
integrated attachment 集成附件
integrated automatic control system 综合自动控制系统;综合自动化系统
integrated automation 综合自动化;整体自动化;集成自动化;全盘自动化
integrated ballast bed 整体式道床
integrated barge 组合式驳船;分节驳船
integrated barge fleet 分节驳船队
integrated barge train 分节驳船队
integrated bathroom 整体卫生间
integrated belt system 组合胶带运输系统;整套输送带系统;整套皮带系统
integrated biochemical process 生物化学联用法
integrated biochemical reactor 一体式生化反应器
integrated biological pond system 一体式生物塘系统
integrated biology-chemistry equipment for sewage treatment 一体式生化污水处理装置
integrated biomass separation membrane bioreactor 一体式生物质分离膜生物反应器
integrated bridge system 组合驾驶台系统;组合船桥系统
integrated brightness 累积亮度
integrated budget formula 综合预算方案
integrated building system 综合建筑系统;建筑体系
integrated cancellation ratio 积累抵消系数;积累抵消率
integrated cartridge 集结药包
integrated case studies 综合个例研究
integrated catchment management 流域综合管理
integrated ceiling 整体式综合吊顶;合成顶棚
integrated ceiling system 复合顶棚体系
integrated channel 整体地沟;集成电路通道
integrated chemical biological flocculation-suspended media blanket process 一体式化学生物絮凝悬浮介质床工艺
integrated chopper 集成斩波器
integrated circuit 集成电路;积分电路

integrated circuit amplifier 集成电路放大器
integrated circuit block 集成电路块
integrated circuit board 集成电路板
integrated circuit breadboard 集成电路试验板
integrated circuit card 集成电路卡
integrated circuit chip 集成电路片
integrated circuit communication data processor 集成电路通信数据处理机
integrated circuit die 集成电路晶片
integrated circuit distance relay device 集成电路距离保护装置
integrated circuit double waveguide laser 集成双波导激器
integrated circuit element 集成电路元件
integrated circuit model 集成电路模型;集成电路模式组件
integrated circuit package 集成电路组件;集成电路封装
integrated circuit resistor 集成电路电阻器
integrated circuit simulator 集成电路模拟器
integrated circuit technique 集成电路技术
integrated circuit test clip 集成电路测试夹
integrated circuit test system 集成电路测试系统
integrated civil engineering 综合土木工程
integrated civil engineering system 综合土木工程系统
integrated coal gasification combined cycle 煤气化联合循环发电
integrated coil 整体式线圈
integrated colo(u)r index 累积色指数
integrated communication 综合通信;联合通信
integrated communication adapter 综合通信适配器;整体通信衔接器;集中通信适配器;集成通信转接器
integrated communication centre 联合通信中心
integrated communication instruction 联合通信指令
integrated communication network 综合通信网
integrated communications activity 联合通信站
integrated communications agency 联合通信局
integrated communication system 综合通信系统
integrated complex 联合企业
integrated component 集成元件;集成器件
integrated component circuit 整体元件电路;集成元件电路;积分组件电路
integrated computer system 集成计算机系统
integrated concentration and flow indicator 组合式浓度流量指示器
integrated console 综合控制台;联控台;集中控制台
integrated container service 集装箱联运业务;集装箱服务公司
integrated container terminal 集装箱联运站
integrated contractor 综合承包商;综合承包者;综合承包人
integrated control 协调防治;综合治理;综合控制;综合防治;综合调节
integrated control in plant 厂内综合治理
integrated control of atmospheric pollution 大气污染综合防治
integrated control of biologic(al) pollution 生物污染综合控制;生物污染防治
integrated control of environment(al) pollution 环境污染综合治理;环境污染综合防治
integrated control of groundwater pollution 地下水污染综合防治
integrated control of noise pollution 噪声污染综合防治
integrated control of urban environment 城市环境综合整治
integrated control of water pollution 水污染综合控制;水污染综合防治
integrated control system 综合控制系统;综合调解系统
integrated control technique 综合控制技术;综合防治技术
integrated conveyor system 成套输送系统
integrated cross-section 积分截面
integrated curb (与路面整体连接的)整体路缘
integrated curve 累积曲线;积分曲线;累积流量曲线
integrated cyclic(al) process 总体循环法
integrated data bank 综合数据库
integrated data base 综合数据库;集中数据库
integrated data handling file 综合数据处理文件
integrated data management system 综合数据管理系统
integrated data network 综合数据网
integrated data presentation and control system 综合数据显示及控制系统
integrated data processing 综合数据处理(法);整体数据处理;集中数据处理;统一数据处理
integrated data processing centre 综合数据处理中心
integrated data processing system 集总数据处理制度;集中数据处理系统
integrated data retrieval system 集中数据检索系统
integrated data storage 综合数据库;集成数据存储
integrated data store 集中数据库;集中数据存储
integrated data system 综合数据系统
integrated deep-water navigation system 深海组合导航系统
integrated demand 平均需气量;平衡负荷;总需求(量)
integrated demand meter 平均累计最大需量计
integrated design 总体设计
integrated development 综合开发;全面发展
integrated development environment 集成开发环境
integrated device 集成器件;集成半导体器件
integrated digital communication network 集成数字通信网
integrated digital logic circuit 集成数字逻辑电路
integrated digital network 综合数字网
integrated digital transmission and switching 综合数字传输和交换
integrated digital volt meter 集成数字伏特计
integrated display system 综合显示系统;航海数据综合显示系统
integrated distribution floor system 综合分布的楼面系统
integrated distribution network system 综合布线网络系统
integrated Doppler 积分多普勒
integrated drainage 合成水系
integrated economy 一体化经济
integrated electromagnetic velocity/radioactive density measuring instrument 电磁流速/放射性浓度综合测量仪
integrated electronic processor 综合电子信息处理机
integrated electronics 集成电子学
integrated electrooptics 集成电光学
integrated element circuit 集成元件电路
integrated emission 累积发射
integrated emulation 集中仿真
integrated emulator 集中仿真程序;集成仿真器;积分模拟程序;统一仿真程序
integrated energy density 积分能量密度
integrated enterprise 综合(性)企业
integrated environmental control 综合环境控制
integrated environment(al) pollution control 环境污染综合控制
integrated equipment 综合设备
integrated evaluation 综合评价
integrated evaluation test 综合评定试验
integrated excitation intensity 累积激发强度
integrated extrusion type 集成挤出型
integrated factor 集结系数
integrated-feed back laser 集成反馈激光器
integrated fiber drawing furnace 玻璃拉丝组合炉
integrated file adapter 整体文件存储衔接器
integrated fire control system 联合火控系统
integrated fixed-film activated sludge process 一体式固着膜活性污泥法
integrated fleet 组合式驳船队
integrated floating dock 组合式浮船坞
integrated flow curve 综合流量曲线;径流累积曲线;径流积分曲线;累积流量曲线
integrated flow-sheet 综合流程
integrated flux 累积流量;积分通量
integrated forest utilization 综合森林利用;森林综合利用
integrated format 整数形式
integrated fuel control carburetor 综合控制供油式化油器
integrated function 被积函数
integrated fund 综合基金
integrated geologic(al) data base 综合地质数据库
integrated geometric(al) factor 积分几何因子
integrated global navigation and surveillance system 组合全球导航与监视系统;综合全球导航监视系统
integrated global ocean services system 综合性全球海洋服务系统
integrated global ocean station system 综合全球海洋站系统
integrated groundwater quality model 综合地下水水质模型
integrated heating system 组合采暖系统;整体采暖系统
integrated hollow fiber membrane bioreactor 一体式中空纤维膜生物反应器
integrated hybrid biofilm reactor 一体式复合式生物膜反应器
integrated hydrolytic acidification-biological aerobic filter 一体式水解酸化好氧生物滤池
integrated identification system 识别综合系统
integrated incident light 累积入射光
integrated index of the anthropogenic load on river 河流人为负荷综合指数
integrated inductor 集成电感器
integrated inertial navigation system 集成式惯性导航系统
integrated information processing 综合信息处理;集中信息处理
integrated information system 综合信息系统;综合情报系统
integrated injection logic 集成注入逻辑
integrated injection logic circuit 集成注入逻辑电路
integrated injection type 集成注入型
integrated instrument panel 综合仪表盘
integrated intensity 累积强度
integrated intercom set 综合内部通信设备
integrated interferometric reflector 集成干涉反光镜
integrated investigation of river basin 流域综合查勘
integrated iron and steel plant 钢铁联合工厂;综合钢铁厂
integrated iron and steel works 综合性钢铁厂;钢铁联合企业
integrated lake-watershed acidification study 湖泊流域酸化综合研究模型
integrated level 集成度
integrated light intensity 集束光强度;积分光强度
integrated line 综合生产流程
integrated line network 综合式路网
integrated load curve 荷载累积曲线
integrated logging 综合性采伐;集成伐木;木材综合(利用)采运;木材综合利用
integrated logic circuit 集成逻辑电路
integrated magnitude 累积星等
integrated management 综合管理
integrated management information 集中管理信息
integrated management information system 综合管理信息系统
integrated management method 综合管理方法
integrated management of coal mines 煤矿综合经营
integrated management of water quality and quantity 水质水量综合管理
integrated management planning 整体化管理计划
integrated manager 结合型经理
integrated manufacturing system 综合生产系统;集成制造系统;设计制造一体化
integrated mapping system 集成制图系统
integrated marketing system 综合销售系统
integrated material handling 综合的材料处理
integrated membrane bioreactor 一体式膜生物反应器
integrated membrane bioreactor system 一体式膜生物反应器系统
integrated membrane-composite bioreactor 一体式膜复合生物反应器
integrated membrane process 一体式膜工艺
integrated membrane-sequencing batch bioreactor 一体式膜序批间歇式生物反应器
integrated membrane system 一体式膜系统
integrated mica 结合云母
integrated microwave/ultraviolet-illumination method 一体式微波/紫外光照法
integrated mill 大型工厂

integrated mode 组合模式
integrated modem 集成调制解调器
integrated monitoring 综合监测
integrated morphology 集成形态学
integrated motor 整体式电动机;积分直流电动机;积分马达;机内电动机
integrated motorists information system 驾驶员综合信息系统
integrated moving average process 结合移动平均数过程
integrated multipurpose water control system 综合多目标水管理系统
integrated national telecommunication system 国家综合电信系统
integrated navigation 组合导航
integrated navigation and autopilot system 组合导航与自动驾驶系统
integrated navigation and collision avoidance system 组合导航及避碰系统
integrated navigation computer 组合导航计算机
integrated navigation display 综合导航显示器
integrated navigation equipment 组合导航设备
integrated navigation system 组合导航系统;综合航行系统;综合导航系统
integrated network 积分网络;综合网络;综合规划;总体网络;集成管网;复杂管网
integrated neutron flux 积分中子通量
integrated noise 累积噪声;积分噪声
integrated noise temperature 累积噪声温度
integrated observation system 综合观测系统
integrated ocean pollution control 海洋污染综合控制
integrated ocean surveillance system 海域环境容量;海洋综合监视系统
integrated oil company 综合性石油公司;大型石油(联合)公司
integrated operating system 集中操作系统
integrated operation 整体操作;联合操作
integrated operational amplifier 集成运算放大器
integrated operation training system 综合运营培训系统
integrated optic(al) bolometer for radiation 集成光学辐射热测量计
integrated optic(al) circuit 集成光学电路;集成光路
integrated optic(al) fiber system 综合光纤系统
integrated optic(al) spectrum analyzer 集成光谱分析器
integrated optic(al) switch 集成光学开关
integrated optic(al) waveguide coupler 集成光学波导耦合器
integrated optics 集成光学;集成光路
integrated ozone value 臭氧总值
integrated package 集成块
integrated peak 积分峰(值)
integrated pest control 病虫害综合防治
integrated pesticide management 农药综合治理
integrated pest management 有害生物综合治理;害物综合防治;虫害综合治理
integrated photoelasticity 积分光(测)弹性(学)
integrated photographic magnitude 累积照相星等
integrated photometer 积分光度计
integrated plan(ning) 统筹计划;综合规划;整体计划;整体规划
integrated planning of water pollution control 水污染综合控制规划
integrated pollution control 污染综合控制
integrated pollution control program(me) 污染综合控制规划
integrated positioning 组合定位
integrated power grid 综合电力网
integrated precision testing method 积分精度测试法
integrated prevention and treatment of pest 害虫综合防治
integrated production line 综合生产线
integrated program(me) for commodities 商品综合方案
integrated program(me) for rail transport 铁路运输综合作业方案
integrated project 综合计划;综合工程
integrated public transport system 一体化公共交通系统
integrated puff model 积分烟团模式

integrated radiance 积分辐射率
integrated radiant emittance 总辐射能;总辐射度
integrated radiation 累积辐射
integrated railway radio system 集成铁路无线电通信系统
integrated real-time contamination monitor 实时污染综合监测仪
integrated refinery 整体化炼厂;联合炼油厂
integrated reflecting power 累积反射率
integrated reflection 累积反射
integrated reflectivity coefficient 积分反射系数
integrated regional cooperation 一体化区域合作
integrated research vessel 综合调查船
integrated resistor 集成电阻
integrated reuse-recycle treatment process 综合利用循环处理过程
integrated risk assessment index system 综合风险评价指标体系
integrated river-basin development 江河流域综合开发;流域综合开发
integrated Ro/Ro system 轮渡式综合运输系统
integrated running test 综合运行测试
integrated rural development 农村综合发展
integrated sample 综合样品
integrated satellite navigation equipment 综合卫星导航设备
integrated satellite terrestrial network 卫星地面整体网络
integrated selective key 综合选择样片
integrated semi-conductor circle 半导体集成电路
integrated sensor 集成传感器
integrated service 集成运营
integrated service digital network 综合业务数字网
integrated service digital network access capability 综合业务数字网的接入能力
integrated service digital network connection 综合业务数字网的连接
integrated service digit network 综合服务数字网(络)
integrated service network 综合业务(通信)网
integrated sewage membrane bioreactor 一体式污水膜生物反应器
integrated sewage treatment 污水综合处理
integrated ship instrumentation system 综合船舶仪表系统
integrated signal 综合信号;积分信号
integrated simulation panel 综合仿真板
integrated software 集成软件
integrated software package 集成软件包
integrated sonar system 综合声呐系统
integrated spectrum 累积光谱
integrated square 平方的积分
integrated standardization 综合标准化
integrated station 综合车站
integrated steam power plant 总装式蒸气动力装置
integrated steam turbine plant 总装式汽轮机装置
integrated steelworks 钢铁联合企业
integrated storage 综合库存;积累存储器
integrated streamer 集成电缆
integrated structure 结合式结构
integrated structure dam 组合式坝
integrated structure design philosophy 总体结构设计原理
integrated study 综合研究(报告)
integrated survey 综合考察
integrated system 综合系统;集中式系统;集成(化)系统;统一系统;成套装置;成套系统
integrated system for the management of agricultural pollution 综合农业污染管理系统
integrated system of liquid effluent treatment 液态污水综合处理系统
integrated task index 综合任务指标;综合的任务指标
integrated terminal equipment 综合型终端设备
integrated test 联调
integrated test requirement 综合测试要求大纲
integrated test system 综合测试系统;综合测试设备;成套测试设备
integrated three-dimensional frame 整体式三维框架
integrated total ocean surveillance 全海洋综合监视
integrated tower 有反应器的蒸馏塔
integrated tows 组合式驳船队;分节驳船队

integrated toxicity 综合毒性
integrated toxicity index 综合毒性指数
integrated tracking 综合跟踪
integrated tracking system 综合跟踪系统;积分跟踪系统
integrated train 专用整体化列车
integrated-transfer technique 整体传输技术
integrated transmission system 综合性传输系统;联合输电系统
integrated transportation 综合运输
integrated transport network 综合运输网
integrated transport protocol 综合传输协议
integrated transport system 综合运输系统
integrated treatment 综合处理
integrated treatment of water pollution 水污染综合治理
integrated tug-barge 组合顶推驳船
integrated type membrane separation 一体式膜分离
integrated unit 整体化装置;联合装置;集成元件
integrated unit flip-flop 集成单元触发器
integrated urban transport system 一体化城市交通系统
integrated use 综合利用
integrated utilization 综合利用
integrated utilization of wood 木材综合利用
integrated vertical-flow constructed wetland 复合垂直流人工湿地
integrated vertical-flow infiltration system 复合垂直流渗流系统
integrated vessel management system 综合船舶管理系统
integrated wastewater discharge standard 污水综合排放标准
integrated wastewater treatment 污水综合处理
integrated wastewater treatment plant 污水综合处理厂;废水综合处理厂
integrated water pollution control 水污染综合控制
integrated water quality data 综合水质数据
integrated water quantity and quality planning 水量水质综合规划
integrated water quantity-quality modeling 水量水质综合模拟
integrated water resource management 水资源综合管理
integrated water resource planning 水利资源综合利用规划
integrated wideband communication system 综合宽带通信系统
integrating acceleration 综合加速度
integrating accelerometer 综合加速测量仪;积分仪;积分加速度表
integrating ampere hour meter 累计安时计
integrating amplifier 累计放大器;积分放大器
integrating apparent energy meter 累计表观功率计
integrating bellows 积分膜盒
integrating block 积算部件
integrating cavity 累积腔
integrating circuit 积分网络;积分电路
integrating comb 汇集排管
integrating comb method 冲量法
integrating densitometer 积分式浓度比色计
integrating depth recorder 自记积深计;积分式自记测深仪
integrating detector 积分探测器;积分检波器
integrating device 累计装置;累计器;累器;积量装置;积分器
integrating divider 脉冲积分选择器
integrating dose meter 积分剂量计
integrating electric(al) timer 累计电流计时器
integrating element 积分回路;积分环节
integrating excess power meter 积算超功率表
integrating exposure meter 积分曝光计
integrating factor 积分因子;积分因数
integrating filter 积分滤波器
integrating float 积深浮标
integrating flowmeter 累计流量计;累积(式)流量计;积分流量计
integrating frequency meter 积量频率计
integrating galvanometer 积分电流计
integrating gear 积分传动装置;积分传动齿轮;积分齿轮
integrating gyroscope 积分陀螺(仪);积分回旋器

integrating gyroscopic accelerometer 积分陀螺加速表
integrating gyrounit 陀螺积分环节
integrating instrument 累计仪器;累计曲线;积算仪;积分仪;积分器;求积仪
integrating ionization chamber 累积电离室
integrating kilowatt-hour meter 积算千瓦时计;积算电度表
integrating light detector 积分光探测器
integrating luxmeter 积算照度表
integrating manometer 累计压力计
integrating measuring instrument 积算累计式测量仪表
integrating mechanism 积分机构;积分机
integrating meter 累计式电表;积量器;积量计;积分器;积分(计算)仪;积分计
integrating method 积分法
integrating motor 积分(直流)电动机;积分马达
integrating multiramp converter 积分式多级变换器
integrating nephelometer 积(分)比浊计
integrating network 积分网络;积分回路
integrating pad 积分衰减器
integrating photometer 累计光度表;累积(式)光度计;积算光度表;积分(式)光度计
integrating photon counter 积分光子计数器
integrating range 积分区间
integrating reactive watthour meter 积算无功瓦时计
integrating relay 积算继电器;积分型继电器
integrating sampler 积分取样器;累积式采样器;积分采样器
integrating sound level meter 积分声级计
integrating sphere 集光球;积算球;积分球
integrating-sphere photometer 积分球光度计
integrating system 积分装置;积分系统
integrating tachometer 积分转数表
integrating time constant 积分时间常数
integrating water flow meter 积分水流量计;累计水流量计
integrating water sampler 综合采水器;积深采水样器;全水层采样器
integrating wattmeter 累计式电度表;积算瓦特计
integrating wheel 积分轮;求积仪导轮
integration 综合;整体化;整合(作用);累计;集成(化);平等待遇
integration accelerometer 综合加速仪;积分加速度仪;积分加速度表
integration by decomposition 分解求积分法
integration by partial fraction 部分分数积分法
integration by parts 分部积分法;分步积分法
integration by reduction 渐化法
integration by substitution 换元积分法;代换积分法
integration by successive reductions 递演积分法;递推积分法;递归积分法
integration center 整合中枢
integration circuit 积分电路
integration coefficient of rock mass 岩体完整性指数
integration colo(u)rimetry 积分比色法
integration constant 积分常数
integration deficient mutant 整合缺陷型
integration efficiency 整合效率
integration method 积分方法;积分法
integration method of velocity measurement 流速积深测量法;垂线综合测流速法
integration of a matrix 矩阵的积分
integration of operation 联合作业
integration of signals 信号积累
integration signal 集积信号
integration step 积分步长
integration subroutine 积分子程序
integration time 积分时间
integration variable 积分变量
integrative action 整合动作
integrative immobilized biological activated carbon bioreactor 一体式固定化生物活性炭生物反应器
integrative levels 整合层次
integrative organization 综合组织
integrative oxidation ditch 一体式氧化沟
integrative six tanks activated sludge reactor 一体式六箱活性污泥反应器
integrative suppression 整合校正
integrator 积累器;积分装置;积分元件;积分网络;积分器;积分(描图)仪;积分机;积分电路;求积仪;求积(分)器
integrator-amplifier 积分放大器
integrator drift 积分漂移
integrator method 积分电路法
integrity 正直;整体性;领土完整;完整性
integrity basis 整基
integrity constraint 完整性约束条件
integrity of core 岩芯的完整度
integrity of sewer system 管道系统完整性
integrity test(ing) 完好性检验(桩工);完整性试验
integrometer 惯性矩面积仪;矩求积仪;惯性矩求积仪
integronics 综合电子设备
intellect 智能;智力
intellectual economy 知识经济
intellectual faculty 智能
intellectual property 知识产权
intellectual rights 知识产权
intellectual technology 智能技术;智能工艺学
intellectual work 脑力劳动
intelligence 消息;智能;智力;情报
intelligence activity 情报活动
intelligence agency 智能代理;情报机关;情报机构
intelligence agent 情报人员
intelligence annex 情报附件
intelligence bureau 情报局
intelligence card reader 智能卡片阅读器
intelligence data 情报资料
intelligence data base 情报数据库
intelligence data system 情报资料系统
intelligence department 情报部(门)
intelligence index 智力指数
intelligence information 情报资料
intelligence-intensive industry 智能密集型产业
intelligence machine 智能机
intelligence office 职业介绍所;情报部(门)
intelligence personnel 情报人员
intelligence photography 侦察摄影
intelligence platform 侦察站
intelligence quotient 智商;智能商数
intelligencer 情报员
intelligence receiver 信息接收机;情报接收机
intelligence requirement 信息获取要求
intelligence robot 智能机器人
intelligence science 智能科学
intelligence signal 载波信号;情报数据信号
intelligence system 信息系统;智能系统;情报系统;探测和识别系统
intelligence terminal 智能终端
intelligence test 智能检查
intelligent amplifier 智能放大器
intelligent building 智能大厦
intelligent cable 灵巧电缆
intelligent calcium-ferrite analyzer 智能钙铁分析仪
intelligent ceramics 智能陶瓷
intelligent composite material 智能复合材料
intelligent computer-aided design 智能计算机辅助设计
intelligent concrete 智能混凝土
intelligent control 智能控制
intelligent controller 智控制器
intelligent data acquisition 智能资料收集;智能数据采集
intelligent data entry terminal 智能数据输入端
intelligent display 智能显示器
intelligent flowmeter 智能流量计
intelligent graphic 智能图
intelligent graphic display 智能(终端)图形显示
intelligent graphic system 智能图形系统
intelligent instrument 智能仪器
intelligent integration 智能积分
intelligent keyboard system 智能键盘系统
intelligent network 智能网络
intelligent pig 管道内检测器
intelligent processing 智能加工
intelligent residence district 智能住宅小区
intelligent robot 智能机器人
intelligent science 智能科学
intelligent sensor 智能传感器
intelligent terminal 智能终端
intelligent ultrasonic detector 智能超声波探伤仪
intelligent ultrasonic prober 智能超声波探伤仪
intelligent use 合理使用
intelligent vehicle highway system 智能车路系统
intelligibility 可理解性;可辨度
intel mobile module 英特尔移动模块
INTELSAT 国际电信卫星
INTELSAT communication network 国际电信卫星通信网
INTELSAT control centre 国际电信卫星组织控制中心
INTELSAT Management Division 国际电信卫星组织管理部门
INTELSAT operation centre 国际电信卫星组织操作中心
INTELSAT system records 国际电信卫星系统记录
INTELSAT-type commercial satellite earth station 国际电信卫星型商用卫星地球站
intemperate 激烈的
intemperate weather 恶劣天气
intemperate wind 烈风
intemperate zone 热带
intend 打算
intendant 经理人;监督人(员)
intended bottom 设计河底;设计底高程
intended course 计划航向
intended gap 预留间隙
intended investment 集约投资
intended performance 预期使用期
intended size 公称尺寸;额定尺寸
intended stress 预计应力
intended target 指定目标
intended track 预计航迹;计划航线;计划航迹
intended voyage 预定航次
intense agitation 强力搅拌
intense bioactivity 强生物活性
intense care ward 重病病房
intense cold 严寒
intense colo(u)r 亮色;浓色;强色
intense cooling plant 强化冷却设备;强化冷冻装置;深度冷冻厂
intense current 强电流
intense depression 强低压
intense earthquake 强烈地震
intense fall 暴雨
intense field 强场
intense gamma radiation 强伽马辐射
intense heat 高温;酷暑;酷热
intense industrialization 高度工业化
intense ion beam 强离子束
intense laser radiation 强激光辐射
intense light pulse 强光脉冲
intense light source 强光源
intensely bunched ion source 强聚束离子源
intense magnetic field 强磁场
intense radiation 强辐射
intense(rain)fall 暴雨
intense shock wave 强激波
intense sinking stage 强烈沉降阶段
intense storm 暴雨;强(烈)暴雨
intense tornado 强龙卷
intense tunable infrared source 强可调红外源
intense ultrasonic wave 强超声波
intense zone 强烈(地震)区
intensification 增强;加深;加强;加厚;强化
intensification modulation 亮度调制
intensification of negative 负片加厚
intensification of radiation 辐射强度
intensification of traffic movement 强化交通流
intensification pulse 加亮脉冲
intensified activated sludge process 强化活性污泥法
intensified activated sludge system 强化活性污泥系统
intensified charging 增强充电
intensified image 增亮图像
intensified mining 强化开采
intensified program(me) 重点计划
intensified response 增强反应
intensified sintering 强化烧结
intensified test 强化的试验
intensifier 照明装置;增器器;增强剂;增器器;增厚剂;扩大器;加厚剂;强化因子;倍加器
intensifier circuit 增强电路;增光电路;放大器级间谐振电路
intensifier coil 增强线圈

intensifier driven hydraulic press 增压器传动液压机
intensifier effective gain 增强器有效增益
intensifier electrode 增强电极；增光(电)极
intensifier gain 增强器增益
intensifier potential 增光电位
intensifier pulse 照明脉冲；增强脉冲
intensifier ring 增强环
intensifier stage 放大级
intensifier target 增强靶
intensifier type 增强器类型
intensify(ing) 增强；激化
intensifying electrode 增强电极
intensifying foil 箔制增光屏；增感箔
intensifying gate 照明脉冲
intensifying pulse 增辉脉冲
intensifying ring 增光环
intensifying screen 增光屏；增感屏；加亮荧光屏
in tension 拉伸状态
intension for claim 索赔意向
intensitometer 强度计
intensity 烈度；集度；密(集)度；强度
intensity abnormal 烈度异常
intensity adjustment 烈度调零
intensity alternation 强度更迭
intensity amplification 强度放大
intensity anomaly 烈度异常
intensity armature 高欧姆电枢
intensity assessment 烈度评定
intensity attenuation 烈度衰减
intensity bridge ratio 强度分路比率
intensity calculation 强度计算
intensity calibrating device 强度校准装置
intensity characteristic 烈度特征
intensity circuit 亮度控制电路
intensity contrast 强度对比度
intensity control 亮度调整；亮度调节；强度控制
intensity control amplifier 亮度调节放大器
intensity-controlling electrode 强度控制电极
intensity convolution integral 强度卷积积分
intensity curve 强度曲线
intensity determination 烈度鉴定
intensity difference 强度差
intensity dip 强度坑；强度降落
intensity discrimination 强度鉴别
intensity distribution 烈度分布；亮度分布；强度分布
intensity-duration curve 强度时间曲线；强度历时曲线
intensity-duration formula 强度时间公式；强度历时公式
intensity envelope function 强度包线函数
intensity exceeding probability 烈度超过概率
intensity exceeding rate 烈度超过率
intensity factor 强度因子；强度因素
intensity function 强度函数
intensity gate 亮度闸；强度(控制)闸
intensity grade 烈度；强度
intensity grid 调制栅
intensity-hue-saturation 强度－色度－饱和度
intensity index 强度指数
intensity interferometer 光强干涉仪；强度干涉仪
intensity interferometry 强度干涉测量法
intensity level 亮度级；亮度电平；辉度水平；强度级；场强电平
intensity level of sound 声力级
intensity line 强度线
intensity map of neotectonic movement 新构造运动强度图
intensity measuring device 强度测定仪
intensity-modulated indicator 亮度调制指示器；强度调制指示器；调幅指示器
intensity-modulated indicex 亮度调制标志
intensity modulation 亮度调制；强度调制
intensity modulation scan 亮度调制显示器
intensity modulator 强度调制器
intensity of absorption bands 吸收带强度
intensity of background value 背景值强度
intensity of back-washing 冲洗强度；反冲洗强度
intensity of bed-load transport 推移质输移强度；推移质输沙强度
intensity of blending 搅拌强度；混合强度
intensity of breaking 拉伸应力
intensity of burning power 燃烧(力)强度
intensity of centrifugal field 离心力场强度

intensity of colo(u)r 色强度
intensity of combustion 燃烧强度
intensity of compression 压缩强度；压缩程度
intensity of compressive stress 压缩应力强度
intensity of cooling 冷却强度
intensity of current 水流强度；电流强度
intensity of cut 采伐强度
intensity of development 土地开发强度；发展密度
intensity of draft 抽力强度
intensity of drag 牵引强度
intensity of draught 牵引强度；牵引力；通风强度
intensity of earthquake 地震强度
intensity of effective rainfall 有效降雨强度；净雨强度；净雨率
intensity of electric(al) field 电场强度
intensity of evapo(u)ration 蒸发强度；蒸发率
intensity of exposure 曝光强度
intensity of fishing 捕捞强度
intensity of gravitational field 引力场强度
intensity of gravity 重力强度
intensity of gravity field 重力场速度
intensity of ground shaking 地震动强度
intensity of hearth 炉底强度
intensity of heat 热强度
intensity of heat dissipation in workshop 车间散热强度
intensity of illumination 照明强度；照度；光照(强)度
intensity of incandescence 灼热力；灼热度
intensity of induced magnetization 磁化强度
intensity of inflation pressure 充气压强
intensity of irradiation 光渗强度；辐照强度
intensity of labo(u)r 劳动强度
intensity of light 照度；光(强)度；光强
intensity of light reflection 反光强度
intensity of light source 光源强度
intensity of load(ing) 荷载强度
intensity of magnetic field 磁场强度
intensity of magnetization 内禀磁感应强度；磁化强度
intensity of oscillation 振荡强度
intensity of peak value 峰值强度
intensity of photosynthesis 光合作用强度
intensity of polarization 极化强度
intensity of pollution 污染强度
intensity of precipitation 降雨强度；降水强度
intensity of pressure 压强；压力强度
intensity of radiation 辐射强度
intensity of radioactivity 放射性强度；放射性活度
intensity of radioactivity per unit mass 比活度
intensity of rain(fall) 降雨率；单位时间降雨量；雨量强度；降雨强度
intensity of rain storm 暴雨强度
intensity of reflected light 反射光强度
intensity of relief 地面起伏强度
intensity of restraint 拘束度
intensity of return current 回流强度
intensity of roasting 焙烧强度
intensity of runoff zone 径流带强度
intensity of selection 选择强度
intensity of shaking 震动强度；振动强度
intensity of shear 剪力强度
intensity of soil erosion 土壤流失强度
intensity of solar radiation 太阳辐射强度；日辐射强度
intensity of sound 声音强度；声强
intensity of spectral lines 谱线强度
intensity of spontaneous polarization 自发极化强度
intensity of storms 暴雨强度
intensity of stress 单位面积的应力；应力强度
intensity of tectonic contrast 构造反差强度
intensity of tectonic movement 构造运动强度
intensity of tension 拉力强度
intensity of traffic 交通强度；运输强度；交通密度
intensity of transpiration 蒸腾强度
intensity of turbulence 阵风分量；紊流强度；紊动强度；湍流强度
intensity of uniform loading 均布荷载强度
intensity of vibration 振动强度
intensity of washwater 冲水强度；冲洗强度
intensity of water 用水强度
intensity of wave 波强度；波的强度
intensity of wave pressure 波压强度

intensity of wind pressure 风压强度
intensity of wind velocity 风速强度
intensity of work 劳动强度
intensity pattern 辐射图形
intensity per unit time 单位时间强度
intensity-rainfall curve 强度雨量关系曲线
intensity rating 烈度评定值
intensity ratio 强度比
intensity receiver 场强接收机
intensity recurrence period 烈度重现周期
intensity reducer 减强器
intensity relative quantities 强度相对数
intensity return period 强度重现期
intensity run on shore 故意搁浅
intensity scale 烈度等级；烈度表；强度等级；强度标度；地震烈度表；调光制
intensity-scale calibration method 强度标法
intensity scale of Japan 日本烈度表
intensity spectrum 强度谱
intensity statistics 烈度统计量
intensity target method 强度指标法
intensity transfer function 强度传递函数
intensity transformation 强度变换
intensity transmission coefficient 光强透射系数
intensity value of being measured element 待测元素强度值
intensity-voltage dependence 强度电压关系
intensity vs time curve 强度时间曲线
intensive 增强的；加强器；加强剂；密集；强烈的
intensive agriculture 集约(化)农业；农村集约化
intensive amount 内含量
intensive analysis 强度分析
intensive and meticulous farming 精耕细作
intensive arc lamp 强弧(光)光灯
intensive care unit 重病监护室；重病护理单元；监护病房
intensive characteristics of rock 岩石的强化特性
intensive course 短期学术讲座
intensive cultivation 精耕细作；集约耕种
intensive culture 集约栽培
intensive depressing region 强烈下降区
intensive dispatch of port cargo 疏港
intensive drying 充分干燥
intensive exploitation 高度开发
intensive farming 集约(农业)经营；集约农业；集约耕作；强化农业
intensive filtration 强化过滤
intensive forestry 集约林业
intensive grassland farming 集约草地农业
intensive grazing system 集约式放牧
intensive index of faulting 构造断裂强度指数
intensive instrumentation arrays 密集仪器台阵
intensive investigation 深入调查
intensive investment 集约投资
intensively farmed agricultural plots 精耕细作农田
intensive management 集约经营
intensive method 精测法
intensive mixer 高功率炼胶机；高功率混合机
intensive mixing 强烈混合；充分搅拌；充分混合
intensive movement 强烈运动
intensive noise 强烈噪声
intensive non-fluxing blender 强力非塑化掺和机
intensive observation area 加强观测区
intensive observation period 加强观测期
intensive parameter 强度参数
intensive production 精耕细作；集约生产
intensive property 强烈性
intensive quantity 强度量
intensive recording mode 集中记录方式
intensive reflector 加强反光器；强光反射镜；探照反射器
intensive sampling 密集抽样
intensive stage 强烈阶段
intensive support 密集支架
intensive survey method 精测法
intensive system 密集整枝法
intensive training course 速成训练班
intensive uplifting region 强烈上升区
intensive use of land 土地高度利用
intention agreement 意向协议书；意向性协定；意向书
intentional additives 特定添加剂
intentional breaching 扒口(堤口)
intentionally left blank 空白页

intention letter 意向书
intention movement 意向运动
intent propagation 意图传播;集中传播
interact 相互感应;相互作用
interactant 相互作用物;反应物
interacting 联动
interacting activity 相互制约活动;交互式活动
interacting arches 交叉拱
interacting computation 交互计算
interacting element 互作用型元件
interacting energy 相互作用能
interacting field 相互作用用场
interacting goal 相互关联目标
interacting nuclei 相互作用核
interacting parallel processing 交互并行处理
interacting particles 相互作用粒子
interacting simulator 交互模拟器;人机对话模拟器
interacting space 相互作用空间
interacting state 相互作用状态
interacting subsystem 相互作用的分支系统
interacting wave train 相干波列
interaction 相互制约;相互影响;相互配合;交互作用;互相作用
interactional inhibitory system 二元互抑体系
interaction among individuals 个体间的相互关系
interaction analysis 相互影响分析;交叉分析
interaction balance 关联平衡;交叉平衡
interaction balance principle 关联平衡原理;交互平衡原理
interaction chamber 互作用腔
interaction circuit 相制电路
interaction coefficient 相互作用系数
interaction cross section 相互作用截面
interaction diagram 干扰图;相互作用图表
interaction effect 相互作用效应;相互的影响
interaction energy 相互作用能;作用能
interaction factor 交互作用因子;互作用因数;互作用系数
interaction fault 交互故障
interaction force 相互作用力
interaction formula 交接公式
interaction gap 互作用(间)隙
interaction Hamiltonian 相互作用哈密顿函数
interaction impedance 互作用阻抗
interaction in social group 社会集团的相互作用
interactionism 相互作用说;交互作用论
interaction loss 互作用损耗
interaction matrix 相互影响矩阵
interaction matrix analysis forecasting 相互影响矩阵分析预测法
interaction mean free path 相互作用平均自由程
interaction method with variable weights 选权迭代法
interaction of atmospheric contaminants 大气污染物的相互作用
interaction of atmospheric pollutants 大气污染物的相互作用
interaction of dislocation 位错的交互作用
interaction of footing 基础的相互影响
interaction of light and temperature 光合温度的相互作用
interaction of molecules 分子相互作用
interaction of multiphase medium 多相介质相互作用
interaction of passing vessel 船吸
interaction of soil and water 土与水的相互作用
interaction parameter 相互作用参数;交互作用参数
interaction partners 相互作用对
interaction phase 相间作用
interaction picture 相互作用图像;相互作用图景
interaction potential 相互作用势
interaction prediction 关联预测;交互预测
interaction prediction approach 关联预测法;交互作用预测法
interaction prediction principle 关联预测原理;交互预估原理
interaction process 相互作用过程;交互作用过程
interaction rate 相互作用率
interaction region 互作用区
interaction representation 相互作用图像
interaction space 互作用空间
interaction stress 相互作用应力;相互作用模型
interactions with abiotic environment 无生命环境的相互作用

interaction time 相互作用时间;交互作用时间
interaction volume 相互作用范围
interaction width 共同作用宽度
interaction with interelectrode 板间互作用
interaction zone 相互作用区域
interactive 交互式;交互(的);重复
interactive action 人机交互动作
interactive advertising 交互式广告
interactive batch processing 交互成批处理
interactive behavio(u)r 相互影响行为
interactive cartographic(al) editing 人机对话制图编辑
interactive computer 交互型计算机
interactive computer-aided design 交互式计算机辅助设计
interactive computer graphic(al) system 人机联系计算机制图系统
interactive computer graphics 交互式计算机图形学;交互式计算机制图
interactive computer graphics system 交互式计算机制图系统
interactive computer mapping system 人机联系计算机制图系统
interactive computer system 交互式计算机系统
interactive computing and control facility 交互计算和控制设备
interactive controller 人机联系控制器
interactive data edit system 人机联系数据编辑系统
interactive debug 交互调试
interactive debugging 交互调试;交互程序调整
interactive design system 交互式设计系统
interactive digital image manipulation system 人机对话数字图像处理系统
interactive digital plotter 交互式数字绘图机
interactive display 相互作用式显示器;交互显示器
interactive display and edit facility 人机联系显示编辑
interactive display control 人机联系显示控制
interactive display controller 人机联系显示控制器
interactive display system 交互显示系统
interactive display terminal 人机交互式显示终端
interactive editing 人机对话编辑
interactive editing system 人机联系编辑系统
interactive environment 交互式环境;交互环境
interactive feature extraction system 人机联系要素抽出系统
interactive forecasting 交互性预测
interactive formatting system 交互格式形成系统
interactive graphic input 交互式图形输入
interactive graphic processing 人—机对话图形处理;交互式图形处理
interactive graphic processing system 交互式图形处理系统
interactive graphics 交互式制图(软件);交互图形;交互式图形(学);人机对话图
interactive graphics system 交互式图形系统;交互式绘图系统;人机联系绘图系统
interactive graphics terminal 交互图形终端
interactive graphy 人机联系自动制图系统;人机联系制图
interactive handler 人机联系处理机
interactive image processing 交互图像处理;图像人机联系处理
interactive instruction 交互指令
interactive interpretation 人机联系解释
interactive keyboard 交互键盘
interactive light pen 交互光笔
interactive man-machine system 人机联系系统
interactive mapping system 人机联系绘图系统
interactive method 反复计算法
interactive mode 交互方式;对话方式
interactive model 相互作用模型;相互影响模型
interactive modelling 人机联系模型法
interactive modification 人机联系修改
interactive multimedia 交互式多媒体
interactive operation 交互操作;人机联系操作
interactive pointer 人机联系光标
interactive problem control system 交互问题控制系统
interactive process control 交互过程控制
interactive processing 交互处理;人机交互处理
interactive program(me) 人机对话程序;交互程序
interactive property 相互用性质
interactive query 交互查询;人机对话式查询

interactive restoration 交互式复原;交互复原
interactive routine 交互式例行程序
interactive searching 交互查找
interactive session 交互式进程
interactive simulation 交互仿真
interactive simulator 交互式模拟程序
interactive skills 相互影响技能
interactive software system 相互作用软件系统
interactive solid design 交互式实体设计
interactive system 交互系统;交互式系统
interactive television 交互式电视
interactive terminal 交互终端设备;交互(式)终端
interactive terminal interface 交互终端接口
interactive type communication 交互式通信
interactive user 交互式用户
interactive utility billing 公用事业相互开票
interactive way 人机联系方式
interactivity 相互作用
interact on each other 互相作用
interadaptation 相互适应
interagency 跨部门的;部门之间的
Inter-Agency Committee on Water Resources 机构间水资源委员会
inter-agency loan 机构间贷款
inter-agency subcontract 机构间分包合同
inter-air space 气隙空间
interalloy 中间合金
interambulacrum 间步带
Inter-American Accounting Association 泛美会计协会
Inter-American Association of Sanitary Engineering 美洲国际卫生工程协会
Inter-American Development Bank 泛美开发银行
Inter-American Housing and Planning Center 美洲国际住宅及城市规划中心
Inter-American Planning Society 美洲国际城市规划学会
inter-and-subtropic(al) convergence zone 热带和亚热带辐合区
interannealed wire 中间退火钢丝
interanneal(ing) 中间退火
interannual change 年际变化
interannual correlation 年际相关;年际关系
interannual regulation 跨年调节;多年调节
interannual shoal variation 浅滩年际变化
interannual variability 年际变率
interannual variation 年际变化
interannual variation of scour and fill 年际冲淤变化
interannual variation of shoal 浅滩年际变化
interaquifer flow 含水层间水流;层间流
interarc basin 弧间盆地
interarrival time 到达(时间)间隔;到达间隔时间
interassembler 交(式)汇编程序
interassimilation 粒间同化作用
interasteric 星点间的
interation 交互影响
interatomic bonding force 原子键力
interatomic diamagnetic current 原子间抗磁性电流
interatomic distance 原子间距(离)
interatomic force 原子间力
interatomic interference 原子间干涉
interatomic spacing 原子间距(离)
interatrial septum 房隔
interatrial tract 房间束
interattraction 相互吸引
interauricular bundle 房间束
interauricular septrm 房间隔
interavailability 相互利用
interaxis 中间轴;两轴之间;两轴线之间的空间
interaxle differential 轴间差速器
interbaluster 栏杆之间;栏杆空当
interband 带间
interbanded coal 夹矸煤
interbank 管束间
interbank exchange rate 同业汇率
interbank loan 同业拆款
interbank rate 银行同业汇价
interbank space 管束间隙
interbank transactions 联行往来
interbank transfers 银行间转账;联行往来
interbasinal development planning 跨流域开发计划;跨流域开发规划
interbasinal diversion 跨流域引水;跨流域调水

interbasinal water diversion 跨流域引水;跨流域调水
interbasin area 流域间面积;流域间地区;河道间地区;区间面积
interbasin diversion 河道间引水
interbasin diversion project 跨流域引水工程
interbasin transfer of lakes 湖泊间水转移
interbasin transfer of water 流域(间)水迁移
interbasin(water) diversion 跨流域引水
interbasin water transfer 跨流域引水;跨流域调水
interbasin water transfer project 跨流域调水工程
inter-bead lack of fusion 焊道间未熔合
interbed 夹层
interbedded 间歇的;夹地层;夹层的;混合的;互层的;层间的
interbedded layer 夹层
interbedded sands 互层砂土
interbedded strata 互层;间层;夹层
interbedded water 层间水
interbedding 互层;中间层
interbed multiples 层间多次反射
interbed tide 逆潮(流)
interbehavio(u)r 相互行为
interblend 相混;互层掺混;混合
interblock 信息记录组
interblock analysis 区组间分析
interblock gap 组间间隔;块间隔
interbonding 交织砌合
interborough 市镇间(的)
interboundary pore 界面孔
interbrace 窗台木和窗头木之间的横撑(窗框中)
inter-branch 分行往来
inter-branch account 分支店间往来账户;联行往来账户;联号往来账
inter-branch excess freight 分店间超支运费
inter-branch transaction 分店间交易;分支店间往来业务
inter-branch transfer 分店间转账事项;分店间往来
interbubble channel 气泡间沟道
interburner 中间补燃加力燃烧室
interbus 联络母线;旁路母线
interbus transformer 联络母线变压器
intercable crosstalk 电缆间串扰
intercalary 插入的
intercalary attachment 中间附着
intercalary cycle 闰周
intercalary day 闰日
intercalary deletion 中间缺失
intercalary growth 节间生长
intercalary inversion 中间倒位
intercalary month 闰月
intercalary strata 地质夹层;夹层
intercalary year 闰年
intercalate 插入
intercalated 间生的;间层的;夹层的;插入的
intercalated bed 夹石;夹层
intercalated beds of sandstone 砂岩夹层
intercalated disc 闰盘
intercalated tapes 叠式绝缘带
intercalated texture 穿织结构
intercalated zone 夹层带
intercalating agent 嵌入剂
intercalation 置闰;隔行扫描;夹杂;夹层;闰期;嵌入;添加;插入作用
intercalation bed 间层;夹层
intercalation chemistry 插入化学
intercalation compound 插入化学物;插层化合物
intercalibration 相互校准
intercalibration of observational instrumentation 观测仪器的相互校准
intercall telephone 内部电话机
intercapillary 毛细管间的
intercardinal 基点间的(方位)
intercardinal heading 象限航向
intercardinal plane 主间平面
intercardinal point 象限点;中间基本方位;中央方位点(即4个象限点的中间点)
intercarrier 载波差拍
intercarrier distance 载波间距离;载波间隔
intercarrier noise suppression 内载波噪声抑制
intercarrier receiving system 内载波接收系统
intercarrier signal 内载波信号
intercarrier sound reception 内载波伴音接收法
intercarrier sound system 内载波伴音系统

intercarrier system 载波差拍制;内载波接收方式
intercavernous 腔间的
intercell 注液电池
intercell communication 单元间通信
intercensal figures 人口调查数字
intercensal study 人口调查研究
intercept 高度差;拦截;截线;截听;截取;截距;截断;截段
intercepted cross road 截断型交叉路;丁字交叉口
intercepted drain system 截流式排水系统
intercepted flow 截流流量
intercepted green 隔离绿地
intercepted length 截取长度
intercepted matter 阻截物
intercepted quantity of rainfall 截流降雨量
intercepted resource 截听资源;被截资源
intercepted river 间断性河流
intercepted station 被截站
intercepted stream 间断性河流
intercepted terminal 截听终端;被截终端
intercepted water 蓄积水;拦蓄水;截留水
intercept equation 截段方程
intercepter 中间收集器;遮断物;冷凝罐;拦截机;扼流板;隔断器;拦截器;截沙阱;截沙池;截流器;截流管;截流存水弯;截击器
intercepter plate 截流板
intercepter sewer system 截流式下水道系统
intercepter valve 遮断阀
intercept form 截距式
intercept ground-based optic(al) recorder 监视地面光学记录仪
intercepting basin 截水池
intercepting chamber 窨井(分开排水系统与阴沟);截流检查井
intercepting channel 截水沟;截流渠;天沟
intercepting conduit-type sewer 截流污水管
intercepting dike 截土堤(公路保护挖坡的挡土墙)
intercepting ditch 天沟;拦水沟;截水沟
intercepting drain 截水(盲)沟;截流(排水)沟;截流排水道
intercepting drain system 截流式排水管系统
intercepting effect 截流效果
intercepting green 隔离绿地
intercepting gutter 集水沟
intercepting inclined plank settling tank 拦截斜板式沉淀池
intercepting inclined plankt sedimentation 拦截斜板式沉淀
intercepting installation 截流设备
intercepting layout 截流式布置
intercepting of water 截水
intercepting pipe 截污管道
intercepting plate 截流板;扰流板
intercepting pollution 截污
intercepting pump 截流泵
intercepting pumping station 截流泵站
intercepting screen 截光屏;承截幕
intercepting sedimentation 拦截沉淀
intercepting sewer 截污管道;截流(沟)管;截流下水道;截流(污水)管;截污沟渠
intercepting sewer network 截污管网
intercepting steam chest 中间汽柜
intercepting subdrain 地下截水管;地下截水沟
intercepting system 截污管网;截流系统;截流式系统;排水系统布置—截流式系统
intercepting trap 截流存水弯;排水防气瓣;水封
intercepting trunk 截听干线;截取干线
intercepting valve 截流阀;启动调节阀;三通阀
intercepting valve head connection 中间阀盖接头
intercepting well 暴雨溢流井
interception 阻;拦截;截流
interception arrangement 截流布置
interception by crops 作物截留(量)
interception by forest 森林截流
interception by grass 草被截留(量)
interception by vegetation 植被截留
interception circuit 强折电路
interception cut 截水沟
interception drain 截水沟
interception effect 折射作用
interception loss 阻截耗损;截留(损)失;截流损失
interception of call 截接服务
interception of rays 光线的折射作用
interception of upper reaches of river 上游河道

截流
interception of water 截水;截断水流;水流截断;水的截流
interception operator 截接话务员
interception point 截流点
interception probability 拦截概率
interception ratio 截流倍数
interception storage 拦截蓄水;截留储蓄
intercept method 高度差法;截线法;截距法;截断法
interceptometer 阻留雨量计;截留仪;截留测定计;截流测定计
intercept operator 截听操作员
interceptor chest 截止阀气室
interceptor drain 截流排水;截水沟
interceptor for sewage 污水截流管
interceptor jar 中间集乳器
interceptor plate 截流板
interceptor sewer 截污管道;截流污水管
interceptor trap 截流存水弯
interceptor valve 阻止阀;截断阀
intercept point 拦截点;截(击)点
intercept receiver 截听接收机
intercept refractivity 比折光度
intercept station 侦听站;截听站;监听台
intercept trunk 暂用中继线
intercept unit 截断设备
intercept velocity 截留速度
intercerebral fissure 半球间裂
intercessor 调解人
interchain fixing 跨链定位
interchange 交替;交换;交叉站台;换接;换乘;互通式立(体)交(叉);互换;反演
interchangeability 可互换性;互换性
interchangeability and replacement 可互换性与置换
interchangeability of gas 燃气互换性
interchangeability of parts 零件的互换性;部件互换表
interchangeable 可交换的;可互换的
interchangeable anti-siphonage device 互换式反虹吸装置;可互换的反虹吸装置
interchangeable anvil pallet 可互换砧面
interchangeable assembling 可互换装配
interchangeable assembly 互换性装配
interchangeable attachment 可互换的附件
interchangeable bit 可换式钻头;可更换钻头
interchangeable body 可互换车身;标准车身
interchangeable bond 可换债券
interchangeable camera cone 可更换的摄影机镜筒
interchangeable card feed 可换卡片输入
interchangeable cassette 更换式暗盒
interchangeable cavity mo(u)ld 可换腔模具
interchangeable connector 互换接头
interchangeable cutter 互换刀片
interchangeable design 可互换的设计;可互换的版本;通用设计
interchangeable equipment 通用装备
interchangeable equipment item 可互换的设备项目
interchangeable European pallet 可互换的欧洲托盘
interchangeable fabricated parts 互换配件
interchangeable facing 可换式牙面
interchangeable front equipment 可互换的前部装置
interchangeable front rig 可互换的前部装置
interchangeable gear 可互换齿轮
interchangeable grain sieve 互换式谷粒筛
interchangeable ground 标准玻璃磨口
interchangeable item 可交换项目
interchangeable lens 可换透镜;可互换镜头
interchangeable magazine 互换式暗盒
interchangeable manufacture 零部件可换的生产;可互换的制造;互换性生产
interchangeable manufacturing 互换性制造
interchangeable material 可互换的材料
interchangeable mirror 可互换反射镜
interchangeable mount 可换支座
interchangeableness 互换性
interchangeable objective 可换物镜;可互换物镜
interchangeable oil filters 互换滤油器
interchangeable parts 可替换零件;可交换部件;可互换零件;可互换部件;互换零件;互换部件;通用配件

interchangeable prism 可互换棱镜
interchangeable program(me) tape 可互换程序带
interchangeable slide 可互换滑板
interchangeable syringe 互换注射器
interchangeable term 通用条款
interchangeable type bar 可换字锤
interchangeable wheel 互换齿轮
interchange agreement 交换协定
interchange box 交换盒
interchange bridge 立交桥
interchange channel 交换通道
interchange circuit 交换线路；通信线路
interchange coefficient 互换系数
interchange cross autostair(case) 互通式立体交叉自动楼
interchange cross-platform 换乘站台；联运站台；互通式立体交叉平台
interchange graph 交换图
interchange instability 槽纹不稳定性
interchange line 交换线
interchange-loading station 换装站
interchange-loading track 换装线【铁】
interchange-loading yard 换装站；换场【铁】
interchange method 替换法(桁架分析用)
interchange of air 空气交换；空气互换
interchange of cause and effect 因果转化
interchange of energy 能量交换
interchange of express highway 高速公路互通式立交
interchange of heat 热量交换；热交换
interchange of matter 物质交换
interchange of momentum 动量交换；动量互换作用
interchange of sea and land 沧桑变化
interchange of vehicles 车辆互通
interchange opinion 交换意见
interchange point 交换点；货物转运站
interchange power 换接功率；互换功率
interchanger 交换机
interchange ramp 互通匝道；立交匝道；互通(式交)坡道
interchange reaction 交换反应；互换作用；代换反应；代换反力
interchange relationship 换乘关系
interchanger of heat 热互换器
interchange service 海陆接转服务
interchange space 交接场
interchange station 枢纽站；交换站；换站；换乘枢纽
interchange track 在转运时货车的运行轨道；交换线【铁】；交换车停留线；车辆交接线
interchange trunk 互换中继线
interchange with special bicycle track 分隔式立体交叉【道】
interchange with weaving section 交织型互通式立交
interchange yard 交换车场【铁】
interchanging 交替；交换；互换
interchannel 信道间
interchannel area 间间地区
interchannel crosstalk 信道间串音
interchannel-grained meander belt deposit 河间湿地沉积
interchannel interference 信道间干扰
interchannel noise 信道噪声
intercision 侧移袭夺【地】
intercity 市际的；城市之间的
intercity bus 市际公共汽车
intercity circuit 长途通信线路；城市间通信电路
intercity communication 城市间通信；市际通信；市际交通
intercity commuting 市际通勤交通
intercity freight traffic 城市间货物运输
intercity line 长途线路【铁】
intercity network 城市间电视网
intercity passenger traffic 城市间旅客运输
intercity rail 城际铁路
intercity relay system 城市间中继系统
intercity road 市际道路
intercity television system 城市间电视传输系统
intercity traffic 市际交通；城市间运输
intercity train 城市间列车
intercity trucking 市际汽车货运
inter-class correlation 组间相关

inter-class correlation coefficient 组间相关系数
inter-class variance 组间方差
intercloud discharge 云间放电
intercoagulation 相互渗透；相互凝聚(作用)；相互凝结(作用)
intercoastal 两海岸间的；两岸间的
intercoastal transportation 国内海港间运输
intercoat adhesion 涂层间附着力
intercoat contamination 涂层间污染
intercoating 中间涂层
intercoil insulation 线圈间绝缘
intercolpar thickening 沟间加厚
intercolumn 柱间空隙；栏间空隙
intercolumnar 柱间的
intercolumnar screen 柱间间断
intercolumniation 柱距；柱间距(定比)【建】；塔间距离定比；分柱法
intercom 内部通话制；对讲通信系统；对讲机；船内通信
inter-combination 相互组合
intercom cable 船内通信电缆
intercom loudspeaker 内部通话扬声器
intercommunicability 相互传递性
intercommunicating line 内部通信线路
intercommunicating pore 连通气孔；连通孔
intercommunicating porosity 互通孔
intercommunicating room 互通的房间
intercommunicating set 内部对话装置
intercommunicating system 内部通话制；内部通话系统
intercommunicating(tele)phone 内部(对讲)电话；对讲电话
intercommunicating telephone set 对讲电话机
intercommunicating telephone system 对讲电话系统
intercommunication 内部通信(联系)；互相联系；相互通信；连通；多向通信；对讲电话装置；船内通信
intercommunication circuit 内部通信电路
intercommunication installation 内部通信装置；内部通话装置
intercommunication master set 内部通话主机
intercommunication panel 船内通信控制板
intercommunication plug switchboard 内部通信插头式交换机
intercommunication set 内部通信设备
intercommunication speaker 内部通话扬声器
intercommunication switch 内部通话开关
intercommunication system 内部通信制；内部通信系统；飞机通信系统；机内通信系统；相互通信制；对讲电话系统
intercommunication system with telephone 电话内线
intercommunication telephone 内部电话
intercommunication telephone set 对讲机
intercommunicator 内部通信机
intercommunity highway 社区间公路
inter-company loan 公司内部贷款
inter-company receivable and payable 公司内部资金的转拨；公司间应收及应付款项
inter-company transactions 联营公司间会计事项
intercomparison 相互比较；互相比较
intercompilation 编译
intercomputer communication system 计算机间通信系统
intercom set 内部通信设备
intercom system 船内通信设备
intercom telephone 内部对讲电话；船内通信电话
intercondenser 中间冷凝器；中间凝汽器
interconnect 连通；互联
interconnectability 可相互连接性
interconnected amplifier 内部互联放大器
interconnected control 相互联系控制机构；联锁控制
interconnected delta connection 互相三角形连接
interconnected electric(al) power system 互联电力系统
interconnected estuarine channels 河口网
interconnected grouting duct 互连灌浆导管
interconnected network 互联电网；互连管网
interconnected pipe system 互连管道系统
interconnected pore 连通孔
interconnected signal system 互连式信号系统
interconnected star connection 星形曲折接法；曲折接法

interconnected star winding 曲折连接绕组
interconnected structure 连通结构
interconnected synchronous generator 互连同步发电机；并网同步发电机
interconnected system 相联结的电力系统；环状管网；互连系统
interconnected transmission system 互连输气系统
interconnected voids 相互连通的孔隙
interconnecting 互连；交互连接
interconnecting cable 连接电缆
interconnecting device 转接设备
interconnecting elbow 反焰管
interconnecting feeder 联络馈线；连络馈线；互联馈路
interconnecting linkage 连接杆
interconnecting network 互连网路
interconnecting nut 中间连接螺母
interconnecting piece 连接部
interconnecting pipe 连通管
interconnecting plug 连接插头
interconnecting road 联络道路
interconnecting roadway 互连道路
interconnecting steam main 互联蒸汽总管
interconnecting switch 联络开关
interconnecting taxiway 互连飞机滑行道；互联汽车道
interconnecting ties 电网联络线
interconnecting transformer 联络变压器
interconnecting wiring diagram 接线图
interconnection 相互耦合；相互连接；内连(接)；互相联系；互相连结；互(相)连(接)
interconnection circuit 互连电路
interconnection constraint 关联约束
interconnection crossover 互连交叉
interconnection diagram 接线图；相互联系图
interconnection film 互连薄膜
interconnection flexibility 互连的灵活性
interconnection list 互连表
interconnection matrix 互连矩阵
interconnection of power system 电力系统互连
interconnection restriction 联结限制
interconnection switch 联络开关
interconnection system 互连系统
interconnection tie 联络馈线
interconnection time delay 互连时间延迟
interconnection vector 关联向量
interconnector 连通管；联络线路
interconsole message program(me) 控制台间信息程序
intercontinental 洲际的；陆间的
intercontinental airport 洲际航空站；洲际航空港
intercontinental base 洲际基线
intercontinental basin 陆内盆地
intercontinental circuit 洲际电路
intercontinental connection 洲际连接
intercontinental geodetic connection by satellite observation 卫星大地测量洲际联测
intercontinental maintenance 洲际维护
intercontinental network 洲际网
intercontinental rift system 陆间裂谷系
intercontinental sea 大陆间海
intercontinental service 洲际业务
intercontinental subduction 陆间俯冲
intercontinental survey by satellite 卫星洲际联测
intercontinental synchronization 洲际同步
intercontinental through-shipment 洲际直达货物
intercontinental transit exchange 洲际经转交换台
intercontinental transmission 洲际传送
intercontinental trunking 洲际中继
intercontour space 地貌模型的阶梯空间
interconversion 相互转换；互相换算
interconversion of directions 向全换算
interconvertibility 相变性
interconvertible 可互换
intercoolant 中间冷却剂
intercooled cycle 中间冷却循环
intercooled diesel engine 中冷增压柴油机
intercooler 中间散热器；中(间)冷(却)器；中间冷却剂
intercooler unit 中间冷却装置
intercooling 中间冷却
intercooling device 中间冷却设备
intercooling of exhaust gas 废气的中间冷却
intercooling stage 中间冷却级

intercoordination 相互协调;相互耦合
intercorrelation 交互相关
intercostal 肋间的;加强肋
intercostal angle 间断角钢
intercostal floor 间断肋板
intercostal girder 间断纵桁
intercostal keelson 间断内龙骨
intercostal member 间断构件
intercostal plate 间断板
intercostal side stringer 间断边纵骨
intercoupling 相互作用;相互耦合;寄生耦合;互耦合
intercractonic basin 陆间盆地
intercretion concretion 填隙结核
inter-crew communication 飞机内通信
inter-crimp 钢丝网中附加皱折波状丝
intercropping 间混作
intercropping corns in woodland 林粮间作
intercross 互相贯通;互交
intercrossing population 互交集团
intercrystalline 晶粒间的;晶间的
intercrystalline attack 晶间侵蚀
intercrystalline barrier 晶间势垒
intercrystalline brittleness 晶粒间脆性
intercrystalline corrosion 晶间侵蚀;晶间腐蚀;结晶粒间腐蚀
intercrystalline crack 晶间裂纹
intercrystalline cracking 沿晶界破裂;晶界断裂
intercrystalline dissolution pore 晶间溶孔
intercrystalline failure 晶界破坏;晶界断裂;晶间破坏
intercrystalline fracture 晶界断裂
intercrystalline pore 晶间孔隙
intercrystalline rupture 粒间断裂
intercrystalline segregation 晶间偏析
intercrystalline water 层间结晶水;内结晶水
inter-cultivation 中耕
intercupola (圆屋顶的)中间夹层;小圆屋顶间;圆顶夹层
intercycle 循环区间;循环间隔;中间周期;中间循环;内(部)周期
intercycle cooler 中间循环冷却器
interdata processor 中间数据处理机
interdeep 山间坳陷
interdelta bay deposit 三角洲间湾沉积
interdeltaic 三角洲间的
interdendritic 枝晶间
interdendritic attack 枝晶间腐蚀
interdendritic corrosion 显微腐蚀;枝晶间腐蚀
interdendritic graphite 枝晶间石墨
interdendritic porosity 枝晶间缩松
interdendritic segregation 枝晶间偏析
interdendritic shrinkage 枝晶间缩松;枝晶间疏松
interdendritic shrinkage porosity 枝晶间缩松
interdentil 齿饰间距
interdependence 相互依赖;相互依存;相互关系;相互关联;互相依赖;相互耦合
interdependence among commodities 商品间的相互依赖
interdependence factor 相倚因素
interdependent 相互依赖的
interdependent function 相依函数
interdependent variable 相依变量
interdialling 局间拨号
interdict 禁令
interdiction countermeasure 反干扰措施
interdiction fire 拦阻射击
interdiction of commerce 禁运
interdiffuse 互相扩散
interdiffusion 相互扩散;互扩散
interdiffusion coefficient 互扩散系数
interdigital 指状组合型
interdigital circuit 交叉指形电路
interdigital electrode 叉指电极
interdigital field linac 交叉指波直线加速器
interdigital line 叉指慢波线
interdigital magnetron 叉指磁控管
interdigital pause 间歇
interdigital transducer 叉指式换能器
interdigitated element 交叉指状单元
interdigitate junction 相嵌连接
interdigitation 指状联结;交叉联结;连锁;交错结合;交错接合
interdigited capacitor 片状分层电容器;叉指状电容器

interdisciplinary 多学科的
interdisciplinary 学科之间的;跨学科的;多种学科的;边缘学科的
interdisciplinary cooperation 跨学科合作;多科性协作
interdisciplinary economics 跨学科经济学
interdisciplinary study 多学科研究
interdisciplinary subject 边缘学科
interdiscipline 跨学科
interdiscipline subject 交叉学科
inter-disciplining 各学科间的
interdistributary area 支流区间;支流面积;分流河道间地区
interdistributary bay 分流间(海)湾
interdistributary bay deposit 分流间湾沉积
interdistrict settlement fund 区间清算资金
interdiurnal pressure variation 气压的日际变化
interdiurnal shoal variation 浅滩日际变化
interdiurnal temperature variation 温度日际变化
interdiurnal variability 日际变化率
interdiurnal variation of shoal 浅滩日际变化
interdiurnal water-level change 水位日际变化
interdiurnal water-level variation 水位日际变化
interdome 圆屋中间隔层;圆屋顶间;圆顶夹层
interdorsal 背间的
inter downward filtration 间歇过流
inter-drain 间沟
interdrive 中间平巷
interduce (窗台木和窗头木之间的)横撑
interdynode voltage 倍增极极间电压
interelectric(al) 电子间的
interelectrode 极间的
interelectrode conductance 极间电导
interelectrode leakage 极间漏泄
interelectrode space 极间空间
interelectrode transadmittance 极间跨导
interelement 单元间
inter-element interference 元素相干扰
inter-element traction 单元间的牵引
inter-enterprise credit 企业间信贷
interest 本金、利息、税金和保险费;保险主体
interest account 利息账户
interest accruing from land funds 土地资金利息
interest advance 预付利息
interest and amortization charges 还本付息支出
interest and tax 利和税
interest arbitrage 利息套戥
interest based on each installment 按每期等额付款计息
interest based on uncollected balance 按未收余额计息
interest-bearing account 计息账户
interest-bearing assets 计息资产
interest-bearing bank debenture 附息银行债券
interest-bearing capital 生息资本;借贷资本
interest-bearing eligible liabilities 附息合格负债
interest-bearing load 生息贷款
interest-bearing note 附息票据
interest-bearing securities 附息证券
interest bill 附息汇票;附利息票据
interest bill of exchange 附息汇票
interest charges 利息费用;收取的利率
interest charged to cost 计入成本利息
interest claim 利息债权
interest clause 利息条款
interest clause bill 附利息条款票据
interest concessions 让利
interest cost 利息成本
interest coupon 息票
interest cover 利息偿付
interest coverage ratio 盈利对利息的倍数;利息偿付比率
interest date 付息日
interest differential 息差年率;利息差额
interest due 到期利息
interest during construction 建筑期利息;施工期利息
interest earned 已取得的利息
interested administration 有关主管机关
interested parties 有关方(面)
interested party 有关当事人;利害关系人
interesterification 酯交换(作用)
interest escalation 利率自动上调
interest exchange 利息支出

interest expenses 支付利息;利息支出;利息费用
interest-extra loan 增息贷款;分期还本付息贷款
interest extra note 利息另计借据
interest factor 利息因素;本利(和)系数
interest field 权利范围
interest for delay 滞付利息
interest for delinquency 滞付利息
interest formula 计息公式
interest free 不计息的
interest free credit 无息信贷;无息贷款
interest free deposit 无息存款
interest free investment 无息投资
interest free loan 无息信贷;无息贷款
interest-free loan fund 免息贷款基金
interest fund 利息基金
interest groups 利益集团
interest in arrears 滞付利息
interest in block 应收利息
interest income 利息收入
interest induced note 利息合计借据
interest-in-general depreciation method 一般利息折旧法
interest in red 应付利息
interest margin 利差幅度
interest method of account current 金额分类计息法
interest money 附息货币;附息的货币
interest on a loan 借款利息
interest on arrears 远期利息;延期利息;欠款利息
interest on borrowing 借款利息
interest on call 活期借款利息
interest on capital 资本利息
interest on condition 附有条件的权益
interest on consumer debt 消费者债务利息
interest on debit balance 借方应付利息
interest on interbranches accounts 联行往来利息
interest on investment 投资利息
interest on loans 借入款利息
interest only mortgage 只分期付息而到期后一次偿还本金的抵押方式
interest on money 息钱
interest on overdue payments 逾期付款利息
interest on payment for purchases 进货款利息;进货付款利息
interest on tax underpaid or postponed 欠缴或滞纳税款利息
interest on the principal 本金的利息
interest paid 已付利息
interest passbook 息折
interest payable 应付利息
interest payable reserve 应付利息准备
interest paying date 付息日
interest payment 支付利息
interest payment and amortization 分期还本付息
interest payment date 付息日
interest payment of foreign loan 国外借款利息支付
interest payment period 利息偿付期;付息期
interest per diem 日息
interest period 利息期
interest policy 固定利益保单;利息政策
interest prepaid 预付利息
interest rate 利率
interest rate cap 最大利率
interest rate differential 利率差异
interest rate differential subsidy 利率差别补贴
interest rate effect 利率效应
interest rate futures 利率期货
interest rate on borrowing 借款利率
interest rate option 利率期权
interest rate risk 利率风险
interest rebate 利息回扣;退息
interest receivable 应收(未收)利息
interest received 已取得的利息;收入利息
interest reserve 利息准备(金)
interest restriction 利息限制
interests 利益
interests of the whole 全局利益
interest spread 利息差额
interest statement 息单
interest subsidization 利息津贴;利息补贴
interest subsidy fund 利息补贴基金
interest subsidy home 利息补贴住宅
interest suspense 利息暂记

interest table 利率表
interest tax 利息税
interest tax deduction certificate 扣除利息税证明书
interest test 兴趣测验
interest to maturity 满期前都附利息
interestuarine 两河口之间的
interest warrant 付息单
interest with holding tax 利息预扣税
interface 相互作用面;联系装置;连接设备;连接电路;离合面;界面;接口;交界面;交接层;交接面;内界面;地震间断面;分界面
interface adapter 接口衔接器
interface between horizons 层接面;层间接触面
interface board 接口板
interface bus 接口总线
interface cable 接口电缆
interface card 接口电路板;接口插件
interface channel 连接通道;接口通道
interface charge 界面电荷
interface checker 接口检查程序
interface chemistry 界面化学
interface chip 接口(芯)片
interface circuit 接口电路
interface clear 接口清除
interface command 接口指令
interface communication 界面传送;接口通信;接口联系装置;人机联系;人机对话
interface compatibility 相互关系的一致性
interface computer 接口计算机
interface condition 交接条件;交接条件
interface configuration 界面形状;界面构形
interface connection 穿通线
interface contact resistance 界面接触电阻
interface control 界面控制
interface control check 接口控制检验
interface control document 接口控制文件
interface controller 接口控制器
interface control module 接口控制模块
interface coupler 界面耦合器
interface cutting 界面切割
interface data 接口数据
interface data unit 接口数据部件
interface debugging 接口调试
interface design 接口(部件)设计
interface detection 界面探测
interface detector 油水界面探测仪;油水界面测定器;界面探测器
interface device 界面装置;接口设备
interface distribution coefficient 界面分配系数
interface drainage 界面排水
interface driver 接口驱动器
interface effect 界面效应
interface element 转换元件;界面元件;界面单元;接口元件
interface energy 界面能;分界面能
interface equipment 接口设备
interface error control 接口错误控制
interface event 交界事项
interface exchanging process 界面交换过程
interface flag 接口标志
interface flexibility 接口灵活性
interface flow velocity 界面流速
interface function 接口功能
interface handling 换装作业
interface latch chip 接口锁存器芯片
interface layer 中间层
interface layer resistance 内面阻;层间电阻
interface level 界面高度
interface level indicator 界面指示器
interface level settling velocity 界面沉降速度
interface location 中间站;联络站;界面定位
interface locator 界面定位器
interface logic 接口逻辑
interface loss 间隙损失
interface management plan 界面管理方案
interface message 接口信息;接口消息
interface message processor 接口信息处理器;接口(信息处理)机;接口报文处理器
interface message processor throughput 接口机吞吐量
interface migration rate 界面迁移率;分界面徙动率
interface mixing 界面混合
interface mix(ture) 界面混染

interface mobility 界面迁移率
interface module 接口组件;接口模块;接口模件
interface of beds 层间界面
interface of host computer 主机接口
interface parameter 接口参数
interface pressure 界面压力;接触面压力;交界面压力
interface procedure 交接手续;换装手续
interface processor 接口处理机
interface reaction 界面反应
interface region 界面区;分界面区
interface relative moisture content 分界相对含水量
interface requirement 接口要求
interface resis 面间电阻
interface resistance 界面电阻
interface routine 接口例行程序;接口程序
interface scattering 界面散射
interface shape 界面形状
interface shear 界面受剪
interface signaling format 接口信号格式
interface software 接口软件
interface stability 界面稳定性;分界面稳定性
interface standard 接口标准
interface state 界面态
interface status flag 联结状态标记;接口状态标记
interface strength 黏结强度;界面黏滞强度;接口强度;接触面强度
interface subsystem 接口子系统
interface switch 接口开关
interface tension 界面张力
interface tension ring method test 界面张力环法试验
interface termination 接口终端
interface tester 接口测试仪
interface turbidity monitoring 界面浊度监测
interface unit 连接器件;接口装置;接口部件
interface vector 连接向量
interface vehicle 界面运动器;冲翼艇
interface velocity 界面速度
interfacial active agent 界面活性剂;表面活性剂;表面活化剂
interfacial activity 界面活动;界面活性
interfacial adhesion 界面黏结;界面黏合
interfacial adsorption 界面(间)吸附
interfacial agent 界面活性剂
interfacial angle 晶面角;(面)(交)角;面间角
interfacial area 界面面积
interfacial bond 界面黏接
interfacial bond strength 界面黏结强度
interfacial composition 界面成分
interfacial contact 面间接触;层面间接触;分界处表面接触
interfacial diffusion 界面扩散
interfacial electrochemistry 界面电化学
interfacial energy 界面能
interfacial exchanging process 界面交换过程
interfacial film 界(面)膜;界面层;面间膜
interfacial flow 界面流
interfacial force 表面张力
interfacial friction 层间摩擦;不平整摩擦
interfacial geology 界面地质
interfacial layer 夹层
interfacial level interfacial height 界面高度
interfacial medium 界面介质
interfacial oscillation 分界面振荡
interfacial phenomenon 界面现象
interfacial polarization 界面极化
interfacial polycondensation 界面缩聚
interfacial potential 界面位能;界面势(能)
interfacial region 界面区
interfacial shear 界面剪切力
interfacial shear strength 界面剪切强度
interfacial spacing 晶面间距
interfacial stress 界面应力
interfacial structure 界面结构
interfacial surface 交界面
interfacial surface tension 面间表面张力
interfacial tensimeter 界面张力计
interfacial tension 界面(间)张力;面际张力;面积张力;分界面上的表面张力;表面张力
interfacial tension of air mercury system 空气汞系统界面张力
interfacial tension of air oil system 空气油系统

界面张力
interfacial tension of air water system 空气水系统界面张力
interfacial tension of oil-water system 油水系统界面张力
interfacial tension of water-gas system 水一天然气系统界面张力
interfacial turbulence 界面团流
interfacial velocity 界面速度
interfacial viscosity ratio 界面黏度比
interfacial wave 界面波;内波
interfacial weld 摩擦面焊合(点)
interfacial zone 界面区
interfacing 衬头;衬布
interfacing device 分界接合装置
interfacing discipline 相邻学科
interfan depression 扇间洼地
interfan depression deposit 扇间洼地沉积
interfan depression of alluvial fan 冲积扇扇间洼地
interfenestral 窗间的
interfenestration 窗间墙宽度;窗间布置;窗距离布置
interference 干涉;干扰;结论;串扰
interference absorber 干涉吸收体;干扰吸收器
interference absorption band 干涉吸收带
interference amplitude 干扰振幅
interference area correlation 干涉区对比
interference atmosphere 干扰大气层
interference band 干涉条纹;干涉光带;干扰谱带;干扰频带
interference blanker 干扰消除器
interference body bolt 高强螺栓
interference characteristic 干涉特性;干扰特性
interference checking 干扰检测
interference circumstances 干扰情况
interference coefficient 干涉系数
interference colo(u)r 干涉色;干扰色
interference colo(u)r chart 干涉色表
interference colo(u)r filter 干涉滤色镜
interference colo(u)rs order 干涉色级序
interference comparator 光干涉比长器;干涉比长仪
interference condition 干涉条件
interference contrast 干涉相衬
interference contrast microscopy 干涉相衬显微术
interference correction 干扰修正
interference current 干扰电流
interference demonstration 干涉演示
interference detection 干扰侦察;干扰检测
interference detector 干扰检测仪
interference diagram 叶片的频谱图;频率图
interference dilatometer 干涉膨胀仪
interference dissociation 干扰分离
interference drag 干涉阻力
interference due to adjacent station 邻台干扰
interference edge filter 干涉截止滤光片
interference effect 干涉效应;干扰作用;干扰效应
interference element 干扰元素
interference elimination 清除干扰
interference eliminator 干扰消除器
interference eyepiece 干涉目镜
interference factor 干扰因素;干扰系数
interference fading 干扰(性)衰落
interference field 干涉场;干扰场
interference-field intensity tester 干扰场强测试仪
interference figure 干涉图形;干涉图(案);干扰图(像)
interference film 干涉膜
interference filter 干涉滤光片;干涉滤光镜;干涉滤波器;干扰滤色片;干扰滤光片;干扰滤波器
interference filter monochromator 干涉滤光片单色仪
interference finding instrument 干扰寻迹器
interference fit 过盈配合;干涉配合;静坐配合;静配合;紧配合
interference-free 抗干扰(的)
interference frequency 干扰频率
interference fringe 干涉图样;干涉条纹;干涉带;干扰带
interference fringe method 干涉条纹法
interference fringe modulation 干涉条纹调制
interference from adjacent channel 邻道干扰
interference from adjacent channel signals 邻道信号干扰
interference from neighbo(u)ring line 邻线干扰
interference function 干涉函数

interference generator 干扰发生器
interference grating 干涉光栅
interference guard band 抗干扰保护频带
interference image 干扰像
interference instrument 干涉仪
interference intake 干扰进气
interference in the co-located station 同站干扰
interference inverter 噪声限制器；杂波抑制器；干扰倒相器；干扰补偿器
interference level 干扰级；干扰电平
interference light filter 干涉滤光器
interference lobe 干涉波瓣
interference locator 干扰定位器
interference loss 干扰损耗
interference magnifier 干涉放大镜
interference meter 干涉流量计
interference method 干涉法
interference metrology 干涉度量学
interference microscope 干涉显微镜；干扰显微镜
interference microscopy 干涉显微术
interference mirror 干涉镜
interference mode 干扰形式
interference modulation 干涉调制
interference modulator 干涉调制器
interference near field pattern 近场干涉图形
interference noise 干扰噪声；干扰杂波
interference objective 干涉物镜
interference of alternating current 交流电干扰
interference of equal inclination 等倾角干涉
interference of equal thickness 等厚度干涉
interference of light 光电比色；光(的)干扰
interference of sound 声扰；声(干扰)；声(的)干扰
interference of sound wave 声波的干扰
interference of tooth 轮齿干涉；齿(的)干涉
interference of waves 波浪干涉；波(的)干扰
interference of wells 井群互扰；井间干扰
interference order 干涉级
interference path difference 干涉程差
interference pattern 干涉图样；干涉图式；干涉图(案)；干扰特性；干涉图形；干涉图(像)
interference phase contrast microscope 干涉相衬显微镜
interference phenomenon 干涉现象；干扰现象
interference photography 干涉摄影
interference photometer 干涉光度计
interference photometry 干涉光度测量法
interference plane 干涉平面
interference point 干涉点
interference polarizing filter 干涉偏振滤光器
interference power 噪声功率；干扰功率
interference prediction 干扰预测
interference preventer 干扰防护装置；防干扰装置
interference prism 干涉棱镜；干扰棱镜
interference range 干扰区；干扰区
interference rangefinder 干涉无线电测距仪
interference receiver 干扰接收机
interference reducer 干扰抑制器
interference reduction 干扰减少
interference reflectometer 干涉反射仪
interference reflector 干涉反射器
interference refractometer 干涉折射计；干扰折射计
interference region 干扰区域；干扰区
interference rejection 干扰抑制；抗(干)扰性；抗干扰能力；抗干扰度；反干扰能力
interference rejection capability 干扰抑制能力
interference response 干扰响应；干扰反应
interference ring 干涉圈；干涉环；干涉光环
interference ripple mark 干涉波痕；交错波痕；扰动波痕
interference search gear 干扰定位器
interference settlement 干扰(性)沉陷；干扰(性)沉降
interference signal 干扰信号
interference source 干扰源
interference source suppression 干扰源抑制
interference spectroscope 干涉分光镜；干扰分光镜
interference spectroscopy 干涉分光(光谱)学
interference spectrum 干涉(光)谱；干扰频谱；干扰光谱
interference spot 干涉斑
interference suppression 干扰抑制；干扰减少
interference suppression filter 干扰抑制滤光器
interference suppressor 干扰抑制器
interference susceptibility 干扰灵敏度

interference system 干扰系统
interference term 干涉项
interference test 干扰试验
interference theory 干涉理论
interference time 干扰时间；空转时间
interference trap 干涉滤波器
interference tube 干涉管
interference unit 干扰器
interference wave 干涉波；干扰波
interference well 干扰井
interference with reception 接收干扰
interference with tests on completion 对竣工检验的干扰
interference zone 干扰区
interferent 干扰物(质)
interferential 干扰的
interferential vibration 干扰振动
interfere with navigation 妨碍航行
interfering 干扰性的；干扰的
interfering background 干扰背景
interfering beam 干涉束光
interfering carrier 干扰载波
interfering component 干扰组分
interfering energy 干扰能量
interfering float time 干扰宽裕时间；干扰浮动时间
interfering impulse 干扰脉冲
interfering ion 干扰离子
interfering material 干扰物质
interfering mode 竞争模式
interfering noise 干扰噪声；干扰杂波
interfering pumping from large-scale well groups 大型井群干扰抽水
interfering signal 干扰信号
interfering singing 干扰啸声
interfering space 干扰区
interfering substance 干扰物质
interfering test pattern 干涉测试图
interfering transmitter 干扰发射机
interfering wave 干扰波
interfering well formula of unsteady flow 非稳定流干扰井公式
interfering wells pumping test 群孔干扰抽水试验
interferogram 干涉图样；干涉图(案)
interferogram method 干涉图法
interferogram technique 干涉图技术
interferography 干涉摄影术
interferometer 干涉仪；干涉计
interferometer antenna 干涉仪天线
interferometer base 干涉仪基线长度
interferometer coating 干涉膜
interferometer effect 干涉仪效应
interferometer measuration system 干涉测量系统
interferometer measurement 干涉仪测定
interferometer metering technique 干涉测量法
interferometer method 干涉仪法
interferometer micrometer 干涉仪式测微计
interferometer microscope 干涉仪显微镜
interferometer modes 干涉仪类型
interferometer modulation 干涉调制
interferometer optics 干涉仪光学
interferometer pattern 干涉仪方向图；干涉仪测得的图样；干涉条纹
interferometer phase 干涉仪相位
interferometer plate 干涉片；干涉仪玻璃板
interferometer polar diagram 干涉仪极坐标方向图
interferometer radar 干涉雷达
interferometer spectrometer 干涉光谱仪
interferometer spectroradiometer 干涉光谱辐射计；干涉分光辐射计
interferometer strain ga(u)ge 应变干涉仪；干涉应变仪；干涉式应变计
interferometer system 干涉仪系统
interferometric astronomy 干涉天体测量
interferometric binary 干涉双星
interferometric deflection 干涉偏转
interferometric fiber optic(al) current sensor 干涉式光纤电流传感器
interferometric fiber optic(al) accelerometer 干涉型光纤加速度计
interferometric fiber optic(al) gyroscope 干涉型光纤陀螺仪
interferometric filter 干涉滤光片；干涉滤光镜
interferometric fringe 干涉条纹
interferometric manometer 干涉真空计；干涉压力计
interferometric microscopy 干涉测量显微术
interferometric null method 干涉衡消法
interferometric optic(al) fiber acoustic(al) sensor 干涉型声光纤声传感器
interferometric radiometer 干涉辐射计
interferometric sensor 干涉型传感器
interferometric spectrometer 干涉分光计
interferometric temperature monitor 干涉计量式温度监视器
interferometric test 干涉法试验
interferometry 干涉量度学；干涉度量学；干涉测量法
interferon 干扰素
interferoscope 干涉镜；干扰显示器
inter-fiber bond 纤维间黏合
interfibrous 纤维间的
interfield cut 场间切换
interfield disparity 场间差异
interfilamentous matrix 细丝间基质
interfile 文件间的
interfile relationship 内部文件关系
interfinger 楔形夹层；相互楔交；指状交错；相互贯穿
interfingering 相互贯穿；楔形夹层；指状交通；相互穿插
interfingering member 相互贯穿的构件；楔形夹层部件；楔形夹层构件
interfix 相互确定；中间定位
interflare matter 耀斑际物质
interflectance method 预定照度法
interflood period 间洪期；洪水间隔期
interfloor operation 上下楼作业
interfloor stair(case) 楼面间楼梯
interfloor traffic 楼层间交通；楼面间交通
interfloor travel 不同层次间的交通(指楼面或桥面)；不同层次的行程；楼层间交通
interflorescence 内部晶化
interflow 中间流；过渡流量；交流；间层流；混流；合流；壤中流；土内水流；地下径流；层间流；表层流
interflow of commodities 物资交流
interflow turbidity current 层间浊流
interfluent 混流的；汇合的
interfluent flow 异重水流；(不同密度的)层间水流
interfluent lava flow 内流熔岩流；混合岩流；地下熔岩流
interfluve 江河分水区；河间岭；河间地分水区；泉间分水区；泉间地；两相同流向水流间的地区
interfluve landform of loessland 黄土沟间地貌
interfluvial 两河之间的；河间的
interfold 交互折叠
interfolding 交错褶皱；交插褶皱
interformational 层间的
interformational bed 层面夹层
interformational breccia 层间角砾岩
interformational conglomerate 层组间砾岩；层间砾岩
interformational sheet intrusion 建造间层状贯入体
interformational sill 层间岩床；内部构造(底)槛
interformational sliding 层间滑动
interformational unconformity 层间不整合
interform reliability coefficient 形式间可靠性系数
interframe 帧间
interframe coding 帧间编码
interframe encode 帧间编码
interfuel substitution 不同燃料相互取代
inter-fund and inter-government transactions 基金直接和政府内部转发
inter-fund balance payable 应付账款款项
inter-fund balance receivable 应收转账款项
inter-fund loan 基金间贷款
inter-fund receivable and payable 基金间应收应付项目
inter-fund transfer 基金间转账
intergelisol 残冻层
intergenerational mobility table 两代间变动率表
interglacial epoch 间冰期；间冰阶
interglacial period 间冰期
interglacial period deposit 间冰期沉积
interglacial rising of sea level 间冰期海面上升
interglacial stage 间冰期
interglaze decoration 釉中装饰
interglobular areas 球间区
interglyph 槽陇间的空间
interglyphe 柱槽间
Inter-Governmental Maritime Consultative

Organization 政府间海事协商组织
Inter-Governmental Oceanographic (al) Commission 政府间海洋委员会
Inter-Governmental Standing Committee on Shipping 政府间常设航运委员会
intergradation 间渡
intergrade 中间形式;中间粒级;中间级配;过渡性(土壤);渐次混合;渐次变迁
intergrain slip 粒间滑移
intergrant 构成部分
intergranular 粒间的;晶(粒)间的;晶界
intergranular attack 粒间侵蚀;晶间腐蚀
intergranular cement 粒间胶结
intergranular contact pressure 粒间接触压力
intergranular corrosion 粒间腐蚀;晶粒间腐蚀;晶间腐蚀
intergranular crack 沿晶开裂;沿边开裂
intergranular cracking 粒间裂缝;颗粒间开裂
intergranular force 颗粒间力;土颗粒间力
intergranular fracture 晶界断口
intergranular friction 颗粒间摩擦
intergranular martensite 粒间马氏体;粒状马氏体钢
intergranular penetration 粒间穿透
intergranular porosity 粒间孔隙(度)
intergranular precipitation 弥散沉淀
intergranular pressure 有效应力;粒间压力;颗粒间压力;土颗粒间压力
intergranular slip-plane 颗粒间滑动面
intergranular space 粒间孔隙;颗粒间空隙;晶间距
intergranular stress (土壤的)颗粒间的应力;粒间应力;晶间应力
intergranular texture 粒间结构;间粒结构
intergranular volume 粒间体积
intergranule 粒间;晶粒之间
intergreen interval delay 绿灯间隔延误
intergreen interval (time) 绿灯间隔时间
intergreen period 绿灯信号间隔时间
intergrind 相互研磨;共同碾磨;互磨
intergrinding 互磨
intergrinding of bentonite 膨润土互磨
interground 相互研磨
interground addition 磨碎时的加入物;磨细掺和剂;研磨添加剂;研磨(时的)添加料;破碎杂料;粉磨添加料
interground additive 磨碎添加剂
interground fuel process 黑生料法
intergrout 中速凝结浆
intergrow 共生;夹石层
intergrow crystal 连生晶体
intergrown along the facies 沿地质相共生
intergrown knot 连生节;活节;混生节;连生树节
intergrowth 连晶;交生;互生
intergrowth cement texture 连生胶结物结构
interhalogen 卤间化合物
interhalogen compound 卤素互(间)化物
interheater 加热器;中间加热器;重新加热器;重热炉;中间过热器;再热器
interhemispheric integration 两半球间的整合作用
interhourly variability 时际变率
interim acceptance certificate 中期验收证书;初步验收证(明)书
interim account 临时账单
interimage effect 像间效应
interim agreement 临时协议;临时协定
interim agreement for a project 建设项目施工临时协议书
interim arrangement 临时办法
interim audit 中期审计;期中审计
interim authorization 暂时批准
interim award 临时裁决
interim beam 内纵梁
interim borrowing 期中借贷
interim budget 中期预算;期中预算
interim certificate 临时(付款)证书;中期证书
interim charges 临时开支
interim closing 期中结账
interim codes 临时编码
interim combination technique 暂时组合法
interim compilation 过渡阶段编绘
interim credit 临时贷款
interim criterion 暂行准则;暂行标准
interim dates 分阶段工期

interim determination of extension 临时的延期决定;中期延期决定
interim disposal site 中间抛泥区
interim dividend 期中股息;暂定股利
interim-earnings statement 期中盈利报表
interim engineering order 临时工程指标
interim estimate 概估;暂估;暂时估算;暂时估价;暂时估计;毛估;期中报表
interim expenses 临时开支;临时费用
interim financial statement 期中财务报表
interim financing 临时(性)筹措资金;临时赊贷;临时集资;期中集资;中间性筹款
interim fixing 中间固定;临时固定
interim garage sign 内部车库标志
interim improvement 临时改善措施;临时改进措施
interim income and loss statement 期中损益计算书
interim income statement 期间损益表
interim index 临时指数
interim inspection 中间检查
interim interest 临时利息
interim loan 临时贷款
interim means 暂行措施;过渡性措施;临时措施;临时办法
interim measure 过渡办法;暂时措施
interim meteorologic (al) satellite 临时气象卫星
interim orbit 中间轨道
interim payment 进度(付)款;中期付款;暂付款
interim payment certificate 中期付款证书
interim port 临时港区
interim price 暂时价格;临时价格
interim primary drinking water regulations 暂行初级饮用水规程
interim primary regulation 暂行基本规则;暂行法规
interim procedure 暂行办法
interim provisions 暂行规定
interim record 临时记录
interim regulations 暂行条例
interim reinforcement 临时钢筋
interim remedial measure 临时补救措施
interim reparation 临时赔偿
interim replacement 中期更新;中间性更新
interim report 中期报告;临时报告;临时报告;阶段报告;单记录速取报告
interim reversion station 中间折返站
interim revision 临时修订版
interim scheme 过渡方案
interim spare parts list 临时备件单
interim specification 暂行规范
interim stage report 中间阶段报告
interim statement 期中报表
interim status 暂时状态
interim stock 中间库存
interim storage 中间储[贮]水池;暂时储[贮]存;暂时储[贮]藏;临时储[贮]存;临时储[贮]藏
interim summary 小结
interim survey 中期测量
interim treatment guide 暂行处理指南
interim use 过渡用途
interim valuation 临时估价
interim work sheet 期中工作报表
interindustry analysis 产业关联分析
interindustry economics 部门间经济学
interindustry emission control 工业排污控制;工业间排放控制
interindustry equilibrium 部门间平衡
interindustry relation analysis 工业关系分析法
interionic attraction 离子间引力
interionic distance 离子际距;离子间距
interior 里面;内面的;内部(的);室内
interior adhesive 室内用黏合剂;内部黏着力
interior air 室内空气
interior air cooler 室内冷气机
interior and exterior finishes 内外装修
interior angle 内角
interior angles of the same side 同傍内角
interior arch 内切拱
interior architecture 室内建筑(学)
interior area 内部楼面面积
interior arrangement 内部装置
interior awning 室内遮篷
interior basin 内陆盆地;内港池
interior bending moment 内部弯矩
interior board 内装板

interior bonding agent 室内用粘合剂;内部粘结剂
interior breakwater 内防波堤
interior building board 室内建筑板
interior building panel 室内建筑板
interior buoyancy 内浮力
interior casing 内框
interior cement (ing agent) 室内黏结剂
interior center 内心
interior chimney 内部烟囱;壁内烟囱
interior chlorinate rubber paint 内部氯化处理的橡胶漆
interior circulation three-phase biological fluidized bed reactor 内循环三相生物流化床反应器
interior city 内地城市
interior climate 室内气候
interior collocation 内部配置
interior column 内柱
interior combustion 浸没燃烧;内部燃烧
interior comfort level 室内舒适水平
interior communication 内部通信(联系)
interior communication cable 船内通信电缆
interior communication of ship 舰船内部通信
interior communication system 内部通信制;内部通信系统
interior communication system of ship 舰船内部通信系统
interior concrete column 室内混凝土柱
interior conduit 内管道;室内管道;暗线(用)导管
interior contact 内切
interior container 内容器
interior continent 内陆
interior convection 内运流
interior core 内部型芯
interior corner 室内角隅;内墙角
interior corner reinforcement 内墙角加固件;内墙角增强筋
interior corridor 内廊;室内走廊
interior-corridor type building 内廊式房屋;内走廊型建筑物
interior corrosion 内部腐蚀
interior court 内天井;内庭;内院;室内庭院
interior craton basin 克拉通内盆地
interior crystalline 内结晶
interior decorating 室内装饰;内部装饰;内部装潢
interior decoration 内部装修;内部装饰;内部装潢;室内装饰;室内布置
interior decorator 室内设计师;室内装饰师
interior delta 内陆三角洲
interior design 室内设计
interior designer 室内设计师
interior décor 室内装饰
interior diaphragm 内横隔板
interior differential needle valve 中差针形调压阀;内差式针形阀
interior diffusion 内扩散
interior diffusion of water 水分的内扩散
interior dimension 室内尺寸
interior distance between brick courses 格孔间距
interior distribution 室内配电
interior door 房门;内门
interior door jamb 内门侧柱;房门框
interior door jamb block 内部门窗侧壁砌体;内部门窗侧壁砌块
interior door jamb tile 内门门窗侧壁砖
interior downpipe 内水落管
interior drainage 内落水;内陆(排)水系;内(部)排水;屋顶内排水
interior drain pipe 室内排水管
interior economy 内地经济
interior electrolysis 内电解法
interior electrolytic-anaerobic-oxic process 内电解厌氧好氧工艺
interior element 内定向元素
interior elevation 室内立面图
interior emulsion paint 室内用乳化涂料
interior enamel 内用瓷漆
interior escape stair (case) 室内安全梯;室内太平梯;室内安全电梯
interior extrapolation method 交叉法;内外插法
interior face 室内顶面;室内墙面
interior facing 内墙饰面
interior fiberboard finish 内部纤维板装修;内部纤维板装饰
interior finish 内装饰;内(部)装修;内部饰面;室

内装修;室内装饰
interior-finish board 室内用装修材料
interior finishing 内装(修);内部修饰
interior finishing in tunnel 隧道内装修
interior finishing varnish 内用罩光清漆;内用末道清漆
interior finish work 内檐装修;(内)(部)装修工程
interior fire protection 内部防火
interior fixtures 室内装置(尤指灯具);内部装饰物;内部装饰品
interior floor area 内部楼面面积
interior flooring 室内地面(材料);内楼面
interior flow 内流
interior focusing 内对光;内调焦
interior focusing system 内对光系统
interior focusing telescope 内聚焦望远镜;内对光望远镜;内调(焦)望远镜
interior furnishing 内部装饰;内部装潢
interior furnishing glass 室内陈设玻璃
interior gallery 内部长廊
interior gallery apartment building 内廊(道)式公寓(建筑)
interior garage sign 内部车库标志;车库内部标志
interior garden 内园;室内花园
interior gate 内部闸
interior gate valve 内部(高压)闸阀
interior gear 内齿轮
interior glass door 室内玻璃门
interior glazed (玻璃的)内部安装;内装玻璃的
interior glazed window 内装玻璃窗
interior glazing 室内玻璃安装;室内玻璃窗
interior gloss varnish 室内上光清漆
interior glue 室内用胶着剂
interior handrail 内部扶手
interior harbo(u)r 内港
interior heat 室内热量;室内(供)热
interior heat gain 室内热增量;室内增热
interior heating system 建筑供暖系统
interior height 车内净高
interior humidity 室内湿度
interior hung scaffold 室内悬吊脚手架;内部悬挂式脚手架;内挂脚手架
interior illumination 室内照明
interior ingress 内初切
interior insulation 室内隔热隔声;内部绝缘
interior joinery 内部细木工作
interior joint 内部节点
interior label 内数据说明符;内部标号;内标
interior lake 内陆湖
interior lamp 内部指示灯
interior landscape design 室内景观设计
interior latex gloss paint 室内胶乳光泽涂料
interior latex semigloss paint 室内胶乳半光泽涂料
interior layout 室内规划;室内布置
interior leak 内部泄漏
interior leakage 内部泄漏量
interior leg braces 内斜撑
interior leg braces of drilling tower 钻塔内斜撑
interior liabilities 内部负债;对内负债
interior lighting 内部照明;室内照明
interior lining 内衬;内部衬砌
interior lining panel 内部护板
interior load (混凝土路面的)中部荷载;板中荷载
interior load stress 中载应力
interior lot 内地段
interior mapping 内映象
interior masonry dividing wall 室内砌筑分隔墙
interior measure 内测度
interior metal stud 内墙用金属立筋;内墙金属龙骨
interior mirror (车身的)内部后视镜
interior moisture 室内湿度;室内潮气
interior moment 内力矩
interior mounting height 室内照明安装高度
interior nodal point 内节点
interior node 内节点
interior noise 室内噪声
interior noise criterion curve 室内噪声标准曲线
interior non-bearing walls 内部非承重墙
interior of a set 集合的内部
interior of building 房屋内部;楼房内部;建筑内部
interior of dry dock 干船(坞)坞室;坞室
interior oil stain 室内用油性着色剂
interior opposite angle 内对角
interior orientation 内方位;(内)(部)定向

interior orientation elements 像片内方位元素
interior packing 内部包装
interior paint 内用漆;室内(用)油漆;室内(用)涂料
interior paint coat 内部涂料
interior panel 室内护墙板
interior panelling 内侧镶板
interior partition 内部隔断;内隔墙
interior partitioning 室内间隔
interior passage 车内通道
interior perspective 室内透视
interior perspective centre 内透视中心
interior picture 室内画
interior piping 屋内管道;内部管线;室内管道
interior plain 内陆平原
interior planet 内行星
interior plantscape 室内植物景观
interior plumbing system 室内排水系统
interior plywood 内部(装修)胶合板;内用胶合板;室内用胶合板
interior point 内点
interior point intermodal 一贯运输
interior port 内港
interior post type container 内柱式集装箱
interior primer 打底涂料;内墙饰面底涂;核心雷管
interior profile 内剖面
interior rate of financial revenue 财务内部收益率
interior relative humidity 室内相对湿度
interior relative moisture 内部相对含水量
interior reveal 室内外抱
interior sap stain 边材内部变色
interior sea 内(陆)海
interior ship tracking system 舰船光学跟踪装置;舰船光学跟踪系统
interior side yard 内部侧院
interior signal lighting 车内信号照明灯
interior single cycle basin 内陆单旋回盆地
interior solution 内解
interior space 内部空间;室内空间
interior span 内跨梁;中间跨;内跨(度);内部跨度
interior spring 内弹簧垫子
interior stain 室内涂色;内部色彩
interior stair (case) 室内楼梯(间)
interior stair rail 室内楼梯栏杆
interior stanchion 室内柱子;室内支柱
interior stop 内装窗玻璃镶条;内部压条;内部定位条;内部挡块
interior storm system 内排水系统
interior street 小区内部街道
interior stress 内应力
interior stringer 内纵梁
interior structure 内部结构
interior structure of the earth 地球内部结构
interior stucco 内装修用灰泥
interior stud 室内壁骨
interior styling (汽车车身的)内部造型
interior supply water 内水源
interior support 内部支柱;室内支柱;内部支承
interior surfacing 室内表面装饰
interior temperature 内部温度;室内温度
interior temperature control 室内温度控制;室内温度调节
interior thickness 中部厚度
interior tile 室内砖
interior tissue 内部组织
interior tracery 挂落【建】;室内花格饰;内部窗(花)格
interior trade 国内贸易
interior traverse 内角导线【测】
interior trim 内部装修;室内装饰件;内镶边
interior trim colo(u)r 内部装饰色彩;车室内器具彩色
interior trim(ming) 内部装饰;内部装潢;内部修饰;内部装修;室内装修
interior-type plywood 内用胶合板;室内用胶合板
interior upset pipe 管端部内加厚管子;端部内加厚管
interior valley 灰岩盆地
interior valve 内部阀;内阀(门)
interior varnish 内用(清)漆;内皮漆
interior view 内视图;内部图
interior wall 内墙
interior wall surface 室内墙面
interior water areas 港内水域
interior waters 港内水

interior waterway 内陆水道;内河水道;内河航道
interior wet standpipes 室内供水管系
interior window 室内窗
interior window cill 内窗槛
interior window frame 内窗框(架)
interior window sill 内窗槛
interior wiring 内线;室内线路;室内路线;室内布线
interior wood stain 室内木材涂色
interior works 室内工作;屋内装修和设施;室内施工;室内工程
interior yard 里院;内院
interior zone 内部区段
interirrigation 灌溉间期
inter-island gap 岛间峡谷
inter-island platform 岛间台地
interjacent 居间的;夹层的;插入的
interjectional 插入的
interjoist 跨度;跨距;格栅(空)档;格栅间
interknot 连结在一起
interlaboratory 实验室间
interlaboratory study 协作试验研究
interlaboratory test comparisons 实验室间的试验比较
interlaboratory trial 实验室间试验
interlace 组合;隔行析象;隔行(扫描);交织;交错存储;交错(操作);交叉存储;夹层
interlaced 编织状的
interlaced arches 交叉拱(门)
interlaced channel 游荡河槽
interlaced fencing 交织围篱;交织围栏
interlaced field 隔行扫描场
interlaced handle 绞花手柄
interlaced island 交织江心洲
interlaced line 套轨线路
interlaced memory 交错存储器
interlaced scanning 隔行扫描;交替扫描
interlace operation 交叉操作
interlace scan 隔行扫描
interlacing 隔行;交错存储;交错;间行
interlacing arcade 交织拱廊;交叉拱廊;内叉拱廊
interlacing arches 柱顶交织拱;交织拱;交织拱门;交叉拱门;内叉拱
interlacing arrangement 编织状排列
interlacing diagram 交叉图(交通)
interlacing drainage 游荡水系
interlacing earth-scraping process 交错铲土法
interlacing memory 交叉存取存储器
interlacing of wires 电缆交织;电线交织
interlacing ornament 绠;交叉饰
interlacing pattern 交织图案
interlacing ratio 隔行比
interlacunar margina 隙间边
interlacustrine overflow stream 湖间河流
interlacustrine transfer of water 湖泊间水转移
interlamellar deformation 层间形变
interlamellar spacing 层间距
interlaminar bonding 夹层黏结;层间黏结
interlaminar erosion strength 层间剥离强度
interlaminar fracture 层间破裂
interlaminar shear strength 层间剪切强度
interlaminar strength 层间强度
interlaminated 纹层状互层
interlaminated resistance 中间电阻
interlamination 层间
interlamination resistance 层间电阻
interlap 相互重叠;内覆盖
interlattice point distance 阵点间距
interlay 夹层;垫衬
interlayer 隔层;芯层;界(面)层;间层;夹层;层间的
interlayer adhesion 层间附着力
interlayer bonding 层间联结;层间连接
inter layer cleavage 层间劈理
interlayer contact 层间接触
interlayer continuity 层间连接
interlayer crush 层间裂隙
interlayer crush belt 层间破碎带
interlayer detachment 层间剥离
interlayer dislocation plane 层间错动面
interlayered bedding 互层层理【地】
interlayer effect 界层效应
interlayer gliding fault 层间滑动断裂
interlayer gliding fracture 层间滑动断裂
interlayering 间层理;互层理;层间排列

interlayer ion 层间离子
interlayer mixture 层间混合物
interlayer of spun glass 叠层玻璃布;玻璃纤维夹层
interlayer rock 内部围岩
interlayer slip surface 层间滑动面
interlayer spacing 层间距
interlayer stress field 层间应力场
interlayer swelling 层间膨胀
interlayer temperature 层间温度
interlayer water 夹层水;层间水
interlayment 屋面油毡层
interleaf 插入纸;中间层;夹层
interleaf friction 板片间摩擦力;板片摩擦
interleave 隔行;交织;交替;交错;交插;间隔因子
interleaved additional channel 插入波道
interleaved array 交错数组
interleaved coil 纠结式线圈;交错线圈
interleaved luminance signal 亮色交错信号
interleaved memory system 交叉存储系统
interleaved model intersection mode 交叉方式
interleaved reduction gear 嵌入式减速齿轮箱
interleaved scanning 隔行扫描
interleaved subscript 交错下标
interleaved transmission 交错传输
interleaved winding 纠结式绕组;交错绕组
interleaver 衬纸机
interleave type core 交叠式铁芯
interleaving 隔行;插页
interleaving access 交叉访问;交叉存取
interleaving memory 交叉存取存储器
interleaving paper 衬(垫)纸;衬底纸
interlibrary loan 馆际互借
inter lighting 室内照明
interlimb angle 翼间角
interline 线之间虚线
interlinear space 行间空白
interlinear spacing 行间间距
interline freight 联运货物;铁路联运货物
interliner 衬里;衬布
interline waybill 联运货物路程单
interlining material 衬料
interlink 联锁;连环;互连
interlinkage 连动装置;交链;互连
interlinkage flux 交链磁通
interlinked 互连的
interlinked hierarchy 互连次结构
interlinked leakage 漏磁链;交链漏磁
interlinked tension 相间电压
interlinking 互连
interloan agreement 相互出借协定
interlobate 尖间
interlobate moraine 中分碛;冰舌间碛
interlobe 叶片间的
interlobular moraine 冰舌间碛
interlock 联系;联锁设备;联锁(器);联动;连结;交替工作;闭锁;安全开关
interlock alarm 联锁报警器
interlock assembly 连锁装置;连锁装配
interlock circuit 联锁回路;联锁电路
interlock coupling 联锁耦联
interlock cylinder 联锁缸
interlocked circuit breaker 联锁断路器
interlocked crossover 联锁渡线
interlocked derail 联锁脱轨器
interlocked film wind and shutter mechanism 卷片和快门联锁机构
interlocked grain 倾斜木纹;扭曲木纹;斜交木纹;交错纹理;交错木纹
interlocked-grain wood 交错纹理木材
interlocked operation 联锁操作;互锁操作
interlocked oscillator 联锁振荡器;内同步振荡器
interlocked pile 联锁桩
interlocked poidometer 联锁自动秤
interlocked rock 沿接触面啮合的岩石
interlocked signal 联锁信号
interlocked socket with switch 带开关的连接插座
interlocked switch 联锁开关;联锁分道叉;联锁道岔【铁】
interlocked texture 镶嵌碎裂结构
interlocked turnout 联锁道岔【铁】
interlocked type waveguide 联锁波导管
interlocked zone 联锁区
interlocker 联锁装置;联锁器
interlock force 联锁力

interlock friction 联结摩擦(钢桩墙)
interlock gap 组间间距
interlock 联动;联锁效应;联锁(法);可联动的;交锁;互锁;咬合作用
interlocking angle 锁角
interlocking apparatus 联锁装置
interlocking area 联锁区
interlocking assembly 联锁装置
interlocking block 联锁闭塞区间【铁】;联锁块(体);联锁扣搭块;扣搭块;咬接砌块
interlocking block pavement 联锁块铺面
interlocking block system 联锁闭塞装置;联锁闭塞系统
interlocking board 企口板;嵌锁(式)板
interlocking building panel 企口墙板;锁结式建筑板
interlocking by electric(al) locks 电锁器联锁
interlocking by electric(al) locks with electric(al) semaphore 电动臂板电锁器联锁
interlocking by electric(al) locks with light signal 色灯电锁器联锁【铁】
interlocking by electric(al) locks with semaphore 臂板电锁器联锁
interlocking by point detector 联锁箱联锁;道岔锁闭检查器联锁
interlocking chart 联锁图表
interlocking clay roof(ing) tile 搭扣黏土瓦
interlocking computer box 联锁计算机柜
interlocking concrete block 混凝土砌块的嵌锁;混凝土砌块的闭锁;路面混凝土砌块
interlocking concrete block pavement 嵌锁式混凝土路面
interlocking concrete pipe 锁锁式混凝土管
interlocking cone 加固锥体
interlocking contact 联锁触头;联锁触点
interlocking control system 联锁控制系统
interlocking cutter 联锁刀具;联锁刀
interlocking data 连锁数据
interlocking device 联动装置;联锁装置
interlocking electromagnet 联锁电磁铁
interlocking equipment 联锁设备
interlocking equipment for main line 正线联锁设备
interlocking equipment in depot 车辆段联锁设备
interlocking frame 互联机构
interlocking gear 联锁机构;闭锁机构
interlocking grain 斜纹;斜交木纹
interlocking joint 联锁接合;联锁缝;连续缝;接头管;交错接缝;企口(嵌)缝;锁口缝;石砌体交叉缝;交错砌缝
interlocking latch 联锁销
interlocking line 联锁(用)线路
interlocking machine 联锁机;号志联锁机
interlocking mechanism 联锁机构;闭锁机构
interlocking metal tape armo(u)r 芦席纹金属铠包
interlocking milling cutter 交齿铣刀;组合错齿槽铣刀
interlocking of tracks 混轨道
interlocking panel apparatus 操纵台设备
interlocking particle 连生颗粒;咬合颗粒
interlocking paving 镶锁铺砌(路面)
interlocking paving stone 联锁铺地砌块
interlocking pipe 锁口管
interlocking plant 联锁装置;联锁设备
interlocking point 互换点
interlocking point coordinates 互换点坐标
interlocking pore 连通孔
interlocking property 联锁性
interlocking relay 连锁继电器;联锁继电器
interlocking resistance 嵌锁力
interlocking revetment 联锁护坡
interlocking ring structures 联结环形构造
interlocking roofing 联锁屋面
interlocking roofing tiles 联锁屋面瓦
interlocking round ridge tiles 联锁式圆形屋脊瓦
interlocking section 锁口截面
interlocking seepage prevention 锁口防渗
interlocking shaft 联锁轴
interlocking sheeting 扣锁板桩;搭扣板桩;联锁钢板桩
interlocking shingle 扣搭屋面板
interlocking side mill 交齿侧铣刀
interlocking signal 联锁信号
interlocking spur 交错山嘴

interlocking station 联锁站
interlocking steel sheet piles 联锁钢板桩
interlocking surface 嵌锁式锁面;嵌挤式面层
interlocking switch 联锁开关
interlocking switch group 联锁组合开关
interlocking system 联锁系统
interlocking table 联锁表
interlocking test 联锁试验
interlocking tile 搭扣瓦;联锁式屋面槽瓦;联结的大型屋面槽瓦;平搭瓦;单搭接瓦;咬接瓦;联锁瓦;槽瓦;搭接瓦
interlocking tile roofing 联锁瓦(片)屋面
interlocking tile with snowrib 挡雪瓦片
interlocking tooth 交错齿;交齿;错齿
interlocking with locally worked points 非集中联锁
interlock interrupt 联锁中断
interlock line 联锁线
interlock of centralization 集中联锁
interlock pore 通孔
interlock protection 连锁保护(装置)
interlock resistance 锁口阻力
interlocks 联动装置
interlock shaft 联动杆
interlocks of steel sheet pile 钢板桩锁口
interlock space 组间间距
interlock system 连锁装置
interlock tension 锁口拉力
interlock tension force 锁口张力
interlocutory decree 临时判决
interloper 无牌营业者;无船舶执照的船;非友船
interlude 中间程序;插算;辅助子程序
interlunation 无月期间
intermarginal line 缘间线
intermarginal sulcus 缘间沟
inter marker group sender 标识群互通发送器
intermate individuals 入选的个体
intermediary 中间状态;中间形态;中间人
intermediary agent 媒介
intermediary character 中间性状
intermediary institution 中介机构
intermediary paper 隔层纸
intermediary solution 中间溶液
intermediary trade 居间贸易
intermediary water mass 中层水团
intermediate 中间物;中间片;中间的;中等木;中等的;居间的
intermediate acceptance 中间交工验收
intermediate access storage 中间存取存储器
intermediate acidity rock 中酸性岩
intermediate agent 中间剂
intermediate aggregate 中间骨料
intermediate air filter 中间空气过滤器
intermediate alkalinity 中等碱度
intermediate alloy 中间合金
intermediate altitude communication satellite 中高空通信卫星
intermediate altitude meteorological satellite 中高度气象卫星
intermediate altitude satellite 中高度卫星
intermediate amplifier 中间放大器
intermediate angle 中间边帮角
intermediate annealing 中间退火
intermediate arbor support 中间刀杆支架
intermediate arch 中间发券
intermediate area 缓冲地带
intermediate assembly point 中间装配地点
intermediate assignment 中间赋值
intermediate axle 中间轴;中轴
intermediate axle bevel gear and pinion 中桥主动及被动伞齿轮
intermediate axle housing (汽车的)中轴壳
intermediate axle left wheel brake tube 中桥左轮制动管路
intermediate band photometry 中带测光
intermediate base 中形管座;中间基
intermediate base crude(oil) 中间基原油;中间基石油
intermediate basin 区间面积
intermediate battery supply board 中间电池供给装置
intermediate bay 中间跨;中间板;桥的中间跨;梢间【建】
intermediate beam 中间梁;辅助梁

intermediate bearer 中间托梁;中间承托;中间承木
intermediate bearing 中间轴承
intermediate bearing casing 中间轴承罩
intermediate bellcrank 中间双臂曲柄;中间直接形杠杆
intermediate belt 中电路;中间(地)带;中间层;中环带;过渡层;过渡(地)带;充气带
intermediate belt gearing 中间皮带传动装置
intermediate benchmark 中间水准点
intermediate bend 中间弯曲;中间弯头
intermediate bin 中间(料)仓
intermediate biofilm organism 中间生物膜生物
intermediate bleeding chamber 中间抽气室
intermediate block 中间锤座;中间程序块;砧面托
intermediate block checking 中间程序块检查
intermediate block post 线路所【铁】
intermediate bloomer 中间开坯机座;中间初轧机
intermediate board frame 中间配线架
intermediate body 中间体
intermediate boson 中间玻色子
intermediate box 中间箱
intermediate box pile 中间组合箱桩
intermediate breaker 中间轧碎机;中间破碎机
intermediate breaking 中等程度轧碎
intermediate brickwork 中间砖墙
intermediate bridge 中承式桥
intermediate buffer 中间缓冲器
intermediate bulkhead 中间舱壁
intermediate bunker 中间料仓
intermediate cable 中间电缆
intermediate cable terminal box 中间电缆盒
intermediate caisson 船坞中部闸门
intermediate caisson groove 船坞中部闸门门槽
intermediate cam 中三角
intermediate capacity rail transit system 中等运量轨道交通系统
intermediate car 中型小客车
intermediate care facility 中等护理疗养设施
intermediate carrier 中间运输工具;中间接运工具
intermediate casing 中间套管;技术套管
intermediate catalyst film 中间催化剂薄膜
intermediate category 中级晶族
intermediate ceiling 假天顶
intermediate center 中转中心
intermediate chafing block 中间防擦块
intermediate chamber (破碎机的)中粒度破碎室
intermediate channel 中间渠道
intermediate character 中间性状
intermediate chemicals 中间化学品
intermediate circuit 中间电路
intermediate circuit sub-control station 中间电路控制分站
intermediate circulation 中层环流
intermediate clarifier 中间澄清池
intermediate class pavement 中级路面;过渡式路面
intermediate clinker breaker 中间熟料破碎机
intermediate clinker crusher 中间熟料破碎机
intermediate coal 中等亮度煤
intermediate coat 中间涂层;中涂层;二道底漆
intermediate coating 中层涂料
intermediate code 中间(代)码;半成码
intermediate colo(u)r 过渡色;间色
intermediate column 中(间)柱;中型柱
intermediate command 中间指令
intermediate commercial organizations 居间性的商业组织
intermediate comminution 中间粉碎
intermediate communication level 中间水平
intermediate computer 中间计算机
intermediate condenser 中间冷凝器
intermediate condition 中密状态
intermediate conductor 中间导线
intermediate connecting rod 中间连杆
intermediate connector 中间连接器
intermediate constituent 中间组成物
intermediate consumer 中间消费者
intermediate consumption 中间消耗
intermediate contact 中间触点
intermediate container 中间罐
intermediate contour 中间等高线;基本等高线;首曲线;非注字等高线
intermediate contrast 中等反差
intermediate control 中间控制
intermediate control change 中间控制变换

intermediate control data 中间控制数据
intermediate controlling dimension 中间控制尺寸
intermediate control office 中途调度站
intermediate conversion 中间转换
intermediate converter 中间变流器
intermediate conveyer 中间传送机
intermediate coolant 中间载热剂;中间冷却剂
intermediate cooler 中间冷却器
intermediate cooling 中间冷却
intermediate coordinates 中介坐标
intermediate copy 中间拷贝
intermediate corbel-bracket set 平身科斗拱
intermediate corona 中介日冕
intermediate correction 中间层校正;中间层改正(量)
intermediate counter 中间计算器
intermediate coupling 中介耦合;中间耦合;居间耦合
intermediate-coupling theory 中介耦合理论
intermediate course 中间夹层;中间层
intermediate coverage range 中间有效范围
intermediate cracking 板中开裂;(混凝土路面的)中间裂缝;板中裂开
intermediate crank pin 中间曲柄销
intermediate crimp 中等折褶;中等卷曲;中等皱缩
intermediate cross girder 中间横梁
intermediate cross-section 中间截面
intermediate crude 中间基石油
intermediate crusher 中(级破)碎机;中间破碎机
intermediate crushing 中等程度轧碎;中间压碎;中间破碎;中级压碎;次级粉碎
intermediate crushing chamber (碎石机的)中碎室
intermediate crushing vibrating rod mill 中间破碎振动棒磨
intermediate crust 过渡壳
intermediate culture 补植
intermediate current 中间值电流;中电流;中层流
intermediate current plasma arc welding 中电流等离子弧焊
intermediate curve 计曲线;首曲线
intermediate cut 中间馏分
intermediate cutting 间伐
intermediate cycle 中间周期;过渡循环
intermediate cycling load 中间荷载
intermediate cylinder 中压汽缸;中间筒
intermediate data set 中间数据集
intermediate decay 中期腐朽
intermediate deck 中间甲板;中承式桥面;中层(桥面)
intermediate dense 中密(的)
intermediate depot 中间站
intermediate depth wave 中度深水波
intermediate developed country 中度开发国家
intermediate diagonal 再分杆
intermediate diameter 中径
intermediate diaphragm 中间(横)隔板;中间挡板;隔仓板
intermediate differential 中间微分
intermediate distillate 中间馏分
intermediate distributing [distribution] frame 中间配线架
intermediate document 阶段成果
intermediate door rail (门的)中冒头;中门横档
intermediate draft 中区牵伸
intermediate draw bar 中间拉杆
intermediate draw bar pin 中间拉杆销
intermediate drawing 过渡标描
intermediate drift burst 中介漂移暴
intermediate drill 中间钻杆;中间钎子
intermediate drivage 中间巷道掘进
intermediate drive 中间传动
intermediate drying 中间烘燥;中间烘干
intermediate dust collector 中间集尘器
intermediate duty fireclay brick 中等耐火(黏土)砖
intermediate earthquake 中深地震
intermediate echo suppressor 中点回波抑制器
intermediate efficiency filter 中效过滤器
intermediate electrode 中间电极
intermediate energy region 中能区
intermediate entry 中间平巷
intermediate equipment 中间设备
intermediate evolute 中间渐屈线
intermediate facies 中间相【地】;中成相
intermediate fastening 中间扣件

intermediate feed 中间喂给装置
intermediate feed inlet 塔中部液体入口
intermediate felling 间伐
intermediate fiberboard 中密度纤维板
intermediate fiberboard sheathing 中间墙衬纤维板
intermediate field 中间场;居间场
intermediate filament 中间丝
intermediate file 中间文件【计】
intermediate film 中间胶片
intermediate fire(d) glaze 中温釉;中火釉
intermediate flange 对接法兰;中间接合盘
intermediate flask 中间砂箱
intermediate float 中间浮子;领水浮子
intermediate floor 中间楼面
intermediate floor beam 中间楼面间梁;中间楼板梁
intermediate floor joist 中层楼板格栅
intermediate floor slab 中间楼面;中间楼板
intermediate flue 中间烟道
intermediate flux 中间熔剂
intermediate foam layer 中间泡沫层
intermediate focal length tracking telescope 中焦距光学跟踪器
intermediate focus 中间焦点
intermediate focus earthquake 中(震)源地震;中深源地震
intermediate focus shock 中源地震
intermediate form 中间形式;中间类型;中等起伏地形;过渡型
intermediate formation 中间形成
intermediate fraction 中间馏分
intermediate frame 中间肋骨;中间构架;辅助肋骨;二道粗纱机
intermediate franchise 不定期特许
intermediate frequency 中间频率
intermediate frequency amplifier 中频放大器
intermediate frequency automatic gain control setting 中频自动增益调整
intermediate frequency ceramic filter 中频陶瓷滤波器
intermediate frequency channel 中频通路
intermediate frequency combining 中频合并
intermediate frequency distortion 中频失真
intermediate frequency feed-through 中频直通
intermediate frequency furnace 中频感应炉
intermediate frequency interconnection 中频互连
intermediate frequency interference 中频干扰
intermediate frequency interference ratio 中频干扰比
intermediate frequency noise 中频噪声
intermediate frequency oscillator 中频振荡器
intermediate frequency output 中频输出
intermediate frequency passband 中频通带
intermediate frequency power amplifier 中频功率放大器
intermediate frequency preamplifier 中频前置放大器;前置中频放大器
intermediate frequency rejection ratio 中频抑制比
intermediate frequency relay 中距离频率继电器
intermediate frequency signal 中频信号
intermediate frequency strip 中放部分
intermediate frequency sweep signal generator 中频扫频仪
intermediate frequency switching 中频转换
intermediate frequency transformer 中频变压器
intermediate fuel oil 中间燃料油;中级燃油
intermediate function 中间功能
intermediate gap 中间隙
intermediate gasoline 中级汽油
intermediate gate 中门;中间闸门;中间坞门;中部闸门
intermediate gate post 门的中间柱;中间门柱
intermediate gear 中速齿轮;中间齿轮;第二速度齿轮;二挡齿轮
intermediate gearbox 中间减速器
intermediate gear bracket 中间齿轮托架
intermediate gearing 中间传动
intermediate gear sleeve 中间齿轮套筒
intermediate geologic(al) report 中间地质报告
intermediate girder 中间梁;中间大梁
intermediate glass 中间玻璃;过渡玻璃
intermediate glass former 中间体玻璃形成物
intermediate glue 中级(耐水)胶黏剂
intermediate goods 中级商品;中间产品;半成品
intermediate gradation 中间粒级

intermediate grade 中间坡(度);中间等级(的);中等坡度
intermediate grade billet 中级钢杆
intermediate grade span 中等跨度
intermediate grade steel 中级钢
intermediate gradient 中间坡(度)
intermediate grain 中间颗粒;中级颗粒
intermediate granular texture 中粒结构
intermediate grinding 中碎;中磨;中间研磨;中级研磨
intermediate grinding compartment 中间粉磨仓
intermediate groin 中间丁坝
intermediate groove 船坞中部槽
intermediate ground water 过渡层地下水;中间地下水
intermediate group 子夜星组;中间群
intermediate guide blade 中间导叶
intermediate guide key 中间底图
intermediate handling 中间处理
intermediate haulage conveyer 中间运输机
intermediate head 中间引水闸门
intermediate heading 中间平巷
intermediate heater 中间加热器
intermediate heat exchanger 中间热交换器
intermediate heating 中间预热;中间加热
intermediate hole 中间炮孔
intermediate hopper 中间储棉箱
intermediate horizon 中间直视距离;中间视距;中间地平圈
intermediate host 中间主体;中间寄主
intermediate hue 间色
intermediate igneous rock 中性火成岩
intermediate image 中间图像
intermediate-image spectrometer 中间成像能谱仪
intermediate impregnation 中间浸渍
intermediate improvement cutting 改进间伐;抚育间伐
intermediate inflow 区间入流;区间来水
intermediate input 中间投入
intermediate input in railway transport industry 铁路运输业中的间投入
intermediate inspection 中期检验;中间检验
intermediate integral 中间积分
intermediate in the scale of development 中度开发
intermediate inventory 半制成品存量
intermediate isotope 中等原子量同位素
intermediate jack 中间接片
intermediate jack rafter 中间小椽
intermediate joint 耙中【疏】
intermediate jointing box 中间接线盒
intermediate joist 中间托梁
intermediate journal 中间轴颈
intermediate junction 中间连接
intermediate junction box 中途接线盒
intermediate lamella 中间板
intermediate landing 中间平台
intermediate landing-station 中间水平井底车场
intermediate lava 中性熔岩
intermediate lava flow 中性熔岩流
intermediate layer 中间层;过渡层;硅镁带;硅镁层;夹层
intermediate layer correction value 中间层校正值
intermediate lens 中间透镜
intermediate level 中间平巷;中等能级
intermediate level cave 中等放射性工作室
intermediate level cell 中等放射性工作室
intermediate level radiation 中能级辐射
intermediate level radioactive waste 中等强度的放射性废料
intermediate level throat 半深度的流液洞
intermediate level waste 中放废物;中等(强度)放射性废物
intermediate lightning arrester 中间避雷器
intermediate line 中间线
intermediate line of fire 中间射向
intermediate lining material 中间衬料
intermediate link 中间连接板;流通环节;流转环节
intermediate liquid 中间液体
intermediate load 腰荷;中间负荷
intermediate load unit 中间负荷机组
intermediate lock head 中间闸首
intermediate longitudinal girder 中间纵梁
intermediate loop 中间回路

intermediate magnitude earthquake 中等震级地震
intermediate main bearing 中间主轴承
intermediate maintenance 中修
intermediate market 中间市场
intermediate mass 中间块
intermediate mast (架空索道的)区间支架
intermediate means 中间设备
intermediate member 中间构件
intermediate memory 中间存储(器);暂时存储
intermediate memory storage 中间存储保持器
intermediate metabolite 中间代谢物
intermediate metallic sealing 中间涂覆金属封接
intermediate mill 中间轧机
intermediate moraine 中分碛
intermediate mullion 中间窗梃
intermediate multiple 中间倍数
intermediate navigation channel 中间航道(船闸)
intermediate negative carry 中间负进位;中间反向进位
intermediateness 中间性
intermediate network node 中间网络节点
intermediate neutron reactor 中能中子反应堆
intermediate nozzle 中间喷嘴
intermediate object program (me) 中间目标程序;中间结果程序
intermediate observation 中间观测;插点观测;附加观测
intermediate ocean current 中间洋流
intermediate octane rating 中间辛烷值
intermediate oil 中间油品
intermediate(oil) tank 中间油罐
intermediate oil/water segregator 中间油水分离器
intermediate orbit 中间轨道
intermediate original plan 过渡原图
intermediate overhaul 中修
intermediate oxide 中间体氧化物
intermediate pack 中间填充;中间装填
intermediate pack strip 中间填充带
intermediate pan 中间盘
intermediate panel 中间镶板
intermediate parallel pack 中间平行填充带
intermediate part 中间部
intermediate partition 中间壁
intermediate pass 预轧孔型;中间轧制孔型
intermediate pass sorting 中间扫描分类法
intermediate peeling 层间脱落
intermediate periodic(al) duty 间断周期性负荷
intermediate period seismograph 中周期地震仪
intermediate phase 中间相
intermediate pier 斗板石;中间(桥)墩
intermediate pile 中间桥桩;中间桩
intermediate pinion 中间齿轮
intermediate pintle 中间舵栓
intermediate pit design 中间露天采场设计
intermediate pivot 中间支点;中间枢轴
intermediate plan 中间阶段规划
intermediate plane 中间平面
intermediate plant 中期植物;中间型植物
intermediate plant house 中温温室
intermediate plate 中间板块;中间板;夹片
intermediate platform 中间站台;中间平台
intermediate point 中途站;中间点;插点;三字点磁罗经
intermediate polar 中介漂移
intermediate pool 运河中间消涌池;闸门河段;闸间水库
intermediate port 中途(停泊)港;中途口岸;中间港
intermediate position 中间位置
intermediate post 中间(支)柱;中间立柱
intermediate potential 中间电位
intermediate power amplifier 中间功率放大器;中等功率放大器
intermediate preheater 中间预热器
intermediate pressure 中压;中间压力;中等压力
intermediate pressure centrifugal blower 中压离心式鼓风机
intermediate pressure chamber 中间压力室
intermediate pressure compressor 中压压缩机;中压压气机
intermediate pressure cylinder 中压(气)缸
intermediate pressure section 中压段;中压部分
intermediate pressure turbine 中压涡轮机;中压透平;中压汽轮机
intermediate principal plane 中主平面;中间主应力面
intermediate principal purlin(e) 中金檩
intermediate principal strain 中(间)主应变
intermediate principal stress 中(间)主应力;第二主应力
intermediate principal stress value 中间主应力值
intermediate processing element 中间处理部件
intermediate product 中间产物;中间产品;中间商品;部分乘积
intermediate product theory 中间产物理论
intermediate proof 中间证明
intermediate propeller shaft 中间传动轴
intermediate publication 过渡版
intermediate pulley 中(间)滑轮;中间皮带轮
intermediate pump 中真空泵
intermediate pumping station 中间泵站
intermediate purlin(e) 上中平衡;金檩;金桁;中间檩
intermediate quantity 中间量
intermediate radical 中间游离基;瞬间游离基
intermediate radius 中间半径
intermediate rafter 花架椽;中(间)椽
intermediate rail (门的)中档;伸缩调整轨;中间冒头;中等栏杆
intermediate range 中间距离;中间范围;中程
intermediate range sonar 中程声呐
intermediate rate biological filter 中负荷生物滤池
intermediate rays 居间射线
intermediate reaction tower 中间反应塔
intermediate reactor 中能中子反应堆;中能堆
intermediate receiver 中间接收器
intermediate reduction 中间磨细;中间粉碎;中间归约
intermediate reduction pinion 中间减速过齿轮
intermediate reduction wheel 中间减速轮
intermediate reflux 中间回流
intermediate region 中能量区;中间区
intermediate relay 中间继电器
intermediate repair of track 线路中修
intermediate repeater 中转站;中间中继器
intermediate repeater station 中间转播站
intermediate report 中间报告
intermediate reservoir 中间热储
intermediate resistance 中间电阻
intermediate resolution 中等清晰度
intermediate resolution spectrometer 中分辨率分光计
intermediate resonance 中间共振
intermediate result 中间结果
intermediate retarder 中间缓行器
intermediate revision 中间检定
intermediate rib 中间肋;居间肋
intermediate ring 接合镜筒
intermediate ringer 中间局振铃器
intermediate ring road 中环路
intermediate rinse 中间冲洗
intermediate road wheel 中间负重轮
intermediate rock 岩石夹层;中性岩
intermediate rock mass 过渡性岩体
intermediate rocks 中性岩类
intermediate roll 中间轧辊;中辊
intermediate rolling 中间轧制
intermediate rolling mill 中间轧机
intermediate rolling train 中间轧机组;粗轧机组
intermediate row teeth 牙轮中间一排齿
intermediates 中间产物;光泽辊
intermediate safety chain 中间保险链
intermediate scale 中间尺度;过渡比例尺
intermediate scale experiment 中间试验
intermediate scale map 中比例尺图
intermediate scale problem 中型问题
intermediate scattering function 中间散射函数
intermediate screening 中间筛(分)
intermediate sea 中海
intermediate sealing glass 封接玻璃;过渡封接玻璃
intermediate section 中型构件;中型型材;中截面;中间部分
intermediate sedimentation 中间沉降;中间沉淀
intermediate sedimentation tank 中间沉淀池;中间沉淀槽
intermediate setting wheel 中间过轮
intermediate settling tank 中间沉淀池
intermediate shaft 中间传动杆;中间(转动齿轮)轴;中间竖井

intermediate shaft bearing 中间(轴)轴承
intermediate shaft bearing sleeve 中间轴轴承套筒
intermediate shaft cap 中间轴盖
intermediate shaft supporting bracket 中间轴支架
intermediate shape 中间形状
intermediate ship 中型船;客货船
intermediate shock 中深地震
intermediate short range photography 中近距摄影
intermediate siding 区间岔线;中间照准【测】;中间读数【测】
intermediate sight 中(间)视;中间观测
intermediate sill 中间槛;中层帽木;中层间托座
intermediate sintering 中间烧结
intermediate size car 中型小客车
intermediate size homogeneous reactor 中型均匀反应堆
intermediate size reduction 中碎;第二级破碎
intermediate slide 中刀架
intermediate slope angle 中间边坡角
intermediate soil 中间土壤;过渡土壤
intermediate solid 中等固体量
intermediate solid solution 中间固溶体;次生固溶体
intermediate solution 中间溶液
intermediate spacing 适中的植距
intermediate span 中跨;中间跨(度)
intermediate span bridge 中跨桥梁
intermediate species 中间物质;瞬间物质
intermediate spectrum reactor 中能中子谱堆
intermediate speed 中速;中等速度
intermediate speed pinion 中间小齿轮
intermediate spindle 中间轴
intermediate stage 中间平台;中(间)期;中间阶段;中间级
intermediate stage of decay 中期腐朽阶段
intermediate stair(case) 中间楼梯
intermediate stanchion 中间支柱
intermediate stand 中间机座
intermediate state 中间状态
intermediate station 中途站;中间台;中间(车)站;插站【测】
intermediate steam 中间蒸汽
intermediate steering arm 中间转向臂
intermediate step 中间级
intermediate stiffener 中间加强肋;中间加强杆;中间加劲肋;中间加劲杆
intermediate stiffener of steel web 钢腹板中间加劲肋
intermediate stiffness 中级刚度
intermediate stock 居间砧木
intermediate stone 中等骨料;中等石块
intermediate stop valve 中间截止阀
intermediate storage 中间储存;中间储[贮]存;中间器材储备;中间存储(器)
intermediate storage tank 中间储罐
intermediate store 中间存储(器)
intermediate storm deposit 中部风暴沉积
intermediate straight line 夹直线
intermediate strain axis 中间应变轴
intermediate strength 中等强度;中期强度
intermediate stress 中间应力
intermediate strike wheel 中间打点轮
intermediate string 中间套管柱;技术套管
intermediate strip pack 中间充填带
intermediate strong motion instrument 中强震仪
intermediate strut 中间支柱;中间支撑;中间横撑
intermediate studdle 中间支柱(井框间)
intermediate sub-control station 中间控制分站
intermediate subsystem 中介次系
intermediate super abrasion furnace black 中超耐磨炉黑
intermediate super abrasion furnace (carbon) black 中超耐磨炉法炭黑
intermediate superheater 中间过热器;再热器
intermediate support 插座;中间支座;中间支架;中间支点;中间支承;中间支撑
intermediate support spar 中间架杆
intermediate surface 中间面
intermediate survey 中间检验;期中检验
intermediate switch 中间开关
intermediate switching station (输电线的)中间开关站
intermediate synoptical observation 辅助天气观测
intermediate system 中间系统

intermediate tank 中间罐
intermediate tankage 中间罐区储液能力
intermediate tank farm 中间站
intermediate tap 中间丝锥;中间螺丝攻;中间抽头;二道螺丝攻
intermediate tax 中间税
intermediate technical examination 中间技术检验
intermediate technical examination position 中检台位【铁】
intermediate technology 中间技术;中等科技
intermediate tectonic level 中间构造层次
intermediate temperature deposits 中温沉积物
intermediate temperature setting adhesive 中温硬化胶着剂;中温固化胶黏剂;中温凝固胶黏剂
intermediate temperature setting agent 中温硬化黏结剂
intermediate temperature setting glue 中温凝固胶合剂
intermediate temperature sludge 中温淤渣
intermediate temperature varnish 中温清漆
intermediate term 中期
intermediate term financing 中期资金筹措
intermediate terminal 中途站;中途港
intermediate thaw(ing) period 中间解冻时期;中间融化时期
intermediate theory states 中间理论州(对于抵押贷款,美国各州立法不同,一些州立法机构认为抵押时所有权归属受押方(称所有权理论州),另一些州认为受押方只有留置权(称留置权州);还有一些州持"中间理论",即一旦借款人未能按期还款,所有权便自动归属受押方)
intermediate thermal medium tank 热载体中间储[贮]槽
intermediate throwoff 中间卸料
intermediate tide 中间潮
intermediate tie 中间系杆
intermediate toll center[centre] 长途电话中心局
intermediate total 中间计数
intermediate trade 中介贸易;中间贸易
intermediate train distance point with siding 有配线分界点
intermediate train distancing point 分界点【铁】
intermediate train spacing point 中间分界点【铁】
intermediate transfer point 中间转运点
intermediate transformer 中间变压器
intermediate transit point 中转点
intermediate transmission 中间传动
intermediate transport 中间传送
intermediate transport technology system 中间运输工艺系统
intermediate transverse frame 中间横向框架
intermediate transverse girder 中间横梁
intermediate treatment 中间处理;中级处理
intermediate trench 中间壕
intermediate trunk distributing frame 中间中继线配线架;中间干线配线架
intermediate truss 中间桁架
intermediate tubular column 中型管柱
intermediate type 中级;中等;中(间)型
intermediate type arbo(u)r support 中间刀杆支架
intermediate type pavement 过渡型路面;中间型路面
intermediate type surface 中级路面
intermediate underdrive gear 中间轴惰轮
intermediate unit 中间单位
intermediate unlocking wheel 中间释放轮
intermediate uranium fluoride 中间铀氟化物
intermediate vadose water 中间渗流水;中间带上层滞水
intermediate value 中(间)值;介值
intermediate value theorem 介值定理;中值定理
intermediate valve 中间阀
intermediate valve chamber 中间阀室
intermediate variable 中间变量
intermediate vector boson 中间矢量玻色子
intermediate velocity cloud 中速云
intermediate wall 中隔墙
intermediate washer 中间垫片
intermediate wastewater 中段废水
intermediate water 岩层间水;中(间)水;中层水;过渡带水;层间水
intermediate water content 中等含水量
intermediate water facility 中水装置
intermediate water layer 过渡带水层

intermediate water mass 中间水质量
intermediate water reuse 中水回用
intermediate water supply 中水供水;中水道
intermediate water supply station 中水给水站
intermediate water supply system 中水供水系统
intermediate wave 深水波与浅水之间的波浪;中(短)波
intermediate wave energy 中波能
intermediate wave transmitter 中波发射机
intermediate wax 中间蜡
intermediate weave 中等格栅;中间织纹
intermediate weight 中等重量
intermediate wheel 空转轮
intermediate wheel and pinion 中心齿轴
intermediate wheel lock 中间轮锁杆
intermediate wheel reverser bridge 中间轮换向夹板
intermediate winding wheel 中间上条轮
intermediate wood 中间材
intermediate wrap drive 中间双滚筒驱动(胶带机)
intermediate yield 中间收获;中等威力
intermediate zone 中间(地)带;中间层;过渡区;过渡(地)带;充气带;二次燃烧区
intermediation 居间业务;中间业务;居间调停
intermediation effect 媒介效应
intermediator 调停人
intermedio-lateral cell column 侧灰柱
intermedium 中间物;中间体
interment 埋葬
intermeshed scanning 间隔扫描
intermeshing of tracks 轨道相互结合
intermeshing rolls (磨碎机的)咬合对辊
intermetallic 金属间(化合)的
intermetallic coating 金属间化合物涂层
intermetallic compound 中间组成物;金属间化合物;金属互化物;电子化合物
intermetallic metal 中间金属
intermetallic phase 金属间相
intermetallic phase particle 中间金属相微粒
intermetallics 金属间化合物;金属互化物
intermetium 竞技场中标柱间的长栏(古罗马)
intermicellar equilibrium 胶束间平衡
intermicellar swelling 胶束间溶胀
intermicrotubular bridge 微管间桥
intermigration 相互迁移
intermingle 混栖;混合;掺杂
intermingled aggregate 混合集料;混合骨料
intermiscibility 相互混(溶)性
intermission 中断;中止(时间);工间休息
intermitosis 分裂间期
intermitotic 分裂间期的
intermittence 周期性;中止;间歇性;间歇现象;间断
intermittence type 间歇式
intermittency 间歇性;间歇现象
intermittency effect 间歇效应
intermittent 间歇机构;间歇的;间断的
intermittent-acting mill 周期式轧机
intermittent aeration 间歇曝气
intermittent aeration tank 间歇曝气池
intermittent agitation 间歇搅动
intermittent automatic block 间歇式自动闭塞
intermittent beacon 间歇式信号灯;间歇式标志;间歇式灯塔
intermittent biological filter 间歇式生物滤池
intermittent biological sand filter 间歇生物砂滤池
intermittent blowdown 定期排污
intermittent bucket type elevator 间歇式提降机;间歇吊斗式升降机
intermittent burning 间断燃烧
intermittent cave river 间歇洞穴河流
intermittent charging 间断加料
intermittent chlorination 间歇加氯;间歇氯化消毒
intermittent contact 断续接触;断续串线
intermittent contact bed 间歇接触床
intermittent continuous wave 断续等幅波
intermittent control 间歇控制;间断调节
intermittent controller 不连续作用调节器
intermittent cooling 间歇供冷
intermittent countercurrent rinsing 间歇式逆流清洗
intermittent creep 间歇性蠕变
intermittent curing 间歇式养护
intermittent current 断续(电)流

intermittent cut 间歇切削
intermittent cyclic extended aeration system 间歇循环延时式曝气法
intermittent defect 间歇性故障
intermittent depressing region 间歇性下降区
intermittent discharge 间歇放电;间隙放电
intermittent disconnection 间歇切断;瞬断;时断时续
intermittent downward filtration 间歇向下过滤
intermittent downward infiltration 间歇下流式过滤池
intermittent drainage extended aeration system 间歇排水延时式曝气法
intermittent drainage lake 间歇性排水湖
intermittent drainage way 间歇式排水道
intermittent dredge(r) 间歇式挖泥机;间歇挖泥机
intermittent dryer 周期干燥器;间歇干燥器
intermittent drying 间歇性晒田;间歇干燥;间隙干燥
intermittent dust removal 定期除灰
intermittent duty 间歇工作(状态);间歇工作制;间歇负载;断续使用;断续工作方式
intermittent earth 间歇接地
intermittent effect 间歇效应
intermittent equipment 间歇工作设备
intermittent error 间歇(性)错误;间发错误
intermittent eruption 间歇喷发
intermittent excavator 间歇式开凿机;间歇式挖土机;间歇性挖土机
intermittent external wall 透水外堤
intermittent failure 间歇失效;间断性故障
intermittent fault 间歇故障
intermittent feed 间歇喂料;间歇投料
intermittent fillet weld 间断角隙焊接;断续角焊缝
intermittent fillet welding 断续角焊
intermittent film projector 断续式放映机
intermittent filter 间歇(式)滤池;间歇过滤器
intermittent filtration 间歇过滤作用;间歇过滤(法)
intermittent firing 间歇燃烧
intermittent-firing duct engine 脉动式喷气发动机
intermittent-flame-exposure test 周期性受灼试验;间歇式受灼试验
intermittent flooding 间歇灌水;间歇溢流
intermittent flow 交替流;间歇性自喷;间歇性水流;间歇性流动;间歇流;季节性流水;季节性河流
intermittent flow sedimentation tank 间歇流沉淀池
intermittent flow settling basin 间断流动沉淀池
intermittent flush 间歇冲洗;断续冲砂
intermittent force 脉动力
intermittent fractionation 间歇分馏
intermittent freezing 间歇凝固;间歇冻结
intermittent gas-lift 间歇举气;间歇气升;间歇气举
intermittent ga(u)ging 间测
intermittent geyser 间歇温泉
intermittent grading 间歇级配;升降交替的坡度
intermittent grazing 间歇性放牧;间歇放牧
intermittent grinding 间歇研磨;分批研磨
intermittent grinding and polishing 单机磨光
intermittent grounding fault 间歇性接地故障
intermittent heat 断续加热
intermittent heating 间歇加热;间歇供暖;间歇采暖
intermittent icing 插封
intermittent ignition 断续点火;间歇点火
intermittent indexing 逐齿分度法;间断分度法
intermittent injection 间歇注入
intermittent input 脉动输入
intermittent integration 间歇累计;间歇积分
intermittent interrupted 断续的
intermittent interrupted river 间歇性河流
intermittent interrupted stream 喀斯特河;间歇中断河;间歇性河流
intermittent irradiation 间歇照射;断续照射
intermittent kiln 间歇作业窑;间歇窑
intermittent lake 间歇湖
intermittent leakage 间歇性渗漏
intermittent lehr 间歇式退火窑
intermittent life 间歇寿命
intermittent light 间歇式信号灯;间歇式标志光;明暗灯标;断续光;等明暗光
intermittent line 虚线;断续线
intermittent load(ing) 断续加载;间歇(性)荷载;间歇负载;脉冲负荷;断续短期负载;断续荷载;定时荷载

intermittently 断断续续地
intermittently aerated biofilter 间歇曝气生物滤池
intermittently aerated membrane bioreactor 间歇曝气膜生物反应器
intermittently aerated submerged membrane bioreactor 淹没式间歇曝气膜生物反应器
intermittently attached sheet roofing 点式连接片材屋面
intermittently decanted extended aeration 间歇倾注延时曝气
intermittently operated activated sludge plant 间歇运作活性污泥处理设备
intermittent manufacture 间断制造
intermittent mixer 间歇式拌和机
intermittent mixing 间断拌和(法)(指水泥土混合料加水后并不立即拌和);间歇搅拌(法);间歇拌和(法)
intermittent mixing plant 间断搅拌工场;间断搅拌设备;间断搅拌车间
intermittent motion 间歇运动
intermittent motion film transport 胶片断续传输装置
intermittent movement equipment 间歇运动设备
intermittent movement machinery 周期性运输机械
intermittent movement projector 断续传动式放映机
intermittent noise 间断性噪声
intermittent oiling 间断油润滑
intermittent operation 间歇操作;间歇运转;间歇运行;间隙操作
intermittent oscillation 间歇振荡
intermittent overhead spray nozzle 间隙式顶喷嘴
intermittent period 间歇周期;间歇期
intermittent periodic(al) spring 间歇周期泉
intermittent period of transmission 传播休止期
intermittent pilot 间歇点火器
intermittent pit 间断坑
intermittent point welding 间隔焊点;断续点焊
intermittent pollution 间歇污染;断续污染
intermittent positive pressure ventilation 间歇正压换气
intermittent pressure filter 间歇式压滤机
intermittent production 间断生产
intermittent production process 断续性生产过程
intermittent projector 断续式放映机
intermittent pulse 脉结代
intermittent quick flashing light 断续急闪光
intermittent quick flash light 等间隙急闪光
intermittent rain 间歇雨;间断雨
intermittent reaction 间歇反应
intermittent recharge 间断回灌
intermittent recorder 间歇记录器;打点式记录仪;打点式记录器
intermittent region 断续区
intermittent river 间歇河(流);时令河
intermittent rolling process 间歇压延法;间歇滚压法
intermittent run 间歇运行
intermittent sampling 间歇采样
intermittent sand filter 间歇砂滤器;间歇砂滤池
intermittent sand filtration 间歇砂滤(作用)
intermittent scanning 间歇扫描
intermittent sedimentation 间歇沉降(作用);间歇沉积(作用);断续沉淀(法)
intermittent service 间歇工作;间歇使用;间歇运行
intermittent servo mechanism 间歇作用伺服机构
intermittent shoulder 间断设有避车道的路肩;间断式路肩
intermittent spring 间歇涌水井;间歇(喷)泉;潮汐泉
intermittent stationary Gaussian process 间歇平稳高斯过程
intermittent step test 间歇性分级试验
intermittent sterilization 间歇消毒;间歇灭菌(法)
intermittent storm 间歇性暴雨
intermittent stream 间歇溪流;间歇河(流);季节性流水;季节性河流;时令河
intermittent stream flow 间歇河流;间歇水流;间歇水道
intermittent stress 周期应力
intermittent submerged sill 间断潜水坝
intermittent subsiding movement 间歇性下降运动
intermittent surge 间断涌流
intermittent system 间歇式系统

intermittent test 间歇负载试验
intermittent thickener 间歇浓缩机
intermittent tidal stream 间歇潮流
intermittent treatment 间歇处理
intermittent tunnel dryer 间歇式隧道干燥器
intermittent type blending silo 间歇式搅拌库
intermittent type cab signal(l)ing 点式机车信号
intermittent type counter current rinsing 间歇式逆流清洗
intermittent type meter 间歇式测量仪表
intermittent type track shifting 间歇式移道法
intermittent type track shifting machine 间断性移轨机
intermittent underground lake 间穴地下湖
intermittent unit 断续装置
intermittent uplifting area in East China 中国东部间歇性隆升区【地】
intermittent uplifting region 间歇性上升区
intermittent upwarping movement 间歇性上升运动
intermittent visibility 间歇通视
intermittent weigh-batch(mixing) plant 间歇式配料搅拌设备;间歇式配料搅拌车间
intermittent weighing 间歇称重
intermittent weld 断续焊缝
intermittent weld at interlock 锁口处断续焊
intermittent welding 间断焊;断续焊(接)
intermittent wiping 刮水器的间歇擦拭
intermittent working 隔年作业
intermittent working equipment 间歇工作设备
intermittent working excavator 间歇工作的挖土机;间歇工作的挖掘机
intermitter 间歇性涌水孔;间歇调节器
intermix 相互混合;搅拌;交杂;掺和;拌和
intermixable 可互拌的
intermixing 相互拌和;搅拌;混合
intermixture 混合料;混合物;掺料;掺混物;掺和料
intermodal 联合的
intermodal agreement 联合运输协定
intermodal cargo 联合货物
intermodal cargo movement 货物联运;货物换装
intermodal carrier 国际联合运输人
intermodal Cokriging method 协同克立格法
intermodal container 联运集装箱
intermodal container terminal 集装箱联运装卸网;集装箱联运站
intermodal container transfer facility 联运集装箱换装设施;集装箱联运中转设施
intermodal coupler 模间耦合器
intermodal distortion 模间畸变
intermodal facility 联运设备;换装设备
intermodalism 联运制;联运化
intermodal movement of freight 货物联运
intermodal paperwork 联运单据
intermodal road 协调联运公路
intermodal service centre 联运业务中心
intermodal system 联运制
intermodal tariff system 联合票价制
intermodal transport 协调联运;多式联运
intermodal transportation 协同一贯联运;协同联运;联(合)运(输);多式联运
intermodal transportation system 联运系统;组合运输系统
intermodal transport document 多式联运单证
intermodal transport operator 联运经营人
intermodal vessel 多式联运船舶
intermode beats 模间拍
intermode spacing frequency 模间间隔频率
intermodillion 檐托座间;斗拱间
intermodulated fluorescence 内调荧光
intermodulation 相互调制;交(叉)调(制);互(相)调(制)
intermodulation distortion 互调失真
intermodulation distortion meter 互调失真仪
intermodulation distortion percentage 互调失真百分比
intermodulation disturbance 互调干扰
intermodulation effect 交调效应;交叉调制作用
intermodulation frequency 互调差频
intermodulation interference 互调干扰
intermodulation product 相互调制分量;交调产物;互调产物
intermolecular 分子间的
intermolecular attraction 分子间(吸)引力

intermolecular bonding energy 分子间键合能
intermolecular cohesion 分子间内聚力
intermolecular condensation 分子间缩合
intermolecular cross-linking 分子间交联
intermolecular distance 分子间距离
intermolecular force 分子间力
intermolecular interaction 分子间相互作用
intermolecular labelling 分子间标记
intermolecular ligation 分子间连接
intermolecular linkage 分子间键
intermolecular migration 分子间转移作用
intermolecular orbital 分子间轨道
intermolecular ordering 分子间有序化
intermolecular oxidation 分子间氧化作用
intermolecular polymerization 分子间聚合
intermolecular potential energy 分子间势能
intermolecular reaction 分子相互反应
intermolecular rearrangement 分子(间)重排
intermolecular repulsion 分子间斥力
intermolecular transfer 分子间转移
intermolecular transposition 分子间重排作用
intermonsoonal rain 间季风雨季
intermont 山间洼地;山间的;山间凹地;山谷
intermontane 山间的
intermontane area 山间地区
intermontane basin 山间盆地
intermontane coal-bearing formation 山间含煤建造
intermontane deep 山间坳陷
intermontane depression 山间洼地;山间凹陷;山间坳地
intermontane glacier 山麓汇合冰川;山间冰川;山谷冰川
intermontane hollow 山间洼地
intermontane plain 山间平原
intermontane plateau 山间高原
intermontane space 山间地带
intermontane trough 山间槽地
intermontane valley 山间(河)谷
intermont glacier 山间盆地冰川
intermonthly variability 月际变异率
intermonthly variation 月际变异;月际变化
intermountain basin 山间盆地
intermountain deep 山间坳陷
intermountainous 山间的
intermountainous basin 山间盆地
intermountainous plain 山间平原
intermountain plain 山间平原
intermountain seismic belt 山间地震带
intermural 墙间的;内外墙中层;墙壁间(距)的;壁间
intermuscular needling 分刺
intermutule 挑檐托块的间距
internal 内在的;内部的;分散相
internal ablation 内部烧蚀
internal abnormal voltage 内部反常电压
internal absorbent method 内吸收法
internal absorptance 内吸收系数;内吸收比;内吸光率
internal absorption 内吸收
internal absorption factor 内吸收因数;内吸收率
internal access 内部沟
internal accessories of tank 油罐内部零件
internal account 境内账户
internal accounting 内部会计
internal accounting control 内部会计控制
internal action 内部作用
internal activity 内在活动
internal addition 内位加成
internal address 内部地址
internal adjustment 内面校正;内部平差
internal adjustment compensator 内调整补偿器
internal adjustment scope 内调式瞄准镜
internal administrative control 内部管理控制
internal admission (蒸气机的)内进气
internal adsorption 内吸附
internal aerodynamics 内流空气动力学
internal agent 内在动因;内因
internal agreement 内部符合
internal air 内部空气
internal air baffle 内导风板
internal air cooler 内部空气冷却器
internal air cooling 内部空气冷却
internal air duct 内部通风沟
internal ambience 室内气氛;室内环境

internal analog(ue) loop 内部模拟环路
internal analysis 内部分析
internal and external equilibrium 内外平衡
internal and external ga(u)ge 内外径规
internal and external load(ing) 内外荷载
internal and external pressure 内外压力
internal and external snap ga(u)ge 内外径规
internal angle 内角;凸墙内角
internal angle bead 阴条砖(釉面砖的配件砖);阴角砖
internal angle to cove skirting 阴角砖;壁脚弯砖阴角(釉面砖的配件砖);壁脚弯砖
internal annular shake 木心环裂
internal antenna 内天线
internal anti-stat 内用抗静电剂
internal applied tanking 内防水套
internal arch 内拱
internal area 内域
internal area ratio 内面积比
internal arithmetic 机内运算;内运算;内部计算
internal arrangement 内部装置;内部协议
internal atmosphere 室内气氛
internal attachment 内加工装置
internal attribute 内部属性;内部表征
internal audit 内部审计;内部稽核
internal auditing 内部审计
internal auditing manual 内部审计规程
internal auditing standard 内部审计标准
internal audit of internal control system 铁路内部审计
internal auditor 内部审计员
internal audit service 内部审计处
internal audit system 内部审计制度
internal band block 内带滑车
internal band brake 内带闸
internal barrier 内势垒
internal beam 内束
internal beam current 内束流
internal beam energy 内束能量
internal beamsplitting layer 内射束分离层
internal behavior 内部性状
internal bending crack 弯曲内裂
internal bending moment 内弯矩
internal benefit 内部效益
internal bevel gear 内锥齿轮
internal bias 内部偏移
internal bird's beak 阴条三角;阴条角尖嘴砖
internal bisector 内分角线
internal blast loading 内燃机喷流荷载
internal block 内分程序;内部分程序
internal block brake 内蹄外张式制动器
internal bond 内胶结强度;内部黏结力;国内债券;内部债券
internal bonding energy 内结合能
internal boundary 内边界(线)
internal boundary condition 内边界条件
internal boundary layer 内边界层
internal bound block 内带滑车
internal bracing 内部拉条
internal brake 内制动力;内闸;内蹄制动器
internal brake shoe 内制动蹄
internal branch 内支
internal break 内裂
internal breakdown 内部损坏
internal breeder 增殖反应堆;内部再生区
internal breeding 内增殖
internal breeding ratio 内增殖比;内部再生系数
internal bridge connection 内桥连接
internal broach 内拉刀
internal broacher 内拉床
internal broaching 内拉削
internal broaching machine 内拉床
internal broken 内破裂
internal bud 内芽
internal buffer 内缓冲区;内部缓冲区
internal building board 内部建筑板
internal building panel 室内建筑用板
internal building sheet 内部建筑(薄)板
internal bull-nose brick 内凹砖
internal bunching 内聚束
internal buoyancy 内浮力;内浮体
internal buoyancy appliance 内部浮力装置
internal cable method 预应力钢缆法
internal caliber ga(u)ge 内径卡尺

internal calibration 内校准;内部校准
internal calibration source 内校准源
internal cal(l)iper ga(u)ge 内径卡规
internal cal(l)ipers 内卡钳
internal catalyst bed 内催化剂床
internal cathodic protection 内阴极保护
internal cause 内因
internal cavity 内部孔穴
internal cell 内室
internal center drilling 内中心钻孔
internal centerless grinder 内圆无心磨床
internal centerless grinding machine 内圆无心磨床
internal chaining 内链接
internal characteristic 内(部)特性
internal characteristic curve 内(部)特性曲线
internal charge 内部装药(爆破)
internal check 内裂;内部牵制;内部核查;内部干裂
internal check system 内部牵制制度
internal check valve (油罐的)内单向阀
internal chiasma 内(部)交叉
internal chill 内冷铁;反白口
internal circular planing 刨圆孔法
internal circulation 内循环
internal circulation anaerobic reactor 内循环厌氧反应器
internal circulation moving bed biofilm reactor 内循环移动床生物膜反应器
internal circulation reactor 内循环反应器
internal circumferential lamella 内环骨板
internal cladding 内挂板
internal classification (磨机的)内部分级
internal classifier 内装式选粉机
internal clearance (滚动轴承等的)内部间隙
internal clock 内时钟;内时标
internal clock generator 内时钟发生器
internal clocking 内部时钟;内部计时
internal clock pulse 内部时钟脉冲
internal clock pulse signal 内部时钟脉冲信号
internal clock source 内部时钟源
internal closing bracket 内闭括号
internal clutch gear 内离合齿轮
internal coagulant 内施凝固剂
internal coating 内部涂层;室内涂层;室内涂料
internal coconut piece 阴三角砖;阴三角
internal code 内部码;内部代码
internal cohesion (土或材料的)内黏力;内凝力
internal coil evapo(u)rator 内盘管蒸发器
internal column 内柱;室内柱
internal combustion 内燃烧;内燃
internal combustion burner 内燃嘴
internal combustion compactor 内燃夯实机
internal combustion engine 内燃式发动机;内燃机
internal combustion engine driven generator 内燃机拖动的发电机
internal combustion engine plant 内燃机厂
internal combustion engine repair ship 内燃机修理船
internal combustion loco(motive) 内燃机车
internal combustion locomotive haulage 内燃机车牵引
internal combustion piston engine 内燃活塞式发动机
internal combustion power station 内燃机发电站
internal combustion pump 内燃泵
internal combustion rammer 内燃夯实机
internal combustion turbine 燃气轮机
internal command 内部命令
internal commerce 国内贸易
internal common tangent 内公切线
internal common tangent plane 内公切面
internal communication 内部通信(联系)
internal communication circuit 内部通信线路
internal communication system 内部通信制;内部通信系统
internal compensation 内消旋作用;内部补偿
internal composition 内部合成
internal compressive safe factor 抗内压安全系数
internal compressive strength 抗内压强度
internal compressor 内部压气机
internal concrete vibrator 插入式混凝土振动器;混凝土内插振捣器;混凝土内部振捣器;插入式混凝土振捣器
internal condensation 内缩合(作用)

internal condition 内部条件
internal conditions for design 设计的内部条件
internal conduction 内导
internal conductor 内部导体
internal configuration 内部构形
internal conical refraction 内锥形折射
internal connection 内部连接
internal consistency 内部相容性;内部符合
internal consistency criterion 内部一致性准则
internal consistency index 内部一致性指数
internal constant 内部常数
internal consultant 内部顾问师
internal consumption 内部消耗;本体消耗
internal contact 内接触
internal contour sawing 内轮廓锯法
internal contracting brake 抱闸
internal contraction 内收缩
internal control 内部控制
internal control analysis 内检分析;内检
internal controller 燃气舵
internal control point 内部控制点
internal control questionnaire 内部控制调查表
internal control report 内部控制报告
internal control standard 内部控制标准
internal control system 内部控制系统;内部管理制度
internal control system audit of accounts payable 应付账款内部控制制度审计
internal control system audit of cash management and accounting 现金管理核算内部控制制度审计
internal convection 内运流
internal conversion 内(部)转换
internal conversion coefficient 内转换系数
internal conversion factor 内转换因子
internal conversion neutron detector 内转换中子探测器
internal conversion ratio 内转换比
internal conversion routine 内部转换程序
internal conversion source 内转换源
internal conversion spectrometer 内转换谱仪
internal cooling 内冷却;内部冷却
internal cooling blow pipe 内冷式吹管
internal cooling generator 内冷发电机
internal cooling grinding 内冷却磨削
internal coordinate 内坐标
internal coordination 内部协调
internal core 内部核心
internal core trade 内部核心贸易
internal cork 内木栓层
internal corner 内角隅;阴角;内角
internal corridor 内廊;内走廊
internal corrosion 内(部)锈蚀;内部腐蚀
internal cost 内部成本
internal counter 内计数管
internal court 内天井;内院;内庭;内部庭园;内天井;室内庭院
internal crack 内部裂纹;内部裂缝
internal cross bracing 内部交叉拉条
internal cross-section 净截面;内截面
internal crucible 本床
internal cubic(al) capacity 内部容积;船舶容积
internal curve hydraulic motor 内曲线液压马达
internal cutting off 内切断
internal cycle hydrolytic acidification 内循环水解酸化
internal cylindric(al) ga(u)ge 柱形测孔规
internal damping 内阻尼
internal damping coefficient 内阻尼系数
internal damping curve 内阻尼曲线
internal damping losses 内阻尼损失
internal data 内部资料;内部数据
internal data field 内数据域;内数据段
internal data structure 内数据结构
internal debt 内债
internal decision making 内部决策
internal decorating 内部装饰;内部装潢
internal decoration 内部装饰;内部装潢
internal defect (铸件橡胶制品等的)内部缺陷
internal defocusing 内散焦
internal deformation 内部变形
internal degree of freedom 内自由度
internal delustring 内部消光
internal design 内部设计

internal designer 内部设计师
internal design pressure 设计内压
internal desuperheater 内部减热器
internal diagnostics 内路诊断
internal dial ga(u)ge 内径千分表
internal diameter 内直径;内径
internal diameter method 内径法
internal diameter of pipe 管子内径;管(的)内径
internal diameter of screen tube 滤水管内径
internal diameter of transporting material tube 输料管内径
internal diameter of well pipe 井管内径
internal dielectric(al) loss 内部介电损耗
internal differential compensating pinion 内差速补偿小行星齿轮
internal diffusion 内扩散
internal dimension 内部尺寸
internal direct sum 内直和
internal disc (接合器的)内圆盘
internal discharge 内卸
internal discharge bucket elevator 中间卸料式提升机;内卸斗式提升机
internal discontinuity 内部间断
internal diseconomy 内部的不经济;内部不经济性
internal dispersion 内色散
internal displacement coordinate 内位移坐标
internal dissipation inequality 内耗不等式
internal distribution 室内配电
internal distribution pattern 内分布型
internal diversity 内部差异
internal division 内分
internal door 内门
internal dormer 斜屋顶内凹窗;内天窗;(斜屋顶的)内凹式老虎窗;斜坡屋顶内窗
internal dose 内剂量
internal double bond 内双键
internal double peak 内双峰
internal downpipe 内部水落管
internal drag 内部阻力
internal drag curve 内阻力曲线
internal drain 内部消耗(指排水);内部排水;内部流失
internal drainage 内泄油;内陆水系
internal drainage layer 内部排水层
internal drainage system 内部排水系统
internal drift current 内部漂流
internal drilling 内钻孔
internal drop 内压降
internal drop hammer 管内落锤
internal drum filter 内滤鼓式过滤机
internal drying test 内部干燥试验
internal dry patch 内部干斑
internal dust circulation 粉尘内循环
internal dynamics 内动力学
internal ear 内耳
internal economic rate of return 内部经济收益率;内部经济回收率;内部经济报酬率
internal economy 内部经济性;内部的节约
internal economy of scale 厂内规模经济
internal education 内部教育
internal efficiency 内效率
internal efficiency ratio 相对内效率
internal elastic membrane 内弹性膜
internal electrically energizable heater element 内供电式加热器元件
internal electric(al) source 内部电源
internal electrode position control 内部电极位置控制
internal electrogravimetry 内电重量法
internal electrolysis 内电解法
internal elevation 室内立面;内部标高;内部高程
internal emergency unlocking handle 车内紧急解锁手柄
internal emission line 内发射线
internal encrustation 内部结垢
internal energy 内能
internal energy balance 内部能量平衡
internal energy per unit mass 单位质量内能
internal engine control 发动机内部控制
internal entropy 内部平均信息量
internal environment 内(在)环境;内部环境
internal equilibrium 内部平衡
internal equipment 内部设备
internal erosion 内部腐蚀;(土壤的)隙缝侵蚀;内侵蚀;层内侵蚀;暗蚀

internal error 内部错误
internal ester 内酯
internal examination 内部检验;内部检查
internal examining relative errors 内检相对误差
internal exhaust gas recirculation 内部排气再循环
internal expanded coupling 内胀式管接头
internal expanding 内涨
internal expanding brake 内胀式制动器
internal expanding shoe brake 内蹄外张式制动器
internal exposure 内照射;内暴露量
internal external gear arrangement 内外齿轮装置
internal external gear clutch 内外齿轮离合器
internal external rotary pump 内外转子泵
internal external upset 内外加厚
internal external-upset drill pipe 内外加厚钻杆
internal facing 内部面饰;内表面修装
internal factor 内在因子
internal failsafe timer 内保险计时器
internal fan kiln 内装风扇式干燥;内燃风扇式干燥窑
internal Faraday rotation 内禀法拉第旋转
internal fault 内部故障
internal feedback 内反馈
internal feed pipe 内给水管
internal fiber board finish 内部纤维板装修;内部纤维板装饰
internal field 内场
internal field alternator 内极式同步发电机
internal field distribution 田间配水
internal finance 内部资金(收益留存加折旧)
internal financial rate of return 内部财务收益率;内部财务回收率;内部财务报酬率
internal financing 内部资金提供;内部筹款;内部财务
internal finish 室内装修
internal finish(ing) paint 内部装修用漆
internal firebox boiler 内火室锅炉
internal fired furnace 内燃烧炉
internal fire-extinguishing system 内部消防系统;内部灭火系统
internal firing of brick 内燃烧砖法
internal fissure 内部裂纹;内部裂缝
internal fittings 内部装置;内部设备;内部配件
internal fittings of tank 油罐内部设备
internal fixation with screws 螺钉内固定术
internal fixation with stainless plate 不锈钢板内固定术
internal fixation with steel plate 钢板内固定术
internal fixture 内部装置
internal floating-head exchanger 内浮盖式换热器
internal flooding 边内注水
internal flow 内流;内部水流;(金属轧断时的)内变形
internal flow characteristics 内部流动特性
internal flue 内烟管;室内烟道
internal flue boiler 内焰管锅炉
internal fluid friction 液体内摩擦
internal fluorescent lighting 内设荧光灯
internal flush 内平的
internal flush drill pipe 内平钻杆
internal flush jointed coupling 贯眼式钻杆接头;内平钻杆接头
internal flush thread 内平扣
internal flush tool joint 贯眼式钻杆接头;内平钻杆接头
internal focusing 内对光;内调焦
internal focusing telescope 内对光望远镜;内调望远镜;内聚焦望远镜
internal force 应力;内力
internal force distribution 内力分布
internal form 内部形式;内部格式
internal fracture 内部破碎
internal fracture test 内部破裂试验;(桩的)拔出试验
internal fragmentation 内部存储碎片;内部存储残片
internal frame(work) 内部模板工程;内模
internal friction 内摩擦(力);内耗;内部摩擦
internal frictional force 内摩擦力
internal frictional losses 内摩擦损失
internal frictional resistance of the locomotive 机车本身内部阻力
internal friction angle (土壤的)内摩擦角
internal friction coefficient 内摩擦系数

internal friction of soil 土壤内摩擦力;土的内摩擦力
internal friction twist principle 内摩擦加捻原理
internal frosting 内面磨砂
internal function register 内操作寄存器;内部功能寄存器
internal furnace 炉胆;内炉膛
internal furnace ion source 内炉离子源
internal gallery 内廊道
internal gallery apartment building 内廊道式公寓建筑
internal gas counter 内气体计数管
internal gas drive 内部气驱
internal gas heated retort 内热釜
internal gas pressure 内气压
internal gate 内部闸
internal gate valve 内部闸阀
internal gating 内选通
internal ga(u)ge 内径规
internal gear 内(啮合)齿轮
internal gear clutch 内齿离合器
internal gear drive 内齿轮传动
internal geared axle 内齿圈轮边减速式驱动桥
internal gear grinder 内齿轮磨床
internal gearing 内啮合;内齿啮合
internal gear pump 内齿轮(水)泵
internal gear rotary pump 内齿轮回转泵
internal gear tester 内齿轮检查仪
internal geometry 内部几何条件
internal geosyncline 内地槽【地】
internal gland 内压盖
internal glass door 内部玻璃门
internal glazing (玻璃的)内部安装;内部玻璃安装;内镶玻璃
internal gloss(clear) varnish 内部上光清漆
internal granular layer 内粒层;内颗粒层
internal graticule (阴极射线管的)内标度
internal grid 内格
internal grid point 内格点
internal grinder 内圆磨床
internal grinding 内圆磨削;内表面研磨
internal grinding action 内磨作用
internal grinding attachment 内圆磨削附件;内磨附件
internal grinding fixture 内磨削夹具
internal grinding head 内圆磨头
internal grinding machine 内圆磨床
internal groove 内槽
internal groove sidewall 内槽壁
internal handrail 内扶手(栏杆)
internal hardboard finish 内部硬质纤维板装修
internal harmonics 内调和函数
internal hazard 隐患;内在危险
internal haze 内混浊
internal head cover 支持盖;内顶盖
internal heat 内部热量;内(能)热
internal heat dryer 内部加热式干燥机
internal heat exchange 内部热交换
internal heat gain 内部热增量;内部取暖
internal heat generation 内热生成;内发热;内部产热
internal heating 内加热
internal heating method 内热法
internal heating method for rock fragmentation 内加热法破岩
internal heating surface 内受热面
internal heat source 内热源
internal hoist 内部卷扬机
internal honing 内圆珩磨
internal honing machine 内圆珩床
internal horizontal drive 内行推动
internal humidity 内部湿度;内湿度
internal hydraulic pressure 内周水压
internal hydrogen cooling 氢内冷
internal idle time 内空闲时间
internal ignition 内部燃烧
internal illumination 内部照明(度);室内照明
internal illumination sign plate 内部照明式标志板
internal impedance 内阻抗
internal incentive mechanism 内部刺激机制
internal indicator 内指示剂;内部指示剂
internal-indirect lighting 内部无影照明;内部反射照明;内部间接照明
internal information 内部信息;内部情报
internal information processing system 内信息处理系统
internal infrastructure 内部基础结构
internal inhibition 内抑制
internal inner ring 内缓冲器
internal input circuit 内部输入电路
internal inspection 内部检验
internal installation 内部安装
internal installation work 内部安装工作
internal instrument installation 仪器埋设
internal insulation 室内隔热隔声;内绝缘;内保温
internal interference 内干扰
internal interrupt 内中断;内部中断
internal interruption 内部中断
internal irradiation 内照射;内辐照
internal irradiation dosimetry 内辐照剂量学
internal irradiation hazard 内照射危险
internal irradiation protection 内照射防护
internal irregularity 内生紊乱
internal irregular point 内非正则点
internal isochron 内部等时线
internality of external diseconomy 外部不经济性内部化
internalization 内在化;内化作用
internalization of environment(al) cost 环境成本内在化
internalized as relations 内在化为关系
internal joinery 内部细木工作
internal junction station 国内联轨站
internal Kelvin's wave 开尔文内波
internal key 内键
internal keyboard 内键盘
internal keyway 内键槽
internal kiln fittings 回转窑内热交换器装置
internal kinetic energy 内在动能
internal label 内部标号
internal labeling 内标记
internal labor market 内部劳动力市场
internal lamination 内部层裂
internal lap 内研磨杆
internal latent heat 内潜能
internal layout 内部规划图;内部布置图
internal leader 暗雨水管;暗水落管
internal leaf (空芯墙的)内堵;(空芯墙的)内层;内部门扉
internal leak 内部泄漏
internal leakage 内部泄漏量;内部漏泄
internal least square 内部最小平方法
internal lebensspuren 内部生物遗迹
internal lesion 内部损伤
internal lighting 室内采光;内部采光
internal limit ga(u)ge 内径极限规
internal limiting membrane 内界膜
internal linear transformation 内部线性变换
internal liner 内衬
internal lining 内衬;内部衬砌
internal linking 内部链接
internal lintel 内过梁
internal lip 内唇
internal liquid cooling 内液冷;内冷水
internal liquidity 内在流量
internal load 内部荷载
internal load bearing wall 内承重墙
internal loading (集装箱的)内载负荷;(集装箱等的)内荷载
internal loan 国内贷款;内债
internal lobe 内叶
internal locking mechanism 内部锁紧机构
internal lock signal 内锁信号
internal loss 内损失;内部损失
internal loss coefficient 内损耗系数
internal loss torque 内损转矩
internal lubrication 内部润滑法
internally compensated operational amplifier 内补偿运算放大器
internally cooled supercharging 内冷增压
internally diffuse(d) junction 内扩散结
internally driven 内传动的
internally fired boiler 内燃锅炉
internally fired furnace 内燃炉;内部燃烧炉
internally frosted bulb 内磨砂灯泡
internally geared motor 内装减速器的电动机
internally generated fund 自有资金
internally generated time signal 内生时间信号
internally heated gas apparatus 内加热气体装置
internally heated oven 直接式烤箱;内热式烤箱
internally heated pressure vessel 内加热高压釜
internally irregular point 内非正则点
internally lubricated reinforced thermoplastics 内润滑增强热塑料
internally lubricated wire rope 内润滑钢丝绳
internally piloted valve 内控阀
internally piloting 内控
internally program(me)ed computer 内程序计算机
internally reflected scattered light 内部反射的散射光
internally scanned laser 内扫描激光器
internally specified index 内部规定指标
internally stored program(me) 内(部)存(储)程序;常驻内存程序
internally tangent 内切
internally tangent circle 内切圆
internally thin set 内薄集
internally ventilated motor 内通风式电动机
internally vibrated concrete 插入式振捣混凝土
internal machine code 内部码
internal machine representation 机内表示
internal machining operation 内部切削加工
internal macro instruction 内部宏指令
internal magnet 内磁铁
internal magnetic field 内磁场
internal magnetic recording 内部磁记录
internal manipulation instruction 内加工指令;内操作指令;内部操纵指令
internal mark 内标
internal masonry dividing wall 内部圬工分隔墙
internal measurement 内部丈量
internal measurement of tank 油罐内体积
internal measuring instrument 孔径测量仪器;内径测量仪
internal medicine 内科
internal member 内套件;被套件;被包容件
internal memory 内存储器;内储存器;内存(储[贮]器)
internal memory-address register 内存地址寄存器
internal method 内标法
internal micrometer 内测微计;内径千分尺;内径千分表
internal migration 内迁移;内部移民
internal milieu 内环境
internal milling cutter 内面铣刀
internal miter 内斜榫
internal mix atomizer 内混雾化器
internal mixer 内部混合机;密炼机;密闭式炼胶机;密闭式混炼机;密闭式混合机
internal mixing 内混合
internal mixing blade 内部搅拌叶片;内拌和翼
internal mixing multiple-component spray nozzle 内混合型多组分喷嘴
internal mixing spray gun 内混合喷枪
internal mobility 内流动性
internal modulation 内调制
internal modulator 内调制器
internal module 内部模块
internal moisture 内部湿度
internal moment 内力矩;内矩
internal moraine 冰川内碛
internal motion resistance 内部行驶阻力
internal mo(u)ld 内模
internal name 内部名字
internal navigation 内河航运;内河航行
internal nodal point 内节点
internal node 内节点
internal noise 内(部)噪声
internal noise bandwidth 内噪声带宽
internal noise source 内噪声源
internal normalization 内标归一化法
internal number 内部数;内部编号
internal number system 内部计数制
internal object 内部目标;内部对象
internal observation 内部观测
internal octagon brick 内八角砖
internal olefin 内烯烃
internal oleyl sulfate 内油醇硫酸酯
internal operation 内表面加工
internal operation ratio 内部操作比
internal optic(al) density 内光学密度;内光密度
internal optic(al) parametric oscillation 内光学

参量振荡
internal optic(al) parametric oscillator 内光学参量振荡器
internal optic(al) path 内光路
internal order 内部秩序
internal organization of the network layer 网络层内部结构
internal orientation 内方位
internal orifice of cochlear aqueduct 蜗水管内口
internal oscillator 内振荡器
internal output circuit 内部输出电路
internal oxidation 内氧化
internal packing 内部气封
internal paint 内部涂料
internal panel 室内建筑用板;内部护墙板
internal partitioning 内部分隔
internal passage 内部通道
internal peak 内峰
internal pedestrian traffic 屋内人行交通
internal performance 内部性态
internal phase angle 内相位角
internal phase ratio 内相比
internal phase shift 内部相移
internal phasing 内部定相
internal phosphorus loading 内磷负荷
internal photoelectric(al) effect 内光电效应
internal pinion 内接小齿轮
internal pipe diameter 内管径
internal pipeline 内部管线
internal pipe size 管的内径
internal pipe thread 管的内螺纹;管内螺纹
internal piping 内部管道;内部管涌
internal plant 内部通信设备
internal plant haulage 工厂内部搬运
internal plastering 内部抹灰
internal plasticization 内增塑(作用)
internal plate 内板
internal plumber 装管工人;室内装管工人
internal plumbing 内部水暖安装;内部卫生管道工程
internal point (螺丝攻的)空心孔;定心孔
internal point of division 内分点
internal poker 插入式振动器
internal polarization modulation 内部偏振调制
internal pore 内空隙
internal porosity 内缩松;内孔隙率;内孔隙度
internal porosity of the aggregate 骨料松隙度;骨料的内部孔隙度
internal position isomer 内异构体
internal potential vibration 内部电位振动
internal power 内部功率
internal power source 内部能源
internal pressure 内压强;内压(力);内部压力
internal pressure current 内压流
internal pressure of pipe 管道内压
internal pressure strength 内压强度
internal pressure strength test 内压强度试验
internal pressure test 内压试验
internal pressure tester 内压试验机
internal prestress 内(部)预(加)应力
internal preventer 钻柱防喷回压阀
internal primer 内底漆
internal primitive water 深层原生水
internal priority 内部优先数;内部优先级
internal probe 内探针
internal probe target 内靶
internal procedure 内过程;内部过程
internal processor error 内部处理机错误
internal processor register 处理器内部寄存器
internal product 内积
internal program(me) 内程序
internal projection 内心投影
internal proportional counter 内源正比计算器
internal Q-switch 内腔 Q 开关
internal quality block 内侧粗劣砌块
internal quality brick 内侧粗劣砖
internal radiation 内照射
internal radiation dose 内部辐射剂量
internal rail 内部铁路(线)
internal railroad system 港内铁路系统
internal rail stress 钢轨内应力
internal railway system 港内铁路系统
internal ramp 内匝道;内坡道
internal rate of return 内在报酬率;内部收益率;内部报酬率;回收率法

internal rate of return method 内部收益率计算法;内部利率法
internal rate of returns on investment 投资的内部收益率
internal ratio 内部比率;内比
internal reactance 内电抗
internal reamer 内孔铰刀
internal reaming 内铰孔
internal rearrangement 内部重排
internal rebunching 内部再聚束
internal receptor 内部感受器
internal recessing 内开槽
internal recycle 内循环
internal recycling 内循环;内部循环
internal redecoration 内部重新装修;内部重新装饰
internal reference (核磁共振的)内标率
internal reference electrode 内参考电极;内参比电极
internal reference generator 内基准发生器
internal reference line 内参比线
internal reference method (磁共振的)内标准法
internal reference muting 内基准抑制
internal reflectance spectroscopy 衰减全反射比
internal reflection 内反射;内部反射
internal reflection colo(u)r 内反射色
internal reflection observation method 内反射观察法
internal reflectivity 内反射率
internal reflux 内回流
internal reflux ratio 内回流比
internal register 内部寄存器
internal regression 内回归
internal relation 内在关系;内部联系;内在联系
internal release agent 内脱模剂
internal report for management 内部管理报告
internal reporting 内部报告
internal report system 内部(的)报告制度
internal representation 内部表示法
internal research 内部研究;内部调查
internal resilience 弹性变形内力;内回弹能
internal resistance 内阻(力);内电阻;内部抵抗
internal resistance curve 内阻力曲线
internal resistance electric(al) melting 内热法;内电阻熔融法
internal restraint 惯性试验(集装箱等)
internal reticular apparatus 内网器
internal return 内部反合
internal reveal 内部外抱
internal revenue 国家税收
internal revenue rate 内含报酬率
internal revenue tax 内地税
internal reversing gear 内反向齿轮;内部回动装置
internal rewards 内部奖励
internal rigging 内部装配;内部吊索
internal ring 内环
internal riser pipe 内立管;内部立管
internal rope thread 内圆牙螺纹
internal rotation 内旋转;内旋
internal rubber mixer 密闭式混胶机
internal sapwood 边材内部
internal scaffold(ing) 室内脚手架;内脚手架;里脚手架
internal scanned laser 内扫描激光器
internal schema 内模式;内部模式
internal scour 内部淘洗;内部冲刷;潜蚀
internal screening 内屏蔽
internal screw 阴螺钉;内螺旋
internal screw cutting tool 内螺纹车刀
internal screw finish 内螺纹瓶口;内螺纹抛光
internal screw ga(u)ge 内螺纹规
internal screw pump 内螺旋泵
internal screw thread 内螺纹;阴螺纹
internal screw thread cock valve 内螺纹旋塞阀
internal screw thread finish 内螺纹瓶口
internal screw thread stop valve 内螺纹截止阀
internal sea 内海
internal sealing 内部填缝;内密封
internal searching 内部查找
internal segment 内节
internal seiche 内假潮;内部湖震
internal self-climbing tower crane 内爬塔式起重机
internal self-focusing 内自聚焦
internal sensitivity speck 内部感光点
internal sensor 内部传感器

internal separation 内分离;分离
internal sequence configuration mode 层序内部结构模式
internal sequence number 内部顺序号
internal serration 内细齿
internal service 内部业务
internal settlement measurement 内部沉陷量测
internal settlement price 内部结算价格
internal shake 内轮裂;内裂;木材内环裂
internal shaping 内成型
internal shell 内壳层
internal shield 内屏蔽;内部屏蔽
internal shield ring 内屏蔽圈
internal shift 内移位
internal short path 内光路
internal shoulder angle 阴角扇形砖
internal shoulder boring 内肩镗孔
internal shoulder turning 内肩车削
internal shrinkage 内部缩孔;内部收缩
internal shroud 内盖
internal shuttering 内模板;内模壳
internal signal 内部信号
internal sizing 内施胶
internal skeleton 内骨骼
internal skin 内表层;内层墙
internal skip-hoist 内部侧卸式起重机;内吊斗提升
internal slatted blind 内石板百叶窗
internal sleeve 内套(管)
internal slope resistance 内电阻率
internal slot 内槽
internal soil drainage 土壤内部排水
internal solitary wave 孤立内波
internal sort 内分类;内部排序
internal sorting 内部分类
internal sort phase 内部分类阶段
internal sound insulation 内隔声
internal soundness 内部质量
internal source 内放射源
internal source of revenue 内部财源
internal source program(me) 内源程序
internal source program(me) pointer 内部源程序指示字
internal source spectrometer 内源谱仪
internal space 内部空间
internal span 内跨度;内跨
internal spatial structure of town 城市内部空间结构
internal spiral 内螺旋
internal spline 内花键
internal spring safety relief valve 内弹簧安全阀
internal spur 内距
internal spur gear 内正齿轮
internal square 内直角尺;内正方形
internal stability 内固;内部稳定性
internal stability number 内固数
internal stability of reinforced earth 加筋土内部稳定性
internal stagnant water 内部积水;内滞水
internal stair(case) 室内楼梯;内楼梯
internal stair(case) rail 内楼梯栏杆
internal standard 内(部)标准;内标物;内标法;内标
internal standard area 内标面积
internal standard element 内标元素
internal standard line 内标线
internal standard method 内标法
internal standard normalization 内标归一化法
internal standard ratio 内部标准比率
internal standard sample 内标试样
internal standing wave 内驻波
internal state 内状态
internal state variable 内态变量
internal stay 内撑
internal steam pipe 内集气管
internal steam work 蒸气内能
internal stop 内部光阑
internal storage 内存(储器);内储存器;内部储[贮]存;内部存储器
internal storage capacity 内存(储器)容量
internal storage location 内存单元
internal storage structure 内存结构
internal storage system 内存储装置;内存储系统
internal stored program(me) 内存储程序
internal straight boring 内直线镗孔
internal straight turning 内直线车削

internal strain 内应变
internal strain-ga(u)ge unit 内应变计
internal strapped block 内带滑车
internal stress 自(重)应力;内应力;剩余应力
internal stress relaxation 内应力松弛
internal stretcher 内拉幅装置
internal stripping section 内部汽提段
internal structure 内(部)结构;内部构造;层内构造
internal structure of city 城市内部结构
internal structure of town 城镇内部结构
internal stud 隔墙竖筋;内部壁骨
internal styling (汽车车身的)内部造型
internal supercharger 内增压器
internal supply and sale expenses 内部供应和销售支出
internal supply and sale income 内部供应和销售收入
internal surface 内表面
internal surface area 内表面积
internal surface filter 内滤式过滤器;内表面过滤器
internal surfacing 室内墙面处理;内粉刷;内衬;内表面修装
internal surge 内涌浪
internal survey 内部检查;内部勘测
internal switch 内部开关
internal symbol 内符号;内部符号
internal symbol form 内符号形式
internal symmetry 内对称性
internal symmetry group 内部对称群
internal symmetry principle 内部对称原理
internal synchronization mode 内同步方式
internal system result 内部系统结果【计】
internal table 内部表
internal taper 内锥度
internal taper boring 内锥形镗孔
internal tapered hole 内锥形孔
internal taper turning 内锥形车削
internal tapping 内攻丝
internal target 内靶
internal target irradiation 内靶辐照
internal-teeth gear 内齿轮
internal-teeth spur gearing 内齿轮传动(装置)
internal telecommunication department 国内电信部门
internal(tele)phone 内部电话
internal telephone system 内部电话系统
internal temperature 内部温度
internal temperature control 内部温度调节;内部温(度)控(制)
internal tension 内张力
internal terrestrial heat 地热
internal texture of ore body 矿体内部结构
internal thermal 内热
internal thermal conductance 内部热传导
internal thermal resistance 内热阻
internal thermocouple 内热电偶
internal threaded bit 内螺纹钻头
internal threaded drill rod 内螺纹钻杆
internal thread finish 内螺纹抛光
internal thread ga(u)ge 内螺纹规
internal thread grinder 内螺纹磨床
internal thread grinding machine 内螺纹磨床
internal threading 内螺纹车削
internal threading tool 内螺纹车刀
internal thread 内螺丝(齿纹);内螺纹
internal tide 内潮
internal tile 内砖
internal time base stability 内部时基稳定度
internal time loop 内时循环
internal timer 内计时器
internal tongue 内舌片
internal tooth 内齿
internal toothed friction plate 内齿摩擦片
internal toothing 内啮合
internal tooth lock washer 内齿锁紧垫圈
internal torque 内扭矩
internal tracery 内窗(花)格
internal trace table 内部跟踪表
internal traction force 内部牵引力
internal trade 国内贸易
internal traffic 国内运费;内部交通;场内交通
internal trans 内转移;内超过
internal transactions 内部会计事项
internal transfer price 内部转让价格
internal transfer profit 内部转让利润
internal transfer system 场内搬运系统
internal transfer vehicle 集装箱场内专用车;场内专用车;场内运输车;场内集装箱专用平板车;场地搬运车
internal transfer vehicle system 场内搬运车方式
internal transformation 内点变换
internal transmission factor 内透射率
internal transmittance 内透射系数;内透射比;内部透射度
internal transport(ation) 内部运输;矿内运输
internal transportation expenses 场内运费
internal transportation system 内部运输系统
internal trap 内自陷;内捕俘
internal treatment 内部处理
internal trimming 室内装修
internal trip 境内出行;市内乘车出行
internal turning tool 内圆车刀
internal type idle limiter 内装式怠速浓度调程限制器
internal unit 内部设备;内部零件
internal unwinding 内退解
internal upset drill pipe 内部加厚的钻管;内加厚钻杆
internal upset(ting) 内加厚
internal use 内部使用
internal vacuum 内部真空度
internal validity 固定效力
internal valuation 内部估价
internal valve 内阀(门)
internal variable 内变量
internal variance 内部方差
internal velocity section 层速度剖面
internal verification testing 内校验测试
internal version 内倒转术
internal vertical drive 内场推动
internal vibrating machine 插入式振捣机
internal vibration 内振动;内部振捣;插入式振捣
internal vibrator 插入式(混凝土)振捣机;插入式振动器;内插式振捣器;棒式内部振捣器
internal viscosity 内黏滞
internal viscous damping 内黏滞阻尼
internal void 内空隙
internal void fraction 内部空隙分率
internal voltage 内电压
internal volume 内体积;容积(集装箱)
internal wage differentials 内部工资差别
internal wall(ing) 内墙;壁内部
internal wall(ing) panel 内壁板;内墙(镶)板
internal wall surface 内墙面
internal wash 内部冲洗
internal waste disposal system 内部废物处理系统
internal water 内水;内陆水;内部水
internal water circulation 内水分循环;内部水循环;水文内循环;水分内循环
internal water cooling 水内冷
internal water course 国内水路;国内水道
internal water deficit 内水分欠缺;内部缺水量
internal water-feed machine 中心供水凿岩机;内部供水式钻机
internal water pressure 内水压力
internal waterproofing lining of basement 内防水套
internal water quality 内部水质
internal waters 国内水域;内水(域);内陆水域;内河;内部水域
internal waterway 国内水路;国内水道
internal wave 内波
internal wave motion 内波运动
internal wave of ocean 海洋内波
internal wear 内部磨损;内部磨板
internal wedge adjustment 内光楔调整
internal weight 内部权
internal welding 内部焊接
internal wheel brake 内胀式车轮制动器
internal wheel spindle 内圆磨initial砂轮轴
internal window 内窗
internal window frame 内窗框(架)
internal window sill 内窗台
internal wiring 内部线路
internal wiring diagram 内部接线图
internal work 内圆加工;内功;内表面加工
internal worm of inlet trunnion 进料端空心轴内螺旋叶片
internal yield pressure 内屈服压力(深钻工艺)
internal zone 内部区域
intern architect 见习建筑师;实习建筑师
international 国际的
International Abstracts in Operations Research 国际运筹学文摘
international accounting standards 国际会计标准
international account settlement payable 应付国际联清算款
International Active Sun Year 国际太阳活动年
International Advisory Committee on marine Sciences 国际海洋科学咨询委员会;国际海事科学委员会
international aerial navigation 国际航运
international affairs 国际事务
international agency 国际机构
International Agency for Research on Cancer 国际癌症研究所
international agreement 国际协定
international aid 国际援助
international airfield 国际机场
International Air Pollution Protection Association 国际空气污染保护协会
international airport 国际机场
International Air Transport Association 国际航空运输协会
international airway 国际航空线
international algebraic language 国际代数语言
international algorithmic language 国际算法语言
International Amateur Radio Union 国际业余无线电爱好者协会
international ampere 国际安培
international analysis code 国际分析电码
international and local shopping 国际和国内采购
international angstrom 国际埃
International Annealed Copper Standard 国际退火铜标准
International Antarctic Meteorological Research Committee 国际南极气象研究委员会
international application satellite 国际应用卫星
international aquifer 国际含水层
international arbitration 国际仲裁
International Arbitration Commission 国际仲裁委员会
International Arbitration League 国际仲裁联盟
international architecture 国际式建筑
International Association Against Noise 国际反噪声协会
International Association for Bridge and Structural Engineering 国际桥梁和结构协会;国际桥梁与结构工程协会
International Association for Earthquake Engineering 国际地震工程协会
International Association for Ecology 国际生态学协会
International Association for Hydraulic Research 国际水力学研究协会
International Association for Shell and Spatial Structure 国际薄壳结构与空间结构协会
International Association for Shell Structures 国际壳体结构协会
International Association for Testing Materials 国际材料试验协会
International Association for the Rhine Ships Register 国际莱茵河船舶注册协会
International Association for Time-keeping 国际计时协会
International Association for water Law 国际领海法协会
International Association of Assessing Officers 国际房地产税务评估者协会;国际估价员协会
International Association of Classification Societies 国际船级社协会
International Association of Consulting Engineers 国际咨询工程师联合会
International Association of Dredging Companies 国际疏浚公司协会
International Association of Drilling Contractors 国际钻井契约者协会;国际钻井承包商协会
International Association of Engineering Geology 国际工程地质协会
International Association of European General Average Adjusters 国际欧洲共同海损理算师协会
International Association of Geodesy 国际大地

测量协会
International Association of Great Lake Ports 大湖区港口国际协会
International Association of Hydrogeologist 国际水文地质学家协会;国际水文地质工作者协会
International Association of Hydrology 国际水文协会
International Association of Hydrology Sciences 国际水文科学协会
International Association of Independent Tanker Owners 国际独立油船船东协会
International Association of Institutes of Navigation 国际航行学会联合会;国际航海学会联合会
International Association of Lighthouse Authorities 国际航标协会;国际灯塔管理协会
International Association of Meteorological and Atmospheric Sciences 国际气象和大气科学协会
International Association of Meteorology and Atmospheric Physics 国际气象和大气物理协会
International Association of Physical Oceanography 国际自然海洋协会
International Association of Port and Harbo(u)rs 国际港湾协会;国际港口协会
International Association of Producers of Insurance and Reinsurance 国际保险与分保协会;国际保险与分保险公司协会
International Association of Refrigerated warehouses 国际冷藏库协会
International Association of Seismology 国际地震协会
International Association of Seismology and Physics of the Earth's Interior 国际地震和地球内心物理学协会
International Association of Shell Structures 国际薄壳结构协会
International Association of Theoretic(al) and Applied Limnology 国际理论和应用湖沼学协会
International Association of the Physical Sciences of the Ocean 国际海洋物理协会
International Association on Water Pollution control 国际污染控制协会
International Association on Water Pollution Research 国际污染研究协会;国际水污染研究学会
International Association on Water Pollution Research and Control 国际水污染研究与控制协会
International Astronautical Federation 国际宇航联合会
International Astronomical Union 国际天文学联合会
International Atomic Energy Agency 国际原子能机构
International Atomic Time 国际原子时(间)
international atomic weight 国际原子量
international atomic weight table 国际原子量表
international auction 国际拍卖
international balance of payments 国际收支平衡
International Bank for Reconstruction and Development 国际复兴开发银行;世界银行
International Bank for Reconstruction and Development Loan 国际复兴开发银行贷款
international bidding 国际招标
international bill of exchange 国际汇票
International Biodiversity Day 国际生物多样性日
International Biological Program(me) 国际生物学计划;国际生物计划
international body 国际机构
international borrowing 国际借贷
international boundary 国境线;国界;国际疆界
International boundary and Water Commission 国际边界和水委员会
international bridge 国境桥
International Bridge Tunnel and Turnpike Association 国际桥隧栈道协会
international broadcasting 国际广播
international broadcast relay 国际广播中继
International broadcast station 国际广播电台
International Bulk Chemical Code 国际散装化学品规则
International Bureau for Rock Mechanics 国际岩石力学局
International Bureau of Weights and Measures 国际计量局;国际度量衡局
international business 国际商业;国际企业
International Business Machine Corporation [IBM] 国际商用机器公司

International Cable Protection Committee 国际电缆保护委员会
international call 国际通话;国际呼叫;国际长途电话
international call letters 国际通话
international call sign 国际通话;国际呼号
international candle 英国烛光;国际(旧)烛光
international carat 国际克拉(金刚石重量单位)
International Cargo Handling Coordination Association 国际货物装卸协会;国际货物装卸协调联合会
international Cartel 国际卡特尔;国际协议书;国际机构
International Cartographic(al) Association 国际制图协会
International Cement Seminar 国际水泥讨论会
International Center for Environment 国际环境中心
International Center or Settlement of Investment Disputes 解决投资争端的国际中心
International Centre for Agricultural Research on Dry Areas 国际干燥区域农业研究中心
International Centre for Industry and Environment 国际工业和环境中心
International Centre for Integrated Mountain Development 国际山地开发中心
International Centre for Research in Agroforestry 国际农林研究中心
International Centre for Training and Education in Environmental Science 环境科教中心
International Chamber of Commerce 国际商会
International Chamber of Shipping 国际航运协会;国际航运公会;国际航业协会;国际航业公会;国际海运联盟;国际海运公会
international check 国际支票;旅行支票
international chemical symbols for the element 国际化学元素符号
International City Management Association 国际城市管理协会
International Civil Aviation Organization 国际民航组织
international claim 国际债权
International classification 国际分类
international classification of diseases 国际疾病分类
international classification of mineral resources 矿产资源国际分类
International classification system 国际分类法
international classification system of soil fractions 国际制土粒分级
international clearing 国际清算;国际结算
international climate convention 国际气候公约
international climate funds 国际气候基金
International Cloud Atlas 国际云图
international code 国际通用旗号;国际电码
international code flag 国际信号旗
International Code for State of Sea 国际海况电码
international code of conduct 国际行为准则;国际行为标准
International Code of Signals 国际码语言;国际信号(代)码;国际电码信号
international code user 国际电码用户
international collaboration 国际合作
International Collaborative Pesticides Analytical Committee 国际农药合作分析委员会
international colo(u)r index 国际色指数
international commerce 国际贸易
international commerce arbitration 国际商事仲裁
international commercial exchange 国际商品交易所
international commercial loans 国际商业贷款
international commercial radio 国际商业无线电
international commercial terms 国际商业条款
International Commision on Oceanography 国际海洋学委员会
International Commission of Agricultural Engineering 国际农业工程委员会
International Commission of Snow and Ice 国际冰雪委员会
International Commission on Cloud Physics 国际云物理学委员会
International Commission on Glass 国际玻璃协会
International Commission on Illumination 国际照明委员会
International Commission on Illumination chromaticity coordinate 国际照明委员会色度坐标
International Commission on Illumination colo(u)r difference equation 国际照明委员会色差方程
International Commission on Illumination daylight illuminant(s) 国际照明委员会日光照明体
International Commission on Large Dams 国际大坝委员会
International Commission on Radiological Protection 国际辐射防护委员会;国际放射性辐射防护委员会
International Commission on Water Quality 国际水质委员会
International Commission Signal Company 国际信号公司委员会
International Committee for the Organization of Traffic at Sea 国际海上交通组织委员会
International Committee of Ground Radio Stations 国际陆地无线电台委员会
International Committee on Acoustics 国际声学委员会
international commodity 国际商品
international commodity agreement 国际商品协定
international commons 国际公有物;国际公地
international communication 国际通信
International Communication Association 国际通信协会
international communication satellite 国际通信卫星
international communication satellite earth terminal 国际通信卫星地面终端
International Communications Satellite Consortium 国际通信卫星联合组织
international comparison 国际比较
international competition 国际竞赛
international competitive bidding 国际竞争性招标;国际竞争性投标
International Computation Center 国际计算中心
international computer 国际制式计算机
International Computer Center 国际计算机中心
International Computer's Limited 国际计算机有限公司
International Confederation of Free Trade Unions 国际自由工会联合会
international conference 国际会议
International Conference for the Safety of Life at Sea 国际海上人命安全会议
International Conference of Building Officials 国际建筑官员大会
International Conference of Free Trade Unions 国际自由贸易联盟会议
International Conference on Coastal Engineering 国际海岸工程会议
International Conference on Computer Communication 国际计算机通信会议
International Conference on Environmental Future 国际环境未来会议
International Conference on Environmental Sensing and Assessment 国际环境自动检测与估价会议
International Conference on marine Pollution 国际海洋污染会议
International Conference on Soil Mechanics and Foundation Engineering 国际土力学与基础工程学会
International Conference on Water and Environment 国际水与环境会议
International Conferences on Engineering in the Ocean Environment 国际海洋环境工程会议
International Conference to Discipline the Use of Radar in Maritime Navigation 规定海上使用雷达的国际会议
International Congress for Modern Architecture 国际现代建筑会议
International Congress of Ecology 国际生态学大会
International Congress of Pesticide Chemistry 国际农药化学大会
International Congress of Soil Science 国际土壤会议
international consortium 国际财团
international construction engineering contract 国际建筑工程承包
International Container Association 国际集装箱协会

International Container Bureau 国际集装箱局
International Container Transport Company 国际集装箱运输公司
international contract 国际承包
international control frequency band 国际管理频带
international control station 国际控制电台
International Convection for the Prevention of Pollution from Ship 国际船舶防污染公约
international convention 国际惯例;国际公约
International Convention for Prevention of Oil Spill Pollution of the Sea 国际防止海上油污染公约
International Convention for Protection of Birds 国际鸟类保护公约
International Convention for Regulation of Whaling 国际捕鲸公约
International Convention for Safe Container 国际集装箱安全公约;国际安全集装箱公约;集装箱安全公约
International Convention for the Prevention of Pollution from Ships 防止船舶引起污染的国际公约
International Convention for the Prevention of Pollution of the Sea by Oil 国际防止海上油污染公约
International Convention for the Safety of Life at Sea 国际海上人命安全公约
International Convention on Civil Liability for Oil Pollution Damage 国际油污损害民事责任公约
International Convention on Environmental Protection 国际环境保护公约
International Convention on Maritime Search 国际海上搜寻救助公约
International Convention on Patents 专利权的国际公约
International Convention on Salvage 国际救助公约
International Convention on Standards for Training 国际船员训练、发证和值班标准公约
International Convention Relating to Intervention on High Seas in Oil Spill Accident 发生油污事故时在公海上进行干涉的国际公约
International Convention Relating to Intervention on the High Seas in Cases of Oil Pollution Casualties 国际干预公海油污染事件公约
international conventions into which China has accessed 中国已加入的国际公约
international convergence 热带辐合区;热带辐合带
international cooperation 国际协作;国际合作
International Cooperation on Marine Engineering 轮机工程系统国际协作委员会【船】
International Cooperation on Marine Engineering System 国际海洋工程协作委员会
international coordinated solar observations 国际太阳联合观测
international coordination group for tsunami warning system in the Pacific 太平洋海啸警报组
international corporation law 国际公司法;跨国公司法
International Cost Engineering Congress 国际造价工程联合会
International Council for Building Research 国际建筑研究与文献委员会
International Council for Environmental Law 国际环境法理事会
International Council for Exploration of the Sea 国际海洋开发理事会
International Council for Protection of Birds 国际鸟类保护理事会
International Council for the Exploration of the Sea 国际海洋考察理事会
International Council of Environment Law 环境法理事会
International Council of Graphic Design Association 图形设计协会国际理事会
International Council of Marine Industry Association 国际海洋工业协会理事会
International Council of Pressel Vessel Technology 国际压力容器学会理事会
International Council of Scientific Unions 国际科学协会理事会;国际科学联合会理事会
International Council on Monuments and Sites 国际名胜古迹理事会

International Court of Justice 国际法庭
International Credit Association 国际信贷协会
International Critical Tables 国际科技常数
international currency 国际通行货币;国际货币
international custom 国际惯例
international custom union 世界关税同盟
International Danube Commission 多瑙河国际委员会
International Data Co-ordination Centre 国际数据协调中心
International Data Rescue Coordination Center 国际资料抢救协调中心
international date line 国际日期变更线;国际日界线;国际换日线;日界线
international debt 国际债务
International Decade for Natural Hazards Reduction 1990 to 2000 国际减灾十年(1990—2000年)
international decade of ocean exploration 国际海洋考察十年
international decade station 国际水文十年测站
international deluxe type hotel 国际豪华型旅馆
International Development Association 国际开发协会
International Development Association credit 国际开发协会信贷
international die 公制螺丝板牙
international digital data service 国际数字数据业务
international direct dialing 国际直接拨号
international direct dial telephone 国际直拨电话
international direct distance dialing 国际直通长途电话
International Directory 国际指南
international distance 节点间距
international distress code 国际遇险求救电码;国际遇险呼救电码;国际遇险呼叫电码
international distress frequency 国际遇险求救频率;国际遇险呼叫频率
international division of labo(u)r 国际分工
International Docker's Federation 国际码头工人联合会
International Documentation Center 国际文献资料中心
international drain basin 国际排水盆地
international dredging market 国际疏浚市场
International Drinking Water Regulations or Standards 国际饮用水规程或标准
International Drinking Water Standards 国际饮用水标准
International Drinking Water Supply and Sanitation Decade 国际饮用水供应与环境卫生十年
International Earth Rotation Service 国际地球自转服务
international economic cooperation 国际经济合作
international economy 国际经济
international electric(al) units 国际电学单位
International Electronic Commission 国际电子委员会
International Electronics Association 国际电子协会
International Electrotechnical Commission 国际电子技术委员会;国际电工委员会
international ellipsoid 国际椭球(体)
international ellipsoid of reference 国际参考椭圆;国际参考椭球
international ellipsoid of rotation 国际旋转椭球
International Energy Agency 国际能源局;国际能源机构
International Energy Programme 国际能源计划
International Environmental Bureau 国际环境局
International Environmental Education Program(me) 国际环境教育计划;国际环境教育方案
International Environmental Information System 国际环境资料系统
International Environmental Law 国际环境法
International Environmental Policy 国际环境政策
international environmental relation 国际环境关系
international environmental specimen bank 国际环境标本库
international environment(al) standard 国际环境标准
international exchange 国际汇兑
international exchange crisis 国际外汇危机
international exhibition 国际展览会
international express train 国际联运旅客特别快车
international fair 国际博览会;博览会

international farad 国际法拉
International Federation for Housing and Planning 国际住宅和城市规划协会
International Federation for Information Processing 国际信息处理联合会
International Federation of Automatic Control 国际自动控制联合会
International Federation of Building and Wood Workers 国际房屋及木工协会
International Federation of Computer Sciences 国际计算机科学联合会
international Federation of Forwarding Agent's Association 国际货运代理人联合会
International Federation of Landscape Architects 国际园林设计师协会
International Federation of National Standardization Association 国家标准化协会国际联盟;国际统一规格协会
International Federation of Operations Research Societies 国际运筹协会联合会
International Federation of Prestressing 国际预应力混凝土协会
International Federation of Shipmaster's Association 国际船长协会联合会;船长协会国际联合会
International Federation of Surveyors 国际测量师联合会;国际测量工作者联合会
international ferry freight 国际渡船运费
International fertilizer Industry Association 国际化肥工业协会
International Finance Corporation 国际金融公司
international finance organization loans 国际金融组织贷款
international financial center 国际金融中心
international financial institution 国际金融机构
International Flashing Light Signaling 国际闪光通信信号
International Freight Conference 国际运费工会
international freight forwarder 国际货运代理人
international frequency assignment plan 国际频率分配计划
International Frequency List 国际频率表
International Frequency Registration Board 国际频率登记委员会
international funding sources 国际资金筹措机构
International Generalization Joint Commission 国际标准化联合委员会
international geodynamic(al) project 国际地球动力计划
International Geographic(al) Corporation 国际地理协会
International Geologic(al) Commission 国际地质协会
International Geologic(al) Congress 国际地质会议;国际地质大会
international geologic(al) correlation program(me) 国际地质相关方案;国际地质对比方案
International Geologic(al) Society 国际地质学会
international geomagnetic reference field 国际地磁参考场
International Geophysical Year 国际地球物理年
International Geosphere-Biosphere Program(me) 国际陆界生物圈方案
International Geotextile Society 国际土工织物学会
International Global Atmospheric Chemistry Program(me) 国际全球大气化学计划
international goods through transport 国际货物联运
International Gravimetric Commission 国际重力测量委员会
international gravity formula 国际重力公式
international gravity standard 国际重力基准
International Gravity Standardization Net 1971 1971年国际重力基准图
international gravity station network 国际重力基点网
International Group of Protection and Indemnity Associations 国际保赔协会集团
international guarantee 国际担保
International Habitate and Human Settlement Foundation 国际生境和人类住区基金
International Harbo(u)r Conference 国际港口会议
international henry 国际亨利
international horse power 国际马力

International House 国际文化馆(位于日本东京)
International Housing and Planning Committee 国际住宅和城市规划委员会
International Hydrographic(al) Bureau 国际水文局;国际水道测量局;国际海道测量局
International Hydrographic(al) Conference 国际水文会议
International Hydrographic(al) Organization 国际航道测量组织;国际海道测量组织
International Hydrological Decade 国际水文十年
international hydrological program(me) 国际水文计划;国际水文(学)规划;国际水文学方案
International Hydrologic Organization 国际水文组织
International Hydrologic Seminar 国际水文研讨会
International Ice Patrol 国际海水巡逻队;国际冰情侦察;国际冰情巡逻
International Ice-patrol Service 国际冰情侦察机构
international indebtedness 国际借贷
international index number 国际指数;国际区站号
international Indian ocean expedition 国际印度洋考查
International Industrial Exhaust Control 国际工业排气控制
International Industrial Television Association 国际工业电视协会
international information network 国际信息网
International Institute for Advanced System Analysis 国际高级系统研究源
International Institute for Applied System Analysis 国际应用系统分析研究院
International Institute for Educational Planning 国际教育规划研究所
International Institute for Environmental and Development 国际环境与发展学会;国际环境和发展研究所
International Institute for Environmental Development 国际环境发展研究所
International Institute for Environment and Development 环境与发展国际研究所
International Institute for the Unification of Private Law 国际私法统一协会
International Institute of Communications 国际通信协会
International Institute of Environmental Affairs 国际环境事务(研究)所
International Institute of Noise Control Engineering 国际噪声控制工程学会
International Institute of Refrigeration 国际制冷学会
International Institute of Tropic(al) Agriculture 国际热带农业研究所
International Institute of welding 国际焊接学会
international investment 国际投资
International Investment Bank 国际投资银行
international investment position 国际投资状况
international investment trust 国际投资信托;国际投资托拉斯
internationality 国际性
internationalization 国际化
internationalized capital 资本的国际化
International Joint Committee for Tall Buildings 国际高层建筑联合委员会
International Journal of Rock Mechanics and Mining sciences 国际岩石力学与矿业工程杂志
international kilocalorie 国际千卡
international knot 国际节(速度单位,1 国际节 = 1.852 千米/小时)
International Labour Association 万国劳动协会
international labo(u)r compensation facility 国际劳动补偿办法
International Labo(u)r Organization 国际劳工组织
International Lake Environment Committee 国际湖泊环境委员会
international landline charges 国际陆线费
International Latitude Service 国际纬度站;国际纬度服务
International Latitude Station 国际纬度站
International Law 国际法;国际公法
International Law Association 国际法(律)协会
International Law Commission 国际法委员会
International Law of Natural Resources Protection 国际自然资源保护法
international law of sea 国际海洋法

International Law of the Sea 国际海法
International Legal Conference on Marine Prevention Law 海洋污染损害问题国际法律会议
international lending institution 国际贷款机构
international level 世界水平
International Life-boat Federation 国际救生艇联合会;国际救生艇会议
International Lifesaving Appliance manufacturers' Association 国际救生设备制造商协会
international limited bidding 国际有限投标
international liquidity unit 国际结算单位
International List of Ship Stations 国际船舶无线电台一览表
International List of Stations 国际电台表
international lithosphere project 国际岩石圈研究计划
international load line certificate 国际船舶载重线证书;国际载重线证书
International Loadline Convention 国际(船舶)载重线公约
International Loadline Record 国际载重线勘定记录
international loan 国际贷款;国际借款
international low water 国际低潮位;国际低潮面
International Lumen 国际流明
internationally accepted methods of absorbing foreign investment 国际通行的吸引外资方式
internationally harmonized standard 国际协调标准
internationally important ecosystem 国际性重要生态系统
internationally recognized 国际公认的
internationally significant ecosystem 有国际意义的生态系统
international mailgram service 国际电报信函业务
international maintenance center 国际维护中心
International Maize and Wheat Improvement Centre 国际玉米小麦改良中心
international map 国际地图
international map of the World 世界地图
International Marine and Oil Development Company 国际海洋和石油开发公司
International Marine and Shipping Conference 国际海事与海运会议
International marine highway 国际海洋航线
international marine radio aids to navigation 国际船用无线电的导航设备
international maritime Bureau 国际海事局
International Maritime Committee 国际海运委员会;国际海事委员会
International Maritime Dangerous Goods Code 国际危险物品海运规则;国际海上危险货物运输规则
International maritime domain 国际海洋领域
International Maritime Law 国际海上法;国际海洋法;国际海事法
International Maritime Organization 国际海事组织
International Maritime Organization Search and Rescue System 国际海事组织搜救系统
International Maritime Pilots' Association 国际引航员协会;国际海上引航协会
international maritime satellite 国际海事卫星
international maritime satellite communication 国际海事卫星通信
International Maritime Satellite Organization 国际海事卫星组织
international maritime satellite system 国际海事卫星系统
international market 国际市场;世界市场
international market price 国际市场价格;世界市场价格
International Mathematical Union 国际数学联合会
international maxwell 国际麦;克斯韦
International Medical Guide 国际船用医药指南
International Meeting on Radio Aid to Marine Navigation 国际无线电助航设备会议
International Meridian 国际子午线
International Meteorological Center 国际气象中心
international meteorological code 国际气象符号;国际气象电码
International Meteorological Organization 国际气象组织
International Meteorological Telecommunication Network in Asia and Pacific 亚洲和太平洋国际气象电信网

international metric system 国际米制;国际公制
International Microwave Power Institute 国际微波功率协会
international migration 国际迁移
international mobile telecommunication 国际移动通信
international modern 国际式建筑
International Modular Group 国际模数组织
international monetary arrangement 国际货币协定
International Monetary Fund 国际金融组织;国际货币基金
international monetary fund credit 国际货币基金组织信贷
International Monetary Fund Organization 国际货币基金组织
international monetary market 国际金融市场
international monetary reserve 国际货币储备
international monetary system 国际货币制度
international money order 国际汇款单;国际汇票
international monitoring measures 国际监督措施
international monopoly 国际垄断
international Morse code 国际莫尔斯电码;大陆电码
international multilateral loans 国际多边贷款
international nautical mile 国际(海)里(1 国际(海)里=1.852 千米/小时)
international navigation 国际沿海航行
international navigational route 国际航道
International Navigation Conference 国际航运会议;国际航海会议
International Navigation Congress 国际航运会议;国际导航会议
International Navigation Simulator Lecturers Conference 国际航海教师模拟器会议
international navigation system 国际导航系统
international NAVTEX service 国际 NAVTEX 业务
international network 国际网络
international network management 国际网络管理
International Network of Basin Organization 国际流域组织站网
International Nitrogen Unit 国际氮素小组
international nomenclature 国际命名法;国际分幅编号
international normal gravity formula 国际正常重力公式
international norms and standards 国际规范和标准
international north Atlantic polar front program(me) 国际北大西洋极峰研究计划
International Nuclear Information System 国际核资料系统
international number 国际号码;国际分幅编号
international numbering scheme 国际编号方案
international number of geologic(al) map 国际地质图编号
international number of light 国际航行灯号码
international obligation 国际债务
International Occupational and Health Information Center 国际职业和卫生资料中心
International Ocean Institute 国际海洋学会
International Oceanographic(al) Commission 国际海洋委员会
International Oceanographic Data Exchange 国际海洋资料交换所
International ocean space 国际海洋空间
international ohm 国际欧姆
International Oil industry 国际石油工业
International Oil Pollution Compensation Fund 国际油污赔偿基金(组织)
International oil Pollution Prevention certificate 国际防止油污染证书
International Oil Pollution Prevention Exhibition and Conference 国际石油污染防护展览与会议
International Oil Tank Commission 国际油轮委员会
International Oil Tankers and Terminals Safety Guide 国际油船及码头安全指南
international one-in-a-million map 国际百分之一世界地图
international one-in-a-million map numbering 国际百分之一世界地图图幅编号
international on-line information retrieval service 国际联机信息检索服务机构

International Organization 国际(性)组织
International Organization for Migration 国际移民组织
International Organization of Legal Metrology 国际法定计量组织
International Ozone Commission 国际臭氧委员会
International Ozone Institute 国际臭氧学会
international packet switching service 国际分组交换业务
international packet switch stream 国际分组交换流(业务)
international pallet 通用集装箱;国际集装箱
international parity price 国际同等价格
international passenger station 国际客运站
international passenger terminal building 国际客运站屋;国际旅客度假村
international passenger train 国际旅客列车
international patent 国际专利
International Petroleum Industry Environment (al) Conservation Association 国际石油工业环境保护协会
international phototelegraph call 国际传真呼叫
international phototelegraph network 国际传真网络
international phototelegraph position 国际传真座席;国际传真座台
international phototelegraph service 国际传真业务
international phototelegraph terminal center 国际传真终端中心
International Phycological Society 国际海藻学会
International pipe standard 国际管材标准
International Planned Parenthood Federation 国际计划生育联合会
International Plant Protection Convention 国际植物保护公约
International Polar Motion Service 国际极移服务
International Polar Year 国际极年;国际地(球)极年
international policy 国际政策
International Pollution Abatement Conference 国际消除污染大会
international Pollution Prevention certificate for the carriage of Noxious Substance in Bulk 国际防止散装运输有毒物质污染证书
international port 国际港(口)
International Post Telecommunication Office 国际邮政电信局
international practical system of units 国际实用单位制
international practical temperature scale 国际温标;国际实用温标
international practice 国际惯例
International Press Telecommunications Committee 国际新闻电信委员会
international price level 国际价格水平
international price of products 产品国际价格
international procedure of frequency assignment 国际频率指配程序
international program(me) change relay 国际中断
International Program(me) for the Improvement of Working Conditions and Environment 国际改进工作条件和环境方案
international program(me) of ocean drilling 国际海洋钻探计划
International Program(me) pm Chemical Safety 国际化学物质安全性计划
International Programs in Atmosphere 国际大气科学和水文计划
international project contract 国际工程发包;国际工程承包
international prototype kilogram 国际千克原型;国际公斤原型;国际标准千克
international public law 国际公法
international public telegraph 国际公众电报
international public telegraph service 国际公众电报业务
international quarantine 国际检疫
International Quiet Sun Year 国际太阳宁静年;国际宁静太阳年
international quotation of price 国际报价
International radiation unit 国际辐射单位
international radio 国际无线电
International Radio Advisory Committee 国际无线电咨询委员会
International Radio Call Sign 国际无线电呼号
International Radio Consultative Committee 国际无线电咨询委员会;国际电信咨询委员会
International Radio Frequency Board 国际无线电频率委员会
International Radiological Commission 国际辐射委员会
International Radio-Maritime Committee 国际海事无线电(通信)委员会;国际海上无线电(通信)委员会
International Radio Scientific Union 国际无线电科学联合会
International Radio Service of Call Signs 国际无线电呼号分类表
international radio signals 国际式无线电时号
International Radio Silence 国际无线电停歇规定;国际无线电静寂时间
International Radio Watch-keeping Period 国际无线电工作时间表
International Railway Congress Association 国际铁路协会
international railway container through transport 国际铁路集装箱联运
International Railway Union 国际铁路联盟
International Railway Union mark 国际铁路联盟标记
International Railway Union Standard of Container 国际铁路联盟集装箱标准
International Rapid Latitude Service 国际纬度快速服务
international rated horsepower 国际额定马力
International Real Estate Federation 国际房地产联合会
international record carrier 国际文件传输业务;国际记录载波
International Red Cross 国际红十字会
international reference atmosphere 国际参考大气
International Reference Center for Waste Disposal 国际废物处置咨询中心
International Reference Centre for Community Water Supply 村社供水国际参考资料中心
International Reference Centre for Waste Disposal 废物处置国际咨询中心
international reference ellipsoid 国际参考椭圆;国际参考椭球
International Referral System 国际查询系统
International Referral System for Sources of Environmental Information 环境资料来源国际查询;国际环境资料调查系统
International Registry of Potentially Toxic Chemicals 潜在有毒化学品国际登记中心;可能有毒化学品国际登记中心;国际潜在有毒化学品注册处;国际潜在有毒化学品登记中心
International Regulation for Tonnage Measurement 国际船舶吨位丈量规则
International Regulations for Preventing Collisions at Sea 国际海上避碰规则
international relations 国际关系
international-relations relentingly 国际关系缓和
international-relations relentlessly 国际关系紧张
International Rescue and First Aid Association 国际救援与急救协会
international reserve 国际储备
International Rhythmic 国际科学式【无】
International Rice Research Institute 国际稻米研究所
International Right of Way Association 国际房地产购买协会
international river 国际河流
international river basin 国际流域;国际河流盆地
International River Improvements Act 国际河流整治法令
International Road Congress 国际道路会议
International Road Federation 国际道路联合会
international road network 国际道路网
International Road Research Documentation 国际道路研究文献
international road transport approval plate 国际公路运输海关公约批准单
International Road Transport Union 国际公路运输联盟
international roughness index 国际平整度指数
international route 国际航线;国际航空公司
international routine maintenance measurement 国际例行维护测量
International Rubber Hardness Degree 国际橡胶硬度(等级)
international rule 国际原木板积表(英尺材)
International Rules for the Interpretation of Trade Terms 国际贸易条件解释通则;贸易术语解释的国际通则
International rules of the road 国际海上避碰规则;国际海上航道规则
international safety certificate 国际船舶安全证书
international safety net 国际安全网
International Salvage Union 国际救捞联盟;国际救捞联合会;国际海难救助联盟
international sanction 国际制裁
International Sanitary Conventions 国际卫生公约
International Sanitary Regulations 国际卫生规则;国际卫生法规
international satellite 国际卫星
International Satellite Cloud Climatology Project 国际卫星云气候学计划
international satellite for ionospheric research 国际电离层研究卫星
International Science Organization 国际科学组织
international scientific vocabulary 国际科技词汇
international screw die 公制螺丝钢板
international screw pitch ga(u)ge 国际螺距规
international screw tap 公制丝锥
international screw thread 国际标准螺纹
international sea 国际海域
international sea area 国际水域;国际海域
international seabed area 国际海底区域
international sea-borne shipping 国际海运船舶吨数
international seaborne trade 国际海运贸易
International Sea Commission 国际海洋委员会
International Seaman's Union 国际海员联盟;国际海员联合会
International Sea Scale 国际浪级表
international security 国际证券
International Seismological Summary 国际地震资料汇编
International Service Coordination Centre 国际业务协调中心
international settlement 国际结算
international seven-unit error detecting code 国际七单元检错电码
international sheet of geologic(al) map 地质图国际分幅
International Shipbuilding Research Association 国际造船研究会
International Ship Operating Service 国际船舶营运服务公司
International Shipowners' Association 国际船东协会
International Shipping Federation 国际海运联合会;国际船舶运输联合会
International Ship Structure Congress 国际船舶结构会议
International Ship Suppliers Association 国际船舶(材料)供应协会
international shopping 国际询价采购;国际采购
international shore connection 国际通岸消防接头
international sieve unit 国际筛规
International Sight-seeing and Tours Association 国际旅游观光协会
international signal code 国际信号(代)码
international signal flag 国际信号旗
international signal system 国际信号系统
international signs 国际符号
international situation 国际形势
International Snow Classification 国际雪分类
International Society for Photogrammetry 国际摄影测量学会
International Society for Photogrammetry and Remote Sensing 国际摄影测量与遥感学会
International Society for Rock Mechanics 国际岩石力学学会
International Society for Testing Materials 国际材料试验协会
International Society of Ecological Restoration 国际生态修复学会
International Society of Hydrological Sciences 国际水文科学学会
International Society of Medical Hydrology and

climatology 国际医疗水文和气候学会
International Society of Mine Surveying 国际矿山测量学会
International Society of Soil Mechanics and Foundation Engineering 国际土力学与基础工程学会
International Society of Soil Science 国际土壤学会
international soil classification 国际土壤分类
International Soil museum 国际土壤博物馆
international soil texture grade 国际土壤质地分级
international spheroid 国际椭球(体)
international standard 国际标准
international standard annealed copper 国际标准退火铜
International Standard Atmosphere 国际标准大气(压)
International Standard Book Number[ISBN]国际标准书号
international standard candle power 国际标准烛光
International Standard Conference 国际标准会议
international standard depth 国际标准深度
International Standardization Association 国际标准化协会
international standardization of geographic(al) names 国际地名标准化
International Standardization Organization [ISO]国际标准化组织
International Standardization Organization code 国际标准组织编码
International Standardization Organization (freight) container 国际标准集装箱
International Standardization Organization pallet 国际标准托盘
International Standardization Organization paper sizes 国际标准纸张尺寸
International Standardization Organization Recommendation 国际标准化组织推荐标准
International Standardization Organization standards for container 国际标准集装箱规格
International Standard Meter 国际标准米;国际米原器
international standard metric thread 国际标准公制螺纹
international standard of water quality 国际水质标准
International Standard Organization[ISO]国际标准组织
international standards 国际水平;国际规格
International Standard Serial Number 国际连续出版物号码
international standard sieves 国际标准筛
international standards method 国际标准化方法
International Standards Organization code 国际标准机构编码
International Standard thread 国际标准螺纹
International Standard Time 国际标准时
International Standard Unit 国际标准单位
international standing 国际地位
international steam table 国际蒸气表
international strait 国际海峡
international stream 国际河流
international style 国际式建筑
International subcontracting 国际分包(办法);国际分包合同
International subcontractor 国际分包商
International Switching and Testing Centre 国际电信交换与测试中心
International Switching Maintenance Centre 国际电信交换维护中心
international symbol mark of the disabled 国际无障碍物通用标志
international symbol of other orientation 晶胞取向不同的国际符号
international symbols 国际符号;国际标志
International Symposium on River Sedimentation 国际河流沉积作用学术讨论会
International Symposium on Water Tracing 国际水示踪学术讨论会
international syndicated bank loans 国际银行贷款
international synoptic(al) analysis code 国际天气分析电码
international synoptic(al) code 国际天气电码
international system 国际系统;国际网络

International System of Electric(al) Units 国际电工单位制;国际电气单位制
international system of units[SI]国际(计量)单位制
International Table of Calorie 国际热量表
International Tanker Owner's Association 国际油船船东会
International Tanker Owners' Pollution Federation 国际油船污染控制联合会
International Tar Conference 国际沥青会议
international tariff 国际联运运价表
international technical control centre 国际技术控制中心
international telecommunication 国际电信
International Telecommunication Congress 国际电信会议
International Telecommunication Convension 国际无线电通信会议;国际电信公约
International Telecommunication Coordination Centre 国际电信业务协调中心
International Telecommunication Division 国际电信司
International Telecommunication Monitoring System 国际电信监测系统
International Telecommunication Satellite 国际电信卫星
International Telecommunication Satellite Organization 国际通信卫星组织;国际电信卫星组织
International Telecommunication Satellite system Management 国际电信卫星系统管理部门
international telecommunication service 国际通信业务;国际电信业务
International Telecommunications Union 国际电信联盟;国际电信联合会
International Telecommunications Union Radio Conference 国际电信联盟无线电会议
International Telecommunications Union Regulations 国际电信联盟规则
international telegraph alphabet 国际电报字母表
International Telegraph Alphabet No.2 第二号国际电报字母表
International Telegraph Alphabet No.3 第三号国际电报字母表
International Telegraph Union 国际电报联盟
international telephone and telegraph communication system 国际电话电报通信系统
International Telephone and Telegraph Corporation 国际电话电报公司
International Telephone Centre 国际电话中心
International Telephone Directory 国际电话号码簿;国际电话号码本
International Telephone Exchange 国际电话局
International Telephone Office 国际电话局
international teleprinter exchange service 国际电传打字机交换业务
international teletype code 国际电传打字机电码
international television broadcasting 国际电视广播
International Television centre 国际电视中心
international telex directory information 国际电传号码簿信息
international telex service 国际用户电报业务
international temperature scale 国际温标
international tender 国际性投标
international tendering company 国际承包公司
international terminal 国际航空终点站;国际交通终点站
international test sieve series 国际成套试验筛(包括19种筛号);国际标准筛系列
international textural grade 国际制土壤质地等级
international thread 国际螺纹
international through goods traffic plan for imports and exports 国际联运进出口运输计划
international through traffic of railway 国际铁路联运
international through traffic passenger shipping documents 国际联运旅客乘车票据
international through traffic station 国际联运站
international ticket 国际联运客票
International Time Bureau 国际时间局
international time signals 国际无线电对时信号
International tolerance standard 国际公差标准
International Tonnage Certificate 国际吨位证书
International Tonnage Convention 国际吨位丈量公约

international tourist hotel 国际旅游宾馆
International Towing Tank Conference 国际船模水池试验会议
international trade 国际贸易
international trade agreement 国际贸易协定
International Trade Centre 国际贸易中心
international trade charter 国际贸易宪章
international trade mark 国际商标
International Trade Organization 国际贸易组织
international trade policy 国际贸易政策
international trade system 国际贸易系统
international trading partner 国际贸易伙伴
international train 国际列车
International Training Centre for Water Resources Management 国际水资源培训中心
international train passenger waiting-room 国际列车候车室
international transfer account telegraph service 国际转账电报业务
international transfer of technology 国际技术转让
international transit 国际过境
International Transit Strait 国际通航海峡
international transmission center 国际传输中心
International Transportation Service 国际运输服务公司
International Transport Committee 国际运输委员会
International Travel Service 国际旅行社
international treaties on environmental protection 环境保护国际条约
international treaty 国际条约
International Tropic(al) Timber Agreement 国际热带木材协定
International Tropic(al) Timber Organization 国际热带木材组织
International Tsunami Information Centre 国际海啸情报中心
International Tunnelling Association 国际隧道学会;国际隧道协会
international undefined boundary 未定国界
international understanding 国际谅解
International Union for Conservation of Nature and Nature Resources 国际自然和自然资源保护联合会
International Union for Pure and Applied Chemistry 国际理论和应用化学联合会
International Union for the Conservation of Nature and Living Resources 保护大自然及生物资源国际联合会
International Union for the Protection of Nature 国际自然保护联合会
International Union for the Scientific Study of Population 国际人口科学研究联合会
International Union for Water Research 国际水事研究会
International Union of Air Pollution Prevention Association 国际防止空气污染协会联合会
International Union of Architects 国际建筑师联合会
International Union of Biological Science 国际生物科学协会;国际生物科学联合会
International Union of Building Centers 国际建筑中心联合会
International Union of Building Societies and Loan Associations 国际建房信贷联合会
International Union of Chemistry 国际化学联合会
International Union of Forestry Research Organization 国际森林研究组织协会
International Union of Geodesy and Geophysics 国际大地测量学与地球物理联合会
International Union of Geologic(al) Sciences 国际地质科学联合会
International Union of Geology and Geophysics 国际地质与地球物理学联合会
International Union of Marine Insurance 国际海上保险联盟;国际海事保险联盟;国际海事保险联合会
International Union of Microbiological Societies 国际微生物学会联合会
International Union of Public Transport 公共交通国际联合会
International Union of Railway 国际铁路联盟
International Union of Surveying and Mapping 国际测绘联合会

International Union of Testing and Research Laboratories for Materials and Structures 国际建筑材料及结构试验及研究实验所联合会
International Union on marine Science 国际海洋科学联合会
international unit 国际单位
international universal galactic coordinate system 国际通用银道坐标系
international universal test chart 国际通用视力表
international upper mantle project 国际上地幔计划
international usage 国际惯例
International Visibility code 国际能见度码
international visual storm warning signal 国际风暴警报目视信号
international volt 国际伏特
international voyage 国际航行
international waiting room 国际候船室;国际候机室;国际候车室
international water 公海
International Water Association 国际水协会
International Water Conference 国际水会议
international watercourse 国际水道
international watercourse system 国际水道系统
International Water Engineering Center 国际水工程中心
International Waterfowl Research Bureau 国际水鸟研究局
International Waterfowl Trust 国际水禽托拉斯
international water law 国际水法
International Water Resources Association 国际水资源协会
international waters 国际水域;国际海域;公海
International Water Supply Association 国际给水(工程)协会
international waterway 国际水道;国际航道
International Whaling Commission 国际捕鲸委员会
International Wireless Certificate 国际无线电证书
International Workshop on Precipitation Measurement 国际降水
International Yacht Racing Union 国际游艇联盟
International Yearbook of Agricultural Statistics 国际农业统计年鉴
International Yearbook of Cartography 国际制图年鉴
internegative 中间底片
intern engineer 实习工程师
internet 互联网【计】;交互网络;互联网(络)
internet browser 互联网浏览器
internet content provider 互联网信息内容供应商
internet explorer (微软公司的)互联网浏览器
internet protocol[IP] 因特网协议;网际协议
internet service provider 互联网服务供应商
internetwork communication 互网络通信
internides 内构造带
internodal cutting 节间插条
internodal destination queue 节点间目的地队列
internodal distance 节点距
internodal tract 结间束
internode 结间部;节间部;节间
internuclear 核间的
internuclear distance 核间距离
internuclear potential 核间互作用势
interoceanic 大洋之间的
interoceanic canal 洋际运河
interoffice account 分支店账户
interoffice cable 局内电缆
interoffice communication 局间通信
interoffice traffic 办公室之间交通
interoffice trunk 局间中继线;局间干线;机关间直接干线
inter-office voucher 列账通知传票
intero-inferiorly 向内下
interoperability 互通性
interorder distance 干涉带间距
interparticle attraction 粒间吸力
interparticle attractive force 粒间引力
interparticle attrition 颗粒间摩擦
interparticle bond 粒间键
interparticle bonding 粒间黏结
interparticle bridging 颗粒间连接
interparticle cohesion 粒间黏合力
interparticle collision 颗粒碰撞

interparticle comminution 粒间粉磨
interparticle compress 粒间压缩
interparticle crushing 粒间破碎
interparticle diffusion 粒子内部扩散
interparticle dissolution pore 粒间溶孔
interparticle distance 粒子间距
interparticle force 粒间作用力
interparticle friction 粒间摩擦;颗粒间摩擦
interparticle porosity 粒间孔隙;颗粒间多孔性
interparticle repulsion 细颗粒间相互排斥作用(土内);粒间斥力
interparticle space 粒子内部空间
interparticle structure 粒间结构
interparticle swelling 粒间膨胀
interparticle track 颗粒内扩散
interparticulate force 粒间作用力
interpass 层间的
interpass temperature 层间温度
interpellation 质询
interpendent 中间吊饰;中间吊灯;中间挂饰
interpenetrate 互相贯通
interpenetrating mo(u)lding 交截线脚;交截线条(后期哥特式)
interpenetrating network 互穿网络
interpenetrating network of samples 交叉样本网
interpenetrating polymer network 相互穿透聚合物网络;互穿聚合物网络
interpenetrating subsampling 交叉子抽样
interpenetration 相互渗透;互相渗透;互相贯穿;互相穿透;穿插
interpenetration blastic texture 穿插变晶结构
interpenetration of folded cylindrical surfaces 折叠圆柱面的(交)互贯(穿)
interpenetration of internal and external space 内外空间互贯
interpenetration of spaces 空间交贯
interpenetration of two parallel barrel vaults 平行筒拱的互贯
interpenetration twin 互穿孪晶;穿插双晶
interperiod allocation 跨期分摊(指在前后两个会计期间分摊)
interperiod allocation account 跨期摊配账户
interperiodic(al) line 周期间线
interperiod inspection 中间技术检查
interperiod tax 跨期税款
interphase 相界面;相间的;中间相;界面相;间期;分界面;中间状态;中间周象
interphase boundary 相间边界
interphase boundary energy 相间边界能
interphase boundary junction 相间边界结
interphase connecting rod 相间连杆
interphase connection 相间连接线
interphase coupling 相间耦合
interphase death 间期死亡
interphase engineering 界面层工程
interphase interaction 相间相互作用
interphase interface 相间分界面
interphase mass transfer 相界面传质
interphase nucleus 间期核
interphase polycondensation 相间缩聚
interphase potential 相间电位;界面电势
interphase precipitate 相际沉淀
interphase quick-break overcurrent protection 相间速断过流保护
interphase reactor 相间电感器
interphase transformer 相间变压器
interphone 内部通信装置;内线自动电话机;内部电话;互通电话机;对讲电话(机)
interphone amplifier 直通电话增音器;内部对讲电话增音机
inter phone communication 飞机内通信
interphone equipment 内部通话设备
interphone radio 内部无线电话
interphone system 内部(对讲)电话系统;对讲电话系统
interpier sheeting 柱间横挡板;柱桩间板桩;柱间支撑
interpilaster 壁柱空当;壁柱净距;壁柱间距
inter-pile-sheeting 桩间挡土板;柱桩间板桩墙;插板桩;桩间水平支撑;柱间水平支撑
interpiracy of water sources 水源互相袭夺
interpit sheeting(sheathing) 水平板桩;横向板桩(混凝土桩支撑);坑内柱间横挡板
interplanar 晶面间的;平面间

interplanar crystal spacing 晶面距离
interplanar distance 晶面间距;平面间距
interplanar spacing 面网间距;平面间距
interplanar spacing d value 面网间距 D 值
interplanetary dust 行星际尘埃
interplanetary field 行星际场
interplanetary flight 宇宙航行
interplanetary magnetic field 行星际磁场
interplanetary magnetic storm 行星际磁暴
interplanetary measurement satellite 行星际测量卫星
interplanetary monitoring platform 行星际监视台;星际监视台
interplanetary monitor satellite 星际监视卫星
interplanetary navigation 行星际航行;星际航行;星际导航
interplanetary shock wave 行星际冲击波
interplant 工厂之间的;间作;间植;套种(的作物)
interplantation 内植
inter-plant bridge 车间联系桥
inter-plant handling 厂内运输
interplanting 林间栽植
interplanting of another crop 套种
interplant operations directive 工厂之间的操作指令
interplant shipping notice 厂际装运通知
interplant transportation 厂际运输
interplate 板块间
interplate earthquake 板块间地震;板际地震
interplatform carbonate rock association 台间碳酸盐岩组合
inter-platform subway 站台间的地下通道
interplay 相互作用
interplay of forces 力的相互作用
interplay of light and shade 光荫交映
interpleader 互讼
interpleural suture 侧板间缝
interpluvial period 间雨期
interpolant 内插式
interpolar 极间的;极间
interpolar axis 极间轴线
interpolate 内插;插入
interpolated contour 内插深线;内插等高线
interpolated data and speech transmission 内插数据与语言传输
interpolated error 内插误差
interpolated interval 内插间距
interpolated point 内插点
interpolated point between contours 内插高程点
interpolated premature beat 插进性过早搏动
interpolated section 插补断面
interpolated synthetic(al) observation 内插同步观测
interpolated value 内插值
interpolater 插入器;分数计算器;补插器
interpolating 内插
interpolating coding area 插值编码区域
interpolating function 插值函数
interpolating multiplier 插值乘法器
interpolating oscillator 内插振荡器
interpolating potentiometer 相间电位;界值电位计
interpolating property 插入性质
interpolating splines 内插样条函数
interpolating to halves 折半内插
interpolation 估读误差;内推法;内插(法);内插表;填写;添改;插值(法);插补
interpolation algorithm 插值算法
interpolational growth 插入生长
interpolation and extrapolation program(me) 内插及外插程序
interpolation by central difference 中差插值法
interpolation by proportional 比例插值法
interpolation by proportional parts 比例内插法;比例插值
interpolation by rate of change 变率内插
interpolation design 内插式设计
interpolation error 估读误差;内插误差;插入法误差;插补误差
interpolation error-filter 内插误差滤波器
interpolation error of gravity-anomaly 重力异常内插误差
interpolation factor 内插因子
interpolation formula 内插公式;插值公式
interpolation for photogrammetry 摄影测量内插

interpolation function 内插函数
interpolation gravity 内插重力值
interpolation line 内插行
interpolation method 内插(值)法;插入法;插值法
interpolation of contour lines 等高线内插
interpolation of contours 等高线内插
interpolation of gravity 重力内插
interpolation of spline function 样条函数插值
interpolation oscillator 内插振荡器
interpolation polynomial 内插多项式;插值多项式
interpolation table 内插表
interpolation well 插值井
interpolative picture coding 内插图像编码
interpolative relation analysis 内插法相对分析
interpolator 转发器;校对机;内插器;插补器;分类器;分类机
interpolator of plotter 绘图机插补器
interpole 中间极;整流极;极间极;换向极;附加(磁)极;辅助(整流)极;补偿磁极
interpole core 附加磁铁芯
interpole generator 带中间极发电机;辅极发电机
interpole shoe 整流磁极极靴;换向极极靴
interpole space 间极间空隙;极间空隙
interpole winding 间极绕组;换向极绕组;附加极绕组
interpolymer 互聚物;异分子聚合物;共聚物;共聚体;内聚体
interpolymerization 共聚作用
interporosity flow coefficient 窜流系数
interpose 介入;插入式(安装);插入
interposed vault 楼盖梁间小穹顶
interposer 插入式选样
interpose type formwork 承插式支架
interposing relay 插入式继电器
interposition 中间位置;截接;间位;插入
interposition growth 侵入生长
interpositive 过度阳片
interpositum 中间帆
interpressure 内压强;内压(力)
interpretation 解释;判译;判读;条款解释;实验结果整理;实验结果分析;成果判读
interpretation and deduction map of geochemistry 地球化学异常解释推断图
interpretation at meeting 会议口译
interpretation based matching 基于解译影像匹配
interpretation by standard procedure 由标准程序的说明
interpretation criterion 解释推断准则
interpretation-estimated result map 解释推断成果图
interpretation exactitude 解释精度
interpretation execution 解释执行
interpretation key 解译标志
interpretation map 推断解释图
interpretation method 译码方法;解释方法;判读方法
interpretation method of borehole electromagnetic wave method 井中电磁波法解释方法
interpretation method of magnetic anomalies 磁异常解释方法
interpretation of air photographs 航空照片判读;航空照片识别
interpretation of a reduction rule program(me) 归约规则程序的解释
interpretation of coal log 煤田测井结果解释
interpretation of contract 解释合同
interpretation of echogram 声图判译;声图判读
interpretation of gravity anomaly 重力异常解释
interpretation of hydrogeologic(al) log 水文测井结果解释
interpretation of law 法律解释;法律的解释
interpretation of logs 测井曲线解释
interpretation of magnetic anomaly 磁异常解释
interpretation of military object 军用目标判读
interpretation of mining log 金属测井结果解释
interpretation of petroleum log 石油测井结果解释(方法)
interpretation of radioactive anomaly 放射性异常解释
interpretation of remotely sensed images 遥感图像解译
interpretation of results 结果分析;成果整理
interpretation of sample means 样品平均数的解释
interpretation of satellite photograph 卫星像片判读
interpretation of scheme 模式解释;模式的解释
interpretation of seismogram 地震图解释
interpretation of topography 地形图解译
interpretation parameter 解释推断参数
interpretation process 解释过程
interpretation program(me) 解释推断程序
interpretation time 解释时间
interpretation way of log 测井结果解释方式
interpretative classification 解释分类
interpretative code 解释(编)码
interpretative execution 解释执行
interpretative log 钻探解译剖面
interpretative method 解释法
interpretative mode 解释方式
interpretative order 解释指令
interpretative process 解释过程
interpretative program(me) 解释程序
interpretative routine 解释的例行程序;解释程序
interpretative rule 解释规则
interpretative simulation 解释模拟
interpretative stratigraphy 解释地层学
interpretative structural modelling 解析结构模型
interpretative subprogram(me) 解释子程序
interpretative subroutine 演算子程序;解释子程序
interpretative system 解释系统;翻译方式
interpretative trace program(me) 解释跟踪程序
interpretative tracing 解释追踪
interpretative version 运算方案;解释方案
interpretator 解译者
interpreted active fault 解译活动断层
interpreted aquifer type 解译含水层类型
interpreted coal-accumulating structure type 解译聚煤构造类型
interpreted coal basin-controlling structure type 解译控煤盆构造类型
interpreted coal bed 解译煤层
interpreted coal-series strata 解译煤系地层
interpreted comprehensive ore guides 解译综合找矿标志
interpreted fault activity of engineering foundation 解译工程基础断层活动程度
interpreted geologic(al) condition of engineering foundation 解译工程基础地质条件
interpreted geologic(al) conditions of mineralization 解译成矿地质条件
interpreted geologic(al) cross section of magnetotelluric method 大地电磁法地质断面图
interpreted geomorphic condition of engineering foundation 解译工程基础地貌条件
interpreted geomorphologic(al) condition 解译地貌条件
interpreted hydrogeologic(al) condition 解译水文地质条件
interpreted hydrogeologic(al) conditions of geothermal field 解译地热田水文地质条件
interpreted impounding structure type 解译蓄水构造类型
interpreted indirect indicators for coal prospecting 解译找煤间接标志
interpreted key bed 解译标志层
interpreted lithologic-stratigraphic(al) unit of engineering foundation 解译工程基础岩性地层单位
interpreted magmatic condition 解译岩浆条件
interpreted metamorphic condition 解译变质条件
interpreted mineralization pattern 解译矿化类型
interpreted oil-gas-bearing basin type 解译油气盆地类型
interpreted oil-gas reservoir structure type 解译油气储集构造类型
interpreted old mine traces 解译老矿遗迹
interpreted quaternary deposit type of engineering foundation 解译工程基础第四系沉积类型
interpreted regional geotectonic background 解译区域大地构造背景
interpreted rock and soil type of engineering foundation 解译工程基础岩土类型
interpreted sedimentary condition 解译沉积条件
interpreted soil indicators 解译土壤标志
interpreted stratigraphic(al) condition 解译地层条件
interpreted structural condition 解译构造条件
interpreted structural condition of engineering foundation 解译工程基础构造条件
interpreted structural conditions of geothermal field 解译地热田地热构造条件
interpreted vegetation indicators 解译植物标志
interpreted wall rock alteration type 解译围岩蚀变类型
interpreter 译码机;转换机;解说员;判读员;翻译者;翻译员;翻译器;翻译机
interpreter code 象征码;翻译(程序)代码
interpreter for reduction rules 归约规则的解释程序
interpreter language 翻译语言
interpreter operation 解释操作
interpreter program(me) 解释程序
interpreter programming 翻译程序编制
interpreter routine 解释子程序;解释例(行)程(序)
interpreting 解释执行;解释过程
interpreting core loss 岩芯遗失解释
interpreting device 资料分享仪器;资料整理仪器
interpretion of result 成果判读
interpretion of satellite image 卫星图像判读;卫星图像解译;卫星图像解释
interpretive classification 描述分类
interpretive element 翻译因素
interpretive language 翻译语言
interpretive map 派生地图
interpretive order 翻译指令
interpretive programming 翻译程序法
interpretive routine 翻译程序
interpretive scheme 翻译模式
interpretive system 翻译系统
interpretive tracing 翻译跟踪法
interpretoscope 译释显示器;判读仪
interpret table 解释表;翻译表
interprocess 工序间
interprocess annealing 工序间退火;中间退火
interprocess communication 进程间通信
interprocess communication facility 进程(间)通信设备
interprocessor 中间处理机
interprocessor communication 处理机间通信
interprocessor interference 处理机间干扰
interprocessor interrupt 处理机间中断
interprogram(me) communicating statement 程序间通信语句
interprogram(me) communication 程序通信;程序间通信
interprovincial expressway 省际高速公路
interprovincial freeway 省际公路
interprovincial highway 省际公路
interprovincial power networks 跨省电网
interproximal cavity 邻面洞
interproximal finishing bur 邻面精修钻
interpulse 中介脉冲
interpulse period 脉冲间歇时间
interpupillary adjustment 眼基线调节【测】
interquartile range 四分位数间距;四分位差
interradium 间辐区
interradius 间辐
interray 间辐条
interreaction 互相反应
interreaction matrix 互作用矩阵
interrecord(ing) gap 记录间隙;记录间隔;记录间断
interrecord(ing) gap length 记录间隔长度
interrecord(ing) relationship 记录间关系
interrecord(ing) structure 记录结构;记录间结构
inter reflection 交互反射
inter-regional 地区内的
inter-regional cooperation 区域间合作
inter-regional freight express train 地区间货物特别快车
inter-regional highway 区域间公路
inter-regional trade 区际贸易
inter-regional trip 区间出行
interrelated 相互联系的
interrelated control 连锁控制
interrelated data 有关资料
interrelated issues 相关问题
interrelation 相互联系;相互关系
interrelationship 相互影响;相互关系;干扰;内在联系
inter-repair time 修理间隔期
inter-republic water transfer 跨国调水
interrod substance 釉柱间质

interrogated time offset frequency agile racon 时间延迟鉴频雷达应答器
interrogate generator 询问信号发生器
interrogate interrupt 询问中断
interrogating processor 询问处理机
interrogating range 询问距离
interrogating signal 询问信号
interrogation antenna 询问天线
interrogation coding 询问编码
interrogation frequency 询问频率
interrogation hole 询问孔
interrogation link 询问信道
interrogation pulse 询问脉冲
interrogation pulse spacing 询问脉冲间隔
interrogation recording and location system 询问记录和定位系统
interrogation signal 询问信号
interrogation spacing 询问间隔
interrogation tele-signaling system 遥示查询制
interrogation unit 询问机
interrogation winding 询问绕组
interrogator 询问应答机;询问器;询问程序
interrogator delay 询问延迟
interrogator-responder 询问应答器;问答器;问答机
interrogator-responder system 询问应答方式
interrogator-responsor 询问应答器;询问应答机
interrogatorset 询问器
interrogator-transmitter 询问器
inter-row space 行距
interrupable instruction 可中断指令
interrupt 中断;遮断
interruptable gas 可中断供应的瓦斯;可中断供应的煤气
interruptable power 可间断供电
interruptable process industry 中断程序作业;中断程序工业
interruptable state 中断状态;可中断状态
interrupt acknowledge 批准中断
interrupt address 中断地址
interrupt address vector 中断地址向量
interrupt analysis 中断分析
interrupt arch mo(u)lding 间断拱线脚
interrupt assignment 中断指定
interrupt assignment strategy 中断指定策略
interrupt bus 中断总线
interrupt button 中断电钮
interrupt class 中断级
interrupt cycle 中断周期
interrupt disable 禁止中断
interrupted 中断的
interrupted acoustic(al) ceiling 不连通的吸声顶棚;间断式吸声吊顶
interrupted ag(e)ing 断续时效;分级时效
interrupted arch 中断的三角檐饰;为装体所中断的拱;(门窗过梁的)弧形檐饰;(山墙的)拱形三角檐饰;中断式拱;间断拱
interrupted arch mo(u)lding 中断拱的装饰线脚
interrupted circuit 被阻断的电路
interrupted continuous waves 断续等幅波;断续波;间断等幅波
interrupted current 断续电流
interrupted cut 断续切削
interrupted cycle of erosion 中断侵蚀旋回;中断的侵蚀旋回
interrupted dc tachometer 断续直流转速计
interrupted discharge of traffic 交通中断
interrupted-drive device 断续驱动装置
interrupted earth 断续接地
interrupted elution 中断洗提
interrupted fault 不连续断层
interrupted fire 断续点火
interrupted fold(ing) 不连续褶皱
interrupted galvanofaradic apparatus 间断直流感应仪
interrupted gear 断续机构
interrupted hardening 阶段淬火;双液淬火;分级淬火
interrupted identification 中断识别
interrupted instruction 中断指令
interrupted lighting row 间断排灯
interrupted line 虚线
interrupted load(ing) 间歇荷载;断续加载;断续荷载;断续负荷
interruptedly pinnate 参差羽状的

interrupted mode 中断方式
interrupted oil supply 间歇供油
interrupted oscillation 间歇振荡
interrupted period of test 试演间断时间
interrupted pour 间断浇注
interrupted production 间歇生产
interrupted profile 中断剖面
interrupted projection 分瓣投影
interrupted protection river banks 间断保护河岸
interrupted quenching 分级淬火
interrupted quick flashing light 断续快速闪光灯;断续急闪光
interrupted return control word 中断返回控制字
interrupted return instruction 中断返回指令
interrupted ringing 断续振铃
interrupted river 中断河;间断河
interrupted screw 带槽螺钉
interrupted signal 断续信号
interrupted solution 不连续解
interrupted speech 断续语言
interrupted spot welding 断续点焊
interrupted stream 中断河;间断河;断续河流
interrupted stroke 羽状条纹
interrupted water table 中断潜水面;间断潜水面
interrupted wave 断续电波
interrupt enable 允许中断
interrupt enable and disable 中断允许与禁止(指令)
interrupt enable flag 允许中断标志
interrupt enable flip-flop 允许中断触发器
interrupter 引信保险隔板;中断器;障碍物;斩波器;断续器
interrupter contact 断续器触点
interrupter duty 断续工作
interrupter gear 射击电动协调装置;断续齿轮
interrupt error 中断错误
interrupter type vibrator 断续式振动器
interrupter vibrator 断续式振动器
interrupt feed-back signal 中断反馈信号
interrupt flag 中断标记
interrupt flip-flop 中断触发器
interrupt freeze mode 中断冻结状态
interrupt handling 中断处理
interrupt handling logic 中断处理逻辑
interrupt handling routine 中断处理程序
interruptibility 可中断性
interruptible customer 间歇性用户;缓冲用户
interruptible jet sensor 遮断射流传感器
interruptible sensor 遮断式传感器
interrupt identification 中断标志
interrupt indicator 中断指示器
interrupting capacity 中断能力;遮断功率;断路容量;断开容量
interrupting current 断续电流
interrupting device 中断装置
interrupting drainage well 间歇排水井
interrupting rating 分断容量
interrupting relay 断续继电器
interrupting spring 断续弹簧
interrupting time 断路时间
interrupt initialization sequence 中断初始化序列
interruption 中止;中断期;遮断物;间断
interruption arc 断流电弧
interruption cable 应急电缆;替换电缆
interruption code 中断码
interruption frequency 断续频率
interruption governor 断续调速器
interruption handling 中断处理
interruption handling routine 中断处理例行程序
interruption loss 截留(雨量)损失【植物】
interruption mask 中断屏蔽
interruption masked status 中断屏蔽状态
interruption of a right 权利的中断
interruption of circuit 断路
interruption of light beam 光束受阻
interruption of operation 运行受阻;运算中断;操作中断
interruption of service 运行障碍;供水中断
interruption of shipping 断航
interruption of throughput 输水中断
interruption on track 线路中断
interruption pending 中断保留
interruption queue 中断排队;中断队列
interruption source 中断原因

interruption status 中断状态
interruption status word 中断状态字
interruption subroutine 中断子(例行)程序
interruption supervisor 中断管理程序
interruption supply 间断供应
interrupt isochronous 等时中断
interrupt latency 中断等待
interrupt lockout 封锁中断
interrupt logging 中断记录
interrupt mask 中断屏蔽
interrupt mask bit 中断屏蔽位
interrupt mask flag 中断屏蔽标志
interrupt mask out 中断屏蔽输出
interrupt mode 中断状态;保持状态
interrupt module 中断模件
interrupt nesting 中断嵌套
interrupt on overflow 上溢中断
interrupt on underflow 下溢中断
interrupt operation 中断操作
interrupt-oriented system 中断用系统;面向中断的系统
interrupt phase 中断状态
interrupt priority 中断优先权
interrupt priority chip 中断优先片
interrupt priority signal 中断优先级信号
interrupt priority system 中断优先系统
interrupt priority table 中断优先表
interrupt priority threshold 中断优先阈值
interrupt prior level 中断优先级
interrupt procedure 中断过程
interrupt processing 中断处理
interrupt processing routine 中断处理程序
interrupt program(me) signal 中断程序信号
interrupt push switch 中断按钮开关
interrupt reassignment 中断重指定
interrupt request 中断请求
interrupt request clearing 清除中断请求
interrupt request handling 处理中断请求
interrupt request signal 中断请求信号
interrupt response 中断响应
interrupt road 中断路
interrupt routine 中断程序;断开程序
interrupt run 中断运行
interrupt service routine 中断服务程序
interrupt servicing 中断服务
interrupt signal feedback 中断信号反馈
interrupt singal 中断信号
interrupt source 中断源
interrupt stacking 中断堆叠;中断保存
interrupt storage area 中断存储区
interrupt system 中断系统
interrupt trap 中断捕获
interrupt/trap acknowledge transaction 中断/陷阱应答处理
interrupt trigger 中断触发器
interrupt trigger signal 中断触发信号
interrupt vector 中断向量;间断向量
interrupt word register 中断字寄存器
interscan 中间扫描
interscendental curve 半超越曲线
inter-scorers reliability 信度
interseasonal variability 季节变率
intersect 横切;相交;交叉;横断
intersected country 丘陵地区;地形起伏地区
intersected terrace 交切阶地
intersecting 相叉的
intersecting angle 相交角;交会角
intersecting arcade 交叉拱廊
intersecting axis 交叉轴线
intersecting barrel 交叉筒拱;交叉圆拱顶
intersecting body 相贯体
intersecting cross-over 交叉渡线
intersecting dislocations 相交位错
intersecting fault 交切断层;交错断层
intersecting fracture 交叉切穿的断裂
intersecting frame 交叉式针梳机
intersecting gill set frame 重针制条机
intersecting gill spreader 重针延展机
intersecting line 相交线;交线;切线
intersecting plan 相交平面
intersecting point 相交点;转角点;交汇点;交(会)点
intersecting relation of joint 节理交切关系
intersecting ribs 交叉穹肋
intersecting roads 相交道路;交叉道路

intersecting roof 交叉屋顶
intersecting routes 相交路线;交叉路线
intersecting runway 交叉跑道
intersecting runway configuration 交叉跑道系统
intersecting section 交叉路段
intersecting shaft 相交轴
intersecting slip bands 相交滑移带
intersecting surface 交叉面;横截面
intersecting tracery 交叉花格窗
intersecting vault 交叉拱顶
intersecting wave 干涉波;三角浪
intersecting zone 交汇区
intersection 路口;截距;交接点;交会;交叉(口);平交道口;道路交叉口
intersectional country 切割地区
intersectional elements 相交因素
intersection angle 相交角;转角;交(会)角;交叉角
intersection approach 交叉口引道;交叉口进入段;交叉路口进口;交叉驶入道路;交叉口进口道
intersection area 交叉区
intersection at angle 平面交叉
intersection at grade 平面交叉
intersection block post 交叉线路所
intersection capacity 交叉口通行能力;交叉口通过能力
intersection census 交叉口交通调查
intersection chart 交织图;网络图
intersection classification 交叉分类
intersection control routine 交互控制程序
intersection control system 交互控制系统(交通管理)
intersection cost 勘察费用
intersection count 交叉口交通调查
intersection cross-walk 交叉口人行横道
intersection curve 横断曲线
intersection data 相交数据
intersection design 交叉口设计
intersection diagram 交叉略图
intersection directional volume flow diagram 交叉流量流向图;交叉口分向流量图
intersection entrance 交叉口进口
intersection exit 交叉点出口(道路);交叉口驶出道路;交叉口出口
intersection-factor method 截面因数法
intersection fault 交错断层
intersection flow 交叉流(量)
intersection fractures of fault with anticline 断裂与背斜交接地段
intersection friction 交叉摩阻
intersection function 交集函数
intersection gate 与门
intersection grade 相交坡度
intersection graph 交图
intersection height 交会高度;交点高度
intersection interference 交叉干扰
intersection legs 相交路段;交叉路段
intersection line 相交线;交线;交会线
intersection lineation 交面纹理
intersection method 交会法
intersection multiway 多路交叉(口)
intersection number 相交数
intersection of barrier 间壁交集
intersection of cross-drift 横巷交会(处)
intersection of faults 断层交叉点
intersection of favo(u)rable bed with fractures 有利岩层与断裂交接地段
intersection of favo(u)rable bed with intrusion 有利岩层与侵入体交接地段
intersection of fuzzy sets 模糊集的交
intersection of ring structure with fault zone 环状构造与断裂带交切地段
intersection of river 河谷交叉
intersection of sets 集合的相交;集的交
intersection of shock waves 冲击波(的)交会
intersection of solids 相贯体
intersection of tangents 切线交叉口
intersection photogrammetry 交会摄影测量学
intersection plan 交叉口平面图
intersection plane 交会平面;交面
intersection point 交会点;交汇点;交换点;交(叉)点
intersection point of diagonals 对角线交叉点
intersection point with angle measurements 测角交会点

intersection rate of return 交叉盈利率;交叉报酬率
intersection roads 交叉道路
intersection roof 带斜沟屋顶;交叉屋顶
intersection speed controller 交叉口车速控制机
intersection station 交会点;交会测站
intersection status 路口状态
intersection street 交叉街道
intersection theorem 相交定理
intersection theory 相交理论
intersection theory method 相交理论法
intersection time 截距时间
intersection treatment 交叉口处理(法)
intersection way 交叉口的相交路段
intersection with widened corners 加宽转角式交叉口
intersector curve 横断曲线
intersect peneplain 交切准平原
intersect point 交会点
interseeding intersowing 添播
intersegment 节间
intersegmental 段间的
intersegment reference 段间引用
interselling 交叉销售
intersertal 间隐的;填隙的;斑状晶间的
intersertal texture 间片结构【地】;填隙结构;间隐结构
intersheathes 金属间层
intersheet bonding 片间黏结
intershield 中间屏蔽
intership radio telegraph 船舶间无线电报
intershop trucking expensee 二场间的搬运费
intersite user communication 各地用户间通信
intersity limitation 强度限
intersity meter 强度测量计
intersity modulated beam 强度调制光束
intersity modulated current 强度调制电流
intersity modulating electrode 强度调制电极
intersity modulation circuit 强度调制电路
intersity modulation scan 强度调制扫描
internsnubber 中间缓冲器
Inter-Society colo(u)r Council of American 美国学会间颜色委员会
intersolubility 互溶性
interspace 用间隔隔开;中间;割间间隔带;留空隙;空隙;空间;净空;间隙;间距;星际
interspace loss 间隙损失
interspace of double glazing 双层玻璃空隙
interspatial texture 填间结构
interspecies correlation estimation 物种间相关估算
intersperse 交替;散置
interspersed carbide 熔焊碳化钨粉末
interspersed clay layer 散置的黏土层;嵌置的黏土层
interspersed matter 铺撒物;铺撒料
interspersion 混布
interspicular region 针状物际区
interspike interval 峰电位间隔
interspinal line 棘间线
interstade 间冰段
interstage 级间的;级间;级际滤波器
interstage amplifier 中间放大级
interstage amplifier section 中间放大级
interstage annealing 中间退火
interstage attemperator 级间减温器
interstage cooler 中间冷却器;级间冷却器
interstage cooling 中间冷却
interstage coupling 管子间耦合;级间耦合
interstage coupling method 级间耦合法
interstage gland 级间气封
interstage line 奇偶行
interstage load resistance 级间负载电阻
interstage network 级间耦合网络
interstage punch 奇偶行穿孔
interstage punching 隔行穿孔;奇数行穿孔
interstage section 接口段;级间部分
interstage shielding 极间屏蔽;级间屏蔽
interstage stabilizing network 级间稳定网络
interstage structure 级间结构
interstage surge tank 中间缓冲器
interstage transformer 级间变压器
interstage voltage 级间电压
interstall pillar 中间矿柱
interstand 中间机座

interstar winding 曲折绕组
interstate 州际的
interstate airways communications station 州间航空通信电台
Interstate Commerce Commission 州际商务委员会;联邦州际商务委员会
interstate compact 州际协定
interstate highway 州际公路
interstate highway system 州际公路系统
interstate land sales 州际土地出售
interstate line 州际管线
interstate park 州际公园
interstate quarantine 洲际检疫
interstate road 州际公路
interstate traffic 州际交通(美国)
interstate trucking 州际货运
interstate waters 国家水道
interstation connection 站间联系
interstation cooperation 站际合作
interstation correlation 站间相关
interstation interference 局间干扰;电台间互扰;电台间干扰
interstation telephone 站间电话
inter station train operation telephone 站间行车电话【铁】
interstellar absorption 星际间吸收
interstellar dust 星际尘埃
interstep 级间
interstep regulating transformer 级间调节变压器
interstice 小间隙;空隙裂缝;空隙;间隙;第二气隙
interstice bin (由三个以上筒仓的外墙构成的)间格料仓
interstices of soil 土壤空隙;土壤裂缝
interstirrup space 箍筋间距
interstitial 填隙的;填隙子;填隙式
interstitial alloy 填隙式合金
interstitial anion 填隙阴离子
interstitial area 城市边缘地区
interstitial atom 填隙原子
interstitial cation 填隙阳离子
interstitial compound 间隙化合物;填隙化合物
interstitial condensation 缝隙结露;隙间结露;缝隙凝结(水)
interstitial configuration 间隙位形
interstitial content 节间密度
interstitial cracking 穿晶断口
interstitialcy 结点间;节间
interstitialcy mechanism 推填机理
interstitial defect 杂质缺陷;间歇式缺陷
interstitial diffusion 填隙式扩散
interstitial embrittlement-sensitive material 间隙脆化敏感材料
interstitial flow 间隙(渗)流;渗流
interstitial fluid 间液;间隙流体
interstitial fraction 隙间部分
interstitial gas 间隙气体
interstitial going 间隙操作
interstitial hydraulic pressure 空隙水压力
interstitial ice 空隙冰;间隙冰;地下冰;缝隙冰
interstitial ion 填隙离子
interstitial lamella 间骨板
interstitial liquid 隙间液体
interstitial material 中间体;填隙物质;填隙(材)料
interstitial matter 填隙物质
interstitial membrane 间隙膜
interstitial nucleus 间位核
interstitial operation 间隙操作
interstitial phase 间歇相;间隙相
interstitial position 晶格中节点间隙;节点间隙
interstitial pressure 裂隙间压力;孔隙压力
interstitial rate 填隙率
interstitial running 间隙操作
interstitial site 间隙位置;间位位置
interstitial solid solution 浸渍固体溶液;填隙固熔体
interstitial space 间隙;组织间隙
interstitial stage 间冰期
interstitial structure 填隙(式)结构
interstitial-substitutional model 填隙替位模型
interstitial surface 孔隙面
interstitial surface area 孔隙面面积
interstitial surface of soil 土的孔隙面
interstitial texture 填充结构
interstitial vacancy pair 填隙空位偶
interstitial velocity 孔隙流速

interstitial void 孔隙
interstitial volume 隙缝体积
interstitial water 岩石间隙水;岩缝间水;构造水;裂隙水;孔隙水;空隙水;间隙水;缝隙水
interstitial water saturation 饱和间隙水量
interstitial working 间隙操作
interstitium (十字形平面房屋的)交叉处;(教堂塔楼下的)正中空间;(教堂平面中的)十字交叉
interstock 中间砧木
interstor(e)y drift 层间偏移
interstorm period 雨间期;间雨期
interstratification 间层(作用);层间;中间层
interstratified 间隔的;层间的
interstratified bed 间层;互层
interstratified coal sample 分层煤样
interstratified mineral 层间矿物
interstream 河间的
interstream area 河间(地)区;分水岭区
interstream divide (两水系的)分水岭
interstream groundwater ridge 河间潜水脊;河间地下水分水线;河间地下(水)分水岭;河间地下(水分水)脊
interswitching unit 变换开关装置
intersymbol interference 码间干扰;符号间干扰
inter-sync 内同步
intersystem communications 系统间通信
intersystem comparison 体系间比较
intersystem problem 系统中的问题
intersystem transition 系间跃迁
intersystole 收缩间期
intertank transfer process 倒灌流程
intertelephone 内部电话
inter-temporal consumption baskets 跨时消费群
intertergite 间背片
interterminal capacitance 端子间电容
intertexture 交织物
intertidal 潮区内的;潮间的
intertidal belt 潮间带
intertidal fascia 潮间带
intertidal flat 潮间坪
intertidal flat deposit 潮间坪沉积
intertidalite 潮间带沉积(物)
intertidal marsh 潮间沼泽地
intertidal natural history 潮区自然特性
intertidal pool 潮水坑
intertidal region 潮水浸淹带;潮漫区;潮浸区;潮间带
intertidal sedimentation 潮间沉积(作用)
intertidal shoal deposit 潮间浅滩沉积
intertidal waters 潮间水域;潮间带水域
intertidal zone 潮汐带;潮浸地带;潮间区;潮间带
intertidal zone ecology 潮间带生态学
intertie 交接横木;交叉杆;交叉栏杆;交叉拉杆;水平杆
intertill 中耕
intertillage 中耕;间作
intertilled crop 中耕作物
intertoll trunk 内部长途中继线;长途台间中继线;长途局间中继线;长途电话局间干线
intertonguing 交错相变化;交错沉积;错相变化
intertooth space 齿间
intertown 市际的;长途的
intertown bus 长途公共汽车
intertown traffic 城镇间交通
intertrack area 股道间空地
intertrack bond 轨道间连接线
intertrack crosstalk 道间串扰;串道
intertrack space 股道间空地
intertrade 互惠贸易
intertrade and transregional economic zones 跨行业跨地区的经济区
inter-train pause 脉冲休止间隔
intertriglyph 三槽陇板间的空间
intertrigo 擦烂
intertrip 联动跳闸
intertripping 联锁跳闸
intertrochanteric crest 转子间嵴
intertrochanteric line 转子间线
intertropic(al) 在南北回归线之间的
intertropic(al) confluence zone 热带汇流区
intertropic(al) convergence zone 热带聚集区;热带辐合区;热带辐合带;赤道辐合区
intertropic(al) front 热带锋;赤道锋
intertropics 南北回归线之间地带内的任何地区

interturn 匝间
interturn fault 匝间故障
interturn insulation 匝间绝缘
interturn protection 匝间保护
interturn short circuit 匝间短路
interturn short circuit protection of generator stator 发电机定子匝间短路保护(装置)
interturn tester 线匝间试验器
intertype 整行排铸机
intertype competition 非同行竞争
interurban 城市间的
interurban bus 市间公共汽车
interurban bus service 市间公共汽车交通
interurban interaction 城市间相互作用
interurban long haul service 市间长途运输
interurban migration 城市间人口迁移
interurban railroad 穿城铁道;市间铁路;市际铁路;市际铁道;越野铁道;城市间铁路;城际铁路
interurban service 市间交通
interurban street 市间街道
interurban traffic 市间交通;市际交通;城市间运输
interurban tramway 市区间的有轨电车
interval 音程;间歇;间期;间隔(时间);区间;时间间隔;差别
interval abutting one another 毗连区间
interval acidized 酸化层段
interval aeration 间歇性充气
interval analysis 区间分析
interval arithmetic 区间运算;区间四则运算
interval-automatic 间歇自动的
interval between coating 涂装间隔
interval between current electrode 供电极距
interval between explosions 爆炸间隔
interval between fissure 裂隙间距
interval between graticule wire 十字丝距
interval between inputs 投入间隔期
interval between lines 线距
interval between measuring electrode 测量极距
interval between pumping sections 取水段间距离
interval between strips 航线间距;航线间隔;航向间隔
interval between two trains 列车间隔
interval between well rows 井排间距
interval break 工间休息
interval clock 间隔时钟
interval coded source 间隔式编码震源
interval contraction 区间收缩
interval correlation 间隔对比;间断对比;地层厚度对比
interval data 间隔型数据
interval distribution 间隔分布
interval dividing technique 区间分割技术
intervale 丘陵间的低地
intervaled scale 区间标尺
interval error 间隔误差
interval estimate 区间估值;区间估计量;区间估计
interval estimation 区间推定;区间估计
interval excavating process 间隔开挖法
interval flow injection analysis 间隔流动分析
interval flow injection analysis instrument 间隔流动分析仪
interval forecast 区间预测
interval fractured 压裂层段
interval from high water 高潮间隙
interval from low-water 低潮间隙
interval function 区间函数
interval graph 区间图
interval halving 区间折半;区间分半法
interval integral 区间积分
interval length of logging 测井层段深度
interval linear programming 区间线性规划
intervallum 间隔带;内外墙间隔带
interval mapping 区间映射
interval measurement 区间计量法
interval migration 区间迁移
interval mode 间隔方式;区间方式
interval number 区间数
interval of columns 柱距;柱间距(离)
interval of convergence 收敛区间
interval of distance 等值距
interval of faults 断层间距
interval of grouping 分组间隔
interval of measuring profiles 量测断面间距
interval of observation line 观测路线间距

interval of observation time 观测时间间隔
interval of optimal water level change 最佳水位变动区间
interval of processing 处理井段厚度
interval of rest 静止的间歇
interval of service 换班时间
interval of stations 站间距离
interval of strip hole 条孔间距
interval of survey line 测线间距
interval of tidal current 潮流期
interval of topographic(al) point 地形点间距
interval of uncertainty 不定区间
interval of water hammer 水锤间隔时间
intervalometer 间隔时间计;间隔时间读出仪;间隔调整器;曝光控制器;曝光控制计;曝光节器;时间间隔计;时间间隔(测量)器;定时器;定时曝光控制器;定时计
interval overlapping one another 相叠区间
interval polling timer 间隔查询定时程序;区间轮询计时器;区间定时询问计时器
interval prediction 区间预测
interval printout 间打印输出
interval record 间隔式记录
interval sampling 间隔抽样
interval sampling theory 区间抽样理论
interval scale 区间尺度
interval scan 间隔扫描
interval sequence 信号显示次序
interval service value 区间服务值
interval signal 周期信号;间隔信号
interval solutions 区间解
interval squeezed 挤水泥层段
interval stock 中间库存
interval time between two trains 列车间隔时间
interval time-out 区间超时
interval timer 精确测时计;间隔时钟;间隔计时器;区间计时器;区间定时器;时间间隔调节器
interval timer in darkroom 暗室定时器
interval topology 区间拓扑
interval transit time 间隔传播时间
interval value 区间值
interval valued extensions 区间值扩展
interval valued function 区间值函数
intervalve 中间管;闸阀间;级(间)(的)
intervalve coupling 管子间耦合;管间耦合
interval vector valued function 区间值向量函数
interval velocity 层速度
interval velocity model 层速度模型
interval velocity plan map 层速度平面图
intervalve transformer 管间变压器
interval-zone 间隔带
intervane burner 旋流叶片式喷燃器
intervariable similarity 变量间相似性
intervascular 管间的
intervascular pitting 管间纹孔式
intervene 干预;介入;插入
intervening area 区间面积
intervening area inflow 区间来水;区间入流
intervening atmosphere 居间大气层
intervening block 开采的矿块
intervening boring 中间钻孔;插孔
intervening channel 中间航道(船闸);闸间航道
intervening cooler 中间冷却器
intervening cooling 中间冷却
intervening inflow 区间入流
intervening obstruction 居间障碍(物)
intervening opportunities model 介入机会模型
intervening portion 错位
intervening releveling 穿插复测水准
intervening river 穿过一个地区的河流;穿过市区的河流
intervening statement 插入语句
intervening strata 夹层
intervening thickness 介入厚度
intervening triangulation 补充三角测量
intervening well 补充井
intervening word 插入字
intervention 干涉;介入
intervention button 应急保险开关;应急按钮;紧急保险按钮
intervention level 干预水平
intervention radiation level 干预辐射水平
intervention sequence 应急(指令)序列
intervention switch 应争保险开关;应急开关;紧

急保险开关
intervention vector 干涉向量
interview 口头查询;面询;会见
interview expenses 访问费用
interviewing 面试
interview survey 访谈式研究
intervisibility 通视;对向通视
intervisibility map 通视条件图
intervisibility test 通视试验
intervisible 相互可见的;可通视的
intervisible condition for pilotage 驾驶通视条件
intervisible point 互通视点
intervisible station 互通视点
inter vivos trust 财产授予者生前信托财产所有权
intervolve 互相盘绕
interwave 干涉波
interweave 交织
interweave length 交织长度
interweave section 交织段
interweaving traffic 交织交通
interwind 互相盘绕
interwinding 中间绕组
interwinding capacity 绕组间电容
interwine 缠绕
interword gap 字间隙;字间间隔
interwork 互相配合
interworking 交互工作
interworking function 互连功能
interworking protocol 互通协议
interwork report 内部报告
interwoven 交织;网贯交织
interwoven fabric 交织布
interwoven fencing 交织围篱;交织围栏
interwoven fibrous structure 交织纤维结构
interwoven floor mat 合编地毡
interxylary cork 木间木栓
interzonal trip 区间出行
interzonal vehicle 区间车
interzone 间断带
interzone call 区域间呼叫
interzone spacing 熔区间间距;区间间距
interzone traffic 区间交通
interzone trip 区间行程;区间出行
in-test 试验期中
intestacy 未留遗嘱
intestinal tract disease 肠道疾病
in the aggregate 总共
in the air 空中
in the channel 在途货物
in the clear 净距;净(空)高度;以内宽计算;以内侧尺寸计算;净宽;净空;净高;明文;无阻(碍)
in the controlled environment chamber 在人工气候室
in the different employments of capital 不同用途的资本
in the direction of the arrow 按箭头方向
in the eye of the law 依法律观点
in the eye of the wind 朝着风吹方向
in the fall 在秋季
in the field 在野外;在现场
in-the-hole drill 潜孔钻机
in-the-home water treatment device 家用水处理装置
in the horizontal position 卧置
in the land owner's employ 土地所有者的工作者
in the large 全局的
in the log 原木形式;未锯开的状况
in the mill drying 粉磨兼烘干
in the money 具实值;较现值有利;已到价合约
in the near future programme 已列入近期规划
in the open 在露天
in the open air 露天
in the planning as separate flows 计划中列为单独流程
in the planning stage 在计划阶段
in the present decade 在目前十年内
in the public domain 不受专利权限制;不受版权限制
in the red 负债
in the right of 根据
in the rough 未经修琢的;未加工的
in-the-seam mining 单一煤层开采法
in the sky 空中

in the small 局部的
in the substance 全重(量)
in the substrate 在基质内
in the vertical position 立置
in the vicinity 在附近
in the watery stage 在灌浆期
in the wet (基坑的)湿挖
in the wind 向上风航行;在上风
in the years already spent 已使用年限
intimate 功能相似的;密切的
intimate admixing 紧密掺和
intimate admixture 紧密混合物;紧密掺和物
intimate contact (颗粒的)紧密接触;密切接触
intimate intergrowth 密集共生
intimate knowledge 丰富的知识
intimate mixing 均匀搅拌;均匀拌和;精细混合
intimate mixture 均匀混合料;均匀混合物;紧密混合物
in-time operation 定时操作
intolerable concentration 不可耐浓度
intolerable contamination 超过容许值的污染
intolerable dose 不容许剂量
intolerable risk 不容许的危险性
intolerance 不耐性;不耐
intolerance of shade 需光度
intolerant organism 阳性生物;对环境变化耐力差的生物;不耐阴生物
intolerant species 敏感种;不耐污染品种
intolerant tree 阳性树
Intolox saddle (packing) 英驼洛克斯鞍形填料
into mapping 映入的映射
intonaco (湿壁画的)最后一道细灰泥;掺大理石粉的细面抹灰灰浆
intonation harbo(u)r 国际港(口)
intonation style 国际式
intone 合调
intorsion 内扭转
intort 内扭转
into stages 分阶段的
into tree 放在树中
in-town station 市内车站
intoxicant grain 中毒谷粒
intoxication 中毒;毒物
intra-African trade 非洲内部贸易
intra-arc basin 弧内盆地
intra-array diffusion 内阵列扩散
intraatrial heart block 房间隔性传导阻滞
intraband transition 带内跃迁;带内过渡
intrabasinal rock 盆内岩石
intrabeam viewing 束内观察
intrabiopelmicrite 内碎屑生物球粒泥晶灰岩
intrabranch 内部往来
intrabuilding network cable 楼内网络电缆
intrabundle 绝缘股线间
intrabundle spacing 分裂导线间距
intracavity 内腔式
intracavity electrooptic modulator 内腔式电光调制器
intracavity frequency doubling 腔中倍频
intracavity image converter 内腔式变像管
intracavity resonantor 内腔式共振器
intracavity saturated absorption 腔内饱和吸收
intracell 晶格之内;晶格内部
intra-channel clarifier 河水澄清剂
intra-city 市内的;同城往来;同城的
intra-city commuting 市内通勤交通
intra-city traffic 市内交通
intra-city travel 市内运行
intraclass 同类的
intraclass correlation 组内相关;同类相关
intraclass correlator 同类相关器
intraclass variance 组内方差
intraclast 内碎屑;内成碎屑灰岩
intraclast aluminous rock 内碎屑铝质岩
intraclast-bearing micritic limestone 含内碎屑微晶灰岩
intraclast-bearing pelitomorphic aluminous rock 含内碎屑泥状铝质岩
intraclastic limestone 内碎屑灰岩
intraclastic-micritic limestone 内碎屑微晶灰岩
intraclastic phosphoraite 内碎屑磷块岩
intraclastic texture 内碎屑结构
intraclast pelitomorphic aluminous rock 内碎屑泥状铝质岩

intraclast siliceous rock 内碎屑硅质岩
intraclinal 内斜的
intracloud discharge 云中放电
intracoastal 沿海的;沿(海)岸的;近岸的
intracoastal waterway 沿海水道;沿海岸的内陆水道;沿岸水道;近岸水道;内陆水道;内航道;沿海运河(约平行于海岸);沿岸运河(约平行于海岸)
intracompany 公司内的
intraconnection 内连(接)
intraconnection track 互联轨迹
intracontinental 陆内的
intracontinental basin 陆间盆地
intracontinental composite basin 内陆复合盆地
intracontinental geosyncline 陆间地槽;内陆地槽
intracontinental lowland 内陆低地
intracontinental sea 陆内海;陆间海
intracontinental subduction 陆内俯冲
intracontinent collision orogen belt 陆内碰撞造山带
intracontinetal rift system 陆间裂谷系
intracoronal retainer 冠内固位体
intracratonic basin 克拉通内盆地;平原地槽
intracratonic geosyncline 克拉通内地槽
intracrustal melting 壳内熔融
intracrystal kink 晶内扭折
intracrystalline 晶内;穿晶
intracrystalline attack 晶体内侵蚀
intracrystalline corrosion 晶体内的侵蚀
intracrystalline flow 晶内流变
intracrystalline fracture 晶粒内断裂
intracrystalline gliding 结晶内滑动
intracrystalline rupture 晶粒内断裂
intractability 难解性
intractable 难控制的;难加工
intractable nonlinearity 难处理的非线性
intra-day limit 日内限额
intra-day overdraft 同一日内透支
intradeep 内渊
intradistrict 区域内的
intrados 拱内圈;拱内侧面;拱腹线;拱底面;拱的内表面;内褶皱面;内拱圈
intrados curvature 内弧曲率
intrados of arch 拱腹(线)
intrados springing line 拱内圈起拱线;拱内弧起拱线;拱腹起拱线
intrados width 拱底面宽度
intra-European 欧洲内部的
intrafolial fold(ing) 层内褶皱;片内褶皱
intraformational 层内的
intraformational bed 层内夹层
intraformational breccia 层内角砾岩
intraformational conglomerate 层内砾岩
intraformational contortion 层内扭曲
intraformational fold(ing) 层内褶皱;片内褶皱
intraframe coding 帧内编码
intraframe image coder 帧内图像编码器
intrageosyncline 陆间地槽;内地向斜;内地槽;副地槽【地】
intragrain slip 晶内滑移
intragranular 粒内的;晶粒内;晶间的
intragranular movement 粒内运动
intragranular porosity 粒内孔隙度
intragranular sliding 颗粒内滑动
intra-harbo(u)r 港内的
intra-industrial 工业内部的;行业内部的
intra-industry 工业内部
in train repair 不摘车修理【铁】
intraisland 岛内的
intra-issue 限内发行
intralaboratory 实验室内的
intralake 湖内的
intralaminar 层内的
intralaminar nuclei 板内核群
intralayer communication 层内通信
intralayer crumple 层内揉皱
intralayer crush belt 层内破碎带
intraline distance 线内距离
intralocular 小房内的
intralocus interaction 位点内互作
intramagmatic 岩浆内的
intramagmatic deposit 内岩浆矿床
intramarginal 边缘内的
intramatrical 衬质内生的

intramicarenite 内碎屑泥晶砾屑灰岩
intramicellar swelling 胶束内溶胀
intramicrite 内碎屑微晶灰岩;内碎屑泥晶灰岩;微晶内碎屑灰岩
intramicrudite 内碎屑微晶砾屑灰岩;内碎屑泥晶砾屑灰岩
intramodal distortion 模内失真;模内畸变
intramode 模内
intramolecularly 分子内
intramolecular transfer 分子内转移
intramontane trough 山间海槽
intramural railway 市内铁路
intranet 企业内联网;内(部互)联网;企业网;铁路物资流通企业
in transit 转运途中;在运输途中;串视
in transit mixing 运输过程中搅拌
intranuclear 核内的
intranuclear crystal 核内晶体
intranuclear energy 核内能
intranuclear forces 核内力
intra-oceanic arc 洋内弧
intra-office communications 办公室内部通信
intra-office network 办公室(间)网络
intra-organic matter diffusion 有机物内浸出
intrapair energy 分子内对内的能量
intraparticle dissolution pore 粒内溶孔
intraparticle transport 粒内迁移
intrapearlitic 珠光体内的
intrapermafrost 永冻土中的;永冻带的;永冻层内的
intrapermafrost water 永冻土层内水
intraphase conductor 分导线
intraplant transportation 厂内运输;厂内输送
intraplate 板块内
intraplate block 板内断块
intraplate earthquake 板内地震;板块内部地震
intraplate seismicity 板内地震运动
intraply hybrid 层内混杂
intra-port 港内的
intraprocessor interrupt 处理机内部中断
intrapulse noise 脉内噪声
intrapulse time 脉冲间隔时间
intrarecord(ing) data structure 记录内部数据结构
intrarecord(ing) slack byte 记录间隙字节
intrarecord(ing) structure 记录内结构
intrared analyzer 红内线分析器
intraregional 地区内部的
intrasellar 蝶鞍内的
intrasonic 超低频
intrasparenite 内碎屑亮晶砂屑灰岩
intrasparite 亮晶内碎屑灰岩;内碎屑亮晶灰岩
intrasparudite 内碎屑亮晶砾屑灰岩
intra-specific diversity 种下多样性
intrastate traffic 州内汽车交通
intrastratal 层内的
intrastratal flow structure 层内流状构造
intrastratal solution 层内溶液;层内溶解作用;层间溶液
intra-striate 内条纹
intrasystem 系统内的
intratarget dosage 靶内剂量
intratectal 覆盖层内的
intratelluric 地内生成的;地内的
intratelluric water 原生水;地内水
intra-terminal track 枢纽内线路
intratrade 内部贸易
intratransaction 内部交易
intratransfer price 内部转让价格
intra-urban interaction 市内相互作用
intra-urban migration 市内人口迁移
intra-urban mobility 市内流动
intra-urban transportation 市内运输
intra vane 内叶片
intra-vane pump 内叶轮泵;内叶片泵;内翼泵
intravibrator 插入式振捣器
intrazonal 地区内部的;地带内的
intrazonal community 隐域植物群落
intrazonal formation 隐域植物群系
intrazonal soil 隐域土
intrazonal trip 区内出行
intrazonal vegetation 地带内植被
intra-zone traffic 区内交通
intra-zone trip 区内出行
intrench 挖壕;掘工事;挖槽
intrenched meander 峡谷蜿蜒河道;深切河湾

intrenched relationship 嵌入关系
intrenched river 嵌入河;深切河流;深切河谷
intrenched stream 嵌入河流;嵌入河;深切河流
intrenched terrace 嵌入台地;嵌入阶地
intrenched valley 嵌入河谷
intrenching tools 挖槽工具;挖掘工具;挖沟工具;土工器具
intrenchment 掘壕沟
in triangle connection 按三角形接法
intricacies 错综复杂的事物
intricacy 错综复杂
intricate 复杂的
intricate cored casting 复杂型芯的铸件
intricate cross section 复杂断面;交错截面
intricate mechanism 复杂机构
intricate mo(u)ld 复杂模具
intricate shape 复杂断面(形状)
in trim 匀平状态
intrinsic(al) 固有的;本征;内在的;内蕴
intrinsic(al) absorption 内蕴吸收;内蕴吸收;本征吸收
intrinsic(al) accuracy 固有精度;内在精(确)度
intrinsic(al) acidity 固有酸度
intrinsic(al) activity 固有活性;内在活性
intrinsic(al) admittance 固有导纳;内在导纳;内蕴导纳
intrinsic(al) adsorption constant 固有吸附常数
intrinsic(al) adsorption equilibrium constant 固有吸附平衡常数
intrinsic(al) alkalinity 固有碱度
intrinsic(al) anisotropy 内在各向异性
intrinsic(al) area 内蕴面积
intrinsic(al) articulation 内关节
intrinsic(al) basicity 固有碱度
intrinsic(al) bioremediation 内源性生物修复
intrinsic(al) bistability 本征型光学双稳性
intrinsic(al) breaking energy 内在断裂能
intrinsic(al) brightness 固有亮度;内蕴亮度;本征亮度;本身亮度
intrinsic(al) brilliance 实际亮度
intrinsic(al) brilliancy 内蕴耀度;本征亮度
intrinsic(al) carrier 本征载流子
intrinsic(al) characteristic 内在特性
intrinsic(al) charge 本征电荷
intrinsic(al) coercivity 内矫顽磁力
intrinsic(al) colo(u)r 本身颜色
intrinsic(al) colo(u)ration 本征着色
intrinsic(al) colo(u)r index 本征色指数
intrinsic(al) concentration 本征浓度
intrinsic(al) conduction 本征导电
intrinsic(al) conductivity 固有导电性;内蕴传导性;本征导电性
intrinsic(al) constant 固有常数
intrinsic(al) contact potential difference 固有接触电位差
intrinsic(al) contaminants 固有杂质;内在杂质
intrinsic(al) coordinates 内部坐标;本征坐标
intrinsic(al) crystal 本征晶体
intrinsic(al) curve 特征曲线;禀性曲线;包络(曲)线
intrinsic(al) defect 本征缺陷
intrinsic(al) density 固有密度
intrinsic(al) derivative 内在导数
intrinsic(al) detector 本征探测器
intrinsic(al) differential geometry 内在微分几何
intrinsic(al) differentiation 内在微分
intrinsic(al) dispersion 内蕴弥散度
intrinsic(al) dissymmetry 内在不对称性
intrinsic(al) dynamic(al) viscosity 固有动力黏度
intrinsic(al) dynamics 内在动态
intrinsic(al) efficiency 内蕴效率
intrinsic(al) electro-luminescence 本征电致发光
intrinsic(al) electronic absorption 本征电子吸收
intrinsic(al) energy 固有能(量);能含量;内蕴能;内能;内蕴能;禀能;本征
intrinsic(al) equation 内蕴方程;禀性方程;本征方程(式)
intrinsic(al) equations of a curve 内蕴曲线方程;内在曲线方程;曲线的内蕴方程
intrinsic(al) error 固有误差;内在误差;基本误差
intrinsic(al) excitation 本征激发
intrinsic(al) factor 内(在)因素;内源因素;内因子
intrinsic(al) fault 内在层错
intrinsic(al) fluxdensity 内蕴磁感应强度
intrinsic(al) friction 内摩擦

intrinsic(al) function 固有函数;内在函数;内蕴函数
intrinsic(al) function name 内在函数名
intrinsic(al) function reference 内在函数引用
intrinsic(al) geodesy 本征大地测量学
intrinsic(al) geometry 内蕴几何学
intrinsic(al) geometry of a surface 曲面的内蕴几何
intrinsic(al) germanium photodiode 本征锗光电二极管
intrinsic(al) growth rate 内在增长率
intrinsic(al) heat 固有热;内热
intrinsic(al) homology 内蕴同调
intrinsic(al) hypothesis 内蕴假设
intrinsic(al) impedance 固有阻抗;内(在)阻抗;内禀阻抗;特征阻抗;本征阻抗
intrinsic(al) inductance 内磁感应强度
intrinsic(al) induction 内在磁感应;内禀感应;内禀磁感应强度;本征感应;铁磁感应
intrinsic(al) infrared detector 本征红外探测器
intrinsic(al) inhibition 内在抑制作用
intrinsic(al) inhomogeneity 内在不均匀性
intrinsic(al) instability 固有不稳定性
intrinsic(al) interest 内在利益
intrinsic(al) internal angle of friction 固有内摩擦角;真内摩擦角
intrinsic(al) ionic conduction 本征离子导电
intrinsic(al) ionization 本征电离
intrinsic(al) junction loss 本征连接损耗
intrinsic(al) layer 本征层
intrinsic(al) level 内在电平;本征能级
intrinsic(al) light 固有光
intrinsic(al) linewidth 本征线宽
intrinsic(al) loss 本征损失
intrinsic(al) luminosity 内禀光度;本身亮度;本身光度
intrinsically 本质上
intrinsically linear model 内线性模型
intrinsically safe 固有安全的;本征安全的
intrinsic(al) magnetic moment 内禀磁矩
intrinsic(al) magnitude 本身星等
intrinsic(al) material 本征材料;本征半导体材料
intrinsic(al) mechanical property 内在机械性能
intrinsic(al) metabolic activity 内在代谢活动
intrinsic(al) micro-bioremediation 内源性微生物修复
intrinsic(al) mobility 本征迁移率
intrinsic(al) noise temperature 内禀噪声温度
intrinsic(al) non-flammability 内在的不燃性
intrinsic(al) nucleation 本征核化;本征成核
intrinsic(al) parity 内蕴对;内禀宇称
intrinsic(al) period 内蕴周期
intrinsic(al) permeability 固有透水性;固有导磁率;内渗透性;渗透率;本征磁导率
intrinsic(al) photo-conductivity 本征光电导
intrinsic(al) photo-conductor 本征光电导体
intrinsic(al) photoemission 内禀光电发射
intrinsic(al) phototransistor 本征光电晶体管
intrinsic(al) pollution 内源性污染
intrinsic(al) precision 内在精(确)度
intrinsic(al) pressure 固有压力;内在压力;内压强;内压(力);内部压力
intrinsic(al) probability 固有概率
intrinsic(al) problem 内在问题
intrinsic(al) procedure 固有过程;内在过程;内部过程
intrinsic(al) property 固有特性;内在性能;内在特征;内在特性;内禀性质;本征性质;固有性状
intrinsic(al) property of a surface 曲面的内蕴性质
intrinsic(al) rate of growth 内在增长率
intrinsic(al) rate of increase 顺时人口净增长率;内禀增长率
intrinsic(al) rate of natural increase 内在自然增长率;内禀自然增长率;天然内在增长率
intrinsic(al) redshift 内禀红移
intrinsic(al) region 本征区
intrinsic(al) resistance 固有阻力;固有电阻;本征阻力;本征电阻
intrinsic(al) resisting moment 内抗矩
intrinsic(al) resistivity 本征电阻率
intrinsic(al) resolution 内在分辨率
intrinsic(al) resonance 内禀共振;本征共振
intrinsic(al) retention mechanism 固有保留机制
intrinsic(al) reward 内在报酬

intrinsic(al) safety 内在防爆安全
intrinsic(al) scattering 本征散射
intrinsic(al) seed 本征籽晶
intrinsic(al) semi-conductor 本征半导体
intrinsic(al) shear strength 固有抗剪强度;内在剪切强度
intrinsic(al) shear strength curve 固有抗剪强度曲线;抗剪强度包(络)线;强度包(络)线
intrinsic(al) shear strength envelope 固有抗剪强度包络线
intrinsic(al) sheet 本征晶片;本征层
intrinsic(al) shrinkage 内在收缩;内部收缩
intrinsic(al) silicon 本征硅
intrinsic(al) solubility 固有溶(解)度
intrinsic(al) speed 本征速度;特性速度
intrinsic(al) spheric(al) geometry 内部球面几何学
intrinsic(al) stability 固有稳定性;固有稳定度;内在稳定性
intrinsic(al) stand-off ratio 本征偏离比;本征变位比
intrinsic(al) strain 内应变
intrinsic(al) strength 固有强度;内在强度;本征强度
intrinsic(al) strength curve 内在强度曲线
intrinsic(al) stress 内应力
intrinsic(al) temperature range 本征温度范围
intrinsic(al) time 固有时间;内裏时间;实时;本征时间
intrinsic(al) tracer 内含示踪剂;内部示踪物
intrinsic(al) transistor 本征晶体管
intrinsic(al) utility 内在效用
intrinsic(al) value 固有值;内在价值;实值;实价;潜在价值
intrinsic(al) variable star 内因变星
intrinsic(al) vector 内部向量
intrinsic(al) viscosity 固有黏度;内裏黏滞性;特性黏度;本征黏度
intrinsic(al) viscosity number 特性黏(度)数
intrinsic(al) viscosity of polymer 聚合物特性黏度
intrinsic(al) viscous damping 内黏滞阻尼
intrinsic(al) wavelength 固有波长
intrinsic(al) worth 内在价值
intrinsic(al) Young's modulus 内部杨氏模量
intrinsic-barrier diode 本征势垒二极管
intrinsic-barrier transistor 本征势垒晶体管
intrinsic-extrinsic fault 内外层错
intrinsic function 内部函数
in triplicate 一式三份
introduce 引入;介绍
introduced gas 带入气体
introduced plant 引入植物
introduced variety 引入品种
introduce foreign capital 引进外资;输入外资
introduce new product 介绍新产品
introduce the budget 提出预算案
introducing of foreign capital 引进国外资本
introducing of foreign investment 引进国外投资;利用外资
introducing of foreign technology 引进国外技术
introducing of technology from abroad 技术引进
introduction 序系;引言;注射;概论;简述;简介;入门(指说明书等的入门知识)
introduction commission 介绍费
introduction course 初级入门教程;插入层
introduction of advanced technology 引进先进技术
introduction of air 引入空气
introduction of foreign capital 引进外资
introduction of foreign investment 引进外国投资
introduction of foreign technology 引进国外技术
introduction of petrology 岩石学序言
introduction of the tidal wave 潮波进入
introduction to accounting theory 会计理论入门
introduction valve 注射阀;进样阀
introductory rate 引入率
introductory report 介绍性报告
introductory statement 介绍性说明
introfaction 加速浸泡作用
introflection 向内弯曲
introflexion 向内弯曲;内屈
introjection 内投作用
introne 内含子
intropolis 向内式城市;漏斗形城市

introscope 内腔探视仪;内腔检视仪;内孔窥视仪;内孔检视仪;内壁检验器
introspection 反省
introversion 内向弯曲;内倾;内翻
introverted personality 内向人格
intruded aggregate concrete 预压骨料混凝土;预填骨料混凝土
intruded concrete 压力灌浆混凝土
intruded rock 侵入岩
intruder 盗窃者;盗窃信息者
intruder alarm 防盗警报器;防窃警报器;防盗器
intruding water 侵入水
intrusion 向内突入;(焊点边缘的)根须;接合线伸长;侵入(作用);侵入体【地】
intrusion agent 注入剂;灌入剂;混合料;添加剂
intrusion aid 灌浆添加剂;助注剂
intrusion alarm 防窃警报器
intrusion and extrusion 侵入与喷出
intrusion concrete 预填骨料混凝土;注浆混凝土;预填集料混凝土
intrusion cone of saltwater 盐水上升锥
intrusion contact trap 侵入接触圈闭
intrusion detector 插入式探测器
intrusion displacement 侵入位移
intrusion grouting 侵入式浇灌;侵入灌浆
intrusion mortar of cement 灌入水泥砂浆
intrusion of mortar 灌(注)砂浆;灌浆灰浆
intrusion of salt 盐分侵入
intrusion of saltwater 盐水入侵
intrusion of seawater 海水入侵
intrusion of tide saltwater 咸潮倒灌
intrusion of water 水侵入
intrusion on an area 面状入侵
intrusion pipe (水泥浆的)灌注管;喷射管
intrusion rock 侵入岩
intrusion spring 侵入泉
intrusion structure 侵入构造
intrusion tectonics 侵入构造
intrusion through fissure-karst passage 沿裂隙岩溶通道入侵
intrusion through river course 沿河道入侵
intrusion through the beach of the sea 沿海岸入侵
intrusion tone 打扰声
intrusion water process 浸水法
intrusive assimilation and contamination 侵入同化混染作用
intrusive body 侵入岩体;侵入体【地】
intrusive breccia 侵入角砾岩
intrusive complex 侵入杂岩
intrusive contact 侵入接触
intrusive dike 侵入岩墙;侵入岩脉
intrusive facies 侵入相
intrusive filtration 侵入过滤
intrusive granite 侵入花岗岩
intrusive growing season 侵入生长时间
intrusive growth 侵入生长
intrusive ice 侵入冰
intrusive igneous rock 侵入火成岩
intrusive injection 侵入体贯入
intrusive mass 侵入体【地】
intrusive mountain 侵入山
intrusive rock 侵入岩体;侵入岩
intrusive rock occurrence 侵入岩产状
intrusive sheet 侵入岩片;侵入岩床;侵入岩层
intrusive sill 侵入岩床
intrusive vein 侵入岩脉;侵入脉
intrusive vulcanicity 侵入火山活动
intrusive zone 侵入带
intubating forceps 插管钳
intubation 插管法;插管
intubator 插管器
intuition 直感
intuition in measuring variable 度量变量中的直观知识
intuition in probability 概率中的直观
intuitive ability 良能
intuitive approach 直觉方法
intuitive forecasting 直观预测
intuitive graphic 直观图
intuitive judg(e)ment 直觉判断;直观判断
intuitive manner 直观状态;直观方式
intuitive order quantity 经验订货量
intuitive recovery procedure 直觉校正过程

intuitive thesis of probability 概率的直观论点
intumescence 泡沸;发泡膨胀
intumescent agent 泡沸剂
intumescent coating 发泡型防火涂料
intumescent flame retardant 膨胀阻燃剂
intumescent paint 热胀漆(一种防火漆)
intumescent sealant 膨胀密封膏;膨胀密封胶
inturning screws 内旋双车
intussusception 套迭
intussusception growth 填充生长
Intze tank 英池式水箱
inulic acid 阿兰酸
inuncta 涂擦剂
inunction 涂擦;敷擦法
inunctum 涂擦剂
inundated 洪水淹没的;洪泛的
inundated area 淹没区;淹没面积;洪泛区;受淹面积;泛区;泛滥区(域)
inundated area of reservoir 水库淹没区
inundated cultivated land 淹没农田;受淹耕地
inundated district 淹水地区;泛区;泛滥区(域);泛滥地区
inundated ice 淹没冰
inundated land 淹没(地)区;淹没地;洪水淹没区;洪泛地;受淹地区;泛区;泛滥土地;泛滥区(域);泛滥地区
inundated mine 淹没矿井
inundated plain 洪泛平原;泛滥平原
inundated population 受淹人口
inundated region 淹没地区
inundated sand 水浸泡过的砂;饱水砂
inundated zone 泛滥区(域)
inundating flood 大洪水;泛滥洪水
inundation 淹水;淹没;浸水;洪水泛滥;大水;泛滥
inundation area 洪水淹没区域
inundation bank 泛滥河岸
inundation bridge 淹没桥;高水位桥
inundation canal 洪水泄水道;洪泛渠;泛灌渠
inundation coefficient of underground river 暗河充水系数
inundation fever 洪水热
inundation index 淹没指标
inundation insurance 洪水保险;水灾保险
inundation map 泛滥区地图
inundation method 饱水法;淹灌法;漫灌法
inundation on tracks 轨道淹没;线路浸水
inundation phase 陆静相
inundation protection 防汛;防洪;防淹没
inundative irrigation 淹灌
inundator 混凝土砂中含水量测定器;注水桶(混凝土搅拌用);浸泡器
inundatus 旱涝交替地区
in unit time 在单位时间里
in urgent need of 迫切要求
in-use testing 实用试验
in vacuo 在真空内
invaded zone 侵入地带;侵入带
invaded zone correction 侵入带校正
invader 侵入物
invading air 侵入空气
invading point 侵界点
invading sea 侵入海岸线;侵入海
invading shoreline 侵入海岸线
invading water 侵入水
invading wave 来袭波
invading waves in estuary 河口入侵波
invaginate 凹入
invagination 折入;内陷;套迭;反折;凹入
invalid 作废的;无效的
invalid address 无效地址
invalid anonymous block 不合法的无名块
invalid arguments 无效自变量
invalidated bonds 宣告无效的债券
invalidated ticket box 废票箱
invalidation 无效;使作废;使无效
invalid authorization code 无效特许码
invalid bit 无效位
invalid block name 不合法的块名称
invalid call 非正确呼叫
invalid characters 无效字符
invalid cheques 无效支票
invalid clause 无效条款
invalid code 无效码
invalid command 无效命令

invalid condition detection 无效条件检测
invalid contract 无效合同
invalid destination 非正确目的地
invalid digit 无效数位
invalid input 无效输入
invalid instruction 无效指令
invalid insurance 伤残保险
invalid key 无用键;无效键
invalid key condition 无效键条件
invalid kitchen 营养厨房;病号厨房
invalid layer name 不合法的层名
invalid limits 不合法的界限
invalid linetype name 不合法的线性名称
invalid name 无效名;不正确名称
invalid op code 无效操作码
invalid page time-sharing 无效页面分时
invalid path 无效途经
invalid point 不合法的点
invalid punch 无效穿孔
invalid selection 不合法的选择;选择无效
invalid sequence 无效序列;无效顺序
invalid ticket 废票
invalid transport 无效运输
invalid view name 不合法的视图名称
invalid voucher 无效凭单
invalid window specification 不合法的窗口说明
invar 因瓦(又称殷瓦);因钢;不胀钢
invar baseline wire 殷瓦基线尺
invar base tape measurement 殷瓦基线尺丈量
invar extensometer 殷钢丝伸缩仪
invariability in price 价格不变性
invariable 恒定;常数;不变的
invariable pendulum 定长摆
invariable plane 不变平面
invariable system 不变体系
invariable with time 不随时间变化的
invariance 不变性;不变式
invariance of cost 价值的不变性
invariance of decision problem 决策问题的不变性
invariance of power 功率不变性
invariance property 不变性
invariant 恒定的;无自由度的;常数因子;不变形的;不变式的;不变量
invariant coordinates 不变坐标
invariant cost 不变费用
invariant distribution 不变分布
invariant embedding 不变嵌入(法)
invariant estimation 不变估计
invariant factor 不变因子
invariant imbedding 不变嵌入法
invariant increase 无变异增长
invariant in space-time 时空不变式
invariant integral 不变积分
invariant load 不变荷载
invariant measure 不变测度
invariant metric 不变度量
invariant of central perspective 中心透视不变式
invariant of stress 应力分量不变式
invariant pendulum 定长摆
invariant point method 不变点法
invariant relation 不变关系
invariant routing 固定选径
invariant state 不变状态
invariant subgroup 正规子群
invariant subspace 不变子空间
invariant system 无变量系
invariant value 不变值
in various proportions 按各种比例
invar level(l)ing rod 殷瓦水准尺
invar level(l)ing staff 殷瓦水准尺;殷钢水准尺
invar measurement 殷瓦尺丈量
invaro 殷瓦劳合金钢
invar pendulum 殷瓦摆(又称因瓦摆)
invar(plotting)scale 殷钢尺;殷瓦绘图尺
invar ribbon 殷瓦带尺
invar rod 殷瓦杆;殷瓦标尺【测】
invar stadia rod 殷瓦视距尺
invar steel 铁镍合金钢;殷钢
invar subtense bar 殷瓦模测尺
invar tape 殷钢卷尺;殷钢尺;因钢带尺
invar tape extensometer 殷瓦尺引伸仪;殷瓦尺伸长计
invar wheel 殷瓦轮
invar wire 殷钢线尺;因钢线尺;殷瓦线尺

invar wire extensometer 殷瓦尺引伸仪;殷瓦尺伸长计
in varying degree 不同程度上
invasion 海侵;侵位
invasion by grass 杂草侵入
invasion of grass 杂草丛生
invasion of weeds 杂草丛生
invasiveness 侵入力
invention 创造;发明物;发明
invention-creation 发明创造
invention of information 发明情报
invention problem 创造问题
inventive concept 发明构思
inventor 发明人;发明家
inventor and creator 发明人和创造人
inventoriable cost 可列入存货的成本
inventorial expenses 可归做成本的费用
inventories of economic activity and population 经济与人口资料要目
inventor's certificate 发明人证书
inventor's certificate system 发明人证书制度
inventor's rights 发明权
inventory 详细目录;库存(清单);清单;物品单;物料量;存货清单;存货(量);存储;财产清册;财产目录
inventory account 存货账;库存核算
inventory adjustment 库存调整
inventory analysis 库存分析;存储分析
inventory and control 库存与控制
inventory and production process 存货与生产过程
inventory and stock level 存储与存货水准
inventory assets 库存资产;存货资产
inventory audit 库存审查;存货审计
inventory building 库存增加
inventory calculation 编目计算
inventory carrying cost 库存费用;存货费用;保管费用
inventory change 库存变动;藏量变化
inventory charges 材料的耗费
inventory checker 库存盘点人
inventory control 库存控制;库存管理;盘存管理;物资管理;存货控制;存货管制
inventory control model 库存管理模型
inventory control problem 库存控制问题;商品盘存控制问题;存储控制问题
inventory control system 存货控制系统
inventory control theory 库存控制理论
inventory cost 库存费用;存料成本
inventory costing and control 库存成本计算与管理;库存成本计价与控制;存货成本计算与管理
inventory costing method 存货计价方法;库存成本计算法
inventory cost selection 库存成本的选择
inventory count slip 盘点存料单;盘存点料单
inventory credit 存货贷款
inventory cut-off date 盘存截止期
inventory cut-off voucher 盘存截止凭单
inventory cycle 库存周期;库存循环;存货周期
inventory date 盘存日期
inventory depletion 库存用完;存货耗尽
inventory depreciation method 盘存计价折旧法
inventory equation 盘存公式;盘存等式
inventory file 清单文件;存货文件
inventory filing 编目文件生成
inventory final 期末存货
inventory financing 库存资金融通;存货资金融通
inventory forced filling rate 装料量
inventory groundwater 地下水储量盈亏估计
inventory index 库存指数;存货指数
inventory index number 库存指数
inventory index of raw materials 原材料存货指数
inventory investment 库存投资
inventory investment cycle 存货投资周期
inventory level 存货水准;存货水平;存货水准
inventory liquidating 库存减少
inventory list 盘存清单;财产目录
inventory loss 库存损失
inventory management 库存管理;物资管理
inventory management simulator 库存管理模拟程序
inventory management system 库存管理制度;库存管理系统
inventory method 库存计算法;盘存法
inventory methods contrasted 盘存方法对照

inventory model 库存模型
inventory of fission product radioactivity 裂变产物的放射性总强度
inventory of fixed assets 固定资产存量
inventory of manpower 人才储备表
inventory of non-tariff barriers 非关税壁垒清单
inventory of raw materials on hand 库存原材料的盘存
inventory of repair parts 修理零件存货;库存修理用零件
inventory of the property 财产目录
inventory of water resources 水利资源清账
inventory on hand 库存存货
inventory price 存货计价;库存价格
inventory pricing 库存计价
inventory pricing method 库存计价法
inventory problem 库存问题
inventory profit 盘存盈余
inventory rating 投料量
inventory record 盘存记录
inventory relevant cost 库存关联费用
inventory report 库存报告单;存货报告单
inventory reserve 库存准备;盘存准备金
inventory risk 库存风险
inventory scheduling 库存日程计划
inventory sheet 库存盘存报表;库存单;盘存报表
inventory-shipment ratio 库存装运比
inventory shortage 盘存短耗
inventory shortage and over report 库存盘盈盘亏报告
inventory short and over 盘存损盘;盘存缺溢
inventory statistics 库存统计(表)
inventory stock 盘存货(物)
inventory stock rails 周转轨
inventory survey 现况调查;资源调查;工程调查资料;库存盘点;库存调查;盘存;实物调查
inventory system 库存系统;记录信息系统;盘存制度;存储系统
inventory system for dependent demand 依赖性需求存货系统
inventory tag 盘存标签
inventory taking 盘存
inventory theory 库存(理)论;存储论;编目理论
inventory to receivable ratio 存货对应收款项的比率
inventory turnover 库存周转率;存货周转
inventory turnover period 库存周转期;存货周转期
inventory turnover ratio 库存周转率;存货周转率
inventory valuation 库存估价
inventory valuation adjustment 库存估价调整(额)
inventory-valuation reserves 存货估价准备
inventory value 库存价值;库存额
inventress 女发明家
inverarite 镍硫铁矿
invernite 正斑花岗岩;闪动斑状花岗岩
inverse 相反的;逆的;倒转的;倒相的;反相的;反量;反换式;反(的)
inverse amplifier 逆放大器
inverse analysis 反演分析
inverse analysis function 解析反函数
inverse annealing 反退火
inverse arc 逆弧
inverse assembler 逆汇编程序;反汇编程序
inverse association 反向结合
inverse astigmatism 反规性散光
inverse azimuth 反方位角
inverse bandwidth 逆带宽;带宽倒数;反向带宽
inverse blow 逆送风
inverse cam 反凸轮
inverse channel 逆信道;倒置信道;反信道
inverse chart 横轴投影地图;横轴海图
inverse chill 反激冷;反淬火;反白口
inverse circuit 反演电路
inverse circular function 反三角函数
inverse code 逆代码
inverse coefficient matrix 逆系数矩阵
inverse component 逆分量
inverse Compton effect 逆康普顿效应
inverse computation 反演计算;反算
inverse computation of coordinates 坐标反算
inverse condemnation 征地损失申诉
inverse consecant 反余割
inverse converter 逆变换器

inverse convolution 逆褶积;逆卷积
inverse correlation 逆相关;负相关;反相关
inverse correspondence 逆对应;反对应
inverse cosine 反余弦
inverse cotangent 反余切
inverse crosstalk 倒置串话
inverse current 逆流
inverse current relay 反向电流继电器
inverse curve 反曲线
inverse cycle 逆循环
inverse cylindric(al) orthomorphic chart 横轴墨卡托投影(地)图
inverse cylindric(al) orthomorphic projection 横轴圆柱正形投影
inverse cylindric(al) orthomorphic projection chart 横轴圆柱正形投影图
inversed cone inlet of shaft kiln 立窑喇叭口
inverse definite time limit characteristic 逆时定时限特性
inverse demand pattern 相反需求类型
inverse density dependence 反密度相关
inverse derivative action 反导数作用
inverse derivative controller 反导数调节器
inverse derivative unit 反微分器
inverse development in electric(al) prospecting 电法勘探的反演
inversed grafting 反接
inverse diffusion 反扩散
inverse diffusion length 反扩散长度
inverse dip 反下垂
inverse discrete cosine transform 离散余弦反变换
inverse dispersion 逆频散;反波散
inverse distance 距离倒数;反向距离
inverse distribution 逆分布
inversed knife coater 反刀涂装机
inversed knife coating 反刀涂装
inverse drain current 反漏电流
inverse draught 反拔模斜度
inversed repulsion motor 反推斥电动机;反排斥电动机
inverse dryer 逆流干燥器
inverse duplex circuit 逆双工电路
inverse duplex receiver 回路双工接收机
inversed V-connection 反 V 形接法
inverse edge 逆棱
inverse electric(al) characteristic 反向电特性
inverse electrode current 反(向)电极电流
inverse element 反元素
inverse emulsion 逆乳浊液
inverse emulsion polymerization 反相乳液聚合反应
inverse equator 横轴赤道
inverse estimation 逆估计
inverse estuary 逆向河口;逆河口
inverse factorial 反阶乘
inverse factorial series 反阶乘级数
inverse fast Fourier transformation 反向快速傅立叶变换
inverse feed 倒反馈
inverse feedback 负反授;负反馈
inverse feedback filter 倒反馈滤波器
inverse ferrite 逆铁淦氧
inverse field 逆序场;反向场
inverse figure 逆图形
inverse file 反文件
inverse filter 逆滤波器;反(向)滤波器
inverse filtering 逆滤波;反向滤波
inverse flame 反焰;倒焰
inverse flame calciner 倒焰式煅烧窑
inverse flame calcining kiln 倒焰式煅烧窑
inverse flame kiln 倒焰窑
inverse flange 反向法兰
inverse flow 逆流;回流;倒流
inverse form 逆形式
inverse formula 反演公式
inverse Fourier transform(ation) 傅立叶反变换;逆傅立叶变换;傅立叶变换
inverse-frequency noise 反比例频率噪声;反转频率噪声
inverse-frequency spectrum 反向频谱;反转频谱
inverse function 逆函数;反函数
inverse function amplifier 反函数放大器
inverse function element 反函数元素
inverse function generation 反函(数)发生

inverse function generator 反函数发生器
inverse function rule 反函数规则
inverse function theorem 反函数定理
inverse gain jamming 逆增益干扰
inverse gas chromatography 反气相色谱法
inverse gas liquid chromatograph 反相气液色谱仪
inverse Gaussian distribution 反高斯分布
inverse gear method 倒尺法
inverse geometry 反向几何学
inverse graded bedding 逆序粒层
inverse grading 逆级配
inverse grid current 反栅流
inverse Hadamard transform 哈达马德反变换
inverse heating-rate curves 升温速率倒数曲线
inverse hemming 反卷边
inverse homomorphic function 逆同态函数
inverse homomorphism 逆同态
inverse homotopy 逆同伦
inverse hour 倒时数;反时针
inverse hyperbolic function 反双曲(线)函数
inverse image 逆像;互反成比;反像
inverse impedance 逆阻抗
inverse index 逆指标;反指标
inverse indicator 逆指标;反指标
inverse inductance matrix 逆电感矩阵
inverse initiation 反向装药
inverse input 反相输入
inverse integrator 逆积分器;反演积分器
inverse interpolation 逆插值法;反内插(法);反插值法
inverse isotone mapping 逆保序映射
inverse isotopic dilution 逆同位素稀释
inverse isotopic dilution method 逆同位素稀释法
inverse iteration 逆迭代;反迭代;反代
inverse Landau damping 逆朗道阻尼
inverse Laplace's transform(action) 拉普拉斯逆变换;拉普拉斯反变换;逆拉普拉斯变换;反拉普拉斯变换
inverse latitude 横轴投影纬度
inverse lattice 倒易晶格
inverse least squares estimate 反向最小二乘估计
inverse least squares method 反向最小二乘方法
inverse limit 逆向极限
inverse limiter 反限制器
inverse limit space 逆向极限空间
inverse limit spectrum 逆极限谱
inverse linear optimal control 逆线性最优控制
inverse linear regression 逆线性回归
inverse logarithm(of a number) 反对数
inverse logging section after stack 迭后反演测井剖面
inverse longitude 横轴投影经度
inversely proportional 成反比(例)的
inversely proportional quantities 反比例量
inversely proportional relationship 反比例关系
inversely related 逆相关
inversely varying cost 逆向变换成本
inversely well-ordered set 逆良序集
inverse magnetostriction effect 反磁致伸缩效应
inverse map 横轴投影地图
inverse mapping 逆映象;逆映射
inverse mapping system 逆映射系
inverse maser 逆微波激射
inverse maser effect 逆微波激射效应
inverse matching 逆匹配
inverse matrix 矩阵求逆;矩阵反演;逆(矩)阵;倒置矩阵;反行列式;反矩阵
inverse matrix table 逆矩阵系数表
inverse Mercator chart 横切墨卡托投影图;横切渐长投影图
inverse Mercator projection chart 横轴墨卡托投影(地)图
inverse meridian 横轴子午线;横轴投影子午线
inverse metamorphism 逆变形
inverse method 反演法
inverse mixture interference 倒易混频干扰
inverse modeling 反演模拟
inverse multinomial sampling 逆多项式抽样
inverse network 逆网络;倒置网络;逆量网络;倒电网络;反演电路
inverse noise 反向噪声
inverse nuclear reaction 反向核反应
inverse number 逆数;反数
inverse Nyquist array method 逆奈奎斯特阵列法

inverse Nyquist diagram 逆奈奎斯特图
inverse of a function 反函数
inverse of a matrix 矩阵的逆;逆矩阵
inverse of a number 数的逆;倒数
inverse of a square matrix 方阵的逆矩阵
inverse of multiplication 乘法反运算
inverse operation 逆运算;逆变运行;倒置运行;反运算;反向运行
inverse operator 逆算子;逆算符;反算子
inverse parallel 横轴投影纬线;反并联
inverse parallel connection 反向并联连接
inverse peak voltage 反向峰(值电)压;反峰电压
inverse penetration coefficient 反渗透率
inverse period 逆电势期
inverse permeability tensor 反向渗透率张量
inverse permutation 逆置换
inverse perspective 反透射画法
inverse photo-electric(al) effect 光电倒效应;反(向)光电效应
inverse piezo-electric(al) effect 反压电效应
inverse plummet 倒垂线
inverse plummet observation 倒垂观测;倒垂法
inverse point 反演点
inverse pole figure 反极像图;反极点图
inverse position computation 后方交会计算;位置反算
inverse position problem 大地位置反算问题
inverse power correction 逆幂校正
inverse power factor 反功率因数
inverse power fluid 逆幂型流体
inverse power law 逆幂法则
inverse power method 逆幂法;反幂法
inverse probability 逆概率
inverse problem 逆问题;反演问题;反算问题
inverse problem of gravity anomaly 重力异常反演问题
inverse problem of potential theory 位论反解问题
inverse process 逆过程
inverse projection 横轴投影
inverse projection parallel 横轴投影纬线
inverse proportion 反比率;反比例;反比(关系)
inverse proportional effect 反比例效应
inverse proposition 逆命题【数】
inverse pyramid of habitat 栖息地倒金字塔
inverse quenching 反淬火
inverse Raman scattering 反转拉曼散射
inverse Raman spectrometry 反向拉曼光谱法
inverse ranks 反秩
inverse ratio 反比(例)
inverse-ratio curve 反比曲线;反曲线
inverse reaction 逆反应
inverse refraction diagram 逆折射图
inverse regression 逆回归
inverse regular representation 逆正则表示
inverse relation 逆关系;反比(例)关系;逆联系
inverse relationship 相反关系
inverse relation telemeter 反比式遥测装置;反比式遥测器
inverse residue code 反剩余码
inverse resonance 电流谐振;并联谐振
inverse-response calculation 反算
inverse rhumb line 横轴投影等变形线;横轴恒向线
inverse sample census(ing) 颠倒采样调查
inverse sampling 逆抽样;颠倒采样
inverse sandwich growth 反向夹层生长
inverse scattering problem 反散射问题
inverse segregation 逆偏析;反偏析
inverse semilog regression 逆半对数回归
inverse serial correlation 逆序列相关
inverse shift 反方向移动
inverse shoran problem 肖兰反算问题
inverse signal 返回信号
inverse sine 反正弦
inverse sine transformation 反正弦变换
inverse sludge index 污泥回流指数;倒泥指数
inverse solution 反解
inverse solution of conic(al) projection 圆锥投影反求法
inverse solution of cylindric(al) projection 圆柱投影反求法
inverse solution of map projection 地图投影反求法
inverse-speed motor 反速电动机;串励特性电动机
inverse spinel 反(式)尖晶石

inverse spinel type structure 反尖晶石晶型结构
inverse square law 平方反比(例)定律
inverse square matrix 逆方阵
inverse square root 平方根倒数
inverse standardization 反推法标准化
inverse standardized rate 反推法标准率
inverse Stark effect 逆斯塔克效应
inverse stress 逆应力
inverse stretched velocity 反拉伸速度
inverse stripping voltammetric method 反向溶出伏安法
inverse stripping volt-ampere method 反向溶出伏安法
inverse substitution 逆代换;反代换
inverse surface 反曲面
inverse symmetric(al) tensor 反对称张量
inverse symmetry 反(演)对称性
inverse synchrotron absorption 逆同步加速吸收
inverse system 逆向系统;反向系(统)
inverse test 反虑料试验
inverse theorem 逆定理;反定理
inverse time 反时;反比时间
inverse time circuit breaker 逆时断路器
inverse time current protection 反时限电流保护
inverse time definite time limit relay 反时一定时限继电器;逆时定限继电器
inverse time definite-time relay 定时限反时限继电器
inverse time-delay 逆时延迟;成反比的时延;反时延迟
inverse time-delay over-current release 反时限过电流脱扣器
inverse time-element 反时延
inverse time induction relay 感应式反时继电器
inverse time-lag 反时滞;反时限;反时延;反时限延时;反比例时滞
inverse time-lag apparatus 反时滞装置
inverse time-lag protection 反时限保护
inverse time-lag relay 反时限继电器
inverse time-lag relay with definite minimum release 有限反时限继电器
inverse time limit 逆比时限;反时延;反比时限
inverse time limit characteristic 反比时限特性
inverse time limit impedance relay 反比时限阻抗继电器
inverse time limit relay 反比时限继电器
inverse time of series 级数的反演
inverse time(over) current protection 反时限过流保护(装置)
inverse time relay 反比延时继电器;反比时间继电器
inverse transform 逆变换;反变换式;反变换
inverse transformation 逆变换;反变换
inverse transformation kernel 反向变换核
inverse transformation method 反变换法
inverse transformation of axis 轴逆变换
inverse transitive closure 逆传递闭包
inverse trigonometric(al) function 反三角函数;逆三角函数
inverse tsunami problem 地震海啸逆问题
inverse type of fever 颠倒型热
inverse value 倒数;相反值
inverse variation 逆变分
inverse vector 逆向量
inverse vector locus 逆矢量轨迹
inverse video 反图像
inverse voltage 逆电压;反(向)电压
inverse volume 容积的倒数
inverse wavelet estimation 求反子波
inverse wavelet filtering 子波反演滤波
inverse Zeeman effect 逆塞曼效应;倒塞曼效应
inverse zoning 逆向分带
inverse Z-transform 反 Z 变换
inversion 转位;转化;转换;逆增;逆向;逆位;逆变(流);内翻;换流;求逆;颠倒;逆转;倒置物;倒位;反ési像;反映;反演(变换);反量;反变
inversional curve 反演曲线
inversion axis 旋转倒轴;反演轴
inversion base 逆温层底
inversion break-up 逆温破坏
inversion break-up fumigation 逆增烟云消散
inversion bridge 反位桥
inversion calculation 反演计算
inversion carrier 反转载流子
inversion centre 反演中心
inversion channel 倒置沟道
inversion chart 反演图
inversion cloud 逆温云
inversion condition 逆温状况
inversion constant 反演常数
inversion converter 变流机
inversion curve 转化曲线;反转曲线
inversion density 反转密度
inversion dip 反凹陷
inversion doubling 反转分裂
inversion energy 反转能量
inversion formula 反演公式;反絮公式;反推公式
inversion formula for Fourier transform 傅立叶变换式的反变换
inversion frequency 转换频率
inversion fumigation plume 逆温下沉型烟羽
inversion geometry 反演几何学
inversion group 反演群
inversion haze 逆温霾
inversion height 逆温层高度
inversion in valley 河谷中逆温;河谷逆温
inversion Laplacian transform 拉普拉斯逆变换
inversion law 反演律
inversion layer 逆转层;逆温层;倒转层;反型层;反变层
inversion layer forming 逆温层形成
inversion layer of temperature 逆温层
inversion level 反转能级
inversion lid 逆温层顶
inversion limitation 反转极限
inversion log 逆温雾
inversion mechanism 反转机制
inversion method 反演法
inversion mode 倒转方式
inversion of an integral 积分的反演
inversion of bit 比特反转
inversion of curves 曲线的反演
inversion of displacement 剪胀点;位移反向点
inversion of kinematic chain 运动链换向
inversion of large matrix 大矩阵反演
inversion of multiple density interface 多层密度介面反演
inversion of precipitation 雨量逆减;降水逆减
inversion of rainfall 雨量逆增
inversion of relief 地形倒转;地形倒置;反立体
inversion of seawater temperature 海水逆温
inversion of single density interface 单层密度介面反演
inversion of stress 应力转换;应力倒向
inversion of temperature 逆温;温度转换;温度逆化;温度逆增
inversion of the image 图像转换
inversion of transform 反变换
inversion of travel time 走时反演
inversion of zonation 逆转成带现象
inversion operation 逆变运行;倒置运行;反运算;反向运行
inversion point 转换点;转化点;转变点;逆转点;倒转点;反演点;反向点
inversion point criterion 剪胀点准则;回胀点准则;反向点准则
inversion problem 反演问题
inversion profile 反转分布
inversion pump rate 反转泵速率
inversion reduction 逆改正
inversion region 反转区
inversion saturation 反转饱和
inversion set 逆转装置
inversion spectrum 转换光谱;反演谱;反向光谱
inversion symmetry 反演对称性;反向对称
inversion technique 反演技术
inversion temperature 转换温度;转变温度
inversion theorem 反演定理
inversion threshold 反转阈
inversion transformation 逆转换
inversion transition 反演跃迁
inversive property 反演性质
inversor 控制器;反演器;反演器
invert 仰拱;转化;管道仰拱;管道内底;内底;倒拱;翻转
invert a linked list 颠倒一个链接表
invert arch 反窟;仰拱;倒拱
invert block (管道内的)底块;倒拱砌块;凹陷砌块
invert center 翻转中心
invert content 转化物含量
invert corner (瓷砖的)阴条
invert drop 反滴水
invertebrate 无脊椎动物;(的)
invertebrate fauna 无脊椎动物区系
inverted 倒转的;倒立的;反相的
inverted air current 逆风流
inverted amplifier 倒相放大器
inverted anaerobic-anoxic-aerobic activated sludge process 倒置厌氧缺氧好氧活性污泥工艺
inverted anticline 倒转背斜
inverted arc 倒弧
inverted arch 倒发券;底拱;反拱;仰拱;倒拱
inverted arch floor 仰拱底板;反拱底板;倒拱底板
inverted arch form 仰拱模板
inverted arch foundation 仰拱基础;倒拱基础;反拱基础
inverted asphalt emulsion 油包水型乳化沥青;油包水(型)沥青乳液;沥青倒乳液
inverted ballast 倒相镇流器
inverted bandwidth 逆带宽
inverted beam 上翻梁
inverted beam method 倒梁法
inverted biologic microscope 倒置生物显微镜;翻转式生物显微镜
inverted block 倒立卷筒;倒拱砌块
inverted bow and chain girder 鱼形桁架;鱼腹式大梁;倒置弓形大梁
inverted-brush contact 反用电刷触点
inverted bucket type steam trap 倒吊桶式疏水器
inverted cam 从动凸轮
inverted camber 倒路拱;反拱(度)
inverted capacity 吸收容量;吸收能力;回渗能力;回灌能力;倒灌容量
inverted capacity of well 井的吸收容量
inverted carbureter[carburetor] 倒置气化器
inverted catenary 倒悬链曲线
inverted cavetto 倒凹饰;倒拱修圆
inverted-channel shaped beam 反槽形梁(板)
inverted choke 倒滤层;反嵌料
inverted compass 倒置罗经
inverted cone 倒圆锥体
inverted cone bur 倒锥形钻
inverted cone depression 倒锥漏斗
inverted connection 倒置连接
inverted converter 逆换流器;反向变换器;反向变换机
inverted converter inverter 反向变流器
inverted crown 路拱
inverted crucible 倒置的坩埚
inverted cup 反向皮碗
inverted curb 倒置缘石;倒置式路缘石;倒拱路边石
inverted crucible 倒置坩埚
inverted cusp 漩涡底谷;(指漩涡的)底谷;倒三角尖顶;倒峰;反尖角;反尖端;反尖顶
inverted cyclotron 反回旋加速管
inverted cylinder 倒置汽缸
inverted cyma recta 倒正反曲线饰
inverted cyma reversa 倒反曲线饰【建】
inverted die 异向模具;倒置模具
inverted dip 倒倾角
inverted dipper 反向铲
inverted direct act engine 倒置直接发动机
inverted directline 倒相直通行
inverted Dirirklet distribution 逆狄利克雷分布
inverted draft 逆向气流;反向气流
inverted drainage well 吸收排水井;逆向排水井;回渗井;回灌井;反渗排水井;吸水井
inverted draught 反向气流
inverted draw cut 顶槽(采矿);顶部掏槽
inverted echo-sounder 反向回声测深仪;反回声测深仪;倒置回声测深仪
inverted emulsion 逆乳液;倒乳液
inverted emulsion drilling fluid 反相乳化钻井液
inverted emulsion mud 反乳化泥浆
inverted engine 倒置发动机;倒立式发动机;倒缸发动机
inverted escapement 反装式擒纵机构
inverted evapo(u)ration 向下蒸发;反向蒸发
inverted extrusion 反挤压
inverted field pulse 场频倒脉冲;反相场脉冲
inverted file 索引文件;倒置的文件;倒序检测文件;倒排文件;反序文件
inverted file model 倒置文件方式

inverted filler 反滤物料
inverted filter 倒滤层；反向滤池；反滤层；反沪层；反过滤器
inverted filter layer 倒滤层
inverted flank 倒转翼；侧转翼
inverted flat roof 倒置平面；倒置平屋顶
inverted fold 倒转褶皱
inverted frame pulse 帧频倒脉冲
inverted frequency converter 换频机
inverted ga(u)ge 倒式压力计；倒式测压管
inverted gravel layer 反滤层
inverted hour 逆时针；反时针(的)
inverted image 倒影；倒像；倒立像
inverted impulse 倒脉冲
inverted in line engine 倒height直列式发动机
inverted input 反相输入；反输入
inverted J curve 反J曲线
inverted joint 反转接头管件
inverted kerb 倒置路缘石；倒拱路边石
inverted king post beam 倒单柱梁
inverted king post girder 倒单柱大梁
inverted king post truss 倒单柱桁架
inverted ladder digital/analog(ue) converter 颠倒梯形数/模转换器
inverted Lamb dip 倒兰伯凹陷；反兰伯凹陷
inverted-L antenna 倒L形天线
inverted layer 地下水区水构筑物反滤层
inverted left-to-right 左右颠倒的
inverted level 仰拱标高；仰拱高程
inverted lining 逆衬砌法
inverted lining method 先拱后墙法
inverted list 倒引表；倒排表
inverted L-type network 倒L形网络
inverted loop 倒飞筋斗
inverted lotus 垂莲饰
inverted L-type calender 倒L形辊压机
inverted L-type portal crane 半门式起重机(双门柱不等长)
inverted machine 反用电机
inverted magnetron 倒置磁控管；反磁控管
inverted market 反向市场
inverted metallurgic microscope 倒置金相显微镜
inverted microscope 倒置显微镜
inverted model 倒模型
inverted motor 反结构电动机
inverted nine-spot pattern 反九点井网
inverted operation of trains 反向运行【铁】
inverted ophthalmoscope 倒像检眼镜
inverted orbitron 反轨旋管
inverted order 倒序(排列)；反逆次序
inverted osmometer 倒置渗压计
inverted output 反相输出
inverted parabola 倒抛物线
inverted pendulum 悬挂摆；倒立摆；倒锤(线)；倒摆
inverted pendulum seismograph 倒置地震仪
inverted pendulum structure 倒摆式结构
inverted penetration 倒提入施工法(一种先浇沥青后撒石料的路面施工方法)；倒铺路面法；倒贯入(法)
inverted penetration macadam 倒贯入式碎石路
inverted penetration macadam pavement 倒贯入式碎石路面
inverted penetration pavement 倒贯入式路面
inverted penetration surface treatment 反向渗透法表面处理
inverted plasticity 搅胀性；反塑性
inverted plumb line 倒垂线
inverted plunge 倒转倾伏
inverted population 反转粒子数；反转分布
inverted population density 反转粒子数密度
inverted position 倒转位置；反位
inverted position telescope 倒镜
inverted pulse 反极性脉冲
inverted pyramid 倒方锥体
inverted pyramid antenna 漏斗形天线；倒角锥天线
inverted queen-post beam 倒双柱梁
inverted queen-post girder 倒双柱大梁
inverted queen-post truss 倒双柱桁架
inverted ram press 反压式压力机；下压式压机
inverted rectifier 逆整流器(直流变交流)；反用整流器；反用换流器
inverted relief 倒地形；倒置地形
inverted repulsion motor 反用推斥电动机；反常推斥式电动机

inverted river 逆流河；倒淌河；倒流河
inverted roller leveller 反压辊式矫直机
inverted roof 倒置(式)屋顶；倒置(式)屋面；倒铺屋面；凹面屋顶
inverted roof membrane assembly 倒置屋面膜装配体
inverted rotary converter 旋转式逆变机；反向旋转式变流器；反向旋转式变流器
inverted rotary torque converter 反转变矩器
inverted running 反向运行
inverted sand drain filter 砂质排水反滤层
inverted sequence 逆序；反序
inverted signal 反信号
inverted siphon[syphon] 倒虹吸(管)；倒虹管；反向虹吸管；倾虹吸
inverted siphon culvert 倒虹吸桶洞；倒虹吸涵洞
inverted siphon pipe 倒虹吸管
inverted siphon sewer 倒虹吸污水管；倒虹吸下水道
inverted slips 反向扳牙
inverted slope 倒坡
inverted spectral term 倒光谱项
inverted spider 反向卡瓦
inverted spin 反螺旋
inverted steam 倒流河
inverted steam trap 倒流隔汽具
inverted Stepanov technique 下导模法
inverted stereo 反立体
inverted stereoscope 反立体镜
inverted stratum 倒转岩层；倒转地层；倒层
inverted stream 逆流河；倒淌河
inverted synchronous converter 反向同步变流器
inverted syncline 倒转向斜
inverted takeover 反向接管企业
inverted talon 反爪饰
inverted T beam 倒T形梁
inverted tee section 倒丁字形断面
inverted telephoto lens 反远摄镜头
inverted telescope 倒像望远镜；倒镜；长望远镜
inverted T-formed budding 倒丁字形芽接
inverted throat 倒凹弧饰
inverted tide 逆潮(流)；倒潮
inverted tooth chain 反齿链
inverted transport 逆向输送
inverted trap 倒疏液器
inverted triangular truss 倒三角形桁架
inverted T section 倒T形断面型材；倒丁字形断面
inverted T-section precast floor slab 倒T形预制楼板
inverted T-shaped footing 倒T形底脚
inverted T-slot 倒T形槽
inverted T type retaining wall 倒T形挡土墙
inverted tube 倒像管；倒虹管
inverted-turn transposition 线圈端部换位
inverted umbrella-shaped aeration machine 倒伞形曝气机
inverted unconformity 倒转不整合
inverted U-shaped precast concrete unit 倒U形预制混凝土单元
inverted U yoke 倒U形卡箍；倒U形轭铁
inverted valve 止回阀；逆止阀
inverted V antenna 倒V形天线
inverted vault 倒拱顶；倒拱形顶
inverted V brace 叉手
inverted-V corbel bracket 人字拱(唐朝的一种斗拱)
inverted V distribution 反V形分布
inverted vee guidance 人字导
inverted vee(slide) way 人字形导轨
inverted vee slip 人字形滑槽
inverted vertical engine 倒立式发动机
inverted V-lag pattern 反V形滞后模式
inverted V-shaped brace 由戗；叉手
inverted weir 倒堰
inverted weld(ing) 仰焊(接)；仰拱焊接；凹面焊(缝)
inverted well 吸水井；吸收井；灌水井；回灌井；倒(流)井；反渗井
inverted Y tower 倒Y形索塔
invert elevation 地下管道内底高程；负高程；管道内径底点高程；仰拱高程；堰顶高程；闸槛高程；闸板高程；管(内)底标高(程)；管道内底高程；渠槽底坡(管道内底)高程；底拱高程；底板高程
invert emulsion 转化乳剂
invert emulsion mud 反相乳化泥浆
invertendo 反比定理

inverter 逆变器；电流换向器；倒相器；反用换流器；反演器；反向换流器；反向变流机；变流器；变流电路
inverter amplifier 放大变流器
inverter buffer 倒相缓冲器；反相(器)缓冲器
inverter circuit 非电路
inverter for power supply for emergency ventilation 强制通风电源转换器
inverter loop 反相回路
inverter matrix 反相器矩阵；反相器矩阵
inverter open collector 收集极开路的反相器
inverter oscillator 倒相振荡器
inverter protection unit 逆变器保护单元
inverter stage 倒相级；倒相极
inverter transformer 逆变器用变压器；倒相变压器；反流变压器
inverter transistor 倒相晶体管
inverter tube 倒相管；变换管
inverter unit 逆变单元
invert form 仰拱模板
invert gate 反相门
invert grade 管道内底坡度
invert hull-building 反模造船法
invertibility 可逆性
invertible 可逆的
invertible function 可逆函数
invertible law 可逆律
invertible matrix 可逆矩阵
inverting action 翻作用
inverting amplifier 倒相放大器；反相放大器
inverting ballast 电流变换器
inverting buffer 反相缓冲器
inverting converter 倒频变换器；反相变换器
inverting element 变换元件
inverting eyepiece 倒像目镜
inverting function 倒相作用
inverting input 倒向输入；倒相输入
inverting lens 倒像透镜
inverting matrix 逆矩阵
inverting parametric device 倒相参量器件；反向参量器件
inverting prism 转像棱镜；倒相棱镜
inverting stratified development 逆梯度发展
inverting system 倒像系统
inverting telescope 倒像望远镜
inverting terminal 倒相输入端
invert instruction 倒数指令
invert level 管道内底高程；(管道的)内底高；倒转水平；倒底高程
invert-lift 底板隆起
invert masonry block 倒拱砌块
invert of lock chamber 闸室底部
invert of manhole 人孔内部；人孔内底
invert on zero 零反相；零翻转
invertor 逆变器；电流换向器
invert pendulum 逆立摆
invert range finder 倒影式测距仪；倒像(式)测距仪
invert Schmitt trigger 反相施密特触发电路
invert siphon[syphon] 倒虹吸
invert soap 逆化皂
invert strut 反装支撑
in vessels subjected to external pressure 在承受外压的容器
invest … with power of attorney 授予委托书
invested assets 投入资产
invested capital 基本工程投资；原有资本；投入资本
invested firm 投资公司；投入资本的商号
invested funds 投入资金
invested in plant 投入产业
invested mo(u)ld 熔模铸型
invested with full authority 授予全权
investee 接受投资者
investigate 调查研究
investigated length 调查长度
investigating method of following the trend 走向追索法
investigation 勘察；勘查；研究；考察；调查(研究)；探讨；踏勘调查；审查
investigation and study 调查研究
investigation area 勘察区；调查区
investigation area of basic geology 基础地质研究区
investigation boat 调查船
investigation bore 勘察孔
investigation by geologic(al) observation point

地质点调查
investigation cost 勘察费用;调查研究费用
investigation degree of economic conditions 经济条件研究程度
investigation degree of engineering-geologic(al) conditions of ore deposit 矿床工程地质条件研究程度
investigation degree of geology of ore deposit 矿床地质研究程度
investigation degree of hydrogeologic(al) conditions of ore deposit 矿床水文地质条件研究程度
investigation degree of mining-technical conditions 开采技术条件研究程度
investigation degree of technologic(al) conditions 加工技术条件研究程度
investigation diving 调查潜水
investigation drift 调查导坑
investigation during construction 施工勘察
investigation from correspondent method 通信观察法
investigation groundwater hazard 地下水有害作用调查
investigation group 考察团;调查团
investigation hole 探洞;探孔;勘探钻孔
investigation material 考察资料
investigation method 勘测方法;调查方法
investigation method for stratified population 分层群体调查法
investigation methodical system in railway statistics 铁路统计调查方法体系
investigation of abandoned mine 废窑调查
investigation of accuracy 精度研究
investigation of bridge 桥址勘察
investigation of building material 建材调查
investigation of contaminative sources 污染源调查
investigation of economic condition 经济条件研究
investigation of engineering-geologic(al) conditions of ore deposit 矿床工程地质条件研究
investigation of festival days and holidays 节假日调查
investigation of foundation 地基察勘;地基查勘
investigation of foundation condition 地基土壤调查
investigation of geobotany 地植物调查
investigation of geologic(al) observation point 观测点研究
investigation of geologic(al) route 地质路线研究
investigation of geologic(al) structure 地质构造调查
investigation of geology of ore deposit 矿床地质研究
investigation of groundwater point 地下水点调查
investigation of hydrogeologic(al) conditions of ore deposit 矿床水文地质条件研究
investigation of karst cave 溶洞调查
investigation of linear road 线路勘察
investigation of mine 矿井调查
investigation of mining-technical conditions 开采技术条件研究
investigation of new tectonic movement 新构造运动调查
investigation of physico-geologic(al) phenomena 物理地质现象调查
investigation of productive mine 生产矿井调查
investigation of railway transportation economics 铁路运输经济调查
investigation of slop deformation 边坡变形调查
investigation of small coal mine 小窑调查
investigation of station site 站址勘察
investigation of subway 地下铁道勘察
investigation of surface water body 地表水体调查
investigation of technologic(al) condition 加工技术条件研究
investigation of tunnel 隧道调查
investigation of underground garage 地下车库勘察
investigation of underground oil store 地下油库勘察
investigation of underground powerhouse 地下厂房勘察
investigation of underground reservoir 地下水库勘察
investigation of underground river 暗河调查

investigation of underground shop 地下商店勘察
investigation of underground storehouse 地下仓库勘察
investigation of underground theatre 地下剧场勘察
investigation of water quality 水质调查
investigation of well 水井调查
investigation on mineral deposits 矿产地质调查
investigation phase 勘察阶段;调查研究阶段
investigation pit 检查坑;探坑
investigation planning 调查工作设计
investigation radiation level 调查辐射水平
investigation radius 调查半径
investigation record of spring 泉水调查记录
investigation report 调查报告
investigation shaft 探井;勘探井
investigation stage 调查阶段
investigation team in railway statistics 铁路统计调查队
investigation test 研究性试验
investigation work 已进行过的调查工作
investigative diving 调查潜水
investigative test 探索性试验
investigator 调查者
investigatory reflex 探究反射
invest in an enterprise 投资于企业
investing 熔模铸造
investing formula 投资公式
investing responsibility system 投资包干责任制
investing temperature 结壳温度
investing time 结壳时间
invest in stock 投资于股票
investiture 装饰;覆盖物
invest material 覆盖材料
investment 围模料;投资;包埋法
investment ability 投资能力
investment abroad 国外业务会计;国外投资;对外投资(额)
investment accelerator 投资加速因素
investment account 投资账
investment act 投资法
investment activity 投资活动
investment adviser 投资顾问
investment adviser act 投资顾问法
investment agency 投资机构
investment allocated by senior administrative agency 上级拨入投资
investment allocation 投资分摊
investment allowance 投资补贴;投资减税;投资抵减
investment along proper lines 投资方向
investment analysi 投资分析
investment-analysis worksheet 投资分析工作表
investment analyst 投资分析师
investment and effect rate 投资效益率
investment and enterprise corporation 投资企业公司
investment appetite 投资喜好
investment appraisal 投资评价;投资估价
investment association 投资协会
investment bank 投资银行
Investment Bank of Sweden 瑞典投资银行
investment base 投资基准;投资基础
investment behavio(u)r 投资动态
investment benefit 投资效益
investment bill 投资汇票
investment bin 翻斗
investment bond 投资性债券
investment boom 投资景气;投资高涨;投资繁荣
investment broker 投资经纪人
investment budget 投资预算
investment budget constraint 投资预算约束
investment burn out furnace 燃烧脱蜡炉
investment buying 投资购买
investment capital 投资资本
investment casting 蜡模铸造法;精密铸造;熔模铸造(法);失蜡造形(铸造)
investment casting method 蜡型精密铸造法;熔模铸造法
investment casting wax 熔模铸造用蜡
investment center 投资中心
investment certificate 投资证书;投资证明;出资证明书
investment certification 出资证明书

investment climate 投资气候;投资环境
investment coefficient 投资系数
investment commitment 投资承诺
investment company 投资公司
Investment Company Act 投资公司法案(美国)
investment company shares 投资公司股票
investment compound 蜡模铸造用耐火材料;熔模铸造涂料
investment consultancy corporation 投资咨询公司
investment consultancy service center 投资咨询服务中心
investment consultant corporation 投资咨询公司
investment consumption cost 投资消耗费
investment contract 投资合同
investment contractural plan 投资契约计划
investment controlling of construction 工程建设投资控制
investment corporation 投资公司
investment cost 投资额;投资费(用);投资成本
investment counselor 投资顾问
investment credit 投资信用(长期信用);投资信托;投资信贷;投资减税
investment credit rating 投资信用等级
investment criterion 投资准则
investment currency 投资货币
investment cycle 投资周期
investment decision 投资决策
investment decision of firms 厂商投资判断
investment demand 投资需求
investment demand schedule 投资需求表
investment direction 投资方向
investment discount 投资折价
investment dislocation 投资失当
investment diversification 投资分散;分散投资
investment dollars 投资额
investment-earnings ratio 投资收益比
investment effect 投资效果
investment effectiveness 投资效果
investment environment 投资环境
investment error 投资误差
investment estimate 投资估算;投资概算
investment estimating 投资估算
investment estimation 投资估算;投资估计
investment estimate 投资估计
investment evaluation 投资评价
investment expenditures 投资支出;投资费(用)
investment expert 投资专家
investment failure 投资失败
investment feature 投资特性
investment financed 筹款进行的投资
investment financing 投资资金筹措
investment firm 投资公司
investment fits the needs of funds 投资需求配套
investment follow-up 投资后续行动
investment for non-productive purpose 非生产目的的投资
investment function 投资函数
investment funds 投资基金
investment funds needed 需要经费
investment goods 投资设备;投资财
investment grant 投资补助金;投资补贴;投资拨款
investment growth rate 投资增长速度
investment guarantee 投资保证
investment guaranty program(me) 投资保证方案
investment house 投资公司
investment impairment allowance 备抵投资亏折
investment in additional current assets 追加流动资产投资
investment in capital construction 基建投资;基本建设投资
investment in capital construction projects for commercial and productive purposes 生产经营性基建投资
investment incentive 投资优待;投资奖励;投资刺激
investment incentive act 奖励投资条例
investment income 投资收益;投资收入
investment income ratio 投资收入比率
investment income surcharge 投资收入附加税
investment index number 投资指数
investment inflation 投资膨胀
investment information bank 投资信息库
investment in general fixed assets 一般固定资产投资
investment in housing 住房投资;房产投资

investment in human resources 人力投资
investment in kind 实物投资
investment in land 土地投资
investment in money 现金投资
investment in plant and machinery 厂房及机械设备投资
investment in property 房地产投资
investment in public sector 公共部门投资
investment institution 投资机构
investment in stock 股票投资
investment in subsidiary 对子公司的投资
investment insurance 投资保险
investment intention 投资意向
investment interest 投资权益
investment in the key areas designated by the state 国家确定的重点领域投资
investment lag 投资时滞;投资的时间延滞
investment law 投资法
investment ledger 投资分类账
investment level movement 投资水平变动
investment limitation 资本限额
investment loan 投资贷款
investment loss 投资损失
investment lump-sum contracting 投资包干
investment makeup 投资结构
investment management 投资管理
investment manager 投资经营者
investment manual 投资手册
investment market 投资市场
investment matrix 投资矩阵
investment medium 投资媒介
investment middleman 投资中间人
investment model 投资模型
investment mo(u)ld 失蜡铸型
investment mo(u)lding 熔模铸造
investment mo(u)lding pattern 熔模
investment multiplier 投资增益乘数;投资乘数
investment multiplier effect 投资乘数作用
investment-net income ratio 投资纯收入率
investment norm 投资限额
investment objective 投资目标
investment of civil work 土建工程投资
investment of engineering 工程投资
investment of fixed assets 固定资产投资
investment of pile plant 反应堆装置投资
investment of reserves 准备金投资
investment of sinking fund 偿债基金投资
investment of water resources developing engineering 水资源开发工程投资
investment on securities 证券投资
investment opportunity 投资机会
investment opportunity schedule 投资机会表
investment orientation 投资方向
investment oriented project 以投资为目标的项目
investment outlay 投资支出
investment outlay coefficient 投资支出函数
investment outside the budget 预算外投资
investment outside the plan 计划外投资
investment pattern 熔模
investment per capita 平均每人投资
investment performance requirements 投资助销
investment per unit product 单位(产品)投资额
investment per unit recharge water volume 单位回灌水量的投资
investment per unit water volume 单位水量的投资
investment phase of a project 项目投资(时)期
investment plan 投资计划
investment plan of real assets 固定资产投资计划
investment policy 投资政策
investment political risks insurance 投资保险
investment pool 联合投资组织
investment portfolio 投资组合;投资证券组合;投资业务
investment portfolio management 投资组合管理
investment portfolio risk 投资组合风险
investment potential 投资潜力
investment precoat 面层涂料
investment premium 投资溢价;投资升水
investment priority 投资重点
investment program(me) 投资计划
investment programming 投资规划
investment project 投资项目;投资计划;投资方案
investment promotion 促进投资
investment propensity 投资倾向

investment property 投资产业
investment proposal 投资方案
investment protection agreement 投资保护协议;投资保护协定
investment rate 投资率
investment rate for unproductive construction 非生产性建设投资率
investment rating 投资评级;投资分级
investment recovery 投资回收;投资还本
investment recovery period 投资回收期
investment research 投资研究
investment reserve 投资准备金
investment resources 投资资源
investment result 投资效益
investment return 投资利润;投资效益
investment revenue 投资收入
investment risk 投资风险
investment risk analysis 投资风险分析
investments belong to current assets 可归属流动资产类的投资
investment schedule 投资表
investments classifiable as current assets 可归属流动资产类的投资;投资归属流动资产
investment securities 投资证券
investment securities traded 投资证券买卖
investment security 投资性证券
investment service 投资服务
investment shares 投资股份
investment spending 投资支出
investment statistics 投资统计
investment stocks 投资股份
investment strategy 投资策略
investment structure 投资结构
investment support activity 投资支持活动
investment surcharge 投资收入附加税
investment tax 投资税
investment tax credit 投资税收抵免;投资(宽)减税额;投资赋税优惠;投资的减税率
investment trust 投资信托;投资托拉斯(即合股投资公司);投资公司
investment trust company 投资信托公司
investment trust on securities 证券投资信托
investment turnover 投资周转率;投资周转额
investment turnover rate 投资周转率
investment value 投资价值
investment value model 投资价值模型
investment willingness 投资意愿
investment within the plan 计划内投资
investment worth 投资价值
investment worthy of compensation 应得报酬得投资
investment yield 投资所得;投资收益(值);投资收入率
investor 客商;投资者
investor from a parent company 母公司投资人
investor-owned utility 归投资者所有的公共事业
investor project 投资者进行的项目
investor's demand 投资的需求
investor's method 投资者方法
investor-sponsor 保证人;倡议人;投资者
investor-sponsored cooperation 投资者支持的协作
investor's yield 投资者的利益
investor tax credit 投资减税额
invest shell casting method 熔模壳型铸造法
invest... with power of attorney 发给委任书
in view of the fact 考虑到实际
invigilator 监视器
invigorated river 多水河流;丰水河
invigorated stream 丰水河
invigorating effect 强化作用
invigorating premodial energy 补元气
invigorating the economy 搞活经济
invigorating water 微冷水
invincible 屋顶天窗玻璃格条
inviscid 无韧性的;非黏性的
inviscid flow 理想流;无黏性流;无黏流动;非黏滞流;非黏性流
inviscid fluid 无黏性流体;非黏(性)流体;不黏滞(性)流体
inviscid stability criterion 非黏性稳定性准则
invisibility 不可见性
invisible 不可见的
invisible assets 账外资产;无形资产
invisible axis 隐轴;暗轴

invisible balance 无形差额
invisible capital 无形资本
invisible chromatogram 不可见色谱图
invisible damage 不可见伤害
invisible dehydration 不显性失水
invisible earnings 无形收益
invisible export 无形出口;非商品输出
invisible exports and imports 无形出口和进口
invisible fat 不可见脂
invisible flame operation 暗火操作
invisible frame 无形框架
invisible gain 无形利益
invisible glass 隐形玻璃;无反光玻璃
invisible glazing 无反光商店橱窗玻璃
invisible green 暗绿色;浓绿;墨绿色;深绿色
invisible heat 光谱的红外部分;不可见热
invisible hinge 暗铰链;暗合页;隐蔽铰链
invisible image 不可见像
invisible import 无形进口;非商品输入
invisible injury 不可见损伤
invisible item 无形收支项目;无形项目(用于国际收支表)
invisible light 不可见光
invisible light filter 不可见光滤光镜
invisible line 虚线;隐线
invisible loss 蒸发损失;蒸发损耗;无形损耗;无形的损失
invisible matter 不可见物质
invisible mending 精密织补;织补
invisible radiation 不可视辐射
invisible ray 不可见射线;不可见光线;不可见辐射
invisible return flow 回归水流
invisible shrinkage 库存无形短缺
invisible spectrum 不可见光谱
invisible stitch 暗缝法
invisible stock 无形股份
invisible supply 无形供给;不可见供应
invisible transaction 无形交易
invisible waste 无形损失
invisible writing ink 隐显墨水
invited bidders 被邀投标人;被邀投标商;被邀投标者
invitation 邀请
invitation delay 邀请延迟
invitation for bid 招标;报价邀请;招标通告
invitation for bidding 招标文件;招标书;投标邀请书;招标通告
invitation for offer 要约引证;邀请开价;邀请发盘;邀请出价;邀请报价;招标
invitation for subscription 邀请捐款;邀请参与认购;邀请参与捐款;认购请帖
invitation for tender 招标单;招标
invitation for tendering 招标文件;招标通告;招标书;投标邀请书
invitation issuing 发标
invitation list 邀请表
invitation of offer 报价邀请
invitation to bid 邀请投标(书);议标;招标文件;招标通告;招标书;投标邀请书;报价邀请
invitation to bid for a project 工程招标
invitation to send 发出邀请
invitation to subscribe 邀请认购
invitation to tender 邀请投标(书);招标文件;招标通告;招标书;报价邀请
invitation to tender for a project 工程招标
invitation to treat 邀请投标人做交易;邀请人谈判;邀请投标人商议;邀请投标人洽谈
invite 聘请
invite applications for a job 招聘
invite bids 征求承包;招标
invited bidder 受邀投标者;被邀投标人;特邀投标者
invited tendering 有限招标
invite for tender 招标
invite public bidding 公开招标
invite subscription for a loan 邀请参与贷款;申请批准放款
invite tenders 征求承包;招标
invite to tender 招标
inviting bid 邀请投标(书)
inviting bids audit 招标审计;投标审计
inviting bids of project construction 工程施工招标
inviting bids of project development 项目开发招标
inviting bids of special engineering contracting

专项工程承包招标
inviting bids of whole procedure 全过程招标
inviting intellectuals from overseas 智力引进
inviting tender 公开招标
inviting tenders for building and outfitting project 建筑安装工程招标
in vitro 在试管内;试管区
invocation 启用;调用
invocation privacy lock 秘级调用锁
invocation of procedure 引用过程;过程调用
invoice 货价票;货价单;配货单;发票;发货单
invoice adjustment 发票(的)调整
invoice amount 发票(金)额
invoice approval 发票的核准
invoice book 发货单存根;进货发票;发票存根;发票簿
invoice-book outwards 销货发票簿;出货簿
invoice clerk 发票管理员
invoice copying book 发票存根簿
invoice cost 发票价;发票成本
invoice cost and charges 发票价值和费用
invoice date 发票日期
invoice discount 发票折扣
invoice duplicate 发票副本
invoice for purchase 进货发票
invoice for sales 销货发票
invoice for value added tax 增值税发票
invoice guard book 发票防护装置簿
invoice method 发票法
invoice number 发票号
invoice of transfer 拨付凭单
invoice of withdrawals 领料单
invoice outward 售出发票
invoice outward book 销货发票簿
invoice price 发票价格;发票价
invoice register 发票登记簿
invoice requisition 请购单
invoice specification 发票明细单;发票明细表
invoice stub 发票存根
invoice value 发票价值;发票价
invoice value plus freight 发票价加运费
invoice weight 毛重;发票注明的重量;发票(所开)重量;发货清单重量
invoice with documents 附提货单的发票
invoice with documents attached 附有证书的发货单
invoicing 开出发票
invoicing currency 发票通货
invoke 调用
invoked block 被调(用)分程序
invoked procedure 调用过程;调用程序;被调用过程
invoking block 调用分程序;调用(程序)块
involatile sample 不挥发性试样
involatile substance 不挥发性物质
involuntary 非出于自愿的
involuntary alienation 财产强制转让
involuntary bankruptcy 强制性破产清理;被动破产
involuntary conversion 因征用引起房地产向货币的强制转换
involuntary conversion of plant and equipment 厂房和设备的强制更换
involuntary conveyance 非自愿房地产所有权转换(如征用、离婚等)
involuntary insolvency 强制破产
involuntary interrupt 意外中断;偶然中断
involuntary inventory 非自愿库存
involuntary investment 非自愿投资
involuntary lien 强行留置权
involuntary loan 非故意的损失;非自愿借款
involuntary loss or gain 非人力所能影响的损益
involuntary movement 不自主运动
involuntary muscular fiber 无横纹肌纤维
involuntary relocation 非自愿移民;非自愿搬迁
involuntary resettlement 非自愿移民
involuntary risk 非自发危险
involuntary stop 强迫停止
involuntary transfer 强制性转让
involute 卷起;渐伸(开)线;内旋的;内卷的
involute and helix measuring instrument 齿轮渐开线及螺旋角测量仪
involute compasses 渐开线规
involute contact ratio 渐开线重叠系数;渐开线接触比
involute cutter 渐开线铣刀
involute cyclone 螺旋式旋流分离器

involute equalizer 螺旋式均压线
involute flank 渐开线齿面
involute gear 渐开线齿轮
involute gear cutter 渐开线齿轮铁刀
involute gear hob 渐开线齿轮滚刀
involute heart cam 渐开线形凸轮
involute inlet 内旋式入口
involute motion 渐开线运动
involute profile 渐开线齿形
involute profile tester 齿轮渐开线仪
involute pump 渐开线(齿轮)泵;螺旋泵
involute rack 渐开线齿条
involute reticle 渐开线形调制盘
involute reticle blade 内旋调制盘叶片
involute spline 渐开线花键
involute spline broach 渐开线花键拉刀
involute spline hob 渐开线花键滚刀
involute tooth 渐开线齿
involute toothing 渐开线齿轮啮合
involute worm 渐开线蜗杆
involution 卷入作用;内转;内卷;乘方;包卷作用
involution form 退化型
involution of high order 高阶对合
involution period 回旋期
involutorial anti-automorphism 对合反自同构
involutory collineation 对合直射
involutory correlation 对合对射
involutory matrix 对合矩阵
involutory transformation 对合变换
involve 自乘;卷入
involved area 受影响的区域
involved in debt 债台高筑
involved strata in fold 褶皱卷入地层
inwale 内护舷纵材;舭板侧舷缘板
inwall 内衬墙;内衬;舭板侧舷缘板
inwall brick 炉衬耐火砖
inward 向内的;进口的;内部;内壁
inward and outward 进入和驶出口;进出港
inward batter of rail 钢轨内倾角
inward bound 向内行驶;来港;进站的;进港;返港
inward-bound light 内向光
inward cant of rail 钢轨内倾角
inward cargo 进口货(物)
inward cash remittance 现金汇入
inward cash transfer 现金转入
inward chamfered window 内向削边窗
inward chance variation 内向机会变差
inward charge 入港费
inward clearance 到港报关
inward clearance certificate 到港报关证书
inward clearing bill 进口检查证;海关检验单
inward collection 进口托收;进口代收;内向代收
inward correspondence 内向对应
inward curve 内弯
inward documentary bills 进口跟单汇票;进口押汇
inward electric(al) current 内向电流
inward flange 内向折边;内凸缘;内边缘
inward flow 内向(水)流
inward flow radial turbine 向心式透平
inward flow turbine 向心式涡轮;辐流式涡轮机;辐流式水轮机;内流式涡轮机;内流式透平;内流式水轮机
inward flux 进入通量;输入通量
inward freight 进口运费
inward freight and cartage 运入运费;进口水陆运费
inward leakage 内向漏失
inward-looking development policy 内向开发政策
inwardly 内里
inwardly illuminated license plate 内部照明的牌照板
inwardly opened casement 内开扇;内开(式)窗扇
inwardly opened door 内开门
inwardly opened window 内开(式)窗
inwardly projecting orifice 内部喷嘴全收缩注孔;内部喷嘴
inward manifest 进口货单;进口舱单;进口报关清单
inwardness 内质
inward normal 内向法线
inward opening 向内开(窗);内开
inward opening door 内开门
inward opening window 内开内窗
inward osmosis 内向渗流
inward-outward cargo tally sheet 进出口理货计

数单
inward-outward dialing system 内外自动电话系统;内外拨号制
inward passage 进口航道
inward pilotage 进港领航费
inward port charges 入港费
inward pressure 内向压力
inward reinsurance 分入再保险
inward remittance 汇入款项
inward running 内向运转
inwards 进口货
inward seepage 向内渗漏
inward slope 内坡
inward substantial variation 内向实质变差
in ward-swinging (门窗的)向内平开窗
inward tipping 向内倾斜的
inward-toll board 长途接收台
inward trade flow 进口贸易流量
inward transit 内运过境
inward turning 内旋式;内向旋转
inward variation 内向变差
inward voyage 返航航程
inward window 内部窗
inwash 巨厚冲积层;冰川边缘沉积;岸边淤积物;岸边淤积
inwashed sediment 泥沙淤积物;泥沙冲积层;冲入泥沙
in-water maintenance 不进坞检修
in-water maintenance facility 不进坞检修设备
in-water-stressed conditions 在缺水情况下
in-water survey 水下检验;不进坞验船
in welded joints 在焊缝上
inwelling 海水倒灌
in white and blue 唐太宗青花瓶(瓷器名)
in-wintering 舍内越冬
Inwood factor 英伍德系数(用于计算年金的系数)
in work 在业
inwrought 豪华装饰的;紧密结合的
inyoite 板硼钙石
iobenzamic acid 碘苯扎酸
iocetamic acid 碘酸胺酸
iochroite 电气石
Ioco 橡胶无缝地板
io collector 离子收集器
iodargyrite 碘银矿
iodate 碘酸盐
iodate ion 碘酸盐离子
Iodate oxygen demand 碘酸盐耗氧量
iodatimetric titration 碘酸盐滴定
iodatimetry 碘酸盐滴定;碘酸盐滴淀法
iodating agent 碘化剂
iodazide 叠氮化碘
iodcarnallite 碘光卤石
iodchromate 碘铬钙石
iodethylene 二碘仿
iodetryl 碘硬酯
iodic acid 正碘酸;碘酸
iodide 碘化物
iodide flux 碘溶剂
iodide goiter 高碘性甲状腺肿
iodide of potassium 碘化钾
Iodide of sodium 碘化钠
iodide process 碘化物法
iodide process metal 碘化物法生产的金属
iodimetric 碘滴定的
iodimetric analysis 定碘量分析
iodimetric method 碘量滴定法
iodimetric titration 碘氧化滴定
iodimetry 碘氧化滴定;碘量法
iodination 碘化作用
iodine air monitor 碘气监测器
iodine bisulfide 碘化硫
iodine chloride test 氯化碘试验
iodine compound 碘化物
iodine consumed 碘消耗量
iodine consumption 耗碘量
iodine-containing substance 含碘物质
iodine coulomb meter 碘电量计
iodine cyanide 氰化碘
iodine deficiency 缺碘;碘缺乏
iodine deposit 碘矿床
iodine disulfide 碘化硫
iodine flask 碘瓶
iodine flask with ground-in glass stopper 带磨

砂玻塞的碘瓶
iodine glycerin 碘甘油
iodine green 碘绿
iodine ion 碘离子
iodine lamp 碘(素)灯
iodine monochloride 一氯化碘
iodine monochloride test 氯化碘试验
iodine number 碘值
iodine number and saponification number factor 碘值与皂化值因数
iodine number calculating 碘值计算
iodine number flask 碘值烧瓶
iodine oxide 氧化碘
iodine-rich product 富碘产物
iodine soap 碘皂
iodine solubilization method 碘加溶法
iodine sorption value 碘吸附值
iodine-starch colo(u)rimetry 碘淀粉比色法
iodine test 碘试验
iodine tincture 碘酒
iodine-tungsten arc lamp 碘钨弧灯
iodine-tungsten lamp 碘钨灯
iodine value 碘值
iodine water 碘水
iodine yellow 碘黄粉
iodipin 碘油
iodism 碘中毒
iodized activated charcoal 含碘活性炭
iodized oil 碘油;碘化油
iodized paper 碘纸
iodized poppy-seed oil 碘化罂粟油
iodized salt 碘化盐;碘化食盐
iodized starch 碘化淀粉
iodoacetamide 碘乙酰胺
iodoacetic acid 碘乙酸
iodoacetone 碘丙酮
iodoalkane 碘代烷
iodobenzene 碘代苯
iodobromite 卤银矿
iodobutane 碘丁烷
iodochlorhydroxy-quinoline 碘氯羟基喹啉
iodo-chlorobromide paper 碘氯溴化银相纸
iodochlorol 氯碘油剂
iodo compounds 碘化合物
iodocyanin 花青染料
iodododecane 癸基碘
iodoeosin test 碘曙红试验
iodoethane 乙基碘;碘乙烷
iodoethanol 碘乙醇
iodoform 碘仿
iodoform gauze 碘仿纱布
iodogorgoic acid 碘珊氨酸
iodohydrocarbon 碘代烃
iodomethane 碘代甲烷
iodometric chlorine test 滴定碘法氯试验
iodometric scale method 碘比色法
iodometric titration 碘滴定
iodometry 碘滴定
iodonaphthalene 碘代萘
iodonitrotetrazolium 碘硝基四唑
iodooctadecane 碘代十八烷
iodopentane 碘戊烷
iodophor 碘消灵;碘载体
iodophthalein sodium 碘酞钠
iodopropane 碘丙烷
iodopropylene 碘丙烯
iodopropylidene glycerol 碘丙甘油
iodoquinoline 碘代喹啉
iodoxamic acid 碘沙酸
iofendylate 碘芬酯
ioglycamic acid 碘甘卡酸
iolite 堇青石
ion 离子
ion accelerating voltage 离子加速电压
ion adsorb 离子吸附
ion adsorption 离子吸附
ion altimeter 电离层高度表
ion antigenism 离子拮抗作用
ionarc process 等离子弧方法
ion balance 离子平衡
ion beam 离子束
ion beam mass spectrometer 离子束质谱仪
ion beam width 离子束宽度
ion bombardment mass analyzer 离子轰击质量分析器
ion burn 离子斑点
ion chamber 离子室;电离室
ion-chamber detector 电离室型探测器
ion characteristic 离子特性
ion-charged bentonite 带电荷离子膨润土
ion chromatograph 离子色谱仪
ion chromatography 离子色谱法;离子层析
ion cloud 离子云
ion cluster 离子簇
ion collection system 离子收集系统
ion concentration 离子浓度
ion conductor 离子导体
ion contour map of groundwater 地下水离子等值线图
ion counter 离子计数器;电离计数器
ion current 离子电流
ion density ratio 离子密度比
ion desorption rate 离子脱附率
ion detection 离子检测
ion detector 离子探测器;离子检测器
ion dialysis 离子渗析
ion diffusion 离子扩散
ion doping 离子掺杂
ion doping technique 离子掺杂技术
ion electrode 离子电极
ion emitting 离子发射
ion energy spread 离子能量发散
ion engine 离子发动机
ion exchange 离子交换
ion exchange bead 离子交换微珠
ion exchange process in wastewater treatment 废水处理的离子交换法
ion exchange strengthening of glass 玻璃的离子交换钢化法
ion exchange treatment wastewater 废水离子交换处理注
ion exponent 离子指数
ion-expulsion ultrafiltration 离子驱动超滤
ion flo(a)tation 离子浮选
ion flo(a)tation method 离子选矿法
ion flow 离子流
ion-free water 无离子水
ion ga(u)ge 电离真空计;电离压力计
ion hydration 离子水化作用
Ionian entablature 爱奥尼克柱顶盘
Ionian Sea 伊奥尼亚海
ionic acceleration 离子加速
ionic accelerator 离子加速器
ionic activity 离子活度;离子酸度
ionic activity coefficient 离子活度系数
ionic adsorbate 离子吸附质
Ionic architectural order 爱奥尼克建筑柱型
Ionic architecture 爱奥尼克建筑
ionic association 离子缔合(作用)
Ionic base 爱奥尼克底座
ionic beam 离子束
ionic binding 离子交换接合
ionic bond 离子键
ionic bonding 离子耦合;离子结合;离子键联;离子键合
ionic bonding force 离子键力
ionic centrifuge 磁控电子管
ionic chamber 电离箱
ionic charge 离子电荷
ionic charge number 离子电荷数
Ionic colonnade 爱奥尼克柱廊
Ionic column 爱奥尼克柱
ionic composition 离子成分
ionic compound 离子化合物
ionic concentration 离子含量
ionic conduction 离子导电;离子传导
ionic conductivity 离子导电性
ionic crystal 离子型结晶
ionic crystal lattice 离子晶格
ionic current 离子电流
Ionic cyma(tium) reversa 爱奥尼克反曲线脚
ionic dispersion 离子分散体
ionic dissociation 离子离解
ionic emulsifying agent 离子乳化剂(泥浆添加剂)
Ionic entablature 爱奥尼克柱顶盘
ionic equation 离子方程式
ionic equilibrium 离子平衡
ionic equivalent conductance 离子当量电导
ionic exchange 离子交换
ionic exchange action 离子交换作用
ionic exchange adsorption 离子交换吸附
ionic exchange agent 离子交换剂
ionic exchange anti-corrosive pigment 离子交换型防锈颜料
ionic exchange bed expansion 离子交换床层膨胀率
ionic exchange capacity 离子交换容量
ionic exchange cellulose 离子交换纤维素
ionic exchange chromatography 离子交换色谱法;离子交换色层分析法;离子交换色层法;离子交换层析
ionic exchange column 离子交换柱
ionic exchange constant 离子交换常数
ionic exchange electrodialysis 离子交换电渗析
ionic exchange equilibrium 离子交换平衡
ionic exchange film 离子交换膜
ionic exchange flame atomic absorption method 离子交换火焰原子吸附法
ionic exchange material 离子交换剂;离子交换材料
ionic exchange membrane 离子交换膜
ionic exchange membrane electrolysis 离子交换膜电解法
ionic exchange membrane method 离子交换膜法
ionic exchange method 离子交换法
ionic exchange paper 离子交换纸
ionic exchange pre-process 离子交换法预处理
ionic exchange process 离子交换过程
ionic exchange property 离子交换性能
ionic exchange purification 离子交换净化
ionic exchanger 离子交换器;离子交换剂;离子交换机
ionic exchange regeneration 离子交换剂再生
ionic exchange resin 离子交换树脂
ionic exchange resin photometry 离子交换树脂光度法
ionic exchange resin regeneration 离子交换树脂再生
ionic exchange separation 离子交换分离
ionic exchange separation method 离子交换分离法
ionic exchange softener 离子交换水软化器
ionic exchange softening method 离子交换软化法
ionic exchange system 离子交换系统
ionic exchange technique 离子交换技术;离子交换法
ionic exchange thin-layer chromatography 离子交换薄层色谱法;离子交换薄层层析
ionic exchange tower 离子交换塔
ionic exchange treatment 离子交换处理
ionic exchange wastewater 离子交换废水
ionic exchange water 离子交换水
ionic exclusion chromatography 离子排斥色谱法
ionic fixation 离子固定作用
ionic formula 电子公式
ionic hydration 离子水化作用;离子水合(作用)
ionic interaction 离子相互作用
ionicity 电离度
ionic link 离子键
ionic micelle 离子胶体
ionic mobility 离子淌度;离子迁移率
ionic mobility spectrometry 离子淌度光谱法
ionic modulation 电离调制
Ionic order 爱奥尼克柱型;爱奥尼克柱式
Ionic order of the Asia Minor type 小亚细亚型爱奥尼亚柱式
ionic osmotic pressure 离子渗透压
Ionic pilaster 爱奥尼克式壁柱
ionic polarization 离子极化
ionic polymerization 离子聚合
Ionic portico 爱奥尼克式门廊
ionic potential 离子电位;电离(电)势
ionic propulsion 离子推进
ionic radius 离子半径
ionic ratio 离子比(率);离子比值
ionic rays 离子射线
ionic reaction 离子反应
ionic regulation 离子调节
Ionic scroll 爱奥尼克式漩涡;爱奥尼克盘蜗
ionic sensitive electrode 离子敏感电极
ionic size 离子大小
ionic source 离子源
ionic species 离子物种

ionic state 离子状态
ionic strength 离子强度
ionic strength adjustment agent 离子强度调节剂
Ionic style 爱奥尼克式
ionic substitution 离子替换
ionic surfactant 离子型表面活性剂
Ionic temple 爱奥尼克式寺院
ionic volume 离子容积
Ionic volute 爱奥尼克式涡涡
ion inverter 离子变频管
ion-ion emission 离子离子发射
ionisation energy 电离能(量)
ionisation layer 电离层
ionite 富硅高岭石
ionitriding treatment equipment 离子氮化处理设备
ionium deficiency method 锾亏损法
ionium excess method 锾过剩法
ionium-protactinium method 锾镤法
ionium-uranium method 锾铀法
ionizability 电离度
ionizable adsorbate 可电离吸附质
ionizable organic species 可电离有机物种
ionization 离子化作用;离子电渗作用;电离(作用)
ionization arc-over 电离闪络
ionization buffer 消电离剂
ionization by light 光致电离
ionization by step 分级电离
ionization cell 电离测定池
ionization chamber 电离箱;电离室
ionization chamber detector 电离室型辐射控测器
ionization chamber method 电离室法
ionization chamber type emanation apparatus 电离室射气仪
ionization characteristic 电离特性曲线
ionization chemical interference 电离化学干扰
ionization chromatographic determination 电离色谱测量
ionization coefficient 电离系数
ionization compensation method 电离补偿法
ionization constant 解离常数;电离常数
ionization constant of base 碱电离常数
ionization constant of salt 盐电离常数
ionization counter 电离计数器;电离计数管
ionization cross section 电离截面
ionization cross section detector 电离截面控测器
ionization current average value 电离电流平均值
ionization curve 电离曲线
ionization degree 电离度
ionization delay 电离延迟
ionization density 电离密度
ionization detector 电离探测器;电离检测器
ionization device 电离装置
ionization dose-meter 电离式辐射剂量计
ionization electrode 电离电极
ionization energy 电离能(量)
ionization equilibrium 电离平衡
ionization fire alarm 电离火警警报器
ionization fire detector 电离式火警探测器
ionization fission chamber 电离裂变室
ionization gap 电离隙
ionization ga(u)ge 电离真空计;电离(压力)计;电离规管
ionization ga(u)ge detector 电离压力计探测器
ionization heat 电离热
ionization in depth 深度电离
ionization interference 电离干扰
ionization layer 电离层
ionization level 电离级
ionization loss 电离损失
ionization manometer 电离压力计
ionization measurement 电离测量
ionization meter 电离强度测量计
ionization of air 空气电离(作用)
ionization of water 水的电离
ionization point 电离点
ionization potential 电离电位;电离(电)势
ionization process 电离过程
ionization product constant of water 水的离子积常数
ionization pulse 电离脉冲
ionization pump 离子泵
ionization puncture 电离击穿
ionization radiation 电离辐射
ionization rate 电离速度

ionization reaction 电离反应
ionization sensor 离子式探测器
ionization series 电离序
ionization smoke detector 电离感烟探测器
ionization solution 电离溶液
ionization space 电离区域
ionization spectrometer 电离谱仪;布拉格分光计
ionization technique 电离技术
ionization temperature 电离温度
ionization tendency 电离倾向
ionization time 电离时间
ionization vacuum ga(u)ge 电离真空计;电离真空规
ionization voltage 电离电压
ionization yield 电离率
ionize 电离
ionized 电离的
ionized calcium 离子钙
ionized-cluster beam epitaxy 电离粒子团束外延
ionized gas 电离气体
ionized gas detector 电离气体探测器
ionized gas laser 电离气体激光器
ionized layer 电离层
ionized molecule 分子离子
ionized state 电离状态
ionized stratum 电离层
ionizer 稳弧剂
ionizing 电离
ionizing collision 电离碰撞
ionizing efficiency 电离效率
ionizing electrode 电离电极
ionizing energy 电离能(量)
ionizing event 电离(作用)
ionizing nuclear radiation meter 电离核辐射测量仪
ionizing radiation 离子型辐射;电离辐射
ionizing radiation detector 电离辐射探测器
ionizing solvent 离子化溶剂
ion life 离子寿命
ion mean life 离子的平均寿命
ion meter 电离压力表
ion microprobe 二次离子质谱仪
ion microprobe analysis 离子微探针分析
ion microprobe mass spectrometer 离子微探针质谱计
ion microscope 场致离子显微镜
ionocolo(u)rimeter 氢离子比色计
ionogemi 电离的
ionogen 电离质
ionogram 电离图
ionography 载体电泳图法
ionomer 离子交联聚合物;离聚物
ionomer resin 离子键树脂;离聚物树脂
ionometer 氢离子浓度计
ionopause 电离层顶
ionophone 阴极送话器
ionoscope 存储摄像管
ionosonde 电离层探测装置;电离层探测器
ionosphere 游离层;离子层;电离圈;电离层
ionosphere error 电离层误差
ionosphere scatter(ing) 电离层散射
ionosphere wave 天空电波
ionospheric absorption 电离层吸收
ionospheric control points 电离层控制点
ionospheric cross modulation 电离层交叉调制
ionospheric disturbance 电离层骚扰;电离层扰动;电离层干扰
ionospheric eclipse 电离层食
ionospheric effect 电离层效应
ionospheric error 电离层(传播)误差
ionospheric forecast 电离层预报
ionospheric height error 电离层高度误差
ionospheric irregularities 电离层不均匀体
ionospheric path error 电离层行程误差;电离层传播途径误差
ionospheric physics of equator 赤道电离层物理学
ionospheric probe 电离层探针
ionospheric propagation 电离层传播
ionospheric propagation sudden disturbance 电离层传播突然扰动
ionospheric radio propagation 电离层无线电传播;电离层电波传播
ionospheric ray 电离层反射波
ionospheric recorder 电离层记录器
ionospheric recording equipment 电离层记录设备

ionospheric reflection 电离层反射
ionospheric refraction 电离层折射
ionospheric refraction correction 电离层折射校正;电离层折射归正;电离层折射改正
ionospheric regions 电离层区
ionospheric satellite 电离层卫星
ionospheric scatter(ing) 电离层散射
ionospheric scatter propagation 电离层散射传播
ionospheric scintillation 电离层闪烁
ionospheric sounding satellite 电离层探测卫星;电离层观测卫星
ionospheric storm 电离层暴
ionospheric wave 天波;电离层反射波;电离层波
ionotron 静电消除器
ionotropy 互变异构现象
ion pair 离子对
ion pair chromatography 离子对色谱法;离子对层析
ion phenylformic acid 离子苯甲酸
ion plasma 等离子区
ion plating 离子电镀法;离子电镀
ion plating equipment 离子镀膜机
ion plating film 电离镀膜
ion-plating handicraft article and machine part 离子镀工艺品和机械零件
ion-precipitation purification 离子沉淀净化
ion probe 离子探针;二次离子质谱仪
ionprobe microanalyzer 离子探针显微分析;离子探针微量分析仪
ion product 离子乘积
ion product constant 离子积常数
ion pump 离子泵
ion-pump mass spectrometer 离子泵质谱仪
ion recombination channel 离子复合通道
ion reflection 离子反射
ion reflection coefficient 离子反射系数
ion resonance mass spectrometer 离子共振型质谱仪
ion runoff 离子径流量
ion scattering analysis 离子散射分析
ion selective electrochemical sensor 离子选择电化学传感器
ion selective electrode 离子选择电极;选择性离子电极
ion selective electrode analysis 离子选择电极分析
ion selective electrode method 离子选择电极法
ion selective membrane 离子选择膜
ion sheath 离子层
ion sieve 离子筛
ionsilicate minerals 链状硅酸盐矿物
ions of the chalcophile type 铜型离子
ions of the inert gas type 惰性气体型离子
ions of the siderophile type 过渡型离子
ion source life 离子源寿命
ion source mass spectrometer 离子源质谱仪;火花源质谱仪;火花源质谱计
ion specific electrode 离子特定电极
ion specific membrane 离子特定膜
ion spectrometry 离子谱法
ion spot 离子斑点;离子斑
ions specifically adsorb 离子特性吸附
ion stick 离子棒
ion stuffing 挤塞效应
ion substitution 离子置换
iontophoresis 电离子透入疗法
ion trajectory 离子轨道
ion-transit-time measurement 离子运行时间测定
ion transmission 离子传输
ion transport 离子移动;离子迁移
ion trap 离子阱
ion trap detector 离子阱检测器
ion trap detector system 离子阱检测器系统
ion trap mass spectrometry 离子阱质谱法
ion triplet 三重离子
ion type 离子类型
ion wave 离子波
iophendylatum 碘苯酯
iophenoic acid 碘芬酸
iopodic acid 碘泊酸
iopronic acid 碘普罗酸
iosefamic acid 碘西法酸
ioseric acid 碘丝酸
iotendylate 碘苯酯
iotroxic acid 碘托西酸

iowaite 水氯铁镁石
lowan glaciation 艾俄瓦冰川作用
I owe you 借条;借据
iozite 方铁矿
IP[initial point] 初始点
ipazine 草怕津
ipentity 完全相同
ipil(e) 太平洋铁木
Iporka 艾波卡低温绝缘材料;保冻塑料袋
Ipro brick 铺路砖
I-prop 工字撑
ipso jure 依法
ipso jure termination of a contract 依法终止合同
ipsonite 残沥青
iraesite 硫砷依矿
I-rail 工字轨
Iranian architecture 伊朗建筑
Iranian Consortium 伊朗国际财团
Iranian Oil Participants Ltd 伊朗石油合资者有限公司
iranite 水铬铅矿
iraqite 硅稀土钙石
Iraq National Oil Company 伊拉克国家石油公司
Iraq Petroleum Company 伊拉克石油公司
iraser 红外(线)激射;红外(线)激光器
iraurite 铱金
ircon 远红外控制器
irdome 红外(线)整流罩
ireko (摇橹船的)橹钉眼;(摇橹船的)橹钉脐
irenine 温和碱
irestone 角岩;角闪岩;硬黏土(质)页岩
irhtemite 斜砷镁钙石
iridarsenite 砷铱矿
iridectomy hook 提勒耳氏钩
iridescence 晕色;钢化应力斑;虹色;虹彩(天);虹彩光;彩虹光
iridescent 荧光性(忽)绿(忽)紫的颜色;虹色的;闪光色
iridescent cloud 虹彩云;彩虹云
iridescent colo(u)r 彩虹色
iridescent decoration 彩虹装饰
iridescent glaze 虹彩釉;彩虹釉
iridescent mirror block 彩虹黑釉
iridescent paper 电光纸
iridescent scintillating effect 虹彩闪烁效应
Iridex 艾利德克斯纺前染色黏胶长丝
iridic chloride 四氯化铱
irid(i)oplatinum 铱铂合金
iridite process 浸镀铬法
iriditing 浸镀铬
iriditing process 浸镀铬法
iridium 自然铱
iridium chloride 四氯化铱
iridium fountain pen with stainless cap 不锈钢套铱金笔
iridium ores 铱矿
iridium oxide 氧化铱
iridium tetrachloride 四氯化铱
iridium wire 铱丝
iridization 虹视
iridizing 上虹膜;上彩虹
iridosmine 铱锇矿;铱锇合金
Iridye 艾利代黏胶染色纱
iriginite 黄钼铀矿
iripatarsenite 砷铂铱矿
iris 窗孔;虹彩色;光圈
iris action 可变光阑作用;膜片作用
iris aperture 光阑孔径;可变光阑孔径;膜片孔径
irisated 虹彩的;虹彩色的
irisation 虹彩(天)
iris blade 可变光阑(叶)片
iris canna 垂花美人蕉
iris change mechanism 可变光阑变化机制
iris-coupled cutoff attenuator 膜孔耦合截止衰减器
iris-coupled filter 膜孔耦合滤波器
iris-coupled wavemeter 膜孔耦合波长计
iris damper 光圈风门
iris diaphragm 膜片;锁光圈;光圈;可变光阑
iris diaphragm photometer 光瞳光度计
irised 彩虹色的
iris ensata 马莲
Irish architecture 爱尔兰建筑
Irish bridge 过水路面;石砌明水沟
irish car(r)ageenin 鹿角菜胶

Irish confetti 砖石碎片
Irish diamond 水晶
Irish fan 铲
Irish gum 爱尔兰胶;鹿角菜胶;角叉菜胶
Irish hurricane 无风而下微雨
Irish mail 坑道运工人车辆
Irishman's sidewalk 人行道
Irish moss 爱尔兰苔;鹿角菜;角叉菜胶
Irish pennant 松飘的束帆索
Irish splice 缩插接
iris in 圈入
irising 彩虹
irising from tempering 钢化彩虹;彩虹
iris mauve 彩虹紫色
iris microphotometer 可变光阑测微光度计
iris mount 可变光阑套
iris opening 膜孔
iris out 圈出
iris paper 虹彩纸
iris photometer 光瞳光度计;可变光阑光度计;虹彩色度计
iris quartz 虹水晶
iris ring 可变光圈;锁光(圈)环
iris setting 装定光圈
iris setting indicator 光阑调整指标器
iris stop equipment 膜片式堵管装置
iris wave-beam device 可变光阑波束装置
iris wave-beam guide 可变光阑波导
iris wave-beam structure 可变光阑波束结构
iris wipe 圈划
Irkutsk polarity superchron 伊尔库茨克极性超时
Irkutsk polarity superchronzone 伊尔库茨克极性超时间
Irkutsk polarity superzone 伊尔库茨克极性超带
iroko 伊罗科木
iron accumulator 碱铁蓄电池
iron acetate 乙酸亚铁
iron acetate liquor 醋酸铁液
Iron Age 铁器时代
iron aggregate 铁屑集料;铁屑骨料
iron aggregate concrete 铁屑集料混凝土;铁屑骨料混凝土;铁(屑)混凝土
iron aggregate joint 铁屑水泥密接头
iron-air cell 铁空气电池
iron-air trap 铁阻气
iron alloy 铁合金
iron-alloy pipe 铁合金管
iron alum 铁明矾;铁矾
iron-alumina ratio 铁铝率;铁铝比(率)
iron-aluminum alloy 铁铝合金
iron-aluminum coating 铁铝包层
iron amalgam 铁汞齐
iron ammonium sulfate 硫酸亚铁铵
iron and steel base 钢铁基地
iron and steel casting 钢铁铸件
iron and steel furnace 黑色金属冶金炉
iron and steel industry 钢铁工业
iron and steel manufacture 钢铁制造
iron and steel material shaped 钢铁型材
iron and steel plant 钢铁厂;炼钢铁厂
iron and steel scrap 废钢铁
iron and steel sheet and plate 钢铁板材
iron and steel sheet piling 铁及钢板桩法凿井
iron and steel wire 钢铁丝材
iron and steel works 钢铁厂;炼钢铁厂
iron angle 角钢;角铁
iron antigorite 铁叶蛇纹石
iron-antimony compound 铁锑化合物
iron-arc 铁弧拱;铁弧
iron arch 铁拱
iron arch bridge 铁拱桥
iron arch(ed) girder 铁拱主梁
iron architecture 铁构建筑;铁建造
iron arc lamp 铁弧灯
iron arsenate 砷酸亚铁
iron article 铁制品
iron-aventuring glass 铁金星玻璃
iron-back 壁炉背部的铁板
iron bacteria 铁细菌;嗜铁细菌
iron balance weight 铁坠陀
iron ball 铁球
iron ballast 铁石碴;压铁;铁碇
iron band 铁箍
iron bar 铁条;铁栓;铁杆;铁棒

iron bark 硬木;铁皮木;铁皮桉;澳洲橡木;桉树(木材)
iron bars 铁栅栏
iron-basalt 铁玄武岩
iron-base bearing 铁基轴承;铁基粉末冶金轴承
iron-base bushing 铁基粉末冶金轴套
iron-based material 含铁材料;铁基材料
iron-base powder 铁基合金粉末
iron-base superalloy 铁基高温合金
iron-base super-strength alloy 铁基超强度合金
iron beam 铁梁
iron-bearing 铁支承;含铁的
iron-bearing aggregate 含铁集料;含铁骨料
iron-bearing alloy 含铁合金
iron bearing formation 含铁岩层;含铁建造
iron-bearing material 含铁材料;含铁料
iron-bearing sludge 含铁污泥
iron-bearing water 含铁水
iron beidellite 铁贝得石
iron billet 短铁条
iron-binding capacity 铁结合力
iron binding wire 铁扎线
iron black 黑铁粉;铁黑;锑黑
iron blast furnace slag 炉渣砂
iron block 支挡铁件;铁滑车
iron block pavement 铸铁块铺面;铸铁块路面;铁块路面
iron blue 青灰色;铁青色;铁蓝色;铁蓝
iron body swing check valve 铁壳摆动式防逆阀
iron body valve 铁体阀
iron bolt 铁栓;铁螺栓
iron boracite 铁方硼石
ironbound 岩石围绕的;坚硬;多岩石;包铁的
iron bound block 铁带滑车
ironbound coast 多岩(石)海岸(无锚地)
ironbound shore 多岩(石)海岸(无锚地)
Ironbox 铁皮黄杨
iron box switch 铁壳开关
iron bracket 铁托
iron brick 矿渣砖;熔渣砖;铁板料
iron brick paving 铸铁砖铺砌;矿渣砖铺砌;铁砖铺砌
iron bridge 铁桥
iron briquette 铁屑团块
iron brown 铁褐色
iron brown pigment 铁棕颜料
iron brucite 铁水镁石
iron buff 铁浅黄
iron cake 含铁滤饼;铁渣
iron cap 铁帽
iron carbide 碳化铁;渗碳体
iron-carbon aeration micro-electrolysis 铁—碳曝气微电解
iron-carbon alloy 铁碳合金
iron-carbon diagram 铁碳平衡图
iron-carbon equilibrium diagram 铁碳平衡图
iron-carbon hydrolysis reactor 铁—碳水解反应器
iron-carbon micro-electrolysis 铁—碳微电解(法)
iron-carbon phase diagram 铁碳相图
iron-carbon ratio 铁碳比
iron carrier 铁件装卸机
iron carrying framing construction 铁框架构造
iron carryover 铁携带
iron case 铁壳
iron cased 加固的;装甲的
iron case meter 铸壳煤气表
iron cash case 保险柜
iron casting 铸铁件;铁铸件;生铁铸造;生铁铸件
iron catalyst 铁催化剂
iron catalyst reaction 铁催化剂反应
iron cathode 铁阴极
iron cell 铁电池
iron cement 含铁水泥;铁质胶结剂;铁屑水泥;铁水泥;铁胶结料;铁胶合剂
iron-cemented carbide 铁结硬质合金
iron-cemented tungsten carbide 铁钨硬质合金
iron-cementite diagram 铁渗碳体平衡图
iron chain bridge 铁链桥
iron chair 铁垫板
iron channel 槽形铁;槽钢
iron chill 冷铁
iron chimney connector 烟囱铁连接件
iron chip internal electrolysis process 铁屑内电

解法
iron chip micro-electrolysis 铁屑微电解法
iron chipping-activated carbon interior-electrolysis 铁屑活性炭内电解法
iron chipping-H$_2$O$_2$ oxidation process 铁屑双氧水氧化法
iron chipping interior-electrolysis 铁屑内电解法
iron chippings and shavings 铁屑
iron chips 铁屑
iron chloride 氯化铁
iron chromate 铬酸铁
iron-chromium binary diagram 铁—铬二元图
iron circuit 铁磁路；磁路铁芯部分
iron citrate 柠檬酸铁铵
iron clad 装甲的；铠装；铁甲；包铁；金属覆盖层；铁壳
iron-clad agreement 铁甲协定（绝对不能违反的协定）
iron-clad battery 铠装蓄电池；铁壳电池
iron-clad cable 金属包覆电缆
iron-clad coil 铁壳线圈
iron-clad contract 铁甲合同
iron-clad cutout 铁壳断流器
iron-clad distribution equipment 铁壳配电装置
iron-clad electromagnet 铠装电磁体
iron-clad galvanometer 铁壳电流计
iron-clad motor 铁壳电动机
iron-clad switch 铁壳开关
iron-clad switchgear 铁壳开关装置
iron-clad transformer 铁壳变压器
iron clamp 铁夹
iron clamping ring 腰铁
iron clamp ring 铁夹环
iron claw 撬棍；铁爪
iron clay 赭土；泥质铁矿；黄黏土；铁黏土
iron clip 铁夹
iron coagulant 铁混凝剂
iron cobalt liquid colo(u)r scale 铁钴比色计
iron coke 铁焦
iron-coke ratio 铁焦比
iron-colo(u)red glass 铁着色玻璃
iron-compass 铁指南针；铁南针
iron compound 铁化合物
iron concretion 铁质结核
iron conduit 铁导管
iron connector 铁制连接件；铁接线盒
iron-constantan 铁康铜
iron-constantan couple 铁康铜热电偶
iron-constantan thermocouple 铁康铜温差电偶；铁康铜热电偶
iron construction 铁结构
iron-consuming bacteria 耗铁细菌
iron-container rectifier 铁壳整流器
iron containing alloy 含铁合金
iron content 含铁量；铁含量
iron-copper 铁铜合金
iron-copper alloy 铁铜合金
iron-copper catalyst 铁铜催化剂
iron core (连接扶手与栏杆的)扁铁；铁芯
iron-core coil 铁芯线圈
iron-core long-stator linear motor 铁芯长定子线性电动机
iron-core material 铁芯材料
iron-core transformer 铁芯变压器
iron corrosion 铁腐蚀
iron cover 舷窗铁盖
iron covering 铁盖
iron cover switch 铁壳开关
iron cross law 铁十字律
iron crucible 铸铁坩埚
iron cutter 铁切割机
iron deficiency 铁质缺乏；铁的不足
iron density 铁的密度
iron deposit 铁沉积物
iron disconnecting (air) trap 铁制隔断拦截器
iron disulfide 二硫化铁
iron-dog 铁夹；铁钳；狗头钉
iron double blocks 双饼铁滑车
iron dowel 铁暗壁
iron dryer 铁催干剂；铁皂；铁干料
iron driver 铆钉工；毛铁工
iron drum 铁桶
iron dust 铁屑；铁末；铁涵洞；铁粉
iron-dust coil 铁粉线圈
iron-dust core 铁粉芯

iron-dust core coil 铁粉芯线圈
iron earth 菱铁矿
iron electric(al) connector 铁接线盒
iron elimination 除铁
iron embedded in concrete 混凝土内(埋)钢筋
ironer 熨平机；烫平机
iron expansion shield 能伸缩活动铁架；金属伸缩棚罩
iron eye 铁环
iron eyebar suspension bridge 铁眼杆悬索桥
iron ferrocyanide 亚铁氰化铁
iron fighter 扎钢筋工；扎铁工
iron-filing mortar 铁屑砂浆
iron filings 铁屑；铁锉屑
iron filled epoxy pattern 铁粉填料环氧树脂模
iron filler 硬质腻子；铁质填料；铁腻子
iron filling 灌铁水
Iron film 铁质薄膜
iron firedog 壁炉炉架；铁柴架壁炉
iron fittings 铁屑
iron flask 铁砂箱
iron fluoride 氟化铁
iron fluosilicate 铁氟硅酸盐
iron foil 铁箔
iron foot 铁足(青瓷的底脚)
iron foot valve 铁底阀
iron for concaved planes 凹刨刨刀
iron formation 铁质建造；铁建造
iron forms 铁模板
iron form(work) 铁模板(工程)
iron founding 生铁铸造
iron-foundry(shop) 铸铁(车)间；铸工车间；铸铁厂；翻砂厂
iron-foundry waste 铸铁厂废物
iron-foundry waste water 铸铁厂废水
iron framework 铁构架；铁框架
iron framing 铁构架
iron framing construction 铁框架结构
iron-free solution 无铁溶液
iron front 铸铁制的建筑物正面
iron gate 铁门
iron gauze 铁丝网；铁纱
iron girder 钢梁
iron girder bridge 钢梁桥；铁梁桥
iron glance 赤铁矿
iron glass 铁玻璃
iron glazing bar 镶玻璃铁条；铁窗樘；铁制嵌玻璃条
iron gossan 铁帽
iron granules 铁微粒
iron-graphite diagram 铁石墨平衡图
iron gray 灰白色；铁灰色
iron grill(e) 铁丝格子；铁花格
iron grill(e) door 铁花栅门
iron group elements 铁族元素
iron gymnite 铁水蛇纹石
iron hammer 铁锤
iron hammer scale 锻鳞
iron-hand 机械手
iron handle screw driver 铁柄螺丝刀
iron hat 钢盔；铁帽
iron heating 烙铁加热
iron hinge 铰链
iron hook 铁钩
iron hoop 铁箍
iron hoop lacing 铁箍方格子构造
iron horse 铁马
ironic hydroxide 氢氧化铁
ironic hydroxide desulfurizing 氢氧化铁法脱硫
Ironier's bronze 艾昂尼尔青铜
iron industry 炼铁工业
ironing 熨平；压平；烙边；拉伸；聚压；减径挤压；挤拉法；烫平；打薄；抽薄；变薄拉深
ironing board 熨衣台；熨衣台；熨衣板；烫衣板
ironing die 拉薄模
ironing machine 熨衣机；熨压机；熨平机；熨衣器
ironing room 熨衣室；烫衣室
ironing-screed 墁平准条；粉平准条；熨烫样板
ironing table 熨衣台
iron intercepting air trap 铁制空气拦截器；铁制阻截空气用存水弯
iron interceptor 铁制拦截器；铁隔断
iron ion 铁离子
iron-iron carbide equilibrium diagram 铁碳化铁平衡图

iron-iron carbide system 铁碳化铁系
iron key to connect the stone blocksand arch ring 石拱圈石块间的铁楔
iron ladle car 铁水罐车
iron lath(ing) 抹灰铁丝网；铁制抹灰底；铁板条
iron law of wages 铁的工资规律
iron-lazulite 铁天蓝石
iron leg 段铁
ironless alternative current servo motor 无铁交流伺服电动机
ironless armature 无铁电枢
ironless magnetic system 无铁的磁系统
iron lintel 铁楣
iron liquor 醋酸铁液
iron lithomarge 铁密高岭土
iron-loaded concrete 铁混凝土
iron lock 铁锁
iron loss 金属烧损；铁芯损失；铁(芯损)耗；铁损；磁芯损耗
iron loss factor 铁损系数；铁耗系数
iron loss per unit weight 单位铁损
iron loss test 铁损试验
iron machine 铁电机
iron magnetic property 铁磁性
iron magnetic spectrometer 铁磁谱仪
iron making 炼铁
iron man 钢铁工人；炼铁工人
iron-manganese colo(u)red glass 铁锰着色玻璃
iron-manganese concretion 铁锰结核
iron manufacture industry sewage 制糖工业污水
iron master 铁器制造商
iron melt 铁水
iron-melting furnace 化铁炉；熔铁炉
iron melting pot 铁制熔锅
iron metal 铁金属
iron metavanadate 钒酸铁
iron meteorite 陨铁；铁陨石
iron mica 含铁云母；云母状铁锈屑；铁云母
iron mica paint 铁云母涂料
iron mica schist 铁云母片岩；铁云片岩
iron mill 铁磨
iron mine 铁矿
iron mineral 铁矿物
iron minium 赤铁；赭土
iron modulus 铁指数
iron modulus of cement (水泥的)铁率
iron-monger 金属器具商；铁器商；小五金商
iron monger for building trades 建筑五金商
ironmongery 五金店；铁器(业)；五金(器具)
ironmongery finishes 小五金饰面
ironmongery materials 小五金材料
iron mongery work 五金安装工程
iron monosulfide 一硫化铁
iron-monticellite 铁钙橄榄石
iron mordant 铁媒染剂
iron mottling 铁锈斑
iron mo(u)ld 锈痕；铁锈迹；铁模；铁斑
iron munting 铁窗格条
iron nail 铁钉
iron natrolite 铁钠沸石(钠沸石与绿泥石的混合物)
iron-nickel 铁镍合金
iron-nickel accumulator 铁镍蓄电池
iron-nickel alloy 铁镍合金
iron-nickel-aluminum magnet 铁镍铝磁铁
iron-nickel-chromium alloy 铁镍铬合金
iron-nickel-copper-molybdenum alloy 铁镍铜钼合金
iron-nickel storage battery 铁镍蓄电池
iron nitrate 硝酸铁
iron nitride 一氮化二铁
iron notch 出铁口；出铁槽
iron ocher 铁赭色；铁赭石
iron ochre 铁赭色
iron oil drum 铁制油桶
iron-oilite 多孔铁
iron oil syringe 铁制润滑油注射器
iron oleate 铁油酸酯
iron openwork 铁花
iron ore 铁矿石；铁矿
iron-ore aggregate 铁矿石集料
iron-ore carrier 铁矿运载工具；铁矿运载车
iron-ore cement 矿渣水泥；铁矿水泥
iron-ore concentrate 铁精矿；精矿
iron-ore concrete aggregate 铁矿石混凝土集料

iron-ore deposit 铁矿床
iron-ore flo(a)tation 铁矿浮选(法)
iron-ore grinding 铁矿研磨
iron-ore mine 铁矿(山)
iron-ore pellet 铁矿石球团;铁矿颗粒
iron-ore processing 铁矿处理
iron-ore road surface 铁矿石路面
iron-ore separator 铁矿分选机
iron-ore slurry pump 铁矿砂浆泵
iron-ore stripe mine 铁矿露天采矿场
iron organisms 铁微生物
iron out 铁滚轧平
iron-oxalate complex 铁—草酸盐络合物
iron oxide 氧化铁;铁矿石
iron oxide adsorbent 氧化铁吸附剂
iron oxide basic 碱性氧化铁
iron oxide black 氧化铁黑;铁黑;四氧化三铁
iron oxide brown 氧化铁棕;漆工棕;铁棕(色)
iron oxide cement 氧化铁胶结物;含铁质水泥
iron oxide-coated sand 涂氧化铁砂
iron oxide coating 氧化铁涂层
iron oxide ore 氧化铁矿
iron oxide paint 氧化铁涂料
iron oxide particle 氧化铁粒子
iron oxide pigment 氧化铁颜料
iron oxide process 氧化铁法
iron oxide purple 氧化铁紫
iron oxide red 铁红
iron oxide residue 氧化铁残渣
iron oxide type electrode 氧化铁型焊条
iron oxide yellow 铁黄
iron oxide yellow pigment 氧化铁黄颜料
iron-oxidizer 铁氧化剂
iron-oxidizing bacteria 铁氧化细菌
iron-oxygen 铁氧系
iron-oxygen battery 铁氧电池
iron oxyhdroxide solid 氧氢氧化铁固体(即针铁矿)
iron oxyhdroxide 氧氢氧化铁
iron pad lock 铁挂锁
iron pail 铁桶
iron paint mill 铁粉漆磨坊
iron pan 铁质地层;铁盘;铁锅;硬岩盘;硬土层
iron pattern 铸铁模型;铁模
iron pavement 铸铁块(铺砌)路面;铁板铺面
iron paving 铸铁铺面
iron peak 铁峰
iron period 铁器时代
iron phosphate coating 磷酸铁膜;磷酸铁层
iron phosphide eutectic 磷共晶
iron picture 铁画
iron pike 铁镐
iron pile 钢桩;铁桩
iron pipe 铁管
iron pipe conduit 铁管道
iron pipe culvert 铁管涵洞
iron pipeline 铁管线
iron pipe mark 铁管标志
iron pipe size 铁管尺寸;铁管内径
iron plate 钢板;铁板
iron plate chimney 铁烟囱
iron plated concrete 钢板衬砌混凝土;包铁混凝土
iron podzol 铁质灰壤
iron pole 铁柱
iron pole piece 软铁极片
iron Portland cement 含铁硅酸盐水泥;矿渣硅酸盐水泥;矿渣波特兰水泥;铁渣水泥;铁硅酸盐水泥;铁波特兰水泥
iron-pot furnace 铁锅熔化炉
iron-powder 铁粉
iron-powder cement 铁粉水泥(防水用)
iron-powder coating 铁粉涂料
iron-powder electrode 铁粉涂料焊条
iron-powder process 铁粉切割法
iron-powder type electrode 铁粉焊条
iron printing process 印铁像;印铁法
iron product 铁制品
iron prop 铁柱
iron protoxalate 草酸亚铁
iron protoxide 氧化亚铁
iron purification 铁的净化
iron putty 含铁油灰;铁油灰
iron pyrite 黄铁矿
iron railing 铁扶手;铁栏杆
iron rake 铁耙

iron ramp 铁夯
iron red (含氧化铁的)红颜料;铁红
iron red glaze 铁红釉
iron red pigment 铁红颜料
iron red plastering 铁锈红涂料
iron reinforcement 铁筋
iron removal 除铁作用;除铁
iron removal contact-oxidation 除铁接触氧化法
iron removal filter 除铁滤池
iron removal plant 除铁车间;除铁装置;除铁厂
iron resinate 树脂酸铁
iron resistance 铁电阻
iron-retention agent 铁络合剂
iron-rich powder process 铁粉氧块切割法
iron ridging 骑马钉;铁屋脊
iron ring 铁圈;铁环
iron-ring support 铁圈支架
iron rivet 铁铆钉
iron rod 钢条;铁棍;铁杆
iron roller 铁碾;铁滚筒
iron roof 铁屋面
iron roof cladding 铁皮屋顶盖板
iron roof cover(ing) 铁皮屋面板;屋顶铁覆盖层
iron roofer 铁皮屋面工
iron roofing 铁皮屋顶;铁皮屋面
iron roof sheathing 铁皮屋面板
iron room 熨衣室
iron rope stayed bridge 铁斜拉桥;铁索斜拉桥
iron rope suspension bridge 铁悬索桥
iron rubber 铁质橡胶
iron runner 流铁槽;铁水沟;出铁槽
iron rust 铁锈
iron rust cement 铁胶合料;含铁水泥
iron rust glaze 铁锈色釉
irons 铁粉
iron safe clause 铁柜条款
iron saffron 印度红
iron salt 铁盐类
iron sand 铁矿砂;铁砂;铸铁砂
iron sandstone 铁砂石;铁砂岩
iron sash bar 钢窗型材
iron scaffold 金属脚手架;铁脚手架
iron scale 氧化铁皮;锅垢;铁氧化皮;铁锈
iron schefferite 铁锰钙辉石
iron scrap 废铁
iron scraper 刮铁刀
iron screw with square head 方头铁螺钉
iron scurf 铁屑
iron-sealing glass 铁封玻璃
iron section 铁制型材;型钢
iron separation 铁质分离
iron shackle 铁钩环
iron shavings 铁棉;铁屑
iron shears 铁剪
iron sheet 铁板;铁片;铁皮
iron sheeting 铁格条
iron sheet paving 铁板铺面
iron sheet pile 铁制薄板桩;铁板桩
iron sheet roof cover(ing) 铁皮屋顶覆盖层
iron sheet roofing 铁皮屋顶
iron sheet roof sheathing 铁皮屋顶夹衬板
iron sheet shears 铁皮剪子;铁皮剪刀
iron-shod 包铁的
iron shot 钢球;铁珠;铁丸;铁砂;铁质斑彩
iron shot concrete 铁丸混凝土
iron shuttering 铁模板
iron sick 铁材腐蚀
iron sight 机械瞄准具;普通瞄准具
iron silicate 硅酸铁
iron-silicate gel 硅酸铁胶
iron-silicon alloy 硅铁合金;硅钢合金;铁硅合金
iron-silicon modulus 硅铁模数
iron-silver composition 铁银制品
iron single block 单饼铁滑车
iron single jack chain 光身铁链
iron skeleton framing construction 铁框架结构
iron skull 铁路锅炉工
iron-skutterudite 铁方钴矿;方砷铁矿
iron slag 铁熔渣
iron slag aggregate 炼铁炉渣骨料;铁矿炉渣粒料
iron slag built-up roof(ing) 炼铁炉渣复合屋面
iron slag concrete 炼铁炉渣混凝土
iron slag fiber 炼铁炉渣纤维
iron slag filler 炉渣填料

iron slag sand concrete 炉渣砂混凝土
iron sleeper 铁轨枕
iron sliding bolt 铁插销
iron slips 铁挡边
iron-smeltery wastewater 炼铁厂废水
iron-smelting factory 熔铁厂;炼铁厂
iron-smelting furnace 炼铁炉
iron-smelting plant 炼铁厂
ironsmith 铁匠;铁工;锻工
iron smoke tube 铁烟囱
iron soap 铁皂;铁干料
iron socket 铁承窝
iron solder 铁焊料
iron soldering 烙铁钎焊
iron solution value 铁溶出值
iron spar 菱铁矿;球菱铁矿
iron speck 铁斑
iron split pin 铁开尾销
iron sponge powder 海绵铁粉
iron spot 铁点;铁斑点;铁斑
iron spring 铁泉
iron square head screw 方头铁螺钉
iron square hinge 方铰链
iron square nut 方头铁螺母
iron stain 铁锈(污)点;铁锈(污染);铁斑
iron stair(case) 铁楼梯;铁扶梯;铁制楼梯
iron stand 铁架
iron stay 铁拉条;铁拉杆
iron stay bolt 铁撑螺栓
iron stearate 硬脂酸铁
iron still 铁釜
ironstone 含铁矿石;铁石;氧化铁;泥铁矿;褐铁矿;铁岩;铁矿(石)
ironstone china 硬质陶器;硬质高强精陶
ironstone hardpans 褐铁矿硬土层
ironstone solder 硬焊料
ironstone ware 硬质陶器
iron storage battery 铁极蓄电池
iron stove 铁炉
iron stove pipe 铁烟囱
iron strap 铁箍条;铁狭条;铁皮条
iron strap loosened 铁皮松开
iron strapped block 铁带滑车
iron strapping 铁皮条打包
iron stud 铁双头螺栓
iron sulfate 硫酸铁
iron sulfide 硫化亚铁;硫化铁
iron sulphide 硫化铁
iron support 支铁
iron tallate 松浆油酸铁
iron tapping hole 出铁口
iron teeth 铁齿
iron tie 铁系杆
iron tie rod 铁系杆
iron timber connector (木结构的)铁制连接件
iron tin funnel 白铁漏斗
iron tire 铁轮
iron-tired cart traffic 铁轮车交通
iron-tired vehicle 铁皮轮车
iron-tired wheel 铁皮轮车
iron titanium oxide 氧化钛铁
iron tourmaline 铁电气石
iron tower 铁塔
iron tramp 铁杂质
iron trap 铁制拦截器;铁存水弯
iron tree 铁木
iron-triangle 铁三角架
iron tube 铁管
iron tube tunnel 铁管隧道
iron turbidity 铁混浊
iron uranite 铁铀云母
iron utilization 铁的利用效率
iron vane instrument 铁叶式仪表;铁片式仪表
iron vane type ammeter 铁片式安培表
iron vane type voltmeter 铁片式伏特计
iron violet 铁紫
iron vitriol 青矾;七水(合)硫酸亚铁;铁绿矾;铁矾
iron ware 铁制品;铁(制)器
iron washer 铁垫圈
iron waste 碎铁片;废铁
iron weight 压铁
iron whiskers 铁金属须
iron window 铁制窗
iron window bar 铁窗挡;铁窗条

iron winds 铁风
iron wire 钢丝;铁丝;铁导线;包装钉
iron wire armo(u)ring 铁丝铠装
iron wire brush 钢丝刷;铁丝刷
iron wire cloth 铁丝布
iron wire coil 铁丝卷
iron wire mill rinse water 铁线材轧机废水
iron wire sieve 铁丝筛
ironwood 坚硬的木料;硬木;坚硬木(材);铁木(结构)
ironwood screw 铁制木螺钉
iron wool 铁棉
iron worker 扎钢筋工;铁工;钢筋工;钢铁工人;铁架构架工;铁器工人
iron worker's shop 铁工场
ironworking 铁加工
ironworks 炼铁厂;铁制零件;铁工(厂);钢铁厂;铁制品;铁制部分;铁件;铁活;锻铁;锤铁
irony 铁似
iron yellow(pigment) 铁黄颜料
iron-zirconium pink 锆铁(粉)红
irradiance [irradiancy] 光辉;辐照度;辐射通量密度;发光
irradiance distribution 辐照度分布
irradiance level 辐照度级
irradiance pattern 辐照图
irradiance responsivity 辐照反应度
irradiant 光照的
irradiate 辐照
irradiated area 照射面积;曝光面积
irradiated coating 照射固化涂层
irradiated crystal 辐照过的晶体
irradiated diamond 受辐射的金刚石
irradiated gas 受辐照气体
irradiated plastic(material) 光渗塑料
irradiated plastics 照射塑料
irradiated polyethylene 照射聚乙烯
irradiated surface 受辐照表面
irradiation 照射杀菌;照射;光渗;辐照量;辐照(度)
irradiation capsule 照射盒
irradiation capsule opener 照射盒开盒装置
irradiation chimera 辐射性嵌合体
irradiation container 照射容器
irradiation correction 照射校正;太阳高度光渗差改正量
irradiation damage 照射损伤;辐照损伤
irradiation dosage 照射剂量
irradiation dose 照射剂量
irradiation drum 辐照筒
irradiation effect 照射效应;辐照效应
irradiation error 光渗差
irradiation field 照射野;辐照场
irradiation fuel 辐照燃料
irradiation hole 照射孔道
irradiation injury 照射伤害;射线杀伤;辐照伤害
irradiation life 辐照寿命
irradiation machine 辐照器
irradiation of inhibition 扩散性抑制
irradiation of solid wastes 固体废物辐射处理
irradiation plant 辐照装置
irradiation-proof glass 防辐照玻璃
irradiation reactor 辐照用反应堆
irradiation-resistant electric(al) glass fiber 耐辐射电绝缘玻璃纤维
irradiation-resistant optic(al) glass 耐辐射光学玻璃
irradiation rig 辐照装置
irradiation-saturation current 辐照饱和电流
irradiation service 照射业务
irradiation stability 辐照稳定性;辐射稳定性
irradiation storage 辐射储藏
irradiation time 照射时间;辐照时间
irradiation treatment of municipal refuse 城市垃圾辐射处理
irradiation treatment of refuse 垃圾辐照处理
irradiation unit 照射装置
irradiation with ultraviolet light 用紫外线辐照
irradiatometer 辐计
irradiator 照射器;辐射器;辐射体
irradome 红外(线)整流罩
irrational 非理性的;不合理的
irrational dispersion 不正常色散;不规则色散
irrational equation 无理数方程;无理方程

irrational error 不合理误差
irrational exponent 无理指数
irrational expression 无理式
irrational extraction of resources 破坏性开采
irrational function 无理函数
irrational invariant 无理不变式
irrational involution 无理对合
irrationalism 反理性主义【建】
irrationality 不合理
irrational layout 不合理布置
irrational location 不合理布局
irrational monomial expression 无理单项式
irrational number 无理数
irrational part 无理部分
irrational quantity 无理量
irrational real number 无理实数
irrational root 无理根
irrational stock of goods 不合理库存
irrational system of units 无理单位制
irrational transport 不合理运输
irreconcilable 不可调和的
irrecoverable 不能恢复的
irrecoverable compliance 不可(回)复柔量
irrecoverable creep 不可复蠕变
irrecoverable deformation 塑性变形
irrecoverable error 不可恢复的错误
irrecoverable read error 不可恢复读出误差
irrecoverable set 不可恢复的凝结
irrecoverable strain 不可恢复应变
irredeemable 不能兑现的;不可兑现的
irredeemable bonds 不规定到期日的债券
irredeemable currency 不能兑现的通货;不兑换的货币
irredeemable loan stock 不兑现公司债券
irreducibility 不可约性
irreducibility criterion 不可约(性)判别准则
irreducible 不可约的;不能降低的;不能减缩;不能还原的;不能复位的;不可约(的)
irreducible algebraic equation 不可约代数方程
irreducible bound water saturation 束缚水饱和度
irreducible cluster integral 不可约集团积分
irreducible component 不可约成分
irreducible correspondence 不可约对应
irreducible curve 不可约曲线
irreducible electrodynamics 不可约电动力学
irreducible equation 不可约方程
irreducible field equation 不可约场方程
irreducible fraction 不可约分数
irreducible function 不可约多项式
irreducible group 不可约群
irreducible image 不可约映像
irreducible integral 不可约积分
irreducible invariant subgroup 不可约不变子群
irreducible Markov chain 不可约马尔可夫链
irreducible matrix 不可约矩阵
irreducible minimum 最小限
irreducible phase 不可约相
irreducible polynomial 不可约(的)多项式
irreducible quadratic irrational number 不可约二次无理数
irreducible representation 不可约表示
irreducible representation of a group 群的不可约表示
irreducible saturation 最低残留饱和度;残余饱和率(岩芯分析)
irreducible system of equations 不可约方程组
irreducible tensor 不可约张量
irreducible tensor operator 不可约张量算符
irreducible water 残余水
irreducible wave equation 不可约波动方程
irredundant disjunctive form 非冗长或形式
irreflexive 漫反射的;反自反
irreflexive relation 漫射关系;非自反关系;反自反关系
irregular 无规律的;紊乱的;未按规定的;复层的;非正式的;不(整)齐的;不匀;不规则的
irregular aggregate 不规则集料;不规则骨料;非正规的集料;非正规的集料;不合规定的骨料
irregular anagalactic nebula 不规则河外星云
irregular annihilation rate 不规则湮没率
irregular arrangement 不规则排列
irregular arrangement method 不规则布置法
irregular array 不规则阵列

irregular astigmatism 不规则像散;不规则散光
irregular beading 边釉不齐
irregular bedding 不整合层理;不均匀基床;不规则层理
irregular bend 不规则河湾
irregular blotches 不规则的斑点
irregular borehole 不规则井眼
irregular bottom 不规则河底;不规则底板
irregular boudin 不规则石香肠
irregular break 不平整断口
irregular brick 找平砖
irregular carrier 不定期船
irregular channel 不规则渠道
irregular chattering 不规则反跳;不定反跳
irregular checking 不规则细裂
irregular close packing 不规则密堆积
irregular coagulation 不规则混凝
irregular coast 沉溺(海)岸;不规则海岸
irregular coastline 不规则海岸线
irregular collective 不规则集合
irregular combustion 不规则燃烧
irregular component 不规则分量
irregular concrete block 异形混凝土块体
irregular concrete block armo(u)r 不规则混凝土块体护面
irregular concrete unit 异形混凝土块体
irregular configuration 不规则形状
irregular contact face 不规则接触面
irregular contour 不规则外形;不规则轮廓
irregular convection 不规则对流
irregular corolla 不整齐花冠
irregular course 乱砌层;不规则砌层
irregular-coursed rubble 乱石墙;乱石砌体;乱石层;乱砌毛石;不规则层乱石;不成层乱石毛石;石组;不规则成层毛石;乱石层
irregular-coursed wall 乱砌层墙
irregular crack 不规则裂纹;不规则裂缝;不规则开裂
irregular cross-section 不均匀横断面;不规则横断面
irregular crystal 不规则晶体;不规则冰晶
irregular curve 折线;曲线板;不规则曲线(规)
irregular cycle 不规则周期
irregular deformation 不规则变形
irregular dense fibrous connective tissue 不规则致密纤维结缔组织
irregular deposit 不定期存款
irregular depression 不规则形坑
irregular depth 不规则入土深度
irregular deviation 非正规离差;非正规离差;不规则偏转;不规则偏差;不定差
irregular dislocation 不规则位错
irregular distortion 不规则失真;不规则畸变;不规则变形
irregular distribution 不规则分布
irregular diurnal tide 不正规日潮;不规则日潮
irregular dominance 不规则显性
irregular drift 不规则漂移
irregular edge 不规则边(缘)
irregular edge dislocation 不规则型位错;不规则刃形错位
irregular endorsement 非正式背书;非正规背书;非正规背签
irregular endorser 不规则背书人
irregular entropy increase 不规则熵增加
irregular error 偶然误差;不规则误差
irregular excitation curve 不规则激发曲线
irregular fetch 不规则风区
irregular fever 不规则热
irregular figure 不规则图形
irregular fissure 不规则裂隙
irregular flagstone 不规则铺面石板
irregular flood pattern 不规则注水井网
irregular floor 不平整台面;不平整桥面;不平整楼面;不平整底板;不平整层面;不规则床面
irregular flow fold 不规则流褶皱
irregular fluctuation 不规则起伏;不规则脉动;不规则变动
irregular fold 不规则褶皱
irregular folding 不规则折叠
irregular fracture 不规则断口;不平断口
irregular frame 不规则框架
irregular French curve 曲线板;不规则(曲)线规
irregular fringe 不规则条纹
irregular galaxy 不规则星系
irregular garden 不规则庭园;不规则花园

irregular geomagnetic field 不规则地磁场
irregular grain 不规则纹理
irregular grain gravure screen 不规则砂目凹版网屏
irregular grid drainage 不规则格状水系
irregular growth 不正常生长
irregular growth rate 不规则生长率
irregular harmonic vibration 不规则谐振动
irregular heir 非正式继承人
irregular hole 不规则井眼
irregular iceberg 尖塔形冰山；峰形冰山
irregular index 不规则指数
irregular indorsement 非正规背书；非正规背签
irregular inquiry 不定期调查
irregular installation 非正款装置
irregular integral 非正则积分
irregular intergrowth 不规则共生
irregular interior node 不规则内结点；不规则内节点
irregular interval 不规则区间
irregularities in vessel surface 容器表面缺陷
irregularity 谬误；奇异性；奇点；参差不齐；不匀度；不齐；不平整(度)；不规则(性)；凹凸不平
irregularity coefficient 不平整系数；不平度系数；不规则系数
irregularity control 不匀率控制
irregularity degree 不平整度
irregularity distribution load 不均匀分布荷载
irregularity index (路面的)不平整指数；不均匀指数；不规则性指数
irregularity of alignment 轨向不平顺
irregularity of cross level 水平不平顺
irregularity of interference fringe 干涉条纹不规则性
irregularity of longitudinal level 高低不平顺
irregularity of track 轨道变形
irregular labor pains 不规则阵痛
irregular lake 畸变湖
irregular length 不规则长度
irregular letter of credit 不可撤消信用证
irregular light 不规则灯光
irregular liner 不定期班轮
irregular load 不规则荷载
irregular loading 不规则加载
Irregular loadlng sequence 不规则倚载序列
irregular lower 曲折下降
irregularly distributed load 非匀布荷载；非均布荷载；不平均分布荷重；不平均分布荷载；不规则分布荷载
irregularly distribution load 不均匀分布荷载
irregularly shaped steel 异形钢材
irregularly varying service 不规则变化使用荷载
irregularly wave axis guide 不规则轴波导
irregular magnetic micropulsation 不规则地磁微脉动
irregular map 不正规图幅
irregular market 不正常市场
irregular meson decay 不规则介子衰变
irregular meteorological disturbance 不规则气象干扰
irregular micropulsation 不规则微脉动
irregular miter 不规则斜角缝【建】
irregular motion 不均匀运动
irregular motion of flowing fluid 不规则流体运动
irregular movement 不规则运动
irregular natural stream 不规则天然河流
irregular nebulae 不规则星云
irregular net 非正则网络
irregular network 不规则网络
irregular node 不规则节点
irregular nuclear reaction 不规则核反应
irregular oil-water interface 不规则的油水界面
irregular operating 不正常运行
irregular ore body 不规则矿体
irregular orientation 不规则取向
irregular oscillation 不规则振荡
irregular particle 不规则颗粒
irregular particle shape 不规则颗粒形状
irregular particle shape of aggregate 不规则粒状集料；不规则粒状骨料
irregular parting 不规则分离
irregular pattern 不规则型式；不规则图案
irregular pattern type cracking 不规则形开裂
irregular paving rubble 不规则铺面毛石
irregular paving sett 不规则铺路小石
irregular periodicity 不规则的周期性

irregular phase change 不规则相位变化
irregular pitch (屋面的)不匀称斜坡；不规则坡度
irregular point 非规则点
irregular pool 不规则油藏
irregular pore 不规划孔
irregular porosity 不规则孔隙度
irregular powder 不规则粉末
irregular profile 不规则纵断面；不规则外形
irregular pulsation 不规则脉动；不规则的脉冲形成
irregular pulse 不规则脉
irregular pumping 不规则泵送
irregular quadrilateral 不规则四边形
irregular reference 乱反射
irregular reflection 不规则反射
irregular refraction 不规则折射
irregular region 非正则域
irregular remainder 不规则剩余
irregular resonance absorption 不规则共振吸收
irregular resonance scattering 不规则共振散射
irregular respiration 呼吸不规则
irregular river bottom 不平整河底
irregular rock formation 不规则岩层
irregular rock stratum 不规则岩层
irregular rotation 不规则自转
irregular rotation of the earth 地球不规则自转
irregular rubble 不规则毛石
irregular rubble gable 不规则毛石山墙
irregular rubble masonry 不规则毛石砌体
irregular rubble paving 不规则毛石铺路；不规则毛石路面
irregular running 不均匀行程
irregular satellite 不规则卫星
irregular scattering 不规则散射
irregular scattering cross-section 不规则散射截面
irregular section 不规则截面；不规则断面
irregular semi-diurnal tide 不正规半日潮；不规则半日潮
irregular service 非定期运输线路；不定期运行；不定期检修；不定期航线；不定期航次；不定期供应；不定期服务
irregular sett 不均匀小方石
irregular settlement 不均匀沉陷；不规则沉陷；不规则沉降
irregular shape 不规则形状
irregular-shaped discontinuity 不规则状结构面
irregular-shaped nodule 不规则状结核
irregular-shaped orebody 不规则状矿体
irregular shed 不整齐梭道
irregular shelterwood system 不规则整渐伐
irregular shoreline 不规则岸线
irregular short-term fluctuation 不规则短期波动
irregular singular point 不规则奇点
irregular slab 异型板
irregular slip surface 不规则滑动面
irregular sounding 不定点测深
irregular spectral line shift 不规则谱线移
irregular spiking 不规则尖峰
irregular splitting 不规则的裂纹
irregular square stone 不规则方石
irregular stacking 不规则堆积
irregular stocking 不正规立木度
irregular structure 不规则结构
irregular style 不整齐式
irregular subsidence 不规则沉陷；不规则沉降
irregular subsoil 不规则基土
irregular surface 不规则表面
irregular surface reaction 不规则表面反应
irregular tax 杂税
irregular temperature variation 不规则温度变化
irregular topography 不规则地形
irregular trend 不规则趋势
irregular type 不规则型
irregular variable star 不规则变星
irregular variation 不规则偏差；不规则变化；不规则变动
irregular vibration 不规则振动
irregular wave 不规则波
irregular wave generator 不规则造波机；不规则波发生器
irregular waveguide 异形波导
irregular wear 非正常磨损；不均匀磨损；不规则磨损
irregular weir 不规则堰；不规则小堰
irregular winding 不规则线圈
irregular working 不正常运行；不正常工作；不规则采区

irregulation 不规则
irregulation error 不规则误差
irrelative 无关系的；非相对的
irrelevance 不相关性；不恰当组合
irrelevance rule 不相关法则
irrelevancy 不相关
irrelevant 无关系的
irrelevant cost 无关成本；非相关成本
irrelevant image 不相关像
irrelevant substance 不相关物质
irremediable 难以改正的；无法补救的；不能修复的；不可换回的
irremediable defect 永久性损坏
irremediable disease 不治之症
irrepairable 无法挽救的；无法挽回的；无法补救的
irreparable 不能修理的
irreplaceable 无法替换的；不能调换的
irreplaceable use value 不可替代的使用价值
irreproachable 无瑕疵的；无以非议的
irreproachable quality 优质
irreproducibility 非再生性
irreproducible 不能再得的
irresistible 不可抗矩的
irresistible beyond control 不可抗矩的灾害
irresistible force 不可抗力
irresistible incident 不可抗拒的事件
irresolvable 不能分解的
irrespective of percentage 不受损失百分比限制；不论损失程度；不论百分比是多少
irrespective of size 统级
irrespective of weather conditions 不拘气候条件
irresponsibility 不负责任
irretrievable 无法弥补的；不能恢复的；不可挽回的
irretrievable loss 不可弥补的损失
irreversibility 不逆演化；不可逆状态；不可逆性；不可回溯性
irreversibility for a production set 生产集的不可逆性
irreversible 不能恢复的；不可逆的
irreversible action 不可逆作用
irreversible adiabatic expansion 不可逆绝热膨胀
irreversible adiabatics 不可逆绝热线
irreversible adiabatic wire 不可逆绝热线
irreversible adsorption 不可逆吸附
irreversible agglomeration process 不可逆附聚过程
irreversible agglutination 不可逆性凝集
irreversible ballasting system 不可逆压载系统
irreversible boundary movement 不可逆畴壁移动
irreversible breaking 不可逆破坏
irreversible cell 不可逆电池
irreversible change 不可逆变化
irreversible circulation 不可逆循环
irreversible coagulation 不可逆凝固
irreversible colloid 不可逆性胶体；不可逆胶体
irreversible compression 不可逆压缩
irreversible controls 不可逆操纵；无回力操纵系统
irreversible control system 不可逆控制系统
irreversible conversion 不可逆变化
irreversible conveyer 单向运输机
irreversible creep 单向蠕动；不可逆(转)徐变；不可逆蠕动
irreversible cycle 不可逆(性)循环
irreversible deformation 永久变形；塑性变形；不可逆变形；不可回复变形
irreversible denaturation 不可逆变性
irreversible diffusion 不可逆扩散
irreversible displacement 永久位移；不可逆位移；不可恢复的位移
irreversible domain wall displacement 不可逆磁畴壁位移
irreversible effect 不可逆效应
irreversible electrode 不可逆电极
irreversible energy loss 不可逆能量损失
irreversible engine 不可逆转发动机
irreversible flow 不可逆流动
irreversible fouling rate 不可逆污着速率
irreversible gear 不可逆传动装置
irreversible gel 不可逆凝胶
irreversible heat cycle 不可逆热循环
irreversible heat engine 不可逆热机
irreversible indicator 不可逆指示剂
irreversible inhibition 不可逆(性)抑制

irreversible isothermal expansion 不可逆等温膨胀
irreversible laser damage 不可逆激光损伤
irreversible magnetic process 单向磁化过程
irreversible magnetization 不可逆磁化
irreversible mercury poisoning 不可逆汞中毒
irreversible operation 不可逆运算;不可逆工作状态
irreversible optical damage 不可逆光损伤
irreversible permeability 不可逆磁(导)率
irreversible phase transition 不可逆相转变
irreversible phenomenon 不可逆现象
irreversible process 单向磁化过程;不可逆过程
irreversible reaction 不可逆反应
irreversible resistance 不可逆阻力
irreversible shrinkage 不能恢复的收缩;不可逆收缩
irreversible state 不可逆状态;不可逆态
irreversible steel 不可逆钢
irreversible steering 不可逆转向(机构)
irreversible steering assembly 自锁转向装置
irreversible steering gear 自锁转向装置;不可逆转向器
irreversible swelling 不可逆溶胀
irreversible tectonic deformation 不可逆构造变形
irreversible temperature-indicating paint 不可逆示温漆;不可逆示温颜料
irreversible thermodynamics 非平衡态热力学;不可逆热力学
irreversible throttling procedure 不可逆节流过程
irreversible toxic effect 不可逆性毒作用
irreversible transformation 单向转变;不可逆转变;不可逆变形
irreversible volatility 不可逆挥发性
irreversible water 不可逆水
irreversible work 不可逆功
irreversible worm 自锁蜗杆;止回蜗杆
irrevocable and confirmed credit 不可撤销保兑信用证
irrevocable condition 不可取消的条件
irrevocable control strategy 不可撤回的控制策略
irrevocable credit 不可撤销的信用证
irrevocable letter of credit 不可撤销的信用证
irrevocable letter of guarantee 不可撤销保函
irrigable 可灌的
irrigable area 可灌溉区;可灌溉面积;可灌溉地;水浇地面积
irrigable bench 可灌台地
irrigable land 有水源的适耕地;可灌溉地
irrigable terrace 可灌台地
irrigate 灌溉;浇灌
irrigated 已灌溉的
irrigated acreage 灌溉英亩数;灌溉亩数
irrigated area 灌区;灌溉区(域);灌溉面积;灌溉地区
irrigated crop 灌溉作物
irrigated electrostatic precipitator 湿式静电集尘器
irrigated farmland 灌溉农田
irrigated farms 灌溉农田
irrigated fields 水田;水浇地
irrigated filter 湿式过滤器
irrigated land 灌溉地;水浇地
irrigated nursery 灌秧田;水秧田
irrigated pasture 灌溉牧地;灌溉牧场
irrigated plantation 灌溉种植(园)
irrigated plot 灌溉地区;灌溉地段
irrigated region 灌溉区域
irrigated sewage field 污水灌溉田
irrigated takyr 灌溉龟裂土
irrigated water 灌溉水
irrigate farmland 沃田
irrgate the fields 浇地
irrigate water quota 灌水定额
irrigating dressing box 冲洗敷料盒
irrigating farming 灌溉农业
irrigating frequency 灌水次数
irrigating head 灌溉水头
irrigating needle 冲灌针
irrigating net(work) 灌溉网
irrigating of borehole 钻孔冲洗
irrigating of drill hole 钻孔冲洗
irrigating river 灌溉河流
irrigating shovel 灌溉开沟铲
irrigating stream 灌溉用水(源);灌溉水流;灌溉河流
irrigating system 灌溉系统
irrigating unit 灌溉单元
irrigating water 灌溉(用)水
irrigating water quota 灌水定额;灌溉定额
irrigation 灌水;灌溉;冲洗法;分畦淹灌
irrigation agriculture 灌溉农业
irrigational facility 灌溉设施
irrigation and drainage engineering 农田水利工程
irrigation and drainage network 排灌网
irrigation and drainage pumping station 排灌站
Irrigation and Power 灌溉与动力(期刊)
irrigation application efficiency 灌溉效率;灌溉水有效利用率
irrigation area 灌溉面积
irrigation area planning 灌区规划
irrigation area ratio 灌溉面积比
irrigation basin 灌溉盆地
irrigation before seeding 播前灌水;播前灌溉
irrigation bonds 水利债券
irrigation border 灌水渠埂;灌溉畦;灌溉(挡水)田埂
irrigation bucket 戽斗
irrigation by channelling water 水灌溉
irrigation by electric(al) power 电力灌溉
irrigation by flooding 淹灌;漫灌
irrigation by furrow 沟灌
irrigation by gravity 自流灌溉
irrigation by infiltration 浸润灌溉;渗流灌溉
irrigation by mechanical power 机械灌溉
irrigation by pump 用泵抽水灌溉;泵灌
irrigation by pumping 扬水灌溉;提水灌溉;抽水灌溉
irrigation by sprinkling 喷灌
irrigation by stages 分散灌水
irrigation by surface flooding 地面淹灌
irrigation by trickling 滴灌
irrigation canal 灌渠;灌溉渠(道);灌溉区;灌溉(水)渠
irrigation canal network 灌溉渠道网
irrigation canal system 灌溉(渠道)系统
irrigation capacity 灌水量
irrigation channel 灌溉渠(道);灌溉(水)渠
irrigation coefficient 灌溉系数
irrigation condition 灌溉条件
irrigation crop 灌溉作物
irrigation cycle 灌溉周期
irrigation dam 灌溉用坝
irrigation date 灌水日期
irrigation demand 灌溉需水量
irrigation design 灌溉设计
irrigation district 灌区;灌溉区
irrigation ditch 灌(水)沟;灌溉支渠;灌溉(明)沟;渠道
irrigation ditcher 灌渠挖掘机
irrigation ditch outlet 灌渠出水口
irrigation diversion 灌溉引水
irrigation draft 灌溉渠水
irrigation draught 灌溉取水
irrigation efficiency 灌溉效率
irrigation engineer 灌溉工程师
irrigation engineering 灌溉工程
irrigation engineering and maintenance 灌溉工程施工与维修
irrigation equipment 灌溉设备
irrigation experiment 灌溉试验
irrigation facility 灌溉设施
irrigation farming 灌溉农业;灌溉耕作
irrigation field 灌溉地;灌溉场
irrigation for frost protection 进行防霜冻灌溉;防霜冻灌溉
irrigation frequency 灌水周期;灌水频次;灌溉次数
irrigation furrow 灌溉沟槽;灌水沟;灌溉(水)沟
irrigation grease 喷雨枪
irrigation head 灌溉水头;灌溉水舌
irrigation hydrant 灌溉水栓;灌溉龙头
irrigation installation 灌溉装置;灌溉设备
irrigation intensity 灌溉强度
irrigation interval 灌水间距;灌溉周期
irrigation in their given order 依次灌溉
irrigationist 水利专家
irrigation junction 灌溉枢纽
irrigation land 灌溉地
irrigation lateral(canal) 灌溉(支)渠
irrigation layout plan 灌区平面布置图;灌区布置规划
irrigation level 灌溉水位
irrigation line 灌溉管道;灌溉线
irrigation main 灌溉干管;灌溉干渠
irrigation method 灌水方式;灌水方法;灌溉方法
irrigation methods and practices 灌溉方法及应用
irrigation month 灌溉月份
irrigation mouth 灌水口
irrigation net(work) 灌溉网;排灌网
irrigation norm 灌溉规范;灌溉定额;灌溉标准
irrigation normal 灌溉准则
irrigation number 灌水次数
irrigation of farmland 农田灌溉
irrigation of orchards 果园灌溉
irrigation operation 灌溉作业
irrigation operation waste 灌溉用水损失
irrigation organization 灌水组织
irrigation pattern 灌溉方式
irrigation period 灌水一次的时间(作物耗水高峰阶段);灌水时间;灌溉期;灌溉季节
irrigation pipe 灌溉水管;灌溉管
irrigation pipe coupler 灌溉水管接头
irrigation pipeline 灌溉管道
irrigation pipe trailer 灌溉管拖车
irrigation plan 灌溉计划
irrigation plant 灌溉站;灌溉设施;灌溉设备
irrigation pool 灌溉水池
irrigation practice 灌溉制度;灌溉措施
irrigation principle 灌溉原则
irrigation procedure 灌溉制度;灌溉程序
irrigation program(me) 灌溉流程图;灌溉方案;灌溉程序;灌溉制度
irrigation project 灌溉计划;灌溉工程
irrigation pump 灌溉用泵;灌溉水泵;灌溉泵
irrigation pumping 灌溉扬水;灌溉抽水
irrigation pumping station 灌溉抽水站;灌溉泵站
irrigation purpose 灌溉目的
irrigation rate 灌溉标准;灌溉(用水)率
irrigation regime 灌溉制度
irrigation requirement 灌水定额;灌溉需水(总)量;灌溉定量;灌溉定额
irrigation reservoir 灌溉水库
irrigation return flow 灌溉回(归)水
irrigation rig 灌溉装置;喷灌装置
irrigation rotation 灌溉轮换;轮灌
irrigation runoff water quality 灌溉径流水质
irrigation schedule 灌溉进度表
irrigation scheme 灌溉方案;灌溉网;灌溉流程图
irrigation science 灌溉科学
irrigation season 灌溉季节;农田用水季节
irrigation sewage disposal 灌溉污水处理;污水灌溉处置;污染灌溉处置
irrigation sewage disposal fields 污水灌溉地
irrigation shovel 灌溉开沟铲
irrigation siphon 灌溉虹吸管
irrigation station 灌水站;灌溉站
irrigation storage 灌溉蓄水;灌溉水库;灌溉库容
irrigation stream 灌溉水流
irrigation strip 灌溉条
irrigation structure 灌溉建筑物;灌溉沟筑物
irrigation syringe 冲洗器
irrigation system 灌溉系统
irrigation system of sewage 水灌溉系统
irrigation technique 灌水技术;灌溉水技术
irrigation test 灌溉试验
irrigation time 灌水时间
irrigation timing and length 灌溉时度和长度
irrigation tobacco 灌溉烟草
irrigation tool 灌水工具
irrigation treatment 灌溉处理
irrigation trench 灌溉沟渠
irrigation tunnel 灌溉引水隧洞;灌溉隧洞;灌溉隧道;水利隧道
irrigation type 灌溉型
irrigation valve 灌水栓;灌溉阀
irrigation water 灌溉用水;灌溉水
irrigation water demand 灌溉需水量;田间需水量
irrigation water leakage 灌溉水渗入
irrigation water management 灌水管理;灌溉用水管理
irrigation water outlet works 灌溉用水泄水工程
irrigation water quality 灌溉(用水)水质
irrigation water quality assessment 灌溉用水水质评价
irrigation water quality standard 灌溉(用水)水质标准;农田灌溉水质标准
irrigation water quota 灌水定额

irrigation water requirement 灌溉需水量；田间需水量
irrigation way 灌水方式
irrigation weir （低坝的）灌溉堰；灌溉用堰
irrigation well 灌溉水井；灌溉井
irrigation with ditch 渠灌
irrigation with ditch and drainage with well 渠灌井排
irrigation with flood water 引洪淤灌；引洪淹灌；引洪漫灌；引洪漫地
irrigation with sewage 污水灌溉；沲灌
irrigation with wastewater 污灌；废水灌溉
irrigation with well 井灌
irrigation with well and ditch 井渠联灌
irrigation work 灌溉工程
irrigator 灌溉者；灌注器；灌水员；灌溉设备；灌溉器具；灌溉车；喷灌机；施灌人员；冲洗器
irritant 刺激物
irritant gas 刺激性气体
irritant poison 刺激性毒物
irritated epiphora with cold tear 迎风冷泪
irritating agent 刺激剂
irritating compound 刺激性化合物
irritating compound smog 刺激性化合物烟雾
irritating pollutant 刺激性污染物
irrometer 土壤湿度计
irrotational 非旋光的；不旋转的
irrotational binding 无旋键
irrotational deformation 无旋转形变；无旋形变；非转动变形
irrotational disturbance 不旋转扰动
irrotational field 无旋场；非旋场
irrotational flow 无旋流（动）；无涡流；非旋转流
irrotational flow model 无旋流模型
irrotationality 无旋（性）
irrotational motion 无旋运动；无旋流动；无涡运动；非旋运动
irrotational strain 无旋应变；非旋应变
irrotational translation wave 无旋转推进波
irrotational type of motion 无旋转型运动
irrotational vector 无旋向量
irrotational vector field 无旋（转）向量场；无旋矢量场
irrotational velocity field 无旋流速场
irrotational vibration 无旋振动
irrotational vortex 无漩涡流；非旋转涡流
irrotational wave 无旋波
irruption 急速增加；侵袭
irruption vein 侵入岩脉
irruptive rock 火成岩；侵入岩
irtron 红外（线）光电管
Irwin consistometer 艾文稠度计
Irwin slump test 艾文下沉试验
irythermal 广温的
isa 锰铜
Isaacs high-speed sampler 艾萨克斯高速采样器
Isaacs Kidd water sampler 艾萨克斯·基德采水样器
isabellin 锰系电阻材料
isabelline 灰黄色的
isacoustic 同强震声；等响
isacoustic(al) curve 等响度曲线
isalea 等日射线
isallobar 等变压线；大气压等变化线
isallobaric analysis 等高压分析
isallobaric chart 等变压图
isallobaric gradient 等变压梯度
isallobaric high 正变压中心【气】
isallobaric low 负变压中心
isallobaric wind 等变压风
isallohypsic wind 等变高风
isallolypse 等变高线
isallotherm 等变温线
isallothermic chart 等温线图
isallothermic line 等变温线
isametral 等偏差线
isamic acid 衣氨酸
isanabat 等上升速度线
isanakatabar 等气压较差线
isanamorphic line 等变形线
isanemone 等风速线
isanomal 等偏差线；等偏差线；等距（离）常线
isanomaly 等异线；等距平线
Isano oil 衣散油

isaphenin 二乙酰酚靛红
isarithm 地形等高线；等值线
isarithmic line 等值线
isarithmic map 人口密度图
isasteric 等容的
isasteric surface 等容面
isatin 靛红
isatron 质谱仪
isaurore 极光等频线
isba 俄罗斯木屋
Iscorex 艾斯科勒克斯装饰用金银线
isdmun 整块端砌
I-section 工字形剖面；工字（形）断面；工形截面
I-section joist 工字形小梁；工字钢
I-section spar 工字形翼梁
I-section steel pile 工字形钢桩
isenerg(e) 等内能线
Isenthal automatic voltage regulator 振荡型自动稳压器；爱生塔尔自动稳压器
isenthalp 等焓线
isenthalpic 等热函线；等焓线；等焓的
isentrope 等熵线；等熵面
isentrope line 等熵线
isentropic 等熵的；等熵
isentropic analysis 等熵面分析
isentropic change 绝热变化；等熵变化
isentropic chart 等熵线图；等熵图
isentropic compression 等熵压缩
isentropic condensation level 抬升凝结高度
isentropic condition 等熵条件
isentropic efficiency 等熵效率
isentropic expansion 等熵膨胀
isentropic expansion nozzle 等熵膨胀喷管
isentropic flow 等熵流动
isentropic heat drop 等熵热降
isentropic index 等熵指数
isentropic line 等熵线
isentropic procedure 等熵过程
isentropic process 等熵过程
isentropic recompression 等熵再压缩
isentropic sheet 等熵层
isentropic stagnation heating 等熵滞止加热
isentropic stream 等熵流
isentropic streamline 等熵流线
isentropic surface 等熵面
isentropic thickness chart 等熵厚度图
isentropic weight chart 等熵重量图
isentropic work 等熵功
isentropity 等熵性
iseoric line 等年温较差线
isepire 等降水大陆度线
iserine 低铁金红石
isethionate 羟乙磺酸酯
isethionic acid 磺基乙醇；羟乙（基）磺酸
I-shape 工字形
I-shaped 工字形的
I-shaped beam 工字钢
I-shaped cross-section 工字形截面；工字形断面
I-shaped plan 工字形平面
I-shaped prop 工字撑
I-shaped suspension shaft 工字悬挂轴
Isherwood frame 纵肋骨；伊舍伍德；框架
Isherwood framing 纵肋结构系统；纵肋架式；纵结构式
Isherwood system 纵结构式；伊舍伍德；体系
Ising coupling 伊辛耦合
isinglass 鱼胶；鳔胶；白云母薄片
isinglass gold-size 鱼胶贴金漆
isinglass stone 云母
Ising model 伊辛模型
Ising-Stevens distribution 艾辛—史蒂文斯分布
iskymeter 现场土壤剪切仪；现场剪切触探仪
Islamic academy 伊斯兰教经学院
Islamic architecture 伊斯兰教建筑
Islamic Development Bank 伊斯兰开发银行
island 舰台；舰桥；栖件柱；岛
island arc 弧形列岛；岛弧
island arc and trench system 弧沟系
island arc chain 岛弧链
Island arc-Continent collision 岛弧大陆碰撞
island-arc gravity anomaly 岛弧重力异常
island arc in East Asia 东亚岛弧
Island arc-island arc collision 岛弧岛弧碰撞
island-arc sea 岛弧海

island-arc system 岛弧系
island arc-trench system 岛弧—海沟体系
island barrier 拱门岛
island berth 岛式泊位
island biogeology 岛屿生物地质学
island breakwater 独立（式）防波堤；岛式防波堤；岛堤
island chain 列岛；岛链
island chart 岛屿图
island code 岛码
island continent 岛洲
island counter 岛式柜台
island country 岛国
island developing country 发展中岛国
island diabasic texture 岛状辉绿结构
island ecology 岛屿生态
island ecosystem 岛屿生态系统
island endemic species 岛屿特有种
islander 岛上居民
island fauna 岛屿动物区系
island flora 岛屿植物区系
island frozen soil 岛状冻土
island garden 岛园
island group 群岛；岛群
island harbo(u)r 岛式港（口）；岛港
island hill 孤基岩丘；岛（状）丘；岛山
island ice 岛冰
island kitchen 岛式厨房（炉灶等布置在中央部位的厨房）
island lighting column 安全岛上的照明灯柱（道路）
island line 岛际航线
island-mainland connection survey 岛陆联测
island mesa 岛状平顶山
island method 岛筑开挖法
island mole 岛状防波堤；岛式防波堤；岛堤
island mountain 岛状山；岛山
island of stability 稳定岛
island platform 岛式站台；岛式平台
island platform roof 岛式平台屋顶
island port 岛港
island relic species 岛屿残留种
island roller 凸纹轧辊（钢板防滑）；格纹轧辊（钢板防滑）
islands and islets 岛屿
island sea 岛海
island serving shelf 平台式工作架
islands for science 科研用的岛屿
island shelf 岛架
island short 岛压点短路
island site 指居住街坊；孤立场地；独立基地
island slope 岛坡
island station 岛式候车站；岛式车站
island station platform 岛式站台
island station roof 单桩支承的屋架（如车站月台上采用的）；单柱屋顶
island structure 岛状结构
island survey(ing) 岛屿测量
island tanker berth 岛式油船泊位
island terminal 岛式码头
island touch 岛压点接触
island-trench type geosyncline 岛弧海沟型地槽
island-type 岛弧型
island-type construction 独立（式）建筑；岛式建筑
island-type cooking range 岛式炉灶
island-type geosyncline 岛弧型地槽
island-type wharf 岛式码头
island universe 岛宇宙
island volcano 岛火山
isle(t) 小岛；岛状地带；陆连岛
iso abnormal 等异常线
isoabnormal line 等异常线
isoabsorptive point 等吸光点
isoacetylene 异乙炔
isoacoustic(al) 等响线
isoacoustic(al) chart 等声强线图
isoacoustic(al) curve 等响线
isoactivity method 等活度法
iso-agglutination 同种凝集
isoagglutinin 同种凝集素
isoallyl 丙烯基
iso amplitude 等振幅线；等振幅的；等变幅线
isoamyl acetate 乙酸戊酯
isoamyl benzoate 苯甲酸异戊酯
iso-amyl butyrate 丁酸异戊酯

isoamylene 异戊烯
isoamyl ketone 二异戊基甲酮
isoamyl propionate 丙酸异戊酯
isoanabaric center 等升压中心
isoanabase line 等沉压线
isoanomaly 等偏差线；等距平线；等常线
isoanomaly curve 等异常曲线
isoatmic line 等蒸发线
isoaxial magnetic anomaly 等轴磁异常
isoazimuth 等方位线；等方位角
isoballast 等压载
isobar 同质异位素；同量异序素；同量异位素；等主应力线；等压线；等权；等气压线
isobar decay 同质异位素衰变
isobaric 等压的；等气压的
isobaric abundance 同量异位素的丰度
isobaric analogue 同量异位素相似态
isobaric analysis 等压分析
isobaric atom 同量异序原子
isobaric chamber 恒压舱
isobaric change of gas 气体的等压变化
isobaric channel 等压线通道
isobaric chart 等压线图；等压面图
isobaric combustion 等压燃烧
isobaric configuration 等压面形势
isobaric contraction 等压性收缩
isobaric control 等压调节
isobaric convergence 等压线辐合
isobaric cooling 等压冷却
isobaric divergence 等压线辐散
isobaric diving equipment 恒压潜水装置；常压潜水装具
isobaric element 同量异位素
isobaric expansion 等压膨胀
isobaric gas counter-diffusion 等压气体逆向扩散
isobaric heat capacity 恒压热容
isobaric heterotope 同量异位素
isobaric line 等压线；等气压线
isobaric line chart 等压线图
isobaric map 等压线图
isobaric mass-change determination 等压质量变化测定
isobaric model 同量异位素模型
isobaric motion 等压运动
isobaric multiplet 同量异位多重态
isobaric nucleus 同量异位核
isobaric procedure 等压过程
isobaric process 恒压过程；等压过程
isobaric section 等压载面；等压截面
isobaric slope 等压面坡度
isobaric space 同量异位素空间；等压空间
isobaric state 同质量位态
isobaric surface 等压面；等气压面
isobaric topography 高度型式
isobaric transformation 同量异位转化
isobaric triad 同量异位三重态
isobaric triplet 同量异位三重态
isobar model 同量异位模型
isobarometric 等气压的
isobarometric(al) line 等压线
isobar polynomial 等权多项式
isobar spacing 等压线间距
isobase contour 等基线图
isobases 等基线
isobath 等值；等(水)深线；等海深线；等高线
isobath chart 等深线海(底地形)图
isobathic 等深(的)
isobathic chart 等深线图；等深线海(底地形)图
isobathic line 等深线
isobath interval 等深距
isobath map 等深线图；等海深线图
isobath map of surface runoff 地表径流深度等值线图
isobath map of underground runoff 地下径流深度等值线图
isobath of water table 潜水位等值线；地下水位等深线；地下水位等高线
isobaths current 等深线流
isobaths of structure in the top of oil layer 油层顶构造等深图
isobathye 等深(的)
isobathye line 等深线
isobathytherm 海水等深等温线；等温深度线；等温深度面

isobiochore 等生态域线
isobitateral leaf 等面叶
isobits 等同二进制位组
isoborneol 异冰片
isobornyl acetate 醋酸异冰片酯
isobornylene 异冰片烯
isobornyl thiocyanoacetate 异冰片酯
isobront 雷暴等时线【气】；等雷暴日数线；初雷等时线
isobrontal line 等雷暴日数线
isobutanol 异丁醇
isobutene 异丁烯
isobutene polymer 聚异丁烯
isobutene rubber 异丁橡胶；异丁(烯)橡胶；丁烯(合成)橡胶
isobutyl acetate 乙酸异丁酯
isobutyl acetylene 异丁基乙炔
isobutyl acrylate 丙烯酸异丁酯
isobutyl alcohol 异丁醇
isobutyl aminobenzoate 氨基苯甲酸异丁酯
isobutyl benzoate 苯甲酸异丁酯
isobutyl bromide 异丁基溴
isobutylene 异丁烯
isobutylene-isoprene 异丁烯橡胶
isobutylene-isoprene copolymer 异丁橡胶
isobutylene-isoprene rubber 异丁橡胶；异丁烯-异戊(间)二烯橡胶
isobutylene resin 聚(异)丁烯树脂
isobutyl isobutyrate 异丁酸异酯
isobutyl phenylacetate 苯乙酸异丁酯
isobutyl propionate 丙酸异丁酯
isobutyl ricinoleate 蓖酸异丁酯
isobutyl urethane 异丁氨基甲酸乙酯
isobutyric 异丁
isobutyric acid 异丁酸
isobutyrone 二异丙基甲酮
isocaloric ration 等能量日粮
isocals 等发热量线
isocamphor 异樟脑
isocandela 等烛光；等光强
isocandela diagram 等烛光线图；等光强图
isocarb 等含碳量线；等固定碳线
isocarbon couple 同碳偶合
isocarbophos 水胺硫磷
isocatabase 等下沉线；等降线；等高线
isoceles spheric(al) triangle 等腰球面三角形
isoceles trapezoid 等腰梯形
isocenter 航摄失真中心；等角点
isocentre plot 等角点辐射三角测量
isocenter triangulation 等角点像片三角测量
isocephalic 有同等高度的头部(浅浮雕中)
isoceraunic 等雷雨(次数和强度)线；等雷频率线
isocercal 等尾的
isochasm 极光等频线
isocheim 冬季等温(平均)线；等冬温线
isochemical indication 同化学性质指示
isochemical metamorphism 等化学变质作用
isochime 等冬温线
isochion 等雪线；等雪深线；等雪量线；等雪厚线
isochlor (地下水的)等含氯量线
isochore 恒容线；等体积线；等容线；等层厚线
isochoric 等体的
isochoric change 等容变化
isochoric deformation 等体积变形；等容积变形
isochoric process 等容过程
isochroism 等时值
isochromate 等色线
isochromatic 同色的；等色的；单色的
isochromatic curve 等色线；等色曲线
isochromatic fringe 等色边纹；单色边纹
isochromatic fringe method 等色条纹法(光弹性力学用)；同色条纹法
isochromatic fringe pattern 等色条纹图
isochromatic interference belt 等色干涉带
isochromatic line 等色线
isochromatic line of sea 海水等水色线
isochromatic pattern 等色线图(偏光试验)
isochromatic photograph 等色照片；等色图像
isochromatics (光弹性分析的)等色线
isochromatic stimuli 等色刺激
isochromatic zone 等色带
isochrome 单压线
isochrome map of photoelastic 光弹实验等色线图
isochromism 同步性

isochromosome 等臂染色体
isochron 等时线
isochronal 等时
isochronal annealing 均时退火；等时退火
isochronal error 等时性误差
isochronal line 同震线；等时线
isochronal map 等时线图
isochronal method 等流时线法
isochronal slices 等时切片
isochronal spiral regulator 等时螺线调节器
isochronal test 等时法测试；等时测试法
isochronal vibration 等时振动
isochron(e) 交通等时区；同时(力)线；同时水坡线；等稳线；等时(曲)线；等时区；等时差线；等年龄线；等流时线；瞬压曲线
isochron(e) age 等时年龄
isochron(e) chart 等时线图；等流时线图
isochroneity 同时性；等时性
isochron(e) line 等龄线
isochron(e) method 同时线法；等(流)时线法
isochronia 等时值
isochronic 等时的
isochronism 等时性
isochronism speed governor 同步调速器
Isochronograf 等时仪
isochronograph 等时图；等时计
isochronous 等时的
isochronous circuit 等时电路
isochronous control 无差调节
isochronous controller 同步控制器
isochronous correspondence 等时对应
isochronous cyclotron 等时性回旋加速器
isochronous digital signal 等时数字信号
isochronous distortion 等时性畸变
isochronous error 等时性误差
isochronous governor 恒速调速器；同步调节器
isochronous modulation 同步调制；等时调制
isochronous multiplexer 等时复用器
isochronous oscillation 同步振荡；等时振荡
isochronous rolling 等周期摆动；等时摆动
isochronous scanning 同步扫描；等时扫描
isochronous signal distortion tester 等时信号失真测试器
isochronous stratigraphic(al) unit 等时性地层单位
isochronous surface 等时面
isochronous transmission 等时传输
isochronous vibration 等时振动
isochron surface method 等时面法
isocitric acid 异棕檬酸
isoclasite 水磷钙石
isoclimate line 气候等值线
isoclimate zone 同气候带
isoclimatic line 气候等值线；等气候线
isoclimatic zone 气候相同地带
isoclinal 均斜的；等斜的；等磁倾线；等磁偏线
isoclinal chart 等倾线图
isoclinal fault 等斜断层
isoclinal fold 等斜褶皱
isoclinal fold in rock beds 岩层等斜向褶皱
isoclinal line 等磁倾线；等磁偏线
isoclinal method 等倾线法
isoclinal valley 同斜屋谷；等斜谷
isocline 等斜线；等倾(斜)线；等磁偏线
isocline chart of photoelastic test 光弹实验等倾线图
isoclined method 等倾法
isocline equation 等倾线方程
isocline line 等倾(斜)线
isocline method 等倾法
isocline planes 等斜平面
isoclinic 等斜的；等倾的；等磁偏线
isoclinic chart 等磁倾线图
isoclinic equator 等磁倾赤道
isoclinic line 等斜线；等倾向线；等倾(斜)线；等磁倾线；等磁偏线
isocolloid 同质胶体
iso-component 异构化组分
isocompound 异构化合物
isocon 分流直象管；分流式正析像管
isoconcentration line 等浓度线
isoconcentration map 等浓度图
isoconcentration point 等浓度点
isoconcentration reduction 等浓度削减

isoconncentration 等浓度
isocorrelation line 等相关线
isocost curve 等成本曲线;等费用曲线
iso-cost line 等成本线
isocotidal 等同潮的
isocount 等放射性线
isocrackate 异构裂化物;加氢裂化产物
isocracking 异构裂化
isocrotyl bromide 异丁烯基溴
isocryme 最冷期等水温线
isocurlus 等漩涡强度线
isocyanate 异氰酸酯;异氰酸盐;封闭型异氰酸酯
isocyanate adhesive 异氰酸酯黏合剂
isocyanate plastics 异氰酸盐塑料
isocyanate resin 异氰酸盐树脂;异氰酸树脂
isocyanate varnish 异氰酸酯清漆;异氰酸盐清漆
isocyanic acid 异氰酸
isocyanide 异氰化物
isocyanilic acid 怪氰酸
isocyanurate foam 异氰脲酸酯泡沫;异三聚氰泡沫
isocyanuric acid 异氰脲酸
isocycle 等原子环
isocycle nucleus 等环核
isocyclic compound 碳环化合物
isocylindric(al) 等圆柱的
isocylindric(al) projection 等圆柱投影
isodamage curve 等震害曲线
Isoda metal 铅基巴氏合金
isodapane 等总运费线
isodef 等少量百分率线
iso-deflection 等挠度
iso-deflection diagram 等挠度图
isodense 等密线
isodense compact 等致密压坯
isodense process 等密度线法
isodensitometer 等密度计
isodensitracer 等密度描绘仪
isodensity 等密度
isodensity diagram 等密度图
isodensity pseudocolo(u)r encoding image 等密度假色编码图像
isodiameter 等直径
isodiametric 等直径的
isodiametric crystal 等侧轴晶体
isodiazotate 反重氮酸盐
isodibromosuccinic acid 异二溴丁二酸
isodiff 等改正线;等差线
isodimorphism 类质二象;同质二形
isodimorphous 同质二形的
isodirectional distribution 同向分布
isodispersoid 等相胶质
isodomon 整块石面砌;整块端砌;块石端砌
isodomum 顺砖错砌;等高层砌石块;整块(石)端砌;块石端砌
isodose curve 等剂量曲线
isodose line 等剂量线
isodose recorder 等剂量记录器
isodrome governor 等速调速器
isodromic 恒值的;等速的
isodromic controller 等速控制器
isodrosotherm 等露点线
isodynam(e) 等(风)力线;等磁力线;等磁力图
isodynamic(al) 等能的;等磁力;等磁动力的
isodynamic(al) chart 等磁力图
isodynamic(al) chart of X X 等值线图
isodynamic(al) chart of Y Y 等值线图
isodynamic(al) chart of Z Z 等值线图
isodynamic(al) line 等强磁力线;等力线;等磁线;等磁强度;等磁力线
isodynamic(al) magnetic separator 等磁力分离仪
isodynamic(al) map 等磁力线图;等磁力地图
isodynamic(al) separator 等磁力分选器;等磁力分离仪;等磁力磁选机
isodyne 等力线;等法向应力线
isoefficiency curve 等效率线;等效率曲线
Isoelastic 等弹性弹簧合金;高镍弹簧钢;等弹性(的)
isoelasticity 等弹性
isoelectric(al) 等电位的
isoelectric(al) edge 等电面;等电边
isoelectric(al) focusing 等电聚焦;等电点聚焦
isoelectric(al) fractionation 等电点分离
isoelectric(al) heating 等电位加热
isoelectric(al) pH 等电 pH

isoelectric(al) pH value of mineral 矿物等电 pH 值
isoelectric(al) point 等电(离)点
isoelectric(al) precipitation 等电沉淀作用
isoelectric(al) zone 等电区
isoelectrofocusing 等电聚焦
isoelectronic compound 等电子化合物
isoelectronic doping 等电子掺杂
isoelectronic focusing 等电子聚焦
isoelectronic ion 等电子离子
isoelectronic recombination center 等电子复合中心
isoelectronic sequence 等电子序;等电子数序
isoelectronic spectrum 等电子光谱
isoenergetic 等能的;等能
isoenergetic acceleration 等能加速度;绝热加速度
isoenergetic displacement 等能位移
isoenergetic stability 等能稳定性
isoenthalpic flow 等焓流动量
iso-enthalpy 等焓线
isoentropic change 等熵变化
isoentropic expansion 绝热膨胀
isoentropic nozzle 等熵喷管
isoepitaxial growth 同质外延生长
iso-epitaxy 同质外延
Isoetes sinensis Palmer 中华水韭
isofacial rock 等相岩
isofacies map 等相图
isoferroplatinum 等轴铁铂矿
isoflavanone 二氢异黄酮
Isoflex 埃索富力斯(一种塑性绝缘材料)
isoflux 等通量
isofoot candle 等勒克斯线面
isofoot candle line 等照(度)线
isoforming 低温形变(热)处理
isofrequency 等频率
isofrequency curve 相等频数曲线
isofrequency pseudocolo(u)r encoding image 等频率假彩色编码图像
isofrequent 等频的
isofrigid 等寒的
isofronts-preiso code 等锋等压电码
isogal 等重力线
isogam 重力等差线;等磁场强度线
isogamme 等磁场强度线
isogel 同构异量质凝胶
isogen 等偏角线
isogenesis 同源发生
isogenetic 同源的;同期成的;同成因的
isogeny 同源
isogeotherms 等地温线;地下等温线
isogon 正多边形【数】;同风向线;同方向线;等角多角形;等磁偏
isogonal 等角偏线;等方位线;等磁偏的
isogonal chart 等磁偏角线图
isogonal conjugate points 等角共轭点
isogonality 等角变换;保角变换
isogonal line 等角线
isogonal mapping 等角映射
isogonal trajectory 等角曲线系;等角轨线;等磁偏角轨线
isogonal transformation 等角变换
isogonic 等偏角;等角的;等磁偏线;等磁偏(的)
isogonic chart 等偏(线)图;等磁偏角线图;等磁差图;D 等值线图
isogonic curve 等方位线
isogonic line 同向线;等斜面线;等偏(角)线;等方位线;等磁倾线;等磁偏线;等磁差线
isogonic map 等磁倾线图
isogonism 等角
isogony 同速生长;等比生长
isogor 等油气比
isogor map 等油气比图
isograd 等变线
isogradal 等温等压线的;等变质级的;等变线的;等变度的
isograde 等温等压线的;等坡的;等变质级的;等变线的;等变度的
isograde rock 等变岩
isogradient 等梯度线
isograft 同系移植物
isogram 等值线图;等值线;等高线图
isogramme line 等磁场线
isograms 等重力线

isogranular 等粒的
isogranular texture 等粒结构
isograph 求根仪;万能尺;等值线图
isograph of erosion and deposition 冲淤等值线图
isography 摹绘
isogrid 地磁等偏线;地磁等变线
isogrid chart 地磁等偏线图
isogriv 等坐标磁偏角线
isogyre 同消色线;等施干涉条纹
isogyre pump turbine 同轴水泵水轮机;水轮泵
isogyres 消光影
isohaline 等盐线;等含盐量线;等盐度线;等盐变线;等盐(度)的
isohaline map 等盐度图
isohalsine 等盐(度)的
iso-hardness diagram 等硬度图
iso-hardness value 等硬度值
isohedral 等面的
isoheight 等高线
isohel 等日照线
isohelic line 等日照线
isohion 等雪线高度线;等雪日线;等雪厚线
isohronal 同步的
isohronon 精密时计
isohume 等水分线;等湿(度)线
isohydric 等碱硬度的;等氢离子的;等 pH 值线
isohydric concentration 等氢离子浓度
isohydric indicator solution 等氢离子指示剂溶液
isohydric technique 等氢离子技术
isohydric welder 等氢离子弧焊机
isohydrocarbon 异构烃
isohyet 等沉陷线;等雨量线;等降雨(量)线;等降水量线;等沉淀线
isohyetal 等沉陷线;等降雨量的
isohyetal chart 降雨分布图
isohyetal gradient 等雨量线梯度
isohyetal line 等雨量线
isohyetal line chart 等雨量线图
isohyetal map 雨量分布图;降雨量等值线图;等(降)雨量线图
isohyetal method 等雨量线法
isohygromen 等湿(度)线
isohygrometric line 等湿(度)线
isohygrotherm 等水温线
isohyperthermic 等过热的
isohypse 等高线
iso-income 均等收入
isoindigo 异靛
isoindole 异吲哚
isoindole metal complex pigment 异吲哚金属络合物颜料
isoindolinone pigment 异吲哚啉酮系颜料
isoinhibitor 同效抑制剂
iso-intensity curve 等降雨(量)强度曲线
iso-ionia 等离子浓度
iso-ionic area 等离子区
iso-ionic point 等离子点
iso-jet burner 高速等温烧嘴
isokatabaric center 等降压中心
isokatabase 等下沉线
isokeraunic 等雷雨(次数和强度)线;等雷频率线
isokinesis sampling 流态采样
isokinetic 等风速线;等动能线
isokinetic flow 等动力流
isokinetic focusing 等动力聚焦法
isokinetic sample 等动态样品
isokinetic sample introduction 等动力进样
isokinetic sampling 同流态取样;等速取样;等速采样;等动能采样(法);等动力取样(法)
isokite 氟磷钙镁石
isokom 等黏线
isolabel(l)ing 同标记
Isolament 埃索蜡蒙(一种表面硬化剂,可以防止水泥地面起尘)
isolantite 高频绝缘陶瓷
isolate 隔离;绝缘的;抽数
isolate bus 绝缘母线
isolated 隔离
isolated actuated control 单点感应控制
isolated actuated signal control 独立感应信号控制
isolated adaptive routing 隔离式自适应路由选择
isolated aeration bioreactor 隔油曝气生物反应器
isolated amplifier 隔离放大器
isolated area 隔离区;隔离带;偏僻地区;独立区

isolated base 隔离底座
isolated basin 孤立盆地;孤立海盆
isolated bath tub 独立浴盆
isolated beam 独立梁
isolated bell-tower 独立钟塔;独立钟楼
isolated block 独立区段
isolated body 隔离体;分离体;分隔体
isolated brake 独立制动
isolated breakwater 岛式防波堤;岛堤
isolated building 独立房屋;独立建筑物
isolated burst 孤立暴
isolated chapel 独立小礼拜堂
isolated chimney 独立(大)烟囱;孤立烟囱
isolated circular cell 单个圆形格体
isolated column 自立柱;独立柱
isolated component 分立元件
isolated consultation room 隔离诊室
isolated control 独立控制
isolated controller 独立控制器;独立控制机;单点控制器;单点控制机
isolated corral 隔离畜圈
isolated culture 分离培养法
isolated danger 孤立障碍物
isolated danger buoy 孤立障碍(浮)标;孤立危险物标
isolated danger mark 单独险标;孤立险标
isolated digital output module 隔离数字输出组件
isolated distribution 孤立状分布
isolated ditch 分离沟
isolated double bond 孤立双键;隔离双键
isolated electric(al) power station 孤立电厂
isolated electrode 隔离电极
isolated event 孤立事件;独立事件
isolated feature 独立地物
isolated fixed cycle signal 单点定周期信号
isolated footing 隔离基础;独立基脚;独立基础;独立底脚
isolated foundation 隔离基础;独立基础;单独基础;防振基础
isolated galaxy 孤立星系
isolated gate 隔离栅
isolated gate bipolar transistor 绝缘门双极晶体管
isolated generating plant 自备发电厂;独立发电厂;单独运转发电站;单独运转发电厂
isolated generating station 自备发电站
isolated grid 独立格网
isolated hall 独立大厦
isolated hill 孤山
isolated house 独立房屋
isolated indication 单个显示
isolated intersection 单独交叉口
isolated intersection control 独立交叉口信号控制;单点交叉口控制
isolated interstice 孤立裂缝;孤立孔隙;孤立间隙;单独裂隙
isolated lagoon 隔离粪尿池
isolated layer 隔层
isolated lift tower 独立提升塔
isolated load 孤立荷载;集中荷载;集中负载;单个荷载
isolated local minimizer 孤立局部最小化
isolated location 隔离位置;隔离单元区
isolated masonry wall 独立砖石墙;独立圬工墙
isolated motor room 隔离电机间
isolated network 孤立网络;独立网(络)
isolated neutral system 中性点绝缘制;中性点不接地系统;不接地系统
isolated node 孤立结点;孤立节点
isolated obstruction 孤立障碍物
isolated obstruction buoy 孤立障碍(浮)标
isolated operation 单机运行;单独运行
isolated operation power plant 单独运转电站
isolated pacing response 隔离整步响应;分离定步响应
isolated panel 隔离节间
isolated partition 隔离间
isolated peak 孤峰
isolated phase bus 离相母线;分相封闭式母线;分析母线
isolated phase switchgear 隔离相开关设备;分离相位开关设备
isolated pier 独立桥墩
isolated piller crane 回转起重机
isolated plant 专用设备;专用发电装置;孤立发电装置;独立发电装置;单个发电装置
isolated planting 孤植
isolated plot 隔离区
isolated point 孤立点;孤点
isolated power system 独立电源系统;独立电力系统;单独电力系统
isolated reaction 孤立反应
isolated reef 孤立礁;孤礁
isolated ripple 孤立波痕
isolated rock 孤立岩石;孤立礁
isolated seed garden 隔离种子园
isolated shear wall system 独立抗剪墙系统
isolated signal 独立信号
isolated signal control 点控制
isolated solar gain 分离式太阳能采集器
isolated solution 孤立解
isolated source 孤立源
isolated state 孤立状态
isolated station 孤立电厂
isolated storm 局部暴雨;孤立暴雨
isolated subsystem 孤立子系统
isolated support 独立支承
isolated system 孤立系统;孤立体系;隔振系统;隔离系统
isolated T-beam 独立T梁
isolated thunderstorm 孤立雷暴
isolated tree 孤丘树;独立树
isolated value 孤值;孤立值
isolated valve 隔离阀
isolated vehicle 隔离车
isolated ventilation 隔绝式通风
isolated vertex 孤立顶点;孤立点
isolated ward 隔离病房
isolated word 孤立字
isolate house 独立住宅
isolate process 绝热过程
isolater 绝缘子;绝缘体
isolateral leaf 等面叶
isolating amplifier 缓冲放大器;分隔放大器
isolating and maintaining 分开保留
isolating body 隔离物件
isolating cable 绝缘电缆
isolating capacitor 隔流电容器
isolating circuit 隔离电路
isolating cock 隔离塞门;塞门
isolating condenser 隔流电容器
isolating geothermal fluid 孤立地热流体
isolating integral 孤立积分
isolating layer 隔离层;绝缘隔层
isolating link 隔离开关
isolating matter 绝缘体
isolating mechanism 隔离机制
isolating membrane 隔离薄膜;隔层
isolating method 隔阻法
isolating non-return valve 隔离止回阀
isolating oil pool 隔油池
isolating oil pool of horizontal flow 平流式隔油池
isolating-oil pool of same-direction flow 同向流式隔油池
isolating partition 隔断;绝缘隔板;隔离间壁
isolating pig house 隔离猪舍
isolating rail seam and linking line 轨道绝缘轨缝和连接线
isolating resistor 隔绝电阻器
isolating shut-off valve 隔离断开阀
isolating switch 隔离开关;切断开关;断路器;断路开关
isolating test coating 涂料防护性试验(层)
isolating test paint coating 隔离式试验涂层
isolating transformer 隔离变压器;分隔变压器;防护变压器;安全变压器
isolating valve 关闭阀;隔离阀;截止阀;切断阀;分离保护阀
isolation 析离;隔振;隔离作用;隔绝;离析;离伐
isolation, selection and evaluation 分离选择和评价
isolation amplifier 隔离放大器
isolation barrier 隔离屏障
isolation belt 隔离带
isolation board 冻结板
isolation booth 隔声室
isolation capability 查找故障能力
isolation control 隔离控制
isolation diagnostic 分离诊断
isolation diffusion 隔离扩散
isolation ditch 分离沟
isolation effect 孤立作用
isolation effectiveness 隔声效率;隔声效果;绝缘效率;绝缘效果
isolation hospital 隔离医院
isolation information system 分割信息系统
isolation in hospital 住院隔离
isolation joint 隔离接头;隔离接缝;隔离缝;绝缘接头;绝缘接缝;施工缝;分隔缝
isolation layer 隔热层
isolation leakage 隔离漏流
isolation medium 绝缘媒质;绝缘介质;分离培养基
isolation membrane 隔离薄膜
isolation method 隔离法;分离计算法;分离法
isolation moat 隔离壕
isolation mounting 隔振安装;隔能装置;隔振装置
isolation network 隔离网络
isolation of blunders 查出错误
isolation of noise 隔声;噪声(的)隔绝
isolation of pollutant 污染物分离
isolation of pure culture 纯培养分离
isolation of risks 风险隔离
isolation of the brake 制动隔离
isolation period 隔离期限;隔离期
isolation position 中立位置
isolation reed mat 防护苇垫
isolation region 隔离区
isolation room 隔离室
isolation strip 隔离条;隔离条;隔离带;绝缘条;保护行;膨胀缝条;隔离地带
isolation technics 分离技术
isolation technique 分离技术
isolation technology 隔离工艺
isolation test 隔离检验
isolation tester 绝流试验仪
isolation transect 隔离样条
isolation transformer 分隔变压器;隔离变压器
isolation trench 隔振沟
isolation valve 安全阀门;隔离阀;分离阀
isolation ward 隔离病房
isolator 隔振子;隔振体;隔离器;隔离开关;绝缘体;单向器;单面波导管;分离器
isolead curve 等提前量曲线
isoleucine 异亮氨酸
isolimb 等月缘线
isoline 等值线;等斜褶皱;等向线;等位线;等量线;等价线;等高线
isoline map 等值线图;等值线地图
isoline method 等值线法
isolit 绝缘胶纸板;绝缘胶木纸
isolite 艾索莱特层压电木绝缘场物
isolith 等岩性线;等岩(性的)
isolith map 等岩性图
isolog(ue) 同族体
isolong 等经度校正线
isoluminance curve 等照度曲线
isoluminance transform 等照度变换
isolux 等照度;等勒克斯度面;等光强度
isolux curve 等照(度)线;等照度曲线
isolux diagram 等照度图;照度分配图
isolux line 等照(度)线
isolvency 无清偿能力
isomagnetic 等磁偏线;等磁力的;等磁的;等磁
isomagnetic chart 等磁力线图;等磁要素图
isomagnetic chart of non-dipole field 非偶极子场等值线图
isomagnetic line 等磁线;等磁偏线;等磁力线
isomagnetic map 等磁线图;等磁力线地图
isomarte 等成分线
isomenal (特my气温的)月平均等值线
isomer 异构体;同分异构体;等雨率线;等降水线
isomercaptobutyric acid 异巯基丁酸
isomeric 同质异能的
isomeric colo(u)r 同谱异色;非条件等色;同色同谱色
isomeric compound 异构物;异构化合物
isomeric fission 同质异能裂变
isomeric function 分节机能
isomeric hydrocarbon 异构烃
isomeric polymer 异构聚合物
isomeric shift 同质异能位移
isomeric transition 同质异能跃迁
isomeride 异构体

isomerism 异性;异构现象;同质异能性;同质异构性;同分异构(作用)
isomerization 异构化作用
isomerization catalyst 异构化催化剂
isomerization heat 异构热
isomerization polymerization 异构化聚合
isomerization process 异构化过程
isomerization reaction 异构化反应
isomerization risk 掺杂险
isomerized drying oil 异构化干性油
isomerized linsed oil 异构化亚麻籽油
isomerized oil 异构化油
isomerized rubber 异构化橡胶
isomeromorphism 同质多形现象
isomers 等雨率线;等比值线
isomertieite 等轴砷锑钯矿
isomesical deposit 同煤沉积
ISO method for cement strength test 国际水泥强度测试法
isometric(al) 等轴的;等体积的;等容积的;等距(离)的;等径的;等角的
isometric(al) axis 等角轴(线)
isometric(al) block diagram 等轴断块图
isometric(al) contraction 等长性收缩
isometric(al) coordinates 等量坐标
isometric(al) coordinate system 等量坐标系
isometric(al) correspondence 等距(离)对应
isometric(al) crystal 等轴晶
isometric(al) (crystal) system 等轴晶系
isometric(al) diagram 等轴测图
isometric(al) dipole antenna 同轴偶极天线
isometric(al) drawing 等视图;等轴测图;等量图;等距(离)绘图;等距(离)画法;等角图;等角投影图;透度图;正等轴测图
isometric(al) family of curve 等距(离)曲线族
isometric(al) fence diagrams 等间栅栏图
isometric(al) grain 等积形颗粒
isometric(al) heat of adsorption 等量吸附热
isometric(al) latitude 等量纬度
isometric(al) line 等容线;等值线
isometric(al) mapping 等距映象
isometric(al) orebody 等轴状矿体
isometric(al) orthogonal net 等距正交网
isometric(al) parallel 等量纬圈;等比线
isometric(al) parameter 等距(离)参数
isometric(al) perspective 等角透视
isometric(al) plotting 等距(离)绘图
isometric(al) pressure 等渗压
isometric(al) process 等体积过程;等容过程
isometric(al) projection 等值投影;等距(离)投影(法);等距(离)射影;等角投影
isometric(al) projection method 等角投影法
isometric(al) rainfall map 等降水率图;等降水百分率图
isometric(al) representation 等距(离)表示
isometric(al) sketch 等轴测图
isometric(al) space 等距(离)空间
isometric(al) surface 等距(离)曲面
isometric(al) surface coordinate 等距(离)曲面坐标
isometric(al) system 等距(离)晶系;立方系;立方晶系
isometric(al) system of surface 等距(离)曲面系
isometric(al) three-dimensional projection 等角三维投影图
isometric(al) training 等长性训练
isometric(al) value 降水百分线
isometric(al) view 等轴测图;等距(离)投影图;等角图;轴测图
isometrics 管道安装图;等体积线;等容线
isometrography 平行线规;等角线规
isometropal 等秋温线
isometropia 等折光性
isometry 等轴现象;等容的
isometry group 等长群
isomoment curve 等力矩线
isomorph 类质同像;类质同晶型体;同形体;同晶(型)体
isomorphic 同形的;同晶型
isomorphic graph 同构图
isomorphic image 同构图形
isomorphic mapping 同构映射
isomorphic space 同构空间
isomorphic substitution 同形代替

isomorphic transformation 同态变换
isomorphism 类质同像;类质同晶型(现象);类质同晶;同(形)性;同形现象;同晶型(现象)
isomorphism of algebra 代数同构
isomorphous 同形的
isomorphous addition 类质同像混入物
isomorphous admixture 类质同像混合物
isomorphous compounds 同晶化合物
isomorphous crystal 同形晶体
isomorphous layer 同形层
isomorphous mixture 类质同晶型混合物;同晶型混合物
isomorphous pair 类质同像对;同形偶
isomorphous polymer 同晶多型聚合物
isomorphous replacement 类质同像置换;同形置换;同晶型置换;同晶替换
isomorphous series 类质同像系列;同晶型系
isomorphous substitution 同形代替;同晶置换;同晶形取代
isomorphous system 同晶系
isomrophous crystal 同型晶体
isomultiple 同位多重性
isomultiplet 同位多重态
isoneph 等值线;等云量线
isonephelic line 等云量线
isonif 等雪量线
isonitroso 异亚硝基
isonitrosoacetone 肟基丙酮
isonitroso-antipyrine 异氰化物
isonitrosocamphor 异亚硝基樟脑
isonival 等雪量的
isonival line 等雪线;等雪深线;等雪量线
isonomalis 磁力等差线
isonym 同名
isooctyl isodecyl phthalate 邻苯二甲酸异辛基异癸酯
isooleic acid 异油酸
isoombre 等蒸发线
isoorthicon 分流正析像管
iso-orthoclase 正钾长石
iso-orthotherm 等正温线
isoosmotic 等渗压的
iso-outlay lines 等费用线
isopach 等厚度
isopachic stress pattern 等厚应力条纹图
isopach map 等厚线图;等厚图
isopach map of aquifer 含水层等厚线图
isopach map of bottom sandstone body thickness 底沙体等厚图
isopach map of coal formation 煤系等厚图
isopach map of ore body 矿体等厚线图
isopach map of quaternary sediment 第四纪沉积物等厚线图
isopach map of Quaternary system 第四系厚度等值线图;第四系等厚图
isopach map of sand 砂的等厚图
isopach map of strata 地层等厚图
isopach map of total coal seam 煤层累计等厚图
isopach map of total sandstone body thickness 砂体累计等厚图
isopachous cement texture 等厚胶结物结构
isopachous fold 等厚褶皱
isopachous line 等厚线
isopachous map 等厚(地质)图
isopachous map of orebody 矿体等厚线图
isopachy map of coal seam 煤层等厚图
isopachyte 等厚线
isopag 等冻期线;等冬季线
isopanchromatic 等全色的
isoparaffin 异构烷烃
isoparametric 等参数的;等参数
isoparametric element 等(参)元;等参数单元
isoparametric formulations 等参数表示
isoparametric method 等参数法
isoparametric type 等参数型
isopectic line 等冻期线;等冰时线;等冰冻线;冰冻等时线
isopen 同相线
isopentanize 航空汽油中加异戊烷
isopentene 异戊烯
isopentylcyclopentane 异戊基环戊烷
isopercental map 等百分比雨量图
isopercentil 等面分数线
isopercental 等百分数的

isopercentral method 等百分数法
isoperianomaly map 等重力异常图
isoperimetric 等周
isoperimetric figure 等周数字
isoperimetric constant 等周常数
isoperimetric curve 等周曲线
isoperimetric line 无变形线
isoperimetric polygon 等周多边形
isoperimetric problem 等周问题
isoperm 等渗透率线;等渗透率型;恒导磁铁镍钴合金
isophase 等相的
isophase bus 同相母线
isophase light 等明暗光
isophase line 等相线
isophase metamorphism 等相变质作用
isophasm of pressure 变压等值线
isophene 等物候线
isophenogamy 同表型配合
isophonic contour 等音感曲线
isophorone 异佛尔酮
isophorone diisocyanate 异佛尔酮二异氰酸酯
isophot 等勒克斯线面
isophotal wavelength 等光波长
isophot curve 等照(度)线;等幅透线
isophote 等照(度)线;等光强线
isophotometer 等光度线记录仪;等光度计
isophotometric atlas 等光度图
isophotometry 等光度测量
isophthalic acid 间苯二甲酸
isophthalic alkyd resin 间苯二甲酸醇酸树脂
isophthalonitrile 异酞腈;间苯二腈
isophthaloyl chloride 间苯二酰氯
isophysical series of metamorphic rocks 变质岩的等物理系列
isopic 同相的
isopical deposit 同类沉积
isopiestic 等压线;等压的;等测压水位的
isopiestic contour line of groundwater 地下水等水压线图
isopiestic level 等压面
isopiestic line 等压线;等水位线;等势线;等测压水位线
isopiestic line of aquifer 含水层等压线
isopiestic map 等压线图【地】
isopiestic method 等压法
isopiestic point 等压点
isopiestic pressure 等压
isopiezometric level map of confined water 承压水等水压线图
isopipteses 同时出现线
isoplanar 同平面;等平面的
isoplanar bipolar memory 等平面双极存储器
isoplanar device 等平面器件
isoplanar integrated injection logic 等平面注入逻辑电路
isoplanar isolation 等平面隔离
isoplanar process 等平面工艺
isoplanar transistor 等平面晶体管
isoplane 等轴平面;等平面
isopleric line 不等体积线
isopleth 等值线;等浓(度)线;等成分面
isopleth graph 等值线统计图
isopleth map 等值图
isopleth of position function 位置函数等值线
isopleth of rainfall 等雨量线
isopleth of temperature 等温线
isopleth of wind speed 等风速线
isopleth radiation 辐射等值线
isopleths drawing 等值线绘制
isopluvial chart 等雨量图
isopluvial line 等雨量线
isopluvial map 等雨累频图;等雨线图
isopolyester 间苯二甲酸聚酯
isopolymolybdate 同多钼酸盐
isopolymorphism 同多形性;同多形现象
isopolytungstate 同多钨酸盐
isopor 周年等磁变线;地磁等年变线;磁要素等年变线
isoporic 等磁变线
isoporic chart 地磁等年变线图;等磁变线图
isoporic chart of DD 等年变率图
isoporic chart of XX 等年变率图
isoporic chart of YY 等年变率图
isoporic chart of ZZ 等年变率图

isoporic line 周年等磁变线;地磁等年变线;等磁变线
isopotential 等位势线
isopotential line 等位线;等势线
isopotential map 等产量图
isopotential point 等电位点
isopotential surface 等势面;等电位面
isopreference curve 等优先曲线;等偏好曲线
isoprene 异戊间二烯;异戊二烯
isoprene rubber 异戊(间)二烯橡胶
isoprenoid alkane 异戊间二烯型烷烃
isopressing 等静压(成型);等静力成型
isoprofit curve 同等利润曲线
isoprofit line 同等利润线
isoprol acetate 乙酸异丙酯
isopropanol 异丙醇
isopropanol test 异丙醇试验
isopropenyl acetate 醋酸异丙烯基酯
isopropoxide 异丙氧化物
isopropoxy benzene 异丙氧基苯
isopropyl acetate 醋酸异丙酯
isopropyl amine 异丙胺
isopropyl benzene 异丙(基)苯;枯烯
isopropyl benzene hydroperoxide 异丙苯过氧化氢
isopropyl benzene oxidation unit 异丙苯氧化装置
isopropyl benzene sulfonate 苯磺酸异丙酯
isopropyl benzoate 苯甲酸异丙酯
isopropyl butyrate 丁酸异丙酯
isopropyl cellosolve 异丁基溶纤剂
isopropyl ether 异丙醚
isopropylnitrate 硝酸异丙酯
isopropyl propionate 丙酸异丙酯
isopsophic index 噪声指数
isopulse 恒定脉冲(的);等脉冲(线)
isopulse system 衡定脉冲系统
isopurpunin 蒽紫红素
isopycnal 等密线;等密度线
isopycnal mixing 等密度混合
isopycnal nature 等密度特性
isopycnal surface 等密度面
isopycnic 等密度面;等密线;等密的
isopycnic line 等密度线
isopycnic slurry 等密度淤浆
isopycnic surface 等密度面
isopygous 等尼节
isopyknic 等体积的;等容的
isopyknic line 等密度线
isoquant 等产量曲线
isoquant diagram 等量图
isoquinoline 异喹啉
isoradical line 等角点辐射线
isorapid trellis casing 等斜度的格构拱道顶盖
isoreagent 同型物
isoreliable seismic design 等可靠性抗震设计
isorheic 等黏液;等黏的
isorithm 等值线
isorotation 共转;等旋光度
isorubber 异构化橡胶
isosalinity line 等盐线
isosalinity map of groundwater 地下水等矿化度图
isosbestic point 等吸光点
isosceles 等腰的;等边(的)
isosceles quadrilateral 等边四边形
isosceles triangle 等腰三角形
isosceles triangular solution 等腰三角形解
isoscope 同位素探伤仪
isoscript 等时仪
isoseism 等震线;等震
isoseismal 等震线(的);等震度线;等震(度)的
isoseismal curve 等震曲线
isoseismal line 等震线;等震度线
isoseismal map 等震线图
isoseismic 等震线(的);等震(度)的
isoseismic line 等震线;等震度线
isosensitivity curve 等敏感度曲线
isoshear 等切变线;等抗剪强度
iso-slopes 等势线
isosmotic 等渗压的
isosmotic pressure 等渗压
isosmotic pressure line 等渗压线
isosmotic solution 等渗压溶液;等渗溶液
isosol 同物异质溶胶
isospace 同空间
ISO speed standard 国际感光度标准
isostasy 压力均衡(现象);均衡现象;地壳均衡说;地壳均衡
isostasy gravity anomaly 重力均衡异常
isostath 等密度线
isostatic(al) 均衡的(地壳);等压的
isostatic(al) adjustment 均衡调整;均衡调节;静力均衡调整
isostatic(al) balance 压力均衡(现象)
isostatic(al) gravity anomaly map 均衡重力异常图
isostatically pressed chrome oxide 等静力氧化铬
isostatically reduced gravity data 均衡归正的重力值
isostatic anomaly 均衡异常;均衡失常;地壳均衡异常
isostatic balance 地壳均衡
isostatic buoyancy 均衡浮力
isostatic calculation 均衡计算
isostatic cogeoid 补偿大地水准面
isostatic compaction 等静压成形
isostatic compensation 均衡补偿
isostatic compensation surface 均衡补偿面
isostatic computation 均衡计算
isostatic cool pressing 冷等静压
isostatic correction 均衡改正
isostatic curve 等压线;等压曲线
isostatic depression 均衡下降
isostatic equilibrium 均衡平衡;地壳均衡
isostatic forming 等静力成形
isostatic frame 静定构架
isostatic geoid 均衡大地水准面
isostatic gravity anomaly 均衡重力异常;均衡重力分异
isostatic line 均衡等力线;静水力平衡线;等压线;等静力线
isostatic mo(u)lding 等静压成型;等静力成形法
isostatic movement 均衡运动
isostatic press 等静压机;均衡力机;等静力机
isostatic pressed brick 等静力成形砖
isostatic pressed refractory 等静压成形耐火材料
isostatic pressing 等压成型;等静力压制;等静压
isostatic pressing machine 等静压机
isostatic pressing treatment 等静压处理
isostatic pressure 等静压力
isostatic pressure vessel 等静压容器
isostatic prestressed concrete 均匀预应力混凝土
isostatic reduction 地壳均衡改正
isostatics 倾度线
isostatic sintering 等静压烧结
isostatic spheroid 均衡椭球
isostatic static warping 均衡静态翘曲
isostatic structure 静定结构;等压结构;等静力结构
isostatic subsidence 均衡沉陷
isostatic surface 均衡面
isostatic table 均衡计算表
isostatic theory 均衡学说;均衡说;地壳均衡理论
isostatic tooling 等压模具;等静力刀具加工
isostearic acid 异硬脂酸
isostension 等张力
isoster 等容线
isostere 等比容线
isosteric 电子等排的;等比容的
isosteric surface 等体积度面;等比容面
isostrain 等应变
isostrain diagram 等应变图
isostrain line 等应变线
isostratification map 等地层图
iso-strength interval 等强线间距
isostress 等应力;等胁强
isostructural 等结构
isostructuralism 同结构性;等结构性
isostructure 同结构;同构造;等(同)结构
isosuccinic acid 异丁二酸
isosulf 异构硫
isosulfur map 等含硫值图
iso-surface 等面;等值面
isotac 同时解冻线;等解冻线
isotach(n) 等(风)速线;等速度曲线;等流速线
isotachoelectrophoresis 等速电泳
isotachoelectrophoresis apparatus 等速电泳仪
isotachophoresis 等速电泳法;等速电泳
isotach(yl) 等速度曲线
isotactic 全同立构的;等规的
isotacticity 全同立构规整度;全同规整度;等规(立构)性
isotactic polymer 全同立构聚合物;等规聚合物
isotactic polymer plastics 等规聚合物塑料
isotactic polypropylene 等规聚丙烯
isotactic polypropylene fiber 全同聚丙烯纤维
isotactic propylene membrane 等规丙烯膜
isotaxy 全同立构
isotemperature line 等色温线
isoteniscope 蒸气(静)压力计;等张力计
isotensoid head contour 平衡型封头曲面;等张力封头曲面
isotest device 绝流试验仪
isothene 等气压平衡线
isothere 等夏温线
isotherm 恒温线;等温线;等势线
isothermal 等温(线)的;等温
isothermal absorption 等温吸附
isothermal absorption equation 等温吸附方程
isothermal absorption model 等温吸附模型
isothermal adsorption 等温吸收
isothermal air supply 等温送风
isothermal analysis 等温分析
isothermal analyzer 等温分析仪
isothermal annealed 等温退火的
isothermal annealing 等温退火
isothermal atmosphere 等温大气层
isothermal bubble 等温气泡;等温空泡
isothermal burning 等温煅烧
isothermal burning technology 等温烧成技术
isothermal calorimeter 等温热量计;等温量热器
isothermal cavity 等温腔
isothermal change 恒温变化;等温转变;等温变化
isothermal chart 等温线图;等温图
isothermal compressibility 等温压缩率
isothermal compression 等温压力;等温压缩
isothermal compression efficiency 等温压缩效率
isothermal compressor 等温压缩机
isothermal condition 等温条件;等温情况
isothermal conductivity 等温电导率
isothermal-conjugate system of curves 等温共轭曲线系
isothermal cooling 等温冷却
isothermal cooling transformation diagram 等温冷却转变图
isothermal core 等温核心
isothermal curing 恒温养护;等温养护;等温固化
isothermal curve 等温(曲)线
isothermal desorption 等温解吸
isothermal diagram 等温图
isothermal efficiency 等温效率
isothermal enclosure 等温闭合罩
isothermal equilibrium 等温平衡
isothermal expansion 等温膨胀
isothermal flow 等温流(动)
isothermal forging 等温锻造
isothermal gas sphere 等温气体球
isothermal hardening 等温淬火
isothermal heating 等温加热
isothermal heat treatment 等温热处理
isothermal horsepower 等温马力;等温功率
isothermal humidification 等温加湿
isothermal incompressibility 等温不可压缩性系数
isothermal jet 等温射流
isothermal jump 等温跳变
isothermal latitude 等温纬度
isothermal layer 同温层;等温层
isothermal level 等温面
isothermal line 恒温线;等(夏)温线;等势线
isothermal linearity 等温线性
isothermally asymptotic surface 等温渐近曲面
isothermal map of groundwater 地下水等温线图
isothermal martensite 等温马氏体
isothermal normalizing 等温正火
isothermal nozzle 等温喷管
isothermal operation 等温操作
isothermal ordering process 恒温有序化过程
isothermal orthogonal system of curves 等温正交曲线系
isothermal overall efficiency 等温总效率
isothermal parameter 等温参数
isothermal phase diagram 等温线图
isothermal plasma 等温等离子体
isothermal precipitation 等温析出
isothermal procedure 等温过程
isothermal process 等温过程

isothermal quenching 等温淬火
isothermal reaction 恒温反应；等温反应
isothermal remanent magnetization 等温剩余磁化(强度)
isothermal rigidity 等温刚性模量
isothermal section 等温区；等温截面
isothermal separation 等温分离
isothermal sintering 等温烧结
isothermal sound speed 等温声速
isothermal spheroidizing 等温球化退火
isothermal storage 等温储存
isothermal storage vessel 等温储存容器
isothermal strain 等温应变
isothermal surface 等温面
isothermal swelling 等温膨胀性
isothermal tempering 等温回火
isothermal TMT 等温形变热处理
isothermal transformation 等温转变；等温变化
isothermal transformation curve 等温转变曲线
isothermal transformation diagram 等温转变图
isothermal trans-stressing 等温形变热处理
isothermal treatment 等温处理
isothermal zone 地下恒温带
isothermic 等温的
isothermic circulation method 恒温流动循环法
isothermic coordinate 等温坐标
isothermic equation for absorption 吸附等温式
isothermic expansion 等温扩张
isothermic-shell calorimeter 恒温量热器
isothermic surface 等温面
isothermic transition 等温转变
isothermic wagon 保温车
isothermobath 断面等水温线；等温槽；等水温线
isotherm ribbon 等温带
isothernal transformation case hardening 等温转变表面硬化处理
isothiocyanate 异硫氰酸酯；异硫氰酸盐
isothiocyanic acid 异硫氰酸
isothrausmatic 等环状的
isothyme 等蒸发量线
isotim 等运费线
isotime line 等时线
isotimic 等值线；等值的
isotomeograph 地球自转测试仪
isotone 等压性；等渗性；保序
isotone mapping 保序映射
isotonic 等张的；等异位
isotonic concentration 等(渗)压浓度；等渗浓度
isotonic contraction 等张收缩
isotonic contration 等渗收缩
isotonicity 等张(力)性；等渗性；保序性
isotonic pressure 等渗压
isotonic regression 保序回归
isotonic regression function 保序回归函数
isotonic solution 等渗溶液
isotonic training 等张性训练
isotope 同位素
isotope abundance 同位素丰度
isotope addition method 同位素加(入)法
isotope analysis 同位素分析
isotope analysis method 同位素分析方法
isotope analysis of inclusion 包体同位素分析
isotope analysis of inclusion solution 包裹体溶液同位素分析
isotope carrier 同位素载体
isotope chart 同位素表
isotope chemistry 同位素化学
isotope composition 同位素组分；同位素成分
isotope contaminant 同位素污染物
isotope contamination 同位素污染；同位素混
isotope content 同位素含量
isotope contour 同位素等值线
isotope correlation 同位素相关性
isotope correlogram 同位素相关图
isotope density probe 同位素密度探测仪
isotope derivation method 同位素衍生法
isotope determination of carbon dioxide inclusion 包裹体二氧化碳同位素分析
isotope dilution analysis 同位素稀释分析
isotope dilution mass spectrometry 同位素稀释质谱法
isotope dilution method 同位素稀释法
isotope dilution spark source mass spectrometry 同位素稀释火花源质谱法

isotope discrimination 同位素辨别法
isotope distribution 同位素分布
isotope distribution function 同位素配分函数
isotope effect 同位素效应
isotope effect between molecular 分子间同位素效应
isotope equilibrium 同位素平衡
isotope equilibrium between minerals 矿物同位素平衡
isotope equilibrium between rocks 岩石同位素平衡
isotope equilibrium constant 同位素平衡常数
isotope equilibrium temperature 同位素平衡温度
isotope evolution 同位素演化
isotope exchange 同位素交换
isotope exchange equilibrium 同位素交换平衡
isotope exchange method 同位素交换法
isotope exchange rate 同位素交换速率
isotope exchange reaction 同位素交换反应
isotope external thermometry 同位素外部测温法
isotope fractionation 同位素分馏；同位素分离
isotope fractionation effect 同位素分馏效应
isotope fractionation equation 同位素分馏方程
isotope fractionation factor 同位素分馏系数
isotope fractionation mechanism 同位素分馏机理
isotope frequency curve 同位素频率曲线
isotope generator 同位素电源
isotope geochemistry 同位素地球化学
isotope geologic(al) age 同位素地质年龄
isotope geology 同位素地质学
isotope geothermometer 同位素地质温度计
isotope handling 同位素操作
isotope handling calculator 同位素操作计算器
isotope heterogeneity 同位素不均性
isotope homogenization 同位素均质化
isotope hydrology program(me) 同位素水文方案
isotope internal thermometry 同位素内部测温法
isotope isogram 同位素等值线图
isotope-label(l)ing 同位素标记
isotope-label(l)ing method 同位素标记法
isotope laboratory 同位素实验室
isotope locator 同位素检查器；同位素检测器
isotope mass ratio 同位素质量比
isotope mass spectrometer 同位素质谱计
isotope measurement of groundwater 地下水同位素测定
isotope method 同位素方法
isotope mineralogy 同位素矿物学
isotope mixing 同位素混合作用
isotope neutralizer 同位素电中和器
isotope partition coefficient 同位素配分系数
isotope pattern curve 同位素类型曲线
isotope peak 同位素峰
isotope pollutant 同位素污染物
isotope pollution 同位素污染
isotope radiation source 放射性同位素源
isotope ratio 同位素比(值)
isotope ratio analysis 同位素比分析
isotope ratio mass spectrometer 同位素比质谱计；同位素比值质谱计
isotope-ratio-recording mass spectrometer 同位素比值记录质谱仪
isotope ratio tracer 同位素比示踪法
isotope ratio tracer method 同位素比示踪物法；同位素比示踪剂法
isotope scanner 同位素扫描仪
isotope sediment concentration ga(u)ge 同位素测沙仪
isotope separation 同位素分离
isotope separation apparatus 同位素分离器
isotope separation factor 同位素分离因数
isotope separation method 同位素分离方法
isotope separation plant 同位素分离工厂
isotope separator 同位素分离器
isotope shift 同位素移动
isotope shift of molecular spectrum 分子光谱同位素移动
isotope specific activity 同位素比活度
isotopes standard 同位素标准
isotopes types 同位素类型
isotope substitute 同位素代替物
isotope tag 同位素标志
isotope temperature scale 同位素温标
isotope thermometer equation 同位素温度方程

isotope thermometry 同位素测温法
isotope tracer 同位素示踪器；同位素示踪剂
isotope tracer technique 同位素示踪技术
isotope tracing 同位素示踪
isotope-velocity method 同位素测流法
isotope water level ga(u)ge 同位素水位计
isotopic abundance 同位素分丰度
isotopic abundance measure 同位素分丰度测量
isotopic age 同位素年龄；同位素法测定的年代；同位素法测定的年代
isotopic age determination 同位素年代测定；放射性测定年代
isotopic age gradient 同位素年龄梯度
isotopically exchangeable phosphat 同位素代换的磷酸盐
isotopically pure 同位素纯
isotopic analysis 同位素分析
isotopic assay 同位素测定(法)
isotopic carrier 载体
isotopic chemistry 同位素化学
isotopic composition 同位素组分；同位素成分
isotopic composition of oil field water 油田水同位素组成
isotopic constituency 同位素成分
isotopic contamination 同位素污染
isotopic content 同位素含量
isotopic correction 同位素校正
isotopic dating 同位素年代年龄测定；同位素测定年龄
isotopic decay method 同位素衰变法
isotopic deposit 同区沉积
isotopic determination 同位素测量
isotopic differentiation 同位素差异
isotopic distribution 同位素分布
isotopic effect 同位素效应
isotopic element 同位(元)素
isotopic enrichment 同位素浓缩
isotopic foil 同位素箔
isotopic geochemical method 同位素化探
isotopic geochemical parameter 同位素地球化学参数
isotopic geochronologic(al) scale 同位素地质年代表
isotopic geochronology 同位素地质年代学
isotopic geothermometer 同位素地热温标
isotopic hydrogeology 同位素水文地质
isotopic input function 同位素输入函数
isotopic isomer 同位素的同质异能素
isotopic mass measuring 同位素质量测量
isotopic maturation curve diagram 同位素熟化曲线图
isotopic measurement 同位素测定(法)
isotopic method 同位素方法
isotopic mixture 同位素混合物
isotopic molecule 同位素分子
isotopic multiplicity 同位素多重性
isotopic number 同位数
isotopic output function 同位素输出函数
isotopic oxygen exchange 同位素氧交换
isotopic poison 同位素毒物
isotopic power 同位素动力
isotopic purity 同位素纯度
isotopic ratio 同位素比
isotopic reactivity coefficient 同位素反应性系数
isotopic sensitivity 同位素灵敏度
isotopic separation 同位素分离
isotopic specific activity 同位素比放射性
isotopic standard 同位素标准
isotopic standard samples 同位素标准样品
isotopic target 同位素靶
isotopic temperature 同位素温度
isotopic thermometer 同位素温度计
isotopic transformation 共线变换
isotopic turbidity meter 同位素含砂量计
isotopic type curve diagram 同位素类型曲线图
isotopic variable 同位素变数
isotopy 合痕
isotranslation regulation 恒位移式调节器
isotransplant system 同系移植系统
isotrimorphism 同质三形；同三晶形
isotron 同位素分析器；同位素分离器
isotrope 各向同性晶体；均质体；均质
isotropic 各向同性的；均质的；均质板
isotropic(al) cone 迷向锥面

isotropic(al) continuous skew(ed) plate 各向同性连续斜交板
isotropic(al) curve 迷向曲线
isotropic(al) cylindrical shell roof 各向同性筒形薄壳屋顶
isotropic(al) line 迷向(直)线
isotropic(al) line of aquifer 含水层等压线
isotropically consolidated drained test 各向等压固结排水试验
isotropically consolidated undrained test 各向等压固结不排水试验
isotropic(al) plate 各向同性板
isotropic(al) radiator 全向辐射器;全方位辐射器
isotropic(al) surface 迷向曲面
isotropic anomaly 均向异常
isotropic antenna 各向同性天线;无方向性天线
isotropic aquifer 各向同性含水层;均质含水层
isotropic body 各向同性天体;各向同性体;均质体
isotropic cell pressure 各向同性室压
isotropic compressible plasma 各向同性压缩等离子体;各向同性可压缩等离子体
isotropic compression 各向同性压缩;各向均匀压缩
isotropic consolidation 各向同性固结作用;各向同性固结;各向等压固结
isotropic consolidation state 等向固结状态
isotropic crystal 各向同性晶体
isotropic deep water 各向同性的深层水
isotropic deformation 各向同性变形
isotropic deposit 各向同性沉积物;各向均匀沉积物;各向均匀沉积物;均质沉积;地质沉积
isotropic dielectric 各向同性电介质
isotropic dielectric(al) medium 均质电介质
isotropic diffusive ultrafilter 各向同性超滤器
isotropic distribution 各向同性分布
isotropic disturbance 均质扰动
isotropic elastic soil 均质弹性土
isotropic emulsion 各向同性乳胶
isotropic fabric 各向同性组构;均质组构
isotropic fiber 各向同性纤维;均质纤维
isotropic fluid 各向同性流体
isotropic flux 各向同性通量
isotropic formation 均质建造;均质地层
isotropic gravity anomaly 均匀重力异常
isotropic half-space 各向同性半空间
isotropic hardening 各向同性硬化;各向等硬化;等向硬化
isotropic hardening theory 各向同性硬化理论;等向硬化理论
isotropic illuminated surface 各向同性受照面
isotropic instrument 均质仪
isotropic inversion 各向同性反演
isotropic layer 各向同性层
isotropic line 空间对角线;极小(直)线
isotropic linear elasticity 各向同性线(性)弹性
isotropic mass 各向同性体
isotropic material 匀质材料;各向同性材料;等向性材料
isotropic medium 各向同性媒质;各向同性介质;均质介质
isotropic medium filter 均质介质滤池
isotropic mineral 均质矿物
isotropic noise 各向同性噪声
isotropic normal compression line 各向同性等压线
isotropic plane 迷向面
isotropic plasma 各向同性等离子体
isotropic plasma slab 各向同性等离子体层
isotropic plate 各向同性板
isotropic point 各向同性点;无向性点
isotropic porosity 无向孔隙
isotropic radiation 各向同性辐射
isotropic radiation field 各向同性辐射场
isotropic radiator 各向同性辐射体
isotropic rig 均质平台
isotropic rock 均质岩石
isotropic saturated soil 各向同性饱和土
isotropic scatter 各向同性散射体
isotropic scattering 各向同性散射(射);均质散射
isotropic seismic source 各向同性震源
isotropic semi-infinite solid 匀质半无限体;各向同性半无限体
isotropic soil 各向同性土;均质土(壤)
isotropic source 各向同性源
isotropic space 各向同性空间
isotropic stress 各向同性应力;各向等应力
isotropic stress state 各向同性应力状态
isotropic structure 各向同性结构
isotropic substance 各向同性物质;均质物质
isotropic surface noise 各向同性表面噪声
isotropic symmetry 各向同性对称;均质对称
isotropic temperature factor 各向同性温度系数
isotropic thermal vibration 各向同性热振动
isotropic thickness ga(u)ge 同位素测厚仪
isotropic tracer 同位素示踪剂
isotropic tracer technique 同位素示踪技术
isotropic transparent material 各向同性透明材料
isotropic turbulence 各向同性湍流(度);各向同性湍动;各向同性湍流;各向同性湍动;均匀素流;均匀紊动;均匀扰动;湍流
isotropic universe 各向同性宇宙
isotropism 各向同性现象;各向同性;均质性
isotropization 各向同性化;均质化
isotropy 各向同性;均质性
isotype 同号标本;等型;反映统计数字的象征性图表
isotype substance 同型物
isotypic 同型的
isotypic specificity 同型特异性
isotypic variation 同型变异
isotypism 同型制;等型性
isovalent 等价的
isovalent colo(u)r 等价色
isovalent conjugation 等价共轭
isovalent hypercon jugation 等价超共轭
isovelocity 等速线
isovels 等速线;等流速线
iso version 催化异构化
isoviscous 等黏度的
isoviscous state 等黏态
isoviscous temperature 等黏温度
isovols 等体积线;等挥发(分)线
isovolumetric(al) 等容的
isovolumetric(al) heat capacity 恒容热容
isovolumetric(al) particle diameter 等容粒径
isovolumetric(al) process 等容过程;恒容过程
isowarping 等挠曲(的)
isowarping curve 等挠曲线
iso-wet bulb temperature line 等湿球温度线
isozonide 异臭氧化物
issuance 颁发
issuance fee 发证费
issuance for enforcement 颁行
issuance of a note 签发票据
issuance of certificate 颁发证书
issuance of certified emission reductions 核准排放量的签发
issuance of materials 发料
issuance of materials on a norm basis 限额发料
issuance of regulation 颁发规章
issue 狭水道的开口处;结局(指网络模型);期号;发料;颁布
issue a certificate 签发证书
issue a drawback 发还退税
issue an order 发布命令
issue a policy 出立保单
issue a summons 发出传票
issue a warrant 签发凭单;出票;发出认购书
issue bidding document 发行招标文件
issue bill of lading 出提单
issue bonds 发行债券
issue bonds at discount 折扣发行
issue by tender 招标发行
issued capital 发行资本
issue debentures 发行债券
issue department 发行部
issued shares 已发行股票
issued stock 已发行股票
issue guarantee 出具保函
issue house 证券发行公司
issue note 发料单
issue note of store transfer 调拨材料发料单
issue of a bill of exchange 开出汇票;签发汇票
issue of bidding documents 发标
issue of check 开支票
issue of licences[licenses] 颁发许可证
issue of securities 发行有价证券
issue payment receipt 填开纳税凭证
issuer 开证人;发行者;发行人
issues 债券
issue share 发行股票
issue slip 发料单
issues of environment(al) impact 环境影响问题
issue the official calendar 授时
issue voucher 发行凭证
issuing authority note 发行说明
issuing bank 开证(银)行;发行银行
issuing company 出单公司
issuing corporation 出单公司
issuing credit 信贷发行;开立信用证
issuing date 开证日期;签发日期
issuing house 发行证券的公司;发行银行
issuing invoice 签发账单;签发清单;签发货单
issuing jet 喷出口
issuing office 出版单位
issuing opening bank 开证(银)行
issuing service 签发处
issuing stock 发行股票
issuing store 发放库存
issuing transaction 发行业务
issuing velocity 涌水速度;冒水速度;射出速度;出流速度
I-steel 窄边工字钢;工字形钢;工字钢
I-steel column 工字钢组合柱
I-steel pile 工字形钢桩
Isteg 埃斯塔(一种扭绞钢筋的牌号)
isthmian 峡的;地峡的
Isthmian Canal 地峡运河
isthmic 峡的;地峡的
isthmus 峡部;地峡;地图制图精度
isthmus armature 细颈形衔铁
istikan with handle 带把小茶杯
istisuite 斜硅钠钙石;伊硅钙石
istle 龙舌兰纤维
I-strut 工字铁支撑;工字钢支撑
isulated conduction 绝缘性传导
iswas 简单计算装置;简单的计算装置
itabaryte 铁英岩
itabirite 铁英岩
itacolumice 可弯砂岩
itaconic(al) acid 衣康酸;亚甲基丁二酸
itaconic(al) anhydride 衣康酸酐;亚甲基丁二酸酐
itai-itai disease 痛痛病
Italian architecture 意大利式建筑
Italianare 意大利形式
Italian asbestos 意大利石棉;透闪石棉
Italian blind 滑槽卷帘;意大利(式)卷帘
Italian chrysolite 符山石
Italian cut 意大利式掏槽
Italian excavation method 意大利式开挖法
Italian ferret 意大利丝带
Italian flexible pavement design method 意大利柔性路面设计法
Italian gardens 意大利园林
Italian Gothic 意大利哥特式
Italianized roofing 意大利式屋顶
Italian modern style 意大利新艺术运动派;意大利现代式
Italian mosaic 意大利镶嵌花砖;意大利式马赛克;意大利嵌镶物
Italian ochre 意大利赭石;意大利赭色
Italian order 意大利式柱型;意大利(式)柱式
Italian pink 荷兰粉红
Italian red 意大利红
Italian Renaissance 意大利文艺复兴
Italian Renaissance style garden 意大利式庭园
Italian Romanesque 意大利罗马风格;意大利罗马式
Italian roof 四坡屋顶;意大利屋顶
Italian roof tile 意大利屋瓦(同时使用平、弧两种瓦);意大利式瓦
Italian round trowel 意大利圆形泥刀
Italian square trowel 意大利方形泥刀
Italian stone pine 意大利五针松
Italian tile 意大利筒瓦;意大利(式)瓦;意大利面砖
Italian tiling 意大利瓦;意大利贴瓦砖法(彩色瓷砖贴法);意大利瓦屋面
Italian window 意大利式窗
Italic capital 大写体体字
italic lettering 斜体字
Italic order 意大利柱型;古意大利柱型
italite 粗白榴岩
italsi 意大利硅铝合金
Italy 意大利
itamalic acid 衣苹酸;羟甲基丁二酸

itatartaric acid 衣酒石酸
item 项目;项;品目;位号;条款;保价邮件
item advance 项目转移;按项序处理;按项目前进
item analysis 项目分析
item-by-item method 逐项比较法
item-by-item plan 计件管理方案
item-by-item sequential sampling plan 逐项连续取样
item code 项码
item compared 比较项目
item cost 项目成本
item counter 操作次数计数器;分项操作计数器
item description 项目说明
item design 项目组成;项目设计
item file 项目文件
item forecast 单项预测
item identification number 项目识别号
item in the tender 投标项目
itemize 逐条列举;分项列举;分条
itemize bill 分别开账
itemized 分项的
itemized account 分项明细账;详细账目;明细账
itemized appropriation 分项拨款
itemized deduction 分项扣减数;法定扣减项目
itemized list 详细项目单
itemized record 明细记录
itemized schedule 项目一览表;明细表
item key 单项判读标志
item line 项目行
item list 明细栏
item master record 库存档案记录
item method 项目法
item number 项目号;项次号
item obtainment time 领料时间
item of balance 平衡项目
item of business 营业项目
item of corrosion test 溶蚀试验项目
item of cost 成本项目
item of evaluation 评价项目
item of expenditures 支出项目;支出要素
item of groundwater behavio(u)r observation 地下水动态观测项目
item of information 信息项
item of information in a table entry 表格项目中的信息栏
item of investigation 调查工作项目
item of payment 支付项目
item of receipt 收入项目
item of repair job 修理项目
item of stock 存料项目
item of tax 税目
item of taxation 税种
item of water treatment 水处理项目
item processing 项目处理
item record 项目记录
item refunded 偿还款项
item repairing order 项目修理通知单
item separation symbol 项目分隔符
items for export as part of foreign aid 出口援外产品
items in tariff 运价表项目
items in transit 在途物品
item size 项目规模;项(的)大小
items master file 库存主档
items not requiring or providing cash 无须动用现金的项目
items on display 展览品
items on display 展览品
items received in advance 预收项目;预收款项
items sent for collection 托收项目;托收款项
item study listing 项目研究编目
item-test regression curve 项目检验回归曲线
item top 项目上端
item under construction 在建项目;施工项目
item-weighted averages 分项加权平均数
item weighting 项目加权
iterate 累接;重复

iterated algorithm 迭代算法
iterated contraction 多重收缩
iterated electric(al) filter 链形滤波器;多节滤波器
iterated extension 多重扩张
iterated fission probability 反复裂变几率
iterated function 累函数;迭代函数;重复函数
iterated game 迭代对策
iterated integral 累积分;迭积分
iterated interpolation method 叠加插值法;迭代插值法
iterated line graph 迭线图
iterated logarithm 迭对数
iterated network 滤波网络;链形网络;累接网络
iterated series 累级数;迭级数
iterated value 迭代值
iterate-interpolation method 迭代内插值法
iteration 累接;重复;迭代法
iteration according to Gauss-Seidel 高斯—赛德尔迭代法
iteration algorithm 迭代法
iteration factor 项迭代因子;迭代因子;重复因子
iteration function 迭代函数
iteration in a program(me) 程序中的迭代法
iteration method 迭代法;递代法
iteration process 迭代过程;迭代法
iterative 迭接的;迭代的;重复的
iterative addition 迭代加法
iterative algorithm 迭代法
iterative analog computer 迭代模拟计算机
iterative analysis 反复分析
iterative array 迭代阵列
iterative attenuation 累接衰减
iterative attenuation constant 累接衰耗常数
iterative automatic stepwise cluster 迭代自组逐步聚类
iterative balance 迭代平衡
iterative calculation 迭代计算
iterative calculation method 迭代计算法
iterative circuit 累接电路
iterative computation 迭代计算
iterative computing method 迭代计算机;迭代计算法
iterative design 迭代设计
iterative division 迭代除法
iterative dO-group 重复循环组
iterative earthing 重复接地
iterative evolution 迭代进化
iterative factor 迭代因子
iterative filter 链形滤波器;累接滤波器
iterative impedance 累接阻抗
iterative implicit method 迭代隐式法
iterative integral 迭代积分
iterative least square method 迭代最小二乘法
iterative loop 迭生循环;迭代循环
iteratively faster 快速迭代
iterative method 迭代(渐近)法;重复法;反复法
iterative nature of decision analysis 决策分析的迭代特性
iterative network 累接网络
iterative noise 迭代噪声
iterative operation 迭代操作
iterative phase constant 累接相位常数
iterative placement algorithm 迭代布局算法
iterative procedure 迭代程序;递次求近法
iterative process 迭代过程;反复过程
iterative program(me) 迭代程序
iterative programming 迭代程序设计
iterative reallocation 迭代实位置
iterative repetition 迭代反复
iterative routine 迭代例行程序;迭代程序
iterative search method 迭代搜索法
iterative sequence 迭代时序
iterative solution 迭代解;重复求解
iterative sonar 迭代声纳
iterative statement 迭代语句
iterative structure 迭合结构;迭代结构

iterative technique 迭代技术(有限元计算法之一)
itinerant inspection 流动检验
itinerant merchant 行商
itinerant peddler 行商
itinerant trader 行商
itinerary 旅行指南;旅行路线;旅行计划;旅程;航海日程表
itinerary chart 海程图
itinerary lever 电锁闭控制杆
itinerary log 旅行日志;旅行日程
itinerary map 旅行路线图;路线图;航线图;海程图
itinerary mission 巡回检查任务
itinerary personnel 巡回检查员
itinerary pillar 路标
itinera versurarum 剧院舞台两翼侧门(罗马)
itoite 伊托石;羟锗铅矾
ltos satellite[improved TIROS operational satellite] 艾托斯卫星
it-plate 石棉橡胶板
itron 伊管
itsindrite 微斜霞石正长岩
ittnerite 变蓝方石
iumper boring bar 冲击钻孔器
ivaarite 钛榴石
lvanium 依瓦尼姆铝合金
lvanof's flexible pavement design method 伊万诺夫柔性路面设计法
ivanovite 水硼氯钙钾石
ivernite 二长斑岩
lverson notation 埃威逊记号;埃威逊记法
ivory 象牙;厚光纸;乳白色
ivory black 象牙(炭)黑;象牙墨;牙黑
ivory black pigment 象牙黑颜料
ivory board 象牙(白)纸板;白卡纸板
ivory bristol 象牙白光泽纸板
ivory china 象牙瓷
ivory chip(ping)s 象牙屑
ivory-colo(u)red 象牙色的
ivory finish 象牙白装饰
ivory glaze 象牙色釉
ivory glazed coat(ing) 象牙色釉表层
ivory glazed finish 象牙色表层
ivory paint 乳白色油漆
ivory palm 象牙椰子树
ivory porcelain 象牙瓷
ivory sculpture 象牙雕刻品
ivory shell 象牙螺
ivory spoon 象牙匙
ivory tint 象牙色
ivory tower 象牙之塔
ivorytype 象牙色浸蜡印像法
ivory wares 牙雕
ivory white 象牙白(一种釉色品)
ivory wood 象牙木(产于南美,淡白带褐黄色有时微绿色)
ivory work 象牙工艺;象牙制品
ivory yellow 象牙黄(一种釉色品)
ivy 常春藤;长春藤
ivy buttercup 常春藤毛茛
lvy glorybind 打碗花
ivy-leaf 常春藤叶
iwakiite 四方锰铁矿
iwan (伊斯兰教清真寺的)穹顶门廊
ixiolite 锡铁矿;锰钽矿
ixionolite 锰钽矿
ixolyte 红蜡石
lzod impact machine 悬臂梁式冲击机;埃左德冲击试验机
lzod impact test 悬臂梁式碰撞试验;悬臂梁式冲击试验;埃德冲击试验;氏冲击试验
lzod impact value 悬臂冲击试验值;埃佐德冲击值
lzod notch V 形缺口
lzod number 埃左德冲击值
lzod test 悬臂式冲击试验;埃左德冲击试验
lzod value 摆式冲击量;埃佐德冲击值
Izu-Ogasawara trench 伊豆一小笠原海沟

J

J.[joule's equivalent]焦耳氏当量
jab 突然冲击
jaborandi oil 美洲毛果(芸香)油
jabot 带褶边的帐帘头；花边绉劈
jacal 小茅屋(墨西哥及美国西南印第安人的杆柱、泥草茅屋)
jacinth 锆石；红锆石；红锆石色；橘红色
jack 支撑物；拉钩连杆；举重器；千斤顶；起重三角架；起重器；弹簧开关；手持风锤；闪锌矿；塞孔；船首旗；插口【电】；插孔
jack adjustment 千斤顶调整
jack and circle 冲击钻具紧扣装置
jack and pinion rack 齿条千斤顶
jack arch 平碹；平拱(顶)；等厚拱；单砖拱
jack arm 起重平衡臂
jackass 锚链孔塞；锚链孔防水塞
jack back 平衡杠杆
jack bar 千斤顶垫木；钻机支柱
jack base 插孔板；塞孔板
jack beam 主梁；托梁；大梁
jackbit 岩芯钻头；凿岩机钎子；可拆式钻头；活(络)钻头；千斤顶帽；手持式风钻钎子
jackbit insert 刃口
jack block 顶升构件法；千斤顶垫块
jack block construction 顶升施工法
jack board 千斤顶枕木；插孔板；插口板
jack bolt 千斤顶螺栓；起重螺杆；调整螺栓；定位螺栓
jack boom 起重杆；绳轮托臂；(挖掘机的)辅助支架；支撑臂；起重吊杆
jack boot 过膝长统靴
jack-boot off (钢丝绳的)接头拉断
jack box 插接箱
jack brick (玻璃坩埚的)垫底平砖
jack carrying frame 千斤顶载架
jack cart 起重机车
jack catch 卡爪
jack chain 起重链；起重机链；八字形链
jack chuck 活络卡盘
jack clutch 牙嵌离合器
jack column 可伸缩立柱
jack contact 插孔接触
jack cord 塞绳
jack creel 活动筒子架
jack cylinder 千斤顶油缸
jack cylinder range 千斤顶油缸行程；起重机油缸行程
jack down 降下(用千斤顶)
jack drill 凿岩钻
jacked-in bridge or culvert 顶进桥涵
jacked-in pile 顶入桩；压入式桩
jacked pile 千斤顶入桩；压入(式)桩；托换基础桩；顶压桩；顶入桩
jacked pile underpinning 压入桩托换
jack end 顶端
jack-ended 插线
jack engine 小型机车；小号发动机；辅助引擎；辅助发动机；小型蒸气机
jack equipment 千斤顶设备
jacket 铸坑；罩布；护套；外罩；外套管；套(筒)；夹套；外套
jacket band screw 套带螺钉
jacket boiler 夹套锅炉
jacket casing 套筒套
jacket closure 夹套盖板
jacket-cooled engine 水套冷却式发动机
jacket cooling 夹套冷却；护套冷却；套管冷却
jacket-cooling system 汽缸套冷却系统
jacket-cooling water 汽缸冷却水
jacket core 水套型芯
jacket cylinder 有套汽缸
jacket drain cock 水套放水阀
jacket drain valve 汽缸套泄水阀
jacketed 加套；套层的；备有夹套的
jacketed cooler 夹套冷却器
jacketed counter 带壳计数器；带壳计数器
jacketed crystallizer 夹套结晶器；套层结晶器
jacketed cylinder 带冷却水套的汽缸；有散热套汽缸
jacketed evapo(u)rator 套层蒸发器；带夹套的蒸发器
jacketed heat exchanger 夹套式换热器
jacketed kettle 套锅；带夹套的锅
jacketed lamp 双层灯
jacketed pipe 蒸气套管
jacketed-pipe high-pressure steam preheater 高压蒸汽套管预热器
jacketed portion 汽缸有水套部分
jacketed pump 夹套泵；外套泵；有套泵
jacketed specimen 封装样品
jacketed still 套层蒸馏器；带夹套的釜
jacketed stove 加外壳的炉子
jacketed syringe 套层注射器
jacketed vessel 夹套式容器
jacketed wall 套墙；水套墙；双层墙
jacket face 片夹面
jacket for pump 泵套
jacket ga(u)ge 汽缸套压力表(冷却水)；片基量规
jacket head 夹套封头
jacket heating 夹套加热；汽缸加热；水套加热；水套供暖
jacketing 蒙套；外加套；套；包壳；加套保温；夹套；外套
jacketing cool 套式冷却
jacketing heat 套式加热
jacketing machine 套管机；包护封机
jacket inlet 水套进口
jacket layer 套层
jacket leak 护套渗漏
jacket leg 导管架腿柱
jacket loss 水套损失
jacket of burner 燃烧器套管
jacket of concrete 混凝土外壳
jacket of water 水夹套
jacket outside cylinder 夹套外筒
jacket passage 水套通道
jacket pipe 套管；水套管
jacket rib 套肋
jacket safety relief valve 保温套安全阀
jacket side 夹套侧
jacket space 护套空间；水套空间
jacket steam 套管蒸汽
jacket temperature 水套内冷却水温度(发动机汽缸)
jacket test pressure 夹套试验压力
jacket tube 转向轴外管；转向轴外管；管状外罩；套筒炮身
jacket type cooling 水套冷却
jacket valve 套阀；套层阀
jacket wall 套壁
jacket water 冷却水；夹套水；汽缸冷却水；水套冷却水
jacket water cock 水套放水阀
jacket water cooler 汽缸套冷却水冷却器；水套冷却器
jacket water pump 汽缸套冷却水泵
jacket water tank 汽缸套冷却水柜
jacket water temperature 水套温度
jack fastener 插口线夹
jack flag 船首旗
jack for prestressed concrete 预应力钢筋张拉器
jack frame 举重架；绞车架；末道粗纱机
jackfruit 木菠萝
jackfurnace 修钎炉
jack hammer 凿岩锤；气锤；手提凿岩机；手持(式)凿岩机；手持式风钻；锤击式凿岩机；冲击凿孔机；风镐
jack hammer drill 撞钻；手持式凿岩机；冲击式凿岩机；冲钻
jack hammer man 手持式凿岩机工
jack-hammer-pusher leg combination 风钻和自动给进气腿联合装置
jack handle 千斤顶手柄
jack-handling tower 挤压塔；冲压塔
jackhead pump 随动泵；井下附属水泵；副泵
jack hole 石门
jack horse 台架
jack houses 升降室
jacking 用千斤顶；张拉；套料；顶托；顶起(用千斤顶)；(隧道、管道等的)顶进施工法；顶进法；顶管施工
jacking accessible cable connections 张拉斜缆接合
jacking anchorage 张拉锚碇；(预应力钢筋的)张拉锚具
jacking apparatus 千斤顶
jacking base 顶推基座；顶管顶推基座
jacking beam 顶管顶梁；反作用梁；顶梁
jacking block 螺旋千斤顶；千斤顶张拉；千斤顶(木)垫块；顶块；顶管顶铁
jacking bolt 顶举螺栓
jacking bracket 千斤顶支座
jacking control stress 张拉控制应力
jacking cycle 顶进周期；顶管顶进周期
jacking delivery motion 走车牵伸输出运动
jacking device 顶托设备；顶升装置；顶起设备；顶进设备；顶管设备
jacking disc 千斤顶垫块
jacking down hull 降台
jacking end 张拉端
jacking end of a bed 预应力钢筋层张拉端头
jacking engine 转车机；盘车机
jacking equipment 千斤顶设备；升降设备；顶管机具
jacking force 张拉力；千斤顶举升力；顶推力；顶管顶推力
jacking frame 顶管托架；顶托架
jacking gear 升降装置
jacking header 千斤顶顶头
jacking jetting process 压入冲洗法；顶推水冲法
jacking load test 顶荷载试验
jacking-machine 起皱机
jacking method 压入法；顶升法
jacking motion 走车牵伸运动
jacking of foundation 基础抬升
jacking of pile 千斤顶顶桩
jacking oil 顶轴油
jacking oil pump 顶轴油泵
jacking operation 升降合作业
jacking pad 起重机支垫
jacking pile 顶进桩；顶压桩；顶入桩
jacking plate 举升凸耳；举升垫板；千斤顶垫板；顶推垫板；提升垫板
jacking platen 压板；千斤顶压板
jacking pocket 千斤顶座孔；顶管顶座孔
jacking pressure 千斤顶压力
jacking pump for bearings 顶轴油泵
jacking recess 顶托用凹槽
jacking rod 千斤顶压杆；顶升钢筋
jacking roll calender 卷布式轧光机
jacking screw 螺旋千斤顶；顶起螺钉
jacking stress 张拉应力；顶抬应力；顶应力；顶管顶应力；最大预应力筋应力
jacking strut 顶抗；顶进支杆；顶管顶进支杆
jacking system 升降系统；顶起系统；顶管系统
jacking test 顶托试验
jacking through anchorage 拉锚式锚
jacking unit 顶压装置
jacking up 抬价；千斤顶顶起；顶起
jacking up hull 升台
jacking up price 抬高市价
jacking weight 升台重量
jacking yoke 千斤顶支架
jack-in method 顶入法
jack-in pile 顶入桩
jack in the basket 笼标(浅滩立标)
jack-in-the-box 螺旋(式)千斤顶
jack in unit 插换部件
jack iron 千斤顶加力杆
jack-king flip-flop 主从触发器
jackknife 水手刀；大折刀；拖挂死角；吊杆屈折
jackknife bridge 折叠(式)桥
jackknife cantilever mast 折叠悬臂式桅杆
jackknife derrick 折叠式钻塔
jackknife door 折叠(刀)式门；折门
jackknife drilling mast 折叠式桅杆(钻探用)
jackknife rig 折叠式钻机

jack ladder 木踏板绳梯；软梯；索梯；绳梯
jack lagging 支拱板条；砌筑壳体用的模板；承重木构件
jack lamp 安全灯
jack latch 带栓打捞钩
jack leg 轻型钻架；起重顶杆；千斤顶支柱；风动钻架
jack lever 千斤顶加力杆；起重杠杆；顶重杠杆
jack lift 起重吊车；千斤顶小车；起重托架
jack lift point 起重器支承点
jack light 篝火
jack line 千斤绳
jack load cell 千斤顶压力盒
jack-loading method 千斤顶加载法
jack loss 千斤顶损失
jackmanizing 深渗碳处理；深度渗碳
jack marking 塞孔符号
jack mate 船友
jack mechanism 起重机构
jackmill 锻钎机
jackmill for hot-milling jackbits 热锻钎头锻钎机
jack nut 支杆螺母；起重螺母
jack off 引导绕纱
jack-o-lantern 夜班守望员；值夜人员；圆灯笼；夜班守卫员
jack pad 千斤顶垫座；千斤顶垫；起重器垫；承压盘
jack pad assembly 千斤顶垫
jack pair 千斤对顶；双千斤顶；双缸液压千斤顶
jack panel 接线盘；塞孔盘；塞孔板；插口板；插孔（面）板；转插板；接线板
jack per line 同号
jack per station 异号
jack pile 支撑桩；顶入桩
jack pile puller 起重拔桩机
jack pine 黑松；北美短叶松
jack plane 大刨；粗刨
jack post 绞车卷筒轴架；撑杆；轴柱
jack prop 顶柱；支柱
jack propelling speed 千斤顶顶进速度
jack pump 油矿泵；抽水泵
jack rack 千斤顶齿柱
jack rafter 小椽；（四坡屋顶的）面坡椽；短椽；顶木
jack rail bender 轨道弯曲器；弯轨器；弯轨机
jack ram 千斤顶提升杆
jack ratio (传动摇臂的)臂长比
jack rib 小弯椽；小肋材；短肋
jack ring 支撑环；塞孔圈
jack rod 系天幕杆；千斤顶提升杆；千斤顶杆；钎杆
jackrod furnace 锻钎炉
jack rod sleeve 滑杆套
jack roll 人力绞盘；人力绞磨；手绞盘
jacks 碎铁片；吊综杆
jack saw 钢锯；横锯
jack screw 螺旋起重机；螺旋顶重器；螺旋顶升器；千斤顶螺杆；螺杆千斤顶；千斤顶螺钉；起重螺钉；起重螺旋
jack-scrowing 抬高市价
jackshaft 溜井；副轴；中间轴；增速轴；凿岩机顶柱；曲柄轴；起重轴；变速（器）传动轴；变速机构传动杆；暗井
jackshaft differential 中桥差速器
jackshaft differential carrier 中桥差速器
jackshaft drive pinion 中间轴主动齿轮
jackshaft housing 中桥箱
jackshaft propeller shaft 中间轴的万向轴
jack's shore 顶撑；背撑；小撑；套管支柱；单柱支架
jack's land 远方雾区
Jackson alloy 铜锡锌合金
Jackson candle method 杰克逊烛光烛度法
Jackson candle turbidimeter 杰克逊光度浊度计；杰克逊烛光浊度计
Jackson factor 杰克逊因子
Jackson theory 杰克逊双层界面理论；粗糙界面理论
Jackson turbidimeter 杰克逊浊度计
Jackson turbidity unit 杰克逊浊度单位
jack spring 塞孔接触弹簧
jack staff 船首旗杆
jack stand 立顶
jackstay 挂物索；桅桁支索；撑杆；扶手杆；分隔索；帆桁支索
jack string 成串权架
jack stringer 小纵梁；外纵梁
jack strip 插口排；插口簧片；插孔簧片
jack switch 插接开关；插塞开关
jack system 升高转子装置

jack tar 水手
jack tar fashion 合乎水手标准
jack test 千斤顶试验
jack timber 小椽；支木；短椽；短撑木；顶木
jack to position 举到位置上；举升就位
jack truss (四坡屋顶的)屋顶半桁架；小(尺寸)桁架；次(要)桁架
jack truss member 次要桁架杆件
jack-up 顶起(来)
jack-up crane 自升式起重机
jack-up drilling platform 自升式钻探平台；自升式钻井平台
jack up expenditures 增加开支
jack-up partition 顶升隔墙
jack-up pile 静压桩
jack-up pile driver 顶起式打桩机
jack-up platform 自升式平台；升降式平台
jack-up rig 自升式钻探平台；自升式(钻井)平台；自升式井架；顶升装置；升降式钻机
jack up the price 抬价；抬高价格
jack vault 平拱顶
jack well 抽水井
jack with handle 手摇千斤顶
jack wood 细纹木(材)
jacky mouite 硅钢铀矿
jaclbsite 锰铁矿
Jacob alloy 铜硅锰合金
Jacobean style 英国雅各布式；(17世纪英国的)雅各宾式
Jacobian curve 雅可比曲线
Jacobian determinant 雅可比行列式；函数行列式；导数行列式
Jacobian determination 雅可比行列式
Jacobian elliptic function 雅可比椭圆函数
Jacobian function 雅可比函数
Jacobian ladder 雅可比软梯
Jacobian matrix 雅可比矩阵；导数矩阵
Jacobian polynomial 雅可比多项式
Jacobian transformation 雅可比变换
Jacobian variety 雅可比簇
Jacobian well formula 雅可比井公式
Jacobi's alloy 雅可比合金
Jacobi's bracket 雅可比括号
Jacobi's condition 雅可比条件
Jacobi's differential equation 雅可比微分方程
Jacobi's ellipsoid 雅可比椭球
Jacobi's elliptic function 雅可比椭圆函数
Jacobi's equation 雅可比方程
Jacobi's formula 雅可比公式
Jacobi's identity 雅可比恒等式
Jacobi's imaginary transformation 雅可比虚变换
Jacobi's integral 雅可比积分
Jacobi's matrix 雅可比矩阵
Jacobi's matrix method 雅可比矩阵法
Jacobi's method 雅可比法
Jacobi's polynomial 雅可比多项式
Jacobi's second method of solution 雅可比第二解法
Jacobi's series 雅可比级数
Jacobi's symbol 雅可比记号
Jacobi's system 雅可比系
Jacobi's theorem 雅可比定理
Jacobi's transformation 雅可比变换
Jacobi's transformation determinant 雅可比变换行列式
Jacobite glasses 雅可白特时代的玻璃
Jacob's drill chuck 茄克波扎头
Jacobsen direction meter 雅可布森流向仪
Jacobsen rearrangement 雅可布森重排
Jacobshaven Glacier 雅可布港冰川
jacobsite 锰铁尖晶石
Jacob's ladder 索梯；雅可比软梯；木踏板绳梯；软梯；绳梯；链梯；链斗提升机
Jacob's membrane 杆status锥体层
Jacob's method for determining 雅可比法
Jacob's staff 测量仪器的支架
Jacob's taper 雅可布锥
jaconet 细薄防水布
jacot tool 枢轴研磨工具
jacquard 提花机
jacquard attachment 提花装置
jacquard board 茶版纸
jacquard card lacer 编绞板机
jacquard cord 纸板串连带

jacquard device 提花装置
jacquard drum 提花滚筒
jacquard embroidering 提花吊线
jacquard fabric 提花织物
jacquard flat knitting machine 提花横机
jacquard lever 提花杆
jacquard lock 提花三角装置
jacquard mechanism 提花装置
jacquard prism racking mechanism 提花棱柱撑动装置
jacquard ribbon 提花带
jacquard weave carpet 提花地毯
jacquard weaves 提花组织；提花织物
Jacquet's method 杰奎特电解抛光法
jacupirangite 钛铁霞辉岩
Jacuzzi 波浪式浴盆
jad 石面截槽；底部掏槽
jadder 割石机；割饥机
jade 玉(石)；绿玉色；翠；翡翠色
jade article 玉器
jade carving 玉雕
jade deposit 玉石矿床
jade glass 玉色玻璃
jade green 玉色；绿玉色；翠绿；玉绿色
jade-inlaid 红木嵌玉石制品
jadeite 硬玉；翡翠
jadeite-diopside 硬玉透辉石
jadeite glaucophane schist 硬玉蓝闪片岩
jadeitite 硬玉岩
jade object 玉器
jade sculpture 玉雕
jade stone 玉(石)；硬玉
jadete 玉(石)
jadeware 玉器
jade workshop 玉器工厂
Jaeger blower 耶格鼓风机
Jaeger method 耶格法
Jaeger's chart 近距测视力标型
Jaeger's test type 近距测视力标型
jae metal 铜镍合金(含铜30%、镍70%)
jaff 复式干扰
jag 锯齿状缺口；锯齿状断口；塞缝
jag bolt 棘爪栓；棘地脚螺栓
jaged terrain 多裂缝地带
jagged 锯齿状的
jagged bit 齿形钻头
jagged casing 齿缘套管
jagged core bit 齿缘(岩芯)钻头；齿形(岩芯)钻头；凿孔空心钻头
jagged edge 锯齿状边缘；不平坦边缘
jagged line 锯齿形线
jagged rocks 山岩
jagged shoreline 锯齿状(海)岸线
jagged terrain 崎岖地形；凹凸不平的地势
jagged ware-eroded coast 崎岖浪蚀岸
jagger 砂钩；齿凿；花边刀；滚刀
Jagger table 雅各振动台(一种混凝土振动台)
jaggies 锯齿线
jagging 开槽口；开凹口；参差
jagging board 洗选矿泥坡面板
jagoite 氯硅铁铅矿
jagowerite 羟磷铝钡石
jag resistance 抗切口性能
jag spike 棘钉
jahnsite 磷铁镁锰钙石
jai-alai court 回力球场
jai-alai hall 回力球场
jail 监狱
jail farm 监狱农场
Jain architecture 耆那式建筑(耆那教是起于印度的宗教)
Jain temple 耆那庙宇
jaipurite 块硫钴矿；灰钴矿
jake (加拿大的)双桅纵帆船；室外厕所
Jak-tung truck 手拉车
jalap resin 紫茉莉树脂
jal-awning window 翻窗组；百叶遮阳窗
jali panel 雕刻镶板
jalousie 遮窗；固定百叶窗；玻璃百叶窗
jalousie door 百叶门
jalousie framing 百叶窗框架
jalousie glass 百叶窗玻璃
jalousie screen 百叶窗遮板
jalousie window 玻璃百叶窗；百叶窗

jalpaite 辉铜银矿
Jalten 贾尔坦锰铜低合金钢;锰铜低合金钢
jam 拥挤交通;压紧;阻塞(交通);卡片堵塞;挤紧;虎钳;卡住;失真胶片;塞住;堵片;插梁
Jamaica ebony 美洲乌木
jam angle tracking 干扰角跟踪
jamb 门窗框;门窗侧框;门窗侧边;门窗侧壁;门窗边框;燃烧室侧壁;洞口侧壁
jamb anchor (门窗的)边框锚固件
jam bar 操作铁钩
jamb block 门窗边框圆角砌块;洞口砌块;侧墙砖;门窗侧柱混凝土垫块
jamb brick 圆边圆角砖;炉头砖;牛鼻砖;门窗边框圆角砖;侧墙砖
jamb brush 掸灰刷
jamb casing 门窗边框厢
jamb cleaner 门框清扫机
jamb depth 边框厚度
jamb duster 扫尘刷
jamb extension 加深的侧膛;侧柱盖板
jamb footing 门边框踢脚台
jamb guard 侧柱护铁;门窗框边框;门窗侧壁护铁
jamb horn 边框冒头
jamb joint 窗框接头
jamb liner 窗边框筒子板
jamb lining 门窗边框筒子板;门窗口筒子板;门窗侧板;门窗边框衬板;端头饰板
jamb lock 边框锁;门梃锁
jamb mo(u)lding 门窗边框线条;门窗边框线脚
jamb nut 锁紧螺母;锁紧螺帽
jambo 凿岩机车
jam bolt 防松螺栓
jamb on door or window 抱框
jamborite 水羟镍石
jamb patch 炉头修补
jamb plate 窗框筒子板
jamb post 门(窗)边框柱;石门窗侧壁
jamb-shafts 门窗立柱;门窗框侧柱
jamb stone 门窗侧壁石;边框墙石;门窗侧石柱;门窗边框石;石门窗侧壁
jamb stone pillar 门窗边框石柱
jamb stone post 门窗边框石柱
jamb wall 阈墙;闸门侧墙;门窗侧墙;门窗边框侧墙;门窗边框(侧)墙;侧墙;侧壁
jamb wall roof 边墙屋顶
jam cleat 甲板上挽压钩
jam density 堵塞密度
jam detection device 阻塞探测装置
James concentrator 詹姆斯摇床
jamesite 砷铁铅锌石
jamesonite 羽毛矿;脆硫锑铅矿
Jame's powder 詹姆斯粉
James table 詹姆斯型摇床
jami 伊斯兰大教堂
Jamin compensator 雅明补偿器
Jamin effect 雅明效应
Jamin fringes 雅明条纹
Jamin interference microscope 雅明干涉显微镜
Jamin intreferometer 雅明干涉仪
Jamin prism 雅明棱镜
Jamin refractometer 雅明折射计
Jamin's chain 雅明氏链
jam(med) density 阻塞密度
jammed pinion 卡住的齿轮
jammer 支柱;杆式集装机;干扰发射机;人为干扰台;人为干扰;爬犁起重架;电波干扰;车载起重机;车载吊车
jammer finder 干扰源探测雷达
jammer transmitter 干扰发射机
jamming 抑制;阻塞;干扰噪声;干扰(杂波);接收时他台来的干扰;人为干扰;堵塞;电波干扰
jamming coverage 干扰影响区
jamming effectiveness 干扰效能
jamming equipment 干扰设备
jamming handle 夹紧装置的手柄
jamming intensity 干扰强度
jamming in the hole 孔内阻塞卡钻
jamming of a drilling tool 卡钻
jamming of drill 卡钻
jamming pattern 干扰图形;干扰图(像)
jamming station 干扰(电)台
jamming vulnerability 抗扰性不良;低抗扰性
jam nut 制动螺母;制动螺帽;止动螺母;防松螺母;扁螺母;保险螺母;保险螺帽;安全螺母

jam nut lock 防松螺帽紧定装置
jam of logs 木材拥塞
jam of traffic 交通拥塞;交通阻塞
jampack 挤满;塞满
jam-packed 挤紧;塞得紧紧的
jam plate 螺丝板;锁板
jam pot cover 阀封
Jamrao open type outlet 贾姆拉奥明渠式泄水口
Jamrao type orifice module outlet 贾姆拉奥方孔自动恒定流量泄水口
jam rivet(t)er 窄处铆(接)机;气动铆枪;气动铆机;伸缩套筒铆锤
jam sensor 压紧检测器;压紧传感器
jam socket machine 管头成型机;污水管管头成形机
jam to signal ratio 干扰信号比
jam type packing 压紧式填料
jam up 筛眼堵塞
jam weld 塞焊;对头焊接;搭头焊
Janbu method of slope stability analysis 詹布法
Jander's equation 扬德尔方程
Janecke coordinate 詹内科坐标;简克标(液—液萃取用)
janeckeite 硅酸三钙石
janggunite 羟黑锰矿
jangipure jute 低级印度黄麻
janiceps 双面联胎
janitor 房屋管理员;看门(工)人;照管房屋的工友
janitor's closet 房屋管理(员)室
janitor's office 房屋管理(员)室
janitor's room 门房;传达室;房屋管理(员)室;门卫室
Janney coupler 詹尼式车钩
Janney flo(a)tation cell 詹尼型浮选槽
Janney mechanical-air machine 詹尼型压气机械搅拌浮选机
Janney motor 轴向回转柱塞液压马达;轴向回转柱塞式液压电动机
Janney pump 轴向回转柱塞泵
Jannin method 贾宁磨损检测法
Janssen's burner 詹森喷灯
Janssen's equation 詹森方程
Janssen system 詹森系统
janua 沿街的前门(古罗马)
January thaw 一月解冻
Janus 双向天线
Janus black 贾纳斯黑
Janus green 贾纳斯绿
jap-A-lac 日本大漆 A
japan 亮漆;漆;天然漆
Japan Accounting Association 日本会计研究学会
japan ager 琼脂
Japan Air Society 日本航空协会
Japan art paper 雕刻凹版用A纸
Japan Association of Corrosion Control 日本防腐协会
Japan Atomic Energy Committee 日本原子能委员会
Japan Atomic Energy Insurance Pool 日本原子能保险组织
Japan Atomic Energy Research Institute 日本原子能研究所
Japan Atomic Fuel Corp. 日本核燃料公司
Japan basin 日本海盆
japan black 黑沥青(清)漆;黑亮漆
Japan/Caribbean route 日本—加勒比海航线
japan cedar 柳杉
Japan central standard time 日本中央标准时
Japan Chamber of Commerce and Industry 日本工商会
japan colo(u)r 日本天然漆色浆
Japan Commercial Arbitration Association 日本商事仲裁协会
Japan Construction Consultants Association 日本建筑咨询业者协会;日本建设咨询业者协会
Japan current 黑潮;日本海(暖)流
Japan Development Bank 日本开发银行
japan drier 日本大漆催干剂
Japan/East Africa route 日本—东非航线
Japan Environmental Agency 日本环境厅
Japan Environmental Association 日本环境协会
Japanese Agricultural Standard 日本农业标准
Japanese alder 赤杨
Japanese apricot 梅
Japanese apricot flower 梅花

Japanese architecture 日本式建筑;日本建筑
Japanese Association of Groundwater Hydrology 日本地下水协会
Japanese Association of Petroleum Technologists 日本石油技术协会
Japanese bell tower 日本钟楼
Japanese birch 白桦
Japanese cedar 柳杉
Japanese cherry 日本樱花
Japanese-east Asia bivalve subprovince 日本—东亚双壳类地理亚区
Japanese Electric(al) Committee Standard 日本电工委员会标准
Japanese Electrotechnical Committee 日本电气学会
Japanese elm 春榆【植】
Japanese Engineering Standard 日本工业标准;日本工程标准
Japanese Federation of Employers Associations 日本经营者团体联盟
Japanese flowering cherry 日本樱花
Japanese fret 日本回纹饰
Japanese frit 白玉
Japanese horse-chestnut 日本七叶树
Japanese Industrial Standard[JIS] 日本工业规格;日本工业标准
Japanese Industrial Standard flat pallet 日本工业标准托盘
Japanese Institute of Certified Public Accountants 日本公认会计师协会
Japanese intensity scale 日本地震烈度表
Japanese ivy 爬山虎【植】
Japanese lacquer 大漆;日本天然漆;深黑漆;亮漆
Japanese larch 日本落叶松
Japanese law 日本律
Japanese Lease International Corporation 日本租赁国际公司
Japanese Marine Corporation 日本海事协会
Japanese mat 日本式褥垫(榻榻米)
Japanese National Railways 日本国有铁路;日本国家铁路公司
Japanese oceanographic(al) Data Center 日本海洋资料中心
Japanese organizations 日本型企业组织
Japanese Overseas Economic Cooperation Funds 日本海外经济协力基金
Japanese Paint Industrial Association 日本涂料工业协会
Japanese Painting Standard 日本涂装(工业协会)标准
Japanese porcelain 日本瓷器
Japanese quince 日本贴梗海棠;日本海棠
Japanese Railway Engineering Association 日本铁路工程协会;日本铁道工程学会
Japanese red 日本红(含硅氧的氢氧化铁)
Japanese roof truss 日式屋架;日本式屋架
Japanese Society of Limnology 日本湖沼学会
Japanese Society of Mechanical Engineers 日本机械工程师学会
Japanese Society of Phycology 日本藻类学会
Japanese standard cone 日本标准锥
Japanese standard time 日本标准时
Japanese style 日本式
Japanese tanoak 日本石柯
Japanese temple 日本神殿
Japanese tile 日式瓦
Japanese Trading Company 日本商社
Japanese tung oil 罂子桐油;日本桐油
Japanese yen 日元;日本货币单位
Japanese yew 紫杉
Japan/Europe route 日本—欧洲航线
Japan External Trade Organization 日本贸易振兴会
Japan Federation of Employers' Association 日本雇主协会联合会
japan finish 日本罩面漆
Japan geosyncline 日本地槽
japan gold size 快干清漆
Japan helwingia 青荚叶
Japan/Hong Kong route 日本—香港航线
japanic acid 日本酸
Japan International Cooperation Agency 日本国际协力事业团
Japan Ionosphere Sounding Satellite 日本电离

层探测卫星
japanite 叶绿泥石
japan lacquer 日本天然漆
Japan Management Association 日本管理学会
Japan Maritime Disaster Prevention Centre 日本海上灾害防止中心
Japan Maritime Safety Agency 日本海上保安厅
Japan/Mediterranean East route 日本—地中海东部航线
Japan/Mediterranean West route 日本—地中海西部航线
Japan Meteorologic(al) Agency 日本气象厅
Japan/Micronesia route 日本—密罗克尼亚航线
Japan/Middle America West Coast route 日本—中美西海岸航线
Japan National Oil Corp 日本国石油公团
Japan National Railways Standards 日本国(家)铁(道)标准
japanned 漆过的;使发黑光;使发黑亮;涂硬质清漆的铁器;水亮漆的;涂漆的
japanned finish 日本罩面黑漆
japanned leather 漆皮
japanner 漆工;油漆工
japanners' brown 漆工棕
Japan/New Yoke route 日本—纽约航线
japanning 沥青清漆烘烤装饰法;涂漆层;涂漆;上漆;烘漆;烤瓷;烧瓷
japanning kettle 漆锅
japanning oven 涂漆用炉;涂漆炉;上漆炉
japanning room 油漆间;漆房
japanning works 油漆工厂
Japan Oriental Leasing Company 日本东方租赁公司
japan pagoda-tree 槐树
Japan Patent Centre 日本专利中心
Japan Petroleum Institute 日本石油学会
Japan Printing Ink Maker's Association Colo(u)r 日本油墨者协会色谱
Japan Productivity Center 日本生产本部
Japan Railway Technical Service 日本海外铁道技术协力会
Japan Sea Cable 日本海水线
Japan sea type geosynclines 日本海型地槽
Japan Sewage Works Agency 日本污水工程局
Japan Shipping Exchange Inc 日本海运集会所
Japan Society of Mechanical Engineering 日本机械工程学会
Japan Society on Water Pollution Research 日本水污染研究学会
Japan/South Africa route 日本—南非航线
Japan/South America East Area route 日本—南美东海岸航线
Japan/South America West Area route 日本—南美西海岸航线
Japan/South Pacific route 日本—太平洋南部航线
Japan standard JLPA 日本的 JLPA 标准
Japan standard time 日本标准时
Japan/Strait Area route 日本—海峡地区航线
Japan stream 黑潮
Japan system of environmental management 日本环境管理体制
Japan/Taiwan route 日本—台湾航线
japan tallow 野漆树蜡;日本蜡
Japan Tax Association 日本租税研究协会
Japan Transport Consultants Association 日本海外运输咨询业者协会
Japan trench 日本海沟
japan varnish 日本生漆
Japan warm current 日本暖海流
japan wax 野漆树蜡;漆蜡;日本蜡
japonica 木蠹;日本山茶
Japonica style 日本式建筑
japopinic acid 日本松节油酸
jar 缸;容器罩;瓮;大口瓶;加耳(电容单位)
Jaraktogenos kurzii 晃模属
Jaramillo event 贾拉米洛事件
Jaramillo normal nopolarity subzone 贾拉米洛正向极性亚带成
Jaramillo normal polarity subchron 贾拉米洛正向极性亚时
Jaramillo normal polarity subchronzone 贾拉米洛正向极性时间带
jar block 冲击锤
jar bumper 振击器(打捞被卡钻具)

jar collar 打头
jar coupling 振击器;冲击活环
jar crotch socket 震击打捞筒
jardiniere (装饰用的)花盆架;花瓶
jargon 黄锆石
jar hammer 冲击锤
jar knocker 振击器(打捞被卡钻具);打捞工具
jarlite 氟铝钠锶石
jar mill 撞击式球磨机;罐形球磨机;罐磨机;缸式磨机;瓷制球磨罐
jar mo(u)lding machine 振动造形机
Jarno taper 贾诺锥度
jaroschite 镁水绿矾
jarosite 黄钾铁矾
jarosite process 钾黄铁矿方法
jar piece 加重管(钻探用);导向杆(冲击锤)
jar-proof 防振的
jarrah 桉树;茄勒木;边缘桉;赤桉
jar-ramming machine 振动造形机
jar ram mo(u)lding machine 振动制模机
jarred 赤桉
jarred loose ground 震松地
jarred loose land 震松地
jarring 震动
jarring action 振击作用
jarring effect 振动效应;振动效应
jarring machine 震击机;震动机;振动机;冲击机
jarring mark 横纹
jarring motion 振动
jarring mo(u)lding machine 振动成形机;振实(式)造形机
jarring of ingot 振动铸锭法
jarring piece 冲击锤导向架
jarring vehicle 颠簸的车辆
jar rod 加重冲击钻杆;传动钻杆
jars 振击器(打捞被卡钻具);冲击锤
jar-shaped kiln 罐形窑
jar socket 震击打捞筒;振击打捞筒
jar squeezer 振压(式)造形机
jar staff 冲击锤导杆
jar test 杯罐试验;(废水的)悬浮物体分离试验;瓶式检验;烧杯试验
jar tests for coagulation 振动凝结试验
jar tests for flocculation 振动絮凝试验
jar tong socket 打捞活环上折断部分的捞筒
jar-to-signal ratio 噪声信号比
jar washer 冲击垫箍
jar weight 冲击锤
jar work off 活环行程
Jasil water bar 雅西尔金属棒(一种检查进水情况的金属棒)
jaskelskiite 硫锑铋铜矿
jasmine(e) 淡黄色
jasmundite 硫硅钙石
Jason clause 杰森条款(遇险时抛弃货物的条款)
jaspachate 玛瑙碧玉
jaspagate 玛瑙碧玉
jaspe 表面彩点
jaspe carpet 碧玉花纹地毯;影纹地毯
jaspe cloth 贾斯佩绸(用于帐帘和室内装饰品);影纹布
jaspe lino(leum) 仿碧玉油地毡;流纹油地毡
jasper 墨绿色;碧玉
jasper agate 碧玉玛瑙
jasper carving 碧玉雕刻品
jasper glass 碧玉玻璃
jasperiod 碧玉状
jasperite 碧玉
jasper marble 碧玉大理石
jasperoid 碧玉
jasperoidization 似碧玉化
jasperoid rock 碧玉质岩
jasper rock 碧玉岩
jasper stone 碧玉石
jasper stoneware 碧玉炻器;钡质炻器
jasper ware 碧玉细炻器
jaspé cloth 影纹布
jaspilite 碧玉铁质岩
jaspilyte 碧玉铁质岩
jaspis 碧玉
jaspoid 玄武玻璃
jasp opal 蛋白碧玉;碧玉蛋白石
jaspure 碧玉大理岩;碧玉大理石
jassa 大螯蜚

jatex 浓缩橡浆;浓浆
Jauncey's scattering formula 姜基散射公式
jaune brilliant 亮黄
jaunt 短途游览
Java black-rot 爪哇黑腐病
javaite 爪哇熔融石
java para 爪哇白拉胶
Java ratio 爪哇比率
Java Sea 爪哇海
Java trench 爪哇海沟
Javelle water 亚维耳水;次氯酸钠消毒液;次氯酸钾消毒水
javellization 消毒净水(法);次氯酸钠消毒净水物
Javel water 亚维尔水;次氯酸盐消毒液
jaw 卡爪;接杆;夹片;夹紧装置;虎钳牙;钳口;凸轮;叉头
jawab 配称建筑,对称房屋
jaw angle 钳角;颚板角
jaw bolt 叉头螺栓;颚夹螺栓
jaw box 包锌的木水槽
jaw brake 爪闸;爪式制动器
jaw breaker 颚形轧碎机;颚式轧碎机;颚式碎石机;颚式破碎机
jaw chuck 爪式卡盘;齿卡盘;颚式夹头
jaw clutch 牙嵌离合器;爪式离合器;颚形离合器;颚式离合器
jaw coupler 爪盘联结节
jaw coupling 爪形联轴器;爪形联轴节;爪形联接器;爪形离合器;爪式联轴器;爪式联轴节;爪式联接器;爪式离合器;爪盘联轴节
jaw crusher 颚形轧碎机;颚式轧碎机;颚式压碎机;颚式碎石机;颚式破碎机
jaw crusher plate 碎石机颚板
jaw crushing plate 颚式破碎板
jaw grab 爪式抓斗;颚式抓斗
jaw gyratory crusher 旋颚式破碎机;颚旋式破碎机
jaw liner 颚衬板
jaw mandrel 爪式心轴
jaw nut 防松螺帽;保险螺帽;锁紧螺帽
jaw of coupling head 接轴铰接叉头
jaw of pile 桩靴
jaw of spanner 扳手钳口
jaw of the chair 轨座颚
jaw of the vice 虎钳口
jaw opening 开度;颚板间距(破碎机入口)
jaw plate 侧板;颚(夹)板;碎石机颚板
jaw setting 颚式破碎机的卸料口间隙宽度;颚板间距(破碎机出口);颚板净空(破碎机小室底部削料点至颚板之间的距离)
jaw shears 颚式剪床
jaw socket 虎钳口
jaws of vise 虎钳子
jaw spanner 爪形扳手
jaw speed 狭口流速
jaw stick 球形握把
jawstock 摆动颚
jawstock shaft 摆动颚偏心轴
jaw-type hydraulic rail brake 钳式液压缓行器
jaw-type jump clutch 波纹面滑跳式离合器
jaw-type pneumatic rail brake 钳式风动缓行器
jaw-type rail brake 钳式轨道减速器;钳式轨道缓行器;钳式轨道制动器
jaw vise 夹具;虎钳
jaw wedge 车轴楔;立式导承调整楔
jaw weld 对接焊
jayrator 移相段
jay-walking 不守规则乱穿街道
jazz the motor 强化发动机
J-bend J 形弯管
J-bolt 弯钩螺栓;钩头螺栓
jealous glass 不透明玻璃
jeanbandyite 津羟锡铁矿
jeanette 细斜纹布
Jeans criterion 金斯判据
Jeans instability 金斯不稳定度
Jeans length 金斯长度
Jeans spheroid 金斯球体
Jean's theorem 金斯定理
Jeans velocity 金斯速度
Jeans wavelength 金斯波长
jears 吊横桁的绞辘
jecoleic acid 介考裂酸
jecolelic acid 十九碳烯酸
jedding axe 尖斧锤;鹤嘴斧;石工斧

jeep trailer 吉普拖车;吉普车的拖车;双轴拖挂货车
jeer 吊横桁的绞辘
jeer block 桁索滑车;桁索
jeers 帆索
Jeffcott tachometer 直读式转速计
jefferisite 水蛭石
Jeffrey air operated jig 杰弗雷型气动跳汰机
Jeffrey diaphragm jig 杰弗雷型隔膜跳汰机
jeffreyite 羟硅铍钙石
Jehol 避暑山庄
jehu 出租汽车驾驶员
jel 胶化;凝胶体;冻胶
jell 凝胶;胶质体
jelled cement 凝胶状水泥
Jellet halfshade 耶雷半荫仪
Jelley-type refractometer 吉莱折射仪
jellied 成胶状
jellied gasoline 凝固汽油
jellies 凝胶剂
Jellif 镍铬电阻合金
jellification 胶凝作用;凝胶作用
jelling 胶凝;变稠
jelling time 凝结时间
jellous 冻状的
jelly 胶状物;胶质;胶冻;浆状体;凝胶;透明冻胶;冻胶;胶质体
jelly consistency 凝胶稠度;冻胶稠度
jelly filled cable 全填充电缆
jellyfish 海面浮标;水母
jellyfish soup with parsley 海底松
jellyfish stage 水母期
jelly fungi 胶质菌
jelly grade 凝胶化度
jellygraph 胶版;胶板
jellying power 胶凝能力
jellying strength 胶凝强度
jelly-like 胶状(的)
jelly mo(u)ld 胶模;吉利冻模型
jelly paint 凝胶漆
jelly powder 果胶粉;明胶粉
jelly print 胶纸印像
jemchuznikovite 草酸镁纳石
jemmy 起模杆;短铁橇;短橇棍
Jena glass 耶拿玻璃
Jena(light) glass 耶拿光玻璃
Jena planetarium 耶拿天象馆(1922年设计的第一个薄壳混凝土圆屋顶)
Jena ware 耶拿玻璃器
jenkin cracking 一种液相热裂化过程
Jenkin's filter 詹金斯过滤器
jenkinsite 铁叶蛇纹石
jennite 羟硅钠钙石
jenny 移动起重机;卷扬机;滑车;单轮滑轮
Jenny Linde 活动林德式窗(一种产品窗型);简妮·林德窗(窗型的一种)
jenny scaffold 活动脚手架;滑动脚手架
jenny wheel 单滑轮起重机
Jenssen cooler 表面冷却器
Jenssen exhaust scrubber 詹森废水洗涤器
Jenssen's classification 詹森分类法
Jenssen's inequality 詹森不等式
jeopardise 遭受危险;危及
jeopardize 遭受危险;危及;受危害
jeopardy clause 危害条款(欧洲货币协定中的一项条款)
jeppeite 钾钛石
jeremejevite 硼铝石
jeremejewrite 硼酸铝石
jerk 震动冲击;急撞;急停;急跳;急摔;急牵;急扭;急顿
jerker 登船检查员
jerker rod 拉杆
jerking 海关搜查
jerking motion 跃动;强烈颤动;冲撞;冲击运动
jerking movement 急动
jerking stress 撞击应力
jerking table 摇床;震淘台
jerkin head (呈斜坡的)山墙尖;山墙尖(呈斜坡的)屋顶;半山头【建】;斜坡山墙尖屋顶;两坡式屋顶(自山墙半高处突起尖顶)
jerkin head roof 两坡式屋顶;半山头屋顶;(山墙尖呈斜坡的)斜山墙尖两坡式屋顶;山墙尖两坡式屋顶
jerk limit 冲击极限

jerk line (一种钻机用的)启动绳;拉紧线
jerkmeter 加速度计
jerk note 清查证
jerk pump 高压燃油喷射泵;脉动(作用)泵
jerk-pump fuel system 脉动泵供油系统
jerk shoe 张紧工作绳的工具
jerk test 振动冲击试验;急冲试验
jerkwater 铁路专用支线
jerky 急拉;冲击的;不平稳的
jerky pulse 急冲脉
jeroboam 大酒瓶
jeromite 晒硫砷矿
jerque note 进口检查证;结关单;海关检验单;清查证
jerquer 最后的巡查;海关缉私官员
jerribag 可折叠桶
jerrican 金属制液体容器;五加仑装的汽油罐
jerried 成胶状
jerry 权宜之计的;偷工减料(的);碳质页岩;粗劣的(工程)
jerry-build 偷工减料
jerry-builder 偷工减料的制造商;偷工减料的营造商
jerry-building 粗糙房屋;低劣房屋;偷工减料的建筑;简陋的房屋
jerry-building construction 偷工减料的建筑
jerry-built 简陋搭建;违章建造的
jerry-built construction 偷工减料建筑
jerry can 简便油桶;汽油罐;长方形白铁罐
jerrygibbsite 羟硅锰石
jerry iron 修缝凿
Jersey barrier 约瑟夫主撞栏
Jersey fireclay brick 高硅黏土砖
Jerusalem pine 耶路撒冷松
jerusalem sage 糙苏
jervisite 钪霓辉石
jesse 烛台(教堂用,呈树枝形)
Jesse window 教堂用窗(树杆状窗格,用彩色玻璃嵌成圣象)
jesting beam 装饰梁;假梁
Jesuit church 耶稣会教堂
Jesuitical style (拉丁美洲的)耶稣教会式
Jesuit Seismologic(al) Association 危险地震协会
Jesus pin (针形的)紧锁销
jet 煤玉;煤精;化油器喷注;黑色大理石;喷注;喷气发动机;喷溅;喷射;贝褐碳
jet abrader 耐磨仪;冲沙试验机
jet action filter 喷气作用滤池
jet-action trencher 射流作用挖沟机
jet action valve 射流管阀
jet adjuster 喷嘴调节装置
jet aerater 喷射式充气器
jet aeration 喷射曝气
jet aeration system 射流式喷气系统
jet aerator 瀑布式喷嘴暖气池;射流曝气器;喷射曝气器
jet aeroplane 喷气式飞机
jet agitation 喷射搅拌
jet agitator 喷射搅拌机;喷射混合器
jet aircraft 喷气式飞机
jet aircraft fuel 喷射式飞机用燃料;喷气飞机用的燃料
jet air exhauster 喷射式排风机
jet airfield 喷气机机场
jet airliner 喷气式客机;喷(气)式班机
jet airplane 喷气式飞机
jet air pump 喷射泵;喷气泵
jet air separator 空气脉冲选粉机
jet alloy 火箭合金
jet analyzer 喷射式分析器
jet-anvil type fluid energy mill 单喷式气流粉碎机;靶板式气流粉碎机
jet apex angle 喷射顶角
jet apparatus 喷射装置
jet arch 喷雾器拱形喷杆
jet area 喷嘴面积;喷嘴出口截面;喷管出口截面积;射流截面
jet area contraction coefficient 喷管收敛系数
jet assist 喷射加速
jet assisted take-off unit 喷气起飞助推器
jet atomizer 喷射雾化器
jetavator 喷流偏转器;射流偏转舵
jet axial velocity 射流轴心速度
jet axis 射流轴(心)
jet band 喷射带
jet bar 横箱带

jet barrier 喷气阻拦屏
jet beck 喷射式绳状染槽
jet bit 喷射钻头;喷射钻(机);喷射式钻头
jet bit drilling 喷射式钻头钻进;射流穿孔
jet black 烟黑;深黑色;深黑的;乌黑发亮的;漆黑(的);煤玉似的
jet-black measuring mark 黑测标
jet blade 喷气发动机叶片
jet blasting 射流吹洗
jet blocking 射流风机阻挡
jet blocking coefficient 射阻系数
jet blocking length 射阻长度
jet blow 射流冲击
jet blower 喷射嘴;喷射式通风机;喷气鼓风机;射流风机
jet boat 喷水推进艇;喷气快艇
jet-bomber runway 射流式袭炸机跑道
jet boundary 射流边界
jet brake 喷射制动;射流制动器
jet bubbler 喷射装置
jet bulking 喷射法膨化变形
jet burner 高速喷射烧嘴;喷射口;喷灯;喷嘴燃烧器;火口
jet calibration 量孔校准;射流参数校准
jet carburet(t)or 喷雾式汽化器;喷雾式化油器
jet casing cutter 喷射切管器
jet cement 特快硬水泥
jet centrifugal pump 喷射式离心泵
jet chamber 喷雾室
jet chemical polishing 喷射化学抛光
jet chimney 蒸气管道
jet city 航空港城市
jet classifier 喷嘴喷射空气分级机
jet-clean 喷嘴清洗器
jet cleaner 喷嘴清洗器(清洁喷嘴的细丝);喷射清釜装置
jet cleaning 喷水清洗;喷射洗涤;喷射清洗;喷射清理;喷气清洗;反吹风清灰
jet cleaning precipitation 喷射洗涤除尘
jet coal 长焰煤;煤玉
jet collector 喷射式集尘器
jet compression 喷射压缩
jet compressor 喷射压气机
jet condensation 喷水冷凝
jet condenser 喷水凝汽器;喷水冷凝器;喷射(式)冷凝器;喷射凝汽器
jet condenser pump 喷水凝汽式泵;喷水凝汽泵
jet cone 喷射流扩展锥
jet contraction 喷射流收缩;射流收敛;水舌收缩;水流收缩
jet control 喷流调节
jet control system 射流控制系统
jet conveyer 抛掷式输送器
jet cooker 喷射式煮浆锅
jet cooling 喷射冷却;射流冷却;射流吹风冷却
jet-core region 射流中心
jet crater 喷丝板沉积物
jetcrete 喷射水泥(砂)浆;喷射混凝土;喷枪喷水泥浆;喷浆
jetcreting 混凝土喷射
jet crop drier 热风喷射式农产品干燥
jet current 急流;射流
jet curtain 喷流幕;射流幕
jet cutter method 喷射挖掘法(冲水打井、用泵吸泥);喷浆挖掘法
jet cutting 喷射切削;水力开凿;射流切割
jet cutting car 移动式水枪
jet-cyclo flo(a)tation cell 喷射旋流式浮选机
jet-cyclo flotator 喷射旋流式浮选机
jet damping 喷流阻尼;射流阻尼
jet dean 喷泉
jet deckle 喷射式定边装置
jet deep-well pump 喷射式深井泵
jet deflection 射流偏斜
jet deflection system 喷流偏转系统
jet deflector 折流器;射流转向器
jet deflexion 喷流偏转
jet deviator 偏流器
jet diffuser 喷流扩散器
jet diffusion 喷射扩散;射流分散
jet-diffusion pump 喷射扩散泵
jet diffusor 射流扩散器
jet dilution 射流稀释
jet direction 喷流方向

jet disperser 射流消能装置;射流扩散装置;射流分散器;射流分散剂
jet dispersion 喷注的雾化;射流分散
jet distance 射程
jet divergence angle 射流扩散角
jet douche 喷射冲洗
jet drainage 喷射泵排水
jet dredge(r) 喷水采泥船;喷射(式)挖泥船;射流挖泥船;冲沙船
jet drier 喷射干燥机;喷气干燥炉
jet drilling 水力钻进;射流穿孔
jet drilling mud 喷射钻进泥浆
jet drilling rig 喷流钻孔机;喷射钻井设备
jet drive 喷气推进;喷气推动;喷气传动
jet dryer 喷射干燥器;喷气干燥机
jet dryers sizing machine 喷气烘燥式浆纱机
jet dust counter 喷射式计尘器
jet dy(e)ing machine 喷射染色机
jet dynamics 射流动力学
jet edge 射流边界
jet-edge generator 喷注边棱发声器
jet-edge-resonator whistle 喷气侧壁共振器哨
jet-edge ultrasonic generator 喷注棱超声发生器
jet eductor 喷射器
jet-effect wind 急流效应风
jet efflux 喷流
jet ejector 喷射器;喷射泵
jet electrode 喷电极
jet electrolysis-plating method 喷射式电镀法
jet electro-plating 喷射电镀
jet elevator 喷水器;水力提升器
jet-enamelled ware 喷涂装饰瓷
jet engine 喷气(式)发动机
jet engine compressor 喷气发动机压气机
jet engine fuel 喷气燃料;喷气发动机燃油
jet engine lubricant 喷气发动机润滑剂
jet engine parts 喷气发动机零件
jet entrance point 射流进入点
jet envelope 喷油柱外层
jetevator 转动式喷管;喷气流偏转器
jet exhaust 喷气口;排气口
jet exhauster 引射器;喷射真空泵;喷射抽气机
jet exit 气流排出;喷嘴出口;尾喷管;射流胀大;射流出口;出口气流
jet expansion 喷嘴出口
jet-fan 射流风机
jet filter 喷吹清灰的袋式吸尘器
jet fire 立窑喷火
jet-flame drill 火焰钻机
jet flame process 热焰喷射法
jet flap rudder 喷射舵;喷射舱
jet flo(a)tation 射流气浮法;射流曝气气浮法
jet flow 急流;喷水流;喷(射)流;喷气流;射流
jet flow fan 射流风机
jet flow gate 射流式闸门
jet flow pump 射流泵
jet fluid 射流
jet foam monitor 架式喷射器
jet force 射流力;射流冲击力
jet-form 喷油柱形状
jet-formed product 喷射制品
jet foundation applicator 喷射涂布装置
jet freezing 喷注冻结法
jet front 喷油柱前锋
jet fuel(oil) 航空煤油;喷气燃料;喷气发动机燃料
jet-fuel-resistant coating 抗喷气机燃料油侵蚀面层(混凝土跑道)
jet galaxy 喷流星系
jet gas turbine 喷气式燃气轮机;喷气燃气透平
jet-generating system 喷气发生系统
jet generator 喷注式发生器;喷注式超声波发生器
jet geometry 射流几何
jet granulation 炉渣喷水粒化;喷水水淬法
jet grinding 喷射研磨
jet-grouted pile 旋喷桩(即喷柱)
jet grouting 旋喷灌浆;旋喷(法);注浆灌浆;高压喷射灌浆;加固土体的旋喷注浆法;喷射注浆;喷射灌浆
jet grouting method 喷射灌浆法
jet hardening 喷水淬火;喷气淬火
jet head 喷油柱前锋;喷(射)头;喷气头
jet height 喷水高度;射流高度
jet holder 喷丝头室

jet hole 喷水孔;水力喷射钻孔
jet hose 喷射软管;射流软管
jet humidification 喷射加湿
jet humidifier 喷水增湿器;喷射加湿器
jet hydraulic horsepower 射流水马力;射流水功率
jet hydrofoil 喷气推进水翼艇;喷气水翼艇
jet ignition 喷气燃烧炉
jet ignition system 喷射点火系统
jet impact 射流冲击
jet impact area 射流冲击范围
jet impact mill 喷射(冲击)式磨机
jet impact range 射流冲击范围
jet-induced circulation control 喷流诱导的环流控制
jet injection 射流喷注;射流喷射
jet injection pipe 喷射器喷管
jet injector 射流泵
jet interaction amplifier 射流管式放大器
jet interaction element 射流互作用型元件
jet lance 喷枪
jet landing barrier 喷气机着陆拦阻装置
jet leaching 喷射浸出
jet length 射流长度
jet lift dredge(r) 喷射泵挖泥船;喷射(式)挖泥船
jet lifter method 提升喷射法(用电动泵将钻井中泥沙喷出)
jetlike structure 喷流状结构
jet line 喷嘴供液管路
jetliner 喷气式客机;喷气(式)班机
jet lubrication 喷射润滑
jet mantle 喷油柱外层
jet method of drilling 水力钻进法
jet method of pile-driving 射水沉桩法
jet micronizer 喷气式微粉磨机
jet mill 喷射式磨机;喷气碾机;喷气磨
jet milling 气流磨;喷射研磨;喷射干燥;喷射粉磨
jet mixer 喷射混合器
jet mixing 喷注入混合;喷射混合;射流搅拌
jet mixing flow 混射流
jet mixing system 喷注混合系统;喷射混合系统
jet motor 喷气式发动机;喷气发动机
jet mo(u)lding 喷射模塑法;喷射模法;喷射成型;射塑喷射模塑法;射塑
jet mo(u)lding nozzle 注射成形喷嘴;喷射模塑喷嘴
jet multilayer dryer 多层热风烘呢机
jet needle 喷油针
jet noise 喷注噪声;喷射噪声;喷射(器)气流噪声;喷气噪声;喷流噪声
jet noise suppressor 喷气发动机消声器
jet nozzle 喷(射)嘴;喷射管;射流喷嘴
jet nozzled rock bit 喷射式牙轮钻头
jet number 量孔符号
jetocopter 喷气式直升机
jet of flame 火焰喷射
jet of water 喷射水;水注;水压喷射
jetometer 润滑油腐蚀性测定仪
jet-O-Mizer 喷射式微粉磨机
jet orifice 喷射口;喷口;喷出口;排气口;射口;排气喷孔
jet orifice separator 喷嘴式分离器
jet out 喷射出
jet overflow dy(e)ing machine 喷射溢流染色机
jet pack 喷气发动机组件
jet parameter calibration 射流参数校正
jet part 喷气部分
jet particle 喷射粒子;喷射粒流
jet passage 喷气通道;喷射管道
jet passage plug 喷口通道塞
jet path 射流轨迹
jet perforating process 射流钻进作业;射流穿孔作业
jet perforation 射流穿孔
jet perforator 喷射式射孔器;喷射式穿孔器;射流穿孔器
jet perforator charge 聚能射孔药包;喷射式穿孔药包
jet phenomenon 喷射现象
jet pierce drill 火焰喷射钻机;热力钻机
jet-pierce drilling 火焰喷射钻机
jet-pierce machine 热力穿孔机;喷焰钻机
jet piercer 火钻
jet-piercing 喷焰钻进;热力穿孔;热焰喷射钻孔
jet-piercing burner 喷焰钻燃烧室

jet-piercing drill 火力钻机;火力喷射式穿孔机;火焰喷射钻具;熔化穿孔机;喷烧穿孔机
jet-piercing drilling 火焰喷射钻进
jet-piercing machine 热力穿孔机
jet-piercing method 热焰喷射钻孔法
jet pile 水冲桩;射水打入的桩
jet pipe 注射管;喷(射)管;喷口管;射流管
jet-pipe oil-operated controller 喷气液压控制器
jet pipe regulator 射流管调节器
jet pipe shroud 喷管隔热层
jet pipe temperature controller 喷管温度调节器
jet pipe thermometer 喷管排气测温计
jet pipe tip 射流管喷嘴;喷管喷嘴
jet pipe valve 喷流管阀;射流管阀
jet piping 喷管
jet plane 喷气式飞机
jet plating 喷镀
jet polishing 喷射抛光
jet port 喷气机场;喷气式飞机场;喷气飞机航空站
jet powder pile 喷粉桩
jet power 喷气射力;喷气推动功率;喷气发动机推动;喷气动力;反冲力
jet-powered car 喷气式汽车
jet-powered junk retriever 孔底反循环冲捞器
jet power ring 喷气功率环
jet power unit 气动力装置;喷气动力装置
jet pressure 喷射压力
jet priming 射流起动
jet printer 喷墨打印机
jet printing 喷射印刷
jet printing ink 喷墨印刷油墨
jet printing system 喷射印花装置
jet probe 射水探测
jet probing 喷射探测;喷射控制;射水杆(土壤)探测;喷射探查
jet problem 射流问题
jet progress 喷气技术进展
jetproof 耐水冲的
jet propellant 喷气发动机燃料
jet-propelled 喷气推进的;喷气发动机推动的
jet-propelled aeroplane 喷气式飞机
jet-propelled aircraft chart 喷气式飞机航空图
jet-propelled boat 喷气推进艇
jet-propelled carriage 喷气推进牵引车
jet-propelled carrier 喷气式运载工具
jet-propelled sprayer 喷气式喷雾器
jet-propelled vessel 喷气推进式船舶
jet propeller 喷射推进器;喷气(式)推进器;射流推进器;喷气螺旋桨
jet-propeller aeroplane 螺旋桨喷气飞机
jet-propeller engine 喷气螺旋桨发动机
jet propulsion 喷气推进;喷射驱动;喷气推进
jet propulsion engine 喷气推进发动机
jet propulsion fuel 喷气式飞机用燃料
jet propulsion laboratory 喷气推进实验室
jet propulsion unit 喷气推进装置
jet propulsive efficiency 喷气推进效率
jet pulse filter 喷气脉冲吸尘器
jet pulverizer 喷射碾磨机;喷气磨;喷射粉碎机;喷射磨粉机
jet pulverizing 喷射粉磨;喷射磨碎;喷射研磨
jet pump 喷吸泵;喷射(式)泵
jet pump dredge(r) 喷射泵挖泥船;射流(泵)挖泥船
jet pump pellet drilling 喷射泵钻粒钻井
jet-pump pellet impact drill bit 射流泵散弹钻进用钻头
jet pump suction dredge(r) 冲沙船
jet purifier 喷洗器;喷气净化器
jet quenching 喷射冷却;喷射淬火
jet ratio 射流比
jet reaction 射流反应
jet-reaction unit 喷流推进装置
jet recompression 射流再压缩
jet refueling vehicle 喷气式飞机用的加油车
jet regulator 喷气调节器;射流调节器
jet reheat temperature control 喷流重热温度控制
jet relay 射流继电器
jet resistant pavement 抗喷气路面
jet reverse circulation drilling 射流反循环钻进
jet ring 喷丝帽浆环
jet-ring-plate 环缝式算板
jet room 喷丝头室
jet rose 喷水莲蓬头;多孔喷嘴
jet route 喷气式飞机航线

jet runway 喷气机用的跑道
jetsam 投弃货物(船舶遇险时投弃的货物)
jet scour 挑流冲刷
jet scrubber 喷射洗涤器;喷射洗涤机;喷射涤气器
jet separation 喷嘴分离;射流分离(现象)
jet separator 射流分离器
jet set cement 喷射水泥;快凝水泥
jet shale 煤玉页岩
jet shower 喷淋浴
jet silencer 喷气发动机消声器
jet size 喷口尺寸;喷管直径
jet slice 喷嘴式堰板
jet space 喷嘴前空间
jet speed 喷射气流速度;喷流速度
jet spray dryer 喷射喷雾干燥机
jet sprayer 喷射式喷雾机;喷气式喷雾器
jet-spray processor 喷雾式洗片机
jet spread 喷射分散;射流分散
jet-spread angle 喷射分散角
jet stack 射流线
jet starter 喷气起动机
jet static pressure 喷流静压(力)
jet steering 喷气操纵
jet stem 喷嘴杆
jet stone 黑色电气石
jet streak 急流
jet stream 急流;喷(水)流;喷气流;射水;射流;喷射水流
jet stream velocity 射流速度
jet-stream whirl 喷流涡流
jet-stream wind 喷射气流
jet stretch 喷头拉伸
jet stretch ratio 喷头拉伸比
jet strip pump 喷射清舱泵
jet strip system 喷射清舱系统
jet structure 喷流结构
jet suction dredge(r) 射流吸泥船;射流挖泥船
jet swelling effect 射流膨胀效应
jet switching 射流转换
jet system 喷流偏斜控制系统
jet tab 喷射舌片
jettage 码头税
jet tail cone 喷管锥体
jet teau 喷水口
jetted 用水冲的;冲出的;喷出的
jetted particle drilling 喷射钢粒钻进
jet ted pile 水冲桩;射水沉桩;射水打入的桩
jetted screw pile 射水螺旋桩;水冲螺旋桩
jetted sidewall craft 侧壁式喷射气垫艇
jetted well 水力钻井
jet temperature indicator 喷流温度指示器;喷流温度计
jetter 喷洗器;喷洗装置
jetter assembly 喷洗装置
jet thrust 喷射推力;喷气发动机推力;反推力
jet thrust measurement 射流推力测定
jet thrust reverser 喷气发动机推力反向器
jettied construction 悬挑式建筑
jetting 注射;喷射法;喷射;喷吹;水力钻井;水力喷射钻井;水力法钻探;射水;射流钻井;冲孔(钻孔的);水力沉桩法
jetting action 射流作用
jetting angle 喷射角(度)
jetting bit 高压射流钻头
jetting chain 冲洗链
jetting control 喷射控制
jetting cutter rod 喷枪杆;射流切割器
jetting device 喷射装置;水力冲洗装置;水力喷射装置
jetting distance 抛射距离(喷式运输机);喷射距离
jetting drill 喷射钻(机);热焰钻孔装置;水力钻(机);射水钻井;射流冲击式钻机
jetting drilling 喷射钻进;喷射钻进;水力钻进;水力钻探
jet(ting) effect 射流效应
jetting fill 水冲填土法;水冲填土;射流填土;吹泄阀
jetting gear 高压射流矿机;水枪;高压射流挖掘机;水力挖掘机
jetting height 喷射高度
jetting hose 喷注水管;喷射软管
jetting in basin 池内喷射
jetting lance 喷枪;射水管(水冲沉桩)
jetting machine 高压清洗机;喷射机
jetting method 水冲沉桩法;(井点的)射水沉没法

jetting model 喷射模型
jetting nozzle 高压喷嘴;喷射管;喷嘴
jetting orifice 井点射水口;(井点的)射水口
jetting out 射出;挑出(如托臂等由墙突出)
jetting piles into place 喷水沉桩法
jetting piling 水力沉桩(法);水冲沉桩(法);射水沉桩法;射水沉桩
jetting pipe 射水管;射流管;喷射管
jetting process 喷射法;水冲法;射水法
jetting pump 冲洗泵;喷射泵
jetting pump transfer 喷射法输送;喷射泵输送
jetting rod 喷射杆;冲洗杆
jetting sled 高压射流挖掘机(汽车式);射流冲击式滑车
jetting suction dredge(r) 冲吸挖泥船
jetting tip 喷口
jetting tool 喷砂工具
jetting transfer 喷射法输送
jetting type well-point 喷射(型)井点
jetting water 喷洗水;喷射机
jetting wave absorber 喷水式防波堤
jettison 弃货(船舶遇险时);抛弃(在遇难时减轻船重的一种措施);抛货(船遇险时为减载);投弃货物(船舶遇险时投弃的货物);投弃(海上航行遇紧急情况时);丢弃(货物);被迫把货物投入海中
jettisonable auxiliary fuel tank 补助油箱
jettisonable ballast 可抛压载
jettisonable pylon 可分离挂梁
jettisonable weight 可抛重量
jettison and washing overboard 抛弃和浪打落海货物
jettison device 弹射器;弹射装置
jettison gear 投弃装置
jettisoning system 抛放系统
jettison launcher 弹射式发射装置
jettison of cargo 抛舱面货物
jet trajectory length 射流挑出长度
jet tray 喷射(形)塔盘;喷射分馏塔盘
jettron 气动开关
jet tub 喷淋式浴缸
jet turbine 喷射式涡轮
jet turbine oil 喷气式透平油
jetty 栈桥;港口外堤;建筑物的突出部分;码头;突码头;突堤;浮码头;防波堤
jetty bent pile 码头排架桩
jetty clause 码头检验条款
jetty dock 堤式船坞
jetty harbo(u)r 突堤港;有突堤的港口
jetty head 栈桥头;突堤前沿;突出堤;(防波堤的)堤头;导流堤头;坝头
jetty-head berth 突堤码头前缘泊位
jetty hook 码头系缆钩
jetty mounted radar 码头雷达
jetty mounted sonar 码头声呐
jet type abrasion test 喷砂磨耗试验
jet type bit 喷射式钻头
jet type carbureter 喷射式气化器
jet type deaerating heater 喷射式除氧加热器
jet-type deaerator 喷射式除气器
jet type dust collector 脉动除尘器;脉冲反吹袋式除尘器;喷射型除尘器
jet type humidifier 喷射式加湿器;喷射加湿器
jet type impulse turbine 水斗式水轮机;冲击式水轮机
jet type reverse circulation tool 喷射式反循环钻具
jet type tricone bit 喷射式三牙轮钻头
jet type washer 喷射式清洗机
jet type water course 喷射式钻头水眼
jetty pier 栈桥式码头;突堤码头;突堤
jetty structure of high level platform supported on pile 高桩码头结构
jetty system 突堤式码头系统
jetty wharf 指形码头;突堤(式)码头
jet unit 喷水推动装置
jet vacuum pump 喷射式真空泵
jet vacuum type hydro-percussive tool 射吸式冲击器
jet valve 喷射阀
jet vane 喷气舵;喷气导流控制片
jet velocity 喷射速度;喷气速度
jet veneer dryer 喷气单板干燥器;喷气单板干燥机
jet viscometer 喷射黏度计

jet wash 喷射冲洗;射流冲洗
jet washer 喷射洗涤器;喷射洗涤机
jet washing 喷射洗涤
jet water 喷射水
jet wave 射流波;自由射流波
jet-wave rectifier 射波整流器
jetway 喷气式飞机跑道
jet-well pump 喷射式深水泵;喷射式深井泵
jewel 宝石轴承;宝石
jewel bearing 宝石轴承
jewel bearing manufac-turing equipment 宝石轴承生产设备
jewel block (船的)信号吊绳滑车;球滑车;单饼小滑车
jeweler's enamel 珍宝珐琅
jeweler's file 宝石匠锉
jewelers lathe 玉工车床
jeweler's piercing saw 珠宝工的弓锯
jeweler's screw 修表螺旋
jeweler's screw-driver 钟表用螺丝刀
jeweler's shop 珠宝店;首饰店
jewel hole 宝石轴承孔
jewelled 装有宝石的
jewelled end-plate 托钻
jewelled ware 宝石光泽瓷器
jeweller 珠宝商
jewel(l)ery alloy 装饰用合金
jewelling 压铝
jewel post 钻石柱
jewelry 宝石
jewelry blue 宝石蓝
jewelry bronze 装饰青铜
jewelry enamel 景泰蓝(装饰品)
jewelry insurance 珠宝保险
jewel sintering machine 宝石烧结机
jewel-tipped 宝石头的
Jewish architecture 犹太(式)建筑
Jewish bema 犹太教堂中的讲坛
Jew's harp 锚环
jew solder 屋顶水泥(俚语);屋顶黏合料
Jew's pitch 半固态沥青
Jew stone 犹太石层;白铁矿
jezekite 红磷钠矿
J-groove 丁形凹沟
J hook J形吊钩
Jiaji 夹脊
Jian-an ware 建安窑
Jiangnan private gardens 江南园林(中国)
jianhaugite 硅钕锰钠石
Jian white ware 建白窑
Jiaotan xia official kiln 效坛下官窑
Jiape drawdown 加坡系数
jib 镶夹条;起重机臂;扒杆;吊臂;副臂;扁栓
jib adjusting gear 起重机悬臂升降机构
jib and cotter 合楔
jib angle indicator 起重臂角度指示器;臂架倾角指示器
jib arm 悬臂;机臂;起重机机臂
jib arm of boom 起重机臂
jib balanced without counterweight 无配重平衡的臂架
jib barrow 支梁式手推车;支架式手推车
jib boom 机臂;后斜帆桁;起重机旋臂;起重杆;起重臂;挺杆;辅助延伸臂
jib-boom crane 装配吊车
jib bracket 起重臂托座;悬臂托座;起重木托座架
jib brake 起重臂制动闸
jib cable 起重臂缆
jib component 起重臂部件
jib crab 吊杆车;悬臂起重机
jib crane 旋臂吊车;旋臂(式)起重机;悬臂(式)起重机;悬臂吊机;悬臂吊车;摇臂起重机;转臂起重机;扒杆(式)起重机;挺杆式起重机;伸臂式起重机;伸臂式固定起重机;动臂起重机;臂式吊车;臂架起重机
jib crane charger 回转式吊车加料机
jib door 起重臂门;隐门;墙面齐平门
jib drum 吊臂卷绕筒
jib-end (输送机卸料端安装的)卸料悬臂
jib extension 辅助伸臂
jib extension ram 吊臂外伸作动筒
jib fold mechanism 臂架折叠机构
jib foot spool 起重臂脚鼓
jib head 截头;截盘座;吊机臂上端;吊臂上端

jib head radius 悬臂活动半径;挖掘机作用半径
jib height 吊臂高度
jib hoist 吊臂卷扬机
jib hoist limit switch 吊臂提升机止限开关
jib-in 插入截盘
jib inclination 吊臂倾斜度
jib inclination angle 吊臂倾斜角
jibing 改变帆的方向
jib length 悬臂长度;起重臂长度;吊臂长度
jib lift cable 起重臂提升缆
jib lift valve 吊臂提升阀
jib loader 旋臂装料机,旋臂装货机;悬臂装料机;摇臂装料机
jib luffing line 吊臂变幅索
jib motion 挺杆运动
jib-mounted drill 转臂式钻机
jib operating radius 吊杆作业半径
jib pin 吊臂脚销
jib point 吊臂顶端
jib point pin 吊臂顶销
jib point sheave 吊臂顶滑轮
jib saw 竖线锯
jib sheave 挺杆滑车
jib shipper shaft 吊臂移动轴
jib side sheave 吊臂侧缆滑轮
jib stopper 起重机旋臂制动器
jib support 臂架搁置架
jib swing angle 吊臂旋转角
jib tooth 截盘齿
jib transit 工具经纬仪
jib type crane 臂架型起重机
jib type tractor loader 转臂式拖拉机装载器
jib wharf 突堤式码头
jib winch 吊臂卷扬机
jib winch inner bushing 吊杆绞盘内衬套
jiffy stand 停车架
jig 小绞辘,钻模;规尺;矿筛;夹具装架;夹紧装置;焊接平台;衰减波群;钓钩;导尺;淘簸筛
jig-adjusted 粗调的
jig assembly 夹具装配
jig back 双索索道(一来一往);往复索道;双线缆道;高架电车道;双线索道
jig bed 跳汰床层
jig borer 坐标镗床;钻模镗床
jig-borer microscope 坐标镗床显微镜
jig boring 坐标镗削;细钻
jig boring machine 坐标镗床
jig bush(ing) 钻件夹具的导套;钻套
jig concentrate 跳汰精矿
jig drill 细钻;钻模钻床
jig feeder 摆动式给料机
jig force 跳汰力
jig for testing equivalent cube 等效立方体试验夹具
jigged bed 跳汰床
jigged-bed adsorption column 跳汰床吸附塔
jigged-bed ion-exchange 跳汰床离子交换
jigged-bed resin-in-pulp process 跳动床树脂矿浆吸附法
jigger 小绞车,小滑车;辘轳,镂花锯;可变耦合变压器;浸染机;减幅振荡变压器;耦合器;盘车;往复摇动式运输机;跳汰机工;跳汰机;水库拦鱼器;衰减波变压器;筛矿器;淘簸筛
jigger bars 搓板带(路面)
jigger block 带索滑车
jigger body 旋坯用泥料
jigger conveyer 振动输送机,振动运送机
jigger coupling 电感耦合
jiggering 旋坯成型;盘车拉环
jigger lifting piston 提升活塞
jigger man 拉坯工
jigger operator 旋坯工
jigger pin 顶料销
jigger-pin die 带顶料装置的模具
jigger saw 往复式线锯;往复式竖线锯
jigger tackle 轻便绞辘
jigging 跳汰选矿法;跳汰选;簸析法
jigging area 跳汰面积
jigging block 样板(机械加工用)
jigging box 跳汰选槽;跳汰机洗箱;跳汰机
jigging cell 跳汰分室
jigging chamber 跳汰室
jigging chute 振动溜槽
jigging compaction 振动压密

jigging conveyer 簸动输送机;震动输送机;振动式斜槽;振动输送机;往复摇动式运输机;振动送机
jigging cycle curve 跳汰周期曲线
jigging down 钻杆冲击钻进
jigging grade 簸动炉排;簸动炉箅
jigging grate 振动式炉箅
jigging machine 簸析机
jigging method 跳汰法
jigging motion 摇摆运动;颠簸运动(向上下或左右)
jigging platform 振动台;倾卸台
jigging plunger 跳汰机活塞
jigging screen 跳汰筛;跳汰机筛板;跳汰筛;簸动筛;振动输送槽
jigging sieve 摇筛机;振动筛;跳汰机筛板
jigging stenter 簸动拉条机
jigging test 簸析试验
jigging washer 振动洗涤器
jiggle 轻摇;摇动;摆动
jiggle bar 摇杆;搓板带;摇手柄
jiggled frame 摇动架
jiggler (提升绳上的)夹车器
jiggly 不平稳的
jig grinder 坐标磨床
jig grinding machine 坐标磨床
jig key 钻模键
jig mill 坐标镗床
jig plate 夹具板;钻模板
jig plunger 跳汰机排渣器
jig point 基点
jig saw 锯曲线机;线锯;细(竖)锯;窄锯条机锯;曲线机锯;竖线锯;往复式竖线锯
jig-saw puzzle of continents 大陆拼合
jig-saw shape 锯齿形
jig stenter 摆动式拉幅机;摆动拉幅机
jig table 跳汰摇床
jig tank 跳汰箱;跳汰机选槽
jig washer 跳汰洗选机;跳汰洗煤机
jig with side bed 侧鼓式跳汰机
Jihaerte formula 吉哈尔特公式
Jilong Port 基隆港
jim 搬运小车;横ударный小车(桥式起重机上)转向架
jimboite 锰硼石
jim crow 轨条挠器;轨道挺直器;人工弯轨机;弯轨器;挺器
jimmer 两页不能分离的铰链
jimmy 煤车;铁橇棍;短橇棍
jimthompsonite 镁川石
Jimu triangle method 济姆三角法
jin 市斤(1市斤=0.5千克)
jin chain hoist 葫芦
jingle bell 门铃;信号铃
Jinjin rug 津锦地毯
jink 车钩
jinnie wheel 链式起重机(俗称神仙葫芦)
jinniwink A形(塔式)起重机;A形杆式起重机;人字动臂起重机
jinny 制动ска;自重滑行坡;固定绞车
jinrikisha 人力车
jitney 小型公共汽车
jitter 歧离;跳动;速度偏差;抖动;不稳定性;图像跳动
jitterbug 图像不稳定故障;混凝土表面压平器;图像跳动;手动摆柄;手工压浆棒子;混凝土铁栅捣棒;蛙式夯实器
jitter gain 抖动增益
jitter tester 抖动测试仪
jitter time 分散时间
jitter tolerance 抖动容限
jitty 联络巷道;联络小巷道;短通道
jiyumill 微粉机
J-joint 丁形接头
joaquinite 硅钠钡钛石
job 职业;职务;职位;工作项目;工作;加工件;任务;程序作业
job abort 作业的中止
job account 工作账;工作分项账户
job accounting routine 作业算帐程序
job action command 作业处理命令
job analysis 作业分析;职业分析;职务分析;工作(项目)分析;工作程序的分析
job applicant 求职人员
job assessment 工作评价

job assignment 分派工作
job assignment notice 工程任务单
jobber 小宗批发商;中间商;杂工;临时工;批发中间商;批发商;半熟练工;分包工;房屋修缮工
jobber market 分包市场;包工市场
Jobber's reamer 机用精铰刀
jobbing 碎修
jobbing casting 零批铸造;零星铸件;零星浇铸
jobbing foundry 铸造厂;零星铸造车间
jobbing gardener 临时园林工(人)
jobbing house 批发店
jobbing market 中间市场
jobbing mill 小型钢轧机;中小型型钢轧机;零批轧机
jobbing sand foundry 零星小件翻砂厂
jobbing sheet 中(型)板;中薄板
jobbing sheet-rolling mill 中板轧机;零批薄板轧机
jobbing shop 特殊订货加工车间;(工地现场的)修理车间
jobbing work 单件小批生产;散工;临时工;包工(指工作);短工;零星作业
jobbing worker 合同工
job breakdown 工作细分
job breakdown method 工序解体分析法
job-by-job 单项(工程)承包;逐项承包(工程等);单项工程聘任
job capacity 工作容量
job captain 设计制图组长;现场设计负责人;工地设计负责人;工地负责人;工程负责人;设计项目负责人
job card 工作卡;工作进度卡;作业卡片;作业登记卡
job change 更换品种;改变品种;工作变更
job change analysis 工作变更分析
job change request 工作变更要求
job characteristic 工作因素
job class 作业类别;作业分类
job cleanup 工地清理;工地完工清扫;建筑工地清理;建筑场地清理工作
job code 工作编号
job completion 工程的扫尾竣工
job condition 工作条件
job content 作业内容;工作内容
job control 作业控制;施工控制
job control block 作业控制块;作业控制块
job control command 作业控制命令
job control communication 作业控制通信
job control information 作业控制信息
job control language 作业控制语言
job control program(me) 作业控制程序
job control statement 作业控制语句
job control stream 作业控制流
job control table 作业控制表
job conversion 职业转换
job cost 分批成本
job cost accounting 分批成本会计
job cost estimating 工作成本估计
job costing 分批控制法;分批成本计算
job cost journal 工程成本日记账
job cost ledger 工程成本分类账;分批成本分类账
job cost method 分批成本法
job cost record 工程成本记录;工程成本账
job cost sheet 分批成本计算单;分批成本单
job cost system 分批成本制(度);工程成本系统
job cost voucher 分批成本凭单
job cover plan 人力安排计划
job creation 开创就业门路;提供就业机会;提供职业岗位
job creation potential 就业潜力
job-cured 现场养护的
job-cured concrete 工地养护混凝土;现场养护混凝土
job-cured cylinder 现场养护的(混凝土)圆柱试体
job-curing 现场养护
job-data 作业数据
job data sheet 作业数据表
job date 作业日期
job deck 作业卡片组
job definition 作业定义
job department 工作部门
job depreciation method 分批折旧法
job description 作业说明(书);工作项目规程;工作说明
job description sheet 工作说明单
job design 作业设计;工作设计
job-designed air-actuated spreader 特制气动摊

铺机;特制气动铺料机
job development credit 扩大就业机会优惠
job displacement 裁员;就业转移
job duplication 兼职
job efficiency 制品合格率
job engineering 作业工程
job enlargement 作业扩展;扩大就业;扩大工作内容
job entry 作业记入;作业输入
job entry subsystem 作业输入子系统;作业入口子系统
job entry subsystem 3 第三类作业输入子系统
job entry system 作业调入系统
job evaluation 工作项目评价;工作评价;工作估计;工作报酬;工种评价;评工法
job-evaluation plan 工作评价计划
job-evaluation program(me) 工作评价方案
job evaluation system 岗位工作评价制
job experience training 工作经验训练
job facility 工程设施
job factor 工作因素;工作项目因素
job family 同类工作
job file control block 作业文件控制块;作业文件控制分程序
job file index 作业文件索引
job flow control 作业流程控制
job flow-time 工作流程时间
job foreman 工作队长
job for life 终身职务
job handling routine 作业管理程序
job history card 记事卡片
job hopper 频繁变换工作的人
job hopping 跳厂换职业
job hunter 找工作的人;求职者;求职人员
job hunting 求职
job initiation 作业起始
job in process 在建工程
job input device 作业输入装置
job input file 作业输入文件
job input queue 作业输入排队
job input stream 作业输入流
job instruction sheet 工作指导卡;工作说明单
job key 工作键
job laboratory 工地试验室;工地实验室
job layout 工程放线
jobless rate 失业率
job library 作业库;工作程序库
job load 作业装入
job location 施工现场;施工地点;施工场所
job logging 作业注册;作业定时记录;工作记录
job lot 小批;零星物品;零星数量;分批数量;分批
job lot contract 小批合同
job lot control 分批控制;分批进料管理;批量控制
job lot costing 分批成本计算
job lot method 分批法
job lot method of cost calculation 分批成本计算法
job lot production 单件生产
job lot system 工作批量法
job made 就地完工;现场完成工作(项目);专门定制
job management 作业控制;作业管理;工作处理控制
job management program(me) 作业管理程序
job mechanic 工地钳工
job method 作业方法
job mix 工地拌和;现场拌和;作业配合;工地拌和混和料
job mix design 现场配合比设计
job-mix(ed) concrete 现拌混凝土;现场搅拌混凝土;现场拌和混凝土;现场拌制混凝土
job-mix(ed) paint 现场调制涂料
job mix formula 现场拌和配方;现场配合公式
job monitor 作业监督程序
job name 作业名
job note 工作单
job number 作业号码;作业编号;在修船只代号;工(作)号;工程(编)号;订货号
job numbering 工程编号
job off 赚钱卖出
job offer 现场提供
job office 工程处;工务处
job operation 加工方法
job opportunity 工作机会;就业机会
job order 作业顺序;工作顺序;工作授权;工作次序;工程协议书;工程任务单;加工顺序;任务书;任务单;派工单;通知单;分批工作通知单;分批订单

job order cost 订单成本;分批成本
job order cost accounting 分批成本会计
job order cost cycle 分批成本循环
job order costing 分批成本计算法;分批成本计算;分批订货成本
job order cost method 分批成本法
job order cost sheet 分批通知成本单;分批成本(计算)单
job order cost system 分批成本制(度)
job order sheet 作业顺序单
job organization 工程组织机构
job orientation 职前指导
job orientation and training 职业前训练;安职训练
job-oriented language 面向作业的语言
job-oriented terminal 面向作业终端;作业终端
job out 分包出去
job output device 作业输出装置;作业输出设备
job output file 作业输出文件
job output queue 作业输出排队;作业输出队列
job output stream 作业输出流
job overhead 工程间接费;工程管理费;业务管理费用
job overhead cost 施工管理费
job pack area 作业装填区;作业装配区
job pack area queue 作业装填区队列
job parameter 作业参数
job parts list 工作零件目录
job-place 现场浇注
job-placed 现场铺设的;现场浇筑(的);现场浇灌;现场浇捣混凝土;现场灌筑
job-placed concrete 现浇混凝土;现场浇筑混凝土;现浇注混凝土;现场浇灌混凝土
job-placed forming 现场灌筑的
job plan 施工进度表;工地平面图;施工计划
job planing 刨削工
job planning 施工计划
job plaster 现场拌制灰泥
job-poured concrete 现场浇筑混凝土;现场浇注混凝土;现场浇灌混凝土
job practice 作业方法;施工方法
job pricing 工作报酬
job priority 作业预先权;作业优先权
job processing 作业处理
job processing control 作业处理控制;工作处理控制
job processing control program(me) 作业处理控制程序
job processing monitor 作业处理监督(程序)
job processing monitor system 作业处理监督系统
job processing monitor termination 作业处理监督终止
job processing unit 作业处理部件
job processing word 作业处理字
job processor 作业调度程序
job production 分批生产
job production order 分批生产通知
job program(me) 加工程序;作业程序;工作计划;工作程序
job program(me) mode 作业程序方式;程序工作方式
job progress 工程进度
job queue 作业排队
job rate 作业定额;职务工资率(指某一具体职务的最低工资率);生产定额;包工制
job rating 作业定额核定;工作评级
job rating sheet 工作(职位)考核表
job ration 生产定额
job record 工作记录;工地记录
job reference 作业基准
job report 施工报告;工作报告
job request 作业请求
job request control routine 作业请求控制程序
job request selection 作业请求选择
job research 作业搜索
job review 工作检查
job rotation 职务轮流;职位轮调;工作换班
job route sheet 工作规划单
job schedule 作业进度表;作业调度程序;工作日程表;工作进度(表);工程进度;施工进度计划
job scheduler task 作业调度程序任务
job scheduling 作业进度;作业调度;排作业时间表;编制工作进度表
job scheduling routine 作业调度程序
job search channels 求职途径

job security 作业保障;职业保障;工作保障
job selection 作业选择
job selector 工作规范选择器;负荷选择器
job selector dial 负荷选择器
job sequence 作业序列;加工指令序列;加工序列;加工程序
job-sequencing module 作业定序模块
job-sequencing system 作业定序系统
job-shaped 施工成型
job sheet 加工单;派工单;施工单
job shop 加工车间;单件小批生产车间
job shop automation 车间自动化
job shop operation 作业安排操作
job shopper 经常调换工作者
job shop scheduling 车间作业进度表;作业安排调度
job shop sequencing problem 作业安排定序问题
job shop simulation 作业安排模拟
job shop simulator 作业安排模拟器
job's housekeeping 工地管理;工地组织
job's housekeeping made 在施工现场制造的
job's housekeeping' mechanic 建筑现场机修工;施工现场机修工;建筑现场机械工;施工现场机械工
job's pound 监狱 job site 现场;工作地段;工作地点;工地;工程现场;工程地点;工地;施工现场
job site check 现场检验
job site installation 施工现场安装;工地设备
job site prestressing 现场预加应力
job site safety checklist 现场安全检验单
job site schedule 现场进度计划
job site scheduler 现场进度计划师
job site stretching 现场拉伸
job site tensioning 现场张拉
jobs of downhole operation 井下作业项目
job specification 作业说明(书);作业规范;工作说明;施工规范
job-splitting 作业分解;一工分做制;全日工改为两个半日工
job spoiler 破坏团体者
job stack 作业堆栈;加工序列
job stacking 作业堆积;成批作业
job-stack system 作业堆栈系统
job standardization 工作标准化
job status 作业状况
job stealer 破坏行规
job step 作业段;作业步骤;加工步骤
job step control 作业步控制
job step control block 作业段控制块;作业步控制分程序
job step task 作业步任务
job-stored 工地储存的
job stream 作业流
job stream input 作业流输入
job stretching of tension 现场预应力张拉
job studies 对工件研究
job study 作业研究;工作研究
job subsidies 岗位津贴
job superintendent 现场监督员;工程管理员
job system 分批制度
job task analysis 工作任务分析
job tensioning 现场预应力张拉
job theodolite 工具经纬仪
job throughput 作业吞吐能量;作业(处理)能力;工作处理能力
job ticket 工作卡;工作进度卡;工票
job ticket box 分配箱
job time card 工作时间卡
job time limit 作业时间限
job time report 工作时间报告;工时报表
job time study 工作时间研究
job time ticket 工作时间卡
job title 工作名称
job-to-job hauling 工地搬迁;建筑场地之间的搬运
job training 工作训练
job training standards 工作训练标准
job vacancy 职位空缺
job vacancy rate 空位率
job wanted 求职
job watch 秒表(代替天文钟测天用)
job welding shop 加工焊接工场
job work 散工;包工
job working shop 现场单件生产车间
job work shop 现场加工车间
jockey 移动;振动膜;张紧轮;连接装置;连接夹具;

jockey pulley 支持轮;张力惰轮;滚轮;导(向)轮;导向滚轮;辅轮
jockey roller 张力辊;卷筒纸导辊;导向滚轮
jockey sprocket 导向链轮
jockey stick 钢绳夹具阻止器
jockey valve 先开阀;辅助阀
jockey weight 活动砝码;微动砝码
jockey well 拉绳滑轮
jockey wheel 支托轮;导向滚轮
Jodelite 周得来(一种木材防腐剂,对病虫害有效)
joesmithite 铅铍闪石
jog 精密送料;慢走刀;慢进给;轻撞;嵌合;粗(糙)面;凹凸部;墙的凹凸面;轻摇
jog bolt 地脚螺栓
jog down 嵌入
jogged dislocation 割价位错
jogged frame 嵌合肋骨
jogged joint 连孔连接
jogger 撞纸机;摆动码纸器
jogging 慢速的;微动;电动机的频繁反复起动;冲动状态;风浪逐渐缓和;轻摇
jogging control 微动控制
jogging speed 慢速度
joggle 折曲;折接;啮合扣;啮合机接件;弯合;突出物;榫;定位销钉
joggle beam 啮合梁;拼(接)梁;拼合梁;榫接梁
joggle column 齿合柱
joggled 啮合的;榫接的
joggled beam 镶衬梁;镶合梁;榫接梁;啮合梁
joggled butt joint 啮合对接
joggle die 折曲模;阶梯(成型)模
joggled joint 拼接接头;弯合;折接;榫接;啮合接
joggled lap joint 啮合搭接
joggled lintel 啮合(小)过梁;拼接小过梁;砖石啮合小过梁
joggled piece 啮合件;啮合柱
joggled post 啮合柱
joggled single lap joint 啮合单面搭接
joggled stone 企口石块
joggled timber 啮合木;斜头原木
joggled work 镶接工作
joggle frame 折接肋骨;榫接肋骨
joggle joint 啮合接(头);啮合接榫;折接;啮合镶接;榫接接头
joggle jointing 夹板接合
joggle piece 齿合柱;支撑件;啮合件;正同柱
joggle plating 折接板
joggle post 桁架中柱;啮合木柱;吊杆柱;齿合柱;受拉柱;榫接柱;正同柱
joggler 榫接操作工;折曲机
joggle shackle 长而微弯形卸扣
joggle tenon 凹凸相咬榫中的凸榫
joggle timbers 折接肋骨;榫接肋骨
joggle truss 拼接架;拼接桁架
joggle work 拼接工作;拼接砌合
joggling 阶梯成型;凹凸接合
joggling die 锻轧模
joggling machine 折曲机;榫接机
joggling table 振动台
joggling test 折曲试验
joggy 锯子
jog method 迂回航行;缓慢进场法
jog strainer 平筛
jog-trough conveyer 振动槽式输送器;振动槽式输送机
johachidolite 硼铝钙石;水氟硼石
Johann crystal geometry 约翰晶体几何学
johannite 铀铜矾;铜铀矾
johannsenite 钙锰辉石;锰钙辉石
Johansen's method 约翰森法(丹麦工程师约翰森的屈服线理论,用于钢筋混凝土板的极限荷载计算)
Johansson block 约翰逊块规
johnboat 平底船;驳船;平底舢板
John bull 约翰色粉(一种粉刷墙壁用的可洗的含油色粉)
Johnny ball 绝缘瓷瓶
Johnny house 户外厕所
Johnson concentrator 约翰逊精选机
Johnson counter 约翰逊计数器
Johnson effect 约翰逊效应(热噪效应)
Johnson ga(u)ge(block)约翰逊规块;量块
Johnson noise 约翰逊噪声;热噪声

Johnson noise thermometry 热噪声测温术
Johnson noise voltage 约翰逊噪声电压;热噪声电压;热噪电压
Johnson-Nyquist noise 热噪声
Johnson power meter 微分功率表
Johnson regulator 差动式调压箱;约翰逊式调压塔
Johnson's bronze 约翰逊轴承青铜
Johnson space center 约翰逊空间中心
Johnson valve 约翰逊阀(高落差水轮机阀);高落差水轮机阀
Johnsten process 庄司顿土壤灌浆加固法
johnstrupite 氟硅铈矿
join 衔接;联结;接合处;交接;参加;保联
join a ship 上船任职
join assembly 连接汇编
join by fusion 熔接焊
join colo(u)r 榫砌砖缝颜料
join dependency 连接依赖
joined and erected timber scaffold(ing)连接安装的木脚手架
joined-label-definition 标号定义组
joined mark 连码
joiner 细木工;装修木工;接合物;木工(工)人
joiner bulkhead 木或金属板舱壁
joiner door (非水密的)轻便门
joiner for concrete moulds 制混凝土板的细木工
joiner glue 接榫胶
joiner plan 舱室木工工程图
joiner's adhesive 接榫胶
joiner's belt 木工工台;木工工作台
joiner's bench 木工台;木工工作台
joiner's chisel 木工凿
joiner's compasses 细木工用的圆规
joiner's cross-cut saw bench 细木横切锯齿;细木横割锯齿
joiner's finish 木装修
joiner's ga(u)ge 木工用墨线序;划线规尺;木工规尺
joiner's glue 木工胶;水胶;细木用胶水
joiner shackle 连接卸扣;锚链卸扣
joiner's hammer 木工锤;羊角锤
joiner's kit 细木工工具箱
joiner's labo(u)rer 木工助手
joiner's mallet 细木工木槌
joiner's plane 木工长刨
joiner's shop 细木工场;细木车间;木工车间
joiner's work 木装修;细木作
joiner's workshop 细木车间
joiner work 细木工程
joinery 细木工作;细木工车间;木工(指工作)
joinery component 细木工件;榫口;接榫件
joinery making 细木工艺
joinery member 细木工件
joinery timber 细木工用材;榫接件
joinery unit 细木工件;榫接件
joinery wood 细木工用材;装修木料
joinery work 木装修;小木作;装修木工
join-homomorphism 保联同态
joining 接合
joining balk 系梁;联系木梁
joining beam 系梁;联系梁;联结杆
joining by mortise and tenon 榫眼接合;榫槽接合
joining by rabbets 槽舌接合;企口接合
joining by screws 螺丝接合
joining indicator 连接指示器
joining iron 连接铁件
joining nipple 接口螺管;接合(短)螺管
joining of microcrack 微裂相接
joining on butt 对头接(合);对接
joining on skew 斜向接合
joining piece 连接片
joining plane 修边刨
joining pointer 连接指示器
joining polyline 封闭接线
joining seed 拼接籽晶
joining shackle 锚链连接卸扣
joining shop 拼接工场
joining-up 连接;咬合
joining-up differentially 差接
joining-up exposure 高光曝光
joining-up in parallel 并接
joining-up in series 串接
joining with keypiece 键接连接
joining with passing tenon 穿榫接头;穿榫接合

joining with peg-shoulder 直榫接头;直榫接合
joining with swelled tenon 扩榫接头;扩榫接合
joining work shop 装修木工车间
joining yard 拼接场;装修木工场
join-irreducible 保联不可约的
join monolithically 无缝连接
join object 要素连接
join on skew 斜向接
join operation 联合运算
join operations centre 联合运算中心
join point 连接点
joint 结点;关节;联合的;节点;接头;接榫;接合(点);接缝;拼接;缝;节理
joint access 连接入口
joint account 共同计账;(银行的)联名账户;联合账户;联合财产
joint action 联合作用;联合行动;接头作用;接合作用;接合动作
joint action of chemicals 化学物质联合作用
joint action og gases 气体联合作用
joint(ad)venture 联合企业;合资企业;合股公司;共同投资;短期合伙
joint ag(e)ing time 胶接老化时间;黏合期;接合期
joint agenda for action 联合行动议程
joint agent 联合代理商;联合代理人
joint agreement 共同契约
joint airport 军民用机场
joint allowance 接缝允许间隙;接缝高差;缝隙
joint alteration coefficient 节理面蚀变程度系数
joint among three-members 三连接头
joint analysis 节点分析(法)
joint and insert 接缝及其填充
joint and several liability 连带负债;共同责任及各自责任;连带责任
joint angle 连接角;接头角度;连接角度
joint annuity 联合年金
joint applicator 填缝机;拼缝器;拼缝机
joint appraisal 联合评估
joint architect 合作建筑师
joint area 接缝面积;胶黏面;接合面
joint arrangement 接缝布置
joint a ship 到船上工作
joint assembly 接合汇编;接缝组装件;集中汇编
joint at acute angle 锐角接头
joint audit 联合审计
joint bar 鱼尼板;连接杆;连接板;接头销钉;接头夹板(鱼尾板)【铁】;接插件
joint beading 接缝凸出;接缝隆起
joint beam 接合梁
joint bearing 关节轴承
joint bed 接合床座
joint belt 节理带
joint between box caisson 沉箱接头
joint between three-members 三联接头
joint bid(ding)合作投标;联合要价;联合投标;联合递价;合伙投标;共同投标
joint binding tape 接缝黏结纸带;嵌缝纸带
joint block 凿石块;连接砌块;接头凸爪;接头块;接块;接合块
joint bolt 连接螺栓;接头螺栓;接合螺栓;插销螺栓;扶手螺栓
joint bond 共同债务
joint-box 接线箱;接线盒;电缆接线箱;分线箱
joint-box compound 接线盒材料;电缆套管填充剂
joint-box cover 电缆连接盒盖
joint branch 分叉接头;分叉接管
joint brazing 接头钎焊
joint bridging 未焊满的节点;跨接接合
joint buildup sequence 焊道熔敷顺序
joint business 联合营业
joint buying office 联合采购处
joint cap 接缝盖;密封盖;闭锁盖;接头密封盖
joint capacity 总合容量
joint capital 共同资本;联合资本;合股资本
joint capital investment 共同投资
joint cargo system 联合货运制
joint catalog(ue)联合目录
joint category 接头类型
joint cement 接缝胶泥;胶黏剂;填缝胶料;接缝水泥
joint center 接缝中心;十字头
joint chair 钢轨接座;接座;铁轨接座
joint characteristic curve of hydraulic coupling and prime motor 液力耦合器原动机联合特性曲线

joint characteristic curve of hydraulic torque converter and prime motor 液力变矩器原动机联合特性曲线
joint charges 联合运输费用
joint checkup 会审
joint circuit 联合线路
joint clamp 接头箍
joint cleaner 清缝机;清缝器
joint cleaning 清缝
joint cleaning machine 混凝土路面清缝机;清缝机
joint clearance 接头间隙;接合间隙;钎缝间隙
joint clearance ga(u)ge 轨隙规
joint clearance of brazing 钎焊间隙
joint clevis for twin tube 双腕臂底座
joint close 密接;合龙
joint closure 路面填缝;接缝填充;填缝;路面填塞接缝
Joint Commission on Atomic Energy 原子能联合委员会
joint committee 联合委员会
joint common risks agreement 联合共同保险协议
joint communication facility 联合通信设备
joint company 联营公司
joint compound 接头填料;接缝剂;胶接剂
joint compound adhesive 接头胶黏剂
joint compound cooker 填缝料熔炉
joint compound melting furnace 填缝料熔炉
joint compound-taping 接头胶带
joint compound-topping 接缝覆面材料
joint conditioning time 接头调理时间;接合养护期
joint condition time 接头期
joint connection 结点连接
joint construction 接头施工;合作施工
joint continuity 节理贯通度
joint contour diagram 节理等密图
joint contribution 共同出资
joint core 节点核心
joint cost 共有成本;共用费用;共同费用;联合费(用);联产品成本
joint coupling 接头连接器;偶接;活节连接器;活接连接器;电缆接头套管;方向接头
joint coupling of cable 电缆接头套管
joint cover(ing) 接缝搭盖;盖缝板;压缝板;接头互相搭接
joint covering strip 接缝盖条
joint credit 共同债权;联合信贷;连带债务
joint creditor 共同债权人
joint cross (万向接头的)十字头;万向十字接头;十字形接合件
joint cross pinblock 十字头
joint current 总电流
joint custody 联名保管(户)
joint cutter 切缝机;连接绞刀
joint-cutting 切缝法;切割制榫;(整体大面积的)切缝;混凝土路面切缝
joint-cutting equipment 切缝机(用于切割混凝土路面)
joint-cutting machine 混凝土路面切缝机;切缝机(用于切割混凝土路面)
joint debt 共同债务
joint declaration 联合声明
joint deformation 节点变形
joint deformity 关节变形
joint demand 共同需求;复合需求
joint density 节理密度【地】
joint depth 接缝深度;接头深度
joint design 接头设计;接头结构;接缝设计;合作设计;接缝尺寸;接点设计;联合设计
joint detailing 接缝细部做法
joint details 接合详图;节点图
joint diagram 节理图(解)
joint dimension 接头尺寸;接缝尺寸
joint displacement 节点移动;接头位移
joint disposal 综合处置;(废水污水的)共同处理;联合排放
joint distribution 联合分配;联合分布
joint distribution function 联合分布函数
joint dovetail 楔形榫接合
joint dowel 定缝销;暗榫
joint draft resolution 联合决议草案
joint drag 扭结带
joint-drag kink bands 节理牵引膝折带
joint drainage 开缝阴沟
joint drawboard 加梢榫接

joint drive 万向节传动
joint dynamometer 万向节式测功仪
jointed arm 摇杆
jointed boarding 接缝望板
jointed brake 木闸瓦的条状制动带
jointed build 接笋船壳板
jointed connecting rod 组合连杆
jointed core 接缝芯板;拼缝芯板
jointed coupling 绞链联轴器
jointed fiber 接续光纤
jointed edge 接合边;黏合侧面;板边拼接
jointed harrow 组合耙
jointed inner plys 拼缝内层单板
jointed rock mass 裂隙岩体;节理岩体
jointed rod 组合钻杆;组合钎子;连接钻杆
jointed rule 英制比例尺;测杆;曲尺;折尺
jointed shaft 铰接轴
jointed spline 连接样条
jointed track 有缝线路
joint efficiency 结合效率;接头系数;接合效率
joint efforts 共同努力
joint elbow 接合弯头
joint element (有限元分析中的)连接单元
joint endorsement 联合背书
joint enterprise 共同事业;联合企业;合资企业;合营企业;合办企业
joint entropy 相关熵;相关平均信息量
joint equation 节点平衡方程
joint equilibrium 节点平衡
joint equity venture 联合投资企业;合股经营企业
jointer 压缝器;管子工;勾缝器;联络员;连接器;连接件;开缝器;接口工;接缝器;接缝镘;涂缝镘;填缝器
jointer ga(u)ge 长刨量规(木工用)
jointer plane 木工长刨;长刨
jointer saw 修边锯;石材细缝锯
joint estate 共有财产;联合租赁房产
joint estimate 联合估计(值)
joint estimation 联合估计
joint evil 关节痛
joint expenses 共同营业费;联运费用;连带费用
joint export agent 联合出口代理商;联合出口代理人
joint export department 联合出口部
joint face 连接面;节理面;接合面;接缝面;分型面
joint facility 联合设备
joint facility income 联合设备收益
joint factor 接合系数
joint failure 接缝破损;接头破坏;(深钻技术中的)脱缝;接缝损坏
joint family 复合家庭
joint fastener 接合片;接合金属片;波纹扣片(连接木构件用)
joint fastening 钢轨连接板;接头扣件
joint faulting 错缝;连接变形;连接断裂
joint file 刃锉
joint filler 接缝(填)料;膨胀填料;填料;油灰填料;嵌缝条
joint filler tape 嵌缝条;接缝填充带
joint fillet 填缝板;嵌缝条
joint fillet cover strip 压缝板
joint filling 节理充填;接缝填充;填缝;堵缝;嵌缝
joint filling composition 填缝混合物
joint filling compound 填缝料;嵌缝混合料
joint filter 嵌缝料
joint financing 联合通融资金;联合融资;联合投资
joint finish 填缝涂料
joint fished 夹板接头
joint fissure 节理裂缝;接头裂缝
joint fissure line 接头缝合线
joint fissure statistics 节理裂隙统计
joint fitting 对接配件
joint fit-up 接头装配
joint five-way turnout 关节型五开道岔
joint fixing 接缝固定
joint flange 接头凸缘;接口法兰;接合凸缘;接合法兰;凸缘结合
joint flash 型芯飞边;披缝
joint float 联合浮动
joint flood control and irrigation use 防洪灌溉联合使用
joint floor(ing) 拼缝地板;企口地板
joint fluid 关节液
joint for allowance 误差缝
joint for attaching instrument connection 连

仪表的接头
joint for balance piston 平衡活塞杆接头
joint for fittings with internal threads 具有内螺纹的接头
joint formation 节理构造
joint-forming machine 接缝成型机
joint-forming metal strip 接缝金条盖缝
joint for shear resistance 抗剪接头
joint free 无缝的
joint frequency 节理频数;节理频率
joint frequency distribution 联合频率分布
joint frequency function 联合频率函数;联合次数函数
joint-friction-resistance 节理缝摩阻力
joint funding 联合筹资
joint gap 轨缝间隙(铁轨);接合缝隙;接缝宽度;接缝间隙;钎缝间隙
joint gate 分型面浇口
joint ga(u)ge 测缝仪;测缝计
joint glue 接合胶;接缝胶
joint glued 接合缝;胶合缝
joint grease 接头润滑剂;丝扣(润滑)油
joint grid 接缝网;接合网络
joint grinding 联合碾磨(花岗岩和沥青)
joint group 联合集团
joint group number 节理组类
joint group of expert on the scientific aspects of marine pollution 海洋污染问题联合专家小组
joint grouting 灌缝;接缝灰浆;接缝灌浆;灰缝灌浆;填缝灌浆;灰浆灌浆
joint-grouting groove 接缝灌浆沟槽
joint grouting work 接缝灌浆工作
joint guarantee 连带担保
joint halved 对搭接头;对接接头;搭接接头
joint heater 接缝加热器
join the two sections of a bridge 合龙(桥梁)
joint hiding 隐藏的接缝
joint hinge 接合铰链
joint histogram 节理直方图
joint holding moment 节点夹持弯矩;节点力矩
joint hoop 节点箍筋
joint housed 暗榫接;暗接榫
joint impedance 结点阻抗
joint imperfection 接缝缺陷
joint implementation 联合实施
joint income 共同收益
joint information content 联合信息量
jointing 原浆勾缝;折叠;拼接修整;微磨刀刃;填缝;刀头修整;单板整边;齿顶修整;齿侧修整;节理;接合
jointing arrangement 接缝排列(布置)
jointing at inside corner 阴角接缝做法
jointing at outside corner 阳角接缝做法
jointing-bonding-adhesive compound 结黏石膏
jointing by mortise and tenon 榫槽接合
jointing cast units by prestressing 用预应力结合的预制件
jointing chamber 电缆交接箱
jointing clamp 接线钳;接线夹
jointing compound 堵缝剂;结缝剂;胶接剂;密封剂;接缝密封料
jointing edge 接触点
jointing element 连接件;连接构件
jointing machine 接合机
jointing material 勾缝料;接合密封材料;接缝材料;密封材料;填缝料;填密垫圈
jointing medium 接合剂
jointing method 连接法;接合方法
jointing mortar 接缝砂浆;接缝灰泥
jointing of concrete structure 混凝土结构接缝
jointing of pipes 管封;管子连接
jointing of sett paving 块料铺砌的再填缝;石块铺砌(面)连接
jointing of tubes with tube sheet 管子与管板的连接
jointing paste 堵缝膏泥;灌浆缝;灌缝膏;接缝膏泥
jointing pattern 接缝图案
jointing piece 连接件
jointing plane 修边刨;合缝刨
jointing press 板材压合机;板材压合机
jointing refractory cement 耐火填缝灰浆;耐火填缝灰泥
jointing ring 连接环
jointing rivet 连接铆钉;铆钉

jointing rule 接榫划线尺;接榫(量)规;砌砖规尺;砌砖长尺;(砌砖抹灰工用的)长直尺;勾缝尺
jointing shackle 连接卸扣
jointing shoe 顶梁(托)座
jointing sleeve 连接套筒
jointing slip 黏结用泥浆
jointing stage 伸长期
jointing strip 压缝条;嵌缝条
jointing surface 接合面
jointing system 接合系统
jointing tool 整齿工具;接缝器;勾缝工具;接榫工具;接缝工具;填缝器
jointing washer 接头垫圈;接口垫圈
joint inspection 联检;联合检查
joint inspection boat 联检船
joint inspection ship 联合检查船
joint installation plan 联合安装计划
joint installing machine 接缝安装机(混凝土路面)
joint integrity 节点整体性
joint intensity 节理密度
joint interpolated motion 接点插入移位
joint investment 联营;合资
joint investment railway 合资铁路
joint invitation to tender 联合招标
joint iron 接缝铁
joint lap 互搭接合
joint leak 接头漏水
joint leakage 接头空隙;接缝渗漏
joint length 节理长度
jointless 无接头的;无(接)缝的;无法兰连接
jointless acid resistant flooring 整体耐酸地面
jointless construction 无缝构造
jointless deck 连续桥面
jointless floor 无缝地板
jointless flooring 菱苦土地面;无缝地面做法;无缝地板;无缝楼板
jointless plastics skirting 整体塑料踢脚
jointless skirting 整体踢脚线
jointless structure 无缝结构
jointless track 无缝线路
jointless track circuit 无(绝)缘轨道电路
joint liability 共同责任
joint line 联系测线;连线;模缝线;合缝线;分型线
joint liner 接合衬垫
joint listing metal ferrule with O-ring 金属包头与O形环的组合接头
joint load 节点荷载
joint loading 节点加荷
joint loan 联合贷款
joint-locked warehouse 连锁仓库;进出货栈
jointly and severally liable 负有连带责任的
jointly assess 评议商定
jointly dependent variable 联合依变数
jointly ergodic random process 联合各态历经随机过程;联合遍历随机过程
jointly-operate 联营
jointly operated 合营的
jointly random process 联合随机过程
jointly run business 联营企业
jointly stationary random process 联合平稳随机过程
joint management 联合管理
joint mapping with plane-table and transit 小平板仪与经纬仪联合测图
joint marginal distribution (三个随机变数联合分布的)联合边缘分布
Joint Maritime Committee 联合海事委员会
joint mark 接合限度;接合痕迹
joint marketing 合作销售
joint masking tape 接缝覆盖带;遮缝带;贴缝带
joint-matching shoe 接缝调高靴(沥青混合料摊铺机自动控制高程的装置)
joint material 嵌缝材料;封填胶
joint maximization 联合最大化
joint measurement 测缝;缝的量测;共同尺寸;总尺寸
joint mechanism 节点原理;节点机构
Joint Meteorologic(al) Board 联合气象局
Joint Meteorologic(al) Committee 联合气象委员会
Joint Meteorologic (al) Radio Propagation Subcommittee 气象无线电传播联合分委员会
Joint Meteorologic(al) Satellite Advisory Committee 气象卫星联合咨询委员会

joint meter 节点计;测缝仪;测缝计
joint metre 测缝仪;测缝计
joint mobility 节点位移(率);节点灵活性
joint moment 节点弯矩
joint mortar 接缝灰浆
joint mortised 榫接合
joint mo(u)ld 成形模板;型模;(预制灰质饰件的)模板
joint movement 缝宽变量;接缝位移;接缝伸缩
joint-mullion 节理式窗棂
joint nature 节理性质
Joint Navigation(al) Satellite Committee 导航卫星联合委员会
joint needle-handle file 铰链针锉
joint nonparametric confidence interval 联合非参数区间
joint note 联名票据;联合票据
joint nut 接头螺母
joint observation 联合观测;接缝观测
joint observation value 联合观测值
joint occupancy date 联合占用日期;联合日期
joint occurrence 节理产状
Joint Oceanic Association 海洋学联合会
joint of casing 一段套管柱
joint of concrete pavement 混凝土路面接缝
joint of drill pipe 一段钻杆柱
joint of framework 构架节点;框架节点;(桁架的)节点
joint offsets 错开
joint of retreat 收缩节理
joint of rupture 断裂缝
joint of segment 管片接头
joint of trigonal frame 三脚架的节点
joint on butt 平接
joint on square 直角接合
joint opening 土隙;裂隙开度;节理裂开;缝张开度;缝隙爆破;缝口
joint operating device 并车装置
joint operating procedure 联合操作程序
joint operation 联营;联合调度
joint operation centre 联合操作中心
joint orientation 节理方向;节理方位
joint-outing 切制制榫;混凝土路面切缝;(整体大面积的)切缝
joint overseas switchboard 联合海外交换台
joint owner 合股船所有人
joint ownership 共同所有(权)
joint packing 接合填密;接合密封;填充垫圈
joint parts 接合配件
joint pattern 节理型;节理图形;节理模式;节理格式
joint penetration 接头熔深;接头焊透层
joint permeability 接缝渗透性;节点渗透性
joint persistence 节理贯通度
joint photograph expert group 联合图像专家组
joint pin 连接柱;连接销;连接针
joint pipe 连接管;接合管
joint placing machine 接缝机
joint plan 接缝平面图;接头平面图
joint plane 接缝面;节理面
joint plant 节理平面
joint plate 连接板;节点板;接合板;接缝板
joint pliers 铰钳;扁钳
joint plug 接缝堵塞物
joint point 接合点
joint pointing 勾缝;平缝;嵌缝
joint pole 同架电杆
joint pole diagram 节理极点图
joint pooling of imports 进口联营
joint pouring 接缝灌浆;灌缝;浇缝
joint pouring machine 接缝灌浆机
joint preparation 接头坡口准备
joint pressing-in method 压缝法
joint probability 联合概率;结合概率;接合概率
joint probability density 联合概率密度
joint probability density function 联合概率密度函数
joint probability distribution 联合概率分布
joint probability distribution function 联合概率分布函数
joint probability function 联合概率函数
joint procedures analysis 联合工序分析
joint product 联产品;复合产品
joint product cost 联产品成本
joint product costing 联产品成本分摊法

joint production cost 联合生产成本
joint product method 联产品法
joint product production 联产品生产
joint products costing 联产品成本计算
joint profile 接缝断面
joint project 联合规划;合办项目
joint protection policy 共同保障保险单
joint purification plant 联合净化厂
joint rail and water transportation 陆海联运
joint rail motor rate 铁路、汽车联运运价率
joint railway 联合铁路
joint raker 清缝耙;接缝支托;接缝斜撑
joint random variable 联合随机变数;联合随机变量
joint rate 联(运)运价;联运运费;联运费率;联合运价(率);联合收费
joint reinforcement 接缝配筋;接缝钢筋;接头钢筋;嵌缝配筋
joint repair heater 接缝维加热器
joint resealing 重新封缝
joint research 合作研究
joint reserve 公共准备金
joint residue 封缝前应清除的杂质;接头杂质
joint resistance 结点电阻;合成电阻
joint responsibility of railway 铁路的连带责任
joint ridging 接缝凸起;接缝隆起
joint rigidity 节点刚度
joint ring 接合密封圈;接合(垫)圈;密封圈;接合环
joint rod 斜接尺;连接杆
joint roller 接缝压辊
joint rose(diagram) 节理玫瑰图
joint rosette 节理玫瑰图
joint rotation 节点旋转
joint rotation angle 节点转动角;节点旋转角
joint roughness 节理粗糙度
joint roughness coefficient 节理(面)粗糙(度)系数
joint roughness value 节理糙度数值
joint rub 上胶密闭接头
joint rule 连接尺;接榫量规;斜角尺;砌砖长尺;斜角接缝尺
joint runner 管接头外缠绳;接口灌口;接缝填料;填缝索;填缝浇口;填缝条
joint saving 联合投资节约金
joint sawing 接缝磨光;锯缝
joint sawing machine 锯缝机;切缝机
joints between blocks 块体间的接缝
joints between caisson 沉箱间的接缝
joints between monoliths 沉井间接缝
joints between tubes 沉管接缝
joints between units 构件间接缝
joint scarfed 嵌接榫
Joint Science Committee 联合科学委员会
joint scientific research project 联合科研项目
joint sea-and-rail transportation 海陆联运
joint seal 接缝止水;封缝;密接
joint sealant 接缝止水剂;接缝密封胶;密封胶;填缝胶;封缝料
joint seal compound 封缝混合料
joint sealed by cover strip 压缝条封缝
joint sealer 敛缝料;接缝封闭剂;填缝料;填缝机;封缝机
joint sealing 封缝止水;封缝
joint sealing compound 抗喷气和燃油的(路面)填缝料;封缝(材)料;密封料;封缝混合料
joint sealing filler 封缝填料
joint sealing lime 密封填缝石灰
joint sealing machine 封缝剂;灌缝机;封缝机
joint sealing mastic 填缝玛琋脂
joint sealing material 接头缝填料;接缝密封材料;填缝料;封缝料
joint sealing strip 封缝条;接缝密封条
joint seam 接缝
joint section 共线区段;合页铰页片
joint section of line 铁路线联合使用区段
joint separation 接头分离
joint set 节理组
joint sheet 接合垫片;接缝垫片;填充垫片
joint sheet packing 接合填密片
joint shield 接缝盖板;接缝盖板;防渗接缝
joint shingle 搭接木瓦;对接的屋面木板瓦
joint signature 联合签署;合签
joint single crossover turnout 单渡线关节型道岔
joint single-way turnout 单开关节型道岔
joint size 连接尺寸

joint slack 联轴器;连接套筒
joint sleeper 接接垫枕;接合小格栅;接缝处小梁;轨枕
joint slip 节点滑动
joint smoothing iron 接缝用烙铁
joints of sewers 管道接缝
joints of the body on the whole 百节
joint space 接点空间
joint spacer 接缝隔条;接缝隔片;接缝插物;接缝间距块
joint spacing 节理间距;接缝间隙;接缝间距;接缝间隔
joint spalling 缝合碎裂;缝口低陷;缝口碎裂
joint spider (万向接头的)十字叉
joint spring 节理泉
joint stages dividing 节理分期
joint-staggering brick 错缝砖
joint statement 联合声明
joint state-private enterprise 公私合营企业
joint station 共用车站;联合车站
joint step 钢轨接头错牙
joint stepped 踏步对搭接头
joint stiffness 节理刚度;接缝刚度
joint stock 共同资本;合资
joint-stock bank 股份银行
joint-stock company 合股公司;股份公司
joint-stock enterprise 合股公司
joint stock limited partnership 联合股份有限公司
joint stool 折叠(式)椅子;榫接凳
joint strap 带状接头;带状结点
joint strength 接缝强度;连接强度
joint strength characteristic 节理强度特性
joint strike 同盟罢工
joint strip 压缝条;嵌缝胶条
joint stub 节点短截头
joint supply 联合供给
joint surface 节理面;接缝面;分型面
joint survey 联合检查;节理勘测
joint survey data 节理勘测资料
joints using welds for sealing 焊接密封连接
joint swing 接缝锯法
joint system 裂隙组;裂隙系;节理系
joint system dividing 节理配套
joint tape 连接带;接缝带;胶纸带;加强筋;加强带;黏缝带
Joint Technical Committee 联合技术委员会
Joint Technical Planning Committee 联合技术规划委员会
joint tee 接头丁字铁件
joint template 锁接头量规
joint tenancy 共有不动产权;共同租赁;联合租赁
joint tenant 不动产共有人;联合租赁者
joint tender 联合投标
joint tendering 合作投标;合伙投标
joint terminal 联合枢纽
joint the map 图幅接边
joint three-way turnout 关节型三开道岔
joint tie 接头轨枕;拼接板
joint tongue 连接榫;榫舌;滑键
joint tongue and groove 企口接缝
joint topical workshop 专题研究班
joint toxicity 联合毒性
joint toxicity action 联合毒性作用
joint trackage 共用线路
joint traffic revenue 联运收入
joint trajectory 节点轨迹;节点移动轨迹
joint translation 节点移动;节点位移
joint treatment 综合处理;联合处理
joint trunk exchange 连接长途电话局
joint tube 锁口管
joint type 节理类型
Joint Typhoon Warning Center 联合台风警报中心(美国海空军)
joint undertaking 共同事业;联合企业;合资企业;合股公司
joint-up exposure 连接曝光
joint use 一线多用;综合利用;共用;联合运用;同杆架设
joint users group 联合用户组
joint valley 节理谷
joint variable 联合变量
joint variation 连变分
joint veins 节理脉
joint velocity 节点移动速度

joint venture 共同风险投资;联营体;联营企业;联合经营;联合承包方式;合资经营;合资;合资企业;合股公司;联合企业;合包(工程)
joint venture company 合资公司
joint venture corporation 合资公司
joint venture firm 合资企业;联合经营商行
joint venture for project 工程项目合资
joint venture investment 合资经营投资;(短期的)合伙投资
joint venture of government and private citizen 公私合营
joint venture project 中外合资项目
joint ventures bidder 联合投标
joint venture work 合作工作
joint washer 密封垫圈
joint wastewater treating facility 废水综合处理设施
joint water-and-rail transportation 水路铁路联运
joint water pressure 节理水压力
joint water-rail-motor rate 水道、铁路、汽车联运价率
joint water-rail rate 水道、铁路联运价率
joint waviness 节理面起伏度
joint weld 焊缝
joint welding 搭焊
joint welding procedure 接头焊接工艺
joint width 接缝宽度;微斜面宽度;缝宽
joint with butt strap 带盖板的平接缝
joint with double cavities 双空腔缝
joint with dovetail groove 燕尾榫接合;鸠尾榫(槽)接合
joint with sealing plate 填缝板接缝
joint with single butt strap 单搭板铆钉对接
joint with single cavity 单空腔缝
joint with single cover plate riveting butt 单搭板铆钉对接
joint with single strap 单搭板铆钉对接
joint with staggered rivets 间行错双面铆钉搭接
joint working group 联合工作组
jointy 成层的;多接缝的;充满极细裂缝的
joint yoke 叉槽
joint zone 接缝区
join up 连接起来
joist 小梁;格棚;托梁;地楞;底托板;副梁;安装托梁
joist anchor 格栅锚栓;墙锚;墙锚固钢板;连接锚;格栅锚垫
joist and plank 梁板两用材料;厚板龙骨
joist bar 格栅钢筋;T形梁钢筋;梁钢
joist bearing 格栅承座;托梁支承;托梁承座
joist bridging 格栅的剪刀形撑;格栅撑
joist ceiling 格栅平顶
joist chair T形梁支座;格栅支座
joist connector 托梁连接器
joisted floor 格栅楼盖
joisted sub-floor 格栅支承的毛地板
joisted underfloor 格栅支承的毛地板
joist floor 格栅楼盖;格栅楼面(仅由格栅承载的木楼板);格栅楼板(仅由格栅承载的木楼板);格栅地板
joist grillage 格栅排格;成列格栅
joist hanger 格栅锚件;格栅吊钩;格栅;梁托;格栅承座
joist head 格栅枕
joist header 格栅枕
joist lumber 方材
joist nail 格栅钉
joist pass 工字钢孔型
joist plate 垫木板
joists and planks 格栅与厚板
joist shearing machine 工字铁剪机
joist shears 工字剪切机;型钢剪切机;钢梁剪切机
joist spacing 格栅间距
joist steel 工字钢;梁钢
joist trimmer 格栅(金属)配件
joist under collar beam 系梁下的格栅
Jojoba wax 西蒙得木蜡
jokokuite 五水锰矾
jokula 冰流;冰川河流
jokulhlaup 冰川(消融)洪水
joliotite 炭铀矿
jolite ceramics 堇青石陶瓷
jolley 辘轳成形
jolleying 阴模旋坯成型
jolli panel 雕刻镶板
Jolly 耐火砖成型机;陶器转盘

Jolly balance 测密(实)度平秤;比重天平
jolly boat 小帆船;杂用艇
jolt 意外的挫折;摇动;震摇;振撼;振动;颠簸;震击
jolt-and-jumble test 振动试验
jolt capacity 振击能力
jolter 震实造型机;震实台;震动器;振实制型机
jolt impact tester 震击试验仪;震击式韧性试验机
jolting 震实;震击
jolting cylinder 震击缸
jolting machine 震实造型机;震击造型机
jolting mo(u)lding 振动成型;振动机制
jolting-mo(u)lding machine 振动成型机
jolting of ingots 震动铸锭法
jolting piston 震击活塞
jolting rate 震动速率
jolting table 振动台
jolting vibrator 震实器;冲击振捣器
jolt knee valve 震实膝形阀
jolt knock-out grid 振击落砂架
jolt mo(u)lding 震实造型;振动机制;振动成型;捣固成型法
jolt mo(u)lding machine 震实造形机;振动造型机;振实造型机;振动成型机
jolt(-packed) 震实
jolt-packed liner 震筑衬里;振筑衬里
jolt-packing 震动填料;振动填料
jolt-ram 震实
jolt rammer 震实造型机
jolt ramming 振动夯实;振动夯击;振动捣实;震实
jolt-ramming machine 振动压型机;振动夯实机;振动夯击机;振动压型机;振动捣打机;振动夯击机;震实造型机;冲击机
jolt-ram pattern draw machine 顶箱压实式造型机
jolt rock-over mo(u)lding machine 翻台震实造型机
jolt rock-over pattern draw mo(u)lding machine 翻箱震实式造型机
jolt rollover draw mo(u)lding machine 翻箱震实式造型机
jolt rollover pattern draw machine 翻箱震实式造型机
jolt-squeeze 震压;振实挤压
jolt-squeeze mo(u)lding machine 振压(式)造型机;震压式造型机
jolt squeeze pattern drawing machine 起模式震压造形机
jolt-squeezer 震压式造型机
jolt-squeezer machine 震压式造型机
jolt squeeze rollover pattern drawing machine 翻转起模式震压造型机
jolt-squeeze rotalift mo(u)lding machine 翻转式震压造型机
jolt-squeezer stripper 顶箱震压式造型机
jolt-squeezer stripper mo(u)lding machine 顶箱震压(式)造型机
jolt stripper mo(u)lding machine 顶箱震实造型机
jolt stroke 震击行程
jolt table 震动台
jolt toughness 震动韧性
jolt toughness tester 震击式韧性试验仪
jolt-type mo(u)lding machine 震实式造型机
jolt vibration 撞击式振动
jolt vibrator 振动式振捣器;颠震振捣器
Joly balance 焦利天平
Joly's steam calorimeter 焦利蒸汽量热器
Joly wax block photometer 焦利腊块光度计
Jominy curve 顶端淬火曲线
Jominy distance 顶端淬火距离
Jominy end quench test 顶端淬透性试验
Jominy test 顶端淬(透性)试验
Jomon style 一种传统的日本建筑风格;日本绳纹文化建筑式样
Jomon-type pottery 绳纹型取土器(日本古土器)
jonah trip 不幸的一次航行;不成功计划
Jona's method 分段绝缘法
Jones Berkshire refiner 琼斯大锥度精磨机
Jones-Bertrams beater 琼斯帕特兰斯打浆机
Jonesboro 琼斯波罗花岗岩(一种产于美国缅因州的浅红色花岗岩)
Jones calculus 琼斯计算法
Jones furnace 琼斯炉
jonesite 硅钛钡钾石
Jones matrix 琼斯矩阵
Jones-Mote reaction 琼斯—莫特反应

Jones noise equivalent power 琼斯等效噪声功率
Jones pulper 琼斯高低速双转盘水力碎浆机
Jones reagent 琼斯试剂
Jones riffle 琼斯格槽缩样器
Jones splitter 琼斯缩分器
Jones thermocouple 琼斯温差电偶
Jones vector 琼斯矢量
Jones zone 琼斯区
jonquil 淡黄色;红黄色;长寿花;黄水仙
Jonval type turbine 章伐耳水轮机(平行或轴向水流水轮机)
Jonval wheel 章伐耳水轮
Joosten chemical solidification process 周斯坦化学溶液凝固掘进法
Joosten consolidation process 周斯坦加固法
Joosten process 周斯坦固土法;周斯坦式(溶液固土)掘进法;周斯坦化学加固土壤法
Joosten two fluid chemical consolidation process 周斯坦双液化学加固法(以水玻璃和氯化钙两种液体注入土壤)
jordan 低速磨浆机
Jordan bearing 推力套筒轴承
jordanite 约硫砷铅矿
jordan mill 低速磨浆机
jordisite 佼硫钼矿
josephinite 镍铁矿
joss house 庙;寺院
joss powder 柏木粉
Jothian series 约特斯统(前寒武纪)
jot stream 热水溪
jotter 拍纸簿
jotting 简短的笔记
joule 焦(耳)(能量,热量功的国际单位)
Joule cycle 布雷顿循环
Joule heating 欧姆加热
Joule/Kelvin 焦耳/开
Joule-Kelvin effect 焦耳—开尔文效应
Joule/kilogram-Kelvin 焦耳/千克·开
Joule-Lenz's effect 焦耳—楞次效应
Joule-Lenz's law 焦耳—楞次定律
Joule magnetostriction 正磁致伸缩
joulemeter 焦耳计
Joule's calorimeter 焦耳热量计
Joule's dissipation 焦耳耗散
Joule's effect 焦耳效应
Joule's energy 焦耳能量
Joule's equivalent 焦耳(热功)当量
Joule's heat 焦耳热
Joule's heating 焦耳加热
Joule's law 焦耳定律
Joule's magnetostriction expansion 焦耳磁致伸缩性膨胀
Joule-Thomson coefficient 焦耳—汤姆逊系数
Joule-Thomson cooler 焦耳—汤姆逊制冷器
Joule-Thomson cooling 焦耳—汤姆逊效应冷却
Joule-Thomson effect 液化效应;焦耳—汤姆逊效应
jounary radius 出行半径
jour-a-jour enamel 透光珐琅
jouravskite 硫碳钙锰石
Jourawski's method 茹拉夫斯基法(俄国工程师,静定桁架的节点分析法)
journal 学报;止推轴颈;杂志;辊颈;航海日志;日志;日记(簿);顺序记录;定期刊物
journal account 流水账;日记簿
journal bearing 主轴承;轴颈轴承;径向轴承;滑动轴颈
journal bearing sleeve 轴颈轴承套
journal bearing surface 轴颈轴承面
journal book 流水账;日记帐;日记簿;分录簿
journal box 轴箱;轴颈箱;轴承箱
journal-box bolt tie bar 轴箱螺栓系杆
journal-box cover 轴颈箱盖
journal-box guide 轴箱导框
journal-box lid 轴颈箱盖
journal-box lid hinge 轴颈箱盖铰链
journal-box lid spring 轴颈箱盖弹簧
journal-box spring seat 轴箱弹簧座
journal-box wedge 轴箱楔
journal-box yoke 轴箱架
journal brass 轴颈铜衬;轴承铜铜
journal brass alloy 轴颈黄铜合金
journal bronze 轴颈轴承青铜;轴颈黄铜铜;减磨青铜
journal buffer 日志缓冲器

journal bush 轴颈衬套
journal cash book 现金出入账簿;现金出纳备查簿
journal collar 轴颈环
journal compound 轴颈润滑脂
journal day book 分录日记账;分录日记簿;日用流水账
journal entry 流水分录;日记账分录;日记簿分录;分录
journal file 日志文件
journal folio 日用流水账;日记账页数;分录账页码
journal for axial load 轴向载荷轴承;轴向负载轴颈
journal for radial load 径向负载轴颈
journal friction 轴颈摩擦
journal housing 轴箱
journaling 记日志;记录数据集中的变动;报表
journalize 分录;编制分录
journalizing 流水分录;日记账分录;分录
journalizing of closing accounts 结算的分录
journal ledger 分类账
journal lid 轴箱盖;轴盖
journalling sleeve 轴颈套筒
journal load 轴颈负荷
journal metal 轴颈合金
journal name 刊物名称
journal neck 轴颈
Journal of Animal Science 畜牧学报
Journal of Applied Mechanics 应用力学杂志
journal of fluid mechanics 流体动力杂志
Journal of Food Science 食品科学学报
Journal of Morphology 形态学报
Journal of Range Management 草原管理学报
Journal of Research of the National Bureau of Standards 美国国家标准局研究杂志
Journal of Soil, Water Conservation 水土保持杂志(美国)
Journal of Soil Science 土壤科学杂志(半年刊)
Journal of the American Concrete Institute 美国混凝土学会杂志
Journal of the British Grassland Society 草地学会学报
Journal of the Construction Division 建筑杂志(美国土木工程师学会)
Journal of the Institute of Water Engineers 给水工程师学会杂志
Journal of the Institute of Sewage Purification 污水净化学会杂志
Journal of the Institution of Electric(al) Engineers 电气工程师学会杂志
Journal of the Institution of Water Engineers 给水工程师学会杂志
Journal of the Soil Mechanics and Foundation Division 土力学与基础工程(期刊名)
Journal of the Water Pollution Control Federation 水污染控制联合会会刊
Journal of the Waterway and Harbo(u)r Division 水道及港口分会会刊
Journal of Water Pollution Control 水污染控制杂志
journal oil 轴颈油
journal output 日志输出
journal packing 轴颈填密;轴颈填料;轴颈密封垫
journal pin 轴颈销
journal pressure 轴颈压力
journal printer 记录打印机
journal reader 报刊杂志读取机
journal resistance 轴颈阻力
journal rest 轴颈支承
journal roll 轴颈辊
journal shaft 轴颈支承的大轴
journal sheet 报表
journal spring 轴箱弹簧
journal spring device 轴箱弹簧装置
journal stirrup 支耳
journal teleprinter 日志电传打印机;日记账电传打印机
journal velocity 轴颈速度
journal voucher 分录凭证
journal with collar 有环轴颈
journey 旅程;路程
journey back to the earth 返回地球的路程
journey-man 短工;计日工(人);职工;满师徒工;雇工;散工
journeyman electrician 熟练电工
journeyman gas fitter 临时煤气用具安装工

journeyman painter 熟练油漆工
journey report 出车报单
journey speed 行程速度;行程车速
journey time 行程时间
journey to shop 购物旅程
journey-to-work 工作旅程;上班旅程;上班交通
journey-to-work travel time 上班行程时间
journey work 雇佣工作;短工工作;计时工;计日工;散工;短工
joustifiable expenditures 合理支出
Joy continuous miner 乔伊连续采煤机;乔伊型连续挖掘机
joy-riding plane 游览机
joystick 远距离驾驶杆;控制杆;变速操纵杆;排挡杆;远距离操作手柄;远距离操纵手柄;驾驶杆;操纵杆
joystick lever 球端杆
joystick pointer 操纵杆光(光标)指示器
joystick signal 操纵杆信号
j polyether plasticizer 聚醚增塑剂
J seal J形止水结构;J形密封结构;J型密封材料
Juan de Fuca plate 胡安德福卡板块【地】
Juan de Fuca ridge 胡安德富卡海岭
juanite 水黄长石
jubbled rock 混杂岩;混合岩
jube 隔拦;屏障【建】;教堂十字架围屏(圣坛前);阁楼(教堂放十字架坛上的);唱诗班屏隔(教堂)
jubilee clip 连接螺旋夹
jubilee exhibition 周年纪念展览会
jubilee skip 小型轻轨翻斗车;小型翻斗车
jubilee track system 窄轨铁路系统
jubilee truck 小型侧卸式货车;小型矿车(侧卸式);小型货车;倾卸式货车;轻轨料车
jubilee wagon 小型铁道车;小型侧卸车;小型(侧翻式)货车;倾卸车;翻斗车
jubo burner 特大废气燃烧器
jud 石面板槽
judas 监视孔;(门,墙上的)外窥孔
judas window 监视窗;窥视孔
Judd colo(u)r difference unit 贾德色差单位
judder 强振动声;强烈振动;图像不稳定;震颤声
Judd graph 贾德黑白格纸
Judd hiding power chart 贾德遮盖力试验纸
juddite 锰钠铁闪石
judge 衡量;判定;审理;裁定;法官
judge a colo(u)r 鉴定颜色纯度
judge distance 目测距离
judge lamp 裁判灯
judgement 鉴定;判定
judgemental identification 经验认定法
judgement creditor 判决(确定)债权人
judgement currency clause 判决通货条款
judgement debt 判决确定的债务
judgement debtor 判决确定债务人;判决债务人
judgement forecast 判断预测
judgement in decision 决策判断
Judgement Jugoslav Resister of Shipping 南斯拉夫船舶登记局
judgement lien 判决留置权;判决抵押;判决留置财产
judgement payable 判决应付款项;判定应付款项
judgement proof 没有资产因而不能对他实行资产扣押者
judgement rates 判决确定的费率;判决费率
judgement receivable 判决应收款项
judgement record 判决记录
judgement sample 判断样本;判定抽样
judgement sampling 鉴定性抽样;判断抽样
judgement selection 判断抽选
judgement summons 法院传讯
judgement theory 判决理论
judgement time 判断时间
judges room 审讯室
judging device 识别装置
judging of grassland condition 草地诊断;草地评定
judging technique 鉴定技术
judgment 判决;判断
judgment currency clause 审理货币条款
judgment lien 判决留置财产;法院判决的留置权
judgment sampling 标准抽样
Judicature Act 司法权法案
judicial 公正的;司法的
judicial action 诉讼
judicial allocation 通过司法裁定确定水量分配;司

法分配
judicial arbitration 法院仲裁
judicial authority 司法机关
judicial decision 司法判断;司法裁决;法院判决
judicial department 司法部门
judicial foreclosure 取消抵押品赎回权的法律手续
judicial guarantee 法律保证
judicial investigation 司法调查
judicial lawsuit 司法诉讼
judicial litigation 司法诉讼
judicial mortgage 法院判定的抵押
judicial person 法人
judicial precedent 司法制例
judicial proceeding 司法手续;审判程序;法律诉讼
judicial program(me) 司法程序
judicial remedies 司法补偿
judicial sale 法院判决的拍卖
judicial security 法院判决的担保
judicial settlement 司法解决
judicial trustee 司法受托人
Judicial Trustees Act 司法受托人法案
Juell pressure flo(a)tation cell 求勒型压力浮选机
jug 有把水罐;汽缸;地震检波器;带柄水罐;粗糙面
jugate 连环体
jugged 锯齿形;齿形(的)
juggle 锯开的木块
juggler 斜柱
juggling 摇动
jug-handle 操作用管线;壶柄
jug-handle interchange 壶柄型立体交叉
jughandle ramp 壶柄式匝道;壶把匝道【道】
jug hustler 放线员
Juglans 胡桃壳
jug line 大线
Jugoslav Resister 南斯拉夫船级社
juice 高利
juicer 电工;榨汁机
juice straining carrier 除屑机
juicy runway 湿的跑道
juienite 硫氰钠钻石
jujube 枣树
jujube tree 枣树
jukebox storage 盘式存储器
juke joint (有自动电唱机的)音乐酒吧间
julgoldite 复铁绿纤石
Julian calender 儒略历
Julian century 儒略世纪
Julian date 儒略日期;儒略日
Julian day 儒略日
Julian day calendar 儒略日历
Julian day number 儒略日数
Julian ephemeris century 儒略历书世纪
Julian ephemeris date 儒略历书日期
Julian epoch 儒略历元;儒略纪元
Julian era 儒略纪元
Julian number 儒略日数
Julian period 儒略周期
Julian year 儒略年
julienite 天青钴矿;毛氰钴矿
Julius Caesar calendar 儒略历
jumble sale 旧货廉卖(英国)
jumble shop 廉价杂货店
jumbo 运木滑板;移动式钻机台;钻孔台车;钻架;钻车;渣口冷却器;空心粘土大砖;巨型设备;巨型喷气机;巨型;巨大的;前支撑帆;特大的;隧道钻车;隧道支模车;隧道运ލ车;隧道孔机;隧道铠框;隧道盾构;大型喷气客机
jumbo air liner 大型喷气式班机
jumbo arm 钻车的钻机托臂
jumbo barge carrier 大型载泊船
jumbo base 大号管基
jumbo bolt 手揿锁的插销
jumbo boom 重(型)吊杆;隧道钻车臂
jumbo brick 大型砖;(尺寸超过规定的)次砖;尺寸超过标准的大砖
jumbo brick structure 大型砖结构
jumbo column section 巨型钢柱
jumbo drill 车装式钻机
jumbo fiber 大号光纤;粗光纤
jumbo for rail service 有轨钻车
jumbo for shotcrete sprayer 喷射机台车
jumboisation 切断接长;切断加长
jumboization 切断接长;切断加长
jumbo jet 巨型喷气机;大型喷气机
jumbo loader 钻装机;凿岩装车联合机
jumbolter (安装锚栓和打眼用的)钻车;杆柱钻机;锚杆钻机
jumbos 撬
jumbo ship 巨型船舶
jumbo size 特大号包装;大(型)尺寸
jumbo-sized 巨型的;巨大的
jumbo-size sheet stacker 大玻璃片码垛机
jumbo stern ramp 大型船尾斜吊桥
jumbo truck 钻车
jumbo tire 大车胎;超低压轮胎
jumbo windmill 巨型风力发动机
jumbo work 船体截断加长工程
jume 板岩瓦片夹钳
jump 矿脉断层;突跃;跳跃;跳变;水跃
jump address 转移地址;跳变地址
jump a leg 窜相位
jumpbo group 大组;大群
jump butt 截根端
jump characteristic 跳跃特性
jump condition 转移条件;跳跃条件
jump continuation 跳变延拓
jump control 转移控制
jump counter 跳跃式脉冲计数器;跳进式脉冲计数器
jump coupling 跳合联轴节
jump dam 跳越坝
jump discontinuity 跳跃间断点;跳跃不连续性
jump distance 跳动距离;水跃距离
jump-down 下落断层
jump drill 钢绳冲击(式)钻机
jump drilling 撞钻;冲击;冲击钻井;索钻
jumped up people 暴发户
jumper 制轮爪;工作服;钢钎;跨接线;开眼钎子;横牵索;桥形接片;钎子;跳线;跳杆;手工钎子;穿孔凿;冲击钻杆;长钻;长凿手钻;长凿
jumper bar 凿岩钎;冲击钎;冲杆;撞钎
jumper bit 落锤锤;钎子;冲击钻(头);冲击锥;冲击锤
jumper boring bar 冲击式凿岩机;冲击式钎子;冲钻杆
jumper bridge for jumping second wheel 跳秒轮用压板
jumper cable 应急电缆;跨接电缆;桥接电缆;分号电缆
jumper cable coupling 跨接电缆接头
jumper clamp 跳线夹
jumper connection 跳接
jumper detacher 卸钎器
jumper drill 跳秒钻;手钻;长钻
jumper extractor 拔钎器
jumper field 跨接线排
jumper hammer 撞击锤;冲孔器
jumper hose coupling 跨接软管联轴节
jumper indicator 分号表
jumper lift 跳秒抬针
jumper line 连接管线
jumper list 配线表
jumper locate 跳杆定位
jumper locker 作业箱
jumper motion 跳跃运动
jumper plug 插座
jumper stay 制动支索;临时支索;横牵索;桅间索
jumper steel 钎杆钢;长凿钢
jumper terminal 跳线端子
jumper tester 跨接测试器
jumper-top blast pipe 顶阀鼓风管;顶阀吹管
jumper tube 跨接管;旁通管
jumper wire 跨接线;跳线;跳接线
jump feed 周期性继续给进;快速越程;跳跃进给
jump field 转移信息组
jumpform 滑升模板;滑模;升板
jumpforming 阶升式模板法;升模施工;提升模板;滑升模板
jump frequency 跳动频率;跳变频率
jump function 跃变函数;阶跃函数;跳跃函数
jump grading 间断级配;跳越级配;跳跃级配
jump hammer 夹板锤;摩擦锤
jump height 跳跃高度;水跃高度
jump if not 条件转移
jump in brightness 亮度跃迁;亮度突变;亮度落差
jump index 跳越分度
jumping 突变现象;跳变
jumping bar 跳杆
jumping chisel 长钻
jumping circuit 跳线路
jumping correspondence 跳动对应
jumping dial 跳字电表盘
jumping drill 跳动钻;手钻;长钻
jumping fire 飞火
jumping form 升模
jumping formwork 滑升模板
jumping frog 蛙式打夯机;跳跃式打夯机
jumping hours finger 跳时拨针
jumping hours indicator 跳时指示盘
jumping hours jumper cover 跳时定位杆
jumping hours jumper maintaining plate 跳时杆压板
jumping hours jumper rivet 跳时定位杆铆钉
jumping hours jumper spring 跳时定位簧
jumping hours jumper spring rivet 跳时定位杆簧铆钉
jumping hours star support 跳时星轮座
jumping jack 手持夯实器
jumping ladder 绳梯
jumping matrix 跳跃矩阵
jumping-meter 垂直跳跃计
jumping mill 浮动式轧机
jumping net 救生网
jumping-off point 起点
jumping phenomenon 跳跃现象;跳动现象
jumping pit 沙坑
jumping probability 跳变概率
jumping relay 跳越继电器;跳动继电器
jumping second detent 跳秒钉拉挡
jumping second driving wheel 跳秒针传动轮
jumping second pallet jewel 跳秒针叉瓦
jumping second whip 跳秒轮
jumping shoe 起重机提升踏板
jumping shuttering 升模施工;提升式模板
jumping three-high stand 跳式三辊机座
jumping up 倾覆
jumping variation 跳跃变化
jumping vault 跃式拱顶(一种具有西里西亚地方特色的建筑形式);跳跃拱顶(古波兰西利亚的地方形式)
jumping wheel jumper 跳轮定位杆
jumping wire 跨接线;跳线
jump in potential 电位跃变;电位跳变
jump in pressure 压力突变
jump instruction 转移指令;跳转指令;跳越指令
jump joint 对头接(合);对接接合
jump lock 转移封锁
jump-off 跳离起飞
jump on rail 跳轨
jump operation 转移操作
jump operator 跳跃算子
jump order 转移指令;跳越指令
jumpover 跳跃轨道;跳过;连接管线;迂回管;绕行管;连续迂回管
jumpover connection 跨越连接管
jump-over distance 跳越距离
jump reaction 突变反应
jump relation 跳变关系
jump relative instruction 相对转移指令
jump resonance 跳跃谐振
jump ring 扣环
jump roll 扁钢轧辊
jump roughing mill 浮动式粗轧机
jump routine 跳转程序
jump saw 升降圆锯机
jump scrape 跳动刮板
jump seat 折叠座(位);活动座位
jump second 跳秒(每隔一秒钟跳动一次的秒针)
jump set 加强支撑;跳指标;跳变器
jump ship 非正式离船;背弃
jump space 跳间隔
jump spark 跳跃火花机;跳跃电火花;跳火
jumpspark coil 高压线圈
jump spark ignition 跳跃火花点火
jump spark system 火花防护罩;跳跃火花制
jump statistic 跳跃统计量
jump steepness 阶跃陡度
jump subroutine 转移子程序
jump suit 连衣裤工作服
jump table 转移表
jump test 可锻性试验;顶锻试验;顶锻检验
jump the rails 出轨
jump the track 脱轨

jump transfer 转移
jump-up 由平巷向上开掘的小天井;逆断层
jump-up base 前进机场
jump-up fault 上投断层
jump valve 回跳阀;跳动阀门
jump vector 转移向量
jump weld 直角对焊接;T 形焊接头;撞击焊接;平头焊接
jump welded pipe 对缝焊管;焊缝管
jump welded tube 焊缝管;对缝焊管
jump welding 丁字形焊
junction 主连点;中继线;连接点;联合;联轨站;连锁交易;接头(段);接界;联轨(处);接点;交叉口;面结;会流点;枢纽(联轨)站;枢纽
junctional complex 连接物;连接复合体
junctional epithelium 附着上皮
junctional membrane 联结膜
junction arm 交叉臂;连续杆
junction at equilibrium 平衡点
junction attachment 接线;连接
junction barrier 结势垒
junction battery 结型电池
junction benchmark 连测水准基点;结点【测】
junction between tunnel and station 隧道与车站接头
junction block 连接头;连接段;接线块;接头管(户内下水道与街道排水管的接口);(排水管与污水管间的)接合处
junction board 接线台
junction bolt 接合螺栓
junction box 联轴器;接线箱;接线盒;交叉口方框;集管箱;分向盒;分线箱;分线匣;分线盒;套管
junction box header 换热器的管束箱
junction buoy 洲尼分汊浮标;中沙内端浮标;航道汇合点浮标
junction canal 联系运河;联渠;连渠;汇合渠道
junction capacitance 结电容
junction capacity 交叉口通行能力;交叉口通过能力
junction center 汇接局;中心站
junction chamber 连接室;接头室;交汇井;汇流井
junction circuit 中继电路;联系电路;连接电路
junction closure 接点闭合
junction compensator 冷端补偿器
junction condition 连接条件
junction conduction 结电导
junction cord 连接塞绳
junction couple 结热偶
junction crossover 结跨越
junction current 结电流
junction depth 结深
junction detail 交接大样
junction detail of level(l)ing 水准结点图
junction diode 结式二极管
junction dock 汇合点港池
junction efficiency 结效率
junction exchange 中继局
junction field 结场
junction field effect 面结型场效应
junction figure 接合图形;接合投影
junction flying 接线飞行
junction frequency 聚合频率
junction gallery 连接廊(道)
junction gate 结型栅
junction gusset(plate) 角接板;接点板
junction house 输送机房;分类处
junction interface 结界面
junction laser 结型激光器
junction leakage 结漏
junction line 连接管线;联轨线;结合线;渡线
junction manhole 管路交叉处人孔;管路交叉处检查井;接头人孔;交汇处人孔(管路、电缆);交汇处检查井(管路、电缆);汇流管道连接检查井
junction marker 路口指示标;交叉口指示标
junction marks 中沙内端标
junction mode 接轨方式
junction of channels 河道汇合点;航道汇合处;水道汇合点
junction of metal for pipes 管道金属接头
junction of sewers 管道交接
junction of the star type 星形枢纽(道路)
junction of three-roads 三岔路口;三叉路口
junction of two streams 河流汇合点;合流点
junction of veins 脉接合
junction photodiode 结型光电二极管

junction piece 连接短管;接合块
junction piece without saddle 无法兰盘接合件
junction pile 转;联结板桩;角桩
junction pipe 连接短管
junction plane 结平面
junction plate 结合板;接合片;接合板
junction point 相交点;联结点;联测点;连接点;连测点;接(合)点;接轨点;会合点;分支点
junction point method 接点法
junction point of level(l)ing line 水准结点
junction point of seismic fault 地震断裂交汇点
junction point of traverses 导线结点
junction pole 接线杆;分线柱
junction pole transposition 分线杆交叉
junction port 转口港;中转港(口);通过港
junction potential 接界电势
junction potential method 接点电势法
junction rail 接合轨
junction railway 联络铁路(线);联轨铁路
junction resistor 结型电阻
junction roundabouts 道路枢纽环道;交叉口环道
junction service 短程通信
junction signal 枢纽信号
junction station 联轨站;接轨站;枢纽站
junction stripe laser 结型条状激光器
junction temperature 结温;结区温度;结点温度
junction terminal district station 枢纽区段站
junction terminal transfer train 枢纽小运转列车
junction tetrode transistor 结型四极晶体管
junction transistor 结式晶体管;面结型晶体管
junction transposition 接线换位
junction triode 结型晶体三极管
junction type 结型
junction type field effect transistor 结型场效应晶体管
junction type photodetector 结型光电探测器
junction type transistor 面结型晶体管
junction valve 连接阀
junction well 交汇井;会井
junctor 联络线;连接机
juncture 接合点;接缝;交界处
June solstice 夏至
jungite 磷铁锌钙石
jungle 密林;热带植丛;未耕作地;丛林;稠密居住区;稠密工厂区
jungle airstrip 丛林简易机场
jungle circuit 稠密电路
jungle land 丛林地
jungle terrain 丛林地
Jungner battery 铁镍蓄电池
junior 次的
Junior Achievement 青年商业社
junior beam 轻型(钢)梁;次(要)梁;副梁;小梁
junior board of directors 初级董事会
junior boards 初级董事会
junior bonds 初级债券
junior capital 初级资本
junior channel 小型槽钢;轻型槽钢
junior counter 简易计数器
junior department store 小型百货商店
junior engineer 初级工程师
junior first engineer's room 副大管轮室
junior inspector 一般检查员
junior keypunch operator 初级穿孔操作员
junior machine 新式机床;小型机车
junior middle school 初级中学
junior partner 新合伙人
junior programmer 初级程序员
junior range circuit 距离辅助电路
junior securities 非优先证券
junior security 非优先抵押权
junior systems analyst 初级系统分析员
juniper 黄绿色
juniper berry 杜松子
juniper berry oil 杜松子油;刺柏油
juniper gum 桧(树)胶;杜松树胶
juniperic acid 桧酸
juniper oil 杜松(籽)油
juniper rod 杜松油
juniper tar oil 杜松油
junitfoite 水硅锌钙石
junjor keypuncher 初级穿孔操作员
junk 金属片;金属废料;假货;(中国式的)木帆船;碎片;舢板船;大号木帆船;废绳;废旧物资

junk a hole 封孔【岩】
junk and sail-boat model 木船模型
junk art 废料艺术
junk basket 抓筒
junk bit 端面铣刀
junk bottle 深(绿)色厚玻璃瓶
junk classification 废品分类
junk damage 钻头外侧磨损
junked and abandoned 打捞失败而报废;废弃的
junked automobile 废汽车
junked plastic cleaning wastewater 废旧塑料清洗废水
junked tire 废旧轮胎
junked well 报废井
Junker's calorimeter 容克量热器
Junker's engine 容克式发动机
Junker's gas calorimeter 容克气体量热计
Junker's mo(u)ld 容克式水冷铤模
Junker's pump 容克型喷油泵
Junker's water dynamometer 容克水力测功器
Junker's water flow calorimeter 容克水流量热器
junket 野餐
junk head 汽缸头;汽缸盖
junk man 船工
junk market 旧货市场
junk pipe 等外管;非标准化管道
junk price 赔本价(钱)
junk retriever 小物品打捞器(钻孔内);小物件打捞器(钻孔内)
junk ring 填料压盖;压圈;压(密封)环;压环;密封环;活塞顶密封环;填隙圈;填料(盒)压环;填料涵压盖;衬圈
junk sculpture 废料雕塑
junk shop 旧货摊;旧货店
junk store 旧货摊
junk tube 等外管
junk value 残值;残余价值;废品值;废料价值
junk wind 行船风;亚洲的东南季风
junk yard 废品清理场
Juno cathode 卷状阴极
junoite 硒硫铋铜铅矿
Jun purple 茄皮紫
junta 琼泰局
Jupiter 木星
Jupiter cycle 岁星纪年
jupper 手眼钢钎
Juraku zuchi 日本红土
Jurassic lime 侏罗纪石灰
Jurassic limestone 侏罗纪石灰岩
Jurassic period 侏罗纪
Jurassic reef 侏罗礁;侏罗礁石
Jurassic sandstone 侏罗纪砂岩;侏罗纪砂岩石
Jurassic system 侏罗系
Jurassic type of folding 侏罗式褶皱
Jura-Trias 侏罗—三叠纪
Jura-type folds 侏罗式褶皱
jurbanite 斜铝矾
juridical person 当事人
juridical person of an association 社团法人
jurisdiction 主权;管辖权;权限;司法权
jurisdictional data 司法材料
jurisdictional dispute 司法争议
jurisdictional issues 司法问题
jurisdictional limit 管辖范围
jurisdiction area 管辖区(域)
jurisdiction within territorial waters 领航管辖权
jurupaite 硅钙镁石
jury 应急构件;应急的;临时的;审查委员会;备用的
jury anchor 应急锚
jury mast 应急桅;正头支杠;临时桅杆
jury of executive opinion method 经理人员集体审定法
jury pump 应急泵;备用泵
jury repair 应急修理;临时应急修理;临时应急修复;紧急抢修
jury rig 应急索具;应急帆;临时应急装置
jury-rigged 临时配备的
jury rudder 应急舵
jury sail 应急帆
jury steering gear 临时操舵装置
jury strut 辅助支柱;应急支柱
jury water pump 备用水泵
jus 法律制度
just and reasonable 公平合理

just-as-good 代用的
just bias 最佳偏置
just compensation 正当补偿;合理赔偿;合理补偿;合理报酬
just compromise 适当调和
just critical 正好在临界状态下
just-freezing solid 刚凝固体
justice 公正
justice of exchange 交易公平
justifiable complaint 正当的申请
justifiable expenditures 正当费用
justification 证实;证明正确;证明;整理位置;整版;论证;码速调整;合理性;齐行;对齐;版面调整
justification control digits 调整控制数字
justification digit 码速调整数字
justification of project 工程论证
justification rate 码速调整率
justification routine 证明程序
justification service digit 码速调整服务数字
justified copy 排字样张
justified line 齐行
justified margin 合理余量;边缘调整
justified price 合理价格
justified raise 合理上涨
justified selling price 鉴定售价
justifier 装版工人;辩释者;证明者;装板材料
justify 证明正确;整版
justify control cost 证明管理费的合理使用
justifying bail 适当担保
justifying digit 调整数字
justifying space 齐行楔
justify margin 匀行裕度;排列余量
just intonation 正确声调
justite 锌铁黄长石
just noticeable difference 恰可识别的色差;辨差阈值
just perceptible difference 最小可辨差异
just perceptible noise 最小可辨噪声
just perceptible visual colo(u)r difference 最小可辨彩色差
just price 公平价格
just size 正确尺寸;正常尺寸
just tolerable noise 最大容许噪声
just ton 短吨(美国);公正吨(1英国公正吨 = 1.0165千克)

just tuning 自然调谐
jut 尖端;突起部;突出部;伸出
jute 黄麻;电缆黄麻包皮
jute and plastics mixed bag 麻塑混编袋
jute backing 麻靠垫;黄麻底背
jute bag 麻袋
jute bagiute bag 黄麻袋(河工抢险用砂袋)
jute bag marking ink 麻袋墨
jute baling press 黄麻打包机
jute board 麻片(止水用);黄麻板;厚硬纸板
jute border 黄麻装饰花边
jute burlap 粗麻帆布;黄麻布袋
jute burlap mat 麻布垫;黄麻垫;黄麻毡
jute calender 黄麻轧光机
jute canvas 棉麻帆布;粗黄麻帆布
jute carder 黄麻梳麻机
jute core 黄麻绳芯
jute cotton blended yarn 黄麻棉混纺纱
jute doubler 黄麻并拈机
jute factory 麻加工厂;黄麻工厂
jute felt 黄麻纤维毡
jute fiber 黄麻纤维
jute hessian(canvas) 黄麻(制的麻)布
jute hessian(canvas) mat 黄麻布毡垫
jute hessian drag 黄麻布拖网;黄麻布捕捞器
jute hessian drag finish (混凝土路面的)麻布刮平抹光;(混凝土的)粗麻布表面处理
jute insertion 麻衬垫;黄麻加固
jute insulated cable 黄麻绝缘电缆
jute lamination 麻叠片;麻叠层
jute mat 麻袋片
jute matting 黄麻地毯
jute mill 黄麻工厂
jute moisture tester 黄麻水分测试仪
jute opener 黄麻开包机
jute packing 黄麻填缝
jute paper 麻浆纸;黄麻纸
jute products 黄麻产品
jute-protected cable 黄麻护层电缆
jute ribboner 黄麻碎茎机
jute rope 麻绳
jute roving frame 黄麻粗纱机
jute sacking 黄麻织造
jute seed oil 黄麻籽油
jute softener 黄麻软麻机

jute spinning and weaving machine 黄麻纺织机械
jute spinning frame 黄麻精纺机
jute spreading machine 黄麻栉梳机;黄麻延展机
jute twine 黄麻绳
jute waste 黄麻回丝;废黄麻;黄麻废料;麻丝
jute webbing 光面宽带
jute wrap(ping) 麻包装皮;黄麻布包装
jute yarn 电缆黄麻包皮线;黄麻线
jut out 突出
jutter 抖纹(螺纹缺陷)
jutting-off-pier 悬臂桥墩;伸壁桥墩
jutty 建筑物上层突出部分;突码头;防波堤;建筑物的悬挑部分
jut window 突出窗;凸窗
juvenile 新生的;岩浆源的
juvenile basalt 初生玄武岩
juvenile drainage 初生水系
juvenile ejecta 初生喷出物
juvenile form 幼年型
juvenile gas 岩浆气(体)
juvenile growth 幼年生长
juvenile mantle 新生地幔;原生地幔
juvenile motion 岩浆运动
juvenile period 幼年期
juvenile relief 幼年地形
juvenile river 初生河流
juvenile soil 幼年土;初生土
juvenile spring 岩浆泉
juvenile stage 幼年期
juvenile water 新生水;原生水;岩浆水;初生水
juvenile water closet 少年坐便器
juvenile wood 幼龄木材;幼龄材;中心木质部
juvey 少年教养院
juvie 少年教养院
juvite 正霞正长岩
juxta-epigenesis 近后期成岩作用
juxtaglomerular apparatus 球旁器
juxtaposed arch 交叉重叠暄;三铺暄
juxtaposed blocks 并列断块
juxtaposed ice stream 并置冰川
juxtaposition 斜接;接触双晶;交叉重叠法;并列;并置
juxtaposition-twin 并置双晶
juxtaposition type of dam and lock 闸坝并列式
J-value 总角动量值

K

K-40AR isochron 钾氩—40 等时线
Kaaba (麦加大寺院中的)穆斯林圣堂
K-A ages 钾氩年龄
Kaapvaal nucleus 卡普瓦尔陆核
kaatialaite 水砷氢铁石
kaavie 卡维雪
Kabah 穆斯林圣堂(麦加大寺院中的)
kabaite 阴地蜡
kabuskiki kaisha 株式会社(日本的股份有限公司)
Kacelite-B 硅藻土耐火砖载体
Kadaya gum 卡拉雅胶
kadmoselite 镉硒矿
Kady mill 高速桨叶式碎土机;卡迪(式)研磨机
Kaena event 喀纳事件
Kaena reversed polarity subchron 卡伊纳反向极性亚时
Kaena reversed polarity subchronzone 卡伊纳反向极性亚时间带
Kaena reversed polarity subzone 卡伊纳反向极性亚带
kaersutite 钛闪石;钛角闪石
kafehydrocyanite 黄血盐
kafir corn 高粱
kafiroic acid 高粱酸
Kahikatea 新西兰白松
kahlerite 铁砷铀云母
Kahne's method 卡恩法
Kahn iron 康恩铁
Kahn system 康恩混凝土加筋系统
Kahn test 康氏试验
kaing good 修补
kainite 钾盐镁矾
kainitite 钾盐镁矾岩
kainolithe 新喷出岩
kainosite 钙钇铈矿;碳硅铈钙石
kainotype 新相
kainotype rock 新相岩
kainozoic basin 新生代盆地
Kainozoic era 新生代【地】
Kainozoic group 新生界
Kaiser dome 凯撒公司生产的半球形圆屋顶
Kaiser floor 凯撒公司生产的楼板
Kaiser wall 凯撒墙壁
Kaiserworth at Goslar 戈斯拉尔的凯撒沃斯住宅
kakemono 条幅;字画
kaki 柿色炻器釉
kakirite 角砾破碎岩
kaki-shibu 柿油;柿浆;生柿汁
kakogeusia 恶味
kakor tokite 条纹霞正长岩
kakotopia 令人厌恶社会;极不理想的居住地
Kakutani theorem 角谷定理
kal 粗铁
kal'a (阿拉伯建筑上的)城堡
kala-azar 黑热病;阿萨姆热
kalaite 含铜绿松石
kalamein door 金属包门;包金属门;铁包门
kalamein fire 金属包皮防火门
kalamein fire door 包金属门;包薄钢板防火门
kalamein sheathing 包金属皮;金属皮;防蚀合金涂层;防腐合金覆层木材
kalasa 印度式建筑的瓶形顶部装饰
kal Baisakhi 卡贝萨基尘飑
kalborsite 氯硼硅铝钾石
kalchstein 方解石
Kaldor-Hicks test 卡尔多—希克斯检验
kaleidophone 示振器
kaleidoscope 万花筒
kaleidoscope prism 四面(体)棱镜
kaleidoscopic packaging 附赠品包装
kalema 几内亚大激浪;海滨激浪
kalfax 卡尔发克斯感光乳剂
kalia 氧化钾
kaliborite 硼镁石
kalicinite 重碳钾石
kalidium 盐爪爪
kalifeldspath 正长石;钾长石;微斜长石;透长石;冰长石

kalimeter 碳酸定量器;碱量计;碱度计;碱定量器
kalimetry 碱定量法
kalinite 纤(维)钾明矾
kalinor 碳酸钾氯化钾混合物
kaliophilite 钾霞石
kalipyrochlore 钾烧绿石
kali salt 钾盐
kalisaltpeter 钾硝石
kalistrontite 钾锶矾
kalium alum 铵明矾
kalium carbonate 碳酸钾
kalium o-phenyl phenate 邻苯基苯酚钾
kalk 石灰
kalkbauxite 钙质铝土矿
kalkowskite 铈钇钛铁矿
kalk saltpeter 钙硝石
kallirotron 卡利罗管
Kalman filter 卡尔曼滤波器
Kalman filtering 卡尔曼滤波
Kalman filtering system 卡尔曼滤波系统
Kalman filter theory 卡尔曼滤波原理
kalopanax 栓木
Kalsbeek counting net 卡斯贝克计数网
kalsilite 原钾霞石;六方钾霞石
kalsilite-bearing ultrabasic extrusive rock 含钾霞石超基性喷出岩
kalsomine 可赛银粉;刷墙水粉;刷墙(粉)料;刷墙鬃刷;石灰浆底;刷墙粉渣
kalunite 钾明矾石
kaluszite 钾石膏
Kalvar 卡尔瓦尔光致散射体
Kalvar film 卡尔瓦胶片;微泡法胶片
Kalvar process 微泡法
Kalzium metal 铝钙合金
kamacite 锥纹石;铁纹石
kamaisshilite 科羟铝黄长石
Kambalda-type nickel deposit 卡姆巴尔达式镍矿床
kame 冰碛阜;冰砾阜
kame moraine 冰砾碛
Kamensiji difference method 卡门斯基差分法
kame plain 冰碛平原
kame terrace 冰碛阶地;冰砾(阜)阶地;冰阜阶地
kame terrain 冰碛地貌
Kaminsty approximation formula 卡明斯基近似公式
kamiokite 钼铁矿
kamitugaite 卡米图加石
kamlolenic acid 粗糠柴酸
kammererite 铬绿泥石
kammgranite 斑闪花岗岩
Kampfe process 肯普费法
Kampfe viscometer 肯普黏度计
kampometer 热辐射计
kamptulicon 橡胶地毯
kam-tap screen 垂直振动筛
Kamyr filter 卡米尔过滤机
kanaekanite 碱硅钙铀钍矿
Kanai's semi-empiric(al) formula 卡赖半径经验公式
kanat 坎儿井;帷幕墙;地下暗渠;暗渠
Kanavec plastometer 卡氏塑性计;卡氏塑度计
Kanawha series 卡纳华统【地】
kanbara clay 蒲原黏土
kanch 扩大巷道;挑顶卧底
Kandatawood 坎达塔木(镶木地板用)
kandite 高岭矿
k-and-k machine 转子型高落式浮选机
kanemite 水硅钠石
kangaroo crane 带斗式起重机;袋鼠式起重机;带斗门座式起重机
kangaroo ship 袋鼠形船
kangaroo system 袋鼠装卸方式
kangaroo tower crane 袋鼠型塔式起重机
kangri 炭火蓝
kangri burn 怀炉灼伤
Kanimbran orogeny 卡宁伯拉造山运动
Kani-method 卡泥法
Kanji characters 汉字字符

kankar 灰质核
kankite 水砷铁石
kankrinite 钙霞石
kanoite 锰辉石
kanonaite 锰红柱石
kanr well 坎尔井(新疆地区的一种灌溉工程)
Kansan's glacial epoch 堪萨斯冰期
Kansan's glacial stage 堪萨斯冰期
Kansan's glaciation 堪萨斯冰川作用
Kansas city standard 堪萨斯城标准
Kansas method of flexible pavement design 堪萨斯柔性路面设计法
Kansas triaxial method 堪萨斯州三轴试验法(设计柔性路面厚度的方法)
kansui clay 寒水石粉
Kanthal 铬铝钴耐热钢;坎萨尔其铬铝电热丝
Kanthal alloy 铬铝钴铁合金;坎瑟尔合金
Kanthal super 高级(堪塔尔)电阻丝
Kanthal wire 铁铬铝电阻丝
kantharos [kantharus] 双柄高脚薄釉酒杯
kantographic(al) block diagram 折坡断块图
Kantorovich method 康托洛维奇法
kaolin 白陶土;高岭土
kaolin brick 高岭土砖
kaolin catalyst 高岭土催化剂
kaolin clay 白陶土;瓷土;高岭土
kaolin crystal 高岭石晶体
kaolin dynamic membrane 高岭土动力膜
kaoline 瓷土
kaoline powder 高岭土粉
kaolin fiber 高岭土纤维
kaolinfon 高岭黏土
kaolin glycerol paste 高岭土甘油酰浆(一种偶联剂)
kaolin group 高岭土族
kaolinic 高岭土的;白陶土的
kaolinic shale 高岭石页岩
kaolinisation 高岭石化;高岭土化(作用)
kaolinised 高岭土化的
kaolinite 高岭石;纯高岭土;纯粹高岭土
kaolinite autoflocculation 高岭土自动絮凝作用
kaolinite clay 高岭土黏土;高岭土黏土
kaolinite deflocculation 高岭土散凝作用
kaolinite group 高岭石类;高岭土属
kaolinite mudstone 高岭石泥岩
kaolinite ore 高岭石岩矿石
kaolinite structure 高岭石结构
kaolinitic clay 高岭石黏土;高岭土质黏土
kaolinitic sand 高岭土砂
kaolinization 高岭土化(作用);高岭石化
kaolinized 高岭土化的
kaolin mineral 高岭土
kaolin ore of halloysite-kaolinite type 埃洛石高岭石型高岭土矿石
kaolin ore of halloysite type 埃洛石型高岭土矿石
kaolin ore of hydromica-halloysite type 水云母埃洛石型高岭土
kaolin ore of hydromicall-kaolinite type 水云母高岭石型高岭土矿石
kaolin ore of kaolinite-halloysite type 高岭石埃洛石型高岭土
kaolin ore of kaolinite type 高岭石型高岭土矿石
kaolin ore of organic-kaolinite type 有机质高岭石高岭土矿石
kaolinosis 高岭土尘肺
kaolin porcelain 瓷器
kaolin powder 陶土粉;高岭土粉末
kaolin sand 高岭砂
kaolin-sandstone 高岭砂岩
kaolinsol 高岭化土;高岭质土
kaolin suspension 高岭土悬浊液
kaolin wool 高岭土质耐火纤维;高岭土棉
kaowool 高岭土质耐火纤维;高岭棉
kapitza method 单结晶形成法
Kaplan runner 转桨式转轮;卡普兰式转轮
Kaplan turbine 转桨式水轮机;卡普兰水轮机
Kaplan wheel 卡普兰式水轮
kapok 木棉

kapok cushion 木棉垫
kapok lifejacket 木棉救生衣
kapok oil 吉贝油
kapor 龙脑香樟树
kappameter 磁化率测定仪
Kappa number 卡伯值
Kapp line 卡普线
Kapp phase advancer 卡普进相机;卡普相位超前补偿器
Kapp vibrator 卡普振子;卡普相位超前补偿器
kapur 龙脑香樟树
kar 冰坑;冰堆;凹地
karaburan 黑风暴(中亚东北风);风沙尘
Karagwec-ankole cycle 卡拉圭—安科勒旋回
Karahari manganese deposit 卡拉哈里锰矿床
karajol 卡拉乔尔风
Karakum massif 卡拉库姆地块
Karakum old land 卡拉库姆古陆
Karamali-Moqinwula fold-fault belt 克拉麦里—莫钦乌拉断裂带
Karameili geodome series 克拉美丽地穹列
Karaoke hall 卡拉OK厅
Kara Sea 喀拉海
karat 克拉(宝石重量单位,1 克拉 = 200 毫克);开(纯金含量单位);赤金
karaya gum 刺梧桐树胶
Karbate 无孔碳
karbogel 碳胶(脱水剂)
kardex 索引卡片柜
karelianite 三方氧钒矿
Karelian orogeny 卡累利阿运动
karema 卡雷马风
karez 灌溉暗渠;坎儿井;地下灌溉渠道
Karginsky interglacial stage 喀尔金斯基间冰期
Karginsky intrestade 卡金斯基间冰阶
kar herbage 山区高草地
Karhunen-Loeve transformation 卡洛变换
karibibite 铁砷石
karidium 氟化钠
karif 卡里夫风
kariz 灌溉暗渠;坎儿井
Karl-Fischer's method 卡尔—费希尔水分测定法
Karl Fisher's titration 费歇尔滴定法
karlite 卡硼镁石
Karlsbad twin 卡斯巴双晶
Karlsbad twin law 卡斯巴双晶律
karlsteinite 高钾碱性花岗岩
Karma 卡马
karmalloy 高电阻镍铬合金
karma month 世间月
Karman analogy 卡曼比拟
Karman constant 卡曼常数
Karman equation (测定水泥比表面积的)卡门公式
Karman number 卡曼数
Karman vortex 卡曼(漩)涡
Karman vortex flow 卡门涡流;卡曼涡流
Karman vortex street 卡门涡列;卡曼涡列
Karman vortex train 卡门涡列
Karmash alloy 锑铜锌轴承合金
Karnak membrane system 凯尔奈克薄膜防水系统
karnauba wax 加洛巴拉;巴西棕榈蜡
Karnuagh map 卡诺图
karoo 雨季草原
karoptrite 硅铝锑锰矿
karpatite 黄地蜡
karpawood 卡帕木(镶木地板用)
karpholite 硅酸铁锰矿
karrah 卡里桉木(坚韧耐久且耐火)
karren 溶沟
karri 卡里桉树;变色桉(一种澳洲硬木)
karroo 雨季草原
Karroo system 卡路系
karst 岩溶;喀斯特
karst accumulation landform 岩溶堆积地貌
karst air-explosion 岩溶气爆
karst aquifer 岩溶含水层;喀斯特含水层
karst area 喀斯特地区
karst base level 岩溶基准面;喀斯特(侵蚀)基准(面)
karst basin 岩溶盆地
Karst-bauxite 喀斯特铝土矿
karst breccia 岩溶角砾岩;喀斯特角砾岩
karst bridge 岩溶桥;天生桥
karst cave 岩溶洞;喀斯特溶洞;溶洞

karst cave characteristics 岩溶洞穴特征
karst cave connection test 岩溶连通试验
karst channel 岩溶通道
karst coast 喀斯特海岸
karst collapse 岩溶陷落;岩溶地面塌陷
karst collapse-column 岩溶陷落柱
karst corridor 岩溶通道;溶蚀通道
karst cycle 岩溶旋回;喀斯特旋回
karst denudation plane 岩溶剥蚀面
karst deposit 岩溶矿床;岩溶堆积物;岩溶沉积物
karst deposits outside cave 洞外岩溶堆积物
karst depression 岩溶洼地;岩溶盆地;喀斯特盆地;溶蚀洼地
karst developing zone 岩溶发育带
karst development range 岩溶发育区范围
karst ditch 溶沟
karsten 岩溶沟
karstenite 硬石膏;无水石膏
karst erosion 岩溶侵蚀;喀斯特侵蚀
karst evolution 岩溶演化
karst factor counted along a line 线岩溶率
karst factor counted on plane 面岩溶率
karst factor of core 岩芯岩溶率
karst feature 岩溶地形;喀斯特地形
karst fen 岩溶沼泽
karst fissure 岩溶裂隙;喀斯特裂隙;溶隙
karst flow passage 岩洞落道
karst form 岩溶形态
karst funnel 岩溶漏斗;喀斯特漏斗
karst geomorphology profile 岩溶地貌剖面图
karst geomorphy 岩溶地貌;喀斯特地貌
karst grotto 岩溶洞穴
karst hall 岩溶大厅
karst hole 喀斯特落水洞
karst hydrogeologic(al) survey 岩溶水文地质调查
karst hydrogeology profile 岩溶水文地质剖面图
karst hydrographic(al) map of underground rivers 岩溶暗河水系图
karstic 岩溶的;喀斯特的
karstic basin 喀斯特盆地
karstic cave 溶穴
karstic collapse 岩溶坍陷;岩溶塌陷;喀斯特塌陷
karstic collapse breccia 岩溶塌陷角砾岩
karstic earth cave 土洞
karstic feature 岩溶形态;岩溶特征;岩溶地形;喀斯特形态;喀斯特特征
karst(ic) formation 喀斯特地层;岩溶地层
karst(ic) geology 岩溶地质(学);喀斯特地质(学)
karst(ic) groundwater 岩溶地下水;喀斯特地下水
karst(ic) hydrogeology 岩溶水文地质(学);喀斯特水文地质(学)
karst(ic) hydrology 喀斯特水文学;岩溶水文学;灰岩区水文学
karstic network 洞穴网
karstic phenomenon 岩溶现象
karstic spring 岩溶泉
karstic water 岩溶水;喀斯特水
karstifiable 可岩溶化的;可喀斯特化的
karsticated rock 岩溶化岩石;喀斯特化岩石
karstification 岩溶作用;岩溶化;喀斯特作用;喀斯特化
karstification model 岩溶化模式
karstification zone 岩溶作用带
karstified 岩溶化的;喀斯特化的
karstified limestone 岩溶化石灰岩
karstified rock 岩溶化岩石;喀斯特化岩石
karst investigation 岩溶调查
karst karren shape 洞穴溶沟类
karst lake 岩溶湖;喀斯特湖
karst land feature 岩溶地貌
karst landform 岩溶地形;岩溶地貌;喀斯特地形
karst landscape 岩溶景观;喀斯特景观
karst limestone 岩溶石灰岩
karstology 岩溶学
karst passage(way) 岩溶通道;喀斯特通道
karst peneplain 岩溶准平原;喀斯特准平原
karst phenomenon 岩溶现象;喀斯特现象
karst physical geology phenomenon 岩溶物理地质现象
karst pit 岩溶井;喀斯特井
karst pit depth 岩溶竖井深度
karst pit diameter 岩溶竖井直径
karst plain 岩溶平原;喀斯特平原;溶蚀平原

karst plateau 岩溶高原;喀斯特高原
karst pond 溶潭
karst process 岩溶作用;喀斯特作用
karst region 岩溶区;岩溶地区;喀斯特地区
karst river 岩溶河(流);喀斯特河;喀斯特暗河
karst shape measure 岩溶形态测量
karst source 岩溶泉;喀斯特(源)泉
karst spring 岩溶泉;喀斯特泉(水)
karst stage 岩溶期
karst stream 岩溶河(流)
karst street 岩溶通道;溶蚀通道
karst subsidence 岩溶坍陷
karst system 岩溶系统;喀斯特系统
karst terrain 岩溶地区;喀斯特地区
karst thermal water 岩溶热水
karst topography 岩溶地形;喀斯特地形
karst tower 灰岩残丘
karst trench 溶槽
karst type 岩溶类型
karst type map 岩溶类型图
karst valley 岩溶谷地;溶谷
karst water 岩溶水;喀斯特水
karst water regime 喀斯特水动态
karst well 岩溶井;喀斯特井;天然井
karst window 喀斯特天窗;天窗
kart 赛车
Kartel 卡特尔
karton 纸板箱;厚纸
karyoclasis 核破裂
karyogamy 核融合
karyokinetic phase 核分裂期
karyolemma 核膜
karyolysis 核溶解
karyomicrosome 核微粒
karyomorphism 核形
karyonide 核系
karyopycnosis 核浓缩
karyorrhexis 核碎裂;核破裂
karyosome 核微粒;核体
karyosphere 核球
karyotheca 核膜
karyotype 核型
kas 衣柜;食橱
kasba 城堡
Kaschin-Beck disease 骨骼氟中毒;卡辛-贝克氏病;氟骨病
kasenite 霞辉碳酸盐岩
kasenite pegmatite 霞辉碳酸盐伟晶岩
Kashi gulf 喀什海湾
kasolite 硅铅铀矿;硅铅铀矿
kasparite 钴镁明矾
kasr 阿拉伯宫殿
Kassai nucleus 开赛陆核
kassite 羟钙钛矿
kasten corer 盒式取样管
katabaric 负变压的
katabasis 希腊正教会教堂
katabatic 下降的
katabatic wind 下降风;下吹风;重力风;山风
katabolism 分解代谢
katachthonous formation 原地建造
kata-condensed rings 渺位缩合环
katadromous fishes 降海产卵鱼类
katadromy 降海繁殖;降海产卵
kata factor 降幂因数
katafront 下滑锋【气】
katagneiss 深变片麻岩
kata-imponite 深脆沥青
katakinesis 放能作用
katakinetomeres 非活化分子
kataklastic rock 碎裂岩
kataklastic structure 碎裂结构
katallobar 负变压中心;负变压线
katallobaric 负变压的
katalysis 催化(作用);触媒(作用)
katametamorphism 深变质(作用)
katamorphic rock 破碎变质岩
katamorphic zone 破碎变质带;碎裂变质带
katamorphism 分化变质;破碎变质现象;风化变质(作用)
Katanga marine trough 加丹加海槽
Katanga orogeny 加丹加运动
kata-orthoclase gneiss 深变正长片麻岩
kataphoresis 阳离子电泳

kataphorite 红钠闪石
kata plagioclasegneiss 深变斜长片麻岩
kata-rock 破碎岩(石);破碎带变质岩;深变质带岩石
kataseism 向震中
kataseismic 向震中的
katastatic stress 弛存应力
katatectic layer 溶蚀残余层
katathermal 深成高温热液的
katathermal ore deposit 深成热液矿床
katathermal solution 高温热液;深成热液
Kata thermometer 卡塔温度计;空调湿度计;下降温度计;(低温的)冷却温度计;(低温的)冷却温度表;低温温度计
katavothra lake 落水洞湖
Katayama disease 日本血吸虫病
katazonal 深变质带的
katazonal metamorphism 深带变质(作用)
katazone 深变质带
kate-isallobar 等负变压线
katergol 火箭燃料
Kater's reversible pendulum 卡特可逆摆
katharobe 清水生物
katharobic zone 清水带
katharobiont 清水生物
katharometer 热导检测器;热导计;导热析气计
katharometry 热导率测量术
kathetometer 高差计
kathode 阴极
kathode ray tube 阴极射线管
kation 阳离子;正离子
katio oil 烛果油
Katmaian type 卡特迈型
Katmanian 卡特迈式
katogence 破坏作用
katogene rock 水成岩;沉积岩
katogenic metamorphism 热液变质作用;深成变质(作用)
katogeno-dynamoetamorphism 深成动力变质
katoite 扣藤石
katolysis 不完全分解
katophorite 红钠闪石
katopohorite 红钠闪石
katoptric system 逆屈光系统
katoptrite 黑硅锑锰矿
katsura tree 连香树
katungite 白榄黄长岩
katydid 高轮运材挂车
katzengold 云母;黑云母
katzensilver 云母;白云母
Katz funnel 卡茨漏斗
Kauertz engine 考尔茨发动机
Kaufman method 螺钉镦锻法
Kaufmann iodine value 考大曼碘值
kauka site 透长花岗岩
kaurene 贝壳杉烯
kaurenic acid 贝壳杉烯酸
kauri 考里树胶;贝壳杉
kauri butanol test 贝壳杉脂丁醇(溶解)试验
kauri-butanol value 贝壳杉脂丁醇(溶液溶解)值
kauri copal 贝壳杉松脂
Kauri gum 栲利胶脂;贝壳杉脂;贝壳杉胶
Kaurinic acid 栲利酸;贝壳杉酸
kauri reduction value 贝壳杉脂稀释值
Kauri resin 栲利树脂;贝壳杉脂
kaurit 尿素树脂接合剂
kauri tolerance value test 贝壳杉脂稀释试验
kauri varnish 贝壳杉清漆
kauroresene 贝壳杉脂素
kaus 考斯风;卡活斯风
kaver 卡浮风
Kavraisky's pseudocylindric(al) arbitrary projection 卡夫拉伊斯基任意伪圆柱投影
kawine 卡法树脂
kawo kawo fibre 木棉纤维
kay 小礁岛
kayak 皮筏
Kayakalan blue 卡雅卡兰蓝
Kayanol blue 卡雅诺尔蓝
Kayanol red 卡雅诺尔红
Kayarus brown 卡雅鲁斯棕
Kaye disk centrifuge 凯氏盘式离心机
Kaye's patent (安装门把手方法)凯氏专利
Kaylo insulation 硅酸钙质高温绝缘材料
Kayser 凯塞

kayser hardness test 卡锁硬度试验
kayserite 片铝石
Kazakh-kashmir old land 哈萨克—克什迷古陆【地】
Kazakhstanian old land 哈萨克斯坦古陆【地】
Kazakhstanian plate 哈萨克斯坦板块【地】
Kazantzevo interglacial stage 喀占特涅夫间冰期
Kazantzevo interstade 卡札采夫间冰期
kbar 千巴(压力单位)
K bentonite 钾盐班脱土
K-braced frame K形支撑框架
K-bracing K形连结杆
kcal/second 千卡/秒(热功率单位)
K corona 连续日冕
keatingine 锌锰辉石
keatite 热液石英
keblah (穆斯林礼拜时的)朝向
Kebrit deep 克卜里特海渊
keckite 磷铁锰钙石
keckle 钢丝索缆包扎捆缠物品;绕绳的碰垫
keckle meckle 贫铅矿石
keckling 防磨卷
Kedafu formula method 柯达夫公式计算法
kedge 小锚;拉锚移船;牵引锚;抛小锚移船
kedge anchor 小锚
kedge rope 小锚绳
kedging 抛小锚移船
keekwilee-house 土房
keel 龙骨突;龙骨(船脊骨);平底煤驳;船龙骨;标准煤驳(重量单位)
keelage 寄港税;入港税;入港费;停泊税;停泊费;入港停泊税
keel-and-bilge block 垫木船;垫船木块;龙骨垫及侧垫;盘木及侧盘木
keel angle 龙骨角铁;龙骨角钢
keel arch 葱头形拱;内外四心桃花拱;葱形拱;反弯尖券;龙骨券
keel batten 内龙骨
keel beam 龙骨墩地基梁
keel bender 龙骨弯曲机
keel block 铸锭;龙骨形小铸锭;龙骨墩(木);龙骨垫(木);基尔试块;艇座;艇架
keel block level 龙骨墩高程
keel block loading 龙骨墩荷载
keel block of dry dock 干坞龙骨墩
keel block tier 龙骨垫木台;盘木台
keelboat 龙骨船
keel boundary line 龙骨外板线
keel clearance 龙骨下富裕水深;龙骨下富余水深
keel condenser 龙骨边凝水管
keel cooler 龙骨冷却机;龙骨式冷却器
keel date 龙骨铺设期
keel dome 船底声呐导流罩
keel draft 龙骨吃水
keeler 小木板;浅木桶;平底浅盆
keele series 基勒统【地】
keel flat plate 龙骨板
keel flying height 龙骨飞高
keel grade 龙骨斜度;龙骨坡度
keel haul line 过底绳
keel laid 动工造船
keel line 龙骨线;开骨线;首尾线;艏艉线
keel metacenter 稳心至龙骨距离
keel mo(u)lding 龙骨线脚;葱形线脚;桃尖线脚;哥特式装饰线脚
keel piece 龙骨构件;尾框底部
keel plate 龙骨板
keel rabbet 龙骨镶口;龙骨槽
keel sagging 龙骨重度
keel scarf 龙骨嵌接
keel shape 龙骨形
keel shore 龙骨支撑;假龙骨【船】
keelson 内龙骨
keelson angle 内龙骨角钢
keelson lug 内龙骨与肋板连接的短角钢
keelson plate 内龙骨(翼)板
keel staple 假龙骨栓钉【船】
keel stop 龙骨尾铁
keel water lift 龙骨水车
keen 尖锐
keen alloy 铜基合金
keen-edged 刀口锐利的
Keenekote 肯尼寇特绝热、隔声和防火材料
Keene's cement 金氏胶结料;金氏(干固)水泥;硬石膏胶结料;干固水泥
Keene's marble cement 金氏干固白色胶结料;金氏(干固)水泥
keenness 锐敏度
keen price 减价;低廉价格
keen to purchase 渴望购买
keep aboriginal 保持原始价值
keep account 记载;计算账面借款;管账;记账;入账
keep a close control of overtime 严格控制加班时间
keep additional accounts in addition to the authorized ones 账目设账
keep a full stand of seedlings 保苗
keep a good offing 保持离地远些航行
keep-alive 点火电极
keep-alive anode 保弧阳极
keep-alive cathode 保弧阴极
keep-alive circuit 维弧电路;保活电路;保弧电路
keep-alive contact 电流保持接点;保弧触点
keep-alive current 保活电流
keep-alive discharge 保活放电
keep-alive electrode 维弧阳极
keep alive grid 保活栅
keep-alive relaxation 保活弛张
keep-alive voltage 激励电压;维弧电压
keep an appointment 履约
keep and bailey castle 带主塔楼城堡;带望楼外墙的城堡
keep and offing 保持离岸航行
keep a spell 接班
keep a tab on the expenses 记录各项费用
keep a tight control on revenue and expenditures 统收统支
keep away 不得接近
keep back 止住;往后退
keep balance 保持平衡
keep bolt 盖螺栓
keep clear 让开
keep confidential 保守秘密
keep consort 保持在一起航行
keep contract 遵守合同
keep cool 放置冷处;保持阴凉;保持冷藏
keep course 保持航向
keep course and speed 保向保速航行
keep down price 平抑价格
keep dry 勿受潮湿;勿放湿处;防湿
keep ecologic(al) balance 保持生态平衡
keeper 衔铁;止动螺母;止动螺帽;看管者;看管人;卡箍;夹;切断销;卫铁;锁紧螺母;锁(紧)螺帽;(机械的)制动;定位器;定位螺钉;定位件;柄;保管人;保磁铁片;保持器;保持件
keeper current 保持电流
keeper electrode 定位电极
keeper hook 扣牢门窗用的S形钩子
keeper of a magnet 保磁用衔铁
keeper plate 锁紧木栓;锁紧木片
keeper voltage 保持器电压
keep flat 平放
keep free of weeds 洁无杂草(指土壤)
keep from heavy weed growth in the field 防治杂草丛生
keep head above water 不欠债
keep head on 保持原航向
keep in a cool place 在冷处保管
keep in a dry place 在干处保管
keep in check 制止
keeping 保管
keeping and handing expenses 保管及处理费用
keeping bench income 保管基准收益
keeping current account 流水账的登记
keeping dry 保持干燥
keeping equipment records 保管设备记录;保管设备的登记
keeping expenses 保管费(用)
keeping from ill weeds 防除恶性杂草
keeping ga(u)ge 砌墙标杆;砌墙高度规准尺;标杆(砌墙用);皮数杆
keeping parts 备用件
keeping perpend 清水对缝
keeping property 耐久性
keeping quality 耐储[贮](藏)性;耐久性
keeping records 保存记录
keeping records and orderliness 保存记录与工作条理

keeping room 家庭起居室
keeping sample 保留样品
keeping spare parts 备件
keeping the centre of tree open 使中间透光(剪树)
keeping the ga(u)ge 砌墙标杆;按层砌筑;砌墙高度规准尺;皮数杆
keeping the prepends 错缝对直;清水对缝;垂直缝每隔一皮保持互相垂直(砌砖法)
keeping the roots from freezing 防止根部伤冻
keeping the soil from puddling 防止土壤胶粘
keeping time 产品有效期;保存期
keeping under pressure 保持压力
keeping up the side 挡边
keep in order 保持秩序;保持整齐
keep in repair 保持良好性能;保持检修完善
keep in sight 保持在视界范围内
keep in storage 储[贮]存
keep intact 保持原样
keep in touch with 保持联络
keep-like tower 城堡主楼形塔;城堡高塔
keep of castle 城堡主垒
keep off 不让接近;不得接近
keep off median sign 中央分隔带禁入标志
keep one's bearings 守住某物标方位
keep one's weather eye open 注意观察天气变化情况
keep one's words 遵守诺言
keep on the go 维持运转
keep on the transit 保持在叠标线上
keep open 保持畅通;保持开放
keep-out area 禁用区
keep out of debt 不借债
keep out of the sun 避免阳光;避免日晒
keep out of the way 让开航路;让开
keep pace with the predicted growth in through-put 跟上预计的吞吐量增长
keep part of the foreign exchange earnings 外汇分成
keep plate 制动板
keep plate fork 压板货叉
keep relay 止动继电器;保护继电器
keep right sign 右行标志
keeps 罐座
keep site clear 保持现场整洁
keep soil from packing together 防止土壤板结
keep standing 存版
keep strictly to the terms of the contract 严格遵守合同条款
keep the land aboard 保持沿海岸航行;保持望见陆地
keep the royal touching 尽量靠上风航行
keep the sea 保持海上航行
keep the sounding-pole going 篙测航驶
keep the vacuum constant 保持真空度不变
keep the wind 尽量守住上风(防下风有险礁)
keep tight 保持密封
keep to instruction 按照指示
keep to leeward 保持下风位置
keep to program(me) 固定施工进度计划;按照计划;按照程序;按计划完成,按计划进行
keep to schedule 按照计划;按计划进行
keep tower 城堡主塔;防守塔
keep under the lee of land 保持海岸挡风航线上
keep up 保持不跌落(价格)
keep up pressure 保持压力
keep up regular supply 保持经常供应
Keep upright 勿倒置;切勿倒置;不要翻倒;保持直立
keep-way vessel 直航船
keepwell 必要条款
keep wire line in tension 拉紧钢丝绳
keep with surroundings 适合环境
keep your quotation 维持你方报价
Keesom force 取向力
keeve 缸;盆;大桶
Keewation 基瓦丁群【地】
keg 小桶;圆铁桶;厚漆铁桶
keg buoy 木桶浮标;捕鱼用木桶浮标
kegelite 硫硅钾锌铅石
keg float 桶浮标;浮筒标;浮桶(标)
kehoeite 土磷锌铝矿;水磷锌铝矿
keister 手提皮箱
ke-it (浇混凝土前放在模板上的)小橡胶条
keithconite 碲钯矿

keith's rule 凯蒂规则(计算热水管辐射面积的规则)
keiviite 硅镱石
Keketongtao basin 科克同套盆地
Kek mill 圆盘式离心粉碎机;凯克磨机
Kelcaloy 凯尔卡洛伊复合钢
Kelcaloy method 高级合金钢冶炼法
keld 河流静水区;喷泉
keldyshite 硅钠锆石
Kelebe hammer 凯利皮锤(捣实试模中砂浆专用)
Kel F 凯尔 F 型树脂
Kellaways rock 凯拉维斯岩
kelleg 小锚
Keller furnace 凯勒式电弧炉
kellering 仿形铣
kellerite 铜镁矾
Keller machine 自动机械雕刻机
Keller process 开勒法
Keller system 凯勒成堆运砖法
Kelley filter 凯利型过滤器
Kelley filter press 凯利型压滤机
kellick 小锚
Kelling's test 凯林试验
Kellner eyepiece 克尔纳目镜
Kellner's eyepiece 开涅尔目镜
Kellner's representation 开涅尔表示法
kellock 小锚
kellogg chart 凯洛格曲线图
Kellogg crossbar system 凯洛格纵横自动交换制;凯洛格交叉式
Kellogg equation 凯洛格方程
Kellogg's oak 加利福尼亚黑栎
Kellog hot-top method 钢锭顶部电加热保温法
kellstone 黏石粉刷
Kelly 主动钻杆
Kelly ball 凯利球体贯入仪
Kelly ball penetration 凯利球体贯入度
Kelly ball(penetration) test 凯利球(体)贯入试验
kelly bar 方钻杆;方形钻杆;凯利(方)钻杆
Kelly bar bushing 钻杆套管;钻杆衬套
Kelly bar drive 钻杆驱动
kelly bar slips 方钻杆卡瓦
kelly bar sub 方钻杆异径接头
kelly board 钻塔台板
kelly bushing 方补心;方钻杆补心;钻杆衬套
kelly cock 方钻杆上部阀
Kelly colo(u)r chart 凯利色品图
Kelly core sampler 凯利原状取土器
kelly crowd 方钻杆压入装置
kelly diagonal size 方钻杆对角尺寸
kelly drive 方钻杆驱动
Kelly filter 凯利叶片式压滤机
Kelly guided grab 凯利导杆抓斗
kelly hole 方钻杆鼠洞
kellyite 锰铝蛇纹石
kelly joint 方钻杆
kelly left-handed joint 方钻杆上接头
kelly overstand 机止余尺【岩】
kelly packer 方钻杆防喷封隔器
kelly platform 钻塔台板
kelly right-handed joint 方钻杆下接头
kelly safety valve 方钻杆安全阀
kelly saver 方钻杆防油罩;方钻杆防油器
kelly saver sub 方钻杆安全接头
kelly spinner 方钻杆旋转器
kelly's rat hole 方钻杆用鼠洞
kelly stem 方钻杆
kelly stop cock 方钻杆回流阀
kelly straightener 方钻杆矫直器
kelly valve 方钻杆防喷阀
kelly wiper 方钻杆刮泥器
Kelmet 铜铅轴承合金
kelp 巨形海草
kelp ashes 海草灰
kelp oil 海草油
kelp salt 海草灰盐
Keltic architecture 凯尔特式建筑
Keluo tectonic knot 科洛构造结
Kelvin 开尔文
Kelvin body 弹黏性固体
Kelvin effect 集肤效应
Kelvin-Helmholtz contraction 开尔文—亥姆霍兹收缩
Kelvin-Hughes echo sounder 开尔文—休斯式回声测深仪

Kelvinometer 开尔文温度计;绝对温度计
Kelvin's absolute scale 开氏绝对温标
Kelvin's balance 开尔文电天平
Kelvin's body 开尔文体
Kelvin scale 开氏温标;绝对温标
Kelvin's chemical tube 开尔文化学测温管
Kelvin's circulation theorem 开尔文环量定理
Kelvin's degree 开尔文绝对温度
Kelvin's effect 开尔文效应
Kelvin's equation 开尔文方程
Kelvin's guard-ring capacitor 开尔文保护环电容器
Kelvin's integrator 开尔文积分器
Kelvin skin effect 趋肤效应
Kelvin's minimum energy theorem 开尔文最小能量定理
Kelvin's model 开尔文模型
Kelvin's rheologic(al) model 开尔文流变模型
Kelvin's solid 开尔文固体
Kelvin's solid model 开尔文固体模型
Kelvin's sounder 开尔文测深仪
Kelvin's substance 开尔文体
Kelvin's temperature 开氏温度;开尔文温度
Kelvin's temperature scale 开氏温标
Kelvin's thermometric scale 开氏温标
Kelvin's tide ga(u)ge 开尔文验潮仪
Kelvin's tuber 开氏测深管
Kelvin's visco-elastic stress-strain relation 开尔文黏弹性应力应变关系
Kelvin's wave 开尔文波
Kelvin temperature 绝对温度;开氏温度
Kelvin temperature scale 绝对温标
Kelvin-Varley slide 开尔文—瓦利尔滑动电阻
Kelvin-Voigt model 开尔文—伏格特模型
kelyanite 揸莱安矿
kelyphytic border 次变边【地】
kelyphytic rime 次变边【地】
kelyphytic-rim texture 次变边结构
Kemble beds 肯布勒层
kemidol 细石灰粉
Kemidoln 凯米多尔石灰
kemmlitzite 砷锶铝矾
Kemnay 阿伯丁(澳、英、加、美、南非产的银灰色有云母黑点的花岗岩)
Kempirsay chromium deposit 肯皮尔赛铬矿床
kempite 氯羟锰矿
ken 视野
Kendall effect 假像效应
Kendall's coefficient 肯德尔系数
Kendall's effect 肯德尔效应
Kendall's tau coefficient 肯德尔等级相关系数
kenematic oily pollutant transport model 运动油质污染物输移模型
kenesthesia 存在感觉
Kennametal 肯纳硬质合金;钴碳化钨硬质合金
Kennedy extractor 肯尼迪萃取器
kennedyite 黑镁铁钛矿;钛镁铁矿
kennedy key 方形切向键
Kennedy's critical velocity 肯尼特临界流速
kennel 下水道;阴沟;烛煤;沟渠;路边水沟;坚硬砂岩;峒窟
Kennelly-Heaviside layer 肯涅利—海维赛层
kennel-type piggery 矮小型猪舍
kennet 大型缆墩;大型缆耳
kenning 海上视界(约 20 海里)
Kennison flow nozzle 肯纳逊流量喷嘴
Kennison nozzle 肯纳逊喷嘴
kenotron 高压整流二极管;大型热阴极二极管;二极整流管
kenotron rectifier 高压二极整流管
kent 制图纸;绘图纸
kentallenite 橄榄二长岩
Kent axe 肯特篙斧
Kent claw hammer 肯特拔钉锤
Kenter shackle 双件式连接卸扣
Kenter type anchor shackle 肯特型锚卡环
Kenter type joining shackle 肯特型连接卡环
Kenter type shackle 肯特型卡环
kentish plow 转壁双向犁
Kentish rag 肯特郡硅质砂岩
Kentish tracery 肯特郡花格窗(英国)
kentite 铵硝
kentledge 压舱料;压载铁(块);压载货;平衡重;配重铁块;铁压头

kentledge goods 压载货;压舱料
Kenton 淡褐色砂岩(诺森伯兰产的)
kentrolite 硅铅锰矿
kentsmithite 黑钒砂岩
Kentucky bluestone 肯塔基州青石(一种蓝灰色砂石)
Kentucky design method 肯塔基州(柔性路面)设计法
Kentucky method of flexible pavement design 肯塔基州柔性路面设计法
Kentucky-rock asphalt 肯塔基岩石与地沥青混合料(一种特制的薄层防滑面层)
kenyaite 水羟硅钠石
Kenyon interceptor 拦截器(人孔中的,有阀可清除阻塞)
kenyte 霓橄粗面岩
kep 罐托;门扣;窗钩;拉手
kepel 脉壁黏土
Kepes model 开派斯模型
kep interlock 罐座联动装置;罐托联锁系统
Keplerian element 开普勒要素
Keplerian ellipse 开普勒轨道
Keplerian ellipsoid 开普勒椭圆
Keplerian equation 开普勒方程
Keplerian law 开普勒定律
Keplerian motion 开普勒运动
Keplerian orbit 开普勒轨道
Keplerian system 开普勒系统
Keplerian telescope 开普勒望远镜
Kepler's element 开普勒要素
Kepler's ellipse 开普勒轨道
Kepler's ellipsoid 开普勒椭圆
Kepler's equation 开普勒方程
Kepler's law 开普勒定律
Kepler's motion 开普勒运动
Kepler's orbit 开普勒轨道
Kepler's second law 等面积定律
Kepler's system 开普勒系统
Kepler's telescope 开普勒望远镜
keps 罐座
kerabitumen 油母岩(质)
keralite 英云角岩
keramic 陶器的
keramics 陶器
keramite 莫来石的别名
Keramonite 金属网为骨架的陶瓷材料
keramovitrons 玻璃黏结陶瓷制品
keramzite 陶结块;素烧黏土填料
keramzite concrete 膨胀陶粒混凝土(一种苏联产品)
kerargyrite 角银矿
keratan 角质素
keratin 革质;角质;石灰浆缓凝剂
keratin compound 角朊化合物
keratinize 角质化
keratohyaline layer stratum granulosum epidermidis 表皮粒层
keratol 涂有硝棉的防水布
keratophyre 角斑岩
keratophyre glass 角斑岩质玻璃
keraunograph 雷暴仪
keraunophone 闪电预示器
kerb 路缘石;路缘;路牙;路边石;护轮槛;道牙石;道牙;侧石;非正式交易所
kerb alignment 路缘准直
kerb bed footway 路缘石人行道
kerb below ground 埋入式路缘;地面下路缘石
kerb brick 路缘砖;路边砖
kerb curb 缘石
kerbed footway 带路缘石的人行道;镶侧石的步行道
Kerber eyepiece 克贝尔目镜
kerb form 路缘模板
kerb grab(bing) equipment 路缘开挖机;路缘石抓取设备
kerb inlet 路边窨井(下水道的路边进口);路边雨水口
kerb lane 外侧车道
kerb line 路边线
kerb loading 路边装卸
kerb machine 路缘石压制机;铺设侧石机械;铺设路缘石机械
kerb market 场外债券市场;交易所外市场
kerb mo(u)ld 路缘模板;路缘模子
kerb parking 路边停车;路旁停车

kerb parking place 路边停车区
kerb press 路缘石压制机
kerb press machine 路缘压路机
kerb shoe (沉井带刃口的)边脚
kerb shoe of caisson 沉箱刃脚
kerbside loading 靠人行道上车或卸货
kerbside parking 侧石边停车
kerbstone 路缘石;路牙;路边石;道牙石
kerbstone of pavement 路边石
kerb tile 弯角面砖
kerb tool 路缘模板
kerchenite 纤磷铁矿
Kerch iron deposit 刻赤铁矿床
Kerchsteel 克尔期砷铁钢
kerenes 煤油烯
kerf 切缝;割缝;锯痕;锯缝;锯槽;截口;划痕;切口宽度;切口;切开;切割(缝);切断沟;劈痕;掏槽;[复] kerve
kerf cutter 切割滚刀
kerfed beam 有截口的梁;底有截口的梁
kerfing 开槽;切沟;割缝
kerf stone 底刃金刚石
Kerguelen plateau 凯尔盖朗海台
kerilite 英云角岩
keriotheca 蜂巢层
kerites 煤油沥青
kerkis (古希腊楔形或梯形的)座席分区
kerlite 蜡蛇纹石
kerma 比释动能
Kermadec trench 克马德克海沟
kerma rate 比释动能率
kermesinus 暗红色
kermesite 桔红硫锑矿;红锑矿
kermes lake 胭脂虫色淀;虫红色淀
kermess 户外定期集市;(美国慈善性质的)义卖集市
Kermet 克美特合金
kermis 墟集市;定期集市
kern 古地块;核心
kern area 核心面积
kernbut 断层侧外丘
kern concrete 柱芯混凝土;柱中心混凝土
kern counter 尘粒计数器;计核器
kern cross-section 柱芯(混凝)断面;核心横断面
kern distance 核心距
kern edge 核心边界
kernel 内核;核(心);枪眼;环孔;炮眼;炮眼城垛;门窗凹进处
kernel-based system 以核心为基础的系统
kernel-decomposition 核心分解(斜长石)
kernel estimator 核估计量
kernel function 核函数
kernel mode 核方式
kernel module 核心模块
kernel normal form 核范式
kernel of a homomorphism 同态的核
kernel of a linear transformation 线性变换的核
kernel of an integral equation 积分方程的核
kernel of an integral transform 积分变换的核
kernel of transformation 变换核
kernel oil 核油
kernel polynomial 核多项式
kernel principal component regression 核主成分回归
kernel program(me) 核心程序
kernel rot 粒霉
kernel stack not valid abort 内核栈无效失败
kernel structure 粒结构
kernel technology 内核技术
kernel test program(me) 核心检测程序
kerne tipple 喷射卸料槽
kernite 科方硼砂;贫水硼砂;四水硼砂
kern language 核心语言
kern limit 截面核心周界;核心边界
kern line 截面核心线
Kern method 克恩涂料黏附性试验法
kern of section 截面核心
kern point 截面核心点(截面主轴上的截面核心周界点)
kern point moment 截面核心点矩
kern-stone 粗粒砂岩
Kern strength 柱芯混凝土强度;核心强度
kerogen 油母岩(质)
kerogen shale 油母(页)岩;沥青质页岩
kerogen thermal degradation theory on origin of petroleum 干酪根热降解成油说
kerosene 火油;灯油
kerosene burner 煤油灯
kerosene burning quality 煤油燃烧性质
kerosene burning test 煤油燃烧试验
kerosene coal 油页岩
kerosene colo(u)r test 煤油颜色测定;煤油色度试验
kerosene cutback 用煤油轻制的沥青
kerosene cutback asphalt 煤油稀释沥青
kerosene cutting torch 煤油切割器
kerosene degreasing 煤油脱脂
kerosene distillate 煤油馏分
kerosene emulsion 石油乳液;石油乳剂
kerosene engine 煤油引擎;煤油发动机
kerosene flash point 煤油闪点
kerosene flo(a)tation 煤油浮选
kerosene-fluxed 用煤油助溶的
kerosene fuel burner 煤油燃烧炉
kerosene glass chimney 煤油灯罩
kerosene lamp 煤油灯
kerosene method 煤油法(测密度用);石油法(测土密度用)
kerosene number 煤油号数;(油毡的)煤油吸收量
kerosene oil 煤油
kerosene oil engine 煤油发动机
kerosene oil-gas 煤油气
kerosene-oil mixture 煤油滑油混合物
kerosene paint 煤油涂料
kerosene poisoning 煤油中毒
kerosene propellant 煤油喷气燃料
kerosene raffinate 精制煤油
kerosene resistance 石油稳定性;石油耐久性;抗煤油性
kerosene sediment test 煤油(中)沉积物(的)测定
kerosene shale 煤油页岩
kerosene smoke 煤油烟
kerosene standard colo(u)rs 煤油的标准颜色
kerosene stock 煤油燃料
kerosene stove 煤油炉
kerosene stripper 煤油用气提塔
kerosene sulfur test 煤油中硫的检测;煤油中硫的测定
kerosene type jet fuel 煤油型喷气燃料
kerosene-type turbine fuel 煤油型燃气轮机燃料
kerosene value (油毡的)煤油吸收量
kerosene wick 煤油灯芯
kerosine 煤油
kerosine absorption of felt 原纸吸油量
kerosine distillate 煤油馏出物
kerosine number 煤油值
kerosine shale 块煤
kerotenes 焦化沥青质
Kerr cell 克尔盒
Kerr cell camera 克尔盒照相机
Kerr constant 克尔常数
Kerr effect 克尔效应;电光克尔效应
Kerr-effect liquid 克尔效应液体
Kerr-effect material 克尔效应材料
Kerr effect self-focusing 克尔效应自聚焦
Kerr electro optic(al) law 克尔电光定律
kerrite 黄绿蛭石
kerriwood 克瑞木(硬木地板用)
Kerr magneticoptic(al) effect 克尔磁光效应;磁光克尔效应
Kerr modulator 克尔调节器
Kerr nonlinearity 克尔非线性
kerrolic acid 开醇酸
kersantite 云斜煌岩
kerstenite 黄硒铅石
Kerus 克鲁司钨钢
kerve 拉槽;掏槽;底槽;kerf 的复数形式
kerylbenzene 煤油烷基苯
kerzinte 镍褐煤
Keshan disease 克山病
Kessener brush 克森勒刷
Kessener brush aeration 凯森纳转刷曝气
Kessener brush aerator 凯森纳转刷曝气器
Kessener revolving brush 凯森纳(循环)旋转刷(活性污泥中用以维持循环和供氧的圆筒形金属刷)
Kessler abrasion tester 凯斯勒耐磨试验机
Kessler mill 周期式带钢轧机;凯斯勒轧机
kesterite 锌黄锡矿
Kesternich test 耐蚀试验

kestner cement 水硬性水泥(耐高温达1300℃); 耐火水泥(耐高温达1300℃)
ket 刃
ketal 酮缩醇;缩酮
keten(e) 乙烯酮;烯酮
ketene lamp 乙烯酮灯
ketene process 乙烯酮法
ketimine 酮亚胺
ketipic acid 草酰二乙酸
ketoacidosis 酮酸中毒
keto-acid soap grease 酮酸皂基润滑脂
ketoacyl 酮脂酰
keto-alcohol 氧基醇;酮醇
ketoalkylation 酮烷基化
keto-amine 酮胺
keto-carboxylic acid 酮酸
ketocoumaran 苯并二氢吡喃酮
keto-enol system 酮烯醇系
keto form 酮式
ketoimine 酮亚胺
ketoisocaproate 酮异己酸
ketoisocaproic acid 酮异己酸
ketoisovalerate 酮异戊酸
ketoisovaleric acid 酮异戊酸
ketol 乙酮醇;酮醇
keto-lactol 内缩酮
ketone 酮(类);缩酮
ketone acetal 酮缩醛
ketone acid 酮酸
ketone alcohol 酮醇
ketone-aldehyde 酮醛
ketone-benzol-dewaxing process 酮苯脱蜡法
ketone condenser 丙酮冷凝器
ketone ester 酮酯
ketone ether 烷氧基酮;酮醚
ketone form 酮式
ketone-formaldehyde resin 酮类甲醛树脂
ketone fractionator 丙酮精馏塔
ketone group 酮基
ketone oil 酮油
ketone oxidation 酮氧化
ketone rancidity 酮败
ketone resin 酮(缩)醛树脂;酮(类)树脂
ketone-sulphoxylate 次硫酸铜
ketonic acid 酮酸
ketonic ether 酮醚
ketonic form 酮式
ketonic group 酮基
ketonic hydrolysis 成酮水解作用;成酮分解
ketonic type 酮式
ketonization 酮化
ketonize 酮化
ketos 凯托斯铬锰钨钢
ketostearate 酮硬脂酸
ketostearic acid 酮硬脂酸
ketotrithion 连三硫酮
ketovinylation 酮乙烯化作用
ketoxime 酮肟
k'ette 小厨房
kettle 锅状的;锅;水(槽)壶;釜(体)
kettle back 穹形顶板
kettle bodied oil 锅熬聚合油
kettle boiled linseed oil 炼制的亚麻油;亚麻子油清漆
kettle-boiled oil 清油;熟油
kettle chain 锅状洼地带
kettle cure 罐中硫化
kettle depression 锅状陷落;锅形陷落
kettledrum 定音鼓;铜鼓(釜状)
kettledrum frame 定音鼓架
kettle fault 锅状断层;锅形断层
kettle for heating joint filler 填缝料熔炉;填缝料加热炉
kettle furnace 倾动式保温炉
kettleholder 勺柄;水壶柄
kettle hole 锅穴;锅形陷洞,壶穴;地表空穴
kettle lake 锅穴湖;锅形湖
kettle moraine 锅形冰碛
Kettleness bed 凯特尔内斯层
Kettler-Drude dispersion equation 凯特勒—德鲁德色散方程
kettle reboiler 锅形再沸器
kettle refining 锅精炼(低熔点金属)
kettle room 开水间;热水供应室

kettlestitch 束结装订
kettle-type reboiler 釜式重沸器
kettnerite 氟碳铋钙石
Ketton cement 拉特兰郡硅酸盐水泥
Keuper-chinle polarity chron 考侬波—钦利极性时
Keuper-chinle polarity chronzone 考侬波—钦利极性时间带
Keuper-chiule polarity zone 考侬波—钦利极性带
Keuper marl 考侬波泥灰岩;斑纹泥灰岩(美国)
Keuper(series) 考侬波统
kevel 尖斧锤;鹤嘴锤;鹤嘴锄;盘绳栓
kevil 石工斧;盘绳栓
kew barometer 定槽气压表
Keweenawan 基维诺期
kew garden 英国皇家植物园;邱园
kew-pattern barometer 寇乌式气压表;定槽式气压表
key 楔片;关键(的);解答;键;机器键;密钥;栓体;电键;封顶
key access 键存取
key-actuating machine 键控机
keyage 码头费
key aggregate 嵌缝填料;嵌缝碎石;嵌缝集料;嵌缝骨料;填缝骨料;塞缝骨料
key-airport 枢纽机场
key and bolt connected timber beam bridge 键结合木梁桥
key architectural characteristics 主要体系特性
key area 关键地区;标题区
key area control 要项控制
key band 特征谱带;电键组
key bar (非受力的)桁架封闭杆;定位键
key beam 楔形梁;键接梁
key bed 键座;基准面;标准层;标志层(位);分界层
key block 拱顶石;封顶板;封顶
key board 键盘;键板;电键板;键块
keyboard control key 键盘控制键
keyboard encoder 键盘编码器
keyboard entry 键盘输入
keyboard function key 键盘功能键
keyboard input 键盘输入
keyboard input printout 键盘输入打印输出
keyboard input simulation 键盘输入模拟
keyboard keyer 键控器
keyboard oscillator 按钮调谐频率振荡器;按钮调谐频率的振荡器
keyboard perforator 键盘穿孔机;键孔穿孔机
keyboard printer 键盘打印机
keyboard programming 键盘编程器
keyboard punch 键盘穿孔机
keyboard select routing 键盘路由选择
keyboard send/receive terminal 键盘收发终端
keyboard typing reperforator 键盘式复凿孔机
key body 主体
key bolt 螺杆销;键螺栓
key break 键断
key brick 拱顶砖;辐射砖;楔砖
key buffer 电键缓冲器
key-buoy 桶形浮标
key button 电键;按钮
key cabinet 电话控制盒;电键箱;钥匙柜
key card puncher 卡片穿孔机
key chain 钥匙链
key change 关键码改变;钥匙齿型;钥匙齿形变化
key changing lock 用可变钥匙开启的锁
key chuck 主夹盘
key city 中心城市
key click 电键声响;电键卡嗒声;电键喀哒声
key click filter 电键浪涌脉冲滤波器
key coat 下涂
key code 关键码;键控代码
key code table 关键码表
key colo(u)r 基本色(调)
key columns 关键列
key column values 键数位位值
key component 主要组分;主要组成(部分);关键组件;关键组分
key compression 关键码压缩
key computer printer 计算机的键盘打印机
key connection 键结合
key console (充当拱顶石的)凸挑饰件
key construction 主要建筑物;重点工程;关键工程
key construction joint 键接施工缝
key control system 钥匙控制系统;键控系统

key count 主要交通站点数;控制站计数
key course 楔形砌层;拱顶石层;锁接砌层;控制层
key cut 基本挖掘;超前沟
key cut hole 掏槽炮眼
key dam 筑主坝;主坝
key data 主要数据
key day 控制日
key department 要害部门
key diagram 原理草图;概略原理图;解说图;索引图;总图
key dimension 主要尺寸
key discharge locations 主要排污点
key domain 关键域
key drawing 纲要图;例图;解释图;接图表;描轮廓线;索引图
key drive 键驱动;键控;键传动
key-driven calculator 拨键计算器
key driver 键起子;销起子
key drop 钥匙孔盖(片);盖孔板
keyed 锁着
keyed-alike cylinders 相同钥匙的圆筒弹子锁
keyed amplifier 键控放大器
keyed attribute 信息标号属性;关键属性
keyed beam 楔合梁;键合梁;加键梁;拼梁;拼合梁;榫接梁;叠合梁;键接梁
keyed block 加键方块
keyed brick 槽形砖
keyed bush 加键轴衬
keyed clamp(ing) 定时箝位
keyed composite girder 加键叠合梁;键接组合梁
keyed compound beam 键接组合梁
keyed compound girder 键接组合梁
keyed construction joint 楔形施工缝;楔形工作缝;企口构造缝;键接施工缝
keyed-differently cylinders 不相同钥匙的圆筒弹子锁
keyed dowel 开槽木销;槽榫
keyed end 锁定端;锁定点
keyed girder 加键离合梁;加键叠合梁;木键合梁;键合梁
keyed-in frame 嵌墙橙子(部分墙体小于洞口)
keyed into 嵌进
keyed joint 键连接
keyed jointing 嵌缝;键接(合);加键接缝;半圆凹缝;凹圆灰缝;勾缝;楔形缝;凹圆弧灰浆缝
keyed miter joint 嵌键斜角缝
keyed mortise and tenon 加键镶榫接合;嵌键斜角缝;牙尖雌雄榫
keyed mortise and tenon joint 键接雌雄榫
keyed-on crank 楔装曲柄
keyed original trace 分色原图;分色清绘原图
keyed pointing 嵌键勾缝;凹圆形灰浆勾缝;凹圆形砂浆勾缝
keyed rainbow signal 键控彩虹信号
keyed scarf 楔形嵌接;锁嵌
keyed soffit 斩平的拱腹面
keyed strutting 键控(格栅)横撑
key element 要素;关键元素
key element of curve 曲线要素
key ended trunk line 端接电缆中继线
key engineering project 关键工程项目
key equipment package 关键设备包
keyer 键控器;调制器;定时器;电键器
keyer adapter 电键匹配附加器
keyer multivibrator 键控多谐振荡器
key error 关键性误差
keyer tube 键控管
key escutcheon 钥匙孔板;键纹板;锁孔盖
Keyes equation 凯斯方程
key event 关键事件;关键结点
key factor 主要因素;主导因子;重要因素;关键因素
key file 组锉;什锦锉
key files set 组挫
key fixing plate 固键板
key flat 拼版晒版
key floating 抹灰底层;基础的整平层
key folding 键叠合
key footing 基础的整平层
key for identification 鉴定检查表
key for plaster 粉刷施工要点
key for searching 查找键
key fossil 标准化石
keyframe 键架
key function 主要功能
key generator 密钥产生器

key groove 楔形角;楔形槽;键槽;插销槽
key group 指示组
key hinge 拱冠铰;拱顶铰
key hog 拱顶加高量
key hold 重点货舱
key holder 持键器
key hole 钥匙孔;基准孔;基准井;电键插孔;标准井;控制钻孔【岩】;键孔;椭圆弹孔;锁眼;栓孔
keyhole Charpy impact test specimen 钥匙形缺口冲击试样
keyhole cover 钥匙孔盖
keyholed back plate 带锁眼底板;带孔的后板
keyhole effect 小孔效应
key hole escutcheon 钥匙孔罩
keyhole-mode welding 小孔型等离子弧焊;穿透型焊接法
keyhole neck 钥匙孔式领圈
keyhole plate 钥匙孔盖板
keyhole saw 栓孔锯;斜形狭圆锯;钢丝锯;开孔锯;键孔锯;细木锯;圆弧锯
keyhole shaped orifice 钥匙孔形喷孔
keyhole slot 销槽;键孔槽;锁眼形长孔
keyhole specimen 有刻痕的冲击试块;钥匙孔形切口试样;冲击弯曲试验钥匙孔试样
keyhole type notch 钥匙式槽口;锁孔式槽口
key horizon 主要土层;基准层;标准层;标志层(位);基准水平线
key-in 嵌入
key industry 主要工业;主导工业;支柱产业;关键工业;基础工业
key industry duty 重要工业税;基础工业保护关税
key industry investment fund 基础工业投资基金
key industry investment trust 基础工业投资信托
keying 楔圈;键控;锁上;锁结
keying action 连接键作用;锁结作用;楔住
keying action of aggregate 骨料的锁结作用;集料锁结作用
keying agent 增黏剂
keying amplifier 电子开关
keying chirps 电键啁啾声
keying circuit 键控电路
keying device 键控装置
keying frequency 键控频率;电键频率
keying-in 新老砖墙咬接;砖石榫接;接搓;键接
keying interval 电键间隔
keying line 遥控线路
keying mix 砌筑法(其中之一种);键接合
keying monitor 键控监测器
keying of entries 分录线索
keying off curve 曲线测设
keying pin 键销
keying ring 嵌铅环
keying signal 键控信号
keying speed 按键率
keying strength 咬合强度
keying unit 键控器
keying wave 键控波;传号波
keying waveform 键控波形
keying wedge 固定楔
key-in knob(door) lock 匙孔在执手中的门锁;匙孔在球形把手内的门锁;门柱锁
key-in region 通过区
key instruction 引导指令【计】;主要指示;主导指令;关键指令
key instrument 主要仪器
key intermediate 关键中间体
key intersection 关键交叉口
key intersection control 关键交叉口控制
key investment project 投资重点
keyite 砷锌镉铜石
key items 重点项目
key job 关键工作
key joint 键接合
key joint pointing 凹形勾缝;关键字检索;拱顶式勾缝;凹圆形砂浆勾缝
key lampholder 开关灯头
key lamp-socket 开关灯头
key layer 标准层;标志层(位)
key ledger 主要分类账
key length 键的长度
keyless chuck 无键的夹具
keyless coupling 无键联轴节
keyless lock 无钥匙锁
keyless ringing 无钥信号;插塞式自动振铃

keyless socket 无开关灯口;无键插座
keyless watch 无钥匙表;不上发条表
keyless work 无键机构
key letter 编码键
key level 键电平
key lever 主控握柄;主控手柄;键柄
key-lifting switch 扳键开关
key light 主光;主灯(光);基本灯光;主照明光
key lighting 主照明
key line 浮标绳
key line plan 基本线型图
key link 中心环节;主要环节
key lock 钥匙锁(闭器);键封锁
key lock apparatus 钥匙锁装置
key locker 钥匙壁箱
keylock switch 钥匙锁定开关
key machine 主机;主导机械
key manufacturing operation 关键工序
key map 总图;纲要图;接图表;索引图
key mat 键垫;键标帽
key metal 母体金属;母合金
key method 电键法
key milestone 主要里程碑
key model 基本模型
key modulation 键控调制;键控
key module 密钥组件
Keynesian growth theory 凯恩斯增长理论
Keynes plan 凯恩斯计划
keynote 基调
keynote address 主题报告
keynote presentation 主题报告
keynoter 主题报告人
keynote speaker 主题发言人;主题报告人
keynote speech 主题报告
key number 钥匙号码;片边号码
key nut 键螺母
key-off 切断
key of joint 接缝企口
key of reference 引用关键字
key of scaffolds 脚手架的固定键
key on 用键固定;接通
key-operated brake 楔形制动器
key operation 主要作业;关键操作
key out 切断;断开
key paper 重要文章
key paste (由黑色糖浆与石墨组成的)润滑油
key pattern 回纹饰;曲折回纹(饰)
key person 关键人员
key personnel 主要人员
key petroleum source bed 标准生油层
key pile 主桩;枢桩;板桩中最后一根桩
key plan 平面布置总图;索引图;索引平面;总平面图
key plate 主版;键板;锁定挡板;地物版;钥匙孔(金属护)板
key point 要点;重点;制高点;战略据点;关键(点)
key point in construction 关键工点
key point investigation 重点调查
key point of control 防治关键点
key point products 重点产品
key pollutant 主要污染物;关键性污染物
key pollution 主要污染
key pollution source 主要污染源
key position 要点;制高点;关键位置;关键点;枢纽(位置)
key post 主要岗位;打印器;坑道支柱;短柱
key project 重点项目;重点工程;关键工程;枢纽工程
key property 主要特性;基本性能
key protrusion 榫舌
key puller 拔键器
key pulley 拨键器
key pulse 键控脉冲;按键脉冲
key pulser 电键发送器
key punch 键(控)穿孔;键孔穿孔机
key puncher 键控穿孔机;穿孔(机操作)员
key punch error 键控穿孔错误
key punch machine 键控穿孔机;键孔穿孔机
key rack 钥匙架
key range 键控范围
key ratio 主要比率
key reaction 主要反应;关键反应
key relaxation 特征松弛
key relay 键控继电器
key rendering 抹灰打底;底涂;粉刷底涂
key resources 关键资源

key resources type industry 关键资源的工业
key ring 键环
key rock 标准岩层;标志岩层
key roll over 键控倒转
key sag 拱顶下垂度;拱中下垂
key sample 主要标本;标准样品
key screw 螺旋键;螺丝把;螺帽扳手
key search 关键字检索
key seat 键槽;销座;栓座;销槽;铣键槽;插键槽
keyseat broach 键槽拉刀
key seated hole 钻具卡槽钻孔
keyseater 键槽机;铣键槽机;键槽铣刀;键槽插床
key seating (孔壁形成的)拉槽;键槽;键;销槽
key-seating machine 开键槽机
key-seating milling machine 键槽铣床
key-seating sticking 键槽卡钻
key-seat rule 铣键槽尺
key section 标准剖面
key sector 关键部门
key segment 刹尖石;封闭管片
key sender 电键发送器;按钮电键;键控发射机
key-sending 电键选择
keysent 电键发送器拨号
key set 电键(组);按(钮电)键;转接板;键集
keyshelf 键座;键架
key shut-out feature 钥匙锁闭功能
key slot 键槽
key slotting machine 键槽插床
key socket 旋钮灯口;开关灯头;键插口;电键插座
key sort 关键码分类
key sorting 按关键字分类
key source 密码索引
key spacer 电键隔板
key staff 钥匙路牌
key state 关键状态
key station 主(要)台站;主(控)台;中心(控制)站;控制(测)站;基本测站
key-station volume count 主要站交通量计数;控制站交通计数
keystone 填缝石;用拱顶石支承;拱心石;拱冠石;拱顶石;嵌缝石;锁砖;塞缝石;(一种用于新水泥工程的)油漆;锁石
keystone at the crown of stone bridge 石拱桥拱顶石
keystone correction 梯形失真校正;梯形畸变校正
keystone distortion 梯形失真;梯形畸变
keystone effect 楔效应;梯形失真效应;梯形畸变效应
keystone layer 拱顶石层;拱心石层
keystone piston ring 梯形断面活塞环
keystone plate 波纹钢板;瓦垅(波)纹钢板;瓦楞钢板;封顶板
keystone ring 梯形环
keystone scanning 梯形扫描
keystone strand wire rope 楔表面钢丝绳;表面钢丝绳
keystone-type piston ring 梯形活塞环
keystone wave 梯形波(形)
keystoning 梯形失真
keystoning correction 梯形失真校正
keystroke 击键
keystroke verification 击键验证
key structure 主要建筑物
key switch 钥匙开关;键(式)开关;电键开关;按键开关
key switch lampholder 带按钮开关灯头
key switch lamp-socket 带按钮开关灯座
key system parameter 关键系统参数
key table 键表
key tablet 钥匙路牌
key tag 钥匙牌;钥匙标签
keytainer 钥匙夹套
key tape load 键带信息输入
key technique 关键技术
key technology 要素技术;关键技术
key tenant 有吸引力的承租人
key the contract 遵守合同
key to legend 图例说明
key top 键顶
key to symbol 图例说明
key touch selector 触键选择器;按键选择器
key to varieties 品种索引表;品种鉴定说明
key to variety 品种说明;品种检索表
key traffic 主要运输

key trench 坝基截水墙槽
key type socket 楔形承插口
key unit 电键部件
key up 激励
key utilization species 主要利用指标种
key value 色调特征
key valve 扳手阀门；键阀；钥匙阀门
key verify 键校对；键盘式验孔
key wall 齿墙；刺墙
key water-control project 水利枢纽工程
keyway 键槽；栓道；销座；销槽；凹凸缝；榫槽；混凝土接合处凹槽
keyway attachment 键槽加工装置
keyway planer 销槽刨床；键槽刨床；刨槽机
keyway planing machine 键槽刨床
keyway slotting machine 键槽机
keyway slotting tool 键槽插刀
key well 基准（钻）孔；基准（钻）井；注入井
key well system 控制钻井法
key word 关键字；主题词；索引字；关键词
keyword-from-title index 标题关键词索引
key word in context 文件标题字；上下文内关键字
key word in title 标题关键字
key word out of context 文献检索字；文件标题字；上下文外关键字
keyword out-of-context index 上下文外关键词索引
key words sorting 分类标记
key works 关键工程
keywrench 活动扳手；活动扳钳
K factor 基线与高度比；K 因子；K 系数；增殖系数
K-factorial connection K 阶乘联络
K-feldspar-cordierite-hornfels facies 钾长石—堇青石—角页岩相
K-G model K—G 模型
Khadar polarity superchron 克哈达极性超时
Khadar polarity superchronzone 克哈达极性超时间带
Khadar polarity superzone 克哈达极性超带
khademite rostite 斜方铝矾
khaki 卡其；黄色的；黄褐色的；土黄色；开利姆地毯
khaki engobe 黄褐色化妆土
khaki glaze 黄褐色釉
khan 篷车旅店；商栈；简陋驿站
khanat 坎儿井
khannesheite 碳钡钠石
Kharktan-Bogda deep fracture zone 哈里克套—博格达深断裂系
khaya 非洲桃花心木
khibinite 粒霞正长岩；褐硅铈石；褐硅铈矿
khibinskite 希宾石
khinite 碲铅铜石
khlellin 开林
Khmer architecture 高棉建筑
khorasan 喀拉逊地毯
khotan 和田地毯
Kiama reversed polarity hyperchron 基亚曼反向极性亚时
Kiama reversed polarity hyperchronzone 基亚曼反向极性巨时间带
Kiama reversed polarity hyperzone 基亚曼反向极性巨带
Kiamitia shales 基奥瓦页岩
Kibaran cycle 基巴兰造山旋回
kibble 凿井用吊桶；提升大吊桶；竖井吊桶；修琢石块（采石场）；吊桶（凿井用）
kibble cage 出碴用吊笼
kibble hoist 吊桶提升机
kibbler 粉碎机
kibble rope 吊桶钢丝绳
kibbler roll 辊式破碎机
kibbler roll-crusher 齿槽辊破碎机
kibbling 粉碎力
kibbling mill 筒式磨碎机；粗磨机
kibleh （伊斯兰教徒的）遥拜方向
kibosh clay 木节黏土
kibushi clay 木节黏土
kibusi clay 有机碳黏土
kich 结晶石墨
kick 根端跳起；后坐；汽油发动机；突跳；反应力；反击力；反冲力；凹底；凹槽砖
kick ahead 前冲
kick a hole 钢绳冲击法钻孔
kick-atomizing 冲击雾射；冲击雾化（的）
kick-atomizing pile driver 冲击喷射打桩机；冲击雾化内燃打桩机
kick-atomizing pile hammer 柴油机打桩锤
kickback 回弹；回跳；退赔；回扣；溢出；反冲；反程
kickback dump 端转翻笼；逆转翻笼；倒车翻车器
kickback line 回扫线
kickback power supply 回扫脉冲电源
kickback transformer 回扫描变压器
kickback type of supply 脉冲电源
kick circuit 急冲电路；脉冲电路；突跳电路
kick copy 分批标记本
kickdown 换低挡
kickdown buffer 高低机缓冲器
kickdown switch 自动跳台开关；低速挡开关；加速系统自动开关
kickdown switch fixed terminal 加速系统自动开关接线柱
kickdown switch plunger 加速系统自动开关柱塞
kicked upstair (case) 明升暗降
kicker 舷外挂机；印件分堆器；加固木块（框架的）；抛勘器；艇外推进器；弹踢器；把套器；施工缝处混凝土的堵头；抖动器；冲击磁铁；侧刀；发泡催化剂；拨壳挺
kicker baffle 导向隔板
kicker coil 冲击线圈
kicker cylinder 回程缸
kicker light 强聚光
kicker magnet 冲击磁铁
kicker mill 锤式缩绒机
kicker plate 加强横木（加固楼梯与混凝土连接）；（楼梯的）防滑条
kicker port 排气孔
kicker ram 辅助柱塞
kicker stone 侧刃金刚石
kick(e)y 弯管器
kick-in arm 进料拨杆
kicking coil 扼流线圈
kicking down 钢绳冲击钻法
kicking out 外挑
kicking piece 横挡撑；腰梁垫木；垫木；撑杆；（门的）下横档护板
kicking plate 门脚护板；防踏板；踢脚板；门的下横挡板
kicking post 马厩隔墙外端的柱子（内端为饲料槽旁的柱子）
kicking recess 柜下防踢的凹进处；踢脚凹空；踢脚凹进处
kicking shunting 溜放调车法
kicking stops 防动绳制
kicking strap 脚蹬金属板；拉紧木板的（金属）条
kicking strip 踢脚板
kicking tackle 缓冲绞辘
kick motor 加速发动机
kick-off 出油（油井）；断开分离装置；甩出机；拨料机
kick-off arm 出料推杆
kick-off mechanism 卸料机；解脱装置；推出机
kick-off point 造斜点
kick-off temperature 引发温度；生效温度；起始分解温度
kick-off valve （空气升液器的）起动阀；解脱旋钮；出料阀门
kick of rudder 浪击舵叶
kick of wheel 舵轮跳动
kickout 龙口（墙体挑出排水口）；斜撑（或拉杆）脱断；斜撑意外松动；支柱意外松动
kickout arm 出料推杆
kickout latch 弹簧插销
kickout lever 断路器杠杆；分离杆；止动杆
kickout master cylinder 断路器活塞杆；止动总泵缸；关掉主汽缸
kickover tool 造斜工具
kick-pedal 脚蹬起动踏板
kickpipe 防踢板（保护外露电缆）
kick plate 刮板；踢板；防滑条（楼梯）；门脚护板；楼梯防滑条；门踢板
kick plate and drip 踢板和滴水槽
kick pleat 外开褶
kick point 变坡点；转折点
kick rail 踢脚栏
kick recess 踢脚凹进处
Kick's law 基克压研定律
kicksort 脉冲幅度分析
kicksorter 脉冲高度分析仪；脉冲高度分析器；选分仪；振幅分析器；脉冲振幅分析器；脉冲高度分析器
kicksort(ing) 振幅分析
kicksorting technique 振幅分析技术
kick spring 反冲弹簧
kickstand 撑脚架
kickstarter 反冲式起动器；反冲式起动机；脚踏起动器；跳动式起动器
kickstarter bearing 冲式起动器的曲柄轴承
kickstarter casing 反冲式起动机箱
kickstarter casing cover 冲蚀起动机箱盖
kick strip 加强板；加固木块；门脚护条；踢脚；挡板
kick time 井涌时间
kick transformer 急冲变压器；脉冲变压器
kick-up 向上弯曲；预留空余；翻车器
kick-up block 阻车器
kick-up frame 降低车辆重心的车架；上弯式支架；凹型车架
kick water pump 反冲式水泵
kick wheel 脚动陶轮
kid 大手餐盘；树枝束；梢捆；埽；柴排；柴捆；舱面鱼池；小木桶；柴把
kiddcreekite 硫钨锡铜矿
kidding 埽工护岸；柴捆作业；填梢护岸；打捆
kiddle 鱼梁；（河中捕鱼用的）拦河栅
kidney 结块；小圆石；小卵石
kidney basin 弯盘
kidney desk 腰圆形书桌
kidney fat 板油
kidney (iron) ore 肾矿石
kidney joint 挠性接头；气隙耦合器
kidney oil 肾色油
kidney ore 肾铁矿
kidney shaped stockpile 腰肾形料堆
kidney stone 黏土中硫化铁团块
kidney stone nephrite 软玉
kidwellite 羟磷铁钠石
Kienlis functionality theory 基恩利斯官能度理论
Kienmayer's amalgam 基尼迈耶汞合金
kier 煮布锅
kier-boiling 煮布锅精练
kier-boiling kiering 煮布锅精练
kier liquor 漂煮液；漂煮锅废液；漂洗液
kier piler 煮布锅堆布器
kier scouring 煮布锅精练
kier stain 煮炼斑渍
kier waste 漂煮废料；漂洗废料
kier wastewater 漂煮锅废水；漂洗废水
kies 黄铁矿
kieseguhur Merck 白色硅藻土载体
kieselguhr 硅藻土
kieselguhr as a rapid filter medium 硅藻土滤料
kieselguhr brick 硅藻土砖
kieselguhr concrete 硅藻土混凝土
kieselguhr covering cord 硅藻土包线
kieselguhr slab 硅藻土板
kieserite 硫镁矾；水镁矾
kieseritite 石盐镁矾岩
kieserohalocarnallite 硅杂盐
kievite 透淡闪石
kikekunemalo 漆用树胶
Kikuchi band 菊池谱带
Kikuchi line 菊池线
kil 黑海白黏土
Kilauean type 基拉韦厄型
kilckoanite 斜方硅钙石
kilderkin 小桶；大桶（颗粒状物容积单位）
kilfoam 抗泡剂
kilhig 推杆
Kilim 机织地毯；基里姆地毯
kilkenny coal 无烟煤
kill 小海峡；阻断
killagh 小艇锚；石锚；代用锚
killalaite 斜水硅钙石
kill and choke line 压井液输送管线
Killarmeyan orogeny 基拉尔尼造山运动
killas 片岩；板岩
killed alloy 安定合金；全脱氧合金
killed carbon steel 镇静碳钢
killed colo(u)r 饱和色
killed ingot 镇静钢锭
killed lime 消石灰；死石灰；失效石灰
killed spirit 焊酸；氯化锌
killed steel 镇静钢；全脱氧钢；脱氧钢
killed wire 去弹性钢丝
killer 抑制器；钢脱氧剂；断路器；扼杀剂

killer boat 捕鲸快艇;捕鲸船
killer circuit 熄灭电路;抑制(器)电路
killer detector-amplifier 抑制器检波放大器
killer pulse 抑制脉冲
killer restrainer 限制器
killer ship 捕鲸快艇
killer smog 有毒烟雾
killer stage 消色电路;抑制级
killer switch 断路器开关
killer tube 抑制管
killer well 压制失控定向井
killer winding 消磁绕组;灭磁绕组
killesse 边沟;雨水槽;沟渠;渠道;(舞台两侧布景之间的)空隙;檐槽;滑槽(板)
kill fish 死鱼
kill(i)ck 代用锚;石锚;小锚;大石块
killi(i)ck hitch 圆材结加半结
killing 切断电流;涂抹节疤
killing agent 脱氧剂;镇静剂
killing back 死亡
killing curve 杀菌曲线
killing freeze 严霜;严寒
killing frost 严霜
killing knot 除掉木节
killing line 切断线路;空泵管道
killing mud weight 压井泥浆比重
killing period 镇静期
killing point of temperature 致死温度
killing roll 矫形滚筒
killing smog 致死烟雾
killing substance 毁灭性物质;致死物质
killing temperature 致死温度
killing the well 压井
killing time 致死时间
Killing vector 基林矢量
killing wells 压制井喷
killing work 屠宰场
killinite 块云母
kill line 压井管线
kill-line manifold 压井管汇
killock 小锚;大石块
kill order 停止供应定货;取消定货
kill radius 杀伤半径
kill ratio 射杀率
kill string 压井管柱
kill strip 死条纹
kill the bill 否决议案
kill the sea 抑制海浪
kill the well 压制井喷
kiln 大窑;窑(炉);干燥炉;炉干燥器;烘干炉;窑烧
kiln alignment with laser 激光调窑
kiln annealing 窑中退火
kiln atmosphere 窑内气氛
kiln attendant 看火工
kiln axis 窑轴线
kiln basin 料浆池
kiln bottom 窑底
kilnboy 干燥程序自动记录
kiln bracing 箍窑
kiln brick 窑烘砖
kiln brown stain 窑干褐变
kiln building 炉窑车间
kiln bunk 窑(干)装材车架
kiln burn 窑干褐变
kiln-burned 窑烧的;窑烘的
kiln burnt brick 窑烘砖
kiln car 窑车
kiln cart 窑车
kiln chamber 窑室
kiln charge 窑装载量;窑干装载量
kiln collapse 倒窑
kiln control logic 回转窑控制逻辑
kiln crackle 窑缝
kiln crown 窑拱;窑顶
kiln degrade 窑干贬值
kiln dirt 落渣
kiln discharge 出窑
kiln-dried 窑室烘干的;烘干的;窑杆木材
kiln-dried lumber 窑杆材
kiln-dried wood 窑干木材;烘干木材
kiln dry (木材的)窑内干燥;窑中烘干
kiln dryer 炉式干燥器
kiln drying 窑(内)烘干;窑中烘干
kiln drying curve 烘窑曲线

kiln-dry lumber 烘干木材;窑内烘干木材
kiln dust 窑灰
kiln entrance 窑门
kiln evapo(u)rator 烘干炉;窑式脱水器
kilneye 窑孔;看火孔;出灰孔;窑口
kiln factory 窑场
kiln feed 入窑生料
kiln feed dust 生料灰
kiln feeder 装窑机;炉窑进料器
kiln feeding unit 生坯进窑机
kiln-feed material 入窑物料
kiln-feed slurry pump 入窑料浆泵
kiln-feed thickener 入窑料浆浓缩机
kiln filler 满窑工
kiln filling 满窑
kiln-fired 窑烘的;窑烧的
kiln firing 烧窑
kiln for large pieces 大器窑
kiln for small pieces 小器窑
kiln-fresh brick 刚出炉的砖;新出窑砖
kiln furniture 窑具;烧窑辅助设备;窑内陶瓷坯盒
kiln gas 窑尾废气;窑气;窑炉烟气
kiln gas scrubber 炉气洗涤器
kiln gate 炉门
kiln glaze 窑釉
kiln hole 窑门洞;窑洞
kiln hood 窑罩
kilning 窑烘(干);烘窑;窑烧;窑烧
kiln insulation 窑隔热层
kiln in the shape of steamed bun 馒头窑
kiln jacking 盘窑
kiln liner 窑内衬垫;窑衬
kiln lining 窑内衬砌;窑里层;窑衬
kiln log 烧窑记录
kiln man 烧窑工人
kiln mark 烧成压痕
kiln mark brick 窑斑砖
kiln-marked brick 窑斑砖
kiln mill 窑炉转磨;综合干燥磨碎机;联合干燥磨碎机
kiln monitoring 看火
kiln operation 炉窑运转;窑炉作业
kiln operator 看火工
kiln pier 基座(转窑);转窑(支)座
kiln placing 装窑
kiln process 窑烧法;回转窑法
kiln refractory brick 窑烘砖;窑制耐火砖
kiln ringing 结窑圈
kiln roasting 窑炉烤烘
kiln roasting kiln 烤窑
kiln roof 窑顶
kiln run 窑干燥过程;炉子工况;未分类的砖或瓦
kiln sample 窑干试样
kiln sand 窑砂;垫砂
kiln scum 窑霜;窑生盐迹;吐渣(表面析出白色浮渣)
kiln seasoned 烘干的
kiln setter 装窑窑具
kiln shaft 立窑筒体
kiln shelf 装窑棚架
kiln shell 窑壳;炉壳;窑壁
kiln shell section between tyres 二轮带间窑筒体段节
kiln shell section under a tyre 轮带下筒体段节
kiln-shell temperature 窑壁温度;窑壳温度
kiln shut-down 停窑
kiln site 窑址
kiln size 炉窑尺寸;窑尺寸
kiln slag 窑渣
kiln stain 窑干材色变;火刺
kiln stone 炉料;窑炉石料
kiln structure 窑体结构
kiln sweat 窑汗
kiln tar 窑焦油
kiln transformation 窑换
kiln truck 装材车
kiln unloading unit 卸窑机;出窑机
kiln ware 窑器
kiln wash 窑墙涂料
kiln waste heat 烧窑排出的余热;窑废热
kiln white (黏土砖的)窑生盐迹;白霜;釉面无光
kiln with enlarged discharge end 热端扩大型窑
kiln with enlarged feed end 冷端扩大型窑
kiln with movable floor 吊顶窑
kiln with(suspension) preheater 带悬浮预热器窑
kiln with travel(l)ing grate 移动算式窑

kiloampere 千安(培)
kilobar 千巴(压力单位)
kilobarn 千靶
kilobaud 千波德
kilobecquorel 千贝克勒尔
kilobit 一千位;千二进制位
kilobits 千比特
kilobyte 一千个字节;千字节(1024字节)
kilocalorie 千卡;大卡
kilocalories per hour 千卡/小时
kilo-character 千字符
kilocoulomb 千库(仑)
kilo cubic feet per hour 千立方英尺小时
kilocurie 千居里
kilocurie source 千居里源
kilocycle 千周;千赫
kilocycles per second 每秒千周;千赫
kilodyne 千达因
kiloelectron volt 千电子伏特
kilogauss 千高斯
kilogradient ratio 坡度千分率
kilograin 千克冷
kilogram-calorie 千(克)卡;大卡
kilogram-equivalent 千克当量;公斤当量
kilogram-equivalent weight 千克当量
kilogram force 公力;千克力
kilogram force meter 千克力·米
kilogram gross weight 毛重千克数
kilogram legal weight 法定千克数
kilogram-mass 千克·质量
kilogram(me) 公斤;千克
kilogram-meter 千克·米(功的单位);公斤·米
kilogram net weight 净重公斤数
kilograms per cubic(al) meter 每立方米千克数;千克/立方米
kilograms per minute 每分钟千克数
kilograms per second 每秒千克数
kilogram-weight 公斤重;千克重
kilohertz 千周;千赫
kilohertz thousand cycles per second 千赫
kilojoule 千焦耳
kiloline 千磁力线
kiloliter 千升;千公升
kilolumen 千流明
kilolumen-hour 千流明·时
kilom 千米;公里
kilomega 千兆;十亿
kilomegabit 千兆位;十亿位
kilomega cycle 千兆周;吉周;千兆周
kilomega hertz 千兆赫
kilometer 公里;千米
kilometerage 公里里程
kilometerage charges 按公里数收费
kilometer grid 公里网
kilometer interval for vehicle maintenance 车辆维护车间里程
kilometer interval of vehicle overhaul 车辆大修间隔里程
kilometer marker 里程碑
kilometer marking 里程碑
kilometer per hour 千米/小时;每小时千米数
kilometer post 公里标;里程标;里程碑
kilometer post of waterway 航道里程标
kilometer recorder for vehicle 车辆计程仪
kilometer scale 公里(图)尺
kilometer sign 公里标
kilometers of track maintained 维修轨道公里
kilometers of track operated 营业线路公里数
kilometers per wagon per day 货车日车公里
kilometer stone 公里标;里程桩;里程石;里程碑
kilometer-ton 千米·吨;公里·吨
kilometer wave 千米波
kilometrage 在公路、航道竖立公里里程碑;公里里程;公里表
kilometrage of double-line railway 铁路复线里程
kilometrage of track operated 运营线路公里
kilometres interval of running responsible accident 行车责任事故间隔里程
kilometre stone 里程石;里程碑
kilometric allowance 每公里税金
kilometric scale 公里(图)尺
kilometric wave 千米波
kilomol(e) 千摩尔
kilonewton 千牛顿

kilonewton meter 千牛顿米
kilooersted 千奥
kilo-ohm 千欧(姆)
kiloparsec 千秒差距
kilopascal 千帕(斯卡)(压力单位)
kilopond 千克力
kilopost 公里标
kilorad facility 千棒设施
kilorad program(me) 千棒计划
kiloroentgen 千伦琴
kilosecond 千秒
kilostere 千立方米;千公秉
kiloton 千吨
kilotron 整流管
kilounit 千单位
kilovar 千乏
kilovar-hour 千乏时
kilovar-hour meter 千乏时计
kilovar instrument 千乏表
kilovolt 千伏(特)
kilovoltage 千伏电压
kilovolt-ampere 千伏安
kilovolt-ampere-hour 千伏安(小)时
kilovolt-ampere-hour meter 千伏安时计
kilovolt-ampere rating 千伏安额定容量;额定千伏安
kilovolt meter 千伏计;千伏(电压)表
kilovolt peak 千伏峰值
kilovolt peak value 千伏峰值
kilowatt 千瓦
kilowatt-day 千瓦日
kilowatt-hour 度【电】;千瓦小时;千瓦时
kilowatt-hour meter 千瓦小时电度表;千瓦时计;电度计;电表
kilowatt input 输入千瓦
kilowatt loss 千瓦损耗
kilowatt-meter 千瓦计;电力(千瓦)计;电力千瓦表
kilowatt of eletctric energy 千瓦电能
kilowatt output 输出千瓦;输出功率
kilowatt-year 千瓦年
kilrr-attenuation 失真衰减器
kilt 石级风化
kimberlite 金伯利岩;角砾云橄岩
kimberlite diamond-bearing formation 金伯利岩含金刚石建造
kimberlite group 金伯利岩类
kim coal 油页岩
Kimeridge clay 启莫里奇黏土
Kimeridge coal 启莫里奇煤
kimitotype 界限层型
Kimmerian(stage) 基米里阶【地】
Kimmeridgian 启莫里阶【地】
Kimoloboard (一种由矿物(硅藻土)组成的)耐火墙板
kimono 和服
kimzeyite 锆榴石;钙锆榴石
kin 日本重量单位
kinaesthetic 动力感
kinckebocker yarn 彩点纱
kind 种类
kind band 扭折带
kindergarten 幼儿园
Kinderhook bed 肯德胡克层【地】
Kinderhookian 肯德胡克群【地】
kindling 燃烧;升炉
kindling point 着火点,燃(烧)点
kindling temperature 着火温度;着火点;燃点
kindling wood 引火木(柴)
kindly 易采的岩石
kindly remit 请即付款
kindly remit by check 请汇支票付款
Kind modulus 金德系数
kind of accounting statements 会计报表的种类
kind of analogy medium 模拟介质的类型
kind of anomaly 异常种类
kind of benefits 福利支付种类;保险支付种类
kind of business 业务种类
kind of canal 运河类型
kind of cavitation 空穴类型;空化类型;汽蚀类型
kind of crystal system 晶系的种类
kind of diving 潜水种类
kind of drill tool 钻探钻具种类
kind of drive 传动方式
kind of economic organization 经济组织种类
kind of formation 地层种类

kind of geothermal anomaly 地热异常种类
kind of land use 土地利用种类;土地利用方式
kind of liquid 液体种类
kind of matrix 胎体种类(金刚石)
kind of precipitation 沉淀类型
kind of reaming shell 扩孔器种类
kind of simple form 单形的种类
kind of special form 特殊形的种类
kind of stabilizer 稳定剂种类
kind of support 支架种类
kind of symmetry type 对称型的类型
kind of tectonic element 构造单元种类
kind of test 试验类别
kind of traction 牵引种类【铁】
kind of wastewater 废水类型
kind of works 工种(名称)
kinds of aids layout 航标配布类别
kinds of cargo 货物分类
kinds of cave sediments 洞穴沉积物种类
kinds of earth heat 地热类型
kinds of flammability gas 可燃气类别
kinds of forest fires 森林火灾的种类
kinds of gas 煤气种类
kinds of heat source 热源种类
kinds of hydrobase-exchange 水化阳离子种类
kinds of mortar 灰浆种类
kinds of nuclear energy 核能种类
kinds of remote sensing applications for geologic(al) mapping 遥感地质制图类型
kinds of root 根的种类
kinedensigraphy 运动密度测定法
kinegraphic(al) control 远距离控制
kinegraphic(al) control unit 远距离控制装置
kinema 电影院
kinematically acceptable solution 满足运动条件的解
kinematically admissible field 动性容许力场
kinematically admissible multiplier 动性容许乘数
kinematically admissible velocity field 动可容速度场
kinematically determinate structure 超静定结构
kinematically indeterminate structure 机动上超静定结构
kinematically non-linear analysis 动力学非线性分析
kinematically rigid 机动上刚性
kinematically similar 运动相似的
kinematic analysis 运动学分析;动力分析
kinematic boundary condition 运动边界条件
kinematic bunching 动态聚束
kinematic capillarity 运动毛管率
kinematic chain 运动链系
kinematic coefficient 运动系数
kinematic coefficient of viscosity 黏度动力系数
kinematic construction 浮动差胀结构
kinematic design 机动设计;动力学设计
kinematic ductility 运动延性;动黏度;动延性
kinematic eddy viscosity 涡动黏度;涡动动力黏滞度;动涡流黏滞度
kinematic eddy viscosity coefficient 动涡流黏滞系数
kinematic elasticity 运动弹性;动弹性
kinematic energy 动能
kinematic envelops 运动包络线
kinematic equilibrium 动态平衡
kinematic fluidity 运动流度;运动流动性
kinematic hardening 运动硬化;随动硬化;动力硬化
kinematic indeterminacy 运动不确定性
kinematic link 运动链
kinematic matrix 运动矩阵
kinematic model 动力模型
kinematic oceanography 运动海洋学;海洋运动学
kinematic operator 运动算子
kinematic parameter 运动参数
kinematic relation 运动关系
kinematic routing 运动波演算
kinematics 运动学
kinematic scheme 传动系统图
kinematic similarity 运动相似性;动力学相似;动力学模拟
kinematic similitude 运动相似性
kinematics of machinery 机械运动学
kinematics of machines 机械运动学

kinematic solution 机动方法求解
kinematic surge 运动波涌
kinematic theorem 机动理论
kinematic theory of framework 结构运动理论;框架的机动理论
kinematic to Saybolt universal viscosity 换算运动黏度成赛波特通用黏度
kinematic type of lineation 线理运动类型
kinematic velocity 运动速度
kinematic viscosity 运动黏滞性;运动黏滞率;运动黏(滞)度;动黏滞性;动黏滞度;动态黏(滞)度
kinematic viscosity coefficient 流动黏度;动黏滞系数
kinematic viscosity for sediment-laden flow 挟沙水流动黏滞度
kinematic viscosity of water 水的运动黏度;水的动黏滞性;水的动黏滞度
kinematic viscosity scale 动力黏度表
kinematic-viscous coefficient 运动黏滞系数
kinematic wave 运动波(浪)
kinematic wave theory 运动波理论;运动波浪说
kinemograph 流速坐标图;转速图表
kinemometer 感应式转速表;流速计;灵敏转速计;灵敏转速表;转速计
kinemometering radar 测水速雷达
Kinepanorama 克尼潘诺拉玛系统
kinescope 显像管;电子显像管
kinescope bulb 显像管外壳
kinescope factory 显像管厂
kinescope recorder 屏幕录像机
kinesimeter 运动测量器
kinetheodolite 电影经纬仪;摄影跟踪经纬仪
kinetic 能动的;动力学的;动(力)的
kinetic aerobic compost 动态好氧堆肥
kinetic air pump 动力空气泵
kinetic air/vacuum 动力空气/真空
kinetic analogy 运动相似性;动力模拟
kinetic analysis 动力学分析;动力分析
kinetic art 活动艺术
kinetic boundary condition 动力边界条件
kinetic chain 动力学链
kinetic characteristic 运动特性;动态特征;动态特性;动力学特性
kinetic characteristic curve 运动特性曲线
kinetic chemicals 致冷机;冷却剂
kinetic coefficient of friction 动力摩擦系数;摩擦动力系数
kinetic colo(u)rimetry method 动力学比色法
kinetic consideration 动力学研究
kinetic constant 动力学常数
kinetic control 动态控制
kinetic control system 运动控制系统;动态控制系统;动力学控制系统
kinetic decomposition 动力学分解
kinetic design 动态设计
kinetic eddy viscosity 涡流动力黏滞度
kinetic eddy viscosity coefficient 涡流动力黏滞系数
kinetic energy 动能;动力能
kinetic energy coefficient 动能系数
kinetic energy correction coefficient 动能改正系数
kinetic energy correction factor 动能改正系数
kinetic energy expenditures 动能消耗
kinetic energy gradient 动能坡度
kinetic energy head 动能高
kinetic energy of glacier 冰川的动能
kinetic energy of impact 冲击动能
kinetic energy of waves 波浪动能
kinetic energy rejection 排出动能
kinetic equation 动能方程;动力学方程;动力论方程
kinetic equilibrium 动态平衡;动力平衡
kinetic factor 动力系数
kinetic filter 动态滤波器
kinetic flow factor 运动水流因数;流动因数;动水系数;动力水流因素;弗劳德数
kinetic force 动力动能
kinetic friction 动摩擦
kinetic frictional stress 动摩擦应力
kinetic friction coefficient 动摩擦系数
kinetic fumigation 动式熏气
kinetic gradient 动力梯度
kinetic head 流速水头;动压水头;动(力)水头;速位差;速度头
kinetic heat effect 动热效应

kinetic heating 动力加热
kinetic height 动压水头
kinetic hypothesis 分子运动假说
kinetic index 动力学指数
kinetic instability 微观不稳定性;动力学不稳定性
kinetic isotope effect 动力学同位素效应
kineticist 动力学家
kineticity 动性;动率
kinetic load of wagon 货车动载重
kinetic masking 活动掩蔽;动力学掩蔽
kinetic metamorphism 侵入动力变质;动力变质(作用)
kinetic method 动力学方法
kinetic micelle 动力学胶束
kinetic model 动力学模型
kinetic modeling 动力学模拟
kinetic molecular theory 分子运动理论
kinetic moment 动力矩
kinetic neutralization test 动态中和试验
kinetic oceanography 动力学海洋学
kinetic oscillogram 动态波形图
kinetic parameter 动力学参数;动力参数
kinetic potential 运动势;动势(的);动力势
kinetic power theorem 动态功率定理
kinetic pressure 速压;动压强;动压力;动(力)压
kinetic pressure tensor 动力压张量
kinetic property 动力学性质
kinetic pump 动力泵
kinetic reaction 动反作用
kinetic resolution 动态分辨率
kinetics 动力学
kinetic simulator 动态特征模拟;动态特性仿真器;动态特性模拟(器)
kinetics of aeration 曝气动力学
kinetics of biochemical treatment of wastewater 废水生物处理动力学
kinetics of carbonate rock dissolution 碳酸盐岩溶解动力学
kinetics of coal gasification 煤炭气化动力学
kinetics of combustion 燃烧动力学
kinetics of comminution 破碎动力学;磨碎动力学
kinetics of condensation 凝结动力学
kinetics of filtration 过滤动力学
kinetics of flame 火焰动力学
kinetics of flo(a)tation 漂浮动力学;浮动动力学
kinetics of gas transfer 气体传递动力学
kinetics of high temperature 高温动力学
kinetics of hydration 水化动力学
kinetics of hydrologic(al) reactions 水文反应动力学
kinetics of ion formation 离子形成动力学
kinetics of natural purification 自然净化动力学
kinetics of reaction 反应动力学
kinetics of sedimentation 沉淀动力学
kinetics of sludge digestion 污泥消化动力学
kinetics of sludge production 污泥生长动力学
kinetics of treatment system 处理系统动力学
kinetic solvent effect 动力溶剂效应
kinetic sound screen 动力学声屏
kinetic spectroflurometric method 动力学荧光测定法
kinetic spectrophotometry 动力学分光光度法
kinetic stability 动力稳定(性);动力稳定(度);动力平衡
kinetic study 动力学研究
kinetic system 动态系统
kinetic tank 活动油箱;移动式油箱
kinetic theory 动力学理论;分子运动学说
kinetic theory of gases 气体动力理论
kinetic thermal coefficient 热动力系数
kinetic transfer 动力学变换
kinetic tremor 动时震颤
kinetic viscosity 动力黏滞率;动(力)黏(滞)度
kinetogenesis 动力生成学说;促动作用
kinetogenic 促动的
kinetonucleus 动核
kinetoplast 动质体
kinetostatics 动静法;运动静力学
kinety 动体
king anchor 将军锚
king-and-queen post 桁架的中柱与双柱
king-and-queen post and wind filling 中柱、双柱与风撑
king-and-queen post truss 立字桁架;立式(字)桁架
king-and-queen roof truss 有中柱与双柱的屋顶桁架
king beam 舱口大梁
king bolt 主销;主(螺)栓;中心立轴;中枢销;大螺栓
king bolt and socket 主轴球窝
king bolt ball and socket 主销球窝
king bolt bush 主销衬套
king bolt truss 钢吊杆桁架
king brick 中心流钢砖;中心流钢砖
king bridge 龙门桅上桁架
king bulkhead 主隔(舱)板;主舱壁
kingbury thrust bearing 单圈推力轴承
king closer 砍角砖;去角(纵剖)砖;超半砖的接砖;楔形砖;七分头;四分之三砖
kingfisher 鱼狗
kingfisher colo(u)red glaze 翡翠釉
kingfisher feather glaze 翠毛釉
kingfisher's feather flower 翠花
King floor 耐火地板(其中的一种)
king girder 主梁
king head 中柱的大头;中柱大小头;桁架中柱大头
kingite 白水磷铝石
king journal 主销;枢轴
king lever 主握柄
king light 主灯
king master mason 圬工领班
king oscillator 主振荡器
king piece 桁架中柱;主柱;主梁
king pile 中心(心)桩;主桩;导桩;定位桩
kingpin 主(轴)销;转向销;转向立轴;中心销;中心立轴;中枢销;滚针
kingpin angle 主销倾角
kingpin bearing 主销座;主销轴承;主销倾角止推销轴承;止推销轴承;止推辊轴承;凸轮止推回转轴承
king pin bush 主销衬套
kingpin cap 主销盖
kingpin cover 中心立轴盖
kingpin cover gasket 中心立轴盖衬
kingpin draw key 中心立轴活动键
kingpin felt 中心立轴毡
king pin inclination 主销倾角;主销内倾(角)
kingpin slewing crane 枢转起重机
kingpin steering assembly 主销式转向机构
kingpin stop screw 主销定位螺钉
kingpin thrust bearing 中心立轴止推轴承
king plank 甲板中心纵板;主板条
kingpost 主柱;主梁;主杆;龙门吊杆柱;将军柱;脊瓜柱;桁架中柱;吊杆柱;船舶吊杆中柱
kingpost antenna 主轴式天线
kingpost antenna mount 主轴式天线座架
king-posted beam 单柱托梁
kingpost girder 单立柱下撑式大梁;单柱大梁
kingpost joint 桁架中柱节点
kingpost rafter 单支柱桁架
kingpost roof 单柱屋架;单柱屋顶;单柱桁架;单(立)柱桁架屋顶
kingpost system 主轴式
kingpost timber truss bridge 单柱木桁架桥
kingpost truss 柱撑式三角桁架;单柱桁架;单立柱(倒三角)桁架;中柱式桁架;单立柱桁架
kingpost truss bridge 单柱桁架桥
kingpost wooden truss 单柱木桁架
kingpost wooden truss bridge 单柱木桁架桥
king rod 钢制拉杆;桁架中吊杆;腹杆;主螺栓;螺栓式中柱;(桁架的)中柱
king roller 主轧辊
king's blue 氧化钴;钴蓝(色)
king's blue glass 钴蓝玻璃
Kingsbury bearing 金斯伯里轴承
kingsbury bearing 金斯伯里轴承
Kingsbury spheric(al) thrust bearing 金斯伯里式止推轴承;球面支柱式止推轴承
Kingsbury-type thrust bearing 楔块式推力轴承;金斯伯里式推力轴承;倾斜瓦块式推力轴承
king's chamber 国王的墓室(金字塔中)
king's glazing 镶配屋面采光玻璃;装人行道采光玻璃
king's highway 主要道路
king size 特大号;超过标准长度的
king-size bed 阔型床;特大床
king-sized 特大的;特别的;大型的;超长的
king-size tanker 大型油船(3万~4万吨以上);大型油船
king's master mason 王宫建筑师
kingsmountite 磷铝锰钙石
king's picture 钱币
king's piece 中柱
king spoke 上舵轮柄
king's post 中柱
king's rod 中柱
king's silver 纯银
king's tomb 王陵
Kingston Agreement 金斯敦协议
Kingston valve 金氏通海阀;主吸入阀;通海阀
king's yellow 雌黄
king-table 水平装饰带(阳台栏杆下的,中世纪建筑)
king tower (塔式起重机的)主塔;中支柱;塔式起重机的主塔;承重柱;吊机塔中塔
king truss 有中柱的桁架;主构架;单柱屋架;单柱桁架
king valve 总阀;主阀
king wood 紫色木;紫木
Kingwood stone 金伍石(产于美国西弗吉尼亚州的一种石英石)
kinichilite 巾碲铁石
kink 鱼骨形断裂;折弯;绞结;纽结;扭折;活套;(设计或施工中的)缺陷;纬缩;(钢丝绳的)死扣;绳索扭结;边卿浪(一种薄板缺陷);平面弯曲
kink band 膝折带;扭结带;漆折带;弯结带
kinked 波状的;起伏的;扭曲的
kinked chain 扭结链
kinked demand curve 扭折的需求曲线
kinked equilibrium orbit 扭曲平衡轨道
kinked region (结构方面的)缺陷区
kinker (打结器的)扭结轴;扭结器
kinker shaft 扭结器轴
kink fold 膝褶皱;尖角褶皱;人字形褶皱
kink folding 膝折褶皱
kinking 扭结(钢丝绳);弯折;扭接;缠线
kinking instability 扭曲不稳定性;扭折不稳定性;回线型的不稳定性
kinking limestone 金陵灰岩(早石炭世)
kinking of hose 软管扭接
kink mode 扭曲模
kink of a curve 曲线的弯折
kink of curvature 曲线弯折部分
kink of curve 曲线扭折
kinky 绞结
Kinnison-Colby's flood-flow formula 肯尼森—库毕洪峰流量公式
kino 吉纳树胶;吉纳胶;电影院;胺树胶
kino eucalyptus 树胶桉
kinoform 显像形式;开诺全息照片
kino gum 桉树胶
kinoin 热带非洲紫檀树脂;奇诺树脂
kinoite 水硅铜钙石
kinoshitalite 钡镁脆云母
kinotheodolite 电影经纬仪
kinotoxin 疲劳毒素
kinscord 花式沟纹
kinship family 大家庭
kintal 百公斤;百千克;百公升
kintic temperature 动力学温度
kinzel test piece 焊接弯曲试验片
kinzigite 榴云岩
K-ion exchange capacity 钾离子交换容量
kiosk 小亭;公用电话室;公共电话亭;凉亭;亭子;书报亭;室内配电亭;电话亭
Kiowa shale 基奥瓦页岩
kip bag 长形帆布工具袋
kipopound 千磅
kipp phenomenon 跳跃现象
kipp relay 冲息多谐振荡器
Kipp's apparatus 基普仪;基普发生器
kips per square inch 千磅/平方英寸
kipushite 羟磷锌铜石
kir 含沥青岩;岩沥青
Kirchhoff's boundary condition 基尔霍夫边界条件
Kirchhoff's current law 基尔霍夫电流定律
Kirchhoff's diffraction 基尔霍夫衍射
Kirchhoff's diffraction formula 基尔霍夫衍射公式
Kirchhoff's diffraction integral 基尔霍夫衍射积分
Kirchhoff's diffraction theory 基尔霍夫衍射理论
Kirchhoff's equation 基尔霍夫方程
Kirchhoff's formulation 基尔霍夫表达式

Kirchhoff's hypothesis 基尔霍夫假设
Kirchhoff's integral 基尔霍夫积分
Kirchhoff's law 基尔霍夫定律
Kirchhoff's law of radiation 基尔霍夫辐射定律
Kirchhoff's migration 基尔霍夫偏移
Kirchhoff's principle 基尔霍夫原理
Kirchhoff's second law 基尔霍夫电压定律
Kirchhoff's theory 基尔霍夫理论
Kirchhoff's voltage law 基尔霍夫电压定律
kirk 苏格兰礼拜堂；十字镐
Kirkby moor flags 基尔克比荒野板岩层
Kirklington sandstone 克尔克林顿砂岩
kirksite 锌合金（模具用）
Kirkup table 基尔库卜型风力摇床；风力淘汰盘
Kirk-wood-Brinkely's theory 柯克伍德—布里恩克赖理论
Kirkwood gaps 柯克伍德空隙
kirner 手工冲击钻
kirovite 镁水绿矾
kirschsteinite 钙铁橄榄石
Kirsten propeller 柯尔斯顿螺旋桨
Kirunatype iron deposit 基鲁纳式铁矿床
kirunavaarite 磁铁岩
kirve 陶槽；截槽
Kiselgel A 细孔硅胶
kish 凝壳；漂浮石墨；铁水上的漂浮石墨
kish carbon 碳鳞
kish collector 石墨捕捉器
kish graphite 集结石墨；高炉石墨；漂浮石墨
kish lock 石墨集结
kish slag 含石墨渣；石墨渣
kisk graphite 初生石墨
Kisol 基索尔（一种蛭石混凝土，轻质，保温，耐火）
KIS process 高效应埋弧自动焊接法
kiss 陶瓷器皿粘连；缝口
kiss and ride 迎面接送乘；家庭接送；送别接乘
kiss coater 轻触涂料机；吻涂机
kiss coating 轻压涂装；轻触涂装；吻（合）涂（布）；单面涂胶层；单面给胶涂层法
kiss core 预埋型蕊
kisser 氧化铁皮斑点
kiss gate 压边内烧口
kissing gate 单人（通行）转门
kiss mark 接触痕；烧结痕
kiss plate 轻压印版
kiss printing 凸纹辊筒印花
kiss roll coating 轻触辊涂
kist 支架工具箱；石砌窖；石砌墓；圣物箱
kistvaen 凯尔特（式）人的石墓室
kit 小桶；鱼笼；用具箱；一组仪器；一套工具；整套（工具）；工具箱；工具包；木桶；配套元件；配套零件；成套组件；成套零件；成套工具；成套部件；背囊
Kitahara type water bottle 北原式采水器
kit assembler 成套汇编程序
kit a vehicle 车辆上配备有成套附件
kit bag 工具袋
kitchen 炊事间；厨房
kitchen appliance 厨房用具
kitchen board 案板
kitchen buffet car 餐车（铁路）；便餐车
kitchen building block module 装配式厨房
kitchen building block unit 厨房单元
kitchen burning appliance 厨房灶具；厨房炉
kitchen cabinet 餐具柜；厨房橱柜；厨房壁柜
kitchen car 厨车
kitchen central heating 厨房集中供暖；厨房中心采暖
kitchen chopping board 案板
kitchen closet 厨房壁柜
kitchen court 厨房天井
kitchen cupboard 碗柜
kitchen/dining room 厨房兼餐厅；连餐厅一起的厨房
kitchen equipment 厨房设备；厨房设施
kitchen equipment layout 厨房设备布置
kitchener 灶灶；厨师
kitchenet(te) 室内厨房；小厨房
kitchen exhaust system 厨房排气系统
kitchen fitments 厨房装备品；厨房设备
kitchen fittings 厨房设备
kitchen fixtures 厨房设备
kitchen floor 布置厨房的楼层
kitchen furnace 厨房炉灶
kitchen furniture 厨房家具；厨房器具

kitchen garbage grinder 厨房垃圾磨碎机
kitchen garden 家庭菜园；菜园
kitchen hatchet with handle 带柄厨房斧
kitchen hood 厨房拔气罩；车篷；遮光板；天棚；出檐
kitchen(ing) car 厨房车
kitchen installation 厨房设备
kitchen midden 贝冢
kitchen of works canteen 工厂食堂的厨房
kitchen rack 厨房餐具架
kitchen range 炉灶；厨灶
kitchen refuse 厨房垃圾
kitchen shelf 厨房餐具架
kitchen sink 洗涤盆；洗涤池；厨房洗碗槽；厨房洗涤盆
kitchen stor(e)y 布置厨房的楼层；厨房楼层
kitchen stove 厨灶
kitchen stove range 厨灶
kitchen system 厨房设备
kitchen table 案桌
kitchen tackle （船上的）厨房用品
kitchen top 案桌
kitchen tower 厨房水塔
kitchen trailer 厨房拖车
kitchen truck 炊事车
kitchen utensil 炊具；厨（房用）具
kitchen ventilation 厨房通风
kitchenware 厨（房用）具；炊具
kitchen waste 厨房废水；厨房垃圾
kitchen waste disposal pipe 厨房下水管
kitchen waste disposal shaft 厨房下水道窨井；厨房垃圾管道
kitchen wastewater 厨房废水
kitchen yard 厨房院子；厨园；菜园
kite 空头支票；空头票据；通融票据；风筝；最高帆；轻帆
kite-air ship 系留气艇
kite balloon 系留气球
kite bill 空头票据
kite camera 俯瞰图照相机；鸟瞰图照相机
kite drag 帆布海锚
kite float 鸢式浮子
kite-flying 发行通融票据；开空头支票
kit element 过滤元件
kite-mark 风筝标志
kite-meteorograph 风筝气象记录器
kite observation 风筝观测
kite reel 风筝线轴
kite winder 楼梯转向斜踏步；转向斜踏步；风筝形踏步【建】；风筝形扇步；鸢式转角踏步；楼梯扇步
kiting 移挪补空；冒空；提现挪用
kiting cheque 通融支票
kiting stocks 空头股票
kitkaite 硒碲镍矿
kit locker 用具储[贮]藏室；水手（用）箱
kitoon 系留气球；无人控制观测气球
kit processor card 成套处理器插件板；成套处理机插件
kit shortage notice 成套器材短缺通知
kit software 成套软件
kittatinnyite 水硅钙锰石
Kittel centrifugal tray 离心式克特尔塔板
Kittel polygonal tray 多角形克特尔塔板
Kittel standard tray 标准克特尔塔板
kittel tray 斜孔网状克特尔塔板
kitten-ball diamond 垒球场
kitting cheque 空头支票
kit tools 成套工具
kit utility 成套用具；成套应用设备；成套实用程序
Kiulungshan series 九龙山统【典】
kiva 基瓦（美国印第安人开会等用的大圆屋）
Kiwi nuclear reactor 基威反应堆
Kizilye tectonic segment 克什尔构造段
Kjeldahl flask 基耶达烧瓶；基耳达测氮瓶；长颈烧瓶
Kjeldahl method for nitrogen determination 基耶达测氮法
Kjeldahl nitrogen determination 基耶达定氮测定
Kjeldahl's method 基耶达定氮法
Kjellmann-Franki cardboard drain method 基尔漫—弗兰开纸板排水法
Kjellmann-Franki machine 排水沟机械；基尔漫—弗兰开（纸板）排水机
kjerulfine 氟磷镁石
kladnoite 酞酰亚胺石；铵基苯石

klaprothite 脆硫铜铋矿
klatsch 谈话会
Klauskran mobile 克劳斯—克兰流动起重车
klaw notch 开槽榫接
klaxon 电喇叭
klaxon device 喇叭；电报警器；电喇叭装置
klebelsbergite 基锑矾
kleberite 水钛铁矿
kleemanite 水羟磷铝锌石
Kleene closure 克林闭包
Kleene's star operator 克林星号算子
Kleene's theorem on fixpoint 克林不动点定理
Kleene's theorem on regular set 克林正则集定理
Kleine block 克赖恩纳空心耐火黏土砖
Kleine floor（先在空心砖缝间置钢筋然后用混凝土浇铸成整体的）楼板
Kleine hollow-brick floor 克赖恩纳空心砖地板
Kleine hollow floor 克赖恩纳空心地板
Kleine hollow floor(clay) brick 克赖恩纳空心楼板（用）砖
Klein-Goldberger model 克莱因—戈德伯格模型
Klein-Gordon equation 克莱因—戈登方程
kleinite 氯氮汞矿
Kleinlogel metal(lic) aggregate 克莱恩洛戈尔金属骨料
Kleinman symmetry condition 克莱曼对称条件
Klein-Nishina formula 克莱因—仁科公式
Kleinpflaster 嵌花式砌块；嵌花式块石路面；嵌花式砌石；小方石块；嵌花式石块
Klein-Rydberg construction 克莱因—里德伯尔结构
Klein-Rydberg method 克莱因—里德伯方法
Klein's hypothesis 克莱因假说
Klein's reagent 克莱因试剂
kleit 高岭土
klementite 镁鳞绿泥石
Klemm glue 克莱姆胶
kleptoscope 潜望镜
Klett-Summerson colo(u)rimeter 克莱特—萨默森比色计
kliegshine 溢光灯（光）；白炽溢光灯；弧光灯；克利灯；强弧（光）光灯
Kliene floor 克利纳楼板
Klingenberg clay 克林根别尔格黏土
Klingenberg permeability（与流体的压力和性质无关的）克林根别尔格岩石真实渗透率
Klingerite 一种橡胶石棉垫料
Kling-type ladle 梨形铁水罐；克林式盛铁桶
klinker brick 过烧砖；缸砖
klinkstone 响岩
klinokinesis 偏动态
klinostat 回转器
klint 硬岩礁；高原陡线；陡岸
klinta klipt 的复数形式
klintite 硬礁岩
klip 悬崖
klippe 孤残层；构造外露层；飞来峰
klipt 陡崖；[复]klintar
klirr-attenuation 失真衰减量
klirr factor 波形失真因数；波形失真系数；非直线性畸变系数
Klischograph 电子刻版机
klockmannite 硒铜蓝；六方硒铜矿
klong 水道
kloof 峡谷；深峡谷；山坡陡谷；冲沟
kloof wind 克洛夫风
Klopfner ga(u)ge 克洛普弗真空计
Klotz equation 克劳兹方程
kluf 山口；峡谷；沟
klydonogram 脉冲电压记录图
klydonograph 浪涌电压记录器；脉冲电压记录器；电涌记录器
klystron 调速管；速度调制管
klystron generator 速调管振荡器
klystron mount 速调管座
klystron oscillator 速调管振荡器
klystron repeater 速度调制管放大器
klystron(tube) 速调管
kmaite 绿云母
K-matrix K 矩阵
K meson 重介子
KMH-meter 千瓦小时电度表
knack 技巧；门路
knacker 旧屋收买人；废船
knag 木节；木瘤

knag gasket 塞垫(一种推拉窗垫套)
knaggy 多节的
knaggy wood 树瘤材；多节木(材)；多节材
knap 丘顶；小山
knapen system 干燥墙体的散湿孔
Knapp bottom pressure ga(u)ge 纳普底压计
knapped flint 敲碎的燧石
knapper 凿石工人；击碎器；破碎器；碎石工；碎石锤；碎石机
knapping hammer 凿石锤；碎石锤
knapping machine 凿石机；破碎机；碎石机
Knapp's forceps 转轴镊
knapsack 背囊；背罐；背负式喷雾器体；背包
knapsack abutments 反重桥台
knapsack air-blast sprayer with engine drive 背负鼓风式机动弥雾器
knapsack algorithm 渐缩算法
knapsack compressed air type 背负式气动喷雾器
knapsack compressed air type sprayer 背负式压气喷雾机；背负式气力喷雾机
knapsack duster 背负式喷粉机；背负式喷雾器
knapsack lever type sprayer 背负式杠杆喷雾机
knapsack mist-duster 背包式喷粉机
knapsack pest control machine 背负式除虫器
knapsack power duster 背负式动力喷雾机
knapsack power mistduster 背负式动力弥雾喷粉机
knapsack problem 背包问题
knapsack pump 背负式喷雾器
knapsack seeder 背负式稀播机；背负式播种机
knapsack sprayer 背负式喷雾器；背包式喷雾器
knapsack sprayer with piston pump 背负式活塞泵喷雾器
knapsack sprayer with pressurized cylinder 背负式压力筒喷雾器
knapsack station 轻便台；背囊式电台
knapsack type sprayer 背负式喷雾器
knar 木结节；木节
K-Na ratio 钾钠比
knar clay 木节黏土
knarled 多节的
knaur 木节
kncading machine 捏练机
knead 搓
kneadable 可搓捏的
kneaded eraser 橡胶擦
kneaded gravel 泥砂砾；泥流砾；泥流搬运砾石
kneaded mass 捏合物质
kneaded rubber 橡胶擦
kneaded structure 捏和结构；捏合结构
kneader 叶片混砂机；捏练机；捏和机；捏合机；黏土拌和器；黏土拌和机；混砂机、和灰工；碎纸机
kneader arm 碎纸机搅拌臂
kneader bar 碎纸机刀片
kneader blade 碎纸机刀片
kneader extruder batch mixer 捏合挤出分批混合机
kneader-mixer 叶片混砂机
kneader type machine 搓(揉)式混砂机
kneader type mixer 搅拌式混砂机；捏合型搅拌机；搓(揉)式混砂机
kneading 练泥；捏制；捏合；揉捏法；揉揉；搓揉法
kneading action 搓揉作用
kneading action of traffic 交通对路面的揉搓作用
kneading and mixing machinery 捏炼混合机
kneading and twisting method 挤拧法
kneading compaction 羊足碾压实；搓揉压实；用羊足碾压实塑性土壤
kneading compactor 揉压机；揉和机
kneading energy 搓揉能
kneading extruder 捏合挤压机
kneading machine 捏土机；捏合机；揉搓式混砂机；碎纸机；搓(揉)式混砂机
kneading method 压法；搓条法
kneading mill 搅拌机；捏合机；混砂机
kneading model 捏塑
kneading pulper 纸浆捏合机
kneading trough 黏土揉混槽；揉合槽；碎浆机槽
knee 膝状物；膝部；肋材；合角铁；曲条；弯曲处；车柱；扶手弯头
knee action 膝形杆动作
knee action suspension 独立悬挂
knee action wheel 肘动轮
knee-and-column 升降台
knee-and-column milling machine 升降台铣床
knee batten 角撑板(条)

knee bend 肘形弯管；弯头；弯管接头；扶手弯头
knee bend 90 degrees 90度弯头
kneeboard 矮墙(赛马场周围)；护膝板；矮墙
knee brace 斜撑；膝形拉条；隅撑；角撑；撑条
knee-braced bracket 斜弯托座；膝形悬臂托架
knee-brace roof 斜撑屋架
knee-brace roof truss 隅撑屋架；有隅撑的屋架
knee bracing 角撑；隅撑
knee bracket 膝形托架
knee bracket plate 弯角连接板；连接板
knee brake 曲柄制动器；弯带闸
knee butt 膝状铰链
knee cap 护膝
knee clamp 膝形夹头
knee colter 膝形犁刀
knee compression 膝状曲线弯曲压缩
kneed 揉；弯的，关节状的；瘤多的
knee-deep 深到膝的
knee desk 写字台
kneed hinge 成直角的铰链(用于箱盖和可启闭的天窗上)
knee fold 膝状褶皱【地】
knee frame 拐弯式构架
knee frequency 拐点频率
knee girder 肘形梁；肘状梁
knee-girder bascule 肘形梁式竖旋桥
knee grip 曲柄支托
knee guard 护膝
knee head loss 弯头水头损失；弯管水头损失
knee hole 容膝孔；弯曲孔
kneehole desk 容膝空当书桌
knee-hole table 两边带屉的写字桌
knee iron 斜角(补强)铁；隅铁；角铁；角补强铁
knee joint 肘接；肘环接接；弯头接合；臂接
knee-jointed 肘连接的
kneeler 斜交石；山墙角石
knee level 膝状曲线弯曲电平
knee lever 曲杆；弯头手柄；膝杆；直角形杠杆
kneeling board 跪板；斜的木垫(教堂跪拜用)
kneeling figure 祈祷雕像；下跪人像
knee loss 弯头(水头)损失
knee of a curve 曲线折点
knee of characteristic 特性曲线最大弯曲处；特性曲线拐点
knee of curve 曲线拐点
knee of head 船头悬伸木
knee operated faucet 膝开式洗脸盆
kneepan-shaped 小盘状的
knee piece 木肘材；曲形部件；曲橼；曲块；角橼；曲片；曲块
knee pipe 直角弯管；曲管；弯头；弯管
knee plate 肘板；角垫板；角撑板
knee point 拐点；挂竞点；曲线弯曲点
knee rafter 曲橼；曲块；角橼
knee roof 折线屋面；复折屋顶；复斜屋顶
knee sensitivity 拐点灵敏度
knee shaped twin 啼状双晶
kneestone 斜交石；山墙角石
knee structure 拐弯式构造
knee strut 抗压支柱
knee table 三角桌
knee timber 拐弯木材；木肘材；曲肘形木材；多节材
knee tool 弯头刀架
knee tube 直角弯管
knee twin 膝状双晶
knee type 升降台型
knee-type structure 膝型构造
knee wall 支撑墙；顶层间壁墙
knickknack 小摆设；小装饰品；小家具
knickpoint 纵坡降点；裂点；河床纵坡陡降点；河床陡坎；侵蚀交叉点；坡折点；陡坎
knick zone 扭结带
knife 小刀
knife applied 刮涂；手工刮涂
knife arbor 圆盘刀片的心轴
knife bar 纵切刀；切割器刀杆；切割器；涂胶刀
knife blade 刀片
knife-blade cover 刀铲式覆土器
knife-blade switch 刀形开关
knife block 刀片滑块
knife board 磨刀架；刀板
knife bracket 压刀板
knife check 刀页裂隙；单板背面裂隙；裁纸刀颊板
knife clip 切割器压刀板

knife coater 刮刀涂层机；刮刀涂布；刮刀(式)涂胶机
knife coating 刮涂(法)；刮刀涂布法；刀片涂布
knife colter 直犁刀
knife consistency (油灰的)刮刀稠度
knife contact 刀形触头；刀形触点
knife coverer 刀式覆土器；覆土板
knife-curl test 漆膜刮卷试验
knife cutter 直犁刀；(制鞋用的)切鞋机；锋刀刮蜡器
knife-cut veneer 刀切成层板；刨刀单板；刀切薄片
knife cylinder 切碎滚筒；刀齿滚筒
knife discharge 刮刀卸泥量
knife disk 圆盘刀
knife-dog 楔形钻杆夹持器
knife drive crank 割刀传动曲柄
knife drum 切碎滚筒；刀齿滚筒
knife edge 锐缘；刃形支承边；刀口；刃脚；刃锋；刀片式刮路机；刀口；刃缘；刀刃
knife-edge bearing 刀形支承；刃形支承；刀口支承；刀形支承
knife-edge contact 刀口触点；闸刀式接点；刀形触点；闸刀式接触；闸刀式开关
knife-edge crest 刃脊
knife-edge cutter 尖刃模
knife-edge cutting machine 刀刃切割机
knife-edged 极锋利
knife edge die 切断模
knife-edged weir 锐缘堰
knife-edge follower 尖顶从动杆；刀形随动件
knife-edge liner 闸刀衬垫
knife-edge load(ing) 刀口荷载；集中线荷载；线(性)荷载
knife-edge magnetometer 刀口式磁力仪
knife-edge method 刀口检验法；刀边法
knife-edge pivot 刀形枢轴；刃形枢轴；刀口支轴
knife-edge pointer 刃形指针
knife-edge refraction 刃形折射
knife-edge relay 刃式继电器；刀形继电器；刀口继电器
knife-edge relay contact 刃形继电器触点
knife-edge scanning 刀形扫描
knife-edge support 刃型支承；刃形支承；刀口支承
knife-edge surface visco(si)meter 刀刃面黏度计
knife-edge suspension 刀口支悬；刀口悬架；刀口吊架
knife-edge switch 闸刀开关；刀形开关
knife-edge test 刀刃试验；刀口检验；刀锋检验
knife edge tread 刃脚踏面
knife-edge valve 刀形阀
knife-edge weir 锐顶堰；刀刃堰；薄壁堰
knife-equipped 装有割刀的
knife-equipped plunger 装有切草刀的柱塞
knife extension blade 伸缩划料刀
knife face 刀面
knife file 刀座；刀形锉；有刀锋的锉刀
knife flywheel 刀盘；刀轮(装金刚石的切裁玻璃用)
knife folding machine 刀式折页机
knife gate 楔形浇口；压边浇口；铡刀式闸门
knife gin saw file 三角尖锉
knife grade 刮刀稠度
knife grinder 磨刀装置；磨刀石；磨刀砂轮；磨刀机；磨刀工；刀刃磨床
knife-grinding attachment 磨刀附加装置
knife grinding machine 磨刀机
knife guide 割刀导向板；切碎装置导向架
knife handle 刀柄
knife head guide 刀头导向板
knife height 刀光高度；刀高
knife hog 枝叶切削机
knife holder 刀夹；刀杆
knife holder ship 水底整平船
knife lanyard 水手刀系带
knife lever 割刀摇臂
knife-line corrosion 刀蚀
knife machine 磨刀器；磨刀机
knife mark 刮刀条花；条痕；刀线；刀痕
knife mill 切碎机
knife of switch 开关闸刀
knife-on-blanket coater 刮刀垫带涂布机
knife-over-roll coater 辊衬刮刀涂布机
knife plane 刨
knife plate 切割器定刀片
knife pleat 顺风裥
knife pressure 剪刀剪切力
knifer 搅土器；剖土机

knife reaper file 刀形锉;单向齿锉
knife register 切割器对心
knife-repair anvil 刀片修理铁砧
knife rest 拒马;(餐桌上的)刀架
knife reversing switch 倒顺刀开关
knife-roller gin 刀辊式轧花机
knife saw 刀锯
knife-scratch test 刀片划痕试验;刀割痕(硬度)试验
knife seal 刀口封接;薄边封接
knife section 动刀片
knife-shaped coin 刀币
knife-shaped lightning arrester 刃形避雷器
knife-shaped needle 刃形指针
knife-shaped pointer 刃形指针
knife shaping tool 刀头刨头
knife sharpener 磨刀机
knife slicker 磨刀器
knife stone 磨刀石
knife strap 割刀刀杆
knife stroke 割刀行程
knife structure electrode 刀形电极
knife support 切割器支撑滑脚
knife switch 闸控开关;闸刀(开关);刀(形)开关
knife test 刀挠检验法;刀割试验法
knife test for plywood 胶合板刀齿试验
knife tool 修边刀具;刀具
knife-tooth harrow 齿刀式耙路机
knife turning tool 刀头车刀
knife type clay crusher 刀式黏土破碎机
knife type rotary harrow 刀式转地
knife type scraper 刮刀式清管器;刀形刮管器
knife type shredder 甩刀式茎秆切碎机
knife varnish 刮涂用高黏度清漆
knife with retractable blades 有伸缩刀刃的刀子
knifing 装木地板作业;(切深孔型中的)切深;切割
knifing filler 细填料;刮涂腻子;适于用灰刀填塞的填塞物(有别于用刷的填缝物)
knifing socket 插刀式插头座
knight 系缆柱
knight engine 套阀发动机
knish 搅棍
knit 接合;密接
knit article 针织品
knit goods 针织物;针织品
knit goods mill 针织厂
knit line 接合线
knit-stitch machine 缝编机
knitted carpet 针织地毯
knitted fabric 针织物;编成的织物
knitted geotextile 编织型土工织物
knitted glove 针织手套
knitted gunning hose 编织喷砂浆管
knitted rubber hose 编织胶管
knitted spraying hose 编织喷管
knitting 涂层翻验
knitting action 交织作用(路面上层混合料);(路面上层混合料的)网结作用
knitting frame 针织机架
knitting layer (新旧混凝土之间的)接合层;(新旧混凝土之间的)水泥浆层
knitting-machine 编织机
knitting mill 针织厂
knitting needle (抽筒钢丝绳的)联结工具
knitting wool 毛线
knittle 小麻绳;缩帆索
knitwear 针织品
knives 练泥机刀片
knob 旋钮;旋扭调节器;执手;开关旋钮;球形雕饰;球形柄;球形把手;撅钮键;桅杆或旗杆上的顶球;雕球饰;把手;按钮
knob-and-basin topography 丘状盆地;凹凹地形;凸凹地形
knob-and-kettle topography 凹凸地形;丘洼地形;凸凹地形
knob-and-tube wiring (装在瓷柱上或管内的)暗线;瓷珠瓷管布线;瓷柱瓷管布线;穿墙布线(法)
knobbing 粗琢石;粗琢块石;球形把手;雕球饰
knobble 削节;小节;削砍石块;节瘤
knobbled iron 熟铁
knobbling 制成把手形;压平突出物;削节;球形把手;雕球饰;制铁坯;破碎块石;石块粗加工
knobbling fire 搅铁炉
knobbling roll 开坯轧辊
knob bolt 球拗门栓;球拗门闩

knob boring machine 冲击钻机
knobboss 球形捏手
knob controller 按钮式操纵器
knob cylinder lock 执手弹子锁
knob dial 鼓形刻度盘
knob door fitting 门把配件
knob door furniture 门把配件
knob door hardware 门把五金配件
knob door lock 圆把手门锁
knob down 按下按钮
knob fittings 圆把手配件
knob hardware 门执手五金配件
knob insulator 鼓形绝缘子;瓷珠;瓷珠;绝缘旋钮
knob latch 球把插锁;球拗碰锁
knob lock 球把门锁;球拗门锁
knob of key 钥匙
knob-operated control 旋钮控制
knob rose 球把垫圈
knob shank 球形把手柄
knob switch 旋钮开关
knob top (把手的)球形端;(把手的)球形头
knob up 拔出按钮
knob wiring 瓷柱布线;瓷珠布线
knock 撞击;敲击;敲打;敲成;打击;爆震
knockability 出砂性
knock agent 抗爆剂
knock bit off 震简冲击杆与钻头连接销
knock-boring machine 冲击钻探机;冲击式钻机
knock characteristic 抗震性;抗爆特性
knock-compound 抗震剂
knock down 卸下;解体;拍板成交;拆开;撞倒;减价;击落数;拆除
knock down chain 可拆卸链
knock down export 散件输出;散件出口;部件出口;半成品出口
knock down fitting 拆装五金连接件
knock down parts 组装零件
knock down test 可锻性试验;顶锻试验
knock down the oil 从水中分出石油
knock down tower 拆卸(式)栈标
knocked down 降低价格;建筑构件或机械部位错位;倾倒;船在大风中横倾状态;组装的房屋构件;(易于拆卸的)预制构件
knocked down building components 桥下建筑部件
knocked down frame 用预制件装配的门框
knocked-down in carload 以拆卸状态装车
knocked down packing 拆卸包装
knocked down price 降低售价
knocked down shipment 分运费
knocked down shipping 拆开运输;分装运输
knocker 信号锤;引爆雷管;岩块;巨石;门环;敲击器;爆震燃料;爆震剂;雷管;起爆钮;信号铃锤
knocker-out 反向撞角
knock-free fuel 无爆震燃料
knock-free operation 无爆震操作
knock free power 无爆震功率
knock-free region 无爆震区
knock hole 定位销孔;顶销孔
knock-in 敲入
knock indication instrument 爆震指示器;爆震器
knock indicator 爆震器;爆震测量(指示)仪
knock inducer 爆震诱导物
knocking 撞击水锤;击碎;碰磕掉釉;捶击法;爆震燃烧;爆震
knocking behaviour 爆震燃烧
knocking-bucker 采石器
knocking combustion 爆燃
knocking explosion 爆炸
knocking gear 撞击机构
knocking of the valve 阀门碰撞
knocking-out 碰撞位移;打ához出;拆除
knocking-out grid 振动落料栅架
knocking over 脱圈
knocking-over action 脱圈动作
knocking-over bar 脱圈板
knocking-over bit 脱圈片
knocking-over comb 脱圈板
knocking-over row 脱圈横列
knocking-over segment 脱圈镰条
knocking-over sinker 脱圈沉降片
knocking-over verge 脱圈栅状齿口
knocking-over wheel 脱圈轮
knocking test 敲击试验;爆震试验;爆击试验

knocking up 重新搅拌(法);重新拌和;拌和
knock intensity 爆震强度
knock intensity indicator 爆震强度指示仪;爆震强度指示器;爆震强度
knock-knee 内弧;叉形腿
knock-limited density index 爆震限制的密度指数
knockmeter 爆震仪;爆燃仪;爆震计
knock-off 歇工;下班;中止工作;减低(价钱);敲去;敲落;收工;自动停机;自动断开;碰撞自停装置;停工时间
knock-off bit 可卸钻头
knock off block 泵管柱悬挂装置
knock-off cam 停机凸轮
knock-off core 易割冒口芯片
knock-off device 撞击自停装置
knock-off head 易割冒口
knock-off joint 脱钩装置;可拆卸接头;杆接头;停动连接
knock-off post 脱钩器杆
knock-off shaft 停车轴
knock-off shower 纸幅切边水针
knock-off the crhistmas tree 采油树破裂(由于高压)
knock-on collision 迎面碰撞;直接碰撞
knock-on displacement 碰撞位移
knockout 压坑;敲落;抛出;出坯;拆模(板);压床打料棒;落砂;击出;抛掷器;脱模;塑物脱膜;顶出器;倒出铸件和壳;打泥芯;打出;出砂;拆卸器;拆卸工具;凝聚器;分液器
knockout actuated stripper 打料机构带动的卸料板
knockout attachment 弹射器
knockout bar 钎子
knockout barrel 滚筒冷却落砂机
knockout box 气体分离器;气体分离箱
knockout bracket 打料横杆的撞击架
knockout chamber 降尘室
knockout coil 水冷分凝器;分凝盘管;分离盘管
knockout core box 脱落式芯盒
knockout cylinder 顶件油缸
knockout drum 分凝罐;分离鼓
knockout grating 落砂栅;落砂架
knockout grid 落砂栅
knockout latch 脱模销紧销
knockout machine 落砂机;取出机;脱模机
knockout measurement 击出测量法
knockout mechanism 顶出机构;出坯装置;出坯机构
knockout pad 推件盘
knockout pin 顶杆;顶出杆
knockout plate 脱模板;推板;甩板;顶击板;打料板
knockout press 脱模力;甩力
knock out property 出砂性
knockout resonance 击出共振
knockout stroke 出坯冲程
knockout tower 分离塔
knock-out zone 分离区
knock over 连接管道;固定管道(其中的一种方法)
knock pin 顶销;止顶;定位销
knock producer 爆震(发生)器
knock property 爆震性
knock rating 震率;爆震率;爆击率;防爆率
knock reducer 减振器;抗震剂
knock-sedative 抗震的
knock stick 螺杆调节工具(钻紧时)
knock suppressor 抑爆剂;抗爆剂
knock tendency 爆震性
knock-test engine 测爆机;爆震试验机
knock tester 爆震试验机
knock test(ing) 爆震(性)试验;抗(爆)震性试验
knock together 急促做成
knock value 抗震值
knock wave 冲击波;敲击波
knoll 圆丘;海丘;堤岸;土墩;小山
knolllite 叶沸石
knoll reef 圆丘礁
knoll spring 圆丘泉;丘泉
Knoop hardness 克氏硬度
Knoop's hardness 努普(显微)硬度
Knoop's hardness number 努普硬度值
Knoop's hardness tester 努普硬度计
Knoop's indentation 努普压痕
Knoop's(indentation) microhardness test 努普(刻痕)微硬度试验
Knoop's indentation test 努普压痕试验

Knoop's indenter 努普硬度仪;努普压头(硬度试验用)
Knoop's method 努普法
Knoop's microhardness test 努普微硬度试验
Knoop's number 努普硬度值
Knoop's scale 努普(硬度)标
Knoop's theory 努普学说
knop 圆形把手;拉手【建】;突出雕饰;雕球饰
knopite 铈钙钛矿
knop yarn 点节线
knorringite 镁铬榴石
knosp 雕球饰
knot 结点;节疤;节(航海单位);纽结;纽点;扭结;木节;每小时航行里;海里/小时;树瘤纹;绳结;打结
knot area ratio 节疤面积比率
knot bed 花结花坛
knot breaker 碎呢开松机;打回丝机
knot brush 分节刷;画笔;涂料刷;球形刷
knot catcher 除节机
knot cluster 节群;节丛;复节
knot diagram 纽结图
knot equivalence 纽结等价
knot facility 捆轧装置;捆轧设备
knot garden 错综复杂的花园设计
knot hole 节孔;木节孔
knothol mixer 隔膜混合器
knot installation 捆轧设备
knotless 无结的
knot-mode digitizing 非连续数字化
knot nozzle propeller 导流管式螺旋桨
knot of pipeline 管线节点
knot plant 捆轧装置;捆轧设备;捆轧厂
knot polynomial 纽结多项式
knot problem 纽结问题
knot projection 纽结射影
knot sealer 木节封闭剂;封闭剂
knots of tectonic network 构造网结
knot sphere 纽结曲面
knots sealing 木节封闭
knot strength (钢丝绳芯的)打结强度
knot table 纽结表
knotted 瘤状;打结的
knotted bar iron 节钢
knotted columns 缠结柱
knotted list 打结的列表
knotted over 遮盖木节
knotted pile carpet 手结栽绒地毯;栽绒地毯
knotted pillar 雕刻成绳结饰的柱
knotted rug 手结栽绒地毯
knotted schist 瘤斑片岩
knotted sheepshank 牢固缩绳结
knotter 结筛;打结器;除节机
knotter assembly 打结器总成
knotter blower 打结器清理风扇
knotter cam 打结器凸轮
knotter cam wheel 打结器凸轮
knotter cleaner 打结器清理装置
knotter disk 打结器压绳盘
knotter drive gear 打结器主动齿轮
knotter hook pinion 打结器卡嘴小齿轮
knotter jaw 打结嘴
knotter jaw spring 打结器卡嘴弹簧
knotter shaft 打结器轴
knotter spring 打结器弹簧
knotter stripper 打结器紧绳钢杆
knot theory 纽结理论
knotting 结绳;木节刮油(木节眼做刮油处理,以便油漆);涂木节疤;塞木节孔;打绳结;封节剂;在木板上打腻子
knotting bill 打结嘴
knotting cycle 打结工作循环
knotting machine 结网机;打绞机
knotting mechanism 打结器
knotting needle 结绳针
knotting strength (钢丝绳绳芯的)打结强度
knotting varnish 木节封闭清漆
knot torus 扭结环面
knott's clay 含碳黏土(用于制砖可省燃料)
knotty 难解;多节的
knotty and knaggy wood 多节木(材)
knotty clay 板结黏土
knotty dolomite 结瘤状白云石;结瘤状白云岩
knotty glass 有疙瘩的玻璃

knot-tying device 打结器
knot tying machine 打结机
knot type 扭结型
knotty pine 多节松木
knotty problem 难题
knotty wood 多节木
knot wood 有节木料
knot work 编织细工;交错线饰
know-how 专有技术;专门知识;专门技术;专门技能;专长;经验;技术指导费;技术秘密;技巧;实际知识;保密情报
know-how agreement 专有技术协议;技术转让协议;技术交流协议
know-how commission 技术转让费
know-how information 专有技术情报;技术信息;技术秘密情报
know-how license 技术许可证
know-how market 技术知识市场
knowledge acquisition 知识获取
knowledge base 知识库
knowledge base consultative system 知识库咨询系统
knowledge base management system 知识库管理系统
knowledge base system 知识库系统
knowledge discovery 知识发现
knowledge engineer 知识工程师
knowledge factory 教育机构
knowledge industry 知识工业
knowledge information processing system 知识信息处理系统
knowledge intensive 知识密集型
knowledge obtain system 知识获得系统
knowledge of result 结果知识
knowledge process language 知识处理语言
knowledge professional 知识专家
knowledge representation 知识表示
Knowles' dobby motion 诺尔斯多臂装置
Knowle's free flow 诺尔斯畅流(一种管子接头商品名)
Knowles' head 诺尔斯多臂装置
Knowles' positive dobby 诺尔斯多臂装置
Knowles's cell 诺尔斯电解池
known 已知数
known address 已知地址
known angle 已知角
known base line 已知基线
known component 已知组分
known condition 已知条件
known direction 已知方向
known economic reserves in places 已探明藏量
known error 已知误差
known error condition 已知错误条件
known function 已知函数
known future value and present value-compute unknown 由已知终值和现值求算未知
known liability 已知负债
known loss 已知(的)损失
known number 已知数
known point 已知点
known quantity 已知数;已知量
known reserves 已知储量;探明储量
known sample 已知样品;已知试样
known segment 已知段
known segment table 已知段表
known side 已知边
known solution 已知溶液
known state 已知状态
known substance 已知物(质)
known universe 已知整体
known value 已知值
know the rope 熟悉船上帆缆业务
know-why 技术原理
Knox cracking 诺克斯裂化法(气相热裂化法)
Knox's and Oxborne furnace 诺克斯—奥克斯本焙烧炉
Knox's process 诺克斯气相裂化过程
Knox's true vapour phase process 诺克斯气相裂化过程
Knox's unit 诺克斯装置
knuckle 转向节;肘节;过渡圆角;关节;钩爪;钩舌;炉颈;炉节;坡度突变;膨隆;万向接头;叉节
knuckle-and-socket joint 链式球形连接;活节连接;铰链节支接头

knuckle arm 羊角臂;转向节臂;方向臂;关节(杆)臂
knuckle assembly 转向节总成
knuckle bearing 关节轴承;铰(式支)座;球形支座
knuckle bearing cone 关节轴承锥
knuckle bend 小圆弯
knuckle bushing 转向节衬套
knuckle buster 螺母扳手;板钳;扳手
knuckle centre 万向接头十字轴
knuckle cone 肘形锥
knuckle drive 铰节传动
knuckle drive flange 关节传动凸缘
knuckle end 转向节臂接头
knuckle flange felt spring 关节凸缘毡弹簧
knuckle gear(ing) 圆齿齿轮(装置);速径齿轮装置
knuckle guard 铰链罩
knuckle guide 铰链式导向装置
knuckle joint 肘形接合;肘形铰接;肘形关节;肘节形接头;折向节点;关节接合;铰链接合;活节接合;胴接头;叉形接头;叉形铰链接合;肘形接头;肘形连接;肘形接缝(双折屋面中两个斜面的交线);肘接(头);铰接
knuckle jointing 肘环套接
knuckle joint press 肘接式压力机;肘杆式压力机
knuckle length 铰轴长度;转向节长度
knuckle-lever coining press 肘杆式压下压力机
knuckle-lever drive 肘杆传动
knuckle-lever press 肘杆压力机
knuckle line 折角线
knuckle man 挂钩工
knuckle mast 可折桅;铰式桅杆
knuckle opener 钩爪钥
knuckle-pin 转向节销;肘销;关节销;钩销;钩舌销;万向接头插销;副连杆销
knuckle-pin angle 关节销内倾角
knuckle-pin center 转向主销轴线与路面的交叉点
knuckle pin center line 转向销轴线
knuckle-pin cover 关节销盖
knuckle-pin nut 关节销螺母
knuckle pivot 转向节销;转向节;枢轴
knuckle pivot center[centre] 转向节旋转中心线
knuckle post 转向节柱
knuckle press 肘杆压机;肘板压机
knuckle radius 转角半径;过渡半径
knuckle rail 肘形轨
knuckle rod 活节杆
knuckle screw thread 圆螺纹
knuckle soldered joint 管子工接头(位于两根铅管成直角处);肘形焊接
knuckle spindle 转向节轴;转向节销;前轮轴
knuckle stern 折角船尾
knuckle strake 船尾折角列板
knuckle support 转向节支架;关节支架
knuckle thrower (机车的)钩舌锁铁
knuckle thrust bearing 转向节止推轴承;关节推力轴承
knuckle tooth 圆弧顶面齿;圆齿
knuckle trunnion 转向节枢销;关节耳轴
knuckling 拉坯;焊接收缩;焊接疤痕;突聚;肘形连接;链环扣连接;(电缆的)扭结
knuckling line slipway 折线变坡滑道【船】
Knudsen cell 努森室
Knudsen cosine law 努森余弦定律
Knudsen flow 自由分子流
Knudsen formula 努森公式
Knudsen ga(u)ge 努森真空计
Knudsen layer 努森层
Knudsen number 努森数
Knudsen sampler 努森采样器
Knudsen's equation 努森方程
Knudsen's table 努森表
Knudsen vacuum ga(u)ge 努森真空计
knulling 周边滚花的纹饰;微凸出的断开圆线脚
knur 木材节疤
knurl 隆起饰;滚花
knurl dies 滚花模
knurled 滚花的
knurled acoustic(al) tile 滚花天花板
knurled deflector 凸形导向器
knurled disc 旋盘
knurled grip center punch 滚花中心冲头
knurled head 滚花头
knurled knob 滚花旋盘;滚花旋钮;滚花捏手
knurled nut 滚花螺母;隆起螺帽
knurled piston 压花修复活塞;滚花活塞

knurled ring 旋环
knurled roll 滚花轧辊(筒);网纹轧辊
knurled screw 滚花螺杆;滚花螺钉
knurled thin nut 滚花扁螺母
knurled thumb screw 滚花螺杆
knurled wheel 滚花轮
knurling 压花纹;滚压周边花;滚花
knurling cutter 滚花刀
knurling machine 压花机
knurling tool 压花机;压花滚轮;压花工具;滚花工具;滚花刀具
knurlizer 压花机
knurlizing machine 滚花机;活塞修复机
knurl mark 辊印
knurl roll 压花辊
knurl roller 压花滚筒
knurls 辊式拉边器
knurr (木材等的)硬节
koa 寇阿相思木
koalmobile 自行式井下无轨矿车;无轨自行矿车
koashvite 硅钛钙钠石
kobeite 钛稀金矿
kobellite 硫锑铋铅矿;硫铋锑铅矿
Kobe Port 神户港【日本】
kobitalium 科毕塔铝合金
kobstone 瘤石
Kochab 北极二
kochenite 琥珀树脂
Koch flask 考科瓶
Koch ring 考科环
Koch's freezing process 科赫冻结凿井法
Koch's microphotometer 科赫测微光度计
Koch's phenomenon 科赫现象
Koch's resistance 科赫电阻
Kodachrome 柯达彩色胶片;柯达彩色胶卷
Kodachrome film 柯达彩色胶卷
Kodacolo(u)r 柯达彩色胶卷
Kodak camera 柯达照相机
Kodak relief plate 柯达光敏树脂凸版
koechlinite 钼目铋矿
Koehring process 凯林热压法
koembang 科厄姆班风
Koenen floor 柯能肋形板
koenenite 氯羟镁铝石
Koenen plate 柯能肋形板
Koeningsberger ratio 科尼斯贝格比
koenlite 重碳地蜡
Koepe hoist 戈培式提升机
koepe hoisting 摩擦式提升
Koepe mine hoist 戈培式矿井提升机
Koepe pulley 摩擦轮;戈培(提升滑)轮
Koepe pulley groove lining 戈培滑轮槽内衬
Koepe reel 戈培轮
Koepe system of winding 戈培氏滑轮卷扬系统
Koepe wheel 戈培轮
Koepe winder 戈培式提升机;戈培卷扬机
Koepe winder brake 戈培式提升机制动
Koepe winding 戈培式提升
koettigite 水砷锌石;水红砷锌石
Kogasin 合成洗涤剂原料油
kogel process (路面防滑的)热处理
Kohinoor 贵重大钻石
Kohinoor test 科努耳试验
Kohinoor value 科努耳值
Kohler coater 气刀涂布机
Kohler illumination 柯勒照明
Kohler illuminator 柯勒照明器
Kohlrausch method 科尔劳施方法
Kohlrausch's law 科尔劳旋定律
Kohnstamm's phenomenon 后继性运动
koilonychia 匙状形甲
koinomatter 正常物质
Koipato group 科帕托群【地】
koji bed 曲床
koji tray 曲盘
kokkite 粒状岩(石)
Ko-kneader 混练挤压机
koktaite 铵石膏
kolbeckite 硅磷锆石;水磷锆石
Kolbe hydrocarbon synthesis 科尔伯烃合成法
Kolbe-Schmitt synthesis 科尔伯—施密特合成
koldflo 冷挤压成型法
kolfanite 阿磷钙铁石
Kolim old land 科里姆古陆

kolinsky 貂皮
Koliuma massif 科累马地块
kolk 紊动漩涡上升流动
kollage 固体润滑油;固体润滑剂
kollanite 硅结砾岩
kollergang 轮碾机;碾砂机
kollermill 轮碾机
Kollsman window 高度表气压调定窗
kolm 含铀煤结核
Kolmergorov-Smirnov filter 科尔默格罗夫—米尔洛夫滤波器
Kolmer test 科尔默试验
Kolmogorov axioms for probabilities 柯尔莫哥罗夫概率
Kolmogorov consistency condition 柯尔莫哥洛夫相容性条件
Kolmogorov criterion 柯尔莫哥罗夫准则
Kolmogorov inequality 柯尔莫哥洛夫不等式
Kolmogorov-Sinal invariant 柯尔莫哥洛夫—希奈不变量
Kolmogorov-Smirnov distribution 柯尔莫哥罗夫—斯米诺夫分布系
Kolmogorov-Smirnov test 柯尔莫哥罗夫—斯米尔诺夫检验法;K-S 检验法【道】
Kolmogorov's zero-one law 柯尔莫哥罗夫 0-1 定律
kolovratite 钒镍矿
kolwezite 钴孔雀石
kolymite 科汞铜矿
Komamura model 考马姆拉模型
komarovite 硅铌钙石
komatiite 科马提岩
kominuter 磨矿机;粉碎机
Kommerell bead bend test 科默雷尔熔接(料)弯曲试验(德国工业标准)
kona 科纳风
kona cyclone 科纳气旋
Konal alloy 康纳尔镍钴合金
Kona storm 背风面风暴;科纳气旋
kondo 日本佛教庙宇的主屋
Kondo alloy 康多合金
Kondo temperature 康多温度
Kondratieff cycle 康德拉捷夫长期波动周期
Kondratieff wave 康德拉捷夫循环波
konel 考涅尔代用白金
Konel metal 康涅尔代用白金
kong 缸;(无锡砾石下的)无矿基岩
Kongque island 孔雀岛
kongsbergite 汞银矿
Konig hardness 柯尼希硬度
Konigsberg bridges problem 哥尼斯堡七桥问题
Konig's theorem 科尼希定理
Konik(e) 镍锰钢
konilite 粉状石灰;粉石英
konimeter 计尘器;尘量计;尘度计;尘埃计算器;空气尘量计;灰尘计数器
koninckite 针磷铁矿
koniogravimeter 计尘器
koniology 微尘学;尘埃学
koniosis 矿工肺尘病
koniscope 检尘器;计尘仪;计尘器;尘粒镜;尘埃镜
konisphere 尘圈
konistra (古希腊剧院中的)合唱队席
konite 镁白云石
konitest 计尘检验
konnarite 硅镍矿
konometer 尘埃计算器;大气尘金计
konoscope 锥光仪
konoscopic observation 锥光观察
konstantan 康铜
konstruktal 康斯合金
konyaite 孔钠镁矾
Kooman's array 库曼天线阵
Kootenai series 库特奈统
kopfe 小山;孤丘
kophthalic acid 异苯二(甲)酸
kopje 丘陵(南非)
Koppen's climate classification 柯本气候分类法
Koppers horizontal-flue oven 考柏斯水平烟道炼焦炉
koppie 小山丘
koppite 重烧绿石
korahinskite 科硼钙石
Kordic algorithm 科迪克算法

Kordofan gum 柯道方胶
Korduct 库尔管
Korea-augite 朝鲜辉石
Korean moonstone 朝鲜月光石
Korean pine 海松【植】;红松
Korean wares 高丽窑
Korean Won 朝鲜圆
Korea Strait 朝鲜海峡
Korimann power loader 科曼强力截装机
korina 李姆巴缆仁木;伦巴木(非洲果仁树)
koris 干涸谷(北非)
koritnigite 科水砷锌石
kornelite 斜红铁矾
kornerupine 柱晶石
kornish boiler 水平单火管锅炉
koronit 柯罗炸药
koroseal 氯乙烯树脂
Korowkin anchorage system 柯罗金锚固装置
korshunovskite 柯羟氯镁石
Kort nozzle 科特导流管
Kort nozzle rudder 科特导流管舵
Korzhinsky phase rule 柯尔斯基相律
Korzhinsky's mineralogic(al) phase rule 柯尔仁斯基矿物相律
kosometsuke 芙蓉手
Kossakin formula 库沙金公式
kossava 科萨瓦风
Kossel theory 完整光滑面理论(晶体生长)
kostovite 针碲金铜矿
kostylcvite 水硅锆石
kotoite 小藤石;粒硼镁石;粒镁硼石;镁硼石
kotulskite 黄碲钯矿
koum 沙质沙漠;纯砂沙漠
koutekite 六方砷铜矿
Koval 考瓦尔铝合金
Kovar 铁镍钴合金;铁钴镍合金
Kovar alloy 科瓦铁镍钴合金;科瓦合金
Kovats index 科瓦茨指数;保留指数
kovdorskite 科碳磷镁石
Koware blue on white vase with purple colo(u)r 哥瓷青花加紫瓶
Koware blue on white water jar 哥瓷青花水盂
Koware globular shape vase 哥瓷铁箍天球瓶
Koware vase with black dragon design 哥瓷铁龙瓶
Koware vase with double animal ears 哥瓷铁箍兽耳瓶
Koyck geometric lag scheme 科伊克几何滞后型式
Koyck transformation 科伊克变换
Kozeny-Carmen equation 柯兹尼—卡曼方程
Kozeny's equation 柯兹尼方程式
kozulite 铁锰钠闪石
k-point 弹性极限
K-polynomial 扭结多项式
K-probe 凯氏探针
K-pump 钾泵
kraal 牛栏;茅舍(南非);家畜栏
krablite 透长凝灰岩
Kraemer-Sarnow mercury pipeline 克莱迈尔—沙尔塔水银吸移管
Kraemer-Sarnow's method 克莱迈尔—沙尔诺(树脂软化点)测定法
Kraemer-Sarnow softening point 克莱迈尔—沙尔塔软化点
Kraemer-Sarnow softening point test 克莱迈尔—沙尔塔软化点试验
Kraemer-Sarnow test 克莱迈尔—沙尔塔沥青熔化点试验
kraft bag 牛皮纸袋
kraft bag paper 牛皮袋纸
kraft bleach effluent 牛皮纸漂白污水;牛皮纸漂白废水
kraft bleaching 牛皮浆漂白
kraft bleach plant effluent 牛皮纸漂白厂污水;牛皮纸漂白厂废水
kraft bleach plant effluent chromophores 牛皮纸漂白厂污水生色团
kraft bleach wastewater 牛皮纸漂白废水;牛皮纸漂白(厂)污水
kraftboard 纸板
kraft-faced building insulation 牛皮纸贴面建筑防潮板
kraft insulating tissue 电绝缘牛皮薄纸
kraft liner 牛皮纸板;牛皮卡纸

kraft liquor 硫酸化液;牛皮(纸)液
kraft machine 牛皮纸造纸机
kraft mill 牛皮纸浆工厂
kraft mill effluent 牛皮纸厂废水
kraft mill waste 牛皮纸废水
kraft paper 牛皮纸;包装纸
kraft paper bag package 牛皮纸袋包装
kraft paper foil 牛皮纸箔
kraft paper free dry 不粘牛皮纸干
kraft process 硫酸盐制浆法;牛皮纸浆制法
kraft pulp 牛皮纸浆
kraft pulp mill 牛皮纸浆厂
kraft pulp-mill wastewater 牛皮纸浆厂污水
Kraft temperature 克拉夫特温度
kraft test liner 牛皮纸箱纸板
kraftwood 装饰性胶合板(內墙面用)
kraft wrapping 牛皮包装纸
kraisslite 砷硅锌锰石
Krakatoan type 克拉卡托型
Krakatoan wind 克拉卡托风
Kramer-Sarnow mercury pipe-line 克拉茂—沙诺水银吸移管(确定沥青材料的软化点用)
Kramers-Kronig relation 克拉茂—克朗尼希(色散)关系
Kramer's modulus 克拉茂模数
Kramer's rule 克拉茂法则
Kramer's theorem 克拉茂法则;克拉茂定理
Kramer's uniformity factor 克拉茂均匀系数
Kranenburg method 直接水压式成型法
Kranz Triplex method 可锻铸铁制造法
Krarup cable 克拉均匀加感电缆;连续加感电缆;均匀加感电缆
krarupization 连续加感;均匀加感
krasnozem 红土;红壤
Krasovsky ellipsoid 克拉索夫斯基椭球
Krasovsky spheroid 克拉索夫斯基椭球
Krassowski ellipsoid of 1938 克拉索夫斯基椭球体(1938年)
Kratky process 克拉基热压法
kratochvilite 芴石
Kratogen 克拉通区
kratometer 棱镜矫视器
kraurosis 干皱
Kraus cost 克劳斯法
Krause mill 钢板冷轧机;克劳兹轧机
Krause rolling mill 克劳兹轧机
krausite 钾铁矾
krauskopfite 水硅钡石
Kraus process 克劳斯法(结构力学);克劳斯法(一种活性污泥法的改型)【给】
Kraut cell 克劳特浮选机
krautite 淡红砷锰石
Krebs-Stormer's visco(si)meter 克雷布斯—史托摩尔型旋转式黏度计
Krebs unit 克雷布斯黏度单位
Kredietbank NV 信贷银行(比利时)
Krein-Milman theorem 克莱因—米尔曼定理
kremastic water 含气渗水;通气层水;渗流水
kremersite 红铵铁盐
Kremlin 克里姆林宫
Kremnitz white 克雷姆尼茨白
Kremser formula 克莱姆塞公式
krennerite 斜方碲金矿;白蹄金银矿
kreosote 杂酚油
krepidoma (希腊古庙柱列的)多阶台座
kribergite 硫磷铝石
Kriging estimators 克立格估计量
Kriging matrix 克立格矩阵
Kriging method 克立格法
Kriging plan 克立格方案
Kriging variance 克立格方差
Kriging weighting coefficient 克立格权系数
krill 鲸食磷虾
krinvoite 硅铬镁石
krith 克瑞(1克瑞=0.0986克)
kroeberite 磁黄铁矿
kroehnkite 柱钠铜矾
krokidolite 青石棉
Kroll corrosive liquid 氢氟酸腐蚀液
Kroll reactor product 克罗尔海绵金属
Kromal 克劳马尔钼高速钢
Kromarc 可焊不锈钢
Kromax 克劳马科镍铬合金
Kromayer lamp 克罗迈尔灯

kromogram 彩色图
Kromore 克劳莫尔镍铬合金
kromscope 彩色图像观察仪
Kronbergs green 茶绿色
Kronecker delta 克朗纳克尔德尔塔
Kronig sum 克罗内克和
Kronig's doublet formula 克罗尼格双谱线公式
Kronstein's number 克隆斯坦值
Krovan 克劳凡铬钒钢
krugite 镁钾钙矾
krummholz 高山矮曲林
krupkaite 库辉铋铜铅矿
krupp austenite steel 奥氏体铬镍合金钢
Krupp ball mill 克虏伯球磨机
Krupp furnace 粒状炭电阻炉
Krupp guard 克鲁伯防护装置
Krupp mill 克鲁普球磨机
Krupp-Renn method 克鲁普转炉炼钢法
Krupp section 克鲁普型材;克鲁普横断面
Krupp suspension preheater 克鲁普悬浮预热器
Krupp triple steel 克鲁普铬钼钒高速钢;克鲁普高速钢
Kruskal limit 克鲁斯凯极限
Kruskal statistic 克鲁斯凯统计量
krutaite 方硒铜矿
krutovite 等轴砷镍矿
kryogen 冰晶
kryogen blue 冰精蓝
kryogen brown 冰精棕
kryogenne 冷却剂
kryolithionite 锂冰晶石
kryometer 低温计
kryoscope 凝固点测定计
kryotron 冷子管
kryptoclimatology 室内小气候学
Kryptol 粒状炭;克利普托尔电阻材料
Kryptol furnace 炭粒炉;炭粒电阻炉
kryptolith 独居石
Kryptol stove 炭棒电阻炉
kryptomagmatic deposits 隐岩浆矿床
kryptomere 隐晶岩
krypton-bromine method 氪—溴法;热发光法
krypton lamp 氪灯
kryptoseismic 隐式地震的
kryptoseismology 隐式地震
kryptotil 纤柱晶石
krystic 冰雪的
krystic geology 冰雪地质学
krytol 炭棒
kryzhanovskite 羟磷铁锰石
kshoal of fish 鱼群
K-slump test K形坍落度试验
ksomarc 可焊接的不锈钢(铬16%,镍2%,其余是铁)
K-space 波矢量空间
K-S system K-S方式
K-strut 双叉式支撑
K-system critical excitation voltage of elements 元素的K系列临界激发电压
ktenasite 基铜矾
ktesiphon arch 悬链曲线拱
K-T gasification method 开—梯气化法
K-theory 扭结理论
K-truss K形桁架
k-type anchorage K式锚
K-type expansive cement K形膨胀水泥
Kuala Lumpur 吉隆坡
kuanite 磷铝铁钡石
Kubelka-Munk equation "a" 库贝尔卡—蒙克等式
Kubelka-Munk scattering coefficient 库贝尔卡—蒙克散射系数
Kubelka-Munk theory 库贝尔卡—蒙克理论
Ku brick 库砖(一种卷折形砖,用于插接新的混凝土梁与老混凝土工程上)
kuchersite 油页岩
Kuck classification schema 库克分类法
kudampro 防潮层铜片
Kuder preference record 库德优先选择记录
Kufil alloy 铜银合金
Kuhl cement 石膏矿渣水泥;库尔水泥
Kuhn-Tucker condition 库恩—塔克条件
Kuhn-Tucker multiplier method 库恩—塔克乘子法
Kuhn-Tucker theorem 库恩—塔克定理
kukersit 库克油页岩
kukersite 含藻岩

Kukkersite 库克油页岩
Kukui oil 烛果油
kulaite 闪霞粒玄岩
Kula plate 库拉板块
kulkeite 绿泥间滑石
Kullberg's balance 库尔伯格摆
kullerudite 斜方硒镍矿
kullgren lignin 低磺酸化木素
Kumamoto-Minamata disease 熊本水俣病(日本)
Kumanal(alloy) 库曼纳尔合金;铜锰铝标准电阻合金
Kumishi basin 库米什盆地
Kumium alloy 高热导率铜铬合金;高电导率铜铬合金
Kumukuli basin 库木库里盆地
kunckle line 棱角线
kunckle press 曲柄连杆式压力机
kunckle tooth 圆顶齿
Kundt rule 孔脱定则
Kundt's constant 孔脱常数
Kundt tube 孔脱管
Kungurian 空谷尔阶【地】
kunheim metal 稀土金属与镁的合金
Kunial 含铝镍铜合金
Kunial brass alloy 库尼尔黄铜合金
Kunifer alloy 铜镍铁合金
Kunkun mountain 昆仑山脉
Kunnifer 库尼镍铜
Kunzel sounding rod 昆策尔测杆
kunzite 紫锂辉石
kupaphrite 铜泡石(天蓝石)
kupfelsilumin 硅铝明合金
kupfernickel 红镍矿
kupferphosphoruranit 铜铀云母
kupferschiefer 铜页岩
kupletskite 锰星叶石
Kuplex 一种铜钝化剂
kupola 穹顶;岩钟
kuppel horizon 穹隆布景;穹隆布置
kupper solder 铅焊料
kupramite 防氨面罩
Kuprodur alloy 库普罗德合金
kuramite 硫锡铜矿
kuranakhite 碲锰铅石
Kuratowski's lemma 库拉托夫斯基引理
kurb-to-kurb crossing distance 侧石到侧石距离
kurchatovite 硼镁钙石
kurgantaite 水硼钙锶石
kurhaus 疗养所;矿泉水疗养所
Kuril basin 千岛海盆
Kuril current 千岛海流
Kurile cold current 亲潮
Kurile current 亲潮
Kuril trench 千岛海沟
kurkar 凝砂块(砂丘砂和碳酸钙胶凝砂的混合料)
kurled tool 滚花工具
Kurllov formula 库尔洛夫式
kurnakovite 库水硼镁石;富水镁硼石
kuroko ore 黑矿
kuroko-type deposit 黑矿型矿床
kuromore 镍铬耐热合金(因科镍)
Kuron's hygroscopicity 康龙吸湿性
Kuroshio countercurrent 黑潮逆流
Kuroshio(current)黑潮
Kuroshio extension 黑潮续流
Kuroshio system 黑潮系系
kursaal (在疗养区的)公共大厅;(在疗养区的)娱乐中心
kurskite 氟钠磷灰石
Kursk-type iron deposit 库尔斯克型铁矿床
kurtosis 尖峰状态;尖峰值;尖度;峭度;峰态;峰度
kurtosis test 峰度检验
Kuruktag fault uplift 库鲁克塔格断隆
Kuruktag marine trough 库鲁克塔格海槽
Kuruktag tectonic segment 库鲁克塔格构造段
Kuruktag upwarping faulted region 库鲁克塔克隆断区
kurumsakite 硅钒锌铝石
kusuite 钒铅铈矿
Kusunoki-damashi oil 桉樟油
Kuteera gum 刺槐树胶
kutinaite 方砷铜银矿
kutnahorite 镁锰方解石;镁菱锰矿
kutnohorite 锰白云石

Kuttern 铜碲合金
Kutter roughness coefficient 库特粗糙系数
Kutter's formula 库特公式
Kuwait 科威特
kuwait Fund 科威特基金会
Kuznets cycle 库茨涅兹循环(指平均周期为 15～25 年幻景气循环)
kuznetsovite 氯砷汞石
Künzel sounding rod 孔泽尔测深杆
K-value 黏度值;K 值;增殖系数黏度值;材料导热性能值
kvanefjeldite 羟硅钙钠石
Kveim test 卡万试验
kvellite 橄闪歪煌岩
Kwanting Reservoir 官厅水库(日本)
k-w-level tube K-W 水准管
k-word 千字【计】
kyak 皮筏
kyanising 升汞浸渍处理;水银防腐法;升汞防腐
kyanite 蓝晶石(岩)
kyanite deposit 蓝晶石矿床
kyanite schist facies 蓝晶石片岩
kyanization 水银防腐法;升汞防腐(法)
kyanize 用升汞浸渍木材;用氯化汞浸渍(木材);升汞防腐
kyanized wood 升汞防腐木材
kyanizing 氯化汞冷浸处理(木材防腐);水银防腐法;升汞防腐(法)
Kyan's process 升汞防腐(法)
kybernetics 控制论
kydex 基迭克斯板材
kylite 辉橄霞斜岩
kylix 高脚双柄大酒杯
kyljack 水管防冻塑性材料
kymatology 脉波学;波浪学;波动学
kymogram 记录图;记波图
kymograph 转筒记录器;转筒波纹记录器;滚筒记录器;记纹鼓;记波器;描波器;曲线描记器;飞机转动记录器;波形自(动)记(录)器;波形记录器
kymographion 波形示波器
kymograph paper 记纹纸
kymography 滚筒记录法;记波术;记波摄影;波形自动测量法
kymokinoscopy 记波运动描记法
Kynal 铝质防滑踢脚板
Kyodai type extractor 搅拌式提取器;搅拌式萃取塔
Kyoto Protocol 京都协议书
kyr 小山
kyrock (美国制造的)沥青(砂)岩
kyrtom 弓形褶皱
kyrtometer 曲面测量计;曲度计
kyschtymite 刚玉黑云钙长岩
kytoon 系留气球;人控制观测气球;风筝式系留气球
kyzylkumite 库钒钛矿
Kyzylkum massif 克齐尔库姆地块

L

laager 装甲车停车处;临时防御营地
laagte 宽平谷地
laavenite 锆钽矿
lab 试验室;实验室
Labarraque's solution 拉巴腊溶液
labbé 拉贝风
lab coat 实验工作服
Labdanum(gum) 劳丹胶
lab data 试验(室)资料
lab dyer 试验室染色机
label 纸条;记录单;签条;披水石;厂牌;标题;标记;标签;标牌;标明;标记(物);标号
label address table 标号地址表
label area 标签区;标记区;标号区
label assignment statement 标记赋值语句
label attaching machine 钉商标机
label attribute 标号属性
label band 贴标签部位
label block 标号信息组
label check 标记检验;标号检验
label clause 标签条款
label cloth 标签布
label coating 标签涂料
label coding 标证编码;标记编码;标号编码
label constant 标号常数
label-corbel table 托臂滴水石
label course 门窗披水线条;出缘线
label-cutting press 标签裁切机
label cylinder 标号柱面
label data 标号数据
label definition 标号定义
label delimiter 标号定义符
label descriptor 标号描述符
labeled plants park 植物公园
label expression 标号表达式
label field 标号字段;标号栏
label format record 标号格式记录
label-gluing machine 标签上胶机
label group 标号组
label handling 标号处理
label handling routine 标号处理程序
label identifier 标号标识符
label incarnation 标号实体
label information 标记信息
label information area 标号信息区
labeling 标志;标贴;标签费
labeling charges 标纸费
labeling machine 制标签机
labeling method 标号计算法;标号法
labeling requirement 标签需要
labeling response 标定反应
labeling scheme 代码电路;标号方案
labeling technique 标号法
label insurance 标签(保)险
label-knotting machine 标签捆包机
label lacquer 标签漆
label(l)ed 标记的
label(l)ed amount 标示量;标记量
label(l)ed atom 标记原子;示踪原子
label(l)ed atom method 标记原子法;示踪原子法
label(l)ed cargo 危险品;贴有危险品标签的货物;特别标注货
label(l)ed common 带标号的公用区
label(l)ed common block 有标号公用块
label(l)ed compound 标记化合物
label(l)ed content 标记含量
label(l)ed door 带有防火检验合格签条的门
label(l)ed element 标记元素
label(l)ed field 标号字段;标号域;标号区段
label(l)ed fire dampers 标定过的防火挡板
label(l)ed frame 带有防火检验合格签条的框架
label(l)ed graph 标记图;标记图;标记数;标号图
label(l)ed isotope 显踪同位素;示踪同位素;标记同位素
label(l)ed molecule 标记分子
label(l)ed network 标号网络
label(l)ed node oriented directed graph 标定的面向节点的有向图

label(l)ed notation 标号标记
label(l)ed ordered tree 带标号有序树
label(l)ed panel 耐火夹芯板
label(l)ed particle 示踪颗粒
label(l)ed phosphorus 标志磷
label(l)ed plant 标记植物
label(l)ed pod 标记盆栽
label(l)ed precursor 标记前体
label(l)ed price 标明价格;标价
label(l)ed series-parallel network 标定(的)序列并行网络
label(l)ed size 标码
label(l)ed statement 有标号语句
label(l)ed tracer 加标志示踪物
label(l)ed transitive digraph 标定可迁有向图
label(l)ed variant 标记变式
label(l)ed window 带有防火检验合格签条的窗;防火规范窗
label(l)er 贴标签机
label(l)ing 加标签;贴商标;贴标签;示踪
label(l)ing act 标签法
label(l)ing machine 贴标签机
label(l)ing method 标示法;标记法
label(l)ing of core box 岩芯盒标记
label(l)ing reader 标记卡阅读器;标号(卡)阅读器
label(l)ing regulation 标签条例
label(l)ing technique 标记技术
label list 标号列表;标号表
label list of a label variable declaration 标号变量说明的标号表
label map 标号变换
label marking machine 打商标机
label mo(u)ld (哥特式建筑特有的)方形滴水石;(门窗上边的)出檐线脚
label mo(u)lding 披水线脚
label number 标号数
label of railway tariff 铁路运价分类表
label paper 商标纸;标签纸
label paper tape 纸条
label prefix 标号前缀
label printer 标记印字机
label printing 标签印花
label printing machine 标签印刷机
label processing 标记处理;标号加工;标号处理
label processor 标号加工程序;标号处理器;标号处理程序
label rack 标签架
label record 标号记录
label routine 标号程序
label save area 标号保存区
label set 标号组
label-sewing machine 缝标签机
labels for packages of dangerous goods 危险货物包装标志
label standard deviation 标准差标记
label standard level 标号标准电平
label stop 披水石端饰;滴水石(末端)浮雕;滴水罩端部浮雕;滴水罩端饰
label subscript 标号下标
label symbol 标记符号
label table 标号表
label terminations 出缘限
label terminator 标号终结符
label trace 标号追踪
label-tying machine 标签系结机
label variable 标号变量
lab-gown 实验服
labial 风琴管;唇状的
labiate ladle 转包;倾动式浇包;带嘴浇包
labile 不稳定的;滑动的
labile acid 不稳定(的)酸
labile area 不稳定区
labile balance 不稳定平衡
labile bond 不稳定键
labile complexes 不稳定配合物
labile coordination compound 活性配位化合物
labile element 不稳定成分
labile equilibrium 不稳(定)平衡

labile factor 不稳定因素
labile factor deficiency 不稳定因子缺乏
labile flow 不稳定流
labile form 不稳形
labile hydrogen 不稳定氢
labile nitrocellulose 不稳定硝化棉
labile oscillator 易变振荡器
labile phosphate 活性磷酸酯;活性磷酸盐
labile phytocoenosium 不稳定植物群落
labile protobitumen 易分解原沥青;易分解有机质
labile pulse 不稳定脉
labile region 易变区;不稳定区
labile shelf 不稳定大陆架
labile state 易变(状)态;不稳(定)状态;不稳定态
labile zone 不稳定区
lability 不稳定性;不耐性;不安定性
lability coefficient 不稳定性系数;不稳定度系数
lability due to sliding 由滑移导致的不稳定性
lability due to uneven settlement 由不均匀沉降导致的不稳定性
lability effect 不稳定性效应
lability number 不稳定性系数;不稳定度系数
lability of emulsion (沥青的)乳液的不稳定性;(沥青的)乳液的不安定性
labite 拉巴石
labium mediale 内侧唇
lable number 标记号码
lab mixer 试验室用混合器
laborage 工资
laboratorian 化验员
laboratories microscope 实验室用显微镜
laboratory 化验室;试验室;实验室
laboratory accreditation 实验室资格证明
laboratory analytical method 室内分析方法
laboratory and green-house 实验室和温室
laboratory apparatus 实验仪器;试验室设备
laboratory appliance 实验室仪器
laboratory assessment 实验室评定
laboratory assessor 实验室评定者
laboratory assistant 化验员
laboratory automation system 实验(室)自动化系统
laboratory bench 试验台;实验工作台
laboratory bench tile 实验室桌面砖;实验室台面砖
laboratory block 实验楼
laboratory boiling glass 玻璃烧器
laboratory building 实验楼
laboratory certification 实验室认证
laboratory channel 实验河槽
laboratory clothing 实验室工作服
laboratory coefficient of permeability 实验室渗透系数
laboratory computer 实验室计算机
laboratory computer hierarchy 实验室计算机体系
laboratory condition 实验室条件
laboratory consolidation test 实验室固结试验
laboratory course 实验学科
laboratory crucible 技工用坩埚
laboratory data 实验数据
laboratory database 实验(室)数据库
laboratory data record 实验数据记录
laboratory decantation method 试验室湿筛分法
laboratory device 化验设备
laboratory digester 实验室蒸煮锅
laboratory diving tank 实验室潜水舱
laboratory drilling 实验室钻探试验
laboratory-dry aggregate 实验干集料
laboratory dyeing machine 实验室染色机
laboratory engine 实验室试验研究用发动机
laboratory engine test 实验室内燃机试验
laboratory environment 化验室环境;实验室环境
laboratory equipment 试验室设备;实验室设备
laboratory evaluation 实验室评定
laboratory examination 化验检查
laboratory exercises 实验手册
laboratory experiment 室内试验;实验室试验;实验室实验
laboratory exposure 实验室气蚀法;实验室风化法

laboratory filter 实验室过滤机
laboratory findings 化验结果;检验所见;化验所见;化验检查所见;实验资料;实验数据
laboratory flume 试验水槽
laboratory flume hood 试验用排烟橱
laboratory for application of remote sensing 遥感应用实验室
laboratory for atmospheric and solar physics 大气和太阳物理实验室
laboratory for external examination 外检实验室
laboratory for internal examination 内检实验室
laboratory for materials testing 材料试验室
laboratory for photoelasticity 光弹性实验室
laboratory for umpire analysis 仲裁实验室
laboratory frame 实验室系
laboratory fume hood 试验室排烟橱
laboratory furnace 实验室用窑;实验室用炉;实验室窑炉
laboratory furniture 实验室设备
laboratory glass 仪器玻璃
laboratory glassware 实验室玻璃器皿
laboratory-grade water 实验室用水
laboratory grinder 技工室磨削机;实验室用磨
laboratory inspection 实验室检验
laboratory installation 实验室装置
laboratory instrument 实验仪器;实验室仪表
laboratory instrument computer 实验设备用计算机
laboratory investigation 实验室研究
laboratory jordan 实验室锥形磨浆机
laboratory knock-testing method 实验室辛烷值测定法
laboratory life 实验室试验寿命
laboratory manual 实验手册
laboratory maximum dry density 实验最大干密度
laboratory measurement 实验室测量
laboratory melts 实验室熔炼产品
laboratory method 实验室方法
laboratory microgrinder 技工室用微型磨削机
laboratory microscope 化验显微镜
laboratory mill 实验型磨矿机;实验室用磨
laboratory minicomputer 实验室小型计算机
laboratory mix design 试验室配合比设计
laboratory mixer 实验室拌和器
laboratory model 实验模型
laboratory model analysis 实验模型分析
laboratory model construction 实验室模型结构
laboratory mortar grinder 实验室研钵研磨机
laboratory muller 实验室用混砂机
laboratory octane number 实验室辛烷值
laboratory of materials strength 材料强度实验室
laboratory of sample analysis 样品分析实验室
laboratory of sample identification 样品鉴定实验室
laboratory of sample testing 样品测试实验室
laboratory oscillator 实验室振荡器
laboratory paper machine 实验室纸机
laboratory paraphernalia 实验用用具
laboratory performance 实验室表现
laboratory pilot plant study 实验室中间生产研究
laboratory pliers 技工钳
laboratory porcelain 化学瓷(器)
laboratory practice 实验室实习
laboratory procedure 化验室研究方法;实验室研究(方)法;实验(室)程序;实验步骤
laboratory proofing 实验室验证
laboratory proportioning 试验室配合比;试验配合
laboratory qualification 实验室鉴定
laboratory quality control 实验室质量控制
laboratory reactor 实验室用炉
laboratory reagent 试验室试剂;实验室试剂
laboratory reference standard 实验室参考标准器
laboratory refiner 实验室用磨浆机
laboratory reliability test 实验室可靠性试验
laboratory report 化验报告;实验报告
laboratory research 试验室研究
laboratory result 实验结果
laboratory rocket 实验室用火箭发动机
laboratory rotor mill 实验室转子磨
laboratory sample 实验样品;实验室试样
laboratory sample divider 实验室样品分样机
laboratory sample number 实验室样号
laboratory sandbath 实验室(用)砂浴
laboratory scale 试验室规模;实验室规模
laboratory-scale apparatus 实验室级器械

laboratory service 实验室业务;实验室的辅助管线
laboratory sheet 化验单
laboratory shop 试制车间;试造车间
laboratory sieve-mesh 实验室标准筛网目
laboratory sieving equipment 实验室筛选设备
laboratory sifter 振动筛分机
laboratory signal processing instrument 实验室信号处理设备
laboratory signal processor 实验室信号处理机
laboratory-simulated degraded imagery 实验室模拟降质像
laboratory simulation 实验室模拟
laboratory simulation test 实验室模拟试验
laboratory sink 化验室洗盆
laboratory size extruder 试验用压出机
laboratory size reactor 化验室型反应器;实验室型反应器
laboratory soil and rock test 岩土室内试验
laboratory soil block cultures 实验室土壤木块培养基
laboratory soil test 室内土工试验
laboratory sole 炉底;炉床
laboratory spectral measurement instrument 实验室波谱测量仪器
laboratory staff 试验室人员
laboratory standard 实验室标准
laboratory standard microphone 实验室标准传声器
laboratory standard of frequency 实验室的频率标准
laboratory stirring device 实验室搅拌器;实验室搅拌装置
laboratory strength 实验强度
laboratory study 试验室研究
laboratory suction filter 实验室吸滤机
laboratory surveying 实验室测量法
laboratory system 实验室系
laboratory technician 化验员;实验员;实验室技术员
laboratory technique 实验室技术
laboratory test 化验;室内试验;实验室试验;实验室测试
laboratory-tested soil specimen 实验室试验土样
laboratory test engine 试验台试验的发动机;实验室试验的发动机
laboratory test engine method 实验室发动机试验法
laboratory testing rig 室内试验装置;实验室试验装置
laboratory test item 化验项目
laboratory test report 化验单
laboratory test scale 实验室试验规模
laboratory tests of stability 稳定性的实验室检验
laboratory tile 实验室铺面砖;实验室瓷砖
laboratory tower 实验塔
laboratory trailer 实验室拖车;实验室挂车
laboratory tweezers 技工镊
laboratory-type instrument 实验室仪器
laboratory use 实验室使用
laboratory vane test 室内十字板试验
laboratory vehicle 移动式试验室;试验车;实验(室用)车;车载试验室
laboratory velocity 实验室系速度
laboratory water 实验室用水
laborer's quality 劳动者素质
labor-exchange team 变工队
labor-intensive goods 需要大量劳动力的货物
labor-intensive industrial product 需要大量劳力的工业产品
labor-saving device 节约人工措施;人工节约法
labor-saving investment 节约人工的投资
labor-saving machinery industry 节约人工的机械工业
labo(u)r 生产;纵横摇摆;工人;劳务;劳力;劳动者;人工
labo(u)r accident 劳动事故
labo(u)r accounting 人工会计
labo(u)r agreement 劳资协议;劳务协议
labo(u)r allocation system 劳动制度
labo(u)r and capital 劳资;劳方和资方
labo(u)r and capital resources 劳动与资本资源
labo(u)r and management 劳资;劳方和资方
labo(u)r and material 工料
labo(u)r and material analysis 工料分析
labo(u)r and material payment bond 工料费合约;人工材料费保单;工料付款承诺书

labo(u)r and salary office 劳资室
labo(u)r and wage plan 劳动工资计划
labo(u)r and welfare expenditures 劳工及福利支出
labo(u)r arbitration 劳动纠纷仲裁
labo(u)r arbitration system 劳动仲裁制度
labo(u)r association 劳工协会
labo(u)r-augmenting technical change 劳力增加型技术变化
labo(u)r benefit 劳动效益
labo(u)r bolding economic indicators 劳动占用经济指标
labo(u)r budget 人工预算
labo(u)r camp 劳工驻地;工人营地
labo(u)r capacity 工率;劳动生产率
labo(u)r-capital dispute 劳资纠纷
labo(u)r-capital relations 劳资关系
La Bour centrifugal pump 拉布离心泵
labo(u)r charges 工资费用
labo(u)r conciliation 劳工争议调解
labo(u)r condition 劳动条件
labo(u)r constant 劳力常数;劳动(力)定额
labo(u)r content 加工工作量
labo(u)r contract 劳务合同;劳工合同;劳动合同;固定劳动量
labo(u)r contractor 承包者;承包商;承包人;劳务承包商
labo(u)r control 劳动管理;人工管理
labo(u)r cost 工人成本;劳务价格;劳力费用;劳动价值;劳动成本;人工费(用);人工成本;劳动工资费
labo(u)r cost account 劳务费账户
labo(u)r cost differential 劳力成本差异
labo(u)r cost in railway enterprise 铁路企业人工成本
labo(u)r cost percentage method 人工成本百分比法
labo(u)r cost ratio 人工成本比率
labo(u)r cost saving plan 人工成本节约计划
labo(u)r cost sheet 人工成本单
labo(u)r court 劳资争议法庭
labo(u)r day 劳动日
labo(u)r discipline 劳动纪律
labo(u)r dispute 劳资纠纷;劳资争议;劳动纠纷
labo(u)r distribution 人工分配
labo(u)r distribution sheet 人工分配单
labo(u)r disturbance 工潮
labo(u)r economic effectiveness indicators 劳动经济效益指标
labo(u)r economics 劳动经济学
labo(u)r efficiency 工率;劳动效率;劳动生产率;人工效率
labo(u)r efficiency of productive worker 生产工人劳动效率
labo(u)r efficiency report 人工效率报告
labo(u)r efficiency variance 人工效率差异
labo(u)r employment 劳务雇佣
labo(u)r emulation 劳动竞赛
labo(u)r environmental quality assessment 劳动环境质量评价
labo(u)r equilibrium 劳动平衡
labo(u)r exchange 产品交换;职业介绍所;劳力交换;劳动力调配
labo(u)r famine 劳动力缺乏
labo(u)r feather bedding 强迫雇佣劳动
labo(u)r flow table 劳动流程表
labo(u)r flux 劳动力流动
labo(u)r force 劳动力
labo(u)r force participation rate 劳动力参加率
labo(u)r forces 劳动资源
labo(u)r force shortage 劳动力短缺
labo(u)r force statistics 劳动统计
labo(u)r grade 劳动等级
labo(u)r health 劳动卫生
labo(u)r hour 工时;劳动小时数
labo(u)r hour basis for overhead application 制造费用分配的人工小时法
labo(u)r hour method 人工小时法
labo(u)r hour rate 工时率;人工小时率
labo(u)r housing 工人住房
labo(u)r hygiene 劳动卫生学;劳动卫生
labo(u)r hygiene standard 劳动卫生标准
labo(u)r income 劳动收入

labo(u)r index 劳动指标
labo(u)r information 劳工资料
labo(u)ring class 工人阶级
labo(u)r in general 一般劳动
labo(u)ring man 劳动者
labo(u)ring-saving 可节省劳动力的
labo(u)r in process 分步人工账户;分步人工;在产工人;在制人工
labo(u)r in progress 在制人工
labo(u)r input 劳动量投入;投工量
labo(u)r instrument 劳动手段
labo(u)r insurance 劳动保险
labo(u)r insurance and welfare 劳保福利
labo(u)r insurance cost 劳保费
labo(u)r insurance expenditures 劳保支出
labo(u)r insurance fund 劳保基金
labo(u)r insurance regulations 劳保条例
labo(u)r insurance system 劳动保险制度
labo(u)r intensity 劳动强度;劳动集约程度
labo(u)r-intensive 劳务密集型;劳务密集的;劳动密集型;劳动密集的
labo(u)r-intensive enterprise 劳动密集型企业
labo(u)r-intensive industry 劳工密集工业;劳动密集型工业
labo(u)r-intensive production 劳动密集生产
labo(u)r-intensive project 劳动密集型项目
labo(u)r-intensive technique 劳动密集技术
labo(u)r in the aggregate 全体劳力
labo(u)r law 劳动法;劳工法
labo(u)r liquidity 劳力流动性;劳动流动性
labo(u)r management 劳动(力)管理;人工管理
labo(u)r management of railway 劳动定额管理
labo(u)r-management relations 劳资关系
labo(u)r market 劳力市场;劳动力市场
labo(u)r market with a hierarchical mobility 分层次流动的劳动力市场
labo(u)r measurement 人工测定
labo(u)r migration 劳动力移动
labo(u)r mix variance 人工搭配差异
labo(u)r mobility 劳工流动性;劳动流动性
labo(u)r of low productivity in agriculture 生产率低的农业劳力
labo(u)r-only subcontracting 只包人工的承包
labo(u)r only subcontractor 承包劳动
labo(u)r organization 工人组织;劳工组织;劳动组织
labo(u)r-oriented industry 劳工导向工业
labo(u)r-output ratio 劳动产量比率
labo(u)r overturn 工作人员变动
labo(u)r personnel quota standard in railway transport enterprise 铁路运输企业劳动定员标准
labo(u)r planned use 计划用工
labo(u)r planning 劳务计划
labo(u)r pool 劳力供应源
labo(u)r population 劳动人口
labo(u)r power 劳动力;人力
labo(u)r process 劳动过程
labo(u)r productivity 工率;劳动(力)生产率
labo(u)r productivity index 劳动生产指数
labo(u)r productivity of stevedore 装卸工人生产率
labo(u)r productivity plan 劳动生产率计划
labo(u)r protect and safety devices 防护设备配置
labo(u)r protecting standard 劳动保护标准
labo(u)r protection 劳动保护
labo(u)r protection act 劳动保护法
labo(u)r protection medical care 劳保医疗
labo(u)r protection payment 劳保支出
labo(u)r quantity standard 人工工时标准
labo(u)r quantity variance 人工工时差异
labo(u)r quota 劳动份额;劳动定额
labo(u)r rate 劳务费;劳动率;劳务价格
labo(u)r rate standard 工资率标准;人工工资率标准
labo(u)r rate variance 人工工资率差异
labo(u)r realizes itself in commodities 劳动物化在商品中
labo(u)r-reducing 可节省劳动力的
labo(u)r regulation 劳动纪律
labo(u)r-related cost 人工相关成本
labo(u)r relation 劳资关系
labo(u)r relief 失业工人救济金
labo(u)r requirement 劳动定员;所需劳动力
labo(u)r requirements per square meter 单方用工
labo(u)r resources 劳力资源;人力资源
labo(u)r reward 劳动报酬
labo(u)r right 劳动用工权
labo(u)r room 待产室
labo(u)r safety device 劳动保护设施
labo(u)r safety measure 劳动安全措施
labo(u)r saving 节约劳动力;人力节约;省工;省力
labo(u)r-saving device 机械装置;省工设备;省工法
labo(u)r-saving equipment 节省劳动设备
labo(u)r-saving invention 节省劳动发明
labo(u)r-saving investment 节省劳动投资
labo(u)r-saving machinery 劳力节省机械
labo(u)r-saving machinery industry 劳力节省机械工业
labo(u)r-saving ratio 节省劳动比率
labo(u)r saving technique 节约劳动力的技术
labo(u)r security 劳动保护
labo(u)r service charges 劳动服务费;劳务费
labo(u)r service company 劳动服务公司
labo(u)r service export 劳务出口
labo(u)r sheet 计工表
labo(u)r shortage 劳动力缺乏
labo(u)r slowdown 惰工;怠工
labo(u)rsome 费力的
labo(u)r specialization 劳动专业化
labo(u)r stability index 人工稳定性指数
labo(u)r standard 人工标准
labo(u)r standard cost 人工标准成本
labo(u)r statement 人工表
labo(u)r statistics 劳动统计
labo(u)r statistics of railway enterprise 铁路劳动统计
labo(u)r statute 劳工条例
labo(u)r strength 劳动强度
labo(u)r stringency 劳力不足
labo(u)r surplus 劳力过剩
labo(u)r system 劳动制度
labo(u)r time variance 人工效率差异;人工时间差异
labo(u)r trouble 劳资纠纷
labo(u)r turnover 劳工周转率;劳动力流动(率);人工周转率
labo(u)r union 工会
labo(u)r union expenditures 工会经费
labo(u)r unit 劳力单位
labo(u)r unrest 劳资争议
labo(u)r usage variance 人工用量差异
labo(u)r variance 人工差异
labo(u)r wage rate 劳动工资率
labo(u)r weekly cost reports 工人费用周报表
labo(u)r working on his own 个体劳动者
labradite 钙钠斜长石;拉长岩
labradophyre 拉长斑岩
Labrador current 拉布拉多海流
labradorescence 拉长晕彩
labradorite 拉长石;浅色辉长岩;闪光拉长石;曹灰长石;富拉玄武岩;钙钠长石
labradorite anorthosite 拉长斜长岩
Labrador Sea 拉布拉多海
labrador spar 拉长石
La Brea sandstone 拉布雷亚砂岩
labrobite 钙长石
labryrinth oil retainer 迷宫式护油圈
lab-scale 实验室规模
lab-size equipment 试验用设备
lab study 实验室研究
lab-type turbidimeter 实验室型浊度计
labuntsovite 硅钛钾钡矿;水硅铌钛矿
Labuotes effect 拉波毕效应
L abutment L 形桥台;L 形坝座
labware 实验室器皿
labyrinth 螺旋管;曲径式密封;曲径(迷宫);曲径;复杂结构;封严圈
labyrinth algorithm 迷宫算法
labyrinth baffle 迷宫式障板
labyrinth bearing 迷宫轴承
labyrinth box 迷宫(式)密封箱;曲径密封箱
labyrinth casing 迷宫气封体
labyrinth cave karst 迷宫式溶洞
labyrinth classifier 迷宫式分级机
labyrinth clearance 迷宫间隙;气封间隙
labyrinth collar 迷宫式密封圈;迷宫密封环
labyrinth compressor 迷宫式压缩机
labyrinth dust collector 迷宫式集尘器
labyrinth fin 迷宫气封疏齿;曲径气封片
labyrinth fret 曲折回纹(饰);迷宫曲径
labyrinth gland 迷宫式压盖;迷宫式(气)封;迷宫式密封;迷宫密封装置;曲径式密封套
labyrinth gland loss 气封损失
labyrinth gland packing 迷宫气封
labyrinth-grease seal 迷宫式润滑脂密封器;迷宫式润滑脂封闭器
labyrinth hedge 迷宫绿篱
labyrinth loudspeaker 迷宫式扬声器
labyrinth nut 迷宫密封螺母
labyrinth of Crete 克里底王的迷宫
labyrinth oil retainer 迷宫式集油器
labyrinth packing 迷宫填料;迷宫式密封(件);迷宫式密封;曲折阻漏;曲折轴垫;曲径轴封;堵漏密封料;曲折式密封
labyrinth packing edge 迷宫密封梳齿边
labyrinth piston 卸荷活塞;迷宫式活塞
labyrinth piston compressor 迷宫式活塞压缩机;迷宫活塞式压缩机
labyrinth pressure ratio 迷宫密封压力比
labyrinth pump 迷宫泵
labyrinth ring 迷宫式密封圈;迷宫式密封环;迷宫环;曲折密封圈;曲径环;气封环
labyrinth seal 拉别令密封;密封迷路;迷宫密封;曲径气封;曲径密封;曲径密封
labyrinth seal device 蓖齿式密封装置
labyrinth seal gland 迷宫式密封装置
labyrinth sealing 迷宫式密封;迷宫式充填物;螺旋止水;迷宫(式)止水
labyrinth seal ring 螺旋止水环;迷宫(式)止水环
labyrinth shaft seal 迷宫式轴封
labyrinth strip 气封片
labyrinth teeth 迷宫齿
labyrinth valve 迷宫阀
labyrinth viewing device 迷宫式窥视装置
labyrinth waterstop 迷宫式填缝条;迷宫内的挡路水池
lac 洋干漆;紫胶;假漆;天然胶质;虫(胶)漆;虫胶
lac acid 紫胶酸
lacality mark 产地号印
Lacasitan(stage) 拉卡赛特阶【地】
lacca 虫漆;虫胶;紫胶虫
lacca coerula 石蕊
laccaic acid 紫胶酸;虫胶酸(色素)
laccal 紫胶酚;虫漆酚
laccase 漆酶
lacceroic acid 紫胶蜡酸;虫胶蜡酸;虫胶醋酸
laccolite 岩盘;岩盖;菌形穹隆
laccolith 岩株;岩盖;菌形穹隆
laccolith mountain 岩盖山
lac dye 紫胶染料;虫胶染料
lace 镶花边;用带穿边;整列穿孔;结带子;花边;全(区)穿孔;束带
lace bar 缀条
lace bark pine 白皮松
lace card 整列穿孔卡片
lace curtain 挑花窗帘
laced beam 缀合梁;空腹梁;花格梁
laced belt 接头带
laced card 全穿孔卡片
laced column 缀合柱;花边柱;束带柱
laced column of built channels 槽钢缀合柱
laced corner 交错砌合的墙角
laced fall 复滑轮绳索
laced nappe 花边状推覆体
laced paper cutter 花边纸刀
laced valley 搭瓦天沟;(瓦片犬牙交错铺砌的)天沟
lace leaf 带花纹的叶子
lace-like spire 花边状塔顶;透孔塔顶;楼空花饰的塔尖顶
lace paper 花边纸
lace punch 全穿孔
lacer 系紧用具;束紧用具
lacerable 可撕裂的
lacerate 划破;撕碎
lacerated 撕裂状的
lacerating machine 拉力试验机;切碎装置
laceration 划破;撕裂;挫裂创
laceration index 破伤指数
lace screen 花边式滤网;辫带(式)滤网
lacet 盘山(道)路

lace trimmings 花边装饰
lacet road 盘山(道)路;旋回道路
lace-up 交织;缀合
lacewood 单球悬铃木
lacework 花纹边;镶花边;网眼针织物
Lacey's formula 雷西公式
Lacey's silt factor 雷西泥沙系数
lac ferri 沉降磷酸铁
laches 不可原谅的不履行合同义务
Lacid 克列酸
lacing 系索;拱圈结合层;联缀;联条;配力钢筋;单缀;分编
lacing angle 角铁牵条
lacing bar 系杆;缀条;拉筋
lacing board 系紧板;缀板;绳索模板;电缆模板;布缆板
lacing course 挂结层;拱圈结合层;拉结层;带层
lacing eye 穿绳眼环;系索环
lacing grommet 系索环;穿绳眼环
lacing hole 拉筋孔;纹板串连孔;穿绳孔
lacing leather 束紧皮带
lacing line 系帆索;固缚索;穿缚绳
lacing of flat bars 铺扁铁条
lacing stand 绕丝架
lacing system 缀板与缀条系统;连缀系统
lacing wire 拉筋;编缀的金属线;束紧的金属线
lack a quorum 不足法定人数
lack contents 内含物不多
lack copper 电解铜
lacker 真漆;亮漆;蜡克;漆(器)
lacker work 打蜡克
lack for underground water 地下水不足
lacking number 缺号
lacklustre 无光泽的
lack of alignment 准线偏斜;中心线偏斜
lack of balance 缺乏平衡;非均衡
lack of biologically mature trees 缺乏生物学上成熟的树木
lack of capital 缺少资金
lack of conformity 缺乏一致性;不符合
lack of definition 不清晰
lack of equipment 设备缺乏;设备不足
lack of filling power 枯瘦;木器漆塌渗
lack of fit 不相称;不适合
lack of fuel 燃料缺乏
lack of fusion 熔化不良;熔合不良;未熔合;熔化不全
lack of hiding 遮盖力弱
lack of interpenetration 层间未焊透
lack of inter-run fusion 层间未熔合
lack of joint penetration 接头未焊透
lack of lubrication 缺乏润滑
lack of manpower 缺乏人力
lack of materials 缺乏存货;缺乏材料;材料短缺
lack of offer 缺少报盘
lack of order 缺少订货
lack of penetration 未熔透;未焊透
lack of rainfall 缺雨
lack of registration 配准不佳
lack of resolution 鉴别力损耗;清晰度欠佳;分辨率不足
lack of root fusion 根部未焊透
lack of sharpness 不清晰
lack of side fusion 侧面未熔合
lack of space 缺少空间
lack of supplies 缺乏供应品
lack of tolerance 缺乏耐性
lack of uniformity 品质欠均匀;不均衡性
lack of uniformly optimal plan 一致最优方案的不足
lack of water 缺水
lac lake 紫胶色淀;虫漆染料
lacmoid 间苯二酚蓝
laconicum 古罗马浴室中的热气浴室;热气室
Lacoste and Romberg 拉科斯特—隆伯格
Lacoste pendulum 拉科斯特摆
Lacoste-Romberg gravimeter 拉科斯特—隆伯格重力仪
lacotile (背衬防水压缩木的)大张仿制屋瓦
Lacour converter 串级变换器
Lacour motor 拉库尔电动机
La Court des Comptes of France 法国审计法院
lacquer 硝基纤维素漆;真漆;亮漆;蜡克;胶片;挥发性漆;喷漆蜡克;喷漆
lacquer brush 清漆刷;漆刷

lacquer cable 漆包线
lacquer carving 漆雕
lacquer coat 漆涂层
lacquer-coated steel sheet 涂漆薄钢板
lacquer curtain coating(method)喷漆法;幕帘涂清漆法
lacquer deposit 漆沉积
lacquer diluent 稀漆剂;油漆稀料;漆料稀释剂
lacquer disc 蜡克盘;胶盘
lacquered board 涂漆板
lacquered hardboard 涂(清)漆(的)硬质木纤维板
lacquered plate 涂漆镀锡薄钢板
lacquered sheet iron 镀漆铁皮
lacquered steel 涂漆钢材
lacquered ware 漆器
lacquer enamel 硝基瓷漆;挥发性瓷漆;搪瓷漆;瓷漆;珐琅
lacquer engraving 雕漆
lacquerer 漆工;油漆工
lacquer factory 油漆厂
lacquer film 挥发性漆膜;清漆薄膜
lacquer for floor 地板漆
lacquer formation 漆生成
lacquering 髹漆;漆涂层;漆沉积;喷清漆;涂(清)漆;刷漆;上漆;打蜡克;成漆;喷漆
lacquering japanning 涂漆
lacquering machine 喷漆机
lacquer lifting 漆咬底;漆起皱
lacquer-like protecting layer 类漆防腐蚀保护
lacquer man 漆工
lacquer master 蜡克主盘
lacquer oil 喷漆用油
lacquer original 蜡克原盘
lacquer paint 喷漆涂料
lacquer painting 漆画
lacquer petroleum 漆用汽油
lacquer plant 漆树
lacquer putty 喷漆底层用油灰;饰面用油灰
lacquer relief 漆片雕
lacquer-removing preparation 除漆制剂
lacquer sealer 硝基(纤维)封闭底漆;漆封剂
lacquer sheathing 漆复盖层
lacquer softener 漆用软化剂
lacquer solution 漆溶液
lacquer solvent 油漆溶剂;助溶剂;溶漆剂;漆用溶剂
lacquer surfacer 硝基(纤维)二道浆
lacquer thinner 硝基(漆)稀料;硝基(纤维)漆稀释剂;香蕉水;稀漆剂;挥发性漆稀释剂;漆(料)稀释剂;漆冲淡剂;喷漆稀料
lacquer tree 漆树
lacquer type organic coating 树脂漆
lacquer varnish 亮(清)漆;亮漆;凡立水;喷涂清漆;喷漆用清漆
lacquer vehicle 亮漆媒液
lacquerware 漆器
lacquer wax 漆蜡
lacquer with gold design 描金漆器
lacquer work 漆细工;涂漆;打蜡克;漆器
lac resin 虫胶树脂
lacroixite 锥晶石
lacrosse 军事测距系统
lacrosse ground 长曲棍球场
lactam 内酰胺;乳胺
lactamic acid 丙氨酸
lactary 奶场
lactate 乳酸酯
lacte 乳白色
lactescence 乳状;乳汁色
lactescent 乳状的;乳色的
lactic acid 乳酸;丙醇酸
lactide 交酯类;丙内酯;丙交酯
lactim 内酰亚胺
Lactobacillus acidophilus 嗜酸乳杆菌
lactocrit 测乳脂汁
lactolite 乳酪塑料
lactone colo(u)ring matter 内酯染料
lactone isomerism 内酯异构现象
lactone process 内酯法
lactone rule 内酯规则
lactonic acid 内酯酸
lactonic ketone 内酯
lactonic ring 内酯环
lactoprene 聚酯橡胶;乳胶

lactose 乳糖
lacuna 缺失(地层);空白区;地层缺失;镶嵌装饰槽;洼地;花格平顶;顶棚镶嵌板;[复] lacunae
lacunae lacuna 的复数形式
lacunar 藻井天花板;藻井平顶;花格平顶【建】;凹格顶棚;镶板顶棚
lacunaria 井式平顶;井式顶棚;坑式平顶(古希腊建筑);(庙宇内殿外走廊的)天花板;花格顶棚
lacunaris 井式平顶
lacunarity 缺顶
lacunary 缺顶的;多小孔的
lacunary function 缺顶函数;缺顶函数
lacunary series 缺顶级数;缺顶级数
lacunary space 缺顶空间;缺顶空间
lacunary structure 缺顶结构
lacunate mode 陷窝状水系模式
lacunosus 网状云
lacus aestatis 夏湖
lacus autumn 秋湖
lacus mortis 死湖
lacus somniorum 梦湖
lacuster 湖泊中心区
lacustrine 湖成的;湖(沉)积的;湖泊的
lacustrine algal facies 湖泊藻质相
lacustrine basin 湖成盆地
lacustrine beach 湖滩
lacustrine clay 湖沼黏土;湖积黏土;湖底黏土;湖成黏土;湖泊沉积黏土
lacustrine-clay soil 湖积黏土土壤
lacustrine conglomerate 湖成砾岩
lacustrine deposit 湖(沼沉)积物;湖相沉积(物);湖成堆积;湖成沉积物;湖沉积;湖沼沉积(物)
lacustrine environment 湖环境;湖泊环境
lacustrine facies 湖相;湖泊相
lacustrine fishes 湖泊鱼类
lacustrine formation 湖积层;湖相沉积层
lacustrine herbaceous facies 湖泊草本相
lacustrine landform 湖成地形
lacustrine limestone 介壳灰岩
lacustrine muck 湖积腐泥;湖底腐泥;湖成腐泥
lacustrine nodule 湖相结核
lacustrine peat 湖积泥炭;湖泊沉积泥煤
lacustrine plain 湖积平原;湖成平原
lacustrine plant 湖沼植物
lacustrine sediment 湖(沼沉)积物;湖相沉积(物);湖泊沉积物
lacustrine sedimentation 湖泊沉积(作用)
lacustrine sedimentation model 湖积沉积模式
lacustrine sedimentology 湖泊沉积学
lacustrine soil 湖积土(壤)
lacustrine structure 湖成结构
lacustrine succession 湖泊序列
lacustrine system 湖泊体系
lacus veris 春湖
lac varnish 紫胶清漆;光(清)漆;清漆;虫胶清漆
lac wax 紫胶蜡;虫胶蜡
lac without sticks 无黏性的紫胶;无黏性的生漆
ladang 移动性农业;迁移农业
ladar 光雷达
ladder 阶梯;码头爬梯;棚架;爬梯;梯(子);梯状物;梯形裂缝;斗架
ladder access 梯道;扶梯;用梯子上达
ladder and cage 梯子和护罩
ladder apes-ear ring 梯形猴耳环
ladder attenuator 梯形衰减器
ladder-back chair 梯形靠背椅
ladder board 梯子板
ladder borehole 梯架式钻孔
ladder bracing 格条;桁架格条;梯形拉条;杵架格条
ladder bucket dredge(r) 多斗挖泥船;链斗式挖泥船
ladder cable through 阶梯式电缆架;阶梯式电缆沟
ladder car 斗式运输车
ladder chain 钩environchain;梯状链;梯形链条
ladder chain carrying bucket 链斗式提升机;运载链斗的梯形链条
ladder chart 梯形图
ladder circuit 梯型线路;梯形电路
ladder cleat 扶梯横档
ladder company 云梯消防队
ladder control 梯式控制
ladder core 板条门芯
ladder cradle 梯座(配合空气腿用)
ladder diagram 阶梯图形;梯形图;顺序控制图
ladder ditcher 梯形挖沟机;多斗式挖沟机;链斗

(式)挖沟机
ladder dredge(r) 链索斗挖泥机;联斗挖泥船;多斗(式)挖泥机;斗梯式挖泥船;梯形链斗式挖泥机;链斗式挖泥机;链斗式挖泥船;链斗式挖泥机
ladder drill 阶梯形钻探
ladder drill hole 梯形式钻孔
ladder drilling 阶梯式钻孔;梯式钻眼法;梯式钻孔(法);梯(架)式钻进(法)
ladder drilling method 梯架式钻孔法;梯架式钻井法
ladder effect 梯形效应
ladder elevator 多斗式挖泥船
ladder escape 安全梯
ladder excavator 链斗挖土机;链式挖掘机;梯式挖掘机;多斗挖土机;斗梯式挖土机
ladder for repairing 修理梯
ladder frame 链斗框架;梯架
ladder fuel element 分级燃料元
ladder gasket 梯形垫块
ladder grid numbers 制图格网数字注记
ladder guard 爬梯安全装置
ladder guide 斗架导轨;斗架导板
ladder hoist 多斗升降机;梯式起重机
ladder hoisting gear (挖泥船的)铰刀架起落装置;斗架起落装置;(挖泥船的)臂铰刀架起落装置
ladder hoist winch 起桥绞车;斗桥提升车
ladder hoist wire 铰刀架缆;斗架缆
ladder hook 梯钩
ladder index 阶梯指数
ladder jack scaffold 油漆工脚手架;梯承脚手架;台架脚手架;轻型脚手架
ladder landing[platform] 梯子平台;(钻塔的)梯台
ladder lodes 梯状矿脉
ladder marking 梯形蚀痕
ladder method 阶梯法
ladder-mounted drill 链斗式钻机
ladder niche 梯槽
ladder of cascades 梯级跌水
ladder of management 管理手段
ladder operator 梯算符
ladder pivot 桥架枢轴
ladder polymer 梯形聚合物
ladder purchase 绞架滑轮组;斗架滑轮组
ladder rack 阶梯形齿条;梯级形齿轨
ladder rail 舷梯杆;舷梯扶手;梯栏杆
ladder recess 爬梯槽
ladder road 梯子道
ladder roadway 梯子间
ladder rod 绳梯棍
ladder roller 斗架滚筒;导梯托辊
ladder rung 梯子横木;梯踏步;绳梯棍;梯阶
ladders 梯状波皮
ladder scaffold(ing) 梯台架;梯式脚手架;台架脚手架;梯承脚手架
ladder scraper 皮带装料铲运机;链斗式括土机
ladder screen 舷梯围布;梯下安全网
ladder secondary linear floating motor 梯状次极直线悬浮电机
laddershaped distribution method 梯形分配法
ladder shell 梯螺
ladder sign 阶梯征
ladders in painting 刷涂梯(形条)纹
ladder sollar 梯子平台
ladder stair(case) 踏板式楼梯;梯级
ladder stand 可移式舷梯架
ladder stay 梯子架
ladder step 扶梯梯级
ladder stile 梯的竖桩
ladder stitch 梯缝
ladder string 梯子侧板
ladder tilt factor 斗架倾角系数
ladder to manhole 人孔梯子
ladder track 梯子形轨道;梯形线;梯线[铁]
ladder trencher 链式挖沟机;多斗式挖土机;多斗式挖沟机
laddertron 梯形管
ladder truck 云梯消防车;梯架汽车
ladder trunnion 桥架耳轴
ladder type 梯形
ladder-type circuit 梯型电路;梯形电路
ladder-type decoupling filter 梯型去耦滤波器
ladder-type ditcher 梯式挖沟机
ladder-type fender 舷梯型橡皮护舷
ladder-type filter 梯形网络;梯形滤波器

ladder-type frame 梯形车架
ladder type network 梯形网络
ladder-type support 梯形棚子
ladder-type trenching machine 梯式挖沟机
ladder unloader 链式卸货机
ladder variable 阶梯变量
ladder veins 梯状矿脉
ladder wall 梯式加筋锚定墙
ladder way 人行梯子;带梯子的垂直矿井通道;梯子道;梯口
ladder-way compartment 梯子间
ladder weave 梯形组织
ladder web 梯状织物;(百叶窗的)梯形带
ladder well 桥档;挖泥槽;梯形裂缝凹槽;竖梯井孔
ladder-well gantry 桥档龙门
ladder winch 梯斗式起重绞车
ladder winch speed controller 桥架绞车速度控制装置
ladder work 梯工
laddic 多孔磁芯
lade 汲取;汲出
laden 装着;载满;荷载的
laden belt 潜水压铅带
laden draft 满载吃水
laden earhead 满浆穗
laden height 满载状态下的高度
laden hull 满载船舶
laden in bulk 散装满载;散装的
laden load 装载量
laden speed 满载航速
laden weight 装载重量;毛重;满载重量;车辆总重(量)
laden weight ratio on front wheel 装车前轮荷载比
laden weight ratio on tyre 装车时轮胎荷载比
Ladestar 北极星
ladies' changing room 女更衣室
ladies' drawing room 女客厅;女休息室
ladies' hair dressing shop 女子理发室;女理发店
ladies' room 女卫生间;女厕(所);女盥洗室
ladies' shop 妇女用品商店
ladies' toilet 女盥洗室;女厕(所)
ladies wash closet 女厕所
ladies' water-closet 女厕(所)
ladies' WC 女厕(所)
lading 荷载;船货
lading door 装料门
lading order for free goods 免税品起货批准单
ladinian regression 拉丁期海退
Ladinian(stage) 拉丁阶【地】
ladkin 硬木刮槽工具(撬起窗上铅条的硬木工具)
ladle 铸桶;铸杓;钢包;铁水包
ladle addition 罐内加料;炉前料;桶内加料;包内添加剂
ladle analysis 桶样分析;包样分析
ladle bail 铸钢桶挂钩
ladle barrow 铸钢桶手推车;铸锭车
ladle body 桶身
ladle bowl 罐壳;桶壳
ladle brick 铸钢桶衬砖;钢包衬砖;盛钢桶衬砖
ladle capacity 桶容量
ladle car 钢水罐车;桶车;铁水包运输车
ladle carriage 铸桶车
ladle carrier 铸桶叉形夹
ladle casing 桶壳
ladle chemistry 桶样成分;包样化学成分
ladle clay 造型黏土;浇(注)包火泥
ladle cooling pot 料勺冷却盘
ladle cover 包盖
ladle crane 铁水包吊车
ladle crane trolley 钢包吊车;铁水包吊运车;盛钢桶吊运车
ladle degassing 包内去氧
ladle deoxidation 包内脱氧
ladle deoxidizing 包内脱氧
ladle dryer 浇包烘炉;烘包器
ladle forehearth 倾动式前炉
ladleful 满桶量
ladle heater 烘包器
ladle heel 包底剩铁
ladle inoculation 包中孕育处理
ladle liner 预制包衬;浇包衬
ladle lining 浇铸时内衬;桶衬
ladle lip 浇包嘴;桶唇
ladleman 铸锭工

ladle mixing 桶内混合
ladle of the bottom pour type 底流式桶
ladle pit 出钢坑
ladle pot 铸罐;浇桶;烧桶;铁水包
ladler 玻璃浇注工
ladle refining 桶中精炼
ladle sample 桶样;包样;包内取样
ladle scull 桶内结壳
ladle shank 抬包
ladle spout 浇包嘴
ladle stand 钢包支座
ladle test 桶样试验;桶形试验
ladle tilter 铁水包倾注装置;盛钢桶倾注装置
ladle-to-ladle 倒包
ladle treatment 包内处理
ladle trunnion 浇包轴颈
ladle wash 浇包涂料
ladle well 桶底虹吸池
ladling 金属装桶
Ladoga precision depth recorder 拉多加型精密回声测深仪
lady 小石板;探照灯控制设备
Lady Chapel 圣母堂;圣母院;小礼拜堂;圣母玛利亚礼拜堂
lady cracker 吨边炮
lady's slipper 美国杓兰属
laevis 光滑的
laevorotatory quartz 左旋石英
Lafarge cement 拉法吉水泥;大理石粉石灰石膏水泥
Lafferty ga(u)ge 热阴极电离计
laffittite 硫砷汞银矿
Lafond's Tables 拉丰表【气】
lag 滞留物;滞后调整;滞后;隔热套;落后;互交车头时距;汽缸隔热层格;汽缸保温套;迟滞;残留粗屑沉积;延迟;套板;防护套
lag adjustment 滞相调整
lagan 系浮标的投弃货物
lag-and-route method 滞后加演算法;迟滞演算法
lag angle 连接角钢;结合角钢
lag behavio(u)r 滞后特性
lag behind 落后
lag bolt 木螺栓;短螺栓;方头螺栓
lag characteristic 延迟特性;落后特性
lag coefficient 滞后系数;后延系数;时间常量
lag compensation 滞后补偿
lag compensator 滞后补偿器
lag correlation 滞后相关;后延相关
lag covariance 落后协方差
lag curve 滞后特性曲线
lag damper 滞后阻尼器
lag deposit 滞留沉积;滞积质;滞积层;滞后淤积;滞后沉积;残留沉积(物);粗化沉积
lag distance 滞积距离;沉积距离
lag domain 时滞域
lag element 滞后元件
lagena 壶
lagend 系浮标的投弃货物
lagengneiss 层状片麻岩
lageniform 瓶形的
lag error 滞后误差;迟滞(误)差
lag-error correction 迟滞误差修正
lag fault 滞后断层;大余角正断层
lag felt 矿渣毡
laggard 落后者
lagged-demand meter 时滞需量计
lagged liner 槽底挡料衬板
lagged pile 加套桩;加套管的桩;套桩;绝热管道;护热管道
lagged pulley 带套鼓轮
lagged type 滞后型
lagged variable of unanticipated inflation 非预期的价格上涨中滞后变量
lagging 支拱木料;护壁(板);横挡板;外罩;外套;挡木板;迟延;保温(外)套;延迟;套板;防护套
lagging angle 滞后角
lagging board 背板(打桩)
lagging capacity 滞后容量
lagging casing 绝热外壳;保温外壳
lagging circuit 滞后电路
lagging cloth 防护布
lagging coil 滞后线圈
lagging commutation 延迟换向;滞后换向
lagging cover 粗镀;粗涂
lagging current 滞后电流

lagging curve 滞后曲线
lagging device 滞相装置;滞后装置
lagging edge 下降边;延迟反馈;脉冲后沿;后缘;后沿;随边
lagging effect 滞后效应
lagging feedback 迟滞反馈
lagging filter 迟后滤波器
lagging gauze 保护层金属膜
lagging gravel 滞后砾石
lagging green signal phase 后期绿灯信号相
lagging half axle 横半轴
lagging index 滞延指数
lagging indicator 滞后(经济)指标;落后指标;拉后指标;拖后指标
lagging jack 拱架;拱鹰架
lagging jacket 汽缸保温套
lagging load 滞后负荷(电流);电流滞后的负荷;电感性负荷
lagging material 绝缘材料;绝热材料;保温材料
lagging of arch-center 拱架模板
lagging of neap 小潮迟后
lagging of phase 相滞;相位滞后
lagging of pile 桩箍
lagging of the tide 潮时落后(每天迟约49分钟)
lagging of tide 月潮间隙;迟潮时间;潮汐滞后;潮汐滞滞;潮时滞后;潮时延后
lagging on the tide retardation 潮时落后(每天迟约49分钟);潮汐迟滞
lagging phase 滞后相位
lagging phase angle 滞后相位角
lagging phase feeding section 滞后相供电臂
lagging phase operation 滞相运行
lagging pile 横挡板;支拱板框;支拱板条
lagging power factor 滞后功率因数
lagging section 预制管道隔热块;保温罩;预制保温罩
lagging section for cold protection 预制防冻保护层
lagging section for heat protection 预制放热保护层
lagging shadow 落后阴影
lagging truck 落后履带
lagging-type equations 滞后型方程
lagging voltage 滞后电压
lagging water 缓流水
lagging wattless load 无功滞后负荷
lag gravel 滞留砾石;残积砾石;残积砂石
lag ignition 迟点火
lag in investment 延缓投资
lag in phase 相位滞后;相滞
lag intake 迟进气
lag in technology 技术落后
lag interval 滞后间隔后区间
lag knock 滞后燃烧爆震
lag lead compensation 滞后超前补偿
lag lead compensator 滞后超前补偿器
lag method 滞时法;滞后法
lag network 滞后网络
lagniappe 免费赠品
lago 湖泊
lag of basin 流域滞时;流域汇流时间
lag of boiler 锅炉外套
lag of compass 罗盘迟滞
lag of controlled plant 调节对象滞后
lag of culmination 中天迟滞
lag of higher order 高次谐波滞后
lag of ignition 发火落后
lag of the gyrocompass 陀螺罗经迟滞
lag of the tide 潮汐滞后
lag of time 滞后时间;时间滞后;时间滞差
lag oil zone 残油带
lagoon 泻湖;近海湖;环礁湖;海湾湖;海岸边咸水湖;污水塘
lagoonal deposit 泻湖沉积
lagoonal formation 泻湖建造
lagoon area 泻湖区
lagoon beach 泻湖滩
lagoon channel 环形珊瑚岛水道
lagoon cliff 泻湖坡
lagoon complex 泻湖复合体
lagooned water 泻湖水
lagoon facies 泻湖相
lagoon for holding sludge 淤泥泻湖;淤泥(沉淀)池;污泥塘

lagoon harbo(u)r 礁湖港;湖港;河口浅水港;河口湖港;浅水河口港;泻湖港湾;近海湖港
lagooning 自然池净化;湖存污;污泥池塘法;塘处理
lagooning of sludge 污泥池蓄处理
lagoon island 泻湖岛;内岛;环形珊瑚岛;环礁
lagoon margin 泻湖边缘
lagoon marsh associes 泻湖沼泽演替系列群丛
lagoon mouth 泻湖口;礁湖河口;浅水河口
lagoon mud 泻湖淤泥
lagoon port 礁湖港;浅水河口港
lagoon pretreatment 塘预处理
lagoon process 氧化塘工艺;氧化塘法;池塘法
lagoon reef 环形珊瑚岛
lagoon sand 泻湖砂
lagoon sand-bar deposit 泻湖沙坝沉积
lagoon scarp 泻湖坡
lagoon sedimentary soil 泻湖沉积土
lagoon shoreface deposit 泻湖滨面沉积
lagoon-side 泻湖岸边土地
lagoon slope 泻湖坡
lagoon water 咸水湖
lagoriolite 硅酸铝钠
lag period 停滞期
lag phase 延迟期;滞后阶段;缓慢期;停滞期;停滞阶段;迟滞期
lag phenomenon 迟滞现象
lag pile 套桩;加套桩
lag position 随后位置
Lagrange's equation 拉格朗日方程
Lagrange's interpolation formula 拉格朗日插值公式
Lagrange's problem in calculus of variations 变分法中的拉格朗日问题
Lagrangian approach 拉格朗日求解法
Lagrangian coordinates 拉格朗日坐标
Lagrangian current measurement 拉格朗日测量法
Lagrangian equation 拉格朗日方程
Lagrangian equation of motion 拉格朗日运动方程
Lagrangian function 拉格朗日函数
Lagrangian generalized velocity 广义速度
Lagrangian identity 拉格朗日恒等式
Lagrangian interpolation 拉格朗日插值
Lagrangian method 拉格朗日法
Lagrangian method of determinate 拉格朗日待定系数
Lagrangian model 拉格朗日模型
Lagrangian multiplier 拉格朗日乘子
Lagrangian-Newton method 拉格朗日一牛顿法
Lagrangian operator 拉格朗日算子
Lagrangian parameter 拉格朗日参数
Lagrangian wave 拉格朗日波
lag ratio 滞积比
lag regression 落后回归;后延回归
lags 延迟支付进口货款
lag sand 滞积砂;残积砂;风蚀留沙
lag screw 长螺钉;方头螺钉;板头尖端木螺钉;方头尖螺钉
lags in response 反应迟钝
lag spike 螺丝钉
lag stage 缓慢阶段(污泥沉降)
lag time 延迟时间;滞时;滞后时间;汇流时间;时间差距
Laguerre polynomial 拉盖尔多项式
Laguerre's differential equation 拉盖尔微分方程
laguna 泻湖;小湖;短暂浅湖
lag unit 滞后部件
lag window 滞后窗
lag wood screw 木螺钉
lahar 泥石流;火山碎屑滑坡;火山泥流(物)
lahar deposit (火山型的)泥石流沉积;泥流沉积(物)
laid 横置
laid antique 条纹仿古纸
laid bare 揭开;裸露的
laid cold 冷铺的;冷铺
laid concealed 暗敷设
laid contrary to the stratum 递层铺砌
laid deck 弯曲状木甲板
laid down cost 敷设成本
laid dry 干砌的;干砌
laid finish 条纹装饰
laid hot 热铺的
laid-in effect 衬热效应
laid-in mo(u)lding 镶入线条;镶嵌线条

laid in panels 格铺砌;按格铺砌
laid-in thread 衬垫纱线
laid length (管道的)铺设长度;敷设长度;敷管长度;安装长度
laid line 直纹线条
laid-off work 解雇;停工
laid on 安装
laid-on-edge course 立砌砖层;侧砌
laid on end 立砌
laid-on mo(u)lding 贴附线脚;安装线条;安装线脚【建】;预制装配式线脚
laid-on stop 钉子樘子上的止门条;钉子樘子上的止窗条
laid-on thread 外嵌凹引线条(装饰用);贴玻璃线
laid paper 直纹纸
laid paving mixture 嵌置的铺面混合料;嵌入的铺面混合物;嵌入的铺面混合料
laid rope 六股七丝钢索
laid touching (砌石的)咬接
laid up 闲置;系船;搁置;卧置;停泊(指锚泊作业);待修;拆卸修理
laid-up capital 闲置资金
laid-up fleet 停止使用的船队
laid-up in port 停泊在港
laid-up return 停泊退费
laid-up ship 闲置船只;停航闲置船;闲置不用的船
laid-up tonnage 闲置不用的船;停航船泊;闲置吨位
laid-up vessel 闲置船舶
laid-up vessel anchorage 闲置船锚地;待修船锚地
laid wire 直纹网
laid wire rope 普通扭纹钢丝索
laigh lift 泵组中最下面的水泵
laihunite 莱河矿
Lainer's reducer 莱纳减薄液
laing's floor 梁氏地板(由预制梁与砖块构成的防火地板)
Laingspan 拉英斯攀式体系建筑(英国建筑体系之一,主要用于学校建筑)
lair 泥潭
lairages 到岸牲口临时栏圈
laissez-faire 放任;不干涉
laissez-passer 免验证;通行证
laitakarite 硫硒铋矿
laitance 乳剂;水泥浮浆;水泥浮沫;浮浆皮;浮浆(混凝土或水泥表面因操作不当,水分过多等原因而产生的乳白色薄膜,凝结后称为浆皮);灰浆;翻沫(由泌水引起的混凝土表面的沫);表面水泥浆;白霜
laitance coating 浆皮;水泥浮浆层;浮浆(表)层(在混凝土表面)
laitance layer 浆皮;水泥翻沫层;浮浆(表)层(在混凝土表面);翻涂层;翻沫层
laitance on surface 表面翻沫;表面浮浆
laitier 浮渣
laity 外行人;门外汉
lakao 中国绿;绿胶;绿膏
lakarpite 钠闪正长岩
lake 媒色颜料;湖泊;色淀
lake-accumulated bank 湖积浅滩
lake acidification 湖泊酸化
lake acidification mitigation 湖泊酸化减弱
lake acidity monitoring network 湖泊酸度监测网
lake aeration 湖水复氧;湖泊曝气
lake algae 湖泊藻类
lake and reservoir floodproofing 湖库防洪
lake and river mud 泥地
lake area 湖区;(湖)泊)面积
lake arm 湖汊
lake asphalt 湖沥青;湖地沥青;湖池沥青;天然沥青
Lake Balkash, Ebi Nor wind 尔伯风
lake bank 湖堤;湖岸
lake-bank slope 湖岸岸坡
lake basin 湖区;湖地;湖(泊)盆地;湖(泊)流域
lake basin bog 湖沼;湖沼
lake beach deposit 湖滩沉积
lake bed 湖积层;湖底;湖床;湖泊沉积矿
lake-belt 湖泊地带
lake biocoenosis 湖泊生物群落
lake blue 色淀蓝
lake bog 湖泊沼泽;湖泊泥塘
lake bordeaux 色淀枣红
lake bottom 湖底
lake bottom mud 湖底淤泥
lake bottom reclamation 开垦湖地

lake breeze 湖风
lake brine 湖盆盐水
lake carrier 湖船
lake center 湖心
lake channel 湖泊航道
lake circulation 湖水混合作用;湖泊环流
lake clay 湖黏土;湖泥
lake cliff 湖蚀岸
lake-cliff trap 湖岸圈闭
lake cold water fish habitat 湖泊冷水鱼类生境
lake colo(u)rs 色淀颜料;色淀染料;沉淀色料
lake conservation 湖泊调蓄;湖泊保护
lake copper 湖铜
lake current 湖流
lake delta 湖三角洲
lake delta deposit 湖泊三角洲沉积
lake delta facies 湖泊三角洲相
lake deposit 湖成泥沙;湖(泊)沉积(物)
lake depth 湖深
lake district 湖区
lake drainage 湖泊排泄;湖泊流域
lake drainage basin 湖盆;湖泊流域;湖泊集水区
lake due to erosion 侵蚀湖
lake due to landslide 山崩湖
lake due to warping 翘曲湖
lake dwelling 湖上木屋;湖边房屋;水上屋;湖上居住
lake dyes 色淀染料
lake eco-environment 湖泊生态环境
lake ecologic(al) system 湖泊生态系统
lake ecosystem 湖泊生态系统
lake ecosystem pollution 湖泊生态系统污染
lake ecosystem restoration 湖泊生态系统修复
lake effect 湖泊效应
lake effect storm 湖上雪暴
lake elevation 湖面高程
lake environment 湖泊环境
lake environmental capacity 湖泊环境容量
lake environmental quality assessment 湖泊环境质量评价
lake-eroded bank 湖蚀浅滩
lake-eroded cave 湖蚀穴
lake eutrophication 湖泊富营养化
lake eutrophication analysis procedure 湖泊富营养化分析程序
lake eutrophic criterion 湖泊富营养标准
lake evaluation index 湖泊评价指标
lake facies 湖相
lake fishery 湖泊渔业
lake fleet 湖运船队
lake floor 湖床
lake-floor plain 湖底平原
lake flora 湖泊植物系
lake flow 湖流
lake fluctuation 湖面波动;湖面变化
lake forecast 湖泊(水文)预报
lakefront 湖滨;湖边平地;湖滨游憩沙滩;湖滨避暑山庄
lake geomorphology 湖泊地貌(学)
lake-george diamond 水晶
lake harbo(u)r 湖港
lakehead 湖尾
lake hydrobios 湖泊水生生物
lake hydrology 湖泊水文学
lake ice 湖冰
lake inlet 湖口;湖(泊)入口
lake intake 湖水引水;湖水进入口
lake investigation 湖泊调查
lakeland 多湖(泊)地区;湖泊地区
lake landform 湖泊地貌
Lakeland walling 湖泊地区围墙
lake length 湖泊长度
lakelet 小湖
lake level 湖水位;湖水面;湖平面;湖面水准
lake loess 湖积黄土;湖泊沉积黄土
lake management 湖泊管理
lake marl 沼灰泥;湖成泥灰岩;湖灰泥;湖灰土
lake marsh 湖沼
lake marsh reclamation 湖沼围垦
lake moor 湖沼
lake morphology 湖泊形态学
lake mud 湖泥
lake natural purification 湖水自净(化)
lake nourishment 湖泊营养情况
lake obliteration 湖泊消失

lake of acid yellow 色淀酸性黄
lake of cold air 冷空气湖
lake oil 琥珀油
lake orange 色淀橙
lake ore 沼铁矿;褐铁矿
lake outflow 湖泊出流量
lake outlet 湖(泊)出口;出湖口
lake-outlet channel 湖泊出(水口)水道;湖泊出水道;出湖口水道
lake peat 湖泥炭;湖成泥炭
lake phytoplankton 湖泊浮游植物
lake pigment 色淀颜料
lake pitch 湖沥青
lake plain 湖(成)平原
lake pollution 湖泊污染
lake pollution control 湖泊污染控制
lake ponded up by lava 熔岩堵成的湖
lake port 湖港
lake pumping station 湖水泵送站
laker 湖船;大湖轮
lake reclamation 湖泊改良
lake red 色淀红
lake region 湖区
lake remover 沉淀色料去除剂
lake resources 湖泊资源
lake restoration 湖泊修复
lake retention 湖泊滞洪
lake rim swamp 湖缘林沼
lake-river oil spill simulation model 湖泊河流溢油模拟模型
lake round scenic spot 湖泊风景区
lake sand 湖砂
lake sediment 湖泊底泥;湖泊沉积(物)
lake sediment anomaly 湖积物异常
lake sediment sample 湖积物样品
lake self-purification 湖水自净(化)
lake shoal 湖岸浅滩
lakeshore 湖滨;湖岸;湖滩
lakeshore development 湖滨开发
lakeshore landform 湖岸地貌
lake shoreline 湖岸线
lakeshore zone 湖岸带
lakeside 湖滨
lakeside channel 滨湖航道
lakeside development 湖滨开发
lakeside hotel 湖滨旅馆
lakeside park 滨湖公园
lakeside zone 湖滨带
lake stage 湖水位;湖泊发育阶段
lake steamer 大湖轮
lake stone 湖石
lake storage 湖泊蓄水(量)
lake strandline 湖岸线
lake stratification 湖水分层;湖泊分层
lake surface 湖面
lake survey 湖泊调查;湖泊测量
lake terrace 湖成阶地;湖(泊)阶地
lake thermohydrodynamics 湖泊热流体动力学
lake toner 色淀调色剂
lake toxic analysis 湖水毒性分析
lake-traffic management 湖上交通管理
lake-traffic regulations 湖上交通管理条例
lake trophic level 湖泊营养水平
lake type 湖型;湖泊类型
lake valley 湖谷
lake view 湖泊景观
lake-village 水上住家;湖上桩屋村落
lake volume 湖容
lake wall 湖墙;湖边低砂砾堤
lakeward 向着湖泊的
lake water 湖水
lake water level 湖水位
lake water quality 湖泊水质
lake water quality mode 湖泊水质模式
lake water quality model 湖泊水质模型
lake water sampling 湖水采样
lake watershed 湖泊流域
lake watershed management 湖泊流域管理
lake water temperature 湖水温度
lake water temperature condition 湖水温度情况
lake with marine vestiges 海迹湖
lake with outlet 有出口湖;湖出流;出口湖
lake without outflow 无出流湖;无口湖
lake without outlet 无口湖;无出口湖

lake zoobenthos 湖泊底栖动物
lake zooplankton 湖泊浮游动物
Lako-type iron deposit 拉科式铁矿床
laky 深红色的;湖状的;多湖泊的
lalique 模制装饰耐热玻璃
Lalique glass 浮雕图案玻璃
Lallemantia oil 拉曼油
lally column 空心圆钢柱;混凝土填心圆筒形钢结构柱
lam 拉综梁;砂质黏土;沙质黏土
lama 矿脉泥
lamagal 镁铝耐火材料
Lamaism 喇嘛教
lamaist pagoda 喇嘛塔
Lamaist temple 喇嘛寺院
lamalginite 薄层藻类体
Lamarck's theory 拉马克学说
lamasery 喇嘛寺院;喇叭庙
Lamassu 人头飞牛雕像(美索不达米亚宫殿庙宇门口)
Lama temple 喇嘛寺院
lambda 兰姆达海洋水文定位系统;人字缝尖;人字点
lambda-type 人字形
Lambda-type structure 人字形构造
lambdoidal 三角形的;人字形的
lambdoidal suture 人字缝
lambdoid margin 人字缘
lamber-core construction 板条芯细木结构
Lambert 朗伯(亮度单位)
Lambert-Beer law 朗伯—比尔定律
Lambert-Garland mooring buoy 朗伯—嘎兰式系船浮标
Lambertian distribution 朗伯分布
Lambertian extended source 朗伯扩展光源
Lambertian radiator 朗伯辐射体
Lambertian reflector 朗伯反射体
Lambertian source 朗伯源
Lambertian surface source 朗伯表面光源
lambertite 斜硅钙铀矿
Lambert's albedo 朗伯反照率
Lambert's azimuthal equal area projection 朗伯特等积方位投影
Lambert's bearing 朗伯特等角投影海图方位;朗伯航向
Lambert's conformal chart 朗伯特正形投影图
Lambert's conformal projection 朗伯特正形投影
Lambert's cosine law 朗伯余弦定律
Lambert's cosine law of incidence 朗伯入射余弦定律
Lambert's cosine relation 朗伯余弦关系式
Lambert's course 朗伯航程
Lambert's emission law 朗伯发射定律
Lambert's equation 朗伯方程
Lambert's grid 朗伯特投影格网
Lambert's law 朗伯定律
Lambert's law of absorption 朗伯吸收定律
Lambert's law of reflection 朗伯反射定律
Lambert's law radiator 朗伯律辐射体
Lambert's projection 朗伯特投影;天顶等积投影
Lambert's surface 朗伯表面
Lambert's zenithal equal area projection 朗伯特等积方位投影
lambing shed 产棚
lamboanite 暗色混合岩
lambrequin (门窗的)贴脸板;门窗装饰性短帘;装饰性挂帘;盔饰盖布;门窗垂饰;边花饰
lamb roller 羊皮加工辊
Lamb's dip 兰姆凹陷
Lamb's dip frequency stabilization 兰姆稳频
lambskin 劣质无烟煤;羊皮纸;羔羊皮
Lamb's problem 兰姆问题
Lamb's semiclassical theory 兰姆半经典理论
Lamb's shift 兰姆移动;拉姆移位
lamb's tongue 羊舌饰【建】
Lamb's wave 兰姆波
lamb's wool cylinder 羊羔毛加工滚筒
Lambton flight 刮板装载机
Lamb wave 板波
lame 金属薄板;金银丝锦缎
lamel 薄板
lamel cathode 薄膜阴极
lamella 联方格网;壳层;片层;薄片(层);薄层
lamella arch 叠层拱
lamella bundle manufacture 板束的制造

lamella clutch 片式离合器
lamella construction 叠层构造
lamella cupola 肋格半球(圆屋)顶
lamella dome 肋格穹顶
lamella grid 叠层网格
lamella heat exchanger 板壳式换热器
lamella mat 网格薄毡;叠层垫;层压板;保温带;矿棉毡
lamella network 叠层网络
lamellar 叶片状;片状的;成薄层;层状(的);层纹状;薄片状的
lamellar air-heater 片式热风器;片式热风机
lamellar body 片层体;片状小体
lamellar bone 板层骨
lamellar boundary 层流边界
lamellar brake 多片闸
lamellar compound 层状化合物
lamellar compound of graphite 石墨层间化合物
lamellar corneal grafting 层间角膜移植
lamellar corneal transplantation 层间角膜移植
lamellar crystal 片状晶体
lamellar crystallization 层状结晶
lamellar deformation 片晶变形
lamellar displacement 片晶位移
lamellar eutectic 层状共晶
lamellar extrusion technique 层状挤压成型技术
lamellar fibril 片状原纤
lamellar field 层流场
lamellar filter 片式滤清器
lamellar flow 片流;层流
lamellar fracture 层状断裂面
lamellar gametophyte 片状配子体
lamellar graphite 片状石墨
lamellar grating interferometer 层状光栅干涉仪
lamellar growth 片状生长;层状生长
lamellar lattice 片晶晶格
lamellar magnet 多层薄片磁铁
lamellar magnetization 薄片磁化
lamellar martensite 片状马氏体
lamellar material 薄片材料
lamellar membrane 片状膜;片膜层;片层膜
lamellar morphology 片状形态
lamella roof 网格屋顶;叠层屋顶;网格屋盖;叠层薄板屋面
lamella roof structure 联合层状结构;网架屋顶结构;叠层屋顶结构
lamellar pearlite 片状珠光体;层状珠光体
lamellar phase 层状相
lamellar pigment 片状颜料
lamellar(pit) prop 格状坑道支撑
lamellar pyrite 白铁矿
lamellar roof 叠层屋顶;叠层式屋顶;网格屋顶;网格屋盖
lamellar roof structure 叠层屋顶结构;网架屋顶结构
lamellar shutter 叶片式快门
lamellar spacing 层间隙
lamellar structure 片晶结构;片层结构;纹层状构造;叠层结构;层状组织;层状构造;薄层(状)构造
lamellar tearing 层状撕裂;层状拉裂;分层撕裂
lamellar thickening 片晶增厚
lamellar vector 片式矢量
lamellar vector field 无旋向量场
lamellasome 片核体;层片体
lamella system 格网结构系统;联方体系
lamellate 片状的
lamellated fracture 成层断裂
lamellated granule 板层颗粒
lamellated plate 层板
lamellate placentation 片状胎座式
lamellation 薄片;薄层;纹理
lamella truss 叠层屋架;叠层桁架
lameller micelles 层状胶束
Lamellibranchiata 瓣鳃动物纲
Lame's constant 拉梅常数
Lame's ellipsoid coordinates 拉梅椭球面坐标
Lame's formula 拉梅公式
Lame stress-director surface 拉梅应力指示面
Lame stress ellipsoid 拉梅应力椭球
Lamex process 拉米克斯法
Lamgar eccentric mooring buoy 兰嘎式偏心系船浮筒
lamina 层状体;薄片
lamina affixed 附着板

lamina basalis chorioideae 玻璃层
laminable 易展性;可展的;可成为薄层的
lamina boards 薄片木心夹板
lamina body 迭聚体
laminac 泡沫塑料;成型用聚酯树脂
lamina clay 薄层黏土
lamina cribrosa 筛板;筛状板
lamina design 铺层设计
lamina dorsalis 背板线
laminae 纹层状;纹层
lamina explosion proof machine 窄隙防爆式电机
laminagraph 断层照相机
laminagraphy 断层照相术
laminal 成层的
laminal filter 片状填料
laminal placentation 薄层胎座式
laminal structure 薄层结构
lamina membranacea 膜板
lamina perpendicularis 铅直板
lamina posterior 后层
lamina propria 固有层
lamina quadrigemina 四叠板
laminar 由薄板组成的;成层的;层状的;层流的;层理的;薄片状的
laminar alumin(i)um particle 片状铝粒
laminar bedding 纹状层理;纹层
laminar boundary 层流边界
laminar boundary layer 片状边界层;片流(运动)边界层;成层边界层;层流附面层;层流边界层
laminar cavitation 层状空泡
laminar composite 层状复合材料
laminar convection 层状对流;分层对流
laminar crack 层状裂纹
laminar current 平流;层流
laminar damping 层流衰减
laminar density flow 层状异重流;分层异重流
laminar diffusion 层流扩散
laminar displacement 薄层位移
lamina reticularis 网状层;网层
laminar film 片状膜;片流层;层流膜;层流薄膜;薄片;薄层
laminar film in pipe 管壁层流膜层
laminar flame 层流焰;层流火焰
laminar flame stability 层流火焰稳定性
laminar flow 滞流;平流;片流;纹流;层流;层流(流动)
laminar flow burner 层流燃烧器
laminar flow cupboard 水平层流通风橱
laminar flow dye laser 层流染料激光器
laminar flow extent 层流区;层流段范围;层流程度
laminar flow hood 层流净化罩
laminar flow in pipe 管中层流
laminar flow resistance 层流阻力
laminar flow separation 层流分离
laminar flow station 层流净化台
laminar flow structure 层状流动构造
laminar fracture 成层断裂
laminar free-surface flow 层状自由面流
laminar friction 层流摩擦
laminar grating 层晶格;层状点阵
laminar heat convection 层流中热传递
laminarian belt 海带分布带
laminarian zone 海带分布区
laminarization 层状;化
laminar layer 层流层;薄层
laminar mechanism 层流机制
laminar mixing 层流混合;分层拌和
laminar model 片流模型;层流模型
laminar motion 片流运动;层状运动;层流(运动);层流
laminar plastic flow 片型塑性流动;片型范性流变
laminar region 层流区
laminar restrictor 层流流阻器
laminar sand-flood 片状积沙
laminar scale 鳞状锈
laminar separation 层流分离现象;层流分离
laminar skin friction 层流表面摩擦;层流表面摩擦
laminar structure 层状结构;层状构造;层性结构
laminar sublayer 层流次层;近壁底层流;河床底滞流层;片流底层;层流亚层
laminar surface flow 层状表面流;表面流
laminar surface of flow 水面层流
laminar symmetry 分层对称性
laminar texture 层流状结构

laminar theory 层流理论
laminar tissue 薄板组织
laminar-turbulent transition 层流—紊流之间的过渡带;层流—紊流过渡;层流—湍流(之间的)过渡段;层流—湍流(之间的)过渡带
laminar vector 片式矢量
laminar velocity 层流速度;层流流速
laminar velocity distribution 层状流速分布;层流速度分布
laminar wake 层流尾涡
laminar wall 层流壁
laminar wing 层流翼
laminar zone 片流区;层流区
laminaset 纹层组
laminate 层压制品;层压塑料;层压;分成薄片;分层
laminate asbestos fabric 层状石棉织物
laminate assemblies 层积材
laminate coordinates 层合板坐标
laminated 页状;叠层的;分层的;薄层组成的
laminated acoustic(al) glass 夹层隔声玻璃
laminated adhesive 覆膜胶
laminated arch 叠(层)拱;层积拱;层板拱
laminated arch beam 层积拱梁
laminated arch bridge 夹板拱桥
laminated arch(ed) girder 层黏土拱梁
laminated armature 叠片电枢
laminated armature conductor 叠片电枢导体
laminated article 叠层制品
laminated bar 多层杆
laminated barrel 薄片卷成的枪管
laminated basalt 板状玄武岩
laminated beam 叠层梁;层积(拱)梁
laminated bearing 夹板胶木轴承;夹板支座;层压材料支承;层叠片材轴承
laminated bedding 细层;纹层
laminated belt 多层带
laminated bending 弯曲层积;层积弯曲
laminated board 贴面板;叠层板;层压板;板条芯胶合板
laminated-boom spar 层压伸臂梁
laminated bowstring truss 叠层弓弦桁架
laminated brush 叠片电刷;分层电刷
laminated-brush switch 叠片刷型开关;刷形开关;叠片(刷触点)开关
laminated bus plate 层压汇流条
laminated calculus 分层结石
laminated chain 无声链;叠片链
laminated clay 季候泥;片状黏土;纹层(又称季候泥);纹层黏土;成层黏土;层状黏土;层积泥;纹状黏土
laminated cloth 多层黏合布;层压布
laminated coal 泥煤;叠层煤
laminated coating 夹布包层;多层涂料;多层涂层
laminated compact 层状坯块;分层坯块
laminated composite structure 层叠式组合结构
laminated conductor 叠片导体;分层导体
laminated conglomerate mudstone 层纹状砾质泥岩
laminated construction 多层薄板构造;叠层结构;叠层构造
laminated contact 叠片触点;分层片触点
laminated core 叠片芯子;叠片铁芯;叠片磁芯;叠木铁芯
laminated cover(ing) 片状覆盖体;叠层盖板
laminated cross 交向层压
laminated curved beam 弧形叠层梁;叠黏木曲梁
laminated decking 层积铺面板
laminated dielectric 薄片介质
laminated disc 叠层盘
laminated elastic spike with single shaft 多层单杆弹性道钉
laminated fabric 叠层织物;层压网纸板;叠层屋面卷材
laminated-fabric plate 层压纤维板
laminated feeder 叠层式喂料机
laminated ferrite memory 叠片铁氧体存储器;层压铁氧体存储器
laminated ferrocement 层压钢丝网水泥
laminated fiber board 层压纤维板
laminated fiber wallboard 叠合纤维板;层压纤维墙板;叠层纤维壁板
laminated film 层压膜
laminated film bag 层压薄膜袋
laminated filter 叠层滤片层地板;叠层式滤器

laminated floor 叠层地板;夹层地板
laminated flow 分层流动
laminated foil 层压箔
laminated formwork board 叠层模板
laminated fracture 层状断口
laminated frame 胶合叠层框架;叠片定子骨架;叠层框架
laminated girder 胶合叠层梁;叠黏木大梁
laminated glass 夹层玻璃;叠层玻璃;层压玻璃;安全玻璃;塑胶夹层玻璃
laminated glassing paper 叠层玻璃纸
laminated glass partition wall 层压玻璃隔墙
laminated glued arch 胶合叠板拱
laminated glued timber arch 胶合叠板拱
laminated hook 片式单钩
laminated insulating board 叠层绝缘板
laminated insulating glass 叠层隔热玻璃
laminated insulating glazing 多层绝缘玻璃片
laminated insulating sheet 叠层绝缘板
laminated insulating slab 叠层隔热板;叠层隔声板
laminated insulation 多层板绝缘;叠层绝缘;层状绝缘;分层绝缘
laminated insulator 分层绝缘体
laminated iron core 叠片铁芯
laminated joint 马牙榫接【建】;叠接角缝(木工);燕尾榫接;层压接合
laminated jointing 板状节理;叠片状节理【地】
laminated lattice girder 叠层格构梁;叠层的网格梁
laminated layer vessel 多层容器
laminated lead sheet 多层铅片板;叠合铅皮
laminated leaf-spring 叠板簧;多片式钢板弹簧
laminated lightweight building slab 建筑用轻型叠层板
laminated lime 叠层石灰
laminated limestone 叶片状灰岩;叠层石灰岩;层纹状石灰岩
laminated magnet 叠片磁铁;叠层磁铁
laminated material 层压材料;多层材料
laminated melamine resin board 蜜胺树脂层压板;叠层三聚氰胺树脂板
laminated metal 层状金属
laminated mica 云母片
laminated mor 层状的森林粗腐殖质
laminated mo(u)lding 多层模压;层压模制
laminated nozzle 层压喷管
laminated overlay 叠层覆盖物
laminated panel 夹层胶压镶板;叠层镶板
laminated paper 层压纸;多层纸
laminated paperboard 层压纸板;叠层石膏板;叠合纸板
laminated paper sheet 层压纸板
laminated pattern lumber 制模叠层木料
laminated plaster board 贴面纸面石膏板;叠层石膏板
laminated plastic board 热压树脂塑(料)胶板;叠层塑料板;层压塑料板
laminated plastic board picking stick 层压塑料板打梭棒
laminated plastic board side lever 层压塑料板侧杆
laminated plastic panel 多层塑料板
laminated plastic panel with decorative surface 层压塑料装饰面板
laminated plastics 积层塑料;塑料层板;多层塑料;层压塑料;层积塑料
laminated plastic sheet 透明塑料片
laminated plastic veneer 多层塑料板;叠合塑料板
laminated plate 叠层板
laminated plywood 胶合叠层板;层积胶合板
laminated pole 叠片极;叠片磁极
laminated pole-shoe 叠层极靴
laminated polyethylene film 分层聚乙烯片
laminated polyvinyl chloride cover(ing) 聚氯乙烯层罩面料;叠层聚氯乙烯覆盖
laminated porcelain 多层瓷器;叠层瓷
laminated portal frame 叠层门架
laminated products 复合制品;叠层产品
laminated rafter 层叠椽
laminated resin 层压树脂
laminated rib 片肋;叠层肋
laminated rigid insulation board 层合硬质绝热板
laminated rock 片岩;纹层岩
laminated rod 层压杆
laminated roof 叠层屋顶
laminated rotor 叠层式转子;叠合式转子

laminated rubber 层压橡胶
laminated rubber bearing 板式橡胶支座;叠层板橡胶支座
laminated rubber bearing for bridge 桥梁板式橡胶支座
laminated safety glass 夹层安全玻璃;叠层(胶合)安全玻璃;层压安全玻璃;薄片胶合安全玻璃
laminated safety plate glass 层叠胶合安全平板玻璃
laminated safety sheet glass 层安全平板玻璃
laminated sampling 分层取样
laminated sapropelic coal 层积腐泥煤
laminated security glass 防弹玻璃;叠层安全玻璃
laminated sheet 层压片材;层压板(材)
laminated sheet glass 层叠平板玻璃
laminated sheet iron 成层薄钢板
laminated shell roof 叠层薄壳屋顶
laminated shield 层状屏蔽
laminated shim 叠层垫片;叠层薄片;叠层薄垫
laminated shuttering board 叠层模壳板;叠层模板
laminated slab 叠合板【港】
laminated soil 层状土(壤)
laminated spring 钢板弹簧;叠层弹簧;叠板(式)弹簧;板(弹)簧;叠板式弹簧钢板
laminated spring hammer 夹板锤
laminated state 层压状态
laminated striation 层纹
laminated structure 页状结构;胶合木结构;胶合板结构;纹层(状)构造;叠层结构;层状结构;层状构造;分层结构;层压结构
laminated suspension spring 叠板弹簧悬挂吊
laminated system 叠层结构;叠层体系
laminated tarpaulin 层压帆布
laminated texture 页片状结构;片状结构
laminated thermosetting plastics 层压热固塑料
laminated timber 胶合木材;叠层木(材);层积材;层叠木板;层积木
laminated timber arch 夹板拱
laminated timber centering 叠板拱架
laminated timber floor 立板木桥面
laminated timber truss 叠层木桁架
laminated tin 叠层洋铁
laminated truss 叠层桁架
laminated V-belt 活络三角带
laminated veneer 层板单板
laminated veneer lumber 单板层积材
laminated vessel 多层式容器
laminated wall component 叠层墙构件
laminated water-proof fabric 叠层防水织物
laminated windscreen bending furnaces 夹层风挡玻璃热弯炉
laminated windshield glass 夹层挡风玻璃
laminated wire(d) glass 钢丝叠层保险玻璃;夹丝夹层玻璃
laminated wood 胶合木;胶合板结构;叠合板【材】;叠层木板;层压木材(材);层压木(板);层压板;层积材;层(合)板
laminated yoke 叠片磁轭;叠层磁轭
laminate-faced fire door 层压板面防火门
laminate insulating board 成层绝缘板;叠层绝缘板
laminate insulation 绝热层压板
laminate joint 层压接缝
laminate material 层合材料;层压材料
laminate mo(u)lding 层压制模;层(压模)塑法
laminate orientation 层板取向
laminate ply 层压层
laminate power bus 多层电源配线电路
laminater 积层装置
laminate structure 层理构造;层状结构;片状构造
laminate technique 层压工艺
laminating 层压(法);层积;分成薄层
laminating cloth 层压布
laminating composition 胶合料;叠层构造
laminating film 层压薄膜
laminating machine 塑料薄膜粘贴机;层压机
laminating press 层压压机
laminating process 层压法
laminating resin 层压树脂
laminating roller 薄膜轧辊
laminating rolling mill 贵金属薄片轧机;箔片轧机
laminating sheeting 层压薄膜;叠层片材
laminating varnish 层压用清漆
lamination 交替片组;纹理;叠片;叠合;叠层(状);成层;层压(法);层压起鳞;层理【地】;层合;层叠;分层作用;分层(现象);薄层

lamination coating 硅钢片漆;层压涂层;薄涂层
lamination coupling 圆盘联轴节
lamination crack 夹层裂纹;层裂
lamination detector 分层缺陷探测仪;分层缺陷探测仪
lamination factor 层叠系数;分层系数
lamination insulation 叠钢片绝缘;冲片绝缘
lamination of maps 地图贴面
lamination of pole 磁极叠片;磁极(冲)片;磁板叠片
lamination stock 层积毛坯料
lamination synthetic(al) leather 层压合成革
laminator 折叠型压片机;胶合机;涂布机;贴膜机;层压机;层合机
laminboard 夹芯板;多层木心夹板;薄木片层夹板;板条芯细木工板;薄层板
laminectomy retractor 单侧椎板拉钩
laminiferous 薄板的
laminoazo-benzene 对二甲氨基偶氮苯
laminography 断面X射线照相术;层折X射线照相法;分层X线照相术
laminoid-fenestral limestone 纹理灰岩
laminwood 胶合木;叠层木(材);层压木(板)
laminwood slide track 胶合滑道;叠层木滑道
Lamipore 层压多孔金属材料
Lami theorem 拉米定理
lamiwood 叠层木(材)
lammerite 拉砷铜石
lammie 膨胀砖
La Mont boiler 拉蒙式锅炉
Lamont's law 拉蒙特定律
lamp 照明器;灯(泡)
lamp, lantern 灯类
lamp adapter 电灯泡附件;灯座适配器;灯头
lampadite 铜锰土
lampan 地沟冲洗砂矿
lamp annunciator 灯示呼唤器
lamp ballast 镇流器
lamp bank 负载灯排;变阻灯排;白炽灯组
lamp bank signal 灯列信号
lamp base 灯座;灯头
lamp base adapter 灯座连接器
lamp black 软质炭黑;灯烟;灯墨;灯黑;烟墨
lampblack pigment 松烟;灯烟
lamp blown glass 灯工玻璃
lampblown glassware 灯工玻璃制品
lamp body 灯壳
lamp bracket 桅灯灯座;灯架;壁灯(托)架
lamp bridge 灯桥【电】
lamp brightness selector 灯光亮度调节器
lamp brush 灯刷
lamp bulb 灯泡
lamp bulb for optic(al) pyrometer 光学高温计用灯泡
lamp bulb vacuum pump 灯泡真空抽气泵
lamp cap 管帽;灯头
lamp carrier 持灯管
lamp changer 换泡器;换灯器
lamp characteristic 电子管特性曲线
lamp characteristic curve 灯泡特性曲线
lamp-charging rack 矿灯充电架
lamp check 灯光检验
lamp check circuit 指示灯检查电路
lamp check test 指示灯检查试验
lamp chimney (油灯的)灯罩
lamp circuit 电灯线路
lamp compartment 灯房
lamp condenser lens 光源聚光透镜
lamp cord 柔性连接线;电灯(软)线;灯绳
lamp cover 灯罩
lamp cup 灯座
lamp current 电灯电流
lamp depreciation 减光补偿
lamp dimmer 灯光调控器;灯光明暗调节器;灯光衰减器
lamp display panel 灯光显示牌
lamp door 灯框
lamp driver 指示灯驱动器;灯驱动器
lamp efficiency 电灯效率;灯光效率;发光效率
Lampen mill 单球磨
lamp extractor 取灯泡器
lamp failure alarm 灯光熄灭信号
lamp filament 灯丝
lamp fittings 电灯组件;电灯配件;灯具附件
lamp globe 圆灯罩

lamp hanger 灯架
lampholder 灯座;灯头;壁灯插座
lampholder plug 电灯插头
lampholder switch 灯头开关
lamp hole 灯孔;灯井;灯管
lamp house 灯罩;灯室
lamp housing 灯罩;灯光屏蔽罩
lamp inrush current (通电瞬间的)冲击电流;电灯的合闸电流
lampion 节日彩色灯泡
lamp jack 灯插口
lamp jacket 玻璃球灯罩
lamp kerosene 灯用煤油
lamp kerosine 点灯用煤油
lamp lens 灯玻璃;车灯玻璃
lamp life 灯泡的平均寿命
lamp light 灯光;灯火
lamplight colo(u)r film 灯光型色彩片
lamp load 电光负载;电光负荷;电灯负载
lamp locker 灯具室;灯具间
lamp lumen depreciation factor 灯光度折减因数;灯光度折减系数;灯照度折减因数;灯照度折减系数
lampman 矿灯管理员
lamp-matrix sign 灯泡方阵型标志
lamp oil 火油;灯油
lamp outlet box 灯头盒
lamp panel 灯泡板;灯盘
lamp pendant 灯垂饰
lamp pens (汽车的)灯玻璃
lamp pH-meter 灯式酸碱计
lamp plug 电灯插头;灯插头;灯塞线
lamp pole 路灯柱;灯柱
lamppost 灯柱;路灯柱;灯杆
lampprophyllite 闪叶石
lamp receptacle 灯座
lamp reflector 灯光反射器;灯管反射器;车灯反射镜
lamp regulator 电灯调节器;电灯电路自动电压调整器
lamp resistance 灯泡电阻
lamp resistor 灯泡电阻器;变阻灯
lamprobolite 玄武岩角闪石;玄武岩闪石
lamproom 灯具室;灯具间;灯房;矿灯管理室;矿灯管理房
lamprophane 闪光矿
lamprophanite 闪光矿
lamprophyllite-lujavrite 闪叶异霞正长岩
lamprophyre 煌斑岩
lamprophyre group 煌斑岩类
lamprophyric texture 煌斑结构
lamproschist 煌斑片岩;片状变质煌斑岩
lamprostibian 闪锑铁锰矿
lamps 灯具
lamps and lanterns 灯光信号灯
lamp screen 灯罩
lampshade 灯罩
lamp shade and reflector 灯罩及反射器
lampshade trim 灯罩边饰
lamp shielding angle 遮光角
lamp signal 灯光信号
lamps indicator 航行灯指示器
lamp socket 灯台;灯(插)座;电灯插座
lamp socket cable contact 灯座接线头
lampstand 灯座;路灯柱;灯柱;灯台;三脚架;三角底座
lamp-stand greening 灯柱绿化
lamp store 灯具室;灯具间
lamp sulfur test 燃灯含硫量试验
lamp switch 电灯开关;灯开关
lamp switch escutcheon 灯开关片
lamp synchronizer 灯泡式同步指示器
lamp target 灯光靶标
lamp test 灯泡试验
lamp trimmer 矿灯清理工
lamp-voltage regulator 白炽灯稳压器;灯电压调整器
lamp wick 灯芯;灯捻
lamp wire 灯线
lamp wire grommet 灯线接头
lamp working 灯工操作
lamson tube 气动输送管(用于传送文件资料)
lanai 外廊;凉台(美国夏威夷);门庭;游廊
Lanaperl blue 拉纳佩勒蓝
lanarkite 黄铅矾;硫酸铅矿

Lanasyn green 拉纳新绿
Lancashire boiler 兰开夏锅炉
Lancashire boiler with corrugated flues 有波形烟道的兰开夏锅炉
Lancashire booster 兰开夏升压器
Lancashire brick 兰开夏砖(一种密实不透水的优质红色面砖)
Lancashire pattern distempering brush 兰开夏式平刷
Lancashire two-flue boiler 兰开夏双烟道锅炉
Lancaster mixer 兰开斯特混合机;逆流(盘式)混合机
Lancastrian(stage) 兰开斯特阶【地】
lance 柳叶刀
lance cutting 氧炬切割
lanced arch 尖拱
lance door 吹灰门
Lancefield precipitation test 兰斯菲尔德沉淀试验
lanceolate 柳叶刀形的
lance pipe 钻管
lance point 钻点;钎尖
lance pointer 矛形指针
lancet 柳叶刀;尖顶窗;提钩;矢状饰【建】;折角饰;矢状饰;长窄尖顶窗;尖顶拱窄窗(英国哥特式建筑)
lancet arch 尖拱;尖顶拱;锐尖拱;复合拱;二心外心桃花尖拱
lancet architecture 尖拱式建筑;锐拱式建筑
lanceted 尖顶拱窄窗的
lancet style 尖拱式建筑
lancet window 尖头窗;尖顶窗;矢形尖卷窗;长尖头窗;长窄尖顶窗
lance type burner 喷枪式燃烧器
lancewood 枪木
Lanchester balancer 兰彻斯特平衡器
Lanchester damper 兰彻斯特减振器
Lanchester equation 兰彻斯特方程
lanciform 尖顶形的
lancing 氧断;切口;切缝
lancing die 切缝模
lancing door 清洁孔
land 国土;陆地;连接盘;刃棱面;刃带;土地;钝边;槽脊
land abutment 陆上桥台;桥头;桥台;岸墩
land abutment of dam 坝岸连接结构
land access 土地的获得
land access benefit 田庄通道受益(道路筑成后田庄的得益)
land access road 进路;田庄出入道路
land accretion 土地围垦;填筑土地;填筑;冲积地;土地改良
land acquirement 土地征用
land acquisition 征购土地;征地;土地征用(贷款)
land acquisition cost 土地征用费
land administration department 土地管理部门
Land Administration Law of the People's Republic of China 中华人民共和国土地管理法
land agency 土地买卖办公所(美国);土地管理处
land agent 土地管理人;地产商;土地经纪人
land aggradation by reclamation 吹填造地
land allocation 土地分配
land allotment 土地核配
land allotment system 土地分配制度
land amelioration 土壤改良
land and building 土地及建筑物;房地产
land and building sinking fund 房地产偿债基金
land-and-sea breeze 海陆风;陆风和海风
land-and-water coordinated transport 水陆联运
land-and-water development 水土开发
land and water excavator 水陆两用挖掘机
land animal 陆生动物
land annexation 土地兼并
land application 土地处理(法);土地处置
land application of sludge 土地施加污泥
land application satellite system 陆地应用卫星系统
land appraisal 土地鉴定;土地估价
land appropriation 土地的分配
land archives 土地档案
land area 陆地部分;土地面积
land area covered with trees 森林覆盖率
land area of the globe 全球大陆地带
land arrangement 土地整理
land asphalt 陆产(地)沥青;土沥青
land assembly 土地合并

land assignment 土地划拨
landau 敞篷汽车;活顶小汽车
Landau damping 朗道阻尼
Landau fluctuation 朗道涨落
landauite 朗道矿
landaulet 四门轿车
Landau level 朗道能级
Landau length 朗道长度
Landau theory 朗道理论
land availability 可利用土地
land axle 地轮轴
land bank 土地银行;不动产银行;土地库
land banking 地产储备
land barrier 陆障
land-based 设在陆上的;岸基地
land-based activity 陆地活动
land-based gas turbine engine 固定式燃气轮机
land-based pollution 陆源污染
land-based pollution source 陆上污染源
land based power plant 固定式动力装置
land-based prototype reactor 陆上模式反应堆
land-based sources of marine pollution 陆源性海洋污染源
land-based sources of ocean pollution 陆源性海洋污染源;陆地来源的海洋污染
land-based sources of pollution 陆源污染源;陆地来源的污染
land-based water quality station 岸基水质监测站
land basin 陆盆
land batture 河滩高地;河漫滩;河岸淹没区;堤前河滩
land bedding 土地分层耕作
land-berm 台地
land betterment 土壤改良;土地改良
land beyond dike[dyke] 堤外滩地;堤外(土)地
land blink 陆映光
land block 陆块;地块
land board 舱口装卸货垫板;土地局
land boiler 陆地锅炉
land bond 土地债券
land-borne 陆运的
land-borne cargo 陆运货物
land-borne trade 陆上贸易
land boundary 地界;基地界线
land boundary map 地界图
land boundary survey 地界测量
land-bound crane 陆地吊
land-bound equipment 陆地设备
land breeze 陆风(从陆地吹向海洋的风);陆地风
land bridge 陆桥;陆地桥
land-bridge movement 陆桥联运
land-bridge route 陆地桥梁路线
land-bridge service 陆桥运输;陆桥联运
landbridge transport(ation) 大陆桥运输
land cable 地面电缆
land caisson 陆上沉箱;陆上沉井
land capability 土地(生产)能力;土地容量;地力
land capability class 地力分级
land capability classification 地力分类
land capability map 地力图;地力(分类)规划图
land capacity 土地生产能力;地力
land capital 土地资本
land carriage 陆运;陆上运输;陆路运费
land carrier 陆运工具
land carrying capacity 土地负荷能力;土地承载容量;土地承受能力
land casing 下套管;把套管放入井内
land certificate 土地证(书);由联邦政府发给的土地证书
land chain 测链(长66英尺);土地测链
land charge 土地负担
land classification 土地分等;土壤分类;土地分类
land-classification map 土地分类图
land clearance 土地上房屋拆除;清楚地面树木;贫民窟的拆除;周刃隙角;清理土地;清理场地
land clearing 清理场地;土地清理;地面清理;场地清理
land-clearing and rock rake 清除树根与岩块的耙子
land-clearing blade 灌木清除机;除荆机
land-clearing machine 地面清理机
land-clearing rake 场地清理耙
land compass 测量罗盘(仪)
land compensating fee 土地补偿费
land compensation and rehabilitation 土地补偿

和安置费
land compensation fee 土地补偿费
land configuration 地形
land confiscation 土地没收
land confiscation on reservoir area 水库征地
land connection 岸边连接
land conservation 土地保存;土地资源保持;土地保护
Land Conservation and Development Commission 土地保护和发展委员会
land conservation policy 土堤保护政策
land consolidation 土地整理;土地固结
land consolidation area 土地整理区;土地加固区
land construction 陆地建造;土地保存
land contour 陆地轮廓线
land contract 房地产转让分期付款合同;土地契约;地契
land control 土地管制
land controlled climate 大陆性气候
land control measures 土地管制措施
land corner 台肩棱边
land cost 土地费用;地价
land cover 地面覆盖(层)
land cover analysis 地面覆盖分析
land covered with forest 森林地带
land craft 陆运工具
land creep 土地蠕变
land cruiser (城市间的)长途汽车
land cut canal 陆上运河
land damage 装船前货损
land deed 土地证
land deposit 陆相沉积(物)
land depot 储[贮]木场
land-derived organic matter 陆源有机物
land desertification 土地沙漠化
land deterioration 土地退化
land developer 土地开发业者
land developing value along the line 沿线土地开发价值
land development 土地开发;土地开垦
land development loan 土地开发贷款
land disposal 掩埋处置;陆地处置;埋入地下;土地处置;地面处置
land disposal needs (废物的)土地处置需求
land disposal of sewage sludge 下水污泥的地面处理
land disposal site 地面处理场
land disposal system 土地处置系统
land divider 路面分道线;分车岛
land drain 土地排水沟;土地排水道;农田排水(沟);地面排水(沟)
land drainage 土地排水;地面排水;大地排水
land drainage act 土地排水法
land drainage pipe 地面排水管
land drainer 排气沟挖掘机;排水沟挖掘机
land-drain hook 地面排水(岩)层
land-drain pipe 农田排水管(道)
land dredge(r) 陆上挖土机;挖泥机;挖土机
land driver 陆上打桩机
lande 荒地
land easement 地役权
land economics 国土经济学;土地经济学
landed 起岸
landed cost 卸岸成本;起岸成本;搬运成本
landed duty paid 起岸税捐付讫条件
landed estate 地产;不动产权
landed price 抵岸价(格);卸岸价格
landed property 土地所有权;地产
landed proprietor 土地所有者
landed quality terms 卸卖品质条件;起卸后货物质量品质条款
landed quantity terms 卸货数量条件;起岸数量条件
landed terms 货物起岸费用条款;起岸费用条款;岸上交货价格
landed weight 卸货重量;卸船重量
landed weight final 卸货重量为准;以起岸重量为准
landed weight/quality 以岸品质/重量
landed weight terms 卸货重量条件
Lande factor 朗德因子
land effect 陆地效应;海岸线效应
Lande g factor 朗德 g 因子
land end of groyne 丁坝接岸段
Landenian(stage) 兰登阶【地】
lander 着陆器;装舱管系;罐座;块石装载工;登陆器;登陆车;登陆舱;出铁槽;出钢槽;登陆艇;罐托(混凝土吊罐);把钩工(混凝土);司罐工
lander-carriage (飞机的)起落架
lander(discharge) valve 装舱阀
landerite 蔷薇榴石
land erosion 土地侵蚀;水土流失
land erosion control 土地侵蚀治理
land eruption 陆地喷发
landes 荆棘地
landesite 褐磷锰铁矿
land-ethics 大地伦理学
land evaluation 土地评价
landevanite 淡红蒙脱石
land evapo(u)ration 陆面蒸发;土壤蒸发;地面蒸发
land excavator 陆上挖掘机
land expropriation 土地征用
land expropriation law 土地征用法
land expropriation right 土地征用权
land extensive industry 广占土地工业
land facies 陆相
land fall 土塌;土崩;山崩;地滑;地崩;返航过程;崩塌;接近陆地;(地层结构上的)塌方;初见陆地【航海】
landfall buoy 近陆浮标
landfall light 近陆灯标;驶近陆地初见灯光;初见灯标
landfall mark 近陆标志;近岸标
land farming 土地耕作
landfast 系缆柱;岸上系缆柱
landfast ice 岸冰;沿岸固定冰;定着冰
land feature 地面特征;地貌
land fertility 土地肥力
land fever 陆地热;河床热
land files 土地台账
landfill 垃圾垫土;掩埋;垃圾土地填埋;埋填入土;土地填埋;土地填筑;填埋;填坑;填地;废碴填埋
landfill compactor 垃圾压实机
landfill contamination 垃圾填埋污染
landfill disposal 垃圾填埋处理;土壤填埋处理;填埋处理
landfill dumping 倾弃垃圾填地
landfill gas 垃圾填埋气;填埋气体;掩埋场气体
landfilling 倾弃垃圾填埋
landfill leachate 垃圾土地填埋淋滤液;垃圾渗滤液
landfill of municipal refuse 城市垃圾填埋
landfill of refuse 垃圾填埋
landfill of solid wastes 固体废物填埋
landfill site 垃圾填埋占地
landfill treatment 垃圾填埋处理
landfill waste in 杂填土
landfill waste site 垃圾填筑场
land filtration 土地浸润;土地过滤
land filtration irrigation 浸润灌溉;土壤浸润灌溉
land floe 厚浮冰块;贴岸冰
land flood(ing) 地面泛流;地面泛滥;陆地淹水;洪水泛滥;内涝
land-flora 土性植物
land flow slide 地面滑移
land fog 陆雾;陆地雾
land following shoe 仿形滑板
land for building 建筑用地
land for external transport 对外交通用地
land for future extension 预留(土)地
landform 地形
landform and geomorgraphy investigation 地形地貌调查
land formation 陆相(地)层;陆地建造【地】
landform colo(u)ration 分层设色法
landform element 地形单元
landform features 地貌形态特征
land forming 土壤整治
landform line 地形线;地表形态线
landform map 地形图;地势图
landform mapping 地势图制图
landform of diapir structure 底辟构造地貌
landform of domal structure 穹隆构造地貌
landform of fold structure 褶曲构造地貌
landform of horizontal stratigraphic(al) structure 水平岩层构造地貌
landform of monoclinal structure 单斜构造地貌
landform plane 地形面
landform remote sensing 地貌遥感
landform symbol 地貌形态图符号
land for public facility 公共设施用地
land for public use 公共用地
land for sale 出售地
land for sale in lots 分售地
land freeze 土地交易冻结
land freight 陆(上)运费;陆路运费;陆地运费
land fund benefit 土地资金收益
land goods without permit 私自起货
land grader 整地机;灌木清除机;除荆机
land grading 平整土地;平地;土地平整
land grant 土地许用证
Landgrant College 联邦政府赠地建立学院(美国)
land gravity survey 陆地重力测量
land held in demesne 领有地
land hemisphere 陆半球;大陆半球
landholder 土地所有者;土地所有人
land holdings per capita 人均占地
land holding tax 土地闲置税
Landholt fringe 兰德霍尔特条纹
land ice 陆冰;岸(缘)冰
land improvement 土壤改良;土地改良
land improvement district 土地改良区域
land improvement loan 土地改良贷款
land increment 土地增值
land increment value duty 土地增值税;土地增价税
land information system 土地信息系统
landing 卸货上岸;卸岸;靠泊;降落;集运点;起岸;停车装卸台;停车平台;梯台;上岸;登陆;登岸处;出车台
landing abutment 岸墩
landing accommodation 降落设施;上岸设施;登陆设施
landing account 卸货记录簿;卸货报告;起货栈单
landing agent 卸货代理商;运输代理行;货物到港代理业务
landing aids 着陆辅助设备;着陆导航设备
landing and delivery 卸货与交付
landing and take-off cycle 起飞着陆循环
landing apron 活动登船梯
landing area 降落区域;降落场(地);起落飞行区;飞机(着陆)场;飞机降落场
landing area floodlight system 机场照明系统
landing barge 登陆艇
landing base 底盘
landing beacon 着陆信标
landing beam 平台梁;系梁;着陆导航波束;楼梯平台梁;着陆信标射束
landing beam transmitter 跑道定位标发射机
landing bearer 楼梯平台托架;楼梯梁板支撑
landing binder 楼梯梁板支撑
landing binding joist 平台联结格栅;楼梯格栅;登陆栅;登陆缚柱
landing book 卸岸货账册;起岸清单
landing bottom 井门;马头门
landing bridge 引桥;栈桥;登陆(栈)桥;登岸桥
landing brow 搭板
landing-call push 楼梯呼叫按钮
landing card 卸货证;登陆证
landing carriage 楼梯平台格栅;平台格栅;登岸跳板
landing casing 套管下放到孔内台阶上
landing ceiling joist 平台顶板格栅;登陆屋顶架;楼梯格栅
landing certificate 卸货证明;起岸证明书
landing charges 卸货费(用);着陆费;起货费;起岸费;上岸费;飞机起降费
landing chassis (飞机的)起落架
landing circle 圆形着陆航线
landing clerk 码头旅客接待员
landing collar 套管的连顶接箍;吊环(吊套管型)
landing compass 陆用罗经
landing craft 坦克登陆艇;登陆艇;登陆舰
landing craft tank 登陆驳船(海洋钻探联络供应用)
landing cross 着陆跑道;着陆十字路
landing curve 降落线迹
landing deceleration aids 着陆减速设备
landing deck 降落甲板
landing depth 下套管深度
landing direction indicator 着陆方向指示器
landing direction light 降落方向灯
landing distance 着陆距离;飞机着陆距离
landing distance available 降落距离
landing door (住宅楼的)楼门;电梯门口;住宅门
landing edge 纵向接头;纵边接缝
landing elevator 电梯入口平台;休息平台
landing entrance 电梯口

landing error 落点误差
landing expenses 卸货费(用);起岸费
landing field 着陆地;降落场(地);飞机(降落)场
landing flap 着陆襟翼
landing flare 着陆照明弹;着陆用照明弹
landing gear 着陆装置;起落装置;(飞机的)起落架;撑角;撑地装置;防收锁;降落装置
landing gear post 起落架支柱
landing gear shock absorber 起落架减震器
landing gear torque arm 起落架防扭臂
landing girder 平台梁
landing ground 着陆场;降落场(地);飞机场
landing head 油管头
landing hire 卸货费(用);起岸费
landing joint 联顶节
landing joist(of stairs)楼梯(休息)平台格栅;平台(格)栅
landing leg 支腿
landing length 楼梯平台长度
landing level 楼梯平台标高;楼梯平台高程
landing light 着陆灯;降落信号灯光
landing load 着陆荷载
landing mat 飞机降落甲板;装拆式飞机起降跑道;金属路面板
landing newel 楼梯扶手转弯支柱;楼梯平台起柱;楼梯扶手角柱
landing noise of aircraft 着陆噪声
landing of beam 电子束射击点
landing of cargo 卸货
landing of stair(case)楼梯休息平台
landing opening 电梯口
landing order 卸货单;出货单
landing permit 卸货证明
landing pier 栈桥码头;靠岸码头;桥台;起货码头;登陆码头;登岸码头;岸墩
landing pitch 路面
landing place 卸货(地)点;卸货处;着陆地;降落场(地);登陆地点;登岸处;飞机降落场;浮码头
landing placing 登陆处
landing platform 卸货平台;下车月台;楼梯平台;浮码头
landing pontoon 码头趸船;趸船码头
landing port 着陆机场
landing procedure 着陆程序
landing quality terms 卸货品质条件
landing quay 靠岸码头
landing ramp 上岸跳板;登陆跳板
landing ring 调节环;定位环
landing run 着陆滑跑;降落滑程
landing runway 着陆跑道;降落跑道;起飞降落跑道;跑道;飞机降落跑道
landing runway lighting 着陆跑道照明
landing seat 支座(套管着陆处)
landing shaft 出煤立井
landing ship 登陆艇;登陆舰
landing ship dock 船坞登陆舰;船尾吊驳母船
landing ship medium 中型登陆艇(海洋钻探用)
landing shock absorption 着陆减震
landing shoulder 固定座;固定台肩
landing site 着陆场;登陆地点
landing skid 装货跳板;降落橇
landing slab 楼梯平台板;平台板
landing speed 着陆速度
landing stage 着陆阶段;轮船码头;简易码头;码头;起货码头;突码头;上岸码头;趸船(码头);承接平台;浮码头;登船;(脚手架上的)材料放料台
landing stage of scaffold 脚手架平台
landing stair(case)登陆跳板
landing station 降落场(地)
landing step 平台踏步;平台式踏步
landing strake 舷缘第二列板
landing strip 着陆跑道;着陆带;简易飞机场;跑道
landing strip landing runway 飞机跑道
landing surveyor 码头检验员
landing swell 向岸涌浪
landing-switch 层位开关
landing system 着陆辅助设备
landing T T 字布;T 形着陆标志
landing tank 坦克登陆艇
landing tax 入国境税
landing tee 丁字风向标
landing top 悬挂点
landing tread 楼梯平台踏板
landing troopship 两栖运输舰

landing troop transport 两栖运输舰
landing type vessel 登陆型货船
landing waiter 海关起卸货监督员
landing weight 卸货重量;到岸重量
landing wharf 卸货码头;靠岸码头;上岸码头
landing width 楼梯平台宽度
landing zone 楼梯转弯处;靠岸区;停机区
land installation 陆用装置
land intensification 土地集约经营;土地集约化
land intensive 土地密集
land intensive industry 土地集约工业
land irrigation 土地灌溉(法)
landis chaser 切向螺纹梳刀
Landis type grinder 莱迪斯式磨床
land jobber 土地投机商
land judging 土地评价
land laid 船在陆地可见范围
land lane 冰区中间通向海岸的水道;冰间水道
land law 土地法(规)
land lead 冰区中间通向海岸的水道
land lease agreement 租地契约
land leasing 土地出让
land leg 陆运段
land legislation 土地立法;土地法(规)
land lending 借地
land level 地面高程;地轮操纵杆
land leveler 平土机;平地机
land-level(l)ing 土地平整;土壤平整
land-level(l)ing operation 平整土地
land-level(l)ing project 土地平整工程
land liable to flood 易淹地区;漫滩;泛滥界;易涝洼地
land license 土地执照
land line 用地界线;路面分道线;路界线;道路用地边线;岸边线;道路征地线;陆上运输线;陆上线路;陆上通信线
land-line adjustment 标定地界
landline charges 陆线费用
landline facility 陆上通信线(路)设施
land line subscriber 地面连线用户
land loan 土地借款;土地贷款
land lock (双线船闸中的)靠岸船闸;内侧船闸;堤防闸门;堤防闸口;岸侧船闸
landlocked 陆地围绕的;陆封的;被陆地包围的;内海;陆围海
landlocked bodies of waters 闭合水域;闭合水区
landlocked body of water 环抱水区;闭塞水体
landlocked country 内陆国(家);内陆地区;内地国家
landlocked drainage 内陆水系
landlocked harbo(u)r 陆封港;陆地环抱港;内陆港(口);天然港(湾)
landlocked inlet of sea 陆地环抱海湾
landlocked lake 陆围湖(泊);内陆湖;封闭湖
landlocked parcel 无路可通的土地
landlocked sea 陆围海;内陆海
landlocked species 内陆种(类)
landlocked waters 闭合水域;闭合水区;内陆水域
landlord 土地所有人;地主
landlordism 地主所有制
landlord's warrant 房东特许状(许可房东强行收取无力交租的房客的个人财产拍卖抵租)
landlord waiver 业主弃权声明书
landlubber 新水手;对船陌生的人
land mammal 陆地哺乳动物
land management 土地管理
land management monopoly 土地经营的垄断
land management right 土地经营权
land map 地形图
landmark 接岸标;里程碑;地面方位标;地面标记;用地标;陆标;界标;划时代的事;文物建筑;地物;地界标;地标;岸标
landmark architecture 历史标志性建筑
landmark beacon 界(址)标灯;地形灯标
landmark feature 方位物
landmark planting 地标栽植
landmass 地块;大(片)陆地;地块
landmass denudation 地块剥蚀
landmass simulator plate 地面情况模拟板
land measure 土地丈量单位
land measurement 土地测量
land melioration 土地改良
land mile 法定英里(1 法定英里=1690 米)
land mine 地雷

land mobile service 陆上移动通信;地面移动通信业务
land mobile station 地面移动电台;地面流动电台
land monument 铁路用地标牌
land movement 土体移动;土体位移;土体滑动
land nationalization 土地国有化
land navigation 地面导航
land net 地界
land-ocean-climate satellite 陆地海洋气候卫星
land-ocean interactions 陆地海洋相互作用
land of cutting tool 刀刃棱面
land office 土地管理所;土地管理局
land oil exploitation 陆上石油开发
Landolt's band 兰多尔特带
Landolt's fiber 兰多尔特纤维
land on fallow rotation 轮荒地
land on which the enterprise are situated 企业占用的土地
land owner 土地所有者;土地所有人
land ownership 土地所有(权);土地所有制
land ownership map 地产图
land ownership rights 土地所有权;田权
landowner's royalty 土地所有人的矿业使用费
land pan 陆地蒸发器;陆地蒸发皿
land parcel 地块
land parcelling plan 土地划分图
land patent 土地专利
land phosphate 磷灰土
land photography method 陆地摄影法
land pier 地墩;码头;河岸防波堤;桥台
land pipeline 陆地管线
land pitch 地沥青;土地倾斜度
land plane 陆上飞机;码头;土地整平机
land planer 大型平地机
land planner 土地规划师
land planning 土地规划
land planning survey 土地规划测量
land plant 陆生植物
land plaster 石膏粉
land policy 土地政策
land pollutant 陆地污染物
land pollution 土地污染;大地污染
land poor 土地贫乏
land preparation 整地
land preparing 整地
land price 土地价格;土地征购价
land price per floor area 楼面地价
land property 地产
land property right 土地产权
land prospecting 土地探测
land protection law 土地保护法
land purchase 土地收购;土地购置
land purchasing price 土地征购价格
land quarantine 陆地检疫
land quay 登陆码头
land readjustment 土地区划整理;土地调整
land readjustment project 土地调整规划
land readjustment work 土地(重新)调整工作;土地重划
land receiver 地面接收装置;地面接收器;地面接收机
land reclaimed from a lake 湖田
land reclamation 造陆;垦荒;开垦(土)地;开垦;围垦;拓垦;土地开垦;土地改造;土地改良;填筑(土);填筑;吹填造地;吹填土;吹泥造地;采区复田;土地整治;垦拓土地
land reclamation and improvement 土地垦殖与改良
land reclamation by enclosure 筑堤造地;筑堤围垦
land reclamation by filling 填土造地;填土围垦
land reclamation by warping 放淤造田
land reclamation machinery 垦殖机械;土地开垦机械
land reconversion 土地复原
land reform 土地改良;土地改革
land reform and tenure 土地改革与土地占有
land register 土壤登记;土地登记;地政局;地籍图;地籍册;地籍(簿)
land-register fee 地产登记费;土地登记费
land registration 土地登记
Land Registration Act 土地登记法例
land registry office 土地登记处
land registry survey 地籍测量
land regulation 土地条例

land rent 地租
land rental 土地租费
land requisition criterion 土地征用标准
land requisition(ing) 土地征用;土地取得
land requisitioning compensation 征地补偿
land requisition line 土地征用线
land reservation for public facility 公共设施保留地
land residual technique 从房屋价值评土地价值的方法(一种土地残值评估方法)
land resource area 土地资源区域
land resources 陆地资源;土地资源
land resources conservation 陆地资源保护;土地资源保护
land resources survey 土地资源调查
land resting 土地休耕
land restoration 土地恢复;地面复原
land retirement 土壤退化;土壤浸染;土地休耕;土地退化
land return 地面(杂乱)回波;地面反射(波)
land revenue 土地收入
land riding 台面引导
land rights 土地权
land risk 陆上风险
land roller 镇压器;平地滚压器;轻便越野小汽车
land royalty 土地使用费
land run-off 陆地径流;地面径流
landsat 陆地卫星;大陆卫星
land satellite 陆地卫星
landsat ground station 陆地卫星地面接收站
landsat imagery 陆地卫星成像
landsat operation controlling center 陆地卫星运转控制中心
landsat TM image 陆地卫星 TM 图像
Landsberger apparatus 兰茨伯格沸点计
landscape 陆上风景;景色;景观;地形;风景;布置园林
landscape aesthetics 景观美学
landscape architect 园林设计师;园艺家;园林建筑师;造园家;景观建筑师
landscape architecture 园林学;园艺建筑学;园林建筑(学);造园(学);造景学;景观建筑学;风景建筑学
landscape art 园林艺术
landscape bureau 园林管理局
landscape cell 景观单元
landscape changes 地貌变化
landscape character 地貌特征
landscape characteristic 景观特征
landscape city 风景城市
landscape climatology 景观气候学
landscape conservation 园林保护;景光保护;环境绿化保护
landscaped area 风景区
landscape design 园林设计;景观设计;风景设计;造园;风景规划;景色设计;城市景观设计;街景设计
landscape development 环境美化;园林建设;风景设计
landscape development of highway 公路绿化
landscape development plan 景观开发计划
landscaped freeway 具有景观的高速公路
landscaped interior 有景观室内
landscaped median(strip) 风景中心地带
landscaped office room 有氯化布置的办公室
landscaped strip 风景地带
landscape ecology 景观生态学
landscape ecosystem 景观生态系统
landscape engineer 园林工程师;造园工程师
landscape engineering 园林工程学;造园工程;绿化工程;景观工程;环境绿化工程;风景工程学;造景工程;庭园风景工程
landscape esthetics 景观学
landscape evaluation 景观评价
landscape feature 景观特征
landscape forest 风景林
landscape garden 风景园
landscape gardener 园艺美化师;园艺家;造园家;造景园艺师;景园艺师;景观园艺家;造园师;风景园艺师
landscape gardening 线路园林化;园艺(学);造园(学);观园艺(学);风景式庭园化
landscape geochemical flow 景观地球化学流
landscape geochemical regionalization 景观地球化学区划
landscape irrigation 景观灌溉

landscape lens 取景镜头;风景镜头
landscape map 景观图
landscape marble 园林造景大理石;美景大理石;风景大理石
landscape of fumarole 喷汽孔景观
landscape painter 风景画家
landscape painting 山水画;风景画
landscape panel 横纹镶板;风景木纹板;园林造景镶板壁
landscape park 天然公园;景观公园
landscape plan 风景规划;景观规划;园林平面布置
landscape planning 景观规划;风景规划
landscape planting 造景栽植;景观绿化;风景造林
landscape preservation 景观保护;环境维护;风景保存
landscape preservation area 景观保护区
landscape prism 景观棱柱
landscape project 土地规划图
landscape protected area 风景保护区
landscape protection 景观保护
landscape protection area 景观保护区
landscaper 景观设计师;庭园设计师;造园家
landscape renovation 景观改造
landscape reservation 景观保护区;天然保护区;风景保护(区)
landscape restoration 恢复景观;风景恢复
landscape screen 风景规划;(分隔办公室的)活动景色围屏
landscape sketch 景观略图
landscape square 景观广场
landscape style 园林式;风景式
landscape treatment 园林设计
landscape type 园林型
landscape unit 景观单位
landscape water treatment 景观水处理
landscape work 环境美化工作;庭园工程
landscape zone 风景区
landscaping 园林设计;造园林;绿化(工程);景观设计;环境美化;庭园布置;设景;风景设计;布置园林;自然美化工作;风景布置
landscaping and erosion control 风景布置和控制冲刷
landscaping design for environmental purposes 美化环境
landscaping office 景观式办公室
landscaping urban district 城市风景区
landscapist 庭园设计师;风景画家;园林设计师
landschaft 自然景色;自然景观;景观
land scraper 平地机
land scurvy 陆地坏血病
land-sea breeze 陆风和海风
land service function 土地使用功能
land-service road 地方道路
land settlement project 土地垦殖计划;土地垦殖规划;土地垦殖工程
landshaft 景观
land shaper 筑埂机
land shaping 土地整形
land shield 陆上隧道建筑防护板
landside 地面边;犁沟壁;犁侧板;背水面;背水侧
landside area 陆域
landside area of port 港口陆域
landside(d) bridge span 岸间的桥梁跨度
landside of terminal building 候机楼对陆侧
landside share 侧板犁铧
landside slope 内坡;(堤的)背水坡
land sky 陆照云光
landslide 滑坡;土壤滑坡;坍坡;坍方;地面滑移;(地层结构上的)塌方;山崩;地滑;地崩;大塌方;崩塌;坐崩(滑崩)
landslide accumulation soil 滑坡堆积土
landslide analysis 滑坡分析
landslide area 滑坡区
landslide axle 滑坡轴
landslide body 滑坡体
landslide cliff 滑坡壁
landslide control 滑坡治理;滑坡防治;坍方控制;滑坡控制
landslide correction 滑坡整治
landslide dam 崩塌形成的坝
landslide deposit 滑坡泥砂
landslide embankment 滑坡堤
landslide factor 滑坡要素
landslide flow 山崩流

landslide investigation 滑坡调查
landslide lake 塌方湖;山崩湖;地滑阻塞湖
landslide length 滑坡长度
landslide location 滑坡位置
landslide loss 滑坡损失
landslide mass 滑坡体
landslide monitoring 滑坡监视
landslide mound 滑动鼓丘
landslide of compound structural plane 复合结构面滑坡
landslide of hard-hemihard rock group 坚硬—半坚硬岩组滑坡
landslide of rock group of hard-soft interlacing 软硬互层岩组滑坡
landslide of soft rock group 软弱岩组滑坡
landslide of unload structural plane 卸荷结构面滑坡
landslide perimeter 滑坡周界
landslide protection 防塌方
landslide protection wall 防坍墙
landslide protection works 坍方防护工程
landslide rapids 滑坡急滩
landslide spring 滑坡泉;山崩泉
landslide surge 地滑激浪
landslide surveillance 滑坡监视
landslide survey 滑坡测量
landslide terrace 滑坡台阶
landslide tongue 滑坡舌
landslide type 滑坡类型
landslide width 滑坡宽度
landslip 坍坡;塌坡;(地层结构上的)塌方;滑坡;大塌方
landslip area 滑坡区
landslip-induced rapids 滑坡急滩
landslip protection 防塌方
land slope 土地坡度;地面坡度
landsman 见习船员
land smoothing 土壤平整;土地平整
land-sourced pollutant 陆源污染物
land sources 土地资源
land speculation 土地投机;地皮投机;地产投机
landspout 陆上旋风;陆上冒海水的天然石洞;陆(上)龙卷(风)
land spreading 地面铺设
land spring 滑坡底泉;平地泉;表层泉
land station 地球站;地面站
land storage tank 地上储[贮]罐;地上油罐
land stream 入海河流
Lands Tribunal 土地裁判庭
land strip 起落区;剥除表土
land structure 陆上建筑物;陆地结构;土地结构
land subdivision 土地细分;园地划分;土地划分
land subleased 转租土地
land subsidence 路缘排水口;陆(地)沉(陷);土地下沉;土地沉陷;地(面)陷(落);地面下陷;地面下沉;地面沉陷;地面沉降;地基沉陷;地表陷落;地表沉陷
land subsidence induced by geothermal field exploitation 热田开发所诱发的地面沉降
land-suitability 土地适用性
land suitability classification 土地适用性分类(指宜农、宜林、宜牧、宜耕等而言)
land surface 地面
land surface area 幅员面积
land surface nuclear burst 地面核爆炸
land surface pyrometer 表面温度计
land surface temperature 地面温度
land survey 土地调查;土地丈量
land surveyer 土地测量员
land surveying 陆地测量;土地测量;大地测量
land swamping 土地沼泽化
land swell 拍岸浪;拍岸涌;拍岸巨浪
land system 土地制度
land tax 钱粮;土地税;田赋;地产税
land tax collect in kind 实物土地税
land tenancy 土地租赁
land tenure 土地占有制度;土地条例;土地使用权;土地使用期;土地(的)租用;土地占用权
land the casing 套管下至孔底
land tie 着地拉杆;锚定拉杆;地锚;地锚拉杆
land-tied island 陆连岛;沙颈岬
land tile 地面排水瓦管;透水陶管
land title 地契
land topography 地形;地貌

land transfer 土地转让
land transit insurance 陆运险
land transport(ation) 陆运;陆上运输;陆地运输;陆路运费
land transportation frequency bands 地面运输频带
land transportation radio services 地面运输无线电业务
land transport operator 陆路运输行业;陆运经营者
land trash (漂动的)冰区岸边碎冰块
land treatment 土壤渗滤处理污水;土地渗滤处理;土地处理(法);田间处理(污水)
land treatment method 土地处理法
land treatment of wastewater 废水陆地处理法
land treatment system 土地处理系统
land trust 土地信托
land trust certificate 土地信托证
land turbine 陆用汽轮机
land type 土地类型
land-type map 土地类型图
Landtz process 垃圾干馏法
land up 淤积;淤高;用土填满;堵塞
land upheaval 地面隆起
land usage 土地利用
land use 土地类别分区(城市区域规划);土地利用(计划);土地使用
land-use act 土地使用法
landuse-activity ratio 土地使用活动率
land-use adjustment 土地利用调整;用地调整
land-use analysis 土地利用分析
land-use and transportation plan 土地使用与交通运输规划
land-use and transport optimization 土地利用及交通最优选择
land-use assessment 城市用地评价
land-use balance 用地平衡
land-use capability 土地利用率
land-use certificate 土地使用证书
land-use claim 用地申请
land-use classes 土地使用分类;土地利用级别;土地利用分类
land-use classification 土地使用分类
land-use control 土地使用控制
land use control planning for line network 线网用地控制规划
land-use cost 土地利用费
land-use data 用地资料
land-use economics 土地利用经济学
land-use fee 征地费;土地使用费
land-use forecasting techniques of transportation planning 土地使用预测技术【道】
land-use intensity 土地使用强度;土地利用密度
land-use inventories 土地使用资料要目
land-use map 用地图;土地利用图
land-use pattern 土地利用形式
land-use permit 建设用地规划许可证
land-use plan 土地利用(规划)图;用地计划
land-use planning 土地利用规划;土地使用规划;土地利用计划
land-use program(me) 土地规划
land-use rate 土地利用率
land-use ratio 土地利用率
land-use regulations 土地利用规则;土地使用规章
land-use right 土地使用权
land-use statistics 土地利用统计
land-use status 土地利用情况
land-use structure 土地利用结构
land-use study 土地利用研究
land-use survey 土地利用调查;土地利用状况调查;土地利用测量
land-use type 土地利用类型
land-use zone 土地用途区
land-use zoning 土地用途分区制
land utilization 土地利用
land utilization intensity 土地利用强度
land utilization rate 土地利用率
land valuation 土地评价
land value 土地(的)价值;地价
land value increment tax 土地增值税
land value map 各块土地财产价值图
landwaiter 装卸监督员;海关人员;税务员
land wall 靠岸(闸)墙;靠岸(船闸)闸墙;(船闸的)岸侧闸墙;岸壁
landward wind 向陆风;海风

land warrant 土地使用证
landwash 高低潮线间海滨;波浪对岸冲刷;高潮线;漫滩地;海滨高潮线
land wastage 土地浪费
land waste 蛮岩;砂砾;风化石;岸屑;蛮石
landwaste slope 流沙坡
land-water 陆源水(河口处);陆地水;(沿岸冰面上的)水沟;地面水
land-water interface 陆桥水界面
land-water ratio 公园水陆面积比率
land waters 沿岸冰面开裂水域
land wave 地形起伏;地面起伏
land way 陆上通道;陆路
land wheel 地轮
land wheel driven bale loader 地轮驱动的装草捆机
land wheel driven rear mounted distributor 地轮驱动的后悬挂式撒布机
land wheel driven spreader 地轮驱动的撒布机
land wind 向海风;陆风(从陆地吹向海洋的风)
land zoning regulations 土地分区原则;土地分区计划
lane 里弄【建】;胡同;航道;点双曲线;车道
lane allocation 车道布置
lane arrangement 车道布置
lane arrangement-phasing system 车道布置—信号定相(位)系统
lane-at-a-time construction 路面按车道施工法;按车道建筑法
lane-at-a-time placement 按车道铺设法
lane balance 平衡车道;车道平衡;车道调剂
lane blockage 车道阻塞
lane bolter 长方孔筛
lane-by-lane mixing 逐条拌和【道】
lane capacity 车道通行能力;车道通过能力
lane centre line 车行道中心线
lane change (道路建筑时的)车道转换;行车道变更;行车道变动
lane changing (道路建筑时的)车道转换
lane choice 车道选择
lane-control signal 车道控制信号;车道管理信号;车辆管理信号
lane count 巷计数
lane design capacity 车道设计通行能力
lane-direction control signal 车道方向控制信号
lane distribution 交通量车道分布;车道分布
lane divider 分道线;分车岛
Lane-Emden equation 埃姆登方程
Lane-Emden function 埃姆登函数
lane flow 车道车流
lane for turning traffic 转弯行车的车道
lane gate 巷门;里弄门
lane guide 车道引导标志;车道指示标志;车道导引设置
lane identification 巷识别
lane indicating signal 车道指示信号;车道分线信号
lane joint 道路纵向缝;纵接头;纵缝
lane line 车道线;车道边界线;分道线
lane line in same direction 车道境界线
lane load(ing) 车辆荷载;车道荷载
lane marker 车道标志设置;车道划线机
lane marking 车道标线
Lane method 直接离心浮集法
lane move 车道变动
lane moving (道路建筑时的)车道转换
lane occupancy 车道占用率;车道占有率
lane occupation 车道占用率;车道占有率
lane-route 海洋航线
lane separator 车道分隔带;分车带
Lane's law 莱恩定律
lane slip 滑巷
Lane's weighted creep theory 莱恩加权徐变理论
lane tortuous line 车道迂回线
lane-use control 车道使用控制
lane-use control marking 车道控制使用标志
lane-use control sign 车道使用指定标志;车道控制使用标志
lane utilization ratio 车道利用比
lane width 船行道宽度;车道宽度
Langaloy 高镍合金
langbanite 硅锑锰矿
langbeinite 无水钾镁矾
Lange gloss meter 朗格光泽仪
Langelier index 朗格利尔指数;饱和指数
Langelier saturation index 朗格利尔饱和指数

langential velocity(component) 切向分速度
Lange photoelectric(al) colo(u)rimeter 朗格光电比色计
Langer beam 朗格尔梁
Langer girder 朗格尔(大)梁
Langerhans layer 表皮粒层
land-water 朗之万函数
Langevin function 朗之万函数
Langevin ion 朗之万离子;大离子
Langevin radiation pressure 朗之万辐射压力
Langevin theory of diamagnetism 朗之万抗磁性理论
Langevin transducer 朗之万换能器;盖板换能器
Langevin vibrator 朗之万振动片
Lang factor 兰氏比率
langisite 砷镍钴矿
langite 蓝铜矾
lang lay 同向捻(法);顺捻
lang-lay line 直捻钢丝绳
lang-lay rope 兰氏捻钢丝绳;平行捻钢丝绳;同(向)捻钢丝绳;顺捻缆
lang-lay wire 顺捻钢丝绳;顺搓钢丝绳
langley 兰利(太阳辐射测量单位)
Langmuir-Hinshelwood kinetic model 朗格缪尔-欣谢尔伍德动力学模型
Langmuir-Hinshelwood mechanism 朗格缪尔—欣谢尔伍德机理
Langmuir's adsorption isotherm 朗格缪尔吸附等温线
Langmuir's dark space 朗格缪尔暗区
Langmuir's effect 朗格缪尔效应
Langmuir's equation 朗格缪尔方程
Langmuir's frequency 朗格缪尔频率;等离子体频率
Langmuir's isotherm equation 朗格缪尔等温方程
Langmuir's law 朗格缪尔定律
Langmuir's model 朗格缪尔模型
Langmuir's oscillation 朗格缪尔振荡
Langmuir's probe 朗格缪尔探针;朗格缪尔测量仪
Langmuir's single level adsorption model 朗格缪尔单层吸附模型
Langmuir's theory 朗格缪尔学说
Langmuir's trongh 朗格缪尔型槽
Langmuir's turbulence 朗格缪尔湍流
Langmuir's wave instability 朗格缪尔波不稳定性
Langmuir's wave turbulence 朗格缪尔波湍动
Langmuir-type adsorption 朗格缪尔型吸附
Langmuir visco(si)meter 朗格缪尔黏度计
Langmuir viscosity 朗格缪尔黏度
langoustier 捕虾船
Langport beds 朗波特层
lang raising screw 地轮起落螺杆
lang-range transport(ation) 长程输送
Lang's left hand lay 左搓钢丝绳
Lang's right-hand lay 右搓钢丝绳
lang traverse 大地测量
language ability 语言能力
language laboratory 语言试验室
language of arbitration 仲裁的语言
language of the agreement 协议书的语言
language of the bid 标书语言;标书文字
language translation 语言翻译
languid 簧舌
Langweiler charge 随行装药
laniard 小绳;拉火绳;收紧索
Lanina 拉尼娜("圣女")现象
lanital (酪素纤维制造的)人造纤维;人造羊毛
Lankford value 兰克福特值
lannonite 氟水铝镁钙矾
lanolin(e) 精制羊毛脂;羊毛脂
lanostane 羊毛甾烷
Lanping tectonic knot 兰坪构造结
lansan 兰桑风
lansfordite 五水碳镁石;多水菱镁矿
lantern 信号台;泥芯架;幻灯机;航标灯;塔式天窗;手灯;灯塔上的灯室;灯笼;灯壳;灯(具)
lantern candle 灯笼蜡烛
lantern clock 灯钟
lantern coelom 提灯腔
lantern column of welded sheet construction 焊板建造的灯柱
lantern cross 穹顶十字架(教堂)
lantern fish 灯笼鱼
lantern gear 针轮大齿轮;针齿轮;灯笼齿轮
lantern gland 填料环
lantern glass 灯用玻璃

lantern lens 幻灯镜头
lantern light 天窗;提灯;顶窗;灯笼式天窗;挂灯
lantern light roof 灯笼式屋顶
lantern mantle 灯头纱罩
lantern opening 灯孔;灯笼式天窗圆屋顶孔
lantern pane 灯用玻璃
lantern pinion 针轮小齿轮;滚柱(或小)齿轮;灯笼式小齿轮
lantern plate 幻灯片
lantern ring 圆顶天窗圈梁;液封环;(水泵的)密封环;套环;灯笼式环
lantern ring for liquid seal 封液环
lantern ring groove 环槽
lantern roof 灯笼式屋顶
lantern roof light 灯笼式屋顶天窗
lantern room 灯房
lantern slide 幻灯片
lantern(slide) projector 幻灯放映机
lantern tower 灯楗;灯塔;穹窿形顶塔楼
lantern-wheel gearing 针轮传动
lanthanite 碳镧石
lanthanite-(Nb) 碳铌石
lanthanon element 镧族元素
lanthanum barium method 镧—铈法
lanthanum boride 硼化镧
lanthanum boride ceramics 硼化镧陶瓷
lanthanum cerium method 镧—铈法
lanthanum chromite ceramics 铬化镧陶瓷
lanthanumdoped lead zirconate-lead titanate 钛酸锆酸镧铅
lanthanum fluoride 氟化镧
lanthanum fluoride active medium 氟化镧激活媒质
lanthanum fluoride infrared transmitting ceramics 氟化镧透红外陶瓷
lanthanum-impregnated silica gel 充满镧的硅胶
lanthanum iodate 碘酸镧
lanthanum iodite 碘化镧
lanthanum ores 镧矿
lanthanum oxalate 草酸镧
lanthanum oxide 氧化镧
lanthanum silicate 硅酸镧
lanthanum titanate ceramics 钛酸镧陶瓷
Lanthar lens 兰泰尔镜头
Lantiron 高硅耐热耐酸铸铁
Lantz process 兰茨法
Lanvirnian(stage) 兰维尔阶【地】
lanyard 小绳;拉火绳;收紧索;短索;牵索(系于安全带上的绳索)
lanyard knot 绳头结
lanyard stopper 制动索;锚链制
lanyard stuff 四股油绳
lanyard thimble 镀锌三角形绳环
Lanyl blue 拉尼尔蓝
Lanz cast iron 珠光体铸铁
Lanz-pearlite process 铸型预热浇注法
Lanz's line 兰茨线
Lanz's point 兰茨点
laoding factor 储备系数
Laomedia astacina 泥虾
lap 研磨模;皱折;折叠;鳞比;互搭;漆膜的搭接复盖;围裙;刷路搭接处;山间凹地;搭接量;搭接部分;搭边;叠接;余面;研磨
lap and lead lever 滑阀控制杆
lap belt (车座位上的)安全带
lap blister 折叠气泡
lap butt 搭接连接
lap calking 搭边捻缝
lap cement 搭接胶泥;搭接胶合料;搭接黏结材料;搭接胶黏剂
lap circle 余面圆
lap coil winding 叠绕组
lap connection 搭接接头
lap dissolve 慢转换
lap dissolve shutter 慢转换装置;慢转换器
lap dovetail 互搭鸠尾榫;搭接鸠尾榫
lap drag 叠板刮路机;叠板路刮
lapel bias 卜头斜度
lapel microphone 佩带式小型话筒;小型话筒
lapel roll line 卜头线
lap evenness tester 棉卷均匀度试验机
lap fillet weld 搭接角焊缝
lap fillet welding 搭接角焊
lap former 成卷装置;成卷机

lap gate 压边浇口
lap grinder 搭头研磨机;清理焊缝磨床
lap guide 导卷架
lapicide 石工
lapidarist 宝石家
lapidary 宝石商;宝石鉴识家;宝石工(艺);宝石雕琢术;宝石收集者
lapie 石灰岩沟;岩沟
lapieite 硫锑镍铜矿
lapilli tuff 火山砾凝灰岩
lapillus 火山砾;[复] lapilli
lapirium chloride 吡胺月酯
lapis 青金石;路程标桩(罗马)
lapis cal aminaris 异极矿
lapis divinus 铜矾
lapis lazuli 天青石(色);杂青金石;琉璃璧;青金石;天然群青
lapislazuli blue 天青石蓝
lapislazuli ware 青金石器皿
lap joint 搭接连接;搭(接)接头;重接;叠接;搭头搭接
lap jointed sheeting 互搭铺板;搭接板
lap-joint fillet weld 搭接焊缝
lap-joint flange 活套法兰;自由回转法兰;搭接凸缘
lap jointing 互搭接头
lap joint of deck 甲板搭结边
lap joint pliers 搭接钳
lap joint sheet(ing) 搭接板
lap joint stud 活套环圈
Laplace's azimuth 拉普拉斯方位角
Laplace's distribution 拉普拉斯分布
Laplace's equation 拉普拉斯方程;调和方程
Laplace's expansion 拉普拉斯展开式
Laplace's expansion theorem 拉普拉斯展开定理
Laplace's formula 拉普拉斯公式
Laplace's hydrodynamic(al) theory 拉普拉斯流体动力学说
Laplace's integral 拉普拉斯积分
Laplace's irrotational motion 拉普拉斯无旋运动
Laplace's law 拉普拉斯定律
Laplace's law of succession 拉普拉斯连续律
Laplace's linear equation 拉普拉斯线性方程
Laplace's measure of dispersion 拉普拉斯离差
Laplace's model 拉普拉斯模型
Laplace's operator 调和算子;拉普拉斯算子
Laplace's partial differential equation 拉普拉斯偏微分方程
Laplace's plane 拉普拉斯平面
Laplace's point 拉普拉斯点
Laplace's station 拉普拉斯点
Laplace's theory 拉普拉斯学说
Laplace's transform(ation) 拉普拉斯变换
Laplace's vector 拉普拉斯矢量
Laplacian azimuth 拉普拉斯方位角
Laplacian coefficient 拉普拉斯系数
Laplacian condition 拉普拉斯条件
Laplacian control 拉普拉斯检核
Laplacian curve 拉普拉斯曲线
Laplacian equation 拉普拉斯方程
Laplacian hydrodynamic(al) theory 拉普拉斯流体动力学说
Laplacian inversion theorem 拉普拉斯反演定理
Laplacian nebular hypothesis 拉普拉斯星云假说
Laplacian operator 拉普拉斯算符;拉普拉斯算子;调和算子
Laplacian point 拉普拉斯点
Laplacian speed of sound 拉普拉斯声速
Laplacian station 拉普拉斯点
Laplacian surface harmonics 拉普拉斯面谐函数
Laplacian transform(ation) 拉普拉斯变换
lap lagging 搭接挡板
laplandite 硅钛铈钠石
La Plata sandstone 拉普拉塔砂岩
lap length of bar splicing 钢筋搭接长度
lap link 连接两段链条的接头;活口链环
lap mark 接头痕迹
lap marking 搭接印记
lap of bucket trace 叠斗;斗迹重叠
lap of coil 曲管卷
lap of length 搭接长度
lap of splice 互搭长度;搭接长度;搭接部分;板搭接
lap of valve 阀余面
La Pointe picker 拉波安特型拣矿机
Laporte selection rule 拉波特选择定则

lapout 侧向尖灭
lap over 重叠;搭接
lap-over effect 包边效应
lap-over seam 搭接缝;重叠矿层
lapped 搭接的;重叠的
lapped butt 端搭接;搭接横缝
lapped corner joint 转角搭接;搭接角接头;搭角接合
lapped dovetail 抽屉(燕尾)榫;搭接鸠尾榫;转角搭接盖面燕尾榫
lapped face 研磨表面
lapped finish 研磨加工;研磨的表面光洁度
lapped joint 搭接法兰;互嵌接合;搭接(焊)缝;互搭接头连接;搭接接头
lapped length 搭接长度
lapped pulp 稀薄纸浆
lapped seam 搭接(纵)缝
lapped seam attachment 叠缝装置
lapped slice 研磨的片子
lapped splice 互拼接合;互搭接合;搭接
lapped steel plate joint 搭接钢板接缝
lapped tenon 互搭榫接;搭接榫
lapped type 重叠式(指混凝土沉排形式)
lapped-type flange 搭接法兰
lapped welding joint 搭接焊缝
lapper 研具;磨床;研磨机
lappet 下垂物;垂花机;浮纹织物;浮经装置
lapping 精研研磨;磨合;打枝;搭叠;重叠;擦光;余面;研磨
lapping agent 抛光剂;研磨剂
lapping allowance 研磨加工留量
lapping burn 研磨烧伤
lapping cloth 衬布
lapping compound 研磨剂;研磨膏;研剂
lapping finish(ing) 研磨
lapping fluid 磨料研磨液
lapping gear 精研齿轮
lapping-in 研合;对研
lapping joint 搭接缝;搭头搭接
lapping liquid 研磨液
lapping lubricant 研磨润滑剂;擦准润滑剂
lapping machine 研磨机;研光机;精研机;磨片机;缠绕机
lapping material for monocrystalline silicon 单晶硅研磨材料
lapping mechanism 轻敲机构
lapping of reinforcing steel 钢筋的搭接
lapping oil 研磨油;研磨液
lapping paste 研磨膏
lapping plate 研磨板;精研板;搭接板
lapping position 垫纱位置
lapping powder 研磨剂;研磨粉
lapping rejects 研磨屑
lapping ring 精研圈
lapping stick 研磨条
lapping surface 研磨面
lapping time 喷路搭接流平时间;搭接时间
lapping tool 研磨工具
lapping wheel 研磨轮
lap plank drag 叠板刮土机
lap position 遮闭位置;滑阀重叠位置;重叠位置
lap resistance welding 搭接电阻焊
lap rivet 搭接铆钉;互搭钉
lap-riveted 铆接的;互搭铆接的
lap riveting 铆搭接;互搭铆(接);搭接铆;搭叠铆接
lap roller 棉卷辊
lap scarf 互搭楔接;互搭榫接;互搭嵌接
lapse 期满失效;时间推移;失效;失误;垂直梯度
lap sealant 搭接密封膏
lap sealant joint 搭接密封接缝
lap seam 重叠岩层;搭接缝
lap seam cathode 叠缝阴极
lap seam welder 搭(缝)焊机
lap seam welding 搭接(缝)焊
lapsed policy 失效保险单
lapse limit 对流层顶
lapse line 递减线
lapse of temperature 温度递减
lapse of time 时间消逝;时间的推移
lapse period 气温递减时期
lapse provision 失效条款
lapse rate 下降比;直减率;减率;气温递减率;温度直减率;温度递降率;温度递减率;递减率;大气温度垂直梯度
lapse rate of air temperature 气温直减率

lap shaver 刮皮刀
lap siding 交叠式会让线;纵列式会车站;互搭壁板;鱼鳞板(墙面);互搭板壁;披叠墙板;搭接面板;护墙板
lapsing schedule 逐期折旧明细表;固定资产增减明细表
lap splice 互搭接头;互搭接合;搭接板;重叠拼接;重叠胶接
lapstreak planking 盖瓦式叠板
lap switch 交叠式三开道岔
Laptev Sea 拉普帖夫海
lap tool 研具
laptop 便携式计算机
lap turnout 交叠式三开道岔
lap-welded 搭焊的;搭焊
lap welded pipe 搭接法焊接的管子;搭焊管
lap welded seam 搭焊缝
lap welded steel pipe 搭焊钢管
lap welded steel tube 搭焊钢管
lap welded tube 叠式搭焊缝管
lap welder 搭焊机
lap weld(ing) 搭焊;叠式焊接;接焊;接头焊接
lap welding machine 接头焊接机
lap width (屋面瓦的)搭接宽度
lap winding 叠式卷绕;叠绕组;叠绕(法)
lap wood 枝条材
lap(work) 搭接
lapworker 搭接工
lap wound 叠绕
lap wound armature 叠绕电枢
laquear 格子天花板;天花板的镶板
laquer 光漆
Laramian cycle 拉拉米旋回【地】
Laramian sand 拉拉米砂层
Laramide orogeny 拉拉米运动【地】
lararium 小型家庭神位(罗马)
Laray viscometer 莱雷黏度计
larboard 左舷
larceny 盗窃;非法侵占财产
larch 落叶松(木)
larch fir 松杉木材
larch pine 南欧黑松
larch shingle 落叶松板瓦
larch-tree 日本落叶松
larch turpentine 落叶松松节油
lardalite 歪霞正长岩
larder 储藏间;食品室
Lard Howe rise 豪勋爵海隆
lard ice 油脂状冰;泥泞冰
lard oil 猪油
lard peat 猪脂泥煤;猪脂泥炭
lareactors 激光聚变堆
large acreage 大面积
large aggregate concrete 大粒料混凝土;大集料混凝土;大骨料混凝土;粗集料混凝土;粗骨料混凝土
large air compressor 大型空气压缩机
large alternate current three-phase synchronous motor 大型交流三相同步电动机
large amount of sediment 巨量泥沙
large amounts 大批
large amplitude 大振幅
large amplitude cyclic(al) deformation 大振幅循环变形
large amplitude flexural vibration 大振幅弯曲振动
large amplitude non-linear condition 大振幅非线性条件
large amplitude shock wave 大波幅激波
large amplitude wave 大振幅波
large and medium direct current motor 大中型直流电动机
large and medium-sized harbo(u)r 大中型港口
large and medium-sized port 大中型港口
large and medium-sized projects 大中型建设项目
large and medium-sized project to be continued 续建大中型项目
large and small-diameter cutter 大小直径铣刀
large and thick meniscus 肥大板根
large and vigorous pulse 脉大有力
large angle 大角度
large-angle boundary 大角度间界;大角度晶间界
large-angle scanner 宽角扫描器;大偏转角扫描器
large-angle scanning 广角扫描
large annual rings 大年轮

large aperture 大口径;大孔径(的)
large aperture interferometer 大孔径干涉仪
large aperture optic(al) system 大孔径光学系统
large aperture seismic array 大孔径地震台阵
large aperture seismic experiment 大窗口地震试验
large area blasting 大面积爆破
large area colo(u)ring 大面积着色
large area flood 大面积洪水
large area foils 大张金属箔
large area of lacustrine soils 大片湖积土
large area photomultiplier 大面积光电倍增器
large area rectifier 大面积整流器
large area surface load 大面积地面荷载
large area survey 大面积测量
large arm caliper 大臂井径仪
large associative processor 大型关联处理机
large astronomical satellite 大型天文卫星
large automatic navigation(al) buoy 大型自动导航浮标
large autonomous submersible 大型自动主潜水器
large ballast 大块石渣
large batch 大批量的
large batch production 大批生产
large bell angle 大料钟角
large bell beam 大钟平衡杆
large bin 大型储仓;大料仓
large-block 大型砌块
large-blocked structure 大型砌块结构;大型砌块构造;大块状结构
large-block masonry 大型砌块圬工
large-block unit 大型砌块部件
large bodies 大型菌体
large bodies of water 大水体
large bolt rope needle 缝帆边绳大针
large bore 大直径探坑
large bore engine 大缸径发动机
large bore furnace 大口径炉
large borehole 大孔径钻孔
large bore hose 大口径胶管
large bore tunnel 大断面隧洞
large brickwork 大型砖块
large bubble aeration 大气泡曝气
large bubble aerator 大气泡曝气池
large building 大厦
large building block 大砌块
large building slab 大板
large bulb ship 大球鼻(首)船
large-bus 大型公共汽车
large caisson 大型沉箱
large calorie 千卡;大卡
large capacitor checking instrument 大电容校验仪
large capacity 大容量的
large-capacity air raid shelter 大型防空掩蔽所;大容量掩体;大容量防空洞
large-capacity belt conveying plant 大容量皮带传送装置
large-capacity car 大号轻便汽车
large-capacity communication 高容量通信
large-capacity flat car 大型平板车
large-capacity grinding mill 大容量研磨机
large-capacity machine units 大型机组
large-capacity memory 大容量存储器
large-capacity meter 大流量计
large-capacity mixer 大容量搅拌机
large-capacity number control turning machine 大型能量数字控制车床
large-capacity refrigerated centrifuge 大容量冷冻离心机
large-capacity stone spreader box 大容量箱形撒石机
large-capacity storage(rain) ga(u)ge 大容量累积雨量器
large-capacity wagon 大型货车
large cast concrete panel 大型预制混凝土板
large center 大型中心
large chemical complex 大型化工联合企业;大化学联合企业
large chemical plant 大型化工厂
large cherry 大樱桃
large city 大城市
large clearing 大面积清除;大面积砍伐森林
large clod 大土块
large coal 块煤
large concrete block 大型混凝土砌块;大型混凝

土块(体)
large conical gear 大圆锥齿轮
large conical gear bolt 大圆锥齿轮螺栓
large contour projector 大型轮廓投影仪
large contract 大型承包工程
large core and large numerical aperture fiber 大芯径大数值孔径光纤
large core fiber[fibre] 大芯径光纤;粗芯光纤
large core memory 大容量磁芯存储器
large correction 大改正
large cross section excavation 大断面开挖
large cylinder 大容积汽缸
large cylinder wharf structure 大圆筒码头结构
large dam 大坝
large dam concrete plant 大坝混凝土搅拌站
large data file 大容量文件
large deflection 大挠度
large deflection theory 大挠度理论
large deformation 大变形
large deformation analysis 大型变形分析
large deformation theory 大形变理论
large delta association 大型三角洲组合
large destructive earthquake 大破坏性地震
large deviation 大偏差
large-diameter 大直径(的);大号
large-diameter borehole 大口径钻孔;大孔径钻孔
large-diameter boring 大口径钻探
large-diameter design core barrel 大直径标准岩芯管(美国4~8英寸)
large-diameter double tube 大直径双层管
large-diameter drill hole 大直径钻孔
large-diameter drilling machine 大孔径钻机;大直径钻机
large-diameter evapo(u)ration pan 大口径蒸发器;大口径蒸发皿
large-diameter hole 大直径钻孔;大直径圆孔
large-diameter optic(al) fiber 大直径光纤
large-diameter pier 大直径支墩;大直径支柱
large-diameter piles in prebored holes 大直径钻孔灌注桩
large-diameter pipe 大(直)径管;大口径管(材)
large-diameter polybutylene plastic fitting 大直径聚丁烯塑料管件
large-diameter polybutylene plastic pipe 大直径聚丁烯塑料管
large-diameter prestressed concrete cylinder pile 大直径预应力混凝土管桩
large-diameter sample 大直径试样
large-diameter source 大角径源
large dipping angle migration 大倾角偏移
large discharge pump 大排量泵
large dish with underglaze blue 青花大盘
large-displacement holography 大位移全息照相术
large-displacement matrix 大位移矩阵
large domestic incinerator 大型民用焚化炉
large drill 重型凿岩机
large-duty 高生产率的
large-duty loader 大型装载机
large dyne 牛顿
large earthquake 大地震
large eccentric compression 大偏心受压
large eccentricity 大偏心
large elastic deformation 大弹性变形
large end bell 大头承口
large end bell reducer 承口大小头
large-engined vessel 大功率船舶;装有大型发动机的船舶
large engineering project 大型工程建设
large equilibrium constant 大平衡常数
large extension table 大号伸长台
large face 大面
large face plate 大花盘;称量托盘
large finisher 大型路面抹面机;大型路面整修机
large fixed loop field 大定源场
large flat head nail 大头钉
large flat head slotted screw 大平槽头螺钉
large floe 巨大浮冰
large flow sampler 大流量采样器
large fly-cutting disc 大型快速切削盘
large forest region 密林带
large form 大模板
large format 大开本;大尺寸;大规模;大型号
large format board 大规格板
large format camera 大像幅摄影机

large form construction 大模板施工
large fracture 大冰隙(超过500米宽)
large-framed 体格大的
large gap spark chamber 大隙火花室
large garden 大花园
large gas engine 大型气体发动机
large gas field 大气田
large gear cutting machine 大型齿轮加工机床
large gear ring 大齿轮圈
large-general-purpose computer 大型通用计算机
large grab pontoon dredge(r) 大型平底船型挖泥船
large grain 大颗粒;粗粒的
large groundwater project 大型地下水工程
large groundwater supply 大型地下水供水
large group connection 多组合连接
large handcart 大板车
large hand sketch 放大图样
large headed nail 大头钉
large high speed machine 大型高速电机
large hole drilling 大口径钻进
large housing estate 大型居民点
large hydraulic project 大型水利工程
large ice field 大冰原
large ice pieces 大冰块
large imperial 大王裁
large induction motor 大型感应电机
large inside micrometer cal(l)ipers 大内径测微卡尺;大内侧螺旋测微器
large ion 大离子
large ion lithophile element 大离子亲石元素
large job 大型工地;大量工作
large junk 沙船
large knot 大(树)节;大节疤;大木节(直径大于1.5英寸的树节子)
large land areas 大地区
large landslide 大型滑坡
large laying and finished machine 大型路面铺设整修机
large-leaved variety 大叶品种
large linear structure 大型线状构造
large linear system 大线性组
large log 大圆材
large-lot production 大批量生产
large low-speed machine 大型低速电机
large magnitude earthquake 大震级地震
large main valve piston 主阀大活塞
large main valve piston ring 主阀大活塞环
large male and female face 大凸凹式密封面
large manned submersible 大型载人潜水器
large marine ecosystem 大海洋生态系统
large matrix problem 大矩阵问题
large matrix store 大容量矩阵存储器
large media 大介质
large-mesh 大网眼的;大网眼;大筛眼
large metal 前轴承
large mixer 大型搅拌机
large mobile wet masher 大型移位式湿饲料搅碎机
large module gear 大模数齿轮
large moored ship 系泊大型船
large motorway restaurant 高速公路餐馆
large mo(u)ld 大型模具
large multi-road crossing 多条道路交汇的广阔交叉口
large navigable canal 大型通行运河
large navigable river 大型通行河流
large navigation(al) buoy 大型助航浮标;大型导航浮标
large number hypothesis 大数假说
large numbers 大批
large numerical control EDM machine 大型数控电火花机床
large numerical control wire-cut EDM machine 大型数控电火花线切割机床
large object salvage system 大型目标营救系统;大目标救捞系统
large offset marine seismic 大偏移海洋地震
large oil field 大油田
large opening 大开孔
large opening flow 大孔口泄流
large open lamp with stand 带灯座大口灯
large open well 大口井
large optic(al) cavity 大光学共振腔
large optic(al) reflector 大型反光镜

large orifice 大孔口
large output 大输出
large oval head nibbed bolt 大扁圆头带榫螺栓
large oval head rivet 大扁圆头铆钉
large oval head socket shank rivet 大扁圆头半空心铆钉
large oval head square neck bolt 大扁圆头方预螺栓
large over-voltage operation 大过电压运行
large paddle-type stirring mechanism 大浆叶搅拌机构
large panel 大组装件;大型墙板;大型板材
large panel building 大板建筑
large panel construction 大型格板结构;大型预制板施工;预制大板构造;大型墙板施工(法);大板建筑
large panel form(work) 大型板材模板
large panelled structure 大板结构
large panel roof slab 大型屋面板
large panel shuttering 大块模板
large panel structure 大型板材建筑(物);大板结构;大节间结构
large panel system 大板结构体系
large panel wall 大型墙板
large panel wall structure 大板墙体结构
large panel wall system 大板墙体系
large parts 大型零部件
large paving sett 大毛石(铺路用)
large photography 大像幅摄影
large pile 大桩
large pitch pocket 大树脂囊
large plant 大型设备
large platform elevator 大平台升降机
large pole 大圆木料;大径材
large pore 大孔
large port 大港
large post 大邮裁
large pour 大量连续浇注
large power motor 大功率电动机
large power transistor 大功率晶体管
large praying chamber 祈祷室
large pressurized vessel 大型压力容器
large primary jaw crusher 大型一级颚式破碎机;大型颚式轧碎机
large program(me) 大型程序
large project 大项目
large projector 大型投影仪
large punch 大冲子
large quantities 大批
large quantities of lime 大量石灰
large quantity 大批量;大量
large radius curve 大半径曲线
large rafter 大椽木
larger curved meander 大弯度曲流;大弯度河曲
large reading dial (自记压力仪的)大刻度盘
large regional system 广域系统;大区域系统
large region yield 大范围屈服
larger erectional constituent(of tide) 大出差分潮
large research microscope 大型研究用显微镜
large reservoir 大水库
large reverse body bush 回动体大衬套
large riprap 大块抛石护坡;大块乱石护面
larger lunar elliptic(al) diurnal constituent 大太阴椭圆日分潮
larger lunar elliptic(al) semidiurnal constituent 大太阴椭圆半日分潮
larger lunar evectional semidiurnal constituent 大太阴出差半日分潮
large road finisher 大型路面平整机;大型路面整修机
larger observed 明显地震
large rocket 大型火箭
large rock mass 大块岩石
large roof slab 大型屋面板
large round-headed nail 大圆头钉
larger solar elliptic(al) semidiurnal constituent 大太阳椭圆半日分潮
large rubble 大型块石;大型毛石
large rubble fill 大型毛石填料
large salt concentration 粗盐浓度
large sample 大子样;大样品;大样本
large-sample distribution 大样分布
large sample method 大样法
large saw 大锯

large scale 大比例尺;大型;大批的;大规模(的);大尺度;大尺寸;大秤;大比例尺(的)【测】
large-scale aerophoto-grammetry 大比例尺航空摄影
large-scale army air style map 大比例尺军用导航图
large scale bench blasting 台阶大爆破
large-scale blasting 大爆破
large-scale building 大型建筑(物)
large-scale building of water conservancy projects 大兴水利
large-scale cartography 大比例尺制图
large-scale characteristic 大尺度特征
large-scale chart 大比例尺海图
large-scale computation 大规模计算
large-scale computer 大型计算机
large-scale concrete block 大尺寸混凝土块
large-scale construction 大规模施工;大规模建筑工程;大规模建设
large-scale consumer 消费量大用户
large-scale convection 大规模对流;大尺度对流
large-scale design drawing 大比例尺设计图
large-scale detail plan 大比例尺详图
large-scale digital computer 大型数字计算机
large-scale digital computing system 大型的数字计算系统
large-scale digitizer 大型数字化转换仪;大型数字化转换器
large-scale direct shear test 大面积直剪试验
large-scale drawing 大比例尺图
large-scale dredging 大规模疏浚
large-scale drilling program(me) 大规模钻探工程;大比例尺钻探网
large-scale drilling system 大口径钻探设备
large-scale earthquake 大规模地震
large-scale electrolytic cell 大型电解槽
large-scale element 大尺寸构件
large-scale enterprise 大型企业
large-scale equipment 大型设备
large-scale erection works 大规模安装工程
large-scale excavation 大规模挖方工程;(大规模)开挖
large-scale excavator 大型挖掘机
large-scale experiment 大型试验;大型实验;大规模试验
large-scale facility 大型设施
large-scale feature 大型特征
large-scale field loading test 大型野外荷载试验
large-scale field test 大型野外试验
large-scale forest damage survey in Europe 欧洲大规模森林损害调查
large-scale form 大型模板
large-scale genetic soil map 大比例尺发生学土壤图
large-scale geologic(al) survey 大比例尺地质测量
large-scale harbo(u)r 大型港口
large-scale hydraulic model 大(比例)尺水工模型
large-scale hydraulic model test 大(比例)尺水工模型试验
large scale industrial district 大型工业区
large-scale industry 重工业;大规模工业
large-scale integrated circuit 大规模集成电路
large-scale integration 大规模集成
large-scale interaction 大尺度相互作用
large-scale knowledge system 大规模知识系统
large-scale landform 大型地貌
large-scale linear programming 大规模线性规划
large-scale map 大比例尺海图;大比例尺(地)图
large-scale mapping 大比例尺测图;大规模测绘地图
large-scale master sheet 大比例尺底图
large-scale mechanized construction 大规模机械化施工
large-scale metrology 大尺寸测量
large-scale model 大尺度模型;大比例尺模型
large-scale model test 大比例尺模型试验
large-scale of bursting water 大型突水
large scale outbreak 大规模暴发
large-scale photograph 大比例尺照片
large-scale pilot plant 大型实验厂
large-scale plan 大比例尺平面图
large-scale planning 大型规划
large-scale plantation 大规模种植园
large-scale pollution 大面积污染;大范围污染

large-scale port 大型港口
large-scale pour 大规模灌注
large-scale prefabricated element 大尺寸预制件
large-scale problem 大型问题
large-scale production 大量成批生产；大规模生产
large-scale programming 大型规划
large-scale project 大型项目；大型规划；大型工程（项目）；大规模建设项目；大规模计划
large-scale project site 大规模工程现场
large-scale public building 大型公共建筑
large-scale regional geologic(al) surveying 大比例尺区域地质调查
large-scale restitution 大比例尺测图
large-scale series 大比例尺系列
large-scale setting out 定线；大规模放样；大规模定线
large-scale shaking table 大型振动台
large-scale shuttering 大型模板
large-scale sociotechnical system 大型社会技术系统
large-scale soil map 大比例尺土壤图
large-scale soil test 大型土工试验；大尺寸土工试验
large-scale structure 大尺度结构
large-scale survey 大面积调查；大比例尺测图；大比例尺测量
large-scale surveying map 大比例尺测图
large-scale system 大(规模)系统
large-scale temporary facility 大型临时设施
large-scale test 大型试验；大规模试验
large-scale time sharing system 大型分时系统
large-scale topographical drawing 大比例尺地形图
large-scale topographic(al) map 大比例尺地形图
large-scale topographic(al) mapping 大比例尺地形测图
large-scale transportation 大规模运输
large-scale trial 工业性试验；大型试验
large-scale turbulence 大尺度紊动；大尺度湍流
large-scale use 大规模利用
large-scale user 大用户
large-scale utilization 大规模利用
large-scale variation 大规模变化
large-scale velocity field 大尺度速度场
large-scale water source 大型水源地
large-scale weather situation 大(范围)天气形势
large-scale works 大型工程
large-scale yard 大型调车场
large-scale yielding 大范围屈服
large-screen 大屏幕
large screen display 大屏幕显示(器)
large screen projector 大屏幕投影器
large screen radar indicator 大屏幕雷达指示器
large-screen television 大屏幕电视
large screen video projector 大型屏幕录像投影机
large self-contained production platform 大型独立生产平台
large sett paving 大型小方石路面
large shaft sleeve 大轴套
large shape 大型钢材
large shed 大库棚
large shock 大地震
large signal 强信号；大信号
large-signal analysis 大信号分析(法)
large-signal conductance 大信号电导
large-signal current gain 大信号电流增益
large-signal performance 大信号性能
large silo 大型筒仓
large size 大尺寸
large-size brick 大型砖
large-size coconut fibre brush 大号椰子纤维刷
large-size collapsible container 大型折叠式集装箱
large-sized 大型的；大型的；大规模的
large-sized cable 大型电缆；大容量电缆
large-sized coal 大块煤
large-sized diesinker 大型锻模仿型铣床
large-sized floor slab 大尺寸地板
large-sized generator rotor slotting machine 大型发电机转子插铣床
large-size(d) high frequency hydraulic vibration table 大型高级液压振动台
large-size diamond 大颗粒金刚石
large-sized long arm 大型长臂挖掘机
large-sized model 大模型
large-sized panel 大组装件
large-sized photographic microscope 大型照相显微镜
large-size drill 大型钻头
large-size floor joist 楼板大龙骨
large-size folding container 大型折叠式集装箱
large-size folding wood dryer 大型折叠木晾衣架
large-size ore deposit 大型矿床
large-size passenger bus 大客车
large-size planer 大型刨床
large-size precision thread ring and plug ga(u)ge 大型精密螺纹环塞规
large-size reverberatory furnace 大型反射炉
large-size sheet stacker 大规格玻璃板堆垛机；大玻璃片码垛机
large Soderberg cell 大型连续自焙阳极电解槽
large space enclosure 大容积密闭罩
large space telescope 大型空间望远镜
large span 大跨度；大跨距的
large span beam 大跨度梁
large span bearing system 大跨度承重结构；大跨度承重体系
large span floor slab 大跨度楼板
large span frame 大跨度构架
large span (pre)cast beam 大跨度预制梁
large span roof 大跨度屋顶
large span shell 大跨度薄壳
large span shell vault 大跨度薄壳拱顶
large span structure 大跨度结构
large span trussed roof 大跨度桁架屋顶
large spinning disk distributor 大圆盘撒布机
large spoon 大匙
large square 大方木
large square bland continuous-casting machine 大方坯连铸机
largest birefringence 最大双折射率
largest crane 最大起重机
largest derricks 最大吊杆
large steam turbine-generator 大型气轮发电机组
large-step method 大步长方法
largest extinction angle 最大消光角
large-stone bit 粗粒金刚石钻头（粒度为1克拉8颗以上）
largest peak discharge 最大洪峰流量
large strain amplitude 大应变幅
large structure 大型结构(物)
large structure testing laboratory 大型结构试验室
largest value guideline 最大准则
largest value theory 极值理论
largest vessel 特大型船；超重型船
large supply main 给水总管；供电主干线
large surface water project 大型地表水工程
large-swath width scanner 大幅宽度扫描器
large system 大系统
large system of equations 大型方程组
large tanker 大型油船
large tank wagon 大型罐车
large taper 大角度的锥体
large tea spoon 大茶匙
large test 大规模试验
large-thick coal seam 巨厚煤层
large throat with connecting large pore 粗孔大喉型
large-tonnage 大吨位；大产量
large-tonnage mine 大型矿山
large-tonnage product 大量产品
large-tonnage ship 大吨位船舶
large-tonnage vessel 大吨位船舶
large toolmaker's microscope 大型工具制造工用显微镜
large tools box 大件工具箱
large-toothed aspen 大齿白杨【植】
large-toothed saw blade 大齿形锯刀
large-town citizen 大城市居民
large transport airplane 大型运输机
large tug 大型拖轮
large-type diagnostic X-ray machine 大型X光诊断机
large-type dozer 大型推土机
large-type drill 大型钻机；大型凿岩机
large-type excavator 大型挖掘机
large-type horizontal metallurgical microscope 大型卧式金相显微镜
large-type sinterer 大型烧结机
large-type tractor 大型拖拉机
large value capacitor 大容量电容器
large vice 大虎钳
large volume 吞吐量大的；大体积的
large volume concrete block 大体积混凝土块体
large volume cylinder 大容量油缸
large volume data exchanger 大容量数据交换机
large volume harbo(u)r 吞吐量大港口
large volume item 大量生产制品
large volume of fire 猛烈的火力
large volume sprayer 大容量喷雾机
large volume spring 大流量温泉
large volume sprinkler 大流量洒水车；大流量人工降雨器；大流量喷洒器；大流量喷灌器
large wall panel 大型墙板
large wall panel construction 大型墙板构造
large wastewater treatment plant 大型废水处理厂
large water heater 大型(浴用)热水器
large water range wharf 高水位差码头
large wave 巨浪
large wave tank 大型(模拟)波浪水池
lariciresinol 松树脂醇
laricis cortex 落叶松皮
Larimer column 工字钢组合柱
Larix decidua 欧洲落叶松
Larkin cementing shoe 拉尔金注水泥法套管鞋
Lark's head 双合套；双合结
larkspur 飞燕草
larmatron 拉马管
larmier 滴水槽；飞檐
Larmor formula 拉莫尔公式
Larmor frequency 拉莫尔频率
Larmor orbit 拉莫尔轨道
Larmor precession 拉莫尔旋进；拉莫尔进动
Larmor precession frequency 拉莫尔进动频率
Larmor radius 拉莫尔半径
Larmor theorem 拉莫尔定理
larmotron 拉莫管
larnite 斜硅酸石；斜硅钙石；甲型硅灰石
larosite 硫铅铜矿
Larrea divaricata 石炭酸灌木
larried up 用拌浆锄拌浇石灰砂浆
Larrson pile 拉逊板桩
larry 矿车用推车；手推车；上坡牵引车；斗底车；电葫芦；秤量车；薄浆雾（英国特恩河口的浓雾）；薄浆；拌浆锄；拌浆铲；半液态砂浆
larry bin 装煤车煤斗
larry car 底卸式矿车
larrying 灌浆；薄浆砌筑
larrying up 薄浆砌砖(法)；浇薄浆（砌砖时）
larsenite 硅铅锌矿
Larssen-Nielson 拉森—奈尔逊体系建筑（丹麦建筑体系之一，构件尺寸与房间相等，装配成箱形，外墙覆面为隔热层夹心板）
Larssen's ledge finder 拉尔森表土层钻探装置（边钻边下套管）
Larssen's pile 拉森型柱桩；拉森前探钢桩
Larssen's piling 拉森桩；拉森钢板桩
Larssen's section 拉森厚木板；拉森型材
Larssen's sheet piling 拉森式钢板桩
Larssen's spile 超前伸梁
Larssen's steel sheet pile 拉森式钢板桩
Larssen's steel sheet pile wall 拉森式钢板桩墙
larvae 幼虫
larvae of oyster 牡蛎幼体
larval phase 幼虫期
larval plankton 浮游幼体
larval population 虫口
larval stage 幼虫期
larvascope 幼虫检查镜
larvikite 歪碱正长岩
lasable dye 可激射染料
lasanum 产科椅
Lasater's bubble-point pressure correction 拉沙特尔泡沸点压力校正
lascar 东印度水手
Lascar closet 拉斯卡式厕所（一种蹲式水冲厕所）
Laschamp event 拉斯线普事件
Laschamp reversed polarity subchron 拉尚反向极性亚时
Laschamp reversed polarity subchronzone 拉尚反向极性亚时间带
Lascnamp reversed polarity subzone 拉尚反向极性亚带
lase 激光辐射
lasecon 激(射)光转换器

laser 激光(器)
laser absolute gravimeter 激光绝对重力仪;激光绝对重力计
laser accelerometer 激光加速仪;激光加速器;激光加速计
laser accumulation 激光能量聚集
laser acoustic(al) signal 激光声学信号
laser acquisition 激光探测
laser acquisition device 激光截获装置
laser acquisition system 激光搜索系统
laser action 激光作用;激光应用
laser activation 激光引发
laser activity 激光性能
laser adjoint application 激光组合应用
laser aerocamera 激光航空照相机
laser aid 激光装置
laser aiming 激光引导
laser airborne dust monitor 激光法飘尘监测仪
laser aligner 激光准直仪;激光准直器
laser alignment 激光准直;激光指向;激光对准
laser alignment deflection measurement 激光准直挠度测量
laser alignment error 激光准直误差
laser alignment method 激光找中心法
laser alignment survey 激光准直测量
laser alignment system 激光准直系统
laser alignment telescope 激光准直望远镜;激光导向仪
laser alignment with zone plate 波带板准直;波带板激光准直
laser altimeter 激光高度计;激光测高仪
laser altitude ga(u)ge 激光测高计
laser amplifier bandwidth 激光带宽
laser anemometer 激光风速仪;激光风速计
laser anemometer signal 激光风速计信号
laser angular rate sensor 激光角速率传感器
laser annealing 激光退火
laser antenna 激光天线
laser aperture 激光器孔径;激光孔径
laser arrangement 激光装置
laser array 激光阵列
laser array axis 激光阵列轴
laser array source package 激光阵列源部件
laser attenuator assembly 激光衰减器
laser automatic leveling device 激光自动找平装置
laser automatic tracking system 激光自动跟踪系统
laser avoidance device 激光报警器
laser backscatter device 激光后向散射装置
laser back scattering 激光反向散射
laser band pass filtering image 激光带通滤波图像
laser bandwidth 激光带宽
laser bar 激光棒
laser-based airborne measurement system 机载激光测量系统
laser-based light-scattering mass detector 激光光散射质量检测器
laser basic mode 激光器基模
laser beacon 激光信标
laser beam 激光束
laser beam acquiring 激光瞄准;测光束瞄准
laser beam controller 激光束控制器
laser beam cutting 激光束切割
laser beam danger 激光束损伤危险
laser beam deflection 激光束偏转
laser beam deflection sensor 激光束偏转传感器
laser beam deflection system 激光束偏转系统
laser beam deflection technique 激光束偏转法
laser beam drilling machine 激光钻孔机
laser beam flying 激光束扫描
laser beam illumination 激光束照明
laser beam image recorder 激光束图像记录器
laser beam image reproducer 激光束图像重现器
laser beam instrument 激光射线仪
laser beam projector 激光束投射器
laser beam recorder 激光束记录器
laser beam setup 激光照准器
laser beam splitter 激光分光镜
laser beam steering instrument 激光束控制器
laser beam tracker 激光束跟踪器
laser beam transit 激光经纬仪
laser beam weld 激光焊
laser beam welding machine 激光焊接机
laser bedding 泥波层理
laser bistable device 双稳态激光装置

laser blasting 激光引爆
laser bleaching 激光漂白
laser bonding 激光焊接
laser bounce 激光反射
laser breakdown 激光击穿
laser burst 激光脉冲
laser calibrated 激光校准的
laser calorimeter 激光量热器
laser camera 激光照相机;激光摄像机
laser camera line-scanning system 激光行扫描摄影系统
laser cascade connection 激光级联
laser cavity 激光(谐振)腔
laser ceilometer 激光云高计;激光测云仪;激光测高仪
laser cell 激光光电元件
laser ceramics 激光陶瓷
laser channel capacity 激光信道容量
laser channel marker 激光航道标志
laser chaos 激光混沌
laser circuit 激光电路
laser coagulation 激光凝固
laser coagulator 激光凝聚器
laser coating 激光镀膜
laser coelostat 激光定向仪
laser coherent 激光相干性
laser collimator 激光准直仪;激光照准仪
laser colo(u)r recorder 激光彩色记录器
lasercom 激光通信
laser communication 激光通信
laser communication device 激光通信装置
laser communication engineering 激光通信工程
laser communication ground system 激光通信地面系统
laser communication link 激光通信线路
laser communication system 激光通信系统
laser communicator 激光通信装置
laser computer 激光计算机
laser computer-output microfilm 激光计算机输出缩微胶片
laser cone calorimeter 激光锥形量热器
laser connector 激光器连接器
laser conversion efficiency 激光转换效率
laser coordination 激光定位
laser coordinator 激光定位器
laser correlation spectroscopy 激光相关光谱学
laser cross-beam velocimeter 交叉激光束速度计
laser crystal 激光晶体
laser crystal contamination 激光晶体污染
laser crystal growing 激光晶体生长
laser crystal orientating instrument 激光晶体定向仪
laser curing 激光固化
laser current transformer 激光变流器
laser cutter 激光切断机
laser cutting 激光切割
laser damage 激光损伤
laser damage of crystal 晶体激光损伤
laser damage of glass 玻璃激光损伤
laser damage of optic(al) coatings 光学镀层激光损伤
laser data display 激光数据显示器
laser data transmission link 激光数据传输线路
laser defect detection 激光缺陷检测
laser deflection 激光偏转
laser deflection modulation 激光偏转调制
laser deflector 激光偏转器
laser density probe 激光媒质密度探针
laser designator 激光指示器
laser detection 激光探测
laser detection system 激光探测系统
laser detector 激光探测器
laser detonator 激光雷管
laser device 激光装置;激光器
laser differential level(l)ing 激光微差水准测量
laser diode 激光二极管
laser directional filtering image 激光方向滤波图像
laser displacement bar 激光束偏移指示器
laser display 激光显示(器)
laser display panel 激光显示板
laser display system 激光显示系统
laser distance measurement instrument 激光测距仪
laser distance measuring instrument 激光测距仪

laser distance ranging 激光测距
laser doping 激光掺杂
laser Doppler anemometer 激光多普勒风速计
laser Doppler anemometry 激光多普勒风速测量法;激光多普勒风速和风向测定法
laser Doppler homodyne detection 激光多普勒零拍检测
laser Doppler radar 激光多普勒雷达
laser Doppler system 激光多普勒系统
laser Doppler velocimeter 激光多普勒速度计;激光多普勒测速仪
laser Doppler velocimetry 激光多普勒速度计测定;激光多普勒测速法
laser Doppler velocity measurement 激光多普勒测速
laser dot 激光点
laser drill 激光钻机
laser drilling 激光钻孔;激光凿岩
laser drilling machine 激光钻孔机
laser drilling system 激光钻孔系统
laser dynamic(al) balancing 激光动平衡
laser earthquake alarm 激光地震报警器
laser effect 激光影响
laser efficiency 激光效率
laser electronic data processing setup 激光电子数据处理装置
laser electro-optic(al) measurement and alignment instrument 激光测准直仪
laser electrooptic technology 激光电光技术
laser element 激光器单元
laser emulsion storage 激光(感光)乳胶存储器
laser end reflector 激光谐振腔端面反射体
laser energized detonation system 激光引爆系统
laser energized explosive device 激光引爆装置;激光激励爆炸装置
laser energy monitor 激光能量监控器
laser engineering 激光工程
laser engraving 激光刻图
laser enhanced images 激光增强图像
laser enhanced multiband composite image 激光增强多波段合成图像
laser enrichment 激光浓缩
laser etalon 激光标准具
laser excitation 激光激发
laser excitation efficiency 激光激发效率
laser exciter 激光激发源
laser extensometer 激光延伸计;激光伸长仪
laser eyepiece 激光目镜
laser eyewear 激光护目镜
laser facility 激光装置
laser fiber 激光玻璃纤维
laser fiber fineness distribution 激光纤维细度分布分析仪
laser fiber optics 激光纤维光学
laser fiber-optic transmission system 激光光纤传输系统
laser filter 激光滤片
laser fire control system 激光火控系统
laser flash tube 激光闪光管
laser flowmeter 激光流量计
laser flow transducer 激光流量传感器
laser fluctuations 激光辐射起伏;激光辐射波动
laser fluorescence 激光荧光;激光发光
laser focal shift monitor 激光焦点偏移监测器
laser focusing system 激光聚焦系统
laser frequency 激光频率
laser frequency correcting element 激光频率校正元件
laser frequency doubling 激光倍频
laser frequency stability 激光(器)频率稳定度
laser frequency stabilization 激光稳频
laser frequency stabilizing system 激光稳频系统
laser frequency switch 激光频率开关
laser fusion propulsion 激光聚变推进器
laser fuze 激光引信
laser gas 激光气体
laser gasification 激光气化
laser gasification spectrum analyser 激光气化光谱分析仪
laser gated 激光选通
laser gated night vision sight 激光选通夜视瞄准器
laser gear 激光设备
laser generated second harmonic 激光辐射二次谐波

laser generation 激光振荡
laser generator 激光发生器
laser geodimeter 激光测距仪
laser geodynamic(al) satellite 激光动力测地卫星;激光地球动力学卫星
laser geodynamic(al) satellite program(me) (lageos) 激光地球动力学卫星计划
laser glass 激光玻璃
laser glaze 激光上釉
laser goggles 激光防护镜
laser grade control 激光坡度控制
laser granulometer 激光粒度测定仪
laser gravimeter 激光重力仪;激光重差计
laser grooving 激光刻槽
laser grooving and scribing 激光刻划
laser ground mapper 激光地面测绘器
laser ground mapping system 激光地面测绘系统
laser guide instrument 激光导向仪
laser guide of vertical shaft 竖井激光指向
laser gun 激光枪
laser gyro axis 激光陀螺轴
laser gyro package 激光陀螺装置
laser gyro(scope) 激光陀螺(仪);激光回转仪
laser harmonic 激光谐波
laser head 激光头
laser head assembly 激光头装置
laser head block 激光头部件
laser heating of plasma 激光加热等离子体
laser height accuracy 激光测高准确度;激光测高精度
laser heterodyne acoustic(al) sensor 激光外差式声学传感器
laser heterodyne measurement 激光外差测量
laser heterodyne system 激光差拍系统
laser higher frequency filtering image 激光高通滤波图像
laser hole 激光钻的孔
laser hole drilling 激光钻孔
laser hole drilling system 激光钻孔系统
laser hologram 激光全息图
laser hologram playback 激光全息图再现
laser hologram vibration measurement 激光全息测振
laser holographic(al) interferometer 激光全息干涉仪
laser holographic(al) interferometry 激光全息干涉(测量)术
laser holography memory 激光全息存储器
laser homing 激光寻的
laser homing control 激光寻的控制
laser homing equipment 激光寻的装置
laser host material 激光基质材料
laser ignited fusion 激光点火聚变
laser illuminated target 激光照明目标
laser illumination 激光照明
laser illuminator 激光照明器
laser image converter 激光图像转换器
laser image processing 图像处理激光扫描机
laser imagery recorder 激光图像记录器
laser image-speckle interferometer 激光影像-斑点干涉仪
laser induced 激光致;激光引发的
laser induced atmospheric breakdown 激光大气击穿
laser induced chemical reaction 激光致化学反应;激光诱导化学反应
laser induced crack 激光致裂纹
laser induced damage 激光致损伤
laser induced damage threshold 激光损伤阈值
laser induced emission 激光致发射
laser induced filamentary damage 激光丝状损伤
laser induced fluorescence 激光诱导荧光
laser induced fusion 激光致聚变
laser induced inclusion damage 激光杂质损伤
laser induced spark 激光引发火花
laser induced spectrometry 激光诱导光谱法
laser induced surface damage 激光表面损伤
laser inertial navigation system 激光惯性导航系统
laser information display system 激光信息显示系统
laser infrared radar 激光红外雷达
lasering 激光作用
lasering safety 激光防护
laser initiated 激光引发的

laser initiation 激光起爆;激光触发
laser instrumentation 激光计测
laser intelligence data 激光信息数据
laser interferometer 激光干涉仪
laser interferometer camera 激光干涉仪照相机
laser interferometer manometer 激光干涉气压计
laser interferometry 激光干涉量学;激光干涉测量
laser intrusion-detector 激光入侵探测器
laser irradiated layered target 激光辐照分层靶
laser irradiated surface 激光辐照面
laser irradiation 激光辐照
laser isotope separation 激光同位素分离
laserium 激光天象仪
laser jamming 激光干扰
laser Kerr cell 激光克尔盒
laser knife 激光刀
laser large screen display 激光大屏幕显示
leading marks 激光导标
laser length measuring machine 激光测长机
laser length standard 激光长度基准
laser level 激光水准仪
laser leveler 激光水准器
laser level ga(u)ge 激光水准器
laser level(l)ing 激光校平;激光水准测量;激光操平;激光找平
laser level meter 激光水平仪
laser levels 激光能级
laser level sensor 激光物位传感器
laser lever 激光器杠杆
laser levitation 激光悬浮
laser light 激光
laser light demodulating system 激光解调系统
laser lighthouse 激光灯塔
laser lighting 激光照明
laser light-scattering by plasma 等离子体的激光散射
laser light-scattering probe 激光散射探头
laser light sensing head 激光灵敏头
laser linear comparator 激光线纹比较仪
laser line following 激光线迎跟踪
laser line scan camera system 激光行扫描照相系统
laser line scanner 激光行扫描仪
laser line-scanning sensor 激光行扫描传感器
laser linewidth 激光线宽
laser local oscillator 激光本机振荡器
laser location 激光定位
laser location observation 激光定位观察法
laser location surveying 激光定位测量
laser locator 激光探测器;激光定位器
laser locking 激光同步
laser lookon 激光跟踪
laser lower frequency filtering image 激光低通滤波图像
laser lower level 激光下能级
laser low-light level television 激光光色显示
laser lunar ranging 激光月球测距;对月激光测距
laser machine 激光加工机
laser machining 激光(机械)加工
laser mapping 激光成像;激光测绘
laser mapping equipment 激光测绘仪
laser mapping system 激光测绘系统
laser marker 激光刺点仪
laser marking system 激光标志系统
laser mass spectrometer 激光质谱仪
laser mass spectrometry 激光质谱法
laser master oscillator 激光主控振荡器
laser material 激光材料
laser material processing 激光材料加工
laser measured height 激光测高
laser medium 激光媒质
laser medium gain curve 激光介质增益曲线
laser melting method 激光熔融法
laser memory 激光存储器
laser micro-analysis 激光显微分析;激光微量分析
laser micro-calorimeter 激光微量热计
laser micro-emission spectrometer 激光显微发射光谱仪
laser micro-machining 激光微型加工
laser micro-probe 激光(显)微探针;激光显微刀;激光微探子;激光微探测器
laser microprobe mass analyzer 激光探针质量分析器
laser micro-probe mass spectrometry 激光微探针质谱分析
laser microprobe mass spectroscopy 激光探针质谱
laser micro-probe spectrochemical analysis 激光显微光谱分析
laser micro-spectral analyzer 激光微光谱分析仪
laser micro-spectrographic(al) analysis 激光微区光谱分析
laser micro-spectroscopy 激光微量光谱分析
laser micro-strain ga(u)ge 激光微应变计
laser micro-welder 激光微件熔焊机
laser mirror 激光器反射镜;激光镜
laser mode control 激光模控
laser modulation 激光调制
laser modulator 激光调制器
laser monitor for pollution 激光污染监测仪;激光测污仪
laser monitoring 激光监测(术)
laser multilateration 激光多边测量
laser multimode operation 激光器多模工作
laser navigation 激光导航
laser navigation(al) equipment 激光导航仪
laser navigation gear 激光导航设备
laser nephelometer 激光烟雾计;激光散射浊度计;激光比浊计
laser nephelometry 激光测浑法;激光比浊法
laser obstacle avoidance sensor 激光防撞传感器
laser obstacle avoidance system 激光防撞系统
laser operation 激光器运转
laser ophthalmoscope 激光检眼镜
laser optic(al) bench 激光器光具座
laser optic(al) demonstration instrument 激光光学演示仪
laser optics 激光光学装置;激光光学
laser optoacoustic(al) detection 激光光声探测
laser optoacoustic(al) method 激光光声法
laser orientation 激光指向
laser oscillation 激光振荡
laser oscillation condition 激光(器)振荡条件;激光器激励条件
laser oscillator 激光振荡器
laser oscillator-amplifier 激光振荡放大器
laser oscillator-amplifier system 激光振荡放大系统
laser oscillator modulator 激光振荡调制器
laser output 激光输出
laser output characteristic 激光输出特性
laser output efficiency 激光输出效率
laser output frequency 激光输出频率
laser output spectrum 激光输出光谱
laser output wavefront 激光输出波前
laser phased array 激光相控阵列
laser phase noise 激光相位噪声
laserphoto 激光照片传真
laser photochromic display 激光光色显示
laser photocoagulation 激光凝结;激光凝固
laser photocoagulator 激光凝聚器;激光凝结器;激光凝固器
laser photolysis 激光光解
laser photometer 激光光度计
laser photometry 激光光度学
laser phototypesetting system 激光照排系统
laser physics 激光物理学
laser pickoff 激光接收(器)
laser pick-up system 激光拾音器系统
laser piercing power 激光穿透能力
laser pipe fibre optics 激光管纤维光学装置
laser planetarium 激光天象仪
laser plotter 激光绘图机
laser plumbing 激光投点
laser plummet 激光垂准仪
laser plummet apparatus 激光铅垂仪
laser point transfer device 激光转点仪
laser point transfer machine 激光转点仪
laser powder 激光粉
laser power 激光功率
laser preamplifier 激光前置放大器
laser precision length measurement 激光精密测距;激光精密测长
laser pressure ga(u)ge 激光压力计
laser printer 激光打印机
laser probability 激光跃迁概率
laser probe 激光探针
laser probe mass spectrometry 激光探针质谱法
laser probing 激光探测

laser processing 激光加工
laser processing system 激光加工系统
laser profile 激光剖面图;激光断面图
laser profilometer system 激光地形纵断面平整度仪(分析未出来场地)
laser projection microscope 激光投影显微镜
laser protective eyewear 激光护目镜
laser protective housing 激光器安全罩
laser pulse control 激光脉冲控制
laser pulse length 激光脉冲长度
laser pulse width 激光脉冲宽度
laser pump(ing) 激光泵
laser pyrolysis 激光高温分解
laser pyrolysis gas chromatography 激光热解气相色谱法
laser pyrolyzer 激光热解器
laser quasimode 激光准模式
laser quenching 激光淬火
laser radar 光探测与测距装置;光雷达;激光雷达
laser radar cross-section 激光雷达截面积
laser radar encoder 激光雷达编码器
laser radar equation 激光雷达方程
laser radar range 激光雷达作用距
laser radar technique 激光雷达技术
laser radiation 激光照射
laser radiation damage 激光辐射损伤
laser radiation detector 激光辐射探测器
laser radiometer 激光辐射计
laser Raman radar 激光拉曼雷达探测器
laser Raman spectrometer 激光拉曼分光光度计
laser Raman spectroscopy 激光拉曼分光学
laser Raman spectrum 激光拉曼光谱
laser range finder 激光测距仪;激光测距机;激光定位仪
laser rangefinder theodolite 激光测距经纬仪
laser ranger 激光测距仪
laser ranging and tracking system 激光测距和跟踪系统
laser ranging apparatus for satellite 卫星激光测距仪
laser ranging device 激光测距仪
laser ranging retroreflector 激光测距后向反射器
laser ranging sensor 激光测距传感器
laser ranging sight 激光测距瞄准具
laser ranging station 激光测距站
laser ranging theodolite 激光测距经纬仪
laser ray nondestructive tire tester 激光轮胎无损检验仪
laser receiver 激光接收器
laser receiver system 激光接收系统
laser recorder 激光录制器;激光记录仪;激光记录器
laser recording head 激光记录头
laser recording system 激光记录系统
laser reflectometer test 激光发射计试验
laser reflector 激光反射器
laser resonator 激光谐振器;激光(器)共振腔
laser rifle 激光枪
laser road surface tester 路面激光测试仪
laser rod 激光棒
laser rod-cooling jacket 激光棒冷却套
laser rod grinding 激光棒研磨
laser rotational sensor 激光转动传感器
laser route device 激光航线测定仪
laser safe level 激光安全水平
laser safety goggle 激光安全护目镜
laser safety standard 激光安全标准
laser satellite 激光卫星
laser satellite ranging 对卫星激光测距
laser satellite tracking installation 激光卫星跟踪装置
laser satellite tracking station 激光卫星跟踪站
laser saturation spectroscopy 激光饱和光谱学
laser scan filtering 激光滤波
laser scanner 激光扫描仪
laser scanner technique 激光扫描技术
laser scanning 激光扫描
laser scanning system 激光扫描系统
laser scatterometer 激光散射计
laser schlieren apparatus 激光纹影仪
laser schlieren method 激光纹影照相法;激光条纹照相法
laser schlieren photography 激光纹影照相术
laser scope 激光显示器;激光观察器
laser scriber 激光划线器;激光划片器

laser scribing 激光划线
laser search apparatus 激光搜索装置
laser searchlight 激光探照灯
laser section 激光剖面图;激光断面图
laser seeker 激光寻找器;激光寻的器
laser seismometer 激光地震仪
laser semiactive homing system 激光半主动寻的系统
laser sensor 激光传感器
laser service connection 激光器连接件
laser shielding eye glass 激光护目镜
laser shoot simulator 激光射击模拟器
laser shot 激光发射
laser side looking radar 激光侧视雷达
laser sight 激光瞄准具
laser signal 激光信号
laser signal device 激光信号装置
laser's intensity 激光强度
laser slicing machine 激光切片机
laser solution 激光溶解
laser sonar 激光声呐
laser sources 激光源
laser speckle 激光散斑
laser speckle field 激光散斑场
laser speckle interferometry 激光斑干涉量度术
laser speckle method 激光斑纹法
laser spectral output 激光光谱输出
laser spectrogram 激光光谱谱图
laser spectrometry 激光光谱法
laser spectrophotometer 激光分光光度计
laser spectroscopy 激光光谱学
laser spectrum 激光光谱
laser spectrum analysis 激光光谱分析
laser spiking 激光尖峰
laser spot 激光焦点
laser spot tracker 激光光斑跟踪器
laser squib 激光引爆器
laser stability 激光器稳定性
laser stadia ranging 激光测距
laser Stark spectroscopy 激光斯塔克光谱学
laser station 激光站
laser storage 激光存储(器)
laser strainmeter 激光应变仪
laser strain seismograph 激光应变地震仪
laser streak velocimeter 激光条纹速度计
laser super-heterodyne receiver 超外差(式)激光接收机
laser surface modification 激光表面改性
laser surface treatment 激光表面处理
laser surgery 激光外科
laser surveillance 激光监视
laser surveillance system 激光监视系统
laser surveying system 激光测量系统
laser switch 激光开关
laser system 激光系统
laser target 激光靶
laser target designator 激光目标指示器
laser target indicator 激光指示器
laser target plasma 激光等离子体
laser target positioner 激光目标位置测定装置
laser target recognition system 激光目标识别系统
laser technique 激光技术
laser technology 激光技术
laser telescope 激光望远镜
laser television 激光电视
laser television camera 激光电视摄像机
laser termination 激光器终端
laser terrain avoidance sensor 激光防撞传感器
laser terrain avoidance system 激光防撞系统
laser terrain-clearance indicator 激光测高计
laser terrain following radar 激光测高雷达
laser terrain profile recorder 激光地形剖面仪;激光地形断面仪
laser terrain profiling equipment 激光地形剖面仪;激光地形断面仪
laser test console 激光测试控制台
laser theodolite 激光经纬仪
laser thermal wave technique 激光热波技术
laser threshold 激光阈值;激光临界值
laser threshold power density 激光阈值功率密度
laser tip 激光尖端
laser tomography 激光层析法
laser topographic(al) instrument 激光地形仪
laser topographic(al) position finder 激光地形仪

laser tracker 激光跟踪器
laser tracking 激光跟踪
laser tracking axis 激光跟踪轴
laser tracking head 激光跟踪头
laser tracking positioning system 激光跟踪定位系统
laser tracking system 激光跟踪系统
laser transit 激光经纬仪
laser transit instrument 激光经纬仪
laser transition 激光跃迁
laser transition frequency 激光跃迁频率
laser transition probability 激光跃迁概率
laser transmission 激光传输
laser transmitter 激光发射机
laser transmitting telescope 激光发射管
laser triad 三激光器陀罗
laser triggered spark gap 激光触发火花隙
laser triggered switch 激光触发开关
laser trimmer 激光微调器
laser trimming 激光微调
laser tube 激光管
laser tube cavity 激光管谐振腔;气体激光器
laser ultrasonic holography 激光超声全息照相
laser underwater communication 水下激光通信
laser vector velocimeter 激光矢量速度计
laser velocimeter 激光速度计;激光测速仪
laser velocimetry 激光速度学
laser vessel 激光器容器
laser vibration probe 激光振动探针
laser voice link 激光通话线路
laser warning receiver 激光警戒接收机
laser watchdog 激光监视
laser waveform generator 激光波形发生器
laser wavefront analyzer[analyser] 激光波前分析仪
laser wave length 激光波长
laser welder 激光焊机
laser welder head 激光焊接头
laser welding 激光焊接
laser welding chamber 激光焊接室
laser welding head 激光焊接头
laser welding machine 激光焊接器
laser welding system 激光焊接系统
laser writer 激光显示器
laser zenith meter 激光垂直仪
lash 捆绑;冲击;缚
lash about 飞荡绳尾
LASH barge 载驳船的船;拉希(子)驳
LASH barge loader 拉希(子)驳装卸机
LASH dock 载驳货船码头
lashed fiber optic(al) cable 缠挂式光缆
lasher 蓄水池;系索;溢流堰;溢洪道;装石工;拦河坝;放石工
lasher-on 挂车链工
lashing 系绳;拉紧;捆索;绳套;绑扎;绑绳
lashing chain 捆绑用的链条;无极绳挂链
lashing device 系紧装置
lashing eye 系索眼环;系孔;绑绳扎成的绳眼
lashing for on deck container 甲板集装箱绑缚
lashing line 捆绑线
lashing net 捆货网兜
lashing operation 紧固作业
lashing post 系绳柱
lashing ring 系(索)环;系缆环;触线圈
lashing ring pad 系绳环垫板
lashing scheme on deck 甲板集装箱绑缚系统
lashing triangle 三角环
lashing wire 系索;拉筋;拉金;捆轧钢丝;捆扎线;束缚线;绑轧钢丝绳
lashing wire coil 捆扎钢丝线圈
LASH lighter 拉希(子)驳
lash method 震动法;振动法
lash rail 系物横杆
lash rope 绑扎绳(子)
LASH system 载驳船系统;拉希方式;激光半主动寻的系统
lash the tiller 缚住舵柄
lash type inertia vibrating shakeout 冲击式惯性振动落砂机
LASH vessel 拉希载驳货船
lasing 产生激光;发射激光
lasing area 激光面
lasing beam diameter 激光束直径
lasing fiber 激光(光学)纤维 激光玻璃纤维

lasing gas 激光气体
lasing light emitter 激光源；激光发射体
lasing medium 激光作用媒质
lasing mode 激光模式
lasing surface 激光作用面
lasing threshold 激光阈限
lasing time 激光振荡时间
lasionite 银星石
Lasky law method 拉斯基律法
Lassaigne's test 拉萨涅试验
lasso 套索
last 延续；最后的
last account of contractor 施工单位竣工决算
last account of project 建设项目竣工决算
lastage 市场税；使用码头税（英国）
last amplifier 终端放大器；末级放大器
last bid 最后报价
last card 最新卡片；最后卡片
last collision 末次碰撞
last column 最后一行
last come first served 后先出法；后到先服务
last contact 复圆月蚀；复圆日
last cry 最新流行品
last cry first-out 最新流行品
last current state 最后当前状态
last cut 最后留分
last cycle 终止周期
last date 最后期限；终日
last detector 末级检波器
last element of chain 链的最末元素
laster plan 长远规划图
last executed instruction 最后执行指令
last exit edge 最后出口边
last finishing pass 终轧孔型
last fix 上一次定位
last groove 精轧孔型；成品孔型
last ice 终冰
last ice date 终冰期
lastics 弹塑体
last in 逆序换算
last-in-chain 链中最末元素
last-in first-out 后进先出（法）；后到先服务
last-in first-out method 后进先出法
last-in first-out stack 后进先出栈
last-in first-out store 后进先出存储器
lasting 厚实斜纹织物
lasting cementing machine 帮角刷胶机
lasting date of investigation 调查工作进行时间
lasting effect 耐久效果
lasting equipment 永久性设备；坚固设备
lasting food 耐久存的食品
lastingness 耐久
lasting of ice cover 封冻历时；封冰历时
lasting of snow cover 积雪历时
lasting property 耐久性能
lasting quality 耐久性能
lasting rain 持续（降）雨；长时间降雨
lasting seal 耐久嵌缝
lasting water system 长期给水系统
last in last out 后进后出（法）
last in on hand 后进现存（法）
last-in still-here 后进还存
last invoice price 最后进价法
last location 最后单元
last macro instruction 最后宏指令
last moment emergency shut-down 应急停堆系统
last multiplier 最后乘子
last notice day 最后交割通知日
last number redial 最末号码重拨
last number(s) 最后号数
last-of-chain 链尾；链的最末元素
last of ebb 退潮末
last-off 最后退卷
last office of destination 最后的收报局
last page 结束页号
last-party release 双方话终拆线
last pass 终轧道次；精轧型；末遍
last pass own code 末遍扩充工作码
last pass phase 最后通过阶段
last pay certificate 最后付款证明书
last-period forecast 近期预测
last phase 终相；残余相
last port 最后到达港；终点港
last quarter 下弦（天文）

last record of a set 集的最末记录
last resort 最后手段
last retarder 末级缓行器
last ride bonus 最后乘次优惠
last schedule 末次会晤
last scratch 最后基准线
last sector 最后扇段
last significant figure 末位有效数字
last snow 终雪
last stage 末级
last stage blade 末级叶片
last statement 结算
last step 最后一档；最后步骤；最高梯级
last-subscriber release 双方话终拆线
last subscript 最后下标
last-survivor annuity 最后生存年金
last symbol 最后符号
last term 最后项；末项
last terminal 最后终端符
last test 最后测试
last test step 最后测试步
last thread failure 接头最末扣断裂
last-to-freeze liquid 残留金属液
last trading day 最后交易日
last trunk capacity 终端中继线容量
last turn 最终轨道
last word 最新发明；最新成就；最新产品；最先进的事物；最高权威；决定性意见
lasureous 暗蓝色的
lasurite 青金石
lat 印度表柱
latacoria 侧膜
latallae 侧膜片
latasuture 侧缝
latch 矿山测量；卡齿；门闩锁；碰锁；握柄键；弹键；弹簧锁；锁存器；门锁；锁住
latch bar 门闩
latch bit test 闩锁位测试
latch body 闩卡体（绳索取芯工具）；弹卡体（绳索取芯工具）
latch bolt 弹键栓；弹簧闩；锁栓；插销闩；门窗闩
latch bracket 弹簧托；闩托
latch bracket screw 弹键托螺钉
latch catch 弹键扣
latch circuit 锁住电路；闩锁电路
latch clutch 掣子离合器
latch cylinder 排料启闭器汽缸
latch decoder 闩锁译码器
latch dog 闩钩；闭锁抓取器
latchdown 锁定
latch drive 闭锁驱动
latched-in contact 锁销接触器
latched position 锁定位置
latched system 密码系统
latch elevator under tool joint 接头下闭锁提引器
latch enable 允许闩
latches 门闩线路
latches circuit 门闩线路
latchet 线垫
latch fitting 锁件；闭锁装置
latch gear 弹键装置；锁销机构
latch gear casing 弹键装置箱
latch handle pin 弹键柄销
latch head 弹簧插销
latch hook 挂钩；卡子；弹键钩；锁钩；挡舌吊沟；掣子
latching 锁住；碰锁；自动关闭设备
latching cam 自锁式凸轮
latching circuit 自锁电路；闭锁电路
latching circulator 自锁循环器
latching current 闭锁电流
latching full adder 闭锁全加器
latching network 自锁网络
latching register 自保持寄存器
latching relay 自锁继电器；闩锁继电器；闭锁继电器
latching strength（车门的）锁栓强度
latching valve 自锁阀
latching voltage 闭锁电压
latch-in relay 闩锁继电器
latch jack（捞砂筒的）带销捞筒；打捞器；门闩插孔
latch-jack dump bailer（带铰链弹簧挡销的）灌注抽筒
latch jamb 装锁门梃
latch key 碰锁钥匙；弹键（门）锁；弹簧锁钥匙
latch lock 弹键锁；弹键闭锁；掣子门锁

latch locking 插销锁定
latch lug jaw 闩上突出的键；闩上突出的锁
latch mode 闩锁状态
latch needle 钩针
latch nut 锁紧螺母；锁紧螺帽；防松螺母
latch on 抓住管子（提引器）；扣石吊卡
latch out 插销脱出
latch out tabulator key 开锁制表键
latch pin 掣子销
latch plate 弹簧板；插销
latch register 闩锁寄存器
latch-release cylinder 开闭汽缸
latch retainer 控制掣子；吸持塞孔；闩承座；闩的定位器
latch retracting case 弹卡活动套筒
latch screw 门栓大螺丝；弹键螺钉
latch shoe 闭止把弹簧架
latch spring 碰锁弹簧；锁扣弹簧；锁紧弹簧；插销锁内弹簧
latch stop 扣铁挡；闩式挡料装置
latchstring 门闩拉线；栓锁带
latch support 弹卡支座
latch-type front head（钻机的）闩锁式前机头；带活销夹钎器
latch-type mechanism 锁闩型机构
latch unit 闩锁器
latch-up 封闭
latch-up protection 闭锁保护
late 延迟
late Alpine geosyncline 晚阿尔卑斯期地槽
late art nouveau 后期新艺术风格；晚期新艺术风格；晚期新艺术运动
late bark 硬皮
late Baroque arch 后期巴洛克建筑形式
late Baroque style 后期巴洛克艺术风格
late-bearing 增阻力的；滞后阻力的
late bids 迟到标书
late burning 晚期用火
Late Caledonian geosyncline 晚加里东期地槽【地】
Late Cambrian epoch 晚寒武世【地】
late Cambrian extinction 晚寒武世绝灭【地】
Late Carboniferous epoch 晚石炭世【地】
late charges 延期罚款；拖延债务的罚款
late come first served 后先出法；后到先服务
late commitment 晚期支付
late commutation 延迟换向
late contact 后动触点
late cretaceous climatic zonation 晚白垩世气候分带
Late Cretaceous epoch 晚白垩世【地】
Late cretaceous extinction 晚白垩世绝灭【地】
Late cretaceous regression 晚白垩世海退【地】
late curing 后期养生；后期养护
late delivery 迟交货
Late Devonian epoch 晚泥盆世【地】
late diagenesis 晚期成岩作用
late effect 远期影响；远期效应；延迟效应；晚期效应；晚发效应
late-fall irrigation 晚秋灌溉
late fallow 生草休闲
late-Frankish architecture 法兰克晚期建筑艺术；晚期法兰克式建筑
late frost 晚霜
late frost damage 晚霜害
late gate 后闸门
late geometric(al) 晚期几何形体式（装饰）
late geometric(al) lattice 晚期几何形窗花格
late-geosyncline stage 晚地槽阶段
late-glacial 后冰期的
late glacial deposit 后期冰川沉积物
late glacial epoch 晚冰期
late glacial period crust movement 晚冰期地壳运动
late Gothic 后期尖拱【建】；后期哥特式（装饰）风格
late Gothic hall church 后期哥特式大厅教堂；晚期哥特式教堂
late Gothic royal chapel 后期哥特式王家礼拜堂
late Gothic tracery 后期哥特式几何窗格
late Gothic vault 后期哥特式尖拱
late Hercynian geosyncline 晚海西期地槽
late ignition 延迟点火；迟点火
late impoundment 后期蓄水
late-kinematic granite 晚构造期花岗岩

late magmatic mineral deposit 晚期岩浆矿床
late magmatic ore deposit 晚期岩浆矿床
late mature 晚成年期
late mature valley 晚成年谷
late maturity river 晚成熟期河流
late-mediaeval architecture 中世纪晚期建筑艺术;后期中古时代建筑
late-model 新型(的);现代形式
late modern architecture 后期现代建筑
late modern style 后期新式艺术风格;后期现代派
latency 潜伏状态;等数时间
latency energy 潜水能
latency period 潜伏期
latency time 等待时间;空转时间;潜在时间
Late Neocathaysian system 晚期新华夏系【地】
laten heat calorimeter 量潜热器
latensification 潜像增强;潜像处理
latent 潜伏的
latent activation site 潜在活化部位
latent aerophilic quality 潜在亲气性(指浮选粒子)
latent ambiguity 潜在含糊性
latent capacity 潜力
latent catalyst 潜催化剂
latent condition 潜伏状态
latent consequence 潜在后果
latent content 潜含量
latent corrosion 潜蚀
latent crimp 潜在卷曲
latent curing agent 潜固化剂;潜伏性固化剂
latent danger 隐患
latent defect 隐蔽事故;隐蔽故障;内在缺陷;潜在瑕疵;潜在事故;潜在缺陷;隐伤
latent defect clause 潜在瑕疵条款
latent demand 潜在需求
latent demand for funds 潜在资金需求
latent development 潜伏发育
latent dishonour 潜在拒付
latent economical geothermal field 亚经济性地热田
latent elastic deformation 潜弹性形变
latent energy 潜能
latent epidemiogenesis 潜伏流行过程
latent evapo(u)ration 潜蒸发
latent force 潜力
latent fusion 潜熔
latent geothermal area 潜在地热区
latent glide system 潜在的滑移系统
latent growth period 潜伏生长期
latent heat 潜热
latent heat calorimeter 潜热量器
latent heat flux 潜热通量
latent heat load 潜热负荷
latent heat of condensation 冷凝潜热
latent heat of evapo(u)rization 蒸发潜热;气化潜热
latent heat of freezing 冰冻潜热
latent heat of fusion 融解潜热;融化潜热;熔融潜热;熔解热;熔化潜热
latent heat of liquefaction 液化潜热
latent heat of melting 熔化潜热
latent heat of solidification 凝固潜热
latent heat of sublimation 升华热;升华潜热
latent heat of vapo(u)rization 气化潜热
latent heat quantity 潜热量
latent heat release 潜热释放
latent Herschel effect 赫歇尔效应
latent hydraulic binding agent 潜在水硬性胶结剂;潜伏水硬性胶合剂
latent hydraulicity 潜在水硬性
latent hydraulic power 水化潜能
latent hydraulic substance 潜伏水硬性物质
latent hydrophobic quality 潜在疏水性(指浮选粒子)
latentiation 潜伏化作用
latent image 潜像
latent image memory 潜像存储器
latent-image stability 潜像稳定性
latent infection 潜伏侵染
latent instability 潜在不稳定(性)
latent light 潜光
latent load 潜热荷载
latent loss 潜亏
latent magma 潜岩浆
latent nuclear energy 潜在的核能

latent of freezing 结冰潜热
latent overpopulation 潜在过剩人口
latent parameter 特征参数
latent period 潜伏期
latent phase 潜伏期
latent picture 潜像
latent polarity 潜极性
latent pollutant 潜在污染物;潜隐污染物;潜水隐污染物
latent pollutant source 潜在污染物源
latent power demand 潜在电能需要
latent preimage 预期潜像;初期潜像
latent productive capacity 生产潜力
latent productive capital 潜在的生产资本
latent property 潜在财产
latent radiation effect 潜伏辐射反应
latent reactivity 潜在活性
latent root 潜伏本征根;特征根;本征根
latent science 潜科学
latent solvent 潜溶剂;助溶剂
latent stress 潜在应力;潜应力
latent stress field 潜在应力场
latent structure 潜结构
latent sub-image 亚潜像
latent time 潜伏时间
latent unemployment 潜在失业
latent variable 潜在变量
latent vector 特征向量;本征向量;本征矢量
late opus 晚期艺术作品;近期作品
Late Ordovician epoch 晚奥陶世【地】
Late Ordovician extinction 晚奥陶世灭【地】
late origin theory of petroleum 晚期生油学说【地】
late-orogenic 晚造山期的
late-orogenic basin 晚造山期盆地
late orogenic phase 晚造山相
late payment 逾期付款
late period 晚期
Late Permian epoch 晚二叠世【地】
Late Permian extinction 晚二叠世绝灭【地】
Late Permian regression 晚二叠世海退【地】
Late Pleistocene epoch 晚更新世【地】
Late pointed style 英国哥特式建筑风格之后期垂直线建筑风格;后期尖拱式;后期哥特式
Late potassium biphthalate 苯二甲酸氢钾
Late Proterozoic extinction 晚元古代绝灭【地】
Late Proterozoic glacial stage 晚元古代冰期【地】
Late Proterozoic transgression 晚元古代海浸【地】
late pruning 晚期修剪
late radiation effect 远期辐射效应;延迟辐射效应;辐射远期效应
lateral 支管渠;抗风支撑;污水旁管;侧向的;侧(面)的;分支管道
lateral aberration 横像差;横向像差
lateral abrasion 横向冲刷;侧蚀
lateral acceleration 横向加速度;横向加速度;侧向加速度
lateral accelerograph 侧向加速自记器
lateral accelerometer 侧向加速表
lateral accretion 横向淤积;侧向加积作用;侧向堆积【地】
lateral accretion deposit 侧向加积沉积
lateral accretion pattern 侧积型
lateral adaptation 侧适应
lateral adhesion 侧向附着
lateral adjustment 横向校正;侧面调节
lateral adjustment of roll 轧辊的轴向调整
lateral aggregation 侧向聚集作用
lateral alignment 侧面准直;横向对准
lateral anomaly 侧部异常
lateral arch 侧向拱;侧跨拱
lateral area 侧面积
lateral area above water 水线上侧面积
lateral areas 侧区
lateral arrangement of station tracks 横列式到发线
lateral attitude 横向位置
lateral axis 横轴线
lateral axle 横轴
lateral balance 侧平衡
lateral ballast resistance 道床横向阻力
lateral bar 横向沙洲;横向沙坝;侧向沙滩;侧向砂坝
lateral-bar screen 横条帘幕
lateral bars tread 横向条纹

lateral basal branch 外底支
lateral basal segment 外基底段
lateral bearing pressure 侧向承压力
lateral bend(ing) 侧向弯曲;横向弯曲;测弯;侧弯
lateral bending test 横向弯曲试验
lateral binder 横向箍筋
lateral binding 横向箍筋;侧向连接
lateral bine 侧蔓
lateral blast 侧向爆破
lateral blow 侧吹风
lateral blue convergence assembly 蓝侧位会聚装置
lateral body 侧体
lateral body axis 联系坐标系横轴
lateral bond 横向黏结
lateral border 外侧界
lateral boundary of aquifer 含水层侧向边界
lateral bow 侧向弯曲;弓弯
lateral bracing 横系杆;(横向)支撑;横向联结系;水平支撑;侧向支撑;侧向加撑
lateral bracing system 横向支撑系统;侧向支撑系统
lateral branch 支流;旁支管;侧枝
lateral buckling 横向压屈;横向翘曲;侧向压屈;侧向翘曲;侧向压曲;侧向挠曲;侧向折
lateral buckling of girders 大梁横向弯曲
lateral bud 侧芽
lateral bulging 侧向凸出;侧向膨胀
lateral burning 侧面燃烧的
lateral bursting water 侧向突水
lateral calcaneal branches 跟外侧支
lateral canal 支运河;支渠;横向渠道;旁支运河;旁支渠道;斗渠;侧运河;侧向渠道
lateral capacitance 侧电容
lateral capacity 支渠过水能力
lateral carinae 侧隆线
lateral caving 侧向淘空
lateral center of gravity 横重心
lateral cerari 侧三角蜡孔
lateral chain 侧链
lateral-chain theory 侧链学说
lateral change of channel 水道横向变化
lateral channel 铁水支沟
lateral chapel 旁侧礼拜堂
lateral chemical potential 侧化学势
lateral chromatic aberration 横向色差
lateral circuit 侧电路
lateral clearance 侧向余宽;横向净空;横动量;侧向裕宽;侧向净空
lateral clear distance of curve 横净距(曲线)
lateral clear distance of horizontal curve 平面线横净距
lateral cleavage 侧劈理
lateral clinometer 侧向倾斜仪;侧向测斜仪
lateral cofferdam 横向围堰;(平行于河道的)侧向围堰
lateral colonnade 侧柱廊
lateral column 侧柱
lateral combination 横向联合
lateral compliance 横向顺性
lateral component 侧向分力
lateral compression 横压力;横向压缩;侧压(力);侧(向)压缩;侧向挤压
lateral compression boring 旁压钻孔
lateral compression coefficient 侧压系数
lateral compression curve 旁压线
lateral compressive force 侧向压力
lateral conductor 斜拉导线;横向导线
lateral conduit 横向管渠
lateral cone 寄生火山锥
lateral confinement 侧向限制;侧限
lateral consequent river 侧面顺向河
lateral consequent stream 侧面顺向河
lateral consolidation pressure 侧向固结压力
lateral contact pin 侧插棒
lateral continuation 旁侧延拓
lateral continuity 侧向延伸
lateral contraction 横(向)收缩;缺口缩颈;侧(向)收缩;侧面收缩
lateral contraction coefficient 侧收缩系数
lateral controller 横向操纵机构
lateral convergence 横向会聚;侧面会聚
lateral cord 侧索
lateral correction 横位校正;侧位校正
lateral correction magnet 横向校正磁铁

lateral corrosion 侧向侵蚀;侧(向)磨蚀;侧向腐蚀
lateral coverage 侧向覆盖
lateral crack 横向裂纹;横向裂缝
lateral crater 侧火山口
lateral creep 横向蠕变;横向徐变
lateral crevasse 侧向裂缝;边缘冰裂缝;侧侧裂隙
lateral crystal growth 横向晶体生长
lateral current 横流
lateral curvature 侧弯
lateral damping 滚动阻尼;倾侧运动阻尼
lateral daylighting 侧面采光
lateral declination 旁向偏差
lateral decubitus film 侧卧水平摄影片
lateral deflection 压侧;横向偏转;横向挠曲(度);横偏系数;侧向偏转;侧向偏离;侧(向)挠度;侧向变位
lateral deflection angle 方向提前角
lateral deformation 横向变形;旁向形变;侧向形变;侧(向)变形
lateral degrees of freedom 侧向自由度;侧向运动自由度
lateral dehiscence 侧面开裂
lateral deviation 左右偏差;横向偏移;横(向)偏差;(钻孔的)水平偏移;侧向偏移;侧向偏差;侧偏(位)
lateral diagonal 横向斜撑;侧向斜撑
lateral difference in water level 横向水位差
lateral diffusion 侧向扩散
lateral dimensions 横向尺寸;侧向尺寸;侧面尺寸
lateral dip angle 横向倾角值
lateral discharge 侧向排泄量;侧向流量
lateral discontinuous unit 侧向不连续单位
lateral dislocation 侧方脱位
lateral dispersion coefficient 横向弥散系数
lateral displacement 横向位移;侧向位移;侧向变位;侧位移;侧方移位
lateral distance 风机边柜
lateral distribution 横向分布
lateral distribution of load 荷载横向分布
lateral distribution of sediment 泥沙的横向分布
lateral distributor 横向配水渠
lateral disturbing 横向干扰
lateral ditch 支沟;外侧沟;边沟
lateral drain 横向沟渠;侧向排水沟;横向排水(沟)
lateral drainage 横向排水;侧向排水
lateral drainage ability 侧向排水能力
lateral drainage system 侧向排水系统
lateral drape coefficient 侧向悬垂系数
lateral drift 水平偏移;侧向偏移
lateral drilling 侧向钻井
lateral duct 支管通;支管道
lateral dune 侧条沙垄
lateral earth movement 横向地面运动
lateral earth pressure 横向土压力;侧向土压力
lateral earth pressure against piles 桩的侧向土压力
lateral earth pressure coefficient 侧土压力系数
lateral earth resistance 侧向土抗力
lateral echo sounder 侧视回声测深仪
lateral economic tie 横向经济联系
lateral effective stress 侧向有效应力
lateral electric(al) measuring curve 侧向电测曲线
lateral electrode 侧电极
lateral element 侧体
lateral elevation 侧视图;侧立面
lateral enrichment 侧富集
lateral entrance 侧门;旁入口
lateral erosion 横向冲刷;旁蚀;侧向侵蚀(作用);侧向冲刷;侧蚀(作用);侧面侵蚀
lateral erosion by water action 水的侧向冲刷作用
lateral error 横向误差;侧向误差
lateral error of traverse 导线横向误差
lateral eruption 侧方喷发
lateral escape 侧向出口
lateral exhaust at the edge of a batch 槽边排风罩
lateral expansion 横向膨胀;侧(向)膨胀
lateral extensometer 横向应变计;横向伸长计;横向膨胀计
lateral facade 侧立面
lateral face 侧面
lateral fan 测移扇
lateral fault 横向断层;横错断层;平移断层;侧移断层;侧向断层
lateral feed 横向进刀

lateral filament 侧丝
lateral fillet 侧面填角焊(缝)
lateral fissure 外侧裂
lateral flexibility 横向塑性;横向挠性
lateral flexion 横向曲率
lateral flexure 横(向弯)曲;侧弯
lateral flow 横向流(动);横流;侧向流(动);侧流
lateral flow buoy 横流浮标
lateral flow of mixture 混合料的侧向流动
lateral flow spillway 侧槽式溢洪道;侧流(式)溢洪道
lateral flow velocity 横向流速
lateral-flux coil 磁悬浮横向磁场线圈
lateral force 侧(向)力
lateral force coefficient 横(向)力系数;侧力系数
lateral force density 侧向密度
lateral-force design 横向力设计;横力设计
lateral force factor 横向水流力系数
lateral force resisting system 抗侧力系统;抗侧力体系
lateral freedom 侧向运动自由度
lateral friction 侧向摩擦(阻力);侧摩阻;侧面摩擦(力)
lateral frontal organ 侧额器
lateral gasification 水平方向气化
lateral geotropism 横向地理;侧向淋溶
lateral grade of bridge deck 桥面横坡
lateral gradient 横(向)坡降;横向坡度;横比降
lateral group 外侧组;侧基
lateral growth rate 横向生长率
lateral guidance 防偏摆横向导框
lateral guide 侧导承
lateral halo 测移晕
lateral headgate 支渠进水闸(门)
lateral heat-transfer 侧向传热
lateral hood 侧吸(风)罩
lateral hydraulic assistance 横向水力辅助
lateral illumination 横向照明
lateral impact 侧向冲击
lateral incaving 侧向淘刷
lateral incision 侧切开
lateral incisor 侧切牙
lateral inclined tomography 侧位倾斜体层照相术
lateral inclinometer 横向倾斜仪;横向倾斜计
lateral inertia 横向惯量
lateral inertia force 侧向惯性力
lateral inflow 横向入流;旁侧入流;旁侧来水;侧向入流;侧向流入;侧向来水
lateral inflow of second kind boundary 第二边界侧向流入量
lateral inhibition 侧抑制
lateral inhomogeneity 侧向不均匀性
lateral instability 横向不稳定性
lateral intake 横进水管
lateral interaction 侧向互作用
lateral intersection 侧方交会(定位法)【测】
lateral inversion 横向倒置;图像左右倒置;侧向反演
lateral irrigation 侧灌
lateral jump 水平定起角
lateral keel 侧向龙骨
lateral lap 旁向重叠;侧向重叠
lateral leaching 侧向淋溶;侧向淋滤;侧向浸出
lateral length scale 横向长度比尺
lateral levee lake 天然堤后湖
lateral level(l)ing 侧向调平
lateral lighting 侧窗采光
lateral line 侧线;歧管线
lateral-line system 侧线系统
lateral linkage 横向联系;水平式联系
lateral load 横向荷载;横向负载;横向负荷;冲力;侧向荷载;侧负载
lateral loaded pile 侧向荷载桩
lateral load equivalent frame 侧向荷载等效框架
lateral loading 侧向荷载
lateral load resisting member 抗侧力构件
lateral load test 侧向荷载试验
lateral lobe 侧叶
lateral locator 侧向定位器;侧面(定)位器
lateral log 侧向测井(图)
lateral longitudinal stria 外侧纵纹
laterally confined compression test 受侧限压缩试验;有侧限的抗压试验
laterally confined specimen 受侧限试样;侧限试样
laterally loaded 受侧向荷载

laterally loaded pile 承受侧向荷载桩
laterally reversed 横向翻转的
laterally stable 横向稳定的
lateral magnification 横向引长;横向扩张;横(向)放大率
lateral magnifying power 线性放大系数
lateral maneuvering range 横向机动范围
lateral mark 侧面标志
lateral marking 侧边标志;侧面标志
lateral mass 侧块
lateral member 横构件
lateral mesosome 侧间体
lateral microscale 横向微比例尺
lateral migration 侧向运移;侧向位移;侧向迁移
lateral migration along source rock 沿母岩侧向运移
lateral migration in the reservoir 在储层内侧向运移
lateral migration of river 河流的横向迁移
lateral mirage 侧现蜃景
lateral misalignment 横偏
lateral modulus of subgrade reaction 侧向基床反力模量
lateral moment 侧向弯矩
lateral momentum 侧向动量
lateral moraine 侧斜冰碛层;侧碛;侧(面)冰碛
lateral moraine bar 侧碛堤
lateral motion 横向运动
lateral movement 沿层横向错动;横向移动;横向(水平)运动;平行层理错动;侧向运动;侧向移动
lateral movement knob 载物台横向运动螺旋
lateral movement stake 横移轴桩;侧向移动标桩【测】
lateral nasal process 外侧鼻突
lateral oblique position 侧斜位
lateral obstructions 侧向障碍物
lateral opening 侧孔;边孔
lateral order distribution 侧序分布
lateral oscillation 横向振荡
lateral oscillatory characteristic 侧向振动特性
lateral outlet brick 出口流钢砖
lateral overlap 横向重叠;旁向重叠;侧向重叠;侧面覆盖
lateral parametric(al) excitation 侧向参数激振
lateral parity 横向奇偶
lateral part 侧部
lateral partition 横隔板
lateral-pattern array 侧测线
lateral photo-effect 侧向光电效应
lateral photoelectric(al) effect 侧向光电效应
lateral pickup 横向传感器
lateral pilacerores 侧蜡板
lateral pile friction resistance 侧壁桩摩擦阻力
lateral pile load test 桩的侧向荷载试验;入土桩承载试验
lateral pile resistance 桩侧壁阻力
lateral pin 横销
lateral pipe 支管
lateral(pipe)line containing sprinklers 有洒水设备的横向管道
lateral placement 路边布局;侧向布置;侧向构筑物
lateral plan 侧视图;侧面视图
lateral planation 旁夷作用;侧夷;侧(向均)夷作用
lateral plane 侧向平面
lateral-plane underwater 水下侧面面积
lateral-plate 侧板;侧蜡板
lateral play 横向移动量
lateral point 侧点
lateral pointing correction 水平瞄准修正
lateral porosity variations from deviated well VSP data 由斜井 VSP 取得孔隙度横向变化
lateral portion 外侧部
lateral position 侧卧位
lateral pressure 横(向)压力;旁压强;旁压力;侧压(力);侧向压力
lateral pressure apparatus 旁压仪;侧压仪
lateral pressure coefficient 侧压力系数
lateral pressure difference 横向压差
lateral pressure of soil 侧土压力
lateral pressure ratio 侧压系数
lateral pressure test 旁压试验
lateral process system 侧突系统
lateral profile 横断面图
lateral profile level(l)ing 横断面水准测量

lateral projecting structure 横向伸出建筑物
lateral projection 侧射影
lateral promotion 侧面升迁
lateral pyramidal tract 锥体侧束
lateral ramp 侧向匝道;侧向坡道
lateral range 侧射程
lateral range function 侧射程函数
lateral reaction 横向反力;侧(向)反力
lateral recharge coefficient 侧向补给系数
lateral recharge increment 侧向补给量
lateral recumbent position 侧卧位
lateral recumbent posture 侧卧式
lateral reducing on branch 分叉缩径分叉管
lateral reducing on one run 一头缩径分叉管
lateral reducing on one run and branch 一头和分叉缩径分叉管
lateral reflection 侧反射
lateral refraction 侧向反射声;旁折光;旁(向)折射;侧向折射
lateral reinforcement 箍筋;横(向)钢筋;侧向钢筋
lateral resilience 横向弹性
lateral resistance 横向阻力;横向抗力;侧向阻力;侧向抵抗力
lateral resistance of pile 桩的横向抗力
lateral resisting capacity 抗侧力能力
lateral resisting structure 抗侧力结构
lateral resolution 横向分辨率
lateral response 侧向干扰运动
lateral restraining pressure 侧限压力
lateral restraint 侧向约束;侧向限制
lateral restraint reinforcement 侧限加固
lateral reversal 镜像
lateral ridge 侧面线脚【建】
lateral rigidity 侧向刚度
lateral rigidity of bridge 桥梁横向刚度
lateral rock seal 侧向岩层封闭
lateral root 外侧根;侧根
lateral run-out 横向偏离
laterals 侧根
lateral safe distance 横向安全距离
lateral scale 侧鳞突
lateral scan 横向扫描
lateral scouring 侧向冲刷
lateral scroll (楼梯扶手终端的)平面涡卷形装饰
lateral secretion 侧分泌
lateral secretion theory 侧分泌说
lateral section 横截面;横断面;剖面;断面;侧(向)截面
lateral sediment transport 横向泥沙输移
lateral seepage 侧向渗流;侧渗(水)
lateral segment 外侧段
lateral semicircular canal 外半规管
lateral sensitivity 偏航敏感度;偏航灵敏度
lateral sensor 侧壁传感器
lateral separation 横向间隔
lateral septum 侧隔壁
lateral service piping 消费管线;供水支管;(横向通往用户的)供煤气管道;(横向通往用户的)供水管道
lateral sewer 污水支管;污水旁管;侧向下水道;侧向污水管;侧向暗沟;侧向沟管
lateral shear 横剪力;侧向剪力;侧向剪力;侧面剪切;侧面剪力;侧断层
lateral shelter 侧方庇荫
lateral shift 横移;横向位移;侧向位移;侧向变位;表面隆起(深挖土层);水平位移(岩石断层)
lateral shifting of centre of gravity of goods 货物重心的横向位移
lateral shift of channel 河槽横向位移
lateral shower 侧向簇射;侧簇射
lateral shrinkage 横向收缩(率);侧向收缩率
lateral sight distance 侧向视距
lateral silo entrance 库侧入口
lateral sinus line 安伯格线
lateral slide 侧向滑坡;侧向滑动
lateral slide mo(u)ld 旁模式模
lateral slipping 横向滑移;横向滑动
lateral slope of ground 地面横坡
lateral slope of top of subgrade 路基顶面横向坡度
lateral-slot screen 横向槽孔筛
lateral soil load 侧向土压力
lateral soil pressure 侧向土压力
lateral sonar 旁视声呐;侧视声呐
lateral sonde 梯度电极系

lateral spacing 侧向间距
lateral spillway 侧溢洪道
lateral spillway channel 旁支溢洪道;旁侧溢洪道;岸边溢洪道
lateral spit 侧沙嘴
lateral spread(ing) 侧向扩张;侧向扩展;横向摊铺
lateral springing of pile 桩的横向弹动;桩的横向窜动
lateral squeezing-out of soft soil 软土侧向挤出
lateral stability 横向稳定(性);横稳性【船】;侧向稳定(度);侧向稳定(性)
lateral stability coefficient 横向稳定系数
lateral stability test 侧向稳定性试验
lateral stabilizer 侧向稳定装置
lateral stay 侧面稳索;侧拉条
lateral stiffness 横向刚性;横向刚度;侧向刚度;侧向劲度
lateral stiffness of track 轨道横向刚度
lateral stopping area 路侧停车处
lateral stops 水平止动装置
lateral storage 沿岸调蓄;支流储蓄(水)
lateral strain 横向应变;侧(向)应变
lateral strain indicator 横向应变指示器;侧向应变指示器
lateral stream 横向水流;侧向水流
lateral strength 横向强度
lateral stress 横向应力;侧(向)应力
lateral structure function 侧向结构函数
lateral strut 横支柱;横向支撑;侧向支撑
lateral support 横向支承;侧向支座;侧向支承;侧向支撑;边缘支承(路面);侧撑;横撑
lateral supporting bar 横底座杆
lateral support system 横向支承系统;横向支撑系统
lateral surface 外侧面;侧面
lateral suspension motor 横向悬挂电机;牵引电机架式悬挂
lateral sway 横向摇摆
lateral sway force of train 列车摇摆力
lateral sweep 横向扫描
lateral swelling 侧向溶胀
lateral swing 侧向摆动
lateral system 支渠系统;水上航道浮标系统;侧墙输水系统(船闸);侧标系统;旁支系统;横向支撑系统
lateral system of braces 横撑系统;横支撑系统;横向支撑系统
lateral tensile stress 侧拉应力
lateral tension 侧向张力
lateral tentative stress 侧控应力
lateral terrace 侧向阶地
lateral thinning 侧向变薄
lateral thrust 横向推力;横向推进;横向推动;横推断层;切向逆断层;侧压(力);侧向推力;侧推(力)
lateral thruster 横向推进器;侧推(力)器
lateral-thrust unit 侧向推力装置
lateral tie 横向栏杆;横向拉条;横向拉杆;横向箍筋;横拉杆;侧面拉条;钢箍
lateral tie beam 水平拉杆
lateral ties of column 钢筋柱箍;钢筋混凝土柱的横箍筋;横向柱箍
lateral tile 旁向倾斜
lateral tilt 旁向倾角;侧向倾斜
lateral-torsional buckling 侧向挠曲;横向扭曲
lateral track model 横向轨道模型
lateral track stability 轨道横向稳定性
lateral transport 横向输移;横向运输;侧向输送
lateral transport of sediment 横向输沙
lateral traverse 侧向横动
lateral trolley 侧滑车
lateral truss 横(向)桁架;侧向桁架;抗风桁架
lateral tubercle 外侧结节
lateral tunnel 支洞;侧洞
lateral turnout 单开道岔【铁】
lateral underpinning 侧向托换
lateral underwater area 水线下船体纵剖面面积
lateral unloading valve of cement silo 水泥库侧卸料器
lateral unloading valve of silo 库侧卸料阀
lateral variation 横向变动;横向变化
lateral vein 侧脉
lateral vein of the first order 第一级侧脉
lateral velocity 横向速度
lateral-vertical ratio 横载—垂载比

lateral vibration 横(向)振动;侧向振动
lateral vibration of shafting 轴系的横振动
lateral view 侧视图;侧视;侧面图
lateral voltage 横向电压
lateral wall 侧帮
lateral wall infarction 侧壁梗死
lateral water transport culvert 横向输水涵洞
lateral water transport gallery 横向输水廊道
lateral wave 横向波;横波;侧波
lateral wear 侧磨;侧面磨损;侧面磨耗
lateral wear of rail head 轨头侧面磨损
lateral wedge 侧向楔体
lateral weld 侧面焊缝;侧焊
lateral wheel runout 车轮横向偏离
lateral wind 侧风
lateral wind bracing 侧向风撑;侧向防风拉条
lateral wind pressure 侧向风压力
lateral wing 侧翼
lateral yield 侧向屈服
lateral zoning 侧向分带
Lateran 拉特兰大教堂(罗马);拉特兰宫(罗马)
later childhood 少年期
late reaction 迟发反应
late release 延迟关闭
late Renaissance 后期文艺复兴式【建】
late Renaissance architecture 后期文艺复兴建筑
latericeous 土红色的;红砖灰状的
latericumbent 侧卧的
laterite 砖红壤(土);红土;铁矾土
laterite chip(ping)s 红土碎片;铁矾土碎屑
laterite clay 红土质黏土
laterite deposit 红土沉积
laterite for cement burden 水泥配料用红土
laterite gel 红土凝胶
laterite gravel 铁矾土砾石
laterite karst 红层岩溶
laterite soil 砖红壤性土;砖红壤(土);铝红土
laterite type of soil formation 砖红壤成土作用
lateritic 红土的
lateritic clay 砖红黏土
lateritic crust 砖红土结壳
lateritic gravel 铁矾土砾石;砖红土卵石;铁矾土卵石
lateritic loam 红壤土
lateritic loamy 红壤
lateritic soil 砖红壤性土;砖红壤(土);红(黏)土;铝红土
lateritious 土红色的;红砖灰状的
lateritite 再造铝红壤;红土碎屑岩;次生红土
lateritization 砖红壤化
lateritoid 准红土
laterization 砖红壤化;红土化(作用)
laterized-type iron deposit 红土化型铁矿床
laternating load 交变荷载
latern(ring) 润滑环
laterogenesis 侧向成因;侧分泌学说
laterolog 横向测井;侧向测井地层
laterolog 3 三电极侧向测井仪;三侧向测井
laterolog 3 curve 三电极屏障电流法测井曲线;三侧向测井曲线
laterolog 6 六侧向测井
laterolog 6 curve 六侧向测井曲线
laterolog 7 七侧向测井
laterolog 7 curve 七侧向测井曲线
laterolog 8 八侧向测井
laterolog 8 curve 八侧向测井曲线
laterolog calibration value 侧向测井刻度值
laterolog curve 侧向测井曲线
laterologger 侧向测井仪
laterolog ging 侧向测井(图)
laterolog resistivity 侧向测井电阻率
laterolog sonde 侧向测井电极系
laterology 钻孔电阻记录
late Roman 晚期罗马风格
late Romanesque 晚期罗马式建筑
late Romanesque church 仿罗马式教堂
lateropulsion 侧步
laterotergite 侧背片
later rest 休眠后期
later sonde at bottom 底部梯度电极系
later sonde at top 顶部梯度电极系
later strength 后期强度
later stress field 后期应力场
later style 后期风格
Later Temple of Artemis at Ephesus 爱菲苏司的

后期"月女神"神庙
late shift 夜班
late slope detection 迟斜检测法
late sowing 迟播
late spark 迟火花
late stage 晚期
latest allowable date 最迟允许日期
latest allowance date 最迟许可时间
late start 迟启
latest breakup 最晚解冻日
latest complete freezing 最晚封冻日
latest entry 最后计算值
latest event occurrence time 最后如期完工时间；建筑最后完工期限
latest fashion flower 最时新花样
latest final ice clearance 最晚终冰日；最晚解冻日
latest finish date 最晚完工时间；最晚完成时间；竣工日期；建筑最后完工日期；最后完工日期
latest finishing time 最迟完工时间；最迟完成时间
latest finish time 最晚完工时间；最晚完成时间；最迟竣工时间；最迟结束时间；竣工时间；竣工日程
latest frost 终霜
latest information 最新资料
latest market report 最近市场报告
latest node time 最迟结点时间
late strength 后期强度
latest snowfall 终降雪
latest start date 最晚开始时间；最晚开工时间；建筑最晚开工时间；最晚开工日期
latest starting time 最晚开始时间；最晚开工时间；最迟开始时间；最迟开工时间
latest starting time of activity 作业最迟开始时间
latest style 新风格
latest technical standard 最新版本技术标准
latest technology 最新工艺
latest time 最迟时间
late timing 延时
Late Triassic climatic zonation 晚三叠世气候分带
Late Triassic epoch 晚三叠世【地】
Late Triassic extinction 晚三叠世绝灭【地】
late variety 晚熟品种
Late Wisconsin 晚维斯康辛【地】
latewood 晚材；夏木材；夏材
late wood ratio 秋材率
late work 后期作品
latex [复] latices 或 latexes 胶乳；乳汁；乳液；乳胶
latex acrylic sealant 胶乳丙烯酸密封膏
latex adhesive 乳胶黏结剂；胶乳胶合剂
latex agglutination inhibiting test 胶乳凝集抑制试验
latex agglutination inhibition reaction 乳胶凝集抑制反应
latex agglutination reaction 胶乳凝集反应
latex agglutination test 胶乳凝集试验；乳胶凝集反应
latex anti-septic 乳胶防腐剂
latex base 乳胶基；胶乳基底
latex bitumen 乳胶沥青
latex bitumen mixture 橡胶沥青混合物；橡浆沥青混合物
latex-bound 乳胶黏结
latex ca(u)lk 胶乳嵌缝膏；胶乳嵌缝料；胶乳嵌缝剂
latex cement 胶乳水泥；胶乳结合剂；胶乳接合剂；橡胶水泥
latex-cement-aggregate mix(ture) 胶乳水泥骨料混合剂
latex-cement mortar 胶乳水泥砂浆
latex-cement sealing compound 乳胶水泥密封复合料
latex-coated fabric 胶乳布
latex coated fabric manufacturing aggregate 胶乳布机组
latex coated glass fabric 胶乳涂复玻璃布
latex coating 乳胶涂料
latex composition 乳胶配合物
latex compound 乳胶配合物
latex compounding 乳胶配合工艺；配料乳胶的制备
latex concrete 乳胶混凝土
latex content 含胶量
latex emulsion 橡乳胶；乳胶乳；乳胶（液）；胶乳液
latex emulsion adhesive 橡胶胶黏合剂；乳胶胶黏剂；胶乳胶黏剂；乳胶黏合剂
latex enamel 乳胶瓷漆
latex examination gloves 检查手套

latex fixation test 胶乳凝集试验
latex flocculation test 胶乳絮状试验
latex floor(ing) finish 乳胶地面涂层；乳胶地板面层
latex foam 胶乳泡沫；泡沫乳胶；泡沫胶乳
latex foam rubber 胶乳泡沫橡胶；乳液泡沫橡胶
latex glue 乳胶黏合料；胶乳；胶水
latex house 建筑用乳胶漆
latex house paint 建筑（用）乳胶涂料；民用乳胶涂料
latex liquid 橡胶乳液
latex mastic 乳胶黏合剂；胶乳玛琋脂
latex mastic compound 乳胶油灰混合物
latex mixer 胶浆搅拌机
latex-modified cement paste 乳胶水泥浆；掺橡胶水泥浆
latex-modified concrete 掺橡胶混凝土
latex mortar 乳胶砂浆
latexometer 胶乳比重计
latex paint 橡浆涂料；乳胶涂料；乳胶漆
latex particle agglutination test 胶乳颗粒凝集试验
latex preservative 乳胶防凝剂；乳胶防腐剂；乳胶保护剂
latex primer-sealer 乳胶封闭底漆
latex proofing 乳胶涂覆
latex resin 乳胶树脂
latex rubber 乳胶橡胶
latex rubber bag 乳胶袋
latex screed 乳胶找平层
latex sealant 胶乳密封膏
latex separator 胶液分离机
latex sprayed rubber 乳液喷雾橡胶
latex stabilization 胶乳稳定性
latex thickener 胶乳增稠剂；乳胶增稠剂
latex wall paint 胶乳（墙面）涂料
latex wastewater 橡浆废水
latex water paint 乳胶漆
latex water-thinned acrylic 水乳型丙烯酸涂料
latex water-thinned polyvinyl acetate 水乳型聚醋酸乙烯涂料
latex water-thinned polyvinyl chloride 水乳型聚氯乙烯涂料
Late Yanshanian geosyncline 晚燕山期地槽【地】
Late Yanshanian subcycle 晚燕山亚旋回【地】
lath 木板条；灰条；灰板条；条板；板条
lath, plaster, float and set 三道粉刷
lath, plaster and set 两道粉刷
lath and a half 1/4 英寸厚板条
lath and plaster 板条抹灰
lath and plaster ceiling 板条抹灰顶棚
lath and plaster partition 条板抹灰隔墙；板条抹灰隔墙；灰板条抹灰隔墙
lath and plaster shed 灰面屋
lath and plaster wall 板条泥镘墙
lath brick 条（形）砖；条形砌窑砖
lath ceiling 板条抹灰天花板
lath clip 板条夹
lath cutter 割板机；切板机
lathe 车床
lathe accessories 车床附件
lathe bed 车（床）身
lathe bit 车刀
lathe carriage 车床拖板；车床刀架
lathe carrier 车床鸡心夹头
lathe center 车床顶尖
lathe center point 车床顶尖端
lathe chuck 车床卡盘；车床夹盘
lathe control system 车床控制系统
lathe cutter 割板机；车床切削刀具
lathe cutting tool 车床切削刀具
lathed ceiling 板条平顶；板条顶棚
lathe design 车床设计
lathed fence 板条围栏
lathe dog 制动爪；车床轧头
lathe drill 卧式车床
lathed system 译密码装置
lathe dynamometer 车床测力计
lathe equipment 车床附件
lathe for copper roller 车花筒机
lathe frame 车床架；车床床座
lathe grinding 车床研磨；车床磨削
lathe head 车头箱
lathe lapping 车床研磨
latheman 车工
lathe mandrel 车床心轴

lathe operation 车床操作
lathe operator 车工
lather 起泡(沫)；起泡；车床工；板条工
lather booster 泡沫促进剂
lather collapse 泡沫破裂
lather end point 泡沫终点
lathe rest 车床支座
Latherometer 泡沫仪
lather quickness 起泡速度
lather-type soap dispenser 起泡式肥皂配出器
lather value 泡沫值
lather volume 泡沫体积
lathe saddle 车床刀架
lathe spanner 车床螺丝扳手
lathe spindle 车床轴
lathe tool 车刀；车床刀具
lathe tool-setting microscope 车床对刀显微镜
lathe treadle 车座踏板
lathe turning 车床切削；车床车削
lathe turret 车床转塔刀架
lathe type winding machine 卧式缠绕机
lathe work 车削加工；车工工作；彩纹；车床工作
lathe worker 车工
lath fence 板条围栏
lath floor 板条地板
lath hammer 板条槌；钉板条斧；钉抹灰板条用锤；带刃刀的拔钉锤；钉锤
lathhouse 板条房屋；板条花房；板条种植房
lathing 钉板条；板条筛网
lathing board 钉木板条；墙上钉板条的板
lathing ceiling 板条抹灰泥天花板；板条抹灰泥顶棚
lathing for stucco 抹灰板条
lathing hammer 板条锤；钉板条锤；钉抹灰板条用锤
lathing hatchet 锤斧；板条斧
lathing insulating mat 抹灰板条墙隔热垫层
lathing mesh 抹灰钢丝网
lathing nail 板条钉
lathing of roof 钉屋面板条
lath insulating mat 板条抹灰泥绝热层
lath laid-and-set 板条打底抹面
lath-like habit 叶片状习性
lath martensite 板条马氏体
lath mesh 板条抹灰网
lath nail 板条钉
lath paper 建筑用纸；防水纸
lath scratcher 板条刮粗器；底灰刮毛板条；粉刷刮毛帚
lath screen 板条帘
lath-shaped 板条状（的）
lath shutter 板条百页
lath timber 板条木材
lath wood 板条木材
lathwork 灰底板条工；钉板条
latialite 蓝方石
latiandesite 安粗安山岩
latibasalt 安粗玄武岩
latices 乳状液；latex 的复数形式
laticometer 胶乳比重计
latifundium 大庄园；大地产；大领地
Latin alphabet 拉丁字母（表）
Latin American Common Market 拉丁美洲共同市场
Latin American Free Trade Area 拉丁美洲自由贸易区
Latin American Free Trade Association 拉丁美洲自由贸易协会
Latin American Integration Association 拉丁美洲综合协会
Latin American Shipowner's Association 拉丁美洲船东协会
Latin architecture 拉丁式建筑；拉丁建筑
Latin cross 拉丁式十字架；十字交叉点；纵长十字架
Latin language 拉丁语；拉丁文
Latin matrix randomization 拉丁方阵随机化
Latin square 拉丁方(矩)
Latin square design 拉丁方设计
Latin square experiment 拉丁方试验
Latin square method 拉丁方格法
latite 熔岩；安粗岩
latite group 安粗岩类
latite porphyry 安粗斑岩
latitude 纬度
latitude and departure 经纬距
latitude and longitude 黄纬和黄经

latitude and speed correction mechanism 纬度和船速差订正器
latitude arc 纬度弧
latitude by (dead) reckoning 积算航纬度;船位推算纬度
latitude by sun's maximum altitude 太阳最大高度求纬度
latitude by Talcott's method 太尔各特法纬度
latitude circle 黄纬圈;纬度圆;纬(度)圈
latitude compensator 纬度补偿器
latitude correction 纬度校正(量);纬度改正(量)
latitude correction coefficient 纬度改正系数
latitude correction value 纬度改正值
latitude corrector 纬度校正器
latitude cutoff 纬度截止
latitude data computer 纬度数据计算机
latitude determination 纬度测定
latitude difference 纬差
latitude distribution 纬度分布
latitude effect 纬度影响;纬度效应
latitude error 纬度误差
latitude factor 纬度因素;纬度因数;纬度系数
latitude from 起程纬度
latitude in 到达点纬度
latitude left 起程纬度
latitude level 纬度水准器
latitude level corrector 水银器纬度修正盘
latitude level(l)ing dial 水银器纬度修正盘
latitude line 纬(度)线;东西向位置线
latitude longitude grid 经纬度格网;经纬网
latitude method 纬度法(测天球船位)
latitude micrometer 纬度测微器
latitude of a heavenly body 天体黄纬
latitude of a place 某地纬度
latitude of epicenter 震中纬度
latitude of exposure 曝光时限
latitude of meridian altitude 中天高度求纬度
latitude of observer 测者纬度
latitude of pedal 底点纬度;垂足纬度
latitude of the well position 钻孔所在位置的纬度
latitude of vertex 顶点纬度
latitude poise 纬度平衡
latitude range 纬度范围
latitude rider 纬度差订正器
latitude scale (海图上的)纬度尺(度)
latitude service 纬度服务
latitude south 南纬
latitude station 纬度站
latitude value 纬度值
latitude variation 纬度变化
latitude zones of climate 纬度气候带
latitudinal band 纬度带
latitudinal dispersivity 横向弥散度
latitudinal distance between the borehole centers 孔心横距
latitudinal distribution 纬向分布;纬度分布
latitudinal extent 纬线跨度
latitudinal grade 横向坡度
latitudinal structural system 纬向构造体系
latitudinal tectonic system 纬向构造体系
latitudinal transgressions and regressions 纬度性海水进退
latitudinal type 纬线型
latitudinal vegetational zonation 植被水平地带性
latitudinal zone 纬度地带
latitudinal zoning 纬向分带性;纬度分带
latiumite 硫硅石;硫硅碱钙石
Latona 便携式油散热器
latorex 红土砖块;板条帘;苗床格子
latosol 砖红土;砖红壤性土;砖红壤(土);淋滤土
latrappite 钙钛铌矿;铌钙钛矿
latrine 沟厕;公共厕所
latrine fitments 旱厕设备
latrine pit 茅坑
latrobe 一种采暖炉;取暖火炉
latrobite 淡红钙长石
latten 金属薄板
latten brass 薄黄铜;黄铜薄片
Lattens 拉丁锌铜合金
latten(-tin) 黄铜片
latter end 末段
latterkin 硬刮槽工具;刮窗嵌槽木;硬木刮槽工具
latter math 后割草
latter testing 未筛试验

lattic dynamics of metal 金属晶格动力学
lattice (格(栅);斜条格构;斜角格构;支承桁架;栅格;格子;(格(式);格;;格架;格构;经纬度格网;网格;图网;图格;点阵;花格
lattice absorption 晶格吸收;点阵吸收
lattice array 点阵排列;点阵列
lattice asymmetry 点阵不对称
lattice bar 格条;格构缀条;格构杆
lattice beacon 笼标(浅滩立标)
lattice beam 格(子)梁;格构梁;花格梁
lattice beam bridge 花格梁桥
lattice bending 点阵弯曲
lattice binding 点阵耦合
lattice board 格子板;网络板
lattice bond 点阵键
lattice boom 格构挺杆;格构式起重臂;格构伸臂;笼格(式)吊杆;桁架臂架;格子吊杆
lattice brace 格构缀条
lattice braced tapered steel mast 锥形桁架钢柱
lattice braced tapered steel pole 锥形桁架钢柱
lattice bracing 格子撑架;格型支撑;格构撑架
lattice brick 格子砖;空心砖;花格砖
lattice bridge 格构(梁)桥
lattice-building 造建晶格;建造晶格
lattice bulwarks 格子舷墙
lattice ceiling 格子顶棚
lattice cell 晶泡
lattice charge 晶格电荷
lattice chart 网格图
lattice circuit 网格电路
lattice clay 网纹黏土
lattice coil 网格线圈;梳形线圈;多层线圈;点阵式线圈
lattice column 格(式)柱;格式框架;缀合柱;格构柱
lattice column mast 格形钻塔;格形桅杆
lattice complex 点阵复合
lattice compliance 晶格顺服
lattice constant 晶格常数;点阵常数
lattice construction 桁架结构
lattice contraction 点阵收缩
lattice coordination number 晶格配位数;点阵配位数
lattice correspondence 点阵对应
lattice curvature 点阵曲率
latticed 缀合的;结构的;网格状的
latticed bar 缀合杆
latticed bracing 格构支撑
latticed column 格构柱
latticed cupola 格构穹顶
latticed derrick crane 格子吊杆起重机
latticed dike [dyke] 格坝
latticed door 格子门
lattice defect 晶格缺陷;点阵缺陷
lattice defect scattering 晶格缺陷散射;点阵缺陷散射
lattice deformation 点阵变形
lattice derrick 笼格(式)吊杆
lattice derrick crane 格梁动臂起重机
latticed fence 网格子栅栏
latticed framework 格构框架
latticed girder 花格大梁;格构(大)梁
lattice diagram 点阵图
lattice dilatation 晶格膨胀
lattice dislocation 晶格位错;点阵畸变
lattice dislocation defect 晶格错位缺陷
lattice disorder 点阵无序
lattice distance 点阵间距
lattice disturbance 晶格结构破坏
latticed member 缀合杆
lattice door 格门
latticed partition wall 缀合隔墙;花格隔墙;花格隔断
latticed perforation plate 梯形穿孔板
lattice drainage 格状排水系(统)
lattice drainage pattern 长方水系型
lattice drier 帘子式烘燥机
lattice dryer 帘子式烘燥机
latticed shell 网壳;网格壳体
latticed stanchion 缀合支柱
latticed steel column 格构钢柱
latticed steel girder 格构钢梁
latticed steeple 网格尖顶
latticed structure 格构式结构;格构结构
latticed strut 格构支撑;缀合支撑
latticed web 格构腹板

latticed wing 翼栅
lattice dynamics 点阵动力学
lattice dynamics of molecucrystals 分子结晶晶格动力学
lattice effect coefficient 方格影响系数
lattice emission 晶格发射
lattice energy 晶格能;点阵能
lattice feeder 帘子式喂给机
lattice fence 格构围栏
lattice filter 桥形网络滤波器
lattice flow 叶栅流动
lattice force 点阵力
lattice form 格构式
lattice frame 格子框架;格式构架;格构框架;格(构)架
lattice framework 格构框架
lattice framing 支条;帽儿梁
lattice gate 格状闸门;网格闸门
lattice girder 格子梁;格构(大)梁;花格大梁;桁架大梁;网构大梁
lattice group 点阵组
lattice heat capacity 点阵热容
lattice heterogeneity 点阵异质
lattice homing 格网导向
lattice homomorphism 格同构
lattice hypothesis 格子假说
lattice imperfection 晶格缺陷;点阵缺陷;点阵不完整性
lattice impurity 点阵杂质
lattice irregularity 点阵缺陷
lattice isomorphism 格同构
lattice jib 格构挺杆;格构吊杆;桁架臂
lattice mast 格子桅;格构抱杆;网架式铁塔;格构桅柱
lattice matching 晶格匹配;点阵匹配
lattice member 格构部件
lattice misfit 点阵错合
lattice model 点阵模型
lattice mo(u)lding 网格饰的凸出方形木线脚
lattice network 格子网络;格状网络;桥形网络;网格状线路;网格形线路
lattice of rutile type 金红石型晶格
lattice operation 格运算
lattice optics 晶格光学
lattice-ordered group 格序群
lattice-ordered ring 格序环
lattice orientation 点阵取向
lattice parameter 晶格参数;点阵参数
lattice parameter method 点阵参数法
lattice pattern 网格图型
lattice plane 格子平面;晶面;晶格(平)面;点阵平面
lattice plate 屋盖网格板;网格板
lattice point 阵点;格点;晶格结点;网点;点阵点
lattice point formula 格点公式
lattice polarization 晶格极化;点阵极化
lattice pole 格构抱杆;格构(式)桅杆
lattice portal 格构桥门
lattice position 点阵位置
lattice purlin(e) 格构檩条;格构桁条
lattice pylon 格构牌楼;格构标塔
lattice reactor 栅格反应堆
lattice reinforcement 格构钢筋网;格构配筋
lattice relaxation 晶格弛豫
lattice resolution 晶格分辨率
lattice resonance 晶格共振
lattice retaining wall 网格式挡土墙
lattice rod 格条;格构杆件
lattice roof 格构屋顶
lattice scattering 晶格散射;点阵散射
lattice screen 格构式拦污栅;网格拦污栅
lattice search 格点寻优法;格点查查;格点搜索(法)
lattice search technique 格点搜索技术;格点搜索法
lattice self-diffusion 晶格自扩散
lattice shell 格构薄壳
lattice-site 晶格点;点阵点
lattice spacing 栅格间距;晶格距离;点阵间隔
lattice square 格子方;网格形方孔
lattice stanchion 格条
lattice steel girder 格构钢梁
lattice steel mast 钢格构桅杆
lattice steel tower 格状结构金属塔
lattice strain 晶格应变;点阵应变
lattice structure 栅格结构;格状构造;格(形)结

构;晶体结构;晶格结构;晶格构造;网格结构;点阵构造
lattice strut 缀材;格构支撑;格条
lattice substitution 点阵的替位
lattice swing bridge 格子旋桥
lattice symmetry 点阵对称
lattice system 格状系统
lattice texture 格子结构;点阵结构
lattice theory 格论;格构理论;晶体(点阵)理论;点阵理论;网络理论
lattice tightness 晶格紧密度
lattice tower 格子形杆架;格架塔;格构形塔;网格塔楼
lattice town truss 多斜杆桁架
lattice tracing 格构缀条
lattice transformation point 晶格转变点
lattice translation 点阵平移
lattice translation vector 晶格平移矢量
lattice truss 格子桁架;格构桁架;网构桁架
lattice type 格构式;点阵类型
lattice-type concave 栅格式凹板
lattice-type filter 桥式滤波器;格子形滤波器
lattice-type frame 格构框架
lattice-type portal frame 格构式框架;格构型框架
lattice-type quantity equip-ment 格式定量器
lattice undercarriage 格构起落架
lattice unit 晶胞;晶架单位
lattice vacancy 晶体空位;晶格空缺;点阵空位
lattice vector 点阵矢量
lattice vibration 晶格振动;点阵振动
lattice vibration quantum 晶格振动量子
lattice wall (指用木或金属砌成的)花格墙
lattice water 晶格水
lattice wave 晶格波;晶格波;点阵波
lattice web 花格腹板;花腹板
lattice winding 斜格形绕组;斜格式绕组;栅格型绕组;篮式线圈
lattice window 花格窗;斜条线(构)窗;格子窗;格构窗;玻璃格条窗;花饰铅条窗
lattice work 构架工作;格子细工;格构细工;格构构件;格构工程;晶格构造;网格结构;花格
lattice-work fence 格构细工栅栏
lattice-work partition 花格隔墙;花格隔断
lattice-wound coil 斜格式线圈;蜂房线圈
latticing 格构构件;双缀
latticing bar 格条
latus 侧部
latus rectum 正焦弦;正交弦;通径
lauan 柳安(木)
laubanite 白沸石
laubmannite 劳磷铁矿
laueite 劳埃石;黄磷锰铁矿
Laue's background 劳厄背底
Laue's breadth 劳厄宽度
Laue's camera 劳厄(型 X 射线)照相机
Laue's class 劳厄对称型
Laue's condition 劳厄条件
Laue's diagram 劳厄图
Laue's diffraction 劳厄衍射
Laue's diffraction diagram 劳厄衍射图(样)
Laue's diffraction equation 劳厄衍射方程
Laue's diffraction pattern 劳厄衍射图(样);劳厄衍射花样
Laue's effect 劳厄效应
Laue's equation 劳厄方程
Laue's group 劳厄群
Laue's index 劳厄指数
Laue's interference 劳厄干涉
Laue's method 劳厄法;晶体法
Laue's monotonic scattering 劳厄单调散射
Laue's pattern 劳厄斑;劳厄图
Laue's photograph 劳厄照片;劳厄图
Laue's photography 劳厄照相(术)
Laue's picture 劳厄照片
Laue's plane 劳厄平面
Laue's point 劳厄点
Laue spot 劳厄斑点
Laue's spike 劳厄尖峰
Laue's spot 劳厄斑(点)
Laue's spot deformation 劳厄斑点的畸变
Laue's symmetry 劳厄对称
Laue's threelemella interferometer 劳厄三片式干涉仪
Laue theory 劳厄理论

laugenite 奥长闪长岩
Laughlin filter 磁铁矿过滤器
laumonite facies 浊沸石相
laumonite-prehnite pumpellyite lawsonite facies group 浊沸石—葡萄石绿纤石—硬柱石相组
laumontite 浊沸石
launayite 劳硫锑铅矿
launch 小汽艇;坐船出去;开出;开办;滑曳;汽艇;弹射;使下水【船】;发射;下水
launch a ship 船下水
launch control system 发射控制系统
launch date 下水日期
launched ship 已下水船
launcher 发射架;发射装置;发射器
launch facility foundation 滑曳设施基础(如船坞中设置)
launch hire 小汽艇租费
launching 投射;滑曳(桥梁架设);下水
launching accuracy 发射准确度
launching aircraft 母机
launching bar 发射樟
launching barrel 发射阱
launching beam 起动杆
launching boost 发射助推器
launching booster 起动助推器
launching cable 牵索
launching calculation 下水计算
launching catapult 升空弹射器
launching ceremony 下水仪式;下水式
launching clause 下水条款
launching compartment 发送阱
launching complex 全套发射设备
launching cradle 斜架下水车【船】;下水(支船)架;滑道承船架;弹射架;随船架;船舶下水滑架;船舶下水滑道;导弹发射架
launching curve 下水曲线
launching dart 发射标
launching drawing 下水图纸
launching erection 滑曳装置
launching facility 下水设备
launching fiber 发射光纤
launching gear 下水装置;放艇装置
launching girder 滑曳梁
launching grade 斜坡度
launching grease 下水滑脂;船用下水润滑脂;船用润滑脂
launching installation 下水装置
launching nose 滑曳导梁;顶推导梁;导梁
launching of a trial meteorologic(al) satellite 发射试验性气象卫星
launching of caisson 沉箱下水
launching of pipeline section 管线下水
launching of submersible 吊放潜水器
launching operation 下水作业
launching pad 发射台
launching platform 下水仪式台;下水(船)台;发射台
launching position indicator 发射位置指示器
launching power 起动功率
launching procedure 下水程序
launching ramp 管线下水坡道;船舶下水滑道
launching shaft 始发井
launching shoe 发射塔
launching site 发射场设备;发射场
launching slide 下水坡道;起飞滑行
launching speed 下滑速度
launching stability 下水稳定性
launching staff 吊放工作人员
launching stand 管线下水坡道;发射台
launching system 起动系统
launching tower 垂直发射装置;发射塔
launching trolley 下水车;滑曳空中吊车;起动车
launching truss 滑曳桁架
launching unit 发射装置
launching valve 发送阀
launching vehicle 两栖车辆;发射车
launching velocity 下滑速度
launching way 滑道;下水滑道;下水船道;船台滑道;船舶下水滑道
launching weight 下水重量;下水时船舶重量;投放重量
launching width 船舶下水的船台面宽度
launch interval 发射间隔
launch-latch 挡弹扣

launch numerical aperture 发射数值孔径
launch out 下水
launch pad 发射坪;发射台
launch-phase tracking laser 发射相位跟踪激光器
launch retrieval apparatus 吊放回收装置
launch ring 发射环
launch shaft and arrival shaft 出发与到达竖井
launch simulator 发射模拟器
launch tug 港内拖轮;拖轮;船舶下水拖轮
launch-type work boat 汽艇型工作船
launch vehicle 运载工具
launch way 船舶下水滑道
launch window 有利发射时机;发射最佳时间
launch airplane 火箭运载机
launder 洗盐槽;洗涤槽;流槽洗涤;流槽;渡槽;出钢槽;槽选机;天沟(俗称)
launder classifier 洗涤分级器;洗涤分级机
launderer 洗槽;流槽;洗衣工;洗涤(机)
launderette 自助洗衣店
laundering 洗烫;湿洗
laundering and dry cleaning 浆洗和干洗
laundering and dyeing shop 洗染店
launder jacket 流槽水套
launderometer 耐洗牢度试验仪
launder sand slicer 槽式分砂机
launder screen 流槽筛
launder separation process 流水槽比重分选法
launder-type hindered-settling hydraulic classifier 流槽式干扰下沉水力分级机
launder-type tandem Menzies hydroseparator 流槽式串联孟席斯型水力分选机
launder-type vortex classifier 流槽式涡流分级机
launder washer 流槽选煤机;槽式洗磨机
launder washing 流槽选矿;槽洗
laundromat 自助洗衣店;自动洗衣店;自助洗衣场
laundromat waste 自动洗衣店废水
laundry 洗衣室;洗衣店
laundry blue 洗衣蓝;洗涤蓝;合成天青石蓝
laundry chute 脏衣(物)溜槽
laundry club 公共洗衣店;共用洗衣房
laundry detergent 洗涤剂
laundry drying ground 晾衣场(地)
laundry equipment 洗衣房设备
laundry machine 洗衣机
laundryman 洗衣工
laundry room 洗衣间;洗衣房
laundry sink 洗衣池;洗衣水槽
laundry soap powder 洗衣粉
laundry solid soap 洗衣皂
laundry stove 洗衣房热风炉
laundry tray 洗衣盆;洗衣池;洗衣槽
laundry tub 洗衣盆;洗衣池
laundry wash house 洗衣房
laundry wastes 洗衣废水
laundry wastewater 洗衣(房)废水
laura (修道士住的)单间小房群体
Lauraian crustal cupola 劳拉壳块
lauraldehyde 十二醛
laurane 月桂烷
Laurasia 劳拉古陆
Laurasia breakup 劳拉大陆解体
Laurasia-Gondwana separation 劳拉大陆与冈瓦纳大陆分离
Laurasia paleocontinent 劳拉古陆
Laurasia-type platform 劳拉型地台
laurate 月桂酸盐
laurdalite 歪霞正长岩
laurel 月桂叶油;月桂树
laurel forest 温带雨林
laurel leaf 月桂叶
laurel-leaf swag 荣誉花环;月桂叶花环;月桂叶饰品
laurel oil 月桂脂
laurelor 月桂
laurel pink 月桂粉红(淡红色)
laurel tallow 月桂脂
laurel wax 月桂蜡
laurel wood 月桂木
laurence 闪烁景
laurene 月桂烯
Laurent expansion 劳伦特展开
Laurent half-shade plate 劳伦半遮片;劳朗半影板
laurentia 劳伦大陆
Laurentia-Baltica collision 劳伦大陆与波罗的大

陆碰撞
Laurentia-Gondwana collision 劳伦大陆与冈瓦纳大陆碰撞
Laurention orogeny 劳伦运动【地】
Laurent plate 劳伦板
Laurent polarimeter 劳伦偏振计
Laurer's canal 劳勒尔管
Laurer-Stieda canal 劳勒尔—斯蒂尔达管
lauric acid 月桂酸；十二烷酸
lauric aldehyde 十二醛
lauric alkyd resin 月桂酸醇酸树脂
lauricidin 十二烷酸二羟丙酯
laurifruticeta 阔叶常绿灌木群落；常绿灌木群落
laurilignosa 亚热带雨林；阔叶常绿木本群落；常绿木本群落
laurionite 羟氯铅矿
laurisilva 温带雨林
laurisilvae 阔叶乔木群落；阔叶林
laurite 硫钌矿
Lauritsen detector 劳里森检测器
Lauritsen electroscope 劳里森验电器
lauroleic acid 月桂烯酸；十二烯酸
laurone 月桂酮
lauronic acid 月桂酮酸
laurophenone 月桂苯酮
lauroyl 月桂酰
laurus 月桂
laurusia 劳拉大陆
Laurusia-Africa collision 劳拉大陆与非洲大陆碰撞
Laurusia-Asia collision 劳拉大陆与亚洲大陆碰撞
Laurus nobilis 月桂
laurvikite 歪碱正长岩
lauryl 月桂酰；十二基
lauryl alcohol 月桂醇；十二烷醇；十二醇
lauryl amine 十二胺
laurylamine hydrochloride 十二烷胺盐酸盐
laurylene 十二烯
lauryl sodium sulfate 十二烷基硫酸钠
lauryl sulfate 十二烷基硫酸盐
laurylsulfoacetate 十二烷基磺化乙酸酯
laury sulfate 硫酸月桂酯
lausenite 六水铁矾
Lauson engine 劳逊发动机
Lautal 劳塔尔铝硅铜合金
lautarite 碘钙石
lauter 过滤
Lauth's mill 劳思式轧机
Lauth's violet 劳思紫
lautite 辉砷铜矿
lav 盥洗室；厕所
lava 流岩；火山熔岩；熔岩
lava aleuritic texture 熔岩砂状结构
lava ash 火山灰；熔岩灰
lava basalt 玄武岩；熔岩玄武岩
lava bed 熔岩层；熔岩床
lava blister 熔岩泡
lavabo（固定在墙上的）盥洗器；（固定在墙上的）盥洗盆；盥洗室
lava boulder texture 熔岩巨砾结构
lava breccia 火山角砾岩
lava brick 熔岩砖
lava-cake agglomerate 熔岩饼集块岩
lava-cake breccia 熔岩饼火山角砾岩
lava cauldron 熔岩火山湖
lava cave 熔岩洞（穴）
lava cone 熔岩锥；熔岩丘
lava desert 熔岩荒野
lava discharge 熔岩喷发
lava dome 岩丘；熔岩穹丘；盾状火山
lava effusion 熔岩喷溢
lava fall 熔岩瀑布
lava field 熔岩原；熔岩荒野
lava flow 火山熔岩流；熔岩流
lava flow dome 溢流丘
lava foam glass 熔岩泡沫玻璃
lava fountain 熔岩喷涌
lavage 洗出法；灌洗（法）
lavage solution 灌洗剂
lava glass 熔岩玻璃
lava lake 熔岩湖
lava lake activity 熔岩湖活动
lava landforms of large extent 大型熔岩地貌
lava levee 熔岩堤
Lavalier microphone 项链式麦克风

Laval nozzle 拉瓦尔喷管
Lavandine oil 雷文定油；杂熏衣草油；熏衣类油
Lavandulol 雷文杜醇；熏衣草醇
lava phase 熔岩相
lava pit 熔岩坑
lava plain 熔岩平原
lava plateau 熔岩台地；熔岩高原
lava plug 熔岩岩颈
lava pore and cavern 溶岩孔洞
lava psephitic texture 熔岩砾状结构
lava sheet 熔岩席
lava slag 熔岩渣
lava soap 去油垢洗手皂（含熔岩粉）；去油垢皂（含浮石粉）
lava soil 熔岩土壤
lava spine 熔岩棘
lava stalactite 熔岩钟乳
lava stalagmite 熔岩石笋
lava streak 熔岩脉
lava terrace 熔岩阶地
lavation 洗出法；灌洗法；冲洗法
lavatory 洗手盆；洗手间；盥洗室；盥洗间；厕所
lavatory basin 洗手盆；洗面器；盥洗盆；盥洗槽
lavatory basin range 盥洗盘
lavatory basin with pedestal 带座洗手盆
lavatory bowl 马桶
lavatory equipment 盥洗室设备
lavatory fan 盥洗室排气扇
lavatory flush(ing) 厕所冲水；盥洗室冲洗
lavatory flushing cistern 盥洗室冲洗水箱
lavatory pan 大便器；冲洗式便桶；冲洗式坐式便桶
lavatory pit 茅坑
lavatory seat 抽水大便器座；马桶座
lavatory seat lid 马桶盖
lavatory soap dispenser 厕所肥皂配出器
lavatory tray 盥洗室水盘
lavatory waste 面盆出水管
lavatory waste pipe 盥洗室排水管；盥洗排水管
lava tube 熔岩管（熔岩流空的管状孔道）
lava tuffaceous texture 熔岩凝灰结构
lava tumulus 熔岩鼓包
lava tunnel 熔岩隧道；熔岩暗道
lava volcano 熔岩火山
lava ware 仿熔岩花纹的炻器
lava wedge 熔岩楔
lave 冲洗
lavender 淡紫色；薰衣草
lavender grey glaze 粉青釉
lavender jar 香料缸
lavender oil 熏衣草油
lavender print 翻正片
lavender spike oil 薄荷油
lavendulan 氯砷钠铜石
lavenite 纳钙锆石；褐锰锆矿
Laves' phases 拉夫斯相
lavishly ornamented 豪华装饰的
Lavitas point 拉维塔斯点
Lavite process 拉维特法
lavoflux 水力淘析器
law 规律；定律；法则；法规
Lawaczek viscometer 拉瓦捷克黏度
law against unfair competition 反对不正当竞争法
law and principle of chemistry 化学定律和原理
law and regulation of commerce and industry 工商法规
lawazulite 硒碲铋矿
lawbreaker 违法者
law-breaking capitalist 不法资本家
law court 法院
law day 法日（记录在纸押物或纸条上的还债日期）；法定还债日
law enforcement 法律(的)实施
Law for Preventing Collisions at Sea 海上避碰规章；海上避碰规则
lawful age 法定年龄
lawful condition 法定条件
lawful damages 法定的损害赔偿
lawful interests 合法利益
lawful measurement of protecting water resources 水资源保护的法律措施
lawful merchandise 合法商品
lawful money 法定货币
lawful occupier 法定占有人
lawful operations 合法经营

lawful property 合法财产
lawful reserve 法定准备金
lawful rights 合法权利
lawful rights and interests 合法权益
lawful tender 法定支付手段
lawful trade 合法贸易
law in force 现行法律
law in operation 现行法律
law merchant 商人惯常法
lawn 细麻布；菌苔；上等细布；草坪
lawn-and-garden equipment 草坪和园艺作业用具
lawn and planting area 绿化区
lawn belt 草坪带；草地带
lawn care business 草地保护企业
lawn clippers 草坪用剪刀
lawn drizzle 草坪喷水
lawn edger 草坪边器；草坪边修剪机
lawn grass 草坪植物
lawn ley 草场
lawn mower 割草机；剪草器；地毯剪毛机；草坪修剪机；草坪割草机；草坪剪草机
lawn mulcher 草坪覆盖器
lawn planting 草坪栽植
lawn roller 草坪辗压器；草坪滚压器
lawn sand 草坪砂
lawn sign 草坪标志牌
lawn sower 草坪撒播机
lawn sprinkler 草坪喷水器；草坪喷洒器；草地喷水器；草地喷灌器
lawn sprinkler system 草坪喷水系统；草地喷洒系统
lawn sweeper 草坪清理机
lawn top mulch 草地覆盖物
lawn trimmer 草坪修剪器
lawn trimmer and edger 草坪修剪边两用机
lawn umbrella 草坪遮阳伞
law observance study 遵守交通规则调查
law of ambient noise pollution prevention and control 环境噪声污染防治法
law of angle 角度定律
law of Archimedes 阿基米德定律
law of area 面积定律
law of association 结合律
law of audit 审计法
law of average 平均律
law of average profit 平均利润法则
law of average rate of profit 平均利润率规律
law of bankruptcy 破产法
law of basin area 流域面积定律
law of Beer 比尔定律
law of belting 带的配置定律
law of Biot-Savart-Laplace 毕奥—萨伐尔—拉普拉斯定律
law of Brewster 布儒斯特定律
law of capillarity 毛细作用定律
law of carriage of goods 货物运输法
law of causality 因果律
law of causation 因果律
law of chance 机遇率；机遇律；随机率；随机定律
law of chemical change 化学变化定律
law of colo(u)r mixing 色混合规律
law of combination of errors 误差合成定律
law of combining proportions 化合比例定律
law of combining volumes 化合体积律
law of combining weight 化合量定律
law of commercial transactions 商业交易法
law of commutation 交换律
law of company 公司法
law of comparative advantage 比较优势法则
law of comparative cost 比较成本定律
law of compatibility 相容性定律
law of compensation 补色法则；补偿法则
law of competition 竞争法则
law of concentration effect 浓度效应律
law of condenser 电容器容量变化律
law of connected vessels 连通器定律
law of conservation and conversion of energy 能量守恒和转换定律
law of conservation of angular momentum 角动量守恒定律
law of conservation of aquatic resources 水产资源保护法
law of conservation of mass 质量守恒定律；质量不灭定律
law of conservation of mass and energy 质能守

恒定律
law of conservation of matter 物质守恒定律;物质不灭定律
law of conservation of momentum 动量守恒定律
law of conservation of unemployed 失业守恒规律
law of constancy of interfacial angle 面(间)角守恒定律
law of constant angles 面间角守恒定律
law of constant angular momentum 等角动量定律
law of constant proportion 定比定律
law of construction of energy 能量守恒定律
law of consumption imitation 消费模拟法则
law of contingent punishment 权变惩罚律
law of contingent reinforcement 权变强化律
law of continuity 连续(性)定律
law of contract 合同法
law of convection 对流定律
law of cooling 冷却定律
law of correlation 相关(定)律;对比定律
law of corresponding states 相应状态定律
law of cosine 余弦定律
law of cost 生产成本法则
law of decay of turbulence 紊流衰减定律
law of decreasing cost 费用递减法则
law of decreasing returns 报酬递减法则
law of definite composition 固定成分律;定比定律
law of definite proportion 定比定律
law of degradation of energy 能量退降律
law of demand 需求(定)律
law of density turbulence 紊流衰减定律
law of development 发展规律
law of dilution 稀释定律
law of diminishing demand 需求递减法则
law of diminishing land fertility 土地肥力递减规律
law of diminishing marginal productivity 边际生产力递减律
law of diminishing marginal utility 边际效用缩减律;边际效用递减(规)律
law of diminishing returns 土地报酬递减律;返回缩减律;享受递减律;利润递减率;减少补偿定律;回收递减律;收益递减律;报酬递减律
law of diminishing returns of land 土地收益递减律
law of diminishing utility 效用递减律
law of distribution 分配(定)律;分布(定)律
law of distribution according work done 按劳分配法则
law of distribution of errors 误差分布定律
law of distribution of velocities 速度分布律
law of divide into two 两分法
law of double logarithm 叠对数定律
law of Dulong and Petit 杜郎和伯替定律
law of dynamic(al) similarity 动态相似律;动(力相)似定律
law of economic contract 经济合同法
law of economic development 经济发展规律
law of economic order of consumption 消费的经济顺序法则
law of effect 效果律
law of elastic demand 弹性需求法则
law of elasticity 弹性原则;弹性定律
law of electromagnetic induction 电磁感应定律;法拉第电磁感应定律
law of electrostatic attraction 库仑定律
law of environmental protection 环境保护法
law of equal ampereturns 等安匝定律
law of equal area 等面积定律
law of equal volumes 体积等同定律
law of equilibrium in connected vessels 连通器平衡定律
law of equi-marginal productivity 同等边际生产率的法则;边际生产均等法则
law of error propagation 误差传播定律
law of errors 误差定律
law of exchange 互换律
law of excluded middle 排中律
law of exponents 指数定律
law of extreme light path 极端光程律
law of extreme path 极端路律
law of faunal succesion 化石层序律
law office 律师事务所
law of force 力律
law of foreign capital 外资法
law of friction 摩擦定律

law of frontality 正面构图法则
law of fund movement 资金运动规律
law of gas diffusion 气体扩散定律
law of geometric(al) crystallography 几体结晶构造定律
law of granulometry 粒度分析定律
law of greatest depth 最大深度定律
law of great numbers 大数定律
law of Guldberg and waage 古德柏和瓦治定律(质量作用定律)
law of heating action of current 电流热作用定律
law of hydraulic similitude 水力相似律
law of hydrodynamic(al) similarity 流体动力学相似律
law of identity 同一律
law of imagery 成像定律
law of import 输入法规
law of incidence 赋税负担法则;负担法则
law of increasing cost 成本费递增律;成本递增规律
law of increasing returns 报酬递增法则
law of independent propagation of light 光的独立传播律
law of indestructibility of matter 物质守恒定律;物质不灭定律
law of indices 指数定律
law of inertia 惯性定律
law of isochronism 等周期定律
law of iterated logarithm 迭对数定律
law of labo(u)r safety and health 劳动安全卫生法
law of land management 土地管理法
law of land subsidence prevention 防止地面沉陷法
law of large numbers 大数定律
law of learning efficacy 学习效果律
law of least square 最小二乘律
law of least time 最短时间定律
law of life 生活规律
law of light 光吸收定律
law of limiting factor 限制因子律
law of linear permeability 线性渗透定律
law of linkage and exchange 连锁和变换规律
law of logarithmic velocity distribution 对数流速分布定律
law of logarithms 对数定律
law of magnetism 磁性定律
law of marine environment protection 海洋环境保护法
law of maritime commerce 海商法
law of maritime prize 海上捕获法
law of marque 海上捕获法
law of mass action 质量作用定律
law of mean 均值定律;平均律
law of mechanical similarity 力学相似定律
law of minimum 最小(值)定律;最低限度律
law of minimum potential energy 最小位能原理;最小位能定律
law of mixtures 混合定律
law of mobile bed sediment model 动床泥沙模型律
law of molecular concentration 分子浓度定律
law of motion 运动定律
law of multiple proportions 倍比定律
law of nations 国际公法;国际法
law of natural protection 自然保护法
law of natural resources 自然资源法
law of nature 自然界(定)律;自然规律
law of negotiable instruments 票据法
law of non-proportional returns 报酬非比例法则
law of non-specialization 非特(殊)化定律
Law of of the People's Republic of China on Wild Animal Protection 中华人民共和国野生动物保护法
law of one price 一价定律
law of overall sediment model 全沙模型律
law of parallel solenoids 平行力管定律
law of partial pressure 气体分压定律;分压(力)定律
law of partition 分配定律
law of perfect gas 理想气体定律
law of period 周期定律
law of periodicity 周期律;周期定律
law of photochemical absorption 光化吸收定律
law of photochemical equivalence 光化当量定律
law of photochemistry 光化学定律

law of photochemistry equivalence 光化学当量定律
law of photoelectricity 光电律;光电定律
law of planning 规划法
law of plastic flow 塑流定律
law of prevention and elimination of radioactive pollution 放射性污染防治法
law of priority 先取权;优先权
law of prize 捕获法
law of probability 概率定律;或然率定律
law of propagation of errors 误差传播率;误差传播定律
Law of Property Act 物业法法案
law of proportional effect 比例效应定律
law of proportionality 比例定律
law of protection of international rivers and lakes 国际河流与湖泊保护法
law of radiation 辐射定律
law of radioactive decay 放射性衰变定律
law of rational index[复] indices 有理指数定律
law of rationality 有理交截定律;整数定律
law of recapitulation 重演律
law of reciprocal demand 相互需求法则
law of reciprocal proportions 互比定律
law of reciprocity 可逆定理;互易性定律;互反律;反比定律
law of rectilinear diameters 密度中线定律
law of reflection 反射定律
law of refraction 折射定律
law of reinforcement size 强化规模律
law of relativity 效力递减律
law of resistance 电阻定律
law of responsibility and authority 权责法
law of restitution 赔偿法
law of scarcity 稀缺规律
law of sea 海洋法
law of segregation 分异定律;分离定律
law of signs 记号律;符号律
law of similarity 相似律;相似(定)律
law of similarity for dredge pump 泥泵相似定律
law of similitude 相似(定)律;同比律
law of sine 正弦定律
law of size distribution 粒度分布定律
law of small numbers 小数定律
law of soil and water conservation 水土保持法
law of solid friction 固体摩擦定律
law of special environmental protection 特殊环境保护法
law of specific heat 比热定律
law of statistic(al) constancy 统计恒性法则
law of statistics 统计法
law of storms 风暴运行定律;风暴律
law of stream gradient 河流坡降定律;河流坡度律
law of stream length 河长定律
law of substitution 代入定律
law of superimposition 叠加定律
law of superposition 叠加定理;叠覆律;叠复律;地层层序律;重叠律;叠加定律;叠加原理
law of supply and demand 供求规律;供求定律
Law of Surveying and Mapping of the People's Republic of China 中华人民共和国测绘法
law of suspended load model 悬沙模型律
law of suspended sediment model 悬沙模型律
law of symmetry 对称性定律;对称定律
law of tangent 正切定律
law of ten percent 10%定律
law of territorial resources 国土资源法
law of territory 国土法
law of the falling tendency of the rate of profit 利润率下降法则;利润率递减法则
law of the flag 船舶国籍法
law of the iterated logarithm 反对数定律
law of the mean 中值定律;平均值定律
law of the open seas 公海法
law of the parallel solenoids 平行网络定律
law of the People's Republic of China on environmental impact assessment 中华人民共和国环境影响评价
Law of the People's Republic of China on Prevention and Control of Water Pollution 中华人民共和国水污染控制法;中华人民共和国水污染防治法
Law of the People's Republic of China on Pro-

tection of Cultural Relics 中华人民共和国文物保护法
Law of the People's Republic of China on Water and Soil Conservation 中华人民共和国水土保持法
law of the place of dispatch 起运地法;启运地法
law of thermal equilibrium 热量平衡定律
law of thermodynamics 热力学定律
law of thermoneutrality 热中和定律
law of the sea 海上法(规)
law of the situation 形势规律
law of tolerance 耐性定律;(环境因素或条件的)容许极限定律
law of tort 民事侵权法
law of total current 全电流定律
law of total sediment transported model 全沙模型律
law of trade 贸易法
law of unequal slope 山坡不对称定律
law of universal gravitation 万有引力定律
law of value 价值规律
law of value without competition 非竞争的价值规律
law of variable proportions 比例变动规律;不定比例律;变动比例法
law of virtual displacement 虚位移定律;虚位移原理
law of virtual velocity 虚速度定律
law of virtual work 虚功定律;虚功原理
law of wages 工资定律
law of water and soil conservation 水土保持法
law of water pollution prevention 水污染防治法;防止水污染法
law of well building application 建井申请法
law of wild plant protection 野生植物保护法
law of zero or unity 零一律
law on transfer of technology 技术转让法
Lawrence tube 三色色标管
lawrencite 陨氯铁
lawrencium 铹
laws and regulations 法规
laws and regulations of energy 能源法规
laws covering pressure vessels 涉及压力容器的法规
laws governing special economy 特有经济规律
laws of development of national economy 国民经济发展规律
laws of Oleron 奥利伦法
laws of space and colo(u)r 空间与色彩定律;空间与色彩原理
laws of the currency of money 货币流通规律
lawsonbalumite 羟锌锰矾
Lawson chair 劳逊椅(软垫矮背附属椅)
Lawson criterion 劳逊判据
lawsonite 硬柱石
lawsonite chlorite epidote schist 硬柱石绿泥石绿帘片岩
lawsonite glaucophane albite schist 硬柱石蓝闪石钠长片岩
lawsonite glaucophane quartz schist 硬柱石蓝闪石英片岩
lawsuit 起诉;诉讼
lawyer 律师
laxar house 麻风病院
laxative effect of water 水的缓泻效应
lax business management 经营管理松弛
laxity 疏密度
lay 绞距;花纹方向;砌法;索股纹路;索股股数;草地轮作;放置;安放
lay aback 逆帆;使船后退
lay a course 定航向
lay a foundation 奠基
lay aft 到船尾就位
lay a keel 安放龙骨
lay alongside 横靠;停靠
lay-aside 储[贮]备;搁在一旁;路侧停车处;放置;备用车道
lay a switch 铺设道岔【铁】
layaway plan 先行定金;逐月付款的累积购买法
lay-back skip 背装式料斗
lay bar 平窗标;水平玻璃压条;水平玻璃压条条;玻璃横格条
lay barge 敷设管线驳船;布管驳船
lay bench 休息长凳

lay berth 闲置船舶停靠的码头;闲置泊位
lay board 托板
lay boat 闲置船舶;长驻船;长期停泊不用的船舶
layboy 自动折纸机
lay brethren (修道院里做杂役的)僧侣
lay brick 砌砖
lay-by 支线;增宽河段;路旁停车处;让船处;停航;迎风停泊;路侧停车处;会船段;单轨独头岔道;避车道;备用车道;漂流流(不抛锚)
lay-by footage 钻探报表外进尺
lay-by materials 放于路侧的备用材料;备用材料
lay-by track 让车道(道路加宽部分)
lay/cancel days 搁置/注销期
Laycock overdrive 莱科克超速器
lay corporation 非营利法人
lay crosswise 交叉向放置;交叉向布设
lay days 装卸期间;装卸货日期;窝工日;停工日;停港时间;停泊时间;停泊日期;受载期间;装修时间;港口耽搁日;搁置时间
lay day statement 货物装卸作业报告书
laydown 制定;敷设;槽孔模片铺设组
laydown arm 放板臂
laydown conveyer 放板输送机
laydown job 钻杆立根拧卸作业
laydown machine 铺设机(沥青混凝土);铺洒碾压机;摊铺机(沥青混凝土)
laydown rack 钻杆摆放架;摆管架
laydown rate 沉淀速率;沉积速率
laydown the pipe 放倒钻杆;放倒管子
laydown thickness 铺设厚度;松铺厚度
laydown true bearings 标出真方位
lay edges 空白边
lay edgewise 沿边铺砌;边对边铺砌
layer 压条枝;进货批次;焊层;铺放者;铺放机;涂层;地层;层(数);敷设机
layer against bomb 遮弹层
layerage 压条法
layerage by girdling 环状剥皮压条机
layer anisotropic media 层状各向异性介质
layer-block structure 层块构造
layer-block tectonics hypothesis 层块构造说
layer board 排水槽支承板;天沟托板
layer-build cell 分层电池
layer-built 分层铺筑的;分层
layer-built cell 成层电池
layer-built embankment 分层铺筑的(路)堤
layer-built mark 分层标
layer-by-layer manuring 分层施肥
layer-by-layer winding 逐层绕法
layer cable 分层绞合电缆
layer-cake-corrosion 夹层腐蚀
layer charging 分装料
layer chromatography 层色谱
layer closure 层位闭合
layer coefficient 层厚系数
layer colo(u)red 测高的
layer-colo(u)red edition 分层设色版
layer colo(u)ring 分层着色法
layer colo(u)rs 分层设色表
layer-colo(u)r series 分层设色系列
layer combustion 层状燃烧
layer construction 分层填土;分层建筑;分层施工
layer corrosion 片状腐蚀;地层腐蚀
layer crystal 层状晶体;层晶
layer curing 分层固化
layer cuttage division 压条扦插区
layer depth 混合层深度;层深
layered 成层的;层状;分层的
layered anorthosite 层状型斜长岩
layered bathymetric(al) chart 分层设色海底地形
layered bitumen 层状沥青
layered clay 层状黏土;分层黏土
layered cloth 垫布(研磨玻璃用)
layered colo(u)red edition 分层设色版
layered colo(u)r series 分层设色系列
layered column chromatography 层叠式柱形色谱法
layered construction 分层施工(法);多层结构;分层堆筑
layered effect 成层体系影响
layered foundation 成层地基
layered igneous rock body 层状火成岩体
layered intaking 分层取水
layered intrusion 层状侵入

layered intrusion body 层状侵入体
layered intrusive body 层状侵入体
layered laser 层状激光器
layered magnesium-aluminum hydroxide 层状氢氧化镁铝
layered map 分层着色地(形)图;分层(设色)图
layered medium 层状介质;分层媒质;分层介质
layered operating system 分层操作系统
layered pavement 成层路面
layered permafrost 层隔永冻层
layered polymer membrane 多层的聚合物膜
layered pool 层状油藏
layered reverberation 分层混响
layered sedimentation 成层沉积;分层沉淀
layered shape detector 层状探测器
layered silicates 层硅酸盐类
layered soil 成层土(壤);层状土(壤)
layered sound insulator 分层隔声材料
layered stratum 层状地层
layered structure 成层结构;成层构造;分层结构;层状结构;层状构造
layered style 分层设色表示法
layered style map 分层设色图
layered system 层状体系
layered tablets 层片
layered target 层状靶
layered upflow carbon adsorption 分层升流炭吸附
layered velocity model 分层速度模型
layered vessel 多层式容器
layer equivalency 层厚均等
layer group lithologic(al) phase chart 分岩系的岩性岩相图
layer growth 层状生长
layer impedance 阴极界面阻抗
layer induce blast 起爆层
layering 压条(法);成层;层次化;层次表示;分层(作用)
layering nodule 层状结核
layering of firedamp 瓦斯分层
layering plot 压条区
layer in slot 槽内绕组层
layer insulation 层间绝缘
layer insulation test 层间绝缘试验
layer interface 层次接口
layer ionization 电离层
layer lattice 层状晶格;层形晶格;层形点阵
layer lattice mineral 层状晶架矿物
layer lattice structure 层状晶格结构
layer line 层线
layer loading 煤炭分层装车法
layer location of anomaly emerged 异常现象出现层位
layer method 成层法
layer mining 分层开采
layer mining ploughing 深层耕作
layer misclosure 层位闭合差
layer motif(pattern) 层状基型
layer mo(u)lding 分层压制模制
layer network element 层网络单元
layer number 层号
layer number of radial pipe 辐射管层数
layer number of setting casing 套管下入层数
layer of a distributed winding 分布绕组层
layer of air 空气层
layer of ashes 灰层
layer of chippings 石屑层;碎片层
layer of clay 黏土层
layer of cloth 布绝缘层
layer of coarse gravel 粗石层
layer of compensation 补偿层
layer of dirt 污物层
layer of discontinuity 不连续层
layer of earth 土层
layer of fine sand 细砂层
layer of foam 泡沫层
layer of humus 腐殖质层
layer of insulation 隔离层;绝缘层;隔热层;防水层
layer of insulation of bridge deck 桥面防水层
layer of light liquid 轻液层
layer of no motion 无流层
layer of non-conducting materials 隔离层
layer of nuburnt gas 不燃气体层
layer of oxide 氧化层
layer of oxidem 氧化膜

layer of oxygen deficient 缺氧层
layer of oxygen minimum 最低含氧层
layer of rods and cones 杆体锥体层
layer of sand 砂层
layer of scum 泡沫层；浮渣层
layer of sediments 泥沙层
layer of strands 钢绞线层
layer of surface retention 地表滞留层
layer of suspended water 悬着水层
layer of transition 过渡层
layer of vine 插蔓
layer of wire 钢丝层(预应力混凝土)；电缆涂层
layer of wood 木质层
layer of wood(en) blocks 木垫板
layer optics 薄膜光学
layer out 画样工
layer-over-half-space 半空间上土层
layer ploughing 分层耕作
layer position hole 定位钻孔
layer press 层压机
layer protocol 层次协议
layer-replaced stack 分层替代叠加
layer resistance 层电阻
layer resolution densilog 层分辨率密度测井
layer resolution density log 层分辨率密度测井
layer roller track 分层辊道
layers 层次
layer short circuit 层间短路
layer silicate 层(状结构)硅酸盐；硅酸盐层
layer silicate structure 硅酸盐层状结构
layers of wood(en) blocks 木垛(砖)层
layer-sphere structure 层圈构造
layer splice 层编接
layer spread method 层铺法
layer steps 分层设色表
layer stripping ratio 分层剥采比
layer structure 层状组织；层状结构；层状构造
layer test 层间绝缘试验
layer thickness 层厚(度)
layer thickness ga(u)ge 膜层测厚仪
layer thickness meter 膜层测定仪；镀层测厚仪
layer thickness of dobie blasting 裸爆炸层厚度
layer thickness of mass concrete 大体积混凝土浇筑层厚度
layer-tinted map 分层设色图
layer tinting 分层着色法
layer tints 分层设色表
layer to layer 逐层
layer to layer signal transfer 层间信号传递
layer to layer transfer 层间转换
layer topology 分层拓扑学
layer type cable 分层式电缆
layer water 间层水
layer winding 层状晶格；分层绕组；分层绕法
layerwise summation method 分层总和法(计算地基沉降)
layer-wound solenoid 层绕螺线管
lay-fit method 嵌铺法
lay-flat 平折性
lay-flat adhesive 平服黏合剂
lay-flat tube irrigation 平铺管喷灌
lay in 储[贮]藏；收起
lay in a stock of merchandise 进货
lay in ceiling 嵌入式顶棚
laying 瞄准；埋设；铺设；砌筑；衬垫层；敷设
laying and finishing machine 摊铺修整机；铺洒碾压机
laying arch by sections 分段砌拱
laying arch continuously 连续砌拱
laying area 砌筑面积；敷设面积；铺设面积
laying attachment 管道铺设附属设备；铺管道附件
laying barge 铺管船；铺管驳船
laying battery 多层产蛋鸡笼
laying blocks 砌石
laying bricks 砖砌；巧工砌造；砖工砌造
laying buoy 铺管浮筒
laying cat 铺管履带拖拉机
laying cloth 铺地板
laying corner 套印规矩线
laying cost 铺设费用；安装费用
laying crawler tractor 铺管履带拖拉机
laying crew 铺管班
laying depth 埋置深度；铺筑厚度；敷设深度
laying down 造船几何画法；放样

laying down of keel 龙骨安置
laying drawing 铺设图；图面布置；布置图(管道、道路等)
laying embargo on ships 禁止船只出入
laying end 装玻璃的研磨台；铺放
laying error 测设误差
laying gang 铺管班
laying guide (印在卷材上的)搭接线；铺设(导)线
laying in a switch 道岔铺设
laying in knitted fabric 衬垫针织物
laying in stocks 采办货物
laying in thread 衬垫纱
laying length 铺设长度；铺管长度；敷设长度
laying line (印在卷材上的)搭接线；铺设(导)线
laying loaded 负荷层
laying machine 码纸机；铺路机；搓绳机；敷设机械；铺设机械
laying material 铺设材料；底层材料
laying mode of route 线路敷设方式
laying of cables 敷设电缆；电缆敷设
laying off 下料；造船几何画法；造材检量；掸刷；放样；停工；下料；擦勺(在新刷油漆表面)；拉平
laying off an angle 角度测设
laying of mosaics 铺马赛克
laying of pipe 管道铺设；铺管；埋管
laying of pipeline 管道敷设
laying of roadway 路面铺筑
laying of secondary track 复线铺设
laying of second track 修建第二线；铺复线；铺第二线
laying of setts 块料铺砌(路面)
laying of third track 修建第三线
laying on cloth 垫布(研磨玻璃用)
laying on plaster 上石膏垫料
laying on trowel 镘刀；泥刀
laying operation 铺管作业
laying out 画样划线；画样工；号料；敷设(线路)；放样；定线；工程放线；布置；规划
laying-out a line 划线
laying-out bench 划线台；定位台
laying-out land 规划土地；土地布置
laying-out table 划线台
laying-out the tunnel axis 隧道轴线之标定
laying-out wire 定线概尺
laying party 铺管班
laying pattern 铺砌用模型板；铺设格式
laying pipe 铺设管道；敷设管道
laying plan 铺管计划；铺设计划
laying point 测设点；铺设地点
laying rate 铺砌速度；浇筑速度；铺设速度
laying reel 中心出料式卷线机；铺料式卷线机
laying ship 管道铺设船；铺管船
laying site 铺管工地；铺管现场
laying subbase 铺设基础底层
laying technique 铺砌技术；铺设方法；浇筑方法；砌筑方法；铺设技术；浇筑技术；砌筑技术
laying temperature (道路油面的)铺筑温度；铺设温度；摊铺温度；浇筑温度
laying tile 砌砖；敷设面砖
laying time of observation net 观测网设置时间
laying to bond 整砖砌合
laying tool 铺管工具
laying tractor 吊管机
laying trowel 刮刀；砌砖泥刀；大灰抹子；瓦刀
laying up 铺放；衬垫；铺网
laying-up a rivet 敲进铆钉
laying-up basin 停船港池
laying-up berth 停船港池
laying-up machine 合绳机
laying-up reel 合绳机工字轮
laying-up returns 退还停泊保险费
laying vehicle 铺管车
laying width 铺设宽度
laying winch 铺设绞车(管道)；铺管绞车；管道敷设绞盘
laying work 铺管作业；砌筑工作；浇筑工作
laying yard 装玻璃的研磨台；铺放
lay-in panel 镶嵌墙板
lay-in section 入口断面
lay-in timber connector 暗藏木接点
lay-in trim 入口装饰
lay-in unit 入口单元
lay-in wood connector 暗藏木接点
lay land 生荒地；处女地；撂荒地

lay lea 轮作草地
lay light 间接采光；平面采光；顶棚采光；吊平顶窗；天花板上水平窗；顶棚窗；平顶灯
layman 门外汉；外行人；外行；非专业人员
lay marks 规矩线
lay off 暂时解雇；临时解雇；停在港外；停工期间
lay-off in winter 冬季停工
lay off rate 暂时解雇率；解雇率
lay of land 地形
lay of line 路线
lay of rope 钢丝绳的搓捻法
lay of wire 芯线绞距
lay on 征税；涂(油漆、颜料、水泥等)；安装(水电、煤气、电话等)
lay-on a given compass 按磁方位角定向
lay-on-air dryer 热风气垫式烘燥机
lay-on roller 压带轮
lay-on the line 筹足(款项)
lay open 暴露；揭开
layout 规划；划线；放样；形式；总体布置；总布置图；格式；建筑布置图；号料；配置(图)；电路设计的配置；场地布置规划；敷设线路；布置；布局(配置)；版面编排
layout and location 配置和布局
layout arrangement of turnout 道岔配列
layout block 划线平台
layout board 设计板
layout by fixed position 按固定位置布置
layout by operation 操作布置
layout by process 分类布置
layout character 格式控制字符；划分字符；排位置字符；打印格式符号；布置字符；布局控制字符
layout chart 线路图；观测系统图；流程图；平面布置图；施工图；布置图
layout check subsystem 格式检查子系统
layout circle 配置圆
layout constant 格式常数；数位分配常数
layout data 格式数据
layout design 平面设计；图纸设计；电路图设计；草图设计；布置设计
layout drawing 配线图；草图；布置图
layout flow chart 布置流程图
layout for artificial harbo(u)r 人工港布置
layout for drilling 钻孔划线
layout for road and transportation in plant area 总图运输
layout grid 坐标网；方格网；布局
layout in field 现场定测
layout line 区划线；配置线
layout location 定线
layout machine 展示机；定线仪；测绘缩放仪
layout man 下料工；划线工
layout map 布置图
layout of aids-to-navigation 航标配布
layout of airport 机场布置
layout of boreholes 钻孔布置
layout of bridge 桥梁布置
layout of bridge style 桥式布置
layout of chart of aids 航标配布图
layout of circular curves 圆曲线放样
layout of column grid 柱网布置
layout of construction work 施工场地布置
layout of control 控制网布置
layout of curve 曲线测设
layout of details 细节布置
layout of drilling equipment 机场布置【岩】
layout of equipment 设备配置图；设备布置(图)
layout of foundation 基础放样
layout of furniture 家具布置
layout of hydrojunction 枢纽布置(图)
layout of lighting 照明平面图
layout of lintel arrangement 过梁平面布置图
layout of pipeline 管线布置
layout of plan 设计方案；编制设计书
layout of port 港口布局
layout of powerhouse 电站厂房布置
layout of power station 电站布置
layout of production line 生产线布置
layout of regulating line 整治线布置
layout of regulation line 整治线布置
layout of reinforcement 加固方案；钢筋布置
layout of road 道路布置；道路规划；道路布设方案
layout of round 一组炮眼布置；炮眼组布置
layout of route 放线

layout of sewerage system 排水管道布置
layout of station 选址;车站(线路房屋)布置图;车站布局;测站设计;测站布置
layout of steps 梯级布置
layout of warehouse 仓库布置
layout of waters 理水
layout plan 总图(布置)设计;总平面图;规划;设计;建筑布置图;平面布置总图;平面(布置)图;平面布置设计;定线图;场地布置规划;布置图;布局图;发展蓝图
layout plan of station 车站股道配制图;车站布置图
layout procedure 配置过程;设计程序
layout road 道路布置;道路规划
layout sheet 总规划图;总布置图;平面图
layout study 布置研究
layout survey 施工放样测量;定线测量
layout table 刻图桌
layout(work) 设计图案
layover (公共交通的)终点停车处;(旅行中的)逗留期间;(旅行中的)中断期间
layover time 终点站停车时间
lay panel 门心板;横纹镶板;平纹镶板;水平镶板
lay plate 分线板
lay ratio 绞距系数;绞距比;扭绞系数
Layrub universal joint 莱鲁布万向接头
layshaft 副轴;变速器副轴;中间轴;侧轴;并置轴
layshaft bearing 中间轴承
layshaft drive gear 中间轴主动齿轮
layshaft gear cluster 中间轴齿轮组
layshaft of geared head 车头箱中间轴
lay sheet 开印试纸
lay ship 闲置船舶
lay stall 垃圾堆
lay sword 筘座脚
lay the grain 整光
lay the keel of a ship 安放龙骨
lay the land 航行到看不见岸的地方
lay the line 敷设管线
laytime 停泊时间;装卸期间;装卸货时间
lay to 掉头向风行船
lay tongs 长柄管钳
lay turn-out 设置岔道(道路);设置分岔(道路)
lay under contribution 强迫缴税
lay up 搁置不用;树脂浸渍增强材料;长期航,敷层;铺砌;铺叠;铺层;绞合;树脂浸渍增强材料;接头;接合处
lay-up mo(u)lding 铺层成型
layup procedure 增强塑料敷层方法
lay up refund 停航退费
lay up return 停泊退费
lay-up room 合片室
lay vessel 闲置船舶
lay waterproof skin 铺防水(面)层
lay wire 模网
lazaret(to)检疫船;传染病医院;近船尾储[贮]藏室;检疫所;麻疯病院;船上的传染病室;甲板间储[贮]藏室;检疫站;隔离病院;隔离病房
lazarevicite 方辉砷铜矿
lazar house 传染病医院
laze flame 弱焰
lazies 点纹花边
laziness of the bubble 气泡迟滞
lazuli 杂青金石
lazuline 暗蓝色的
lazulite 天蓝石
lazurfeldspar 蓝正长石
lazurite 青金石;天青石
lazy arm 吊臂
lazy bar 火炉工具挂杆
lazy board 木制支架;垫板(铺管用)
lazy element 惰性元素
lazy flow 低流动性
lazy guy 吊杆稳索
lazy guy method 锤出吊货法
lazy halyard 小升降绳
lazy H antenna 双偶极子H天线
lazy jack 屈伸起重机;补偿联动器;系列杠杆
lazy line 轻便缆
lazy painter 轻便(的)艇首缆
lazy pinion 空转小齿轮;空载小齿轮
lazy stream 缓流
lazy strike 怠工
lazy Susan 转动餐盘;餐桌转盘;橱柜旋转台
lazy thermometer 惰性温度计;惰性温度表

lazy tongs 多铰式伸缩钳;打捞工具;同步机构;惰钳
lb 磅
L-band radar L波段雷达
L-band radiometer L波段辐射计
L-bar 角铁;角钢;角板
L-beam L形梁;不等边角钢
Lbe wind 尔伯风
L-block separation L形耐火砖分隔
LCL team yard 零担货场
L-column L形柱(预制混凝土框架)
LD50 test 致命(死)量50%试验
LD model 赖特—邓肯模型
l-dodecanol 十二醇
lea 缕;草地轮作
Lea and Nurse permeability apparatus 李和诺瑟渗透性测定装置
Lea and Parker lime saturation factor 李和派克石灰饱和系数
lea body 草地犁体
leach 滤灰槽;浸析;浸提液;浸出(液);淋溶
leachability 可溶出性;可浸出性;可淋溶性
leachable 可滤取
leachant 溶浸剂
leachate 沥滤液;沥滤产物
leachate characteristic 沥滤液特征
leachate circulation 沥滤液循环
leachate collection system 沥滤液收集系统
leachate concentration 沥滤液浓度
leachate polluted groundwater 沥滤液污染的地下水
leachate re-circulation 沥滤液再循环
leachate re-cycling 沥滤液再循环
leachates 滤出物
leachate treatment 沥滤液处理
leach chemistry 溶浸化学
leached alkali soil 淋溶碱土
leached brown earth 强淋溶土;淋溶出的褐色土
leached chernozem 淋溶黑钙土
leached glass 浸渍过的玻璃
leached glass fiber[fibre] 酸沥滤玻璃纤维;高硅氧玻璃纤维
leached horizon 淋溶层
leached image guide 酸溶传像束
leached layer 淋溶层;淋滤层
leached-replacement pool trend of carbonate rock 溶浊交代型碳酸盐岩油气藏趋向带
leached salt cavern 溶盐空穴
leached soil 淋溶土;土余土
leached surface 溶滤面;渗漏面;沥滤表面
leached trap 溶滤圈闭
leached zone 淋溶带;淋滤带;溶滤带
leacher 浸取器
leachine 淋溶
leaching 滤出;淋洗;淋滤;沥取;浸析作用;溶滤作用;浸滤;浸沥;浸出
leaching agent 助滤剂;浸出剂;溶浸剂;助沥滤剂
leaching agitator 溶浸搅拌机
leaching basin 过滤池;滤水池
leaching bed 污水渗床;渗沟;渗床
leaching by agitation 搅溶法
leaching cesspool 沥滤污水坑;污水渗坑;污水渗井;渗漏污水池
leaching concentrate 浸出精矿
leaching corrosion time 淋溶时间
leaching-crystalization process 溶浸结晶法
leaching effect 淋溶作用
leaching erosion 淋溶侵蚀
leaching estimation and chemistry model-pesticide 农药溶化估算与化学模型
leaching evaluation of agricultural chemical methodology 农用化学物溶化评估法
leaching experiment 淋滤实验
leaching experiment of dissolvable salt of soil 土壤可溶盐淋滤实验
leaching factor 淋洗因数
leaching field 沥滤场;浸湿范围;渗水场地
leaching fraction 溶化分数
leaching halo 淋滤晕范围
leaching horizon 淋溶层
leaching into a soil 渗入土壤
leaching layer 淋溶层
leaching liquor 沥滤液
leaching loss 淋失;沥滤损失
leaching materials of local environmental science 乡土环境教材
leaching of calcium 钙的淋溶
leaching of cement paste 水泥浆滤出;水泥浆渗漏(由酸溶液等使水化物分解)
leaching of glass 玻璃的浸析作用
leaching of nitrate 硝酸盐的淋溶
leaching operation 淋洗作业
leaching-out 洗出;沥滤出;浸出
leaching pit 污水坑;渗水井坑;渗滤坑
leaching process 沥滤法
leaching prosess 溶浸工序
leaching rate 渗毒率;渗出率;毒料渗出率
leaching ratio 淋溶率
leaching requirement 沥滤条件;淋溶溶水量;沥滤要求;(土壤的)冲洗需水量
leaching solution 沥滤液;沥滤溶液;浸提液;浸出液
leaching test 流失试验;淋溶试验
leaching trench 渗滤沟
leaching water 淋洗水;溶滤水
leaching well 沥(滤)水井;渗(水)井
leaching zone of groundwater 熔滤潜水带
leach liquor 浸提废液
leach material 可沥滤矿物;溶浸物
leach pile 沥滤法回收矿物堆
leach pit 污水渗坑;渗坑
leach pit for sewage 污水渗坑
leach-precipitation 溶浸沉淀
leach-precipitation-flo(a)tation 沥滤-沉淀浮选选矿法
leach residue 浸取残渣;浸出残渣
lead 自然铅;领先;前置;铅测锤;牵引方向;提前量;膛线周径;水砣;三注意定位及了望;三注意测深;导引;导向柱;导联;导路;导线;导距;导架;导程;传爆元件;超前角;引线;超前
lead absorber (船尾的)铅吸收体
lead accumulation 铅累积
lead accumulator 铅蓄电池
lead acetate 乙酸铅;醋酸铅
lead acetate medium 醋酸铅培养基
lead acetate method 乙酸铅法
lead acetate poisoning 醋酸铅中毒
lead-acetate test 乙酸铅试验
lead-acid battery 铅酸电池;铅酸电池
lead-acid cell 铅酸蓄电池;铅酸电池
lead-acid cobat battery 铅酸钴电池
lead acid storage battery 铅酸蓄电池
lead acid storage battery for navigation aids 航标用铅酸蓄电池
lead acoumulator 铅蓄电池
lead additive 铅添加剂
lead aerosol 铅质气溶胶
lead ages 铅法年龄
lead alkali glass 铅碱玻璃
lead alloy 铅合金
lead-alloy cable sheath 铅合金电缆包皮
lead-alloy pipe 铅合金管
lead-alloy sheath 铅合金包皮
lead-alloy-sheathed cable 铅合金包皮电缆
lead-alloy sheathing 铅合金皮
lead-alloy tube 铅合金管
lead-alpha age method 铅—阿发粒子年代测定法
lead-alpha method 阿尔法铅法
lead aluminate 铝酸铅
leadamalgam 汞铅矿
lead and lag 超前和滞后(国际收支)
lead and oil 铅白与亚麻籽油(作白油漆用)
lead-and-oil paint 油铅油漆
lead and taper ga(u)ge 扣齿啮合计
lead angle 螺旋升角;前置角;导前角;导程角;超越角;超前角;投弹角
lead-antiknock additive 铅基抗爆震添加剂
lead antimonite 锑酸铅
lead apron 铅护裙;铅防护围裙
lead arsenate 砷酸氢铅
lead aryl 芳基铅
lead arylide 芳基铅
lead ash 铅灰;铅砂
lead attachment 引线焊接;导引附件;导向附件
lead azide 氮化铅;叠氮化铅
lead azide alumin(i)um detonator 叠氮化铅铅雷管
lead back 铅背滨鹬
lead back lath 铅梢条板
lead ballast 铅压载

lead bank 牵头银行
lead bar 铅条
lead barium glass 铅钡玻璃
lead base alloy 铅基合金
lead-base anti-friction alloy 铅基抗摩擦合金
lead-base babbit metal 铅基轴承合金
lead-base babbitt 铅基巴氏合金
lead-based additive 含铅添加剂
lead-based coatings 铅基涂层
lead-base grease 铅皂润滑脂
lead-base paint 铅基油漆
lead-base(priming) paint 铅基漆;铅基底漆
lead-base white metal linings 铅基白合金衬层
lead bat 铅楔;墙上固定铅泛水的铅楔块
lead bath 铅淬火槽;铅淬火;镀铅槽
lead battery 酸性蓄电池组
lead battery metal 蓄电池铅合金
lead beam 引导光束
lead-bearing 含铅
lead-bearing enamel 含铅釉
lead-bearing steel 含铅钢
lead bell type flush(ing) tank 铅按钮式冲水箱
lead-belt screen 惯性圆盘式振动筛
lead biologic(al) shied 生物防护铅屏;铅生物防护屏蔽
lead biologic(al) shielding wall 铅生物防护墙
lead bit 定向钻头;导向钻头
lead blade of bit 钻头的导向翼
lead block 导向滑轮;铅柱;铅块;水砣滑车
lead block compression test 铅柱压缩试验
lead block expansion test 铅柱膨胀试验
lead blue 带灰色的碱式硫酸铅蓝
lead bonding 引线焊接;引线接合
lead borate 硼酸铅
lead borate glaze 铅硼釉;硼酸铅釉
lead borosilicate 硼硅酸铅
lead borosilicate glaze 硼硅酸铅釉
lead bottom 测斜仪打印器
lead box 出线盒【电】
lead box rainwater gutter 铅槽雨水沟
lead brass 铅黄铜
lead bronze 铅青铜
lead bronze wrapped bush 包铅青铜衬套
lead bullion 铅锭;粗铅
lead burned joint 铅焊接头
lead burner 管(子)工;铅(管)工
lead burning 熔铅结合;铅焊
lead bushing 引线孔板;引线导管;引线导板
lead button 铅销;铅质垫片;铅制间隔物;铅制定位物;铅制垫片
lead cable 铅包电缆;引(出)线电缆
lead-cable automatic guidance system 引导电缆式自动导向系统
lead cable press 电缆铅包压机;电缆铅包皮压力机
lead cable sheath 电缆铅包皮
lead came 铅棂条;镶嵌窗玻璃有檐铅条
lead capacitance 引线电容;导线电容
lead-capped nail 铅帽钉;铅头钉
lead capping 铅帽盖;铅压(封)顶
lead carbonate 碳酸铅
lead casing 铅包装
lead cast 重铸铅板;纸板重铸铅板;铅铸法
lead castle 铅堡
lead ca(u)lking 铅封;铅填缝;灌铅
lead cesspool 铅雨水斗
lead chamber 铅室
lead chamber crystals 铅室结晶
lead chamber gases 铅室气
lead chamber pan 铅室底盘
lead chamber process 铅室法
lead chamber space 铅室容积
lead charge 传爆药
lead chart 提前量图表
lead chloride 氯化铅
lead chromate 铬酸铅;铬黄;铅铬黄
lead chromate coated pigment 铬酸铅包膜颜料;铬酸铅包核颜料
lead chromate pigment 铬酸铅颜料
lead chrome green 铅铬绿
lead chromes 铅铬(黄橙色颜料)
lead-circuit 超前电路
lead clamp 引线夹
lead closet flush(ing) pipe 厕所冲水管;厕所铅下水管

lead-clothed 包铅的
lead-clothed glazing bar 铅包玻璃窗格条
lead-coated 包铅的;涂铅的;镀铅的
lead-coated copper flashing(piece) 包铅的铜防水片;包铅的铜泛水板
lead-coated copper sheet 涂铅铜板;铅皮铜板;铅包铜板
lead-coated copper strip 涂铅铜带
lead-coated metal 涂铅的金属
lead-coated pipe 包铅管;涂铅管
lead-coated sheet 镀铅薄钢板
lead-coated steel plate 镀铅钢板
lead-coating 铅皮;包铅
lead coil 铅旋管;铅盘管
lead collar 铅环;出线套
lead collimator 铅准直器
lead compass 铅笔圆规
lead compensation 超前校正;超前补偿
lead compensator 超前补偿器
lead compound 先导化合物;铅化合物
lead computer 提前量计算尺
lead concentrate 精炼铅
lead connection 引线连接
lead content 含铅量
lead control 导向控制;导向调节;导数调节;超前控制
lead coping 铅帽盖;铅顶盖
lead copper matte 铅铜锍;铅冰铜
lead core door 铅皮门
lead corrosion 铅腐蚀
lead corrosion test 铅腐蚀试验
lead coulombmeter 铅极电量计
lead-covered 铅的;镀铅的;包铅的
lead-covered armo(u)red cable 铅包加强的电缆
lead-covered cable 铅皮电缆;包铅电缆;铅包电缆;引出线电缆
lead-covered cupola 铅覆盖的穹顶
lead-covered dome 铅覆盖的穹隆
lead-covered glazing bar 铅包玻璃窗格条
lead-covered insulated cable 铅包绝缘电缆
lead-covered wire 铅皮线;铅包线
lead covering 铅包皮
lead crystal(glass) 铅晶质玻璃
lead cupola 包铅圆屋顶
lead curtain of chamber 铅墙
lead curve 导向曲线;导曲线
lead cutting edge 主切削刃
lead cyanamide 氰氨化铅
lead cyanamide anti-rust paint 氰氨化铅防锈漆
lead cyanamide rustproof paint 氰氨化铅防锈漆
lead cyanide 氰化铅
lead cycle 铅循环
lead cylinder 铅圆筒;铅汽缸
lead damp course 铅皮防水层;铅皮防潮层
lead data 前导数据
lead deposit 铅沉积
lead dialkyl 二烷基铅
lead dibromide 二溴化铅
lead dichloride 二氯化铅
lead dichromate 重铬酸铅
lead dimethide 二甲基铅
lead dimethyl 二甲基铅
lead dioxide 二氧化铅
lead dioxide method 二氧化铅法
lead discharge pipe 铅排水管
lead disilicate 二硅酸铅
lead dome 包铅圆屋顶
lead dot 铅栓;铅销;铅铸销(固定屋瓦用)
lead dowel 铅销钉;铅榫
lead draining pipe 排水铅管;铅排水管
lead drier 铅干料;铅催干剂
lead dross 熔渣
lead dust 铅灰;铅尘
leaded 加铅的;含铅;填铅;包铅的;镀铅的
leaded alloy 铅合金;加铅合金
leaded babbitt alloys 巴比铅合金
leaded beveled plate (铅条镶的)斜边彩画玻璃
leaded brass 加铅黄铜
leaded bronze 加铅青铜;含铅青铜
leaded fuel 加铅燃料;含铅燃油;含铅燃料
leaded gasoline 加铅汽油;含铅汽油
leaded glass 镶铅玻璃;铅条玻璃;(铅条镶嵌的)窗玻璃
leaded gun metal 含铅炮铜

leaded joint 铅封接
leaded light 花饰铅条窗;铅条(镶嵌)玻璃窗
leaded low-phosphorus steel 含铅低磷钢
leaded paint 含铅涂料
leaded panel 彩色玻璃板
leaded screen wire brass 筛网铅黄铜
leaded tin bronze 低铅锡青铜(铅低于0.5%)
leaded zinc 含铅氧化锌;含铅锌白
leaded zinc oxide 含铅氧化锌;铅锌白
lead elbow 铅管弯头;铅弯头
leaden 铅制的;铅色的;天色灰暗
lead encased 铅包外壳的
lead encasing of hose 软管铅接合
lead engineer 首席工程师
leaden washer 铅制垫圈
leader 向导船;引头;引导部分;主干;指挥员;落水管;领舰;领机;胶片前导;特价品;首项;首部记录;首部;导鱼堤;导引部分;导水管;导脉;导杆;引线
leader bud 主轴芽
leader cable 导行电缆;导线电缆;导航电缆(在港道水底);引线(电缆)
leader cam 靠模凸轮
leader card 标题卡(片)
leader drain 主排水沟;主排水管
leader drum 水落管圆筒
leader extension 水落管接长部分
leader film 引导胶片;牵引片
leader head 雨水斗;落水斗;水落斗;排水管水头
leader hook 水落管箍
leader hopper 水落斗
leader label 带帖标记
leader line 引线;带箭头指引线;旁注线;引出线
leader mill 精整轧机
leader of crew 井队长
leader pass 精轧前孔型;成品前孔
leader pilot 引导
leader pin 导销
leader pipe 水落管
leader record 引导记录;标题记录
leader rolls 成品前机座的轧辊
lead error 螺距误差;导程误差
leader rotate type pile driver 导架回转式打桩机
leader-rubber gloves 铅橡胶手套
leadership 领导能力
leader shoe 雨水管出水弯头;水落管鞋
leaders in their chosen field of learning 学科带头人
leaderstone 脉壁泥
leader strap 雨水管卡子;落水管卡;水落(管)固定带
leader stroke 先导闪击
leader vehicle 领头车辆;前车
lead facing 铅涂面层;铅贴面
lead fatigue test 引线疲劳试验
lead file 铅锉
lead-filled epoxy resin 铅环氧树脂
lead filler 铅衬垫;铅垫板;铅填充物
lead finder 泄漏探测器
lead flake 铅片
lead flashing(piece) 铅泛水片;铅披水片;铅皮泛水;铅泛水条
lead flat 铅皮平屋顶
lead flat roof 铅皮平屋顶
lead float file 扁铅锉
lead fluoride 氟化铅
lead fluorosilicate 氟硅酸铅;铅氟硅酸盐
lead flushing pipe 冲水铅管;铅冲水管;铅下水管
lead foil 铅箔
lead foil screen 铅屏蔽;铅箔增感屏
lead fouling 铅沉积
lead foundry 铅铸造厂
lead frame bond 引线框式键合
lead-free 无铅的;不含四乙铅的
lead-free automobile fuel 无铅车用燃料
lead-free fuel 无铅燃料
lead free gasoline 无铅汽油
lead-free glaze 无铅釉
lead-free paint 无铅油漆(适用于存放食物等且易受铅沾污之处);无铅涂料;无铅漆;无铅涂漆
lead-free pigment 无铅颜料
lead free toy coating 无铅玩具涂料
lead frit 铅熔块
lead fuse(wire) 铅熔丝;铅保险丝;保险铅丝
lead gasket 铅封;铅垫圈(防水表壳内用)

lead ga(u)ge 导程检查仪
lead glance 方铅矿
lead glance pulp 方铅矿纸浆
lead glass 含铅玻璃;铅玻璃;铅条玻璃窗
lead glass fiber 铅玻璃纤维
lead glaze(d finish)铅釉(条镶嵌玻璃)
lead glazing 铅条(镶嵌)玻璃;上铅釉
lead gloves 铅手套
lead gray 铅灰色
lead grease 铅皂润滑脂
lead grey 铅灰色
lead grid 电池铅板
lead grill 电池铅板
lead gutter 铅檐沟;铅皮天沟;铅排水沟
lead halide 卤化铅
lead hammer 铅锤
lead hat 铅帽
lead head nail 铅头钉
leadhillite 硫碳铅矿
lead hinge 铅铰链
lead hip 铅斜屋脊;铅屋脊
lead holder 铅芯架
lead hole 导向孔;导孔
lead hydrate 氢氧化铅
lead hydrogan sulfate 硫酸氢铅
lead hydroxide 氢氧化铅
lead hydroxyl carbonate 碳酸羟铅
lead hypophosphite 次磷酸铅
lead impedance 引线阻抗
lead-in 引入线;首字;输入端;引入
lead-in-air indicator 空气含铅量指示器
lead-in brush 引入电刷
lead-in bushing 引入套管
lead-in clamp 引入线夹
lead-in conductor 引入导线;引入线
lead-in dolphin 绞缆墩;导墩
lead inductance 引线电感;导线电感
leading 加铅(条);加空铅;前移量;导行;导前;导向的;超前
leading ahead 拎头
leading anchorage device 墩头锚具
leading and lading 应收账款的挪后补前
leading article 特价品;社论
leading axle 引导轴;前轴;导轴
leading band 仰斜矿房
leading beacon 定向标;导向标;导航标;导标;导航叠标
leading black 超前黑色
leading block 导向滑轮;导向滑车;导(向)块
leading block eyeplate 导向滑车眼板
leading bogie 前转向架
leading brush 引前电刷
leading buoy 导航浮标
leading cadre of float-and-sink 浮沉试验负责人
leading cadre of laboratory 化验室负责人
leading cadre of sizing test 筛分试验负责人
leading chain 领头链
leading channel 进口水道
leading coach 引导客车
leading coefficient 首项系数
leading collector 进水干管;进口干道
leading commodities 主要货物
leading control 打头控制
leading crank 导曲柄
leading crop 主要作物
leading current 导前电流;超前电流
leading decision 前导判定
leading diabolo roller 导向凹面镇压轮
leading diagonal 主对角线
leading dimension 主要尺寸;当前维
leading down wire 引上线
leading draughtsman 主要绘图员
leading edge 进气边;前缘;前沿;上升边;导边;穿孔卡前端;门窗扇开关侧边;门窗锁安锁的一边
leading edge during driving 导向边(打桩时)
leading edge locus 前缘轨迹
leading edge of a shield 盾构前沿;盾构前缘
leading edge overshoot 前沿过冲
leading edge pulse time 脉冲前沿时间;脉冲前沿上升时间
leading edge septum 前沿切割板
leading edge time 前沿时间
leading edge tracking 前沿跟踪
leading effect 超前效应

leading electrode 引导电极
leading end 前端;引导端
leading-end acceleration 前端加速度
leading-end clamp 前端夹板
leading exploration line 主导勘探线
leading exporter 主要出口国
leading face 迎水面(叶片);正面;工作面;超前工作面
leading factor 主导因子
leading fairway of lock 船闸引航道
leading feature 主要特征;主要特性
leading flange 导轮缘
leading fossil 主导化石;标准化石
leading frame 主构件;槽板(施工放线用);主(导)构架
leading ghost 超前重影;超前重影像
leading graphic 标题图形;前导图形字符;引导图形符
leading green signal phase 前期绿灯信号显示
Leading Group Environment(al) Protection under the State Council of China 中国国务院所属环境保护领导小组
leading half axle 前进半轴(拖拉机转弯时)
leading hand 第一把手;领班;工长
leading heading 超前平巷;超前煤巷
leading hook 前端弯钩
leading-in 引入
leading-in box 引入箱;进线盒
leading-in bracket 进线架
leading-in cable 引入电缆;引线电缆;进局电缆
leading indicator 先导指标数字;先行指数;先行经济指标;超前指标
leading industry 先导产业
leading-in end 引入端
leading-in equipment 引船进坞设备
leading-in gear 引船进坞装置
leading-in girder 导梁
leading-in insulator 引入绝缘管
leading-in line 进站线(路);进场线路
leading-in phase 超前项;超前相
leading-in pole 引入杆
leading-in terminal 引入端子
leading-in track 进站线(路)
leading-in tube 引入绝缘管
leading-in wire 引入线
leading ion 主要离子
leading-isotope age 铅同位素年代
leading jetty 船闸导航墙;导堤
leading knitting system 主成圈系统
leading levee 导堤
leading light 引导灯(桩);后桅灯;导向灯标;导航(叠)标灯;导(航)灯
leading line 叠标导航线;导向线;导索;导航线
leading line of bearing from an objective beyond chart 方位引示线
leading line of navigation 导航线
leading load 电容性负荷;电流超前的(负)载;超前负载
leading locomotive 本务机车【铁】
leading manufacturer 主要厂商
leading mark 认识货物的标志;道路标志;方向标;叠标(导航);导(航)标
leading market 主要市场
leading marks method 导标法;导标断面测深法;导航标法
leading minor 主子式
leading-off rods 引出杆
leading ore 铅矿
leading-out 引出
leading-out end 引出端
leading-out terminal 引出端(子);出线端
leading-out wire 引出线
leading oval pass 成品前椭圆孔型
leading pass 终轧前孔型
leading peak 前沿峰;谱峰伸前;谱带伸长
leading pedestrian phase 行人先走信号显示
leading phase 超前项;超前相
leading phase angle 相位超前角;超前角
leading phase feeding section 引前相供电臂
leading phase operation 进相运行
leading pile 定位桩;导桩
leading pipe 导管
leading place 超前工作面
leading plug 前置火花塞
leading pole 导磁极

leading pole tip 磁极前端
leading port authorities 主要港务管理机构
leading portion 前段;前导部分
leading power factor 超前功率因数
leading radiator 前缘散热器
leading ramp 引入滑轨
leading ratio 铅比
leading roller 导轮
leading room 主要洞室;急救洞室
lead-in groove 引入螺线;引入槽;导入螺线
leading schedule 主要附表
leading screw 导杆
leading screw lathe 丝杠车床
leading screw support 导螺杆轴承
leading sheave 导向滑轮;导轮
leading ship 领导船;导航船
leading shoot 主梢
leading snow fence 防雪栅
leading spiral 导入螺线
leading spurious signal 超前乱真信号
leading stone 极磁铁矿
leading support 超前支护
leading surface 先导工作面
leading tap 螺母丝锥
leading tape 引带
leading tectonic system 主导构造体系
leading time 超前时间
leading transient 前沿瞬变
leading truck 前转角架;前进履带;前导转向架;导向架
leading turn 绿前转弯
leading up cable 引上电缆
leading uranium age method 铅铀年代测定法
leading uranium ratio 铅铀比
leading variety 当家品种
leading vehicle 前导车;头车
leading voltage 超前电压
leading wagon 引导货车
leading wall 前壁
leading wattless load 无功电容性负荷
leading wave 前缘波;头波
leading wheel 主动轮;驱动轮;导轮
leading wheelsets 前导轮组
leading white 超前白色
leading wind direction 主导风向
leading wire 引线;导线【电】;爆破母线
leading zero 先行零;引导零;前导零
leading zero suppress 消去前补零;删去前导零
lead in hair 发铅
lead in inductance 引线电感
lead in insulator 引入(线)绝缘子;穿墙绝缘子
lead in jetty 导堤
lead in light 引进灯(机场)
lead in nipple 引入线短接管
lead in oil 白铅油
lead in petrol 汽油中铅
lead in plate 引入板
lead in porcelain tube 引入线瓷管
lead in roll 引入辊
lead in screw 引入螺钉
lead in section 入口段
lead insert(ion)铅插入物;铅插入层;铅垫片;铅垫板
lead insulation 铅绝缘层
lead intensifier 铅加厚液
lead in wire 引入线;引药线;联络线;输入线
lead iodide 碘化铅
lead ion 铅离子
lead iron oxide 氧化铁铅
lead iron oxide anti-corrosive paint 含铅氧化铁防锈涂料;含铅氧化铁防锈漆
lead isotope single-stage model 铅同位素单阶段模式
lead joint 浇铅接合;铅接口;填铅接合(缝);充铅接头;充铅接缝;灌铅接合
lead jointed pipe line 铅接口管线
lead kiln operator 看火工长
lead knife 铅刀
lead-lag 超前滞后
lead-lag compensation 超前滞后补偿
lead-lag network 超前滞后网络
lead-lag relationship 不同统计数列间的时滞变化关系
lead lanthanum zirconate titanate 锆钛酸铅镧;钛酸锆酸镧铅

lead lap 用铅重叠
lead lavatory flush(ing) pipe 厕所内嵌冲水管
lead layer 铅层
lead-lead acid cell 铅蓄电池;铅极酸性蓄电池
leadless 无铅的
leadless colo(u)r 无铅色料
leadless enamel 无铅釉;无铅搪瓷
leadless glaze 无铅釉
leadless inverted device 无引线变换器
leadless paint 无铅漆
leadless piezoelectric(al) ceramics 无铅压电陶瓷
lead light 导航灯;(用铅条镶嵌的)玻璃窗
lead-light glazier 花饰铅条窗玻璃工
lead limit switch 行程限位开关;引纹限位开关
lead line 蓝线;接受管;铅线;铅锤线;铅锤索;铅垂线;水砣绳;出油管线;测深线;测深索;测深绳;测铅绳;伯顿氏线
lead-lined 铅衬的;衬铅的
lead-lined cistern 衬铅储[贮]水器;衬铅容器
lead-lined door 衬铅门;防射线门;防辐射门;铅门;防X射线的铅门
lead-lined frame 防射线门框;铅衬门框;防X射线的包铅门楹
lead-lined launder 衬铅流槽
lead-lined pipe 衬铅管;铅衬管
lead-lined wooden tank 衬铅木箱
lead line mark 测深绳标记;水砣绳标记
leadline method 铅锤索测深法
lead lining 铅衬里;铅衬;镀铅
lead-lining of box roof gutter 屋面箱形下水沟的铅衬里
lead link marks 水砣长度记录;水砣长度记号
lead linoleate 亚油酸铅
lead load 铅负荷
lead lubrication process 熔铅润滑拉丝法
lead magnesio-niobate 铌镁酸铅
lead magnesio-niobate ceramics 铌镁酸铅陶瓷
lead magnesio tungstate ceramics 钨镁酸铅陶瓷
leadman 测深锤手;锤手
lead manager 牵头经理
lead manganese drier 铅锰干燥剂
lead marcasite 闪锌矿
lead mark 导航标志
lead matte 铅锍;粗铅;冰铅
lead melting equipment 熔铅设备
lead melting furnace 熔铅炉
lead melting kettle 熔铅锅
lead metasilicate 硅酸铅;偏硅酸铅
lead method of age determination 铅法地质年代测定法
lead mill 铅片
lead mine 铅矿
lead mineral 铅矿
lead molybdate 钼酸铅
lead monoxide 一氧化铅;氧化铅;蜜陀僧(氧化铅);密陀僧
lead mo(u)lding 铅铸型法
lead nail 铅头钉;铜铅合金钉;大头钉;铅屋面钉;(固定屋顶铅皮的)小铜钉
lead naphthenate 环烷酸铅
lead network 微分网络;超前网络
lead niobate 铌酸铅
lead niobate-nickelate ceramics 铌镍酸铅陶瓷
lead niobiummagnesium zirconate titanate ceramics 铌镁锆钛酸铅陶瓷
lead niobiumzine zirconate titanate ceramics 铌镁锆钛酸铅陶瓷
lead nitrate 硝酸铅
lead number 车床特性
lead of crossing 辙叉导距【铁】
lead off dike 导流坝
lead of helix 螺旋导程
lead of screw 螺旋导程
lead of valve 阀导柱
lead oleate 油酸铅
lead oleate plaster 油酸铅硬膏
lead ore 铅矿石;铅矿
lead orthoplumbate 四氧化三铅
lead out 领头;输出;导出;引出端;管线
lead-out groove 引出纹槽;盘尾纹
lead-out terminal 输出端子
lead out wire 孔外电雷管导线
lead-overgroove 过渡纹
lead oxalate 草酸铅

lead oxide 一氧化铅;氧化铅;密陀僧
lead oxide photoconductive layer 氧化铅光电导层
lead oxide photoconductive tube 氧化铅光电导管
lead oxide powder 氧化铅粉
lead oxide red 红铅粉;四氧化三铅
lead oxide vidicon 氧化铅光电管
lead pad 铅垫
lead paint 含铅漆;铅漆;铅丹;白铅漆;铅油
lead pan 铅盘
lead-paper cable 空气纸绝缘铅包电缆
lead para-aminobenzoate 对氨基苯甲酸铅
lead patented wire 铅淬钢丝
lead patenting 铅浴淬火;铅浴索氏化处理
lead pattern 引线图案
lead paving 铅板铺面
lead pellet (卸扣的)铅粒填封
lead pencil scratch test 耐划试验;铅笔刮痕硬度试验
lead perchlorate 高氯酸铅
lead perchloride 过氯化铅
lead peroxide candle 二氧化铅管
lead peroxide method 过氧化铅法
lead peroxide red 红色过氧化铅;红色二氧化铅
lead peroxydisulfate 过二硫酸铅
lead phosphate 磷酸铅
lead phosphite 亚磷酸铅
lead phosphosilicate 磷硅酸铅
lead pig 铅块;铅锭
lead pigment 铅颜料
lead pipe 铅管
lead pipe elbow 铅管弯头
lead pipe joint 铅管接头
lead pipe press 铅管压力(试验)机
lead pipe rigidity 铅管样强直
lead pipe soldering 铅管焊接
lead pipe support 铅管支架
lead pipe work 铅管工
lead piping 铅管
lead plaster 铅膏
lead plate 铅板
lead plate test 铅板试验
lead plating 镀铅;包铅
lead plug 钻销;铅塞;扒锯子;铅抓钉
lead plumb 铅垂线;铅锤
lead plumbate 高铅酸铅;铅酸铅
lead plummet 铅垂球;垂球
lead poisoning 铅中毒;铅毒
lead polarity of a transformer 变压器引线极性
lead pollutant 铅污染物
lead pollution 铅污染
lead pourer's mo(u)ld 铅浇注模
lead powder 金属铅粉;铅末;铅粉
lead powder impulse filler 脉冲铅粉机
lead powder primer paint 铅粉底漆
lead prediction 提前量预测
lead pressure pipe 铅质压力管
lead primer 铅底漆
lead priming paint 铅质底层漆
lead profile press 铅质造型床
lead proof 注铅检测
lead protoxide 一氧化铅
lead pursuit 迎面导航方式;引导跟踪方式
lead radiation shield 铅质热辐射罩具
lead radiation shielding wall 热辐射铅质遮热墙板
leadrail 合拢轨
lead rainwater gutter 铅檐沟;铅雨水管
lead red 红丹(又名红铅)
lead regulus 铅块;锑铅合金
lead resinate 树脂酸铅
lead resistance 引线电阻
lead-restricted paint 铅限量(色)漆;低铅(油)漆;少铅漆(含氧化铅少于5%)
lead return launder 返铅流槽
lead ridge 铅皮屋脊
lead ring 铅环;导环
lead riser 引线头;铅冒口
lead rivet 铅销
lead rod 铅棒
lead roof 铅皮屋顶
lead roof gutter 铅屋檐沟;铅皮屋顶天沟
lead roofing 铅皮屋面
lead rope 铅绳
lead rubber 含铅橡胶(辐射防护);铅橡皮
lead-rubber apron 铅橡胶围裙

lead-rubber sheet 铅橡胶布
leads 两个滑车中间的辘绳部分
lead safe 铅座;铅托架;铅底板;铅盘;铅垫
lead salt 铅盐
lead screen thickness 铅屏厚度
lead screw 引入槽;丝杠;丝杆;导螺杆
lead-screw drive 导螺杆驱动
lead screw for cross-slide 横向进给螺杆
lead screw for saddle 床鞍螺杆
lead screw for top slide 刀架螺杆
lead-screw gear 丝杠交换齿轮
lead-screw monocomparator 导杆螺旋单像比长仪
lead screw tester 丝杠检查仪
lead seal 引线封焊;输入端封接
lead seal casting machine 灌铅机;铅封压铸机
lead-sealed sheets 罩铅板
lead sealing 铅封;引线焊接
lead sealing pliers 铅封钳
lead section press 铅断面冲压机
lead sharpener 磨铅心机
lead sharping machine 铅磨削器
lead sheath 铅护套;铅鞘;铅壳;(玻璃窗格条的)铅皮外层
lead-sheathed 包铅的;铅包皮的
lead-sheathed cable 铅包电缆
lead sheathed cotton-insulated distributing cable 棉纱绝缘铅包配线电缆
lead-sheathed steel-taped 铅包钢带
lead-sheathed triple core cable 铅包三芯电力电缆
lead-sheathed wire 铅包线
lead-sheathing 铅包皮;包铅
lead sheet 铅皮;薄铅板;青铅;铅板
lead-sheet roof 铅皮屋顶
lead-sheet roofing 铅皮屋面
lead shield 铅屏;铅护套;铅套;铅罩
lead shield compartment 铅屏蔽室;铅屏蔽舱
lead shielding 铅板屏;铅屑板;铅屑蔽
lead shielding wall 铅屏蔽墙
lead shield thickness of radioactive sampling 辐射取样铅屏厚度
lead shoe 导向板
lead shoot (平屋顶,通过墙身的)排水铅槽
lead shot 铅珠;铅粒
lead shot for hunting 打猎铅弹
lead silicate 硅酸铅
lead silicate glass 硅酸铅玻璃
lead silicate white 硅酸铅白
lead silico-chromate 硅铬酸铅
lead silicofluoride 硅氟化铅
lead silico-titanate 硅钛酸铅
lead sinker 铅锤;铅测深锤
lead slag 铅渣
lead slate (管道穿过屋顶处的)铅泛水;铅板(瓦);铅皮管脚泛水
lead sleeve 铅压接套;铅套筒;铅皮管脚泛水
leadsman 铅手;水砣手;水砣测深手;锤手;测深员
leadsman gripe 水砣手安全带
leadsman's apron 水砣手防湿围裙
leadsman's gripe 水砣手安全带
leadsman's platform 水砣手台;锤测台;测深台
lead smelting 铅熔炼
lead smelting slag 铅矿渣
lead soaker 块状铅泛水板;防漏铅板;铅皮立墙泛水
lead soap 铅皂
lead-soap lubricant 铅基润滑剂
lead soil pipe 铅污水管
lead solder 铅焊料
lead soldering 铅焊接
lead spar 硫酸铅矿;白铅矿
lead spindle 主电动机联轴器
lead spitter 铅水嘴;铅皮套管(落水管的一段排水套管);雨水管接口;引线曳放器
lead spitter fuse lighter 铅管导爆索点火器
lead splash lap 熔铅喷雾盖板
lead sponge 海绵铅
lead spreader 引线展延装置
lead stabilizer 导向稳定器
lead stannate 锡酸铅
lead stearate 硬脂酸铅
lead storage battery 铅蓄电池
lead strain 引线拉力
lead strip 铅条
lead suboxide 一氧化二铅
lead suboxide anti-corrosive paint 黑铅粉防锈

漆;一氧化二铅防锈漆
lead sugar 乙酸铅;醋酸铅
lead sulfate 硫酸铅
lead sulfide 硫化铅
lead sulfide cell 硫化铅光电元件
lead sulfide detector 硫化铅探测器
lead sulfide detector cell 硫化铅试池
lead sulfide film 硫化铅薄膜
lead sulfide photoconductive cell 硫化铅光电导元件
lead sulfide photodetector 硫化铅光电探测器;硫化铅光电检测器
lead sulfochromate 硫代铬酸铅
lead sulfur isotope correlation 铅—硫同位素相关性
lead sulphate 硫酸铅
lead sulphite 硫化铅
lead surfacing 铅贴面
lead susceptibility 感性铅
lead-suspended chip 引线悬挂的片子
lead table 提前量表
lead tack 铅垫块平头钉;平头铅条;定位铅条;铅(块)平头钉;线垫(砌砖用);补漏垫;小钉
lead tank 铅衬槽
lead tantanate 钽酸铅
lead tempering bath 回火铅槽
lead terminal 引线端子
lead tester 导程检查仪
lead tetrachloride 四氯化铅
lead tetr(a)oxide 四氧化三铅
lead thiocyanate 硫氰化铅
lead thiosulfate 硫代硫酸铅
lead through 引入
lead time 预订时间;准备时间;交付周期;交付时间;前置时间;前期工作时间;投产前需时;调达时间;提前周期;生产准备时间;超前时间;产品从设计至投产间的时间;发展周期;研制周期;更换模具的时间;投产准备阶段;订货至交货时间;订货与交货间隔
lead-time bias 领先时间偏差
lead time for capacity expansion 扩大生产能力所需的时间
leadtime offset 研制周期补偿
lead-tin alloy 铅锡合金
lead-tin solder 铅锡焊料
lead-tin-telluride 铅锡碲化物
lead-tin-telluride crystal 铅锡碲晶体
lead-tin-telluride detector 铅锡碲探测器
lead-tin telluride detector arrays (热成像用的)铅锡碲化物探测器阵列
lead titanate 钛酸铅
lead titanate ceramics 钛酸铅陶瓷
lead titanate piezoelectric(al) ceramics 钛酸铅压电陶瓷
lead to 导致
lead toilet flush(ing) pipe 厕所铅质冲水管
lead toilet waste water pipe 厕所铅质污水管
lead tolerance 容许含铅量
lead tongs 上管钳(拧开下管钳固定接头时使用)
lead-tong man 管钳操作工
lead torque 引线扭力
lead tower 指向塔;打桩架
lead track 驼峰溜放线
lead track of special section 特殊断面牵出线
lead trap 铅阻汽具;铅存水弯
lead tray 铅盘
lead treated steel 含铅易切削钢
lead tree 铅树
lead trimming 引线修齐
lead trim press 铅修饰物的压制
lead trough 铅屋檐沟;铅天沟;铅槽
lead truck 前转向架
lead tube 铅管
lead tubing 铅管
lead tungstate 钨酸铅
lead type letter 白体字
lead underwriter 牵头承保人
lead unit press 铅构件压制
lead up gasoline 加铅汽油
lead valley(gutter) 铅斜沟槽
lead vanadate 钒酸铅
lead-vehicle 头车
lead vein 细脉;铅矿脉;导脉
lead vitriol 硫酸铅矿;铅矾

lead washer 铅垫片;铅衬垫;铅垫圈
lead waste(pipe) 铅污水管
lead wastewater 含铅废水
lead waterbar 挡水条
lead weathering 铅板防水条(利用铅作为密封材料)
lead wedge 铅楔
lead weight 铅坠;铅锤;吊锤
lead welding 铅焊
lead white 碱式碳酸铅;铅粉;铅白
lead window 铅窗
lead wing 铅翼
lead wire 母线(爆破用);入线;铅线;铅丝;出线;引出线
lead wire compensation 引线补偿;引导线补偿
lead wire compensator 导线影响补偿器
lead wire stone cage 填石铅丝笼
lead wolframate 钨酸铅
lead wool 铅毛;细小铅条;铅纤维;铅绒
lead wool ca(u)lking 铅棉缝堵;铅毛堵缝
lead works 铅制品;铅衬;制铅(工)厂;铅矿熔炼工厂
leady 含铅;似铅
lead yarn 铅毛
leady matte 含铅冰铜
lead-zinc deposit in skarn 矽卡岩型铅锌矿床
lead zinc oxide 氧化铅锌
lead zinc primer 铅锌底漆
lead zirconate 锆酸铅
lead zirconate titanate 锆钛酸铅
leaf 叶饰;叶(瓣);快门叶片;节流门;门扇;门扉
leaf actuator 闸刀开关;刀形断路器
leafage 叶饰;树叶
leaf analysis 叶片分析
leaf and canopy temperature 叶片和冠层覆盖的温度
leaf and dart 叶饰和箭头饰;叶箭饰;叶瓣与箭头花纹装饰线脚
leaf-and-dart mo(u)lding 叶饰与针刺饰;叶镖装饰线脚
leaf and square 叶瓣与方块花纹装饰线脚
leaf and tongue 叶饰和舌饰;叶与舌饰
leaf apex 叶端
lea farming 草田轮作;草地农业
leaf-bearing forest 阔叶树林
leaf-bearing tree 阔叶树
leaf blade 叶片
leaf bridge 仰开桥;开合桥;铰链式仰开桥;竖旋桥;吊桥
leaf-by-leaf intrusion 层层侵入
leaf canopy 林冠
leaf-canopy inhibition germination 绿荫效应
leaf cast 落叶病
leaf chain 无声链
leaf clarifier 叶片澄清机;滤叶
leaf clay 页状黏土;叶状黏土
leaf coal 纸煤;页状煤;叶状煤
leaf-comb 单冠
leaf cover 枯枝落叶层
leaf crop 产叶作物
leaf crown 叶饰皇冠
leaf dam 活瓣式闸门坝;选择桁架木板坝
leaf door 摺门;双扇门;折门
leaf driving motor 过滤叶片驱动电机
leaf driving unit 过滤叶片驱动装置
leafed alumin(i)um powder 片状铝粉;铝粉片
leafed iron oxide 片状氧化铁
leafed pigment 片状颜料
leafed powder 悬浮金属粉末;片状粉末
leaf electrometer 箔片静电计;箔静电计
leaf fall 落叶
leaf fastener 叶片紧固器
leaf-fat 板油
leaf filter 叶片式过滤器;叶滤机;片页式过滤器;板式过滤器
leaf-floating plant 浮叶植物
leaf gate 开合门;人字门
leaf-gilding 贴金箔;包金
leaf gneiss 贯入片麻岩
leaf gold 叶片;叶状层;金箔;金叶
leaf index 叶形指数
leafing 叶展;金属粉末悬浮现象;漂浮;出叶
leafing agent 叶展剂;漂浮剂
leafing agitator 叶片式搅拌器
leafing alumin(i)um 铝箔

leafing retention 叶展持久性;漂浮持久性
leafing stability 叶展稳定性;漂浮稳定性
leafing-type alumin(i)um powder 叶展型铝粉;漂浮型铝粉
leafing value 漂浮值
leaflet 小叶;广告
leaflet crystal 叶片状结晶
leaf-like decoration 叶饰;叶状装饰
leaf-like ornamentation 叶状饰
leaf-like stem 叶状枝
leaf-like texture 叶片状结构
leaf-like tissue 叶状组织
leaf litter 落叶层
leaf mining 分层开采
leaf mo(u)ld 腐叶土;腐殖叶;腐殖土;腐叶色
leaf necrosis 树叶枯死
leaf of a laminated spring 叠板弹簧
leaf of bascule bridge 衡重式仰开桥翼;竖旋桥翼;竖旋桥的翼
leaf of blocks 砖外壳(空芯墙);空心砖外壳
leaf of Descartes 笛卡尔蔓叶线
leaf of diaphragm 光圈瓣;光阑薄片
leaf of grass 草叶
leaf of hinge 合页片;合叶片
leaf optical system 薄片光学系统
leaf ornamentation 叶饰
leaf pattern 叶形
leaf pattern decoration 木叶纹饰
leaf peat 页泥炭;叶泥炭
leaf photosynthesis 叶片光合作用
leaf rake head 草耙头
leaf roll 卷叶
leaf scar 叶痕
leaf shale 片页岩
leaf-shaped curve 叶状窗花格(哥特式);叶状曲线
leaf sight 瞄准标尺
leaf spring 翼簧;钢板弹簧;簧片;片弹簧;叠板(弹)簧;板(弹)簧;叶片式弹簧
leaf spring center bolt 钢板弹簧中心螺栓
leaf-area index 叶面指数
leaf-spring suspension 片弹簧悬挂
leaf support 过滤叶片支架
leaf suspension spring 扁簧弹簧悬挂
leaf temperature 叶片温度
leaf test 滤叶试验
leaf through 翻阅
leaf trace 叶迹
leaf tree 阔叶树
leaf type 叶型
leaf-type damper 空气瓣阀
leaf-type filter 圆盘式过滤器
leaf-type shutter 叶片式快门
leaf-type vacuum filter 真空叶滤器
leaf valve 叶片阀;簧片阀;瓣状活门;舌阀
leaf water potential 叶片水势能
leaf wood 阔叶树木材;阔叶林(木);硬材(乔木);阔叶树木材
leafwork 叶饰;叶饰细雕;叶饰雕刻;雕花工
leafy 叶状的;叶状
leafy frieze 叶状饰带;叶状檐壁
leafy liverwort 叶状地线
leafy powder 薄片状粉末
league 里格(长度单位,1 里格≈3 英里)
League of Arab States 阿拉伯国家联盟
League of Nations Building 国际联盟大厦
lea harrow 草地耙
Leahy screen 凸轮冲击筛
leak 漏失;漏扎;漏洞;漏出量;流失;跑料;漏损(量)
leakage 许可的漏损率;泄漏;泄出量;漏泄;漏损(量);漏失(量);漏出(量);渗水量;渗漏
leakage air 漏入空气;漏出空气
leakage and breakage 漏损与破损
leakage and drip 泄漏和滴失
leakage and/or thief 漏水和/或盗窃
leakage anomaly 渗漏异常
leakage apron 防渗铺盖
leakage area 渗漏范围
leakage around dam abutment 绕坝渗漏量
leakage basin 滤水池
leakage chamber 漏气凝水柜
leakage check 密封检查
leakage coefficient 越流系数;漏泄系数;漏损系数;漏电系数;漏磁系数;渗漏系数

leakage coil 泄放线圈
leakage concentration magnitude 泄放浓度值
leakage condenser 漏电凝水柜
leakage conductance 漏泄电导;电漏
leakage conductivity 泄漏电导率
leakage conductor 线路避雷器
leakage control 泄漏控制
leakage current 泄漏电流;泄放电流;漏泄电流;漏风;漏电
leakage current detector 泄漏电流检测器
leakage detector 管漏水探测器;漏电指示器;检漏器;接地指示器;探漏仪
leakage discharge 泄漏放电
leakage distance 泄漏距离;漏电距离
leakage efficiency 密封效率
leakage end 漏磁端
leakage factor 越流因素;越流因数;漏泄系数;漏损系数;漏水因数;漏水系数;漏失系数;渗漏因素;渗漏系数
leakage field 漏磁场;漏磁场
leakage finder 管漏水探测器
leakage flow 漏流
leakage flow rate 泄漏流量;渗漏流量
leakage fluid dram 排液口
leakage flux 泄漏通量;漏磁通(量)
leakage form adjacent aquifer 相邻含水层越流补给
leakage from reservoir 库岸渗漏;水库渗漏(量)
leakage halo 渗透晕;渗滤晕;渗漏晕
leakage impedance 泄漏阻抗;漏泄阻抗
leakage indicator 泄漏检测器;检漏计;渗漏指示器
leakage indicator hole 泄漏指示孔
leakage inductance 漏电感;磁漏电感
leakage in fault zone 断层破碎带渗漏
leakage intake 泄漏进风
leakage intake system 双风巷通风系统
leakage interrupter leakage breaker 漏电断路器
leakage layer 越流层
leakage line 泄漏管路;磁漏线
leakage location 检漏
leakage locator 渗漏定位器;渗漏探测器
leakage loss 泄漏损失;漏泄损失;漏电损耗;渗漏损失;漏失量
leakage magnetic flue 漏磁通管
leakage magnetic flux 漏磁通(量)
leakage measurement 渗漏量观测
leakagemeter 检漏仪
leakage method 漏磁法
leakage neutron 泄漏中子
leakage of current 电流漏泄
leakage of dam abutment 坝肩渗漏
leakage of dam foundation 坝下渗漏;坝基渗漏量
leakage of electricity 漏电
leakage of grout 漏浆;冒浆
leakage of interfluve 河间地块渗漏
leakage of light 漏光
leakage of mortar 漏浆
leakage of pipe 管道渗漏
leakage of piping 管系渗漏量
leakage of rain 雨水渗透;漏雨
leakage of water 漏水
leakage over subdivide 分水岭渗漏
leakage passage 漏油流道
leakage path 泄漏通道;泄漏电路;漏水线路;漏路线;漏磁路径;渗径
leakage permeance 漏磁导率
leakage point 泄漏点
leakage position 漏处
leakage power 漏泄功率;耗散功率
leakage-proof 漏水防止;避漏
leakage protection 堵漏;漏泄保护;防漏
leakage protective system 漏泄保护系统;防漏系统
leakage radiation 漏辐射
leakage radiation dose 泄漏辐射剂量
leakage rate 泄漏率;泄漏量;漏气率;漏气率
leakage reactance 漏抗;漏磁电抗
leakage recharge 越流补给
leakage reduction 泄漏抑制;渗漏抑制
leakage resistance 泄漏电阻;漏(泄)电阻;流电阻
leakage scene 泄漏现场
leakage spectrum 泄漏光谱
leakage spot displacement 光点的电位移
leakage stoppage 堵漏
leakage survey 检漏

leakage tank 泄漏罐;泄漏槽
leakage test 泄漏试验;漏气试验;检漏;密封性试验;密封(性)检验;气密性试验;气密性实验;真空式试验;漏电试验;检漏试验;渗漏试验
leakage tester 密封性测试器;检漏器
leakage through karst cave 岩溶洞穴渗漏
leakage through the old channel 古河道渗漏
leakage transformer 磁漏变压器
leakage voltage 漏电压
leakage water 越流水;漏失水;渗滤水;渗漏水;漏量
leakage water outlet 漏水出口
leakage well 漏失井
leak air 漏空气
leak alarm 渗漏报警器;漏水警报
leakance 泄漏系数;漏泄系数;漏泄电导;漏电;漏磁;电漏
leak check 泄漏检验;检漏
leak circuit 泄放电路
leak clamp 修理夹;止漏管箍;防漏(箍)夹
leak coil 泄放线圈
leak collector 积泄器
leak detecting 检漏
leak detection 泄漏检查;紧密性检查;检漏;检查漏泄;渗漏检查;渗漏检测
leak detection by listening 停漏
leak detection equipment 检漏装置
leak detection method 检漏法
leak detector 泄漏探测器;泄电指示器;漏失检验器;检漏仪;检漏器;燃料元件破裂检测器;探漏器;渗漏探测仪;检漏器;测漏器;短路指示器;电漏指示器;检测器;接地指示器;探漏仪
leak detector for pipeline 管道检漏器
leaked 漏的
leaker 漏泄构件;漏气构件;漏钢;渗水铸件;渗漏铸件
leak finder 寻漏器
leak-free 密闭;不漏的
leak hunting 检漏;测漏
leak-in 漏风
leak indicator 渗漏显示器;示漏器
leakiness 泄漏程度;漏泄程度;不密实
leakiness check 密封检查
leakiness detector 探漏仪;探漏器
leakiness free 密封的
leaking 滤出;溢出;泡漏;泄漏
leaking blow-off valve 漏水排放阀
leaking brine 淋滤型卤水
leaking discharge 越流排泄
leaking dose 泄漏剂量
leaking in 漏入水量
leaking joint 漏节;渗漏接缝
leaking tuyere 漏气风口
leaking well 渗漏井
leakless 不漏的
leak level 泄漏电平
leak localizer 检漏器;探漏器
leak location assembly 检漏装置
leak locator 测漏器
leakly 漏的
leak-off chute 漏水槽
leak-off connection 放泄接头
leak-off pipe 泄流管;漏泄管
leakoff-type shaft seal 迷宫式轴封;曲径式轴封
leak oil line 漏油管
leak out 泄露;漏出
leak pinpointing 漏处定位
leak plugging 堵塞漏水
leak preventer 防漏剂
leak prevention 防漏
leak preventive 防漏剂
leakproof 不漏的;防漏(的);不漏电
leakproof additive 防漏失剂
leakproof battery 全封闭电池
leakproof fit 推合座;紧密配合;密配合的
leakproof fuel cell 密封油箱
leakproof grab 防漏抓斗
leakproof grab for bulk fertilizer 散粒肥防漏抓斗
leakproof grab for bulk grain 散粮防漏抓斗
leakproofing material 止水材料;防漏材料
leakproof material 防漏材料
leakproof motor pump 防漏式电动泵
leakproofness 密封性;密闭度
leakproof pump 密封泵

leakproof roof 防漏屋顶
leakproof seal 防漏密封
leak protection 防漏
leak rate 漏损率;漏失率;漏气率
leak resistance 泄放电阻;防漏性能
leak source 漏点
leak stoppage 防漏
leak stopper 塞漏材料
leak stopping outfit 堵漏工具
leak test 漏气试验;漏电试验;密封性检验;真空式试验;检漏试验;渗漏试验
leak test by filling water 水封试验
leak-tested 密封度试验;防漏试验
leak-testing apparatus 密封性检验仪
leak through 渗透;滴漏
leak through liquid 渗流
leak tight 密封的;气密的;不透气的;不漏的;真空密封;防漏的
leak tightness 密封性;密闭度
leak-tight system 气密系统
leak to ground 接地漏电
leak valve 泄放阀;针阀;漏气阀
leak water 渗漏水
leaky aquifer 越流性含水层;漏水含水层;渗漏蓄水层
leaky coaxial cable 漏缆
leaky depression 渗水洼地
leaky duplex 漏顶式双冲构造
leaky feeder 漏缆
leaky foundation 漏水地基;渗漏地基
leaky gasket 漏油衬垫
leaky joint 渗漏接缝
leaky lawn 渗漏草场
leaky mode 漏模
leaky mode buried-hetero-structure 漏模掩埋式异质结构
leaky mutant 渗漏突变体
leaky pipe 渗漏管
leaky-pipe antenna 波导隙缝天线;波导开缝天线
leaky recharge 有越流补给
leaky recharge element 有越流补给单元
leaky reservoir 漏库
leakring 瓶口泄漏
leaky riveting 松铆(接)
leaky seam 漏水缝
leaky shallow well 渗漏浅井
leaky ship 渗漏的船
leaky system 越流系统
leaky theory 越流理论
leaky transform fault 渗漏转换层
leaky valve 泄漏的阀
leaky wave 漏泄波;漏波
leaky wave antenna 漏波天线
leaky waveguide 漏隙波导;开槽波导(管)
leaky waveguide antenna 开缝波导天线
leam 沼泽地排水(沟)
lea management 草地经营
Leam Chabang Port 林差纲港(泰国)
lean 倾斜;贫料;偏向;低可塑性的
lean absorption oil 贫吸收油
lean alloy steel 低合金钢
lean bow 尖形船首
lean burden 贫装料
lean-burn 稀燃
lean-burn emission controlled engine 稀混合气式净化排气发动机
lean-burn engine 稀混合气发动机
lean cement concrete 贫混凝土;少灰水泥混凝土;少水泥混凝土
lean clay 贫黏土;瘦黏土;低塑性黏土;低可塑性黏土
lean clay of low plasticity 低塑性瘦黏土
lean coal 贫煤;瘦煤;低质煤;低级煤
lean coking coal 瘦焦煤
lean concrete 贫混凝土;贫灰混凝土;少灰混凝土;低强度等级混凝土
lean concrete base 贫混凝土基底;贫混凝土基层;贫混凝土底层
lean concrete pad 贫混凝土垫层
lean concrete test cube 贫混凝土试块
lean concrete test cylinder 贫混凝土试验用圆柱体
lean concrete underlayer 贫混凝土垫层
lean developer 不饱和显影剂
leaner air fuel ratio 更稀的空气燃油比
lean fresh concrete 新拌的贫混凝土

lean freshly mixed concrete 新拌的贫混凝土
lean fuel-air ratio 稀混气比
lean fuel mixture 稀混合物;贫燃料混合气
lean gas 贫气;低发热量气
lean gas riser 贫气提升管
leaning 倾斜;倾向;贫油;使贫化
leaning against the wind 凭借风势
leaning device 倾斜机构
leaning out 倾向于稀的(混合气)
leaning thread 倾斜螺纹;梯形螺纹
leaning tower 斜塔
leaning tower of Pisa 比萨斜塔
leaning wheel 倾斜轮
leaning wheel grader 倾斜轮式平地机;车轮可倾式平地机;车辆可倾式平地机
lean lime 贫石灰
lean-limit (混合气燃烧的)稀薄极限;贫油极限
lean material 选矿后的废石;贫矿;废矿石
lean metering 贫油调节
lean mix 贫混合料;贫拌和;少灰合料;少灰混合
lean(mix) concrete 少水泥混凝土;贫混凝土;少灰混凝土
lean mixed mortar 少灰拌和砂浆
lean mixing 贫搅拌
lean mixture 稀混合物;稀混合气;贫搅拌;贫(灰)混合料;少油混合料;少水泥混合料;少灰泥;贫拌和料
lean mixture limit 稀混合气极限
lean mixture maximum power 贫燃料混合物最大功率
lean mixture rating method 贫油混合物抗爆性评定法
lean mixture strength 稀混合气强度
lean mortar 贫砂浆;贫灰浆;少灰砂浆;低标号砂浆
lean mo(u)lding sand 瘦型砂;瘦砂
leanness 贫乏
lean oil 解吸油;贫油
lean oil shale 贫油页岩
lean open-type mixture 多孔贫混合料
lean ore 贫矿石;贫矿
lean paint 薄油漆;稀油漆
lean phase 稀相;疏相
lean phase bed 稀相床
lean phase conveying 稀流输送
lean position 稀混合气的位置
lean rolled concrete 碾压贫混凝土;少碾压混凝土
lean sand asphalt 贫砂沥青混合料
lean slag 贫渣;瘦渣
lean solution 贫液;贫溶液;废溶液
lean solution flash drum 贫液内蒸槽
lean stick 拐杖
lean-to 一面坡屋;单坡披屋;单坡棚;单坡顶;单坡的;披屋
lean-to dormer 单坡(屋)顶老虎窗;单坡屋顶窗;单坡老虎窗
lean-to heat bed 单斜面温床
lean-to house 单斜面温室
lean-to ladder 梯子
lean-to mansard roof 单面折线形屋顶;单面复折屋顶
lean-to ring roof 单坡环形屋面
lean-to-roof 单坡屋顶;披屋
lean-to-roof purlin(e) 单坡屋架檩条;单坡屋面檩条
lean-to roof strut 单坡屋顶支柱
lean-to roof tile 单坡屋顶瓦
lean-to roof truss 单坡屋顶桁架
lean-to skylight 单坡的天窗
lean-to trussed strut 单坡桁架支撑
Lea-Nurse method 利纳斯(比表面积测定)法
lean year 歉收年
leaper 小浪花;跳跃者;船首溅沫
leap forward 跃进
leapfrog 蛤蟆夯;蛙式夯土机;用动力夯夯;机动夯;蛙式夯;跳步前进;动力夯
leapfrog cascade 跳阶级联
leapfrog fashion 跳蛙式排水;多级明坑排水
leapfrogged 跳背式
leapfrogging 交互跃进;跳蛙式
leapfrogging tactics 跳蛙战术
leapfrog method 跳法法;跳背法;蛙法法;时间中心差法
leapfrog routine 蛙跳式路线
leapfrog switching 跳步开关
leapfrog test 跳步检验;跳步测试

leaping formwork 升模;提升(式)模板
leaping frog 跳蛙式打夯机
leaping legs 跳跃足
leaping motion 跳跃运动
leaping of divide 分水界移动
leaping organ 跳跃器
leaping shuttering 滑升模板;提升式模板
leaping weir 下水道溢流堰;跃水分水堰;跳越堰
leaping-weir overflow 跳越堰溢流
lea plow 草地犁
leap month 闰月
leap ore 贫锡矿
leap second 闰秒
leap second adjustment 闰秒调整
leap-weir 跳越堰
leap year 闰年
lear 玻璃退火窑
lear board 排水槽支承板;排水沟支承板;雨水槽支承板;天沟托板
Lear engine 利尔发动机
learl foot 纵向长度
learned characteristics 学习特性
learned organization 学术团体
learned report 学术报告
learned society 学术团体;学会
learner 学习者
learner driver 学习驾驶员
learnership 培训协议
learner's pool 初学者游泳池
learning 学问学习
learning activities 学习活动
learning algorithm 学习算法
learning and adapting capability 学习适应能力
learning capacity 学习容量
learning control 学习控制
learning control system 学习控制系统
learning curve 学习曲线;研讨地线
learning-employment sequence 学习就业序列
learning identification 学习辨识
learning machine 学习机
learning matrix 学习矩阵
learning mechanism 学习机制
learning network 学习网络
learning program(me) 学习程序
learning resource center 学习资料中心
learning system 学习系统
learning theory 学习理论
leasable area 可出租面积
lease 租赁;租借;租地;分经;分绞;租约
lease agreement 租赁契约;租赁合同;租约
lease automatic custody transfer 井区自动转输站
leaseback 租回已售出产业;返租;反租赁
lease banded hank 分绞绞纱
lease by estoppel 禁止推翻租借
lease charges 租用费
lease contract 租赁契约;租赁合同
lease cord 分绞绳
leased channel 租用信道;租用线路
leased circuit 专用路线
leased circuit connection 租用电路链接
leased circuit data transmission service 租用电路数传业务
leased construction equipment 租用施工机械
leased data communication service 专用数据通信服务网
leased dedicated channel 租用信道
leased facility 租用设施;租用设备
leased freight car 租用货车
leased group link 租用基群线路
leased housing 出租房屋
lease-distribution system 井区配电系统
leased line 租用线(路);专用线路;租用线路
leased line annunciator 专线报警器
leased line channel 租用线路信道
leased line network 租用线路网;专线通信网;租用专线通信网
leased network service 租用网路业务
leased private channel 租用的专用信道
leased satellite 租用卫星
leased territory 租借地
lease equipment 矿场设备
lease finance company 租赁财务公司
lease financing 租赁筹资
lease fixed assets 租赁固定资产

leasehold 租来的;租借物;租赁权;租期(限);租(借)地;土地租用权;租赁权;租约;租赁
leaseholder 承租人;租赁人;租赁者;租借人;租户;租地人;持证者
leasehold interest insurance 租赁权益保险
leasehold mortgage 租赁物抵押贷款
leasehold property 租赁财产;租借物业
leasehold ship 租来的船
leasehold value 租赁价值
lease in batches 批租
lease-lend 平等租借交换
lease making 分经
lease pin 分经筘
lease purchase 租借购买
lease-purchase agreement 租赁购置协议
lease rate quotes 租金计算率
lease reed 分绞筘
lease rent 租赁费
lease rod 分经筘;分绞棒
lease tank 井区储罐;油矿油罐
lease term 租期;租赁期间;租借期限
lease trade 租赁贸易
lease with option to purchase 有购买权的租约
leasing 租赁;租金
leasing comb 分径筘
leasing land in batches 土地批租
leasing machine 分经机
leasing of equipment 设备出租
leasing system for fixed assets 固定资产租赁制
leasing trade 租赁贸易
least 极微
least action 最小作用量
least action principle 最小作用量原理
least amount of solid 最小固体量
least average cost 平均成本最低
least circle of aberration 最小像差圆
least-coat estimating and scheduling system 最低成本估算与调度系统;最低成本估算与调度法
least common denominator 最小公分母
least common multiple 最小公倍数
least comparative disadvantage 最小的比较劣势
least cost 最小成本;最低成本;低费用;低造价
least cost analysis 最小费用分析
least cost combination 最低成本组合
least cost estimating 最低成本估算
least cost expansion 费用最小的发展方案
least cost input combination 最小成本的投入组合
least cost method 最小成本法;最小费用法
least cost model 最小费用模型
least cost operation 最低费用运转
least cost option 最少费用方案
least cost point 最小费用点
least cost route 最小费用路由
least cost solution 最小费用解
least-cost-to-build 建筑费用最小
least count 最小计数;最小读数
least count of vernier 游标盘最小读数
least depth 最小深度
least dimension 最小尺度
least distance 最小距离
least distance classification 最小距离分类
least distance of distinct vision 明视最小距离;明视最短距离
least effective dose 最小有效剂量;最小有效量
least energy principle 最小能量原理
least energy theory 最小能量理论
least error 最小误差
least extinction angle 最小消光角
least factor-cost for producing 生产要素得最低成本
least fatal dose 最小致死剂量
least field capacity 最小田间持水量
least fixed-point 最小定点
least fixpoint 最小不动点
least fixpoint of functionals 泛函的最小不动点
least harmonic majorante 最小调和优函数
least-integral-square error 最小积分平方误差
least limit 最小极限
least maneuverable ship 操纵不灵敏的船
least mean square error 最小均方误差
least mean square optimization 最小均方最佳化
least mean square velocity 最小均方速度
least moment 最小力矩
least navigable depth 最小通航水深

least navigation depth 最小通航水深
leastone 层状砂岩
least operator bias 最小操作误差
least payment 最低限额缴款数
least perceptible chromaticity 最低可见色度差
least perceptible chromaticity difference 最小可辨色(度)差
least perceptible difference 最小可辨差别
least play 最小间隙(机械)
least precise factor 最小精密因数
least price distortion 最少价格影响
least principal stress 最小主应力
least quantity of rock powder 最低岩粉用量
least radius 最小半径
least radius of gyration 最小回转半径
least reading 最小读数
least readout 最小直读
least recently used 刚使用过的
least recorded depth 记录最小水深
least residue 最小剩余
least resistance 最小阻力;最小抵抗力
least resistance line 最小抵抗线
least significant bit 最小有效位;最低有效位;最低位
least significant character 最低有效位组;最低位字符
least significant difference 最小显著数;最小显著差异;最小显著差(数);最少显著差
least significant digit 最低有效数位;最小有效数;最低有效位(数);最低位数字;最低(位)有效数字;最低(数)位
least significant end 最低端
least significant position 最低有效位
least splitting field 最小可分域
least square 最小平方;最小二乘(方)
least square adjustment 最小二乘(法)平差
least-square analysis 最小二乘方分析
least square approximation 最小二乘方逼近;最小二乘逼近(法)
least square collocation 最小二乘(法)配置
least square correlation 最小二乘法相关
least square criterion 最小二乘判别准则;最小二乘方准则
least square error approximation 最小二乘方逼近
least square estimate 最小二乘估值
least square estimation 最小二乘估计
least square estimation process 最小二乘推估法
least square filtering 最小二乘法滤波
least square fit(ting) 最小二乘(法)拟合;最小二乘法选配
least square inverse filtering 最小平方反滤波
least square linear relationship 最小平方线性关系
least square mapping technique 最小平方映射技术
least square method 最小平方法;最小二乘方法;最小二乘法
least square method in quantitative analysis 定量分析中最小二乘法
least square polynomial fit 最小二乘多项拟合
least square prediction 最小二乘推估
least square procedure 最小二乘法程序;最小二乘法
least square regression 最小二乘回归
least squares 最小二乘法
least squares calculation 最小二乘方计算
least squares deconvolution 最小平方反褶积
least squares filtering 最小二乘滤波;最小平方滤波
least squares solution 最小二乘;最小二乘法解
least squares prediction 最小二乘方估算
least squares straight line fit 最小方差直线拟合
least square theory 最小二乘法理论
least strain axis 最小应变轴
least stream power 最小水流功率
least thermal stability coal 热稳定性极差煤
least time path 最小时程;最速降线;捷线;最短时程
least-time principle 最短时间原理;费马原理
least total cost method 最低总成本法
least upper bound 最小上界【数】;上确界
least water-holding capacity 最小持水量
least wetted perimeter 最小湿周边
least work 最小功
least-work principle 最小功原理
least-work theorem 最小功定理
least-work theory 最小功理论
leat 夹矸的煤层;泥炭堆;人工水沟;渠道;水道;水车引水道
leat board 天沟托板
lea tester 缕纱强力试验机
leather 皮革制品;皮革
leather adhesive 皮革黏合剂
leather apron 皮制防渗帷幕
leather bellows 皮网箱;皮腔;皮老虎;皮风箱
leather belt 皮带
leather board 仿皮革纸板;革制板;皮制板
leather boat-body-rest pads 牛皮靠把
leather bowl 皮碗
leather bucket 皮碗活塞;皮碗
leather case 皮套
leather cement 皮件胶合料
leather cloth 漆布;油布;人造革;防水布
leather collar 皮圈;皮垫圈
leather cup 皮碗;皮胀圈
leather cup washer 皮碗垫圈
leather diaphragm 皮料薄膜
leather door 皮革门
leather dope 皮革漆
leather dye 皮革染料
leatherette 假皮;人造革;纸革
leatherette paper 充皮纸
leather fabric 皮革织物
leather gasket 皮密封垫片;皮垫圈;皮衬垫
leather glue 革胶;皮革胶
leather goods 皮革制品
leather goods and furs 毛皮
leather hand pad 皮革手垫
leather-hanging 皮革挂帘
leather-hard 半干状(指坯体干燥程度);半干
leather-hard-hardness 半干状态硬度(窑业制品毛坯的)
leather hollows 凹形皮革件
leather hose 皮软管
leather imitation paper 仿革纸
leather industry 皮革工业
leather industry sewage 制革工业废水
leather industry wastes 制革废水
leather lacquer 皮革喷漆
leather-like coating 仿皮革涂层
leather measuring tape 皮卷尺
leather nailing machine 钉皮机
leatheroid 薄钢纸;纸皮;假皮;人造革
leather package 密封皮碗
leather packing 皮填料;皮革填充物;皮垫
leather packing collar 皮垫圈
leather packing for balance piston 平衡活塞皮垫
leather palm working gloves 皮工作手套
leather polishing wheel 牛皮抛光轮
leatherpulp board 皮革纸浆板
leather sealing ring 皮密封圈
leather sealing strip 皮带止水条;皮革衬条
leather seat 皮鞍座
leather sheet 皮衬
leather strap 皮带
leather strop 革砥
leather substance 革质
leather tape condenser 皮带分条搓条机
leather tar 皮革焦油
leather varnish 皮革清漆
leather washer 皮垫圈;皮衬垫
leather wastes 制革废物;制革废水
leather wastewater 制革废水
leatherwood 沼泽革木
leather working gloves 劳保手套
leave 离开;模锻斜度
leave break 超过回船期
leave cost 休假费
leave draft 出港吃水
leave from the deciduous trees 落叶树树叶
leave-in-place form 留置原位的(混凝土)模板
leave in trust 委托
leave leave 超过假期
leave leeway 留有余地
leave on full salary 付全部工资的休假
leave open 不加遮盖
leave-out (路面施工暂时留出的)空当
leave outstanding 搁置不管;搁置不付;不予偿还的
leave port 驶离港;出港
leave something as security for a loan 押账
leave strip 保留带
leave surface time 入水时间
leave the land uncultivated 土地撂荒
leave the pipe hanging 让管道悬挂
leave tree 保留树木
leave unused 闲置
leave without pay 留职停薪;无薪假;停薪留职
leave with pay 薪资照付的假期
leaving 离开
leaving air 污浊空气;排出空气
leaving area 排气面积
leaving bell 出钟
leaving dock 出坞
leaving draft 离港吃水
leaving energy 出口能量
leaving lock in a curvilinear way 曲线出闸
leaving loss 余速损失
leaving momentum 输出动量
leavings 剩余;剩货;残渣;渣滓
leaving signal 出发信号
leaving speed 出口速度
leaving speed at retarder 缓行器出口速度
leaving speed from retarder 缓行器出口速度
leaving the dock 驶出船闸
leaving the land 离开陆地
leaving velocity 余速
leaving-velocity loss 余速损失
Lebanon cedar 西南亚雪杉
lebbek-tree 山槐
Lebedeff polarising interferometer 列别捷夫偏光干涉仪
Lebesgue integral 勒贝格积分
Lebesgue measure 勒贝格测度
Lebesgue number 勒贝格数
Leblanc connection 勒布朗克接法;二相变三相的接法
Leblanc connexion 勒布朗克连接法
Leblanc phase advancer 勒布朗克进相器
Leblanc process 勒布朗
Leblanc system 勒布朗克系统
leca 黏土陶粒
leca block 黏土陶粒块
lech 拱顶石
Le Chatelier apparatus 勒夏特列埃仪(用于测定水泥的安定性)
Le Chatelier soundness test 检查水泥安定性的试验
Le Chatelier's principle 勒夏特列埃原理
Le Chatelier's thermocouple 勒夏特列埃热电偶
Le Chatelier test 勒夏特列埃衰减试验;勒夏特列埃水泥安定性试验
lechatellierite 焦石英
Lecher 勒谢尔线
Lecher line 勒谢尔线
Lecher wire 勒谢尔线
Lecher wire wavemeter 勒谢尔线
leck 硬黏土;石状黏土
Leclanche cell 勒克朗歇电池
lecontite 钠铵矾
lectern 写字台;桌面倾斜的讲台(寺院);讲坛;读经台;(放讲稿的)小台架
lectin 外源凝集素
lectostratotype 选定标准地层剖面;选层型
lectotype 选型;选模式;选模标本
lecture 讲座
lecture area 教学区
lecture building 教学楼
lecture experiment 演示实验
lecture hall 演讲厅;讲演厅;大讲堂
lecture room 讲堂
lecture theater 阶梯教室;梯级教室;演讲厅;讲堂
Leda clay 莱达黏土
ledatron 莱达管
ledbit 有铅垫板的沥青防水层;(有铅心的)沥青防潮层
Ledbury Shales 莱德伯里页岩
Leddel alloy 莱登锌合金
ledeburite eutectic 莱氏体共晶体
Lederberg technique 莱德伯格技术
ledge 檐;岩崖;岩石架;岩礁;中梁;靠墙板;礁脉;横档;含矿岩层;突出部(分);凸缘;凸耳;石梁;石脊;舱内横围板;壁架;暗礁;岸(边)礁
ledge bar 舱口盖承
ledge batten (门的)上下冒头
ledged and braced door 直拼斜撑门;直拼Z形撑门;拼板斜撑门(门扇不带框)
ledged door 直拼撑门;(不带框的)拼板门;实拼门;

直拼斜撑框构门;带斜撑拼板门;无框门;板条门
ledge excavation 岩石开挖;岩面开挖
ledge joint 搭接焊缝;搭接(接合)
ledge jointing 搭接焊缝
ledge matter 脉质
ledg(e)ment 横线脚;上口方板;横线条
ledge mustard 篱芥(植物)
ledge of rocks 石梁;岩礁
ledger 总账;账簿;矿脉底部;模板檩条;横木;卧木;大横杆;分类账;分户账
ledger accounts 分类账账户;分类账户
ledger analysis 分类账分析
ledger analysis sheet 分类账分析表
ledger assets 账面资产;分类账上的资产
ledger balance 总账平衡;分类账余额;分类平衡器
ledger beam 花篮梁
ledger board 圈梁;脚手板;滑道护木;篱笆;长板条;栅栏顶板;栏杆扶手;栏顶板;脚手架横杆;木架隔层横木;格栅横托;围栏板;滑道护木
ledger books 分类账簿
ledger budget 法定总预算书
ledger card 总账卡片
ledger control 分类账控制;分类统制账
ledger folio 分类账页数;分类账页次
ledger forms 分类账表格
ledger for track maintenance 线路维修登记簿
ledger journal 分类日记账
ledge rock 岩脉;岩床;礁石;基岩;含矿岩石;凸出岩架;坚灯岩;层岩;岸礁
ledger paper 账簿纸
ledger plate 横板;卧材;固定板;格栅横托木
ledger rock 岩床;突岩;礁岩;坚硬岩;层岩;岸礁
ledger sheet 分类账账页;分类账页次;分类账
ledger strip 格栅横托木;大梁底部镶条;横木条
ledger transfer 分类账转账
ledger transfer method 分类账结转法
ledge waterstop 止水边条
ledgment 砖石装饰腰线;砖石装饰横线脚
Ledian 列德期【地】
Ledian(stage) 莱弟阶【地】
ledikite 伊利石
ledloy 莱德洛伊易切削钢;含铅钢
ledloy free cutting steel 加铅易切削钢
ledmorite 榴霞正长岩
Ledoux bell meter 勒杜克浮钟压力计
Ledrit 铅黄铜
Leduc current 勒杜克电流
Leduc effect 磁致温差效应
lee 下风;背风面;背风;背冰川的
lee anchor 下风锚;惰锚
lee armo(u)r 内侧护面
lee bank 海底堆积阶地;背风岸
lee board 防横漂板;下风板;横漂抵板(平底船的下风边)
lee breakwater 下风防波堤;背风(侧)防波堤
lee cable 惰链
leech 小型开口沉井;纵帆下风缘;横帆两侧边缘;水蛭;帆缘
lee chain 惰链
leech dermatitis 水蛭皮炎
leechee 荔枝
leech rope 天幕边绳;帆缘绳
lee current 顺风潮流
lee depression 背风坡低压
lee dune 背风沙丘
lee eddy 背风漩涡;下风漩涡
lee face 下风面;背流面;背风面
leegte 宽平谷地
lee helm 上风舵
Leeman "doorstopper" borehole strainmeter 利曼"门塞"钻孔应变计;"doorstopper" technique 利曼"门塞"法;利曼"挡器"法
Leeman's multi-component borehole strain meter 利曼多分量钻孔应变计
Lee-MeCall 李麦柯后张法(一种预应力混凝土钢筋后张法)
Lee-MeCall bar 李麦柯式高强粗钢筋(预应力混凝土中)
Lee-MeCall system 李麦柯式高强钢筋法(预应力混凝土中)
Leendertse block 里思德兹式混凝土护坡块
lee of the shore 岸的下风侧
lee-persistent pesticide 低残留农药
leer 缓冷炉;退火窑;退火炉;玻璃退火容;玻璃退火炉

leer pan 退火盘
Lee's algorithm 李氏算法
lees black 葡萄酒渣炭黑
Lee's hologram 李氏全息图
lee shore 下风岸;背风海岸;背风岸
lee side 下风舷;下风面;下风侧;背流面;背风面
lee slope 下风坡;背风坡
Lee's model 李氏模型
Lee's wave 李氏波
lee tide 下风潮;顺风潮
lee tide current 顺风潮流
lee trough 动力槽;背风槽
lee vortex behind hills 山地背风涡旋
leeward 下风的;下风;在下风的;背风的
leeward bank 下风岸;背风岸
leeward breakwater 下风防波堤
leeward chord 下风弦(杆)
leeward end of fetch 风区前沿;背风区前沿
leeward(ly) 向下风
leeward side 下风面;下风侧;背风面;背风侧
leeward slope 背风坡
leeward squall 迎面大风雨
leeward tide 下风潮;顺风潮
leeward tide current 顺风潮流
lee wave 顺风波浪;背风波
leeway 机动性;活动余地;漂移;偏航(角);风压漂移;风压角;风压差;风吹移;允许误差;时间损失;安全裕度;漂流角
leeway angle 漂流角;风流压差
leeway drift 漂流角;风流压差
leeway indicator 风压差指示器
lefkoweld 环氧树脂类黏结剂;环氧树脂类黏合剂
left 左侧的;左边
left adjoint 左伴随
left alignment 左边对齐
left angle 左角
left anterior 左前
left anterior descending branch 左前降支
left-averted photography 左偏离摄影术
left bank 左岸【地理】;左岸(面向下游)
left bank tributary 左岸支流
left border 左缘;左侧缘
left bracket 左括号
left-component 左侧数
left conveyor reverse cover 左侧输送机回动盖
left corner bottom-up 左角自底向上
left curly bracket 左波形括号
left-cut tool 左切刀
left-distributor cover 左侧分配器盖
left driving 左侧行驶;靠左(侧)行驶
left elevation 左视图
left end marker 左端标记
left entry 左侧记入
left exit turn 驶出左转弯
left fold 左褶皱
left front electrical equipment mounting 左前方电气设备安装
left gable tile 左山墙瓦
left half axle 左半轴
left half axle housing 左半轴壳
left-hand 左向;左手的;左面的
left-hand adder 左移加法器;高位加法器
left-hand airscrew 左旋螺桨
left-hand auger 左向旋转螺旋
left-hand border 左图廓
left-hand component 左侧制分量
left-hand continuity 左方连续
left-hand control car 左转向盘车辆;左边驾驶汽车;左座驾驶车辆
left-hand corner 左转
left-hand crankshaft 左式曲轴
left-hand crossing 左渡线
left-hand crusher 左式破碎机;左侧型破碎机
left-hand curve 左向曲线
left-hand derivative 左导数
left-hand designation 左向标志
left-hand digit 左边的数字;高位数位
left-hand door 左开门
left-hand drive 左座驾驶;左御(车)
left-hand drive car 左转向盘车辆
left-handed 向左旋转;反时针;左手方向的
left-handed coordinate system 左旋坐标系;左手坐标系
left-handed door 左开门

left-handed engine 左转发动机
left-handed form 左形
left-hand edge 西图廓
left-handed helix 左向螺旋线
left-handed machine 反装
left-handed moment 左转力矩
left-handed multiplication 左乘法
left-handed nut 左旋螺母
left-handed opening 左开关
left-handed polarization 左旋偏振;左旋极化
left-handed polarized wave 左旋极化波
left-handed propeller 左旋螺旋桨;反时针转螺旋桨
left-handed quartz 左旋石英
left-handed reference frame 左手参考系
left-handed rope 左搓绳
left-handed rotation 左旋;逆时针旋转;左转
left-handed rule 左旋定则
left-handed screw 左旋螺杆
left-handed separation 左形移动;左行离距
left-handed system 左手定则;左旋定则
left-handed thread 倒牙
left-handed twist 左扭转;左旋螺旋线
left-handed vessel 左旋推进器船
left-handed watch 左表
left-handed window 左开窗
left-hand end 左端
left-hand engine 左转发动机
left-hand facing tool 左削平面车刀
left-hand fault 左旋断层
left-hand fusee 左向发条轮
left-hand helicity 左螺旋
left-hand helix 左侧螺旋线;左向螺旋线
left-hand hinge 左向铰链
left-hand lang-lay 同向左捻
left-hand lay 左转扭绞;左捻
left-hand limit 左旋极限
left-hand lock 左手门锁
left-hand loom 左手织机
left-hand lower derivate 左方下导数
left-hand machine 左侧操作机床
left-hand magneto 左转磁电机
left-hand mark 左侧标
left-hand member 左边部分
left-hand movement 左传动机芯
left-hand nut 左旋螺纹螺母
left-hand opening 左开关
left-hand opening valve 左旋阀
left-hand ordinary lay 交叉左捻
left-hand pedal 左踏板
left-hand photograph 左像片
left-hand polarization 左向偏振
left-hand reverse door 左手外开门
left-hand rotating fan 左转(通)风机
left-hand rotation 左向旋转
left-hand rotation reamer 左切铰刀
left-hand round edge external corner 左侧圆角阳角砖
left-hand rule 左手定则
left-hand running 左侧行车
left-hand screw 左旋螺钉;左螺旋
left-hand screw drill pipe 反扣钻杆
left-hand screw thread 左旋螺纹
left-hand side 左面;左边
left-hand side back link 左端回指连线
left-hand side outlet elbow 左手测流肘管
left-hand side outlet tee 左手测流三通
left-hand side tool 左削侧刨刀
left-hand spin 左螺旋
left-hand spiral bevel gear 左旋螺旋伞齿轮
left-hand staggered junction 左方错位式交叉口
left-hand stairway 左扶手楼梯
left-hand steering 左座转向
left-hand stone 左侧进站
left-hand switch 左向道岔
left-hand system 左手系
left-hand table 左侧型摇床
left-hand tap 反扣公锥
left-hand taper 左旋螺锥形特性
left-hand thread 左(旋)螺纹;反螺纹
left-hand-thread rod 左丝扣钻杆
left-hand thread tap 左旋螺丝锥
left-hand three finger rule 左手三指定则
left-hand tool 左切刀;反手刀具
left-hand tool bit 左削刀头

left-hand tool joint 反扣钻杆
left-hand tool joint thread 反扣接头螺纹
left-hand tools 左削刀具
left-hand traffic 左侧行车的交通
left-hand turn 左转弯;左转;左向旋转
left-hand turning 左向转弯;左向旋转
left-hand turnout 左开道岔;左侧道岔
left-hand turn prohibited 禁止左转(弯)
left-hand twist rail 左扭钢轨
left-hand unit 左侧装置
left-hand upper derivative 左方上导数
left-hand variable 左侧变量
left-hand winding 左向绕组
left-hand wing rail 左翼轨
left-hand zero 左边零
left hemisphere 左半球
left-in 留守
left-in-place formwork 留在原处的模板
left justified 左侧调整
left justified text 左对齐文本
left justify 左对位;左调整
left-laid 左捻
left laid rope 左搓绳
left-lane 左边车道;左车道
left-lateral fault 左旋断层;左行断层;左推断层;左侧断层
left-lateral motion 向左侧运动
left-lateral position 左侧卧位;左侧位
left-lateral strike-slip fault 左行走向滑动断层
left lay 左顺捻;左捻
left lay rope 左捻钢(丝)绳
left limit 左极限
left line 左线
left linear form production 左线性形式产生式
left longitude 左经度
left-luggage office 行李寄存处
left main exhaust port 左侧排气主口
left margin 左缘;左端
left marginal bank 左边岸
left marginal branch 左缘支
left-most bit 最左位
left-most cell 最左单元
left-most derivation 最左导出
left-most terminal set 左端终止集
left-most tree 最左树
left-normal-slip fault 左行正向滑动断层
left-off movement 左转驶出行驶
leftover 剩余物;剩余的;废屑料;废料;屑物;下脚料
leftover area 荒地
leftover bits and pieces 下脚料;边角料
leftover bits and pieces of raw material 边角余料
left pallet stone 擒纵叉左叉瓦
left part 左部
left part list 左部表
left part of a rule 规则左部
left rail 左股钢轨
left rear wheel 左后轮
left recursion 左递归
left-recursion problem 左循环问题
left recursive definition 左递归定义
left recursive rule 左递归规则
left recursive way 左递归方式
left-reverse-slip fault 左行逆向滑动断层
left-right 左-右
left-right asymmetry 左右不对称
left-right control 左右调整
left-right deviation 左右偏差
left-right guidance 左右制导
left-right indicator 左右方向;零位指示器
left-right needle 左右向指针;左右方向指针
left-right parse 左到右分析
left-right staggered junction 左右错位式交叉(口)
left rudder 左舵
left screw 左螺旋;左侧螺旋桨
left shift 向左移(位);左移(位)
left-shift times of multiplier 乘法器左移次数
left side 左侧;反面
left side bank 左岸(面向下游)
left sideline of edge-cut 挖槽左边线
left side of the figure 图的左侧
left slip 左滑
left slope angle 左坡角
left square bracket 左方括号

left stair(case) 左转楼梯
left steering trapezoid connection 左转向梯形臂
left substitutability 左置换
left subtree in a binary tree 二叉树中的左子树
left surface 左侧面
left swing line 左向旋转绳索
left tapered plate 左转弯楔形衬砌环
left tile 左瓦
left-to-right parser 左到右分析算法
left-to-right reading 正像
left-to-right scan 左到右扫描
left track 左侧履带
left translation 左平移
left turn 左转弯;左转变
left-turn ban 禁止向左转
left-turning traffic 左转行车
left-turn lane (交叉口外的)左转(弯)车道;左转专用车道
left-turn leg 左旋转撑脚;左旋转支柱
left turnout 左开道岔
left-turn ramp 左转匝道
left-turn slot 左转弯专用道
left-turn stand 左转站立;左转支座
left turn switch 左开道岔
left verge tile 左边缘瓦
left view 左视图
leftward skew slab 左向斜板
leftward welding 左向焊法
left-wheel 左转弯走
leg 桩腿;支柱;支腿;犁柱;航段;焊脚;气压排液管;棚子支托
legacy 遗址;遗赠;遗产
legacy duty 分遗产税;遗产税
legal 法定的;法定
legal address 法定地址
legal administrator 法定财产管理人
legal advice 法律咨询
legal adviser 法律顾问
legal age 法定年龄
legal agency 法定代理
legal agent of owner 法定船东代表
legal aid certificate 法律援助证书
legal ampere 法定安培
legal appropriation 法定用途;法定拨款
legal arbitration 法律仲裁;按法律进行仲裁
legal aspect 法定概念
legal aspects of pollution 污染的法律问题
legal assets 法定资产
legal assignment 法定转让
legal authority 法律认可
legal authority for air pollution 空气污染法律依据
legal authorization 法律许可
legal binding force of economic contract 经济合同的法律约束力
legal body 法人;法律机构
legal bond 法定证券;法定债券
legal broker 合法经纪人
legal budget 法定预算
legal cap 大八开(图)纸
legal capacity 权利能力;法定身分;法定资格
legal capacity of juristic person 法人的权利能力
legal capital 法定资本
legal compensation 法律补偿
legal consciousness of environment 环境法律意识
legal consolidated budget 法定总预算
legal constituted authority 法统
legal constraint 法定约束
legal construction of railway statistics 铁路统计法制建设
legal consultation 法律咨询处;法律顾问处
legal control 合法控制
legal cost 诉讼费
legal creditor 法定债权人
legal currency 法定货币
legal custody 法定监护;法定保管
legal day 一日(到午夜12点为止);法定日
legal debt limit 法定债务限额
legal debt margin 法定债务限额
legal debtor 法定债务人
legal delay period 法定延误期间
legal delay time 法定延误时间
legal demi-gross weight 法定半毛重
legal department 法律部门
legal documents 法定单证

legal domicile 法定住所
legal duty 法定义务
legal earned income 法定收入
legal earned reserve 法定盈利准备金
legal earned surplus reserve 法定盈余公积金;法定已获盈余准备金
legal effect 法律效力
legal entity 法人实体;法人;法律实体;法定实体;法定单位
legal estate 法定不动产
legal evidence 法定证据
legal expenses 诉讼费;法律费
legal formalities 法律程式
legal fruits 法定果实(即利息地租等)
legal gas 法定天然气
legal guardian 法定监护人
legal heir 合法继承人;法定继承人
legal holder 法定持有人
legal holiday 法定假日;法定假期
legal hour 法定时间;法定工作时间
legal immigration 合法迁移
legal income from property 法定财产收益
legal institutions 法制
legal instrument 法律文件;法定证书
legal interest 法定利息;法定权益
legal investment 合法投资;法定投资
legal investment list 法定投资清单
legality 合法性;合法
legality control 合法性控制
legalization 合法化
legalize 签证
legalized invoice 签证发票
legal jurisdiction 法定管辖
legal keeping 法定持有
legal lending limits 法定贷款限额
legal liability 法律责任
legal lien 法定留置权;法定财产留置权
legal limit (汽车的)法定速率限制
legal limits of time 法定时限
legal liquidation 法定清算
legal list 法定投资项目一览表;可上市证券;合法证券;合法投资
legally binding agreement 有法律效力的协议
legal man 法人
legal margin 法定保证金
legal maximum of interest rates 法定最高利率
legal measure 法律措施
legal measurement unit 法定计量单位
legal measuring instruments 法定测量器具
legal merchandise 合法商品
legal metrology 法制计量学
legal minimum rate 法定最低利率
legal minimum reserve 法定最低准备金
legal minimum wage 法定最低工资
legal money 法币
legal monoply 法定专营
legal monopoly 法定垄断;法定独占
legal name 依法登记的名称;正规名称;名称;法定名称
legal net weight 法定净重
legal norm of environment 环境法律规范
legal notice 法律公告
legal obligation 法律责任;法定债务;法定义务
legal occupation 法定职业
legal occupier 法定占有人
legal ohm 法定欧姆
legal oil 法定石油
leg along 排绳
legal organization 法定组织
legal or judicial security 法定的或法院判决的担保
legal owner 法定所有人;合法业主;合法所有人
legal payload 容许载重量;法定有效荷载
legal penalty 法律处罚
legal person 当事人;法人
legal personality 法人资格
legal personal representative 法定代理人
legal personalty 法定动产
legal possession 法定占有
legal price 法定价格
legal private property 法定私有财产
legal procedure 诉讼程序;法律程序;法定程序
legal proceedings 法律程序;法定程序
legal process 法律程序
legal professional privilege 法定的职业权力

legal profit 法定利润
legal program(me) 合法程序
legal property 法定财产
legal protection 法律保障;法律保护
legal public loan fund 法定公债基金;法定公共贷款基金
legal qualification 法定资格
legal quay 特许码头(英国)
legal rate 法定汇率
legal rate of interest 法定利率
legal ratification 法律认为
legal ratio 法定率
legal regulation 法规
legal relation of environment 环境法律关系
legal remedy 法律补救方法
legal remedy of air pollution 空气污染的法律制裁
legal representation letter 审计客户律师表白书
legal representative 法人代表;法定代理人;法定代表人
legal required field intensity 指定场强;规定场强
legal requirement on deposit 法定存款准备金标准
legal requirements of banks 法定银行准备金
legal reserve 法定准备金;法定存款;法定储备金
legal reserve fund 法定准备基金;法定基金;法定储备金
legal reserve policy 法定准备政策
legal reserve requirements 法定存款准备金比率
legal residence 依法登记的住所;法定住所
legal responsibility 法律责任;法定的责任
legal retriction 法律约束
legal retrieval 合法检索
legal rights 合法权利;法定权利
legal sanction 法律制裁
legal security 法定担保
legal service area 广播区域
legal settlement 法律清标
legal size 法定规格纸
legal speed 法定最高速度;法定速度
legal speed limit 法定时速限制
legal standard 法定标准
legal standard of value 法定价格(标准);法定价值标准
legal subject 法律主体
legal successor 合法继承人
legal surplus 法定公积金
legal system 法制
legal system of environment 环境法制
legal system of science and technology 科技法制
legal tare 法定皮重
legal tax cut 合法减税
legal tender 法定债款;法定货币;法偿;法币
legal tender bonds 法偿货币债券
legal tender money 法偿货币
legal tender paper money 法偿纸币
legal tender powers 法偿能力
legal term 法定支付期;法定期限
legal time 法定时间;法定工作时间
legal title 合法所有权;所有权;法定所有权;法定权利
legal unit 法定单位
legal unit of measurement 法定计量单位
legal valuation 法定评价;法定估价
legal valuation method of inventory 法定库存估价法
legal valve 法定价值
legal volt 法定伏特
legal warranty 法定捣牵
legal water level 法定水位
legal weight 法定重量
legal welfare expenses 法定福利费
legal wheel load 法定轮载
legal year 法定年度
legatee 遗产承受人;受遗赠者
legatee of inheritance 遗产继承人
legation 公使馆
leg bridge 以立柱作支承的梁式桥;柱支承梁桥;支柱桥(台桥);立柱式桥
leg cal(l)ipers 通用卡钳
leg clearance 门架净空
leg drop 舞台两侧的窄幕
legena 瓜状体
legend 图例;图的说明;图表符号;图标符号(插图说明)
legendary 图廊外
legendary data 图廊外资料

legend of commands 指令图例
legend of symbols 图例符号;常用符号表
legend plate 图例板;图标板
Legendre's coefficient 勒让德系数
Legendre's condition 勒让德条件
Legendre's contact transformation 勒让德变换
Legendre's equation 勒让德方程
Legendre's function 勒让德函数
Legendre's polynomial 勒让德多项式
Legendre's polynomials 勒让德多项式系
Legendre's rule 勒让德法则
Legendre's series 勒让德级数
Legendre's theorem 勒让德定理
Legendre's transformation 勒让德变换
leggatt 木签子
legget 整理茅草的工具
leggings 绑腿
leggy 多相位
leg-hammer 支架式凿岩机
Le Ghatelier apparatus 理查德里仪(用于测量水硬性水泥安定性的仪器)
leghorn 麦杆黄(色)
legibility 易读性;明视度;明了性;清晰性;清晰度
legibility distance 可读距离
legibility time 判读时间
legionnaires's disease 军团病
legislation 法规
legislation for city planning 城市立法规划法规
legislature 立法机关
legitimacy 合法性
legitimate authority 法定职权
legitimate business 正当业务
legitimate claim 合理的索赔
legitimate cost 正当成本
legitimate house 营业性戏院
legitimate income 正当收入
legitimate interests 合法利益
legitimate name 正统命名;合法命名
legitimate power 法定力
legitimate procedure 正规手续;合法手续
legitimate rights 合法权利
legitimate smoke 合法用火烟
legitimate sovereign 法定主权
leg length 焊脚长度
leglet 腿饰
leg level(l)ing screw 脚安平螺旋;安平脚螺旋
leg members (钻塔的)支柱;塔腿构件
leg-mounted pneumatic rock drill 风动支架凿岩机
legnum 翅瓣缘
leg of a bridge 电桥臂
leg of a fillet weld 角焊缝焊脚;角焊缝
leg of angle 角钢肢;角边
leg of bit 巴掌
leg of circuit 电路臂
leg of compasses 圆规脚
leg of derrick 钻塔腿
leg of fillet weld 圆角焊肉厚
leg of frame 构架支柱;框架支柱;框架边
leg of mutton 三角软帆;羊腿形的;三角形的
leg of pitch 支柱间距
leg of the bit 钻头翼
leg of tower 塔脚
leg pathway 路径
leg piece 撑住;(钻塔的)支柱;立柱
leg pipe 短铸铁支风管;冷凝器气压管;大气冷凝管
leg-pulling 扳腿推拿手法
legrandite 基性砷锌矿;水羟砷锌石
leg reinforcing 塔腿加固
leg rest 搁脚
leg roller 车脚小轮
leg room 伸腿的地方(车、飞机等);伸脚空间(座位train)
leg row boat 脚划船
leg-secured type 支脚固定式
legs of frame 刚架支腿
leg supporter 搁脚
leg type 支腿式;支脚式
legume crop 豆类作物
leguminous 豆荚状
leguminous cover 豆科覆盖植物
leguminous structure 豆荚状构造
leg vice 虎钳;台虎钳;长腿虎钳
leg wire 脚线
leg wore material 脚线材料

legwork 新闻采访工作;跑腿活
Lehigh bead bend test 焊道弯曲试验
Lehigh jig 理海型跳汰机
Lehigh restraint cracking test 里海裂纹试验;拘束抗裂试验
Lehigh specimen 里海试样(焊接试验用)
Lehigh type cracking test 里海裂纹试验
lehiite 磷钙碱铝石;白磷碱铝石
Lehmann discontinuity 勒曼不连续面
Lehmann process 勒曼选煤法
Lehmann's rule 勒曼规则(后方交会试错法)
lehnerite 板磷铁矿
lehr 退火窑;退火炉;长隧道窑;玻璃退火窑
lehr assistant 退火辅助工
lehr attendant 退火工
lehr belt 退火窑输送网带
lehr breakage 退火破损量
lehr crack 退火破裂
lehr end serviceman 退火辅助工
lehr loader 退火窑送料装置
lehr man 退火工
lehr mat 退火窑输送网带
lehr minder 退火工
lehr operator 退火工
lehuntite 钠沸石;豆科植物
Leibnitz's rule 莱布尼兹法则
Leibnitz's test 莱布尼兹判别法
Leidenfrost point 莱顿弗罗斯特点
Leidenfrost's phenomenon 莱顿弗罗斯特现象
leidleite 流安松脂岩
leifite 白针柱石
Leighton Buzzard sand (英国供水泥试验的)莱顿布扎尔德砂
Leighton Buzzard silver sand 莱顿布扎尔德银砂
leightonite 钾钙铜矾
leiocome 絮凝剂
leiocome glue 絮凝胶
leiomyoblastoma 成平滑肌瘤
Leipzig market halls 莱比锡商场
Leipzig yellow 莱比锡黄;铅铬黄;铬黄
leisure area 休养区;游览区
leisure center[centre] 消闲中心;休假中心
leisure facility 休息设施;闲暇消遣设施
leisure industry 闲暇工业
leisure service 空闲服务
leisure traffic 闲时交通
leisure zone 游憩地带
leiteite 亚肿锌石
Leithner's blue 莱兹纳尔蓝
Leitz interference microscope 莱茨干涉显微镜
Leitz lens 丽斯镜头(德国)
Leitz sector shutter 莱茨扇形快门
l electric(al) circuit plan 电路图
Lemac 聚醋酸乙烯酯系列
Lemaitre cosmo logic(al) model 勒梅特宇宙模型
Lemaitre regularization 勒梅特正规化
Leman prism 莱曼棱镜
Lemarquand 铜锌基锡镍钴合金
lemma 辅助定理
Lemnaceae 浮萍科
lemna polyrhiza L 浮萍(属)
lemnasite 钠磷锰矿
lemnian earth 水磨土
lemniscate 双纽线
lemniscate coordinates 双纽线坐标
lemniscate curve 双纽曲线
lemniscate function 双纽线函数
lemniscate of Bernoulli 伯努利双纽线
lemniscate reception 双纽线方向性图接收
lemniscate transition curve 双纽型缓和曲线
lemon 柠檬;柠檬树;淡黄色
lemon brown rot 柠檬褐腐病
lemon chrome yellow 铬酸钡黄颜料;柠檬黄(色);柠檬铬黄
lemon grating 柠檬皮条片
lemon lac 柠檬胶
lemon oil 柠檬油
lemon pale 柠檬苍白色
lemon pips oil 柠檬子油
lemon salt 柠檬盐
lemon scented gum 柠檬桉
lemon seed oil 柠檬子油
lemon shellac 柠檬黄级紫胶;柠檬虫漆
lemon spline 柠檬塞缝片

lemon wood 柠檬木
lemon yellow 柠檬黄(颜料);柠檬黄(色)
lemon yellow pale 浅柠檬黄
lemoynite 水钠钙锆石
lenad 似长石类
Lenard ray 勒纳射线
Lenard tube 勒纳德管
Lenard window 勒纳德窗
lend 借出;贷款;贷
lendable 可借贷的
lend a hand 协助
lend at interest 计息贷款;放出有息贷款
lender 出借人;贷方;出借者
lender of last resort 最后贷款人
lending 放贷
lending and borrowing business 借贷业
lending at call 即期贷款
lending bank 贷款银行
lending department 出纳处
lending foreign exchange account 出借外汇账户
lending institution 贷款机构
lending library 图书馆租书处;公共图书馆;租书处;收费图书馆
lending margin 借贷边际
lending of development funds 开发基金贷款
lending operation 贷款业务
lending rate 贷款利率
lending sand 配料砂
lend-lease 平等租借交换
lend-lease act 租借法案
lend money on interest 附息贷款
lend money on security 凭抵押品贷款
lend money on usury 放债;放高货利
L-endwall L形端墙
lenetic 静水群落(的)
lenetic community 静水群落
lengenbachite 辉砷银铅矿
length 长宽高;长度
length adjustment platform 长度调节板
length adjustment stop 长度调节板
length at crest 坝顶长度
length attribute 长度属性
length-azimuth-frequency histogram analysis of lineament 线性体长度方位频数直方图分析
length bar 量棒
length between buffers 缓冲器保险杠间距
length between coupler heads 车钩钩头间距
length between perpendicular of ship 船舶垂线间长
length between perpendiculars 两柱间长;垂线间距;垂线间长(度)
length-breadth ratio 长宽比(率)
length cargo 长件
length change 长度变化
length character 长度代号
length check 长度检验
length coefficient 长度系数
length condition 长度条件
length correction 长度校正;长度改正
length cutting 纵切
length depth ratio 长深比
length diameter ratio 长度直径比
length draft ratio 长度吃水比
lengthen 延长;拉长
lengthen drilling line 由绞车放钢绳
lengthened chain 长链
lengthened code 延长码
lengthened coupling 加长联轴器
lengthened dipole 加长偶极子
lengthened motor lorry 加长载货汽车
lengthener 伸长器
lengthening 延长;拉长;加长;延伸
lengthening bar 延伸杆(件);接长杆件;加长杆
lengthening coefficient 延伸系数;伸长系数
lengthening coil 加长线圈
lengthening crossing loop 延长会让站
lengthening inductance 加感线圈
lengthening joint 接长接头;增长接头;加长连接
lengthening of ship body 船身加长法
lengthening of timber 木料接长
lengthening piece 接长料;接长杆件;底链
lengthening reaction 伸长反应
lengthening rod 钻杆接杆;加长钻杆;接杆(钻井钻杆接长用)

lengthening structural timber 接长结构木料
lengthen label 加长标号
lengthen symbol 加长符号
length equation 基线方程;长度方程(式)
length extension vibration mode 长度伸缩振动模式
length factor 长度系数
length fast 延性负;负延性
length feed 纵向进给;纵向进刀
length field 长度字段
length for butt seam inspection 对接焊缝探伤长度
length-frequency histogram analysis of lineament 线性体长度频数直方图分析
length ga(u)ge 长度规;长度计
length-height index 长高指数
length-height ratio 长高比
length increment 钎子组每根长度差
length limit 长度极限
length machined 机切长度
length mean diameter 长度平均径
length measurement 长度量算;长度量测
length-measuring instrument 测长仪;长度计量仪器
length-measuring interferometer 测长干涉仪
length-measuring machine measuring machine 测长机
length-measuring numeral display meter 测长数显表
length meter 测长器
length modifier 长度修饰语;长度修改因子
length modulation 长度调制
length of 4 way pipe 四通长度
length of action 啮合长度
length of active fault 活断层长度
length of a curve 曲线的长度
length of advancement 前移距
length of a knot 节长
length of an interval 区间的长
length of anomaly body 异常体长度
length of a particle 长径
length of aperiodicity stretch 非周期延伸长度
length of arc 弧长
length of armature 电转子长
length of artificial recharge engineering 回灌工程长度
length of asbestos fiber 石棉纤维长度
length of a scale division 标度分格间距
length of backwater reach 回水距离
length of bailer 提捞筒长度
length of base 基线长度
length of berth 泊位长度
length of body 体长
length of boundary wall 围墙长度
length of bridge 桥长
length of budget period 预算期限的长度
length of burning zone 燃烧带长度
length of caisson 沉箱长
length of cantilever 悬臂梁长度;悬臂长度
length of car body underframe 车体底架长度
length of casing 套管长度
length of casing section 套管段长
length of channel 河道长度
length of closed loop 闭合环线周长
length of closed section 隧道净长
length of common normal 公法线长度
length of conductor 导管长度
length of conduit 通道长度
length of constructed line 线路建筑长度
length of construction line 线路建筑长度
length of core 岩芯长度
length of core barrel 岩芯管长度
length of corner cut 转角切除长度
length of correlation time window 相关时窗时间
length of cracks 裂缝长度;地裂缝长度
length of crest 坝顶长度
length of curves 曲线长(度);弯道长度
length of cut 切削长度;切割尺寸
length of cylinder 圆柱长度
length of dam 坝长;坝顶长度
length of day 日照长度
length of deconvolution operator 反褶积算子长度
length of deformation band 变形带长度
length of directed-path 有向路长度
length of dispersion band 弥散带长度

length of dock 坞长度
length of dominant fiber[fibre] 主纤维长度
length of down-comer 降液管长度;下液管长度
length of draw 拉伸长度
length of drilling 钻探进尺
length of drill rod 钻杆长度
length of dripstone 滴石长度
length of dry dock 干坞长度
length of earth fracture 裂缝长度【地】
length of echelon array 雁列带长度
length of electric(al) prospecting section 电测剖面长度
length of embankment 入土长度(指桩)
length of embedment 埋入长度
length of entrance 进流段长度
length of equal effect pipeline 等效管线长度
length of equivalent pendulum 等效摆长度
length of exposure 曝光时间;曝光持续时间
length of faulting 断层作用长度
length of feed 给进长度
length of fence 围墙长度
length of field 字段长度
length of fish 落鱼长度
length of fit 螺纹旋入长度;配合长度
length of flame 火焰长度
length of flow mark 流痕长度
length of fold 褶皱长度
length of forward movement 前移距
length of freight loading-unloading 货物装卸有效长度
length of frost free period 无霜期间
length of frost free season 无霜期间
length of geologic(al) section line 地质剖面线长度
length of glide mass 滑体长度
length of grade section 坡段长度
length of gradient 坡段长度
length of half car 半节车厢长度
length of handling 铁路装卸长度
length of haul 运距
length of height line 高程路线长度
length of height-tide coastline 高潮海岸线长度
length of horse 断片长度
length of identifier 标识符长度
length of inclined well 斜井的长度
length of increment 焊段长度
length of inner diameter arc 内径弧长
length of jaspe sheet 影纹布的长度
length of jump 跃长;水跃长度
length of karstic stony column 石柱长度
length of keel 龙骨长度
length of landslide crack 滑坡裂隙长度
length of lay 捻距;捻长;(绳的)摆动长度;捻距长;接线长度
length of leg wire of electric(al) detonator 电雷管脚线长度
length of life 寿命;使用寿命;使用期限
length of lifting bridge 吊桥长度
length of lifting bridge beam 吊桥托梁长度
length of lifting drawbridge 吊桥长度
length of line 线路长度
length of lineament 线性体长度
length of line constructed 线路建筑长度
length of line in operation 运营线路长度;铁路通车长度
length of line open to traffic 运营线路长度
length of line operated 运营线路长度
length of line operation 运营线路长度
length of liner 尼管长度
length of loading/unloading siding 铁路装卸长度
length of load waterline 满载水线长
length of lowered pantograph 受电弓降低后长度
length of low-tide coastline 低潮海岸线长度
length of machine 机长
length of magnetic path 磁路长度
length of maximum grade 最大坡道之长度
length of mean-tide coastline 平均潮位海岸线长度
length of mean turn 匝的平均长度
length of measurement 计量长度
length of melt period 融雪持续时间;解冻持续时间
length of milling 套铣长度
length of model 模型长度
length of mooring line 锚链长度
length of mud-stone flow body 泥石流体长

length of normal 法线长度
length of observation line for test 试验观测线长度
length of observation tunnel 观测平硐长度
length of observing line 观测线长度
length of offset 被错动长度
length of oil 油的延展长度
length of one round 一次(掘进)循环进度(长度)
length of orebody 矿体长度
length of outer diameter arc 外径弧长
length of overland flow 地面漫流长度
length of pendulum 摆长
length of penetration 贯入度;渗透深度
length of perpendiculars 垂直距离
length of pipe 管子长度;管长
length of pipeline 管线长度
length of pipe section 管段长度;管段
length of piston stroke 活塞冲程
length of plugging 堵塞长度
length of pole arc 极弧长度
length of port railway 港口铁路线长度
length of profile 剖面长度
length of protection 保障期限
length of pumping sections 取水段长度
length of quay 码头长度
length of queue 队长
length of radial pipe 辐射管长度
length of railroad lines in service 铁路营业里程
length of railway line 铁路线路长度
length of railway special line 铁路专用线长度
length of reach 河段长度
length of record 记录期限;记录长度
length of regime observation line 动态观测线长度
length of residence 居住期限
length of restraint 嵌固长度;镶接长度
length of river 河流长度
length of road 道路长度
length of roll 胶片长度;卷材长度
length of round 一周长度;一圈长度
length of run 运转时间;运(程)距;运程;游尼距;展开长度;推进长度;船尾端长;游程长度
length of runoff zone 径流带长度
length of sampler 取样器长度
length of sampling 采样长度
length of sand settling tube 沉沙管长度
length of saw blade 锯条长度
length of scale 标尺长度
length of scan line 扫描线长度
length of screen tube 滤水管长度
length of search 搜索长度
length of seismic prospecting section 地震剖面长度
length of service 工龄
length of service bonus 服务年限奖金
length of shakes (木材的)裂纹长度
length of sheet flood 漫流长度
length of sheet flow 漫流长度
length of shift 值班时间
length of ship 船的长度
length of side 边长
length of single well pipe 单节井管长度
length of skirt 裙座长度
length of slope 斜坡长度;边坡长度
length of soil sample 土样长度
length of span 跨距长度
length of splice 搭接长度
length of spread 传播长度;穿插长度
length of spur line 支线长
length of stalactite 钟乳石长度
length of station 车站长度
length of station site 站坪长度
length of straight flange 直边长度
length of straight pipe 管子展开长度
length of strike for sheet 板体走向长度
length of string 拉丝长度;拔丝长度
length of strip hole 条孔长度
length of stroke 冲程长度;冲程
length of structure 建筑物长度
length of support 支承长度;支座长度
length of switching area 咽喉区长度
length of switch rail 尖轨长度
length of tangent(ial)(line) 切线长(度);切距
length of tectonic region 构造区的长度
length of temporarily operating railway 铁路临时营业长度

length of test sector 实验段长度
length of the cycle 循环长度
length of the day 日长
length of the interfacial wave 界面波长
length of the intervals 间隔时间
length of the largest vessel 最大船舶长度
length of the limitation period 时效期
length of the magnetic body 磁性体长度
length of the month 月长
length of the smallest vessel 最小船舶长度
length of the year 年长
length of thread 螺纹长度
length of thread engagement 螺纹啮合长度
length of thread ring 造扣长度
length of time 时期;持续时间
length of time employed 工龄
length of time window of deep layer 深层时窗长度
length of time window of shallow layer 浅层时窗长度
length of tooth 齿高
length of track laid out 铺轨长度
length of track on ground 履带着地长度
length of train 列车长度
length of transfer 传力长度;传输距离;传递长度
length of transporting water line 输水线路长度
length of trap 圈闭长度
length of travel 行程
length of tray pitch spacer 塔板定距管杆长度
length of trolley wire network 触线网长度
length of tube 管子长度
length of tunnel proper 隧道净长
length of turnout 道岔长度
length of TV under well 井下电视段长度
length of underground river 暗河长度
length of underground stream 暗河长度
length of vane 十字板头长度
length of vector 向量长(度)
length of vitality 使用期限
length of wagon 车身长度
length of wagon floor surface for bearing goods weight 车辆负重面长度
length of warranty 保险期
length of wave 波长
length of wave crest line 波峰长度
length of wave group 波群长度
length of weight sequence 权序列长度
length of weld 焊缝长度
length of well rows 井排长度
length of wharf 码头长度
length of wheel base 轴距离长度
length on face 辊身长度
length on water line 水线长;吃水线长
length overall 船舶全长;总长(度)
length per bar 每根长
length per member 每块构件长
length ratio 长度比
length rod 测杆
length scale 长度规;长度比例(尺);长度比
length shear ga(u)ge 定尺剪切挡板
length slow 延性正;正延性
lengthsman 长度测量员
length specification 长度说明
length stop 纵向制动器;纵向止动器
length to beam ratio 长高比【船】
length-to-diameter ratio 长度直径比;长径比
length to height ratio 长高比
length tolerance 长度公差
length to width ratio 长宽比
length travel 纵向位移;纵向进给
length tube vertical evapo(u)rator 垂直长管型蒸发器
length unit 长度单位
length variation 长度变化;长度偏差
lengthways milling 纵铣
length width ratio 长宽比(率)
lengthwise 长向的;纵向的;纵排(货物装舱)
lengthwise direction 纵向
lengthwise fold 纵向折叠
lengthwise movement 纵向运动
lengthwise oscillation 纵向振荡
lengthwise position 纵向位置
lengthwise seat 纵长座
lengthwise section 纵断面

lengthwise shrinkage crack 纵向缩缝
lengthwise streaking 纵向条痕
lengthwise tear 纵向撕破
lengthwise travel rate of table 工作台纵向移动速度
lengthy cargo 特长货物;超长件;超长货物;长件货
lengthy cargo charges 超大货物运费
lengthy charges 超长费;长件货加价费
lengthy shut down periods 长期停机时间
Lenham beds 伦哈姆层
lennilenapeite 淡硬绿泥石
leno 纱罗织物
lenoblite 二水矾石
leno cloth 纱罗织物
leno edge 纱罗布边
leno-fastening 纱罗边
leno loom 纱罗织机
leno reed 纱罗筘
leno-selvedge 绞边;纱罗布边
leno weave 纱罗织法
Lenox method 伦诺克斯法
lens 局部过厚;镜头;镜片;透镜体;透镜;电子透镜;扁平矿体;扁豆状体
lens aberration 透镜像差
lens abnormality 晶状体异常
lens adapter 附加透镜
lens-adapter ring 透镜适配圈
lens angle 镜头视场角;透镜角
lens antenna 透镜天线
lens aperture 镜头光圈(刻度);透镜孔径
lens aperture indicator 镜头光圈指数器
lens aperture number 透镜孔径数
lens aperture ratio 针孔透镜比
lens array imaging 透镜阵列成像
lens assembly 透镜组
lensatic compass 透镜磁罗盘;读数放大罗盘
lens axis 透镜光轴
lens barrel 透镜筒;透镜镜筒
lens bending 透镜配曲调整
lens blank 透镜毛坯
lens blooming 透镜薄膜
lens board 镜头板;透镜柜;透镜板
lens bracket 镜头支架
lens brush 镜头毛刷
lens calibration 镜头校准
lens cap 镜头盖;物镜盖;透镜罩;透镜盖
lens carrier 透镜架;透镜柜
lens carrier slide 镜头滑动架
lens case 镜头罩
lens cell 镜头元件;镜框
lens center[centre] 镜头中心;透镜中心
lens changing 镜头更换
lens channel 透镜孔道
lens cleaner 抹镜水
lens coating 镜头涂层;透镜镀膜
lens collar 镜头环
lens combination 透镜系统
lens compass 透镜罗盘
lens component 透镜部件
lens condenser 聚光镜
lens cone 镜筒;物镜框
lens control 透镜控制
lens convergence 透镜会聚
lens convergence plane 透镜会聚面
lens-corrected horn 透镜校正角
lens-coupled viewfinder 透镜耦合瞄准器
lens coupling 透镜耦合
lens cover 透镜盖
lens coverage 镜头视界;镜头拍摄范围;透镜视角
lens covering a small angle of field 窄视角透镜
lens covering a wide angle of field 宽视角透镜;广视角透镜
lens curvature 透镜曲率
lens diameter 透镜直径
lens diaphragm 镜头光阑
lens diaphragm opening 透镜光阑孔;透镜光阑孔
lens diaphragm scale 物镜光圈刻度
lens disc 透镜盘
lens disk 透镜盘
lens distortion 镜头畸变;透镜畸变
lens doublet 透镜偶极子;双透镜物镜
lens drum 透镜筒;透镜鼓
lens drum scanner 透镜轮扫掠器
lens edge 镜片外圆

lens efficiency 透镜分辨能力;透镜分辨率
lens element 透镜元件
lens equation 透镜方程
lens face 透镜面
lens factor 透镜因数(透镜光学分辨能力);透镜分辨周数;透镜分辨能力
lens field 透镜像场;透镜视场
lens field illumination 透镜视场照明
lens flare 镜头眩光;透镜反射光斑
lens focus 透镜聚焦;透镜焦点
lens formula 透镜公式
lens gasket 透镜垫片
lens grinding 透镜磨光
lens holder 镜头支架;镜头架;透镜架;透镜柜
lens hood 遮光罩;镜头遮光罩;物镜遮光(罩);透镜遮光罩
lens hood case 遮光罩
lensing 透镜状地层
lens iris 透镜(锁)光圈;透镜可变光阑
lens jacket 镜头套;透镜层
lens key 木材肉眼识别检索表
Len's law 锷次定律
lens light 透镜光
lens light guide 透镜光导管
lens-like 类透镜的
lens-like medium 类透镜媒质
lens loop 晶状体环
lens loss 透镜损耗
lensmaker's formula 薄透镜公式
lens matrix 透镜矩阵
lens measure 检镜仪
lens measure ga(u)ge 透镜表
lens meter 透镜检查仪
lens mount 镜头座;镜头筒;透镜(框)架
lens mount adapter 透镜接头
lens mounting 物镜框
lens nucleus 晶状体核
lens of clay 黏土透镜体
lens of extreme aperture 最大相对孔径物镜;临界孔径物镜
lens of sand 砂透镜
lens of silt 淤泥透镜
lens of variable focal length 可变焦距透镜
lens of wide aperture 大相对孔径物镜
lensoid 透境状
lens(o)meter 焦度计;镜度计
lens opening 镜头孔径;镜头开度;透镜孔(径)
lens optic(al) length 镜头光学长度
lens panel 物镜框;透镜圆片;镜头板
lens paper 镜头纸;拭镜头纸
lens particles 晶状体微粒
lens performance 镜箱性能
lens periphery 透镜边缘
lens pit 晶状体凹
lens plane 透镜面
lens plate 晶状体板
lens power 透镜光学能力
lens processing 透镜加工
lens profile 透镜剖面
lens pyrometer 透镜高温计
lens radial distortion 透镜径向失真
lens rate 透镜速率
lens register 镜面配准
lens response 透镜响应
lens retainer spring 透镜护圈簧
lens rim 透镜边缘
lens ring 透镜圈;透镜环
lens ring washer 透镜垫
lens scanning disk 透镜扫描盘
lens screen 透镜遮光片
lens seismic facies unit 透镜状地震岩相单元
lens selection 透镜选择
lens shade 遮光罩;镜箱罩
lens shaped 透镜状(的);扁豆状;透镜形的
lens-shaped stratified deposit 透镜状成层矿床
lens shield 遮光罩;镜头挡光板
lens shim 透镜垫片
lens shutter 中心快门;快门;镜头快门;透镜光闸
lens shutter drum 镜头快门转鼓
lens slide 镜头滑座
lens spectrometer 透镜分光计
lens speed 镜头速率;镜速率;透镜速度
lens stereoscope 立体镜;透镜式立体镜
lens stop 透镜光圈;透镜光阑

lens strength 焦度;透镜光焦度
lens surface 透镜面
lens system 透镜系统
lens telescope 透镜望远镜
lens test 透镜检验
lens tester 检镜仪;透镜检验仪
lens testing chart 透镜测试表
lens thickness 透镜厚度
lens tissue 镜头(薄)纸;抹镜纸;擦镜头纸
lens-to-screen distance 透镜对屏间距
lens transmission 透镜透光率
lens transmission efficiency 镜头透过率
lens transmission factor 透镜透光因数
lens truss 鱼形桁架;叶形桁架;棱形桁架
lens tube 镜头筒
lens turret 镜旋转头;透镜旋转台;透镜回转头
lens turret matting shot 镜头转轮遮摄
lens-type trap 透镜型圈闭
lens velocity 透镜速度
lens washer 透镜垫
lens watch 透镜深度规
lens wavebeam guide 透镜光导管
lens wiping paper 擦镜头纸
lens with automatic diaphragm 自动光阑透镜
lenthionine 蘑菇香精
lentic ecosystem 静水生态系统
lentic evnironment 死水环境
lenticle 透镜体;扁豆状体;扁豆体
lenticle of clay 黏土透镜
lenticonus 圆锥形晶状体
lenticular 两面凸的;晶状体;透镜状(的);透镜的;双凸透镜状;饼状;扁豆状
lenticular arch 双叶拱
lenticular arch bridge 双叶拱桥
lenticular beam 鱼腹式梁;组合梁;扁豆形(组合)梁
lenticular bed 透镜体层
lenticular bedding 透镜(状)层理
lenticular bedding structure 透镜状层理构造
lenticular bob 透镜形摆锤
lenticular cloud 夹状云
lenticular cross-bedding structure 透镜状交错层理构造
lenticular deposit 透镜状沉积
lenticular deposit sand 透镜状沉积砂
lenticular domain 透镜域
lenticular film 双凸透镜状胶片;凹凸式胶片
lenticular film process 微透镜胶片法
lenticular girder bridge 鱼腹式梁桥;透镜式梁桥
lenticular intercalation 扁豆夹层
lenticular interlayer 扁豆夹层
lenticular loop 豆状核袢
lenticular martensite 透镜(状)马氏体
lenticular migmatite 扁豆状混合岩
lenticular nucleus 豆状核
lenticular ore body 透镜状矿体;扁豆状矿体
lenticular pit aperture 透镜形纹孔口
lenticular plate 柱面光栅板
lenticular plate camera 双凸板摄影机
lenticular pseudofluidal structure 透镜状假流纹构造
lenticular screen 柱面光栅
lenticular structure 凸镜状结构;透镜状构造
lenticular truss 鱼形桁架;鱼腹式桁架;叶形桁架
lenticular truss bridge 鱼腹式桁架桥
lenticular unit 透镜部件
lenticular void 透镜状空隙
lenticular wire 扁线
lenticulated screen 柱面光栅
lenticulation 透镜光栅膜制造方法;透镜光栅
lenticule 微透镜;扁豆夹层
lentic water 死水;静水
lentiform 透镜状的;饼状的;扁豆状
lentiform beam 鱼形梁;鱼腹式梁;组合梁;扁豆形组合梁
lentiform nucleus 豆状核
lentiform ore body 透镜状矿体
lentiginose 着色斑的
lentiginous 着色斑的
lentigo stain 着色斑
lentil 小扁豆层(体);扁豆状夹层
lentil-headed screw 扁头铆钉;扁头螺钉
lentoid 透镜状结构;透镜状(的)
Lentz valve gear 伦兹式阀装置
Lenz's law 楞次定律

Lenz's rule 楞次定则
Leonard control 发动机电动机控制
Leonard converter 发电机电动机转换器
Leonard dynamo 变速用直流发电机
Leonardian 伦纳德统【地】
Leonardo's band 节制索
Leonard system 伦纳德方式
Leon firedamp tester 莱昂型瓦斯检定器
Leonhardt prestressing system 莱昂哈特预应力体系
Leonhardt system 莱昂哈特法(预应力钢筋张拉系统)
leonite 钾镁矾
Leontief inverse 莱昂惕夫倒数
leontine 牡丹草亭
leopard cat 豹猫
leopards skin limestone 豹皮灰岩
leopardwood 豹斑木(产于圭亚那)
Leopold related matrix method of environmental impact assessment 利奥波德环境影响评价相关矩阵法
Leopold underdrain system 滤砖配水系
lepadidae 茗荷儿
lepas 茗荷儿
Lepel discharger 勒佩尔放电器
lepel quenched spark-gap 盘形猝熄火花放电器
lepers hospital 麻风病医院
lepersonnite 莱普生石
leper's squint 教堂圣坛右面的小矮窗
leper window 低侧窗;(中世纪教堂外墙上的)小窗
lepidic 鳞屑的
lepidoblastic 鳞片变晶状
lepidoblastic texture 鳞片变晶结构
lepidocrocite 纤铁矿
lepidocrocite rock 纤铁矿岩
lepido granoblastic texture 鳞片花岗变晶结构
lepido granularblastic texture 鳞片粒状变晶结构
lepidolite 鳞云母;锂云母
lepidolite rock 锂云母岩
lepidophytotelinite 鳞木结构镜质体
lepisphere 鳞球
lepitophyto-vitrite 鳞木微镜煤
Lepocinclis 鳞孔藻属
lepoidolite 红云母
Lepol grate process preheater 立波尔窑炉算子预热机
Lepol kiln 立波尔(水泥回转)窑
Lepol travel(l)ing grate 立波尔移动式炉算子
Lepospondyli 壳椎亚纲
leprosarium 麻疯病院
leprosery 麻疯病院
leptite 长英麻粒岩
leptochlorite 鳞绿泥石
leptodermous 薄壁的
leptogeosyncline 瘦地槽;薄地槽
leptokurtic distribution 尖峰态分布;凸峰态分布
leptokurtosis 尖峰态;凸峰态;峰态
leptometer 比粘计
Leptomitus lacteus 水节霉
lepton 轻子;轻粒子
leptonomorphology 膜形态学
leptosporangiate 薄囊的
leptothermal deposit 亚中温热液矿床
lepto-zygenema 细偶线期
leptynolite 片状角岩
lermontovite 稀土磷铀矿;水铈铀磷钙石
Leroux formation 勒罗克斯组
lerrite 黄绿云母
Lerwick Series 莱威克统【地】
lesbian cyma(tium) 古希腊波状花边;带叶和箭饰的鸟混线脚;双弯曲线(脚)
lesbian leaf 水叶装饰
lesbian rule 铅制线脚型板
lesche 古希腊宴会厅
lesene 无帽壁柱;壁柱
less-cohesive soil 少黏性土
less container load 并柜卸载
less developed area 不发达地区
less developed countries 欠发达国家
less easily falling roof 中等冒落顶板
lessee 租户;承租人
lessen 缩小
lessening 缩减
lessening fold 缩减的褶皱

lesser-calorie 小卡
lesser circulation 小循环
lesser civet cat 小灵猫
lesser ebb 小落潮流;最小退潮流;最小落潮流(速度);(一天两次落潮中的)较弱落潮流;低低潮
lesser flood 最小涨潮流(速度);低高潮
lesserite 多水硼镁石
lesser river 小河;次要河流
lesser road 简易道路;低级道路
less favorable currency 软币
less important work 辅助作业
Lessing ring packing 勒辛环填料
Lessing rings 勒辛环
lessive brown earth 白浆化棕壤
less noble metal 次贵金属
lessor 出租者;出租人
lessor of small plots 小土地出租者
lessor's interest 出租人权利
less permeable layer 弱透水层
lesspollution 无公害
less privileged country 条件差的国家
less prominent 不突出
less reinforced concrete 少筋混凝土
less saline water 低盐水
less simple morphologic(al) 形态较简单的
less soluble gas 非溶性气体
less suitable water 不太适用的水
less than average 中下
less-than-carload 沿途零担车;铁路运输零担;零担的;拼车货
less-than-carload freight store 沿线零担仓库【铁】
less-than-carload freight train 沿途零担货物列车
less-than-carload freight transshipment station 零担货(物)
less-than-carload lot 零担货(物)
less-than-carload rate 包裹运价;零担运价(率)
less-than-carload team yard 零担货场
less-than-carload traffic 零担运输
less than condition 小于条件
less than contained load 拼箱装载
less than container load 零担的;拼箱货;未满载货柜
less than container load cargo 非整箱货
less than container lot 零担货(物)
less than match 小于符合
less than normal refraction 小于正常折射
less than or equal to 小于或等于
less than part-load transshipment station 零担货(物)
less than truckload 零担的;拼车货;不足一车零担货
less-than-truckload lot 载货车零担货物
less thermal stability coal 热稳定性较差煤
less-trafficked area 交通稀少地区
less turbid 微浊
less-well calcined 煅烧不良的
lestage 市场税
leste 累斯太风
let 路面排水沟
let a contract 订立合同;发包订约
let a mulch top of the soil 残茬留在地面
let contract 发包
let down 配漆;调漆;放料;程序下降
let-down heat exchanger 下泄热交换器
let-down stream 下行水流
let-down tank 排放槽
let-down vessel 放电容器
let engine test cell 喷气发动机试车间
let fall 丢下;倒下
let go 解缆;河岸崩决;放手
let go anchor 抛锚
let go a rope 解缆
let go the stern line 解尾缆
lethal 致死的
lethal agent 致死剂
lethal chamber 屠宰场
lethal concentration 致死浓度
lethal concentration fifty percent 半致死浓度
lethal concentration low 低致死浓度
lethal concentration required to kill 50% of the test animals 致死中浓度
lethal dosage 致死剂量
lethal dose 致死量
lethal dose 50% 半致死量;半致死剂量

lethal dose fifty percent 半致死剂量
lethal dryness 致死干(燥)度
lethal effect 致死效应
lethal equivalent 致死当量
lethal exposure 致死照射
lethal factor 致死因素
lethal gas 致死毒气
lethal index 致死指数;致死积
lethal irradiation 致死辐射
lethality 致死率
lethality rate 致死率
lethal mutation 致死突变
lethal radiation dose 致死辐射剂量
lethal range 致死限
lethal ray 致死光线
lethal substance 致死物质
lethal synthesis 致死性合成
lethal tange 致死范围
lethal temperature 致死温度
lethal threshold 致死阈(值);致死限值
lethal time 致死时间
lethal tools 凶器
lethal toxin 致死毒素
lethal weapon 凶器
lethal zone 致死区
lethargy 对数能量损失
let-in-braces 柱斜撑杆;插入式斜撑
let in concrete 放进混凝土
let-in flap 嵌入挡板
let it work itself 任其自己运行
let-off gear(stand) 导出装置
let-off pipe 出水管
let out 路肩排水沟
let out screw 松开螺钉
letovicite 氢铵矾;酸性铵矾
lettable area 可出租的面积
letter 信函;字母;铅字
letter agreement 书信合约;书面契约;协议书
letter box 信箱;供料机通道口
letter box blue 信箱蓝
letter-box company 信箱公司
letter-box plate 信箱板(安装在大门上的信槽孔板)
letter character 字母符号
letter chute 信件滑槽
letter code 字母代码
letter contract 书面合同
letter decoration 字母装饰;字体装饰
letter drop-off 邮箱;信箱
letter drop plate 信箱投口金属板;投信口遮板
lettered grid 注记坐标网
lettered message 文字通告
lettered square 有注记坐标网格
letter enrichment 以字体增添装饰
letter error 印刷错误
letterhead 图廓整饰样图
lettering 压花;注记(编排);地图注记;打号;编字码
lettering foil 透明注记
lettering in topographic(al) maps 地形图注记
lettering of chart 海图注记
lettering of light 灯标注记
lettering of map 图面排字
lettering on bottom 瓶底印记
lettering on drawing 图纸上的文字书写
lettering on map surface 图面注记
lettering plate 注记板
letter of abandonment 委付书
letter of acceptance 中标通知书;中标函;接受信;接受投标函;承兑函;投标接受书
letter of advice 汇款通知书;发货通知书;起运通知;通知书;通知单;汇款通知单
letter of advise 通知书
letter of agreement 协议书;协定书
letter of allotment 分配书
letter of application 申请书
letter of appointment 聘书;委任书
letter of assignment 转让书
letter of assurance 航运许可证
letter of attorney 委托书;委任状;授权(文)书;代理证书
letter of attornment 转租通知书
letter of authority 委托书;委托拨款证;授权证书
letter of authority from manufacturer 生产厂家的授权信

letter of authorization 委托书;授权书
letter of award 决标书;裁决书;判决书
letter of cancellation 解约书
letter of capacity 能力判决书
letter of comfort 同意协助书
letter of commitment 委托书;承诺书
letter of confirmation 证实书;确认书
letter of credence 介绍信
letter of credit [L/C]信用证;银行现金保证函;使用证;银行信用证
letter of credit charges 信用证费
letter of credit issued ledger 开除信用证分类证
letter of credit ledger 信用证总账
letter of credit of government to government 政府间信用证
letter of credit opened by brief cable 简电开证
letter of credit payment 凭信用证交货
letter of credit terms 信用证条件
letter of credit without recourse 不可追索的信用证
letter of credit with telegraphic(al) transfer reimbursement clause 电报索汇条款信用证
letter of endorsement 背书信
letter of guarantee 担保书;保证书;保函;信用保证书
letter of guarantee for bid 投标保证书
letter of hypothecation 质押书;抵押证书
letter of identification 签字式样证明书;身份证书;身份介绍信
letter of indemnity 认赔书;赔偿保证书;损失赔偿保证书;保结书
letter of indication 签字式样证明书
letter of inquiry 询价信;询价函件
letter of instruction 押汇信用指示书
letter of insurance 卖方出具给买方的确认已保险信函
letter of intent 交付书;申述意图函件;契约条款释意信;意向书
letter of intention 意向书
letter of interest 意向书
letter of introduction 介绍信
letter of invitation 邀请信
letter of licence 延期索偿同意书
letter of lien 留置权书;扣押权信
letter of offer and acceptance 交货验收单
letter of patent 特许证书;专利证(书)
letter of ratification 批准书
letter of recommendation 介绍信;推荐信;推荐书
letter of reference 介绍信;查询信
letter of remittance 信汇
letter of representation 审计客户表白书
letter of subrogation 权益转让书
letter of undertaking 担保书;承诺书
letter orders 信函订单
letter ornament 字母装饰;字体装饰
letter patent 特许证(指专利)
letter plate 信箱投信口开缝板;街门投信口
letter press 铅印;凸版印刷机(的总称)
letter press ink 铅印油墨;凸版油墨
letterpress paper 凸版印刷纸
letterpress print coater 多辊转印涂布机
letter press printing 凸版印刷
letterpress typographic(al) printing 凸版印刷
letter printing 铅印
letter punch 印记冲模;钢字码
letter report 通信报告
letters and lettering 字体和写法
letter-set 凸版胶印
letter sizes 代表尺寸的字母
letter slot 信槽孔
letters of administration 遗产管理委任状
letter sorting machine 分信机
letters patent 特许证书
letters testamentary 法庭授予遗嘱执行人权利的命令
letter stock 库存未登记股票
letter string 字母串
letter symbol 文字符号
letter transfer 信汇
letter-type combination lock 字母锁;字母式暗码锁
letterwood 豹斑木(圭亚那)
let the seller beware 卖方自担的风险
letting 公开开标;公布标价
letting down 降低硬脆性(一种钢热处理法);下滑松弛回火

letting of a contract 发包
letting of bid 开标
letting of contract 合同书
letting tint 回火色
Leucaena glauca 银合欢【植】
leucanterite 淡铁矾
leucargyrite 银勒银矿
leucaugite 白辉石
leucine 亮氨酸;氨基己酸
leucismus 白变
leucite 白榴石
leucite-basalt 白榴石玄武岩
leucite basanite 白榴碧玄岩
leucite-bearing ultrabasic extrusive rock 含白榴石超基性喷出岩
leucite kulaite 白榴闪霞粒玄岩
leucite melilitite 白榴黄长岩
leucite nosean sodalite phonolite 白榴石勒方石方钠石响岩
leucite nosean sodalite phonolite porphyry 白榴石勒万石方钠石响岩斑岩
leucite phonolite 白榴(石)响岩
leucite syenite 白榴正长岩
leucite tephrite 白榴(石)碱玄岩
leucite tuff 白榴石凝灰岩
leucitite 白榴岩
leucitophyre 白榴斑岩
leuco 无色的
leucoandesite 浅色安山岩
leucobasalt 浅色玄武岩;浅色玄武岩
leucochalcite 橄榄铜矿
leuco-compound 无色化合物
leucocrate 淡色岩
leucocratic 浅色的
leucocratic dike 淡色岩脉
leucocratic rock 淡色岩
leucodiorite 浅色闪长岩;淡(色)闪长岩
leuco dye 染料隐色基;无色染料
leucofluorescein 白荧光素
leucogabbro 浅色辉长苏长岩
leucogabbro norite 浅色辉长苏长岩
leucogarnet 淡榴石;淡钙铅榴石
leuco granoblastite 浅粒岩
leucohornblende gabbro 浅色角闪辉长岩
leucomalachite green 无色孔雀石绿
leucomb 吊车上挑出的遮盖
leucomethylene blue 隐色亚甲基蓝;无色亚甲基蓝
leucomicrodiorite 浅色微晶闪长岩
leucon 复沟型;碱的可塑性材料,不溶于水不渗透酸
leuconorite 浅色苏长岩
leuconoritegabbro 浅色苏长辉长岩
leucooivine gabbro 浅色橄榄辉长岩
leucoolivine norite 浅色橄榄苏长岩
leucophane 白皮石
leucophanite 白铍石
leucophite 白环蛇纹岩
leucophoenicite 淡硅锰石
leucophosphite 淡磷钾铁矿
leucophyllite 淡云母
leucophyre 糟化辉绿岩
leucopigment 隐色颜料;无色颜料
leucoplast 白色体
leuco-sapphire 白蓝宝石
leucoscope 光学高温计;色光(光)度计;白色偏光镜
leucosphenite 淡钡钛石
leucosyenite 浅色正长岩
leucothionine 白硫堇
leucotroctolitic norite 浅色长苏长岩
leucotroctrolite 浅色橄长岩
leuco vat dye 隐色还原染料
leucoxene 白钛石
leukasmus 白皮斑
leukeran 苯丁酸氮芥
leuko 白色
leukocytic crystal 莱登晶体
leukonic acid 白酮酸
leukoplast 白色体
leukorrhea with bloody discharge 赤带
leukous 白的
leval alloy 铜银共晶合金
levant cotton 草棉
levante 累范特风
levantera 累范太腊风
levecel 耳房【建】;厢房;附属房

leveche 累韦切风
levecon 电平调节
levee 天然堤;天然冲积堤;堤防;堤(坝);冲积堤
levee back 堤防背(水)面;堤背
levee bank 堤岸;堤
levee base 堤底
levee body 堤体;堤身
levee breach (堤坝的)决口;堤防决口;堤岸决口
levee break (堤坝的)决口;破堤
levee construction 筑堤;堤防工程;堤的建筑;防波堤施工
levee core wall 堤芯墙;堤核芯墙
levee crest 堤顶
levee crown 堤顶
leveed bank 冲积滩;河海滩岸;筑堤护岸
leveed channel 有堤水道;有堤渠道
levee delta 堤状三角洲
levee deposit 天然堤沉积
leveed pond 堤成池;筑堤围湖;筑堤围池
leveed river 有堤防河道
levee extension 延长堤线
levee facies 天然堤相
levee failure 崩堤
levee fill 堤岸填筑
levee footing 堤防底脚
levee foundation 堤基
levee free-board 堤顶超高
levee gate 堤防闸门;堤坝闸门
levee grade 堤顶纵坡;堤顶高程;堤顶高;堤(的)坡度
levee maintenance 堤防维修;堤防维护
levee management 堤防管理
levee muck ditch 堤脚泥沟
levee of settlement 堤的沉降
levee opening 堤防闸口
levee patrol 巡堤员
levee protection 堤防保护
levee raising 堤防加高;堤坝加高
levee ramp 堤上坡道;堤岸斜坡道
levee reinforcement 堤防加固
levee restoration 堤防整修;堤防修复
levee revetment 护堤壁;堤面护坡;堤防护岸
levee ridge 冲积堤脊线
levee road 堤路
levee slide 坍堤;堤身滑坡
levee slope 堤坡
levee sloughing 堤身成边;堤身崩塌
levee sluice 堤上泄水闸
levee spacing 堤间距
levee strengthening 堤防加固
levee-surrounded field 围田
levee system 堤防系统;堤系
levee toe 堤脚
levee undermining 堤底淘刷
levee widening 堤防加宽;堤坝培厚;堤坝加宽
levee width 堤宽
level 校平;级别;能级;平田;平面;平舱;水准(仪);水平(面);定坡度;单排触点;程度;层次;标高
level above threshold 阈上水平;阈上级
level a building with the ground 拆除建筑物
level adjusting attenuator 电平调整衰减器
level adjustment 水平调准
level advance line 平面钻进线(采矿)
level alarm 液位报警(器);液体报警(器)
level alarm high low 液面高下限警报器
level allowance 允许液面
level and smooth 平滑的
level annuity plan 逐年平摊偿还计划
level attitude 水平位置
level axis 水准轴
level bar 水准尺;水平尺
level bar screen 扁平条杆筛
level-based system 分层系统
level bed 平床
level block 标号信息组
level boards 活动百叶窗
level body length 水平体长
level bolt 通天插销
level book 水准手册;水准手簿;水准记录簿;水准测量手簿
level border irrigation 淹灌
level bottom environment 平底环境
level bottom zone 平坦海底带

level breath 水面宽(度)
level bridge 平桥
level bubble 水准仪;水准器;水准气泡
level canal 平水运河
level capacity 级容量;满平容积;标志容积
level chamber 水准气泡室
level charges plan 平衡费用计划
level check 料位检测;电平检验
level clamp 电平固定
level compander 电平压缩扩展器
level comparator 水平仪式比测仪;电平比较器
level compensator 水准(仪)调节器;电平补偿器;分层补偿器
level-compound excitation 平复激励;平复励
level-compound excited motor 平复激电动机
level concrete pad 平整水泥垫
level condition 水平状态
level constant 水准仪常数;水准常数
level control 信号电平控制;液面控制;料位控制;级位控制;钳位电平调整;水准测量控制;水位控制;水平面调节;水平控制;电平控制;电平调整;电平调节;水准控制
level control circuit 电平控制电路
level control element 水面控制元件
level controller 液面控制器;液面调整器;液面调节器;位面控制器;水位控制器
level control network 水位控制网
level control station 电平调整台
level control table 级控制表;级别控制表;分级控制表
level control valve 水位调节阀
level converter 电平变换器
level conveyer 水平运输机
level correction 水准(仪)校正;水平校正
level country 平坦地区
level course 整平层;水平巷道
level cross country 平原不平地区
level crossing 平面交叉;平交道(口);平交;水平交叉;道口;铁路与公路平交道;铁路道口
level crossing alarm device 道口报警装置
level crossing controlled at site 就地控制道口
level crossing gate 平交道拦;平交道拦路木;平交道拦路门(栅)
level crossing protection 平交道口防护
level crossing protection installation 道口防护装置
level crossing safety installation 道口安全装置
level crossing signal 平面交叉信号
level crossing spectroscopy 能级交叉光谱学
level crossing warning device 道口报警装置
level crossing watchman 道口看守员
level crossing with automatic barrier 自动栏木道口
level crossing with barriers worked locally 栏木就地操纵道口
level crossing with barriers worked remotely 栏木遥控操纵道口
level crossing with signal 有信号防护道口
level cross-section 水平横截面;水平断面
level cut carpet 平绒地毯
level cutting 平巷掘进;水平掏槽;水平开挖
level deconvolution 展平反褶积
level deexcitation 能级去激活
level density 能级密度
level-depth relation curve 水位水深关系曲线
level detection 电平检测
level detection device 料位测定装置
level detector 料位检测器;料位高度测定仪;电平探测器;电平检测器;仓面指示器;水平探测器
level determination 水平测定
level diagram 电平图;分层流图
level difference 高(程)差;能级差;声级差;电平差
level difference in beam top surface 梁顶面水平偏差
level difference meter 电平差测试仪(计量器)
level difference of sieve tray floor 筛板面水平差
level displacement table 层位移表
level distribution 电平分布
level drift 横洞;水平坑道
level drive 沿走向掘进
level dyeing 匀染;均匀染色
level east 盘东【测】
leveled time 正常时间
level elevation 中段标高;中段高程
level ellipsoid 水准椭球

level ellipsoid of revolution 水准旋转椭球
level error 水准误差;水平差;地平差
level face 水准面;水平面
level feeder 水平喂料机
level finding 找平【测】
level fixer 电平固定器
level flight 水平飞行
level flight indicator 水平飞行指示器
level fold 非倾伏褶曲
level-full capacity 总容积;满槽容量;(水库的)总库容
level gallery 平巷
level ga(u)ge 水准指针;液位计;液面指示器;液面计;料面计;料面测量仪表;平导轨;水准仪;水准器;水位计;水平计;水准轨
level ga(u)ge for absorber 吸收器液面计
level ga(u)ge for evapo(u)rator 蒸发器液面计
level glass 液面视镜
level governor 水平调节器
level grade 平坡度;平坡
level grade between opposite gradients 分坡平段
level-gradient changeover brake 平道坡道可调式制动机
level-gradient changeover device 平道坡道转换装置
level graduation 水准器划分
level graphic instrument 电平图records
level grizzly 平格筛;扁平条格筛
level ground 平地;水平地面
level gutter 平檐沟;水平排水沟
level hydrograph 水位过程线
level ice 平坦海冰;平滑冰
level indicated control 液面指示控制
level indicating controller 液面指示控制器
level indicator 液位计;液面指示器;料位指示器;料面指示器;级指示器;能级指示器;水平指示器;水平规;电平指示符;层指示符;层效指示符;水准仪
level indicator with rotating paddles 旋转叶轮式料位指示器
level instrument 位面计;水准仪;水平仪;测量水平仪
level interval 中段间距;水平间距
level inverted 水平倒飞
level landslide 深层滑坡
level(l)ed bucket capacity 平斗斗容
level(l)ed reference error 水平参考误差
level(l)er 整平者;整平器;校直机;校平器;平整机;水准测量员;水准测量(人)员;钢板矫平机;调平器
level(l)er roll 矫直辊
level life 能级寿命
level limit switch 料位限位开关
level line 平坡线路;平道线路;平道;水准线;水平线;等高线
level line and surface 水平线及水平面
level-line repayment 按固定比额分期偿还
level(l)ing 流平;矫直;矫平;平整;调平;水准测量;水平调节;抄平【测】;测平(检验墙身垂直度);夷平作用;平层
level(l)ing action 均涂作用;均涂效应
level(l)ing adjustment 调平装置;水准测量平差
level(l)ing agent 均化剂;均染剂;流平剂
level(l)ing along the line 路线水准测量
level(l)ing and smoothing the bank line 平整岸线
level(l)ing attachment 水准测量附加装置
level(l)ing auger 分配螺旋
level(l)ing base 整平的基底;基准面;水准基线;水准面;水准基点
level(l)ing beam 刮尺;修平尺;修平板
level(l)ing bedding 水平层理
level(l)ing bench 平整台;水平平台
level(l)ing binder course 找平砂浆层
level(l)ing blanket 找平层;平整层
level(l)ing block 校平垫块;水平压块;水平校正块
level(l)ing board 找平板
level(l)ing bolt 调平螺钉
level(ling) bottle 平液水准瓶
level(l)ing branch line 支水准路线
level(l)ing bubble 水准仪;水准气泡
level(l)ing bulb 平液球管;水准球管
level(l)ing bulb reservoir 校平储[贮]液球
level(l)ing by means of barometer 气压表测高
level(l)ing charges 平舱费;平堆费

level(l)ing coat 找平层
level(l)ing compound 均涂合成剂;匀涂合成剂
level(l)ing concrete 整平混凝土;找平混凝土层;混凝土整平层;抄平棍凝土层;混凝土整平
level(l)ing control lever 水平调节手柄
level(l)ing course 整平层;找平层;灰浆平整层;平整层;水平层
level(l)ing crossing 水平交叉口
level(l)ing culvert 灌水廊道;(船闸、船坞的)灌水涵洞;水面平衡涵洞
level(l)ing culvert intake 水面平衡涵洞进水口
level(l)ing culvert outlet 水面平衡涵洞出水口
level(l)ing cylinder 调平油缸
level(l)ing data 水准资料
level(l)ing device 液面控制装置;调整装置;平装置;水平调节装置;电梯自动平层装置
level(l)ing discrepancy 水准测量不符值
level(l)ing drag scraper 平地拖曳刮土机;平地拖铲
level(l)ing dryer 平幅烘燥机
level(l)ing dye 均染染料
level(l)ing effect 均化效应
level(l)ing equipment 水平找平设备;水平调节设备
level(l)ing factor method 系数修正法
level(l)ing factors 评定因素
level(l)ing finisher 水平整修机
level(ling) flask 平液水准瓶
level(l)ing(foot) screw 安平脚螺旋
level(l)ing ga(u)ge 水平仪
level(l)ing gear 调平机构
level(l)ing ground 铲平
level(l)ing head 三角架头
level(l)ing increment 水准测量高差增量;水准测量高差
level(l)ing information 水准资料
level(l)ing instrument 水准仪(器);水准测量仪器;水平仪;平尺
level(l)ing jack 找平千斤顶;调平(用)千斤顶
level(l)ing layer 找平层;整平层
level(l)ing lever 水平调节杆
level(l)ing lift rod 可提升斜拉杆;可提升杆;水平调节起落杆
level(l)ing line 水准线;水准路线
level(l)ing loop 水准环;水准回线
level(l)ing machine 整平机;矫正机;矫平机;平土机;平路机;调平机;平整机
level(l)ing marker 水准标志
level(l)ing mass 整平填料;找平填充料
level(l)ing mechanism 调平机构;水平调节机构
level(l)ing mirror 调平镜
level(l)ing mortar 找平灰浆;找平砂浆
level(l)ing net 水准网
level(l)ing network 调整网;水准网
level(l)ing of capital gains 资本盈利水平
level(l)ing off 找平
level(l)ing of ground 场地平整;平整场地
level(l)ing of instrument 仪器置平
level(l)ing(of) model 模型置平
level(l)ing of production 均衡生产
level(l)ing of soft shoulders 软弱路肩的找平;软弱路肩的平整
level(l)ing of the track 轨道调直
level(l)ing operation 找平作业
level(l)ing operation 水准测量工作
level(l)ing origin 水准原点
level(l)ing pad 调平垫片;水准基座
level(l)ing party 水准测量队
level(l)ing peg 平栓;水准测量桩;水准桩;水平基桩
level(l)ing pillar 水准标石
level(l)ing pipe 调平连通管
level(l)ing plan 高程透写图
level(l)ing plank 找平板
level(l)ing plate 测量底座(钢垫板);检定平板;尺垫【测】;水准板
level(l)ing plummet 测平垂球
level(l)ing point 水准点
level(l)ing pole 塔尺【测】;水准标尺;水平(标)杆;水平标尺
level(l)ing practice 水准测量作业
level(l)ing process 整平横断面
level(l)ing production 评定生产
level(l)ing property 流平性;均涂性能
level(l)ing ram 调平油缸

level(l)ing range 调平范围
level(l)ing ring 整平圈;水准圈(道路排水口的)
level(l)ing rod 塔尺【测】;水准杆;水平标杆;水准(标)尺;水准(标)尺
level(l)ing rod error 水准标尺误差
level(l)ing route map 水准路线图
level(l)ing rule 平尺;准直尺;水准尺
level(l)ing screed(material) 找平砂浆层;整平板;找平板
level(l)ing screw 校平螺旋;校平螺钉;平准螺钉;调平螺钉;水准螺旋;水准螺钉
level(l)ing shoe 调平支脚;水平稳定器
level(l)ing solvent 均化溶剂;平准溶剂
level(l)ing spindle 水平主轴
level(l)ing spring plate 弹簧底板
level(l)ing staff 塔尺【测】;水准(标)尺;水准标杆
level(l)ing support 水准架;校平架
level(l)ing survey(ing) 水准测量
level(l)ing tape 水准卷尺;水准带尺
level(l)ing tool 整平工具;找平工具
level(l)ing tripod 水准仪三脚架
level(l)ing tube 平液管
level(l)ing underlay 找平层
level(l)ing-up 整平;拉平
level(l)ing valve 高度阀;高度控制阀;平衡阀;(空气悬架的)自动调平阀
level(l)ing vessel 平液容器;水准容器
level(l)ing washer 校平板
level(l)ing with theodolite 经纬仪水准法
level(l)ing work 找平;填平地坪;整平工作;水准测量工作
level(l)ing zone 平层调整区(电梯)
level loader 水平式装载机
level loading 平装
level loop 水准环
level luffing 水平变幅;水平(吊杆)起落
level-luffing boom 水平升降吊杆
level-luffing cable 水平升降吊杆缆索
level-luffing crane 平行运送旋臂起重机;平臂起重机;水平(位移)变幅起重机;平等运送旋臂起重机;水平俯仰起重机;鹅头式伸臂起重机;水面俯仰起重机
level-luffing gantry crane 水平变幅门座式起重机
level-luffing gear 水平位移变幅起重装置;水平位移变幅机构;水平起落杆滑车;水平起落杆齿轮
level-luffing jib 水平起落吊杆支架
level-luffing jib crane 水平变幅臂架式起重机
level-luffing line 水平起落吊杆索
level-luffing ram 水平起落吊杆顶杆
level-luffing slewing crane 水平旋转起重机起落吊杆
level-luffing tower crane 水平起落吊杆塔式起重机
level man 水准测量(人)员
level mark 水准(面)点;水平标志;水准面标记;基准面(标记);高程点
level mark on column 柱上水准点
level measurement 水平面测量;标高测量
level measurement meter 料位测量仪表
level measuring set 水平仪;电平测试器;电平测量器
level meter 液位指示器;液位(指示)计;液体计;液面计;料位计;水平仪;电平计;电平表;声级计;电平指示器;水位指示器;水平指示器
level meter of material 料位计
level model 水平模型
level monitor 液面监视器
level monitoring 料面监控
level multiple 弧层复联线束
levelness 水平度
levelness of top surface of shuttle 货叉上表面水平度
level net 水准网
level notes 水准手簿;水准记录簿
level number 中段号;层数;层次号码;层号
level of abstraction 抽象阶段;抽象度
level of acceptable risk 可接受风险安全度
level of access 存取级;存取层
level of addressing 定址级数
level of aspiration 期望水平
level of atmospheric inversion 大气逆温层高度
level of audibility 听度级
level of automation 自动化程度
level of background noises 背景噪声级
level of base plate and support 底板与支座高程

level of base rock 岩面高程
level of bedrock surface 基岩顶高程
level of capital gains 资本盈利水平
level of cavitation 空蚀度;气蚀等级;气蚀程度
level of commodity price at wholesale 批发商品价格水平
level of compensation 补偿水平(面)
level of competency 能力水平
level of compliance 遵守的水准
level of concrete mix 混凝土拌和物等级
level of confidence 置信水准;置信水平;置信级
level of consumption 消费水平
level of control 管理水平
level of cooperation 合作程度
level of damage 损坏的程度
level of development 发展水平
level of drive 激励电平
level of education 教育等级
level of effort 工作规模
level of environment(al) pollution load 环境负荷水平
level off 整平;刮平;改平;矫直;恢复水平;取得平衡;稳定;调(成)水平
level of factor 因子指标
level of farming mechanization 农业机械化水平
level of fixed number of labo(u)r 定员水平
level of floor(ing) 地板面标高;地板面高程
level of foundation 基础水平;地基高程;基础高程
level-off position 改平位置
level of free convection 自由对流高度
level of ga(u)ge zero 水尺零点高程
level of gravity potential 重力位水准面
level of gray tone 灰度级
level of ground 地面标高;地面高程
level of groundwater 潜水水位
level of illumination 照明水平;照明等级;照度等级
level of instrument (测量时的)仪器高程
level of integrity 配套程度;完整程度
level of liquid 液面高度;液位高度
level of loading 加料量
level of loudness 响度级
level of maintenance 车辆维护等级
level of management 管理水平
level of minimum sound velocity 最小声速级
level of nesting 嵌套层(次);程序的嵌套层次
level of noise 噪声值;噪声级
level of no motion 无运动面
level of nondivergence 无辐散高度
level of no strain 无应变范围
level of nozzle face 接管面标高;接管面高程
level of objective 目标层次
level of optimization 最优化水平
level of orders 订货范围
level of organic metamorphism 有机变质程度
level of organization 组织结构
level of original bed 原地面高程
level of pollutant 污染物量;污染水平
level of pollution control 污染控制水平
level of position 职别
level of premium 平均保(险)费
level of price 物价水准;物价水平
level of prices at wholesale 批发商品价格水平
level of reference 参考标高;参考高程
level of residue 渣量;残渣量
level of risk 危险性程度;危险程度
level of saturation 潜水面(指地下水);地下水面;饱和水面
level of scholarship 学术水平
level of sea 海平面
level of sensitivity 感觉程度
level of service 道路服务水平;服务水准;服务水平
level of significance 显著性水准;显著性水平;显著性有效指标;有效水平;有效级;重要性程度
level of skill 技能水平
level of solidification 固化程度
level of solids in reactor 反应器中固相高度
level of sound 声级
level of standard 标准的级别
level of standardization 标准化水平;标准化目的
level of storage 蓄水高度
level of stress 应力水平(等级);应力密度
level of subsoil water 潜水面(不指地下水);地下(水)水位
level of support ring 支承圈标高;支承圈高程

level of tail water 下游水位;尾水水位
level of technology 工艺水平
level of the bottom 底面标高;底面高程
level of the de-polarized component 去极化分量电平
level of the hydraulic pressure 水压水位
level of the trench bottom 沟底标高;沟底高程
level of toxic pollutant 有毒污染物水平
level of treatment 处理等级
level of tremied concrete 水下浇筑混凝土顶面;水下灌注混凝土顶面
level of uncertainty 不确定性程度
level of upper pond 上游库水位;上池水位
level of water 水位;水平水;水位高度
level of water being sucked 吸水程度
level of wave-base 浪蚀基面
level of zero of ga(u)ge 水尺零点高程
level one file 一级文件
level-one variable 一级变量;第一级变量
level oscillator 电平振荡器
level over-load 操作定额过载
level party 水准组;水准测量队
level-payment mortgage 平衡摊还抵押(每月等额偿付的抵押)
level peg 水平基桩;水平标桩
level pillar 水平矿柱
level plan 水平平面图
level plane 水准面;水平面
level plane surface 平坦的水准表面
level plug 水准塞;调平塞;调平栓;校平塞;校平栓
level pole 水准标尺
level position 水平状态;水平位置
level premium 平均保(险)费
level press 水平压力机
level pressure 定压
level pressure coefficient 常压力系数
level pressure control 基准压力调节
level priority interrupt 级优先中断
level probe 液位探针
level railroad line 平坦铁路线
level railway crossing 铁路与公路平面交叉口
level rainwater gutter 水平雨水沟
level raised by wind 风吹水位升高
level range 电平测量范围
level reach 河段;水道区段
level reading 高程读数
level recorder 液面记录器;自记水位仪;电平记录计;水平记录仪;电平记录器
level recorder ga(u)ge 自记水位仪;自记水位计
level recording instrument 电平记录仪
level recording meter 电平记录仪
level reference 水平基准(面)
level regulating course 水准校准层;整平层
level regulating device 电平调节装置
level regulator 液面调节器;水准调节器;电平调节器
level reservoir 静水库
level ridge 平垄
level ridge vaults 拱脊同高的交叉拱顶;穹脊在同一水平上的两个穹
level rise velocity 水位升高速度;水位上涨速度;水面上升速度(船闸灌水时)
level road 缓坡道路;主平巷;平路
level road horsepower 水平道路行驶阻力功率
level rod 量油杆;塔尺【测】;水准尺;水准测杆
level roller runway 水平辊道
level roof gutter 水平屋面天沟
level routing 水位演算
level ruler 水平尺
level saturation method 级饱和法
level scale of alidade 照准部水准器格值
level scheme 能级图
level screen 水平筛
level screen of bars 杆式平面筛;平格筛
level seam 水平矿脉
level section 水平横断面;水平段
level sensibility 水准器灵敏度
level-sensing device 水位记录装置;料位计;水位传感装置
level-sensitive scan design 电平相关扫描设计(法)
level sensitivity 水准灵敏度
level sensor 料位探测器;水位控制器;水平传感器;电平探测器
level set 水平集合

level setting 电平设定;电平调整;电平调定
level shift amplifier 电平漂移放大器
level shift diode 电平移动二极管
level shifter 电平转移电路
level shifting 电平移动
level shoe 调平支脚;水准尺端
level sight glass 液面指示玻璃管;玻璃液位表;玻璃示油规
level signal 电平信号
level signal generator 电平信号发生器
level slicing 分层分割
levels of fecal coliform 大肠菌水平
levels of prevention 分级预防
level soil 平整土地
level spacing 能级间隔;水平间距;电平间隔
level spheroid 等位椭球
level stability 水平稳定性
level staff 水准标尺
level stage 水位
level stake 水平桩
level status block 级别状态块
level storage 静水容
level storage yard 平货位堆货场
level surface 液面;位势面;水准面;水平面;等势面
level-surface method 水准面法
level survey 高程测量;水准测量;水平测量
level switch 料位开关;箱位电平开关;钳拉电平转换;水平开关;电平开关
level table 平台
level table guide 平台导承
level tangent track 平直线路
level tank 液位槽
level team 水准组
level tendering 摆平投标
level terrace 水田梯田;水平梯田
level terrain 平坦地形;平坦地区;平坦地(带)
level tester 校水准器;水准器检验仪;水准管检定器
level testing instrument 水准检验仪;水准管检定器
level theodolite 水准经纬仪
level throat 水平式流液洞;水平流液洞
level-to-level administration 分级管理
level tracer 电平图示仪
level tracing receiver 电平直观仪
level track 平路;水平轨道;缓坡道路
level transducer 水位发送器;水位传送器
level transit 水准经纬仪;测平镜
level translation 电平变换
level translator 电平转移器;电平变换器
level trier 验表器;检平器;水准仪灵敏度测量仪;水准器检验器;水准(管)检定器
level-trimming 平舱
Level-Trol 特罗尔液位调节器
level tryer 水平式试验器
level tube 管状水准器;校平管;气泡水准管;水准仪管;水准器;水准(测)管
level tube axis 水准管轴
level tube bubble 水准管气泡
level tube housing 水准管框
level tunnel 平坡隧道
level type angle ga(u)ge 水平式测角仪
level type flo(a)tation machine 直流槽型浮选机
level untying 平面疏解【铁】
level-up 找平【测】;拉平
level-up course 平整层
level-up wedge 填平补齐用楔
level valve test bench 高度调整阀试验装置
level vessel 平液容器
level vial 水准仪气泡;水准器;水准气泡;水平仪气泡
level vibrating screen 水平振动筛
level west 盘西
level width 能级宽度
level-wind 水平绞线器
level winding 均缠
level-winding device 尺度绕平装置
level with compass 带罗盘的水准仪
level work area 级工作区
level zero interrupt vector 零级中断向量
lever 杠杆;控制杆
lever action 杠杆作用
lever action air pump 杠杆抽气泵
lever action bolt 暗插销;杠杆式插销
lever action pump and hose 杠杆泵连软管
lever action strut 杠杆作用支柱
lever-actuated die 杠杆动作的模具

lever-actuated pusher type stock guide 杠杆动作推杆式导料板
leverage 杠杆作用;杠杆系;杠杆利益;杠杆机构;杠杆传动;杠杆臂长比;杠杆比率;投机能力;财务杠杆作用;浮动股息
leverage action 杠杆作用
leveraged buyout 借债谋利
leveraged lease 杠杆租赁
leverage effect 杠杆效应;举债经营效果;外贸杠杆作用
leverage factor 杠杆系数
leverage fund 杠杆资金
leverage lease 杠杆租赁
leverage of force 力的杠杆作用
leverage principle 财务调度杠杆原理
leverage ratio 杠杆比率
leveraging 杠杆作用
lever amplification strain ga(u)ge 杠杆式放大应变仪;杠杆放大式应变计
lever and weight safety valve 杠杆重锤式安全阀
lever and weight spring swing check valve 杠杆弹簧摆动式止回阀
lever and weight swing check valve 杠杆和锤摆动式止回阀
lever arm 杠杆臂;杆臂;作用力臂
lever-arm deflection indicator 杠杆臂挠度计
leverarm of stability 稳性力臂
lever balance 杠杆天平;天平
lever bell pull 杠杆式铃拉手
lever block 杠杆式拉力滑车;手扳葫芦;手摆式拉力滑车;闭塞杆
lever boards 活动气窗;活动百叶(窗);百叶窗板
lever bolt 杠杆式插销
lever box 联动柄箱
lever brace 挺穿孔器
lever bracket 杆托;杠杆支架
lever bracket bolt and nut 杆托螺栓及螺母
lever brake 杆闸;杆式制动器;手制动器
lever change 变换手柄位置;变换手柄;变速杆
lever chuck 带臂夹盘
lever clamp 杠杆夹具;偏心夹具
lever collar 手柄套管
lever connection 杠杆连接
lever contact 杆式接点
lever control 杠杆操纵;杠杆控制
lever controller 活砧式掣链器
lever crank mechanism 摆杠杆曲柄连杆机构
lever cut-out 杠杆切断
lever deflectometer 杠杆弯沉仪;杠杆挠度仪(即贝克曼弯度仪)
lever dial indicator 杠杆式千分表;杠杆式百分表
lever distribution method 杠杆分配法
lever dolly 杠杆支持
lever draw bridge 杠杆式活动桥;升降桥;杠杆操纵开合桥
lever drill 杠杆钻
lever dynamometer 杠杆测力计
levered suspension 杠杆式悬置
lever effect 杠杆作用
lever-equal beam balance 杠杆式等臂天平
lever escapement 杠杆式擒纵机构;叉式擒纵机构
Leverett function 莱弗里特函数
lever feed 杠杆进给
lever feedback servovalve 杠杆反馈伺服阀
lever feed core drill 手把给进岩芯钻机
lever floor weigh bridge 杠杆式地上衡
lever fork 叉形杠杆头
lever for pan position 固定锅体手把
lever for throwing the machine in-and-out-of gear 机器传动的离合杆
lever fulcrum 杠杆支点
lever ga(u)ge 杠杆检查量规
lever gear door 提升门;提开门
lever grease gun 润滑脂枪
lever gun 杠杆式焊枪
lever gun welding head 杠杆式点焊钳
lever hammer 杠杆锤
lever handle 杠杆手柄;门把手;横(把)执手;搬把式门拉手;弯把执手
lever handle fittings 杠杆式门拉手小五金;杠杆式门拉手设备
lever handle hardware 杠杆式门拉手附件
lever handle pin 杆手柄销
lever handle-plate 杠杆式拉手板

lever harrow 杠杆把;杆式耙;杆齿耙
lever head 杠杆头
lever hook 杠杆钩
lever indicator 杠杆指示器;杠杆指示计;杠杆百分表
lever jack 杠杆(式)千斤顶;杠杆起重机
lever key 杠杆键
lever latch 杠杆锁;杠杆掣柄
lever latch cap screw 手柄弹键盖螺钉
lever law 杠杆定律
lever lift 杠杆提升机;杠杆起重机
lever lift cultivator 杠杆提升式耕种机
lever lock 杠杆锁;杆锁;握柄锁闭;弹子锁
lever locking 柄式锁闭
lever mechanism 杠杆机理;杠杆机构
lever micrometer 杠杆千分尺
lever motion 杠杆运动
lever movement 锚式机芯;擒纵机构
lever nail pulley 拔钉杆
lever of crane 起重机臂
lever of first order 第一类杠杆
lever of force 力臂
lever of stability 回复力臂
lever of the first kind 第一类杠杆
lever of the third order 第三类杠杆
lever-operated 手把操作
lever-operated valve 杠杆操纵阀
lever operating 杠杆操纵
lever operation 杠杆传动
lever pantograph 杠杆缩放仪
lever pin 杠销;杆销
lever principle 杠杆原理;杠杆法则;杠杆定理
lever punch 杠杆(式)冲压机;杠杆式冲床
lever punching machine 杠杆式冲压机
lever punch press 杠杆式压力机;杠杆式压床
lever pusher 杠杆式推钢机
lever ratio 杠杆比
lever relationship 杠杆关系
lever release valve 杠杆放松阀
lever relief valve 杠杆泄放阀
lever reversing gear 杠杆回动装置
lever riveter 杠杆式铆接机
lever roller 杠杆滚轮
lever rope 杠杆索
lever rule 杠杆规则;杠杆定则;杠杆定理
levers 杠杆系
lever safety valve 杠杆安全阀
lever saw 杠杆锯
lever scale 杠秤;杆秤
lever-set punch 杠杆穿孔机
lever shaft 杠杆轴
lever shaft arm 杠杆轴臂
lever shaft nut 杠杆轴螺母
lever shaft packing 杠杆轴填密
lever shaft plug 杠轴塞
lever shear 杠杆剪力
lever shears 杠杆式剪切机;杠杆式剪(床)
lever shears alligator shears 杠杆式剪切机
lever sleeve 杠杆套筒
lever spike extractor 长柄拔钉钳
lever sprayer 杠杆式喷雾机
lever spring 杠杆式弹簧
leverstand 联动柄座
lever steerer 操舵杆
lever steering 杠杆式转向机构;杠杆控制
lever stop 杠杆制动器
lever stopper 活砧式掣链器
lever-suspension wheel 摇臂悬置轮
lever switch 杠杆开关;杠杆操纵开关;手柄开关
lever test 杠杆式测微仪;杠杆试验
lever test bar 杠杆试棒;杠杆测试棒
lever testing machine 杠杆式试验机
lever tongs 杠杆夹具;杠杆钳
lever tumbler 制栓杆
lever-tumbler lock 杠杆锁;制栓杆锁
lever-type 杠杆型;杠杆式(的)
lever-type barbed wire strainer 杠杆式倒刺铁丝拉紧器
lever-type brush-holder 杠杆式电刷握持器
lever-type clamp 杠杆夹钳
lever-type feedback 杠杆反馈
lever-type gate lifting device 杠杆式启门设备
lever-type grease gun 杠杆式润滑脂枪
lever-type hand grease gun 杠杆式润滑脂枪

lever-type independent suspension 杠杆式独立悬架
lever-type indicator 杠杆式指示表
lever-type jack 手把型千斤顶
lever-type operator 杠杆式窗开关
lever-type pressure grease gun 杠杆式注压枪
lever-type safety valve 杠杆式安全阀
lever-type starter 杠杆式起动器
lever under-ground weigh bridge 杠杆式地中衡
lever-under-screen operator 纱窗下开关器
lever unequalbeam balance 横杆式不等臂天平
lever valve 杠杆阀
lever weighting 杠杆加压
lever with ovoid grip 卵形手柄杆
lever-wood 美洲铁木
leviathan 大型洗衣机;巨型远洋轮
leviathan washer 大洗涤机
leviathan wool 刺绣毛线
leviating melting apparatus 悬熔设备
levigate 淘选
levigated abrasive 细磨磨料
levigating 磨细
levigation 悬浮分级;研碎;研末;磨细;淘选;水磨;沉淀法选矿
levigelinite 均匀凝胶体;均匀腐植体
Levi graph 勒维图
levitate 漂浮的;漂浮
levitated spherator 漂浮球形器
levitated super-conducting magnet train 超导磁悬浮列车
levitated superconducting ring 漂浮超导环
levitated vehicle 悬浮车
levitating 漂起
levitating bubble 浮起气泡
levitation 悬浮;漂浮(感);浮置
levitation chassis 磁悬浮车底盘
levitation coil 磁悬浮系统线圈
levitation control 磁悬浮控制
levitation effect 悬浮效应;漂浮效应
levitation evapo(u)rator 漂浮蒸发器
levitation guidance system 磁悬浮导向系统
levitation heating 悬空加热法;悬浮加热法
levitation magnetic 磁悬浮系统磁体
levitation melting 悬熔法;悬浮熔融
levitation rail 磁悬浮系统轨道
levitation regulator 磁悬浮系统调节器
levitation stator 磁悬浮系统定子
levitation stator ga(u)ge 磁悬浮定子标距
levitron 漂浮器
levoform 左式
levoversion (眼)左转
levulinate 乙酰丙酸盐
levulinic 乙酰丙
levulinic acid 乙酰丙酸
levulinic aldehyde 乙酰丙醛
levy 征收
levy duties on 征税;收税
levyine 插晶菱沸石
levying of fines 征收罚款
levy in kind 征实
levyite 插晶菱沸石
levyne 插晶菱沸石
levynite 插晶菱沸石
levy on environmentally harmful consumption 对环境有害的消费征税
Levy's criterion (重力坝设计的)利维准则
levy tax 征税
levy tax on 对……抽税
levy toll on 征税
lew 避雨棚;砖坯帽盖
lewatit M2 强碱性阴离子交换树脂
lewis 起重爪;吊楔
Lewis acid 路易斯酸
lewis anchor 起重爪;吊楔;吊石块用的锚具
lewis bar 起重楔;吊楔杆
lewis bolt 地脚螺栓;楔形地脚螺栓;棘螺栓;带眼螺栓;大头螺栓
lewis hole 楔形槽;吊楔孔;大块岩石劈开孔;岩石劈开孔;石块表面的凹陷
Lewisian(stage) 利维斯阶(苏格兰)【地】
lewisite 降落伞式雷达干扰发射机
lewis of crane 起重爪
Lewis's acid 路易斯酸
Lewis's acid site 路易斯酸性点

Lewis's bolt 路易斯螺栓
lewis sheeting (做百叶窗的)薄钢片
Lewis's metal 路易斯易熔合金
lewisson 起重爪;吊楔
lewisson bolt 吊石栓
Lewis's theory 路易斯理论
Lewistonian(stage) 利维斯顿阶(美国)
lexical 词法的;词典的
lexical ambiguity 词法多义性;词法的歧义性
lexical analyser 词法分析程序
lexical analysis 词法分析
lexical analysis phase 词法分析阶段
lexical analyzer 词法分析器
lexical conversion 词法转换
lexical function 词法功能
lexical processor 词法处理程序
lexical routine 词汇程序
lexical symbol 词法符号
lexical transformation 词汇转换
lexical unit 词法单位
lexicographic order 辞典式顺序;词典顺序;词典编辑次序
lexicon 专门词汇;辞典;词汇;词典
lex loci rei sitae 所在地法律
lex loci solutionis 债务履行地法
lex situs 所在地法律
ley 锡铅轴承合金;混合草地;草地轮作
Leyden blue 莱登蓝
Leyden jar 莱登瓶
Leyden's crystals 莱登晶体
ley farming 草田轮作
lgneous facies 火成相
Lhasa block 拉萨地块
L-head L形顶撑端头;L形(支)撑头
L-head cylinder 侧阀汽缸
L-headed jetty 曲尺形栈桥;L形栈桥
L-head engine 侧置气门发动机
L-head groin 勾头丁坝
L-head jetty L形栈桥
L-head spur dike 勾头丁坝
L-hoe blade 单刀平切锄铲
liabilities 负债
liabilities accounting 负债核算
liabilities out of book 账外负债
liabilities to capital ratio 负债对资本比率
liability 易感性;义务;债务;责任
liability accident 责任事故
liability account 负债类账户;负债账户
liability and responsibility 负债与责任
liability balance 负债余额对照表
liability between the parties 双方之间的责任
liability certificate 负债证明书
liability clause 责任条款
liability composition 负债结构
liability dividend 负债股利
liability for damages 损害赔偿责任;赔偿责任;损害赔偿责任
liability for delay 延迟负责;误期责任
liability for endorsement 背书责任
liability for environment(al) damage 环境损害的责任
liability for guarantee 保证责任
liability for injury 人身事故责任
liability for light and power 应付电费
liability for loss 损失所负责任
liability for nuclear damages 核破坏的责任
liability for payroll tax 应付工薪税
liability for two years 两年责任
liability fraying 磨损倾向
liability insurance 责任险;责任保险;个人义务保险;对第三者负责的保险
liability insurance for two years 两年责任险
liability insurer 责任保险承保人;赔偿责任承保人
liability management 债务管理;负债管理
liability method 负债法(用于迟延所得税)
liability mix report 负债混合报告表
liability of acceptance 承兑责任
liability of client 业主责任
liability of cracking 易裂性
liability of the carrier 承运人的责任
liability of the consultant 咨询人的责任;咨询工程师的责任
liability of the insurer 承包者责任;承包商责任;承包人责任;保险公司的责任

liability on guaranties 保证责任
liability outstanding 未清偿债务
liability payable on demand 即付债务
liability reserve 负债准备
liability risk 负债风险;负责风险
liability salvage 责任救助
liability system 责任制(度)
liability to accidents 事故发生;事故责任
liability to cracking 易裂性
liability to frost damage 易冻坏性;易冻性
liability to pay compensation 赔偿责任
liability to shoaling 易淤浅性
liability to weathering 易风化性
liability with interest 附息负债
liability without fault 无过错责任
liable jointly and severally 负共同连带责任
liable to a tax 征税的
liable to attack 易受攻击
liable to duty 应付税
liable to infection 容易感染的
liable to insect injury 容易受虫害的
Li abundance 锂丰度
Liacopoulos phenomenon 莱科波洛斯现象
liaison 联络;传输连接
liaison aeroplane 联络飞机
liaison between technical committees 技术委员会的联络
Liaison Committee of International Technical Associations of Civil Engineering 国际土木工程技术团体联络委员会
liaison man 联络员
liaison meeting 联络会议
liaison office 联络处
liaison plan 协调计划(资源计划之一)
liana 胶藤;藤本植物
liana rubber 藤胶
liandratite 铌钽铀矿
lianoid form 藤本型
Lianr 光学镜头
Liao trichromatic decoration 辽三彩
Liapunov's function 李亚普诺夫函数
Liapunov's method 李亚普诺夫法
Liapunov's second method 李亚普诺夫第二方法
Liapunov's stability 李亚普诺夫稳定(性)
Liapunov's theorem 李亚普诺夫定理
Liapunov's vector function 李亚普诺夫向量函数
Liar 光学物镜
liardite 冻蛋白石
Lias 里亚斯统【地】;蓝色石灰岩
Lias brick 里亚斯黏土砖
Lias clay 里亚斯黏土
Lias lime 蓝色石灰
Liassic clay 里亚斯黏土
Liassic limestone 里亚斯石灰石(烧石灰用的黏土质石灰石)
Liassic sandstone 里亚斯黏土砂岩
Liassic(series) 里亚斯统【地】
Lias stone 里亚斯石
Libby counter 炭黑计数器
Libby detector 利比探测器
Libby-Owens 利比法(一种拉制玻璃方法)
Li-bearing brine 含锂卤水
libeccio 利比锡奥风
liberal factor of safety 安全裕度;安全余裕;安全余量
liberalization of economy 经济自由化
liberalization of trade 贸易自由化
liberalize the condition of loans 放宽贷款条件
liberal market access 自由进入市场
liberal style 写意
liberate 逸出
liberated capital 行动自由的资本
liberated heat 放出的热量
liberating size 解离粒度
liberation 游离;解离;释出
liberation coefficient of diamond 金刚石解离系数
liberation mesh 解离网目
liberation of gases 气体的逸出;放出气体
liberation of heat 热量释放;意大利新艺术形式的复兴;放热
liberation of intergrown constituents 连生组分的解离
liberation of reserved storage 保留存储的释放
liberator cell 脱铜槽

liberhenite 磷铜矿
liberite 锂铍石
liberty 自由;短假期
liberty day 放假日期
liberty hall 便厅
liberty man 放假上岸的船员
liberty mutual automotive crash simulator 自由相互碰撞模拟装置
liberty of trading 贸易自由
liberty ship 自由轮
liberty to touch and stay 驻泊条款
Libolite 西非天然沥青
libollite 暗沥青
libra 磅
librarian 库管理程序;收编程序;使用操作系统的程序;程序库生成程序;程序库管理程序
librarian programming 库管理程序设计
librarian system 库管理程序系统;库程序系统
library 图书馆
library allocation 库分配
library allocator 库分配程序;程序库分配程序
library automotion 程序库自动化
library binding 图书馆装帧
library command 库命令
library control statement 程序库控制语句
library directory 程序库目录
library editor 程序库编辑程序
library facility 库功能;程序库功能
library file 库文件
library file descriptor 程序库文件描述符
library function 库函数;集合函数
library function routine 库函数例行程序;库函数程序
library information 程序库信息
library maintenance 程序库维护
library maintenance routine 程序库维护程序
library management 库管理;程序库管理程序;程序库管理
library member 库成员;程序库成员
library module 库模块
library negative mo(u)ld 馆藏立体地图印膜
library of standard components 标准组件库
library of symbols 符号库
library package 库程序包
library postlude 程序库尾部
library prelude 库序部;程序库序部
library procedure 库过程
library program(me) 程序库程序
library programming 库程序编制
library reference programming 引用库的程序设计;参考库的程序设计
library routine 库存(例行)程序;程序库(例行)程序
library searching 库搜索
library service 程序库服务
library software 库存软件
library steps 书库便梯
library structure 程序库结构
library subprogram(me) 程序库子程序
library subroutine 库存子程序;程序库子程序
library support 程序库供应
library text 库文
library track 参考道
library update 程序库更新
library work area 库工作区
libration 天平动
librational axis 天平动轴
libration deviation 天平动偏异
libration effect 天平动效应
libration ellipse 天平动椭圆
libration in latitude 纬秤动
libration in longitude 经秤动
libration on latitude 纬天平动
libration point 秤重点
libration theory 天平动理论
libraty file designator 程序库文件标志符
Libya architecture 利比亚建筑
licanic acid 里坎酸;巴西果油酸
licence = license
licence agreement 许可证协议;许可证协定
licence for the export of commodities 商品出口许可证;出口货物许可证
licence-issuing authority 发证机关
licence-number-matching method 汽车牌号对照法
licence revoked 驾驶执照吊销

licence suspended 执照暂时吊销或扣押
licence system of producing of hazardous chemicals 化学危险品生产许可证制度
licence system of using water 取水许可制度
license 许可证;许可;执照;接受方;特许证;特许
license and permit bond 许可证及执照;执照及许可证
license and quota system 许可配额混合制
license bracket 车照架
license carrier 车照架
license condition 许可证条件
licensed architect 注册建筑师;领有执照的建筑师
licensed contractor 注册承包商;领有执照的承包商
licensed deposit takers 特许收储机构
licensed engineer 注册工程师;领有执照的工程师
licensed lender 小额贷款公司
licensed material 特许材料
licensed pilot 有执照的引水员
licensed premises 注册房地产
licensed process 获得许可证的生产过程
licensed(public) accountant 注册会计师
licensed technology 许可转让的技术
licensee 许可证接受人;许可证接受方;许可证持有者;引进方;技术引进方;获得许可的人;持证者;被许可方;受方
licensee estoppel 许可证接受方不得违约
license expenses 许可证费(用);执照费;发给许可证费用
license fee 牌照费;许可证费(用)
license in place 许可证有效
license number 执照号码
license plate 汽车牌照;牌照;牌照板
license production 许可证生产;特许生产
licenser 批发执照者;出让方;发给许可证者;发放许可证者
license system 许可证制度
license system of using water 取水许可制度
license tax 执照税;牌照税
license to discharge 倾卸许可证
license to trade 许可通商;通商许可证
license to trade ordinance 贸易许可证条例
license trade 许可贸易
license trade union 许可证贸易联盟
license transactions 许可证贸易
licensing 给许可证;许可证交易
licensing agreement 许可证协定;许可证(贸易)协议
licensing bank 签证银行
licensing examination 执照考试
licensing of export 出口许可证
licensing of import 进口许可
licensing operations 许可证业务
licensing procedure 许可证程序
licensing program(me) 许可程序
licensing requirement 申请执照的条件;申请许可证的条件
licensing system 许可证制度
licensing system for multimodal transport operator 发给联运人执照的制度
licensor 许可方;技术输出方;认可者;批发执照者;出让方;发照照者;发证人;发许可证者;核准人
licentiate 领有开业证书人员
lichee 荔枝
lichen 青苔(绿藻类);苔藓;地衣
lichen albus 白苔癣
lichen blue 石蕊蓝;地衣蓝
lichen desert 地衣类荒漠
lichen ecology 地衣生态学
Lichenes Imperfecti 半知地衣
lichen fungus 地衣真菌
lichenology 地衣学
lichen vegetation 地衣植被
lich gate 教堂墓地前面有顶盖的门;墓地廊道;停枢门道
lich-house 殡舍;停尸房;太平间
Lichner's blue 李希纳蓝
lich-stone 架棺石
Lichtenberg's alloy 利登彼格铅锡铋易熔合金
lick 盐沼(泽)
licker-in grinder 磨刺辊机
licker-in grinding and covering machine 刺辊包磨机
licker-in roller 刺毛辊
Lickey quartzite 利基石英岩

lick roller 舔袍辊;冲洗滚筒
li coast 淤泥质海岸
lid 闸口;盖头;凸缘;大气温度逆增
lidar 激光雷达;激光红外雷达
lidar light detection and ranging 激光探测和测距
lidar meteorology 激光雷达气象学
liddicoatite 钙锂电气石
lid embossing machine 罐盖压花机
lido (远洋轮上的)露天游泳池;(豪华的)海滨浴场
lido deck 游泳池甲板
lid off 盖子失落
lid on price 限价
lid printing machine 罐盖印字机
liduid electrooptic(al) cell 液体电光元件
lie 座落;横卧;倒向
Lie algebra 李氏代数
lie along the land 沿岸航行
lie at anchor 锚泊
lie at anchor in a harbor 停泊(指系泊作业)
lie athwart 横风或横流停泊
lie athwart the wind 横风停泊
lie at single anchor 单锚停泊
liebenbergite 镍橄榄石
liebenerite 白假霞石
Liebermann-Burchard test 利伯曼—布查尔德试验
Liebermann-Storch test 利伯曼—斯托希试验
Liebermann test 利伯曼试验
Liebermann test for phenols 苯酚的利伯曼试验
Liebig condenser 李比希冷凝器
liebigite 铀钙石;绿碳钙铀矿;碳铀钙石
Liebig's law of minimum 李比希最低因子定律
Liebmann effect 利布曼效应
Lie bracket 李氏括号
Lie derivative 李氏导数
lie detector 测伪器
lie idle 闲置
lie in berth 靠码头停泊
lieingite 利硫砷铅矿
lien 留置权
lien and hypothecation 财产的留置权及抵押权
lien clause 留置权条款
lien for freight 运费留置权
lien for taxes 纳税留置
lien mechanic's (手续上的)扣押权;款项尚未给付的通知
lien of broker 经纪人留置权
lien on cargo 货物留置(权)
lien on goods 货物留置(权)
lien release 工料付讫;(用以解除留置权)付款证明
lien structure 外界构筑物
lien theory 留置权理论(认为受甲方只有留置权而无法定所有权)
lien waiver 扣押权;留置权;扣押权;放弃声明书;留置权的放弃
lier 玻璃退火窑;长隧道窑;退火炉
lierne 穹隆副肋;枝肋
lierne rib 哥特式穹顶的装饰性肋;枝肋;穹顶枝肋
lierne vault 枝肋穹隆;枝肋拱;扇形肋穹顶
lierne vaulting 穹隆副肋作业;扇形穹顶;枝肋穹顶;用装饰性肋的穹顶
Liesegan ring 间歇沉淀环
Lie series 李氏级数
lie the interred course 在预定的航线上
lie time 停工时间
Lie transformation 李氏变换
lieu 场所
lie up 卧冬
Liewen evapo(u)rator 列文式蒸发器
life acceleration factor 加速老化系数
life activity 生活活动力
life and force pump 提升与压力泵
life annuity 终身年金
life area 生活场所
life-arena 生活场所
life arrow 救生抛缆的引头
life assemblage 生活组合
life assurance 人寿保险
life belt 救生带;生物带;(车座位上的)安全带
lifebench 救生凳
lifeblood 命脉;生命线
lifeboat 救生艇;救生舢板;救生船;救护船
lifeboat capacity 救生艇容量;救生艇容积
lifeboat compass 救生艇罗经
lifeboat drag 救生艇浮锚

lifeboat equipment 救生艇属具
lifeboat falls 吊救生艇的辘绳
lifeboat hook 救生艇挽钩;救生艇钩篙
lifeboat launching gear 救生艇降落装置
lifeboatman 救生艇(艇)员
lifeboat station 救生艇站
lifeboat tackle 救生艇吊杆绞辘
life-boost cathode 耐久阴极
life buoy 救生圈
life buoy flare 救生圈烟火
life buoy light 救生圈信号;救生圈浮灯
life buoy self-igniting light 带自亮浮灯的救生艇
life buoy signal 救生圈信号
life circle 生活周期;生活圈
life compression chamber 救生加压舱
life condition 生存条件
life cradle 救生吊框
life cycle 寿命周期;世代;生命周期;生活周期;生存期
life cycle analysis method 寿命周期分析法
life cycle cost(ing) 寿命周期成本;全寿命费用计算;寿命周期价格;全寿命期费用
life cycle cost method 全寿命费用计算法
life cycle diagram 生活史表解
life cycle management of product 产品从出厂到最后消费管理
life cycle of product 产品寿命周期
life cycle test 交变荷载耐久试验;重复荷载耐久试验
life detector 使用期限测试器
life duration 耐用期
life estate 非世袭的终身财产
life expectancy 预регу(使用)寿命;预期使用年限;预计使用期限;预计使用年限;预计可用年限;预测使用期限;预测使用年限;估计寿命;概率寿命;期望寿命;寿命预期
life factor 耐久性因素;耐久性系数;寿命系数;使用年限因素
life float 救生浮具;救生浮艇
life-force 生命力
life form 生活型式;生活方式
life-form class 生活型级
life-form dominance 生活型优势
life-form spectrum 生活型谱
life-form system 生活型系统
life formula 寿命计算公式
life goal 生活目标
life grab 救生抓索
life grease lubrication 长效润滑油(脂)
life guard 救生员;排障器
life habit 生活习性
life history 生活史
life instinct 生活本能
life insurance 人寿(保)险;死亡保险;寿命保险;使用期限保险
life insurance contract 人身保险合同
life insurance corporation 人寿保险公司
life insurance for groups 团体人寿险
life insurance policy 人寿保险单
life insurance premiums 人寿保险费
life insurance trust 人寿保险信托
life insurance with dividend 人寿保险费
life intensity 生活强度
life interest 终身财产所有权(指非世袭的财产);终身权益
life jacket 救生衣;救生背心
life kite 救生发报机风筝
life layer 地球生物圈
life leasehold 终身租赁
life length 寿命
lifeless 非生命的
lifeless rubber 无弹力橡胶
life level 寿命水平
life light 救生灯
lifelike colo(u)r 逼真颜色
life line 安全线;生命线;安全索;安全栏绳;(车座位上的)安全带;救生绳;救生索
lifeline earthquake engineering 生命线地震工程
lifeline game 生存线对策
life line gun 救生索发射炮;救生抛绳枪
life line rocket apparatus 救生绳火箭发射器
life line-throwing appliance 救生索发射器;救生抛绳设备
life line-throwing gun 救生索抛射枪;救生索发射炮;救生抛绳枪

life-load stress 活载应力
life loss 生命损失
life lubrication 长效润滑
life mortar 救生索抛射枪
life net 救生网;救护网;安全网
life of a depreciable assets 固定资产使用年限
life of blast furnace 高炉寿命
life of contract 期货合同有效期
life of equipment 设备寿命
life of face 工作面开车年限(采矿)
life office 人寿保险业
life of furnace 炉的使用寿命
life of loan 借款期限
life of pavement 路面使用年限;路面(铺面)寿命
life of project 工程使用年限
life of reservoir 水库寿命;水库使用年限
life of stock 库存期限
life of the die 模子寿命;模转寿命
life of well 井的开采期限
life performance 寿命性能
life performance curve 光源特性与寿命的关系曲线;寿命性能曲线
life period 生活时期;存在时期;存在时间
life policy 人寿保险单
life power 持久功率
life preserver 救生用具;救生器;救生船
life process 生活过程
life raft 救生筏
life raft autoreleasing 救生筏自动释放
life rail 桅杆栅栏;安全栏杆
life repair cost 全使用期内修理费;全部使用期的修理费;使用期修理费用
life requirement 耐久性要求;寿命要求
life ring 救生艇舷握系环;救生圈
life rope 安全索;(车座位上的)安全带
life salvage 救生
life saver (塔上工人的)安全带;救生员
lifesaving 救生
life-saving appliance 救生设备
life-saving appliances certificate 救生设备证书
life-saving equipment 救生设备
life-saving float 救生浮具
life-saving raft 救生筏
life-saving rocket 救生火箭
life-saving service 救生站;救生机构;海上救助业务;海事救援队
life-saving ship 救生船
life-saving signal 救生信号
life-saving station 救生站;救生艇站
life size 同实物一样大小;原物尺寸;原大小
life span 有效期限;生存时间;生存期间;升降孔;平均生命期;寿命;使用期限;存在时间
life-span of oil 油使用期限
life-span taper 寿命锥体
life-span toxicity test 终生毒性试验
life-span triangle 寿命三角形
life's prime want 生活的第一需要
life stage 生活期
life style 生活方式
life suit 救生服
life-support back pack 救生背包
life support system 救生系统;生命支持系统
life system 生命系统
life table 耐用年限表;寿命表;使用年限表(固定资产)
life tenant 终身受益人;土地终身占有人;终身租户;永久租户
life test(ing) 使用寿命试验;使用耐久性试验;工作期限试验;耐用性试验;寿命试验;使用期限试验
life-test rack 寿命试验台
life thermal protective aids 救生保温用具
life-threatening 危急生命的
life time 活性期;使用年限(设备或建筑物);连续操作时间;使用寿命;生存期;使用期限;寿命
lifetime dilation 寿命延长
lifetime dose 终身剂量
lifetime employment 终身就业
lifetime employment system 终身雇佣制;职务终身制
lifetime for satellite 卫星寿命
lifetime limitation 寿命极限
lifetime lubricated 永久润滑的;一次加油润滑的;全程润滑;润滑使用期
lifetime lubrication 一次加油润滑;永恒润滑;长效润滑
lifetime of energy level 能级寿命
lifetime of energy state 能态寿命
lifetime of satellite 卫星寿命
lifetime of set and reset cycles 反转寿命
lifetime of the state 能态寿命
lifetime prediction 寿命预测
lifetime probability 使用期限概率
lifetime service 长期使用
lifetime warranty 终身保证期
life tools 工具寿命
life-up roller 过渡辊台
life vest 救生衣;救生圈
life waistcoat 救生背心
life zone 生物带
life zone ecology 生物带生态学
LIFO dollar-value method 按价值计算的后进先出法
lift 载货电梯;掘起;举起;阶段高度;(混凝土等的)浇筑层;起箱;起落;起吊;提升跨度;提起;升起;电梯;搭车;垂直电梯;升降机
liftability 起模性
liftable and lowerable belt conveyer 可升降胶带输送机;升降式胶带运输机
lift accumulator 起落蓄能器
lift agitator 提升搅拌器
lift-and-carry mechanism 升高并移动零件的机构
lift and carry transfer 提升移送装置
lift and conveyer 升降输送机
lift angle 升角(叉摆)
lift apartment house 有电梯的公寓房子
lift arm 起落臂;提升臂
lift arm extension 提升臂延伸
lift arm support pin 提升臂支撑销钉
lift attendant 电梯操纵员;电梯服务员;提升机司机
lift away shutter door 提升门;提升百叶门
lift bank 电梯组
lift bar 提升杆
lift beam furnace 升降杆送料炉
lift between two levels 水平间垂高
lift block 起重滑轮
lift bolt 吊栓;起重螺杆;环首螺栓
liftboy 开电梯工人
lift bracing 升力拉条
lift bracket 升降滑架
lift bridge 升降桥;升降式开合桥;吊桥
lift by lift 分层(浇筑)
lift by the stem 尾浮【船】
lift cable 提升索
lift capacity 电梯容量;提升机容量
lift car 电梯厢;提升罐笼;电梯笼
lift-car door 提升罐笼门;电梯笼门
lift cargo 承运货物
lift carriage 梯厢
lift chains 提升链
lift chamber 升船箱;升船厢
lift check valve 提升止回阀;升降式止回阀
lift clutch 起落自动器;起落离合器
lift coefficient 升力系数
lift component 升力分量
lift control 升力控制;升降控制;电梯控制
lift controller 电梯控制器
lift conveyer 升运器
lift counter 往返行程计数器
lift crank 起落弯臂;提升曲柄
lift curve 升力曲线
lift-curve-slope parameter 升力曲线斜率参数
lift cycles 疲劳寿命循环数
lift cylinder 起重油缸;提升油缸
lift cylinder arm 提升油缸臂
lift cylinder trunnion 提升机活塞旋转耳轴
lift device 起重装置
lift-device nozzle 升力喷管
lift divergence 升力减小
lift dock 升降式船坞;浮船坞
lift door 电梯门
lift-drag ratio 举阻比;升阻比;升力阻力比;升力力比
lift dump 升力卸减
lift dwelling block 有电梯的住房街坊
lift dwelling house 有电梯的居住房子
lifted block 地垒;断块;隆起
lifted load 起升荷载
lifted root 挖出块根
lifted throat 上升式流液液洞
lift efficiency 升力特性;升举性能
lift engager 提升装置接合器;升降机衔接器
lift engine 腾空机
lift entrance 电梯入口
lifter 起重机(构);启门机;提升装置;提升机;提钩;抬刀机构;升降机;砂钩;电梯司机;底板眼;升降机
lifter arm 提升器臂
lifter bevel wheel (高低牙的)升降伞形齿轮
lifter blade 扬料叶片
lifter board 提升台
lifter bracket 升降托脚
lifter cam 升降凸轮
lifter change wheel 升降变换齿轮
lifter flight 扬料板;升举刮板
lifter guide 挺杆导管;顶杆导轨
lifter hole 底炸孔;辅助炮眼
lifter-loader 升运装载机;升运装载车
lifter motion (移ényi片的)提升运动
lifter motion cam shaft 升降凸轮轴
lifter pin 升降销
lifter rod 升降杆
lifter roller 挺杆滚轮
lifter roof 升降顶;储罐浮顶;浮顶
lifter roof tank 升降式浮顶罐
lifter slide 升降滑板
lifter wheel 挖掘轮
lifter winch 起重绞车;升降绞车
lift fan 升力风扇;(气垫车的)升降风扇
lift flap plain gate 升卧式平面闸门
lift-floor 顶升楼板
lift force 升力;上举力
lift fork 叉车;叉式升降机
lift frame 举升车架;提升框架;吊运架
lift full-deck jumbo 升降平台式钻孔台车
lift gas 提升用气体;提升气体
lift gas return 提升气体返回口
lift gate 吊门;拦(路)木;提升闸门;提升式门;升降(闸)门
lift gate feeder 提升闸门加料机
lift gate weir 提升闸门堰
lift gear 起重装置
lift guide rail 电梯导轨
lift hall 电梯前的大厅
lift hammer 落锤;提升锤
lift handle 提升柄;提柄
lift head 扬程
lift height 浇筑层高度;分层厚度;浇筑层段高度;提升高度
lift hook 吊钩
lift indicator 升力指示器
lifting 离焰;纵高;起重;起出;尾浮【船】;提升;升起;升高;上升;吊装;吊起
lifting acceleration 起升加速度
lifting airscrew 上升螺旋桨
lifting altitude 起升高度
lifting and inspecting installation 升高检查设备
lifting and operating key 提升与操纵键
lifting and traversing mechanism 提升转向机构
lifting apparatus 卷扬装置;起重装置
lifting appliance 起重装置;提升装置;提升设备;升降设备
lifting appliance quadrennial certificate 吊货四年一次检验证
lifting appliances test record 船舶起重和起货设备试验证书
lifting arm 升降臂;提升杆
lifting attachment 起重附加装置;提升附加装置
lifting bail 提吊钩;吊环
lifting bar 提升杆
lifting barrel 提桶
lifting barrier 提升限制;升降式栏木
lifting beam 吊起梁;吊装托梁;起重天秤;起重梁;起重横梁;启门梁;吊重梁
lifting beam with claws 料耙横梁
lifting beam with electromagnets 电磁吸盘横梁
lifting beam with hooks 吊钩横梁
lifting beam with tongs 夹钳横梁
lifting bed valve 升降式底阀
lifting belt 升运带
lifting blade 掘起铲;起针刀片
lifting block 起重机;提升滑轮组;起重滑车
lifting body 穿梭机;升力体
lifting-body reentry 升力体再入
lifting bolt 吊环;起重螺杆;起重螺杆;吊栓;吊环螺栓;环首螺栓

lifting bracket 起重支架
lifting brake 起重制动(器);起重制动闸
lifting bridge 升降桥;提升桥;吊桥
lifting bridge for inclined shaft 斜井吊桥
lifting bucket wheel 斗轮式提升机
lifting cable 起重索;吊索
lifting cam 升降凸轮
lifting cap 起钻护箍
lifting capacity 载重能力;起重(容)量;起重(机起重)能力;起吊能力;提升量;水泵扬程;起举能力
lifting capacity of crane 起重机起重量;吊车起重能力
lifting capacity of derrick 钻塔起重量
lifting capacity of feed mechanism 给进机构上举能
lifting capacity of floating dock 浮船坞举力
lifting capacity of free-on-wheels 轮上不受载时的起重能力
lifting capacity on outriggers 在支架上的起重能力
lifting car dumper 提升式翻车机
lifting chain 起重链;升降铰链;吊链;滑车链
lifting channel 提升槽钢
lifting charge 升扬装药
lifting chute 槽形闸门
lifting clamps 提升钳
lifting component 上向分力
lifting condensation level 抬升凝结高度
lifting construction technique 提升的施工技术
lifting crane 起重吊车
lifting curve 提升曲线;升程曲线(凸轮)
lifting cylinder 悬挂装置油缸;液力起重缸;提升油缸
lifting deck 吊车台;吊车台面
lifting desk 升降台
lifting device 卷扬装置;启门设备,启闭设备;提升装置;提升设备;升降装置;起重机;扬料装置;起重装置
lifting dipping 船首突然下倾现象
lifting dock 升船浮坞;单坞墙式浮(船)坞;岸边浮坞
lifting dog 提升抓;提升器;提升环
lifting dog assembly(in overshot head)（在打捞筒端的）提升环总成
lifting dog spring 提升卡钳弹簧
lifting door 提升(式)门
lifting door fittings 升降门配件;电梯门配件
lifting door furniture 升降门设备,电梯门设备
lifting door hardware 升降门附件;电梯门附件
lifting drum 提升卷筒;提升鼓
lifting drum brake 提升卷筒制动器;提升鼓制动器
lifting drum shaft 提升卷筒轴;提升鼓轴
lifting efficiency 提升效率
lifting electromagnet 起重电磁铁
lifting equipment 起重机械;提升装置;抬船设备;提升设备;起重设备
lifting equipment for bridge maintenance 桥梁维修顶升设备
lifting eye 吊眼;吊环;吊耳
lifting eye bolt 带吊环的提升螺栓;吊眼
lifting eyelet 吊孔
lifting eye nut 吊眼螺帽;吊环
lifting eyes 集装箱配件
lifting fall 起重索
lifting fan 升力风扇
lifting finder 翻转装置
lifting finger 升降指;翻钢回转装置
lifting fitting 起重用部件
lifting flat gate 提升平板闸门
lifting flight 扬料板;提料翼板
lifting flights lifter 带提升刮板的提升机
lifting floor and sliding formwork combined construction 升板滑模联合施工
lifting fog 抬升雾;上升雾
lifting force 起升力;提升力;升举力;上举力
lifting force moment 升力力矩
lifting fork 叉式起重机;提升叉
lifting form(work)提升模板
lifting from fork pocket test 叉槽试验
lifting from the top test 顶品试验
lifting gate 升降闸门;升闸门;提升式闸门;栏水门
lifting-gate feeder 提板式加料机;提闸门式加料机;升闸闸板式加料机;升板式装料机
lifting gear 起吊设备;启门机构;提升装置;提升机械;提升机构;提传动装置;升降装置;升降起重联动装置;提起重机;起重装置
lifting gearing 起落机构

lifting grab 钳式带卷吊具
lifting guard 井口防护栏
lifting guide pillar 起重导杆;升降导杆
lifting guiding runways 起升道轨
lifting guiding wheel device 起升导向轮装置
lifting handle 起重手柄;提升把手
lifting head sheet-handling system 升端送纸系统
lifting height 扬程(指起重机升降范围);提升高度;拔高度[铁]
lifting height curve 起升高度曲线
lifting height limiter 起升高度限位器;上极限位器[机]
lifting hoist 提升绞车
lifting hole 起吊孔;吊孔
lifting hook 起重钩;提升钩;吊钩
lifting hook-type gate 钩式提升闸门（双层平板闸门的上门扇）;提升式闸门;钩钩式闸门
lifting hydraulic system 提升液压系统
lifting injector 吸引喷射器;吸液喷射器
lifting inserts 起吊用预埋件
lifting jack 起重机;千斤顶;起重器;举重机
lifting jet 喷压提升机;喷气提升机
lifting key 起吊开关
lifting knife 提刀;升降刀
lifting lever 提升杆;拨叉
lifting limit switch 提升极限开关
lifting-line 升力线;承重线
lifting link 提升联杆
lifting load 上举荷载
lifting lock 提升闸
lifting lock gate 提升式闸门
lifting-lowering of load 荷载升降
lifting lug 吊扣;吊耳;挂耳
lifting lug for equipment installation 设备安装吊具
lifting magnet 吸铁吊具;起重吸盘;起重(机)电磁铁;起重磁铁;起重磁盘;升降磁铁;电磁吸(铁)盘;磁运器;磁力起重机
lifting magnet crane 电磁吊盘;电磁吊车
lifting magnet with tines 有尖叉的起重电磁铁;有尖齿的起重电磁铁
lifting mechanism 起重机械装置;提升机制;提升机构
lifting method 起吊方法
lifting method not to be used 禁止使用起吊法
lifting moment 倾覆力矩;上托力矩;上升力矩
lifting motor 起重电动机
lifting movement 上升运动
lifting of a girder 大梁起吊
lifting of concrete 混凝土剥落（拆模时）
lifting-off of rail 钢轨上扬
lifting of pipe lines 起出输送管
lifting of roll 提举辊子
lifting of water 水的提升;抽水;扬水;提水
lifting pad 吊板
lifting performance curve 起重特性曲线
lifting piece 提升装置;提升部件
lifting pin 顶升杆;吊销;月楔;起重爪
lifting pipe 引上线用管;提升管;提升分线管;上升分线管
lifting piping line 提升管线
lifting plan 纹钉图
lifting plane 升降舵
lifting plant 起重机械
lifting plate 起模板;平台升降台
lifting platform 升降(平)台;提升平台
lifting plow 掘起犁
lifting plunger 回程柱塞
lifting poker 升降杆
lifting position 上升位置
lifting power 起重量;起重力;启门力;提升(能)力;提升量;升举力;上举力
lifting press 提升压紧装置
lifting pressure 升压
lifting prong 叉形挖掘铲
lifting propeller 升力螺旋桨
lifting puffer 提升小绞车
lifting pump 提升泵;抽吸泵
lifting pump with bucket valve piston 有活塞止回阀的提升泵
lifting pump with hollow plunger 有空心柱塞的提升泵
lifting rack 提升齿条;爬梯
lifting range 举升行程;起升范围【机】
lifting reentry 利用升力再入

lifting repair 起道修理
lifting rig 起重设备;起吊设备;提升装置
lifting rod 起重杆;提升杆;升降杆
lifting rope 提升索;吊索;吊绳
lifting screw 螺旋千斤顶;螺旋起重机;千斤顶螺杆;起模螺钉;升降螺杆;螺旋起重器
lifting set 起重装置
lifting shaft 升降轴;升降轴
lifting shaft arm 升降轴臂
lifting shaft bracket 升轴托
lifting share 掘起铲
lifting shelf 扬料架
lifting ship capacity 升船能力
lifting ship power 升船能力
lifting shovel 掘起铲
lifting shutter 提升式百叶(门);提升式开闭器;提升式百叶窗
lifting sliding door 提升滑动门
lifting sliding door furniture 起吊滑门设备
lifting sliding door hardware 提升滑门附件
lifting sliding window 提升滑窗
lifting sling 卷扬索套;起重吊索;升降索套
lifting sling under bottom of sunken vessel 沉船底千斤
lifting span 提升式桥孔
lifting speed 举升速度;起重速度;提升速度
lifting spindle 起重丝杠;升降螺杆
lifting spreader 集装箱专用吊具;吊架
lifting stage 提升阶段
lifting stem 提杆
lifting stirrup 起重框架
lifting strength 起吊强度
lifting stroke 升举行程
lifting surface area 升力面面积
lifting system 升力产生装置
lifting table 平行升降台
lifting tackle 起重葫芦;起重滑轮;起重滑车;升降机;起重装置;起重滑车
lifting tail 升力尾翼
lifting test by bottom fitting 底吊试验
lifting test by fork pockets 叉槽试验
lifting test by up top fitting 顶吊试验
lifting the bottom test 底吊试验
lifting time 起升时间
lifting tongs 块石提升夹钳;提升夹钳
lifting tool 提引工具【岩】
lifting tower 竖井井架
lifting truck 起重卡车;提升小车
lifting truck of high altitude working 高空作业车
lifting trunnion 起重轴颈
lifting turbine cover 上汽缸起吊
lifting unit 提升装置;升力装置
lifting up of vines 提蔓
lifting valve 提升阀
lifting valve gear 提升阀装置
lifting velocity 升起速度;提升速度
lifting walkway 提升式通道;提升式过道
lifting wall 提升门墙
lifting web 升运带
lifting weight 提升重量
lifting winch 起重卷扬机;提升绞车;卷扬机
lifting window 上拉窗;起吊窗
lifting window fitting 起吊窗五金;起吊窗设备
lifting with dry chamber 干运
lifting with floating crane 起重船打捞法
lifting without water 干运
lifting with water 湿运
lifting with wetted chamber 湿运
lifting work 起重工作
lifting worker 起重工
lifting yoke 提手
lift inland lock 单级船闸;提升式单级船闸
lift installation 起吊设备;电梯装置;电梯安装
lift irrigated area 提水灌溉区(域);提水灌溉面积
lift irrigation 扬水灌溉;提水灌溉;抽水灌溉
lift irrigation area 提水灌溉区(域)
lift jacking form 顶升模板
lift jet 喷压提升机;喷气提升机
lift joint 浇筑层段;水平张节理;释重节理;层间接缝;浇筑层段缝;升降机回合面
lift kick-out 动臂升程定位器
lift ladder 升降梯
lift landing 升降机平台;电梯平台
lift landing entrance 电梯平台入口

lift latch 回转门闩;转动门闩;起吊机插锁;手揿插销;提闩
lift leg 提升腿
lift lever 提升操作手柄
lift limiter 升程限制器
lift limit switch 提升限位开关
lift line 提升管线;提升管路;(浇筑层的)层面线;承载线;升力线
lift line man 潜水员助手
lift link 提升杆
lift linkage 提升机构杆系
lift link screw 提升杆调整螺杆
lift loading 升力分布
lift lobby 电梯厅
lift lock 升船闸;升船机;单级船闸;船闸;升降闸
lift lock arm 提升式舱口盖扳手
lift machine 电梯驱动主机
lift machine room 电梯机房;起吊机房
lift magnet 起重电磁铁
lift maker 升降机制造厂
liftman 开电梯工人
lift mast 起吊柱
lift method 提升法
lift motor 电梯用电动机
lift movement 上下运动;升降运动
lift multiple dwelling 有电梯的多层住宅
lift navigation lock 单级船闸;提升式单级船闸
lift of a lock 船闸升程;船闸水头
lift of a valve 阀升程
lift of concrete 混凝土浇筑层;混凝土的升运送高;混凝土的剥落(拆模时)
lift of dredge pump 泥泵扬程
lift-off 起飞;发射;弹射;升空
lift-off a seat 离开阀座
lift-off attitude 起飞仰角
lift-off butt 活络合页;活脱铰链;抽心铰链
lift-off container truck 吊装集装箱运货车
lift-off hinge 活脱铰链;套芯铰链
lift-off poverty 脱贫
lift-off pressure 离地张力(预应力混凝土构件);离地张拉压力(预应力混凝土构件)
lift-off seal 提升密封
lift-off site 起飞场
lift of hook above ground 吊钩离地高度
lift of lock 船闸水头;船闸上游水位差
lift of pump 水泵扬程;泵压头(高度);泵(的)扬程
lift of pump suction 泵的抽吸高度
lift of the pump suction 泵的吸高
lift of valve plate 阀片升程
lift-on/lift-off 起重机装卸
lift-on/lift-off bulk tipping container 吊上吊下散装倾卸货箱
lift-on/lift-off container ship 吊装式集装箱船
lift-on/lift-off system 吊装装卸方式;吊上吊下装卸系统;吊进吊出法
lift-on the sea 海上起吊
lift-on the water 水上起吊
liftout attachment 推出装置;顶出装置
liftout bolt 顶出螺栓
liftout crucible furnace 坑式坩埚炉
liftout crucible type furnace 坑式坩埚炉
liftout plate 提升板;顶出板
liftout plunger 提升柱塞
liftout roller 过渡辊台;提升辊
lift park 升降式停车库;升降式停车场
lift pin 起模针;起模顶杆;升降机竖井
lift pin stripper mo(u)lding machine 顶箱式造型机
lift piston 起落油缸活塞;提升油缸活塞
lift pit 电梯井底坑;升降机竖井;电梯井坑
lift-plate floor 升板楼面
lift platform 电梯地板;提升台;升降机平台
lift platform truck 有提升平台的装卸车;有提升台的载货车
lift pod 升力发动机吊舱
lift pot 提升罐;提升斗
lift pressure 扬压力
lift pump 吸扬泵;扬水泵;扬液泵;升水泵;抽扬泵;提升泵
lift pumping station 提升泵站
lift rail 升降轨
lift ram 起落油缸;提升油缸
lift range (悬挂装置的)升程
lift rate 提升速率

lift residence block 有电梯的住宅区段;有电梯的住宅街坊
lift residence building 有电梯的住宅建筑
lift residential building 有电梯的居住建筑
lift rod adjuster 提升杆调节器
lift rod pin 提升杆销
lift rod stroke 顶杆行程
lift roller 提升滚轮
lift-rolling hatchcover 层叠式舱盖装置
lift rope 提升钢丝绳
lifts 层数(脚手架在垂直方向)
lift screw 螺旋起重器
lift set 升高组
lift shaft 升降机轴;升降机竖井;升降机井(道);电梯井
lift shaft arm 提升轴臂
lift shutter 提升式开闭器
lift-slab 顶升楼板;升板(法)
lift-slab collar 升板卡圈
lift-slab column 升板支柱;顶升楼板的柱
lift-slab concrete floor 升板混凝土楼面板
lift-slab construction 升板施工法;升板法结构;吊板施工
lift-slab method 升板法
lift-slab method of construction 升板施工法
lift-slab roof 顶升板屋面
lift-slab span 顶升板跨度
lift-slab structure 顶升板结构
lift-slab system 升板施工法;升板体系
lift-slab technique 升板技术
lift slack-cable-switch 电梯行索松断保护开关
lift sling here 此处吊起
lift span 提升式桥孔;升降式桥孔;提升式桥跨
lift speed 提升速度
lift stage 升降舞台
lift/stair(case) core 电梯/楼梯中心
lift station 扬水站;提升站;水泵站;升液站(污水);抽水站;泵站
lift stop 提升限制器
lift stroke 提升行程
lift sub 吊卡
lift surface 冲面
lift table 升降台
lift tackle (吊杆的)升降绞辘;起重滑车
lift technique 提升技术
lift the embargo 解除禁运
lift thickness 浇筑层厚度;铺土厚度;层厚(度);夯实铺土厚度
lift thrust 升力发动机推力
lift time 提升时间
lift tower 提升井筒;起重塔
lift trailer 起重装卸拖车
lift transducer 升力换能器
lift transporter 拖头牵引的起重运输机
lift truck 自动装卸车;起重车辆;提升车;升降式搬运车;升降车;装有提升机的卡车;升降式装卸车;升降叉车搬运机;起重车
lift trunk 升降机围壁;升降机通道;升降机井道
lift tube 升水管
lift-type 升式
lift-type car park 升降机车库;升降机停车场
lift-type disk 升式盘
lift-type valve 升式阀
lift-type valve cock 升式阀旋塞
lift unit 升力装置
lift unit frame 大型成组托架
lift unit frame ship 成组托架专用船
lift unit frame trailer 成组托架拖车;成组托架挂车
lift-up drawbar 起落式牵引装置
lift-up method 顶升法
lift valve 升水管道阀;升(举)阀;支撑阀
lift van 装卸箱;装卸箱货物
lift velocity 提升速度
lift vessel 浮吊
lift wall 闸首墙;闸室墙
liftway 电梯井道
lift well 升降机竖井;升降机井孔;升降机井(道);电梯井
lift well door 电梯井门
lift wire 升力线
lift with automatic push-button control 自动按钮控制的电梯;自动按钮控制的升降机
lift with axial floater 轴向浮筒式升船机;轴向浮筒式举船机
lift with counterweight 平衡重式举船机(由低水面部分一边举船过闸至高水一边);平衡重式升船机
lift with crane 用起重机起吊
lift yoke 提升吊架
ligament stress 孔桥带应力;带状应力
ligamentum bifurcatum 分歧韧带
ligamentum quadratum 方形韧带
ligamentum reflexum 反转韧带
ligand 配位体;配位基;配价体
ligand chromatography 配位体色谱(法);配位色谱法
ligand concentration 配位体浓度
ligand exchange 配位体交换
ligand exchange chromatography 配位体交换色谱法;配位体交换层析
ligand field stabilization energy 配位场稳定化能
ligand field theory 配位场理论
ligand ion 配位体离子
ligand number 配位体数
ligand polymer 配位体聚合物
ligand-promoted dissolution process 助配位体溶解过程
ligand solvent 配位体溶剂
ligands share 配位体份额
ligasoid 液气悬胶
ligature grid 绑扎格网;绑扎网格
ligger 拌灰板;小搅拌板(苏格兰);横托木;(茅屋的)屋脊板;(盖茅草屋顶用的)红褐色胶黏料
light 轻度;轻便的;灯;淡
light aboard ship 龙门吊式载驳货船
light absorbency 吸光性
light absorbent 吸光材料
light-absorbing 吸收光线的
light-absorbing aerosols 光吸收烟雾剂
light-absorpting shade 采光罩
light absorption 光吸收;光的吸收(作用)
light absorption line 光吸收线;光吸收谱线
light absorption spectrometry 光吸收光谱法
light abutment 轻型桥台
light accumulation of snow 轻微积雪
light-activated element 光敏元件
light activated silicon-controlled rectifier 光激可控硅闸晶体管
light activated silicon switch 光激硅开关
light activated switch 光敏开关;光激开关
light activated thyristor 光激可控硅闸晶体管
light active switch 光敏开关
light adaptability 光适应
light adaptation 光适应(性);亮适应;明适应;明视适应性
light-admitting 透光的
light-admitting board 透光板
light-admitting grill(e) 透光格子窗;透光格栅
light-admitting plastic material 透光塑料
light-admitting plastic sheet(ing) 透光塑料薄板
light-admitting quality 透光性
light-admitting sheet 透光薄板
light adobe brick 轻质黏土砖;轻质风干砖坯
light aeroplane 小型飞机
light ag(e)ing 光致老化;光时效;光老化
lightage steel joist 轻钢龙骨
light aggregate 轻质骨料;轻质集料
light air 稀薄空气;一级风;高层大气;软风;轻风
light aircraft strip 轻型飞机简易跑道
light airplane 轻型飞机
light alkylate 轻质烷基化物
light alloy 轻质合金;轻合金
light alloy casting 轻合金铸造
light alloy girder 轻合金梁
light alloy girder bridge 轻合金梁桥
light alloy housing 轻合金缸体
light alloy metallurgy 轻合金冶金学
light alloy metal plate 轻合金板
light alloy piston 轻合金活塞
light alloy structure 轻合金结构
light amplification 光放大
light amplifier 光增强器;光放大器
light amplitude 光振幅
light amplitude modulation 光振幅调制
light amplitude modulator 光振幅调制器
light analyzer[analyser] 光分析器;光检偏器
light and air easement 通风采光权
light and beacon list 灯标及航标表
light and bell buoy 闪光钟响雪标;灯光警钟浮标
light and bulk freight 轻浮货物

light and horn buoy 闪光号笛浮标
light and power cost 取暖照明及动力成本
light and shade 明暗(表现)
light and shade contrast 明暗对比
light and shade face 明暗面
light angle 光入射角;灯光角度
light annealing 光亮退火
light antenna 光天线;光束导向天线
light aperture 光孔径
light apparatus 灯具;发光设备
light application ratio 光照比
light application time 光照时间
light architecture 精巧建筑
light area 照明区(域)
light area defense 轻型面防御
light armored car 轻型装甲汽车
light armo(u)red protection 轻装甲防护
light armo(u)ring 轻装甲
light ash 轻质苏打
light attenuation 光衰减
light axle 轻轴;轻型轴
light axle attached by gluing 用胶结法压装轮对的轻型轴
light baffle 挡光板
light-balancing filter 光平衡滤光镜;中性滤光片
light band 照明带;光带
light barium crown glass 轻钡冕玻璃
light barium flint glass 轻钡火石玻璃
light barrier 光势垒
light basis weight plate 轻型板
light bath 光浴
light beacon 色灯信号;灯桩(上设灯标);灯光信号;灯(光导)标
light beacon system 光信标系统
light beam 光柱;光注
light beam acquiring 光束瞄准
light beam deflector 光束致偏器;光束偏转器;光束偏折器
light beam diaphragm 光束隔膜
light beam divergence angle 光束发散角
light beam oscillograph 光束示波器
light beam pick-in 光束拾音器
light beam pick-up 光注拾声器;光束拾声器
light beam propagation error 光束传播误差
light beam recording oscillograph 光束记录示波器
light beam remote control 光束式遥控
light beam scanning 光束扫描
light beam splitter 光束分光镜;分束镜;分光镜
light beam welding machine 光束焊机
light beam width 光束宽度
light beat 光拍
light beating bucket receiver 光拍斗形接收器
light bed load movement 薄的底砂运动;薄的底质运动
light bending 轻型钢筋弯曲
light bias 偏光;光线偏向;轻微漏光
light bias lamp 背景光灯
light bill 灯标税单
light bituminous carriageway 浅色沥青路面
light blading 整修轻型;轻刨;轻刮
light blast(ing) 小爆破
light block 预制轻混凝土块
light block masonry 轻型圬工结构
light blue 浅蓝(色);淡青色;淡蓝色;淡蓝的
light blue-black 浅蓝黑色
light blue-drown 浅蓝褐色
light blue glass 浅绿色玻璃
light blue glaze 浅蓝釉
light blue-green 浅蓝绿色
light blue-grey 浅蓝灰色
light blue pull 浅蓝印样
light blue-violet 浅蓝紫色
light bluish green 玉色
light board 轻型板材;轻质板
light boat 灯船;灯标船
light body 低稠度
light boom 吊光桁
light box 航行识别灯控制箱;透写台
light bracket 轻型托架;灯架
light break reaction 闪光反应
light breeze 轻风;二级风
light brick 轻质砖;轻质砖;轻型砖
light bridge 灯光平台;灯光渡桥;灯光调整电桥;舞台安装天桥;舞台安灯天桥

light bridge train 轻型架桥梁辐重队
light brightness control 光亮色控制;光亮度控制
light brown 浅褐色;浅棕色
light brown-black 浅褐黑色
light brown-blue 浅褐蓝色
light brown glaze 铁锈花釉
light brown-green 浅褐绿色
light brown-grey 浅褐灰色
light brown matter 淡褐色物质
light brown-red 浅褐红色
light brown-violet 浅褐紫色
light brown-yellow 浅褐黄色
light building block 轻型建筑砌块;轻型建筑砌块
light building board 轻型建筑板;轻质建筑板材;轻质建筑板材
light bulb 灯热灯泡;灯泡
light bulb-lens combination 灯泡透镜组合
light bulb torus 灯泡环型安全壳
light bulky cargo 轻泡货
light buoy 挂灯浮标;灯光浮标;灯浮标;浮灯标;浮标灯;发光浮标
light burden 轻负荷
light-burned 轻烧的
light-burned bauxite 轻烧矾土熟料
light-burned clinker 轻烧熟料
light-burned dolomite 轻烧白云石
light-burned impregnated brick 轻烧油浸砖
light-burned magnesia 轻质煅烧氧化镁;轻燃氧化镁;轻烧镁砂
light-burned refractory ware 轻烧耐火制品
light-burning 轻烧
light-burnt lime 轻烧石灰
light bus 中型客车;中型公共汽车;旅行车;轻便客车
light button 光钮;光按钮
light cable bushing 灯线衬套;灯丝衬套
light cable percussion rig 轻型钢绳冲击钻架
light calcined dolomite 轻烧白云石
light cap 轻压盖
light car 小轿车;轻型小客车;轻型汽车;小(排量)汽车
light cargo 轻(泡)货;轻量货品;松泡货
light cargo carrier 轻型运货车
light carrier 光载波
light carrier bundle 光导束
light carrier gas 轻载气
light carrier injection 光载波注入
light-case 薄壁的
light-case bomb 薄壳炸弹
light casing 灯罩壳
light casting 轻型铸件
light cast member 轻质预制件;轻型预制构件
light cast stone 浅色铸石
light-catalysed 光催化的
light C bag 轻型C字麻袋
light cell 光电池
light center 发光中心
light-center length 光源中心长度;发光中心长度
light chamot(te) brick 轻质耐火砖
light channel steel 轻型槽钢
light character 光学性质
light characteristic 亮度特性;灯(光性)质;灯光特征
light chart 照明图表
light chestnut 淡栗色
light chopper 遮光器;截光器;光线断续器
light cinder aggregate 轻煤渣集料;轻煤渣骨料
light circle 光圆
light clay 轻质黏土;轻黏土
light clay brick 轻型黏土砖;轻黏土砖
light clay(ey) loam 轻亚黏土;轻黏壤土
light climate 光照气候;不同气候的灯光能见度
light climate factor 光照气候系数
light clock 光钟
light closing weld 珠焊
light closing welding 轻连续焊接
light coal 轻煤;气煤
light coarse aggregate 轻质粗集料;轻质粗骨料
light-coated electrode 敷料电极;薄皮焊条
light coating 薄药皮焊条;薄皮焊条;薄涂层
light cobalt violet 淡钴紫
light cold rolled sheet 冷轧薄板
light collapsible 轻微湿陷
light collector 聚光器
light colo(u)r 灯色;光亮颜色;淡色(的)
light colo(u)r asphalt tile 淡色沥青砖

light-colo(u)red 浅色(的);轻度着色的;淡色的
light-colo(u)red earth 浅色土
light-colo(u)red granite 浅色花岗岩
light-colo(u)red intrusive rock 浅色侵入岩
light-colo(u)red mineral 浅色矿物
light-colo(u)red rock 浅色岩
light-colo(u)red soil 浅色土
light column 轻型柱
light commutation circuit 光开关电路
light-compass reaction 补偿向性
light compensation point 光补偿点
light component 轻型建筑构件;轻质的构成体
light concrete 轻混凝土
light concrete aggregate 轻质混凝土集料;轻混凝土骨料
light concrete beam 轻型混凝土梁;轻质混凝土梁
light concrete block 轻混凝土砌块
light concrete building 轻型混凝土建筑;轻质混凝土建筑
light concrete building block 轻型混凝土砌块;轻质混凝土砌块
light concrete chimney pot 轻型混凝土烟囱筒体;轻质混凝土烟囱筒体
light concrete column 轻型混凝土柱;轻质混凝土柱
light concrete structure 轻(型)混凝土结构;轻质混凝土结构
light concrete wall panel 轻混凝土墙板
light condenser 照明灯聚光镜
light condition 压载状态;空载状态;轻载状态;轻载条件;照明环境
light conditioning 光调;照明调节
light conduction 光传导
light conduit 照明电缆管道
light cone 光锥(体)
light connector cable 照明开关导线
light construction 轻型结构;轻型建筑;轻型构造
light construction material 轻质建筑材料
light construction method 轻型结构方法;轻型建筑方法
light consumer 电灯用户;减光器
light continuance welder 轻型连续焊
light continuous welding 轻连续焊接
light control 灯光控管;照明控制;光(量)控(制);灯头控制;灯光控制;灯光调节
light control-console 灯光控制台
light control device 光度控制装置
light-controlled pedestrian crossing 信号灯控制人行横道
light control room 灯光控制室
light control tape 照明控制带;灯光控制带
light control technique 光束控制技术
light converter 光换能器
light core 轻心材
light correction filter 光校正滤光片
light counter 光计量器
light-coupled semi-conductor switch 光耦合半导体开关
light coupler 光耦合器
light court 采光天井;天井
light cross 光柱
light crossing signal 交叉口灯光标志
light cross-section method 光切断法
light crown glass 轻冕(光学)玻璃
light crude 轻质原油
light cruiser 轻巡洋舰;轻型巡洋舰
light cube cargo 轻泡货
light-cupola 上部采光的圆屋顶;轻穿顶;透光穹顶
light current 光电流;弱电流
light current engineering 弱电工程学
light-current system 弱电控制系统
light-curve 光变曲线;光曲线
light-curve in two colo(u)rs 两色光变曲线
light cut 光截面;浅切削
light cutting 浅挖
light cycle oil 轻循环油
light cylinder 光速圆柱面
light dark adaptation 亮暗适应
light day 光日
light decay 光衰减
light-decay characteristic 余辉特性;光度衰减特性
light deflecting technique 光偏转技术
light deflection 光偏转
light deflector 光偏转器
light delay 光延迟

light delivery truck 轻型送货车
light demanding tree 阳性树
light demolition work 轻型拆除工作
light densimeter 光密度计
light densitometry 光密度测定法
light-density railroad 低输送密度铁路
light dependentresistor 光敏电阻器
light detection 光探测
light detection and ranging 光测定法
light diaphragm 光阑
light diesel fuel 轻质柴油燃料
light diesel oil 轻柴油
light diffuser 光扩散器;灯光扩散器
light-diffusing ceiling 散光顶棚
light-diffusing ceiling panel 光漫射板
light-diffusing ceiling system 光漫射系统;漫射光天花板体系
light-diffusing ceiling wall 光漫射墙(面)
light-diffusing glass 光漫射玻璃
light-diffusing window 光漫射窗
light diffusion 光(的)漫射;光散射
light dimmer 调光器;遮光器;灯罩
light-directing block 透光玻璃砖
light director 照明监督
light disk 光盘
light displacement of vessel 空船排水量;船舶空载排水量
light distortion 轻微畸变
light distribution 照明配电;光强分布;配光;光(线)分布
light distribution box 轻配电箱
light distribution curve 光强分布曲线;光分布曲线
light-distribution photometer 光度分布计
light disturbance 光扰(动)
light ditch 灯光槽
light-dividing device 分光装置;分光设备
light dope 轻微掺杂
light dosimeter 曝光控制器;曝光控制计
light draft 空载吃水
light-draft bottom 低阻力犁体
light draft craft 浅水船
light drag 光曳;光牵引
light draught 空载吃水;轻载吃水
light draught steamer 浅水轮船
light drawn 轻度冷拔
light drilling 轻型钻机钻进
light drilling equipment 轻型凿岩设备
light driver 光驱动器
light due 灯塔费
light dues 灯标费;灯标、航标、浮标费
light duration 光照延续时间
light duty 轻型(的);轻工作制;轻荷载;轻负载;灯塔费
light-duty axle 轻载驱动桥
light-duty cast-iron well cover seating 轻型铸铁井盖
light-duty coating 轻防腐蚀涂料
light-duty compressor 小型压缩机
light-duty crane 轻型起重机
light-duty crusher 轻型破碎机
light-duty derrick 轻型吊杆装置
light-duty engine 轻型发动机
light-duty flexible swing door 轻型灵活旋转门
light-duty floor(ing) 轻型地板;轻荷载地板;轻型楼面
light-duty lathe 轻型车床
light-duty lorry 轻型载货汽车
light-duty machine 小功率机床;轻型机械
light-duty oil bath air cleaner 轻型油浴式空气滤清器
light-duty pile driver 轻型打桩机;轻荷载打桩机
light-duty pneumatic drill 轻型风钻
light-duty power manipulator 轻型电动机械手
light-duty power roller 轻型电动压路机
light-duty regulation 轻型整治
light-duty regulation works 轻型整治工程
light-duty road 轻载公路
light-duty safety clamps 轻型钻杆安全夹器
light-duty scaffold 轻型脚手架
light-duty series 轻型系列
light-duty service 驳负荷运行
light-duty test 轻载(荷)试验;轻负载试验;轻(负)荷试验
light-duty trailer 轻型挂车
light-duty truck 轻型载货汽车
light-duty vehicle 轻型(载货)汽车

light dynamic(al) sounding 轻型动力触探
light earth 疏松土(壤)
light E bag 轻型E字麻袋
light echo 回光
light economic structure 轻型经济结构
lighted 装灯的
lighted beacon 灯桩(上设立标);灯立标;灯标
lighted bell buoy 装灯钟铃浮标
lighted buoy 灯光浮标
lighted fiber 木化纤维
lighted float 发光浮子
lighted marks 灯光标志
lighted sound buoy 音响灯(浮)标;声光浮标;灯光音响浮标
lighted whistle buoy 装灯笛声浮标
light effect 光效应;灯光效果
light effect of ultra-sound 超声波光效应
light electric(al) coupler 光电耦合器
light electric(al) tachometer 光电测速仪
light electric(al) transducer 光电换能器
light elevation 标灯高程
light ellipse 光椭圆
light emission 光放射;光发射
light-emission detector 光射式检验器
light emitter 光发体;光发射器;灯光信号发送器
light-emitting area 发光面积;发光面
light emitting bitumen logging 发光沥青测井
light-emitting capillary analysis 发光毛细分析
light-emitting component 发光元件
light emitting correlation analysis 发光对比分析
light-emitting diode 发光二极管;光发射二极管
light emitting drop-out point analysis 发光点滴分析
light-emitting junction 发光结
light-emitting material 光发射材料
light-emitting screen 发光荧光屏;发光屏
lighten 照亮;减轻;变轻
lighten a ship 减轻船载;减轻船荷
light-end products 轻质产品
lightened floor 空心肋板
lightened plate frame 空心肋板
lightened ship 减载船
lightener core 简化泥芯
light energy 光能量;光能;发光能
light energy distribution 光能分布
light-energy distribution curve 光能分布曲线
light engine 单机;没有挂列车的机车
light engine attaching with vehicles 单机附挂
light engine oil 轻质发动机油
light engine running 单机运行
light engraving 浅刻
lightening 发光;闪光;闪电
lightening admixture 减轻剂
lightening a ship 卸货减轻船的负载
lightening-caused fire 偶然雷击火
lightening core 减轻芯
lightening hole 减重孔;(零件的)减少自重孔;减轻孔(金属零件减轻重量的孔);发光孔
lightening on water 水上减载
lightening plate frame 空心肋板
lightening power 消色力
lighten load 减轻荷载;减轻负载;减轻负荷
lighten the burden on enterprises 减轻企业负担
light equation 光行时差;光(行)差;光变方程
light equipment 轻型装备;轻型设备;轻便设备
lighter 引火器;照明装置;照明器;港内驳船;货驳;发光器;驳船
lighter aboard ship 子母船;拉希船;拉希驳;母船;浮水货框;载驳(子)母船;载驳(子)船;普通载驳货船
lighter aboard ship system 载驳船系统
lighterage 过驳;驳运(费);驳船装卸;驳船费
lighterage anchorage 过驳锚地
lighterage limit 驳运费限额;驳运范围
lighterage on water 水上过驳
lighterage operation 过驳作业
lighterage pier 车船间货物驳运码头
lighterage port 过驳港
lighterage service 驳运业务
lighterage volume 驳运量
lighter berth 驳船碰泊处;驳船泊位
lighter body 轻质基(石油)
lighter charges 驳运费;驳船费
lighter clerk 计算驳船到离码头的助理员

lighter/container ship 载驳/集装箱船
lighter demurrage 驳船过期停泊费
lighter due 驳运费
lighter flint 火石
lighter freight station 码头型货станции;驳船货运站
lighter freight station cargo 驳船货运站货
lightering 驳运
lighter loaded cargo 整驳货
lighterman 驳船工人
lighter note 驳船纳税证明书
lighter pool 驳船水域
lighter risk 驳船险
light error 微小误差
lighter's berth 驳船码头
lighter sealing charges 驳船封仓费
light(er) section 小型型钢
lighter service 驳船运输
lighter's wharf 卸货码头;卸货场;驳船码头
lighter-than-water non-aqueous phase liquid 比水轻的非水相液体
lighter traffic 驳船运输
light excitation 光激发;发光激励
light excited current 光激电流法
light exciting pulse 光激发脉冲
light expanded clay aggregate 轻质膨胀粘土骨料
light exposure 曝光量
light exposure test 耐光性试验
light extinction meter 消光计
light extinction method 消光法
light extinguished 灯光熄灭
light fabric 轻薄织物
light facade 淡色的立面;轻快的立面
light facade element 淡色立面构件;轻质立面构件
light face 白体活字
light face letter 细线体字
light facility 轻型设备
lightfast 晒不褪色的
lightfast colo(u)r 不褪色的色彩
lightfast material 耐光材料
light-fastness 耐光度;耐晒性;耐光性;不褪色性
light fast traffic 轻型快速交通;轻车快速交通
light fiber 明纤维
light field 光场;(光弹性力学试验的)亮区
light field image 光场像
light field method 光场法
light fill 低填土;矮路堤
light filler 小填角焊缝;轻质填角;填满角焊缝
light filler block 浅填角焊缝
light fillet 浅角焊缝
light fillet weld 小填角焊;轻型填角焊;凹形角焊
light filter 滤色片;颜色滤色片;滤光器
light filtered cylinder oil 轻质发动机油
light filter factor 滤光器因数
light filtering 滤光
light filtering medium 轻滤料
light filter material 轻滤料
light filtration 滤光
light finishing 轻浆整理
light fire brick 轻烧砖
light fire-clay insulating brick 轻质耐火黏土砖;轻质耐火黏土保温砖
light fitting 灯光;灯具;照明配件;照明装置;灯光设备;灯具设备
light fitting with integral mounting 带整套支架的灯光设备
light fixture 轻型设备;灯具
light flare 闪光
light flash 闪光
light flash excitation 闪光激发
light flashing signal 通信闪光灯
light flint 轻火石玻璃
light flint glass 轻燧石玻璃
light float 灯标;灯标船;船状灯浮;船形灯标
light floe 薄浮冰块;薄冰块
light flux 光通量
light flux meter 光通量计
light flux ratio 光通量比
light flux surface density 光通表面密度
light-focusing glass fiber 聚光玻璃纤维
light-focusing plastic fiber 光聚焦塑料纤维
light fog 光学灰雾;光灰雾;轻雾(霾)
light force fit 轻压配合
light forest 疏林
light fraction 轻馏分

light fraction of oil 原油的轻馏分
light framing 轻型框架木料;轻型骨架
light freeze 轻冻
light freight 轻货
light frequency 光频率;光频
light frequency amplification 光频放大
light frequency modulation 光频率调制
light frost 轻雾(霾)
light frother 轻油起泡剂
light fuel 轻质燃料
light fuel oil 轻质燃油
light fugitive 不耐光的
light-fuse 灯用熔断丝
light gain 光增益
light gap 光隙
light gap method 光隙法
light gap micrometer 光隙测微器
light gas oil 轻质瓦斯油;轻气体油
light gasoline 轻质汽油
light gate 光闸
light gathering 聚光能力
light-gathering device 聚光器
light-gathering optics 聚光器;聚光光学系统
light-gathering power 聚光本领
light gating 光选通
light-gating cathode-ray tube 光闸阴极射线管
light-ga(u)ge 小断面的
light ga(u)ge carriageway 曲线车行道
light-ga(u)ge cold-formed member 薄板冷成型件
light-ga(u)ge cold-formed steel member 冷轧轻钢;冷轧的轻型钢材
light-ga(u)ge cold-formed steel structural member 薄的冷加工成型钢结构构件
light-ga(u)ge copper tube 薄壁铜管;薄铜管
light-ga(u)ge metal mo(u)ld 轻金属模板;薄钢模板
light-ga(u)ge plate 薄钢板;薄板
light-ga(u)ge railroad 窄轨铁路;轻轨铁道
light-ga(u)ge railway 窄轨铁路;轻轨铁道;轻轨铁路
light-ga(u)ge section steel 轻型型钢
light-ga(u)ge sheet 薄板;薄钢板
light-ga(u)ge steel 轻型型钢;轻量型钢
light-ga(u)ge steel construction 轻钢薄壁结构;轻型钢结构
light-ga(u)ge steel framing 轻型钢构架;轻型钢结构
light-ga(u)ge steel member 用轻型钢制作的钢构件;轻钢构件
light-ga(u)ge steel section 薄壁型钢
light-ga(u)ge steel structure 轻型钢结构
light-ga(u)ge structure 曲线结构
light-ga(u)ge track 窄轨铁路;轻轨铁道
light-ga(u)ge welding 薄板焊接
light-ga(u)ge wire 细号线
light generation 光振荡
light globe 圆球灯罩;白炽灯;灯泡;灯罩
light-gold bronze finish 亮金青铜饰面
light grab 轻型抓斗
light grade 小坡度;低标号
light gradient 小坡度
light grain 轻谷物
light gray 浅灰色;浅灰
light green 浅绿色;品绿;淡绿色
light green-black 浅绿黑色
light green-brown 浅绿褐色
light green glaze 浅绿釉
light green-grey 浅绿灰色
light green-yellow 浅绿黄色
light grey 浅灰色;浅灰
light grey-black 浅灰黑色
light grey-blue 浅灰蓝色
light grey-brown 浅灰褐色
light grey-green 浅灰绿色
light grey-red 浅灰红色
light grey-violet 浅灰紫色
light grey-white 浅灰白色
light grey-yellow 浅灰黄色
light ground 流岩;明亮的背景;冲刷下的岩土
light guide 光学纤维;光导
light guide bundles 传光束
light guide cart 光导向小车
light guide efficiency 光导效率
light guide refractometer 光导折射计

light guide rod 导光棒
light guide tube 投影管
light guideway transit system 轻型导向交通系统
light-guiding film 光导膜
light gum veins 弥合缝隙的树胶
light gun 光枪;光电子枪;光笔
light halo 光晕
light(hammer) drill 轻型凿岩机
light hardening 光坚膜;光照锻练;光硬化
light hardening agent 光硬化剂
light harmonic 光谐波
light harvesting pigment 聚光色素
light head 光电传感头;曝光头
light-heat ag(e)ing 光热老化
light helicopter 轻型直升机
light hilly area 微丘区
light hoist 轻型提升机
light hoisting gear 轻型起重机
light hoisting tackle 轻型起重滑车
light hole 落水洞;水冲穴;地面塌陷洞
light homer 光寻的头
light homing 光自动跟踪
light homing guidance 光寻的制导
light homing head 光自动寻的头
light hook speed 空吊速率
light hopper car 轻便漏斗车
light hopper wagon 混凝土斗车;轻型漏斗车;轻便漏斗车;底开车;漏斗车
light horse 轻型马
light hours 轻负荷时
lighthouse 曝光台;灯台;灯塔;灯室
lighthouse cavity 灯塔式空腔
lighthouse diode 灯塔二极管
lighthouse lantern 灯塔灯笼
light house list 灯塔表
light house tender 灯塔补给船;灯塔供应船
lighthouse tube 盘封管;灯塔管
light hydrocarbon 轻质烃
light hydrocarbon diagenesis stage 轻烃成岩作用阶段
light hydrogen 轻氢
light ice 薄冰
light ice floe 薄冰块
light illuminating factor 照明率
light image converter 光像变换器;光变像器
light impulse 光脉冲
light index 光照指数
light indicator 指示灯;灯光指示器
light-induced modulation 光感应调制
light-induced tanning 光硬化作用
light industrial district 轻工业区
light industrial machinery plant 轻工业机械厂
light industrial products 轻工业(产)品
light industry 轻工业
light industry products 轻工业产品
lighting 照明;灯光;采光
lighting and heating 照明供暖
lighting and ventilating shaft 照明通风两用筒
lighting aperture 照明窗
lighting apparatus 照明装置;照明器具;感光设备
lighting apparatus for vehicle 车辆照明器
lighting arrangement 照明布置图;采光装置
lighting art 照明艺术
lighting atrium 采光中庭
lighting-at-sea 海上过驳
lighting attachment 照明装置
lighting batten (舞台吊灯具和布景的)钢管
lighting booth 灯光控制室;舞台吊灯具和布景的钢管;布景灯光控制;剧院灯光控制室
lighting branch circuit 照明用支路;照明用分支电路
lighting bridge 照明天桥
lighting bulb 照明灯泡
lighting burner 照明灯
lighting by roof-light(s) 用屋顶光照明
lighting by skylight 天窗采光
lighting cable 电灯线;照明线路;照明电缆
lighting cable drum 照明电缆盘
lighting calculation 照明计算
Lighting cement 莱丁水泥(商品名)
lighting changeover box 照明切换箱
lighting circuit 照明线路;照明电路
lighting column (路灯的)灯柱
lighting component 照明构成部分
lighting condition 照明条件;光照条件

lighting console 调光台
lighting consumer 照明用户
lighting consumption 照明耗量
lighting contrast 照明对比
lighting contrast ratio 照明对比度
lighting control 照明控制
lighting control box 照明控制箱
lighting control characteristic 照明控制特性
lighting controller 照明控制器
lighting control panel 照明控制盘;灯光控制板
lighting-control relay 照明控制继电器
lighting control room 照明控制室
lighting cost 照明费用
lighting culture 光照培养法
lighting current 照明电流
lighting current motor 照明电流发动机
lighting demand 照明需量
lighting design 照明设计;采光设计
lighting device 照明装置;照明灯具
lighting diffuser 漫射光照明装置;柔光器;光漫射器
lighting director 照明监督
lighting display 照明景象
lighting distribution box 照明配电箱
lighting distribution panel 照明配电盘
lighting distribution room 照明配电室
lighting dynamo 照明用发电机;照明发电机
lighting effect 照明效应
lighting efficiency 照明效率
lighting engineer 照明工程师
lighting engineering 照明技术;照明工程(学)
lighting equipment 照明设备;灯光器具
Lighting Equipment Manufacturer's Association 照明器具制造商协会(英国)
lighting facility 照明装置;照明设施;照明设备
lighting facility of road 道路照明设施
lighting feeder 照明馈路
lighting fitting 照明设备;装灯配件;照明配件
lighting fixture 照明装置;照明器材;灯具
lighting fixture raceway 照明装置电缆管道
lighting gas 照明气;燃用煤气;灯用煤气
lighting generator 照明(用)发电机
lighting glass 照明玻璃;漫射光玻璃
lighting glass ware 照明玻璃制品
lighting grid 灯具悬吊格栅
lighting hole 点火孔;点火口
lighting hours 照明时间
lighting inspection 光检验
lighting installation 照明装置
lighting instrument 照明设备
lighting layout 照明设计
lighting leads 照明用导线
lighting level 照明等级;照度级
lighting line 照明线
lighting loading 轻负载;轻荷载;轻负荷;照明负载;照明负荷;船空载;放电承受能力
lighting mains 照明网络;照明干线;照明电源
lighting man 照明员
lighting mast 照明桅;照明塔;路灯柱
lighting mast with arm 带支臂的采光柱
lighting meter 照明用户电度表;照明电度表
lighting motor-generator set 照明用电动发电机组
lighting network 照明网络
lighting-off 灭灯;停止照明
lighting-off torch 点火火炬;点火棒
lighting of long tunnel 长隧道照明
lighting of site 工地照明
lighting outlet 照明出线口;照明电源引出线;照明电源插座
lighting panel 照明配电盘;照明格板;照明配线板;照明配线盘
lighting peak 照明尖峰;照明峰荷
lighting plant 照明装置;照明设备
lighting plant depreciation apportion and overhaul charges 照明设备折旧摊销及大修费
lighting platform 灯光平台
lighting point 照明点
lighting pole 灯柱
lighting post 路灯柱;灯柱
lighting power 照明功率;光能;亮度;发光能力
lighting power box 照明电源箱
lighting power distribution panel 照明配电箱
lighting proof 防避闪电
lighting protector 闪电防护器
lighting-reflecting curb 反光缘石

lighting regulator 照明调节器
lighting row 照明排灯
lighting set 照明装置
lighting set-up 照明设计
lighting sight glass 照明视镜
lighting slab 发光板
lighting staff 灯柱
lighting standard 照明标准;照度标准;灯柱;电灯杆
lighting sub-distribution 照明配电
lighting supply source 照明电源
lighting switch 照明开关
lighting switchboard 照明开关箱
lighting switch bushing 灯开关衬套
lighting switchgear 照明开关装置
lighting switch handle 照明开关手柄
lighting switch operating spider 灯开关操作十字架
lighting system 照明系统;照明方式;灯光系统
lighting tariff 照明电价表;照明费率;照明电费表
lighting technique 照明技术
lighting time 照明时间
lighting tower 照明塔
lighting transformer 照明变压器
lighting tube 荧光管
lighting unit 照明装置;照明器;照明单位
lighting-up 点燃
lighting-up time 照明时间
lighting-up tuyere 点火口
lighting voltage regulator 照明网络稳压器
lighting window 采光窗
lighting wire 照明电线
lighting wiring 照明布线
light injury 光损伤
light input 光流输入
light-inspection car 轻型检查车
light installation 照明装置
light instrument 轻便工具
light integral meter 积分光度计
light-integrating meter 光量积分仪
light-integrating switch 聚光开关
light integrator 光学积分器;曝光积分器
light intensifier 光增强器
light intensity 照度;光强;光(谱)强度;亮度;发光强度
light intensity controller 光强控制器
light intensity detector 光强检测器
light intensity direct determination method 光强直接测定法
light intensity equalizer 光强度调节器
light intensity fluctuation 光强起伏;光强波动
light intensity indicator 光强指示器
light intensity measurement 光强测定
light intensity meter 光强计
light intensity modulation 光强调制
light intensity modulator 光强调制器
light intensity variation 光强波动
light intercepting 火光自动警报器
light intercepting curtain 遮光幕
light interference 光干涉
light interference comparator 光干涉比较仪;光干涉比长器
light interference control technique 光干涉测量控制技术
light intermittent duty 轻间隙负载
light intermittent test 轻负载断续试验
light interstitial material 白色中间体
light interstitial phase 白色中间相
light in the night sky 夜天光
light in weight 重量轻
light irradiation stability 光照稳定性
light irrigation 轻灌;浅灌
lightish 较轻的;淡色的
light isotope 轻同位素
light kaolin 轻质高岭土
light Karuni 轻质卡路尼(一种坚硬地板材料)
light keeper 司光员;灯塔管理员
light kinemometer 光测速仪
light knock 轻微爆震
light lantern 航标灯
light lathe 轻型车床
light leak(age) 漏光
light-leakage loss 光漏损失
light level 亮度级
light lime concrete 轻质石灰混凝土;轻石灰三合土
light lime-sand brick 轻质灰砂砖

light line 亮纹条;灯光导航线;导航标灯线
light line antenna 电灯用天线
light-line contour 等高线
light line speed 空吊速率
light liquid 轻液
light liquid dispersion pipe 轻液分布管
light liquid inlet(outlet) 轻液入出口
light liquid paraffin 轻液状石蜡
light liquid petrolatum 轻质液状石蜡
light list 航行灯表;灯塔;灯标(位置)表
light list number 航行灯号码表;灯塔表号码;灯标表号码
light load 轻荷载;压载【船】;照明负荷
light load adjustment 轻负载调节;轻载调整
light load compensating device 轻载补偿装置
light load compensation 轻负载补偿
light load condition 空载状态
light load displacement 空载排水量;压载排水量
light loaded district 轻负荷区
light loaded network 轻负载网络
light load line 空载水线;空载吃水线
light load period 轻载期间
light load test 轻载试验
light loam 轻(质)亚黏土;轻(质)壤土;轻质垆坶
light lock 防火门;暗室口避光装置
light locomotive 轻型机车;轻便机车
light logic 光逻辑
light loss 光损失;照明损耗
light loss factor 光损失系数;减光系数
light lubricant 轻润滑油
light lubricating oil 轻质润滑油
light luminous flux 光通量
lightly coated electrode 轻药焊条;薄药皮焊条;薄涂(料)焊条
lightly covered 轻度覆盖的
lightly damped ga(u)ge 小惯性压力表;灵敏压力计
lightly over-consolidated clay 轻微固结黏土
lightly pigmented skin 浅色素皮肤;浅肤色皮肤
lightly reinforced pavement 少量配筋(混凝土)路面
lightly rusted 轻度锈蚀
lightly toughened glass 轻度钢化玻璃
lightly used line 清淡运输线(路)
light machine oil 轻质机械油
light magnesia 轻质镁氧;轻烧氧化镁
light magnesium carbonate 轻质碳酸镁
light magnesium oxide 轻质氧化镁
light maintenance 小修(理)
light maintenance and installation truck 小修和安装
light man 照明工作人员
light marker 灯标
light mark galvanometer 光标检流计
light mark microammeter 光点微安表
light maroon 浅栗色
light mast crane 轻(型)桅杆(式)转臂重机;桅杆(式)转臂起重机
light material 小截面材料;薄钢材;照明材料
light material bucket 轻型料斗;轻质材料的料斗;轻质材料的铲斗
light mattress work 轻排工;轻褥工;氢沉排
light measurement 光度学
light measuring system 光测系统
light-mesh steel fabric 细眼钢丝网
light metal 轻金属
light-metal alloy 轻金属合金;轻合金
light-metal blind slat 轻金属百叶板窗
light-metal builder's furniture 轻金属建筑工人用的家具
light-metal builder's hardware 轻金属建筑工人用的五金
light-metal cap 轻金属帽盖
light-metal connecting element 轻金属连接构件
light-metal construction 轻金属建筑;轻金属结构
light-metal construction section 轻金属结构部件
light-metal door closer 轻金属关门装置
light-metal elevating belt conveyer 轻金属皮带提升输送机
light-metal facade 轻金属立面
light-metal fittings 轻金属小五金;轻金属配件
light-metal frame 轻金属框架
light-metal furniture 轻金属家具
light-metal glazing 轻金属玻璃窗

light-metal hardware 轻金属五金制品
light-metal jalousie 轻金属百叶窗
light-metal lattice(d) girder 轻金属格构桁架;轻金属网格梁
light-metal lever handle 轻金属杠杆式(门)拉手
light-metal louvers 轻金属百叶板
light-metal ore 轻金属矿石
light-metal partition(wall) 轻金属隔墙
light-metal plain webbed beam 轻金属带腹板的梁
light-metal plain webbed girder 轻金属平腹板大梁
light-metal profile 金属型材断面
light metal rim 轻金属轮辋
light-metal roof cladding 轻金属屋面表层
light-metal roof truss 轻金属屋架
light-metals commodities 轻金属矿产
light-metal shape 轻金属模型
light-metal shutter door 轻金属百叶门
light-metal slated blind 轻金属条百叶窗
light-metal slated roller blind 轻金属卷帘式百叶窗
light-metal solid webbed beam 轻金属实心腹板梁
light-metal solid webbed girder 轻金属实心腹板大梁
light-metal space load bearing structure 轻金属空间承重结构
light-metal structural engineering 轻金属结构工程
light-metal structural section 轻金属建筑(结构)型材
light-metal trim 轻金属装饰
light-metal trussed girder 轻金属桁架大梁
light-metal tube 轻金属管材
light-metal tubular scaffold(ing) 轻金属管脚手架
light-metal unit 轻金属(结构)单元
light-metal window 轻金属窗
light-metal window sill 轻金属窗台
light meter 曝光测定;测光表;照度计;光度计
light microphone 光敏传声器
light microscope 光显微镜
light microscope photon microscope 光学显微镜
light microscope technique 光学显微镜技术
light microscopy 光学显微镜(检查)法
light microsecond 光微秒
light mild sand 轻亚砂土
light mineral 轻矿物
light mineral aggregate 轻质矿物骨料;轻质矿物集料
light mobile machine shop 轻型活动修理车
light modulated gas discharge tube 光调制气体放电管
light modulation 光载波注入;光调制;光调节
light modulation detector 光调制探测器;光调制检测器
light modulator 光调(制)器;光调节器;调光器
light money 灯塔附加费
light mono tower crane 轻型单塔式起重机
light mounted 24-disc harrow 24片悬挂轻耙
light mud brick 轻质泥土砖
light mud pumping 冒浆轻微
light navigation 灯标导航
light-negative 负光电导性
lightness 颜色明度;光亮度;明亮度;轻微
lightness correction 亮度校正
lightness hue 色调和谐;色彩柔和;色彩和谐
lightness scale 明度标
light neutron method 光中子法
light nilas 浅暗冰
lightning arrester 过压放电器;电涌放电器;防雷接地;避雷装置;避雷针;避雷器;照明过压保险装置
lightning arrester for power supply 电源防雷装置
lightning arrester terminal box 避雷器接线盒
lightning ball 闪电火球
lightning call 闪急通报
lightning cement 闪光水泥
lightning chain 闪电链
lightning channel 闪道
lightning conductor 接地极;避雷装置;避雷器的接地极;避雷(导)线;避雷针(导线);避雷器
lightning coverage 雷电保护
lightning-current meter 雷电流计
lightning current steepness 雷电流陡度
lightning damage 雷电损害
lightning discharge 雷闪放电;闪电放电
lightning disturbance 雷电事故
lightning-diverting cable 分雷电缆
lightning fault 雷电故障

lightning flash 电闪
lightning fork 闪电叉
lightning gap 避雷器放电(空)隙;避雷器放电间隙
lightning generator 人造闪电发生器
lightning grounding 闪电接地
lightning guard 避雷针;避雷器
lightning impulse 闪电脉冲
lightning impulse residual voltage 雷电冲击残压
lightning impulse withstand voltage 雷电耐压;雷电冲击耐受电压;承受雷电冲击电压
lightning impulse withstand voltage test 雷击冲击试验
lightning interference 雷电干扰
lightning mark 雷击电纹;雷击斑
lightning performance 雷电性能
lightning plate needle 板式避雷针
lightning plate protector 板式避雷器
lightning prevention design 避雷设计
lightning proof 防雷击装置
lightning-protecting engineering 避雷技术
lightning protection 闪电保护;防雷保护装置;防雷;避雷
lightning protection cable 避雷线
lightning protection device 避雷装置
lightning-protection engineering 避雷工程(学)
lightning protection equipment 雷电保护装置
lightning protection grounding layout 建筑物防雷接地平面图
lightning protection system 避雷系统;避雷装置
lightning protector 避雷器
lightning protector plate 避雷板
lightning recorder 雷电记录器;闪电记录器
lightning ring 闪电轮
lightning rod 避雷装置;避雷针;避雷器
lightning rod support 避雷针支架
lightning-safe 避雷
lightning shake 发射形环裂;闪电状裂纹;放射形环裂
lightning spike 避雷针
lightning storm 雷雨;雷暴;闪电暴雷
lightning streak 雷电闪光
lightning stroke 雷击;电击
lightning-stroke recorder 雷击记录器
lightning surge 雷电电涌;雷电过电压
lightning switch 避雷开关
lightning tooth cross-cut saw 闪齿横割锯
lightning trip-out rate 雷电跳闸率
lightning trouble 雷电事故
light noncombustible construction 轻型耐火结构
light occluder 光防护
light off 点火
light of sight 瞄准线
light oil 轻质石油;轻油
light oil constituents 轻油组分
light oil cracking 轻油裂化
light oil distillate 轻油馏分
light oil engine 轻油机
light oil heater 轻油加热器
light oil maximizing method 轻油最大限度提取法(洛马克斯法)
light oil tank 轻油柜
light on 照明开始
light on reading 光照读数
light opening 透光口
light-operated switch 光控开关
light operation 轻作业
light optic(al) flint 轻火石光学玻璃
light-order (航道标灯的)光级
light oscillograph 光线示波器
light outlet 电灯头
light output 光辐射;发光效率;光(能)输出
light output ratio 光输出比
light overhaul 小修(理)
light overhead contact line 轻型接触导线
light overtone 光谐波
light packet 光脉冲群
light-painted 淡色油漆的
light panel 场致发光板;轻型方框
light partition 轻隔墙
light partition wall 轻隔墙;轻质隔墙
light-passing board 透光板
light-passing plastic 透光塑料
light-passing plastic material 透光塑性材料
light-passing plastic sheet(ing) 透光塑料薄板

light-passing sheet(ing) plastic 透光塑料片
light paste 软膏
light path 光路
light path length 光程长
light path selector 光路选择杆
light pattern 光图像;光发射图样
light pattern needle holder 轻型持针钳
light peat 浅色泥炭
light pen 光电笔;光笔
light pen attention 光笔注意装置;光笔中断
light pencil 光锥(体)
light pen control 光笔控制
light pen detect 光笔检出
light pen detection 光笔检出;光笔检测
light pen device 光笔装置
light pen display 光笔显示器
light penetration 光透射;光透入;光穿透;浅贯入
light pen hit 光笔指点;光笔命中;光笔检测
light pen interrupt 光笔中断
light pen register 光笔寄存器
light pen strike 光笔检测;光笔触碰;光笔触击
light pen tracking 光笔跟踪
light pen unit 光笔装置
light perception 光感觉
light period 光照期;灯光周期
light permeability 透光性;光渗透性
light petrol 轻汽油
light phase 发光方式
light phase modulation 光相应调制
light photon 可见光子
light photon beam 可见光子束
light pile hammer 轻型(打)桩锤
light pillar 光柱;日柱
light pipe 光导向装置;(舞台吊灯具和布景的)钢管;导光管;光(导)管
light-pipe optics 光管光学(装置)
light-pipe oscillation 光管振荡
light pitch 缓倾斜
light placement 空载排水吨
light plant 活动发电厂;电灯厂;电站;发电机组
light plastic spar 发光塑料标桩
light plate 轻质板;中钢板;轻质板材;轻磅镀锡薄钢板;薄钢板
light plate mill 中板轧机
light platinum metal 铂类轻金属
light plethysmograph 光体积描记器
light plot 照明图表
light pneumatic drilling machine 轻型风动式钻机
light point 照明插座;光点
light polarization demodulation 光偏振解调制
light polarization demodulator 光偏振解调器
light polarization modulation 光偏振调制
light polarization modulator 光偏振调制器
light pole 灯杆;路灯杆
light pollution 光污染;轻(度)污染
light porous clay brick 轻质多孔黏土砖
light positive 正光电导性
lightpot 舞台照明法
light powdered cement 轻质粉状水泥
light power 光功率
light power meter 光功率计
light power motor 小型电动机
light prefabrication 轻型预制(产品)
light press fit 轻压配合
light pressure 光压;轻微压力
light pressure fit 轻压配合
light pressure measurement 光压测量
light printing 晒印
lightprinting device 光绘装置;晒印装置
light probe 光探针
light projection 光投射
light projector 投光器
light-proof 不透光的;遮光(的);耐光
light-proof blind 遮光卷帘
light-proof enclosure 暗箱
light-proof louver 遮光百叶窗
light-proof material 不透光材料
light-proofness 耐光性;遮光性;遮光效果
light-proof paper 防光纸
light-proof vent 不透光通气孔
light propagation vector 光传播矢量
light protable conveyer 轻型移动式输送器
light protective cone 遮光罩
light pruning 轻修;轻度修剪

light pulse 光脉冲
light pulse bonding 光脉冲焊接
light pulse generator 脉冲光源
light pulse modulation 光脉冲调制
light pulse propagation 光脉冲传播
light pulse repetition rate 光脉冲重复频率
light pump 光泵
light-pumped solid state laser 光抽运固体激光器
light pumping semiconductor laser 光抽运半导体激光器
light quality 灯质
light quantity 光通量;光(度)量
light quantum 光量子
light quantum maser 光量子激射器
light radiation from sea-surface 海面向上光辐射
light rail 轻轨铁路;轻轨
light rail car 轻油车(铁路用)
light rail line 轻轨线路
light rail motor tractor 轻型机车
light rail rapid transit system 快速轻轨交通系统
light railroad 轻轨铁道;轻便铁路;轻便铁道
light railroad system 窄轨铁路系统;轻轨铁路系统
light rail system 轻轨系统
light rail track bolting wrench 轻轨螺栓扳手
light rail traffic 轻轨交通
light rail transit 轻轨交通
light rail transit system 轻轨交通系统
light rail transport 轻轨运输
light rail vehicle 轻轨车辆
light railway 轻便铁路;轻便铁道;轻便轨道
light railway bridge 轻便铁道桥梁
light railway material 轻轨铁路器材;窄轨铁路器材
light railway system 窄轨铁路系统;轻轨铁路系统
light railway track 轻便铁轨
light railway truck (窄轨的)轻便铁道车
light railway van 轻便轨车
light rain 小雨;毛毛雨;微雨
light range 灯光照距;灯光射程
light-ranging 光测距
light rapid transit 轻型快速交通
light rapid transit system 轻型快速交通系统
light rare earth 轻稀土
light rare earth element deficiency 轻稀土亏损
light rare-earth ores 轻稀土矿
light ratio 照明比;光比
light ray 光线;光射线
light ray bending 光线转折
light ray brazing 红外(线)钎焊
light ray control 光电控制
light ray pointer galvanometer 光线指针式检流计
lightray propagation 光信号传播
light ray welding 光束焊接
light reaction 光(化)反应;明反应
light receiver 光接收器
light recovery section 轻型装备抢救组
light recycle stock 轻质循环进料
light red 耐晒红;浅红色
light reddish chestnut 浅红粟子色
light red-drown 浅红褐色
light red-grey 浅红灰色
light-red silver ore 淡红银矿;硫砷银矿
light-reducing glass 减光玻璃
light reduction 小压下量
light red-violet 浅红紫色
light red-yellow 浅红黄色
light reflectance 光反射率
light-reflectance apparatus 反光仪
light reflecting characteristic 反光特性
light reflecting curb 反光路缘(石);反光侧石
light reflection 光反射
light reflection value 反光量
light-reflective glass 反射玻璃;反光玻璃
light reflectivity 光反射性;光反射率
light reflector 灯光反射器
light reflex 光反射;光发射
light refraction 光(的)折射
light regime 光照制度
light region 明区
light regulation 光调节
light regulator 照明调节器;灯光调节器;光稳定器
light relay 光继电器
light relay rack 电码继电器架
light repair 小修(理);临时修理
light repair shop for hauled stock 搬运设备小修

车间
light requirement 需光量;需光度
light resistance 光电阻;抗光性;耐光性;耐光的
light-resistant 耐光;阻光的;耐光的
light-resistant paper 防光纸
light resistor 光能电阻器;光敏电阻器
light revolving tower crane 轻型旋转塔式起重机
light rig 轻型钻探设备
light rigid conduit 轻质硬导线管
light ring 光圈
light road oil 轻质铺路油;轻质路油
light rock boring machine 轻型凿岩机
light rock drill 轻型凿岩机
light rocket 照明火箭
light rod 细盘条
light roller check 轻碾裂缝
light roof 轻型屋顶
light roof truss 轻型屋架
light rubber cot 薄胶指套
light-ruby silver 淡红银矿
light running 小负荷运转;轻载运行
light-running fit 轻转配合;轻动配合
lights 轻磅镀锡薄钢板
light samples 光脉冲
light sand 松砂
light sand-lime brick 轻质灰砂砖
light sandy clay 轻砂黏土
light sandy loam 轻亚砂土
light saturated point 光饱和点
light saturation point 光饱和点
light scaffold(ing) 轻型脚手架
light scaling 轻微剥落
light scantling vessel 轻结构船
light scatter 光漫射
light scattering 光散射
light scattering apparatus 光散射仪
light scattering diagram 光散射表图解
light scattering glassware 光散射玻璃器皿
light scattering in solids 固体内光散射
light scattering meter 光散射仪
light scattering method 光散法
light scattering photometer 光散射光度计;散光计;散射曝表;散射光度计
light scattering table 光散射表
light scattering technique 光散射技术;光散射法
light scissors truss 轻型剪式屋架
light screed 轻质括板
light screen 遮光板
light screen planting 遮光栽植
light scribing 光刻
light scribing device 光刻装置
light section 小断面;小型钢;轻型(型)钢
light section(al) steel 轻质型钢
light section car 轻型轨道车
light section method 光切法;光切断法
light section microscope 光切型显微镜
light section mill 小型型材轧机;小型钢轧机
light section steel structure 轻型钢结构
light sector 照明区(域);光照扇形区
light-seeking 光自动寻的;光定位
light selecting lever 光路选择杆
light self-trapping 光自陷
light sensation 光敏;光感
light sense 光感
light sensitive 感光的;光敏的
light-sensitive aquatic organism 光敏水生物
light-sensitive array 光敏阵列
light-sensitive cathode 光敏阴极
light-sensitive cell 光敏电池;光电探测器
light-sensitive cell pyrometer 光敏元件高温计
light-sensitive ceramics 光敏陶瓷
light-sensitive coating 光敏涂料;光敏涂层
light-sensitive compound 光敏化合物
light-sensitive detector 光敏检测器;光敏探测器;光电探测器
light-sensitive device 光敏仪器;光敏装置
light-sensitive diode 光敏二极管
light-sensitive effect 光敏效应
light-sensitive electron tube 光敏电子管
light-sensitive element 光敏元件
light-sensitive emulsion 光敏胶;感光乳剂
light-sensitive engraving 光敏刻图
light-sensitive flag switch 光电旗形开关
light-sensitive glass 光敏玻璃

light-sensitive imaging tube 光敏成像管
light-sensitive lacquer 光刻胶
light-sensitive layer 光敏层;感光层
light-sensitive layer lamination proof 色层叠合打样法
light-sensitive material 光敏材料;感光材料
light-sensitive metal 光敏金属
light-sensitive mosaic 光敏镶嵌面
light-sensitiveness 感光性
light-sensitive paper 光敏纸
light-sensitive polymer 光敏聚合物
light-sensitive relay 光敏继电器
light-sensitive resin 感光性树脂
light-sensitive semiconductor material 光敏半导体材料
light-sensitive surface 光敏面;感光膜;感光面
light-sensitive tube 光电管
light-sensitive vehicle detector 光感检车器
light sensitivity 光敏度;感光性;感光度
light sensitization 光敏感作用
light sensor 光电探测器
light service track 工地轻便轨道
light shade 遮光物;浅色相;淡色调;不饱色;遮光帘
light shaft 采光竖井;采光井
light shale 轻页岩
light shape 小型钢材
light shear line 轻型剪切机作业线
light sheet 薄钢皮;薄钢板;薄钢片
light sheet metal 轻金属薄板
light shell 照明弹
light shield 遮光罩
light ship 浮灯标;空载船;灯(标)船
light ship condition 空载状态
lightship mooring 灯船碇泊设备
light shot 小爆破
light shoulder 瓶肩过薄
lightshow 光显示
light shower 小阵雨
light shutter 轻百叶窗
light-shutter tube 光快门管
light signal 光(示)信号;闪光信号;色灯信号;灯(光)信号
light-signal transfer characteristic 光信号转换特性
light sign board 照明标志牌;灯光招牌
light silicate concrete 轻质硅酸盐混凝土
light silty loam 轻粉壤土
lights in line 界限叠标灯
light sized 小尺寸的;小号的
light sizing 轻浆
light skein 小绞丝
light-skinned structure 薄蒙皮结构
light slag 轻质炉渣
light slag concrete 轻质炉渣混凝土
light slag powder 轻质矿渣粉末
light slash 细采伐剩余物
light slide 遮光板
light slit 光隙;光缝
light slot 灯槽
light slushing oil 铸模用润滑油
lightsman 灯船船员
light smooth wheel roller 轻型平碾
light snow 小雪
light socket 电灯插座
light soda ash 轻质纯碱
light soil 轻质土(壤);松质土;砂土
light solids 轻岩粉
light solid tile 轻质实心砖
lights on land 陆上灯标
light sounding test 轻便触探试验
light sounding test blow count 轻便触探试验锤击数
light source 照明源;光源
light source colo(u)r 光源颜色;光源色
light source condenser 光源聚光镜
light source coupler 光源耦合器
light source driver 光源激励器
light source efficacy 光源功效
light source efficiency 光源效率
light source for aids 航标光源
light source from laser 激光光源
light source moniter 光源监视器
light source of gas-discharge lamp 气体放电光源
light source power 光源功率
light source quantum effeciency 光源量子效率

light source stabilizer 光源稳压器
light spectrograph 光摄谱仪
light spectrum 光谱
light sphagnum peat 浅色水藓泥炭
light spill 光漏失
light spirit 轻油
light split 漏斗
light splitter 分光器;分光镜
light splitting 分光
light-splitting electrooptical system 分光电光系统
light-splitting optic(al) system 分束光学系统;分光光学系统
light spot 光点;亮点
light spot direct current ammeter 光标式直流电流表
light spot direct current voltmeter 光标式直流电流表
light spot galvanometer 光点微电计;光点检流计;光(点)电流计
light spot millivoltmeter 光标式千分电压表
light spot projector 光点投影器
light spot recorder 光点记录器
light spot scanner 光点扫描器
light spot scanning 光点扫描(法)
light spot type 镜测读数型
light spot type meter 光点指示式仪表;光标式测量仪表
light stability 光稳定性;耐光性
light stabilizer 光稳定剂
light stabilizing(agent) 耐光物质;光稳定物质
light-stable unsaturated polyester resin 光稳定不饱和聚酯树脂
light staff 灯杆;灯柱;灯竿
light-stained sapwood 轻斑边材;轻斑材料
light standard 灯杆
light starching 轻浆
light starting duty 轻载启动负载
light station 灯桩;灯标站
light status 光线状况
light steel girder 轻钢大梁
light steel pallet 轻钢货架;轻钢集装箱
light steel structure 轻钢结构
light steel unit 轻钢构件;轻钢单元
lights tending and inspection 航标补给检查
light step 光阶
light stimulus 光激励
light stone 浅石色
light stopper 轻便油灰刀;轻便刮铲
light storage 光存储
light storage device 光存储装置;光存储器件
light straw 淡黄色
light stream 光流
light stroke 光射病
light-struck 光照的;漏过光
light structural concrete 轻质结构混凝土
light stucco 轻型拉毛粉刷
light suit 轻型潜水服
light sum 光和
light supply 给光量
light surface mulch 路面细覆盖料;表面薄层覆盖料
light surfacing 简易路面;铺筑简易路面
light switch 光开关;电灯开关;照明开关
light switch toggle 照明开关肘节
light system 电灯线路体系
light table 透明桌;修版台;映绘台;透写台;灯塔表;灯标表
light-taking window 采光窗
light tar 轻质焦油沥青
light target 灯光靶标
light tar oil 轻质柏油
light telegraph 灯光信号设备
light tender 航标工作船
light terracing grader 轻便台地整平机
light test 光隙试验
light-textured soil 轻质土(壤);砂质土(壤)
light-tight 防光的;不透光(的);不漏光
light-tight blind 不透光的百叶窗;不透光的遮帘
light-tight box 暗箱
light-tight compartment 挡光隔板
light-tight curtain 不透光布幕
light-tight door 不透光的门
light-tight jalousie 遮光(固定)百叶窗;不透光的百叶窗
light-tight louvers 不透光的百叶窗

light-tight slatted blind 不透光的板条百叶窗
light-tight system 遮光系统
light-tight tube 不透光管
light-tight window 遮光窗
light tile 轻(砖)瓦;轻瓦
light tillage 浅耕
light time 光行时;光时
light-time curve 光通量时间曲线
light tint 淡色
light tint paint 浅色漆;淡色漆
light-to-current conversion 光电变换
light-to-dark current ratio 光流与暗流比率
light-tonnage vessel 小吨位船舶
light top roller 轻质辊
light tormentor (舞台口竖边安装灯具的)管柱
light totating tower crane 轻型旋转塔式起重机
light tower 灯塔
light tower crane 轻型塔式起重机
light tower hoist 轻型塔吊
light tows 空载驳船队
light-tracing paper 描图纸;透明纸
light tractor 小功率拖拉机
light traffic 交通不繁忙;轻量交通
light traffic line 清淡运输线(路)
light traffic period 运输淡期
light traffic road 低交通量道路;轻轨交通量道路
light traffic route 清淡运输线(路)
light traffic volume 轻交通量
light transducer 光传感器
light-transfer characteristic 光电转换特性;光(的)传输特性
light transmission 光透射;光透过率;光传输;透光(性);透光量
light-transmission coefficient 透光系数
light-transmission value 透光率
light transmittance 光透过率;透射率;透光性;透光率;透光度
light-transmittance ceramics 透光陶瓷
light-transmittance efficiency 光效能
light-transmittance meter 透光度计
light transmitter 光发射器
light-transmitting capacity 光导能力
light-transmitting ceramics 透光陶瓷;透明陶瓷
light-transmitting filament 光导丝
light-transmitting medium 光传输媒质
light-transmitting rod 导光棒
light-transmitting specimen 光透射抽样
light trap 陷波器;灯光捕虫器;挡光装置;阻光通道;避光装置
light-trapped opening 光闸孔
light-trapped ventilation 不透光通风
light trapping 光陷阱
light traveling modulator 光行波调制器
light treatment 光照处理
light-triggered alarm 光通报警器;光触发报警器
light trim 轻型装饰
light trough 灯槽
light truck 轻型货车;轻型运汽车;轻便货车;轻型载货汽车
light truck with long wheel base 长轴距轻型载货汽车
light tube 光管;灯管
light tug 小型拖轮
light type abutment 轻型桥台
light type frame pile driving plant 轻型框架式打桩机
light type letter 白体字
light type pier 轻型桥墩
light type pile frame 轻型桩架;轻型打桩架
light-type truck 轻型货车
light type winch 轻型卷扬机;轻型绞车
light unit 光度单位
light up 照亮;理清缠绳
light use 轻度利用
light value 光量值
light valve 光阀
light valve array 光阀阵列
light valve shutter 光闸
light valve tube 效应光阀管
light-van 小型货车;客货两用车;轻型厢式送货车
light vault 轻型拱;轻型拱顶
light vector 光矢(量)
light vehicle 轻型载货车;轻型车辆
light vehicle lane 轻型轿车道

light vehicle traffic 轻型车辆交通
light-vehicular traffic 轻型车辆的运输(公路上)
light velocity 光速
light vessel 航标灯船;灯(标)船
light vibrating plate 轻微振动板
light vibrating screed 轻微振动整平板
light vibration 光振动
light violet 浅紫色
light violet-black 浅紫黑色
light violet-blue 浅紫蓝色
light violet-brown 浅紫褐色
light violet-grey 浅紫灰色
light violet-red 浅紫红色
light viscosity oil 低黏度油
light voyage 空载航行
light wagon drill 轻型钻车
light wall 轻质墙;薄壁
light-wall conduit 薄壁管
light wall pipe 薄壁管
light washing 轻度洗涤
light water 轻水;普通水
light-water breeder reactor 轻水增殖堆
light(water) line 空载水线;空载吃水线;轻船水线
light-water moderated reactor 轻水慢化堆
light-water moderator 轻水慢化剂
light-water reactor 轻水反应堆
light watt 光瓦特
light wave 光波
light wave communication 光波通信
light wave fields 光波场
light wavelength 光的波长
light wave rangefinder 光波测距仪
light wave regenerator 光波再生器
light wave retardation 光波延时(作用)
light wave spectrum analyzer 光波光谱分析仪
light wave train 光波列
light wave vector 光波矢量
light-way divider 光分路器
lightweight 空重;空载排水量;空载吃水;轻质;轻型;轻量化;轻便
lightweight adobe brick 轻质泥土砖
lightweight aggregate 轻(质)集料;轻(质)骨料
lightweight aggregate block 轻集料预制块;轻骨料预制块
lightweight aggregate concrete 轻(质)集料混凝土;轻(质)骨料混凝土
lightweight aggregate concrete block 轻集料混凝土砌块;轻骨料混凝土砌块
lightweight aggregate concrete building block 轻骨料混凝土建筑砌块
lightweight aggregate plant 轻集料厂;轻骨料厂
lightweight aggregate poured concrete 轻(质)骨料浇混凝土
lightweight angle steel 轻型角钢
lightweight base 轻质基座
lightweight base system 轻质基座体系
lightweight beam 轻质梁
lightweight block 轻(质)砌块;轻(质混凝土)块
lightweight block masonry(work) 轻质块材圬工
lightweight block partition(wall) 轻质块材隔墙
lightweight body 轻重量车身
lightweight bottle 轻量瓶
lightweight brick 轻砖;轻质砖
lightweight bridge deck(ing) 轻型钢桥面板
lightweight building block 轻质建筑块材
lightweight building felt 轻质建筑油毡
lightweight building insulation (grade) board 轻质建筑绝缘板材
lightweight building member 轻质建筑构件
lightweight building panel 轻质建筑预制板
lightweight building sheet 轻质建筑薄板
lightweight building slab 轻质建筑板材
lightweight building unit 轻质建筑单元
lightweight cable 轻电缆
lightweight calcium silicate brick 轻质硅酸钙砖
lightweight camera 轻便型照相机
lightweight car 轻型小客车
lightweight case 轻质表壳
lightweight castable 轻质无定形材料;轻质火泥
lightweight cast cavity tile 轻质空心砖
lightweight cast cellular expanded concrete screed 轻质膨胀混凝土找平层
lightweight cast component 轻质铸造零件
lightweight cast concrete component 轻质混凝土浇筑构件
lightweight cast concrete component unit 轻质混凝土浇制复合构件
lightweight cast concrete member 轻质混凝土浇制构件
lightweight cast concrete unit 轻质混凝土浇制单元
lightweight cast concrete ware 轻质混凝土浇铸制品
lightweight cast member 轻质浇制构件
lightweight cast unit 轻质浇制单元
lightweight cast wall slab 轻质浇制墙板
light weight cement 低密度水泥
lightweight center strength cable 轻型中心增强电缆
lightweight chamotte brick 轻质黏土砖
lightweight chip 轻量木片
lightweight cinder aggregate 轻质熔渣骨料
lightweight clay brick 轻质黏土砖
lightweight coarse aggregate 轻质粗骨料
lightweight coaxial cable 轻型同轴电缆
lightweight concrete 低密度混凝土;轻(质)混凝土;轻量混凝土
lightweight concrete aggregate works 轻质混凝土骨料工程
lightweight concrete beam 轻质混凝土梁
lightweight concrete block 轻质混凝土砌块
lightweight concrete bridge 轻质混凝土桥
lightweight concrete building component 轻质混凝土浇制构件
lightweight concrete building member 轻质混凝土建筑构件
lightweight concrete cast(ing) 轻质混凝土浇铸
lightweight concrete cast member 轻质混凝土预制件
lightweight concrete cavity block 轻质混凝土空心砌块
lightweight concrete cavity tile 轻质混凝土空心砖
lightweight concrete chimney 轻质混凝土烟囱
lightweight concrete chimney pot 轻质混凝土烟囱套
lightweight concrete column 轻质混凝土柱
lightweight concrete core 轻质混凝土核心
lightweight concrete facade 轻质混凝土立面
lightweight concrete filler slab 轻质混凝土填充板
lightweight concrete filler tile 轻质混凝土填充砖
lightweight concrete floor filler 轻质混凝土楼板填充块
lightweight concrete floor slab 轻质混凝土楼板
lightweight concrete frame 轻质混凝土框架
lightweight concrete garage 轻质混凝土车库
lightweight concrete girder 轻质混凝土大梁
lightweight concrete hollow block 轻质混凝土空心砌块;轻混凝土空心块(体)
lightweight concrete hollow filler 轻质混凝土空心填充块
lightweight concrete hollow tile 轻质混凝土空心砖
lightweight concrete lintel 轻质混凝土过梁
lightweight concrete load bearing frame 轻质混凝土承重框架
lightweight concrete mixer 轻质混凝土搅拌机
lightweight concrete panel 轻质混凝土嵌板
lightweight concrete plank 轻质混凝土板
lightweight concrete pot 轻质混凝土罐
lightweight concrete product 轻质混凝土产品
lightweight concrete roof(ing) slab 轻质混凝土屋面板
lightweight concrete screed material 轻质混凝土找平材料
lightweight concrete screed(topping) 轻质混凝土找平层
lightweight concrete slab 轻质混凝土楼板
lightweight concrete solid block 轻质混凝土实心块
lightweight concrete solid floor 轻质混凝土实心楼板
lightweight concrete structure 轻质混凝土结构
lightweight concrete supporting frame 轻质混凝土结构框架
lightweight concrete topping slab 轻质混凝土顶板
lightweight concrete wall 轻质混凝土墙
lightweight concrete wall slab 轻质混凝土墙板
lightweight concrete weight carrying frame 轻质混凝土承重框架
lightweight concrete with corrugated steel sheeting 带瓦楞钢板的轻质混凝土

lightweight construction 轻质结构;轻质构造;轻型结构;轻便结构
lightweight constructional material 轻质构造材料
lightweight constructiond method 轻质构造方法
lightweight container 轻瓶
lightweight core 轻质核心
lightweight core-out floor unit 轻质空心楼板单元
lightweight core-out plastic floor unit 轻质空心塑料楼板单元
lightweight corer 轻型取芯管
lightweight corundum brick 轻质刚玉砖
lightweight deck(ing) 轻质钢面板
lightweight deck(ing) system 轻质钢板体系
lightweight diesel engine 轻型柴油机
lightweight dive haven 轻潜水员安全所
lightweight diving 轻潜水
lightweight diving apparatus 轻潜水设备
lightweight diving equipment 轻潜水设备
lightweight diving gear 轻潜水设备
lightweight diving outfit 轻潜水设备
lightweight drill tubing 轻钻井管
lightweight expanded clay aggregate 黏土陶粒轻集料;膨胀黏土轻集料;膨胀黏土骨料;陶粒轻集料;陶粒轻骨料
lightweight expanded concrete aggregate 轻质膨胀混凝土骨料
lightweight expaned plastics 轻质膨胀塑料
lightweight exposed concrete block 轻质外露混凝土砌块
lightweight exposed concrete tile 轻质外露混凝土瓦;轻质外露混凝土片砖
lightweight external plaster 轻质外粉刷墙
lightweight external rendering 轻质外粉刷
lightweight extrusion 轻质挤压型材
lightweight fabric 轻质织物
lightweight fabrication 轻量化制造法
lightweight facade 轻质建筑立面
lightweight fair-faced concrete 轻质清水混凝土
lightweight fair-faced concrete block 轻质清水混凝土砌块
lightweight fair-faced concrete tile 轻质清水混凝土砖
lightweight filler 轻质填充料
lightweight filler block 轻质填充块材
lightweight filler slab 轻质填充板材
lightweight fine grain 轻质细粒
lightweight fine grained sand 轻质细粒砂
lightweight fines 轻质地板
lightweight fire brick 轻质耐火砖
lightweight fireclay brick 轻质黏土砖
lightweight floor 轻质填充楼板
lightweight floor filler tile 轻质地板填充花砖
lightweight floor girder 轻质地板梁
lightweight foamed plastics 轻质泡沫塑料
lightweight frame 轻框架
lightweight gas turbine 轻型燃气轮机
lightweight girder 轻质大梁
lightweight grader 轻型平地机
lightweight grain 轻量谷类
lightweight gypsum partition tile 轻质石膏隔墙板
lightweight gypsum plaster 轻质石膏灰泥;轻质石膏粉刷
lightweight hand rock drill 手动轻型岩石钻
lightweight helmet 轻装头盔
lightweight hide 轻磅皮
lightweight high alumina brick 轻质高铝砖
lightweight hollow block 轻质空心砌块
lightweight hollow concrete block 轻质空心混凝土砌块
lightweight hollow filler 轻质空心填充块
lightweight I-beam 轻型工字钢
lightweight inertial navigation system 轻型惯性导航系统
lightweight insulating brick 轻质保温砖
lightweight insulating concrete 轻质绝缘混凝土;轻质隔热混凝土;轻量绝热混凝土
lightweight insulating screed 轻质绝缘层
lightweight insulating screed material 轻质绝缘层材料
lightweight insulation (grade) board 轻质绝缘分度板
lightweight iron 轻便式熨斗
lightweight laser 轻便型激光器
lightweight lime concrete 轻(质)石灰三合土;轻三合土;轻质石灰混凝土砌块
lightweight lime concrete tile 轻质石灰混凝土砖
lightweight loading 轻荷载
lightweight loam 轻质亚黏土
lightweight loam slab 轻质亚砂土板
lightweight loam wall 轻质黏土墙
lightweight mask 轻型面罩
lightweight masonry(work) 轻质圬工
lightweight material 轻质材料
lightweight material partition 轻质隔墙
lightweight member 轻质构件
lightweight metal 轻质金属
lightweight metal alloy 轻质合金
lightweight metal blind slat 轻金属百叶窗条板
lightweight metal coach 轻金属客车
lightweight metal construction 轻质金属结构
lightweight metal construction section 轻质金属结构断面
lightweight metal construction trim 轻质金属结构装饰
lightweight metal construction unit 轻质金属结构单元
lightweight metal door closer 轻质金属门闭合装置
lightweight metal facade 轻质金属立面
lightweight metal fittings 轻质金属设备零件
lightweight metal frame 轻质金属框架
lightweight metal furniture 轻质金属家具
lightweight metal glazing 轻质金属钢框玻璃窗
lightweight metal hardware 轻质金属器具;轻质金属五金
lightweight metal jalousie 轻金属百叶窗
lightweight metal lattice(d) girder 轻金属格构大梁结构
lightweight metal lever handle 轻质金属杠杆手柄
lightweight metal louvers 轻质金属百叶窗
lightweight metal partition(wall) 轻质金属隔墙
lightweight metal plain web(bed) beam 轻质金属宽板横梁
lightweight metal plain webbed girder 轻质金属宽板大梁
lightweight metal plank 轻质金属板
lightweight metal profile 轻质金属断面
lightweight metal roll-up overhead door 轻金属上卷门
lightweight metal roof cladding 轻质金属屋面覆盖板
lightweight metal roof covering 轻质金属屋面盖板
lightweight metal roof lattice truss 轻质金属网格桁架
lightweight metal roof sheathing 轻质金属屋面覆盖层
lightweight metal section 轻质金属型材
lightweight mineral aggregate 轻质矿物集料;轻质矿物骨料
lightweight mirror 轻型反射镜
lightweight mixed clacined gypsum building plaster 轻质混合煅烧施工建筑灰泥
lightweight monolithic box 轻型整体盒子结构
lightweight motorcycle 轻型摩托车
lightweight mud 低比重泥浆;轻泥浆
lightweight mud brick 轻质黏土砖
light weight of car 车辆自重
lightweight panel 轻质板
lightweight paper 轻磅纸
lightweight partition wall 轻质隔墙
lightweight pe(a)rlite curtain wall 轻质珍珠岩幕墙
lightweight pipe 轻质管子;轻量管子
lightweight piston 轻金属活塞
lightweight plaster 轻质粉刷;轻质灰浆
lightweight plastic ball 轻质塑料球
lightweight plastic foam 轻质塑料泡沫
lightweight plywood 轻质胶合板
lightweight porous clay brick 轻质多孔黏土砖
lightweight pot 轻质罐
lightweight power unit 轻便动力设备
lightweight precast concrete block 预制轻质混凝土槽形板;预制轻质混凝土块
lightweight precast concrete channel slab 预制轻质混凝土槽形板
lightweight precast concrete comound unit 轻质预制混凝土复合构件
lightweight precast concrete component 轻质预浇混凝土构件
lightweight precast concrete member 轻质预制混凝土构件
lightweight precast concrete product 轻质预制混凝土产品
lightweight precast concrete ware 轻质预制混凝土制品
lightweight precast featherweight concrete channel slab 预制轻质混凝土槽形板
lightweight precast unit 轻质预制单元
lightweight precast wall slab 轻质预制墙板
lightweight prefabricated concrete compound unit 轻质预制混凝土复合单元
lightweight prestressed concrete member 轻质预应力混凝土构件
lightweight prestressed concrete unit 轻质预应力混凝土单元
lightweight profile 轻质型材
lightweight refractory 轻质耐火材料
lightweight refractory brick 轻质耐火砖
lightweight refractory material 轻质耐火材料
lightweight refractory product 轻质耐火产品
lightweight reinforced concrete 轻质钢筋混凝土
lightweight rendering 底涂层
light weight rescue apparatus 简易救生器
lightweight rock drill 轻便钻岩机
lightweight roofing 轻质屋面;轻质屋顶
lightweight roofing section 轻质屋顶断面
lightweight roofing shape 轻质屋顶形式
lightweight roofing solid web girder 轻质屋面实心腹板大梁
lightweight roofing spatial loading bearing structure 轻质屋面空间承重结构
lightweight roofing structural engineering 轻质屋面结构工程
lightweight roofing three-dimensional load bearing 轻质屋面三向荷载承重
lightweight roofing tile 轻质屋面瓦
lightweight roofing trim 轻质屋面装饰
lightweight roofing trussed girder 轻质屋面桁架大梁
lightweight roofing tube 轻质屋面安装管架
lightweight roofing tubular scaffold 轻质屋面安装管子脚手架
lightweight roofing unit 轻质屋面单元
lightweight roofing window 轻质屋顶窗
lightweight roofing window sill 轻质屋面窗台
lightweight sand 轻砂;轻质砂
lightweight sand-lime brick 轻质黄砂石灰砖
lightweight scaffold(ing) 轻质脚手架
lightweight screed 轻质找平层;轻质括板
lightweight secondary crusher 轻型二次破碎机
lightweight section 轻型(型)钢;小型钢;型钢;轻质型材
lightweight section(al) steel 轻型钢材;轻质型钢
lightweight sediment 轻质泥沙
lightweight shape 轻质型材
lightweight sheet 轻质板
lightweight sheet piling 轻质板桩
lightweight shipment 轻量货运输
lightweight shoes 轻装潜水鞋
lightweight silica brick 轻质硅砖
lightweight sintered aggregate 轻质陶粒
lightweight slag 轻质熔渣
lightweight slag block 轻质熔渣块
lightweight slag building block 轻质熔渣建筑砌块
lightweight slag concrete 轻质熔渣混凝土
lightweight slag concrete block 轻质熔渣混凝土砌块
lightweight slag concrete tile 轻质熔渣混凝土砖
lightweight slag powder 轻质熔渣粉末
lightweight slag tile 轻质熔渣砖
lightweight sloping roof 轻质斜坡屋顶
lightweight soild block 轻质实心砌块
lightweight solid tile 轻质实心砖
lightweight steel beam 轻质钢梁
lightweight steel component 轻质钢构件
lightweight steel construction 轻(质)钢结构
lightweight steel floor 轻质钢地板
lightweight steel frame 轻质钢框架
lightweight steel girder 轻质钢梁
lightweight steel girder floor 轻质钢梁地板
lightweight steel member 轻型钢构件
lightweight steel section 轻质钢材断面
lightweight steel shape 轻型钢材;冷弯型钢

lightweight steel structure 轻型钢结构
lightweight steel unit 轻钢单元
lightweight stopper 轻型向上式凿岩机；轻型伸缩式凿岩机；轻型挡块
lightweight structural clay product 轻质结构用黏土产品
lightweight structural concrete 轻质结构混凝土
lightweight structure 轻质结构；轻型建筑物；轻型结构；轻钢结构
lightweight stucco 轻质拉毛粉刷
lightweight substrate system 轻质衬底制
lightweight substructure system 轻质下部结构体系
lightweight tile 轻质砖瓦
lightweight tile masonry(work) 轻质砖瓦圬工
lightweight tile partition(wall) 轻质砖隔墙
lightweight tonnage 轻量吨位
lightweight trim 轻质装修
lightweight trussed girder 轻质桁架大梁
lightweight type water swivel 轻型水龙头；轻型水接头
lightweight unit 活动房屋；轻质单位
lightweight variable-depth towed sonar 轻型可变深度拖曳式声呐
lightweight vault 轻质拱顶
lightweight vehicle 轻型车辆
lightweight vehicle construction 轻型车辆结构
lightweight vertical coring(clay) brick 轻质竖向空心黏土砖
lightweight wall 轻质墙
lightweight wall slab 轻质墙板
lightweight warning radar station 轻型警戒雷达站
light weld 轻焊缝
light welding 轻焊接；浅填焊接
light welding structure 薄壁焊接结构
light well 小天井；采光井；通风竖井
light wheel tractor 轻型轮(胎)式拖拉机
light whistle buoy 音响灯(浮)标
light white pocket 轻微囊状鳞朽
light wind 轻风
light wire 光导束
light wood 易燃干木料；明子林；轻(质)木材；轻木；多脂木材；易燃木材
light wood fibre board 轻质木材纤维板
light wood oil 轻木油；汽馏松节油
light work 轻作业
light writer 光刻字机
light year 光年
light yellow 浅黄色；淡黄色；鹅黄
light yellow-brown 浅黄褐色
light yellow filter 浅黄色滤光镜
light yellow-green 浅黄绿色
light yellow-grey 浅黄灰色
light yellow-red 浅黄红色
light yield 光输出
light zone 光弧区；亮度区
light-zone depth 光亮带深度
Lignacite 木萨特(一种轻质骨料)
lignaloe oil 沉香木油
lignaloes 沉香木
ligne ga(u)ge 等内规
ligneous 木材的；木制的；木质的
ligneous plant 木本植物
lignification 木(质)化作用；木质化；惰性化作用
lignified cork 木质化木栓
lignified fiber 木纤维
lignified tissue 木质化组织
lignified wood 改质木材；木质素胶合板
lignifying 木质化；木素化
lignin-dissolving fungi 木质素溶解菌
lignin(e) 木素；木质素
lignin(e) acid 木素酸
lignin(e) adhesive 木质素黏胶剂
lignin(e) building unit 木素结构单元
lignin(e) content 木素含量
lignin(e) flocculant 木质素絮凝剂
lignin(e) liquor 纸浆残液
lignin(e) organic matter 木质素有机物
lignin(e) paste 木质素胶
lignin(e) phenol formaldehyde resin 木素酚醛树脂；木素苯酚甲醛树脂
lignin(e) plastics 木质纤维塑料；木(质)素塑料
lignin(e) reaction 木素反应
lignin(e) removal 除木质素
lignin(e) resin 木质素树脂
lignin(e) resin fixed biofilm anaerobic fluidized bed 木质素树脂固着生物膜厌氧流化床
lignin(e) sulfonate 木素磺酸酯；木素磺酸盐
lignin(e) sulphonate 木素磺酸盐
lignin(e) sulphonic acid 木素磺酸；木素磺酸
lignin(e) tar 木质素焦油
lignin(e) tar pitch 木质素柏油膏；木质素硬沥青；木质素焦油脂
lignin-humus complex 木素腐殖质复合体
lignin-sulphonate 磺化木质素
lignite 褐煤
lignite A 黑褐煤
lignite B 棕褐煤
lignite benzine 褐煤汽油
lignite bitumite 褐性烟煤
lignite char 碳化褐煤
lignite coal tar 褐煤焦油
lignite mine 褐煤矿
lignite oil 褐煤油
lignite paraffin(e) 褐煤石蜡
lignite pitch 褐煤沥青
lignite process 褐煤加工法
lignite resin 褐煤树脂
lignite tar 褐煤焦油
lignite tar pitch 褐煤焦油沥青；褐煤硬沥青；褐煤柏油脂
lignite wax 蒙丹蜡；褐煤蜡
lignitic coal 褐煤
lignitous coal 褐煤
lignocellulose 木素纤维素；木质纤维素
lignocellulose derived dissolved organic carbon 从木质纤维素中得到的溶解有机碳
lignocellulosic 木质纤维的
lignocellulosic anthracite 木质纤维无烟煤
lignocerane 二十四烷
lignoceric acid 木焦油酸；二十四烷酸；巴西棕榈酸
ligno-chrome gel 木铬胶
ligno-concrete 木筋混凝土
ligno-humus 木(质)素腐殖质
lignone 木纤维质
lignosa 木本植被
lignose 木纤维素；木本群落
lignosite 木素磺酸钙
lignosol 木浆
lignosulfonate 磺化木质素；木素磺化盐
lignosulfonic acid 木质(素)磺酸；木素磺酸
lignosulphonate 木质(素)磺酸盐；磺化木质素
lignosulphonate mud 木质素磺酸盐泥浆
lignosulphonic acid 木(素)磺酸
lignotuber 木块茎
lignum 木材；木
lignum shotcrete 木支撑
lignum vitae 镰形树胶；铁梨木
lignumvitae bearing 铁梨木轴承；层压胶木轴承
lignumvitae bush 铁梨木轴衬
lignumvitae strip 铁梨木嵌环；铁梨木条
ligrified fiber 木质纤维
ligroin(e) 里格若英；挥发油；轻石油；石油轻溶剂油；石油醚；粗汽油
liguid media 液态介质
ligulate 舌状体
liguliformcolulus 舌状体
likasite 羟磷硝铜矿
like attribute 相似属性
like electricity 同号电
like-grained 单一粒径的；单纹理
like-grained concrete 单一粒径骨料混凝土；相似颗粒混凝土；均质混凝土
like-grained gravel 等粒径卵石；同纹理砾石
like in kind property 同类财产
likelihood 相似性；可能发生的事情；可能成功的迹象；似然率
likelihood bedded 似层状
likelihood criterion 似然准则
likelihood detection 概率性探测
likelihood estimate 似然估值；似然估计量
likelihood estimation 相似估计；似然估计
likelihood function 概率函数；似然(率)函数
likelihood method 相似法
likelihood ratio 概率比；似然比
likelihood ratio functional 似然比泛函
likelihood ratio method 相似检定法；相似比值法
likelihood ratio test 概率比检验法；似然比值检验；似然比值检定
likelihood ratio test threshold 概率比测试阈值
likeness 相似性；类似
like numbers 同名数；同类数
like parity 同奇偶性
like polarity 同极性
like pole 同性极；同名极
like products 相同产品
like terms 同类项
likin 厘金
liking 亲合性；亲合力
lilac 紫丁香；丁香(花)
lilac colo(u)r 浅紫色
lilac colo(u)r lustre 浅紫光泽彩；淡紫光泽彩
lilac jade 丁香紫玉
lilac pale 淡紫藤色
lilianite 硫铋铅矿
liliquoid 乳状胶体
Lilly controller 李利控制器
lily 指北花纹；百合；洁白的
lily bulb 百合
lily-flower 白花花
lily gilding 镀白金
lily magnolia 木兰
lily pad ice 饼状冰
lily root 百合
lily root flour 百合粉
lily white 百合白色
limacon 蜗牛形曲线
liman 小湾或河口；淤泥浅湾；淤泥底的海边湖；江河入海的港湾；碱沼；溺谷；泥湾；河口沼泽；河口淤泥沉积；河口(湾)
liman coast 溺谷海岸
liman irrigation 雪水蓄灌；洪漫灌溉；春水蓄灌
limatura 锉屑
limb (树的)大枝；测角器；分度盘
limb brightening 临边增亮
limb darkening 周边减光；临边昏暗
limber board 污水道盖板；舱底污水道盖板
limber hole 流水孔
limber kentledge 压载铁(块)；压载货
limberoller conveyer 柔性液轴输送机
limber passage 污水道
limber(s) 通水孔
limber space 污水道
limber's plate 舱底污水道盖板
limbers-rope 舱底去污绳
limber strake 内龙骨翼板
limbic 边缘的
limbic system 边缘系
limb infrared monitoring of the stratosphere 平流层临边红外监测
limbo 中间过渡地带；垃圾场；监狱
limb of electromagnet 电磁(铁)铁芯
limb of fold 褶曲翼
limb radiance infrared radiometer 临边辐射红外线辐射仪
limb scanning 临边扫描
limburgite 玻基辉橄榄岩
limbus 明显边缘
limbus fossae ovalis 卵圆窝缘
limbus spiralis 螺旋缘
lime 氧化钙；酸橙；石灰
lime accumulation 石灰累积(作用)
lime addition 石灰添加料；石灰添加(量)
lime aggressive 石灰侵蚀的
lime aggressivity 石灰侵蚀性
lime agitator 石灰搅拌机；石灰搅拌器
lime algae 灰藻
lime alkali glaze 石灰—碱釉
lime aluminous cement 石灰矾土水泥
lime and breeze 石灰煤渣
lime and breeze slag 石灰煤渣矿渣
lime and cement mortar 水泥石灰砂浆；石灰水泥砂浆；白灰水泥砂浆
lime and cement mortar plaster 水泥石灰砂浆粉刷
lime and cement spreader 石灰和水泥撒料器
lime and earth foundation 灰土基础
lime and fly-ash 石灰粉煤灰
lime and fly-ash slag 石灰粉煤灰矿渣
lime and fly-ash soil 石灰粉煤灰土
lime and soda process 石灰苏打法
lime application 石灰施用
limear selection method 线选法

lime ash 石灰粉
lime-ash flooring 灰渣石地面；灰渣铺面；灰渣石（灰）铺面
lime bag 石灰袋
lime banker 石灰池
lime base 石灰基底
lime basecoat 石灰基底层
lime based 石灰打底的
lime base grease 钙基润滑脂；石灰脂；石灰基底油脂
lime base mud 灰基泥浆
lime base process 石灰法
lime bath 石灰浴
lime-bearing claystone 含灰质黏土岩
lime-bearing dolomite 含灰质白云岩
lime-bearing sandstone 含灰质砂岩
lime-bearing waste 含石灰废物
lime bin 化石灰池；石灰库；石灰仓
lime binding capacity 石灰浆黏结能力
lime biologic(al) sludge 石灰生物污泥
lime biologic(al) treatment 石灰生物处理
lime blast 石灰斑
lime bloom 混凝土表面起霜；石灰花
lime blowing 石灰胀裂（建筑砖缺陷）
lime blue 铜蓝；石灰蓝
lime boil 石灰水煮炼
lime bonded silica brick 石灰结合的硅砖
lime bonded silica refractory 石灰结合的硅质耐火材料
lime borate glass 硼酸钙玻璃
lime-borosilicate glass 硼硅酸钙玻璃
lime-bound 用石灰结合的；石灰结合的
lime-bound aerated concrete 石灰胶结多孔混凝土
lime-bound building component 石灰结合的建筑构件
lime-bound macadam 石灰(黏)结碎石(路)
lime brick 石灰砖；硅酸盐砖
lime bright annealed wire 中性介质中退火的钢丝
lime brush 石灰浆刷
limeburner 石灰窑；烧石灰工人
lime burning 窑烧石灰；石灰煅烧；锻烧石灰；煅烧石灰
lime burning kiln 石灰窑
lime burning plant 石灰(窑)厂
lime burnt clay cement 石灰烧黏土水泥
lime cake 石灰饼
lime cake waste 石灰滤渣；石灰(滤)饼废物
lime cake wastewater 石灰滤渣废水
lime carbonate 碳酸钙
lime carrier 石灰载体
lime cartridge 石灰包(炸药)
lime caustic precipitation 石灰苛性沉淀
lime caving 石灰崩落开采法
lime caving bin 石灰开采漏斗
lime cement 石灰水泥；石灰胶结料
lime cemented sandstone 石灰结合的砂岩
lime cement finish 石灰水泥灰浆抹面；石灰水泥面层
lime cement flyash concrete 石灰水泥粉煤灰(三合)混凝土
lime cement mixed plaster 石灰水泥粉刷
lime cement mortar 石灰水泥砂浆
lime-cement stucco 石灰水泥拉毛粉饰
lime chamotte mortar 石灰陶渣砂浆；石灰陶渣灰泥
lime chloride 氯化钙
lime cinder mixture 石灰煤渣混合料
lime classifier 石灰消化分离器
lime clay 白灰黏土；石灰黏土
lime claystone 灰质黏土岩
lime clinker 氧化钙熟料
lime coagulation 石灰凝结(法)；石灰混凝(法)
lime coagulation sedimentation 石灰凝结沉淀
lime coarse aggregate concrete 石灰粗集料混凝土；石灰粗骨料混凝土
lime coat 石灰涂层
lime coating 涂石灰；(钢丝的)石灰处理
lime column 石灰桩；石灰柱
lime combing capacity 石灰化合能力
lime compatibility 石灰的相容性
lime concrete 石灰三合土；石灰混凝土(石灰、砂、砾石混合物)；三合土
lime concretion 钙质结核；石灰(质)结核
lime conditioning 石灰熟化
lime content 石灰(质)含量

lime content determination 石灰含量测定
lime cream 石灰乳
lime crown glass 钙冕玻璃
lime crucible 石灰坩埚
lime crusher 石灰压碎机
limed 刷石灰；石灰处理的；水泥砌合的；刷了石灰的
lime decarbonization 石灰脱碳酸化(作用)
lime deposit 碳酸钙沉积物；水垢；石灰质沉积物；碳酸钙沉淀物；锅垢
lime-depositing 碳酸钙沉淀的
lime deviation method 石灰误差法
limed hide 石灰鞣革
lime dinas 石灰硅石
lime disintegration 石灰碎裂；石灰风化；石灰分解
lime distributor 石灰撒布机
limed oak 修饰过的橡木家具；浅灰色橡木(面)
limed oil 钙质松浆油；钙脂油
lime dosage 石灰剂量
lime dose 石灰投量
limed paint oil 钙脂(松浆油)清漆
limed poly-pale 石灰聚合松香
limed rosin 钙脂(松香)；松香钙皂；石灰松香；苛化树脂；经石灰作用的树脂
limed rosin enamel-interior 钙脂内用瓷漆
limed rosin varnish 钙脂清漆；石灰松香清漆
lime dust 石灰粉(尘)
lime earth 灰土
lime earth-broken brick concrete 碎砖三合土
lime earthenware 石灰质精陶器；石灰精陶
lime earth mix 灰土配合比
lime earth rammed 灰土夯实
lime earth wall 灰土墙
lime efflorescence 石灰引起的起霜现象；石灰风化；粉化石灰
lime emulsion 石灰乳(浊)液
lime encrusted 钙化的；固结石灰
lime external plaster 石灰外表面抹灰泥
lime external rendering 石灰外粉刷
lime factor 石灰因素；石灰系数
limefast 耐石灰的；耐碱的
limefast pigment 耐石灰颜料；耐碱颜料
limefast pigment for colo(u)ring cement 水泥着色耐碱颜料
lime feeder 石灰加料器
lime feldspar 钙(质)长石
lime ferritic electrode 碱性焊条
lime finish coat 石灰抹面层
lime flax fibered mortar 白灰膏麻刀灰
lime flax fibred mortar 白灰膏麻灰刀
lime-fly ash block 粉煤灰硅酸盐砌块
lime fly ash cement 石灰粉煤灰水泥
lime-flayash-crushed stone base (course) 石灰粉煤灰碎石基层
lime-flayash-crushed sand gravel base (course) 石灰粉煤灰砂砾基层
lime flyash (mixture) 石灰粉煤灰(混合料)
lime flyash stabilization 石灰粉煤灰稳定；石灰粉煤灰炼加固(法)
lime gas purification 石灰气纯化
lime glass 钙玻璃；氧化钙玻璃；石灰玻璃
lime glaze 石灰釉
lime granulated-slag (mixture) 石灰水淬渣(混合料)
lime gravel 灰砾；石灰砾石
lime gravel island 灰砾岛
lime grease 石灰脂
lime green 绿土；石灰绿；深绿色；酸橙绿(暗黄绿色)
lime grout(ing) 石灰灌浆
lime gypsum mortar 石灰石膏灰浆
lime gypsum process 石灰石膏工艺
lime-handling equipment 石灰运输设备
lime hardpan 石灰硬磐
lime-harmotome 钙交沸石
lime hydrate 消石灰
lime hydrating machine 石灰水化机；石灰消化机；石灰熟化机
lime hydration 石灰的消化作用
lime-improved soil 石灰改良土壤
lime injection 石灰粉喷射；打石灰桩(稳定土壤法)
lime insulating plaster 石灰隔热抹灰层
lime in the soil 土壤里的石灰
lime iron-concretion 石灰质铁质结核

lime-Keene's cement 石灰基恩水泥(高温焙烧白石膏粉与石膏粉拌成的面层粉刷材料)
lime kiln 灰窑；石灰(煅烧)窑
lime kiln gas 石灰窑气
lime knot 石灰结节
lime lead glass 钙铅玻璃
lime lee still 石灰拘酸蒸馏器
lime light 强白光；注意点；灰光(灯)；石灰灯
limelight flap 面光
lime liquor 石灰液
lime-magnesia ratio 石灰镁氧比例；钙镁率；钙镁比(率)
lime marl 灰质泥灰岩
lime matt 钙质无光釉；石灰无光釉
lime matt glaze 钙质无光釉
lime mica 珍珠云母
lime-mica schist 钙云母片岩
lime milk 石灰乳；石灰浆；石灰液
lime mill 石灰石粉碎机
lime mixer 石灰拌和机
lime-modified soil 石灰改善土壤
lime modulus 石灰模数
lime mortar 石灰砂浆；白灰砂浆
lime mortar flooring 石灰砂浆地板
lime mortar plastering 石灰砂浆粉刷
lime mortar undercoat 白灰砂浆打底
lime mud 石灰渣；石灰泥浆
lime mud disposal 石灰渣处理；石灰泥处理
lime mud filter 白泥过滤机
lime mudrock 灰泥岩
limene 橙油倍半萜
lime nitrogen 石灰氮
lime nodule 石灰结核
lime oil 酸橙油
lime (or cement) stabilized ash 石灰(或水泥)稳定渣
lime paint 石灰油漆；石灰涂料
lime pan 石灰硬盘
lime paste 石灰膏；石灰浆刷；石灰胶
lime paste handling pump 灰浆输送泵
lime pile 石灰桩；石灰柱；石灰堆
lime pile with sand 石灰砂桩
lime pillar 灰岩柱
lime pit 石灰坑；石灰池
lime plant 石灰厂
lime plaster 石灰涂层；石灰粉刷；石灰砂浆；石灰膏；石灰粉饰；掺砂石灰膏
lime plaster emulsified asphalt 石灰膏乳化沥青
lime plastering 石灰粉刷；石灰粉饰
lime plaster with hemp cut 麻刀灰
lime plaster with straw pulp 纸筋灰
lime popping 石类爆裂
lime pops 石灰爆
lime potential 石灰位减少表示酸性增加；石灰位测定降水酸碱度的标准
lime powder 石灰粉末；石灰粉；潮解石灰
lime pozzolan 石灰火山(灰)混合料
lime pozzolanic cement 石灰火山灰(质)水泥
lime-precipitating 碳酸钙沉淀的
lime precipitation 石灰沉淀(法)
lime process 石灰法；水化石灰法
lime producer 石灰岩生产者
lime product 碳酸钙制品；石灰制品
limeproof paint 防石灰油漆
lime pulverizer 石灰碾碎机
lime purification 石灰净化
lime purifying 石灰净化
lime putty 石灰腻子；石灰密封油背；石灰膏
lime quarry 石灰矿；采石灰石场
lime raker 石灰搅拌器；石灰耙
lime ratio 石灰比(例)
lime reactivity 灰石活化性
lime recalcining 石灰再(煅)烧；石灰二次煅烧
lime recovery 石灰回收
lime recycling system 石灰循环系统
lime red 石灰红
lime refractory 钙质耐火材料；石灰耐火材料
lime requirement 石灰需要量
lime resistance 耐钙性
lime-resistant surfactant 耐钙表面活性剂
lime roasting 石灰培烧
lime rock 钙质岩；未固结石灰岩；石灰岩；碳酸钙；石灰石
lime rock tower 石灰石塔

lime rosin ready mixed paint 钙脂调合漆
lime rubble rock 石灰碎砾岩
lime rust-coating 石灰防锈涂层
lime salt 石灰盐
lime sand 灰砂;石灰(质)砂
lime sand brick 灰砂砖;石灰砂砖;白灰砂砖;硅石砖
lime sand brick machine 制灰砂砖机
lime sand-broken brick concrete 灰砂碎砖混凝土;石灰三合土
lime sand island 灰沙岛
lime sand mortar 石灰砂砂浆
lime sand pile 石灰砂桩
lime sand plaster 灰砂粉刷
lime sand rock 石灰(泡)砂岩
lime sandstone 灰质砂岩;石灰砂岩;石灰砂石
lime-saturated 石灰饱和的
lime saturation 石灰饱和度
lime saturation factor 石灰饱和系数
lime saturation factor of cement 水泥的石灰饱和系数
lime saturation ratio 石灰饱和比
lime saturation value 石灰饱和值
lime saturator 石灰饱和器
lime screen 石灰筛
limes death 致死界量
lime-secreting 碳酸钙沉淀的
lime set 凝液
lime shaft kiln 石灰竖窑
lime silica ratio 钙硅比
lime-silicate brick 钙硅酸盐砖
lime-silicate concrete 硅酸盐混凝土
lime-silicate concrete block 硅酸盐混凝土砌块
lime-silicate glass 石灰玻璃
lime-silicate rock 钙硅岩;硅酸钙岩石
lime silica-water system 石灰氧化硅水系统
lime silo 石灰储[贮]存槽;石灰筒仓
lime sink 落水洞
lime slag cement 石灰矿渣水泥
lime slaked in the air 气化石灰
lime slaker 石灰消化器;石灰熟化器
lime slaking 石灰消化(作用);石灰熟化
lime slaking drum 石灰消化筒
lime slaking machine 石灰消化机
lime slaking pit 石灰消化坑
lime slaking tank 石灰消化槽
lime slaking trough 石灰消化槽
lime slate 石灰质板岩
lime sludge 石灰污泥
lime sludge disposal 石灰污泥处置
lime sludge treatment 石灰污泥处理
lime slurry 石灰泥;石灰浆
lime slurry treatment for acid waste 废酸石灰处理
limes margins 沟边
lime soap 钙皂;石灰皂
lime soap dispersing agent 钙皂分散剂
lime soda ash process 灰渣苏打法水处理
lime soda ash softening 石灰苏打灰软化(法)
lime-soda-feldspar 钙钠长石
lime soda glass 钙镁玻璃;钙钠玻璃
lime soda method 石灰苏打法
lime soda process 石灰苏打法
lime soda softening 石灰苏打软化(法)
lime soda softening method 石灰苏打软水法;石灰苏打软化法
lime soda softening process 石灰苏打软水法
lime soda treatment 石灰苏打处理
lime-sodium carbonate softening method 石灰碳酸钠软化法;石灰纯碱软化法
lime softening 石灰软化(法)
lime softening plant 石灰软化设备
lime soil 钙质土;灰土;石灰土
lime soil base (course) 石灰土基层
lime soil column 灰土井桩
lime soil compaction pile 灰土桩挤密;灰土挤密桩
lime soil cushion 灰土垫层
lime soil pavement 石灰土路面
lime soil pile 灰土桩
lime soil well 灰土井
lime solidification 石灰固化
lime sower 石灰撒施机;石灰撒布机
lime spray drying 喷洒石灰干燥法
lime spreader 石灰撒布机;石灰撒布机;施石灰机
lime stabilization 石灰稳定(法);石灰加固法;石灰稳定土

lime stabilization of sludge 污泥石灰稳定;污泥石灰处理
lime-stabilized 石灰稳定的
lime stabilized ash 石灰稳定灰渣
lime stabilized sand pavement 石灰稳定砂土路面
lime stabilized soil 石灰加固土
lime stain 石灰斑
lime stake 石灰桩
lime status 石灰含量
lime stimulation 石灰激发
lime stirrer 石灰搅拌器
limestone 灰岩;石灰石;碳酸钙;石灰岩(质)
limestone addition 石灰石混合料;石灰石掺和料
limestone aggregate 石灰石集料;石灰石骨料
limestone ballast 石灰石道砟
limestone-based process 石灰石法
limestone block 石灰石块
limestone breaker 石灰石破碎机
limestone brick 石灰岩砖
limestone brown loam 石灰石褐土
limestone cave 灰岩洞;石灰岩溶洞
limestone cavern 石灰岩洞
limestone chip(ping)s 石灰石屑
limestone coal group 边缘煤群
limestone coarse aggregate concrete 石灰石粗集料混凝土;石灰石粗骨料混凝土
limestone concretion 石灰石凝块
limestone conglomerate 石灰岩砾岩
limestone crusher 石灰石破碎机
limestone crushing plant 石灰石破碎设备
limestone debris 石灰石岩屑
limestone deposit 石灰岩矿床;石灰石沉积
limestone dust 石灰石粉末
limestone-faced precast panel 预制石灰岩面板
limestone facies 灰岩相;石灰石相
limestone filler 石灰石屑填料
limestone flour 石灰石粉
limestone flux 石灰石助熔剂
limestone for alkali industry 制碱用灰岩
limestone for carbide 电石用灰岩
limestone for cement 水泥用灰岩
limestone for chemical firtiliger 化肥用灰岩
limestone for construction 建筑石料用灰岩
limestone for flux 熔剂用灰岩
limestone for glass 玻璃用灰岩
limestone formation 石灰岩建造
limestone gravel 石灰质砾石
limestone kiln 石灰窑
limestone log 石灰岩测井
limestone marble 石灰石大理石;石灰石大理石
limestone masonry (work) 石灰石圬工
limestone mastic 石灰石玛琋脂
limestone needs 石灰石的需要
limestone neutralization 石灰石中和
limestone neutralization treatment 石灰石中和处理
limestone pebble conglomerate 石灰质卵石砾岩
limestone phyllite 石灰石千枚岩
limestone plant 石灰土植物;石灰石厂
limestone powder 石灰石粉
limestone putty 石灰石油灰
limestone quarry 石灰石矿;石灰石采石场
limestone quarry operator 石灰石采石工
limestone reactor 石灰石反应器
limestone red earth 石灰石红土
limestone reservoir 灰岩储集层
limestone ripper 石灰石土松土机
limestone rock 石灰石岩
limestone rock asphalt 石灰岩(地)沥青
limestone rubble 石灰岩碎石
limestone sand 石灰岩砂
limestone scrubber sludge 石灰石清洗污泥
limestone scrubbing (process) 石灰石清洗法
limestone sink 落水洞;灰岩陷坑;石灰岩渗坑;石灰岩落水洞
limestone slate 灰质板岩;石灰石页岩
limestone soil 石灰土;石灰发育的土壤
limestone sonde 灰岩电极系
limestone storage 石灰石仓库
limestone store 石灰石仓库
limestone stratum 石灰石地层
limestone tar macadam 石灰石—柏油碎石路
limestone terrain 石灰岩地形
limestone tower 石灰石吸收塔

limestone treatment for acid wastewater 酸废水石灰处理法
limestone type 石灰岩类型
limestone wash 粉刷石灰
limestone washer 石灰石清洗机
limestone weighing feeder 石灰石计量喂料机
limestone wet scrubber 石灰石湿式洗涤器
lime storage 石灰库
lime sulfur 石硫合剂
lime sulphur (solution) 石灰硫黄合剂
lime suspension 石灰悬浊液;石灰悬浮液
limes zero 无毒界量
lime tallow wash 油脂石灰水涂刷
limetitania type electrode 钙钛型焊条
lime trass mortar 石灰火山灰浆
lime treated aggregate 石灰处理的集料;石灰处理的骨料
lime treated mud 灰基泥浆;石灰处理的泥浆
lime treated straw 碱化秸秆
lime treatment 石灰加固;石灰处理
lime tree 菩提树;欧椴树
lime trough 石灰槽
limette oil 白柠檬油
lime uranite 钙铀云母
lime vat 石灰还原浴
lime wash 涂白;刷石灰水;刷白;石灰刷白;石灰浆涂刷;白灰浆;涂墙石灰乳;石灰水;石灰刷浆
limewash brush 刷石灰(浆)刷排刷
lime washing 石灰水浆涂料;用石灰刷白(石材的临时保护)
lime wash nozzle 石灰溶液喷嘴
lime water 石灰水
lime water softening plant 石灰软水装置;石灰软水厂;石灰软化设备
lime water softening test 石灰软水试验
lime weeping (混凝土裂缝处分泌出的)石灰乳
lime white 熟石灰;石灰浆;石灰白
lime whitewash 石灰刷白;石灰水粉刷
lime whiting 刷石灰浆;石灰刷白;石灰粉刷
limewood 北美椴木
lime yard 石灰间;石灰场
lime yard waste 石灰(堆)场废物
lime yard waste (water) 石灰场废水
limey dolomite 灰质白云岩
lime yellow 石灰黄
limey marl 灰质泥灰岩
liminal contrast 阈值反衬;阈值对比度
liminal stimulus 最低限刺激
liminal value 最低极限值
liming 钙化;浸灰;浸(法);加石灰;涂石灰;灰处理;施(用)石灰;撒石灰
liming apparatus 石灰熟化器
liming chamber 石灰乳室
liming material 石灰浸物质
liming of a soil 土壤施加石灰
liming out 灰析
liming process 灰浸(法)
liming stain 石灰斑
liming still 石灰乳槽
liming tank 石灰乳槽
liminimeter (测定湖面水位变位的)精密水位仪
liminometer (测定湖面水位变位的)精密水位仪
limit 限额;限度;极限(的);额度
limit up or down 停板价
limit acceleration factor 极限过载;极限过荷
limit-access project 保密工程
limit accurate method 极限精确法
limit address 界地址
limitaiton on credit creation 信用产生的界限
limit analysis 极限分析
limit analysis method 极限分析法
limit analysis method of reinforcement 极限分析补强法
limit analysis solution 极限分析解
limit angle 极限角;临界角
limit angle of slope under water 水下极限坡角
limitation 限制;限止;限度;局限性(极限);界限;极限
limitational production function 限制生产函数
limitation clauses 责任
limitation in maintenance operations 养路作业安全规定
limitation of access 工作范围限制证
limitation of action 诉讼时效

limitation of a method 方法使用范围
limitation of climate change 气候变化的限度
limitation of curvature 曲度限制
limitation of delivery 流量限制
limitation of length 长度极限
limitation of length of path 冲程限度
limitation of length of stroke 路程限度;冲程限度
limitation of liability 责任限制;职责范围
limitation of nonproductive input 限制非生产性投入
limitation of over-length 超长界限
limitation of Panama Canal passable ship 巴拿马运河通行限制
limitation of shipowner's liability 船东责任限制
limitation of ship size 船型限制
limitation of space 空间限制
limitation of speed 速度限制
limitation of the period for tax exemption 免税期限
limitation of velocity 极限速度;临界速度
limitation on length 长度限制
limitation on shelf-life 存放期限
limitation on usage 使用上的限制
limitations of brazed vessels 钎焊容器限制
limitations of seismic investigation 地震勘测限制
limitations on welded vessels 焊接容器的限制
limitations on welding 焊接限制
limitation to stream's self-purification 河流自净(作用的)限度;河水自净限度
limitation velocity 限制速度
limitation zone of building height 建筑高度限制区
limitator 限制器
limit at pulsating stress 脉冲应力极限
limit blade 限止叶片
limit bridge 窄量程电桥
limitcator 电触式极限传感器
limit check 极限校验;极限检验;极限检查
limit checking trap 界限校验自陷
limit collar 限制环;限定环
limit condition 极限状态
limit condition design method 极限状态设计法
limit conductance 极限电导
limit control 限额控制;极限(位置)控制;限量控制;极限(值)调整;安全控制;限度控制;限位控制
limit-control system 极限位置控制系统
limit curvature 极限曲率
limit cycle 极限周值;极限环
limit cycle of a differential equation 微分方程的极限环
limit cycles oscillations 极限环振荡
limit deflection 极限偏转量
limit deformation 极限变形
limit deposit velocity 极限沉积速度
limit design(ing) 最大强度设计法;极限(荷载)设计;极限强度设计;极限(状态)设计;极限负荷设计
limit design method 极限设计法
limit dextrin 有限糊精
limit dilution passage 限度稀释传代
limit dimension 限制尺寸
limit dimensioning method 定极限尺寸法
limit dose 极限剂量;极量
limit drainage amount of polluted water 限制污水排放量
limit drainage concentration of polluted water 限制污水排放浓度
limited 受限制
limited access bridge 限制进入的桥梁
limited access data 有限访问数据
limited access highway 限制进入的公路;快速公路
limited access road 限制进入的道路
limited access roof 有限通行屋顶
limited access route 全封闭线路
limited air pressure 限制空气压力
limited amount 限制额
limited apparent flow 有限表观流动
limited area fine mesh model 有限域细网格模式
limited area model 有限域模式
limited area sprinkler system 有限面积喷淋系统
limited audit 局部审计
limited availability 有限利用度
limited bandwidth 限定的频带宽度
limited bending stress 极限弯曲应力
limited bid 不公开招标
limited carcinogenicity test 短期致癌实验
limited check 限额支票;限定金额及支付期的支票
limited circular diameter of tube layout 布管限定圆直径
limited clearance sign 净空限制标
limited coasting trade 沿海短程运输
limited common area 有限的公共面积
limited company 有限责任公司;有限公司
limited competitive bidding 有限竞争性招标
limited concentration ratio 极限浓缩倍数
limited configuration interaction 有限构型相互作用
limited consumption 消费不足;限制消费
limited converse domain 有限后域
limited coverage range 有限射程范围
limited cracking 有限裂缝
limited credit 信用限额
limited creep stress 蠕变极限应力;蠕变极限
limited cross-country mobility 有限的越野机动性
limited data transfer and display equipment 有限数据传递及显示装置
limited deformation 有限形变;有限变形
limited distribution 有限散发
limited element response spectrum 有限元反应谱分析
limited express 特快列车
limited filtration velocity 极限滤速
limited flow 有限流动
limited fluid flow 有限流体流动
limited form wave 极限波
limited fraction 有界分数
limited frequency range 有限频率范围
limited function 囿函数
limited gain amplifier 有限增益放大器
limited grain size 临界颗粒尺寸
limited-height 高度有限的
limited highway 机动车专用公路
limited hump height at summer 夏季限制峰高
limited image number 有限次映射
limited inquiry 有限询价
limited inspection 局部检测
limited integrator 限量积分器;有限积分器
limited international bidding 有限国际招标;国际有限招标
limited in years of market forecasting 市场预测年限
limited laser pulse by Fourier transform 傅立叶变换限制的光脉冲
limited level 极限高程;极限高度
limited liability 限制责任;有限责任
limited liability co. 股份有限公司
limited liability company 股份有限公司
limited life component 有限寿命部件
limited life structure 寿命短建筑物
limited mutual solubility 有限互溶性
limited offer 限定报价
limited optimization 有限优化
limited order 限价订单
limited orifice 限流孔口
limited output 限定输出功率
limited output power 极限输出功率
limited oxygen demand 最终需氧量;极限需氧量
limited oxygen demand index 极限需氧量指数
limited partner 有限股东;有限合伙
limited partnership 有限合伙;两合公司
limited-payment life insurance 限期缴费人寿保险
limited-payment life policy 限期缴费人寿保险单
limited period 期限
limited plastic flow 有限塑性流动
limited pot life 有限活化寿命;有限活化期
limited prestressing 有限预应力
limited prestressing concrete bridge 有限预应力混凝土桥梁
limited price order 限制价格订货
limited price store 限价商店
limited production 有限生产
limited proportionality region 有限正比区
limited queue length 排队线长度受限制
limited range 限界;有限范围
limited recursion 有限递归法
limited reserves of foreign exchange 外汇资金短缺
limited resources 有限资金
limited road 限制通过能力的道路
limited-rotation hydraulic actuator 有限转动液力促动器
limited-run trial production 小批试制
limited safe 有限安全
limited sample bioassay 有限样本生物测定
limited-scan spot beams 有限扫掠点波束
limited service route 限制航线
limited set of displacement coordinates 位移坐标的有限集合
limited signal 限幅信号
limited slip 止滑
limited slip bolted joint 有限滑动栓接缝
limited slip differential 防滑差速器
limited space-charge accumulation mode 限累模式
limited speed 限制速度
limited speed road 限速路
limited stability 有限稳定性
limited steerable antenna 有限可控天线
limited subgrade frost penetration design 限制冰冻深度的路基(高程)设计
limited subgrade frost penetration method 限制路基冰冻深度设计法
limited submission 有限的公开招标行动
limited swelling 有限泡胀
limited tender 限制性投标
limited tendering 有限招标
limited traverse 有限方向限;横向限制
limited value assignment 限值排布
limited value of discharge standard 排放标准极限值
limited variation 有界变更
limited visibility 有限能见度
limited way 限制进出的道路
limit equilibrium analysis 极限平衡分析
limit equilibrium condition 极限平衡条件
limit equilibrium method 刚体极限平衡法;极限平衡法
limiter 限制器;限(制)幅器;限位器;限动器;限动件
limiter characteristic 限制器特性
limiter circuit 限制电路;限幅电路
limiter electrode 限制器电极
limiter shunt 限流分路器
limiter stage 限制级
limit failure 极限破坏
limit for arsenic in water 水中砷的极限
limit for production and consumption 生产消费限制
limit frame for balance weight 坠砣限制架
limit frequency 截止频率
limit gas 界限(燃)气;极限燃气;极限煤气
limit ga(u)ge 限制量计;极限(量)规;定限雨量器
limit gauging 极限测量法
limit governor 极限调速器
limit grade 极限坡度
limit indication 极限指示
limit indicator 限流指示器;极限指示器
limit inferior 下极限
limiting 限幅
limiting address 极限地址
limiting altitude (飞行的)极限高度
limiting ambient temperature 极限环境温度
limiting amplifier 限制放大器;限幅放大器
limiting angle 极限角
limiting angle of friction 极限摩擦角;摩擦极限角
limiting angle of resolution 极限分辨角
limiting aperture 限制孔径
limiting apparent magnitude 极限视星等
limiting availability 极限有效度
limiting bearing capacity of surroundings 羽岩极限承载力
limiting bed 界限层
limiting bending moment 极限弯矩
limiting capacity 极限生产能力
limiting case 极限情况
limiting circle 中心圆;极限图
limiting circuit 限制电路;限幅电路
limiting concentration 极限允许浓度;极限浓度
limiting condition 极限状况;极限条件;极限状态;限制条件
limiting condition for operation 极限操作条件
limiting conductivity 极限电导率
limiting control action 极限控制作用;极限控制动作
limiting creep 极限蠕变
limiting creep stress 极限徐变应力;极限蠕变应力

limiting current 极限电流
limiting current density 极限电流密度
limiting curve 限制曲线;控制弯段;极限曲线
limiting danger line 危险线;危险界限;碍航物界线
limiting date 限制日期
limiting deflection 极限挠曲量;极限挠度
limiting density 极限密度
limiting deposit velocity 极限沉积速度
limiting depth 限制水深;极限深度
limiting design value 设计限值
limiting device 限制(幅)器;限位器
limiting device cover 限止器盖
limiting dextrin 极限糊精
limiting diaphragm 限制膜片
limiting diaphram 极限光栏
limiting dilution 极限稀释法
limiting dimension 极限尺寸;净空;限制尺寸
limiting discharge 限制流量;极限排放
limiting distribution 极限分布
limiting distribution of the Haga's test 芳贺检验极限分布
limiting draft marks 极限吃水标志
limiting drawing ratio 极限拉延比
limiting effluent-quality 出水水质极限值
limiting equilibrium 极限平衡
limiting equilibrium mechanics 极限平衡力学
limiting error 极限误差
limiting factor 限制因子;限制因素;极限因数;起限制作用的因素
limiting-fall rating 正常落差水位流量关系曲线
limiting feedback 限制反馈;有限反馈
limiting feedforward 限制前馈;有限前馈
limiting field 限制场;抑制场
limiting fissure 界裂
limiting-flow device 限流设施
limiting-flow shower head 限流淋浴器
limiting-flow valve 限流阀
limiting frequency 极限频率
limiting friction 极限摩擦
limiting function 极限函数
limiting fuse 限流熔断器
limiting grade 限制坡度;最大坡降;最大坡度
limiting grade of ores 矿石最低品位;矿石极限品位
limiting grade of ore drawn 出矿极限品位
limiting gradient 极限坡度;限制坡度;最大坡降;最大坡度;限制纵坡
limiting grading curve 极限粒度曲线;极限级配曲线
limiting grain density 极限粒度;最大粒度
limiting grizzly 限粒格筛
limiting head 极限扬程;极限水头
limiting height 极限高度
limiting height of hump 限制峰高【铁】
limiting hottest-spot temperature 极限最热点温度
limiting infiltration rate 稳定入渗率
limiting information capacity 极限信息容量
limiting insulation temperature 极限绝缘温度
limiting intensity 极限强度
limiting law 极限定律
limiting layer 限制层
limiting level 限幅电平
limiting level control 极限电平控制
limiting level range 限制电平范围
limiting line 极限线;边界线;限止线;相界线;分界线
limiting load design method 极限荷载设计法
limiting load method 极限荷载法
limiting Mach number 极限马赫数
limiting magnification 极限放大率
limiting magnitude 极限星等
limiting margin 限幅边际
limiting maximum stress 最大限制应力;最大极限应力;极限最大应力
limiting member 限动件
limiting membrane 界膜
limiting mesh 筛孔极限尺寸
limiting mesh of the grind 粉磨粒度极限值
limiting minable thickness 极限可采厚度
limiting minimum stress 最小限制应力;最小极限应力;极限最小应力
limiting moisture capacity 极限含水能力
limiting noise emission 噪声发射极限
limiting number 极限数
limiting of resolution 分辨力限度
limiting penetration 极限贯入度
limiting photographic magnitude 极限照相星等

limiting plane 界面
limiting plate 界板
limiting point 极限点;边界点
limiting polarization 极限极化
limiting pollutional load 极限允许污染量
limiting position 极限位置
limiting potential 极限电位
limiting pressure 极限压力;极限压强
limiting pressure closing valve 超压切断阀
limiting pressure combustion 限压燃烧
limiting probabilities 极限概率
limiting quality 极限质量
limiting quantity 极限量
limiting radii 极限半径
limiting range of stress 应力极限;应力(的)极限范围
limiting reactor 限流电抗器
limiting resistance 极限阻力
limiting resistor 限流电阻器
limiting resolution 极限分辨能力;极限分辨率
limiting resolution angle 极限分辨角
limiting resolving angle 最小分辨角
limiting resolving power 极限分辨能力;极限分辨率
limiting retention factor 极限保留因子
limiting retention volume 极限保留体积
limiting rudder 最大偏度方向舵
limiting scanning power 极限扫描力
limiting scour depth 最大冲刷深度;极限冲刷深度
limiting screen 限粒筛;有限筛选
limiting screen aperture 极限筛孔眼
limiting sscreen size 极限筛尺寸
limiting section 限制区间
limiting service temperature 极限使用温度
limiting sighting error angle 极限瞄准误差角
limiting size 限制尺寸;极限尺度;极限尺寸
limiting slenderness 极限细长比
limiting slit 限制缝
limiting solubility 极限溶解度
limiting spatial frequency 极限空间频率
limiting speed 限制速度
limiting state 极限状态
limiting state design 极限状态设计
limiting state of equilibrium 极限平衡状态
limiting strain 极限应变
limiting stratum 界限层
limiting strength 极限强度
limiting stress 极限应力;临界应力
limiting structure 限界建筑物;控制建筑物
limiting sulcus 界沟
limiting surface 界面;限制表面
limiting surface of yielding 塑性变形的极限表面
limiting switch 行程开关
limiting temperature 临界温度;极限温度
limiting test 限度试验
limiting the growth of the money supply 紧缩银根
limiting the number of fishing vessels 限制渔船数
limiting thickness of rock intercalation 夹石剔除厚度
limiting time 限定时间
limiting total amount of fishing 限制总捕捞量
limiting tractive force 限界推移力;极限牵引力
limiting tractive power 最大挟砂力
limiting value 极限值;限阈
limiting valve 限制阀
limiting valve body 限制阀体
limiting valve cap 限制阀盖
limiting valve spring 限制阀簧
limiting variety 有限种类;有限变化
limiting velocity 流速极限;极限(俯冲)速度;极限流速
limiting vertical resolution 最高垂直分解力;极限分解力
limiting viscosity 特性黏度
limiting viscosity number 特性黏(度)数;本征黏度
limiting volume 极限体积
limiting wage increases 限制工资增长
limiting water depth 控制水深;极限水深
limiting wave 极限波
limiting wave direction 波浪射线的极限方向
limiting wave height 极限波高
limiting wave number 极限波数
limiting zone 限心;极限区
limit-in-mean 均值极限

limitless 无限的
limit level measurement 极限料位测定
limit line 极限线
limit load 限制荷载;最大使用荷载;极限荷载;极限负荷
limit-load design 极限荷载设计
limit loading 极限负载
limit load method 极限荷载法
limit load of pile 桩的极限荷载
limit loop 极限环
limit mark 界限标
limit material requisition 限额领料单
limit matrix 极限矩阵
limit moisture content 界限含水量
limit moment 极限弯矩
limit of accommodation 调节极限
limit of accuracy 精确限度;精确极限;精密限度;精度界限;精度极限;精度范围
limit of admissible error 容许误差极限
limit of allowable error 容许误差极限;容许误差范围
limit of allowance for wear 磨损留量极限
limit of a sequence 序列的极限
limit of audibility 能听极限
limit of backrush 回溃界限;(波的)回卷界限
limit of backwash 回溃界限;(波的)回卷界限
limit of backwater 壅水界限;回水界限;回水极限;回水范围
limit of bearing capacity 承载能力极限
limit of building area 建筑面积限定;建筑面积限制
limit of compatibility 相容极限
limit of compensation 赔偿限额
limit of concentration 浓度限度
limit of consistency 稠度限度;稠度界限;稠度极限
limit of convergence 收敛极限
limit of cooling temperature 冷却温度极限
limit of credit 信用额度
limit of creep 徐变极限;滑动极限;蠕变极限
limit of deformation 变形极限
limit of detection 检出限(界);检测极限
limit of detection of impurities 杂质试验极限;杂质检测极限
limit of determination 测定限(度)
limit of diffusion 扩散极限
limit of drift ice 浮冰界
limit of driving a pile 打桩的极限深度
limit of economic haul 经济运距范围;经济运距限度
limit of elasticity 弹性限度;弹性极限
limit of endurance 持久限度;持久极限
limit of endurance test 持久性试验极限
limit of equilibrium 平衡极限
limit of error 限差(范围);最大(容许)误差;误差限度;误差极限;误差范围
limit of excavation 开挖范围(界线);挖掘界限
limit of explosion 爆炸极限
limit of fatigue 疲劳界限;疲劳极限;疲劳强度
limit of fatigue in tension 受拉疲劳强度
limit of fineness 精细度限度
limit of flocculation 絮状沉淀单位;絮凝限度;絮凝极限
limit of free haul 最大免费运距;土方的最大免费运距
limit of friction 摩擦极限
limit of function 函数极限
limit of ice-shelf 浮架界
limit of identification 最低检出量;鉴定限度
limit of impurities 杂质限度
limit of inflammability (混合气燃烧的)稀薄极限;可燃性极限;可燃极限
limit of integration 积分水下限;积分域;积分范围
limit of kernel 核心范围
limit of liability 责任限额
limit of liquidity 液性极限;液限
limit of map 图廓
limit of maximum financial expenditures 财政支出的最高数量界限
limit of maximum financial revenue 财政收入最大限量
limit of national fishing zone 国家渔区界线
limit of navigation channel 航道界限
limit of oscillation 摆动限度
limit of overdrawn account 透支限额
limit of pack-ice 浮冰界

limit of passive earth pressure 极限被动土压力
limit of permissible error 容许误差极限
limit of plastic flow 塑流极限
limit of plasticity 塑性极限;塑限
limit of precision 最大精度;精确极限;精密限度
limit of profitable haul 最大盈利运距
limit of prohibited area 禁区界线
limit of proportion 比例限度;比例限
limit of proportionality 比例限度;比例界限;比例极限
limit of quantitation 定量限
limit of release 排放极限
limit of resolution 分辨限度;分辨极限
limit of rollability 可轧制性极限
limit of rupture 破裂极限;破坏极限
limit of saturation 饱和限度
limit of scour(ing) 冲刷限制;冲刷界限
limit of sector 光弧界线;光弧极限
limit of seed size 晶种粒度范围
limit of sensitivity 灵敏度极限
limit of sequence 序列极限
limit of sheet 图廓
limit of shunt 调车限界
limit-of-shunt sign 调车限界标
limit of size 极限尺寸
limit of space 空间限度
limit of stability 稳定限度;稳定极限;安定极限
limit of stress 应力极限;应力限度
limit of swash 波浪上爬冲界限;波浪上冲界线
limit of territorial water 领海基线
limit of the atmosphere 大气极限
limit of tidal current 潮流界限;潮汐流限
limit of tidal influence 潮汐影响界限
limit of tidal river reach 感潮河口段界限
limit of tide 潮汐界限;潮界
limit of tolerance 公差极限;耐性界限;容许量;公差界限
limit of toxic concentration 中毒浓度极限
limit of transmission 传输限度
limit of ultimate strength 极限强度;极限强度范围
limit of undermining 潜挖限制
limit of underwashing (河床的)冲刷极限
limit of unsurveyed area 未勘测地区界线
limit of uprush 涌浪界限;上涌界限;爬坡界限
limit of vacuum pull 真空吸力极限
limit of variable 变数极限
limit of variation 变化范围
limit of visibility 可见极限;能见度;能见(度)界限;能见(度)极限;能见度范围;视见限度
limit of visible spectrum 可见光谱范围
limit of wash 冲刷限度;冲刷界限
limit of wear 磨损极限;磨耗极限
limit of yielding (抽地下水的)极限出水量;屈服极限;屈服点;沉陷限度
limit on the left 左极限
limit on the right 右极限
limit option 界限选择
limit order 限价订单;限定性指令;停板订单
limitotype 区域层型
limit output 极限输出;极限出力
limit permissible concentration 极限容许浓度
limit pile load 桩极限荷载
limit plane 临界面;极限面;分界面
limit plane wave 界面波
limit plug 限制塞
limit point 极点
limit power 极师;功率
limit pressure 极限压力
limit pretightening sealing load 界限预紧密封比压
limit prices 限价
limit pricing 限制性定价
limit priority 限制优先级;极限优先权
limit process 极限法;极限步骤;求极限过程
limit profile 机车车辆限界(美国)
limit profitable haul 最大有利运距
limit range 限制范围;极限范围
limit ratio 限幅比
limit register 界(限)寄存器;上下界寄存器
limit resolution 极限分辨能力
limit rise 升达限幅
limit rod 限制杆
limits and fits 极限与配合
limits-checking 界限检查
limit screen size 有限的筛网尺寸;有限的筛孔尺寸

limit screw 止动螺钉
limit screw pitch ga(u)ge 螺距极限规
limits file 界限文件;极限文件
limit shock 极限地震
limit size 限定尺寸;极限尺寸
limit slenderness ratio of boom 吊杆极限长细比
limit slope 临界坡度;极限坡度
limits of authority 权限
limits of effective current range 有效电流极限
limits of financial capacity 财力负担限度
limits of functions and powers 职权范围
limits of integration 积分限
limit solid solution 有限固溶体
limit stage 极限阶段
limit state 极限状态
limit-state design 极限状态设计;塑性设计
limit state design method 极限状态设计法
limit-state design philosophy 极限强度设计原理
limit state design theory 极限状态设计原理;极限状态设计理论
limit state equation 极限状态方程
limit state for normal use 正常使用极限状态
limit state of bearing capacity 承载力极限状态
limit state of cracking 开裂极限状态
limit state of crack width 裂缝宽度极限状态
limit state of deformation 变形的极限状态
limit stop 行程限制器;限制器;止动器;极限开关;限位挡;限动挡
limit stop valve 限止阀
limit strength 极限强度
limit suction lift 极限吸程
limit superior 上极限
limit surface of yielding 软化极限曲面
limit surface wave 界面波
limit switch 限制开关;限位开关;终端开关;极限开关;停止开关;行程开关
limit switches guarding progressive die 限位开关保护的连续模
limit switch for hoist motion 提升运动限制开关;升降运动的限制开关
limit switch for luffing motion 俯仰运动的限制开关
limit switch for travel motion 运行限制开关
limit system 公差制;极限制
limit table 限额表
limit theorem 极限理论
limit to growth 增长的限度;生长极限
limit up 升达限幅
limit value 阈限值;界限值;极限值
limit valve 限位阀
limit voltage 极限电压
limit weight 极限重量
limmer 沥青石灰岩;沥青石灰石
limn 描画
Limnaea stagnalis 静水锥实螺
limnetic 湖沼的
limnetic algae 湖藻
limnetic coal deposit 湖沼煤系
limnetic ecosystem 湖沼生态系统
limnetic facies 湖沼相;湖相
limnetic sedimentary type 静水沉积型
limneticum 湖水盐度(范围);淡水盐度范围
limnetic zone 湖沼区;湖沼带;湖沼透光层
limnic 湖沼的;湖泊的
limnic coal deposit 湖沼煤系
limnicole 湖沼动物
limnicolous 栖湖沼的
limnigram 自记湖泊水位图
limnigraph 自记湖泊水位计;自记录水位计
limnimeter 湖泊水位仪;潮位计
limnium 湖沼群系;湖沼群落
limnocorral 凸纹围栏
limnocryptophyte 沼泽植物
limnodic 盐土沼泽群落
limnodium 沼泽群落
limnogenic rock 湖沼沉积岩
limno-geotic 淡水环境
limnograph 湖面水质;自记水位计
limnologic(al) meteorology 湖沼气象学
limnologic(al) region 湖沼区;湖区
limnologic(al) transformation 湖沼演化;湖泊演替;湖泊演化;湖泊演变
limnologist 湖沼学家
limnology 湖沼学;湖沼水质学;湖沼生物学;湖泊水文学;淡水生物学

limnophilous 喜沼泽的
limnophilus 沼泽种类
limnoplankton 湖沼浮生物;湖沼浮游生物;淡水浮游生物
limnoquartzite 淡水石英岩
limnoria 蛀船虫;木蠹;船蛆
limnoria lignorum 蛀木水虱
limo 柠檬
limoge 绘画珐琅
limoid 消化石灰;熟石灰
limonite 褐铁矿;褐铁石
limonite cement 褐铁矿胶结物
limonite concrete 褐铁矿混凝土
limonite-oolite 褐铁矿鲕状岩
limonite ore 褐铁矿砂;褐铁矿矿石
limonite rock 褐铁岩
limonium suffrulicosum 补血草
limous 混浊的
limousine 小客(轿)车;大型高级轿车
limousine terminal 客车终点站火枢纽站
limp 软封面书
limp base 半硬基地;地下基地
limp cloth 软布面
limp-diaphragm 柔软膜片
limpet 开口小沉井;蠓
limpet asbestos 喷涂石棉
limpet mine 水下爆破弹
limpet washer 锥度垫圈;锁紧垫圈;波纹边垫板
limpidity 清澈度;透明度;清晰度
limp leather goods 软皮片
limpness 柔软性
Limpopo cycle 林波波造山旋回
limp state 柔软态
limp wall 柔性墙
limy 含石灰的;似石灰;石灰质;石灰质(的)
limy clay 灰质黏土;石灰质黏土;含石灰的黏土
limy concretion 石灰质凝结物;石灰质团块
limy gravel 黏结灰质砾石
limy sandstone 石灰质砂岩
limy soil 灰质土(壤);石灰性土壤;石灰质土壤
linac 线性电子加速器;直线(电子)加速器
linac acceptance 直线加速器接收度
linac adjustment 直线加速器调试
linac axis 直线加速器轴线
linac-cyclotron combination 直线加速器回旋加速器组合
linac duty factor 直线加速器负载因子
linac focusing 直线加速器聚焦
linac gun 直线加速器枪
linac injector 直线加速器注入器
linac length 直线加速器长度
linac operation group 直线加速器运行组
linac output 直线加速器输出
linac stage 直线加速器级
linac tank 直线加速器加速腔
linaloe wood oil 沉香木油
linalool 沉香醇;芳樟醇
linarite 青钳矾
linasec 纸孔带加工装置
linatex 防锈乳胶
linbay 棚屋(英国)
Linblad suction thickener 林卜拉德真空浓缩机
linburgite 玻基辉橄岩
linch 田埂坎;梯田崖
linchet 田埂坎;梯田崖
linchpin 楔(栓);销;制轮楔;开口销;锚杆销;车轴销
Lincolnshire limestone 林肯郡灰岩
Lincoln type milling machine 林肯式铣床
Lincoln weld 埋弧自动焊
lincrusta 油毡纸;凸纹墙纸;浮雕墙纸
lincrusta-walton 软木屑与氧化的亚麻籽油制成的浮雕壁纸
lindackerite 水砷氢铜石;砷镍铜矾
Lindapter 固定接合器(可避免在钢构件上的钻孔)
Linde copper sweetening 林德氯化铜脱硫法
Linde drill 熔化穿孔机
Lindelof space 林德洛夫空间
Lindemann's electrometer 林德曼静电计
Lindemann's glass 林德曼玻璃;透紫外线玻璃
Lindemann's joint 林德曼接合
Lindemann's method 林德曼法
linden 欧洲椴;菩提树
Lindenmayer system 林登梅耶系统【地】

linden oil 椴树油
Linde's apparatus 林德液化器
Linde welding 林德钢管对焊法
lindgrenite 钼铜矿
lindinosite 钠闪正花岗岩
Lindsay's surface fabric 林德舍地面加劲条(一种耐磨水泥地面中的加劲钢条)
lindslengite 钛钡铭石
Lindstedt equation 林德斯塔方程
lindstromite 辉铋铜铅矿
line 线条;线路【电】;线路;线道;路线;行(业);行列;品系;排;磁致伸缩延迟线;船队;衬
line 线(1 线 = 1/12 英寸)
linea alba 白线
line abreast 横队
line absorption 线吸收
line access point 线路通路点
line accumulator 线路蓄电池
line activities 线路业务
line acutance 线划清晰度
line adapter 线路衔接器;线路适配器
line adaptor 线路转接器
linea dentata 齿状线
line admittance 线路电纳;线路导纳
line advance 移行【计】
lineae ablicantes 白纹
lineage 线状腐蚀坑;系谱
lineage segment zone 谱系枝带
lineage structure 脉理构造;嵌晶结构;嵌晶构造
lineage zone 谱系带
linea intertrochanterica 转子间线
lineal change 线性变化
lineal expansion 线性膨胀
lineal pressure 线压
lineal scale length 直线刻度长度
lineal shrinkage 线收缩;纵缩(沿长度方向的收缩)
lineal shrinkage test 线缩试验
lineal source of water pollution 水污染线源
lineal town 线状延伸的城市;带形城市
linea mediana 正中线
lineament 线状特征物;线性构造;线型构造;区域断陷线;地貌轮廓线
lineament along drainage system 水系线性体
lineament map 线性构造图
lineament related to deep faults 深断裂线性体
lineaments along original structure of rock 岩石原生构造线性体
lineaments along stratigraphic(al) unconformity 地层不整合线性体
lineament structure 区域线性构造;棋盘格式构造
lineament tectonics 地貌构造学
line amplifier 线路放大器
line amplifier bay 线路放大器架
line amplitude 行幅度
line amplitude adjustment 行幅度调整
line amplitude control 行宽控制;行幅度控制
line analysis 线分析(法)
line and arc intersect 线与弧相交
line and bridge building depreciation expenses 线桥建筑折旧费
line and dot interlace 行点交织
line-and-grade stakes 放样桩
line-and-half toning 等高线加晕渲表示法
line and level 线路与标高;线路与高程
line and pin 砌砖挂线
line and trunk group 线路和中继线群
line angle 线角
linea nuchae inferior 下项线
linea nuchae suprema 最上项线
linea pectinata 耻状线
linear 线状的;线形的;线条状;直线性的;直线(型)的;区域断陷线
linear absorption coefficient 线(性)吸收系数
linear acceleration 线(性)加速度;直线加速度
lineae acceleration method 线性加速度法
linear accelerator 线性加速器;直线(性)加速器
linear accelerator injector 直线加速器注入器
linear-accelerator meson factory 直线加速器介子工厂
linear-accelerator structure 直线加速器结构
linear accelerograph 线性加速自记仪;线性加速自记器
linear accelerometer 线性加速表
linear accumulative damage criterion 线性累积损坏准则;线性积累破坏准则
linear action disk valve 平板滑阀
linear activity 线性(核)放射性
linear actuator 线性致动器;线性操作机构;直线运动液压缸;直线运动促动器;往复缸
linear adjoint group 线性伴随群
linear admissible function 线性容许函数
linear adsorption 线性吸附
linear adsorption coefficient 线性吸收系数
linear aerodynamic characteristics 线性空气动力特性
linear after contraction 线性残余收缩;残余收缩;线性再收缩
linear after expansion 线性残余膨胀
linear air distribution 线性气流分布
linear algebra 线性代数
linear algebraic equation 线性代数方程
linear algebraic group 线性代数群
linear alkylate sulfonate 直链烷化磺酸盐
linear alkylbenzene 直链烷基苯
linear alkylbenzene sulfonate 直链烷基苯磺酸盐
linear alteration zone 线状蚀变带
linear alternative gradient system 线性交变梯度系统
linear amorphous polymer 线性无定形聚物
linear amplification 线性放大(倍数)
linear amplifier 线性放大器;直线性放大器
linear analogue control 线性模拟控制
linear analysis 线性分析;线分析(法)
linear-and-angular-movement pickup 线位移和角位移传感器
linear angle encoder 线性角编码器
linear angular intersection 边角交会(法);边角联合交会
linear annealing 线性退火
linear anode current 线性输出电流
linear anomaly 浅状异常
linear antenna array 线式天线阵
linear anticline zone 线状背斜带
linear approximation 线性近似;线性逼近
linear arch 抗力线拱;线性拱
linear arrangement 线性排列;直线排列
linear array 线性阵列;线性台床;线性排列;线型阵列;线形组合;直线组列;直线(式)天线阵
linear array of simple sound source 单声源线列阵
linear array of simple source 线列声源
linear array receiver 线列(阵)接收机
linear array scanner 线性阵列扫描仪
linear array scanning 线阵扫描
linear asynchronous structure 线性异步结构
linear attenuation 线性衰减
linear attenuation coefficient 线性衰减系数;线吸收系数
linear axiom 直线公理
linear ball bearing 直线球轴承
linear base 线性时基;直线基阵
linear beam density 线性束流密度
linear beam tube 直线注管
linear bearing 直线轴承
linear bearing system 直线支承体系
linear behavio(u)r 线性性质
linear bending theory 线性弯曲理论
linear birefringence 线性双折射
linear block code 线性块码;线性分组码
linear Boolean recursion 线性布尔递归
linear boundary operators 线性边缘算子
linear boundary value problem 线性边值问题
linear bounded automat 线性有界自动机
linear bounded automation 线性有限自动机
linear branch 一次枝线
linear buckling theory 线性压曲理论
linear building 直线形建筑
linear bundle 线丛
linear burning rate 线性燃烧速率
linear burn up 线性燃耗
linear calibration curve 线性校准曲线
linear car-following model 线性随车模型;线性车辆跟随模型
linear carriage head-positioning actuator 车载直线移动式磁头定位驱动器
linear chain 线链;直链
linear chain of triangulation 线形三角链(锁)
linear change 线性变化;(混凝土路面的)线变形
linear change of reheating 重烧线性变化
linear-channel method 刻线法
linear character 线性特征员
linear characteristic 线性特性;直线特性
linear charge concentration 线装药密度
linear charge density 线性电荷密度
linear chart 线图;直线图
linear chroma distortion 线性色度失真
linear chromatography 线性色谱(法)
linear circuit 线性网络;线性电路
linear circuit test 线性电路试验
linear city 狭长城市;带形城市;长形城市
linear classifier 线性分类器
linear clavate 长棒状的籽粒
linear cleavage 线状劈理;直线劈理
linear clipper 线性削波器
linear closure 线性闭包;线闭合差
linear closure principle 拉直线关闭规律
linear code 线性码
linear coefficient of thermal expansion 热线膨胀系数;热膨胀线性系数
linear colloid 线状胶体
linear colo(u)r demodulation 线性彩色解调
linear colo(u)rimetric method 线性比色法
linear combination 线性组合
linear combination of atomic orbitals 原子轨道函数线性组合
linear commutation 直线换向
linear comparator 线性比较器
linear complex 线性线丛
linear compliance 线性柔度
linear compound 直链化合物
linear computation 线性计算
linear condensation 线型缩合
linear condensation polymer 线形缩聚物
linear condensed rings 线型缩合环
linear condition 线性条件
linear conductor 线性导体
linear conductor antenna 直线导体天线
linear congruence 线性同余
linear connection 线性联络
linear conservation system 线性守恒系统
linear constraint 线性约束
linear construction 直尺作图法
linear contact lay wire rope 线接触钢丝绳
linear content 线尺度
linear continuous set 线性连续集
linear continuum 线性连续统
linear contraction 线收缩
linear control 线性控制
linear control electromechanism 线性控制电动机构
linear control process 线性控制过程
linear control system 线性控制系统
linear control theory 线性控制理论
linear convergence 线性收敛
linear coordinates 线坐标;直线坐标
linear correlation 线性相关;直线相关
linear correlation coefficient 直线相关系数
linear cost function 线性费用函数
linear coupled dynamic theory 线性耦合动态论
linear coupler 线性耦合装置;线性耦合器
linear cover-degree 直线盖度
linear creep 线性蠕变
linear crystallizer 线式结晶器
linear current 线性电流
linear current density 线性电流密度
linear current range 线性电流范围
linear cut-off 线性截止
linear cut-off low-pass filter 线性截止低通滤波器
linear cutting machine 线切割机床
linear damage law 线性损伤法则
linear damping 线性阻尼
linear decision model 线性决策模型
linear decision rule 线性决策律;线性决策法则
linear decoration 线状装饰;线形装饰
linear decrease 直线下降
linear decrease voltage 线性下降电压
linear decrement 线性衰减量
linear deflection 线变位
linear deformation 线性形变;线(性)变形;纵向变形;拉伸伸长;长度变形
linear delay circuit 线性延伸电路
linear delay distortion 线性延迟失真
linear demand function 线性需求函数

linear demodulator circuit 线性解调器电路
linear densimeter 线性密度计
linear densitometer 线性光密度计
linear dependence 线性相关
linear dependence relation 线性相关性关系
linear depletion 线型损耗
linear detection 线性检波
linear detector 线性检波器
linear device 直线装置
linear diagram 线形示意图;只表示结构各构件中心线的图
linear diameter 线径
linear dielectric 线性电介质
linear difference equation 线性差分方程
linear differential-difference equation 线性差分微分方程
linear differential equation 线性微分方程
linear differential equation of high order 高阶线性微分方程
linear differential operator 线性微分算子;线性微分算符
linear diffuser 条形散流器
linear diffusion 线性扩散
linear digitizer 线划数字化器
linear dilatation 线膨胀
linear dilatometer 线膨胀测试仪
linear dilatometry 线膨胀测量法
linear dimension 线性维数;线性尺寸;线尺度;一维
linear dimensional stability 线性尺寸稳定性
linear dimensioning 长度标注;长度尺寸标注
linear Diophantine equation 线性迪奥番廷方程
linear direct current saturation control 线性直流饱和控制
linear direct reading thermometer 线性直读式温度计
linear discontinuities 线状缺陷
linear discrepancy 线量误差;长度闭合差
linear discrete programming 线性离散规划
linear discrete valued system(s) 线性离散值系
linear discriminant function 线性判别函数
linear discriminator 线性鉴频器
linear disjointedness 线性不相交性
linear disjointure 线性无缘性
linear dispersion 线性色散;线性分散;线色散率
linear displacement 线性位移;线位移
linear displacement ga(u)ge 线位移测量计
linear distance 直线距离
linear distorted 线性失真的
linear distortion 线性失真;线性畸变
linear distribution 线性分布;线形分布
linear doubling time 线性倍增时间
linear drift 线性漂移
linear drilling 直线钻孔
linear drive 线性驱动
linear dune 线状沙丘;线形沙丘
linear dynamic(al) property 线性动力特性
linear dynamic(al) range of amplifier 放大器线性动态范围
linear dynamic(al) system 线性动态系统;线性动力系统
linear dynamic model 线性动态模型
linear dynamic range 线性动态范围
linear dynamics 线性动力学
linear earthquake 线状地震;直线地震;线型地震;线性地震
linear eccentricity 偏心律;偏心距
linear elastic design response spectrum 线弹性设计反应谱
linear elastic fracture mechanics 线性弹性断裂力学;线弹性断裂力学
linear elastic gravity-turn-on analysis 一次加荷重力弹性分析;一次重力加荷线性弹性分析
linear elasticity 线性弹性;线性弹性变形
linear elastic manner 线性弹性方式
linear elastic modulus 线弹性模量
linear elastic theory 线弹性理论
linear elastodynamics 线弹性动力学
linear elastoplastic fracture mechanics 线弹塑性断裂力学
linear electric(al) constant 线性电气常数;线路电气常数
linear electric(al) current loop 线性电流环
linear electric(al) motor 直线电动机
linear electrical parameter 传输线参数

linear electron accelerator 线性电子加速器
linear electronic matrixing 线性电子矩阵化
linear electro-optical effect 线性电光效应
linear element 线状要素;线元;线性元素;线性元件;线性环节;线性单元;微弧;长度元
linear elongation rate 线性伸长率
linear eluant strength gradient 线性洗脱液强度梯度;线性洗提力梯度
linear encoder 线性编码器
linear energy transfer 线性能量传递;线能量转移;阻止本领;传能线密度
linear equalizer 线性均衡器
linear equation 线性方程(式);线性方差;一次方程(式);直线方程(式)
linear equation solver 线性方程解算装置
linear equation system program(me) 线性方程组程序
linear equilibrium line 线性平衡线
linear equivalence classes 线性等价类
linear erosion 线状侵蚀;线性侵蚀
linear error 线性误差;线长误差;长度误差
linear error of closure 长度闭合差
linear eruption 线状喷出;裂缝喷溢
linear ester 线型酯
linear estimate 线性估计
linear estimation 线性估计
linear etching 线形蚀刻
linear expanding 线性扩展
linear expansibility 线胀性;线膨胀系数
linear expansion 线胀率;线性展开;线(性)膨胀;直线膨胀
linear expansion coefficient 线性膨胀系数;线膨胀系数
linear expansion coefficient of rail steel 轨钢的线膨胀系数
linear expansion factor linear 线膨胀系数
linear expansion rate 线膨胀率
linear expansion relay 线膨胀型继电器
linear expansivity 线胀性
linear expectation 长度增加
linear expenditures system 线性支出系统
linear explanatory economic model 线性解释经济模型
linear expression 线性表示
linear extension 线延伸
linear extinction coefficient 线性衰减系数
linear extrapolation 线性外延(法);线性外推(法);直线外插法
linear extrapolation distance 线性外推距离
linear extrapolation length 直线外推长度
linear fabric 线状组构
linear factor 线性因子
linear failure rate distribution 线性失误率分布
linear feature 线性特征
linear feature extraction 线性特征提取
linear feedback control 线性反馈控制
linear feedback system 线性反馈系统
linear figure 线性图形
linear filter 线性滤波器;线性过滤器
linear filtering 线性滤波
linear filter model 线性过滤模型
linear fitting 线性拟合
linear flame propagation 火焰线性传播
linear float guide 直线浮筒导轨
linear flow 线(性)流;线性进程;直线流动
linear flow characteristic 线性流量特性
linear flow rate 线性流速
linear flow structure 线状流动构造;流线构造;片板状流动构造;层流构造
linear fluidity 线性流动
linear focus 直线聚焦
linear focusing 直线聚焦
linear focus lens 线性聚焦透镜
linear fold 线形褶皱
linear foliation 线状叶理
linear foot 延英尺;纵(英)尺
linear form 线性形式
linear fractional function 线性分式函数
linear fractional group 线性分式群
linear fractional programming 线性分式规划
linear fractional transformation 线性分式变换;茂比乌斯变换
linear fraction transformation 线性分式变换
linear fracture density 裂缝线密度

linear fracture mechanics 线性断裂力学
linear-free-energy-related 线性自由能相关
linear free-energy relationship 线性自由能相关法;线性自由能关系
linear frequency modulation 线性调频
linear frequency response 线性频率响应
linear friction 线性摩擦
linear function 线性函数;线型函数;一次函数;直线函数
linear functional 线性泛函
linear functional differential equation 线性泛函微分方程
linear fundamental figure 线性基本图形
linear gain 线性增益
linear gain receiver 线性增益接收机
linear gas velocity 气体线速
linear gate 线性门
linear generator 线性发电机
linear genus 线性亏格
linear gradient 线性梯度
linear gradient junction 线性梯度结
linear grading 线性梯度
linear graduate scale 分划直尺
linear graduation 直线刻划
linear graph 线状图;线性图
linear-graphic(al) method 直线图解法
linear gravimeter 弦线重力仪
linear group 线性群
linear growth 线性增长
linear growth rate 线性生长速率
linear hachure 晕滃线
linear hard core pinch 直线硬芯箍缩
linear hash 线性散列
linear heat generation rate 线功率
linear heating power 线功率
linear heat rating 线功率
linear high polymer 线型高聚物
linear high-speed test bench 线性高速试验台
linear homogeneous equation 线性齐次方程;齐次一次方程
linear homogeneous function 线性均匀功能
linear homogeneous group 线性齐次群
linear homogeneous production function 线性均匀生产函数
linear homogeneous system of equations 线性齐次方程组
linear homopolymer 线性均聚物
linear hopper 直线型料斗
linear hydrocarbon 直链烃
linear hypothesis 线性假设
linear hypothesis model 线性假设模型
linear hypothesis testing 线性假设的检验
linear hysteretic 线性滞后
linear ideal chromatography 线性理想色谱
linear image pick-up device 线性摄像器件
linear imaging scanner system 线性成像扫描系统
linear imbedding 线性嵌入
linear impedance 线性阻抗
linear-impedance relay 线性阻抗继电器
linear impervious boundary 直线隔水边界
linear impulse 力的冲量
linear inch 线性英寸
linear increase voltage 线性上升电压
linear increasing load 线性增加荷载
linear independence 线性无关(性);线性不相关;线划无关
linear indication 线性显示
linear induction accelerator 直线感应加速器
linear induction motor 线性感应电动机
linear induction pump 线性感应泵
linear inductive coupling storage(store) 线性电感耦合存储器
linear inductosyn 线位移感应式传感器;直线式感应同步器;直线式感应传感器
linear inequalities 线性不等式组
linear inequality 线性不等式
linear inequality constraint 线性不等式约束
linear information processing language 线性信息处理语言
linear infrastructure 线性基础结构
linear injection system 线型注入系统
linear-input form strategy 线性输入形式策略
linear input tape 线性输入带
linear instantaneous value 线性瞬时值

linear integral equation 线性积分方程
linear integrated circuit 线性集成电路
linear interpolater 线性插补器
linear interpolation 线性内插(法);线性插值(法);直线性内插法;比例内插
linear interpolation method 线性内插法;线性插值法
linear intersection(method)边交会(测量)法;直线交会法;长度交会法;测变交会法
linear intrusion 线状浸入
linear inversion method 线性反演法
linear involution 线性对合
linear ion density 线离子密度;离子的线密度
linear ionization 线电离
linear irrotational flow 线性无旋流
linearisation 线性化
linear isometry 线性等距
linear isotherm 线性等温线
linearity 线性(度);直线性
linearity accelerator 直线性加速器
linearity check 线性比较
linearity checker 线性测试器;直线性检查仪
linearity circuit 线性化电路;线性电路
linearity coil 直线性线圈
linearity compensator 线性补偿器
linearity control 线性控制
linearity control circuit 线性控制电路
linearity controller 线性控制器
linearity correcting amplifier 线性校正放大器
linearity correction 线性校正
linearity correction circuit 线性校正电路;直线性校正电路
linearity curve 线性曲线
linearity dependence 线性相关
linearity distortion 线性失真
linearity error 线性误差
linearity factor 线性因数;线性系数
linearity independence 线性无关
linearity measuring wave form 直线性测量波形
linearity operator 线性算子;线性算符
linearity pattern 线性图案
linearity potentiometer 线性电位器
linearity range 线性段
linearity region 直线区域
linearity sector 直线区
linearity staircase signal 线性梯级信号
linearity test card 线性测试卡
linearity test generator 线性测试振荡器;直线性测试振荡器
linearity test signal 线性测试信号
linearizable curve 可化为直线的曲线
linearization 线性化;直线化
linearization concept 线性化概念
linearization of function 函数线性化
linearization of non-linear system 非线性系统的线性化
linearize 线化
linearized aerodynamics 线化空气动力学
linearized condition 线性条件
linearized equation 线性化方程
linearized extension of map 映射的线性化扩充
linearized flow 线性化流动
linearized function 线性化函数
linearized maximum likelihood estimator 线性化极大似然估计量
linearized method 线化法
linearized network 线性化网络
linearized observation equation 线性化观测方程
linearized polynomial 线性化多项式
linearized system 线性化系统
linearized theoretical models 线性理论模型
linearized theory 线性理论;线化理论
linearizer 线性化电路;线性化单元
linearizing resistance 直线化电阻
linear junction 线性结
linear lag 线性滞后
linear large scale integrated circuit 线性大规模集成电路
linear latitude 线纬度
linear lattice 线性格
linear law 线性定律;直线定律
linear law of seepage flow 线性渗透定律
linear leaf 长叶片
linear least square method 线性最小平方法

linear left-to-right scan 自左至右线扫描
linear length 线性长度
linear level 线性电平
linear light 长光;光束;光线;线(形)光;直线性光
linear lighting source 线光源
linear lightning 线状闪电
linear light source 线形光源;线开光源
linear light system 线性光学系统
linear line complex 线性线丛
linear line congruence 线性线汇
linear list 线性表
linear load 线(性)荷载;单位长度线荷载;单位长度负荷;线负载
linear load bearing system 线性荷载体系
linear load-carrying system 线性负荷体系
linear loaded system 线性负荷体系
linear loading 线性荷载
linear logarithmic amplifier 线性对数放大器
linear log quantizer 线性对数量化器
linear longitude 线经度
linear low density polyethylene 线性低密度聚乙烯
linearly compact subcategory 线性紧子范畴
linearly dependent coefficient 线性相关系数
linearly dependent vector 线性相关的向量
linearly elastic 线性弹性变形的
linearly equivalent divisor 线性等价因子
linearly graded impurity distribution 线性分级杂质分布
linearly graded junction 线性分级结
linearly independent 线性无关的
linearly independent elements 线性无关元
linearly independent family 线性无关族
linearly independent integral 线性无关积分
linearly independent quantities 线性无关量组
linearly independent set 线性无关集
linearly independent solutions 线性无关解
linearly independent vector 线性无关的向量
linearly ordered chain 线性有序链
linearly ordered module 线性有序模
linearly ordered set 线性有序集
linearly polarized light 线(式)偏振光;线偏光;平面偏振光
linearly polarized light beam 线偏振光束
linearly polarized light output 线性偏振光输出
linearly polarized mode 线性偏振极化模
linearly polarized output 线性极化输出
linearly polarized wave 线性极化波;线偏振波;直线偏振波
linearly semi ordered space 线性半序空间
linearly separable 线性可分离
linearly separable function 线性可分函数
linearly unstable motion 线性不稳定运动
linearly variable resistance 线性可变电阻
linearly varying load 线性变化荷载
linearly varying strain 线性受力变形
linearly varying stress field 线性变化应力场
linear macromolecule 线型大分子
linear magnetic anomaly 线性磁异常
linear magnetostriction factor 线性磁致伸缩系数
linear magnification 线性放大(倍数);单向放大(率)
linear magnification coefficient 线性放大系数
linear manifold 线性流形【数】
linear mapping 线性映射
linear material behavio(u)r 线性材料性能
linear matrix 线性矩阵
linear maximum likelihood method 线性极大似然法
linear mean diameter 线性平均直径;线平均径
linear measure 线性测度;直线量测;直线度量;尺度;长度(单位);长度计量
linear measurement 长度测量;线性测量
linear measuring 线测标
linear measuring assembly 线性测量装置
linear memory 线存储器
linear metal productivity 线金属量
linear meter 线性仪表;延米
linear method 直线法
linear metre of radiograph 射线照相的延米
linear metric space 线性度量空间
linear microcircuit 线性微电路
linear microdensitometer 线性微密度计
linear microsystem 线性微系统
linear migration 线性回游;直线迁移

linear migration of large earthquake 大地震直线迁移
linear minimization 线性极小化
linear minimization problem 线性极小化问题
linear mixed estimation method 线性混合估计法
linear mode 线性模式
linear model 线性型;线性模型
linear modulation 线性调制
linear moment 线性矩;线性力矩
linear momentum 线(性)动量;冲量
linear motion 线性运动;直线运动
linear motion actuator 直线电机
linear motion electric(al) drive 线性电气传动
linear motion slides 线性运动滑块
linear motion type seat 直线移动式座椅
linear motor 线性电动机;直线电(动)机
linear motor principle 直线驱动原理
linear motor table 直线电机绘图机
linear movement pick-up 线位移传感器
linear multiparameter system 线性多参数系统
linear multiple accelerator 直线倍增加速器
linear multiple regression model 线性多元回归模型
linear multistep method 线性多步法
linear multivariable control 线性多变量控制
linear multivariable system 线性多变量系统
linear net 线性网
linear network 线性网络
linear network method 线性网络法
linear network theory 线性网络理论
linear noise-cancelling microphone 线性抗噪声传声器
linear non-ideal chromatography 线性非理想色谱法
linear non-stationary geostatistics 线性非平稳地质统计学
linear normalization 线性归一化
linear normal strain 线性正应变
linear number 线性数
linear objective 线性目标
linear objective function 线性目标函数
linear oligomer 线形低聚物
linear opening 火焰管开孔
linear operation 线性运算
linear operational element 线性运算部件
linear operations on the sample data 样本数据线性运算
linear operator 线性运算器;线性算子;线性算符
linear optical system 线性光学系统
linear opties 线性光学
linear optimal stochastic system 线性最佳随机系统
linear optimal system 线性最佳系统
linear optimization 线性(最)优化
linear order 线性序;直线序列;全序
linear ore ratio 线含矿率
linear organization 线性结构
linear oscillation 线性振荡
linear oscillator 线性振子;线性振荡器
linear output 线性输出
linear output feedback 线性输出反馈
linear output horizon sensor 线性输出地平仪
linear oxidation rate 线性氧化速率
linear pantograph 线性缩放仪
linear parabolic equations 线性抛物型方程
linear paraffin 直链烷烃
linear parallax 线性视差
linear parallelism 线状排列
linear-parallel to dominant line array 平行线组合
linear parameter 线性参数
linear parametric amplifier 线性参数放大器
linear park 带形公园
linear partial differential equation 线性偏微分方程
linear partial differential operator 线性偏微分算子
linear partition isotherm 线性分配等温线
linear passive coupling network 线性无源耦合网络
linear path 直线形轨道;单线小路
linear-path screen 直线轨迹振动筛
linear path system 线性分支系统
linear pattern 直线型
linear pattern shooting 直线型爆破
linear pencil 线性束
linear permutation 线性排列

linear perspective 线性透视;直线透视(图)
linear perturbation 线性摄动;线性扰动
linear perturbation analysis 线性微扰分析
linear perturbation theory 小扰动法;线性扰动理论
linear pervious boundary 直线透水边界
linear phase 线性影像相值
linear phase comparator 线性相位比较器
linear phase distortion 线性相位失真
linear phase filter 线性相位滤波器
linear phase operation 线性相位运用;线性相位特性工作
linear phase shift 线性相移
linear phase sweep 线性相箝位
linear phase system 线性相位特性系统
linear phase system function 线性相位系统函数
linear phenomenon 线性现象
linear pinch 直线箍缩
linear pinch-machine 直线箍缩机
linear pit 线状纹孔
linear pitch 直线节距
linear planimeter 线性求积仪
linear planting 列植
linear plastic optical fiber 线性塑料光纤
linear plastic theory 线形塑性理论
linear point set 线性点集
linear-polar coordinate plot 线性极坐标图
linear polarization 线(性)极化;线偏振
linear polarization correlation measurement 线性极化关联测量
linear polarization method 线性极化法
linear polarized light 线性偏振光
linear polyamide 线型聚酰胺
linear polyester 线形聚酯
linear polyethylene 线型聚乙烯
linear polymer 线型聚合物
linear polynomial 线性多项式
linear potential 线性电位
linear potentiometer 线性电位计;线性电位器
linear power 线性额定功率
linear power amplifier 线性功率放大器
linear power rating 线功率
linear precedence function 线性优先函数
linear precision 测边精度
linear prediction 线性预测
linear prediction theory 线性预测理论
linear predictive coding 线性预测编码
linear predictor 线性预报值
linear pressing 直线压制
linear pressure 线压力
linear pressure loss 沿程压力损失
linear prestressing 长线预应力法;预应力直线强拉;线性预加应力(钢筋混凝土)
linear probability model 线性概率模型
linear process 线性过程;直线性过程
linear products 定尺商品钢材
linear program(me) 线性规划;线性程序
linear program(me) part 程序直线部分
linear programming 线性规划;线性程序设计;直线式程序
linear programming analysis 线形规划分析
linear programming deconvolution 线性规划反褶积
linear programming distribution method 线性规划分配法
linear programming method 线性规划法
linear programming model 线性规划模型
linear programming problem 线性规划问题
linear programming program(me) 线性规划程序
linear programming technique 线性规划法
linear programming theory 线性规划法
linear programming under uncertainty 不确定情况下的线性规划
linear progression of stresses 应力直线变化
linear propagation 直线传播
linear propulsion 线性推进
linear pseudometric space 线性伪度量空间
linear pulse 线性调频脉冲
linear pulse amplifier 线性脉冲放大器
linear pulse filter 线性脉冲滤波器
linear pulse motor 直线脉冲电动机;直线步进电动机
linear-quadratic-Gaussian problem 高斯线性二次问题
linear quantizer 线性脉冲调制器;线性量化器;线性分层器;线性变换器
linear radial sweep 线性径向扫描
linear radiation source 线状辐射源
linear radiative stopping power 线辐射阻止本领
linear radiator 线形辐射器
linear radio-frequency mass spectrometer 直线射频质谱仪
linear ramp 线性斜坡电压
linear range 线性范围
linear rank 线性秩
linear rate 线速率
linear rate constant 线性速率常数
linear rate law 线性速率定律
linear rate of flow 直线流速
linear rating 线功率
linear ratio 线性比;直线比
line array 直线阵
linear ray 单列射线
line-array sensor 线列传感器
linear reactance 线性电流
linear reactor 线性电抗器
linear reciprocating sweep 线性往复扫描
linear recombination 线性复合
linear rectification 线性检波
linear rectifier 线性检波器
linear recurrence 线性递归
linear recurrence relations 线性递推关系
linear recurrent sequence 线性循环序列
linear recurring sequence 线性递归序列
linear reduction of tariffs 线性关税减让法
linear reef pool trend 线状礁油气藏趋向带
linear regeneration sampling 线状更新抽样法
linear regenerative extraction 直线再生引出
linear regression 线性回归;直线迴归;直线回归
linear regression analysis 线性回归分析
linear regression equation 直线回归方程;直接回归方程
linear regression equation of bi-variable 二元直线回归方程
linear regression equation of one variable 一元线性回归方程
linear regression estimate 线性回归估计值
linear regression method 线性回归法
linear regression model 线性回归模型
linear regression technique 线性回归技术
linear regulation control law 线性管理法规
linear regulator 线性调节器
linear regulator problem 线性调节器问题
linear rehash method 线性再散列法
linear relation(ship) 线性关系;直线关系
linear reluctance motor 直线式磁阻电动机
linear repayment 线性偿付
linear repeater 线性转发器
linear representation 线性表示
linear reservoir 线性水库
linear reservoir model 线性水库模型
linear reservoir system 线性水库系统;线性水库群
linear residuum 线状风化壳
linear resistance 线性电阻
linear-resistance flowmeter 线性电阻流量计
linear resistive element 线性电阻元件
linear resistor 线性电阻器;线性电阻
linear resolution 线形分解;行分解
linear resolver 线性旋转变压器
linear resolving power 线分辨率
linear resonance 线性共振
linear resonant system 线性共振系统
linear resonator accelerator 线性谐振加速器
linear response 线性响应;线性反应
linear responsibility chart 线性责任图
linear restriction 线性约束
linear rigidity 线性刚度
linear rocker bearing 线性摇臂轴承
linear rotational flow 线性旋转流
linear rotational spring 线性旋转弹簧
linear ruler 纵尺
linear sales organization 直线式销售组织
linear salvation energy relationship 线性溶剂化能相关法
linear sample bias 线性试样偏差
linear sample plot 线上标(准)地;线形样方
linear saw-tooth wave 线性锯齿波
linear scale 线性刻度;线性标度;线性尺;线性比例尺;直线比例尺;长度比例(尺)
linear scale factor 线性比例系数
linear scale ohmmeter 线性刻度欧姆表
linear scale ratio 线性比尺
linear scan(ning) 线性扫描;直线扫描
linear scanning speed 线扫描速度
linear scan voltammetry 线性扫描伏安法
linear scattering 线性散射
linear schistosity 线状片理
linear sealing 线性密封;线密封
linear search 线性查找;一维搜索
linear seasonal variation 线性变化型季节变动
linear segment 线性部分
linear selection 线性选择
linear selection storage 线选存储器
linear selection switch 线选开关
linear selection system 线选法
linear semilogrithmic trend line 线性半对数趋势线
linear sensitivity measures 线性敏感度测量法
linear separable function 线性可分函数
linear sequence circuit 线性时序电路
linear sequential machine 线性序列机
linear servo actuator 线性伺服执行机构
linear servomechanism 线性伺服机构
linear servosystem dynamics 线性伺服系统动力学
linear settlement 带状聚落
linear settling rate 线性沉降速率
linear shape function 线性形状函数
linear shear apparatus 直线剪切装置
linear shell theory 线性薄壳理论
linear shift register 线性移位寄存器
linear ship loader 直线装船机
linear shrinkage 线性收缩;线收缩;纵向收缩率
linear shrinkage rate 线缩率
linear signal 线性信号
linear simple group 线性单群
linear simple lattice 线性简单点阵
linear simultaneous equation 线性联立方程
linear single-equation model 线性单一方程式模式
linear single-loop control system 线性单回路控制系统
linear size 线性尺寸
linear slip surface 直线型滑动面
linear slit 线缝
linear slope 直线型边坡
linear-slope delay filter 线性斜率延迟滤波器
linear-slot array 直线缝隙天线阵
linear smoothing 线性修匀;线性校平;线性平滑;线性光滑
linear smoothing algorithm 线性平滑算法
linear smoothing with five point 五点线性平滑
linear smoothing with three-point 三点线性平滑
linear sound 线性声源
linear source 线源
linear source of water pollution 水污染线源
linear space 线性空间
linear space-invariant optical system 线性空间不变光学系统
linear space truss 线性空间桁架
linear spanning 线性生成
linear spark counter 直线火花计数管
linear spatial filter 线性空间滤波器
linear specific power 线性比功率
linear spectral model 线性谱模型
linear spectral space 线性谱空间
linear spectrum 线性谱
linear speed 线(性)速度;线速率
linear speed-concentration model 线性速度—密度模型
linear speed-density model 线性车速-密度模型
linear speed method 线性速度测量法;直线速度测算法
linear split system 线性分流系统;线性分离系统
linear spring 线性弹簧
linear stability analysis 线性稳定性分析
linear staircase function generator 线性阶式函数发生器
linear Stark effect 线性斯塔克效应
linear state of stress 应力的线性状态
linear stationary geostatistics 线性平稳地质统计学
linear statistical sampling model 线性统计抽样模型
linear stator 直线定子
linear stator machine 直线定子电机
linear step motor 直线步进电动机

linear stepping motor 直线步进电动机
linear stiffness 线性刚度
linear stochastic hypothesis 线性随机假设
linear stochastic regression model 线性随机回归模型
linear stochastic system 线性随机系统
linear stopping power 线性阻止能力；阻止能力
linear store 线存储器
linear strain 线应变；线性应变
linear strain diagram 线性应变图
linear strain rate 线应变率
linear strain ratio 线应变率
linear strain seismograph 线性应变地震仪
linear strategy rule 线性策略规则
linear stratification 线状层理
linear stream 线性流
linear stress 线应力；线性应力
linear stress distribution 应力的线性分布
linear stress field 线性应力场
linear stressing 加线性应力；线性应力
linear stretching 线性伸长；线性扩展
linear structural equation 线性结构方程
linear structural equation system 线性结构方程组
linear structure 线性结构；直线型结构；直线结构
linear structure system 直线型结构体系
linear style of ornamentation 直线形式装饰
linear subgraph 线性子图
linear submanifold 线性子流形
linear subspace 线性子空间
linear substitution 线性代换
linear superpolymer 线型高聚物
linear superposition 线性叠加；线型叠加
linear supply function 线性供应承数
linear supporting system 线形支承体系
linear survey 线状调查
linear sweep 线性扫描；直线扫描
linear sweep delay circuit 线性扫描延迟电路
linear sweep generator 直线扫描发生器；线性扫描振荡器
linear sweep rate 线性扫描率
linear sweep voltammetry 线性扫描伏安法
linear switch 线性开关
linear switching 线性开关；断路
linear synchronous machine 线性同步电机
linear synchronous motor 线性同步电机
linear system 线性系(统)；线性体系
linear system analysis 线性系统分析
linear systematic statistic 线性系统统计量
linear systematic statistics function 线性系统统计函数
linear system of algebra 线性代数系
linear system of complex 线性线丛系
linear system of curves 线性曲线系
linear system of equations 线性方程组
linear system of point groups 线性点集系
linear system of surfaces 线性曲面系
linear system theory 线性系统理论
linear system with constant coefficient 常系数线性系统
linear system with variable coefficients 变系数线性系统
linear taper 线性电阻分布特性
linear target 线状目标
linear temperature gradient 线性温度梯度
linear tempering 线性回火
linear temporal filtering 线性瞬时滤波
linear tensioning 直线张拉
linear tensor 线性张量
linear term 线性项
linear test 线性检验
linear tetrad 单列四分体
linear theory 线性理论
linear theory method 线性理论法
linear thermal expansibility 线性热膨胀率
linear thermal expansion 线性热膨胀
linear thermal resistance 线热阻
linear thickness 线性厚度
linear three-dimensional technique 线性三维技术
linear thrust misalignment 推力向量线性位移
linear time algorithm 线性时间算法
linear time base 线性时基；直线时基
linear time-base oscillator 线性时基振荡器
linear time base sweep 线性扫描
linear time delay circuit 线性延时电路

linear time invariant filter 线性非时变滤波器
linear time-invariant system 线性时间不变系统；常参数线性系统
linear time-quantized control system 线性时间整量化系统
linear time-variant channel 线性时变信道
linear time-varying network 线性时变参数网络
linear titration 线性滴定
linear-to-log converter 线性对数变换器
linear total shrinkage 直线总收缩
linear town 线性延伸的城市；带形城市；带形城镇
linear trace 线性扫描
linear tracking 线性跟踪
linear track load 每延米轨道压力
linear transducer 线性换能器
linear transfer 直线式自动线
linear transform 线性变换式
linear transformation 线性变换；线性位移；一次变换
linear transformer 线性变压器
linear transient analysis 线性瞬态分析
linear translation 线性位移
linear traverse microscope 直线移侧显微镜
linear traverse technique (测定混凝土空隙特征等的)直线移测技术
linear trend 线性趋势；线性倾向
linear triangulation chain 线形锁；线形三角链(锁)
linear triangulation network 线形三角网
linear truss 线形桁架
linear tube 线性摄像管
linear turbine 线性涡轮机
linear two-dimensional structure 线性二维结构
linear type 线性型；直线型；直线式
linear-type concrete hinge-bearing 线式混凝土铰支座
linear unbiased estimation 线性无偏估计
linear unequality 线性不等式
linear unit 线性装置；线性器件
linear unit hydrograph theory 线性单位过程线理论
linear unseparable function 线性不可分函数
linear variable 线性变量
linear variable differential transformer 线性(变换)差动变压器；直线位移差动变压器
Linear variable displacement transducers 线性差动传感器
linear variable resistor 线性可变电阻
linear variation 线性变化
linear variation load 线性变化荷载
linear variation parameter system 线性变参数系统
linear vector function 线性向量函数；线性矢量函数；一次矢函数
linear vector space 线性向量空间；线性矢量空间
linear vehicle capacity 线性车辆容量
linear velocity 线(性)速度；直线速度
linear vernier 线性游标
linear-vertical to dominant line array 垂直线性组合
linear vibration 线性振动
linear vibrations of elastic bodies 弹性体的线性振动
linear viscoelastic behaviour 线性黏弹特性
linear viscoelasticity 线性黏弹性；线形黏弹性
linear viscoelastic material 线形黏弹性材料
linear viscoelastic medium 线形黏弹性介质
linear viscoelastic micropolar material 线性黏弹微极性材料
linear viscous damping 线性黏性阻尼
linear viscous damping coefficient 线性黏性阻尼系数
linear viscous dashpot 线性黏滞阻尼器
linear voltage 线性电压
linear voltage differential transformer 线性电压差分传感器
linear voltage displacement transducer 线性电压位移传感器
linear voltage wave form 线性电压波形
linear water body 线状水体
linear waters 线状水体
linear wave 线性波；直线波
linear waveform 线性波形
linear waveform distortion 线性波形失真
linear waveform response 线性波形响应
linear waveguide accelerator 线性波导加速器
linear wave theory 线性波理论

linear weight-carrying system 线形载重系统
linear well rows 直线井排
linear whole body scanner 线性全身扫描机
linear width of peak at high-speed 高速度处峰的线宽
linear width of peak at low-speed 低速度处峰的线宽
linear wind-wave interaction 线性风浪相互作用
linear Z-pinch 直线Z箍缩
linea semi-lunaris 半月线
line astern 纵队
line asymmetry 谱线不对称性
lineation 线状构造；线形结构；线条；勾画轮廓；画线
lineational orientation 区域构造线方向
lineation structure 线状构造；线理构造
linea trapezoidea 斜方线
line attachment 接入线路
line attendant 巡线员
line attenuation 线路衰减
line balance 线路平衡度；线路平衡
line balance converter 线路平衡变换器；跨越设备
line balancing 线路平衡；人工投入线性平衡法；生产线平衡
line bank 接线排
line bank contact 线弧接头；触排接头
line-bar resaw 旋臂锯；悬臂再锯机
line battery 线路电池
line bearing 导向轴承；线支承
line bend 图像行畸变
line bias 线路偏置
line blanketing 谱线覆盖
line blanketing index 谱线覆盖指数
line blend 配料混合试验法
line-blending 在管道中掺和；管道混合
line blind 挡风闸板
line block 线条凸版；线路码组
line blocking 谱线覆盖
line blow 直线性强风；强风(六级风)
line boat 系缆工作船；送缆船；带缆船
line booster pump 管线增压泵；管道补压
line borer 直线镗床
line boring 直线镗孔；直线式布钻孔；共轴孔
line boring bar 同轴多孔镗杆
line boring machine 钻轴机
line boss 扑火总指挥；生产指挥人员
line brattice 纵向隔风墙
line break 小检修；输送管道断裂
line breaker 线路开关；断路器；线路断路器
line break signal 断线信号
line bridge 跨线天桥
line broadening 线展宽；谱线变宽
line broadening by turbulence 湍流致宽
line broadening method 线宽化
line buffer 线路缓冲器
line building 直列式建筑物；行列式建筑
line building-out network 线路附加网络
line bundle 线丛
line burner 塔顶拔杆【岩】；棒状燃烧器；管状燃烧器
line-busy tone 线路忙音
line-by-line clamping 逐行箝位
line-by-line horizontal read out 逐行水平读出
line-by-line mapping 逐行扫描成像
line-by-line scan 逐行扫描
line-by-line synchronization 逐行同步
line-by-pass device 旁路式装置
line-by-pass system 旁路式系统
line camp 火险营部
line capacitance 线路电容
line capacitor 线路电容器
line capacitor voltage transducer 线路电容器电压变换器
line capacity 线路载客量；线路容量；线路能力；作业线能力
line carrying capacity 线路通过能力
line carrying heavy traffic 繁忙线路
line carrying light traffic 清淡运输线(路)
line casting machine 整行铸排机；铸条机
line center 谱线中心
line centering 对行
line char 线路图
line characteristic 线路特性
line characteristic distortion 线路特性失真
line charge 线电荷
line charge model 线荷模型；磁荷模型

line charge voltage 线路充电电压
line charging capacity 线路充电容量
line charging current 线路无载电流
line charging power 线路充电容量
line chart 线图；单线图
line check 小检修；输送管线断裂
line checking 线形细裂
line check valve 管路单向阀；管道止回阀
line choke 线路节流阀
line choking coil 线路扼流圈
line-chromatic number 线色数
line circuit 有线电路；有线电话；外线电路；用户线电路；天线电路
line circuit-breaker 线路断路器；线路断流器
line classification 线路分类；线路等级；铁路等级
line cleaning 使线圈光滑
line clear 线路开通；线路出清
line clearance with water 用水隔油
line clear procedure 线路出清程序
line clinometer 线性测斜仪
line clipping 线段裁剪
line clipping algorithm 线段裁剪算法
line clogging 管路阻塞
line closed down 封闭的线路；封闭的铁路线
line closed to traffic 停止营业线路
line closure 线路封锁
line cloud 线状云
line code 行代码；线路码；一行代码
line code pattern 线路码型
line coiler 钓线架
line-colo(u)r 线条彩色印件
line colo(u)ring 线着色
line communication 有线通信
line commutated inverter 线换向变流器
line complex 丝丛
line composition 对行
line compression 单位线压力(压路机)
line computation 线计算
line concentrating unit 线路集中单元
line concentrator 线路集中器；用户集线网；集线装置；集线器
line conditioning 线路调节
line conductance 线电导
line conductor 线路导线；线路导体；导线【电】
line-cone 线圆锥
line configuration 网络配置
line congruence 线汇
line-connected graph 线连通图
line connection equipment 线路连接设备
line-connectivity 线连通度；边连通度
line connector 线路连接器；接线器
line constant 线路常数；输电线路常数
line construction 线路架设
line contact 线接触
line contact bearing 线接触轴承
line contactor 线路接触器
line contact pressure 线接触压力
line contour 谱线轮廓
line control 线路控制；线控
line control block 线路控制分程序
line control computer 线路控制计算机
line control routine 线路控制程序
line conversion 连续转换线条
line coordinates 线坐标
line coordinates equation 线标方程
line copy 线划原图
line copy plate 线划板【测】
line cord 线路电缆；电源软线
line-cord resistor 电源软线电阻器
line corona 线状电晕；线路电晕
line coupling 线耦合
line coupling transmission system 线间耦合传播方式
line covering 线覆盖
line covering number 线覆盖数；边覆盖数
line crawl buffer 爬行缓冲器
line crawl cancel 爬行清除
line crawl inverter 爬行倒相器
line crew 线路工程队
line cross 穿线法
line cross point coordinates 测线交点坐标
line cross point stake number 测线交点桩号
line current 线路电流；线电流
line current tester 线路电流计

line current transducer 线路电流变换器
line cut 线条凸版；绳割；线雕(刻)
line cutting 浅刻技术
line cutting technique 线刻技术
line-cutting wastewater 线切割废水
lined 有护板的；有衬里的；衬砌的；衬里的
lined and coated with bitumen 衬砌和涂沥青
lined and cotton mixture 棉麻混合织物
lined bearing 轴承衬里金属
lined bend 消声弯头
lined borehole 有套管钻孔；衬壁钻孔
lined canal 有衬砌的渠道；衬砌渠道；衬砌明沟；砌面明沟
lined crucible 涂层坩埚；衬里坩埚
lined ditch 有衬砌沟渠；有衬层的明沟
lined duct 加衬管；带衬里的风道；衬砌暗渠
line defect 线状缺陷；线(形)缺陷；位错
line definition 行清晰度；行分辨度
line delay 导线延迟
lined elbow 消声弯头
line delete command 删行命令
line delete symbol 删行符(号)
line deletion character 删行字符
line density 线密度
line depot 线路上油库
line detection 线条检测
line development 线路改进；展线
lined excavation 有支护的开挖
lined formwork 平直模板
line diagram 线形示意图；线路图；单线图；构件中心线的图
line differential protection 纵差保护
line digraph 线有向图
line disconnecting switch 线路隔离开关
line disconnection 切断线路
line-disjoint paths 线不相交道路
line dislocation 线位错
line displacement 谱线位移
line display range 行显示区段
line distance 线距
line distortion 线路失真；线路畸变
line division 线分划
line down 回线故障
line drawing 线路图；线条画；线描；线拉伸；线画图；轮廓线图；两维轮廓图；画线；三维轮廓图
line drawing brush 画线笔刷
line drawing display 绘线图形显示；线形图显示
line drawing machine 画线机
line drawing processing 线图处理
line drawn in ink 线画清绘
line drilling 线性钻进；沿直线钻孔；沿同一勘探线上钻孔钻进；直线钻孔
lined ring 衬环
line driver 线路激励器；长线驱动器
line driving amplifier 线路激励放大器
line drop 线路(电)压降；线终表示器
line drop compensation 线路电压降补偿
line drop compensator 线路压降补偿器
line drop contour chart 根据连续线勾绘的等高线图
line dropped control 断续控制
line dropped controller 断续控制器
line drop signal 线路吊牌信号
line drop voltmeter compensator 线电压降电压表补偿器
line drum 绳索卷筒
lined sight glass 带衬里视镜
lined sight glass with nozzle 带颈衬里视镜
lined tunnel 衬砌隧洞
lined vessel 衬里容器
lined with aluminium foil in the wood(en) case 木箱内衬铝箔纸
lined with glue 衬胶蝶阀
line editor 行编辑程序
line effect 线状花纹
line element 线(元)素；线路单元；线画要素
line element method 线元法
line element vector 线索向量；线素向量
line emission 线发射
line emission cloud 谱线发射云
line end 线端
line ending 线端端正
line-end station 线路终点站
line engineer 线路工程师；管线工程师
line engraving 线划刻图；线雕(刻)

line equalizer 线路均衡器
line equipment 线路设备；管线设备
line error 线路差错；分划误差
line etching 线型蚀刻；线形蚀刻；线条刻蚀
line execution control block 线路执行控制功能块
line executive 业务管理人员；幕前执行主管
line expansion 线性膨胀
line expansion coefficient 线膨胀系数
line facility 线路设备
line fail detection set 线路故障探测装置
line failure 线路故障
line fault 线路故障
line fault anticipator 线路故障探测装置
line feature 线划要素
line-fed motor 直接馈电电动机
line feed 线路馈电；电路馈电
line feed character 换行字符
line feed code 换行码
line feeder 线路馈电表；加强线【电】
line feed key 换行键
line fetch 线风区
line figure 线画图形
line fill 线路占用率；管线内物品
line filler 线画填料
line filter 线路滤波器
line filter balance 线路滤波器平衡
line finder 寻线器；寻线机；线寻找器；找线器；行定位器
line finder circuit 寻线机电路
line-finder shelf 寻线器架
line-finder switch 寻线器开关
line-firing 线状烧焰
line fishing 绳钩钓法
line-fit 线段拟合
line-fitting 线段拟合
line fittings 线路配件
line flow 线路流通量
line flow equation 线路潮流方程式
line flux 线路通量
line flying 测线飞行
line-focus 线状焦点
line-follower 线划跟踪器
line-following 线划跟踪；线跟踪
line-following device 线划跟踪装置
line font 线形
line for acids 耐酸管
line force 缆索拉力；缆拉力
line for locomotive to shed 入段线【铁】
line for locomotive to station yard 出段线【铁】
line form 线状
line formation 谱线形成
line freeing circuit 闭塞解除电路
line frequency 线路频率；行扫描频率；行频
line-frequency coreless induction furnace 无芯工频感应电炉
line-frequency furnace 工频炉；工频感应电炉
line-frequency induction 工频感应
line function 生产线职能
line gale 二分点风暴
line gamma ray survey 路线伽马测量
line gap 线隙；线路避雷器
line generation 向量产生；线划综合
line generator 线产生器；直线产生器
line geometry 线素几何学
line graduation 分划线
line graph 线图；直线图
line guy 线路拉线
line-halftone combination 线条网目混合版
line handling 缆索操纵(靠码头用)；撇缆带缆
line hanger 尾管悬挂器
line hardware 线路器具
line haul 沿线路装运；长途运输
line haul cost for transport 运输的路途费用
line-haul trucking 货车长途运输
lineheader 自由断面隧道掘进机
line heat detector 线路感温器
line heating 带钢加热
line heat source 线性热源
line height 行高
line hit 线路干扰；线路扰扰
line hold 行同步
line-hole drilling 沿巷道周边钻进
line hook 系链钩；系船钩
line hydrophone 线列水听器

line identification 谱线证认;管道标记;管线标记
line idle 线路空闲
line image 线画图形;线画图像
line impedance 线路阻抗;传输阻抗
line in 排列成行;对正直线
line inclusion 线状夹渣;线状夹杂物;链状夹杂物
line independence number 线独立数;边独立数
line-indices 反射线指数
line inductance 线路电感
line influence 线路影响
line inlet 线路入口
line input 线路输入
line insert command 行插入命令
line insertion 行插入
line inspection 巡线;线路检查
line inspection car 线路测验车
line installed in plaster 抹灰内装暗线
line installed on insulators 装在绝缘子上的电缆
line installed on plaster 安装在抹灰面上的电缆
line insulation 线路绝缘
line insulator 线路绝缘子;管线安装工
line integral 线积分
line intensity 线压力;谱线强度
line intensity ratio 线强度比
line-intercept 线截
line-intercept method 样线法
line interface 线路接口
line interface controller 线路接口控制器
line interface hardware 线路接口硬件
line interface unit 线路接口部件
line interlace 隔行扫描
line interrupter 线条断续器
line interruption 线路中断
line inversion 线反演
line involution 线束对合
line isolation relay 线路隔离继电器
line item 限额项目;行式项目;排列项
line item appropriation 分项拨款;分项经费
line item budget 分项预算;明细支出预算
line-item cost 分项成本
line iteration 线迭代
line iteration method 线迭代法
line jack 线路塞孔
line label 行标号
line lapped 零遮盖
line lattice 直线晶格
line layout 测线布置
line length 线长度;测线长度
line length between higher-class points 高级点间线长
line length between junction points 结点间线长
line lengthener 线路加长器
line level 线水准仪;气泡水准器;测定水平高程(房屋基础施工);行电平;线路电平;沿传输线电平;拉线找平
line level(l)ing 线水准测量
line light 聚光灯
line lightning arrestor 线路避雷针
line lightning performance 线路雷电特性
line light source 线光源
line like a bulb 灯泡线
line link 线路环节
line-link controller 线链控制器
line link control station 线路链路控制站
line-link frame 线路塞绳架
line link sub-control station 线路链路控制分站
linellae 线条
line load(ing) 线荷重;线荷载
line load intensity 线路负荷强度
line location 定线;线位;线路选位;线路定位
line lock 行锁定
linelock playback mode 行锁定重放方式
line log amplifier 线性对数(特性)放大器
line log receiver 线性对数接收机
line longitudinal differential protection device 线路纵向差动保护装置
line longitudinal profile 线路纵断面(图)
line loop 线路回路;线路环路
line loop resistance 环线电阻;环路电阻
line loss 线损;线路损失;线路损耗;管路损失;输气损耗
line lubricator 管路给油器
line maintenance 线路维修
line maker 画线员;画线机;画线工

line man 线路工;铁路养路工人;线路检修工;测线员
line manhole 管线检查井
lineman's detector 携带式检电器
line map 线路图;路线图
line mapping 制作线画图
line mark 行标记
line marker 画标线机;路面画线机
line marker connector 线路识别连接器
line marking 直线标志
line marking of expropriated land 征地红线
line marking paint 路面画线漆;画线漆
line material 线路器材
line measurement 谱线测量
linemen's pliers 电工钳
line meter 电源电压表
line microphone 线列传声器
line milling 直线铣削
line misclosure 测线闭合差
line mixer 流动混合器
line modulator 线调制器
line monitor 线路监视器
line monitoring equipment 线路监控设备
line monitoring tube 行频信号监视管
line-mounted valve 管式连接阀
linen 亚麻布
line name 测线名
linen-backed 裱糊的
linen-backed map 裱布地图
linen-backed paper 裱布纸
linen bakelite 亚麻电木
linen bank 亚麻布纹高级书写纸
linen button 布钮扣
linen canvas 亚麻帆布
linen checks 亚麻条格布
linen chute 待洗床单运送槽
linen closet 被服室
linen cloth 亚麻织物;亚麻布
linen cloth grinding wheel 亚麻布碾磨轮
linen crash 亚麻粗布
linen cupboard 待洗衣物柜;床单橱
linen drapery 亚麻布帐帘
linen duck 亚麻帆布
line negative 刻图阴片
line net 线网
line net adaptability 线网适应性
line net length 线网长度
line net planning 线网规划
line net total length 线网总长(度)
line network 线路网;管道网
line network execution planning 线网实施规划
line network execution program(me) 线网实施规划
line network filter 线网滤波器
line network management system 线网管理系统
line network passenger volume 线网客流
line network plan 线网方案
line network scheme 线网方案
line network scale 线网规模
line network structure 线网结构
linen fiber 亚麻纤维
linen finish 亚麻布纹饰面;麻布纹整饰
linenfold 折叠亚麻布饰面镶板;作对称的布褶或布卷花饰的雕刻版
linen-fold panel 仿折叠亚麻布饰面镶板;褶布饰镶板
linen goods 亚麻布料
linen hose 亚麻布水龙带
linen industry 亚麻工业
linen locker 被服库
linen look 麻型
linen mesh 亚麻网眼布
line noise 线路噪声;线路杂音
line non-linear coefficient 线路非直线系数
line non-linear factor 线路曲折因数
line not operated 未运营线路;未营业的铁路线
linen paper 亚麻布纸;布纹纸
linen pot 亚麻联匹
linen rags 亚麻破布
linen reed count 亚麻织机筘号
linen room 被服室;洗床单房(旅馆);床单储存室
linen rope 亚麻绳
linen roughs 帆布型亚麻布
linen rug 亚麻地毯底布
linen sheeting 亚麻被单布
linen store 被服库

linen-tape 布卷尺;亚麻带;麻布卷尺;亚麻布卷尺
linen thread 麻线
line number 行数;线数;行号;行编号;测线号
line numbered file 行编号文件
line number of the working area 工区测线数
line numbers in one circuit 每周圆孔行数
linen yarn 亚麻纱
line objective 业务部门目标
line occupation for works 施工封闭线路
line of action 作用线
line of action of force 力的作用线
line of aim 瞄准线
line of alignment 线路走向
line of application 作用线;力的作用线
line of approach 方针
line of apsides 拱线长度;拱线
line of arch 拱线
line of arch center 拱中心线
line of arch pressure 拱压力线
line of auditorium rake 观众厅坡度线
line of balance 平衡作业线;平衡线(图)
line of balance technique 平衡路线法
line of bearing 走向线;方向线;方位线
line of beauty 曲线美
line of best fit 最小二乘法线;最佳拟合线
line of breakers 碎浪线
line of buckets 料斗行(列);吊斗绳索
line of building 房基(线);建筑界线
line-of-business reporting 分部财务报告
line of capacity 线路容量
line of cells 电解槽系列
line of centers[centers] 拱的中心轨迹线;中心轨迹线;中心联线;中心连线
line of chord 弦长比例尺
line of cleavage 开裂线;解理线;破坏线;劈裂线;劈理线
line of collimation 对准线;准直线;透视线;视准轴线;视直线
line of columns 柱列
line of connection 连接线
line of constant bearing 等方位线
line of constant dip 恒定倾斜线
line of constant scale 等比线
line of contact 接触线;啮合线(齿轮传动);切线
line of corresponding discharge 流量相关曲线
line of corresponding stage 水位相关曲线
line of credit 信用线;信用额度;信贷限额;贷款限额;信贷额度
line of creep 蠕流线;蠕动线;蠕变线;爬径;塑流线;渗流线
line of cultivation 耕作线
line of curvature 曲率线
line of cut 切割线
line of declination 赤纬圈
line of deflection 挠曲线
line of demarcation 限制线;分界线;边界线
line of departure 纵坐标轴;抛落线;搜寻起始线;出发线【铁】;发射线
line of descent 系谱
line of dip 斜向线;倾斜线;倾向线
line of discontinuity 间断线;不连续线
line of discount 贴现额度
line of divergence 辐散线
line of electric(al) force 电力线(路)
line of elevation 标高线;高程线
line of energy head 能高线
line of engagement 啮合线(齿轮传动)
line of equal density 等密度线
line of equal distortion 等变形线
line of equal parallax 等视差线
line of equal pressure 等压线
line of equal principal stress 等主应力线
line of equal scale 等比线
line of equal sediment concentration 等沉积物浓度线
line of equal settlement 等沉陷线;等沉降线
line of equal shear 等剪力线
line of equal subsidence 等沉陷线
line of equal temperature 等温线
line of equal variations 等磁偏线
line of equal wave height 等波高线
line of equidistance 等距(离)线
line of excavation 开挖线
line of excavation construction 基坑开挖线

line of existing ground 原地面线
line off 用支划(分)开
line of fall 落地线
line of fastest flow 最大流速线;中泓线
line of fastest velocity of flow 最大流速线
line of feedback coupling 回授耦合线
line of firn 永久积雪线;万年雪线
line of flo(a)tation 吃水线;浮力曲线
line of flow 流(迹)线;迹线
line of flux 通量线
line of force 力线
line of fracture 破裂线;断裂线
line of fusion 熔合线
line of gravity 重力线
line of greatest slope 最大坡度线
line of holes 穿孔线
line of horizontal stress 水平应力线;横向受力线
line of hyper-fine structure 超精细结构谱线
line of impact 击中线
line of influence 影响线;感应线
line of inside framing 肋骨内缘线
line of intensity 强度线
line of intersection 交汇线;交会线;交叉线
line of least pressure 最小压力线
line of least resistance 最小阻力线;最小阻抗线;最小抗滑线;最小抵抗线
line of least squares 最小二乘方线
line of level(l)ing 水准路线
line of levels 水准线;标高线
line of loading 荷载线
line of longitude 经度线
line of magnetic force 磁力线
line of magnetic induction 磁感应线
line of magnetization 磁化线
line of main curvature 主曲率线
line of maximum depth 最大水深线;水流动力轴线;深水线;深泓线
line of maximum pressure 最大压力线
line of maximum velocity 最大流速线
line of max stream 最大流速线
line of movement data 迁移基准线
line of no distortion 无畸变线
line of normal curvature 法向曲率线
line of nosing 楼梯级突出线
line of no variations 零磁偏线;零磁差线
line of ornament 线形装饰
line of outcrop 露头线
line of percolation 透水线;渗流线
line of phase transformation 相转换线
line of piles 桩位线;排桩
line of pipes 输送管线
line of policy 方针
line of position 位置线;船位线
line of position run 转移船位线
line of posts 柱位线
line of pressure 压力线;啮合线(齿轮传动)
line of principal curvature 主曲率线
line of principal stress 主应力线;主应力迹线
line of production 流水生产线;生产线
line of projection 投影线
line of quickest descent 最快速下降线
line of radius vector 矢径线
line of rail 铁路;轨道
line of raised water level 壅高水位线
line of reference 基准线;参考线
line of regression 归算线;脊线;回归线
line of resistance 阻力线;抗力线
line of resultants of reservoir empty 水库空库时合力线
line of resultants of reservoir full 水库满库时合力线
line of return 返回线路
line of right ascension 赤经圈
line of right-of-way 道路红线
line of roll forming 辊锻线
line of rupture 破裂线;破坏线
line of same dip 等倾(斜)线
line of saturation 浸润线;饱和线;饱和水位(潜水面)
line of section 截交线
line of seepage 浸润线;潜水面(指地下水);渗漏线;渗流线
line of separation 分离线
line of setback 避车线

line of shade 阴线
line of shadow 影线;阴影线
line of shear(ing) stress 剪应力线
line of shortest length 最短线
line of sight 瞄准线;视准线;视线;视距通信系统
line-of-sight component 视向分量
line-of-sight course 目视航向
line-of-sight coverage 直视可达范围
line-of-sight geometry 瞄准几何学
line-of-sight link 视线链路
line-of-sight propagation 视线传播
line-of-sight velocity 视向速度
line of singularity 奇异线
line of slide 滑动线;坍方线;滑裂线
line of sliding 滑移线;滑动线;坍方线;滑裂线
line of slip 滑移线
line of slope 坡线
line of solidification 凝固线
line of sounding 探深线;测深线
line of space 空间轮廓
line of spectrum 光谱线
line of springs 泉群分析
line of stratification 层线;分层线【地】
line of strike 走向线;走向【地】
line of support 支承线
line of symmetry 对称线
line of thrust 推力线
line of tidal wave 潮波线
line of total head 总压头线;总水头线
line of traffic 交通线路
line of transference 转移线
line of travel 楼梯行走线(离扶手48厘米处);行走线;轨迹线
line of trend 倾向线
line of turning tool travel 车刀移动线
line of vault 拱顶线
line of vehicles 车辆线
line of vision 视向
line of water area 水涯线
line of water-level decline 水位下降线
line of water probing 探水线
line of wave 波线
line of weakness 最小强度线;最小抵抗线;弱线;脆质线
line of wells 油井群;油井线
line of zero moment 零矩线
line of engagement 接触线
lineograph 描线规;画线器;画线规
lineoid 超平面
line oiler 自动注油器;管路注油器;管路给油器
line open to traffic 开始运营的铁路线;投入运营线路
line operated to traffic 运营线路
line organization 分级式组织
line original 线画原图
line or retirement 退回线
line oscillator 线振荡器
line outage 线路停电
line-outage calculator 线路停电率计算器
line outlet 线路出口
line out of service 线路停用
line out of use 不再使用的路线
line output 线路输出
line oven 线炉
line overhead 线路开销
line overlap factor 线路重复因数
line overrelaxation 线超松弛
line packing 管道密封;管道储气
line pad 线性衰减器;线路衰减器
line pair 线耦;线对
line parameter 线路参数;传输线参数
line passenger circulation 线路客流周转量
line pattern 线条样式;线条图形;直线图形
line pen 直线笔
line period 线扫描周期
line perpendicular to projection plane 投影面垂直线
line photograph 划线复照片
line pilot alarm 线路导频报警
line pin 挂线销钉;挂线标杆
line pipe 管线;干线用管;总管;管道
line piping 管系
line plan 线形图;线路平面图;线路平面
line plan and profile 线路平面及纵断面

line planting 列植
line plate 线划板【测】
line-plot survey 沿线标地调查
line-plotter 线绘图机
line plume 线状烟缕
line pointer 行指示字;行指针
line polar or a curve 曲线的极线
line pole 线路电杆
line pole foundation depth 行柱基础深度
line pollution source 线污染源
line polygon 连线多边形;索多边形
line post 贴面;两梁间的横梁;护面;定期轮船;定期班轮;行间立柱;行柱
line-post insulator 线路柱形绝缘子
line power 线路功率
line power spectrum 线路功率谱
line pressure 线压;管线压力;管线压力;管道压力;气压系统操作压力;输送管压力
line-pressure-competent tunnel 衬砌承压隧洞
line print 线画原图
line printer 行式印刷装置;线条印刷机;宽行打印机;划线机;行式打印机;行录制器
line procedure specification 控制通道程序清单
line production 流水作业;流水生产;连续作业法
line production method 流水生产法;线路生产法
line profile 线路纵断面(图);谱线轮廓
line program(me) 线性规划
line prompt area 行提示符区
line protection 输电线路保护装置;输电线路保护
line protection breaker 断路器;线路保险开关
line protocol 线路协议
line pull 钢丝绳拉力;绳索拉力
line pulsing 仿真线脉冲调制
line pump 管线泵
line pumping 管线上泵送
line purging 扫线
liner 邮船;轴箱衬垫;轴瓦;直线规;隔离器;加设护顶梁;内担;内衬层;画线工具;套筒;定期客轮;定期班轮;衬圈;衬砌;衬里;衬垫;班轮;低电离核区
line radiation 线路辐射
line rail 基准轨
line ranger 定线器
line raster 行光栅
line rate 线速率
liner backing 轴承衬支座
liner band 衬带
liner bill of lading 班轮提单
liner blank 汽缸套毛坯
liner boat 定期船
liner bushing 衬套
liner cargo 客轮货物;班轮货物;班机货物
liner centralizer 衬管扶正器
liner company 邮船公司;班轮公司
liner completion 衬管完井
liner depth 尾管深
line-reactor model 线堆模型
line reaming 铰同心孔
line receiver 线路接收机;线接收机;谱线接收机;长线接收器
line reconstruction 线路改建
line record 线路记录
line rectifier 线性检波器
line reflection 线发射;直线反映
line regenerative repeater 线路再生中继机
line regrading 线路坡度整修;线路改坡
line-regular graph 线正则图
line regulation 线路压差(率);线路调整率;线路电压调整;电源(电压)调整(率)
line regulator 管线减振器
line relaxation 线松弛
line relaxation method 线性张弛法
line relay 线路继电器;有线中继
line relaying equipment 有线中继设备
line relay system 有线中继制
line release 线路断开;工料付讫
line relief mapping 制作立体地图
line repair 线路修复
line repeater 线路增音机
line repeating coil 转ценной线圈
line replace command 换行命令
line reserve method 线储量法
line residual current 线路残余电流;大地回流
line resistance 线路电阻;管道阻力
line resistance compensation 线性电阻补偿

line resolution 线条分辨检验图;行分辨
line retirement 搜寻边缘线
line retry parameter 线路重执参数
line-return period 行扫描回程期间
line reversal 谱线反转;谱线变换
line reversal method 回线法(测温用)
line revision 改线
liner flange 汽缸套凸缘
liner fold panel(l)ing 带内衬的护壁板
liner goods 班轮货物
liner hanger 衬管悬挂器
line rider 管线警卫
line ridership 线路乘客总量
liner lane 内侧车道
liner lead 汽缸套导向表面
liner man 样板工
liner material 衬料;衬里材料
liner neck 汽缸套的上定位环带
liner negligence clause 标准班轮附加险条款
line rocket gun 救生索火箭发射器
line rod 花杆【测】
liner of cylinder 汽缸套筒
liner off 画样工
line-rooted graph 线有根图
line roulette 线转迹线
line-route map 线路图
liner plate 支护板;垫板;衬(砌)板
liner plate cofferdam 衬砌板围堰
liner port 定期船港
liner rate 班轮费率
liner ring 衬环
liner service 班轮业务
liner's freight tariff 班轮运费表
liners of high carbon medium-chromium alloy steel 高碳中铬合金钢衬板
liner space 垫板间风
liner terms 泊拉条件;船方不承担装卸费;班轮条款;班轮条件
liner transport 班轮运输
liner tube 衬管
line rule 线路维护规程;线路维修规则
line runner 带缆水手
line rupture protection valve 管路破坏保护阀
liner vessel 班轮
liner wear 衬套磨损;衬板磨损
liner with pressing bar 压条衬板
lines 线形图;横剖面型线图【船】
line sampling 线性采样;直线抽样法
linesboat 带线船
line scale 直线标度
line-scan camera 行扫描摄影机
line scanner 行扫描仪;行扫描器
line scanning 线路扫描;行扫描
line scanning imaging device 行扫描成像装置
line scanning period 线扫描周期
line scanning spot 线性扫描点
line scan transformer 行扫描变压器
line scan tube 行扫描仪;行扫描器
line scattering coefficient 线散射系数
line-screen 线条网屏;网线板;条形屏
line scriber 刻线仪
lines drawing 船体线型
line search 线搜索
line section 线段
line sectionalizing 线路分段
line seepage 浸润线
line segment 线段;搅拌机弓形内衬;弓形炉衬
line selection 线选择;纯系选择
line selection unit 线路选择器
line selector 线路选择器;终接器
line selector oscilloscope 分行示波器
line separating filter 线路频带分隔滤波器
line series 线级数
line series ruled surface 直纹面的线列
line service 线路保养
line shaft 总轴;主传动轴;轴心线;中间轴;天轴;动力轴
line shaft bearing 支撑轴承;船舶传动轴轴承
line shaft hanger 主轴吊架
line shafting 总轴系;天轴;传动轴系
line shape 线形
line shaped track 一字线
line shape function 线形函数
line-sharing system 分路传输系统

line sharpness 线画光洁度
line sheet 限额表
line shift 光谱线的位移;谱线位移
line shipping 定期航运
line shock 管路冲击;水击
line shortage 不满行
line showing finishing allowance 线示加工留量
line shrinkage 线性收缩
line side 线路侧
lineside isolator 轨旁隔离开关
lineside signal 线路信号
line-sight distance 视线距离
line signal 振铃信号
line signal code 线路信号编码
line signal element 线路信号单元
line signal sender connector 线路信号发送连接器
line simulator 线路仿真器
lines intersect 线段相交
line size 管道尺寸
line skipping 跳行
line slope 行斜率
line(s)man 巡线工;线务员;线路工;养路工;架线工;放线员;放样工(人)
linesmen's pliers 钢丝钳;电工钳
line smoothing 线性光滑;线光顺;曲线光滑
lines of constant time difference 等时差线
lines of curvature 曲率线
lines of equal magnetic variation 等偏差线;等磁差图
lines of equal potential 等势线
lines of equal variation 等偏差线
lines of maintenance 维修类别范畴
lines of occlusion 咬合线
line softening 连续直流软化
line source 线污染源;线(光)源
line source heater 线型源加热器
line source model 线污染源模型;线(声)源模型
line source of air pollution 空气污染线源
line source of earthquake 线状震源
line source of pollution 线污染源
line source of sound 线声源
line source structure 线源结构
line spacing 线宽;测线间距;测深线间距
line span 线路挡距
line spectrum 线状光谱;线状(频)谱;线型谱带;线(光)谱;水平扫描
line spectrum test 线谱鉴定
line speed 线速度;线路速率;传送带速度
lines per inch 每英寸行数
lines per picture 每帧行数
line sphere transformation 线球变换
line spinning 旋绕管子的线
line splitter 线路分割器
line splitting 线分裂;谱线分裂
line squall 线状风暴;线飑
lines superior 上极限
line stabilization 线稳定
line-stabilized oscillator 线稳振荡器
line stake 线路桩;路线(标)桩
line standard 线标准;分划式标尺
line starter 全电压起动器
line-start motor 线路起动电动机;直接起动电动机
line station 线路停车站
line stochastic system 线性随机系统
line stopping 管线断水
line storm 分点风暴;二分点风暴
line straightness 线画平直度
line strainer 管道滤网;管路网式过滤器
line strength 谱线强度
line stretcher 线耦合装置;线路伸延器;线扩充器;线长延(伸)器
line-stretcher phase shifter 拉线器移相器
line string 线划串
line-strip punching 穿孔记录
line strobe monitor 选行监视器;线扫描示波器;线路选通监视器
line structure 流水结构;工厂生产组织;线伏组织
line style 线条类型;线型
line subsystem 线路子系统
line support 线支承
line supported 用线支承的
line surcharge load 线超载
line survey(ing) 线路测量

line susceptance 线路电纳
line switch 寻线机;预选器;线路开关
line switch board 预选器架
line switching 线路开关;线路交换
line switching concentrator 线路交换集中器;线交换集中器
line switching system 线路交换制;线路交换系统
line switching technique 线路转换技术
line switching type 线路交换型
line-switch ring 线路倒换环
line symbol 线状符号;线画符号
line symmetric graph 线对称图
line symmetry 线对称
line synchronization 线路同步
line synchronizing pulse 水平同步脉冲
line system control block 线路系统控制功能块
line tank farm 线路上油库
line tap 线路分支
line tape 卷尺
line target 线状目标
line telecommunications 有线电信
line telephone 有线电话
line templete 线样板
line tension 线张力
line terminal 线路终端;主引线端子
line terminal equipment 线路终端设备
line terminal switchboard 线路末端交换台
line terminating equipment 线路终端设备
line terminating network 线路终端网络
line termination 线路终端
line termination capacity 线路端接容量
line termination controller 线路终接控制器
line termination equipment 线路终端设备
line termination unit 有线终端设备
line term of estimation variance 线估计方差
line test 线测验
line test-access point 线路测试入口点
line tester 线路试验器;试线器
line testing subsystem 线路测试子系统
line test trunk 线路试验中继线
line thinning 按垄疏伐
line through center 中心连线
line throughput capacity 线路通过能力【铁】
line thrower 抛缆器
line throwing appliance 抛缆设备
line throwing gun 轻便抛绳枪;抛绳枪
line tilt 行倾斜
line timber 排列支架
line-time base valve 直线时基管
line-to-earth fault 线路接地事故
line-to-earth voltage 线(路)对地电压
line-to-ground 线路接地;线路对地
line-to-ground fault 线路接地故障
line-to-ground shock hazard 线路接地冲击危险
line-to-ground short circuit 线(路)对地短路
line-to-ground voltage 线地电压
line-to-line 间间;两线间
line-to-line fault 线对线故障
line-to-line noise influence voltage 线间噪扰电压
line-to-line short circuit 线间短路
line-to-line valve 零遮盖阀;零叠量阀
line-to-line voltage 线(间)电压
line-to-neutral 从导线到中性点的
line-to-neutral voltage 线中电压
line topology 线画拓扑结构
line tower 线路支架
line trace mill 轮廓仿形铣床
line tracer 线画跟踪器
line tracing 线画跟踪
line traffic 线路业务量;线路通信量;定期船舶交通;分线运输;分线交通
line traffic coordination 线路业务调度
line transect 线带;线形样条
line transect sample 线断面样本
line transfer 送缆
line transformer 线转换器;线路变压器;行变换器
line translation 行转换;行变换
line transmission 线路传输
line transmission error 线路传输误差
line transmission frequency band 线路传输频带
line transmission frequency spectrum 线路传输频谱
line transmission group 线路传输群
line trap 线路阻波器;线路陷波器;线路隔波器;线

列陷波电路;塞流线圈;导线插头
line trawler 拖钓渔船
line tree 界标树
line tuning 线路调谐
line tying 线捆扎
line type 测线类型
line type centrifugal separator 管道型离心式分离器
line type cracking 线型开裂
line type definition 线型定义
line type fire detector 线型火灾探测器
line type modulator 线型调制器
line type pulser 线式脉冲器
line type stacking 波浪形堆料法
line type wire-drawing machine 直线式钢筋拔丝机
line unit 接线盒
line up 校直;校正;排成一行;调成一直线;调整;对管;垫整
line-up clamp 对管器
line-up gang 对管班
line upkeep 线路维护
line-up of numeric(al) control machine tools 数字控制机床群
line-up party 对管班
line-up team 对管班
line-up test 校准试验;校准设备;校直试验
line-up with range marks 对准导标
line used by section 段管线
line-use ratio 线路使用系数;扫描行使率
line utilization 线路使用程度
line valve 管线上普通阀
line variator 变线仪
line vignetting 行遮光
line voltage 线(路)电压;电网电压
line voltage controller 线电压控制
line voltage distribution 线电压分布
line voltage regulator 线路电压调整器;电源电压调节器
line voltage start 线电压起动
line voltage transducer 线路电压变换器
line volume 线涡流
line vortex 线涡流;线涡
line walker 巡线员
line walking 巡线
line weight 线画粗度;线粗
line weld(ing) 滚焊;缝焊;焊(接);直线焊缝焊接
line width 线宽(度);线画宽度;行宽
line width characteristic 线宽特性
line width control 线宽控制
line width control technique 线宽控制技术
line width ga(u)ge 线径测量计
line width scale factor 线宽比例因子
line width variator 变线仪
line winding 交流网测绕组
line wiper 线路弧刷
line wire 线路导线
line with bricks 砖石支护;砌砖(墙)
line with distributed parameter 分布参数线
line with flat gradients 缓坡线
line with lead 衬铅
line with rubber 衬以橡胶;衬胶
line with steep gradients 陡坡线
line with wicker-work 用篱笆排列;用编枝排列
line work 线路工程;线画原图;线路工作;线画清绘
linework positive 线画阳片
line zero pointer 第零行指针
Lingbi stone 灵璧石
lingering 慢燃
ling field 移动场
ling rupture 管路破坏
lingua geographica 地图样舌
lingual arch 舌弧
lingual bar 舌状沙洲;舌形沙坝
linguale 舌点
lingual groove 舌槽
linguall-bar deposit 舌形砂坝沉积
lingui-form 舌状
Lingula flags 舌形贝板层
linguloid bar 似舌形沙坝
linguoclination 舌侧倾斜
linguoid bar 舌状沙洲
linguoid ripple mark 尖лиз形波痕;舌状大波痕
linhay 棚屋(英国)
liniment 涂抹油;涂抹glim;搽剂;擦剂
linimentum 搽剂
linimentum intertrigo 搽剂
linimentum volatile 挥发性搽剂
lining 窑衬;砖砌内衬;砖衬;炉衬;里层;加衬;内衬;筒子板;绳钩法;垫片;砌砌;衬(里)层;衬里;衬垫;衬壁
lining a canal bottom with clay 用黏土加固渠道底槽
lining alloy 轴承衬合金;衬里合金;衬合金
lining arch 衬砌拱
lining bar 拔道撬棍;拔道棍
lining belt 衬带
lining board 镶板;内壁板;垫板;衬(里)板
lining board of dredge pump 泥泵衬板
lining brake 衬面闸
lining break-in 摩擦衬片磨合
lining brick 衬砌砖;衬(里)砖;砌壁砖;面砖
lining brick for converter 转炉炉衬砖
lining burn-back 炉衬熔蚀
lining coefficient of friction 衬层的摩擦系数
lining component 内衬部件
lining concrete 衬砌混凝土;衬里混凝土;混凝土衬层
lining concrete slab 混凝土衬板
lining cross-section 衬砌断面
lining fabric 衬里织物
lining felt 衬毡
lining foil 衬箔
lining form 脚手架
lining form platform truck 衬砌模板台车
lining for support of rock load 岩石荷载支承衬砌
lining for water-level ga(u)ge 水平尺调整
lining glaze 摩擦片表面釉光化
lining in 加粗成图;加深草图上的线条
lining leather 衬里革
liningless cupola 无炉衬冲天炉
lining life 炉衬寿命;衬料寿命
lining machine 渠道铺筑机;衬砌铺砌机;衬里铺砌机;整道机
lining mass 衬料
lining material 衬砌材料;衬里材料;衬垫(材)料;衬(材)料
lining member 衬里构件
lining metal 衬金属
lining mo(u)ld 衬砌模
lining nail 半圆头钉
lining of a pipe 管衬砌
lining of bearing 轴承衬层
lining of canal 渠道护面
lining of door casing 门窗衬垫;门框衬垫
lining of door frame 门框衬垫
lining of fire brick 耐火砖衬里
lining of friction 摩擦衬面
lining of operations 衬砌工作
lining of rock mass 岩体衬砌支护
lining of shaft 炉身内衬;竖井支护
lining of slope 坡面铺砌
lining of smooth face 光滑面衬砌
lining of the side wall(s) 边墙衬里
lining of tunnel 隧道衬砌
lining of well 井衬砌
lining out 锯木前画线;成行栽植;板上画线(锯木前)
lining over refractory 耐火材料上加衬
lining pad 摩擦衬块
lining panel 衬板
lining paper 衬纸;壁纸衬纸;环衬;廉价壁纸(作衬纸用)
lining peg 路界桩;路界;界标
lining pen 直线笔
lining piece 衬料
lining pipe 衬砌管
lining plank 衬垫板;衬砌木板
lining plate 衬板
lining plate with rib 凸棱衬板
lining pole 花杆【测】;测杆;标杆
lining products 衬里产品
lining ring 衬圈;衬砌环
lining rubber 橡胶衬里;衬里橡胶
lining rubber tanker 衬胶槽车
lining section 衬砌剖面
lining segment 衬砌块
lining service group 整套衬砌工具
lining sheet 衬里薄板
lining sheeting 横板;衬板;封檐板
lining sight 简单瞄准器
lining size 衬里尺寸;覆盖尺寸;挡板尺寸
lining slab 外墙加劲板;衬里板
lining split 衬砌破损
lining stone 衬石;衬砌石块
lining stoneware 衬面陶瓷
lining strip 内墙板条
lining strutting 衬砌支架
lining stud 内墙板间柱
lining surface 摩擦衬片表面
lining thickness 衬砌厚度
lining tile 衬砖;衬瓦
lining tool 画线器;修网线刻刀;画线工具;小扁漆刷
lining tube 衬砌管
lining turf 草皮栽植;植草皮
lining type 衬砌类型
lining-up 规正;砌衬;以物物砌;在薄的材料边上加狭条(起加强作用);排齐;衬砌;内衬
lining wall 内隔墙;衬(砌)墙;衬砌壁
lining ware 衬里
lining wear 炉衬侵蚀;衬里磨损
lining winch 衬里绞车
lining with column-type sidewalls 柱式边墙衬砌
lining with continuous arched sidewalls 连拱墙衬砌
lining work 砌筑内衬工作
linishing machine 砂带磨光机
link 固定配线;链合;链杆;链环;连通的连接片;连接设备;连接(符);连接部件;矿车连接器;接线;环节;无线电通信线路;网络连线;网络节
linkable program(me) 可连接程序
link access attribute 连接访问属性
link access procedure 链路接入规程
link access procedure balanced 平衡型链路接入规程
link access procedure for modems 调制解调器连接访问协议
link address 连接地址
link adjusting gear 连杆调节装置
linkage 匝连;杠杆机构;杆系;链系;链接;链合;联锁;连锁;离子键;键结;键合;耦合;磁通匝连数
linkage area 连接区
linkage bar 连系钢筋
linkage beam 连系梁
linkage check chain 悬挂装置限位链
linkage check rod 悬挂装置锁定杆
linkage coefficient 磁链系数
linkage computer 联动计算机
linkage convention 连接约定;连接匹配
linkage device 联动装置
linkage drawbar 牵引杆;牵引板
linkage editing 连接编辑
linkage editor 连锁编辑程序;连续编辑程序
linkage fault 连接故障
linkage flag 连接标记
linkage function 连接功能
linkage group 链组;连锁群
linkage guide 联动导轨
linkage hitch 悬挂装置
linkage in cross-impact studies 交互影响研究中的连接
linkage information 连接信息
linkage instruction 连接指令
linkage interrupt 连接中断
linkage in three dimensions 三维空间联动装置
linkage law 连锁定律
linkage levelling screw 悬挂装置;调平螺杆
linkage map 连锁图
linkage mounting 悬挂装置
linkage of coroutine 共行程序的连接
linkage of risks 风险连锁
linkage parameter 杆系参数
linkage pitman 悬挂装置铰接轴
linkage point 铰接点
linkage ratio 链系比
linkage section 连接节;连接段
linkage segment 连接段
linkage stabilizer 悬挂装置的锁定器
linkage system 联动装置;连锁装置;连锁制
linkage typology analysis 经济联系型分析
linkage unit 联动组件;连接装置;连接单元
linkage value 连锁值
linkage voltage test 连续加压试验
linkage winch 悬挂式绞车;车悬挂式绞车

link allotter 线路分配器;链式分配器
link and pin coupler 杆销联结器
link arm 连杆臂;斜拉杆
link ball 拉杆球端
link bar 连接(钢)筋;铰接杆件;铰接杆(采矿);铰连杆(采矿)
link belt 链条;链带
link belt conveyer furnace 环链式炉
Link-Belt drum type concentrator 林克-拜尔特型滚筒式精选机
link-belt feeder 链带给料机
link-belt PD screen 偏心式振动筛
link-belt UP screen 不平衡轮式振动筛
link bit 链接位;连接位
link block 林克(消波)混凝土块体;连节滑块;连接模块;连接(滑)块;连接分程序;密封销;导块
link bolt 链环插销;铰链螺销;铰接螺栓
link bond 链条联结
link box 活节连接器
link bridge 链桥;链式悬桥
link-by-link signal 逐段联络信令【铁】
link canal 连接渠道
link center[centre] 联结中心
link chain 节链;节链链;扁节链;链条;链环
link checksum 链路检验和
link chute adapter 链槽接口
link circuit 链式电路;链路;联络线路
link connection 链路连接
link control 连接控制
link control message 通信线路控制信息
link control module 连接控制模块;连接控制块
link control procedure 链路控制规程;连接控制过程
link control program(me) 连接控制程序
link coupling 环节耦合
link curve 索状多边形;连接曲线
link data unit 链路控制数据装置
link definition 连元定义
link dormer 连接的老虎窗;大屋顶窗
linked 偶合的;连接的
linked allocation 链接分配;连接分配
linked allocation of stack 堆栈的链接分配
linked allocation of table 表格的链接分配
linked arch 铰接拱
linked arched girder 连连拱梁
linked bar 连接的钢筋
linked bar chart program(me) 连续条线图
linked circular kiosk 双环亭
linked connection 铰接
linked control system 联动控制系统
linked deal 连锁交易
linked entry office 网络进口局
linked exit office 网络出口局
linked file 连接文件
linked industry 关连工业
link-edit 连接编辑
link editing 连接编辑
link edition 连接编辑
link editor 连接编辑程序
linked linear list 连接线性表
linked list 链(接)表;连接表
linked memory allocation 耦合存储器分配
linked numbering scheme 网络编号方案
linked office 网络局
linked paired comparison designs 相联成对比较设计
linked pool 串联水库
linked purlin(e) 连接的檩条
linked queue 链式队列;链接队列
linked reaction 耦合作用
linked reservoir 串联水库
linked roof bar (金属支架的)组合顶梁
linked routine 闭型例行程序;闭型程序
linked square kiosk 套方亭
linked switch 联动开关;连接开关;耦合开关;联锁开关
linked system 联动系统;交通信号联动系统
linked trip 连续出行;全程出行
linked veins 链状矿物
Linkenbach table 林肯巴哈型圆形固定淘汰盘
link encryption 链路加密;通信线路保密
linker 连接程序
Linke scale 林克标【气】
link exchange 链路交换
link expansion band 弹性钢表带

link face 链环面
link feed tooth 送链齿
link field 连接域
link for steady arm 定位器连接环
link frame 链节架
link fulcrum 链杆支轴
link fuse 链熔线;带接线片熔丝片
link gear 联动机构
link gearing 连杆传动装置
link group 连接组;接线组
link hanger 连杆吊架;连杆吊架
link hinge 搭襻铰链
link indicator 链接位指示器;连接(位)指示器
link information 连接信息
linking 联锁;套口
linking agent 黏结剂
linking anchor 连环锚
linking automation 连接自动化
linking beam 连系梁
linking course 套口横列
linking jig 套口架
linking loader 链装程序;连接负载
linking machine 合缝机;套口机
linking machine needle 套口针
linking member 联系杆件
linking module 连接模块
linking point of track 轨道连接点
linking route 连接线(路);连接路线
linking seam 套口缝
linking station 中继站;中继台
linking system 连锁制
linking taxiway 连接滑行道
linking taxiway connecting 联络道
linking terminal for roadbed structural reinforcement 道床结构钢筋连接端子
linking terminal for tunnel structural reinforcement 隧道结构钢筋连接端子
linking-up 操纵杆加速操作;接上
linking-up road 联络线
linking-up ship 联络船
linking-up station 中继(电)台
link insulator 串式绝缘子
link joint 链纹;联连接
link lever 连杆
link library 连接库
link line 联络线;连接线;连(接)线
link linkage 连接程序
link list 连接表
linkload 区间客流负荷
link mechanism 连杆机构
link model 联结模式
link motion 链节运动;链动机;连杆运动
link-motion drive 连杆运动机构传动
link motion valve gear 连杆配汽机构;连杆阀动装置
link of chain 链环
link off 断开的连接片
link of levels 水准测量区段
link order 连接指令;耦合指令;返回指令
link or pointer variable 连接或指针变量
link overflow 连接溢出
link pack area 连接装配区
link pack area queue 连续装配区排队
link pack update area 连接装配更新区
link pin 折页销;联杆销;连接销;连杆销
link plate 链节板;连接板
link polygon 索多边形
link press 连杆式压力机
link protocol 链路协议
link pumping station 支管泵站
link ratio critical speed 杆系传动比临界速率
link receiver 接力接收机
link relative ratio 环比
link resonance 网络节共振
link road 连接(道)路;连接线
link rod 连杆
links 高尔夫球场;海岸沙滩;海岸草原
link saddle 滑动鞍
link shoe strap 链节衬垫簧片
links-links design 双反面图案
links-links fabric 双反面织物
links-links flat bar knitting machine 双反面横机
links-links jacquard cam 双反面提花三角
links-links knitting 双反面编织
links-links pattern drum 双反面提花滚筒

links-links patterning mechanism 双反面提花装置
links-links stitch 双反面线圈
link sort 连接分类
link span 连接翼板;连接架
link station 链路站
link stop 链制动
link stretch 链节弹性伸展
link support 链节托
link-supported clump fender 链条支承的重块护舷
link system 支管系统;联结系统
link test 链路测试
link to node 对节点的连接
link to the left 对左边的连接
link transmitter 接力发射机;强方向发射束发射机
link type chain 环接链条
link unit 连接单元
link-up 连贯;衔接
link-up pontoon 垫档趸船
link-up the networks 联网
link variable 链路变量
linkway 联系通道
link winch 悬挂式绞车
linkwork 连杆运动;杆系;链系;联动装置;铰接机构
linn 瀑布;溪谷
linnaeite 硫钴矿
linocut 油毡板;亚麻油毡浮雕版
Linofelt 利诺弗尔特(一种墙隔热材料)
linoleate of lead 铅亚油酸盐
linoleate of manganese 锰亚油酸盐
linoleic acid 蓼酸;亚麻仁油酸
linoleic acid erol ester 甘油三亚油酸酯
linoleic acid glycerol ester 甘油十八(碳)烯酸酯
linol(e)in 甘油三亚油酸酯;亚麻油精;亚麻油脂
linolek acid 亚油酸
linolelaidic acid 反亚油酸
linolenate 亚麻酸盐
linolenate dryer 亚麻油盐催干剂
linolenic acid 亚麻酸;次亚麻(仁)油酸
linolenic acid glycerol ester 甘油亚麻酸酯
linolenic and linoleic 亚麻酸和亚油酸
linolenin 亚麻精;甘油三亚麻酸酯
lino(leum) 油地毯;油漆布;油地毡;亚麻(籽)油(漆)布;亚麻油毡;漆布;地漆布(铺地用)
lino(leum) base 油毡底层
lino(leum) (bonding) adhesive 油地毡胶黏剂;油毡黏结剂
lino(leum) cement 油毡黏结剂;油(地)毡胶黏剂
lino(leum) cover 油毡护面
lino(leum) deck 油毡甲板;油毡床面
lino(leum) felt 油地毡块
lino(leum) floor 油毡铺地
lino(leum) floor cover(ing) 油毡铺地面层
lino(leum) flooring 油毡铺地;油地毡(铺)地面;油地毡楼地面;漆布地面
lino(leum) ga(u)ge 油毡规格
lino(leum) knife 油毡刀;割毡刀
lino(leum) laying 铺油毡
lino(leum) lining 油毡衬里
lino(leum) oil 油(地)毡(用)油;亚麻油;漆布油
lino(leum) oil cloth varnish 油布清漆
lino(leum) print paint 油毡印刷漆
lino(leum) surface 油地毡毡面
lino(leum) tar 油地毡焦油
lino(leum) tile 油地毡面砖;油(地)毡块
lino(leum) varnish 油毡清漆
linoleyl alcohol 亚油基醇
linolic acid 蓼酸
linolin 亚油精
linolk acid 亚油酸
linophyre 线状斑岩
linophyric 线斑状
linotape 浸漆绝缘布带
Linotile 林诺泰(一种无缝合成地板);亚麻油毡铺块
linotron 莱诺管
Linotype 整行铸排机
linoxanthine 亚麻黄质
linoxyn 氧化亚麻(仁)油;干亚麻子油
linseed 亚麻籽
linseed cake 亚麻籽饼
linseed fatty acid 亚麻仁脂肪酸
linseed meal 亚麻籽粉
linseed oil 亚麻(籽)油;亚麻仁油;胡麻油
linseed oil acid 亚麻籽油酸
linseed oil alkyd 亚麻籽油醇酸树脂

linseed oil base	亚麻籽油基
linseed oil-bearing	含亚麻籽油(的)
linseed oil combination test	亚麻仁油混合试验
linseed oil drying	亚麻籽油干燥
linseed oil emulsion	亚麻籽油乳剂
linseed oil fatty acid	亚麻仁油脂肪酸
linseed oil for paints	涂料用亚麻籽油;油料用亚麻籽油
linseed oil paint	亚麻籽油油漆;亚麻籽油涂料
linseed oil priming agent	亚麻籽油打底剂
linseed oil putty	亚麻籽油灰
linseed oil soap	亚麻仁油皂
linseed oil solution	亚麻籽油溶液
linseed oil stand oil	调墨用熟油;叠合油料;亚麻仁油;亚麻籽熟油
linseed oil varnish	亚麻清漆;亚麻漬漆;亚麻籽油清漆;亚麻油(清)漆;亚麻油凡立水;亚麻仁油清漆
linseed oil without foots	无油脚亚麻籽油
linseed roller	亚麻脱粒辊
linseed stand oil	亚麻籽熟油;亚(麻油)定油
linseed-tung stand oil	亚桐定油;亚麻油桐定油
Linseis plastometer	林氏可塑性测定仪
linsen bedding	透镜层理
linsey	亚麻羊毛交织物;带条纹的砂岩;混杂粗呢布
lint	毛绒;亚麻布;纤维屑
linted latex paint	棉绒胶乳涂料
lin-Tee beam	单T板
lintel	过梁;炉壁托圈;楣;大梁砖
lintel beam	楣;过梁
lintel block	过梁砌块;盖砖
lintel brick	过梁砌砖;门门顶砖石
lintel course	(门窗的)过梁层;门窗过梁层
lintel-damp-course	过梁缓冲层;过梁防潮层
linteled	有过梁的
lintel-falseblock	过梁模块;过梁支架
lintel-falseblock reinforcement	过梁模块增强物
lintel girder	炉腰支圈
lintel machine	过梁机械
lintel mo(u)lding	门楣饰
lintel of a door	门楣
lintel of doors and windows	门窗过梁
lintel soffit	过梁底面
lintel stone	石过梁
lintel-tol	水平横楣
linter	剥绒纤维;剥绒机
linter gin	剥棉短绒机
linter pulp	棉绒浆
linters	棉绒纤维
Lintile	林泰(一种地毯)
lint paper	棉绒纸
lint retainer	纤维夹持器
Linville truss	林维尔桁架
liny	似线
Lion bearing alloy	莱昂轴承合金
lion frieze	雕狮饰檐壁
lion gate	狮大门
Lion indicator	跨规
lion-mask	狮面具;狮面装饰
Lion metal	拉昂锡基轴承合金
lion tape	浸漆绝缘布带
Lion Tomb at Cnidos	克尼达斯的狮墓
liotex	镁质地板的总称
liottite	利钙霞石
lip	缘边;铸嘴;突缘;凸缘;凸出部(分);导脂器;刀刃;玻璃液流料槽
lipa	油脂
lip angle	楔唇角;楔角
liparite	萤石;硅长雀石;流纹岩;铁纹石;石英粗面岩
liparite tuff	流纹凝灰岩
lip block	支撑垫块;顶板法
lip bore	唇形钻孔
lip casting	压边浇铸
lip clearance	钻缘隙角;背角
lip curb	唇状路缘;缓坡缘石
lip elevation	坎唇高程
lip grinding	挡边磨削
lip-guided	挡边引导;挡边导向
lip hat section steel	带缘帽型钢
lip height	切削刃高度
lip hollow shape	带缘空心型材
lipid	类脂
lipid-containing wastewater	含类脂废水
lipide	脂质;脂类;类脂(化合)物
lipid resinite	类脂树脂体
lipid solubility	类脂可溶性
lip joint	赤陶砌块间的唇状接合;陶土构件的企口
Lipkin's pycnometer	利普金比重计
lipless guard	无唇护刀器
lip loss	进口边缘损失
lip mask	喷漆唇形罩
lip microphone	唇式传声器
lip mo(u)ld	凸缘线脚;唇瓣装饰
lipochromous	无色的
lip of the jet	喷管边缘
lipoid	类脂
lipoidal matter	类脂物(质)
lip on tyre	轮箍唇
lipophile liquid	亲油液体
lipophilic dispersant	亲油分散基
lipophilic hydromicelle	亲油极性芯胶束
lipophilicity	亲脂性;亲油性
lipophilic nature	亲油性
Lipotes vexillifer	白鳍豚
Lipowitz alloy	利玻维兹(低温易熔)合金
lip packing	带唇边环形密封
lipped	只在砖的周边用灰浆(指在圬工接缝处)
lippped angle	卷边角钢
lipped channel	卷边槽钢
lipped floor brick	带突缘的铺地砖
lipped joint	半搭接;唇形接合;凸缘接合;舌形接头;舌形接口
lip(ped) sieve	鱼鳞筛
lipper	小浪花;微波海面;跳跃者;船首溅沫;飞溅浪花
Lippich polarizer	李皮奇偏振器
Lippich prism	李皮奇棱镜
lip-piece	支撑加important木
lipping	支撑垫块;(平板门的)贴边木条;唇状突出;半搭接;唇形结合;镶边;木条镶边
lipping of rail head	轨头肥边
lip plate	碟形边
Lippman fringes	李普曼条纹
Lippmann colo(u)r process	李普曼彩色法
Lippmann effect	李普曼效应
Lippmann emulsion	李普曼乳胶;李普曼乳剂
Lippmann hologram	李普曼全息图
Lippmann holography	李普曼全息术
Lippmann process	李普曼(方)法
Lippmann's colo(u)r photography	李普曼彩色摄影术
lip pour ladle	转包
lip pressure	边部压力
lip relief angle	钻缘后角
lip ring	炉头钢圈;炉顶钢圈
lipscombite	四方复铁天蓝石
lip screen	分极筛
lip seal	唇形密封
lip seal fitting	带唇形密封的接头
lip seal(ing) ring	唇形密封环
lip streak	拉引线道;槽口线道
lip strike	锁舌孔板;(锁舌孔板的)撞唇
lip surface	前面
lip tile	唇砖
liptinite	壳质煤素质
liptinite kerogen	脂质组干酪根
liptite	微稳定煤;微壳质煤
liptobiolite	残烛煤;残殖煤;残留生物岩
liptobiolite forming process	残植化作用
liptobiolith	残烛煤;残殖煤;残留生物岩
liptocoenosis	残体群落
liptodetrinite	壳质体;碎屑稳定体;碎屑壳质体
liptotriolith	残生物岩
lip trap	(老式的)唇形存水弯
lip-type bear	唇型油封;唇型密封环
lip type bearing	带挡边轴承
lip union	内突环式连接管;凸缘活接头
lip Z steel	带缘Z形钢
liquable	可液化的
liqualino	抹灰黏结剂(一种水泥砂浆)
liquamatic	水力驱动
liquate	熔析
liquated copper	熔析铜
liquated lead	熔析铅
liquated-lead kettle	熔析铅锅
liquate out	熔析出
liquating apparatus	熔析设备
liquating kettle	熔析锅
liquation	液析;液化作用;固相分离;离析性;熔析;熔融;熔离;熔解分析
liquation bath	熔析槽
liquation deposit	熔离矿床
liquation furnace	熔化炉
liquation hearth	熔析炉
liquation lead	熔析铅
liquation point	液化点
liquation refining	熔析精炼
liqudbath furnace	熔炉
liquefacient	解凝剂
liquefactant	熔解物
liquefaction	液化(作用);振动液化
liquefaction category	液化等级
liquefaction coal	液化用煤
liquefaction effect	液化效应
liquefaction failure	土壤液化塌坍;液化破坏
liquefaction foundation	液化地基
liquefaction grade	地基液化等级
liquefaction index	地基液化指数;液化指数
liquefaction of coal	煤液化;煤的液化
liquefaction of cohesiveless soil	无黏性土液化
liquefaction of sand(soil)	矿土流化;砂土液化
liquefaction of sandy soil	沙性土壤液化
liquefaction of saturated soil	饱和土液化
liquefaction of soil	土的液化
liquefaction point	液化点
liquefaction potential	液化(潜)势;液化可能性;砂土潜在液化能力
liquefaction potention	液化潜在能力
liquefaction probability	液化概率
liquefaction process	液化法
liquefaction ratio	液化比率
liquefaction resistant factor	抗液化系数
liquefaction strength	液化强度
liquefaction stress ratio	液化应力比
liquefaction susceptibility	液化灵敏度
liquefaction value	液化值
liquefaction with limited strain potential	具有有限应变势的液化
liquefactive degeneration	液化变性
liquefiable	可液化的
liquefied air	液态空气;液化空气
liquefied ammonia tank vehicle	液氨槽车
liquefied bituminous material	液态沥青
liquefied butane	液态丁烷
liquefied carbolic acid	液态石炭酸
liquefied carbon dioxide tank vehicle	液态二氧化碳槽车
liquefied chlorine gas	液化氯气
liquefied cohesionless-particle flow	液化松散颗粒流
liquefied compound	液态化合物
liquefied compressed gas	液化气压缩气体
liquefied gas	液体气体;液化气(体);液化惰性气体
liquefied gas aerosols	液化气雾剂
liquefied gas carrier	液化气(运输)船
liquefied gas pipeline	液化气管道
liquefied gas pump	液化气泵
liquefied gas tank	液化气体储[贮]罐
liquefied gas tanker	液化气运输船;液化气体槽车
liquefied gas truck	煤气筒汽车
liquefied gas vapo(u)rizer	液态气体气化装置
liquefied inflammable gas	液化易燃气
liquefied methane	液化甲烷
liquefied methane gas	液化沼气;液化甲烷气
liquefied methane gas tanker	液化甲烷运输船
liquefied natural gas	液化天然气;液态天然气
liquefied natural gas carrier	液化天然气船
liquefied natural gas tank	液化天然气罐;液化气罐
liquefied nitrogen truck	液氮车
liquefied noble gas	液化惰性气体
liquefied oxygen tank vehicle	液氧槽车
liquefied particles	液化颗粒集合体
liquefied petrolatum	液化石蜡
liquefied petroleum gas	液化石油气
liquefied petroleum gas additive	液化石油气添加剂
liquefied petroleum gas automobile	液化石油气汽车
liquefied petroleum gas bottle	液化石油气钢瓶
liquefied petroleum gas buses	液化石油气公共汽车
liquefied petroleum gas carrier	液化石油气船
liquefied petroleum gas container	液化石油气钢瓶
liquefied petroleum gas cylinder	液化石油气

（钢）瓶
liquefied petroleum gas distribution station 液化气储配站
liquefied petroleum gas engine 液化石油气发动机；液化气发动机
liquefied petroleum gas equipment 液化石油气设备
liquefied petroleum gas fuel 液化石油气燃料
liquefied petroleum gas high pressure holder 液化石油气高压气罐
liquefied petroleum gas odorant 液化石油气增味剂
liquefied petroleum gas platforming 液化石油气铂重整
liquefied petroleum gas powered car 液化气动力车
liquefied petroleum gas regulator 液化气调节器
liquefied petroleum gas relaying station 液化气转输站
liquefied petroleum gas retail depot 液化气零售点
liquefied petroleum gas tank 液化气罐
liquefied petroleum gas tank vehicle 液化石油气槽车
liquefied petroleum gas tank-wagon 液化石油气铁路槽车
liquefied petroleum gas terminal 液化石油气码头
liquefied petroleum gas truck 液化石油气机车
liquefied phenol 液态酚；液化苯酚
liquefied propane(gas) 液化丙烷
liquefied propane gas tanker 液化丙烷运输船
liquefied refinery gas 液化炼厂气；精制液化气
liquefied region 液化区
liquefied resin 液化树胶；液化树脂
liquefied soil 液化砂土
liquefier 液化器；液化剂
liquefy 液化；变成液态
liquefying 液化
liquefying agent 液化剂
liquefying air 液化空气
liquefying gas 液化气(体)
liquefying plant 液化装置
liquefying point 液化点
liquefying process 液化过程
liquefying sand layer 液化砂层
liquefying stress 液化应力
liquefying time 液化时间
liquescent 可液化的；变液的
liquid 液体；汁
liquid absorbent 液体吸收剂
liquid absorption 吸液量
liquid accelerator 液体促进剂
liquid accumulation 液态储存
liquid acetylene 液态乙炔
liquid acrylic polymer system 液态丙烯酸聚合物系统
liquid adhesive 液体胶；液体黏合剂；黏合液
liquid admix(ture) 液体混合物；液态附加剂
liquid adsorption method 液体吸附法
liquid aerosol 液体气溶胶
liquid agent 液态附加剂
liquid aggregate state 液体凝集状态
liquid air 液态空气
liquid air bottle 液态空气瓶
liquid air container 液态空气容器
liquid air-cooled detector 液态空气致冷探测器
liquid air cycle engine 液态空气循环发动机
liquid air-entaining agent 液体加气剂
liquid air temperature 液态空气温度
liquid air trap 液态空气阱
liquid alloy 液态合金
liquid alumin(i)um 液态铝
liquid alumin(i)um globule 液态铝珠
liquidambar 枫香树香脂
liquidambar formosana slice 枫香片
liquid ammonia 液态氨；液氨；氨液；氨水
liquid ammonia carrier 液态氨运输船
liquid ammonia seperator 液态氨分离器
liquid ammonia storage tank 液氨储[贮]槽
liquid and paste mixer 液体和浆体混合器
liquid anion 液态阴离子
liquid annealing 液体退火
liquid antifreezer 防冻液剂
liquid application 液施法；液施
liquid applied chlorosulfonated polyethylene 氯磺化聚乙烯涂料
liquid applied neoprene 氯丁橡胶涂料
liquid applied waterproofing system 防水涂料系统
liquid ash extraction 液体灰分提取法
liquid asphalt 轻制沥青；液体(地)沥青；液态沥青；残油
liquid asphaltic material 液体(地)沥青材料；液态沥青材料
liquid asphaltic products 液态沥青
liquid assets 易变现金的财产；流动资产
liquid assets rations 流动资产限额
liquidate 熔化分离；清算；清偿
liquidate appropriation 清算拨款
liquidate assets 清算资产
liquidated account 已清算账户
liquidated corporation 已清理的公司
liquidated damages 误期赔偿费；日罚款额（承包合约中延误工期每日定额罚款）；已清偿(了结)的损失赔偿金；违约罚金；损害赔偿金；确定的赔偿金；清偿损失额；违约金；违约罚款
liquidated damages for delay 延期损失赔偿额；误期罚款；误期赔偿费
liquidated damages for process performance failure 工艺性能未保证赔偿费
liquidated debt 已清偿债务
liquidated demand 清偿债务之要求
liquidated obligation 已清偿(的)债务
liquidated sum 实得还款数目
liquidated unit price method 清算单价法
liquidating account 清算账户
liquidating partner 清算人
liquidation 液化；熔析；熔解；清算；清理
liquidation account 清算变现表；清理账户；清理变产表
liquidation affairs 清算事务
liquidation balance sheet 清算资产负债表
liquidation dividends 清算分摊额
liquidation expenses of fixed assets 固定资产清理费用
liquidation gain or loss 清算损益；清理损益
liquidation income 清算收入
liquidation in installments 分期摊还的清算；分批摊还款项的清算；分批摊还的清算
liquidation in instalments 分期还款的清算；分期偿付或摊还款项的清算
liquidation management of international through-transport 国际联运清单管理
liquidation of accomplished task 完成工作清算
liquidation of a judicial person 法人清算
liquidation of claim 债务清理
liquidation of debts 清场债务；清理债务
liquidation of fixed assets 固定资产清理
liquidation of property and appraisal of assets 清产核资
liquidation or bankruptcy proceedings 破产诉讼
liquidation price 清算价值(破产企业资产的)
liquidation procedure 清理程序
liquidation profit and loss 清算损益
liquidation reserve 清偿准备金
liquidation sale 停业清理(大)拍卖
liquidation unit price 清算单价
liquidation value 清算价值；清理价值
liquidation value per share 每股清算价值
liquid atomization 液体雾化
liquidator 清算人；清理人；财产清算人
liquid balance 现金结余；液体平衡
liquid ballast 液体压舱；液体压载
liquid bath 液浴；液池；熔渣池
liquid bath furnace 液态排渣炉膛
liquid bed biologic(al) reactor 液化床生物反应器
liquid bed reactor 液化床反应器
liquid binder 液体黏合剂
liquid bitumen 液体(地)沥青；液态沥青
liquid bituminous material 液体(地)沥青材料；液态沥青材料；液状沥青材料
liquid blade cooling 叶片液冷
liquid blast cleaning 液体抛丸清洗
liquid blasting 液体喷砂处理；液体喷砂
liquid bleach 液体漂白剂
liquid blending system 液体掺和系统；液化掺和系统
liquid blocking 液封
liquid bond-breaker 液态脱模剂
liquid bonding agent 液态黏结剂
liquid brake 液压制动器；液力制动(器)；液动闸
liquid bright gold 液体亮金；亮金水
liquid bright platinum 亮白金水
liquid brush rectifier 电解电刷整流器
liquid bubble tracer 液滴示踪法
liquid bulk cargo 液体散货；散装液体
liquid bulk container 液体散货集装箱；罐式集装箱；散液集装箱
liquid buoyancy material 液体浮力材料
liquid calorimeter 液态热量计；液体量热计
liquid capacitor 液体电容器
liquid capacity 液体容量；液体流量
liquid capital 流动资本
liquid carbon dioxide 液态二氧化碳
liquid carbonitriding 液体碳氮共渗
liquid carburizing 液体渗碳处理；液体渗碳
liquid cargo 液体货(物)
liquid cargo container 液体货物集装箱；散液集装箱
liquid cargo filling height 液货罐装高度
liquid cargo ship 液化船
liquid cash funds 流动现金
liquid casting resin 液体铸塑树脂
liquid caustic soda 液体烧碱
liquid cell 液体电池；液体池；液槽
liquid cement 液体掺碳剂；胶液
liquid center bending 液心顶弯
liquid charging stock 液体原料
liquid chemical dosimeter 液体化学剂量计
liquid chemical rocket 液体化学燃料火箭
liquid chemicals tanker 液体化学品船
liquid chemistry analysis from compressive soil 土壤压出液分析
liquid chiller 液体冷却器
liquid chlorine 液态氯；液氯
liquid chlorine diffuser 液氯扩散器
liquid chlorine injector 液氯喷射器
liquid chlorine metering tank 液氯计量槽
liquid chlorine section 液氯工段
liquid chlorine storage tank 液氯储[贮]罐
liquid chromatogram 液相色谱；液体色谱
liquid chromatograph 液相色谱仪
liquid chromatograph/Fourier transform infrared spectrometer 液相色谱傅立叶红外光谱联合测定仪
liquid chromatograph/mass spectrometer 液相色谱质谱仪联用
liquid chromatograph/mass spectrometer computer 液相色谱质谱联合测定仪
liquid chromatography 液相色谱(法)；液相层析；液体色谱法；液体层析
liquid chromatography-mass spectrometry 液相色谱法质谱法联用
liquid circulating temperature control 液体循环温度控制
liquid circulation chlorine compressor 液环式氯气压缩机
liquid cladding 液体包层
liquid clutch 液压离合器；液体离合器；液体联轴器
liquid coating(material) 液态涂面料；液体涂料
liquid cold-tar creosote 冷用杂酚油；常温液态杂酚油
liquid colo(u)rant 液体颜料
liquid column 液柱
liquid column barometer 液柱气压表
liquid column chromatogrpahy 液相柱色谱法；液相柱层析
liquid column ga(u)ge 液柱压力计
liquid column manometer 液柱压力计
liquid combination 液体混合燃料
liquid commodities 液体矿产
liquid compass 液体罗盘；液体罗经；湿式罗盘；充液罗盘
liquid-compressed steel 液态挤压钢；加压凝固钢
liquid compressibility factor 液体压缩系数
liquid concrete 液体混凝土；液状混凝土；流态混凝土
liquid concrete admix(ture) 混凝土的液体外加剂；液态混凝土附加剂
liquid concrete agent 混凝土的液体外加剂
liquid concrete floor hardener 混凝土楼板的硬化液剂
liquid condensed film 液体凝聚膜；液态凝聚膜
liquid condenser 液体(介质)电容器

liquid condition 液态;液体状态
liquid conduction 液体导电
liquid containing binder 液体黏结剂
liquid content 液相含量
liquid contraction 液态收缩
liquid controller 液体控制器;液体控制阀
liquid control valve 液体控制阀
liquid coolant 液体冷却剂;冷却液
liquid cooled 液冷式
liquid cooled brake 液冷式制动装置
liquid cooled engine 液冷(式)发动机
liquid cooled gas-turbine blade 液冷式燃气轮机叶片
liquid cooled nozzle 液冷式喷管
liquid cooled reactor 液体冷却反应堆
liquid cooled rotor 液冷转子
liquid cooled stator winding 液冷定子绕组
liquid cooled transformer 液冷变压器
liquid cooler 液体冷却器
liquid cooling 液体冷却;液冷
liquid cooling generator 液冷发电机
liquid cooling medium 液冷介质
liquid cooling system 液冷却系统
liquid core 液相穴
liquid core fiber 液芯纤维;液芯光纤
liquid core fiber-optic(al) waveguide 液芯光纤波导
liquid core optic(al) fiber 液芯光纤
liquid corrosion 液体腐蚀;液蚀
liquid counter 液体计数器
liquid counting 液体计数
liquid crucible technique 液体坩埚技术
liquid crystal 液晶体;液晶
liquid crystal box 液晶盒
liquid crystal cell 液晶元件;液晶盒
liquid crystal compound 液晶混合物
liquid crystal display 液晶显示(器)
liquid crystal display device 液晶显示器
liquid crystal display glass 液晶显示玻璃
liquid crystal display material 液晶显示材料
liquid crystal display operation 液晶显示操作
liquid crystal image converter 液晶像转换器
liquid crystal imaging 液晶成像
liquid crystal infrared device 液晶红外器件
liquid crystal injection 液晶注入
liquid crystal layer 液晶层
liquid crystal light valve 液晶光阀
liquid crystalline phase 液晶相
liquid crystalline state 液晶态
liquid crystal memory 液晶存储器
liquid crystal microscope 液晶显微镜
liquid crystal non-destructive testing 液晶无探伤法
liquid crystal optics 液晶光学
liquid crystal pictorial display 液晶显像
liquid crystal polymer 液晶聚合物
liquid crystal state 液晶态
liquid crystal stationary phase 液晶固定相
liquid crystal watch 液晶表
liquid crystal waveguide 液晶波导
liquid cubic 液货舱容
liquid culture 液体培养
liquid curing 液体硫化
liquid cutting of wood 木材水力切割;木材水力采伐
liquid cyaniding 液体氰化法
liquid cycle 液体循环
liquid cyclone 液体旋风器;液体分流器;液体分尘器;水力旋流(分离)器
liquid damped compass 液体阻尼罗盘
liquid damper 油压缓冲器;液压减振器;液体阻尼器;液体减振器
liquid damping 液体阻尼
liquid damping device 液体阻尼装置
liquid damp-proofing and permeability reducing agent 防潮防渗液
liquid deformation 液体形变
liquid dehumidifying 液体干燥(作用);液体防潮(作用);液体除湿(作用)
liquid delay line 液体延迟线
liquid densifier 液态浓缩剂
liquid densifying admix(ture) 液体浓缩混合物
liquid densifying agent 液态浓缩剂
liquid densimeter 液体密度计
liquid density 液体稠度
liquid depolarizer 液体去极化剂

liquid deposit 流动存款
liquid deposition 液相沉积
liquid desiccant 液体干燥剂
liquid desiccation 液体干燥剂脱水
liquid detergent 液体洗涤剂
liquid development 液体显影
liquid dielectric 液体电介质
liquid dielectric(al) capacitor 液体介质电容器;液态电解质电容器
liquid differentiation 液态分异作用
liquid diffusion factor 液体扩散因子
liquid diffusity 液体扩散率
liquid dimmer 液体减光器
liquid discharge 排液
liquid discharging operation 卸液作业
liquid disintegrated cast iron 液碎铸铁
liquid disintegrated powder 液碎铁粉
liquid dispenser 液体喷雾器
liquid displacement 液体置换
liquid distribution 淋水装置
liquid distributor 液体分配头;喷液机;喷液车;布液管
liquid dividing head 液体取样头
liquid dominated system 液态为主的地热系统
liquid dosimeter 流体剂量计
liquid dosing apparatus 液剂量计;流体剂量计
liquid dryer 液体干燥剂;液体干料;液体催干剂;燥液
liquid drop 液滴
liquid drop medium 液滴媒质
liquid drop model 液滴模型
liquid drop model of nucleus 核的液滴模型
liquid drop separator 液滴分离器
liquid dryer 液体干燥剂
liquid dump bailer 灌浆桶
liquid dung 厩水;粪液;粪水
liquid duplicating 湿法复制
liquid dye laser 液体激光器
liquid dye penetrant examination record 液体着色渗透探伤记录
liquid dynamics 液体动力学
liquid effluent 液体流出物
liquid elastic deformation 液体弹性形变
liquid electrode 液体电极
liquid electrolyte 液体电解质
liquid electrolytic capacitor 电解液电容器
liquid element 液态元素
liquid elevating valve 液面阀
liquid elution chromatography 液相洗脱色谱
liquid emulsion 液体容器液体乳胶
liquid encapsulate Czochralski technique 液封直拉法
liquid encapsulated Czochralski growth 液封提拉法
liquid encapsulation technique 液相覆盖技术;液体掩盖技术
liquidensitometer 液体密度计
liquid entrainment 液体雾沫
liquid envelope 液体包封;薄塑料涂层
liquid ethylene 液态乙烯
liquid expanded film 液态扩张膜
liquid expansion thermometer 液体(膨胀)温度计
liquid explosive 液体炸药;液态炸药
liquid explosive drill 液体炸药钻机
liquid extract 液体提出物;流浸膏(剂);萃取液
liquid extraction 液体提取
liquid face 液面
liquid fat 液态油脂
liquid feed pump 液体进料泵
liquid fertilizer 液体肥料
liquid fertilizer mixer-proportioner 液肥混合调配器
liquid-filled acoustic(al) lens 充液声透镜
liquid-filled fiber 充液纤维
liquid-filled fuse unit 液体熔断器;充油熔断器
liquid-filled membrane pressure ga(u)ge 充液隔膜压力计
liquid-filled motor 充液式电动机
liquid-filled porosity 充液的孔隙性
liquid-filled prism 充液棱镜
liquid-filled thermometer 液体温度计;液体温度表;充液式温度计
liquid filler 液体填孔剂;液态腻子
liquid filling machine 液体分装机;灌装机
liquid film 液体膜;液膜

liquid film coefficient 液态膜系数;液膜(传递)系数
liquid film control(ling) 液膜控制
liquid film lubrication 液膜润滑
liquid film resistance 液膜阻力
liquid film resonance 液膜共振
liquid film seal 液膜密封
liquid film separation process 液态膜分离工艺
liquid film type 液膜式
liquid filter 液体(过)滤器;铝合金过滤器
liquid filtering cartridge 液体过滤筒
liquid filtration 液体过滤
liquid fire 液体燃烧剂火焰
liquid fire gun 喷火器
liquid Fisher-Tropschhydrocarbons 费托液体烃
liquid flame hardening 液体火焰淬火
liquid floated gyroscope 液浮陀螺仪
liquid floated pendulous accelerometer 液浮摆式加速度仪
liquid flooding 液体溢流
liquid floor soap 液体洗地板皂
liquid flow 液体流(动);液流
liquid flow counter 流液计数器
liquid flow equation 流体流动方程
liquid flow indicator 液体流动指示器
liquid flow measurement 液体流量测量
liquid flowmeter 液体流量计
liquid flow rate 液体流率
liquid flow sight glass 液体流动看窗
liquid fluidized bed 液体流化床
liquid flux 液体溶剂
liquid flux cover 液态熔剂盖
liquid forging 模压铸造
liquid for self-curing resin 自凝树脂液
liquid for zinc-phosphate cement 磷酸锌粘固粉液
liquid friction 液体摩擦
liquid fuel 液体燃料;液态燃料;流体燃料
liquid fuel burner 液体燃料燃烧器
liquid fuel burning appliance 燃油设施
liquid fuel commodities 液体燃料矿产
liquid fuel motor 液体燃料发动机
liquid fuel oil 液体燃料油
liquid fuel reactor 液体燃料堆
liquid fuel sustainer 液体燃料主发动机
liquid full try cock 液位检查旋塞
liquid funds tax 流动资金占用税
liquid funds tax rate 流动资金占用税率
liquid furnace 浴炉
liquid fuse 液体熄弧保安器;液体熔丝;液体熔断丝;充油熔断丝
liquid fuse unit 油熔断器;液体熔断器;液体保安器;液浸保安器
liquid gallon 液加仑
liquid gas 液化气(体)
liquid gas distributor 液气分布器
liquid gas equilibrium region 液—气两相平衡区
liquid gas fuel 液化气体燃料
liquid gasket 液体密封(料)
liquid-gas ratio 液气比
liquid-gas system 液气系统
liquid-gas transfer compressor 液化气输送压缩机
liquid-gas vehicle 液化气体燃料车
liquid gate 液体片门;液门
liquid ga(u)ge 液体负压计
liquid gel 液态凝胶
liquid glass 液体玻璃;液态玻璃;硅酸钠
liquid glue 液态胶;液胶
liquid gradient 液体梯度
liquid gravity control 液体重力控制
liquid grease 液体油脂;液体润滑脂
liquid growth 液相生长
liquid gum 液态胶质
liquid hammer 液击;液锤
liquid hand compass 手持液体罗盘
liquid handling pump 液压泵
liquid hardener 液态硬化剂(混凝土表面处理用)
liquid hardening 液体淬火
liquid hauler 液体牵引机
liquid hauling truck 液体运输车
liquid head 液柱;液(体)压头
liquid header 液体收集器;液体收集管;排液总管
liquid heater 液体加热器
liquid heat exchanger 液体热交换器
liquid heating fuel 采暖燃油
liquid helium 液态氦

liquid hold-up 液体滞留(器);液体藏量
liquid honing 液体研磨;液体打磨;喷气清理;水砂抛光;水磨
liquid honing machine 液体研磨机
liquid horse-power 液体马力;液体功率
liquid hourly space velocity 液体空间速度
liquid hydrocarbon 液烃;液体烃;液态烃
liquid hydrocarbonceous fuel 液烃燃料
liquid hydrogen 液氢
liquid hydrogen bath 液氢槽
liquid hydrogen bubble chamber 液氢气泡室
liquid hydrogen dewar 液态氢的绝热容器
liquid hydrogen plant 液氢站
liquid hydrogen pump 液氢泵
liquid hydrogen sulfide 液态硫化氢
liquid-immersed reactor 液浸电抗器;充油电抗器
liquid-immersed transformer 液浸变压器
liquid-immersible 液体密封的;液体可浸入的
liquid immersion test 液体浸入法试验
liquid immiscibility 液体不混溶性;液态分离作用;熔离作用
liquid in clinker 熟料中熔融物
liquid inclusion 液体包裹体;液态包裹物
liquid index 液性指数
liquid indicator 液体流动指示器;液面指示器;液体指示器;液面计
liquid inductance 液感
liquid inelastic deformation 液体非弹性形变
liquid-in-glass thermograph 玻管液体温度计
liquid-in-glass thermometer 液体温度计;玻璃(管)液体温度计;玻管液体温度计
liquid injection valve 液体注射阀
liquid ink 墨汁
liquid inlet 液体入口
liquid insulating material 液体绝缘材料
liquid insulation 液体绝缘
liquid insulator 液体绝缘体
liquid interface indicator 液体界面指示计
liquid interferometer 液体干涉仪
liquid inventory 熔液流量;溶液流量
liquid investment 流动投资;包埋料液
liquid investment audit 短期投资审计
liquid investment income audit 短期投资收益的审计
liquid investment source audit 短期投资来源的审计
liquid investment transfer audit 短期投资转让的审计
liquid investment valuation audit 短期投资计价的审计
liquid ion exchange 液体离子交换
liquid ion exchange membrane electrode 液体离子交换膜电极
liquid ion exchanger 液体离子交换剂;液态离子交换剂
liquid ion exchange technique 液态离子交换技术
liquid iron 铁水
liquid isostatic(al) pressing 液等静压成型
liquidity 液性;流动程度;变现能力
liquidity analysis 流动性分析
liquidity balance 清偿差额
liquidity basis 清偿基础
liquidity cost 流动资金的费用
liquidity dilemma 清偿中的困境
liquidity diversification 流动性多样化
liquidity factor 液性指数
liquidity index 液性指数;流性指数;流动性指标
liquidity limit 液性极限;流动性极限
liquidity preference demand 流动偏好需求
liquidity preference function 流动偏好函数
liquidity premium 流动性升水
liquidity rate 变现率
liquidity ratio 流动(性)比率;清偿能力比率
liquidity report 流动性报告
liquidity risk 流动性风险;清算风险
liquidity shortage 清偿力不足
liquidity squeeze 流动性紧缩
liquidizing point 液化点
liquid jet 液体喷射
liquid jet air pump 喷吸气泵
liquid junction 液体接头
liquid junction potential 液体接界电势;液体接触电位;液界电位差;液结电势
liquid koji 液体曲

liquid lamella 液体薄膜
liquid laser 液体激光器
liquid latex 液态胶乳
liquid layer 液层
liquid layer column 液层柱
liquid lens 液体透镜;液态透镜
liquid level 液位;液面
liquid level alarm 液位报警器;液面警报器
liquid level coefficient of reactivity 反应性水位系数
liquid level control 液位调节;液面控制
liquid level controller 液位控制器;液面控制器;液面调节器
liquid level control range 液面控制范围
liquid level depth after job 举空深度
liquid leveler 液面计;液面高度计
liquid level ga(u)ge 液位(流量)计;液体水平压力计;液面指示器;液位仪
liquid level ga(u)ge against frost 防霜液面计
liquid level ga(u)ge of medium pressure 中压液面计
liquid level ga(u)ge with float 浮子液面计
liquid level ga(u)ge with jacket 夹套型液面计
liquid level ga(u)ge with lining 带衬里液面计
liquid level ga(u)ge with magnetic buoyage 磁性浮标液面计
liquid level ga(u)ge with nozzle 带颈液面计
liquid level ga(u)ge with remote controlled buoyage 浮标遥测液面计
liquid level in casing 套管内液体水平
liquid level indicating transmitter 液面指示变送器
liquid level indicator 液位指示器;液位指示计;液面指示器;液面(指示)计;液面仪;反射式玻璃液面表
liquid level inspection cock 液位检查旋塞
liquid level measurement 液面测定
liquid level meter 液位计
liquid level method 液面水平测量法
liquid level recorder 液位记录器;液面记录器;自记液面计;自动液位记录器
liquid level recording controller 液面记录控制器
liquid level regulator 液位调节器
liquid level relay 液面浮动继电器
liquid level sensor 液位传感器;液面传感器
liquid level signal 液位指示信号
liquid level switch 液位开关
liquid level transmitter 液位遥测仪;液面发送器
liquid liabilities audit 流动负债审计
liquid liability 流动负债
liquid limie 液态极限
liquid limit 液性限度;液性限界;液性界限;液限;黏点;流(性)限(度)
liquid limit apparatus 液限仪;流限仪
liquid limit determination 液限测定
liquid limit device 液限仪
liquid limiting centrifugation 液体极限离心分离
liquid limit machine 液限仪
liquid limit of soil 土壤液限;土的液限
liquid limit test 液限试验
liquid line 液(相)线;液体管线;液体管(路)
liquid-liquid chromatography 液一液色谱法
liquid-liquid chromtagraphy 液一液层析
liquid-liquid column chromtagraphy 液液柱状色谱法;液液柱状层析
liquid-liquid electrode 液一液电极
liquid-liquid extraction 液一液萃取法
liquid-liquid extraction gas chromatography 液液萃取气相色谱法;液液萃取气相层析
liquid-liquid extraction separation 液液萃取分离
liquid-liquid extraction system 液液提取系统
liquid-liquid immiscibility 液液不混溶性
liquid-liquid partition chromatography 液液分配色谱法
liquid-liquid phase separation 液液相分离
liquid live microorganism 流动活微生物
liquid load calorimeter 流体负载量热计
liquid lubricant 液体润滑剂;液态润滑剂
liquid lubrication 液体润滑
liquid macrogol 液体聚乙二醇
liquid magma 液态岩浆
liquid magnesium 液态镁
liquid magnesium chloride 液态氯化镁
liquid manometer 液体压力计;液拉压力表
liquid manure 粪液

liquid mass stability 液体质量稳定性
liquid mature 液体肥料
liquid mature gutter 厩液肥沟
liquid mature pit 液肥坑
liquid measure 液体计量;液体测量器;液量(度量);液量单位
liquid measurement 液体计量
liquid media penetration rate 液体介质渗透速率
liquid medium 液体培养基;液体介质;液态介质
liquid medium sonic delay line 液体介质声延迟线
liquid melt 熔融物
liquid membrane 液膜
liquid membrane concrete curing 混凝土液膜养护
liquid membrane curing 液膜养护
liquid membrane curing compound 液态养护膜
liquid membrane electrode 液态薄膜电极;液膜电极
liquid membrane extraction technique 液膜萃取技术
liquid membrane forming compound 液体薄膜形成剂(养护混凝土用)
liquid membrane isolation 液膜分离
liquid membrane isolation wastewater separation 液膜分离废水处理
liquid membrane process 液膜工艺
liquid membrane separation 液膜分离
liquid membrane separation technology 液膜分离技术
liquid memory 液体存储器
liquid mercury 液态汞
liquid mercury lift 液汞提升
liquid metal 液态金属;熔融金属
liquid metal breeder reactor 液态金属增殖堆
liquid metal brush 液态金属电刷
liquid metal cathode 液态金属阴极
liquid metal charge 液态金属装料
liquid metal coolant 液态金属冷却剂;液态金属冷却介质
liquid metal-coolant circuit 液态金属冷却剂回路
liquid metal-cooled reactor 液态金属冷却堆
liquid metal extraction 液态金属萃取
liquid metal fast breeder reactor 液体金属快速增殖反应堆
liquid metal forging 模压铸造
liquid metal fuel 液态金属燃料
liquid metal fuel cell 液态金属燃料电池
liquid metal fuel reactor 液态金属燃料堆
liquid metal heat exchanger 液态金属热交换器
liquid metal heattransfer fluid 液体金属传热流体
liquid metal heat-transfer system 液态金属传热系统
liquid metal infiltration 液态金属渗透
liquid metal magnetohydrodynamic generator 液态金属磁流体发电机
liquid metal nuclear fuel 液态金属核燃料
liquid metal reactor 液态金属反应堆
liquid metal slip-ring 液态金属集电环
liquid metal spray column 液态金属喷雾塔
liquid metal steam turbine 液态金属汽轮机
liquid metal welding 浇注补焊
liquid meter 液体计量器
liquid metering vessel 液体计量器
liquid methane 液体甲烷
liquid method 液施法;泥水稳定液施工法(防护钻孔壁面用)
liquid milk wastewater 液态奶废水
liquid mill 液体磨机
liquid mirror 液体镜;液体反射镜
liquid moderator reactor 液体慢化剂(反应)堆
liquid monomer plastic 液态单体塑胶
liquid monomer process 液态单体工艺
liquid mortar densifier 液态砂浆增浓剂;液态灰浆稠化剂
liquid motor fuel 液体动力燃料
liquid mud-stone flow 稀性泥石流
liquid natural gas 液化天然气
liquid natural gas carrier 液化天然气船
liquid natural gas terminal 液化天然气码头
liquidness 液态
liquid nitrogen 液态氮;液(化)氮
liquid nitrogen(cold) trap 液氮冷阱
liquid nitrogen container 液氮储[贮]存罐
liquid nitrogen converter 液氮气化器
liquid nitrogen-cooled detector 液氮致冷探测器

liquid nitrogen-cooled generator 液氮冷却(超导)发电机
liquid nitrogen cooling 液氮致冷;液氮制冷;液氮冷却
liquid nitrogen cooling crystal 液氮致冷晶体
liquid nitrogen cooling system 液氮冷却系统
liquid nitrogen engine 液态氮发动机
liquid nitrogen freezing system 液态氮冷冻装置
liquid nitrogen temperature 液氮温度
liquid nitrous oxide 液态氧化二氮
liquid nutrient medium 液体培养基
liquid of high density 高密度液体
liquid of silicate cement 硅黏固粉液
liquidoid 液相;固溶线
liquid oil in cavity 孔洞油苗
liquid oil in fossil cast 化石印膜中油苗
liquid oil in vug 晶洞油苗
liquidometer 液位计;液面计
liquid opaque 修改液
liquid optic(al) adhesive 液体光学胶黏剂
liquid ore 液态矿
liquid outlet 液体出口;滤液出口
liquid oxidation fat acid 液态化脂肪酸
liquid oxygen 液氧;液态氧
liquid oxygen-alcohol unit 液氧酒精发动机
liquid oxygen cartridge 液氧爆破筒
liquid oxygen explosive 液氧爆炸;液(态)氧炸药
liquid oxygen-gasoline unit 液氧汽油发动机
liquid oxygen generator 液态氧发生器
liquid oxygen pump 液氧泵
liquid oxygen storage tank 液态氧储[贮]存柜
liquid pack 液体密封;水封
liquid packing 液力密封;液封;水力压紧;水封器
liquid paper 修正液
liquid paraffin 液状石蜡;液体石蜡;石蜡油
liquid parting 脱模液
liquid partition chromatography 液相分配色谱法
liquid penetrant examination 液体渗透探伤
liquid penetrant inspection 液体着色探伤;液体渗透检验法;渗液探伤;渗液检验(法)
liquid penetrant test 渗液探伤试验
liquid penetrant testing 液体渗透探伤
liquid penetration test 液体渗透试验
liquid petrolatum 凡士林;液状石蜡;液体矿脂;石蜡油
liquid petrolatum quality test 液体矿脂质量试验
liquid petroleum asphaltic bitumen 液体石油沥青
liquid petroleum gas 瓶装液化(石油)气
liquid petroleum gas ship 液态石油气运输船
liquid petroleum gas system 液化器系统
liquid phantom 液相
liquid phase 液相
liquid phase activated carbon treatment 液相活性炭处理法
liquid phase adsorption method 液相吸附法
liquid phase air oxidation 液相空气氧化
liquid phase burner 液相燃烧器
liquid phase chromatography 液相色谱
liquid phase combustion 液相燃烧;液态燃烧
liquid phase cracking 液相裂化
liquid phase cracking process 液相裂化过程;液相分解过程
liquid phase deposition 液相沉淀
liquid phase epitaxial growth 液相外延生长
liquid phase epitaxial method 液相外延法
liquid phase epitaxy 液相外延
liquid phase extraction 液相抽提
liquid phase fluorination 液相氟化
liquid phase granular activated carbon 液相颗粒活性炭
liquid phase gum 液相胶
liquid phase hot pressing 液相热压
liquid phase hydrogenation 液相氢化
liquid phase immiscibility 液相不混溶
liquid phase interference 液相干扰
liquid phase loading 液相荷载量
liquid phase mode 液相模
liquid phase nitration 液相硝化
liquid phase of soil 土壤液相
liquid phase operation 液相操作
liquid phase oxidation 液相氧化(作用)
liquid phase polymerization 液相聚合
liquid phase reaction 液相反应
liquid phase refining 液相精制

liquid phase region 液相区
liquid phase separation 液相分离
liquid phase sintering 液相烧结
liquid phase sintering mechanism 液相烧结机理
liquid phase surface 液相面
liquid phase suspension process 液相悬浮过程
liquid phase thermal cracking 液相热裂化
liquid phase volume 液相量
liquid piezometer 液体测压计
liquid piston compressor 液体活塞式压缩机;液环式压缩机;水环式压缩机
liquid piston rotary compressor 液体活塞式旋转压气机
liquid plasticizing aid 液体增塑剂
liquid polishing agent 液体抛光剂;抛光液
liquid polymer 液相聚合物
liquid polymerization 液相聚合
liquid potential 液面电位差
liquid potential function 流势函数
liquid power 液体动力燃料
liquid prefilming agent 液体预膜剂
liquid pressing 液压模锻
liquid pressure 水压头;液压;液体压力
liquid pressure drive 液体传动
liquid pressure ga(u)ge 液体压力计
liquid pressure nitriding 加压液体氮化法
liquid pressure pick-up 液体压力传感器
liquid pressure pump set 液压泵组
liquid pressure scales 液压秤
liquid pressure transducer 液压转换器
liquid pressurized gas 液态压缩气体
liquid prism 液体棱镜
liquid propane 液态丙烷;液化丙烷
liquid propane engine 液化丙烷燃料发动机
liquid propellant 液体燃料;液体火箭燃料
liquid propellant injection system 液体燃料喷射系统
liquid propellant rocket engine 液体推进剂火箭发动机
liquid propellant starter 液体推进剂启动机
liquid propellant sustainer 液体推进剂主发动机
liquid proportioner 液体比例混合器
liquid pump 液体泵;泵
liquid pump seal 液泵密封
liquid purification 液体精制法;湿法脱硫
liquid purification process 液相精制过程
liquid quantity meter 液体流量计
liquid quart 液量夸脱
liquid quench hardening glass 液冷钢化玻璃
liquid ratio 流动比率
liquid reagent 液体试剂
liquid receiver 储液器
liquid recorder 液位记录器
liquid rectifier 液体整流器;电解整流器
liquid recycle model 液体再循环模型
liquid re-distributor 液体再分布器
liquid regulating resistor 液体可变电阻器
liquid regulator 液体调节器
liquid reservoir 储[贮]液器
liquid residue 液状残渣;液体残渣
liquid resin 树胶;液体松脂;液态树脂
liquid resistance 液体电阻;液体变阻器
liquid resistor 液体介质电阻器;液体电阻器
liquid resistor rating plant (牵引动力装置的)水阻试验装置
liquid resistor testing plant (牵引动力装置的)水阻试验装置
liquid rheostat 液体变阻器;液浸变阻器
liquid rigidity 液体刚性
liquid rime 液态霜淞
liquid ring compressor 液环式压缩机;水环式压缩机
liquid ring pump 液体环式泵
liquid ring vacuum pump 液环式真空泵
liquid roofing 液态屋面材料;屋面涂料
liquid roofing material 液态屋面材料
liquid root canal filling 根管塑化
liquid rosin 松香;液体树脂;妥尔油;松浆油
liquid rotary compressor 液环式压缩机
liquid rotary pump 液环泵
liquid rubber 液态橡胶
liquid sample counter 液体试样计数管
liquid sampler 液体取样器
liquid scintillation 液体闪烁现象;液体闪烁(法)

liquid scintillation activity meter 液体闪烁测量仪
liquid scintillation counter 液体闪烁计数器
liquid scintillation counting 液体闪烁计数
liquid scintillation detector 液体闪烁探测器
liquid scintillation radioassay 液体闪烁放射性分析
liquid scintillation spectrometer 液晶闪烁分光计
liquid scintillation spectrometry 液体闪烁光谱测定
liquid scintillation spectrum 液体闪烁光谱
liquid scintillation technique 液体闪烁技术
liquid scintillator 液体闪烁体;液体闪烁谱
liquid scrubber 液体洗涤器
liquid scrubbing 液体洗涤
liquid seal 油封;液压封闭;液封;水封(器)
liquid sealant 液状密封剂
liquid sealed discharge 液封放卸
liquid sealed pump 液封泵
liquid sealing 液体密封
liquid sedimentation balance 微粒沉积分析天平
liquid self-hardening sand 流态自硬砂
liquid semiconductor 易变半导体;液体半导体
liquid separation membrane 液体分离膜
liquid separator 液体分离器
liquid sewage sludge 液态污水污泥
liquid shear 液体切力
liquid sheet 液膜
liquid sheet disintegration 液膜破裂
liquid ship 液化货船
liquid shrinkage 液体收缩;液态收缩
liquid shut-down system 液体停堆系统
liquid siccative 干燥剂;液体脱水剂
liquid sight glass 液体观察角
liquid simulating bed 液体模拟床
liquid sintering 液相烧结
liquid slag 液体炉渣;液态渣;熔融渣
liquid slag producer 液体排渣发生炉
liquid slip-regulator 液体转差调节器
liquid sludge 液(体)状污泥;液态污泥;湿污泥
liquid sludge pump 液体厩肥泵;泥浆肥料泵
liquid slugging 液击
liquid smoke 分馏木材的首馏分
liquid soap base 液体皂基
liquid soap dispenser 皂水器(洗手用)
liquid sodium 液体钠
liquid solid adsorption 液固吸附
liquid solid adsorption chromatography 液固吸附色谱法
liquid solid boundary 液固界面
liquid solid chromatography 液固色谱(法)
liquid solid contraction 凝固收缩
liquid solid cyclone 水力旋流器
liquid solid equilibrium 凝固平衡
liquid solid extraction 凝固萃取
liquid solid separation 液固分离
liquid solid transition point 液固(态)转化点
liquid solution 液体溶液
liquid sonic cell 液体声池
liquid sorbent dehumidifier 液体吸附减湿器
liquid source diffusion 液态源扩散
liquid source radium content 液体源镭含量
liquid space velocity 液体空间速度
liquid sprayer 液体喷雾器
liquid-spreader 喷液机
liquid spring 液体弹簧
liquid springing 液压减振
liquid spring unit 液压悬架装置;液体减震器
liquid squeeze 液体压缩
liquid starter 液体起动器
liquid state 液状;液态
liquid state bioconversion process 液态生物转化工艺
liquid state diffusion 液态扩散
liquid steel 钢液;钢水
liquid storage 储[贮]液罐
liquid storage sphere 储液球罐
liquid storage vessel 储[贮]液容器
liquid stream 液流
liquid strength 液体强度
liquid sulfur dioxide 液态二氧化硫
liquid sulfur dioxide-benzene process 苯液体二氧化硫法
liquid sulfur dioxide extraction process 液态二氧化硫抽提过程
liquid surface 液体界面;液面
liquid surface acoustic(al) holography 液面声

全息(术)
liquid surface adsorption 液面吸附
liquid surface interferometer 液面干涉仪
liquid surface level 液面标高;液面高程
liquid surface load 液面负荷
liquid surface waterproofing agent 液体表面防水剂
liquid surfactant membrane 液表面活性剂膜
liquid surge baffle 液体缓冲挡板
liquid suspension 液体悬浮液
liquid suspension reactor 液体悬浮燃料堆
liquid sustainer 液体推进剂主发动机
liquid tach(e)ometer 液体转速计
liquid tar product 液体焦油产物;液体柏油制品
liquid thermal diffusion 液相热扩散法
liquid thermometer 钢管水银土壤强度计
liquid thermostat 液体节温器;液体恒温器
liquid-tight 液体不能透过的;液封的;液密的;不透液的
liquid-tight dustguard joint 水封防尘连接
liquid tightness 液体紧密性
liquid-tight rubber mo(u)ld 防水橡胶模
liquid-tight waste 不透液废物
liquid-tight waterrepeller 液密的防水剂
liquid-tight wax 液密的蜡
liquid-tight workability agent 液密增塑剂
liquid trailer 液罐挂车
liquid trap 液体阱;液体分离器;液体捕集器;集液器;疏液器
liquid tubular manometer 液体管式压力表
liquid upper limit 流性上限
liquidus 液相线;液线
liquidus curve 液相(曲)线
liquidus line 液相曲线;液线
liquidus phase 液相面
liquidus sintering 液相线烧结
liquidus temperature 液(相)线温度
liquid valve 液体阀
liquid vapo(u)r heat exchanger 液汽热交换器
liquid vapo(u)r interface 液气界面
liquid vapo(u)r mixture 液汽混合物
liquid vapo(u)r surface 液气分界面;水气分界面
liquid viscosimeter 液体黏度计
liquid viscosity 液体黏性;液体黏度
liquid viscosity theory 液体黏度理论
liquid volume 液体体积
liquid volume hourly space velocity 液体体积空间速度
liquid volume measurement 液体容积计量
liquid volume viscosity 液体体积黏度
liquid vortex contactor 液体涡动接触器
liquid wall 液壁
liquid wall ionization chamber 液壁电离室
liquid waste 废液;液状污物;液体废物;液体废污;废水
liquid waste and solid waste disposal and treatment 液体废物和固体物的弃置和处理
liquid waste disposal 液体废物处理;废液处理;废水废液处理系统
liquid waste disposal and treatment 液体废物的弃置和处理
liquid waste disposal technique 液体废物处理技术
liquid waste from nuclear reactor 核反应堆废液
liquid waste incineration 液状废料焚化;液体废料焚化
liquid waste incinerator 液状废料焚化炉;废液焚烧炉
liquid waste processing system 废液处理系统
liquid water 液体水;液态水
liquid water content 液态水含量
liquid water content of snow 雪水含量
liquid wax 蜡水;液体石蜡;液体蜡
liquid weigh process 液称量法
liquid whistle 液体警笛
liquid yield 液体产率
liquid zone 液区
liquid zoning 液体区域化
liquification 液化
liquified acetylene 液态乙炔
liquified air 液化空气
liquified air container 液态空气容器
liquified chlorine 液态氯
liquified chlorine gas 液态氯气
liquified ethylene 液态乙烯

liquified fuel gas 液化气体燃料
liquified gas 液化气体
liquified gas carrier 液化石油气船
liquified hydrogen 液态氢
liquified nitrogen consumption 液氮消耗量
liquified oxygen 液态氧
liquifier 液化剂;稀释剂
liquilizer 液化器
liquimetric autopour 定量自动浇注
liquogel 液状凝胶;液胶体
Liquon 澄清池(其中的一种)
liquor 液体;液剂;流体;料液;溶液(剂)
liquor ammoniae fortis 浓氨溶液
liquor analysis 溶液分析
liquor condensate 冷凝液
liquor(feed liquor) inlet 料液进(入)口
liquor finish 液体抛光
liquor finish draw 湿法拉制
liquor finishing 钢丝染红处理
liquor gutta percha 马来乳胶溶液
liquor iodine phenolatus 酚制碘溶液
liquor-level regulator 液面调节器
liquor of the same tap 同样品质的酒
liquor plumbi subacetatis dilutus 稀碱式醋酸铅溶液
liquor plumbi subacetatis fortis 浓碱式醋酸铅溶液
liquor pump 液体泵
liquor ratio 液比
liquorrhea 液溢
liquor sample 液体试样;溶液试样
liquor separation 液体分离
liquor separator 液体分离器;分液器
liquor store 酒库;液体储藏
liquor stream 液流
liquor-to-wood ratio 液比
liquostriction 液浸变形
lira 里拉
liroconite 水砷铝铜矿
L-iron 角铁;角钢
lisena (罗马式的)壁柱条带
Lishi loess 离石黄土
lisimeter 土壤渗透仪;土壤渗水仪;土壤渗水计;测渗计
liskeardite 砷铁铝石;砷铁铝矿
Lissamine green 利萨明绿
Lissamine violet 利萨明紫
list 一览表;列举;链表检索;目录表;名单;名册;清单;倾侧;上市的全部股票;表格;表册;边材木条;列表;横倾【船】
List and Index of Admiralty Chart Folio 英版海图图夹索引图
list area 表域;表区
list building 开列名单
list cell 表元
list compacting 表格压缩
list data structure 表(格)数据结构
list-directed 表式
list-directed data field length 表式数据域长度
list-directed data specification 表式数据说明部分
list directed input 表式输入;表控输入
list-directed input list 表控(制)输入表
list directed input-output 表(列)式输入输出;表控输入输出;表格型输入输出
list-directed input-output statement 表式输入输出语句
list-directed length of field 表式域长度
list-directed output 表式输出
list-directed transmission 表式传输;直接表格传送;表格式传输
listed brand (指在交易所的)上场商品
list edge 锡鳞;毛翅
list editor program(me) 表编辑程序
listed price 牌价;明码
listed ships 在册船舶
listed stock (指在交易所挂牌的)上市股票
listed substance 列入清单物质
listel 柱槽突面;平缘;平板脚;扁带饰
list element 列表元(素)
listen continuously 连续监听值班
listener 收听器
listening appliance 收听装置;收听器
listening device 助听器;窃听器;听音器;收听装置;收听器
listening-in device 窃听装置

listening-in line 监听线
listening jack 应答塞孔
listening key 监听键
listening method 听声法(检查地下管道漏水)
listening mine 声响水雷
listening mode 监听方式;收听方式
listening plug 听音插头
listening post 监听站;潜听哨
listening set 被动声呐
listening sonar 噪声定位声呐;停测声呐;听声声呐;被动声呐
listening station 侦察站
listen only sonar 噪声定位声呐;被动声呐
lister 制表人;双壁开沟犁
lister furrow 犁沟
listerine 防腐溶液
Lister lens 利斯式透镜
Lister objective 利斯式物镜
Lister's anti-septic 氰化锌汞
lister shovel 双壁开沟铲
list event 表事件
list file 表文件
list file structure 表文件结构;编目文件结构
list for investigation 调查表
list form 表格形式
list frame sampling 名单框抽样
list handling statement 表处理语句
list head in thread tree 穿线树中的表头
listing 列表(编目);锉饰边缘;侧倾;修边;边材;横倾【船】
listing agent 上市代理人(房地产)
listing angle 侧倾角
listing basin 鱼鳞坑;围垄池;田间蓄水池
listing contract 房地产商合伙契约
listing control 编表控制
listing method 列表法
listing of a ship 船身倾斜
listing on the exchange 在交易所挂牌;在交易所登记
listing program(me) 上市程序
listing ship 倾斜船
listing technique 列举法
listing unit 名册单位
list inserting 表插入法
list insertion 表插入
list language 表语言
list linearization 表格线性化
list manipulation 表处理
list notation 表标记
list of anchor bolts for the equipments 设备地脚螺栓一览表
list of architects 建筑师索引;建筑师目录
list of articles 物品清单
list of available space 可利用空间表
list of award 决标单
list of balance 结余清单
list of bidding items 招标项目
list of bidding items and quantities 投标项目与数量清单
list of bolts and tie rods 螺栓及拉杆表
list of building materials 建筑材料表
list of buildings and structures 建筑物构造物名称表
list of cargo 装货清单
list of colo(u)r indices 彩色索引表
list of conspicuous object 显著物标一览表;显著目标一览表
list of constructions 建筑结构索引
list of contents 图纸目录
list of coordinates 坐标表
list of cost estimate 估价单
list of currency 货币及金银清单
list of dam sites 坝址一览表
list of direction 方向表
list of directive radio beacon 无线电指向标表
list of distribution 分配表
list of documents 合同文件目录;文件目录
list of drawings 图纸清单;图纸目录
list of element 元素表
list of equipment 设备清单;设备目录
list of equipment and materials 设备及材料明细表;设备材料清单
list of errata 正误表;勘误表
list of expenses 费用清单

list of fixtures 备品目录
list of forwarded traffic 发送货物(运输)清单
list of furniture and fixture 设备装置目录
list of geographic(al) names 地名表
list of hiring charges 租费清单
list of illustration 插图目录
list of import commodities 进口商品目录
list of instruments 文件目录
list of internal finishes 房间做法表
list of labo(u)r and materials 工料单
list of leeway 风压差表
list of lights 灯塔表;灯标表
list of lintels 过梁表
list of machinery and materials 设备材料清单
list of main equipment 主要设备(明细)表
list of materials and fittings 材料及设备明细表;材料及配件明细表
list of materials and labo(u)r 工料单
list of materials and main equipment 主要设备材料清单;主要设备材料表
list of modification 更改清单
list of notion 概念表
list of open points 待定项目清单
list of parts 零件目录(表)
list of pattern indices 图形索引表
list of payment 付款表
list of piping erection 管道安装览表
list of piping supports 管道支架一览表
list of pollutant 污染物清单
list of polluting substance 污染物清单
list of polyline indices 折线索引表
list of preliminary determination 初定震中表
list of prices 价目表;价格表
list of property 财产清单;财产目录
list of quantity of doors and windows 门窗数量表
list of quotations 牌价
list of radio beacon 无线电指向标表
list of radio signals 无线电信号表
list of railway goods tariff rate 铁路货物运价率表
list of reinforcement 钢筋表
list of sailings 船期表
list of schedule 钢筋表
list of ship's stores 船舶备用物品
list of signs 图式符号;图例;符号表
list of spare parts 备件清单
list of standard drawings 标准图纸目录
list of technical characteristic 技术特性表
list of text indices 正文索引表
list of the comprehensive materials 综合材料表
list of the equipment 设备一览表
list of views 对景图目录
list of working drawings 施工图(图纸)目录
Listomatic camera 照相复印机
Liston chopper 利斯顿机械换流器
list price 价格目录价;基本价格;目录价(格);定价
list price method 标准价格法
list procedure 列表程序
list processing 表处理
list processing condition 表处理条件
list processing function 表处理功能;表处理操作
list processing language 表(格)处理语言;报表处理语言
list processing routine 表处理程序
list processing structure 表处理结构
list processing system 列表处理系统;表加工系统;表格处理系统
list processing technique 表处理技术
list processor 表(格)处理机;编目处理机
list pump 横倾平衡泵
list reclamation 表的回收
list record(ing) content 记录内容列表
list record(ing) headers 记录标头列表
list representation 表格表示法
listric fault 犁式断层;铲形断层
listric surface 铲形断裂面;凹形断裂面
list sampling 名单抽样
list scanner 表扫描程序
list schedule 表格调度法
list scheduling 表(格)调度
list separator 表分隔符
list sorting method 表分类法
list storage 表的存储
list structure 链表结构;条状结构;带状结构;表结构
list structure form 表结构形式

list table 编目表
list to port 左舷倾侧
listvenite 滑石菱镁片岩
litany desk 牧师跪祷桌
lit-by-lit 一层一层的
litchfieldite 钠云霞正长岩
litchi 荔枝
lite 窗玻璃片;玻璃窗;平板玻璃切裁尺寸
lite limestone 鲕状石灰岩
liter 立升;升
literage limits 驳船作业区
literal 程序文字
literal code 文字编码
literal coefficient 文字系数
literal constant 文字常数
literal contract 成文合同;书面合同
literal equation 文字方程
literal expression 文字表达式;文字表达法
literal node 基本式节点;文字节点
literal notation 文字注记;文字记号
literal operand 文字操作数
literal pool 文字库
literal register 常数寄存器
literal translation 逐字翻译;直译
literary language 正式语言
literary property 版权
literature 著作;文献
literature cited 引用文献;参考文献
literature reference 参考文献
literature research 文献检索
literature review 文献评述
literature search 文献检索
literature survey 文献调查
liter capacity 公升容积
liter per second 升每秒
liters of water 公升水
liter weight 升重
liter weight of clinker 熟料升重
liter weight test of clinker 熟料的升重测定
lithamide 氮化锂
litharenite 岩屑砂屑岩
litharge 一氧化铅;氧化铅;密陀僧;铅黄
litharge-glycerin cement 铅黄甘油胶合剂
litharge-glycerine mortar 一氧化铅与甘油灰泥
litharge stock 一氧化铅混合剂
Lithargyrum 密陀僧
lithargysmus 一氧化铅中毒;密陀僧中毒
Lithcote 里得寇特(一种涂料)
lithe board 易变形板
litheness 柔韧性
Lithenia granite (产于美国乔治亚州的)青灰色细质花岗岩
lithergol 液固混合推进剂
lithfilm reprography 干片复照
lithia 氧化锂;锂氧
lithia ceramics 氧化锂陶瓷
lithia granite 锂云母花岗岩
lithian muscovite 锂白云母
lithia porcelain 氧化锂瓷
Lithic 石器时代
lithical 石质的
lithic arenite 岩屑砂屑岩
lithic arkose 岩屑长石砂岩
lithic arkosic wacke 岩屑长石砂岩质瓦克岩
lithic contact 母质层;土壤岩面接触层;石质层(土壤岩石接触层)
lithic-crystalloclastic psammitic texture 岩晶屑砂状结构
lithic drainage 地下水系
lithic era 原始岩代
lithic feldspathic graywacke 岩屑长石杂屑岩
lithic graywacke 岩屑杂屑岩;石质杂砂岩
lithic ore 石质矿石
lithic psammitic texture 岩屑砂状结构
lithic quartz graywacke 岩屑石英杂砂岩
lithic quartz sandstone 岩屑石英砂岩
lithic sandstone 岩屑砂岩;石质砂岩
lithic subarkosic wacke 岩屑亚长石砂岩质瓦克岩
lithic sublabile arenite 岩屑亚稳定砂屑岩
lithic tuff 岩屑凝灰岩;石质凝灰岩
lithic tuff lava 岩屑凝灰熔岩
lithic-vitroclastic tuff 岩玻屑凝灰岩
lithic-vitroclastic tuffaceous texture 岩玻屑凝灰结构

lithic wacke 岩屑瓦克岩
lithification 岩化(作用);石化(作用)
lithin 彩色水泥(砂浆)涂料
lithin coating 彩色水泥砂浆(墙壁)涂装
lithinit 里季尼特(硬质合金)
lithiocarmine 锂卡红
lithionite 锂云母
lithiophilite 磷锰锂矿;锰磷锂矿
lithiophorite 锂硬锰矿
lithiophosphate 块磷锂矿
lithistid 压缝石条
lithistida 石海绵目
lithite 平衡石
lithium acetate 醋酸锂
lithium-air battery 锂空气电池
lithium aluminium hydride reduction 氢化铝锂还原
lithium aluminohydride 铝氢化锂
lithium amalgam 锂汞齐
lithium amide 氨基锂;氨化锂
lithium base grease 锂基润滑脂
lithium battery 锂电池组
lithium bichromate 重铬酸锂
lithium borate 硼酸锂
lithium borohydride 硼氢化锂
lithium borosilicate 硼硅酸锂
lithium bromide 溴化锂
lithium bromide absorption-type refrigerating machine 溴化锂吸收式制冷机
lithium bromide chiller 溴化锂制冷机
lithium bromide refrigerating machine 溴化锂制冷机
lithium bronze 锂青铜
lithium cell 锂电解槽;锂电池
lithium chloride 氯化锂
lithium citrate 柠檬酸锂
lithium dichromate 重铬酸锂
lithium dihydrogen phosphate 磷酸二氢锂
lithium drift detector 锂漂移探测器
lithium drifted germanium crystal 锂漂移锗晶体
lithium drifted silicon detector 锂漂移硅探测器
lithium drift germanium detector 锂漂移锗探测器
lithium filter 锂滤色镜
lithium fluoride 氟化锂
lithium fluoride dosimetry 氟化锂剂量测定法
lithium fluoride single crystal 氟化锂单晶体
lithium fluoride thermoluminescence dosimeter 氟化锂热释光剂量计
lithium fluoride whisker 氟化锂晶须
lithium fluoroborate 氟硼酸锂
lithium fluorosilicate 氟硅酸锂
lithium grease 锂润滑脂
lithium halide 卤化锂
lithium hypobromite 次溴酸锂
lithium hypochlorite 次氯酸锂
lithium iodate 碘酸锂
lithium iodate single crystal 碘酸锂单晶体
lithium iodide 碘化锂
lithium iodine battery 锂碘电池
lithium ion adsorption 锂离子吸附
lithium metasilicate 硅酸锂;偏硅酸锂
lithium niobate 铌酸锂
lithium niobate stress ga(u)ge 铌酸应力计
lithium niobate-tantalate 铌钽酸锂
lithium nitride 一氮化三锂
lithium ore 锂矿(石)
lithium oxalate 草酸锂
lithium oxide 氧化锂
lithium perchlorate 高氯酸锂
lithium potassium niobate 铌酸钾锂
lithium propoxide 丙醇锂
lithium salts 锂盐
lithium silicate 硅酸锂
lithium silicide 硅化锂
lithium soap grease 锂基润滑脂
lithium spectrum 锂光谱
lithium star 锂星
lithium stearate 硬脂酸锂
lithium sulfur battery 锂硫蓄电池
lithium tetraborate 四硼酸锂
lithium titanate 钛酸锂
lithium tritoxide 氚氧化锂
lithium-type pressure gun grease 锂皂型压力枪用润滑脂

lithium vanadate 钒酸锂
lithium water 锂水
lithium zirconate 锆酸锂
lithizone 岩性带
litho 岩石
litho-blue 蓝色平印油墨
lithocarbon 石碳
lithocenosis 碎石清除术
lithochemical survey 岩石化学测量
lithochromy 彩色石印
lithoclase 岩裂;破裂面
lithoclast 碎石器
lithoclastic 碎石的
lithoclasty 碎石术
Lithococcus 石球藻属
lithocon 硅存储管
lithoconion 碎石器
lithocrete 沥青地面(其中的一种)
lithodensity log 岩性密度测井
lithodensity log curve 岩性密度测井曲线
lithodomous 钻石的
litho-facies 岩性相;岩相
lithofacies analysis 岩相分析法
lithofacies association 岩性相组合
lithofacies change trap 岩相变化圈闭
lithofacies gone anomaly of intrusive body 侵入岩体相带异常
lithofacies gone anomaly of metamorphic rock 变质岩系相带异常
lithofacies gone anomaly of sedimentary rock formation 沉积岩系相带异常
lithofacies gone anomaly of volcanic rock formation 火山岩系相带异常
lithofacies map 岩相图
lithofacies-paleotopographic(al) map 岩相古地理图
lithofalt paving 沥青铺路块(一种用于车辆繁忙的道路)
litho felt 石印毡
lithofraction 岩裂作用
lithogeneous process 造岩过程
lithogenesis 岩石成因学;造岩
lithogenesis fissure water 成岩裂隙水
lithogenesis pore 成岩孔隙
lithogenesy 岩石成因学;石的形成
lithogenetic unit 岩层单位
lithogenous 造岩的
lithogenous component 造岩组分
lithogenous process 造岩作用
lithogeny 岩石成因学
lithogeochemical method 岩石地球化学方法
lithogeochemical survey 岩石地球化学测量
lithoglyph 石刻
lithograph 石印
lithographic(al) 石印的
lithographic(al) chalk 石印石
lithographic(al) crayon 石印蜡毛
lithographic(al) film 平版制版胶片
lithographic(al) ink 石印墨;石版油墨
lithographic(al) limestone 石印灰岩;印版石灰岩
lithographic(al) offset duplicater 胶印复印机
lithographic(al) oil 平版用清漆;石印油
lithographic(al) plate 平版
lithographic(al) press 平板印刷机
lithographic(al) printing 石版印刷
lithographic(al) printing plant 地图印刷厂
lithographic(al) representation 胶印
lithographic(al) slate 石印板(岩)
lithographic(al) stone 石印灰岩;石印石
lithographic(al) texture 石印结构
lithographic(al) varnish 平版用清漆;石印清漆
lithograph stone 石印石
lithography 平版印刷术;石印术;石版印刷(术)
lithography roll 石印胶滚
lithogration texture 石印石结构
lithoherm 岩礁
lithoidite 隐晶流纹岩
lithoid tufa 石质华
litholapaxy 碎石洗出术
lithologic(al) analysis diagram 岩性分析图
lithologic(al) and pelaeogeographic(al) condition 岩相古地理条件
lithologic(al) boundary 岩性边界
lithologic(al) character 岩性性质

lithologic(al) character of roof rock 盖层岩性
lithologic(al) characters of apical plate of aquifer 含水层顶板地层岩性
lithologic(al) characters of aquifer 含水层岩性
lithologic(al) characters of aquifer where tracer is put into 示踪剂投放含水层岩性
lithologic(al) characters of aquifuge 隔水层岩性
lithologic(al) characters of blasting segment 爆破段岩性
lithologic(al) characters of bottom layer of aquifer 含水层底板地层岩性
lithologic(al) characters of experimental part 试验段岩性
lithologic(al) characters of waste disposal stratum 排污层岩性
lithologic(al) characters of water-beating section 含水段岩性
lithologic(al) characters profile of aeration zone 包气带岩性剖面
lithologic(al) classification of source rock 烃源岩的岩石类型
lithologic(al) composition 岩石组成
lithologic(al) contact zone 岩性接触带
lithologic(al) control 岩性控制
lithologic(al) correlation 岩性对比
lithologic(al)-geologic(al) map of quaternary system 第四系岩性地质图
lithologic(al) identity 岩性相等
lithologic(al) log 岩性录井(图)
lithologic(al) map 岩石(分布)图;岩性图;岩相图
lithologic(al) oil-gas field 岩性油气田
lithologic(al) order 岩性阶
lithologic(al) pattern 岩性模式
lithologic(al) pool 岩性油藏
lithologic(al) relative thickness 岩性相对厚度
lithologic(al) section 岩性断面图
lithologic(al) similarity 岩性相似(性)
lithologic(al) state 岩性状态
lithologic(al) thickness 岩性厚度
lithologic(al) transition frequency 岩性转移频数
lithologic(al) triangle 岩相三角图
lithologic(al) unit 岩性(地层)单位
lithologic character 岩性特征;岩性
lithology 岩性学;岩石学
lithology analysis 岩性分析
lithology factor 岩性因素
lithology of fault rock 断层岩性质
lithology of lubricating layer 滑润层岩性
lithology of parting 夹矸岩性
Lithol red 入漆朱;立索尔红
Lithol 立索尔颜料
litholyte 溶石液灌注器
litholytic 溶石的
lithomarge 密高岭土
litho master 石印原版;石印盘
lithometeor 大气尘粒
lithonite ore 锂云母锂矿石
lithontriptic 碎石剂
litho oil 平版用清漆;石印油
litho-paper 石印纸
Lithoperl 利索派森里粉刷液
lithophane 暗花装饰
lithophile 亲岩的;亲石的
lithophile element 在地壳集中的元素;亲岩元素;亲岩元素;亲硅元素
lithop(h)one 立德粉
lithophotography 光刻照相术
lithophotogravure 石印照相制版
lithophtes 石生植物
lithophylic property 亲硅酸盐性
lithophysa 石泡
lithophysa structure 石泡构造
lithophysa welded tuff 石泡熔结凝灰岩
lithoplaxy 碎石术
lithopone 硫化亚铅;锌钡白(一种用于油漆的白色颜料)
lithopone grade 锌钡白级
litho printings 石版印刷纸
litho-relict 岩石残余物
lithosere 石生演替系列
lithosis 石屑肺;尘肺病
Lithosite 利索派里砂浆;水硅铝钾石
lithosized blue 石印蓝
lithosoil 石质土

lithosol 固结岩屑土;石质土
lithosol fast pink 立托索尔坚牢桃红
lithosolid compressibility 岩石固体压缩系数
lithosome 岩(性)体;岩石体
lithosphere 锌钡白;岩(石)圈;岩石层;地圈;陆圈;陆界;地球岩石圈
lithospheric bulge 岩石圈膨胀
lithospheric earthquake 岩石圈地震
lithospheric fracture 岩石圈断裂
lithospheric isostasy 岩石圈均衡
lithospheric plate 岩石圈板块
lithospheric signal 岩石圈信号
lithospheric slab 岩石圈板块
lithospheric splitting 岩石圈分裂
lithospheric subplate 岩石圈下板块;岩石圈次板块
lithosporic 石斑
lithostatic(al) 地压的
lithostatic(al) loading 岩石静态负荷
lithostatic(al) pressure 岩压;岩石静压力;静岩压力;积土压力;地面压强;地面压力;地静压力
lithostatic(al) stress 静岩应力;地压应力;地静应力
lithostone 石版
lithostratic unit 岩性地层单位
lithostratigraphic(al) boundary 岩性地层界线
lithostratigraphic(al) framework 岩石地层格架
lithostratigraphic(al) unit 岩石地层单位
lithostratigraphic(al) zone 岩性地层带;岩石地层带
lithostratigraphy 岩相层序;岩石地层(学)
lithostratigraphy method 岩性地层学方法
lithostrome 均质岩层
lithostrotum opus 装饰性铺地(古希腊和罗马住宅)
lithostyle 石针
lithothrypty 碎石术
lithotint 彩色石印
lithotome 切石刀
lithotomous 钻石的
lithotomy 切石术
lithotope 岩石沉积环境
lithotripsia 碎石术
lithotripsy 碎石术
lithotriptor 碎石器
lithotrite 碎石钳;碎石器
lithotrity 碎石术
lithotype 煤岩类型
lithotypes of coal by luster 按光泽分类的肉眼煤岩类型
lithoxyl(e) 石化木
lithoxylite 石化木
lithozone 岩性带
lithpone 锌钡白
lith process 平版工艺
litidionite 硅碱铜矿
litigant 诉讼当事人
litigate 提出诉讼
litmus neutral test paper 中性石蕊试纸
litmus paper 石蕊试纸
litmus red test paper 红色石蕊试纸
litnogeochemical anomaly 岩石圈地球化学异常
litophere 外壳
litosilo 镁氧水泥甲板敷料
lit-par-lit gneiss 间层片麻岩
lit-par-lit injection 间层注入;间层贯入
lit-par-lit intrusion 间层侵入
lit-par-lit schistosity 间层片理
lit-par-lit structure 间层构造
litre =liter
litter 死地被物
litter basket 果皮箱;污物桶
litter-bin (街上的)垃圾箱;保洁箱;垃圾桶
litter fence 活动铁丝网
litter horizon 枯枝落叶层
litter layer 枯枝落叶层
litter magnet 磁力清理垃圾器
litter receptacle 杂物箱;垃圾箱;杂物间
litter stretcher 担架
litter weight 窝重
little beakiron 小台钻
little Bessemer steel 小型酸性转炉钢
little by little 逐渐地
little consolidation 略固结
little covering plate 小盖板
little drop technique 微滴术
little end 小头
little-end bearing 小端轴承

little end of connecting rod 连杆小头
little-expansive fusion caking 微膨胀熔融黏结
little fissure 少缝隙
little-fuse bolometer 熔断丝式测辐射热计
little gas escape 少量气体逸出
little giant 小型水枪
little hard water 微硬水
little heap 小堆
little hydraulic lime 低水凝性石灰
little ice age 小冰期
little inch 小尺寸管子(指直径24英寸及以下的管子)
little joiner 小木楔;小木块
little leaf disease 小叶病
little monsoon 小季节风
little open void ratio 小开空隙率
little piece checkroom 小件寄存处
little poisonous 低毒的
little red wagon 活动厕所
Little's alloy 利特尔合金
little school 教养院
little slate 小石板(瓦);小页岩
little spring 小弹簧
little theater 小剧场
Littleton-point 利特尔顿点(玻璃黏点为 $10^{7.6}$ 泊的温度)
little tree willow 小树柳
little window 小窗
littoral 沿海的;沿岸;涨潮淹没地带;潮湿间的;滨海的
littoral accumulation 沿岸淤积;沿岸堆积;沿岸沉积(物)
littoral area 沿岸地区;沿岸区;湖滨地带;海滨地带;海岸区;潮汐区(域);潮滩区;潮间区;滨海区
littoral barrier 拦沙埂
littoral beach ridge 滨岸堤
littoral belt 浅潮地带;滨海地带
littoral berm 海滩阶地
littoral bottom 滨海底
littoral clastic sedimentation model 滨海碎屑沉积模式
littoral cliff 滨海悬岸
littoral climate 海滨气候;海岸气候
littoral community 浅海群落
littoral condition 沿岸条件
littoral cone 岸边火山锥
littoral conglomerate 滨岸砾岩
littoral country 沿海国(家)
littoral current 沿海潮流;沿岸(海)流;近岸流;海岸径流;顺岸流;沿岸(潮)流
littoral deposit 沿岸沉积(物);岩沉积;湖滨沉积;湖滩潮间带淤积(物);湖滩潮间带堆积(物);湖滩潮间带沉积(物);海滩沉积;海滩潮间带沉积(物);潮间带沉积;潮间带淤积(物);潮间带堆积(物);潮滩沉积;潮间带沉积;潮滩沉积;滨海沉积(物)
littoral deposition 沿岸沉积作用
littoral drift 沿岸物质流;沿岸输沙;沿岸漂移;沿岸漂沙;沿岸(泥沙)流;海岸漂沙;滨海漂积物
littoral drift belt 沿岸漂沙带
littoral drift coast 有漂砂的沿岸地带
littoral dune 沿岸沙丘;湖滨沙丘;滨海沙丘
littoral environment 沿岸环境;海滨环境;滨海环境
littoral facies 沿岸相;潮滩相;潮间带相;海岸相;滨海相
littoral fauna 沿岸动物区系
littoral flat 滨海平台
littoral fringe 海带状边界
littoral industrial area 沿海工业区
littoral industrial zone 沿海工业地带
littoral lagoon deposit 滨海泻湖沉积
littoral land 沿岸(土)地
littoral landform 沿岸地形;海洋地形;海岸地形;滨海地形
littoral marsh deposit 滨海沼泽沉积
littoral movement 沿岸运动;沿岸移动;沿岸输移;沿岸输沙;沿岸漂沙
littoral mud and sand flow 沿岸泥沙流
littoral nourishment 沿岸淤滩;海滩养护;补沙护滩
littoral placer 滨海砂矿
littoral plain 沿岸平原;湖滩平原;湖滨平原;海滩平原;海滨平原;海岸平原;堆积平原;潮滩平原;滨海平原
littoral plain deposit 滨海平原沉积
littoral plain swamp 海岸平原沼泽
littoral process 海岸演变过程;岸滩演变过程

littoral region 沿岸地带;海岸区;潮汐区(域);滨海区
littoral right 海岸(用水)权;湖岸(用水)权;沿岸使用权;海岸使用权;海滨产权;岸线使用权
littoral rock 潮滩岩;岸滩岩
littoral Sabkha sedimentation model 滨海萨巴哈沉积模式
littoral sand bar deposit 滨海沙坝沉积
littoral sand dune deposit 滨海沙丘沉积
littoral sand movement 沿岸泥沙运动
littoral sediment 沿岸泥沙;沿岸沉积(物);滨海沉积(物)
littoral sedimentary soil 滨海沉积土
littoral sedimentation 沿岸沉积作用
littoral sediment transport 沿岸输沙
littoral sheeted sand deposit 滨海席状沙沉积
littoral shelf 浪成湖滨台地;海岸浅滩
littoral state 沿海国(家);沿岸国
littoral system 沿岸系
littoral terrace 海洋阶地;海岸阶地;海滨阶地;海岸阶地
littoral topography 沿岸地形;滨海地形
littoral tracer 沿岸示踪沙;沿岸流示踪剂
littoral transport 沿岸搬运;沿岸运输;沿岸输移;沿岸漂沙
littoral vegetation 海岸植被
littoral water 沿岸水
littoral wind 沿岸风
littoral zone 沿岸区;沿岸(地)带;开采陷落带;近岸带;海岸带;潮汛带;潮汐带;潮滩带;潮浸地带;潮间带;滨海(地)带;浅岸带
littorideserta 海滨荒漠群落
Littrow's condition 利特罗条件
Littrow's grating spectrograph 利特罗光栅摄谱仪
Littrow's mirror 利特罗反射镜
Littrow's monochromator 利特罗单色仪
Littrow's mounting 利特罗(棱镜)装置
Littrow's prism 利特罗棱镜
Littrow's prism spectrograph 利特罗棱镜摄谱仪
Littrow's quartz spectrograph 利特罗石英摄谱仪
Littrow's spectrograph 利特罗摄谱仪
Littrow's spectrometer 利特罗分光计
Littrow's type collimator lens 利特罗式准直透镜
lituate 外伸
liturgical water vessels 宗教仪式用水盆
litus 海滨群落
lituus 连锁螺线
lityy white 纯白
litzendraht wire 编织线
Litzmann's obliquity 利兹曼倾斜
Litz wire 利兹线;绞合线;辫编线;编组线
Liumogen 琉莫根磷光体
livability 居住适用性;居住适应性
livability space 可居住空间
livability space ratio 居住面积比值
livable floor area 适于居住的楼面积
livable room 适于居住的房间
live 居住;活的;带电(的)
live animal 活动物
live ark shell 活赤贝
live axle 行走轮轴;驱动轴;驱动轮轴;动轴;传动(力的)轴
live beam pass 开口梁形轧槽
live bed scour 动床冲刷
live birth index 出生存活率
live bolt 弹簧门
live boom 独立变幅的动臂(挖掘机)
live boom of excavator 挖土机活动伸臂
live-bottom furnace 单电极电弧熔炼炉
live box 活鱼箱;活水笼;网箱;饲养场
live brake lever 浮装闸杆
live broadcast 现场广播;直播
live cable test cap 带电电缆试验盖头
live camera 现场摄像机
live capacity 有效库容;活库容
live cargo 活货
live center 活顶尖;回转顶尖
live center of roller 辊子有效顶端
live chassis 带电机壳
live circuit 有电电路;带电电线
live conductor 有电导线;火线;带电导线;带电导体
live crack 活裂缝
live crude 充气原油
live deck 出河台

live edge 活性边(油漆未干的边);湿边
live enclosure 矮树栏;树篱
live end 有效端;交混回响壁;加电压端;活跃端;带电端
live entry 带电进(电泳)槽
live equipment 带电设备
live farming 畜牧业
live fence (用树木组成的)绿篱;灌木篱
live fish carrier 运活渔船
live form 现代地形
live freshwater crab 大闸蟹
live front 盘面接线
live front switchboard 盘面接线式配电盘
live gas 新鲜气体
live glacier 现代冰川;流动冰川;活冰川
live graphite 含铀块石墨;放射性石墨
live gravity load 重力活荷载
live guy 活动牵索
live hole 点火孔
livehood economy 民生经济
livehood space 生活空间
live hopper 活动料箱
live hydraulic pump 独立传动的液压泵
live hydraulic system 独立液压系统
live ice 流动冰川;活冰川
live input 常通输入
live knot 连体节;连生节;活络钻头;活节
live landform 现代地形
live lever 浮装杆
livelihood 生活资料
live line 装满的管线;火线
live line tester 试电器
live line work 带电作业;带电工作
live load 有效负载;有效负荷;工作负载;活(荷)载;活负载;活(动)荷载;使用荷载;使用负载;实用负载;动(力)负载;动(力)负荷;动荷
live load moment 活载力矩;活载弯矩
live load on deck 面板上活荷载
live load stress 活载应力;活荷载应力
lively coal 易碎块煤
live machine 可使用的机器
live market 牲畜市场
liveness 住房的声学质量
live oak 常绿橡树
live open steam 新汽;直接蒸汽
live parking 短暂停车;不离车停放
live part 运转中的部件;带电部件;有电部(零)件
live pass 工作孔型
live peat 层状腐殖质泥煤
live pick-up 电视室内摄制;广播室广播;直接录音;室内摄影
live pit 新灰坑
liveplate 通风孔式板状炉排
live power drive 连续功率驱动
live power take-off 独立式功率输出轴
live program(me) 实况广播
live question 当前问题
liver 油品变稠结块
live rail 载电轨;地磅活动轨;导电轨
live recording 现场录音
livered thickened oil 稠硬油
live ring 滚圈;轮带;动圈
livering 硬化(指涂料油漆的变质);膏化;泥浆加稠;泥浆稠化
live ring migration 轮带位移
livering of slurry 泥浆结团
live riser 热冒口
live-roll conveyer 滚轴运输机;自动滚动式滚柱输送机;电动滚轴输送机
live roller 辊式输送器;辊式输送机;传动辊(道);转动辊
live-roller bed 传动辊道
live-roller conveyer 滚轴运输机;滚轴输送机;辊道输送机;电动滚轴输送机;驱动式滚柱输送机
live-roller cradle (升船用的)缆车
live-roller gear 传动辊齿
live-roller system 联珠式滚轮系统
live-roller type spillway 缆车滑道
live-roll feeder 辊式进料器
live-roll grizzly 滚轴筛;多轴辊轴筛
live-roll table 传动辊道
live room 活跃室;混响室;交混回响室
liver opal 硅乳石
Liverpool head 通风管护罩(防海浪溅入)

Liverpool quay terms 利物浦码头条款(英国)
liver rock (取自矿中最佳部位的)优质石材
liver spotting 硅砖上的肝色斑点
live rubber 高碳性橡胶;高弹性橡胶
live runway 现用跑道
liverwort 地钱(藓类)
livery 胶结的;板结的;转让产权;转让资产;车辆出租行;财产所有权让渡
livery of seisin 所有权的让渡;财产所有权的让渡
livery stable 出租马车处;马房
live screw bottomed ice day tank 底部设有活动螺旋的日调节冰箱
live shaft 转动轴
live silo 自流筒仓;活筒仓;卸料仓
live silo capacity 活动料箱容量;自由调节的仓储[贮]容量
livesite 埃洛高岭石
live species 现有物种
live spindle 旋转心轴;旋车主轴
live stage 真实级
live steam 新蒸汽;高压蒸汽
live steam pipe 新汽管;主汽管
live steam piping 主汽管道
live steam valve 新汽阀
livestock 牧业;牲畜
livestock and plant 活动植物
livestock and quaculture industries 畜牧、水产业
livestock breeding 畜牧业;养殖业;家畜繁殖;牲畜饲养
livestock building 饲养家畜建筑物
livestock container 装牲畜用的集装箱;家畜集装箱;动物集装箱
livestock conversion coefficient 牲畜折合系数
livestock enclosure 牲口栏
livestock farm 畜牧场
livestock farming 畜牧业
livestock husbandry 畜牧业
livestock hydrant 牲畜给水栓
livestock insurance 牲畜保险
livestock judge 家畜鉴定
livestock judging 牲畜评价
livestock loading platform 牲畜装车站台
livestock loading ramp 牲畜装车斜台
livestock management 家畜管理
livestock number 牲畜数量
livestock pool 饮畜池
livestock product 畜产品
livestock production 牧业生产
livestock production structure 牲畜结构
livestock production system 畜牧生产体系
livestock raising 饲养业
livestock ramp 斜面装车台
livestock ranching 牲畜大农业
livestock rearing 饲养业
livestock reservoir 饮畜池
livestock shed 牲口棚
livestock spray 家用消毒剂
livestock transit insurance 牲畜运输保险
livestock wagon 家畜车;牲畜车
livestock waste collection 禽畜废物收集
livestock water supply 家畜给水
live storage 活用蓄水量;有效储备;有效库容;有效储料库;周转容量;活库容;仓库场;活堆场
live storage bin 活储[贮]仓;有效储仓;活存仓
live storage capacity 有效库容
live stress 活载应力
live stub axle 驱动短轴
live studio 活化播音室;实况广播室;长混响播音室
live talent studio 直接表演演播室
live terminal 带电端子
live time 使用期限
live transmission 实时传输
live truck lever 转向架移动杠杆
live up to one's name 名副其实
live vaccine 活菌苗
live volume 有效容积
liveweight gain 增重
liveweight increase 增重
liveweight increment 增重
live weight loss 减重
live wire 有电线;载电线;火线;活线;通电铁丝网;通电电线;输电线;输电电缆;带电电线;带电导线
live wire entanglement 电网
live wood 新伐木材;潮湿木材

live zone 拌缸净体积【道】
livid 青灰色的;铅色的;青黑色的
livid brown 青黑棕色
lividity 青灰色;铅色;青黑色
livid pind 青黑粉红色
livid purple 青黑紫红色
livid violet 青黑紫蓝色
living 生活
living accommodation 居住处所;居住舱室;生活舱室
living allowance 助学金;生活资料;生活津贴
living aquatic resource 水生生物资源
living area 起居室;居住(使用)面积;生活区
living boat 住宿船
living body 生物体
living car 生活车
living chamber 居住处所;居住舱室;生活舱
living clock 生物钟
living collection 标本园
living compartment 居住处所;居住舱室
living computer 运转中计算机
living condition 生活状况;生活条件
living cost 生活费(用)
living deck 居住甲板
living dining 起居室兼餐室
living dining kitchen 起居室—餐室—厨房
living-dwelling zone 生活居住区
living environment 生活环境;生存环境
living environmental condition 生活环境条件
living environmental conditions of organisms 生物的生存环境条件
living environment of coral reef 珊瑚礁生活环境
living example 实例
living expenses 生活费(用)
living fence 绿篱
living filter 生物滤池
living fire break 防火植物带
living floor area 居住面积
living floor space 居住面积
living form 生活型
living fossil 活化石
living glacier 现代冰川;流动冰川;活冰川
living hut 居住用简屋;居住用茅舍;临时居住木板房
living insecticide 生物杀虫剂
living in the body of another animal 生活在其他动物体内
living kitchen 住房兼厨房;起居室兼厨房
living level 生活水平
living light 生活照明
living marine resources 海洋生物资源
living material 生活物质
living monitor 活的监测器(一般指海洋生物)
living nature 生物界
living needs 生活需要
living organism 活性有机物;生物机体;生活有机体
living polymer 活性高聚物
living quarter 住宅(区);住宿船;居住(小)区;生活区;居住处所;居住舱室;宿舍(区)
living quarters for staff and worker 职工宿舍
living requirement 生活需要
living resources 生物资源
living room 起居室
living room-entrance hall 起居室门厅
living sand 浮沙
living soil 地面土壤;地表土壤;表土
living space 居处;居所;居住面积;居住空间;生存空间
living space of urban population 城市居民居住面积
living space per capita 人均居住面积
living sphere 生活圈
living standard 生活水平;生活标准
livingstonite 硫汞锑矿
living substance 活质;生活物质
living terrace 生活用平台
living-tree pergola 树棚
living unit 居住单元
living unit floor space 生活单元楼板面积
living wage 能维持生活的工资;生活工资
living water use 生活用水
living wire 输电线
living zone 生活区;生存带
liwan 穹顶门廊(清真寺)

lixiviant 浸出溶液;浸出剂
lixiviate 浸析;浸滤
lixiviation 淋洗;浸滤(作用);溶滤
lixivium 浸滤液;碱液
lizardite 鳞石;利蛇纹石;板蛇纹石
ljardite 冻石
L joint 原生平缓节理;L节理
L joint of igneous rock 火成岩水平节理
Ljungstrom steam turbine 容克斯川式汽轮机;辐流式汽轮机
Ljungstrom turbine 辐流式气轮机
Llanberis Slates 蓝贝里斯板岩
Llandeilian(stage) 蓝代洛阶【地】
Llandeilian flags and limestone 蓝代洛板层和灰岩
Llandellian series 蓝代洛统【地】
Llandoverian 蓝达夫里阶【地】
llano 南美洲大草原
Llanvirn series 蓝维尔恩统【地】
Llanvirnian(stage) 蓝维尔恩阶【地】
Lloyd's, companies 劳氏公司;劳埃德公司
Lloyd's agent 劳氏(船级社)代理人;劳氏保险协会代理人;劳氏保险公司代理人;劳埃德船级社代理人;劳埃德保险协会代理人;劳埃德保险公司代理人
Lloyd's association 劳氏船级协会
Lloyd's average bond 劳氏海损契约
Lloyd's Brokers Association 劳氏经纪协会;劳埃德经纪协会
Lloyd's certificate 劳氏船级证书;劳埃德(船级)证书
Lloyd's class 劳氏(合社)船级
Lloyd's form of average bond 劳氏海损协议格式;劳埃德海损协议格式;劳氏合社海损担保格式
Lloyd's Insurance Brokers Association 劳氏保险经纪商协会;劳氏保险经纪人协会;劳埃德保险经纪商协会;劳埃德保险经纪人协会
Lloyd interferogram 劳埃德干涉图
Lloyd's List 劳氏船名录;劳氏船舶年鉴;劳埃德船名录;劳埃德船舶年鉴
Lloyd's machinery certificate 劳氏机械合格证书;劳埃德机械合格证书
Lloyd's Maritime Information Service 劳氏海运信息服务处;劳埃德海运信息服务处
Lloyd's Market 劳氏市场(即伦敦保险市场)
Lloyd's mirror 劳埃德镜
Lloyd's number 劳氏数;劳埃德数
Lloyd's open form 劳氏公开格式;劳埃德公开格式
Lloyd's Refrigerating Machinery Certificate 劳氏冷冻机械证书;劳埃德冷冻机械证书
Lloyd's register 劳埃德船舶年鉴;劳氏(船级协会)船舶年鉴
Lloyd's Register of Shipping 劳氏船级社;劳埃德船级社
Lloyd's Report 劳氏报告
Lloyd's rule 劳氏(合社)船(级)规范;劳埃德船规范
Lloyd's Shipping Index 劳氏船舶动态日报
Lloyd's survey 劳氏验船协会;劳埃德验船协会
Lloyd's Surveyor 劳氏公证行
Lloyd's underwriter 劳氏保险公司(劳氏又译劳合社);劳埃德保险公司;劳氏承保人
LL-shaped entrance and exit 死巷式出入口
LM Ericsson Telefonaktinbolaget 埃里克森电话工业公司
LMG tanker 液化甲烷运输船
LNG[Liquid Natural Gas] 液化天然气
India proof paper 版画印样用薄纸
loacked reverse 反位锁闭
load 装入指令;装模料;重物;载重;量木材单位;加入量;荷重;荷载情况;荷载;取数;送入;充填;负重;负担;负载;负荷;搬运物;装填
loadability 荷载能力
loadable 可受载;适于承载
load accumulator 负载累加器
load accumulator with magnitude 负载数值累加器
load acting point 荷载点
load action 负荷作用
load address 装入场址
load adjuster 荷载调整器
load adjustment 负荷调整;负荷调节
load admittance 负载导纳
loadage 装载量;载重量;积载重;货重
load allocation 负载分配;负荷配置

loadamatic control 负载变化自动控制；负荷变化自动控制
load analysis 负载分析
load analyzer 负荷分析器
load and carry dumper 铲运自卸车
load-and-carry equipment 载运机
load-and-carry machine 装运机
load-and-go 装入立即执行(程序)；装配立即执行；装入并执行；程序装入立即执行
load and linkage editor 装入连接编辑程序
load and resistance factor design 荷载与抗力系数设计
load and strength histogram 荷载强度进程曲线图
load and transport 装运
load and unloading and discharging task revenue 装卸排作业收入
load and weigh cell 测力称重传感器
load angle 负荷角
load anticipator 负荷预测器
load application 施加荷载
load applying unit 加载设备；加荷设备
load area 受压面积；负荷区域
load arm 负载臂
load a ship with cargo 载货上船
load at failure 破损载荷；破坏载荷；断裂负荷
load at first crack 开裂载重；开裂荷载；始裂荷载；初现裂缝荷载；初裂荷载
load at fracture 断裂荷载；断裂负荷
load at rupture 断裂荷载；断裂负荷；破坏荷载
load axis 负荷轴线
load backfeed 负载反馈
load-back method 还馈法；负载还馈法
load balance 负载平衡
load balancing 荷载平衡；负载均衡
load balancing design method 负荷平衡设计法
load balancing group 均分负载组；负载均衡组
load-balancing method 荷载平衡法
load ball 负荷球
load barrel 索筒
load bearing 承载(能力)；受荷；承重(的)
load-bearing aggregate 承重集料；承重骨料
load-bearing area 承重面积
load-bearing bogie 承载转向架
load-bearing brick 承重砖
load-bearing brickwork construction 砖承重结构
load-bearing capability 承载能力
load-bearing capacity 荷载容量；地基支承力；承载(能)力
load-bearing characteristic 负荷特性；荷载特征；荷载特性；负载特性
load-bearing cross wall 承重横墙
load-bearing cross wall unit 承重横墙单元
load-bearing diaphragm 承重隔膜
load-bearing external wall unit 承重外墙单元
load-bearing face work 承重的(镶)面层
load-bearing facing 承重面层
load-bearing fixing 支承固定件；承重配件
load-bearing frame 承重支架；承重框架；承重构架；承载构架
load-bearing hollow brick 承重空心砖
load-bearing in longitudinal direction 纵向承重
load-bearing in transverse direction 横向承重
load-bearing layer 持力层
load-bearing masonry 承重圬工；承重砌体
load-bearing mechanism 承重机构
load-bearing member 承重构件；承载构件；承荷构件
load-bearing metal stud 金属承载立筋；承重金属龙骨
load-bearing panel 承重墙板
load-bearing panel construction 承重板墙结构
load-bearing partition 承重隔墙
load-bearing pavement 承重路面
load-bearing pile 承重桩
load-bearing plane 承重平面
load-bearing plate 承压板
load-bearing power 承重能力
load-bearing property 承载性能
load-bearing rib 承重肋
load-bearing ring 承重环
load-bearing skeleton 承重骨架；承重构架
load-bearing skeleton construction 承重框架
load-bearing skeleton member 承重骨架构件

load-bearing skeleton rigidity 承重骨架刚度
load-bearing skeleton structure 承重骨架结构
load-bearing stiffener 承载加劲肋；承载加劲杆
load-bearing stratum 持力层
load-bearing strength 承载强度
load-bearing structural insulating material 承重保温材料
load-bearing structural tile 承重瓦；预应力槽瓦
load-bearing structure 支承结构；承重结构
load-bearing structure system 承重结构系统
load-bearing surfacing 承重(路)表面
load-bearing tape 承载带
load-bearing test 承载试验
load-bearing tile 承重砖
load-bearing unit 承重单元
load-bearing wall 承重墙
load-bearing wall construction 承重墙建造；承重墙构造
load-bearing wall of block 承重砌块墙
load-bearing (wall) panel 承重墙板；承载板墙
load-bearing wall structure 承重墙结构
load bending stress 荷载弯拉应力
load binder 杠杆提升器；紧索具
load block 提升滑轮组；加载块；加荷块
load boom 荷载吊杆
load bounding 界限荷载
load box 负荷电阻箱；荷载箱；测力盒
load brake 重锤闸；超载制动器；超载制动器
load-break rating 遮断容量
load-break switch 负载断连开关
load bridging resistor 负载转移电阻
load bucket capacity 载重机铲斗堆装容量
load buffer memory 加载缓冲存储器
load button 加载按钮；起重按钮；输入按钮
load by human crowd 人行荷载；人群荷载
load cable 载重钢缆；工作钢丝绳；牵引钢丝绳
load calibrating device 负载检验器
load capacitor 负载电容器；负荷电容器
load capacity 荷载(容)量；承载能力；载重容量；负载能力；负荷容量
load capacity limiter 过载保护装置
load capacity of crane 起重机起重能力
load capacity on fifth wheel 第五轮负荷(汽车)
load capacity test 承载能力试验
load car 缆车；空中吊运车
load carrier 货船
load carry duty 承载率
load carrying 负荷的；载重的；带负荷运行的
load-carrying ability 荷载容量；载重量
load-carrying capability 承重能力
load-carrying capacity 承载能力；车辆运装能力；负荷容量；承载能力；负载容量；承载能力
load-carrying construction 承力载结构；承力结构
load-carrying member 承重构件；承载构件；承荷结构
load-carrying power 承载能力；负载能力
load-carrying property 承载性能；承载能力
load-carrying structure 承重结构；承力结构
load-carrying test track 负荷试验线【铁】
load-carrying vehicle 载重车辆；载重车
load case 荷载的类型
load cast 重荷模；荷重模；负载铸型
load casted 负载铸型的
load-casting 负载铸型作用
load cell 荷重计；压力盒；压力传感器；拉压传感器；荷重筒(石料压力试验用)；加载块；加荷块；荷传感器；荷载计；测压元件；测力计；测力盒；测力(传感)器；负载传感器；荷载传感器；加载盒
load cell hopper scale 料斗电子秤
load cell transducer 负荷容器传感器
load cell weighing equipment 压力传感式定量自动秤
load center 负载重心；负荷中心
load-center substation 枢纽变电站
load chain 载重链；承重链
load change 负荷变动
load change test 变荷载试验；变负载试验
load channelling 荷载转移途径
load characteristic 负载特性曲线；负载特性
load characteristics of internal combustion engine 内燃机负荷特性
load characteristic test 荷载特征试验；荷载特性试验；负荷特性试验；负载特性试验
load chart 载重曲线图；负荷分布图；负荷曲线图；负荷变动图
load circuit 负载电路；负荷电路
load circuit efficiency 负载电路效率
load class 负荷等级
load classification number 荷载等级数；荷载分类指数
load classification number method 荷载分级数法
load classification number system 荷载分类代号制
load coefficient 荷载常数；荷载系数
load coefficient of explosive charged 炸药装填系数
load coil 感应加热线圈
load column 承重柱；载重柱
load combination 荷载组合
load compensation 荷载调整；负载补偿；负荷补偿
load compilation 荷载分类
load component 荷载组成部分
load-compression diagram 负载压缩图；荷载压缩图
load concentrated at a point 集中在一点上的荷载
load condenser 负载电容器
load condition 重载状态；载重状态；荷载条件；承载状态；负载状态；负荷状态；负荷情况
load consolidation curve 荷载—固结曲线
load constant 荷载常数
load container 整装货柜；整装集装箱
load control 负荷控制
load-controlled braking 荷载控制式制动
load controller 控制装载装置；负荷调节器
load control word instruction 装入控制字命令
load conversion factor 荷载换算系数
load conveyer 负载传输机
load couple 炉料热电偶(测量炉料温度)
load crab 侧向夹向器
load criterion 荷载准则
load cross line 荷载交叉线(连续梁图解法)；负荷交叉线
load current 负载电流；负荷电流
load-current meter 负载电流计；负荷电流表
load curve 载重曲线；荷载曲线；承重特征曲线；载荷曲线
load cycle 荷载重复的周期；负荷循环
load cycling 周期性荷载
load cylinder 加载汽缸；加载油缸；负载油缸
load-data base 输入数据库
load data counter 送入数据计数器
load deck 大间隔炮眼群
load decreasing device according to frequency 按频率减负荷装置
load decrement 每级卸荷
load deflection 荷载挠度；加载位移
load-deflection curve 荷载(挠)变形曲线；荷载挠度曲线；负载挠度曲线
load-deflection diagram 荷载挠度(曲线)图
load-deflection plot 荷载挠度区
load-deflection relation 荷载挠度关系
load deflexion 载荷挠度
load deformation 负荷变形；荷载变形
load-deformation characteristics 荷载变形特性
load-deformation curve 过载畸变图；应力应变图；荷载变形曲线
load-deformation diagram 荷载变形图；负荷变形曲线图
load demand 负荷需要量
load demand comparator 负荷需要比较仪
load density 装载密度；荷载密集度；负载密度；负荷密度
load dependent 按荷载而定的
load dependent brake 荷载控制式制动器
load dependent relay valve 载控继动阀
load determination 荷载的确定
load device 荷载装置
load diagram 装舱记录曲线；荷载(简)图；负荷图；负荷曲线
load diffusion 负荷分布；负荷分散
load-discharge of river 河水含沙量
load-discharge ratio 输沙率流量比
load-discharge relationship 泥沙流量关系
load dispatch center 负荷调度中心
load dispatcher 配电员
load-dispatching board 供电配电盘
load-dispatching office 配电所
load displacement 满载排水量；载重排水量；积载排水量；荷载排水量

load-displacement curve 荷载—位移曲线
load distance 运距;载运距离
load-distributing curve 荷载分布曲线
load-distributing foundation 分布荷载的基础
load distribution 荷载分布;配电;负载分散;负荷分配;负荷分布
load distribution ability 荷载分布能力
load distribution action 荷载分布作用
load distribution characteristic 荷载分布特性
load distribution curve 配电(曲)线;荷载分布(曲)线;负荷分布曲线
load distribution factor 荷载分布系数
load distribution in a combined power system 复合电力系统中的负荷分配
load distribution line 荷载分布线;负载分配曲线
load distribution plate 承压板
load distribution power 荷载分布能力
load disturbance 负荷扰动
load diversity 荷载不同时率;负荷差异
load divider 负载分配器
load-dividing valve 荷载分配阀
load division 负荷分区
load double precision 倍精度寄存
load down 降负荷
load draftload draught 满载吃水
load draught 载重吃水;负荷吃水
load drop 荷载减小
load-dropping test 减负荷试验
load due to wind pressure 风压荷载;风(力)荷载
load dump test 甩负荷试验
load duration 荷载持续时间;荷载历时
load duration curve 荷载历时曲线;负荷历时曲线;负荷持续时间曲线
loaded 载重的;加载的;荷载的
loaded antenna 加载天线;加感天线
loaded area 荷载区;荷载面积;承载面积
loaded barge storage 重载驳船停泊区
loaded beam 承载的梁;承重梁
loaded brick 承载砖
loaded cable 加感电缆;负载电缆
loaded car 重车
loaded cars to be delivered at junction station 移交车
loaded cavity 加载谐振腔
loaded chord 受载弦杆;承载弦
loaded circuit 加载电路
loaded column 荷载柱
loaded concrete 重混凝土
loaded condition 满载状态
loaded container 重箱;载货集装箱;载货货柜;实货柜
loaded container spreader 重箱吊具
loaded core 已装料的堆芯
loaded density of sample 样品充填密度
loaded diffused optic(al) waveguide 加载扩散光波导
loaded displacement 满载排水量
loaded displacement tonnage 满载排水吨位
loaded draught 满载吃水;满负载时的下沉量
loaded filter 反滤层;堵塞的过滤器;压载滤水体;阻塞的滤清器;护脚倒滤层;承载反滤层
loaded float 加载浮子;荷载浮筒;水下浮标;双子浮标;深水浮子;深水浮标
loaded frame 受荷载的框架
loaded freeboard 满载干舷
loaded hole 装药炮孔
loaded impedance 加载阻抗;承载阻抗
loaded in longitudinal direction 纵向加载
loaded in transverse direction 横向加载
loaded inverted filter 加荷的倒置过滤器
loaded length 装药长度;荷载长度;荷载长度
loaded line 装满的管线;荷载线;加感线路
loaded magazine 已装片暗盒
loaded masonry(work) 承重圬工(结构)
loaded mechanism 受载机理;受载的机理
loaded motional impedance 动态阻抗;负荷动态阻抗
loaded mud 高比重泥浆
loaded part 有负荷的零件
loaded plane 荷载作用的平面;承载平面
loaded price 附加费用价格
loaded Q(value) 负载时的 Q 值
loaded radius 负载半径
loaded region 承载区

loaded resin 荷载树脂
loaded river 挟沙河流;含沙河流
loaded rubber 填料橡胶
loaded running speed 重载运行速度
loaded sheet 填料纸
loaded skeleton 受荷载的骨架
loaded space structure 受荷载的空间结构
loaded speed 满载航速
loaded spring 荷载簧
loaded stream 挟沙河流;含沙河流
loaded surface 加载面
loaded swelling 有荷载膨胀
loaded swelling rate 有荷载膨胀率
loaded three-dimensional structure 受荷载的三维结构
loaded to capacity 满载
loaded to collapse 加载到破坏
loaded track 受载轨道
loaded-up condition 荷载状态
loaded valve 荷载阀
loaded wagon 重车
loaded wagon direction 重车方向
loaded wagon flow table 重车车流表
loaded wagon kilometers 重车走行公里
loaded wagons delivered at junction station 移交重车
loaded waterline length 满载吃水线长度
loaded weight 装载重量
loaded wheel 塞料磨轮
loaded work piece 负载工件
load effect 荷载效应
load efficiency 负载效率
load ejection 装货顶推(机)
load-elongation curve 荷载伸长曲线;负载伸长曲线
load-elongation diagram 荷载伸长曲线图;负载伸长图
load equalization 负荷平衡
load equivalency 荷重当量
load equivalent 等效载重;荷载当量
loader 装载机;装卸机;装填机;装料机(械);装货人;装砖机;装砖车;载重车;加载装置;加载器;贷款机构;承载器;搬运设备
loader arm 装载机起重臂
loader attachment 加料器附件
loader backhoe 前装反铲挖掘机
loader block 装入程序块
loader boom 装载机转臂;横梁式装料机
loader-compatible form 适于装人程序的形式
loader conveyer 装载机的运输机;自装式运输机
loader-digger 挖掘装载机;挖掘装载两用机
loader discharge boom 装载卸料悬臂;装载机卸料悬臂
loader discharge conveyer 装载机的卸载运输机
loader-dozer 装载两用推土机;装载推土两用机
loader drift 装载平巷
loader end man 装载机后座管理员
loader engineer 装载机司机
loader equipment depreciation apportion and overhaul charges 装岩设备折旧摊销及大修费
loader for loading flat stored grain 平台存粮装载机
loader for loading ground stored grain 地上存粮装载机
loader front conveyer 装载机的前部运输机
loader gate road 装载工平巷(采矿)
loader gathering head 装载机集料机头;装载机骨料机头
loader-harvester 收获运装机
loader island 装卸岛
loader itself 拌和机装料斗(混凝土)
loader mechanism 装载机构
loader-mounted jumbo 装在装载机上的钻车
loader-mounted shovel 装载斗
loader raising gear 装料器提升机构
loader rear(ward) conveyer 装载机的尾部运输机
load error 负载误差
loader runway 装载机导轨;装载机滑道
loader scale 装料器秤
loader screen 筛分装载机
loader's door race 装填机门座圈
loader shovel 装载铲;铲斗车
loader skip 装料器斗
loader slide 装车台
loader tramming lever 装载机行走操纵杆

loader transporter 装运车
loader truck 起重汽车
loader tyre 装载车轮胎
loader-unloader 装卸机
loader wharf 装载码头
load estimate 荷载估computing;荷载估计;负荷估算
load-extension curve 拉伸应力应变图;荷载延伸曲线;荷载伸长曲线;荷载拉伸曲线
load factor 流量系数;装载因子;装载因数;荷载因子;荷载因数;荷重系数;负载系数;负载率;负载因素
load factor design 荷载系数设计;荷载系数法设计;荷载因数设计
load factor design method 荷载系数设计法
load factor method 破损阶段法;荷载因数法;荷载系数法;破坏阶段法
load factor of intersection 交叉口负荷系数
load factor of locked ship 过闸船舶装载系数;过泊船舶装载系数
load factor rate 负荷因素比率
load feeder 荷载馈线
load file 装入文件
load flow 上网潮流;潮流(电力)
load flow calculation 负荷量计算
load-flow structure 负荷流动构造
load fluctuation 荷载波动;荷载变化;负荷波动
load fold 负载褶皱
load following 负荷跟踪
load forecast 负荷预测
load form crowd 人群荷载
load-free speed 无负荷速度
load frequency 输出频率
load frequency control 输出频率调节;负荷频率控制
load frequency design factor 荷载频率设计系数
load from crowd 人行荷载
load ga(u)ge 载重计;测载计;测力计
load gear 起落架;荷载齿轮
load-generating system 加荷系统;产生压力的装置
load governor 荷载调整器;负载调节器
load growth 负荷增长
load handling 负载装卸
load handling device position indicator 吊具位置指示器
load haul dump 装运卸砟联合机
load haul-dump unit 装运机
load hauled 托运荷重;牵引总重;牵引重量;牵引质量
load holding device 载重保持装置;载重把持法;捆扎荷载装置;加载装置;荷载保持装置;保持荷载装置
load hook 起重吊钩;吊钩
load hopper 装料漏斗
load hour factor 负荷时间率
load image 装入图像
load immediate instruction 立即装入指令
load impact 负荷冲击
load impact allowable 容许冲击负荷
load impact allowance 容许冲击荷载
load impedance 终端阻抗;负载阻抗;负荷阻抗
load impedance modulation 荷载阻抗法调制
load in bulk 散装装运
load inclination 荷载倾(斜)角
load inclination correction factor 荷载倾斜因素;荷载倾斜校正系数
load increase 荷载增加;负载增长
load increment 载重增加;每级荷载;荷载增量;负荷增量
load increment ratio 荷载增量比
load index 荷载指数
load index register instruction 负荷索引寄存器指令
load-indicating bolt 荷载指示螺栓;示压螺栓;高强安全螺栓
load indicator 载重指示器;量测荷载计;荷载指示器;起重量指示器;测载计;测力计;测载计;负荷指示仪
load indicator of cutter motor 铰刀电机负荷指示器
load-induced earthquake 荷载诱发地震
load-induced stress 附加应力
load influence zone 荷载影响区(域)
loading 吸收量;装炸药;装载;装岩;装入;装料;装货;装车;荷载作用;加重;加重(载);加荷(载);寄存[计];集装箱装箱作业;浓度;货物装箱;含

量;弃渣处理;配入量;存放;采购;附加成本;装填;负荷
loading airlock 装载用气闸
loading airlock chamber 装载用气闸室
loading analysis 荷载分析
loading and discharging 装卸
loading and haulage in tunnel exploration 坑道装岩运输
loading and over-flowing dredging 装舱溢流
loading and unloading 装卸;加载与卸载;加载与卸荷;加荷与卸载;加荷与卸荷
loading and unloading accident 装卸事故
loading and unloading bay 装卸港池
loading and unloading capability 装卸能力
loading and unloading capacity 吞吐量
loading and unloading cargo 装卸货物
loading and unloading crane 装卸起重机
loading and unloading expenses 装卸费(用)
loading and unloading facility 装卸设备
loading and unloading machine 装卸机械
loading and unloading mechanization 装卸机械化
loading and unloading operation 装卸作业
loading and unloading platform 装卸平台
loading and unloading point 装卸地点
loading and unloading ramp 装卸(斜)坡道
loading and unloading risks 货物装卸险
loading and unloading road 装卸运输路线
loading and unloading station 装卸站
loading and unloading time 装卸时间
loading and unloading track 装卸线【铁】
loading angle 装填角
loading aperture 装料孔
loading apparatus 负载装置;荷载仪
loading appliances 装载器械
loading arcade 登机廊道
loading arch 承载拱
loading area 载货面积;装载(地)区;装载场;装入区;加载面积;码头装卸区;承重面积;负载面积;负荷区
loading arm 装料悬臂;装料臂
loading arm tube 装料悬臂管
loading arrangement 装载安排
loading attachment 装载装置
loading auger 装料螺旋
loading back method 反馈法
loading bank 装货台
loading bay 装卸湾;装卸港池;装货场;装车车位;进料台;安装现场;装料间;装载场;装货间;装车跨;进料场
loading belt 装料皮带(机);装料输送带;装料皮带带式装载机
loading belt conveyer 装载用皮带输送机
loading belt jib 装料皮带托架
loading berm 加压戗台;反压台;反压护道
loading berth 装车车位;装货泊位;装船泊位;散装码头
loading bin 填料筒;装料仓;装载器;装料(存)仓
loading boom 悬臂装载吊杆;装料横梁;装货臂架;吊货杆
loading bridge 装载桥;装卸桥;装卸起重机;载重桥;高架桥;桥式装料机;桥式装船机;载重机桥(架)
loading bridge hoist 装载桥式吊车
loading brigade 装载组
loading broker 装货经纪人
loading bucket 铲斗;装载斗;装岩斗
loading by crab's paws loader 蟹爪式装岩机装岩
loading by electric(al) scraper 电耙装岩
loading by hand 人工装岩
loading by load-haul-dump equipment 装运卸设备装岩
loading by shovel-haul equipment 铲运机装岩
loading by shovel loader 铲斗装岩机装岩
loading cable 承载索
loading calculation 负荷计算
loading calculator 负荷计算器
loading capacity 功率;起重量;装载量;起重能力;装载能力;载重能力;载重量;荷重能力;承载能力;负荷能力;负荷量
loading capacity of bridge 桥梁承载能力
loading capacity of car 车辆载重(量)
loading capacity of column 柱荷载量
loading capacity of hopper barge 泥驳装载量
loading capacity of land treatment 土地处理负荷量
loading capacity of vehicle 车辆装载量
loading capacity of the magazine 暗匣装片量
loading cartridge 定量装载器
loading case 负荷线圈盒;装货箱;荷载情况
loading cask 装料罐
loading certificate 装货证明书
loading chain 木材拖链
loading charges 装卸费(用);装货费(用);装船费(用)
loading chart 荷载分布图;装料图表
loading chute 装载溜槽;装载槽;装畜通道;装料槽;装货滑槽;供料滑槽;供料槽;进料槽;散装货装货筒;散装装货槽
loading clamp 装料夹钳
loading clearance 装载限界
loading coat 加荷板;压止层(地下室柔性防水层的混凝土覆盖层);抗水板;重压板(地下室抗水压地板);加荷板
loading code 荷载规范
loading coefficient 荷载系数;负荷系数;装载系数;加载系数;加感系数
loading coil 加感线圈
loading coil spacing 加感线圈间距;加感节距
loading combination 荷载组合
loading concrete 铺设的混凝土面层(地下室沥青地面上);抗水板
loading condition 荷载分布情况;荷载状态;负荷条件;装载情况;负荷条件
loading conductivity box 负荷导电箱
loading control 加载控制器
loading conveyer 装载运输机;装料输送机
loading cradle 成品收集筐;成品筐架
loading crane 装载起重机;装料吊车;装货吊车;装吊起重机
loading criterion 加载准则;加载标准;加荷标准;负荷标准
loading curve 荷载曲线;负荷曲线;加载曲线;加荷曲线;荷重曲线
loading cycle 加载周期;加载循环;加荷周期;加荷循环;荷载循环
loading cycles to failure 装载周期至破损阶段
loading cylinder 装载汽缸
loading data 装载数据
loading data of machines 机器的负荷数据
loading data sheet 负荷数据单
loading days 装货天数
loading decrement 每级卸荷
loading-deformation curve 荷载变形曲线
loading-deformation relationship 荷载变形关系
loading degree 装载程度
loading density 装药密度;装填密度;装料密度;加载密度;加荷密度
loading depot 装载站
loading device 加料装置;负载装置;装载设备;装料装置;加载装置
loading diagram 装载分布图;荷载图;载载图(示)
loading disc 满载吃水线标图
loading disk 载重线线圈;加感圆盘
loading distribution 装载分布;荷载分布
loading dock 装船码头;装卸场;货物中转站;仓库装卸月台;装载支架;装卸站台;装卸平台;装货平台
loading dock shelter 装卸平台棚;装卸台雨棚
loading door 侧台大门;舞台后运布景门
loading dose 负荷剂量
loading dredging 装舱作业;装舱施工
loading drift 装载平巷
loading edge 装岩斗
loading effect 负载作用
loading elevator 装载提升机
loading embankment 加载填方;加荷填方
loading end 加载端;承截面
loading equipment 加载设备;装载设备;装岩设备
loading error 装配错误;加载错误;加感误差;输入误差;输入差错
loading estimate 载额估计
loading facility 装额设备;装料设施;装料设备;装货设备
loading factor 负荷系数;加载因数;积载因素;积载因数;荷载系数;荷载因数;负载因数
loading factor of container 箱载(重)利用率
loading flexible conduit 装料软管
loading fold 负荷褶皱
loading foot slope 压坡脚
loading force 加载力
loading fork 装料叉
loading for maximum stress 最大应力的荷载
loading frame 荷载架;加载(载)架;加载框架;加荷载架;加荷框架;荷重架
loading function 荷载函数
loading gang memo 装货跳板
loading gantry 门式装船机
loading ga(u)ge 装载界限;限界架;装载限界;载重吃水标尺;载重标准;量载(荷)规;测载器
loading grab 装岩斗
loading grabbing crane 抓斗装载起重机装料抓斗
loading grade of lateral pressure test 旁压试验加压等级
loading groove 装料口
loading ground 承载土
loading hat 加感线圈罩
loading hatch 装卸舱口
loading head 装船机机头;集料器
loading height 装载高度;(汽车的)底板高度
loading history 加载历史;加载过程;加荷历史;加荷过程;荷载历史
loading hopper 装料(漏)斗;装卸散装货料斗;装载斗(仓);进料漏斗
loading hose 装药软管;装料软管
loading hydraulic system 装载液压系统;加荷液压系统
loading in bulk 散装发运;散装
loading increment 荷载增量;每级荷载
loading in daylight 白光下装片
loading index 荷载指数
loading induced stress 荷载引起的应力
loading inductance 加感
loading influence zone 荷载影响区(域)
loading in increments 分阶(增值)加荷;分阶段加荷
loading inlet 装料口
loading installation 装料设施;加荷设备;装车装置;装载起重设施;装料设备
loading instruction 载重规定;载重定额
loading instrument 配载仪
loading intensity 装载强度;加载强度;负荷强度
loading interval 装填时间
loading in turn 依次装货;按次序装货
loading island 装载岛;公交车站安全岛;停车岛;岛式站台;岛式平台;岛式公交车站
loading jack 装载索具;装车台;装车起重器
loading lag 装填滞后
loading leg 溜送槽
loading lifting capacity 额定起重量
loading lifting mechanism 起重机构
loading limit 荷载极限;装载限度;载重限度
loading limitation 载重限制
loading line 装载吃水线;装载高度;充气管路
loading list 装货清单;入港报关单
loading location 装载站
loading machine 装载机;装料器;装料机(械);装货机;带载机组
loading-machine runner 装载机司机
loading machine with chain buckets 链斗装载机
loading manhole 装加感线圈的进入孔
loading manifold 灌装集流管
loading mark 滴料折痕
loading material 加感材料;填入料;填充剂
loading mechanism 加载装置;加载机构;加荷机构
loading member 装卸工人
loading method 装载方式;装载方法;装岩方法
loading motor 加载电机
loading net 装卸网兜;(装卸用的)网兜;吊货网(兜)
loading note 装料单
loading of bridge 桥梁载重量
loading of concentrated weight 集重装载
loading of effluent weir 出水堰负荷
loading of ground 地基加荷
loading of hole 炮眼装药
loading of pipe 管内填塞(弯管时)
loading of river 河流负荷
loading of source 装源
loading of stream 河流负荷
loading of timber 木材装载
loading of water 水体荷载
loading on the berth 船在泊位装货中

loading operation 装料作业
loading operator 装料作业人员;装料工
loading out facility 装船设备
loading overall 单一货种的装运
loading pad 加载垫板;加荷垫板
loading pan 吊桶装载斗
loading parameter 荷载参数
loading path 加载途径;加荷途径;荷载途径;荷载路径
loading pattern 装货方式;装料方式
loading period 荷载周期;载荷周期
loading permit 装船许可证
loading pigment 颜料填充剂;体质颜料
loading pipe 填装管;预填物管子(填有沙子);装填料管;装货溜管;装货滑槽;填料管(弯管时管内填料以防变形)
loading pipeline 装料管线
loading piston 加载活塞;加荷活塞
loading place 装载货物处;装货地点
loading plan 装载计划;装料计划;配载图
loading plan by stations 站别装车计划
loading plant 装载设备;装货设备
loading plate 荷载板;装载板;装盘;加荷板;加载板
loading plate test 荷载试验;承载板试验
loading platform 装卸站台;装车台;装料台架;仓库装卸站台;装载台;装油平台;装料平台;装料(平)台;装货站台;装货平台;荷载台;上下客站台
loading platform track 站台铁路线
loading pocket 装载仓
loading point 加感点;装载(地)点;装料点;装驳站;加载点;加有稳定剂的地基加载点;荷载作用点;荷载点
loading pole 加载杆;加载导柱;加荷杆;爆炸点
loading pollution sources 负荷污染源
loading port 装运港;装货港;载货港(口)
loading port and destination 装运口岸及目的地
loading port survey 装卸港检查
loading position 负荷位置;荷载位置;装填位置;装料位置;滴料位置
loading pressure 箱内气体压力;增压压力
loading procedure 加载过程;加荷过程
loading process 装带过程;加载过程
loading production quota 装岩生产定额
loading productivity 装岩生产率
loading program(me) 装入程序;装料计划
loading property 负荷特性;荷载特性
loading pump 装料泵
loading quay 装料码头;装货码头;装船码头;载装码头
loading rack 装料台;灌装栈桥
loading ram 加载活塞;加荷活塞
loading ramp 装货跳板;装卸场;斜坡装卸台;装载坡台;装载坡道;装料(斜)台;装车斜坡台
loading rate 装载速度;装货率;荷载率;装货效率;加载速率;加荷速率;升负荷速度;负荷率
loading rating 荷载速率
loading ratio 装载比;装药率(单位体积岩石的炸药量);装填比;装货比
loading regime 带载方式
loading regulations 载重线规则
loading report 装货报表
loading resistor 负载电阻器;负载电阻
loading reversal 反复荷载
loading rhythm 负荷韵律
loading rig 加载设备;加荷设备
loading ring 装货环;装料环;加载环;加带环
loading roller 负荷滚轮
loading routine 装入程序;输入程序
loading run 装载循环
loading schedule 装料图;装料计划
loading scow 大型货盘
loading screen 装料筛;加料筛
loading section 加感段;加感部分
loading sequence 装载顺序
loading shaft 进料口
loading shed 仓库前装卸车棚;装卸棚
loading sheet 装料板
loading shift 装料班
loading shock 加载波动;加荷突变;冲击加载
loading shovel 装料铲;装卸铲;装料机铲斗;装料铲;铲装车;铲斗车
loading shovel bucket 装载铲斗(车)
loading side 负载的一侧;负载的一边;加载的一侧;加载的一边

loading siding 装载(侧)线;装卸线【铁】;装车线(路)
loading silo 装料筒仓
loading site 装载地区;装载地点
loading skip 装载料车;装料翻斗(混凝土);装载斗仓;装料斗
loading slab 浮式地板(地板下有橡胶衬垫);重压板;抗水板;加载板;加荷板
loading soil 承载土层
loading space 装载面积;装料室;载料空间;船货装载空间
loading speed 装货速度;加载速率;加载速度;加荷速度;负载速度
loading spot 装货地点
loading spout 装货溜管;装货滑槽
loading spring 负荷弹簧
loading stage 装货台;装载提升机;装料台;装卸栈桥;卸料台;装料平台;加载时间;加荷时间
loading standard 加载标准;加荷标准
loading standard for design vehicle 车辆荷载标准
loading state 负荷状态
loading station 装载装置;装载站;装料站;装货站;起运站
loading stick 装药棍
loading strain 荷载应变
loading-strain curve 荷载应变曲线;荷载变形曲线
loading-strain diagram 荷载应变图
loading-strain property 荷载应变性能
loading strength 承载强度
loading stress 荷载应力
loading surface 装载面
loading surplus 附加利益
loading survey 装载检查;装舱检查
loading system 承载系统;装载设备;载带系统;(试桩)加载装置;加载系统;(试桩的)加荷装置;加感制
loading table 装料图表;装载表
loading tackle 装卸滑轮
loading team 装载队
loading temperature 装模料温度
loading term 装载期限
loading terminal 装料站;装料码头;装货码头;装装码头;货运终点站
loading test(ing) 负荷试验;载重试验;加载试验;荷载试验;加荷试验
loading time 加载时间;装载时间;装岩时间;装填持续时间;装货时间;加荷时间
loading time allowed 容许装载时间
loading time used 实用卸货时间
loading to capacity 装满
loading to collapse 加载至破坏
loading to failure 加载至破坏;加荷至破坏
loading to normal capacity 额定负荷
loading tower 装载塔;装卸塔;装货塔架
loading tower operator 装载塔司机
loading track 装载轨道;装卸线【铁】;装车线(路)
loading tray 装零星货物的吊盘;装料盘
loading tripod 装运三脚架
loading trolley 起重绞车
loading trough 装载槽
loading truck 装药车
loading turn 装货次序
loading type 装载方式
loading unit 装入机头;加载单元
loading-unloading 装卸
loading-unloading facility 装卸车设施
loading-unloading line 装卸线
loading-unloading platform 装卸站台
loading-unloading siding 装卸线
loading-unloading time of rail car 铁路车辆装卸作业时间
loading up 加负荷;加负荷;负荷升高
loading vessel 装药器
loading voucher 装料单
loading warranties 货运种类或数量限制保证
loading waveguide 加载波导
loading way 仓库前装卸车棚
loading weigher 装料称量装置
loading weight 荷载重量;装船重量;装重;荷载重
loading well 供料室;供料井
loading wharf 装货码头;装料码头;装货码头;装船码头;载装码头
loading wheel 负重轮

loading width 装载宽度
loading winch 装料卷扬机;装卸绞车;起重绞车;起货机
loading yard 装载场;装货场;装车场
loading zone 装载区;加载区域;加荷区域
load initial table 初始装入表
load input 荷载输入
load insensitive device 负载不敏感元件
load instruction 取指令;取数指令
load intensity 荷载强度;负荷强度
load interest in constructional period 建设期贷款利息
load interrupt mask 装入中断屏蔽
load isolator 负载隔离器
load jaws 负载夹紧装置
load jib 承重挺杆
load key 装载键;输入键;打入键
load lamp power meter 灯泡负荷功率计
load lateral distribution 荷载横向分布
load lateral distribution coefficient 荷载横向分布系数
load leads 负载引线
load level 负荷等级
load-leveler shock absorber 车身调平减振器;负荷调平式减震器
load level(l)ing 负载均衡;负载调整;负载平衡
load-level(l)ing relay 平衡负荷继电器;负载限制继电器
load-lift height 起升高度【机】
load lifting limiter 荷载限制器;起重量限制器
load lifting speed 荷载提升速度;起重速度;起升速度【机】
load limit 荷载极限;负载极限
load limitation 负载限度;荷载限度;负载极限
load limit changer 负载极限变换器
load limiter 荷载限制器
load limiting device 荷载限制器;负载限制器
load limiting ga(u)ge 最大装载界限
load limiting resistor 负载限制电阻器
load limiting system 荷载限制系统
load limiting winch 自动调整张力绞缆机
load limit motor 负荷限制电机
load limit operation 限负荷运行
load limit switch 负载极限开关
load limit valve 负载限制阀
loadline 载重吃水线;载重(吃水)线;荷载线;负载线;负载
loadline assignment 载重线勘定
loadline certificate 干舷证书;载重线证书
loadline convention 载重线公约
loadline disc 载重线标圈
loadline guide block 载重线导向轮;载重线导向块
loadline mark 干舷标志;载重线标志;满载吃水线;载重吃水标志;载重吃水标记
loadline regulations 载重线规则
load line rules, zone, areas and seasonal periods 载重线规则、地带、区域及季节期
loadline ship 勘有载重线标志的船
loadline survey 载重线检验
load location detector 货位异常检测装置
load loss 负荷损耗;荷载损失;负荷损失;负载损耗;负载损失
load-lowering height 下降高度【机】
load-lowering height indicator 下降深度指示器
load-lowering speed 载重下降速度;下降速度【机】
load magnitude 负荷数量
load maintainer 加载稳定器
load maintainer system 静载维持系统
load map 装入配置图;装配图
load mark 载重标志;干舷标志
load masonry building 承重砖石房屋
loadmaster 装卸长
load matching 负载匹配;负载平衡
load matrix 荷载矩阵
load measurement 荷载测量;负荷测定
load measuring baseplate 测力垫板
load measuring cell 负载传感器;测力计
load measuring device 负荷测定装置;负载测定装置
load measuring recorder 荷重记录器
load memory lockout register 负载存储锁定寄存器
load metamorphism 承载变质;承载变形;承载变性;负荷变质(作用)
load meter 测荷仪;荷载计;落地磅秤

load meterage 设备负荷测量
load micro-program(me) 输入微程序
load mode 装载式;装入方式;传送方式
load module 装载程序;装入模块;装配组件;装配模块;寄存信息块;输入程序(片)
load module library 装配模块库;输入模块库
load moment 荷载力矩;起重力矩;负载力矩;荷载转矩
load moment limiter 载重力矩限制器
load monitoring system 载货监视系统
load mo(u)ld 负荷模
load new program(me) 装入新程序
load of earth pressure 土荷载
load off 卸荷;卸载
load of glacier 冰川携带物
load of grinding media 研磨体装填量
load of loosen bedrock 松弛荷载
load of loosen ground 松弛荷载
load of map content 地图容量
load of river 河流泥沙
load of roadway 线路荷载
load of roof 顶部荷载
load of rotary table 转盘负荷
load of taxation 赋税重压
loadometer 轮载测定器;(称量载重汽车的)落地(磅)秤;轮压测定器;荷载(测力)计;测载仪;测荷仪
load on 加载;载上
load on axle 轴心加载;轴上荷载
load on call 调用装入
load on pile 桩荷载
load on roadway 车道荷载
load on section 路段载重;区段荷载
load on spring 簧上荷载
load on the wheel 轮上负荷
load on top 油脚不清装油法;最大负荷;最大负荷;顶装法
load on top system 油脚不清装油制度
load-operated brake 荷载自制式制动器
load operation 加载操作;取操作
load outreach indicator 荷载幅度指示器
load-out tunnel 装载隧道
load over the entire area 全面积负荷
load pacer 均匀加载装置
load pad 承载垫板
load path 荷载传布途径
load pattern 荷载模式;负荷特性;负荷曲线图
load peak 峰(值)荷;高峰负荷;负荷峰值;负荷尖峰;负荷峰值
load-penetration curve 荷载—贯入曲线;负荷—贯入曲线
load-penetration test 荷载贯入试验
load per axle 轴载重
load per unit area 单位面积负荷;单位面积产量(池窑);池窑单位面积产量
load per unit length 单位长度荷载
load per unit of length 单位长度的负荷
load per unit of mass 单位质量含量
load per unit of volume 单位体积含量
load phase 带载相位;带载相
load pipe tester 管道荷载检验仪
load piston 举重活塞;加压柱塞(土壤承载力试验)
load pocket 负荷囊
load point 荷载作用点;信息起始点;输入点
load point indicator 信息起始点标记
load point marker 信息起始点标记
load polygon 荷载多边形
load position indicator 加载位置指示器;加料位置指示器
load pouch 负荷囊
load power 负载功率;负荷功率
load prediction 负荷预测
load prediction curve 负荷预测曲线
load pressure 荷载压力;负荷压力
load pressure brake 负荷压迫制动器
load pressure feedback 负荷压力反馈
load pressure retarder 重力式缓行器
load profile 负荷曲线
load-proportional braking equipment 正比于荷重的制动装置
load-proportional valve 荷载均衡阀
load proportioner 负荷量调节器
load proportioner system 负荷按比例调整系统
load pulley block 载重滑车组;载重滑车架
load quantity 荷载量

load radius 装载半径;起重机装载半径
load range 荷载幅度;负荷范围
load rate 单位负载;额定负荷;荷载率
load rating 设计负荷;设计负载;额定装载量;额定荷载;额定负载;荷载率定
load ratio 荷载比;负荷率;负荷比
load ratio adjuster 带载电压调整装置
load ratio control 带载电压调节
load ratio voltage regulator 带载电压调整器;负载电压调整器
load reactance 负荷电抗
load reaction 负荷反应
load-reaction brake 荷载反作用制动器
load reading 荷载读数
load reapplication 重新加载;重新加荷
load record 负荷记录
load recrystallization 重压再结晶
load redistribution 荷载重分布
load-reducing influence 负荷降低影响
load reduction 负荷减低
load register 装入寄存器
load register instruction 存入寄存器指令
load regulation 装载规程;负载调整率;负载调节;负荷调节;加载规程;加荷规程
load regulator 负载调整器;负载调节器;负荷调整器
load rejection 供电暂停;弃负荷;抛负荷;甩负荷;负荷中断
load relay 负荷继电器
load removal 卸载;卸(负)荷
load removal age 卸荷龄期
load repeat counter 负荷重复计数器
load repetition 重复负载;负载循环
load resistance 荷载阻力;负荷电阻;负载电阻
load-resistance photocell 负载电阻光电管;负荷电阻光电管
load respondent valve 荷载感应阀
load response 负荷反应
load responsible control 荷载调节
load responsible controller 力调节器
load-responsive control 反应负荷变化控制
load retention test 持载试验
load reversal 荷载变号;荷载反向
load-reversal test 反复荷载试验
load rig 装卸设备;加荷设备
load ring 量力杯;集索圈
load river 含泥沙河流
load rope 重载钢丝绳;吊重索
load running 满载运转;负载运行;负载运转
load sampling 泥沙取样
load saturation curve 负载饱和曲线;负荷饱和曲线
load scheme 负荷计划
loadscreen 筛分装载机
load sediment entrainment of river 河流挟沙(也称河流挟砂)
load segment 装配段
load selector 负荷选择器
load selector switch 负载选择开关;负荷选择开关
load sensing torque converter 荷载传感式变扭器;负载传感式扭矩变换器
load sensitive device 负载敏感元件;负荷敏感元件
load sensitivity 负荷灵敏度
load sensor 荷载传感器;测力器;负载检测器
load separation point 负荷分离点
load setting gear 负荷给定装置
load-settlement curve 荷载沉降曲线
load-settlement diagram 荷载沉降曲线图
load-settlement graph 荷载沉降曲线图
loads for tunnel 隧道荷载
load sharing 均分负载(法);负载均分;负荷分配;分载
load-sharing between substations 变电所间可负载复用
load sharing magnetic switch 均分负载磁开关
load-sharing matrix switch 均分负载矩阵开关;负荷分配矩阵开关
load sharing switch 负荷分配开关
load-sharing system 负荷分配系统
load sheave 吊货滑车
load-shedding 能使荷载扩散的;分区停电;减荷;切断若干线路的电流供应(电源过载时);甩负荷;分区切断负荷;局部停电(防止电站超荷)
load-shedding equipment 减载装置;甩负荷装置
load shifting 移载
load shifting resistor 负荷转移电阻

load shock 冲击荷载;负荷冲击
load shut-down test 甩负荷试验
load side 负荷端
load signal conditioner 负载信号调节器
load skeleton 承重骨架
load-slip curve 负载滑动曲线
load spacing 荷载间距
load/span factor 荷载跨度系数
load spectrum 荷载谱
load speed 负载速度
load spread 荷载分布
load spreader 负载分散器
load-spreading 荷载分布
load-spreading ability 荷载分布能力
load-spreading curve 荷载分布曲线
load spreading property 荷载分布特性
load stage 加载时阶;加载时期;荷载阶段
loadstone 磁石;极磁铁矿;天然磁石
load store instruction 装入存储指令
load strain 荷载应变
load-strain curve 荷载应变曲线
load-strain diagram 负荷应变图;荷载应变曲线图
load stream 含泥沙河流
load strength 荷载强度
load stress 荷载应力
load strip 荷载带
load structure 负载构造;负荷构造;负荷构成
load-supply situation 负载及供电情况
load support 承料座;承料支撑
load supporting 承载
load supporting characteristic 承载特征;承载特性
load supporting girder 承载梁
load suppression gear 负荷限制器
load surface 承重面;装载面
load susceptance 负荷电纳
load-swelling curve 荷载—膨胀曲线
load swing 负载波动;负荷波动;负荷变化;负荷摆动
load switch 负载开关;负荷开关
load table 装载表;反压台
load tension curve 负荷张力曲线
load term 装载期限
load test 加载试验;带负荷试验;负载试验
load testing machine 负载测试机
load testing of structures 结构荷载试验;建筑物荷载试验
load test method 荷载试验法
load test of bridge 桥梁载重试验
load test on pile 桩载试验;桩载试验
load thrown off 负荷卸除
load thrown on 负荷带上
load throw off 供电暂停;弃负荷;抛负荷;负荷中断;甩负荷
load throw-off test 甩负荷试验
load time 装入时间;加荷时间;存入时间;负载时间
load-time deflection curve 负载时间挠度曲线;载荷时间与挠度的关系曲线;载重时间挠曲线;荷载时间挠曲线;挠度与时间关系曲线;荷载时间挠度(关系)曲线;受载时间挠度曲线
load-time displacement chart 荷载时间位移关系图
load tipping moment 倾覆力矩
load to collapse 载重崩溃;加载(至)破坏;加荷破坏;破坏荷载
load-to-load 逐批结算
load torque 荷载力矩;负载转矩;负载力矩;负荷扭矩
load train 轮压系列;荷载系列
load transducer 荷载传感器
load transfer 轴重转移;轴载转移;荷载转移;荷载传递;传送装入;负荷转移
load transfer analysis mechanism 荷载传递分析机构
load transfer assembly 货物转载设施;传力杆定位器;荷载传递设备
load transfer characteristic 荷载传递特征
load transfer device 载重转移装置;荷载转移装置;荷载传送装置;荷载传递装置;输送带;传送带;(混凝土路面的)传力装置
load transference 负荷转移;荷载转移
load transference joint (混凝土路面的)传力式接缝;传力接缝
load transfer mechanism 荷载传递机理;传力机理
load transfer method 荷载传递(方)法
load transfer path 荷载转移途径
load transferring area 荷载传递区

load transfer to rear axle during starting 起动时轴重向后轴转移
load transformer 负载变压器;负荷变压器
load transmission 荷载传递
load transport 泥沙输送
load type 负载类型
load uniformly distributed over span 跨度上均布荷载
load unit 负荷单位
load-unload cycle 加载卸载环;加载卸荷环;加荷卸载环;加荷卸荷环
load unload point 装卸料点
load up 装入;升负荷
load-up condition 负荷状态
load uprise 负荷上升率
load up variation 负荷上升率
load valley 负荷曲线低谷
load variation 负荷波动;负荷变动
load vector 装入向量;荷载向量
load vehicle 荷载车;载重车
load velocity relation 负荷速度关系
load verification 负荷校准
load voltage 负载电压;负荷电压
load wall 承重墙
load water-line 载重(水)线;载重线标志;载重吃水线;满载吃水线
load water-line coefficient 满载水线面面积系数;水线面系数
load water mark 载重吃水标志;载重吃水标记
load water plane 满载水线面
load water plane area 满载水线面积
load wave 负荷波纹
load weighing device 称重设备
load weigh scheme 车辆称重系统
load weight of car 车辆装载重量
load weight signal 荷载信号
load with ink 着墨
load-yield curve 荷载屈服曲线
load-yield recorder 荷载下沉记录仪
load zone 负荷范围
loaf 凸起图案
loafing shed 散放棚
loaf-like jointing 圆面包状节理【地】
loam 亚黏土;垆坶;烂砂;黏土砂泥;壤土;沃土;肥土
loam beater 揉麻泥工具
loam board 刮板
loam brick 垆坶砖;壤土砖;夯土砖;泥砖;烂砂砖
loam casting 黏土制型法;麻泥(型)铸造;砌砖铸造;砖铸;泥铸
loam clay 壤黏土
loam concrete 黏土混凝土
loam core 垆坶芯;壤土芯;黏土型芯;泥芯
loam core wall 壤土芯墙
loam cutter 夯砌面砖
loam desert 壤土荒漠地;垆坶荒漠地
loam fill(ing) 砂黏土填料;垆坶填料;垆坶填筑;回填亚黏土
loam glaze 天然土釉
loam ground 壤土质底
loam heath 壤土荒野地;垆坶荒野地
loamification 亚黏土化(作用);壤质化
loam kneader 黏土砂捏和机
loam kneader masticator 立式黏土砂泥捏和机
loam mill 烂砂碾磨机;黏土砂捏和机
loam mortar 砂黏土灰泥;垆坶灰泥
loam mo(u)ld 烂泥型;麻泥型;黏土砂泥型;泥型;黏土铸造(物);泥塑;泥模
loam mo(u)lding 黏土造形;泥型铸造
loam road 垆坶路;壤土路
loam sand 亚砂土;高黏土型砂;烂泥砂;壤土质砂
loam sand mo(u)ld 黏土砂泥型
loam seal(ing) 垆坶密封;壤土密封;亚黏土截水墙
loam wall(ing) 垆坶(筑)墙;壤土(筑)墙;夯土墙;土墙;板筑土墙
loam wedge 垆坶楔(块);壤土楔(块)
loamy 垆坶质的;壤土的
loamy bottom 壤土墙
loamy clay 垆坶质黏土;壤质黏土
loamy coarse sand 壤质粗砂
loamy earth 垆坶质土
loamy fine sand 壤质细砂
loamy fine soil 壤质细土
loamy gravel 带黏土砾石;带壤土砾石
loamy ground 垆坶质土

loamy light sand 亚壤土砂
loamy marl 壤土(质)灰泥
loamy sand 垆坶砂;壤(质)砂土
loamy sandstone 壤土砂石
loamy soil 壤(质)土;垆坶(质)土;黏土地;肥沃土
loamy texture 壤质构造;壤质
loan 借款;贷款;出借
loanable capital 可贷资本
loanable fund 可贷资金
loan account 贷款账户
loan administration 贷款管理机构
loan against collateral 抵押贷款
loan agreement 借款协议;借款契约;借款合约;借款合同;贷款协议;贷款协定;放贷合同
loan amount 贷款额
loan and advance 贷款与预付
loan and trust company 信托公司
loan applicant 申请贷款人;申请借款人
loan application 贷款申请书;借款申请书;申请借款
loan approved 已批准的贷款
loan authority 贷款管理机构;放款机构;放款当局
loan bank 放款银行
loan bill 贷款汇单
loan bond 债券
loan broker 贷款经纪人
loan business 贷款业务
loan calling 提前偿付债款
loan capital 借入资本;借贷资本;贷款资本
loan ceiling 贷款最高限额;放款限度
loan closing 贷款收盘;贷款结束;贷款结账
loan commission 贷款佣金
loan commitments 承付贷款
loan committee 贷款委员会
loan competition 信贷竞争
loan consortium 借款财团;贷款财团
loan contract 借款契约;贷款契约;放款合同
loan delinquency 贷款拖欠
loan-deposit ratio 贷款与存款比率
loan desk 借书处
loan disbursed 已拨款的贷款
loan document 贷款文件
loaned flat 免息借券
loaned labo(u)r 借用劳动力
loanee 债务人;借入者
loaner 债权人;借用物;借出者;贷款人;贷款部门
loan extended 展期贷款
loan financing 贷款筹措
loan for a short-time 短期贷款
loan for consumption 消费信贷;消费贷款
loan for exclusive use 专用贷款
loan for export 出口贷款
loan for technical facility 小型技术措施贷款
loan for the settlement of accounts 结算贷款
loan for use 使用贷款
loan from foreign powers 外债
loan fund 贷款资金;借贷资金;贷放基金
loan guarantee 借款保证函
loan guaranteed 有担保的贷款
loan guarantee program(me) 贷款担保计划
loan holder 押款的受押人;债券持有者;债券持有人
loan implementation 贷款拨付
loan in excess of the quota 超定额贷款
loan interest 借款利息;借贷利息;贷款利息
loan interest rate 贷款利率
loan in trust 贷款信托
loan ledger 借款总账;借款分类账
loan loss 贷款损失
loan loss allowance 备抵贷款损失
loan market 借款市场;贷款市场
loan money 借款
loan No. 贷款编号
loan note 借款证;借款凭证
loan of development project 开发项目放款
loan office 当铺;贷款处
loan officer 信贷员
loan of project 项目贷款
loan of taxation 赋税的重压
loan on credit 信用贷款
loan on favo(u)rable terms 优惠贷款
loan on guarantee 保证贷款
loan on interest 息债
loan on notes 票据抵押贷款
loan on security 押款

loan outstanding 未偿贷款
loan participation 参与贷款
loan payable 应付贷款
loan policy 贷款保险
loan portfolio 贷款业务量
loan pricing 贷款计息
loan principal repayment 贷款本金偿还额
loan proceed 借入资金;所得贷款;贷款收入
loan race 信贷竞争
loan rate 贷款利率
loan rate of interest 放款利率
loan redemption 贷款偿付
loan repay capital with interest statement 借款还本付息计算方案
loan repayment 偿还借款
loan repayment period 贷款偿还期
loan request 贷款申请
loans and discounts 借款与贴现
loans due to over-stocking 超储积压借款
loan secured by account receivable 应收款作担保的借款
loan service 贷款业务
loan servicing 贷款业务;借款的付息及还本
loan shark 高利贷者
loan society 信用合作社
loans on bills 票据抵押贷款
loans payable 应付借款
loans receivable 应收贷款
loan stock 借款股票
loans to subsidiaries 对附金属企业放款;附属公司借款
loan syndicate 借款财团;贷款财团
loan teller 信贷员
loan term 借款条件
loan tied to an index 按指数计值偿还的贷款
loan to subsidiaries 附属机构贷款
loan-to-value ratio 贷款与价值比
loan transaction 贷款业务
loan under hire purchase agreement 分期付款购货合同放款
loan value 押借额度
loan worthy 有偿付能力
loathing of cold 憎寒
Lobachevsky space 罗巴切夫斯基空间
lobate 舌状的
lobate coast 舌状海岸
lobate delta 桨叶形三角洲;舌状三角洲;朵状三角洲
lobate rill mark 舌状流痕;朵状细流痕
lobate scarp 叶状悬崖
lobby(area) 走廊;休息室;接待室;门厅;门廊;前厅;前室
lobby banking 门厅银行业务
lobby man (戏院、剧场的)收票员
lobe 叶型;叶形轮;叶片;正弦的半周;浅片;凸起;凸角;朵体;插接瓣;波瓣;冰川舌
lobe amplitude 波瓣幅度
lobe comparison method 波瓣比较法
lobed arch 有花饰的拱
lobed dome 扁圆屋顶
Lobed element pump 罗茨泵
lobed impeller 瓣叶形转子(罗茨鼓风机)
lobed impeller meter 叶形转子式流量计
lobed rotor compressor 罗茨压缩机
lobed rotor motor 罗茨电动机
lobed wheel 叶形轮
lobe frequency 波瓣频率
lobe-half-power width 波瓣半功率宽度
lobe pattern 波瓣图
lobe penetration 波瓣穿透
lobe penetration ratio 波瓣穿透率
lobe plate 凸轮板
lobe pump 罗茨泵;凸轮泵
lobe region 波瓣区
lobe rotation 波瓣旋转
lobe shaping 波瓣成型
lobe sweep(ning) interferometer 扫瓣干涉仪
lobe swing 波瓣摆动;瓣摆动
lobe switch 波瓣转换开关;波瓣变换开关;瓣变换开关
lobe switching 射束转换法;波瓣转换
lobe switching method 波瓣转换法;波瓣晃动法
lobe switch oscillator 波瓣转换开关振荡器
lobe width 波瓣宽度

lobing antenna 等信号区转换天线；波束可控天线
loblolly pine 火炬松
lob mechanic 工地机械工
lobster boat 捕虾船
lobster shift 夜班
lobulus parietalis inferior 顶下叶
lobulus quadrangularis 方形小叶
lobus parietalis 顶叶
local 现场的；局限性；局部的；地区的；地方（的）；本地的
local acceleration 局部加速度；当地加速度
local acceptance 限地承兑
local access 地方支路入口；专线入口；必停站口（指公共汽车逢站必停的路口）
local access and transport area 本地接入和传送地区
local action 自放电（电池）；局地作用；局部作用（量）；局部腐蚀；当地诉讼
local activity 局部活动性
local address 局部地址
local addressing 局部编址
local adjustment 局部平差【测】；局部调整；测站平差
local aeronautical chart 区域航空图
local agency 地方机构；地方办事处；当地分理处；当地代理行；当地代理商；当地代理人；当地代理处；当地代办所
local air conditioning 局部区域空气调节
local airline 地方航线
local allocation tax 地方分配税
local allocation tax grant 地方分配税收补贴
local amplification 局部放大
local amplifier 本机放大器
local and general independence 局部总体不相依
local annealing 局部退火
local anomaly 局部异常
local apparent noon 地方视（正）午
local apparent solar time 当地视太阳时
local apparent time 地方视时；当地视时
local application 局部应用
local aquifuge 局部隔水层
local architecture 当地建筑；地方建筑
local area 局部面积；局部地区网；区间面积
local area network 局部区域网络；局部地区网络；区域网络；地区网络；本地网；局域网
local area network simulation 局域网仿真
local area network system 区域网络系统
local array 地方台阵
local assistance 局部协助；地区协助
local astronomical time 地方天文时
local attraction 局部引力；局部磁场干扰；地方性磁干扰；地方磁干扰
local authority 地方政府；地方政权；地方管理局；地方当局；本地主管机关
local authority apartment unit 地方当局建造的公寓单元
local authority dwelling 地方当局建造的住宅
local authority estate 地方当局建造的公共住房
local automatic message accounting 局内自动信息记账；局部自动信息计算
local automation system 站内自动化系统；局部自动化系统；测站自动化系统；本机自动化系统
local availability 当地具备
local background 局部背景
local backup 局部后备（保护）；局部备用设备；局部备用方案
local base level 局部基准面；当地零点；当地基准面
local base vector 局部基向量
local batch job 本地成批作业
local batch processing 就地成批处理；本地成批处理
local battery 自给电池组；本机电池
local bearing strength 局部受压强度
local best-fitting ellipsoid 局部定位椭球
local bill of lading 国内运输提单；局部提单
local B/L 国内运输提单
local bond stress 局部黏结力；局部约束应力
local borrow 就地取土坑；就地取料坑
local branch 地方支线
local branched railroad 地方铁路支线
local branched railway 地方铁路支线
local branch road 局部支路
local brand 区域性品牌
local break 局部开裂

local breeding 地方品种
local buckling 局部纵弯（失稳）；局部弯曲；局部压屈；局部弯曲；局部屈曲；失稳
local buckling damage 局部压屈损坏（指钢梁）
local buckling of steel pile 钢桩的局部压屈
local buffer storage 局部缓冲存储器
local building 当地建筑
local building code 地方建筑规范；地方建筑法规；当地建筑规范
local building materials 地方建筑材料
local building regulation 地方建筑规程
local bulging 局部凸出
local burst mode 局部分段式
local bus 局部总线
local busy 局部占线；局部电话线不空
local bypass 局部分路
local cable 局内电缆；局部电缆；市话电缆；本机电缆
local call 局部呼叫；市内呼叫；市（内电）话；本地呼叫
local canalization 局部渠化
local capacity 地方能力
local car 货物作业车；本站作业车
local carburizing 局部渗碳
local cargo 当地货物
local carrier 地方运输行；地方船队；本地载波
local cars and empty cars dispatching plan 管内工作车输送和空车配送计划
local cars dispatching plan in district 区段管内车辆输送计划
local cars to be unloaded 管内工作量
local Cartesian coordinates 局部笛卡尔坐标
local cause 局部原因
local cell 局部电池
local cell corrosion 局部电池腐蚀
local cement industry 地方水泥工业
local cement plant 小水泥厂；地方水泥厂
local central office 市内中心局；市内中继线路；市话局；地区中心局；分局；本地中心局；本地电话局
local change 局地变化；局部变化
local change of velocity with time 局部速度变化
local channel 局部通道；部分信道；本地信道；本地通道
local channel loopback 本地信道环路
local character 分区编号
local charges 当地费用
local checker 本机振荡器检验器；本机检验器
local chequer 本机振荡器检验器
local chronicles 地方志
local circuit 局部环路；局部电路；本机电路
local circulation 局地环流；局部循环；局部环流；地方性环流
local circulation ventilation 局部循环通风
local circumstance 局部环境；当地情况
local city 地方城市
local civil time 地方民用时（间）
local clause 地区条款
local clearing 同城票据交换
local climate 小气候；局地气候；局部气候；地区气候；地方（性）气候
local climate in city 城市小气候
local clock 测站钟
local cluster of galaxies 本星系团
local cluster of stars 本星团
local coastal wind 地方性陆风
local code 局部编码；地方法则；地方法规
local code and regulation 地方法则；地方法规
local coherence 局部相干性
local coil 本机振荡线圈；本地线圈
local colo(u)r 地方色彩；局部光彩；本色
local communication 局部通信；地区通信
local community 区域群落
local commutativity 局部可换性
local competitive bidding 国内竞争性投标
local composite loopback 本地复合环回
local comprehensive equilibrium 地区综合平衡
local computer 本地计算机
local computer network 局部计算机网络；局部地区计算机网；本地计算机网
local condition 地方状况；定位条件；当地条件
local conditioning 局部调节
local conditions and customs 风土人情
local congestion 局部拥挤
local connection 市内通话；地方支线；地方连接线
local connectivity 局部连通度

local construction materials 当地建筑材料
local contact effect of cracked section 裂面局部接触反应
local contraction 局部束窄；局部缩窄
local contraction reach 局部狭窄河段；局部束窄河段
local control 现场控制；直接操纵；局部控制；局部防治；局部调整；局部调节；就地控制；本机控制
local control circuit 局部控制电路
local control equipment 就地级控制设备
local controller 单点控制器；单点控制机；本地控制器
local control panel 就地控制盘
local control survey 局部控制测量
local control system 直接操作装置
local convergence 局部收敛
local coordinates 当地坐标；局部坐标
local coordinate system 局部坐标系；地方坐标系；当地坐标系（统）
local correlation 局部对比
local corrosion 局部腐蚀
local cost 当地费用
local cost waiver 当地费用豁免
local couple 局部电偶
local cracking 局部开裂
local criterion 局部性准则；局部判据
local critical flux 局部临界通量
local crowning 局部凸起
local crush injury 局部挤压伤
local crystal oscillator 本机晶体振荡器
local currency 国内货币；地方货币；当地通货；当地货币
local current 局部水流；局部电流
local curve of a tide 当地潮汐曲线
local customs 当地风俗习惯
local damage 局部损坏
local database 局部数据库
local data base management system 局部数据库管理系统
local data transmission 局部数据传输
local datum 地方基准；当地零点；当地基准面
local decomposition 局部分解
local defect 局部缺陷
local deflection 局部偏转；局部偏斜；局部偏位；局部偏差；局部挠曲；局部变形
local deformation 局部变形；局部变态
local density approximations 局部密度近似求解法
local density of seawater 海水当场密度
local dent 局部凹陷
local dependence 局部相依
local depression 局部沉降；局部沉陷；局部凹陷；局部陷落
local depth structure map 局部深度构造图
local derivative 局部导数；偏导数
local desiccation 局部干燥
local destination 本地目的地；本地接收站
local destructiveness 地区性破坏
local devanning 当地拆箱（指集装箱）
local development city 地方开发城市
local device 本地设备
local differentiability 局部可微性
local digital loopback 本地数字环回
local dike [dyke] 民堤；地方堤防
local dip 局部扰动；局部倾斜
local distant control 近程—远程控制
local distant switch 本地远区转换开关
local distortion 局部扭曲；局部畸变；局部变形
local distribution 地域分布
local distribution of elements 元素地方分布
local distributor 地方性分流道路
local disturbance 局部扰动
local divide 局部分水界
local dose 局部剂量
local drying underwater welding 局部干燥水下焊接
local duty 地方税
locale 地点；场所
local earthquake 局部地震；地方震
local earthquake intensity 场区烈度
local effect 局部作用（量）；局部效应
local effect array 场地影响阵列；场地影响台阵
local element 局部电池
local elongation 局部伸长
local elongation at fracture 断裂处局部伸长

local emission source 本地排放来源
local employment 就近就业
local employment act 市内就业条例;地方就业条例;本地就业条例
local enlarged pile 局部加粗桩
local environmental protection 乡土保护
local erosion 局部侵蚀;局部冲刷
local erosion base level 地方侵蚀基准面
local excavation 局部开挖
local exchange 市内交换;市内电话局;地区内部交换
local exchange center 本地交换中心;地区交换中心;市话局
local exchanger 本地交换机
local exhaust system 局部排风系统
local exhaust ventilation 局部通风装置;局部排气通风;局部排风
local expert 土专家
local extension 局部伸长
local extension at fracture 断裂处局部伸长
local extra observation 加点观测
local extremum 局部极值
local factor 局部因素
local failure 局部故障;局部破坏
local fallout 局部沉降物;局部沉降
local fan 国产风机;局部通风机
local fatigue 局部疲劳
local fault 局部故障
local feedback 本机回授;本机反馈
local file 局部文件
local fire alarm control panel 区域报警器
local fisheries inspector 地区渔业检查员
local fishery 地区渔业
local flag 局部标记
local flow rate 局部流速
local flow velocity 局部流速
local focal length 局部焦距
local forecast 局地预报;地方预报;当地(天气)预报
local freight 局部运费;第二船费;地方货运
local freight agent 当地运输代理商
local frequency 本振频率
local frequency control 本机频率控制
local freshet 局部发水
local friction drag 局部摩擦阻力
local galaxy 本星系
local general 局部通用
local geochemical survey 局部地球化学测量
local geoidal map 局部大地水准面图
local geology 局部地质学
local geothermal gradient 局部地热梯度
local global connection 局部整体连接
local goods transportation 管内货物运输
local government 地方政府
local grant-in-aid 地方补助金
local gravity anomaly 局部重力异常
local gravity circulation 局部重力循环
local groundwater 区域地下水
local ground watering 区域地下水
local ground water survey 区域地下水调查
local ground water system 局部地下水流系统
local group 本星系群
local growth factor 地方增长系数
local habitat 局部生境
local harbo(u)r 地方港(口)
local hardening 局部硬化;局部淬硬;局部淬火
local head loss 局部水头损失
local health agency 地方卫生机关;地方卫生机构
local heat flux 局部热通量
local heating 局部增温;局部升温;局部加热
local heat source 局部热源
local height system 地方高程系
local highway 地方性道路;地方公路
local holidays 地方节日
local horizon 当地地平面
local hospital 地方医院
local hot spot 局部过热点
local hour angle 地方时角
local housing authority 地方住房管理机构
local illumination 局部照明
local improvement 局部整治;局部加固;局部改造;局部改善
local improvement fund 地方改良基金
local improvement reserve 地方改良准备金
local inclination 局部倾斜

local independence dependence in general 局部不相依总体相依
local indicator system 局部指标体系
local industry 地方工业
local inertial frame 局域惯性架
local inertial system 局域惯性系统
local infection 局域传染;局部传染
local infiltration 局部浸润
local inflow 局部进水量;区间水量;区间入流;区间流入;区间来水;地方来水;当地流入水量;当地来水
local influence 地方影响
local information 地方情报;本地信息
local information center 地区情报中心
local instability 局部失稳;局部不稳定
local instability failure 局部失稳破坏
local intelligence 局部智能
local intensification 局部加厚
local interaction 局部相互作用
local interest 局部利益;当地利益
local interference 当地干扰
local inter-urban traffic 当地市际交通
local inversion 局部逆温
local ionization equilibrium 局部电离平衡
local irradiation 局部照射;局部辐射
local irregularity 单个微观不平度
local irrigation work 地方灌溉工程
localism 地方观念
local isostasy 局部均衡
localities of joint measurements 节理统计点位置
localities of strain measurement point 应变测量点位置
locality 局部(性);群落地;位置;地方级
locality disinfection 区域消毒
locality factor 地点因素
locality of high natural stress area 高地应力区位置
locality of program's memory reference 程序的存储引用位置
locality protection 区域保护
localizability 可局限性;可定域性
localization 国产化;局限;局部化;局部富集作用;定域;地方化
localization data of target 目标定位数据
localization economy 局部经济;地方经济
localization method 定位法
localization of boundary stress 边缘应力的局部性
localization of faults 故障位置测定;确定故障点
localization of perspective region 远景区定位
localization of sound source 声源定位
localization of target 目标定位
localization of wave 波定域性
localizator 抑制剂;指向标;探测器;定位信标;定位器
localized air supply 局部送风
localized air supply system 局部送风系统
localized cooling 局部降温
localized corrosion 局部腐蚀
localized crushing 局部破碎;局部压碎;局部破碎
localized deposition 局部沉积
localized drop in temperature 局部降温
localized error 局部误差;定域误差
localized general lighting 局部一般照明
localized lighting 分区一般照明
localized model 局部模型
localized negative charge 局部负电荷
localized network 局部网(络)
localized purchasing 分权采购
localized raw materials 地方原料
localized recommendation 分区推荐
localized regulation 局部调节
localized release 局部泄水
localized scour 冲刷坑
localized scouring 局部冲刷
localized shear failure 局部剪切破坏;局部剪力破坏
localized source 局部源
localized state 局部态;定域态
localized states transition 局部态跃迁
localized target 局部目标
localized tempering 局部回火
localized tetanus 局限性强直
localized vector 局限矢量;束缚向量
localizer 抑制剂;指向标;检漏器;探测器;定位信标;定位器
localizer antenna 着陆航向信标台天线;定位器天线

localizer beacon 着陆航向信标
localizer(beam) radio transmitter 定位定向无线电信号发射机
localizer on course line 航线定位信标
local junction circuit 市内中继线路
local key 局部键
local knowledge 局部知识
local laboratory array 地方实验室台阵【地震】;本地实验室台阵
local labo(u)r 地方劳力;当地劳工;当地劳动力;当地工人
local lacuna 局部缺失
local latitude 当地纬度;测者纬度
local lesion 局部损害
local level 当地水准面;当地水平面;当地级
local level control 就地级控制
local-level plan(ning) 地方规划
local level post 当地雇用人员名额
local lighting 局部照明
local lighting lamp 局部照明灯
local limit theorem 局部极限定理
local line 局部线路;市内线路;本局电话线;本地线路
local line system 市内线路制
local load 局部负荷
local loading and unloading rate 管内装卸率
local loans fund 地方信贷基金
local-local link 本地到本地链路
local lock 局部锁
local loop 用户回线;用户环路;局部循环;局部回路;内部回路;本地循环;本地环路
local loss 局部损失;局部损耗
local lubricator 局部润滑器
local lunar time 地方太阴时
locally-administered state enterprise 地方国营企业
locally asymptotically most powerful test 局部渐近最大功效检验
locally attached 局部连接的;本地连接的
locally attached station 本地连接站;本地连接的站
locally attached terminal 本地连接终端;本地连接的终端
locally available material 地方(可用)材料;当地(可用)材料
locally available soil 当地可用土壤
locally catenative sequence 局部连接序列
locally compact 局部紧致
locally controlled power station 当地控制电站;部分受控电站
locally convex 局部凸的
locally most powerful rank order test 局部最大功效等级顺序检查
locally operated turnout 非集中道岔
locally optimal plan 局部最优方案
locally optimal solution 局部最优解
locally produced map 野战制印图
locally produced raw materials 就地取材
locally stationary process 局部平稳过程
locally supplied equipment 国内提供的设备;地方提供的设备;当地(提供的)设备;本国提供的设备
locally thinning 局部变薄
locally vectorizable 可局部向量化
local Mach number 当地马赫数
local magnetic anomaly 局部磁异常;局部磁力异常
local magnetic anomaly chart 局部磁异常图
local magnetic disturbance 局部磁(场)干扰;地方(性)磁干扰
local magnetic field 局部磁场
local magnitude 局部震震级;近震(体波)震级;地区震震级;地方震震级
local manual control 就地手动控制
local map 邮政地图
local market 生产地市场
local marshalling station 地方性编组站【铁】
local material 地方材料;当地材料;本地材料
local materials for buildings 当地建筑材料
local maximum 局部最大值;局部极大(值)
local mean moon 地方平(正)午
local mean sea level 当地平均海面
local mean time 地方平时;地方标准时;当地平均时间;地方平均时(间)
local mechanism ventilation 局部机械通风
local membrane stress 局部薄膜应力
local memory 局部存储区;本机存储器
local meridian 地方子午线;地方子午圈;当地子午圈;测者子午圈

local metamorphism 局部变质(作用);接触变质(作用)
local method 土办法
local microfield 局部微观场
local microfield intensity 局部微观场强
local minimization problem 局部极小化问题
local minimum 局部最小值;局部极小(值)
local miscellaneous taxes 地方杂税
local mode 本机方式;本地(传送)方式
local-mode coverage 区域方式覆盖
local model 局部模型
local modem 本地调制解调器
local mounted 就地安装
local movable bed 局部动床
local movement 局部性运动
local multiple access 本地多重访问
local multiple reflection 局部多次反射
local multipoint distribution service 本地多点分配业务
local municipal authorities 当地市政机关
local name 车间编号
local natural resources 地方自然资源
local navigation(al) warning 地区航海警告
local network 局部网(络);本地通信网
local network control program(me) 局地网控制程序;局部网络控制程序;本地网络控制程序
local networking 局部联网
local network of stations 地方台网
local node 本地节点
local nomenclature 地区命名
local noon 地方正午
local notice to mariners 地区性航行通告
local numbering 分区编号
local number of geologic(al) map 地方地质图编号
local observation net 局部性观测网
local operating cost 地方业务费用
local operation mode 局控模式
local operator workstation 本地操作工作站
local optimization 局部最优化
local optimum 局部最优值;局部最优(点);局部最佳状态;局部优化
local order 局域有序
local order track 车辆编组线
local origin 当地来源
local or own type 局部或固有类型
local or relative minimum 局部或相对减小
local oscillation 局部振荡;本地振荡;本(机)振(荡)
local oscillator 局部振荡器;本机振荡器;本地振荡器
local oscillator drive 本机振荡器激励
local oscillator filter 本振滤波器
local oscillator injection 本机振荡器注入
local oscillator radiation 本振辐射;本机振荡器辐射
local oscillator tube 本机振荡管
local overheating 局部过热
local packet 本地传输包
local panel 就地表盘
local parameter 局部参数
local parasite 当地寄生菌
local passenger flow 管内客流
local passenger flow diagram 管内客流图
local passenger train 管内旅客列车
local pattern matching 局部模式匹配
local permeability 局部渗透率
local personnel 当地人员
local perturbation 局部扰动
local phenomenon 局部现象
local phenomenon of flow 局部水流现象
local plan(ning) 地方规划
local plug 局部转换开关;局部插头
local pneumatic process 局部气压法
local police station 派出所
local politicians 当地政界
local pollutant discharge standards 地方污染排放标准
local pollution 局部(地区)污染
local pollution source 局部污染源
local polynomial model 局部多项式模型
local populace 当地居民
local port 地方港湾;地方港(口);当地港湾
local port surcharge 小港运费附加费
local postweld heat treatment 局部焊后热处理
local potential 局部电位
local power source 局部电源

local power supply 局部电源;地方电源
local precipitation 局部降水(量);地方性降水(量);当地降水(量)
local preheating 局部预热
local pressure 局部压力;局部受压;定点压强
local price 当地价格
local printer 本地打印机
local procedure error 本地程序错
local processing unit 现场处理单元
local processor 局部处理器;局部处理机;本地处理机
local product 土产(品)
local project 地方项目
local project or dent 局部凸出或凹陷
local protecting grazing 局部保护放牧
local public finance 地方财政
local public utility loan 地方业务费用
local pumping station 区域泵房;区域泵站
local radiation effect 局部辐射效应
local railroad 地方铁路
local railway 铁路支线;地方性铁路;地方铁路
local rain(fall) 地形雨;局部降雨
local rate 管内运价率;当地价格;当地单价;当地比率;当地比价;本线路运价率
local raw material 本地性原料
local reaction 局部反应
local readjustment 局部调整
local reception 本地接收
local reduction 局部减薄
local reference-beam holography 局部参考光束全息术
local reference coordinate system 当地参考坐标系(统)
local reference system 地方参考坐标系
local reflex 局部反射
local refraction 局部折射;地方大气折射
local register 局部寄存器
local regression 局部海退
local regularization 局部正规化
local regulation 局部整治;局部调节;地方规范
local regulations of environment 地方性环境法规
local reinforcement 局部加强;局部补强
local relative fall of sea level 局部海面相对下降
local relative rise of sea level 局部海面相对上升
local release 本节车缓解
local-relief method 分层设色法
local-remote link 本地到远地链路
local-remote relay 本地远地切换继电器
local representative 本地代表(商)
local resistance 局部阻力
local resistance factor 局部阻力系数;局部阻抗系数;局部强度系数;局部电阻系数
local resonance 局部共振;本机谐振
local resource 本地资源;地方资源
local restriction 局部限制
local result 局部结果
local river basin 局部河流流域
local road 地方性道路;支路
local road network 地方道路网
local rounding error 局部舍入误差
local round-off error 局部舍入误差
local route 地方路线
local rules of environment 地方环境规章
local runoff 涝水;内水;区间径流;当地径流
local satellite computer 本地卫星计算机
local scale 局部规模;局部比例尺
local scaling factor 区域比例系数
local scour 局部冲刷
local scour hole 局部冲刷坑
local scour near pier 桥墩局部冲刷
local scour pit 局部冲刷坑
local screening plant 当地筛选场
local script 地方名
local section 局部剖面
local segment 局部段
local seismicity 局部地震活动性
local seismic network 地方地震台网
local seismic requirement 地方性抗震要求
local selection 局部选择
local selector 市内选择机
local sender 本地发射机
local separation 局部分离
local service 支线运输;局部服务;市内公共汽车;短途运输(线);地方性服务交通

local service area 局部服务区;本地服务区
local service urban bus 市内公共汽车
local settlement 局部沉降
local shared resources 局部共享资源;本地共享资源
local shear 局部剪切;局部剪力
local shear failure 局部剪切破坏;局部剪力破坏
local shield(ing) 局部屏蔽
local shock 地方震
local shopping 国内采购;当地采购
local side 本机端;本机侧
local side friction 局部侧摩阻;局部侧摩擦力
local sidereal noon 地方恒星(正)午
local sidereal time 地方恒星时
local siding 管内工作车停留线;货物作业车停留线
local signal 本局信号;本地信号
local site 本地站
local site condition 局部场地条件
local site geology 局部场地地质(情况)
local soil 局部土壤
local soil type 土组
local solar time 地方太阳时
local solution 局部解
local speciality 土特产
local species 当地种
local specification 地方规范
local squeeze 局部挤压伤
local stability 局部稳定性
local stabilization 局部稳定作用;局部加固
local standard 本地标准
local standard time 地区标准时;地方标准时;当地标准时(间)
local star stream 本星流
local star system 本星团
local station 本地(电)台;市内分局;市内电话局;地区站;本地台;本地广播台
local statistic 局部统计量
local stellar system 本星群
local stock 当地库存
local storage 局部存储器
local store 地区库存
local strain 局部应变;局部变形
local stream velocity 局部流速;当地流速
local street 地方街道
local strength 局部强度
local stress 局部应力
local stress relief heat treatment 消除局部应力热处理
local stress relieving 局部消除应力
local structural discontinuity 局部结构不连续性
local structural map 局部构造图
local structural map in time domain 时域局部构造图
local style 地方风格
local subcarrier 本机副载波
local supercluster 本超星系团
local supergalaxy 本超星系
local supervisory system 局部监理系统
local supplies 地方资源;当地资源;当地材料
local surcharge 地方附加费
local surrounding 局部环境
local survey 局部测量
local switch 局部转换;局部开关
local switch board 市内交换机
local symbol 局部符号
local synchronous signal 本机同步信号
local system 局部系统;本地系统
local system queue area 局部系统队列区;本地系统队列区
local tax 地方税
local taxation 地方税
local tax law 地方税法
local tax payment 地方税款缴纳
local tectonics 局部地质构造
local tectonic stress field 局部构造应力场
local telephone exchange 市内电话局
local telephone network 地区电话网
local telephone office 市话局;地区电话所
local telephone switching system 市内电话局
local telephone system 市内电话
local temperature 局部温度
local tempering 区域强化
local term 地方地理通名
local terminal 本地终端设备;本地终端

local term 当地交货条件
local terrain correction 局部地形改正
local test 局部试验;局部测试
local theoretic(al) datum 当地理论基准面
local thermal runaway 局部热击穿
local thermodynamic(al) equelibrium 局部热力学平衡
local thermodynamic(al) equilibrium 局域平衡;局部热动平衡
local threshold 局部异常下限
local through road 当地过境道路
local thunderstorm 地方性雷暴
local time 地方时(间);当地时间
local time offset 地区时差
local title 本地题目
local topography 局部地形
local toxic effect 局部毒作用
local tracking 局部跟踪
local traffic 当地交通;境内交通;市区通话(业务);市内交通;短途运输;地方交通;本地交通
local traffic only sign 禁止过境交通标志
local traffic revenue 管内运输收入
local train 区间列车;普通列车;慢车;管内列车;区段列车
local training 就地培训
local transaction 本地事项;本地事务处理;本地交往
local transfer tax 地方转移税;地方传递税
local transformation 局部变换
local transgression 局部海浸
local transit 地方运输车辆;当地中天
local transmitter 本地发射机
local transportation 区域性运输;地方交通
local triangulation 局部三角测量
local trigger 局部触发
local true time 真当地时间;地方真时;地方夏时
local trunk 市内中继线
local unconformity 局部不整合
local uniformizer 局部单值化参数
local update procedure 局部更新过程
local user terminal 地区用户终端;本地用户终端
local value 局部值
local vanning 当地装箱(指集装箱)
local variable 静态变量;局部变数;局部变量
local variable declaration 局部变量说明
local variable symbol 局部变量符(号)
local variation 局地变化;局部变化;局部变动;地方变化
local variety 地方品种
local velocity 局地速度;局部速度
local vent 排气道(便池房的);局部排气口;(在存水弯前的)通风管道;局部通气管;局部通风孔;局部排气管
local ventilating pipe 局部通气管;局部通风管
local ventilation 局部通风
local ventilation resistance 局部通风阻力
local vertical 地方垂直;当地铅垂方向;当地垂线;本点垂直
local vibration 局部摆动;局部振动
local view 局部视图
local virtual protocol 本地虚拟协议
local volume table 地方材积表
local wagon flow 地方车流
local water blocking 局部堵水
local water discharge pump room 局部排水泵房
local water law 地方水法
local waterway 地方航道
local weather report 地方天气报告
local weldability 局部可焊性
local wind 局部地区风;地方性风;地域风
local wind direction 局部风向;区域性风向;地方性风向
local workable coal seam 局部可采煤层
local X-ray apparatus 定位 X 射线照射装置
local yard 管内工作车场
local yield(ing) 局部屈服;局部折屈;局部凹陷
local zone time 地区时;地方区时
locant 位次
locate 放样;定位;测定位置;定线;设置
locate a wreck with towing grapnel 四爪拖轮扫沉船
located frequency for railway use 铁路专用频率
located in welded joint 位于焊接接头中
located object 定线目标;地物;方位物;方位点
located station 已知坐标的地面点
located tectonic unit 所在的构造单元

locate function 定位功能
locate mode 配位式;定位(方)式
locater 探测程序
locate statement 定位语句
locating 放样;定位;测定位置;定线
locating axis 定位轴线
locating back 定位板;压平板
locating blasting 定位爆破
locating block 定位块
locating bush 定位管
locating by direct locating intersection point 直接定交点定线
locating cement channeling 定位水泥沟渠工程
locating center punch 定位中心冲(头)
locating collar 紧固套
locating cone 定位锥
locating detent 定位爪
locating device 定位装置
locating engineer 选线工程师
locating file 定位文件;查找文件
locating grade 定线坡度
locating hole 定位孔
locating map 位置图
locating method by statistic(al) graph 定位系统图表法
locating of mineral prospective district 成矿远景区圈定
locating of root 勘根法
locating pad 工艺卡头
locating parameter 定位参数;位置参量
locating piece 定位块
locating pin 定线测针;定位销;挡料销
locating plate 定位板
locating plug 定位塞套
locating point 观察点;基准点;定(位)点;地质点
locating ring 定位环
locating rod 定位杆
locating screw 定位螺钉
locating shaft 定位轴
locating signal 定位信号
locating slide rod 定位滑杆
locating slot 定位槽
locating sounding by vertical angle method 垂直角法水深定位
locating soundings by stretching wire 断面索水深定位
locating stuck point 确定(井内)被卡点
locating stud 定位双头螺栓
locating surface 基准面;定位面
locating system 定位系统;定位方式
locating tab 定位销
locating tab assembly 定位销总成
locating valve seat assembly 定位阀座总成
location 位置;特定区域;所在地;地址;地点;存储单元;场所;测位;部位;布点
location above water 水上定位
location adder 地址寄存器
location address data 定位地址数据
locational analysis 区位分析
locational factor 区位因素
locational integration 区位一体化
locational interdependence 区划中的相互制约关系
locational parameter 定位参数
locational quotient 区位商(数)
locational requirement 区位需求
locational rent 区位租
locational theory 区位理论
locational triangle 区位三角形
location and lookout 三注意定位及瞭望;三注意测深
location and navigation device-passive 无源定位与导航装置
location anomaly of geochemistry 局部地球化学异常
location beacon 示位信标(机场);定位信标
location bearing 定位轴承
location between the exterior and interior 半表半里
location buoy 定位浮标
location by cross-section 横断面定线
location by cross section with rope 断面索法
location by forward intersection 散点交会定位法
location by setting out angle 拨角定线
location case 地点角色
location center line 定位中心线

location center 定位中心
location clause 堆积地点条款
location clerk 货物理货员
location coefficient 位置系数
location condition 选址条件;选点条件;定位条件
location constant 位置常数;单元常数
location counter 指令(位置)计数器;位置计数器;定位计数器;单元计数器;存储单元计数器;程序计数器
location damages 地表的损害赔款
location decision method 布局决策方法
location delimiter 存储区定界符
location detection 位置检测
location deviation 定位偏差
location diagram 接图表
location dimension 定位尺寸
location drawing 地盘图;定位图;位置图
location engineer 选线工程师
location factor 位置因子;定位因素
location fit 定位配合
location flag 定位旗
location for installation 安装位置图
location-free procedure 浮动过程
location hole 定位孔
location identifier 存储区标识符
location index 位置索引;图幅接合表;索引图
location information 位置信息;存储单元信息
location in plain region 平原地区选线
location layout 定位布置;场地布置
location map 建筑物现状图;位置图;索引图;定线图;定位图
location map of monitoring point 监测点位置图
location mark 装配标记
location mode 定位方式;单元状态;存放方式;分布方法
location mode data item 定位状态数据项目;定位方式数据项;存放方式数据项;分布方法的数据项
location monument 矿区标志;界石;界标
location network spectrum survey 网格定点能谱测量
location number 位置号码
location of anchorage 锚定位置
location of apparatus 仪器配置
location of barrage 堰址;闸址;坝址;坝的位置
location of borehole 定孔位【岩】
location of boring 定孔位【岩】
location of boundary 勘界
location of bridge abutment 桥墩定位
location of bridge pier 桥墩定位
location of bursting water 突水点位置
location of components 零件配置
location of dam 坝址;坝的位置
location of dangerous cargo container stowage 危险货物集装箱位置
location of diapiric structure 底辟分布位置
location of discharge area 排泄区位置
location of distribution 分布地点
location of driving unit 传动装置位置
location of factory 厂址
location of fault inflexion 断层拐点位置
location of fault intersection 断层交叉点位置
location of faults 故障探测;损坏处测定
location of fault scarp 断层崖位置
location of fish school 鱼群位置探索
location of flat 断坪位置
location of foci 震源位置
location of fold 褶皱的位置
location of fold axle 褶皱轴位置
location of fold distribution 褶皱分布位置
location of gliding tectonics 滑动构造位置
location of impact structure 撞击构造位置
location of impressions 模膛的布排
location of industrial districts 工业区布局
location of industry 选厂址;作业划区;工业位置;工业所在地点
location of jump 水跃位置
location of klippe 飞来峰位置
location of leak 漏水地点
location of line 定线
location of line in mountain 越岭选线
location of line in mountain and valley region 山区河谷定线;山区河谷地区选线
location of line on hilly land 丘陵地区选线
location of magnetic survey area 磁测区位置

location of manhole 检查井位置
location of manufacturing 工业位置;工业地点
location of marshalling yard 编组场配置
location of mistake 错误勘定;错误定位
location of mountain line 越岭选线
location of nappe rootzone 推覆体根带位置
location of observation project 观测工程位置
location of order 指令位置
location of ore body in space 矿体空间位置
location of pal(a)eoland 古陆位置
location of parking area 停车场位置
location of parking facility 停车场选址
location of pier and abutment 桥梁墩台定位
location of pilot tunnel 导洞位置
location of pipe 管道位置
location of plate 板块位置
location of point source dischargers 点污染源排污装置位置
location of production 生产布局
location of profile 剖面位置
location of prospective district 远景区位置
location of pumphouse 泵房位置
location of pumping station 泵站位置
location of railway route selection 铁路选线
location of ramp 断坡位置
location of root 寻根法;根的位置
location of route 路线位置
location of sample 取样地点
location of sewage treatment plants 污水厂位置
location of sound sources 声源定位
location of spoil 弃渣位置
location of spring 泉水位置
location of stress concentration zone 应力集中区范围
location of table at coordinates 工作台坐标定位
location of tank 水池位置
location of testing site 试验场位置
location of the ground stress measuring hole 地应力测量孔位置
location of the ground stress measuring point 地应力测量地区名称
location of the machine to be installed 设备安装地点
location of(the) well 定井位;井位
location of water-level 水位位置
location of window 构造窗位置
location of work piece 对准工件位置
location on a slope 斜坡上位置
location over mountain 越岭选线
location parameter 位置参数
location plan 平面布置图;总平面图;现场位置平面图;工程布置图;矿山开拓位置图;位置图;地盘图
location-plate 定位板
location pointer 位置指示符
location problem 定位问题;布局问题
location procedure 定线程序
location quotient 区位系数
location reagent 定位剂;斑点显色剂
location register 定位寄存器
location reinforcement 定位钢筋
locations 定位件
location selecting for dam basis 坝基选址
location selection 位置选择
location sketch 位置略图
locations passed by axial-trace 褶皱轴迹经过地点
locations passed through by fault 断层经过地点
location sub 定位短节
location survey 定(位)测(量);定线测量;定线
location test 定位试验
location theory 区位理论;选址论;选择厂址论
location variable 位置变量
location with cross-section method 横断面选线
locator 探测器;探测程序;搜索装置;示位信标(机场);定位子;定位仪;定位器;定位程序;定位仪;测位器
locator argument 定位自变数
locator beacon 定位器信标
locator card 定位卡
locator data 定位数据
locator device 定位设备
locator device state 定位设备状态
locator dual collimating line 双定位器准直线
locator mode 定位设备的操作方式
locator of pipe 摆管定位

locator pins 定位器插头
locator qualification 定位(器)限定
locator qualifier 定位修饰符
locator's hand level 勘测用手提水准仪
locator slide 定位滑道
locator system 定位仪系统
locator variable 定位(符)变量
loch 矿脉中裂缝;矿脉中空穴;环山澳;海湾;长窄的海湾;滨海湖
lochan 小湖;小滨海湖
lochmium 植丛群落
lociation 超亚植物群丛
lock 制动楔;闸;固定;锁合装置;锁定器;水闸;上保险;闭锁锻造;闭锁;保险栓
lockable 可锁的
lockable compartment 门顶可锁室
lock a file 封锁文件
lockage 闸程;水闸通行税;水闸用材(料);过闸(费);水闸高低度;水闸高差;船闸系统;船闸通行费
lockage arrangement 过闸排档
lockage basin 闸池
lockage bay 闸池
lockage button 同步按钮
lockage chamber 闸段(河运两闸门之间)
lockage despatching[dispatching] 过闸调度
lockage facility 过闸设施;船闸
lockage gate 闸门
lockage in regular order 按序过闸
lockage mode 过闸方式
lockage-on 锁定
lockage technique 过闸技术
lockage time 过闸时间
lockage water 闸内水量;过闸用水;过闸耗水(量);船闸用水量
lockage well cup 可锁井帽
lock alloy 保险合金
lock and block 联锁闭塞
lock-and-block system 联锁闭塞制
lock and bolt 锁加闩
lock and dam 闸坝
lock-and-dam method 闸坝改善航运法;闸坝法(改善航道)
lock-and-dam obstructive to navigation 碍航闸坝
lock-and-follow scanner 同步跟踪扫描装置
lock and lock lift 锁销与锁销提臂
lock and weir 闸坝
lock approach 船闸引航道;船库引航道
lock approach channel 船闸引渠
lock attendant 船闸工作人员
lock automatization 船闸自动化
lock auxiliary wall 辅助引航墙
lock backset 锁心后退量;锁中心到门边距离
lock ball 锁紧球
lock banking eccentric 锁桨制动偏心销
lock banking stop 锁止片
lock bar 锁杆
lock-bar pipe 锁条管
lock barrel 锁心柱
lock bar steel pipe 箍锚钢管;加箍钢管
lock basin 闸室(船闸);船闸室
lock bay 闸室容水量;运河阔幅段;闸室(船闸);船闸室(水舱)
lock beading 卷边结合
lock bevel 锁舌斜面
lock block 嵌锁木块(门上);装锁木块;门上装锁板;闩块
lock bolt 锁紧螺栓;止动销;锁紧螺杆;定位螺栓;船闸螺栓
lock bolt support with shotcrete 喷锚支护
lock bottom (船闸的)闸底;船闸底
lock bridge 闸桥
lock butt joint 锁底对接接头
lock canal 设闸运河;船闸运河
lock capacity 船闸通过能力
lock catch 锁扣;闭锁掣子
lock chain 梯级(式)船闸
lock chamber 船闸室墙;闸室(船闸);升船箱;船闸(闸)室
lock chamber gate 闸门
lock chamber wall 闸室岸壁;船闸室墙;船闸墙
lock check gate 船闸节制闸门;止逆门;闸门;防逆闸门
lock circuit 同步电路
lock clasp 销轴锁紧

lock clip 门锁定位卡子
lock cock 带锁旋塞
lock complex 船闸区
lock corner 啮合拐角;锁紧角;锁折角
lock counter 过闸计数器
lock cover 锁紧压片
lock crew 船闸人员
lock culvert 船闸排水涵洞;船闸排水渠;船闸输水涵洞
lock cut 船闸引渠;船闸引河;(人工开挖的)船闸引航道
lock door cam 门锁杆凸轮;锁杆凸轮(集装箱)
lock door rail 装锁门冒头
lock down 船闸向下游驶去;(木排的)扎锁木
lock drop 船闸上游水位差
lock-due 过闸(收)费
locke coil wire rope 密封钢丝绳
locked 封闭的
locked basin 有闸港池
locked-bell system 锁闭式信号铃系统
locked cable 锁紧缆
locked chain 固定链系
locked coil conductor 吸持线圈导线
locked coil construction (钢丝绳的)锁丝结构
locked coil rope 铠装电缆;锁丝钢丝绳;封闭式钢索;封闭式钢丝绳;包缠式钢丝绳;箍环缆索;铠装缆索
locked coil strand 封闭圈钢绞线;封闭钢丝索
locked-coil wire rope 包缠式钢缆
lockedcover switch 锁盖开关
locked die 扣合模
locked fault 锁止断层
locked file 锁定文件
locked groove 同心槽;锁槽;闭纹
locked harbo(u)r 受浅滩或暗礁包围的港
locked-in capital 已占用资本;闭锁资本
locked-in line 锁定线
locked-in oscillator 同步振荡器
locked-in stress 内应力
locked joint 咬口接缝
locked market 封锁市场
locked mode 锁模;锁定模式
locked name 被锁名字;被封闭的名字
locked normal 定位锁闭
locked normal and reverse 定反位锁闭
locked oscillator 锁相振荡器;锁定振荡器
locked-oscillator detector 锁定振荡器检波器
locked page 被锁页面;被封闭的页
locked point 锁止点
locked position 锁闭位置
locked press 锁定压烫机
locked record 被锁记录;被封锁的记录;被封闭的记录
locked resource 受保护资源;带密码资源;被锁资源
locked river 有闸河流;渠化河流
locked rotor 锁定转子
locked rotor current 止转转子电流
locked rotor torque 制动转子力矩
locked rotor voltage 转子停转电压
locked scarf 锁紧嵌接
locked-seam 潜缝
locked stream 渠化河流
locked subcarrier 受锁副载波
locked test 加料循环试验
locked train reduction gear 功率分支减速齿轮
locked-up stress 锁紧应力
locked vessel 闸内船只
locked waterway 有闸水道
locked wheel 止动的车轮(制动时)
locked-wheel braking 卡紧轮子的制动
locked-wire rope 锁丝钢丝绳
locked-wire strand cable 含有钢丝心线的多股电缆
lock emptying 船闸泄水;船闸放水
lock engineering 船闸工程
lock entrance 闸口
lock equalization time 船闸输水时间;船闸泄水时间;船闸充水时间
lock equipment 气闸设备
locker 机架;可锁容器;小碗柜;柜;冷藏间;锁扣装置;锁夹;储物柜;储物舱;存衣柜(运动员的小型柜)
locker-er-plant 抽屉式冷柜
locker frame 设备柜支架
locker lighting 设备柜照明

locker paper 冷藏包装纸
locker plant 冷藏厂；抽屉式冷柜；小室冷库
locker ring 锁扣环；衣物柜环
locker room 衣物间；生活间；更衣室；衣帽间
Locker two-piece ring 洛克金属双环木材连接件
locket 落水洞
lock exchange flow 船闸交换流
lock face 锁面
lock facility 封锁
lockfast 锁牢的
lock file 齐头六边形锉
lock-filers clamp 虎钳夹
lock filling 船闸灌水；船闸充水
lock filling and emptying system 船闸充水与泄水系统；船闸充泄系统
lock filling branch manifold 船闸充水分总管
lock filling system 船闸充水设备
lock filling time 船闸充水时间
lock-fit lead 锁紧引线
lock fitted 装锁的
lock flights 阶梯(式)船闸；多级船闸；船闸梯级；梯级(式)船闸
lock flights in continuous arrangement 连续布置的多级船闸
lock flights in individual arrangement 分离式多级船闸；分开布置的多级船闸
lock flights with intermediate 分离式多级船闸
lock flights with intermediate canal 分开布置的多级船闸
lock floor 闸(室)底板；船闸地板；船闸底(板)
lock-following strobe 同步选通脉冲
lock former 铁板开槽机；薄铁板折板机
lock-free stretch 无闸河段
lock-free system 无锁定方式
lock front 闸门前面；前闸；锁端
lock front bevel 锁端斜面
lock full lost 闸门灌水损失
lock full of water 一闸水
lock furniture 装锁家具
lock gate 闸门；(船闸的)闸门；船闸闸门
lock-gate hatch 闸(板)门；排洪闸
lock group 船闸群
lock guide 船闸引水渠；船闸(引)航道
lock head (船闸的)闸首；船闸闸首；船闸水头
lock head wall 闸首墙；闸墙
lock hook 弹簧舌小钩；防脱钩
lock hopper 闭锁料斗
lock-hopper system 闭锁料斗系统
lock-in 锁定；锁紧
lock-in amplifier 锁定放大器
lock-in feature 锁定特征
locking 制动；吸持；联锁；同步跟踪；锁定；锁闭
locking air pressure 锁闭风压
locking and unlocking rail (索道的)挂结和脱开轨道
locking angle 啮合角；锁角
locking apparatus 锁紧装置
locking arm 锁臂
locking arrangement 锁紧装置
locking assembly 锁定组件
locking bail 锁紧钩
locking ball 定位钢球
locking bar 门闩；锁杆；门栓；锁扣门闩；锁紧杆
locking batten bar 带锁封舱铁条
locking bed (火车车架上的)锁紧座；锁簧床；锁床
locking blade 锁紧叶片
locking bolt 锁紧螺栓；锁定螺栓；防松螺栓
locking box 闭锁盒
locking buckle 搭扣
locking cam 锁紧凸轮
locking cap 止动螺帽；固定帽
locking circuit 吸持电路；同步电路；锁闭电路
locking clamp 锁紧装置
locking clip 锁籍
locking clutch 锁紧离合器
locking cock 锁闭开关
locking coil 保持线圈
locking collar 锁圈
locking cone 止棱；锥形锁销；锁紧锥
locking contact 联锁触头；联锁触点
locking control 同步控制；锁定控制
locking control voltage 同步控制电压
locking course 锁紧横列
locking crank 关闭曲柄

locking cross bar 锁合横杆
locking cycle time 船闸作业周期
locking cylinder 锁紧缸
locking device 脚手斜撑；剪刀撑；防松装置；制动装置；止动装置；固定器；锁扣装置；锁紧装置；锁紧机构；锁定装置；封闭装置；闭锁装置；闭塞装置；保险设备
locking dog 锁定爪
lock in gear 闭锁机构
locking element 闭锁装置
locking escape 封锁换码
locking face 锁紧面
locking fastener 闭锁紧固件
locking fill 锁定充填
locking force 合型力；锁紧力；锁闭力
locking frequency 锁定频率
locking gasket 保险垫片
locking gear 停止机构；锁紧装置；锁定装置
locking grip 夹具
locking handle 闭锁(手)柄；锁紧把手；止动柄；锁(紧)柄
locking head 锚机滚筒接合栓
locking hole 锁孔
locking in 闭塞；关进；锁定；同步
locking in geologic(al) fault 断层闭锁作用
locking key 止动键；制动按钮；制定键；锁(紧)键；锁定开关；锁定键
locking latch 插接锁定；锁紧装置
locking length of bolt 钳杆锁固长度
locking lever 止动杠杆；止动柄；联锁杆；锁定(连)杆
locking link 联锁杆；锁定连杆
locking lip 制动唇；锁闭盖
locking loop 锁环
locking lug 闭锁卡铁；闭锁突榫
locking magnet 吸持磁铁；锁定磁铁
locking mandrel 锁定芯轴
locking maneuver 进出船闸的操纵
locking material 制锁材料；联锁件；锁扣材料
locking mechanism 止动机构；锁闭机构
locking moment 约束弯矩
locking nut 接线螺母；锁紧螺母；锁紧螺帽；锁紧(环)；锁定螺母；防松螺母
locking of a lever 止动手柄
locking of forceps 合拢锁扣
locking of wheels 轮对抱死
locking on 闭锁
locking operation 船闸过闸作业；过闸作业
locking out 封闭工厂；同步损失；失步
locking out cock 止气塞
locking out of the spoil 弃土分离；废石料分离
locking pawl 锁定爪；制动爪
locking phase 固定相；同步相(位)；锁定相(位)
locking piece 锁定片
locking pin 插销；制动销；定位销；制动栓；锁销；防松栓
locking pintle 紧锁舵栓
locking plate 锁紧木栓；锁紧木片；定锁盘；防松板
locking-plate pinion 锁板齿轴
locking pliers 大力钳
locking plug 闭锁塞
locking plunger 止动杆
locking portion 导锁部分
locking press button 锁定按钮
locking process 锁定过程
locking profile 锁住侧面
locking pulse 锁定脉冲
locking quadrant 止动扇形齿轮；止动扇形板
locking ram 制动夯具
locking range 同步范围；锁定范围
locking recess 锁合凹座
locking relay 锁定继电器；闩锁继电器
locking requisition 锁闭条件
locking rim 锁环轮辋
locking ring 紧箍；锁圈；密封圈；锁紧环
locking ring mount 固定环支座
locking rod 锁杆；锁止杆；锁闭杆
locking screw 制动螺钉；锁紧螺钉；锁紧螺旋；防松螺丝；防松螺钉
locking sheet 联锁表
locking shift character 封闭移位符号
locking signal 禁止信号；锁定信号
locking spring 锁紧弹簧；锁簧
locking steel wire 锁紧钢丝
locking stile 装锁门梃；门锁边梃；门梃(装锁用)

locking strip 锁条；止动带；锁定带
locking stud 止动销(钉)
locking switch 锁盖开关
locking system 锁固体系；锁闭系统
locking tappet 凸轮推杆锁定
locking thread form 防松螺纹样式
locking time 船闸过闸时间；过闸时间；锁定时间
locking transfer contact 锁定切换触点
locking-type plug 固定插头；锁式插头
locking valve 锁紧阀
locking washer 防松垫圈；防松垫片；锁紧垫圈；锁紧用钢丝
locking water demand 过闸需水量
locking wedge 锁定楔；锁紧楔
lock-in mixer 同步混频器；锁定混频器
lock-in oscillator 同步振荡器
lock-in period 不能预付的时期
lock input 同步输入
lock-in range 牵引范围；同步范围；锁定范围
lock-in relay 闭塞继电器
lock-in synchronism 牵入同步；锁定同步
lock-in time 同步时间
lock-in tube 锁式管
lock jamb 装锁舌盒的门框边梃
lock-jaw 锁口风
lock joint 紧边互锁接缝(柔性金属屋顶的接缝)；榫接；锁口(接头)；扣接合；锁口；锁接合；锁接头
lock keeper 锁舌盒；船闸管理员；水闸管理员；船闸工作人员；船闸操纵人员
lock key 封锁关键码
lock key replacer 锁键替换装器
lock knitting 锁缝编织
lock knob 固定钮；锁(定)钮
lock lay-bay 船闸会船段
lock lever stud 锁紧桩
lock lever yoke 锁紧杆
lock lift 船闸水头；船闸升程；船闸上游水位差；船闸提升高度
lock lifter 锁销提臂
lock light (防击击的)不锈钢窗；耐火窗
lock list 封锁表
lock location 闸址(船闸)；船闸闸址
lock magnet 锁定磁铁
lockmaking 制锁
lockman 船闸工作人员
lock master 船闸负责人
lock mechanism 制动机构；锁定机构
lock miter 交叉榫斜角接；互锁斜角接；斜接封闭
lock nut 制动螺母；止动螺母；锁紧螺母；保险螺母；自锁螺帽；防松螺帽；锁紧螺母；锁紧螺帽；防松螺母
lock nut washer 扣定螺帽垫片；锁紧螺母垫圈
lock of metal for bag 包用金属锁
lock of metal for vehicle 车辆用金属锁
lock-on 自动跟踪；锁住；锁定
lock-on boundary 锁定边界
lock-on circuit 强制同步自动跟踪电路；强制同步自保持电路；同步电路；受锁电路
lock-on connection 锁紧接头
lock-on counter 同步计数器；同步计量器
lock-on digitizer 跟踪数字化器
lock-on digitizing 跟踪数字化
lock-on range 捕捉距离
lock-on relay 同步继电器
lock-on signal 锁定信号；捕获信号
lock operating lever 锁紧操作杆
lock operation 船闸作业；船闸运行
lock operation(al) water 船闸作业用水
lock option 锁定选择
lockout 解雇；休业；锁定；封锁(器)；封闭工厂；闭锁；闭厂；保护
lockout circuit 闭塞电路；保持电路；闭锁电路
lockout pulse 整步脉冲
lockout relay 连锁继电器；闭锁继电器
lockout switch 联锁开关
lockover 翻转锁定
lockover circuit 双稳态电路
lock paddle 船闸输水口；船闸输水阀；船库输水道；充水闸
lock pawl 止动爪；锁定爪
lock piece 锁止件
lock pillar 锁扣支柱
lock pin 固定销；锁(紧)销；保险销
lock pin housing 锁定销壳体

lock pintle 锁紧舵栓
lock pit 船闸基坑
lock plate 锁舌盒;锁舌板;锁定板
lock plateau 船闸平台
lock platform 船闸平台
lock plunger 锁紧栓
lock position 锁止位置
lock positively 联锁
lock radius 锁紧半径
lock rail 装锁舌板横档;装锁冒头;装锁横档;装门锁横挡;中冒头;门锁横挡
lockrand 锁边
lock reinforcement 门锁金属垫板
lock reinforcing unit （金属门上的）锁匣
lock resolution 锁分解
lock ring 紧箍
lock rise 船闸提升高度;船闸上下游水位差
lock rod 锁杆
lock room 隔声室;锁定区
lock rotor frequency 锁定转子频率
locksaw 曲线（截）锯;圆锯;装锁锯
lock screw 锁紧螺钉;止动螺钉;锁定螺钉;船闸工作人员
lock seam 卷边锁缝;咬口接缝;卷（边）接缝
lock seam door 面板两侧作互锁缝的门
lock seaming 锁口接缝
lock seaming dies 卷边接合模
lock seaming operation 锁缝操作
lock seam pipe 直卷边接缝管
lock seam tube 直卷边接缝管
lock-servosystem 锁相伺服系统
lockset 装锁夹具;成套门锁;锁具
locks for alumin(i)um door and window 铝门窗锁
lock sheet piling 企口钢板桩
lock sheet piling bar 带锁口的钢板桩;带锁的钢板桩
lockshield valve 屏蔽闸（便于暖气片拆修）;平衡阀
lock side wall 闸室边墙;闸室墙
lock significant depth 门槛水深
lock signal 船闸信号
lock sill 人字门槛;船闸门槛;船闸闸址;闸（门）槛
lock site 闸址（船闸）;船闸闸址;船闸位置
lock sleeve 扣定套筒;锁(紧)套
locksman 船闸管理人员;水闸管理员;船闸操纵人员
locksmith 修锁工人;制锁工;锁匠
lock smith's clamp 小虎钳
lock smith's hammer 锁匠的锤
lock smith's work 锁匠工作;锁工
lockspit 挖掘标线
lock spring 固定锁弹簧;锁紧簧
lock spring ring 锁紧弹簧环
lock station 闸站【铁】;枢纽闸站
lock-step procedure 因循守旧过程
lock steps 梯级（式）船闸
lock stile 装锁门梃;装锁竖框;门锁挺
lockstitch blindstitch 锁式暗缝
lockstitch button holder 锁眼机
lockstitch course 锁缝线圈横列
lockstitch indexing cam 锁缝分度凸轮
lock stop 停止锁
lock strike 锁栓眼板;锁舌盒
lock-strip cavity 锁紧带槽;锁紧条空腔
lock-strip gasket 锁紧填隙片;门闩圈;加嵌条窗玻璃垫圈;锁条式密封垫
lock support 锁紧杆座
lock tender 气闸操作员;水闸看守人;船闸管理人
lock test 拘束试验;束缚试验
lock test by low-frequency 低频牵引试验
lock thimble 互衔心环
lock-tile floor system 互锁型地砖系统;锁固地板系统
lock time 击发间隙
lock traffic signal 船闸通行信号
lock tube 锁套
lock-type hose coupling 锁紧式软管接头
lock uint 锁合装置
lock unit 整步器;同步单元
lock-up 锁住;装版;船过闸向上游驶去;上锁保管;拘留所;锁上行动;顶托;封孔蓄水
lock-up certificate of deposits 封锁定期存单;封锁存款单
lock-up clutch 锁止离合器;闭锁离合器
lock-up cock 锁闭开关

lock-up device 锁定装置
lock-up relay 锁定继电器;闪锁继电器
lock-up stage 结束阶段
lock-up time 锁定时间
lock valve 带锁阀;船闸输水阀;保险阀
lock valve gate 截止阀
lock valve globe 球形截止阀
lock wall 闸墙;船闸闸墙;船闸岸壁
lock wall culvert 闸墙内输水涵洞;闸墙输水廊道
lock washer 防松垫圈;止动垫圈;弹簧垫圈;弹簧垫片;锁紧垫圈
lock washer type spring 定片式弹簧
lockwire 安全锁线
lock with high lift 高水头船闸;竖井式船闸
lock with storage chamber 具有蓄水室的船闸;储［贮］水船闸
lockwork 锁的机构
lock-woven mesh 机织钢丝网
lock-woven steel fabric 机织钢丝网
lock wrench 大力钳
lock yarn 毛圈线
loco 机车;火车头;交易地点（商品市场术语）;卖方当地交货价格;牵引机车;当地交货
loco driver 机车驾驶员
locofoco 黄磷火柴
loco haulage 机车运输
locomobile 自动机车;锅驼机
locomorphic stage 原地成岩变化阶段;硬结期
locomotion 运转;移动
loco motive 机车【铁】;机动的;火车头;牵引机车
locomotive and car repair shop 机车车辆修理厂
locomotive axle load 机车轴重
locomotive bed 机车试验台
locomotive boiler 机车锅炉
locomotive cab 火车驾驶室
locomotive car 机车【铁】
locomotive changing point 机车折返站
locomotive converted running kilometers 机车换算走行公里
locomotive cowcatcher 机车排障器
locomotive crane 机车吊机;蒸汽回转起重机;轨道式起重机;机车起重机;铁路起重机
locomotive crank axle 机车曲柄轴
locomotive cylinder 机车汽缸
locomotive daily output 机车日产量
locomotive daily working plan 机车日计划
locomotive day 机车日
locomotive depot 机务段
locomotive depot driver 机务段驾驶员
locomotive depot track 机务段管线
locomotive depreciation apportion and overhaul charges 机车折旧摊销及大修费
locomotive dispatching 机车调度
locomotive drawing dead weight-ton-kilometre 机车牵引载重吨公里
locomotive drive 机车拖动
locomotive engine 火车头;机车
locomotive equipment 机车设备
locomotive facility 机务设备
locomotive fireman 机车司炉
locomotive for negotiating curve 转向机车【铁】
locomotive gas turbine 机车燃气轮机
locomotive gradient 矿用机车坡度
locomotive gross ton kilometers 机车总重吨公里
locomotive haul adapter 机车车钩
locomotive holding track 机待线【铁】
locomotive holdtrack 机待线【铁】
locomotive inductor 机车感应器
locomotive in operation 运用机车
locomotive in reserve 备用机车【铁】
locomotive jib crane 机车悬臂起重机;机车起重机
locomotive kilometer 机车行走公里;机车公里
locomotive kilometers on the road 机车沿线走行公里
locomotive load cell test stand 机车牵引力测定装置
locomotiveness 变换位置方法;移动性能;位置变换性能
locomotive noise 机车噪声
locomotive oil 汽缸油
locomotive operation 铁路机车运用
locomotive operation and maintenance 机务;机车业务
locomotive panel 机车控制板

locomotive pilot 机车导航仪
locomotive power control 机车调度
locomotive providing total adhesion 全黏着机车
locomotive rating 机车牵引力检定;机车额定吨位
locomotive regulator 机车调节器
locomotive remote control 机车遥控
locomotive repairing depot 检修机务段【铁】
locomotive routing 机车线路;机车交路
locomotive running depot 运用机务段;机车运用段
locomotive running kilometerage between two repairs 检修公里
locomotive run-round track 机车迂回线
locomotive runs 机车线路;机车交路
locomotive-screw changing station 机务换乘所
locomotive sector expenses 机务部门支出
locomotive service and temporary rest line 整备待班线
locomotive service and temporary rest track 整备待班线
locomotive servicing point 机务整备所
locomotive serving track 机车整备线
locomotive shed 机车库【铁】;机车房
locomotive speed characteristics 机车速度特性
locomotive temporary rest 机车待班
locomotive terminal 机务段;区段站【铁】
locomotive testing 机车试验
locomotive testing bed 机车试验台
locomotive testing plant 机车试验装置
locomotive tonnage rating 机车牵引吨数
locomotive total mass 机车整备质量
locomotive track 机车走行线
locomotive tractive characteristic curve 机车牵引特性曲线
locomotive tractive force 机车牵引力
locomotive turnaround depot 机务折返段
locomotive turnaround point 机务折返所
locomotive type boiler 机车式锅炉
locomotive type fire box 机车式燃烧室
locomotive underframe 机车底架
locomotive under repair 再修机车
locomotive valve gear 机车的阀动装置
locomotive vehicle plant 机车车辆厂
locomotive waiting track 机待线【铁】
locomotive weight in working order 机车重量
locomotive with alternating current/direct current motor converter set 交直流转换机车
locomotive with commutator motors 单向整流子电动机电力机车
locomotive with single phase/three phase converter set 异步牵引电动机动力机车
locomotive working diagram 机车周转图
locomotive working district 机车牵引区段
locomotive working plan 机车工作计划
locomotive working program(me) 机车工作方案
loco price 当地交货价格
locor （静力图解中的）力矢量
loctal 锁式
loctal socket 锁式管座
locular 小腔的
loculation 小腔形成
loculus 古墓中放骨灰瓮的凹处
locus 座位;空间位置;几何轨迹;位点;所在地
locus ceruleus 蓝斑
locus function 轨迹函数
locus of an equation 方程的轨迹
locus of centre of buoyancy 浮心曲线
locus of center 中心轨迹线;拱圈的中心轨迹线
locus of discontinuity 断裂面线;断裂面位置;不连续区;不连续轨迹
locus of high-speed flow 高速水流轨迹线
locus of high velocity flow 高速水流轨迹线
locus of high water 高水位轨迹
locus of low water 低水位轨迹
locus of metacenter 稳心曲线
locus of metacentric radii 稳心半径轨迹
locus of points 点(的)轨迹
locus sigilli 盖章处;盖印处;封印处
locust 洋槐;刺槐
locust control 治理蝗灾
locust tree 槐树;刺槐
locutorium 修道院会客室
locutory 修道院会客室
lodar 罗达远程精确测位器;方位测定器
lode 排水阴沟;引水业务;脉矿;水路;水沟

lode apex 采矿露头
lode deposit 脉状矿床
loded iron 脉铁
Lode-Duncan model 罗德—邓肯模型
lode gold 脉金
lode mining 开采矿脉
lode ore 脉矿石
Lode's angle 罗德角
Lode's number 罗德数
Lode's parameter 罗德参数
lodestone 吸铁石;极磁铁矿;脉石;天然磁铁;磁石
lodestuff 脉内矿物
lode tin 脉锡
lodge 小旅馆;搭挂;传达室
lodge a claim 索赔
lodge-book 旅馆登记簿
lodge claim 提出赔偿要求
lodgement 沉积;寓所;交存;堆积
lodg(e)ment till 底碛
lodgepole 棚屋支柱
lodgepole pine 红松【植】
lodger 副支梁材;房客
lodging 住处;倒伏;住房;寄宿处
lodging house 私人房屋;公寓;宿舍
lodging knee 梁后水平肘材
lodging room 分间出租的房间
lodging-rooming house 分间出租的住房;分间出租的宿舍;分间出租的房间
lodging system 调休制【铁】
lodgment for investor 投资者存放
lodgment of water 涝;积水;存积水
lodgment table 墙基脚;勒脚;横线脚
lodicator 货油分布指示器
lodos 洛多斯风
lodranite 古铜辉石橄榄石铁陨石;橄榄古铜陨铁
loellingite 斜方砷铁矿
loess 黄土
loessal 黄土质的
loessal loam 黄土类壤土
loessal soil 黄土质土
loess area 黄土区
loess bridge 黄土桥
loess child 僵石
loess chile 黄土结核
loess clay 黄色钙质黏土
loess collapsibility 黄土湿陷
loess collapsible slumping 湿陷性黄土
loess column 黄土柱
loess concretion 僵石;黄土结核;黄土僵
loess conservation 黄土保持
loess deposit 黄土沉积
loess deposit slide 黄土沉积滑坡
loess dish 黄土碟
loess doll 黄土结核;僵石
loess erosion 黄土侵蚀
loess flat 黄土坪
loess flow 黄土流
loess for cement burden 水泥配料用黄土
loess formation 黄土层
loess geomorphology 黄土地貌
loess gully 黄土冲沟
loess highland 黄土高原
loess hillock 黄土崩
loess(i)al 黄土的
loess(i)al nodule 黄土钙结核
loess(i)al plateau 黄土高原
loess(i)al soil 黄土;黄土状土;黄土性土(壤);黄土类土
loessic 黄土的
loessic stage 黄土期
loessification 黄土化(作用)
loessite 黄土岩
loess Jian 黄土塥
loess kindchen 黄土结核
loessland 黄土地(区)
loess landform 黄土地貌
loess landslide 黄土滑坡
loess land use 黄土利用
loess-like loam 黄土类壤土
loess-like soil 黄土状土
loess-lime underlay 灰背
loess loam 黄土性壤土;黄土垆姆
loess nodule 黄土结核
loess plain 黄土平原

loess plateau 黄土高原
loess region 黄土区
loess relief 黄土地貌
loess ridge 黄土梁
loess sinkhole 黄土陷穴
loess soil 黄土质土(壤);黄土性土(壤);黄土;大孔性土
loess tableland 黄土塬
loess topography 黄土地形
loess-type granular deposit 黄土质粒状沉积
loess vale 黄土珊
loess wedge 黄土冰楔
loess well 黄土井
loeweite[loewite] 钠镁矾
loferite 鸟眼灰岩
loft 教堂中的楼座;舞台前部空间;楼厢;鸽舍;厩楼;谷仓顶棚;屋顶下储藏空间;悬杆;阁楼;膨起;输送机道;顶楼层
loft antenna 屋顶天线;顶棚天线
loft barn 双层畜舍
loft beam 屋顶层横梁
loft building 统间式建筑;大空间房屋
loft-dried 风干的;悬挂干燥
loft dryer 干燥箱;箱式干燥器;干燥棚;悬挂干燥器
loft drying 悬挂干燥
lofting 支架上的垫料;大样;放样
lofting plume 高耸的烟羽;高升空中的烟羽
lofting room 放样间
loft insulation 顶层绝热
loft ladder 阁楼爬梯;隐梯;阁楼扶梯;暗藏梯
loft paper 放样纸
loft plan office 统间式办公室
loft room 阁楼房间
loftsman 放样员;放样工(人)
loft stair(case) 折梯
log 运行记录;圆木;圆材;原木;钻孔记录;值班簿;载人日志簿;记录簿;计程仪;木头;木材;航海日志;登记;大木料;测井记录;测程器;记录;测井曲线;测井;测程仪【航海】
log amplifier 对数放大器
Loganian system 洛甘宁系【地】
Logan slabbing machine 洛根型片式截煤机
Logarcheve tangential method 罗加杰夫切线法
logarithm 对数
logarithmic 对数的
logarithmically periodic(al) dipole 对数周期式偶极子
logarithmic amplifier 对数放大器
logarithmic amplitude spectrum 对数振幅谱
logarithmic attenuator 对数衰减器
logarithmic barrier function 对数障碍函数
logarithmic calculator 对数计算器
logarithmic chart 对数图
logarithmic circuit 对数电路
logarithmic computation 对数计算
logarithmic computer 对数计算机
logarithmic coordinate paper 对数坐标纸
logarithmic coordinates 对数坐标
logarithmic cosine 余弦对数
logarithmic creep 对数蠕变
logarithm(ic) criterion 对数判定(法)
logarithmic curve 对数曲线
logarithmic curve type 对数曲线型
logarithmic cycle 对数循环
logarithmic decrease 对数递减
logarithmic decrement 对数减缩率;对数减(缩)量
logarithmic decrement coefficient 对数衰减系数
logarithmic decrement rate 对数衰减率
logarithmic demodulation 对数解调器
logarithmic derivative 对数导数
logarithmic detector 对数检波器
logarithmic deviation 对数偏差
logarithmic differentiation 对数微分法;对数微分
logarithmic diluter 对数稀释器
logarithmic diode 对数二极管
logarithmic distribution 对数分布
logarithmic divided meter 对数刻度计
logarithmic divider 对数除法器
logarithmic double spiral 对数双螺线
logarithmic energy change 能量的对数变化
logarithmic energy spectrum 对数能谱
logarithmic exposure range 对数曝光量范围
logarithmic exposure scale 对数曝光量范围
logarithmic fast time constant circuit 对数短时

间常数电路
logarithmic flow-concentration model 对数型流量密度模型
logarithmic formula 对数公式
logarithmic frequency sweep rate 对数频率扫描率
logarithmic function 对数函数
logarithmic function model 对数函数模型
logarithmic graph 对数图像
logarithmic graphic instrument 对数特性图示仪
logarithmic growth 指数生长
logarithmic growth phase 对数生长期
logarithmic increase in population 人口的对数增长
logarithmic increment 对数增量
logarithmic integral 对数积分
logarithmic law 对数律;对数法则
logarithmic least square method 对数最小二乘法
logarithmic-likelihood ratio 对数似然比
logarithmic mean particle diameter 对数平均粒径
logarithmic mean temperature 对数平均温度
logarithmic mean temperature difference 对数平均温(度)差
logarithmic mean value 对数平均值
logarithmic megohmmeter 对数律刻度高阻计
logarithmic multiplier 对数(式)乘法器
logarithmic normal 对数正常
logarithmic normal distribution 对数正态分布
logarithmic normal probability paper 对数标准概率纸
logarithmic optic(al) wedge tach(e)ometer 对数光楔视距仪
logarithmic oscilloscope 对数示波器
logarithmic paper 对数坐标(记录)纸;对数(格)纸
logarithmic phase 对数期
logarithmic plot 对数坐标图
logarithmic potential 对数势
logarithmic potentiometer 对数律电位计
logarithmic probability 对数概率
logarithmic probability paper 对数概率坐标纸;对数概率格纸
logarithmic profile of velocity 对数速度剖面图
logarithmic property 对数性质
logarithmic reproduction curve 对数再生曲线
logarithmic resistance 对数电阻
logarithmic scale 对数刻度;对数计算尺;对数分度;对数度尺;对数尺(度);对数标度;对数比例尺
logarithmic-scale noise meter 对数刻度噪声计
logarithmic scales 对数表
logarithmic search 对数检索
logarithmic search method 对数查找法
logarithmic sensitivity 对数灵敏度
logarithmic series 对数级数
logarithmic sheet 对数坐标(记录)纸
logarithmic sine 正弦对数
logarithmic skew curve factor 对数偏斜曲线系数
logarithmic solution 对数解法;对数解
logarithmic speed 对数感光度
logarithmic speed-concentration model 对数型速度—密度模型
logarithmic spiral 对数螺(旋)线
logarithmic spiral curve 对数螺线曲线
logarithmic spiral method 对数螺线法
logarithmic sprayer 对数型喷雾器
logarithmic strain 有效应变;自然应变;真实应变;对数应变
logarithmic survivor curve 对数生存曲线;对数存活者曲线;对数残存物曲线
logarithmic table 对数表
logarithmic tach(e)ometer staff 对数视距尺
logarithmic temperature scale 对数温标
logarithmic time intervals 对数时间间隔
logarithmic titration 对数滴定法
logarithmic-to-linear converter 对数函数线性函数变换器
logarithmic transformation 对数变换
logarithmic trigonometric(al) function 三角函数对数表;对数三角函数
logarithmic utility function 对数效用函数
logarithmic vacuum-tube voltmeter 对数刻度真空管伏特计
logarithmic variogram 对数变差函数
logarithmic velocity distribution 对数流速分布
logarithmic velocity profile 对数速度廓线;对数速度分布图;对数流速分布;风速对数廓线

logarithmic viscosity 比浓对数黏度
logarithmic viscosity number 对数黏度值;比浓对数粘度植
logarithmic voltage quantizer 对数电压分层器
logarithmic voltmeter 对数(刻度)伏特计
logarithmic wind(speed) profile 对数风速廓线;风速对数廓线
logarithm of abundance 对数丰度
logarithm of exposure 曝光量对数
logarithm of time fitting method 时间对数拟合法;对数时效法
logarithm-of-time method 时间对数法
logarithmoid 广对数螺线
log band sawing machine 原木带锯
log barge 木材驳船
log-base 对数底(数)
log blue 原木蓝变
log board 航海记事牌
log body 原木运输车厢
logbook 值班簿;记事簿;记录册;航行日记簿;航海日志;报务日记簿
log boom 计程仪支杆;木排拦河埂;木浮排;木材堰;护栏木栅;(河口或港口的)木排浮坝
log brand 原木号印
log bridge 圆木桥
log cabin 圆木(墙)小屋;井干式构造木屋;小木屋;木屋
log cabin construction 井干式构架
log cabin siding 圆木屋的外墙披叠板;视如圆木屋的外墙披叠板
log carriage 原木运转车
log carrier 运木船
log chip 测速板
log chute 原木溜槽;原木滑道;木材筏道;滑木道;放木斜槽;放木滑槽
log circular saw 原木圆锯
log class 原木等级
log classifier 流槽式分级机
log clock 里程指示器;计程仪航程计数器
log cofferdam 木围堰
log column 圆木柱
log construction 井干式构造;圆木结构;原木结构
log control table 记录控制表
log conversion voltmeter 对数变换伏特计
log conveyer 运木机;木料输送机;圆木输送机
log-crib 木笼;圆木叠框
log-crib abutment 木笼桥台;叠框式岸墩;木笼式岸墩;圆形木笼岸墩
log-crib dam 木笼坝
log-crib pitching 木笼护坡
log-crib revetment 木笼护岸;木笼护坡
log culvert 圆木涵洞
log curing 木料养护
log curve and plot 测井曲线及成果图件
log cycle 对数循环
log dam 木材流放坝;木坝
log data composite interpretation 测井资料综合解释
log data set 记录数据集
log deck 原木楞台
log decrement(rate) 对数衰减率
log deflector 木筏转向设备;木筏转向设备
log deviation 对数偏差
log device 记录装置
log-diagram 对数曲线图;记录曲线
log digester 原木汽蒸锅
log drag 圆木刮路器;圆木路刮;原木刮路器
log driver 木筏工人;木材流放工(人);放筏工
log drum 记录鼓
log dump 储[贮]木场;集材场
loge 前部楼座;剧院包厢
logeion 说白区(古希腊剧场舞台上);希腊剧院舞台
log elevator 圆木提升机;原木提升机
log equipment 测井仪器设备
log-exposure curve 曝光量对数曲线
log extract 航海日志摘要;航海日记摘要
log-face ga(u)ge 航程指示器
log file 记录文件
log flume 原木流送槽
log fork 圆木装卸叉
log frame 多锯机;垂直锯木架;垂直锯架
log frame saw 原木框锯;框式锯木机;大型直锯
logged 沼泽的;半淹没的
logged-crib abutment 木笼式桥台;木龙桥台

logged crib pier 木笼式桥台
logged on time 连接时间
logged pile 圆木桩;加箍桩
logger 采运木材工;伐木人;测井记录者;锯木工;记录器;计算机记录器;登记器;测井仪;采运工;采伐机械;伐木者;伐木工
logger system 记录系统
logger task 记录任务
loggia 凉廊;(建筑物一侧俯临庭院的)走廊;敞廊;凹阳台
logging 存入;记录;工作记录;溜放木材;记入;集运作业;木材采运(作业);木材采伐;事件记录;采运(作业);浮运木材;伐区作业;伐木(业);测井(曲线);测井记录法;[复]loggias 或 loggie
logging arch 集材拱架
logging cable 电测用电缆
logging calibration 测井刻度
logging chain 木材捆运链
logging clearing 集材
logging cost 测井费
logging crane 伐木起重机
logging current 测井供电电流
logging cycle 采伐周期
logging dam 木材流放坝
logging data 录入数据;记录数据
logging data on tape 记录带
logging depot 采运管理站
logging depth 测井深度
logging desk 值班台
logging device 记录装置;测井装置
logging disturbed factor 测井时的手扰因素
logging end depth 测井终止深度
logging film 记录胶片
logging frame 曳索机架
logging industry 木材采运工业
logging in stratum 地层检测
logging instrument 测井仪
logging locomotive 运木材机车
logging machine 记录机
logging method 集木材方法;测井方法
logging method of electric(al) prospecting 电测深法
logging module 记录模块
logging motion arrestor 测井升沉制动器
logging number of electric(al) prospecting 电测深点数
logging oscillograph 测井示波仪
logging pass 筏道
logging path 筏道
logging program(me)记录程序
logging road 筏道;集材道路;放木道
logging saw 伐木锯
logging scissors 原木铁具
logging service facility 记录服务程序
logging service request 记录服务申请
logging show 采运作业面
logging speed correction 测速校正
logging start depth 测井起始深度
logging station 伐木场
logging succession 采伐演替
logging synthesis curve plot 测井综合曲线图
logging television 井下电视机
logging temperature 测井温度
logging time 测井时间
logging tongs 带钩夹具
logging tool calibration 测井仪器刻度
logging truck 测井工程车
logging typewriter 记录打字机
logging unit 测井装置
logging wheels 高轮运材挂车
logging winch 绞木机
log glass 计时沙漏
log gorge 木材阻塞
log governor 计程仪运速轮
log grade 原木等级
log grading curve 原木等级曲线
log grapple 抓木器;原木抓扬机;木材抓钩
log growth phase 对数增殖期
log habitat 原木生境
log harrow 圆木耙
log haul 木材拖运
log haul-up 原木起装置;曳木机;木材拖运机
log hoisting apparatus 原木提升设备
log house 木房(屋)

log hut 圆木小屋
logic(al) access level 逻辑存取层
logic(al) action 逻辑作用;逻辑操作
logic(al) add 逻辑加
logic(al) adder 逻辑加法器
logic(al) addition 逻辑加法
logic(al) address 逻辑地址
logic(al) algebra 逻辑代数
logic(al) algebraic equation with two unknowns 二元逻辑代数方程式
logic(al) analog 逻辑模拟
logic(al) analyser 逻辑分析仪;逻辑分析器
logic(al) analysis 逻辑分析
logic(al) analysis device 逻辑分析装置
logic(al) analysis system 逻辑分析系统
logic(al) analyzer[analyser]逻辑分析仪;逻辑分析器
logic(al) analyzer for maintenance planning 逻辑模拟分析系统
logic(al) arithmetic unit 逻辑运算器
logic(al) array 逻辑数组
logic(al) axiom 逻辑公理
logic(al) beauty 合理美
logic(al) block 逻辑块;逻辑单元;逻辑部件
logic(al) box 逻辑环
logic(al) calculation 逻辑运算;逻辑演算
logic(al) capability 逻辑能力
logic(al) card 逻辑卡片;逻辑插件;逻辑板
logic(al) channel 逻辑信道
logic(al) channel number 逻辑通道号
logic(al) chart 逻辑(流程)图
logic(al) choice 逻辑选择
logic(al) circuit 逻辑线路;逻辑电路
logic(al) circuit card 逻辑电路板
logic(al) circuit for self-adaptation 自适应逻辑线路
logic(al) clamp 逻辑钳位
logic(al) combination 逻辑组合
logic(al) comparator 逻辑比较器
logic(al) comparison 逻辑比较
logic(al) compatibility 逻辑兼容性
logic(al) complement 逻辑补
logic(al) composite variable 逻辑组合变量
logic(al) computer 逻辑运算计算机
logic(al) conception of probability 概率论的逻辑概念
logic(al) condition 逻辑条件
logic(al) connected terminal 逻辑连接终端
logic(al) connection 逻辑连结;逻辑连接
logic(al) connection terminal 逻辑连接终端
logic(al) connective 逻辑连接符;逻辑连接词
logic(al) connector 逻辑连接符
logic(al) connect terminal 逻辑连接终端
logic(al) consequence 逻辑结果
logic(al) constant 逻辑常数
logic(al) contouring 合理描绘等高线
logic(al) control 逻辑控制
logic(al) conversion 逻辑转换;逻辑变换
logic(al) conversion matrix 逻辑变换矩阵
logic(al) converter 逻辑变换器
logic(al) core 逻辑磁芯
logic(al) correction 逻辑对比
logic(al) curve 逻辑曲线
logic(al) data 逻辑型数据
logic(al) data base 逻辑数据库;逻辑数据基
logic(al) data base definition 逻辑数据库定义
logic(al) data base design 逻辑数据库设计
logic(al) data base record 逻辑数据库记录;逻辑数据基记录
logic(al) data independence 逻辑数据独立性
logic(al) data link 逻辑数据链路
logic(al) data number 逻辑数据码
logic(al) data path 逻辑数据通路
logic(al) data structure 逻辑数据结构
logic(al) data transfer 逻辑数据传送;逻辑数据传输
logic(al) decision 逻辑判断;逻辑判定;逻辑决策
logic(al) decision process 逻辑决策过程
logic(al) decision table 逻辑判定表
logic(al) deduction by reasoning 逻辑推演
logic(al) descripto 逻辑描述符
logic(al) design 逻辑设计
logic(al) design automation 逻辑设计自动化
logic(al) designer 逻辑设计者

logic(al) design system 逻辑设计系统
logic(al) detector 逻辑检测器
logic(al) determinant 逻辑行列式
logic(al) development system 逻辑开发系统
logic(al) device 逻辑设备;逻辑器件
logic(al) device address 逻辑设备地址
logic(al) device number 逻辑设备号
logic(al) device order 逻辑设备指令;逻辑设备命令
logic(al) device table 逻辑输入输出设备表
logic(al) diagram 逻辑图(框);逻辑框图
logic(al) diagram of reliability 可靠性逻辑图
logic(al) difference 逻辑异;逻辑差
logic(al) division 逻辑分类
logic(al) drawing 逻辑图
logic(al) edit 逻辑编辑
logic(al) editing symbol 逻辑编辑符
logic(al) element 逻辑元件;逻辑单元
logic(al) end 逻辑结尾
logic(al) entity 逻辑实体
logic(al) equation 逻辑方程(式)
logic(al) equation generator 控制逻辑生成器
logic(al) equation simulation 逻辑方程模拟
logic(al) equipment table 逻辑设备表
logic(al) equivalence 逻辑等价
logic(al) error 逻辑错误
logic(al) escape symbol 逻辑转义符
logic(al) event 逻辑事件
logic(al) expander 逻辑扩大器
logic(al) expression 逻辑表达式
logic(al) factor 逻辑因子
logic(al) family 逻辑系列
logic(al) fault 逻辑故障
logic(al) fault condition 逻辑出错条件
logic(al) file 逻辑文件
logic(al) file name 逻辑文件名
logic(al) file system 逻辑文件系统
logic(al) flexibility 逻辑灵活性
logic(al) flow 逻辑流程
logic(al) flow chart 逻辑流程图
logic(al) form 逻辑形式
logic(al) formula 逻辑公式
logic(al) function 逻辑作用;逻辑函数
logic(al) gate 逻辑选择器开关;逻辑门
logic(al) grammar 逻辑文法
logic(al) ground 逻辑接地
logic(al) group 逻辑组
logic(al) group number 逻辑组号
logic(al) hazard 逻辑险态;逻辑冒险
logic(al) image 逻辑图像
logic(al) implication 逻辑隐含
logic(al) indicator 逻辑指示器
logic(al) inequality 逻辑不等性
logic(al) inference 逻辑推理
logic(al) inference engine 逻辑推理机
logic(al) information method 逻辑信息法
logic(al) in memory 逻辑存取器;内存储逻辑
logic(al) inspection 逻辑检查
logic(al) instruction 逻辑指令
logic(al) instruction set 逻辑指令系统
logic(al) integrated circuit 逻辑集成电路
logic(al) interface 逻辑接口
logic(al) inverter 逻辑倒相器
logicality 逻辑性
logicalized memory 逻辑化存储器
logic(al) leading end 逻辑前端
logic(al) level 逻辑级;逻辑电平;逻辑层次
logic(al) level of a structure member 结构成员的逻辑层次
logic(al) level of programming 程序设计逻辑级
logic(al) light 逻辑指示
logic(al) line 逻辑行
logic(al) line end symbol 逻辑行结束符
logic(al) line group 逻辑线组
logic(al) link 逻辑连接
logic(al) link layer 逻辑链路层
logic(al) link path 逻辑连接通路;逻辑连接路径
logic(al) logging 逻辑记入
logically complete 逻辑完全性
logically connected terminal 逻辑连接的终端
logically true 逻辑正确
logic(al) machine 逻辑机
logic(al) map 逻辑图
logic(al) mathematics 逻辑数学
logic(al) matrix 逻辑矩阵

logic(al) matrix equation 逻辑矩阵方程式
logic(al) message 逻辑消息
logic(al) mistake 逻辑错误
logic(al) model 逻辑模型
logic(al) module 逻辑组件;逻辑模块
logic(al) multiplication 逻辑乘(法);逻辑操作
logic(al) multiply 逻辑积;逻辑乘
logic(al) name 逻辑名字
logic(al) name table 逻辑名表
logic(al) network 逻辑网络
logic(al) network concept 逻辑网络概念
logic(al) number 逻辑编号
logic(al) of events 事件的必然
logic(al) of modality 模态逻辑
logic(al) of relations 关系逻辑
logic(al) operand 逻辑操作数
logic(al) operation 逻辑运算;逻辑操作
logic(al) operation instruction 逻辑运算指令
logic(al) operator 逻辑(运)算符;逻辑算子;逻辑操作符
logic(al) order 逻辑指令
logic(al) organization 逻辑组织;逻辑结构
logic(al) oscilloscope 逻辑示波器
logic(al) outcome 逻辑结果
logic(al) overlap coefficient 逻辑重叠系数
logic(al) page number 逻辑页编号
logic(al) paging 逻辑分页
logic(al) partition(ing) 逻辑划分
logic(al) pattern 逻辑图
logic(al) pointer 逻辑指示字
logic(al) polarity 逻辑极性
logic(al) primary 逻辑初等量
logic(al) probe 逻辑探针;逻辑探头
logic(al) probe indicator 逻辑探头指示器
logic(al) procedure diagram 逻辑流程框图
logic(al) product 逻辑乘积;逻辑积
logic(al) product gate 逻辑乘门
logic(al) program(me) 逻辑程序
logic(al) programming 逻辑程序设计
logic(al) protocol 逻辑协议
logic(al) pulser 逻辑脉冲发生器
logic(al) record 逻辑记录
logic(al) record interface 逻辑记录接口
logic(al) relation 逻辑关系(式)
logic(al) relay 逻辑继电器
logic(al) roll-out 暂时逻辑转出;逻辑转出
logic(al) rule 逻辑规则;逻辑尺
logic(al) schematic 逻辑框图
logic(al) scope board 逻辑显示板
logic(al) sequence 逻辑序列;逻辑顺序;逻辑次序;逻辑程序
logic(al) shift 逻辑转换;逻辑移位
logic(al) shift operation 逻辑移位操作
logic(al) shift right 逻辑右移
logic(al) simulation 逻辑模拟
logic(al) specification 逻辑规格
logic(al) spectrum 逻辑系列
logic(al) state 逻辑状态
logic(al) state analyzer 逻辑状态分析器
logic(al) state indicator 逻辑状态指示器
logic(al) state table 逻辑状态表
logic(al) storage 逻辑存储
logic(al) structure 逻辑结构
logic(al) sum 逻辑和
logic(al) sum gate 逻辑和门
logic(al) swing 逻辑摆幅
logic(al) switch 逻辑开关
logic(al) synthesis 逻辑综合
logic(al) system 逻辑系统
logic(al) system design 逻辑系统设计
logic(al) tab 逻辑制表
logic(al) table 逻辑表
logic(al) term 逻辑项
logic(al) terminal 逻辑终端
logic(al) terminal pool 逻辑终端组
logic(al) testing 逻辑测试
logic(al) timer 逻辑时钟;逻辑计时器
logic(al) tracing 逻辑追踪
logic(al) tree 逻辑树
logic(al) trigger generator 逻辑触发生成器
logic(al) truth 逻辑永真(式)
logic(al) twins 逻辑孪生
logic(al) type 逻辑型;逻辑类型
logic(al) unit 逻辑设备;逻辑单元;逻辑部件

logic(al) unit connection test 逻辑部件连接测试
logic(al) unit services 逻辑设备服务;逻辑单元服务
logic(al) unit type 逻辑单元类型
logic(al) value 逻辑值
logic(al) variable 逻辑变数;逻辑变量
logic(al) verification system 逻辑检验系统
logic(al) verify 逻辑核对
logic(al) volume 逻辑卷
logic(al) wiring 逻辑布线
logic(al) with shift 移位逻辑
logic(al) word 逻辑词
logic(al) zero clock 逻辑零时钟
logicor 逻辑磁芯
log-in 进入系统;记入;请求联机;登记;注册;登录
log-in and log-out 录入与退出
login dialogue box 登录时对话框
log indicator 计程仪记录器;计程仪指示器;测程仪指示器
log-initiated checkpoint 记录启动检查点
log interpretation 测井成果解释
logistic 计算术;数理符号逻辑;对数的
logistical airfield 后勤机场
logistical base 供应基地
logistical vessel 供应船
logistic base 补给基地
logistic curve 增加曲线
logistic curve fitting 曲线拟合
logistic delay 后续服务滞后
logistic management 后勤管理
logistic planning 后勤规划
logistic probability figure of radon 氡气对数概率图
logistic probability figure of radon value 氡值对数概率图
logistics 后勤;输给系统
logistic ship 后勤船
logistics of distribution 分配后勤
logistics system 货物的流程系统
logistic supplies 后勤供应
LOGISTRIP automatic system 油船自动装卸压载系统
logit 分对数
logitron 磁性逻辑元件
log jack 造材搬钩
log jam 原木堵塞;流木拥塞;木块堵塞;木材堵塞
log kicker 原木推送机;踢木器
log landing 木材装车场
log leveler 半圆木平路器;半圆木平路机
log-likelihood ratio 对数似然比
log line 计程仪索;计程(仪)绳;探测索;测速绳;测量仪绳;测程仪绳;测程线;测程索
log-linear recurrence relationship 对数直线重复关系
log line marks 测速绳标记
log loader 圆木装载机;原木装载机
log lock 过木闸;放木闸
loglog 重对数
log-log coordinates 双对数坐标
log-log diagnostic plot 双对数诊断图版法
log-log graph paper 重对数纸
log-log grid 双对数坐标系统
log-log paper 重对数(坐标)纸;双对数坐标纸;复对数坐标纸
log-log plot 双对数坐标图;双对数标绘
log-log range 重对数范围
log-log scale 双对数坐标;重对数图尺;重对数比(例)尺;重叠对数图尺;复对数标尺
log-log selection 重对数选择性
log-log slope 双对数斜率
log-log spectral ratio 双对数频谱比
log-log transformation 重对数变换
log-log weighting coefficient 重对数加权系数
log mark 原木号印
log market 木材市场
log marking paint 原木标记漆
log mean temperature difference 对数平均温(度)差
lognormal 对数正态;对数正常
log-normal distribution 正态对数分布;对数正态分布
log-normal frequency distribution 对数正态频率分布
log-normal probability 对数正态概率图

log-normal probability paper 对数正态概率坐标纸
log of borehole 钻孔柱状图;钻孔记录;测井记录
log of boring number 孔号记录
log of drill hole 钻孔柱状图
logoff 注销;关机;断开链路
log of freight train performance 货物列车性能日记
log of test pit 试坑记录
logograph 略字;标识符号
logometer 电流比计;测电流比计;比率表;测程仪【航海】
log-on 进入系统;记入;登记;开机
log-on data 启动数据
log-on message 启动信息
logon procedure 注册过程
logotype 标识符号
logout 注销;运行记录;事件记录
log paper 对数纸;对数格纸
log pass 放木道;圆木桩;伐道
log pass structure 过木建筑物
log path 放木道
log Pearson Type 3 distribution 对数皮尔逊 III 型分布
log-periodic(al) antenna 对数周期天线
log-periodic(al) dipole 对数周期式偶极子
log pile 圆木桩
log pond 原木池;漂木存放水区
log-probability 对数概率
log-probability law 对数概率定律
log-probability paper 对数概率坐标纸;对数概率格纸
log public word 测井通用词
log raft 圆木堆;木排;木筏
logrank test 时序检验
log reading 测井曲线量值
log register 计程仪航程指示器;计程记录仪;测程仪记录器
log revetment 原木护岸
logrolling 滚木头;搬运木材
log rotator 计程仪转子
log rule 原木测杖;原木材积表;材积计算式;材积计算表;材积尺
log run 原木制材量
log saw 原木锯;头锯;大木锯
log scale 原木(检)尺;原木材积表
log scaling 原木检尺
log screw 木螺钉;方头木螺钉
log seal 原木止水;止水木条;木止水条
log sensitivity 对数灵敏度
log's error record book 计程仪误差记录簿
log sheet 运转日期记录;钻孔柱状图;记录卡(片);记录表;对数坐标(记录)纸;对数纸;地质剖面图
log-ship 手用测程器;扇形计程器;原木运输船
log skidder 圆木拖运机
log slide 滑道
log slip 木材拖运;木材滑道
log sluice 原木流放渠
log spiral 对数螺旋线;对数螺线
log's reading 计程仪读数
log stain 原木变色
log stop 木闸门;叠梁闸门
log storage 原木场
log survey 原木库存量
log table 对数表
log tally 原木计数
log task 日志任务
log tool 测井仪
log track 拖木材的线路
log trailer 木料拖车
log training wall 原木导堤;木栏导流堤
log transcriber 计程记录仪
log-trash boom 污物栅栏
log treatment 原木处理
log truck 原木运输车
log type entry 登录类型表目
logu 弹性不足
loguat 枇杷树
log unified number 测井统一编号
log unit 对数单位
log volume 原木材积
log volume table 原木材积表
log wagon 圆木装运车
log walled sauna but 原木盖的桑拿浴室小屋;原木盖的塞汽浴室小屋

log washer 洗矿槽;洗矿机;倾斜洗选槽;分级槽
logway 木材放流槽;集材道;放木道;筏道
log weir 原木堰;木堰;木材流放堰;筏堰
log well 观测井
logwood 苏方树;长胶合板面层;苏木
logwood extract 苏木提取物;苏木精
log yard 木材码垛;木材堆(置)场
log yarder 圆木码垛机
log zero 测井零线
Lohmann hologram 罗门全息图
Lo-Hsing-To zero datum 罗星塔零点
Lok'd bar 霍普钢窗(一种专利钢窗)
Loke hand level 罗克手提水准仪
lokkaite 水碳钇石
Lok-strength internal fracture test (桩的)拔出试验
Lok-strength test 劳克试验
loktal base 锁式管座
loll 静止角;丧失稳性
lolly 片冰
lolly ice 海上浮冰;松冰团;疏松冰;冰晶
lolly sludge lump 松冰团
Lo/Lo system 吊装装卸方式;吊上吊下装卸系统
Lomanche trench 罗曼什海沟
Lomas nut 伦孟螺帽;孟形枢帽
Lomax method 洛马克斯法;轻油强化法
Lombard architecture 伦巴第建筑;意大利北部建筑
lombarde 伦巴德风
Lombard popler 伦巴第白杨木
Lombard rate 证券抵押;短期贷款利息;伦巴第利率
lombardy polar 钻天杨
lomomite 浊沸石
lomonosovite 磷硅钛钠石
Lomonosov ridge 罗蒙诺索夫海岭
lomontite 浊沸石
Londinian(stage) 浪丁阶【地】
London basin 伦敦盆地
London Building Acts 英国建筑法规;伦敦建筑法规
London bullion market 伦敦黄金市场
London clay 伦敦盆地河流沉积黏土(供制砖用);伦敦黏土
London Commodity Exchange 伦敦商品交易所
London Convention 伦敦公约
London Court of Arbitration 伦敦仲裁法庭
London equation 伦敦方程
London force 伦敦力
London foreign exchange market 伦敦外汇市场
London gold market 伦敦黄金市场
London International Financial Futures Exchange 伦敦国际金融期货交易所
London landed clause 伦敦卸货条款
London landed terms 伦敦到岸价格条件
London Metal Exchange 伦敦金属交易所
London pattern trowel 伦敦型泥刀
London penetration depth 伦敦穿透深度
London Port 伦敦(蒂尔伯里)港
London price 伦敦价格(指美国股票在伦敦的价格)
London screwdriver 伦敦旋凿
London smog incident 伦敦烟雾事件
London smoke 伦敦烟色;暗灰色;灰黑色;红黄色
London standard 伦敦标准
London stock 白垩黏土砖;伦敦砖;普通砖
London stock brick 伦敦手制砖
London Stock Exchange 伦敦证券交易所
London Transport 伦敦交通局
London-type smog 伦敦型(烟)雾
London unit 伦敦单位
London weighting 伦敦加权
London white 白铅;重白铅;伦敦白
lonealing 低温退火
lone-wolf prospector 单干户勘探工作者
long 长的
long acceleration 持续加速度
long acting 长效的
long acting preparation 长效制剂
long addendum 长齿顶高
long addendum gear 径向正变位齿轮;长齿顶齿轮
long addendum tooth 加高齿
long afterglow 长余辉
long aggregate 长骨料
longan 龙眼
long and base line buoy 长索浮标
long and heavy coating 长厚窑皮
long and heavy goods 超长超重货物;长(大)件和重件货物;长大笨重货物

long and nodding 长而下垂
long and short 总之;简括
long and short addendum system 长短齿顶高齿轮系
long and short(angle) quoin 长短交替突角
long and short ridge tile 长短脊瓦
long and short technique 长短交替技术
long and short-term objectives 长期目标与短期目标
long and short work 长短交替砌石;长短砌合
long arc 大摆幅
long arc xenon lamp 长弧氙灯
long arm (启闭门上的)有钩长臂;长臂;万拱;慢拱
long armed door 机动门(水密滑动舱壁门)
long arm jack 长臂千斤顶
long arm mirror 长臂镜
long arm paint brush 长柄漆刷
long auger 槽钻头
long auger drill 长螺旋钻机
long axial azimuth of circular features 环形体长轴方向
long axis 长轴
long axis anticline 长轴背斜
long axis plotting 长轴绘图
long axle direction of settlement funnel 沉降漏斗长轴方向
long bar burner 长管燃烧器
long base 长基线
long base broom drag 长底盘刮路刷
long base drag 长底盘路刮;长刀刮路机
long base line acoustic(al) positioning system 长基线水声定位系统
long base-line system 长基线系统
long base-multiple-blade drag 长底盘多刃刮路机
Long Beach Harbor 长滩港(美国)
long beam cutting machine 长杆式裁切机
long bed for pretensioning 先张法长线台座
long bend 长弯头
long bend calking tool 长弯捻缝凿
long bent point screw 长弯头画线针
long bill 长期汇票
long blade 长叶片
long blast 汽笛长声;长声
long blast drilling 深孔凿岩
long blast hole work 深孔崩落开采法
long-boat 大艇
long bolt 长螺杆
long bolt support 长螺栓支撑
long bone 长管骨
long boom(ed) dragline 长臂索铲;长臂吊铲
long boom loader 长转臂式装载机;长臂式装载机
long boom quay crane 长臂码头起重机
long boom stripping shovel 长臂剥离挖掘机;长臂剥离挖掘铲
long boom wharf crane 码头(用)长臂起重机;长臂码头起重机
long borehole 深孔
long borer 槽钻
long borer auger 长钻
long branched 长枝的
long branch routine 长转移程序
long branch sequence 长转移顺序
long bridge 长桥楼
long bridge house 长桥楼
long bridge vessel 长桥楼船
long burning time booster 长期工作助推器
long burst 长时间脉冲
long butt 长截头
long cane pruning 长梢修剪
long card grinding roller 长磨辊
long center conveyer 长线输送机
long centumweight 长担(1 长担 = 112 磅)
long chain(age) 长链
long chain branching 长链支化
long chain diacid 长链二酸
long chain dicarboxylic acid 长链二羧酸
long chain fatty acid 长链脂肪酸
long chain molecule 长链分子
long chain polymer 长链聚合物
long charge 整孔装药;柱状装(炸)药
long chord 长弦
long chord winding 长节距绕组
long church 长列式教堂
long clay 高塑性黏土

long cloak 斗篷
long code word 长码字
long column 长柱(长过20倍宽的柱)
long-columnar habit 长圆柱状晶体形
long column testing machine 长柱试验机
long compartment mill 多仓长磨
long cone cyclone 长锥体旋风除尘器
long connecting hose 长连接软管
long continued load(ing) 长期负载;长期负荷
long coordinate string 长坐标数字串
long cord 长考
long core machine 长铁芯电机
long corridor 长廊
long counter 全波计数管;长计数管
long coupling pin 长联结销
long crested sea 长峰波
long crested wave 长脊波;长峰波
long crested weir 宽顶堰;长顶堰
long cross garnet 长十字形螺铰
long crown glass 长冕玻璃
long-cultivated field 熟地
long cultivated land 久耕地;熟地
long cut 长路堑
long cut wood 纵截木;长原木;长木材
long cycle life 长周期寿命
long cylinder 长圆筒
long cylindric(al) roller bearing 长圆柱滚子轴承
long cylindric(al) shell 长圆柱形薄壳
longdal log washer 回转板洗矿机
long dash 长划
long dash line 长虚线;长画虚线
long day 长日照
long day crop 长日照作物
long day plant 长日照植物
long day short-night plant 长日照短夜植物
long delay blasting cap 长(期)延时引爆雷管
long denotation 长标志
long diagonal of indentation 压痕的长对角线
long diamond mesh 长菱形网眼;长菱形筛孔
long dike 长堤
long discharge 长放电
long distance 远距离;长距离的;长距离
long-distance bus 长途(公共)汽车
long-distance bus station 长途(公共)汽车站
long-distance bus stop 长途公共汽车站
long-distance cable 长途电缆
long-distance call 远距离呼叫;长途呼叫;长途电话;长距离通话
long-distance carrier cable 长途载波电缆
long-distance communication 远距离通信;长途通信
long-distance commuting 长距离通勤交通
long-distance conveyance 远距离输送
long-distance conveyer 长距离胶带运输机;长距离(胶带)输送机
long-distance driver 长途驾驶员
long-distance goods traffic 长途货运
long-distance guided communication 远距波导通信
long-distance haulage work 长距离运输
long-distance heat 远距离热源
long-distance heat intake 远距离热源引进
long-distance heat supply pipeline 长距离供暖管线
long-distance lens 远摄镜头
long-distance light 远光
long-distance line 长途线路;长距离铁路;长距离管线;长距离线路
long-distance local 本地长距离站
long-distance loop 长途回路;长途环路
long-distance mobile phone 长途移动电话
long-distance motor-coach train 远程电动旅客列车
long-distance navigation system 远距离导航系统
long-distance of migration 长距离运移
long-distance operation 长距离遥控
long-distance passenger traffic 长途客运
long-distance pipeline 长距离输水管线;长距离(输气)管线
long-distance(power) transmission 长距离输电
long-distance radio station 远距离无线电站
long-distance railcar 远距离轨道车
long-distance railway 长距离铁路
long-distance recorder 远距离记录仪;远程记录仪器;遥测仪器
long-distance remote control 长距离遥控
long-distance ship-shore radio communication 船岸远距离无线电通信;岸船远距离无线电通信
long-distance shot 远距离摄影
long-distance signaling 远距离信号
long-distance sound propagation 远程声传播
long-distance stage transmitter 远传水位计;遥测水位计
long-distance supply 远距离供能;远距离供电
long-distance symmetric(al) cable 长途对称电缆
long-distance telephone(call) 长途电话;长途通话
long-distance telephone exchange 长途电话局
long-distance telephone exchanger 长途电话交换机
long-distance telephone office 长途电话局
long-distance telephoto lens 长距离远摄镜头
long-distance terminal 长途终端
long-distance traffic 长途通信业务;长途通信;远距离交通
long-distance transmission line 长途输电线;长距离输电线路
long-distance transmission system 长距离输电系统
long-distance transport 长途运输;长距离输送;远距离输送
long-distance trunk 长途中继线
long-distance water level recorder 远距水位记录仪;远传自记水位计;遥测水位计
long-distance water pipeline 远距离供水管
long-distance water supply 远距离供水
long-distance xerography 长距静电摄影术;远距离静电印刷
long division 长除法
long dolly (桩的)长垫盘;替灯;送桩;(打桩时用的)垫桩;长顶桩;长垫桩
long-dotted line 短画(虚)线
long down grade section 长大直线地段
long downhill grade 长大下坡道
long dozen 13个
long drill 长钻
long dry process kiln 干法长窑
long dummy 调直工具(管子工用)
long duration 长(时)期;长历时;长(持续)时间
long duration deep dive 持续深潜水
long duration employment 经常性失业
long duration load 持久荷载;长时荷载;长期荷载
long duration static test 蠕变试验;长期静力试验
long duration test 耐久试验;长期试验;连续(负荷)试验
long ears 长耳子
long end interest 较长期的利息
long end link 长型末端链环
long epoch environmental effect 远期环境影响;远期环境效应;长期接触;长期感受;长期暴露
longer 纵堆装法;较长
longeron 翼梁;纵梁;桁梁
longer rail 长钢轨
longest 最长的
long-established station 长期观测站;常设测站
longest operation sequence 最长作业顺序
longest-operation-time heuristic method 最长作业时间探索法
longevity 资历长的;资力;寿命;长期任职
longevity of service 长使用寿命
longevity pay 按工作年资增加工资
longevity test 寿命试验
long extension 长远开拓;长期扩展
long-eye auger 深眼木钻;深孔钻
long face 走砖面;顺砖面;长的一面;长边面
long face place 长工作面;长壁工作面
long fake 并列排绳法
long fallow land 长期休闲地
long fascine 梢龙;长梢捆;长柴笼
long fatigue life 耐疲劳性
long feed 纵向进给
long-feed drill jumbo 长进给钻车;深(给)进钻车
long fiber 长纤维
long fiberd 长纤维的
long fiberd asbestos 长绒石棉
long fibered paper 长纤维纸
long fibered pulp 长纤维纸浆
long fiber glass cloth 长丝玻璃布
long fiber grease 长纤维润滑脂
long fiber reinforced thermoplastic 长纤维增强热塑料
long figure 昂贵
long firm 皮包公司
long flame 长焰;长火焰
long flame burner 长焰烧嘴;长火焰燃烧器
long flame coal 长焰煤
long flame gas burner 长焰烧嘴
long flashing light 长闪光
long flat car 长大平车
long flat nose pliers 尖嘴钳;长尖嘴钳
long flat nose pliers with side cutting 长尖嘴扁口钳
long flat nose side cutting pliers 长尖嘴偏口钳
long flax 长亚麻
long float 杠尺;长抹子;长镘板
long floating point 长浮点数
long flowering period 长花期
long fluorescent lag 长余辉
long focal length camera 长焦距照相机
long focal length camera lens 长焦距照相镜头
long focal length lens 长焦距物镜;长焦距透镜
long focal length telemicroscope 长焦距望远显微镜
long focus 长焦距
long focus lens 长焦距透镜
long focus magnetic lens 长焦距磁透镜
long focus objective 长焦距物镜
long forceps 长镊
long forecastle 长首楼
long form 详细格式
long form bill of lading 详式提单;全式提单
long form report 详细审计报告
long fruit branch 长果枝
long gallery 长廊
long glass 航海望远镜;长性玻璃;长型漏砂计
long glass fiber[fibre] 长玻璃纤维
long grating linear movement sensor 长光栅线性位移传感器
long gravity wave 长重力波
long green 现钞(尤指美钞)
long grinder 纵向碎木机
long grooving plane 长开槽刨
long hafted billhook 修树枝用的长柄刀
long half-lift 长半衰期
long half paper 长半张
long half round nose side cutting pliers 长半圆嘴偏口钳
long half-round slide valve 长半圆滑阀
long handled float 长柄镘板;长柄镘;长柄抹子
long handled hammer 长柄锤
long handle float 长柄镘刀;长柄抹子
long handle grass shears 长柄刈草铰剪
long handle rust scraper 长柄铲锈刀
long handle shovel 长柄铲
long handle wire brush 长柄钢丝刷
longhand method 笔算法
long hand shovel 长柄铲
long haul(age) 长距离运输;长运距;远程送运;长途运输;长距离输送
long haul carrier system 长距离载波系统
long haul communication 长距离通信
long haul material 远运距材料
long haul optic(al) link 长距离光线路
long haul over-the highway truck 长途货运汽车
long haul radio 长距离无线电通信
long haul telemetry 远程遥测
long haul toll transit switch 长途电话转换开关
long header 厚墙中顺砖横砌法;连续丁砖
long heavy down grade 长大下坡道
long heavy grade 长大(上)坡道【铁】
long heavy swell 长狂涌;八级涌浪
long hemp 长大麻
long high swell 长狂涌
long hole benching 深孔梯段掘进法;深孔梯段回采法;深孔梯段采矿法
long hole bit 深眼嘴子;深眼钻
long hole blasting 深孔凿岩;深孔爆破
long hole blasting technique 深孔爆破技术
long hole boring apparatus 钻深孔装置
long hole drill 长孔钻机;深眼钻
long hole drilling 深孔钻进;深孔凿岩
long hole drilling machine 深孔钻机
long hole grouting 深孔灌浆

long hole infusion coal mining method 长孔水封爆破采煤法
long hole jetting 长孔水力采煤法
long hole method 深眼爆破法；深孔爆破法
long hole percussion machine 冲击式深孔钻机
long hole percussive machine 冲击式深孔钻机
long hole shrinkage method 深孔留矿法
long holing 深孔钻进
longhorn beetle 天牛
long hour motor 持续运行电动机
long hundred 一百多
long hundredweight 长担（1 长担 = 112 磅）
longifolic acid 长叶酸
longimanous 长手的
longimetry 长度测量
long inks and short inks 长丝墨和短丝墨
long instruction 长指令
long instruction format 长指令格式
long internal wave 长内波
long international voyage 国际长途航线
long irradiated 长时间照射的
longitude 经线；经度
longitude and altitude 经纬度
longitude and latitude of station 台站经纬度
longitude by chronometer 时间求经度
longitude by (dead) reckoning 积算航经度；船位推算经度
longitude by equal altitude 近午等高度天体经度
longitude by observation 观测经度
longitude by transit of star 恒星中天求经度
longitude circle 子午圈；经（度）圈；黄经圈
longitude correction 经度校正
longitude determination 经度测定
longitude difference 经距；经差
longitude distribution 经度分布
longitude east 东经
longitude effect 经度效应
longitude equation 经度方程
longitude error 经度误差
longitude factor 经度因数；经度系数
longitude from 起程经度
longitude in 到达经度
longitude in arc 用角度表示经度
longitude in time 用时间单位表示经度
longitude-latitude division 经纬线分幅
longitude left 起程经度
longitude method 经度法
longitude of ascending 升交点黄经
longitude of descending node 降交点黄经
longitude of epicenter 震中经度
longitude of node 交点黄经
longitude of periastron 近星点经度
longitude of perigee 近地点黄经
longitude of perihelion 近日点黄经
longitude of the well position 钻孔所在位置的经度
longitude of vertex 顶点经度
longitude plane 子午面
longitude scale 经度比例尺
longitude station 经度站
longitude term 经度项
longitude zone 经度地带
longitudinal 纵行的；纵向的；轴向的
longitudinal aberration 纵向象差
longitudinal acceleration 纵向加速度
longitudinal acceleration sensor 纵向加速传感器
longitudinal action 纵向动作
longitudinal adjustment 纵向调整
longitudinal air flow 纵向气流
longitudinal aisle 纵过道
longitudinal alignment 纵向校准；纵向定位
longitudinal antenna fuze 纵向天线引信
longitudinal apparent resistivity curve 纵向视电阻率曲线
longitudinal arch 纵向拱券；纵向拱
longitudinal arch kiln 纵拱窑
longitudinal arrangement 纵轴排列；纵列式（站线）布置
longitudinal arrangement of standard heavy trucks 标准车队的纵向布置
longitudinal arrangement of station tracks 纵列式到发线
longitudinal arrangement type junction terminal 顺列式枢纽
longitudinal attenuation 传输衰耗

longitudinal attitude 纵向姿态
longitudinal automatic single spindle lathe 纵切单轴自动车床
longitudinal axis 纵轴；纵向轴
longitudinal axis circular current 纵轴环流
longitudinal axis of station 车站纵轴
longitudinal baffle 纵向隔板；纵向挡板；纵向折流板
longitudinal balance 纵平衡
longitudinal ballast resistance 道床纵向阻力
longitudinal bar 纵向沙洲；纵向钢筋；纵沙洲；沙滩；纵筋
longitudinal-bar deposit 纵向沙坝沉积
longitudinal bead bend test 纵缝弯曲试验；焊道纵向弯曲试验
longitudinal bead notched bend test 纵缝缺口弯曲试验
longitudinal bead test 纵向可焊性试验；纵向珠焊试验；纵向熔珠试验
longitudinal beam 纵向梁；纵梁
longitudinal beam section 纵向束流截面
longitudinal beam technique 纵波法（探伤）
longitudinal bearing wall construction 纵墙承重结构
longitudinal bed 长方形堆场
longitudinal bend(ing) 纵向弯曲；总纵弯曲
longitudinal bending moment 纵向弯矩
longitudinal bending strength 纵向弯力；纵向弯曲强度
longitudinal bending stress 纵向受弯应力
longitudinal bent 纵向构架；纵向构架
longitudinal binary fission 纵二分裂
longitudinal blast 纵向爆破
longitudinal blending bed 长方形预均化堆场
longitudinal blowing process 立吹法；垂直喷吹法
longitudinal bond 纵向拉结砌法；纵向多层顺砖砌墙法；纵向多层砌体；纵向多层砌合；纵列砌合法（美国）
longitudinal bonding of rotary kiln lining 回转窑纵向交错砌砖法
longitudinal bottom slope of sedimentation tank 沉淀池底纵坡
longitudinal bow 纵向弯曲；轴向弯曲
longitudinal brace system 纵向连接系统
longitudinal bracing 纵向支撑；纵向竖联
longitudinal bracket 纵梁拉杆
longitudinal bulkhead 纵向舱壁；纵隔堵；纵舱壁【船】
longitudinal (bullfloat) finishing machine 大抹子整平机；长柄镘板整平机；纵向镘板整平机
longitudinal cable 纵向钢丝束（预应力配筋结构）
longitudinal cable force 纵向系缆力（船闸）
longitudinal carrier cable 纵向受力绳；纵向承力索
longitudinal cavity 纵向空腔；（黏土带孔板的）孔洞
longitudinal center joint 纵向中线（焊）接缝；纵向中线
longitudinal center line 轴向中心线；纵向中线；纵向中（心）线
longitudinal center of buoyancy 浮心纵向位置；纵（向）浮心
longitudinal center of buoyancy from after perpendicular 浮心距尾垂纵距离
longitudinal center of buoyancy from forward perpendicular 浮心距首垂纵距离
longitudinal center of gravity 纵重心；重心纵向位置
longitudinal chain mobilization 纵向链流动化
longitudinal channel wave 纵通道波
longitudinal characteristic 纵向运动特性
longitudinal check 纵向检验
longitudinal chromatic aberration 纵向色差
longitudinal church 平面教堂；长形教堂
longitudinal circuit 纵向电路
longitudinal clearance 纵向余隙
longitudinal clinometer 纵向倾斜计；俯仰角指示器
longitudinal coast 纵向海岸；纵式海岸；纵长海岸
longitudinal coastline 纵向海岸线；纵式海岸线
longitudinal coefficient 圆柱形系数；棱形系数
longitudinal cofferdam 纵向围堰
longitudinal commissure 纵连合索
longitudinal comparator 纵动比较仪
longitudinal compensation 纵向补偿
longitudinal component 纵向分量；纵向部分
longitudinal compression 纵向压缩；纵向挤压力
longitudinal compression process 顺纹压缩法

longitudinal compressive force 纵向压力
longitudinal conductivity 纵向传导率
longitudinal connection system 纵向连接系统
longitudinal consequent river 纵顺向河
longitudinal consequent stream 纵顺向河
longitudinal construction 纵向结构
longitudinal construction joint 纵向施工缝；纵向建筑缝；纵向工作缝
longitudinal continuous band illuminant 纵向连续带光源
longitudinal contraction 纵向收缩
longitudinal controller 纵向操纵机构
longitudinal conveyer 纵向传送装置
longitudinal coordinate 纵坐标
longitudinal copying 纵向仿形切削；纵向仿形
longitudinal corner crack 角部纵裂
longitudinal correction 齿端倒角
longitudinal correlation coefficient 纵相关系数
longitudinal counterbalancing 纵向平衡补偿
longitudinal crack 纵（向）裂纹；纵（向）裂缝；纵向开裂
longitudinal cracking 纵向裂痕；纵向裂缝
longitudinal crane girder 纵向起重机梁；起重机纵梁
longitudinal crevasse 纵向裂纹；纵（向）裂缝；纵向冰隙；纵向冰缝；纵劈理；纵裂隙
longitudinal crista 纵行嵴
longitudinal cross-section 纵剖面；纵向截面
longitudinal crown bar 纵顶杆
longitudinal crown stay 纵顶撑
longitudinal culvert 纵向输水廊道；纵向（输水）涵洞
longitudinal current 纵向水流；纵向电流
longitudinal current force 纵向水流力；纵向潮流力
longitudinal current force coefficient 纵向水流力系数
longitudinal current velocity 纵向流速
longitudinal curvature 纵向弯曲
longitudinal curved slipway 弧形滑道【船】
longitudinal cutter 纵向切割机
longitudinal cutting 纵（向）切（削）；纵（向）割
longitudinal dam 纵向围堰；顺坝；导流坝
longitudinal damping 纵向阻尼；纵向振动减振
longitudinal debunching 纵向散聚
longitudinal deceleration 纵向减速
longitudinal deformation 纵向形变；纵向变形
longitudinal degrees of freedom 纵向自由度
longitudinal deviation 纵偏差
longitudinal differential 纵向差动
longitudinal differential protection 纵联差动保护
longitudinal diffusion 纵向扩散
longitudinal diffusion coefficient 纵向扩散系数
longitudinal diffusivity 纵向扩散系数
longitudinal dike 导流堤；纵向围堰；顺堤；顺坝
longitudinal dike at downstream end of island 洲尾顺坝
longitudinal dike at end 洲尾顺坝
longitudinal dike at head 洲头顺坝
longitudinal dilatation 纵向膨胀
longitudinal dip component versus depth diagram 纵向倾角分量-深度图
longitudinal direction 纵向流变；纵向
longitudinal dispersion 纵向弥散；纵向分散
longitudinal dispersion coefficient 纵向弥散系数；纵向分散系数；弥散系数
longitudinal dispersivity 纵向弥散度
longitudinal displacement 纵向位移
longitudinal distance 纵向距离；纵距
longitudinal distance between the centres of two boreholes 孔心纵距
longitudinal distance of photography 摄影纵距
longitudinal distortion 纵向畸变
longitudinal distribution 纵向分布
longitudinal disturbed voltage 纵向扰动电压
longitudinal divergence 纵向发散
longitudinal divide 纵向分水岭；纵分水界
longitudinal divider 纵向分隔板
longitudinal division 纵裂
longitudinal drainage 纵向水系；纵向排水
longitudinal drainage for bridge surface 桥面纵向排水
longitudinal drainage pipe 纵向排水管
longitudinal dredging 纵挖【疏】
longitudinal dredging method 纵挖法

longitudinal ductility 纵向延性
longitudinal dune 纵向沙丘;纵长沙丘
longitudinal earthquake 纵向地震
longitudinal edge 纵向边
longitudinal edge beam 纵向边(缘)梁
longitudinal effect 纵效应
longitudinal efflux 顺向冲采
longitudinal elasticity 纵向弹性
longitudinal electric(al) field 纵向电场
longitudinal electromagnetic force 纵向电磁力
longitudinal electromechanical coupling factor 纵向机电耦合系数
longitudinal electromotive force 纵向电动势
longitudinal elevator 纵向升降机
longitudinal embankment 纵向堤岸;顺堤
longitudinal erosion 纵侵蚀作用
longitudinal erosion valley 纵向侵蚀谷
longitudinal error 纵向误差;经度误差
longitudinal error of traverse 导线纵向误差
longitudinal expansion 纵向膨胀;纵向拉伸
longitudinal expansion joint 纵向伸缩缝;纵向膨胀缝
longitudinal extension 纵向伸缩
longitudinal extent 经线跨度;经度相距
longitudinal facial crack 面部纵裂
longitudinal fall (混凝土溜槽的)纵向坡度;纵向坡降
longitudinal fault 纵向断层;纵断层
longitudinal fault belt 纵向断层带
longitudinal feed 纵向送进;纵向进给
longitudinal feed gear 纵向进刀装置
longitudinal fiber 纵向纤维
longitudinal field 纵向场
longitudinal field lens 纵向场透镜
longitudinal field type 纵向场式
longitudinal fillet weld 纵向(贴)角焊
longitudinal filter 纵向滤波器
longitudinal finisher 纵向修整机
longitudinal fish plate 纵向鱼尾板;纵向接合板
longitudinal fission 纵裂
longitudinal fissure 纵裂
longitudinal flame 纵火焰
longitudinal flame furnace 纵火焰窑
longitudinal flame tank furnace 纵火焰池窑
longitudinal floor culvert 纵向底部廊道
longitudinal floor culvert system 底部纵向廊道充泄水系统
longitudinal floor manifolds 底部纵向多支廊道
longitudinal flow 纵向流动
longitudinal flow reactor 纵向连续反应器
longitudinal fluting 纵向凹槽
longitudinal flux coil 纵向磁通线圈
longitudinal flux magnet 纵向磁通磁体
longitudinal focusing 纵向聚焦
longitudinal fold 纵向折叠;纵褶皱
longitudinal force 纵向力
longitudinal force coefficient 纵向力系数
longitudinal form error 齿向误差
longitudinal fracture 纵向裂纹;纵向裂缝;纵(向)断裂
longitudinal frame 纵向构架;纵肋骨;纵构架
longitudinal framed ship 纵向构架船
longitudinal framed vessel 纵肋架式船
longitudinal framing 纵向构架;纵肋结构系统;纵肋架式;纵结构式;纵骨架式
longitudinal framing system 纵骨架结构系统
longitudinal friction 纵向摩擦力
longitudinal frictional force 纵向摩擦力
longitudinal friction factor of pavement 纵向路面摩擦系数
longitudinal fundamental optic(al) frequency 纵向光学基频率
longitudinal furrows mark 纵向沟脊
longitudinal geologic(al) section 地质纵剖面(图)
longitudinal girder 纵(向)大梁;纵梁;纵桁
longitudinal girder and cross beam connection 纵横梁连结
longitudinal girt 纵支杆;纵向围梁;纵向加劲条梁
longitudinal grade 纵坡度;纵向坡度
longitudinal grade of bridge 桥梁纵坡
longitudinal gradient 纵顺梯度;纵坡(度);纵比降;经向梯度
longitudinal grain 纵向木纹;纵纹

longitudinal grain board 纵向木纹板
longitudinal grinding 纵向磨
longitudinal groove 纵向沟纹;纵向槽
longitudinal growth 纵向生长
longitudinal hair crack 纵向细微裂纹;纵向发裂
longitudinal heating 纵向加热
longitudinal horizontal bracing 纵向水平支撑
longitudinal hump profile 驼峰纵向断面(铁路)
longitudinal hydraulic assistance 纵向水力助推
longitudinal illumination 纵向照明
longitudinal impedance 纵向阻抗
longitudinal imposed load 纵向外加荷载
longitudinal incision 纵切开
longitudinal inclined building berth 纵向倾斜船台
longitudinal inclined shiplift 纵向斜升升船机
longitudinal inclinometer 纵向倾斜计
longitudinal instability 纵向不稳定性
longitudinal interference 纵向干扰
longitudinal interval 纵向间隔;纵向间隔宽度
longitudinal irregularity 纵向不平顺
longitudinal jack rafter 纵向小椽
longitudinal joint 纵向接缝;纵向接头;纵结合;纵缝;砖墙横缝;轴向接头
longitudinal joint efficiency 纵向接头效率
longitudinal joint groove former 纵缝开槽机
longitudinal jointing 纵向接缝;纵(向)接头;纵节理【地】;纵焊缝
longitudinal judder 纵向
longitudinal key 长方键
longitudinal knife 纵切刀
longitudinal launching 纵向下水
longitudinal layer 纵层
longitudinal length extension vibration mode 纵向长度伸缩振动模式
longitudinal length modes 纵向长度振动模式
longitudinal level 纵向水准器;高低水准器
longitudinal level(l)ing 纵向调平;纵断面水准测量
longitudinal level of rail 轨道前后高低
longitudinal lie 纵产式
longitudinal load(ing) 纵(向)荷载;纵向负载;纵向负荷
longitudinal locking 纵式锁闭
longitudinal locking bar 纵向锁条
longitudinal log haul-up 纵向拉木机
longitudinally acting force of concrete 混凝土纵向作用力
longitudinally corrugate 纵向波状的
longitudinally frame 纵骨架
longitudinally framed hull 纵构架式船体
longitudinally framed vessel 纵骨架式船
longitudinally mounted engine 纵向布置的发动机
longitudinally split rim 对开式轮辋
longitudinally stable 纵向稳定的
longitudinally swing migration 纵向摆动迁移
longitudinally traversable 纵向移动的
longitudinal-made delay line 纵向延迟线
longitudinal magnetic dipole 纵向磁偶极
longitudinal magnetic field 纵向磁场
longitudinal magnetic flux 纵向磁通
longitudinal magnetization 纵向磁化
longitudinal magnetomotive force 纵向磁通势
longitudinal magnetoresistance 纵向磁致电阻
longitudinal magnetostriction 纵向磁致伸缩
longitudinal magnification 纵性放大率;纵向放大(率);轴向放大率
longitudinal manifold 纵向集合管(道);纵向汇管
longitudinal mass 纵向质量
longitudinal member 纵向构件;纵向杆件;纵构件
longitudinal metacenter 纵稳定(中)心
longitudinal metacentric height 纵稳心高度
longitudinal metacentric radius 纵向稳心半径
longitudinal microscopic dispersion 纵向微观弥散
longitudinal milling 纵向铣削
longitudinal mixing 纵向混合;纵顺混合
longitudinal mode 纵向振型
longitudinal mode delay line 纵向振荡模延迟线
longitudinal mode of motion 纵向运动
longitudinal modulus 纵向模数;纵向模量
longitudinal moment 纵向力矩
longitudinal moment of inertia 纵摇纵向惯性矩;纵向惯矩
longitudinal momentum 纵向动量
longitudinal motion 纵向运动;纵向移动;纵向驱动
longitudinal movement 纵向运动;纵向移动;轴向移动

longitudinal movement of air 纵向空气运动
longitudinal multibucket excavator 纵向挖掘多斗挖掘机
longitudinal number 纵向数;纵向排号
longitudinal numeral 纵向数;纵向排号
longitudinal of geometric(al) model 几何模型航向倾斜
longitudinal order 纵次序
longitudinal oscillation 纵向振荡;纵向摆动
longitudinal overlap 纵向重叠;航向重叠
longitudinal overturned car 纵向
longitudinal parallax 纵视差
longitudinal parallax screw 纵向视差螺旋;纵视差螺旋
longitudinal parity 纵向奇偶性;纵向奇偶位
longitudinal parity check 纵向奇偶性检验;纵向奇偶检验
longitudinal parking 平行停靠;平行停车;顺列式停车
longitudinal partition 纵隔板
longitudinal pass partition plate 纵向隔板
longitudinal pavement line 纵向路面线
longitudinal periodicity 纵向周期性
longitudinal permeability 纵向透水性
longitudinal phase 纵相
longitudinal phase curve 纵向相位曲线
longitudinal phase space 纵向相空间
longitudinal photographic(al) distance 摄影纵距
longitudinal photomagnetoelectric(al) effect 纵向光磁电效应
longitudinal piezoelectric(al) effect 纵向压电效应
longitudinal pipe 纵顺管
longitudinal pitch 纵向中心距;纵向节距;纵向间距
longitudinal plan 纵剖面图;前视图
longitudinal plane 纵向平面
longitudinal plane of symmetry 纵向中心铅垂面
longitudinal plane wave 纵向平面波
longitudinal play 纵向游隙
longitudinal polarization 纵向极化
longitudinal polarization wave 纵偏振波
longitudinal polarized beam 纵向极化束
longitudinal pressure 纵(向)压力
longitudinal prestress(ing) 纵向预(加)应力
longitudinal profile 纵向轮廓;路线中心线坡度图;纵(向)剖面;纵剖面图;纵断面(图)(尤指道路)
longitudinal profile of depth 深度纵剖面
longitudinal profile of double incline 双峰(地形)的纵断面
longitudinal profile of river 河流纵剖面;河道纵断面(图)
longitudinal profile of riverbed 河底纵断面;河床纵剖面;河床纵断面
longitudinal profile of sewers 管道纵断面
longitudinal profile of terrace 阶地纵剖面
longitudinal projected area 侧投影面积
longitudinal propagation 纵向传播
longitudinal pulling method 纵向拖拉法
longitudinal quadrupole 纵向四极
longitudinal railway slip 纵向船排滑道
longitudinal rapidity 纵向快度
longitudinal recording 纵向记录
longitudinal redundance 纵向冗余
longitudinal redundancy 纵向冗余
longitudinal redundancy check 纵向冗余检验;纵向多余数位检验
longitudinal redundancy check character 纵向冗余检验符号
longitudinal refractive index 纵向折射率
longitudinal register 纵向记录
longitudinal registration 纵向记录
longitudinal reinforcement 纵加强筋;纵列钢筋;纵向配筋;纵向钢筋
longitudinal reinforcement of the collecting net 收集网横向钢筋
longitudinal reinforcing 纵向增强
longitudinal relaxation 纵向弛豫
longitudinal removal 纵向迁移
longitudinal repeat 纵向重复
longitudinal repulsion 纵向排斥
longitudinal resistance 纵向阻力
longitudinal resistance seam welding 纵向电阻缝焊
longitudinal resolution 纵向分辨(率)

longitudinal resolving power 纵向分辨力
longitudinal resonance 纵向共振
longitudinal resonator transducer 纵向振子换能器
longitudinal response characteristic 纵向运动频率特性
longitudinal restoring moment 纵向复原力矩
longitudinal rib 纵肋;纵向肋条
longitudinal rib of vault 穹隆纵肋
longitudinal ridge 纵向脊
longitudinal ridge-rib 纵向脊肋
longitudinal ridge roofed shed 锥形尖顶储料棚
longitudinal righting moment 纵向恢复力矩
longitudinal rigidity 纵向刚度
longitudinal river 纵向河
longitudinal river slope 纵比降(河流);河道纵比降
longitudinal road marking 纵向标线[道]
longitudinal rod 纵向钢筋;纵向杆
longitudinal rolling 纵轧
longitudinal roof bar 纵顶杆
longitudinal sand-ridge 纵向沙垄
longitudinal scale lofting 纵向缩尺放样
longitudinal scan 纵向扫描
longitudinal scoring machine 纵切机
longitudinal screen tension 纵向筛网张力
longitudinal seal 纵向密封;纵向封垫;纵向密封
longitudinal seam 纵向缝;纵缝;纵(边)接缝
longitudinal seam welding 纵向滚(线)焊接;纵向缝焊;纵缝焊接
longitudinal seat 纵向长座椅
longitudinal section 纵剖面;纵切面;纵截面;纵断面(图)
longitudinal section area 纵截面积
longitudinal section measurement 纵断面测量
longitudinal section of grotto 溶洞纵剖面图
longitudinal section of groundwater body 地下水体纵截面
longitudinal section of pipe trench 管沟纵剖面图
longitudinal section of riverbed 河底纵断面;河床纵断面
longitudinal section profile 纵断面图
longitudinal segmentation 纵裂
longitudinal seiche 纵向假潮;纵向湖震
longitudinal separation 纵向距离;纵向间隔;前后距离
longitudinal sequence 纵向焊接顺序;纵向次序
longitudinal shaft 纵轴式干燥窑;纵轴
longitudinal shape 纵向形状
longitudinal shear 纵向剪切;纵向剪力
longitudinal shear crack 纵向剪切裂隙
longitudinal shearing flow 纵向剪切流
longitudinal shearing stress 纵向剪切应力
longitudinal shearing test 纵向剪切试验
longitudinal shift 纵移;纵向移动
longitudinal shifting of centre of gravity of goods 货物重心的纵向位移
longitudinal shift of focus 焦点纵向移位
longitudinal ship incline 纵向斜面升船机
longitudinal shrinkage 纵向收缩
longitudinal shrinkage crack 纵向收缩裂纹
longitudinal shrinkage rate 纵向线缩率
longitudinal shrinking 纵向收缩
longitudinal sill 纵槛
longitudinal sleeper 纵(向)轨枕
longitudinal sleeper track 纵向轨枕线路
longitudinal slide 纵向滑板
longitudinal sliding motion 纵向滑动
longitudinal slip 纵向滑道
longitudinal slipway 纵向修船滑道;纵向滑道
longitudinal slipway and wedged chassis 纵向斜架滑道
longitudinal slipway with rail tracks 纵向船排滑道
longitudinal slope 纵向坡度;纵坡(度);纵比降
longitudinal slope line 纵向坡线
longitudinal slope of dike top (crest) 坝顶纵坡
longitudinal slope of dock floor 坞底板纵坡
longitudinal sloping hearth furnace 斜底式炉
longitudinal slot 纵向槽
longitudinal sound velocity 纵声速
longitudinal space charge effect 纵向空间电荷效应
longitudinal spacing 纵向间距;纵距
longitudinal spheric(al) aberration 纵向球(面像)差
longitudinal split 纵裂
longitudinal spring 纵簧;纵弹簧
longitudinal stability 纵向稳定(性);纵稳定性
longitudinal stability criterion 纵向安全性准则
longitudinal stability test 纵向稳定性试验
longitudinal stabilizer 纵向稳定器
longitudinal stabilizing nozzle 纵向稳定喷口
longitudinal stack 纵向叠加
longitudinal static stability 纵向静态稳定性;纵向静力稳定性
longitudinal stay 纵向拉线;锅炉纵向撑条
longitudinal stiffener 纵向加劲杆
longitudinal stiffness 纵向劲度;纵向刚度
longitudinal stop 纵向止动器
longitudinal strain 线应变;纵向应变;纵向变形
longitudinal strain of given direction 给定方位的长度应变
longitudinal stream 纵向流;纵向河;后会河
longitudinal strength 纵向强度
longitudinal strength member 纵向强力构件
longitudinal stress 纵向应力
longitudinal stress in flange hub 法兰颈部轴向应力
longitudinal stretching force 纵向延伸力;纵向拉力;纵向张力
longitudinal striation 纵纹
longitudinal strike-off (blade) 纵向整平板
longitudinal stringer 纵材
longitudinal structure 纵向(受力)结构
longitudinal strut 纵支杆;纵向(支)撑杆;纵联杆
longitudinal study 纵向研究;纵向调查(研究)
longitudinal surface 纵断面
longitudinal surface velocity 纵向表面流速
longitudinal survey 纵向面调查
longitudinal suspension 链型悬挂
longitudinal system 纵肋结构;纵构架式
longitudinal system of framing 纵骨架式
longitudinal table travel 纵向工作台移动
longitudinal tangent 纵向切线
longitudinal tectonic system 纵向构造体系
longitudinal tendon 纵向预应力腱;纵向预应力钢筋束
longitudinal tensile force 纵向拉力
longitudinal tensile stress 纵向拉应力
longitudinal tension stress 纵向拉应力
longitudinal testing 纵向试验
longitudinal thermal conduction 纵向热传导
longitudinal thickness modes 纵向厚度模式
longitudinal thrust 纵向推力
longitudinal thrustfault 纵向冲断层
longitudinal tie 纵(向)系杆;纵向联(结)杆;纵向轨枕;纵拉杆
longitudinal tierod 纵向拉杆;长向拉条
longitudinal tilt 纵向倾斜;航向倾斜;航向倾角
longitudinal tilting wheel 航向倾斜手轮
longitudinal timber 纵向轨枕;木纵梁
longitudinal tool rest 纵向刀架
longitudinal track 纵向磁迹
longitudinal transmission check 纵向传输校验
longitudinal transport 纵向迁移
longitudinal travel 纵向位移
longitudinal traverse 纵向移动
longitudinal traverse gear 纵向自走机构
longitudinal trim 纵倾[船]
longitudinal trimmer 纵向割边机
longitudinal truss 纵向桁架
longitudinal tube 纵管
longitudinal tube spacing 纵向管距
longitudinal tubule 纵小管
longitudinal turbulence 纵向漩涡;纵向涡流
longitudinal type 纵型
longitudinal type district station 纵列式区段站
longitudinal type of shop 纵列式车间
longitudinal type passing station 纵列式会让站
longitudinal underdrain 纵向暗沟
longitudinal undulator 纵向波荡器
longitudinal valley 纵向河谷;纵谷
longitudinal vegetational zonation 植被经度地带性
longitudinal vein 纵脉
longitudinal velocity 纵向速度;纵波速度
longitudinal ventilation 纵向(式)通风;轴向通风
longitudinal ventilation system 纵向通风系统
longitudinal ventilation with jet-blower 射流通风
longitudinal vertical bracing 纵向垂直支撑
longitudinal vertical plane 纵直(纵)平面
longitudinal vibration 纵向振动
longitudinal vibration characteristics 纵向振动特性
longitudinal vibration transducer 纵振换能器
longitudinal vibrator 纵向振子;纵向振动器
longitudinal view 纵视图
longitudinal viscosity 纵向黏滞性
longitudinal void 纵向空穴
longitudinal voltage 纵向分裂电压
longitudinal wall 纵墙;纵挡板;纵壁;导墙
longitudinal wall hanger bearing 墙托架轴承
longitudinal wall load bearing 纵墙承重
longitudinal warping 纵向翘曲;纵向变形
longitudinal warping joint 纵向伸缩缝;纵向温度缝
longitudinal wave 压缩波;纵波;地震纵波;P波
longitudinal wave motion 纵波运动
longitudinal web stiffener 纵向梁腹加劲杆;纵向腹板加劲肋
longitudinal weld 纵(焊)缝;纵(向)焊
longitudinal welding 纵(向)焊接;纵缝焊接
longitudinal welding machine 纵向焊接机
longitudinal welding seam 纵焊缝
longitudinal weld seam 纵向焊缝
longitudinal wind force 纵向风力
longitudinal wind force coefficient 纵风力系数
longitudinal winding 纵向缠绕
longitudinal winding loop tenacity 纵向缠绕环扣强度;环扣强度
longitudinal wire of the cloth 金属丝布的经线(筛网)
longitudinal zoning 纵向分带;经向分带(性);经度分带
long jawed 绳索松股
long-jump pit 跳远沙坑
long keeping 耐储[贮]藏的
long lag phosphor 长余辉磷光体;长余辉发光物质
long laid rope 软搓绳
long lasting 持久的;耐久的;长效的
long lath 长条板;石膏板条
long lay 软搓
long lead 分立式引线
long lead item 合同前已订货货物
long lead screw 长丝杠
long leaf 长叶
longleaf pine 长生树;长叶松
longleaf pine oil 长叶松油
longleaf tobacco 长叶烟草
long leg chamber cross 长腿室四通
long leg chamber tee 长腿室三通
long leg frame section 长腿框架构件
long legged 续航力大的舰船
long length additional 超长附加费
long length logging 长材集运
long lengths charges 超长费
long life 长寿命(的);长使用寿命;长命
long life anti-fouling coating 长效型防污漆
long life grease 长寿润脂;长使用期润滑脂
long life isotope 长寿同位素
long life lamp 长寿灯泡
long life machine 耐用机械
long life operating data 长期运转数据
long life parts 建筑物使用寿命最大的部分
long life top coat 耐久面漆;长效面漆
long life tube 长寿命管
long lift build 长程式成型
long line 长线法(预应力);长线;长途的
long-linearity control 远距线性调整器
long line azimuth survey 长边方位角测量
long line concrete 长线张拉(预应力钢筋)混凝土
long line current 长线电流
long line drawing frame 长亚麻并条机
long line effect 长线效应
long line fishing 延绳钩
long line frequency control 长线频率控制
long line method 长线法
long line method of pretensioning 长线先张法
long line platform mo(u)lding 长线台座成型
long line prestressed concrete 长线法预应力混凝土
long line prestressed concrete process 长线预应力混凝土张拉法
long line pretensioning 长线先张法(预应力)
long line pretensioning banks 长线预应力张拉台座
long line process 长线台座工艺
long line production process 长线法;长线预应力

张拉法
long liner 拖钓船；长线钓鱼船
long lines 长距离回线
long lines engineering 长线路通信工程
long line system （预应力混凝土的）长线系统；长线法
long line theory 长线理论
long line traveling-wave antenna 长线行波天线
long lining 延绳钩法
long link chain 长链条；长环节链
long lip strike 长锁舌碰扣
long lived 长寿命的；长使用期的；长存的
long lived assets 长期资本
long lived coating 耐久涂层
long lived facility 耐用设备
long lived iso tope 长寿命同位素
long lived toxic substance 长命有毒物质
long lived tracer 长效示踪剂
long lived waste 长寿命废物
long load 长件货物
long lobed pointed trefoil 长瓣三尖叶饰
long locomotive runs 长交路
long loop detector 长环框式检测器
long low swell 长轻涌；二级涌浪
long mesh fabric 长眼钢丝网
long mesh wire cloth 长网眼金属丝布
long mesh woven wire screen 长网眼金属丝织筛网
long moderate swell 长中涌
Longmyndian Series 隆敏德统【地】
longn 长大梁
long narrow table 条案
long neck 长颈烧瓶
long necked flask 长颈烧瓶
long neck funnel 长颈漏斗
long neck pliers 长颈钳
long needle-like crystal 长针状晶体
long nipple 带螺母外丝接头；长螺纹接头；长管接头；双外螺丝；带螺帽外丝接头
long nosed pliers 长尖嘴钳
long nose-like 长鼻状
long nose locking wrench 长尖嘴大力钳
long nose pliers 长嘴钳；长头钳
long nozzle heel nailing machine 长尖嘴钉跟机
long nozzle oil can 长嘴油壶
long number 多位数；长数
long nut tap 长柄螺母丝锥
long offset addressing 长偏移量寻址
long oil 长油度
long oil alkyd 长油度醇酸树脂；长油醇酸树脂（含油量60%）
long oil alkyd resin 长油度醇酸树脂
long oil gloss paint 长油度有光涂料
long oil skin 长油布雨衣
long oil type 长油性
long oil varnish 油性清漆；长油（性）清漆；长油度清漆
long oval 长椭圆形
long over-hung shaft 长外伸轴
long pair twist quad 长线对搓合四芯线组；长对四芯绞线组
long pass filter 长通滤波器
long past due 早已过期；早已到期
long path air spectrum 长光程空气光谱
long path infrared gas detector 长距离红外气体检测器
long path infrared system 长距离红外系统
long path interferometer 长光程干涉仪
long path monitoring 长光程监控
long path multiple reflection 全程多次反射（地震勘探）
long peak 长链抛锚
long peck line 长划虚线
long pennant 长三角旗
long period 长周期；长期（的）
long period array 长周期台阵
long period average discharge 多年平均流量
long period average flow 长期平均流量
long period base line distortion 长周期基线畸变
long period changing temperature 长周期变化温度
long period constituent 长（周）期分潮
long period containment 长期封闭
long period creep test 长期蠕变试验

long period dynamic(al) stability 长周期运动稳定性
long period fluctuation 长周期波动；长期波动
long period force 朔望生潮力；长周期引力潮；长周期生潮力
long period forecast(ing) 长期预报
long period gravity wave 长周期重力波
long period ground motion 长周期地面运动
long periodicity 长周期性
long period instability 长周期不稳定性
long period microtremor 长周期微力振动；长周期脉动
long period oscillation 长周期振荡
long period oscillation damping 长周期振动阻尼
long period pendulum 长周期摆
long period perturbation 长周期摄动
long period rain recorder 长期雨量记录仪
long period runoff relation 长期径流关系
long period seiche 长周期假潮
long period seismogram 长周期地震图
long period seismograph 长周期地震仪；长周期地层仪
long periods of wet soil 土壤长期潮湿
long period storage reservoir 长期蓄水水库；长期调节水库；多年调节水库
long period structure 长周期结构
long period superlattic 长周期超点阵
long period supply price 长期供给价格
long period surge 长周期（涌）波
long period swell 长周期涌
long period term 长周期项
long period test 长周期试验
long period tide 长周期潮（汐）
long period tide-generating force 长周期生潮力
long period transport plan 运输计划
long period variation 长周期变量；长期变化
long period variation in earth's temperature 地温长期变化
long period vibration 长周期振动
long period wave 长周期波
long persistence 长余辉
long persistence oscilloscope 长余辉示波器
long persistence screen 长余辉荧光屏
long pile 长桩
long-pillar working 留大煤柱开采法
long pin 长销
long pipe 长管
long pitch corrugated sheet 大波纹板
long pitch winding 长节距线组
long pitman mower 长连杆式割草机
long plane 长刨
long play 密纹唱片
long playing rocket 长期运行的卫星
long point rail 长心轨
long poop 长尾楼
long position 超买
long price 高价；昂贵；昂价
long primer 长点火管
long prismatic 长棱形的
long production run 大批生产；大量生产
long profile of river 全河剖面
long progressive wave 长前进浪
long pruning 长剪
long pug mill 单臂捣泥机；单臂揉捏机
long pulse 长脉冲
long pulse dye laser 长脉冲染料激光器
long pulse laser 长脉冲激光器
long pulse radar 长脉冲雷达
long purse 富裕；富有
long rachis segments length 长穗轴节
long radius 长半径；大半径
long radius 45° bend 45°大半径弯头
long radius bend 大半径弯管；大半径弯曲
long radius curve 长半径曲线；大半径曲线
long radius elbow 大弯曲半径弯管；大半径弯头；长半径弯头；月弯
long radius grating 长半径光栅
long radius system 长半径系统
long rail 长轨
long rail string 长轨条
long range 远程（的）；远期的；长程的；长射程；长期；长距离（的）
long-range accuracy 远程精度
long-range aids 远程助航设备

long-range air-navigation chart 远程导航图
long-range bias 大范围偏压
long-range budget 长期预算
long-range cash planning 长期资金计划
long-range chart 远程航图
long-range conception 长期设想
long-range control 远程控制
long-range data 远程数据
long-range data terminal 远程数据终端
long-range design 长远设计
long-range development 长期发展计划
long-range development plan 远景发展规划；长期开发计划
long-range disorder 长程无序
long-range elasticity 广范围弹性；高弹性
long-range excavator conveyer[conveyor] 远程挖土运输机
long-range force 长程力
long-range forecast(ing) 长期预报；大范围预报
long-range fossil 长延限化石
long-range indicator element 远程指示元素
long-range input 远程输入
long-range interaction 远程相互作用
long-range laser radar 远程激光雷达
long-range line network 远景线网
long-range meteorologic(al) forecast(ing) 长期气象预报
long-range navigation[loran] 远程导航；远程航行；罗兰（一种远距离导航系统）
long-range navigation aids 远程导航设备；罗兰导航系统
long-range navigation computer 远航计算机
long-range navigation system 远距离导航系统；脉冲时差双曲线导航系统；远程导航系统；罗兰
long-range navigation zone 远程导航区
long-range operation 长期作业
long-range order 长程有序；长程序
long-range perspective 远景
long-range photograph 远摄像片
long-range plan(ning) 远景规划；长远计划；长远规划；长期（经营）计划；长期规划
long-range plan of coal industry 煤炭工业远景规划
long-range position determination system 远程定位系统
long-range prediction 远期预测；长期预测；长期预报
long-range product 粒度不均匀的产品
long-range profit planning 长期获利计划
long-range prognosis 长期预测
long-range program(me) 长期规划
long-range radar 远程雷达
long-range radio beacon 远程无线电信标
long-range radio location land station 远程定位岸台
long-range scan 长距扫描
long-range ship 长航程船
long-range shovel 长臂电铲
long-range sonar 远程声呐
long-range sprinkler 远射程喷灌机
long-range stress field 长程应力场
long-range sweep 长距扫描
long-range telemetering 远程遥测
long-range telemetering buoy 远程遥测浮标
long-range tracking mode 远距跟踪方式
long-range transmission 远距离输送；远距离输电；远距离传送；远程输送
long-range transport 长距离输送
long-range transport of pollutant 污染物远程输送
long-range trawler 远程拖网船
long-range version 远程型
long-range weather forecast(ing) 长期天气预报
long-reach 伸距长的
long reach plug 长型火花塞
long rear-pivoting boom 后枢接式长转臂
long rectangular wave generator 长矩形波发生器
long relative stability stage 长期稳定阶段
long residual action 长效性；长期残留作用
long residuum 久沸残渣
long response 慢反应
long response time 慢反应时间
long reverberation time 长时间混响
long reversed S-shape 长反S形
long rod 长水准尺

long rod insulator 棒形绝缘子
long rolling sea 涌浪;长涌;长浪;长滚浪;长波
long rotational grassland 长期轮作草地
long round nose pliers 圆尖咀钳;长圆嘴钳
long routing 长交路
long rubber seal 长橡胶封
long run 长时间操作;长期运转;长期使用
long run average 平均状况
long run average cost 远航程平均价格;长期平均成本
long run cost variation 长期成本变化
long run frequency 长期频率
long run marginal cost 长航程限界价格
long running 连续运行;连续旋转;长期运行;长期旋转
long run production cost schedule 长期生产费用表
long run test 连续(式)试验;长期试验
long saw 长锯
long scan 慢扫描;长扫描
Long's coefficient 朗氏系数
long scope 长链抛锚
long scraper 长漆铲
long screw 长螺钉;管道连接螺钉(用于水暖卫生工程);长螺纹(管)
long screw nipple 长螺纹套筒
long sea outfall 污水远海排污口
long season crop 长期作物
long section line 长剖面测线
long sections 超长条
long service life 长使用寿命
long shaft crusher 长轴(圆锥)破碎机
long shaft pendulum tool 长柄砂舂
long shank auger 长柄木工钻
long shank carborundum needle 长柄砂针
long shank machine tap 长柄机用丝锥
long shell 长薄壳;长壳
long shift 长移位
long-shift control 远距位移调整器
long shift suspended spindle gyratory crusher 长悬轴式旋回破碎机
long shoot 长茎
longshore 沿海岸;在海岸工作的;海岸边的
longshore bar 沿岸沙洲;沿岸沙埂;沿岸沙堤;沿岸沙坝;海岸砂坝;海岸沙洲
longshore current 沿岸(海)流;近岸(海)流;顺岸(海)流
longshore drift 沿岸漂砂;沿岸物质流;沿岸漂移;沿岸漂流;沿岸沉积物;海岸边流;远岸漂移
longshore fishery 沿岸渔业
longshoreman 沿岸工作居民;装卸工(人);港口工人;码头装卸工(人);码头工人
longshoreman's gang 装卸组
longshoremen's shelter 装卸工人休息棚
longshore movement of material 泥沙纵向运动
longshore transport 沿岸运输;沿岸输送;沿岸漂沙
longshore trough 沿岸洼槽;沿岸海沟;沿岸沟谷;沿岸凹槽
longshore union 码头工会
longshore wind 海岸风
longshore work 港口作业;码头装卸工作
longshoring work 码头装卸工作
long short day plant 长短日植物
long shot 远景
long shunt 长分流器
long shunt compound generator 长并激复励发电机
long shunt compound machine 长并复励电机
long shunt compound winding 长分流复励绕组
long shunt winding 长分路绕组;长并联绕组;长并励绕组
long side round edge opposites (釉面砖的)对长圆边条砖
long side round edge tile 长边圆角釉面砖
long-sighted 远视的
long silk waste 长废丝
long skip 长跳越
long ski system 长撬方式
long slab 长板
long slag 长渣
long slope 长坡
long slot burner 长缝式燃烧器
long slot coupler 长缝耦合器
long spaced count rate meter 长间距计数率计

long spaced detector 长间距探测器
long spaced neutron log 长源距中子测井
long spaced sonde 长电极系
long spaced sonic logger 长源距声波测井仪
long spacing 长间距
long spacing densilog 长源距密度测井
long spacing densilog curve 长源距密度测井曲线
long spacing density log 长源距密度测井
long spacing neutron-neutron log curve 长源距中子—中子测井曲线
long span 长距堆放(原木的);大跨度;长跨(度)
long-span arch 大跨度拱
long-span bridge 大跨度桥(梁);长跨(度)桥(梁)
long-span structure 大跨度结构
long-span land smoother 长跨距平地机
long-span line 长杆距线路
long-span precast gypsum slab 预制大跨石膏板
long-span rib 大跨度肋
long-span roof 大跨度屋顶
long-span shell 大跨度薄壳
long-span steel framing 大跨钢构架
long-span structure 大跨度结构
long-span trussed beam 大跨度桁架梁
long-span tunnel 大跨度隧道
long spark 长放电
long splice (麻绳的)长绞接头;(麻绳的)长捻接;长插接(绳结)
long split 长劈裂
long spout oil can 长嘴油壶
long S-shape 长S形
long stage airline 长途直飞航线
long stalked 长茎的
long stalked pear 长柄梨
long standing 可长期存在;经久的;长期存在的
long standing business dealings 长期经营
long staple 长纤维
long stapled 长纤维的
long-stator cable 长定子用电缆
long-stator linear motor 轨道直线电机;长定子直线电机
long-stator linear motor with iron core 长定子带铁芯直线电机
long-stator section 长定子分段
long stay 长链抛锚
long steel product 线状钢材
long steel tape 钢卷尺
long steel wire saw 长钢丝锯
long steep grade 长大坡长;长大坡道【铁】
long steep up-gradient 长大上坡道;长陡上坡道
long stem cutting 长枝插
long stem funnel 长柄漏斗
long stem insulator 长形绝缘子
long stemmed nozzle 长杆喷嘴
long stick 长码尺
long straight reach 长直河段
long stretch of straight line 长大直线地段
long string 采油套管
long stringed gutter tile 连锁式屋面槽瓦
long strip footing 长条形基脚;条形底脚;长条形基础
long strip foundation 长条形基础
long stroke 长冲程
long-stroke brake 长程制动器
long-stroke distribution 长冲程分配
long-stroke friction press 长冲程摩擦压(砖)机;大冲程摩擦压(砖)机
long-stroke jack 长行程千斤顶;长冲程千斤顶
long-stroke pump 长行程泵;长冲程泵
long-stroke ram 长行程液压顶杆;长冲程液压顶杆;长行程液压活塞;长冲程液压活塞
long-stroke riveting hammer 长行程铆钉锤;长冲程铆钉锤
long-stroke shock absorber 长冲程吸振器
long-stroke steam engine 长冲程蒸汽机
long-stroke strut 长冲程减振支柱
long stud link 长型末端链环
long surf 长激浪
long surf clam 蛤蜊
long suspension insulator 长悬式绝缘子
long sustained load 持久荷载
long sustained load test 持久荷载试验
long-sweep ell 远拂肘管;大曲率半径弯管
long-sweeping curve 大半径曲线
long swell 长涌浪

long swell wave 长涌浪
long symbol 长符号
long tackle 长绞辘
long tackle block 提琴式滑车;长绳滑车组
long tail block 姐妹滑车
long tailed anchovy 凤尾鱼
long tailed engine 顿钻发动机
long tailed pair amplifier 差动放大器
long tailed pheasant 长尾雉
long tangent 长夹直线
long taper 长锥体
long taper die tap 板牙丝锥
long tapering 长尖削的
long tee hinge 长T形铰链
long telescope 倒相型远镜;长望远镜
long-term 长期(的);远期的
long-term adaptation 长期适应
long-term aging 长期老化
long-term agreement 长期协议
long-term allowable bearing capacity 长期允许承载力
long-term allowable stress 长期许用应力
long-term and expanded program(me) of oceanic exploration and research 海洋勘探研究长期扩大方案
long-term aseismic slip 长期无振滑动;长期抗震滑动
long-term assets 固定资产
long-term audit 长期借款审计
long-term average 长期平均
long-term average annual energy 多年平均发电量
long-term average annual power output 多年平均发电量
long-term average depth of rainfall 长期平均降雨量
long-term average discharge 多年平均流量;长期平均流量
long-term balance 长期平衡
long-term balance ratio 长期平衡系数
long-term behavio(u)r 长期(运转)性状;长期(运转)性能
long-term benefit 长远利益;长期受益
long-term bill 长期汇票
long-term biochemical oxygen demand 长期生化需氧量
long-term blanket contact 长期一揽子合同
long-term budget 长期预算
long-term capital 长期资本资产销售收益;长期资本
long-term care 长期管理
long-term change 长期变化
long-term channel trend 长期河槽趋势
long-term climatic cycle 长期气候循环
long-term climatic trend 长期气候趋势
long-term cohesion of frozen soil 冻土的长期内聚力
long-term constituent 长期分潮
long-term construction contract accounting 长期建筑合同会计处理
long-term construction contracts 长期承包工程;长期承包合同
long-term containment 长期封闭
long-term contracts 长期合同
long-term cooperation 长期合作
long-term corrosion resistance coating 长效防腐蚀涂料
long-term corrosion test 长期腐蚀试验
long-term cost 固定开支
long-term credit 长期信贷;长期贷款
Long-Term Credit Bank of Japan 日本长期信贷银行
long-term creep 长期蠕变
long-term creep test 长期徐变试验;长期蠕变试验
long-term crop 长期作物
long-term cyclic(al) load 长周期性荷载
long-term decision making 长期决策
long-term deflection 长期挠度;长期变形【测】
long-term degradation 长期退级
long-term deposit 长期存款
long-term development 远景发展;长期发展
long-term development plan 远景发展规划;长期发展计划
long-term drift 长期漂移
long-term dues accounting audit 长期应付款会计核算的审计

long-term dues audit 长期应付款的审计
long-term dues authenticity and legality audit 长期应付款真实性
long-term dues valuation audit 长期应付款计价的审计
long-term earthquake prediction 长期地震预报
long-term economy 长期节约
long-term effect 长时作用;长时后效;长期影响;长期效应;长期时效
long-term effects of pollutant 污染物质远期效应;污染物的长期效应;化学品的长期影响
long-term environmental effect 长期环境影响;长期接触;长期感受;长期暴露;远期环境影响;远期环境效应;远期环境终端
long-term environmental impact 远期环境终端
long-term equilibrium 长期平衡
long-term evolution 长期演变
long-term evolvement 长期演变
long-term exposure 长期暴露
long-term fading 慢衰落
long-term film 长寿胶片
long-term financing 长期融资;长期贷款;长期资金融通;长期提供资金
long-term fire danger 长期火险性
long-term fix 长期固定
long-term fix area 长期固定区
long-term flood predication 长期洪水预测
long-term flow prediction 长期预测流量
long-term forecast(ing) 长期预报
long-term forward contract 远期合同
long-term frequency drift 长期频率漂移
long-term frequency stability 长期频率稳定度
long-term ga(u)ging 长期测流
long-term ga(u)ging of flow 长期流量测量
long-term genetic adaptation 长期发展适应
long-term global trend 全球长期趋势
long-term goal 远期目标;长期目标
long-term goods transport plan 长远货物运输计划
long-term government bonds 长期国债
long-term hydrostatic(al) strength 长期静水压强度
long-term income 长期收入
long-term insurance 长期保险
long-term intake 长期摄入
long-term interest-free loan 长期无息贷款
long-term interest rate 长期利率
long-term interest 长远利益
long-term investment 长期投资
long-term investment audit 长期投资审计
long-term investment decision-making 长期投资决策
long-term investment projects 长期投资项目
long-term irradiation 长期辐照
long-term lease 长期租赁
long-term lease contract 长期使用契约
long-term load action 长期荷载作用
long-term load(ing) 长期荷载;长期负载;长期负荷
long-term loading test 长期荷载试验
long-term loan 长期贷款
long-term maintenance 长期维修
long-term material plan 长期物资计划
long-term memory 长期记忆;长期存储器
long-term monitoring 长期监测
long-term non-toxicity test 长期无毒试验
long-term objective 长期目标
long-term observation 长期观测
long-term observation of engineering geology 工程地质长期观测
long-term observation record 长期观测记录;长系列观测资料
long-term operation 长期作业
long-term outdoor exposure 长期户外曝晒
long-term outline plan 远景规划纲要;长期规划纲要
long-term passenger transportation plan 旅客运输长远计划
long-term performance 长期性工作性能
long-term perspective 长远观点
long-term plan 长期计划
long-term planning 远期规划;长远发展;长远计划;长远规划;长期规划
long-term poisoning 长期中毒
long-term pollution 长期污染
long-term potential 长期潜力;发展前景

long-term prediction 长期预测;长期预报;远期预测
long-term prediction of concentration of contaminants 长期污染浓度预报
long-term prepayment 长期预付款
long-term productivity 长期生产能力
long-term program(me) 长期规划
long-term project 长期方案;长期计划
long-term quality 长期特性
long-term rate of interest 长期利率
long-term record 长期观测记录
long-term reference 长周期运动分量
long-term regulation 多年调节
long-term report 详细查账报告
long-term reproducibility 长期再现性
long-term research 长远研究
long-term residents 常住居民
long-term ring bending strain 长期弯环应变
long-term rotation 长轮伐期
long-term running test 长期运转试验
long-term sampling 长时间采样
long-term saving 长期存款
long-term sedimentation 长期沉积
long-term sediment yield 长期淤积量;长期沉积量;长期产沙量
long-term settlement 长期沉降
long-term shear strength 长期抗剪强度
long-term solvency 长期偿付能力
long-term stability 长时间稳定性;长期稳定性;长期稳定度
long-term station 长期观测站
long-term stiffness 长期刚度
long-term storage 长期蓄水;长期库容;长期储存;长期保存;多年调节库容
long-term strain 长期应变
long-term strategy 长期战略
long-term strength 后期强度;长期强度
long-term sustainability 长期可持续性
long-term sustained loading 长期持续荷载
long-term test 长期试验;连续试验;连续负荷试验
long-term testing 长期测试
long-term time average 长期时间平均数
long-term tolerability test 长期耐受性检验
long-term trade gap 长期贸易逆差
long-term trend 长期趋势
long-term unexpired cost 长期未耗用成本
long-term unfavo(u)rable balance of trade 长期贸易逆差
long-term value 长期价值
long-term variation 长期变化
long-term wastewater 长期废水
long-term water system 长期给水系统
long terne sheet 镀铅锡箔铁板;镀铅锡箔钢板
long thermal neutron detector 长寿命热中子探测器
long T hinge 长T形铰链
long thread casing 长扣套管
long threaded bell joint 长螺纹钟口接头
long thrust 冲刺
long time 长期荷载;长期
long-time annual flow 多年平均流量
long-time average annual flow 正常径流(量);多年平均径流(量)
long-time average annual value 多年平均值
long-time average discharge 多年平均流量
long-time base 长时基
long-time burning oil 久燃煤油(信号灯用);长期燃烧油
long-time creep test 长期徐变试验;长期蠕变试验
long-time deflection 持久挠度
long-time earthquake prediction 长时间地震预报;地震长期预报
long-time exposure 长时间曝光
long-time load(ing) 长期荷载;长期负载;长期负荷;持久荷载
long-time load(ing) test 持久荷载试验;长期荷载试验;长时间负荷试验
long-time performance of concrete 混凝土长期性能
long-time pumping test 长期抽水试验
long-time quality 长期使用后的特性
long timer 长时间计时器
long-time sediment yield 长期沉积量
long-time series 长时间系列
long-time service testing 长期荷载试验

long-time settlement 长期沉降
long-time solution 长时间溶解
long-time stability 长期稳定性;长期稳定度
long-time strength 持久强度;长期强度
long-time test 连续试验;持久试验;长期试验
long-time trend 长期趋势
long toggle plate 长时板
long toll call 长途通话
long Tom 长汤姆;长形倾斜淘金槽
long ton 英吨;重吨;长吨
long tooth 长顶齿
long tooth stone rake 长齿石耙
long trade 远洋贸易
long transient effect 长瞬态效应
long trip 长途旅行;长行程
long trunk call 长途呼叫
long tube 长管
long tube absorption cell method 长吸收管法
long tube chime 长管音调门铃
long tube evapo(u)rator 长管蒸发器
long tube falling film evapo(u)ration 长管垂直降膜式蒸发器
long tube nozzle 长管嘴
long tube vertical evapo(u)rator 长管竖式蒸发器;长管立式蒸发器
longulite 联珠晶子;长联雏晶
long varnish 长油性清漆;长油度清漆
long vegetable fiber 麻类纤维
long vernier 长游标
long vincula 长纽
long walk bridge 长步桥(油船上)
longwall 长壁法回采
longwall advancing 长壁前进式开采(采矿);前进式回采工作面
longwall advancing to the dip 沿倾向向下前进式长壁开采法
longwall advancing to the strike 沿倾向前进式长壁开采法
longwall coal cutter 长壁工作面截煤机;长工作面用截煤机
longwall face 长(壁)工作面(采矿)
longwall face conveyer 长壁工作面输送机(采矿)
longwall face mining 长壁工作面开采(采矿)
long wall generator (of underground gasification) 长壁炉法
longwall mining 长壁开采(法)
long wall mining method 长壁式开采法
longwall retreating 长壁后退式回采(采矿);后退式回采工作面
longwall-shortwall coalcutter 万能截煤机
longwall undercutter 长壁工作面用截煤机
longwall working 长壁采矿法
long wave 久远波动;长浪;长波;比水深长得多的波浪
long wave antenna 长波天线
long wave band 长波段
long wave broadcasting transmitter 长波广播发射机
long wave communication 长波通信
long wave FM radio telephone 长波调频无线电话
long wave infrared afocal zoom telescope 长波红外远焦可调式望远镜
long wave infra red detector 长波红外探测器
long wave infrared region 长波红外波段
long wavelength fiber 长波长光纤
long wavelength laser 长波长激光器
long wavelength laser diode 长波长激光二极管
long wavelength low-amplitude positive anomaly at ocean side of trench 海沟洋侧长波低幅正异常
long wavelength pass filter 长波长通过滤光片
long wavelength pass optic(al) filter 长波长光通滤波器
long wavelength region 长波区
long wavelength semiconductor laser 长波长半导体激光器
long wavelength thermal detector 长波长热探测器
long wavelength threshold 红限;长波限
long wave lighting diode 长波发光二极管
long wave pass filter 长波通滤光片
long wave radar 长波雷达
long wave radiation 长波辐射
long wave radio 长波收音机
long wave receiver 长波接收机

long wave theory 长波理论
long wave transmitter 长波发射机
long wave ultravoilet radiation 长波紫外辐射
long wave undulation 长波浪钢轨磨耗
long wearing 耐磨
long weight 长形重锤
long welded rail 焊接长钢轨
long welded rail transporting and working train 长钢轨运输作业车
long wet process kiln 湿法长窑
long wet-process rotary kiln 湿法长回转窑
long wet spells 长期潮湿天气
long wheelbase grader 长车架平地机
long wing sweep 长翼箭形铲
long wire antenna 长线天线
long wire field 长导线场
long wire-lock loop 长线锁环路
long wire method 钢丝法;长导线法
longwood 长胶合板面层
long word 长字
long yoke 长偏转线圈
Lonja de la Seda at Valencia 巴伦西亚的丝绸交易所(15世纪)
lonnealing 低温回火
lonsdaleite 六方碳;六方金刚石
lont-time diffusion 长时间扩散
loo 洗手间;洛风
loof 船首尖部
look ahead 向前看;预取;超前屏蔽;超前
look ahead adder 先行加法器
look ahead carry generator 先行进位发生器;超前进位产生器
look ahead carry output 先行进位输出
look-ahead control 超前控制;先行控制
look-ahead facility 先行装置
look ahead strategy 先行策略
look ahead system 超前式
look ahead unit 先行控制部件
look-alike 表面相似的
look angle 视角
look aside buffer 后备缓冲器;监视缓冲器
look aside memory 后备存储器
look-at-me function 中断功能
look-back test 回送检查
look box 观察孔;观察检验孔
look direction 观测方向
looker 查找程序
look for coal 找煤
look forward to 期待
looking glass 窥水镜;镜子
looking-glass finish 镜面抛光;抛光到镜面
looking signal 同步信号
look on net 侦察鱼群的网
lookout 观景处;(挑出城山墙屋顶的)椽;(挑出城山墙屋顶的)檩;挑檐支架;悬挑支架;望楼;瞭望台;瞭望
lookout angle 视角
lookout assist device 瞭望辅助装置
lookout bridge 瞭望桥楼
lookout deck 瞭望台;瞭望平台
lookout floor 瞭望层
lookout gallery 监视廊
lookout gondola (观光塔的)瞭望舱
lookout station 瞭望台
lookout telegraph 瞭望传信器
lookout tower 瞭望台;瞭望塔;望楼
look over 翻阅
look see 一般检查
lookum (行车或起重滑轮上的)罩盖;披屋;(覆盖卷扬机的)棚子
look-up 探求;查找;查出;查阅
look-up command 查找指令;查找命令
look-up instruction 查找指令;查表指令;探查指令
look-up table 查(对)表;对照表;查表法
look-up technique 查表技术
loom 上现晨景【气】;非金属软管
loom arch 织机上梁
looming 幽影;海市蜃楼;蜃景
looming healding 穿综筘
looming of the land 隐约看见陆地
looming of the light 地平以下的灯塔光弧
loom mounting 织机上轴
loom of an oar 桨柄
loom side 织机墙板

loom stare 坯布
loom state fabric 坯布
Loonen model 劳农模型
loony bin 精神病院;疯人院
loop 线圈;匝;活套轧制;活套;回(转)线;回路;环行路;环道;旁通导管;网孔;套状沙洲;绳回环
loop actuating signal 回路作用信号;环路动作信号
loop admittance matrix 回路导纳矩阵
loop aerial 环状天线;环形天线
loop analysis 回路分析法
loop anchorage 环状锚固
loop-and-trunk layout 网路位置图
loop antenna 环形框形天线;回路天线;环形天线
loop around 循环
loop back 回送;返回
loop back function 回送功能
loop back roll 活套辊
loop back test 线路循环试验;回送试验
loop bar 环形沙洲;环形沙滩;活头杆;套状沙洲;套状沙石;套杆
loop-battery pulsing 环路电池脉冲发生
loop belt system 环形夹皮带系统
loop body 循环节;循环(本)体
loop box 循环专用单元;环路箱
loop brake 环形天线固定螺钉
loop break 末端钩环
loop cable 环网电缆
loop capacitance 回路电容
loop chain 环链;花环链幕
loop checker 循环检查程序
loop checking 循环线检验
loop checking system 循环检查控制;循环检查系统;回送校验系统
loop circuit 环形电路;环路【电】;封闭线;闭回路;闭合回路
loop classifier 回路分级机;环形分级机;环式分级机
loop clause 循环子句
loop closure 环线闭合差
loop cloth 毛圈织物
loop-coil 螺环
loop command 循环指令
loop compensator 弧形伸缩器
loop compilation 循环编译
loop computing function 循环计算操作
loop connection 回线连接
loop construction 线圈结构
loop control 环路控制
loop control algorithm 循环控制算法
loop counter 循环计数器
loop-coupled 回路连接的
loop coupling 回线耦合
loop culvert 跨闸涵洞;(船闸闸首的)环形廊道;短廊道
loop current 回路电流
loop curve 环形(率定)曲线;闭合曲线
loop cut 纽形剖线
loop cutoff 狭颈裁弯;环形裁弯
loop cut-off characteristic 回线截止频率特性
loop cutting machine 剪绒机
loop depot 环形车站;环道车站
loop detector 环线式测车器
loop development 环形展线;环形展开
loop diagram 环形图
loop diallings 环路断开脉冲发生
loop diameter 缠绕直径
loop difference signal 回路差信号
loop direction finder 环状天线测向仪;框形测向器
loop-disconnect pulsing 环路断开脉冲发生
loop distribution 循环分配
loop distribution algorithm 循环分配算法
loop drier 悬挂式烘布机;悬挂式烘布机
loop economizer 盘管省煤器
looped bar 拐角环筋
looped barrier 环形沙洲
looped carpet 毛圈地毯
looped ends diamond 菱角菱形
looped fabric 毛圈织物
looped lace 起圈花边
looped link 钩环
looped network 环状管网
looped pile 毛圈
looped pipe 环状管
looped (pipe) line 环状管道;环状管线

looped power supply 环状供电
looped rug 毛圈地毯
looped system 反馈调节系统
looped wire 带钩导线
looped yarn 毛圈线
loop ends 循环结束
loop equation 回路方程;环(路)方程
looper 捆束机;防折器;(屋面卷材铺设机中的)卷圈器;活套挑;停圈机;套口机;打环机;撑套器
looper arm 撑套杆
looper clip 套口机剪线装置
looper control 套口控制器
looper course 套口横列
looper gear 活套支持器;活套挑
looper point 套口针片
looper roll 活套辊
loop error 循环错误
loop-excitation function 回线激发函数
loop expansion bend 膨胀弯管
loop expansion pipe 膨胀管;涨力弯管
loop feedback signal 回路反馈信号;环路反馈信号
loop feeding 环形供电
loop filament 环形灯丝
loop-film 循环片
loop filter 环路滤波器
loop formation 环路形成;毛圈形成
loop forming 成圈
loop-forming element 成圈机件
loop free 无循环
loop-free algorithm 无循环算法
loopful 钚环量
loop function 循环操作
loop gain 回路增益;回路放大(系数);环路增益
loop-gain method 环路增益法
loop galvanometer 回线检流计;回线电流计;环形电流计;环路检流计;卡表
loop head 循环入口
loop heating system 循环加热系统
loophole 环眼;箭窗;箭孔;窥孔;透光孔;观察孔;换气孔;漏洞;瞭望孔;枪眼;透光孔
loopholed gallery 有透光孔的画廊
loop-hole door (舞台上的)地板门;通气门;活板门;有狭孔门
loop identification 环的识别
loop impedance 回线阻抗
loop-impedance matrix 闭路阻抗矩阵
loop impulse 回路脉冲
loop inductance 回线电感;回路电感;环线电感;环路电感
looping 环链形;环;成圈;成环
looping bed 活套台
looping car 活套车
looping channel 卷取槽;卷料槽;(圆盘的)活套槽
looping compensator 环形补偿器
looping execution 循环执行
looping floor 转环地面
looping in 形成回路;形成环路;环形安装;环接
looping machine 套口机
looping merchant mill 活套式型钢轧机
looping method 环线法
looping mill 线材(滚)轧机;活套式轧机
looping needle 套口针
looping network 环状管网
looping observation 闭合导线观测
looping out 电灯接熔断的常法
looping pipe 管式围盘;环(形)管
looping plume 圈形烟缕;波浪形烟
looping routine 循环程序
looping seam 套口缝
looping test 打结试验
loop initialization 循环预置;(初始状态的)循环初置;循环初始化
loop input signal 环路输入信号
loop invariant 循环不变式
loop inversion 循环反演
loop joint 环状接点;back环接头
loop jump 循环转移
loop knot 环结
loop lake 牛轭湖;环形湖
loop length 线圈长度
loop lifter 撑套器
loop lifting device 毛圈起圈装置
loop line 圈线;回线;环线线(路);环(绕)线(路);

盘旋路
loop-line flow counter 循环流动计数器
loop locked 闭环的
loop meter 电表回路
loop method 环路法
loop-mile 环线英里
loop motor 环流电动机
loop network 环网
loop of retrogression 逆行圈
loop of strain stress curve 应力应变曲线回环
loop optimization 循环优化
loop optimization method 循环优化法
loop-oriented equation 按环列的方程
loop oscillograph 回路输出信号;振子式示波器;回线示波器;回路示波仪;环线示波器
loop output signal 回路输出信号;环路输出信号
loop parameter 循环参数
loop pile 起圈绒头;毛圈绒头;圈绒面
loop-pile carpet 毛圈线地毯;圈绒地毯
loop program(me) 循环程序
loop programming 环形程序设计
loop pulsing 封闭脉冲发送
loop radiator 环形辐射器
loop railway 环行铁道;环行铁道;环线铁路
loop raised fabric 毛圈织物
loop ramp 环形匝道【道】;环形坡道
loop rating 绳套曲线
loop rating curve 水位流量关系环线
loop ratio 环路传递函数
loop reactance 回线电抗
loop reactor 环形反应器
loop receiver 探向接收机
loop receiving antenna 环形接收天线
loop rectangularity 回线矩形性
loop regulator 活套垂度调节器
loop resistance 回线电阻;环线电阻
loop resistance measurement 回路电阻测量
loop restriction 环路限制
loop reversal 循环倒换
loop road 绕越道路;环道;环线;环(行)路
loop rod 环头杆
loop rug 毛圈地毯
loop scavenged cylinder 回流换气汽缸
loop scavenging 回流换气法
loop seal 环形水封;环形管水封;盘封
loop selvedge 毛圈边
loop sense antenna 环形辨向天线
loop sensor 环形传感器
loop service 环路供电
loop siding 会让线
loop space 闭路空间
loop speed-up 循环加速
loop speed-up hierarchy 循环加速层次
loop spit 环形沙嘴
loop stage-discharge relation 水位流量关系环线
loop state 循环状态
loop statement 循环语句【计】
loop station 环形车站
loop step at the pocket sliding door car end 2 二号车端头滑动门处踏步
loopstick antenna 铁氧体棒(形)天线
loop stop 循环停机
loop street 半环路;环状街道
loop strength 互扣强度
loop structure 线圈结构
loop table 循环表
loop table element 循环表元素
loop termination 循环结束
loop test 环形试验;环线试验;环路试验
loop-test bar 回路试验铁条;环线试验铁条
loop tester 循环检查程序;环线测试器
loop track 环行道;铁路环线
loop track of hump 驼峰迂回线
loop traffic volume 匝道交通量
loop transfer function 环路传递函数;闭合传递函数
loop transmission 回线输电;环线制传输
loop transmittance 环路透射比
loop traverse 环形导线;闭合导线【测】
loop turn-back track 环行折返线
loop-type characteristic 闭合回线特性
loop-type junction terminal 环形枢纽
loop-type of locomotive routing system 循环交路
loop-type pit bottom 环行式井底车场
loop variable 循环变量

loop vent 排气孔;循环通风;环形通气管;环式通气管
loop voltage 回路电压
loopway 环形岔道;环形车道转盘;错车岔道
loop web sling 环形带扣
loop wheel machine 台车;吊机
loop winding 环向缠绕
loop window 环形窗;环孔窗;竖直狭通气孔
loop wire 回线;环线
loop yarn 圈纱
loop yarn twister 起圈花线并捻机
loose 解开;松(散)的;疏松的
loose aggregate 松散集料;松散骨料;疏松(集)料;疏松骨料
loose alluvium 松散冲积层
loose and open surface 松散透水面层;松散透水路面
loose anti-coupling 弱反耦合
loose apron 松散石料护坦;松散石料护板;松散护坦
loose ashes 飞灰;松散飞灰
loose ash pond 飞灰池
loose ash pond water 飞灰池水
loose axle 空载车轴
loose barn 散放牛舍
loose base 松底层;松基层
loose beam heads 松套边盘
loose black 粉末炭黑
loose blade 可拆卸螺旋桨叶
loose blasting 松动爆破;松爆破
loose blasting charge 松动爆破药包
loose blasting crater 松动爆破漏斗
loose blocks 疏松块堆;疏松块堆
loose bolt 松动螺柱
loose boss 自由轮毂
loose bottom 活动底板;疏松底板
loose-boundary model 疏松界面型
loose-box (牲畜的)分隔栏;无栓分隔栏;单厩间
loose brake backing plate 松动制动盘
loose brush 疏铺梢料
loose brush mattress 松(散)梢褥;疏铺柴排
loose brush revetment 疏铺梢护岸
loose bulk density 松散容重
loose bush 可换衬套;活动衬套
loose butt hinge 活叶铰链;活脱铰链;活络合页;抽芯铰链
loose cargo 散装货;单件货物
loose cavity plate 带空腔活动模板
loose ceiling 可动的舱底板
loose cement 松散水泥;散装水泥
loose center 随转顶尖
loose change gear 可互换变速齿轮
loose charge 松动药包
loose circuit 松耦合电路
loose coal 疏松煤
loose collar 松紧环
loose colo(u)r 浮色
loose combination 松散联合(体);疏松结合
loose compaction 疏松压制
loose condition 松散状态
loose connection 松动连接;不良连结
loose constraint 松的约束;松约约束;非严格约束
loose contact 接触不良;松动触点;不良触点
loose contamination 松散污染
loose core 松散岩芯;(混凝土的水泥浆流失及骨料离析)松散夹芯;粗集料离析(混凝土);粗骨料离析(混凝土);抽芯
loose-coupled type pipe 松接式管道
loose coupling 可拆卸联轴节;松耦合;松联轴节;松开联轴节;松动结合
loose-crank operator 可卸柄操纵器
loose cubic(al) meter 松散立方米
loose cubic(al) yard 松散立方码
loose cut veneer 切割镶面板
loosed end 绳索散端
loose density 松装密度
loose deposit aquifer 松散含水层
loose deposit of quaternary 第四系松散堆积物
loose deposits 疏松沉淀物
loose depth 松(散)厚度;松铺厚度
loose dry density 松杆密度
loose dust 飞扬的粉尘
loose earth 松土
loose eccentric 游动偏心器
loose-eccentric wheel 游动偏心轮

loose edge 脱边;松边;掉边
loose fiber 散状纤维
loose fiber pollution 松散纤维污染
loose filament 散丝
loose fill 松散填充;松填土;松填方;松散充填料;疏松充填;坝料干填
loose filler 疏松填料
loose-fill insulation 松散填充绝热材料;松散料保温
loose fill type insulant 松散绝热材料
loose fit 松装配;宽装;粗装配;粗糙加装;转动配合;间隙配合;松配(合);松动配合
loose fitting 松动零件;散配件
loose fitting piston 松配合活塞
loose flange 活套法兰;松套凸缘;松套法兰(盘)
loose flange joint 松套法兰盘接头
loose flange type 活套式法兰
loose fold 松折
loose-footed roadway arch 疏松底部的平巷拱(采矿)
loose formations 疏开队形;散开队形
loose foundation 松散地面;松散地基
loose frame type 活套框架式
loose frame type core box 脱框式芯盒
loose framing 疏成帧
loose frozen soil 松散冻土
loose fund 游资
loose gland 松压盖
loose glass 玻璃屑
loose goods 松散货物;散装货(物)
loose grade mix 松级配混合料
loose grain 自由冲击;松面
loose grasses 稀流草层
loose gravel 松散砾石;松砾石
loose-grazed 无人放牧的
loose grid (用系绳法操作的)舞台平衡锤系统
loose ground 松散岩体;松散土(壤);松散地(面);松散地基;松软地面;松软地层;松地;松散地面
loose headstock 后顶针座;床尾
loose headstock center 头座随转顶尖
loose heel switch 活尖头跟端转辙器
loose hemp fiber 麻絮
loose housing shed 散放饲养棚
loose housing system 散放牛舍
loose hub 松衬套
loose insulation 松散料隔热层;松散保温材料
loose interconnection 松散互连
loose joint 活接;松套连接;松接头;松接头
loose-joint butt 可拆活页;可拆铰链;活络合页
loose-joint hinge 可拆铰链;可拆合页;活关节合页;插销铰链
loose justification 不精确齐行
loose knot 木材上松节疤;脱落节;松(木)节;疏松节疤(木材)
loose laid and ballasted roofing 松铺压顶屋面
loose laid roofing 松铺屋面
loose laid tiles 松铺地砖
loose laying depth 松铺厚度
loose-leaf 活页装订;活页
loose-leaf format 活页本
loose-leaf map 多幅地图
loose-lifting piston 松动活塞;大间隙活塞
loose liner 滑动衬垫;活动衬管
loose lintel 临时过梁;活动过梁;无连接的过梁;松弛过梁
loose list 松弛表
loose lock 活动锁口;活动的屋面板接口
loose-loop model 松散线圈型
loosely bound 结合松散的;松弛结合
loosely bound water 弱结合水
loosely coupled core 松散耦合线芯
loosely coupled interprocessor 松联结处理机
loosely coupled processor 松联结处理机
loosely filled vermiculite 松填蛭石
loosely grained rock 松散岩石
loosely packed 疏堆积
loosely spread 未捣实的(混凝土)
loosely spread concrete 振捣不足的混凝土;振捣不实的混凝土;未压实混凝土;未捣实混凝土;松布混凝土
loosely stowed 堆装松散
loosely wound roll 松卷纸卷
loosely wound spiral 松卷螺旋
loose main diagonal 松散的主斜杆

loose masonry 干砌圬工;干砌瓦工;干砌体;干砌石砌体;干堆圬工
loose material 化学不稳定材料;松散物料;松散(材)料;散粒料;松软物质
loose material grab 松散材料抓斗
loose measure 未压实体积;松散状态下测量;松散状态下测定的;松散体积;量;松方;粗测
loose measurement 粗测
loose measure volume 松方体积;松散土方
loose membrane 疏松膜
loose metal 松石
loose mix 松级配混合料
loose mo(u)ld 可卸式压模;可拆玻璃模脚
loose mo(u)lding (木质的)可拆玻璃压条
loosen 松开;放松
loosened 分散的;松散的;疏松的
loosened cake 松饼
loosened carbon 松的炭渣
loosened concrete 已损坏混凝土;已损害混凝土
loosened grain 疏松纹理;粗松纹理
loosened layer 疏松层
loosened rock 松石
loosened rock mass 松散岩体
loosened zone 松碎带;松动区;不胶结带
loose needle survey 罗盘粗测;罗盘测量
loosener 松土机
looseness 松散性;松劲;松度;松弛;不牢固性
looseness factor 放宽系数
looseness of soil 土壤松散
looseness of structure 结构松散
loosen grain 松散纹理
loosening 松开;松散
loosening bar 敲棒
loosening coefficient 松散系数
loosening depth 松散深度;松土深度
loosening earthwork 松土工(作)
loosening equipment 松土工具
loosening leverage 松杆率
loosening machine 松土机
loosening of coupling 松开万管接头
loosening of rock mass 地层松动
loosening of the tube 电子管松动
loosening pressure 松散压力;松动压力;松弛土压
loosening pressure design 松动压力设计
loosening sleeper bed 扒松轨枕槽
loosening strength 松弛强度;卸荷强度
loosening wedge 拆卸楔
loosening work 松解工作;松动工作
loosening zone 松动圈
loosen texture 散体结构
loosen the brake 松开制动踏板
loosen up price 放开价格
loosen zone 免压圈
loose of buoyancy method 损失浮力法
loose ore 松散状矿石
loose organic sand 松散有机质砂
loose overburden 松散覆盖层
loose packed 散装
loose packed density 松装密度
loose pack ice 稀疏冰
loose packing 疏松填充;疏松积
loose packing unit weight 松堆容重
loose pack rolling 松叠轧制
loose part 可拆部分;松动部分
loose parts monitor 零件松动监测器
loose pattern 单体模样;粗制模型
loose pavement 松散路面
loose pe(a)rlite 松散的珍珠岩
loose penstock 松接式压力钢管
loose pick 松纬
loose piece 木模活块;活络块
loose pile 松纬起绒
loose pin 活动定位销;导向销
loose-pin butt 活销铰链
loose pin butt hinge 活络销铰链;抽芯合页;抽芯铰链
loose pin hinge 抽芯铰链;枢轴铰链
loose pinion 空转小齿轮
loose placement 松铺
loose planting 散植
loose-plate transfer mo(u)ld 活板式传递塑模
loose powder 松粉;疏松粉末
loose powder sintering 松散的烧结粉
loose product 散装产品;散装货物

loose propeller blade 可拆卸螺旋桨叶
loose protection course 松散保护层
loose pulley 游(滑)轮;拉绳滑轮;空转皮带轮;惰轮;独立滑车
loose putty 松散油灰
loose rein 单缰控缰
loose rib 平挡圈
loose ring 松圈
loose rivet 未铆紧铆钉;松铆钉
loose riveting 松铆(接)
loose rock 松散碎石块;未胶结岩石;危石;松石;松散岩石;松软岩石;疏松岩石;浮石
loose rock check dam 碎石谷坊
loose rock dam 堆石坝
loose rock dump 抛石;堆石(体)
loose rock fall 松石坍落
loose rock-fill 抛石填充;抛石充填;松散废石充填(料);疏松填石;堆石
loose rock mass 松散岩体
loose-rod screen deck 活动棒条筛面
loose roller 中间轮
loose roof 松软顶板
loose running 无载运行;松弛在轴上
loose running fit 松转配合
loose saline soil 疏松盐土
loose sand 松(散)砂;散砂
loose sand compaction 松散砂压实
loose sand compaction by vibration 松散砂振密
loose scale 疏松氧化皮;疏松水垢;疏松的氧化皮
loose selvedge 松边
loose shunting 溜放调车法
loose side 镶板的切割面;胶合板木片层的面;皮带松边;松面;单板松面;单板背面
loose sintering 松装烧结
loose sleeper 轨枕空吊板
loose snow avalance 松散雪崩
loose socket 平滑离合器
loose soil 松土;松散土(壤);疏松土(壤)
loose soil compaction 松散土压实
loose spiral end 松螺旋
loose stall 方形圈
loose standard 轻定额;保守定额
loose state 松散状态;疏松状态
loose stock 散装货
loose-stone 干砌石;松石;松动金刚石
loose-stone dam 干砌石坝
loose-stone pitching 干砌石护坡
loose stop 活的(门窗)阻止条
loose stowage 不紧凑装载
loose straw 疏松稿秆
loose structure 松散结构;多孔结构
loose stuff 松散岩层
loose surface 松散面层
loose-surface road 松散面层的道路
loose texture 结构疏松;疏松质地
loose texture of single particle 松散单粒结构
loose thickness 松铺厚度;松散密度;松厚度
loose tie 轨道暗坑;松动的轨枕
loose to ga(u)ge 松动轮规
loose tongue 固紧钉;合板钉;横舌榫;嵌销;嵌入榫
loose tongue joint 抽芯接头
loose tongue miter 活榫舌斜角接
loose track 松弛的履带
loose tube optic(al) fiber 松套光纤
loose tube splicer 松管接头器
loose twist 松捻
loose type 松开轮箍
loose type flange 松式法兰
loose unit weight 松散(料)单位重量;干松容重
loose valve 截止阀
loose veneer 松单板
loose volume 松方;松体积;松散体积;松散容积;疏松体积
loose volume batching 松散体积配料(称量)法
loose volume weight 松堆容重
loose wedge 松散楔体
loose weight 松重(量);松散(料)重量;疏松重量
loose weld 脱焊;断丝
loose wheel 游滑轮;松带
loose winding 松卷
loose wire gripper 松线夹
loose wool 散棉;原棉;散棉线;短纤维玻璃棉
loose yards 石方;松散土方量;松散(料)容量;(土的)松方量;松散体积

loose zone 松动区
loos from lockage 过闸用水量耗损
loosing a track 松脱履带
loosing equipment 松土机
lop 剪短;下垂的;树木剪枝;悬垂;修剪;碎波
lop and top 枝梢材
loparite 铈铌钙钛矿
lophaite 蠕绿泥石
lophodont 脊齿型
lophodont teeth 脊型齿
Lophophyllidudae 顶柱珊瑚科
lopolith 岩盆;告盘
lopper 剪枝剪刀;剪枝工;斩波器;砍除器
lopping 截枝;截短
lopping shears 剪枝刀;钳工台长柄剪;长柄树枝大剪刀
lopraite 铬铅矿
loprotron 整流射线管
lopside 倾向一方
lopsided 倾斜边的;偏重;不均称;一边高一边低的;不平衡的
lopsided arch 不对称拱
lopwood 枝材
loquat 枇杷
Lorac(Long-Range Accuracy System) 罗拉克导航系统;精密无线电导航系统;罗拉克导航制
Lorad(Long-Range Detection) 罗拉德远距离探测系统
Lorain type shell liner 劳雷因型衬板压条
Loran 罗兰远程导航系统
Loran A 罗兰 A
Loran A/C receiver 罗兰 A/C 接收机
Loran A system 罗兰 A 导航系统
Loran C 罗兰 C
Loran C alarm 罗兰 C 警告
Loran chain 罗兰链
Loran chart 罗兰(海)图;罗兰导航图
Loran chart line 罗兰位置线
Loran C-inertial integrated navigation 罗兰 C 惯性导航系统
Loran C monitor station 罗兰 C 监控台
Loran coordinates 罗兰坐标
Loran coverage 罗兰覆盖区
Loran C signal 罗兰 C 信号
Loran C system 罗兰 C 系统
Loran curve 罗兰曲线
lorandite 红铊矿
Loran fix 罗兰定标;罗兰船位
Loran fixing 罗兰定位
Loran indicator 罗兰指示器;罗兰显示管;双曲线定位仪用显示管
Loran inertial navigation system 罗兰惯性导航系统
Loran line 罗兰线
Loran location 罗兰定位
Loran method 罗兰法
Loran monitor station 罗兰监测站
Loran navigator 罗兰导航仪
Loran network 罗兰导航网
Loran/Omega course and track equipment 罗兰奥米伽航向航迹装置
Loran position line 罗兰位置线
Loran receiver 罗兰(导航系统)接收机
Loran repetition rate 罗兰重复频率
Loran scope 罗兰显示器
Loran simulator 劳兰模拟器
loranskite 钇钽矿
Loran station 罗兰发射台
Loran system 罗兰导航制;罗兰(导航)系统
Loran table 罗兰(导航)表
Loran timer 罗兰定时图
Loran triad 罗兰台链
Loran triplet 罗兰台链
lordosis 鞍背状脊柱
Lorentz broadening 碰撞变宽
Lorentz coil 洛伦兹线圈;笼形线圈
Lorentz equation 洛伦兹方程
Lorentz force 洛伦兹力
Lorentz force density 洛伦兹力密度
Lorentz four-vector 四元矢量
Lorentz frame 洛伦兹框架
Lorentz ga(u)ge 洛伦兹规范
Lorentz group 洛伦兹群
Lorentz line-splitting theory 洛伦兹谱线劈裂理论

Lorentz line type 洛伦兹线型
Lorentz matrix 洛伦兹矩阵
Lorentz relation 洛伦兹关系(式)
Lorentz theory of light sources 洛伦兹光源理论
Lorenz apparatus 洛伦兹仪
Lorenz bored pile 洛伦兹螺旋钻孔桩;洛伦兹混凝土钻桩
Lorenz curve 洛伦兹曲线
lorenzenite 硅钠钛石
Lorenz/Fehlmann method 劳费氏沉箱沉降法(沉箱做成上小下大,上部灌膨润土稀浆)
Lorenz-Lorentz's formula 洛伦兹—洛伦兹公式
Lorhumb line 罗伦导航网时间变更线;两族双曲线中等变化值连线
lorica 兜甲
loricated pipe 内部涂有沥青的管子
loricula (教堂内墙上的)斜形小窗洞
lorop photography 远距离摄影
Lorraine cross 双横档十字架
lorry 小车;运料车;运货车;载货汽车;平台四轮车;推料车;手车;斗底车
lorry agitator 搅拌车(混凝土)
lorry body 货车车厢;货车车身
lorry-borne road heater 装在货车上的路面加热器
lorry crane 汽车(式)起重机;汽车吊
lorry-dumped riprap 汽车倒下的碎石
lorry dump piler 倾倒堆垛车;自行散料抛掷码垛机
lorry entrance 货车入口
lorry excavator 挖土车
lorry haul(ing) 货车运输
lorry lane 货车车道;货车通道
lorry loader 自动装卸机;装卸车;汽车式装载机
lorry loading 货车装车处
lorry loading crane 装车起重机;随车(式)起重机;随车(式)吊
lorry-loading list 载货汽车装载清单
lorry mixer 搅拌车;汽车式搅拌机
lorry-mounted bucket(elevator) loader 斗式提升装载车
lorry-mounted crane 汽车式起重机;汽车吊;汽车式钻机;起重汽车;随车(式)起重机;随车(式)吊
lorry-mounted ditcher 挖沟车
lorry-mounted excavator 挖土车
lorry mounted hydraulic loader 装在汽车上的液压装载机
lorry-mounted(mechanical) auger 装有机械螺旋钻的卡车
lorry-mounted mixer 车载混凝土搅拌机
lorry mounted spreader 装在载货汽车上的撒布机
lorry-mounted tipping unit 装在汽车上的翻斗装置
lorry-mounted winch 安装在汽车上的绞盘式起重机
lorry rail 手推车轨道;手推料车轨道
lorry rim 运货汽车轮辋
lorry scale 货车地磅;汽车秤
lorry terminal 运货汽车终点站
lorry tire 载货汽车用轮胎
lorry track 手推车轨道
lorry tyre 载货车用轮胎
lorry 运货汽车(英国)
lorry wash(down) 冲洗车
lorry winch 装有绞盘的载货汽车;载货汽车绞盘
lorry with trailer 带拖车的运货汽车
lorum 轴节间片
lorymer 飞檐;滴水槽;压顶线脚【建】;墙顶挑出的凸带
Los Angeles abrasion machine 洛杉矶磨损试验机;洛杉矶磨蚀试验机
Los Angeles abrasion test 洛杉矶磨蚀试验
Los Angeles abrasion tester 洛杉矶磨损试验机;洛杉矶磨蚀试验机
Los Angeles abrasion testing machine 洛杉矶磨耗试验机
Los Angeles basin 洛杉矶盆地
Los Angeles machine 洛杉矶骨料磨损试验机;洛杉矶集料磨耗试验机
Los Angeles Port 洛杉矶港
Los Angeles rattler 洛杉矶石料磨耗机
Los Angeles test 磨耗机试验;洛杉矶试验
Los Angeles-type photochemical smog 洛杉矶光化学型雾
Los Angeles-type smog 洛杉矶型雾
lose a bid 未中标;未得标
lose ability to control the market 失去控制市场能力

lose an election 落选
lose from natural calamity 自然灾害损失
lose ground 处于不利地位;被流走
lose line 失效控制线
lose money 亏本
lose money in a business 蚀本生意
lose one's capital 赔本
lose one's market 失去买卖机会
lose one's outlay 赔本
loser 亏损企业;亏本者;损失物
lose return 泥浆漏失
loser of property 财产失主
lose speed 失去速度
lose their colo(u)r or turn yellow 褪色或变黄
lose time 延误
lose-time accident 停工事故
lose turgor 失去膨压
lose water 泥浆漏失
lose way 减少航速
loseyite 蓝锌锰矿;碳锌锰矿
losing bid 未中标
losing business 亏损企业
Losinger system 洛辛格系统
losing lock 失锁
losing party 败诉方
losing proposition 亏损企业
losing river 隐入河
losing step 失步
losing stream 隐入河;(补给地下水的)亏水河;损流河;渗流河
losing synchronism 失步
losing tender 未中标
loss 亏损;损失,损耗;负效益
loss accident 丢失事故
loss account 亏损账户
loss adjuster 保险赔偿估定员
loss advice 损失通知书;出险通知
loss analysis 亏损分析
loss and damage 损失与损伤
loss and gains account 损益账
loss and gains on retirement of fixed assets 固定资产拆除损益
loss and profit accounting system 损益核算制度
loss and profit carried forward 结转下期损益
loss angle 损失角;损耗角;衰减角
loss angle technique 损耗角技术
loss apportionment 损失分摊
loss area 亏损区
loss assessment 估定损失;损失估计;损失额评定
loss at sea 海损
loss at total reflection 全反射损耗
loss behind the shock 激波后压力损失
loss bordereaux 损失明细表
loss by evapo(u)ration 蒸发损失;蒸发损耗
loss by lockage 过闸水量损失;过闸耗水量;船闸过船时的流量损失
loss by percolation 渗漏损失
loss by reflection 反射损耗
loss by roasting 焙烧损失
loss by run-off of soil 水土流失
loss by solution 溶失量;溶解损失(量)
loss by solution test 溶失量测定
loss by swirls 涡流损失
loss capital 资金亏损;亏本;蚀本
loss carried forward from the last term 前期结转损失
loss carried forward to the following term 结转下期损失
loss carried forward to the next term 结转下期损失
loss carry back 回计亏损;亏损抵前
loss carrying-forward 计入后期亏损
loss ceiling 损失上限
loss characteristic 损耗特性
loss code rate 漏码率
loss coefficient 漏电系数;损失系数;损耗系数
loss compensator 损耗补偿器
loss compliance 损耗柔量
loss conductance 损耗电导
loss cone 漏逸锥面;逃逸锥(面);损失锥
loss-cone angle 损失锥角
loss-cone instability 损失锥不稳定性
loss contour 损失线;等损失线
loss control 损耗控制

loss cost 损失性费用
loss credit 失信用
loss credit standing 失去信用
loss current 损耗电流
loss-delay system 混合(排队)系统
loss density 损耗密度
loss distance 损耗距离
loss draft 为损失支付的费用
loss due to bouncing 反弹损失
loss due to condensation 凝结损耗
loss due to creep 徐变的损失;蠕变的损失
loss due to frequency conversion 变频损耗
loss due to friction 磨损;摩擦损失
loss due to leakage 漏耗
loss due to obstruction 阻塞损失;故障损失
loss due to project scrapped 报废工程损失
loss due to shock 震动损失
loss due to sudden contraction 骤然收缩损失;断面骤小损失
loss due to sudden enlargement 骤然扩大损失;断面骤大损失
loss due to valve 开关损失;阀门损失
loss effect 损耗效应
lossening of macroscopic planned management 放松宏观计划管理
losser 损耗元件
losser circuit 非周期振荡电路
losses and gains brought forward 前期滚存损益
losses in a year 年度亏损
losses in soil 土壤流失
losses outstanding 未决赔款
losses paid 已付赔款
losses permitted by policy 政策性亏损
loss estimate 损耗估计
loss evaluation 损耗估算
Lossev effect 注入式电致发光
loss experience 灭失记录(航运)
loss factor 损失因数;损失率;损失因子;损耗因素;损耗因数;损耗系数
loss failure period 老化失效期;老化故障期
loss forwarded 结转损失
loss free condition 无损失条件
loss-free dielectric 无损耗电介质
loss-free line 无损耗线
loss-free medium 无损耗介质
loss-free reflectivity mechanism 无损耗反射机理
loss-free wall 无损耗壁
loss from accident 偶然事故的损失
loss from devaluation 贬值损失
loss from erosion 剥蚀损失
loss from obsolescence 废弃损失
loss from reappraisal of fixed assets 固定资产重估损失
loss from scrap disposition 废料处理损失;废旧资产处理损失
loss from suspension 停业损失
loss function 损失函数;损耗函数
loss head 压头损失;水头损失
loss head due to enlargement 断面扩大产生的水头损失
Lossiemouth beds 洛西默思层【地】
loss in definition 清晰度降低
loss index 损耗指数
loss in energy 能量损失
lossing normal colo(u)r 颜色失常
lossing proposition 亏本生意
loss in gradient 坡度损失
lossing solution 溶失量
loss in head 压头损失
loss in head in entrance 进水水头损失
loss in level 水头损失
loss in production 生产损失
loss in revenue 收益损失
loss in rigidity 刚度损失
loss in smelting 熔融损失
loss in stiffness 劲度损失
loss intensity 漏失强度
loss in the suspension of work 停工损失
loss in turbine blade 涡轮叶片绕流损失
loss in weight 重量损失;重量不足;短重
loss-in-weight batch blender 减重分批掺和机
loss-in-weight feeder 减重喂料机;失重补偿给料器
loss-in-weight under heating 加热失重
loss in yield 产量损失

loss leader 亏本出售的商品
lossless cable 无耗电缆
lossless cavity 无损耗共振器
lossless channel 无损耗通道
lossless code 无损失电码
lossless compression 无损压缩
lossless element 无耗元件
lossless filter 无耗滤波器
lossless junction 无损耗结
lossless line 无损耗线
lossless material 无耗损材料
lossless medium 无损耗介质
lossless network 无耗网络
lossless waveguide 无损耗波导
loss limit 损耗限制
loss maintenance component 损耗维护部件
loss maker 亏本企业;蚀本生意
loss measurement 损耗测量
loss measuring instrument 损耗测量仪
loss mechanism 损耗机理
loss meter 损耗电度表;损耗测度表
loss modulation 损耗调制
loss modulus 损耗系数;损耗模量
loss money in a business 贴本
loss multiplier 损耗倍增器
loss of accuracy 精(确)度损失;降低精度
loss of adhesion 黏结力损失;黏着破坏
loss of adjustment 失调
loss of advanced profits insurance 预期利润损失保险
loss of bond 握裹耗损;黏结耗损
loss of calcination 煅烧损失
loss of cash 现金损失
loss of charge 充电损失
loss of circulation 循环(钻)液损失;泥浆循环损失;循环损失;钻孔漏失
loss of containment 限制使用失去效率
loss of contrast 对比度损失;反差损失
loss of control 失去操纵
loss of control at macroscopic level 宏观失控
loss of controllability 失去操纵性
loss of coolant 冷却剂损失
loss-of-coolant accident 冷却剂流失事故;失水事故
loss-of-coolant experiment 冷却剂流失试验;失水试验
loss of core material 堤芯石损失
loss of cross section 横截面损失
loss of current term 本期损失
loss of(drill) core 钻孔岩芯损耗
loss of dry inhibitor 干性下降阻抑剂
loss of dynamic(al) head due to obstructing object 障碍产生的水头损失
loss of dynamic(al) head due to sudden change of cross-section(al) area 横断面突然变化引起的动力水头损失
loss of efficiency factor 效率降低系数
loss of energy 能(量)损失
loss of energy head 能头损失
loss of epidermis 失去表皮
loss of excitation 失磁
loss of excitation protection 失磁保护(装置)
loss of fall 落差损失;压头损失;跌落损失
loss of field 失磁;磁场损耗
loss of fluid yield 流体产量减损
loss of frame 帧丢失
loss of freight 运费损失
loss of friction 摩擦损失
loss of friction head 摩阻水头损失
loss of gloss 光泽损失
loss of goods 货物的灭失
loss of hardness 硬度损失
loss of head 压头损失;落差(水头)损失;水头损失;失头
loss of head due to contraction 收缩管的压头;收缩段水龙头损失;断面收缩产生的水头损失
loss of head due to enlargement 扩大段水头损失
loss of head due to entrance 管进口段的压头损失
loss of head due to exit 管出口段压头损失
loss of head due to obstructions 阻障产生的水头损失;障碍段水头损失
loss of head due to pipe fitting 管道配件的压头损失
loss of head ga(u)ge 压头损失仪;压头损失计;水头损失(量测)计

loss of head ga(u)ge of filter 滤池水头损失计
loss of head in bends 管道转弯产生的水头损失
loss of head indicator 水头损失指示器
loss of head in entrance 进口(阻力产生的)水头损失
loss of head in friction 摩擦(产生的)水头损失
loss of head of filter 滤池水头损失
loss of heat 热损失;热(量)损失;失热
loss of heat through wall 筒体热损失;壁面热损失
loss of human life 人命损失
loss of ignition 燃烧损失;点火失效
loss of import duty risk 进口关税险
loss of indication 道岔失去表示
loss of information 信息丧失
loss of interlocking 失去联锁
loss of inventory valuation 库存降价损失
loss of kinetic energy due to impact 冲击动能损失
loss of life 使用寿命降低
loss of light 光损失
loss of littoral drift 漂砂损失
loss of load expectation 负荷减少期望值
loss of load probability 负荷损失概率;负荷减少概率
loss of machine 机械损耗;掉炉;电机损耗
loss of macroeconomic control 宏观失控
loss of magnetic reversals 反复磁化损耗
loss of market 行情下跌;市价跌落;市场损失;失去市场;商品货物跌价
loss of mobility 失去机动性
loss of momentum 动量损失
loss of natural structure 自然结构的丧失
loss of nitrogen 氮素损失
loss of package 整件遗失;整件缺失
loss of phase 失相;断相
loss of pointer 指针丢失
loss of polarity 极性消失
loss of power 动力损失;电源损耗
loss of pressure 压力下降;压力损失;压力损耗;压力降低;失压
loss of prestress 预应力损失
loss of prestress friction 摩擦引起的预应力损失
loss of pretension 预张力损失
loss of profit insurance 营业中断收益损失保险
loss of radio control 失去无线电控制
loss of returns 循环损失;井口无(泥浆)返出;吸浆损失
loss of rigidity 刚度损失
loss of self-stress 自应力损失
loss of sheet 掉炉
loss of signal 信号丢失
loss of significance 精度损失
loss of soil 土壤流失;水土流失;失土
loss of soil moisture 损耗土壤水分
loss of species 绝种;物种消失
loss of speed 速度损失
loss of stability 失去安定性
loss of stiffness 降低刚度
loss of strength 降低强度;强度损失
loss of stress 应力损失
loss of suction 失吸现象
loss of synchronism 同步性破坏
loss of synchronism protection 失步保护
loss of synchronization 失去同步
loss of tension 张力损失
loss of time 时间损失
loss of total pressure 总压力损失
loss of total stress 总应力损失
loss of traction 牵引力损失;打滑损失
loss of turgidity 丧失饱满度
loss of underground water 地下水损失
loss-of-use insurance 价值损失保险
loss of vacuum 失去真空
loss of velocity head 速度头损失
loss of voltage 电压损失
loss of volume 体积减少
loss of voyage 航程损失
loss of water 水分损失
loss of weight 重量损失;减重;失重
loss of weight feeder 减量秤喂料机
loss of well 井漏
loss of workability 和易性丧失
loss of working time 误工
loss on acid washing 酸洗损失
loss on capital assets 固定资产损失

loss on clearing warehouse 清仓损失
loss on defective products 废品损失
loss on devaluation 贬值损失
loss on disposal 非流动资产出售损失
loss on drying 干燥失重
loss one's capital 亏本
loss on head due to obstruction 障碍物引起的动水头损失
loss on head in bends 弯头水头损失;管道转弯产生的水头损失
loss on heat 热能损失
loss on heating 加热损失(量)
loss on heating test 加热损失试验
loss on idle time 停工损失
loss on ignition 点火损失量;灼减量;烧灼损失;灼烧失量;火灾损失;强热失量;烧灼减量;烧失量
loss on insurance claim 保险损失索赔;保险金差损
loss on melting 熔化损失
loss on property abandoned 财产废弃损失
loss on property destroyed 财产毁损失
loss on property devaluation 财产跌价损失
loss on property retired 资产废置损失;财产退废损失;财产废置损失;财产报废损失
loss on realization of assets 资产变价损失;变卖资产损失
loss on retirement 退废损失
loss on retirement of fixed assets 固定资产报废损失
loss on weight 失重
loss opportunity for irrigation 错过灌水机会
loss order 赔款支付命令
loss paid 已付赔款
loss per pass 单程损耗
loss prevention 防损
loss probability 呼损率;损耗概率
loss quantity 流失量
loss queue system 损失制排队系统
loss rate 损失率
loss rate of rock strength 岩石强度损失率
loss rate of rock weight by frost 岩石冻失率
loss ratio 亏本率;赔款率;赔付率;损耗比
loss recovery 损失回收
loss reinstatement clauses 赔款后保险金额复原条款
loss report 损失报告书
loss reserve 赔款准备金
loss resistance 损耗电阻
loss soda ash percentage 纯碱飞散率
loss source 损失源
loss spread over a number of years 若干年内分头损失
loss-summation method 损失汇总法;损耗相加法
loss surveying 测漏
loss sustained during loading and unloading 装卸期间损失
loss system 立即拒绝体制;立即服务方式
loss tangent 损失角正切值;损耗角正切
loss tangent test 损耗角正切试验
loss test 损耗试验
loss through shock wave 激波中压力损失
loss through standing 储存损失;储存时损失
loss time 空载时间;损耗时间
loss torque 损耗转矩
loss water 失水
lossy 有损耗的衰减器
lossy attenuator 有损耗衰减器
lossy cable 有损耗电缆
lossy coaxial cable 有损耗同轴电缆
lossy compression 有损压缩
lossy dielectric 有耗介质
lossy line 高损耗线
lossy material 有损耗材料
lossy medium 有损耗媒质;有损耗介质
lossy termination 有耗终端负载
lost area 无功面积;损失面积
lost argon 氩丢失量
lost articles accompanying passengers 旅客遗失物品
lost balloon 测风气球
lost buoyancy method 损失浮力法
lost call 呼损
lost check 遗失的支票
lost circulation 泥浆漏失;全漏失
lost circulation interval 漏水段

lost circulation material 堵漏材料
lost circulation of drill hole 钻孔漏失
lost circulation pressure 漏失压力
lost clause 灭失条款
lost-colo(u)r process 落色工艺
lost condition 迷失状态;丢失状态
lost core 孔内残留岩芯
lost cores during lifting 提升时脱离岩芯;提升时损失岩芯;提升时丢失岩芯
lost count 漏失计数;漏读数
lost data 损失数据
lost drill pipe 磨损的钻杆
lost energy 能量损失
lost ground 失利的;漏失的(材料);落后的;失败的;处于劣势;残留岩芯
lost head 压头损失;落差(水头)损失;切头;水头损失;废品
lost-head nail 小头钉;埋头钉
lost heat 损失热(量)
lost hill 孤立残山;孤立残丘
lost labo(u)r 虚耗人工
lost material 消耗的材料;磨损(的)材料;损耗的材料
lost money in business 亏蚀
lost motion 无效运动;滞差运动;隙动差;滞后运动;空转(动);空载行程;空运转;空动;无效运转;废料
lost motion of micrometer 测微器隙动差
lost motion of screw 齿隙差
lost mountain 孤(立残)山;孤立残丘
lost of gloss 失光
lost of not lost clause 灭失或不灭失条款
lost-oil lubrication 无效油润滑
lost or not lost clause 货物损失运费照付条款;不管损失已否发生的条款
lost pass 空轧道次
lost pattern casting 失模铸造
lost policy holder 投保人失踪;失踪被保险人
lost river 隐入河;干河;流失河;潜入河
lost round 失效炮孔组
lost sales 丧失销售
lost ship 失踪船舶
lost strata 地层缺失;层序间断
lost stream 隐入河;干河;流失河;潜入河
lost surface 磨损的面层
lost surfacing 磨损面层;磨耗的面层
lost time 虚耗时间;消耗工时;损失时间
lost update 丢失修改
lost usefulness 效用消失
lost water 水损耗
lost wax casting 蜡模铸造法;熔模铸造;失蜡铸造
lost wax mo(u)lding 熔模铸造
lost wax process 熔模铸造过程;失蜡铸造法;失蜡铸造法
lost work 无效功
lost working hours unaccountable to workers 非工人责任造成的损失时间
lot 一批;一块;一堆;工区;建筑用地;合同包;批量;地皮;地段;份额
lot-acceptance sampling 批量接受抽样;认可抽样检验;分批认可抽样
lot accounting 分批会计;分批核算
lot and building account 房屋账户
lot boundary 地界
lot card 批号卡
lot cargo 大宗货物;成批货(物)
lot commission 拍卖佣金;大量手续费
lot cost system 分批成本核算制
lot depth 地基深度
lot diameter variation 批量直径的变化值
lot division 工区划分
lotharmeyerite 钙锰锌矿
lotic community 激流群落
lotic ecosystem 激流生态系统
lotic water 激流水;活水
lot identification mark 批号;产品批号
lotiform 莲花瓣形的;莲苞形式的
lotio 洗剂
lotio alba 白色洗液
lotio flava 黄色洗剂
lotion 洗剂;洗液
Lotka's law 洛特卡定律
lot line 区段边界线(楼群);房地产边界线;土地边界线;地界;基地(边)线;地(盘)界线

lot-line wall 地界墙
lot mean diameter 批量平均直径
lot measurement 地块测量;地皮测量
lot-method 分批法
lot note 分批票据
lot number 货物批号;签号;批数;批号
lot of ground 地块
lot of land 一块地皮
lotos 莲饰;荷花饰
lotos bud capital 蓓蕾饰柱头
lotos column 荷花圆柱
lotos glossom capital 莲花饰柱头;荷花饰柱头
lotos ornament 荷花饰
lot plan 地段(平面)图
lot planning 基地规划
lot plot method 分批测定方法
lot production 成批生产
lot-product production 分批产品的生产
lot quality protection 分批质量保护
lotrite 绿纤石
lot sample 批样
lots cost system 分批成本计算
lot separation 分票隔垫
lot serial production 成批生产
lot size 批量大小;批量;订购数量
lot-size inventory management interpolation technique 批量货单管理内插技术
lots of record 在册土地
lottery 抽签法
lot test 抽样试验
lotting 安达标签
lot tolerance 逐批容差
lot-tolerance failure rate 批内允许故障率;批内插允许破损率
lot-tolerance percentage defective 容忍疵品比率
lot-tolerance percent defective 批内插允许次品率;缺点允许率;批量允许不良率
lotus 莲饰;莲花(池);荷花饰
Lotus alloy 洛特斯铅锑锡轴承合金
lotus base 莲瓣柱基
lotus canopy 莲花宝盖
lotus capital (古埃及的)莲饰柱头;荷花饰柱头
lotus column 莲饰柱;荷花饰柱子
lotus flower 莲花
lotus-form structure 莲花状构造
lotus lamp 荷叶灯
lotus leaf 荷莲叶
lotus pendant 垂莲柱
lotus pond 荷花池
lotus-root like 藕状的
lotus-rootlike boudin 藕节形石吞肠
lotus-root-shaped 藕状的
lotus-root-shaped channel 藕节状河道
lotus seed 莲籽
lotus-shaped grab bucket 荷花抓斗
lotus tree 忘忧树
lot weight tolerance 批重容差
lot zone 拟建房屋的土地
loud 高声的
loud-break switch 高压大电流开关
loud hailer 扩音器
loud hailer control 船内通信控制
loudness 响度;音量;高声
loudness analyzer[analyser] 响度分析仪
loudness balance 响度平衡
loudness contours 等响线
loudness control 响度控制
loudness control circuit 响度控制电路
loudness control switch 响度控制开关
loudness density 响度密度
loudness directivity index 响度指向性指数
loudness efficiency factor 响度效率因子
loudness equalization control 响度均衡控制
loudness evaluation 响度鉴定
loudness function 响度函数
loudness index 响度指数
loudness level 响度级
loudness of sound 音度
loudness pattern 响度图;等响图
loudness rating 响度评定
loudness recruitment 响度重振;响度复原
loudness reduction 响度降低
loudness scale 响度等级;响度标度
loudness sensation 响度的感觉

loudness test 响度试验
loudness unit 响度单位
loudounite 水硅锆钠钙石
loudspeaker 扬声器;广播;喇叭;扩音器
loudspeaker cabinet 扬声器箱
loudspeaker enclosure 扬声器箱
loudspeaker group 扬声器组
loudspeaker hole 扬声器孔
loudspeaker housing 扬声器箱
loudspeaker in driver's cab 驾驶室广播喇叭
loudspeaker set 广播喇叭
loudspeaker system 广播系统;扬声器系统
loudspeaking telephone 扩音电话
loudspeaking telephone set 扩声电话机
loud spot 屋内声响焦点;响区
loud trailer 大功率指向性扬声器
lough 海湾;环山湖;湖泊
loughlinite 纤钠海泡石;硅镁石
Louis-seize style 路易十六风格
lounge 休息厅;休息室;公共休息室;起居室;大休息室
lounge antenna 室内垂直偶极天线
lounge car (火车的)休息车棚
lounge chair 安乐椅;躺椅
lounge hall 休息厅
loup 熟铁块;不定形铁块
loupe 接目放大镜;简单放大镜;寸镜
Loup river series 卢普河统【地】
loury 天气阴沉
louse cage (伐木工的)山中小屋
lousicide 灭虱剂
Loutorick system 劳托维克方式
Loutorick transfer system 劳托维克集装箱移放方式
louver 遮荫棚;叶栅;装饰栅;遮光罩;遮光格栅;固定百页窗;气窗;通气窗;条板;天窗;百叶板石;百叶窗
louverall ceiling 间接照明式顶棚;百叶窗式顶棚
louver board 天窗板;开孔散热片;散热片;散热板;百叶(窗)板
louver damper 百叶窗挡板
louver door 百叶门
louver dryer 百叶窗式干燥器
louvered (air) intake 百叶式进气口
louvered air outlet 透气百叶窗;百叶式出气口;百叶窗出气口
louvered awning 遮阳篷
louvered awning blind 百叶遮阳;活动遮篷(窗帘);遮阳百叶
louvered battens 百叶窗叶板
louvered blade 百叶片
louvered board 百叶窗板;百叶板;百叶式挡板
louvered café door 百叶式摇门(半高的,饮食店用)
louvered ceiling 百叶窗式顶棚;遮光栅格天棚
louvered combustion chamber 百叶窗式燃烧室
louvered die 切口模具
louvered door 百叶门;百叶式门(通风用)
louvered drum sand cooler 砂子冷却滚筒
louvered intake 百叶窗进气口;百叶式进气口
louvered lighting 百叶窗式照明
louvered overhang 鱼鳞板;遮阳百叶;百叶吊帘
louvered screen 百叶箱
louvered slats 百叶窗式条板
louver fan 百叶窗式电扇
louvering die 百叶窗模具
louver lighting 天窗采光照明;散光照明;百叶窗散光照明
louver mo(u)lding 缀模
louver of upper impervious layer 隔水顶板天窗
louver screen 可卷百叶窗;百页纱窗
louver separator 百叶窗式分离器;百叶窗除尘器;百叶窗分离器
louver shielding angle 百叶窗遮蔽角;百叶栅遮蔽角
louver shutter 鱼鳞板;活动百叶窗;活百叶窗;翻板闸门;百叶窗式快门;百叶窗式航摄快门;百叶窗快门
louver sieve 鱼鳞筛
louver slide 通风筒口开关
louver system 压入逸散式(通风)
louver vane 百叶导流栅;百叶窗式导流栅
louver ventilated 百叶孔通风的
louver window 气窗;百叶窗
louver window frame 百叶窗架

Louvre 卢浮美术馆(法国)
louvre = louver
louvre blade 百叶窗片
louvre tablet 百叶窗叶片
louvre ventilation 百叶窗通风
lovangar 林中草地
Lovar tape 洛瓦带尺
lovchorrite 褐硅铈石;褐硅铈矿
lovdarite 铍硅钠石
love arrows 发金红石
love-lies-bleeding 栽培苋
Lovelock detector 拉夫洛克检测器
lovely white glaze 甜白釉
Love number 乐甫数
loveringite 钛铈钙矿
love seat 恋人沙发;双人座椅;双人沙发
Love's number 洛夫数
Love wave 洛夫波
Lovibond colo(u)rimeter 洛维邦色度计
Lovibond colo(u)r system 洛维邦颜色体系
Lovibond comparator 洛维邦调比较计
Lovibond glass standard 洛维邦德玻璃标准
Lovibond tintometer 洛维邦色辉计;洛维邦色调计;洛维邦测色计
lovozerite 基性异性石
low aberration deflection yoke 低像差偏转系统
low abnormal pressure 低异常压力
low absorption spectrophotometry 低吸光度法
low access 慢加速存取
low-access memory 慢速存取存储器
low accuracy 低精确度
low-activity 低放射性水平
low activity data processing 低活动率数据处理
low air 供加压舱的空气
low alarm 低位报警器
low albite 低钠长石
low alkali borosilicate glass fiber 低碱硼硅酸盐玻璃纤维
low alkali cement 低碱(度)水泥
low alkali ceramics 低碱(陶)瓷
low alkali clinker 低碱熟料
low alkalinity 低碱度
low alkalinity waste(water) 低碱度废水
low alkali Portland cement 低碱硅酸盐水泥;低碱波特兰水泥
low alkali silica glass 低碱石英玻璃
low alkali sodium silicate 低碱硅酸钠
low alkali sulphate-resisting cement 低碱抗硫酸盐水泥
low alloy 低合金
low alloy cast iron 低合金铸铁
low alloyed 低合金的
low alloy high-strength steel 低合金高强度钢
low alloy high tensile 低合金高强度
low alloy high-tensile steel 低合金高强钢
low alloy high-tensile structural steel 低合金高强度结构钢
low alloy rail steel 低合金钢轨钢
low alloy steel 低合金钢
low alloy steel covered arc welding electrode 低合金钢焊条
low alloy steel high side wagon 低合金钢敞车
low alloy steel plate 低合金钢板
low alloy structural steel 低合金结构钢
low alloy threading steel 低合金螺纹钢
low alloy tool steel 低合金工具钢
low altitude 低空
low altitude aerial photograph 低空摄影像片;低空航照片;低空航摄像片
low altitude aerial photography 低空摄影
low altitude aircraft 低空飞机
low altitude antenna 低仰角天线
low altitude defence system 低空防御系统
low altitude dish 低辐射天线;低仰角天线
low altitude indicator 低空指示器
low altitude passage 低仰角通过
low altitude radiosonde 低空探空仪
low altitude satellite 低高度卫星
low alumina cement 低铝水泥
low alumina fireclay refractory 低铝黏土耐火材料
low amino baking enamel 低氨基烘漆
low amplitude vibration 低幅振动
low and deep 低沉
low and intermediate tensile strength 低中抗拉强度
low-and-moderate-income housing 中低收入住房
low-and-nonwaste technology 低废和无废技术
low angle 低角度边界;低角度;低角
low-angle dip 缓倾斜;缓倾角
low-angled steel block plane 单手小钢刨
low-angle fault 缓倾(角)断层;缓角断层;平倾断层;低角(度)断层
low-angle faulting 缓角断层作用
low-angle grain boundary 小角晶粒间界;低角度晶界;低角晶粒间界
low-angle light scattering 小角光散射
low-angle plane 低角刨
low-angle scattering 小角散射(法)
low-angle smooth bottom steel bench plane 低角平底钢刨
low-angle spray 低角度喷洒
low-angle stacking fault plane 低角度堆垛层错面
low-angle sunlight 低角度日光
low-angle thrust 低角度冲断层
low angular sheen 低角度光泽
low annual precipitation 低年雨量
low apsis 近地点
low area 低地
low area storm 低压区风暴
low aromatic white spirit 低芳烃石油溶剂
low-ash 低灰分的
low-ash coal 低灰煤
low ash fuel 低灰燃料
low atmosphere 低层大气
low atmosphere layer 低层天气
low atmosphere ozone 低层大气臭氧
low atmosphere ozone pollution 低层大气臭氧污染
low atmospheric pressure 低气压
low-attenuation range 低衰减范围
low-back chair 低背椅
low-background counter 低本底计数器
low bake curing alkyd/melamine-formaldehyde system 低温固化三聚氰胺甲醛/酸树脂系统
low-bake finish 低温烘烤(表面)处理;低温烘漆;低温烘烤面漆
low band 低波段
low-band boost 低频段响应提升
low band filter 低通滤波器
low bandwidth 低带宽
low bank 低岸
low bank of earth between field 田埂
low base-drag nozzle 小底部阻力喷管
low bay 低高度的工业厂房
low beam 暗光灯;小光灯;(车头的)低焦距光束
low bearing oil pressure governor 轴承低油压遮断装置
low bed 低座挂车;低平板拖车
low bed loader 低底盘装料车
low bed semitrailer 低车架半牵引车;低架式半拖车
low bed trailer 低架拖车;短平底拖车;平底拖车
low bid 低价标;低标
low binding 弱结合;松弛耦合
low birch thicket 矮灌丛
low birefringent fiber 低双折射光纤
low-blast furnace 低压鼓风炉;低压高炉
low block 低矮建筑(群);矮平房
low blower 低增压器
low bog 低地沼泽
low boiler 低沸(点)化合物;低沸剂
low boiling 低沸点的
low boiling liquid 低沸点液体
low boiling naphtha 低沸点石脑油
low boiling point 低沸点
low boiling point liquid 低沸点液体
low boiling point solvent 低沸点溶剂
low-boom 下弦杆;(桁架的)下弦
low boy 低座挂车;短平底拖车;低底盘载重汽车;低底盘混凝土吊罐运输车;矮(多斗)柜;短衣柜
low brake 低速制动器;低速闸
low brass 下半轴瓦;软黄铜;低锌黄铜
low brightness 低明度
low-browed 入口低矮的
low building 低矮建筑(群)
low-built 低构架的
low-built car 低重心车辆
low-built chassis 低车身盘
low bush 矮灌木林
low calorific value coal 低发热量煤
low calorific value gas 低热值煤气
low-capacitance cable 低电容电缆;低分布电容电缆
low-capacitance tube 低极间电容电子管
low-capacity 小容量;低电容
low-capacity cable 小容量电缆
low-capacity hydraulic system 小容量液压系统
low capped pile foundation 低桩承台;低承台桩基
low-carbon 低碳的;低碳
low-carbon cast iron 低碳铸铁
low-carbon cast steel 低碳铸钢
low-carbon cold drawn steel wire 冷拔低碳钢丝
low-carbon content 低含碳量
low-carbon ferrochromium 低碳铬铁
low-carbon ferromanganese 低碳锰铁
low-carbon ferronickel 低碳镍铁
low-carbon martensite 低碳马氏体
low-carbon martensitic steels 低碳马氏体钢
low-carbon pig iron 低碳生铁
low-carbon steel 软钢;低碳钢
low-carbon steel electrode 低碳钢焊条
low-carbon steel pipe 低碳钢管
low-carbon steel welding electrode 低碳钢电焊条
low-carbon tar 低碳焦油(沥青)
low-carbon welding wires 低碳焊接线材
low-carbon wire 低碳钢丝
low carrier modulation 低载频调制
low ceiling 低平顶
low-ceilinged room 低平顶房间;低顶棚房间
low-ceiling room 低平顶房屋
low cement castable 低水泥浇注料
low cement content 低水泥用量
low cement refractory castable 低水泥含量的耐火浇注料
low channel 低频道
low channel antenna 低频道天线
low channel seismogram 低频道地震图
low-chrome steel 低铬钢
low chromium alloy steel 低铬合金钢
low circle 恒隐圈
low circulation 低倍循环
low cistern 低水箱
low-class barroom 低级酒吧
low class residential zone 低等阶层住宅区
low cloud 低云
low coal seam 薄煤层
low coast 低平海滨;低平(海)岸;低海岸
low coastline 低岸线
low-coefficient glass 低膨胀系数玻璃
low column 低料柱
low-compaction applicator 低压轮胎式施肥机
low compressibility 低压缩性
low compressible soil 低压缩性土
low-compression 低压缩的
low-compression motor 低压缩发动机
low compression ratio 低压缩比
low concentrated ammonia-containing wastewater 含低浓度氨废水
low concentrated domestic sewage wastewater 低浓度生活污水处理
low concentrated domestic wastewater 低浓度生活废水
low concentrated methanol wastewater 低浓度甲醇废水
low concentrated municipal wastewater 低浓度城市废水
low concentrated organic wastewater 低浓度有机废水
low concentrated saline wastewater 低浓度盐质废水
low concentrated soluble wastewater 低浓度溶性废水
low concentrated wastewater 低浓度废水
low concentration 低浓度
low conductivity coating 低导热涂层
low conductivity water 低电导水;纯水
low-consistency plaster 低稠度石膏灰
low consumption engine 低耗油量发动机
low-contact correlation 低度接触相关
low-contract use 低度接触用途
low content alloy 低合金
low contrast 低反差
low contrast developer 软调显影剂
low convex area 低凸起区

low copper nickel matte anode 低铜镍锍阳极
low copper silica glass 低铜石英玻璃
low correlation 弱相关；低度相关
low cost 低造价；低费用；低成本
low-cost automation 低价自动化
low-cost bituminous surface 低价沥青路面
low-cost compacting method 简易击实法
low-cost computer 简易型计算机
low-cost credit 低息贷款
low-cost die 低值模具
low-cost dredging 简易疏浚
low-cost dwelling house 廉价住宅
low-cost goods 廉价值货
low-cost highway 低(造)价公路；低级公路
low-cost homes 低价住宅；低价住房
low-cost housing 低价住房；低(造)价住宅；低标准房屋
low-cost pavement 简易路面；低造价路面
low-cost road 低价公路；简易道路；低(造)价道路；低级道路
low-cost road surface 简易路面
low-cost sorbent 廉价吸附剂
low-cost surface 低级路面；低价路面
low counter gear 低副机构
low country 低洼地
low course 下游段
low coverage 小视界；近距(离)
low crack edge 低绽边
low crest breakwater 低堤顶防波堤
low cristobalite 低温型方石英
low crown 横坡小的路拱；低路拱
low current 小电流；低电流
low-current beam 弱流束
low-current digital circuit 弱电流数字电路
low-current heater 弱电流热丝；低电流热子
low-current plasma arc welding 小电流等离子弧焊
low-current target 弱流靶
low-cut filter 低灵敏度滤波器；低阻滤波器
low-cut frequency 低截频率
low-cut frequency slope 低截频徒度
low-cycle fatigue 低循环疲劳；低周疲劳
low-cycle fatigue meter 低循环疲劳强度计
low-cycle operation 低周波运行
low-cyclic(al) loading 低循环荷载
low-cyclic(al) loading constant amplitude test 低循环等荷载幅值试验
low-cyclic(al) loading constant strain rate test 低循环应变速率试验
low dam 潜坝；低坝
low-damped oscillator 小阻尼振子
low dark current 弱暗电流
low dark decay 低惰性
low data-rate input 低数据率输入
low deck of bridge 双层桥的下层桥面
low deck trailer 低架拖车
low definition 低清晰度
Lowden drier 劳登干燥机
low density 密度低的
low-density ablation material 低密度烧蚀材料
low-density alloy 低密度合金
low-density beam 弱流束；低密度粒子束
low-density board 低密度板
low-density cement 低密度水泥
low-density charge 低密度炉料
low-density concrete 低密度混凝土
low-density data system 低密度数据系统
low-density development 低密度发展
low-density dynamite 低密度黄色炸药
low density fiberboard 低密度纤维板；软质纤维板
low-density filler 低密度填料
low density gas 稀薄气体
low-density layer 低密度层
low-density light non-aqueous phase liquid 低密度轻质非水相液体
low density line 低运输密度线路
low-density packaging 低密度打包
low-density particle board 低密度刨花板
low-density passenger line 低运输密度客运线路
low-density pickup baler 低密度捡拾压捆机
low-density point of interest 低密度特征点
low-density polyethylene 低密度聚乙烯
low-density press baler 低密度压捆机
low density railroad 低运输密度铁路
low-density recording mode 低密度记录方式
low-density sampling 低密度采样
low-density wood 低密度木材
low-density wood chipboard 低密度碎木板；低密度刨花板
low depression 低俯角
low-depth bridge 低高度桥
low-depth girder 低高度梁
low dielectric(al) loss glass 低介电损耗玻璃
low dielectric(al) loss glass fiber 低介电损耗玻璃纤维
low-differential saturator 低差压式饱和器
low dilution 低度稀释
low dip(angle) 缓倾角；低倾角
low discharge 小流量；低水平排料；低(水)流量
low dispersion optic(al) glass 低色散光学玻璃
low dispersion spectrum 低色散光谱
low dissolved oxygen 低溶解氧
low dissolved oxygen concentration activated sludge system 低溶解氧浓度活性污泥系统
low distillation thermometer 低温蒸馏温度计
low distortion 轻度失真
low dose irradiation 小剂量照射
low dose of exposure 小剂量照射
low dose tolerance 低量耐受性
low down closet 低式冲水厕所
low-down manure spreader 低架厩肥撒布机
low-drag body 小阻力物体
low-drag cowl 减阻罩
low-drag profile 低阻叶型
low drive power modulator 低激发功率调制器
low dry strength 干强度低
low ductility 韧性低
low-ductility steel 低塑性钢
low-duty 小功率工作状态；小功率的；轻型
low-duty-cycle switch 脚时转换开关；短时工作开关
low-duty fireclay brick 低级耐火黏土砖；轻质耐火砖
low-duty of water 低效用水率；低灌溉率
low-duty pulsed tube 低工作比的脉冲管
low-duty supercharger 小力增压器
low-early-strength cement 低早期强度水泥；早期低强(度)水泥
low earth barrier 低土围堤；矮土(围)墙
low earth orbit 近地轨道
low earth orbit satellite 低轨道卫星
low efficiency 低效率；低利用率
low elasticity 低弹性
low electric(al) conductivity 低导电率
low electrode 低电极
low-elevation 低标高；低高程
low embankment 潜堤
low emission 弱放射；弱发射
low-emission coating 溶剂挥发量低的涂料；低散发涂料
low-emission equipment 低散发设备
low-emission material 低散发量材料
low-emission production 低(散发)污染生产
low-emission source 低污染排放源
low-emission technology 低污染(排放)技术
low emissivity 低发射率
low-emissivity coating 低发射涂料
low-emissivity glass 低辐射率玻璃
low end 低档的
low energy 低能的
low energy accelerator 低能量加速器
low energy building 低能耗建筑
low energy charged particle 低能带电粒子
low energy coast 低能海岸
low energy component 低能组分；低能成分
low energy consumption 低耗能
low energy consumption building 低耗能建筑
low energy content cement 低能量含量水泥
low energy electron-beam monitor 低能电子束监测器
low energy electron diffraction 低能电子衍射
low energy electron microscopy 低能电子显微镜
low energy environment 低能量环境；低能环境
low energy geothermal field 低能位地热田
low energy ion scattering spectroscopy 低能离子散射谱
low energy membrane bioreactor 低能膜生物反应器
low energy orbital 低能轨道
low energy physics 低能物理学
low energy proton spectrometer 低能质子能谱仪
low energy relay 低(耗)能继电器
low energy resonance 低能谐振
low energy scrubber 低能耗洗涤器
low energy ultrasound assisted bioreactor 低能超声波辅助生物反应器
Lowenhertz thread 卢温赫兹螺纹
low enriched fuel cycle 低浓燃料循环
low enrichment reactor 低浓缩铀反应堆
low enrichment uranium 低浓缩铀
low enthalpy fluid 低焓流体
low enthalpy well 低焓井
low entropy outlook 低熵世界观
lower 下托牙；下部；较低；放低
lower acceptance value 下限接受值
lower accumulator 下限累加器
lower acid 低级酸
lower adjusting valve 下端调整阀
lower air cylinder gasket 下气筒垫密片
lower alcohol 低级醇
lower alternate depth 低共轭水深
lower altitudes 低海拔地区
lower anchor bracket 下承锚底座
lower anchors 下连接子卡
lower angle brace 下隅撑
lower apex of fold 褶皱底
lower appendage 船体水线下附属物；船体水线下附属体
lower approach 下游引航道
lower approach channel 下游引航道
lower approximate value 下近似值；偏小近似值
lower approximation value 偏小近似值
lower apron 迎水坡水下护坦；水下护坦；水下护脚；防波堤迎水斜坡水下部分
lower apse 近星点
lower aquifer 下部含水量
lower arch bar 下拱杆
lower atmosphere layer 低层大气层
lowerator conveyer 包装件输送机
lower auxiliary steam port 下部副气口
lower away 松下
lower back 下背
lower bainite 下贝氏体
lower ball cover 下轴承盖
lower band 下边；低频带
lower bank interest rates for savings deposits and loands 降低银行存贷款利率
lower bar 下鼠笼条；下层线棒(电机)
lower base figure 压低基数
lower basement 底层地下室；第二层地下室
lower basin (船闸的)下游河段
lower beach 前滨
lower beacon light 下信标灯
lower beam 下横梁
lower bearing 下轴瓦；下(部支承)轴承
lower bearing bracket 下机架
lower bearing housing 下轴承箱
lower bearing housing cap 下轴承箱盖
lower bearing spider 下轴承架
lower bed 下层；下层
lower bell 大料钟
lower bench 下梯段；下台阶；下段(地层露头)
lower berth 下铺
lower block 下滑车
lower bolster 下模板；下梁；下垫板
lower boom 下弦杆；下弦
lower boom junction plate 下弦接板
lower boom longitudinal bar 下弦纵杆
lower boom member 下弦杆部件
lower boom plate 下弦板
lower boom rod 下弦杆
lower border 下图廓；下边线
lower bosh line 下炉腹线；炉腹底线
lower bound 下限；下界
lower boundary 下边界
lower bound depth of the water bearing formation 含水层底界面深度
lower bound expression 下界表达式
lower bound solution 下限解
lower bound theorem 下限原理；下限定理
lower box 下箱
lower bracket 下轴承支架
lower bracket for cantilever 腕臂下底座
lower bracket for double track cantilever 双腕臂下底座

lower branch 下子午圈;下半圈
lower branch of Greenwich 格林威治子圈
lower branch of the meridian 下子午线
lower branch of the observer's meridian 测者子圈
lower brass 下轴瓦
lower bridge 下(层)桥楼;下层驾驶台
lower bridge deck 下层桥面
lower bubble 下部气泡式鼓包(机身);照准部水准器
lower building 低层建筑(物)
lower bunk 下铺
lower cage 下鼠笼
lower calorific power 低卡值;低发热量
lower calorific value 低位热值;低卡值;低发热量
lower camber 下弦
lower camber side 下弧面
Lower Cambrian 早寒武世【地】
Lower Cambrian series 下寒武统【地】
lower camshaft drive gear 下凸轮轴驱动齿
lower camshaft worm gear 下凸轮轴蜗轮
lower canal reach 下运河段
lower canopy 下冠层
lower cap 底盖
Lower Carboniferous series 下石炭统
lower cardan 下端万向节
lower carriage 低车架;下部机械滑动部分
lower case 小写体
lower case alphabetic character 小写字符
lower case letter 小写字母
lower case type 小写字体
lower categories 低级分类单位
lower center 下死点
lower chain 下链条
lower chamber 下闸室
lower change point 下临界点
lower cheek plate (破碎机的)下(颚)板
lower chord 下弦杆;下弦
lower chord bar 下弦杆
lower chord junction plate 下弦联板
lower chord member 下弦构件
lower-chord panel point 下弦节点
lower chord rod 下弦杆
lower chroma sideband 色度信号下边带
lower chromosphere 色球低层
lower circle 下盘
lower citadel 下层城堡
lower clamp 下盘制动螺旋
lower class 下类
lower class limit 组下限
lower class treatment of wastewater 污水一级处理
lower cloud point 下浊点
lower coat 下层(面层)
lower coefficient of resource abundance 资源丰度系数低
lower coil 下部线圈
lower colonnade 下柱列
lower confidence interval 下置信区间
lower confidence limit 下置信限;可信区间下限
lower confining bed of aquifer 含水层下部不透水层;含水层下部隔水层
lower contact 下接头
lower control arm 下控制杆
lower control limit 下控制限;下管理限;控制下限
lower core support barrel 下部堆芯围筒
lower core(support) plate 下部堆芯板
lower corona 日冕低层
lower cost 降低成本
lower-cost region 成本较低的区域
lower couch(roll) 下覆辊
lower counter 船尾突出下部
lower coupling 连接器(汽车)
lower course 下游河段;下层
lower course of river 河流下游
lower course of stream 河流下游
lower cover 下机壳;下盖
lower crankcase 下曲轴箱
lower Cretaceous 早白垩世【地】
lower Cretaceous series 下白垩统【地】
lower critical cooling rate 下临界冷却速度
lower critical stress 低临界应力
lower critical velocity 下限临界流速;低临界流速;次临界速度
lower crown height 下冠高
lower crown length 下冠长

lower crust 下地壳层
lower crustsima 下地壳(硅镁)
lower culmination 下中天
lower curtate 下区段;低部
lower curve 下曲线
lower cut 下部掏槽;粗切削
lower-cut-off frequency 下限截止频率
lower cutting position 下切削位置
lower cylinder 下汽缸
lower cylinder bore wear 下止点附近汽缸壁的磨损
lower cylinder half 下半汽缸
lower cyma 下裂
lower dead center 下止点
lower dead-center indicator 下止点指示器
lower dead point 下止点
lower deck 下层桥面;下(层)甲板
lower decking level 下部装罐水平
lower decomposition 下半分解
lower deep 下深槽
lower delta-plain deposit 下三角洲平原沉积
lower density 下密度
lower derivative 下导数
lower derived function 下导函数
lower development 下游梯级
lower deviation 下偏差
Lower Devonian 早泥盆世【地】
Lower Devonian series 下泥盆统【地】
lower diaphragm plate and valve stem 下部隔膜板及阀杆
lower die 下模
lower differential coefficient 下导数
lower discharge tunnel 下部泄水隧道;泄流底涵
lower distribution plate 下分配板
lower door rail (门的)下冒头
lower drum 下气包
lower edge 下图廓;底边
lower edge board 下边板
lowered ground water level 降低地下水位
lowered position 落下位置
lowered superficial resistance 表气不固
lowered water level 降落水位
lower emulsion 下层乳剂
lower end 下端
lower end buoy 中沙下端浮标
lower end of dredge-cut 挖槽下端
lower end of the mast 桅杆下端
lower end shield 下半端罩
lower envelope curve 下包络线
lower envelope principle 下包络原理
lowerer 下楼机
lower expansion chamber 下扩大室(调压井)
lower explosive limit 爆炸下限
lower extreme 下限;下端
lower extreme point 下端点
lower failure plane 下层破裂面
lower falsework 下部脚手架
lower fan 下扇
lower fascia 柱顶过梁最下一条
lower fiber 下缘纤维
lower fill 路堤下层
lower fillet and fascia 下枋
lower fitting radius 后回转半径
lower fixed jaw face 下固定颚板面
lower flange 下翼缘;下法兰
lower flange of girder 梁下翼(缘);梁的下翼缘
lower flange plate 下翼缘板
lower flat pole 低层磁极
lower floor 下层楼
lower flow regime(n) 缓流状态;下部水流动态;低流态
lower frame 下框架
lower frame bracing 下部框架撑条
lower frame rail 车架下梁
lower frequency component 低频分量
lower frequency limit of audibility 可听声频下限
lower frigid zone 下寒带
lower funnel 下层烟囱
lower gasket 下衬
lower gate 下游(船闸)闸门;下(首)闸门
lower gate bay 下闸首
lower gate block 下闸首墩
lower gate recess 下游闸门门槽
lower ga(u)ge 下水尺;下端水位尺
lower ga(u)ge cock 下端试水位旋塞

lower-grade metamorphic soft coal 低变质烟煤
Lower Greensand 下绿砂层(白垩纪);下海绿石砂
lower grid 低格栅
lower grid plate 下栅板
lower groundwater table 降低的地下水面
lower guard sill 下游防护槛
lower guard wall 下游护墙
lower gudgeon 闸门底枢
lower guide 下导轨
lower guide bearing 下导轴承
lower guide metal 下导轴瓦
lower guide rail 下导轨
lower guide track 下导轨;(门的)底导轨
lower guide vane ring 下导叶环
lower half 下面部分;下半部
lower half assembly 下半组装
lower half bearing 轴承下瓦
lower half casing 下半汽缸
lower half crankcase 下半曲轴箱
lower half nut 对开螺母下部
lower half-plane 下半平面
lower half-power frequency 下半功率频率
lower half-space 下半空间
lower harbo(u)r 与外海接近的港口
lower hazard limit 危害下限
lower head 下闸首;底盖
lower header 下水箱
lower head-water section 下游船闸的上游河段
lower heating value 低位发热值;低位发热量;低(发)热值
lower height tree 低层树
lower high water 低高潮
lower high water interval 低高潮间隙
lower-horizon soil 下层土
lower house 下议院
lower housing 下部壳体
lower hybrid wave 低温杂波
lower in 下沟
lower-income group 较低收入组别
lower-income housing 较低收入者住房
lowering 下降;降低;低落;低沉;下沉
lowering brake control 下降制动器控制(装置)
lowering brake switching 下降制动开关
lowering conveyer 下放物体的垂直运输机
lowering device 降弓装置
lowering funnel 可放倒烟囱
lowering furnace 下移烧结炉
lowering girder 落梁
lowering groundwater table 地下水位下降
lowering height limiter 下降极限限位器【机】
lowering in (管道的)下入沟内;下沟
lowering jack 放下千斤顶
lowering limb 下降段;亏水曲线;退水(曲)线
lowering limiter 下降深度限位器
lowering line 堵漏砧吊索
lowering mast 可放倒式桅杆
lowering mechanism 下落机构
lowering melt process 坩埚拉丝法
lowering motion 下降运动;降落运动
lowering of amplitude 振幅降低
lowering of bucket ladder 斗架下放
lowering of freezing point 冰点下降
lowering of groundwater 地层降水
lowering of groundwater level 地下水位下降;降低地下水位
lowering of hardness and desalination 降硬脱盐
lowering of horizon 水平线降低
lowering of pressure 压力降低
lowering of roadbed 落底
lowering of salvage pontoon 打捞浮筒溜缆
lowering of temperature 温度下降
lowering of the groundwater 地下水位降低
lowering of the inner rail half elevation 降低内轨半超高
lowering of the water level by wind 风退水
lowering of the water table 地下水位降低
lowering of track 落道
lowering of underground water 降低地下水位
lowering of vapo(u)r pressure 蒸汽压下降
lowering of viscosity 黏滞度降低
lowering of water by effect of wind 风降水位
lowering of water-level 水位下降;水面下降
lowering of water table 降低地下水位

lowering position 落下位置
lowering precast tunnel sections 下降预制(混凝土)隧道筒管部分
lowering rate 下落速度;下降速度
lowering salvage pontoon 沉放打捞浮筒
lowering speed 下降速度;松吊速度
lowering speed of the hook 吊钩下降速度
lowering temperature crystallization 降温结晶法
lowering time 下落时间
lowering wedges 拆除架楔
lower inlet for medium 介质下部入口管
lower in quotation 报价较低
lower integral 下积分
lower interlock relay 下联锁继电器
lower jaw 下颚
Lower Jurassic 早侏罗世【地】;下侏罗纪【地】
Lower Jurassic series 下侏罗统【地】
lower lateral 下部侧撑;底部侧撑
lower lateral bracing 下弦横向水平支撑;下(弦)横撑;纵向下平联
lower latitude 下纬度
lower layer 下一层;下层
lower leaf 下游门叶;下游门扉;下分层
lower leg 泄水管(道)
lower level intake 深式进水口
lower level interrupt 低级中断
lower level loading 下装
lower level node 下层节点
lower level objective 低水平目标
lower lift drum 下部提升筒
lower lift lock 低水头船闸
lower light 下面光
lower limb 下翼;下(边)缘
lower limestone group 下灰岩群
lower limit 下限限额;下限尺寸;最小极限;最小尺寸;低限
lower limit function 下限函数
lower limiting filter 高通滤波器
lower limiting frequency 下限截止频率
lower limit of crystallization temperature 析晶温度下限
lower limit of detectability 最低检出量;可检出的最低限值
lower limit of detection 检测下限
lower limit of flammability 着火浓度下限;爆炸下限
lower limit of organic carbon 有机碳下限含量
lower limit of separation 分选下限
lower limit of stability 下限稳定性;安定下限
lower limit of ultimate strength 极限强度下限;低限强度极点
lower limit of variation 偏差下限
lower limit of yield strength 屈服强度下限
lower limit on the right 右下限
lower limit register 下限寄存器
lower limit value 下限值
lower link 下拉杆
lower lip (进气道的)前下缘;(虹吸道的)出口屋斗;(虹吸道的)出口反弧段
lower load block 吊钩夹套;取物滑轮匣(起重机吊具)
lower load factor 低负荷因素
lower lock approach 下游引航道;船闸下游引航道
lower lock gate 下游(船闸)闸门;下闸门
lower low-water 低低潮
lower low-water datum 低低潮基准面
lower low-water interval 低低潮间隙
lower low-water line 低低潮水位线
lower-lying 低洼土壤
lower main steam port 下部主汽口
lower management 基层管理
lower mantle 下地幔;地幔下部
lower margin 下缘;下图廓
lower marginal area 下部边缘区
lower mast 下桅
lower mature condensated gas 低成熟凝析气
lower mature source rock 低成熟源岩
lower mature stage 低成熟阶段
lower member 低级物
lower-method of mining 矿业开发技术低
Lower Mississippian 早密西西比世【地】
lower miter gate 下闸首人字门
lower mode vibration 低次波振动
lower molecular weight hydrocarbon 低分子量烃类
lower-molecular weight organic solution 低分子量有机液
lower-most point 最低点
lower motion 下盘转动
lower mo(u)ld half 下模;凹模
lower nappe profile 水舌下缘线
lower new red sandstone 下层新红砂岩
lower nicol 下偏光镜
lower node 下节点
lower noise circular saw 低噪声手动进料木工圆锯机
lower nut 下螺母
lower of historic(al) cost or net realizable value 历史成本与可变现净值孰低
lower of total cost or total market 总成本与总市价孰低
lower oil header 下集油箱
lower oil pan 下油盘
lower operating range control 下限控制
lower order bias estimator 较低阶有偏估计量
lower-order point 测图点
lower order structure 低级构造
lower-order triangulation 低等三角测量
lower-order ordinate set 下纵标集
Lower Ordovician 早奥陶世【地】;下奥陶纪【地】
Lower Ordovician series 下奥陶统【地】
lower orifice 阴窍
lower output 较低输出
lower oxygen consumption water 耗氧量稍低的水
lower paid bracket 低工资等级
lower pair 面接对偶;低副
lower panel 下翼片
lower part 下面部分;下半部;深部
lower particle size limit 粒度下限
lower part of crankcase 曲轴箱的下面部分
lower pass 下轧道;下孔道
Lower Pennsylvanian 早宾夕法尼亚世
Lower Permian 早二叠世【地】
Lower Permian series 下二叠统
lower pile cap 下承台
lower pilot step 下排障器踏级
lower pintel 闸门底枢
lower pintle 下枢轴
lower pintle casting 底轴浇铸
lower pivot 闸门底枢
lower pivot casting 下枢轴铸模;下枢轴铸件;下枢轴浇铸;下枢轴铸造
lower plane 底面
lower plane-bed facies lamellar structure 下部平底相纹理构造
lower plant 低等植物
lower plastic limit 下塑限;塑性下限;塑限下限(土壤含水量指标)
lower plastic moisture content limit 塑性下限含水量
lower plate 下盘;下夹板;下底板;下部板块;主夹板;底壁
lower platen 下压板
lower Pleistocene 早更新世
lower plunger 下冲杆
lower polar stratosphere 极地平流层下部
lower pole 下极
lower pole face 下极面
lower pole piece 下极头
lower polyolefins 低聚烯烃
lower pond (船闸的)下游河段;船闸下游水库
lower pool 尾水池;下游塘;(船闸的)下游河段;下深槽;船闸下游水库;船闸后池
lower pool elevation 下游水位
lower pool of lock 船闸下游水面
lower portal leg 门腿【机】
lower powered 低功率的
lower pressure 低压力
lower pressure and production stage 低压低产期
lower-pressure limit 低压极限
lower prestressing level 下层预应力标高
lower price 降价
lower principal purlin(e) 下金檩;下金桁
lower priority group 低优先群
lower priority interrupt 低优先中断
Lower Proterozoic era 下元古代【地】
lower punch 下凸模;下模冲;下冲杆
lower punch pressure 下冲头压力
lower purlin(e) tiebeam 下金枋
lower quartile length 下四分位长度
lower race ring 下座圈
lower rail 下导轨;下横木;里轨
lower ram 下锤头
lower range 低量程
lower range value 下限值
lower reach 下游(河)段;下游(边)
lower reaches of stream 河流下游
lower red 暗红色
lower reservoir 下游水库
lower residual pesticide 低残留农药
lower resistance pH glasselectrode 低阻 pH 玻璃电极
lower retaining wall 下挡墙
lower return nozzle for medium 介质下部回流管
lower rigging 下桅的左右支索
lower ring 下部温带板
lower river 下游
lower-river course 下游河道
lower roll 下辊
lower roller 支重轮
lower rope 底部绳索
lower row 下列
lower runner boss cover 转轮泄水锥下段
lower sample 下层试样
lower scaffold 下部脚手架
lower scale 刻度下段
lower screen 下层筛
lower seal 下密封
lower seat 下承座
lower seat connecting arm 下座椅连杆
lower section 下面部分
lower segment 下节
lower semi-continuity 下半连续性
lower semi-continuous function 下半连续函数
lower semi-lattice 下半格
lower semi-modular lattice 下半模格
lower semisphere 下半球
lower separator 下横梁
lower shaft 下轴
lower shed 下梭口
lower shell 下筒体
lower shield 下半端罩
lower shoe 下底板
lower shoe bearing 导向轴承
lower shore 前演
lower shrouds 下桅的左右支索
lower side 下面;下侧(面)
lower sideband characteristic 下边带特性
lower sideband spectrum 下边带频谱
lower sideband upconverter 下边带上变频器
lower side flat 下边滩
lower side panel 下侧板
lower side rail 下桁材
lower sieve 下筛;草籽筛
Lower Silurian 早志留世【地】;下志留【地】
Lower Silurian series 下志留统【地】
lower sleeve 下套筒
lower slice 下分层
lower sluice gate 下游泄水闸门
lower socket 下承座
Lower Sonoran life zone 下北美生物带
lower span wire 软横跨下定位索
lower specification limit 下规定限
lower speed 减低速度
lower speed drying 降速干燥
lower spider 下机架
lower stack 下层烟囱
lower stack limit 堆栈下限
lower standard 低标准
lower state 低能态
lower station 下游电站
lower steam cylinder gasket 下汽缸垫密片
lower steam inlet port 下进汽口
lower stem seal 下阀杆密封
lower stock 下舵杆
lower storage basin 下游蓄水池
lower storage yard 低货位堆货场;低货位
lower storey of forest 下层林
lower stream course 下游河道
lower stretch 下游河段
lower strings 下弦
lower subalpine 下亚高山
lower sublittoral zone 下浅海地带

lower support of bucket ladder 斗架下支承
lower surface camber 下曲面
lower surface depth of shale in coal formation 夹矸底界面的深度
lower surge basin 下调压井；下调压池
lower surge chamber 下调压室
lower suspension arm 下悬挂臂
lower swing jaw face 下摆动颚板面
lower tank 下箱
lower tank block 池壁下部砖
lower-techniques of mineral commodity processing 矿产加工技术低
lower tectonic level 下部构造层次
lower temperate zone 下温带
lower tension carriage 下部拉紧装置
lower terrigenous formation 下部陆源建造
lower the duty 课税从轻
lower the land 远远离开陆地
lower the price 削价；降低价格
lower the temperature 降温
lower thrust wall 下游承推墙
lower tolerance limit 下容差限
lower tooth 下齿
lower toxic limit 耐毒下限；毒性下限
lower tracer 下扫描线
lower track 下道
lower transit 下中天
lower transverse strut 下横撑
lower tread roller 支重轮
lower tree layer 低树层
lower triangular matrix 下三角形矩阵
Lower Triassic 早三迭世【地】
Lower Triassic sandstone 早期三叠系砂岩
lower Triassic series 下三叠统【地】
lower troposphere 对流层下部
lower tumbler 下导轮
lower tween deck 下二层舱；下的中甲板
lower value seat 下端阀座
lower valve 下方值
lower variation 下变差
lower variation of tolerance 下偏差
lower wall 下盘；下底板；主夹板；断层下盘；下壁
lower wall halo 下盘晕
lower water course 河流下游
lower-water indicator 低水位指示器
lower water line 低潮线
lower water mark 低潮线(痕迹)；低潮线标志
lower weight 坠重
lower whorl 下轮
lower wig-wag 下摇摆轮
lower window edge 窗口的下界
lower wing 下翼
lower wing wall 下闸首翼墙
lower wishbone link 下叉连杆
lower yield point 下屈服点；屈服点下限
lowery pressure process 半细胞加压法(木材防腐处理)
lowest 最低的
lowest acceptable rate of return 能接受的最低收益率
lowest achievable emission rate 最低可达排放(速)率
lowest assured discharge 最小保证流量
lowest assured natural stream flow 最小可靠天然径流
lowest astronomical tide 最低天文潮
lowest astronomical tide level 最低天文潮位
lowest atmospheric layer 最低大气层；最低层大气；贴地大气层
lowest average 最低平均数
lowest bid 最低报价；(指标的)竞买的最低价；出价最低的投标
lowest bidder 最低价投标人
lowest bid price 最低标价
lowest common denominator 最小公分母；最低公分母
lowest common multiple 最小公倍数；最低公倍数
lowest critical point 最小临界点
lowest critical value 最小临界值
lowest critical velocity 最低临界速度
lowest discharge 最小流量
lowest dose level 最低剂量
lowest effective power 最低有效功率
lowest elevation 最低高程

lowest energy 最低能量
lowest evaluated bid 最低评标价投标；估价最低投标
lowest evenness 最低均匀
lowest ever-known discharge 最小枯水流量；历史最小流水量
lowest ever known water 最低枯水位
lowest ever-known water level 历史最低水位；历史最低潮位
lowest ground water 最低地下水
lowest guaranteed discharge 最小保证流量
lowest height of embankment 路堤最低高度
lowest high water 最低高水位
lowest ignition point 最低着火点
lowest ineffective concentration tested 测试的最小无影响浓度
lowest interest rate 最低(银行)利率
lowest level 最低标高；最低高程
lowest level allowing water quality standard 最低水平容许水质标准
lowest level node 最低层节点
lowest limit 最小尺寸
lowest low tide 最低低潮
lowest low water 最低低潮
lowest low water datum 最低水位基准面
lowest low water level 最低低水位；最低低潮位；历史最低低潮位
lowest low water springs 大潮最低低潮位；大潮最低低潮面
lowest low water spring tide 朔望大潮最低水位；大潮最低水位
lowest mode of vibration 最低振型
lowest navigable discharge 最低通航流量
lowest navigable stage 最低通航水位
lowest normal low water 最低正常低潮；正常低潮
lowest normal tide 最低正常水位
lowest observable adverse effect level 有毒性反应的最低浓度
lowest observed adverse effect level 最小觉察到的有害效应水平；最低可见有害作用水平
lowest observed effect concentration 最小觉察到的效应浓度；最低有影响浓度
lowest observed effect level 最小觉察到的效应水平；最低可见作用水平
lowest observed frequency 最低观测频率
lowest offer 最低价的报盘
lowest operating frequency 最低工作频率
lowest order 最低位
lowest oxygen consumption water 耗氧量很低的水
lowest passenger deck 最低客舱甲板
lowest percentage extractability 最小可提取性百分率
lowest permissible temperature for welding 焊接允许的最低温度
lowest point 最低点
lowest point of foundation 基础最低点
lowest possible frequency 最低可用频率
lowest possible low water 最低可能低潮位；可能最低水位；可能最低低潮位
lowest possible price 最低可；尽可能的最低价
lowest price limit 最低限价
lowest production cost 最低生产成本
lowest qualified bidder 信誉可靠最低价的投标商；最低合格投标人
lowest quotation 最低报价
lowest recorded discharge 实测最小流量
lowest recorded level 已记录的最低水平面；记录最低水位；实测最低水位
lowest recorded stage 已记录的最低水位；实测最低水位
lowest rejected concentration tested 最小排污浓度测定值
lowest required radiating power 最低要求辐射功率
lowest reservoir level 最低库水位
lowest resonance 最小能量时的共振
lowest responsible bid 合格最低标；信誉可靠最低标价
lowest responsible bidder 信誉可靠最低标价的投标人；最低合格投标人；信誉可靠最低标价的投标商
lowest responsive bid 信誉可靠最低标价；最低合格标单

lowest safe waterline 最低安全水位
lowest self-ignition point 最低自燃温度
lowest slump 最小塌落度；最干硬稠度(即最小坍落度)
lowest speed 最低速度
lowest stage of groundwater table 最低地下水位；地下水最低水位
lowest storage level 最低库水位
lowest streamflow 最小径流(量)
lowest taxable limit 最低课税限度
lowest temperature 最低温度；最低温差
lowest tender 最(低的)投标；最低报价；出价最低的投标
lowest term 最低项；最低条件
lowest tide water mark 最低潮位标志
lowest unfilled molecular orbital 分子最低空余轨道
lowest upper elevation 上游最低水位
lowest upper pool elevation 上游最低库水位
lowest usable frequency 最低可用频率
lowest useful frequency 最低使用频率
lowest useful high frequency 最低可用高频
lowest value of measured pressure 测量压力最低值
lowest value on record 最低记录值
lowest water level 最低水位；最低潮位
lowest wave trough 最低波谷
low excess air 低过量空气
low-expansion 低膨胀系数的
low-expansion alloy 低膨胀合金
low-expansion coefficient 低膨胀系数
low-expansion glass 低膨胀玻璃
low-expansion glass-ceramics 低膨胀微晶玻璃
low-expansion material 低膨胀系数材料；低膨胀材料
low-expansion steel 低膨胀钢
low expansion cement 低膨胀水泥
low explosive 低效炸药；低级炸药；低爆速炸药
low-explosive forming 低爆炸成型
low-factor line 低功率因数线路
low-fall installation 低水头(水力发电)设备；低水头电站
low fat fishes 低脂鱼类
low fat marine product 低脂海产品
low fidelity 低保真度
low field layer 低草层
low field loss 低场损耗
low finish 低光泽加工
low finned tube 低翅片管
low fire 文火
low-fired porcelain 低火度瓷器；低温瓷(器)
low firing enamel 低温搪瓷
low flammability 不易燃性
low-flash 低温闪蒸；低温发火；低闪点(的)
low flood plain 低河漫滩
low-floor 低地板
low-floor bus 低地板式公共汽车
low floor charger 低台装料机
low flow 低流动性；枯水流量；枯水；低流量
low-flow augmentation 枯水期水源；枯水期水量；枯水期补充；枯水流量增加
low-flow channel 枯水河槽
low-flow forecast 低水预报；枯水预报
low-flow frequency 枯水位频
low-flow frequency curve 枯水位频曲线；枯水流量频率曲线；低流量频率曲线
low-flow navigation depth 枯水通航深度
low-flow period 枯水期；枯季；低水期
low-flow plumbing fixture 低水量卫生设备
low-flow season 枯水期；枯水季节；枯水季
low-flow shower 低流量喷水头
low-flow shower head 低流量喷水头
low-flow-water level 低水水位
low-flow year 枯水年；少水年；旱年
low fluid loss cement 低失水水泥
low flush tank 低水箱
low flush toilet 低冲洗水量马桶
low-flutter magnet 小调变度磁铁；低颤磁铁
low-flux reactor 低通量反应堆
low-flying aircraft 低空飞行飞机
low foaming 低泡的
low foaming surfactant 低泡表面活性剂
low foam metal cleaner 低泡金属清洗剂

low forest 低乔林;矮林
low form 低温型(晶体);低等类型
low-freeboard ship 低干舷船
low-freezing 低凝固点
low-freezing explosive 难凝炸药;低温炸药
low-freezing liquid 低凝液体
low-frequency 低周波的;低频率的;低频率;低频;频低
low-frequency absorption section 低频吸收剖面
low-frequency acquisition 低频显示
low-frequency alternate current instrument for resistivity 低频交流电阻率仪
low-frequency amplification 低频放大
low-frequency amplification stage 低频放大级
low-frequency amplifier 低频放大器
low-frequency analyser and ranging sonobuoy 低频分析器与测距声呐浮标
low-frequency antenna 低频天线
low-frequency background 低频背景
low-frequency band 低频带
low-frequency beacon 低频信标
low-frequency cable 低频电缆
low-frequency ceramics filter 低频陶瓷滤波器
low-frequency characteristic oscillograph 低频特性图示仪
low-frequency choke 低频扼流圈
low-frequency communication cable 低频通信电缆
low-frequency compensation 低频补偿
low-frequency compensator 低频补偿器
low-frequency correction 低频校正
low-frequency current 低频电流
low-frequency cutoff 低频截止
low-frequency cycle 低周周期;低频循环荷载
low-frequency cycle constant-load-amplitude test 低频循环等荷载幅值试验
low-frequency cycle constant-strain-rate test 低频循环等应变速率试验
low-frequency dielectric(al) separation method 低频介电分离法
low-frequency dielectric(al) separator 低频介电分离仪
low-frequency direction finder 低频测向器
low-frequency distortion 低频失真
low-frequency dry-flashover voltage 低频干闪络电压
low-frequency electric(al) furnace 低频电炉
low-frequency electric(al) porcelain 低频电瓷
low-frequency electric(al) smelting furnace 低频电气熔铁炉
low-frequency electromagnetometer 低频电磁仪
low-frequency field 低频场
low-frequency filter 低频滤波器
low frequency filtering image with partial coherent light 部分相干光低通滤波图像
low-frequency fluctuation 低频起伏
low-frequency furnace 低频炉
low-frequency gain 低频增益
low-frequency ga(u)ge 低频仪
low-frequency generator 低频发生机
low-frequency geophone 低频检波器
low-frequency head 低频部件
low-frequency heating 低频加热
low-frequency impedance corrector 低频阻抗校正器
low-frequency induction furnace 低频感应电炉;低频电感加热炉
low-frequency industrial oscilloscope 低频工业示波器
low-frequency iron core inductance 低频铁芯电感
low-frequency limit 低频极限
low-frequency limit circuit 低频限制电路
low-frequency magnetic anomaly 低频磁异常
low-frequency modulation 低频调制
low-frequency navigation(al) system 低频导航系统
low-frequency noise 低频噪声
low-frequency notch filtering 低频陷波速过滤
low-frequency omnidirectional range 低频全向作用距离
low-frequency oscillator 低频振荡器
low-frequency oscillograph 低频示波器
low-frequency padder 低频微调电容器;低频垫整电容器

low-frequency parameter 低频参数
low-frequency power amplifier 低频功率放大器;低频功率放大电路
low-frequency preamplifier 低频前置放大器
low-frequency propagation 低频传播
low-frequency pulp screen 低频筛浆机
low-frequency puncture voltage 低频击穿电压
low-frequency quasi-static dynamic(al) load 低频准静态动力荷载
low-frequency radio astronomy 低频射电天文学
low-frequency radio telescope 低频射电望远镜
low-frequency relay 低频率继电器;低频继电器
low-frequency resistance 低频电阻
low-frequency response 低频响应
low-frequency ringer 低频振铃器
low-frequency shock transducer 低频振动传感器
low-frequency signal generator 低频信号发生器
low-frequency sound insulation 低频隔声
low-frequency spark array 低频电火花组合
low-frequency spectrum 低频光谱
low-frequency stage 低频段
low-frequency start 低频起动
low-frequency stress 低频率应力
low-frequency substitution error 低频替代误差
low-frequency time and frequency dissemination 低频时频发播
low-frequency transconductance 低频跨导
low-frequency transduction 低频传导
low-frequency transformer 低频变压器
low-frequency transistor 低频晶体管
low-frequency tube 低频管
low-frequency vibration 低频振动
low-frequency vibratory mixing 低频振动搅拌
low-frequency wave 低频波
low-frequency wet-flashover voltage 低频湿闪络电压
low friction bearing 低摩擦轴承
low friction operation 低摩擦操作
low fuel consumption petrol engine 节油改进型汽油机
low-fume and harmfulness electrode 低尘低毒焊条
low-fusible ash 低熔灰分
low-gain 低增益
low-gain channel 低增益通道
low-gallonage spraying 低容量喷雾
low-gantry cable bulldozer 低架钢索平路机;低架钢索推土机
low-g centrifuge 小重力加速度离心机
low gear 低(速)挡;低速齿轮;低挡
low-grade 低级的;低等的;劣等;轻度的;低质量;低质的;低品位的;低劣(的)
low-grade antimony metal 低品位金属锑
low-grade asbestos 低级石棉
low-grade cement 低标号水泥;低强度等级水泥
low-grade cement mortar 低标号水泥砂浆;低强度等级水泥砂浆
low-grade coal 低质煤;低品位煤
low-grade concrete 低强度混凝土;低标号混凝土
low-grade copper 低级铜
low-grade deposit 低品位矿床
low-grade energy 低能
low-grade fuel 劣质燃料;低品位燃料;低级燃料;低质燃料
low-grade goods 劣货
low-grade heat consumer 低位热能用户
low-grade level 低级
low-grade lignite coal 低质褐煤
low-grade lime 低品级石灰;低(品位)石灰
low-grade material 低品位原料
low-grade metal 低品位金属
low-grade metamorphosed 低变质
low-grade mineral 低品位矿石
low-grade ore 低级矿;贫矿(石)
low-grade regional metamorphism 低级区域变质作用
low-grade silicate cotton 低品位炉渣棉;低品位矿渣棉
low-grade silicate wool 低品位炉渣绒;低品位矿渣绒
low-grade tectonic unit 低级构造单元
low-grade wolfram 低品位钨矿
low gradient 小坡度;缓坡;平缓梯度;平缓坡度;平缓坡道;低梯度

low-gravity 低比重
low ground 洼地;低(洼)地
low growing plant 矮生植物
low growing vegetation 矮生植物;矮生植被
low hardness circulating cooling water system 低硬度循环冷却水系统
low hardness water 软水;低硬度水
low hazard contents 不易燃建筑物料;低火灾隐患
low-hazard industrial buildings 低危险工业建筑物;低危险工业房屋
low-hazard storage buildings 低危险储[贮]藏建筑物
low-head 低压头;低水头;低式;低落差;低架式
low-head centrifugal pump 低扬程离心泵;低水头离心泵
low-head cyclonic-type separator 低压旋流式分离器
low-head development 低水头开发
low-headed training 矮干整枝;低干整枝
low-head hydroelectric(al) plant 低水头水力发电厂;低水头电站
low-head hydroelectric(al) power plant 低水头水电厂
low-head hydroelectric(al) power station 低水头水电站
low-head hydro-junction 低水头枢纽
low-head installation 低水头(水力发电)设备
low-head power plant 低水头发电厂;低水头电站
low-head pump 低压头泵
low-headroom grinding plant 低净空粉磨车间
low-head scheme 低水头工程;低水头方案;低水头电站
low-head storm sewer 低水头雨水管
low-head turbine 低压头涡轮机;低水头水轮机
low-head(vibrating) screen 低头振动筛
low-head water power station 低水头水电站
low-head water turbine 低水头水轮机
low-hearth 精炼炉床
low heat appliance 低热灶具
low heat cement 低(水化)热水泥
low heat concrete 低热混凝土
low heat duty clay 低熔点黏土;低品位耐火黏土
low heat energy kiln 低热耗窑
low heat expansive slag cement 低热微膨胀矿渣水泥
low heat flow zone 低热流带
low heating value 低(发)热值
low heating value gas 低热值气体
low heat method 低热法
low heat of hydration 水化低热
low heat of hydration cement 低热水泥;低水化热水泥
low heat oil well cement 低热油井水泥
low heat Portland(blast furnace) slag cement 低热硅酸盐矿渣水泥;低热波特兰矿渣水泥;低热水泥
low heat Portland cement 低热硅酸盐水泥;低热(波特兰)水泥
low-hedge 低树篱;矮树篱
low-height car 矮车身车辆
low-helix drill 平螺旋钻
low-high junction 低高结
low-high plane nutrition 低高营养水平
low-high speed change 低高速变换
low hold 底舱
low humic acids coal 低腐殖酸煤
low hump 低驼峰
low hydration heat cement 低水化热水泥
low hydrogen alloy rod 低氢合金焊条
low hydrogen electrode 低氢焊条
low hydrogen type covered electrode 低氢型涂料电弧焊条
low hydrogen welding electrode 低氢型(电)焊条
low hydrogen welding rod 低氢型(电)焊条
low hysteresis silicon steel 低磁滞硅钢
low hysteresis steel 低磁滞钢
low-ice soil 轻度冻土
low idle 小油门慢车;低速空转
low-impedance 低阻抗的
low-impedance coupling 低阻抗耦合
low-impedance manganin stress ga(u)ge 低阻抗锰铜应力计
low-impedance measurement 低阻抗测量
low-impedance path 低阻通路;低阻通道;低阻抗

回路
low-impedance sampling oscilloscope 低阻抗采样示波器
low-impedance switching tube 低阻抗开关管
low incidence 小冲角
low incinerator 平型炉;低火层焚化炉
low-income block 低租金住宅
low-income housing 低收入者住房
low-income housing project 低收入者住房建设项目
low-income level 低收入水平
low-income project 低收入者住房设计项目
low-income residence 低收入者住宅
low-index material 低折射率材料
low indicator 低指示器
low-inductance synchro 低电感自动同步机
low inflammability 低易燃性
lowing phreatic level 降低潜水位
lowing the salt location in soil 压盐
low input/output program(me) 低速输入输出程序
low input voltage converter 低输入电压变换器
low in saturation 低饱和状态
low insulation 低级绝缘
low intensity 低强度;低烈度
low intensity atomizer 低强度雾化器
low intensity beam 低流强束
low intensity drum 弱磁转筒
low intensity field 弱磁场
low intensity magnetic separation 弱磁选
low intensity(magnetic) separator 弱磁场磁选机
low intensity radiation 低强度辐射
low intensity reciprocity failure 低强度倒易律失效
low intensity storm 雨量小的阵雨;小雨
low intensity ultrasound 低强度超声波
low intensive 低强度的
low interest credit 低息贷款
low interest fund 低息资金
low interest loans 低息贷款
low interest policy 低利政策
low interest rate 低利率
low interpretability 可判程度低
low ion 低电离离子
low ionization nuclear emission-line region 低电离核区
low irrigated emulsion 低含水量乳化液
low island 低岛
lowitz 日珥
low jetty 潜堤;潜坝
low joint 钢轨低接头;低接头
low-key 暗色调;低调
low-key(ed) image 软调图像
low-key(ed) picture 软调图像
low-key gradation 暗调层次
low-key image 低调图像
low-key lighting 阴暗色调照明;有节制的照明
low-key mask 低调蒙片
low-key photograph 低调照片
low-key picture 暗色调相片
low-key tone 暗调屏
low-K tholeiite 低钾拉斑玄武岩
low labo(u)r cost 低劳动成本
lowland 洼地;低(洼)地
lowland area 低地区
lowland deposit 低地沉积层
lowland forest 低温林;低湿林
lowland lake 低地湖
lowland meadow 低地草原
lowland moor 低位沼泽;低位泥炭土;低地沼泽
lowland reach 低平原河段
lowland rice 水稻
lowland river 平原河流;低地河流
lowland zone 低地河段
low latitude 低纬度
low-latitude climate 低纬度气候
low-latitude desert climate 低纬度沙漠气候
low-latitude disturbance 低纬度扰动
low-latitude region 低纬度区
low-layer tester 底层测试仪
low-laying land 低洼地
low lead 低铅
low lead crystal glass 低铅玻璃
low lead fuel 低含铅汽油
low-lead gas 低铅汽油
low-level 下部中段的;低水平面;低水平(的);低能

级;低空的;低级(的);低电平;低高程;高程低的
low-level air 大气低层
low-level air temperature 低气温
low-level amplification 低电平放大
low-level amplifier 低电平放大器
low-level amplitude limiter 低电平限幅器
low-level audible alarm 低音音响警报
low-level bog 低沼;低洼沼泽;低洼湿地;低洼泥沼地
low-level bridge 低水位桥;漫水桥
low-level bus 低底盘公共汽车
low-level canvas cover 低空帆布篷
low-level cistern 低水箱;低位水箱
low-level cloud 低云
low-level code 低级代码
low-level code continuity check 低级代码和连续性校验;低级编码和连续性校验
low-level condenser 低位凝气器
low-level counting 低水平测量
low-level district 低水位区
low-level dosimeter 低剂量率剂量计
low-level exposure 低剂量接触
low-level flat 低洼地
low-level flushing cistern 低水箱(抽水马桶)
low-level flush toilet 低水箱冲水厕所
low-level grate 低水平格条
low-level groin 河底潜丁坝;低丁坝
low-level ground 低地
low-level groyne 河底潜丁坝
low-level input jack 低电平信号输入塞孔
low-level intake 深进水口
low-level irradiation 低水平照射
low-level jet condenser 低位喷水凝气器;低位喷射凝气器
low-level land 低洼地
low-level language 初级语言
low-level laterite 低含量铝红土
low-level lighting system 低照度照明系统
low-level light source 弱光源
low-level logic 低电平逻辑
low-level logic circuit 低电平逻辑电路
low-level machine 低电面浮选机
low-level makeup 低水位补给
low-level mill 低液面排矿磨碎机
low-level mixing plant 低位(设置的)拌和站;低位(设置的)拌和机
low-level modulation 低功率调制;低电平调制
low-level multiplexer 低电平倍增器
low-level multiplexing 低电平放大
low-level network 低级网络
low-level of cave water 洞穴低水位
low-level of radiation 低能级辐射
low-level of yield 低产量
low-level outlet 泄水底孔;深排水口
low-level outlet tunnel 底部泄水隧洞
low-level overflow gate 低位溢流门
low-level ozone 低空臭氧
low-level packet 低级包
low-level parametric amplifier 小功率信号参量放大器;低电平参量放大器
low-level platform supported on bearing piles 低桩承台
low-level pulser 低电平脉冲发生器
low-level radiation 低强度放射
low-level radioactive laundry wastewater 低强度放射性浆洗废水
low-level radioactive waste 低含量放射性废弃物;低水平放射性废物;低强度放射性废物
low-level radioactive wastewater 低放射性废水
low-level radioactivity 低强度放射性
low-level railway 地下铁道
low-level receiver 低电平接收器
low-level redundant development 低水平重复建设
low-level relieving platform supported on bearing piles 低桩承台
low-level schedule and high level schedule 低级调度与高级调度
low-level scheduling 低级调度
low-level season 枯水季
low-level slewing tower crane 下回转塔式起重机
low-level software 低级软件
low-level sonde 低空探空仪
low-level source 小功率源
low-level sprinkler irrigation 地面小孔细流喷灌

low-level(stack-gas) economizer 低温省煤器
low-level storage bunker 低料位储仓
low-level suite 低水平套房
low-level technology 低水平技术
low-level waste 低水平废物;低放(射性)废物
low-level water closet 低水箱抽水马桶
low-level water tank 低(水)位水箱;底部水箱
low-level whole body counter 低强度全身放射性污染计数器
low-lever granite 深部花岗岩
low lift 低压的;低扬程
low-lift blade grader 低举刃式平地机
low-lift construction (大体积混凝土的)薄层施工
low-lift construction for bulk concrete 大体积混凝土薄层施工
low-lift construction for mass concrete 大体积混凝土薄层施工
low-lift grouting 低砌筑层灌浆(空心砌块墙)
low-lift lock 低水位差船闸;低水头船闸
low-lift platform truck 低升程平板式载重汽车
low-lift pump 低扬程(水)泵;低压泵
low-lift pumping station 低扬程泵站
low-lift pump station 一级泵站;低扬程泵站
low-lift tidal lock 低水头潮汐船闸
low-lift truck 低举升车
low light 低照度的
low light indicator 微光警告信号
low light intensity 弱光强度
low light level 低照度;低亮度
low-light-level imaging 低照度成像
low-light level microspectrophotometer 低光级显微分光光度计
low-light level night vision device 微光夜视仪
low-light level television 微光电视
low-light sensitivity emulsion 低感光度乳剂
low-light sextant 微光夜视六分仪
low lime aluminate refractory cement 低钙铝酸盐耐火水泥
low-limed cement 低(氧化)钙水泥;低石灰水泥
low lime process 低石灰法
low lime rock 低钙石灰石
low limit 下限
low limit of tolerance 下限公差
low limit of water stable area 水稳定场下限
low limit register 下限寄存器
low limit value of mercury anomaly 汞异常下限值
low line 低压线路
low load adjustment 低载调整装置
low loader 低架装载机;轻型装载机
low loading amplifier 低阻加载放大器
low loading floor 低装载底盘;低装载车台
low loading two wheeled spreader 低架双轮撒肥车
low load-lifting speed 微升速度【机】
low load-lowering speed 微降速度【机】
low load period 低负荷期
low load trailer 低车厢底板挂车;低架挂车
low located peat 低位泥炭
low loss 低损耗(的)
low-loss cable 低耗电缆
low-loss coil 低耗线圈
low-loss dielectric 低损耗介电体;低损耗电介质
low-loss ferrite 低损耗铁氧体
low-loss filter 低耗滤波器
low-loss fluorinated ethylene propylene clad silica fiber 低损耗氟化乙丙烯包层石英光纤
low-loss glass 低损失玻璃;低介电损耗玻璃
low-loss modulator 低损耗调制器
low-loss optic(al) fiber 低损耗光纤
low-loss power transformer 低损耗电力变压器
low-loss silica fiber 低损耗石英光纤
low-loss waveguide 低损耗波导管
low-luminosity image 低亮度图像
low-lying area 低洼地区;低地
low-lying country 低地国
low-lying district 低洼地区
low-lying ground 低洼地区
low-lying ground low-lying land 低地
low-lying land 低矮地区;洼地;低地
low-lying level 低能级
low-lying marsh land 低洼沼泽地
low-lying sand 河谷砂
low-lying swamp 低位沼泽;低地沼泽
low-lying swamp land 低洼地带的沼泽地

lowly process（木材的）半定量浸注法
low machinability 低切削性
low machinability steel 难切削钢；难加工钢材
low-magnification seismograph 低倍率地震计
low magnitude earthquake 低震级地震
low-maintenance battery 少保养蓄电池
low manganese rail steel 低锰钢轨钢
low manganese steel 低锰钢
low manganese structural steel 低锰结构钢
low-mark cement 低标号水泥
low-mass panel 轻型板
low-mass particle 小质量质点
low material input-output ratio 低材料输入输出比
low meadow 低洼草地
low-medium mountain 低中山（绝对高度 1000～3500 米）
low-melting 低熔点的；低熔点的
low-melting alloy 低熔点合金；易熔合金
low-melting component 低熔点组分
low-melting compound 低熔点化合物
low-melting constituent 低熔点组分
low-melting copper-soluble metal 低熔点铜溶金属
low-melting enamel 低熔搪瓷
low-melting glass 低熔（点）玻璃
low-melting glaze 低温釉；低熔点釉
low-melting heavy metal 低熔点重金属
low-melting metal ingredient 低熔点金属组分
low-melting metal phase 低熔点金属相
low-melting phase 低熔相
low-melting point 低熔点的
low-melting point alloy 低熔点合金
low-melting point alloy die 低熔点合金模具
low-melting point metal 低熔点金属
low-melting point slag 低熔点炉渣
low-melting point solder 低熔点焊料
low-melting sealing frit 低熔点焊료玻璃
low-melting solder 软钎料
low-melting solids 低熔点固体物质
low-melt point air-tight binder 低熔点密封胶结料
low methane mine 低沼气矿井
low middling 下中级品
low-mid level 枯水中位
low mineralization 低矿化
low-modulation track 浅调制磁迹
low-modulus resin 低模量树脂；低弹模树脂
low-moisture silage 低水分储[贮]料
low molecular weight 低分子量
low molecular weight chlorinated hydrocarbon 低分子量氯化烃
low molecular weight volatile fatty acid 低分子量挥发性脂肪酸
low moor 低位沼泽；低位湿原；低洼沼泽
low moor bog 低位泥炭沼泽
low moor peat 低位泥炭；低位原泥炭；低沼泥炭
low moor wood peat 低洼地沼泽森林泥炭
low mountain 低山区；低山（绝对高度 500～1000 米）
low mountain range 低矮山脉
low-mounted draw bar 拉杆低位安装
low-negative relief 低负突起
low nitrogen oil 低氮石油
low-noise 低噪声
low-noise air screw compressor 低噪声螺杆空气压缩机
low-noise amplifier 低噪声放大器
low-noise antenna 低噪声天线
low-noise axial fan 低噪声轴流式风机
low-noise bearing 低噪声轴承
low-noise cable 低噪声电光缆
low-noise channel 低噪声信道
low-noise characteristic 低噪声特性曲线
low-noise crystal oscillator 低噪声晶体振荡器
low-noise high precision calculating amplifier 低噪声高精度运算放大器
low-noise klystron 低噪声速调管
low-noise lamp 低噪声（白炽）灯
low-noise level transformer 低噪声变压器
low-noise line 低噪声线路
low-noise machine 低噪声机器
low-noise measuring amplifier 低噪声测量放大器
low-noise nozzle 低噪声喷口
low-noise preamplifier 低噪声前置放大器
low-noise receiver 低噪声接收机
low-noise reflector antenna 低噪声反射器天线
low-noise transistor 低噪声晶体管

low-noise travel(l)ing wave tube 低噪声行波管
low-noise ventilator 低噪声通风机
low NO_x burner 低氮氧化物排放量的燃烧器；氮化物排放量低的燃烧器
low NO_x combustion 低氮化氧燃烧
low NO_x emission combustion technology 低氮氧化物燃烧技术
low-NO_x technology 排出氮氧化物少的技术
low-NO_x technology 排出氮氧化物少的技术
low nutrient 低营养
low-nutrient water 含低养分的水
low-oblique 浅倾的
low-oblique aerial photograph 浅倾航片
low oblique photography 浅倾摄影
low-odo(u)r 淡气味
low of wild animals protection 野生动物保护法
low ohmmeter 低电阻计
low oil 轻质油
low oil alarm 低油位警报器；低油位警报；低油位报警器
low-oil alarm system 少油警报
low-oil-content circuit-breaker 少油开关
low-oil content measuring transformer 少油式测量用变压器
low oil level 低油位
low oil pressure 低油压；油压低
low optic(al) absorption 少量光吸收
low order 低位（数）；低次；低阶
low-order add circuit 低位加法电路；低阶相加线路
low-order burst 低效率爆炸
low-order connection supervision 低阶连接监控
low-order digit 最低位数字；低位数（位）；低阶数字
low-order end 最右端；低位端
low-order function 低级功能
low-order goods 低级商品
low-order interface 低阶接口
low-order manufacturing 初级工业
low-order memory area 小地址存储区域
low-order merge 低阶合并
low-order path adaption 低阶通道适配
low-order path connection 低阶通道连接
low-order path overhead monitoring 低阶通道开销监视
low-order path termination 低阶通道终端
low-order path unequipped generator 低阶通道未装载发生器
low-order position 低位（位置）；低价位置
low-order virtual container 低阶虚容器
low ordinary 次等的
low organic carbon wastewater 低有机碳废水
low output laser system 低输出激光系统
low oxide 低价氧化物
low oxygen consumption water 耗氧量低的水
low-paid workers 低工资职工
low paper detection 缺纸检测
low paper indicator 缺纸指示灯
low parallel bar 低双杠
low parapet 矮女儿墙
low parry arc 下偏耳
low partition 半截隔断；矮隔墙；矮隔断
low-pass 低通
low-pass circuit 低通电路
low-pass filter 高阻滤波器；低通滤光片；低通滤波器；低梯度滤波器
low-pass filtering 低通滤波
low pass main frequency 低通主频
low peak hour 低峰期
low-penetration 低针入度（的）
low-penetration asphalt 低针入度地沥青
low penetration electrode 小熔深焊条
low-performance equipment 低性能设备
low-performance sealant 低功能密封膏
low permeability 低透水性；低渗透性
low-persistent pesticide 低残留农药
low perviousness 低透水性
low phosphate content 低磷酸盐含量
low phosphate detergent 低磷去污垢剂
low phosphoric water treatment agent 低磷水处理剂
low phosphorous pig iron 低磷生铁
low phosphorus 低磷
low phosphorus coal 低磷煤
low phosphorus pre-film 低磷预膜
low pH value cement 低碱（度）水泥；低级水泥

low-pilling variant 低起球变性纤维
low pitch 缓和坡度
low pitch cone roof 低倾度锥形顶盖
low pitched glazing 缓轻度玻璃窗
low pitched roof 缓坡顶；缓坡层顶
low pitch gable roof 缓双坡屋顶
low pitch roof 低坡屋顶
low plain 底平原（绝对高度 0～200 米）；低平原
low plastic clay 低塑性黏土
low-plastic clay soil 低塑性黏质土
low-plasticity 低塑性
low-plastic silt 低塑性粉土
low platform 低站台
low platform trailer 低平台拖车
low point 低点
low point of the slope 坡度的低点
low point of vertical curve 竖曲线最低点
low-pole wood 矮秆树干；矮干树林
low pollution 低污染度
low pollution coating 低污染涂料
low pollution emission 低污染排放
low pollution fuel 低污染燃料
low pollution heat engine 低污染热机
low pollution power source 低污染动力源
low pollution source 低污染源
low pollution technique 低污染技术
low pollution vehicle 低污染车辆
low pollution waters 低污染水域；低污染水体
low polymer 低聚合物
low polymerized polyurethane leaking agent 低聚合度聚氨酯堵漏剂
low-polymerized polyurethane leakproofing agent 氰凝防渗剂；氰凝堵漏剂
low population area 人口稀疏地区
low population density 低人口密度
low-porosity 低孔隙度
low porosity fired clay brick 低气孔黏土砖
low-porosity graphite rod 致密石墨棒
low potential 低电势
low-power 小功率；低功率；低倍（镜）
low-power amplifier 小功率放大器
low-power circulation 低倍率循环
low-power data retention 低功率数据保持
low-power direct current electromagnetic relay 低功率直流电磁继电器
low-power direct current wet-type electromagnet for valve 低功率直流湿式阀用电磁铁
low-powered 小功率的
low-powered engine 小功率发动机
low-powered fan marker 小功率扇形指点标
low-powered money 弱力货币
low-powered non-directional radio beacon 小功率全向无线电信标
low-powered reaper 低功率收割机
low-powered ringer 小功率振铃器
low-powered vessel 小功率船；低速船
low-power factor circuit 小功率因数电路
low-power factor mercury lamp ballast 低功率因数水银灯镇流器
low-power factor transformer 高电抗变压器
low-power factor wattmeter 低功率因素瓦特表
low-power field 低倍视野
low-power gear 小功率齿轮
low-power interrogator 小功率询问器
low-power lamp 小功率灯
low-power laser rangefinder 低功率激光测距机
low-power lens 低倍镜
low-power level signal 小功率信号
low-power light source 弱光源
low-power load 小功率负载
low-power logic 低功率逻辑
low-power machine 小功率电机
low-power measurement 小功率测量；低功率测量
low-power microscope 低倍显微镜
low-power modulation system 小功率调制系统
low-power motor 小型电动机；小功率电动机
low-power objective 低倍物镜
low-power output reactor 小功率反应堆
low-power range 低功率区段
low-power range system 低输出系统
low-power repeater station 小功率中继转播台
low-power research reactor 小功率研究性反应堆
low-power run 低功率运行
low-power station 小功率电台

low-power telescope 低倍率望远镜
low-power video transmitter 小功率图像发射机
low-power voltage regulator 低功率调压器
low-power water boiler reactor 小功率沸腾式反应堆
low-power waveguide termination 小功率波导终端负载
low-power winding 小功率绕组
low-pressure 低压强;低压(的);低电压
low-pressure accumulator 低压蓄力器
low-pressure acetylene 低压乙炔
low-pressure acetylene generator 低压乙炔发生器
low-pressure admission 低压进气
low-pressure air atomization 低压空气喷雾法
low-pressure air burner 低压鼓风燃烧器
low-pressure air compressor 低压空气压缩机
low-pressure air equipment 低压风动设备
low-pressure air floating fender 低气压浮动式护舷
low-pressure air receiver 低压空气瓶
low-pressure air starter 低压空气起动机
low-pressure air system 低压空气系统
low-pressure ammonia receiver 低压氨储存器
low-pressure area 低(气)压区
low-pressure asbestos rubber jointing sheet 低压石棉橡胶板
low-pressure baler 低密度捡拾压捆机
low-pressure barometer 低压气压表
low-pressure bell-type recording flowmeter 低压钟罩形记录式流量计
low-pressure boiler 低压锅炉
low-pressure bubble 低压漩涡
low-pressure burner 低压烧嘴;低压燃烧室;低压燃烧器
low-pressure cabinet 低压箱
low-pressure capacity 低压能力
low-pressure casing spray 低压缸喷雾装置
low-pressure center 低压中心
low-pressure centrifugal blower 低压离心式鼓风机
low-pressure centrifugal pump 低压离心泵
low-pressure chamber 低压室
low-pressure charging 低压增压
low-pressure chemical vapo(u)r deposition 低压化学气相沉积
low-pressure cleaning 低压清洗
low-pressure coarse oil filter 低压粗油过滤器
low-pressure combination of gases 低压气体混合物
low-pressure combustion chamber 低压燃烧室
low-pressure compressor 低压压缩机;低压气机
low-pressure condensing turbine 低压凝汽式汽轮机
low-pressure connector 低压接头
low-pressure controller 低压控制器
low-pressure cross head 低压十字头
low-pressure cutout 低压断路(制冷机)
low-pressure cutter 低压切割器;低压割炬
low-pressure cylinder 低压汽缸;低压(汽)缸
low-pressure detector 低压测试器
low-pressure diamond 低压金刚石
low-pressure die casting 低压铸造
low-pressure diecasting machine 低压铸造机
low-pressure direct current plasma 低压直流等离子体
low-pressure disc 低压轮
low-pressure distribution lamp 低压配电盘
low-pressure district 低压区
low-pressure-drop preheater 低压损预热器
low-pressure drum 低压气包
low-pressure eccentric 低压偏心轮
low-pressure end 低压端
low-pressure engine 低压力发动机
low-pressure equipment 低压设备
low-pressure expansion machine 低压膨胀机
low-pressure feed heater 低压给水加热器
low-pressure filter 低压滤油器
low-pressure fixed spray nozzle 低压固定式喷嘴
low-pressure flame 低压火焰
low-pressure flange 低压法兰
low-pressure float valve 低压浮子阀;低压浮头阀;低压浮球阀
low-pressure fluid flow 低压流体流
low-pressure formation 低压岩层;低压地层;低压结构层
low-pressure fuel oil pressure ga(u)ge 低压燃油压力表
low-pressure gallery 低压坑道
low-pressure gas 低压煤气;低压气体
low-pressure gas burner 低压燃烧器;大气式燃烧器;低压煤气喷烧器
low-pressure ga(u)ge 低压压力表;低压水表;低压计;吸入压力表
low-pressure generator 低压发生器
low-pressure grout hole 低压灌浆孔
low-pressure grouting 低压灌浆
low-pressure grouting method (水泥混凝土的)低压灌浆法
low-pressure gun 低压油枪
low-pressure heater 低压加热器
low-pressure heating 低压采暖;低压加热
low-pressure heating boiler 低压加热锅炉
low-pressure high volume pump 低压力大排量泵
low-pressure hose 低压软管
low-pressure hot spray 低压热喷涂
low-pressure hot water boiler 低压热水锅炉
low-pressure hot water heating 低压热水供暖
low-pressure hydraulic control 低压液压操纵;低压液压控制
low-pressure hydraulic excavator 低压液压挖土机
low-pressure hydraulic gear pump 低压液压齿轮泵
low-pressure hydraulics 低压液压系统;低压水力学;低压液压装置
low-pressure hydrogenation unit 低压加氢装置
low-pressure injection mo(u)lding 低压注射模塑;低压注入模塑
low-pressure intake 低压进水口;低压进水装置
low-pressure intake device 低压进水装置;低压进水设备
low-pressure intake unit 低压进水装置;低压进水设备
low-pressure laminate 低压层压板;低压叠层板;低压层压塑料;低压层合板
low-pressure laminating 低压层压法
low-pressure laminating resin 低压积层脂;低压层压用树脂
low-pressure lamination 低压层压法
low-pressure lamp 低压灯
low-pressure light source 低压光源
low-pressure load reducing fork 低压减荷叉
low-pressure load reducing fork diaphragm 低压减荷叉皮膜
low-pressure load reduction fork spring 低压减荷叉弹簧
low-pressure melting 低压熔化
low-pressure mercury lamp 低压水银灯
low-pressure mercury vapo(u)r lamp 低压水银蒸汽灯;低压汞灯
low-pressure microwave plasma 低压微波等离子体
low-pressure mo(u)lding 低压制模法;低压模塑法;低压模(塑成型)
low-pressure nozzle 低压喷嘴
low-pressure overlay 低压塑料印纹板
low-pressure pipe 低压管(道)
low-pressure(pipe)line 低压管线
low-pressure piping 低压管道
low-pressure piston 低压活塞
low-pressure plasma 低压等离子体
low-pressure pneumatic conveying system 低压气动输送系统
low-pressure pneumatics 低压气动技术
low-pressure polyethylene 低压聚乙烯
low-pressure polyethylene fibre 低压聚乙烯纤维
low-pressure preheater 低压预热器
low-pressure process 低压法
low-pressure pulse ultraviolet treatment unit 脉冲式低压紫外线治疗机
low-pressure pump 低压泵
low-pressure purge 低压气体吹洗
low-pressure refrigeration 低压冷冻
low-pressure regional metamorphic facies series 低压区域变质相系
low-pressure regional metamorphism 低压区域变质作用
low-pressure resin 低压(固化)树脂
low-pressure rubber hose 低压橡胶软管
low-pressure rubber sheet 低压胶板
low-pressure safety cut-out 低压安全断流器
low-pressure safety valve 低压安全阀
low-pressure safety valve assembly 低压安全阀总成
low-pressure safety valve spring 低压安全阀弹簧
low-pressure saturated steam 低压饱和蒸汽
low-pressure section 低压部分
low-pressure sensor 低压传感器
low-pressure separator 低压分离器
low-pressure shaft 低压轴
low-pressure side 低压侧
low-pressure slide valve 低压滑阀
low-pressure sodium lamp 低压钠灯
low-pressure sodium vapo(u)r lamp 低压钠汽灯
low-pressure sprayer 低压喷雾器
low-pressure spraying unit 低压喷射装置;低压喷雾装置
low-pressure sprinkler 低压喷淋机;低压(头)喷灌机
low-pressure stage 低压段
low-pressure stainless steel hose 低压不锈钢软管
low-pressure steam 低压蒸汽
low-pressure steam boiler 低压蒸汽锅炉
low-pressure steam-cured 低压蒸汽养护的
low-pressure steam curing 低压蒸汽养护
low-pressure steam engine 低压蒸汽机
low-pressure steam generator 低压蒸汽发生器
low-pressure steam heating 低压蒸汽供暖;低压蒸汽采暖
low-pressure steam heating system 低压蒸汽供暖系统
low-pressure steam pipe 低压蒸汽管
low-pressure steam system 低压蒸汽系统
low-pressure steam turbine 低压蒸汽涡轮机;低压蒸汽透平;低压蒸汽轮机;低压汽轮机
low-pressure synthesis 低压合成
low-pressure synthesis of super-hard coating 超硬涂层的低压合成
low-pressure system 低压热水系统;低压系统
low-pressure tank 低压储[贮]罐
low-pressure tap 低压水龙头
low-pressure test 低压试验
low-pressure torch 低压焊炬;低压吹管
low-pressure transfer mo(u)lding 低压塑成形;低压模塑传递法
low-pressure tube 低压管(道)
low-pressure tubing 低压管道
low-pressure tunnel 低压隧道
low-pressure turbine 低压涡轮(机);低压汽轮机
low-pressure turbine stage 低压涡轮级
low-pressure turbocharger 低压涡轮增压器
low-pressure type burner 低压型燃烧器
low-pressure tire 低压轮胎
low-pressure vacuum pump 高(级)真空泵
low-pressure valve 低压阀门
low-pressure vapo(u)r heating system 低压蒸汽供暖系统
low-pressure vessel 低压容器
low-pressure warm air sprayer 低压温风喷涂机;低压暖风喷涂机
low-pressure water system 低压供水系统
low-pressure well 低压井
low-pressure wet steam 低压湿蒸汽
low-pressure wheel 低压轮
low-pressure zone 低压区
low priced and consumable supplies 低值易耗品
low priced goods 下价货
low-print tape 低迹带
low priority 低级优先权;低级优先级
low priority ready queue 低优先就绪队列
low probability ground movement 低概率地面运动
low-producing 低产量的
low production 小批生产;小量生产
low productive soil 低产土壤
low productivity in agriculture 农业生产率低
low profile 低型的;低断面的
low-profile additive 低收缩添加剂
low-profile batching plant 自走式分配料设备
low-profile central plant 自走式集中配拌和机
low-profile design concept 低外形设计概念
low-profile electronic platform scale 薄型电子台秤
low-profile hot melt granular 低型面热熔成粒机
low-profile mixing plant 自走式搅拌设备
low-profile plant 自行式拌制混凝土装置;低平式

拌和厂
low-profile pulpwood fork 低纸浆木材叉(具);低位纸浆木材叉(具)
low-profile staddle carrier 低型跨式载运车
low pulse frequency 低脉冲频率
low-purity 低纯度
low quality 下等品;质量不高;低品质度;劣等;低质量(的)
low quality airdry roughages 低质量干粗饲料
low quality fuel 低质燃料
low radiation plastic welder 低辐射热合机
low radioactivity glass 低放射性玻璃
low radioactivity waste 低放射性废物
low radio frequency 无线电低频;低射频
low rail 曲线内轨;低轨(曲线内轨)
low-rainfall area 低雨量地区
low range 低量程;低挡;低倍率
low-range flowmeter 低量程流量计
low-range oxygen analyzer[analyser] 低氧量分析器
low-range pressure ga(u)ge 低量程气压表
low rank anthracite 低级无烟煤
low-rank fuel 低级燃料
low-rank graywacke 低级杂砂岩
low-rank hydrothermal alteration 低级水热蚀变
low-rank metamorphism 低级变质作用;低度变质作用
low rate 低速的
low-rate aerobic pond 低负荷生物塘
low-rate biologic(al) filter 普通生物滤池;低负荷生物滤池
low-rate filter 低速过滤池;低负荷滤池;慢滤池;低速度滤池
low-rate filtration 低负荷过滤
low-rate of reusability 资源再生利用率低
low-rate silver zinc battery 低速率银锌蓄电池
low-rate trickling filter 低负荷滴滤池;低滤率滴滤池
low ratio of modulus 低模量比岩石
low-reactance grounding 低电抗接地
low reactance-resistance ratio 低电抗电阻比
low reaction force 低反力型
low reaction force fender 低反力护舷
low reading thermometer 低温温度计
low red heat 次红热
low-reflection film 低反射膜;防反射膜;防反射胶片
low refractive and high dispersion optic(al) glass 低折射率高色散光学玻璃
low refractive high dispersive glass 低折射高色散玻璃
low-reinforced beam 少筋梁
low relative humidity 低相对湿度
low release fertilizer 缓释肥料
low relief 低地势;浅雕;浅浮雕;低浮雕
low relief area 起伏小的地区
low relief carving 浅浮槽
low relief frieze 浅浮雕檐壁
low relief terrain 丘陵区;低山区
low-rent apartment 低租金公寓
low-rent apartment unit 低租金公寓单元
low-rent building 低租金房屋
low-rent demonstration program(me) 低租金住房示范(建设)计划
low-rent house 低租金房屋
low-rent house-building 低租金住房建造
low-rent housing 简易楼;低租金房屋建设
low-rent project 低租金住房设计项目
low-rent public housing 低公共住房;低房租公共住房
low-rent unit 低租金公寓
low-residual 低残留
low-residual-phosphorous copper 低磷铜;低残留磷铜
low-resistance 低电阻(的)
low-resistance damping winding 低电阻阻尼绕组
low-resistance glass oil-trap 低阻玻璃捕油器
low-resistance potentiometer 低阻电位计
low-resistance precision potentiometer 低阻精密电位计
low-resistance thermometry 低阻温度测量术
low-resistance transmitter 低阻发射机
low-resistance vapo(u)rizer 低阻汽化器
low-resistive interesting point 低阻交点
low-resistive shielding disturbance 低阻屏蔽干扰
low resistive shielding layer 低阻屏蔽层

low resistivity shielding 低阻屏蔽
low resolution 低分辨率
low resolution infrared radiometer 低分辨率红外辐射计
low resolution mass spectrometer 低分辨率质谱仪
low resolution spectrometer 低分辨率分光计
low resolution spectrometry 低分辨率光谱法
low responder 低应答者
low-response 响应慢;低响应
low retaining wall 矮挡土墙
low return 低反射
low-ripple generator 低脉动发电机
low-rise 层数少且不设置楼梯的楼房
low-rise block 低层公寓
low-rise building 低层建筑;低层房屋
low-rise development 低建发展
low-rise dwelling 低层住宅
low rise/high density 低层高密度
low-rise/low-density 低层低密度
low-rise roof 低(坡)屋顶;矮屋顶
low-rise spheric(al) shell 低曲率壳结构
low-rise(storied) building 低层建筑(物)
low-rise structure 低层建筑(物)
low-risk technology 低风险技术
low river basin 低洼盆地
low runaway speed 低飞逸转速
low runoff 枯水径流;低水径流
low salinity characteristic 低盐特性
low salinity reservoir 低盐度热储
low salinity water 低盐渍度水
low saturation 饱和度小
low scale 基本刻度;低读数
low seam 薄矿层
low-seam conveyer 薄煤层运输机;薄矿层输送机
low section-height tire 扁平轮胎(汽车用)
low seismicity region 低地震活动区
low-sensitivity terminal unit 低灵敏度终端装置
low service 低压配水
low service district 低压供水区
low servo 低速加力
low-set-clay 低沉降性黏土
low setting point oil 耐寒性润滑油
low-shaft blast furnace 矮高炉
low-shaft furnace 低身竖炉;矮高炉
low shear dispersion 低切变分散
low shear visco(si)meter 低剪切黏度计
low shear viscosity 低剪切黏度
low shore 低岸
low shrink additive 低收缩添加剂
low shrinkage 低收缩
low-shrinkage concrete 低缩性混凝土;低收缩混凝土
low-shrinkage fiber 低收缩纤维
low shrink resin 低收缩树脂
low side 低端
low side band 下边带
low side charging 低压侧充注
low sided car 低边车
low sided open wagon 低边敞车
low sided ship 低舷船
low sided wagon 低边车
low side float valve 低压浮子阀;低压浮头阀;低压浮球阀
low side gondola car 低边敞车
low side mill 低边多辊式磨机
low side roller mill 低面íé磨机
low side window (老教堂内高坛墙上的)小窗;低侧窗
low silicate product 低硅产品
low silicon 低硅的
low silicon bronze 低硅青铜
low silicon bronze alloy 低硅青铜合金
low silicon cast iron 低硅铸铁
low silicon pig iron 低硅生铁
low-sill structure 下平巷底板结构;下底梁结构
low sintered composite contact metal 低温烧结的复合电触头合金
low sintering temperature 低温烧结(的)
low-slag cement 少熔渣水泥;低矿渣水泥;低熔渣水泥;低掺量矿渣(硅酸盐)水泥
low slope 低边坡
low-slump 低坍落度的
low-slump concrete 干硬(性)混凝土;低坍落度混凝土

low-slung modernistic building 低矮式建筑
low-slung structure 低矮式建筑
low smelting heat 低温熔炼
low smoke 低发烟
low-smoke fiber optic(al) cable 低烟阻燃光缆
low-smoke halogen free 低烟无卤
low-smoke resin 低挥发树脂
low-soda alumina 低碱氧化铝
low sodium water 低钠水
low soil moisture 低土壤湿度
low solid 低固体分
low-solid non-dispersed mud 低固相非分散性泥浆
low solids mud 低固相泥浆
low solubility glaze 低铅溶出釉
low solvency solvent 低溶解力溶剂
low solvent paint 低溶剂含量漆
Lowson technique 高频设备调谐技术
low-sounding horn 低音喇叭
low specific speed 低比速
low speed 慢速(度);低速度;低速(的)
low-speed adjustment 低速调整
low-speed aerator 低速曝气器
low-speed agitator 低速搅拌器
low-speed air explosive bomb 低速空爆弹
low-speed analogue computer 低速模拟计算机
low-speed anemometer 低速风速仪
low-speed autobalancing centrifuge 低速自动平衡离心机
low-speed balancing 低速动平衡
low-speed band 低速带
low-speed characteristic 低速特性
low-speed cluster bus 低速群集器总线
low-speed coal mill 钢球磨煤机
low-speed condition 低速条件
low-speed conical refiner 低速锥形精磨机
low-speed current meter 低流速仪
low-speed data rate 低速数据传输率
low-speed down travel 微降值【机】
low-speed dump 低速转储
low-speed engine 低速发动机
low-speed film 低感光度胶片
low-speed flow 低速流
low-speed gear 低速挡;低速齿轮
low-speed handling 低速操作
low-speed heavy cut 低速重切削
low-speed high torque hydraulic motor 低速大转矩液压马达
low-speed high torque hydraulic pump 低速大转矩液压泵
low-speed high-torque shredder 低速高转矩撕碎机
low-speed high torque turbodrill 低速大转矩涡轮钻具
low-speed hydraulic motor 低速液压马达
low-speed jet 低速射流
low-speed lifting 微提升【机】
low-speed line 低速线路
low-speed lowering 微下降【机】
low-speed machine 低速机
low-speed memory 低速记忆装置;低速存储器
low-speed mill 慢速轧机;低速碎矿机
low-speed motor 低速电动机
low-speed needle valve 低速针阀
low-speed Pitot static head 低(流)速毕托管静水头
low-speed pump 低速(水)泵
low-speed register 低速寄存器
low-speed regulation 低速调节
low-speed rolling 低速轧制
low-speed running 低速运行
low-speed sand filter 慢速砂滤池
low-speed sand filtration 低速砂滤
low-speed section of hump yard 驼峰编组站低速区段;驼峰编组场低速区段
low-speed shaft 低速轴
low-speed stability characteristic 低速稳定性
low-speed storage 低速存储器
low-speed surface aeration system 低速表面曝气系统
low-speed synchromotor 低速同步电机
low-speed synchronous motor 低速同步电机
low-speed tracking 低速跟踪
low-speed turbine aerator 低速涡轮曝气器;低速透平曝气器
low-speed up travel 微升值【机】